AP 3270

SIGHT REDUCTION TABLES FOR AIR NAVIGATION

VOLUME 1

SELECTED STARS

EPOCH 1995·0

CORRECTION FOR PRECESSION AND NUTATION FOR SURFACE NAVIGATION

Although designed for use in the air, this volume is being increasingly used for the reduction of astronomical sights at sea.

The altitudes and azimuths of stars as tabulated in this volume are calculated for the mean equinox of 1995·0. For strict accuracy it is necessary to apply to a position line or fix, deduced from these tables, a correction for the effects of precession and nutation. Table **5** gives such corrections, but only to the nearest minute of arc for use in air navigation.

The accompanying tables give the corrections for the years 1991–1999, to the nearest 0'.1 in distance and 1° in true bearing; they follow the design of Table **5** and should be used in the same way. It is suggested that they be used instead of Table **5** whenever the additional accuracy is required.

CORRECTION FOR PRECESSION AND NUTATION

LHA ♈	North latitudes							South latitudes						LHA ♈
	N 80°	N 70°	N 60°	N 50°	N 40°	N 20°	0°	S 20°	S 40°	S 50°	S 60°	S 70°	S 80°	
							1991							
°	′ °	′ °	′ °	′ °	′ °	′ °	′ °	′ °	′ °	′ °	′ °	′ °	′ °	°
0	**1·2** 204	**1·4** 220	**1·6** 230	**1·9** 237	**2·2** 241	**2·5** 245	**2·7** 247	**2·5** 245	**2·1** 240	**1·9** 235	**1·6** 228	**1·3** 217	**1·1** 200	**0**
30	**1·3** 228	**1·6** 237	**1·9** 242	**2·2** 246	**2·4** 248	**2·6** 250	**2·6** 250	**2·3** 247	**1·7** 239	**1·4** 232	**1·2** 219	**0·9** 198	**0·9** 171	**30**
60	**1·4** 250	**1·8** 254	**2·1** 256	**2·3** 258	**2·5** 259	**2·7** 259	**2·5** 259	**2·0** 256	**1·4** 249	**1·0** 240	**0·6** 219	**0·5** 174	**0·7** 134	**60**
90	**1·5** 272	**1·8** 271	**2·1** 271	**2·4** 271	**2·5** 271	**2·7** 271	**2·4** 271	**1·9** 271	**1·2** 272	**0·8** 273	**0·3** 279	**0·2** 074	**0·6** 086	**90**
120	**1·4** 294	**1·8** 289	**2·1** 286	**2·3** 284	**2·5** 283	**2·7** 282	**2·5** 283	**2·1** 286	**1·4** 294	**1·0** 303	**0·7** 322	**0·6** 001	**0·7** 039	**120**
150	**1·3** 316	**1·6** 306	**1·9** 300	**2·2** 296	**2·4** 293	**2·6** 291	**2·6** 291	**2·3** 294	**1·8** 301	**1·5** 308	**1·2** 320	**1·0** 338	**0·9** 004	**150**
180	**1·1** 340	**1·3** 323	**1·6** 312	**1·9** 305	**2·1** 300	**2·5** 295	**2·7** 293	**2·5** 295	**2·2** 299	**1·9** 303	**1·6** 310	**1·4** 320	**1·2** 336	**180**
210	**0·9** 009	**0·9** 342	**1·2** 321	**1·4** 308	**1·7** 301	**2·3** 293	**2·6** 290	**2·6** 290	**2·4** 292	**2·2** 294	**1·9** 298	**1·6** 303	**1·3** 312	**210**
240	**0·7** 046	**0·5** 006	**0·6** 321	**1·0** 300	**1·4** 291	**2·0** 284	**2·5** 281	**2·7** 281	**2·5** 281	**2·3** 282	**2·1** 284	**1·8** 286	**1·4** 290	**240**
270	**0·6** 094	**0·2** 106	**0·3** 261	**0·8** 267	**1·2** 268	**1·9** 269	**2·4** 269	**2·7** 269	**2·5** 269	**2·4** 269	**2·1** 269	**1·8** 269	**1·5** 268	**270**
300	**0·7** 141	**0·6** 179	**0·7** 218	**1·0** 237	**1·4** 246	**2·1** 254	**2·5** 257	**2·7** 258	**2·5** 257	**2·3** 256	**2·1** 254	**1·8** 251	**1·4** 246	**300**
330	**0·9** 176	**1·0** 202	**1·2** 220	**1·5** 232	**1·8** 239	**2·3** 246	**2·6** 249	**2·6** 249	**2·4** 247	**2·2** 244	**1·9** 240	**1·6** 234	**1·3** 224	**330**
360	**1·2** 204	**1·4** 220	**1·6** 230	**1·9** 237	**2·2** 241	**2·5** 245	**2·7** 247	**2·5** 245	**2·1** 240	**1·9** 235	**1·6** 228	**1·3** 217	**1·1** 200	**360**
							1992							
0	**0·8** 201	**0·9** 218	**1·1** 229	**1·3** 236	**1·4** 240	**1·7** 245	**1·8** 247	**1·7** 245	**1·5** 241	**1·3** 236	**1·1** 229	**0·9** 219	**0·8** 202	**0**
30	**0·9** 225	**1·1** 235	**1·3** 241	**1·5** 245	**1·6** 247	**1·8** 249	**1·8** 249	**1·6** 246	**1·2** 239	**1·0** 232	**0·8** 220	**0·7** 200	**0·6** 174	**30**
60	**1·0** 248	**1·2** 252	**1·4** 255	**1·6** 257	**1·7** 258	**1·8** 258	**1·7** 258	**1·4** 255	**0·9** 247	**0·7** 238	**0·5** 218	**0·4** 177	**0·5** 139	**60**
90	**1·0** 270	**1·2** 270	**1·4** 270	**1·6** 270	**1·7** 270	**1·8** 270	**1·6** 270	**1·3** 270	**0·8** 269	**0·5** 269	**0·2** 268	**0·1** 094	**0·4** 091	**90**
120	**1·0** 291	**1·2** 287	**1·4** 284	**1·6** 283	**1·7** 282	**1·8** 281	**1·7** 282	**1·4** 285	**0·9** 292	**0·7** 301	**0·5** 321	**0·4** 004	**0·5** 043	**120**
150	**0·9** 314	**1·1** 304	**1·3** 298	**1·5** 295	**1·6** 292	**1·8** 290	**1·8** 291	**1·6** 293	**1·2** 301	**1·0** 308	**0·8** 321	**0·7** 340	**0·6** 007	**150**
180	**0·8** 338	**0·9** 321	**1·1** 311	**1·3** 304	**1·5** 299	**1·7** 295	**1·8** 293	**1·7** 295	**1·4** 300	**1·3** 304	**1·1** 311	**0·9** 322	**0·8** 339	**180**
210	**0·6** 006	**0·7** 340	**0·8** 320	**1·0** 308	**1·2** 301	**1·6** 294	**1·8** 291	**1·8** 291	**1·6** 293	**1·5** 295	**1·3** 299	**1·1** 305	**0·9** 315	**210**
240	**0·5** 041	**0·4** 003	**0·5** 322	**0·7** 302	**0·9** 293	**1·4** 285	**1·7** 282	**1·8** 282	**1·7** 282	**1·6** 283	**1·4** 285	**1·2** 288	**1·0** 292	**240**
270	**0·4** 089	**0·1** 086	**0·2** 272	**0·5** 271	**0·8** 271	**1·3** 270	**1·6** 270	**1·8** 270	**1·7** 270	**1·6** 270	**1·4** 270	**1·2** 270	**1·0** 270	**270**
300	**0·5** 137	**0·4** 176	**0·5** 219	**0·7** 239	**0·9** 248	**1·4** 255	**1·7** 258	**1·8** 259	**1·7** 258	**1·6** 257	**1·4** 256	**1·2** 253	**1·0** 249	**300**
330	**0·6** 173	**0·7** 200	**0·8** 219	**1·0** 232	**1·2** 239	**1·6** 247	**1·8** 249	**1·8** 250	**1·6** 248	**1·5** 245	**1·3** 242	**1·1** 236	**0·9** 226	**330**
360	**0·8** 201	**0·9** 218	**1·1** 229	**1·3** 236	**1·4** 240	**1·7** 245	**1·8** 247	**1·7** 245	**1·5** 241	**1·3** 236	**1·1** 229	**0·9** 219	**0·8** 202	**360**
							1993							
0	**0·4** 194	**0·5** 213	**0·6** 225	**0·7** 234	**0·8** 239	**0·9** 245	**1·0** 246	**0·9** 246	**0·8** 242	**0·7** 238	**0·6** 232	**0·5** 223	**0·4** 209	**0**
30	**0·5** 219	**0·6** 230	**0·7** 237	**0·8** 242	**0·9** 245	**1·0** 248	**1·0** 248	**0·9** 245	**0·7** 238	**0·6** 232	**0·5** 221	**0·4** 205	**0·4** 182	**30**
60	**0·5** 242	**0·6** 247	**0·8** 251	**0·8** 253	**0·9** 254	**1·0** 255	**0·9** 255	**0·8** 251	**0·5** 243	**0·4** 234	**0·3** 216	**0·2** 184	**0·3** 149	**60**
90	**0·5** 264	**0·7** 265	**0·8** 266	**0·9** 266	**0·9** 266	**1·0** 266	**0·9** 266	**0·7** 265	**0·4** 262	**0·3** 258	**0·1** 242	**0·1** 135	**0·2** 105	**90**
120	**0·5** 285	**0·7** 282	**0·8** 281	**0·9** 279	**0·9** 279	**1·0** 278	**0·9** 279	**0·7** 281	**0·5** 288	**0·3** 296	**0·2** 318	**0·1** 015	**0·3** 055	**120**
150	**0·5** 308	**0·6** 300	**0·7** 295	**0·8** 292	**0·9** 290	**1·0** 288	**0·9** 289	**0·8** 292	**0·6** 300	**0·5** 308	**0·4** 323	**0·3** 346	**0·3** 016	**150**
180	**0·4** 331	**0·5** 317	**0·6** 308	**0·7** 302	**0·8** 298	**0·9** 294	**1·0** 294	**0·9** 295	**0·8** 301	**0·7** 306	**0·6** 315	**0·5** 327	**0·4** 346	**180**
210	**0·4** 358	**0·4** 335	**0·5** 319	**0·6** 308	**0·7** 302	**0·9** 295	**1·0** 292	**1·0** 292	**0·9** 295	**0·8** 298	**0·7** 303	**0·6** 310	**0·5** 321	**210**
240	**0·3** 031	**0·2** 356	**0·3** 324	**0·4** 306	**0·5** 297	**0·8** 289	**0·9** 285	**1·0** 285	**0·9** 286	**0·8** 287	**0·8** 289	**0·6** 293	**0·5** 298	**240**
270	**0·2** 075	**0·1** 045	**0·1** 298	**0·3** 282	**0·4** 278	**0·7** 275	**0·9** 274	**1·0** 274	**0·9** 274	**0·9** 274	**0·8** 274	**0·7** 275	**0·5** 276	**270**
300	**0·3** 125	**0·1** 165	**0·2** 222	**0·3** 244	**0·5** 252	**0·7** 259	**0·9** 261	**1·0** 262	**0·9** 261	**0·9** 261	**0·8** 259	**0·7** 258	**0·5** 255	**300**
330	**0·3** 164	**0·3** 194	**0·4** 217	**0·5** 232	**0·6** 240	**0·8** 248	**0·9** 251	**1·0** 252	**0·9** 250	**0·8** 248	**0·7** 245	**0·6** 240	**0·5** 232	**330**
360	**0·4** 194	**0·5** 213	**0·6** 225	**0·7** 234	**0·8** 239	**0·9** 245	**1·0** 246	**0·9** 246	**0·8** 242	**0·7** 238	**0·6** 232	**0·5** 223	**0·4** 209	**360**

The above table gives the correction to be applied to a position line or a fix for the effects of precession and nutation from the mean equinox of 1995·0. Each entry consists of the distance (in bold type) in nautical miles, and the direction (true bearing) in which the position line or fix is to be moved. The table is entered firstly by the year, then by choosing the column nearest the latitude and finally the entry nearest the LHA♈ of observation; no interpolation is necessary, though in extreme cases near the beginning or end of a year (but not the end of 1994 or the beginning of 1995 when the corrections are zero) values midway towards those of the previous or following years may be taken.

Example. In 1992 a fix is obtained in latitude N 23° when LHA♈ is 71°. Entering the table with the year 1992, latitude N 20°, and LHA♈ 60° gives **1′·8** 258° which indicates that the fix is to be transferred 1·8 miles in true bearing 258°.

TABLE 8—REFRACTION

TO BE *SUBTRACTED* FROM SEXTANT ALTITUDE

R	(a) Height in thousands of feet												*R*
	0	5	10	15	20	25	30	35	40	45	50	55	
′	°	°	°	°	°	°	°	°	°	°	°	°	′
	90	90	90	90	90	90	90	90	90	90	90	90	
0													0
	63	59	55	51	46	41	36	31	26	20	17	13	
1													1
	33	29	26	22	19	16	14	11	10	10	10	10	
2													2
	21	19	16	14	12	10	10	10					
3													3
	16	14	12	10	10								
4													4
	12	11	10										
5													5
	10	10											

R	(b) Height in thousands of metres																				*R*
	0	1	2	3	4	5	6	7	8	9	10	11	12	13	14	15	16	17	18	19	
′	°	°	°	°	°	°	°	°	°	°	°	°	°	°	°	°	°	°	°	°	′
	90	90	90	90	90	90	90	90	90	90	90	90	90	90	90	90	90	90	90	90	
0																					0
	63	61	58	55	53	50	46	43	40	37	33	30	26	23	20	17	15	13	11	10	
1																					1
	33	31	28	26	24	21	19	17	16	14	12	11	10	10	10	10	10	10	10		
2																					2
	21	20	18	16	15	13	12	11	10	10	10	10									
3																					3
	16	14	13	12	11	10	10	10													
4																					4
	12	11	10	10	10																
5																					5
	10	10																			

Choose the column appropriate to height, in units of 1000 feet in table 8(a) or in units of 1000 metres in table 8(b), and find the range of altitude in which the sextant altitude lies; the corresponding value of *R* is the refraction to be subtracted from the sextant altitude.

TABLE 9—CORIOLIS (*Z*) CORRECTION

STANDARD DOME REFRACTION

To be *subtracted* from sextant altitude when using sextant suspension in a perspex dome.

Alt.	Refn.	Alt.	Refn.
°	′	°	′
10	8	50	4
20	7	60	4
30	6	70	3
40	5	80	3

This table must not be used if a calibration table is fitted to the dome, or if a flat glass plate is provided, or for non-standard domes.

Ground speed knots	Latitude										Ground speed knots
	0°	10°	20°	30°	40°	50°	60°	70°	80°	90°	
	′	′	′	′	′	′	′	′	′	′	
50	0	0	0	1	1	1	1	1	1	1	50
100	0	0	1	1	2	2	2	2	3	3	100
150	0	1	1	2	3	3	3	4	4	4	150
200	0	1	2	3	3	4	5	5	5	5	200
250	0	1	2	3	4	5	6	6	6	7	250
300	0	1	3	4	5	6	7	7	8	8	300
350	0	2	3	5	6	7	8	9	9	9	350
400	0	2	4	5	7	8	9	10	10	10	400
450	0	2	4	6	8	9	10	11	12	12	450
500	0	2	4	7	8	10	11	12	13	13	500
550	0	3	5	7	9	11	12	14	14	14	550
600	0	3	5	8	10	12	14	15	16	16	600
650	0	3	6	9	11	13	15	16	17	17	650
700	0	3	6	9	12	14	16	17	18	18	700
750	0	3	7	10	13	15	17	18	19	20	750
800	0	4	7	10	13	16	18	20	21	21	800
850	0	4	8	11	14	17	19	21	22	22	850
900	0	4	8	12	15	18	20	22	23	24	900

Apply by moving the position line a distance *Z* to starboard (right) of the track in northern latitudes, and to port (left) in southern latitudes.

BUBBLE SEXTANT ERROR

Sextant No.

Alt.	Corr.
°	′

TABLE 1.—Altitude Correction for Change in Position of Observer

Correction for 4 Minutes of Time

Rel. Zn	Ground Speed in Knots																		Rel. Zn
	50	100	150	200	250	300	350	400	450	500	550	600	650	700	750	800	850	900	
°	′	′	′	′	′	′	′	′	′	′	′	′	′	′	′	′	′	′	°
000	+3	+7	+10	+13	+17	+20	+23	+27	+30	+33	+37	+40	+43	+47	+50	+53	+57	+60	000
005	3	7	10	13	17	20	23	27	30	33	37	40	43	46	50	53	56	60	355
010	3	7	10	13	16	20	23	26	30	33	36	39	43	46	49	53	56	59	350
015	3	6	10	13	16	19	23	26	29	32	35	39	42	45	48	52	55	58	345
020	3	6	9	13	16	19	22	25	28	31	34	38	41	44	47	50	53	56	340
025	3	6	9	12	15	18	21	24	27	30	33	36	39	42	45	48	51	54	335
030	+3	+6	+9	+12	+14	+17	+20	+23	+26	+29	+32	+35	+38	+40	+43	+46	+49	+52	330
035	3	5	8	11	14	16	19	22	25	27	30	33	35	38	41	44	46	49	325
040	3	5	8	10	13	15	18	20	23	26	28	31	33	36	38	41	43	46	320
045	2	5	7	9	12	14	16	19	21	24	26	28	31	33	35	38	40	42	315
050	2	4	6	9	11	13	15	17	19	21	24	26	28	30	32	34	36	39	310
055	2	4	6	8	10	11	13	15	17	19	21	23	25	27	29	31	33	34	305
060	+2	+3	+5	+7	+8	+10	+12	+13	+15	+17	+18	+20	+22	+23	+25	+27	+28	+30	300
065	1	3	4	6	7	8	10	11	13	14	15	17	18	20	21	23	24	25	295
070	1	2	3	5	6	7	8	9	10	11	13	14	15	16	17	18	19	21	290
075	1	2	3	3	4	5	6	7	8	9	9	10	11	12	13	14	15	16	285
080	1	1	2	2	3	3	4	5	5	6	6	7	8	8	9	9	10	10	280
085	+0	+1	+1	+1	+1	+2	+2	+2	+3	+3	+3	+3	+4	+4	+4	+5	+5	+5	275
090	0	0	0	0	0	0	0	0	0	0	0	0	0	0	0	0	0	0	270
095	−0	−1	−1	−1	−1	−2	−2	−2	−3	−3	−3	−3	−4	−4	−4	−5	−5	−5	265
100	1	1	2	2	3	3	4	5	5	6	6	7	8	8	9	9	10	10	260
105	1	2	3	3	4	5	6	7	8	9	9	10	11	12	13	14	15	16	255
110	1	2	3	5	6	7	8	9	10	11	13	14	15	16	17	18	19	21	250
115	1	3	4	6	7	8	10	11	13	14	15	17	18	20	21	23	24	25	245
120	2	3	5	7	8	10	12	13	15	17	18	20	22	23	25	27	28	30	240
125	−2	−4	−6	−8	−10	−11	−13	−15	−17	−19	−21	−23	−25	−27	−29	−31	−33	−34	235
130	2	4	6	9	11	13	15	17	19	21	24	26	28	30	32	34	36	39	230
135	2	5	7	9	12	14	16	19	21	24	26	28	31	33	35	38	40	42	225
140	3	5	8	10	13	15	18	20	23	26	28	31	33	36	38	41	43	46	220
145	3	5	8	11	14	16	19	22	25	27	30	33	35	38	41	44	46	49	215
150	3	6	9	12	14	17	20	23	26	29	32	35	38	40	43	46	49	52	210
155	−3	−6	−9	−12	−15	−18	−21	−24	−27	−30	−33	−36	−39	−42	−45	−48	−51	−54	205
160	3	6	9	13	16	19	22	25	28	31	34	38	41	44	47	50	53	56	200
165	3	6	10	13	16	19	23	26	29	32	35	39	42	45	48	52	55	58	195
170	3	7	10	13	16	20	23	26	30	33	36	39	43	46	49	53	56	59	190
175	3	7	10	13	17	20	23	27	30	33	37	40	43	46	50	53	56	60	185
180	−3	−7	−10	−13	−17	−20	−23	−27	−30	−33	−37	−40	−43	−47	−50	−53	−57	−60	180

Interpolation for Altitude Correction for Less Than 4 Minutes of Time

Interval of Time	Value from Tables 1 and 2 (For values greater than 60′ see opposite page)																				Interval of Time
	3	6	9	12	15	18	21	24	27	30	33	36	39	42	45	48	51	54	57	60	
m s	′	′	′	′	′	′	′	′	′	′	′	′	′	′	′	′	′	′	′	′	m s
0 00	0	0	0	0	0	0	0	0	0	0	0	0	0	0	0	0	0	0	0	0	0 00
10	0	0	0	1	1	1	1	1	1	1	1	2	2	2	2	2	2	2	2	3	10
20	0	1	1	1	1	2	2	2	2	3	3	3	3	4	4	4	4	5	5	5	20
30	0	1	1	2	2	2	3	3	3	4	4	5	5	5	6	6	6	7	7	8	30
40	1	1	2	2	3	3	4	4	5	5	6	6	7	7	8	8	9	9	10	10	40
0 50	1	1	2	3	3	4	4	5	6	6	7	8	8	9	9	10	11	11	12	13	0 50
1 00	1	2	2	3	4	5	5	6	7	8	8	9	10	11	11	12	13	14	14	15	1 00
10	1	2	3	4	4	5	6	7	8	9	10	11	11	12	13	14	15	16	17	18	10
20	1	2	3	4	5	6	7	8	9	10	11	12	13	14	15	16	17	18	19	20	20
30	1	2	3	5	6	7	8	9	10	11	12	14	15	16	17	18	19	20	21	23	30
40	1	3	4	5	6	8	9	10	11	13	14	15	16	18	19	20	21	23	24	25	40
1 50	1	3	4	6	7	8	10	11	12	14	15	17	18	19	21	22	23	25	26	28	1 50
2 00	2	3	5	6	8	9	11	12	14	15	17	18	20	21	23	24	26	27	29	30	2 00
10	2	3	5	7	8	10	11	13	15	16	18	20	21	23	24	26	28	29	31	33	10
20	2	4	5	7	9	11	12	14	16	18	19	21	23	25	26	28	30	32	33	35	20
30	2	4	6	8	9	11	13	15	17	19	21	23	24	26	28	30	32	34	36	38	30
40	2	4	6	8	10	12	14	16	18	20	22	24	26	28	30	32	34	36	38	40	40
2 50	2	4	6	9	11	13	15	17	19	21	23	26	28	30	32	34	36	38	40	43	2 50
3 00	2	5	7	9	11	14	16	18	20	23	25	27	29	32	34	36	38	41	43	45	3 00
10	2	5	7	10	12	14	17	19	21	24	26	29	31	33	36	38	40	43	45	48	10
20	3	5	8	10	13	15	18	20	23	25	28	30	33	35	38	40	43	45	48	50	20
30	3	5	8	11	13	16	18	21	24	26	29	32	34	37	39	42	45	47	50	53	30
40	3	6	8	11	14	17	19	22	25	28	30	33	36	39	41	44	47	50	52	55	40
3 50	3	6	9	12	14	17	20	23	26	29	32	35	37	40	43	46	49	52	55	58	3 50
4 00	3	6	9	12	15	18	21	24	27	30	33	36	39	42	45	48	51	54	57	60	4 00

Time of fix (tab 1) or computation (tab 2)	Sign from 4-min. Table	To observed altitude	To tabulated altitude	To intercept
Later than observation	+	Add	Subtract	Toward
	−	Subtract	Add	Away
Earlier than observation	+	Subtract	Add	Away
	−	Add	Subtract	Toward

ALTERNATIVE TABLE 1—Altitude Correction for Change in Position of Observer

CORRECTION FOR 1 MINUTE OF TIME

Rel. Zn	GROUND SPEED IN KNOTS																		Rel. Zn
	50	100	150	200	250	300	350	400	450	500	550	600	650	700	750	800	850	900	
°	′	′	′	′	′	′	′	′	′	′	′	′	′	′	′	′	′	′	°
000	+ 0.8	+ 1.7	+ 2.5	+ 3.3	+ 4.2	+ 5.0	+ 5.8	+ 6.7	+ 7.5	+ 8.3	+ 9.2	+ 10.0	+ 10.8	+ 11.7	+ 12.5	+ 13.3	+ 14.2	+ 15.0	000
002	0.8	1.7	2.5	3.3	4.2	5.0	5.8	6.7	7.5	8.3	9.2	10.0	10.8	11.7	12.5	13.3	14.2	15.0	358
004	0.8	1.7	2.5	3.3	4.2	5.0	5.8	6.7	7.5	8.3	9.1	10.0	10.8	11.6	12.5	13.3	14.1	15.0	356
006	0.8	1.7	2.5	3.3	4.1	5.0	5.8	6.6	7.5	8.3	9.1	9.9	10.8	11.6	12.4	13.3	14.1	14.9	354
008	0.8	1.7	2.5	3.3	4.1	5.0	5.8	6.6	7.4	8.3	9.1	9.9	10.7	11.6	12.4	13.2	14.0	14.9	352
010	+ 0.8	+ 1.6	+ 2.5	+ 3.3	+ 4.1	+ 4.9	+ 5.7	+ 6.6	+ 7.4	+ 8.2	+ 9.0	+ 9.8	+ 10.7	+ 11.5	+ 12.3	+ 13.1	+ 14.0	+ 14.8	350
012	0.8	1.6	2.4	3.3	4.1	4.9	5.7	6.5	7.3	8.2	9.0	9.8	10.6	11.4	12.2	13.0	13.9	14.7	348
014	0.8	1.6	2.4	3.2	4.0	4.9	5.7	6.5	7.3	8.1	8.9	9.7	10.5	11.3	12.1	12.9	13.7	14.6	346
016	0.8	1.6	2.4	3.2	4.0	4.8	5.6	6.4	7.2	8.0	8.8	9.6	10.4	11.2	12.0	12.8	13.6	14.4	344
018	0.8	1.6	2.4	3.2	4.0	4.8	5.5	6.3	7.1	7.9	8.7	9.5	10.3	11.1	11.9	12.7	13.5	14.3	342
020	+ 0.8	+ 1.6	+ 2.3	+ 3.1	+ 3.9	+ 4.7	+ 5.5	+ 6.3	+ 7.0	+ 7.8	+ 8.6	+ 9.4	+ 10.2	+ 11.0	+ 11.7	+ 12.5	+ 13.3	+ 14.1	340
022	0.8	1.5	2.3	3.1	3.9	4.6	5.4	6.2	7.0	7.7	8.5	9.3	10.0	10.8	11.6	12.4	13.1	13.9	338
024	0.8	1.5	2.3	3.0	3.8	4.6	5.3	6.1	6.9	7.6	8.4	9.1	9.9	10.7	11.4	12.2	12.9	13.7	336
026	0.7	1.5	2.2	3.0	3.7	4.5	5.2	6.0	6.7	7.5	8.2	9.0	9.7	10.5	11.2	12.0	12.7	13.5	334
028	0.7	1.5	2.2	2.9	3.7	4.4	5.2	5.9	6.6	7.4	8.1	8.8	9.6	10.3	11.0	11.8	12.5	13.2	332
030	+ 0.7	+ 1.4	+ 2.2	+ 2.9	+ 3.6	+ 4.3	+ 5.1	+ 5.8	+ 6.5	+ 7.2	+ 7.9	+ 8.7	+ 9.4	+ 10.1	+ 10.8	+ 11.5	+ 12.3	+ 13.0	330
032	0.7	1.4	2.1	2.8	3.5	4.2	4.9	5.7	6.4	7.1	7.8	8.5	9.2	9.9	10.6	11.3	12.0	12.7	328
034	0.7	1.4	2.1	2.8	3.5	4.1	4.8	5.5	6.2	6.9	7.6	8.3	9.0	9.7	10.4	11.1	11.7	12.4	326
036	0.7	1.3	2.0	2.7	3.4	4.0	4.7	5.4	6.1	6.7	7.4	8.1	8.8	9.4	10.1	10.8	11.5	12.1	324
038	0.7	1.3	2.0	2.6	3.3	3.9	4.6	5.3	5.9	6.6	7.2	7.9	8.5	9.2	9.9	10.5	11.2	11.8	322
040	+ 0.6	+ 1.3	+ 1.9	+ 2.6	+ 3.2	+ 3.8	+ 4.5	+ 5.1	+ 5.7	+ 6.4	+ 7.0	+ 7.7	+ 8.3	+ 8.9	+ 9.6	+ 10.2	+ 10.9	+ 11.5	320
042	0.6	1.2	1.9	2.5	3.1	3.7	4.3	5.0	5.6	6.2	6.8	7.4	8.1	8.7	9.3	9.9	10.5	11.1	318
044	0.6	1.2	1.8	2.4	3.0	3.6	4.2	4.8	5.4	6.0	6.6	7.2	7.8	8.4	9.0	9.6	10.2	10.8	316
046	0.6	1.2	1.7	2.3	2.9	3.5	4.1	4.6	5.2	5.8	6.4	6.9	7.5	8.1	8.7	9.3	9.8	10.4	314
048	0.6	1.1	1.7	2.2	2.8	3.3	3.9	4.5	5.0	5.6	6.1	6.7	7.2	7.8	8.4	8.9	9.5	10.0	312
050	+ 0.5	+ 1.1	+ 1.6	+ 2.1	+ 2.7	+ 3.2	+ 3.7	+ 4.3	+ 4.8	+ 5.4	+ 5.9	+ 6.4	+ 7.0	+ 7.5	+ 8.0	+ 8.6	+ 9.1	+ 9.6	310
052	0.5	1.0	1.5	2.1	2.6	3.1	3.6	4.1	4.6	5.1	5.6	6.2	6.7	7.2	7.7	8.2	8.7	9.2	308
054	0.5	1.0	1.5	2.0	2.4	2.9	3.4	3.9	4.4	4.9	5.4	5.9	6.4	6.9	7.3	7.8	8.3	8.8	306
056	0.5	0.9	1.4	1.9	2.3	2.8	3.3	3.7	4.2	4.7	5.1	5.6	6.1	6.5	7.0	7.5	7.9	8.4	304
058	0.4	0.9	1.3	1.8	2.2	2.6	3.1	3.5	4.0	4.4	4.9	5.3	5.7	6.2	6.6	7.1	7.5	7.9	302
060	+ 0.4	+ 0.8	+ 1.3	+ 1.7	+ 2.1	+ 2.5	+ 2.9	+ 3.3	+ 3.8	+ 4.2	+ 4.6	+ 5.0	+ 5.4	+ 5.8	+ 6.3	+ 6.7	+ 7.1	+ 7.5	300
062	0.4	0.8	1.2	1.6	2.0	2.3	2.7	3.1	3.5	3.9	4.3	4.7	5.1	5.5	5.9	6.3	6.7	7.0	298
064	0.4	0.7	1.1	1.5	1.8	2.2	2.6	2.9	3.3	3.7	4.0	4.4	4.7	5.1	5.5	5.8	6.2	6.6	296
066	0.3	0.7	1.0	1.4	1.7	2.0	2.4	2.7	3.1	3.4	3.7	4.1	4.4	4.7	5.1	5.4	5.8	6.1	294
068	0.3	0.6	0.9	1.2	1.6	1.9	2.2	2.5	2.8	3.1	3.4	3.7	4.1	4.4	4.7	5.0	5.3	5.6	292
070	+ 0.3	+ 0.6	+ 0.9	+ 1.1	+ 1.4	+ 1.7	+ 2.0	+ 2.3	+ 2.6	+ 2.9	+ 3.1	+ 3.4	+ 3.7	+ 4.0	+ 4.3	+ 4.6	+ 4.8	+ 5.1	290
072	0.3	0.5	0.8	1.0	1.3	1.5	1.8	2.1	2.3	2.6	2.8	3.1	3.3	3.6	3.9	4.1	4.4	4.6	288
074	0.2	0.5	0.7	0.9	1.1	1.4	1.6	1.8	2.1	2.3	2.5	2.8	3.0	3.2	3.4	3.7	3.9	4.1	286
076	0.2	0.4	0.6	0.8	1.0	1.2	1.4	1.6	1.8	2.0	2.2	2.4	2.6	2.8	3.0	3.2	3.4	3.6	284
078	0.2	0.3	0.5	0.7	0.9	1.0	1.2	1.4	1.6	1.7	1.9	2.1	2.3	2.4	2.6	2.8	2.9	3.1	282
080	+ 0.1	+ 0.3	+ 0.4	+ 0.6	+ 0.7	+ 0.9	+ 1.0	+ 1.2	+ 1.3	+ 1.4	+ 1.6	+ 1.7	+ 1.9	+ 2.0	+ 2.2	+ 2.3	+ 2.5	+ 2.6	280
082	0.1	0.2	0.3	0.5	0.6	0.7	0.8	0.9	1.0	1.2	1.3	1.4	1.5	1.6	1.7	1.9	2.0	2.1	278
084	0.1	0.2	0.3	0.3	0.4	0.5	0.6	0.7	0.8	0.9	1.0	1.0	1.1	1.2	1.3	1.4	1.5	1.6	276
086	0.1	0.1	0.2	0.2	0.3	0.3	0.4	0.5	0.5	0.6	0.6	0.7	0.8	0.8	0.9	0.9	1.0	1.0	274
088	0.0	0.1	0.1	0.1	0.1	0.2	0.2	0.2	0.3	0.3	0.3	0.3	0.4	0.4	0.4	0.5	0.5	0.5	272
090	0.0	0.0	0.0	0.0	0.0	0.0	0.0	0.0	0.0	0.0	0.0	0.0	0.0	0.0	0.0	0.0	0.0	0.0	270
092	− 0.0	− 0.1	− 0.1	− 0.1	− 0.1	− 0.2	− 0.2	− 0.2	− 0.3	− 0.3	− 0.3	− 0.3	− 0.4	− 0.4	− 0.4	− 0.5	− 0.5	− 0.5	268
094	0.1	0.1	0.2	0.2	0.3	0.3	0.4	0.5	0.5	0.6	0.6	0.7	0.8	0.8	0.9	0.9	1.0	1.0	266
096	0.1	0.2	0.3	0.3	0.4	0.5	0.6	0.7	0.8	0.9	1.0	1.0	1.1	1.2	1.3	1.4	1.5	1.6	264
098	0.1	0.2	0.3	0.5	0.6	0.7	0.8	0.9	1.0	1.2	1.3	1.4	1.5	1.6	1.7	1.9	2.0	2.1	262
100	0.1	0.3	0.4	0.6	0.7	0.9	1.0	1.2	1.3	1.4	1.6	1.7	1.9	2.0	2.2	2.3	2.5	2.6	260
102	− 0.2	− 0.3	− 0.5	− 0.7	− 0.9	− 1.0	− 1.2	− 1.4	− 1.6	− 1.7	− 1.9	− 2.1	− 2.3	− 2.4	− 2.6	− 2.8	− 2.9	− 3.1	258
104	0.2	0.4	0.6	0.8	1.0	1.2	1.4	1.6	1.8	2.0	2.2	2.4	2.6	2.8	3.0	3.2	3.4	3.6	256
106	0.2	0.5	0.7	0.9	1.1	1.4	1.6	1.8	2.1	2.3	2.5	2.8	3.0	3.2	3.4	3.7	3.9	4.1	254
108	0.3	0.5	0.8	1.0	1.3	1.5	1.8	2.1	2.3	2.6	2.8	3.1	3.3	3.6	3.9	4.1	4.4	4.6	252
110	0.3	0.6	0.9	1.1	1.4	1.7	2.0	2.3	2.6	2.9	3.1	3.4	3.7	4.0	4.3	4.6	4.8	5.1	250
112	− 0.3	− 0.6	− 0.9	− 1.2	− 1.6	− 1.9	− 2.2	− 2.5	− 2.8	− 3.1	− 3.4	− 3.7	− 4.1	− 4.4	− 4.7	− 5.0	− 5.3	− 5.6	248
114	0.3	0.7	1.0	1.4	1.7	2.0	2.4	2.7	3.1	3.4	3.7	4.1	4.4	4.7	5.1	5.4	5.8	6.1	246
116	0.4	0.7	1.1	1.5	1.8	2.2	2.6	2.9	3.3	3.7	4.0	4.4	4.7	5.1	5.5	5.8	6.2	6.6	244
118	0.4	0.8	1.2	1.6	2.0	2.3	2.7	3.1	3.5	3.9	4.3	4.7	5.1	5.5	5.9	6.3	6.7	7.0	242
120	0.4	0.8	1.3	1.7	2.1	2.5	2.9	3.3	3.8	4.2	4.6	5.0	5.4	5.8	6.3	6.7	7.1	7.5	240
122	− 0.4	− 0.9	− 1.3	− 1.8	− 2.2	− 2.6	− 3.1	− 3.5	− 4.0	− 4.4	− 4.9	− 5.3	− 5.7	− 6.2	− 6.6	− 7.1	− 7.5	− 7.9	238
124	0.5	0.9	1.4	1.9	2.3	2.8	3.3	3.7	4.2	4.7	5.1	5.6	6.1	6.5	7.0	7.5	7.9	8.4	236
126	0.5	1.0	1.5	2.0	2.4	2.9	3.4	3.9	4.4	4.9	5.4	5.9	6.4	6.9	7.3	7.8	8.3	8.8	234
128	0.5	1.0	1.5	2.1	2.6	3.1	3.6	4.1	4.6	5.1	5.6	6.2	6.7	7.2	7.7	8.2	8.7	9.2	232
130	0.5	1.1	1.6	2.1	2.7	3.2	3.7	4.3	4.8	5.4	5.9	6.4	7.0	7.5	8.0	8.6	9.1	9.6	230
132	− 0.6	− 1.1	− 1.7	− 2.2	− 2.8	− 3.3	− 3.9	− 4.5	− 5.0	− 5.6	− 6.1	− 6.7	− 7.2	− 7.8	− 8.4	− 8.9	− 9.5	− 10.0	228
134	0.6	1.2	1.7	2.3	2.9	3.5	4.1	4.6	5.2	5.8	6.4	6.9	7.5	8.1	8.7	9.3	9.8	10.4	226
136	0.6	1.2	1.8	2.4	3.0	3.6	4.2	4.8	5.4	6.0	6.6	7.2	7.8	8.4	9.0	9.6	10.2	10.8	224
138	0.6	1.2	1.9	2.5	3.1	3.7	4.3	5.0	5.6	6.2	6.8	7.4	8.1	8.7	9.3	9.9	10.5	11.1	222
140	0.6	1.3	1.9	2.6	3.2	3.8	4.5	5.1	5.7	6.4	7.0	7.7	8.3	8.9	9.6	10.2	10.9	11.5	220
142	− 0.7	− 1.3	− 2.0	− 2.6	− 3.3	− 3.9	− 4.6	− 5.3	− 5.9	− 6.6	− 7.2	− 7.9	− 8.5	− 9.2	− 9.9	− 10.5	− 11.2	− 11.8	218
144	0.7	1.3	2.0	2.7	3.4	4.0	4.7	5.4	6.1	6.7	7.4	8.1	8.8	9.4	10.1	10.8	11.5	12.1	216
146	0.7	1.4	2.1	2.8	3.5	4.1	4.8	5.5	6.2	6.9	7.6	8.3	9.0	9.7	10.4	11.1	11.7	12.4	214
148	0.7	1.4	2.1	2.8	3.5	4.2	4.9	5.7	6.4	7.1	7.8	8.5	9.2	9.9	10.6	11.3	12.0	12.7	212
150	0.7	1.4	2.2	2.9	3.6	4.3	5.1	5.8	6.5	7.2	7.9	8.7	9.4	10.1	10.8	11.5	12.3	13.0	210
152	− 0.7	− 1.5	− 2.2	− 2.9	− 3.7	− 4.4	− 5.2	− 5.9	− 6.6	− 7.4	− 8.1	− 8.8	− 9.6	− 10.3	− 11.0	− 11.8	− 12.5	− 13.2	208
154	0.7	1.5	2.2	3.0	3.7	4.5	5.2	6.0	6.7	7.5	8.2	9.0	9.7	10.5	11.2	12.0	12.7	13.5	206
156	0.8	1.5	2.3	3.0	3.8	4.6	5.3	6.1	6.9	7.6	8.4	9.1	9.9	10.7	11.4	12.2	12.9	13.7	204
158	0.8	1.5	2.3	3.1	3.9	4.6	5.4	6.2	7.0	7.7	8.5	9.3	10.0	10.8	11.6	12.4	13.1	13.9	202
160	0.8	1.6	2.3	3.1	3.9	4.7	5.5	6.3	7.0	7.8	8.6	9.4	10.2	11.0	11.7	12.5	13.3	14.1	200
162	− 0.8	− 1.6	− 2.4	− 3.2	− 4.0	− 4.8	− 5.5	− 6.3	− 7.1	− 7.9	− 8.7	− 9.5	− 10.3	− 11.1	− 11.9	− 12.7	− 13.5	− 14.3	198
164	0.8	1.6	2.4	3.2	4.0	4.8	5.6	6.4	7.2	8.0	8.8	9.6	10.4	11.2	12.0	12.8	13.6	14.4	196
166	0.8	1.6	2.4	3.2	4.0	4.9	5.7	6.5	7.3	8.1	8.9	9.7	10.5	11.3	12.1	12.9	13.7	14.6	194
168	0.8	1.6	2.4	3.3	4.1	4.9	5.7	6.5	7.3	8.2	9.0	9.8	10.6	11.4	12.2	13.0	13.9	14.7	192
170	0.8	1.6	2.5	3.3	4.1	4.9	5.7	6.6	7.4	8.2	9.0	9.8	10.7	11.5	12.3	13.1	14.0	14.8	190
172	− 0.8	− 1.7	− 2.5	− 3.3	− 4.1	− 5.0	− 5.8	− 6.6	− 7.4	− 8.3	− 9.1	− 9.9	− 10.7	− 11.6	− 12.4	− 13.2	− 14.0	− 14.9	188
174	0.8	1.7	2.5	3.3	4.1	5.0	5.8	6.6	7.5	8.3	9.1	9.9	10.8	11.6	12.4	13.3	14.1	14.9	186
176	0.8	1.7	2.5	3.3	4.2	5.0	5.8	6.7	7.5	8.3	9.1	10.0	10.8	11.6	12.5	13.3	14.1	15.0	184
178	0.8	1.7	2.5	3.3	4.2	5.0	5.8	6.7	7.5	8.3	9.2	10.0	10.8	11.7	12.5	13.3	14.2	15.0	182
180	0.8	1.7	2.5	3.3	4.2	5.0	5.8	6.7	7.5	8.3	9.2	10.0	10.8	11.7	12.5	13.3	14.2	15.0	180

Time of fix or computation	Sign from 1-min Table	To observed altitude	To tabulated altitude	To intercept
Later than observation	+	Add	Subtract	Toward
	−	Subtract	Add	Away
Earlier than observation	+	Subtract	Add	Away
	−	Add	Subtract	Toward

ALTERNATIVE TABLE 2—Altitude Correction for Change in Position of Body

CORRECTION FOR 1 MINUTE OF TIME

TRUE Zn	LATITUDE IN DEGREES 0	5	10	15	20	25	30	35	40	45	50	55	60	65	70	75	80	85	TRUE Zn
°	′	′	′	′	′	′	′	′	′	′	′	′	′	′	′	′	′	′	°
090	+15.0	+14.9	+14.8	+14.5	+14.1	+13.6	+13.0	+12.3	+11.5	+10.6	+ 9.6	+ 8.6	+ 7.5	+ 6.3	+ 5.1	+ 3.9	+ 2.6	+ 1.3	090
092	15.0	14.9	14.8	14.5	14.1	13.6	13.0	12.3	11.5	10.6	9.6	8.6	7.5	6.3	5.1	3.9	2.6	1.3	088
094	15.0	14.9	14.7	14.5	14.1	13.6	13.0	12.3	11.5	10.6	9.6	8.6	7.5	6.3	5.1	3.9	2.6	1.3	086
096	14.9	14.9	14.7	14.4	14.0	13.5	12.9	12.2	11.4	10.5	9.6	8.6	7.5	6.3	5.1	3.9	2.6	1.3	084
098	14.9	14.8	14.6	14.3	14.0	13.5	12.9	12.2	11.4	10.5	9.5	8.5	7.4	6.3	5.1	3.8	2.6	1.3	082
100	+14.8	+14.7	+14.5	+14.3	+13.9	+13.4	+12.8	+12.1	+11.3	+10.4	+ 9.5	+ 8.5	+ 7.4	+ 6.2	+ 5.1	+ 3.8	+ 2.6	+ 1.3	080
102	14.7	14.6	14.4	14.2	13.8	13.3	12.7	12.0	11.2	10.4	9.4	8.4	7.3	6.2	5.0	3.8	2.5	1.3	078
104	14.6	14.5	14.3	14.1	13.7	13.2	12.6	11.9	11.1	10.3	9.4	8.3	7.3	6.2	5.0	3.8	2.5	1.3	076
106	14.4	14.4	14.2	13.9	13.5	13.1	12.5	11.8	11.0	10.2	9.3	8.3	7.2	6.1	4.9	3.7	2.5	1.3	074
108	14.3	14.2	14.0	13.8	13.4	12.9	12.4	11.7	10.9	10.1	9.2	8.2	7.1	6.0	4.9	3.7	2.5	1.2	072
110	+14.1	+14.0	+13.9	+13.6	+13.2	+12.8	+12.2	+11.5	+10.8	+10.0	+ 9.1	+ 8.1	+ 7.0	+ 6.0	+ 4.8	+ 3.6	+ 2.4	+ 1.2	070
112	13.9	13.9	13.7	13.4	13.1	12.6	12.0	11.4	10.7	9.8	8.9	8.0	7.0	5.9	4.8	3.6	2.4	1.2	068
114	13.7	13.7	13.5	13.2	12.9	12.4	11.9	11.2	10.5	9.7	8.8	7.9	6.9	5.8	4.7	3.5	2.4	1.2	066
116	13.5	13.4	13.3	13.0	12.7	12.2	11.7	11.0	10.3	9.5	8.7	7.7	6.7	5.7	4.6	3.5	2.3	1.2	064
118	13.2	13.2	13.0	12.8	12.4	12.0	11.5	10.8	10.1	9.4	8.5	7.6	6.6	5.6	4.5	3.4	2.3	1.2	062
120	+13.0	+12.9	+12.8	+12.5	+12.2	+11.8	+11.3	+10.6	+10.0	+ 9.2	+ 8.4	+ 7.5	+ 6.5	+ 5.5	+ 4.4	+ 3.4	+ 2.3	+ 1.1	060
122	12.7	12.7	12.5	12.3	12.0	11.5	11.0	10.4	9.7	9.0	8.2	7.3	6.4	5.4	4.4	3.3	2.2	1.1	058
124	12.4	12.4	12.2	12.0	11.7	11.3	10.8	10.2	9.5	8.8	8.0	7.1	6.2	5.3	4.3	3.2	2.2	1.1	056
126	12.1	12.1	12.0	11.7	11.4	11.0	10.5	9.9	9.3	8.6	7.8	7.0	6.1	5.1	4.2	3.1	2.1	1.1	054
128	11.8	11.8	11.6	11.4	11.1	10.7	10.2	9.7	9.1	8.4	7.6	6.8	5.9	5.0	4.0	3.1	2.1	1.0	052
130	+11.5	+11.4	+11.3	+11.1	+10.8	+10.4	+10.0	+ 9.4	+ 8.8	+ 8.1	+ 7.4	+ 6.6	+ 5.7	+ 4.9	+ 3.9	+ 3.0	+ 2.0	+ 1.0	050
132	11.1	11.1	11.0	10.8	10.5	10.1	9.7	9.1	8.5	7.9	7.2	6.4	5.6	4.7	3.8	2.9	1.9	1.0	048
134	10.8	10.8	10.6	10.4	10.1	9.8	9.3	8.8	8.3	7.6	6.9	6.2	5.4	4.6	3.7	2.8	1.9	0.9	046
136	10.4	10.4	10.3	10.1	9.8	9.4	9.0	8.5	8.0	7.4	6.7	6.0	5.2	4.4	3.6	2.7	1.8	0.9	044
138	10.0	10.0	9.9	9.7	9.4	9.1	8.7	8.2	7.7	7.1	6.5	5.8	5.0	4.2	3.4	2.6	1.7	0.9	042
140	+ 9.6	+ 9.6	+ 9.5	+ 9.3	+ 9.1	+ 8.7	+ 8.4	+ 7.9	+ 7.4	+ 6.8	+ 6.2	+ 5.5	+ 4.8	+ 4.1	+ 3.3	+ 2.5	+ 1.7	+ 0.8	040
142	9.2	9.2	9.1	8.9	8.7	8.4	8.0	7.6	7.1	6.5	5.9	5.3	4.6	3.9	3.2	2.4	1.6	0.8	038
144	8.8	8.8	8.7	8.5	8.3	8.0	7.6	7.2	6.8	6.2	5.7	5.1	4.4	3.7	3.0	2.3	1.5	0.8	036
146	8.4	8.4	8.3	8.1	7.9	7.6	7.3	6.9	6.4	5.9	5.4	4.8	4.2	3.5	2.9	2.2	1.5	0.7	034
148	7.9	7.9	7.8	7.7	7.5	7.2	6.9	6.5	6.1	5.6	5.1	4.6	4.0	3.4	2.7	2.1	1.4	0.7	032
150	+ 7.5	+ 7.5	+ 7.4	+ 7.2	+ 7.0	+ 6.8	+ 6.5	+ 6.1	+ 5.7	+ 5.3	+ 4.8	+ 4.3	+ 3.8	+ 3.2	+ 2.6	+ 1.9	+ 1.3	+ 0.7	030
152	7.0	7.0	6.9	6.8	6.6	6.4	6.1	5.8	5.4	5.0	4.5	4.0	3.5	3.0	2.4	1.8	1.2	0.6	028
154	6.6	6.6	6.5	6.4	6.2	6.0	5.7	5.4	5.0	4.6	4.2	3.8	3.3	2.8	2.2	1.7	1.1	0.6	026
156	6.1	6.1	6.0	5.9	5.7	5.5	5.3	5.0	4.7	4.3	3.9	3.5	3.1	2.6	2.1	1.6	1.1	0.5	024
158	5.6	5.6	5.5	5.4	5.3	5.1	4.9	4.6	4.3	4.0	3.6	3.2	2.8	2.4	1.9	1.5	1.0	0.5	022
160	+ 5.1	+ 5.1	+ 5.1	+ 5.0	+ 4.8	+ 4.6	+ 4.4	+ 4.2	+ 3.9	+ 3.6	+ 3.3	+ 2.9	+ 2.6	+ 2.2	+ 1.8	+ 1.3	+ 0.9	+ 0.4	020
162	4.6	4.6	4.6	4.5	4.4	4.2	4.0	3.8	3.6	3.3	3.0	2.7	2.3	2.0	1.6	1.2	0.8	0.4	018
164	4.1	4.1	4.1	4.0	3.9	3.7	3.6	3.4	3.2	2.9	2.7	2.4	2.1	1.7	1.4	1.1	0.7	0.4	016
166	3.6	3.6	3.6	3.5	3.4	3.3	3.1	3.0	2.8	2.6	2.3	2.1	1.8	1.5	1.2	0.9	0.6	0.3	014
168	3.1	3.1	3.1	3.0	2.9	2.8	2.7	2.6	2.4	2.2	2.0	1.8	1.6	1.3	1.1	0.8	0.5	0.3	012
170	+ 2.6	+ 2.6	+ 2.6	+ 2.5	+ 2.4	+ 2.4	+ 2.3	+ 2.1	+ 2.0	+ 1.8	+ 1.7	+ 1.5	+ 1.3	+ 1.1	+ 0.9	+ 0.7	+ 0.5	+ 0.2	010
172	2.1	2.1	2.1	2.0	2.0	1.9	1.8	1.7	1.6	1.5	1.3	1.2	1.0	0.9	0.7	0.5	0.4	0.2	008
174	1.6	1.6	1.5	1.5	1.5	1.4	1.4	1.3	1.2	1.1	1.0	0.9	0.8	0.7	0.5	0.4	0.3	0.1	006
176	1.0	1.0	1.0	1.0	1.0	1.0	0.9	0.9	0.8	0.7	0.7	0.6	0.5	0.4	0.4	0.3	0.2	0.1	004
178	0.5	0.5	0.5	0.5	0.5	0.5	0.5	0.4	0.4	0.4	0.3	0.3	0.3	0.2	0.2	0.1	0.1	0.0	002
180	0.0	0.0	0.0	0.0	0.0	0.0	0.0	0.0	0.0	0.0	0.0	0.0	0.0	0.0	0.0	0.0	0.0	0.0	000
182	− 0.5	− 0.5	− 0.5	− 0.5	− 0.5	− 0.5	− 0.5	− 0.4	− 0.4	− 0.4	− 0.3	− 0.3	− 0.3	− 0.2	− 0.2	− 0.1	− 0.1	− 0.0	358
184	1.0	1.0	1.0	1.0	1.0	1.0	0.9	0.9	0.8	0.7	0.7	0.6	0.5	0.4	0.4	0.3	0.2	0.1	356
186	1.6	1.6	1.5	1.5	1.5	1.4	1.4	1.3	1.2	1.1	1.0	0.9	0.8	0.7	0.5	0.4	0.3	0.1	354
188	2.1	2.1	2.1	2.0	2.0	1.9	1.8	1.7	1.6	1.5	1.3	1.2	1.0	0.9	0.7	0.5	0.4	0.2	352
190	2.6	2.6	2.6	2.5	2.4	2.4	2.3	2.1	2.0	1.8	1.7	1.5	1.3	1.1	0.9	0.7	0.5	0.2	350
192	− 3.1	− 3.1	− 3.1	− 3.0	− 2.9	− 2.8	− 2.7	− 2.6	− 2.4	− 2.2	− 2.0	− 1.8	− 1.6	− 1.3	− 1.1	− 0.8	− 0.5	− 0.3	348
194	3.6	3.6	3.6	3.5	3.4	3.3	3.1	3.0	2.8	2.6	2.3	2.1	1.8	1.5	1.2	0.9	0.6	0.3	346
196	4.1	4.1	4.1	4.0	3.9	3.7	3.6	3.4	3.2	2.9	2.7	2.4	2.1	1.7	1.4	1.1	0.7	0.4	344
198	4.6	4.6	4.6	4.5	4.4	4.2	4.0	3.8	3.6	3.3	3.0	2.7	2.3	2.0	1.6	1.2	0.8	0.4	342
200	5.1	5.1	5.1	5.0	4.8	4.6	4.4	4.2	3.9	3.6	3.3	2.9	2.6	2.2	1.8	1.3	0.9	0.4	340
202	− 5.6	− 5.6	− 5.5	− 5.4	− 5.3	− 5.1	− 4.9	− 4.6	− 4.3	− 4.0	− 3.6	− 3.2	− 2.8	− 2.4	− 1.9	− 1.5	− 1.0	− 0.5	338
204	6.1	6.1	6.0	5.9	5.7	5.5	5.3	5.0	4.7	4.3	3.9	3.5	3.1	2.6	2.1	1.6	1.1	0.5	336
206	6.6	6.6	6.5	6.4	6.2	6.0	5.7	5.4	5.0	4.6	4.2	3.8	3.3	2.8	2.2	1.7	1.1	0.6	334
208	7.0	7.0	6.9	6.8	6.6	6.4	6.1	5.8	5.4	5.0	4.5	4.0	3.5	3.0	2.4	1.8	1.2	0.6	332
210	7.5	7.5	7.4	7.2	7.0	6.8	6.5	6.1	5.7	5.3	4.8	4.3	3.8	3.2	2.6	1.9	1.3	0.7	330
212	− 7.9	− 7.9	− 7.8	− 7.7	− 7.5	− 7.2	− 6.9	− 6.5	− 6.1	− 5.6	− 5.1	− 4.6	− 4.0	− 3.4	− 2.7	− 2.1	− 1.4	− 0.7	328
214	8.4	8.4	8.3	8.1	7.9	7.6	7.3	6.9	6.4	5.9	5.4	4.8	4.2	3.5	2.9	2.2	1.5	0.7	326
216	8.8	8.8	8.7	8.5	8.3	8.0	7.6	7.2	6.8	6.2	5.7	5.1	4.4	3.7	3.0	2.3	1.5	0.8	324
218	9.2	9.2	9.1	8.9	8.7	8.4	8.0	7.6	7.1	6.5	5.9	5.3	4.6	3.9	3.2	2.4	1.6	0.8	322
220	9.6	9.6	9.5	9.3	9.1	8.7	8.4	7.9	7.4	6.8	6.2	5.5	4.8	4.1	3.3	2.5	1.7	0.8	320
222	−10.0	−10.0	− 9.9	− 9.7	− 9.4	− 9.1	− 8.7	− 8.2	− 7.7	− 7.1	− 6.5	− 5.8	− 5.0	− 4.2	− 3.4	− 2.6	− 1.7	− 0.9	318
224	10.4	10.4	10.3	10.1	9.8	9.4	9.0	8.5	8.0	7.4	6.7	6.0	5.2	4.4	3.6	2.7	1.8	0.9	316
226	10.8	10.8	10.6	10.4	10.1	9.8	9.3	8.8	8.3	7.6	6.9	6.2	5.4	4.6	3.7	2.8	1.9	0.9	314
228	11.1	11.1	11.0	10.8	10.5	10.1	9.7	9.1	8.5	7.9	7.2	6.4	5.6	4.7	3.8	2.9	1.9	1.0	312
230	11.5	11.4	11.3	11.1	10.8	10.4	10.0	9.4	8.8	8.1	7.4	6.6	5.7	4.9	3.9	3.0	2.0	1.0	310
232	−11.8	−11.8	−11.6	−11.4	−11.1	−10.7	−10.2	− 9.7	− 9.1	− 8.4	− 7.6	− 6.8	− 5.9	− 5.0	− 4.0	− 3.1	− 2.1	− 1.0	308
234	12.1	12.1	12.0	11.7	11.4	11.0	10.5	9.9	9.3	8.6	7.8	7.0	6.1	5.1	4.2	3.1	2.1	1.1	306
236	12.4	12.4	12.2	12.0	11.7	11.3	10.8	10.2	9.5	8.8	8.0	7.1	6.2	5.3	4.3	3.2	2.2	1.1	304
238	12.7	12.7	12.5	12.3	12.0	11.5	11.0	10.4	9.7	9.0	8.2	7.3	6.4	5.4	4.4	3.3	2.2	1.1	302
240	13.0	12.9	12.8	12.5	12.2	11.8	11.3	10.6	10.0	9.2	8.4	7.5	6.5	5.5	4.4	3.4	2.3	1.1	300
242	−13.2	−13.2	−13.0	−12.8	−12.4	−12.0	−11.5	−10.8	−10.1	− 9.4	− 8.5	− 7.6	− 6.6	− 5.6	− 4.5	− 3.4	− 2.3	− 1.2	298
244	13.5	13.4	13.3	13.0	12.7	12.2	11.7	11.0	10.3	9.5	8.7	7.7	6.7	5.7	4.6	3.5	2.3	1.2	296
246	13.7	13.7	13.5	13.2	12.9	12.4	11.9	11.2	10.5	9.7	8.8	7.9	6.9	5.8	4.7	3.5	2.4	1.2	294
248	13.9	13.9	13.7	13.4	13.1	12.6	12.0	11.4	10.7	9.8	8.9	8.0	7.0	5.9	4.8	3.6	2.4	1.2	292
250	14.1	14.0	13.9	13.6	13.2	12.8	12.2	11.5	10.8	10.0	9.1	8.1	7.0	6.0	4.8	3.6	2.4	1.2	290
252	−14.3	−14.2	−14.0	−13.8	−13.4	−12.9	−12.4	−11.7	−10.9	−10.1	− 9.2	− 8.2	− 7.1	− 6.0	− 4.9	− 3.7	− 2.5	− 1.2	288
254	14.4	14.4	14.2	13.9	13.5	13.1	12.5	11.8	11.0	10.2	9.3	8.3	7.2	6.1	4.9	3.7	2.5	1.3	286
256	14.6	14.5	14.3	14.1	13.7	13.2	12.6	11.9	11.1	10.3	9.4	8.3	7.3	6.2	5.0	3.8	2.5	1.3	284
258	14.7	14.6	14.4	14.2	13.8	13.3	12.7	12.0	11.2	10.4	9.4	8.4	7.3	6.2	5.0	3.8	2.5	1.3	282
260	14.8	14.7	14.5	14.3	13.9	13.4	12.8	12.1	11.3	10.4	9.5	8.5	7.4	6.2	5.1	3.8	2.6	1.3	280
262	−14.9	−14.8	−14.6	−14.3	−14.0	−13.5	−12.9	−12.2	−11.4	−10.5	− 9.5	− 8.5	− 7.4	− 6.3	− 5.1	− 3.8	− 2.6	− 1.3	278
264	14.9	14.9	14.7	14.4	14.0	13.5	12.9	12.2	11.4	10.5	9.6	8.6	7.5	6.3	5.1	3.9	2.6	1.3	276
266	15.0	14.9	14.7	14.5	14.1	13.6	13.0	12.3	11.5	10.6	9.6	8.6	7.5	6.3	5.1	3.9	2.6	1.3	274
268	15.0	14.9	14.8	14.5	14.1	13.6	13.0	12.3	11.5	10.6	9.6	8.6	7.5	6.3	5.1	3.9	2.6	1.3	272
270	15.0	14.9	14.8	14.5	14.1	13.6	13.0	12.3	11.5	10.6	9.6	8.6	7.5	6.3	5.1	3.9	2.6	1.3	270

Time of fix or computation	Sign from 1-min table	To observed altitude	To tabulated altitude	To intercept
Later than observation	+	Add	Subtract	Toward
	−	Subtract	Add	Away
Earlier than observation	+	Subtract	Add	Away
	−	Add	Subtract	Toward

CORRECTION FOR PRECESSION AND NUTATION

LHA ♈	North latitudes							South latitudes						LHA ♈
	N 80°	N 70°	N 60°	N 50°	N 40°	N 20°	0°	S 20°	S 40°	S 50°	S 60°	S 70°	S 80°	
	1994													
°	′ °	′ °	′ °	′ °	′ °	′ °	′ °	′ °	′ °	′ °	′ °	′ °	′ °	°
0	**0·1** 136	**0·1** 152	**0·1** 177	**0·1** 202	**0·1** 221	**0·1** 239	**0·2** 246	**0·2** 249	**0·2** 249	**0·2** 249	**0·2** 247	**0·2** 244	**0·2** 241	**0**
30	**0·1** 169	**0·1** 184	**0·1** 199	**0·1** 211	**0·2** 220	**0·2** 231	**0·2** 236	**0·2** 237	**0·2** 235	**0·2** 232	**0·2** 228	**0·2** 223	**0·1** 215	**30**
60	**0·1** 198	**0·1** 209	**0·2** 217	**0·2** 223	**0·2** 228	**0·2** 232	**0·2** 233	**0·2** 231	**0·2** 224	**0·2** 218	**0·1** 210	**0·1** 200	**0·1** 188	**60**
90	**0·1** 225	**0·2** 231	**0·2** 235	**0·2** 238	**0·2** 240	**0·2** 241	**0·2** 239	**0·2** 233	**0·1** 219	**0·1** 207	**0·1** 191	**0·1** 174	**0·1** 157	**90**
120	**0·2** 250	**0·2** 253	**0·2** 254	**0·2** 255	**0·2** 256	**0·2** 255	**0·2** 253	**0·1** 247	**0·1** 227	**0·1** 202	**0·1** 164	**0·1** 136	**0·1** 122	**120**
150	**0·2** 275	**0·2** 274	**0·2** 274	**0·2** 274	**0·2** 273	**0·2** 274	**0·2** 274	**0·1** 276	**0·1** 284	**0·0** —	**0·0** —	**0·1** 078	**0·1** 082	**150**
180	**0·2** 299	**0·2** 296	**0·2** 293	**0·2** 291	**0·2** 291	**0·2** 291	**0·2** 294	**0·1** 301	**0·1** 319	**0·1** 338	**0·1** 003	**0·1** 028	**0·1** 044	**180**
210	**0·1** 325	**0·2** 317	**0·2** 312	**0·2** 308	**0·2** 305	**0·2** 303	**0·2** 304	**0·2** 309	**0·2** 320	**0·1** 329	**0·1** 341	**0·1** 356	**0·1** 011	**210**
240	**0·1** 352	**0·1** 340	**0·1** 330	**0·2** 322	**0·2** 316	**0·2** 309	**0·2** 307	**0·2** 308	**0·2** 312	**0·2** 317	**0·2** 323	**0·1** 331	**0·1** 342	**240**
270	**0·1** 023	**0·1** 006	**0·1** 349	**0·1** 333	**0·1** 321	**0·2** 307	**0·2** 301	**0·2** 299	**0·2** 300	**0·2** 302	**0·2** 305	**0·2** 309	**0·1** 315	**270**
300	**0·1** 058	**0·1** 044	**0·1** 016	**0·1** 338	**0·1** 313	**0·1** 293	**0·2** 287	**0·2** 285	**0·2** 284	**0·2** 285	**0·2** 286	**0·2** 287	**0·2** 290	**300**
330	**0·1** 098	**0·1** 102	**0·0** —	**0·0** —	**0·1** 256	**0·1** 264	**0·2** 266	**0·2** 266	**0·2** 267	**0·2** 266	**0·2** 266	**0·2** 266	**0·2** 265	**330**
360	**0·1** 136	**0·1** 152	**0·1** 177	**0·1** 202	**0·1** 221	**0·1** 239	**0·2** 246	**0·2** 249	**0·2** 249	**0·2** 249	**0·2** 247	**0·2** 244	**0·2** 241	**360**
	1995													
0	**0·3** 045	**0·4** 053	**0·5** 059	**0·5** 063	**0·6** 065	**0·6** 067	**0·6** 067	**0·5** 063	**0·4** 054	**0·3** 046	**0·3** 032	**0·2** 013	**0·2** 349	**0**
30	**0·4** 068	**0·4** 072	**0·5** 074	**0·5** 076	**0·6** 077	**0·6** 077	**0·6** 076	**0·4** 073	**0·3** 063	**0·2** 051	**0·1** 026	**0·1** 344	**0·2** 313	**30**
60	**0·4** 091	**0·4** 091	**0·5** 090	**0·6** 090	**0·6** 090	**0·6** 090	**0·5** 090	**0·4** 091	**0·2** 091	**0·1** 092	**0·0** —	**0·1** 267	**0·2** 269	**60**
90	**0·4** 113	**0·4** 109	**0·5** 106	**0·5** 105	**0·6** 104	**0·6** 103	**0·6** 105	**0·4** 108	**0·3** 118	**0·2** 130	**0·2** 154	**0·1** 194	**0·2** 224	**90**
120	**0·3** 137	**0·4** 128	**0·4** 122	**0·5** 118	**0·5** 116	**0·6** 113	**0·6** 114	**0·5** 117	**0·4** 126	**0·3** 134	**0·3** 147	**0·2** 166	**0·2** 189	**120**
150	**0·3** 162	**0·3** 146	**0·4** 136	**0·4** 128	**0·5** 124	**0·6** 118	**0·6** 117	**0·6** 118	**0·5** 123	**0·4** 128	**0·4** 135	**0·3** 145	**0·3** 160	**150**
180	**0·2** 191	**0·2** 167	**0·3** 148	**0·3** 134	**0·4** 126	**0·5** 117	**0·6** 113	**0·6** 113	**0·6** 115	**0·5** 117	**0·5** 121	**0·4** 127	**0·3** 135	**180**
210	**0·2** 227	**0·1** 196	**0·1** 154	**0·2** 129	**0·3** 117	**0·4** 107	**0·6** 104	**0·6** 103	**0·6** 103	**0·5** 104	**0·5** 106	**0·4** 108	**0·4** 112	**210**
240	**0·2** 271	**0·1** 273	**0·0** —	**0·1** 088	**0·2** 089	**0·4** 089	**0·5** 090	**0·6** 090	**0·6** 090	**0·6** 090	**0·5** 090	**0·4** 089	**0·4** 089	**240**
270	**0·2** 316	**0·1** 346	**0·2** 026	**0·2** 050	**0·3** 062	**0·4** 072	**0·6** 075	**0·6** 077	**0·6** 076	**0·5** 075	**0·5** 074	**0·4** 071	**0·4** 067	**270**
300	**0·2** 351	**0·2** 014	**0·3** 033	**0·3** 046	**0·4** 054	**0·5** 063	**0·6** 066	**0·6** 067	**0·5** 064	**0·5** 062	**0·4** 058	**0·4** 052	**0·3** 043	**300**
330	**0·3** 020	**0·3** 035	**0·4** 045	**0·4** 052	**0·5** 057	**0·6** 062	**0·6** 063	**0·6** 062	**0·5** 056	**0·4** 052	**0·4** 044	**0·3** 034	**0·3** 018	**330**
360	**0·3** 045	**0·4** 053	**0·5** 059	**0·5** 063	**0·6** 065	**0·6** 067	**0·6** 067	**0·5** 063	**0·4** 054	**0·3** 046	**0·3** 032	**0·2** 013	**0·2** 349	**360**
	1996													
0	**0·6** 035	**0·8** 047	**0·9** 055	**1·1** 060	**1·2** 063	**1·3** 066	**1·3** 067	**1·2** 064	**1·0** 057	**0·8** 051	**0·7** 042	**0·6** 027	**0·5** 006	**0**
30	**0·7** 058	**0·9** 065	**1·0** 068	**1·2** 071	**1·3** 072	**1·3** 074	**1·3** 073	**1·1** 069	**0·8** 061	**0·6** 051	**0·5** 034	**0·4** 006	**0·4** 334	**30**
60	**0·8** 080	**0·9** 082	**1·1** 083	**1·2** 084	**1·3** 084	**1·3** 085	**1·2** 084	**1·0** 083	**0·6** 078	**0·4** 071	**0·2** 049	**0·2** 325	**0·3** 292	**60**
90	**0·8** 102	**0·9** 100	**1·1** 099	**1·2** 098	**1·3** 097	**1·3** 097	**1·2** 098	**1·0** 099	**0·6** 105	**0·4** 113	**0·2** 137	**0·2** 207	**0·3** 243	**90**
120	**0·7** 125	**0·9** 118	**1·0** 113	**1·1** 111	**1·2** 109	**1·3** 108	**1·3** 108	**1·1** 112	**0·8** 120	**0·6** 129	**0·5** 145	**0·4** 171	**0·4** 202	**120**
150	**0·6** 148	**0·8** 135	**0·9** 127	**1·0** 121	**1·2** 118	**1·3** 114	**1·3** 114	**1·2** 116	**1·0** 122	**0·9** 128	**0·7** 137	**0·6** 151	**0·5** 171	**150**
180	**0·5** 174	**0·6** 153	**0·7** 138	**0·8** 129	**1·0** 123	**1·2** 116	**1·3** 113	**1·3** 114	**1·2** 117	**1·1** 120	**0·9** 125	**0·8** 133	**0·6** 145	**180**
210	**0·4** 206	**0·4** 174	**0·5** 146	**0·6** 129	**0·8** 119	**1·1** 111	**1·3** 107	**1·3** 106	**1·3** 108	**1·2** 109	**1·0** 112	**0·9** 115	**0·7** 122	**210**
240	**0·3** 248	**0·2** 215	**0·2** 131	**0·4** 109	**0·6** 102	**1·0** 097	**1·2** 096	**1·3** 095	**1·3** 096	**1·2** 096	**1·1** 097	**0·9** 098	**0·8** 100	**240**
270	**0·3** 297	**0·2** 333	**0·2** 043	**0·4** 067	**0·6** 075	**1·0** 081	**1·2** 082	**1·3** 083	**1·3** 083	**1·2** 082	**1·1** 081	**0·9** 080	**0·8** 078	**270**
300	**0·4** 338	**0·4** 009	**0·5** 035	**0·6** 051	**0·8** 060	**1·1** 068	**1·3** 072	**1·3** 072	**1·2** 071	**1·1** 069	**1·0** 067	**0·9** 062	**0·7** 055	**300**
330	**0·5** 009	**0·6** 029	**0·7** 043	**0·9** 052	**1·0** 058	**1·2** 064	**1·3** 066	**1·3** 066	**1·2** 062	**1·0** 059	**0·9** 053	**0·8** 045	**0·6** 032	**330**
360	**0·6** 035	**0·8** 047	**0·9** 055	**1·1** 060	**1·2** 063	**1·3** 066	**1·3** 067	**1·2** 064	**1·0** 057	**0·8** 051	**0·7** 042	**0·6** 027	**0·5** 006	**360**

The above table gives the correction to be applied to a position line or a fix for the effects of precession and nutation from the mean equinox of 1995·0. Each entry consists of the distance (in bold type) in nautical miles, and the direction (true bearing) in which the position line or fix is to be moved. The table is entered firstly by the year, then by choosing the column nearest the latitude and finally the entry nearest the LHA♈ of observation; no interpolation is necessary, though in extreme cases near the beginning or end of a year (but not the end of 1994 or the beginning of 1995 when the corrections are zero) values midway towards those of the previous or following years may be taken.

Example. Early in 1996 a fix is obtained in latitude S 46° when LHA♈ is 247°. Entering the table with the year 1996, latitude S 50°, and LHA♈ 240° gives **1′·2** 096° as compared with **0′·6** 090° for 1995 which indicates that the fix is to be transferred 0·9 miles in true bearing 093°.

CORRECTION FOR PRECESSION AND NUTATION

LHA ♈	N 80°	N 70°	N 60°	N 50°	N 40°	N 20°	0°	S 20°	S 40°	S 50°	S 60°	S 70°	S 80°	LHA ♈
	North latitudes							South latitudes						
							1997							
°	′ °	′ °	′ °	′ °	′ °	′ °	′ °	′ °	′ °	′ °	′ °	′ °	′ °	°
0	1·0 031	1·2 044	1·4 053	1·6 058	1·8 062	2·0 066	2·1 067	1·9 064	1·6 059	1·4 053	1·2 044	1·0 031	0·8 012	0
30	1·1 054	1·3 062	1·6 066	1·8 069	1·9 071	2·1 072	2·0 072	1·7 068	1·3 060	1·0 052	0·8 037	0·6 011	0·7 341	30
60	1·1 076	1·4 079	1·7 081	1·8 082	2·0 082	2·1 083	1·9 082	1·5 080	1·0 074	0·7 066	0·4 044	0·3 340	0·5 301	60
90	1·2 098	1·4 097	1·7 096	1·9 095	2·0 095	2·1 095	1·9 095	1·5 096	0·9 100	0·6 106	0·3 125	0·2 218	0·5 251	90
120	1·1 120	1·4 114	1·6 110	1·8 108	1·9 107	2·1 105	2·0 106	1·7 109	1·2 118	0·9 127	0·7 144	0·6 174	0·6 208	120
150	1·0 143	1·2 131	1·4 124	1·6 119	1·8 116	2·0 113	2·1 113	1·9 115	1·5 122	1·3 128	1·1 138	0·9 154	0·8 175	150
180	0·8 168	1·0 149	1·2 136	1·4 127	1·6 121	1·9 116	2·1 113	2·0 114	1·8 118	1·6 122	1·4 127	1·2 136	1·0 149	180
210	0·7 199	0·6 169	0·8 143	1·0 128	1·3 120	1·7 112	2·0 108	2·1 108	1·9 109	1·8 111	1·6 114	1·3 118	1·1 126	210
240	0·5 239	0·3 200	0·4 136	0·7 114	1·0 106	1·5 100	1·9 098	2·1 097	2·0 098	1·8 098	1·7 099	1·4 101	1·1 104	240
270	0·5 289	0·2 322	0·3 055	0·6 074	0·9 080	1·5 084	1·9 085	2·1 085	2·0 085	1·9 085	1·7 084	1·4 083	1·2 082	270
300	0·6 332	0·6 006	0·7 036	0·9 053	1·2 062	1·7 071	2·0 074	2·1 075	1·9 073	1·8 072	1·6 070	1·4 066	1·1 060	300
330	0·8 005	0·9 026	1·1 042	1·3 052	1·5 058	1·9 065	2·1 067	2·0 067	1·8 064	1·6 061	1·4 056	1·2 049	1·0 037	330
360	1·0 031	1·2 044	1·4 053	1·6 058	1·8 062	2·0 066	2·1 067	1·9 064	1·6 059	1·4 053	1·2 044	1·0 031	0·8 012	360
							1998							
0	1·3 028	1·5 042	1·8 052	2·1 058	2·4 062	2·7 066	2·8 067	2·6 065	2·2 059	1·9 054	1·6 046	1·3 034	1·2 015	0
30	1·4 051	1·8 060	2·1 065	2·4 068	2·6 070	2·8 071	2·7 071	2·4 068	1·8 060	1·4 052	1·1 038	0·9 015	0·9 346	30
60	1·5 074	1·9 077	2·2 079	2·5 080	2·7 081	2·8 081	2·6 080	2·1 078	1·4 072	1·0 063	0·6 042	0·4 347	0·7 307	60
90	1·6 095	1·9 094	2·3 094	2·5 093	2·7 093	2·8 093	2·6 093	2·0 094	1·3 097	0·8 100	0·4 115	0·2 229	0·7 257	90
120	1·5 117	1·9 112	2·2 108	2·4 106	2·6 105	2·8 104	2·7 105	2·2 108	1·6 116	1·2 125	0·9 143	0·7 177	0·8 212	120
150	1·4 140	1·7 129	2·0 122	2·2 118	2·5 115	2·8 112	2·8 112	2·5 115	2·0 122	1·7 128	1·4 139	1·1 156	1·0 179	150
180	1·2 165	1·3 146	1·6 134	1·9 126	2·2 121	2·6 115	2·8 113	2·7 114	2·4 118	2·1 122	1·8 128	1·5 138	1·3 152	180
210	0·9 194	0·9 165	1·1 142	1·4 128	1·8 120	2·4 112	2·7 109	2·8 109	2·6 110	2·4 112	2·1 115	1·8 120	1·4 129	210
240	0·7 233	0·4 193	0·6 138	1·0 117	1·4 108	2·1 102	2·6 100	2·8 099	2·7 099	2·5 100	2·2 101	1·9 103	1·5 106	240
270	0·7 283	0·2 311	0·4 065	0·8 080	1·3 083	2·0 086	2·6 087	2·8 087	2·7 087	2·5 087	2·3 086	1·9 086	1·6 085	270
300	0·8 328	0·7 003	0·9 037	1·2 055	1·6 064	2·2 072	2·7 075	2·8 076	2·6 075	2·4 074	2·2 072	1·9 068	1·5 063	300
330	1·0 001	1·1 024	1·4 041	1·7 052	2·0 058	2·5 065	2·8 068	2·8 068	2·5 065	2·2 062	2·0 058	1·7 051	1·4 040	330
360	1·3 028	1·5 042	1·8 052	2·1 058	2·4 062	2·7 066	2·8 067	2·6 065	2·2 059	1·9 054	1·6 046	1·3 034	1·2 015	360
							1999							
0	1·6 026	1·9 041	2·2 051	2·6 057	3·0 061	3·4 065	3·6 067	3·4 065	2·8 060	2·5 055	2·1 047	1·7 035	1·5 018	0
30	1·8 049	2·2 058	2·6 063	3·0 067	3·3 069	3·6 071	3·5 070	3·0 067	2·3 059	1·9 052	1·5 039	1·2 017	1·2 349	30
60	1·9 072	2·4 075	2·8 078	3·2 079	3·4 080	3·6 080	3·3 079	2·7 077	1·8 070	1·3 061	0·8 040	0·6 351	0·9 311	60
90	2·0 093	2·5 093	2·9 092	3·2 092	3·4 092	3·6 092	3·3 092	2·6 093	1·6 094	1·0 097	0·4 106	0·2 242	0·8 262	90
120	1·9 115	2·4 110	2·8 107	3·1 105	3·4 104	3·6 103	3·4 104	2·8 107	1·9 115	1·5 124	1·0 143	0·8 179	1·0 216	120
150	1·7 138	2·1 127	2·5 121	2·9 117	3·2 114	3·5 111	3·5 112	3·1 114	2·5 121	2·1 128	1·7 139	1·4 157	1·3 181	150
180	1·5 162	1·7 145	2·1 133	2·5 125	2·8 120	3·4 115	3·6 113	3·4 115	3·0 119	2·6 123	2·2 129	1·9 139	1·6 154	180
210	1·2 191	1·2 163	1·5 141	1·9 128	2·3 121	3·0 113	3·5 110	3·6 109	3·3 111	3·0 113	2·6 117	2·2 122	1·8 131	210
240	0·9 229	0·6 189	0·8 140	1·3 119	1·8 110	2·7 103	3·3 101	3·6 100	3·4 100	3·2 101	2·8 102	2·4 105	1·9 108	240
270	0·8 278	0·2 298	0·4 074	1·0 083	1·6 086	2·6 087	3·3 088	3·6 088	3·4 088	3·2 088	2·9 088	2·5 087	2·0 087	270
300	1·0 324	0·8 001	1·0 037	1·5 056	1·9 065	2·8 073	3·4 076	3·6 077	3·4 076	3·1 075	2·8 073	2·4 070	1·9 065	300
330	1·3 359	1·4 023	1·7 041	2·1 052	2·5 059	3·1 066	3·5 068	3·5 069	3·2 066	2·9 063	2·5 059	2·1 053	1·7 042	330
360	1·6 026	1·9 041	2·2 051	2·6 057	3·0 061	3·4 065	3·6 067	3·4 065	2·8 060	2·5 055	2·1 047	1·7 035	1·5 018	360

The above table gives the correction to be applied to a position line or a fix for the effects of precession and nutation from the mean equinox of 1995·0. Each entry consists of the distance (in bold type) in nautical miles, and the direction (true bearing) in which the position line or fix is to be moved. The table is entered firstly by the year, then by choosing the column nearest the latitude and finally the entry nearest the LHA♈ of observation; no interpolation is necessary, though in extreme cases near the beginning or end of a year (but not the end of 1994 or the beginning of 1995 when the corrections are zero) values midway towards those of the previous or following years may be taken.

Example. In 1999 a fix is obtained in latitude N 57° when LHA♈ is 95°. Entering the table with the year 1999, latitude N 60°, and LHA♈ 90° gives **2·9** 092° which indicates that the fix is to be transferred 2·9 miles in true bearing 092°.

TABLE 2.—Altitude Correction for Change in Position of Body

Correction for 4 Minutes of Time

True Zn	Latitude in Degrees																		True Zn
	0	5	10	15	20	25	30	35	40	45	50	55	60	65	70	75	80	85	
°	′	′	′	′	′	′	′	′	′	′	′	′	′	′	′	′	′	′	°
090	+60	+60	+59	+58	+56	+54	+52	+49	+46	+42	+39	+34	+30	+25	+21	+16	+10	+5	090
095	60	60	59	58	56	54	52	49	46	42	38	34	30	25	20	15	10	5	085
100	59	59	58	57	56	54	51	48	45	42	38	34	30	25	20	15	10	5	080
105	58	58	57	56	54	53	50	47	44	41	37	33	29	24	20	15	10	5	075
110	56	56	56	54	53	51	49	46	43	40	36	32	28	24	19	15	10	5	070
115	54	54	54	53	51	49	47	45	42	38	35	31	27	23	19	14	9	5	065
120	+52	+52	+51	+50	+49	+47	+45	+43	+40	+37	+33	+30	+26	+22	+18	+13	+9	+5	060
125	49	49	48	47	46	45	43	40	38	35	32	28	25	21	17	13	9	4	055
130	46	46	45	44	43	42	40	38	35	33	30	26	23	19	16	12	8	4	050
135	42	42	42	41	40	38	37	35	33	30	27	24	21	18	15	11	7	4	045
140	39	38	38	37	36	35	33	32	30	27	25	22	19	16	13	10	7	3	040
145	34	34	34	33	32	31	30	28	26	24	22	20	17	15	12	9	6	3	035
150	+30	+30	+30	+29	+28	+27	+26	+25	+23	+21	+19	+17	+15	+13	+10	+8	+5	+3	030
155	25	25	25	24	24	23	22	21	19	18	16	15	13	11	9	7	4	2	025
160	21	20	20	20	19	19	18	17	16	15	13	12	10	9	7	5	4	2	020
165	16	15	15	15	15	14	13	13	12	11	10	9	8	7	5	4	3	1	015
170	10	10	10	10	10	9	9	9	8	7	7	6	5	4	4	3	2	1	010
175	+5	+5	+5	+5	+5	+5	+5	+4	+4	+4	+3	+3	+3	+2	+2	+1	+1	+0	005
180	0	0	0	0	0	0	0	0	0	0	0	0	0	0	0	0	0	0	000
185	−5	−5	−5	−5	−5	−5	−5	−4	−4	−4	−3	−3	−3	−2	−2	−1	−1	−0	355
190	10	10	10	10	10	9	9	9	8	7	7	6	5	4	4	3	2	1	350
195	16	15	15	15	15	14	13	13	12	11	10	9	8	7	5	4	3	1	345
200	21	20	20	20	19	19	18	17	16	15	13	12	10	9	7	5	4	2	340
205	25	25	25	24	24	23	22	21	19	18	16	15	13	11	9	7	4	2	335
210	30	30	30	29	28	27	26	25	23	21	19	17	15	13	10	8	5	3	330
215	−34	−34	−34	−33	−32	−31	−30	−28	−26	−24	−22	−20	−17	−15	−12	−9	−6	−3	325
220	39	38	38	37	36	35	33	32	30	27	25	22	19	16	13	10	7	3	320
225	42	42	42	41	40	38	37	35	33	30	27	24	21	18	15	11	7	4	315
230	46	46	45	44	43	42	40	38	35	33	30	26	23	19	16	12	8	4	310
235	49	49	48	47	46	45	43	40	38	35	32	28	25	21	17	13	9	4	305
240	52	52	51	50	49	47	45	43	40	37	33	30	26	22	18	13	9	5	300
245	−54	−54	−54	−53	−51	−49	−47	−45	−42	−38	−35	−31	−27	−23	−19	−14	−9	−5	295
250	56	56	56	54	53	51	49	46	43	40	36	32	28	24	19	15	10	5	290
255	58	58	57	56	54	53	50	47	44	41	37	33	29	24	20	15	10	5	285
260	59	59	58	57	56	54	51	48	45	42	38	34	30	25	20	15	10	5	280
265	60	60	59	58	56	54	52	49	46	42	38	34	30	25	20	15	10	5	275
270	−60	−60	−59	−58	−56	−54	−52	−49	−46	−42	−39	−34	−30	−25	−21	−16	−10	−5	270

Interpolation for Altitude Correction for Less Than 4 Minutes of Time

Interval of Time	Value from Tables 1 and 2 (For values less than 60′ see opposite page)																				Interval of Time
	63	66	69	72	75	78	81	84	87	90	93	96	99	102	105	108	111	114	117	120	
m s	′	′	′	′	′	′	′	′	′	′	′	′	′	′	′	′	′	′	′	′	m s
0 00	0	0	0	0	0	0	0	0	0	0	0	0	0	0	0	0	0	0	0	0	0 00
10	3	3	3	3	3	3	3	4	4	4	4	4	4	4	4	5	5	5	5	5	10
20	5	6	6	6	6	7	7	7	7	8	8	8	8	9	9	9	9	10	10	10	20
30	8	8	9	9	9	10	10	11	11	11	12	12	12	13	13	14	14	14	15	15	30
40	11	11	12	12	13	13	14	14	15	15	16	16	17	17	18	18	19	19	20	20	40
0 50	13	14	14	15	16	16	17	18	18	19	19	20	21	21	22	23	23	24	24	25	0 50
1 00	16	17	17	18	19	20	20	21	22	23	23	24	25	26	26	27	28	29	29	30	1 00
10	18	19	20	21	22	23	24	25	25	26	27	28	29	30	31	32	32	33	34	35	10
20	21	22	23	24	25	26	27	28	29	30	31	32	33	34	35	36	37	38	39	40	20
30	24	25	26	27	28	29	30	32	33	34	35	36	37	38	39	41	42	43	44	45	30
40	26	28	29	30	31	33	34	35	36	38	39	40	41	43	44	45	46	48	49	50	40
1 50	29	30	32	33	34	36	37	39	40	41	43	44	45	47	48	50	51	52	54	55	1 50
2 00	32	33	35	36	38	39	41	42	44	45	47	48	50	51	53	54	56	57	59	60	2 00
10	34	36	37	39	41	42	44	46	47	49	50	52	54	55	57	59	60	62	63	65	10
20	37	39	40	42	44	46	47	49	51	53	54	56	58	60	61	63	65	67	68	70	20
30	39	41	43	45	47	49	51	53	54	56	58	60	62	64	66	68	69	71	73	75	30
40	42	44	46	48	50	52	54	56	58	60	62	64	66	68	70	72	74	76	78	80	40
2 50	45	47	49	51	53	55	57	60	62	64	66	68	70	72	74	77	79	81	83	85	2 50
3 00	47	50	52	54	56	59	61	63	65	68	70	72	74	77	79	81	83	86	88	90	3 00
10	50	52	55	57	59	62	64	67	69	71	74	76	78	81	83	86	88	90	93	95	10
20	53	55	58	60	63	65	68	70	73	75	78	80	83	85	88	90	93	95	98	100	20
30	55	58	60	63	66	68	71	74	76	79	81	84	87	89	92	95	97	100	102	105	30
40	58	61	63	66	69	72	74	77	80	83	85	88	91	94	96	99	102	105	107	110	40
3 50	60	63	66	69	72	75	78	81	83	86	89	92	95	98	101	104	106	109	112	115	3 50
4 00	63	66	69	72	75	78	81	84	87	90	93	96	99	102	105	108	111	114	117	120	4 00

Time of fix (tab 1) or computation (tab 2)	Sign from 4-min. Table	To observed altitude	To tabulated altitude	To intercept
Later than observation	+	Add	Subtract	Toward
	−	Subtract	Add	Away
Earlier than observation	+	Subtract	Add	Away
	−	Add	Subtract	Toward

TABLE 3.—Conversion of Arc to Time

°	h m	°	h m	°	h m	°	h m	°	h m	°	h m	′	m s	″	s
0	0 0	60	4 0	120	8 0	180	12 0	240	16 0	300	20 0	0	0 0	0	0.00
1	0 4	61	4 4	121	8 4	181	12 4	241	16 4	301	20 4	1	0 4	1	0.07
2	0 8	62	4 8	122	8 8	182	12 8	242	16 8	302	20 8	2	0 8	2	0.13
3	0 12	63	4 12	123	8 12	183	12 12	243	16 12	303	20 12	3	0 12	3	0.20
4	0 16	64	4 16	124	8 16	184	12 16	244	16 16	304	20 16	4	0 16	4	0.27
5	0 20	65	4 20	125	8 20	185	12 20	245	16 20	305	20 20	5	0 20	5	0.33
6	0 24	66	4 24	126	8 24	186	12 24	246	16 24	306	20 24	6	0 24	6	0.40
7	0 28	67	4 28	127	8 28	187	12 28	247	16 28	307	20 28	7	0 28	7	0.47
8	0 32	68	4 32	128	8 32	188	12 32	248	16 32	308	20 32	8	0 32	8	0.53
9	0 36	69	4 36	129	8 36	189	12 36	249	16 36	309	20 36	9	0 36	9	0.60
10	0 40	70	4 40	130	8 40	190	12 40	250	16 40	310	20 40	10	0 40	10	0.67
11	0 44	71	4 44	131	8 44	191	12 44	251	16 44	311	20 44	11	0 44	11	0.73
12	0 48	72	4 48	132	8 48	192	12 48	252	16 48	312	20 48	12	0 48	12	0.80
13	0 52	73	4 52	133	8 52	193	12 52	253	16 52	313	20 52	13	0 52	13	0.87
14	0 56	74	4 56	134	8 56	194	12 56	254	16 56	314	20 56	14	0 56	14	0.93
15	1 0	75	5 0	135	9 0	195	13 0	255	17 0	315	21 0	15	1 0	15	1.00
16	1 4	76	5 4	136	9 4	196	13 4	256	17 4	316	21 4	16	1 4	16	1.07
17	1 8	77	5 8	137	9 8	197	13 8	257	17 8	317	21 8	17	1 8	17	1.13
18	1 12	78	5 12	138	9 12	198	13 12	258	17 12	318	21 12	18	1 12	18	1.20
19	1 16	79	5 16	139	9 16	199	13 16	259	17 16	319	21 16	19	1 16	19	1.27
20	1 20	80	5 20	140	9 20	200	13 20	260	17 20	320	21 20	20	1 20	20	1.33
21	1 24	81	5 24	141	9 24	201	13 24	261	17 24	321	21 24	21	1 24	21	1.40
22	1 28	82	5 28	142	9 28	202	13 28	262	17 28	322	21 28	22	1 28	22	1.47
23	1 32	83	5 32	143	9 32	203	13 32	263	17 32	323	21 32	23	1 32	23	1.53
24	1 36	84	5 36	144	9 36	204	13 36	264	17 36	324	21 36	24	1 36	24	1.60
25	1 40	85	5 40	145	9 40	205	13 40	265	17 40	325	21 40	25	1 40	25	1.67
26	1 44	86	5 44	146	9 44	206	13 44	266	17 44	326	21 44	26	1 44	26	1.73
27	1 48	87	5 48	147	9 48	207	13 48	267	17 48	327	21 48	27	1 48	27	1.80
28	1 52	88	5 52	148	9 52	208	13 52	268	17 52	328	21 52	28	1 52	28	1.87
29	1 56	89	5 56	149	9 56	209	13 56	269	17 56	329	21 56	29	1 56	29	1.93
30	2 0	90	6 0	150	10 0	210	14 0	270	18 0	330	22 0	30	2 0	30	2.00
31	2 4	91	6 4	151	10 4	211	14 4	271	18 4	331	22 4	31	2 4	31	2.07
32	2 8	92	6 8	152	10 8	212	14 8	272	18 8	332	22 8	32	2 8	32	2.13
33	2 12	93	6 12	153	10 12	213	14 12	273	18 12	333	22 12	33	2 12	33	2.20
34	2 16	94	6 16	154	10 16	214	14 16	274	18 16	334	22 16	34	2 16	34	2.27
35	2 20	95	6 20	155	10 20	215	14 20	275	18 20	335	22 20	35	2 20	35	2.33
36	2 24	96	6 24	156	10 24	216	14 24	276	18 24	336	22 24	36	2 24	36	2.40
37	2 28	97	6 28	157	10 28	217	14 28	277	18 28	337	22 28	37	2 28	37	2.47
38	2 32	98	6 32	158	10 32	218	14 32	278	18 32	338	22 32	38	2 32	38	2.53
39	2 36	99	6 36	159	10 36	219	14 36	279	18 36	339	22 36	39	2 36	39	2.60
40	2 40	100	6 40	160	10 40	220	14 40	280	18 40	340	22 40	40	2 40	40	2.67
41	2 44	101	6 44	161	10 44	221	14 44	281	18 44	341	22 44	41	2 44	41	2.73
42	2 48	102	6 48	162	10 48	222	14 48	282	18 48	342	22 48	42	2 48	42	2.80
43	2 52	103	6 52	163	10 52	223	14 52	283	18 52	343	22 52	43	2 52	43	2.87
44	2 56	104	6 56	164	10 56	224	14 56	284	18 56	344	22 56	44	2 56	44	2.93
45	3 0	105	7 0	165	11 0	225	15 0	285	19 0	345	23 0	45	3 0	45	3.00
46	3 4	106	7 4	166	11 4	226	15 4	286	19 4	346	23 4	46	3 4	46	3.07
47	3 8	107	7 8	167	11 8	227	15 8	287	19 8	347	23 8	47	3 8	47	3.13
48	3 12	108	7 12	168	11 12	228	15 12	288	19 12	348	23 12	48	3 12	48	3.20
49	3 16	109	7 16	169	11 16	229	15 16	289	19 16	349	23 16	49	3 16	49	3.27
50	3 20	110	7 20	170	11 20	230	15 20	290	19 20	350	23 20	50	3 20	50	3.33
51	3 24	111	7 24	171	11 24	231	15 24	291	19 24	351	23 24	51	3 24	51	3.40
52	3 28	112	7 28	172	11 28	232	15 28	292	19 28	352	23 28	52	3 28	52	3.47
53	3 32	113	7 32	173	11 32	233	15 32	293	19 32	353	23 32	53	3 32	53	3.53
54	3 36	114	7 36	174	11 36	234	15 36	294	19 36	354	23 36	54	3 36	54	3.60
55	3 40	115	7 40	175	11 40	235	15 40	295	19 40	355	23 40	55	3 40	55	3.67
56	3 44	116	7 44	176	11 44	236	15 44	296	19 44	356	23 44	56	3 44	56	3.73
57	3 48	117	7 48	177	11 48	237	15 48	297	19 48	357	23 48	57	3 48	57	3.80
58	3 52	118	7 52	178	11 52	238	15 52	298	19 52	358	23 52	58	3 52	58	3.87
59	3 56	119	7 56	179	11 56	239	15 56	299	19 56	359	23 56	59	3 56	59	3.93
60	4 0	120	8 0	180	12 0	240	16 0	300	20 0	360	24 0	60	4 0	60	4.00

AP 3270

SIGHT REDUCTION TABLES
FOR
AIR NAVIGATION

VOLUME 1
SELECTED STARS
EPOCH 1995·0

LONDON: HMSO

UNITED KINGDOM EDITION
© *Crown copyright* 1992
Applications for reproduction should be made to HMSO
First published 1992

HMSO
on behalf of the Defence Mapping Agency
Hydrographical/Topographic Center, U.S.A., by whose permission
these tables are reproduced in the United Kingdom

HMSO publications are available from:

HMSO Publications Centre
(Mail, fax and telephone orders only)
PO Box 276, London, SW8 5DT
Telephone orders 071-873 9090
GENERAL ENQUIRIES 071-873 0011
(queuing system in operation for both numbers)
Fax orders 071-873 8200

HMSO Bookshops
49 High Holborn, London, WC1V 6HB (counter service only) 071-873 0011 Fax 071-873 8200
258 Broad Street, Birmingham, B1 2HE 021-643 3740 Fax 021-643 6510
Southey House, 33 Wine Street, Bristol, BS1 2BQ 0272 264306 Fax 0272 294515
9-21 Princess Street, Manchester M60 8AS 061-834 7201 Fax 061-833 0634
16 Arthur Street, Belfast, BT1 4GD 0232 238451 Fax 0232 235401
71 Lothian Road, Edinburgh, EH3 9AZ 031-228 4181 Fax 031-229 2734

HMSO's Accredited Agents
(See Yellow Pages)

and through good booksellers

Obtainable in Canada and U.S.A. from:

Bernan Associates, 9730 E. George Palmer Highway, Lanham, MD 20706

UNITED STATES EDITION

For sale by
authorized sales agents of the
Defence Mapping Agency
Hydrographic/Topographic Centre,
Washington, D.C. 20315

ISBN 011 772731 8

Printed in the United Kingdom for HMSO
Dd295567 2/93 C100 G531 10170

FOREWORD

The *Sight Reduction Tables for Air Navigation* consist of three volumes of comprehensive tables of altitude and azimuth designed for the rapid reduction of astronomical sights in the air. The present volume (Volume 1) contains tables for selected stars for all latitudes, calculated for the epoch of 1995·0, and replaces the previous edition calculated for the epoch of 1990·0; it is intended for use for about 5 years, when a new edition based on a later epoch will be issued. Volume 2 for latitudes 0°–40° and Volume 3 for latitudes 39°–89° contain tables for integral degrees of declination and provide for sights of the Sun, Moon and planets; these tables are permanent.

The time argument in the examples is denoted by UT (Universal Time). It is also known as GMT (Greenwich Mean Time).

Sight Reduction Tables for Air Navigation are published in the USA as Pub. No. 249 and in the UK as AP 3270. The Defense Mapping Agency Hydrographic/Topographic Center is responsible for the compilation of these tables. The Nautical Almanac Office of the U.S. Naval Observatory and H.M. Nautical Almanac Office have co-operated in their design and preparation. Copy for photographic reproduction of the tabular pages was prepared on a Videocomp photosetter automatically controlled by a magnetic tape produced by computer at the U.S. Naval Observatory.

The content and format of these three volumes may not be changed without the approval of Working Party 70 of the Air Standardization Coordinating Committee.

Users should refer corrections, additions, and comments for improving this product to DIRECTOR, DEFENSE MAPPING AGENCY, Attn: PR 8613 Lee Highway, Fairfax, VA 22031-2137.

NAVIGATIONAL STARS, EPOCH 1995·0

Alphabetical Order

Name	No.	Magnitude Visual	Magnitude S−4	SHA	Dec
				° ′	° ′
Acamar	**7**	3·1	3·2	315 29	S 40 19
ACHERNAR	**5**	0·6	0·1	335 37	S 57 16
ACRUX	**30**	1·1	0·5	173 25	S 63 04
*Adhara	**19**	1·6	1·2	255 24	S 28 58
ALDEBARAN	**10**	1·1	3·1	291 06	N 16 30
Alioth	**32**	1·7	1·5	166 33	N 55 59
Alkaid	**34**	1·9	1·5	153 10	N 49 20
*Al Na'ir	**55**	2·2	1·8	28 01	S 46 59
*Alnilam	**15**	1·8	1·3	276 01	S 1 12
Alphard	**25**	2·2	4·4	218 10	S 8 38
Alphecca	**41**	2·3	2·1	126 23	N 26 44
Alpheratz	**1**	2·2	1·8	357 58	N 29 04
ALTAIR	**51**	0·9	1·0	62 22	N 8 51
*Ankaa	**2**	2·4	3·9	353 29	S 42 20
ANTARES	**42**	1·2	3·7	112 43	S 26 25
ARCTURUS	**37**	0·2	1·9	146 09	N 19 12
*Atria	**43**	1·9	4·1	107 58	S 69 01
*Avior	**22**	1·7	3·3	234 24	S 59 30
*Bellatrix	**13**	1·7	1·2	278 47	N 6 21
BETELGEUSE	**16**	0·1–1·2	2·5–3·6	271 16	N 7 24
CANOPUS	**17**	−0·9	−0·8	264 02	S 52 42
CAPELLA	**12**	0·2	1·3	280 55	N 46 00
DENEB	**53**	1·3	1·4	49 41	N 45 16
Denebola	**28**	2·2	2·2	182 48	N 14 36
Diphda	**4**	2·2	3·6	349 10	S 18 01
Dubhe	**27**	2·0	3·4	194 09	N 61 47
*Elnath	**14**	1·8	1·4	278 30	N 28 36
*Eltanin	**47**	2·4	4·6	90 53	N 51 29
Enif	**54**	2·5	4·8	34 01	N 9 51
FOMALHAUT	**56**	1·3	1·3	15 39	S 29 39
*Gacrux	**31**	1·6	4·1	172 17	S 57 05
Gienah	**29**	2·8	2·5	176 07	S 17 31
*HADAR	**35**	0·9	0·3	149 08	S 60 21
Hamal	**6**	2·2	3·8	328 17	N 23 26
*Kaus Australis	**48**	2·0	2·0	84 02	S 34 23
Kochab	**40**	2·2	4·3	137 19	N 74 11
*Markab	**57**	2·6	2·3	13 52	N 15 11
Menkar	**8**	2·8	5·3	314 30	N 4 04
*Menkent	**36**	2·3	3·5	148 24	S 36 21
Miaplacidus	**24**	1·8	1·8	221 43	S 69 42
Mirfak	**9**	1·9	2·4	309 01	N 49 51
Nunki	**50**	2·1	1·9	76 16	S 26 18
Peacock	**52**	2·1	1·7	53 41	S 56 45
POLLUX	**21**	1·2	2·5	243 45	N 28 02
PROCYON	**20**	0·5	0·8	245 14	N 5 14
Rasalhague	**46**	2·1	2·2	96 19	N 12 34
REGULUS	**26**	1·3	1·0	207 58	N 12 00
RIGEL	**11**	0·3	0·0	281 26	S 8 12
RIGIL KENT.	**38**	0·1	0·9	140 11	S 60 49
*Sabik	**44**	2·6	2·5	102 29	S 15 43
Schedar	**3**	2·5	4·1	349 57	N 56 31
Shaula	**45**	1·7	1·3	96 41	S 37 06
SIRIUS	**18**	−1·6	−1·5	258 46	S 16 43
SPICA	**33**	1·2	0·7	158 46	S 11 08
Suhail	**23**	2·2	4·6	223 03	S 43 25
VEGA	**49**	0·1	0·0	80 48	N 38 47
*Zubenelgenubi	**39**	2·9	3·2	137 21	S 16 01

Order of SHA

Name	No.	Magnitude Visual	Magnitude S−4	SHA	Dec
				° ′	° ′
*Markab	**57**	2·6	2·3	13 52	N 15 11
FOMALHAUT	**56**	1·3	1·3	15 39	S 29 39
*Al Na'ir	**55**	2·2	1·8	28 01	S 46 59
Enif	**54**	2·5	4·8	34 01	N 9 51
DENEB	**53**	1·3	1·4	49 41	N 45 16
Peacock	**52**	2·1	1·7	53 41	S 56 45
ALTAIR	**51**	0·9	1·0	62 22	N 8 51
Nunki	**50**	2·1	1·9	76 16	S 26 18
VEGA	**49**	0·1	0·0	80 48	N 38 47
*Kaus Australis	**48**	2·0	2·0	84 02	S 34 23
*Eltanin	**47**	2·4	4·6	90 53	N 51 29
Rasalhague	**46**	2·1	2·2	96 19	N 12 34
Shaula	**45**	1·7	1·3	96 41	S 37 06
*Sabik	**44**	2·6	2·5	102 29	S 15 43
*Atria	**43**	1·9	4·1	107 58	S 69 01
ANTARES	**42**	1·2	3·7	112 43	S 26 25
Alphecca	**41**	2·3	2·1	126 23	N 26 44
Kochab	**40**	2·2	4·3	137 19	N 74 11
*Zubenelgenubi	**39**	2·9	3·2	137 21	S 16 01
RIGIL KENT.	**38**	0·1	0·9	140 11	S 60 49
ARCTURUS	**37**	0·2	1·9	146 09	N 19 12
*Menkent	**36**	2·3	3·5	148 24	S 36 21
*HADAR	**35**	0·9	0·3	149 08	S 60 21
Alkaid	**34**	1·9	1·5	153 10	N 49 20
SPICA	**33**	1·2	0·7	158 46	S 11 08
Alioth	**32**	1·7	1·5	166 33	N 55 59
*Gacrux	**31**	1·6	4·1	172 17	S 57 05
ACRUX	**30**	1·1	0·5	173 25	S 63 04
Gienah	**29**	2·8	2·5	176 07	S 17 31
Denebola	**28**	2·2	2·2	182 48	N 14 36
Dubhe	**27**	2·0	3·4	194 09	N 61 47
REGULUS	**26**	1·3	1·0	207 58	N 12 00
Alphard	**25**	2·2	4·4	218 10	S 8 38
Miaplacidus	**24**	1·8	1·8	221 43	S 69 42
Suhail	**23**	2·2	4·6	223 03	S 43 25
*Avior	**22**	1·7	3·3	234 24	S 59 30
POLLUX	**21**	1·2	2·5	243 45	N 28 02
PROCYON	**20**	0·5	0·8	245 14	N 5 14
*Adhara	**19**	1·6	1·2	255 24	S 28 58
SIRIUS	**18**	−1·6	−1·5	258 46	S 16 43
CANOPUS	**17**	−0·9	−0·8	264 02	S 52 42
BETELGEUSE	**16**	0·1–1·2	2·5–3·6	271 16	N 7 24
*Alnilam	**15**	1·8	1·3	276 01	S 1 12
*Elnath	**14**	1·8	1·4	278 30	N 28 36
*Bellatrix	**13**	1·7	1·2	278 47	N 6 21
CAPELLA	**12**	0·2	1·3	280 55	N 46 00
RIGEL	**11**	0·3	0·0	281 26	S 8 12
ALDEBARAN	**10**	1·1	3·1	291 06	N 16 30
Mirfak	**9**	1·9	2·4	309 01	N 49 51
Menkar	**8**	2·8	5·3	314 30	N 4 04
Acamar	**7**	3·1	3·2	315 29	S 40 19
Hamal	**6**	2·2	3·8	328 17	N 23 26
ACHERNAR	**5**	0·6	0·1	335 37	S 57 16
Diphda	**4**	2·2	3·6	349 10	S 18 01
Schedar	**3**	2·5	4·1	349 57	N 56 31
*Ankaa	**2**	2·4	3·9	353 29	S 42 20
Alpheratz	**1**	2·2	1·8	357 58	N 29 04

The star numbers and names are the same as in *The Air Almanac*.

* Not in tabular pages of Volume 1.

INTRODUCTION

DESCRIPTION OF THE TABLES

These tables, designated as Volume 1 of the three-volume series of AP 3270, *Sight Reduction Tables for Air Navigation*, contain values of the altitude (to the nearest minute) and the true azimuth (to the nearest degree) of seven selected stars for the complete ranges of latitude and hour angle of Aries. The arrangement provides, for any position and time, the best selection of seven of the stars available for observation and, for these seven stars, data for presetting before observation and for accurate reduction of the sights after observation.

In the calculation of the altitudes and azimuths the mean places of the stars for epoch 1995·0 have been used; corrections for precession and nutation are given in Table 5, but their omission will not give rise to a positional error greater than two miles in the years 1992–1997. No correction for refraction has been included in the tabulated altitudes, so that the full correction must be applied to the sextant altitudes.

AP 3270 has been designed for use with *The Air Almanac*, but the tables in this volume may be used with a clock, or other device, giving sidereal time. With the normal procedure of plotting a sight from an assumed position, no interpolation is required for the stars tabulated.

ENTERING ARGUMENTS AND ARRANGEMENT

Latitude. Tabulations are given for every whole degree of latitude from 89° north to 89° south. From 69° north to 69° south all data for a single latitude appear on two facing pages; from 70° to the poles, both north and south, the data for a single latitude appear on one page.

LHA Aries. The vertical argument on each page is the local hour angle of the first point of Aries (LHA ♈). It ranges from 0° to 360°; in general the interval is 1°, but between latitudes 70° and the poles it is increased to 2°.

Selected stars. The tabulated (or computed) altitude (Hc) and the true azimuth (Zn) are given for seven selected stars for each latitude and each entry of LHA ♈. The selection of stars is used unchanged for each group of 15 entries of LHA ♈ (30° for latitudes over 69°, 15° for lower latitudes); within each such group the order of arrangement is that of the azimuths corresponding to the first entry. Of each selection of seven stars, three are marked with a diamond symbol (♦) as being suitable for a three-star fix.

A total of 41 stars is used, of which 19 are of the first magnitude (brighter than magnitude 1·5) and 17 of the second magnitude. The names of first-magnitude stars are given in capital letters. A complete list of the 57 stars selected for astro-navigation is given on page iii and an asterisk is printed beside those stars not used in this volume. The adopted names and numbers agree with those used in *The Air Almanac*. The S–4 magnitudes are applicable to astro-trackers employing S–4 photo-sensitive response.

Many factors were considered in selecting the stars, including azimuth, magnitude, altitude and continuity. Continuity was sought in regard to both latitude and hour angle, particularly for latitude where changes are not immediately evident by inspection.

USE OF THE TABLES

The tables are intended for use for two distinct operations—the planning of observations, and their reduction. It is important that full use should be made of the tables for the planning of observations.

INTRODUCTION

Planning of observations. Since only seven stars are given it is essential to refer to the tables before observation, in order to ensure that data will be available for the reduction of the observations. This is done by estimating latitude and LHA ♈ for the proposed time of observation, from a knowledge of the DR position and GHA ♈ from *The Air Almanac,* or if necessary from Table **4**. On reference to the tables this information gives immediately the seven stars available, together with their approximate altitudes and azimuths. From these seven stars, the observer can select those which best suit his particular purpose and the prevailing conditions; the approximate altitudes and azimuths make identification easy, and enable the sextant to be preset to the approximate altitude.

Example. On 1993 January 1 a navigator proposes to observe at $12^h\ 22^m$ UT in DR position N 54° 17′, E 175° 05′.

		h	m			°	′
From *The Air Almanac,* PM page for January 1 at UT		12	20,	GHA ♈	=	286	10
	flap, increment for		2		=		30
	Sum = GHA ♈ for UT	12	22,	GHA ♈	=	286	40
	Longitude, added if east, subtracted if west				+	175	05
Sum, adjusted as necessary for multiples of 360°				LHA ♈	=	101	45

Reference to page 52 for Lat. 54° N, LHA ♈ 102° shows that the selected stars are *Dubhe* (azimuth 051°), ♦*REGULUS* (117°), *PROCYON* (163°), *SIRIUS* (181°), ♦*RIGEL* (206°), *Mirfak* (284°), and ♦*DENEB* (340°), all being at convenient altitudes between 12° and 59°. No change in the selection will take place for about 50 minutes before or 10 minutes after the time proposed, but if the observations are delayed, *PROCYON* and *DENEB* will be replaced by *Denebola* and *ALDEBARAN*; the same stars are used for latitude 55°, though *Hamal* and *Schedar* replace *Mirfak* and *DENEB* in latitude 53°. The navigator will accordingly plan his programme of observations from among these stars, bearing in mind that the Moon is at first quarter. It should be noted that this preliminary calculation of LHA ♈ may often be modified to serve as a basis for the reduction of the sights, without further reference to *The Air Almanac.*

If observations are made of stars other than those selected, they can be reduced by the use of Volumes **2** and **3** of these tables, provided the declinations are less than 30° north or south. A list of such stars, with their positions, is given in those volumes as well as being indicated in the star lists in *The Air Almanac.* Observations of other stars must be reduced by other methods or tables.

Reduction of sights. GHA ♈ is taken from *The Air Almanac* for the actual time of observation and combined with an assumed longitude, close to the DR longitude, to make LHA ♈ a whole degree, or an even degree for latitudes above 69°. The tables are entered with the whole degree of latitude nearest to the DR latitude, the value of LHA ♈ found above, and the name of the star observed; they give, without interpolation, the tabulated altitude (Hc) and azimuth (Zn). The intercept is found in the usual way by comparing the corrected sextant altitude (Ho) with the tabulated altitude:

towards the star if the sextant altitude is *greater* than the tabulated altitude;
away from the star if the sextant altitude is *less* than the tabulated altitude.

The sextant reading must be corrected for instrument error, dome refraction (if applicable), and refraction (from Table **8**) before being compared with the tabulated altitude. The sight is plotted from the assumed position, defined by the whole degree of latitude and the assumed longitude. This assumed position may previously be adjusted for the effect of Coriolis (see Table **9**), advanced or retarded to another time, and (in extreme cases) shifted to allow for precession and nutation (see Table **5**); alternatively these corrections may be made to the position line or, in the case of the corrections from Tables **5** and **9**, to the final fix. The application of these corrections is considered separately on pages ix and x.

INTRODUCTION

Example. On 1993 January 1, in DR position N 54° 17′, E 175° 46′ at height 9 000 ft. (3 km), an observation of *PROCYON* is obtained at $12^h 21^m 25^s$ UT; the sextant reading is 40° 36′ and the correction for instrument error and dome refraction is −4′.

	h m s			° ′		° ′
From *The Air Almanac*					Sextant altitude	40 36
PM page for January 1 at UT	12 20 00	GHA ♈	=	286 10	Dome refraction, etc.	−4
flap, increment for	1 25		=	0 21	Refraction (Table **8**)	−1
Sum = GHA ♈ for UT	12 21 25	GHA ♈	=	286 31	Corrected sextant altitude (Ho)	40 31
Assumed longitude, added because east,			+	175 29	From the tables (page 52)	
Sum, less 360°		LHA ♈	=	102	Tabulated altitude (Hc)	40 09 Az. (Zn) 163°
					Intercept	22 *towards*

The assumed latitude is N 54°, the assumed longitude is E 175° 29′, and the intercept of 22′ is plotted from this position in true bearing 163°. The position line is drawn perpendicular to this direction.

Usually, sights of several stars will be taken in rapid succession to give a fix. The example below illustrates the use of tables for the reduction of a typical set of observations.

Example. On 1993 January 1, in DR position N 45° 49′, W 25° 47′ (for $23^h 47^m$ UT) at height 3 000 ft. (1 km), sights are taken as follows:

Star	UT	Sextant altitude	Instrument error, etc.
	h m s	° ′	′
Dubhe	23 44 15	37 46	−5
RIGEL	23 47 33	35 55	−5
Alpheratz	23 51 55	33 15	−6

	Dubhe UT	*Dubhe* GHA ♈	*RIGEL* UT	*RIGEL* GHA ♈	*Alpheratz* UT	*Alpheratz* GHA ♈
From *The Air Almanac*	h m s	° ′	h m s	° ′	h m s	° ′
PM page for January 1	23 40	96 38	23 40	96 38	23 50	99 08
flap, increments for	4 15	1 04	7 33	1 53	1 55	0 29
Sum = GHA ♈ for given UT	23 44 15	97 42	23 47 33	98 31	23 51 55	99 37
Assumed longitude, subtracted because west		−25 42		−25 31		−25 37
Sum = LHA ♈		72		73		74

	Dubhe Altitude	*Dubhe* Az.	*RIGEL* Altitude	*RIGEL* Az.	*Alpheratz* Altitude	*Alpheratz* Az.
	° ′		° ′		° ′	
Sextant altitude	37 46		35 55		33 15	
Instrument error and dome refraction	−5		−5		−6	
Refraction (Table **8**)	−1		−1		−1	
Corrected sextant altitude (Ho)	37 40		35 49		33 08	
Tables, p. 68 assumed Lat. 46° N and LHA ♈ as above; Hc and Zn	37 43	037°	35 34	173°	32 30	280°
Intercept	3 *away*		15 *towards*		38 *towards*	

In this example, the assumed longitudes for all observations are taken as close as possible to the DR longitude at $23^h 47^m$; shorter intercepts can often be obtained by relating the assumed position to the DR position at the time of observation. The intercepts are plotted from the respective assumed positions, latitude N 46°, respective longitudes W 25° 42′, W 25° 31′ and W 25° 37′, transferred as necessary for the motion of the aircraft between the time of observation and that of the fix, for the effect of Coriolis acceleration and for precession and nutation. These shifts may be made to the position lines instead of to the assumed positions from which they are constructed, or, for the last two corrections, directly to the fix.

USE OF CORRECTING TABLES

As indicated in the foregoing example, corrections are required for the following, in addition to refraction.

Coriolis acceleration. This correction, given in Table **9** on the inside back cover, may be applied either to each individual observation or to the fix reduced from several observations. When applied to individual observations, either the position line or the assumed position from which it is constructed must be shifted by the distance *Z* miles perpendicular to the track. The rule for applying this correction is given at the foot of Table **9**.

Precession and nutation. The correction in Table **5** on page **322** is normally to be ignored. If, in extreme cases, it is necessary to allow for the change in the positions of the stars, the correction may be treated in the same way as the Coriolis correction and applied to the final fix, or to individual position lines or assumed positions. The correction is applicable only to sights reduced with this volume of tables.

Motion of the observer. If it is desired to get a fix from two or more observations, the resulting position lines must be reduced to a common time, usually the time of one of them. This may be done in two ways: the position lines of observations made earlier or later than this time may be transferred on the plotting chart to allow for the motion of the aircraft in the time-interval concerned, or the corrected sextant altitudes (or intercepts) may be adjusted to allow for the motion of the aircraft in the time-interval concerned.

In the first case, the shift may be applied to the position line or to the assumed position from which it is constructed.

In the second case, the adjustment to corrected sextant altitude may be taken from Table **1** on the inside front cover, interpolating where necessary. Table **1** gives, in the upper part, the correction for a time-interval of 4 minutes, while the lower part enables this to be extended to any time-interval. By reversing the sign of this correction, it may be applied to the tabulated altitude instead of to the corrected sextant altitude, or it may be applied directly to the intercept by the rules given.

Example. In the preceding example on page ix the aircraft was flying at 400 knots on a track 257°T.

From Table **9** the *Z* correction is found to be 8′ and the assumed positions, position lines or the deduced fix must be shifted a distance 8 miles to the starboard (right) of track (for northern latitude), i.e. in direction 347°T.

From Table **5** the correction for precession and nutation is found to be 1 mile in direction 250°T and is to be applied similarly. Both corrections are made by construction on the plotting chart.

Corrections for the change in position of the aircraft will be applied to the corrected sextant altitudes of the first and third stars, so that the fix will be obtained at the time, $23^h\ 47^m\ 33^s$, of the middle observation.

Star	Azimuth	True Track	Relative Azimuth	Table 1 upper part	Time Interval	Correction from lower part of Table 1 to Sext. Alt.	Correction from lower part of Table 1 to Intercept	Adjusted Corr. Sext. Alt.	Adjusted Intercept
	°	°	°	′	m s	′	′	° ′	′
Dubhe	037	257	140	−20	+3 18	−17	17 *away*	37 23	20 *away*
Alpheratz	280	257	023	+24	−4 22	−26	26 *away*	32 42	12 *towards*

where Relative Azimuth = Azimuth − True Track, (adding 360° if necessary).

The above table is largely self-explanatory; the value for the time-interval of $4^m\ 22^s$ is found from the lower part of Table **1** by adding the correction for 4^m to that for 22^s or by doubling that for $2^m\ 11^s$.

INTRODUCTION

Motion of the body. If the time of observation differs from that corresponding to the tabular value of LHA ♈, the entry for this value may still be used if a correction for the motion of the body (due to the rotation of the Earth) in the time interval is applied to the altitude (or intercept). Table **2**, on the page facing the inside front cover, provides for this correction. It enables observations made at different times to be reduced and plotted from the same assumed position, while using the same common value of LHA ♈. Since the time to which this value of LHA ♈ corresponds is usually that at which the fix is desired, it is convenient to combine the corrections for motion of the body with those for the motion of the observer, as the time intervals are the same.

When both the tables for the changes in position of observer and body are used, the quantities taken from the upper parts of Tables **1** and **2** should be summed and the sum used in entering the lower parts of the tables, (values of the sum less than 60′ being used in the lower part of Table 1 and values greater than 60′ in the lower part of Table **2**).

Example. The second example on page ix is reduced using Tables **1** and **2**, assuming that the aircraft was flying at 400 knots on track 257°T, and that the fix is required for $23^h\ 47^m$; the sights are:

	UT	Sextant Altitude
	h m s	° ′
Dubhe	23 44 15	37 46
RIGEL	23 47 33	35 55
Alpheratz	23 51 55	33 15

	UT	GHA ♈
From *The Air Almanac*	h m	° ′
PM page for January 1	23 40	96 38
flap, increment for	7	1 45
Sum = GHA ♈ for	23 47	98 23
Assumed longitude (west, subtract)		−25 23
LHA ♈		73

	Dubhe Altitude	*Dubhe* Az.	*RIGEL* Altitude	*RIGEL* Az.	*Alpheratz* Altitude	*Alpheratz* Az.
	° ′		° ′		° ′	
Sextant altitude	37 46		35 55		33 15	
Instrument error and dome refraction	−5		−5		−6	
Refraction (Table **8**)	−1		−1		−1	
Corrected sextant altitude (Ho)	37 40		35 49		33 08	
Tables, p. 68 assumed Lat. 46° N, LHA ♈ 73°	38 08	037°	35 34	173°	33 11	279°
Intercept	28 *away*		15 *towards*		3 *away*	

The adjustments to these intercepts, for changes in position of observer and body, are found as follows:

Star	Azimuth	True Track	Relative Azimuth	Table 1	Table 2	Sum	Time Interval	Corrections to Intercept	Adjusted Intercept
	°	°	°	′	′	′	m s	′	′
Dubhe	037	257	140	−20	+25	+5	+2 45	3 *towards*	25 *away*
RIGEL	173	257	276	+3	+5	+8	−0 33	1 *away*	14 *towards*
Alpheratz	279	257	022	+25	−41	−16	−4 55	20 *towards*	17 *towards*

INTRODUCTION

POLE STAR TABLES

Table **6** on page 324, facing the inside back cover, gives the *Q* correction to be applied to the corrected sextant altitude of *Polaris*, in the same form as in *The Air Almanac*; the only difference is that it is based on the position of *Polaris* for epoch 1995·0. Refraction is not included. It should be noted that the table in *The Air Almanac* is recalculated each year and is therefore slightly more accurate than Table 6.

Table **7** gives the azimuth of *Polaris*, to 0°·1, for latitudes up to N 70° and for all hour angles; interpolation in LHA ♈ may sometimes be necessary.

Example. On 1993 January 1 at 02ʰ 43ᵐ 32ˢ UT at height 10 000 ft. (3 km), in longitude W 48° 18′, an observation was made of the altitude of *Polaris*, sextant reading 54° 51′, instrument error and dome refraction −4′; the latitude is found as follows:

From *The Air Almanac*,

	°	′
GHA ♈ (02ʰ 40ᵐ)	140	46
Increment (3ᵐ 32ˢ)	0	53
GHA ♈ (02ʰ 43ᵐ 32ˢ)	141	39
Longitude (west, subtract)	−48	18
LHA ♈	93	21

	°	′
Sextant altitude	54	51
Instrument error, etc.		−4
Refraction (Table **8**)		−1
Corrected sextant altitude (Ho)	54	46
(Table **6**, LHA ♈ = 93° 21′)		−25
Latitude	54	21

A correction is theoretically necessary for precession and nutation. Table **5** indicates that the deduced position line (here a parallel of latitude) should be shifted a distance of 1 mile in direction 270°; this leaves the latitude unchanged. The position line should, of course, be shifted for Coriolis acceleration.

Entering Table **7** with the nearest latitude (N 55°) and the value of LHA ♈ (93°), the azimuth of Polaris is found as 358°·9.

SPECIAL TECHNIQUES

The arrangement of the tabulations in this volume lends itself to the use of special techniques of observation and reduction, designed to save calculation and plotting or to allow for precomputation. These techniques are not fully described here, but the principles upon which they are based are given below; users will doubtless develop methods to suit their own requirements.

1. If the interval between observations is four minutes (4ᵐ), or a multiple of 4ᵐ, *The Air Almanac* need only be used to calculate LHA ♈ for one of the observations, since GHA ♈ changes by 1° (to within the accuracy of these tables) in 4ᵐ. For the remaining observations, the same value of LHA ♈ can be used and the intercepts plotted from assumed positions adjusted by the appropriate number of whole degrees of longitude; alternatively the same assumed position can be used and the values of LHA ♈ adjusted by the appropriate number of whole degrees. Since the rate of change of GHA ♈ is not exactly 1° in 4ᵐ these procedures are most accurately used for a three-star fix when LHA ♈ is calculated for the middle observation.

For latitudes greater than 69° (for which LHA ♈ is tabulated in even degrees only) the alternative procedure may be used with an 8ᵐ interval between observations, or with a 4ᵐ interval providing that assumed positions are selected which differ by 1° of longitude and which, together with 1° adjustment to LHA ♈ for the 4ᵐ interval, produce values of LHA ♈ in even degrees.

2. By making the observations at predetermined times ("scheduled shooting"), the tabulated altitudes and azimuths can be extracted beforehand and the same values used both for presetting the sextant and for the subsequent reduction of the sights.

3. All corrections, normally applied to the sextant altitude, may be applied to the tabulated altitude (with reversed signs), or to the assumed position, before an observation is made; similarly, corrections for Coriolis acceleration (Table **9**) and precession and nutation (Table **5**) may be applied to the assumed position, and the respective azimuth and its reciprocal drawn from it before an observation is made, thus enabling the intercept to be measured off (along the azimuth line '*towards*' or its reciprocal '*away*'), and the position line to be drawn (perpendicular to the azimuth line, through the end of the intercept) very quickly after the observation.

4. GHA ♈ may, if necessary, be deduced from Table **4**, on pages 320 and 321, for any date and time in the years 1991-1999 without reference to *The Air Almanac*.

Example. On 1993 January 1 the DR position at $01^h\ 00^m$ UT of an aircraft flying at a height of 18 000 ft., (5 km), on track 345°T, and with a ground speed of 300 knots is S 10° 55′, E 47° 05′. It has been decided to use the alternative procedure given in the first special technique described above, and observations are made with an artificial-horizon sextant having no instrument error, as follows:

	UT	Sextant Altitude
	h m s	° ′
ARCTURUS	00 56 00	30 21
ACRUX	01 00 00	35 24
PROCYON	01 04 00	38 36

	ARCTURUS		*ACRUX*			*PROCYON*	
	UT		UT	GHA ♈		UT	
	h m s		h m s	° ′		h m s	
	00 56 00		01 00 00	115 41		01 04 00	
Assumed longitude (east, add)				+47 19			
LHA ♈		162° 00′		163 00			164° 00′
	Altitude	Az.	Altitude		Az.	Altitude	Az.
	° ′		° ′			° ′	
Sextant altitude	30 21		35 24			38 36	
Refraction (Table **8**)	−1		−1			−1	
Corrected sextant altitude (Ho)	30 20		35 23			38 35	
Tables, p. 182 assumed Lat. 11°S and LHA ♈ as above; Hc and Zn	30 39	060°	35 17		167°	38 22	286°
Intercept		19 *away*	6 *towards*			13 *towards*	

In this example, all observations are plotted from latitude S 11° 00′, longitude E 47° 19′, adjusted for the effect of Coriolis acceleration, precession and nutation, or these corrections may be made to the position lines or to the final fix. The correction to be applied for the effect of Coriolis acceleration (Table **9**) is 1 mile to port (left) of track (for southern latitudes), i.e. in direction 255°T, and that for precession and nutation (Table **5**) is 1 mile in direction 290°T.

A widely used method of precomputation not limited to a specific time-interval between observations, is illustrated in the following example. All observations are made before the desired time of fix, and the corrections from Tables **1** and **2** are applied to the sextant altitudes; thus the signs for these are used as they appear in the tables.

INTRODUCTION

Example. On 1993 January 1 at $03^h\ 00^m$ UT the DR position of an aircraft flying at a height of 30 000 ft., (9 km) is predicted to be S 42° 50′, E 12° 10′. The aircraft is on track 290°T, with a ground speed of 600 knots, and a three-star fix is desired for $03^h\ 00^m$ UT. The following precomputations are made before any observations are taken:

	° ′
GHA ♈ at $03^h\ 00^m$ UT	145 46
Assumed longitude (east, add)	+12 14
LHA ♈	158 00

Entering the tables with an assumed latitude of 43° S, and LHA ♈ 158° (p. 246), it is decided to observe *SPICA*, *ACHERNAR* and *PROCYON*. The respective Hc and Zn for each star is extracted, their respective relative azimuths calculated, and the corrections from Tables **1** and **2** determined for 1 minute of time.

	SPICA	*ACHERNAR*	*PROCYON*
	° ′	° ′	° ′
Tabulated altitude (Hc)	40 53	17 30	27 55
Azimuth (Zn)	063	204	309
Relative Azimuth	133	274	019
Correction for 4^m (Table 1)	−27	+2	+38
Correction for 4^m (Table 2)	+39	−18	−34
Combined correction for 4^m	+12	−16	+4
Combined correction for 1^m	+3.0	−4.0	+1.0

One decimal place is required in the combined correction for 1^m to avoid the introduction of errors when multiplying by the time-interval.

The combined corrections for 1^m may be obtained in a similar manner, but without the division by 4, by use of Alternative Tables **1** and **2**, altitude corrections for change in position respectively of observer and body for 1 minute of time, which are included in this volume as an additional bookmark.

After the precomputations above have been completed, observations are made with an artificial-horizon sextant having no instrument error, as follows:

	UT	Sextant Altitude
	h m s	° ′
SPICA	02 50 00	40 31
ACHERNAR	02 53 00	17 45
PROCYON	02 57 00	28 02

These observations are corrected for refraction (Table **8**), and then for the combined corrections from Tables **1** and **2** to advance each observation to $03^h\ 00^m$ UT. *SPICA* is advanced 10^m, so the correction to be applied to *SPICA* is 10 × (+3′·0) = +30′; *ACHERNAR* is advanced 7^m, so the correction is 7 × (−4′·0) = −28′; *PROCYON* is advanced 3^m, so the correction is 3 × (+1′·0) = +3′. The adjusted, corrected sextant altitude (Ho) is then compared with the tabulated altitude (Hc) for each body, and the fix is plotted in the usual manner from the one assumed position (S 43° 00′, E 12° 14′) which was used to calculate LHA ♈. Corrections for Coriolis acceleration (Table **9**), 11 miles to port (left) of track (for southern latitude), i.e. in direction 200°T, and for precession and nutation (Table **5**), 1 mile in direction 300°T, may be applied to the assumed position before the intercepts are plotted, to the position lines or to the fix obtained.

	SPICA		*ACHERNAR*		*PROCYON*	
UT	h m s 02 50 00		h m s 02 53 00		h m s 02 57 00	
	Altitude	Az.	Altitude	Az.	Altitude	Az.
	° ′		° ′		° ′	
Sextant altitude	40 31		17 45		28 02	
Refraction (Table **8**)	0		−1		−1	
Combined correction (Tables **1** and **2**)	+30		−28		+3	
Adjusted Ho	41 01		17 16		28 04	
Tables, p. 246 assumed Lat. 43°S and LHA ♈ 158°; Hc and Zn	40 53	063°	17 30	204°	27 55	309°
Intercept		8 *towards*		14 *away*		9 *towards*

SIGHT REDUCTION TABLES
FOR
AIR NAVIGATION

TABULATIONS FOR SELECTED STARS EPOCH 1995·0

LAT 89°N

LHA ♈	Hc	Zn	Hc	Zn	Hc	Zn	Hc	Zn	Hc	Zn	Hc	Zn	Hc	Zn
	Dubhe		POLLUX		◆CAPELLA		Alpheratz		◆DENEB		VEGA		◆ARCTURUS	
0	60 48	014	27 36	063	46 10	100	30 04	178	45 54	230	38 56	262	18 23	326
2	60 49	016	27 37	065	46 12	102	30 04	180	45 53	232	38 54	264	18 21	328
4	60 50	018	27 39	067	46 15	104	30 04	182	45 51	235	38 52	266	18 20	330
6	60 50	020	27 41	069	46 17	106	30 04	184	45 49	237	38 50	268	18 19	332
8	60 51	021	27 43	071	46 19	108	30 03	186	45 47	239	38 48	270	18 18	334
10	60 52	023	27 45	073	46 21	110	30 03	188	45 46	241	38 45	272	18 18	336
12	60 53	025	27 47	075	46 23	112	30 03	190	45 44	243	38 43	274	18 17	338
14	60 54	027	27 49	077	46 24	114	30 02	192	45 42	245	38 41	276	18 16	340
16	60 55	029	27 51	079	46 26	116	30 02	194	45 40	247	38 39	278	18 15	342
18	60 56	031	27 53	081	46 28	118	30 01	196	45 38	249	38 37	280	18 15	344
20	60 57	033	27 56	083	46 30	120	30 01	198	45 36	251	38 35	282	18 14	346
22	60 58	035	27 58	085	46 32	122	30 00	200	45 34	253	38 33	284	18 14	348
24	60 59	037	28 00	087	46 34	124	29 59	202	45 32	255	38 31	286	18 13	350
26	61 00	039	28 02	089	46 35	126	29 59	204	45 30	257	38 29	288	18 13	352
28	61 02	041	28 04	091	46 37	128	29 58	206	45 28	259	38 27	290	18 13	354
	Dubhe		POLLUX		◆CAPELLA		Alpheratz		◆DENEB		VEGA		◆ARCTURUS	
30	61 03	043	28 06	093	46 39	130	29 57	208	45 26	261	38 25	292	18 13	356
32	61 05	045	28 08	095	46 40	132	29 56	210	45 24	263	38 23	294	18 13	358
34	61 06	047	28 10	097	46 42	134	29 55	212	45 22	265	38 21	296	18 12	000
36	61 08	049	28 12	099	46 43	136	29 53	214	45 20	267	38 19	298	18 13	002
38	61 09	051	28 14	101	46 45	138	29 52	216	45 18	269	38 17	300	18 13	004
40	61 11	053	28 16	103	46 46	140	29 51	218	45 16	271	38 16	301	18 13	006
42	61 13	055	28 18	105	46 47	142	29 50	220	45 13	273	38 14	303	18 13	008
44	61 14	057	28 20	107	46 49	144	29 48	222	45 11	275	38 12	305	18 13	010
46	61 16	059	28 22	109	46 50	146	29 47	224	45 09	277	38 11	307	18 14	012
48	61 18	061	28 24	111	46 51	148	29 45	226	45 07	279	38 09	309	18 14	014
50	61 20	062	28 26	113	46 52	150	29 44	228	45 05	281	38 07	311	18 15	016
52	61 22	064	28 28	115	46 53	152	29 42	230	45 03	283	38 06	313	18 15	018
54	61 24	066	28 30	117	46 54	154	29 41	232	45 01	285	38 04	315	18 16	020
56	61 25	068	28 32	119	46 55	157	29 39	234	44 59	287	38 03	317	18 17	022
58	61 27	070	28 34	121	46 56	159	29 37	236	44 57	289	38 01	319	18 18	024
	◆ARCTURUS		REGULUS		POLLUX		◆CAPELLA		Alpheratz		◆DENEB		VEGA	
60	18 19	026	11 57	088	28 35	123	46 56	161	29 35	238	44 55	291	38 00	321
62	18 20	028	11 59	090	28 37	125	46 57	163	29 34	240	44 53	293	37 59	323
64	18 21	030	12 01	092	28 39	127	46 57	165	29 32	242	44 51	295	37 58	325
66	18 22	032	12 04	094	28 41	129	46 58	167	29 30	244	44 49	297	37 56	327
68	18 23	034	12 06	096	28 42	131	46 58	169	29 28	246	44 47	299	37 55	329
70	18 24	036	12 08	098	28 44	133	46 59	171	29 26	248	44 46	301	37 54	331
72	18 25	038	12 10	100	28 45	135	46 59	173	29 24	250	44 44	303	37 53	333
74	18 27	040	12 12	102	28 47	137	46 59	175	29 22	253	44 42	305	37 52	335
76	18 28	042	12 14	104	28 48	139	47 00	177	29 20	255	44 40	306	37 52	337
78	18 29	044	12 16	106	28 49	141	47 00	179	29 18	257	44 39	308	37 51	339
80	18 31	046	12 18	108	28 51	143	47 00	181	29 16	259	44 37	310	37 50	341
82	18 32	048	12 20	110	28 52	145	47 00	183	29 14	261	44 36	312	37 49	343
84	18 34	050	12 22	112	28 53	147	46 59	185	29 12	263	44 34	314	37 49	345
86	18 36	052	12 24	114	28 54	149	46 59	187	29 10	265	44 33	316	37 48	347
88	18 37	054	12 26	116	28 55	151	46 59	189	29 08	267	44 31	318	37 48	349
	◆ARCTURUS		REGULUS		POLLUX		◆CAPELLA		Alpheratz		◆DENEB		VEGA	
90	18 39	056	12 28	118	28 56	154	46 58	191	29 06	269	44 30	320	37 47	351
92	18 41	058	12 29	120	28 57	156	46 58	193	29 04	271	44 28	322	37 47	353
94	18 42	060	12 31	122	28 58	158	46 58	195	29 01	273	44 27	324	37 47	355
96	18 44	062	12 33	124	28 59	160	46 57	197	28 59	275	44 26	326	37 47	357
98	18 46	064	12 35	126	28 59	162	46 56	199	28 57	277	44 25	328	37 47	359
100	18 48	066	12 36	128	29 00	164	46 56	201	28 55	279	44 24	330	37 47	001
102	18 50	068	12 38	130	29 00	166	46 55	203	28 53	281	44 23	332	37 47	003
104	18 52	070	12 40	132	29 01	168	46 54	205	28 51	283	44 22	334	37 47	005
106	18 54	072	12 41	134	29 01	170	46 53	207	28 49	285	44 21	336	37 47	007
108	18 56	074	12 43	136	29 02	172	46 52	209	28 47	286	44 20	338	37 47	009
110	18 58	076	12 44	138	29 02	174	46 51	211	28 45	288	44 19	340	37 48	011
112	19 00	078	12 45	140	29 02	176	46 50	213	28 43	290	44 19	342	37 48	013
114	19 02	080	12 47	142	29 02	178	46 49	216	28 41	292	44 18	344	37 49	015
116	19 04	082	12 48	144	29 02	180	46 47	218	28 39	294	44 18	346	37 49	017
118	19 06	084	12 49	146	29 02	182	46 46	220	28 37	296	44 17	348	37 50	019
	VEGA		◆ARCTURUS		REGULUS		POLLUX		◆CAPELLA		Alpheratz		◆DENEB	
120	37 51	021	19 08	086	12 50	148	29 02	184	46 45	222	28 35	298	44 17	350
122	37 51	023	19 10	088	12 51	150	29 02	186	46 43	224	28 34	300	44 16	352
124	37 52	024	19 12	090	12 52	152	29 02	188	46 42	226	28 32	302	44 16	354
126	37 53	026	19 15	092	12 53	154	29 01	190	46 40	228	28 30	304	44 16	356
128	37 54	028	19 17	094	12 54	156	29 01	192	46 39	230	28 28	306	44 16	358
130	37 55	030	19 19	096	12 55	158	29 01	194	46 37	232	28 27	308	44 16	000
132	37 56	032	19 21	098	12 56	160	29 00	196	46 35	234	28 25	310	44 16	002
134	37 57	034	19 23	100	12 57	162	28 59	198	46 34	236	28 23	312	44 16	004
136	37 59	036	19 25	102	12 57	164	28 59	200	46 32	238	28 22	314	44 16	006
138	38 00	038	19 27	104	12 58	166	28 58	202	46 30	240	28 20	316	44 16	008
140	38 01	040	19 29	106	12 58	168	28 57	204	46 28	242	28 19	318	44 17	010
142	38 03	042	19 31	108	12 59	170	28 56	206	46 26	244	28 18	320	44 17	011
144	38 04	044	19 33	110	12 59	172	28 55	208	46 25	246	28 16	322	44 17	013
146	38 05	046	19 35	112	12 59	174	28 54	210	46 23	248	28 15	324	44 18	015
148	38 07	048	19 37	114	12 59	176	28 53	212	46 21	250	28 14	326	44 19	017
	◆DENEB		VEGA		◆ARCTURUS		REGULUS		POLLUX		◆CAPELLA		Alpheratz	
150	44 19	019	38 09	050	19 39	116	12 59	178	28 52	214	46 19	252	28 13	328
152	44 20	021	38 10	052	19 41	118	13 00	180	28 51	216	46 17	254	28 12	330
154	44 21	023	38 12	054	19 42	120	12 59	182	28 50	218	46 15	256	28 11	332
156	44 22	025	38 14	056	19 44	122	12 59	184	28 48	220	46 13	258	28 10	334
158	44 22	027	38 15	058	19 46	124	12 59	186	28 47	222	46 11	260	28 09	336
160	44 23	029	38 17	060	19 48	126	12 59	188	28 46	224	46 09	262	28 08	338
162	44 25	031	38 19	062	19 49	128	12 59	190	28 44	226	46 06	264	28 07	340
164	44 26	033	38 21	064	19 51	130	12 58	192	28 43	228	46 04	266	28 07	342
166	44 27	035	38 23	066	19 53	132	12 58	194	28 41	230	46 02	268	28 06	344
168	44 28	037	38 25	068	19 54	134	12 57	196	28 39	232	46 00	270	28 06	346
170	44 29	039	38 27	070	19 56	136	12 57	198	28 38	234	45 58	272	28 05	348
172	44 31	041	38 29	072	19 57	138	12 56	200	28 36	236	45 56	274	28 05	350
174	44 32	043	38 31	074	19 58	140	12 55	202	28 34	238	45 54	276	28 04	352
176	44 34	045	38 33	076	20 00	142	12 54	204	28 32	240	45 52	278	28 04	354
178	44 35	047	38 35	078	20 01	144	12 53	206	28 30	242	45 50	280	28 04	356

LHA ♈	Hc	Zn	Hc	Zn	Hc	Zn	Hc	Zn	Hc	Zn	Hc	Zn	Hc	Zn
	◆DENEB		VEGA		◆ARCTURUS		REGULUS		POLLUX		◆CAPELLA		Alpheratz	
180	44 37	049	38 37	080	20 02	146	12 52	208	28 29	244	45 48	282	28 04	358
182	44 38	051	38 39	082	20 03	148	12 51	210	28 27	246	45 46	284	28 04	000
184	44 40	053	38 41	084	20 04	150	12 50	212	28 25	248	45 44	286	28 04	002
186	44 42	055	38 43	086	20 06	152	12 49	214	28 23	250	45 42	288	28 04	004
188	44 43	057	38 45	088	20 06	154	12 48	216	28 21	252	45 40	290	28 04	006
190	44 45	059	38 47	090	20 07	156	12 47	218	28 19	254	45 38	292	28 04	008
192	44 47	061	38 49	092	20 08	158	12 45	220	28 17	256	45 36	294	28 05	010
194	44 49	063	38 51	094	20 09	160	12 44	222	28 15	258	45 34	296	28 05	012
196	44 51	065	38 53	096	20 10	162	12 43	224	28 13	260	45 32	298	28 06	014
198	44 53	067	38 56	098	20 10	164	12 41	226	28 11	262	45 30	300	28 06	016
200	44 54	069	38 58	100	20 11	166	12 40	228	28 09	264	45 28	302	28 07	018
202	44 56	071	39 00	102	20 11	168	12 38	230	28 06	266	45 27	304	28 07	020
204	44 58	073	39 02	104	20 12	170	12 36	232	28 04	268	45 25	306	28 08	022
206	45 00	075	39 04	106	20 12	172	12 35	234	28 02	270	45 23	308	28 09	024
208	45 02	077	39 06	108	20 12	174	12 33	236	28 00	272	45 22	310	28 10	026
	Alpheratz		◆DENEB		VEGA		◆ARCTURUS		REGULUS		POLLUX		◆CAPELLA	
210	28 11	028	45 04	079	39 08	110	20 12	176	12 31	238	27 58	274	45 20	312
212	28 12	030	45 07	081	39 10	112	20 12	178	12 29	240	27 56	276	45 18	314
214	28 13	032	45 09	083	39 12	114	20 12	180	12 28	242	27 54	278	45 17	316
216	28 14	034	45 11	085	39 13	116	20 12	182	12 26	244	27 52	280	45 16	318
218	28 15	036	45 13	087	39 15	118	20 12	184	12 24	246	27 50	282	45 14	320
220	28 16	038	45 15	089	39 17	120	20 12	186	12 22	248	27 48	284	45 13	322
222	28 18	040	45 17	091	39 19	122	20 12	188	12 20	250	27 46	286	45 12	324
224	28 19	042	45 19	093	39 21	124	20 12	190	12 18	252	27 44	288	45 10	326
226	28 20	044	45 21	095	39 22	126	20 11	192	12 16	254	27 42	290	45 09	327
228	28 22	046	45 23	097	39 24	128	20 11	194	12 14	256	27 40	292	45 08	329
230	28 23	048	45 25	099	39 26	130	20 10	196	12 12	258	27 38	294	45 07	331
232	28 25	050	45 27	101	39 27	132	20 09	198	12 10	260	27 36	296	45 06	333
234	28 27	052	45 29	103	39 29	134	20 09	200	12 08	262	27 34	298	45 05	335
236	28 28	054	45 31	105	39 30	136	20 08	202	12 06	264	27 32	300	45 04	337
238	28 30	056	45 33	107	39 32	138	20 07	204	12 04	266	27 31	302	45 04	339
	Alpheratz		◆DENEB		VEGA		◆ARCTURUS		REGULUS		POLLUX		◆CAPELLA	
240	28 32	058	45 35	109	39 33	140	20 06	206	12 02	268	27 29	304	45 03	341
242	28 34	059	45 37	111	39 34	142	20 05	208	11 59	270	27 27	306	45 02	343
244	28 35	061	45 39	113	39 36	144	20 04	210	11 57	272	27 25	308	45 02	345
246	28 37	063	45 41	115	39 37	146	20 03	212	11 55	274	27 24	310	45 01	347
248	28 39	065	45 43	117	39 38	148	20 02	214	11 53	276	27 22	312	45 01	349
250	28 41	067	45 45	119	39 39	150	20 01	216	11 51	278	27 21	314	45 00	351
252	28 43	069	45 47	121	39 40	152	20 00	218	11 49	280	27 19	316	45 00	353
254	28 45	071	45 49	123	39 41	154	19 58	220	11 47	282	27 18	318	45 00	355
256	28 47	073	45 50	125	39 42	156	19 57	222	11 45	284	27 16	320	45 00	357
258	28 49	075	45 52	127	39 43	159	19 55	224	11 43	286	27 15	322	45 00	359
260	28 51	077	45 54	129	39 43	161	19 54	226	11 41	288	27 14	324	45 00	001
262	28 53	079	45 55	131	39 44	163	19 52	228	11 39	290	27 13	326	45 00	003
264	28 55	081	45 57	133	39 45	165	19 51	230	11 37	292	27 11	328	45 00	005
266	28 57	083	45 58	135	39 45	167	19 49	232	11 35	294	27 10	330	45 00	007
268	28 59	085	46 00	137	39 46	169	19 48	234	11 33	296	27 09	332	45 00	009
	◆CAPELLA		Alpheratz		◆DENEB		VEGA		◆ARCTURUS		Dubhe		POLLUX	
270	45 01	011	29 01	087	46 01	139	39 46	171	19 46	236	61 31	286	27 08	334
272	45 01	013	29 03	089	46 03	141	39 46	173	19 44	238	61 29	288	27 08	336
274	45 02	015	29 06	091	46 04	143	39 46	175	19 42	240	61 27	290	27 07	338
276	45 02	017	29 08	093	46 05	145	39 47	177	19 40	242	61 25	292	27 06	340
278	45 03	019	29 10	095	46 06	147	39 47	179	19 39	244	61 23	294	27 05	342
280	45 03	021	29 12	097	46 07	149	39 47	181	19 37	246	61 21	296	27 05	344
282	45 04	023	29 14	099	46 08	151	39 47	183	19 35	248	61 19	298	27 04	346
284	45 05	024	29 16	101	46 09	153	39 47	185	19 33	250	61 18	300	27 04	348
286	45 06	026	29 18	103	46 10	155	39 46	187	19 31	252	61 16	302	27 03	350
288	45 07	028	29 20	105	46 11	157	39 46	189	19 29	254	61 14	304	27 03	352
290	45 08	030	29 22	107	46 12	159	39 46	191	19 27	256	61 12	306	27 03	354
292	45 09	032	29 24	109	46 13	161	39 45	193	19 25	258	61 11	308	27 02	356
294	45 10	034	29 26	111	46 13	163	39 45	195	19 23	260	61 09	310	27 02	358
296	45 11	036	29 28	113	46 14	165	39 44	197	19 21	262	61 07	312	27 02	000
298	45 13	038	29 30	115	46 14	167	39 43	199	19 18	264	61 06	313	27 02	002
	POLLUX		◆CAPELLA		Alpheratz		◆DENEB		VEGA		◆ARCTURUS		Dubhe	
300	27 02	004	45 14	040	29 32	117	46 15	169	39 43	201	19 16	266	61 04	315
302	27 03	006	45 15	042	29 34	119	46 15	172	39 42	203	19 14	268	61 03	317
304	27 03	008	45 17	044	29 35	121	46 15	174	39 41	205	19 12	270	61 02	319
306	27 03	010	45 18	046	29 37	124	46 16	176	39 40	207	19 10	272	61 00	321
308	27 04	012	45 20	048	29 39	126	46 16	178	39 39	209	19 08	274	60 59	323
310	27 04	014	45 21	050	29 41	128	46 16	180	39 38	211	19 06	276	60 58	325
312	27 05	016	45 23	052	29 42	130	46 16	182	39 37	213	19 04	278	60 57	327
314	27 05	018	45 25	054	29 44	132	46 16	184	39 36	215	19 02	280	60 55	329
316	27 06	020	45 26	056	29 45	134	46 15	186	39 35	217	19 00	282	60 54	331
318	27 07	022	45 28	058	29 47	136	46 15	188	39 33	219	18 58	284	60 53	333
320	27 07	024	45 30	060	29 48	138	46 15	190	39 32	221	18 56	286	60 53	335
322	27 08	026	45 32	062	29 50	140	46 14	192	39 31	223	18 54	288	60 52	337
324	27 09	028	45 34	064	29 51	142	46 14	194	39 29	225	18 52	290	60 51	339
326	27 10	029	45 36	066	29 52	144	46 13	196	39 28	227	18 50	292	60 50	341
328	27 11	031	45 38	068	29 53	146	46 13	198	39 26	229	18 48	294	60 49	343
	POLLUX		◆CAPELLA		Alpheratz		◆DENEB		VEGA		◆ARCTURUS		Dubhe	
330	27 12	033	45 40	070	29 55	148	46 12	200	39 24	231	18 46	296	60 49	345
332	27 14	035	45 41	072	29 56	150	46 11	202	39 23	233	18 44	298	60 48	347
334	27 15	037	45 43	074	29 57	152	46 11	204	39 21	235	18 42	300	60 48	349
336	27 16	039	45 46	076	29 58	154	46 10	206	39 19	237	18 40	302	60 48	350
338	27 17	041	45 48	078	29 59	156	46 09	208	39 17	240	18 39	304	60 47	352
340	27 19	043	45 50	080	29 59	158	46 08	210	39 16	242	18 37	306	60 47	354
342	27 20	045	45 52	082	30 00	160	46 07	212	39 14	244	18 35	308	60 47	356
344	27 22	047	45 54	084	30 01	162	46 06	214	39 12	246	18 34	310	60 47	358
346	27 23	049	45 56	086	30 01	164	46 04	216	39 10	248	18 32	312	60 47	000
348	27 25	051	45 58	088	30 02	166	46 03	218	39 08	250	18 31	314	60 47	002
350	27 27	053	46 00	090	30 02	168	46 02	220	39 06	252	18 29	316	60 47	004
352	27 28	055	46 02	092	30 03	170	46 00	222	39 04	254	18 28	318	60 47	006
354	27 30	057	46 04	094	30 03	172	45 59	224	39 02	256	18 26	320	60 47	008
356	27 32	059	46 06	096	30 03	174	45 57	226	39 00	258	18 25	322	60 48	010
358	27 34	061	46 08	098	30 04	176	45 56	228	38 58	260	18 24	324	60 48	012

LAT 88°N

LHA ♈	Hc	Zn	Hc	Zn	Hc	Zn	Hc	Zn	Hc	Zn	Hc	Zn	Hc	Zn
	Dubhe		POLLUX		◆CAPELLA		Alpheratz		◆DENEB		VEGA		◆ARCTURUS	
0	59 50	013	27 08	063	46 20	099	31 04	178	46 32	231	39 04	262	17 33	327
2	59 51	015	27 12	065	46 24	101	31 04	180	46 29	233	39 00	264	17 30	328
4	59 52	017	27 16	067	46 28	103	31 04	182	46 25	235	38 56	266	17 28	330
6	59 54	019	27 20	069	46 33	105	31 03	184	46 22	237	38 52	268	17 26	332
8	59 55	021	27 24	071	46 37	107	31 03	186	46 18	239	38 48	270	17 24	334
10	59 57	023	27 28	073	46 41	109	31 03	188	46 15	241	38 43	272	17 23	336
12	59 58	025	27 32	075	46 44	111	31 02	190	46 11	244	38 39	274	17 21	338
14	60 00	026	27 36	077	46 48	113	31 01	192	46 07	246	38 35	276	17 20	340
16	60 02	028	27 40	079	46 52	115	31 00	194	46 03	248	38 31	278	17 18	342
18	60 04	030	27 44	081	46 56	117	30 59	196	45 59	250	38 27	280	17 17	344
20	60 06	032	27 48	083	47 00	119	30 58	198	45 56	252	38 23	282	17 16	346
22	60 08	034	27 52	085	47 03	121	30 56	200	45 52	254	38 19	284	17 15	348
24	60 11	036	27 56	087	47 07	123	30 55	202	45 47	256	38 15	286	17 14	350
26	60 13	038	28 01	089	47 10	125	30 53	204	45 43	258	38 11	288	17 14	352
28	60 16	040	28 05	091	47 14	127	30 51	206	45 39	260	38 07	290	17 13	354
	Dubhe		POLLUX		◆CAPELLA		Alpheratz		◆DENEB		VEGA		◆ARCTURUS	
30	60 19	042	28 09	093	47 17	129	30 49	209	45 35	262	38 03	292	17 13	356
32	60 22	044	28 13	095	47 20	131	30 47	211	45 31	264	37 59	294	17 13	358
34	60 25	046	28 17	097	47 23	133	30 45	213	45 27	266	37 55	296	17 12	000
36	60 28	047	28 22	099	47 26	135	30 43	215	45 23	268	37 51	298	17 13	002
38	60 31	049	28 26	101	47 29	138	30 40	217	45 18	270	37 48	300	17 13	004
40	60 34	051	28 30	103	47 32	140	30 38	219	45 14	272	37 44	302	17 13	006
42	60 37	053	28 34	105	47 35	142	30 35	221	45 10	274	37 41	304	17 14	008
44	60 41	055	28 38	107	47 37	144	30 32	223	45 06	276	37 37	306	17 14	010
46	60 44	057	28 42	109	47 39	146	30 30	225	45 02	278	37 34	308	17 15	012
48	60 48	059	28 46	111	47 42	148	30 27	227	44 58	280	37 31	310	17 16	014
50	60 51	061	28 50	113	47 44	150	30 23	229	44 54	282	37 27	312	17 17	016
52	60 55	063	28 54	115	47 46	152	30 20	231	44 49	284	37 24	314	17 18	018
54	60 59	065	28 57	117	47 48	154	30 17	233	44 45	286	37 21	316	17 20	020
56	61 03	067	29 01	119	47 50	156	30 14	235	44 41	288	37 18	318	17 21	022
58	61 06	069	29 05	121	47 51	158	30 10	237	44 37	290	37 16	320	17 23	024
	◆ARCTURUS		REGULUS		POLLUX		◆CAPELLA		Alpheratz		◆DENEB		VEGA	
60	17 25	026	11 55	088	29 08	123	47 53	160	30 07	239	44 33	292	37 13	322
62	17 27	028	11 59	090	29 12	125	47 54	162	30 03	241	44 30	294	37 11	324
64	17 29	030	12 03	092	29 15	127	47 55	164	29 59	243	44 26	295	37 08	326
66	17 31	032	12 07	094	29 18	129	47 56	166	29 55	245	44 22	297	37 06	328
68	17 33	034	12 12	096	29 22	131	47 57	169	29 52	247	44 18	299	37 04	330
70	17 35	036	12 16	098	29 25	133	47 58	171	29 48	249	44 15	301	37 02	332
72	17 38	038	12 20	100	29 28	135	47 59	173	29 44	251	44 11	303	37 00	334
74	17 40	040	12 24	102	29 31	137	47 59	175	29 40	253	44 08	305	36 58	335
76	17 43	042	12 28	104	29 33	139	47 59	177	29 36	255	44 04	307	36 56	337
78	17 46	044	12 32	106	29 36	141	48 00	179	29 32	257	44 01	309	36 55	339
80	17 49	046	12 36	108	29 39	143	48 00	181	29 28	259	43 58	311	36 53	341
82	17 52	048	12 40	110	29 41	145	47 59	183	29 24	261	43 55	313	36 52	343
84	17 55	050	12 44	112	29 43	147	47 59	185	29 19	263	43 52	315	36 51	345
86	17 58	052	12 48	114	29 46	149	47 59	187	29 15	265	43 49	317	36 50	347
88	18 02	054	12 52	116	29 48	151	47 58	189	29 11	267	43 46	319	36 49	349
	◆ARCTURUS		REGULUS		POLLUX		◆CAPELLA		Alpheratz		◆DENEB		VEGA	
90	18 05	056	12 55	118	29 50	153	47 57	191	29 07	269	43 43	321	36 48	351
92	18 09	058	12 59	120	29 52	155	47 56	193	29 03	271	43 41	323	36 48	353
94	18 12	060	13 03	122	29 53	157	47 55	195	28 58	273	43 38	325	36 47	355
96	18 16	062	13 06	124	29 55	159	47 54	198	28 54	275	43 36	327	36 47	357
98	18 20	064	13 10	126	29 56	161	47 53	200	28 50	277	43 34	329	36 47	359
100	18 23	066	13 13	128	29 57	163	47 51	202	28 46	279	43 32	331	36 47	001
102	18 27	068	13 16	130	29 59	165	47 50	204	28 42	281	43 30	333	36 47	003
104	18 31	070	13 20	132	30 00	168	47 48	206	28 38	283	43 28	335	36 47	005
106	18 35	071	13 23	134	30 00	170	47 46	208	28 34	285	43 26	336	36 48	007
108	18 39	073	13 26	136	30 01	172	47 44	210	28 30	287	43 24	338	36 48	009
110	18 43	075	13 28	138	30 02	174	47 42	212	28 26	289	43 23	340	36 49	011
112	18 47	077	13 31	140	30 02	176	47 40	214	28 22	291	43 22	342	36 50	012
114	18 51	079	13 34	142	30 02	178	47 37	216	28 18	293	43 20	344	36 51	014
116	18 55	081	13 36	144	30 02	180	47 35	218	28 14	295	43 19	346	36 52	016
118	19 00	083	13 39	146	30 02	182	47 32	220	28 10	297	43 18	348	36 53	018
	VEGA		◆ARCTURUS		REGULUS		POLLUX		◆CAPELLA		Alpheratz		◆DENEB	
120	36 54	020	19 04	085	13 41	148	30 02	184	47 29	222	28 07	299	43 18	350
122	36 56	022	19 08	087	13 43	150	30 02	186	47 26	224	28 03	301	43 17	352
124	36 58	024	19 12	089	13 45	152	30 01	188	47 23	226	27 59	303	43 16	354
126	36 59	026	19 16	091	13 47	154	30 01	190	47 20	228	27 56	305	43 16	356
128	37 01	028	19 20	093	13 49	156	30 00	192	47 17	231	27 53	307	43 16	358
130	37 03	030	19 25	095	13 51	158	29 59	194	47 14	233	27 49	309	43 16	000
132	37 05	032	19 29	097	13 52	160	29 58	196	47 11	235	27 46	311	43 16	002
134	37 08	034	19 33	099	13 54	162	29 56	198	47 07	237	27 43	313	43 16	004
136	37 10	036	19 37	101	13 55	164	29 55	200	47 04	239	27 40	315	43 16	005
138	37 13	038	19 41	103	13 56	166	29 54	202	47 00	241	27 37	317	43 17	007
140	37 15	040	19 45	105	13 57	168	29 52	204	46 56	243	27 34	319	43 17	009
142	37 18	042	19 49	107	13 58	170	29 50	206	46 52	245	27 31	321	43 18	011
144	37 21	044	19 53	109	13 58	172	29 48	208	46 49	247	27 29	323	43 19	013
146	37 24	046	19 57	111	13 59	174	29 46	210	46 45	249	27 26	325	43 20	015
148	37 27	048	20 01	113	13 59	176	29 44	212	46 41	251	27 24	327	43 21	017
	◆DENEB		VEGA		◆ARCTURUS		REGULUS		POLLUX		◆CAPELLA		Alpheratz	
150	43 23	019	37 30	050	20 05	115	13 59	178	29 42	214	46 37	253	27 22	329
152	43 24	021	37 33	052	20 09	118	14 00	180	29 39	216	46 33	255	27 20	331
154	43 26	023	37 36	054	20 12	120	13 59	182	29 37	218	46 29	257	27 18	332
156	43 27	025	37 40	055	20 16	122	13 59	184	29 34	220	46 25	259	27 16	334
158	43 29	027	37 43	057	20 19	124	13 59	186	29 31	222	46 21	261	27 14	336
160	43 31	029	37 47	059	20 23	126	13 58	188	29 28	225	46 16	263	27 12	338
162	43 33	031	37 51	061	20 26	128	13 58	190	29 25	227	46 12	265	27 11	340
164	43 35	033	37 54	063	20 29	130	13 57	192	29 22	229	46 08	267	27 10	342
166	43 38	035	37 58	065	20 33	132	13 56	194	29 19	231	46 04	269	27 08	344
168	43 40	037	38 02	067	20 36	134	13 55	196	29 16	233	46 00	271	27 07	346
170	43 43	038	38 06	069	20 39	136	13 54	198	29 13	235	45 56	273	27 06	348
172	43 45	040	38 10	071	20 42	138	13 52	200	29 09	237	45 51	275	27 06	350
174	43 48	042	38 14	073	20 44	140	13 51	202	29 06	239	45 47	277	27 05	352
176	43 51	044	38 18	075	20 47	142	13 49	204	29 02	241	45 43	279	27 04	354
178	43 54	046	38 22	077	20 49	144	13 47	206	28 58	243	45 39	281	27 04	356

LHA ♈	Hc	Zn	Hc	Zn	Hc	Zn	Hc	Zn	Hc	Zn	Hc	Zn	Hc	Zn
	◆DENEB		VEGA		◆ARCTURUS		REGULUS		POLLUX		◆CAPELLA		Alpheratz	
180	43 57	048	38 26	079	20 52	146	13 45	208	28 54	245	45 35	283	27 04	358
182	44 00	050	38 30	081	20 54	148	13 43	210	28 51	247	45 31	285	27 04	000
184	44 03	052	38 34	083	20 56	150	13 41	212	28 47	249	45 27	287	27 04	002
186	44 07	054	38 38	085	20 58	152	13 39	214	28 43	251	45 23	289	27 04	004
188	44 10	056	38 43	087	21 00	154	13 36	216	28 39	253	45 19	291	27 04	006
190	44 14	058	38 47	089	21 02	156	13 34	218	28 35	255	45 15	293	27 05	008
192	44 17	060	38 51	091	21 04	158	13 31	220	28 31	257	45 11	295	27 06	010
194	44 21	062	38 55	093	21 05	160	13 29	222	28 27	259	45 07	297	27 06	012
196	44 25	064	38 59	095	21 07	162	13 26	224	28 23	261	45 04	299	27 07	014
198	44 28	066	39 03	097	21 08	164	13 23	226	28 18	263	45 00	301	27 08	016
200	44 32	068	39 08	099	21 09	166	13 20	228	28 14	265	44 56	303	27 10	018
202	44 36	070	39 12	101	21 10	168	13 16	230	28 10	267	44 53	305	27 11	020
204	44 40	072	39 16	103	21 11	170	13 13	232	28 06	269	44 49	307	27 12	022
206	44 44	074	39 20	105	21 11	172	13 10	234	28 02	271	44 46	309	27 14	024
208	44 48	076	39 24	107	21 12	174	13 06	236	27 58	273	44 43	310	27 16	026
	Alpheratz		◆DENEB		VEGA		◆ARCTURUS		REGULUS		POLLUX		◆CAPELLA	
210	27 18	027	44 52	078	39 28	109	21 12	176	13 03	238	27 53	275	44 40	312
212	27 20	029	44 56	080	39 32	111	21 12	178	12 59	240	27 49	277	44 37	314
214	27 22	031	45 00	082	39 36	113	21 12	180	12 56	242	27 45	279	44 34	316
216	27 24	033	45 05	084	39 39	115	21 12	182	12 52	244	27 41	281	44 31	318
218	27 26	035	45 09	086	39 43	117	21 12	184	12 48	246	27 37	283	44 28	320
220	27 29	037	45 13	088	39 47	119	21 12	186	12 44	248	27 33	285	44 26	322
222	27 31	039	45 17	090	39 51	121	21 11	188	12 40	250	27 29	287	44 23	324
224	27 34	041	45 21	092	39 54	123	21 11	190	12 36	252	27 25	289	44 21	326
226	27 37	043	45 26	094	39 58	125	21 10	192	12 32	254	27 21	291	44 18	328
228	27 40	045	45 30	096	40 01	128	21 09	194	12 28	256	27 17	293	44 16	330
230	27 43	047	45 34	098	40 04	130	21 08	196	12 24	258	27 13	295	44 14	332
232	27 46	049	45 38	100	40 07	132	21 06	198	12 20	260	27 09	297	44 12	334
234	27 49	051	45 42	102	40 10	134	21 05	200	12 16	262	27 06	299	44 11	336
236	27 52	053	45 46	104	40 13	136	21 04	202	12 12	264	27 02	301	44 09	338
238	27 56	055	45 50	106	40 16	138	21 02	204	12 07	266	26 58	303	44 07	340
	Alpheratz		◆DENEB		VEGA		◆ARCTURUS		REGULUS		POLLUX		◆CAPELLA	
240	27 59	057	45 54	108	40 19	140	21 00	206	12 03	268	26 55	305	44 06	342
242	28 03	059	45 58	110	40 22	142	20 58	208	11 59	270	26 51	307	44 05	343
244	28 06	061	46 02	112	40 24	144	20 56	211	11 55	272	26 48	309	44 04	345
246	28 10	063	46 06	114	40 27	146	20 54	213	11 51	274	26 45	311	44 03	347
248	28 14	065	46 10	116	40 29	148	20 52	215	11 47	276	26 42	313	44 02	349
250	28 18	067	46 14	118	40 31	150	20 49	217	11 42	278	26 39	314	44 01	351
252	28 22	069	46 17	120	40 33	152	20 47	219	11 38	280	26 36	316	44 00	353
254	28 26	071	46 21	122	40 35	154	20 44	221	11 34	282	26 33	318	44 00	355
256	28 30	073	46 24	124	40 37	156	20 41	223	11 30	284	26 30	320	44 00	357
258	28 34	075	46 28	126	40 38	158	20 38	225	11 26	286	26 28	322	44 00	359
260	28 38	077	46 31	128	40 40	160	20 35	227	11 22	288	26 25	324	44 00	001
262	28 42	079	46 34	130	40 41	162	20 32	229	11 18	290	26 23	326	44 00	003
264	28 46	081	46 37	132	40 42	164	20 29	231	11 14	292	26 21	328	44 00	005
266	28 50	083	46 41	134	40 43	166	20 26	233	11 10	294	26 18	330	44 00	007
268	28 54	085	46 43	136	40 44	168	20 22	235	11 07	296	26 16	332	44 01	009
	◆CAPELLA		Alpheratz		◆DENEB		VEGA		◆ARCTURUS		Dubhe		POLLUX	
270	44 02	011	28 58	087	46 46	138	40 45	171	20 19	237	61 14	288	26 14	334
272	44 03	012	29 03	089	46 49	140	40 46	173	20 15	239	61 10	290	26 13	336
274	44 04	014	29 07	091	46 52	142	40 46	175	20 12	241	61 06	292	26 11	338
276	44 05	016	29 11	093	46 54	144	40 47	177	20 08	243	61 02	294	26 10	340
278	44 06	018	29 15	095	46 57	147	40 47	179	20 04	245	60 58	295	26 08	342
280	44 07	020	29 19	097	46 59	149	40 47	181	20 00	247	60 54	297	26 07	344
282	44 09	022	29 23	099	47 01	151	40 47	183	19 57	249	60 51	299	26 06	346
284	44 10	024	29 28	101	47 03	153	40 46	185	19 53	251	60 47	301	26 05	348
286	44 12	026	29 32	103	47 05	155	40 46	187	19 49	253	60 44	303	26 04	350
288	44 14	028	29 36	105	47 06	157	40 45	189	19 45	255	60 40	305	26 04	352
290	44 16	030	29 40	107	47 08	159	40 45	191	19 41	257	60 37	307	26 03	354
292	44 18	032	29 44	109	47 09	161	40 44	193	19 36	259	60 33	309	26 03	356
294	44 21	034	29 48	111	47 11	163	40 43	195	19 32	261	60 30	311	26 02	358
296	44 23	036	29 52	113	47 12	165	40 41	197	19 28	263	60 27	313	26 02	000
298	44 25	038	29 55	115	47 13	167	40 40	199	19 24	265	60 24	315	26 02	002
	POLLUX		◆CAPELLA		Alpheratz		◆DENEB		VEGA		◆ARCTURUS		Dubhe	
300	26 03	004	44 28	040	29 59	117	47 14	169	40 39	201	19 20	267	60 21	317
302	26 03	006	44 31	042	30 03	119	47 14	171	40 37	203	19 16	269	60 18	319
304	26 03	008	44 34	044	30 06	121	47 15	173	40 35	206	19 11	271	60 16	320
306	26 04	010	44 37	045	30 10	123	47 15	176	40 33	208	19 07	273	60 13	322
308	26 05	012	44 40	047	30 13	125	47 16	178	40 31	210	19 03	275	60 11	324
310	26 06	014	44 43	049	30 17	127	47 16	180	40 29	212	18 59	277	60 08	326
312	26 07	015	44 46	051	30 20	129	47 16	182	40 27	214	18 55	279	60 06	328
314	26 08	017	44 49	053	30 23	131	47 15	184	40 25	216	18 51	281	60 04	330
316	26 09	019	44 53	055	30 26	133	47 15	186	40 22	218	18 47	283	60 02	332
318	26 11	021	44 56	057	30 29	135	47 15	188	40 20	220	18 43	285	60 00	334
320	26 12	023	45 00	059	30 32	137	47 14	190	40 17	222	18 38	287	59 58	336
322	26 14	025	45 03	061	30 35	139	47 13	192	40 14	224	18 34	289	59 56	338
324	26 16	027	45 07	063	30 38	141	47 12	194	40 11	226	18 31	291	59 55	339
326	26 18	029	45 11	065	30 40	143	47 11	196	40 08	228	18 27	293	59 53	341
328	26 20	031	45 15	067	30 43	145	47 10	198	40 05	230	18 23	295	59 52	343
	POLLUX		◆CAPELLA		Alpheratz		◆DENEB		VEGA		◆ARCTURUS		Dubhe	
330	26 22	033	45 18	069	30 45	147	47 08	200	40 02	232	18 19	297	59 51	345
332	26 25	035	45 22	071	30 47	149	47 07	202	39 58	234	18 15	299	59 50	347
334	26 27	037	45 26	073	30 49	151	47 05	205	39 55	236	18 12	301	59 49	349
336	26 30	039	45 30	075	30 51	153	47 03	207	39 51	238	18 08	303	59 48	351
338	26 32	041	45 34	077	30 53	155	47 02	209	39 48	240	18 05	305	59 48	353
340	26 35	043	45 39	079	30 55	158	46 59	211	39 44	242	18 01	307	59 47	354
342	26 38	045	45 43	081	30 56	160	46 57	213	39 40	244	17 58	309	59 47	356
344	26 41	047	45 47	083	30 58	162	46 55	215	39 36	246	17 55	311	59 47	358
346	26 44	049	45 51	085	30 59	164	46 52	217	39 33	248	17 52	313	59 47	000
348	26 47	051	45 55	087	31 00	166	46 50	219	39 29	250	17 49	315	59 47	002
350	26 51	053	45 59	089	31 01	168	46 47	221	39 25	252	17 46	317	59 47	004
352	26 54	055	46 04	091	31 02	170	46 44	223	39 21	254	17 43	319	59 47	006
354	26 57	057	46 08	093	31 03	172	46 41	225	39 17	256	17 40	321	59 48	008
356	27 01	059	46 12	095	31 03	174	46 38	227	39 13	258	17 37	323	59 48	010
358	27 05	061	46 16	097	31 03	176	46 35	229	39 08	260	17 35	325	59 49	011

LHA	Hc	Zn	Hc	Zn	Hc	Zn	Hc	Zn	Hc	Zn	Hc	Zn	Hc	Zn
♈	Dubhe		POLLUX		◆CAPELLA		Alpheratz		◆DENEB		VEGA		◆ARCTURUS	
0	58 52	013	26 41	062	46 29	098	32 04	178	47 09	232	39 12	263	16 43	327
2	58 53	015	26 46	064	46 35	100	32 04	180	47 04	234	39 06	265	16 39	329
4	58 55	017	26 52	066	46 41	102	32 04	182	46 59	236	38 59	267	16 36	331
6	58 57	018	26 58	068	46 47	104	32 03	184	46 54	238	38 53	269	16 33	333
8	58 59	020	27 04	070	46 54	106	32 03	186	46 48	240	38 47	271	16 30	335
10	59 01	022	27 10	072	47 00	108	32 02	188	46 43	242	38 40	273	16 28	337
12	59 04	024	27 16	074	47 05	110	32 01	190	46 37	244	38 34	275	16 25	338
14	59 06	026	27 22	076	47 11	112	32 00	192	46 32	246	38 28	277	16 23	340
16	59 09	028	27 28	078	47 17	114	31 58	194	46 26	249	38 22	279	16 21	342
18	59 12	029	27 34	080	47 23	116	31 57	196	46 20	251	38 16	281	16 19	344
20	59 15	031	27 40	082	47 28	118	31 55	199	46 14	253	38 09	283	16 18	346
22	59 19	033	27 46	084	47 34	120	31 53	201	46 08	255	38 03	285	16 16	348
24	59 22	035	27 53	086	47 39	122	31 50	203	46 02	257	37 57	287	16 15	350
26	59 26	037	27 59	088	47 44	124	31 48	205	45 56	259	37 51	289	16 14	352
28	59 30	039	28 05	090	47 50	126	31 45	207	45 50	261	37 45	291	16 13	354
	Dubhe		POLLUX		◆CAPELLA		Alpheratz		◆DENEB		VEGA		◆ARCTURUS	
30	59 34	041	28 12	092	47 55	128	31 42	209	45 43	263	37 40	293	16 13	356
32	59 38	042	28 18	094	47 59	131	31 39	211	45 37	265	37 34	295	16 13	358
34	59 42	044	28 24	096	48 04	133	31 36	213	45 31	267	37 28	297	16 12	000
36	59 46	046	28 30	098	48 09	135	31 32	215	45 25	269	37 23	299	16 13	002
38	59 51	048	28 37	100	48 13	137	31 29	217	45 18	271	37 17	301	16 13	004
40	59 56	050	28 43	102	48 17	139	31 25	219	45 12	273	37 12	303	16 14	006
42	60 01	052	28 49	104	48 21	141	31 21	221	45 06	275	37 07	305	16 14	008
44	60 06	054	28 55	106	48 25	143	31 16	223	44 59	277	37 02	307	16 15	010
46	60 11	056	29 01	108	48 29	145	31 12	225	44 53	279	36 57	309	16 16	012
48	60 16	057	29 07	110	48 32	147	31 07	227	44 47	281	36 52	311	16 18	014
50	60 21	059	29 13	112	48 36	149	31 03	229	44 41	283	36 47	313	16 19	016
52	60 27	061	29 18	114	48 39	151	30 58	231	44 35	285	36 42	314	16 21	018
54	60 32	063	29 24	116	48 42	154	30 53	233	44 29	287	36 38	316	16 23	020
56	60 38	065	29 30	118	48 44	156	30 48	235	44 23	289	36 34	318	16 26	022
58	60 44	067	29 35	120	48 47	158	30 43	237	44 17	290	36 30	320	16 28	024
	◆ARCTURUS		REGULUS		POLLUX		◆CAPELLA		Alpheratz		◆DENEB		VEGA	
60	16 31	026	11 52	087	29 41	122	48 49	160	30 37	239	44 11	292	36 26	322
62	16 33	028	11 58	089	29 46	124	48 51	162	30 32	241	44 05	294	36 22	324
64	16 36	030	12 05	091	29 51	126	48 53	164	30 26	243	44 00	296	36 18	326
66	16 40	032	12 11	093	29 56	128	48 55	166	30 21	246	43 54	298	36 15	328
68	16 43	034	12 17	095	30 01	130	48 56	168	30 15	248	43 49	300	36 12	330
70	16 47	036	12 23	097	30 05	133	48 57	170	30 09	250	43 43	302	36 09	332
72	16 50	038	12 30	099	30 10	135	48 58	173	30 03	252	43 38	304	36 06	334
74	16 54	040	12 36	101	30 14	137	48 59	175	29 57	254	43 33	306	36 03	336
76	16 58	041	12 42	103	30 19	139	48 59	177	29 51	256	43 28	308	36 01	338
78	17 03	043	12 48	105	30 23	141	49 00	179	29 45	258	43 23	310	35 58	340
80	17 07	045	12 54	107	30 27	143	49 00	181	29 39	260	43 18	312	35 56	342
82	17 12	047	13 00	109	30 30	145	48 59	183	29 33	262	43 14	314	35 54	343
84	17 16	049	13 06	111	30 34	147	48 59	185	29 26	264	43 09	316	35 53	345
86	17 21	051	13 12	113	30 37	149	48 58	187	29 20	266	43 05	318	35 51	347
88	17 26	053	13 17	115	30 40	151	48 57	189	29 14	268	43 01	320	35 50	349
	◆ARCTURUS		REGULUS		POLLUX		◆CAPELLA		Alpheratz		◆DENEB		VEGA	
90	17 31	055	13 23	117	30 43	153	48 56	192	29 08	270	42 57	322	35 49	351
92	17 36	057	13 29	119	30 46	155	48 55	194	29 01	272	42 53	323	35 48	353
94	17 42	059	13 34	121	30 49	157	48 53	196	28 55	274	42 49	325	35 47	355
96	17 47	061	13 39	123	30 51	159	48 51	198	28 49	276	42 46	327	35 47	357
98	17 53	063	13 45	125	30 53	161	48 49	200	28 42	278	42 42	329	35 47	359
100	17 58	065	13 50	127	30 55	163	48 47	202	28 36	280	42 39	331	35 47	001
102	18 04	067	13 55	129	30 57	165	48 45	204	28 30	282	42 36	333	35 47	003
104	18 10	069	13 59	131	30 58	167	48 42	206	28 24	284	42 34	335	35 47	005
106	18 16	071	14 04	133	30 59	169	48 39	208	28 18	286	42 31	337	35 48	007
108	18 22	073	14 08	135	31 00	171	48 36	211	28 12	288	42 29	339	35 49	008
110	18 28	075	14 13	138	31 01	174	48 33	213	28 06	290	42 26	341	35 50	010
112	18 34	077	14 17	140	31 02	176	48 29	215	28 00	291	42 24	343	35 51	012
114	18 40	079	14 21	142	31 02	178	48 25	217	27 54	293	42 23	344	35 52	014
116	18 46	081	14 25	144	31 02	180	48 22	219	27 49	295	42 21	346	35 54	016
118	18 53	083	14 28	146	31 02	182	48 18	221	27 43	297	42 20	348	35 56	018
	VEGA		◆ARCTURUS		REGULUS		POLLUX		◆CAPELLA		Alpheratz		◆DENEB	
120	35 58	020	18 59	085	14 32	148	31 02	184	48 13	223	27 37	299	42 19	350
122	36 00	022	19 05	087	14 35	150	31 01	186	48 09	225	27 32	301	42 18	352
124	36 03	024	19 11	089	14 38	152	31 01	188	48 04	227	27 27	303	42 17	354
126	36 05	026	19 18	091	14 41	154	31 00	190	48 00	229	27 21	305	42 16	356
128	36 08	028	19 24	093	14 44	156	30 58	192	47 55	231	27 16	307	42 16	358
130	36 11	030	19 30	095	14 46	158	30 57	194	47 50	233	27 11	309	42 16	000
132	36 14	032	19 36	097	14 48	160	30 55	196	47 45	236	27 07	311	42 16	002
134	36 18	034	19 43	099	14 51	162	30 54	198	47 40	238	27 02	313	42 16	004
136	36 21	035	19 49	101	14 52	164	30 51	200	47 34	240	26 58	315	42 17	005
138	36 25	037	19 55	103	14 54	166	30 49	202	47 29	242	26 53	317	42 17	007
140	36 29	039	20 01	105	14 56	168	30 47	204	47 23	244	26 49	319	42 18	009
142	36 33	041	20 07	107	14 57	170	30 44	206	47 18	246	26 45	321	42 19	011
144	36 37	043	20 13	109	14 58	172	30 41	209	47 12	248	26 41	323	42 21	013
146	36 42	045	20 19	111	14 59	174	30 38	211	47 06	250	26 37	325	42 22	015
148	36 46	047	20 25	113	14 59	176	30 35	213	47 00	252	26 34	327	42 24	017
	◆DENEB		VEGA		◆ARCTURUS		REGULUS		POLLUX		◆CAPELLA		Alpheratz	
150	42 26	019	36 51	049	20 30	115	14 59	178	30 31	215	46 54	254	26 30	329
152	42 28	021	36 56	051	20 36	117	15 00	180	30 28	217	46 48	256	26 27	331
154	42 30	023	37 01	053	20 42	119	14 59	182	30 24	219	46 42	258	26 24	333
156	42 33	024	37 06	055	20 47	121	14 59	184	30 20	221	46 36	260	26 22	335
158	42 35	026	37 11	057	20 52	123	14 59	186	30 15	223	46 29	262	26 19	337
160	42 38	028	37 16	059	20 58	125	14 58	188	30 11	225	46 23	264	26 17	339
162	42 41	030	37 22	061	21 03	127	14 57	190	30 07	227	46 17	266	26 14	340
164	42 45	032	37 27	063	21 08	129	14 56	192	30 02	229	46 11	268	26 12	342
166	42 48	034	37 33	065	21 12	131	14 54	194	29 57	231	46 04	270	26 11	344
168	42 52	036	37 38	067	21 17	133	14 52	196	29 52	233	45 58	272	26 09	346
170	42 55	038	37 44	069	21 21	135	14 51	198	29 47	235	45 52	274	26 08	348
172	42 59	040	37 50	071	21 26	137	14 49	200	29 42	237	45 46	276	26 06	350
174	43 03	042	37 56	073	21 30	139	14 46	202	29 37	239	45 39	278	26 05	352
176	43 08	044	38 02	075	21 34	141	14 44	204	29 31	241	45 33	280	26 05	354
178	43 12	046	38 08	076	21 38	143	14 41	206	29 26	243	45 27	282	26 04	356

LHA	Hc	Zn	Hc	Zn	Hc	Zn	Hc	Zn	Hc	Zn	Hc	Zn	Hc	Zn
♈	◆DENEB		VEGA		◆ARCTURUS		REGULUS		POLLUX		◆CAPELLA		Alpheratz	
180	43 17	047	38 14	078	21 41	146	14 38	208	29 20	245	45 21	284	26 04	358
182	43 21	049	38 20	080	21 45	148	14 35	210	29 14	247	45 15	286	26 04	000
184	43 26	051	38 27	082	21 48	150	14 32	212	29 08	249	45 09	288	26 04	002
186	43 31	053	38 33	084	21 51	152	14 28	214	29 02	251	45 03	290	26 04	004
188	43 36	055	38 39	086	21 54	154	14 25	216	28 56	253	44 57	292	26 05	006
190	43 41	057	38 45	088	21 57	156	14 21	218	28 50	255	44 51	294	26 05	008
192	43 47	059	38 52	090	21 59	158	14 17	220	28 44	257	44 45	296	26 06	010
194	43 52	061	38 58	092	22 02	160	14 13	222	28 38	259	44 40	298	26 08	012
196	43 58	063	39 04	094	22 04	162	14 09	224	28 32	261	44 34	300	26 09	014
198	44 03	065	39 11	096	22 06	164	14 04	226	28 26	263	44 29	302	26 11	016
200	44 09	067	39 17	098	22 07	166	13 59	228	28 19	265	44 24	303	26 12	017
202	44 15	069	39 23	100	22 09	168	13 55	231	28 13	267	44 18	305	26 14	019
204	44 21	071	39 29	102	22 10	170	13 50	233	28 07	269	44 13	307	26 16	021
206	44 27	073	39 35	104	22 11	172	13 45	235	28 01	271	44 08	309	26 19	023
208	44 33	075	39 41	106	22 12	174	13 40	237	27 54	273	44 04	311	26 21	025
	Alpheratz		◆DENEB		VEGA		◆ARCTURUS		REGULUS		POLLUX		◆CAPELLA	
210	26 24	027	44 39	077	39 47	108	22 12	176	13 34	239	27 48	275	43 59	313
212	26 27	029	44 45	079	39 53	111	22 12	178	13 29	241	27 42	277	43 55	315
214	26 30	031	44 51	081	39 59	113	22 12	180	13 23	243	27 36	279	43 50	317
216	26 34	033	44 57	083	40 05	115	22 12	182	13 18	245	27 29	281	43 46	319
218	26 37	035	45 04	085	40 10	117	22 12	184	13 12	247	27 23	283	43 42	321
220	26 41	037	45 10	087	40 16	119	22 11	186	13 06	249	27 17	285	43 38	323
222	26 45	039	45 16	089	40 21	121	22 11	188	13 00	251	27 11	287	43 34	325
224	26 49	041	45 23	091	40 27	123	22 10	190	12 54	253	27 05	289	43 31	327
226	26 53	043	45 29	093	40 32	125	22 08	192	12 48	255	26 59	291	43 27	329
228	26 57	045	45 35	095	40 37	127	22 07	194	12 42	257	26 54	293	43 24	330
230	27 02	047	45 41	097	40 42	129	22 05	196	12 36	259	26 48	295	43 21	332
232	27 07	049	45 48	099	40 47	131	22 03	198	12 30	261	26 42	297	43 18	334
234	27 11	051	45 54	101	40 52	133	22 01	201	12 24	263	26 37	299	43 16	336
236	27 16	053	46 00	103	40 56	135	21 59	203	12 17	265	26 31	301	43 13	338
238	27 21	055	46 06	105	41 00	137	21 56	205	12 11	267	26 26	303	43 11	340
	Alpheratz		◆DENEB		VEGA		◆ARCTURUS		REGULUS		POLLUX		◆CAPELLA	
240	27 26	057	46 12	107	41 05	139	21 54	207	12 05	269	26 21	305	43 09	342
242	27 32	059	46 18	109	41 09	141	21 51	209	11 59	271	26 16	307	43 07	344
244	27 37	061	46 24	111	41 13	143	21 48	211	11 52	273	26 11	309	43 05	346
246	27 43	063	46 30	113	41 16	145	21 44	213	11 46	275	26 06	311	43 04	348
248	27 48	064	46 36	115	41 20	147	21 41	215	11 40	277	26 01	313	43 03	349
250	27 54	066	46 41	117	41 23	150	21 37	217	11 34	279	25 57	315	43 02	351
252	28 00	068	46 47	119	41 26	152	21 33	219	11 27	281	25 52	317	43 01	353
254	28 06	070	46 52	121	41 29	154	21 29	221	11 21	283	25 48	319	43 00	355
256	28 12	072	46 57	123	41 32	156	21 25	223	11 15	285	25 44	321	43 00	357
258	28 18	074	47 03	125	41 34	158	21 21	225	11 09	287	25 40	323	43 00	359
260	28 24	076	47 08	127	41 36	160	21 16	227	11 03	289	25 36	325	43 00	001
262	28 30	078	47 13	129	41 38	162	21 12	229	10 57	291	25 33	327	43 00	003
264	28 36	080	47 17	131	41 40	164	21 07	231	10 51	293	25 29	329	43 00	005
266	28 42	082	47 22	133	41 42	166	21 02	233	10 46	295	25 26	331	43 01	007
268	28 49	084	47 27	136	41 43	168	20 57	235	10 40	297	25 23	332	43 02	008
	◆CAPELLA		Alpheratz		◆DENEB		VEGA		◆ARCTURUS		Dubhe		POLLUX	
270	43 03	010	28 55	086	47 31	138	41 44	170	20 52	237	60 55	289	25 20	334
272	43 04	012	29 01	088	47 35	140	41 45	172	20 46	239	60 49	291	25 18	336
274	43 05	014	29 07	090	47 39	142	41 46	175	20 41	241	60 43	293	25 15	338
276	43 07	016	29 14	092	47 43	144	41 46	177	20 35	243	60 37	295	25 13	340
278	43 09	018	29 20	094	47 46	146	41 47	179	20 30	245	60 32	297	25 11	342
280	43 11	020	29 26	096	47 50	148	41 47	181	20 24	247	60 26	299	25 09	344
282	43 13	022	29 32	098	47 53	150	41 47	183	20 18	249	60 21	301	25 08	346
284	43 16	024	29 39	100	47 56	152	41 46	185	20 12	251	60 15	303	25 06	348
286	43 18	026	29 45	102	47 59	154	41 46	187	20 06	253	60 10	305	25 05	350
288	43 21	028	29 51	104	48 02	156	41 45	189	20 00	255	60 05	307	25 04	352
290	43 24	029	29 57	106	48 04	159	41 43	191	19 54	257	60 00	308	25 03	354
292	43 27	031	30 03	108	48 06	161	41 42	193	19 48	259	59 55	310	25 03	356
294	43 31	033	30 09	110	48 08	163	41 40	195	19 42	261	59 50	312	25 02	358
296	43 34	035	30 15	112	48 10	165	41 39	198	19 35	263	59 46	314	25 02	000
298	43 38	037	30 20	114	48 11	167	41 37	200	19 29	265	59 41	316	25 02	002
	POLLUX		◆CAPELLA		Alpheratz		◆DENEB		VEGA		◆ARCTURUS		Dubhe	
300	25 03	004	43 42	039	30 26	116	48 13	169	41 34	202	19 23	267	59 37	318
302	25 03	006	43 46	041	30 32	118	48 14	171	41 32	204	19 17	269	59 33	320
304	25 04	008	43 50	043	30 37	120	48 15	173	41 29	206	19 10	271	59 29	322
306	25 05	009	43 54	045	30 42	123	48 15	175	41 27	208	19 04	273	59 25	323
308	25 06	011	43 59	047	30 48	125	48 16	178	41 24	210	18 58	275	59 22	325
310	25 07	013	44 03	049	30 53	127	48 16	180	41 20	212	18 52	277	59 18	327
312	25 09	015	44 08	051	30 58	129	48 16	182	41 17	214	18 45	279	59 15	329
314	25 11	017	44 13	052	31 03	131	48 15	184	41 13	216	18 39	281	59 12	331
316	25 13	019	44 18	054	31 07	133	48 15	186	41 09	218	18 33	283	59 09	333
318	25 15	021	44 23	056	31 12	135	48 14	188	41 05	220	18 27	285	59 06	335
320	25 17	023	44 29	058	31 16	137	48 13	190	41 01	222	18 21	287	59 03	336
322	25 20	025	44 34	060	31 20	139	48 12	192	40 57	225	18 15	289	59 01	338
324	25 22	027	44 39	062	31 25	141	48 10	194	40 52	227	18 09	291	58 59	340
326	25 25	029	44 45	064	31 28	143	48 09	197	40 48	229	18 03	293	58 56	342
328	25 29	031	44 51	066	31 32	145	48 07	199	40 43	231	17 58	295	58 55	344
	POLLUX		◆CAPELLA		Alpheratz		◆DENEB		VEGA		◆ARCTURUS		Dubhe	
330	25 32	033	44 57	068	31 36	147	48 05	201	40 38	233	17 52	297	58 53	346
332	25 35	035	45 02	070	31 39	149	48 02	203	40 33	235	17 46	299	58 51	347
334	25 39	037	45 08	072	31 42	151	48 00	205	40 28	237	17 41	301	58 50	349
336	25 43	039	45 14	074	31 45	153	47 57	207	40 23	239	17 36	303	58 49	351
338	25 47	041	45 20	076	31 48	155	47 54	209	40 17	241	17 30	305	58 48	353
340	25 51	043	45 26	078	31 50	157	47 51	211	40 12	243	17 25	307	58 48	355
342	25 55	045	45 33	080	31 53	159	47 48	213	40 06	245	17 20	309	58 47	356
344	26 00	047	45 39	082	31 55	161	47 44	215	40 00	247	17 16	311	58 47	358
346	26 05	049	45 45	084	31 57	163	47 40	218	39 54	249	17 11	313	58 47	000
348	26 09	051	45 51	086	31 58	166	47 36	220	39 48	251	17 06	315	58 47	002
350	26 14	053	45 58	088	32 00	168	47 32	222	39 42	253	17 02	317	58 47	004
352	26 19	054	46 04	090	32 01	170	47 28	224	39 36	255	16 58	319	58 48	006
354	26 25	056	46 10	092	32 02	172	47 24	226	39 30	257	16 54	321	58 48	007
356	26 30	058	46 16	094	32 03	174	47 19	228	39 24	259	16 50	323	58 49	009
358	26 35	060	46 23	096	32 03	176	47 14	230	39 18	261	16 46	325	58 50	011

 LAT 86°N

LHA	Hc	Zn	Hc	Zn	Hc	Zn	Hc	Zn	Hc	Zn	Hc	Zn	Hc	Zn
♈	Dubhe		POLLUX		◆CAPELLA		Alpheratz		◆DENEB		VEGA		◆ARCTURUS	
0	57 53	013	26 13	062	46 37	097	33 04	178	47 46	233	39 18	264	15 52	327
2	57 55	014	26 20	064	46 45	099	33 04	180	47 39	235	39 10	266	15 48	329
4	57 57	016	26 28	066	46 53	101	33 04	182	47 32	237	39 02	268	15 44	331
6	58 00	018	26 35	068	47 01	103	33 03	184	47 25	239	38 53	270	15 40	333
8	58 02	020	26 43	070	47 09	105	33 02	186	47 18	241	38 45	272	15 36	335
10	58 05	021	26 51	072	47 17	107	33 01	188	47 10	243	38 37	274	15 33	337
12	58 09	023	26 59	074	47 25	109	33 00	190	47 03	245	38 28	276	15 29	339
14	58 12	025	27 07	076	47 33	111	32 58	192	46 55	247	38 20	278	15 26	341
16	58 16	027	27 15	078	47 41	113	32 56	195	46 47	250	38 12	280	15 24	343
18	58 20	029	27 24	080	47 49	115	32 54	197	46 39	252	38 03	282	15 21	344
20	58 24	030	27 32	082	47 56	117	32 52	199	46 31	254	37 55	284	15 19	346
22	58 28	032	27 40	084	48 04	119	32 49	201	46 23	256	37 47	286	15 18	348
24	58 33	034	27 48	086	48 11	121	32 46	203	46 15	258	37 39	288	15 16	350
26	58 37	036	27 57	088	48 18	123	32 42	205	46 07	260	37 31	290	15 15	352
28	58 42	038	28 05	090	48 25	125	32 39	207	45 59	262	37 23	292	15 14	354
	Dubhe		POLLUX		◆CAPELLA		Alpheratz		◆DENEB		VEGA		◆ARCTURUS	
30	58 48	039	28 14	092	48 32	128	32 35	209	45 50	264	37 16	294	15 13	356
32	58 53	041	28 22	094	48 38	130	32 30	211	45 42	266	37 08	296	15 13	358
34	58 59	043	28 30	096	48 44	132	32 26	213	45 34	268	37 01	298	15 12	000
36	59 04	045	28 39	098	48 51	134	32 21	215	45 25	270	36 53	300	15 13	002
38	59 10	047	28 47	100	48 56	136	32 16	217	45 17	272	36 46	301	15 13	004
40	59 17	049	28 55	102	49 02	138	32 11	219	45 09	274	36 39	303	15 14	006
42	59 23	050	29 03	104	49 08	140	32 06	222	45 00	276	36 32	305	15 15	008
44	59 30	052	29 11	106	49 13	142	32 00	224	44 52	278	36 25	307	15 16	010
46	59 36	054	29 19	108	49 18	144	31 54	226	44 44	280	36 19	309	15 18	012
48	59 43	056	29 27	110	49 23	147	31 48	228	44 35	282	36 12	311	15 20	014
50	59 50	058	29 35	112	49 27	149	31 42	230	44 27	284	36 06	313	15 22	016
52	59 57	060	29 43	114	49 31	151	31 35	232	44 19	286	36 00	315	15 24	018
54	60 05	062	29 50	116	49 35	153	31 29	234	44 11	288	35 54	317	15 27	020
56	60 12	064	29 58	118	49 39	155	31 22	236	44 03	289	35 49	319	15 30	022
58	60 20	065	30 05	120	49 42	157	31 15	238	43 55	291	35 43	321	15 33	024
	◆ARCTURUS		REGULUS		POLLUX		◆CAPELLA		Alpheratz		◆DENEB		VEGA	
60	15 37	026	11 49	087	30 12	122	49 45	159	31 08	240	43 48	293	35 38	323
62	15 40	028	11 58	089	30 19	124	49 48	162	31 00	242	43 40	295	35 33	325
64	15 44	030	12 06	091	30 26	126	49 51	164	30 53	244	43 33	297	35 29	327
66	15 49	031	12 14	093	30 33	128	49 53	166	30 45	246	43 25	299	35 24	328
68	15 53	033	12 23	095	30 39	130	49 55	168	30 37	248	43 18	301	35 20	330
70	15 58	035	12 31	097	30 46	132	49 56	170	30 30	250	43 11	303	35 16	332
72	16 03	037	12 39	099	30 52	134	49 58	172	30 22	252	43 04	305	35 12	334
74	16 08	039	12 48	101	30 58	136	49 59	175	30 14	254	42 57	307	35 08	336
76	16 13	041	12 56	103	31 04	138	49 59	177	30 06	256	42 50	309	35 05	338
78	16 19	043	13 04	105	31 09	140	50 00	179	29 57	258	42 44	311	35 02	340
80	16 25	045	13 12	107	31 14	142	50 00	181	29 49	260	42 38	313	34 59	342
82	16 31	047	13 20	109	31 19	144	49 59	183	29 41	262	42 32	314	34 57	344
84	16 37	049	13 28	111	31 24	147	49 59	185	29 33	264	42 26	316	34 55	346
86	16 43	051	13 35	113	31 28	149	49 58	187	29 24	266	42 20	318	34 53	347
88	16 50	053	13 43	115	31 33	151	49 56	190	29 16	268	42 15	320	34 51	349
	◆ARCTURUS		REGULUS		POLLUX		◆CAPELLA		Alpheratz		◆DENEB		VEGA	
90	16 57	055	13 51	117	31 37	153	49 55	192	29 08	270	42 09	322	34 50	351
92	17 04	057	13 58	119	31 40	155	49 53	194	28 59	272	42 04	324	34 49	353
94	17 11	059	14 05	121	31 44	157	49 51	196	28 51	274	42 00	326	34 48	355
96	17 18	061	14 12	123	31 47	159	49 48	198	28 43	276	41 55	328	34 47	357
98	17 26	063	14 19	125	31 50	161	49 46	200	28 34	278	41 51	330	34 47	359
100	17 33	065	14 26	127	31 52	163	49 43	203	28 26	280	41 47	332	34 47	001
102	17 41	067	14 33	129	31 55	165	49 39	205	28 18	282	41 43	333	34 47	003
104	17 48	069	14 39	131	31 57	167	49 36	207	28 10	284	41 39	335	34 48	005
106	17 56	071	14 45	133	31 58	169	49 32	209	28 02	286	41 36	337	34 48	006
108	18 04	073	14 51	135	32 00	171	49 27	211	27 54	288	41 33	339	34 49	008
110	18 12	075	14 57	137	32 01	173	49 23	213	27 46	290	41 30	341	34 51	010
112	18 20	077	15 02	139	32 02	176	49 18	215	27 38	292	41 27	343	34 52	012
114	18 29	079	15 08	141	32 02	178	49 13	218	27 30	294	41 25	345	34 54	014
116	18 37	081	15 13	143	32 02	180	49 08	220	27 23	296	41 23	347	34 56	016
118	18 45	083	15 18	145	32 02	182	49 03	222	27 15	298	41 21	348	34 59	018
	VEGA		◆ARCTURUS		REGULUS		POLLUX		◆CAPELLA		Alpheratz		◆DENEB	
120	35 02	020	18 53	085	15 22	147	32 02	184	48 57	224	27 08	300	41 19	350
122	35 05	022	19 02	087	15 27	149	32 01	186	48 51	226	27 01	302	41 18	352
124	35 08	024	19 10	089	15 31	152	32 00	188	48 45	228	26 53	304	41 17	354
126	35 11	025	19 19	091	15 35	154	31 59	190	48 39	230	26 47	306	41 16	356
128	35 15	027	19 27	093	15 38	156	31 57	192	48 32	232	26 40	308	41 16	358
130	35 19	029	19 35	095	15 42	158	31 55	194	48 25	234	26 33	310	41 16	000
132	35 23	031	19 44	097	15 45	160	31 53	196	48 18	236	26 27	312	41 16	002
134	35 28	033	19 52	099	15 48	162	31 50	198	48 11	239	26 21	314	41 16	003
136	35 32	035	20 00	101	15 50	164	31 48	201	48 04	241	26 15	315	41 17	005
138	35 37	037	20 08	103	15 52	166	31 45	203	47 57	243	26 09	317	41 18	007
140	35 42	039	20 16	105	15 54	168	31 41	205	47 49	245	26 04	319	41 19	009
142	35 48	041	20 25	107	15 56	170	31 38	207	47 42	247	25 58	321	41 20	011
144	35 53	043	20 32	109	15 57	172	31 34	209	47 34	249	25 53	323	41 22	013
146	35 59	045	20 40	111	15 58	174	31 30	211	47 26	251	25 48	325	41 24	015
148	36 05	047	20 48	113	15 59	176	31 25	213	47 18	253	25 44	327	41 26	017
	◆DENEB		VEGA		◆ARCTURUS		REGULUS		POLLUX		◆CAPELLA		Alpheratz	
150	41 29	018	36 11	048	20 56	115	15 59	178	31 20	215	47 10	255	25 39	329
152	41 32	020	36 18	050	21 03	117	16 00	180	31 15	217	47 02	257	25 35	331
154	41 35	022	36 24	052	21 11	119	15 59	182	31 10	219	46 54	259	25 31	333
156	41 38	024	36 31	054	21 18	121	15 59	184	31 05	221	46 46	261	25 27	335
158	41 42	026	36 38	056	21 25	123	15 58	186	30 59	223	46 37	263	25 24	337
160	41 45	028	36 45	058	21 32	125	15 57	188	30 53	225	46 29	265	25 21	339
162	41 49	030	36 52	060	21 39	127	15 56	190	30 47	227	46 21	267	25 18	341
164	41 54	032	36 59	062	21 45	129	15 54	192	30 41	229	46 12	269	25 15	343
166	41 58	034	37 07	064	21 52	131	15 52	194	30 35	231	46 04	271	25 13	345
168	42 03	035	37 14	066	21 58	133	15 50	196	30 28	234	45 55	273	25 11	346
170	42 08	037	37 22	068	22 04	135	15 48	198	30 21	236	45 47	275	25 09	348
172	42 13	039	37 30	070	22 10	137	15 45	200	30 14	238	45 39	277	25 07	350
174	42 18	041	37 38	072	22 15	139	15 42	202	30 07	240	45 30	279	25 06	352
176	42 24	043	37 46	074	22 21	141	15 38	204	30 00	242	45 22	281	25 05	354
178	42 30	045	37 54	076	22 26	143	15 35	206	29 52	244	45 14	283	25 04	356

LHA	Hc	Zn	Hc	Zn	Hc	Zn	Hc	Zn	Hc	Zn	Hc	Zn	Hc	Zn
♈	◆DENEB		VEGA		◆ARCTURUS		REGULUS		POLLUX		◆CAPELLA		Alpheratz	
180	42 36	047	38 02	078	22 31	145	15 31	208	29 45	246	45 06	285	25 04	358
182	42 42	049	38 10	080	22 35	147	15 27	210	29 37	248	44 58	287	25 04	000
184	42 48	051	38 18	082	22 40	149	15 23	212	29 29	250	44 50	289	25 04	002
186	42 55	053	38 27	084	22 44	151	15 18	215	29 21	252	44 42	291	25 04	004
188	43 02	054	38 35	086	22 48	153	15 13	217	29 13	254	44 34	293	25 05	006
190	43 09	056	38 43	088	22 51	156	15 08	219	29 05	256	44 27	295	25 06	008
192	43 16	058	38 52	090	22 55	158	15 03	221	28 57	258	44 19	297	25 07	010
194	43 23	060	39 00	092	22 58	160	14 57	223	28 49	260	44 12	299	25 09	012
196	43 30	062	39 08	094	23 01	162	14 51	225	28 41	262	44 04	300	25 11	013
198	43 38	064	39 17	096	23 03	164	14 45	227	28 32	264	43 57	302	25 13	015
200	43 45	066	39 25	098	23 05	166	14 39	229	28 24	266	43 50	304	25 15	017
202	43 53	068	39 33	100	23 07	168	14 33	231	28 16	268	43 43	306	25 18	019
204	44 01	070	39 42	102	23 09	170	14 26	233	28 07	270	43 37	308	25 21	021
206	44 09	072	39 50	104	23 10	172	14 19	235	27 59	272	43 30	310	25 24	023
208	44 17	074	39 58	106	23 11	174	14 12	237	27 51	274	43 24	312	25 27	025
	Alpheratz		◆DENEB		VEGA		◆ARCTURUS		REGULUS		POLLUX		◆CAPELLA	
210	25 31	027	44 25	076	40 06	108	23 12	176	14 05	239	27 42	276	43 18	314
212	25 35	029	44 33	078	40 14	110	23 12	178	13 58	241	27 34	278	43 12	316
214	25 39	031	44 41	080	40 22	112	23 12	180	13 51	243	27 26	280	43 06	318
216	25 43	033	44 49	082	40 29	114	23 12	182	13 43	245	27 17	282	43 01	320
218	25 48	035	44 58	084	40 37	116	23 12	184	13 36	247	27 09	284	42 55	321
220	25 53	037	45 06	086	40 44	118	23 11	186	13 28	249	27 01	286	42 50	323
222	25 58	039	45 14	088	40 52	120	23 10	188	13 20	251	26 53	288	42 45	325
224	26 03	041	45 23	090	40 59	122	23 09	190	13 12	253	26 45	290	42 41	327
226	26 09	043	45 31	092	41 06	124	23 07	192	13 04	255	26 37	292	42 36	329
228	26 15	044	45 39	094	41 13	126	23 05	195	12 56	257	26 30	294	42 32	331
230	26 21	046	45 48	096	41 19	128	23 03	197	12 48	259	26 22	296	42 28	333
232	26 27	048	45 56	098	41 26	130	23 00	199	12 40	261	26 15	298	42 24	335
234	26 33	050	46 04	100	41 32	132	22 57	201	12 31	263	26 07	300	42 21	337
236	26 40	052	46 13	102	41 38	134	22 54	203	12 23	265	26 00	302	42 18	338
238	26 46	054	46 21	104	41 44	137	22 51	205	12 15	267	25 53	303	42 15	340
	Alpheratz		◆DENEB		VEGA		◆ARCTURUS		REGULUS		POLLUX		◆CAPELLA	
240	26 53	056	46 29	106	41 50	139	22 47	207	12 06	269	25 46	305	42 12	342
242	27 00	058	46 37	108	41 55	141	22 43	209	11 58	271	25 39	307	42 10	344
244	27 07	060	46 45	110	42 00	143	22 39	211	11 49	273	25 33	309	42 07	346
246	27 15	062	46 53	112	42 05	145	22 35	213	11 41	275	25 26	311	42 05	348
248	27 22	064	47 00	114	42 10	147	22 30	215	11 33	277	25 20	313	42 04	350
250	27 30	066	47 08	116	42 15	149	22 25	217	11 25	279	25 14	315	42 02	352
252	27 38	068	47 15	118	42 19	151	22 20	219	11 16	281	25 08	317	42 01	353
254	27 45	070	47 23	120	42 23	153	22 15	221	11 08	283	25 03	319	42 00	355
256	27 53	072	47 30	122	42 26	155	22 09	223	11 00	285	24 57	321	42 00	357
258	28 01	074	47 37	124	42 30	158	22 03	225	10 52	287	24 52	323	42 00	359
260	28 09	076	47 44	126	42 33	160	21 57	227	10 44	289	24 47	325	42 00	001
262	28 18	078	47 50	128	42 35	162	21 51	229	10 36	291	24 43	327	42 00	003
264	28 26	080	47 57	131	42 38	164	21 44	231	10 28	293	24 38	329	42 00	005
266	28 34	082	48 03	133	42 40	166	21 38	233	10 21	295	24 34	331	42 01	006
268	28 42	084	48 09	135	42 42	168	21 31	235	10 13	297	24 30	333	42 02	008
	◆CAPELLA		Alpheratz		◆DENEB		VEGA		◆ARCTURUS		Dubhe		POLLUX	
270	42 04	010	28 51	086	48 15	137	42 43	170	21 24	237	60 34	291	24 26	335
272	42 05	012	28 59	088	48 21	139	42 45	172	21 17	239	60 26	293	24 23	337
274	42 07	014	29 07	090	48 26	141	42 46	174	21 10	241	60 18	295	24 20	338
276	42 09	016	29 16	092	48 31	143	42 46	177	21 02	243	60 11	297	24 17	340
278	42 12	018	29 24	094	48 36	145	42 47	179	20 55	245	60 04	299	24 14	342
280	42 14	020	29 32	096	48 41	147	42 47	181	20 47	247	59 56	301	24 12	344
282	42 17	021	29 41	098	48 45	150	42 46	183	20 39	249	59 49	302	24 09	346
284	42 21	023	29 49	100	48 49	152	42 46	185	20 31	252	59 42	304	24 08	348
286	42 24	025	29 57	102	48 53	154	42 45	187	20 23	254	59 35	306	24 06	350
288	42 28	027	30 05	104	48 56	156	42 44	189	20 15	256	59 29	308	24 05	352
290	42 32	029	30 13	106	49 00	158	42 42	191	20 07	258	59 22	310	24 04	354
292	42 36	031	30 22	108	49 03	160	42 40	194	19 59	260	59 16	312	24 03	356
294	42 40	033	30 29	110	49 05	162	42 38	196	19 51	262	59 10	313	24 02	358
296	42 45	035	30 37	112	49 08	165	42 36	198	19 42	264	59 04	315	24 02	000
298	42 50	037	30 45	114	49 10	167	42 33	200	19 34	266	58 58	317	24 02	002
	POLLUX		◆CAPELLA		Alpheratz		◆DENEB		VEGA		◆ARCTURUS		Dubhe	
300	24 03	004	42 55	038	30 53	116	49 12	169	42 30	202	19 26	268	58 52	319
302	24 03	006	43 00	040	31 00	118	49 13	171	42 27	204	19 17	270	58 47	321
304	24 04	007	43 06	042	31 07	120	49 14	173	42 23	206	19 09	272	58 42	323
306	24 06	009	43 11	044	31 15	122	49 15	175	42 19	208	19 01	274	58 37	324
308	24 07	011	43 17	046	31 22	124	49 16	178	42 15	210	18 52	276	58 32	326
310	24 09	013	43 23	048	31 28	126	49 16	180	42 11	213	18 44	278	58 27	328
312	24 11	015	43 30	050	31 35	128	49 16	182	42 06	215	18 36	279	58 23	330
314	24 13	017	43 36	052	31 42	130	49 15	184	42 01	217	18 27	281	58 19	332
316	24 16	019	43 43	054	31 48	132	49 14	186	41 56	219	18 19	283	58 15	333
318	24 19	021	43 50	056	31 54	134	49 13	188	41 51	221	18 11	285	58 11	335
320	24 22	023	43 57	057	32 00	136	49 12	190	41 45	223	18 03	287	58 08	337
322	24 25	025	44 04	059	32 05	138	49 10	193	41 39	225	17 55	289	58 05	339
324	24 29	027	44 11	061	32 11	140	49 08	195	41 33	227	17 47	291	58 02	341
326	24 33	029	44 18	063	32 16	143	49 06	197	41 27	229	17 40	293	57 59	342
328	24 37	031	44 26	065	32 21	145	49 04	199	41 21	231	17 32	295	57 57	344
	POLLUX		◆CAPELLA		Alpheratz		◆DENEB		VEGA		◆ARCTURUS		Dubhe	
330	24 41	033	44 34	067	32 26	147	49 01	201	41 14	233	17 24	297	57 55	346
332	24 46	035	44 41	069	32 30	149	48 57	203	41 07	236	17 17	299	57 53	348
334	24 51	037	44 49	071	32 35	151	48 54	205	41 00	238	17 10	301	57 51	349
336	24 56	038	44 57	073	32 38	153	48 50	208	40 53	240	17 03	303	57 50	351
338	25 01	040	45 05	075	32 42	155	48 46	210	40 46	242	16 56	305	57 49	353
340	25 07	042	45 13	077	32 46	157	48 42	212	40 38	244	16 49	307	57 48	355
342	25 13	044	45 22	079	32 49	159	48 37	214	40 31	246	16 43	309	57 47	357
344	25 19	046	45 30	081	32 51	161	48 33	216	40 23	248	16 36	311	57 47	358
346	25 25	048	45 38	083	32 54	163	48 28	218	40 15	250	16 30	313	57 47	000
348	25 31	050	45 46	085	32 56	165	48 22	220	40 07	252	16 24	315	57 47	002
350	25 38	052	45 55	087	32 58	167	48 17	222	39 59	254	16 18	317	57 47	004
352	25 44	054	46 03	089	33 00	170	48 11	225	39 51	256	16 13	319	57 48	005
354	25 51	056	46 12	091	33 01	172	48 05	227	39 43	258	16 07	321	57 49	007
356	25 58	058	46 20	093	33 02	174	47 59	229	39 35	260	16 02	323	57 50	009
358	26 05	060	46 28	095	33 03	176	47 52	231	39 27	262	15 57	325	57 51	011

 LAT 85°N

LHA ♈	Hc	Zn	Hc	Zn	Hc	Zn	Hc	Zn	Hc	Zn	Hc	Zn	Hc	Zn
	Dubhe		POLLUX		◆CAPELLA		Alpheratz		◆DENEB		VEGA		◆ARCTURUS	
0	56 55	012	25 44	061	46 43	096	34 04	178	48 22	234	39 24	265	15 02	327
2	56 57	014	25 54	063	46 53	098	34 04	180	48 13	236	39 14	267	14 57	329
4	57 00	016	26 03	065	47 04	100	34 04	182	48 04	238	39 03	269	14 51	331
6	57 03	017	26 13	067	47 14	102	34 03	184	47 55	240	38 53	271	14 46	333
8	57 06	019	26 22	069	47 24	104	34 02	186	47 46	242	38 42	273	14 42	335
10	57 09	021	26 32	071	47 34	106	34 01	188	47 37	244	38 32	275	14 37	337
12	57 13	023	26 42	073	47 44	108	33 59	191	47 27	246	38 22	277	14 33	339
14	57 17	024	26 52	075	47 54	110	33 57	193	47 18	248	38 11	279	14 30	341
16	57 22	026	27 02	077	48 04	112	33 54	195	47 08	251	38 01	281	14 27	343
18	57 27	028	27 13	079	48 14	114	33 52	197	46 58	253	37 51	283	14 24	345
20	57 32	030	27 23	081	48 23	116	33 48	199	46 48	255	37 41	285	14 21	346
22	57 37	031	27 33	083	48 32	118	33 45	201	46 38	257	37 30	287	14 19	348
24	57 43	033	27 44	085	48 42	120	33 41	203	46 27	259	37 21	289	14 17	350
26	57 48	035	27 54	087	48 51	122	33 37	205	46 17	261	37 11	291	14 15	352
28	57 55	037	28 04	089	48 59	125	33 32	207	46 07	263	37 01	292	14 14	354
	Dubhe		POLLUX		◆CAPELLA		Alpheratz		◆DENEB		VEGA		◆ARCTURUS	
30	58 01	038	28 15	091	49 08	127	33 27	209	45 56	265	36 51	294	14 13	356
32	58 08	040	28 25	093	49 16	129	33 22	212	45 46	267	36 42	296	14 13	358
34	58 14	042	28 36	095	49 24	131	33 16	214	45 36	269	36 33	298	14 12	000
36	58 22	044	28 46	097	49 32	133	33 10	216	45 25	271	36 23	300	14 13	002
38	58 29	046	28 57	099	49 39	135	33 04	218	45 15	273	36 14	302	14 13	004
40	58 37	047	29 07	101	49 47	137	32 57	220	45 04	275	36 06	304	14 14	006
42	58 44	049	29 17	103	49 54	139	32 50	222	44 54	277	35 57	306	14 15	008
44	58 52	051	29 27	105	50 00	142	32 43	224	44 43	279	35 49	308	14 17	010
46	59 01	053	29 37	107	50 07	144	32 36	226	44 33	281	35 41	310	14 19	012
48	59 09	055	29 47	109	50 13	146	32 28	228	44 23	283	35 33	312	14 21	014
50	59 18	056	29 57	111	50 18	148	32 20	230	44 13	285	35 25	314	14 24	016
52	59 26	058	30 07	113	50 24	150	32 12	232	44 03	286	35 18	316	14 27	018
54	59 35	060	30 16	115	50 29	152	32 04	234	43 53	288	35 10	317	14 30	020
56	59 45	062	30 26	117	50 33	155	31 55	236	43 43	290	35 03	319	14 34	022
58	59 54	064	30 35	119	50 38	157	31 46	238	43 33	292	34 57	321	14 38	024
	◆ARCTURUS		Dubhe		POLLUX		◆CAPELLA		Alpheratz		◆DENEB		VEGA	
60	14 42	025	60 03	066	30 44	121	50 41	159	31 37	240	43 23	294	34 50	323
62	14 47	027	60 13	068	30 53	123	50 45	161	31 28	243	43 14	296	34 44	325
64	14 52	029	60 23	069	31 01	125	50 48	163	31 19	245	43 05	298	34 38	327
66	14 57	031	60 33	071	31 10	128	50 51	166	31 09	247	42 56	300	34 33	329
68	15 03	033	60 42	073	31 18	130	50 53	168	31 00	249	42 47	302	34 28	331
70	15 09	035	60 53	075	31 26	132	50 55	170	30 50	251	42 38	304	34 23	333
72	15 15	037	61 03	077	31 34	134	50 57	172	30 40	253	42 29	306	34 18	334
74	15 21	039	61 13	079	31 41	136	50 58	174	30 30	255	42 21	308	34 14	336
76	15 28	041	61 23	081	31 48	138	50 59	177	30 20	257	42 13	309	34 09	338
78	15 35	043	61 34	083	31 55	140	51 00	179	30 10	259	42 05	311	34 06	340
80	15 42	045	61 44	085	32 02	142	51 00	181	29 59	261	41 57	313	34 02	342
82	15 50	047	61 54	087	32 08	144	50 59	183	29 49	263	41 49	315	33 59	344
84	15 58	049	62 05	089	32 14	146	50 58	185	29 38	265	41 42	317	33 57	346
86	16 06	051	62 15	091	32 20	148	50 57	188	29 28	267	41 35	319	33 54	348
88	16 14	053	62 26	093	32 25	150	50 56	190	29 18	269	41 28	321	33 52	349
	◆ARCTURUS		REGULUS		POLLUX		◆CAPELLA		Alpheratz		◆DENEB		VEGA	
90	16 22	055	14 18	117	32 30	152	50 54	192	29 07	271	41 22	323	33 50	351
92	16 31	057	14 27	119	32 35	155	50 51	194	28 57	273	41 16	325	33 49	353
94	16 40	059	14 36	121	32 39	157	50 48	196	28 46	275	41 10	326	33 48	355
96	16 49	061	14 45	123	32 43	159	50 45	199	28 36	277	41 04	328	33 47	357
98	16 58	063	14 54	125	32 46	161	50 42	201	28 25	279	40 59	330	33 47	359
100	17 08	065	15 02	127	32 50	163	50 38	203	28 15	281	40 54	332	33 47	001
102	17 17	067	15 10	129	32 53	165	50 34	205	28 05	283	40 49	334	33 47	003
104	17 27	069	15 18	131	32 55	167	50 29	207	27 55	285	40 44	336	33 48	005
106	17 37	071	15 26	133	32 57	169	50 24	210	27 45	287	40 40	338	33 49	006
108	17 46	073	15 34	135	32 59	171	50 19	212	27 35	289	40 36	339	33 50	008
110	17 56	075	15 41	137	33 00	173	50 13	214	27 25	291	40 33	341	33 52	010
112	18 07	076	15 48	139	33 01	176	50 07	216	27 15	292	40 30	343	33 54	012
114	18 17	078	15 55	141	33 02	178	50 01	218	27 06	294	40 27	345	33 56	014
116	18 27	080	16 01	143	33 02	180	49 54	220	26 56	296	40 24	347	33 59	016
118	18 37	082	16 07	145	33 02	182	49 47	223	26 47	298	40 22	349	34 02	018
	VEGA		◆ARCTURUS		REGULUS		POLLUX		◆CAPELLA		Alpheratz		◆DENEB	
120	34 05	020	18 48	084	16 13	147	33 02	184	49 40	225	26 38	300	40 20	350
122	34 09	021	18 58	086	16 18	149	33 01	186	49 32	227	26 29	302	40 19	352
124	34 13	023	19 09	088	16 24	151	32 59	188	49 25	229	26 20	304	40 17	354
126	34 17	025	19 19	090	16 28	153	32 58	190	49 17	231	26 11	306	40 17	356
128	34 22	027	19 30	092	16 33	155	32 56	192	49 08	233	26 03	308	40 16	358
130	34 27	029	19 40	094	16 37	157	32 53	194	49 00	235	25 55	310	40 16	000
132	34 32	031	19 50	096	16 41	160	32 50	197	48 51	237	25 47	312	40 16	002
134	34 37	033	20 01	098	16 44	162	32 47	199	48 42	239	25 39	314	40 16	003
136	34 43	035	20 11	100	16 48	164	32 44	201	48 33	242	25 32	316	40 17	005
138	34 49	037	20 21	102	16 50	166	32 40	203	48 24	244	25 25	318	40 18	007
140	34 56	038	20 32	104	16 53	168	32 36	205	48 14	246	25 18	320	40 20	009
142	35 02	040	20 42	106	16 55	170	32 31	207	48 05	248	25 11	322	40 21	011
144	35 09	042	20 52	108	16 57	172	32 26	209	47 55	250	25 05	324	40 24	013
146	35 16	044	21 01	110	16 58	174	32 21	211	47 45	252	24 59	325	40 26	014
148	35 24	046	21 11	112	16 59	176	32 15	213	47 35	254	24 53	327	40 29	016
	◆DENEB		VEGA		◆ARCTURUS		REGULUS		POLLUX		◆CAPELLA		Alpheratz	
150	40 32	018	35 31	048	21 21	114	16 59	178	32 09	215	47 25	256	24 48	329
152	40 35	020	35 39	050	21 30	116	17 00	180	32 03	217	47 15	258	24 42	331
154	40 39	022	35 47	052	21 40	119	16 59	182	31 57	220	47 05	260	24 37	333
156	40 43	024	35 56	054	21 49	121	16 59	184	31 50	222	46 54	262	24 33	335
158	40 48	026	36 04	056	21 58	123	16 58	186	31 43	224	46 44	264	24 29	337
160	40 52	027	36 13	058	22 06	125	16 57	188	31 35	226	46 34	266	24 25	339
162	40 57	029	36 22	059	22 15	127	16 55	190	31 28	228	46 23	268	24 21	341
164	41 02	031	36 31	061	22 23	129	16 53	192	31 20	230	46 13	270	24 18	343
166	41 08	033	36 40	063	22 31	131	16 50	194	31 12	232	46 02	272	24 15	345
168	41 14	035	36 49	065	22 39	133	16 48	196	31 03	234	45 52	274	24 12	347
170	41 20	037	36 59	067	22 46	135	16 45	198	30 55	236	45 41	276	24 10	348
172	41 26	039	37 09	069	22 54	137	16 41	200	30 46	238	45 31	278	24 08	350
174	41 33	041	37 19	071	23 01	139	16 37	202	30 37	240	45 21	280	24 07	352
176	41 40	042	37 29	073	23 07	141	16 33	204	30 28	242	45 10	282	24 05	354
178	41 47	044	37 39	075	23 14	143	16 29	207	30 19	244	45 00	284	24 04	356

LHA ♈	Hc	Zn	Hc	Zn	Hc	Zn	Hc	Zn	Hc	Zn	Hc	Zn	Hc	Zn
	◆DENEB		VEGA		◆ARCTURUS		REGULUS		POLLUX		◆CAPELLA		Alpheratz	
180	41 55	046	37 49	077	23 20	145	16 24	209	30 09	246	44 50	286	24 04	358
182	42 02	048	37 59	079	23 26	147	16 19	211	30 00	248	44 40	288	24 04	000
184	42 10	050	38 09	081	23 31	149	16 13	213	29 50	250	44 30	290	24 04	002
186	42 18	052	38 20	083	23 37	151	16 07	215	29 40	252	44 20	292	24 04	004
188	42 27	054	38 30	085	23 42	153	16 01	217	29 30	254	44 11	294	24 05	006
190	42 35	056	38 40	087	23 46	155	15 55	219	29 20	256	44 01	296	24 07	008
192	42 44	058	38 51	089	23 50	157	15 48	221	29 09	258	43 52	297	24 08	010
194	42 53	059	39 01	091	23 54	159	15 41	223	28 59	260	43 43	299	24 10	011
196	43 02	061	39 12	093	23 58	162	15 34	225	28 49	262	43 34	301	24 12	013
198	43 11	063	39 22	095	24 01	164	15 26	227	28 38	264	43 25	303	24 15	015
200	43 20	065	39 33	097	24 03	166	15 19	229	28 28	266	43 16	305	24 18	017
202	43 30	067	39 43	099	24 06	168	15 11	231	28 18	268	43 08	307	24 21	019
204	43 40	069	39 53	101	24 08	170	15 02	233	28 07	270	42 59	309	24 25	021
206	43 50	071	40 04	103	24 10	172	14 54	235	27 57	272	42 51	311	24 29	023
208	43 59	073	40 14	105	24 11	174	14 45	237	27 46	274	42 44	313	24 33	025
	Alpheratz		◆DENEB		VEGA		◆ARCTURUS		REGULUS		POLLUX		◆CAPELLA	
210	24 37	027	44 10	075	40 24	107	24 12	176	14 36	239	27 36	276	42 36	315
212	24 42	029	44 20	077	40 34	109	24 12	178	14 27	241	27 25	278	42 29	316
214	24 47	031	44 30	079	40 44	111	24 12	180	14 18	243	27 15	280	42 22	318
216	24 53	033	44 40	081	40 53	113	24 12	182	14 09	245	27 05	282	42 15	320
218	24 59	034	44 51	083	41 03	115	24 12	184	13 59	247	26 55	284	42 08	322
220	25 05	036	45 01	085	41 12	117	24 11	186	13 49	249	26 45	286	42 02	324
222	25 11	038	45 11	087	41 21	119	24 09	188	13 40	251	26 35	288	41 56	326
224	25 18	040	45 22	089	41 30	121	24 08	190	13 30	253	26 25	290	41 50	328
226	25 25	042	45 32	091	41 39	123	24 06	193	13 20	255	26 15	292	41 45	329
228	25 32	044	45 43	093	41 48	125	24 03	195	13 10	257	26 05	294	41 39	331
230	25 39	046	45 53	095	41 56	128	24 00	197	12 59	259	25 56	296	41 35	333
232	25 47	048	46 04	097	42 04	130	23 57	199	12 49	261	25 47	298	41 30	335
234	25 55	050	46 14	099	42 12	132	23 54	201	12 39	263	25 37	300	41 26	337
236	26 03	052	46 24	101	42 20	134	23 50	203	12 28	265	25 28	302	41 22	339
238	26 11	054	46 34	103	42 28	136	23 45	205	12 18	267	25 20	304	41 18	341
	Alpheratz		◆DENEB		VEGA		◆ARCTURUS		Dubhe		POLLUX		◆CAPELLA	
240	26 20	056	46 45	105	42 35	138	23 41	207	62 45	263	25 11	306	41 15	342
242	26 28	058	46 55	107	42 42	140	23 36	209	62 35	266	25 03	308	41 12	344
244	26 37	060	47 05	109	42 48	142	23 31	211	62 24	268	24 55	310	41 09	346
246	26 46	062	47 14	111	42 54	144	23 25	213	62 14	270	24 47	312	41 07	348
248	26 56	064	47 24	113	43 00	146	23 19	215	62 03	272	24 39	314	41 05	350
250	27 05	066	47 34	115	43 06	149	23 13	217	61 53	273	24 31	316	41 03	352
252	27 15	067	47 43	117	43 11	151	23 07	219	61 42	275	24 24	317	41 02	353
254	27 25	069	47 52	119	43 16	153	23 00	221	61 32	277	24 17	319	41 01	355
256	27 34	071	48 01	121	43 21	155	22 53	223	61 22	279	24 11	321	41 00	357
258	27 44	073	48 10	123	43 25	157	22 45	225	61 11	281	24 04	323	41 00	359
260	27 54	075	48 19	125	43 29	159	22 38	228	61 01	283	23 58	325	41 00	001
262	28 05	077	48 27	128	43 32	161	22 30	230	60 51	285	23 52	327	41 00	003
264	28 15	079	48 35	130	43 35	164	22 22	232	60 41	287	23 47	329	41 01	005
266	28 25	081	48 43	132	43 38	166	22 14	234	60 31	289	23 42	331	41 02	006
268	28 35	083	48 51	134	43 41	168	22 05	236	60 21	291	23 37	333	41 03	008
	◆CAPELLA		Alpheratz		◆DENEB		VEGA		◆ARCTURUS		Dubhe		POLLUX	
270	41 05	010	28 46	085	48 58	136	43 43	170	21 56	238	60 12	293	23 32	335
272	41 07	012	28 56	087	49 06	138	43 44	172	21 47	240	60 02	295	23 28	337
274	41 09	014	29 07	089	49 12	140	43 45	174	21 38	242	59 53	296	23 24	339
276	41 12	016	29 17	091	49 19	143	43 46	177	21 29	244	59 43	298	23 20	341
278	41 15	017	29 28	093	49 25	145	43 47	179	21 19	246	59 34	300	23 17	342
280	41 18	019	29 38	095	49 31	147	43 47	181	21 10	248	59 25	302	23 14	344
282	41 21	021	29 49	097	49 37	149	43 46	183	21 00	250	59 16	304	23 11	346
284	41 25	023	29 59	099	49 42	151	43 46	185	20 50	252	59 08	306	23 09	348
286	41 30	025	30 09	101	49 47	153	43 44	187	20 40	254	58 59	307	23 07	350
288	41 34	027	30 19	103	49 51	156	43 43	190	20 30	256	58 51	309	23 05	352
290	41 39	029	30 30	105	49 55	158	43 41	192	20 20	258	58 43	311	23 04	354
292	41 44	030	30 40	107	49 59	160	43 39	194	20 10	260	58 35	313	23 03	356
294	41 50	032	30 50	109	50 02	162	43 36	196	19 59	262	58 28	315	23 03	358
296	41 55	034	30 59	111	50 06	164	43 33	198	19 49	264	58 21	316	23 02	000
298	42 01	036	31 09	113	50 08	166	43 29	200	19 39	266	58 13	318	23 02	002
	POLLUX		◆CAPELLA		Alpheratz		◆DENEB		VEGA		◆ARCTURUS		Dubhe	
300	23 03	004	42 08	038	31 19	115	50 10	169	43 26	202	19 28	268	58 07	320
302	23 04	006	42 14	040	31 28	117	50 12	171	43 22	205	19 18	270	58 00	322
304	23 05	007	42 21	042	31 37	119	50 14	173	43 17	207	19 07	272	57 54	324
306	23 06	009	42 28	043	31 46	121	50 15	175	43 12	209	18 57	274	57 48	325
308	23 08	011	42 35	045	31 55	124	50 15	177	43 07	211	18 46	276	57 42	327
310	23 11	013	42 43	047	32 04	126	50 16	180	43 01	213	18 36	278	57 36	329
312	23 13	015	42 51	049	32 12	128	50 16	182	42 55	215	18 26	280	57 31	331
314	23 16	017	42 59	051	32 20	130	50 15	184	42 49	217	18 15	282	57 26	332
316	23 19	019	43 07	053	32 28	132	50 14	186	42 43	219	18 05	284	57 21	334
318	23 23	021	43 15	055	32 36	134	50 13	188	42 36	222	17 55	286	57 17	336
320	23 27	023	43 24	057	32 43	136	50 11	191	42 29	224	17 45	288	57 13	338
322	23 31	025	43 33	059	32 50	138	50 09	193	42 22	226	17 35	290	57 09	339
324	23 35	027	43 42	060	32 57	140	50 06	195	42 14	228	17 25	292	57 05	341
326	23 40	029	43 51	062	33 04	142	50 03	197	42 06	230	17 16	294	57 02	343
328	23 45	030	44 00	064	33 10	144	50 00	199	41 58	232	17 06	296	56 59	345
	POLLUX		◆CAPELLA		Alpheratz		◆DENEB		VEGA		◆ARCTURUS		Dubhe	
330	23 51	032	44 10	066	33 16	146	49 57	202	41 50	234	16 57	298	56 57	346
332	23 57	034	44 20	068	33 22	148	49 52	204	41 41	236	16 48	300	56 54	348
334	24 03	036	44 29	070	33 27	151	49 48	206	41 32	238	16 39	302	56 52	350
336	24 09	038	44 39	072	33 32	153	49 43	208	41 23	240	16 30	303	56 50	351
338	24 16	040	44 49	074	33 36	155	49 38	210	41 14	242	16 21	305	56 49	353
340	24 22	042	44 59	076	33 41	157	49 33	212	41 05	245	16 13	307	56 48	355
342	24 30	044	45 10	078	33 45	159	49 27	215	40 55	247	16 05	309	56 47	357
344	24 37	046	45 20	080	33 48	161	49 21	217	40 45	249	15 57	311	56 47	358
346	24 45	048	45 30	082	33 51	163	49 14	219	40 36	251	15 49	313	56 47	000
348	24 53	050	45 40	084	33 54	165	49 08	221	40 26	253	15 41	315	56 47	002
350	25 01	052	45 51	086	33 57	167	49 01	223	40 16	255	15 34	317	56 47	004
352	25 09	054	46 01	088	33 59	169	48 53	225	40 06	257	15 27	319	56 48	005
354	25 17	056	46 12	090	34 01	172	48 46	228	39 55	259	15 20	321	56 49	007
356	25 26	058	46 22	092	34 02	174	48 38	230	39 45	261	15 14	323	56 51	009
358	25 35	060	46 33	094	34 03	176	48 30	232	39 35	263	15 08	325	56 52	010

LHA ♈	Hc	Zn	Hc	Zn	Hc	Zn	Hc	Zn	Hc	Zn	Hc	Zn	Hc	Zn
	Dubhe		POLLUX		◆CAPELLA		Alpheratz		◆DENEB		VEGA		◆ARCTURUS	
0	55 56	012	25 15	061	46 49	095	35 04	178	48 57	235	39 29	266	14 12	327
2	55 59	014	25 27	063	47 01	097	35 04	180	48 46	237	39 17	268	14 05	329
4	56 02	015	25 38	065	47 13	099	35 04	182	48 36	239	39 04	270	13 59	331
6	56 05	017	25 49	067	47 26	101	35 03	184	48 25	241	38 52	272	13 53	333
8	56 09	019	26 01	069	47 38	103	35 02	186	48 14	243	38 39	274	13 47	335
10	56 13	020	26 13	071	47 50	105	35 00	189	48 02	245	38 27	276	13 42	337
12	56 18	022	26 25	073	48 02	107	34 58	191	47 51	247	38 14	278	13 38	339
14	56 23	024	26 37	075	48 14	109	34 55	193	47 39	249	38 02	280	13 33	341
16	56 28	025	26 49	077	48 26	111	34 52	195	47 27	252	37 49	281	13 29	343
18	56 33	027	27 01	079	48 38	113	34 49	197	47 15	254	37 37	283	13 26	345
20	56 39	029	27 13	081	48 49	115	34 45	199	47 03	256	37 25	285	13 23	347
22	56 46	031	27 26	083	49 00	117	34 41	201	46 51	258	37 13	287	13 20	348
24	56 52	032	27 38	085	49 11	119	34 36	203	46 39	260	37 01	289	13 18	350
26	56 59	034	27 51	087	49 22	121	34 31	206	46 26	262	36 49	291	13 16	352
28	57 06	036	28 03	089	49 33	124	34 25	208	46 14	264	36 38	293	13 14	354
	Dubhe		POLLUX		◆CAPELLA		Alpheratz		◆DENEB		VEGA		◆ARCTURUS	
30	57 14	037	28 16	091	49 43	126	34 19	210	46 01	266	36 26	295	13 13	356
32	57 21	039	28 28	093	49 53	128	34 13	212	45 49	268	36 15	297	13 13	358
34	57 29	041	28 41	095	50 03	130	34 06	214	45 36	270	36 04	299	13 12	000
36	57 38	043	28 53	097	50 12	132	33 59	216	45 24	272	35 53	301	13 13	002
38	57 47	044	29 06	099	50 22	134	33 51	218	45 11	274	35 42	303	13 13	004
40	57 55	046	29 18	101	50 30	136	33 43	220	44 59	276	35 32	305	13 14	006
42	58 05	048	29 30	103	50 39	139	33 35	222	44 46	278	35 22	307	13 16	008
44	58 14	050	29 43	105	50 47	141	33 26	224	44 34	280	35 12	308	13 18	010
46	58 24	052	29 55	107	50 55	143	33 17	227	44 22	282	35 02	310	13 20	012
48	58 34	053	30 07	109	51 02	145	33 08	229	44 09	284	34 53	312	13 23	014
50	58 44	055	30 18	111	51 09	147	32 59	231	43 57	285	34 43	314	13 26	016
52	58 54	057	30 30	113	51 16	150	32 49	233	43 45	287	34 35	316	13 30	018
54	59 05	059	30 42	115	51 22	152	32 39	235	43 33	289	34 26	318	13 34	020
56	59 16	060	30 53	117	51 27	154	32 28	237	43 21	291	34 18	320	13 38	021
58	59 27	062	31 04	119	51 33	156	32 18	239	43 10	293	34 10	322	13 43	023
	◆ARCTURUS		Dubhe		POLLUX		◆CAPELLA		Alpheratz		◆DENEB		VEGA	
60	13 48	025	59 38	064	31 15	121	51 37	159	32 07	241	42 58	295	34 02	324
62	13 54	027	59 49	066	31 26	123	51 42	161	31 56	243	42 47	297	33 55	325
64	14 00	029	60 01	068	31 36	125	51 46	163	31 44	245	42 36	299	33 48	327
66	14 06	031	60 12	070	31 46	127	51 49	165	31 33	247	42 25	301	33 41	329
68	14 13	033	60 24	072	31 56	129	51 52	168	31 21	249	42 15	303	33 35	331
70	14 20	035	60 36	073	32 06	131	51 55	170	31 09	251	42 04	305	33 29	333
72	14 27	037	60 48	075	32 15	133	51 57	172	30 57	253	41 54	306	33 24	335
74	14 35	039	61 01	077	32 24	135	51 58	174	30 45	255	41 44	308	33 19	337
76	14 43	041	61 13	079	32 32	137	51 59	177	30 33	257	41 34	310	33 14	338
78	14 51	043	61 25	081	32 41	140	52 00	179	30 21	259	41 25	312	33 09	340
80	15 00	045	61 38	083	32 49	142	52 00	181	30 09	261	41 16	314	33 05	342
82	15 09	047	61 50	085	32 56	144	51 59	183	29 56	263	41 07	316	33 02	344
84	15 18	049	62 03	087	33 04	146	51 58	186	29 44	265	40 58	318	32 58	346
86	15 28	051	62 15	089	33 10	148	51 57	188	29 31	267	40 50	319	32 56	348
88	15 38	053	62 28	091	33 17	150	51 55	190	29 19	269	40 42	321	32 53	350
	◆ARCTURUS		REGULUS		POLLUX		◆CAPELLA		Alpheratz		◆DENEB		VEGA	
90	15 48	055	14 45	117	33 23	152	51 52	192	29 06	271	40 34	323	32 51	351
92	15 58	057	14 56	119	33 29	154	51 49	195	28 54	273	40 27	325	32 49	353
94	16 09	058	15 07	121	33 34	156	51 46	197	28 41	275	40 20	327	32 48	355
96	16 19	060	15 18	123	33 39	158	51 42	199	28 29	277	40 13	329	32 47	357
98	16 30	062	15 28	125	33 43	161	51 38	201	28 16	279	40 07	331	32 47	359
100	16 42	064	15 38	127	33 47	163	51 33	204	28 04	281	40 01	332	32 47	001
102	16 53	066	15 48	129	33 51	165	51 28	206	27 52	283	39 55	334	32 47	003
104	17 05	068	15 58	131	33 54	167	51 22	208	27 39	285	39 50	336	32 48	004
106	17 16	070	16 07	133	33 56	169	51 16	210	27 27	287	39 45	338	32 49	006
108	17 28	072	16 16	135	33 58	171	51 10	212	27 15	289	39 40	340	32 51	008
110	17 40	074	16 25	137	34 00	173	51 03	215	27 04	291	39 36	342	32 53	010
112	17 52	076	16 33	139	34 01	175	50 55	217	26 52	293	39 32	343	32 55	012
114	18 05	078	16 41	141	34 02	178	50 48	219	26 40	295	39 29	345	32 58	014
116	18 17	080	16 49	143	34 02	180	50 39	221	26 29	297	39 26	347	33 01	016
118	18 29	082	16 56	145	34 02	182	50 31	223	26 18	299	39 23	349	33 05	017
	VEGA		◆ARCTURUS		REGULUS		POLLUX		◆CAPELLA		Alpheratz		◆DENEB	
120	33 09	019	18 42	084	17 03	147	34 01	184	50 22	226	26 07	301	39 21	351
122	33 13	021	18 54	086	17 10	149	34 00	186	50 13	228	25 57	303	39 19	352
124	33 18	023	19 07	088	17 16	151	33 59	188	50 04	230	25 46	305	39 18	354
126	33 23	025	19 19	090	17 22	153	33 57	190	49 54	232	25 36	307	39 17	356
128	33 28	027	19 32	092	17 28	155	33 54	193	49 44	234	25 26	308	39 16	358
130	33 34	029	19 44	094	17 33	157	33 51	195	49 34	236	25 16	310	39 16	000
132	33 40	030	19 57	096	17 37	159	33 48	197	49 23	238	25 07	312	39 16	002
134	33 47	032	20 09	098	17 41	161	33 44	199	49 12	240	24 58	314	39 16	003
136	33 54	034	20 22	100	17 45	164	33 40	201	49 01	243	24 49	316	39 17	005
138	34 01	036	20 34	102	17 48	166	33 35	203	48 50	245	24 40	318	39 19	007
140	34 08	038	20 46	104	17 51	168	33 30	205	48 39	247	24 32	320	39 20	009
142	34 16	040	20 58	106	17 54	170	33 24	207	48 27	249	24 24	322	39 23	011
144	34 24	042	21 10	108	17 56	172	33 18	209	48 15	251	24 17	324	39 25	012
146	34 33	044	21 22	110	17 57	174	33 12	212	48 03	253	24 09	326	39 28	014
148	34 42	046	21 34	112	17 59	176	33 05	214	47 51	255	24 02	328	39 31	016
	◆DENEB		VEGA		◆ARCTURUS		REGULUS		POLLUX		◆CAPELLA		Alpheratz	
150	39 35	018	34 51	047	21 45	114	17 59	178	32 58	216	47 39	257	23 56	330
152	39 39	020	35 00	049	21 57	116	18 00	180	32 51	218	47 27	259	23 50	331
154	39 43	022	35 10	051	22 08	118	17 59	182	32 43	220	47 14	261	23 44	333
156	39 48	023	35 20	053	22 19	120	17 59	184	32 35	222	47 02	263	23 38	335
158	39 53	025	35 30	055	22 30	122	17 58	186	32 26	224	46 50	265	23 33	337
160	39 59	027	35 40	057	22 40	124	17 56	188	32 17	226	46 37	267	23 29	339
162	40 05	029	35 51	059	22 50	126	17 54	190	32 08	228	46 24	269	23 24	341
164	40 11	031	36 02	061	23 00	128	17 51	192	31 58	230	46 12	271	23 21	343
166	40 18	033	36 13	063	23 10	130	17 49	194	31 49	232	45 59	273	23 17	345
168	40 24	034	36 24	065	23 19	132	17 45	196	31 39	235	45 47	275	23 14	347
170	40 32	036	36 35	066	23 29	134	17 41	198	31 28	237	45 34	277	23 11	349
172	40 39	038	36 47	068	23 37	137	17 37	201	31 18	239	45 22	279	23 09	350
174	40 47	040	36 59	070	23 46	139	17 33	203	31 07	241	45 10	281	23 07	352
176	40 55	042	37 11	072	23 54	141	17 28	205	30 56	243	44 57	283	23 06	354
178	41 04	044	37 23	074	24 02	143	17 22	207	30 44	245	44 45	285	23 05	356

LAT 84°N

LHA ♈	Hc	Zn	Hc	Zn	Hc	Zn	Hc	Zn	Hc	Zn	Hc	Zn	Hc	Zn
	◆DENEB		VEGA		◆ARCTURUS		REGULUS		POLLUX		◆CAPELLA		Alpheratz	
180	41 13	046	37 35	076	24 09	145	17 16	209	30 33	247	44 33	287	23 04	358
182	41 22	047	37 47	078	24 16	147	17 10	211	30 21	249	44 21	289	23 04	000
184	41 31	049	37 59	080	24 23	149	17 04	213	30 10	251	44 09	291	23 04	002
186	41 41	051	38 12	082	24 29	151	16 57	215	29 58	253	43 58	293	23 05	004
188	41 51	053	38 24	084	24 35	153	16 49	217	29 46	255	43 46	294	23 06	006
190	42 01	055	38 37	086	24 41	155	16 42	219	29 34	257	43 35	296	23 07	008
192	42 11	057	38 49	088	24 46	157	16 33	221	29 21	259	43 24	298	23 09	009
194	42 22	059	39 02	090	24 50	159	16 25	223	29 09	261	43 13	300	23 11	011
196	42 33	061	39 14	092	24 54	161	16 16	225	28 56	263	43 02	302	23 14	013
198	42 44	062	39 27	094	24 58	163	16 07	227	28 44	265	42 52	304	23 17	015
200	42 55	064	39 39	096	25 02	166	15 58	229	28 31	267	42 41	306	23 20	017
202	43 06	066	39 52	098	25 04	168	15 48	231	28 19	269	42 31	308	23 24	019
204	43 18	068	40 04	100	25 07	170	15 39	233	28 06	271	42 21	310	23 29	021
206	43 30	070	40 16	102	25 09	172	15 28	235	27 54	273	42 12	311	23 33	023
208	43 41	072	40 29	104	25 11	174	15 18	237	27 41	275	42 03	313	23 38	025
	Alpheratz		◆DENEB		VEGA		◆ARCTURUS		REGULUS		POLLUX		◆CAPELLA	
210	23 44	027	43 53	074	40 41	106	25 12	176	15 07	239	27 29	277	41 54	315
212	23 50	029	44 05	076	40 53	108	25 12	178	14 56	241	27 16	279	41 45	317
214	23 56	030	44 18	078	41 05	110	25 12	180	14 45	243	27 04	281	41 37	319
216	24 02	032	44 30	080	41 16	112	25 12	182	14 34	245	26 52	283	41 28	321
218	24 09	034	44 42	082	41 28	114	25 12	184	14 22	247	26 40	285	41 21	323
220	24 16	036	44 55	084	41 39	116	25 10	186	14 11	249	26 28	287	41 13	324
222	24 24	038	45 07	086	41 50	118	25 09	188	13 59	251	26 16	289	41 06	326
224	24 32	040	45 20	088	42 01	121	25 07	191	13 47	253	26 04	291	40 59	328
226	24 40	042	45 32	090	42 12	123	25 04	193	13 35	255	25 52	293	40 53	330
228	24 49	044	45 45	092	42 22	125	25 01	195	13 23	257	25 41	295	40 47	332
230	24 57	046	45 57	094	42 33	127	24 58	197	13 11	259	25 29	296	40 41	334
232	25 07	048	46 10	096	42 42	129	24 54	199	12 58	261	25 18	298	40 35	335
234	25 16	050	46 22	098	42 52	131	24 50	201	12 46	263	25 07	300	40 30	337
236	25 26	052	46 35	100	43 01	133	24 45	203	12 33	265	24 57	302	40 26	339
238	25 36	053	46 47	102	43 10	135	24 40	205	12 21	267	24 46	304	40 21	341
	Alpheratz		◆DENEB		VEGA		◆ARCTURUS		Dubhe		POLLUX		◆CAPELLA	
240	25 46	055	46 59	104	43 19	137	24 34	207	62 51	265	24 36	306	40 18	343
242	25 56	057	47 11	106	43 27	140	24 28	209	62 38	267	24 26	308	40 14	344
244	26 07	059	47 24	108	43 35	142	24 22	211	62 26	269	24 16	310	40 11	346
246	26 18	061	47 35	110	43 43	144	24 15	213	62 13	271	24 07	312	40 08	348
248	26 29	063	47 47	112	43 50	146	24 08	216	62 01	273	23 57	314	40 06	350
250	26 40	065	47 59	114	43 57	148	24 01	218	61 48	275	23 49	316	40 04	352
252	26 52	067	48 10	116	44 03	150	23 53	220	61 36	277	23 40	318	40 02	354
254	27 03	069	48 21	118	44 09	152	23 45	222	61 23	279	23 32	320	40 01	355
256	27 15	071	48 32	120	44 15	155	23 36	224	61 11	281	23 24	322	40 00	357
258	27 27	073	48 43	122	44 20	157	23 27	226	60 59	283	23 16	323	40 00	359
260	27 39	075	48 53	125	44 25	159	23 18	228	60 47	285	23 09	325	40 00	001
262	27 51	077	49 03	127	44 29	161	23 09	230	60 35	287	23 02	327	40 00	003
264	28 03	079	49 13	129	44 33	163	22 59	232	60 23	289	22 55	329	40 01	004
266	28 16	081	49 23	131	44 36	166	22 49	234	60 11	291	22 49	331	40 02	006
268	28 28	083	49 32	133	44 39	168	22 39	236	59 59	292	22 43	333	40 04	008
	◆CAPELLA		Alpheratz		◆DENEB		VEGA		◆ARCTURUS		Dubhe		POLLUX	
270	40 06	010	28 41	085	49 41	135	44 42	170	22 28	238	59 48	294	22 38	335
272	40 08	012	28 53	087	49 50	137	44 44	172	22 17	240	59 36	296	22 33	337
274	40 11	014	29 06	089	49 58	140	44 45	174	22 06	242	59 25	298	22 28	339
276	40 14	015	29 18	091	50 06	142	44 46	176	21 55	244	59 14	300	22 23	341
278	40 17	017	29 31	093	50 14	144	44 47	179	21 43	246	59 03	302	22 20	343
280	40 21	019	29 43	095	50 21	146	44 47	181	21 32	248	58 53	303	22 16	345
282	40 25	021	29 56	097	50 28	148	44 46	183	21 21	250	58 42	305	22 13	346
284	40 30	023	30 08	099	50 34	151	44 45	185	21 09	252	58 32	307	22 10	348
286	40 35	024	30 21	101	50 40	153	44 44	187	20 57	254	58 22	309	22 08	350
288	40 40	026	30 33	103	50 46	155	44 42	190	20 45	256	58 13	311	22 06	352
290	40 46	028	30 45	105	50 51	157	44 40	192	20 32	258	58 03	312	22 04	354
292	40 52	030	30 57	107	50 55	159	44 37	194	20 20	260	57 54	314	22 03	356
294	40 59	032	31 09	109	51 00	162	44 34	196	20 08	262	57 45	316	22 03	358
296	41 05	034	31 21	111	51 03	164	44 30	198	19 55	264	57 37	318	22 02	000
298	41 13	035	31 32	113	51 06	166	44 26	201	19 43	266	57 28	319	22 02	002
	POLLUX		◆CAPELLA		Alpheratz		◆DENEB		VEGA		◆ARCTURUS		Dubhe	
300	22 03	004	41 20	037	31 44	115	51 09	168	44 21	203	19 30	268	57 20	321
302	22 04	005	41 28	039	31 55	117	51 11	171	44 16	205	19 18	270	57 13	323
304	22 05	007	41 36	041	32 06	119	51 13	173	44 10	207	19 05	272	57 05	325
306	22 07	009	41 44	043	32 17	121	51 15	175	44 05	209	18 53	274	56 58	326
308	22 09	011	41 53	045	32 28	123	51 15	177	43 58	211	18 40	276	56 51	328
310	22 12	013	42 02	047	32 38	125	51 16	180	43 51	214	18 28	278	56 45	330
312	22 15	015	42 11	048	32 48	127	51 16	182	43 44	216	18 15	280	56 38	331
314	22 19	017	42 21	050	32 58	129	51 15	184	43 37	218	18 03	282	56 33	333
316	22 22	019	42 30	052	33 08	131	51 14	186	43 29	220	17 51	284	56 27	335
318	22 27	021	42 40	054	33 17	133	51 12	189	43 21	222	17 39	286	56 22	336
320	22 31	023	42 51	056	33 26	135	51 10	191	43 12	224	17 27	288	56 17	338
322	22 36	025	43 01	058	33 35	138	51 07	193	43 03	226	17 15	290	56 13	340
324	22 42	026	43 12	060	33 43	140	51 04	195	42 54	229	17 03	292	56 08	342
326	22 48	028	43 23	062	33 51	142	51 01	198	42 44	231	16 51	294	56 05	343
328	22 54	030	43 34	063	33 58	144	50 57	200	42 34	233	16 40	296	56 01	345
	POLLUX		◆CAPELLA		Alpheratz		◆DENEB		VEGA		◆ARCTURUS		Dubhe	
330	23 00	032	43 45	065	34 06	146	50 52	202	42 24	235	16 29	298	55 58	347
332	23 07	034	43 57	067	34 13	148	50 47	204	42 14	237	16 18	300	55 55	348
334	23 14	036	44 08	069	34 19	150	50 42	207	42 03	239	16 07	302	55 53	350
336	23 22	038	44 20	071	34 25	152	50 36	209	41 52	241	15 57	304	55 51	352
338	23 30	040	44 32	073	34 31	154	50 30	211	41 41	243	15 46	306	55 50	353
340	23 38	042	44 44	075	34 36	157	50 23	213	41 30	245	15 36	308	55 48	355
342	23 46	044	44 56	077	34 41	159	50 16	215	41 19	247	15 26	310	55 47	357
344	23 55	046	45 09	079	34 45	161	50 09	218	41 07	249	15 17	312	55 47	358
346	24 04	048	45 21	081	34 49	163	50 01	220	40 55	251	15 08	314	55 47	000
348	24 14	049	45 33	083	34 52	165	49 53	222	40 43	254	14 59	315	55 47	002
350	24 23	051	45 46	085	34 55	167	49 44	224	40 31	256	14 50	317	55 47	003
352	24 33	053	45 58	087	34 58	169	49 35	226	40 19	258	14 42	319	55 48	005
354	24 43	055	46 11	089	35 00	171	49 26	228	40 07	260	14 34	321	55 50	007
356	24 54	057	46 23	091	35 02	174	49 16	231	39 54	262	14 26	323	55 51	009
358	25 05	059	46 36	093	35 03	176	49 07	233	39 42	264	14 19	325	55 53	010

LAT 83°N

LHA	Hc	Zn	Hc	Zn	Hc	Zn	Hc	Zn	Hc	Zn	Hc	Zn	Hc	Zn
♈	Dubhe		POLLUX		◆CAPELLA		Alpheratz		◆DENEB		VEGA		◆ARCTURUS	
0	54 57	012	24 46	061	46 53	094	36 03	178	49 31	236	39 33	266	13 21	327
2	55 00	013	24 59	063	47 07	096	36 04	180	49 18	238	39 19	268	13 14	329
4	55 04	015	25 12	065	47 22	098	36 04	182	49 06	240	39 04	270	13 06	331
6	55 08	017	25 25	066	47 36	100	36 03	184	48 53	242	38 49	272	12 59	333
8	55 12	018	25 39	068	47 51	102	36 01	186	48 40	244	38 35	274	12 53	335
10	55 17	020	25 53	070	48 05	104	35 59	189	48 27	246	38 20	276	12 47	337
12	55 22	022	26 07	072	48 19	106	35 57	191	48 13	248	38 06	278	12 42	339
14	55 28	023	26 21	074	48 33	108	35 54	193	48 00	251	37 51	280	12 37	341
16	55 34	025	26 35	076	48 47	110	35 50	195	47 46	253	37 37	282	12 32	343
18	55 40	026	26 49	078	49 01	112	35 46	197	47 32	255	37 23	284	12 28	345
20	55 47	028	27 03	080	49 14	114	35 42	199	47 18	257	37 09	286	12 24	347
22	55 54	030	27 18	082	49 27	116	35 37	202	47 03	259	36 55	288	12 21	349
24	56 01	032	27 32	084	49 40	118	35 31	204	46 49	261	36 41	290	12 18	350
26	56 09	033	27 47	086	49 53	120	35 25	206	46 34	263	36 27	292	12 16	352
28	56 17	035	28 01	088	50 06	123	35 18	208	46 20	265	36 14	294	12 15	354
	Dubhe		POLLUX		◆CAPELLA		Alpheratz		◆DENEB		VEGA		◆ARCTURUS	
30	56 26	037	28 16	090	50 18	125	35 11	210	46 05	267	36 00	296	12 13	356
32	56 35	038	28 31	092	50 30	127	35 04	212	45 51	269	35 47	298	12 13	358
34	56 44	040	28 45	094	50 41	129	34 56	214	45 36	271	35 35	300	12 12	000
36	56 53	042	29 00	096	50 52	131	34 47	216	45 21	273	35 22	301	12 13	002
38	57 03	043	29 14	098	51 03	133	34 38	219	45 07	275	35 10	303	12 14	004
40	57 13	045	29 29	100	51 14	136	34 29	221	44 52	277	34 58	305	12 15	006
42	57 24	047	29 43	102	51 24	138	34 19	223	44 38	279	34 46	307	12 17	008
44	57 35	049	29 57	104	51 33	140	34 09	225	44 23	281	34 34	309	12 19	010
46	57 46	050	30 12	106	51 42	142	33 58	227	44 09	283	34 23	311	12 22	012
48	57 57	052	30 26	108	51 51	145	33 48	229	43 55	284	34 12	313	12 25	014
50	58 09	054	30 39	110	51 59	147	33 36	231	43 41	286	34 02	315	12 28	016
52	58 21	056	30 53	112	52 07	149	33 25	233	43 27	288	33 51	316	12 33	018
54	58 33	057	31 06	114	52 14	151	33 13	235	43 13	290	33 41	318	12 37	019
56	58 45	059	31 20	116	52 21	154	33 01	237	42 59	292	33 32	320	12 42	021
58	58 58	061	31 33	118	52 27	156	32 48	240	42 46	294	33 23	322	12 48	023
	◆ARCTURUS		Dubhe		POLLUX		◆CAPELLA		Alpheratz		◆DENEB		VEGA	
60	12 54	025	59 11	063	31 45	120	52 33	158	32 35	242	42 33	296	33 14	324
62	13 00	027	59 24	064	31 58	122	52 38	160	32 23	244	42 20	298	33 05	326
64	13 07	029	59 37	066	32 10	124	52 43	163	32 09	246	42 07	300	32 57	328
66	13 15	031	59 51	068	32 22	127	52 47	165	31 56	248	41 54	302	32 50	329
68	13 22	033	60 04	070	32 34	129	52 51	167	31 42	250	41 42	303	32 43	331
70	13 31	035	60 18	072	32 45	131	52 54	170	31 28	252	41 30	305	32 36	333
72	13 39	037	60 32	074	32 56	133	52 56	172	31 14	254	41 18	307	32 29	335
74	13 48	039	60 46	075	33 06	135	52 58	174	31 00	256	41 06	309	32 23	337
76	13 58	041	61 01	077	33 16	137	52 59	176	30 46	258	40 55	311	32 18	339
78	14 07	043	61 15	079	33 26	139	53 00	179	30 32	260	40 44	313	32 13	341
80	14 17	045	61 29	081	33 36	141	53 00	181	30 17	262	40 34	315	32 08	342
82	14 28	047	61 44	083	33 45	143	52 59	183	30 03	264	40 23	316	32 04	344
84	14 39	049	61 58	085	33 53	145	52 58	186	29 48	266	40 14	318	32 00	346
86	14 50	050	62 13	087	34 01	148	52 56	188	29 34	268	40 04	320	31 57	348
88	15 01	052	62 28	089	34 09	150	52 54	190	29 19	270	39 55	322	31 54	350
	◆ARCTURUS		REGULUS		POLLUX		◆CAPELLA		Alpheratz		◆DENEB		VEGA	
90	15 13	054	15 12	116	34 16	152	52 51	193	29 04	272	39 46	324	31 52	352
92	15 25	056	15 25	118	34 23	154	52 47	195	28 50	274	39 37	325	31 50	353
94	15 37	058	15 38	121	34 29	156	52 43	197	28 35	276	39 29	327	31 48	355
96	15 50	060	15 50	123	34 34	158	52 39	199	28 21	278	39 22	329	31 47	357
98	16 03	062	16 02	125	34 40	160	52 34	202	28 06	280	39 14	331	31 47	359
100	16 16	064	16 14	127	34 44	163	52 28	204	27 52	282	39 07	333	31 47	001
102	16 29	066	16 26	129	34 48	165	52 22	206	27 38	284	39 01	335	31 47	003
104	16 42	068	16 37	131	34 52	167	52 15	209	27 23	286	38 55	336	31 48	004
106	16 56	070	16 48	133	34 55	169	52 08	211	27 09	288	38 49	338	31 49	006
108	17 10	072	16 58	135	34 58	171	52 00	213	26 56	290	38 44	340	31 51	008
110	17 24	074	17 09	137	35 00	173	51 52	215	26 42	291	38 39	342	31 54	010
112	17 38	076	17 18	139	35 01	175	51 43	218	26 28	293	38 35	344	31 56	012
114	17 52	078	17 28	141	35 02	178	51 34	220	26 15	295	38 31	345	32 00	014
116	18 07	080	17 37	143	35 02	180	51 24	222	26 02	297	38 27	347	32 03	015
118	18 21	082	17 46	145	35 02	182	51 14	224	25 49	299	38 24	349	32 07	017
	VEGA		◆ARCTURUS		REGULUS		POLLUX		◆CAPELLA		Alpheratz		◆DENEB	
120	32 12	019	18 35	084	17 54	147	35 01	184	51 04	226	25 36	301	38 22	351
122	32 17	021	18 50	086	18 02	149	35 00	186	50 53	229	25 24	303	38 20	353
124	32 22	023	19 05	088	18 09	151	34 58	188	50 42	231	25 12	305	38 18	354
126	32 28	025	19 19	090	18 16	153	34 56	191	50 31	233	25 00	307	38 17	356
128	32 34	026	19 34	092	18 22	155	34 53	193	50 19	235	24 48	309	38 16	358
130	32 41	028	19 48	094	18 28	157	34 49	195	50 07	237	24 37	311	38 16	000
132	32 48	030	20 03	096	18 33	159	34 45	197	49 54	239	24 26	313	38 16	002
134	32 56	032	20 18	098	18 38	161	34 41	199	49 41	242	24 16	315	38 17	003
136	33 04	034	20 32	100	18 43	163	34 36	201	49 28	244	24 05	316	38 18	005
138	33 12	036	20 46	102	18 47	165	34 30	203	49 15	246	23 56	318	38 19	007
140	33 21	038	21 01	104	18 50	168	34 24	206	49 02	248	23 46	320	38 21	009
142	33 30	039	21 15	106	18 53	170	34 18	208	48 48	250	23 37	322	38 24	010
144	33 40	041	21 29	108	18 55	172	34 11	210	48 34	252	23 28	324	38 26	012
146	33 49	043	21 43	110	18 57	174	34 03	212	48 20	254	23 20	326	38 30	014
148	34 00	045	21 56	112	18 58	176	33 55	214	48 06	256	23 12	328	38 34	016
	◆DENEB		VEGA		◆ARCTURUS		REGULUS		POLLUX		◆CAPELLA		Alpheratz	
150	38 38	018	34 10	047	22 10	114	18 59	178	33 47	216	47 52	258	23 04	330
152	38 42	019	34 21	049	22 23	116	19 00	180	33 38	218	47 38	260	22 57	332
154	38 47	021	34 32	051	22 36	118	18 59	182	33 29	220	47 23	262	22 50	334
156	38 53	023	34 43	053	22 49	120	18 58	184	33 19	222	47 09	264	22 44	335
158	38 59	025	34 55	054	23 02	122	18 57	186	33 09	225	46 54	266	22 38	337
160	39 05	027	35 07	056	23 14	124	18 55	188	32 58	227	46 39	268	22 33	339
162	39 12	028	35 20	058	23 26	126	18 53	190	32 48	229	46 25	270	22 28	341
164	39 19	030	35 32	060	23 37	128	18 50	192	32 36	231	46 10	272	22 23	343
166	39 27	032	35 45	062	23 49	130	18 47	194	32 25	233	45 56	274	22 19	345
168	39 35	034	35 58	064	24 00	132	18 43	197	32 13	235	45 41	276	22 16	347
170	39 43	036	36 11	066	24 11	134	18 38	199	32 01	237	45 27	278	22 12	349
172	39 52	038	36 25	068	24 21	136	18 33	201	31 49	239	45 12	280	22 10	351
174	40 01	039	36 38	070	24 31	138	18 28	203	31 36	241	44 58	282	22 08	352
176	40 10	041	36 52	072	24 40	140	18 22	205	31 23	243	44 43	284	22 06	354
178	40 20	043	37 06	074	24 49	142	18 16	207	31 10	245	44 29	286	22 05	356

LHA	Hc	Zn	Hc	Zn	Hc	Zn	Hc	Zn	Hc	Zn	Hc	Zn	Hc	Zn
♈	◆DENEB		VEGA		◆ARCTURUS		REGULUS		POLLUX		◆CAPELLA		Alpheratz	
180	40 30	045	37 20	075	24 58	145	18 09	209	30 56	247	44 15	288	22 04	358
182	40 41	047	37 34	077	25 06	147	18 02	211	30 43	249	44 01	290	22 04	000
184	40 52	049	37 49	079	25 14	149	17 54	213	30 29	251	43 48	292	22 04	002
186	41 03	050	38 03	081	25 22	151	17 46	215	30 15	253	43 34	293	22 05	004
188	41 14	052	38 18	083	25 28	153	17 37	217	30 01	255	43 21	295	22 06	006
190	41 26	054	38 32	085	25 35	155	17 28	219	29 47	257	43 08	297	22 08	008
192	41 38	056	38 47	087	25 41	157	17 19	221	29 32	260	42 55	299	22 10	009
194	41 50	058	39 01	089	25 46	159	17 09	223	29 18	262	42 42	301	22 12	011
196	42 03	060	39 16	091	25 51	161	16 59	225	29 04	264	42 30	303	22 15	013
198	42 15	062	39 31	093	25 56	163	16 48	227	28 49	266	42 18	305	22 19	015
200	42 28	063	39 45	095	26 00	165	16 37	229	28 34	268	42 06	307	22 23	017
202	42 42	065	40 00	097	26 03	168	16 26	231	28 20	270	41 54	308	22 28	019
204	42 55	067	40 14	099	26 06	170	16 14	233	28 05	271	41 43	310	22 33	021
206	43 09	069	40 29	101	26 08	172	16 03	235	27 51	273	41 32	312	22 38	023
208	43 22	071	40 43	103	26 10	174	15 50	237	27 36	275	41 21	314	22 44	025
	Alpheratz		◆DENEB		VEGA		◆ARCTURUS		REGULUS		POLLUX		◆CAPELLA	
210	22 50	026	43 36	073	40 57	105	26 11	176	15 38	239	27 21	277	41 11	316
212	22 57	028	43 50	075	41 11	107	26 12	178	15 25	241	27 07	279	41 01	318
214	23 04	030	44 05	077	41 25	109	26 12	180	15 12	243	26 53	281	40 51	319
216	23 11	032	44 19	079	41 39	111	26 12	182	14 59	245	26 38	283	40 42	321
218	23 19	034	44 33	081	41 52	113	26 11	184	14 46	248	26 24	285	40 33	323
220	23 28	036	44 48	083	42 05	116	26 10	186	14 32	250	26 10	287	40 24	325
222	23 37	038	45 02	085	42 19	118	26 08	189	14 18	252	25 56	289	40 16	327
224	23 46	040	45 17	087	42 31	120	26 06	191	14 04	254	25 42	291	40 08	329
226	23 55	042	45 31	089	42 44	122	26 03	193	13 50	256	25 29	293	40 01	330
228	24 05	044	45 46	091	42 56	124	25 59	195	13 36	258	25 16	295	39 54	332
230	24 15	045	46 01	093	43 08	126	25 55	197	13 22	260	25 02	297	39 47	334
232	24 26	047	46 15	095	43 20	128	25 51	199	13 07	262	24 49	299	39 41	336
234	24 37	049	46 30	097	43 31	130	25 46	201	12 53	264	24 37	301	39 35	338
236	24 48	051	46 44	099	43 42	132	25 40	203	12 38	265	24 24	303	39 30	339
238	25 00	053	46 59	101	43 53	135	25 34	205	12 24	267	24 12	305	39 25	341
	Alpheratz		◆DENEB		VEGA		◆ARCTURUS		Dubhe		POLLUX		◆CAPELLA	
240	25 12	055	47 13	103	44 03	137	25 28	207	62 55	267	24 00	307	39 20	343
242	25 24	057	47 27	105	44 13	139	25 21	210	62 40	269	23 49	308	39 16	345
244	25 36	059	47 41	107	44 22	141	25 13	212	62 25	271	23 37	310	39 13	347
246	25 49	061	47 55	109	44 31	143	25 05	214	62 11	273	23 26	312	39 09	348
248	26 02	063	48 09	111	44 40	145	24 57	216	61 56	275	23 16	314	39 07	350
250	26 15	065	48 23	113	44 48	148	24 48	218	61 42	277	23 05	316	39 04	352
252	26 28	067	48 36	115	44 55	150	24 39	220	61 27	279	22 55	318	39 02	354
254	26 41	068	48 49	117	45 02	152	24 29	222	61 13	281	22 46	320	39 01	355
256	26 55	070	49 02	119	45 09	154	24 19	224	60 59	283	22 37	322	39 00	357
258	27 09	072	49 15	121	45 15	156	24 09	226	60 44	285	22 28	324	39 00	359
260	27 23	074	49 27	124	45 21	159	23 58	228	60 30	287	22 19	326	39 00	001
262	27 37	076	49 39	126	45 26	161	23 47	230	60 16	289	22 11	328	39 00	003
264	27 51	078	49 51	128	45 30	163	23 36	232	60 03	290	22 04	329	39 01	004
266	28 06	080	50 02	130	45 34	165	23 24	234	59 49	292	21 56	331	39 02	006
268	28 20	082	50 13	132	45 38	168	23 12	236	59 35	294	21 50	333	39 04	008
	◆CAPELLA		Alpheratz		◆DENEB		VEGA		◆ARCTURUS		Dubhe		POLLUX	
270	39 06	010	28 35	084	50 24	134	45 41	170	23 00	238	59 22	296	21 43	335
272	39 09	012	28 49	086	50 34	137	45 43	172	22 47	240	59 09	298	21 37	337
274	39 12	013	29 04	088	50 44	139	45 45	174	22 34	242	58 56	299	21 32	339
276	39 16	015	29 19	090	50 53	141	45 46	176	22 21	245	58 44	301	21 27	341
278	39 20	017	29 33	092	51 02	143	45 47	179	22 08	247	58 31	303	21 22	343
280	39 24	019	29 48	094	51 11	145	45 47	181	21 54	249	58 19	305	21 18	345
282	39 29	021	30 02	096	51 19	148	45 46	183	21 41	251	58 07	307	21 15	347
284	39 35	022	30 17	098	51 26	150	45 45	185	21 27	253	57 56	308	21 11	348
286	39 40	024	30 31	100	51 33	152	45 43	188	21 13	255	57 44	310	21 09	350
288	39 47	026	30 46	102	51 40	154	45 41	190	20 59	257	57 33	312	21 06	352
290	39 53	028	31 00	104	51 46	157	45 38	192	20 44	259	57 22	313	21 05	354
292	40 00	030	31 14	106	51 51	159	45 35	194	20 30	261	57 12	315	21 03	356
294	40 08	031	31 28	108	51 56	161	45 31	197	20 16	263	57 02	317	21 03	358
296	40 15	033	31 42	110	52 01	164	45 27	199	20 01	265	56 52	319	21 02	000
298	40 24	035	31 55	112	52 05	166	45 22	201	19 46	267	56 42	320	21 02	002
	POLLUX		◆CAPELLA		Alpheratz		◆DENEB		VEGA		◆ARCTURUS		Dubhe	
300	21 03	004	40 32	037	32 09	114	52 08	168	45 16	203	19 32	269	56 33	322
302	21 04	005	40 41	039	32 22	116	52 11	170	45 10	205	19 17	271	56 24	324
304	21 06	007	40 50	040	32 35	118	52 13	173	45 04	208	19 03	273	56 16	325
306	21 08	009	41 00	042	32 48	120	52 14	175	44 57	210	18 48	275	56 08	327
308	21 11	011	41 10	044	33 00	122	52 15	177	44 49	212	18 33	277	56 00	329
310	21 14	013	41 20	046	33 12	125	52 16	180	44 41	214	18 19	279	55 53	330
312	21 17	015	41 31	048	33 24	127	52 16	182	44 33	216	18 04	280	55 46	332
314	21 21	017	41 42	050	33 36	129	52 15	184	44 24	219	17 50	282	55 39	334
316	21 26	019	41 53	051	33 47	131	52 13	187	44 15	221	17 36	284	55 33	335
318	21 31	021	42 05	053	33 58	133	52 11	189	44 05	223	17 22	286	55 27	337
320	21 36	022	42 17	055	34 09	135	52 09	191	43 55	225	17 08	288	55 21	339
322	21 42	024	42 29	057	34 19	137	52 06	193	43 44	227	16 54	290	55 16	340
324	21 48	026	42 41	059	34 29	139	52 02	196	43 33	229	16 40	292	55 11	342
326	21 55	028	42 54	061	34 38	141	51 58	198	43 22	231	16 27	294	55 07	344
328	22 02	030	43 07	063	34 47	143	51 53	200	43 10	234	16 14	296	55 03	345
	POLLUX		◆CAPELLA		Alpheratz		◆DENEB		VEGA		◆ARCTURUS		Dubhe	
330	22 09	032	43 20	065	34 55	146	51 48	203	42 59	236	16 01	298	55 00	347
332	22 17	034	43 33	066	35 03	148	51 42	205	42 46	238	15 48	300	54 57	349
334	22 26	036	43 47	068	35 11	150	51 35	207	42 34	240	15 35	302	54 54	350
336	22 34	038	44 00	070	35 18	152	51 29	209	42 21	242	15 23	304	54 52	352
338	22 44	040	44 14	072	35 25	154	51 21	212	42 08	244	15 11	306	54 50	354
340	22 53	041	44 28	074	35 31	156	51 13	214	41 55	246	14 59	308	54 48	355
342	23 03	043	44 42	076	35 36	158	51 05	216	41 41	248	14 48	310	54 47	357
344	23 13	045	44 57	078	35 42	161	50 56	218	41 28	250	14 37	312	54 47	358
346	23 24	047	45 11	080	35 46	163	50 47	220	41 14	252	14 26	314	54 47	000
348	23 35	049	45 25	082	35 50	165	50 37	223	41 00	254	14 16	316	54 47	002
350	23 46	051	45 40	084	35 54	167	50 27	225	40 46	256	14 06	318	54 48	003
352	23 57	053	45 54	086	35 57	169	50 16	227	40 31	258	13 56	320	54 49	005
354	24 09	055	46 09	088	35 59	171	50 06	229	40 17	260	13 47	321	54 50	007
356	24 21	057	46 24	090	36 01	173	49 54	231	40 02	262	13 38	323	54 52	008
358	24 34	059	46 38	092	36 03	176	49 43	234	39 48	264	13 29	325	54 54	010

LHA ♈	POLLUX Hc	Zn	◆CAPELLA Hc	Zn	Hamal Hc	Zn	Alpheratz Hc	Zn	◆DENEB Hc	Zn	VEGA Hc	Zn	◆Alioth Hc	Zn
0	24 17	060	46 56	093	30 10	146	37 03	178	50 04	237	39 37	267	48 10	349
2	24 31	062	47 13	095	30 19	148	37 04	180	49 50	239	39 20	269	48 07	350
4	24 46	064	47 29	097	30 28	150	37 03	182	49 35	241	39 03	271	48 05	352
6	25 01	066	47 46	099	30 36	152	37 03	184	49 21	243	38 46	273	48 03	354
8	25 17	068	48 02	101	30 43	155	37 01	187	49 06	245	38 30	275	48 01	355
10	25 32	070	48 19	103	30 50	157	36 59	189	48 50	247	38 13	277	48 00	357
12	25 48	072	48 35	105	30 56	159	36 56	191	48 35	249	37 57	279	47 59	359
14	26 04	074	48 51	107	31 02	161	36 52	193	48 19	252	37 40	281	47 59	000
16	26 20	076	49 07	109	31 07	163	36 48	195	48 03	254	37 24	283	48 00	002
18	26 36	078	49 23	111	31 12	165	36 44	197	47 47	256	37 08	285	48 00	004
20	26 53	080	49 38	113	31 16	167	36 38	200	47 31	258	36 52	287	48 02	005
22	27 09	082	49 53	115	31 19	170	36 32	202	47 14	260	36 36	289	48 04	007
24	27 26	084	50 08	117	31 22	172	36 26	204	46 58	262	36 20	291	48 06	009
26	27 42	086	50 23	119	31 24	174	36 19	206	46 41	264	36 05	293	48 09	010
28	27 59	087	50 37	122	31 25	176	36 11	208	46 25	266	35 49	294	48 12	012

LHA ♈	◆Alioth Hc	Zn	POLLUX Hc	Zn	◆CAPELLA Hc	Zn	Hamal Hc	Zn	Alpheratz Hc	Zn	◆DENEB Hc	Zn	VEGA Hc	Zn
30	48 16	014	28 16	089	50 52	124	31 26	178	36 03	210	46 08	268	35 34	296
32	48 20	016	28 32	091	51 05	126	31 26	180	35 54	213	45 51	270	35 19	298
34	48 25	017	28 49	093	51 19	128	31 26	182	35 45	215	45 35	272	35 05	300
36	48 30	019	29 06	095	51 32	130	31 25	185	35 35	217	45 18	274	34 50	302
38	48 36	021	29 22	097	51 44	133	31 23	187	35 25	219	45 01	276	34 36	304
40	48 42	022	29 39	099	51 56	135	31 21	189	35 14	221	44 45	278	34 23	306
42	48 48	024	29 55	101	52 08	137	31 18	191	35 03	223	44 28	280	34 09	308
44	48 55	026	30 12	103	52 19	139	31 15	193	34 51	225	44 12	282	33 56	310
46	49 03	027	30 28	105	52 30	141	31 10	195	34 39	228	43 56	283	33 44	311
48	49 11	029	30 44	107	52 40	144	31 06	197	34 27	230	43 39	285	33 31	313
50	49 19	031	31 00	110	52 49	146	31 00	200	34 14	232	43 23	287	33 19	315
52	49 28	032	31 15	112	52 58	148	30 55	202	34 00	234	43 07	289	33 08	317
54	49 37	034	31 31	114	53 07	151	30 48	204	33 47	236	42 52	291	32 56	319
56	49 46	036	31 46	116	53 15	153	30 41	206	33 33	238	42 36	293	32 46	321
58	49 56	038	32 01	118	53 22	155	30 33	208	33 18	240	42 21	295	32 35	322

LHA ♈	◆Alioth Hc	Zn	POLLUX Hc	Zn	◆CAPELLA Hc	Zn	Hamal Hc	Zn	Alpheratz Hc	Zn	◆DENEB Hc	Zn	VEGA Hc	Zn
60	50 07	039	32 15	120	53 29	158	30 25	210	33 04	242	42 06	297	32 25	324
62	50 17	041	32 30	122	53 35	160	30 17	212	32 49	244	41 51	299	32 16	326
64	50 29	043	32 44	124	53 40	162	30 07	215	32 34	246	41 37	300	32 07	328
66	50 40	044	32 58	126	53 45	165	29 58	217	32 18	248	41 22	302	31 58	330
68	50 52	046	33 11	128	53 49	167	29 47	219	32 03	250	41 08	304	31 50	332
70	51 04	048	33 24	130	53 53	169	29 37	221	31 47	252	40 55	306	31 42	333
72	51 17	050	33 36	132	53 55	172	29 26	223	31 31	254	40 41	308	31 35	335
74	51 30	051	33 48	134	53 57	174	29 14	225	31 15	256	40 28	310	31 28	337
76	51 43	053	34 00	137	53 59	176	29 02	227	30 58	258	40 16	311	31 22	339
78	51 57	055	34 11	139	54 00	179	28 50	229	30 42	260	40 03	313	31 16	341
80	52 10	057	34 22	141	54 00	181	28 37	231	30 25	262	39 51	315	31 11	343
82	52 24	059	34 33	143	53 59	183	28 24	233	30 09	264	39 40	317	31 06	344
84	52 39	060	34 42	145	53 58	186	28 10	235	29 52	266	39 29	319	31 02	346
86	52 53	062	34 52	147	53 55	188	27 56	237	29 36	268	39 18	321	30 58	348
88	53 08	064	35 01	149	53 53	191	27 42	240	29 19	270	39 07	322	30 55	350

LHA ♈	◆ARCTURUS Hc	Zn	REGULUS Hc	Zn	POLLUX Hc	Zn	◆CAPELLA Hc	Zn	Alpheratz Hc	Zn	◆DENEB Hc	Zn	VEGA Hc	Zn
90	14 38	054	15 38	116	35 09	151	53 49	193	29 02	272	38 57	324	30 52	352
92	14 52	056	15 53	118	35 16	154	53 45	195	28 45	274	38 48	326	30 50	353
94	15 06	058	16 08	120	35 24	156	53 41	198	28 29	276	38 39	328	30 49	355
96	15 20	060	16 22	122	35 30	158	53 35	200	28 12	278	38 30	330	30 47	357
98	15 34	062	16 36	124	35 36	160	53 29	202	27 56	280	38 22	331	30 47	359
100	15 49	064	16 50	126	35 41	162	53 23	205	27 39	282	38 14	333	30 47	001
102	16 04	066	17 03	128	35 46	164	53 15	207	27 23	284	38 07	335	30 47	003
104	16 20	068	17 16	130	35 50	167	53 08	209	27 07	286	38 00	337	30 48	004
106	16 35	070	17 28	132	35 54	169	52 59	211	26 51	288	37 53	338	30 50	006
108	16 51	072	17 41	134	35 57	171	52 50	214	26 35	290	37 48	340	30 52	008
110	17 07	074	17 52	137	35 59	173	52 40	216	26 20	292	37 42	342	30 54	010
112	17 23	076	18 04	139	36 01	175	52 30	218	26 04	294	37 37	344	30 58	012
114	17 39	078	18 14	141	36 02	178	52 20	221	25 49	296	37 33	346	31 01	013
116	17 56	079	18 25	143	36 02	180	52 09	223	25 34	298	37 29	347	31 05	015
118	18 12	081	18 35	145	36 02	182	51 57	225	25 20	300	37 25	349	31 10	017

LHA ♈	VEGA Hc	Zn	◆ARCTURUS Hc	Zn	REGULUS Hc	Zn	POLLUX Hc	Zn	◆CAPELLA Hc	Zn	Alpheratz Hc	Zn	◆DENEB Hc	Zn
120	31 15	019	18 29	083	18 44	147	36 01	184	51 45	227	25 05	302	37 23	351
122	31 21	021	18 45	085	18 53	149	36 00	186	51 33	230	24 51	303	37 20	353
124	31 27	023	19 02	087	19 01	151	35 58	188	51 20	232	24 37	305	37 18	354
126	31 34	024	19 19	089	19 09	153	35 55	191	51 06	234	24 24	307	37 17	356
128	31 41	026	19 35	091	19 16	155	35 51	193	50 53	236	24 11	309	37 16	358
130	31 48	028	19 52	093	19 23	157	35 47	195	50 39	238	23 58	311	37 16	000
132	31 56	030	20 09	095	19 29	159	35 43	197	50 24	240	23 46	313	37 16	001
134	32 05	032	20 25	097	19 35	161	35 38	199	50 10	243	23 34	315	37 17	003
136	32 14	034	20 42	099	19 40	163	35 32	201	49 55	245	23 22	317	37 18	005
138	32 23	035	20 58	101	19 45	165	35 25	204	49 39	247	23 11	319	37 20	007
140	32 33	037	21 15	103	19 49	167	35 18	206	49 24	249	23 00	321	37 22	009
142	32 44	039	21 31	105	19 52	170	35 11	208	49 08	251	22 49	322	37 25	010
144	32 54	041	21 47	107	19 55	172	35 03	210	48 52	253	22 39	324	37 28	012
146	33 05	043	22 03	109	19 57	174	34 54	212	48 36	255	22 30	326	37 32	014
148	33 17	045	22 18	111	19 58	176	34 45	214	48 20	257	22 21	328	37 36	016

LHA ♈	◆DENEB Hc	Zn	VEGA Hc	Zn	◆ARCTURUS Hc	Zn	REGULUS Hc	Zn	POLLUX Hc	Zn	◆CAPELLA Hc	Zn	Alpheratz Hc	Zn
150	37 41	017	33 29	046	22 34	113	19 59	178	34 35	217	48 04	259	22 12	330
152	37 46	019	33 41	048	22 49	115	20 00	180	34 25	219	47 47	261	22 04	332
154	37 52	021	33 54	050	23 04	117	19 59	182	34 14	221	47 31	263	21 57	334
156	37 58	023	34 07	052	23 19	119	19 58	184	34 03	223	47 14	265	21 49	336
158	38 04	025	34 20	054	23 33	122	19 57	186	33 51	225	46 57	267	21 43	337
160	38 12	026	34 34	056	23 47	124	19 55	188	33 39	227	46 41	269	21 37	339
162	38 19	028	34 48	058	24 01	126	19 52	190	33 27	229	46 24	271	21 31	341
164	38 27	030	35 02	059	24 14	128	19 49	192	33 14	231	46 07	273	21 26	343
166	38 36	032	35 16	061	24 27	130	19 45	195	33 01	233	45 51	275	21 21	345
168	38 45	033	35 31	063	24 40	132	19 40	197	32 47	236	45 34	277	21 17	347
170	38 54	035	35 46	065	24 52	134	19 35	199	32 33	238	45 18	279	21 14	349
172	39 04	037	36 02	067	25 04	136	19 30	201	32 19	240	45 01	281	21 11	351
174	39 14	039	36 17	069	25 15	138	19 23	203	32 05	242	44 45	283	21 08	352
176	39 25	041	36 33	071	25 26	140	19 17	205	31 50	244	44 29	285	21 06	354
178	39 36	042	36 49	073	25 37	142	19 09	207	31 35	246	44 12	287	21 05	356

LHA ♈	◆DENEB Hc	Zn	VEGA Hc	Zn	◆ARCTURUS Hc	Zn	REGULUS Hc	Zn	POLLUX Hc	Zn	◆CAPELLA Hc	Zn	Alpheratz Hc	Zn
180	39 48	044	37 05	075	25 47	144	19 01	209	31 19	248	43 57	289	21 04	358
182	40 00	046	37 21	077	25 56	146	18 53	211	31 04	250	43 41	291	21 04	000
184	40 12	048	37 37	079	26 05	148	18 44	213	30 48	252	43 25	292	21 04	002
186	40 24	050	37 54	081	26 14	151	18 35	215	30 32	254	43 10	294	21 05	004
188	40 37	052	38 10	082	26 22	153	18 25	217	30 16	256	42 55	296	21 06	006
190	40 51	053	38 27	084	26 29	155	18 15	219	29 59	258	42 40	298	21 08	007
192	41 04	055	38 43	086	26 36	157	18 04	221	29 43	260	42 25	300	21 11	009
194	41 18	057	39 00	088	26 42	159	17 53	223	29 27	262	42 11	302	21 14	011
196	41 32	059	39 17	090	26 48	161	17 41	225	29 10	264	41 57	304	21 17	013
198	41 47	061	39 33	092	26 53	163	17 29	228	28 53	266	41 43	305	21 21	015
200	42 01	063	39 50	094	26 58	165	17 16	230	28 37	268	41 30	307	21 26	017
202	42 16	065	40 07	096	27 02	167	17 03	232	28 20	270	41 17	309	21 31	019
204	42 31	066	40 23	098	27 05	170	16 50	234	28 03	272	41 04	311	21 36	021
206	42 47	068	40 40	100	27 08	172	16 36	236	27 47	274	40 51	313	21 42	022
208	43 03	070	40 56	102	27 10	174	16 23	238	27 30	276	40 39	315	21 49	024

LHA ♈	Alpheratz Hc	Zn	◆DENEB Hc	Zn	VEGA Hc	Zn	◆ARCTURUS Hc	Zn	REGULUS Hc	Zn	POLLUX Hc	Zn	◆CAPELLA Hc	Zn
210	21 56	026	43 18	072	41 12	104	27 11	176	16 08	240	27 13	278	40 28	316
212	22 04	028	43 34	074	41 28	106	27 12	178	15 54	242	26 57	280	40 16	318
214	22 12	030	43 50	076	41 44	108	27 12	180	15 39	244	26 41	282	40 05	320
216	22 21	032	44 07	078	42 00	111	27 12	182	15 24	246	26 24	284	39 55	322
218	22 30	034	44 23	080	42 16	113	27 11	184	15 08	248	26 08	286	39 45	324
220	22 39	036	44 40	082	42 31	115	27 10	187	14 53	250	25 52	288	39 35	325
222	22 49	038	44 56	084	42 46	117	27 07	189	14 37	252	25 36	290	39 26	327
224	22 59	039	45 13	086	43 01	119	27 05	191	14 21	254	25 21	292	39 17	329
226	23 10	041	45 29	088	43 15	121	27 01	193	14 05	256	25 05	293	39 09	331
228	23 22	043	45 46	090	43 29	123	26 57	195	13 49	258	24 50	295	39 01	333
230	23 33	045	46 03	092	43 43	125	26 53	197	13 32	260	24 35	297	38 53	334
232	23 45	047	46 19	094	43 57	127	26 47	199	13 16	262	24 20	299	38 46	336
234	23 58	049	46 36	096	44 10	130	26 42	201	12 59	264	24 06	301	38 40	338
236	24 10	051	46 53	098	44 22	132	26 35	203	12 43	266	23 52	303	38 33	340
238	24 23	053	47 09	100	44 35	134	26 28	206	12 26	268	23 38	305	38 28	341

LHA ♈	Alpheratz Hc	Zn	◆DENEB Hc	Zn	VEGA Hc	Zn	◆ARCTURUS Hc	Zn	Dubhe Hc	Zn	POLLUX Hc	Zn	◆CAPELLA Hc	Zn
240	24 37	055	47 26	102	44 46	136	26 21	208	62 56	269	23 24	307	38 23	343
242	24 51	057	47 42	104	44 58	138	26 13	210	62 40	271	23 11	309	38 18	345
244	25 05	058	47 58	106	45 09	140	26 04	212	62 23	273	22 58	311	38 14	347
246	25 19	060	48 14	108	45 19	143	25 55	214	62 06	275	22 46	313	38 11	348
248	25 34	062	48 30	110	45 29	145	25 45	216	61 50	277	22 34	315	38 07	350
250	25 49	064	48 45	112	45 38	147	25 35	218	61 33	279	22 22	316	38 05	352
252	26 04	066	49 01	114	45 47	149	25 25	220	61 17	281	22 11	318	38 03	354
254	26 19	068	49 16	116	45 55	152	25 14	222	61 00	283	22 00	320	38 01	356
256	26 35	070	49 31	118	46 03	154	25 02	224	60 44	285	21 49	322	38 00	357
258	26 51	072	49 45	120	46 10	156	24 51	226	60 28	287	21 39	324	38 00	359
260	27 07	074	50 00	123	46 17	158	24 38	229	60 12	288	21 30	326	38 00	001
262	27 23	076	50 14	125	46 22	160	24 25	231	59 56	290	21 21	328	38 00	003
264	27 39	078	50 27	127	46 28	163	24 12	233	59 41	292	21 12	330	38 01	004
266	27 55	080	50 40	129	46 32	165	23 59	235	59 25	294	21 04	332	38 03	006
268	28 12	082	50 53	131	46 36	167	23 45	237	59 10	296	20 56	333	38 05	008

LHA ♈	◆CAPELLA Hc	Zn	Alpheratz Hc	Zn	◆DENEB Hc	Zn	VEGA Hc	Zn	◆ARCTURUS Hc	Zn	Dubhe Hc	Zn	POLLUX Hc	Zn
270	38 07	010	28 28	084	51 05	134	46 40	170	23 31	239	58 55	297	20 49	335
272	38 10	011	28 45	086	51 17	136	46 42	172	23 17	241	58 41	299	20 42	337
274	38 14	013	29 02	088	51 29	138	46 44	174	23 02	243	58 26	301	20 36	339
276	38 18	015	29 18	089	51 40	140	46 46	176	22 47	245	58 12	303	20 30	341
278	38 22	017	29 35	091	51 50	142	46 47	179	22 32	247	57 58	304	20 25	343
280	38 27	018	29 52	093	52 00	145	46 47	181	22 16	249	57 44	306	20 20	345
282	38 33	020	30 08	095	52 09	147	46 46	183	22 00	251	57 31	308	20 16	347
284	38 39	022	30 25	097	52 18	149	46 45	185	21 45	253	57 18	309	20 13	348
286	38 46	024	30 41	099	52 26	152	46 43	188	21 29	255	57 05	311	20 09	350
288	38 53	026	30 58	101	52 34	154	46 40	190	21 12	257	56 53	313	20 07	352
290	39 00	027	31 14	103	52 41	156	46 37	192	20 56	259	56 41	315	20 05	354
292	39 08	029	31 30	106	52 47	159	46 33	195	20 40	261	56 29	316	20 04	356
294	39 16	031	31 46	108	52 53	161	46 29	197	20 23	263	56 18	318	20 03	358
296	39 25	033	32 02	110	52 58	163	46 24	199	20 06	265	56 07	320	20 02	000
298	39 34	034	32 18	112	53 03	166	46 18	201	19 50	267	55 56	321	20 03	002

LHA ♈	POLLUX Hc	Zn	◆CAPELLA Hc	Zn	Alpheratz Hc	Zn	◆DENEB Hc	Zn	VEGA Hc	Zn	◆ARCTURUS Hc	Zn	Dubhe Hc	Zn
300	20 03	004	39 44	036	32 33	114	53 07	168	46 11	204	19 33	269	55 46	323
302	20 05	005	39 54	038	32 48	116	53 10	170	46 04	206	19 16	271	55 36	325
304	20 06	007	40 05	040	33 03	118	53 12	173	45 57	208	19 00	273	55 26	326
306	20 09	009	40 15	042	33 18	120	53 14	175	45 49	210	18 43	275	55 17	328
308	20 12	011	40 27	043	33 32	122	53 15	177	45 40	213	18 26	277	55 09	329
310	20 15	013	40 38	045	33 46	124	53 16	180	45 31	215	18 10	279	55 00	331
312	20 19	015	40 50	047	34 00	126	53 15	182	45 21	217	17 53	281	54 52	333
314	20 24	017	41 03	049	34 13	128	53 15	184	45 11	219	17 37	283	54 45	334
316	20 29	019	41 16	051	34 26	130	53 13	187	45 00	221	17 21	285	54 38	336
318	20 34	020	41 29	053	34 39	132	53 11	189	44 49	224	17 05	287	54 31	338
320	20 40	022	41 42	054	34 51	135	53 08	191	44 37	226	16 49	289	54 25	339
322	20 47	024	41 56	056	35 03	137	53 04	194	44 25	228	16 33	291	54 20	341
324	20 54	026	42 10	058	35 14	139	53 00	196	44 12	230	16 18	293	54 14	342
326	21 02	028	42 24	060	35 25	141	52 55	198	43 59	232	16 02	294	54 10	344
328	21 10	030	42 39	062	35 35	143	52 49	201	43 46	234	15 47	296	54 05	346

LHA ♈	POLLUX Hc	Zn	◆CAPELLA Hc	Zn	Hamal Hc	Zn	Alpheratz Hc	Zn	◆DENEB Hc	Zn	VEGA Hc	Zn	◆Alioth Hc	Zn
330	21 18	032	42 54	064	27 02	115	35 45	145	52 43	203	43 32	236	49 51	323
332	21 27	034	43 09	066	27 17	117	35 54	147	52 36	205	43 18	239	49 41	325
334	21 37	036	43 24	067	27 31	119	36 03	149	52 29	208	43 04	241	49 32	327
336	21 47	037	43 40	069	27 46	121	36 11	152	52 21	210	42 49	243	49 23	329
338	21 57	039	43 55	071	28 00	123	36 19	154	52 12	212	42 34	245	49 14	330
340	22 08	041	44 11	073	28 14	125	36 26	156	52 03	215	42 19	247	49 06	332
342	22 19	043	44 27	075	28 27	127	36 32	158	51 53	217	42 03	249	48 59	334
344	22 31	045	44 43	077	28 40	129	36 38	160	51 43	219	41 47	251	48 51	335
346	22 43	047	45 00	079	28 53	131	36 43	162	51 32	221	41 32	253	48 45	337
348	22 55	049	45 16	081	29 05	133	36 48	165	51 21	224	41 15	255	48 38	339
350	23 08	051	45 33	083	29 17	136	36 52	167	51 09	226	40 59	257	48 32	340
352	23 21	053	45 49	085	29 29	138	36 56	169	50 57	228	40 43	259	48 27	342
354	23 34	055	46 06	087	29 40	140	36 59	171	50 44	230	40 26	261	48 22	344
356	23 48	056	46 23	089	29 50	142	37 01	173	50 31	232	40 10	263	48 18	345
358	24 02	058	46 39	091	30 00	144	37 02	176	50 18	235	39 53	265	48 14	347

LAT 81°N

LHA ♈	Hc	Zn	Hc	Zn	Hc	Zn	Hc	Zn	Hc	Zn	Hc	Zn	Hc	Zn
	POLLUX		◆CAPELLA		Hamal		Alpheratz		◆DENEB		VEGA		◆Alioth	
0	23 47	060	46 58	091	31 00	146	38 03	178	50 37	238	39 39	268	47 11	349
2	24 03	062	47 17	093	31 10	148	38 04	180	50 20	240	39 20	270	47 08	351
4	24 20	064	47 36	095	31 20	150	38 03	182	50 04	242	39 01	272	47 05	352
6	24 37	066	47 54	097	31 29	152	38 02	184	49 47	244	38 43	274	47 03	354
8	24 54	068	48 13	100	31 37	154	38 01	187	49 30	246	38 24	276	47 01	356
10	25 11	069	48 31	102	31 45	156	37 58	189	49 13	248	38 05	278	47 00	357
12	25 29	071	48 50	104	31 52	159	37 55	191	48 55	251	37 47	280	46 59	359
14	25 47	073	49 08	106	31 59	161	37 51	193	48 38	253	37 28	282	46 59	000
16	26 05	075	49 26	108	32 05	163	37 46	195	48 20	255	37 10	284	47 00	002
18	26 23	077	49 44	110	32 10	165	37 41	198	48 01	257	36 52	286	47 01	004
20	26 42	079	50 01	112	32 14	167	37 35	200	47 43	259	36 34	288	47 02	005
22	27 00	081	50 18	114	32 18	169	37 28	202	47 24	261	36 16	289	47 04	007
24	27 19	083	50 35	116	32 21	172	37 21	204	47 06	263	35 59	291	47 07	009
26	27 37	085	50 52	118	32 23	174	37 13	206	46 47	265	35 41	293	47 10	010
28	27 56	087	51 08	121	32 25	176	37 04	209	46 29	267	35 24	295	47 13	012
	◆Alioth		POLLUX		◆CAPELLA		Hamal		Alpheratz		◆DENEB		VEGA	
30	47 18	014	28 15	089	51 24	123	32 26	178	36 55	211	46 10	269	35 07	297
32	47 22	015	28 34	091	51 40	125	32 26	180	36 45	213	45 51	271	34 51	299
34	47 27	017	28 52	093	51 55	127	32 26	182	36 34	215	45 32	273	34 34	301
36	47 33	019	29 11	095	52 10	129	32 25	185	36 23	217	45 13	275	34 18	303
38	47 39	020	29 30	097	52 24	132	32 23	187	36 11	220	44 55	277	34 03	304
40	47 46	022	29 48	099	52 38	134	32 20	189	35 59	222	44 36	279	33 47	306
42	47 53	023	30 07	101	52 51	136	32 17	191	35 46	224	44 18	281	33 32	308
44	48 01	025	30 25	103	53 04	138	32 13	193	35 33	226	43 59	283	33 18	310
46	48 09	027	30 44	105	53 16	141	32 08	195	35 19	228	43 41	284	33 04	312
48	48 18	028	31 02	107	53 28	143	32 03	198	35 05	230	43 23	286	32 50	314
50	48 27	030	31 19	109	53 39	145	31 57	200	34 51	232	43 05	288	32 37	316
52	48 37	032	31 37	111	53 49	148	31 50	202	34 36	234	42 47	290	32 24	317
54	48 47	033	31 55	113	53 59	150	31 43	204	34 20	236	42 30	292	32 11	319
56	48 58	035	32 12	115	54 08	152	31 35	206	34 04	239	42 12	294	31 59	321
58	49 09	037	32 28	117	54 16	155	31 26	208	33 48	241	41 55	296	31 48	323
	◆Alioth		POLLUX		◆CAPELLA		Hamal		Alpheratz		◆DENEB		VEGA	
60	49 20	039	32 45	119	54 24	157	31 17	211	33 32	243	41 39	298	31 36	325
62	49 32	040	33 01	121	54 31	159	31 07	213	33 15	245	41 22	299	31 26	326
64	49 44	042	33 17	123	54 37	162	30 57	215	32 58	247	41 06	301	31 16	328
66	49 57	044	33 33	125	54 43	164	30 46	217	32 40	249	40 50	303	31 06	330
68	50 10	045	33 48	128	54 47	167	30 34	219	32 23	251	40 34	305	30 57	332
70	50 24	047	34 02	130	54 51	169	30 22	221	32 05	253	40 19	307	30 48	334
72	50 38	049	34 17	132	54 55	171	30 09	223	31 47	255	40 04	308	30 40	336
74	50 52	051	34 30	134	54 57	174	29 56	225	31 28	257	39 50	310	30 33	337
76	51 07	052	34 44	136	54 59	176	29 43	228	31 10	259	39 36	312	30 26	339
78	51 22	054	34 56	138	54 59	179	29 29	230	30 52	261	39 22	314	30 19	341
80	51 37	056	35 09	140	55 00	181	29 14	232	30 33	263	39 09	316	30 14	343
82	51 53	057	35 20	142	54 59	184	28 59	234	30 14	265	38 56	317	30 08	345
84	52 09	059	35 31	145	54 57	186	28 44	236	29 56	267	38 43	319	30 04	346
86	52 25	061	35 42	147	54 55	188	28 28	238	29 37	269	38 31	321	29 59	348
88	52 42	063	35 52	149	54 52	191	28 12	240	29 18	271	38 20	323	29 56	350
	◆ARCTURUS		REGULUS		POLLUX		◆CAPELLA		Alpheratz		◆DENEB		VEGA	
90	14 03	054	16 05	116	36 01	151	54 48	193	28 59	273	38 09	325	29 53	352
92	14 18	056	16 22	118	36 10	153	54 43	196	28 41	275	37 58	326	29 51	354
94	14 34	058	16 38	120	36 18	155	54 38	198	28 22	277	37 48	328	29 49	355
96	14 50	060	16 54	122	36 26	158	54 32	200	28 03	279	37 38	330	29 47	357
98	15 06	062	17 10	124	36 32	160	54 25	203	27 45	281	37 29	332	29 47	359
100	15 23	064	17 25	126	36 39	162	54 17	205	27 26	283	37 20	333	29 47	001
102	15 40	066	17 40	128	36 44	164	54 09	208	27 08	285	37 12	335	29 47	003
104	15 57	067	17 55	130	36 49	166	54 00	210	26 50	287	37 05	337	29 48	004
106	16 14	069	18 09	132	36 53	169	53 50	212	26 32	289	36 58	339	29 50	006
108	16 32	071	18 23	134	36 56	171	53 40	215	26 15	290	36 51	340	29 52	008
110	16 50	073	18 36	136	36 59	173	53 29	217	25 57	292	36 45	342	29 55	010
112	17 08	075	18 48	138	37 01	175	53 17	219	25 40	294	36 40	344	29 59	012
114	17 26	077	19 01	140	37 02	178	53 05	221	25 23	296	36 35	346	30 03	013
116	17 45	079	19 12	142	37 02	180	52 52	224	25 06	298	36 30	347	30 07	015
118	18 03	081	19 24	145	37 02	182	52 39	226	24 50	300	36 27	349	30 13	017
	VEGA		◆ARCTURUS		REGULUS		POLLUX		◆CAPELLA		Alpheratz		◆DENEB	
120	30 18	019	18 22	083	19 34	147	37 01	184	52 25	228	24 34	302	36 23	351
122	30 25	021	18 40	085	19 44	149	36 59	186	52 11	231	24 18	304	36 21	353
124	30 31	022	18 59	087	19 54	151	36 57	189	51 56	233	24 02	306	36 19	354
126	30 39	024	19 18	089	20 03	153	36 54	191	51 41	235	23 47	308	36 17	356
128	30 47	026	19 37	091	20 11	155	36 50	193	51 26	237	23 33	309	36 16	358
130	30 55	028	19 55	093	20 18	157	36 45	195	51 10	239	23 18	311	36 16	000
132	31 04	030	20 14	095	20 26	159	36 40	197	50 53	241	23 05	313	36 16	001
134	31 14	031	20 33	097	20 32	161	36 34	200	50 37	244	22 51	315	36 17	003
136	31 24	033	20 51	099	20 38	163	36 27	202	50 20	246	22 38	317	36 18	005
138	31 34	035	21 10	101	20 43	165	36 20	204	50 02	248	22 26	319	36 20	007
140	31 45	037	21 28	103	20 47	167	36 12	206	49 45	250	22 13	321	36 22	008
142	31 57	039	21 46	105	20 51	169	36 04	208	49 27	252	22 02	323	36 25	010
144	32 09	040	22 05	107	20 54	172	35 54	210	49 09	254	21 51	325	36 29	012
146	32 21	042	22 22	109	20 56	174	35 45	213	48 51	256	21 40	326	36 33	014
148	32 34	044	22 40	111	20 58	176	35 34	215	48 33	258	21 30	328	36 38	015
	◆DENEB		VEGA		◆ARCTURUS		REGULUS		POLLUX		◆CAPELLA		Alpheratz	
150	36 43	017	32 47	046	22 57	113	20 59	178	35 23	217	48 14	260	21 20	330
152	36 49	019	33 01	048	23 15	115	21 00	180	35 12	219	47 56	262	21 11	332
154	36 55	021	33 15	050	23 31	117	20 59	182	34 59	221	47 37	264	21 03	334
156	37 02	022	33 30	051	23 48	119	20 58	184	34 47	223	47 18	266	20 55	336
158	37 10	024	33 44	053	24 04	121	20 56	186	34 34	226	47 00	268	20 47	338
160	37 18	026	34 00	055	24 20	123	20 54	188	34 20	228	46 41	270	20 40	339
162	37 26	028	34 15	057	24 36	125	20 51	190	34 06	230	46 22	272	20 34	341
164	37 35	030	34 31	059	24 51	127	20 47	193	33 51	232	46 03	274	20 28	343
166	37 45	031	34 47	061	25 06	129	20 43	195	33 36	234	45 45	276	20 23	345
168	37 55	033	35 04	063	25 20	131	20 38	197	33 21	236	45 26	278	20 19	347
170	38 05	035	35 21	065	25 34	134	20 32	199	33 05	238	45 08	280	20 15	349
172	38 16	037	35 38	066	25 47	136	20 26	201	32 49	240	44 49	282	20 11	351
174	38 28	038	35 55	068	26 00	138	20 19	203	32 33	242	44 31	284	20 09	353
176	38 39	040	36 13	070	26 12	140	20 11	205	32 16	244	44 13	286	20 07	354
178	38 52	042	36 30	072	26 24	142	20 03	207	31 59	246	43 55	288	20 05	356

LHA ♈	Hc	Zn	Hc	Zn	Hc	Zn	Hc	Zn	Hc	Zn	Hc	Zn	Hc	Zn
	◆DENEB		VEGA		◆ARCTURUS		REGULUS		POLLUX		◆CAPELLA		Alpheratz	
180	39 05	044	36 48	074	26 35	144	19 54	209	31 42	248	43 37	290	20 04	358
182	39 18	046	37 07	076	26 46	146	19 44	211	31 24	251	43 19	291	20 04	000
184	39 31	047	37 25	078	26 56	148	19 34	213	31 06	253	43 02	293	20 04	002
186	39 45	049	37 43	080	27 06	150	19 24	215	30 48	255	42 45	295	20 05	004
188	40 00	051	38 02	082	27 15	152	19 13	217	30 30	257	42 28	297	20 06	006
190	40 15	053	38 20	084	27 23	155	19 01	220	30 12	259	42 11	299	20 09	007
192	40 30	055	38 39	086	27 31	157	18 49	222	29 53	261	41 55	301	20 11	009
194	40 45	056	38 58	088	27 38	159	18 36	224	29 35	263	41 39	303	20 15	011
196	41 01	058	39 17	090	27 45	161	18 23	226	29 16	265	41 23	304	20 19	013
198	41 17	060	39 35	091	27 51	163	18 09	228	28 57	267	41 08	306	20 23	015
200	41 33	062	39 54	093	27 56	165	17 55	230	28 38	269	40 53	308	20 28	017
202	41 50	064	40 13	095	28 00	167	17 41	232	28 20	271	40 39	310	20 34	019
204	42 07	066	40 31	097	28 04	169	17 26	234	28 01	273	40 24	312	20 40	020
206	42 24	067	40 50	099	28 07	172	17 10	236	27 42	275	40 10	313	20 47	022
208	42 42	069	41 08	102	28 09	174	16 54	238	27 23	276	39 57	315	20 54	024
	Alpheratz		◆DENEB		VEGA		◆ARCTURUS		REGULUS		POLLUX		◆CAPELLA	
210	21 02	026	42 59	071	41 27	104	28 11	176	16 38	240	27 05	278	39 44	317
212	21 11	028	43 17	073	41 45	106	28 12	178	16 22	242	26 46	280	39 31	319
214	21 20	030	43 35	075	42 03	108	28 12	180	16 05	244	26 28	282	39 19	321
216	21 30	032	43 54	077	42 21	110	28 12	182	15 48	246	26 10	284	39 08	322
218	21 40	034	44 12	079	42 38	112	28 11	184	15 31	248	25 52	286	38 56	324
220	21 50	035	44 30	081	42 56	114	28 09	187	15 13	250	25 34	288	38 46	326
222	22 01	037	44 49	083	43 13	116	28 07	189	14 56	252	25 16	290	38 35	328
224	22 13	039	45 08	085	43 29	118	28 04	191	14 38	254	24 58	292	38 25	329
226	22 25	041	45 26	087	43 46	120	28 00	193	14 20	256	24 41	294	38 16	331
228	22 38	043	45 45	089	44 02	122	27 55	195	14 01	258	24 24	296	38 07	333
230	22 51	045	46 04	090	44 17	124	27 50	197	13 43	260	24 07	298	37 59	335
232	23 04	047	46 23	092	44 33	127	27 44	199	13 24	262	23 51	300	37 51	336
234	23 18	049	46 41	094	44 48	129	27 37	202	13 06	264	23 35	302	37 44	338
236	23 32	050	47 00	096	45 02	131	27 30	204	12 47	266	23 19	303	37 37	340
238	23 47	052	47 19	098	45 16	133	27 22	206	12 28	268	23 03	305	37 31	342
	Alpheratz		◆DENEB		VEGA		◆ARCTURUS		Dubhe		POLLUX		◆CAPELLA	
240	24 02	054	47 37	101	45 29	135	27 14	208	62 56	271	22 48	307	37 25	343
242	24 17	056	47 56	103	45 42	138	27 05	210	62 37	273	22 33	309	37 20	345
244	24 33	058	48 14	105	45 55	140	26 55	212	62 19	275	22 19	311	37 16	347
246	24 49	060	48 32	107	46 07	142	26 45	214	62 00	277	22 05	313	37 12	349
248	25 06	062	48 50	109	46 18	144	26 34	216	61 41	279	21 52	315	37 08	350
250	25 22	064	49 07	111	46 28	146	26 23	218	61 23	281	21 39	317	37 05	352
252	25 39	066	49 25	113	46 39	149	26 11	221	61 04	283	21 26	319	37 03	354
254	25 57	068	49 42	115	46 48	151	25 58	223	60 46	285	21 14	320	37 01	356
256	26 14	069	49 59	117	46 57	153	25 45	225	60 28	286	21 02	322	37 00	357
258	26 32	071	50 15	119	47 05	156	25 32	227	60 10	288	20 51	324	37 00	359
260	26 50	073	50 32	122	47 12	158	25 18	229	59 52	290	20 40	326	37 00	001
262	27 08	075	50 47	124	47 19	160	25 03	231	59 35	292	20 30	328	37 00	003
264	27 26	077	51 03	126	47 25	162	24 49	233	59 18	294	20 20	330	37 01	004
266	27 44	079	51 18	128	47 30	165	24 33	235	59 01	295	20 11	332	37 03	006
268	28 03	081	51 32	130	47 35	167	24 18	237	58 44	297	20 02	334	37 05	008
	◆CAPELLA		Alpheratz		◆DENEB		VEGA		◆ARCTURUS		Dubhe		POLLUX	
270	37 08	010	28 21	083	51 46	133	47 39	169	24 02	239	58 27	299	19 54	335
272	37 11	011	28 40	085	52 00	135	47 42	172	23 46	241	58 11	301	19 47	337
274	37 15	013	28 59	087	52 13	137	47 44	174	23 29	243	57 55	302	19 40	339
276	37 20	015	29 17	089	52 25	139	47 46	176	23 12	245	57 39	304	19 33	341
278	37 25	016	29 36	091	52 37	142	47 47	179	22 55	247	57 24	306	19 28	343
280	37 31	018	29 55	093	52 49	144	47 47	181	22 38	249	57 09	307	19 22	345
282	37 37	020	30 14	095	52 59	146	47 46	183	22 20	251	56 54	309	19 18	347
284	37 43	022	30 32	097	53 09	149	47 45	186	22 02	253	56 39	311	19 14	349
286	37 51	023	30 51	099	53 19	151	47 42	188	21 44	255	56 25	312	19 10	350
288	37 58	025	31 09	101	53 28	153	47 39	190	21 26	257	56 12	314	19 08	352
290	38 07	027	31 28	103	53 36	156	47 36	193	21 07	259	55 58	316	19 05	354
292	38 15	029	31 46	105	53 43	158	47 31	195	20 49	261	55 45	317	19 04	356
294	38 25	030	32 04	107	53 50	160	47 26	197	20 30	263	55 33	319	19 03	358
296	38 34	032	32 22	109	53 56	163	47 20	199	20 11	265	55 21	321	19 02	000
298	38 45	034	32 40	111	54 01	165	47 14	202	19 53	267	55 09	322	19 03	002
	POLLUX		◆CAPELLA		Alpheratz		◆DENEB		VEGA		◆ARCTURUS		Dubhe	
300	19 03	003	38 55	036	32 57	113	54 05	168	47 06	204	19 34	269	54 58	324
302	19 05	005	39 07	038	33 14	115	54 09	170	46 58	206	19 15	271	54 47	325
304	19 07	007	39 18	039	33 31	117	54 12	172	46 50	209	18 56	273	54 36	327
306	19 10	009	39 30	041	33 48	119	54 14	175	46 40	211	18 38	275	54 26	329
308	19 13	011	39 43	043	34 04	121	54 15	177	46 30	213	18 19	277	54 17	330
310	19 17	013	39 56	045	34 20	123	54 16	180	46 20	215	18 00	279	54 08	332
312	19 21	015	40 09	046	34 35	126	54 15	182	46 09	218	17 42	281	53 59	333
314	19 26	017	40 23	048	34 50	128	54 14	184	45 57	220	17 24	283	53 51	335
316	19 32	018	40 37	050	35 05	130	54 13	187	45 45	222	17 05	285	53 43	337
318	19 38	020	40 52	052	35 19	132	54 10	189	45 32	224	16 47	287	53 36	338
320	19 45	022	41 07	054	35 33	134	54 07	192	45 19	226	16 30	289	53 29	340
322	19 52	024	41 22	056	35 46	136	54 02	194	45 05	229	16 12	291	53 23	341
324	20 00	026	41 38	057	35 59	138	53 57	196	44 50	231	15 54	293	53 17	343
326	20 09	028	41 54	059	36 11	140	53 52	199	44 36	233	15 37	295	53 12	344
328	20 18	030	42 10	061	36 23	143	53 45	201	44 20	235	15 20	297	53 07	346
	POLLUX		◆CAPELLA		Hamal		Alpheratz		◆DENEB		VEGA		◆Alioth	
330	20 27	032	42 27	063	27 27	114	36 34	145	53 38	204	44 05	237	49 02	324
332	20 37	033	42 44	065	27 44	116	36 44	147	53 30	206	43 49	239	48 52	326
334	20 48	035	43 01	067	28 00	119	36 54	149	53 22	208	43 33	242	48 41	327
336	20 59	037	43 18	068	28 17	121	37 04	151	53 13	211	43 16	244	48 31	329
338	21 11	039	43 36	070	28 33	123	37 12	153	53 03	213	42 59	246	48 22	331
340	21 23	041	43 53	072	28 48	125	37 20	156	52 52	215	42 42	248	48 13	332
342	21 35	043	44 11	074	29 03	127	37 28	158	52 41	218	42 24	250	48 05	334
344	21 48	045	44 29	076	29 18	129	37 35	160	52 29	220	42 06	252	47 57	336
346	22 02	047	44 48	078	29 33	131	37 41	162	52 17	222	41 48	254	47 49	337
348	22 16	048	45 06	080	29 47	133	37 46	164	52 04	224	41 30	256	47 42	339
350	22 30	050	45 25	082	30 00	135	37 51	167	51 51	227	41 12	258	47 36	341
352	22 44	052	45 43	084	30 13	137	37 55	169	51 37	229	40 54	260	47 30	342
354	22 59	054	46 02	086	30 25	139	37 58	171	51 22	231	40 35	262	47 24	344
356	23 15	056	46 21	088	30 37	142	38 00	173	51 07	233	40 16	264	47 20	346
358	23 31	058	46 39	089	30 49	144	38 02	176	50 52	236	39 58	266	47 15	347

LAT 80°N

LHA ♈	Hc	Zn	Hc	Zn	Hc	Zn	Hc	Zn	Hc	Zn	Hc	Zn	Hc	Zn
	POLLUX		◆CAPELLA		Hamal		Alpheratz		◆DENEB		VEGA		◆Alioth	
0	23 16	060	46 59	090	31 49	145	39 03	178	51 08	239	39 40	269	46 12	349
2	23 35	061	47 20	092	32 01	148	39 04	180	50 50	241	39 20	271	46 09	351
4	23 53	063	47 41	094	32 12	150	39 03	182	50 32	243	38 59	273	46 06	352
6	24 12	065	48 02	096	32 22	152	39 02	184	50 13	245	38 38	275	46 03	354
8	24 31	067	48 22	098	32 31	154	39 00	187	49 54	247	38 17	277	46 01	356
10	24 50	069	48 43	100	32 40	156	38 57	189	49 34	250	37 57	279	46 00	357
12	25 10	071	49 03	102	32 48	158	38 54	191	49 15	252	37 36	281	45 59	359
14	25 30	073	49 23	105	32 55	161	38 49	193	48 55	254	37 16	283	45 59	000
16	25 50	075	49 44	107	33 02	163	38 44	196	48 35	256	36 55	284	46 00	002
18	26 10	077	50 03	109	33 08	165	38 38	198	48 14	258	36 35	286	46 01	004
20	26 30	079	50 23	111	33 13	167	38 31	200	47 54	260	36 15	288	46 02	005
22	26 51	081	50 42	113	33 17	169	38 24	202	47 33	262	35 56	290	46 05	007
24	27 11	083	51 01	115	33 20	172	38 15	205	47 13	264	35 36	292	46 07	008
26	27 32	084	51 20	117	33 23	174	38 06	207	46 52	266	35 17	294	46 11	010
28	27 53	086	51 38	119	33 25	176	37 56	209	46 31	268	34 58	296	46 15	012
	◆Alioth		POLLUX		◆CAPELLA		Hamal		Alpheratz		◆DENEB		VEGA	
30	46 19	013	28 14	088	51 56	122	33 26	178	37 46	211	46 10	270	34 40	298
32	46 24	015	28 34	090	52 14	124	33 26	180	37 35	213	45 50	272	34 21	299
34	46 30	017	28 55	092	52 31	126	33 26	183	37 23	216	45 29	274	34 03	301
36	46 36	018	29 16	094	52 48	128	33 25	185	37 11	218	45 08	276	33 46	303
38	46 43	020	29 37	096	53 04	131	33 22	187	36 58	220	44 47	278	33 28	305
40	46 50	021	29 57	098	53 19	133	33 20	189	36 44	222	44 27	280	33 12	307
42	46 58	023	30 18	100	53 34	135	33 16	191	36 30	224	44 06	282	32 55	309
44	47 07	025	30 38	102	53 49	137	33 11	193	36 15	226	43 46	283	32 39	311
46	47 16	026	30 59	104	54 02	140	33 06	196	35 59	229	43 26	285	32 23	312
48	47 25	028	31 19	106	54 15	142	33 00	198	35 43	231	43 06	287	32 08	314
50	47 35	030	31 39	108	54 28	144	32 53	200	35 27	233	42 46	289	31 54	316
52	47 46	031	31 58	110	54 40	147	32 46	202	35 10	235	42 26	291	31 39	318
54	47 57	033	32 18	112	54 51	149	32 38	204	34 53	237	42 07	293	31 26	320
56	48 08	035	32 37	115	55 01	152	32 29	207	34 35	239	41 48	295	31 12	321
58	48 20	036	32 56	117	55 10	154	32 19	209	34 17	241	41 29	296	31 00	323
	◆Alioth		REGULUS		POLLUX		◆CAPELLA		Alpheratz		◆DENEB		VEGA	
60	48 33	038	11 27	086	33 14	119	55 19	156	33 59	243	41 11	298	30 47	325
62	48 46	039	11 48	088	33 32	121	55 27	159	33 40	245	40 52	300	30 36	327
64	48 59	041	12 09	090	33 50	123	55 34	161	33 21	247	40 35	302	30 25	329
66	49 13	043	12 30	092	34 07	125	55 40	164	33 01	250	40 17	304	30 14	330
68	49 28	045	12 51	094	34 24	127	55 46	166	32 42	252	40 00	306	30 04	332
70	49 43	046	13 11	096	34 40	129	55 50	169	32 22	254	39 43	307	29 55	334
72	49 58	048	13 32	098	34 56	131	55 54	171	32 02	256	39 27	309	29 46	336
74	50 13	050	13 53	100	35 12	133	55 57	174	31 42	258	39 11	311	29 37	338
76	50 30	051	14 13	102	35 27	136	55 59	176	31 21	260	38 55	313	29 30	339
78	50 46	053	14 33	104	35 41	138	55 59	179	31 01	262	38 40	314	29 23	341
80	51 03	055	14 54	106	35 55	140	56 00	181	30 40	264	38 26	316	29 16	343
82	51 20	056	15 14	108	36 08	142	55 59	184	30 19	266	38 11	318	29 10	345
84	51 38	058	15 33	110	36 20	144	55 57	186	29 58	268	37 58	320	29 05	346
86	51 55	060	15 53	112	36 32	146	55 54	189	29 38	270	37 44	322	29 01	348
88	52 14	062	16 12	114	36 43	149	55 51	191	29 17	272	37 32	323	28 57	350
	◆ARCTURUS		REGULUS		POLLUX		◆CAPELLA		Alpheratz		◆DENEB		VEGA	
90	13 27	054	16 31	116	36 54	151	55 46	194	28 56	274	37 20	325	28 54	352
92	13 44	056	16 50	118	37 04	153	55 41	196	28 35	275	37 08	327	28 51	354
94	14 02	058	17 08	120	37 13	155	55 35	198	28 14	277	36 57	329	28 49	355
96	14 19	060	17 26	122	37 21	157	55 28	201	27 54	279	36 46	330	28 48	357
98	14 37	061	17 43	124	37 29	160	55 20	203	27 33	281	36 36	332	28 47	359
100	14 56	063	18 00	126	37 36	162	55 11	206	27 13	283	36 27	334	28 47	001
102	15 15	065	18 17	128	37 42	164	55 02	208	26 53	285	36 18	336	28 47	002
104	15 34	067	18 33	130	37 47	166	54 52	211	26 33	287	36 09	337	28 49	004
106	15 53	069	18 49	132	37 52	169	54 41	213	26 13	289	36 02	339	28 50	006
108	16 13	071	19 04	134	37 55	171	54 29	215	25 53	291	35 54	341	28 53	008
110	16 33	073	19 19	136	37 58	173	54 16	218	25 34	293	35 48	342	28 56	010
112	16 53	075	19 33	138	38 00	175	54 03	220	25 15	295	35 42	344	29 00	011
114	17 13	077	19 47	140	38 02	177	53 50	222	24 56	297	35 37	346	29 04	013
116	17 33	079	20 00	142	38 02	180	53 35	225	24 38	299	35 32	348	29 09	015
118	17 54	081	20 12	144	38 02	182	53 20	227	24 20	300	35 28	349	29 15	017
	VEGA		◆ARCTURUS		REGULUS		POLLUX		◆CAPELLA		Alpheratz		◆DENEB	
120	29 21	019	18 14	083	20 24	146	38 01	184	53 05	229	24 02	302	35 24	351
122	29 28	020	18 35	085	20 35	148	37 59	186	52 49	232	23 44	304	35 21	353
124	29 36	022	18 56	087	20 46	151	37 56	189	52 32	234	23 27	306	35 19	355
126	29 44	024	19 17	089	20 56	153	37 53	191	52 15	236	23 11	308	35 17	356
128	29 53	026	19 37	091	21 05	155	37 48	193	51 58	238	22 54	310	35 16	358
130	30 02	027	19 58	093	21 14	157	37 43	195	51 40	240	22 39	312	35 16	000
132	30 12	029	20 19	095	21 22	159	37 37	198	51 22	243	22 23	314	35 16	001
134	30 22	031	20 40	097	21 29	161	37 31	200	51 03	245	22 08	315	35 17	003
136	30 34	033	21 00	099	21 35	163	37 23	202	50 44	247	21 54	317	35 18	005
138	30 45	035	21 21	101	21 41	165	37 15	204	50 25	249	21 40	319	35 20	007
140	30 57	036	21 41	103	21 46	167	37 06	206	50 05	251	21 27	321	35 23	008
142	31 10	038	22 02	105	21 50	169	36 56	209	49 45	253	21 14	323	35 26	010
144	31 23	040	22 22	107	21 53	172	36 46	211	49 25	255	21 02	325	35 30	012
146	31 37	042	22 42	109	21 56	174	36 35	213	49 05	257	20 50	327	35 35	014
148	31 51	044	23 01	111	21 58	176	36 23	215	48 44	259	20 39	328	35 40	015
	◆DENEB		VEGA		◆ARCTURUS		REGULUS		POLLUX		◆CAPELLA		Alpheratz	
150	35 46	017	32 05	045	23 21	113	21 59	178	36 11	217	48 24	261	20 28	330
152	35 52	019	32 21	047	23 40	115	22 00	180	35 58	220	48 03	263	20 18	332
154	35 59	020	32 36	049	23 59	117	21 59	182	35 44	222	47 42	265	20 09	334
156	36 07	022	32 52	051	24 17	119	21 58	184	35 30	224	47 22	267	20 00	336
158	36 15	024	33 08	053	24 35	121	21 56	186	35 16	226	47 01	269	19 52	338
160	36 24	026	33 25	055	24 53	123	21 53	188	35 00	228	46 40	271	19 44	340
162	36 33	027	33 42	056	25 10	125	21 50	191	34 45	230	46 19	273	19 37	341
164	36 43	029	34 00	058	25 27	127	21 46	193	34 28	232	45 58	275	19 31	343
166	36 53	031	34 18	060	25 43	129	21 41	195	34 11	235	45 38	277	19 25	345
168	37 04	033	34 36	062	25 59	131	21 35	197	33 54	237	45 17	279	19 20	347
170	37 16	034	34 55	064	26 15	133	21 29	199	33 37	239	44 57	281	19 16	349
172	37 28	036	35 13	066	26 30	135	21 22	201	33 19	241	44 36	283	19 12	351
174	37 40	038	35 33	068	26 44	137	21 14	203	33 00	243	44 16	285	19 09	352
176	37 53	040	35 52	069	26 58	139	21 05	205	32 42	245	43 56	287	19 07	354
178	38 07	041	36 12	071	27 11	142	20 56	207	32 23	247	43 36	289	19 05	356

LHA ♈	Hc	Zn	Hc	Zn	Hc	Zn	Hc	Zn	Hc	Zn	Hc	Zn	Hc	Zn
	◆DENEB		VEGA		◆ARCTURUS		REGULUS		POLLUX		◆CAPELLA		Alpheratz	
180	38 21	043	36 32	073	27 24	144	20 46	209	32 03	249	43 16	290	19 04	358
182	38 36	045	36 52	075	27 36	146	20 36	211	31 44	251	42 57	292	19 04	000
184	38 50	047	37 12	077	27 47	148	20 24	214	31 24	253	42 38	294	19 04	002
186	39 06	049	37 32	079	27 58	150	20 13	216	31 04	255	42 19	296	19 05	004
188	39 22	050	37 53	081	28 08	152	20 00	218	30 44	257	42 00	298	19 07	006
190	39 38	052	38 13	083	28 18	154	19 47	220	30 23	259	41 42	300	19 09	007
192	39 55	054	38 34	085	28 26	156	19 34	222	30 03	261	41 24	301	19 12	009
194	40 12	056	38 55	087	28 34	159	19 19	224	29 42	263	41 07	303	19 16	011
196	40 29	057	39 16	089	28 41	161	19 05	226	29 21	265	40 49	305	19 20	013
198	40 47	059	39 36	091	28 48	163	18 49	228	29 00	267	40 32	307	19 25	015
200	41 05	061	39 57	093	28 54	165	18 34	230	28 40	269	40 16	309	19 31	017
202	41 23	063	40 18	095	28 59	167	18 18	232	28 19	271	40 00	310	19 37	018
204	41 42	065	40 39	097	29 03	169	18 01	234	27 58	273	39 44	312	19 44	020
206	42 01	067	40 59	099	29 06	172	17 44	236	27 37	275	39 29	314	19 51	022
208	42 20	068	41 20	101	29 09	174	17 26	238	27 16	277	39 14	316	20 00	024
	Alpheratz		◆DENEB		VEGA		◆ARCTURUS		REGULUS		POLLUX		◆CAPELLA	
210	20 08	026	42 40	070	41 40	103	29 11	176	17 08	240	26 56	279	39 00	318
212	20 18	028	42 59	072	42 01	105	29 12	178	16 50	242	26 35	281	38 46	319
214	20 28	030	43 19	074	42 21	107	29 12	180	16 31	244	26 15	283	38 33	321
216	20 38	031	43 39	076	42 41	109	29 12	182	16 13	246	25 55	285	38 20	323
218	20 50	033	44 00	078	43 00	111	29 11	184	15 53	248	25 35	287	38 08	325
220	21 01	035	44 20	080	43 19	113	29 09	187	15 34	250	25 15	289	37 56	326
222	21 14	037	44 41	082	43 39	115	29 06	189	15 14	252	24 55	290	37 44	328
224	21 26	039	45 01	084	43 57	117	29 02	191	14 54	254	24 36	292	37 34	330
226	21 40	041	45 22	086	44 16	119	28 58	193	14 34	256	24 17	294	37 23	331
228	21 54	043	45 43	087	44 34	122	28 53	195	14 14	258	23 58	296	37 14	333
230	22 08	044	46 04	089	44 51	124	28 47	197	13 53	260	23 39	298	37 05	335
232	22 23	046	46 25	091	45 08	126	28 40	200	13 33	262	23 21	300	36 56	337
234	22 38	048	46 45	093	45 25	128	28 33	202	13 12	264	23 03	302	36 48	338
236	22 54	050	47 06	095	45 41	130	28 25	204	12 51	266	22 46	304	36 41	340
238	23 10	052	47 27	097	45 57	132	28 16	206	12 30	268	22 28	306	36 34	342
	Alpheratz		◆DENEB		VEGA		◆ARCTURUS		Dubhe		POLLUX		◆CAPELLA	
240	23 27	054	47 48	099	46 12	135	28 07	208	62 54	273	22 12	308	36 28	344
242	23 44	056	48 08	101	46 26	137	27 57	210	62 33	275	21 55	309	36 22	345
244	24 01	058	48 28	104	46 40	139	27 46	212	62 12	277	21 40	311	36 17	347
246	24 19	060	48 49	106	46 54	141	27 34	215	61 52	279	21 24	313	36 13	349
248	24 37	061	49 09	108	47 06	144	27 22	217	61 31	281	21 09	315	36 09	350
250	24 56	063	49 28	110	47 18	146	27 09	219	61 11	283	20 55	317	36 06	352
252	25 14	065	49 48	112	47 30	148	26 56	221	60 50	284	20 41	319	36 04	354
254	25 33	067	50 07	114	47 40	150	26 42	223	60 30	286	20 27	321	36 02	356
256	25 53	069	50 26	116	47 50	153	26 28	225	60 10	288	20 14	323	36 00	357
258	26 12	071	50 44	118	47 59	155	26 13	227	59 51	290	20 02	324	36 00	359
260	26 32	073	51 02	121	48 08	157	25 57	229	59 31	292	19 50	326	36 00	001
262	26 52	075	51 20	123	48 15	160	25 41	231	59 12	293	19 39	328	36 00	003
264	27 12	077	51 37	125	48 22	162	25 25	233	58 53	295	19 28	330	36 01	004
266	27 33	079	51 54	127	48 28	164	25 08	235	58 34	297	19 18	332	36 03	006
268	27 53	081	52 11	129	48 33	167	24 50	237	58 16	298	19 09	334	36 06	008
	◆CAPELLA		Alpheratz		◆DENEB		VEGA		◆ARCTURUS		Dubhe		POLLUX	
270	36 09	009	28 14	083	52 27	132	48 38	169	24 33	240	57 58	300	19 00	336
272	36 13	011	28 34	084	52 42	134	48 41	171	24 14	242	57 40	302	18 51	337
274	36 17	013	28 55	086	52 57	136	48 44	174	23 56	244	57 22	304	18 44	339
276	36 22	015	29 16	088	53 11	139	48 46	176	23 37	246	57 05	305	18 37	341
278	36 27	016	29 37	090	53 24	141	48 47	179	23 18	248	56 48	307	18 30	343
280	36 34	018	29 58	092	53 37	143	48 47	181	22 59	250	56 32	309	18 24	345
282	36 40	020	30 19	094	53 49	146	48 46	183	22 39	252	56 16	310	18 19	347
284	36 48	021	30 39	096	54 01	148	48 44	186	22 19	254	56 00	312	18 15	349
286	36 55	023	31 00	098	54 11	150	48 42	188	21 59	256	55 44	313	18 11	350
288	37 04	025	31 21	100	54 21	153	48 38	190	21 39	258	55 30	315	18 08	352
290	37 13	027	31 41	102	54 30	155	48 34	193	21 18	260	55 15	317	18 06	354
292	37 23	028	32 01	104	54 39	158	48 29	195	20 58	262	55 01	318	18 04	356
294	37 33	030	32 21	106	54 46	160	48 23	197	20 37	264	54 47	320	18 03	358
296	37 44	032	32 41	108	54 53	162	48 17	200	20 16	266	54 34	321	18 02	000
298	37 55	034	33 01	110	54 59	165	48 09	202	19 55	268	54 21	323	18 03	002
	POLLUX		◆CAPELLA		Alpheratz		◆DENEB		VEGA		◆ARCTURUS		Dubhe	
300	18 03	003	38 07	035	33 20	112	55 04	167	48 01	204	19 34	270	54 09	325
302	18 05	005	38 19	037	33 39	115	55 08	170	47 52	207	19 14	272	53 57	326
304	18 07	007	38 32	039	33 58	117	55 11	172	47 42	209	18 53	274	53 46	328
306	18 10	009	38 45	041	34 17	119	55 14	175	47 32	211	18 32	276	53 35	329
308	18 14	011	38 59	042	34 35	121	55 15	177	47 21	214	18 11	278	53 24	331
310	18 18	013	39 13	044	34 52	123	55 16	180	47 09	216	17 51	279	53 15	332
312	18 23	015	39 28	046	35 10	125	55 15	182	46 56	218	17 30	281	53 05	334
314	18 29	016	39 43	048	35 27	127	55 14	185	46 43	220	17 10	283	52 56	336
316	18 35	018	39 59	049	35 43	129	55 12	187	46 29	223	16 50	285	52 48	337
318	18 42	020	40 15	051	35 59	131	55 09	189	46 15	225	16 30	287	52 40	339
320	18 49	022	40 31	053	36 14	133	55 05	192	46 00	227	16 10	289	52 33	340
322	18 57	024	40 48	055	36 29	136	55 01	194	45 44	229	15 50	291	52 26	342
324	19 06	026	41 05	057	36 43	138	54 55	197	45 28	232	15 31	293	52 20	343
326	19 16	028	41 23	058	36 57	140	54 48	199	45 11	234	15 12	295	52 14	345
328	19 26	030	41 41	060	37 10	142	54 41	202	44 54	236	14 53	297	52 09	346
	POLLUX		◆CAPELLA		Hamal		Alpheratz		◆DENEB		VEGA		◆Alioth	
330	19 36	031	41 59	062	27 51	114	37 23	144	54 33	204	44 37	238	48 14	325
332	19 47	033	42 17	064	28 10	116	37 34	146	54 24	207	44 19	240	48 02	326
334	19 59	035	42 36	066	28 29	118	37 46	149	54 14	209	44 01	242	47 51	328
336	20 11	037	42 55	068	28 47	120	37 56	151	54 04	211	43 42	244	47 40	330
338	20 24	039	43 15	069	29 05	122	38 06	153	53 53	214	43 23	247	47 30	331
340	20 37	041	43 34	071	29 22	124	38 15	155	53 41	216	43 04	249	47 20	333
342	20 51	043	43 54	073	29 39	126	38 23	158	53 28	218	42 44	251	47 11	335
344	21 06	044	44 14	075	29 56	128	38 31	160	53 15	221	42 25	253	47 02	336
346	21 20	046	44 35	077	30 12	131	38 38	162	53 01	223	42 05	255	46 54	338
348	21 36	048	44 55	079	30 27	133	38 44	164	52 47	225	41 44	257	46 46	339
350	21 51	050	45 15	081	30 42	135	38 49	166	52 31	228	41 24	259	46 39	341
352	22 08	052	45 36	083	30 57	137	38 54	169	52 16	230	41 03	261	46 33	343
354	22 24	054	45 57	085	31 11	139	38 57	171	52 00	232	40 43	263	46 27	344
356	22 41	056	46 18	086	31 24	141	39 00	173	51 43	234	40 22	265	46 21	346
358	22 59	058	46 38	088	31 37	143	39 02	175	51 26	237	40 01	267	46 17	348

LAT 79°N

LHA	Hc	Zn	Hc	Zn	Hc	Zn	Hc	Zn	Hc	Zn	Hc	Zn	Hc	Zn
♈	POLLUX		◆CAPELLA		Hamal		Alpheratz		◆DENEB		VEGA		◆Alioth	
0	22 46	059	46 59	089	32 38	145	40 03	178	51 39	240	39 41	270	45 13	349
2	23 06	061	47 22	091	32 51	147	40 04	180	51 19	242	39 18	272	45 10	351
4	23 26	063	47 45	093	33 03	149	40 03	182	50 58	244	38 55	274	45 06	353
6	23 46	065	48 08	095	33 15	152	40 02	185	50 37	246	38 33	276	45 04	354
8	24 07	067	48 30	097	33 25	154	40 00	187	50 16	249	38 10	278	45 02	356
10	24 28	069	48 53	099	33 35	156	39 57	189	49 55	251	37 47	279	45 00	357
12	24 50	070	49 16	101	33 44	158	39 52	191	49 33	253	37 25	281	44 59	359
14	25 12	072	49 38	103	33 52	160	39 48	194	49 11	255	37 02	283	44 59	000
16	25 34	074	50 00	105	33 59	163	39 42	196	48 49	257	36 40	285	45 00	002
18	25 56	076	50 22	108	34 06	165	39 35	198	48 27	259	36 18	287	45 01	004
20	26 18	078	50 44	110	34 11	167	39 27	200	48 04	261	35 56	289	45 03	005
22	26 41	080	51 05	112	34 16	169	39 19	203	47 41	263	35 35	291	45 05	007
24	27 03	082	51 26	114	34 20	171	39 10	205	47 19	265	35 14	293	45 08	008
26	27 26	084	51 47	116	34 23	174	39 00	207	46 56	267	34 53	295	45 12	010
28	27 49	086	52 07	118	34 25	176	38 49	209	46 33	269	34 32	296	45 16	012
	◆Alioth		POLLUX		◆CAPELLA		Hamal		Alpheratz		◆DENEB		VEGA	
30	45 21	013	28 12	088	52 27	121	34 26	178	38 37	212	46 10	271	34 12	298
32	45 26	015	28 34	090	52 47	123	34 26	180	38 25	214	45 47	273	33 52	300
34	45 32	016	28 57	092	53 06	125	34 26	183	38 12	216	45 24	275	33 32	302
36	45 39	018	29 20	094	53 24	127	34 24	185	37 58	218	45 01	277	33 13	304
38	45 46	019	29 43	096	53 42	130	34 22	187	37 43	220	44 39	279	32 54	306
40	45 54	021	30 06	098	54 00	132	34 19	189	37 28	223	44 16	281	32 35	307
42	46 03	023	30 28	100	54 16	134	34 15	191	37 12	225	43 54	282	32 17	309
44	46 12	024	30 51	102	54 32	137	34 10	194	36 56	227	43 31	284	32 00	311
46	46 22	026	31 13	104	54 48	139	34 04	196	36 39	229	43 09	286	31 43	313
48	46 32	027	31 35	106	55 03	141	33 57	198	36 21	231	42 47	288	31 26	315
50	46 43	029	31 57	108	55 17	144	33 50	200	36 03	233	42 26	290	31 10	316
52	46 54	031	32 19	110	55 30	146	33 41	202	35 44	236	42 04	292	30 55	318
54	47 06	032	32 40	112	55 42	148	33 32	205	35 25	238	41 43	294	30 40	320
56	47 19	034	33 01	114	55 54	151	33 22	207	35 06	240	41 22	295	30 25	322
58	47 32	036	33 22	116	56 04	153	33 11	209	34 46	242	41 02	297	30 11	324
	◆Alioth		REGULUS		POLLUX		◆CAPELLA		Alpheratz		◆DENEB		VEGA	
60	47 45	037	11 23	086	33 43	118	56 14	156	34 25	244	40 42	299	29 58	325
62	47 59	039	11 46	088	34 03	120	56 23	158	34 05	246	40 22	301	29 45	327
64	48 14	040	12 09	090	34 22	122	56 31	161	33 44	248	40 02	303	29 33	329
66	48 29	042	12 32	092	34 41	124	56 38	163	33 22	250	39 43	304	29 22	331
68	48 45	044	12 54	094	35 00	126	56 44	166	33 01	252	39 25	306	29 11	332
70	49 01	045	13 17	096	35 18	129	56 49	168	32 39	254	39 06	308	29 01	334
72	49 17	047	13 40	097	35 36	131	56 53	171	32 16	256	38 49	310	28 51	336
74	49 34	049	14 03	099	35 53	133	56 56	174	31 54	258	38 31	312	28 42	338
76	49 52	050	14 25	101	36 09	135	56 58	176	31 32	260	38 14	313	28 34	340
78	50 09	052	14 48	103	36 25	137	56 59	179	31 09	262	37 58	315	28 26	341
80	50 28	054	15 10	105	36 40	139	56 59	181	30 46	264	37 42	317	28 19	343
82	50 46	055	15 32	107	36 55	142	56 59	184	30 23	266	37 27	319	28 13	345
84	51 05	057	15 53	109	37 09	144	56 57	186	30 01	268	37 12	320	28 07	347
86	51 25	059	16 15	111	37 22	146	56 54	189	29 38	270	36 57	322	28 02	348
88	51 45	061	16 36	113	37 34	148	56 50	191	29 15	272	36 44	324	27 58	350
	◆ARCTURUS		REGULUS		POLLUX		◆CAPELLA		Alpheratz		◆DENEB		VEGA	
90	12 52	054	16 57	115	37 46	150	56 45	194	28 52	274	36 30	325	27 54	352
92	13 10	055	17 17	117	37 57	153	56 39	196	28 29	276	36 18	327	27 51	354
94	13 29	057	17 38	119	38 07	155	56 32	199	28 06	278	36 05	329	27 49	355
96	13 49	059	17 57	121	38 16	157	56 24	201	27 44	280	35 54	331	27 48	357
98	14 09	061	18 17	124	38 25	159	56 15	204	27 21	282	35 43	332	27 47	359
100	14 29	063	18 35	126	38 33	162	56 05	206	26 59	284	35 33	334	27 47	001
102	14 50	065	18 54	128	38 39	164	55 54	209	26 37	286	35 23	336	27 47	002
104	15 10	067	19 12	130	38 45	166	55 43	211	26 15	288	35 14	338	27 49	004
106	15 32	069	19 29	132	38 50	168	55 31	214	25 53	289	35 06	339	27 51	006
108	15 53	071	19 46	134	38 55	171	55 18	216	25 32	291	34 58	341	27 54	008
110	16 15	073	20 02	136	38 58	173	55 04	219	25 11	293	34 51	343	27 57	010
112	16 37	075	20 18	138	39 00	175	54 49	221	24 50	295	34 44	344	28 01	011
114	16 59	077	20 33	140	39 02	177	54 34	223	24 29	297	34 38	346	28 06	013
116	17 21	079	20 47	142	39 02	180	54 18	226	24 09	299	34 33	348	28 11	015
118	17 44	080	21 01	144	39 02	182	54 01	228	23 49	301	34 29	350	28 18	017
	◆VEGA		ARCTURUS		◆REGULUS		POLLUX		◆CAPELLA		Alpheratz		DENEB	
120	28 25	018	18 07	082	21 14	146	39 01	184	53 44	230	23 30	303	34 25	351
122	28 32	020	18 29	084	21 26	148	38 59	187	53 26	233	23 10	305	34 22	353
124	28 40	022	18 52	086	21 38	150	38 56	189	53 07	235	22 52	306	34 19	355
126	28 49	024	19 15	088	21 49	152	38 52	191	52 48	237	22 34	308	34 17	356
128	28 59	025	19 38	090	21 59	155	38 47	193	52 29	239	22 16	310	34 16	358
130	29 09	027	20 01	092	22 09	157	38 41	196	52 09	242	21 59	312	34 16	000
132	29 20	029	20 24	094	22 17	159	38 34	198	51 49	244	21 42	314	34 16	001
134	29 31	031	20 46	096	22 25	161	38 27	200	51 28	246	21 26	316	34 17	003
136	29 43	033	21 09	098	22 32	163	38 19	202	51 07	248	21 10	318	34 18	005
138	29 56	034	21 32	100	22 39	165	38 10	205	50 45	250	20 55	319	34 21	007
140	30 09	036	21 54	102	22 44	167	38 00	207	50 24	252	20 40	321	34 24	008
142	30 23	038	22 17	104	22 49	169	37 49	209	50 02	254	20 26	323	34 27	010
144	30 37	040	22 39	106	22 53	171	37 37	211	49 40	256	20 13	325	34 32	012
146	30 52	041	23 01	108	22 56	174	37 25	213	49 17	258	20 00	327	34 37	013
148	31 07	043	23 22	110	22 58	176	37 12	216	48 55	261	19 48	329	34 42	015
	DENEB		◆VEGA		ARCTURUS		◆REGULUS		POLLUX		◆CAPELLA		Alpheratz	
150	34 48	017	31 23	045	23 44	112	22 59	178	36 58	218	48 32	263	19 36	331
152	34 55	018	31 40	047	24 05	114	23 00	180	36 44	220	48 09	265	19 25	332
154	35 03	020	31 57	049	24 25	116	22 59	182	36 29	222	47 47	267	19 15	334
156	35 11	022	32 14	050	24 46	118	22 58	184	36 13	224	47 24	269	19 05	336
158	35 20	024	32 32	052	25 06	120	22 56	186	35 57	227	47 01	271	18 56	338
160	35 30	025	32 50	054	25 25	122	22 53	188	35 40	229	46 38	272	18 48	340
162	35 40	027	33 09	056	25 44	124	22 49	191	35 23	231	46 15	274	18 40	342
164	35 50	029	33 28	058	26 03	127	22 44	193	35 05	233	45 52	276	18 33	343
166	36 02	031	33 48	060	26 21	129	22 39	195	34 46	235	45 30	278	18 27	345
168	36 14	032	34 08	061	26 39	131	22 33	197	34 27	237	45 07	280	18 22	347
170	36 26	034	34 28	063	26 56	133	22 26	199	34 08	239	44 45	282	18 17	349
172	36 39	036	34 49	065	27 12	135	22 18	201	33 48	241	44 22	284	18 13	351
174	36 53	037	35 09	067	27 28	137	22 09	203	33 27	243	44 00	286	18 10	353
176	37 07	039	35 31	069	27 43	139	22 00	205	33 07	246	43 38	288	18 07	354
178	37 22	041	35 52	071	27 58	141	21 49	207	32 46	248	43 16	289	18 05	356

LHA	Hc	Zn	Hc	Zn	Hc	Zn	Hc	Zn	Hc	Zn	Hc	Zn	Hc	Zn
♈	DENEB		◆VEGA		ARCTURUS		◆REGULUS		POLLUX		◆CAPELLA		Alpheratz	
180	37 37	043	36 14	073	28 12	143	21 38	210	32 24	250	42 55	291	18 04	358
182	37 53	044	36 36	074	28 25	145	21 27	212	32 03	252	42 34	293	18 04	000
184	38 09	046	36 58	076	28 38	148	21 14	214	31 41	254	42 13	295	18 04	002
186	38 26	048	37 20	078	28 50	150	21 01	216	31 19	256	41 52	297	18 05	004
188	38 43	050	37 43	080	29 01	152	20 48	218	30 57	258	41 32	299	18 07	005
190	39 01	051	38 05	082	29 12	154	20 33	220	30 34	260	41 12	300	18 10	007
192	39 19	053	38 28	084	29 21	156	20 18	222	30 11	262	40 53	302	18 13	009
194	39 37	055	38 51	086	29 30	158	20 03	224	29 49	264	40 33	304	18 17	011
196	39 56	057	39 14	088	29 38	161	19 46	226	29 26	266	40 15	306	18 22	013
198	40 16	059	39 37	090	29 45	163	19 30	228	29 03	268	39 56	308	18 27	015
200	40 35	060	40 00	092	29 52	165	19 12	230	28 40	270	39 38	309	18 33	017
202	40 56	062	40 22	094	29 57	167	18 54	232	28 17	272	39 21	311	18 40	018
204	41 16	064	40 45	096	30 02	169	18 36	234	27 54	274	39 04	313	18 48	020
206	41 37	066	41 08	098	30 06	171	18 17	236	27 32	276	38 47	315	18 56	022
208	41 58	068	41 31	100	30 09	174	17 58	238	27 09	277	38 31	316	19 05	024
	Alpheratz		◆DENEB		VEGA		◆ARCTURUS		REGULUS		POLLUX		◆CAPELLA	
210	19 14	026	42 19	069	41 53	102	30 11	176	17 38	240	26 46	279	38 16	318
212	19 25	028	42 41	071	42 15	104	30 12	178	17 18	243	26 24	281	38 00	320
214	19 36	029	43 02	073	42 38	106	30 12	180	16 57	245	26 01	283	37 46	322
216	19 47	031	43 24	075	43 00	108	30 12	182	16 37	247	25 39	285	37 32	323
218	19 59	033	43 47	077	43 21	110	30 11	185	16 15	249	25 17	287	37 19	325
220	20 12	035	44 09	079	43 43	112	30 08	187	15 54	251	24 55	289	37 06	327
222	20 26	037	44 32	081	44 04	114	30 05	189	15 32	253	24 34	291	36 53	328
224	20 40	039	44 54	083	44 24	116	30 01	191	15 10	255	24 13	293	36 42	330
226	20 54	041	45 17	085	44 45	119	29 56	193	14 48	257	23 52	295	36 31	332
228	21 09	042	45 40	086	45 04	121	29 51	195	14 26	258	23 31	297	36 20	334
230	21 25	044	46 03	088	45 24	123	29 44	198	14 03	260	23 11	298	36 10	335
232	21 41	046	46 26	090	45 43	125	29 37	200	13 41	262	22 51	300	36 01	337
234	21 58	048	46 48	092	46 01	127	29 29	202	13 18	264	22 31	302	35 52	339
236	22 15	050	47 11	094	46 19	129	29 20	204	12 55	266	22 12	304	35 44	340
238	22 33	052	47 34	096	46 37	132	29 10	206	12 32	268	21 53	306	35 37	342
	Alpheratz		◆DENEB		VEGA		◆ARCTURUS		Alioth		POLLUX		◆CAPELLA	
240	22 51	054	47 57	098	46 54	134	29 00	208	62 31	242	21 35	308	35 30	344
242	23 10	055	48 19	100	47 10	136	28 48	211	62 10	244	21 17	310	35 24	346
244	23 29	057	48 42	102	47 25	138	28 36	213	61 49	246	21 00	312	35 19	347
246	23 48	059	49 04	104	47 40	141	28 24	215	61 28	248	20 43	313	35 14	349
248	24 08	061	49 26	107	47 54	143	28 10	217	61 07	251	20 27	315	35 10	351
250	24 28	063	49 48	109	48 08	145	27 56	219	60 45	253	20 11	317	35 07	352
252	24 49	065	50 10	111	48 20	148	27 41	221	60 23	255	19 56	319	35 04	354
254	25 10	067	50 31	113	48 32	150	27 26	223	60 01	257	19 41	321	35 02	356
256	25 31	069	50 52	115	48 43	152	27 10	225	59 39	259	19 27	323	35 00	357
258	25 52	070	51 12	117	48 54	155	26 53	228	59 16	261	19 13	325	35 00	359
260	26 14	072	51 32	119	49 03	157	26 36	230	58 53	263	19 00	326	35 00	001
262	26 36	074	51 52	122	49 12	159	26 18	232	58 31	265	18 48	328	35 00	002
264	26 58	076	52 11	124	49 19	162	26 00	234	58 08	267	18 36	330	35 02	004
266	27 21	078	52 30	126	49 26	164	25 42	236	57 45	269	18 25	332	35 04	006
268	27 43	080	52 48	128	49 32	167	25 22	238	57 22	271	18 15	334	35 06	008
	◆CAPELLA		Alpheratz		◆DENEB		VEGA		◆ARCTURUS		Alioth		POLLUX	
270	35 10	009	28 06	082	53 06	131	49 37	169	25 03	240	56 59	273	18 05	336
272	35 14	011	28 28	084	53 23	133	49 41	171	24 43	242	56 36	275	17 56	338
274	35 18	013	28 51	086	53 40	135	49 43	174	24 22	244	56 13	277	17 48	339
276	35 24	014	29 14	088	53 55	138	49 45	176	24 02	246	55 51	279	17 40	341
278	35 30	016	29 37	090	54 10	140	49 47	179	23 40	248	55 28	281	17 33	343
280	35 36	018	30 00	092	54 25	142	49 47	181	23 19	250	55 06	283	17 27	345
282	35 44	019	30 23	094	54 38	145	49 46	183	22 57	252	54 43	284	17 21	347
284	35 52	021	30 46	096	54 51	147	49 44	186	22 36	254	54 21	286	17 16	349
286	36 00	023	31 08	098	55 03	150	49 41	188	22 13	256	54 00	288	17 12	351
288	36 09	025	31 31	100	55 14	152	49 37	191	21 51	258	53 38	290	17 09	352
290	36 19	026	31 53	102	55 25	155	49 33	193	21 29	260	53 16	292	17 06	354
292	36 30	028	32 16	104	55 34	157	49 27	195	21 06	262	52 55	293	17 04	356
294	36 41	030	32 38	106	55 43	159	49 21	198	20 43	264	52 34	295	17 03	358
296	36 52	031	33 00	108	55 50	162	49 13	200	20 20	266	52 14	297	17 02	000
298	37 05	033	33 21	110	55 57	164	49 05	203	19 58	268	51 54	299	17 03	002
	POLLUX		◆CAPELLA		Alpheratz		◆DENEB		VEGA		◆ARCTURUS		Alioth	
300	17 04	003	37 18	035	33 43	112	56 02	167	48 56	205	19 35	270	51 34	300
302	17 05	005	37 31	037	34 04	114	56 07	169	48 45	207	19 12	272	51 14	302
304	17 08	007	37 45	038	34 25	116	56 11	172	48 35	210	18 49	274	50 55	304
306	17 11	009	37 59	040	34 45	118	56 13	175	48 23	212	18 26	276	50 36	306
308	17 15	011	38 14	042	35 05	120	56 15	177	48 10	214	18 03	278	50 18	307
310	17 20	013	38 30	044	35 25	122	56 16	180	47 57	217	17 41	280	50 00	309
312	17 25	015	38 46	045	35 44	124	56 15	182	47 43	219	17 18	282	49 42	311
314	17 31	016	39 02	047	36 03	127	56 14	185	47 28	221	16 56	284	49 25	312
316	17 38	018	39 19	049	36 21	129	56 12	187	47 13	223	16 34	286	49 08	314
318	17 45	020	39 37	051	36 38	131	56 08	190	46 57	226	16 12	288	48 52	316
320	17 54	022	39 55	052	36 55	133	56 04	192	46 40	228	15 50	289	48 36	317
322	18 03	024	40 13	054	37 12	135	55 59	195	46 23	230	15 29	291	48 21	319
324	18 12	026	40 32	056	37 28	137	55 52	197	46 05	232	15 07	293	48 06	320
326	18 22	027	40 51	058	37 43	139	55 45	200	45 47	235	14 47	295	47 52	322
328	18 33	029	41 11	059	37 57	142	55 37	202	45 28	237	14 26	297	47 38	324
	POLLUX		◆CAPELLA		Hamal		Alpheratz		◆DENEB		VEGA		◆Alioth	
330	18 45	031	41 30	061	28 15	113	38 11	144	55 28	205	45 08	239	47 25	325
332	18 57	033	41 51	063	28 36	116	38 24	146	55 18	207	44 48	241	47 12	327
334	19 10	035	42 11	065	28 57	118	38 37	148	55 07	210	44 28	243	47 00	329
336	19 23	037	42 32	067	29 17	120	38 48	151	54 55	212	44 08	245	46 48	330
338	19 37	039	42 53	069	29 36	122	38 59	153	54 42	214	43 47	247	46 37	332
340	19 52	040	43 15	070	29 56	124	39 09	155	54 29	217	43 25	250	46 26	333
342	20 07	042	43 36	072	30 14	126	39 19	157	54 15	219	43 04	252	46 16	335
344	20 23	044	43 58	074	30 33	128	39 27	159	54 00	222	42 42	254	46 07	337
346	20 39	046	44 20	076	30 51	130	39 35	162	53 45	224	42 20	256	45 58	338
348	20 56	048	44 43	078	31 08	132	39 41	164	53 28	226	41 57	258	45 50	340
350	21 13	050	45 05	080	31 24	134	39 47	166	53 11	229	41 35	260	45 42	341
352	21 31	052	45 28	082	31 41	136	39 52	169	52 54	231	41 12	262	45 35	343
354	21 49	054	45 50	084	31 56	139	39 56	171	52 36	233	40 50	264	45 29	345
356	22 07	055	46 13	085	32 11	141	40 00	173	52 17	235	40 27	266	45 23	346
358	22 26	057	46 36	087	32 25	143	40 02	175	51 58	238	40 04	268	45 18	348

LAT 78°N

LHA	Hc	Zn	Hc	Zn	Hc	Zn	Hc	Zn	Hc	Zn	Hc	Zn	Hc	Zn
♈	POLLUX		◆CAPELLA		Hamal		Alpheratz		◆DENEB		VEGA		◆Alioth	
0	22 15	059	46 58	088	33 28	145	41 03	178	52 08	241	39 41	271	44 14	350
2	22 36	061	47 23	090	33 42	147	41 04	180	51 46	243	39 16	273	44 10	351
4	22 58	063	47 48	092	33 55	149	41 03	182	51 24	245	38 51	275	44 07	353
6	23 21	064	48 13	094	34 07	151	41 02	185	51 01	248	38 26	276	44 04	354
8	23 43	066	48 37	096	34 19	153	40 59	187	50 38	250	38 01	278	44 02	356
10	24 06	068	49 02	098	34 30	156	40 56	189	50 14	252	37 37	280	44 00	357
12	24 30	070	49 27	100	34 39	158	40 51	192	49 50	254	37 12	282	43 59	359
14	24 53	072	49 51	102	34 48	160	40 46	194	49 26	256	36 48	284	43 59	000
16	25 17	074	50 16	104	34 56	162	40 39	196	49 02	258	36 24	286	44 00	002
18	25 41	076	50 40	106	35 03	165	40 32	198	48 37	260	36 00	288	44 01	004
20	26 05	078	51 03	109	35 10	167	40 24	201	48 13	262	35 36	290	44 03	005
22	26 30	080	51 27	111	35 15	169	40 14	203	47 48	264	35 13	291	44 05	007
24	26 55	082	51 50	113	35 19	171	40 04	205	47 23	266	34 50	293	44 09	008
26	27 19	083	52 13	115	35 22	174	39 53	208	46 58	268	34 27	295	44 13	010
28	27 44	085	52 35	117	35 25	176	39 41	210	46 33	270	34 05	297	44 17	011
	◆Alioth		POLLUX		◆CAPELLA		Hamal		Alpheratz		◆DENEB		VEGA	
30	44 22	013	28 09	087	52 57	119	35 26	178	39 28	212	46 08	272	33 43	299
32	44 28	014	28 34	089	53 19	122	35 26	180	39 14	214	45 43	274	33 21	301
34	44 35	016	28 59	091	53 40	124	35 26	183	39 00	217	45 19	276	33 00	302
36	44 42	018	29 24	093	54 00	126	35 24	185	38 45	219	44 54	278	32 39	304
38	44 50	019	29 49	095	54 20	128	35 22	187	38 29	221	44 29	280	32 19	306
40	44 58	021	30 14	097	54 39	131	35 18	189	38 12	223	44 05	282	31 59	308
42	45 07	022	30 38	099	54 58	133	35 13	192	37 55	225	43 40	283	31 39	310
44	45 17	024	31 03	101	55 16	136	35 08	194	37 37	228	43 16	285	31 20	311
46	45 28	025	31 27	103	55 33	138	35 02	196	37 18	230	42 52	287	31 02	313
48	45 39	027	31 51	105	55 49	140	34 54	198	36 59	232	42 28	289	30 44	315
50	45 50	029	32 15	107	56 05	143	34 46	201	36 39	234	42 05	291	30 27	317
52	46 02	030	32 39	109	56 19	145	34 37	203	36 18	236	41 42	293	30 10	319
54	46 15	032	33 02	111	56 33	148	34 27	205	35 57	238	41 19	294	29 54	320
56	46 29	033	33 26	113	56 46	150	34 16	207	35 36	240	40 56	296	29 38	322
58	46 43	035	33 48	115	56 58	153	34 04	209	35 14	242	40 34	298	29 23	324
	◆Alioth		REGULUS		POLLUX		◆CAPELLA		Alpheratz		◆DENEB		VEGA	
60	46 57	037	11 18	085	34 11	117	57 09	155	34 51	245	40 12	300	29 09	326
62	47 12	038	11 43	087	34 32	120	57 19	158	34 29	247	39 51	302	28 55	327
64	47 28	040	12 08	089	34 54	122	57 28	160	34 06	249	39 30	303	28 42	329
66	47 44	041	12 33	091	35 15	124	57 35	163	33 42	251	39 09	305	28 29	331
68	48 01	043	12 58	093	35 35	126	57 42	166	33 19	253	38 49	307	28 18	333
70	48 18	045	13 23	095	35 55	128	57 48	168	32 55	255	38 29	309	28 06	334
72	48 36	046	13 48	097	36 15	130	57 52	171	32 30	257	38 10	310	27 56	336
74	48 54	048	14 12	099	36 33	132	57 56	173	32 06	259	37 51	312	27 46	338
76	49 13	049	14 37	101	36 52	135	57 58	176	31 41	261	37 33	314	27 37	340
78	49 32	051	15 01	103	37 09	137	57 59	179	31 17	263	37 15	316	27 29	341
80	49 52	053	15 26	105	37 26	139	57 59	181	30 52	265	36 58	317	27 22	343
82	50 12	054	15 50	107	37 42	141	57 58	184	30 27	267	36 41	319	27 15	345
84	50 32	056	16 13	109	37 57	143	57 56	186	30 02	269	36 25	321	27 09	347
86	50 53	058	16 37	111	38 12	146	57 53	189	29 37	271	36 10	322	27 03	348
88	51 15	059	17 00	113	38 25	148	57 48	192	29 12	273	35 55	324	26 59	350
	◆ARCTURUS		REGULUS		POLLUX		◆CAPELLA		Alpheratz		◆DENEB		VEGA	
90	12 16	053	17 23	115	38 38	150	57 43	194	28 47	275	35 41	326	26 55	352
92	12 36	055	17 45	117	38 50	152	57 36	197	28 23	277	35 27	328	26 52	354
94	12 57	057	18 07	119	39 01	155	57 28	199	27 58	278	35 14	329	26 49	355
96	13 18	059	18 28	121	39 12	157	57 19	202	27 33	280	35 02	331	26 48	357
98	13 40	061	18 50	123	39 21	159	57 10	205	27 09	282	34 50	333	26 47	359
100	14 02	063	19 10	125	39 30	161	56 59	207	26 44	284	34 39	334	26 47	001
102	14 24	065	19 30	127	39 37	164	56 47	210	26 20	286	34 28	336	26 47	002
104	14 47	067	19 50	129	39 44	166	56 34	212	25 57	288	34 18	338	26 49	004
106	15 10	069	20 09	131	39 49	168	56 20	215	25 33	290	34 09	339	26 51	006
108	15 33	071	20 27	133	39 54	170	56 06	217	25 10	292	34 01	341	26 54	008
110	15 57	072	20 45	136	39 57	173	55 50	219	24 47	294	33 53	343	26 58	009
112	16 21	074	21 02	138	40 00	175	55 34	222	24 24	296	33 46	345	27 02	011
114	16 45	076	21 19	140	40 02	177	55 17	224	24 02	297	33 40	346	27 07	013
116	17 09	078	21 34	142	40 02	180	54 59	227	23 40	299	33 34	348	27 13	015
118	17 34	080	21 50	144	40 02	182	54 41	229	23 18	301	33 30	350	27 20	016
	◆VEGA		ARCTURUS		◆REGULUS		POLLUX		◆CAPELLA		Alpheratz		DENEB	
120	27 28	018	17 59	082	22 04	146	40 01	184	54 22	231	22 57	303	33 25	351
122	27 36	020	18 23	084	22 17	148	39 58	187	54 02	234	22 36	305	33 22	353
124	27 45	022	18 48	086	22 30	150	39 55	189	53 41	236	22 16	307	33 19	355
126	27 54	023	19 13	088	22 42	152	39 50	191	53 20	238	21 56	309	33 17	356
128	28 04	025	19 38	090	22 53	154	39 45	194	52 59	240	21 37	310	33 16	358
130	28 15	027	20 03	092	23 04	157	39 39	196	52 37	243	21 18	312	33 16	000
132	28 27	029	20 28	094	23 13	159	39 32	198	52 15	245	21 00	314	33 16	001
134	28 39	030	20 53	096	23 22	161	39 23	200	51 52	247	20 43	316	33 17	003
136	28 52	032	21 17	098	23 30	163	39 14	203	51 29	249	20 26	318	33 19	005
138	29 06	034	21 42	100	23 37	165	39 04	205	51 05	251	20 09	320	33 21	006
140	29 20	036	22 07	102	23 43	167	38 53	207	50 41	253	19 53	322	33 24	008
142	29 35	038	22 31	104	23 48	169	38 41	209	50 17	256	19 38	323	33 28	010
144	29 51	039	22 55	106	23 52	171	38 29	212	49 53	258	19 23	325	33 33	012
146	30 07	041	23 19	108	23 55	174	38 15	214	49 29	260	19 10	327	33 38	013
148	30 23	043	23 43	110	23 58	176	38 01	216	49 04	262	18 56	329	33 44	015
	DENEB		◆VEGA		ARCTURUS		◆REGULUS		POLLUX		◆CAPELLA		Alpheratz	
150	33 51	017	30 41	045	24 06	112	23 59	178	37 46	218	48 39	264	18 44	331
152	33 58	018	30 58	046	24 29	114	24 00	180	37 30	221	48 15	266	18 32	333
154	34 07	020	31 17	048	24 52	116	23 59	182	37 13	223	47 50	268	18 21	334
156	34 16	022	31 36	050	25 14	118	23 58	184	36 56	225	47 25	270	18 10	336
158	34 25	023	31 55	052	25 36	120	23 55	186	36 38	227	47 00	272	18 01	338
160	34 35	025	32 15	054	25 57	122	23 52	189	36 19	229	46 35	274	17 52	340
162	34 46	027	32 35	055	26 18	124	23 48	191	36 00	231	46 10	275	17 43	342
164	34 58	028	32 56	057	26 38	126	23 43	193	35 40	234	45 45	277	17 36	344
166	35 10	030	33 17	059	26 58	128	23 37	195	35 20	236	45 20	279	17 29	345
168	35 23	032	33 39	061	27 18	130	23 30	197	34 59	238	44 56	281	17 23	347
170	35 36	034	34 01	063	27 36	132	23 22	199	34 38	240	44 32	283	17 18	349
172	35 50	035	34 23	064	27 54	135	23 14	201	34 16	242	44 07	285	17 14	351
174	36 05	037	34 46	066	28 12	137	23 04	203	33 54	244	43 43	287	17 10	353
176	36 20	039	35 09	068	28 29	139	22 54	206	33 31	246	43 20	289	17 07	354
178	36 36	040	35 32	070	28 45	141	22 43	208	33 08	248	42 56	290	17 05	356

LHA	Hc	Zn	Hc	Zn	Hc	Zn	Hc	Zn	Hc	Zn	Hc	Zn	Hc	Zn
♈	DENEB		◆VEGA		ARCTURUS		◆REGULUS		POLLUX		◆CAPELLA		Alpheratz	
180	36 53	042	35 56	072	29 00	143	22 31	210	32 45	250	42 33	292	17 04	358
182	37 10	044	36 19	074	29 15	145	22 18	212	32 21	252	42 10	294	17 04	000
184	37 27	046	36 43	076	29 29	147	22 04	214	31 57	254	41 47	296	17 04	002
186	37 45	047	37 08	077	29 42	149	21 50	216	31 33	256	41 25	298	17 05	004
188	38 04	049	37 32	079	29 54	152	21 35	218	31 09	258	41 03	299	17 07	005
190	38 23	051	37 57	081	30 05	154	21 19	220	30 44	260	40 41	301	17 10	007
192	38 43	053	38 21	083	30 16	156	21 03	222	30 20	262	40 20	303	17 14	009
194	39 03	054	38 46	085	30 26	158	20 46	224	29 55	264	40 00	305	17 18	011
196	39 23	056	39 11	087	30 35	160	20 28	226	29 30	266	39 39	306	17 23	013
198	39 44	058	39 36	089	30 43	163	20 09	229	29 05	268	39 19	308	17 29	015
200	40 05	060	40 01	091	30 50	165	19 50	231	28 40	270	39 00	310	17 36	016
202	40 27	061	40 26	093	30 56	167	19 31	233	28 15	272	38 41	312	17 43	018
204	40 49	063	40 51	095	31 01	169	19 11	235	27 50	274	38 23	313	17 51	020
206	41 12	065	41 16	097	31 05	171	18 50	237	27 26	276	38 05	315	18 00	022
208	41 35	067	41 40	099	31 08	174	18 29	239	27 01	278	37 48	317	18 10	024
	Alpheratz		◆DENEB		VEGA		◆ARCTURUS		REGULUS		POLLUX		◆CAPELLA	
210	18 20	026	41 58	069	42 05	101	31 11	176	18 07	241	26 36	280	37 31	319
212	18 31	027	42 21	070	42 29	103	31 12	178	17 45	243	26 12	282	37 14	320
214	18 43	029	42 45	072	42 54	105	31 12	180	17 23	245	25 47	284	36 59	322
216	18 56	031	43 08	074	43 18	107	31 12	182	17 00	247	25 23	286	36 44	324
218	19 09	033	43 33	076	43 41	109	31 10	185	16 37	249	24 59	288	36 29	325
220	19 23	035	43 57	078	44 05	111	31 08	187	16 14	251	24 36	289	36 15	327
222	19 38	037	44 21	080	44 28	113	31 05	189	15 50	253	24 12	291	36 02	329
224	19 53	038	44 46	082	44 50	115	31 00	191	15 26	255	23 49	293	35 50	330
226	20 09	040	45 11	084	45 13	118	30 55	193	15 02	257	23 26	295	35 38	332
228	20 25	042	45 36	085	45 35	120	30 49	196	14 38	259	23 04	297	35 26	334
230	20 42	044	46 00	087	45 56	122	30 41	198	14 13	261	22 42	299	35 16	336
232	21 00	046	46 25	089	46 17	124	30 33	200	13 48	263	22 20	301	35 06	337
234	21 18	048	46 50	091	46 37	126	30 24	202	13 24	265	21 59	303	34 56	339
236	21 36	049	47 15	093	46 57	129	30 15	204	12 59	267	21 38	304	34 48	341
238	21 56	051	47 40	095	47 16	131	30 04	207	12 34	269	21 18	306	34 40	342
	Alpheratz		◆DENEB		VEGA		◆ARCTURUS		Alioth		POLLUX		◆CAPELLA	
240	22 15	053	48 05	097	47 35	133	29 52	209	62 58	243	20 58	308	34 33	344
242	22 36	055	48 30	099	47 53	135	29 40	211	62 36	246	20 39	310	34 26	346
244	22 56	057	48 54	101	48 10	138	29 27	213	62 13	248	20 20	312	34 20	347
246	23 17	059	49 19	103	48 26	140	29 13	215	61 50	250	20 02	314	34 15	349
248	23 39	061	49 43	105	48 42	142	28 58	217	61 26	252	19 44	316	34 11	351
250	24 01	063	50 07	108	48 57	145	28 43	219	61 02	255	19 27	317	34 07	352
252	24 23	064	50 30	110	49 11	147	28 26	222	60 38	257	19 10	319	34 04	354
254	24 46	066	50 54	112	49 24	149	28 09	224	60 14	259	18 54	321	34 02	356
256	25 09	068	51 17	114	49 36	152	27 52	226	59 49	261	18 39	323	34 00	357
258	25 32	070	51 39	116	49 48	154	27 34	228	59 24	263	18 24	325	34 00	359
260	25 56	072	52 01	118	49 58	157	27 15	230	58 59	265	18 10	327	34 00	001
262	26 20	074	52 23	121	50 08	159	26 56	232	58 35	267	17 57	329	34 00	002
264	26 44	076	52 44	123	50 16	161	26 36	234	58 10	269	17 44	330	34 02	004
266	27 08	078	53 05	125	50 24	164	26 15	236	57 45	271	17 32	332	34 04	006
268	27 32	080	53 25	127	50 30	166	25 54	238	57 20	273	17 21	334	34 07	007
	◆CAPELLA		Alpheratz		◆DENEB		VEGA		◆ARCTURUS		Alioth		POLLUX	
270	34 10	009	27 57	081	53 45	130	50 35	169	25 33	240	56 55	275	17 10	336
272	34 15	011	28 22	083	54 04	132	50 40	171	25 11	242	56 30	277	17 00	338
274	34 20	013	28 47	085	54 22	134	50 43	174	24 48	244	56 05	279	16 51	340
276	34 26	014	29 11	087	54 39	137	50 45	176	24 26	246	55 41	280	16 43	341
278	34 32	016	29 36	089	54 56	139	50 47	179	24 03	249	55 16	282	16 35	343
280	34 39	018	30 01	091	55 12	142	50 47	181	23 39	251	54 52	284	16 29	345
282	34 47	019	30 26	093	55 27	144	50 46	183	23 16	253	54 28	286	16 23	347
284	34 56	021	30 51	095	55 41	146	50 44	186	22 52	255	54 04	288	16 17	349
286	35 05	023	31 16	097	55 55	149	50 41	188	22 28	257	53 40	289	16 13	351
288	35 15	024	31 41	099	56 07	151	50 36	191	22 03	259	53 17	291	16 09	352
290	35 25	026	32 05	101	56 19	154	50 31	193	21 39	261	52 54	293	16 06	354
292	35 37	028	32 30	103	56 29	156	50 25	196	21 14	263	52 31	295	16 04	356
294	35 49	029	32 54	105	56 39	159	50 18	198	20 49	264	52 08	296	16 03	358
296	36 01	031	33 18	107	56 47	161	50 09	201	20 24	266	51 46	298	16 02	000
298	36 14	033	33 42	109	56 54	164	50 00	203	19 59	268	51 24	300	16 03	002
	POLLUX		◆CAPELLA		Alpheratz		◆DENEB		VEGA		◆ARCTURUS		Alioth	
300	16 04	003	36 28	034	34 05	111	57 01	167	49 50	205	19 34	270	51 03	301
302	16 06	005	36 43	036	34 28	113	57 06	169	49 39	208	19 10	272	50 42	303
304	16 08	007	36 58	038	34 51	115	57 10	172	49 27	210	18 45	274	50 21	305
306	16 12	009	37 13	040	35 13	117	57 13	174	49 14	213	18 20	276	50 01	306
308	16 16	011	37 29	041	35 35	120	57 15	177	49 00	215	17 55	278	49 41	308
310	16 21	013	37 46	043	35 56	122	57 16	180	48 45	217	17 30	280	49 22	310
312	16 27	014	38 03	045	36 17	124	57 15	182	48 29	220	17 06	282	49 03	311
314	16 34	016	38 21	046	36 38	126	57 14	185	48 13	222	16 42	284	48 44	313
316	16 41	018	38 40	048	36 58	128	57 11	187	47 56	224	16 17	286	48 26	315
318	16 49	020	38 58	050	37 17	130	57 07	190	47 38	226	15 54	288	48 09	316
320	16 58	022	39 18	052	37 36	132	57 03	193	47 20	229	15 30	290	47 52	318
322	17 08	024	39 38	053	37 54	135	56 57	195	47 01	231	15 07	292	47 35	320
324	17 18	025	39 58	055	38 12	137	56 49	198	46 41	233	14 44	294	47 19	321
326	17 29	027	40 19	057	38 28	139	56 41	200	46 21	235	14 21	295	47 04	323
328	17 41	029	40 40	059	38 44	141	56 32	203	46 00	238	13 59	297	46 49	324
	POLLUX		◆CAPELLA		Hamal		Alpheratz		◆DENEB		VEGA		◆Alioth	
330	17 53	031	41 01	060	28 39	113	39 00	143	56 22	205	45 39	240	46 35	326
332	18 07	033	41 23	062	29 02	115	39 14	146	56 11	208	45 17	242	46 21	328
334	18 21	035	41 45	064	29 24	117	39 28	148	55 59	210	44 55	244	46 08	329
336	18 35	037	42 08	066	29 46	119	39 41	150	55 46	213	44 32	246	45 56	331
338	18 50	038	42 31	068	30 08	121	39 53	152	55 32	215	44 09	248	45 44	332
340	19 06	040	42 54	069	30 29	123	40 04	155	55 17	218	43 46	250	45 33	334
342	19 22	042	43 18	071	30 49	125	40 14	157	55 01	220	43 22	253	45 22	335
344	19 39	044	43 41	073	31 09	128	40 23	159	54 45	223	42 58	255	45 12	337
346	19 57	046	44 05	075	31 29	130	40 32	161	54 27	225	42 34	257	45 02	339
348	20 15	048	44 30	077	31 48	132	40 39	164	54 09	227	42 10	259	44 54	340
350	20 34	049	44 54	079	32 06	134	40 46	166	53 51	230	41 45	261	44 45	342
352	20 53	051	45 18	081	32 24	136	40 51	168	53 31	232	41 20	263	44 38	343
354	21 13	053	45 43	082	32 41	138	40 56	171	53 11	234	40 56	265	44 31	345
356	21 33	055	46 08	084	32 57	140	40 59	173	52 51	237	40 31	267	44 25	346
358	21 54	057	46 33	086	33 13	142	41 02	175	52 30	239	40 06	269	44 19	348

LHA ♈	Hc	Zn	Hc	Zn	Hc	Zn	Hc	Zn	Hc	Zn	Hc	Zn	Hc	Zn
	POLLUX		◆CAPELLA		Hamal		Alpheratz		◆DENEB		VEGA		◆Alioth	
0	21 44	058	46 55	087	34 16	144	42 03	178	52 37	242	39 40	271	43 15	350
2	22 07	060	47 22	089	34 32	146	42 04	180	52 13	244	39 13	273	43 11	351
4	22 31	062	47 49	091	34 46	149	42 03	182	51 48	247	38 46	275	43 07	353
6	22 55	064	48 16	093	35 00	151	42 02	185	51 23	249	38 19	277	43 04	354
8	23 19	066	48 43	095	35 12	153	41 59	187	50 58	251	37 52	279	43 02	356
10	23 44	068	49 10	097	35 24	155	41 55	189	50 32	253	37 26	281	43 00	357
12	24 09	070	49 37	099	35 35	158	41 50	192	50 06	255	36 59	283	42 59	359
14	24 34	072	50 03	101	35 45	160	41 44	194	49 40	257	36 33	285	42 59	000
16	25 00	073	50 30	103	35 53	162	41 37	196	49 14	259	36 07	287	43 00	002
18	25 26	075	50 56	105	36 01	164	41 29	199	48 47	261	35 41	288	43 01	003
20	25 52	077	51 22	107	36 08	167	41 20	201	48 20	263	35 16	290	43 03	005
22	26 19	079	51 47	109	36 14	169	41 09	203	47 54	265	34 51	292	43 06	007
24	26 45	081	52 13	112	36 18	171	40 58	206	47 27	267	34 26	294	43 09	008
26	27 12	083	52 38	114	36 22	173	40 46	208	47 00	269	34 01	296	43 13	010
28	27 39	085	53 02	116	36 24	176	40 33	210	46 33	271	33 37	298	43 18	011
	◆Alioth		POLLUX		◆CAPELLA		Hamal		Alpheratz		◆DENEB		VEGA	
30	43 24	013	28 06	087	53 26	118	36 26	178	40 19	213	46 06	273	33 14	299
32	43 30	014	28 33	089	53 50	120	36 26	180	40 04	215	45 39	275	32 50	301
34	43 37	016	29 00	091	54 13	123	36 26	183	39 48	217	45 12	277	32 27	303
36	43 45	017	29 27	093	54 35	125	36 24	185	39 31	219	44 45	279	32 05	305
38	43 53	019	29 54	095	54 57	127	36 21	187	39 14	222	44 19	281	31 43	307
40	44 02	020	30 21	097	55 18	130	36 17	189	38 56	224	43 52	282	31 22	308
42	44 12	022	30 47	099	55 38	132	36 12	192	38 37	226	43 26	284	31 01	310
44	44 22	023	31 14	101	55 58	134	36 06	194	38 17	228	43 00	286	30 40	312
46	44 33	025	31 41	103	56 17	137	35 59	196	37 56	230	42 34	288	30 21	314
48	44 45	027	32 07	105	56 35	139	35 51	198	37 35	232	42 08	290	30 01	315
50	44 57	028	32 33	107	56 52	142	35 42	201	37 14	235	41 43	292	29 43	317
52	45 10	030	32 58	109	57 08	144	35 32	203	36 51	237	41 18	293	29 25	319
54	45 24	031	33 24	111	57 23	147	35 21	205	36 28	239	40 54	295	29 07	321
56	45 38	033	33 49	113	57 38	149	35 09	207	36 05	241	40 29	297	28 51	322
58	45 53	034	34 14	115	57 51	152	34 56	210	35 41	243	40 06	299	28 34	324
	◆Alioth		REGULUS		POLLUX		◆CAPELLA		Alpheratz		◆DENEB		VEGA	
60	46 09	036	11 14	085	34 38	117	58 03	155	35 17	245	39 42	301	28 19	326
62	46 25	037	11 40	087	35 02	119	58 14	157	34 52	247	39 19	302	28 04	328
64	46 42	039	12 07	089	35 25	121	58 24	160	34 27	249	38 56	304	27 50	329
66	46 59	041	12 34	091	35 48	123	58 33	162	34 02	251	38 34	306	27 37	331
68	47 17	042	13 01	093	36 10	125	58 40	165	33 36	253	38 13	308	27 24	333
70	47 35	044	13 28	095	36 32	127	58 47	168	33 10	255	37 52	309	27 12	335
72	47 54	045	13 55	097	36 53	130	58 52	170	32 44	257	37 31	311	27 01	336
74	48 14	047	14 22	099	37 14	132	58 56	173	32 17	259	37 11	313	26 51	338
76	48 34	049	14 48	101	37 33	134	58 58	176	31 51	261	36 51	314	26 41	340
78	48 54	050	15 15	103	37 52	136	58 59	179	31 24	263	36 32	316	26 32	342
80	49 15	052	15 41	105	38 11	138	58 59	181	30 57	265	36 14	318	26 24	343
82	49 37	053	16 07	107	38 28	141	58 58	184	30 30	267	35 56	320	26 17	345
84	49 58	055	16 33	109	38 45	143	58 56	187	30 03	269	35 39	321	26 10	347
86	50 21	057	16 58	111	39 01	145	58 52	189	29 36	271	35 22	323	26 04	349
88	50 44	058	17 23	113	39 16	147	58 47	192	29 09	273	35 06	325	25 59	350
	◆ARCTURUS		REGULUS		POLLUX		◆CAPELLA		Alpheratz		◆DENEB		VEGA	
90	11 40	053	17 48	115	39 30	150	58 41	195	28 42	275	34 51	326	25 55	352
92	12 02	055	18 12	117	39 43	152	58 33	197	28 15	277	34 36	328	25 52	354
94	12 24	057	18 36	119	39 55	154	58 25	200	27 49	279	34 22	330	25 49	356
96	12 47	059	18 59	121	40 07	156	58 15	203	27 22	281	34 09	331	25 48	357
98	13 10	061	19 22	123	40 17	159	58 04	205	26 56	283	33 56	333	25 47	359
100	13 34	063	19 45	125	40 26	161	57 52	208	26 29	285	33 44	335	25 47	001
102	13 58	065	20 07	127	40 35	163	57 39	210	26 03	287	33 33	336	25 48	002
104	14 23	066	20 28	129	40 42	166	57 25	213	25 38	288	33 23	338	25 49	004
106	14 48	068	20 48	131	40 48	168	57 10	215	25 12	290	33 13	340	25 51	006
108	15 13	070	21 08	133	40 53	170	56 53	218	24 47	292	33 04	341	25 55	008
110	15 39	072	21 28	135	40 57	173	56 36	220	24 22	294	32 56	343	25 59	009
112	16 05	074	21 46	137	41 00	175	56 18	223	23 58	296	32 48	345	26 03	011
114	16 31	076	22 04	139	41 02	177	56 00	225	23 34	298	32 42	346	26 09	013
116	16 57	078	22 22	142	41 02	180	55 40	228	23 10	300	32 36	348	26 15	015
118	17 23	080	22 38	144	41 02	182	55 20	230	22 47	302	32 31	350	26 23	016
	◆VEGA		ARCTURUS		◆REGULUS		POLLUX		◆CAPELLA		Alpheratz		DENEB	
120	26 31	018	17 50	082	22 54	146	41 00	184	54 59	232	22 24	303	32 26	351
122	26 39	020	18 17	084	23 08	148	40 58	187	54 37	235	22 02	305	32 23	353
124	26 49	021	18 44	086	23 22	150	40 54	189	54 14	237	21 40	307	32 20	355
126	26 59	023	19 11	088	23 35	152	40 49	191	53 52	239	21 19	309	32 18	356
128	27 10	025	19 38	090	23 48	154	40 43	194	53 28	242	20 58	311	32 16	358
130	27 22	027	20 05	091	23 59	156	40 36	196	53 04	244	20 38	313	32 16	000
132	27 34	028	20 32	093	24 09	158	40 28	198	52 40	246	20 18	314	32 16	001
134	27 48	030	20 59	095	24 19	161	40 19	201	52 15	248	19 59	316	32 17	003
136	28 01	032	21 25	097	24 27	163	40 09	203	51 49	250	19 41	318	32 19	005
138	28 16	034	21 52	099	24 35	165	39 58	205	51 24	252	19 23	320	32 22	006
140	28 31	035	22 19	101	24 41	167	39 46	208	50 58	255	19 06	322	32 25	008
142	28 47	037	22 45	103	24 47	169	39 33	210	50 32	257	18 50	324	32 29	010
144	29 04	039	23 11	105	24 51	171	39 20	212	50 05	259	18 34	325	32 34	011
146	29 21	041	23 37	107	24 55	173	39 05	214	49 39	261	18 19	327	32 40	013
148	29 39	042	24 03	109	24 57	176	38 49	217	49 12	263	18 05	329	32 46	015
	DENEB		◆VEGA		ARCTURUS		◆REGULUS		POLLUX		◆CAPELLA		Alpheratz	
150	32 53	016	29 58	044	24 28	111	24 59	178	38 33	219	48 45	265	17 51	331
152	33 01	018	30 17	046	24 53	113	25 00	180	38 15	221	48 18	267	17 39	333
154	33 10	020	30 37	048	25 18	115	24 59	182	37 57	223	47 51	269	17 27	334
156	33 20	021	30 57	050	25 42	117	24 58	184	37 38	225	47 24	271	17 15	336
158	33 30	023	31 18	051	26 05	120	24 55	186	37 19	228	46 58	273	17 05	338
160	33 41	025	31 39	053	26 29	122	24 51	189	36 58	230	46 31	275	16 55	340
162	33 53	026	32 01	055	26 51	124	24 47	191	36 37	232	46 04	276	16 46	342
164	34 05	028	32 23	057	27 14	126	24 41	193	36 16	234	45 37	278	16 38	344
166	34 18	030	32 46	058	27 35	128	24 35	195	35 54	236	45 10	280	16 31	345
168	34 32	031	33 09	060	27 56	130	24 27	197	35 31	238	44 44	282	16 25	347
170	34 46	033	33 33	062	28 17	132	24 19	199	35 08	240	44 18	284	16 19	349
172	35 01	035	33 57	064	28 36	134	24 09	201	34 44	243	43 51	286	16 15	351
174	35 17	037	34 21	066	28 55	136	23 59	204	34 20	245	43 26	288	16 11	353
176	35 33	038	34 46	068	29 14	138	23 48	206	33 55	247	43 00	289	16 08	355
178	35 50	040	35 11	069	29 31	141	23 36	208	33 30	249	42 35	291	16 06	356

LAT 77°N

LHA ♈	Hc	Zn	Hc	Zn	Hc	Zn	Hc	Zn	Hc	Zn	Hc	Zn	Hc	Zn
	DENEB		◆VEGA		ARCTURUS		◆REGULUS		POLLUX		◆CAPELLA		Alpheratz	
180	36 08	042	35 37	071	29 48	143	23 23	210	33 05	251	42 10	293	16 04	358
182	36 26	043	36 02	073	30 04	145	23 09	212	32 39	253	41 45	295	16 04	000
184	36 45	045	36 28	075	30 19	147	22 54	214	32 13	255	41 21	297	16 04	002
186	37 05	047	36 54	077	30 33	149	22 38	216	31 47	257	40 57	298	16 05	004
188	37 24	048	37 21	079	30 47	151	22 22	218	31 21	259	40 33	300	16 08	005
190	37 45	050	37 47	081	30 59	154	22 05	221	30 54	261	40 10	302	16 11	007
192	38 06	052	38 14	082	31 11	156	21 47	223	30 27	263	39 47	304	16 14	009
194	38 27	054	38 41	084	31 21	158	21 28	225	30 01	265	39 25	305	16 19	011
196	38 49	055	39 08	086	31 31	160	21 09	227	29 34	267	39 03	307	16 25	013
198	39 12	057	39 35	088	31 40	162	20 49	229	29 07	269	38 42	309	16 31	015
200	39 35	059	40 02	090	31 47	165	20 28	231	28 40	271	38 21	311	16 38	016
202	39 58	061	40 29	092	31 54	167	20 07	233	28 13	273	38 01	312	16 46	018
204	40 22	062	40 56	094	32 00	169	19 45	235	27 46	275	37 41	314	16 55	020
206	40 46	064	41 22	096	32 04	171	19 23	237	27 19	277	37 22	316	17 05	022
208	41 10	066	41 49	098	32 08	173	19 00	239	26 52	279	37 04	317	17 15	024
	Alpheratz		◆DENEB		VEGA		◆ARCTURUS		REGULUS		POLLUX		◆CAPELLA	
210	17 26	025	41 35	068	42 16	100	32 11	176	18 37	241	26 26	280	36 46	319
212	17 38	027	42 00	070	42 42	102	32 12	178	18 13	243	25 59	282	36 28	321
214	17 51	029	42 26	071	43 09	104	32 12	180	17 49	245	25 33	284	36 11	322
216	18 04	031	42 52	073	43 35	106	32 12	182	17 24	247	25 07	286	35 55	324
218	18 19	033	43 18	075	44 01	108	32 10	185	16 59	249	24 41	288	35 40	326
220	18 34	035	43 44	077	44 26	110	32 08	187	16 33	251	24 15	290	35 25	327
222	18 49	036	44 10	079	44 51	112	32 04	189	16 08	253	23 50	292	35 11	329
224	19 06	038	44 37	081	45 16	115	31 59	191	15 42	255	23 25	294	34 57	331
226	19 23	040	45 03	083	45 40	117	31 53	194	15 16	257	23 01	296	34 44	333
228	19 40	042	45 30	084	46 04	119	31 46	196	14 49	259	22 37	297	34 32	334
230	19 59	044	45 57	086	46 27	121	31 39	198	14 23	261	22 13	299	34 21	336
232	20 18	046	46 24	088	46 50	123	31 30	200	13 56	263	21 49	301	34 10	338
234	20 37	047	46 51	090	47 13	125	31 20	202	13 29	265	21 27	303	34 00	339
236	20 57	049	47 18	092	47 34	128	31 09	205	13 02	267	21 04	305	33 51	341
238	21 18	051	47 45	094	47 55	130	30 58	207	12 35	269	20 42	307	33 43	343
	Alpheratz		◆DENEB		VEGA		◆ARCTURUS		Alioth		POLLUX		◆CAPELLA	
240	21 39	053	48 12	096	48 16	132	30 45	209	63 24	245	20 21	308	33 35	344
242	22 01	055	48 39	098	48 35	135	30 31	211	63 00	247	20 00	310	33 28	346
244	22 23	057	49 05	100	48 54	137	30 17	213	62 35	250	19 40	312	33 22	348
246	22 46	058	49 32	102	49 12	139	30 02	215	62 09	252	19 20	314	33 16	349
248	23 09	060	49 58	104	49 29	142	29 46	218	61 43	254	19 01	316	33 12	351
250	23 33	062	50 24	106	49 46	144	29 29	220	61 17	256	18 42	318	33 08	352
252	23 57	064	50 50	109	50 01	146	29 11	222	60 51	258	18 25	320	33 04	354
254	24 22	066	51 15	111	50 16	149	28 53	224	60 24	260	18 07	321	33 02	356
256	24 46	068	51 40	113	50 29	151	28 34	226	59 58	263	17 51	323	33 01	357
258	25 11	070	52 05	115	50 42	154	28 14	228	59 31	265	17 35	325	33 00	359
260	25 37	071	52 29	117	50 53	156	27 53	230	59 04	267	17 20	327	33 00	001
262	26 03	073	52 53	119	51 04	158	27 32	232	58 37	269	17 06	329	33 00	002
264	26 29	075	53 16	122	51 13	161	27 11	235	58 10	271	16 52	331	33 02	004
266	26 55	077	53 39	124	51 21	163	26 48	237	57 43	272	16 39	332	33 04	006
268	27 21	079	54 01	126	51 28	166	26 25	239	57 16	274	16 27	334	33 07	007
	◆CAPELLA		Alpheratz		◆DENEB		VEGA		◆ARCTURUS		Alioth		POLLUX	
270	33 11	009	27 48	081	54 23	129	51 34	168	26 02	241	56 49	276	16 15	336
272	33 16	011	28 15	083	54 43	131	51 39	171	25 38	243	56 22	278	16 05	338
274	33 21	012	28 41	085	55 03	133	51 43	173	25 14	245	55 56	280	15 55	340
276	33 27	014	29 08	087	55 23	136	51 45	176	24 49	247	55 29	282	15 46	341
278	33 34	016	29 35	089	55 41	138	51 47	178	24 24	249	55 03	284	15 38	343
280	33 42	017	30 02	091	55 59	141	51 47	181	23 59	251	54 37	285	15 31	345
282	33 50	019	30 29	093	56 15	143	51 46	184	23 33	253	54 11	287	15 24	347
284	34 00	021	30 56	095	56 31	146	51 43	186	23 07	255	53 45	289	15 18	349
286	34 09	022	31 23	096	56 46	148	51 40	189	22 41	257	53 20	291	15 14	351
288	34 20	024	31 50	098	57 00	151	51 35	191	22 15	259	52 55	292	15 10	352
290	34 31	026	32 16	100	57 12	153	51 29	194	21 48	261	52 30	294	15 07	354
292	34 43	027	32 43	102	57 24	156	51 23	196	21 22	263	52 05	296	15 04	356
294	34 56	029	33 09	104	57 34	158	51 15	199	20 55	265	51 41	297	15 03	358
296	35 10	031	33 35	107	57 44	161	51 05	201	20 28	267	51 17	299	15 02	000
298	35 24	032	34 01	109	57 52	164	50 55	203	20 01	269	50 54	301	15 03	002
	POLLUX		◆CAPELLA		Alpheratz		◆DENEB		VEGA		◆ARCTURUS		Alioth	
300	15 04	003	35 39	034	34 26	111	57 59	166	50 44	206	19 34	271	50 31	303
302	15 06	005	35 54	036	34 51	113	58 05	169	50 32	208	19 07	273	50 09	304
304	15 09	007	36 10	037	35 16	115	58 09	172	50 18	211	18 40	275	49 46	306
306	15 13	009	36 27	039	35 40	117	58 13	174	50 04	213	18 13	277	49 25	307
308	15 17	011	36 44	041	36 04	119	58 15	177	49 49	216	17 46	278	49 04	309
310	15 23	013	37 02	043	36 28	121	58 16	180	49 33	218	17 20	280	48 43	311
312	15 29	014	37 21	044	36 51	123	58 15	182	49 15	220	16 53	282	48 23	312
314	15 36	016	37 40	046	37 13	125	58 14	185	48 58	223	16 27	284	48 03	314
316	15 44	018	37 59	048	37 35	127	58 11	188	48 39	225	16 01	286	47 44	315
318	15 53	020	38 20	049	37 56	130	58 06	190	48 19	227	15 35	288	47 25	317
320	16 02	022	38 40	051	38 16	132	58 01	193	47 59	230	15 10	290	47 07	319
322	16 13	024	39 02	053	38 36	134	57 54	196	47 38	232	14 44	292	46 49	320
324	16 24	025	39 23	055	38 55	136	57 47	198	47 17	234	14 20	294	46 32	322
326	16 36	027	39 46	056	39 13	138	57 38	201	46 55	236	13 55	296	46 16	323
328	16 49	029	40 08	058	39 31	141	57 27	203	46 32	239	13 31	298	46 00	325
	POLLUX		◆CAPELLA		Hamal		Alpheratz		◆DENEB		VEGA		◆Alioth	
330	17 02	031	40 31	060	29 02	112	39 48	143	57 16	206	46 09	241	45 45	327
332	17 16	033	40 55	061	29 27	115	40 03	145	57 04	209	45 45	243	45 30	328
334	17 31	035	41 19	063	29 51	117	40 18	147	56 50	211	45 21	245	45 17	330
336	17 47	036	41 43	065	30 15	119	40 32	150	56 36	214	44 56	247	45 03	331
338	18 03	038	42 08	067	30 39	121	40 46	152	56 20	216	44 31	249	44 51	333
340	18 20	040	42 33	069	31 02	123	40 58	154	56 04	219	44 05	251	44 39	334
342	18 38	042	42 58	070	31 24	125	41 09	157	55 47	221	43 40	253	44 27	336
344	18 56	044	43 24	072	31 46	127	41 19	159	55 29	224	43 14	255	44 16	337
346	19 15	046	43 49	074	32 07	129	41 29	161	55 10	226	42 47	258	44 06	339
348	19 35	047	44 15	076	32 28	131	41 37	164	54 50	228	42 21	260	43 57	340
350	19 55	049	44 42	078	32 48	133	41 44	166	54 29	231	41 54	262	43 48	342
352	20 16	051	45 08	080	33 07	136	41 50	168	54 08	233	41 28	264	43 40	344
354	20 37	053	45 35	081	33 25	138	41 55	171	53 46	235	41 01	266	43 33	345
356	20 59	055	46 02	083	33 43	140	41 59	173	53 24	238	40 34	268	43 27	347
358	21 21	057	46 28	085	34 00	142	42 02	175	53 01	240	40 07	270	43 21	348

LHA ♈	Hc	Zn	Hc	Zn	Hc	Zn	Hc	Zn	Hc	Zn	Hc	Zn	Hc	Zn
	POLLUX		◆CAPELLA		Hamal		Alpheratz		◆DENEB		VEGA		◆Alioth	
0	21 12	058	46 52	086	35 05	144	43 03	178	53 04	243	39 38	272	42 16	350
2	21 37	060	47 21	088	35 22	146	43 04	180	52 38	246	39 09	274	42 12	351
4	22 02	062	47 50	090	35 37	148	43 03	182	52 12	248	38 40	276	42 08	353
6	22 28	064	48 19	092	35 52	151	43 01	185	51 45	250	38 11	278	42 05	354
8	22 54	066	48 48	094	36 06	153	42 58	187	51 17	252	37 42	280	42 02	356
10	23 21	067	49 17	096	36 19	155	42 54	190	50 49	254	37 14	282	42 00	357
12	23 48	069	49 46	098	36 30	157	42 49	192	50 21	256	36 46	284	41 59	359
14	24 15	071	50 14	100	36 41	160	42 42	194	49 53	258	36 18	285	41 59	000
16	24 43	073	50 43	102	36 51	162	42 34	197	49 24	260	35 50	287	42 00	002
18	25 11	075	51 11	104	36 59	164	42 26	199	48 56	262	35 22	289	42 01	003
20	25 39	077	51 39	106	37 06	166	42 16	201	48 27	264	34 55	291	42 03	005
22	26 07	079	52 07	108	37 13	169	42 04	204	47 58	266	34 28	293	42 06	006
24	26 36	081	52 34	110	37 18	171	41 52	206	47 29	268	34 01	295	42 10	008
26	27 05	082	53 01	113	37 22	173	41 39	208	47 00	270	33 35	296	42 14	009
28	27 33	084	53 28	115	37 24	176	41 25	211	46 31	272	33 09	298	42 19	011
	◆Alioth		POLLUX		◆CAPELLA		Hamal		Alpheratz		◆DENEB		VEGA	
30	42 25	012	28 02	086	53 54	117	37 26	178	41 09	213	46 02	274	32 44	300
32	42 32	014	28 31	088	54 20	119	37 26	180	40 53	215	45 33	276	32 19	302
34	42 39	015	29 00	090	54 45	122	37 26	183	40 36	218	45 04	278	31 55	304
36	42 47	017	29 29	092	55 09	124	37 24	185	40 18	220	44 36	280	31 31	305
38	42 56	019	29 58	094	55 33	126	37 21	187	39 59	222	44 07	282	31 07	307
40	43 06	020	30 27	096	55 56	129	37 16	190	39 39	224	43 39	283	30 44	309
42	43 16	022	30 56	098	56 18	131	37 11	192	39 18	227	43 11	285	30 22	311
44	43 27	023	31 25	100	56 40	133	37 04	194	38 57	229	42 43	287	30 00	312
46	43 39	025	31 53	102	57 00	136	36 57	196	38 34	231	42 15	289	29 39	314
48	43 51	026	32 22	104	57 20	138	36 48	199	38 12	233	41 48	291	29 19	316
50	44 04	028	32 50	106	57 39	141	36 38	201	37 48	235	41 21	292	28 59	318
52	44 18	029	33 17	108	57 57	143	36 27	203	37 24	237	40 54	294	28 39	319
54	44 33	031	33 45	110	58 13	146	36 15	206	36 59	240	40 28	296	28 21	321
56	44 48	032	34 12	112	58 29	149	36 02	208	36 34	242	40 02	298	28 03	323
58	45 04	034	34 39	114	58 44	151	35 48	210	36 08	244	39 36	299	27 46	325
	◆Alioth		REGULUS		POLLUX		◆CAPELLA		Alpheratz		◆DENEB		VEGA	
60	45 20	035	11 08	085	35 05	116	58 57	154	35 42	246	39 11	301	27 29	326
62	45 37	037	11 37	087	35 31	118	59 09	157	35 15	248	38 47	303	27 13	328
64	45 55	038	12 06	089	35 56	120	59 20	159	34 48	250	38 23	305	26 58	330
66	46 13	040	12 35	091	36 21	123	59 30	162	34 21	252	37 59	306	26 44	331
68	46 32	041	13 05	093	36 45	125	59 38	165	33 53	254	37 36	308	26 31	333
70	46 52	043	13 33	095	37 08	127	59 45	167	33 25	256	37 13	310	26 18	335
72	47 12	045	14 02	097	37 31	129	59 51	170	32 56	258	36 51	312	26 06	337
74	47 32	046	14 31	099	37 53	131	59 55	173	32 28	260	36 30	313	25 55	338
76	47 54	048	15 00	101	38 15	133	59 58	176	31 59	262	36 09	315	25 45	340
78	48 15	049	15 28	103	38 36	136	59 59	179	31 30	264	35 49	317	25 35	342
80	48 38	051	15 56	105	38 55	138	59 59	181	31 02	266	35 29	318	25 27	344
82	49 00	053	16 24	107	39 15	140	59 58	184	30 33	268	35 10	320	25 19	345
84	49 24	054	16 52	109	39 33	142	59 55	187	30 04	270	34 52	322	25 12	347
86	49 47	056	17 19	111	39 50	145	59 51	190	29 34	272	34 34	323	25 06	349
88	50 12	057	17 46	113	40 06	147	59 46	192	29 05	274	34 17	325	25 00	350
	◆ARCTURUS		REGULUS		POLLUX		◆CAPELLA		Alpheratz		◆DENEB		VEGA	
90	11 04	053	18 13	115	40 22	149	59 39	195	28 37	276	34 01	327	24 56	352
92	11 27	055	18 39	117	40 36	151	59 31	198	28 08	278	33 45	328	24 52	354
94	11 52	057	19 05	119	40 49	154	59 21	201	27 39	280	33 30	330	24 50	356
96	12 16	059	19 30	121	41 02	156	59 10	203	27 10	281	33 16	332	24 48	357
98	12 41	061	19 55	123	41 13	158	58 58	206	26 42	283	33 03	333	24 47	359
100	13 07	062	20 19	125	41 23	161	58 45	209	26 14	285	32 50	335	24 47	001
102	13 33	064	20 43	127	41 32	163	58 30	211	25 46	287	32 38	337	24 48	002
104	13 59	066	21 06	129	41 40	165	58 15	214	25 19	289	32 27	338	24 49	004
106	14 26	068	21 28	131	41 47	168	57 58	216	24 51	291	32 17	340	24 52	006
108	14 53	070	21 49	133	41 52	170	57 40	219	24 24	293	32 07	342	24 55	008
110	15 20	072	22 10	135	41 56	173	57 22	221	23 58	295	31 59	343	24 59	009
112	15 48	074	22 30	137	42 00	175	57 02	224	23 31	296	31 51	345	25 05	011
114	16 16	076	22 50	139	42 02	177	56 41	226	23 06	298	31 43	347	25 10	013
116	16 44	078	23 08	141	42 02	180	56 20	229	22 40	300	31 37	348	25 17	014
118	17 13	080	23 26	143	42 02	182	55 58	231	22 15	302	31 32	350	25 25	016
	◆VEGA		ARCTURUS		◆REGULUS		POLLUX		◆CAPELLA		Alpheratz		DENEB	
120	25 33	018	17 41	081	23 43	145	42 00	184	55 35	234	21 51	304	31 27	352
122	25 43	020	18 10	083	23 59	148	41 57	187	55 11	236	21 27	306	31 23	353
124	25 53	021	18 39	085	24 14	150	41 53	189	54 47	238	21 04	307	31 20	355
126	26 04	023	19 08	087	24 28	152	41 48	192	54 22	241	20 41	309	31 18	356
128	26 16	025	19 37	089	24 42	154	41 42	194	53 56	243	20 19	311	31 16	358
130	26 28	026	20 06	091	24 54	156	41 34	196	53 30	245	19 57	313	31 16	000
132	26 42	028	20 35	093	25 05	158	41 25	199	53 03	247	19 36	315	31 16	001
134	26 56	030	21 04	095	25 15	160	41 16	201	52 36	249	19 16	316	31 17	003
136	27 11	032	21 33	097	25 24	163	41 05	203	52 09	252	18 56	318	31 19	005
138	27 26	033	22 02	099	25 33	165	40 53	206	51 41	254	18 37	320	31 22	006
140	27 42	035	22 30	101	25 40	167	40 40	208	51 13	256	18 19	322	31 26	008
142	28 00	037	22 59	103	25 46	169	40 25	210	50 45	258	18 01	324	31 30	010
144	28 17	039	23 27	105	25 51	171	40 10	213	50 17	260	17 45	326	31 35	011
146	28 36	040	23 55	107	25 55	173	39 54	215	49 48	262	17 29	327	31 41	013
148	28 55	042	24 22	109	25 57	176	39 37	217	49 19	264	17 13	329	31 48	015
	DENEB		◆VEGA		ARCTURUS		◆REGULUS		POLLUX		◆CAPELLA		Alpheratz	
150	31 56	016	29 15	044	24 50	111	25 59	178	39 19	219	48 50	266	16 59	331
152	32 04	018	29 35	046	25 17	113	26 00	180	39 00	222	48 21	268	16 45	333
154	32 14	020	29 56	047	25 43	115	25 59	182	38 41	224	47 52	270	16 32	335
156	32 24	021	30 18	049	26 09	117	25 57	184	38 20	226	47 23	272	16 20	336
158	32 35	023	30 40	051	26 35	119	25 55	186	37 59	228	46 54	274	16 09	338
160	32 46	024	31 03	053	27 00	121	25 51	189	37 37	230	46 25	276	15 59	340
162	32 59	026	31 26	054	27 24	123	25 46	191	37 14	233	45 56	278	15 49	342
164	33 12	028	31 50	056	27 48	125	25 40	193	36 51	235	45 28	279	15 41	344
166	33 26	029	32 14	058	28 12	127	25 33	195	36 27	237	44 59	281	15 33	345
168	33 40	031	32 39	060	28 35	129	25 25	197	36 02	239	44 31	283	15 26	347
170	33 56	033	33 04	061	28 57	132	25 15	199	35 37	241	44 03	285	15 20	349
172	34 12	034	33 30	063	29 18	134	25 05	202	35 11	243	43 35	287	15 15	351
174	34 29	036	33 56	065	29 39	136	24 54	204	34 45	245	43 07	289	15 11	353
176	34 46	038	34 23	067	29 58	138	24 42	206	34 19	247	42 40	290	15 08	355
178	35 04	039	34 50	069	30 17	140	24 29	208	33 52	249	42 13	292	15 06	356

LAT 76°N

LHA ♈	Hc	Zn	Hc	Zn	Hc	Zn	Hc	Zn	Hc	Zn	Hc	Zn	Hc	Zn
	DENEB		◆VEGA		ARCTURUS		◆REGULUS		POLLUX		◆CAPELLA		Alpheratz	
180	35 23	041	35 17	071	30 36	142	24 15	210	33 24	251	41 46	294	15 04	358
182	35 43	043	35 44	072	30 53	144	23 59	212	32 57	254	41 19	296	15 04	000
184	36 03	045	36 12	074	31 09	147	23 43	214	32 29	256	40 53	297	15 04	002
186	36 23	046	36 40	076	31 25	149	23 27	217	32 00	258	40 28	299	15 06	004
188	36 45	048	37 08	078	31 39	151	23 09	219	31 32	260	40 03	301	15 08	005
190	37 06	050	37 37	080	31 53	153	22 50	221	31 03	262	39 38	303	15 11	007
192	37 29	051	38 06	082	32 05	155	22 31	223	30 35	264	39 14	304	15 15	009
194	37 52	053	38 34	084	32 17	158	22 11	225	30 06	265	38 50	306	15 20	011
196	38 15	055	39 03	085	32 27	160	21 50	227	29 37	267	38 27	308	15 26	013
198	38 39	056	39 32	087	32 37	162	21 28	229	29 08	269	38 04	309	15 33	014
200	39 04	058	40 01	089	32 45	164	21 06	231	28 39	271	37 42	311	15 41	016
202	39 28	060	40 30	091	32 53	167	20 43	233	28 10	273	37 20	313	15 49	018
204	39 54	062	40 59	093	32 59	169	20 20	235	27 41	275	36 59	315	15 59	020
206	40 20	063	41 28	095	33 04	171	19 55	237	27 12	277	36 39	316	16 09	022
208	40 46	065	41 57	097	33 08	173	19 31	239	26 43	279	36 19	318	16 20	024
	Alpheratz		◆DENEB		VEGA		◆ARCTURUS		REGULUS		POLLUX		◆CAPELLA	
210	16 32	025	41 12	067	42 26	099	33 10	176	19 06	241	26 14	281	36 00	320
212	16 45	027	41 39	069	42 54	101	33 12	178	18 40	243	25 46	283	35 42	321
214	16 58	029	42 06	071	43 23	103	33 12	180	18 14	245	25 18	285	35 24	323
216	17 13	031	42 34	072	43 51	105	33 12	182	17 47	247	24 50	287	35 07	325
218	17 28	033	43 02	074	44 19	107	33 10	185	17 20	249	24 22	288	34 50	326
220	17 44	034	43 30	076	44 46	109	33 07	187	16 53	251	23 55	290	34 34	328
222	18 01	036	43 58	078	45 14	112	33 03	189	16 25	253	23 28	292	34 19	330
224	18 18	038	44 26	080	45 40	114	32 58	191	15 57	255	23 01	294	34 05	331
226	18 37	040	44 55	082	46 07	116	32 52	194	15 29	257	22 35	296	33 51	333
228	18 56	042	45 24	083	46 33	118	32 44	196	15 00	259	22 09	298	33 38	334
230	19 15	043	45 53	085	46 58	120	32 36	198	14 32	261	21 43	300	33 26	336
232	19 36	045	46 22	087	47 23	122	32 26	200	14 03	263	21 18	301	33 15	338
234	19 56	047	46 51	089	47 47	125	32 15	203	13 34	265	20 54	303	33 04	339
236	20 18	049	47 20	091	48 11	127	32 04	205	13 05	267	20 30	305	32 54	341
238	20 40	051	47 49	093	48 33	129	31 51	207	12 36	269	20 06	307	32 45	343
	Alpheratz		◆DENEB		VEGA		◆ARCTURUS		Alioth		POLLUX		◆CAPELLA	
240	21 03	053	48 18	095	48 56	131	31 37	209	63 49	247	19 43	309	32 37	344
242	21 26	054	48 47	097	49 17	134	31 23	211	63 22	249	19 21	311	32 30	346
244	21 50	056	49 15	099	49 38	136	31 07	214	62 55	252	18 59	312	32 23	348
246	22 15	058	49 44	101	49 57	138	30 51	216	62 27	254	18 38	314	32 17	349
248	22 39	060	50 12	103	50 16	141	30 33	218	61 59	256	18 18	316	32 12	351
250	23 05	062	50 41	105	50 34	143	30 15	220	61 30	258	17 58	318	32 08	353
252	23 31	064	51 08	107	50 51	146	29 56	222	61 02	260	17 39	320	32 05	354
254	23 57	065	51 36	109	51 07	148	29 36	224	60 33	262	17 20	322	32 02	356
256	24 23	067	52 03	112	51 21	151	29 15	227	60 04	264	17 03	323	32 01	357
258	24 50	069	52 30	114	51 35	153	28 54	229	59 36	266	16 46	325	32 00	359
260	25 18	071	52 56	116	51 48	156	28 31	231	59 07	268	16 30	327	32 00	001
262	25 45	073	53 22	118	51 59	158	28 09	233	58 38	270	16 14	329	32 00	002
264	26 13	075	53 47	121	52 10	161	27 45	235	58 08	272	16 00	331	32 02	004
266	26 41	077	54 12	123	52 19	163	27 21	237	57 40	274	15 46	332	32 05	006
268	27 10	079	54 36	125	52 26	166	26 56	239	57 11	276	15 33	334	32 08	007
	◆CAPELLA		Alpheratz		◆DENEB		VEGA		◆ARCTURUS		Alioth		POLLUX	
270	32 12	009	27 38	080	55 00	127	52 33	168	26 31	241	56 42	278	15 21	336
272	32 17	011	28 07	082	55 22	130	52 38	171	26 06	243	56 13	280	15 09	338
274	32 23	012	28 36	084	55 44	132	52 42	173	25 39	245	55 45	281	14 59	340
276	32 29	014	29 05	086	56 05	135	52 45	176	25 13	247	55 16	283	14 49	342
278	32 37	016	29 34	088	56 25	137	52 46	178	24 46	249	54 48	285	14 40	343
280	32 45	017	30 03	090	56 45	140	52 47	181	24 18	251	54 20	287	14 33	345
282	32 54	019	30 32	092	57 03	142	52 45	184	23 51	253	53 52	288	14 26	347
284	33 03	020	31 01	094	57 20	145	52 43	186	23 23	255	53 25	290	14 20	349
286	33 14	022	31 30	096	57 37	147	52 39	189	22 55	257	52 58	292	14 14	351
288	33 25	024	31 58	098	57 52	150	52 34	191	22 26	259	52 31	294	14 10	352
290	33 37	025	32 27	100	58 06	152	52 28	194	21 58	261	52 05	295	14 07	354
292	33 50	027	32 56	102	58 19	155	52 20	196	21 29	263	51 39	297	14 04	356
294	34 04	029	33 24	104	58 30	158	52 11	199	21 00	265	51 13	299	14 03	358
296	34 18	030	33 52	106	58 40	160	52 01	201	20 31	267	50 48	300	14 02	000
298	34 33	032	34 20	108	58 50	163	51 50	204	20 02	269	50 23	302	14 03	002
	POLLUX		◆CAPELLA		Alpheratz		◆DENEB		VEGA		◆ARCTURUS		Alioth	
300	14 04	003	34 49	034	34 47	110	58 57	166	51 38	206	19 33	271	49 58	304
302	14 06	005	35 05	035	35 14	112	59 04	169	51 24	209	19 04	273	49 34	305
304	14 09	007	35 22	037	35 41	114	59 09	171	51 10	211	18 35	275	49 11	307
306	14 13	009	35 40	039	36 07	116	59 12	174	50 54	214	18 06	277	48 48	308
308	14 18	011	35 59	040	36 33	118	59 15	177	50 37	216	17 37	279	48 25	310
310	14 24	013	36 18	042	36 58	120	59 16	180	50 20	219	17 09	281	48 03	312
312	14 31	014	36 37	044	37 23	123	59 15	182	50 01	221	16 40	283	47 42	313
314	14 38	016	36 58	045	37 47	125	59 13	185	49 41	223	16 12	285	47 21	315
316	14 47	018	37 19	047	38 11	127	59 10	188	49 21	226	15 44	286	47 01	316
318	14 56	020	37 40	049	38 34	129	59 05	191	49 00	228	15 16	288	46 41	318
320	15 07	022	38 02	050	38 56	131	58 59	193	48 38	230	14 49	290	46 22	319
322	15 18	023	38 25	052	39 17	133	58 52	196	48 15	233	14 22	292	46 03	321
324	15 30	025	38 48	054	39 38	136	58 43	199	47 52	235	13 55	294	45 45	322
326	15 42	027	39 12	056	39 58	138	58 34	201	47 28	237	13 29	296	45 28	324
328	15 56	029	39 36	057	40 17	140	58 22	204	47 03	239	13 03	298	45 11	326
	POLLUX		◆CAPELLA		Hamal		Alpheratz		◆DENEB		VEGA		◆Alioth	
330	16 10	031	40 01	059	29 25	112	40 35	142	58 10	207	46 38	242	44 55	327
332	16 26	033	40 26	061	29 52	114	40 52	145	57 56	209	46 12	244	44 39	329
334	16 42	034	40 51	062	30 18	116	41 09	147	57 41	212	45 45	246	44 25	330
336	16 58	036	41 17	064	30 44	118	41 24	149	57 26	215	45 19	248	44 11	332
338	17 16	038	41 44	066	31 09	120	41 39	152	57 09	217	44 52	250	43 57	333
340	17 34	040	42 10	068	31 34	122	41 52	154	56 51	220	44 24	252	43 44	335
342	17 53	042	42 37	070	31 58	124	42 04	156	56 32	222	43 56	254	43 32	336
344	18 13	043	43 05	071	32 22	127	42 15	159	56 12	225	43 28	256	43 21	338
346	18 33	045	43 32	073	32 45	129	42 25	161	55 51	227	43 00	258	43 10	339
348	18 54	047	44 00	075	33 07	131	42 34	163	55 29	229	42 31	260	43 00	341
350	19 16	049	44 28	077	33 29	133	42 42	166	55 07	232	42 03	262	42 51	342
352	19 38	051	44 57	079	33 50	135	42 49	168	54 44	234	41 34	264	42 43	344
354	20 00	053	45 25	080	34 10	137	42 54	170	54 20	236	41 05	266	42 35	345
356	20 24	054	45 54	082	34 29	139	42 58	173	53 55	239	40 36	268	42 28	347
358	20 48	056	46 23	084	34 47	142	43 01	175	53 30	241	40 07	270	42 22	348

LAT 75°N

LHA ♈	Hc Zn	Hc Zn	Hc Zn	Hc Zn	Hc Zn	Hc Zn	Hc Zn
	◆POLLUX	CAPELLA	Hamal	◆Alpheratz	DENEB	◆VEGA	Alioth
0	20 40 058	46 47 085	35 53 143	44 03 178	53 31 244	39 35 273	41 17 350
2	21 07 060	47 18 087	36 11 146	44 04 180	53 03 247	39 04 275	41 12 352
4	21 34 061	47 49 089	36 28 148	44 03 182	52 34 249	38 33 277	41 08 353
6	22 01 063	48 20 091	36 44 150	44 01 185	52 05 251	38 02 279	41 05 354
8	22 29 065	48 51 093	36 59 152	43 58 187	51 35 253	37 32 281	41 02 356
10	22 58 067	49 22 095	37 13 155	43 53 190	51 05 255	37 01 282	41 00 357
12	23 26 069	49 53 097	37 26 157	43 47 192	50 35 257	36 31 284	40 59 359
14	23 56 071	50 24 099	37 37 159	43 40 195	50 05 259	36 01 286	40 59 000
16	24 25 073	50 55 101	37 48 162	43 32 197	49 34 261	35 31 288	41 00 002
18	24 55 074	51 25 103	37 57 164	43 22 199	49 03 263	35 02 290	41 01 003
20	25 25 076	51 55 105	38 05 166	43 11 202	48 32 265	34 33 292	41 04 005
22	25 55 078	52 25 107	38 11 169	42 59 204	48 01 267	34 04 293	41 07 006
24	26 26 080	52 55 109	38 17 171	42 46 206	47 30 269	33 36 295	41 10 008
26	26 56 082	53 24 111	38 21 173	42 32 209	46 59 271	33 08 297	41 15 009
28	27 27 084	53 52 114	38 24 176	42 16 211	46 28 273	32 41 299	41 20 011
	Alioth	◆POLLUX	CAPELLA	◆Hamal	Alpheratz	DENEB	◆VEGA
30	41 27 012	27 58 086	54 21 116	38 26 178	42 00 213	45 57 275	32 14 301
32	41 34 014	28 29 088	54 48 118	38 26 180	41 42 216	45 26 277	31 47 302
34	41 41 015	29 00 090	55 15 120	38 26 183	41 23 218	44 55 279	31 21 304
36	41 50 017	29 31 091	55 42 123	38 23 185	41 04 220	44 25 281	30 56 306
38	41 59 018	30 02 093	56 08 125	38 20 187	40 43 223	43 54 283	30 31 308
40	42 09 020	30 33 095	56 33 127	38 15 190	40 22 225	43 24 284	30 07 309
42	42 20 021	31 04 097	56 57 130	38 10 192	39 59 227	42 54 286	29 43 311
44	42 32 023	31 35 099	57 20 132	38 03 194	39 36 229	42 25 288	29 20 313
46	42 44 024	32 05 101	57 43 135	37 54 197	39 12 232	41 55 290	28 57 314
48	42 57 026	32 36 103	58 05 137	37 45 199	38 47 234	41 26 291	28 35 316
50	43 11 027	33 06 105	58 25 140	37 34 201	38 22 236	40 57 293	28 14 318
52	43 26 029	33 36 107	58 45 142	37 22 204	37 56 238	40 29 295	27 54 320
54	43 41 030	34 05 109	59 03 145	37 09 206	37 29 240	40 01 297	27 34 321
56	43 57 032	34 34 111	59 20 148	36 55 208	37 02 242	39 34 298	27 15 323
58	44 14 033	35 03 114	59 36 150	36 40 210	36 34 244	39 07 300	26 57 325
	Alioth	◆POLLUX	CAPELLA	◆Hamal	Alpheratz	DENEB	◆VEGA
60	44 31 035	35 31 116	59 51 153	36 24 213	36 06 247	38 40 302	26 39 327
62	44 49 036	35 59 118	60 04 156	36 06 215	35 37 249	38 14 304	26 22 328
64	45 08 038	36 26 120	60 16 159	35 48 217	35 08 251	37 48 305	26 07 330
66	45 27 039	36 53 122	60 27 161	35 29 219	34 39 253	37 23 307	25 51 332
68	45 47 041	37 19 124	60 36 164	35 09 222	34 09 255	36 59 309	25 37 333
70	46 08 042	37 44 126	60 44 167	34 48 224	33 39 257	36 35 310	25 24 335
72	46 29 044	38 09 128	60 50 170	34 26 226	33 08 259	36 11 312	25 11 337
74	46 51 045	38 33 131	60 55 173	34 03 228	32 38 261	35 49 314	24 59 339
76	47 13 047	38 56 133	60 58 176	33 39 230	32 07 263	35 26 315	24 48 340
78	47 36 049	39 18 135	60 59 178	33 15 232	31 36 265	35 05 317	24 38 342
80	47 59 050	39 40 137	60 59 181	32 50 235	31 05 267	34 44 319	24 29 344
82	48 24 052	40 00 140	60 58 184	32 25 237	30 34 269	34 24 320	24 21 345
84	48 48 053	40 20 142	60 55 187	31 58 239	30 03 271	34 05 322	24 13 347
86	49 13 055	40 39 144	60 50 190	31 31 241	29 32 272	33 46 324	24 07 349
88	49 39 056	40 56 146	60 44 193	31 04 243	29 01 274	33 28 325	24 01 350
	Alioth	◆REGULUS	POLLUX	◆CAPELLA	Alpheratz	◆DENEB	VEGA
90	50 05 058	18 38 114	41 13 149	60 37 196	28 30 276	33 11 327	23 56 352
92	50 32 060	19 06 116	41 29 151	60 28 198	27 59 278	32 54 329	23 53 354
94	50 59 061	19 33 118	41 43 153	60 17 201	27 29 280	32 38 330	23 50 356
96	51 26 063	20 01 120	41 56 156	60 05 204	26 58 282	32 23 332	23 48 357
98	51 54 064	20 27 122	42 09 158	59 52 207	26 28 284	32 09 334	23 47 359
100	52 22 066	20 53 124	42 20 160	59 37 209	25 58 286	31 56 335	23 47 001
102	52 51 068	21 18 126	42 29 163	59 21 212	25 28 288	31 43 337	23 48 002
104	53 20 069	21 43 128	42 38 165	59 04 215	24 59 289	31 31 339	23 49 004
106	53 49 071	22 07 131	42 45 168	58 46 217	24 30 291	31 20 340	23 52 006
108	54 18 073	22 30 133	42 51 170	58 27 220	24 01 293	31 10 342	23 56 008
110	54 48 075	22 53 135	42 56 172	58 06 223	23 33 295	31 01 343	24 00 009
112	55 18 076	23 14 137	42 59 175	57 45 225	23 05 297	30 53 345	24 06 011
114	55 49 078	23 35 139	43 01 177	57 22 228	22 37 299	30 45 347	24 12 013
116	56 19 080	23 55 141	43 02 180	56 59 230	22 10 300	30 38 348	24 19 014
118	56 50 082	24 14 143	43 02 182	56 35 232	21 44 302	30 32 350	24 27 016
	◆VEGA	ARCTURUS	◆REGULUS	POLLUX	◆CAPELLA	Alpheratz	DENEB
120	24 36 018	17 32 081	24 32 145	43 00 185	56 10 235	21 18 304	30 27 352
122	24 46 019	18 03 083	24 50 147	42 57 187	55 44 237	20 52 306	30 23 353
124	24 57 021	18 34 085	25 06 149	42 53 189	55 18 239	20 27 308	30 20 355
126	25 09 023	19 05 087	25 21 152	42 47 192	54 51 242	20 03 309	30 18 356
128	25 21 025	19 36 089	25 35 154	42 40 194	54 23 244	19 39 311	30 16 358
130	25 34 026	20 07 091	25 49 156	42 32 197	53 55 246	19 16 313	30 16 000
132	25 49 028	20 38 093	26 01 158	42 22 199	53 26 248	18 54 315	30 16 001
134	26 04 030	21 09 095	26 12 160	42 12 201	52 57 251	18 32 317	30 17 003
136	26 19 031	21 40 097	26 22 162	42 00 204	52 27 253	18 11 319	30 19 005
138	26 36 033	22 11 099	26 30 165	41 47 206	51 58 255	17 51 320	30 22 006
140	26 53 035	22 41 101	26 38 167	41 32 208	51 27 257	17 32 322	30 26 008
142	27 11 037	23 12 102	26 45 169	41 17 211	50 57 259	17 13 324	30 31 010
144	27 30 038	23 42 104	26 50 171	41 01 213	50 26 261	16 55 326	30 36 011
146	27 50 040	24 12 106	26 54 173	40 43 215	49 56 263	16 38 328	30 43 013
148	28 10 042	24 42 108	26 57 176	40 25 218	49 25 265	16 22 329	30 50 014
	DENEB	◆VEGA	ARCTURUS	◆REGULUS	POLLUX	◆CAPELLA	Alpheratz
150	30 58 016	28 31 043	25 11 110	26 59 178	40 05 220	48 54 267	16 06 331
152	31 07 018	28 53 045	25 40 113	27 00 180	39 45 222	48 23 269	15 52 333
154	31 17 019	29 15 047	26 08 115	26 59 182	39 24 224	47 52 271	15 38 335
156	31 28 021	29 38 049	26 36 117	26 57 184	39 02 227	47 21 273	15 25 337
158	31 39 023	30 02 050	27 04 119	26 54 187	38 39 229	46 50 275	15 13 338
160	31 52 024	30 26 052	27 31 121	26 50 189	38 15 231	46 19 277	15 02 340
162	32 05 026	30 51 054	27 57 123	26 45 191	37 50 233	45 48 279	14 52 342
164	32 19 028	31 16 056	28 23 125	26 38 193	37 25 235	45 17 280	14 43 344
166	32 34 029	31 42 057	28 48 127	26 31 195	36 59 237	44 47 282	14 35 346
168	32 49 031	32 09 059	29 13 129	26 22 197	36 33 240	44 17 284	14 28 347
170	33 05 032	32 35 061	29 36 131	26 12 200	36 06 242	43 47 286	14 21 349
172	33 22 034	33 03 063	29 59 133	26 01 202	35 38 244	43 17 288	14 16 351
174	33 40 036	33 31 064	30 22 135	25 49 204	35 10 246	42 47 289	14 12 353
176	33 59 037	33 59 066	30 43 138	25 36 206	34 41 248	42 18 291	14 08 355
178	34 18 039	34 28 068	31 03 140	25 22 208	34 12 250	41 50 293	14 06 356

LHA ♈	Hc Zn	Hc Zn	Hc Zn	Hc Zn	Hc Zn	Hc Zn	Hc Zn
	◆DENEB	VEGA	◆ARCTURUS	REGULUS	◆POLLUX	CAPELLA	Alpheratz
180	34 38 041	34 56 070	31 23 142	25 06 210	33 43 252	41 21 295	14 04 358
182	34 58 042	35 26 072	31 42 144	24 50 213	33 13 254	40 53 296	14 04 000
184	35 20 044	35 55 073	31 59 146	24 33 215	32 43 256	40 26 298	14 04 002
186	35 42 046	36 25 075	32 16 149	24 15 217	32 13 258	39 58 300	14 06 004
188	36 04 047	36 56 077	32 32 151	23 56 219	31 43 260	39 32 302	14 08 005
190	36 27 049	37 26 079	32 46 153	23 36 221	31 12 262	39 05 303	14 12 007
192	36 51 051	37 56 081	33 00 155	23 15 223	30 41 264	38 40 305	14 16 009
194	37 15 052	38 27 083	33 12 157	22 53 225	30 10 266	38 15 307	14 21 011
196	37 40 054	38 58 085	33 24 160	22 31 227	29 39 268	37 50 308	14 27 013
198	38 06 056	39 29 087	33 34 162	22 08 229	29 08 270	37 26 310	14 35 014
200	38 32 058	40 00 088	33 43 164	21 44 231	28 37 272	37 02 312	14 43 016
202	38 58 059	40 31 090	33 51 166	21 19 234	28 06 274	36 39 313	14 52 018
204	39 25 061	41 02 092	33 58 169	20 54 236	27 35 276	36 17 315	15 02 020
206	39 52 063	41 33 094	34 03 171	20 28 238	27 04 278	35 56 317	15 13 022
208	40 20 064	42 04 096	34 07 173	20 01 240	26 33 280	35 35 318	15 25 023
	Alpheratz	◆DENEB	VEGA	◆ARCTURUS	REGULUS	POLLUX	◆CAPELLA
210	15 38 025	40 48 066	42 35 098	34 10 176	19 34 242	26 03 281	35 14 320
212	15 51 027	41 17 068	43 06 100	34 12 178	19 07 244	25 33 283	34 55 322
214	16 06 029	41 46 070	43 36 102	34 12 180	18 39 246	25 02 285	34 36 323
216	16 21 031	42 15 071	44 06 104	34 12 182	18 10 248	24 33 287	34 18 325
218	16 38 032	42 45 073	44 36 106	34 10 185	17 41 250	24 03 289	34 00 327
220	16 55 034	43 15 075	45 06 108	34 07 187	17 12 252	23 34 291	33 43 328
222	17 12 036	43 45 077	45 35 111	34 02 189	16 42 254	23 05 293	33 27 330
224	17 31 038	44 15 079	46 04 113	33 57 192	16 12 256	22 36 294	33 12 331
226	17 51 040	44 46 081	46 32 115	33 50 194	15 42 258	22 08 296	32 58 333
228	18 11 041	45 17 082	47 00 117	33 42 196	15 12 260	21 41 298	32 44 335
230	18 32 043	45 47 084	47 28 119	33 33 198	14 41 261	21 14 300	32 31 336
232	18 53 045	46 18 086	47 55 121	33 22 201	14 10 263	20 47 302	32 19 338
234	19 16 047	46 49 088	48 21 124	33 11 203	13 39 265	20 21 304	32 08 340
236	19 39 049	47 20 090	48 46 126	32 58 205	13 08 267	19 55 305	31 58 341
238	20 02 050	47 51 092	49 11 128	32 44 207	12 37 269	19 30 307	31 48 343
	Alpheratz	◆DENEB	VEGA	◆ARCTURUS	Dubhe	POLLUX	◆CAPELLA
240	20 26 052	48 22 094	49 35 131	32 30 210	62 12 283	19 06 309	31 39 345
242	20 51 054	48 53 096	49 58 133	32 14 212	61 42 284	18 42 311	31 31 346
244	21 17 056	49 24 098	50 20 135	31 57 214	61 12 286	18 19 313	31 24 348
246	21 43 058	49 55 100	50 42 138	31 39 216	60 42 288	17 56 314	31 18 349
248	22 09 060	50 25 102	51 02 140	31 20 218	60 12 289	17 34 316	31 13 351
250	22 36 061	50 56 104	51 22 142	31 01 221	59 43 291	17 13 318	31 09 353
252	23 04 063	51 26 106	51 40 145	30 40 223	59 14 293	16 53 320	31 05 354
254	23 32 065	51 55 108	51 57 147	30 18 225	58 46 294	16 33 322	31 02 356
256	24 00 067	52 25 110	52 14 150	29 56 227	58 18 296	16 15 324	31 01 358
258	24 29 069	52 54 113	52 29 152	29 33 229	57 50 297	15 56 325	31 00 359
260	24 58 071	53 22 115	52 42 155	29 09 231	57 23 299	15 39 327	31 00 001
262	25 27 072	53 50 117	52 55 158	28 45 233	56 56 300	15 23 329	31 01 002
264	25 57 074	54 17 119	53 06 160	28 19 235	56 29 302	15 07 331	31 02 004
266	26 27 076	54 44 122	53 16 163	27 53 238	56 03 304	14 53 333	31 05 006
268	26 57 078	55 10 124	53 25 165	27 27 240	55 37 305	14 39 334	31 08 007
	◆CAPELLA	Alpheratz	◆DENEB	VEGA	◆ARCTURUS	Dubhe	POLLUX
270	31 13 009	27 28 080	55 36 126	53 32 168	27 00 242	55 12 307	14 26 336
272	31 18 010	27 59 082	56 00 129	53 38 171	26 32 244	54 48 308	14 14 338
274	31 24 012	28 29 084	56 24 131	53 42 173	26 04 246	54 23 309	14 02 340
276	31 31 014	29 00 086	56 47 134	53 45 176	25 36 248	54 00 311	13 52 342
278	31 39 015	29 31 088	57 09 136	53 46 178	25 07 250	53 36 312	13 43 343
280	31 47 017	30 02 089	57 30 139	53 47 181	24 37 252	53 14 314	13 35 345
282	31 57 019	30 33 091	57 50 141	53 45 184	24 08 254	52 52 315	13 27 347
284	32 07 020	31 04 093	58 09 144	53 43 186	23 38 256	52 30 317	13 21 349
286	32 18 022	31 35 095	58 27 146	53 38 189	23 08 258	52 09 318	13 15 351
288	32 30 023	32 06 097	58 43 149	53 33 192	22 37 260	51 49 320	13 11 353
290	32 43 025	32 37 099	58 59 152	53 26 194	22 06 262	51 29 321	13 07 354
292	32 57 027	33 08 101	59 13 154	53 18 197	21 36 264	51 10 322	13 05 356
294	33 11 028	33 38 103	59 26 157	53 08 199	21 05 266	50 51 324	13 03 358
296	33 26 030	34 08 105	59 37 160	52 57 202	20 34 268	50 33 325	13 02 000
298	33 42 032	34 38 107	59 47 163	52 45 205	20 03 270	50 16 327	13 03 002
	POLLUX	◆CAPELLA	Alpheratz	◆DENEB	VEGA	◆ARCTURUS	Dubhe
300	13 04 003	33 59 033	35 07 109	59 55 165	52 31 207	19 32 271	49 59 328
302	13 06 005	34 16 035	35 36 111	60 03 168	52 17 210	19 01 273	49 43 330
304	13 10 007	34 34 037	36 05 113	60 08 171	52 01 212	18 30 275	49 28 331
306	13 14 009	34 53 038	36 33 116	60 12 174	51 44 215	17 59 277	49 13 332
308	13 19 011	35 13 040	37 01 118	60 15 177	51 25 217	17 28 279	48 59 334
310	13 25 012	35 33 042	37 28 120	60 16 180	51 06 219	16 57 281	48 45 335
312	13 33 014	35 54 043	37 55 122	60 15 182	50 46 222	16 27 283	48 33 337
314	13 41 016	36 15 045	38 21 124	60 13 185	50 25 224	15 57 285	48 21 338
316	13 50 018	36 38 046	38 47 126	60 10 188	50 03 227	15 27 287	48 09 339
318	14 00 020	37 00 048	39 11 128	60 04 191	49 40 229	14 57 289	47 59 341
320	14 11 022	37 24 050	39 35 131	59 58 194	49 16 231	14 28 290	47 49 342
322	14 23 023	37 48 052	39 58 133	59 50 196	48 51 234	13 59 292	47 40 344
324	14 35 025	38 13 053	40 21 135	59 40 199	48 26 236	13 31 294	47 31 345
326	14 49 027	38 38 055	40 42 137	59 29 202	48 00 238	13 03 296	47 23 346
328	15 03 029	39 03 057	41 03 140	59 17 205	47 33 240	12 35 298	47 16 348
	◆POLLUX	CAPELLA	Hamal	◆Alpheratz	DENEB	◆VEGA	Alioth
330	15 19 031	39 30 058	29 47 111	41 23 142	59 03 207	47 06 243	44 04 328
332	15 35 032	39 56 060	30 16 113	41 41 144	58 48 210	46 38 245	43 48 329
334	15 52 034	40 23 062	30 44 116	41 59 146	58 32 213	46 10 247	43 32 331
336	16 10 036	40 51 063	31 12 118	42 16 149	58 15 215	45 41 249	43 18 332
338	16 29 038	41 19 065	31 39 120	42 31 151	57 56 218	45 12 251	43 03 334
340	16 48 040	41 47 067	32 06 122	42 46 153	57 36 221	44 42 253	42 50 335
342	17 08 041	42 16 069	32 32 124	42 59 156	57 16 223	44 12 255	42 37 337
344	17 29 043	42 45 070	32 57 126	43 11 158	56 54 226	43 42 257	42 25 338
346	17 51 045	43 15 072	33 22 128	43 22 161	56 31 228	43 11 259	42 14 340
348	18 13 047	43 44 074	33 46 130	43 32 163	56 08 231	42 41 261	42 04 341
350	18 36 049	44 14 076	34 09 132	43 40 165	55 43 233	42 10 263	41 54 343
352	19 00 050	44 45 078	34 32 135	43 47 168	55 18 235	41 39 265	41 45 344
354	19 24 052	45 15 079	34 54 137	43 53 170	54 52 238	41 08 267	41 37 346
356	19 49 054	45 46 081	35 14 139	43 58 173	54 26 240	40 37 269	41 30 347
358	20 14 056	46 16 083	35 34 141	44 01 175	53 59 242	40 06 271	41 23 349

LAT 74°N

LHA	Hc	Zn	Hc	Zn	Hc	Zn	Hc	Zn	Hc	Zn	Hc	Zn	Hc	Zn
♈	◆POLLUX		CAPELLA		Hamal		◆Alpheratz		DENEB		◆VEGA		Alioth	
0	20 08	057	46 42	084	36 41	143	45 03	177	53 56	246	39 31	274	40 18	350
2	20 36	059	47 14	086	37 01	145	45 04	180	53 26	248	38 58	276	40 13	352
4	21 05	061	47 47	088	37 19	148	45 03	182	52 55	250	38 26	278	40 09	353
6	21 34	063	48 21	090	37 36	150	45 01	185	52 23	252	37 53	280	40 05	355
8	22 04	065	48 54	092	37 52	152	44 57	187	51 52	254	37 20	281	40 02	356
10	22 34	067	49 27	094	38 07	154	44 52	190	51 20	257	36 48	283	40 00	357
12	23 05	068	50 00	096	38 21	157	44 46	192	50 47	259	36 16	285	39 59	359
14	23 36	070	50 32	098	38 33	159	44 38	195	50 15	261	35 44	287	39 59	000
16	24 07	072	51 05	100	38 44	161	44 29	197	49 42	263	35 13	289	40 00	002
18	24 38	074	51 38	102	38 54	164	44 19	200	49 09	265	34 41	290	40 01	003
20	25 10	076	52 10	104	39 03	166	44 07	202	48 36	267	34 11	292	40 04	005
22	25 43	078	52 42	106	39 10	168	43 54	204	48 03	269	33 40	294	40 07	006
24	26 15	080	53 14	108	39 16	171	43 40	207	47 30	270	33 10	296	40 11	008
26	26 48	081	53 45	110	39 21	173	43 24	209	46 57	272	32 41	298	40 16	009
28	27 20	083	54 16	112	39 24	176	43 07	212	46 24	274	32 12	299	40 21	011
	Alioth		◆POLLUX		CAPELLA		◆Hamal		Alpheratz		DENEB		◆VEGA	
30	40 28	012	27 53	085	54 46	114	39 26	178	42 49	214	45 51	276	31 43	301
32	40 35	014	28 26	087	55 16	117	39 26	180	42 30	216	45 18	278	31 15	303
34	40 43	015	28 59	089	55 45	119	39 25	183	42 10	219	44 46	280	30 47	305
36	40 52	016	29 32	091	56 14	121	39 23	185	41 49	221	44 13	282	30 20	306
38	41 02	018	30 06	093	56 42	124	39 20	187	41 27	223	43 41	283	29 54	308
40	41 13	019	30 39	095	57 09	126	39 15	190	41 04	225	43 09	285	29 28	310
42	41 24	021	31 11	097	57 35	129	39 08	192	40 40	228	42 37	287	29 03	311
44	41 36	022	31 44	099	58 00	131	39 01	195	40 15	230	42 06	289	28 39	313
46	41 49	024	32 17	101	58 25	134	38 52	197	39 49	232	41 35	291	28 15	315
48	42 03	025	32 49	103	58 48	136	38 41	199	39 23	234	41 04	292	27 52	317
50	42 18	027	33 21	105	59 11	139	38 30	202	38 55	237	40 33	294	27 30	318
52	42 33	028	33 53	107	59 32	141	38 17	204	38 27	239	40 03	296	27 08	320
54	42 49	030	34 25	109	59 52	144	38 03	206	37 59	241	39 34	297	26 47	322
56	43 06	031	34 56	111	60 11	147	37 48	209	37 30	243	39 05	299	26 27	323
58	43 23	033	35 26	113	60 28	150	37 32	211	37 00	245	38 36	301	26 08	325
	Alioth		◆POLLUX		CAPELLA		◆Hamal		Alpheratz		DENEB		◆VEGA	
60	43 41	034	35 57	115	60 44	152	37 14	213	36 30	247	38 08	303	25 49	327
62	44 00	036	36 26	117	60 59	155	36 55	215	35 59	249	37 40	304	25 31	329
64	44 20	037	36 56	119	61 12	158	36 36	218	35 28	251	37 13	306	25 15	330
66	44 40	039	37 24	121	61 24	161	36 15	220	34 56	253	36 47	308	24 59	332
68	45 01	040	37 52	123	61 34	164	35 53	222	34 24	255	36 21	309	24 43	334
70	45 23	042	38 19	126	61 42	167	35 31	224	33 52	257	35 56	311	24 29	335
72	45 45	043	38 46	128	61 49	170	35 07	226	33 20	259	35 31	313	24 16	337
74	46 08	045	39 12	130	61 54	172	34 43	229	32 47	261	35 07	314	24 03	339
76	46 32	046	39 36	132	61 58	175	34 18	231	32 14	263	34 44	316	23 52	340
78	46 56	048	40 01	134	61 59	178	33 52	233	31 42	265	34 21	318	23 41	342
80	47 21	049	40 24	137	61 59	181	33 25	235	31 09	267	33 59	319	23 31	344
82	47 46	051	40 46	139	61 58	184	32 57	237	30 36	269	33 38	321	23 23	345
84	48 12	052	41 07	141	61 54	187	32 29	239	30 02	271	33 17	322	23 15	347
86	48 38	054	41 27	144	61 49	190	32 00	241	29 29	273	32 57	324	23 08	349
88	49 05	055	41 46	146	61 43	193	31 31	244	28 56	275	32 38	326	23 02	351
	Alioth		◆REGULUS		POLLUX		◆CAPELLA		Alpheratz		◆DENEB		VEGA	
90	49 33	057	19 02	114	42 04	148	61 34	196	28 23	277	32 20	327	22 57	352
92	50 01	059	19 32	116	42 21	151	61 24	199	27 51	279	32 03	329	22 53	354
94	50 29	060	20 02	118	42 37	153	61 13	202	27 18	281	31 46	331	22 50	356
96	50 58	062	20 31	120	42 51	155	61 00	205	26 46	282	31 30	332	22 48	357
98	51 27	063	20 59	122	43 04	158	60 45	207	26 14	284	31 15	334	22 47	359
100	51 57	065	21 27	124	43 16	160	60 29	210	25 42	286	31 01	336	22 47	001
102	52 27	067	21 54	126	43 27	163	60 12	213	25 10	288	30 48	337	22 48	002
104	52 58	068	22 20	128	43 36	165	59 53	216	24 39	290	30 35	339	22 50	004
106	53 29	070	22 46	130	43 44	167	59 33	218	24 08	292	30 24	340	22 52	006
108	54 00	072	23 11	132	43 50	170	59 12	221	23 37	293	30 13	342	22 56	007
110	54 32	073	23 35	134	43 55	172	58 50	224	23 07	295	30 03	344	23 01	009
112	55 03	075	23 58	136	43 59	175	58 27	226	22 37	297	29 55	345	23 07	011
114	55 35	077	24 20	139	44 01	177	58 02	229	22 08	299	29 47	347	23 13	013
116	56 08	078	24 42	141	44 02	180	57 37	231	21 40	301	29 40	348	23 21	014
118	56 40	080	25 02	143	44 02	182	57 11	234	21 11	303	29 33	350	23 30	016
	◆VEGA		ARCTURUS		◆REGULUS		POLLUX		◆CAPELLA		Alpheratz		DENEB	
120	23 39	018	17 23	081	25 22	145	44 00	185	56 44	236	20 44	304	29 28	352
122	23 50	019	17 56	083	25 40	147	43 56	187	56 16	238	20 17	306	29 24	353
124	24 01	021	18 29	085	25 58	149	43 52	189	55 47	241	19 50	308	29 20	355
126	24 13	023	19 02	087	26 14	151	43 46	192	55 18	243	19 25	310	29 18	357
128	24 27	024	19 35	088	26 29	154	43 38	194	54 49	245	19 00	312	29 16	358
130	24 41	026	20 08	090	26 43	156	43 29	197	54 18	248	18 35	313	29 16	000
132	24 56	028	20 41	092	26 56	158	43 19	199	53 47	250	18 11	315	29 16	001
134	25 11	029	21 14	094	27 08	160	43 07	202	53 16	252	17 49	317	29 17	003
136	25 28	031	21 47	096	27 19	162	42 55	204	52 45	254	17 26	319	29 20	005
138	25 46	033	22 19	098	27 28	165	42 40	206	52 13	256	17 05	321	29 23	006
140	26 04	035	22 52	100	27 37	167	42 25	209	51 40	258	16 44	322	29 27	008
142	26 23	036	23 25	102	27 43	169	42 09	211	51 08	260	16 24	324	29 32	009
144	26 43	038	23 57	104	27 49	171	41 51	213	50 35	262	16 06	326	29 37	011
146	27 04	040	24 29	106	27 54	173	41 32	216	50 02	264	15 47	328	29 44	013
148	27 25	041	25 00	108	27 57	176	41 12	218	49 29	266	15 30	329	29 52	014
	DENEB		◆VEGA		ARCTURUS		◆REGULUS		POLLUX		◆CAPELLA		Alpheratz	
150	30 01	016	27 48	043	25 32	110	27 59	178	40 51	220	48 56	268	15 14	331
152	30 10	018	28 10	045	26 03	112	28 00	180	40 29	223	48 23	270	14 58	333
154	30 20	019	28 34	047	26 33	114	27 59	182	40 06	225	47 50	272	14 44	335
156	30 32	021	28 58	048	27 03	116	27 57	184	39 43	227	47 17	274	14 30	337
158	30 44	022	29 23	050	27 32	118	27 54	187	39 18	229	46 44	276	14 18	338
160	30 57	024	29 49	052	28 01	120	27 49	189	38 52	232	46 11	278	14 06	340
162	31 11	026	30 15	053	28 29	122	27 44	191	38 26	234	45 39	280	13 55	342
164	31 26	027	30 42	055	28 57	124	27 37	193	37 59	236	45 06	281	13 46	344
166	31 41	029	31 10	057	29 24	127	27 29	195	37 31	238	44 34	283	13 37	346
168	31 57	030	31 38	059	29 50	129	27 19	198	37 03	240	44 02	285	13 29	347
170	32 15	032	32 06	060	30 16	131	27 08	200	36 34	242	43 30	287	13 22	349
172	32 33	034	32 35	062	30 40	133	26 57	202	36 04	245	42 58	289	13 17	351
174	32 51	035	33 05	064	31 04	135	26 44	204	35 34	247	42 27	290	13 12	353
176	33 11	037	33 34	066	31 27	137	26 30	206	35 04	249	41 56	292	13 08	355
178	33 31	039	34 05	067	31 49	139	26 14	209	34 33	251	41 26	294	13 06	356

LHA	Hc	Zn	Hc	Zn	Hc	Zn	Hc	Zn	Hc	Zn	Hc	Zn	Hc	Zn
♈	◆DENEB		VEGA		◆ARCTURUS		REGULUS		◆POLLUX		CAPELLA		Alpheratz	
180	33 52	040	34 36	069	32 10	142	25 58	211	34 01	253	40 56	295	13 04	358
182	34 14	042	35 07	071	32 30	144	25 41	213	33 29	255	40 26	297	13 04	000
184	34 36	044	35 38	073	32 49	146	25 22	215	32 57	257	39 57	299	13 04	002
186	34 59	045	36 10	075	33 07	148	25 03	217	32 25	259	39 28	301	13 06	004
188	35 23	047	36 42	076	33 24	150	24 42	219	31 52	261	39 00	302	13 08	005
190	35 48	049	37 14	078	33 40	153	24 21	221	31 20	263	38 32	304	13 12	007
192	36 13	050	37 47	080	33 54	155	23 59	223	30 47	265	38 05	306	13 17	009
194	36 39	052	38 19	082	34 08	157	23 35	226	30 14	267	37 38	307	13 22	011
196	37 05	054	38 52	084	34 20	159	23 11	228	29 41	269	37 12	309	13 29	013
198	37 32	055	39 25	086	34 31	162	22 46	230	29 08	271	36 47	311	13 37	014
200	37 59	057	39 58	088	34 41	164	22 21	232	28 35	272	36 22	312	13 45	016
202	38 27	059	40 31	090	34 49	166	21 55	234	28 02	274	35 58	314	13 55	018
204	38 56	060	41 04	091	34 56	169	21 27	236	27 29	276	35 35	316	14 06	020
206	39 25	062	41 37	093	35 02	171	21 00	238	26 56	278	35 12	317	14 17	021
208	39 54	064	42 10	095	35 07	173	20 31	240	26 23	280	34 50	319	14 30	023
	Alpheratz		◆DENEB		VEGA		◆ARCTURUS		REGULUS		POLLUX		◆CAPELLA	
210	14 43	025	40 24	065	42 43	097	35 10	176	20 03	242	25 51	282	34 28	320
212	14 58	027	40 54	067	43 16	099	35 12	178	19 33	244	25 19	284	34 07	322
214	15 13	029	41 25	069	43 48	101	35 12	180	19 03	246	24 47	286	33 48	324
216	15 30	030	41 56	071	44 21	103	35 12	182	18 33	248	24 15	287	33 28	325
218	15 47	032	42 27	072	44 53	105	35 10	185	18 02	250	23 43	289	33 10	327
220	16 05	034	42 59	074	45 24	108	35 06	187	17 30	252	23 12	291	32 52	329
222	16 24	036	43 31	076	45 56	110	35 01	189	16 59	254	22 42	293	32 35	330
224	16 44	038	44 03	078	46 27	112	34 55	192	16 27	256	22 11	295	32 19	332
226	17 04	039	44 35	080	46 57	114	34 48	194	15 55	258	21 42	297	32 04	333
228	17 26	041	45 08	081	47 27	116	34 39	196	15 22	260	21 12	298	31 50	335
230	17 48	043	45 41	083	47 57	118	34 30	199	14 50	262	20 44	300	31 36	337
232	18 11	045	46 14	085	48 25	120	34 18	201	14 17	264	20 15	302	31 24	338
234	18 34	047	46 47	087	48 54	123	34 06	203	13 44	266	19 47	304	31 12	340
236	18 59	048	47 20	089	49 21	125	33 52	205	13 11	268	19 20	306	31 01	341
238	19 24	050	47 53	091	49 48	127	33 38	208	12 38	269	18 54	308	30 51	343
	Alpheratz		◆DENEB		VEGA		◆ARCTURUS		Dubhe		POLLUX		◆CAPELLA	
240	19 50	052	48 26	093	50 14	130	33 22	210	61 57	285	18 28	309	30 42	345
242	20 16	054	48 59	095	50 39	132	33 05	212	61 26	286	18 03	311	30 33	346
244	20 43	056	49 32	097	51 03	134	32 47	214	60 54	288	17 38	313	30 26	348
246	21 10	057	50 05	099	51 26	137	32 27	217	60 23	290	17 14	315	30 19	350
248	21 39	059	50 37	101	51 48	139	32 07	219	59 52	291	16 51	317	30 14	351
250	22 07	061	51 10	103	52 09	142	31 46	221	59 21	293	16 29	318	30 09	353
252	22 36	063	51 42	105	52 29	144	31 24	223	58 51	294	16 07	320	30 05	354
254	23 06	065	52 13	107	52 48	147	31 01	225	58 21	296	15 46	322	30 03	356
256	23 36	066	52 45	109	53 05	149	30 37	227	57 51	297	15 26	324	30 01	358
258	24 07	068	53 16	111	53 22	152	30 12	230	57 22	299	15 07	326	30 00	359
260	24 38	070	53 47	114	53 37	154	29 47	232	56 53	300	14 49	327	30 00	001
262	25 09	072	54 17	116	53 50	157	29 20	234	56 25	302	14 31	329	30 01	002
264	25 41	074	54 46	118	54 02	160	28 53	236	55 57	303	14 15	331	30 02	004
266	26 12	076	55 15	120	54 13	162	28 25	238	55 29	305	13 59	333	30 05	006
268	26 45	078	55 43	123	54 22	165	27 57	240	55 02	306	13 45	335	30 09	007
	◆CAPELLA		Alpheratz		◆DENEB		VEGA		◆ARCTURUS		Dubhe		POLLUX	
270	30 13	009	27 17	079	56 11	125	54 30	168	27 28	242	54 36	308	13 31	336
272	30 19	010	27 50	081	56 37	128	54 37	170	26 59	244	54 10	309	13 18	338
274	30 25	012	28 22	083	57 03	130	54 41	173	26 29	246	53 45	311	13 06	340
276	30 33	014	28 55	085	57 28	132	54 45	176	25 58	248	53 20	312	12 55	342
278	30 41	015	29 28	087	57 52	135	54 46	178	25 27	250	52 56	313	12 45	344
280	30 50	017	30 01	089	58 15	138	54 47	181	24 56	252	52 32	315	12 37	345
282	31 00	018	30 34	091	58 36	140	54 45	184	24 24	254	52 09	316	12 29	347
284	31 11	020	31 08	093	58 57	143	54 42	186	23 52	256	51 46	318	12 22	349
286	31 23	022	31 41	095	59 16	145	54 38	189	23 20	258	51 24	319	12 16	351
288	31 35	023	32 13	097	59 35	148	54 32	192	22 48	260	51 03	320	12 11	353
290	31 49	025	32 46	099	59 51	151	54 24	195	22 15	262	50 42	322	12 07	354
292	32 03	026	33 19	101	60 07	154	54 15	197	21 42	264	50 22	323	12 05	356
294	32 18	028	33 51	103	60 21	156	54 05	200	21 09	266	50 02	325	12 03	358
296	32 34	030	34 23	105	60 33	159	53 53	202	20 36	268	49 44	326	12 02	000
298	32 51	031	34 55	107	60 44	162	53 39	205	20 03	270	49 25	327	12 03	002
	POLLUX		◆CAPELLA		Alpheratz		◆DENEB		VEGA		◆ARCTURUS		Dubhe	
300	12 04	003	33 08	033	35 27	109	60 53	165	53 25	208	19 30	272	49 08	329
302	12 07	005	33 27	035	35 58	111	61 01	168	53 09	210	18 57	274	48 51	330
304	12 10	007	33 46	036	36 29	113	61 07	171	52 51	213	18 24	276	48 35	332
306	12 15	009	34 06	038	36 59	115	61 12	174	52 33	215	17 51	278	48 19	333
308	12 20	011	34 26	039	37 29	117	61 15	177	52 13	218	17 18	279	48 05	334
310	12 27	012	34 48	041	37 58	119	61 16	180	51 52	220	16 46	281	47 51	336
312	12 34	014	35 10	043	38 26	121	61 15	182	51 30	223	16 13	283	47 37	337
314	12 43	016	35 33	044	38 54	123	61 13	185	51 07	225	15 41	285	47 25	338
316	12 53	018	35 56	046	39 22	126	61 09	188	50 43	228	15 10	287	47 13	340
318	13 03	020	36 20	048	39 48	128	61 03	191	50 19	230	14 38	289	47 02	341
320	13 15	021	36 45	049	40 14	130	60 56	194	49 53	232	14 07	291	46 52	342
322	13 27	023	37 10	051	40 39	132	60 47	197	49 26	235	13 36	293	46 42	344
324	13 41	025	37 36	053	41 03	134	60 37	200	48 59	237	13 06	294	46 33	345
326	13 55	027	38 03	054	41 26	137	60 25	203	48 31	239	12 36	296	46 25	347
328	14 11	029	38 30	056	41 48	139	60 11	205	48 02	241	12 07	298	46 18	348
	◆POLLUX		CAPELLA		Hamal		◆Alpheratz		DENEB		◆VEGA		Alioth	
330	14 27	030	38 58	058	30 09	111	42 10	141	59 56	208	47 33	244	43 14	328
332	14 44	032	39 26	059	30 39	113	42 30	144	59 40	211	47 03	246	42 56	330
334	15 02	034	39 55	061	31 10	115	42 49	146	59 22	214	46 33	248	42 40	331
336	15 21	036	40 24	063	31 39	117	43 07	148	59 03	216	46 02	250	42 24	333
338	15 41	038	40 53	064	32 08	119	43 24	151	58 43	219	45 30	252	42 10	334
340	16 02	039	41 23	066	32 37	121	43 39	153	58 22	222	44 59	254	41 55	336
342	16 23	041	41 54	068	33 05	123	43 54	155	57 59	224	44 27	256	41 42	337
344	16 45	043	42 25	070	33 32	125	44 07	158	57 36	227	43 55	258	41 30	338
346	17 08	045	42 56	071	33 59	128	44 18	160	57 11	229	43 22	260	41 18	340
348	17 32	047	43 27	073	34 25	130	44 29	163	56 45	232	42 49	262	41 07	341
350	17 56	048	43 59	075	34 50	132	44 38	165	56 19	234	42 17	264	40 57	343
352	18 21	050	44 31	077	35 14	134	44 46	168	55 52	237	41 44	266	40 47	344
354	18 47	052	45 03	078	35 37	136	44 52	170	55 24	239	41 11	268	40 39	346
356	19 14	054	45 36	080	36 00	139	44 57	173	54 55	241	40 37	270	40 31	347
358	19 41	056	46 09	082	36 21	141	45 01	175	54 26	243	40 04	272	40 24	349

LHA ♈	Hc	Zn	Hc	Zn	Hc	Zn	Hc	Zn	Hc	Zn	Hc	Zn	Hc	Zn
	◆POLLUX		CAPELLA		Hamal		◆Alpheratz		DENEB		◆VEGA		Alioth	
0	19 36	057	46 35	083	37 29	143	46 03	177	54 20	247	39 27	275	39 19	350
2	20 06	059	47 10	085	37 50	145	46 04	180	53 48	249	38 52	277	39 14	352
4	20 36	061	47 45	087	38 10	147	46 03	182	53 15	251	38 17	278	39 09	353
6	21 07	063	48 20	089	38 28	149	46 01	185	52 41	254	37 42	280	39 05	355
8	21 38	064	48 55	090	38 45	152	45 57	188	52 07	256	37 08	282	39 02	356
10	22 10	066	49 30	092	39 01	154	45 52	190	51 33	258	36 34	284	39 01	358
12	22 42	068	50 05	094	39 16	156	45 45	193	50 59	260	36 00	286	38 59	359
14	23 15	070	50 40	096	39 29	159	45 36	195	50 24	262	35 26	288	38 59	000
16	23 48	072	51 15	098	39 41	161	45 27	198	49 49	264	34 53	289	39 00	002
18	24 22	074	51 49	100	39 52	164	45 15	200	49 14	266	34 20	291	39 01	003
20	24 55	075	52 24	102	40 01	166	45 03	202	48 39	268	33 48	293	39 04	005
22	25 30	077	52 58	104	40 09	168	44 49	205	48 04	270	33 16	295	39 07	006
24	26 04	079	53 31	107	40 15	171	44 33	207	47 29	272	32 44	296	39 11	008
26	26 38	081	54 05	109	40 20	173	44 16	210	46 54	273	32 13	298	39 17	009
28	27 13	083	54 38	111	40 24	176	43 58	212	46 19	275	31 42	300	39 22	010
	Alioth		◆POLLUX		CAPELLA		◆Hamal		Alpheratz		DENEB		◆VEGA	
30	39 29	012	27 48	085	55 10	113	40 26	178	43 39	215	45 44	277	31 12	302
32	39 37	013	28 23	087	55 42	115	40 26	180	43 19	217	45 10	279	30 42	303
34	39 45	015	28 58	088	56 14	118	40 25	183	42 57	219	44 35	281	30 13	305
36	39 55	016	29 33	090	56 44	120	40 23	185	42 34	222	44 01	283	29 45	307
38	40 05	018	30 08	092	57 14	122	40 19	188	42 10	224	43 26	284	29 17	308
40	40 16	019	30 43	094	57 44	125	40 14	190	41 46	226	42 53	286	28 50	310
42	40 28	021	31 18	096	58 12	127	40 07	192	41 20	228	42 19	288	28 23	312
44	40 41	022	31 53	098	58 39	130	39 59	195	40 53	231	41 46	290	27 58	314
46	40 54	023	32 28	100	59 06	132	39 49	197	40 26	233	41 13	291	27 32	315
48	41 09	025	33 02	102	59 31	135	39 38	200	39 57	235	40 41	293	27 08	317
50	41 24	026	33 36	104	59 55	138	39 26	202	39 28	237	40 08	295	26 45	319
52	41 40	028	34 10	106	60 18	140	39 12	204	38 58	239	39 37	297	26 22	320
54	41 57	029	34 44	108	60 40	143	38 57	207	38 28	242	39 06	298	26 00	322
56	42 14	031	35 17	110	61 01	146	38 41	209	37 57	244	38 35	300	25 39	324
58	42 33	032	35 49	112	61 20	149	38 23	211	37 25	246	38 05	302	25 18	325
	Alioth		◆POLLUX		CAPELLA		◆Hamal		Alpheratz		DENEB		◆VEGA	
60	42 52	034	36 22	114	61 37	151	38 04	214	36 53	248	37 35	303	24 59	327
62	43 11	035	36 53	116	61 53	154	37 44	216	36 20	250	37 06	305	24 40	329
64	43 32	037	37 25	119	62 07	157	37 23	218	35 47	252	36 38	307	24 22	330
66	43 53	038	37 55	121	62 20	160	37 01	220	35 13	254	36 10	308	24 06	332
68	44 15	040	38 25	123	62 31	163	36 38	223	34 39	256	35 43	310	23 50	334
70	44 38	041	38 54	125	62 40	166	36 14	225	34 05	258	35 16	312	23 35	335
72	45 01	042	39 22	127	62 48	169	35 48	227	33 31	260	34 50	313	23 21	337
74	45 25	044	39 50	129	62 54	172	35 22	229	32 56	262	34 25	315	23 07	339
76	45 50	045	40 17	132	62 57	175	34 55	231	32 21	264	34 00	316	22 55	341
78	46 15	047	40 42	134	62 59	178	34 27	234	31 46	266	33 36	318	22 44	342
80	46 41	048	41 07	136	62 59	181	33 59	236	31 11	268	33 13	320	22 34	344
82	47 08	050	41 31	138	62 58	184	33 30	238	30 36	270	32 51	321	22 25	346
84	47 35	051	41 54	141	62 54	188	32 59	240	30 01	272	32 29	323	22 16	347
86	48 03	053	42 15	143	62 48	191	32 29	242	29 26	274	32 09	324	22 09	349
88	48 31	054	42 36	145	62 41	194	31 58	244	28 51	275	31 49	326	22 03	351
	Alioth		◆REGULUS		POLLUX		◆CAPELLA		Alpheratz		◆DENEB		VEGA	
90	49 00	056	19 26	114	42 55	148	62 32	197	28 16	277	31 30	328	21 58	352
92	49 29	058	19 58	116	43 13	150	62 21	200	27 41	279	31 11	329	21 53	354
94	49 59	059	20 30	118	43 30	153	62 09	203	27 07	281	30 54	331	21 50	356
96	50 29	061	21 01	120	43 46	155	61 54	205	26 33	283	30 37	333	21 48	357
98	51 00	062	21 31	122	44 00	157	61 38	208	25 58	285	30 21	334	21 47	359
100	51 31	064	22 00	124	44 13	160	61 21	211	25 25	287	30 07	336	21 47	001
102	52 03	065	22 29	126	44 24	162	61 02	214	24 51	288	29 53	337	21 48	002
104	52 35	067	22 57	128	44 34	165	60 42	217	24 18	290	29 40	339	21 50	004
106	53 08	069	23 24	130	44 42	167	60 20	219	23 45	292	29 27	341	21 53	006
108	53 40	070	23 51	132	44 49	170	59 57	222	23 13	294	29 16	342	21 57	007
110	54 14	072	24 17	134	44 55	172	59 33	225	22 41	296	29 06	344	22 02	009
112	54 47	074	24 41	136	44 59	175	59 08	227	22 10	297	28 57	345	22 08	011
114	55 21	075	25 05	138	45 01	177	58 41	230	21 39	299	28 48	347	22 15	012
116	55 55	077	25 28	140	45 02	180	58 14	232	21 09	301	28 41	349	22 23	014
118	56 29	079	25 50	143	45 02	182	57 46	235	20 39	303	28 34	350	22 32	016
	◆VEGA		ARCTURUS		◆REGULUS		POLLUX		◆CAPELLA		Alpheratz		DENEB	
120	22 42	017	17 13	081	26 11	145	45 00	185	57 17	237	20 10	305	28 29	352
122	22 53	019	17 48	082	26 30	147	44 56	187	56 47	240	19 41	306	28 24	353
124	23 05	021	18 23	084	26 49	149	44 51	190	56 16	242	19 13	308	28 21	355
126	23 18	023	18 58	086	27 07	151	44 44	192	55 45	244	18 46	310	28 18	357
128	23 32	024	19 33	088	27 23	153	44 36	195	55 13	247	18 20	312	28 16	358
130	23 47	026	20 08	090	27 38	156	44 27	197	54 40	249	17 54	314	28 16	000
132	24 02	028	20 43	092	27 52	158	44 16	200	54 08	251	17 29	315	28 16	001
134	24 19	029	21 18	094	28 05	160	44 03	202	53 34	253	17 05	317	28 17	003
136	24 37	031	21 53	096	28 16	162	43 49	204	53 00	255	16 41	319	28 20	005
138	24 55	033	22 28	098	28 26	164	43 34	207	52 26	257	16 19	321	28 23	006
140	25 14	034	23 02	100	28 35	167	43 18	209	51 52	259	15 57	323	28 27	008
142	25 35	036	23 37	102	28 42	169	43 00	212	51 17	262	15 36	324	28 32	009
144	25 56	038	24 11	104	28 49	171	42 41	214	50 42	264	15 16	326	28 39	011
146	26 18	039	24 45	106	28 53	173	42 21	216	50 08	266	14 57	328	28 46	013
148	26 40	041	25 19	108	28 57	175	41 59	219	49 33	267	14 38	330	28 54	014
	DENEB		◆VEGA		ARCTURUS		◆REGULUS		POLLUX		◆CAPELLA		Alpheratz	
150	29 03	016	27 04	043	25 52	110	28 59	178	41 37	221	48 57	269	14 21	331
152	29 13	017	27 28	044	26 25	112	29 00	180	41 13	223	48 22	271	14 05	333
154	29 24	019	27 53	046	26 57	114	28 59	182	40 49	226	47 47	273	13 50	335
156	29 36	021	28 18	048	27 29	116	28 57	184	40 23	228	47 12	275	13 35	337
158	29 48	022	28 45	050	28 00	118	28 53	187	39 57	230	46 37	277	13 22	339
160	30 02	024	29 12	051	28 31	120	28 49	189	39 29	232	46 03	279	13 10	340
162	30 17	025	29 39	053	29 01	122	28 43	191	39 01	234	45 28	281	12 58	342
164	30 32	027	30 08	055	29 31	124	28 35	193	38 32	237	44 54	282	12 48	344
166	30 48	029	30 37	056	30 00	126	28 26	196	38 03	239	44 20	284	12 39	346
168	31 06	030	31 06	058	30 27	128	28 16	198	37 32	241	43 46	286	12 31	347
170	31 24	032	31 36	060	30 55	130	28 05	200	37 01	243	43 12	288	12 24	349
172	31 43	033	32 07	062	31 21	132	27 52	202	36 30	245	42 39	289	12 18	351
174	32 02	035	32 38	063	31 46	135	27 38	204	35 58	247	42 06	291	12 13	353
176	32 23	037	33 09	065	32 11	137	27 23	207	35 25	249	41 33	293	12 09	355
178	32 44	038	33 41	067	32 34	139	27 07	209	34 52	251	41 01	295	12 06	356

LAT 73°N

LHA ♈	Hc	Zn	Hc	Zn	Hc	Zn	Hc	Zn	Hc	Zn	Hc	Zn	Hc	Zn
	◆DENEB		VEGA		◆ARCTURUS		REGULUS		◆POLLUX		CAPELLA		Alpheratz	
180	33 06	040	34 14	069	32 57	141	26 50	211	34 19	253	40 30	296	12 04	358
182	33 29	041	34 47	070	33 18	143	26 31	213	33 45	255	39 58	298	12 04	000
184	33 53	043	35 20	072	33 39	146	26 11	215	33 11	257	39 28	300	12 04	002
186	34 17	045	35 54	074	33 58	148	25 50	217	32 36	259	38 57	301	12 06	004
188	34 42	046	36 27	076	34 16	150	25 29	220	32 02	261	38 28	303	12 09	005
190	35 08	048	37 01	078	34 33	152	25 06	222	31 27	263	37 58	305	12 12	007
192	35 34	050	37 36	079	34 49	155	24 42	224	30 52	265	37 30	306	12 17	009
194	36 01	051	38 10	081	35 03	157	24 17	226	30 17	267	37 02	308	12 23	011
196	36 29	053	38 45	083	35 16	159	23 52	228	29 42	269	36 34	310	12 30	012
198	36 57	055	39 20	085	35 28	162	23 25	230	29 07	271	36 08	311	12 38	014
200	37 26	056	39 55	087	35 38	164	22 58	232	28 32	273	35 42	313	12 48	016
202	37 56	058	40 30	089	35 47	166	22 30	234	27 57	275	35 16	314	12 58	018
204	38 25	060	41 05	091	35 55	168	22 01	236	27 22	277	34 52	316	13 09	020
206	38 56	061	41 40	092	36 01	171	21 31	238	26 47	279	34 28	318	13 21	021
208	39 27	063	42 15	094	36 06	173	21 01	240	26 13	280	34 04	319	13 35	023
	Alpheratz		◆DENEB		VEGA		◆ARCTURUS		REGULUS		POLLUX		◆CAPELLA	
210	13 49	025	39 58	065	42 50	096	36 10	175	20 31	242	25 38	282	33 42	321
212	14 04	027	40 30	066	43 25	098	36 12	178	19 59	244	25 04	284	33 20	322
214	14 21	029	41 03	068	44 00	100	36 12	180	19 27	246	24 30	286	32 59	324
216	14 38	030	41 35	070	44 34	102	36 12	183	18 55	248	23 57	288	32 39	326
218	14 56	032	42 09	072	45 08	104	36 09	185	18 22	250	23 23	290	32 19	327
220	15 15	034	42 42	073	45 42	107	36 06	187	17 49	252	22 51	292	32 01	329
222	15 35	036	43 16	075	46 15	109	36 01	190	17 15	254	22 18	293	31 43	330
224	15 56	037	43 50	077	46 48	111	35 54	192	16 41	256	21 46	295	31 26	332
226	16 18	039	44 24	079	47 21	113	35 46	194	16 07	258	21 15	297	31 10	334
228	16 40	041	44 59	080	47 53	115	35 37	196	15 33	260	20 44	299	30 55	335
230	17 04	043	45 33	082	48 25	117	35 26	199	14 58	262	20 13	301	30 41	337
232	17 28	045	46 08	084	48 55	119	35 14	201	14 23	264	19 43	302	30 28	338
234	17 53	046	46 43	086	49 26	122	35 01	203	13 48	266	19 14	304	30 15	340
236	18 19	048	47 18	088	49 55	124	34 47	206	13 13	268	18 45	306	30 04	342
238	18 45	050	47 53	090	50 24	126	34 31	208	12 38	270	18 17	308	29 53	343
	Alpheratz		◆DENEB		VEGA		◆ARCTURUS		Dubhe		POLLUX		◆CAPELLA	
240	19 12	052	48 28	092	50 52	129	34 14	210	61 41	286	17 50	310	29 44	345
242	19 40	053	49 03	094	51 18	131	33 55	212	61 08	288	17 23	311	29 35	346
244	20 09	055	49 38	096	51 44	133	33 36	215	60 35	290	16 57	313	29 27	348
246	20 38	057	50 13	098	52 09	136	33 15	217	60 02	291	16 32	315	29 20	350
248	21 08	059	50 48	100	52 33	138	32 54	219	59 29	293	16 07	317	29 15	351
250	21 38	061	51 22	102	52 56	141	32 31	221	58 57	294	15 44	319	29 10	353
252	22 09	062	51 57	104	53 17	143	32 08	224	58 25	296	15 21	320	29 06	354
254	22 40	064	52 30	106	53 38	146	31 43	226	57 54	297	14 59	322	29 03	356
256	23 12	066	53 04	108	53 57	149	31 17	228	57 23	299	14 38	324	29 01	358
258	23 44	068	53 37	110	54 14	151	30 51	230	56 52	300	14 18	326	29 00	359
260	24 17	070	54 10	112	54 31	154	30 24	232	56 22	302	13 58	327	29 00	001
262	24 50	072	54 42	115	54 45	156	29 55	234	55 53	303	13 40	329	29 01	002
264	25 24	073	55 14	117	54 59	159	29 27	236	55 23	304	13 22	331	29 03	004
266	25 57	075	55 45	119	55 10	162	28 57	238	54 55	306	13 06	333	29 05	005
268	26 31	077	56 15	121	55 20	165	28 27	241	54 27	307	12 50	335	29 09	007
	◆CAPELLA		Alpheratz		◆DENEB		VEGA		◆ARCTURUS		Dubhe		POLLUX	
270	29 14	009	27 06	079	56 45	124	55 29	167	27 56	243	53 59	309	12 36	336
272	29 20	010	27 40	081	57 13	126	55 36	170	27 25	245	53 32	310	12 22	338
274	29 27	012	28 15	083	57 41	129	55 41	173	26 53	247	53 05	312	12 10	340
276	29 34	013	28 50	084	58 08	131	55 45	176	26 20	249	52 39	313	11 58	342
278	29 43	015	29 25	086	58 34	134	55 46	178	25 47	251	52 14	314	11 48	344
280	29 52	017	30 00	088	58 59	136	55 47	181	25 14	253	51 49	316	11 38	345
282	30 03	018	30 35	090	59 22	139	55 45	184	24 40	255	51 25	317	11 30	347
284	30 14	020	31 10	092	59 44	142	55 42	187	24 06	257	51 01	318	11 23	349
286	30 27	021	31 45	094	60 06	144	55 37	189	23 32	259	50 38	320	11 17	351
288	30 40	023	32 20	096	60 25	147	55 30	192	22 58	261	50 16	321	11 12	353
290	30 54	025	32 55	098	60 44	150	55 22	195	22 23	263	49 55	323	11 08	354
292	31 09	026	33 29	100	61 00	153	55 12	198	21 48	264	49 34	324	11 05	356
294	31 25	028	34 04	102	61 16	156	55 01	200	21 13	266	49 13	325	11 03	358
296	31 42	029	34 38	104	61 29	159	54 48	203	20 38	268	48 54	327	11 02	000
298	32 00	031	35 12	106	61 41	162	54 33	206	20 03	270	48 35	328	11 02	002
	POLLUX		◆CAPELLA		Alpheratz		◆DENEB		VEGA		◆ARCTURUS		Dubhe	
300	11 04	003	32 18	033	35 46	108	61 51	165	54 18	208	19 28	272	48 16	329
302	11 07	005	32 37	034	36 19	110	62 00	167	54 00	211	18 53	274	47 59	331
304	11 11	007	32 57	036	36 52	112	62 07	170	53 41	214	18 18	276	47 42	332
306	11 15	009	33 18	037	37 24	114	62 11	173	53 21	216	17 43	278	47 26	333
308	11 21	011	33 40	039	37 56	116	62 14	176	53 00	219	17 08	280	47 11	335
310	11 28	012	34 02	041	38 27	118	62 16	180	52 38	221	16 34	282	46 56	336
312	11 36	014	34 26	042	38 57	121	62 15	183	52 14	224	16 00	283	46 42	337
314	11 45	016	34 50	044	39 27	123	62 13	186	51 49	226	15 26	285	46 29	339
316	11 56	018	35 14	045	39 56	125	62 08	189	51 24	228	14 52	287	46 17	340
318	12 07	020	35 40	047	40 25	127	62 02	192	50 57	231	14 19	289	46 05	341
320	12 19	021	36 06	049	40 52	129	61 54	195	50 29	233	13 46	291	45 54	343
322	12 32	023	36 32	050	41 19	132	61 45	198	50 01	236	13 13	293	45 44	344
324	12 47	025	37 00	052	41 45	134	61 33	200	49 31	238	12 41	295	45 35	345
326	13 02	027	37 28	054	42 10	136	61 20	203	49 01	240	12 09	297	45 27	347
328	13 18	029	37 56	055	42 33	138	61 05	206	48 31	242	11 38	298	45 19	348
	◆POLLUX		CAPELLA		Hamal		◆Alpheratz		DENEB		◆VEGA		Alioth	
330	13 35	030	38 25	057	30 30	110	42 56	141	60 49	209	47 59	245	42 22	329
332	13 54	032	38 55	059	31 02	112	43 18	143	60 31	212	47 27	247	42 05	330
334	14 13	034	39 25	060	31 35	114	43 38	145	60 12	215	46 55	249	41 47	332
336	14 33	036	39 56	062	32 06	116	43 58	148	59 51	217	46 22	251	41 31	333
338	14 54	037	40 27	064	32 37	119	44 16	150	59 29	220	45 48	253	41 16	334
340	15 15	039	40 59	065	33 08	121	44 33	153	59 06	223	45 15	255	41 01	336
342	15 38	041	41 31	067	33 38	123	44 48	155	58 42	225	44 41	257	40 47	337
344	16 01	043	42 03	069	34 07	125	45 02	157	58 16	228	44 06	259	40 34	339
346	16 26	045	42 36	070	34 35	127	45 15	160	57 50	230	43 32	261	40 21	340
348	16 51	046	43 09	072	35 03	129	45 26	162	57 22	233	42 57	263	40 10	342
350	17 16	048	43 43	074	35 30	131	45 36	165	56 54	235	42 22	265	39 59	343
352	17 43	050	44 17	076	35 55	134	45 44	167	56 24	238	41 47	267	39 50	345
354	18 10	052	44 51	078	36 20	136	45 51	170	55 54	240	41 12	269	39 41	346
356	18 38	054	45 25	079	36 44	138	45 57	172	55 24	242	40 37	271	39 33	347
358	19 07	055	46 00	081	37 07	140	46 01	175	54 52	245	40 02	273	39 25	349

LHA ♈	Hc	Zn	Hc	Zn	Hc	Zn	Hc	Zn	Hc	Zn	Hc	Zn	Hc	Zn
	◆POLLUX		CAPELLA		Hamal		◆Alpheratz		DENEB		◆VEGA		Alioth	
0	19 03	057	46 27	082	38 17	142	47 03	177	54 43	248	39 21	276	38 20	350
2	19 34	059	47 04	084	38 39	144	47 04	180	54 08	251	38 45	277	38 14	352
4	20 06	060	47 40	086	39 00	147	47 03	183	53 33	253	38 08	279	38 09	353
6	20 39	062	48 17	087	39 20	149	47 01	185	52 57	255	37 31	281	38 06	355
8	21 12	064	48 55	089	39 38	151	46 56	188	52 21	257	36 55	283	38 03	356
10	21 46	066	49 32	091	39 55	154	46 51	190	51 45	259	36 19	285	38 01	358
12	22 20	068	50 09	093	40 11	156	46 43	193	51 09	261	35 43	286	37 59	359
14	22 54	069	50 46	095	40 25	158	46 34	195	50 32	263	35 08	288	37 59	000
16	23 29	071	51 23	097	40 38	161	46 24	198	49 55	265	34 33	290	38 00	002
18	24 04	073	51 59	099	40 49	163	46 12	200	49 18	267	33 58	292	38 02	003
20	24 40	075	52 36	101	40 59	166	45 58	203	48 41	269	33 24	293	38 04	005
22	25 16	077	53 12	103	41 08	168	45 43	205	48 04	271	32 50	295	38 08	006
24	25 52	079	53 48	105	41 15	171	45 26	208	47 27	273	32 17	297	38 12	007
26	26 29	080	54 24	107	41 20	173	45 08	210	46 50	275	31 44	299	38 17	009
28	27 05	082	54 59	110	41 24	175	44 49	213	46 13	277	31 12	300	38 23	010
	Alioth		◆POLLUX		CAPELLA		◆Hamal		Alpheratz		DENEB		◆VEGA	
30	38 31	012	27 42	084	55 33	112	41 26	178	44 28	215	45 36	278	30 40	302
32	38 39	013	28 19	086	56 07	114	41 26	180	44 06	217	45 00	280	30 09	304
34	38 47	015	28 56	088	56 41	116	41 25	183	43 43	220	44 23	282	29 39	305
36	38 57	016	29 33	090	57 14	119	41 23	185	43 19	222	43 47	284	29 09	307
38	39 08	017	30 10	092	57 46	121	41 19	188	42 53	224	43 11	285	28 39	309
40	39 19	019	30 47	094	58 17	124	41 13	190	42 27	227	42 35	287	28 11	311
42	39 32	020	31 24	096	58 48	126	41 06	193	41 59	229	42 00	289	27 43	312
44	39 45	022	32 01	097	59 17	129	40 57	195	41 31	231	41 25	290	27 16	314
46	39 59	023	32 38	099	59 46	131	40 46	197	41 02	234	40 51	292	26 50	316
48	40 14	025	33 14	101	60 13	134	40 35	200	40 31	236	40 17	294	26 24	317
50	40 30	026	33 50	103	60 39	136	40 21	202	40 00	238	39 43	296	25 59	319
52	40 47	027	34 26	105	61 04	139	40 07	205	39 28	240	39 10	297	25 36	321
54	41 04	029	35 02	107	61 28	142	39 50	207	38 56	242	38 37	299	25 12	322
56	41 23	030	35 37	109	61 50	145	39 33	209	38 23	244	38 05	301	24 50	324
58	41 42	032	36 12	112	62 10	148	39 14	212	37 49	246	37 33	302	24 29	326
	Alioth		◆POLLUX		CAPELLA		◆Hamal		Alpheratz		DENEB		◆VEGA	
60	42 02	033	36 46	114	62 30	151	38 54	214	37 15	249	37 02	304	24 08	327
62	42 22	035	37 20	116	62 47	154	38 33	216	36 40	251	36 32	306	23 49	329
64	42 44	036	37 53	118	63 03	157	38 10	219	36 05	253	36 02	307	23 30	331
66	43 06	037	38 25	120	63 16	160	37 47	221	35 29	255	35 33	309	23 13	332
68	43 29	039	38 57	122	63 29	163	37 22	223	34 53	257	35 04	310	22 56	334
70	43 52	040	39 28	124	63 39	166	36 56	225	34 17	259	34 36	312	22 40	336
72	44 17	042	39 58	127	63 47	169	36 29	228	33 41	261	34 09	314	22 25	337
74	44 42	043	40 28	129	63 53	172	36 01	230	33 04	263	33 42	315	22 11	339
76	45 08	045	40 56	131	63 57	175	35 32	232	32 27	265	33 17	317	21 59	341
78	45 34	046	41 24	133	63 59	178	35 03	234	31 50	267	32 52	318	21 47	342
80	46 01	048	41 50	136	63 59	181	34 32	236	31 13	268	32 27	320	21 36	344
82	46 29	049	42 16	138	63 57	185	34 01	238	30 36	270	32 04	322	21 26	346
84	46 57	051	42 40	140	63 53	188	33 29	240	29 59	272	31 41	323	21 18	347
86	47 26	052	43 03	143	63 47	191	32 57	243	29 22	274	31 20	325	21 10	349
88	47 56	054	43 25	145	63 39	194	32 24	245	28 45	276	30 59	326	21 04	351
	Alioth		◆REGULUS		POLLUX		◆CAPELLA		Alpheratz		◆DENEB		VEGA	
90	48 26	055	19 50	113	43 46	147	63 29	197	28 08	278	30 39	328	20 58	352
92	48 56	057	20 24	115	44 05	150	63 18	200	27 32	280	30 20	330	20 54	354
94	49 28	058	20 57	117	44 23	152	63 04	203	26 55	282	30 01	331	20 50	356
96	49 59	060	21 30	119	44 40	155	62 48	206	26 19	283	29 44	333	20 48	357
98	50 32	061	22 02	121	44 55	157	62 31	209	25 43	285	29 27	334	20 47	359
100	51 04	063	22 33	123	45 09	159	62 12	212	25 07	287	29 12	336	20 47	001
102	51 37	064	23 04	125	45 21	162	61 52	215	24 32	289	28 57	338	20 48	002
104	52 11	066	23 34	127	45 32	164	61 30	218	23 57	291	28 43	339	20 50	004
106	52 45	067	24 03	130	45 41	167	61 06	221	23 23	292	28 31	341	20 53	006
108	53 20	069	24 31	132	45 48	170	60 41	223	22 49	294	28 19	342	20 57	007
110	53 54	071	24 58	134	45 54	172	60 15	226	22 15	296	28 08	344	21 02	009
112	54 30	072	25 25	136	45 59	175	59 48	229	21 42	298	27 58	345	21 09	011
114	55 05	074	25 50	138	46 01	177	59 20	231	21 10	300	27 50	347	21 16	012
116	55 41	076	26 14	140	46 02	180	58 50	234	20 38	301	27 42	349	21 25	014
118	56 17	077	26 37	142	46 02	182	58 20	236	20 06	303	27 35	350	21 34	016
	◆VEGA		ARCTURUS		◆REGULUS		POLLUX		◆CAPELLA		Alpheratz		DENEB	
120	21 45	017	17 03	080	27 00	144	45 59	185	57 49	239	19 36	305	27 29	352
122	21 56	019	17 40	082	27 21	147	45 56	187	57 16	241	19 05	307	27 25	353
124	22 09	021	18 17	084	27 40	149	45 50	190	56 44	243	18 36	309	27 21	355
126	22 22	022	18 54	086	27 59	151	45 43	192	56 10	246	18 07	310	27 18	357
128	22 37	024	19 31	088	28 16	153	45 34	195	55 36	248	17 40	312	27 16	358
130	22 53	026	20 08	090	28 33	155	45 24	197	55 01	250	17 12	314	27 16	000
132	23 09	027	20 45	092	28 47	158	45 12	200	54 26	252	16 46	316	27 16	001
134	23 27	029	21 22	093	29 01	160	44 59	202	53 51	255	16 21	317	27 17	003
136	23 45	031	21 59	095	29 13	162	44 44	205	53 15	257	15 56	319	27 20	005
138	24 05	032	22 36	097	29 24	164	44 28	207	52 39	259	15 32	321	27 23	006
140	24 25	034	23 12	099	29 33	166	44 10	210	52 02	261	15 09	323	27 28	008
142	24 46	036	23 49	101	29 41	169	43 51	212	51 25	263	14 47	324	27 33	009
144	25 08	037	24 25	103	29 48	171	43 30	215	50 49	265	14 26	326	27 40	011
146	25 31	039	25 01	105	29 53	173	43 09	217	50 12	267	14 06	328	27 47	012
148	25 55	041	25 37	107	29 57	175	42 46	219	49 35	269	13 47	330	27 56	014
	DENEB		◆VEGA		ARCTURUS		◆REGULUS		POLLUX		◆CAPELLA		Alpheratz	
150	28 05	016	26 19	042	26 12	109	29 59	178	42 22	222	48 58	271	13 28	332
152	28 16	017	26 45	044	26 47	111	30 00	180	41 57	224	48 20	272	13 11	333
154	28 27	019	27 11	046	27 21	113	29 59	182	41 30	226	47 43	274	12 55	335
156	28 39	020	27 38	047	27 55	115	29 57	184	41 03	228	47 06	276	12 40	337
158	28 53	022	28 06	049	28 28	117	29 53	187	40 35	231	46 30	278	12 26	339
160	29 07	024	28 34	051	29 01	119	29 48	189	40 06	233	45 53	280	12 13	340
162	29 22	025	29 03	052	29 33	121	29 41	191	39 36	235	45 17	282	12 01	342
164	29 39	027	29 33	054	30 04	123	29 33	193	39 05	237	44 40	283	11 50	344
166	29 56	028	30 03	056	30 35	126	29 24	196	38 33	239	44 04	285	11 41	346
168	30 14	030	30 34	058	31 04	128	29 13	198	38 01	242	43 29	287	11 32	348
170	30 33	031	31 06	059	31 33	130	29 01	200	37 28	244	42 53	289	11 25	349
172	30 52	033	31 38	061	32 01	132	28 48	202	36 55	246	42 18	290	11 18	351
174	31 13	035	32 11	063	32 28	134	28 33	205	36 21	248	41 44	292	11 13	353
176	31 35	036	32 44	064	32 55	136	28 17	207	35 46	250	41 10	294	11 09	355
178	31 57	038	33 18	066	33 20	139	28 00	209	35 11	252	40 36	295	11 06	356

LAT 72°N

LHA ♈	Hc	Zn	Hc	Zn	Hc	Zn	Hc	Zn	Hc	Zn	Hc	Zn	Hc	Zn
	◆DENEB		VEGA		◆ARCTURUS		REGULUS		◆POLLUX		CAPELLA		Alpheratz	
180	32 20	039	33 52	068	33 44	141	27 41	211	34 35	254	40 03	297	11 04	358
182	32 44	041	34 26	070	34 06	143	27 21	213	34 00	256	39 30	299	11 04	000
184	33 09	043	35 01	071	34 28	145	27 00	216	33 23	258	38 58	300	11 04	002
186	33 34	044	35 37	073	34 49	147	26 38	218	32 47	260	38 26	302	11 06	004
188	34 00	046	36 12	075	35 08	150	26 15	220	32 10	262	37 55	304	11 09	005
190	34 27	047	36 48	077	35 26	152	25 50	222	31 34	264	37 24	305	11 13	007
192	34 55	049	37 24	079	35 43	154	25 25	224	30 57	266	36 54	307	11 18	009
194	35 23	051	38 01	080	35 58	157	24 59	226	30 20	268	36 25	308	11 24	011
196	35 52	052	38 37	082	36 12	159	24 32	228	29 43	270	35 56	310	11 32	012
198	36 22	054	39 14	084	36 25	161	24 04	230	29 06	272	35 28	312	11 40	014
200	36 52	056	39 51	086	36 36	164	23 35	232	28 28	274	35 01	313	11 50	016
202	37 23	057	40 28	088	36 46	166	23 05	235	27 51	275	34 34	315	12 01	018
204	37 55	059	41 05	090	36 54	168	22 34	237	27 15	277	34 08	317	12 13	020
206	38 27	061	41 42	092	37 01	171	22 03	239	26 38	279	33 43	318	12 26	021
208	38 59	062	42 19	094	37 06	173	21 31	241	26 01	281	33 19	320	12 40	023
	Alpheratz		◆DENEB		VEGA		◆ARCTURUS		REGULUS		POLLUX		◆CAPELLA	
210	12 55	025	39 32	064	42 56	095	37 10	175	20 58	243	25 25	283	32 55	321
212	13 11	027	40 06	066	43 33	097	37 12	178	20 25	245	24 49	285	32 32	323
214	13 28	028	40 40	067	44 10	099	37 12	180	19 51	247	24 13	286	32 10	324
216	13 46	030	41 14	069	44 46	101	37 12	183	19 17	249	23 38	288	31 49	326
218	14 05	032	41 49	071	45 23	103	37 09	185	18 42	251	23 03	290	31 29	328
220	14 25	034	42 24	072	45 59	106	37 05	187	18 07	253	22 28	292	31 09	329
222	14 46	035	43 00	074	46 34	108	37 00	190	17 31	255	21 54	294	30 51	331
224	15 08	037	43 36	076	47 09	110	36 53	192	16 56	256	21 20	296	30 33	332
226	15 31	039	44 12	078	47 44	112	36 44	194	16 19	258	20 47	297	30 17	334
228	15 55	041	44 48	079	48 18	114	36 34	197	15 43	260	20 15	299	30 01	336
230	16 20	043	45 25	081	48 52	116	36 23	199	15 06	262	19 42	301	29 46	337
232	16 45	044	46 01	083	49 24	118	36 10	201	14 29	264	19 11	303	29 32	339
234	17 12	046	46 38	085	49 57	121	35 56	204	13 53	266	18 40	304	29 19	340
236	17 39	048	47 15	087	50 28	123	35 41	206	13 15	268	18 10	306	29 07	342
238	18 07	050	47 52	089	50 59	125	35 24	208	12 38	270	17 40	308	28 56	343
	Alpheratz		◆DENEB		VEGA		◆ARCTURUS		Dubhe		POLLUX		◆CAPELLA	
240	18 35	051	48 29	091	51 29	128	35 05	211	61 24	288	17 11	310	28 46	345
242	19 05	053	49 06	092	51 58	130	34 46	213	60 48	290	16 43	312	28 37	347
244	19 35	055	49 43	094	52 25	133	34 25	215	60 14	291	16 16	313	28 28	348
246	20 05	057	50 20	096	52 52	135	34 03	217	59 39	293	15 49	315	28 21	350
248	20 37	059	50 57	098	53 18	138	33 40	220	59 05	294	15 24	317	28 15	351
250	21 08	060	51 34	100	53 42	140	33 16	222	58 32	296	14 59	319	28 10	353
252	21 41	062	52 10	102	54 05	143	32 51	224	57 59	297	14 35	320	28 06	354
254	22 14	064	52 46	105	54 27	145	32 25	226	57 26	299	14 12	322	28 03	356
256	22 48	066	53 22	107	54 48	148	31 57	228	56 53	300	13 49	324	28 01	358
258	23 22	067	53 57	109	55 07	150	31 29	230	56 22	301	13 28	326	28 00	359
260	23 56	069	54 32	111	55 24	153	31 00	233	55 50	303	13 08	328	28 00	001
262	24 31	071	55 06	113	55 40	156	30 30	235	55 19	304	12 48	329	28 01	002
264	25 06	073	55 40	116	55 55	159	30 00	237	54 49	306	12 30	331	28 03	004
266	25 42	075	56 13	118	56 07	161	29 28	239	54 19	307	12 12	333	28 06	005
268	26 18	077	56 46	120	56 18	164	28 56	241	53 50	308	11 56	335	28 10	007
	◆CAPELLA		Alpheratz		◆DENEB		VEGA		◆ARCTURUS		Dubhe		POLLUX	
270	28 15	009	26 54	078	57 17	123	56 27	167	28 23	243	53 21	310	11 41	337
272	28 21	010	27 30	080	57 48	125	56 35	170	27 50	245	52 53	311	11 27	338
274	28 28	012	28 07	082	58 18	128	56 41	173	27 16	247	52 25	313	11 13	340
276	28 36	013	28 44	084	58 47	130	56 44	175	26 42	249	51 58	314	11 01	342
278	28 45	015	29 21	086	59 15	133	56 46	178	26 07	251	51 32	315	10 50	344
280	28 55	016	29 58	088	59 42	135	56 47	181	25 32	253	51 06	317	10 40	345
282	29 06	018	30 35	090	60 07	138	56 45	184	24 56	255	50 41	318	10 32	347
284	29 18	020	31 12	091	60 31	141	56 41	187	24 20	257	50 16	319	10 24	349
286	29 31	021	31 49	093	60 54	143	56 36	190	23 44	259	49 52	321	10 18	351
288	29 45	023	32 26	095	61 15	146	56 29	192	23 07	261	49 29	322	10 12	353
290	30 00	024	33 03	097	61 35	149	56 20	195	22 30	263	49 07	323	10 08	354
292	30 15	026	33 39	099	61 53	152	56 09	198	21 54	265	48 45	325	10 05	356
294	30 32	027	34 16	101	62 10	155	55 57	201	21 17	267	48 24	326	10 03	358
296	30 50	029	34 52	103	62 25	158	55 43	204	20 40	269	48 03	327	10 02	000
298	31 08	031	35 28	105	62 38	161	55 27	206	20 02	271	47 44	329	10 03	002
	◆CAPELLA		Hamal		Alpheratz		◆DENEB		VEGA		◆ARCTURUS		Dubhe	
300	31 27	032	21 42	081	36 04	107	62 49	164	55 10	209	19 25	273	47 25	330
302	31 48	034	22 19	083	36 39	109	62 58	167	54 51	212	18 48	274	47 06	331
304	32 09	035	22 56	085	37 14	111	63 06	170	54 31	214	18 11	276	46 49	333
306	32 30	037	23 33	086	37 48	113	63 11	173	54 10	217	17 35	278	46 32	334
308	32 53	039	24 10	088	38 22	116	63 14	176	53 47	219	16 58	280	46 16	335
310	33 17	040	24 47	090	38 55	118	63 16	180	53 23	222	16 22	282	46 01	336
312	33 41	042	25 24	092	39 27	120	63 15	183	52 57	225	15 45	284	45 47	338
314	34 06	043	26 01	094	39 59	122	63 12	186	52 31	227	15 10	286	45 33	339
316	34 32	045	26 38	096	40 30	124	63 08	189	52 03	229	14 34	287	45 20	340
318	34 59	047	27 15	098	41 01	126	63 01	192	51 34	232	13 59	289	45 08	342
320	35 26	048	27 51	100	41 30	129	62 52	195	51 05	234	13 24	291	44 57	343
322	35 54	050	28 28	102	41 59	131	62 42	198	50 34	237	12 50	293	44 47	344
324	36 22	051	29 04	104	42 26	133	62 29	201	50 03	239	12 16	295	44 37	346
326	36 52	053	29 40	106	42 53	135	62 15	204	49 31	241	11 42	297	44 28	347
328	37 22	055	30 15	108	43 18	138	61 59	207	48 58	243	11 10	299	44 20	348
	◆POLLUX		CAPELLA		Hamal		◆Alpheratz		DENEB		◆VEGA		Alioth	
330	12 43	030	37 52	056	30 50	110	43 42	140	61 41	210	48 24	246	41 31	329
332	13 03	032	38 23	058	31 25	112	44 06	142	61 22	213	47 50	248	41 12	331
334	13 23	034	38 55	060	31 59	114	44 28	145	61 01	216	47 16	250	40 55	332
336	13 44	036	39 27	061	32 33	116	44 48	147	60 39	218	46 41	252	40 38	333
338	14 06	037	40 00	063	33 06	118	45 08	150	60 15	221	46 05	254	40 21	335
340	14 29	039	40 33	065	33 38	120	45 26	152	59 50	224	45 30	256	40 06	336
342	14 53	041	41 07	066	34 10	122	45 42	155	59 23	227	44 53	258	39 51	338
344	15 17	043	41 41	068	34 41	124	45 57	157	58 56	229	44 17	260	39 38	339
346	15 43	044	42 16	070	35 11	127	46 11	160	58 27	232	43 40	262	39 25	341
348	16 09	046	42 51	071	35 41	129	46 23	162	57 58	234	43 04	264	39 13	342
350	16 36	048	43 26	073	36 09	131	46 34	165	57 27	237	42 27	266	39 02	343
352	17 04	050	44 02	075	36 37	133	46 43	167	56 56	239	41 50	268	38 52	345
354	17 33	052	44 38	077	37 03	135	46 50	170	56 24	241	41 12	270	38 42	346
356	18 02	053	45 14	078	37 29	138	46 56	172	55 51	244	40 35	272	38 34	348
358	18 32	055	45 50	080	37 53	140	47 00	175	55 17	246	39 58	274	38 27	349

LAT 71°N

LHA ♈	Hc	Zn	Hc	Zn	Hc	Zn	Hc	Zn	Hc	Zn	Hc	Zn	Hc	Zn
	◆POLLUX		CAPELLA		Hamal		◆Alpheratz		DENEB		◆VEGA		Alioth	
0	18 30	057	46 18	081	39 04	142	48 03	177	55 04	250	39 15	276	37 21	351
2	19 03	058	46 56	083	39 27	144	48 04	180	54 28	252	38 36	278	37 15	352
4	19 37	060	47 35	084	39 50	146	48 03	183	53 50	254	37 58	280	37 10	353
6	20 11	062	48 14	086	40 11	149	48 00	185	53 13	256	37 19	282	37 06	355
8	20 46	064	48 53	088	40 31	151	47 56	188	52 34	258	36 41	284	37 03	356
10	21 21	065	49 32	090	40 49	153	47 50	190	51 56	260	36 04	285	37 01	358
12	21 57	067	50 11	092	41 06	156	47 42	193	51 17	262	35 26	287	36 59	359
14	22 33	069	50 50	094	41 21	158	47 32	196	50 39	264	34 49	289	36 59	000
16	23 10	071	51 29	096	41 35	161	47 21	198	50 00	266	34 12	291	37 00	002
18	23 47	073	52 08	098	41 47	163	47 08	201	49 21	268	33 36	292	37 02	003
20	24 24	074	52 47	100	41 57	165	46 53	203	48 42	270	33 00	294	37 04	005
22	25 02	076	53 25	102	42 06	168	46 37	206	48 03	272	32 24	296	37 08	006
24	25 40	078	54 03	104	42 14	170	46 19	208	47 24	274	31 50	297	37 12	007
26	26 18	080	54 41	106	42 19	173	46 00	211	46 45	276	31 15	299	37 18	009
28	26 57	082	55 18	108	42 23	175	45 39	213	46 06	277	30 41	301	37 24	010
	Alioth		◆POLLUX		CAPELLA		◆Hamal		Alpheratz		DENEB		◆VEGA	
30	37 32	012	27 36	084	55 55	110	42 26	178	45 17	216	45 27	279	30 08	303
32	37 40	013	28 15	085	56 31	113	42 26	180	44 54	218	44 49	281	29 35	304
34	37 49	014	28 54	087	57 07	115	42 25	183	44 29	220	44 10	283	29 04	306
36	38 00	016	29 33	089	57 42	117	42 22	185	44 03	223	43 32	284	28 32	308
38	38 11	017	30 12	091	58 16	120	42 18	188	43 36	225	42 55	286	28 02	309
40	38 23	019	30 51	093	58 50	122	42 12	190	43 08	227	42 17	288	27 32	311
42	38 36	020	31 30	095	59 23	125	42 04	193	42 38	230	41 40	290	27 03	313
44	38 49	021	32 09	097	59 54	127	41 55	195	42 08	232	41 04	291	26 34	314
46	39 04	023	32 47	099	60 25	130	41 44	198	41 37	234	40 28	293	26 07	316
48	39 20	024	33 26	101	60 54	132	41 31	200	41 05	236	39 52	295	25 40	318
50	39 36	026	34 04	103	61 22	135	41 17	203	40 32	239	39 17	296	25 14	319
52	39 53	027	34 42	105	61 49	138	41 01	205	39 58	241	38 42	298	24 49	321
54	40 12	028	35 20	107	62 15	141	40 44	207	39 24	243	38 08	300	24 25	323
56	40 31	030	35 57	109	62 39	144	40 25	210	38 48	245	37 34	301	24 02	324
58	40 50	031	36 34	111	63 01	147	40 05	212	38 13	247	37 01	303	23 39	326
	Alioth		◆POLLUX		CAPELLA		◆Hamal		Alpheratz		DENEB		◆VEGA	
60	41 11	033	37 10	113	63 22	150	39 44	214	37 36	249	36 28	304	23 18	328
62	41 33	034	37 46	115	63 40	153	39 21	217	37 00	251	35 57	306	22 57	329
64	41 55	035	38 21	117	63 57	156	38 57	219	36 22	253	35 25	308	22 38	331
66	42 18	037	38 55	119	64 13	159	38 32	221	35 45	255	34 55	309	22 19	333
68	42 42	038	39 29	121	64 26	162	38 05	224	35 07	257	34 25	311	22 02	334
70	43 06	040	40 02	124	64 37	165	37 38	226	34 29	259	33 56	313	21 45	336
72	43 32	041	40 34	126	64 46	168	37 09	228	33 50	261	33 27	314	21 30	337
74	43 58	043	41 05	128	64 52	172	36 40	230	33 11	263	33 00	316	21 15	339
76	44 25	044	41 35	130	64 57	175	36 09	232	32 32	265	32 33	317	21 02	341
78	44 52	045	42 05	133	64 59	178	35 38	235	31 53	267	32 07	319	20 50	342
80	45 20	047	42 33	135	64 59	182	35 06	237	31 14	269	31 41	320	20 38	344
82	45 49	048	43 00	137	64 57	185	34 32	239	30 35	271	31 17	322	20 28	346
84	46 19	050	43 26	140	64 53	188	33 59	241	29 56	273	30 53	324	20 19	347
86	46 49	051	43 51	142	64 46	191	33 24	243	29 17	275	30 31	325	20 11	349
88	47 20	053	44 14	144	64 38	195	32 49	245	28 38	277	30 09	327	20 04	351
	Alioth		◆REGULUS		POLLUX		◆CAPELLA		Alpheratz		◆DENEB		VEGA	
90	47 51	054	20 14	113	44 36	147	64 27	198	28 00	278	29 48	328	19 59	352
92	48 23	056	20 50	115	44 57	149	64 14	201	27 21	280	29 28	330	19 54	354
94	48 55	057	21 25	117	45 16	152	63 59	204	26 43	282	29 09	331	19 51	356
96	49 29	059	21 59	119	45 34	154	63 42	207	26 05	284	28 50	333	19 48	357
98	50 02	060	22 33	121	45 50	157	63 23	210	25 27	286	28 33	335	19 47	359
100	50 36	062	23 06	123	46 05	159	63 03	213	24 50	287	28 17	336	19 47	001
102	51 11	063	23 39	125	46 18	162	62 40	216	24 12	289	28 02	338	19 48	002
104	51 46	065	24 10	127	46 29	164	62 17	219	23 36	291	27 47	339	19 50	004
106	52 21	066	24 41	129	46 39	167	61 51	222	23 00	293	27 34	341	19 53	006
108	52 57	068	25 11	131	46 47	169	61 24	225	22 24	295	27 22	342	19 58	007
110	53 34	069	25 40	133	46 54	172	60 56	227	21 49	296	27 11	344	20 03	009
112	54 11	071	26 07	136	46 58	174	60 27	230	21 14	298	27 00	346	20 10	011
114	54 48	073	26 34	138	47 01	177	59 57	233	20 40	300	26 51	347	20 18	012
116	55 25	074	27 00	140	47 02	180	59 25	235	20 06	302	26 43	349	20 26	014
118	56 03	076	27 25	142	47 02	182	58 53	238	19 33	303	26 36	350	20 36	016
	◆VEGA		ARCTURUS		◆REGULUS		POLLUX		◆CAPELLA		Alpheratz		DENEB	
120	20 47	017	16 53	080	27 48	144	46 59	185	58 19	240	19 01	305	26 30	352
122	21 00	019	17 31	082	28 11	146	46 55	187	57 45	242	18 29	307	26 25	353
124	21 13	021	18 10	084	28 32	148	46 49	190	57 10	245	17 59	309	26 21	355
126	21 27	022	18 49	086	28 51	151	46 41	193	56 34	247	17 29	311	26 18	357
128	21 42	024	19 28	087	29 10	153	46 32	195	55 58	249	16 59	312	26 16	358
130	21 59	026	20 07	089	29 27	155	46 21	198	55 21	252	16 31	314	26 16	000
132	22 16	027	20 46	091	29 43	157	46 08	200	54 44	254	16 03	316	26 16	001
134	22 34	029	21 25	093	29 57	160	45 54	203	54 06	256	15 36	318	26 18	003
136	22 53	030	22 04	095	30 10	162	45 38	205	53 28	258	15 10	319	26 20	004
138	23 14	032	22 43	097	30 22	164	45 21	208	52 50	260	14 45	321	26 24	006
140	23 35	034	23 22	099	30 32	166	45 02	210	52 11	262	14 21	323	26 28	008
142	23 57	035	24 00	101	30 40	169	44 41	213	51 32	264	13 58	325	26 34	009
144	24 20	037	24 39	103	30 47	171	44 20	215	50 53	266	13 36	326	26 41	011
146	24 44	039	25 16	105	30 52	173	43 57	217	50 14	268	13 15	328	26 49	012
148	25 09	040	25 54	107	30 56	175	43 32	220	49 35	270	12 55	330	26 57	014
	DENEB		◆VEGA		ARCTURUS		◆REGULUS		POLLUX		◆CAPELLA		Alpheratz	
150	27 07	015	25 35	042	26 31	109	30 59	178	43 07	222	48 56	272	12 36	332
152	27 18	017	26 02	044	27 08	111	31 00	180	42 40	225	48 17	274	12 18	333
154	27 30	019	26 29	045	27 44	113	30 59	182	42 12	227	47 38	275	12 01	335
156	27 43	020	26 57	047	28 20	115	30 56	185	41 43	229	47 00	277	11 45	337
158	27 57	022	27 26	049	28 55	117	30 53	187	41 13	231	46 21	279	11 30	339
160	28 12	023	27 56	050	29 30	119	30 47	189	40 42	234	45 42	281	11 16	340
162	28 28	025	28 26	052	30 04	121	30 40	191	40 10	236	45 04	283	11 04	342
164	28 45	026	28 57	054	30 37	123	30 32	194	39 37	238	44 26	284	10 53	344
166	29 03	028	29 29	055	31 09	125	30 22	196	39 04	240	43 48	286	10 42	346
168	29 22	030	30 02	057	31 41	127	30 10	198	38 29	242	43 11	288	10 33	348
170	29 41	031	30 35	059	32 12	129	29 58	200	37 54	244	42 34	289	10 26	349
172	30 02	033	31 09	060	32 41	132	29 43	203	37 19	247	41 57	291	10 19	351
174	30 24	034	31 43	062	33 10	134	29 27	205	36 43	249	41 21	293	10 14	353
176	30 46	036	32 18	064	33 38	136	29 10	207	36 06	251	40 45	294	10 09	355
178	31 09	037	32 53	066	34 04	138	28 52	209	35 29	253	40 10	296	10 06	356

LHA ♈	Hc	Zn	Hc	Zn	Hc	Zn	Hc	Zn	Hc	Zn	Hc	Zn	Hc	Zn
	◆DENEB		VEGA		◆ARCTURUS		REGULUS		◆POLLUX		CAPELLA		Schedar	
180	31 34	039	33 29	067	34 30	140	28 32	211	34 52	255	39 35	298	37 43	353
182	31 59	041	34 05	069	34 54	143	28 11	214	34 14	257	39 01	299	37 38	354
184	32 24	042	34 42	071	35 17	145	27 49	216	33 36	259	38 27	301	37 35	356
186	32 51	044	35 19	073	35 39	147	27 25	218	32 57	261	37 54	303	37 33	357
188	33 18	045	35 56	074	36 00	149	27 01	220	32 18	263	37 21	304	37 31	359
190	33 47	047	36 34	076	36 19	152	26 35	222	31 40	265	36 49	306	37 31	000
192	34 16	049	37 12	078	36 37	154	26 08	224	31 01	266	36 18	307	37 31	001
194	34 45	050	37 50	080	36 53	156	25 40	227	30 22	268	35 47	309	37 32	003
196	35 16	052	38 29	081	37 08	159	25 11	229	29 43	270	35 17	311	37 35	004
198	35 47	053	39 08	083	37 22	161	24 42	231	29 04	272	34 48	312	37 38	006
200	36 18	055	39 47	085	37 34	163	24 11	233	28 25	274	34 19	314	37 42	007
202	36 51	057	40 26	087	37 44	166	23 39	235	27 46	276	33 52	315	37 47	008
204	37 24	058	41 05	089	37 53	168	23 07	237	27 07	278	33 24	317	37 54	010
206	37 57	060	41 44	091	38 00	171	22 34	239	26 28	280	32 58	319	38 01	011
208	38 31	062	42 23	093	38 05	173	22 00	241	25 50	281	32 33	320	38 09	012
	Alpheratz		◆DENEB		VEGA		◆ARCTURUS		REGULUS		POLLUX		◆CAPELLA	
210	12 00	025	39 06	063	43 02	095	38 09	175	21 26	243	25 12	283	32 08	322
212	12 17	027	39 41	065	43 41	096	38 12	178	20 51	245	24 34	285	31 44	323
214	12 35	028	40 16	066	44 19	098	38 12	180	20 15	247	23 56	287	31 21	325
216	12 54	030	40 52	068	44 58	100	38 12	183	19 39	249	23 19	289	30 59	326
218	13 14	032	41 29	070	45 36	102	38 09	185	19 02	251	22 42	290	30 38	328
220	13 35	034	42 06	072	46 14	105	38 05	187	18 25	253	22 06	292	30 18	330
222	13 57	035	42 43	073	46 52	107	37 59	190	17 47	255	21 30	294	29 59	331
224	14 21	037	43 21	075	47 29	109	37 52	192	17 09	257	20 54	296	29 40	333
226	14 45	039	43 58	077	48 06	111	37 43	195	16 31	259	20 20	298	29 23	334
228	15 10	041	44 37	078	48 42	113	37 32	197	15 53	261	19 45	299	29 06	336
230	15 35	042	45 15	080	49 18	115	37 20	199	15 14	263	19 12	301	28 51	337
232	16 02	044	45 54	082	49 53	117	37 06	202	14 35	264	18 38	303	28 36	339
234	16 30	046	46 32	084	50 27	120	36 51	204	13 56	266	18 06	305	28 22	340
236	16 58	048	47 11	086	51 00	122	36 34	206	13 17	268	17 34	307	28 10	342
238	17 28	049	47 50	088	51 33	124	36 16	209	12 38	270	17 03	308	27 58	344
	Alpheratz		◆DENEB		VEGA		◆ARCTURUS		Dubhe		POLLUX		◆CAPELLA	
240	17 58	051	48 29	089	52 05	127	35 57	211	61 04	290	16 33	310	27 48	345
242	18 28	053	49 08	091	52 36	129	35 36	213	60 27	291	16 03	312	27 38	347
244	19 00	055	49 47	093	53 06	132	35 14	215	59 51	293	15 35	314	27 30	348
246	19 32	056	50 26	095	53 34	134	34 51	218	59 15	294	15 07	315	27 22	350
248	20 05	058	51 05	097	54 02	137	34 26	220	58 40	296	14 40	317	27 16	351
250	20 39	060	51 44	099	54 28	139	34 01	222	58 05	297	14 14	319	27 11	353
252	21 13	062	52 22	101	54 53	142	33 34	224	57 30	299	13 48	321	27 06	354
254	21 47	064	53 01	103	55 16	144	33 06	227	56 56	300	13 24	322	27 03	356
256	22 23	065	53 38	105	55 38	147	32 37	229	56 23	301	13 01	324	27 01	358
258	22 58	067	54 16	107	55 59	150	32 07	231	55 50	303	12 38	326	27 00	359
260	23 35	069	54 53	110	56 18	152	31 36	233	55 17	304	12 17	328	27 00	001
262	24 11	071	55 29	112	56 35	155	31 05	235	54 45	305	11 57	329	27 01	002
264	24 48	072	56 05	114	56 50	158	30 32	237	54 13	307	11 37	331	27 03	004
266	25 26	074	56 41	116	57 04	161	29 59	239	53 42	308	11 19	333	27 06	005
268	26 04	076	57 15	119	57 16	164	29 25	241	53 12	309	11 02	335	27 10	007
	◆CAPELLA		Alpheratz		◆DENEB		VEGA		◆ARCTURUS		Alkaid		Dubhe	
270	27 15	009	26 42	078	57 49	121	57 26	167	28 50	244	54 23	267	52 42	311
272	27 22	010	27 20	080	58 22	124	57 34	170	28 15	246	53 44	269	52 13	312
274	27 29	012	27 58	082	58 54	126	57 40	172	27 39	248	53 05	271	51 44	313
276	27 37	013	28 37	083	59 25	129	57 44	175	27 03	250	52 26	273	51 16	315
278	27 47	015	29 16	085	59 55	131	57 46	178	26 26	252	51 47	274	50 49	316
280	27 57	016	29 55	087	60 24	134	57 47	181	25 49	254	51 08	276	50 22	317
282	28 09	018	30 34	089	60 51	137	57 45	184	25 11	256	50 30	278	49 56	319
284	28 21	019	31 13	091	61 17	139	57 41	187	24 33	258	49 51	280	49 30	320
286	28 35	021	31 52	093	61 42	142	57 35	190	23 55	259	49 13	282	49 06	321
288	28 49	023	32 31	095	62 05	145	57 28	193	23 16	261	48 35	283	48 42	323
290	29 05	024	33 10	097	62 26	148	57 18	196	22 38	263	47 57	285	48 18	324
292	29 21	026	33 49	099	62 46	151	57 06	199	21 59	265	47 19	287	47 56	325
294	29 39	027	34 27	101	63 04	154	56 53	201	21 20	267	46 42	288	47 34	327
296	29 57	029	35 06	103	63 20	157	56 38	204	20 41	269	46 05	290	47 13	328
298	30 16	030	35 44	105	63 35	160	56 21	207	20 02	271	45 28	292	46 52	329
	◆CAPELLA		Hamal		Alpheratz		◆DENEB		VEGA		◆ARCTURUS		Dubhe	
300	30 36	032	21 32	080	36 21	107	63 47	163	56 02	210	19 23	273	46 33	330
302	30 58	033	22 11	082	36 58	109	63 57	167	55 42	212	18 44	275	46 14	332
304	31 20	035	22 50	084	37 35	111	64 05	170	55 21	215	18 05	277	45 56	333
306	31 42	037	23 29	086	38 12	113	64 11	173	54 57	218	17 26	278	45 38	334
308	32 06	038	24 08	088	38 47	115	64 14	176	54 33	220	16 47	280	45 22	336
310	32 31	040	24 47	090	39 22	117	64 16	179	54 07	223	16 09	282	45 06	337
312	32 56	041	25 26	092	39 57	119	64 15	183	53 40	225	15 31	284	44 51	338
314	33 22	043	26 05	094	40 31	121	64 12	186	53 11	228	14 53	286	44 37	339
316	33 49	044	26 44	095	41 04	123	64 07	189	52 42	230	14 16	288	44 24	341
318	34 17	046	27 23	097	41 36	126	64 00	192	52 11	233	13 39	290	44 11	342
320	34 46	048	28 01	099	42 07	128	63 50	196	51 39	235	13 02	291	44 00	343
322	35 15	049	28 40	101	42 38	130	63 39	199	51 07	238	12 26	293	43 49	345
324	35 45	051	29 18	103	43 07	132	63 25	202	50 33	240	11 51	295	43 39	346
326	36 15	052	29 56	105	43 35	135	63 10	205	49 59	242	11 15	297	43 30	347
328	36 47	054	30 33	107	44 02	137	62 52	208	49 24	244	10 41	299	43 22	348
	◆POLLUX		CAPELLA		Hamal		◆Alpheratz		DENEB		◆VEGA		Alioth	
330	11 52	030	37 19	056	31 10	109	44 28	139	62 33	211	48 49	247	40 39	330
332	12 12	032	37 51	057	31 47	111	44 53	142	62 12	214	48 13	249	40 20	331
334	12 33	034	38 24	059	32 23	113	45 17	144	61 49	217	47 36	251	40 02	332
336	12 55	035	38 58	061	32 59	115	45 39	147	61 25	220	46 59	253	39 44	334
338	13 18	037	39 32	062	33 34	117	45 59	149	61 00	222	46 21	255	39 27	335
340	13 42	039	40 07	064	34 08	120	46 19	152	60 33	225	45 43	257	39 11	337
342	14 07	041	40 42	065	34 42	122	46 36	154	60 04	228	45 05	259	38 56	338
344	14 33	042	41 18	067	35 15	124	46 53	157	59 35	230	44 27	261	38 42	339
346	15 00	044	41 54	069	35 47	126	47 07	159	59 04	233	43 48	263	38 28	341
348	15 28	046	42 31	070	36 18	128	47 20	162	58 32	236	43 09	265	38 16	342
350	15 56	048	43 08	072	36 48	130	47 32	164	58 00	238	42 30	267	38 04	344
352	16 25	050	43 45	074	37 17	133	47 41	167	57 26	240	41 51	269	37 54	345
354	16 55	051	44 23	076	37 46	135	47 49	170	56 52	243	41 12	271	37 44	346
356	17 26	053	45 01	077	38 13	137	47 56	172	56 17	245	40 33	273	37 35	348
358	17 58	055	45 39	079	38 39	139	48 00	175	55 41	247	39 54	275	37 28	349

LAT 70°N

LHA ♈	◆POLLUX Hc	Zn	CAPELLA Hc	Zn	Hamal Hc	Zn	◆Alpheratz Hc	Zn	DENEB Hc	Zn	◆VEGA Hc	Zn	Alioth Hc	Zn
0	17 57	056	46 08	080	39 51	141	49 03	177	55 25	251	39 08	277	36 22	351
2	18 31	058	46 48	082	40 16	143	49 04	180	54 46	253	38 27	279	36 15	352
4	19 07	060	47 29	083	40 40	146	49 03	183	54 06	255	37 47	281	36 10	353
6	19 42	062	48 10	085	41 02	148	49 00	185	53 26	257	37 07	283	36 06	355
8	20 19	063	48 51	087	41 23	151	48 55	188	52 46	259	36 27	284	36 03	356
10	20 56	065	49 32	089	41 42	153	48 49	191	52 06	261	35 47	286	36 01	358
12	21 33	067	50 13	091	42 00	155	48 40	193	51 25	263	35 08	288	35 59	359
14	22 11	069	50 54	093	42 16	158	48 30	196	50 44	265	34 29	290	35 59	000
16	22 50	070	51 35	095	42 31	160	48 18	198	50 03	267	33 51	291	36 00	002
18	23 29	072	52 16	097	42 44	163	48 04	201	49 22	269	33 13	293	36 02	003
20	24 08	074	52 56	099	42 55	165	47 48	204	48 41	271	32 35	295	36 05	005
22	24 48	076	53 37	101	43 05	168	47 31	206	48 00	273	31 58	296	36 08	006
24	25 27	078	54 17	103	43 13	170	47 12	209	47 19	275	31 22	298	36 13	007
26	26 08	079	54 57	105	43 19	173	46 51	211	46 38	277	30 46	300	36 19	009
28	26 48	081	55 36	107	43 23	175	46 29	214	45 58	278	30 10	301	36 25	010

LHA ♈	Alioth Hc	Zn	◆POLLUX Hc	Zn	CAPELLA Hc	Zn	◆Hamal Hc	Zn	Alpheratz Hc	Zn	DENEB Hc	Zn	◆VEGA Hc	Zn
30	36 33	011	27 29	083	56 15	109	43 26	178	46 06	216	45 17	280	29 36	303
32	36 42	013	28 10	085	56 54	111	43 26	180	45 41	219	44 37	282	29 01	305
34	36 51	014	28 51	087	57 32	114	43 25	183	45 15	221	43 57	284	28 28	306
36	37 02	016	29 32	089	58 09	116	43 22	185	44 47	223	43 17	285	27 55	308
38	37 13	017	30 13	091	58 46	118	43 17	188	44 18	226	42 38	287	27 23	310
40	37 26	018	30 54	092	59 21	121	43 11	190	43 48	228	41 59	289	26 52	311
42	37 39	020	31 35	094	59 56	123	43 03	193	43 17	230	41 20	290	26 22	313
44	37 53	021	32 15	096	60 30	126	42 53	195	42 45	233	40 42	292	25 52	315
46	38 09	022	32 56	098	61 03	128	42 41	198	42 12	235	40 04	294	25 23	316
48	38 25	024	33 37	100	61 34	131	42 27	200	41 38	237	39 27	295	24 56	318
50	38 42	025	34 17	102	62 04	134	42 12	203	41 03	239	38 50	297	24 29	320
52	39 00	027	34 57	104	62 33	137	41 55	205	40 27	242	38 13	299	24 02	321
54	39 19	028	35 37	106	63 01	140	41 37	208	39 50	244	37 38	300	23 37	323
56	39 38	029	36 16	108	63 26	142	41 17	210	39 13	246	37 03	302	23 13	325
58	39 59	031	36 55	110	63 51	145	40 56	213	38 36	248	36 28	304	22 49	326

LHA ♈	Alioth Hc	Zn	◆POLLUX Hc	Zn	CAPELLA Hc	Zn	◆Hamal Hc	Zn	Alpheratz Hc	Zn	DENEB Hc	Zn	◆VEGA Hc	Zn
60	40 20	032	37 33	112	64 13	149	40 33	215	37 57	250	35 54	305	22 27	328
62	40 43	034	38 11	114	64 33	152	40 09	217	37 18	252	35 21	307	22 06	329
64	41 06	035	38 48	116	64 52	155	39 43	220	36 39	254	34 48	308	21 45	331
66	41 30	036	39 24	119	65 08	158	39 17	222	36 00	256	34 16	310	21 26	333
68	41 55	038	40 00	121	65 23	161	38 49	224	35 20	258	33 45	311	21 08	334
70	42 20	039	40 35	123	65 35	165	38 19	226	34 39	260	33 15	313	20 50	336
72	42 46	041	41 09	125	65 44	168	37 49	229	33 59	262	32 45	315	20 34	338
74	43 13	042	41 42	127	65 52	171	37 18	231	33 18	264	32 16	316	20 19	339
76	43 41	043	42 14	130	65 57	175	36 46	233	32 37	266	31 48	318	20 05	341
78	44 10	045	42 45	132	65 59	178	36 12	235	31 56	268	31 21	319	19 52	343
80	44 39	046	43 15	134	65 59	182	35 38	237	31 15	270	30 55	321	19 41	344
82	45 09	048	43 44	137	65 57	185	35 03	240	30 34	272	30 29	322	19 30	346
84	45 40	049	44 11	139	65 52	188	34 27	242	29 53	273	30 05	324	19 21	347
86	46 11	050	44 38	141	65 45	192	33 51	244	29 12	275	29 41	326	19 12	349
88	46 43	052	45 03	144	65 36	195	33 14	246	28 31	277	29 18	327	19 05	351

LHA ♈	Alioth Hc	Zn	◆REGULUS Hc	Zn	POLLUX Hc	Zn	◆CAPELLA Hc	Zn	Alpheratz Hc	Zn	◆DENEB Hc	Zn	VEGA Hc	Zn
90	47 15	053	20 37	113	45 26	146	65 24	198	27 51	279	28 57	329	18 59	352
92	47 49	055	21 15	115	45 48	149	65 10	202	27 10	281	28 36	330	18 54	354
94	48 22	056	21 52	117	46 09	151	64 53	205	26 30	283	28 16	332	18 51	356
96	48 57	058	22 28	119	46 28	154	64 35	208	25 50	284	27 57	333	18 48	357
98	49 32	059	23 04	121	46 45	156	64 15	211	25 11	286	27 39	335	18 47	359
100	50 07	061	23 39	123	47 01	159	63 52	214	24 31	288	27 22	336	18 47	001
102	50 43	062	24 13	125	47 15	161	63 28	217	23 52	290	27 06	338	18 48	002
104	51 20	064	24 46	127	47 27	164	63 03	220	23 14	291	26 51	340	18 50	004
106	51 57	065	25 19	129	47 38	167	62 36	223	22 36	293	26 37	341	18 54	006
108	52 34	067	25 50	131	47 46	169	62 07	226	21 59	295	26 25	343	18 58	007
110	53 12	068	26 21	133	47 53	172	61 37	229	21 22	297	26 13	344	19 04	009
112	53 50	070	26 50	135	47 58	174	61 05	231	20 45	299	26 02	346	19 11	011
114	54 29	071	27 19	137	48 01	177	60 33	234	20 10	300	25 53	347	19 19	012
116	55 08	073	27 46	139	48 02	180	59 59	237	19 34	302	25 44	349	19 28	014
118	55 47	074	28 12	142	48 02	182	59 24	239	19 00	304	25 37	350	19 39	015

LHA ♈	◆VEGA Hc	Zn	ARCTURUS Hc	Zn	◆REGULUS Hc	Zn	POLLUX Hc	Zn	◆CAPELLA Hc	Zn	Alpheratz Hc	Zn	DENEB Hc	Zn
120	19 50	017	16 42	080	28 37	144	47 59	185	58 48	241	18 26	306	25 31	352
122	20 03	019	17 23	081	29 00	146	47 55	188	58 12	244	17 53	307	25 25	354
124	20 16	020	18 03	083	29 23	148	47 48	190	57 35	246	17 21	309	25 21	355
126	20 31	022	18 44	085	29 44	150	47 40	193	56 57	248	16 49	311	25 18	357
128	20 47	024	19 25	087	30 03	153	47 30	195	56 18	251	16 19	313	25 16	358
130	21 04	025	20 06	089	30 21	155	47 18	198	55 39	253	15 49	314	25 16	000
132	21 22	027	20 47	091	30 38	157	47 05	201	55 00	255	15 20	316	25 16	001
134	21 42	029	21 28	093	30 53	159	46 49	203	54 20	257	14 52	318	25 18	003
136	22 02	030	22 09	095	31 07	162	46 32	206	53 40	259	14 25	319	25 20	004
138	22 23	032	22 50	096	31 19	164	46 14	208	52 59	261	13 59	321	25 24	006
140	22 45	034	23 31	098	31 30	166	45 54	211	52 19	263	13 33	323	25 29	008
142	23 08	035	24 11	100	31 39	168	45 32	213	51 38	265	13 09	325	25 35	009
144	23 32	037	24 52	102	31 46	171	45 09	216	50 57	267	12 46	326	25 42	011
146	23 57	038	25 31	104	31 52	173	44 44	218	50 16	269	12 24	328	25 50	012
148	24 23	040	26 11	106	31 56	175	44 18	220	49 35	271	12 03	330	25 59	014

LHA ♈	DENEB Hc	Zn	◆VEGA Hc	Zn	ARCTURUS Hc	Zn	◆REGULUS Hc	Zn	POLLUX Hc	Zn	◆CAPELLA Hc	Zn	Schedar Hc	Zn
150	26 09	015	24 50	042	26 50	108	31 59	178	43 51	223	48 54	273	39 44	332
152	26 21	017	25 18	043	27 29	110	32 00	180	43 22	225	48 13	275	39 26	334
154	26 33	018	25 47	045	28 07	112	31 59	182	42 53	228	47 32	276	39 08	335
156	26 47	020	26 16	047	28 45	114	31 56	185	42 22	230	46 51	278	38 51	337
158	27 01	022	26 46	048	29 22	116	31 52	187	41 50	232	46 11	280	38 35	338
160	27 17	023	27 17	050	29 59	118	31 46	189	41 17	234	45 31	282	38 20	339
162	27 33	025	27 49	052	30 34	120	31 39	191	40 43	237	44 51	284	38 06	341
164	27 51	026	28 22	053	31 09	122	31 30	194	40 09	239	44 11	285	37 53	342
166	28 10	028	28 55	055	31 44	125	31 20	196	39 33	241	43 31	287	37 41	343
168	28 29	029	29 29	057	32 17	127	31 07	198	38 57	243	42 52	289	37 30	345
170	28 50	031	30 04	058	32 49	129	30 54	201	38 20	245	42 14	290	37 20	346
172	29 11	032	30 39	060	33 21	131	30 39	203	37 42	247	41 35	292	37 11	348
174	29 34	034	31 15	062	33 51	133	30 22	205	37 04	249	40 57	294	37 02	349
176	29 57	036	31 51	063	34 21	135	30 04	207	36 26	251	40 20	295	36 55	350
178	30 22	037	32 28	065	34 49	138	29 44	210	35 47	253	39 43	297	36 48	352

LAT 70°N

LHA ♈	◆DENEB Hc	Zn	VEGA Hc	Zn	◆ARCTURUS Hc	Zn	REGULUS Hc	Zn	◆POLLUX Hc	Zn	CAPELLA Hc	Zn	Schedar Hc	Zn
180	30 47	039	33 05	067	35 16	140	29 23	212	35 07	255	39 07	298	36 43	353
182	31 13	040	33 43	068	35 42	142	29 01	214	34 27	257	38 31	300	36 39	354
184	31 40	042	34 22	070	36 06	144	28 37	216	33 47	259	37 56	302	36 35	356
186	32 08	043	35 01	072	36 30	147	28 13	218	33 06	261	37 21	303	36 33	357
188	32 36	045	35 40	074	36 51	149	27 46	220	32 26	263	36 47	305	36 31	359
190	33 06	046	36 19	075	37 12	151	27 19	223	31 45	265	36 14	306	36 31	000
192	33 36	048	36 59	077	37 31	154	26 51	225	31 04	267	35 41	308	36 31	001
194	34 07	050	37 39	079	37 48	156	26 21	227	30 23	269	35 09	310	36 33	003
196	34 38	051	38 20	081	38 04	158	25 51	229	29 42	271	34 38	311	36 35	004
198	35 11	053	39 00	082	38 18	161	25 20	231	29 01	273	34 07	313	36 38	005
200	35 44	054	39 41	084	38 31	163	24 47	233	28 20	275	33 38	314	36 43	007
202	36 17	056	40 22	086	38 42	166	24 14	235	27 39	276	33 09	316	36 48	008
204	36 52	058	41 03	088	38 51	168	23 40	237	26 58	278	32 40	317	36 54	010
206	37 27	059	41 44	090	38 59	170	23 05	239	26 18	280	32 13	319	37 02	011
208	38 02	061	42 25	092	39 05	173	22 29	241	25 38	282	31 47	321	37 10	012

LHA ♈	Alpheratz Hc	Zn	◆DENEB Hc	Zn	VEGA Hc	Zn	◆ARCTURUS Hc	Zn	REGULUS Hc	Zn	POLLUX Hc	Zn	◆CAPELLA Hc	Zn
210	11 06	025	38 38	062	43 06	094	39 09	175	21 53	243	24 58	284	31 21	322
212	11 23	026	39 15	064	43 47	096	39 12	178	21 16	245	24 18	286	30 56	324
214	11 42	028	39 52	066	44 28	097	39 12	180	20 38	247	23 39	287	30 32	325
216	12 02	030	40 30	067	45 08	099	39 11	183	20 00	249	23 00	289	30 09	327
218	12 23	032	41 08	069	45 49	101	39 09	185	19 21	251	22 21	291	29 47	328
220	12 45	033	41 46	071	46 29	104	39 04	187	18 42	253	21 43	293	29 26	330
222	13 08	035	42 25	072	47 08	106	38 58	190	18 03	255	21 05	294	29 06	331
224	13 33	037	43 05	074	47 48	108	38 50	192	17 23	257	20 28	296	28 47	333
226	13 58	039	43 44	076	48 27	110	38 41	195	16 43	259	19 52	298	28 29	334
228	14 24	040	44 24	078	49 05	112	38 29	197	16 02	261	19 16	300	28 11	336
230	14 51	042	45 04	079	49 43	114	38 16	200	15 22	263	18 40	301	27 55	338
232	15 19	044	45 45	081	50 20	116	38 02	202	14 41	265	18 06	303	27 40	339
234	15 48	046	46 25	083	50 56	119	37 46	204	14 00	267	17 32	305	27 26	341
236	16 18	047	47 06	085	51 32	121	37 28	207	13 19	269	16 58	307	27 13	342
238	16 48	049	47 47	086	52 07	123	37 09	209	12 38	270	16 26	309	27 01	344

LHA ♈	Alpheratz Hc	Zn	◆DENEB Hc	Zn	VEGA Hc	Zn	Rasalhague Hc	Zn	◆ARCTURUS Hc	Zn	Dubhe Hc	Zn	◆CAPELLA Hc	Zn
240	17 20	051	48 28	088	52 40	126	30 40	153	36 48	211	60 43	292	26 50	345
242	17 52	053	49 09	090	53 13	128	30 58	155	36 26	214	60 05	293	26 40	347
244	18 25	054	49 50	092	53 45	131	31 15	157	36 03	216	59 27	294	26 31	348
246	18 59	056	50 31	094	54 16	133	31 30	160	35 38	218	58 50	296	26 23	350
248	19 33	058	51 12	096	54 45	136	31 43	162	35 12	220	58 13	297	26 17	351
250	20 08	060	51 53	098	55 13	138	31 55	164	34 45	223	57 37	299	26 11	353
252	20 44	061	52 33	100	55 40	141	32 06	167	34 17	225	57 01	300	26 07	355
254	21 21	063	53 14	102	56 05	144	32 14	169	33 47	227	56 26	301	26 03	356
256	21 57	065	53 54	104	56 28	146	32 22	171	33 16	229	55 51	303	26 01	358
258	22 35	067	54 33	106	56 50	149	32 27	173	32 45	231	55 17	304	26 00	359
260	23 13	068	55 12	108	57 11	152	32 31	176	32 12	234	54 43	305	26 00	001
262	23 51	070	55 51	111	57 29	155	32 33	178	31 39	236	54 10	307	26 01	002
264	24 30	072	56 29	113	57 46	157	32 34	180	31 04	238	53 37	308	26 03	004
266	25 09	074	57 07	115	58 01	160	32 33	183	30 29	240	53 05	309	26 06	005
268	25 49	076	57 44	117	58 13	163	32 30	185	29 53	242	52 33	310	26 11	007

LHA ♈	◆CAPELLA Hc	Zn	Alpheratz Hc	Zn	◆DENEB Hc	Zn	VEGA Hc	Zn	◆ARCTURUS Hc	Zn	Alkaid Hc	Zn	Dubhe Hc	Zn
270	26 16	008	26 29	077	58 20	120	58 24	166	29 17	244	54 26	268	52 03	312
272	26 23	010	27 09	079	58 55	122	58 33	169	28 40	246	53 45	270	51 32	313
274	26 30	012	27 49	081	59 29	125	58 39	172	28 02	248	53 04	272	51 03	314
276	26 39	013	28 30	083	60 02	127	58 44	175	27 24	250	52 23	274	50 34	316
278	26 49	015	29 11	085	60 34	130	58 46	178	26 45	252	51 42	276	50 05	317
280	27 00	016	29 52	087	61 05	133	58 47	181	26 05	254	51 01	277	49 38	318
282	27 12	018	30 33	088	61 34	135	58 45	184	25 26	256	50 21	279	49 11	320
284	27 25	019	31 14	090	62 02	138	58 41	187	24 46	258	49 40	281	48 44	321
286	27 39	021	31 55	092	62 29	141	58 34	190	24 06	260	49 00	283	48 19	322
288	27 54	022	32 36	094	62 54	144	58 26	193	23 25	262	48 20	284	47 54	323
290	28 10	024	33 17	096	63 17	147	58 16	196	22 44	264	47 41	286	47 30	325
292	28 27	025	33 57	098	63 39	150	58 03	199	22 03	266	47 01	288	47 06	326
294	28 45	027	34 38	100	63 58	153	57 49	202	21 22	268	46 22	289	46 44	327
296	29 04	029	35 18	102	64 15	156	57 33	205	20 41	269	45 44	291	46 22	328
298	29 24	030	35 58	104	64 31	160	57 14	208	20 00	271	45 06	293	46 01	330

LHA ♈	◆CAPELLA Hc	Zn	Hamal Hc	Zn	Alpheratz Hc	Zn	◆DENEB Hc	Zn	VEGA Hc	Zn	◆ARCTURUS Hc	Zn	Dubhe Hc	Zn
300	29 45	032	21 22	080	36 38	106	64 44	163	56 54	210	19 19	273	45 40	331
302	30 07	033	22 03	082	37 17	108	64 55	166	56 33	213	18 38	275	45 21	332
304	30 30	035	22 43	084	37 56	110	65 04	169	56 09	216	17 58	277	45 02	333
306	30 54	036	23 24	086	38 34	112	65 10	173	55 45	219	17 17	279	44 44	335
308	31 19	038	24 05	087	39 12	114	65 14	176	55 18	221	16 37	281	44 27	336
310	31 44	039	24 46	089	39 49	116	65 16	179	54 50	224	15 56	282	44 11	337
312	32 11	041	25 27	091	40 26	118	65 15	183	54 21	226	15 16	284	43 55	339
314	32 38	042	26 08	093	41 02	121	65 12	186	53 51	229	14 37	286	43 41	340
316	33 06	044	26 49	095	41 36	123	65 06	190	53 19	231	13 58	288	43 27	341
318	33 35	046	27 30	097	42 11	125	64 58	193	52 47	234	13 19	290	43 14	342
320	34 05	047	28 11	099	42 44	127	64 48	196	52 13	236	12 40	292	43 02	344
322	34 35	049	28 51	101	43 16	129	64 35	199	51 39	239	12 02	293	42 51	345
324	35 07	050	29 31	103	43 47	132	64 21	203	51 03	241	11 25	295	42 41	346
326	35 39	052	30 11	105	44 17	134	64 04	206	50 27	243	10 48	297	42 31	347
328	36 11	053	30 51	107	44 46	136	63 45	209	49 50	245	10 12	299	42 23	349

LHA ♈	◆POLLUX Hc	Zn	CAPELLA Hc	Zn	Hamal Hc	Zn	◆Alpheratz Hc	Zn	DENEB Hc	Zn	◆VEGA Hc	Zn	Alioth Hc	Zn
330	11 00	030	36 45	055	31 30	109	45 14	139	63 24	212	49 12	248	39 48	330
332	11 21	032	37 18	057	32 08	111	45 40	141	63 01	215	48 34	250	39 28	331
334	11 43	033	37 53	058	32 47	113	46 05	144	62 37	218	47 55	252	39 08	333
336	12 06	035	38 28	060	33 24	115	46 29	146	62 11	221	47 16	254	38 50	334
338	12 30	037	39 04	061	34 01	117	46 51	149	61 43	224	46 36	256	38 32	336
340	12 55	039	39 40	063	34 37	119	47 11	151	61 14	226	45 56	258	38 16	337
342	13 22	041	40 17	065	35 13	121	47 30	154	60 44	229	45 16	260	38 00	338
344	13 49	042	40 54	066	35 48	123	47 48	156	60 12	232	44 35	262	37 45	340
346	14 17	044	41 32	068	36 22	125	48 03	159	59 40	234	43 55	264	37 32	341
348	14 46	046	42 11	070	36 55	128	48 17	161	59 06	237	43 14	266	37 19	342
350	15 16	048	42 49	071	37 27	130	48 30	164	58 31	239	42 33	268	37 07	344
352	15 46	049	43 28	073	37 58	132	48 40	167	57 55	242	41 52	270	36 56	345
354	16 18	051	44 08	075	38 28	134	48 48	169	57 18	244	41 11	272	36 46	347
356	16 50	053	44 47	076	38 57	136	48 55	172	56 41	246	40 30	274	36 37	348
358	17 23	055	45 27	078	39 24	139	49 00	175	56 03	249	39 49	275	36 29	349

LHA ♈	Hc	Zn	Hc	Zn	Hc	Zn	Hc	Zn	Hc	Zn	Hc	Zn	Hc	Zn
	◆POLLUX		CAPELLA		Hamal		◆Alpheratz		DENEB		◆VEGA		Alioth	
0	17 24	056	45 57	079	40 37	141	50 03	177	55 44	252	39 00	278	35 22	351
1	17 42	057	46 18	080	40 51	142	50 03	179	55 23	253	38 39	279	35 19	351
2	18 00	058	46 39	081	41 04	143	50 04	180	55 02	255	38 18	280	35 16	352
3	18 18	059	47 00	081	41 17	144	50 04	181	54 42	256	37 56	281	35 13	353
4	18 36	060	47 21	082	41 29	145	50 03	183	54 21	257	37 35	282	35 11	354
5	18 55	060	47 43	083	41 41	146	50 02	184	54 00	258	37 14	282	35 08	354
6	19 14	061	48 04	084	41 53	148	50 00	185	53 39	259	36 53	283	35 06	355
7	19 33	062	48 26	085	42 04	149	49 57	187	53 18	260	36 32	284	35 05	356
8	19 52	063	48 47	086	42 15	150	49 55	188	52 56	261	36 12	285	35 03	356
9	20 11	064	49 08	087	42 26	151	49 51	189	52 35	262	35 51	286	35 02	357
10	20 30	065	49 30	088	42 36	153	49 48	191	52 14	263	35 30	287	35 01	358
11	20 50	066	49 51	089	42 45	154	49 43	192	51 52	264	35 10	288	35 00	358
12	21 10	067	50 13	090	42 55	155	49 39	194	51 31	265	34 49	288	34 59	359
13	21 29	067	50 34	090	43 04	156	49 33	195	51 10	266	34 29	289	34 59	000
14	21 49	068	50 56	091	43 12	157	49 28	196	50 48	267	34 09	290	34 59	000
	Alioth		◆POLLUX		CAPELLA		Hamal		◆Alpheratz		DENEB		◆VEGA	
15	35 00	001	22 09	069	51 17	092	43 20	159	49 21	198	50 27	268	33 49	291
16	35 00	002	22 29	070	51 39	093	43 28	160	49 15	199	50 05	269	33 29	292
17	35 01	002	22 50	071	52 00	094	43 35	161	49 07	200	49 44	270	33 09	293
18	35 02	003	23 10	072	52 22	095	43 41	162	49 00	201	49 22	270	32 49	294
19	35 03	004	23 31	073	52 43	096	43 48	164	48 52	203	49 01	271	32 29	294
20	35 05	004	23 51	074	53 04	097	43 53	165	48 43	204	48 39	272	32 10	295
21	35 07	005	24 12	074	53 26	098	43 59	166	48 34	205	48 18	273	31 50	296
22	35 09	006	24 33	075	53 47	099	44 04	168	48 25	207	47 56	274	31 31	297
23	35 11	007	24 53	076	54 08	100	44 08	169	48 15	208	47 35	275	31 12	298
24	35 13	007	25 14	077	54 29	101	44 12	170	48 04	209	47 13	276	30 53	299
25	35 16	008	25 35	078	54 50	102	44 15	171	47 54	211	46 52	277	30 34	299
26	35 19	009	25 56	079	55 11	103	44 18	173	47 43	212	46 31	278	30 16	300
27	35 23	009	26 18	080	55 32	104	44 21	174	47 31	213	46 09	279	29 57	301
28	35 26	010	26 39	081	55 53	105	44 23	175	47 19	214	45 48	279	29 39	302
29	35 30	011	27 00	082	56 14	107	44 25	177	47 07	216	45 27	280	29 21	303
	Alioth		◆POLLUX		ALDEBARAN		◆Hamal		Alpheratz		DENEB		◆VEGA	
30	35 34	011	27 21	083	32 11	135	44 26	178	46 54	217	45 06	281	29 03	304
31	35 39	012	27 43	083	32 26	136	44 26	179	46 41	218	44 45	282	28 45	304
32	35 43	013	28 04	084	32 41	137	44 26	180	46 27	219	44 24	283	28 27	305
33	35 48	013	28 26	085	32 55	138	44 26	182	46 14	221	44 03	284	28 10	306
34	35 53	014	28 47	086	33 09	139	44 25	183	46 00	222	43 42	285	27 52	307
35	35 58	015	29 08	087	33 23	140	44 24	184	45 45	223	43 21	285	27 35	308
36	36 04	015	29 30	088	33 37	141	44 22	185	45 30	224	43 01	286	27 18	308
37	36 10	016	29 51	089	33 50	142	44 20	187	45 15	225	42 40	287	27 02	309
38	36 16	017	30 13	090	34 03	144	44 17	188	45 00	227	42 20	288	26 45	310
39	36 22	017	30 34	091	34 16	145	44 14	189	44 44	228	41 59	289	26 29	311
40	36 29	018	30 56	092	34 28	146	44 10	191	44 28	229	41 39	290	26 12	312
41	36 36	019	31 17	093	34 40	147	44 06	192	44 11	230	41 19	290	25 57	313
42	36 43	019	31 39	094	34 52	148	44 01	193	43 55	231	40 59	291	25 41	313
43	36 50	020	32 00	095	35 03	149	43 56	194	43 38	232	40 39	292	25 25	314
44	36 57	021	32 22	096	35 14	150	43 50	196	43 21	233	40 19	293	25 10	315
	Alioth		◆POLLUX		ALDEBARAN		◆Hamal		Alpheratz		DENEB		◆VEGA	
45	37 05	022	32 43	097	35 24	152	43 44	197	43 03	235	39 59	294	24 55	316
46	37 13	022	33 04	098	35 34	153	43 38	198	42 46	236	39 39	295	24 40	317
47	37 21	023	33 26	099	35 44	154	43 31	199	42 28	237	39 20	295	24 25	317
48	37 30	024	33 47	099	35 53	155	43 23	201	42 10	238	39 00	296	24 11	318
49	37 39	024	34 08	100	36 02	156	43 16	202	41 51	239	38 41	297	23 57	319
50	37 48	025	34 29	101	36 10	157	43 07	203	41 33	240	38 22	298	23 43	320
51	37 57	026	34 50	102	36 18	159	42 59	204	41 14	241	38 03	299	23 29	321
52	38 06	026	35 11	103	36 26	160	42 50	206	40 55	242	37 44	299	23 15	322
53	38 16	027	35 32	104	36 33	161	42 40	207	40 36	243	37 26	300	23 02	322
54	38 26	028	35 53	105	36 40	162	42 30	208	40 17	244	37 07	301	22 49	323
55	38 36	028	36 14	106	36 47	163	42 20	209	39 57	246	36 49	302	22 36	324
56	38 46	029	36 34	107	36 53	164	42 09	211	39 38	247	36 31	303	22 24	325
57	38 57	030	36 55	108	36 58	166	41 58	212	39 18	248	36 13	303	22 12	326
58	39 07	030	37 15	109	37 03	167	41 46	213	38 58	249	35 55	304	22 00	326
59	39 18	031	37 35	110	37 08	168	41 34	214	38 38	250	35 37	305	21 48	327
	Alioth		◆POLLUX		ALDEBARAN		◆Hamal		Alpheratz		DENEB		◆VEGA	
60	39 30	032	37 55	112	37 12	169	41 22	215	38 17	251	35 19	306	21 36	328
61	39 41	032	38 15	113	37 16	170	41 10	217	37 57	252	35 02	306	21 25	329
62	39 53	033	38 35	114	37 19	172	40 57	218	37 37	253	34 45	307	21 14	330
63	40 04	034	38 55	115	37 22	173	40 43	219	37 16	254	34 28	308	21 03	330
64	40 17	034	39 14	116	37 25	174	40 30	220	36 55	255	34 11	309	20 53	331
65	40 29	035	39 33	117	37 27	175	40 15	221	36 34	256	33 54	310	20 43	332
66	40 41	036	39 52	118	37 28	176	40 01	222	36 14	257	33 38	310	20 33	333
67	40 54	037	40 11	119	37 29	178	39 46	224	35 53	258	33 22	311	20 23	334
68	41 07	037	40 30	120	37 30	179	39 32	225	35 32	259	33 05	312	20 14	335
69	41 20	038	40 49	121	37 30	180	39 16	226	35 10	260	32 50	313	20 04	335
70	41 33	039	41 07	122	37 30	181	39 01	227	34 49	261	32 34	313	19 56	336
71	41 47	039	41 25	123	37 29	183	38 45	228	34 28	262	32 18	314	19 47	337
72	42 01	040	41 43	124	37 28	184	38 29	229	34 07	263	32 03	315	19 39	338
73	42 14	041	42 00	126	37 26	185	38 12	230	33 45	264	31 48	316	19 31	339
74	42 29	041	42 18	127	37 24	186	37 55	231	33 24	265	31 33	317	19 23	339
	Alioth		◆REGULUS		PROCYON		◆ALDEBARAN		Hamal		Alpheratz		◆DENEB	
75	42 43	042	15 49	098	21 04	137	37 22	187	37 39	233	33 03	266	31 18	317
76	42 57	043	16 11	099	21 19	138	37 19	189	37 21	234	32 41	267	31 04	318
77	43 12	043	16 32	100	21 33	139	37 15	190	37 04	235	32 20	267	30 50	319
78	43 27	044	16 53	101	21 47	140	37 11	191	36 46	236	31 58	268	30 36	320
79	43 42	045	17 14	102	22 01	141	37 07	192	36 28	237	31 37	269	30 22	320
80	43 57	045	17 35	103	22 14	142	37 02	193	36 10	238	31 15	270	30 08	321
81	44 13	046	17 56	104	22 27	143	36 57	195	35 52	239	30 54	271	29 55	322
82	44 28	047	18 17	104	22 40	144	36 51	196	35 33	240	30 32	272	29 42	323
83	44 44	048	18 38	105	22 52	145	36 45	197	35 15	241	30 11	273	29 29	324
84	45 00	048	18 58	106	23 04	146	36 39	198	34 56	242	29 49	274	29 16	324
85	45 16	049	19 19	107	23 16	147	36 32	199	34 36	243	29 28	275	29 04	325
86	45 32	050	19 39	108	23 27	149	36 25	200	34 17	244	29 06	276	28 52	326
87	45 49	050	20 00	109	23 39	150	36 17	202	33 58	245	28 45	277	28 40	327
88	46 05	051	20 20	110	23 49	151	36 09	203	33 38	246	28 24	278	28 28	327
89	46 22	052	20 40	111	24 00	152	36 00	204	33 18	247	28 02	279	28 16	328

LAT 69°N

LHA ♈	Hc	Zn	Hc	Zn	Hc	Zn	Hc	Zn	Hc	Zn	Hc	Zn	Hc	Zn
	Alioth		◆REGULUS		PROCYON		BETELGEUSE		◆ALDEBARAN		Hamal		◆DENEB	
90	46 39	052	21 00	112	24 10	153	28 24	181	35 51	205	32 58	248	28 05	329
91	46 56	053	21 20	113	24 19	154	28 23	183	35 42	206	32 38	249	27 54	330
92	47 14	054	21 40	114	24 29	155	28 22	184	35 32	208	32 18	251	27 44	330
93	47 31	055	21 59	115	24 37	156	28 21	185	35 22	209	31 58	252	27 33	331
94	47 49	055	22 19	116	24 46	157	28 19	186	35 12	210	31 37	253	27 23	332
95	48 06	056	22 38	117	24 54	158	28 16	187	35 01	211	31 17	254	27 13	333
96	48 24	057	22 57	118	25 02	159	28 13	188	34 49	212	30 56	255	27 03	334
97	48 42	057	23 16	119	25 09	160	28 10	189	34 38	213	30 35	255	26 54	334
98	49 01	058	23 34	120	25 16	161	28 06	190	34 26	214	30 14	256	26 45	335
99	49 19	059	23 53	121	25 23	163	28 02	192	34 14	216	29 53	257	26 36	336
100	49 37	060	24 11	122	25 29	164	27 58	193	34 01	217	29 32	258	26 27	337
101	49 56	060	24 29	123	25 35	165	27 53	194	33 48	218	29 11	259	26 19	337
102	50 15	061	24 47	124	25 41	166	27 47	195	33 34	219	28 50	260	26 10	338
103	50 33	062	25 05	125	25 46	167	27 42	196	33 21	220	28 29	261	26 03	339
104	50 52	062	25 22	126	25 50	168	27 36	197	33 07	221	28 08	262	25 55	340
	Alioth		◆REGULUS		PROCYON		BETELGEUSE		◆CAPELLA		Hamal		◆DENEB	
105	51 12	063	25 39	127	25 55	169	27 29	198	63 34	223	27 46	263	25 48	340
106	51 31	064	25 56	128	25 58	170	27 22	199	63 19	224	27 25	264	25 41	341
107	51 50	065	26 13	130	26 02	171	27 15	200	63 04	226	27 04	265	25 34	342
108	52 10	065	26 29	131	26 05	173	27 07	202	62 48	227	26 42	266	25 27	343
109	52 29	066	26 46	132	26 07	174	26 59	203	62 32	229	26 21	267	25 21	344
110	52 49	067	27 01	133	26 10	175	26 50	204	62 16	230	25 59	268	25 15	344
111	53 09	068	27 17	134	26 11	176	26 42	205	61 59	231	25 38	269	25 09	345
112	53 29	068	27 33	135	26 13	177	26 32	206	61 42	233	25 16	270	25 04	346
113	53 49	069	27 48	136	26 14	178	26 23	207	61 25	234	24 55	271	24 59	347
114	54 09	070	28 02	137	26 14	179	26 13	208	61 07	235	24 33	272	24 54	347
115	54 29	071	28 17	138	26 14	180	26 02	209	60 49	237	24 12	273	24 50	348
116	54 50	071	28 31	139	26 14	181	25 52	210	60 31	238	23 50	274	24 45	349
117	55 10	072	28 45	140	26 13	182	25 41	211	60 13	239	23 29	275	24 41	350
118	55 30	073	28 59	141	26 12	184	25 29	212	59 54	241	23 07	275	24 38	350
119	55 51	074	29 12	142	26 11	185	25 18	214	59 35	242	22 46	276	24 34	351
	◆VEGA		ARCTURUS		◆REGULUS		PROCYON		◆CAPELLA		Schedar		DENEB	
120	18 53	017	16 31	079	29 25	143	26 09	186	59 16	243	45 20	312	24 31	352
121	18 59	018	16 52	080	29 38	145	26 06	187	58 57	244	45 04	313	24 28	353
122	19 06	019	17 14	081	29 50	146	26 03	188	58 38	245	44 48	314	24 26	354
123	19 13	019	17 35	082	30 02	147	26 00	189	58 18	247	44 33	315	24 23	354
124	19 20	020	17 56	083	30 14	148	25 57	190	57 58	248	44 18	315	24 22	355
125	19 28	021	18 18	084	30 25	149	25 53	191	57 38	249	44 03	316	24 20	356
126	19 36	022	18 39	085	30 36	150	25 48	192	57 18	250	43 48	317	24 18	357
127	19 44	023	19 00	086	30 46	151	25 43	194	56 58	251	43 33	317	24 17	357
128	19 52	024	19 22	087	30 56	152	25 38	195	56 38	252	43 18	318	24 17	358
129	20 01	024	19 43	088	31 06	153	25 32	196	56 17	253	43 04	319	24 16	359
130	20 10	025	20 05	089	31 16	155	25 26	197	55 56	254	42 50	319	24 16	000
131	20 19	026	20 26	089	31 25	156	25 20	198	55 36	255	42 36	320	24 16	001
132	20 29	027	20 48	090	31 33	157	25 13	199	55 15	256	42 22	321	24 16	001
133	20 39	028	21 09	091	31 42	158	25 06	200	54 54	258	42 09	321	24 17	002
134	20 49	028	21 31	092	31 49	159	24 58	201	54 33	259	41 56	322	24 18	003
	DENEB		◆VEGA		ARCTURUS		◆REGULUS		PROCYON		◆CAPELLA		Schedar	
135	24 19	004	20 59	029	21 52	093	31 57	160	24 50	202	54 12	260	41 42	323
136	24 20	004	21 10	030	22 14	094	32 04	161	24 42	203	53 50	261	41 30	323
137	24 22	005	21 21	031	22 35	095	32 11	163	24 33	204	53 29	262	41 17	324
138	24 24	006	21 32	032	22 57	096	32 17	164	24 24	206	53 08	263	41 04	325
139	24 27	007	21 43	032	23 18	097	32 23	165	24 15	207	52 47	264	40 52	325
140	24 29	007	21 55	033	23 39	098	32 28	166	24 05	208	52 25	265	40 40	326
141	24 32	008	22 07	034	24 01	099	32 33	167	23 55	209	52 04	266	40 28	327
142	24 35	009	22 19	035	24 22	100	32 38	168	23 44	210	51 42	267	40 16	327
143	24 39	010	22 32	036	24 43	101	32 42	169	23 33	211	51 21	267	40 05	328
144	24 43	011	22 44	037	25 04	102	32 45	171	23 22	212	50 59	268	39 54	329
145	24 47	011	22 57	037	25 25	103	32 49	172	23 11	213	50 38	269	39 43	329
146	24 51	012	23 10	038	25 46	104	32 52	173	22 59	214	50 16	270	39 32	330
147	24 56	013	23 24	039	26 07	105	32 54	174	22 46	215	49 55	271	39 21	331
148	25 01	014	23 37	040	26 28	106	32 56	175	22 34	216	49 33	272	39 11	332
149	25 06	014	23 51	041	26 48	107	32 58	176	22 21	217	49 12	273	39 01	332
	DENEB		◆VEGA		ARCTURUS		◆REGULUS		PROCYON		◆CAPELLA		Schedar	
150	25 12	015	24 05	041	27 09	108	32 59	178	22 08	218	48 50	274	38 51	333
151	25 17	016	24 20	042	27 29	109	32 59	179	21 54	219	48 29	275	38 41	334
152	25 23	017	24 34	043	27 49	110	33 00	180	21 40	220	48 08	276	38 32	334
153	25 30	018	24 49	044	28 10	111	32 59	181	21 26	221	47 46	277	38 22	335
154	25 36	018	25 04	045	28 30	112	32 59	182	21 12	223	47 25	278	38 13	336
155	25 43	019	25 19	046	28 50	113	32 58	183	20 57	224	47 04	278	38 05	336
156	25 50	020	25 35	046	29 09	114	32 56	185	20 42	225	46 42	279	37 56	337
157	25 58	021	25 50	047	29 29	115	32 54	186	20 27	226	46 21	280	37 48	338
158	26 05	021	26 06	048	29 48	116	32 52	187	20 12	227	46 00	281	37 40	338
159	26 13	022	26 22	049	30 08	117	32 49	188	19 56	228	45 39	282	37 32	339
160	26 22	023	26 39	050	30 27	118	32 46	189	19 40	229	45 18	283	37 24	340
161	26 30	024	26 55	050	30 46	119	32 42	190	19 24	230	44 57	284	37 17	340
162	26 39	024	27 12	051	31 05	120	32 38	192	19 07	231	44 36	284	37 10	341
163	26 48	025	27 29	052	31 23	121	32 33	193	18 50	232	44 15	285	37 03	342
164	26 57	026	27 46	053	31 41	122	32 28	194	18 33	233	43 55	286	36 56	342
	DENEB		◆VEGA		ARCTURUS		◆REGULUS		POLLUX		◆CAPELLA		Schedar	
165	27 07	027	28 03	054	32 00	123	32 23	195	40 21	241	43 34	287	36 50	343
166	27 17	028	28 20	055	32 17	124	32 17	196	40 02	242	43 13	288	36 44	344
167	27 27	028	28 38	055	32 35	125	32 11	197	39 43	243	42 53	289	36 38	344
168	27 37	029	28 56	056	32 53	126	32 04	199	39 24	244	42 33	289	36 32	345
169	27 47	030	29 14	057	33 10	127	31 57	200	39 04	245	42 13	290	36 27	346
170	27 58	031	29 32	058	33 27	128	31 50	201	38 45	246	41 52	291	36 22	346
171	28 09	031	29 50	059	33 44	129	31 42	202	38 25	247	41 32	292	36 17	347
172	28 21	032	30 08	059	34 00	131	31 34	203	38 05	248	41 13	293	36 12	348
173	28 32	033	30 27	060	34 16	132	31 25	204	37 45	249	40 53	294	36 07	348
174	28 44	034	30 46	061	34 32	133	31 16	205	37 25	250	40 33	294	36 03	349
175	28 56	034	31 05	062	34 48	134	31 07	206	37 05	251	40 14	295	35 59	350
176	29 08	035	31 24	063	35 03	135	30 57	208	36 44	252	39 54	296	35 56	350
177	29 21	036	31 43	064	35 18	136	30 47	209	36 24	253	39 35	297	35 52	351
178	29 34	037	32 02	064	35 33	137	30 36	210	36 03	254	39 16	298	35 49	352
179	29 47	038	32 22	065	35 48	138	30 25	211	35 43	255	38 57	298	35 46	353

LHA ♈	Hc	Zn	Hc	Zn	Hc	Zn	Hc	Zn	Hc	Zn	Hc	Zn	Hc	Zn
	◆DENEB		VEGA		◆ARCTURUS		REGULUS		◆POLLUX		CAPELLA		Schedar	
180	30 00	038	32 41	066	36 02	139	30 14	212	35 22	256	38 38	299	35 43	353
181	30 13	039	33 01	067	36 15	141	30 03	213	35 01	257	38 19	300	35 41	354
182	30 27	040	33 21	068	36 29	142	29 51	214	34 40	258	38 01	301	35 39	355
183	30 41	041	33 41	069	36 42	143	29 38	215	34 19	259	37 42	302	35 37	355
184	30 55	041	34 01	069	36 55	144	29 26	216	33 58	260	37 24	302	35 35	356
185	31 09	042	34 21	070	37 07	145	29 13	218	33 36	261	37 06	303	35 34	357
186	31 24	043	34 41	071	37 20	146	28 59	219	33 15	262	36 48	304	35 33	357
187	31 39	044	35 02	072	37 31	147	28 46	220	32 54	263	36 30	305	35 32	358
188	31 53	044	35 22	073	37 43	149	28 32	221	32 33	264	36 13	305	35 31	359
189	32 09	045	35 43	074	37 54	150	28 18	222	32 11	265	35 55	306	35 31	359
190	32 24	046	36 04	075	38 04	151	28 03	223	31 50	266	35 38	307	35 31	000
191	32 40	047	36 24	076	38 15	152	27 48	224	31 28	267	35 21	308	35 31	001
192	32 55	048	36 45	076	38 24	153	27 33	225	31 07	268	35 04	309	35 31	001
193	33 11	048	37 06	077	38 34	155	27 18	226	30 45	269	34 47	309	35 32	002
194	33 28	049	37 27	078	38 43	156	27 02	227	30 24	270	34 31	310	35 33	003
	Schedar		◆DENEB		VEGA		◆ARCTURUS		REGULUS		◆POLLUX		CAPELLA	
195	35 34	003	33 44	050	37 48	079	38 52	157	26 46	228	30 02	271	34 14	311
196	35 35	004	34 00	051	38 09	080	39 00	158	26 30	229	29 41	271	33 58	312
197	35 37	005	34 17	051	38 31	081	39 08	159	26 14	230	29 19	272	33 42	312
198	35 39	005	34 34	052	38 52	082	39 15	161	25 57	231	28 58	273	33 26	313
199	35 41	006	34 51	053	39 13	083	39 22	162	25 40	233	28 36	274	33 11	314
200	35 43	007	35 08	054	39 35	083	39 28	163	25 23	234	28 15	275	32 56	315
201	35 46	007	35 26	055	39 56	084	39 34	164	25 06	235	27 53	276	32 40	316
202	35 49	008	35 43	055	40 17	085	39 40	165	24 48	236	27 32	277	32 25	316
203	35 52	009	36 01	056	40 39	086	39 45	167	24 30	237	27 11	278	32 11	317
204	35 55	009	36 19	057	41 00	087	39 50	168	24 12	238	26 50	279	31 56	318
205	35 59	010	36 37	058	41 22	088	39 54	169	23 54	239	26 28	280	31 42	319
206	36 03	011	36 56	059	41 43	089	39 58	170	23 35	240	26 07	281	31 28	319
207	36 07	011	37 14	059	42 05	090	40 02	172	23 17	241	25 46	281	31 14	320
208	36 11	012	37 33	060	42 26	091	40 05	173	22 58	242	25 25	282	31 00	321
209	36 16	013	37 51	061	42 48	092	40 07	174	22 39	243	25 04	283	30 47	322
	Schedar		◆DENEB		VEGA		◆ARCTURUS		REGULUS		◆POLLUX		CAPELLA	
210	36 21	014	38 10	062	43 09	093	40 09	175	22 20	244	24 43	284	30 33	322
211	36 26	014	38 29	063	43 31	094	40 11	176	22 00	245	24 22	285	30 20	323
212	36 31	015	38 48	063	43 52	095	40 12	178	21 41	246	24 02	286	30 08	324
213	36 37	016	39 08	064	44 14	096	40 12	179	21 21	247	23 41	287	29 55	325
214	36 43	016	39 27	065	44 35	097	40 12	180	21 01	248	23 20	288	29 43	325
215	36 49	017	39 47	066	44 56	097	40 12	181	20 41	249	23 00	289	29 31	326
216	36 56	018	40 06	067	45 18	098	40 11	183	20 21	250	22 40	289	29 19	327
217	37 02	018	40 26	067	45 39	099	40 10	184	20 01	251	22 19	290	29 07	328
218	37 09	019	40 46	068	46 00	100	40 09	185	19 41	252	21 59	291	28 56	329
219	37 16	020	41 06	069	46 21	101	40 06	186	19 20	253	21 39	292	28 45	329
220	37 23	020	41 26	070	46 42	102	40 04	188	19 00	254	21 20	293	28 34	330
221	37 31	021	41 46	071	47 03	104	40 01	189	18 39	255	21 00	294	28 24	331
222	37 39	022	42 07	072	47 24	105	39 57	190	18 18	255	20 40	295	28 13	332
223	37 47	022	42 27	072	47 45	106	39 53	191	17 57	256	20 21	296	28 03	332
224	37 55	023	42 48	073	48 05	107	39 49	193	17 36	257	20 01	297	27 53	333
	Schedar		◆DENEB		VEGA		◆ARCTURUS		REGULUS		◆POLLUX		CAPELLA	
225	38 04	024	43 08	074	48 26	108	39 44	194	17 15	258	19 42	297	27 44	334
226	38 12	024	43 29	075	48 46	109	39 39	195	16 54	259	19 23	298	27 34	335
227	38 21	025	43 50	076	49 07	110	39 33	196	16 33	260	19 04	299	27 25	335
228	38 31	026	44 11	077	49 27	111	39 27	197	16 12	261	18 46	300	27 16	336
229	38 40	026	44 32	077	49 47	112	39 20	199	15 51	262	18 27	301	27 08	337
230	38 50	027	44 53	078	50 07	113	39 13	200	15 29	263	18 09	302	27 00	338
231	39 00	028	45 14	079	50 26	114	39 05	201	15 08	264	17 51	303	26 52	338
232	39 10	028	45 35	080	50 46	115	38 57	202	14 46	265	17 33	304	26 44	339
233	39 20	029	45 56	081	51 05	116	38 49	203	14 25	266	17 15	304	26 36	340
234	39 31	030	46 17	082	51 24	118	38 40	205	14 04	267	16 57	305	26 29	341
235	39 41	030	46 39	083	51 43	119	38 31	206	13 42	268	16 40	306	26 22	342
236	39 52	031	47 00	084	52 02	120	38 22	207	13 21	269	16 22	307	26 16	342
237	40 04	032	47 21	084	52 21	121	38 12	208	12 59	270	16 05	308	26 09	343
238	40 15	032	47 43	085	52 39	122	38 01	209	12 38	271	15 48	309	26 03	344
239	40 27	033	48 04	086	52 57	123	37 51	211	12 16	272	15 32	310	25 57	345
	Alpheratz		◆DENEB		VEGA		Rasalhague		◆ARCTURUS		Alioth		◆CAPELLA	
240	16 42	051	48 26	087	53 15	125	31 34	153	37 39	212	65 45	261	25 52	345
241	16 59	052	48 47	088	53 32	126	31 43	154	37 28	213	65 23	262	25 46	346
242	17 16	052	49 09	089	53 50	127	31 53	155	37 16	214	65 02	263	25 41	347
243	17 33	053	49 30	090	54 07	128	32 02	156	37 04	215	64 41	264	25 37	348
244	17 50	054	49 52	091	54 24	129	32 10	157	36 51	216	64 19	265	25 32	348
245	18 08	055	50 13	092	54 40	131	32 18	158	36 38	217	63 58	266	25 28	349
246	18 25	056	50 35	093	54 56	132	32 26	159	36 25	219	63 36	267	25 24	350
247	18 43	057	50 56	094	55 12	133	32 33	161	36 12	220	63 15	268	25 21	351
248	19 01	058	51 18	095	55 27	135	32 40	162	35 58	221	62 53	269	25 17	352
249	19 20	058	51 39	096	55 43	136	32 47	163	35 43	222	62 32	270	25 14	352
250	19 38	059	52 00	097	55 57	137	32 53	164	35 29	223	62 10	271	25 11	353
251	19 57	060	52 22	098	56 12	139	32 59	165	35 14	224	61 49	272	25 09	354
252	20 15	061	52 43	099	56 26	140	33 04	166	34 59	225	61 27	273	25 07	355
253	20 34	062	53 04	100	56 39	141	33 09	168	34 43	226	61 06	274	25 05	355
254	20 53	063	53 25	101	56 53	143	33 13	169	34 28	228	60 44	275	25 03	356
	Alpheratz		◆DENEB		ALTAIR		Rasalhague		◆ARCTURUS		Alioth		◆CAPELLA	
255	21 12	064	53 46	102	23 51	133	33 17	170	34 12	229	60 23	276	25 02	357
256	21 32	065	54 07	103	24 06	134	33 21	171	33 55	230	60 02	276	25 01	358
257	21 51	065	54 28	104	24 21	135	33 24	172	33 39	231	59 40	277	25 00	358
258	22 11	066	54 49	105	24 36	136	33 27	173	33 22	232	59 19	278	25 00	359
259	22 31	067	55 10	106	24 51	137	33 29	175	33 05	233	58 58	279	25 00	000
260	22 51	068	55 31	107	25 06	138	33 31	176	32 48	234	58 37	280	25 00	001
261	23 11	069	55 51	108	25 20	139	33 32	177	32 30	235	58 15	281	25 00	001
262	23 31	070	56 12	109	25 34	140	33 33	178	32 12	236	57 54	282	25 01	002
263	23 51	071	56 32	110	25 47	141	33 34	179	31 54	237	57 33	282	25 02	003
264	24 11	072	56 52	111	26 01	142	33 34	180	31 36	238	57 12	283	25 03	004
265	24 32	072	57 12	113	26 14	144	33 33	182	31 18	239	56 51	284	25 05	005
266	24 52	073	57 32	114	26 26	145	33 33	183	30 59	240	56 31	285	25 06	005
267	25 13	074	57 51	115	26 38	146	33 31	184	30 40	241	56 10	286	25 09	006
268	25 34	075	58 11	116	26 50	147	33 30	185	30 21	242	55 49	286	25 11	007
269	25 55	076	58 30	117	27 02	148	33 28	186	30 02	244	55 29	287	25 14	008

LAT 69°N

LHA ♈	Hc	Zn	Hc	Zn	Hc	Zn	Hc	Zn	Hc	Zn	Hc	Zn	Hc	Zn
	◆CAPELLA		Alpheratz		◆DENEB		ALTAIR		Rasalhague		◆ARCTURUS		Alioth	
270	25 17	008	26 15	077	58 49	118	27 13	149	33 25	187	29 43	245	55 08	288
271	25 20	009	26 36	078	59 08	120	27 24	150	33 22	189	29 23	246	54 48	289
272	25 24	010	26 57	079	59 26	121	27 35	151	33 19	190	29 04	247	54 27	289
273	25 27	011	27 19	080	59 45	122	27 45	152	33 15	191	28 44	248	54 07	290
274	25 32	011	27 40	081	60 03	123	27 55	153	33 10	192	28 24	249	53 47	291
275	25 36	012	28 01	081	60 20	125	28 04	154	33 06	193	28 04	250	53 27	292
276	25 41	013	28 22	082	60 38	126	28 13	156	33 01	194	27 44	251	53 07	292
277	25 46	014	28 44	083	60 55	127	28 22	157	32 55	196	27 23	252	52 47	293
278	25 51	014	29 05	084	61 12	129	28 30	158	32 49	197	27 03	253	52 27	294
279	25 56	015	29 26	085	61 29	130	28 38	159	32 43	198	26 42	254	52 08	295
280	26 02	016	29 48	086	61 45	131	28 46	160	32 36	199	26 22	255	51 48	295
281	26 08	017	30 09	087	62 01	133	28 53	161	32 29	200	26 01	256	51 29	296
282	26 15	018	30 31	088	62 17	134	29 00	162	32 21	201	25 40	257	51 10	297
283	26 21	018	30 52	089	62 32	136	29 06	163	32 13	202	25 19	257	50 51	298
284	26 28	019	31 14	090	62 47	137	29 12	165	32 05	204	24 58	258	50 32	298
	◆CAPELLA		Hamal		Alpheratz		◆DENEB		VEGA		◆ARCTURUS		Alioth	
285	26 35	020	16 04	066	31 35	091	63 01	138	59 37	189	24 37	259	50 13	299
286	26 43	021	16 24	067	31 57	092	63 15	140	59 33	191	24 16	260	49 54	300
287	26 50	021	16 44	068	32 18	092	63 29	141	59 29	192	23 55	261	49 36	301
288	26 58	022	17 04	069	32 40	093	63 42	143	59 24	194	23 33	262	49 17	301
289	27 07	023	17 24	070	33 01	094	63 55	144	59 19	195	23 12	263	48 59	302
290	27 15	024	17 44	071	33 23	095	64 07	146	59 13	197	22 51	264	48 41	303
291	27 24	024	18 04	071	33 44	096	64 19	148	59 07	198	22 29	265	48 23	303
292	27 33	025	18 25	072	34 05	097	64 30	149	59 00	200	22 08	266	48 05	304
293	27 42	026	18 45	073	34 27	098	64 41	151	58 52	201	21 46	267	47 47	305
294	27 52	027	19 06	074	34 48	099	64 51	152	58 44	203	21 25	268	47 29	306
295	28 02	027	19 27	075	35 09	100	65 01	154	58 36	204	21 03	269	47 12	306
296	28 12	028	19 47	076	35 30	101	65 10	156	58 27	206	20 42	270	46 55	307
297	28 22	029	20 08	077	35 51	102	65 19	157	58 17	207	20 20	271	46 38	308
298	28 32	030	20 29	078	36 12	103	65 27	159	58 07	208	19 59	272	46 21	308
299	28 43	031	20 50	079	36 33	104	65 34	160	57 57	210	19 37	273	46 04	309
	◆CAPELLA		Hamal		Alpheratz		◆DENEB		VEGA		◆ARCTURUS		Dubhe	
300	28 54	031	21 11	080	36 54	105	65 41	162	57 46	211	19 16	274	44 48	331
301	29 06	032	21 33	081	37 15	106	65 48	164	57 35	213	18 54	274	44 38	332
302	29 17	033	21 54	081	37 35	107	65 53	166	57 23	214	18 33	275	44 28	333
303	29 29	034	22 15	082	37 56	108	65 58	167	57 10	215	18 12	276	44 18	333
304	29 41	034	22 36	083	38 16	109	66 03	169	56 58	217	17 50	277	44 08	334
305	29 53	035	22 58	084	38 36	110	66 07	171	56 45	218	17 29	278	43 59	335
306	30 06	036	23 19	085	38 56	111	66 10	172	56 31	220	17 08	279	43 50	335
307	30 18	037	23 41	086	39 16	112	66 12	174	56 17	221	16 46	280	43 41	336
308	30 31	037	24 02	087	39 36	113	66 14	176	56 03	222	16 25	281	43 32	336
309	30 45	038	24 24	088	39 56	114	66 15	178	55 48	224	16 04	282	43 24	337
310	30 58	039	24 45	089	40 15	115	66 16	179	55 33	225	15 43	283	43 15	338
311	31 12	040	25 07	090	40 35	117	66 16	181	55 18	226	15 22	284	43 07	338
312	31 25	041	25 28	091	40 54	118	66 15	183	55 02	227	15 01	285	42 59	339
313	31 40	041	25 50	092	41 13	119	66 13	185	54 46	229	14 41	285	42 52	340
314	31 54	042	26 11	093	41 32	120	66 11	186	54 30	230	14 20	286	42 44	340
	◆CAPELLA		Hamal		Alpheratz		◆DENEB		VEGA		◆Alphecca		Dubhe	
315	32 08	043	26 33	093	41 50	121	66 09	188	54 13	231	27 54	272	42 37	341
316	32 23	044	26 54	094	42 09	122	66 05	190	53 56	233	27 32	273	42 30	341
317	32 38	044	27 15	095	42 27	123	66 01	192	53 39	234	27 11	274	42 23	342
318	32 53	045	27 37	096	42 45	124	65 57	193	53 22	235	26 50	275	42 17	343
319	33 08	046	27 58	097	43 02	125	65 51	195	53 04	236	26 28	276	42 11	343
320	33 24	047	28 19	098	43 20	126	65 45	197	52 46	237	26 07	277	42 04	344
321	33 40	047	28 41	099	43 37	128	65 39	198	52 28	239	25 45	278	41 59	344
322	33 56	048	29 02	100	43 54	129	65 32	200	52 09	240	25 24	279	41 53	345
323	34 12	049	29 23	101	44 10	130	65 24	202	51 51	241	25 03	280	41 48	346
324	34 28	050	29 44	102	44 27	131	65 16	203	51 32	242	24 42	281	41 42	346
325	34 45	051	30 05	103	44 43	132	65 07	205	51 13	243	24 21	281	41 37	347
326	35 01	051	30 26	104	44 59	133	64 58	207	50 53	244	24 00	282	41 33	348
327	35 18	052	30 47	105	45 14	135	64 48	208	50 34	245	23 39	283	41 28	348
328	35 35	053	31 08	106	45 29	136	64 37	210	50 14	247	23 18	284	41 24	349
329	35 52	054	31 28	107	45 44	137	64 26	212	49 54	248	22 57	285	41 20	349
	CAPELLA		◆Hamal		Alpheratz		◆ALTAIR		VEGA		Alphecca		◆Alioth	
330	36 10	054	31 49	108	45 59	138	26 17	216	49 35	249	22 36	286	38 56	330
331	36 27	055	32 09	109	46 13	139	26 04	217	49 14	250	22 16	287	38 45	331
332	36 45	056	32 29	110	46 27	141	25 51	218	48 54	251	21 55	288	38 35	332
333	37 03	057	32 49	111	46 40	142	25 37	219	48 34	252	21 35	289	38 25	332
334	37 21	058	33 09	112	46 53	143	25 24	220	48 13	253	21 14	290	38 15	333
335	37 39	058	33 29	113	47 06	144	25 10	221	47 53	254	20 54	290	38 05	334
336	37 58	059	33 49	114	47 18	146	24 55	223	47 32	255	20 34	291	37 56	334
337	38 16	060	34 09	115	47 30	147	24 40	224	47 11	256	20 14	292	37 47	335
338	38 35	061	34 28	116	47 42	148	24 26	225	46 50	257	19 54	293	37 38	336
339	38 54	062	34 47	117	47 53	149	24 10	226	46 29	258	19 34	294	37 29	336
340	39 13	062	35 06	118	48 04	151	23 55	227	46 08	259	19 15	295	37 21	337
341	39 32	063	35 25	119	48 14	152	23 39	228	45 47	260	18 55	296	37 12	338
342	39 51	064	35 44	120	48 24	153	23 23	229	45 26	261	18 36	297	37 04	339
343	40 10	065	36 02	121	48 33	154	23 07	230	45 04	262	18 17	297	36 57	339
344	40 30	066	36 20	123	48 43	156	22 50	231	44 43	263	17 58	298	36 49	340
	◆POLLUX		CAPELLA		Hamal		◆Alpheratz		ALTAIR		◆VEGA		Alioth	
345	13 19	043	40 50	066	36 38	124	48 51	157	22 33	232	44 22	264	36 42	341
346	13 34	044	41 09	067	36 56	125	48 59	158	22 16	233	44 00	265	36 35	341
347	13 49	045	41 29	068	37 14	126	49 07	160	21 59	234	43 39	266	36 28	342
348	14 04	046	41 49	069	37 31	127	49 14	161	21 41	235	43 17	267	36 22	343
349	14 19	046	42 09	070	37 48	128	49 21	162	21 24	236	42 56	268	36 15	343
350	14 35	047	42 30	070	38 05	129	49 27	164	21 06	237	42 35	269	36 09	344
351	14 51	048	42 50	071	38 21	130	49 33	165	20 48	238	42 13	270	36 03	345
352	15 07	049	43 10	072	38 38	131	49 38	166	20 29	239	41 52	271	35 58	345
353	15 23	050	43 31	073	38 54	132	49 43	168	20 11	240	41 30	272	35 53	346
354	15 40	051	43 51	074	39 09	134	49 47	169	19 52	241	41 09	273	35 47	347
355	15 57	052	44 12	075	39 25	135	49 51	170	19 33	242	40 47	273	35 43	347
356	16 14	053	44 33	075	39 40	136	49 54	172	19 14	243	40 26	274	35 38	348
357	16 31	053	44 54	076	39 55	137	49 57	173	18 55	244	40 04	275	35 34	349
358	16 48	054	45 15	077	40 09	138	50 00	175	18 35	245	39 43	276	35 30	349
359	17 06	055	45 36	078	40 23	139	50 01	176	18 16	246	39 21	277	35 26	350

LHA ♈	Hc	Zn	Hc	Zn	Hc	Zn	Hc	Zn	Hc	Zn	Hc	Zn	Hc	Zn
	◆POLLUX		CAPELLA		Hamal		◆Alpheratz		DENEB		◆VEGA		Alioth	
0	16 50	056	45 44	078	41 23	140	51 03	177	56 01	254	38 51	279	34 23	351
1	17 09	057	46 06	079	41 38	141	51 03	179	55 39	255	38 29	280	34 20	352
2	17 28	058	46 29	080	41 51	142	51 04	180	55 18	256	38 07	281	34 17	352
3	17 47	058	46 51	080	42 05	144	51 04	181	54 56	257	37 45	281	34 14	353
4	18 06	059	47 13	081	42 18	145	51 03	183	54 34	258	37 23	282	34 11	354
5	18 25	060	47 35	082	42 31	146	51 01	184	54 12	259	37 01	283	34 09	354
6	18 45	061	47 57	083	42 43	147	50 59	186	53 50	260	36 39	284	34 07	355
7	19 05	062	48 20	084	42 55	148	50 57	187	53 28	261	36 17	285	34 05	356
8	19 24	063	48 42	085	43 07	150	50 54	188	53 05	262	35 56	286	34 03	356
9	19 44	064	49 04	086	43 18	151	50 51	190	52 43	263	35 34	287	34 02	357
10	20 05	064	49 27	087	43 29	152	50 47	191	52 21	264	35 13	287	34 01	358
11	20 25	065	49 49	087	43 39	153	50 42	192	51 58	265	34 51	288	34 00	358
12	20 46	066	50 12	088	43 49	155	50 37	194	51 36	266	34 30	289	33 59	359
13	21 06	067	50 34	089	43 58	156	50 31	195	51 13	267	34 09	290	33 59	000
14	21 27	068	50 57	090	44 07	157	50 25	197	50 51	268	33 48	291	33 59	000
	Alioth		◆POLLUX		CAPELLA		Hamal		◆Alpheratz		DENEB		◆VEGA	
15	34 00	001	21 48	069	51 19	091	44 16	158	50 18	198	50 29	269	33 27	292
16	34 00	002	22 09	070	51 42	092	44 24	160	50 11	199	50 06	270	33 06	292
17	34 01	002	22 30	071	52 04	093	44 31	161	50 04	201	49 44	271	32 45	293
18	34 02	003	22 51	071	52 27	094	44 39	162	49 55	202	49 21	272	32 25	294
19	34 03	004	23 13	072	52 49	095	44 45	163	49 47	203	48 59	273	32 04	295
20	34 05	004	23 34	073	53 11	096	44 51	165	49 38	205	48 36	273	31 44	296
21	34 07	005	23 56	074	53 34	097	44 57	166	49 28	206	48 14	274	31 24	297
22	34 09	006	24 17	075	53 56	098	45 02	167	49 18	207	47 51	275	31 04	297
23	34 11	006	24 39	076	54 18	099	45 07	169	49 08	209	47 29	276	30 44	298
24	34 14	007	25 01	077	54 40	100	45 11	170	48 57	210	47 07	277	30 24	299
25	34 17	008	25 23	078	55 02	101	45 15	171	48 45	211	46 44	278	30 05	300
26	34 20	008	25 45	078	55 25	102	45 18	173	48 33	212	46 22	279	29 45	301
27	34 23	009	26 07	079	55 46	103	45 21	174	48 21	214	46 00	280	29 26	302
28	34 27	010	26 29	080	56 08	104	45 23	175	48 08	215	45 38	280	29 07	302
29	34 31	010	26 51	081	56 30	105	45 24	176	47 55	216	45 16	281	28 48	303
	Alioth		◆POLLUX		ALDEBARAN		◆Hamal		Alpheratz		DENEB		◆VEGA	
30	34 35	011	27 13	082	32 53	134	45 26	178	47 42	218	44 54	282	28 29	304
31	34 40	012	27 36	083	33 09	135	45 26	179	47 28	219	44 32	283	28 11	305
32	34 45	013	27 58	084	33 24	136	45 26	180	47 14	220	44 10	284	27 52	306
33	34 50	013	28 20	085	33 40	137	45 26	182	46 59	221	43 48	285	27 34	306
34	34 55	014	28 43	086	33 55	139	45 25	183	46 44	222	43 27	286	27 16	307
35	35 00	015	29 05	087	34 09	140	45 24	184	46 29	224	43 05	286	26 58	308
36	35 06	015	29 28	088	34 24	141	45 22	186	46 13	225	42 43	287	26 41	309
37	35 12	016	29 50	088	34 38	142	45 19	187	45 57	226	42 22	288	26 23	310
38	35 18	017	30 13	089	34 51	143	45 16	188	45 41	227	42 01	289	26 06	310
39	35 25	017	30 35	090	35 05	144	45 13	189	45 24	228	41 39	290	25 49	311
40	35 32	018	30 57	091	35 18	145	45 09	191	45 07	230	41 18	290	25 32	312
41	35 39	019	31 20	092	35 30	147	45 04	192	44 50	231	40 57	291	25 16	313
42	35 46	019	31 42	093	35 42	148	44 59	193	44 32	232	40 36	292	24 59	314
43	35 54	020	32 05	094	35 54	149	44 54	195	44 14	233	40 16	293	24 43	315
44	36 01	021	32 27	095	36 06	150	44 48	196	43 56	234	39 55	294	24 27	315
	Alioth		◆POLLUX		ALDEBARAN		◆Hamal		Alpheratz		DENEB		◆VEGA	
45	36 09	021	32 50	096	36 17	151	44 42	197	43 38	235	39 34	294	24 12	316
46	36 18	022	33 12	097	36 27	152	44 35	199	43 19	237	39 14	295	23 56	317
47	36 26	023	33 34	098	36 37	154	44 27	200	43 00	238	38 54	296	23 41	318
48	36 35	023	33 56	099	36 47	155	44 19	201	42 41	239	38 34	297	23 26	319
49	36 44	024	34 19	100	36 57	156	44 11	202	42 22	240	38 14	298	23 11	319
50	36 53	025	34 41	101	37 06	157	44 02	204	42 02	241	37 54	298	22 57	320
51	37 03	025	35 03	102	37 14	158	43 53	205	41 43	242	37 34	299	22 42	321
52	37 12	026	35 25	103	37 22	159	43 44	206	41 23	243	37 15	300	22 28	322
53	37 22	027	35 47	104	37 30	161	43 34	207	41 02	244	36 55	301	22 15	323
54	37 32	027	36 08	105	37 37	162	43 23	209	40 42	245	36 36	302	22 01	323
55	37 43	028	36 30	106	37 44	163	43 12	210	40 22	246	36 17	302	21 48	324
56	37 54	029	36 52	107	37 50	164	43 01	211	40 01	247	35 58	303	21 35	325
57	38 04	029	37 13	108	37 56	165	42 49	212	39 40	248	35 39	304	21 22	326
58	38 16	030	37 35	109	38 02	167	42 37	213	39 19	249	35 21	305	21 10	327
59	38 27	031	37 56	110	38 07	168	42 24	215	38 58	250	35 02	305	20 57	327
	Alioth		◆POLLUX		ALDEBARAN		◆Hamal		Alpheratz		DENEB		◆VEGA	
60	38 38	031	38 17	111	38 11	169	42 11	216	38 37	252	34 44	306	20 45	328
61	38 50	032	38 38	112	38 15	170	41 58	217	38 15	253	34 26	307	20 34	329
62	39 02	033	38 59	113	38 19	172	41 44	218	37 54	254	34 08	308	20 22	330
63	39 15	033	39 19	114	38 22	173	41 30	219	37 32	255	33 51	309	20 11	331
64	39 27	034	39 40	115	38 24	174	41 15	221	37 11	256	33 33	309	20 00	331
65	39 40	035	40 00	116	38 26	175	41 00	222	36 49	257	33 16	310	19 50	332
66	39 53	035	40 20	117	38 28	176	40 45	223	36 27	258	32 59	311	19 39	333
67	40 06	036	40 40	118	38 29	178	40 30	224	36 05	259	32 42	312	19 29	334
68	40 19	037	41 00	119	38 30	179	40 14	225	35 43	259	32 25	312	19 19	335
69	40 33	037	41 19	120	38 30	180	39 58	226	35 21	260	32 09	313	19 10	335
70	40 46	038	41 38	121	38 30	181	39 41	228	34 59	261	31 52	314	19 01	336
71	41 00	039	41 58	123	38 29	183	39 25	229	34 36	262	31 36	315	18 52	337
72	41 14	039	42 16	124	38 28	184	39 08	230	34 14	263	31 20	315	18 43	338
73	41 29	040	42 35	125	38 26	185	38 50	231	33 52	264	31 05	316	18 35	339
74	41 43	041	42 53	126	38 24	186	38 33	232	33 29	265	30 49	317	18 27	340
	Alioth		◆REGULUS		PROCYON		◆ALDEBARAN		Hamal		Alpheratz		◆DENEB	
75	41 58	041	15 57	098	21 48	137	38 21	187	38 15	233	33 07	266	30 34	318
76	42 13	042	16 20	098	22 03	138	38 18	189	37 57	234	32 44	267	30 19	319
77	42 28	043	16 42	099	22 18	139	38 14	190	37 38	235	32 22	268	30 04	319
78	42 44	043	17 04	100	22 33	140	38 10	191	37 20	236	31 59	269	29 50	320
79	42 59	044	17 26	101	22 47	141	38 06	192	37 01	238	31 37	270	29 35	321
80	43 15	045	17 48	102	23 01	142	38 01	194	36 42	239	31 15	271	29 21	322
81	43 31	045	18 10	103	23 15	143	37 55	195	36 22	240	30 52	272	29 08	322
82	43 47	046	18 32	104	23 29	144	37 49	196	36 03	241	30 30	273	28 54	323
83	44 03	047	18 54	105	23 42	145	37 43	197	35 43	242	30 07	274	28 41	324
84	44 20	048	19 15	106	23 54	146	37 36	198	35 23	243	29 45	275	28 27	325
85	44 36	048	19 37	107	24 07	147	37 29	200	35 03	244	29 22	275	28 15	325
86	44 53	049	19 58	108	24 19	148	37 21	201	34 43	245	29 00	276	28 02	326
87	45 10	050	20 20	109	24 30	149	37 13	202	34 22	246	28 38	277	27 49	327
88	45 27	050	20 41	110	24 42	150	37 04	203	34 02	247	28 15	278	27 37	328
89	45 45	051	21 02	111	24 52	152	36 55	204	33 41	248	27 53	279	27 25	328

LHA ♈	Hc	Zn	Hc	Zn	Hc	Zn	Hc	Zn	Hc	Zn	Hc	Zn	Hc	Zn
	Alioth		◆REGULUS		PROCYON		BETELGEUSE		◆ALDEBARAN		Hamal		◆DENEB	
90	46 02	052	21 23	112	25 03	153	29 24	181	36 46	206	33 20	249	27 14	329
91	46 20	052	21 43	113	25 13	154	29 23	183	36 36	207	32 59	250	27 02	330
92	46 38	053	22 04	114	25 23	155	29 22	184	36 25	208	32 38	251	26 51	331
93	46 56	054	22 25	115	25 32	156	29 20	185	36 15	209	32 16	252	26 40	331
94	47 14	054	22 45	116	25 41	157	29 18	186	36 04	210	31 55	253	26 30	332
95	47 32	055	23 05	117	25 50	158	29 16	187	35 52	211	31 33	254	26 19	333
96	47 51	056	23 25	118	25 58	159	29 13	188	35 40	213	31 12	255	26 09	334
97	48 10	056	23 45	119	26 06	160	29 09	189	35 28	214	30 50	256	26 00	335
98	48 28	057	24 04	120	26 13	161	29 05	191	35 15	215	30 28	257	25 50	335
99	48 47	058	24 24	121	26 20	162	29 01	192	35 02	216	30 06	258	25 41	336
100	49 06	059	24 43	122	26 27	164	28 56	193	34 49	217	29 44	259	25 32	337
101	49 26	059	25 02	123	26 33	165	28 51	194	34 35	218	29 22	260	25 23	338
102	49 45	060	25 21	124	26 39	166	28 45	195	34 21	219	29 00	261	25 15	338
103	50 05	061	25 39	125	26 44	167	28 39	196	34 07	220	28 38	262	25 07	339
104	50 24	061	25 57	126	26 49	168	28 33	197	33 52	222	28 15	263	24 59	340
	Alioth		◆REGULUS		PROCYON		BETELGEUSE		◆CAPELLA		Hamal		◆DENEB	
105	50 44	062	26 16	127	26 53	169	28 26	198	64 17	224	27 53	264	24 51	341
106	51 04	063	26 33	128	26 57	170	28 19	200	64 01	226	27 31	265	24 44	341
107	51 24	063	26 51	129	27 01	171	28 11	201	63 45	227	27 08	266	24 37	342
108	51 44	064	27 08	130	27 04	172	28 03	202	63 28	229	26 46	267	24 30	343
109	52 04	065	27 25	131	27 07	174	27 54	203	63 11	230	26 23	268	24 24	344
110	52 25	066	27 42	132	27 09	175	27 45	204	62 54	232	26 01	268	24 17	344
111	52 45	066	27 58	133	27 11	176	27 36	205	62 36	233	25 39	269	24 11	345
112	53 06	067	28 15	134	27 13	177	27 26	206	62 18	234	25 16	270	24 06	346
113	53 27	068	28 31	135	27 14	178	27 16	207	61 59	236	24 54	271	24 01	347
114	53 48	069	28 46	137	27 14	179	27 06	208	61 41	237	24 31	272	23 56	348
115	54 09	069	29 01	138	27 14	180	26 55	209	61 22	238	24 09	273	23 51	348
116	54 30	070	29 16	139	27 14	181	26 43	211	61 02	240	23 46	274	23 46	349
117	54 51	071	29 31	140	27 13	183	26 32	212	60 43	241	23 24	275	23 42	350
118	55 12	072	29 45	141	27 12	184	26 20	213	60 23	242	23 01	276	23 38	351
119	55 34	072	29 59	142	27 10	185	26 08	214	60 03	243	22 39	277	23 35	351
	◆VEGA		ARCTURUS		◆REGULUS		PROCYON		◆CAPELLA		Schedar		DENEB	
120	17 55	017	16 20	079	30 13	143	27 08	186	59 43	244	44 39	313	23 32	352
121	18 02	018	16 42	080	30 26	144	27 06	187	59 23	246	44 23	314	23 29	353
122	18 09	019	17 04	081	30 39	145	27 03	188	59 02	247	44 06	315	23 26	354
123	18 16	019	17 27	082	30 52	146	26 59	189	58 41	248	43 51	315	23 24	354
124	18 24	020	17 49	083	31 04	148	26 56	190	58 20	249	43 35	316	23 22	355
125	18 32	021	18 11	084	31 16	149	26 51	191	57 59	250	43 19	317	23 20	356
126	18 40	022	18 33	085	31 28	150	26 47	193	57 38	251	43 04	317	23 19	357
127	18 48	023	18 56	085	31 39	151	26 42	194	57 17	253	42 49	318	23 17	357
128	18 57	023	19 18	086	31 49	152	26 36	195	56 55	254	42 34	319	23 17	358
129	19 06	024	19 41	087	32 00	153	26 30	196	56 34	255	42 19	319	23 16	359
130	19 16	025	20 03	088	32 10	154	26 24	197	56 12	256	42 04	320	23 16	000
131	19 25	026	20 26	089	32 19	155	26 17	198	55 50	257	41 50	321	23 16	001
132	19 35	027	20 48	090	32 28	157	26 10	199	55 28	258	41 36	321	23 16	001
133	19 46	027	21 11	091	32 37	158	26 02	200	55 06	259	41 22	322	23 17	002
134	19 56	028	21 33	092	32 45	159	25 54	201	54 44	260	41 08	323	23 18	003
	DENEB		◆VEGA		ARCTURUS		◆REGULUS		PROCYON		◆CAPELLA		Schedar	
135	23 19	004	20 07	029	21 56	093	32 53	160	25 46	202	54 22	261	40 55	323
136	23 21	004	20 18	030	22 18	094	33 01	161	25 37	204	54 00	262	40 41	324
137	23 22	005	20 29	031	22 40	095	33 08	162	25 28	205	53 37	263	40 28	325
138	23 25	006	20 41	031	23 03	096	33 14	164	25 18	206	53 15	264	40 15	325
139	23 27	007	20 53	032	23 25	097	33 21	165	25 08	207	52 53	265	40 02	326
140	23 30	007	21 05	033	23 47	098	33 26	166	24 58	208	52 30	266	39 50	327
141	23 33	008	21 17	034	24 10	098	33 32	167	24 47	209	52 08	267	39 38	327
142	23 36	009	21 30	035	24 32	099	33 36	168	24 36	210	51 45	268	39 26	328
143	23 40	010	21 43	036	24 54	100	33 41	169	24 25	211	51 23	269	39 14	329
144	23 44	010	21 56	036	25 16	101	33 45	171	24 13	212	51 00	270	39 02	329
145	23 48	011	22 09	037	25 38	102	33 48	172	24 01	213	50 38	271	38 51	330
146	23 53	012	22 23	038	26 00	103	33 51	173	23 48	214	50 15	272	38 40	331
147	23 57	013	22 37	039	26 22	104	33 54	174	23 35	215	49 53	272	38 29	331
148	24 03	014	22 51	040	26 44	105	33 56	175	23 22	216	49 30	273	38 18	332
149	24 08	014	23 06	040	27 05	106	33 57	176	23 09	218	49 08	274	38 08	333
	DENEB		◆VEGA		ARCTURUS		◆REGULUS		PROCYON		◆CAPELLA		Schedar	
150	24 14	015	23 20	041	27 27	107	33 59	178	22 55	219	48 46	275	37 57	333
151	24 20	016	23 35	042	27 48	108	33 59	179	22 41	220	48 23	276	37 47	334
152	24 26	017	23 50	043	28 09	109	34 00	180	22 26	221	48 01	277	37 38	335
153	24 32	017	24 06	044	28 31	110	33 59	181	22 11	222	47 39	278	37 28	335
154	24 39	018	24 21	044	28 52	111	33 59	182	21 56	223	47 16	279	37 19	336
155	24 46	019	24 37	045	29 13	112	33 57	184	21 41	224	46 54	279	37 10	337
156	24 54	020	24 53	046	29 33	113	33 56	185	21 25	225	46 32	280	37 01	337
157	25 02	020	25 10	047	29 54	114	33 54	186	21 09	226	46 10	281	36 52	338
158	25 10	021	25 26	048	30 14	115	33 51	187	20 53	227	45 48	282	36 44	339
159	25 18	022	25 43	048	30 35	116	33 48	188	20 36	228	45 26	283	36 36	339
160	25 26	023	26 00	049	30 55	117	33 45	189	20 19	229	45 04	284	36 28	340
161	25 35	023	26 17	050	31 14	118	33 41	191	20 02	230	44 42	285	36 20	341
162	25 44	024	26 34	051	31 34	119	33 37	192	19 45	231	44 21	285	36 13	341
163	25 54	025	26 52	052	31 54	120	33 32	193	19 27	232	43 59	286	36 06	342
164	26 03	026	27 09	052	32 13	121	33 27	194	19 10	233	43 37	287	35 59	343
	DENEB		◆VEGA		ARCTURUS		◆REGULUS		POLLUX		◆CAPELLA		Schedar	
165	26 13	027	27 27	053	32 32	122	33 21	195	40 50	241	43 16	288	35 52	343
166	26 23	027	27 45	054	32 51	124	33 15	196	40 30	242	42 55	289	35 46	344
167	26 34	028	28 04	055	33 09	125	33 08	198	40 10	243	42 33	290	35 40	345
168	26 44	029	28 22	056	33 28	126	33 01	199	39 50	245	42 12	290	35 34	345
169	26 55	030	28 41	057	33 46	127	32 54	200	39 30	246	41 51	291	35 29	346
170	27 07	030	29 00	057	34 04	128	32 46	201	39 09	247	41 30	292	35 23	347
171	27 18	031	29 19	058	34 21	129	32 38	202	38 48	248	41 10	293	35 18	347
172	27 30	032	29 38	059	34 39	130	32 29	203	38 27	249	40 49	294	35 13	348
173	27 42	033	29 57	060	34 56	131	32 20	204	38 06	250	40 28	294	35 09	349
174	27 54	033	30 17	061	35 13	132	32 10	206	37 45	251	40 08	295	35 04	349
175	28 07	034	30 36	061	35 29	133	32 00	207	37 24	252	39 48	296	35 00	350
176	28 19	035	30 56	062	35 45	134	31 50	208	37 03	253	39 28	297	34 56	351
177	28 32	036	31 16	063	36 01	136	31 39	209	36 41	254	39 08	297	34 53	351
178	28 45	036	31 36	064	36 17	137	31 28	210	36 19	255	38 48	298	34 50	352
179	28 59	037	31 56	065	36 32	138	31 17	211	35 58	256	38 28	299	34 47	353

LAT 68°N

LHA ♈	Hc	Zn	Hc	Zn	Hc	Zn	Hc	Zn	Hc	Zn	Hc	Zn	Hc	Zn
	◆DENEB		VEGA		◆ARCTURUS		REGULUS		◆POLLUX		CAPELLA		Schedar	
180	29 13	038	32 17	066	36 47	139	31 05	212	35 36	257	38 08	300	34 44	353
181	29 27	039	32 37	066	37 02	140	30 53	214	35 14	258	37 49	301	34 41	354
182	29 41	039	32 58	067	37 16	141	30 40	215	34 52	259	37 30	301	34 39	355
183	29 55	040	33 19	068	37 30	142	30 27	216	34 30	260	37 11	302	34 37	355
184	30 10	041	33 40	069	37 43	144	30 14	217	34 08	261	36 52	303	34 35	356
185	30 25	042	34 01	070	37 57	145	30 00	218	33 45	262	36 33	304	34 34	357
186	30 40	043	34 22	071	38 09	146	29 46	219	33 23	263	36 14	305	34 33	357
187	30 55	043	34 43	071	38 22	147	29 32	220	33 01	264	35 56	305	34 32	358
188	31 11	044	35 04	072	38 34	148	29 17	221	32 39	265	35 38	306	34 31	359
189	31 26	045	35 26	073	38 45	149	29 02	222	32 16	265	35 20	307	34 31	359
190	31 42	046	35 47	074	38 57	151	28 47	223	31 54	266	35 02	308	34 31	000
191	31 58	046	36 09	075	39 08	152	28 31	224	31 31	267	34 44	308	34 31	001
192	32 15	047	36 31	076	39 18	153	28 16	226	31 09	268	34 26	309	34 31	001
193	32 31	048	36 53	077	39 28	154	27 59	227	30 46	269	34 09	310	34 32	002
194	32 48	049	37 15	077	39 38	155	27 43	228	30 24	270	33 52	311	34 33	003
	Schedar		◆DENEB		VEGA		◆ARCTURUS		REGULUS		◆POLLUX		CAPELLA	
195	34 34	003	33 05	049	37 36	078	39 47	157	27 26	229	30 01	271	33 35	311
196	34 35	004	33 22	050	37 59	079	39 55	158	27 09	230	29 39	272	33 18	312
197	34 37	005	33 40	051	38 21	080	40 04	159	26 52	231	29 16	273	33 02	313
198	34 39	005	33 57	052	38 43	081	40 11	160	26 34	232	28 54	274	32 45	314
199	34 41	006	34 15	052	39 05	082	40 19	161	26 16	233	28 32	275	32 29	314
200	34 44	007	34 33	053	39 27	083	40 26	163	25 58	234	28 09	276	32 13	315
201	34 46	007	34 51	054	39 50	084	40 32	164	25 40	235	27 47	277	31 57	316
202	34 49	008	35 09	055	40 12	084	40 38	165	25 22	236	27 25	277	31 42	317
203	34 53	009	35 28	056	40 34	085	40 44	166	25 03	237	27 02	278	31 27	317
204	34 56	009	35 46	056	40 57	086	40 49	168	24 44	238	26 40	279	31 12	318
205	35 00	010	36 05	057	41 19	087	40 53	169	24 25	239	26 18	280	30 57	319
206	35 04	011	36 24	058	41 42	088	40 57	170	24 05	240	25 56	281	30 42	320
207	35 08	011	36 43	059	42 04	089	41 01	171	23 46	241	25 34	282	30 28	321
208	35 13	012	37 02	059	42 27	090	41 04	173	23 26	242	25 12	283	30 13	321
209	35 18	013	37 22	060	42 49	091	41 07	174	23 06	243	24 50	284	30 00	322
	Schedar		◆DENEB		VEGA		◆ARCTURUS		REGULUS		◆POLLUX		CAPELLA	
210	35 23	013	37 41	061	43 12	092	41 09	175	22 46	244	24 28	285	29 46	323
211	35 28	014	38 01	062	43 34	093	41 10	176	22 26	245	24 07	285	29 32	324
212	35 33	015	38 21	063	43 56	094	41 12	178	22 05	246	23 45	286	29 19	324
213	35 39	015	38 41	063	44 19	095	41 12	179	21 45	247	23 23	287	29 06	325
214	35 45	016	39 01	064	44 41	096	41 12	180	21 24	248	23 02	288	28 53	326
215	35 52	017	39 22	065	45 04	097	41 12	181	21 03	249	22 41	289	28 41	327
216	35 58	017	39 42	066	45 26	097	41 11	183	20 42	250	22 19	290	28 29	327
217	36 05	018	40 03	067	45 48	098	41 10	184	20 21	251	21 58	291	28 17	328
218	36 12	019	40 23	067	46 10	099	41 08	185	19 59	252	21 37	292	28 05	329
219	36 20	019	40 44	068	46 33	100	41 06	186	19 38	253	21 17	293	27 53	330
220	36 27	020	41 05	069	46 55	101	41 03	188	19 16	254	20 56	293	27 42	330
221	36 35	021	41 26	070	47 17	102	41 00	189	18 55	255	20 35	294	27 31	331
222	36 43	021	41 47	071	47 38	103	40 56	190	18 33	256	20 15	295	27 20	332
223	36 51	022	42 08	071	48 00	105	40 52	191	18 11	257	19 55	296	27 10	333
224	37 00	023	42 30	072	48 22	106	40 47	193	17 49	258	19 35	297	27 00	333
	Schedar		◆DENEB		VEGA		◆ARCTURUS		REGULUS		◆POLLUX		CAPELLA	
225	37 09	023	42 51	073	48 44	107	40 42	194	17 27	259	19 15	298	26 50	334
226	37 18	024	43 13	074	49 05	108	40 37	195	17 05	260	18 55	299	26 40	335
227	37 27	025	43 35	075	49 26	109	40 30	196	16 43	261	18 35	299	26 31	336
228	37 36	025	43 56	076	49 48	110	40 24	198	16 21	261	18 16	300	26 22	336
229	37 46	026	44 18	076	50 09	111	40 17	199	15 59	262	17 56	301	26 13	337
230	37 56	027	44 40	077	50 30	112	40 09	200	15 36	263	17 37	302	26 04	338
231	38 06	027	45 02	078	50 50	113	40 01	201	15 14	264	17 18	303	25 56	339
232	38 17	028	45 24	079	51 11	114	39 53	203	14 52	265	16 59	304	25 48	339
233	38 27	029	45 46	080	51 31	115	39 44	204	14 29	266	16 41	305	25 40	340
234	38 38	029	46 08	081	51 52	116	39 35	205	14 07	267	16 22	305	25 33	341
235	38 50	030	46 30	082	52 12	118	39 25	206	13 44	268	16 04	306	25 25	342
236	39 01	031	46 53	082	52 31	119	39 15	207	13 22	269	15 46	307	25 18	342
237	39 12	031	47 15	083	52 51	120	39 04	209	12 59	270	15 28	308	25 12	343
238	39 24	032	47 37	084	53 10	121	38 54	210	12 37	271	15 11	309	25 05	344
239	39 36	033	48 00	085	53 30	122	38 42	211	12 14	272	14 53	310	24 59	345
	Alpheratz		◆DENEB		VEGA		Rasalhague		◆ARCTURUS		Alioth		◆CAPELLA	
240	16 04	050	48 22	086	53 48	123	32 27	152	38 30	212	65 53	264	24 54	346
241	16 21	051	48 45	087	54 07	125	32 37	153	38 18	213	65 30	265	24 48	346
242	16 39	052	49 07	088	54 25	126	32 47	155	38 06	214	65 08	266	24 43	347
243	16 57	053	49 29	089	54 44	127	32 56	156	37 53	216	64 45	267	24 38	348
244	17 15	054	49 52	090	55 01	128	33 05	157	37 40	217	64 23	267	24 33	349
245	17 33	055	50 14	091	55 19	130	33 14	158	37 26	218	64 01	268	24 29	349
246	17 52	056	50 37	092	55 36	131	33 22	159	37 12	219	63 38	269	24 25	350
247	18 10	056	50 59	092	55 53	132	33 30	160	36 58	220	63 16	270	24 21	351
248	18 29	057	51 22	093	56 09	134	33 37	162	36 43	221	62 53	271	24 18	352
249	18 48	058	51 44	094	56 25	135	33 44	163	36 28	223	62 31	272	24 15	352
250	19 07	059	52 07	095	56 41	136	33 51	164	36 12	224	62 08	273	24 12	353
251	19 27	060	52 29	096	56 56	138	33 57	165	35 57	225	61 46	274	24 09	354
252	19 46	061	52 51	097	57 11	139	34 02	166	35 41	226	61 23	275	24 07	355
253	20 06	062	53 14	098	57 26	140	34 07	167	35 25	227	61 01	276	24 05	355
254	20 26	062	53 36	099	57 40	142	34 12	169	35 08	228	60 39	276	24 03	356
	Alpheratz		◆DENEB		ALTAIR		Rasalhague		◆ARCTURUS		Alioth		◆CAPELLA	
255	20 46	063	53 58	100	24 31	133	34 16	170	34 51	229	60 16	277	24 02	357
256	21 06	064	54 20	101	24 48	134	34 20	171	34 34	230	59 54	278	24 01	358
257	21 26	065	54 42	102	25 04	135	34 23	172	34 17	231	59 32	279	24 00	358
258	21 47	066	55 04	103	25 20	136	34 26	173	33 59	232	59 10	280	24 00	359
259	22 07	067	55 26	104	25 35	137	34 29	174	33 41	234	58 48	281	24 00	000
260	22 28	068	55 47	106	25 50	138	34 31	176	33 23	235	58 25	281	24 00	001
261	22 49	069	56 09	107	26 05	139	34 32	177	33 04	236	58 03	282	24 00	001
262	23 10	069	56 30	108	26 20	140	34 33	178	32 46	237	57 42	283	24 01	002
263	23 31	070	56 52	109	26 34	141	34 34	179	32 27	238	57 20	284	24 02	003
264	23 52	071	57 13	110	26 48	142	34 34	180	32 07	239	56 58	285	24 03	004
265	24 13	072	57 34	111	27 02	143	34 33	182	31 48	240	56 36	285	24 05	005
266	24 35	073	57 55	112	27 15	144	34 33	183	31 29	241	56 15	286	24 07	005
267	24 56	074	58 16	113	27 28	145	34 31	184	31 09	242	55 53	287	24 09	006
268	25 18	075	58 36	115	27 40	147	34 29	185	30 49	243	55 32	288	24 11	007
269	25 40	076	58 57	116	27 53	148	34 27	186	30 29	244	55 10	288	24 14	008

LHA ♈	Hc	Zn	Hc	Zn	Hc	Zn	Hc	Zn	Hc	Zn	Hc	Zn	Hc	Zn
	◆CAPELLA		Alpheratz		◆DENEB		ALTAIR		Rasalhague		◆ARCTURUS		Alioth	
270	24 17	008	26 02	076	59 17	117	28 05	149	34 25	187	30 08	245	54 49	289
271	24 21	009	26 23	077	59 37	118	28 16	150	34 21	189	29 48	246	54 28	290
272	24 24	010	26 45	078	59 56	119	28 27	151	34 18	190	29 27	247	54 07	291
273	24 28	011	27 07	079	60 16	121	28 38	152	34 14	191	29 07	248	53 46	291
274	24 33	011	27 30	080	60 35	122	28 48	153	34 09	192	28 46	249	53 25	292
275	24 37	012	27 52	081	60 54	123	28 58	154	34 04	193	28 25	250	53 04	293
276	24 42	013	28 14	082	61 13	125	29 08	155	33 59	195	28 03	251	52 43	294
277	24 47	014	28 36	083	61 31	126	29 17	156	33 53	196	27 42	252	52 23	294
278	24 53	014	28 59	084	61 49	127	29 26	158	33 47	197	27 21	253	52 03	295
279	24 58	015	29 21	084	62 07	129	29 34	159	33 40	198	26 59	254	51 42	296
280	25 04	016	29 43	085	62 24	130	29 42	160	33 33	199	26 37	255	51 22	297
281	25 11	017	30 06	086	62 41	131	29 50	161	33 25	200	26 16	256	51 02	297
282	25 17	017	30 28	087	62 58	133	29 57	162	33 17	202	25 54	257	50 42	298
283	25 24	018	30 51	088	63 14	134	30 03	163	33 09	203	25 32	258	50 22	299
284	25 31	019	31 13	089	63 30	136	30 10	164	33 00	204	25 10	259	50 03	299
	◆CAPELLA		Hamal		Alpheratz		◆DENEB		VEGA		◆ARCTURUS		Alioth	
285	25 39	020	15 40	066	31 36	090	63 46	137	60 36	189	24 48	260	49 43	300
286	25 46	020	16 00	067	31 58	091	64 01	139	60 32	191	24 26	261	49 24	301
287	25 54	021	16 21	068	32 21	092	64 15	140	60 28	192	24 05	262	49 05	302
288	26 03	022	16 42	069	32 43	093	64 30	142	60 23	194	23 41	263	48 46	302
289	26 11	023	17 03	069	33 05	094	64 43	143	60 17	196	23 19	264	48 27	303
290	26 20	023	17 24	070	33 28	095	64 57	145	60 11	197	22 56	265	48 08	304
291	26 29	024	17 45	071	33 50	096	65 09	146	60 04	199	22 34	266	47 49	304
292	26 39	025	18 06	072	34 13	097	65 21	148	59 56	200	22 12	266	47 31	305
293	26 48	026	18 28	073	34 35	098	65 33	150	59 48	202	21 49	267	47 12	306
294	26 58	027	18 49	074	34 57	099	65 44	151	59 40	203	21 27	268	46 54	306
295	27 08	027	19 11	075	35 19	099	65 55	153	59 30	205	21 04	269	46 36	307
296	27 19	028	19 33	076	35 41	100	66 05	155	59 21	206	20 42	270	46 18	308
297	27 29	029	19 54	077	36 04	101	66 14	156	59 11	208	20 19	271	46 01	308
298	27 40	030	20 16	077	36 26	102	66 23	158	59 00	209	19 57	272	45 43	309
299	27 51	030	20 38	078	36 47	103	66 31	160	58 49	211	19 34	273	45 26	310
	◆CAPELLA		Hamal		Alpheratz		◆DENEB		VEGA		◆ARCTURUS		Dubhe	
300	28 03	031	21 00	079	37 09	104	66 38	161	58 37	212	19 12	274	43 55	332
301	28 15	032	21 23	080	37 31	105	66 45	163	58 25	214	18 50	275	43 44	333
302	28 27	033	21 45	081	37 53	106	66 51	165	58 12	215	18 27	276	43 34	333
303	28 39	033	22 07	082	38 14	107	66 57	167	57 59	216	18 05	277	43 24	334
304	28 51	034	22 29	083	38 35	108	67 02	169	57 45	218	17 43	278	43 14	334
305	29 04	035	22 52	084	38 57	109	67 06	170	57 31	219	17 20	278	43 05	335
306	29 17	036	23 14	085	39 18	110	67 09	172	57 17	221	16 58	279	42 55	336
307	29 30	036	23 36	086	39 39	112	67 12	174	57 02	222	16 36	280	42 46	336
308	29 44	037	23 59	086	40 00	113	67 14	176	56 47	223	16 14	281	42 37	337
309	29 57	038	24 21	087	40 20	114	67 15	178	56 31	225	15 52	282	42 28	337
310	30 11	039	24 44	088	40 41	115	67 16	179	56 15	226	15 30	283	42 20	338
311	30 25	039	25 06	089	41 01	116	67 16	181	55 59	227	15 08	284	42 11	339
312	30 40	040	25 29	090	41 21	117	67 15	183	55 42	229	14 46	285	42 03	339
313	30 54	041	25 51	091	41 41	118	67 13	185	55 25	230	14 25	286	41 55	340
314	31 09	042	26 13	092	42 01	119	67 11	187	55 08	231	14 03	287	41 48	340
	◆CAPELLA		Hamal		Alpheratz		◆DENEB		VEGA		◆Alphecca		Dubhe	
315	31 24	042	26 36	093	42 21	120	67 08	189	54 50	232	27 51	273	41 40	341
316	31 39	043	26 58	094	42 40	121	67 04	190	54 32	234	27 29	274	41 33	342
317	31 55	044	27 21	095	42 59	122	67 00	192	54 14	235	27 06	275	41 26	342
318	32 11	045	27 43	096	43 18	123	66 55	194	53 56	236	26 44	276	41 20	343
319	32 27	045	28 06	097	43 37	125	66 49	196	53 37	237	26 22	277	41 13	344
320	32 43	046	28 28	098	43 55	126	66 43	197	53 18	238	25 59	277	41 07	344
321	32 59	047	28 50	099	44 13	127	66 36	199	52 59	240	25 37	278	41 01	345
322	33 15	048	29 12	100	44 31	128	66 28	201	52 39	241	25 15	279	40 55	345
323	33 32	048	29 34	101	44 49	129	66 20	203	52 19	242	24 53	280	40 49	346
324	33 49	049	29 56	102	45 06	130	66 11	204	51 59	243	24 31	281	40 44	347
325	34 06	050	30 18	103	45 23	131	66 01	206	51 39	244	24 08	282	40 39	347
326	34 24	051	30 40	103	45 40	133	65 51	208	51 19	245	23 47	283	40 34	348
327	34 41	052	31 02	104	45 56	134	65 40	209	50 58	247	23 25	284	40 29	348
328	34 59	052	31 24	105	46 12	135	65 29	211	50 38	248	23 03	285	40 25	349
329	35 17	053	31 45	106	46 28	136	65 17	213	50 17	249	22 41	285	40 21	350
	CAPELLA		◆Hamal		Alpheratz		◆ALTAIR		VEGA		Alphecca		◆Alioth	
330	35 35	054	32 07	107	46 43	137	27 05	216	49 56	250	22 20	286	38 03	331
331	35 53	055	32 28	108	46 58	139	26 52	218	49 35	251	21 58	287	37 52	331
332	36 11	055	32 50	109	47 13	140	26 38	219	49 13	252	21 37	288	37 42	332
333	36 30	056	33 11	110	47 27	141	26 24	220	48 52	253	21 15	289	37 31	333
334	36 49	057	33 32	111	47 41	142	26 09	221	48 30	254	20 54	290	37 21	333
335	37 08	058	33 53	113	47 54	144	25 54	222	48 09	255	20 33	291	37 11	334
336	37 27	058	34 13	114	48 08	145	25 39	223	47 47	256	20 12	292	37 02	335
337	37 46	059	34 34	115	48 20	146	25 24	224	47 25	257	19 51	293	36 52	335
338	38 05	060	34 54	116	48 33	147	25 08	225	47 03	258	19 31	293	36 43	336
339	38 25	061	35 14	117	48 44	149	24 52	226	46 41	259	19 10	294	36 34	337
340	38 45	062	35 34	118	48 56	150	24 36	227	46 19	260	18 50	295	36 25	337
341	39 04	062	35 54	119	49 07	151	24 19	228	45 57	261	18 29	296	36 17	338
342	39 24	063	36 14	120	49 17	153	24 02	229	45 34	262	18 09	297	36 09	339
343	39 44	064	36 33	121	49 28	154	23 45	230	45 12	263	17 49	298	36 01	339
344	40 05	065	36 52	122	49 37	155	23 28	231	44 50	264	17 29	299	35 53	340
	◆POLLUX		CAPELLA		Hamal		◆Alpheratz		ALTAIR		◆VEGA		Alioth	
345	12 35	043	40 25	066	37 11	123	49 46	157	23 10	232	44 27	265	35 45	341
346	12 50	044	40 46	066	37 30	124	49 55	158	22 52	233	44 05	266	35 38	342
347	13 06	045	41 06	067	37 48	125	50 03	159	22 34	234	43 43	267	35 31	342
348	13 22	045	41 27	068	38 07	126	50 11	161	22 16	235	43 20	268	35 24	343
349	13 38	046	41 48	069	38 25	127	50 18	162	21 57	236	42 58	269	35 18	344
350	13 54	047	42 09	070	38 42	129	50 25	163	21 38	237	42 35	270	35 12	344
351	14 11	048	42 30	070	39 00	130	50 31	165	21 19	238	42 13	271	35 06	345
352	14 28	049	42 51	071	39 17	131	50 37	166	21 00	239	41 50	272	35 00	346
353	14 45	050	43 13	072	39 34	132	50 42	167	20 41	240	41 28	273	34 54	346
354	15 02	051	43 34	073	39 50	133	50 46	169	20 21	241	41 05	273	34 49	347
355	15 20	051	43 56	074	40 07	134	50 50	170	20 01	242	40 43	274	34 44	348
356	15 37	052	44 17	074	40 23	135	50 54	172	19 41	243	40 21	275	34 39	348
357	15 55	053	44 39	075	40 38	136	50 57	173	19 21	244	39 58	276	34 35	349
358	16 13	054	45 01	076	40 54	138	50 59	174	19 01	245	39 36	277	34 31	350
359	16 31	055	45 23	077	41 09	139	51 01	176	18 40	246	39 14	278	34 27	350

LHA	Hc	Zn	Hc	Zn	Hc	Zn	Hc	Zn	Hc	Zn	Hc	Zn	Hc	Zn
♈	◆POLLUX		CAPELLA		Hamal		◆Alpheratz		DENEB		◆VEGA		Alioth	
0	16 16	056	45 31	077	42 09	139	52 03	177	56 17	255	38 42	280	33 24	351
1	16 36	056	45 54	078	42 24	141	52 03	179	55 54	256	38 19	280	33 20	352
2	16 55	057	46 17	078	42 39	142	52 04	180	55 32	257	37 56	281	33 17	352
3	17 15	058	46 40	079	42 53	143	52 03	181	55 09	258	37 33	282	33 14	353
4	17 35	059	47 03	080	43 07	144	52 03	183	54 46	259	37 10	283	33 11	354
5	17 55	060	47 26	081	43 21	145	52 01	184	54 23	260	36 47	284	33 09	354
6	18 16	061	47 49	082	43 34	147	51 59	186	53 59	261	36 24	285	33 07	355
7	18 36	062	48 13	083	43 46	148	51 57	187	53 36	262	36 02	286	33 05	356
8	18 57	062	48 36	084	43 59	149	51 53	188	53 13	263	35 39	286	33 03	356
9	19 18	063	48 59	084	44 10	150	51 50	190	52 50	264	35 17	287	33 02	357
10	19 39	064	49 23	085	44 22	152	51 45	191	52 26	265	34 54	288	33 01	358
11	20 00	065	49 46	086	44 33	153	51 41	193	52 03	266	34 32	289	33 00	358
12	20 21	066	50 09	087	44 43	154	51 35	194	51 39	267	34 10	290	33 00	359
13	20 43	067	50 33	088	44 53	155	51 29	195	51 16	268	33 48	291	32 59	000
14	21 04	068	50 56	089	45 03	157	51 23	197	50 53	269	33 26	291	32 59	000
	Alioth		◆POLLUX		CAPELLA		Hamal		◆Alpheratz		DENEB		◆VEGA	
15	33 00	001	21 26	068	51 20	090	45 12	158	51 15	198	50 29	270	33 04	292
16	33 00	002	21 48	069	51 43	091	45 20	159	51 08	200	50 06	271	32 43	293
17	33 01	002	22 10	070	52 07	092	45 28	161	51 00	201	49 42	272	32 21	294
18	33 02	003	22 32	071	52 30	093	45 36	162	50 51	202	49 19	273	32 00	295
19	33 03	004	22 54	072	52 53	094	45 43	163	50 42	204	48 55	274	31 39	296
20	33 05	004	23 16	073	53 17	095	45 49	164	50 32	205	48 32	275	31 18	296
21	33 07	005	23 39	074	53 40	096	45 55	166	50 22	206	48 09	275	30 57	297
22	33 09	006	24 01	075	54 03	097	46 01	167	50 11	208	47 45	276	30 36	298
23	33 12	006	24 24	075	54 27	097	46 06	168	50 00	209	47 22	277	30 15	299
24	33 14	007	24 47	076	54 50	098	46 10	170	49 48	210	46 59	278	29 55	300
25	33 17	008	25 10	077	55 13	099	46 14	171	49 36	212	46 36	279	29 34	300
26	33 21	008	25 33	078	55 36	101	46 17	172	49 24	213	46 13	280	29 14	301
27	33 24	009	25 56	079	55 59	102	46 20	174	49 11	214	45 49	281	28 54	302
28	33 28	010	26 19	080	56 22	103	46 23	175	48 57	216	45 26	281	28 35	303
29	33 32	010	26 42	081	56 45	104	46 24	176	48 43	217	45 04	282	28 15	304
	Alioth		◆POLLUX		ALDEBARAN		◆Hamal		Alpheratz		DENEB		◆VEGA	
30	33 36	011	27 05	082	33 34	134	46 26	178	48 29	218	44 41	283	27 56	304
31	33 41	012	27 28	082	33 51	135	46 26	179	48 14	219	44 18	284	27 36	305
32	33 46	012	27 51	083	34 07	136	46 26	180	47 59	221	43 55	285	27 17	306
33	33 51	013	28 15	084	34 24	137	46 26	182	47 44	222	43 33	286	26 58	307
34	33 57	014	28 38	085	34 39	138	46 25	183	47 28	223	43 10	286	26 40	308
35	34 02	014	29 01	086	34 55	139	46 23	184	47 12	224	42 48	287	26 21	308
36	34 08	015	29 25	087	35 10	140	46 21	186	46 55	226	42 25	288	26 03	309
37	34 14	016	29 48	088	35 25	142	46 19	187	46 38	227	42 03	289	25 45	310
38	34 21	016	30 12	089	35 39	143	46 16	188	46 21	228	41 41	290	25 27	311
39	34 28	017	30 35	090	35 53	144	46 12	190	46 03	229	41 19	290	25 09	312
40	34 35	018	30 58	091	36 07	145	46 08	191	45 45	230	40 57	291	24 52	312
41	34 42	018	31 22	092	36 20	146	46 03	192	45 27	232	40 35	292	24 35	313
42	34 49	019	31 45	092	36 33	147	45 58	194	45 09	233	40 13	293	24 18	314
43	34 57	020	32 09	093	36 45	148	45 52	195	44 50	234	39 52	294	24 01	315
44	35 05	020	32 32	094	36 57	150	45 46	196	44 31	235	39 31	294	23 45	316
	Alioth		◆POLLUX		ALDEBARAN		◆Hamal		Alpheratz		DENEB		◆VEGA	
45	35 13	021	32 55	095	37 09	151	45 39	198	44 12	236	39 09	295	23 28	316
46	35 22	022	33 19	096	37 20	152	45 32	199	43 52	237	38 48	296	23 12	317
47	35 31	022	33 42	097	37 31	153	45 24	200	43 32	238	38 27	297	22 57	318
48	35 40	023	34 05	098	37 41	154	45 15	201	43 12	240	38 06	298	22 41	319
49	35 49	024	34 29	099	37 51	156	45 07	203	42 52	241	37 46	298	22 26	320
50	35 58	024	34 52	100	38 01	157	44 57	204	42 31	242	37 25	299	22 11	320
51	36 08	025	35 15	101	38 10	158	44 48	205	42 10	243	37 05	300	21 56	321
52	36 18	026	35 38	102	38 18	159	44 37	207	41 49	244	36 44	301	21 41	322
53	36 29	026	36 01	103	38 27	160	44 27	208	41 28	245	36 24	301	21 27	323
54	36 39	027	36 23	104	38 34	162	44 15	209	41 07	246	36 04	302	21 13	324
55	36 50	028	36 46	105	38 41	163	44 04	210	40 45	247	35 45	303	20 59	324
56	37 01	028	37 09	106	38 48	164	43 52	212	40 24	248	35 25	304	20 46	325
57	37 12	029	37 31	107	38 54	165	43 39	213	40 02	249	35 06	305	20 32	326
58	37 23	030	37 53	108	39 00	166	43 26	214	39 40	250	34 46	305	20 19	327
59	37 35	030	38 16	109	39 05	168	43 13	215	39 18	251	34 27	306	20 07	328
	Alioth		◆POLLUX		ALDEBARAN		◆Hamal		Alpheratz		DENEB		◆VEGA	
60	37 47	031	38 38	110	39 10	169	42 59	216	38 55	252	34 08	307	19 54	328
61	37 59	032	39 00	111	39 14	170	42 45	218	38 33	253	33 50	308	19 42	329
62	38 12	032	39 22	112	39 18	171	42 31	219	38 11	254	33 31	308	19 30	330
63	38 24	033	39 43	113	39 21	173	42 16	220	37 48	255	33 13	309	19 19	331
64	38 37	034	40 05	114	39 24	174	42 01	221	37 25	256	32 55	310	19 07	332
65	38 50	034	40 26	115	39 26	175	41 45	222	37 02	257	32 37	311	18 56	332
66	39 03	035	40 47	116	39 28	176	41 29	224	36 40	258	32 19	311	18 46	333
67	39 17	036	41 08	117	39 29	178	41 13	225	36 17	259	32 02	312	18 35	334
68	39 31	036	41 29	118	39 30	179	40 56	226	35 53	260	31 45	313	18 25	335
69	39 45	037	41 49	120	39 30	180	40 39	227	35 30	261	31 27	314	18 15	336
70	39 59	038	42 09	121	39 30	181	40 22	228	35 07	262	31 11	314	18 06	336
71	40 13	038	42 29	122	39 29	183	40 04	229	34 44	263	30 54	315	17 57	337
72	40 28	039	42 49	123	39 28	184	39 46	231	34 21	264	30 37	316	17 48	338
73	40 43	040	43 09	124	39 26	185	39 28	232	33 57	265	30 21	317	17 39	339
74	40 58	040	43 28	125	39 23	186	39 09	233	33 34	266	30 05	317	17 31	340
	Alioth		◆REGULUS		PROCYON		◆ALDEBARAN		Hamal		Alpheratz		◆DENEB	
75	41 13	041	16 05	097	22 32	136	39 21	188	38 50	234	33 10	267	29 50	318
76	41 28	041	16 28	098	22 48	137	39 17	189	38 31	235	32 47	268	29 34	319
77	41 44	042	16 51	099	23 03	138	39 13	190	38 12	236	32 24	269	29 19	320
78	42 00	043	17 15	100	23 19	140	39 09	191	37 52	237	32 00	270	29 04	320
79	42 16	043	17 38	101	23 34	141	39 04	192	37 33	238	31 37	271	28 49	321
80	42 32	044	18 01	102	23 49	142	38 59	194	37 13	239	31 13	271	28 34	322
81	42 48	045	18 24	103	24 03	143	38 53	195	36 52	240	30 50	272	28 20	323
82	43 05	045	18 46	104	24 17	144	38 47	196	36 32	241	30 26	273	28 06	323
83	43 22	046	19 09	105	24 31	145	38 40	197	36 11	242	30 03	274	27 52	324
84	43 39	047	19 32	106	24 44	146	38 33	199	35 50	244	29 40	275	27 38	325
85	43 56	047	19 54	107	24 57	147	38 25	200	35 29	245	29 16	276	27 25	326
86	44 13	048	20 17	108	25 10	148	38 17	201	35 08	246	28 53	277	27 12	326
87	44 31	049	20 39	109	25 22	149	38 08	202	34 46	247	28 30	278	26 59	327
88	44 49	049	21 01	110	25 34	150	37 59	203	34 25	248	28 07	279	26 47	328
89	45 07	050	21 23	111	25 45	151	37 50	205	34 03	249	27 43	280	26 34	329

LAT 67°N

LHA	Hc	Zn	Hc	Zn	Hc	Zn	Hc	Zn	Hc	Zn	Hc	Zn	Hc	Zn
♈	Alioth		◆REGULUS		PROCYON		BETELGEUSE		◆ALDEBARAN		Hamal		◆DENEB	
90	45 25	051	21 45	112	25 56	152	30 24	181	37 40	206	33 41	250	26 22	329
91	45 43	051	22 07	113	26 07	153	30 23	183	37 29	207	33 19	251	26 10	330
92	46 01	052	22 28	114	26 17	155	30 22	184	37 18	208	32 57	252	25 59	331
93	46 20	053	22 50	114	26 27	156	30 20	185	37 07	209	32 35	253	25 48	332
94	46 39	053	23 11	115	26 36	157	30 18	186	36 55	211	32 12	254	25 37	332
95	46 58	054	23 32	116	26 45	158	30 15	187	36 43	212	31 50	255	25 26	333
96	47 17	055	23 53	117	26 54	159	30 12	188	36 31	213	31 27	256	25 16	334
97	47 36	056	24 14	118	27 02	160	30 08	189	36 18	214	31 04	257	25 05	335
98	47 55	056	24 34	119	27 10	161	30 04	191	36 04	215	30 41	258	24 56	335
99	48 15	057	24 54	121	27 17	162	30 00	192	35 51	216	30 18	259	24 46	336
100	48 35	058	25 14	122	27 24	163	29 55	193	35 37	218	29 55	260	24 37	337
101	48 55	058	25 34	123	27 31	165	29 49	194	35 22	219	29 32	261	24 28	338
102	49 15	059	25 54	124	27 37	166	29 43	195	35 07	220	29 09	261	24 19	339
103	49 35	060	26 13	125	27 42	167	29 37	196	34 52	221	28 46	262	24 10	339
104	49 55	060	26 33	126	27 48	168	29 30	197	34 37	222	28 23	263	24 02	340
	Alioth		◆REGULUS		PROCYON		BETELGEUSE		◆CAPELLA		Hamal		◆DENEB	
105	50 15	061	26 52	127	27 52	169	29 23	199	64 59	226	27 59	264	23 54	341
106	50 36	062	27 10	128	27 57	170	29 15	200	64 42	227	27 36	265	23 47	342
107	50 57	062	27 29	129	28 00	171	29 07	201	64 25	229	27 13	266	23 40	342
108	51 18	063	27 47	130	28 04	172	28 58	202	64 07	230	26 49	267	23 33	343
109	51 38	064	28 05	131	28 07	173	28 49	203	63 49	232	26 26	268	23 26	344
110	52 00	064	28 22	132	28 09	175	28 40	204	63 30	233	26 02	269	23 20	345
111	52 21	065	28 40	133	28 11	176	28 30	205	63 11	235	25 39	270	23 13	345
112	52 42	066	28 57	134	28 13	177	28 20	206	62 52	236	25 15	271	23 08	346
113	53 04	067	29 13	135	28 14	178	28 09	208	62 32	237	24 52	272	23 02	347
114	53 25	067	29 30	136	28 14	179	27 58	209	62 13	239	24 29	273	22 57	348
115	53 47	068	29 46	137	28 14	180	27 47	210	61 52	240	24 05	274	22 52	348
116	54 09	069	30 01	138	28 14	181	27 35	211	61 32	241	23 42	274	22 48	349
117	54 31	070	30 17	139	28 13	183	27 23	212	61 11	242	23 18	275	22 43	350
118	54 53	070	30 32	141	28 12	184	27 10	213	60 51	244	22 55	276	22 39	351
119	55 15	071	30 47	142	28 10	185	26 57	214	60 29	245	22 32	277	22 36	351
	◆VEGA		ARCTURUS		◆REGULUS		PROCYON		◆CAPELLA		Schedar		DENEB	
120	16 58	017	16 09	079	31 01	143	28 08	186	60 08	246	43 58	314	22 32	352
121	17 05	018	16 32	080	31 15	144	28 05	187	59 47	247	43 41	315	22 29	353
122	17 12	018	16 55	081	31 29	145	28 02	188	59 25	248	43 24	315	22 26	354
123	17 20	019	17 18	081	31 42	146	27 59	189	59 03	250	43 08	316	22 24	354
124	17 28	020	17 41	082	31 55	147	27 55	190	58 41	251	42 51	317	22 22	355
125	17 36	021	18 04	083	32 07	148	27 50	192	58 19	252	42 35	317	22 20	356
126	17 44	022	18 28	084	32 19	149	27 45	193	57 56	253	42 20	318	22 19	357
127	17 53	022	18 51	085	32 31	151	27 40	194	57 34	254	42 04	319	22 17	357
128	18 02	023	19 14	086	32 42	152	27 34	195	57 11	255	41 49	319	22 17	358
129	18 12	024	19 38	087	32 53	153	27 28	196	56 49	256	41 33	320	22 16	359
130	18 21	025	20 01	088	33 04	154	27 21	197	56 26	257	41 18	320	22 16	000
131	18 31	026	20 25	089	33 14	155	27 14	198	56 03	258	41 03	321	22 16	001
132	18 42	026	20 48	090	33 23	156	27 06	199	55 40	259	40 49	322	22 16	001
133	18 52	027	21 11	091	33 33	158	26 58	200	55 17	260	40 34	322	22 17	002
134	19 03	028	21 35	092	33 41	159	26 50	202	54 54	261	40 20	323	22 18	003
	DENEB		◆VEGA		ARCTURUS		◆REGULUS		PROCYON		◆CAPELLA		Schedar	
135	22 19	004	19 14	029	21 58	092	33 50	160	26 41	203	54 30	262	40 06	324
136	22 21	004	19 26	030	22 22	093	33 58	161	26 32	204	54 07	263	39 53	324
137	22 23	005	19 38	030	22 45	094	34 05	162	26 22	205	53 44	264	39 39	325
138	22 25	006	19 50	031	23 08	095	34 12	163	26 12	206	53 21	265	39 26	326
139	22 27	007	20 02	032	23 32	096	34 18	165	26 02	207	52 57	266	39 13	326
140	22 30	007	20 14	033	23 55	097	34 24	166	25 51	208	52 34	267	39 00	327
141	22 33	008	20 27	034	24 18	098	34 30	167	25 40	209	52 10	268	38 47	328
142	22 37	009	20 40	034	24 42	099	34 35	168	25 28	210	51 47	269	38 35	328
143	22 41	010	20 54	035	25 05	100	34 40	169	25 16	211	51 23	270	38 22	329
144	22 45	010	21 08	036	25 28	101	34 44	170	25 04	212	51 00	271	38 10	330
145	22 49	011	21 21	037	25 51	102	34 47	172	24 51	214	50 37	272	37 59	330
146	22 54	012	21 36	038	26 14	103	34 51	173	24 38	215	50 13	273	37 47	331
147	22 59	013	21 50	038	26 36	104	34 53	174	24 24	216	49 50	274	37 36	332
148	23 04	013	22 05	039	26 59	105	34 56	175	24 10	217	49 26	275	37 25	332
149	23 10	014	22 20	040	27 22	106	34 57	176	23 56	218	49 03	275	37 14	333
	DENEB		◆VEGA		ARCTURUS		◆REGULUS		PROCYON		◆CAPELLA		Schedar	
150	23 16	015	22 35	041	27 44	107	34 59	178	23 42	219	48 40	276	37 04	334
151	23 22	016	22 51	042	28 07	108	34 59	179	23 27	220	48 16	277	36 53	334
152	23 28	016	23 06	042	28 29	109	35 00	180	23 11	221	47 53	278	36 43	335
153	23 35	017	23 22	043	28 51	110	34 59	181	22 56	222	47 30	279	36 33	336
154	23 42	018	23 38	044	29 13	111	34 59	182	22 40	223	47 07	280	36 24	336
155	23 50	019	23 55	045	29 35	112	34 57	184	22 24	224	46 44	281	36 15	337
156	23 57	020	24 11	046	29 57	113	34 56	185	22 07	225	46 21	281	36 05	338
157	24 05	020	24 28	046	30 18	114	34 53	186	21 51	226	45 58	282	35 57	338
158	24 14	021	24 45	047	30 40	115	34 51	187	21 34	227	45 35	283	35 48	339
159	24 22	022	25 03	048	31 01	116	34 48	188	21 16	228	45 12	284	35 40	339
160	24 31	023	25 20	049	31 22	117	34 44	190	20 59	229	44 49	285	35 32	340
161	24 40	023	25 38	050	31 43	118	34 40	191	20 41	230	44 27	286	35 24	341
162	24 49	024	25 56	050	32 03	119	34 35	192	20 23	231	44 04	286	35 16	341
163	24 59	025	26 14	051	32 24	120	34 30	193	20 04	232	43 42	287	35 09	342
164	25 09	026	26 33	052	32 44	121	34 25	194	19 46	233	43 19	288	35 02	343
	DENEB		◆VEGA		ARCTURUS		◆REGULUS		POLLUX		◆CAPELLA		Schedar	
165	25 19	026	26 51	053	33 04	122	34 19	195	41 18	242	42 57	289	34 55	343
166	25 30	027	27 10	054	33 24	123	34 12	197	40 58	243	42 35	290	34 48	344
167	25 41	028	27 29	054	33 43	124	34 05	198	40 37	244	42 13	290	34 42	345
168	25 52	029	27 48	055	34 03	125	33 58	199	40 15	245	41 51	291	34 36	345
169	26 03	029	28 07	056	34 22	126	33 50	200	39 54	246	41 29	292	34 30	346
170	26 15	030	28 27	057	34 40	127	33 42	201	39 32	247	41 08	293	34 25	347
171	26 27	031	28 47	058	34 59	128	33 33	202	39 11	248	40 46	294	34 20	347
172	26 39	032	29 07	058	35 17	129	33 24	204	38 49	249	40 25	294	34 15	348
173	26 51	032	29 27	059	35 35	131	33 14	205	38 27	250	40 03	295	34 10	349
174	27 04	033	29 47	060	35 53	132	33 04	206	38 05	251	39 42	296	34 05	349
175	27 17	034	30 07	061	36 10	133	32 54	207	37 42	253	39 21	297	34 01	350
176	27 30	035	30 28	062	36 27	134	32 43	208	37 20	254	39 00	297	33 57	351
177	27 43	035	30 49	063	36 44	135	32 32	209	36 57	255	38 40	298	33 54	351
178	27 57	036	31 09	063	37 00	136	32 20	210	36 35	256	38 19	299	33 50	352
179	28 11	037	31 31	064	37 16	137	32 08	212	36 12	257	37 58	300	33 47	353

LHA ♈	Hc	Zn	Hc	Zn	Hc	Zn	Hc	Zn	Hc	Zn	Hc	Zn	Hc	Zn
	◆DENEB		VEGA		◆ARCTURUS		REGULUS		◆POLLUX		CAPELLA		Schedar	
180	28 25	038	31 52	065	37 32	138	31 55	213	35 49	257	37 38	301	33 44	353
181	28 40	038	32 13	066	37 48	140	31 43	214	35 26	258	37 18	301	33 42	354
182	28 54	039	32 34	067	38 03	141	31 29	215	35 03	259	36 58	302	33 39	355
183	29 09	040	32 56	067	38 17	142	31 16	216	34 40	260	36 38	303	33 37	355
184	29 24	041	33 18	068	38 31	143	31 02	217	34 17	261	36 19	304	33 36	356
185	29 40	041	33 40	069	38 45	144	30 47	218	33 54	262	35 59	304	33 34	357
186	29 55	042	34 02	070	38 59	145	30 33	219	33 31	263	35 40	305	33 33	357
187	30 11	043	34 24	071	39 12	147	30 18	220	33 07	264	35 21	306	33 32	358
188	30 27	044	34 46	072	39 25	148	30 02	222	32 44	265	35 02	307	33 31	359
189	30 44	044	35 08	072	39 37	149	29 47	223	32 21	266	34 43	307	33 31	359
190	31 00	045	35 30	073	39 49	150	29 30	224	31 57	267	34 25	308	33 31	000
191	31 17	046	35 53	074	40 00	151	29 14	225	31 34	268	34 06	309	33 31	001
192	31 34	047	36 16	075	40 11	153	28 57	226	31 10	269	33 48	310	33 31	001
193	31 51	047	36 38	076	40 22	154	28 40	227	30 47	270	33 30	310	33 32	002
194	32 08	048	37 01	077	40 32	155	28 23	228	30 23	271	33 13	311	33 33	003
	Schedar		◆DENEB		VEGA		◆ARCTURUS		REGULUS		◆POLLUX		CAPELLA	
195	33 34	003	32 26	049	37 24	077	40 42	156	28 06	229	30 00	272	32 55	312
196	33 35	004	32 44	050	37 47	078	40 51	157	27 48	230	29 37	273	32 38	313
197	33 37	005	33 02	050	38 10	079	41 00	159	27 30	231	29 13	273	32 21	313
198	33 39	005	33 20	051	38 33	080	41 08	160	27 11	232	28 50	274	32 04	314
199	33 41	006	33 38	052	38 56	081	41 16	161	26 53	233	28 26	275	31 47	315
200	33 44	007	33 57	053	39 19	082	41 23	162	26 34	234	28 03	276	31 30	316
201	33 47	007	34 15	053	39 42	083	41 30	164	26 14	235	27 40	277	31 14	316
202	33 50	008	34 34	054	40 06	084	41 36	165	25 55	236	27 17	278	30 58	317
203	33 53	009	34 53	055	40 29	084	41 42	166	25 35	237	26 53	279	30 42	318
204	33 57	009	35 13	056	40 52	085	41 47	167	25 16	238	26 30	280	30 27	319
205	34 01	010	35 32	057	41 16	086	41 52	169	24 55	239	26 07	281	30 11	319
206	34 05	011	35 52	057	41 39	087	41 56	170	24 35	240	25 44	282	29 56	320
207	34 09	011	36 12	058	42 03	088	42 00	171	24 15	241	25 21	282	29 41	321
208	34 14	012	36 32	059	42 26	089	42 04	173	23 54	242	24 58	283	29 27	322
209	34 19	013	36 52	060	42 49	090	42 06	174	23 33	243	24 36	284	29 12	322
	Schedar		◆DENEB		VEGA		◆ARCTURUS		REGULUS		◆POLLUX		CAPELLA	
210	34 24	013	37 12	060	43 13	091	42 09	175	23 12	244	24 13	285	28 58	323
211	34 30	014	37 33	061	43 36	092	42 10	176	22 51	245	23 50	286	28 44	324
212	34 35	015	37 53	062	44 00	093	42 12	178	22 29	246	23 28	287	28 30	325
213	34 41	015	38 14	063	44 23	094	42 12	179	22 08	247	23 05	288	28 17	325
214	34 48	016	38 35	063	44 47	095	42 12	180	21 46	248	22 43	289	28 04	326
215	34 54	016	38 56	064	45 10	096	42 12	181	21 24	249	22 21	289	27 51	327
216	35 01	017	39 17	065	45 33	096	42 11	183	21 02	250	21 59	290	27 38	328
217	35 08	018	39 38	066	45 56	097	42 10	184	20 40	251	21 37	291	27 26	328
218	35 15	018	40 00	067	46 20	098	42 08	185	20 18	252	21 15	292	27 13	329
219	35 23	019	40 21	067	46 43	099	42 06	187	19 55	253	20 53	293	27 01	330
220	35 31	020	40 43	068	47 06	100	42 03	188	19 33	254	20 32	294	26 50	331
221	35 39	020	41 05	069	47 29	101	41 59	189	19 10	255	20 11	295	26 38	331
222	35 47	021	41 27	070	47 52	102	41 55	190	18 48	256	19 49	295	26 27	332
223	35 56	022	41 49	071	48 15	103	41 51	192	18 25	257	19 28	296	26 17	333
224	36 04	022	42 11	071	48 38	104	41 46	193	18 02	258	19 07	297	26 06	334
	Schedar		◆DENEB		VEGA		◆ARCTURUS		REGULUS		◆POLLUX		CAPELLA	
225	36 13	023	42 33	072	49 00	105	41 40	194	17 39	259	18 47	298	25 56	334
226	36 23	024	42 56	073	49 23	107	41 34	195	17 16	260	18 26	299	25 46	335
227	36 32	024	43 18	074	49 45	108	41 28	197	16 53	261	18 05	300	25 36	336
228	36 42	025	43 41	075	50 07	109	41 21	198	16 30	262	17 45	301	25 27	337
229	36 52	026	44 04	076	50 30	110	41 14	199	16 06	263	17 25	301	25 17	337
230	37 02	026	44 26	076	50 52	111	41 06	200	15 43	264	17 05	302	25 08	338
231	37 13	027	44 49	077	51 13	112	40 57	202	15 20	265	16 45	303	25 00	339
232	37 24	028	45 12	078	51 35	113	40 48	203	14 56	266	16 26	304	24 52	340
233	37 35	028	45 35	079	51 57	114	40 39	204	14 33	266	16 07	305	24 44	340
234	37 46	029	45 58	080	52 18	115	40 29	205	14 10	267	15 47	306	24 36	341
235	37 57	030	46 21	081	52 39	116	40 19	207	13 46	268	15 29	307	24 28	342
236	38 09	030	46 44	081	53 00	118	40 08	208	13 23	269	15 10	307	24 21	343
237	38 21	031	47 07	082	53 20	119	39 57	209	12 59	270	14 51	308	24 14	343
238	38 33	032	47 31	083	53 41	120	39 46	210	12 36	271	14 33	309	24 08	344
239	38 46	032	47 54	084	54 01	121	39 34	211	12 12	272	14 15	310	24 01	345
	Alpheratz		◆DENEB		VEGA		Rasalhague		◆ARCTURUS		Alioth		◆CAPELLA	
240	15 26	050	48 17	085	54 21	122	33 20	152	39 21	213	65 58	266	23 55	346
241	15 44	051	48 41	086	54 41	124	33 31	153	39 08	214	65 35	267	23 50	346
242	16 02	052	49 04	087	55 00	125	33 41	154	38 55	215	65 11	268	23 44	347
243	16 21	053	49 27	088	55 19	126	33 51	155	38 41	216	64 48	269	23 39	348
244	16 39	054	49 51	088	55 38	127	34 01	157	38 27	217	64 25	270	23 35	349
245	16 58	054	50 14	089	55 57	129	34 10	158	38 13	218	64 01	270	23 30	349
246	17 18	055	50 38	090	56 15	130	34 18	159	37 58	220	63 38	271	23 26	350
247	17 37	056	51 01	091	56 33	131	34 26	160	37 43	221	63 14	272	23 22	351
248	17 57	057	51 25	092	56 50	132	34 34	161	37 28	222	62 51	273	23 19	352
249	18 16	058	51 48	093	57 07	134	34 42	162	37 12	223	62 27	274	23 15	352
250	18 36	059	52 12	094	57 24	135	34 48	164	36 56	224	62 04	275	23 12	353
251	18 56	060	52 35	095	57 40	136	34 55	165	36 39	225	61 41	276	23 10	354
252	19 17	060	52 58	096	57 56	138	35 01	166	36 22	226	61 17	277	23 07	355
253	19 37	061	53 22	097	58 12	139	35 06	167	36 05	228	60 54	277	23 05	355
254	19 58	062	53 45	098	58 27	141	35 11	168	35 48	229	60 31	278	23 04	356
	Alpheratz		◆DENEB		ALTAIR		Rasalhague		◆ARCTURUS		Alioth		◆CAPELLA	
255	20 19	063	54 08	099	25 12	132	35 15	170	35 30	230	60 08	279	23 02	357
256	20 40	064	54 31	100	25 29	133	35 19	171	35 12	231	59 45	280	23 01	358
257	21 01	065	54 54	101	25 46	134	35 23	172	34 54	232	59 22	281	23 00	358
258	21 22	066	55 17	102	26 02	135	35 26	173	34 35	233	58 59	281	23 00	359
259	21 43	066	55 40	103	26 19	137	35 28	174	34 16	234	58 36	282	23 00	000
260	22 05	067	56 03	104	26 35	138	35 30	176	33 57	235	58 13	283	23 00	001
261	22 27	068	56 25	105	26 50	139	35 32	177	33 38	236	57 50	284	23 00	001
262	22 49	069	56 48	106	27 06	140	35 33	178	33 18	237	57 27	285	23 01	002
263	23 10	070	57 10	107	27 21	141	35 34	179	32 58	238	57 05	285	23 02	003
264	23 33	071	57 33	108	27 35	142	35 34	180	32 38	239	56 42	286	23 03	004
265	23 55	072	57 55	110	27 50	143	35 33	182	32 18	240	56 20	287	23 05	004
266	24 17	072	58 17	111	28 04	144	35 32	183	31 57	241	55 57	288	23 07	005
267	24 39	073	58 39	112	28 17	145	35 31	184	31 37	243	55 35	288	23 09	006
268	25 02	074	59 00	113	28 30	146	35 29	185	31 16	244	55 13	289	23 12	007
269	25 25	075	59 22	114	28 43	147	35 27	186	30 55	245	54 51	290	23 15	007

LAT 67°N

LHA ♈	Hc	Zn	Hc	Zn	Hc	Zn	Hc	Zn	Hc	Zn	Hc	Zn	Hc	Zn
	◆CAPELLA		Alpheratz		◆DENEB		ALTAIR		Rasalhague		◆ARCTURUS		Alioth	
270	23 18	008	25 47	076	59 43	115	28 56	148	35 24	188	30 34	246	54 29	291
271	23 21	009	26 10	077	60 04	117	29 08	150	35 21	189	30 12	247	54 07	291
272	23 25	010	26 33	078	60 25	118	29 20	151	35 17	190	29 50	248	53 45	292
273	23 29	011	26 56	079	60 46	119	29 31	152	35 13	191	29 29	249	53 23	293
274	23 34	011	27 19	079	61 06	120	29 42	153	35 08	192	29 07	250	53 02	293
275	23 39	012	27 42	080	61 26	122	29 52	154	35 03	194	28 45	251	52 40	294
276	23 44	013	28 05	081	61 46	123	30 02	155	34 57	195	28 23	252	52 19	295
277	23 49	014	28 28	082	62 05	124	30 12	156	34 51	196	28 00	253	51 58	296
278	23 55	014	28 52	083	62 25	126	30 21	157	34 44	197	27 38	254	51 37	296
279	24 00	015	29 15	084	62 44	127	30 30	159	34 37	198	27 15	255	51 16	297
280	24 07	016	29 38	085	63 02	128	30 38	160	34 29	199	26 53	256	50 55	298
281	24 13	017	30 02	086	63 20	130	30 46	161	34 21	201	26 30	256	50 34	298
282	24 20	017	30 25	087	63 38	131	30 54	162	34 13	202	26 07	257	50 13	299
283	24 27	018	30 48	088	63 56	133	31 01	163	34 04	203	25 44	258	49 53	300
284	24 35	019	31 12	088	64 13	134	31 07	164	33 54	204	25 21	259	49 33	300
	◆CAPELLA		Hamal		Alpheratz		◆DENEB		VEGA		◆ARCTURUS		Alioth	
285	24 42	020	15 15	066	31 35	089	64 29	136	61 35	190	24 58	260	49 13	301
286	24 50	020	15 36	066	31 59	090	64 45	137	61 31	191	24 35	261	48 53	302
287	24 58	021	15 58	067	32 22	091	65 01	139	61 26	193	24 12	262	48 33	303
288	25 07	022	16 20	068	32 46	092	65 16	140	61 21	194	23 48	263	48 13	303
289	25 16	023	16 41	069	33 09	093	65 31	142	61 15	196	23 25	264	47 54	304
290	25 25	023	17 03	070	33 32	094	65 45	143	61 08	198	23 02	265	47 34	305
291	25 34	024	17 25	071	33 56	095	65 59	145	61 00	199	22 38	266	47 15	305
292	25 44	025	17 48	072	34 19	096	66 12	147	60 52	201	22 15	267	46 56	306
293	25 54	026	18 10	073	34 42	097	66 25	148	60 44	202	21 52	268	46 37	307
294	26 04	026	18 32	074	35 06	098	66 37	150	60 35	204	21 28	269	46 18	307
295	26 15	027	18 55	074	35 29	099	66 48	152	60 25	205	21 05	270	46 00	308
296	26 26	028	19 18	075	35 52	100	66 59	154	60 14	207	20 41	271	45 41	309
297	26 37	029	19 40	076	36 15	101	67 09	155	60 04	209	20 18	272	45 23	309
298	26 48	029	20 03	077	36 38	102	67 18	157	59 52	210	19 54	272	45 05	310
299	27 00	030	20 26	078	37 01	103	67 27	159	59 40	212	19 31	273	44 47	311
	◆CAPELLA		Hamal		Alpheratz		◆DENEB		VEGA		◆ARCTURUS		Dubhe	
300	27 11	031	20 49	079	37 24	104	67 35	161	59 28	213	19 08	274	43 02	332
301	27 24	032	21 12	080	37 47	105	67 42	163	59 14	214	18 44	275	42 51	333
302	27 36	032	21 35	081	38 09	106	67 49	164	59 01	216	18 21	276	42 41	334
303	27 49	033	21 58	082	38 32	107	67 55	166	58 47	217	17 58	277	42 30	334
304	28 02	034	22 21	082	38 54	108	68 00	168	58 33	219	17 34	278	42 20	335
305	28 15	035	22 45	083	39 16	109	68 05	170	58 18	220	17 11	279	42 10	335
306	28 28	035	23 08	084	39 38	110	68 09	172	58 02	222	16 48	280	42 00	336
307	28 42	036	23 31	085	40 00	111	68 11	174	57 46	223	16 25	281	41 51	337
308	28 56	037	23 55	086	40 22	112	68 14	176	57 30	224	16 02	281	41 42	337
309	29 10	037	24 18	087	40 44	113	68 15	178	57 14	226	15 39	282	41 33	338
310	29 24	038	24 42	088	41 06	114	68 16	179	56 57	227	15 16	283	41 24	338
311	29 39	039	25 05	089	41 27	115	68 16	181	56 39	228	14 53	284	41 15	339
312	29 54	040	25 28	090	41 48	116	68 15	183	56 22	230	14 31	285	41 07	340
313	30 09	040	25 52	091	42 09	117	68 13	185	56 04	231	14 08	286	40 59	340
314	30 24	041	26 15	092	42 30	118	68 10	187	55 45	232	13 46	287	40 51	341
	◆CAPELLA		Hamal		Alpheratz		◆DENEB		VEGA		◆Alphecca		Dubhe	
315	30 40	042	26 39	092	42 50	119	68 07	189	55 27	234	27 48	273	40 44	341
316	30 56	043	27 02	093	43 11	120	68 03	191	55 08	235	27 24	274	40 36	342
317	31 12	043	27 26	094	43 31	121	67 58	193	54 48	236	27 01	275	40 29	343
318	31 28	044	27 49	095	43 51	123	67 53	194	54 29	237	26 38	276	40 22	343
319	31 44	045	28 12	096	44 10	124	67 47	196	54 09	238	26 14	277	40 15	344
320	32 01	046	28 36	097	44 30	125	67 40	198	53 49	240	25 51	278	40 09	344
321	32 18	046	28 59	098	44 49	126	67 32	200	53 28	241	25 28	279	40 03	345
322	32 35	047	29 22	099	45 08	127	67 24	202	53 08	242	25 05	280	39 57	346
323	32 52	048	29 45	100	45 26	128	67 15	204	52 47	243	24 42	281	39 51	346
324	33 10	049	30 08	101	45 44	130	67 05	205	52 26	244	24 19	281	39 46	347
325	33 27	049	30 31	102	46 02	131	66 55	207	52 05	245	23 56	282	39 40	347
326	33 45	050	30 54	103	46 20	132	66 44	209	51 43	247	23 33	283	39 35	348
327	34 04	051	31 17	104	46 37	133	66 32	210	51 22	248	23 10	284	39 31	349
328	34 22	052	31 40	105	46 54	134	66 20	212	51 00	249	22 48	285	39 26	349
329	34 40	053	32 02	106	47 11	135	66 07	214	50 38	250	22 25	286	39 22	350
	CAPELLA		◆Hamal		Alpheratz		◆ALTAIR		VEGA		Alphecca		◆Alioth	
330	34 59	053	32 25	107	47 27	137	27 53	217	50 16	251	22 02	287	37 11	331
331	35 18	054	32 47	108	47 43	138	27 39	218	49 54	252	21 40	288	37 00	332
332	35 37	055	33 09	109	47 58	139	27 25	219	49 31	253	21 18	288	36 49	332
333	35 56	056	33 31	110	48 14	140	27 10	220	49 09	254	20 56	289	36 38	333
334	36 16	056	33 53	111	48 28	142	26 55	221	48 46	255	20 34	290	36 28	334
335	36 35	057	34 15	112	48 43	143	26 39	222	48 24	256	20 12	291	36 17	334
336	36 55	058	34 37	113	48 56	144	26 23	223	48 01	257	19 50	292	36 07	335
337	37 15	059	34 58	114	49 10	146	26 07	224	47 38	258	19 28	293	35 58	336
338	37 35	059	35 20	115	49 23	147	25 50	225	47 15	259	19 07	294	35 48	336
339	37 55	060	35 41	116	49 36	148	25 34	226	46 52	260	18 45	295	35 39	337
340	38 16	061	36 02	117	49 48	149	25 16	227	46 29	261	18 24	295	35 30	338
341	38 36	062	36 23	118	49 59	151	24 59	228	46 05	262	18 03	296	35 21	338
342	38 57	062	36 43	119	50 11	152	24 41	230	45 42	263	17 42	297	35 13	339
343	39 18	063	37 04	120	50 21	153	24 23	231	45 19	264	17 21	298	35 04	340
344	39 39	064	37 24	121	50 32	155	24 05	232	44 56	265	17 01	299	34 56	340
	◆POLLUX		CAPELLA		Hamal		◆Alpheratz		ALTAIR		◆VEGA		Alioth	
345	11 51	043	40 00	065	37 44	122	50 41	156	23 47	233	44 32	266	34 49	341
346	12 07	044	40 21	066	38 03	123	50 50	158	23 28	234	44 09	267	34 41	342
347	12 23	044	40 43	066	38 23	125	50 59	159	23 09	235	43 45	268	34 34	342
348	12 40	045	41 04	067	38 42	126	51 07	160	22 50	236	43 22	269	34 27	343
349	12 56	046	41 26	068	39 01	127	51 15	162	22 30	237	42 58	270	34 20	344
350	13 13	047	41 48	069	39 20	128	51 22	163	22 11	238	42 35	271	34 14	344
351	13 31	048	42 09	069	39 38	129	51 29	164	21 51	239	42 12	272	34 08	345
352	13 48	049	42 31	070	39 56	130	51 35	166	21 30	240	41 48	273	34 02	346
353	14 06	050	42 54	071	40 14	131	51 40	167	21 10	241	41 25	273	33 56	346
354	14 24	050	43 16	072	40 31	132	51 45	169	20 50	242	41 01	274	33 51	347
355	14 42	051	43 38	073	40 48	134	51 49	170	20 29	243	40 38	275	33 45	348
356	15 00	052	44 01	074	41 05	135	51 53	171	20 08	244	40 15	276	33 41	348
357	15 19	053	44 23	074	41 22	136	51 56	173	19 47	245	39 51	277	33 36	349
358	15 38	054	44 46	075	41 38	137	51 59	174	19 26	246	39 28	278	33 32	350
359	15 57	055	45 08	076	41 54	138	52 01	176	19 04	247	39 05	279	33 28	350

LHA	Hc	Zn	Hc	Zn	Hc	Zn	Hc	Zn	Hc	Zn	Hc	Zn	Hc	Zn
♈	◆POLLUX		CAPELLA		Hamal		◆Alpheratz		DENEB		◆VEGA		Alioth	
0	15 42	055	45 17	076	42 54	139	53 02	177	56 32	257	38 31	280	32 25	351
1	16 02	056	45 41	077	43 10	140	53 03	178	56 08	258	38 07	281	32 21	352
2	16 23	057	46 05	077	43 26	141	53 04	180	55 44	259	37 43	282	32 18	352
3	16 43	058	46 28	078	43 41	142	53 03	181	55 20	260	37 20	283	32 15	353
4	17 04	059	46 52	079	43 56	144	53 03	183	54 56	261	36 56	284	32 12	354
5	17 25	060	47 16	080	44 10	145	53 01	184	54 32	262	36 32	285	32 09	354
6	17 46	060	47 40	081	44 24	146	52 59	186	54 08	263	36 09	285	32 07	355
7	18 07	061	48 05	082	44 37	147	52 56	187	53 43	264	35 45	286	32 05	356
8	18 29	062	48 29	082	44 50	149	52 53	189	53 19	265	35 22	287	32 03	356
9	18 50	063	48 53	083	45 02	150	52 49	190	52 55	266	34 59	288	32 02	357
10	19 12	064	49 17	084	45 14	151	52 44	192	52 30	267	34 35	289	32 01	358
11	19 34	065	49 41	085	45 26	152	52 39	193	52 06	268	34 12	290	32 00	358
12	19 56	066	50 06	086	45 37	154	52 33	194	51 42	269	33 49	290	32 00	359
13	20 19	066	50 30	087	45 48	155	52 27	196	51 17	269	33 27	291	31 59	000
14	20 41	067	50 55	088	45 58	156	52 20	197	50 53	270	33 04	292	31 59	000
	Alioth		◆POLLUX		CAPELLA		Hamal		◆Alpheratz		DENEB		◆VEGA	
15	32 00	001	21 04	068	51 19	089	46 07	158	52 12	199	50 28	271	32 41	293
16	32 00	002	21 26	069	51 43	090	46 16	159	52 04	200	50 04	272	32 19	294
17	32 01	002	21 49	070	52 08	090	46 25	160	51 56	201	49 40	273	31 57	294
18	32 02	003	22 12	071	52 32	091	46 33	162	51 46	203	49 15	274	31 35	295
19	32 04	004	22 35	071	52 57	092	46 40	163	51 37	204	48 51	275	31 13	296
20	32 05	004	22 58	072	53 21	093	46 47	164	51 26	206	48 27	276	30 51	297
21	32 07	005	23 22	073	53 45	094	46 53	166	51 16	207	48 02	277	30 29	298
22	32 09	006	23 45	074	54 10	095	46 59	167	51 04	208	47 38	277	30 07	298
23	32 12	006	24 09	075	54 34	096	47 04	168	50 52	210	47 14	278	29 46	299
24	32 15	007	24 32	076	54 58	097	47 09	170	50 40	211	46 50	279	29 25	300
25	32 18	008	24 56	077	55 22	098	47 13	171	50 27	212	46 26	280	29 04	301
26	32 21	008	25 20	078	55 46	099	47 17	172	50 14	214	46 02	281	28 43	302
27	32 25	009	25 44	078	56 10	100	47 20	174	50 00	215	45 38	282	28 22	302
28	32 29	010	26 08	079	56 34	101	47 22	175	49 46	216	45 14	282	28 02	303
29	32 33	010	26 32	080	56 58	102	47 24	176	49 31	218	44 50	283	27 42	304
	Alioth		◆POLLUX		ALDEBARAN		◆Hamal		Alpheratz		DENEB		◆VEGA	
30	32 38	011	26 56	081	34 15	133	47 25	178	49 16	219	44 27	284	27 21	305
31	32 42	012	27 20	082	34 33	134	47 26	179	49 01	220	44 03	285	27 01	306
32	32 47	012	27 44	083	34 50	135	47 26	180	48 45	221	43 39	286	26 42	306
33	32 53	013	28 08	084	35 07	137	47 26	182	48 28	223	43 16	287	26 22	307
34	32 58	014	28 33	085	35 24	138	47 25	183	48 11	224	42 53	287	26 03	308
35	33 04	014	28 57	086	35 40	139	47 23	184	47 54	225	42 29	288	25 44	309
36	33 10	015	29 21	086	35 56	140	47 21	186	47 37	226	42 06	289	25 25	310
37	33 17	016	29 46	087	36 12	141	47 18	187	47 19	228	41 43	290	25 06	310
38	33 23	016	30 10	088	36 27	142	47 15	188	47 01	229	41 20	290	24 48	311
39	33 30	017	30 34	089	36 41	143	47 11	190	46 42	230	40 57	291	24 29	312
40	33 37	017	30 59	090	36 56	145	47 07	191	46 23	231	40 35	292	24 11	313
41	33 45	018	31 23	091	37 10	146	47 02	193	46 04	232	40 12	293	23 54	314
42	33 53	019	31 48	092	37 23	147	46 56	194	45 45	234	39 50	294	23 36	314
43	34 01	019	32 12	093	37 36	148	46 50	195	45 25	235	39 27	294	23 19	315
44	34 09	020	32 36	094	37 49	149	46 43	197	45 05	236	39 05	295	23 02	316
	Alioth		◆POLLUX		ALDEBARAN		◆Hamal		Alpheratz		DENEB		◆VEGA	
45	34 17	021	33 01	095	38 01	150	46 36	198	44 45	237	38 43	296	22 45	317
46	34 26	021	33 25	096	38 13	152	46 28	199	44 24	238	38 21	297	22 28	318
47	34 35	022	33 49	097	38 25	153	46 20	200	44 03	239	38 00	297	22 12	318
48	34 44	023	34 13	097	38 35	154	46 11	202	43 42	240	37 38	298	21 56	319
49	34 54	023	34 38	098	38 46	155	46 02	203	43 21	241	37 17	299	21 40	320
50	35 04	024	35 02	099	38 56	156	45 52	204	42 59	243	36 55	300	21 24	321
51	35 14	025	35 26	100	39 05	158	45 42	206	42 37	244	36 34	301	21 09	321
52	35 24	025	35 50	101	39 14	159	45 31	207	42 15	245	36 13	301	20 54	322
53	35 35	026	36 14	102	39 23	160	45 20	208	41 53	246	35 53	302	20 39	323
54	35 45	027	36 37	103	39 31	161	45 08	210	41 31	247	35 32	303	20 25	324
55	35 57	027	37 01	104	39 39	163	44 56	211	41 08	248	35 12	304	20 10	325
56	36 08	028	37 25	105	39 46	164	44 43	212	40 46	249	34 51	304	19 56	325
57	36 19	029	37 48	106	39 52	165	44 30	213	40 23	250	34 31	305	19 43	326
58	36 31	029	38 12	107	39 58	166	44 16	215	40 00	251	34 11	306	19 29	327
59	36 43	030	38 35	108	40 04	168	44 02	216	39 37	252	33 52	307	19 16	328
	Alioth		◆POLLUX		ALDEBARAN		◆Hamal		Alpheratz		DENEB		◆VEGA	
60	36 55	031	38 58	109	40 09	169	43 47	217	39 13	253	33 32	307	19 03	329
61	37 08	031	39 21	110	40 13	170	43 33	218	38 50	254	33 13	308	18 51	329
62	37 21	032	39 44	111	40 17	171	43 17	219	38 26	255	32 54	309	18 38	330
63	37 34	032	40 06	112	40 21	173	43 02	221	38 03	256	32 35	310	18 26	331
64	37 47	033	40 29	113	40 24	174	42 45	222	37 39	257	32 16	310	18 15	332
65	38 00	034	40 51	114	40 26	175	42 29	223	37 15	258	31 58	311	18 03	333
66	38 14	034	41 13	116	40 28	176	42 12	224	36 51	259	31 39	312	17 52	333
67	38 28	035	41 35	117	40 29	178	41 55	225	36 27	260	31 21	313	17 41	334
68	38 42	036	41 57	118	40 30	179	41 37	227	36 03	261	31 04	313	17 31	335
69	38 57	036	42 18	119	40 30	180	41 19	228	35 39	262	30 46	314	17 21	336
70	39 11	037	42 40	120	40 30	181	41 01	229	35 15	263	30 28	315	17 11	337
71	39 26	038	43 01	121	40 29	183	40 43	230	34 51	264	30 11	316	17 01	337
72	39 41	038	43 22	122	40 27	184	40 24	231	34 26	265	29 54	316	16 52	338
73	39 56	039	43 42	123	40 25	185	40 05	232	34 02	266	29 38	317	16 43	339
74	40 12	040	44 02	124	40 23	186	39 45	233	33 38	267	29 21	318	16 34	340
	Alioth		◆REGULUS		PROCYON		◆ALDEBARAN		Hamal		Alpheratz		◆DENEB	
75	40 27	040	16 12	097	23 15	136	40 20	188	39 26	235	33 13	268	29 05	319
76	40 43	041	16 37	098	23 32	137	40 17	189	39 06	236	32 49	268	28 49	319
77	40 59	042	17 01	099	23 48	138	40 12	190	38 45	237	32 25	269	28 33	320
78	41 16	042	17 25	100	24 04	139	40 08	191	38 25	238	32 00	270	28 17	321
79	41 32	043	17 49	101	24 20	140	40 03	193	38 04	239	31 36	271	28 02	322
80	41 49	044	18 13	102	24 36	141	39 57	194	37 43	240	31 11	272	27 47	322
81	42 06	044	18 37	103	24 51	142	39 51	195	37 22	241	30 47	273	27 32	323
82	42 23	045	19 01	104	25 05	143	39 44	196	37 00	242	30 23	274	27 18	324
83	42 40	045	19 24	104	25 20	145	39 37	198	36 39	243	29 58	275	27 03	324
84	42 58	046	19 48	105	25 34	146	39 30	199	36 17	244	29 34	276	26 49	325
85	43 15	047	20 11	106	25 47	147	39 21	200	35 55	245	29 10	277	26 35	326
86	43 33	047	20 35	107	26 00	148	39 13	201	35 32	246	28 46	277	26 22	327
87	43 51	048	20 58	108	26 13	149	39 04	203	35 10	247	28 21	278	26 09	327
88	44 09	049	21 21	109	26 26	150	38 54	204	34 47	248	27 57	279	25 56	328
89	44 28	049	21 44	110	26 38	151	38 44	205	34 25	249	27 33	280	25 43	329

LAT 66°N

LHA	Hc	Zn	Hc	Zn	Hc	Zn	Hc	Zn	Hc	Zn	Hc	Zn	Hc	Zn
♈	Alioth		◆REGULUS		PROCYON		BETELGEUSE		◆ALDEBARAN		Hamal		◆DENEB	
90	44 46	050	22 07	111	26 49	152	31 24	181	38 34	206	34 02	250	25 30	330
91	45 05	051	22 29	112	27 00	153	31 23	183	38 23	207	33 39	251	25 18	330
92	45 24	051	22 52	113	27 11	154	31 22	184	38 11	209	33 15	252	25 06	331
93	45 43	052	23 14	114	27 22	155	31 20	185	37 59	210	32 52	253	24 55	332
94	46 03	053	23 37	115	27 32	157	31 18	186	37 47	211	32 29	254	24 43	333
95	46 22	053	23 59	116	27 41	158	31 15	187	37 34	212	32 05	255	24 32	333
96	46 42	054	24 20	117	27 50	159	31 11	188	37 21	213	31 41	256	24 22	334
97	47 02	055	24 42	118	27 59	160	31 07	190	37 07	214	31 18	257	24 11	335
98	47 22	055	25 03	119	28 07	161	31 03	191	36 53	216	30 54	258	24 01	336
99	47 42	056	25 25	120	28 15	162	30 58	192	36 39	217	30 30	259	23 51	336
100	48 02	057	25 46	121	28 22	163	30 53	193	36 24	218	30 06	260	23 41	337
101	48 23	057	26 06	122	28 29	164	30 47	194	36 09	219	29 42	261	23 32	338
102	48 43	058	26 27	123	28 35	165	30 41	195	35 53	220	29 18	262	23 23	339
103	49 04	059	26 47	124	28 41	167	30 34	196	35 37	221	28 54	263	23 14	339
104	49 25	059	27 07	125	28 46	168	30 27	198	35 21	223	28 29	264	23 06	340
	Alioth		◆REGULUS		PROCYON		BETELGEUSE		◆CAPELLA		Hamal		◆DENEB	
105	49 46	060	27 27	126	28 51	169	30 20	199	65 41	227	28 05	265	22 58	341
106	50 07	061	27 47	127	28 56	170	30 12	200	65 22	229	27 41	266	22 50	342
107	50 28	061	28 06	128	29 00	171	30 03	201	65 04	230	27 16	267	22 42	342
108	50 50	062	28 25	129	29 03	172	29 54	202	64 45	232	26 52	268	22 35	343
109	51 11	063	28 44	130	29 06	173	29 45	203	64 25	233	26 28	269	22 28	344
110	51 33	063	29 02	131	29 09	175	29 35	204	64 05	235	26 03	269	22 22	345
111	51 55	064	29 20	133	29 11	176	29 24	206	63 45	236	25 39	270	22 15	345
112	52 17	065	29 38	134	29 12	177	29 14	207	63 25	238	25 14	271	22 09	346
113	52 39	065	29 56	135	29 14	178	29 02	208	63 04	239	24 50	272	22 04	347
114	53 01	066	30 13	136	29 14	179	28 51	209	62 43	240	24 26	273	21 58	348
115	53 24	067	30 30	137	29 14	180	28 39	210	62 22	241	24 01	274	21 53	348
116	53 46	068	30 46	138	29 14	181	28 26	211	62 00	243	23 37	275	21 49	349
117	54 09	068	31 02	139	29 13	183	28 14	212	61 38	244	23 13	276	21 44	350
118	54 32	069	31 18	140	29 12	184	28 00	213	61 16	245	22 48	277	21 40	351
119	54 54	070	31 34	141	29 10	185	27 47	214	60 54	246	22 24	278	21 36	351
	◆VEGA		ARCTURUS		◆REGULUS		PROCYON		◆CAPELLA		Schedar		DENEB	
120	16 00	017	15 57	078	31 49	142	29 08	186	60 32	248	43 16	315	21 33	352
121	16 08	018	16 21	079	32 03	143	29 05	187	60 09	249	42 58	315	21 30	353
122	16 15	018	16 45	080	32 18	145	29 02	188	59 46	250	42 41	316	21 27	354
123	16 23	019	17 09	081	32 32	146	28 58	189	59 23	251	42 24	317	21 24	354
124	16 31	020	17 33	082	32 45	147	28 54	191	59 00	252	42 08	317	21 22	355
125	16 40	021	17 57	083	32 58	148	28 49	192	58 37	253	41 51	318	21 20	356
126	16 48	022	18 21	084	33 11	149	28 44	193	58 13	254	41 35	318	21 19	357
127	16 58	022	18 46	085	33 23	150	28 38	194	57 50	256	41 19	319	21 18	357
128	17 07	023	19 10	086	33 35	151	28 32	195	57 26	257	41 03	320	21 17	358
129	17 17	024	19 34	087	33 47	153	28 25	196	57 02	258	40 47	320	21 16	359
130	17 27	025	19 59	087	33 58	154	28 18	197	56 38	259	40 32	321	21 16	000
131	17 37	026	20 23	088	34 08	155	28 11	198	56 14	260	40 17	322	21 16	001
132	17 48	026	20 47	089	34 18	156	28 03	200	55 50	261	40 02	322	21 16	001
133	17 59	027	21 12	090	34 28	157	27 55	201	55 26	262	39 47	323	21 17	002
134	18 10	028	21 36	091	34 37	158	27 46	202	55 02	263	39 32	324	21 18	003
	DENEB		◆VEGA		ARCTURUS		◆REGULUS		PROCYON		◆CAPELLA		Schedar	
135	21 19	004	18 22	029	22 01	092	34 46	160	27 37	203	54 38	264	39 18	324
136	21 21	004	18 34	030	22 25	093	34 54	161	27 27	204	54 13	265	39 04	325
137	21 23	005	18 46	030	22 49	094	35 02	162	27 17	205	53 49	266	38 50	326
138	21 25	006	18 58	031	23 14	095	35 09	163	27 06	206	53 25	267	38 36	326
139	21 28	007	19 11	032	23 38	096	35 16	164	26 55	207	53 00	268	38 23	327
140	21 31	007	19 24	033	24 02	097	35 23	166	26 44	208	52 36	268	38 09	327
141	21 34	008	19 37	033	24 26	098	35 28	167	26 32	209	52 12	269	37 56	328
142	21 38	009	19 51	034	24 51	099	35 34	168	26 20	211	51 47	270	37 43	329
143	21 42	010	20 05	035	25 15	099	35 39	169	26 07	212	51 23	271	37 31	329
144	21 46	010	20 19	036	25 39	100	35 43	170	25 54	213	50 58	272	37 19	330
145	21 50	011	20 33	037	26 03	101	35 47	172	25 41	214	50 34	273	37 07	331
146	21 55	012	20 48	037	26 27	102	35 50	173	25 27	215	50 10	274	36 55	331
147	22 00	013	21 03	038	26 50	103	35 53	174	25 13	216	49 45	275	36 43	332
148	22 06	013	21 18	039	27 14	104	35 55	175	24 58	217	49 21	276	36 32	333
149	22 12	014	21 34	040	27 38	105	35 57	176	24 43	218	48 57	277	36 21	333
	DENEB		◆VEGA		ARCTURUS		◆REGULUS		PROCYON		◆CAPELLA		Schedar	
150	22 18	015	21 50	041	28 01	106	35 58	178	24 28	219	48 33	277	36 10	334
151	22 24	016	22 06	041	28 25	107	35 59	179	24 13	220	48 08	278	35 59	335
152	22 31	016	22 22	042	28 48	108	36 00	180	23 57	221	47 44	279	35 49	335
153	22 38	017	22 38	043	29 11	109	35 59	181	23 40	222	47 20	280	35 39	336
154	22 45	018	22 55	044	29 34	110	35 59	182	23 24	223	46 56	281	35 29	336
155	22 53	019	23 12	045	29 57	111	35 57	184	23 07	224	46 32	282	35 19	337
156	23 01	019	23 29	045	30 19	112	35 55	185	22 50	225	46 08	282	35 10	338
157	23 09	020	23 47	046	30 42	113	35 53	186	22 32	226	45 45	283	35 01	338
158	23 18	021	24 05	047	31 04	114	35 50	187	22 14	227	45 21	284	34 52	339
159	23 26	022	24 22	048	31 27	115	35 47	188	21 56	229	44 57	285	34 43	340
160	23 35	022	24 41	049	31 49	116	35 43	190	21 38	230	44 34	286	34 35	340
161	23 45	023	24 59	049	32 10	117	35 39	191	21 19	231	44 10	286	34 27	341
162	23 55	024	25 18	050	32 32	118	35 34	192	21 00	232	43 47	287	34 19	342
163	24 05	025	25 36	051	32 53	119	35 29	193	20 41	233	43 24	288	34 12	342
164	24 15	025	25 56	052	33 15	120	35 23	194	20 21	234	43 00	289	34 04	343
	DENEB		◆VEGA		ARCTURUS		◆REGULUS		POLLUX		◆CAPELLA		Schedar	
165	24 26	026	26 15	052	33 35	121	35 17	196	41 46	243	42 37	290	33 57	344
166	24 36	027	26 34	053	33 56	122	35 10	197	41 24	244	42 14	290	33 51	344
167	24 48	028	26 54	054	34 17	124	35 02	198	41 02	245	41 52	291	33 44	345
168	24 59	028	27 14	055	34 37	125	34 55	199	40 40	246	41 29	292	33 38	346
169	25 11	029	27 34	056	34 57	126	34 46	200	40 18	247	41 06	293	33 32	346
170	25 23	030	27 54	056	35 17	127	34 38	202	39 55	248	40 44	294	33 26	347
171	25 35	031	28 14	057	35 36	128	34 29	203	39 32	249	40 22	294	33 21	348
172	25 48	031	28 35	058	35 55	129	34 19	204	39 10	250	40 00	295	33 16	348
173	26 00	032	28 56	059	36 14	130	34 09	205	38 47	251	39 37	296	33 11	349
174	26 13	033	29 17	060	36 32	131	33 58	206	38 23	252	39 16	297	33 06	350
175	26 27	034	29 38	060	36 51	132	33 47	207	38 00	253	38 54	297	33 02	350
176	26 40	034	29 59	061	37 09	133	33 36	208	37 37	254	38 32	298	32 58	351
177	26 54	035	30 21	062	37 26	134	33 24	210	37 13	255	38 11	299	32 54	351
178	27 09	036	30 42	063	37 43	136	33 12	211	36 49	256	37 50	300	32 51	352
179	27 23	037	31 04	064	38 00	137	32 59	212	36 26	257	37 28	300	32 48	353

LHA ♈	Hc	Zn	Hc	Zn	Hc	Zn	Hc	Zn	Hc	Zn	Hc	Zn	Hc	Zn
	◆DENEB		VEGA		◆ARCTURUS		REGULUS		◆POLLUX		CAPELLA		Schedar	
180	27 38	037	31 26	064	38 17	138	32 46	213	36 02	258	37 07	301	32 45	353
181	27 52	038	31 48	065	38 33	139	32 32	214	35 38	259	36 47	302	32 42	354
182	28 08	039	32 10	066	38 49	140	32 18	215	35 14	260	36 26	303	32 40	355
183	28 23	040	32 33	067	39 04	141	32 04	216	34 50	261	36 06	303	32 38	355
184	28 39	040	32 55	068	39 19	143	31 49	218	34 26	262	35 45	304	32 36	356
185	28 55	041	33 18	068	39 34	144	31 34	219	34 02	263	35 25	305	32 34	357
186	29 11	042	33 41	069	39 48	145	31 19	220	33 37	264	35 05	306	32 33	357
187	29 27	042	34 04	070	40 02	146	31 03	221	33 13	265	34 46	306	32 32	358
188	29 44	043	34 27	071	40 15	147	30 47	222	32 49	266	34 26	307	32 31	359
189	30 01	044	34 50	072	40 28	149	30 31	223	32 24	267	34 07	308	32 31	359
190	30 18	045	35 13	073	40 41	150	30 14	224	32 00	268	33 48	309	32 31	000
191	30 35	045	35 36	073	40 53	151	29 57	225	31 36	269	33 29	309	32 31	001
192	30 52	046	36 00	074	41 04	152	29 39	226	31 11	270	33 10	310	32 31	001
193	31 10	047	36 23	075	41 16	153	29 21	227	30 47	270	32 51	311	32 32	002
194	31 28	048	36 47	076	41 26	155	29 03	228	30 22	271	32 33	312	32 33	003
	Schedar		◆DENEB		VEGA		◆ARCTURUS		REGULUS		◆POLLUX		CAPELLA	
195	32 34	003	31 46	048	37 11	077	41 37	156	28 45	230	29 58	272	32 15	312
196	32 36	004	32 05	049	37 34	078	41 46	157	28 26	231	29 34	273	31 57	313
197	32 37	005	32 23	050	37 58	078	41 55	158	28 07	232	29 09	274	31 39	314
198	32 39	005	32 42	051	38 22	079	42 04	160	27 48	233	28 45	275	31 22	315
199	32 42	006	33 01	051	38 46	080	42 12	161	27 28	234	28 21	276	31 04	315
200	32 44	007	33 20	052	39 10	081	42 20	162	27 08	235	27 56	277	30 47	316
201	32 47	007	33 39	053	39 34	082	42 27	163	26 48	236	27 32	278	30 31	317
202	32 50	008	33 59	054	39 59	083	42 34	165	26 28	237	27 08	278	30 14	318
203	32 54	008	34 19	054	40 23	084	42 40	166	26 08	238	26 44	279	29 58	318
204	32 58	009	34 39	055	40 47	084	42 46	167	25 47	239	26 20	280	29 41	319
205	33 02	010	34 59	056	41 11	085	42 51	169	25 26	240	25 56	281	29 26	320
206	33 06	010	35 19	057	41 36	086	42 56	170	25 05	241	25 32	282	29 10	321
207	33 10	011	35 40	057	42 00	087	43 00	171	24 43	242	25 08	283	28 55	321
208	33 15	012	36 00	058	42 24	088	43 03	172	24 21	243	24 44	284	28 39	322
209	33 20	012	36 21	059	42 49	089	43 06	174	24 00	244	24 21	285	28 24	323
	Schedar		◆DENEB		VEGA		◆ARCTURUS		REGULUS		◆POLLUX		CAPELLA	
210	33 26	013	36 42	060	43 13	090	43 08	175	23 38	245	23 57	285	28 10	323
211	33 31	014	37 03	060	43 38	091	43 10	176	23 16	246	23 34	286	27 55	324
212	33 37	014	37 25	061	44 02	092	43 12	178	22 53	247	23 10	287	27 41	325
213	33 43	015	37 46	062	44 26	093	43 12	179	22 31	248	22 47	288	27 27	326
214	33 50	016	38 08	063	44 51	094	43 12	180	22 08	249	22 24	289	27 14	326
215	33 57	016	38 30	064	45 15	095	43 12	181	21 45	250	22 01	290	27 00	327
216	34 04	017	38 51	064	45 39	095	43 11	183	21 22	251	21 38	291	26 47	328
217	34 11	018	39 14	065	46 04	096	43 10	184	20 59	252	21 15	291	26 34	329
218	34 18	018	39 36	066	46 28	097	43 08	185	20 36	253	20 53	292	26 22	329
219	34 26	019	39 58	067	46 52	098	43 05	187	20 12	254	20 30	293	26 10	330
220	34 34	020	40 21	067	47 16	099	43 02	188	19 49	255	20 08	294	25 58	331
221	34 42	020	40 43	068	47 40	100	42 58	189	19 25	256	19 45	295	25 46	332
222	34 51	021	41 06	069	48 04	101	42 54	191	19 02	256	19 23	296	25 34	332
223	35 00	021	41 29	070	48 28	102	42 50	192	18 38	257	19 01	297	25 23	333
224	35 09	022	41 52	071	48 52	103	42 44	193	18 14	258	18 40	297	25 12	334
	Schedar		◆DENEB		VEGA		◆ARCTURUS		REGULUS		◆POLLUX		CAPELLA	
225	35 18	023	42 15	071	49 16	104	42 39	194	17 50	259	18 18	298	25 02	335
226	35 28	023	42 38	072	49 39	105	42 32	196	17 26	260	17 57	299	24 51	335
227	35 38	024	43 01	073	50 03	106	42 25	197	17 02	261	17 36	300	24 41	336
228	35 48	025	43 25	074	50 26	108	42 18	198	16 38	262	17 15	301	24 31	337
229	35 58	025	43 48	075	50 49	109	42 10	199	16 14	263	16 54	302	24 22	338
230	36 09	026	44 12	075	51 12	110	42 02	201	15 50	264	16 33	303	24 13	338
231	36 19	027	44 35	076	51 35	111	41 53	202	15 25	265	16 13	303	24 04	339
232	36 31	027	44 59	077	51 58	112	41 44	203	15 01	266	15 52	304	23 55	340
233	36 42	028	45 23	078	52 20	113	41 34	204	14 37	267	15 32	305	23 47	341
234	36 53	029	45 47	079	52 43	114	41 23	206	14 12	268	15 12	306	23 39	341
235	37 05	029	46 11	080	53 05	115	41 13	207	13 48	269	14 53	307	23 31	342
236	37 17	030	46 35	080	53 27	116	41 01	208	13 23	269	14 33	308	23 24	343
237	37 30	031	46 59	081	53 49	118	40 49	209	12 59	270	14 14	309	23 17	343
238	37 42	031	47 23	082	54 10	119	40 37	211	12 35	271	13 55	309	23 10	344
239	37 55	032	47 47	083	54 32	120	40 25	212	12 10	272	13 36	310	23 04	345
	Alpheratz		◆DENEB		VEGA		Rasalhague		◆ARCTURUS		Alioth		◆CAPELLA	
240	14 47	050	48 11	084	54 53	121	34 13	152	40 12	213	66 02	268	22 57	346
241	15 06	051	48 36	085	55 13	122	34 24	153	39 58	214	65 37	269	22 51	346
242	15 25	052	49 00	086	55 34	124	34 35	154	39 44	215	65 13	270	22 46	347
243	15 44	053	49 24	086	55 54	125	34 46	155	39 30	217	64 48	271	22 41	348
244	16 04	053	49 49	087	56 14	126	34 56	156	39 15	218	64 24	272	22 36	349
245	16 23	054	50 13	088	56 33	127	35 05	158	39 00	219	64 00	273	22 31	349
246	16 43	055	50 37	089	56 53	129	35 14	159	38 44	220	63 35	273	22 27	350
247	17 03	056	51 02	090	57 12	130	35 23	160	38 28	221	63 11	274	22 23	351
248	17 24	057	51 26	091	57 30	131	35 31	161	38 12	222	62 46	275	22 19	352
249	17 44	058	51 51	092	57 48	133	35 39	162	37 56	224	62 22	276	22 16	352
250	18 05	058	52 15	093	58 06	134	35 46	163	37 39	225	61 58	277	22 13	353
251	18 26	059	52 39	094	58 23	135	35 53	165	37 21	226	61 34	278	22 10	354
252	18 47	060	53 04	095	58 40	137	35 59	166	37 04	227	61 10	278	22 08	355
253	19 08	061	53 28	096	58 57	138	36 04	167	36 46	228	60 45	279	22 05	355
254	19 30	062	53 52	097	59 13	140	36 10	168	36 27	229	60 21	280	22 04	356
	Alpheratz		◆DENEB		ALTAIR		Rasalhague		◆ARCTURUS		Alioth		◆CAPELLA	
255	19 51	063	54 17	098	25 52	132	36 14	169	36 09	230	59 57	281	22 02	357
256	20 13	064	54 41	099	26 10	133	36 19	171	35 50	231	59 33	282	22 01	358
257	20 35	064	55 05	100	26 28	134	36 22	172	35 30	232	59 10	282	22 00	358
258	20 57	065	55 29	101	26 45	135	36 25	173	35 11	234	58 46	283	22 00	359
259	21 19	066	55 53	102	27 02	136	36 28	174	34 51	235	58 22	284	22 00	000
260	21 42	067	56 17	103	27 19	137	36 30	176	34 31	236	57 58	285	22 00	001
261	22 04	068	56 40	104	27 35	138	36 32	177	34 11	237	57 35	285	22 00	001
262	22 27	069	57 04	105	27 51	139	36 33	178	33 50	238	57 11	286	22 01	002
263	22 50	069	57 28	106	28 07	140	36 34	179	33 30	239	56 48	287	22 02	003
264	23 13	070	57 51	107	28 22	142	36 34	180	33 09	240	56 25	288	22 03	004
265	23 36	071	58 14	108	28 37	143	36 33	182	32 47	241	56 01	288	22 05	004
266	23 59	072	58 37	109	28 52	144	36 32	183	32 26	242	55 38	289	22 07	005
267	24 22	073	59 00	110	29 06	145	36 31	184	32 04	243	55 15	290	22 10	006
268	24 45	074	59 23	111	29 20	146	36 29	185	31 42	244	54 52	290	22 12	007
269	25 09	075	59 46	113	29 34	147	36 26	186	31 20	245	54 30	291	22 15	007

LAT 66°N

LHA ♈	Hc	Zn	Hc	Zn	Hc	Zn	Hc	Zn	Hc	Zn	Hc	Zn	Hc	Zn
	◆CAPELLA		Alpheratz		◆DENEB		ALTAIR		Rasalhague		◆ARCTURUS		Alioth	
270	22 19	008	25 32	076	60 08	114	29 47	148	36 23	188	30 58	246	54 07	292
271	22 22	009	25 56	076	60 30	115	29 59	149	36 20	189	30 36	247	53 44	293
272	22 26	010	26 20	077	60 52	116	30 12	150	36 16	190	30 13	248	53 22	293
273	22 30	010	26 44	078	61 14	118	30 24	151	36 11	191	29 50	249	52 59	294
274	22 35	011	27 08	079	61 36	119	30 35	153	36 06	193	29 27	250	52 37	295
275	22 40	012	27 32	080	61 57	120	30 46	154	36 01	194	29 04	251	52 15	295
276	22 45	013	27 56	081	62 18	121	30 57	155	35 55	195	28 41	252	51 53	296
277	22 51	013	28 20	082	62 39	123	31 07	156	35 48	196	28 18	253	51 31	297
278	22 56	014	28 44	082	62 59	124	31 16	157	35 41	197	27 55	254	51 09	297
279	23 03	015	29 08	083	63 19	125	31 26	158	35 34	198	27 31	255	50 48	298
280	23 09	016	29 32	084	63 39	127	31 35	159	35 26	200	27 07	256	50 26	299
281	23 16	016	29 57	085	63 58	128	31 43	161	35 17	201	26 44	257	50 05	299
282	23 23	017	30 21	086	64 17	130	31 51	162	35 08	202	26 20	258	49 44	300
283	23 30	018	30 45	087	64 36	131	31 58	163	34 59	203	25 56	259	49 23	301
284	23 38	019	31 10	088	64 54	133	32 05	164	34 49	204	25 32	260	49 02	301
	◆CAPELLA		Hamal		Alpheratz		◆ALTAIR		VEGA		◆ARCTURUS		Alioth	
285	23 46	019	14 50	065	31 34	089	32 12	165	62 35	190	25 08	261	48 41	302
286	23 54	020	15 12	066	31 59	090	32 18	166	62 30	192	24 44	262	48 21	303
287	24 02	021	15 35	067	32 23	091	32 23	168	62 25	193	24 20	263	48 00	303
288	24 11	022	15 57	068	32 47	092	32 28	169	62 19	195	23 55	264	47 40	304
289	24 20	022	16 20	069	33 12	092	32 33	170	62 12	197	23 31	265	47 20	305
290	24 30	023	16 43	070	33 36	093	32 37	171	62 05	198	23 07	265	47 00	305
291	24 40	024	17 06	071	34 01	094	32 40	172	61 57	200	22 42	266	46 40	306
292	24 50	025	17 29	071	34 25	095	32 43	173	61 48	201	22 18	267	46 20	307
293	25 00	025	17 52	072	34 49	096	32 46	175	61 39	203	21 54	268	46 01	307
294	25 10	026	18 15	073	35 13	097	32 48	176	61 29	205	21 29	269	45 42	308
295	25 21	027	18 39	074	35 38	098	32 50	177	61 19	206	21 05	270	45 22	309
296	25 32	028	19 02	075	36 02	099	32 51	178	61 08	208	20 41	271	45 03	309
297	25 44	028	19 26	076	36 26	100	32 51	179	60 56	209	20 16	272	44 45	310
298	25 56	029	19 50	077	36 50	101	32 51	180	60 44	211	19 52	273	44 26	311
299	26 08	030	20 13	078	37 14	102	32 51	182	60 31	212	19 27	274	44 08	311
	◆CAPELLA		Hamal		Alpheratz		◆ALTAIR		VEGA		◆ARCTURUS		Dubhe	
300	26 20	031	20 37	078	37 38	103	32 50	183	60 18	214	19 03	275	42 09	333
301	26 32	031	21 01	079	38 01	104	32 48	184	60 04	215	18 39	275	41 58	333
302	26 45	032	21 25	080	38 25	105	32 47	185	59 49	217	18 14	276	41 47	334
303	26 58	033	21 49	081	38 48	106	32 44	186	59 34	218	17 50	277	41 36	335
304	27 12	033	22 13	082	39 12	107	32 41	187	59 19	220	17 26	278	41 26	335
305	27 25	034	22 38	083	39 35	108	32 38	189	59 03	221	17 02	279	41 16	336
306	27 39	035	23 02	084	39 58	109	32 34	190	58 47	223	16 38	280	41 06	336
307	27 53	036	23 26	085	40 21	110	32 29	191	58 30	224	16 14	281	40 56	337
308	28 07	036	23 50	086	40 44	111	32 25	192	58 13	225	15 50	282	40 46	337
309	28 22	037	24 15	087	41 07	112	32 19	193	57 55	227	15 26	283	40 37	338
310	28 37	038	24 39	087	41 29	113	32 13	194	57 37	228	15 02	284	40 28	339
311	28 52	039	25 04	088	41 52	114	32 07	196	57 19	230	14 39	284	40 19	339
312	29 07	039	25 28	089	42 14	115	32 00	197	57 00	231	14 15	285	40 11	340
313	29 23	040	25 52	090	42 36	116	31 53	198	56 41	232	13 52	286	40 03	340
314	29 39	041	26 17	091	42 58	117	31 45	199	56 21	233	13 28	287	39 55	341
	CAPELLA		◆Hamal		Alpheratz		◆ALTAIR		VEGA		Alphecca		◆Dubhe	
315	29 55	042	26 41	092	43 19	118	31 37	200	56 02	235	27 44	274	39 47	342
316	30 11	042	27 05	093	43 41	120	31 28	201	55 42	236	27 20	275	39 39	342
317	30 28	043	27 30	094	44 02	121	31 19	203	55 21	237	26 55	276	39 32	343
318	30 45	044	27 54	095	44 23	122	31 09	204	55 01	238	26 31	277	39 25	343
319	31 02	045	28 18	096	44 43	123	30 59	205	54 40	240	26 07	278	39 18	344
320	31 19	045	28 43	097	45 04	124	30 49	206	54 18	241	25 43	278	39 11	345
321	31 36	046	29 07	098	45 24	125	30 38	207	53 57	242	25 19	279	39 05	345
322	31 54	047	29 31	099	45 43	126	30 27	208	53 35	243	24 55	280	38 59	346
323	32 12	048	29 55	099	46 03	128	30 15	209	53 14	244	24 31	281	38 53	346
324	32 30	048	30 19	100	46 22	129	30 03	210	52 51	245	24 07	282	38 47	347
325	32 48	049	30 43	101	46 41	130	29 50	212	52 29	247	23 43	283	38 42	348
326	33 07	050	31 07	102	47 00	131	29 37	213	52 07	248	23 19	284	38 37	348
327	33 26	050	31 31	103	47 18	132	29 24	214	51 44	249	22 55	285	38 32	349
328	33 44	051	31 55	104	47 36	133	29 10	215	51 21	250	22 32	285	38 27	349
329	34 04	052	32 18	105	47 53	135	28 56	216	50 58	251	22 08	286	38 23	350
	CAPELLA		◆Hamal		Alpheratz		◆ALTAIR		VEGA		Alphecca		◆Alioth	
330	34 23	053	32 42	106	48 10	136	28 41	217	50 35	252	21 45	287	36 18	331
331	34 42	053	33 05	107	48 27	137	28 27	218	50 12	253	21 22	288	36 07	332
332	35 02	054	33 28	108	48 44	138	28 11	219	49 48	254	20 59	289	35 55	333
333	35 22	055	33 51	109	49 00	140	27 56	220	49 25	255	20 36	290	35 44	333
334	35 42	056	34 14	110	49 15	141	27 40	221	49 01	256	20 13	291	35 34	334
335	36 02	056	34 37	111	49 30	142	27 23	222	48 37	257	19 50	291	35 23	335
336	36 23	057	35 00	112	49 45	144	27 07	224	48 13	258	19 27	292	35 13	335
337	36 43	058	35 22	113	49 59	145	26 50	225	47 49	259	19 05	293	35 03	336
338	37 04	059	35 45	114	50 13	146	26 32	226	47 25	260	18 42	294	34 53	337
339	37 25	059	36 07	115	50 26	148	26 15	227	47 01	261	18 20	295	34 43	337
340	37 46	060	36 29	116	50 39	149	25 57	228	46 37	262	17 58	296	34 34	338
341	38 07	061	36 51	117	50 52	150	25 39	229	46 13	263	17 36	297	34 25	339
342	38 29	062	37 12	119	51 03	152	25 20	230	45 49	264	17 14	297	34 16	339
343	38 50	062	37 33	120	51 15	153	25 01	231	45 24	265	16 53	298	34 08	340
344	39 12	063	37 55	121	51 26	154	24 42	232	45 00	266	16 31	299	34 00	341
	◆POLLUX		CAPELLA		Hamal		◆Alpheratz		ALTAIR		◆VEGA		Alioth	
345	11 07	043	39 34	064	38 15	122	51 36	156	24 23	233	44 36	267	33 52	341
346	11 23	043	39 56	065	38 36	123	51 46	157	24 03	234	44 11	268	33 44	342
347	11 40	044	40 18	066	38 57	124	51 55	158	23 44	235	43 47	269	33 37	343
348	11 57	045	40 40	066	39 17	125	52 04	160	23 23	236	43 23	270	33 29	343
349	12 15	046	41 03	067	39 36	126	52 12	161	23 03	237	42 58	271	33 23	344
350	12 32	047	41 25	068	39 56	127	52 19	163	22 42	238	42 34	272	33 16	345
351	12 50	048	41 48	069	40 15	128	52 26	164	22 22	239	42 09	273	33 10	345
352	13 09	049	42 11	069	40 34	129	52 33	165	22 01	240	41 45	273	33 03	346
353	13 27	049	42 34	070	40 53	131	52 39	167	21 39	241	41 21	274	32 58	347
354	13 46	050	42 57	071	41 11	132	52 44	168	21 18	242	40 56	275	32 52	347
355	14 04	051	43 20	072	41 29	133	52 49	170	20 56	243	40 32	276	32 47	348
356	14 24	052	43 43	073	41 47	134	52 53	171	20 34	244	40 08	277	32 42	349
357	14 43	053	44 06	073	42 04	135	52 56	173	20 12	245	39 44	278	32 37	349
358	15 02	054	44 30	074	42 21	136	52 59	174	19 50	246	39 19	279	32 33	350
359	15 22	054	44 53	075	42 38	138	53 01	176	19 28	247	38 55	280	32 28	350

LAT 65°N

LHA ♈	Hc	Zn	Hc	Zn	Hc	Zn	Hc	Zn	Hc	Zn	Hc	Zn	Hc	Zn
	◆POLLUX		CAPELLA		Hamal		◆Alpheratz		DENEB		◆VEGA		Alioth	
0	15 08	055	45 02	075	43 39	138	54 02	177	56 45	258	38 20	281	31 25	351
1	15 29	056	45 26	076	43 56	139	54 03	178	56 20	259	37 55	282	31 22	352
2	15 50	057	45 51	076	44 12	141	54 04	180	55 55	260	37 31	283	31 18	353
3	16 11	058	46 16	077	44 28	142	54 03	181	55 30	261	37 06	284	31 15	353
4	16 33	058	46 40	078	44 44	143	54 02	183	55 05	262	36 41	284	31 12	354
5	16 54	059	47 05	079	44 59	144	54 01	184	54 40	263	36 17	285	31 10	354
6	17 16	060	47 30	080	45 13	146	53 59	186	54 14	264	35 52	286	31 07	355
7	17 38	061	47 55	081	45 27	147	53 56	187	53 49	265	35 28	287	31 05	356
8	18 01	062	48 20	081	45 41	148	53 52	189	53 24	266	35 04	288	31 04	356
9	18 23	063	48 45	082	45 54	149	53 48	190	52 59	267	34 40	289	31 02	357
10	18 46	063	49 10	083	46 07	151	53 43	192	52 33	268	34 16	289	31 01	358
11	19 08	064	49 36	084	46 19	152	53 37	193	52 08	269	33 52	290	31 00	358
12	19 31	065	50 01	085	46 31	153	53 31	195	51 43	270	33 28	291	31 00	359
13	19 54	066	50 26	086	46 42	155	53 25	196	51 17	271	33 05	292	30 59	000
14	20 18	067	50 51	086	46 52	156	53 17	198	50 52	272	32 41	293	30 59	000
	Alioth		◆POLLUX		CAPELLA		Hamal		◆Alpheratz		DENEB		◆VEGA	
15	31 00	001	20 41	068	51 17	087	47 03	157	53 09	199	50 27	272	32 18	293
16	31 00	002	21 05	069	51 42	088	47 12	159	53 01	201	50 01	273	31 55	294
17	31 01	002	21 28	069	52 07	089	47 21	160	52 51	202	49 36	274	31 32	295
18	31 02	003	21 52	070	52 33	090	47 30	161	52 42	203	49 11	275	31 09	296
19	31 04	004	22 16	071	52 58	091	47 37	163	52 31	205	48 45	276	30 46	297
20	31 05	004	22 40	072	53 24	092	47 45	164	52 20	206	48 20	277	30 23	297
21	31 07	005	23 04	073	53 49	093	47 51	165	52 09	208	47 55	278	30 01	298
22	31 10	006	23 29	074	54 14	094	47 58	167	51 57	209	47 30	278	29 39	299
23	31 12	006	23 53	075	54 39	095	48 03	168	51 44	210	47 05	279	29 17	300
24	31 15	007	24 17	075	55 05	096	48 08	169	51 31	212	46 40	280	28 55	301
25	31 18	008	24 42	076	55 30	097	48 13	171	51 18	213	46 15	281	28 33	301
26	31 22	008	25 07	077	55 55	098	48 16	172	51 04	214	45 50	282	28 11	302
27	31 26	009	25 31	078	56 20	099	48 20	173	50 49	216	45 25	283	27 50	303
28	31 30	009	25 56	079	56 45	100	48 22	175	50 34	217	45 01	283	27 29	304
29	31 34	010	26 21	080	57 10	101	48 24	176	50 18	218	44 36	284	27 08	305
	Alioth		◆POLLUX		ALDEBARAN		◆Hamal		Alpheratz		DENEB		◆VEGA	
30	31 39	011	26 46	081	34 56	133	48 25	178	50 03	220	44 11	285	26 47	305
31	31 44	011	27 11	081	35 15	134	48 26	179	49 46	221	43 47	286	26 26	306
32	31 49	012	27 36	082	35 33	135	48 26	180	49 29	222	43 23	287	26 06	307
33	31 54	013	28 01	083	35 51	136	48 26	182	49 12	223	42 58	287	25 46	308
34	32 00	013	28 27	084	36 08	137	48 25	183	48 54	225	42 34	288	25 26	308
35	32 06	014	28 52	085	36 25	138	48 23	185	48 36	226	42 10	289	25 06	309
36	32 12	015	29 17	086	36 42	139	48 21	186	48 18	227	41 46	290	24 46	310
37	32 19	015	29 42	087	36 58	141	48 18	187	47 59	228	41 23	291	24 27	311
38	32 26	016	30 08	088	37 14	142	48 14	189	47 40	230	40 59	291	24 08	312
39	32 33	017	30 33	089	37 30	143	48 10	190	47 20	231	40 35	292	23 49	312
40	32 40	017	30 59	089	37 45	144	48 05	191	47 01	232	40 12	293	23 31	313
41	32 48	018	31 24	090	37 59	145	48 00	193	46 40	233	39 49	294	23 12	314
42	32 56	019	31 49	091	38 13	146	47 54	194	46 20	234	39 25	294	22 54	315
43	33 04	019	32 15	092	38 27	148	47 48	195	45 59	236	39 02	295	22 36	315
44	33 12	020	32 40	093	38 41	149	47 41	197	45 38	237	38 39	296	22 18	316
	Alioth		◆POLLUX		ALDEBARAN		◆Hamal		Alpheratz		DENEB		◆VEGA	
45	33 21	021	33 05	094	38 53	150	47 33	198	45 17	238	38 17	297	22 01	317
46	33 30	021	33 30	095	39 06	151	47 25	200	44 55	239	37 54	297	21 44	318
47	33 39	022	33 56	096	39 18	152	47 16	201	44 33	240	37 32	298	21 27	319
48	33 49	022	34 21	097	39 29	154	47 07	202	44 11	241	37 09	299	21 10	319
49	33 59	023	34 46	098	39 40	155	46 57	204	43 49	242	36 47	300	20 54	320
50	34 09	024	35 11	099	39 51	156	46 47	205	43 26	243	36 25	300	20 38	321
51	34 19	024	35 36	100	40 01	157	46 36	206	43 04	244	36 04	301	20 22	322
52	34 30	025	36 01	101	40 10	159	46 24	207	42 41	246	35 42	302	20 06	322
53	34 41	026	36 26	102	40 19	160	46 12	209	42 17	247	35 21	303	19 51	323
54	34 52	026	36 51	103	40 28	161	46 00	210	41 54	248	34 59	303	19 36	324
55	35 03	027	37 16	104	40 36	162	45 47	211	41 31	249	34 38	304	19 21	325
56	35 15	028	37 40	105	40 43	164	45 33	213	41 07	250	34 17	305	19 07	326
57	35 27	028	38 05	105	40 50	165	45 20	214	40 43	251	33 57	306	18 53	326
58	35 39	029	38 29	106	40 57	166	45 05	215	40 19	252	33 36	306	18 39	327
59	35 51	030	38 53	107	41 02	167	44 50	216	39 55	253	33 16	307	18 25	328
	Alioth		◆POLLUX		ALDEBARAN		◆Hamal		Alpheratz		DENEB		◆VEGA	
60	36 04	030	39 17	109	41 08	169	44 35	218	39 30	254	32 56	308	18 12	329
61	36 17	031	39 41	110	41 12	170	44 19	219	39 06	255	32 36	309	17 59	330
62	36 30	031	40 05	111	41 16	171	44 03	220	38 42	256	32 16	309	17 46	330
63	36 43	032	40 29	112	41 20	172	43 47	221	38 17	257	31 57	310	17 34	331
64	36 57	033	40 52	113	41 23	174	43 30	222	37 52	258	31 37	311	17 22	332
65	37 10	033	41 16	114	41 26	175	43 13	224	37 27	259	31 18	312	17 10	333
66	37 24	034	41 39	115	41 28	176	42 55	225	37 02	260	30 59	312	16 58	333
67	37 39	035	42 02	116	41 29	178	42 37	226	36 37	261	30 41	313	16 47	334
68	37 53	035	42 24	117	41 30	179	42 18	227	36 12	262	30 22	314	16 36	335
69	38 08	036	42 47	118	41 30	180	42 00	228	35 47	263	30 04	314	16 26	336
70	38 23	037	43 09	119	41 30	181	41 40	230	35 22	264	29 46	315	16 16	337
71	38 38	037	43 31	120	41 29	183	41 21	231	34 57	264	29 28	316	16 06	337
72	38 54	038	43 53	121	41 27	184	41 01	232	34 32	265	29 11	317	15 56	338
73	39 09	038	44 15	122	41 25	185	40 41	233	34 06	266	28 53	317	15 47	339
74	39 25	039	44 36	124	41 23	187	40 21	234	33 41	267	28 36	318	15 38	340
	Alioth		◆REGULUS		PROCYON		◆ALDEBARAN		Hamal		Alpheratz		◆DENEB	
75	39 41	040	16 20	097	23 58	136	41 19	188	40 00	235	33 16	268	28 20	319
76	39 58	040	16 45	098	24 16	137	41 16	189	39 39	236	32 50	269	28 03	320
77	40 14	041	17 10	099	24 33	138	41 11	190	39 18	237	32 25	270	27 47	320
78	40 31	042	17 35	099	24 50	139	41 07	192	38 56	238	32 00	271	27 31	321
79	40 48	042	18 00	100	25 06	140	41 01	193	38 35	240	31 34	272	27 15	322
80	41 05	043	18 25	101	25 22	141	40 55	194	38 13	241	31 09	273	26 59	323
81	41 22	044	18 50	102	25 38	142	40 49	195	37 50	242	30 44	274	26 44	323
82	41 40	044	19 14	103	25 53	143	40 42	197	37 28	243	30 18	274	26 29	324
83	41 58	045	19 39	104	26 08	144	40 34	198	37 05	244	29 53	275	26 14	325
84	42 16	045	20 04	105	26 23	145	40 26	199	36 42	245	29 28	276	26 00	326
85	42 34	046	20 28	106	26 37	146	40 18	200	36 19	246	29 03	277	25 46	326
86	42 52	047	20 52	107	26 51	148	40 09	202	35 56	247	28 38	278	25 32	327
87	43 11	047	21 16	108	27 05	149	39 59	203	35 33	248	28 12	279	25 18	328
88	43 30	048	21 41	109	27 18	150	39 49	204	35 09	249	27 47	280	25 05	328
89	43 48	049	22 04	110	27 30	151	39 38	205	34 45	250	27 22	281	24 51	329

LHA ♈	Hc	Zn	Hc	Zn	Hc	Zn	Hc	Zn	Hc	Zn	Hc	Zn	Hc	Zn
	Alioth		◆REGULUS		PROCYON		BETELGEUSE		◆ALDEBARAN		Hamal		◆DENEB	
90	44 08	049	22 28	111	27 42	152	32 24	181	39 27	207	34 22	251	24 39	330
91	44 27	050	22 52	112	27 54	153	32 23	183	39 16	208	33 58	252	24 26	331
92	44 46	051	23 15	113	28 05	154	32 22	184	39 04	209	33 33	253	24 14	331
93	45 06	051	23 39	114	28 16	155	32 20	185	38 51	210	33 09	254	24 02	332
94	45 26	052	24 02	115	28 27	156	32 17	186	38 38	211	32 45	255	23 50	333
95	45 46	052	24 25	116	28 36	157	32 14	187	38 25	213	32 20	256	23 39	334
96	46 06	053	24 47	117	28 46	159	32 11	189	38 11	214	31 55	257	23 28	334
97	46 27	054	25 10	118	28 55	160	32 07	190	37 57	215	31 31	258	23 17	335
98	46 47	054	25 32	119	29 04	161	32 02	191	37 42	216	31 06	259	23 06	336
99	47 08	055	25 55	120	29 12	162	31 57	192	37 27	217	30 41	260	22 56	337
100	47 29	056	26 16	121	29 19	163	31 52	193	37 11	218	30 16	261	22 46	337
101	47 50	056	26 38	122	29 26	164	31 46	194	36 55	220	29 51	262	22 36	338
102	48 11	057	27 00	123	29 33	165	31 39	196	36 39	221	29 26	263	22 27	339
103	48 32	058	27 21	124	29 39	166	31 32	197	36 22	222	29 01	264	22 18	340
104	48 54	058	27 42	125	29 45	168	31 24	198	36 05	223	28 35	264	22 09	340
	Alioth		◆REGULUS		PROCYON		BETELGEUSE		◆CAPELLA		Hamal		◆DENEB	
105	49 15	059	28 02	126	29 50	169	31 16	199	66 20	229	28 10	265	22 01	341
106	49 37	060	28 23	127	29 55	170	31 08	200	66 01	231	27 45	266	21 53	342
107	49 59	060	28 43	128	29 59	171	30 59	201	65 41	232	27 20	267	21 45	343
108	50 21	061	29 03	129	30 03	172	30 50	202	65 21	234	26 54	268	21 38	343
109	50 43	062	29 22	130	30 06	173	30 40	204	65 00	235	26 29	269	21 31	344
110	51 06	062	29 42	131	30 09	175	30 29	205	64 39	237	26 03	270	21 24	345
111	51 28	063	30 01	132	30 11	176	30 18	206	64 18	238	25 38	271	21 17	346
112	51 51	064	30 19	133	30 12	177	30 07	207	63 56	239	25 13	272	21 11	346
113	52 14	064	30 38	134	30 13	178	29 56	208	63 34	241	24 47	273	21 05	347
114	52 37	065	30 56	135	30 14	179	29 43	209	63 12	242	24 22	274	21 00	348
115	53 00	066	31 13	136	30 14	180	29 31	210	62 50	243	23 57	274	20 55	349
116	53 23	066	31 31	138	30 14	181	29 18	211	62 27	244	23 32	275	20 50	349
117	53 46	067	31 47	139	30 13	183	29 04	213	62 04	246	23 06	276	20 45	350
118	54 09	068	32 04	140	30 12	184	28 51	214	61 41	247	22 41	277	20 41	351
119	54 33	068	32 20	141	30 10	185	28 36	215	61 17	248	22 16	278	20 37	352
	◆VEGA		ARCTURUS		◆REGULUS		PROCYON		◆CAPELLA		Schedar		DENEB	
120	15 03	017	15 45	078	32 36	142	30 07	186	60 54	249	42 33	315	20 33	352
121	15 10	017	16 09	079	32 51	143	30 04	187	60 30	250	42 16	316	20 30	353
122	15 18	018	16 34	080	33 06	144	30 01	188	60 06	252	41 58	317	20 27	354
123	15 26	019	16 59	081	33 21	145	29 57	189	59 42	253	41 41	317	20 25	355
124	15 35	020	17 24	082	33 35	147	29 53	191	59 17	254	41 23	318	20 22	355
125	15 44	021	17 50	083	33 49	148	29 48	192	58 53	255	41 07	318	20 20	356
126	15 53	021	18 15	084	34 02	149	29 42	193	58 28	256	40 50	319	20 19	357
127	16 02	022	18 40	084	34 15	150	29 36	194	58 04	257	40 33	320	20 18	358
128	16 12	023	19 05	085	34 28	151	29 30	195	57 39	258	40 17	320	20 17	358
129	16 22	024	19 31	086	34 40	152	29 23	196	57 14	259	40 01	321	20 16	359
130	16 32	025	19 56	087	34 51	153	29 16	197	56 49	260	39 45	322	20 16	000
131	16 43	025	20 21	088	35 03	155	29 08	199	56 24	261	39 29	322	20 16	001
132	16 54	026	20 46	089	35 13	156	29 00	200	55 59	262	39 14	323	20 16	001
133	17 05	027	21 12	090	35 23	157	28 51	201	55 34	263	38 59	323	20 17	002
134	17 17	028	21 37	091	35 33	158	28 41	202	55 09	264	38 44	324	20 18	003
	DENEB		◆VEGA		ARCTURUS		◆REGULUS		PROCYON		◆CAPELLA		Schedar	
135	20 19	004	17 29	029	22 03	092	35 42	159	28 32	203	54 44	265	38 29	325
136	20 21	004	17 41	029	22 28	093	35 51	161	28 22	204	54 18	266	38 14	325
137	20 23	005	17 54	030	22 53	093	35 59	162	28 11	205	53 53	267	38 00	326
138	20 26	006	18 07	031	23 19	094	36 07	163	28 00	206	53 28	268	37 46	327
139	20 28	007	18 20	032	23 44	095	36 14	164	27 48	208	53 02	269	37 32	327
140	20 31	007	18 33	033	24 09	096	36 21	165	27 36	209	52 37	270	37 19	328
141	20 35	008	18 47	033	24 34	097	36 27	167	27 24	210	52 12	271	37 05	329
142	20 38	009	19 01	034	24 59	098	36 32	168	27 11	211	51 46	272	36 52	329
143	20 42	010	19 16	035	25 24	099	36 38	169	26 58	212	51 21	272	36 39	330
144	20 47	010	19 30	036	25 49	100	36 42	170	26 45	213	50 56	273	36 27	330
145	20 51	011	19 45	036	26 14	101	36 46	171	26 30	214	50 30	274	36 14	331
146	20 56	012	20 00	037	26 39	102	36 50	173	26 16	215	50 05	275	36 02	332
147	21 02	013	20 16	038	27 04	103	36 53	174	26 01	216	49 40	276	35 50	332
148	21 07	013	20 32	039	27 29	104	36 55	175	25 46	217	49 15	277	35 38	333
149	21 13	014	20 48	040	27 53	105	36 57	176	25 31	218	48 49	278	35 27	334
	DENEB		◆VEGA		ARCTURUS		◆REGULUS		PROCYON		◆CAPELLA		Schedar	
150	21 20	015	21 04	040	28 18	106	36 58	178	25 15	219	48 24	279	35 16	334
151	21 26	015	21 20	041	28 42	107	36 59	179	24 58	220	47 59	279	35 05	335
152	21 33	016	21 37	042	29 06	108	37 00	180	24 42	222	47 34	280	34 54	335
153	21 40	017	21 54	043	29 30	109	36 59	181	24 25	223	47 09	281	34 44	336
154	21 48	018	22 12	043	29 54	110	36 58	182	24 07	224	46 44	282	34 34	337
155	21 56	018	22 29	044	30 18	111	36 57	184	23 50	225	46 20	283	34 24	337
156	22 04	019	22 47	045	30 42	112	36 55	185	23 32	226	45 55	283	34 14	338
157	22 13	020	23 05	046	31 05	113	36 53	186	23 13	227	45 30	284	34 05	339
158	22 21	021	23 23	047	31 29	114	36 50	187	22 55	228	45 06	285	33 56	339
159	22 31	021	23 42	047	31 52	115	36 46	189	22 36	229	44 41	286	33 47	340
160	22 40	022	24 01	048	32 15	116	36 42	190	22 17	230	44 17	287	33 39	341
161	22 50	023	24 20	049	32 38	117	36 38	191	21 57	231	43 53	287	33 30	341
162	23 00	024	24 39	050	33 00	118	36 33	192	21 37	232	43 29	288	33 22	342
163	23 10	024	24 58	051	33 22	119	36 27	193	21 17	233	43 05	289	33 14	343
164	23 21	025	25 18	051	33 45	120	36 21	195	20 57	234	42 41	290	33 07	343
	DENEB		◆VEGA		ARCTURUS		◆REGULUS		POLLUX		◆CAPELLA		Schedar	
165	23 32	026	25 38	052	34 06	121	36 14	196	42 13	244	42 17	291	33 00	344
166	23 43	027	25 58	053	34 28	122	36 07	197	41 50	245	41 53	291	32 53	344
167	23 54	027	26 18	054	34 50	123	35 59	198	41 27	246	41 30	292	32 46	345
168	24 06	028	26 39	054	35 11	124	35 51	199	41 04	247	41 06	293	32 40	346
169	24 18	029	27 00	055	35 32	125	35 43	201	40 41	248	40 43	294	32 34	346
170	24 31	030	27 21	056	35 52	126	35 33	202	40 17	249	40 20	294	32 28	347
171	24 43	030	27 42	057	36 13	127	35 24	203	39 53	250	39 57	295	32 22	348
172	24 56	031	28 03	058	36 33	128	35 14	204	39 30	251	39 34	296	32 17	348
173	25 09	032	28 25	058	36 52	129	35 03	205	39 05	252	39 11	297	32 12	349
174	25 23	033	28 46	059	37 12	131	34 52	206	38 41	253	38 48	297	32 07	350
175	25 37	033	29 08	060	37 31	132	34 40	208	38 17	254	38 26	298	32 03	350
176	25 51	034	29 30	061	37 50	133	34 28	209	37 53	255	38 04	299	31 59	351
177	26 05	035	29 52	061	38 08	134	34 16	210	37 28	256	37 42	300	31 55	352
178	26 20	036	30 15	062	38 26	135	34 03	211	37 03	257	37 20	300	31 51	352
179	26 35	036	30 37	063	38 44	136	33 50	212	36 39	258	36 58	301	31 48	353

LHA ♈	Hc	Zn	Hc	Zn	Hc	Zn	Hc	Zn	Hc	Zn	Hc	Zn	Hc	Zn
	DENEB		◆VEGA		ARCTURUS		◆Denebola		REGULUS		POLLUX		◆CAPELLA	
180	26 50	037	31 00	064	39 01	137	39 34	184	33 36	213	36 14	259	36 36	302
181	27 05	038	31 23	065	39 18	139	39 32	185	33 22	215	35 49	260	36 15	303
182	27 21	038	31 46	065	39 35	140	39 30	186	33 07	216	35 24	261	35 53	303
183	27 37	039	32 09	066	39 51	141	39 27	187	32 52	217	34 59	262	35 32	304
184	27 53	040	32 32	067	40 07	142	39 23	189	32 37	218	34 34	263	35 11	305
185	28 09	041	32 56	068	40 22	143	39 19	190	32 21	219	34 08	264	34 51	306
186	28 26	041	33 19	069	40 37	144	39 15	191	32 05	220	33 43	265	34 30	306
187	28 43	042	33 43	069	40 52	146	39 09	192	31 48	221	33 18	266	34 10	307
188	29 00	043	34 07	070	41 06	147	39 04	194	31 31	222	32 53	266	33 50	308
189	29 17	044	34 31	071	41 19	148	38 58	195	31 14	223	32 27	267	33 30	308
190	29 35	044	34 55	072	41 33	149	38 51	196	30 57	225	32 02	268	33 10	309
191	29 53	045	35 19	073	41 45	151	38 44	197	30 39	226	31 37	269	32 50	310
192	30 11	046	35 43	074	41 57	152	38 36	198	30 20	227	31 11	270	32 31	311
193	30 29	047	36 07	074	42 09	153	38 28	200	30 02	228	30 46	271	32 12	311
194	30 47	047	36 32	075	42 20	154	38 19	201	29 43	229	30 21	272	31 53	312
	DENEB		◆VEGA		ARCTURUS		◆Denebola		REGULUS		POLLUX		◆CAPELLA	
195	31 06	048	36 56	076	42 31	156	38 10	202	29 23	230	29 55	273	31 34	313
196	31 25	049	37 21	077	42 41	157	38 00	203	29 04	231	29 30	274	31 16	314
197	31 44	049	37 46	078	42 51	158	37 49	205	28 44	232	29 05	275	30 57	314
198	32 04	050	38 11	079	43 00	159	37 39	206	28 24	233	28 39	275	30 39	315
199	32 23	051	38 35	079	43 09	161	37 27	207	28 04	234	28 14	276	30 22	316
200	32 43	052	39 00	080	43 17	162	37 16	208	27 43	235	27 49	277	30 04	316
201	33 03	052	39 25	081	43 25	163	37 04	209	27 22	236	27 24	278	29 47	317
202	33 23	053	39 51	082	43 32	164	36 51	210	27 01	237	26 59	279	29 30	318
203	33 44	054	40 16	083	43 38	166	36 38	212	26 39	238	26 34	280	29 13	319
204	34 04	055	40 41	084	43 44	167	36 24	213	26 18	239	26 09	281	28 56	319
205	34 25	055	41 06	084	43 50	168	36 10	214	25 56	240	25 44	282	28 40	320
206	34 46	056	41 31	085	43 55	170	35 56	215	25 34	241	25 19	282	28 24	321
207	35 07	057	41 57	086	43 59	171	35 41	216	25 11	242	24 54	283	28 08	322
208	35 28	058	42 22	087	44 03	172	35 26	217	24 49	243	24 30	284	27 52	322
209	35 50	058	42 47	088	44 06	174	35 10	219	24 26	244	24 05	285	27 37	323
	DENEB		◆VEGA		ARCTURUS		◆Denebola		REGULUS		POLLUX		◆CAPELLA	
210	36 12	059	43 13	089	44 08	175	34 54	220	24 03	245	23 41	286	27 21	324
211	36 33	060	43 38	090	44 10	176	34 38	221	23 40	246	23 17	287	27 07	324
212	36 55	061	44 03	091	44 11	178	34 21	222	23 17	247	22 52	288	26 52	325
213	37 18	061	44 29	092	44 12	179	34 04	223	22 53	248	22 28	288	26 38	326
214	37 40	062	44 54	093	44 12	180	33 47	224	22 30	249	22 04	289	26 24	327
215	38 02	063	45 19	093	44 12	182	33 29	225	22 06	250	21 40	290	26 10	327
216	38 25	064	45 45	094	44 11	183	33 10	226	21 42	251	21 17	291	25 56	328
217	38 48	064	46 10	095	44 10	184	32 52	228	21 18	252	20 53	292	25 43	329
218	39 11	065	46 35	096	44 08	185	32 33	229	20 54	253	20 30	293	25 30	330
219	39 34	066	47 00	097	44 05	187	32 14	230	20 29	254	20 06	294	25 17	330
220	39 57	067	47 25	098	44 02	188	31 54	231	20 05	255	19 43	294	25 05	331
221	40 20	067	47 50	099	43 58	189	31 35	232	19 40	256	19 20	295	24 53	332
222	40 44	068	48 15	100	43 53	191	31 15	233	19 16	257	18 57	296	24 41	333
223	41 08	069	48 40	101	43 48	192	30 54	234	18 51	258	18 34	297	24 30	333
224	41 31	070	49 05	102	43 43	193	30 34	235	18 26	259	18 12	298	24 18	334
	◆DENEB		VEGA		Rasalhague		◆ARCTURUS		Denebola		Dubhe		◆CAPELLA	
225	41 55	071	49 30	103	31 17	134	43 37	195	30 13	236	64 18	291	24 07	335
226	42 19	071	49 55	104	31 35	136	43 30	196	29 51	237	63 54	291	23 57	335
227	42 43	072	50 19	105	31 52	137	43 23	197	29 30	238	63 31	292	23 46	336
228	43 07	073	50 43	106	32 09	138	43 15	198	29 08	239	63 07	292	23 36	337
229	43 32	074	51 08	107	32 26	139	43 07	200	28 46	240	62 44	293	23 26	338
230	43 56	074	51 32	108	32 43	140	42 58	201	28 24	241	62 20	294	23 17	338
231	44 20	075	51 56	110	32 59	141	42 48	202	28 02	242	61 57	294	23 08	339
232	44 45	076	52 20	111	33 15	142	42 39	204	27 40	243	61 34	295	22 59	340
233	45 10	077	52 43	112	33 30	143	42 28	205	27 17	244	61 11	295	22 50	341
234	45 34	078	53 07	113	33 45	144	42 17	206	26 54	245	60 48	296	22 42	341
235	45 59	079	53 30	114	33 59	146	42 06	207	26 31	246	60 26	296	22 34	342
236	46 24	079	53 53	115	34 14	147	41 54	209	26 07	247	60 03	297	22 27	343
237	46 49	080	54 16	116	34 27	148	41 42	210	25 44	248	59 40	298	22 19	344
238	47 14	081	54 39	117	34 40	149	41 29	211	25 20	249	59 18	298	22 12	344
239	47 39	082	55 01	119	34 53	150	41 16	212	24 57	250	58 56	299	22 06	345
	Alpheratz		◆DENEB		VEGA		Rasalhague		◆ARCTURUS		Alioth		◆CAPELLA	
240	14 09	050	48 04	083	55 23	120	35 06	151	41 02	213	66 02	270	21 59	346
241	14 28	051	48 29	084	55 45	121	35 18	153	40 48	215	65 37	271	21 53	347
242	14 48	052	48 55	084	56 06	122	35 29	154	40 33	216	65 12	272	21 47	347
243	15 08	052	49 20	085	56 28	124	35 40	155	40 18	217	64 46	273	21 42	348
244	15 28	053	49 45	086	56 49	125	35 51	156	40 02	218	64 21	274	21 37	349
245	15 48	054	50 11	087	57 09	126	36 01	157	39 46	219	63 56	275	21 32	350
246	16 09	055	50 36	088	57 30	127	36 10	158	39 30	221	63 31	275	21 28	350
247	16 30	056	51 01	089	57 50	129	36 19	160	39 13	222	63 05	276	21 24	351
248	16 51	057	51 27	090	58 09	130	36 28	161	38 56	223	62 40	277	21 20	352
249	17 12	057	51 52	091	58 28	131	36 36	162	38 39	224	62 15	278	21 16	353
250	17 34	058	52 17	091	58 47	133	36 43	163	38 21	225	61 50	279	21 13	353
251	17 55	059	52 43	092	59 06	134	36 50	164	38 03	226	61 25	279	21 10	354
252	18 17	060	53 08	093	59 24	136	36 57	166	37 44	228	61 00	280	21 08	355
253	18 39	061	53 33	094	59 41	137	37 03	167	37 25	229	60 35	281	21 06	355
254	19 01	062	53 59	095	59 58	138	37 08	168	37 06	230	60 10	282	21 04	356
	Alpheratz		◆DENEB		ALTAIR		Rasalhague		◆ARCTURUS		Alioth		◆CAPELLA	
255	19 24	062	54 24	096	26 32	132	37 13	169	36 47	231	59 45	282	21 02	357
256	19 46	063	54 49	097	26 51	133	37 18	171	36 27	232	59 21	283	21 01	358
257	20 09	064	55 14	098	27 09	134	37 22	172	36 07	233	58 56	284	21 00	358
258	20 32	065	55 39	099	27 27	135	37 25	173	35 46	234	58 31	285	21 00	359
259	20 55	066	56 04	100	27 45	136	37 28	174	35 26	235	58 07	285	21 00	000
260	21 18	067	56 29	101	28 03	137	37 30	175	35 05	236	57 42	286	21 00	001
261	21 41	067	56 54	102	28 20	138	37 32	177	34 43	237	57 18	287	21 00	001
262	22 05	068	57 19	103	28 37	139	37 33	178	34 22	238	56 54	288	21 01	002
263	22 28	069	57 43	104	28 53	140	37 34	179	34 00	239	56 30	288	21 02	003
264	22 52	070	58 08	105	29 09	141	37 34	180	33 38	241	56 06	289	21 04	004
265	23 16	071	58 32	107	29 25	142	37 33	182	33 16	242	55 42	290	21 05	004
266	23 40	072	58 56	108	29 40	143	37 32	183	32 54	243	55 18	290	21 07	005
267	24 04	072	59 20	109	29 55	144	37 31	184	32 31	244	54 54	291	21 10	006
268	24 28	073	59 44	110	30 10	146	37 29	185	32 08	245	54 31	292	21 13	007
269	24 53	074	60 08	111	30 24	147	37 26	187	31 45	246	54 07	292	21 16	007

LAT 65°N

LHA ♈	Hc	Zn	Hc	Zn	Hc	Zn	Hc	Zn	Hc	Zn	Hc	Zn	Hc	Zn
	◆CAPELLA		Alpheratz		◆DENEB		ALTAIR		Rasalhague		◆ARCTURUS		Alioth	
270	21 19	008	25 17	075	60 32	112	30 38	148	37 23	188	31 22	247	53 44	293
271	21 23	009	25 42	076	60 55	113	30 51	149	37 19	189	30 59	248	53 21	294
272	21 27	010	26 06	077	61 18	115	31 04	150	37 15	190	30 35	249	52 57	294
273	21 31	010	26 31	078	61 41	116	31 16	151	37 10	191	30 11	250	52 34	295
274	21 36	011	26 56	078	62 04	117	31 28	152	37 05	193	29 48	251	52 12	296
275	21 41	012	27 21	079	62 26	118	31 40	153	36 59	194	29 24	252	51 49	296
276	21 47	013	27 46	080	62 48	120	31 51	155	36 53	195	28 59	253	51 26	297
277	21 52	013	28 11	081	63 10	121	32 02	156	36 46	196	28 35	254	51 04	298
278	21 58	014	28 36	082	63 32	122	32 12	157	36 39	198	28 11	255	50 41	298
279	22 05	015	29 01	083	63 53	124	32 21	158	36 31	199	27 46	256	50 19	299
280	22 11	016	29 26	084	64 14	125	32 31	159	36 22	200	27 22	257	49 57	300
281	22 18	016	29 51	085	64 35	127	32 39	160	36 13	201	26 57	257	49 35	300
282	22 25	017	30 17	085	64 55	128	32 48	162	36 04	202	26 32	258	49 13	301
283	22 33	018	30 42	086	65 14	130	32 56	163	35 54	203	26 07	259	48 52	302
284	22 41	018	31 07	087	65 34	131	33 03	164	35 44	205	25 42	260	48 30	302
	◆CAPELLA		Alpheratz		Enif		◆ALTAIR		VEGA		◆ARCTURUS		Alioth	
285	22 49	019	31 33	088	28 00	133	33 10	165	63 34	190	25 17	261	48 09	303
286	22 58	020	31 58	089	28 18	134	33 16	166	63 29	192	24 52	262	47 48	304
287	23 06	021	32 23	090	28 36	135	33 22	167	63 23	194	24 27	263	47 27	304
288	23 15	021	32 49	091	28 54	136	33 27	169	63 17	195	24 02	264	47 06	305
289	23 25	022	33 14	092	29 11	137	33 32	170	63 10	197	23 37	265	46 45	306
290	23 35	023	33 39	093	29 28	138	33 36	171	63 02	199	23 11	266	46 25	306
291	23 45	024	34 05	094	29 45	139	33 40	172	62 53	200	22 46	267	46 04	307
292	23 55	024	34 30	095	30 01	141	33 43	173	62 44	202	22 21	268	45 44	308
293	24 06	025	34 55	095	30 17	142	33 46	174	62 34	204	21 55	269	45 24	308
294	24 17	026	35 20	096	30 33	143	33 48	176	62 24	205	21 30	270	45 04	309
295	24 28	027	35 46	097	30 48	144	33 49	177	62 12	207	21 05	270	44 45	310
296	24 39	027	36 11	098	31 03	145	33 51	178	62 01	209	20 39	271	44 25	310
297	24 51	028	36 36	099	31 17	146	33 51	179	61 48	210	20 14	272	44 06	311
298	25 03	029	37 01	100	31 31	147	33 51	180	61 35	212	19 49	273	43 47	311
299	25 15	030	37 26	101	31 45	148	33 51	182	61 21	213	19 23	274	43 28	312
	◆CAPELLA		Alpheratz		Enif		◆ALTAIR		VEGA		◆ARCTURUS		Alioth	
300	25 28	030	37 51	102	31 58	149	33 50	183	61 07	215	18 58	275	43 09	313
301	25 41	031	38 15	103	32 10	151	33 48	184	60 52	217	18 33	276	42 50	313
302	25 54	032	38 40	104	32 23	152	33 46	185	60 37	218	18 08	277	42 32	314
303	26 08	032	39 04	105	32 34	153	33 44	186	60 21	220	17 42	278	42 14	315
304	26 21	033	39 29	106	32 46	154	33 41	188	60 05	221	17 17	279	41 56	315
305	26 35	034	39 53	107	32 57	155	33 37	189	59 48	222	16 52	279	41 38	316
306	26 50	035	40 17	108	33 07	156	33 33	190	59 30	224	16 27	280	41 21	317
307	27 04	035	40 41	109	33 17	157	33 28	191	59 13	225	16 02	281	41 03	317
308	27 19	036	41 05	110	33 27	159	33 23	192	58 54	227	15 38	282	40 46	318
309	27 34	037	41 29	111	33 36	160	33 18	193	58 36	228	15 13	283	40 29	318
310	27 50	038	41 53	112	33 44	161	33 11	195	58 17	229	14 48	284	40 13	319
311	28 05	038	42 16	113	33 52	162	33 05	196	57 57	231	14 24	285	39 56	320
312	28 21	039	42 39	114	34 00	163	32 58	197	57 37	232	13 59	286	39 40	320
313	28 37	040	43 02	115	34 07	164	32 50	198	57 17	233	13 35	286	39 24	321
314	28 53	040	43 25	117	34 13	166	32 42	199	56 57	235	13 10	287	39 08	322
	CAPELLA		◆Hamal		Alpheratz		Enif		◆ALTAIR		VEGA		◆Alioth	
315	29 10	041	26 43	091	43 48	118	34 19	167	32 33	200	56 36	236	38 52	322
316	29 27	042	27 08	092	44 10	119	34 25	168	32 24	202	56 15	237	38 37	323
317	29 44	043	27 34	093	44 32	120	34 30	169	32 14	203	55 53	238	38 22	324
318	30 01	043	27 59	094	44 54	121	34 34	170	32 04	204	55 31	240	38 07	324
319	30 19	044	28 24	095	45 15	122	34 38	172	31 54	205	55 09	241	37 52	325
320	30 37	045	28 49	096	45 37	123	34 42	173	31 43	206	54 47	242	37 37	325
321	30 55	046	29 15	097	45 58	124	34 45	174	31 31	207	54 25	243	37 23	326
322	31 13	046	29 40	098	46 19	126	34 47	175	31 20	209	54 02	244	37 09	327
323	31 31	047	30 05	099	46 39	127	34 49	176	31 07	210	53 39	246	36 55	327
324	31 50	048	30 30	100	46 59	128	34 50	178	30 54	211	53 16	247	36 42	328
325	32 09	049	30 55	101	47 19	129	34 51	179	30 41	212	52 52	248	36 28	329
326	32 28	049	31 20	102	47 39	130	34 51	180	30 28	213	52 29	249	36 15	329
327	32 47	050	31 44	103	47 58	131	34 51	181	30 14	214	52 05	250	36 03	330
328	33 07	051	32 09	104	48 17	133	34 50	182	29 59	215	51 41	251	35 50	331
329	33 26	051	32 34	105	48 35	134	34 49	184	29 44	216	51 17	252	35 38	331
	CAPELLA		◆Hamal		Alpheratz		Enif		◆ALTAIR		VEGA		◆Alioth	
330	33 46	052	32 58	106	48 53	135	34 47	185	29 29	217	50 53	253	35 25	332
331	34 06	053	33 23	107	49 11	136	34 44	186	29 14	219	50 28	254	35 14	332
332	34 27	054	33 47	108	49 28	138	34 42	187	28 58	220	50 04	255	35 02	333
333	34 47	054	34 11	109	49 45	139	34 38	188	28 41	221	49 39	256	34 51	334
334	35 08	055	34 35	110	50 01	140	34 34	190	28 25	222	49 15	257	34 40	334
335	35 29	056	34 59	111	50 17	142	34 30	191	28 08	223	48 50	258	34 29	335
336	35 50	057	35 22	112	50 33	143	34 25	192	27 50	224	48 25	259	34 18	336
337	36 11	057	35 46	113	50 48	144	34 19	193	27 32	225	48 00	260	34 08	336
338	36 33	058	36 09	114	51 03	146	34 13	194	27 14	226	47 35	261	33 58	337
339	36 54	059	36 32	115	51 17	147	34 07	196	26 56	227	47 10	262	33 48	338
340	37 16	060	36 55	116	51 30	148	33 59	197	26 37	228	46 45	263	33 39	338
341	37 38	060	37 18	117	51 44	150	33 52	198	26 18	229	46 20	264	33 29	339
342	38 00	061	37 40	118	51 56	151	33 44	199	25 59	230	45 54	265	33 20	340
343	38 22	062	38 03	119	52 08	152	33 35	200	25 39	231	45 29	266	33 12	340
344	38 45	063	38 25	120	52 20	154	33 26	201	25 19	232	45 04	267	33 03	341
	CAPELLA		ALDEBARAN		◆Hamal		Alpheratz		◆ALTAIR		VEGA		◆Alioth	
345	39 07	063	17 29	088	38 47	121	52 31	155	24 59	233	44 38	268	32 55	341
346	39 30	064	17 54	089	39 08	122	52 41	157	24 39	234	44 13	269	32 47	342
347	39 53	065	18 20	090	39 30	123	52 51	158	24 18	235	43 48	270	32 39	343
348	40 16	066	18 45	091	39 51	124	53 00	159	23 57	236	43 22	271	32 32	343
349	40 39	066	19 10	092	40 12	125	53 09	161	23 36	237	42 57	272	32 25	344
350	41 02	067	19 36	093	40 32	127	53 17	162	23 14	238	42 32	273	32 18	345
351	41 26	068	20 01	094	40 52	128	53 24	164	22 52	239	42 06	273	32 12	345
352	41 49	069	20 26	095	41 12	129	53 31	165	22 30	240	41 41	274	32 05	346
353	42 13	069	20 51	096	41 32	130	53 37	167	22 08	241	41 16	275	31 59	347
354	42 37	070	21 17	097	41 51	131	53 43	168	21 46	242	40 51	276	31 54	347
355	43 01	071	21 42	097	42 10	132	53 48	170	21 23	243	40 25	277	31 48	348
356	43 25	072	22 07	098	42 29	133	53 52	171	21 01	244	40 00	278	31 43	349
357	43 49	072	22 32	099	42 47	135	53 55	173	20 38	245	39 35	279	31 38	349
358	44 13	073	22 57	100	43 05	136	53 58	174	20 15	246	39 10	279	31 34	350
359	44 37	074	23 22	101	43 22	137	54 01	175	19 51	247	38 45	280	31 29	351

LHA	Hc	Zn	Hc	Zn	Hc	Zn	Hc	Zn	Hc	Zn	Hc	Zn	Hc	Zn
♈	◆POLLUX		CAPELLA		Hamal		◆Alpheratz		DENEB		◆VEGA		Alioth	
0	14 33	055	44 46	074	44 24	138	55 02	177	56 56	260	38 08	282	30 26	351
1	14 55	056	45 11	075	44 41	139	55 03	178	56 30	261	37 42	283	30 22	352
2	15 17	057	45 36	075	44 59	140	55 04	180	56 04	262	37 17	284	30 19	353
3	15 39	057	46 02	076	45 15	141	55 03	181	55 38	263	36 51	284	30 15	353
4	16 01	058	46 28	077	45 31	142	55 02	183	55 12	264	36 26	285	30 12	354
5	16 24	059	46 53	078	45 47	144	55 01	185	54 46	265	36 01	286	30 10	355
6	16 46	060	47 19	079	46 03	145	54 58	186	54 20	266	35 35	287	30 07	355
7	17 09	061	47 45	079	46 17	146	54 55	188	53 54	266	35 10	288	30 05	356
8	17 32	062	48 11	080	46 32	148	54 51	189	53 27	267	34 45	288	30 04	356
9	17 55	062	48 37	081	46 46	149	54 47	191	53 01	268	34 20	289	30 02	357
10	18 19	063	49 03	082	46 59	150	54 42	192	52 35	269	33 56	290	30 01	358
11	18 42	064	49 29	083	47 12	151	54 36	194	52 08	270	33 31	291	30 00	358
12	19 06	065	49 55	084	47 24	153	54 29	195	51 42	271	33 06	292	30 00	359
13	19 30	066	50 21	084	47 36	154	54 22	197	51 16	272	32 42	292	29 59	000
14	19 54	067	50 47	085	47 47	155	54 14	198	50 50	273	32 18	293	29 59	000
	Alioth		◆POLLUX		CAPELLA		Hamal		◆Alpheratz		DENEB		◆VEGA	
15	30 00	001	20 18	067	51 13	086	47 58	157	54 06	200	50 23	274	31 54	294
16	30 00	002	20 43	068	51 40	087	48 08	158	53 57	201	49 57	275	31 30	295
17	30 01	002	21 07	069	52 06	088	48 17	159	53 47	202	49 31	275	31 06	296
18	30 02	003	21 32	070	52 32	089	48 26	161	53 37	204	49 05	276	30 42	296
19	30 04	004	21 56	071	52 59	090	48 35	162	53 26	205	48 39	277	30 19	297
20	30 06	004	22 21	072	53 25	091	48 42	164	53 14	207	48 12	278	29 55	298
21	30 08	005	22 46	072	53 51	091	48 49	165	53 02	208	47 46	279	29 32	299
22	30 10	006	23 11	073	54 17	092	48 56	166	52 49	210	47 21	280	29 09	299
23	30 13	006	23 37	074	54 44	093	49 02	168	52 36	211	46 55	280	28 46	300
24	30 16	007	24 02	075	55 10	094	49 07	169	52 22	212	46 29	281	28 24	301
25	30 19	007	24 28	076	55 36	095	49 12	171	52 08	214	46 03	282	28 01	302
26	30 23	008	24 53	077	56 02	096	49 16	172	51 53	215	45 37	283	27 39	303
27	30 26	009	25 19	077	56 28	097	49 19	173	51 38	216	45 12	284	27 17	303
28	30 31	009	25 44	078	56 55	098	49 22	175	51 22	218	44 46	284	26 55	304
29	30 35	010	26 10	079	57 21	099	49 24	176	51 05	219	44 21	285	26 34	305
	Alioth		◆POLLUX		ALDEBARAN		◆Hamal		Alpheratz		DENEB		◆VEGA	
30	30 40	011	26 36	080	35 37	132	49 25	178	50 48	220	43 55	286	26 12	306
31	30 45	011	27 02	081	35 56	133	49 26	179	50 31	222	43 30	287	25 51	306
32	30 50	012	27 28	082	36 15	134	49 26	180	50 13	223	43 05	288	25 30	307
33	30 56	013	27 54	083	36 34	136	49 26	182	49 55	224	42 40	288	25 09	308
34	31 01	013	28 20	084	36 52	137	49 25	183	49 37	226	42 15	289	24 48	309
35	31 08	014	28 46	084	37 10	138	49 23	185	49 18	227	41 50	290	24 28	310
36	31 14	015	29 13	085	37 27	139	49 20	186	48 58	228	41 26	291	24 08	310
37	31 21	015	29 39	086	37 44	140	49 17	187	48 39	229	41 01	291	23 48	311
38	31 28	016	30 05	087	38 01	141	49 14	189	48 18	231	40 37	292	23 28	312
39	31 35	016	30 31	088	38 17	142	49 09	190	47 58	232	40 12	293	23 09	313
40	31 43	017	30 58	089	38 33	144	49 04	192	47 37	233	39 48	294	22 49	313
41	31 51	018	31 24	090	38 48	145	48 59	193	47 16	234	39 24	294	22 30	314
42	31 59	018	31 50	091	39 03	146	48 52	194	46 55	235	39 00	295	22 12	315
43	32 07	019	32 17	092	39 18	147	48 46	196	46 33	236	38 37	296	21 53	316
44	32 16	020	32 43	092	39 32	148	48 38	197	46 11	238	38 13	297	21 35	317
	Alioth		◆POLLUX		ALDEBARAN		◆Hamal		Alpheratz		DENEB		◆VEGA	
45	32 25	020	33 09	093	39 45	150	48 30	199	45 48	239	37 49	297	21 17	317
46	32 34	021	33 35	094	39 58	151	48 21	200	45 26	240	37 26	298	20 59	318
47	32 44	022	34 02	095	40 11	152	48 12	201	45 03	241	37 03	299	20 42	319
48	32 53	022	34 28	096	40 23	153	48 02	203	44 40	242	36 40	300	20 25	320
49	33 04	023	34 54	097	40 35	155	47 52	204	44 16	243	36 17	300	20 08	320
50	33 14	023	35 20	098	40 46	156	47 41	205	43 53	244	35 55	301	19 51	321
51	33 24	024	35 46	099	40 56	157	47 29	207	43 29	245	35 32	302	19 35	322
52	33 35	025	36 12	100	41 06	158	47 17	208	43 05	246	35 10	303	19 19	323
53	33 47	025	36 38	101	41 16	160	47 05	209	42 41	247	34 48	303	19 03	323
54	33 58	026	37 04	102	41 25	161	46 52	211	42 16	249	34 26	304	18 47	324
55	34 10	027	37 29	103	41 33	162	46 38	212	41 52	250	34 04	305	18 32	325
56	34 21	027	37 55	104	41 41	163	46 24	213	41 27	251	33 43	305	18 17	326
57	34 34	028	38 20	105	41 48	165	46 09	214	41 02	252	33 21	306	18 03	327
58	34 46	029	38 46	106	41 55	166	45 54	216	40 37	253	33 00	307	17 48	327
59	34 59	029	39 11	107	42 01	167	45 39	217	40 12	254	32 39	308	17 34	328
	Alioth		◆POLLUX		ALDEBARAN		◆Hamal		Alpheratz		DENEB		◆VEGA	
60	35 12	030	39 36	108	42 06	168	45 22	218	39 47	255	32 19	308	17 21	329
61	35 25	030	40 01	109	42 11	170	45 06	219	39 21	256	31 58	309	17 07	330
62	35 38	031	40 26	110	42 16	171	44 49	221	38 56	257	31 38	310	16 54	330
63	35 52	032	40 51	111	42 20	172	44 32	222	38 30	258	31 18	311	16 41	331
64	36 06	032	41 15	112	42 23	174	44 14	223	38 04	259	30 58	311	16 29	332
65	36 20	033	41 39	113	42 25	175	43 56	224	37 39	260	30 38	312	16 17	333
66	36 35	034	42 04	114	42 27	176	43 37	226	37 13	260	30 19	313	16 05	334
67	36 49	034	42 27	115	42 29	178	43 18	227	36 47	261	30 00	313	15 53	334
68	37 04	035	42 51	116	42 30	179	42 59	228	36 21	262	29 41	314	15 42	335
69	37 19	035	43 15	117	42 30	180	42 39	229	35 55	263	29 22	315	15 31	336
70	37 35	036	43 38	118	42 30	181	42 19	230	35 28	264	29 03	316	15 21	337
71	37 50	037	44 01	119	42 29	183	41 59	231	35 02	265	28 45	316	15 10	338
72	38 06	037	44 24	120	42 27	184	41 38	233	34 36	266	28 27	317	15 00	338
73	38 22	038	44 46	122	42 25	185	41 17	234	34 10	267	28 09	318	14 51	339
74	38 38	039	45 09	123	42 22	187	40 56	235	33 44	268	27 52	319	14 42	340
	Alioth		◆REGULUS		PROCYON		◆ALDEBARAN		Hamal		Alpheratz		◆DENEB	
75	38 55	039	16 26	096	24 41	135	42 19	188	40 34	236	33 17	269	27 34	319
76	39 12	040	16 53	097	24 59	137	42 15	189	40 12	237	32 51	270	27 17	320
77	39 29	040	17 19	098	25 17	138	42 11	190	39 50	238	32 25	271	27 00	321
78	39 46	041	17 45	099	25 35	139	42 05	192	39 27	239	31 58	272	26 44	321
79	40 03	042	18 11	100	25 52	140	42 00	193	39 05	240	31 32	272	26 28	322
80	40 21	042	18 36	101	26 09	141	41 54	194	38 42	241	31 06	273	26 12	323
81	40 39	043	19 02	102	26 25	142	41 47	196	38 19	242	30 40	274	25 56	324
82	40 57	044	19 28	103	26 41	143	41 39	197	37 55	243	30 13	275	25 40	324
83	41 15	044	19 53	104	26 57	144	41 31	198	37 32	244	29 47	276	25 25	325
84	41 33	045	20 19	105	27 12	145	41 23	199	37 08	246	29 21	277	25 10	326
85	41 52	045	20 44	106	27 27	146	41 14	201	36 44	247	28 55	278	24 56	327
86	42 11	046	21 10	107	27 42	147	41 04	202	36 19	248	28 29	279	24 41	327
87	42 30	047	21 35	108	27 56	148	40 54	203	35 55	249	28 03	279	24 27	328
88	42 49	047	22 00	108	28 09	149	40 44	204	35 30	250	27 37	280	24 13	329
89	43 09	048	22 25	109	28 22	151	40 33	206	35 06	251	27 11	281	24 00	329

LAT 64°N

LHA	Hc	Zn	Hc	Zn	Hc	Zn	Hc	Zn	Hc	Zn	Hc	Zn	Hc	Zn
♈	Alioth		◆REGULUS		PROCYON		BETELGEUSE		◆ALDEBARAN		Hamal		◆DENEB	
90	43 28	049	22 49	110	28 35	152	33 24	182	40 21	207	34 41	252	23 47	330
91	43 48	049	23 14	111	28 47	153	33 23	183	40 09	208	34 16	253	23 34	331
92	44 08	050	23 38	112	28 59	154	33 21	184	39 56	209	33 51	254	23 21	332
93	44 28	050	24 03	113	29 11	155	33 19	185	39 43	211	33 25	255	23 09	332
94	44 49	051	24 27	114	29 21	156	33 17	186	39 29	212	33 00	256	22 57	333
95	45 09	052	24 51	115	29 32	157	33 14	187	39 15	213	32 34	257	22 45	334
96	45 30	052	25 14	116	29 42	158	33 10	189	39 01	214	32 09	257	22 33	335
97	45 51	053	25 38	117	29 51	159	33 06	190	38 46	215	31 43	258	22 22	335
98	46 12	054	26 01	118	30 00	161	33 01	191	38 30	217	31 17	259	22 11	336
99	46 33	054	26 24	119	30 09	162	32 56	192	38 14	218	30 51	260	22 01	337
100	46 54	055	26 47	120	30 17	163	32 50	193	37 58	219	30 25	261	21 51	337
101	47 16	055	27 09	121	30 24	164	32 44	195	37 41	220	29 59	262	21 41	338
102	47 38	056	27 32	122	30 31	165	32 37	196	37 24	221	29 33	263	21 31	339
103	48 00	057	27 54	123	30 38	166	32 29	197	37 07	222	29 07	264	21 22	340
104	48 22	057	28 16	124	30 44	168	32 22	198	36 49	224	28 41	265	21 13	340
	Alioth		◆REGULUS		PROCYON		BETELGEUSE		◆CAPELLA		Hamal		◆DENEB	
105	48 44	058	28 37	125	30 49	169	32 13	199	66 59	231	28 15	266	21 04	341
106	49 06	059	28 59	126	30 54	170	32 04	200	66 38	232	27 48	267	20 56	342
107	49 29	059	29 20	127	30 58	171	31 55	201	66 17	234	27 22	268	20 48	343
108	49 52	060	29 40	129	31 02	172	31 45	203	65 56	235	26 56	269	20 40	343
109	50 14	061	30 01	130	31 05	173	31 35	204	65 34	237	26 30	270	20 33	344
110	50 37	061	30 21	131	31 08	174	31 24	205	65 12	238	26 03	270	20 26	345
111	51 00	062	30 41	132	31 11	176	31 12	206	64 49	240	25 37	271	20 19	346
112	51 24	062	31 00	133	31 12	177	31 01	207	64 26	241	25 11	272	20 13	346
113	51 47	063	31 19	134	31 13	178	30 48	208	64 03	242	24 44	273	20 07	347
114	52 11	064	31 38	135	31 14	179	30 36	209	63 40	244	24 18	274	20 01	348
115	52 34	064	31 57	136	31 14	180	30 23	211	63 16	245	23 52	275	19 56	349
116	52 58	065	32 15	137	31 14	181	30 09	212	62 52	246	23 26	276	19 51	349
117	53 22	066	32 32	138	31 13	183	29 55	213	62 28	247	23 00	277	19 46	350
118	53 46	066	32 50	139	31 11	184	29 40	214	62 03	249	22 34	278	19 42	351
119	54 10	067	33 07	140	31 09	185	29 26	215	61 39	250	22 07	278	19 38	352
	◆VEGA		ARCTURUS		◆REGULUS		PROCYON		◆CAPELLA		Schedar		DENEB	
120	14 06	017	15 32	078	33 23	142	31 07	186	61 14	251	41 50	316	19 34	352
121	14 13	017	15 58	079	33 39	143	31 04	187	60 49	252	41 32	316	19 31	353
122	14 21	018	16 24	080	33 55	144	31 00	188	60 24	253	41 14	317	19 28	354
123	14 30	019	16 50	081	34 10	145	30 56	190	59 59	254	40 56	318	19 25	355
124	14 38	020	17 16	081	34 25	146	30 52	191	59 33	256	40 39	318	19 23	355
125	14 47	021	17 42	082	34 40	147	30 46	192	59 08	257	40 21	319	19 21	356
126	14 57	021	18 08	083	34 54	148	30 41	193	58 42	258	40 04	320	19 19	357
127	15 07	022	18 34	084	35 07	150	30 35	194	58 16	259	39 47	320	19 18	358
128	15 17	023	19 00	085	35 20	151	30 28	195	57 51	260	39 31	321	19 17	358
129	15 27	024	19 26	086	35 33	152	30 21	196	57 25	261	39 14	321	19 16	359
130	15 38	024	19 53	087	35 45	153	30 13	198	56 59	262	38 58	322	19 16	000
131	15 49	025	20 19	088	35 57	154	30 05	199	56 33	263	38 42	323	19 16	001
132	16 00	026	20 45	089	36 08	155	29 56	200	56 06	264	38 26	323	19 16	001
133	16 12	027	21 11	089	36 18	157	29 47	201	55 40	265	38 10	324	19 17	002
134	16 24	028	21 38	090	36 29	158	29 37	202	55 14	266	37 55	325	19 18	003
	DENEB		◆VEGA		ARCTURUS		◆REGULUS		PROCYON		◆CAPELLA		Schedar	
135	19 19	003	16 36	028	22 04	091	36 38	159	29 27	203	54 48	267	37 40	325
136	19 21	004	16 49	029	22 30	092	36 47	160	29 16	204	54 22	267	37 25	326
137	19 23	005	17 02	030	22 57	093	36 56	162	29 05	206	53 55	268	37 10	326
138	19 26	006	17 15	031	23 23	094	37 04	163	28 54	207	53 29	269	36 56	327
139	19 29	006	17 29	032	23 49	095	37 12	164	28 42	208	53 03	270	36 42	328
140	19 32	007	17 43	032	24 15	096	37 19	165	28 29	209	52 36	271	36 28	328
141	19 35	008	17 57	033	24 41	097	37 25	166	28 16	210	52 10	272	36 14	329
142	19 39	009	18 12	034	25 08	098	37 31	168	28 03	211	51 44	273	36 00	330
143	19 43	009	18 26	035	25 34	099	37 36	169	27 49	212	51 18	274	35 47	330
144	19 48	010	18 41	035	26 00	099	37 41	170	27 35	213	50 51	275	35 34	331
145	19 53	011	18 57	036	26 25	100	37 45	171	27 20	214	50 25	275	35 22	331
146	19 58	012	19 13	037	26 51	101	37 49	173	27 05	215	49 59	276	35 09	332
147	20 03	012	19 29	038	27 17	102	37 52	174	26 50	217	49 33	277	34 57	333
148	20 09	013	19 45	039	27 43	103	37 55	175	26 34	218	49 07	278	34 45	333
149	20 15	014	20 01	039	28 08	104	37 57	176	26 17	219	48 41	279	34 33	334
	DENEB		◆VEGA		ARCTURUS		◆REGULUS		PROCYON		◆CAPELLA		Schedar	
150	20 22	015	20 18	040	28 34	105	37 58	177	26 01	220	48 15	280	34 22	335
151	20 28	015	20 35	041	28 59	106	37 59	179	25 44	221	47 49	280	34 11	335
152	20 36	016	20 53	042	29 24	107	38 00	180	25 26	222	47 23	281	34 00	336
153	20 43	017	21 10	042	29 49	108	37 59	181	25 09	223	46 57	282	33 49	336
154	20 51	018	21 28	043	30 14	109	37 58	182	24 51	224	46 32	283	33 39	337
155	20 59	018	21 46	044	30 39	110	37 57	184	24 32	225	46 06	284	33 28	338
156	21 07	019	22 05	045	31 04	111	37 55	185	24 13	226	45 41	284	33 19	338
157	21 16	020	22 23	046	31 28	112	37 52	186	23 54	227	45 15	285	33 09	339
158	21 25	021	22 42	046	31 52	113	37 49	187	23 35	228	44 50	286	33 00	340
159	21 35	021	23 01	047	32 17	114	37 46	189	23 15	229	44 25	287	32 51	340
160	21 44	022	23 21	048	32 40	115	37 41	190	22 55	230	43 59	288	32 42	341
161	21 54	023	23 40	049	33 04	116	37 37	191	22 35	231	43 34	288	32 33	341
162	22 05	024	24 00	049	33 28	117	37 31	192	22 14	232	43 09	289	32 25	342
163	22 15	024	24 20	050	33 51	118	37 25	194	21 53	233	42 45	290	32 17	343
164	22 26	025	24 40	051	34 14	119	37 19	195	21 32	234	42 20	291	32 10	343
	DENEB		◆VEGA		ARCTURUS		◆REGULUS		POLLUX		◆CAPELLA		Schedar	
165	22 38	026	25 01	052	34 37	120	37 12	196	42 39	244	41 55	291	32 02	344
166	22 49	026	25 22	052	35 00	121	37 04	197	42 16	246	41 31	292	31 55	345
167	23 01	027	25 43	053	35 22	122	36 56	198	41 52	247	41 07	293	31 48	345
168	23 13	028	26 04	054	35 44	123	36 48	200	41 27	248	40 42	294	31 42	346
169	23 26	029	26 25	055	36 06	124	36 39	201	41 03	249	40 18	294	31 35	347
170	23 38	029	26 47	056	36 27	126	36 29	202	40 38	250	39 55	295	31 29	347
171	23 51	030	27 09	056	36 49	127	36 19	203	40 14	251	39 31	296	31 24	348
172	24 05	031	27 31	057	37 10	128	36 08	204	39 49	252	39 07	297	31 18	348
173	24 18	032	27 53	058	37 30	129	35 57	206	39 24	253	38 44	297	31 13	349
174	24 32	032	28 15	059	37 50	130	35 46	207	38 58	254	38 20	298	31 08	350
175	24 47	033	28 38	059	38 10	131	35 34	208	38 33	255	37 57	299	31 04	350
176	25 01	034	29 00	060	38 30	132	35 21	209	38 08	256	37 34	300	31 00	351
177	25 16	034	29 23	061	38 49	133	35 08	210	37 42	257	37 12	300	30 56	352
178	25 31	035	29 47	062	39 08	135	34 54	211	37 17	258	36 49	301	30 52	352
179	25 46	036	30 10	063	39 27	136	34 40	213	36 51	259	36 26	302	30 49	353

LAT 64°N

LHA ♈	Hc	Zn	Hc	Zn	Hc	Zn	Hc	Zn	Hc	Zn	Hc	Zn	Hc	Zn
	DENEB		◆VEGA		ARCTURUS		◆Denebola		REGULUS		POLLUX		◆CAPELLA	
180	26 02	037	30 33	063	39 45	137	40 34	184	34 26	214	36 25	260	36 04	302
181	26 18	037	30 57	064	40 03	138	40 32	185	34 11	215	35 59	261	35 42	303
182	26 34	038	31 21	065	40 20	139	40 29	186	33 56	216	35 33	262	35 20	304
183	26 50	039	31 44	066	40 37	140	40 26	187	33 40	217	35 07	262	34 58	305
184	27 07	040	32 08	066	40 54	142	40 23	189	33 24	218	34 41	263	34 37	305
185	27 24	040	32 33	067	41 10	143	40 18	190	33 08	219	34 15	264	34 16	306
186	27 41	041	32 57	068	41 26	144	40 13	191	32 51	221	33 49	265	33 54	307
187	27 58	042	33 21	069	41 41	145	40 08	192	32 33	222	33 22	266	33 33	308
188	28 16	042	33 46	070	41 56	146	40 02	194	32 16	223	32 56	267	33 13	308
189	28 34	043	34 11	070	42 10	148	39 56	195	31 58	224	32 30	268	32 52	309
190	28 52	044	34 36	071	42 24	149	39 49	196	31 39	225	32 03	269	32 32	310
191	29 10	045	35 01	072	42 37	150	39 41	197	31 20	226	31 37	270	32 12	310
192	29 29	045	35 26	073	42 50	151	39 33	199	31 01	227	31 11	271	31 52	311
193	29 48	046	35 51	074	43 03	153	39 24	200	30 42	228	30 45	272	31 32	312
194	30 07	047	36 16	074	43 14	154	39 15	201	30 22	229	30 18	273	31 12	313
	DENEB		◆VEGA		ARCTURUS		◆Denebola		REGULUS		POLLUX		◆CAPELLA	
195	30 26	048	36 42	075	43 26	155	39 05	202	30 02	230	29 52	273	30 53	313
196	30 45	048	37 07	076	43 37	156	38 55	204	29 42	231	29 26	274	30 34	314
197	31 05	049	37 33	077	43 47	158	38 44	205	29 21	232	29 00	275	30 15	315
198	31 25	050	37 58	078	43 56	159	38 33	206	29 00	234	28 33	276	29 57	315
199	31 45	050	38 24	079	44 06	160	38 21	207	28 39	235	28 07	277	29 38	316
200	32 06	051	38 50	079	44 14	162	38 09	208	28 17	236	27 41	278	29 20	317
201	32 26	052	39 16	080	44 22	163	37 56	210	27 55	237	27 15	279	29 02	318
202	32 47	053	39 42	081	44 30	164	37 43	211	27 33	238	26 49	279	28 45	318
203	33 08	053	40 08	082	44 37	166	37 29	212	27 11	239	26 23	280	28 27	319
204	33 29	054	40 34	083	44 43	167	37 15	213	26 48	240	25 57	281	28 10	320
205	33 51	055	41 00	084	44 48	168	37 00	214	26 25	241	25 32	282	27 53	320
206	34 12	056	41 26	084	44 54	169	36 45	216	26 02	242	25 06	283	27 37	321
207	34 34	056	41 52	085	44 58	171	36 29	217	25 39	243	24 40	284	27 21	322
208	34 56	057	42 18	086	45 02	172	36 13	218	25 15	244	24 15	285	27 04	323
209	35 18	058	42 45	087	45 05	173	35 57	219	24 52	245	23 50	285	26 49	323
	DENEB		◆VEGA		ARCTURUS		◆Denebola		REGULUS		POLLUX		◆CAPELLA	
210	35 41	058	43 11	088	45 08	175	35 40	220	24 28	246	23 24	286	26 33	324
211	36 03	059	43 37	089	45 10	176	35 23	221	24 04	247	22 59	287	26 18	325
212	36 26	060	44 04	090	45 11	178	35 06	222	23 40	248	22 34	288	26 03	326
213	36 49	061	44 30	091	45 12	179	34 48	224	23 15	249	22 09	289	25 48	326
214	37 12	061	44 56	092	45 12	180	34 29	225	22 51	250	21 44	290	25 33	327
215	37 35	062	45 22	092	45 12	182	34 11	226	22 26	251	21 19	291	25 19	328
216	37 58	063	45 49	093	45 11	183	33 52	227	22 01	251	20 55	291	25 05	328
217	38 22	064	46 15	094	45 09	184	33 32	228	21 36	252	20 30	292	24 52	329
218	38 45	064	46 41	095	45 07	186	33 13	229	21 11	253	20 06	293	24 38	330
219	39 09	065	47 07	096	45 04	187	32 52	230	20 46	254	19 42	294	24 25	331
220	39 33	066	47 33	097	45 01	188	32 32	231	20 20	255	19 18	295	24 12	331
221	39 57	067	48 00	098	44 57	190	32 11	232	19 55	256	18 54	296	24 00	332
222	40 21	067	48 26	099	44 52	191	31 51	233	19 29	257	18 31	296	23 48	333
223	40 46	068	48 51	100	44 47	192	31 29	234	19 04	258	18 07	297	23 36	333
224	41 10	069	49 17	101	44 41	194	31 08	235	18 38	259	17 44	298	23 24	334
	◆DENEB		VEGA		Rasalhague		◆ARCTURUS		Denebola		Dubhe		◆CAPELLA	
225	41 35	070	49 43	102	31 58	134	44 35	195	30 46	237	63 56	293	23 13	335
226	41 59	070	50 09	103	32 17	135	44 28	196	30 24	238	63 32	293	23 02	336
227	42 24	071	50 34	104	32 36	136	44 20	197	30 02	239	63 07	294	22 51	336
228	42 49	072	51 00	105	32 54	137	44 12	199	29 39	240	62 43	294	22 41	337
229	43 14	073	51 25	106	33 11	138	44 03	200	29 16	241	62 19	295	22 31	338
230	43 39	074	51 50	107	33 29	140	43 54	201	28 53	242	61 56	295	22 21	339
231	44 05	074	52 15	108	33 45	141	43 44	203	28 30	243	61 32	296	22 12	339
232	44 30	075	52 40	109	34 02	142	43 34	204	28 06	244	61 08	296	22 03	340
233	44 56	076	53 05	111	34 18	143	43 23	205	27 43	245	60 45	297	21 54	341
234	45 21	077	53 29	112	34 34	144	43 11	206	27 19	246	60 21	297	21 45	342
235	45 47	077	53 54	113	34 49	145	42 59	208	26 55	247	59 58	298	21 37	342
236	46 13	078	54 18	114	35 04	146	42 47	209	26 30	248	59 35	299	21 29	343
237	46 38	079	54 42	115	35 18	148	42 34	210	26 06	249	59 12	299	21 22	344
238	47 04	080	55 06	116	35 32	149	42 20	211	25 41	250	58 49	300	21 14	344
239	47 30	081	55 29	117	35 45	150	42 06	213	25 17	251	58 26	300	21 08	345
	Alpheratz		◆DENEB		VEGA		Rasalhague		◆ARCTURUS		Alioth		◆CAPELLA	
240	13 30	050	47 56	082	55 52	119	35 58	151	41 52	214	66 01	272	21 01	346
241	13 50	050	48 22	082	56 15	120	36 11	152	41 37	215	65 35	273	20 55	347
242	14 10	051	48 48	083	56 38	121	36 23	153	41 21	216	65 09	274	20 49	347
243	14 31	052	49 14	084	57 00	122	36 34	155	41 06	218	64 42	275	20 43	348
244	14 52	053	49 41	085	57 22	124	36 45	156	40 49	219	64 16	276	20 38	349
245	15 13	054	50 07	086	57 44	125	36 56	157	40 33	220	63 50	277	20 33	350
246	15 34	055	50 33	087	58 06	126	37 06	158	40 15	221	63 24	277	20 29	350
247	15 56	055	50 59	088	58 27	127	37 15	159	39 58	222	62 58	278	20 24	351
248	16 18	056	51 26	088	58 47	129	37 24	161	39 40	224	62 32	279	20 20	352
249	16 40	057	51 52	089	59 08	130	37 33	162	39 22	225	62 06	280	20 17	353
250	17 02	058	52 18	090	59 28	132	37 41	163	39 03	226	61 40	280	20 14	353
251	17 24	059	52 44	091	59 47	133	37 48	164	38 44	227	61 14	281	20 11	354
252	17 47	060	53 11	092	60 06	134	37 55	165	38 25	228	60 48	282	20 08	355
253	18 10	060	53 37	093	60 25	136	38 01	167	38 05	229	60 23	283	20 06	356
254	18 33	061	54 03	094	60 43	137	38 07	168	37 45	230	59 57	283	20 04	356
	Alpheratz		◆DENEB		ALTAIR		Rasalhague		◆ARCTURUS		Alioth		◆CAPELLA	
255	18 56	062	54 30	095	27 12	131	38 12	169	37 24	231	59 31	284	20 02	357
256	19 19	063	54 56	096	27 31	132	38 17	170	37 04	233	59 06	285	20 01	358
257	19 42	064	55 22	097	27 51	133	38 21	172	36 43	234	58 41	286	20 00	358
258	20 06	065	55 48	098	28 10	134	38 25	173	36 21	235	58 15	286	20 00	359
259	20 30	065	56 14	099	28 28	135	38 28	174	36 00	236	57 50	287	20 00	000
260	20 54	066	56 40	100	28 46	137	38 30	175	35 38	237	57 25	288	20 00	001
261	21 18	067	57 06	101	29 04	138	38 32	177	35 16	238	57 00	288	20 00	001
262	21 42	068	57 32	102	29 22	139	38 33	178	34 53	239	56 35	289	20 01	002
263	22 07	069	57 57	103	29 39	140	38 34	179	34 30	240	56 10	290	20 02	003
264	22 31	070	58 23	104	29 56	141	38 34	180	34 08	241	55 46	290	20 04	004
265	22 56	070	58 48	105	30 12	142	38 33	182	33 44	242	55 21	291	20 06	004
266	23 21	071	59 14	106	30 28	143	38 32	183	33 21	243	54 56	292	20 08	005
267	23 46	072	59 39	107	30 44	144	38 31	184	32 57	244	54 32	292	20 10	006
268	24 11	073	60 04	108	30 59	145	38 28	185	32 34	245	54 08	293	20 13	007
269	24 36	074	60 29	109	31 14	146	38 26	187	32 10	246	53 44	294	20 16	007

LHA ♈	Hc	Zn	Hc	Zn	Hc	Zn	Hc	Zn	Hc	Zn	Hc	Zn	Hc	Zn
	◆CAPELLA		Alpheratz		◆DENEB		ALTAIR		Rasalhague		◆ARCTURUS		Alioth	
270	20 20	008	25 02	075	60 54	111	31 28	147	38 22	188	31 46	247	53 20	294
271	20 24	009	25 27	075	61 18	112	31 42	149	38 18	189	31 21	248	52 56	295
272	20 28	010	25 52	076	61 42	113	31 56	150	38 14	190	30 57	249	52 32	296
273	20 32	010	26 18	077	62 07	114	32 09	151	38 09	192	30 32	250	52 08	296
274	20 37	011	26 44	078	62 30	115	32 21	152	38 03	193	30 07	251	51 45	297
275	20 42	012	27 09	079	62 54	117	32 33	153	37 57	194	29 42	252	51 21	298
276	20 48	012	27 35	080	63 17	118	32 45	154	37 51	195	29 17	253	50 58	298
277	20 54	013	28 01	081	63 40	119	32 56	155	37 43	197	28 52	254	50 35	299
278	21 00	014	28 27	081	64 03	121	33 07	157	37 36	198	28 26	255	50 12	300
279	21 07	015	28 53	082	64 26	122	33 17	158	37 27	199	28 01	256	49 49	300
280	21 13	015	29 19	083	64 48	123	33 27	159	37 19	200	27 35	257	49 27	301
281	21 21	016	29 45	084	65 10	125	33 36	160	37 09	201	27 10	258	49 04	301
282	21 28	017	30 12	085	65 31	126	33 45	161	36 59	203	26 44	259	48 42	302
283	21 36	018	30 38	086	65 52	128	33 53	163	36 49	204	26 18	260	48 20	303
284	21 44	018	31 04	087	66 13	129	34 00	164	36 38	205	25 52	261	47 58	303
	◆CAPELLA		Alpheratz		Enif		◆ALTAIR		VEGA		◆ARCTURUS		Alioth	
285	21 52	019	31 30	088	28 40	133	34 08	165	64 33	191	25 26	262	47 36	304
286	22 01	020	31 57	088	29 00	134	34 14	166	64 27	192	25 00	263	47 14	305
287	22 10	021	32 23	089	29 19	135	34 20	167	64 21	194	24 34	264	46 52	305
288	22 20	021	32 49	090	29 37	136	34 26	168	64 15	196	24 08	264	46 31	306
289	22 29	022	33 16	091	29 55	137	34 31	170	64 07	198	23 42	265	46 10	306
290	22 39	023	33 42	092	30 13	138	34 35	171	63 59	199	23 15	266	45 49	307
291	22 50	023	34 08	093	30 30	139	34 39	172	63 49	201	22 49	267	45 28	308
292	23 00	024	34 34	094	30 47	140	34 43	173	63 40	203	22 23	268	45 07	308
293	23 11	025	35 01	095	31 04	141	34 45	174	63 29	205	21 57	269	44 47	309
294	23 22	026	35 27	096	31 20	142	34 48	176	63 18	206	21 30	270	44 26	310
295	23 34	026	35 53	097	31 36	143	34 49	177	63 06	208	21 04	271	44 06	310
296	23 46	027	36 19	098	31 52	145	34 51	178	62 53	210	20 38	272	43 46	311
297	23 58	028	36 45	099	32 07	146	34 51	179	62 40	211	20 11	273	43 26	312
298	24 10	029	37 11	099	32 21	147	34 51	180	62 26	213	19 45	274	43 07	312
299	24 23	029	37 37	100	32 36	148	34 51	182	62 11	214	19 19	274	42 47	313
	◆CAPELLA		Alpheratz		Enif		◆ALTAIR		VEGA		◆ARCTURUS		Alioth	
300	24 36	030	38 03	101	32 49	149	34 50	183	61 56	216	18 53	275	42 28	313
301	24 50	031	38 29	102	33 03	150	34 48	184	61 40	218	18 27	276	42 09	314
302	25 03	031	38 54	103	33 15	151	34 46	185	61 24	219	18 00	277	41 50	315
303	25 17	032	39 20	104	33 28	153	34 43	186	61 07	221	17 34	278	41 32	315
304	25 31	033	39 45	105	33 40	154	34 40	188	60 49	222	17 08	279	41 13	316
305	25 46	034	40 10	106	33 51	155	34 36	189	60 32	224	16 42	280	40 55	316
306	26 00	034	40 36	107	34 02	156	34 32	190	60 13	225	16 16	281	40 37	317
307	26 15	035	41 01	108	34 12	157	34 27	191	59 54	226	15 51	281	40 19	318
308	26 31	036	41 26	109	34 22	158	34 22	192	59 35	228	15 25	282	40 02	318
309	26 46	037	41 50	110	34 32	160	34 16	194	59 15	229	14 59	283	39 44	319
310	27 02	037	42 15	111	34 41	161	34 09	195	58 55	231	14 34	284	39 27	320
311	27 18	038	42 39	112	34 49	162	34 02	196	58 35	232	14 08	285	39 10	320
312	27 34	039	43 03	114	34 57	163	33 55	197	58 14	233	13 43	286	38 53	321
313	27 51	039	43 28	115	35 04	164	33 47	198	57 52	235	13 18	287	38 37	321
314	28 08	040	43 51	116	35 11	166	33 38	200	57 31	236	12 52	288	38 21	322
	CAPELLA		◆Hamal		Alpheratz		Enif		◆ALTAIR		VEGA		◆Alioth	
315	28 25	041	26 44	091	44 15	117	35 18	167	33 29	201	57 09	237	38 05	323
316	28 42	042	27 10	092	44 38	118	35 23	168	33 20	202	56 47	238	37 49	323
317	29 00	042	27 37	093	45 01	119	35 29	169	33 10	203	56 24	240	37 33	324
318	29 17	043	28 03	094	45 24	120	35 33	170	32 59	204	56 01	241	37 18	325
319	29 36	044	28 29	095	45 47	121	35 38	172	32 48	205	55 38	242	37 03	325
320	29 54	044	28 55	096	46 09	122	35 41	173	32 37	207	55 15	243	36 48	326
321	30 12	045	29 22	096	46 31	123	35 44	174	32 25	208	54 51	245	36 33	326
322	30 31	046	29 48	097	46 53	125	35 47	175	32 12	209	54 27	246	36 19	327
323	30 50	047	30 14	098	47 15	126	35 49	176	31 59	210	54 03	247	36 05	328
324	31 09	047	30 40	099	47 36	127	35 50	178	31 46	211	53 39	248	35 51	328
325	31 29	048	31 06	100	47 57	128	35 51	179	31 32	212	53 14	249	35 37	329
326	31 48	049	31 32	101	48 17	129	35 51	180	31 18	213	52 50	250	35 24	330
327	32 08	049	31 57	102	48 37	131	35 51	181	31 03	214	52 25	251	35 11	330
328	32 28	050	32 23	103	48 57	132	35 50	182	30 48	216	52 00	252	34 58	331
329	32 49	051	32 49	104	49 16	133	35 49	184	30 33	217	51 35	253	34 45	332
	CAPELLA		◆Hamal		Alpheratz		Enif		◆ALTAIR		VEGA		◆Alioth	
330	33 09	052	33 14	105	49 35	134	35 47	185	30 17	218	51 10	254	34 33	332
331	33 30	052	33 39	106	49 54	136	35 44	186	30 00	219	50 44	255	34 20	333
332	33 51	053	34 05	107	50 12	137	35 41	187	29 44	220	50 19	257	34 08	333
333	34 12	054	34 30	108	50 30	138	35 37	189	29 27	221	49 53	258	33 57	334
334	34 33	055	34 55	109	50 47	139	35 33	190	29 09	222	49 27	259	33 45	335
335	34 55	055	35 19	110	51 04	141	35 29	191	28 51	223	49 01	260	33 34	335
336	35 17	056	35 44	111	51 21	142	35 23	192	28 33	224	48 36	261	33 23	336
337	35 39	057	36 09	112	51 37	143	35 17	193	28 15	225	48 10	261	33 13	337
338	36 01	057	36 33	113	51 52	145	35 11	195	27 56	226	47 43	262	33 03	337
339	36 23	058	36 57	114	52 07	146	35 04	196	27 37	227	47 17	263	32 53	338
340	36 45	059	37 21	115	52 21	148	34 57	197	27 17	229	46 51	264	32 43	339
341	37 08	060	37 45	116	52 35	149	34 49	198	26 57	230	46 25	265	32 33	339
342	37 31	060	38 08	117	52 48	150	34 41	199	26 37	231	45 59	266	32 24	340
343	37 54	061	38 31	118	53 01	152	34 32	200	26 16	232	45 33	267	32 15	340
344	38 17	062	38 55	119	53 13	153	34 22	202	25 56	233	45 06	268	32 06	341
	CAPELLA		ALDEBARAN		◆Hamal		Alpheratz		◆ALTAIR		VEGA		◆Alioth	
345	38 40	063	17 27	088	39 17	120	53 25	155	25 35	234	44 40	269	31 58	342
346	39 03	063	17 53	089	39 40	121	53 36	156	25 13	235	44 14	270	31 50	342
347	39 27	064	18 20	090	40 02	123	53 46	157	24 52	236	43 47	271	31 42	343
348	39 51	065	18 46	091	40 24	124	53 56	159	24 30	237	43 21	272	31 34	344
349	40 15	066	19 12	092	40 46	125	54 05	160	24 08	238	42 55	273	31 27	344
350	40 39	066	19 38	093	41 08	126	54 14	162	23 45	239	42 29	273	31 20	345
351	41 03	067	20 05	093	41 29	127	54 22	163	23 23	240	42 02	274	31 13	346
352	41 27	068	20 31	094	41 50	128	54 29	165	23 00	241	41 36	275	31 07	346
353	41 51	069	20 57	095	42 10	129	54 35	166	22 37	242	41 10	276	31 01	347
354	42 16	069	21 23	096	42 30	130	54 41	168	22 14	243	40 44	277	30 55	347
355	42 41	070	21 49	097	42 50	132	54 47	169	21 50	244	40 18	278	30 49	348
356	43 05	071	22 16	098	43 10	133	54 51	171	21 26	245	39 52	279	30 44	349
357	43 30	072	22 42	099	43 29	134	54 55	172	21 03	246	39 26	279	30 39	349
358	43 55	072	23 08	100	43 47	135	54 58	174	20 39	247	39 00	280	30 35	350
359	44 20	073	23 33	101	44 06	136	55 01	175	20 14	248	38 34	281	30 30	351

LHA ♈	Hc	Zn	Hc	Zn	Hc	Zn	Hc	Zn	Hc	Zn	Hc	Zn	Hc	Zn
	◆POLLUX		CAPELLA		Hamal		◆Alpheratz		DENEB		◆VEGA		Alioth	
0	13 59	055	44 29	073	45 08	137	56 02	177	57 06	261	37 55	283	29 27	351
1	14 21	055	44 55	074	45 26	138	56 03	178	56 39	262	37 29	284	29 23	352
2	14 44	056	45 21	074	45 44	139	56 04	180	56 12	263	37 02	284	29 19	353
3	15 06	057	45 47	075	46 02	141	56 03	182	55 45	264	36 36	285	29 16	353
4	15 29	058	46 14	076	46 19	142	56 02	183	55 18	265	36 10	286	29 13	354
5	15 53	059	46 40	077	46 35	143	56 00	185	54 51	266	35 44	287	29 10	355
6	16 16	060	47 07	078	46 52	144	55 58	186	54 24	267	35 18	288	29 08	355
7	16 40	060	47 33	078	47 07	146	55 55	188	53 57	268	34 52	288	29 06	356
8	17 03	061	48 00	079	47 22	147	55 51	189	53 29	269	34 26	289	29 04	357
9	17 27	062	48 27	080	47 37	148	55 46	191	53 02	270	34 00	290	29 02	357
10	17 52	063	48 54	081	47 51	150	55 40	192	52 35	271	33 35	291	29 01	358
11	18 16	064	49 21	082	48 04	151	55 34	194	52 08	271	33 09	291	29 00	358
12	18 40	065	49 47	082	48 17	152	55 27	195	51 40	272	32 44	292	29 00	359
13	19 05	065	50 15	083	48 30	154	55 20	197	51 13	273	32 19	293	28 59	000
14	19 30	066	50 42	084	48 42	155	55 11	199	50 46	274	31 54	294	28 59	000
	Alioth		◆POLLUX		CAPELLA		Hamal		◆Alpheratz		DENEB		◆VEGA	
15	29 00	001	19 55	067	51 09	085	48 53	156	55 02	200	50 19	275	31 29	295
16	29 00	002	20 20	068	51 36	086	49 03	158	54 53	202	49 52	276	31 04	295
17	29 01	002	20 45	069	52 03	087	49 13	159	54 42	203	49 25	277	30 40	296
18	29 02	003	21 11	070	52 30	087	49 23	160	54 31	204	48 58	277	30 15	297
19	29 04	004	21 36	070	52 57	088	49 32	162	54 20	206	48 31	278	29 51	298
20	29 06	004	22 02	071	53 25	089	49 40	163	54 07	207	48 04	279	29 27	298
21	29 08	005	22 28	072	53 52	090	49 47	165	53 55	209	47 37	280	29 03	299
22	29 10	005	22 54	073	54 19	091	49 54	166	53 41	210	47 10	281	28 40	300
23	29 13	006	23 20	074	54 46	092	50 00	167	53 27	212	46 43	281	28 16	301
24	29 16	007	23 46	075	55 14	093	50 06	169	53 13	213	46 17	282	27 53	302
25	29 19	007	24 13	075	55 41	094	50 11	170	52 57	214	45 50	283	27 30	302
26	29 23	008	24 39	076	56 08	095	50 15	172	52 42	216	45 24	284	27 07	303
27	29 27	009	25 05	077	56 35	096	50 19	173	52 26	217	44 57	285	26 44	304
28	29 31	009	25 32	078	57 02	097	50 22	175	52 09	219	44 31	285	26 21	305
29	29 36	010	25 59	079	57 29	098	50 24	176	51 52	220	44 05	286	25 59	305
	Alioth		◆POLLUX		ALDEBARAN		◆Hamal		Alpheratz		DENEB		◆VEGA	
30	29 41	011	26 25	080	36 17	132	50 25	178	51 34	221	43 38	287	25 37	306
31	29 46	011	26 52	080	36 37	133	50 26	179	51 16	223	43 12	288	25 15	307
32	29 51	012	27 19	081	36 57	134	50 26	180	50 57	224	42 47	288	24 53	308
33	29 57	012	27 46	082	37 16	135	50 26	182	50 38	225	42 21	289	24 32	308
34	30 03	013	28 13	083	37 35	136	50 25	183	50 18	226	41 55	290	24 11	309
35	30 09	014	28 40	084	37 54	137	50 23	185	49 58	228	41 30	291	23 50	310
36	30 16	014	29 07	085	38 12	138	50 20	186	49 38	229	41 04	291	23 29	311
37	30 23	015	29 34	086	38 30	140	50 17	188	49 17	230	40 39	292	23 08	311
38	30 30	016	30 02	086	38 48	141	50 13	189	48 56	231	40 14	293	22 48	312
39	30 38	016	30 29	087	39 05	142	50 08	190	48 35	233	39 49	294	22 28	313
40	30 45	017	30 56	088	39 21	143	50 03	192	48 13	234	39 24	294	22 08	314
41	30 53	018	31 23	089	39 37	144	49 57	193	47 51	235	38 59	295	21 49	314
42	31 02	018	31 51	090	39 53	146	49 50	195	47 28	236	38 34	296	21 29	315
43	31 10	019	32 18	091	40 08	147	49 43	196	47 06	237	38 10	297	21 10	316
44	31 19	019	32 45	092	40 23	148	49 35	198	46 42	238	37 46	297	20 51	317
	Alioth		◆POLLUX		ALDEBARAN		◆Hamal		Alpheratz		DENEB		◆VEGA	
45	31 29	020	33 12	093	40 37	149	49 27	199	46 19	240	37 22	298	20 33	318
46	31 38	021	33 39	094	40 51	150	49 18	200	45 56	241	36 58	299	20 15	318
47	31 48	021	34 07	095	41 04	152	49 08	202	45 32	242	36 34	299	19 57	319
48	31 58	022	34 34	095	41 17	153	48 58	203	45 07	243	36 10	300	19 39	320
49	32 08	023	35 01	096	41 29	154	48 47	204	44 43	244	35 47	301	19 21	321
50	32 19	023	35 28	097	41 40	155	48 35	206	44 18	245	35 23	302	19 04	321
51	32 30	024	35 55	098	41 51	157	48 23	207	43 54	246	35 00	302	18 47	322
52	32 41	024	36 22	099	42 02	158	48 10	208	43 29	247	34 37	303	18 31	323
53	32 52	025	36 49	100	42 12	159	47 57	210	43 03	248	34 15	304	18 15	324
54	33 04	026	37 15	101	42 21	161	47 43	211	42 38	249	33 52	305	17 59	324
55	33 16	026	37 42	102	42 30	162	47 29	212	42 12	250	33 30	305	17 43	325
56	33 28	027	38 09	103	42 38	163	47 14	214	41 47	251	33 08	306	17 28	326
57	33 41	028	38 35	104	42 46	164	46 59	215	41 21	252	32 46	307	17 12	327
58	33 53	028	39 02	105	42 53	166	46 43	216	40 55	253	32 24	307	16 58	328
59	34 06	029	39 28	106	42 59	167	46 26	218	40 29	254	32 03	308	16 43	328
	Alioth		◆POLLUX		ALDEBARAN		◆Hamal		Alpheratz		DENEB		◆VEGA	
60	34 20	029	39 54	107	43 05	168	46 09	219	40 02	255	31 41	309	16 29	329
61	34 33	030	40 20	108	43 10	170	45 52	220	39 36	256	31 20	310	16 15	330
62	34 47	031	40 46	109	43 15	171	45 34	221	39 09	257	30 59	310	16 02	331
63	35 01	031	41 11	110	43 19	172	45 16	223	38 43	258	30 39	311	15 49	331
64	35 15	032	41 37	111	43 22	174	44 57	224	38 16	259	30 18	312	15 36	332
65	35 30	033	42 02	112	43 25	175	44 38	225	37 49	260	29 58	312	15 23	333
66	35 45	033	42 27	113	43 27	176	44 19	226	37 22	261	29 38	313	15 11	334
67	36 00	034	42 52	114	43 29	177	43 59	227	36 55	262	29 18	314	14 59	335
68	36 15	034	43 17	115	43 30	179	43 39	229	36 28	263	28 59	315	14 47	335
69	36 30	035	43 42	116	43 30	180	43 18	230	36 01	264	28 39	315	14 36	336
70	36 46	036	44 06	117	43 30	181	42 57	231	35 34	265	28 20	316	14 25	337
71	37 02	036	44 30	118	43 29	183	42 36	232	35 07	266	28 01	317	14 15	338
72	37 18	037	44 54	120	43 27	184	42 14	233	34 40	267	27 43	317	14 05	338
73	37 35	037	45 17	121	43 25	185	41 52	234	34 13	268	27 25	318	13 55	339
74	37 51	038	45 41	122	43 22	187	41 30	235	33 45	269	27 07	319	13 45	340
	Alioth		◆REGULUS		PROCYON		◆ALDEBARAN		Hamal		Alpheratz		◆DENEB	
75	38 08	039	16 33	096	25 24	135	43 18	188	41 07	237	33 18	269	26 49	320
76	38 25	039	17 00	097	25 43	136	43 14	189	40 44	238	32 51	270	26 31	320
77	38 43	040	17 27	098	26 01	137	43 09	191	40 21	239	32 24	271	26 14	321
78	39 00	041	17 54	099	26 20	138	43 04	192	39 58	240	31 56	272	25 57	322
79	39 18	041	18 21	100	26 38	139	42 58	193	39 34	241	31 29	273	25 40	322
80	39 36	042	18 48	101	26 55	140	42 52	195	39 10	242	31 02	274	25 24	323
81	39 55	042	19 14	102	27 12	142	42 44	196	38 46	243	30 35	275	25 08	324
82	40 13	043	19 41	102	27 29	143	42 37	197	38 22	244	30 08	276	24 52	325
83	40 32	044	20 08	103	27 45	144	42 28	198	37 57	245	29 41	276	24 36	325
84	40 51	044	20 34	104	28 01	145	42 20	200	37 32	246	29 14	277	24 21	326
85	41 10	045	21 00	105	28 17	146	42 10	201	37 07	247	28 47	278	24 06	327
86	41 29	045	21 27	106	28 32	147	42 00	202	36 42	248	28 20	279	23 51	327
87	41 48	046	21 53	107	28 47	148	41 49	204	36 17	249	27 53	280	23 36	328
88	42 08	047	22 19	108	29 01	149	41 38	205	35 51	250	27 26	281	23 22	329
89	42 28	047	22 44	109	29 15	150	41 27	206	35 25	251	26 59	282	23 08	330

LAT 63°N

LHA ♈	Hc	Zn	Hc	Zn	Hc	Zn	Hc	Zn	Hc	Zn	Hc	Zn	Hc	Zn
	Alioth		◆REGULUS		PROCYON		BETELGEUSE		◆ALDEBARAN		Hamal		◆DENEB	
90	42 48	048	23 10	110	29 28	151	34 24	182	41 14	207	34 59	252	22 55	330
91	43 08	048	23 36	111	29 41	152	34 23	183	41 02	209	34 33	253	22 41	331
92	43 29	049	24 01	112	29 53	154	34 21	184	40 48	210	34 07	254	22 28	332
93	43 50	050	24 26	113	30 05	155	34 19	185	40 34	211	33 41	255	22 15	333
94	44 11	050	24 51	114	30 16	156	34 16	186	40 20	212	33 15	256	22 03	333
95	44 32	051	25 16	115	30 27	157	34 13	188	40 05	213	32 48	257	21 51	334
96	44 53	052	25 41	116	30 38	158	34 09	189	39 50	215	32 21	258	21 39	335
97	45 14	052	26 05	117	30 47	159	34 05	190	39 34	216	31 55	259	21 28	335
98	45 36	053	26 29	118	30 57	160	34 00	191	39 18	217	31 28	260	21 17	336
99	45 58	053	26 53	119	31 06	162	33 54	192	39 02	218	31 01	261	21 06	337
100	46 20	054	27 17	120	31 14	163	33 48	193	38 45	219	30 34	262	20 55	338
101	46 42	055	27 40	121	31 22	164	33 42	195	38 27	221	30 07	263	20 45	338
102	47 04	055	28 04	122	31 29	165	33 35	196	38 09	222	29 40	264	20 35	339
103	47 26	056	28 27	123	31 36	166	33 27	197	37 51	223	29 13	265	20 26	340
104	47 49	056	28 49	124	31 42	167	33 19	198	37 32	224	28 46	266	20 16	341
	Alioth		◆REGULUS		PROCYON		BETELGEUSE		◆CAPELLA		Hamal		◆DENEB	
105	48 12	057	29 12	125	31 48	169	33 10	199	67 36	233	28 19	266	20 07	341
106	48 35	058	29 34	126	31 53	170	33 00	201	67 14	234	27 52	267	19 59	342
107	48 58	058	29 56	127	31 57	171	32 51	202	66 52	236	27 24	268	19 51	343
108	49 21	059	30 18	128	32 02	172	32 40	203	66 29	237	26 57	269	19 43	344
109	49 44	059	30 39	129	32 05	173	32 29	204	66 06	239	26 30	270	19 35	344
110	50 08	060	31 00	130	32 08	174	32 18	205	65 42	240	26 03	271	19 28	345
111	50 32	061	31 21	131	32 10	176	32 06	206	65 19	242	25 35	272	19 21	346
112	50 55	061	31 41	132	32 12	177	31 54	207	64 54	243	25 08	273	19 15	346
113	51 19	062	32 01	133	32 13	178	31 41	209	64 30	244	24 41	274	19 08	347
114	51 44	063	32 20	135	32 14	179	31 28	210	64 05	246	24 14	274	19 02	348
115	52 08	063	32 40	136	32 14	180	31 14	211	63 40	247	23 47	275	18 57	349
116	52 32	064	32 59	137	32 14	181	31 00	212	63 15	248	23 20	276	18 52	349
117	52 57	064	33 17	138	32 13	183	30 45	213	62 50	249	22 52	277	18 47	350
118	53 21	065	33 35	139	32 11	184	30 30	214	62 24	250	22 25	278	18 42	351
119	53 46	066	33 53	140	32 09	185	30 15	215	61 59	252	21 59	279	18 38	352
	◆VEGA		ARCTURUS		◆REGULUS		PROCYON		◆CAPELLA		Schedar		DENEB	
120	13 08	017	15 20	078	34 10	141	32 07	186	61 33	253	41 07	316	18 34	352
121	13 16	017	15 46	079	34 27	142	32 03	187	61 07	254	40 49	317	18 31	353
122	13 24	018	16 13	079	34 43	143	32 00	189	60 40	255	40 30	318	18 28	354
123	13 33	019	16 40	080	34 59	145	31 55	190	60 14	256	40 12	318	18 25	355
124	13 42	020	17 07	081	35 15	146	31 50	191	59 47	257	39 54	319	18 23	355
125	13 51	020	17 34	082	35 30	147	31 45	192	59 21	258	39 36	320	18 21	356
126	14 01	021	18 01	083	35 45	148	31 39	193	58 54	259	39 18	320	18 19	357
127	14 11	022	18 28	084	35 59	149	31 33	194	58 27	260	39 01	321	18 18	358
128	14 21	023	18 55	085	36 13	150	31 26	196	58 00	261	38 44	321	18 17	358
129	14 32	024	19 22	086	36 26	152	31 18	197	57 34	262	38 27	322	18 16	359
130	14 43	024	19 49	086	36 38	153	31 10	198	57 06	263	38 10	323	18 16	000
131	14 55	025	20 16	087	36 51	154	31 01	199	56 39	264	37 54	323	18 16	001
132	15 06	026	20 43	088	37 02	155	30 52	200	56 12	265	37 38	324	18 16	001
133	15 18	027	21 11	089	37 14	156	30 43	201	55 45	266	37 22	324	18 17	002
134	15 31	028	21 38	090	37 24	158	30 33	202	55 18	267	37 06	325	18 18	003
	DENEB		◆VEGA		ARCTURUS		◆REGULUS		PROCYON		◆CAPELLA		Schedar	
135	18 20	003	15 44	028	22 05	091	37 34	159	30 22	204	54 51	268	36 50	326
136	18 21	004	15 57	029	22 32	092	37 44	160	30 11	205	54 23	269	36 35	326
137	18 24	005	16 10	030	23 00	093	37 53	161	29 59	206	53 56	270	36 20	327
138	18 26	006	16 24	031	23 27	094	38 01	162	29 47	207	53 29	271	36 05	327
139	18 29	006	16 38	031	23 54	094	38 09	164	29 35	208	53 02	272	35 51	328
140	18 32	007	16 52	032	24 21	095	38 17	165	29 22	209	52 35	272	35 37	329
141	18 36	008	17 07	033	24 48	096	38 23	166	29 08	210	52 07	273	35 22	329
142	18 40	009	17 22	034	25 15	097	38 30	167	28 54	211	51 40	274	35 09	330
143	18 44	009	17 37	034	25 42	098	38 35	169	28 40	212	51 13	275	34 55	330
144	18 49	010	17 53	035	26 09	099	38 40	170	28 25	214	50 46	276	34 42	331
145	18 54	011	18 08	036	26 36	100	38 45	171	28 10	215	50 19	277	34 29	332
146	18 59	012	18 25	037	27 03	101	38 49	172	27 54	216	49 52	277	34 16	332
147	19 05	012	18 41	038	27 30	102	38 52	174	27 38	217	49 25	278	34 04	333
148	19 11	013	18 58	038	27 56	103	38 55	175	27 21	218	48 58	279	33 51	334
149	19 17	014	19 15	039	28 23	104	38 57	176	27 04	219	48 31	280	33 39	334
	DENEB		◆VEGA		ARCTURUS		◆REGULUS		PROCYON		◆CAPELLA		Schedar	
150	19 24	015	19 32	040	28 49	105	38 58	177	26 47	220	48 04	281	33 28	335
151	19 31	015	19 50	041	29 15	106	38 59	179	26 29	221	47 37	282	33 16	335
152	19 38	016	20 08	041	29 42	107	39 00	180	26 11	222	47 11	283	33 05	336
153	19 46	017	20 26	042	30 08	108	38 59	181	25 53	223	46 44	283	32 54	337
154	19 54	017	20 44	043	30 34	108	38 58	182	25 34	224	46 18	284	32 43	337
155	20 02	018	21 03	044	30 59	109	38 57	184	25 15	225	45 51	285	32 33	338
156	20 11	019	21 22	044	31 25	110	38 55	185	24 55	226	45 25	285	32 23	339
157	20 20	020	21 41	045	31 50	111	38 52	186	24 35	227	44 59	286	32 13	339
158	20 29	020	22 01	046	32 16	112	38 49	188	24 15	228	44 33	287	32 03	340
159	20 39	021	22 20	047	32 41	113	38 45	189	23 54	229	44 07	288	31 54	340
160	20 49	022	22 40	048	33 06	114	38 41	190	23 33	230	43 41	288	31 45	341
161	20 59	023	23 00	048	33 30	115	38 36	191	23 12	231	43 15	289	31 36	342
162	21 10	023	23 21	049	33 55	116	38 30	193	22 51	233	42 49	290	31 28	342
163	21 21	024	23 42	050	34 19	118	38 24	194	22 29	234	42 24	291	31 20	343
164	21 32	025	24 03	051	34 43	119	38 17	195	22 07	235	41 58	291	31 12	344
	DENEB		◆VEGA		ARCTURUS		◆REGULUS		POLLUX		◆CAPELLA		Schedar	
165	21 44	026	24 24	051	35 07	120	38 10	196	43 05	245	41 33	292	31 04	344
166	21 55	026	24 45	052	35 30	121	38 02	197	42 40	246	41 08	293	30 57	345
167	22 08	027	25 07	053	35 54	122	37 53	199	42 15	247	40 43	294	30 50	345
168	22 20	028	25 28	054	36 17	123	37 44	200	41 50	248	40 18	294	30 43	346
169	22 33	028	25 51	054	36 39	124	37 35	201	41 24	250	39 53	295	30 37	347
170	22 46	029	26 13	055	37 02	125	37 25	202	40 59	251	39 29	296	30 31	347
171	23 00	030	26 35	056	37 24	126	37 14	204	40 33	252	39 04	297	30 25	348
172	23 13	031	26 58	057	37 46	127	37 03	205	40 07	253	38 40	297	30 20	349
173	23 27	031	27 21	057	38 08	128	36 51	206	39 41	254	38 16	298	30 14	349
174	23 42	032	27 44	058	38 29	129	36 39	207	39 15	255	37 52	299	30 09	350
175	23 56	033	28 07	059	38 50	130	36 26	208	38 49	256	37 28	299	30 05	350
176	24 11	034	28 30	060	39 10	132	36 13	210	38 22	257	37 05	300	30 00	351
177	24 26	034	28 54	061	39 30	133	36 00	211	37 56	257	36 41	301	29 56	352
178	24 42	035	29 18	061	39 50	134	35 45	212	37 29	258	36 18	302	29 52	352
179	24 57	036	29 42	062	40 10	135	35 31	213	37 02	259	35 55	302	29 49	353

LHA	Hc	Zn	Hc	Zn	Hc	Zn	Hc	Zn	Hc	Zn	Hc	Zn	Hc	Zn
♈	DENEB		◆VEGA		ARCTURUS		◆Denebola		REGULUS		POLLUX		◆CAPELLA	
180	25 13	036	30 06	063	40 29	136	41 34	184	35 16	214	36 35	260	35 32	303
181	25 30	037	30 30	064	40 47	137	41 32	185	35 00	215	36 08	261	35 09	304
182	25 46	038	30 55	064	41 06	139	41 29	186	34 44	216	35 42	262	34 46	304
183	26 03	039	31 19	065	41 23	140	41 26	187	34 28	218	35 14	263	34 24	305
184	26 20	039	31 44	066	41 41	141	41 22	189	34 11	219	34 47	264	34 02	306
185	26 38	040	32 09	067	41 58	142	41 17	190	33 54	220	34 20	265	33 40	307
186	26 55	041	32 34	067	42 14	143	41 12	191	33 36	221	33 53	266	33 18	307
187	27 13	041	32 59	068	42 30	145	41 07	193	33 18	222	33 26	267	32 57	308
188	27 31	042	33 25	069	42 46	146	41 00	194	33 00	223	32 59	268	32 35	309
189	27 50	043	33 50	070	43 01	147	40 54	195	32 41	224	32 32	269	32 14	309
190	28 08	044	34 16	071	43 15	148	40 46	196	32 21	225	32 04	270	31 53	310
191	28 27	044	34 42	071	43 29	150	40 38	198	32 02	227	31 37	270	31 33	311
192	28 46	045	35 08	072	43 43	151	40 30	199	31 42	228	31 10	271	31 12	312
193	29 06	046	35 34	073	43 56	152	40 20	200	31 22	229	30 43	272	30 52	312
194	29 25	046	36 00	074	44 08	153	40 11	201	31 01	230	30 15	273	30 32	313
	DENEB		◆VEGA		ARCTURUS		◆Denebola		REGULUS		POLLUX		◆CAPELLA	
195	29 45	047	36 26	075	44 20	155	40 00	203	30 40	231	29 48	274	30 12	314
196	30 05	048	36 52	075	44 31	156	39 50	204	30 19	232	29 21	275	29 52	314
197	30 26	049	37 19	076	44 42	157	39 38	205	29 57	233	28 54	276	29 33	315
198	30 46	049	37 45	077	44 52	159	39 27	206	29 35	234	28 27	277	29 14	316
199	31 07	050	38 12	078	45 02	160	39 14	208	29 13	235	28 00	277	28 55	317
200	31 28	051	38 38	079	45 11	161	39 01	209	28 51	236	27 33	278	28 36	317
201	31 49	051	39 05	079	45 20	163	38 48	210	28 28	237	27 06	279	28 18	318
202	32 10	052	39 32	080	45 27	164	38 34	211	28 05	238	26 39	280	28 00	319
203	32 32	053	39 59	081	45 35	165	38 20	212	27 42	239	26 12	281	27 42	319
204	32 54	054	40 26	082	45 41	167	38 05	214	27 18	240	25 46	282	27 24	320
205	33 16	054	40 53	083	45 47	168	37 49	215	26 54	241	25 19	283	27 07	321
206	33 38	055	41 20	084	45 53	169	37 34	216	26 30	242	24 52	283	26 50	322
207	34 01	056	41 47	084	45 57	171	37 17	217	26 06	243	24 26	284	26 33	322
208	34 23	056	42 14	085	46 01	172	37 01	218	25 42	244	24 00	285	26 17	323
209	34 46	057	42 41	086	46 05	173	36 44	220	25 17	245	23 33	286	26 00	324
	DENEB		◆VEGA		ARCTURUS		◆Denebola		REGULUS		POLLUX		◆CAPELLA	
210	35 09	058	43 08	087	46 08	175	36 26	221	24 52	246	23 07	287	25 44	324
211	35 32	059	43 36	088	46 10	176	36 08	222	24 27	247	22 41	288	25 29	325
212	35 55	059	44 03	089	46 11	177	35 50	223	24 02	248	22 15	288	25 13	326
213	36 19	060	44 30	090	46 12	179	35 31	224	23 37	249	21 49	289	24 58	326
214	36 43	061	44 57	091	46 12	180	35 12	225	23 11	250	21 24	290	24 43	327
215	37 06	061	45 25	091	46 12	182	34 52	226	22 46	251	20 58	291	24 28	328
216	37 30	062	45 52	092	46 11	183	34 32	227	22 20	252	20 33	292	24 14	329
217	37 55	063	46 19	093	46 09	184	34 12	229	21 54	253	20 08	293	24 00	329
218	38 19	064	46 46	094	46 07	186	33 52	230	21 28	254	19 43	293	23 46	330
219	38 43	064	47 13	095	46 04	187	33 31	231	21 02	255	19 18	294	23 33	331
220	39 08	065	47 40	096	46 00	188	33 09	232	20 35	256	18 53	295	23 20	332
221	39 33	066	48 07	097	45 56	190	32 48	233	20 09	257	18 28	296	23 07	332
222	39 58	067	48 34	098	45 51	191	32 26	234	19 42	257	18 04	297	22 54	333
223	40 23	067	49 01	099	45 46	192	32 04	235	19 16	258	17 40	298	22 42	334
224	40 48	068	49 28	100	45 39	194	31 41	236	18 49	259	17 16	298	22 30	334
	◆DENEB		VEGA		Rasalhague		◆ARCTURUS		Denebola		Dubhe		◆CAPELLA	
225	41 13	069	49 55	101	32 40	134	45 33	195	31 19	237	63 32	294	22 19	335
226	41 39	070	50 22	102	33 00	135	45 25	196	30 56	238	63 07	295	22 07	336
227	42 05	070	50 48	103	33 19	136	45 17	198	30 33	239	62 43	295	21 56	337
228	42 30	071	51 15	104	33 38	137	45 09	199	30 09	240	62 18	296	21 46	337
229	42 56	072	51 41	105	33 56	138	44 59	200	29 45	241	61 54	296	21 35	338
230	43 22	073	52 07	106	34 14	139	44 50	202	29 21	242	61 29	297	21 25	339
231	43 48	073	52 34	107	34 32	140	44 39	203	28 57	243	61 05	297	21 16	339
232	44 14	074	53 00	108	34 49	141	44 28	204	28 33	244	60 41	298	21 06	340
233	44 40	075	53 25	109	35 06	143	44 17	206	28 08	245	60 17	298	20 57	341
234	45 07	076	53 51	110	35 22	144	44 05	207	27 43	246	59 53	299	20 48	342
235	45 33	076	54 16	111	35 38	145	43 52	208	27 18	247	59 29	299	20 40	342
236	46 00	077	54 42	113	35 53	146	43 39	209	26 53	248	59 06	300	20 32	343
237	46 26	078	55 07	114	36 08	147	43 25	211	26 28	249	58 42	301	20 24	344
238	46 53	079	55 32	115	36 23	148	43 11	212	26 02	250	58 19	301	20 17	345
239	47 20	080	55 56	116	36 37	149	42 56	213	25 36	251	57 55	302	20 10	345
	Alpheratz		◆DENEB		VEGA		Rasalhague		◆ARCTURUS		Alioth		◆CAPELLA	
240	12 51	049	47 47	080	56 20	117	36 51	151	42 41	214	65 57	275	20 03	346
241	13 12	050	48 14	081	56 45	118	37 04	152	42 26	216	65 30	276	19 56	347
242	13 33	051	48 41	082	57 08	120	37 16	153	42 09	217	65 03	276	19 50	347
243	13 54	052	49 08	083	57 32	121	37 28	154	41 53	218	64 36	277	19 45	348
244	14 16	053	49 35	084	57 55	122	37 40	155	41 36	219	64 09	278	19 39	349
245	14 37	054	50 02	085	58 18	124	37 51	157	41 18	221	63 42	279	19 34	350
246	14 59	054	50 29	085	58 40	125	38 02	158	41 00	222	63 15	279	19 29	350
247	15 22	055	50 56	086	59 03	126	38 12	159	40 42	223	62 48	280	19 25	351
248	15 44	056	51 23	087	59 24	128	38 21	160	40 23	224	62 21	281	19 21	352
249	16 07	057	51 50	088	59 46	129	38 30	162	40 04	225	61 55	282	19 17	353
250	16 30	058	52 18	089	60 07	130	38 38	163	39 45	226	61 28	282	19 14	353
251	16 53	058	52 45	090	60 27	132	38 46	164	39 25	228	61 02	283	19 11	354
252	17 16	059	53 12	091	60 48	133	38 53	165	39 04	229	60 35	284	19 08	355
253	17 40	060	53 39	092	61 07	135	39 00	167	38 44	230	60 09	284	19 06	356
254	18 04	061	54 07	092	61 26	136	39 06	168	38 23	231	59 42	285	19 04	356
	Alpheratz		◆DENEB		ALTAIR		Rasalhague		◆ARCTURUS		Alioth		◆CAPELLA	
255	18 27	062	54 34	093	27 51	131	39 11	169	38 01	232	59 16	286	19 03	357
256	18 51	063	55 01	094	28 11	132	39 16	170	37 40	233	58 50	286	19 01	358
257	19 16	063	55 28	095	28 32	133	39 20	172	37 18	234	58 24	287	19 00	358
258	19 40	064	55 55	096	28 51	134	39 24	173	36 56	235	57 58	288	19 00	359
259	20 05	065	56 22	097	29 11	135	39 27	174	36 33	236	57 32	288	19 00	000
260	20 30	066	56 49	098	29 30	136	39 30	175	36 10	238	57 06	289	19 00	001
261	20 55	067	57 16	099	29 49	137	39 32	177	35 47	239	56 40	290	19 00	001
262	21 20	068	57 43	100	30 07	138	39 33	178	35 24	240	56 15	290	19 01	002
263	21 45	068	58 10	101	30 25	139	39 34	179	35 00	241	55 49	291	19 02	003
264	22 10	069	58 36	102	30 42	140	39 34	180	34 36	242	55 24	292	19 04	004
265	22 36	070	59 03	103	31 00	142	39 33	182	34 12	243	54 59	292	19 06	004
266	23 01	071	59 29	104	31 16	143	39 32	183	33 48	244	54 34	293	19 08	005
267	23 27	072	59 56	105	31 33	144	39 30	184	33 23	245	54 09	294	19 11	006
268	23 53	072	60 22	107	31 48	145	39 28	185	32 59	246	53 44	294	19 13	007
269	24 19	073	60 48	108	32 04	146	39 25	187	32 34	247	53 19	295	19 17	007

LAT 63°N

LHA	Hc	Zn	Hc	Zn	Hc	Zn	Hc	Zn	Hc	Zn	Hc	Zn	Hc	Zn
♈	◆CAPELLA		Alpheratz		◆DENEB		ALTAIR		Rasalhague		◆ARCTURUS		Alioth	
270	19 20	008	24 45	074	61 14	109	32 19	147	39 22	188	32 08	248	52 54	296
271	19 24	009	25 12	075	61 40	110	32 33	148	39 18	189	31 43	249	52 30	296
272	19 29	009	25 38	076	62 05	111	32 47	149	39 13	191	31 18	250	52 05	297
273	19 33	010	26 04	077	62 30	112	33 01	151	39 08	192	30 52	251	51 41	297
274	19 38	011	26 31	077	62 55	114	33 14	152	39 02	193	30 26	252	51 17	298
275	19 44	012	26 58	078	63 20	115	33 27	153	38 56	194	30 00	253	50 53	299
276	19 49	012	27 24	079	63 45	116	33 39	154	38 49	196	29 34	254	50 29	299
277	19 55	013	27 51	080	64 09	118	33 51	155	38 41	197	29 08	255	50 06	300
278	20 02	014	28 18	081	64 33	119	34 02	156	38 33	198	28 42	256	49 42	301
279	20 08	015	28 45	082	64 57	120	34 13	158	38 24	199	28 15	257	49 19	301
280	20 15	015	29 12	083	65 20	122	34 23	159	38 15	200	27 49	258	48 56	302
281	20 23	016	29 39	083	65 43	123	34 32	160	38 05	202	27 22	258	48 32	302
282	20 31	017	30 06	084	66 06	125	34 41	161	37 55	203	26 55	259	48 10	303
283	20 39	018	30 33	085	66 28	126	34 50	162	37 44	204	26 28	260	47 47	304
284	20 47	018	31 00	086	66 50	128	34 58	163	37 33	205	26 02	261	47 24	304
	◆CAPELLA		Alpheratz		Enif		◆ALTAIR		VEGA		◆ARCTURUS		Alioth	
285	20 56	019	31 27	087	29 21	132	35 05	165	65 32	191	25 35	262	47 02	305
286	21 05	020	31 55	088	29 41	133	35 12	166	65 26	193	25 08	263	46 40	305
287	21 14	020	32 22	089	30 01	134	35 19	167	65 19	195	24 41	264	46 17	306
288	21 24	021	32 49	090	30 20	135	35 25	168	65 12	197	24 13	265	45 55	307
289	21 34	022	33 16	090	30 39	136	35 30	170	65 04	198	23 46	266	45 34	307
290	21 44	023	33 44	091	30 57	138	35 34	171	64 55	200	23 19	267	45 12	308
291	21 55	023	34 11	092	31 16	139	35 39	172	64 45	202	22 52	268	44 51	309
292	22 06	024	34 38	093	31 33	140	35 42	173	64 35	204	22 25	269	44 30	309
293	22 17	025	35 05	094	31 51	141	35 45	174	64 23	205	21 57	269	44 09	310
294	22 28	025	35 32	095	32 08	142	35 47	176	64 11	207	21 30	270	43 48	310
295	22 40	026	36 00	096	32 24	143	35 49	177	63 58	209	21 03	271	43 27	311
296	22 52	027	36 27	097	32 41	144	35 51	178	63 45	211	20 36	272	43 07	312
297	23 05	028	36 54	098	32 56	145	35 51	179	63 31	212	20 08	273	42 46	312
298	23 18	028	37 21	099	33 12	146	35 51	180	63 16	214	19 41	274	42 26	313
299	23 31	029	37 47	100	33 26	148	35 51	182	63 00	216	19 14	275	42 06	313
	◆CAPELLA		Alpheratz		◆Enif		ALTAIR		VEGA		◆ARCTURUS		Alioth	
300	23 44	030	38 14	101	33 41	149	35 50	183	62 44	217	18 47	276	41 47	314
301	23 58	031	38 41	102	33 55	150	35 48	184	62 27	219	18 20	276	41 27	315
302	24 12	031	39 08	103	34 08	151	35 46	185	62 10	220	17 53	277	41 08	315
303	24 26	032	39 34	104	34 21	152	35 43	187	61 52	222	17 26	278	40 49	316
304	24 41	033	40 01	105	34 33	153	35 40	188	61 34	223	16 59	279	40 30	316
305	24 56	033	40 27	105	34 45	155	35 36	189	61 15	225	16 32	280	40 11	317
306	25 11	034	40 53	106	34 57	156	35 31	190	60 55	226	16 05	281	39 53	318
307	25 26	035	41 19	107	35 08	157	35 26	191	60 35	228	15 39	282	39 35	318
308	25 42	036	41 45	109	35 18	158	35 20	193	60 15	229	15 12	283	39 17	319
309	25 58	036	42 11	110	35 28	159	35 14	194	59 54	231	14 45	283	38 59	320
310	26 14	037	42 36	111	35 37	161	35 07	195	59 33	232	14 19	284	38 41	320
311	26 31	038	43 02	112	35 46	162	35 00	196	59 11	233	13 53	285	38 24	321
312	26 47	038	43 27	113	35 54	163	34 52	197	58 49	235	13 26	286	38 07	321
313	27 04	039	43 52	114	36 02	164	34 44	199	58 27	236	13 00	287	37 50	322
314	27 22	040	44 17	115	36 09	165	34 35	200	58 04	237	12 34	288	37 33	323
	CAPELLA		◆Hamal		Alpheratz		Enif		◆ALTAIR		VEGA		◆Alioth	
315	27 39	041	26 45	090	44 42	116	36 16	167	34 25	201	57 41	239	37 17	323
316	27 57	041	27 12	091	45 06	117	36 22	168	34 15	202	57 17	240	37 01	324
317	28 15	042	27 39	092	45 30	118	36 28	169	34 05	203	56 54	241	36 45	324
318	28 33	043	28 07	093	45 54	119	36 33	170	33 54	204	56 30	242	36 29	325
319	28 52	043	28 34	094	46 18	120	36 37	171	33 42	206	56 05	243	36 13	326
320	29 11	044	29 01	095	46 41	121	36 41	173	33 30	207	55 41	245	35 58	326
321	29 30	045	29 28	096	47 04	123	36 44	174	33 18	208	55 16	246	35 43	327
322	29 49	045	29 55	097	47 27	124	36 46	175	33 05	209	54 51	247	35 28	328
323	30 09	046	30 22	098	47 49	125	36 49	176	32 51	210	54 26	248	35 14	328
324	30 29	047	30 49	099	48 12	126	36 50	178	32 37	211	54 01	249	35 00	329
325	30 49	048	31 16	100	48 33	127	36 51	179	32 23	213	53 35	250	34 46	329
326	31 09	048	31 43	101	48 55	129	36 51	180	32 08	214	53 09	251	34 32	330
327	31 29	049	32 10	101	49 16	130	36 51	181	31 53	215	52 44	252	34 18	331
328	31 50	050	32 36	102	49 37	131	36 50	182	31 37	216	52 18	254	34 05	331
329	32 11	050	33 03	103	49 57	132	36 48	184	31 21	217	51 51	255	33 52	332
	CAPELLA		◆Hamal		Alpheratz		Enif		◆ALTAIR		VEGA		◆Alioth	
330	32 32	051	33 29	104	50 17	133	36 46	185	31 04	218	51 25	256	33 39	332
331	32 53	052	33 56	105	50 37	135	36 44	186	30 47	219	50 59	257	33 27	333
332	33 15	053	34 22	106	50 56	136	36 41	187	30 30	220	50 32	258	33 15	334
333	33 36	053	34 48	107	51 14	137	36 37	189	30 12	221	50 05	259	33 03	334
334	33 58	054	35 14	108	51 33	139	36 32	190	29 54	223	49 39	260	32 51	335
335	34 21	055	35 40	109	51 50	140	36 27	191	29 35	224	49 12	261	32 40	336
336	34 43	055	36 05	110	52 08	141	36 22	192	29 16	225	48 45	262	32 29	336
337	35 05	056	36 31	111	52 24	143	36 16	194	28 57	226	48 18	263	32 18	337
338	35 28	057	36 56	112	52 41	144	36 09	195	28 37	227	47 51	264	32 07	337
339	35 51	058	37 21	113	52 56	145	36 02	196	28 17	228	47 24	264	31 57	338
340	36 14	058	37 46	114	53 12	147	35 54	197	27 57	229	46 57	265	31 47	339
341	36 37	059	38 11	115	53 26	148	35 46	198	27 36	230	46 29	266	31 37	339
342	37 01	060	38 35	116	53 40	150	35 37	200	27 15	231	46 02	267	31 28	340
343	37 24	060	38 59	117	53 54	151	35 28	201	26 54	232	45 35	268	31 18	341
344	37 48	061	39 24	119	54 07	153	35 18	202	26 32	233	45 08	269	31 10	341
	CAPELLA		ALDEBARAN		◆Hamal		Alpheratz		◆ALTAIR		VEGA		◆Alioth	
345	38 12	062	17 25	088	39 47	120	54 19	154	26 10	234	44 41	270	31 01	342
346	38 36	063	17 52	089	40 11	121	54 31	155	25 48	235	44 13	271	30 53	343
347	39 00	063	18 19	089	40 34	122	54 42	157	25 25	236	43 46	272	30 45	343
348	39 25	064	18 46	090	40 57	123	54 52	158	25 03	237	43 19	273	30 37	344
349	39 49	065	19 14	091	41 20	124	55 02	160	24 40	238	42 52	273	30 29	344
350	40 14	065	19 41	092	41 42	125	55 11	161	24 16	239	42 24	274	30 22	345
351	40 39	066	20 08	093	42 04	126	55 19	163	23 53	240	41 57	275	30 15	346
352	41 04	067	20 35	094	42 26	127	55 27	164	23 29	241	41 30	276	30 09	346
353	41 29	068	21 02	095	42 48	129	55 34	166	23 05	242	41 03	277	30 02	347
354	41 54	068	21 30	096	43 09	130	55 40	167	22 41	243	40 36	278	29 56	348
355	42 20	069	21 57	097	43 30	131	55 45	169	22 17	244	40 09	279	29 51	348
356	42 45	070	22 24	098	43 50	132	55 50	171	21 52	245	39 42	279	29 45	349
357	43 11	071	22 51	099	44 10	133	55 54	172	21 27	246	39 15	280	29 40	349
358	43 37	071	23 18	099	44 30	134	55 58	174	21 02	247	38 49	281	29 35	350
359	44 03	072	23 44	100	44 49	136	56 00	175	20 37	248	38 22	282	29 31	351

LHA	Hc	Zn	Hc	Zn	Hc	Zn	Hc	Zn	Hc	Zn	Hc	Zn	Hc	Zn
♈	♦POLLUX		CAPELLA		Hamal		♦Alpheratz		DENEB		♦VEGA		Alioth	
0	13 24	054	44 11	072	45 51	136	57 02	177	57 15	263	37 42	283	28 27	351
1	13 47	055	44 37	073	46 11	137	57 03	178	56 47	264	37 14	284	28 23	352
2	14 10	056	45 04	074	46 30	139	57 04	180	56 19	265	36 47	285	28 20	353
3	14 34	057	45 31	074	46 48	140	57 03	182	55 51	266	36 20	286	28 16	353
4	14 58	058	45 59	075	47 06	141	57 02	183	55 22	267	35 53	287	28 13	354
5	15 21	059	46 26	076	47 23	142	57 00	185	54 54	267	35 26	287	28 10	355
6	15 46	059	46 53	077	47 40	144	56 58	186	54 26	268	34 59	288	28 08	355
7	16 10	060	47 21	077	47 57	145	56 54	188	53 58	269	34 32	289	28 06	356
8	16 34	061	47 48	078	48 12	146	56 50	190	53 30	270	34 06	290	28 04	357
9	16 59	062	48 16	079	48 28	148	56 45	191	53 02	271	33 39	291	28 02	357
10	17 24	063	48 43	080	48 42	149	56 39	193	52 34	272	33 13	291	28 01	358
11	17 49	063	49 11	080	48 57	150	56 32	194	52 05	273	32 47	292	28 00	358
12	18 15	064	49 39	081	49 10	152	56 25	196	51 37	274	32 21	293	28 00	359
13	18 40	065	50 07	082	49 23	153	56 17	197	51 09	274	31 55	294	27 59	000
14	19 06	066	50 35	083	49 36	154	56 08	199	50 41	275	31 29	294	27 59	000
	Alioth		♦POLLUX		CAPELLA		Hamal		♦Alpheratz		DENEB		♦VEGA	
15	28 00	001	19 31	067	51 03	084	49 48	156	55 59	201	50 13	276	31 04	295
16	28 00	002	19 57	068	51 31	084	49 59	157	55 48	202	49 45	277	30 38	296
17	28 01	002	20 23	068	51 59	085	50 09	159	55 37	204	49 17	278	30 13	297
18	28 02	003	20 50	069	52 27	086	50 19	160	55 26	205	48 49	279	29 48	297
19	28 04	004	21 16	070	52 55	087	50 29	161	55 14	207	48 21	279	29 23	298
20	28 06	004	21 43	071	53 23	088	50 37	163	55 01	208	47 54	280	28 58	299
21	28 08	005	22 09	072	53 51	089	50 45	164	54 47	210	47 26	281	28 34	300
22	28 11	005	22 36	072	54 19	090	50 52	166	54 33	211	46 58	282	28 09	300
23	28 13	006	23 03	073	54 48	090	50 59	167	54 18	212	46 31	282	27 45	301
24	28 17	007	23 30	074	55 16	091	51 05	169	54 03	214	46 03	283	27 21	302
25	28 20	007	23 57	075	55 44	092	51 10	170	53 47	215	45 36	284	26 57	303
26	28 24	008	24 24	076	56 12	093	51 15	172	53 30	217	45 09	285	26 34	303
27	28 28	009	24 52	077	56 40	094	51 18	173	53 13	218	44 42	286	26 10	304
28	28 32	009	25 19	077	57 08	095	51 21	175	52 55	219	44 14	286	25 47	305
29	28 37	010	25 47	078	57 36	096	51 24	176	52 37	221	43 47	287	25 24	306
	Alioth		♦POLLUX		ALDEBARAN		♦Hamal		Alpheratz		DENEB		♦VEGA	
30	28 42	010	26 14	079	36 57	131	51 25	177	52 19	222	43 21	288	25 01	306
31	28 47	011	26 42	080	37 18	132	51 26	179	52 00	223	42 54	289	24 39	307
32	28 53	012	27 10	081	37 38	133	51 26	180	51 40	225	42 27	289	24 17	308
33	28 58	012	27 38	082	37 59	134	51 26	182	51 20	226	42 01	290	23 55	309
34	29 05	013	28 06	082	38 19	136	51 24	183	50 59	227	41 34	291	23 33	309
35	29 11	014	28 34	083	38 38	137	51 22	185	50 38	229	41 08	291	23 11	310
36	29 18	014	29 02	084	38 57	138	51 20	186	50 17	230	40 42	292	22 50	311
37	29 25	015	29 30	085	39 16	139	51 16	188	49 55	231	40 16	293	22 28	312
38	29 32	015	29 58	086	39 34	140	51 12	189	49 33	232	39 50	294	22 08	312
39	29 40	016	30 26	087	39 52	141	51 07	191	49 11	234	39 24	294	21 47	313
40	29 48	017	30 54	088	40 09	143	51 02	192	48 48	235	38 59	295	21 27	314
41	29 56	017	31 22	089	40 26	144	50 55	194	48 25	236	38 33	296	21 06	315
42	30 05	018	31 50	089	40 42	145	50 48	195	48 01	237	38 08	297	20 47	316
43	30 14	019	32 18	090	40 58	146	50 41	196	47 38	238	37 43	297	20 27	316
44	30 23	019	32 47	091	41 14	148	50 33	198	47 13	239	37 18	298	20 08	317
	Alioth		♦POLLUX		ALDEBARAN		♦Hamal		Alpheratz		DENEB		♦VEGA	
45	30 32	020	33 15	092	41 28	149	50 24	199	46 49	241	36 53	299	19 49	318
46	30 42	020	33 43	093	41 43	150	50 14	201	46 24	242	36 28	299	19 30	319
47	30 52	021	34 11	094	41 57	151	50 04	202	45 59	243	36 04	300	19 11	319
48	31 02	022	34 39	095	42 10	153	49 53	204	45 34	244	35 40	301	18 53	320
49	31 13	022	35 07	096	42 23	154	49 41	205	45 09	245	35 16	302	18 35	321
50	31 24	023	35 35	097	42 35	155	49 29	206	44 43	246	34 52	302	18 17	322
51	31 35	024	36 03	098	42 46	156	49 16	208	44 17	247	34 28	303	18 00	322
52	31 46	024	36 31	098	42 57	158	49 03	209	43 51	248	34 04	304	17 43	323
53	31 58	025	36 59	099	43 08	159	48 49	210	43 25	249	33 41	304	17 26	324
54	32 10	025	37 27	100	43 18	160	48 34	212	42 59	250	33 18	305	17 10	325
55	32 22	026	37 54	101	43 27	161	48 19	213	42 32	251	32 55	306	16 54	325
56	32 35	027	38 22	102	43 36	163	48 04	214	42 05	252	32 32	307	16 38	326
57	32 47	027	38 49	103	43 44	164	47 47	216	41 38	253	32 10	307	16 22	327
58	33 00	028	39 17	104	43 51	165	47 31	217	41 11	254	31 47	308	16 07	328
59	33 14	029	39 44	105	43 58	167	47 14	218	40 44	255	31 25	309	15 52	328
	Alioth		♦POLLUX		ALDEBARAN		♦Hamal		Alpheratz		DENEB		♦VEGA	
60	33 27	029	40 11	106	44 04	168	46 56	220	40 17	256	31 03	309	15 38	329
61	33 41	030	40 38	107	44 09	169	46 38	221	39 50	257	30 42	310	15 23	330
62	33 55	030	41 05	108	44 14	171	46 19	222	39 22	258	30 20	311	15 09	331
63	34 10	031	41 32	109	44 18	172	46 00	223	38 54	259	29 59	311	14 56	332
64	34 24	032	41 58	110	44 22	173	45 40	225	38 27	260	29 38	312	14 43	332
65	34 39	032	42 24	111	44 25	175	45 21	226	37 59	261	29 17	313	14 30	333
66	34 54	033	42 51	112	44 27	176	45 00	227	37 31	262	28 57	314	14 17	334
67	35 10	033	43 17	113	44 29	177	44 39	228	37 03	263	28 36	314	14 05	335
68	35 25	034	43 42	114	44 30	179	44 18	229	36 35	264	28 16	315	13 53	335
69	35 41	035	44 08	115	44 30	180	43 57	231	36 07	265	27 57	316	13 41	336
70	35 57	035	44 33	117	44 30	181	43 35	232	35 39	266	27 37	316	13 30	337
71	36 14	036	44 58	118	44 29	183	43 12	233	35 11	267	27 18	317	13 19	338
72	36 30	036	45 23	119	44 27	184	42 50	234	34 43	267	26 59	318	13 09	339
73	36 47	037	45 48	120	44 24	185	42 27	235	34 15	268	26 40	318	12 59	339
74	37 04	038	46 12	121	44 21	187	42 04	236	33 47	269	26 21	319	12 49	340
	Alioth		♦REGULUS		PROCYON		♦ALDEBARAN		Hamal		Alpheratz		♦DENEB	
75	37 21	038	16 39	096	26 06	135	44 18	188	41 40	237	33 18	270	26 03	320
76	37 39	039	17 07	097	26 26	136	44 13	190	41 16	238	32 50	271	25 45	321
77	37 57	039	17 35	098	26 45	137	44 08	191	40 52	240	32 22	272	25 27	321
78	38 15	040	18 03	098	27 04	138	44 03	192	40 28	241	31 54	273	25 10	322
79	38 33	041	18 31	099	27 23	139	43 57	193	40 03	242	31 26	274	24 53	323
80	38 51	041	18 59	100	27 41	140	43 50	195	39 38	243	30 58	274	24 36	323
81	39 10	042	19 26	101	27 59	141	43 42	196	39 13	244	30 30	275	24 19	324
82	39 29	042	19 54	102	28 17	142	43 34	197	38 47	245	30 02	276	24 03	325
83	39 48	043	20 21	103	28 34	143	43 25	199	38 22	246	29 34	277	23 47	326
84	40 07	044	20 49	104	28 50	144	43 16	200	37 56	247	29 06	278	23 31	326
85	40 27	044	21 16	105	29 06	146	43 06	201	37 30	248	28 38	279	23 15	327
86	40 47	045	21 43	106	29 22	147	42 55	203	37 04	249	28 10	280	23 00	328
87	41 07	045	22 10	107	29 37	148	42 44	204	36 37	250	27 42	280	22 45	328
88	41 27	046	22 37	108	29 52	149	42 33	205	36 11	251	27 15	281	22 31	329
89	41 47	047	23 04	109	30 07	150	42 20	206	35 44	252	26 47	282	22 16	330

LHA	Hc	Zn	Hc	Zn	Hc	Zn	Hc	Zn	Hc	Zn	Hc	Zn	Hc	Zn
♈	Alioth		♦REGULUS		PROCYON		BETELGEUSE		♦ALDEBARAN		Hamal		♦DENEB	
90	42 08	047	23 30	110	30 20	151	35 24	182	42 07	208	35 17	253	22 02	331
91	42 28	048	23 57	111	30 34	152	35 23	183	41 54	209	34 50	254	21 49	331
92	42 49	048	24 23	112	30 47	153	35 21	184	41 40	210	34 23	255	21 35	332
93	43 11	049	24 49	112	30 59	154	35 19	185	41 26	211	33 56	256	21 22	333
94	43 32	050	25 15	113	31 11	156	35 16	186	41 11	213	33 29	257	21 09	333
95	43 53	050	25 41	114	31 22	157	35 13	188	40 55	214	33 01	258	20 57	334
96	44 15	051	26 07	115	31 33	158	35 09	189	40 39	215	32 33	259	20 45	335
97	44 37	051	26 32	116	31 43	159	35 04	190	40 23	216	32 06	260	20 33	336
98	44 59	052	26 57	117	31 53	160	34 59	191	40 06	218	31 38	261	20 22	336
99	45 21	053	27 22	118	32 03	161	34 53	192	39 49	219	31 10	262	20 10	337
100	45 44	053	27 47	119	32 11	163	34 47	194	39 31	220	30 42	262	20 00	338
101	46 06	054	28 11	120	32 19	164	34 40	195	39 12	221	30 14	263	19 49	339
102	46 29	054	28 35	121	32 27	165	34 32	196	38 54	222	29 46	264	19 39	339
103	46 52	055	28 59	122	32 34	166	34 24	197	38 35	223	29 18	265	19 29	340
104	47 15	056	29 23	123	32 41	167	34 16	198	38 15	225	28 50	266	19 20	341
	Alioth		♦REGULUS		PROCYON		BETELGEUSE		♦CAPELLA		Hamal		♦DENEB	
105	47 39	056	29 46	124	32 47	168	34 06	200	68 11	235	28 22	267	19 11	341
106	48 02	057	30 09	126	32 52	170	33 57	201	67 48	236	27 54	268	19 02	342
107	48 26	057	30 32	127	32 57	171	33 46	202	67 24	238	27 26	269	18 53	343
108	48 50	058	30 54	128	33 01	172	33 36	203	67 00	239	26 58	270	18 45	344
109	49 13	058	31 17	129	33 05	173	33 24	204	66 36	241	26 30	271	18 37	344
110	49 38	059	31 38	130	33 08	174	33 12	205	66 11	242	26 01	271	18 30	345
111	50 02	060	32 00	131	33 10	176	33 00	207	65 46	244	25 33	272	18 23	346
112	50 26	060	32 21	132	33 12	177	32 47	208	65 21	245	25 05	273	18 16	347
113	50 51	061	32 42	133	33 13	178	32 34	209	64 55	246	24 37	274	18 10	347
114	51 15	061	33 02	134	33 14	179	32 20	210	64 29	247	24 09	275	18 04	348
115	51 40	062	33 22	135	33 14	180	32 05	211	64 03	249	23 41	276	17 58	349
116	52 05	063	33 42	136	33 14	181	31 51	212	63 37	250	23 13	277	17 53	349
117	52 30	063	34 01	137	33 13	183	31 35	213	63 10	251	22 45	277	17 48	350
118	52 56	064	34 20	138	33 11	184	31 20	215	62 44	252	22 17	278	17 43	351
119	53 21	065	34 39	140	33 09	185	31 03	216	62 17	253	21 49	279	17 39	352
	♦VEGA		ARCTURUS		♦REGULUS		PROCYON		♦CAPELLA		Schedar		DENEB	
120	12 10	016	15 07	077	34 57	141	33 06	186	61 50	255	40 23	317	17 35	352
121	12 19	017	15 34	078	35 14	142	33 03	187	61 22	256	40 04	318	17 31	353
122	12 27	018	16 02	079	35 31	143	32 59	189	60 55	257	39 46	318	17 28	354
123	12 36	019	16 29	080	35 48	144	32 54	190	60 28	258	39 27	319	17 25	355
124	12 45	020	16 57	081	36 04	145	32 49	191	60 00	259	39 08	319	17 23	355
125	12 55	020	17 25	082	36 20	147	32 44	192	59 32	260	38 50	320	17 21	356
126	13 05	021	17 53	083	36 36	148	32 38	193	59 05	261	38 32	321	17 19	357
127	13 15	022	18 21	083	36 50	149	32 31	194	58 37	262	38 14	321	17 18	358
128	13 26	023	18 49	084	37 05	150	32 23	196	58 09	263	37 57	322	17 17	358
129	13 37	023	19 17	085	37 18	151	32 16	197	57 41	264	37 40	322	17 16	359
130	13 48	024	19 45	086	37 32	152	32 07	198	57 13	265	37 23	323	17 16	000
131	14 00	025	20 13	087	37 44	154	31 58	199	56 45	266	37 06	324	17 16	001
132	14 12	026	20 41	088	37 57	155	31 49	200	56 17	267	36 49	324	17 16	001
133	14 25	027	21 09	089	38 08	156	31 39	201	55 48	268	36 33	325	17 17	002
134	14 38	027	21 38	090	38 20	157	31 28	203	55 20	269	36 17	325	17 18	003
	DENEB		♦VEGA		ARCTURUS		♦REGULUS		PROCYON		♦CAPELLA		Schedar	
135	17 20	003	14 51	028	22 06	090	38 30	159	31 17	204	54 52	269	36 01	326
136	17 22	004	15 04	029	22 34	091	38 40	160	31 05	205	54 24	270	35 45	327
137	17 24	005	15 18	030	23 02	092	38 50	161	30 53	206	53 56	271	35 30	327
138	17 26	006	15 32	030	23 30	093	38 59	162	30 41	207	53 28	272	35 15	328
139	17 29	006	15 46	031	23 58	094	39 07	163	30 27	208	52 59	273	35 00	328
140	17 33	007	16 01	032	24 26	095	39 15	165	30 14	209	52 31	274	34 45	329
141	17 36	008	16 16	033	24 54	096	39 22	166	30 00	211	52 03	275	34 31	330
142	17 40	009	16 32	034	25 22	097	39 28	167	29 45	212	51 35	275	34 17	330
143	17 45	009	16 47	034	25 50	098	39 34	169	29 30	213	51 07	276	34 03	331
144	17 50	010	17 04	035	26 18	099	39 39	170	29 15	214	50 39	277	33 49	331
145	17 55	011	17 20	036	26 46	099	39 44	171	28 59	215	50 11	278	33 36	332
146	18 00	012	17 36	037	27 14	100	39 48	172	28 42	216	49 43	279	33 23	333
147	18 06	012	17 53	037	27 42	101	39 52	174	28 26	217	49 16	279	33 10	333
148	18 12	013	18 11	038	28 09	102	39 54	175	28 08	218	48 48	280	32 57	334
149	18 19	014	18 28	039	28 37	103	39 57	176	27 51	219	48 20	281	32 45	334
	DENEB		♦VEGA		ARCTURUS		♦REGULUS		PROCYON		♦CAPELLA		Schedar	
150	18 25	014	18 46	040	29 04	104	39 58	177	27 33	220	47 53	282	32 33	335
151	18 33	015	19 04	040	29 31	105	39 59	179	27 14	221	47 25	283	32 21	336
152	18 40	016	19 23	041	29 58	106	40 00	180	26 55	223	46 58	283	32 10	336
153	18 48	017	19 41	042	30 25	107	39 59	181	26 36	224	46 30	284	31 59	337
154	18 56	017	20 00	043	30 52	108	39 58	183	26 17	225	46 03	285	31 48	338
155	19 05	018	20 19	043	31 19	109	39 57	184	25 57	226	45 36	286	31 37	338
156	19 14	019	20 39	044	31 46	110	39 55	185	25 36	227	45 09	286	31 27	339
157	19 23	020	20 59	045	32 12	111	39 52	186	25 16	228	44 42	287	31 17	339
158	19 33	020	21 19	046	32 38	112	39 48	188	24 55	229	44 15	288	31 07	340
159	19 43	021	21 39	046	33 04	113	39 44	189	24 33	230	43 48	289	30 58	341
160	19 53	022	22 00	047	33 30	114	39 40	190	24 12	231	43 21	289	30 48	341
161	20 04	022	22 20	048	33 56	115	39 34	191	23 50	232	42 55	290	30 39	342
162	20 15	023	22 41	049	34 21	116	39 28	193	23 27	233	42 28	291	30 31	342
163	20 26	024	23 03	049	34 46	117	39 22	194	23 05	234	42 02	292	30 23	343
164	20 37	025	23 24	050	35 11	118	39 15	195	22 42	235	41 36	292	30 14	344
	DENEB		♦VEGA		ARCTURUS		♦REGULUS		POLLUX		♦CAPELLA		Schedar	
165	20 49	025	23 46	051	35 36	119	39 07	196	43 30	246	41 10	293	30 07	344
166	21 02	026	24 08	052	36 01	120	38 59	198	43 04	247	40 44	294	29 59	345
167	21 14	027	24 30	052	36 25	121	38 50	199	42 38	248	40 19	294	29 52	346
168	21 27	028	24 53	053	36 49	122	38 41	200	42 11	249	39 53	295	29 45	346
169	21 40	028	25 15	054	37 13	123	38 31	201	41 45	250	39 28	296	29 39	347
170	21 54	029	25 38	055	37 36	124	38 20	203	41 18	251	39 02	297	29 32	347
171	22 07	030	26 01	056	37 59	125	38 09	204	40 52	252	38 37	297	29 26	348
172	22 22	030	26 25	056	38 22	127	37 57	205	40 25	253	38 12	298	29 21	349
173	22 36	031	26 48	057	38 44	128	37 45	206	39 58	254	37 47	299	29 15	349
174	22 51	032	27 12	058	39 07	129	37 32	207	39 30	255	37 23	299	29 10	350
175	23 06	033	27 36	059	39 28	130	37 19	209	39 03	256	36 58	300	29 05	351
176	23 21	033	28 00	059	39 50	131	37 05	210	38 36	257	36 34	301	29 01	351
177	23 37	034	28 24	060	40 11	132	36 51	211	38 08	258	36 10	302	28 57	352
178	23 52	035	28 49	061	40 32	133	36 36	212	37 41	259	35 46	302	28 53	352
179	24 09	035	29 14	062	40 52	134	36 21	213	37 13	260	35 22	303	28 49	353

LAT 62°N

LHA ♈	Hc	Zn	Hc	Zn	Hc	Zn	Hc	Zn	Hc	Zn	Hc	Zn	Hc	Zn
	DENEB		♦VEGA		ARCTURUS		♦Denebola		REGULUS		POLLUX		♦CAPELLA	
180	24 25	036	29 38	062	41 12	136	42 33	184	36 05	215	36 45	261	34 59	304
181	24 42	037	30 03	063	41 31	137	42 31	185	35 49	216	36 17	262	34 35	304
182	24 59	038	30 29	064	41 50	138	42 29	186	35 32	217	35 49	263	34 12	305
183	25 16	038	30 54	065	42 09	139	42 25	188	35 15	218	35 21	264	33 49	306
184	25 34	039	31 19	065	42 27	140	42 21	189	34 58	219	34 53	265	33 27	306
185	25 52	040	31 45	066	42 45	142	42 16	190	34 40	220	34 25	266	33 04	307
186	26 10	040	32 11	067	43 02	143	42 11	192	34 21	221	33 57	267	32 42	308
187	26 28	041	32 37	068	43 19	144	42 05	193	34 02	223	33 29	268	32 19	309
188	26 47	042	33 03	068	43 35	145	41 59	194	33 43	224	33 01	268	31 58	309
189	27 06	042	33 29	069	43 51	147	41 51	195	33 23	225	32 33	269	31 36	310
190	27 25	043	33 56	070	44 06	148	41 44	197	33 03	226	32 04	270	31 14	311
191	27 44	044	34 22	071	44 21	149	41 35	198	32 43	227	31 36	271	30 53	311
192	28 04	045	34 49	072	44 35	150	41 26	199	32 22	228	31 08	272	30 32	312
193	28 24	045	35 16	072	44 49	152	41 17	201	32 01	229	30 40	273	30 11	313
194	28 44	046	35 43	073	45 02	153	41 07	202	31 40	230	30 12	274	29 51	313
	DENEB		♦VEGA		ARCTURUS		♦Denebola		REGULUS		POLLUX		♦CAPELLA	
195	29 04	047	36 10	074	45 14	154	40 56	203	31 18	231	29 44	275	29 30	314
196	29 25	047	36 37	075	45 26	156	40 44	204	30 56	232	29 16	275	29 10	315
197	29 46	048	37 04	075	45 37	157	40 33	206	30 33	233	28 48	276	28 50	316
198	30 07	049	37 31	076	45 48	158	40 20	207	30 10	234	28 20	277	28 31	316
199	30 28	050	37 59	077	45 58	160	40 07	208	29 47	235	27 52	278	28 11	317
200	30 50	050	38 26	078	46 08	161	39 54	209	29 24	237	27 24	279	27 52	318
201	31 11	051	38 54	079	46 17	162	39 40	210	29 00	238	26 56	280	27 33	318
202	31 33	052	39 21	079	46 25	164	39 25	212	28 36	239	26 28	280	27 15	319
203	31 56	052	39 49	080	46 33	165	39 10	213	28 12	240	26 01	281	26 56	320
204	32 18	053	40 17	081	46 40	166	38 55	214	27 48	241	25 33	282	26 38	320
205	32 41	054	40 45	082	46 46	168	38 38	215	27 23	242	25 06	283	26 20	321
206	33 03	054	41 13	083	46 51	169	38 22	216	26 58	243	24 38	284	26 03	322
207	33 26	055	41 41	084	46 56	170	38 05	218	26 33	244	24 11	285	25 46	323
208	33 50	056	42 09	084	47 01	172	37 48	219	26 08	245	23 44	285	25 29	323
209	34 13	057	42 37	085	47 04	173	37 30	220	25 42	246	23 17	286	25 12	324
	DENEB		♦VEGA		ARCTURUS		♦Denebola		REGULUS		POLLUX		♦CAPELLA	
210	34 37	057	43 05	086	47 07	175	37 11	221	25 17	247	22 50	287	24 55	325
211	35 01	058	43 33	087	47 10	176	36 53	222	24 51	247	22 23	288	24 39	325
212	35 25	059	44 01	088	47 11	177	36 33	223	24 25	248	21 56	289	24 23	326
213	35 49	059	44 29	089	47 12	179	36 14	225	23 58	249	21 29	290	24 08	327
214	36 13	060	44 57	090	47 12	180	35 54	226	23 32	250	21 03	290	23 53	327
215	36 38	061	45 26	090	47 12	182	35 34	227	23 05	251	20 37	291	23 38	328
216	37 02	062	45 54	091	47 11	183	35 13	228	22 38	252	20 10	292	23 23	329
217	37 27	062	46 22	092	47 09	184	34 52	229	22 12	253	19 44	293	23 08	330
218	37 52	063	46 50	093	47 07	186	34 30	230	21 45	254	19 19	294	22 54	330
219	38 17	064	47 18	094	47 03	187	34 09	231	21 17	255	18 53	295	22 41	331
220	38 43	064	47 46	095	47 00	189	33 46	232	20 50	256	18 27	295	22 27	332
221	39 08	065	48 14	096	46 55	190	33 24	233	20 23	257	18 02	296	22 14	332
222	39 34	066	48 42	097	46 50	191	33 01	234	19 55	258	17 37	297	22 01	333
223	39 59	067	49 10	098	46 44	193	32 38	235	19 28	259	17 12	298	21 48	334
224	40 25	067	49 38	099	46 38	194	32 15	237	19 00	260	16 47	299	21 36	335
	♦DENEB		VEGA		Rasalhague		♦ARCTURUS		Denebola		Dubhe		♦CAPELLA	
225	40 51	068	50 06	100	33 21	133	46 30	195	31 51	238	63 06	296	21 24	335
226	41 18	069	50 34	101	33 42	134	46 23	197	31 27	239	62 41	297	21 13	336
227	41 44	070	51 01	102	34 02	135	46 14	198	31 03	240	62 16	297	21 01	337
228	42 10	070	51 29	103	34 21	136	46 05	199	30 39	241	61 51	298	20 50	337
229	42 37	071	51 56	104	34 40	138	45 56	201	30 14	242	61 26	298	20 40	338
230	43 04	072	52 23	105	34 59	139	45 45	202	29 49	243	61 01	299	20 29	339
231	43 30	072	52 51	106	35 18	140	45 34	203	29 24	244	60 37	299	20 19	340
232	43 57	073	53 18	107	35 36	141	45 23	205	28 59	245	60 12	300	20 10	340
233	44 24	074	53 44	108	35 53	142	45 11	206	28 33	246	59 48	300	20 00	341
234	44 52	075	54 11	109	36 10	143	44 58	207	28 07	247	59 23	300	19 51	342
235	45 19	076	54 38	110	36 27	144	44 45	209	27 41	248	58 59	301	19 43	342
236	45 46	076	55 04	111	36 43	146	44 31	210	27 15	249	58 35	301	19 34	343
237	46 13	077	55 30	112	36 59	147	44 17	211	26 49	250	58 11	302	19 26	344
238	46 41	078	55 56	114	37 14	148	44 02	212	26 22	251	57 47	302	19 19	345
239	47 09	079	56 22	115	37 29	149	43 46	214	25 56	251	57 23	303	19 12	345
	Alpheratz		♦DENEB		VEGA		Rasalhague		♦ARCTURUS		Alioth		♦CAPELLA	
240	12 12	049	47 36	079	56 47	116	37 43	150	43 31	215	65 51	277	19 05	346
241	12 33	050	48 04	080	57 13	117	37 57	152	43 14	216	65 23	278	18 58	347
242	12 55	051	48 32	081	57 37	118	38 10	153	42 57	217	64 55	278	18 52	348
243	13 17	052	49 00	082	58 02	120	38 22	154	42 40	219	64 28	279	18 46	348
244	13 39	053	49 27	083	58 26	121	38 34	155	42 22	220	64 00	280	18 40	349
245	14 02	053	49 55	083	58 50	122	38 46	156	42 04	221	63 32	281	18 35	350
246	14 24	054	50 23	084	59 14	123	38 57	158	41 45	222	63 04	281	18 30	350
247	14 47	055	50 51	085	59 37	125	39 08	159	41 26	224	62 37	282	18 26	351
248	15 11	056	51 20	086	60 00	126	39 17	160	41 06	225	62 09	283	18 22	352
249	15 34	057	51 48	087	60 23	128	39 27	161	40 46	226	61 42	283	18 18	353
250	15 58	057	52 16	088	60 45	129	39 35	163	40 26	227	61 14	284	18 14	353
251	16 21	058	52 44	088	61 07	130	39 44	164	40 05	228	60 47	285	18 11	354
252	16 46	059	53 12	089	61 28	132	39 51	165	39 44	229	60 20	285	18 09	355
253	17 10	060	53 40	090	61 49	133	39 58	166	39 22	230	59 53	286	18 06	356
254	17 34	061	54 08	091	62 09	135	40 04	168	39 00	232	59 26	287	18 04	356
	Alpheratz		♦DENEB		ALTAIR		Rasalhague		♦ARCTURUS		Alioth		♦CAPELLA	
255	17 59	061	54 37	092	28 30	130	40 10	169	38 38	233	58 59	287	18 03	357
256	18 24	062	55 05	093	28 51	131	40 15	170	38 16	234	58 32	288	18 01	358
257	18 49	063	55 33	094	29 12	133	40 20	171	37 53	235	58 05	289	18 00	358
258	19 14	064	56 01	095	29 33	134	40 24	173	37 29	236	57 39	289	18 00	359
259	19 39	065	56 29	096	29 53	135	40 27	174	37 06	237	57 12	290	18 00	000
260	20 05	066	56 57	097	30 13	136	40 30	175	36 42	238	56 46	291	18 00	001
261	20 31	066	57 25	098	30 32	137	40 32	177	36 18	239	56 19	291	18 00	001
262	20 57	067	57 53	099	30 52	138	40 33	178	35 54	240	55 53	292	18 01	002
263	21 23	068	58 21	100	31 10	139	40 34	179	35 29	241	55 27	292	18 02	003
264	21 49	069	58 48	101	31 29	140	40 34	180	35 04	242	55 01	293	18 04	004
265	22 15	070	59 16	102	31 46	141	40 33	182	34 39	243	54 35	294	18 06	004
266	22 42	070	59 44	103	32 04	142	40 32	183	34 14	244	54 10	294	18 08	005
267	23 08	071	60 11	104	32 21	143	40 30	184	33 49	245	53 44	295	18 11	006
268	23 35	072	60 38	105	32 37	145	40 28	186	33 23	246	53 18	296	18 14	007
269	24 02	073	61 05	106	32 54	146	40 25	187	32 57	247	52 53	296	18 17	007

LHA ♈	Hc	Zn	Hc	Zn	Hc	Zn	Hc	Zn	Hc	Zn	Hc	Zn	Hc	Zn
	♦CAPELLA		Alpheratz		♦DENEB		ALTAIR		Rasalhague		♦ARCTURUS		Alioth	
270	18 21	008	24 29	074	61 32	107	33 09	147	40 21	188	32 31	248	52 28	297
271	18 25	009	24 56	075	61 59	108	33 24	148	40 17	189	32 05	249	52 03	297
272	18 29	009	25 23	075	62 26	109	33 39	149	40 12	191	31 38	250	51 38	298
273	18 34	010	25 50	076	62 52	111	33 53	150	40 06	192	31 11	251	51 13	299
274	18 39	011	26 18	077	63 19	112	34 07	151	40 00	193	30 45	252	50 48	299
275	18 45	012	26 45	078	63 45	113	34 20	153	39 54	194	30 18	253	50 24	300
276	18 51	012	27 13	079	64 10	114	34 33	154	39 46	196	29 51	254	49 59	300
277	18 57	013	27 40	080	64 36	116	34 45	155	39 38	197	29 24	255	49 35	301
278	19 03	014	28 08	080	65 01	117	34 57	156	39 30	198	28 56	256	49 11	302
279	19 10	015	28 36	081	65 26	118	35 08	157	39 21	199	28 29	257	48 47	302
280	19 18	015	29 04	082	65 51	120	35 19	158	39 11	201	28 01	258	48 23	303
281	19 25	016	29 32	083	66 15	121	35 29	160	39 01	202	27 34	259	48 00	303
282	19 33	017	30 00	084	66 39	123	35 38	161	38 50	203	27 06	260	47 36	304
283	19 41	017	30 28	085	67 02	124	35 47	162	38 39	204	26 38	261	47 13	305
284	19 50	018	30 56	085	67 26	126	35 55	163	38 27	206	26 10	262	46 50	305
	♦CAPELLA		Alpheratz		Enif		♦ALTAIR		VEGA		♦ARCTURUS		Alioth	
285	19 59	019	31 24	086	30 01	132	36 03	164	66 30	191	25 43	263	46 27	306
286	20 08	020	31 52	087	30 22	133	36 11	166	66 24	193	25 15	264	46 04	306
287	20 18	020	32 20	088	30 42	134	36 17	167	66 17	195	24 47	264	45 42	307
288	20 28	021	32 48	089	31 02	135	36 23	168	66 10	197	24 18	265	45 19	308
289	20 38	022	33 17	090	31 22	136	36 29	169	66 01	199	23 50	266	44 57	308
290	20 49	022	33 45	091	31 42	137	36 34	171	65 51	201	23 22	267	44 35	309
291	20 59	023	34 13	092	32 00	138	36 38	172	65 41	203	22 54	268	44 13	309
292	21 11	024	34 41	092	32 19	139	36 42	173	65 29	205	22 26	269	43 51	310
293	21 22	025	35 09	093	32 37	140	36 45	174	65 17	206	21 58	270	43 30	310
294	21 34	025	35 37	094	32 55	142	36 47	176	65 04	208	21 30	271	43 09	311
295	21 46	026	36 05	095	33 12	143	36 49	177	64 51	210	21 01	272	42 47	312
296	21 59	027	36 33	096	33 29	144	36 50	178	64 36	212	20 33	272	42 26	312
297	22 12	027	37 01	097	33 45	145	36 51	179	64 21	213	20 05	273	42 06	313
298	22 25	028	37 29	098	34 01	146	36 51	180	64 05	215	19 37	274	41 45	313
299	22 38	029	37 57	099	34 17	147	36 51	182	63 49	217	19 09	275	41 25	314
	♦CAPELLA		Alpheratz		♦Enif		ALTAIR		VEGA		♦ARCTURUS		Alioth	
300	22 52	030	38 25	100	34 32	148	36 50	183	63 31	218	18 41	276	41 05	315
301	23 06	030	38 53	101	34 46	150	36 48	184	63 14	220	18 13	277	40 45	315
302	23 21	031	39 20	102	35 00	151	36 46	185	62 55	222	17 45	278	40 25	316
303	23 35	032	39 48	103	35 14	152	36 43	187	62 36	223	17 17	279	40 06	316
304	23 50	032	40 15	104	35 27	153	36 39	188	62 17	225	16 49	279	39 46	317
305	24 05	033	40 43	105	35 39	154	36 35	189	61 57	226	16 22	280	39 27	318
306	24 21	034	41 10	106	35 51	155	36 30	190	61 36	228	15 54	281	39 08	318
307	24 37	035	41 37	107	36 03	157	36 25	192	61 15	229	15 26	282	38 50	319
308	24 53	035	42 04	108	36 14	158	36 19	193	60 53	231	14 59	283	38 31	319
309	25 09	036	42 30	109	36 24	159	36 12	194	60 31	232	14 31	284	38 13	320
310	25 26	037	42 57	110	36 34	160	36 05	195	60 09	233	14 04	285	37 55	321
311	25 43	037	43 23	111	36 43	161	35 58	196	59 46	235	13 37	285	37 37	321
312	26 00	038	43 50	112	36 52	163	35 49	198	59 23	236	13 10	286	37 20	322
313	26 18	039	44 16	113	37 00	164	35 41	199	59 00	237	12 43	287	37 02	322
314	26 35	039	44 42	114	37 07	165	35 31	200	58 36	239	12 16	288	36 45	323
	CAPELLA		♦Hamal		Alpheratz		Enif		♦ALTAIR		VEGA		♦Alioth	
315	26 53	040	26 45	090	45 07	115	37 14	166	35 21	201	58 11	240	36 29	324
316	27 12	041	27 13	091	45 33	116	37 21	168	35 11	202	57 47	241	36 12	324
317	27 30	042	27 41	092	45 58	117	37 26	169	35 00	204	57 22	242	35 56	325
318	27 49	042	28 10	093	46 23	118	37 32	170	34 48	205	56 57	244	35 40	325
319	28 08	043	28 38	094	46 48	119	37 36	171	34 36	206	56 32	245	35 24	326
320	28 28	044	29 06	094	47 12	121	37 40	173	34 24	207	56 06	246	35 08	327
321	28 47	044	29 34	095	47 36	122	37 44	174	34 11	208	55 40	247	34 53	327
322	29 07	045	30 02	096	48 00	123	37 46	175	33 57	209	55 14	248	34 38	328
323	29 27	046	30 30	097	48 23	124	37 48	176	33 43	211	54 48	249	34 23	328
324	29 47	046	30 58	098	48 47	125	37 50	178	33 28	212	54 21	251	34 08	329
325	30 08	047	31 26	099	49 09	126	37 51	179	33 13	213	53 55	252	33 54	330
326	30 29	048	31 53	100	49 32	128	37 51	180	32 58	214	53 28	253	33 40	330
327	30 50	049	32 21	101	49 54	129	37 51	181	32 42	215	53 01	254	33 26	331
328	31 11	049	32 49	102	50 16	130	37 50	183	32 25	216	52 34	255	33 12	332
329	31 32	050	33 16	103	50 37	131	37 48	184	32 08	217	52 07	256	32 59	332
	CAPELLA		♦Hamal		Alpheratz		Enif		♦ALTAIR		VEGA		♦Alioth	
330	31 54	051	33 44	104	50 58	133	37 46	185	31 51	219	51 39	257	32 46	333
331	32 16	051	34 11	105	51 19	134	37 43	186	31 33	220	51 12	258	32 33	333
332	32 38	052	34 38	106	51 39	135	37 40	187	31 15	221	50 44	259	32 21	334
333	33 00	053	35 05	107	51 58	136	37 36	189	30 57	222	50 16	260	32 09	335
334	33 23	053	35 32	108	52 17	138	37 31	190	30 38	223	49 49	261	31 57	335
335	33 46	054	35 59	109	52 36	139	37 26	191	30 18	224	49 21	262	31 45	336
336	34 09	055	36 26	110	52 54	141	37 21	192	29 59	225	48 53	263	31 34	336
337	34 32	056	36 52	111	53 12	142	37 14	194	29 38	226	48 25	264	31 23	337
338	34 55	056	37 18	112	53 29	143	37 07	195	29 18	227	47 57	265	31 12	338
339	35 19	057	37 44	113	53 46	145	37 00	196	28 57	228	47 29	266	31 01	338
340	35 42	058	38 10	114	54 02	146	36 52	197	28 36	229	47 01	266	30 51	339
341	36 06	058	38 36	115	54 17	147	36 43	199	28 14	230	46 33	267	30 41	340
342	36 30	059	39 02	116	54 32	149	36 34	200	27 52	231	46 05	268	30 31	340
343	36 54	060	39 27	117	54 46	150	36 24	201	27 30	232	45 36	269	30 22	341
344	37 19	060	39 52	118	55 00	152	36 13	202	27 08	233	45 08	270	30 13	341
	CAPELLA		ALDEBARAN		♦Hamal		Alpheratz		♦ALTAIR		VEGA		♦Alioth	
345	37 43	061	17 22	087	40 17	119	55 13	153	26 45	234	44 40	271	30 04	342
346	38 08	062	17 50	088	40 41	120	55 25	155	26 22	236	44 12	272	29 55	343
347	38 33	063	18 18	089	41 05	121	55 37	156	25 59	237	43 44	273	29 47	343
348	38 58	063	18 47	090	41 29	122	55 48	158	25 35	238	43 16	274	29 39	344
349	39 23	064	19 15	091	41 53	123	55 58	159	25 11	239	42 47	274	29 32	345
350	39 49	065	19 43	092	42 17	124	56 07	161	24 47	240	42 19	275	29 24	345
351	40 14	065	20 11	093	42 40	126	56 16	162	24 23	241	41 51	276	29 17	346
352	40 40	066	20 39	094	43 02	127	56 24	164	23 58	242	41 23	277	29 10	346
353	41 06	067	21 07	094	43 25	128	56 32	166	23 33	242	40 55	278	29 04	347
354	41 32	068	21 35	095	43 47	129	56 38	167	23 08	243	40 28	279	28 58	348
355	41 58	068	22 03	096	44 09	130	56 44	169	22 43	244	40 00	279	28 52	348
356	42 24	069	22 31	097	44 30	131	56 49	170	22 17	245	39 32	280	28 46	349
357	42 51	070	22 59	098	44 51	133	56 54	172	21 51	246	39 04	281	28 41	350
358	43 17	071	23 27	099	45 11	134	56 57	174	21 26	247	38 37	282	28 36	350
359	43 44	071	23 55	100	45 32	135	57 00	175	20 59	248	38 09	283	28 32	351

LAT 61°N

LHA ♈	Hc	Zn	Hc	Zn	Hc	Zn	Hc	Zn	Hc	Zn	Hc	Zn	Hc	Zn
	◆POLLUX		CAPELLA		Hamal		◆Alpheratz		DENEB		◆VEGA		Alioth	
0	12 49	054	43 52	071	46 34	135	58 02	177	57 21	264	37 27	284	27 28	352
1	13 13	055	44 19	072	46 55	137	58 03	178	56 52	265	36 59	285	27 24	352
2	13 37	056	44 47	073	47 14	138	58 04	180	56 23	266	36 31	286	27 20	353
3	14 01	057	45 15	073	47 34	139	58 03	182	55 54	267	36 03	287	27 17	353
4	14 25	058	45 43	074	47 52	140	58 02	183	55 25	268	35 35	287	27 13	354
5	14 50	058	46 11	075	48 11	142	58 00	185	54 56	269	35 08	288	27 11	355
6	15 15	059	46 39	076	48 28	143	57 57	187	54 27	270	34 40	289	27 08	355
7	15 40	060	47 07	076	48 45	144	57 53	188	53 58	271	34 13	290	27 06	356
8	16 05	061	47 35	077	49 02	146	57 49	190	53 29	271	33 45	290	27 04	357
9	16 31	062	48 04	078	49 18	147	57 44	191	53 00	272	33 18	291	27 02	357
10	16 56	062	48 32	079	49 34	148	57 37	193	52 31	273	32 51	292	27 01	358
11	17 22	063	49 01	079	49 49	150	57 30	195	52 02	274	32 24	293	27 00	358
12	17 48	064	49 29	080	50 03	151	57 23	196	51 33	275	31 57	293	27 00	359
13	18 15	065	49 58	081	50 17	153	57 14	198	51 04	276	31 31	294	26 59	000
14	18 41	066	50 27	082	50 30	154	57 05	199	50 35	276	31 04	295	26 59	000
	Alioth		◆POLLUX		CAPELLA		Hamal		◆Alpheratz		DENEB		◆VEGA	
15	27 00	001	19 07	066	50 55	082	50 42	155	56 55	201	50 06	277	30 38	296
16	27 00	002	19 34	067	51 24	083	50 54	157	56 44	203	49 37	278	30 12	296
17	27 01	002	20 01	068	51 53	084	51 05	158	56 32	204	49 09	279	29 46	297
18	27 03	003	20 28	069	52 22	085	51 16	160	56 20	206	48 40	280	29 20	298
19	27 04	003	20 55	070	52 51	086	51 25	161	56 07	207	48 11	280	28 54	299
20	27 06	004	21 23	070	53 20	086	51 35	163	55 53	209	47 43	281	28 29	299
21	27 08	005	21 50	071	53 49	087	51 43	164	55 39	210	47 14	282	28 04	300
22	27 11	005	22 18	072	54 18	088	51 51	165	55 24	212	46 46	283	27 39	301
23	27 14	006	22 46	073	54 47	089	51 57	167	55 08	213	46 17	283	27 14	302
24	27 17	007	23 13	074	55 16	090	52 04	168	54 52	215	45 49	284	26 49	302
25	27 20	007	23 41	074	55 46	091	52 09	170	54 35	216	45 21	285	26 25	303
26	27 24	008	24 09	075	56 15	092	52 14	171	54 18	217	44 53	286	26 01	304
27	27 28	008	24 38	076	56 44	093	52 18	173	54 00	219	44 25	286	25 36	305
28	27 33	009	25 06	077	57 13	093	52 21	174	53 42	220	43 57	287	25 13	305
29	27 38	010	25 34	078	57 42	094	52 24	176	53 22	222	43 29	288	24 49	306
	Alioth		◆POLLUX		ALDEBARAN		◆Hamal		Alpheratz		DENEB		◆VEGA	
30	27 43	010	26 03	079	37 36	131	52 25	177	53 03	223	43 02	289	24 26	307
31	27 48	011	26 31	079	37 58	132	52 26	179	52 43	224	42 34	289	24 02	308
32	27 54	012	27 00	080	38 19	133	52 26	180	52 22	226	42 07	290	23 40	308
33	28 00	012	27 29	081	38 41	134	52 26	182	52 01	227	41 40	291	23 17	309
34	28 06	013	27 57	082	39 01	135	52 24	183	51 40	228	41 13	292	22 54	310
35	28 13	013	28 26	083	39 22	136	52 22	185	51 18	230	40 46	292	22 32	311
36	28 20	014	28 55	084	39 42	137	52 19	186	50 55	231	40 19	293	22 10	311
37	28 27	015	29 24	084	40 01	139	52 16	188	50 33	232	39 52	294	21 48	312
38	28 34	015	29 53	085	40 20	140	52 11	189	50 10	233	39 25	294	21 27	313
39	28 42	016	30 22	086	40 39	141	52 06	191	49 46	234	38 59	295	21 06	314
40	28 50	017	30 51	087	40 57	142	52 00	192	49 22	236	38 33	296	20 45	314
41	28 59	017	31 20	088	41 14	143	51 54	194	48 58	237	38 07	297	20 24	315
42	29 08	018	31 49	089	41 31	145	51 46	195	48 34	238	37 41	297	20 04	316
43	29 17	018	32 18	090	41 48	146	51 38	197	48 09	239	37 15	298	19 43	317
44	29 26	019	32 47	091	42 04	147	51 30	198	47 44	240	36 49	299	19 24	317
	Alioth		◆POLLUX		ALDEBARAN		◆Hamal		Alpheratz		DENEB		◆VEGA	
45	29 36	020	33 16	091	42 20	148	51 20	200	47 18	241	36 24	299	19 04	318
46	29 46	020	33 46	092	42 35	150	51 10	201	46 52	243	35 59	300	18 45	319
47	29 56	021	34 15	093	42 49	151	50 59	203	46 27	244	35 34	301	18 26	320
48	30 06	022	34 44	094	43 03	152	50 48	204	46 00	245	35 09	301	18 07	320
49	30 17	022	35 13	095	43 16	153	50 35	205	45 34	246	34 44	302	17 48	321
50	30 28	023	35 42	096	43 29	155	50 23	207	45 07	247	34 19	303	17 30	322
51	30 40	023	36 11	097	43 41	156	50 09	208	44 40	248	33 55	304	17 12	323
52	30 51	024	36 39	098	43 53	157	49 55	210	44 13	249	33 31	304	16 55	323
53	31 03	025	37 08	099	44 04	159	49 40	211	43 46	250	33 07	305	16 38	324
54	31 16	025	37 37	100	44 14	160	49 25	212	43 19	251	32 43	306	16 21	325
55	31 28	026	38 06	100	44 24	161	49 09	214	42 51	252	32 20	306	16 04	326
56	31 41	026	38 34	101	44 33	163	48 53	215	42 23	253	31 56	307	15 48	326
57	31 54	027	39 03	102	44 41	164	48 36	216	41 55	254	31 33	308	15 32	327
58	32 07	028	39 31	103	44 49	165	48 19	218	41 27	255	31 10	308	15 16	328
59	32 21	028	39 59	104	44 56	167	48 00	219	40 59	256	30 48	309	15 01	329
	Alioth		◆POLLUX		ALDEBARAN		◆Hamal		Alpheratz		DENEB		◆VEGA	
60	32 35	029	40 27	105	45 03	168	47 42	220	40 31	257	30 25	310	14 46	329
61	32 49	029	40 55	106	45 08	169	47 23	222	40 02	258	30 03	310	14 31	330
62	33 03	030	41 23	107	45 13	171	47 03	223	39 34	259	29 41	311	14 17	331
63	33 18	031	41 51	108	45 18	172	46 43	224	39 05	260	29 19	312	14 03	332
64	33 33	031	42 18	109	45 22	173	46 23	225	38 37	261	28 58	313	13 49	332
65	33 48	032	42 46	110	45 25	175	46 02	226	38 08	262	28 36	313	13 36	333
66	34 04	032	43 13	111	45 27	176	45 41	228	37 39	263	28 15	314	13 23	334
67	34 19	033	43 40	112	45 29	177	45 19	229	37 10	264	27 54	315	13 11	335
68	34 35	034	44 07	113	45 30	179	44 57	230	36 41	265	27 34	315	12 58	336
69	34 52	034	44 33	115	45 30	180	44 34	231	36 12	265	27 13	316	12 46	336
70	35 08	035	45 00	116	45 30	181	44 12	232	35 43	266	26 53	317	12 35	337
71	35 25	035	45 26	117	45 28	183	43 48	234	35 14	267	26 34	317	12 24	338
72	35 42	036	45 52	118	45 27	184	43 25	235	34 45	268	26 14	318	12 13	339
73	35 59	037	46 17	119	45 24	186	43 01	236	34 16	269	25 55	319	12 03	339
74	36 16	037	46 42	120	45 21	187	42 37	237	33 47	270	25 36	320	11 53	340
	Alioth		◆REGULUS		PROCYON		◆ALDEBARAN		Hamal		Alpheratz		◆DENEB	
75	36 34	038	16 45	095	26 48	134	45 17	188	42 12	238	33 18	271	25 17	320
76	36 52	038	17 14	096	27 09	136	45 13	190	41 47	239	32 49	272	24 58	321
77	37 10	039	17 43	097	27 29	137	45 07	191	41 22	240	32 20	273	24 40	322
78	37 29	040	18 12	098	27 49	138	45 01	192	40 57	241	31 51	273	24 22	322
79	37 47	040	18 40	099	28 08	139	44 55	194	40 31	242	31 22	274	24 05	323
80	38 06	041	19 09	100	28 27	140	44 48	195	40 05	244	30 53	275	23 47	324
81	38 25	041	19 38	101	28 46	141	44 40	196	39 39	245	30 24	276	23 30	324
82	38 44	042	20 06	102	29 04	142	44 31	198	39 13	246	29 55	277	23 13	325
83	39 04	042	20 35	103	29 22	143	44 22	199	38 46	247	29 26	278	22 57	326
84	39 24	043	21 03	104	29 39	144	44 12	200	38 19	248	28 57	278	22 41	327
85	39 44	044	21 31	105	29 56	145	44 02	202	37 52	249	28 28	279	22 25	327
86	40 04	044	21 59	105	30 12	146	43 51	203	37 25	250	28 00	280	22 09	328
87	40 24	045	22 27	106	30 28	147	43 39	204	36 58	251	27 31	281	21 54	329
88	40 45	045	22 55	107	30 44	149	43 27	206	36 30	252	27 03	282	21 39	329
89	41 06	046	23 23	108	30 58	150	43 14	207	36 02	253	26 34	283	21 24	330

LHA ♈	Hc	Zn	Hc	Zn	Hc	Zn	Hc	Zn	Hc	Zn	Hc	Zn	Hc	Zn
	Alioth		◆REGULUS		PROCYON		BETELGEUSE		◆ALDEBARAN		Hamal		◆DENEB	
90	41 27	047	23 50	109	31 13	151	36 24	182	43 00	208	35 35	254	21 10	331
91	41 48	047	24 18	110	31 27	152	36 23	183	42 46	209	35 07	255	20 56	331
92	42 09	048	24 45	111	31 40	153	36 21	184	42 32	211	34 38	256	20 42	332
93	42 31	048	25 12	112	31 53	154	36 19	185	42 17	212	34 10	257	20 29	333
94	42 53	049	25 39	113	32 06	155	36 16	186	42 01	213	33 42	257	20 16	334
95	43 15	049	26 06	114	32 17	157	36 12	188	41 45	214	33 13	258	20 03	334
96	43 37	050	26 32	115	32 29	158	36 08	189	41 28	216	32 45	259	19 51	335
97	43 59	051	26 58	116	32 39	159	36 03	190	41 11	217	32 16	260	19 38	336
98	44 22	051	27 24	117	32 50	160	35 58	191	40 53	218	31 48	261	19 27	336
99	44 45	052	27 50	118	32 59	161	35 52	193	40 35	219	31 19	262	19 15	337
100	45 08	052	28 16	119	33 08	162	35 45	194	40 17	220	30 50	263	19 04	338
101	45 31	053	28 41	120	33 17	164	35 38	195	39 57	222	30 21	264	18 53	339
102	45 54	053	29 06	121	33 25	165	35 30	196	39 38	223	29 52	265	18 43	339
103	46 17	054	29 31	122	33 32	166	35 21	197	39 18	224	29 23	266	18 33	340
104	46 41	055	29 56	123	33 39	167	35 12	199	38 57	225	28 54	267	18 23	341
	Alioth		◆REGULUS		PROCYON		BETELGEUSE		◆CAPELLA		Hamal		◆DENEB	
105	47 05	055	30 20	124	33 45	168	35 03	200	68 45	237	28 25	268	18 14	342
106	47 29	056	30 44	125	33 51	169	34 53	201	68 20	238	27 56	268	18 05	342
107	47 53	056	31 07	126	33 56	171	34 42	202	67 55	240	27 27	269	17 56	343
108	48 17	057	31 31	127	34 00	172	34 31	203	67 30	241	26 58	270	17 48	344
109	48 42	058	31 54	128	34 04	173	34 19	205	67 04	243	26 29	271	17 40	344
110	49 06	058	32 17	129	34 07	174	34 06	206	66 38	244	26 00	272	17 32	345
111	49 31	059	32 39	130	34 10	175	33 54	207	66 12	246	25 31	273	17 25	346
112	49 56	059	33 01	131	34 12	177	33 40	208	65 45	247	25 02	274	17 18	347
113	50 21	060	33 23	132	34 13	178	33 26	209	65 19	248	24 32	274	17 11	347
114	50 46	060	33 44	134	34 14	179	33 12	210	64 51	249	24 04	275	17 05	348
115	51 12	061	34 05	135	34 14	180	32 57	212	64 24	251	23 35	276	16 59	349
116	51 37	062	34 25	136	34 14	181	32 41	213	63 57	252	23 06	277	16 54	350
117	52 03	062	34 45	137	34 13	183	32 25	214	63 29	253	22 37	278	16 49	350
118	52 29	063	35 05	138	34 11	184	32 09	215	63 01	254	22 08	279	16 44	351
119	52 54	063	35 24	139	34 09	185	31 52	216	62 33	255	21 39	280	16 40	352
	◆VEGA		ARCTURUS		◆REGULUS		PROCYON		◆CAPELLA		Schedar		DENEB	
120	11 13	016	14 53	077	35 43	140	34 06	186	62 05	256	39 39	318	16 36	352
121	11 21	017	15 22	078	36 01	141	34 02	188	61 36	257	39 20	318	16 32	353
122	11 30	018	15 50	079	36 19	143	33 58	189	61 08	258	39 01	319	16 29	354
123	11 39	019	16 19	080	36 37	144	33 54	190	60 39	260	38 41	319	16 26	355
124	11 49	020	16 47	081	36 54	145	33 48	191	60 11	261	38 23	320	16 23	355
125	11 59	020	17 16	081	37 10	146	33 42	192	59 42	262	38 04	321	16 21	356
126	12 09	021	17 45	082	37 26	147	33 36	193	59 13	263	37 46	321	16 19	357
127	12 20	022	18 14	083	37 42	148	33 29	195	58 44	264	37 28	322	16 18	358
128	12 31	023	18 43	084	37 57	150	33 21	196	58 15	264	37 10	322	16 17	358
129	12 42	023	19 12	085	38 11	151	33 13	197	57 46	265	36 52	323	16 16	359
130	12 54	024	19 41	086	38 25	152	33 04	198	57 17	266	36 35	323	16 16	000
131	13 06	025	20 10	087	38 38	153	32 55	199	56 48	267	36 17	324	16 16	001
132	13 18	026	20 39	087	38 51	155	32 45	201	56 19	268	36 00	325	16 16	001
133	13 31	026	21 08	088	39 03	156	32 34	202	55 50	269	35 44	325	16 17	002
134	13 44	027	21 37	089	39 15	157	32 23	203	55 21	270	35 27	326	16 18	003
	DENEB		◆VEGA		ARCTURUS		◆REGULUS		PROCYON		◆CAPELLA		Schedar	
135	16 20	003	13 58	028	22 06	090	39 26	158	32 12	204	54 52	271	35 11	326
136	16 22	004	14 12	029	22 35	091	39 36	159	32 00	205	54 23	272	34 55	327
137	16 24	005	14 26	030	23 04	092	39 46	161	31 47	206	53 54	273	34 39	328
138	16 27	006	14 40	030	23 33	093	39 56	162	31 34	207	53 25	273	34 24	328
139	16 30	006	14 55	031	24 02	094	40 04	163	31 20	209	52 56	274	34 09	329
140	16 33	007	15 10	032	24 31	094	40 12	165	31 06	210	52 27	275	33 54	329
141	16 37	008	15 26	033	25 00	095	40 20	166	30 51	211	51 58	276	33 39	330
142	16 41	009	15 42	033	25 29	096	40 27	167	30 36	212	51 29	277	33 25	331
143	16 46	009	15 58	034	25 58	097	40 33	168	30 21	213	51 00	277	33 10	331
144	16 50	010	16 14	035	26 27	098	40 38	170	30 04	214	50 31	278	32 56	332
145	16 56	011	16 31	036	26 56	099	40 43	171	29 48	215	50 02	279	32 43	332
146	17 01	011	16 48	036	27 24	100	40 48	172	29 31	216	49 34	280	32 29	333
147	17 07	012	17 06	037	27 53	101	40 51	173	29 13	217	49 05	281	32 16	334
148	17 14	013	17 23	038	28 22	102	40 54	175	28 55	219	48 37	281	32 04	334
149	17 20	014	17 41	039	28 50	103	40 56	176	28 37	220	48 08	282	31 51	335
	DENEB		◆VEGA		ARCTURUS		◆REGULUS		PROCYON		◆CAPELLA		Schedar	
150	17 27	014	18 00	039	29 18	104	40 58	177	28 18	221	47 40	283	31 39	335
151	17 35	015	18 18	040	29 47	104	40 59	179	27 59	222	47 11	284	31 27	336
152	17 43	016	18 37	041	30 15	105	41 00	180	27 39	223	46 43	284	31 15	337
153	17 51	017	18 57	042	30 43	106	40 59	181	27 20	224	46 15	285	31 04	337
154	17 59	017	19 16	042	31 10	107	40 58	183	26 59	225	45 47	286	30 52	338
155	18 08	018	19 36	043	31 38	108	40 57	184	26 38	226	45 19	287	30 42	338
156	18 17	019	19 56	044	32 06	109	40 54	185	26 17	227	44 51	287	30 31	339
157	18 27	019	20 16	045	32 33	110	40 51	186	25 56	228	44 24	288	30 21	340
158	18 37	020	20 37	045	33 00	111	40 48	188	25 34	229	43 56	289	30 11	340
159	18 47	021	20 58	046	33 27	112	40 44	189	25 12	230	43 28	290	30 01	341
160	18 57	022	21 19	047	33 54	113	40 39	190	24 49	231	43 01	290	29 52	341
161	19 08	022	21 40	048	34 21	114	40 33	192	24 26	232	42 34	291	29 42	342
162	19 19	023	22 02	048	34 47	115	40 27	193	24 03	233	42 07	292	29 34	343
163	19 31	024	22 24	049	35 13	116	40 20	194	23 40	234	41 40	292	29 25	343
164	19 43	024	22 46	050	35 39	117	40 13	195	23 16	235	41 13	293	29 17	344
	DENEB		◆VEGA		ARCTURUS		◆REGULUS		POLLUX		◆CAPELLA		Schedar	
165	19 55	025	23 08	051	36 05	118	40 05	197	43 53	247	40 46	294	29 09	344
166	20 08	026	23 31	051	36 30	119	39 56	198	43 27	248	40 20	294	29 01	345
167	20 21	027	23 54	052	36 56	120	39 47	199	42 59	249	39 53	295	28 54	346
168	20 34	027	24 17	053	37 21	122	39 37	200	42 32	250	39 27	296	28 47	346
169	20 47	028	24 40	054	37 45	123	39 27	202	42 05	251	39 01	297	28 40	347
170	21 01	029	25 04	054	38 10	124	39 15	203	41 37	252	38 35	297	28 34	348
171	21 15	029	25 27	055	38 34	125	39 04	204	41 09	253	38 09	298	28 28	348
172	21 30	030	25 51	056	38 57	126	38 52	205	40 41	254	37 44	299	28 22	349
173	21 45	031	26 15	057	39 21	127	38 39	207	40 13	255	37 18	299	28 16	349
174	22 00	032	26 40	057	39 44	128	38 26	208	39 45	256	36 53	300	28 11	350
175	22 15	032	27 04	058	40 07	129	38 12	209	39 17	257	36 28	301	28 06	351
176	22 31	033	27 29	059	40 29	130	37 57	210	38 48	258	36 03	301	28 02	351
177	22 47	034	27 54	060	40 51	132	37 42	211	38 20	259	35 38	302	27 57	352
178	23 03	034	28 19	060	41 13	133	37 27	213	37 51	260	35 14	303	27 54	353
179	23 20	035	28 45	061	41 34	134	37 11	214	37 23	261	34 49	304	27 50	353

LHA	Hc	Zn	Hc	Zn	Hc	Zn	Hc	Zn	Hc	Zn	Hc	Zn	Hc	Zn
♈	DENEB		◆VEGA		ARCTURUS		◆Denebola		REGULUS		POLLUX		◆CAPELLA	
180	23 37	036	29 10	062	41 54	135	43 33	184	36 55	215	36 54	262	34 25	304
181	23 54	037	29 36	063	42 15	136	43 31	185	36 38	216	36 25	263	34 01	305
182	24 11	037	30 02	063	42 35	137	43 28	186	36 20	217	35 56	264	33 38	306
183	24 29	038	30 28	064	42 54	139	43 25	188	36 02	219	35 27	265	33 14	306
184	24 47	039	30 54	065	43 13	140	43 20	189	35 44	220	34 58	265	32 51	307
185	25 05	039	31 21	066	43 32	141	43 15	190	35 25	221	34 29	266	32 28	308
186	25 24	040	31 47	066	43 50	142	43 10	192	35 06	222	34 00	267	32 05	308
187	25 43	041	32 14	067	44 07	144	43 04	193	34 46	223	33 31	268	31 42	309
188	26 02	041	32 41	068	44 24	145	42 57	194	34 26	224	33 02	269	31 19	310
189	26 21	042	33 08	069	44 41	146	42 49	196	34 06	225	32 33	270	30 57	310
190	26 41	043	33 35	069	44 57	147	42 41	197	33 45	226	32 04	271	30 35	311
191	27 01	044	34 02	070	45 12	149	42 32	198	33 24	227	31 35	272	30 13	312
192	27 21	044	34 30	071	45 27	150	42 23	200	33 02	229	31 06	273	29 52	312
193	27 41	045	34 57	072	45 41	151	42 13	201	32 40	230	30 37	273	29 30	313
194	28 02	046	35 25	072	45 55	153	42 02	202	32 18	231	30 08	274	29 09	314
	DENEB		◆VEGA		ARCTURUS		◆Denebola		REGULUS		POLLUX		◆CAPELLA	
195	28 23	046	35 53	073	46 08	154	41 51	203	31 55	232	29 39	275	28 48	315
196	28 44	047	36 20	074	46 21	155	41 39	205	31 32	233	29 10	276	28 28	315
197	29 06	048	36 48	075	46 33	157	41 27	206	31 09	234	28 41	277	28 07	316
198	29 27	048	37 17	075	46 44	158	41 14	207	30 45	235	28 12	278	27 47	317
199	29 49	049	37 45	076	46 55	159	41 00	208	30 21	236	27 43	278	27 27	317
200	30 11	050	38 13	077	47 05	161	40 46	210	29 57	237	27 14	279	27 08	318
201	30 33	050	38 41	078	47 14	162	40 31	211	29 32	238	26 46	280	26 48	319
202	30 56	051	39 10	079	47 23	163	40 16	212	29 08	239	26 17	281	26 29	319
203	31 19	052	39 39	079	47 31	165	40 00	213	28 42	240	25 49	282	26 11	320
204	31 42	053	40 07	080	47 38	166	39 44	215	28 17	241	25 20	283	25 52	321
205	32 05	053	40 36	081	47 44	168	39 27	216	27 52	242	24 52	283	25 34	321
206	32 28	054	41 05	082	47 50	169	39 10	217	27 26	243	24 24	284	25 16	322
207	32 52	055	41 33	083	47 56	170	38 52	218	27 00	244	23 56	285	24 58	323
208	33 16	055	42 02	084	48 00	172	38 34	219	26 33	245	23 27	286	24 41	323
209	33 40	056	42 31	084	48 04	173	38 15	220	26 07	246	23 00	287	24 23	324
	DENEB		◆VEGA		ARCTURUS		◆Denebola		REGULUS		POLLUX		◆CAPELLA	
210	34 04	057	43 00	085	48 07	175	37 56	222	25 40	247	22 32	288	24 06	325
211	34 28	057	43 29	086	48 10	176	37 37	223	25 13	248	22 04	288	23 50	326
212	34 53	058	43 58	087	48 11	177	37 17	224	24 46	249	21 37	289	23 34	326
213	35 18	059	44 27	088	48 12	179	36 56	225	24 19	250	21 09	290	23 18	327
214	35 43	059	44 56	089	48 12	180	36 36	226	23 52	251	20 42	291	23 02	328
215	36 08	060	45 25	089	48 12	182	36 14	227	23 24	252	20 15	292	22 47	328
216	36 33	061	45 55	090	48 11	183	35 53	228	22 57	253	19 48	292	22 31	329
217	36 59	062	46 24	091	48 09	184	35 31	230	22 29	254	19 21	293	22 17	330
218	37 24	062	46 53	092	48 06	186	35 09	231	22 01	255	18 54	294	22 02	330
219	37 50	063	47 22	093	48 03	187	34 46	232	21 33	255	18 28	295	21 48	331
220	38 16	064	47 51	094	47 59	189	34 23	233	21 04	256	18 02	296	21 34	332
221	38 42	064	48 20	095	47 54	190	34 00	234	20 36	257	17 35	296	21 21	333
222	39 09	065	48 49	096	47 49	191	33 36	235	20 08	258	17 09	297	21 07	333
223	39 35	066	49 18	097	47 43	193	33 12	236	19 39	259	16 44	298	20 54	334
224	40 02	067	49 47	098	47 36	194	32 48	237	19 11	260	16 18	299	20 42	335
	◆DENEB		VEGA		Rasalhague		◆ARCTURUS		Denebola		Dubhe		◆CAPELLA	
225	40 29	067	50 15	099	34 02	133	47 28	196	32 23	238	62 39	298	20 30	335
226	40 56	068	50 44	100	34 23	134	47 20	197	31 58	239	62 13	298	20 18	336
227	41 23	069	51 13	101	34 44	135	47 11	198	31 33	240	61 48	299	20 06	337
228	41 50	069	51 41	102	35 04	136	47 02	200	31 08	241	61 22	299	19 55	338
229	42 17	070	52 10	103	35 25	137	46 52	201	30 42	242	60 57	300	19 44	338
230	42 44	071	52 38	104	35 44	138	46 41	202	30 16	243	60 32	300	19 33	339
231	43 12	072	53 06	105	36 03	139	46 29	204	29 50	244	60 07	301	19 23	340
232	43 40	072	53 34	106	36 22	140	46 17	205	29 24	245	59 42	301	19 13	340
233	44 07	073	54 02	107	36 40	142	46 05	207	28 58	246	59 17	301	19 04	341
234	44 35	074	54 30	108	36 58	143	45 51	208	28 31	247	58 52	302	18 54	342
235	45 03	075	54 58	109	37 16	144	45 37	209	28 04	248	58 28	302	18 45	343
236	45 31	075	55 25	110	37 32	145	45 23	210	27 37	249	58 03	303	18 37	343
237	46 00	076	55 52	111	37 49	146	45 08	212	27 10	250	57 39	303	18 29	344
238	46 28	077	56 19	112	38 05	148	44 52	213	26 42	251	57 14	304	18 21	345
239	46 56	078	56 46	113	38 20	149	44 36	214	26 15	252	56 50	304	18 13	345
	Alpheratz		◆DENEB		VEGA		Rasalhague		◆ARCTURUS		Alioth		◆CAPELLA	
240	11 33	049	47 25	078	57 13	115	38 35	150	44 20	216	65 43	279	18 06	346
241	11 55	050	47 53	079	57 39	116	38 49	151	44 02	217	65 14	280	18 00	347
242	12 17	051	48 22	080	58 05	117	39 03	152	43 45	218	64 45	281	17 53	348
243	12 40	052	48 50	081	58 31	118	39 16	154	43 27	219	64 17	281	17 47	348
244	13 03	052	49 19	081	58 57	119	39 29	155	43 08	221	63 48	282	17 41	349
245	13 26	053	49 48	082	59 22	121	39 41	156	42 49	222	63 20	283	17 36	350
246	13 49	054	50 17	083	59 47	122	39 52	157	42 29	223	62 52	283	17 31	351
247	14 13	055	50 46	084	60 11	123	40 03	159	42 09	224	62 23	284	17 26	351
248	14 37	056	51 15	085	60 35	125	40 14	160	41 49	225	61 55	285	17 22	352
249	15 01	056	51 44	085	60 59	126	40 24	161	41 28	226	61 27	285	17 18	353
250	15 25	057	52 13	086	61 22	127	40 33	162	41 06	228	60 59	286	17 15	353
251	15 50	058	52 42	087	61 45	129	40 41	164	40 45	229	60 31	286	17 12	354
252	16 15	059	53 11	088	62 07	130	40 49	165	40 23	230	60 03	287	17 09	355
253	16 40	060	53 40	089	62 29	132	40 56	166	40 00	231	59 35	288	17 06	356
254	17 05	060	54 09	090	62 51	133	41 03	167	39 37	232	59 08	288	17 04	356
	Alpheratz		◆DENEB		ALTAIR		Rasalhague		◆ARCTURUS		Alioth		◆CAPELLA	
255	17 30	061	54 38	091	29 09	130	41 09	169	39 14	233	58 40	289	17 03	357
256	17 56	062	55 07	091	29 31	131	41 14	170	38 51	234	58 13	290	17 01	358
257	18 21	063	55 36	092	29 53	132	41 19	171	38 27	236	57 45	290	17 00	358
258	18 47	064	56 05	093	30 14	133	41 23	173	38 03	237	57 18	291	17 00	359
259	19 14	064	56 34	094	30 35	134	41 27	174	37 38	238	56 51	291	17 00	000
260	19 40	065	57 03	095	30 56	135	41 29	175	37 14	239	56 24	292	17 00	001
261	20 06	066	57 32	096	31 16	136	41 31	177	36 49	240	55 57	293	17 00	001
262	20 33	067	58 01	097	31 36	137	41 33	178	36 23	241	55 30	293	17 01	002
263	21 00	068	58 30	098	31 55	139	41 34	179	35 58	242	55 04	294	17 02	003
264	21 27	068	58 59	099	32 14	140	41 34	180	35 32	243	54 37	294	17 04	004
265	21 54	069	59 27	100	32 33	141	41 33	182	35 06	244	54 11	295	17 06	004
266	22 21	070	59 56	101	32 51	142	41 32	183	34 40	245	53 44	296	17 08	005
267	22 49	071	60 24	102	33 09	143	41 30	184	34 13	246	53 18	296	17 11	006
268	23 16	072	60 53	103	33 26	144	41 28	186	33 47	247	52 52	297	17 14	006
269	23 44	072	61 21	104	33 43	145	41 24	187	33 20	248	52 26	297	17 18	007

LHA	Hc	Zn	Hc	Zn	Hc	Zn	Hc	Zn	Hc	Zn	Hc	Zn	Hc	Zn
♈	◆CAPELLA		Alpheratz		◆DENEB		ALTAIR		Rasalhague		◆ARCTURUS		Alioth	
270	17 22	008	24 12	073	61 49	105	33 59	146	41 21	188	32 53	249	52 00	298
271	17 26	009	24 40	074	62 17	106	34 15	148	41 16	190	32 25	250	51 35	298
272	17 30	009	25 08	075	62 45	108	34 30	149	41 11	191	31 58	251	51 09	299
273	17 35	010	25 36	076	63 13	109	34 45	150	41 05	192	31 30	252	50 44	300
274	17 40	011	26 04	077	63 40	110	35 00	151	40 59	193	31 03	253	50 19	300
275	17 46	012	26 32	077	64 07	111	35 13	152	40 52	195	30 35	254	49 54	301
276	17 52	012	27 01	078	64 34	112	35 27	153	40 44	196	30 07	255	49 29	301
277	17 58	013	27 29	079	65 01	114	35 39	155	40 36	197	29 39	256	49 04	302
278	18 05	014	27 58	080	65 28	115	35 52	156	40 27	199	29 10	257	48 39	303
279	18 12	014	28 27	081	65 54	116	36 03	157	40 17	200	28 42	258	48 15	303
280	18 20	015	28 55	081	66 20	118	36 14	158	40 07	201	28 13	259	47 51	304
281	18 27	016	29 24	082	66 45	119	36 25	159	39 56	202	27 45	259	47 26	304
282	18 36	017	29 53	083	67 10	121	36 35	161	39 45	204	27 16	260	47 03	305
283	18 44	017	30 22	084	67 35	122	36 44	162	39 33	205	26 48	261	46 39	305
284	18 53	018	30 51	085	68 00	124	36 53	163	39 21	206	26 19	262	46 15	306
	◆CAPELLA		Alpheratz		Enif		◆ALTAIR		VEGA		◆ARCTURUS		Alioth	
285	19 02	019	31 20	086	30 41	131	37 01	164	67 29	192	25 50	263	45 52	307
286	19 12	019	31 49	087	31 02	132	37 09	166	67 23	194	25 21	264	45 28	307
287	19 21	020	32 18	087	31 24	133	37 16	167	67 15	196	24 52	265	45 05	308
288	19 32	021	32 47	088	31 45	135	37 22	168	67 07	198	24 23	266	44 42	308
289	19 42	022	33 16	089	32 05	136	37 28	169	66 57	200	23 54	267	44 20	309
290	19 53	022	33 45	090	32 25	137	37 33	170	66 47	202	23 25	268	43 57	309
291	20 04	023	34 14	091	32 45	138	37 37	172	66 36	204	22 56	268	43 35	310
292	20 16	024	34 43	092	33 04	139	37 41	173	66 24	206	22 27	269	43 13	311
293	20 28	024	35 12	093	33 23	140	37 44	174	66 11	207	21 58	270	42 51	311
294	20 40	025	35 41	094	33 42	141	37 47	175	65 57	209	21 29	271	42 29	312
295	20 52	026	36 10	094	34 00	142	37 49	177	65 42	211	21 00	272	42 07	312
296	21 05	027	36 39	095	34 17	143	37 50	178	65 27	213	20 31	273	41 46	313
297	21 18	027	37 08	096	34 34	145	37 51	179	65 11	215	20 01	274	41 25	314
298	21 32	028	37 37	097	34 51	146	37 51	180	64 54	216	19 32	275	41 04	314
299	21 46	029	38 06	098	35 07	147	37 51	182	64 36	218	19 03	275	40 43	315
	◆CAPELLA		Alpheratz		◆Enif		ALTAIR		VEGA		◆ARCTURUS		Alioth	
300	22 00	029	38 35	099	35 23	148	37 50	183	64 18	220	18 35	276	40 22	315
301	22 14	030	39 03	100	35 38	149	37 48	184	63 59	221	18 06	277	40 02	316
302	22 29	031	39 32	101	35 53	150	37 45	185	63 40	223	17 37	278	39 42	316
303	22 44	032	40 01	102	36 07	152	37 42	187	63 20	225	17 08	279	39 22	317
304	22 59	032	40 29	103	36 20	153	37 38	188	62 59	226	16 39	280	39 02	318
305	23 15	033	40 57	104	36 33	154	37 34	189	62 38	228	16 11	281	38 43	318
306	23 31	034	41 25	105	36 46	155	37 29	190	62 16	229	15 42	281	38 23	319
307	23 47	034	41 54	106	36 58	156	37 24	192	61 54	231	15 14	282	38 04	319
308	24 04	035	42 21	107	37 09	158	37 17	193	61 31	232	14 45	283	37 45	320
309	24 21	036	42 49	108	37 20	159	37 11	194	61 08	233	14 17	284	37 27	321
310	24 38	036	43 17	109	37 30	160	37 03	195	60 44	235	13 49	285	37 08	321
311	24 55	037	43 44	110	37 40	161	36 55	197	60 20	236	13 21	286	36 50	322
312	25 13	038	44 12	111	37 49	162	36 47	198	59 56	237	12 53	286	36 32	322
313	25 31	038	44 39	112	37 58	164	36 37	199	59 31	239	12 25	287	36 15	323
314	25 49	039	45 06	113	38 05	165	36 28	200	59 06	240	11 57	288	35 57	323
	CAPELLA		◆Hamal		Alpheratz		Enif		◆ALTAIR		VEGA		◆Alioth	
315	26 08	040	26 45	089	45 32	114	38 13	166	36 17	201	58 41	241	35 40	324
316	26 26	041	27 14	090	45 59	115	38 19	167	36 06	203	58 15	243	35 23	325
317	26 45	041	27 43	091	46 25	116	38 25	169	35 55	204	57 49	244	35 06	325
318	27 05	042	28 12	092	46 51	117	38 31	170	35 43	205	57 23	245	34 50	326
319	27 24	043	28 41	093	47 17	118	38 35	171	35 30	206	56 57	246	34 34	326
320	27 44	043	29 10	094	47 42	120	38 40	172	35 17	207	56 30	247	34 18	327
321	28 04	044	29 39	095	48 07	121	38 43	174	35 03	209	56 03	249	34 02	328
322	28 24	045	30 08	096	48 32	122	38 46	175	34 49	210	55 36	250	33 47	328
323	28 45	045	30 37	097	48 57	123	38 48	176	34 34	211	55 08	251	33 32	329
324	29 06	046	31 06	097	49 21	124	38 50	177	34 19	212	54 41	252	33 17	329
325	29 27	047	31 35	098	49 45	125	38 51	179	34 04	213	54 13	253	33 02	330
326	29 48	047	32 04	099	50 08	127	38 51	180	33 47	214	53 45	254	32 48	331
327	30 10	048	32 32	100	50 31	128	38 51	181	33 31	216	53 17	255	32 33	331
328	30 32	049	33 01	101	50 54	129	38 50	183	33 14	217	52 49	256	32 20	332
329	30 54	049	33 29	102	51 16	130	38 48	184	32 56	218	52 21	257	32 06	332
	CAPELLA		◆Hamal		Alpheratz		Enif		◆ALTAIR		VEGA		◆Alioth	
330	31 16	050	33 58	103	51 38	132	38 46	185	32 38	219	51 52	258	31 53	333
331	31 38	051	34 26	104	52 00	133	38 43	186	32 19	220	51 24	259	31 40	334
332	32 01	052	34 54	105	52 21	134	38 40	188	32 01	221	50 55	260	31 27	334
333	32 24	052	35 22	106	52 41	136	38 35	189	31 41	222	50 26	261	31 14	335
334	32 47	053	35 50	107	53 02	137	38 31	190	31 21	223	49 58	262	31 02	335
335	33 10	054	36 18	108	53 21	138	38 25	191	31 01	224	49 29	263	30 50	336
336	33 34	054	36 45	109	53 40	140	38 19	193	30 41	225	49 00	264	30 39	337
337	33 58	055	37 13	110	53 59	141	38 12	194	30 20	227	48 31	265	30 27	337
338	34 21	056	37 40	111	54 17	142	38 05	195	29 58	228	48 02	266	30 16	338
339	34 46	056	38 07	112	54 34	144	37 57	196	29 37	229	47 33	267	30 05	339
340	35 10	057	38 34	113	54 51	145	37 49	198	29 15	230	47 04	268	29 55	339
341	35 34	058	39 01	114	55 07	147	37 40	199	28 52	231	46 35	268	29 45	340
342	35 59	058	39 27	115	55 23	148	37 30	200	28 30	232	46 06	269	29 35	340
343	36 24	059	39 54	116	55 38	150	37 20	201	28 07	233	45 37	270	29 25	341
344	36 49	060	40 20	117	55 52	151	37 09	202	27 43	234	45 08	271	29 16	342
	CAPELLA		ALDEBARAN		◆Hamal		Alpheratz		◆ALTAIR		VEGA		◆Alioth	
345	37 14	061	17 19	087	40 45	118	56 06	153	27 20	235	44 38	272	29 07	342
346	37 40	061	17 48	088	41 11	119	56 19	154	26 56	236	44 09	273	28 58	343
347	38 05	062	18 17	089	41 36	120	56 32	156	26 32	237	43 40	274	28 50	343
348	38 31	063	18 46	090	42 01	121	56 43	157	26 07	238	43 11	274	28 42	344
349	38 57	063	19 16	091	42 26	123	56 54	159	25 42	239	42 42	275	28 34	345
350	39 23	064	19 45	091	42 50	124	57 04	160	25 17	240	42 13	276	28 26	345
351	39 49	065	20 14	092	43 14	125	57 13	162	24 52	241	41 45	277	28 19	346
352	40 16	065	20 43	093	43 38	126	57 22	164	24 26	242	41 16	278	28 12	347
353	40 42	066	21 12	094	44 01	127	57 30	165	24 01	243	40 47	279	28 05	347
354	41 09	067	21 41	095	44 24	128	57 37	167	23 35	244	40 18	279	27 59	348
355	41 36	068	22 10	096	44 47	129	57 43	168	23 08	245	39 50	280	27 53	348
356	42 02	068	22 39	097	45 09	131	57 49	170	22 42	246	39 21	281	27 48	349
357	42 30	069	23 08	098	45 31	132	57 53	172	22 15	247	38 52	282	27 42	350
358	42 57	070	23 36	099	45 53	133	57 57	173	21 48	248	38 24	283	27 37	350
359	43 24	070	24 05	099	46 14	134	58 00	175	21 21	249	37 56	283	27 32	351

LAT 60°N

LHA ♈	Hc	Zn	Hc	Zn	Hc	Zn	Hc	Zn	Hc	Zn	Hc	Zn	Hc	Zn
	Dubhe		◆CAPELLA		ALDEBARAN		Hamal		◆Alpheratz		ALTAIR		◆VEGA	
0	32 16	008	43 32	070	24 44	100	47 17	135	59 02	177	21 15	250	37 12	285
1	32 20	008	44 00	071	25 14	101	47 38	136	59 03	178	20 47	251	36 43	286
2	32 25	009	44 28	072	25 43	102	47 59	137	59 04	180	20 18	252	36 15	286
3	32 29	010	44 57	072	26 13	103	48 19	138	59 03	182	19 50	253	35 46	287
4	32 34	010	45 26	073	26 42	104	48 38	140	59 02	183	19 21	254	35 17	288
5	32 40	011	45 54	074	27 11	105	48 57	141	59 00	185	18 52	255	34 49	289
6	32 45	011	46 23	074	27 40	105	49 16	142	58 57	187	18 23	255	34 20	289
7	32 51	012	46 52	075	28 09	106	49 34	144	58 53	188	17 54	256	33 52	290
8	32 58	012	47 21	076	28 37	107	49 52	145	58 48	190	17 25	257	33 24	291
9	33 04	013	47 50	077	29 06	108	50 08	146	58 42	192	16 56	258	32 56	292
10	33 11	013	48 20	077	29 34	109	50 25	148	58 36	193	16 26	259	32 28	292
11	33 18	014	48 49	078	30 03	110	50 40	149	58 28	195	15 57	260	32 01	293
12	33 25	014	49 18	079	30 31	111	50 55	151	58 20	197	15 27	261	31 33	294
13	33 33	015	49 48	080	30 59	112	51 10	152	58 11	198	14 58	262	31 06	295
14	33 41	016	50 17	080	31 26	113	51 24	153	58 01	200	14 28	263	30 39	295
	Dubhe		◆CAPELLA		ALDEBARAN		Hamal		◆Alpheratz		DENEB		◆VEGA	
15	33 49	016	50 47	081	31 54	114	51 37	155	57 51	202	49 58	278	30 12	296
16	33 58	017	51 17	082	32 21	115	51 49	156	57 39	203	49 28	279	29 45	297
17	34 06	017	51 46	083	32 48	116	52 01	158	57 27	205	48 59	280	29 18	298
18	34 15	018	52 16	084	33 15	117	52 12	159	57 14	206	48 29	281	28 52	298
19	34 25	018	52 46	084	33 41	118	52 22	161	57 00	208	48 00	282	28 26	299
20	34 34	019	53 16	085	34 08	119	52 32	162	56 46	209	47 30	282	27 59	300
21	34 44	019	53 46	086	34 34	120	52 41	164	56 31	211	47 01	283	27 33	301
22	34 54	020	54 16	087	35 00	121	52 49	165	56 15	212	46 32	284	27 08	301
23	35 04	020	54 46	088	35 25	122	52 56	167	55 58	214	46 03	284	26 42	302
24	35 15	021	55 16	088	35 50	123	53 02	168	55 41	215	45 34	285	26 17	303
25	35 26	021	55 46	089	36 15	124	53 08	170	55 24	217	45 05	286	25 52	304
26	35 37	022	56 16	090	36 40	126	53 13	171	55 05	218	44 36	287	25 27	304
27	35 48	023	56 46	091	37 04	127	53 17	173	54 46	220	44 07	287	25 02	305
28	36 00	023	57 16	092	37 28	128	53 21	174	54 27	221	43 39	288	24 38	306
29	36 12	024	57 46	093	37 51	129	53 23	176	54 07	223	43 10	289	24 14	306
	Dubhe		◆CAPELLA		ALDEBARAN		Hamal		◆Alpheratz		DENEB		◆VEGA	
30	36 24	024	58 15	094	38 15	130	53 25	177	53 46	224	42 42	290	23 49	307
31	36 36	025	58 45	095	38 37	131	53 26	179	53 25	225	42 14	290	23 26	308
32	36 49	025	59 15	096	39 00	132	53 26	180	53 04	227	41 46	291	23 02	309
33	37 02	026	59 45	097	39 22	133	53 26	182	52 42	228	41 18	292	22 39	309
34	37 15	026	60 15	098	39 44	134	53 24	184	52 19	229	40 50	292	22 16	310
35	37 28	027	60 45	099	40 05	136	53 22	185	51 56	230	40 22	293	21 53	311
36	37 42	027	61 14	100	40 25	137	53 19	187	51 33	232	39 55	294	21 30	312
37	37 56	028	61 44	101	40 46	138	53 15	188	51 09	233	39 28	294	21 08	312
38	38 10	028	62 13	102	41 06	139	53 11	190	50 45	234	39 00	295	20 46	313
39	38 24	029	62 43	103	41 25	140	53 05	191	50 21	235	38 33	296	20 24	314
40	38 39	029	63 12	104	41 44	142	52 59	193	49 56	237	38 06	297	20 03	315
41	38 54	030	63 41	105	42 02	143	52 52	194	49 30	238	37 40	297	19 42	315
42	39 09	030	64 10	106	42 20	144	52 44	196	49 05	239	37 13	298	19 21	316
43	39 24	031	64 38	107	42 37	145	52 36	197	48 39	240	36 47	299	19 00	317
44	39 40	031	65 07	108	42 54	147	52 26	199	48 13	241	36 20	299	18 39	318
	◆Dubhe		POLLUX		BETELGEUSE		◆ALDEBARAN		Hamal		Alpheratz		◆DENEB	
45	39 56	032	33 18	091	28 02	129	43 11	148	52 17	200	47 46	242	35 54	300
46	40 12	032	33 48	092	28 25	130	43 26	149	52 06	202	47 20	244	35 28	301
47	40 28	033	34 18	092	28 48	131	43 41	150	51 54	203	46 53	245	35 03	301
48	40 44	033	34 48	093	29 10	132	43 56	152	51 42	205	46 25	246	34 37	302
49	41 01	034	35 17	094	29 32	133	44 10	153	51 30	206	45 58	247	34 12	303
50	41 18	035	35 47	095	29 54	134	44 23	154	51 16	207	45 30	248	33 47	303
51	41 35	035	36 17	096	30 15	135	44 36	156	51 02	209	45 02	249	33 22	304
52	41 52	036	36 47	097	30 36	136	44 48	157	50 47	210	44 34	250	32 57	305
53	42 10	036	37 17	098	30 56	138	45 00	158	50 32	212	44 06	251	32 32	305
54	42 27	037	37 46	099	31 17	139	45 10	160	50 16	213	43 38	252	32 08	306
55	42 45	037	38 16	100	31 36	140	45 21	161	49 59	214	43 09	253	31 44	307
56	43 04	037	38 46	101	31 55	141	45 30	162	49 42	216	42 40	254	31 20	307
57	43 22	038	39 15	102	32 14	142	45 39	164	49 24	217	42 11	255	30 56	308
58	43 41	038	39 44	103	32 32	143	45 47	165	49 06	218	41 42	256	30 33	309
59	43 59	039	40 14	104	32 50	144	45 54	166	48 47	220	41 13	257	30 10	310
	◆Dubhe		POLLUX		BETELGEUSE		◆ALDEBARAN		Hamal		Alpheratz		◆DENEB	
60	44 18	039	40 43	104	33 07	145	46 01	168	48 28	221	40 44	258	29 47	310
61	44 37	040	41 12	105	33 24	146	46 07	169	48 08	222	40 14	259	29 24	311
62	44 57	040	41 41	106	33 41	148	46 13	170	47 47	224	39 45	260	29 01	312
63	45 16	041	42 09	107	33 56	149	46 17	172	47 26	225	39 15	261	28 39	312
64	45 36	041	42 38	108	34 12	150	46 21	173	47 05	226	38 46	262	28 17	313
65	45 56	042	43 06	109	34 26	151	46 24	175	46 43	227	38 16	263	27 55	314
66	46 16	042	43 34	110	34 41	152	46 27	176	46 21	228	37 46	264	27 33	314
67	46 37	043	44 02	112	34 54	153	46 29	177	45 58	230	37 16	264	27 12	315
68	46 57	043	44 30	113	35 08	155	46 30	179	45 35	231	36 46	265	26 51	316
69	47 18	044	44 58	114	35 20	156	46 30	180	45 12	232	36 17	266	26 30	316
70	47 39	044	45 25	115	35 32	157	46 30	182	44 48	233	35 47	267	26 10	317
71	48 00	045	45 52	116	35 44	158	46 28	183	44 24	234	35 17	268	25 49	318
72	48 21	045	46 19	117	35 54	159	46 26	184	43 59	236	34 47	269	25 29	318
73	48 42	046	46 46	118	36 05	161	46 24	186	43 34	237	34 17	270	25 10	319
74	49 04	046	47 12	119	36 14	162	46 21	187	43 09	238	33 47	271	24 50	320
	◆Dubhe		REGULUS		PROCYON		◆BETELGEUSE		ALDEBARAN		◆Alpheratz		DENEB	
75	49 25	047	16 51	095	27 30	134	36 24	163	46 16	188	33 17	271	24 31	320
76	49 47	047	17 20	096	27 52	135	36 32	164	46 12	190	32 47	272	24 12	321
77	50 09	048	17 50	097	28 13	136	36 40	165	46 06	191	32 17	273	23 53	322
78	50 32	048	18 20	098	28 33	137	36 47	167	46 00	193	31 47	274	23 35	323
79	50 54	048	18 50	099	28 53	138	36 54	168	45 53	194	31 17	275	23 17	323
80	51 17	049	19 19	100	29 13	139	37 00	169	45 46	195	30 47	276	22 59	324
81	51 39	049	19 49	101	29 32	141	37 05	170	45 37	197	30 17	277	22 41	325
82	52 02	050	20 18	101	29 51	142	37 10	172	45 28	198	29 47	277	22 24	325
83	52 25	050	20 48	102	30 10	143	37 14	173	45 19	199	29 18	278	22 07	326
84	52 48	051	21 17	103	30 28	144	37 17	174	45 08	201	28 48	279	21 51	327
85	53 12	051	21 46	104	30 45	145	37 20	175	44 57	202	28 18	280	21 34	327
86	53 35	052	22 15	105	31 02	146	37 22	177	44 46	203	27 49	281	21 18	328
87	53 59	052	22 44	106	31 19	147	37 23	178	44 34	205	27 19	281	21 03	329
88	54 22	053	23 13	107	31 35	148	37 24	179	44 21	206	26 50	282	20 47	330
89	54 46	053	23 41	108	31 50	149	37 24	180	44 07	207	26 21	283	20 32	330

LHA ♈	Hc	Zn	Hc	Zn	Hc	Zn	Hc	Zn	Hc	Zn	Hc	Zn	Hc	Zn
	Dubhe		◆REGULUS		PROCYON		BETELGEUSE		◆ALDEBARAN		Mirfak		◆DENEB	
90	55 10	053	24 10	109	32 05	151	37 24	182	43 53	209	65 51	263	20 18	331
91	55 34	054	24 38	110	32 20	152	37 23	183	43 39	210	65 21	264	20 03	332
92	55 59	054	25 06	111	32 34	153	37 21	184	43 23	211	64 51	265	19 49	332
93	56 23	055	25 34	112	32 47	154	37 18	185	43 08	212	64 21	266	19 35	333
94	56 48	055	26 02	113	33 00	155	37 15	187	42 51	214	63 51	267	19 22	334
95	57 12	056	26 30	114	33 12	156	37 12	188	42 34	215	63 21	268	19 09	334
96	57 37	056	26 57	114	33 24	157	37 07	189	42 17	216	62 51	268	18 56	335
97	58 02	056	27 24	115	33 35	159	37 02	190	41 59	217	62 21	269	18 44	336
98	58 27	057	27 51	116	33 46	160	36 56	192	41 40	219	61 51	270	18 32	337
99	58 52	057	28 18	117	33 56	161	36 50	193	41 21	220	61 21	271	18 20	337
100	59 18	058	28 45	118	34 06	162	36 43	194	41 02	221	60 51	272	18 08	338
101	59 43	058	29 11	119	34 15	163	36 36	195	40 42	222	60 21	273	17 57	339
102	60 09	059	29 37	120	34 23	165	36 27	196	40 22	223	59 51	274	17 47	339
103	60 34	059	30 03	121	34 30	166	36 19	198	40 01	225	59 21	274	17 36	340
104	61 00	059	30 28	122	34 38	167	36 09	199	39 40	226	58 52	275	17 26	341
	Alioth		◆REGULUS		PROCYON		BETELGEUSE		◆ALDEBARAN		Mirfak		◆DENEB	
105	46 30	054	30 53	123	34 44	168	35 59	200	39 18	227	58 22	276	17 17	342
106	46 55	055	31 18	125	34 50	169	35 49	201	38 56	228	57 52	277	17 08	342
107	47 19	055	31 43	126	34 55	171	35 37	202	38 33	229	57 22	277	16 59	343
108	47 44	056	32 07	127	35 00	172	35 26	204	38 10	230	56 52	278	16 50	344
109	48 09	057	32 31	128	35 04	173	35 13	205	37 47	231	56 23	279	16 42	345
110	48 34	057	32 54	129	35 07	174	35 00	206	37 24	232	55 53	280	16 34	345
111	48 59	058	33 18	130	35 10	175	34 47	207	37 00	234	55 24	280	16 27	346
112	49 25	058	33 40	131	35 12	177	34 33	208	36 35	235	54 54	281	16 19	347
113	49 50	059	34 03	132	35 13	178	34 18	210	36 11	236	54 25	282	16 13	347
114	50 16	059	34 25	133	35 14	179	34 03	211	35 46	237	53 55	283	16 06	348
115	50 42	060	34 47	134	35 14	180	33 48	212	35 20	238	53 26	283	16 00	349
116	51 08	060	35 08	135	35 14	182	33 32	213	34 55	239	52 57	284	15 55	350
117	51 34	061	35 29	136	35 13	183	33 15	214	34 29	240	52 28	285	15 49	350
118	52 01	062	35 49	138	35 11	184	32 58	215	34 03	241	51 59	285	15 45	351
119	52 27	062	36 09	139	35 09	185	32 40	216	33 37	242	51 30	286	15 40	352
	◆VEGA		ARCTURUS		◆REGULUS		PROCYON		BETELGEUSE		◆CAPELLA		DENEB	
120	10 15	016	14 40	077	36 29	140	35 06	186	32 22	218	62 18	258	15 36	352
121	10 24	017	15 09	078	36 48	141	35 02	188	32 04	219	61 48	259	15 32	353
122	10 33	018	15 38	079	37 07	142	34 58	189	31 45	220	61 19	260	15 29	354
123	10 42	019	16 08	079	37 25	143	34 53	190	31 25	221	60 49	261	15 26	355
124	10 52	019	16 37	080	37 43	144	34 47	191	31 06	222	60 20	262	15 23	355
125	11 02	020	17 07	081	38 00	146	34 41	192	30 45	223	59 50	263	15 21	356
126	11 13	021	17 37	082	38 16	147	34 34	194	30 25	224	59 20	264	15 19	357
127	11 24	022	18 06	083	38 33	148	34 27	195	30 04	225	58 50	265	15 18	358
128	11 35	023	18 36	084	38 48	149	34 19	196	29 42	226	58 20	266	15 17	358
129	11 47	023	19 06	084	39 03	150	34 10	197	29 20	227	57 50	267	15 16	359
130	11 59	024	19 36	085	39 18	152	34 01	198	28 58	228	57 20	268	15 16	000
131	12 11	025	20 06	086	39 32	153	33 51	200	28 35	229	56 50	269	15 16	000
132	12 24	026	20 36	087	39 45	154	33 41	201	28 12	230	56 20	270	15 16	001
133	12 37	026	21 06	088	39 58	155	33 30	202	27 49	232	55 50	271	15 17	002
134	12 51	027	21 36	089	40 10	157	33 19	203	27 25	233	55 20	271	15 18	003
	DENEB		◆VEGA		ARCTURUS		◆REGULUS		PROCYON		BETELGEUSE		◆CAPELLA	
135	15 20	003	13 05	028	22 06	090	40 22	158	33 07	204	27 01	234	54 50	272
136	15 22	004	13 19	029	22 36	090	40 33	159	32 54	205	26 37	235	54 20	273
137	15 24	005	13 34	029	23 06	091	40 43	160	32 41	207	26 13	236	53 50	274
138	15 27	006	13 48	030	23 36	092	40 53	162	32 27	208	25 48	237	53 21	275
139	15 30	006	14 04	031	24 06	093	41 02	163	32 13	209	25 22	238	52 51	276
140	15 34	007	14 19	032	24 36	094	41 10	164	31 58	210	24 57	239	52 21	276
141	15 38	008	14 35	032	25 06	095	41 18	166	31 43	211	24 31	240	51 51	277
142	15 42	009	14 52	033	25 35	096	41 25	167	31 27	212	24 05	241	51 21	278
143	15 46	009	15 08	034	26 05	097	41 32	168	31 11	213	23 39	242	50 52	279
144	15 51	010	15 25	035	26 35	098	41 37	169	30 54	215	23 13	242	50 22	279
145	15 57	011	15 42	035	27 05	098	41 43	171	30 37	216	22 46	243	49 52	280
146	16 03	011	16 00	036	27 34	099	41 47	172	30 19	217	22 19	244	49 23	281
147	16 09	012	16 18	037	28 04	100	41 51	173	30 01	218	21 52	245	48 54	282
148	16 15	013	16 36	038	28 33	101	41 54	175	29 42	219	21 24	246	48 24	282
149	16 22	014	16 55	038	29 03	102	41 56	176	29 23	220	20 57	247	47 55	283
	DENEB		◆VEGA		ARCTURUS		Denebola		◆REGULUS		POLLUX		◆CAPELLA	
150	16 29	014	17 13	039	29 32	103	40 26	144	41 58	177	50 43	231	47 26	284
151	16 37	015	17 33	040	30 01	104	40 44	146	41 59	179	50 20	232	46 57	285
152	16 45	016	17 52	041	30 30	105	41 00	147	42 00	180	49 56	233	46 28	285
153	16 53	016	18 12	041	30 59	106	41 16	148	41 59	181	49 32	234	45 59	286
154	17 02	017	18 32	042	31 28	107	41 32	149	41 58	183	49 07	236	45 30	287
155	17 11	018	18 52	043	31 57	108	41 47	151	41 56	184	48 42	237	45 01	288
156	17 20	019	19 13	044	32 25	109	42 01	152	41 54	185	48 17	238	44 33	288
157	17 30	019	19 33	044	32 54	110	42 15	153	41 51	187	47 51	239	44 04	289
158	17 40	020	19 55	045	33 22	111	42 28	154	41 47	188	47 25	240	43 36	290
159	17 51	021	20 16	046	33 50	112	42 41	156	41 43	189	46 59	241	43 08	290
160	18 01	022	20 38	047	34 17	113	42 53	157	41 38	190	46 33	243	42 40	291
161	18 13	022	21 00	047	34 45	114	43 04	158	41 32	192	46 06	244	42 12	292
162	18 24	023	21 22	048	35 12	115	43 15	160	41 25	193	45 39	245	41 44	292
163	18 36	024	21 44	049	35 40	116	43 25	161	41 18	194	45 12	246	41 17	293
164	18 48	024	22 07	050	36 07	117	43 35	162	41 11	196	44 44	247	40 49	294
	◆DENEB		VEGA		ARCTURUS		◆SPICA		REGULUS		◆POLLUX		CAPELLA	
165	19 01	025	22 30	050	36 33	118	13 12	143	41 02	197	44 16	248	40 22	295
166	19 14	026	22 53	051	37 00	119	13 30	144	40 53	198	43 49	249	39 54	295
167	19 27	026	23 17	052	37 26	120	13 47	145	40 43	199	43 20	250	39 27	296
168	19 40	027	23 40	053	37 52	121	14 04	146	40 33	201	42 52	251	39 01	297
169	19 54	028	24 04	053	38 17	122	14 21	147	40 22	202	42 24	252	38 34	297
170	20 08	029	24 28	054	38 43	123	14 37	148	40 11	203	41 55	253	38 07	298
171	20 23	029	24 53	055	39 08	124	14 52	149	39 58	205	41 26	254	37 41	299
172	20 38	030	25 17	055	39 32	125	15 07	150	39 46	206	40 57	255	37 15	299
173	20 53	031	25 42	056	39 57	126	15 22	151	39 32	207	40 28	256	36 49	300
174	21 08	031	26 07	057	40 21	127	15 36	152	39 19	208	39 59	257	36 23	301
175	21 24	032	26 32	058	40 44	129	15 50	153	39 04	209	39 30	258	35 57	301
176	21 40	033	26 58	058	41 08	130	16 03	154	38 49	211	39 00	259	35 31	302
177	21 57	034	27 24	059	41 30	131	16 16	155	38 33	212	38 31	260	35 06	303
178	22 14	034	27 49	060	41 53	132	16 28	156	38 17	213	38 01	261	34 41	303
179	22 31	035	28 15	061	42 15	133	16 40	157	38 01	214	37 32	262	34 16	304

LAT 60°N

LHA ♈	Hc	Zn	Hc	Zn	Hc	Zn	Hc	Zn	Hc	Zn	Hc	Zn	Hc	Zn
	◆DENEB		VEGA		ARCTURUS		◆SPICA		REGULUS		◆POLLUX		CAPELLA	
180	22 48	036	28 42	061	42 37	134	16 52	158	37 44	215	37 02	263	33 51	305
181	23 05	036	29 08	062	42 58	136	17 02	159	37 26	217	36 32	263	33 27	305
182	23 23	037	29 35	063	43 19	137	17 13	160	37 08	218	36 02	264	33 02	306
183	23 42	038	30 01	064	43 39	138	17 23	161	36 49	219	35 33	265	32 38	307
184	24 00	038	30 28	064	43 59	139	17 32	162	36 30	220	35 03	266	32 14	307
185	24 19	039	30 55	065	44 18	140	17 41	163	36 10	221	34 33	267	31 51	308
186	24 38	040	31 23	066	44 37	142	17 49	164	35 50	222	34 03	268	31 27	309
187	24 57	040	31 50	066	44 55	143	17 57	165	35 30	224	33 33	269	31 04	309
188	25 17	041	32 18	067	45 13	144	18 05	166	35 09	225	33 03	270	30 41	310
189	25 37	042	32 45	068	45 30	145	18 11	167	34 48	226	32 33	271	30 18	311
190	25 57	043	33 13	069	45 47	147	18 18	168	34 26	227	32 03	271	29 55	312
191	26 17	043	33 41	069	46 03	148	18 24	169	34 04	228	31 33	272	29 33	312
192	26 38	044	34 10	070	46 19	149	18 29	170	33 42	229	31 03	273	29 11	313
193	26 59	045	34 38	071	46 34	151	18 34	171	33 19	230	30 33	274	28 49	314
194	27 20	045	35 06	072	46 48	152	18 38	173	32 56	231	30 03	275	28 28	314
	◆DENEB		VEGA		Rasalhague		◆ARCTURUS		REGULUS		◆POLLUX		CAPELLA	
195	27 41	046	35 35	072	21 28	102	47 02	153	32 32	232	29 33	276	28 06	315
196	28 03	047	36 04	073	21 57	103	47 15	155	32 08	233	29 03	277	27 45	316
197	28 25	047	36 32	074	22 26	104	47 28	156	31 44	234	28 33	277	27 24	316
198	28 47	048	37 01	075	22 55	105	47 39	157	31 19	235	28 04	278	27 04	317
199	29 10	049	37 30	076	23 24	106	47 51	159	30 54	236	27 34	279	26 43	318
200	29 32	049	37 59	076	23 53	107	48 01	160	30 29	237	27 04	280	26 23	318
201	29 55	050	38 28	077	24 21	108	48 11	162	30 04	239	26 35	281	26 03	319
202	30 18	051	38 58	078	24 50	109	48 20	163	29 38	240	26 06	281	25 44	320
203	30 42	051	39 27	079	25 18	110	48 28	164	29 12	241	25 36	282	25 24	320
204	31 05	052	39 57	079	25 46	111	48 36	166	28 46	242	25 07	283	25 05	321
205	31 29	053	40 26	080	26 14	112	48 43	167	28 19	243	24 38	284	24 47	322
206	31 53	053	40 56	081	26 42	113	48 49	169	27 53	244	24 09	285	24 28	322
207	32 17	054	41 25	082	27 10	114	48 55	170	27 26	244	23 40	286	24 10	323
208	32 41	055	41 55	083	27 37	115	49 00	172	26 59	245	23 11	286	23 52	324
209	33 06	055	42 25	083	28 04	116	49 04	173	26 31	246	22 42	287	23 35	324
	◆DENEB		VEGA		Rasalhague		◆ARCTURUS		REGULUS		◆POLLUX		CAPELLA	
210	33 31	056	42 55	084	28 31	116	49 07	174	26 04	247	22 14	288	23 17	325
211	33 56	057	43 25	085	28 58	117	49 09	176	25 36	248	21 45	289	23 00	326
212	34 21	058	43 54	086	29 24	118	49 11	177	25 08	249	21 17	290	22 44	327
213	34 47	058	44 24	087	29 51	119	49 12	179	24 40	250	20 48	290	22 27	327
214	35 12	059	44 54	088	30 17	120	49 12	180	24 11	251	20 20	291	22 11	328
215	35 38	060	45 24	088	30 42	122	49 12	182	23 43	252	19 53	292	21 55	329
216	36 04	060	45 54	089	31 08	123	49 11	183	23 14	253	19 25	293	21 40	329
217	36 30	061	46 24	090	31 33	124	49 09	185	22 45	254	18 57	294	21 25	330
218	36 56	062	46 54	091	31 58	125	49 06	186	22 17	255	18 30	294	21 10	331
219	37 23	062	47 24	092	32 22	126	49 03	187	21 48	256	18 03	295	20 55	331
220	37 49	063	47 54	093	32 47	127	48 58	189	21 18	257	17 35	296	20 41	332
221	38 16	064	48 24	094	33 10	128	48 53	190	20 49	258	17 09	297	20 27	333
222	38 43	064	48 54	095	33 34	129	48 48	192	20 20	259	16 42	298	20 14	333
223	39 10	065	49 24	096	33 57	130	48 41	193	19 50	259	16 15	298	20 00	334
224	39 38	066	49 54	096	34 20	131	48 34	195	19 21	260	15 49	299	19 48	335
	DENEB		◆VEGA		Rasalhague		◆ARCTURUS		Denebola		Dubhe		◆CAPELLA	
225	40 05	066	50 24	097	34 42	132	48 26	196	32 55	239	62 10	300	19 35	336
226	40 33	067	50 53	098	35 04	133	48 17	197	32 29	240	61 44	300	19 23	336
227	41 00	068	51 23	099	35 26	134	48 08	199	32 03	241	61 18	300	19 11	337
228	41 28	069	51 53	100	35 47	135	47 58	200	31 37	242	60 52	301	18 59	338
229	41 56	069	52 22	101	36 08	137	47 47	202	31 10	243	60 27	301	18 48	338
230	42 24	070	52 51	102	36 29	138	47 36	203	30 43	244	60 01	302	18 37	339
231	42 53	071	53 21	103	36 49	139	47 24	204	30 16	245	59 36	302	18 27	340
232	43 21	071	53 50	104	37 08	140	47 11	206	29 49	246	59 10	302	18 17	341
233	43 49	072	54 19	105	37 27	141	46 58	207	29 22	247	58 45	303	18 07	341
234	44 18	073	54 48	106	37 46	142	46 44	208	28 54	248	58 20	303	17 57	342
235	44 47	074	55 16	107	38 04	143	46 30	210	28 26	249	57 55	304	17 48	343
236	45 16	074	55 45	109	38 21	145	46 15	211	27 58	250	57 30	304	17 39	343
237	45 45	075	56 13	110	38 39	146	45 59	212	27 30	251	57 05	305	17 31	344
238	46 14	076	56 41	111	38 55	147	45 42	214	27 01	251	56 41	305	17 23	345
239	46 43	077	57 09	112	39 11	148	45 26	215	26 33	252	56 16	305	17 15	346
	DENEB		◆VEGA		Rasalhague		◆ARCTURUS		Denebola		Dubhe		◆CAPELLA	
240	47 12	077	57 37	113	39 27	149	45 08	216	26 04	253	55 52	306	17 08	346
241	47 41	078	58 05	114	39 42	151	44 50	217	25 35	254	55 27	306	17 01	347
242	48 11	079	58 32	115	39 56	152	44 32	219	25 06	255	55 03	307	16 55	348
243	48 40	080	58 59	117	40 10	153	44 13	220	24 37	256	54 39	307	16 48	348
244	49 10	080	59 25	118	40 23	154	43 53	221	24 08	257	54 15	308	16 42	349
245	49 39	081	59 52	119	40 36	156	43 33	222	23 39	258	53 52	308	16 37	350
246	50 09	082	60 18	121	40 48	157	43 13	224	23 10	259	53 28	308	16 32	351
247	50 39	083	60 43	122	40 59	158	42 52	225	22 40	260	53 05	309	16 27	351
248	51 08	083	61 09	123	41 10	159	42 31	226	22 10	261	52 42	309	16 23	352
249	51 38	084	61 34	125	41 20	161	42 09	227	21 41	262	52 18	310	16 19	353
250	52 08	085	61 58	126	41 30	162	41 47	228	21 11	262	51 55	310	16 15	353
251	52 38	086	62 22	127	41 39	163	41 24	229	20 41	263	51 33	311	16 12	354
252	53 08	087	62 46	129	41 47	165	41 01	231	20 12	264	51 10	311	16 09	355
253	53 38	087	63 09	130	41 55	166	40 38	232	19 42	265	50 48	312	16 07	356
254	54 08	088	63 31	132	42 02	167	40 14	233	19 12	266	50 25	312	16 04	356
	Schedar		DENEB		◆ALTAIR		Rasalhague		◆ARCTURUS		Dubhe		◆CAPELLA	
255	37 15	039	54 38	089	29 47	130	42 08	169	39 50	234	50 03	313	16 03	357
256	37 34	039	55 08	090	30 10	131	42 13	170	39 25	235	49 41	313	16 01	358
257	37 54	040	55 38	091	30 33	132	42 18	171	39 01	236	49 19	314	16 00	358
258	38 13	041	56 08	092	30 55	133	42 23	172	38 35	237	48 57	314	16 00	359
259	38 33	041	56 38	093	31 17	134	42 26	174	38 10	238	48 36	314	16 00	000
260	38 53	042	57 08	094	31 38	135	42 29	175	37 44	239	48 15	315	16 00	001
261	39 13	042	57 38	095	31 59	136	42 31	176	37 18	240	47 53	315	16 00	001
262	39 33	043	58 08	095	32 20	137	42 33	178	36 52	242	47 32	316	16 01	002
263	39 53	043	58 37	096	32 40	138	42 34	179	36 26	243	47 12	316	16 03	003
264	40 14	044	59 07	097	33 00	139	42 34	180	35 59	244	46 51	317	16 04	004
265	40 35	045	59 37	098	33 19	140	42 33	182	35 32	245	46 31	317	16 06	004
266	40 56	045	60 07	099	33 38	142	42 32	183	35 05	246	46 10	318	16 09	005
267	41 18	046	60 36	100	33 57	143	42 30	184	34 37	247	45 50	318	16 11	006
268	41 39	046	61 06	101	34 15	144	42 27	186	34 10	248	45 30	319	16 15	006
269	42 01	047	61 35	102	34 32	145	42 24	187	33 42	249	45 11	319	16 18	007

LHA ♈	Hc	Zn	Hc	Zn	Hc	Zn	Hc	Zn	Hc	Zn	Hc	Zn	Hc	Zn
	◆CAPELLA		Alpheratz		◆ALTAIR		Rasalhague		◆ARCTURUS		Alkaid		Dubhe	
270	16 22	008	23 54	073	34 49	146	42 20	188	33 14	250	53 31	282	44 51	320
271	16 26	009	24 23	074	35 06	147	42 15	190	32 46	251	53 02	283	44 32	320
272	16 31	009	24 52	074	35 22	148	42 10	191	32 17	252	52 32	284	44 13	321
273	16 36	010	25 21	075	35 37	150	42 04	192	31 49	253	52 03	284	43 54	321
274	16 42	011	25 50	076	35 52	151	41 57	194	31 20	253	51 34	285	43 35	322
275	16 47	011	26 19	077	36 06	152	41 50	195	30 51	254	51 05	286	43 17	322
276	16 54	012	26 48	078	36 20	153	41 42	196	30 22	255	50 36	286	42 58	323
277	17 00	013	27 18	078	36 34	154	41 33	197	29 53	256	50 08	287	42 40	323
278	17 07	014	27 47	079	36 46	156	41 24	199	29 24	257	49 39	288	42 22	324
279	17 14	014	28 17	080	36 58	157	41 14	200	28 55	258	49 11	288	42 05	324
280	17 22	015	28 46	081	37 10	158	41 03	201	28 25	259	48 42	289	41 47	325
281	17 30	016	29 16	082	37 21	159	40 52	203	27 56	260	48 14	290	41 30	325
282	17 38	016	29 45	083	37 31	160	40 40	204	27 26	261	47 46	290	41 13	326
283	17 47	017	30 15	083	37 41	162	40 28	205	26 56	262	47 18	291	40 56	326
284	17 56	018	30 45	084	37 50	163	40 15	206	26 27	263	46 50	292	40 39	327
	◆CAPELLA		Alpheratz		◆ALTAIR		Rasalhague		◆ARCTURUS		Alkaid		Dubhe	
285	18 05	019	31 15	085	37 59	164	40 01	208	25 57	264	46 22	292	40 23	327
286	18 15	019	31 45	086	38 07	165	39 47	209	25 27	265	45 54	293	40 07	328
287	18 25	020	32 15	087	38 14	167	39 32	210	24 57	265	45 27	294	39 51	328
288	18 36	021	32 45	088	38 21	168	39 17	211	24 27	266	44 59	294	39 35	329
289	18 46	021	33 15	088	38 27	169	39 01	213	23 57	267	44 32	295	39 20	329
290	18 58	022	33 45	089	38 32	170	38 44	214	23 27	268	44 05	296	39 05	330
291	19 09	023	34 15	090	38 37	172	38 28	215	22 57	269	43 38	296	38 50	330
292	19 21	024	34 45	091	38 41	173	38 10	216	22 27	270	43 11	297	38 35	331
293	19 33	024	35 15	092	38 44	174	37 52	217	21 57	271	42 45	298	38 20	331
294	19 46	025	35 45	093	38 47	175	37 34	218	21 27	272	42 18	298	38 06	332
295	19 58	026	36 15	094	38 49	177	37 15	220	20 57	272	41 52	299	37 52	332
296	20 12	026	36 45	095	38 50	178	36 56	221	20 27	273	41 26	300	37 38	333
297	20 25	027	37 14	096	38 51	179	36 36	222	19 57	274	41 00	300	37 25	333
298	20 39	028	37 44	096	38 51	180	36 15	223	19 27	275	40 34	301	37 11	334
299	20 53	029	38 14	097	38 51	182	35 55	224	18 58	276	40 08	302	36 58	334
	CAPELLA		◆Alpheratz		Enif		ALTAIR		◆VEGA		ARCTURUS		◆Dubhe	
300	21 08	029	38 44	098	36 14	148	38 49	183	65 04	221	18 28	277	36 45	335
301	21 22	030	39 13	099	36 29	149	38 48	184	64 44	223	17 58	277	36 33	335
302	21 37	031	39 43	100	36 45	150	38 45	186	64 23	224	17 28	278	36 20	336
303	21 53	031	40 13	101	36 59	151	38 42	187	64 02	226	16 59	279	36 08	337
304	22 09	032	40 42	102	37 14	152	38 38	188	63 40	228	16 29	280	35 57	337
305	22 25	033	41 11	103	37 27	154	38 33	189	63 18	229	16 00	281	35 45	338
306	22 41	033	41 40	104	37 40	155	38 28	191	62 55	231	15 30	282	35 34	338
307	22 58	034	42 09	105	37 53	156	38 22	192	62 31	232	15 01	283	35 23	339
308	23 15	035	42 38	106	38 05	157	38 16	193	62 07	233	14 32	283	35 12	339
309	23 32	035	43 07	107	38 16	158	38 09	194	61 43	235	14 02	284	35 01	340
310	23 49	036	43 36	108	38 27	160	38 01	196	61 18	236	13 33	285	34 51	340
311	24 07	037	44 04	109	38 37	161	37 53	197	60 53	238	13 04	286	34 41	341
312	24 25	037	44 32	110	38 46	162	37 44	198	60 28	239	12 36	287	34 31	341
313	24 44	038	45 01	111	38 55	163	37 34	199	60 02	240	12 07	288	34 22	342
314	25 02	039	45 29	112	39 03	165	37 24	201	59 36	242	11 38	288	34 13	342
	CAPELLA		◆Hamal		Alpheratz		Enif		◆ALTAIR		VEGA		◆Alioth	
315	25 21	040	26 44	089	45 56	113	39 11	166	37 13	202	59 09	243	34 51	324
316	25 41	040	27 14	090	46 24	114	39 18	167	37 02	203	58 42	244	34 34	325
317	26 00	041	27 44	091	46 51	115	39 24	169	36 50	204	58 15	245	34 17	326
318	26 20	042	28 14	092	47 18	116	39 30	170	36 37	205	57 48	246	34 00	326
319	26 40	042	28 44	092	47 45	117	39 35	171	36 24	207	57 20	248	33 44	327
320	27 00	043	29 14	093	48 11	119	39 39	172	36 10	208	56 52	249	33 27	327
321	27 21	044	29 44	094	48 37	120	39 43	174	35 56	209	56 24	250	33 11	328
322	27 42	044	30 14	095	49 03	121	39 46	175	35 41	210	55 56	251	32 56	329
323	28 03	045	30 44	096	49 29	122	39 48	176	35 26	211	55 27	252	32 40	329
324	28 24	046	31 13	097	49 54	123	39 50	177	35 10	212	54 59	253	32 25	330
325	28 46	046	31 43	098	50 19	124	39 51	179	34 54	214	54 30	254	32 10	330
326	29 08	047	32 13	099	50 44	126	39 51	180	34 37	215	54 01	255	31 55	331
327	29 30	048	32 43	100	51 08	127	39 51	181	34 19	216	53 32	256	31 41	332
328	29 52	048	33 12	101	51 31	128	39 50	183	34 02	217	53 03	257	31 27	332
329	30 14	049	33 41	101	51 55	129	39 48	184	33 43	218	52 33	258	31 13	333
	CAPELLA		◆Hamal		Alpheratz		Enif		◆ALTAIR		VEGA		◆Alioth	
330	30 37	050	34 11	102	52 18	131	39 46	185	33 25	219	52 04	259	30 59	333
331	31 00	050	34 40	103	52 40	132	39 43	186	33 05	220	51 34	260	30 46	334
332	31 23	051	35 09	104	53 02	133	39 39	188	32 46	222	51 05	261	30 33	335
333	31 47	052	35 38	105	53 24	135	39 35	189	32 25	223	50 35	262	30 20	335
334	32 11	052	36 07	106	53 45	136	39 30	190	32 05	224	50 05	263	30 08	336
335	32 34	053	36 36	107	54 06	137	39 24	192	31 44	225	49 36	264	29 55	336
336	32 59	054	37 04	108	54 26	139	39 18	193	31 23	226	49 06	265	29 43	337
337	33 23	054	37 33	109	54 45	140	39 11	194	31 01	227	48 36	266	29 32	338
338	33 47	055	38 01	110	55 04	142	39 03	195	30 39	228	48 06	267	29 21	338
339	34 12	056	38 29	111	55 23	143	38 55	197	30 16	229	47 36	268	29 10	339
340	34 37	056	38 57	112	55 40	144	38 46	198	29 53	230	47 06	269	28 59	339
341	35 02	057	39 25	113	55 57	146	38 36	199	29 30	231	46 36	270	28 48	340
342	35 27	058	39 52	114	56 14	147	38 26	200	29 07	232	46 06	270	28 38	341
343	35 53	058	40 19	115	56 30	149	38 16	202	28 43	233	45 36	271	28 28	341
344	36 19	059	40 46	116	56 45	150	38 04	203	28 19	234	45 06	272	28 19	342
	CAPELLA		ALDEBARAN		◆Hamal		Alpheratz		◆ALTAIR		VEGA		◆Alioth	
345	36 44	060	17 16	087	41 13	117	56 59	152	27 54	235	44 36	273	28 10	342
346	37 10	061	17 46	088	41 40	118	57 13	154	27 29	236	44 06	274	28 01	343
347	37 37	061	18 16	088	42 06	120	57 26	155	27 04	237	43 36	275	27 52	344
348	38 03	062	18 46	089	42 32	121	57 38	157	26 39	238	43 06	275	27 44	344
349	38 30	063	19 16	090	42 58	122	57 50	158	26 13	239	42 36	276	27 36	345
350	38 56	063	19 46	091	43 23	123	58 01	160	25 47	240	42 07	277	27 28	345
351	39 23	064	20 16	092	43 48	124	58 10	162	25 21	241	41 37	278	27 21	346
352	39 50	065	20 46	093	44 13	125	58 20	163	24 54	242	41 07	279	27 14	347
353	40 17	065	21 16	094	44 37	126	58 28	165	24 28	243	40 37	279	27 07	347
354	40 45	066	21 46	095	45 01	127	58 35	166	24 01	244	40 08	280	27 01	348
355	41 12	067	22 16	095	45 25	129	58 42	168	23 34	245	39 38	281	26 54	349
356	41 40	067	22 46	096	45 48	130	58 48	170	23 06	246	39 09	282	26 49	349
357	42 08	068	23 15	097	46 11	131	58 53	171	22 39	247	38 40	283	26 43	350
358	42 35	069	23 45	098	46 33	132	58 57	173	22 11	248	38 10	283	26 38	350
359	43 03	069	24 15	099	46 55	133	59 00	175	21 43	249	37 41	284	26 33	351

LAT 59°N

LAT 59°N

LHA ♈	Hc	Zn	Hc	Zn	Hc	Zn	Hc	Zn	Hc	Zn	Hc	Zn	Hc	Zn
	Dubhe		♦CAPELLA		ALDEBARAN		Hamal		♦Alpheratz		ALTAIR		♦VEGA	
0	31 16	008	43 11	069	24 54	099	47 59	134	60 02	176	21 36	250	36 56	286
1	31 21	008	43 40	070	25 25	100	48 21	135	60 03	178	21 06	251	36 27	286
2	31 25	009	44 09	071	25 55	101	48 42	136	60 04	180	20 37	252	35 57	287
3	31 30	009	44 38	071	26 25	102	49 03	138	60 03	182	20 08	253	35 28	288
4	31 35	010	45 08	072	26 56	103	49 24	139	60 02	183	19 38	254	34 58	289
5	31 41	011	45 37	073	27 26	104	49 44	140	60 00	185	19 08	255	34 29	289
6	31 47	011	46 07	073	27 56	105	50 03	142	59 56	187	18 38	256	34 00	290
7	31 53	012	46 36	074	28 25	106	50 22	143	59 52	189	18 08	257	33 31	291
8	31 59	012	47 06	075	28 55	107	50 41	144	59 47	190	17 38	258	33 02	292
9	32 06	013	47 36	076	29 25	108	50 58	146	59 41	192	17 08	258	32 34	292
10	32 13	013	48 06	076	29 54	109	51 15	147	59 34	194	16 38	259	32 05	293
11	32 20	014	48 36	077	30 23	110	51 32	149	59 26	196	16 07	260	31 37	294
12	32 27	014	49 06	078	30 52	111	51 48	150	59 17	197	15 37	261	31 09	295
13	32 35	015	49 36	079	31 21	112	52 03	151	59 08	199	15 06	262	30 41	295
14	32 43	015	50 07	079	31 50	113	52 17	153	58 57	201	14 35	263	30 13	296
	Dubhe		♦CAPELLA		ALDEBARAN		Hamal		♦Alpheratz		DENEB		♦VEGA	
15	32 51	016	50 37	080	32 18	114	52 31	154	58 46	202	49 48	280	29 45	297
16	33 00	016	51 08	081	32 46	115	52 44	156	58 34	204	49 18	280	29 18	297
17	33 09	017	51 38	082	33 14	116	52 56	157	58 21	205	48 48	281	28 50	298
18	33 18	018	52 09	082	33 42	117	53 08	159	58 08	207	48 17	282	28 23	299
19	33 28	018	52 39	083	34 09	118	53 19	160	57 53	209	47 47	283	27 56	300
20	33 37	019	53 10	084	34 37	119	53 29	162	57 38	210	47 17	283	27 29	300
21	33 47	019	53 41	085	35 04	120	53 38	163	57 22	212	46 47	284	27 03	301
22	33 57	020	54 12	085	35 30	121	53 47	165	57 05	213	46 17	285	26 36	302
23	34 08	020	54 42	086	35 57	122	53 54	166	56 48	215	45 47	285	26 10	302
24	34 19	021	55 13	087	36 23	123	54 01	168	56 30	216	45 18	286	25 44	303
25	34 30	021	55 44	088	36 49	124	54 07	169	56 11	218	44 48	287	25 18	304
26	34 41	022	56 15	089	37 14	125	54 12	171	55 52	219	44 18	288	24 53	305
27	34 53	022	56 46	090	37 40	126	54 17	173	55 32	221	43 49	288	24 28	305
28	35 05	023	57 17	090	38 04	127	54 20	174	55 12	222	43 20	289	24 03	306
29	35 17	023	57 48	091	38 29	128	54 23	176	54 51	224	42 51	290	23 38	307
	Dubhe		♦CAPELLA		ALDEBARAN		Hamal		♦Alpheratz		DENEB		♦VEGA	
30	35 29	024	58 19	092	38 53	129	54 25	177	54 29	225	42 22	290	23 13	308
31	35 42	024	58 49	093	39 17	130	54 26	179	54 07	226	41 53	291	22 49	308
32	35 55	025	59 20	094	39 40	132	54 26	180	53 45	228	41 24	292	22 25	309
33	36 08	025	59 51	095	40 03	133	54 26	182	53 22	229	40 55	292	22 01	310
34	36 21	026	60 22	096	40 25	134	54 24	184	52 58	230	40 27	293	21 37	310
35	36 35	026	60 53	097	40 47	135	54 22	185	52 34	231	39 59	294	21 14	311
36	36 49	027	61 23	098	41 09	136	54 19	187	52 10	233	39 30	295	20 50	312
37	37 03	027	61 54	099	41 30	137	54 15	188	51 45	234	39 02	295	20 28	313
38	37 17	028	62 24	100	41 51	139	54 10	190	51 20	235	38 34	296	20 05	313
39	37 32	028	62 55	101	42 11	140	54 04	191	50 54	236	38 07	297	19 43	314
40	37 47	029	63 25	102	42 31	141	53 57	193	50 28	238	37 39	297	19 21	315
41	38 02	030	63 55	103	42 50	142	53 50	195	50 02	239	37 12	298	18 59	316
42	38 17	030	64 25	104	43 09	144	53 42	196	49 35	240	36 45	299	18 37	316
43	38 33	031	64 55	105	43 27	145	53 33	198	49 08	241	36 18	299	18 16	317
44	38 49	031	65 25	106	43 44	146	53 23	199	48 41	242	35 51	300	17 55	318
	♦Dubhe		POLLUX		BETELGEUSE		♦ALDEBARAN		Hamal		Alpheratz		♦DENEB	
45	39 05	032	33 18	090	28 40	129	44 01	147	53 13	201	48 14	245	35 24	301
46	39 21	032	33 49	091	29 03	130	44 18	149	53 02	202	47 46	245	34 57	301
47	39 37	033	34 20	092	29 27	131	44 33	150	52 50	204	47 18	246	34 31	302
48	39 54	033	34 51	093	29 50	132	44 49	151	52 37	205	46 50	247	34 05	303
49	40 11	034	35 22	094	30 13	133	45 03	152	52 23	207	46 21	248	33 39	303
50	40 28	034	35 52	094	30 36	134	45 17	154	52 09	208	45 52	249	33 13	304
51	40 46	035	36 23	095	30 58	135	45 30	155	51 54	209	45 24	250	32 48	305
52	41 03	035	36 54	096	31 19	136	45 43	156	51 39	211	44 54	251	32 23	305
53	41 21	035	37 25	097	31 41	137	45 55	158	51 23	212	44 25	252	31 57	306
54	41 39	036	37 55	098	32 01	138	46 06	159	51 06	214	43 56	253	31 32	307
55	41 57	036	38 26	099	32 22	139	46 17	161	50 48	215	43 26	254	31 08	307
56	42 16	037	38 56	100	32 42	140	46 27	162	50 30	216	42 56	255	30 43	308
57	42 35	037	39 27	101	33 01	142	46 36	163	50 12	218	42 26	256	30 19	309
58	42 53	038	39 57	102	33 20	143	46 45	165	49 53	219	41 56	257	29 55	309
59	43 12	038	40 27	103	33 39	144	46 53	166	49 33	220	41 26	258	29 31	310
	♦Dubhe		POLLUX		BETELGEUSE		♦ALDEBARAN		Hamal		Alpheratz		♦DENEB	
60	43 32	039	40 57	104	33 57	145	47 00	167	49 13	222	40 56	259	29 08	311
61	43 51	039	41 27	105	34 14	146	47 06	169	48 52	223	40 26	260	28 44	311
62	44 11	040	41 57	106	34 31	147	47 12	170	48 30	224	39 55	261	28 21	312
63	44 31	040	42 27	107	34 48	148	47 17	172	48 09	226	39 25	262	27 58	313
64	44 51	041	42 56	108	35 04	150	47 21	173	47 46	227	38 54	262	27 36	313
65	45 11	041	43 26	109	35 19	151	47 24	174	47 24	228	38 23	263	27 13	314
66	45 32	042	43 55	110	35 34	152	47 27	176	47 00	229	37 53	264	26 51	315
67	45 52	042	44 24	111	35 48	153	47 29	177	46 37	230	37 22	265	26 30	315
68	46 13	043	44 53	112	36 02	154	47 30	179	46 13	232	36 51	266	26 08	316
69	46 34	043	45 21	113	36 15	155	47 30	180	45 48	233	36 20	267	25 47	317
70	46 55	044	45 50	114	36 27	157	47 30	182	45 23	234	35 49	268	25 26	317
71	47 17	044	46 18	115	36 39	158	47 28	183	44 58	235	35 18	269	25 05	318
72	47 38	044	46 46	116	36 51	159	47 26	184	44 33	236	34 48	270	24 44	319
73	48 00	045	47 13	117	37 01	160	47 24	186	44 07	237	34 17	270	24 24	319
74	48 22	045	47 41	118	37 11	162	47 20	187	43 41	239	33 46	271	24 04	320
	♦Dubhe		REGULUS		PROCYON		♦SIRIUS		ALDEBARAN		♦Alpheratz		DENEB	
75	48 44	046	16 56	095	28 12	134	11 18	154	47 16	189	33 15	272	23 44	321
76	49 06	046	17 27	096	28 34	135	11 31	155	47 11	190	32 44	273	23 25	321
77	49 29	047	17 57	097	28 56	136	11 44	156	47 05	191	32 13	274	23 06	322
78	49 51	047	18 28	097	29 17	137	11 56	157	46 59	193	31 42	275	22 47	323
79	50 14	048	18 59	098	29 38	138	12 08	158	46 51	194	31 11	275	22 29	324
80	50 37	048	19 29	099	29 58	139	12 19	159	46 43	196	30 41	276	22 10	324
81	51 00	048	20 00	100	30 19	140	12 30	160	46 35	197	30 10	277	21 52	325
82	51 23	049	20 30	101	30 38	141	12 40	161	46 25	198	29 39	278	21 35	326
83	51 46	049	21 00	102	30 57	142	12 50	162	46 15	200	29 09	279	21 17	326
84	52 10	050	21 30	103	31 16	143	12 59	163	46 04	201	28 38	280	21 00	327
85	52 34	050	22 00	104	31 34	145	13 08	164	45 53	202	28 08	280	20 44	328
86	52 57	051	22 30	105	31 52	146	13 16	165	45 41	204	27 38	281	20 27	328
87	53 21	051	23 00	106	32 09	147	13 24	166	45 28	205	27 08	282	20 11	329
88	53 45	051	23 30	106	32 26	148	13 31	167	45 15	206	26 37	283	19 56	330
89	54 10	052	24 00	107	32 42	149	13 38	168	45 01	208	26 07	284	19 40	330

LHA ♈	Hc	Zn	Hc	Zn	Hc	Zn	Hc	Zn	Hc	Zn	Hc	Zn	Hc	Zn
	Dubhe		♦REGULUS		PROCYON		SIRIUS		♦ALDEBARAN		Mirfak		♦DENEB	
90	54 34	052	24 29	108	32 57	150	13 44	169	44 46	209	65 57	265	19 25	331
91	54 58	053	24 58	109	33 12	151	13 50	170	44 31	210	65 26	266	19 10	332
92	55 23	053	25 27	110	33 27	153	13 55	171	44 15	212	64 56	267	18 56	333
93	55 48	054	25 56	111	33 41	154	13 59	172	43 58	213	64 25	268	18 42	333
94	56 13	054	26 25	112	33 54	155	14 04	173	43 41	214	63 54	269	18 28	334
95	56 38	054	26 53	113	34 07	156	14 07	174	43 23	215	63 23	269	18 15	335
96	57 03	055	27 22	114	34 20	157	14 10	175	43 05	217	62 52	270	18 02	335
97	57 28	055	27 50	115	34 31	158	14 13	176	42 46	218	62 21	271	17 49	336
98	57 54	056	28 18	116	34 42	160	14 15	177	42 27	219	61 50	272	17 37	337
99	58 19	056	28 45	117	34 53	161	14 16	178	42 07	220	61 19	273	17 24	337
100	58 45	056	29 13	118	35 03	162	14 17	179	41 47	222	60 48	274	17 13	338
101	59 11	057	29 40	119	35 12	163	14 17	180	41 26	223	60 18	274	17 01	339
102	59 37	057	30 07	120	35 21	164	14 17	181	41 05	224	59 47	275	16 51	340
103	60 02	057	30 34	121	35 29	166	14 17	182	40 43	225	59 16	276	16 40	340
104	60 29	058	31 00	122	35 36	167	14 15	183	40 21	226	58 45	277	16 30	341
	Alioth		♦REGULUS		PROCYON		BETELGEUSE		♦ALDEBARAN		Mirfak		♦DENEB	
105	45 55	053	31 26	123	35 43	168	36 56	200	39 59	227	58 15	278	16 20	342
106	46 20	054	31 52	124	35 49	169	36 44	202	39 36	229	57 44	278	16 10	342
107	46 45	055	32 17	125	35 54	170	36 33	203	39 12	230	57 14	279	16 01	343
108	47 10	055	32 42	126	35 59	172	36 21	204	38 49	231	56 43	280	15 52	344
109	47 36	056	33 07	127	36 03	173	36 08	205	38 24	232	56 13	280	15 44	345
110	48 01	056	33 32	128	36 07	174	35 54	206	38 00	233	55 42	281	15 36	345
111	48 27	057	33 56	129	36 10	175	35 40	208	37 35	234	55 12	282	15 28	346
112	48 53	057	34 19	130	36 12	177	35 26	209	37 10	235	54 42	283	15 21	347
113	49 19	058	34 43	131	36 13	178	35 11	210	36 44	236	54 12	283	15 14	347
114	49 45	058	35 06	133	36 14	179	34 55	211	36 18	237	53 42	284	15 08	348
115	50 11	059	35 28	134	36 14	180	34 39	212	35 52	238	53 12	285	15 01	349
116	50 38	059	35 50	135	36 14	182	34 22	213	35 26	240	52 42	285	14 56	350
117	51 05	060	36 12	136	36 13	183	34 05	215	34 59	241	52 12	286	14 50	350
118	51 32	060	36 33	137	36 11	184	33 47	216	34 32	242	51 42	287	14 45	351
119	51 58	061	36 54	138	36 08	185	33 29	217	34 04	243	51 13	287	14 41	352
	♦Kochab		ARCTURUS		♦REGULUS		PROCYON		BETELGEUSE		♦CAPELLA		DENEB	
120	52 33	026	14 26	077	37 15	139	36 05	186	33 10	218	62 29	260	14 37	353
121	52 46	026	14 56	077	37 35	140	36 01	188	32 51	219	61 59	261	14 33	353
122	53 00	026	15 26	078	37 54	142	35 57	189	32 31	220	61 28	262	14 29	354
123	53 14	027	15 57	079	38 13	143	35 52	190	32 11	221	60 57	263	14 26	355
124	53 28	027	16 27	080	38 31	144	35 46	191	31 50	222	60 27	264	14 24	355
125	53 42	027	16 58	081	38 49	145	35 40	193	31 29	223	59 56	265	14 21	356
126	53 56	027	17 28	082	39 07	146	35 33	194	31 08	225	59 25	266	14 19	357
127	54 10	028	17 59	082	39 23	148	35 25	195	30 46	226	58 54	267	14 18	358
128	54 25	028	18 29	083	39 40	149	35 17	196	30 23	227	58 24	268	14 17	358
129	54 39	028	19 00	084	39 55	150	35 08	197	30 01	228	57 53	269	14 16	359
130	54 54	028	19 31	085	40 10	151	34 58	199	29 38	229	57 22	269	14 16	000
131	55 09	028	20 02	086	40 25	153	34 48	200	29 14	230	56 51	270	14 16	000
132	55 23	029	20 33	087	40 39	154	34 37	201	28 50	231	56 20	271	14 16	001
133	55 38	029	21 03	087	40 52	155	34 26	202	28 26	232	55 49	272	14 17	002
134	55 53	029	21 34	088	41 05	156	34 14	203	28 02	233	55 18	273	14 18	003
	DENEB		♦VEGA		ARCTURUS		♦REGULUS		PROCYON		BETELGEUSE		♦CAPELLA	
135	14 20	003	12 12	028	22 05	089	41 17	158	34 01	205	27 37	234	54 47	274
136	14 22	004	12 26	029	22 36	090	41 29	159	33 48	206	27 12	235	54 16	274
137	14 24	005	12 41	029	23 07	091	41 39	160	33 34	207	26 46	236	53 46	275
138	14 27	006	12 57	030	23 38	092	41 50	161	33 20	208	26 20	237	53 15	276
139	14 30	006	13 12	031	24 09	093	41 59	163	33 05	209	25 54	238	52 44	277
140	14 34	007	13 28	032	24 40	094	42 08	164	32 50	210	25 28	239	52 14	278
141	14 38	008	13 45	032	25 10	094	42 16	165	32 34	211	25 01	240	51 43	278
142	14 42	008	14 01	033	25 41	095	42 24	167	32 18	213	24 35	241	51 12	279
143	14 47	009	14 18	034	26 12	096	42 30	168	32 01	214	24 07	242	50 42	280
144	14 52	010	14 36	035	26 43	097	42 36	169	31 43	215	23 40	243	50 12	281
145	14 58	011	14 53	035	27 13	098	42 42	171	31 25	216	23 12	244	49 41	281
146	15 04	011	15 11	036	27 44	099	42 46	172	31 07	217	22 45	245	49 11	282
147	15 10	012	15 30	037	28 14	100	42 50	173	30 48	218	22 17	246	48 41	283
148	15 17	013	15 49	038	28 45	101	42 54	175	30 29	219	21 48	247	48 11	284
149	15 24	014	16 08	038	29 15	102	42 56	176	30 09	220	21 20	248	47 41	284
	DENEB		♦VEGA		ARCTURUS		Denebola		♦REGULUS		POLLUX		♦CAPELLA	
150	15 31	014	16 27	039	29 45	102	41 15	144	42 58	177	51 21	232	47 11	285
151	15 39	015	16 46	040	30 15	103	41 33	145	42 59	179	50 56	233	46 41	286
152	15 47	016	17 06	041	30 45	104	41 50	146	43 00	180	50 31	234	46 11	286
153	15 56	016	17 27	041	31 15	105	42 07	148	42 59	181	50 06	235	45 42	287
154	16 05	017	17 47	042	31 45	106	42 23	149	42 58	183	49 40	237	45 12	288
155	16 14	018	18 08	043	32 15	107	42 39	150	42 56	184	49 14	238	44 43	289
156	16 23	019	18 29	043	32 44	108	42 54	151	42 54	185	48 48	239	44 14	289
157	16 33	019	18 50	044	33 13	109	43 09	153	42 51	187	48 22	240	43 44	290
158	16 44	020	19 12	045	33 43	110	43 22	154	42 47	188	47 55	241	43 16	291
159	16 55	021	19 34	046	34 12	111	43 36	155	42 42	189	47 27	242	42 47	291
160	17 06	021	19 56	046	34 40	112	43 48	157	42 37	191	47 00	243	42 18	292
161	17 17	022	20 19	047	35 09	113	44 00	158	42 31	192	46 32	245	41 49	293
162	17 29	023	20 42	048	35 37	114	44 11	159	42 24	193	46 04	246	41 21	293
163	17 41	024	21 05	049	36 05	115	44 22	161	42 16	195	45 36	247	40 53	294
164	17 54	024	21 28	049	36 33	116	44 32	162	42 08	196	45 07	248	40 24	295
	♦DENEB		VEGA		ARCTURUS		♦SPICA		REGULUS		♦POLLUX		CAPELLA	
165	18 06	025	21 52	050	37 01	117	14 01	143	42 00	197	44 39	249	39 56	295
166	18 20	026	22 15	051	37 28	118	14 19	144	41 50	198	44 10	250	39 29	296
167	18 33	026	22 39	051	37 55	119	14 37	145	41 40	200	43 40	251	39 01	297
168	18 47	027	23 04	052	38 22	120	14 54	146	41 29	201	43 11	252	38 33	297
169	19 01	028	23 28	053	38 49	121	15 11	147	41 18	202	42 42	253	38 06	298
170	19 16	028	23 53	054	39 15	122	15 28	148	41 06	204	42 12	254	37 39	299
171	19 31	029	24 18	054	39 41	123	15 44	149	40 53	205	41 42	255	37 12	299
172	19 46	030	24 43	055	40 07	125	15 59	150	40 40	206	41 12	256	36 45	300
173	20 01	031	25 09	056	40 32	126	16 14	151	40 26	207	40 42	257	36 18	301
174	20 17	031	25 34	057	40 57	127	16 29	152	40 11	209	40 12	258	35 52	301
175	20 33	032	26 00	057	41 21	128	16 43	153	39 56	210	39 42	259	35 25	302
176	20 50	033	26 26	058	41 46	129	16 57	154	39 41	211	39 12	260	34 59	303
177	21 07	033	26 53	059	42 09	130	17 10	155	39 24	212	38 41	261	34 33	303
178	21 24	034	27 19	059	42 33	131	17 23	156	39 07	214	38 11	262	34 08	304
179	21 41	035	27 46	060	42 56	133	17 36	157	38 50	215	37 40	262	33 42	305

LAT 59°N

LHA ♈	Hc	Zn	Hc	Zn	Hc	Zn	Hc	Zn	Hc	Zn	Hc	Zn	Hc	Zn
	◆DENEB		VEGA		ARCTURUS		◆SPICA		REGULUS		◆POLLUX		CAPELLA	
180	21 59	035	28 13	061	43 18	134	17 47	158	38 32	216	37 09	263	33 17	305
181	22 17	036	28 40	062	43 40	135	17 59	159	38 14	217	36 39	264	32 52	306
182	22 35	037	29 07	062	44 02	136	18 09	160	37 55	218	36 08	265	32 27	307
183	22 54	037	29 34	063	44 23	137	18 20	161	37 36	219	35 37	266	32 02	307
184	23 13	038	30 02	064	44 44	139	18 29	162	37 16	221	35 06	267	31 38	308
185	23 32	039	30 30	064	45 04	140	18 39	163	36 55	222	34 35	268	31 13	309
186	23 52	039	30 58	065	45 24	141	18 47	164	36 35	223	34 05	269	30 49	309
187	24 11	040	31 26	066	45 43	142	18 55	165	36 13	224	33 34	269	30 26	310
188	24 31	041	31 54	067	46 02	144	19 03	166	35 52	225	33 03	270	30 02	311
189	24 52	042	32 23	067	46 20	145	19 10	167	35 30	226	32 32	271	29 39	311
190	25 12	042	32 51	068	46 37	146	19 17	168	35 07	227	32 01	272	29 16	312
191	25 33	043	33 20	069	46 54	148	19 23	169	34 44	228	31 30	273	28 53	313
192	25 54	044	33 49	070	47 10	149	19 28	170	34 21	230	30 59	274	28 30	313
193	26 16	044	34 18	070	47 26	150	19 33	171	33 57	231	30 28	275	28 08	314
194	26 38	045	34 47	071	47 41	152	19 37	172	33 33	232	29 58	275	27 46	315
	◆DENEB		VEGA		Rasalhague		◆ARCTURUS		REGULUS		◆POLLUX		CAPELLA	
195	27 00	046	35 16	072	21 40	102	47 56	153	33 08	233	29 27	276	27 24	315
196	27 22	046	35 46	073	22 10	103	48 09	154	32 44	234	28 56	277	27 02	316
197	27 44	047	36 15	073	22 40	104	48 22	156	32 19	235	28 26	278	26 41	317
198	28 07	048	36 45	074	23 10	105	48 35	157	31 53	236	27 55	279	26 20	317
199	28 30	048	37 15	075	23 40	106	48 46	158	31 27	237	27 24	280	25 59	318
200	28 53	049	37 45	076	24 10	106	48 57	160	31 01	238	26 54	280	25 38	319
201	29 16	050	38 15	076	24 40	107	49 08	161	30 35	239	26 24	281	25 18	319
202	29 40	050	38 45	077	25 09	108	49 17	163	30 08	240	25 53	282	24 58	320
203	30 04	051	39 15	078	25 38	109	49 26	164	29 41	241	25 23	283	24 38	321
204	30 28	052	39 45	079	26 07	110	49 34	166	29 14	242	24 53	284	24 19	321
205	30 52	052	40 15	079	26 36	111	49 42	167	28 47	243	24 23	284	24 00	322
206	31 17	053	40 46	080	27 05	112	49 48	168	28 19	244	23 53	285	23 41	323
207	31 42	054	41 16	081	27 33	113	49 54	170	27 51	245	23 23	286	23 22	323
208	32 07	054	41 47	082	28 02	114	49 59	171	27 23	246	22 54	287	23 04	324
209	32 32	055	42 18	083	28 30	115	50 03	173	26 55	247	22 24	288	22 46	325
	◆DENEB		VEGA		Rasalhague		◆ARCTURUS		REGULUS		◆POLLUX		CAPELLA	
210	32 57	056	42 48	083	28 58	116	50 07	174	26 26	248	21 55	288	22 28	325
211	33 23	056	43 19	084	29 25	117	50 09	176	25 58	249	21 26	289	22 11	326
212	33 49	057	43 50	085	29 53	118	50 11	177	25 29	250	20 56	290	21 54	327
213	34 15	058	44 20	086	30 20	119	50 12	179	25 00	251	20 27	291	21 37	327
214	34 41	058	44 51	087	30 47	120	50 12	180	24 30	252	19 59	291	21 20	328
215	35 07	059	45 22	087	31 13	121	50 12	182	24 01	253	19 30	292	21 04	329
216	35 34	060	45 53	088	31 40	122	50 11	183	23 32	253	19 01	293	20 48	329
217	36 01	060	46 24	089	32 06	123	50 09	185	23 02	254	18 33	294	20 33	330
218	36 27	061	46 55	090	32 32	124	50 06	186	22 32	255	18 05	295	20 18	331
219	36 55	062	47 26	091	32 57	125	50 02	188	22 02	256	17 37	295	20 03	332
220	37 22	062	47 57	092	33 22	126	49 58	189	21 32	257	17 09	296	19 48	332
221	37 49	063	48 28	093	33 47	127	49 52	190	21 02	258	16 41	297	19 34	333
222	38 17	064	48 58	093	34 11	128	49 46	192	20 32	259	16 14	298	19 20	334
223	38 45	064	49 29	094	34 35	129	49 40	193	20 01	260	15 47	299	19 06	334
224	39 13	065	50 00	095	34 59	130	49 32	195	19 31	261	15 20	299	18 53	335
	DENEB		◆VEGA		Rasalhague		◆ARCTURUS		Denebola		Dubhe		◆CAPELLA	
225	39 41	066	50 31	096	35 22	132	49 24	196	33 26	239	61 40	301	18 40	336
226	40 09	066	51 01	097	35 45	133	49 15	198	32 59	240	61 13	302	18 28	336
227	40 37	067	51 32	098	36 08	134	49 05	199	32 32	241	60 47	302	18 16	337
228	41 06	068	52 03	099	36 30	135	48 54	201	32 04	242	60 21	302	18 04	338
229	41 35	068	52 33	100	36 52	136	48 43	202	31 37	243	59 55	303	17 52	339
230	42 03	069	53 04	101	37 13	137	48 31	203	31 10	244	59 29	303	17 41	339
231	42 32	070	53 34	102	37 34	138	48 19	205	30 42	245	59 03	303	17 30	340
232	43 01	071	54 04	103	37 54	139	48 05	206	30 13	246	58 37	304	17 20	341
233	43 31	071	54 34	104	38 14	141	47 51	207	29 45	247	58 12	304	17 10	341
234	44 00	072	55 04	105	38 33	142	47 37	209	29 16	248	57 46	305	17 00	342
235	44 29	073	55 34	106	38 52	143	47 22	210	28 48	249	57 21	305	16 51	343
236	44 59	073	56 03	107	39 10	144	47 06	212	28 19	250	56 56	305	16 42	343
237	45 29	074	56 33	108	39 28	145	46 49	213	27 50	251	56 31	306	16 33	344
238	45 58	075	57 02	109	39 45	147	46 32	214	27 20	252	56 06	306	16 25	345
239	46 28	075	57 31	110	40 02	148	46 15	215	26 51	253	55 41	307	16 17	346
	DENEB		◆VEGA		Rasalhague		◆ARCTURUS		Denebola		Dubhe		◆CAPELLA	
240	46 58	076	58 00	112	40 18	149	45 56	217	26 21	254	55 16	307	16 10	346
241	47 28	077	58 29	113	40 34	150	45 38	218	25 51	255	54 51	307	16 03	347
242	47 58	078	58 57	114	40 49	152	45 18	219	25 22	256	54 27	308	15 56	348
243	48 29	078	59 25	115	41 03	153	44 59	221	24 52	257	54 03	308	15 50	348
244	48 59	079	59 53	116	41 17	154	44 38	222	24 21	258	53 38	309	15 44	349
245	49 29	080	60 20	118	41 30	155	44 17	223	23 51	258	53 14	309	15 38	350
246	50 00	081	60 48	119	41 43	157	43 56	224	23 21	259	52 50	310	15 33	351
247	50 30	081	61 14	120	41 55	158	43 34	225	22 50	260	52 27	310	15 28	351
248	51 01	082	61 41	122	42 06	159	43 12	227	22 20	261	52 03	310	15 23	352
249	51 32	083	62 07	123	42 17	160	42 49	228	21 49	262	51 40	311	15 19	353
250	52 02	084	62 33	124	42 27	162	42 26	229	21 19	263	51 16	311	15 16	353
251	52 33	085	62 58	126	42 36	163	42 03	230	20 48	264	50 53	312	15 12	354
252	53 04	085	63 23	127	42 45	164	41 39	231	20 17	265	50 30	312	15 09	355
253	53 35	086	63 47	129	42 53	166	41 14	232	19 47	266	50 07	313	15 07	356
254	54 05	087	64 11	130	43 00	167	40 50	234	19 16	266	49 45	313	15 05	356
	Schedar		DENEB		◆ALTAIR		Rasalhague		◆ARCTURUS		Dubhe		◆CAPELLA	
255	36 29	038	54 36	088	30 25	129	43 07	168	40 25	235	49 22	313	15 03	357
256	36 48	039	55 07	089	30 49	130	43 12	170	39 59	236	49 00	314	15 01	358
257	37 08	040	55 38	089	31 12	131	43 18	171	39 34	237	48 38	314	15 00	359
258	37 27	040	56 09	090	31 36	132	43 22	172	39 08	238	48 16	315	15 00	359
259	37 47	041	56 40	091	31 58	133	43 26	174	38 41	239	47 54	315	15 00	000
260	38 08	041	57 11	092	32 20	134	43 29	175	38 15	240	47 32	316	15 00	001
261	38 28	042	57 42	093	32 42	136	43 31	176	37 48	241	47 10	316	15 00	001
262	38 49	042	58 12	094	33 04	137	43 33	178	37 20	242	46 49	317	15 01	002
263	39 10	043	58 43	095	33 25	138	43 34	179	36 53	243	46 28	317	15 03	003
264	39 31	043	59 14	096	33 45	139	43 34	180	36 25	244	46 07	318	15 04	004
265	39 52	044	59 45	097	34 05	140	43 33	182	35 57	245	45 46	318	15 06	004
266	40 14	045	60 15	098	34 25	141	43 32	183	35 29	246	45 26	318	15 09	005
267	40 36	045	60 46	099	34 44	142	43 30	184	35 01	247	45 05	319	15 12	006
268	40 57	046	61 17	100	35 03	143	43 27	186	34 32	248	44 45	319	15 15	006
269	41 20	046	61 47	101	35 21	145	43 24	187	34 03	249	44 25	320	15 19	007

LHA ♈	Hc	Zn	Hc	Zn	Hc	Zn	Hc	Zn	Hc	Zn	Hc	Zn	Hc	Zn
	◆CAPELLA		Alpheratz		◆ALTAIR		Rasalhague		◆ARCTURUS		Alkaid		Dubhe	
270	15 23	008	23 36	072	35 39	146	43 19	188	33 34	250	53 18	283	44 05	320
271	15 27	009	24 06	073	35 56	147	43 14	190	33 05	251	52 48	284	43 46	321
272	15 32	009	24 35	074	36 13	148	43 09	191	32 36	252	52 18	285	43 26	321
273	15 37	010	25 05	075	36 29	149	43 02	192	32 06	253	51 48	285	43 07	322
274	15 43	011	25 35	076	36 44	150	42 55	194	31 37	254	51 18	286	42 48	322
275	15 49	011	26 05	076	36 59	152	42 48	195	31 07	255	50 49	287	42 29	323
276	15 55	012	26 35	077	37 14	153	42 39	196	30 37	256	50 19	288	42 10	323
277	16 02	013	27 05	078	37 28	154	42 30	198	30 07	257	49 50	288	41 52	324
278	16 09	014	27 36	079	37 41	155	42 20	199	29 37	258	49 20	289	41 34	324
279	16 16	014	28 06	080	37 53	156	42 10	200	29 07	259	48 51	289	41 16	325
280	16 24	015	28 36	080	38 06	158	41 59	202	28 36	260	48 22	290	40 58	325
281	16 32	016	29 07	081	38 17	159	41 47	203	28 06	261	47 53	291	40 41	326
282	16 41	016	29 37	082	38 28	160	41 35	204	27 35	261	47 24	291	40 23	326
283	16 49	017	30 08	083	38 38	161	41 22	205	27 05	262	46 56	292	40 06	327
284	16 59	018	30 39	084	38 48	163	41 08	207	26 34	263	46 27	293	39 49	327
	◆CAPELLA		Alpheratz		◆ALTAIR		Rasalhague		◆ARCTURUS		Alkaid		Dubhe	
285	17 08	019	31 10	084	38 56	164	40 54	208	26 03	264	45 59	293	39 33	328
286	17 18	019	31 40	085	39 05	165	40 39	209	25 32	265	45 30	294	39 16	328
287	17 29	020	32 11	086	39 12	166	40 24	210	25 02	266	45 02	295	39 00	329
288	17 39	021	32 42	087	39 19	168	40 08	212	24 31	267	44 34	295	38 44	329
289	17 51	021	33 13	088	39 26	169	39 51	213	24 00	268	44 06	296	38 28	330
290	18 02	022	33 44	089	39 31	170	39 34	214	23 29	268	43 39	297	38 13	330
291	18 14	023	34 15	090	39 36	171	39 17	215	22 58	269	43 11	297	37 57	331
292	18 26	023	34 46	090	39 40	173	38 58	217	22 27	270	42 44	298	37 42	331
293	18 38	024	35 16	091	39 44	174	38 40	218	21 56	271	42 16	299	37 28	332
294	18 51	025	35 47	092	39 47	175	38 21	219	21 25	272	41 49	299	37 13	332
295	19 04	026	36 18	093	39 49	177	38 01	220	20 55	273	41 22	300	36 59	333
296	19 18	026	36 49	094	39 50	178	37 41	221	20 24	274	40 56	300	36 45	333
297	19 32	027	37 20	095	39 51	179	37 20	222	19 53	274	40 29	301	36 31	334
298	19 46	028	37 51	096	39 51	180	36 59	224	19 22	275	40 03	302	36 17	334
299	20 00	028	38 21	097	39 51	182	36 38	225	18 51	276	39 37	302	36 04	335
	CAPELLA		◆Alpheratz		Enif		ALTAIR		◆VEGA		Alphecca		◆Dubhe	
300	20 15	029	38 52	097	37 04	147	39 49	183	65 48	223	34 44	265	35 51	335
301	20 30	030	39 23	098	37 21	148	39 47	184	65 27	224	34 14	266	35 38	336
302	20 46	030	39 53	099	37 37	150	39 45	186	65 05	226	33 43	267	35 26	336
303	21 02	031	40 24	100	37 52	151	39 41	187	64 43	227	33 12	267	35 13	337
304	21 18	032	40 54	101	38 07	152	39 37	188	64 20	229	32 41	268	35 01	337
305	21 34	032	41 24	102	38 21	153	39 33	189	63 56	231	32 10	269	34 49	338
306	21 51	033	41 54	103	38 35	154	39 27	191	63 32	232	31 39	270	34 38	338
307	22 08	034	42 24	104	38 48	156	39 21	192	63 08	234	31 08	271	34 27	339
308	22 25	035	42 54	105	39 00	157	39 14	193	62 42	235	30 37	272	34 16	339
309	22 43	035	43 24	106	39 12	158	39 07	195	62 17	236	30 07	273	34 05	340
310	23 01	036	43 54	107	39 23	159	38 59	196	61 51	238	29 36	273	33 55	341
311	23 19	037	44 23	108	39 33	161	38 50	197	61 25	239	29 05	274	33 44	341
312	23 38	037	44 53	109	39 43	162	38 41	198	60 58	240	28 34	275	33 34	342
313	23 56	038	45 22	110	39 53	163	38 31	200	60 31	242	28 03	276	33 25	342
314	24 16	039	45 51	111	40 01	165	38 20	201	60 03	243	27 33	277	33 15	343
	CAPELLA		◆Hamal		Alpheratz		Enif		◆ALTAIR		VEGA		◆Alioth	
315	24 35	039	26 43	088	46 19	112	40 09	166	38 09	202	59 36	244	34 02	325
316	24 55	040	27 14	089	46 48	113	40 16	167	37 57	203	59 08	246	33 45	325
317	25 15	041	27 44	090	47 16	114	40 23	168	37 44	204	58 39	247	33 27	326
318	25 35	041	28 15	091	47 44	115	40 29	170	37 31	206	58 11	248	33 10	327
319	25 55	042	28 46	092	48 12	117	40 34	171	37 17	207	57 42	249	32 53	327
320	26 16	043	29 17	093	48 39	118	40 39	172	37 03	208	57 13	250	32 37	328
321	26 37	043	29 48	094	49 07	119	40 42	174	36 48	209	56 44	251	32 20	328
322	26 59	044	30 19	094	49 34	120	40 46	175	36 33	210	56 15	252	32 04	329
323	27 20	045	30 50	095	50 00	121	40 48	176	36 17	212	55 45	254	31 48	329
324	27 42	045	31 20	096	50 27	122	40 50	177	36 01	213	55 15	255	31 33	330
325	28 04	046	31 51	097	50 52	123	40 51	179	35 44	214	54 46	256	31 18	331
326	28 26	047	32 22	098	51 18	125	40 51	180	35 26	215	54 16	257	31 03	331
327	28 49	047	32 52	099	51 43	126	40 51	181	35 08	216	53 45	258	30 48	332
328	29 12	048	33 23	100	52 08	127	40 50	183	34 49	217	53 15	259	30 33	332
329	29 35	049	33 53	101	52 33	128	40 48	184	34 30	219	52 45	260	30 19	333
	CAPELLA		◆Hamal		Alpheratz		Enif		◆ALTAIR		VEGA		◆Alioth	
330	29 58	049	34 23	102	52 57	130	40 45	185	34 11	220	52 14	261	30 05	334
331	30 22	050	34 54	103	53 20	131	40 42	187	33 51	221	51 44	262	29 52	334
332	30 46	051	35 24	104	53 43	132	40 38	188	33 30	222	51 13	263	29 39	335
333	31 10	051	35 54	105	54 06	134	40 34	189	33 09	223	50 43	263	29 26	335
334	31 34	052	36 23	105	54 28	135	40 29	190	32 48	224	50 12	264	29 13	336
335	31 58	053	36 53	106	54 49	136	40 23	192	32 26	225	49 41	265	29 00	337
336	32 23	053	37 23	107	55 11	138	40 16	193	32 04	226	49 10	266	28 48	337
337	32 48	054	37 52	108	55 31	139	40 09	194	31 42	227	48 39	267	28 36	338
338	33 13	055	38 21	109	55 51	141	40 01	196	31 19	229	48 09	268	28 25	338
339	33 38	055	38 50	110	56 10	142	39 52	197	30 55	230	47 38	269	28 14	339
340	34 04	056	39 19	111	56 29	144	39 43	198	30 32	231	47 07	270	28 03	340
341	34 29	057	39 48	112	56 47	145	39 33	199	30 08	232	46 36	271	27 52	340
342	34 55	057	40 16	113	57 04	147	39 23	201	29 43	233	46 05	271	27 42	341
343	35 21	058	40 45	114	57 21	148	39 11	202	29 18	234	45 34	272	27 32	341
344	35 48	059	41 13	116	57 37	150	39 00	203	28 53	235	45 03	273	27 22	342
	CAPELLA		ALDEBARAN		◆Hamal		Alpheratz		◆ALTAIR		VEGA		◆Alioth	
345	36 14	059	17 13	086	41 40	117	57 52	151	28 28	236	44 32	274	27 12	343
346	36 41	060	17 43	087	42 08	118	58 07	153	28 02	237	44 02	275	27 03	343
347	37 07	061	18 14	088	42 35	119	58 20	154	27 36	238	43 31	276	26 55	344
348	37 34	061	18 45	089	43 02	120	58 33	156	27 10	239	43 00	276	26 46	344
349	38 02	062	19 16	090	43 29	121	58 45	158	26 43	240	42 29	277	26 38	345
350	38 29	063	19 47	091	43 55	122	58 57	159	26 17	241	41 59	278	26 30	346
351	38 56	063	20 18	092	44 21	123	59 07	161	25 49	242	41 28	279	26 23	346
352	39 24	064	20 49	092	44 47	124	59 17	163	25 22	243	40 58	280	26 15	347
353	39 52	065	21 20	093	45 12	125	59 26	164	24 55	244	40 27	280	26 08	347
354	40 20	065	21 50	094	45 37	127	59 34	166	24 27	245	39 57	281	26 02	348
355	40 48	066	22 21	095	46 02	128	59 41	168	23 59	246	39 27	282	25 56	349
356	41 16	067	22 52	096	46 26	129	59 47	169	23 30	247	38 56	283	25 50	349
357	41 45	067	23 23	097	46 50	130	59 52	171	23 02	247	38 26	283	25 44	350
358	42 13	068	23 53	098	47 13	131	59 56	173	22 33	248	37 56	284	25 39	350
359	42 42	069	24 24	099	47 36	133	59 59	175	22 05	249	37 26	285	25 34	351

LHA ♈	Hc	Zn	Hc	Zn	Hc	Zn	Hc	Zn	Hc	Zn	Hc	Zn	Hc	Zn
	Dubhe		♦CAPELLA		ALDEBARAN		Hamal		♦Alpheratz		ALTAIR		♦VEGA	
0	30 17	008	42 49	068	25 04	099	48 40	133	61 02	176	21 56	251	36 40	286
1	30 21	008	43 19	069	25 35	100	49 03	134	61 03	178	21 25	252	36 09	287
2	30 26	009	43 49	070	26 07	101	49 26	136	61 04	180	20 55	252	35 39	288
3	30 31	009	44 19	070	26 38	102	49 47	137	61 03	182	20 25	253	35 09	289
4	30 36	010	44 49	071	27 09	103	50 09	138	61 02	184	19 54	254	34 39	289
5	30 42	010	45 19	072	27 40	104	50 30	140	60 59	185	19 24	255	34 09	290
6	30 48	011	45 49	072	28 11	104	50 50	141	60 56	187	18 53	256	33 39	291
7	30 54	011	46 19	073	28 42	105	51 10	142	60 51	189	18 22	257	33 09	291
8	31 00	012	46 50	074	29 12	106	51 29	144	60 46	191	17 51	258	32 40	292
9	31 07	013	47 20	075	29 43	107	51 48	145	60 40	192	17 20	259	32 11	293
10	31 14	013	47 51	075	30 13	108	52 06	146	60 32	194	16 48	260	31 41	294
11	31 21	014	48 22	076	30 43	109	52 23	148	60 24	196	16 17	261	31 12	294
12	31 29	014	48 53	077	31 13	110	52 39	149	60 15	198	15 46	261	30 43	295
13	31 37	015	49 24	077	31 43	111	52 55	151	60 05	199	15 14	262	30 15	296
14	31 45	015	49 55	078	32 12	112	53 10	152	59 54	201	14 43	263	29 46	296
	Dubhe		♦CAPELLA		ALDEBARAN		Hamal		♦Alpheratz		DENEB		♦VEGA	
15	31 54	016	50 26	079	32 42	113	53 25	154	59 42	203	49 38	281	29 18	297
16	32 02	016	50 57	080	33 11	114	53 39	155	59 29	205	49 07	282	28 50	298
17	32 12	017	51 29	080	33 40	115	53 51	157	59 15	206	48 36	282	28 22	299
18	32 21	017	52 00	081	34 08	116	54 04	158	59 01	208	48 05	283	27 54	299
19	32 30	018	52 31	082	34 37	117	54 15	160	58 45	209	47 34	284	27 26	300
20	32 40	018	53 03	083	35 05	118	54 26	161	58 29	211	47 03	284	26 59	301
21	32 51	019	53 34	083	35 33	119	54 35	163	58 13	213	46 32	285	26 32	301
22	33 01	019	54 06	084	36 01	120	54 44	164	57 55	214	46 01	286	26 05	302
23	33 12	020	54 38	085	36 28	121	54 53	166	57 37	216	45 31	286	25 38	303
24	33 23	020	55 09	086	36 55	122	55 00	168	57 18	217	45 00	287	25 11	304
25	33 34	021	55 41	086	37 22	123	55 06	169	56 58	219	44 30	288	24 45	304
26	33 45	022	56 13	087	37 48	124	55 12	171	56 38	220	44 00	289	24 19	305
27	33 57	022	56 45	088	38 15	125	55 16	172	56 17	222	43 30	289	23 53	306
28	34 09	023	57 16	089	38 40	126	55 20	174	55 56	223	43 00	290	23 27	306
29	34 22	023	57 48	090	39 06	128	55 23	176	55 34	225	42 30	291	23 02	307
	Dubhe		♦CAPELLA		ALDEBARAN		Hamal		♦Alpheratz		DENEB		♦VEGA	
30	34 34	024	58 20	091	39 31	129	55 25	177	55 11	226	42 00	291	22 36	308
31	34 47	024	58 52	091	39 55	130	55 26	179	54 48	227	41 31	292	22 11	309
32	35 00	025	59 24	092	40 20	131	55 26	180	54 25	229	41 01	293	21 47	309
33	35 14	025	59 55	093	40 43	132	55 26	182	54 01	230	40 32	293	21 22	310
34	35 27	026	60 27	094	41 07	133	55 24	184	53 36	231	40 03	294	20 58	311
35	35 41	026	60 59	095	41 30	134	55 22	185	53 11	233	39 34	295	20 34	311
36	35 55	027	61 30	096	41 52	136	55 18	187	52 46	234	39 05	295	20 10	312
37	36 10	027	62 02	097	42 14	137	55 14	189	52 20	235	38 36	296	19 47	313
38	36 24	028	62 34	098	42 36	138	55 09	190	51 53	236	38 08	297	19 24	314
39	36 39	028	63 05	099	42 57	139	55 03	192	51 27	238	37 40	297	19 01	314
40	36 54	029	63 36	100	43 17	140	54 56	193	51 00	239	37 11	298	18 38	315
41	37 09	029	64 08	101	43 37	142	54 48	195	50 33	240	36 43	299	18 16	316
42	37 25	030	64 39	102	43 57	143	54 40	196	50 05	241	36 16	299	17 54	316
43	37 41	030	65 10	103	44 15	144	54 30	198	49 37	242	35 48	300	17 32	317
44	37 57	031	65 41	104	44 34	145	54 20	200	49 09	243	35 20	301	17 11	318
	♦Dubhe		POLLUX		BETELGEUSE		♦ALDEBARAN		Hamal		Alpheratz		♦DENEB	
45	38 13	031	33 18	089	29 17	128	44 51	147	54 09	201	48 40	244	34 53	301
46	38 30	032	33 50	090	29 42	129	45 09	148	53 57	203	48 11	246	34 26	302
47	38 47	032	34 21	091	30 06	130	45 25	149	53 44	204	47 42	247	33 59	303
48	39 04	033	34 53	092	30 30	131	45 41	151	53 31	206	47 13	248	33 32	303
49	39 21	033	35 25	093	30 54	132	45 56	152	53 17	207	46 43	249	33 06	304
50	39 38	034	35 57	094	31 17	133	46 11	153	53 02	209	46 14	250	32 40	304
51	39 56	034	36 28	095	31 40	135	46 25	155	52 46	210	45 44	251	32 14	305
52	40 14	035	37 00	095	32 02	136	46 38	156	52 30	211	45 14	252	31 48	306
53	40 32	035	37 32	096	32 24	137	46 51	157	52 13	213	44 43	253	31 22	306
54	40 50	035	38 03	097	32 46	138	47 02	159	51 56	214	44 13	254	30 56	307
55	41 09	036	38 35	098	33 07	139	47 14	160	51 37	216	43 42	255	30 31	308
56	41 28	036	39 06	099	33 28	140	47 24	162	51 19	217	43 11	256	30 06	308
57	41 47	037	39 38	100	33 48	141	47 34	163	50 59	218	42 41	257	29 41	309
58	42 06	037	40 09	101	34 08	142	47 43	164	50 39	220	42 10	258	29 17	310
59	42 25	038	40 40	102	34 27	143	47 51	166	50 18	221	41 38	259	28 53	310
	♦Dubhe		POLLUX		BETELGEUSE		♦ALDEBARAN		Hamal		Alpheratz		♦DENEB	
60	42 45	038	41 11	103	34 46	145	47 58	167	49 57	222	41 07	260	28 28	311
61	43 05	039	41 42	104	35 04	146	48 05	169	49 35	224	40 36	261	28 05	312
62	43 25	039	42 13	105	35 21	147	48 11	170	49 13	225	40 04	261	27 41	312
63	43 45	040	42 43	106	35 39	148	48 16	171	48 50	226	39 33	262	27 18	313
64	44 05	040	43 14	107	35 55	149	48 20	173	48 27	228	39 01	263	26 55	314
65	44 26	041	43 44	108	36 11	150	48 24	174	48 03	229	38 30	264	26 32	314
66	44 47	041	44 15	109	36 27	152	48 27	176	47 39	230	37 58	265	26 09	315
67	45 08	041	44 45	110	36 41	153	48 29	177	47 15	231	37 26	266	25 47	315
68	45 29	042	45 14	111	36 56	154	48 30	179	46 50	233	36 55	267	25 25	316
69	45 50	042	45 44	112	37 09	155	48 30	180	46 24	234	36 23	268	25 03	317
70	46 12	043	46 13	113	37 22	156	48 29	182	45 58	235	35 51	269	24 41	318
71	46 33	043	46 43	114	37 35	158	48 28	183	45 32	236	35 19	269	24 20	318
72	46 55	044	47 12	115	37 47	159	48 26	184	45 06	237	34 48	270	23 59	319
73	47 17	044	47 40	116	37 58	160	48 23	186	44 39	238	34 16	271	23 38	320
74	47 39	045	48 09	117	38 08	161	48 20	187	44 12	239	33 44	272	23 18	320
	♦Dubhe		REGULUS		PROCYON		♦SIRIUS		ALDEBARAN		♦Alpheratz		DENEB	
75	48 02	045	17 01	095	28 53	133	12 12	154	48 15	189	33 12	273	22 58	321
76	48 24	045	17 32	095	29 16	134	12 26	155	48 10	190	32 41	274	22 38	322
77	48 47	046	18 04	096	29 39	135	12 39	156	48 04	192	32 09	274	22 18	322
78	49 10	046	18 36	097	30 01	137	12 52	157	47 57	193	31 37	275	21 59	323
79	49 33	047	19 07	098	30 22	138	13 04	158	47 49	194	31 05	276	21 40	324
80	49 56	047	19 39	099	30 44	139	13 15	159	47 41	196	30 34	277	21 22	324
81	50 20	048	20 10	100	31 04	140	13 26	160	47 32	197	30 02	278	21 03	325
82	50 43	048	20 41	101	31 25	141	13 37	161	47 22	199	29 31	278	20 45	326
83	51 07	048	21 12	102	31 45	142	13 47	162	47 12	200	28 59	279	20 28	326
84	51 31	049	21 44	102	32 04	143	13 57	163	47 00	201	28 28	280	20 10	327
85	51 55	049	22 15	103	32 23	144	14 06	164	46 48	203	27 57	281	19 53	328
86	52 19	050	22 45	104	32 41	145	14 14	165	46 36	204	27 26	282	19 36	329
87	52 43	050	23 16	105	32 59	146	14 22	166	46 22	206	26 55	282	19 20	329
88	53 08	050	23 47	106	33 16	148	14 30	167	46 08	207	26 24	283	19 04	330
89	53 32	051	24 17	107	33 33	149	14 36	168	45 54	208	25 53	284	18 48	331

LAT 58°N

LHA ♈	Hc	Zn	Hc	Zn	Hc	Zn	Hc	Zn	Hc	Zn	Hc	Zn	Hc	Zn
	Dubhe		♦REGULUS		PROCYON		SIRIUS		♦ALDEBARAN		Mirfak		♦DENEB	
90	53 57	051	24 48	108	33 49	150	14 43	169	45 38	210	66 01	267	18 33	331
91	54 22	052	25 18	109	34 05	151	14 49	170	45 22	211	65 29	268	18 17	332
92	54 47	052	25 48	110	34 20	152	14 54	171	45 06	212	64 58	269	18 03	333
93	55 12	052	26 18	111	34 35	153	14 59	172	44 48	213	64 26	270	17 48	333
94	55 37	053	26 47	112	34 49	155	15 03	173	44 30	215	63 54	271	17 34	334
95	56 02	053	27 17	113	35 02	156	15 07	174	44 12	216	63 22	271	17 20	335
96	56 28	053	27 46	114	35 15	157	15 10	175	43 53	217	62 51	272	17 07	335
97	56 53	054	28 15	115	35 27	158	15 13	176	43 34	219	62 19	273	16 54	336
98	57 19	054	28 44	115	35 38	159	15 15	177	43 13	220	61 47	274	16 41	337
99	57 45	055	29 12	116	35 49	161	15 16	178	42 53	221	61 15	275	16 29	338
100	58 11	055	29 41	117	36 00	162	15 17	179	42 32	222	60 44	275	16 17	338
101	58 37	055	30 09	118	36 09	163	15 17	180	42 10	223	60 12	276	16 06	339
102	59 03	056	30 37	119	36 18	164	15 17	181	41 48	225	59 40	277	15 54	340
103	59 30	056	31 04	120	36 27	165	15 17	182	41 25	226	59 09	278	15 43	340
104	59 56	056	31 31	121	36 34	167	15 15	183	41 02	227	58 37	278	15 33	341
	Alioth		♦REGULUS		PROCYON		BETELGEUSE		♦ALDEBARAN		Mirfak		♦DENEB	
105	45 19	053	31 58	122	36 41	168	37 52	201	40 39	228	58 06	279	15 23	342
106	45 44	053	32 25	123	36 48	169	37 40	202	40 15	229	57 35	280	15 13	343
107	46 10	054	32 51	125	36 53	170	37 28	203	39 51	230	57 03	281	15 04	343
108	46 35	054	33 18	126	36 58	172	37 15	204	39 26	232	56 32	281	14 55	344
109	47 01	055	33 43	127	37 03	173	37 02	205	39 01	233	56 01	282	14 46	345
110	47 27	055	34 09	128	37 06	174	36 48	207	38 36	234	55 30	283	14 38	345
111	47 54	056	34 34	129	37 09	175	36 33	208	38 10	235	54 59	283	14 30	346
112	48 20	056	34 58	130	37 12	177	36 18	209	37 44	236	54 28	284	14 23	347
113	48 46	057	35 22	131	37 13	178	36 02	210	37 17	237	53 57	285	14 16	348
114	49 13	057	35 46	132	37 14	179	35 46	211	36 50	238	53 26	285	14 09	348
115	49 40	058	36 10	133	37 14	180	35 29	213	36 23	239	52 56	286	14 03	349
116	50 07	058	36 33	134	37 14	182	35 12	214	35 56	240	52 25	287	13 57	350
117	50 34	059	36 55	135	37 13	183	34 54	215	35 28	241	51 55	287	13 51	350
118	51 01	059	37 17	137	37 11	184	34 35	216	35 00	242	51 25	288	13 46	351
119	51 29	060	37 39	138	37 08	185	34 16	217	34 32	243	50 54	289	13 41	352
	♦Kochab		ARCTURUS		♦REGULUS		PROCYON		BETELGEUSE		♦CAPELLA		DENEB	
120	51 39	025	14 12	076	38 00	139	37 05	187	33 57	218	62 39	262	13 37	353
121	51 52	026	14 43	077	38 21	140	37 01	188	33 37	219	62 07	263	13 33	353
122	52 06	026	15 14	078	38 41	141	36 56	189	33 17	221	61 35	264	13 30	354
123	52 20	026	15 45	079	39 01	142	36 51	190	32 56	222	61 04	265	13 26	355
124	52 34	026	16 17	080	39 20	144	36 45	192	32 34	223	60 32	266	13 24	355
125	52 48	027	16 48	080	39 38	145	36 38	193	32 12	224	60 00	267	13 21	356
126	53 03	027	17 19	081	39 56	146	36 31	194	31 50	225	59 29	268	13 19	357
127	53 17	027	17 51	082	40 14	147	36 23	195	31 27	226	58 57	268	13 18	358
128	53 32	027	18 22	083	40 31	148	36 14	196	31 04	227	58 25	269	13 17	358
129	53 46	027	18 54	084	40 47	150	36 05	198	30 41	228	57 53	270	13 15	359
130	54 01	028	19 25	085	41 03	151	35 55	199	30 17	229	57 21	271	13 15	000
131	54 16	028	19 57	085	41 18	152	35 44	200	29 53	230	56 50	272	13 15	000
132	54 30	028	20 29	086	41 33	153	35 33	201	29 28	231	56 18	273	13 15	001
133	54 45	028	21 01	087	41 47	155	35 21	202	29 03	232	55 46	273	13 17	002
134	55 01	028	21 32	088	42 00	156	35 09	204	28 38	233	55 14	274	13 18	003
	DENEB		♦VEGA		ARCTURUS		♦REGULUS		PROCYON		BETELGEUSE		♦CAPELLA	
135	13 20	003	11 19	028	22 04	089	42 13	157	34 56	205	28 12	234	54 43	275
136	13 22	004	11 34	028	22 36	090	42 24	159	34 42	206	27 46	235	54 11	276
137	13 25	005	11 49	029	23 08	090	42 36	160	34 28	207	27 20	236	53 39	277
138	13 28	006	12 05	030	23 39	091	42 46	161	34 13	208	26 53	237	53 08	277
139	13 31	006	12 21	031	24 11	092	42 56	162	33 58	210	26 26	238	52 36	278
140	13 35	007	12 37	031	24 43	093	43 06	164	33 42	211	25 59	239	52 05	279
141	13 39	008	12 54	032	25 15	094	43 14	165	33 25	212	25 31	240	51 34	280
142	13 43	008	13 11	033	25 46	095	43 22	166	33 08	213	25 04	241	51 02	280
143	13 48	009	13 28	034	26 18	096	43 29	168	32 50	214	24 36	242	50 31	281
144	13 53	010	13 46	034	26 50	097	43 35	169	32 32	215	24 07	243	50 00	282
145	13 59	011	14 04	035	27 21	097	43 41	170	32 14	216	23 39	244	49 29	283
146	14 05	011	14 23	036	27 53	098	43 46	172	31 55	217	23 10	245	48 58	283
147	14 11	012	14 42	037	28 24	099	43 50	173	31 35	219	22 41	246	48 27	284
148	14 18	013	15 01	037	28 56	100	43 53	175	31 15	220	22 12	247	47 56	285
149	14 25	013	15 20	038	29 27	101	43 56	176	30 55	221	21 43	248	47 25	285
	DENEB		♦VEGA		ARCTURUS		Denebola		♦REGULUS		POLLUX		♦CAPELLA	
150	14 33	014	15 40	039	29 58	102	42 03	143	43 58	177	51 57	233	46 55	286
151	14 41	015	16 00	040	30 29	103	42 22	145	43 59	179	51 32	234	46 24	287
152	14 49	016	16 21	040	31 00	104	42 40	146	44 00	180	51 06	235	45 54	287
153	14 58	016	16 41	041	31 31	105	42 58	147	43 59	181	50 40	236	45 24	288
154	15 07	017	17 02	042	32 02	106	43 15	148	43 58	183	50 13	238	44 53	289
155	15 17	018	17 24	043	32 32	107	43 31	150	43 56	184	49 46	239	44 23	289
156	15 27	018	17 45	043	33 02	107	43 47	151	43 54	185	49 19	240	43 53	290
157	15 37	019	18 07	044	33 33	108	44 02	152	43 50	187	48 51	241	43 24	291
158	15 47	020	18 30	045	34 03	109	44 16	154	43 46	188	48 23	242	42 54	291
159	15 58	021	18 52	045	34 33	110	44 30	155	43 41	189	47 55	243	42 25	292
160	16 10	021	19 15	046	35 02	111	44 43	156	43 36	191	47 26	244	41 55	293
161	16 21	022	19 38	047	35 32	112	44 56	158	43 29	192	46 57	246	41 26	293
162	16 34	023	20 01	048	36 01	113	45 07	159	43 22	193	46 28	247	40 57	294
163	16 46	023	20 25	048	36 30	114	45 19	160	43 15	195	45 59	248	40 28	295
164	16 59	024	20 49	049	36 59	115	45 29	162	43 06	196	45 29	249	39 59	295
	♦DENEB		VEGA		ARCTURUS		♦SPICA		REGULUS		♦POLLUX		CAPELLA	
165	17 12	025	21 13	050	37 28	116	14 49	143	42 57	197	45 00	250	39 30	296
166	17 25	025	21 37	050	37 56	117	15 07	144	42 47	199	44 30	251	39 02	297
167	17 39	026	22 02	051	38 24	118	15 26	145	42 36	200	44 00	252	38 34	297
168	17 54	027	22 27	052	38 52	119	15 44	146	42 25	201	43 29	253	38 05	298
169	18 08	028	22 52	053	39 20	121	16 01	147	42 13	203	42 59	254	37 37	299
170	18 23	028	23 17	053	39 47	122	16 19	148	42 01	204	42 28	255	37 10	299
171	18 38	029	23 43	054	40 14	123	16 35	149	41 47	205	41 58	256	36 42	300
172	18 54	030	24 09	055	40 40	124	16 51	150	41 33	207	41 27	257	36 15	301
173	19 10	030	24 35	055	41 07	125	17 07	151	41 19	208	40 56	258	35 47	301
174	19 26	031	25 01	056	41 32	126	17 22	152	41 04	209	40 25	259	35 20	302
175	19 42	032	25 28	057	41 58	127	17 37	153	40 48	210	39 53	260	34 53	303
176	19 59	032	25 54	058	42 23	128	17 51	154	40 32	212	39 22	260	34 27	303
177	20 16	033	26 21	058	42 48	129	18 05	155	40 15	213	38 51	261	34 00	304
178	20 34	034	26 48	059	43 12	131	18 18	156	39 57	214	38 19	262	33 34	305
179	20 52	034	27 16	060	43 36	132	18 31	157	39 39	215	37 48	263	33 08	305

LAT 58°N

LHA	Hc	Zn	Hc	Zn	Hc	Zn	Hc	Zn	Hc	Zn	Hc	Zn	Hc	Zn
♈	♦DENEB		VEGA		ARCTURUS		♦SPICA		REGULUS		♦POLLUX		CAPELLA	
180	21 10	035	27 43	060	44 00	133	18 43	158	39 21	216	37 16	264	32 42	306
181	21 28	036	28 11	061	44 23	134	18 55	159	39 02	218	36 44	265	32 16	306
182	21 47	036	28 39	062	44 45	135	19 06	160	38 42	219	36 13	266	31 51	307
183	22 06	037	29 07	063	45 07	137	19 16	161	38 22	220	35 41	267	31 26	308
184	22 26	038	29 35	063	45 29	138	19 26	162	38 01	221	35 09	268	31 01	308
185	22 45	039	30 04	064	45 50	139	19 36	163	37 40	222	34 37	268	30 36	309
186	23 05	039	30 32	065	46 10	140	19 45	164	37 18	223	34 06	269	30 11	310
187	23 25	040	31 01	065	46 30	142	19 53	165	36 56	225	33 34	270	29 47	310
188	23 46	041	31 30	066	46 50	143	20 01	166	36 34	226	33 02	271	29 23	311
189	24 07	041	31 59	067	47 09	144	20 09	167	36 11	227	32 30	272	28 59	312
190	24 28	042	32 29	068	47 27	146	20 15	168	35 47	228	31 58	273	28 35	312
191	24 49	043	32 58	068	47 45	147	20 22	169	35 24	229	31 27	274	28 12	313
192	25 11	043	33 28	069	48 02	148	20 27	170	34 59	230	30 55	274	27 49	314
193	25 33	044	33 57	070	48 18	150	20 32	171	34 35	231	30 23	275	27 26	314
194	25 55	045	34 27	070	48 34	151	20 37	172	34 10	232	29 52	276	27 03	315
	♦DENEB		VEGA		Rasalhague		♦ARCTURUS		REGULUS		♦POLLUX		CAPELLA	
195	26 17	045	34 57	071	21 52	102	48 49	152	33 45	233	29 20	277	26 41	316
196	26 40	046	35 28	072	22 24	102	49 03	154	33 19	234	28 49	278	26 19	316
197	27 03	047	35 58	073	22 55	103	49 17	155	32 53	235	28 17	278	25 57	317
198	27 26	047	36 28	073	23 25	104	49 30	157	32 27	236	27 46	279	25 35	318
199	27 50	048	36 59	074	23 56	105	49 42	158	32 00	237	27 14	280	25 14	318
200	28 13	049	37 29	075	24 27	106	49 54	159	31 33	239	26 43	281	24 53	319
201	28 37	049	38 00	076	24 57	107	50 04	161	31 06	240	26 12	282	24 32	320
202	29 01	050	38 31	076	25 28	108	50 15	162	30 38	241	25 41	282	24 12	320
203	29 26	050	39 02	077	25 58	109	50 24	164	30 10	242	25 10	283	23 52	321
204	29 51	051	39 33	078	26 28	110	50 32	165	29 42	243	24 39	284	23 32	322
205	30 15	052	40 04	079	26 58	111	50 40	167	29 14	243	24 08	285	23 12	322
206	30 41	052	40 35	079	27 27	112	50 47	168	28 45	244	23 37	286	22 53	323
207	31 06	053	41 06	080	27 57	113	50 53	170	28 17	245	23 07	286	22 34	324
208	31 31	054	41 38	081	28 26	114	50 58	171	27 48	246	22 36	287	22 15	324
209	31 57	054	42 09	082	28 55	115	51 03	173	27 18	247	22 06	288	21 57	325
	♦DENEB		VEGA		Rasalhague		♦ARCTURUS		REGULUS		♦POLLUX		CAPELLA	
210	32 23	055	42 41	082	29 24	115	51 06	174	26 49	248	21 36	289	21 39	326
211	32 49	056	43 12	083	29 52	116	51 09	176	26 19	249	21 06	289	21 21	326
212	33 16	056	43 44	084	30 21	117	51 11	177	25 49	250	20 36	290	21 03	327
213	33 42	057	44 16	085	30 49	118	51 12	179	25 19	251	20 06	291	20 46	328
214	34 09	058	44 47	086	31 17	119	51 12	180	24 49	252	19 36	292	20 29	328
215	34 36	058	45 19	086	31 44	120	51 12	182	24 19	253	19 07	293	20 13	329
216	35 03	059	45 51	087	32 11	121	51 11	183	23 48	254	18 38	293	19 57	330
217	35 31	060	46 22	088	32 38	123	51 08	185	23 18	255	18 09	294	19 41	330
218	35 58	060	46 54	089	33 05	124	51 05	186	22 47	256	17 40	295	19 25	331
219	36 26	061	47 26	090	33 31	125	51 01	188	22 16	257	17 11	296	19 10	332
220	36 54	062	47 58	091	33 57	126	50 57	189	21 45	258	16 42	296	18 55	332
221	37 22	062	48 30	091	34 23	127	50 51	191	21 14	258	16 14	297	18 40	333
222	37 50	063	49 01	092	34 48	128	50 45	192	20 43	259	15 46	298	18 26	334
223	38 18	064	49 33	093	35 13	129	50 38	194	20 12	260	15 18	299	18 12	334
224	38 47	064	50 05	094	35 38	130	50 30	195	19 40	261	14 50	300	17 59	335
	DENEB		♦VEGA		Rasalhague		♦ARCTURUS		Denebola		Dubhe		♦CAPELLA	
225	39 16	065	50 37	095	36 02	131	50 21	197	33 56	240	61 08	303	17 46	336
226	39 45	066	51 08	096	36 26	132	50 12	198	33 28	241	60 41	303	17 33	337
227	40 14	066	51 40	097	36 49	133	50 01	200	33 01	242	60 15	303	17 20	337
228	40 43	067	52 11	098	37 12	134	49 50	201	32 32	243	59 48	304	17 08	338
229	41 12	068	52 43	099	37 35	136	49 39	202	32 04	244	59 22	304	16 56	339
230	41 42	068	53 14	100	37 57	137	49 26	204	31 35	245	58 56	304	16 45	339
231	42 11	069	53 46	101	38 18	138	49 13	205	31 06	246	58 30	305	16 34	340
232	42 41	070	54 17	102	38 39	139	48 59	207	30 37	247	58 03	305	16 23	341
233	43 11	070	54 48	103	39 00	140	48 45	208	30 08	248	57 38	306	16 13	341
234	43 41	071	55 19	104	39 20	141	48 29	209	29 38	249	57 12	306	16 03	342
235	44 11	072	55 50	105	39 40	143	48 13	211	29 09	250	56 46	306	15 54	343
236	44 41	072	56 20	106	39 59	144	47 57	212	28 39	251	56 20	307	15 44	344
237	45 12	073	56 51	107	40 17	145	47 40	213	28 09	252	55 55	307	15 36	344
238	45 42	074	57 21	108	40 35	146	47 22	215	27 39	252	55 30	307	15 27	345
239	46 13	074	57 51	109	40 53	147	47 03	216	27 08	253	55 04	308	15 19	346
	DENEB		♦VEGA		Rasalhague		♦ARCTURUS		Denebola		Dubhe		♦CAPELLA	
240	46 43	075	58 21	110	41 10	149	46 44	217	26 38	254	54 39	308	15 11	346
241	47 14	076	58 51	111	41 26	150	46 25	219	26 07	255	54 14	309	15 04	347
242	47 45	077	59 21	112	41 42	151	46 05	220	25 36	256	53 50	309	14 57	348
243	48 16	077	59 50	114	41 57	152	45 44	221	25 05	257	53 25	309	14 51	349
244	48 47	078	60 19	115	42 11	154	45 23	222	24 34	258	53 00	310	14 45	349
245	49 18	079	60 47	116	42 25	155	45 01	224	24 03	259	52 36	310	14 39	350
246	49 49	079	61 16	117	42 38	156	44 39	225	23 32	260	52 12	311	14 33	351
247	50 21	080	61 44	119	42 50	158	44 16	226	23 00	261	51 48	311	14 29	351
248	50 52	081	62 12	120	43 02	159	43 53	227	22 29	262	51 24	311	14 24	352
249	51 24	082	62 39	121	43 13	160	43 29	229	21 58	262	51 00	312	14 20	353
250	51 55	082	63 06	123	43 24	161	43 05	230	21 26	263	50 36	312	14 16	354
251	52 27	083	63 32	124	43 33	163	42 41	231	20 54	264	50 13	313	14 13	354
252	52 58	084	63 58	126	43 43	164	42 16	232	20 23	265	49 50	313	14 10	355
253	53 30	085	64 24	127	43 51	165	41 51	233	19 51	266	49 26	313	14 07	356
254	54 01	086	64 49	129	43 58	167	41 25	234	19 19	267	49 03	314	14 05	356
	Schedar		DENEB		♦ALTAIR		Rasalhague		♦ARCTURUS		Dubhe		♦CAPELLA	
255	35 42	038	54 33	086	31 03	129	44 05	168	40 59	235	48 40	314	14 03	357
256	36 01	039	55 05	087	31 28	130	44 12	170	40 33	236	48 18	315	14 01	358
257	36 21	039	55 37	088	31 52	131	44 17	171	40 06	238	47 55	315	14 00	359
258	36 41	040	56 08	089	32 16	132	44 22	172	39 39	239	47 33	316	14 00	359
259	37 02	040	56 40	090	32 39	133	44 26	174	39 12	240	47 11	316	14 00	000
260	37 22	041	57 12	091	33 02	134	44 29	175	38 44	241	46 49	316	14 00	001
261	37 43	041	57 44	091	33 25	135	44 31	176	38 16	242	46 27	317	14 00	001
262	38 04	042	58 16	092	33 47	136	44 33	178	37 48	243	46 05	317	14 01	002
263	38 26	042	58 47	093	34 09	137	44 34	179	37 20	244	45 44	318	14 03	003
264	38 47	043	59 19	094	34 30	138	44 34	180	36 51	245	45 23	318	14 04	004
265	39 09	043	59 51	095	34 51	140	44 33	182	36 22	246	45 01	319	14 07	004
266	39 31	044	60 22	096	35 12	141	44 32	183	35 53	247	44 41	319	14 09	005
267	39 53	044	60 54	097	35 32	142	44 30	185	35 24	248	44 20	320	14 12	006
268	40 15	045	61 26	098	35 51	143	44 27	186	34 54	249	43 59	320	14 15	006
269	40 38	046	61 57	099	36 10	144	44 23	187	34 24	250	43 39	320	14 19	007

LHA	Hc	Zn	Hc	Zn	Hc	Zn	Hc	Zn	Hc	Zn	Hc	Zn	Hc	Zn
♈	♦CAPELLA		Alpheratz		♦ALTAIR		Rasalhague		♦ARCTURUS		Alkaid		Dubhe	
270	14 23	008	23 18	072	36 28	145	44 19	189	33 54	251	53 03	285	43 19	321
271	14 28	009	23 48	073	36 46	146	44 13	190	33 24	252	52 32	285	42 59	321
272	14 33	009	24 19	074	37 03	148	44 08	191	32 54	253	52 02	286	42 39	322
273	14 38	010	24 49	074	37 20	149	44 01	193	32 23	254	51 31	287	42 20	322
274	14 44	011	25 20	075	37 36	150	43 54	194	31 53	255	51 01	287	42 00	323
275	14 50	011	25 51	076	37 52	151	43 46	195	31 22	256	50 31	288	41 41	323
276	14 56	012	26 22	077	38 07	152	43 37	197	30 51	257	50 00	289	41 22	324
277	15 03	013	26 53	077	38 21	154	43 27	198	30 20	257	49 30	289	41 04	324
278	15 10	013	27 24	078	38 35	155	43 17	199	29 49	258	49 00	290	40 45	325
279	15 18	014	27 55	079	38 48	156	43 06	201	29 18	259	48 31	291	40 27	325
280	15 26	015	28 26	080	39 01	157	42 55	202	28 47	260	48 01	291	40 09	326
281	15 34	016	28 57	081	39 13	159	42 42	203	28 15	261	47 31	292	39 51	326
282	15 43	016	29 29	081	39 24	160	42 29	205	27 44	262	47 02	292	39 33	327
283	15 52	017	30 00	082	39 35	161	42 16	206	27 12	263	46 33	293	39 16	327
284	16 02	018	30 32	083	39 45	162	42 02	207	26 41	264	46 03	294	38 59	328
	♦CAPELLA		Alpheratz		♦ALTAIR		Rasalhague		♦ARCTURUS		Alkaid		Dubhe	
285	16 11	018	31 03	084	39 54	164	41 47	208	26 09	265	45 34	294	38 42	328
286	16 22	019	31 35	085	40 03	165	41 31	210	25 37	265	45 05	295	38 25	329
287	16 32	020	32 07	086	40 11	166	41 15	211	25 06	266	44 37	296	38 09	329
288	16 43	021	32 38	086	40 18	167	40 59	212	24 34	267	44 08	296	37 52	330
289	16 55	021	33 10	087	40 24	169	40 42	213	24 02	268	43 40	297	37 36	330
290	17 06	022	33 42	088	40 30	170	40 24	215	23 30	269	43 11	297	37 21	331
291	17 18	023	34 14	089	40 35	171	40 05	216	22 59	270	42 43	298	37 05	331
292	17 31	023	34 46	090	40 40	173	39 46	217	22 27	271	42 15	299	36 50	332
293	17 44	024	35 17	091	40 44	174	39 27	218	21 55	271	41 47	299	36 35	332
294	17 57	025	35 49	091	40 47	175	39 07	219	21 23	272	41 20	300	36 20	333
295	18 10	025	36 21	092	40 49	177	38 47	221	20 52	273	40 52	301	36 05	333
296	18 24	026	36 53	093	40 50	178	38 26	222	20 20	274	40 25	301	35 51	334
297	18 38	027	37 24	094	40 51	179	38 04	223	19 48	275	39 58	302	35 37	334
298	18 53	027	37 56	095	40 51	180	37 42	224	19 16	276	39 31	302	35 23	335
299	19 07	028	38 28	096	40 51	182	37 20	225	18 45	276	39 04	303	35 10	335
	CAPELLA		♦Alpheratz		Enif		ALTAIR		♦VEGA		Alphecca		♦Dubhe	
300	19 23	029	38 59	097	37 55	147	40 49	183	66 32	224	34 50	265	34 56	336
301	19 38	030	39 31	098	38 12	148	40 47	184	66 10	226	34 18	266	34 43	336
302	19 54	030	40 02	098	38 28	149	40 44	186	65 47	227	33 46	267	34 31	337
303	20 10	031	40 34	099	38 44	150	40 41	187	65 23	229	33 14	268	34 18	337
304	20 27	032	41 05	100	39 00	152	40 37	188	64 59	231	32 43	269	34 06	338
305	20 43	032	41 36	101	39 14	153	40 32	190	64 34	232	32 11	270	33 54	338
306	21 01	033	42 08	102	39 29	154	40 26	191	64 08	234	31 39	271	33 42	339
307	21 18	034	42 39	103	39 42	155	40 20	192	63 42	235	31 07	271	33 31	339
308	21 36	034	43 09	104	39 55	157	40 13	193	63 16	237	30 35	272	33 19	340
309	21 54	035	43 40	105	40 07	158	40 05	195	62 49	238	30 04	273	33 09	340
310	22 12	036	44 11	106	40 19	159	39 56	196	62 22	239	29 32	274	32 58	341
311	22 31	036	44 41	107	40 30	160	39 47	197	61 55	241	29 00	275	32 48	341
312	22 50	037	45 12	108	40 40	162	39 38	199	61 27	242	28 29	276	32 37	342
313	23 09	038	45 42	109	40 50	163	39 27	200	60 58	243	27 57	276	32 28	342
314	23 29	038	46 12	110	40 59	164	39 16	201	60 30	245	27 25	277	32 18	343
	CAPELLA		♦Hamal		Alpheratz		Enif		♦ALTAIR		VEGA		♦Alioth	
315	23 48	039	26 41	088	46 41	111	41 07	166	39 04	202	60 01	246	33 13	325
316	24 09	040	27 13	089	47 11	112	41 15	167	38 52	204	59 32	247	32 55	326
317	24 29	040	27 44	090	47 40	113	41 22	168	38 39	205	59 02	248	32 38	326
318	24 50	041	28 16	090	48 09	114	41 28	169	38 25	206	58 33	249	32 20	327
319	25 11	042	28 48	091	48 38	115	41 33	171	38 11	207	58 03	251	32 03	328
320	25 32	042	29 20	092	49 07	117	41 38	172	37 56	208	57 33	252	31 46	328
321	25 54	043	29 51	093	49 35	118	41 42	173	37 41	210	57 03	253	31 29	329
322	26 15	044	30 23	094	50 03	119	41 45	175	37 25	211	56 32	254	31 13	329
323	26 37	044	30 55	095	50 31	120	41 48	176	37 08	212	56 01	255	30 57	330
324	27 00	045	31 27	096	50 58	121	41 50	177	36 51	213	55 31	256	30 41	330
325	27 22	046	31 58	097	51 25	122	41 51	179	36 33	214	55 00	257	30 25	331
326	27 45	046	32 30	097	51 52	124	41 51	180	36 15	216	54 29	258	30 10	332
327	28 08	047	33 01	098	52 18	125	41 51	181	35 56	217	53 58	259	29 55	332
328	28 32	048	33 33	099	52 44	126	41 50	183	35 37	218	53 26	260	29 40	333
329	28 55	048	34 04	100	53 09	127	41 48	184	35 17	219	52 55	261	29 26	333
	CAPELLA		♦Hamal		Alpheratz		Enif		♦ALTAIR		VEGA		♦Alioth	
330	29 19	049	34 35	101	53 34	129	41 45	185	34 57	220	52 23	262	29 12	334
331	29 43	050	35 06	102	53 59	130	41 42	187	34 36	221	51 52	263	28 58	334
332	30 07	050	35 37	103	54 23	131	41 38	188	34 15	222	51 20	264	28 44	335
333	30 32	051	36 08	104	54 47	133	41 33	189	33 53	224	50 49	265	28 31	336
334	30 57	051	36 39	105	55 10	134	41 28	191	33 31	225	50 17	266	28 18	336
335	31 22	052	37 10	106	55 33	135	41 21	192	33 08	226	49 45	266	28 05	337
336	31 47	053	37 40	107	55 55	137	41 15	193	32 45	227	49 14	267	27 53	337
337	32 12	053	38 11	108	56 16	138	41 07	194	32 22	228	48 42	268	27 41	338
338	32 38	054	38 41	109	56 37	140	40 59	196	31 58	229	48 10	269	27 29	339
339	33 04	055	39 11	110	56 57	141	40 50	197	31 34	230	47 38	270	27 18	339
340	33 30	055	39 41	111	57 17	143	40 40	198	31 10	231	47 06	271	27 06	340
341	33 56	056	40 11	112	57 36	144	40 30	200	30 45	232	46 35	272	26 55	340
342	34 22	057	40 40	113	57 54	146	40 19	201	30 19	233	46 03	272	26 45	341
343	34 49	057	41 09	114	58 12	147	40 07	202	29 54	234	45 31	273	26 35	342
344	35 16	058	41 38	115	58 28	149	39 55	203	29 28	235	44 59	274	26 25	342
	CAPELLA		ALDEBARAN		♦Hamal		Alpheratz		♦ALTAIR		VEGA		♦Alioth	
345	35 43	059	17 09	086	42 07	116	58 44	150	29 01	236	44 28	275	26 15	343
346	36 10	059	17 40	087	42 35	117	59 00	152	28 35	237	43 56	276	26 06	343
347	36 38	060	18 12	088	43 04	118	59 14	154	28 08	238	43 24	276	25 57	344
348	37 05	061	18 44	089	43 32	119	59 28	155	27 41	239	42 53	277	25 48	345
349	37 33	061	19 16	090	43 59	120	59 41	157	27 13	240	42 21	278	25 40	345
350	38 01	062	19 47	090	44 27	121	59 53	159	26 46	241	41 50	279	25 32	346
351	38 29	063	20 19	091	44 54	122	60 04	160	26 18	242	41 19	280	25 24	346
352	38 57	063	20 51	092	45 20	123	60 14	162	25 49	243	40 47	280	25 17	347
353	39 26	064	21 23	093	45 47	125	60 23	164	25 21	244	40 16	281	25 10	348
354	39 54	064	21 55	094	46 13	126	60 32	166	24 52	245	39 45	282	25 03	348
355	40 23	065	22 26	095	46 38	127	60 39	167	24 23	246	39 14	283	24 57	349
356	40 52	066	22 58	096	47 03	128	60 46	169	23 54	247	38 43	283	24 51	349
357	41 21	066	23 30	096	47 28	129	60 51	171	23 25	248	38 12	284	24 45	350
358	41 50	067	24 01	097	47 53	131	60 56	173	22 55	249	37 41	285	24 40	351
359	42 20	068	24 33	098	48 17	132	60 59	175	22 25	250	37 10	286	24 35	351

LHA ♈	Hc	Zn	Hc	Zn	Hc	Zn	Hc	Zn	Hc	Zn	Hc	Zn	Hc	Zn
	Dubhe		◆CAPELLA		ALDEBARAN		Hamal		◆Alpheratz		ALTAIR		◆VEGA	
0	29 17	008	42 27	068	25 13	099	49 21	132	62 02	176	22 15	251	36 23	287
1	29 22	008	42 57	068	25 46	099	49 45	134	62 03	178	21 44	252	35 51	288
2	29 27	009	43 27	069	26 18	100	50 08	135	62 04	180	21 13	253	35 20	289
3	29 32	009	43 58	070	26 50	101	50 31	136	62 03	182	20 42	254	34 49	289
4	29 37	010	44 29	070	27 22	102	50 53	137	62 02	184	20 10	255	34 19	290
5	29 43	010	45 00	071	27 54	103	51 15	139	61 59	186	19 39	256	33 48	291
6	29 49	011	45 30	072	28 26	104	51 36	140	61 55	187	19 07	256	33 18	291
7	29 55	011	46 02	072	28 57	105	51 57	141	61 51	189	18 35	257	32 47	292
8	30 02	012	46 33	073	29 29	106	52 17	143	61 45	191	18 03	258	32 17	293
9	30 09	012	47 04	074	30 00	107	52 37	144	61 38	193	17 31	259	31 47	293
10	30 16	013	47 35	074	30 31	108	52 55	146	61 30	195	16 59	260	31 17	294
11	30 23	013	48 07	075	31 02	109	53 13	147	61 22	197	16 27	261	30 47	295
12	30 31	014	48 38	076	31 33	109	53 31	149	61 12	198	15 55	262	30 18	296
13	30 39	015	49 10	076	32 04	110	53 47	150	61 01	200	15 22	263	29 48	296
14	30 47	015	49 42	077	32 34	111	54 03	152	60 49	202	14 50	263	29 19	297
	Dubhe		◆CAPELLA		ALDEBARAN		Hamal		◆Alpheratz		DENEB		◆VEGA	
15	30 56	016	50 14	078	33 05	112	54 18	153	60 37	204	49 26	282	28 50	298
16	31 05	016	50 46	078	33 35	113	54 33	155	60 23	205	48 54	283	28 21	298
17	31 14	017	51 18	079	34 05	114	54 46	156	60 09	207	48 22	283	27 53	299
18	31 24	017	51 50	080	34 34	115	54 59	158	59 54	209	47 51	284	27 24	300
19	31 33	018	52 22	080	35 04	116	55 11	159	59 38	210	47 19	285	26 56	300
20	31 43	018	52 54	081	35 33	117	55 22	161	59 21	212	46 47	285	26 28	301
21	31 54	019	53 27	082	36 02	118	55 33	162	59 03	214	46 16	286	26 00	302
22	32 04	019	53 59	083	36 31	119	55 42	164	58 45	215	45 45	287	25 32	303
23	32 15	020	54 32	083	36 59	120	55 51	166	58 25	217	45 13	287	25 05	303
24	32 26	020	55 04	084	37 27	122	55 58	167	58 06	218	44 42	288	24 38	304
25	32 38	021	55 37	085	37 55	123	56 05	169	57 45	220	44 11	289	24 11	305
26	32 50	021	56 09	086	38 22	124	56 11	171	57 24	221	43 40	289	23 44	305
27	33 02	022	56 42	086	38 49	125	56 16	172	57 02	223	43 10	290	23 18	306
28	33 14	022	57 14	087	39 16	126	56 20	174	56 39	224	42 39	291	22 51	307
29	33 26	023	57 47	088	39 42	127	56 23	175	56 16	226	42 08	291	22 25	307
	Dubhe		◆CAPELLA		ALDEBARAN		Hamal		◆Alpheratz		DENEB		◆VEGA	
30	33 39	023	58 20	089	40 08	128	56 25	177	55 53	227	41 38	292	21 59	308
31	33 52	024	58 52	090	40 33	129	56 26	179	55 29	228	41 08	293	21 34	309
32	34 06	024	59 25	091	40 59	130	56 26	180	55 04	230	40 38	293	21 09	310
33	34 19	025	59 58	091	41 23	131	56 26	182	54 39	231	40 08	294	20 43	310
34	34 33	025	60 30	092	41 48	133	56 24	184	54 13	232	39 38	295	20 19	311
35	34 47	026	61 03	093	42 11	134	56 21	185	53 47	234	39 09	295	19 54	312
36	35 01	026	61 36	094	42 35	135	56 18	187	53 21	235	38 39	296	19 30	312
37	35 16	027	62 08	095	42 58	136	56 13	189	52 54	236	38 10	297	19 06	313
38	35 31	027	62 41	096	43 20	137	56 08	190	52 26	237	37 41	297	18 42	314
39	35 46	028	63 13	097	43 42	139	56 01	192	51 59	239	37 12	298	18 19	315
40	36 01	028	63 46	098	44 03	140	55 54	194	51 31	240	36 43	299	17 56	315
41	36 17	029	64 18	099	44 24	141	55 46	195	51 02	241	36 14	299	17 33	316
42	36 33	029	64 50	100	44 44	142	55 37	197	50 33	242	35 46	300	17 10	317
43	36 49	030	65 22	101	45 04	144	55 27	198	50 04	243	35 18	301	16 48	317
44	37 05	030	65 54	102	45 23	145	55 16	200	49 35	244	34 50	301	16 26	318
	◆Dubhe		POLLUX		BETELGEUSE		◆ALDEBARAN		Hamal		Alpheratz		◆DENEB	
45	37 22	031	33 17	089	29 54	128	45 42	146	55 05	202	49 05	245	34 22	302
46	37 39	031	33 49	090	30 19	129	45 59	148	54 52	203	48 36	247	33 54	302
47	37 56	032	34 22	090	30 45	130	46 17	149	54 39	205	48 06	248	33 27	303
48	38 13	032	34 55	091	31 10	131	46 33	150	54 25	206	47 35	249	32 59	304
49	38 30	033	35 27	092	31 34	132	46 49	152	54 10	208	47 05	250	32 32	304
50	38 48	033	36 00	093	31 58	133	47 04	153	53 55	209	46 34	251	32 05	305
51	39 06	034	36 33	094	32 22	134	47 19	154	53 38	211	46 03	252	31 39	306
52	39 24	034	37 05	095	32 45	135	47 33	156	53 21	212	45 32	253	31 12	306
53	39 43	035	37 38	096	33 08	136	47 46	157	53 03	214	45 00	254	30 46	307
54	40 01	035	38 10	096	33 30	137	47 58	158	52 45	215	44 29	255	30 20	308
55	40 20	035	38 43	097	33 52	138	48 10	160	52 26	216	43 57	256	29 54	308
56	40 39	036	39 15	098	34 14	140	48 21	161	52 06	218	43 26	257	29 29	309
57	40 59	036	39 48	099	34 35	141	48 31	163	51 46	219	42 54	258	29 03	310
58	41 18	037	40 20	100	34 55	142	48 40	164	51 25	221	42 22	259	28 38	310
59	41 38	037	40 52	101	35 15	143	48 49	165	51 03	222	41 50	260	28 13	311
	◆Dubhe		POLLUX		BETELGEUSE		◆ALDEBARAN		Hamal		Alpheratz		◆DENEB	
60	41 58	038	41 24	102	35 34	144	48 57	167	50 41	223	41 18	260	27 49	311
61	42 18	038	41 56	103	35 53	145	49 04	168	50 18	225	40 45	261	27 25	312
62	42 38	039	42 28	104	36 12	146	49 10	170	49 55	226	40 13	262	27 00	313
63	42 58	039	42 59	105	36 29	148	49 15	171	49 31	227	39 41	263	26 37	313
64	43 19	040	43 31	106	36 47	149	49 20	173	49 07	228	39 08	264	26 13	314
65	43 40	040	44 02	107	37 03	150	49 24	174	48 42	230	38 35	265	25 50	315
66	44 01	040	44 33	108	37 19	151	49 26	176	48 17	231	38 03	266	25 26	315
67	44 22	041	45 04	109	37 35	152	49 28	177	47 52	232	37 30	267	25 04	316
68	44 44	041	45 35	110	37 50	154	49 30	179	47 26	233	36 58	268	24 41	317
69	45 05	042	46 06	111	38 04	155	49 30	180	46 59	235	36 25	268	24 19	317
70	45 27	042	46 36	112	38 17	156	49 29	182	46 32	236	35 52	269	23 57	318
71	45 49	043	47 07	113	38 30	157	49 28	183	46 05	237	35 20	270	23 35	319
72	46 11	043	47 36	114	38 43	159	49 26	185	45 38	238	34 47	271	23 14	319
73	46 34	043	48 06	115	38 54	160	49 23	186	45 10	239	34 14	272	22 53	320
74	46 56	044	48 36	116	39 05	161	49 19	188	44 42	240	33 42	273	22 32	321
	◆Dubhe		REGULUS		PROCYON		◆SIRIUS		ALDEBARAN		◆Alpheratz		DENEB	
75	47 19	044	17 05	094	29 34	133	13 07	154	49 14	189	33 09	273	22 11	321
76	47 42	045	17 38	095	29 58	134	13 20	155	49 09	190	32 36	274	21 51	322
77	48 05	045	18 10	096	30 21	135	13 34	156	49 03	192	32 04	275	21 31	323
78	48 28	045	18 43	097	30 44	136	13 47	157	48 55	193	31 31	276	21 11	323
79	48 52	046	19 15	098	31 07	137	13 59	158	48 48	195	30 59	277	20 52	324
80	49 15	046	19 48	099	31 29	138	14 11	159	48 39	196	30 26	277	20 33	325
81	49 39	047	20 20	099	31 50	139	14 23	160	48 29	198	29 54	278	20 14	325
82	50 03	047	20 52	100	32 11	140	14 34	161	48 19	199	29 22	279	19 56	326
83	50 27	047	21 24	101	32 32	142	14 44	162	48 08	200	28 49	280	19 37	327
84	50 51	048	21 56	102	32 52	143	14 54	163	47 56	202	28 17	281	19 20	327
85	51 15	048	22 28	103	33 11	144	15 03	164	47 44	203	27 45	281	19 02	328
86	51 40	049	23 00	104	33 30	145	15 12	165	47 30	205	27 13	282	18 45	329
87	52 04	049	23 32	105	33 49	146	15 20	166	47 16	206	26 41	283	18 28	329
88	52 29	049	24 03	106	34 07	147	15 28	167	47 02	207	26 10	284	18 12	330
89	52 54	050	24 35	107	34 24	148	15 35	168	46 46	209	25 38	285	17 56	331

LAT 57°N

LHA ♈	Hc	Zn	Hc	Zn	Hc	Zn	Hc	Zn	Hc	Zn	Hc	Zn	Hc	Zn
	Dubhe		◆REGULUS		PROCYON		SIRIUS		◆ALDEBARAN		Mirfak		◆DENEB	
90	53 19	050	25 06	107	34 41	150	15 42	169	46 30	210	66 03	269	17 40	331
91	53 44	051	25 37	108	34 57	151	15 48	170	46 13	211	65 30	270	17 24	332
92	54 09	051	26 08	109	35 13	152	15 53	171	45 56	213	64 58	271	17 09	333
93	54 35	051	26 39	110	35 28	153	15 58	172	45 38	214	64 25	272	16 55	334
94	55 00	052	27 09	111	35 43	154	16 03	173	45 20	215	63 52	273	16 40	334
95	55 26	052	27 40	112	35 57	155	16 06	174	45 00	217	63 20	273	16 26	335
96	55 52	052	28 10	113	36 10	157	16 10	175	44 41	218	62 47	274	16 12	336
97	56 18	053	28 40	114	36 23	158	16 12	176	44 20	219	62 15	275	15 59	336
98	56 44	053	29 09	115	36 35	159	16 14	177	43 59	220	61 42	276	15 46	337
99	57 10	053	29 39	116	36 46	160	16 16	178	43 38	222	61 10	276	15 34	338
100	57 36	054	30 08	117	36 57	161	16 17	179	43 16	223	60 37	277	15 21	338
101	58 02	054	30 37	118	37 07	163	16 17	180	42 53	224	60 05	278	15 09	339
102	58 29	054	31 06	119	37 16	164	16 17	181	42 30	225	59 32	279	14 58	340
103	58 55	055	31 34	120	37 25	165	16 17	182	42 07	226	59 00	279	14 47	340
104	59 22	055	32 03	121	37 33	166	16 15	183	41 43	228	58 28	280	14 36	341
	Alioth		◆REGULUS		PROCYON		BETELGEUSE		◆ALDEBARAN		Mirfak		◆DENEB	
105	44 42	052	32 30	122	37 40	168	38 48	201	41 19	229	57 56	281	14 26	342
106	45 08	052	32 58	123	37 47	169	38 36	202	40 54	230	57 24	281	14 16	343
107	45 34	053	33 25	124	37 53	170	38 23	203	40 29	231	56 52	282	14 06	343
108	46 00	053	33 52	125	37 58	171	38 10	205	40 03	232	56 20	283	13 57	344
109	46 26	054	34 19	126	38 02	173	37 56	206	39 37	233	55 48	283	13 48	345
110	46 53	054	34 45	127	38 06	174	37 41	207	39 11	234	55 16	284	13 40	345
111	47 19	055	35 11	128	38 09	175	37 26	208	38 44	235	54 44	285	13 32	346
112	47 46	055	35 36	129	38 12	177	37 11	209	38 17	237	54 13	285	13 24	347
113	48 13	056	36 01	130	38 13	178	36 54	211	37 50	238	53 41	286	13 17	348
114	48 40	056	36 26	131	38 14	179	36 37	212	37 22	239	53 10	287	13 10	348
115	49 08	057	36 50	133	38 14	180	36 20	213	36 54	240	52 39	287	13 04	349
116	49 35	057	37 14	134	38 14	182	36 02	214	36 25	241	52 08	288	12 58	350
117	50 03	058	37 38	135	38 12	183	35 43	215	35 57	242	51 37	288	12 52	350
118	50 30	058	38 01	136	38 10	184	35 24	217	35 28	243	51 06	289	12 47	351
119	50 58	059	38 23	137	38 08	185	35 04	218	34 59	244	50 35	290	12 42	352
	◆Kochab		ARCTURUS		◆REGULUS		PROCYON		BETELGEUSE		◆CAPELLA		DENEB	
120	50 44	025	13 58	076	38 45	138	38 04	187	34 44	219	62 46	264	12 38	353
121	50 58	025	14 30	077	39 06	139	38 00	188	34 23	220	62 13	265	12 34	353
122	51 12	025	15 02	078	39 27	141	37 55	189	34 02	221	61 41	266	12 30	354
123	51 26	026	15 34	079	39 48	142	37 50	190	33 40	222	61 08	267	12 27	355
124	51 40	026	16 06	079	40 08	143	37 44	192	33 18	223	60 36	268	12 24	355
125	51 55	026	16 38	080	40 27	144	37 37	193	32 55	224	60 03	268	12 22	356
126	52 09	026	17 10	081	40 46	145	37 29	194	32 32	225	59 30	269	12 20	357
127	52 23	026	17 42	082	41 04	147	37 21	195	32 09	227	58 58	270	12 18	358
128	52 38	027	18 15	083	41 22	148	37 12	197	31 45	228	58 25	271	12 17	358
129	52 53	027	18 47	083	41 39	149	37 02	198	31 21	229	57 52	272	12 16	359
130	53 07	027	19 20	084	41 55	150	36 52	199	30 56	230	57 20	273	12 16	000
131	53 22	027	19 52	085	42 11	152	36 41	200	30 31	231	56 47	273	12 16	000
132	53 37	027	20 25	086	42 26	153	36 29	202	30 05	232	56 14	274	12 16	001
133	53 52	028	20 57	087	42 41	154	36 17	203	29 39	233	55 42	275	12 17	002
134	54 08	028	21 30	088	42 55	156	36 04	204	29 13	234	55 09	276	12 19	003
	DENEB		◆VEGA		ARCTURUS		◆REGULUS		PROCYON		BETELGEUSE		◆CAPELLA	
135	12 20	003	10 25	028	22 03	088	43 08	157	35 50	205	28 47	235	54 37	276
136	12 22	004	10 41	028	22 35	089	43 20	158	35 36	206	28 20	236	54 04	277
137	12 25	005	10 57	029	23 08	090	43 32	160	35 21	208	27 53	237	53 32	278
138	12 28	006	11 13	030	23 41	091	43 43	161	35 06	209	27 25	238	53 00	279
139	12 31	006	11 29	031	24 13	092	43 53	162	34 50	210	26 57	239	52 27	279
140	12 35	007	11 46	031	24 46	093	44 03	164	34 33	211	26 29	240	51 55	280
141	12 39	008	12 03	032	25 19	093	44 12	165	34 16	212	26 01	241	51 23	281
142	12 44	008	12 21	033	25 51	094	44 20	166	33 58	213	25 32	242	50 51	282
143	12 49	009	12 39	034	26 24	095	44 28	168	33 40	214	25 03	243	50 19	282
144	12 54	010	12 57	034	26 56	096	44 34	169	33 21	216	24 34	244	49 47	283
145	13 00	011	13 15	035	27 29	097	44 40	170	33 02	217	24 05	245	49 15	284
146	13 06	011	13 34	036	28 01	098	44 45	172	32 42	218	23 35	246	48 44	284
147	13 13	012	13 54	037	28 34	099	44 50	173	32 22	219	23 05	246	48 12	285
148	13 20	013	14 13	037	29 06	100	44 53	174	32 01	220	22 35	247	47 40	286
149	13 27	013	14 33	038	29 38	100	44 56	176	31 40	221	22 05	248	47 09	286
	DENEB		◆VEGA		ARCTURUS		Denebola		◆REGULUS		POLLUX		◆CAPELLA	
150	13 35	014	14 53	039	30 10	101	42 51	143	44 58	177	52 33	234	46 38	287
151	13 43	015	15 14	039	30 42	102	43 11	144	44 59	179	52 07	235	46 06	288
152	13 51	016	15 35	040	31 14	103	43 30	145	45 00	180	51 40	236	45 35	288
153	14 00	016	15 56	041	31 46	104	43 48	147	44 59	181	51 12	237	45 04	289
154	14 10	017	16 18	042	32 17	105	44 06	148	44 58	183	50 45	239	44 34	290
155	14 19	018	16 40	042	32 49	106	44 23	149	44 56	184	50 17	240	44 03	290
156	14 30	018	17 02	043	33 20	107	44 39	151	44 53	185	49 48	241	43 32	291
157	14 40	019	17 24	044	33 51	108	44 55	152	44 50	187	49 19	242	43 02	292
158	14 51	020	17 47	044	34 22	109	45 10	153	44 45	188	48 50	243	42 32	292
159	15 02	020	18 10	045	34 53	110	45 24	154	44 40	190	48 21	244	42 01	293
160	15 14	021	18 33	046	35 24	111	45 38	156	44 35	191	47 52	245	41 31	294
161	15 26	022	18 57	047	35 54	112	45 51	157	44 28	192	47 22	247	41 02	294
162	15 38	023	19 21	047	36 25	113	46 03	159	44 21	194	46 52	248	40 32	295
163	15 51	023	19 45	048	36 55	114	46 15	160	44 12	195	46 21	249	40 02	296
164	16 04	024	20 09	049	37 25	115	46 26	161	44 04	196	45 51	250	39 33	296
	◆DENEB		VEGA		ARCTURUS		◆SPICA		REGULUS		◆POLLUX		CAPELLA	
165	16 17	025	20 34	049	37 54	116	15 37	143	43 54	198	45 20	251	39 04	297
166	16 31	025	20 59	050	38 23	117	15 56	144	43 44	199	44 49	252	38 35	297
167	16 45	026	21 24	051	38 52	118	16 15	145	43 33	200	44 18	253	38 06	298
168	17 00	027	21 50	052	39 21	119	16 34	146	43 21	202	43 47	254	37 37	299
169	17 15	027	22 15	052	39 50	120	16 52	147	43 08	203	43 15	255	37 08	299
170	17 30	028	22 41	053	40 18	121	17 09	148	42 55	204	42 44	256	36 40	300
171	17 46	029	23 07	054	40 46	122	17 27	149	42 42	206	42 12	257	36 12	301
172	18 02	029	23 34	054	41 13	123	17 43	150	42 27	207	41 40	258	35 44	301
173	18 18	030	24 01	055	41 41	124	17 59	151	42 12	208	41 08	259	35 16	302
174	18 34	031	24 27	056	42 07	125	18 15	152	41 56	209	40 36	259	34 48	303
175	18 51	032	24 55	056	42 34	126	18 30	153	41 40	211	40 04	260	34 21	303
176	19 09	032	25 22	057	43 00	128	18 45	154	41 23	212	39 32	261	33 54	304
177	19 26	033	25 49	058	43 26	129	18 59	155	41 05	213	38 59	262	33 26	304
178	19 44	034	26 17	059	43 51	130	19 13	156	40 47	214	38 27	263	33 00	305
179	20 02	034	26 45	059	44 16	131	19 26	157	40 28	216	37 54	264	32 33	306

LAT 57°N

LHA ♈	Hc	Zn	Hc	Zn	Hc	Zn	Hc	Zn	Hc	Zn	Hc	Zn	Hc	Zn
	◆DENEB		VEGA		ARCTURUS		◆SPICA		REGULUS		◆POLLUX		CAPELLA	
180	20 21	035	27 13	060	44 40	132	19 39	158	40 09	217	37 22	265	32 07	306
181	20 40	036	27 42	061	45 04	133	19 51	159	39 49	218	36 49	266	31 40	307
182	20 59	036	28 10	061	45 28	135	20 02	160	39 29	219	36 17	267	31 14	308
183	21 18	037	28 39	062	45 51	136	20 13	161	39 08	220	35 44	267	30 49	308
184	21 38	038	29 08	063	46 13	137	20 23	162	38 46	222	35 11	268	30 23	309
185	21 58	038	29 37	063	46 35	138	20 33	163	38 24	223	34 39	269	29 58	310
186	22 19	039	30 06	064	46 56	140	20 43	164	38 02	224	34 06	270	29 33	310
187	22 39	040	30 36	065	47 17	141	20 51	165	37 39	225	33 33	271	29 08	311
188	23 00	040	31 06	066	47 37	142	20 59	166	37 15	226	33 01	272	28 43	311
189	23 22	041	31 35	066	47 57	144	21 07	167	36 52	227	32 28	272	28 19	312
190	23 43	042	32 05	067	48 16	145	21 14	168	36 27	228	31 55	273	27 55	313
191	24 05	042	32 36	068	48 35	146	21 20	169	36 03	230	31 23	274	27 31	313
192	24 27	043	33 06	068	48 52	148	21 26	170	35 38	231	30 50	275	27 07	314
193	24 49	044	33 36	069	49 10	149	21 32	171	35 12	232	30 18	276	26 44	315
194	25 12	044	34 07	070	49 26	150	21 36	172	34 46	233	29 45	277	26 21	315
	◆DENEB		VEGA		Rasalhague		◆ARCTURUS		REGULUS		◆POLLUX		CAPELLA	
195	25 35	045	34 38	070	22 04	101	49 42	152	34 20	234	29 13	277	25 58	316
196	25 58	046	35 09	071	22 36	102	49 57	153	33 54	235	28 40	278	25 35	317
197	26 22	046	35 40	072	23 08	103	50 11	155	33 27	236	28 08	279	25 13	317
198	26 45	047	36 11	073	23 40	104	50 25	156	32 59	237	27 36	280	24 51	318
199	27 09	047	36 42	073	24 12	105	50 38	158	32 32	238	27 04	281	24 29	319
200	27 33	048	37 13	074	24 43	106	50 50	159	32 04	239	26 31	281	24 08	319
201	27 58	049	37 45	075	25 15	107	51 01	160	31 36	240	25 59	282	23 47	320
202	28 23	049	38 16	076	25 46	107	51 12	162	31 07	241	25 28	283	23 26	321
203	28 48	050	38 48	076	26 17	108	51 21	163	30 39	242	24 56	284	23 05	321
204	29 13	051	39 20	077	26 48	109	51 30	165	30 10	243	24 24	284	22 45	322
205	29 38	051	39 52	078	27 19	110	51 38	166	29 40	244	23 52	285	22 25	323
206	30 04	052	40 24	078	27 49	111	51 46	168	29 11	245	23 21	286	22 05	323
207	30 30	053	40 56	079	28 20	112	51 52	169	28 41	246	22 50	287	21 45	324
208	30 56	053	41 28	080	28 50	113	51 57	171	28 11	247	22 18	288	21 26	324
209	31 22	054	42 00	081	29 20	114	52 02	173	27 41	248	21 47	288	21 08	325
	◆DENEB		VEGA		Rasalhague		◆ARCTURUS		REGULUS		◆POLLUX		CAPELLA	
210	31 49	055	42 32	082	29 49	115	52 06	174	27 11	249	21 16	289	20 49	326
211	32 15	055	43 05	082	30 19	116	52 09	176	26 40	250	20 46	290	20 31	326
212	32 42	056	43 37	083	30 48	117	52 11	177	26 09	251	20 15	291	20 13	327
213	33 09	057	44 10	084	31 17	118	52 12	179	25 39	252	19 44	291	19 55	328
214	33 37	057	44 42	085	31 46	119	52 12	180	25 07	252	19 14	292	19 38	329
215	34 04	058	45 15	085	32 14	120	52 12	182	24 36	253	18 44	293	19 21	329
216	34 32	058	45 47	086	32 43	121	52 10	183	24 05	254	18 14	294	19 05	330
217	35 00	059	46 20	087	33 10	122	52 08	185	23 33	255	17 44	294	18 48	331
218	35 28	060	46 53	088	33 38	123	52 05	186	23 02	256	17 14	295	18 33	331
219	35 56	060	47 25	089	34 05	124	52 01	188	22 30	257	16 45	296	18 17	332
220	36 25	061	47 58	089	34 32	125	51 56	189	21 58	258	16 16	297	18 02	333
221	36 54	062	48 31	090	34 59	126	51 50	191	21 26	259	15 46	298	17 47	333
222	37 22	062	49 03	091	35 25	127	51 44	192	20 54	260	15 18	298	17 32	334
223	37 51	063	49 36	092	35 51	128	51 36	194	20 22	261	14 49	299	17 18	335
224	38 21	064	50 09	093	36 16	129	51 28	195	19 49	261	14 20	300	17 04	335
	DENEB		◆VEGA		Rasalhague		◆ARCTURUS		Denebola		Dubhe		◆CAPELLA	
225	38 50	064	50 41	094	36 41	130	51 19	197	34 26	240	60 35	304	16 51	336
226	39 19	065	51 14	095	37 06	132	51 09	198	33 57	241	60 08	305	16 38	337
227	39 49	066	51 46	096	37 30	133	50 58	200	33 29	242	59 41	305	16 25	337
228	40 19	066	52 19	096	37 54	134	50 46	201	33 00	243	59 14	305	16 13	338
229	40 49	067	52 51	097	38 17	135	50 34	203	32 30	244	58 48	305	16 01	339
230	41 19	067	53 24	098	38 40	136	50 21	204	32 01	245	58 21	306	15 49	339
231	41 49	068	53 56	099	39 03	137	50 07	206	31 31	246	57 55	306	15 38	340
232	42 20	069	54 28	100	39 24	138	49 53	207	31 01	247	57 28	306	15 27	341
233	42 50	069	55 00	101	39 46	140	49 37	209	30 30	248	57 02	307	15 16	342
234	43 21	070	55 32	102	40 07	141	49 21	210	30 00	249	56 36	307	15 06	342
235	43 52	071	56 04	103	40 27	142	49 05	211	29 29	250	56 10	307	14 56	343
236	44 23	071	56 36	104	40 47	143	48 47	213	28 59	251	55 44	308	14 47	344
237	44 54	072	57 07	105	41 06	144	48 30	214	28 28	252	55 18	308	14 38	344
238	45 25	073	57 39	106	41 25	146	48 11	215	27 56	253	54 53	309	14 29	345
239	45 56	073	58 10	107	41 43	147	47 52	217	27 25	254	54 27	309	14 21	346
	DENEB		◆VEGA		Rasalhague		◆ARCTURUS		Denebola		Dubhe		◆CAPELLA	
240	46 27	074	58 41	109	42 01	148	47 32	218	26 54	255	54 02	309	14 13	346
241	46 59	075	59 12	110	42 18	149	47 11	219	26 22	256	53 37	310	14 06	347
242	47 31	076	59 43	111	42 34	151	46 50	221	25 50	257	53 12	310	13 59	348
243	48 02	076	60 13	112	42 50	152	46 29	222	25 18	258	52 47	310	13 52	349
244	48 34	077	60 43	113	43 05	153	46 07	223	24 46	258	52 22	311	13 46	349
245	49 06	078	61 13	115	43 19	155	45 44	224	24 14	259	51 57	311	13 40	350
246	49 38	078	61 43	116	43 33	156	45 21	226	23 42	260	51 32	311	13 34	351
247	50 10	079	62 12	117	43 46	157	44 57	227	23 10	261	51 08	312	13 29	351
248	50 42	080	62 41	118	43 58	159	44 33	228	22 38	262	50 44	312	13 25	352
249	51 14	080	63 10	120	44 10	160	44 09	229	22 05	263	50 20	313	13 20	353
250	51 46	081	63 38	121	44 21	161	43 44	230	21 33	264	49 56	313	13 16	354
251	52 19	082	64 05	123	44 31	163	43 18	232	21 00	265	49 32	313	13 13	354
252	52 51	083	64 33	124	44 40	164	42 53	233	20 28	265	49 08	314	13 10	355
253	53 24	083	65 00	126	44 49	165	42 26	234	19 55	266	48 45	314	13 07	356
254	53 56	084	65 26	127	44 57	167	42 00	235	19 23	267	48 21	315	13 05	356
	Schedar		DENEB		◆ALTAIR		Rasalhague		◆ARCTURUS		Dubhe		◆CAPELLA	
255	34 54	038	54 29	085	31 40	128	45 04	168	41 33	236	47 58	315	13 03	357
256	35 14	038	55 01	086	32 06	129	45 10	169	41 06	237	47 35	315	13 02	358
257	35 34	039	55 34	087	32 31	130	45 16	171	40 38	238	47 12	316	13 00	359
258	35 55	039	56 06	087	32 55	131	45 21	172	40 10	239	46 50	316	13 00	359
259	36 16	040	56 39	088	33 20	132	45 25	173	39 42	240	46 27	317	13 00	000
260	36 37	040	57 12	089	33 44	133	45 28	175	39 13	242	46 05	317	13 00	001
261	36 58	041	57 44	090	34 07	135	45 31	176	38 44	243	45 43	318	13 00	001
262	37 19	041	58 17	091	34 30	136	45 33	178	38 15	244	45 21	318	13 01	002
263	37 41	042	58 50	091	34 53	137	45 34	179	37 46	245	44 59	318	13 03	003
264	38 03	042	59 22	092	35 15	138	45 34	180	37 16	246	44 38	319	13 05	004
265	38 25	043	59 55	093	35 37	139	45 33	182	36 46	247	44 16	319	13 07	004
266	38 47	043	60 28	094	35 58	140	45 32	183	36 16	248	43 55	320	13 09	005
267	39 10	044	61 00	095	36 19	141	45 29	185	35 46	249	43 34	320	13 12	006
268	39 33	044	61 33	096	36 39	142	45 26	186	35 15	250	43 13	321	13 16	006
269	39 56	045	62 05	097	36 58	144	45 23	187	34 45	251	42 53	321	13 20	007

LHA ♈	Hc	Zn	Hc	Zn	Hc	Zn	Hc	Zn	Hc	Zn	Hc	Zn	Hc	Zn
	◆CAPELLA		Alpheratz		◆ALTAIR		Rasalhague		◆ARCTURUS		Alkaid		Dubhe	
270	13 24	008	22 59	072	37 17	145	45 18	189	34 14	252	52 47	286	42 32	322
271	13 28	008	23 30	072	37 36	146	45 13	190	33 43	252	52 16	287	42 12	322
272	13 33	009	24 01	073	37 54	147	45 06	192	33 11	253	51 45	287	41 52	322
273	13 39	010	24 33	074	38 11	148	44 59	193	32 40	254	51 13	288	41 32	323
274	13 45	011	25 04	075	38 28	150	44 52	194	32 08	255	50 42	289	41 12	323
275	13 51	011	25 36	075	38 44	151	44 43	196	31 37	256	50 11	289	40 53	324
276	13 58	012	26 08	076	39 00	152	44 34	197	31 05	257	49 41	290	40 34	324
277	14 05	013	26 39	077	39 15	153	44 24	198	30 33	258	49 10	290	40 15	325
278	14 12	013	27 11	078	39 29	155	44 14	200	30 01	259	48 39	291	39 56	325
279	14 20	014	27 43	079	39 43	156	44 02	201	29 29	260	48 09	292	39 37	326
280	14 28	015	28 15	079	39 56	157	43 50	202	28 57	261	47 39	292	39 19	326
281	14 36	016	28 47	080	40 09	158	43 37	204	28 24	262	47 08	293	39 01	327
282	14 45	016	29 20	081	40 20	160	43 24	205	27 52	262	46 38	293	38 43	327
283	14 55	017	29 52	082	40 32	161	43 10	206	27 20	263	46 09	294	38 25	328
284	15 04	018	30 24	083	40 42	162	42 55	208	26 47	264	45 39	295	38 08	328
	◆CAPELLA		Alpheratz		◆ALTAIR		Rasalhague		◆ARCTURUS		Alkaid		Dubhe	
285	15 15	018	30 57	083	40 52	163	42 40	209	26 15	265	45 09	295	37 51	328
286	15 25	019	31 29	084	41 01	165	42 23	210	25 42	266	44 40	296	37 34	329
287	15 36	020	32 02	085	41 09	166	42 07	211	25 09	267	44 10	296	37 17	329
288	15 47	020	32 34	086	41 16	167	41 49	213	24 37	268	43 41	297	37 01	330
289	15 59	021	33 07	087	41 23	169	41 31	214	24 04	269	43 12	298	36 44	330
290	16 11	022	33 40	087	41 29	170	41 13	215	23 31	269	42 43	298	36 28	331
291	16 23	023	34 12	088	41 35	171	40 54	216	22 59	270	42 15	299	36 12	331
292	16 36	023	34 45	089	41 39	173	40 34	218	22 26	271	41 46	299	35 57	332
293	16 49	024	35 18	090	41 43	174	40 14	219	21 53	272	41 18	300	35 42	332
294	17 02	025	35 50	091	41 46	175	39 53	220	21 21	273	40 49	301	35 27	333
295	17 16	025	36 23	092	41 49	177	39 32	221	20 48	274	40 21	301	35 12	333
296	17 30	026	36 56	092	41 50	178	39 10	222	20 15	274	39 54	302	34 57	334
297	17 45	027	37 28	093	41 51	179	38 48	223	19 43	275	39 26	302	34 43	334
298	17 59	027	38 01	094	41 51	180	38 25	225	19 10	276	38 58	303	34 29	335
299	18 14	028	38 33	095	41 51	182	38 02	226	18 38	277	38 31	304	34 15	335
	CAPELLA		◆Alpheratz		Enif		ALTAIR		◆VEGA		Alphecca		◆Dubhe	
300	18 30	029	39 06	096	38 45	146	41 49	183	67 15	226	34 54	266	34 02	336
301	18 46	029	39 38	097	39 03	148	41 47	184	66 51	227	34 21	267	33 48	336
302	19 02	030	40 11	098	39 20	149	41 44	186	66 27	229	33 49	268	33 35	337
303	19 19	031	40 43	099	39 36	150	41 40	187	66 02	231	33 16	269	33 23	337
304	19 35	031	41 15	099	39 52	151	41 36	188	65 36	232	32 43	270	33 10	338
305	19 53	032	41 48	100	40 08	153	41 31	190	65 10	234	32 11	270	32 58	338
306	20 10	033	42 20	101	40 23	154	41 25	191	64 43	235	31 38	271	32 46	339
307	20 28	033	42 52	102	40 37	155	41 18	192	64 16	237	31 05	272	32 35	339
308	20 46	034	43 24	103	40 50	156	41 11	194	63 48	238	30 33	273	32 23	340
309	21 05	035	43 55	104	41 03	158	41 03	195	63 20	240	30 00	274	32 12	340
310	21 23	035	44 27	105	41 15	159	40 54	196	62 52	241	29 27	275	32 01	341
311	21 42	036	44 58	106	41 26	160	40 45	198	62 23	242	28 55	275	31 51	341
312	22 02	037	45 30	107	41 37	161	40 34	199	61 54	244	28 22	276	31 40	342
313	22 21	037	46 01	108	41 47	163	40 24	200	61 25	245	27 50	277	31 30	342
314	22 41	038	46 32	109	41 57	164	40 12	201	60 55	246	27 18	278	31 21	343
	CAPELLA		◆Hamal		Alpheratz		Enif		◆ALTAIR		VEGA		◆Alioth	
315	23 02	039	26 38	087	47 03	110	42 05	165	40 00	203	60 25	247	32 24	326
316	23 22	039	27 11	088	47 33	111	42 13	167	39 47	204	59 54	249	32 06	326
317	23 43	040	27 44	089	48 04	112	42 20	168	39 33	205	59 24	250	31 47	327
318	24 04	041	28 16	090	48 34	113	42 27	169	39 19	206	58 53	251	31 30	327
319	24 26	041	28 49	091	49 04	114	42 32	171	39 04	208	58 22	252	31 12	328
320	24 47	042	29 22	092	49 33	116	42 37	172	38 49	209	57 51	253	30 55	328
321	25 09	043	29 54	092	50 03	117	42 42	173	38 33	210	57 20	254	30 38	329
322	25 32	043	30 27	093	50 32	118	42 45	175	38 16	211	56 48	255	30 21	330
323	25 54	044	31 00	094	51 00	119	42 48	176	37 59	212	56 16	256	30 05	330
324	26 17	045	31 32	095	51 29	120	42 50	177	37 41	214	55 44	257	29 49	331
325	26 40	045	32 05	096	51 57	121	42 51	179	37 22	215	55 13	258	29 33	331
326	27 03	046	32 37	097	52 25	123	42 51	180	37 04	216	54 40	259	29 17	332
327	27 27	047	33 10	098	52 52	124	42 51	181	36 44	217	54 08	260	29 02	332
328	27 51	047	33 42	099	53 19	125	42 50	183	36 24	218	53 36	261	28 47	333
329	28 15	048	34 14	099	53 45	126	42 48	184	36 03	220	53 04	262	28 32	334
	CAPELLA		◆Hamal		Alpheratz		Enif		◆ALTAIR		VEGA		◆Alioth	
330	28 39	048	34 46	100	54 12	128	42 45	185	35 42	221	52 31	263	28 18	334
331	29 04	049	35 18	101	54 37	129	42 41	187	35 21	222	51 59	264	28 04	335
332	29 29	050	35 50	102	55 02	130	42 37	188	34 59	223	51 26	265	27 50	335
333	29 54	050	36 22	103	55 27	132	42 32	189	34 36	224	50 54	266	27 36	336
334	30 19	051	36 54	104	55 51	133	42 27	191	34 14	225	50 21	267	27 23	336
335	30 45	052	37 26	105	56 15	134	42 20	192	33 50	226	49 48	268	27 10	337
336	31 10	052	37 57	106	56 38	136	42 13	193	33 26	227	49 16	269	26 57	338
337	31 36	053	38 29	107	57 00	137	42 05	195	33 02	228	48 43	269	26 45	338
338	32 02	054	39 00	108	57 22	139	41 56	196	32 37	229	48 10	270	26 33	339
339	32 29	054	39 31	109	57 44	140	41 47	197	32 12	231	47 38	271	26 21	339
340	32 55	055	40 02	110	58 04	142	41 37	199	31 47	232	47 05	272	26 10	340
341	33 22	055	40 32	111	58 24	143	41 26	200	31 21	233	46 32	273	25 59	340
342	33 49	056	41 03	112	58 43	145	41 15	201	30 55	234	46 00	273	25 48	341
343	34 16	057	41 33	113	59 02	146	41 03	202	30 29	235	45 27	274	25 38	341
344	34 44	057	42 03	114	59 20	148	40 50	204	30 02	236	44 55	275	25 28	342
	CAPELLA		ALDEBARAN		◆Hamal		Alpheratz		◆ALTAIR		VEGA		◆Alioth	
345	35 11	058	17 04	086	42 33	115	59 36	150	29 35	237	44 22	276	25 18	343
346	35 39	059	17 37	087	43 02	116	59 53	151	29 07	238	43 50	277	25 08	343
347	36 07	059	18 10	087	43 31	117	60 08	153	28 39	239	43 17	277	24 59	344
348	36 35	060	18 42	088	44 00	118	60 22	155	28 11	240	42 45	278	24 50	345
349	37 04	061	19 15	089	44 29	119	60 36	156	27 43	241	42 12	279	24 42	345
350	37 32	061	19 48	090	44 57	120	60 49	158	27 14	242	41 40	280	24 34	346
351	38 01	062	20 20	091	45 25	121	61 00	160	26 45	243	41 08	280	24 26	346
352	38 30	062	20 53	092	45 53	123	61 11	162	26 16	244	40 36	281	24 18	347
353	38 59	063	21 26	093	46 20	124	61 21	163	25 47	245	40 04	282	24 11	348
354	39 28	064	21 58	093	46 47	125	61 30	165	25 17	245	39 32	283	24 04	348
355	39 58	064	22 31	094	47 14	126	61 38	167	24 47	246	39 00	283	23 58	349
356	40 27	065	23 03	095	47 40	127	61 45	169	24 17	247	38 28	284	23 52	349
357	40 57	066	23 36	096	48 06	129	61 50	171	23 47	248	37 57	285	23 46	350
358	41 27	066	24 09	097	48 31	130	61 55	172	23 17	249	37 25	286	23 40	351
359	41 57	067	24 41	098	48 56	131	61 59	174	22 46	250	36 54	286	23 35	351

LHA	Hc	Zn	Hc	Zn	Hc	Zn	Hc	Zn	Hc	Zn	Hc	Zn	Hc	Zn
♈	Dubhe		◆CAPELLA		ALDEBARAN		Hamal		◆Alpheratz		ALTAIR		◆VEGA	
0	28 18	008	42 04	067	25 22	098	50 01	131	63 01	176	22 35	251	36 05	288
1	28 22	008	42 34	067	25 55	099	50 26	133	63 03	178	22 03	252	35 33	288
2	28 27	009	43 05	068	26 28	100	50 50	134	63 04	180	21 31	253	35 01	289
3	28 32	009	43 37	069	27 01	101	51 14	135	63 03	182	20 58	254	34 29	290
4	28 38	010	44 08	069	27 34	102	51 37	137	63 02	184	20 26	255	33 58	291
5	28 44	010	44 39	070	28 07	102	52 00	138	62 59	186	19 54	256	33 27	291
6	28 50	011	45 11	071	28 40	103	52 22	139	62 55	188	19 21	257	32 55	292
7	28 56	011	45 43	071	29 12	104	52 44	141	62 50	190	18 48	258	32 24	293
8	29 03	012	46 14	072	29 45	105	53 05	142	62 44	191	18 15	259	31 53	293
9	29 10	012	46 46	072	30 17	106	53 25	144	62 37	193	17 42	259	31 23	294
10	29 17	013	47 18	073	30 49	107	53 45	145	62 28	195	17 09	260	30 52	295
11	29 25	013	47 51	074	31 21	108	54 04	146	62 19	197	16 36	261	30 22	295
12	29 33	014	48 23	074	31 53	109	54 22	148	62 09	199	16 03	262	29 52	296
13	29 41	014	48 55	075	32 25	110	54 39	149	61 57	201	15 30	263	29 22	297
14	29 49	015	49 28	076	32 56	111	54 56	151	61 45	203	14 57	264	28 52	298
	Dubhe		◆CAPELLA		ALDEBARAN		Hamal		◆Alpheratz		DENEB		◆VEGA	
15	29 58	015	50 00	076	33 27	112	55 12	152	61 32	204	49 13	283	28 22	298
16	30 07	016	50 33	077	33 58	113	55 27	154	61 17	206	48 40	284	27 53	299
17	30 17	016	51 06	078	34 29	114	55 41	156	61 02	208	48 08	284	27 23	300
18	30 26	017	51 39	079	35 00	115	55 55	157	60 46	209	47 35	285	26 54	300
19	30 36	017	52 12	079	35 30	116	56 07	159	60 29	211	47 03	286	26 25	301
20	30 46	018	52 45	080	36 00	117	56 19	160	60 11	213	46 31	286	25 57	302
21	30 57	019	53 18	081	36 30	118	56 30	162	59 53	214	45 59	287	25 28	302
22	31 08	019	53 51	081	37 00	119	56 40	164	59 33	216	45 27	288	25 00	303
23	31 19	020	54 24	082	37 29	120	56 49	165	59 13	218	44 55	288	24 32	304
24	31 30	020	54 57	083	37 58	121	56 57	167	58 52	219	44 23	289	24 04	304
25	31 42	021	55 30	083	38 27	122	57 04	169	58 31	221	43 51	290	23 37	305
26	31 54	021	56 04	084	38 55	123	57 10	170	58 09	222	43 20	290	23 09	306
27	32 06	022	56 37	085	39 23	124	57 15	172	57 46	224	42 48	291	22 42	306
28	32 18	022	57 11	086	39 50	125	57 19	174	57 22	225	42 17	292	22 15	307
29	32 31	023	57 44	086	40 18	126	57 23	175	56 58	227	41 46	292	21 49	308
	Dubhe		◆CAPELLA		ALDEBARAN		Hamal		◆Alpheratz		DENEB		◆VEGA	
30	32 44	023	58 18	087	40 45	127	57 25	177	56 33	228	41 15	293	21 22	309
31	32 57	024	58 51	088	41 11	128	57 26	179	56 08	229	40 44	294	20 56	309
32	33 11	024	59 25	089	41 37	130	57 26	180	55 42	231	40 14	294	20 30	310
33	33 25	025	59 58	090	42 03	131	57 26	182	55 16	232	39 43	295	20 05	311
34	33 39	025	60 32	091	42 28	132	57 24	184	54 49	233	39 13	295	19 39	311
35	33 53	026	61 05	091	42 53	133	57 21	186	54 22	235	38 43	296	19 14	312
36	34 08	026	61 39	092	43 17	134	57 17	187	53 55	236	38 12	297	18 49	313
37	34 22	026	62 12	093	43 41	136	57 12	189	53 27	237	37 43	297	18 25	313
38	34 38	027	62 46	094	44 04	137	57 07	191	52 58	238	37 13	298	18 01	314
39	34 53	027	63 19	095	44 27	138	57 00	192	52 29	240	36 43	299	17 37	315
40	35 08	028	63 53	096	44 49	139	56 52	194	52 00	241	36 14	299	17 13	315
41	35 24	028	64 26	097	45 10	140	56 44	196	51 31	242	35 45	300	16 50	316
42	35 40	029	64 59	098	45 32	142	56 34	197	51 01	243	35 16	300	16 27	317
43	35 57	029	65 33	099	45 52	143	56 24	199	50 31	244	34 47	301	16 04	318
44	36 13	030	66 06	100	46 12	144	56 13	201	50 01	245	34 18	302	15 41	318
	◆Dubhe		POLLUX		BETELGEUSE		◆ALDEBARAN		Hamal		Alpheratz		◆DENEB	
45	36 30	030	33 15	088	30 30	127	46 31	146	56 00	202	49 30	247	33 50	302
46	36 47	031	33 49	089	30 57	128	46 50	147	55 47	204	48 59	248	33 22	303
47	37 05	031	34 22	090	31 23	129	47 08	148	55 33	205	48 28	249	32 54	304
48	37 22	032	34 56	091	31 49	130	47 25	150	55 19	207	47 56	250	32 26	304
49	37 40	032	35 29	091	32 14	131	47 42	151	55 03	208	47 25	251	31 58	305
50	37 58	033	36 03	092	32 39	133	47 58	152	54 47	210	46 53	252	31 31	306
51	38 16	033	36 36	093	33 03	134	48 13	154	54 30	211	46 21	253	31 04	306
52	38 34	034	37 10	094	33 27	135	48 27	155	54 12	213	45 49	254	30 37	307
53	38 53	034	37 43	095	33 51	136	48 41	157	53 53	214	45 17	255	30 10	307
54	39 12	034	38 17	096	34 14	137	48 54	158	53 34	216	44 44	256	29 43	308
55	39 31	035	38 50	097	34 37	138	49 06	159	53 14	217	44 12	257	29 17	309
56	39 50	035	39 23	097	34 59	139	49 18	161	52 53	219	43 39	258	28 51	309
57	40 10	036	39 57	098	35 21	140	49 28	162	52 32	220	43 06	259	28 25	310
58	40 30	036	40 30	099	35 42	141	49 38	164	52 10	221	42 33	260	27 59	311
59	40 50	037	41 03	100	36 03	143	49 47	165	51 47	223	42 00	260	27 34	311
	◆Dubhe		POLLUX		BETELGEUSE		◆ALDEBARAN		Hamal		Alpheratz		◆DENEB	
60	41 10	037	41 36	101	36 23	144	49 55	167	51 24	224	41 27	261	27 09	312
61	41 30	038	42 09	102	36 42	145	50 03	168	51 01	225	40 54	262	26 44	312
62	41 51	038	42 41	103	37 01	146	50 09	170	50 37	227	40 21	263	26 20	313
63	42 12	038	43 14	104	37 20	147	50 15	171	50 12	228	39 47	264	25 55	314
64	42 33	039	43 47	105	37 38	148	50 19	173	49 47	229	39 14	265	25 31	314
65	42 54	039	44 19	106	37 55	150	50 23	174	49 21	231	38 40	266	25 07	315
66	43 15	040	44 51	107	38 12	151	50 26	176	48 55	232	38 07	267	24 44	316
67	43 37	040	45 23	108	38 28	152	50 28	177	48 28	233	37 33	267	24 20	316
68	43 58	041	45 55	109	38 43	153	50 30	179	48 01	234	37 00	268	23 57	317
69	44 20	041	46 27	110	38 58	155	50 30	180	47 34	235	36 26	269	23 35	318
70	44 43	041	46 58	111	39 12	156	50 29	182	47 06	237	35 53	270	23 12	318
71	45 05	042	47 29	112	39 26	157	50 28	183	46 38	238	35 19	271	22 50	319
72	45 27	042	48 00	113	39 38	158	50 26	185	46 09	239	34 46	272	22 28	320
73	45 50	043	48 31	114	39 50	160	50 23	186	45 40	240	34 12	272	22 06	320
74	46 13	043	49 02	115	40 02	161	50 19	188	45 11	241	33 39	273	21 45	321
	◆Dubhe		REGULUS		PROCYON		◆SIRIUS		ALDEBARAN		◆Alpheratz		DENEB	
75	46 36	043	17 10	094	30 15	132	14 01	154	50 14	189	33 05	274	21 24	322
76	46 59	044	17 43	095	30 39	134	14 15	155	50 08	191	32 32	275	21 03	322
77	47 22	044	18 16	096	31 04	135	14 29	156	50 01	192	31 58	276	20 43	323
78	47 46	045	18 50	096	31 27	136	14 42	157	49 54	194	31 25	276	20 23	324
79	48 09	045	19 23	097	31 50	137	14 55	158	49 45	195	30 52	277	20 03	324
80	48 33	045	19 56	098	32 13	138	15 07	159	49 36	197	30 18	278	19 44	325
81	48 57	046	20 30	099	32 35	139	15 19	160	49 26	198	29 45	279	19 25	326
82	49 21	046	21 03	100	32 57	140	15 30	161	49 16	199	29 12	280	19 06	326
83	49 46	047	21 36	101	33 19	141	15 41	162	49 04	201	28 39	280	18 47	327
84	50 10	047	22 09	102	33 39	142	15 51	163	48 52	202	28 06	281	18 29	328
85	50 35	047	22 41	103	34 00	143	16 01	164	48 39	204	27 33	282	18 11	328
86	51 00	048	23 14	103	34 19	145	16 10	165	48 25	205	27 00	283	17 54	329
87	51 24	048	23 47	104	34 39	146	16 18	166	48 10	207	26 28	283	17 37	330
88	51 49	048	24 19	105	34 57	147	16 26	167	47 55	208	25 55	284	17 20	330
89	52 15	049	24 51	106	35 15	148	16 34	168	47 39	209	25 23	285	17 03	331

LAT 56°N

LHA	Hc	Zn	Hc	Zn	Hc	Zn	Hc	Zn	Hc	Zn	Hc	Zn	Hc	Zn
♈	Dubhe		◆REGULUS		PROCYON		SIRIUS		◆ALDEBARAN		Mirfak		◆DENEB	
90	52 40	049	25 24	107	35 33	149	16 41	169	47 22	211	66 02	272	16 47	332
91	53 05	049	25 56	108	35 50	150	16 47	170	47 05	212	65 29	272	16 31	332
92	53 31	050	26 27	109	36 06	152	16 53	171	46 46	213	64 55	273	16 16	333
93	53 57	050	26 59	110	36 22	153	16 58	172	46 28	215	64 22	274	16 01	334
94	54 22	050	27 31	111	36 37	154	17 02	173	46 08	216	63 48	275	15 46	334
95	54 48	051	28 02	112	36 51	155	17 06	174	45 48	217	63 15	275	15 32	335
96	55 14	051	28 33	113	37 05	156	17 09	175	45 28	219	62 42	276	15 18	336
97	55 41	051	29 04	114	37 18	158	17 12	176	45 07	220	62 08	277	15 04	336
98	56 07	052	29 34	114	37 31	159	17 14	177	44 45	221	61 35	278	14 51	337
99	56 33	052	30 05	115	37 42	160	17 16	178	44 23	222	61 02	278	14 38	338
100	57 00	052	30 35	116	37 53	161	17 17	179	44 00	223	60 29	279	14 26	338
101	57 26	053	31 05	117	38 04	162	17 17	180	43 36	225	59 56	280	14 13	339
102	57 53	053	31 35	118	38 14	164	17 17	181	43 12	226	59 22	280	14 02	340
103	58 20	053	32 04	119	38 23	165	17 17	182	42 48	227	58 50	281	13 50	341
104	58 47	054	32 33	120	38 31	166	17 15	183	42 23	228	58 17	282	13 39	341
	Alioth		◆REGULUS		PROCYON		BETELGEUSE		◆ALDEBARAN		Mirfak		◆DENEB	
105	44 05	051	33 02	121	38 39	168	39 44	201	41 58	229	57 44	282	13 29	342
106	44 31	052	33 30	122	38 46	169	39 31	202	41 32	231	57 11	283	13 19	343
107	44 57	052	33 59	123	38 52	170	39 18	204	41 06	232	56 38	284	13 09	343
108	45 24	053	34 26	124	38 57	171	39 04	205	40 40	233	56 06	284	12 59	344
109	45 51	053	34 54	126	39 02	173	38 50	206	40 13	234	55 33	285	12 50	345
110	46 17	054	35 21	127	39 06	174	38 35	207	39 45	235	55 01	285	12 42	345
111	46 45	054	35 48	128	39 09	175	38 19	209	39 18	236	54 29	286	12 34	346
112	47 12	054	36 14	129	39 11	176	38 03	210	38 50	237	53 56	287	12 26	347
113	47 39	055	36 40	130	39 13	178	37 46	211	38 21	238	53 24	287	12 18	348
114	48 07	055	37 06	131	39 14	179	37 28	212	37 53	239	52 52	288	12 11	348
115	48 34	056	37 31	132	39 14	180	37 10	213	37 24	240	52 20	288	12 05	349
116	49 02	056	37 55	133	39 14	182	36 51	215	36 54	241	51 49	289	11 59	350
117	49 30	057	38 20	134	39 12	183	36 32	216	36 25	242	51 17	290	11 53	350
118	49 59	057	38 44	135	39 10	184	36 12	217	35 55	243	50 45	290	11 48	351
119	50 27	058	39 07	137	39 08	185	35 51	218	35 25	244	50 14	291	11 43	352
	◆Kochab		ARCTURUS		◆REGULUS		PROCYON		BETELGEUSE		◆CAPELLA		DENEB	
120	49 50	024	13 43	076	39 30	138	39 04	187	35 30	219	62 51	266	11 38	353
121	50 04	025	14 16	077	39 52	139	39 00	188	35 09	220	62 18	267	11 34	353
122	50 18	025	14 49	077	40 14	140	38 55	189	34 47	221	61 44	268	11 30	354
123	50 32	025	15 21	078	40 35	141	38 49	191	34 25	223	61 11	268	11 27	355
124	50 46	025	15 54	079	40 56	143	38 42	192	34 02	224	60 37	269	11 24	355
125	51 01	025	16 27	080	41 16	144	38 35	193	33 38	225	60 04	270	11 22	356
126	51 15	026	17 00	081	41 35	145	38 27	194	33 14	226	59 30	271	11 20	357
127	51 30	026	17 34	081	41 54	146	38 18	196	32 50	227	58 57	272	11 18	358
128	51 44	026	18 07	082	42 13	147	38 09	197	32 25	228	58 23	273	11 17	358
129	51 59	026	18 40	083	42 30	149	37 59	198	32 00	229	57 50	273	11 16	359
130	52 14	026	19 13	084	42 47	150	37 48	199	31 35	230	57 16	274	11 16	000
131	52 29	027	19 47	085	43 04	151	37 37	201	31 09	231	56 43	275	11 16	000
132	52 44	027	20 20	086	43 20	153	37 25	202	30 42	232	56 09	276	11 16	001
133	52 59	027	20 54	086	43 35	154	37 12	203	30 16	233	55 36	276	11 17	002
134	53 14	027	21 27	087	43 49	155	36 58	204	29 48	234	55 02	277	11 19	003
	DENEB		◆Kochab		ARCTURUS		◆REGULUS		PROCYON		BETELGEUSE		◆CAPELLA	
135	11 20	003	53 30	027	22 01	088	44 03	157	36 44	205	29 21	235	54 29	278
136	11 23	004	53 45	027	22 34	089	44 16	158	36 30	207	28 53	236	53 56	279
137	11 25	005	54 01	028	23 08	090	44 28	159	36 14	208	28 25	237	53 23	279
138	11 28	006	54 16	028	23 41	090	44 40	161	35 58	209	27 57	238	52 50	280
139	11 32	006	54 32	028	24 15	091	44 51	162	35 42	210	27 28	239	52 17	281
140	11 35	007	54 47	028	24 48	092	45 01	163	35 24	211	26 59	240	51 44	281
141	11 40	008	55 03	028	25 22	093	45 10	165	35 07	213	26 30	241	51 11	282
142	11 44	008	55 19	028	25 55	094	45 18	166	34 48	214	26 00	242	50 38	283
143	11 49	009	55 35	028	26 29	095	45 26	167	34 29	215	25 31	243	50 06	283
144	11 55	010	55 51	028	27 02	096	45 33	169	34 10	216	25 01	244	49 33	284
145	12 01	011	56 07	029	27 36	096	45 39	170	33 50	217	24 30	245	49 00	285
146	12 07	011	56 23	029	28 09	097	45 45	172	33 29	218	24 00	246	48 28	285
147	12 14	012	56 39	029	28 42	098	45 49	173	33 08	219	23 29	247	47 56	286
148	12 21	013	56 55	029	29 15	099	45 53	174	32 47	220	22 58	248	47 24	287
149	12 29	013	57 11	029	29 49	100	45 56	176	32 25	222	22 27	249	46 52	287
	DENEB		◆VEGA		ARCTURUS		Denebola		◆REGULUS		POLLUX		◆CAPELLA	
150	12 37	014	14 07	039	30 22	101	43 39	142	45 58	177	53 08	235	46 20	288
151	12 45	015	14 28	039	30 54	102	43 59	144	45 59	179	52 41	236	45 48	289
152	12 54	015	14 49	040	31 27	103	44 19	145	46 00	180	52 13	237	45 16	289
153	13 03	016	15 11	041	32 00	103	44 38	146	45 59	181	51 44	239	44 44	290
154	13 12	017	15 33	041	32 33	104	44 56	147	45 58	183	51 15	240	44 13	291
155	13 22	018	15 55	042	33 05	105	45 14	149	45 56	184	50 46	241	43 42	291
156	13 33	018	16 18	043	33 37	106	45 31	150	45 53	186	50 17	242	43 10	292
157	13 43	019	16 41	044	34 09	107	45 48	151	45 49	187	49 47	243	42 39	293
158	13 54	020	17 04	044	34 41	108	46 03	153	45 45	188	49 17	244	42 08	293
159	14 06	020	17 27	045	35 13	109	46 18	154	45 40	190	48 47	245	41 38	294
160	14 18	021	17 51	046	35 45	110	46 33	155	45 33	191	48 16	246	41 07	294
161	14 30	022	18 15	046	36 16	111	46 46	157	45 27	193	47 45	248	40 37	295
162	14 43	022	18 40	047	36 47	112	46 59	158	45 19	194	47 14	249	40 06	296
163	14 56	023	19 05	048	37 18	113	47 11	160	45 10	195	46 43	250	39 36	296
164	15 09	024	19 30	048	37 49	114	47 23	161	45 01	197	46 11	251	39 06	297
	◆DENEB		VEGA		ARCTURUS		◆SPICA		REGULUS		◆POLLUX		CAPELLA	
165	15 23	025	19 55	049	38 20	115	16 24	143	44 51	198	45 39	252	38 36	298
166	15 37	025	20 20	050	38 50	116	16 44	144	44 40	199	45 07	253	38 07	298
167	15 51	026	20 46	051	39 20	117	17 04	145	44 29	201	44 35	254	37 37	299
168	16 06	027	21 12	051	39 50	118	17 23	146	44 17	202	44 03	255	37 08	299
169	16 22	027	21 38	052	40 19	119	17 42	147	44 04	203	43 31	256	36 39	300
170	16 37	028	22 05	053	40 48	120	18 00	148	43 50	205	42 58	257	36 10	301
171	16 53	029	22 32	053	41 17	121	18 18	149	43 36	206	42 25	258	35 41	301
172	17 09	029	22 59	054	41 46	122	18 35	150	43 20	207	41 52	258	35 12	302
173	17 26	030	23 26	055	42 14	123	18 52	151	43 05	209	41 20	259	34 44	303
174	17 43	031	23 53	055	42 42	125	19 08	152	42 48	210	40 47	260	34 16	303
175	18 00	031	24 21	056	43 09	126	19 24	153	42 31	211	40 13	261	33 48	304
176	18 18	032	24 49	057	43 36	127	19 39	154	42 14	212	39 40	262	33 20	304
177	18 36	033	25 17	057	44 03	128	19 53	155	41 55	214	39 07	263	32 52	305
178	18 54	033	25 46	058	44 29	129	20 07	156	41 36	215	38 34	264	32 25	306
179	19 13	034	26 14	059	44 55	130	20 21	157	41 17	216	38 00	265	31 58	306

LHA ♈	Hc	Zn	Hc	Zn	Hc	Zn	Hc	Zn	Hc	Zn	Hc	Zn	Hc	Zn
	♦DENEB		VEGA		ARCTURUS		♦SPICA		REGULUS		♦POLLUX		CAPELLA	
180	19 32	035	26 43	059	45 20	132	20 34	158	40 57	217	37 27	266	31 31	307
181	19 51	035	27 12	060	45 45	133	20 47	159	40 36	219	36 53	266	31 04	307
182	20 10	036	27 41	061	46 10	134	20 58	160	40 15	220	36 20	267	30 38	308
183	20 30	037	28 11	062	46 33	135	21 10	161	39 53	221	35 46	268	30 11	309
184	20 51	037	28 40	062	46 57	136	21 20	162	39 31	222	35 13	269	29 45	309
185	21 11	038	29 10	063	47 20	138	21 31	163	39 08	223	34 39	270	29 19	310
186	21 32	039	29 40	064	47 42	139	21 40	164	38 45	224	34 06	271	28 54	311
187	21 53	039	30 10	064	48 04	140	21 49	165	38 21	226	33 32	271	28 28	311
188	22 14	040	30 40	065	48 25	142	21 58	166	37 57	227	32 59	272	28 03	312
189	22 36	041	31 11	066	48 45	143	22 06	167	37 32	228	32 25	273	27 38	312
190	22 58	041	31 42	066	49 05	144	22 13	168	37 07	229	31 52	274	27 14	313
191	23 20	042	32 12	067	49 24	146	22 19	169	36 41	230	31 18	275	26 49	314
192	23 43	043	32 43	068	49 43	147	22 25	170	36 16	231	30 45	276	26 25	314
193	24 06	043	33 15	068	50 01	148	22 31	171	35 49	232	30 11	276	26 02	315
194	24 29	044	33 46	069	50 18	150	22 36	172	35 22	233	29 38	277	25 38	316
	♦DENEB		VEGA		Rasalhague		♦ARCTURUS		REGULUS		♦POLLUX		CAPELLA	
195	24 52	045	34 17	070	22 16	101	50 35	151	34 55	234	29 05	278	25 15	316
196	25 16	045	34 49	071	22 49	102	50 50	153	34 28	235	28 32	279	24 52	317
197	25 40	046	35 21	071	23 21	103	51 05	154	34 00	237	27 58	279	24 29	318
198	26 04	046	35 52	072	23 54	103	51 20	156	33 32	238	27 25	280	24 06	318
199	26 29	047	36 24	073	24 27	104	51 33	157	33 03	239	26 52	281	23 44	319
200	26 53	048	36 56	073	24 59	105	51 46	159	32 35	240	26 19	282	23 22	320
201	27 18	048	37 29	074	25 31	106	51 58	160	32 06	241	25 47	283	23 01	320
202	27 43	049	38 01	075	26 04	107	52 09	162	31 36	242	25 14	283	22 39	321
203	28 09	050	38 33	075	26 36	108	52 19	163	31 07	243	24 41	284	22 18	321
204	28 35	050	39 06	076	27 07	109	52 28	165	30 37	244	24 09	285	21 57	322
205	29 00	051	39 39	077	27 39	110	52 37	166	30 07	245	23 37	286	21 37	323
206	29 27	052	40 11	078	28 11	111	52 44	168	29 36	245	23 04	286	21 17	323
207	29 53	052	40 44	078	28 42	112	52 51	169	29 05	246	22 32	287	20 57	324
208	30 20	053	41 17	079	29 13	113	52 57	171	28 35	247	22 00	288	20 37	325
209	30 47	053	41 50	080	29 44	113	53 02	172	28 04	248	21 28	289	20 18	325
	♦DENEB		VEGA		Rasalhague		♦ARCTURUS		REGULUS		♦POLLUX		CAPELLA	
210	31 14	054	42 23	081	30 14	114	53 06	174	27 32	249	20 57	289	19 59	326
211	31 41	055	42 56	081	30 45	115	53 09	175	27 01	250	20 25	290	19 41	327
212	32 08	055	43 29	082	31 15	116	53 11	177	26 29	251	19 54	291	19 22	327
213	32 36	056	44 03	083	31 45	117	53 12	179	25 57	252	19 22	292	19 05	328
214	33 04	057	44 36	084	32 15	118	53 12	180	25 25	253	18 51	292	18 47	329
215	33 32	057	45 09	084	32 44	119	53 12	182	24 53	254	18 20	293	18 30	329
216	34 00	058	45 43	085	33 13	120	53 10	183	24 21	255	17 50	294	18 13	330
217	34 29	058	46 16	086	33 42	121	53 08	185	23 48	256	17 19	295	17 56	331
218	34 58	059	46 50	087	34 10	122	53 05	187	23 16	257	16 49	295	17 40	331
219	35 26	060	47 23	088	34 39	123	53 00	188	22 43	257	16 18	296	17 24	332
220	35 56	060	47 57	088	35 06	125	52 55	190	22 10	258	15 48	297	17 08	333
221	36 25	061	48 30	089	35 34	126	52 49	191	21 37	259	15 19	298	16 53	333
222	36 54	062	49 04	090	36 01	127	52 42	193	21 04	260	14 49	299	16 38	334
223	37 24	062	49 37	091	36 28	128	52 34	194	20 31	261	14 20	299	16 24	335
224	37 54	063	50 11	092	36 54	129	52 26	196	19 58	262	13 51	300	16 10	335
	DENEB		♦VEGA		Rasalhague		♦ARCTURUS		Denebola		Dubhe		♦CAPELLA	
225	38 24	064	50 44	093	37 20	130	52 16	197	34 55	241	60 01	306	15 56	336
226	38 54	064	51 18	093	37 45	131	52 06	199	34 26	242	59 33	306	15 43	336
227	39 24	065	51 51	094	38 11	132	51 54	200	33 56	243	59 06	306	15 30	337
228	39 54	065	52 25	095	38 35	133	51 42	202	33 26	244	58 39	307	15 17	338
229	40 25	066	52 58	096	38 59	134	51 29	203	32 56	245	58 12	307	15 05	339
230	40 56	067	53 32	097	39 23	136	51 16	205	32 25	246	57 45	307	14 53	340
231	41 27	067	54 05	098	39 46	137	51 01	206	31 55	247	57 19	307	14 41	340
232	41 58	068	54 38	099	40 09	138	50 46	208	31 24	248	56 52	308	14 30	341
233	42 29	069	55 11	100	40 31	139	50 30	209	30 52	249	56 26	308	14 19	342
234	43 00	069	55 44	101	40 53	140	50 13	211	30 21	250	55 59	308	14 09	342
235	43 32	070	56 17	102	41 14	141	49 56	212	29 49	251	55 33	309	13 59	343
236	44 03	071	56 50	103	41 35	143	49 38	213	29 18	252	55 07	309	13 49	344
237	44 35	071	57 22	104	41 55	144	49 19	215	28 46	253	54 41	309	13 40	344
238	45 07	072	57 55	105	42 14	145	49 00	216	28 14	253	54 15	310	13 31	345
239	45 39	072	58 27	106	42 33	146	48 40	217	27 41	254	53 49	310	13 23	346
	DENEB		♦VEGA		Rasalhague		♦ARCTURUS		Denebola		Dubhe		♦CAPELLA	
240	46 11	073	59 00	107	42 52	148	48 19	219	27 09	255	53 23	310	13 15	347
241	46 43	074	59 32	108	43 09	149	47 58	220	26 37	256	52 58	311	13 07	347
242	47 15	074	60 03	109	43 26	150	47 36	221	26 04	257	52 33	311	13 00	348
243	47 47	075	60 35	110	43 42	152	47 13	223	25 31	258	52 07	311	12 53	349
244	48 20	076	61 06	112	43 58	153	46 50	224	24 58	259	51 42	312	12 47	349
245	48 52	077	61 37	113	44 13	154	46 27	225	24 25	260	51 17	312	12 41	350
246	49 25	077	62 08	114	44 27	155	46 03	226	23 52	261	50 52	312	12 35	351
247	49 58	078	62 39	115	44 41	157	45 38	228	23 19	261	50 28	313	12 30	351
248	50 31	079	63 09	117	44 54	158	45 13	229	22 46	262	50 03	313	12 25	352
249	51 04	079	63 38	118	45 06	159	44 48	230	22 13	263	49 39	314	12 21	353
250	51 37	080	64 08	119	45 17	161	44 22	231	21 39	264	49 14	314	12 17	354
251	52 10	081	64 37	121	45 28	162	43 55	232	21 06	265	48 50	314	12 13	354
252	52 43	081	65 06	122	45 38	164	43 29	233	20 32	266	48 26	315	12 10	355
253	53 16	082	65 34	124	45 47	165	43 01	235	19 59	267	48 03	315	12 07	356
254	53 49	083	66 01	125	45 55	166	42 34	236	19 25	267	47 39	315	12 05	356
	Schedar		DENEB		♦ALTAIR		Rasalhague		♦ARCTURUS		Dubhe		♦CAPELLA	
255	34 06	037	54 23	084	32 17	128	46 03	168	42 06	237	47 16	316	12 03	357
256	34 27	038	54 56	084	32 43	129	46 09	169	41 38	238	46 52	316	12 02	358
257	34 47	038	55 29	085	33 09	130	46 15	171	41 09	239	46 29	317	12 01	359
258	35 08	039	56 03	086	33 35	131	46 20	172	40 40	240	46 06	317	12 00	359
259	35 29	039	56 36	087	34 00	132	46 25	173	40 11	241	45 43	317	12 00	000
260	35 51	040	57 10	087	34 25	133	46 28	175	39 41	242	45 21	318	12 00	001
261	36 12	040	57 43	088	34 49	134	46 31	176	39 12	243	44 58	318	12 00	001
262	36 34	041	58 17	089	35 13	135	46 33	178	38 41	244	44 36	319	12 01	002
263	36 56	041	58 50	090	35 36	136	46 34	179	38 11	245	44 14	319	12 03	003
264	37 18	042	59 24	091	35 59	137	46 34	180	37 41	246	43 52	320	12 05	003
265	37 41	042	59 58	092	36 22	139	46 33	182	37 10	247	43 30	320	12 07	004
266	38 04	043	60 31	092	36 44	140	46 32	183	36 39	248	43 09	320	12 10	005
267	38 27	043	61 05	093	37 05	141	46 29	185	36 07	249	42 48	321	12 13	006
268	38 50	044	61 38	094	37 26	142	46 26	186	35 36	250	42 27	321	12 16	006
269	39 13	044	62 12	095	37 47	143	46 22	188	35 04	251	42 06	322	12 20	007

LAT 56°N

LHA ♈	Hc	Zn	Hc	Zn	Hc	Zn	Hc	Zn	Hc	Zn	Hc	Zn	Hc	Zn
	♦CAPELLA		Alpheratz		♦ALTAIR		Rasalhague		♦ARCTURUS		Alkaid		Dubhe	
270	12 24	008	22 40	071	38 06	144	46 17	189	34 32	252	52 30	287	41 45	322
271	12 29	008	23 12	072	38 26	146	46 12	190	34 00	253	51 58	288	41 24	323
272	12 34	009	23 44	073	38 44	147	46 05	192	33 28	254	51 26	288	41 04	323
273	12 40	010	24 16	073	39 02	148	45 58	193	32 56	255	50 54	289	40 44	323
274	12 46	011	24 48	074	39 20	149	45 50	195	32 23	256	50 23	290	40 24	324
275	12 52	011	25 21	075	39 37	150	45 41	196	31 51	257	49 51	290	40 04	324
276	12 59	012	25 53	076	39 53	152	45 32	197	31 18	258	49 20	291	39 45	325
277	13 06	013	26 26	077	40 09	153	45 21	199	30 45	259	48 49	291	39 26	325
278	13 14	013	26 58	077	40 24	154	45 10	200	30 12	260	48 17	292	39 07	326
279	13 22	014	27 31	078	40 38	155	44 58	201	29 39	260	47 46	293	38 48	326
280	13 30	015	28 04	079	40 51	157	44 46	203	29 06	261	47 15	293	38 29	327
281	13 39	015	28 37	080	41 04	158	44 32	204	28 33	262	46 45	294	38 11	327
282	13 48	016	29 10	080	41 17	159	44 18	205	28 00	263	46 14	294	37 53	327
283	13 57	017	29 43	081	41 28	161	44 03	207	27 26	264	45 44	295	37 35	328
284	14 07	018	30 16	082	41 39	162	43 48	208	26 53	265	45 13	296	37 17	328
	♦CAPELLA		Alpheratz		♦ALTAIR		Rasalhague		♦ARCTURUS		Alkaid		Dubhe	
285	14 18	018	30 49	083	41 49	163	43 32	209	26 19	266	44 43	296	37 00	329
286	14 28	019	31 23	083	41 58	164	43 15	211	25 46	266	44 13	297	36 42	329
287	14 39	020	31 56	084	42 07	166	42 58	212	25 12	267	43 43	297	36 25	330
288	14 51	020	32 30	085	42 15	167	42 40	213	24 39	268	43 13	298	36 09	330
289	15 03	021	33 03	086	42 22	168	42 21	214	24 05	269	42 44	299	35 52	331
290	15 15	022	33 36	087	42 28	170	42 02	216	23 32	270	42 14	299	35 36	331
291	15 28	022	34 10	087	42 34	171	41 42	217	22 58	271	41 45	300	35 20	332
292	15 40	023	34 43	088	42 39	172	41 22	218	22 25	271	41 16	300	35 04	332
293	15 54	024	35 17	089	42 43	174	41 01	219	21 51	272	40 47	301	34 48	333
294	16 08	024	35 51	090	42 46	175	40 39	221	21 18	273	40 19	301	34 33	333
295	16 22	025	36 24	091	42 49	176	40 17	222	20 44	274	39 50	302	34 18	334
296	16 36	026	36 58	092	42 50	178	39 54	223	20 11	275	39 22	303	34 03	334
297	16 51	026	37 31	092	42 51	179	39 31	224	19 37	276	38 53	303	33 49	335
298	17 06	027	38 05	093	42 51	180	39 08	225	19 04	276	38 25	304	33 35	335
299	17 21	028	38 38	094	42 51	182	38 44	226	18 31	277	37 58	304	33 21	336
	CAPELLA		♦Alpheratz		Enif		ALTAIR		♦VEGA		Alphecca		♦Dubhe	
300	17 37	029	39 12	095	39 35	146	42 49	183	67 56	227	34 58	267	33 07	336
301	17 54	029	39 45	096	39 53	147	42 47	185	67 31	229	34 24	268	32 53	337
302	18 10	030	40 18	097	40 11	148	42 44	186	67 05	231	33 51	269	32 40	337
303	18 27	031	40 52	098	40 28	150	42 40	187	66 39	233	33 17	269	32 27	338
304	18 44	031	41 25	099	40 45	151	42 35	189	66 12	234	32 43	270	32 15	338
305	19 02	032	41 58	099	41 01	152	42 30	190	65 44	236	32 10	271	32 02	339
306	19 20	033	42 31	100	41 16	153	42 24	191	65 16	237	31 36	272	31 50	339
307	19 38	033	43 04	101	41 31	155	42 17	193	64 48	239	31 03	273	31 38	340
308	19 56	034	43 37	102	41 45	156	42 09	194	64 19	240	30 29	273	31 27	340
309	20 15	035	44 10	103	41 58	157	42 01	195	63 50	241	29 56	274	31 15	341
310	20 34	035	44 42	104	42 11	159	41 52	197	63 20	243	29 22	275	31 04	341
311	20 54	036	45 15	105	42 23	160	41 42	198	62 50	244	28 49	276	30 54	342
312	21 14	036	45 47	106	42 34	161	41 31	199	62 20	245	28 16	277	30 43	342
313	21 34	037	46 19	107	42 45	162	41 20	200	61 49	247	27 42	277	30 33	343
314	21 54	038	46 51	108	42 54	164	41 08	202	61 18	248	27 09	278	30 23	343
	CAPELLA		♦Hamal		Alpheratz		Enif		♦ALTAIR		VEGA		♦Alioth	
315	22 15	038	26 35	087	47 23	109	43 03	165	40 55	203	60 47	249	31 34	326
316	22 36	039	27 09	088	47 54	110	43 12	166	40 42	204	60 15	250	31 16	327
317	22 57	040	27 42	089	48 26	111	43 19	168	40 27	206	59 44	251	30 57	327
318	23 19	040	28 16	089	48 57	112	43 26	169	40 13	207	59 12	253	30 39	328
319	23 41	041	28 50	090	49 28	113	43 32	170	39 57	208	58 40	254	30 21	328
320	24 03	042	29 23	091	49 59	114	43 37	172	39 41	209	58 07	255	30 04	329
321	24 25	042	29 57	092	50 29	116	43 41	173	39 24	210	57 35	256	29 46	329
322	24 48	043	30 30	093	50 59	117	43 45	175	39 07	212	57 02	257	29 29	330
323	25 11	044	31 04	094	51 29	118	43 48	176	38 49	213	56 30	258	29 13	330
324	25 34	044	31 37	094	51 58	119	43 50	177	38 31	214	55 57	259	28 56	331
325	25 58	045	32 11	095	52 28	120	43 51	179	38 12	215	55 24	260	28 40	332
326	26 22	046	32 44	096	52 56	121	43 51	180	37 52	216	54 51	261	28 24	332
327	26 46	046	33 17	097	53 25	123	43 51	181	37 32	218	54 18	262	28 09	333
328	27 10	047	33 51	098	53 53	124	43 49	183	37 11	219	53 44	263	27 53	333
329	27 35	047	34 24	099	54 20	125	43 47	184	36 50	220	53 11	264	27 38	334
	CAPELLA		♦Hamal		Alpheratz		Enif		♦ALTAIR		VEGA		♦Alioth	
330	27 59	048	34 57	100	54 48	126	43 45	185	36 28	221	52 38	264	27 24	334
331	28 24	049	35 30	101	55 14	128	43 41	187	36 05	222	52 04	265	27 09	335
332	28 50	049	36 03	101	55 41	129	43 37	188	35 43	223	51 31	266	26 55	335
333	29 15	050	36 36	102	56 06	130	43 32	190	35 19	225	50 57	267	26 41	336
334	29 41	051	37 08	103	56 32	132	43 26	191	34 56	226	50 24	268	26 28	337
335	30 07	051	37 41	104	56 56	133	43 19	192	34 31	227	49 50	269	26 15	337
336	30 33	052	38 13	105	57 21	135	43 11	194	34 07	228	49 17	270	26 02	338
337	31 00	052	38 46	106	57 44	136	43 03	195	33 42	229	48 43	271	25 49	338
338	31 27	053	39 18	107	58 07	138	42 54	196	33 16	230	48 10	271	25 37	339
339	31 53	054	39 50	108	58 29	139	42 44	198	32 50	231	47 36	272	25 25	339
340	32 21	054	40 22	109	58 51	141	42 34	199	32 24	232	47 03	273	25 14	340
341	32 48	055	40 53	110	59 12	142	42 23	200	31 57	233	46 29	274	25 02	341
342	33 15	056	41 25	111	59 32	144	42 11	202	31 30	234	45 56	275	24 51	341
343	33 43	056	41 56	112	59 51	145	41 58	203	31 03	235	45 22	275	24 41	342
344	34 11	057	42 27	114	60 10	147	41 45	204	30 35	236	44 49	276	24 30	342
	CAPELLA		ALDEBARAN		♦Hamal		Alpheratz		♦ALTAIR		VEGA		♦Alioth	
345	34 39	057	17 00	086	42 58	114	60 28	149	30 07	237	44 15	277	24 21	343
346	35 08	058	17 33	086	43 28	115	60 45	150	29 39	238	43 42	278	24 11	344
347	35 36	059	18 07	087	43 58	116	61 01	152	29 10	239	43 09	278	24 02	344
348	36 05	059	18 40	088	44 28	117	61 16	154	28 41	240	42 36	279	23 53	345
349	36 34	060	19 14	089	44 58	118	61 31	156	28 12	241	42 03	280	23 44	345
350	37 03	061	19 47	090	45 27	119	61 44	157	27 43	242	41 30	281	23 36	346
351	37 32	061	20 21	090	45 56	121	61 57	159	27 13	243	40 57	281	23 28	347
352	38 02	062	20 55	091	46 25	122	62 08	161	26 43	244	40 24	282	23 20	347
353	38 31	062	21 28	092	46 53	123	62 18	163	26 13	245	39 51	283	23 13	348
354	39 01	063	22 02	093	47 21	124	62 28	165	25 42	246	39 18	284	23 06	348
355	39 31	064	22 35	094	47 49	125	62 36	167	25 11	247	38 46	284	22 59	349
356	40 01	064	23 09	095	48 16	126	62 43	168	24 40	248	38 13	285	22 53	350
357	40 32	065	23 42	096	48 43	128	62 50	170	24 09	249	37 41	286	22 47	350
358	41 02	065	24 15	096	49 09	129	62 55	172	23 38	250	37 09	286	22 41	351
359	41 33	066	24 49	097	49 35	130	62 59	174	23 06	251	36 37	287	22 36	351

LAT 55°N

LHA ♈	Hc	Zn	Hc	Zn	Hc	Zn	Hc	Zn	Hc	Zn	Hc	Zn	Hc	Zn
	◆CAPELLA		ALDEBARAN		Hamal		◆Alpheratz		ALTAIR		◆VEGA		Alioth	
0	41 39	066	25 30	098	50 40	130	64 01	176	22 53	252	35 46	288	21 32	352
1	42 11	067	26 04	098	51 06	132	64 03	178	22 21	253	35 13	289	21 27	353
2	42 43	067	26 38	099	51 31	133	64 04	180	21 48	254	34 41	290	21 23	353
3	43 14	068	27 12	100	51 56	134	64 03	182	21 15	254	34 09	291	21 19	354
4	43 46	068	27 46	101	52 21	136	64 01	184	20 41	255	33 37	291	21 15	354
5	44 18	069	28 20	102	52 44	137	63 58	186	20 08	256	33 04	292	21 12	355
6	44 51	070	28 53	103	53 07	138	63 54	188	19 35	257	32 33	293	21 09	356
7	45 23	070	29 27	104	53 30	140	63 49	190	19 01	258	32 01	293	21 07	356
8	45 55	071	30 00	105	53 52	141	63 43	192	18 27	259	31 29	294	21 05	357
9	46 28	071	30 33	106	54 13	143	63 35	194	17 53	260	30 58	295	21 03	357
10	47 01	072	31 06	106	54 34	144	63 26	196	17 19	261	30 27	295	21 01	358
11	47 33	073	31 39	107	54 53	146	63 16	198	16 45	261	29 56	296	21 00	359
12	48 06	073	32 12	108	55 12	147	63 05	200	16 11	262	29 25	297	21 00	359
13	48 39	074	32 45	109	55 31	149	62 53	201	15 37	263	28 54	297	20 59	000
14	49 12	075	33 17	110	55 48	150	62 40	203	15 03	264	28 24	298	20 59	000
	Dubhe		◆CAPELLA		ALDEBARAN		Hamal		◆Alpheratz		DENEB		◆VEGA	
15	29 00	015	49 46	075	33 49	111	56 05	152	62 26	205	48 59	284	27 54	299
16	29 09	016	50 19	076	34 21	112	56 21	153	62 11	207	48 26	285	27 23	299
17	29 19	016	50 53	077	34 53	113	56 36	155	61 55	209	47 52	285	26 53	300
18	29 29	017	51 26	077	35 25	114	56 50	157	61 38	210	47 19	286	26 24	301
19	29 39	017	52 00	078	35 56	115	57 03	158	61 20	212	46 46	287	25 54	301
20	29 49	018	52 33	079	36 27	116	57 15	160	61 01	214	46 13	287	25 25	302
21	30 00	018	53 07	079	36 58	117	57 27	162	60 42	215	45 41	288	24 56	303
22	30 11	019	53 41	080	37 28	118	57 37	163	60 21	217	45 08	289	24 27	303
23	30 22	019	54 15	081	37 58	119	57 47	165	60 00	219	44 35	289	23 58	304
24	30 34	020	54 49	081	38 28	120	57 55	167	59 38	220	44 03	290	23 30	305
25	30 45	020	55 23	082	38 58	121	58 03	168	59 16	222	43 31	291	23 02	305
26	30 58	021	55 57	083	39 27	122	58 09	170	58 53	223	42 59	291	22 34	306
27	31 10	021	56 31	083	39 56	123	58 15	172	58 29	225	42 27	292	22 06	307
28	31 23	022	57 05	084	40 25	124	58 19	173	58 04	226	41 55	292	21 39	307
29	31 36	022	57 40	085	40 53	126	58 22	175	57 39	228	41 23	293	21 12	308
	Dubhe		◆POLLUX		BETELGEUSE		RIGEL		◆Hamal		Alpheratz		◆DENEB	
30	31 49	023	24 43	076	23 38	112	15 00	130	58 25	177	57 13	229	40 51	294
31	32 02	023	25 16	077	24 10	113	15 26	131	58 26	179	56 47	231	40 20	294
32	32 16	024	25 50	077	24 41	114	15 52	132	58 26	180	56 20	232	39 49	295
33	32 30	024	26 24	078	25 13	115	16 17	133	58 25	182	55 52	233	39 18	296
34	32 44	025	26 57	079	25 44	116	16 42	134	58 24	184	55 25	235	38 47	296
35	32 59	025	27 31	080	26 14	117	17 07	134	58 21	186	54 56	236	38 16	297
36	33 14	026	28 05	080	26 45	118	17 31	135	58 17	187	54 28	237	37 45	297
37	33 29	026	28 39	081	27 15	119	17 55	136	58 12	189	53 59	238	37 15	298
38	33 44	027	29 13	082	27 45	120	18 19	137	58 06	191	53 29	240	36 44	299
39	34 00	027	29 47	083	28 15	121	18 42	138	57 59	193	52 59	241	36 14	299
40	34 15	028	30 21	083	28 44	122	19 05	139	57 51	194	52 29	242	35 44	300
41	34 31	028	30 56	084	29 13	123	19 27	140	57 42	196	51 58	243	35 15	300
42	34 48	029	31 30	085	29 42	124	19 49	141	57 32	198	51 28	244	34 45	301
43	35 04	029	32 04	086	30 11	125	20 10	142	57 21	199	50 56	245	34 16	302
44	35 21	030	32 38	087	30 39	126	20 31	143	57 09	201	50 25	247	33 46	302
	Dubhe		◆POLLUX		BETELGEUSE		RIGEL		◆Hamal		Alpheratz		◆DENEB	
45	35 38	030	33 13	087	31 06	127	20 51	144	56 56	203	49 53	248	33 18	303
46	35 56	030	33 47	088	31 34	128	21 11	145	56 42	204	49 21	249	32 49	304
47	36 13	031	34 22	089	32 01	129	21 31	146	56 28	206	48 49	250	32 20	304
48	36 31	031	34 56	090	32 27	130	21 50	147	56 12	208	48 17	251	31 52	305
49	36 49	032	35 30	091	32 54	131	22 08	148	55 56	209	47 44	252	31 24	305
50	37 07	032	36 05	092	33 19	132	22 26	149	55 39	211	47 11	253	30 56	306
51	37 26	033	36 39	092	33 45	133	22 43	150	55 21	212	46 38	254	30 28	307
52	37 44	033	37 14	093	34 10	134	23 00	151	55 02	214	46 05	255	30 00	307
53	38 03	034	37 48	094	34 34	135	23 17	152	54 42	215	45 32	256	29 33	308
54	38 22	034	38 22	095	34 58	136	23 32	153	54 22	217	44 59	257	29 06	308
55	38 42	034	38 57	096	35 21	138	23 47	154	54 01	218	44 25	258	28 39	309
56	39 01	035	39 31	097	35 44	139	24 02	155	53 40	220	43 51	259	28 13	310
57	39 21	035	40 05	098	36 07	140	24 16	156	53 18	221	43 18	260	27 46	310
58	39 41	036	40 39	098	36 29	141	24 29	158	52 55	222	42 44	260	27 20	311
59	40 02	036	41 13	099	36 50	142	24 42	159	52 31	224	42 10	261	26 54	312
	◆Dubhe		POLLUX		SIRIUS		◆RIGEL		Hamal		◆Alpheratz		DENEB	
60	40 22	037	41 47	100	10 14	140	24 55	160	52 07	225	41 36	262	26 29	312
61	40 43	037	42 21	101	10 36	141	25 06	161	51 42	226	41 01	263	26 03	313
62	41 03	038	42 54	102	10 57	142	25 17	162	51 17	228	40 27	264	25 38	313
63	41 25	038	43 28	103	11 18	143	25 28	163	50 52	229	39 53	265	25 14	314
64	41 46	038	44 02	104	11 39	144	25 38	164	50 25	230	39 19	266	24 49	315
65	42 07	039	44 35	105	11 59	145	25 47	165	49 59	232	38 44	267	24 25	315
66	42 29	039	45 08	106	12 19	146	25 55	166	49 31	233	38 10	267	24 01	316
67	42 51	040	45 41	107	12 38	146	26 03	167	49 04	234	37 36	268	23 37	317
68	43 13	040	46 14	108	12 57	147	26 11	168	48 36	235	37 01	269	23 13	317
69	43 35	040	46 47	109	13 15	148	26 17	169	48 07	236	36 27	270	22 50	318
70	43 57	041	47 19	110	13 33	149	26 23	171	47 39	238	35 52	271	22 27	319
71	44 20	041	47 51	111	13 50	150	26 29	172	47 09	239	35 18	272	22 05	319
72	44 43	042	48 23	112	14 07	151	26 33	173	46 40	240	34 44	272	21 42	320
73	45 06	042	48 55	113	14 23	152	26 37	174	46 10	241	34 09	273	21 20	321
74	45 29	042	49 27	114	14 39	153	26 41	175	45 40	242	33 35	274	20 59	321
	Dubhe		◆REGULUS		PROCYON		SIRIUS		◆RIGEL		Hamal		◆DENEB	
75	45 52	043	17 14	094	30 55	132	14 54	154	26 43	176	45 09	243	20 37	322
76	46 15	043	17 48	094	31 21	133	15 09	155	26 45	177	44 38	244	20 16	322
77	46 39	044	18 22	095	31 46	134	15 24	156	26 47	178	44 07	245	19 55	323
78	47 03	044	18 56	096	32 10	135	15 37	157	26 47	179	43 36	246	19 35	324
79	47 27	044	19 31	097	32 34	136	15 51	158	26 48	180	43 04	247	19 14	324
80	47 51	045	20 05	098	32 58	137	16 03	159	26 47	182	42 32	248	18 55	325
81	48 15	045	20 39	099	33 21	139	16 15	160	26 46	183	42 00	249	18 35	326
82	48 40	045	21 13	100	33 43	140	16 27	161	26 44	184	41 28	250	18 16	326
83	49 04	046	21 47	100	34 05	141	16 38	162	26 41	185	40 55	251	17 57	327
84	49 29	046	22 21	101	34 27	142	16 49	163	26 38	186	40 23	252	17 38	328
85	49 54	046	22 54	102	34 48	143	16 58	164	26 34	187	39 50	253	17 20	328
86	50 19	047	23 28	103	35 08	144	17 08	165	26 29	188	39 17	254	17 02	329
87	50 44	047	24 01	104	35 28	145	17 17	166	26 24	189	38 44	255	16 45	330
88	51 09	047	24 35	105	35 47	146	17 25	167	26 18	190	38 10	256	16 28	330
89	51 35	048	25 08	106	36 06	148	17 32	168	26 12	192	37 37	257	16 11	331

LHA ♈	Hc	Zn	Hc	Zn	Hc	Zn	Hc	Zn	Hc	Zn	Hc	Zn	Hc	Zn
	Dubhe		◆REGULUS		PROCYON		SIRIUS		◆RIGEL		Mirfak		◆DENEB	
90	52 00	048	25 41	107	36 24	149	17 39	169	26 04	193	65 59	274	15 54	332
91	52 26	048	26 14	107	36 42	150	17 46	170	25 57	194	65 25	275	15 38	332
92	52 52	049	26 47	108	36 59	151	17 52	171	25 48	195	64 51	275	15 22	333
93	53 18	049	27 19	109	37 15	152	17 57	172	25 39	196	64 17	276	15 07	334
94	53 44	049	27 52	110	37 31	154	18 02	173	25 29	197	63 42	277	14 52	334
95	54 10	050	28 24	111	37 46	155	18 06	174	25 19	198	63 08	277	14 37	335
96	54 36	050	28 56	112	38 00	156	18 09	175	25 08	199	62 34	278	14 23	336
97	55 03	050	29 28	113	38 13	157	18 12	176	24 56	200	62 00	279	14 09	337
98	55 29	051	29 59	114	38 26	158	18 14	177	24 44	201	61 26	279	13 56	337
99	55 56	051	30 30	115	38 39	160	18 16	178	24 31	202	60 52	280	13 42	338
100	56 23	051	31 02	116	38 50	161	18 17	179	24 18	203	60 18	281	13 30	339
101	56 50	051	31 32	117	39 01	162	18 17	180	24 04	204	59 45	281	13 17	339
102	57 17	052	32 03	118	39 11	164	18 17	181	23 50	205	59 11	282	13 05	340
103	57 44	052	32 33	119	39 21	165	18 17	182	23 35	207	58 37	283	12 54	341
104	58 11	052	33 03	120	39 29	166	18 15	183	23 19	208	58 04	283	12 43	341
	◆Dubhe		Denebola		REGULUS		◆SIRIUS		RIGEL		ALDEBARAN		◆Mirfak	
105	58 38	053	22 06	096	33 33	121	18 13	184	23 03	209	42 37	230	57 30	284
106	59 05	053	22 40	097	34 02	122	18 11	185	22 46	210	42 10	231	56 57	284
107	59 33	053	23 14	098	34 31	123	18 07	186	22 29	211	41 43	232	56 24	285
108	60 00	053	23 48	099	35 00	124	18 04	187	22 11	212	41 16	234	55 50	286
109	60 28	053	24 22	099	35 29	125	17 59	188	21 53	213	40 48	235	55 17	286
110	60 56	054	24 56	100	35 57	126	17 54	189	21 34	214	40 20	236	54 44	287
111	61 23	054	25 30	101	36 24	127	17 49	190	21 14	215	39 51	237	54 11	287
112	61 51	054	26 04	102	36 51	128	17 42	191	20 54	216	39 22	238	53 39	288
113	62 19	054	26 37	103	37 18	129	17 36	192	20 34	217	38 53	239	53 06	288
114	62 47	054	27 11	104	37 45	130	17 28	193	20 13	218	38 23	240	52 33	289
115	63 15	055	27 44	105	38 11	131	17 20	194	19 52	219	37 53	241	52 01	290
116	63 43	055	28 17	106	38 36	133	17 12	195	19 30	220	37 23	242	51 28	290
117	64 11	055	28 50	107	39 01	134	17 03	196	19 08	221	36 52	243	50 56	291
118	64 40	055	29 23	107	39 26	135	16 53	197	18 46	222	36 21	244	50 24	291
119	65 08	055	29 56	108	39 50	136	16 43	198	18 22	223	35 50	245	49 52	292
	Kochab		◆ARCTURUS		REGULUS		◆SIRIUS		BETELGEUSE		CAPELLA		◆Schedar	
120	48 55	024	13 29	076	40 14	137	16 32	199	36 17	220	62 55	268	35 07	321
121	49 09	024	14 02	076	40 37	138	16 21	200	35 55	221	62 20	269	34 45	321
122	49 23	024	14 36	077	41 00	140	16 09	201	35 32	222	61 46	269	34 24	322
123	49 37	025	15 09	078	41 22	141	15 57	202	35 09	223	61 11	270	34 02	322
124	49 52	025	15 43	079	41 43	142	15 44	203	34 45	224	60 37	271	33 41	323
125	50 06	025	16 17	080	42 04	143	15 30	204	34 21	225	60 03	272	33 21	323
126	50 21	025	16 51	080	42 24	144	15 16	205	33 56	226	59 28	273	33 00	324
127	50 35	025	17 25	081	42 44	146	15 01	206	33 31	227	58 54	273	32 40	324
128	50 50	025	17 59	082	43 03	147	14 46	206	33 05	229	58 19	274	32 20	325
129	51 05	026	18 33	083	43 21	148	14 31	207	32 39	230	57 45	275	32 00	325
130	51 20	026	19 07	084	43 39	150	14 15	208	32 13	231	57 11	276	31 41	326
131	51 35	026	19 41	084	43 56	151	13 58	209	31 46	232	56 37	276	31 22	326
132	51 50	026	20 15	085	44 13	152	13 41	210	31 19	233	56 02	277	31 03	327
133	52 05	026	20 50	086	44 29	153	13 23	211	30 51	234	55 28	278	30 44	327
134	52 21	027	21 24	087	44 44	155	13 05	212	30 23	235	54 54	279	30 26	328
	Kochab		◆ARCTURUS		REGULUS		◆POLLUX		BETELGEUSE		CAPELLA		◆Schedar	
135	52 36	027	21 58	088	44 58	156	59 49	214	29 55	236	54 20	279	30 07	328
136	52 52	027	22 33	088	45 11	157	59 29	216	29 26	237	53 46	280	29 50	329
137	53 07	027	23 07	089	45 24	159	59 09	218	28 57	238	53 13	281	29 32	330
138	53 23	027	23 42	090	45 36	160	58 47	219	28 28	239	52 39	281	29 15	330
139	53 39	027	24 16	091	45 48	162	58 25	221	27 59	240	52 05	282	28 58	331
140	53 54	027	24 50	092	45 58	163	58 03	222	27 29	241	51 31	283	28 41	331
141	54 10	027	25 25	092	46 08	164	57 39	224	26 59	242	50 58	283	28 24	332
142	54 26	028	25 59	093	46 17	166	57 15	225	26 28	243	50 24	284	28 08	332
143	54 42	028	26 34	094	46 25	167	56 50	227	25 57	244	49 51	285	27 52	333
144	54 58	028	27 08	095	46 32	169	56 25	228	25 27	244	49 18	285	27 37	333
145	55 14	028	27 42	096	46 38	170	55 59	229	24 55	245	48 45	286	27 22	334
146	55 30	028	28 16	097	46 44	171	55 33	231	24 24	246	48 12	287	27 07	334
147	55 46	028	28 50	098	46 49	173	55 06	232	23 52	247	47 39	287	26 52	335
148	56 02	028	29 25	098	46 53	174	54 39	233	23 20	248	47 06	288	26 38	336
149	56 19	028	29 59	099	46 56	176	54 11	235	22 48	249	46 33	288	26 23	336
	◆VEGA		ARCTURUS		SPICA		◆REGULUS		POLLUX		◆CAPELLA		Schedar	
150	13 20	038	30 32	100	11 12	129	46 58	177	53 42	236	46 01	289	26 10	337
151	13 41	039	31 06	101	11 39	130	46 59	179	53 14	237	45 28	290	25 56	337
152	14 03	040	31 40	102	12 05	131	47 00	180	52 45	238	44 56	290	25 43	338
153	14 25	041	32 14	103	12 31	131	46 59	181	52 15	240	44 23	291	25 30	338
154	14 48	041	32 47	104	12 56	132	46 58	183	51 45	241	43 51	292	25 18	339
155	15 10	042	33 20	105	13 22	133	46 56	184	51 15	242	43 19	292	25 06	340
156	15 34	043	33 54	106	13 47	134	46 53	186	50 45	243	42 48	293	24 54	340
157	15 57	043	34 27	106	14 11	135	46 49	187	50 14	244	42 16	293	24 42	341
158	16 21	044	35 00	107	14 35	136	46 44	189	49 43	245	41 44	294	24 31	341
159	16 45	045	35 32	108	14 59	137	46 39	190	49 11	246	41 13	295	24 20	342
160	17 09	045	36 05	109	15 22	138	46 32	191	48 39	248	40 42	295	24 09	342
161	17 34	046	36 37	110	15 45	139	46 25	193	48 08	249	40 11	296	23 59	343
162	17 59	047	37 10	111	16 07	140	46 17	194	47 35	250	39 40	296	23 49	344
163	18 24	047	37 42	112	16 29	141	46 08	196	47 03	251	39 09	297	23 40	344
164	18 50	048	38 13	113	16 51	142	45 59	197	46 30	252	38 39	298	23 30	345
	◆DENEB		VEGA		ARCTURUS		◆SPICA		REGULUS		◆POLLUX		CAPELLA	
165	14 28	024	19 15	049	38 45	114	17 12	143	45 48	198	45 58	253	38 08	298
166	14 43	025	19 41	050	39 16	115	17 33	144	45 37	200	45 25	254	37 38	299
167	14 58	026	20 08	050	39 47	116	17 53	145	45 25	201	44 52	255	37 08	299
168	15 13	026	20 34	051	40 18	117	18 13	146	45 12	202	44 18	256	36 38	300
169	15 28	027	21 01	052	40 48	118	18 32	147	44 59	204	43 45	257	36 08	301
170	15 44	028	21 28	052	41 18	119	18 51	147	44 44	205	43 11	257	35 39	301
171	16 00	029	21 56	053	41 48	120	19 09	148	44 29	206	42 38	258	35 09	302
172	16 17	029	22 23	054	42 18	122	19 27	149	44 14	208	42 04	259	34 40	302
173	16 34	030	22 51	054	42 47	123	19 44	150	43 57	209	41 30	260	34 11	303
174	16 51	031	23 19	055	43 15	124	20 01	151	43 40	210	40 56	261	33 43	304
175	17 09	031	23 48	056	43 44	125	20 17	152	43 22	212	40 22	262	33 14	304
176	17 27	032	24 16	056	44 12	126	20 32	153	43 04	213	39 48	263	32 46	305
177	17 45	033	24 45	057	44 40	127	20 48	154	42 45	214	39 14	264	32 18	306
178	18 04	033	25 14	058	45 07	128	21 02	156	42 25	215	38 40	265	31 50	306
179	18 23	034	25 43	058	45 33	130	21 16	157	42 05	217	38 05	266	31 22	307

LAT 55°N

LHA	Hc	Zn	Hc	Zn	Hc	Zn	Hc	Zn	Hc	Zn	Hc	Zn	Hc	Zn
♈	DENEB		◆VEGA		ARCTURUS		◆SPICA		REGULUS		POLLUX		◆CAPELLA	
180	18 42	035	26 12	059	46 00	131	21 30	158	41 44	218	37 31	266	30 55	307
181	19 02	035	26 42	060	46 26	132	21 42	159	41 23	219	36 57	267	30 27	308
182	19 22	036	27 12	060	46 51	133	21 55	160	41 01	220	36 22	268	30 00	309
183	19 42	036	27 42	061	47 16	134	22 06	161	40 38	222	35 48	269	29 34	309
184	20 03	037	28 12	062	47 40	136	22 17	162	40 15	223	35 13	270	29 07	310
185	20 24	038	28 43	062	48 04	137	22 28	163	39 51	224	34 39	271	28 41	310
186	20 45	038	29 13	063	48 27	138	22 38	164	39 27	225	34 05	271	28 15	311
187	21 07	039	29 44	064	48 50	140	22 47	165	39 03	226	33 30	272	27 49	312
188	21 28	040	30 15	064	49 12	141	22 56	166	38 38	227	32 56	273	27 23	312
189	21 50	040	30 46	065	49 33	142	23 04	167	38 12	228	32 21	274	26 58	313
190	22 13	041	31 17	066	49 54	144	23 11	168	37 46	230	31 47	275	26 33	313
191	22 36	042	31 49	066	50 14	145	23 18	169	37 20	231	31 13	275	26 08	314
192	22 59	042	32 20	067	50 33	146	23 25	170	36 53	232	30 39	276	25 43	315
193	23 22	043	32 52	068	50 52	148	23 30	171	36 26	233	30 04	277	25 19	315
194	23 46	044	33 24	069	51 10	149	23 35	172	35 58	234	29 30	278	24 55	316
	◆DENEB		VEGA		Rasalhague		◆ARCTURUS		REGULUS		◆POLLUX		CAPELLA	
195	24 09	044	33 56	069	22 27	100	51 27	151	35 30	235	28 56	278	24 31	317
196	24 34	045	34 29	070	23 00	101	51 44	152	35 02	236	28 22	279	24 08	317
197	24 58	045	35 01	071	23 34	102	51 59	154	34 33	237	27 48	280	23 44	318
198	25 23	046	35 33	071	24 08	103	52 14	155	34 04	238	27 14	281	23 21	319
199	25 48	047	36 06	072	24 41	104	52 28	157	33 34	239	26 41	282	22 59	319
200	26 13	047	36 39	073	25 15	105	52 41	158	33 05	240	26 07	282	22 36	320
201	26 38	048	37 12	073	25 48	106	52 54	160	32 35	241	25 33	283	22 14	320
202	27 04	049	37 45	074	26 21	107	53 05	161	32 05	242	25 00	284	21 53	321
203	27 30	049	38 18	075	26 54	107	53 16	163	31 34	243	24 27	285	21 31	322
204	27 56	050	38 51	075	27 26	108	53 26	164	31 03	244	23 53	285	21 10	322
205	28 22	050	39 25	076	27 59	109	53 35	166	30 32	245	23 20	286	20 49	323
206	28 49	051	39 58	077	28 31	110	53 43	167	30 01	246	22 47	287	20 29	324
207	29 16	052	40 32	078	29 04	111	53 50	169	29 29	247	22 14	288	20 08	324
208	29 43	052	41 05	078	29 36	112	53 56	171	28 57	248	21 42	288	19 48	325
209	30 11	053	41 39	079	30 07	113	54 01	172	28 25	249	21 09	289	19 29	326
	◆DENEB		VEGA		Rasalhague		◆ARCTURUS		REGULUS		◆POLLUX		CAPELLA	
210	30 38	054	42 13	080	30 39	114	54 05	174	27 53	250	20 36	290	19 09	326
211	31 06	054	42 47	080	31 10	115	54 09	175	27 21	251	20 04	291	18 51	327
212	31 34	055	43 21	081	31 42	116	54 11	177	26 48	252	19 32	291	18 32	328
213	32 02	055	43 55	082	32 12	117	54 12	179	26 16	253	19 00	292	18 14	328
214	32 31	056	44 29	083	32 43	118	54 12	180	25 43	253	18 28	293	17 56	329
215	32 59	057	45 03	083	33 13	119	54 12	182	25 10	254	17 57	294	17 38	330
216	33 28	057	45 37	084	33 43	120	54 10	183	24 36	255	17 25	294	17 21	330
217	33 57	058	46 11	085	34 13	121	54 08	185	24 03	256	16 54	295	17 04	331
218	34 26	059	46 46	086	34 42	122	54 04	187	23 30	257	16 23	296	16 47	332
219	34 56	059	47 20	086	35 11	123	54 00	188	22 56	258	15 52	296	16 31	332
220	35 26	060	47 54	087	35 40	124	53 54	190	22 22	259	15 21	297	16 15	333
221	35 55	060	48 29	088	36 08	125	53 48	191	21 49	260	14 51	298	16 00	334
222	36 25	061	49 03	089	36 36	126	53 41	193	21 15	260	14 20	299	15 44	334
223	36 56	062	49 38	090	37 04	127	53 32	195	20 41	261	13 50	299	15 30	335
224	37 26	062	50 12	090	37 31	128	53 23	196	20 07	262	13 20	300	15 15	336
	◆DENEB		VEGA		Rasalhague		◆ARCTURUS		REGULUS		Dubhe		◆CAPELLA	
225	37 56	063	50 46	091	37 58	129	53 13	198	19 32	263	59 25	307	15 01	336
226	38 27	063	51 21	092	38 25	130	53 02	199	18 58	264	58 58	307	14 47	337
227	38 58	064	51 55	093	38 51	132	52 50	201	18 24	265	58 30	308	14 34	338
228	39 29	065	52 30	094	39 16	133	52 38	202	17 50	265	58 03	308	14 21	338
229	40 00	065	53 04	095	39 41	134	52 24	204	17 15	266	57 36	308	14 09	339
230	40 32	066	53 38	096	40 06	135	52 10	205	16 41	267	57 09	308	13 56	340
231	41 03	067	54 12	097	40 30	136	51 55	207	16 07	268	56 42	309	13 45	340
232	41 35	067	54 47	097	40 53	137	51 39	208	15 32	269	56 15	309	13 33	341
233	42 06	068	55 21	098	41 17	139	51 22	210	14 58	270	55 48	309	13 22	342
234	42 38	068	55 55	099	41 39	140	51 05	211	14 23	270	55 22	309	13 12	342
235	43 10	069	56 29	100	42 01	141	50 47	213	13 49	271	54 55	310	13 01	343
236	43 43	070	57 02	101	42 22	142	50 28	214	13 15	272	54 29	310	12 52	344
237	44 15	070	57 36	102	42 43	143	50 08	215	12 40	273	54 02	310	12 42	344
238	44 47	071	58 10	103	43 04	145	49 48	217	12 06	274	53 36	311	12 33	345
239	45 20	072	58 43	104	43 23	146	49 27	218	11 32	274	53 10	311	12 25	346
	Schedar		◆DENEB		VEGA		Rasalhague		◆ARCTURUS		Denebola		◆Dubhe	
240	28 39	029	45 53	072	59 16	105	43 42	147	49 05	219	27 24	256	52 44	311
241	28 56	029	46 26	073	59 49	106	44 00	148	48 43	221	26 51	257	52 19	312
242	29 13	030	46 58	073	60 22	108	44 18	150	48 20	222	26 17	258	51 53	312
243	29 30	030	47 31	074	60 55	109	44 35	151	47 57	223	25 43	258	51 27	312
244	29 48	031	48 05	075	61 27	110	44 51	152	47 33	225	25 10	259	51 02	313
245	30 05	031	48 38	075	62 00	111	45 07	154	47 09	226	24 36	260	50 37	313
246	30 24	032	49 11	076	62 32	112	45 22	155	46 44	227	24 02	261	50 12	313
247	30 42	033	49 45	077	63 03	114	45 36	156	46 18	228	23 28	262	49 47	314
248	31 01	033	50 18	077	63 35	115	45 49	158	45 52	230	22 54	263	49 22	314
249	31 19	034	50 52	078	64 06	116	46 02	159	45 26	231	22 19	264	48 57	314
250	31 39	034	51 26	079	64 36	118	46 14	161	44 59	232	21 45	264	48 33	315
251	31 58	035	51 59	079	65 07	119	46 25	162	44 32	233	21 11	265	48 08	315
252	32 18	035	52 33	080	65 37	120	46 35	163	44 04	234	20 37	266	47 44	315
253	32 38	036	53 07	081	66 06	122	46 45	165	43 36	235	20 02	267	47 20	316
254	32 58	036	53 41	081	66 35	123	46 53	166	43 07	237	19 28	268	46 56	316
	Schedar		DENEB		◆ALTAIR		Rasalhague		◆ARCTURUS		Alkaid		◆CAPELLA	
255	33 18	037	54 15	082	32 53	127	47 01	168	42 38	238	60 32	279	11 03	357
256	33 39	037	54 49	083	33 21	128	47 08	169	42 09	239	59 58	280	11 02	358
257	34 00	038	55 24	084	33 48	129	47 15	170	41 40	240	59 24	281	11 01	359
258	34 21	038	55 58	084	34 14	130	47 20	172	41 10	241	58 51	281	11 00	359
259	34 43	039	56 32	085	34 40	131	47 24	173	40 40	242	58 17	282	11 00	000
260	35 04	039	57 06	086	35 06	132	47 28	175	40 09	243	57 43	282	11 00	001
261	35 26	040	57 41	087	35 31	134	47 31	176	39 38	244	57 10	283	11 00	001
262	35 48	040	58 15	087	35 55	135	47 33	178	39 07	245	56 36	284	11 01	002
263	36 11	041	58 49	088	36 20	136	47 34	179	38 36	246	56 03	284	11 03	003
264	36 33	041	59 24	089	36 43	137	47 34	180	38 04	247	55 29	285	11 05	003
265	36 56	042	59 58	090	37 07	138	47 33	182	37 32	248	54 56	286	11 07	004
266	37 19	042	60 33	091	37 29	139	47 31	183	37 00	249	54 23	286	11 10	005
267	37 43	043	61 07	091	37 52	140	47 29	185	36 28	250	53 50	287	11 13	006
268	38 06	043	61 41	092	38 13	142	47 26	186	35 56	251	53 17	287	11 16	006
269	38 30	044	62 16	093	38 34	143	47 22	188	35 23	252	52 44	288	11 20	007

LHA	Hc	Zn	Hc	Zn	Hc	Zn	Hc	Zn	Hc	Zn	Hc	Zn	Hc	Zn
♈	◆CAPELLA		Alpheratz		◆ALTAIR		Rasalhague		◆ARCTURUS		Alkaid		Kochab	
270	11 25	008	22 21	071	38 55	144	47 16	189	34 50	253	52 12	288	63 24	333
271	11 30	008	22 53	072	39 15	145	47 11	191	34 17	254	51 39	289	63 08	333
272	11 35	009	23 26	072	39 34	146	47 04	192	33 44	255	51 07	290	62 53	333
273	11 41	010	23 59	073	39 53	148	46 56	193	33 11	256	50 34	290	62 37	333
274	11 47	011	24 32	074	40 11	149	46 48	195	32 38	257	50 02	291	62 21	333
275	11 53	011	25 05	075	40 29	150	46 39	196	32 04	257	49 30	291	62 06	333
276	12 00	012	25 38	075	40 46	151	46 29	198	31 30	258	48 58	292	61 50	332
277	12 07	013	26 11	076	41 02	153	46 18	199	30 57	259	48 26	293	61 34	332
278	12 15	013	26 45	077	41 17	154	46 06	200	30 23	260	47 54	293	61 18	332
279	12 23	014	27 18	078	41 32	155	45 54	202	29 49	261	47 23	294	61 02	332
280	12 32	015	27 52	078	41 46	156	45 41	203	29 15	262	46 51	294	60 45	332
281	12 41	015	28 26	079	42 00	158	45 27	204	28 41	263	46 20	295	60 29	332
282	12 50	016	29 00	080	42 13	159	45 12	206	28 07	264	45 49	295	60 13	332
283	13 00	017	29 33	081	42 25	160	44 57	207	27 32	264	45 18	296	59 57	332
284	13 10	017	30 07	081	42 36	162	44 41	208	26 58	265	44 47	296	59 40	332
	◆CAPELLA		Alpheratz		◆ALTAIR		Rasalhague		◆ARCTURUS		Alkaid		Kochab	
285	13 21	018	30 42	082	42 46	163	44 24	210	26 24	266	44 16	297	59 24	332
286	13 32	019	31 16	083	42 56	164	44 07	211	25 49	267	43 46	298	59 08	332
287	13 43	020	31 50	084	43 05	166	43 49	212	25 15	268	43 15	298	58 51	332
288	13 55	020	32 24	084	43 13	167	43 30	214	24 41	269	42 45	299	58 35	332
289	14 07	021	32 58	085	43 21	168	43 10	215	24 06	269	42 15	299	58 19	332
290	14 19	022	33 33	086	43 27	170	42 50	216	23 32	270	41 45	300	58 02	332
291	14 32	022	34 07	087	43 33	171	42 30	217	22 57	271	41 15	300	57 46	332
292	14 45	023	34 41	088	43 38	172	42 09	219	22 23	272	40 46	301	57 30	332
293	14 59	024	35 16	088	43 42	174	41 47	220	21 49	273	40 16	302	57 13	332
294	15 13	024	35 50	089	43 46	175	41 25	221	21 14	273	39 47	302	56 57	332
295	15 27	025	36 25	090	43 48	176	41 02	222	20 40	274	39 18	303	56 41	332
296	15 42	026	36 59	091	43 50	178	40 38	223	20 06	275	38 49	303	56 24	332
297	15 57	026	37 33	092	43 51	179	40 14	225	19 31	276	38 20	304	56 08	332
298	16 13	027	38 08	093	43 51	181	39 50	226	18 57	277	37 52	304	55 52	332
299	16 28	028	38 42	093	43 51	182	39 25	227	18 23	277	37 24	305	55 36	332
	CAPELLA		◆Alpheratz		Enif		ALTAIR		◆VEGA		Alphecca		◆Alioth	
300	16 45	028	39 16	094	40 24	145	43 49	183	68 36	229	35 01	268	35 59	318
301	17 01	029	39 51	095	40 43	147	43 47	185	68 09	231	34 26	268	35 37	319
302	17 18	030	40 25	096	41 02	148	43 43	186	67 42	233	33 52	269	35 14	320
303	17 35	030	40 59	097	41 20	149	43 39	187	67 14	234	33 17	270	34 52	320
304	17 53	031	41 33	098	41 37	150	43 35	189	66 46	236	32 43	271	34 30	321
305	18 11	032	42 07	099	41 54	152	43 29	190	66 17	238	32 09	272	34 08	321
306	18 29	032	42 41	100	42 10	153	43 23	191	65 48	239	31 34	272	33 47	322
307	18 48	033	43 15	100	42 25	154	43 15	193	65 18	241	31 00	273	33 25	322
308	19 06	034	43 49	101	42 40	156	43 07	194	64 48	242	30 25	274	33 04	323
309	19 26	034	44 23	102	42 54	157	42 59	195	64 18	243	29 51	275	32 44	323
310	19 45	035	44 56	103	43 07	158	42 49	197	63 47	245	29 17	276	32 23	324
311	20 05	036	45 30	104	43 19	160	42 39	198	63 15	246	28 43	276	32 03	324
312	20 25	036	46 03	105	43 31	161	42 28	199	62 44	247	28 08	277	31 43	325
313	20 46	037	46 36	106	43 42	162	42 16	201	62 12	248	27 34	278	31 23	325
314	21 07	038	47 09	107	43 52	164	42 03	202	61 40	250	27 00	279	31 04	326
	CAPELLA		◆Hamal		Alpheratz		Enif		◆ALTAIR		VEGA		◆Alioth	
315	21 28	038	26 32	086	47 42	108	44 01	165	41 50	203	61 07	251	30 44	326
316	21 49	039	27 06	087	48 15	109	44 10	166	41 36	205	60 35	252	30 25	327
317	22 11	039	27 41	088	48 47	110	44 18	168	41 22	206	60 02	253	30 07	327
318	22 33	040	28 15	089	49 19	111	44 25	169	41 06	207	59 29	254	29 48	328
319	22 55	041	28 49	090	49 51	112	44 31	170	40 50	208	58 56	255	29 30	328
320	23 18	041	29 24	090	50 23	113	44 36	172	40 33	210	58 22	256	29 12	329
321	23 41	042	29 58	091	50 54	114	44 41	173	40 16	211	57 49	257	28 55	330
322	24 04	043	30 33	092	51 26	116	44 44	174	39 58	212	57 15	258	28 37	330
323	24 27	043	31 07	093	51 56	117	44 47	176	39 39	213	56 42	259	28 20	331
324	24 51	044	31 41	094	52 27	118	44 49	177	39 20	215	56 08	260	28 04	331
325	25 15	045	32 16	095	52 57	119	44 51	179	39 00	216	55 34	261	27 47	332
326	25 39	045	32 50	096	53 27	120	44 51	180	38 40	217	55 00	262	27 31	332
327	26 04	046	33 24	096	53 57	122	44 51	181	38 19	218	54 26	263	27 15	333
328	26 29	046	33 58	097	54 26	123	44 49	183	37 58	219	53 51	264	27 00	333
329	26 54	047	34 33	098	54 55	124	44 47	184	37 35	220	53 17	265	26 45	334
	CAPELLA		◆Hamal		Alpheratz		Enif		◆ALTAIR		VEGA		◆Alioth	
330	27 19	048	35 07	099	55 23	125	44 44	186	37 13	222	52 43	266	26 30	335
331	27 45	048	35 41	100	55 51	127	44 41	187	36 50	223	52 08	267	26 15	335
332	28 10	049	36 14	101	56 18	128	44 36	188	36 26	224	51 34	268	26 01	336
333	28 36	049	36 48	102	56 45	129	44 31	190	36 02	225	51 00	268	25 47	336
334	29 03	050	37 22	103	57 11	131	44 24	191	35 37	226	50 25	269	25 33	337
335	29 29	051	37 55	104	57 37	132	44 17	192	35 12	227	49 51	270	25 19	337
336	29 56	051	38 29	104	58 02	134	44 10	194	34 47	228	49 16	271	25 06	338
337	30 23	052	39 02	105	58 27	135	44 01	195	34 21	229	48 42	272	24 54	339
338	30 50	053	39 35	106	58 51	137	43 52	197	33 55	230	48 08	272	24 41	339
339	31 18	053	40 08	107	59 14	138	43 41	198	33 28	232	47 33	273	24 29	340
340	31 45	054	40 41	108	59 37	140	43 30	199	33 01	233	46 59	274	24 17	340
341	32 13	054	41 13	109	59 59	141	43 19	201	32 33	234	46 25	275	24 06	341
342	32 41	055	41 46	110	60 20	143	43 06	202	32 05	235	45 50	276	23 55	341
343	33 10	056	42 18	111	60 41	144	42 53	203	31 37	236	45 16	276	23 44	342
344	33 38	056	42 50	112	61 00	146	42 39	204	31 09	237	44 42	277	23 33	343
	◆CAPELLA		ALDEBARAN		◆Diphda		Enif		ALTAIR		◆VEGA		Alioth	
345	34 07	057	16 55	085	13 45	155	42 25	206	30 40	238	44 08	278	23 23	343
346	34 36	057	17 29	086	13 59	156	42 09	207	30 10	239	43 34	279	23 13	344
347	35 05	058	18 04	087	14 13	157	41 53	208	29 41	240	43 00	279	23 04	344
348	35 34	059	18 38	088	14 26	158	41 37	210	29 11	241	42 26	280	22 55	345
349	36 04	059	19 13	088	14 39	159	41 19	211	28 41	242	41 52	281	22 46	345
350	36 33	060	19 47	089	14 52	160	41 01	212	28 10	243	41 18	281	22 37	346
351	37 03	060	20 21	090	15 03	160	40 43	213	27 40	244	40 44	282	22 29	347
352	37 33	061	20 56	091	15 15	161	40 24	215	27 09	244	40 11	283	22 21	347
353	38 03	062	21 30	092	15 25	162	40 04	216	26 38	245	39 37	284	22 14	348
354	38 34	062	22 05	093	15 35	163	39 43	217	26 06	246	39 04	284	22 07	348
355	39 04	063	22 39	093	15 45	164	39 22	218	25 35	247	38 31	285	22 00	349
356	39 35	063	23 13	094	15 54	165	39 01	219	25 03	248	37 58	286	21 54	350
357	40 06	064	23 48	095	16 02	166	38 39	221	24 31	249	37 24	286	21 48	350
358	40 37	065	24 22	096	16 10	167	38 16	222	23 58	250	36 52	287	21 42	351
359	41 08	065	24 56	097	16 18	168	37 53	223	23 26	251	36 19	288	21 37	351

LAT 54°N

LHA ♈	Hc	Zn	Hc	Zn	Hc	Zn	Hc	Zn	Hc	Zn	Hc	Zn	Hc	Zn
	◆CAPELLA		ALDEBARAN		Hamal		◆Alpheratz		ALTAIR		◆VEGA		Alioth	
0	41 15	065	25 38	097	51 18	129	65 01	176	23 12	252	35 27	289	20 32	352
1	41 47	066	26 13	098	51 45	131	65 03	178	22 38	253	34 53	290	20 28	353
2	42 19	066	26 48	099	52 12	132	65 04	180	22 04	254	34 20	291	20 23	353
3	42 51	067	27 23	100	52 38	133	65 03	182	21 30	255	33 47	291	20 19	354
4	43 24	068	27 57	101	53 03	135	65 01	184	20 56	256	33 14	292	20 16	354
5	43 56	068	28 32	101	53 28	136	64 58	186	20 22	257	32 42	292	20 12	355
6	44 29	069	29 06	102	53 52	138	64 54	188	19 48	257	32 09	293	20 09	356
7	45 02	069	29 41	103	54 15	139	64 48	190	19 13	258	31 37	294	20 07	356
8	45 35	070	30 15	104	54 38	140	64 41	192	18 39	259	31 05	294	20 05	357
9	46 08	071	30 49	105	55 00	142	64 33	194	18 04	260	30 33	295	20 03	357
10	46 42	071	31 23	106	55 22	143	64 24	196	17 29	261	30 01	296	20 01	358
11	47 15	072	31 57	107	55 43	145	64 13	198	16 54	262	29 29	296	20 00	359
12	47 49	072	32 31	108	56 03	146	64 02	200	16 19	263	28 58	297	20 00	359
13	48 22	073	33 04	109	56 22	148	63 49	202	15 44	263	28 27	298	19 59	000
14	48 56	074	33 38	110	56 40	149	63 35	204	15 09	264	27 55	298	19 59	000
	Dubhe		◆CAPELLA		ALDEBARAN		Hamal		◆Alpheratz		DENEB		◆VEGA	
15	28 02	015	49 30	074	34 11	111	56 58	151	63 20	206	48 44	285	27 25	299
16	28 12	016	50 04	075	34 44	111	57 14	153	63 04	208	48 10	286	26 54	300
17	28 21	016	50 38	075	35 16	112	57 30	154	62 47	210	47 36	287	26 23	300
18	28 31	017	51 12	076	35 49	113	57 45	156	62 30	211	47 02	287	25 53	301
19	28 42	017	51 46	077	36 21	114	57 59	158	62 11	213	46 28	288	25 23	302
20	28 52	018	52 21	077	36 53	115	58 12	159	61 51	215	45 55	288	24 53	302
21	29 03	018	52 55	078	37 25	116	58 24	161	61 30	217	45 22	289	24 23	303
22	29 14	019	53 30	079	37 56	117	58 35	163	61 09	218	44 48	290	23 54	304
23	29 25	019	54 04	079	38 27	118	58 45	164	60 47	220	44 15	290	23 25	304
24	29 37	020	54 39	080	38 58	119	58 54	166	60 24	221	43 42	291	22 56	305
25	29 49	020	55 14	081	39 29	121	59 01	168	60 00	223	43 09	291	22 27	306
26	30 01	021	55 49	081	39 59	122	59 08	170	59 36	225	42 36	292	21 58	306
27	30 14	021	56 24	082	40 29	123	59 14	172	59 11	226	42 04	293	21 30	307
28	30 27	022	56 59	083	40 58	124	59 19	173	58 45	228	41 31	293	21 02	308
29	30 40	022	57 34	083	41 27	125	59 22	175	58 19	229	40 59	294	20 34	308
	Dubhe		◆POLLUX		BETELGEUSE		RIGEL		◆Hamal		Alpheratz		◆DENEB	
30	30 53	023	24 28	075	24 01	112	15 38	130	59 25	177	57 52	230	40 27	294
31	31 07	023	25 02	076	24 33	113	16 05	131	59 26	179	57 24	232	39 55	295
32	31 21	024	25 37	077	25 06	114	16 31	131	59 26	181	56 56	233	39 23	296
33	31 35	024	26 11	078	25 38	115	16 58	132	59 25	182	56 28	234	38 51	296
34	31 50	024	26 46	078	26 10	116	17 24	133	59 23	184	55 59	236	38 20	297
35	32 04	025	27 20	079	26 41	117	17 49	134	59 20	186	55 29	237	37 48	297
36	32 19	025	27 55	080	27 13	117	18 14	135	59 16	188	55 00	238	37 17	298
37	32 35	026	28 30	081	27 44	118	18 39	136	59 11	189	54 29	240	36 46	299
38	32 50	026	29 04	081	28 15	119	19 03	137	59 05	191	53 59	241	36 15	299
39	33 06	027	29 39	082	28 45	120	19 27	138	58 57	193	53 28	242	35 45	300
40	33 22	027	30 14	083	29 16	121	19 50	139	58 49	195	52 57	243	35 14	300
41	33 38	028	30 49	084	29 46	122	20 13	140	58 39	197	52 25	244	34 44	301
42	33 55	028	31 24	084	30 15	123	20 36	141	58 29	198	51 53	245	34 14	302
43	34 12	029	31 59	085	30 45	124	20 58	142	58 17	200	51 21	247	33 44	302
44	34 29	029	32 35	086	31 14	125	21 19	143	58 05	202	50 48	248	33 14	303
	Dubhe		◆POLLUX		BETELGEUSE		RIGEL		◆Hamal		Alpheratz		◆DENEB	
45	34 46	030	33 10	087	31 42	126	21 40	144	57 51	203	50 16	249	32 45	303
46	35 04	030	33 45	088	32 10	127	22 01	145	57 37	205	49 43	250	32 15	304
47	35 22	031	34 20	088	32 38	128	22 21	146	57 21	207	49 09	251	31 46	305
48	35 40	031	34 56	089	33 06	129	22 40	147	57 05	208	48 36	252	31 17	305
49	35 58	031	35 31	090	33 33	130	22 59	148	56 48	210	48 02	253	30 49	306
50	36 16	032	36 06	091	33 59	132	23 18	149	56 30	211	47 29	254	30 20	306
51	36 35	032	36 41	092	34 25	133	23 35	150	56 11	213	46 55	255	29 52	307
52	36 54	033	37 17	092	34 51	134	23 53	151	55 52	215	46 20	256	29 24	308
53	37 13	033	37 52	093	35 16	135	24 10	152	55 31	216	45 46	257	28 56	308
54	37 33	034	38 27	094	35 41	136	24 26	153	55 10	218	45 12	258	28 29	309
55	37 52	034	39 02	095	36 05	137	24 41	154	54 48	219	44 37	259	28 01	310
56	38 12	034	39 37	096	36 29	138	24 57	155	54 26	220	44 03	260	27 34	310
57	38 32	035	40 12	097	36 53	139	25 11	156	54 03	222	43 28	260	27 07	311
58	38 52	035	40 47	098	37 15	140	25 25	157	53 39	223	42 53	261	26 41	311
59	39 13	036	41 22	098	37 37	142	25 38	158	53 14	225	42 18	262	26 14	312
	◆Dubhe		POLLUX		SIRIUS		◆RIGEL		Hamal		◆Alpheratz		DENEB	
60	39 34	036	41 57	099	11 00	140	25 51	159	52 49	226	41 43	263	25 48	313
61	39 55	037	42 32	100	11 22	141	26 03	161	52 24	227	41 08	264	25 22	313
62	40 16	037	43 06	101	11 44	142	26 14	162	51 57	229	40 33	265	24 57	314
63	40 37	037	43 41	102	12 06	143	26 25	163	51 31	230	39 58	266	24 32	314
64	40 59	038	44 15	103	12 27	144	26 35	164	51 03	231	39 23	267	24 07	315
65	41 20	038	44 50	104	12 48	145	26 45	165	50 36	232	38 48	267	23 42	316
66	41 42	039	45 24	105	13 08	145	26 54	166	50 07	234	38 12	268	23 17	316
67	42 04	039	45 58	106	13 28	146	27 02	167	49 39	235	37 37	269	22 53	317
68	42 27	039	46 32	107	13 47	147	27 09	168	49 10	236	37 02	270	22 29	318
69	42 49	040	47 05	108	14 06	148	27 16	169	48 40	237	36 27	271	22 06	318
70	43 12	040	47 39	109	14 24	149	27 22	170	48 10	238	35 51	271	21 42	319
71	43 35	041	48 12	110	14 42	150	27 28	172	47 40	240	35 16	272	21 19	319
72	43 58	041	48 45	111	14 59	151	27 33	173	47 10	241	34 41	273	20 56	320
73	44 21	041	49 18	112	15 16	152	27 37	174	46 39	242	34 06	274	20 34	321
74	44 44	042	49 51	113	15 33	153	27 40	175	46 07	243	33 30	275	20 12	321
	Dubhe		◆REGULUS		PROCYON		SIRIUS		◆RIGEL		Hamal		◆DENEB	
75	45 08	042	17 17	093	31 35	132	15 48	154	27 43	176	45 36	244	19 50	322
76	45 31	042	17 52	094	32 01	133	16 04	155	27 45	177	45 04	245	19 28	323
77	45 55	043	18 28	095	32 27	134	16 18	156	27 47	178	44 32	246	19 07	323
78	46 19	043	19 03	096	32 52	135	16 33	157	27 47	179	43 59	247	18 46	324
79	46 44	044	19 38	097	33 17	136	16 46	158	27 48	180	43 27	248	18 26	325
80	47 08	044	20 13	097	33 42	137	16 59	159	27 47	182	42 54	249	18 05	325
81	47 32	044	20 48	098	34 05	138	17 12	160	27 46	183	42 21	250	17 45	326
82	47 57	045	21 23	099	34 29	139	17 24	161	27 44	184	41 48	251	17 26	327
83	48 22	045	21 57	100	34 51	140	17 35	162	27 41	185	41 14	252	17 07	327
84	48 47	045	22 32	101	35 14	141	17 46	163	27 37	186	40 40	253	16 48	328
85	49 12	046	23 07	102	35 35	143	17 56	164	27 33	187	40 07	254	16 29	329
86	49 37	046	23 41	103	35 57	144	18 06	165	27 29	188	39 33	255	16 11	329
87	50 03	046	24 15	103	36 17	145	18 15	166	27 23	189	38 59	256	15 53	330
88	50 28	047	24 50	104	36 37	146	18 23	167	27 17	191	38 24	257	15 35	331
89	50 54	047	25 24	105	36 57	147	18 31	168	27 10	192	37 50	258	15 18	331

LHA ♈	Hc	Zn	Hc	Zn	Hc	Zn	Hc	Zn	Hc	Zn	Hc	Zn	Hc	Zn
	Dubhe		◆REGULUS		PROCYON		SIRIUS		◆RIGEL		Mirfak		◆DENEB	
90	51 20	047	25 58	106	37 15	148	18 38	169	27 03	193	65 54	276	15 01	332
91	51 46	048	26 32	107	37 34	150	18 45	170	26 55	194	65 19	277	14 45	333
92	52 12	048	27 05	108	37 51	151	18 51	171	26 46	195	64 44	277	14 29	333
93	52 38	048	27 39	109	38 08	152	18 56	172	26 37	196	64 09	278	14 13	334
94	53 04	048	28 12	110	38 24	153	19 01	173	26 27	197	63 34	279	13 58	335
95	53 31	049	28 45	111	38 40	154	19 05	174	26 16	198	63 00	279	13 43	335
96	53 57	049	29 18	112	38 55	156	19 09	175	26 05	199	62 25	280	13 28	336
97	54 24	049	29 51	112	39 09	157	19 12	176	25 53	200	61 50	281	13 14	337
98	54 51	050	30 23	113	39 22	158	19 14	177	25 40	201	61 15	281	13 00	337
99	55 18	050	30 55	114	39 35	159	19 16	178	25 27	202	60 41	282	12 47	338
100	55 45	050	31 27	115	39 47	161	19 17	179	25 13	204	60 06	282	12 34	339
101	56 12	050	31 59	116	39 58	162	19 17	180	24 59	205	59 32	283	12 21	339
102	56 39	051	32 31	117	40 09	163	19 17	181	24 44	206	58 58	284	12 09	340
103	57 06	051	33 02	118	40 19	165	19 16	182	24 28	207	58 23	284	11 57	341
104	57 34	051	33 33	119	40 28	166	19 15	183	24 12	208	57 49	285	11 46	341
	◆Dubhe		Denebola		REGULUS		◆SIRIUS		RIGEL		ALDEBARAN		◆Mirfak	
105	58 01	051	22 12	096	34 03	120	19 13	184	23 55	209	43 15	231	57 15	285
106	58 29	051	22 47	096	34 34	121	19 10	185	23 38	210	42 47	232	56 41	286
107	58 56	052	23 22	097	35 04	122	19 07	186	23 20	211	42 19	233	56 07	286
108	59 24	052	23 57	098	35 33	123	19 03	187	23 02	212	41 51	234	55 34	287
109	59 52	052	24 32	099	36 03	124	18 59	188	22 43	213	41 22	235	55 00	288
110	60 20	052	25 07	100	36 32	125	18 54	189	22 24	214	40 53	236	54 26	288
111	60 47	052	25 41	101	37 00	126	18 48	190	22 04	215	40 23	238	53 53	289
112	61 15	053	26 16	102	37 28	128	18 41	191	21 43	216	39 53	239	53 19	289
113	61 43	053	26 50	102	37 56	129	18 34	192	21 22	217	39 23	240	52 46	290
114	62 12	053	27 25	103	38 23	130	18 27	193	21 01	218	38 53	241	52 13	290
115	62 40	053	27 59	104	38 50	131	18 19	194	20 39	219	38 22	242	51 40	291
116	63 08	053	28 33	105	39 17	132	18 10	195	20 16	220	37 50	243	51 07	291
117	63 36	053	29 07	106	39 43	133	18 01	196	19 54	221	37 19	244	50 34	292
118	64 04	053	29 41	107	40 08	134	17 51	197	19 30	222	36 47	245	50 02	292
119	64 33	053	30 15	108	40 33	135	17 40	198	19 07	223	36 15	246	49 29	293
	Kochab		◆ARCTURUS		REGULUS		◆SIRIUS		BETELGEUSE		CAPELLA		◆Schedar	
120	48 00	023	13 14	075	40 58	137	17 29	199	37 03	220	62 56	270	34 20	321
121	48 14	024	13 48	076	41 22	138	17 17	200	36 40	221	62 21	271	33 58	322
122	48 28	024	14 22	077	41 45	139	17 05	201	36 16	222	61 45	271	33 36	322
123	48 43	024	14 57	078	42 08	140	16 52	202	35 52	224	61 10	272	33 15	323
124	48 57	024	15 31	079	42 30	141	16 39	203	35 28	225	60 35	273	32 54	323
125	49 12	024	16 06	079	42 52	143	16 25	204	35 03	226	60 00	274	32 33	324
126	49 26	025	16 40	080	43 13	144	16 11	205	34 37	227	59 24	274	32 12	324
127	49 41	025	17 15	081	43 33	145	15 56	206	34 11	228	58 49	275	31 51	325
128	49 56	025	17 50	082	43 53	146	15 40	207	33 45	229	58 14	276	31 31	325
129	50 11	025	18 25	082	44 12	148	15 24	208	33 18	230	57 39	277	31 11	326
130	50 26	025	19 00	083	44 31	149	15 07	209	32 51	231	57 04	277	30 51	326
131	50 41	025	19 35	084	44 49	150	14 50	209	32 23	232	56 29	278	30 32	327
132	50 56	026	20 10	085	45 06	152	14 33	210	31 55	233	55 54	279	30 12	327
133	51 12	026	20 45	086	45 22	153	14 15	211	31 27	234	55 19	279	29 53	328
134	51 27	026	21 20	086	45 38	154	13 56	212	30 58	235	54 45	280	29 35	328
	Kochab		◆ARCTURUS		REGULUS		◆POLLUX		BETELGEUSE		CAPELLA		◆Schedar	
135	51 42	026	21 56	087	45 53	156	60 38	215	30 29	236	54 10	281	29 16	329
136	51 58	026	22 31	088	46 07	157	60 18	217	29 59	237	53 35	281	28 58	329
137	52 14	026	23 06	089	46 20	158	59 56	219	29 29	238	53 01	282	28 40	330
138	52 29	026	23 41	090	46 33	160	59 34	220	28 59	239	52 26	283	28 23	330
139	52 45	027	24 17	090	46 44	161	59 10	222	28 29	240	51 52	283	28 05	331
140	53 01	027	24 52	091	46 55	163	58 47	223	27 58	241	51 18	284	27 48	331
141	53 17	027	25 27	092	47 05	164	58 22	225	27 27	242	50 43	284	27 32	332
142	53 33	027	26 02	093	47 15	165	57 57	226	26 56	243	50 09	285	27 15	333
143	53 49	027	26 38	094	47 23	167	57 31	228	26 24	244	49 35	286	26 59	333
144	54 05	027	27 13	094	47 31	168	57 05	229	25 52	245	49 01	286	26 43	334
145	54 21	027	27 48	095	47 37	170	56 38	231	25 20	246	48 28	287	26 28	334
146	54 37	027	28 23	096	47 43	171	56 10	232	24 48	247	47 54	288	26 12	335
147	54 53	027	28 58	097	47 48	173	55 42	233	24 15	248	47 20	288	25 57	335
148	55 09	027	29 33	098	47 52	174	55 14	235	23 43	249	46 47	289	25 43	336
149	55 26	027	30 08	099	47 55	176	54 45	236	23 10	249	46 14	289	25 29	336
	◆VEGA		ARCTURUS		SPICA		◆REGULUS		POLLUX		◆CAPELLA		Schedar	
150	12 32	038	30 43	100	11 49	129	47 58	177	54 16	237	45 40	290	25 15	337
151	12 54	039	31 17	100	12 17	129	47 59	178	53 46	238	45 07	291	25 01	337
152	13 17	040	31 52	101	12 44	130	48 00	180	53 16	240	44 34	291	24 47	338
153	13 39	040	32 27	102	13 10	131	47 59	181	52 45	241	44 02	292	24 34	339
154	14 02	041	33 01	103	13 37	132	47 58	183	52 14	242	43 29	292	24 22	339
155	14 26	042	33 35	104	14 03	133	47 56	184	51 43	243	42 56	293	24 09	340
156	14 49	042	34 09	105	14 28	134	47 52	186	51 11	244	42 24	294	23 57	340
157	15 13	043	34 43	106	14 53	135	47 48	187	50 39	245	41 52	294	23 45	341
158	15 38	044	35 17	107	15 18	136	47 44	189	50 07	246	41 20	295	23 34	341
159	16 02	045	35 51	108	15 43	137	47 38	190	49 35	248	40 48	295	23 23	342
160	16 27	045	36 25	109	16 07	138	47 31	192	49 02	249	40 16	296	23 12	343
161	16 52	046	36 58	110	16 30	139	47 24	193	48 29	250	39 44	297	23 02	343
162	17 18	047	37 31	111	16 53	140	47 15	194	47 56	251	39 13	297	22 52	344
163	17 43	047	38 04	112	17 16	141	47 06	196	47 22	252	38 42	298	22 42	344
164	18 10	048	38 37	113	17 38	141	46 56	197	46 49	253	38 10	298	22 32	345
	◆DENEB		VEGA		ARCTURUS		◆SPICA		REGULUS		◆POLLUX		CAPELLA	
165	13 34	024	18 36	049	39 09	114	18 00	142	46 45	199	46 15	254	37 39	299
166	13 48	025	19 02	049	39 41	115	18 21	143	46 33	200	45 41	255	37 09	300
167	14 03	026	19 29	050	40 13	116	18 42	144	46 21	201	45 07	256	36 38	300
168	14 19	026	19 56	051	40 45	117	19 02	145	46 07	203	44 33	257	36 08	301
169	14 35	027	20 24	051	41 16	118	19 22	146	45 53	204	43 58	257	35 37	301
170	14 51	028	20 52	052	41 47	119	19 41	147	45 39	206	43 24	258	35 07	302
171	15 08	028	21 19	053	42 18	120	20 00	148	45 23	207	42 49	259	34 38	302
172	15 25	029	21 48	053	42 49	121	20 18	149	45 07	208	42 15	260	34 08	303
173	15 42	030	22 16	054	43 19	122	20 36	150	44 50	210	41 40	261	33 38	304
174	15 59	030	22 45	055	43 49	123	20 53	151	44 32	211	41 05	262	33 09	304
175	16 17	031	23 14	055	44 18	124	21 10	152	44 13	212	40 30	263	32 40	305
176	16 36	032	23 43	056	44 47	125	21 26	153	43 54	213	39 55	264	32 11	305
177	16 55	032	24 12	057	45 15	126	21 42	154	43 34	215	39 20	265	31 43	306
178	17 14	033	24 42	057	45 44	128	21 57	155	43 14	216	38 45	265	31 14	307
179	17 33	034	25 11	058	46 11	129	22 11	156	42 53	217	38 10	266	30 46	307

LAT 54°N

LHA	Hc	Zn	Hc	Zn	Hc	Zn	Hc	Zn	Hc	Zn	Hc	Zn	Hc	Zn
♈	DENEB		◆VEGA		ARCTURUS		◆SPICA		REGULUS		POLLUX		◆CAPELLA	
180	17 53	034	25 41	059	46 39	130	22 25	157	42 31	219	37 34	267	30 18	308
181	18 13	035	26 12	059	47 05	131	22 38	158	42 09	220	36 59	268	29 50	308
182	18 33	036	26 42	060	47 32	132	22 51	159	41 46	221	36 24	269	29 23	309
183	18 54	036	27 13	061	47 57	134	23 03	161	41 23	222	35 49	270	28 55	310
184	19 15	037	27 44	061	48 23	135	23 14	162	40 59	223	35 13	270	28 28	310
185	19 36	038	28 15	062	48 47	136	23 25	163	40 34	224	34 38	271	28 02	311
186	19 58	038	28 46	063	49 12	138	23 35	164	40 09	226	34 03	272	27 35	311
187	20 20	039	29 17	063	49 35	139	23 45	165	39 44	227	33 28	273	27 09	312
188	20 42	039	29 49	064	49 58	140	23 54	166	39 18	228	32 52	274	26 43	313
189	21 05	040	30 21	065	50 20	142	24 02	167	38 52	229	32 17	274	26 17	313
190	21 28	041	30 52	065	50 42	143	24 10	168	38 25	230	31 42	275	25 51	314
191	21 51	041	31 25	066	51 03	144	24 17	169	37 57	231	31 07	276	25 26	314
192	22 14	042	31 57	067	51 23	146	24 24	170	37 30	232	30 32	277	25 01	315
193	22 38	043	32 29	067	51 42	147	24 29	171	37 02	233	29 57	277	24 36	316
194	23 02	043	33 02	068	52 01	149	24 35	172	36 33	235	29 22	278	24 12	316
	◆DENEB		VEGA		Rasalhague		◆ARCTURUS		REGULUS		◆POLLUX		CAPELLA	
195	23 26	044	33 35	069	22 37	100	52 19	150	36 04	236	28 47	279	23 47	317
196	23 51	045	34 08	069	23 12	101	52 36	152	35 35	237	28 12	280	23 23	318
197	24 16	045	34 41	070	23 46	102	52 53	153	35 05	238	27 38	281	23 00	318
198	24 41	046	35 14	071	24 21	103	53 08	155	34 35	239	27 03	281	22 36	319
199	25 06	046	35 47	071	24 55	103	53 23	156	34 05	240	26 28	282	22 13	319
200	25 32	047	36 21	072	25 30	104	53 37	158	33 34	241	25 54	283	21 51	320
201	25 58	048	36 54	073	26 04	105	53 50	159	33 04	242	25 20	283	21 28	321
202	26 24	048	37 28	073	26 38	106	54 02	161	32 32	243	24 45	284	21 06	321
203	26 51	049	38 02	074	27 11	107	54 13	162	32 01	244	24 11	285	20 44	322
204	27 17	049	38 36	075	27 45	108	54 24	164	31 29	245	23 37	286	20 22	323
205	27 44	050	39 10	075	28 19	109	54 33	165	30 57	246	23 03	286	20 01	323
206	28 11	051	39 44	076	28 52	110	54 41	167	30 25	247	22 30	287	19 40	324
207	28 39	051	40 18	077	29 25	111	54 49	169	29 52	247	21 56	288	19 20	324
208	29 06	052	40 53	077	29 58	111	54 55	170	29 20	248	21 23	289	18 59	325
209	29 34	053	41 27	078	30 31	112	55 01	172	28 47	249	20 49	289	18 39	326
	◆DENEB		VEGA		Rasalhague		◆ARCTURUS		REGULUS		◆POLLUX		CAPELLA	
210	30 02	053	42 02	079	31 03	113	55 05	174	28 14	250	20 16	290	18 20	326
211	30 31	054	42 36	080	31 35	114	55 08	175	27 40	251	19 43	291	18 00	327
212	30 59	054	43 11	080	32 07	115	55 11	177	27 07	252	19 10	292	17 41	328
213	31 28	055	43 46	081	32 39	116	55 12	179	26 33	253	18 37	292	17 23	328
214	31 57	056	44 21	082	33 11	117	55 12	180	26 00	254	18 05	293	17 04	329
215	32 26	056	44 56	082	33 42	118	55 12	182	25 26	255	17 32	294	16 46	330
216	32 56	057	45 31	083	34 13	119	55 10	184	24 51	256	17 00	295	16 29	330
217	33 25	057	46 06	084	34 43	120	55 07	185	24 17	257	16 28	295	16 11	331
218	33 55	058	46 41	085	35 14	121	55 04	187	23 43	257	15 57	296	15 54	332
219	34 25	059	47 16	085	35 44	122	54 59	188	23 08	258	15 25	297	15 38	332
220	34 55	059	47 51	086	36 13	123	54 53	190	22 34	259	14 54	297	15 22	333
221	35 25	060	48 26	087	36 43	124	54 47	192	21 59	260	14 22	298	15 06	334
222	35 56	060	49 01	088	37 12	125	54 39	193	21 24	261	13 51	299	14 50	334
223	36 27	061	49 37	088	37 40	127	54 30	195	20 50	262	13 21	300	14 35	335
224	36 58	062	50 12	089	38 08	128	54 21	197	20 15	262	12 50	300	14 21	336
	Schedar		◆DENEB		VEGA		Rasalhague		◆ARCTURUS		Denebola		◆Dubhe	
225	24 08	020	37 29	062	50 47	090	38 36	129	54 10	198	35 52	242	58 48	308
226	24 20	021	38 00	063	51 22	091	39 03	130	53 59	200	35 21	243	58 21	309
227	24 33	021	38 31	063	51 58	092	39 30	131	53 46	201	34 50	244	57 53	309
228	24 46	022	39 03	064	52 33	093	39 57	132	53 33	203	34 18	245	57 26	309
229	24 59	022	39 35	065	53 08	093	40 23	133	53 19	204	33 46	246	56 58	309
230	25 13	023	40 07	065	53 43	094	40 48	134	53 04	206	33 13	247	56 31	310
231	25 27	024	40 39	066	54 18	095	41 13	136	52 48	207	32 41	248	56 04	310
232	25 41	024	41 11	066	54 54	096	41 37	137	52 32	209	32 08	249	55 37	310
233	25 56	025	41 43	067	55 29	097	42 01	138	52 14	210	31 35	250	55 10	310
234	26 11	025	42 16	068	56 04	098	42 25	139	51 56	212	31 01	251	54 43	311
235	26 26	026	42 49	068	56 38	099	42 47	140	51 37	213	30 28	252	54 16	311
236	26 41	026	43 21	069	57 13	100	43 10	142	51 17	215	29 54	253	53 50	311
237	26 57	027	43 54	069	57 48	101	43 31	143	50 57	216	29 21	254	53 23	311
238	27 13	027	44 27	070	58 23	102	43 52	144	50 36	217	28 47	255	52 57	312
239	27 30	028	45 01	071	58 57	103	44 13	145	50 14	219	28 13	255	52 30	312
	Schedar		◆DENEB		VEGA		Rasalhague		◆ARCTURUS		Denebola		◆Dubhe	
240	27 46	029	45 34	071	59 31	104	44 32	147	49 52	220	27 38	256	52 04	312
241	28 03	029	46 07	072	60 06	105	44 51	148	49 28	222	27 04	257	51 38	313
242	28 21	030	46 41	072	60 40	106	45 10	149	49 05	223	26 30	258	51 12	313
243	28 38	030	47 15	073	61 13	107	45 27	151	48 40	224	25 55	259	50 47	313
244	28 56	031	47 48	074	61 47	108	45 44	152	48 16	225	25 21	260	50 21	314
245	29 14	031	48 22	074	62 20	109	46 01	153	47 50	227	24 46	261	49 55	314
246	29 33	032	48 56	075	62 54	111	46 16	155	47 24	228	24 11	261	49 30	314
247	29 51	032	49 30	076	63 26	112	46 31	156	46 58	229	23 36	262	49 05	314
248	30 10	033	50 05	076	63 59	113	46 45	157	46 31	230	23 01	263	48 40	315
249	30 29	033	50 39	077	64 31	114	46 58	159	46 04	232	22 26	264	48 15	315
250	30 49	034	51 13	078	65 03	116	47 10	160	45 36	233	21 51	265	47 50	316
251	31 09	034	51 48	078	65 35	117	47 22	162	45 07	234	21 16	266	47 25	316
252	31 29	035	52 22	079	66 06	118	47 33	163	44 39	235	20 41	266	47 01	316
253	31 49	035	52 57	079	66 37	120	47 43	164	44 10	236	20 05	267	46 37	317
254	32 09	036	53 32	080	67 07	121	47 52	166	43 40	237	19 30	268	46 12	317
	◆Schedar		DENEB		ALTAIR		◆Rasalhague		ARCTURUS		◆Alkaid		Kochab	
255	32 30	036	54 06	081	33 29	127	48 00	167	43 10	238	60 22	281	66 02	339
256	32 51	037	54 41	082	33 58	128	48 07	169	42 40	240	59 47	282	65 49	339
257	33 13	037	55 16	082	34 25	129	48 14	170	42 09	241	59 13	282	65 36	338
258	33 34	038	55 51	083	34 53	130	48 19	172	41 39	242	58 38	283	65 23	338
259	33 56	038	56 26	084	35 19	131	48 24	173	41 07	243	58 04	283	65 10	337
260	34 18	039	57 01	084	35 46	132	48 28	175	40 36	244	57 29	284	64 56	337
261	34 40	039	57 36	085	36 12	133	48 31	176	40 04	245	56 55	285	64 42	337
262	35 03	040	58 11	086	36 37	134	48 33	178	39 32	246	56 21	285	64 28	336
263	35 25	040	58 47	087	37 02	135	48 34	179	39 00	247	55 47	286	64 14	336
264	35 48	041	59 22	087	37 27	136	48 34	180	38 27	248	55 13	286	64 00	336
265	36 11	041	59 57	088	37 51	138	48 33	182	37 55	249	54 40	287	63 45	335
266	36 35	042	60 32	089	38 15	139	48 31	183	37 22	250	54 06	287	63 30	335
267	36 58	042	61 08	090	38 38	140	48 29	185	36 48	251	53 32	288	63 15	335
268	37 22	043	61 43	090	39 00	141	48 25	186	36 15	252	52 59	289	63 00	335
269	37 46	043	62 18	091	39 22	142	48 21	188	35 41	253	52 25	289	62 45	334

LHA	Hc	Zn	Hc	Zn	Hc	Zn	Hc	Zn	Hc	Zn	Hc	Zn	Hc	Zn
♈	◆Mirfak		Alpheratz		◆ALTAIR		Rasalhague		◆ARCTURUS		Alkaid		Kochab	
270	18 54	025	22 01	070	39 43	143	48 16	189	35 08	254	51 52	290	62 30	334
271	19 09	026	22 34	071	40 04	145	48 10	191	34 34	254	51 19	290	62 15	334
272	19 24	027	23 07	072	40 24	146	48 03	192	34 00	255	50 46	291	61 59	334
273	19 40	027	23 41	073	40 44	147	47 55	194	33 26	256	50 13	291	61 44	334
274	19 57	028	24 15	073	41 02	148	47 46	195	32 51	257	49 40	292	61 28	334
275	20 13	029	24 49	074	41 21	150	47 36	197	32 17	258	49 08	292	61 12	333
276	20 30	029	25 23	075	41 38	151	47 26	198	31 42	259	48 35	293	60 56	333
277	20 48	030	25 57	076	41 55	152	47 15	199	31 08	260	48 03	294	60 40	333
278	21 05	030	26 31	076	42 11	153	47 03	201	30 33	261	47 30	294	60 24	333
279	21 23	031	27 05	077	42 27	155	46 50	202	29 58	262	46 58	295	60 08	333
280	21 42	032	27 40	078	42 41	156	46 36	204	29 23	262	46 26	295	59 52	333
281	22 00	032	28 14	079	42 55	157	46 21	205	28 48	263	45 54	296	59 36	333
282	22 19	033	28 49	079	43 09	159	46 06	206	28 13	264	45 23	296	59 20	333
283	22 39	033	29 23	080	43 21	160	45 50	208	27 38	265	44 51	297	59 04	333
284	22 58	034	29 58	081	43 33	161	45 34	209	27 03	266	44 20	297	58 47	333
	◆CAPELLA		Alpheratz		◆ALTAIR		Rasalhague		◆ARCTURUS		Alkaid		Kochab	
285	12 24	018	30 33	082	43 44	163	45 16	210	26 28	267	43 49	298	58 31	332
286	12 35	019	31 08	082	43 54	164	44 58	212	25 52	267	43 17	298	58 15	332
287	12 46	019	31 43	083	44 03	165	44 39	213	25 17	268	42 47	299	57 58	332
288	12 58	020	32 18	084	44 12	167	44 20	214	24 42	269	42 16	300	57 42	332
289	13 11	021	32 53	085	44 20	168	44 00	215	24 07	270	41 45	300	57 26	332
290	13 23	022	33 28	085	44 26	169	43 39	217	23 31	271	41 15	301	57 09	332
291	13 37	022	34 03	086	44 32	171	43 17	218	22 56	271	40 44	301	56 53	332
292	13 50	023	34 39	087	44 38	172	42 55	219	22 21	272	40 14	302	56 37	332
293	14 04	024	35 14	088	44 42	174	42 33	220	21 46	273	39 44	302	56 20	332
294	14 18	024	35 49	089	44 46	175	42 10	222	21 10	274	39 15	303	56 04	332
295	14 33	025	36 24	089	44 48	176	41 46	223	20 35	275	38 45	303	55 48	332
296	14 48	026	37 00	090	44 50	178	41 22	224	20 00	275	38 16	304	55 31	333
297	15 03	026	37 35	091	44 51	179	40 57	225	19 25	276	37 47	304	55 15	333
298	15 19	027	38 10	092	44 51	181	40 32	226	18 50	277	37 18	305	54 59	333
299	15 35	028	38 45	093	44 50	182	40 06	228	18 15	278	36 49	306	54 43	333
	CAPELLA		◆Alpheratz		Enif		ALTAIR		◆VEGA		Alphecca		◆Alioth	
300	15 52	028	39 21	093	41 13	145	44 49	183	69 14	231	35 03	268	35 14	319
301	16 09	029	39 56	094	41 33	146	44 46	185	68 46	233	34 27	269	34 51	319
302	16 26	030	40 31	095	41 53	147	44 43	186	68 18	235	33 52	270	34 28	320
303	16 43	030	41 06	096	42 11	149	44 39	187	67 49	236	33 17	271	34 06	320
304	17 01	031	41 41	097	42 29	150	44 34	189	67 19	238	32 42	271	33 43	321
305	17 20	032	42 16	098	42 47	151	44 28	190	66 49	240	32 06	272	33 21	321
306	17 38	032	42 51	099	43 03	153	44 21	192	66 18	241	31 31	273	33 00	322
307	17 57	033	43 26	099	43 19	154	44 14	193	65 47	242	30 56	274	32 38	322
308	18 16	033	44 00	100	43 34	155	44 06	194	65 16	244	30 21	275	32 17	323
309	18 36	034	44 35	101	43 49	156	43 56	196	64 44	245	29 46	275	31 55	324
310	18 56	035	45 10	102	44 02	158	43 47	197	64 12	247	29 11	276	31 35	324
311	19 16	035	45 44	103	44 15	159	43 36	198	63 39	248	28 36	277	31 14	325
312	19 37	036	46 18	104	44 27	161	43 24	200	63 06	249	28 01	278	30 54	325
313	19 58	037	46 52	105	44 39	162	43 12	201	62 33	250	27 26	279	30 34	326
314	20 19	037	47 26	106	44 49	163	42 59	202	62 00	251	26 51	279	30 14	326
	CAPELLA		◆Hamal		Alpheratz		Enif		◆ALTAIR		VEGA		◆Alioth	
315	20 41	038	26 28	086	48 00	107	44 59	165	42 45	204	61 26	253	29 54	327
316	21 02	039	27 03	087	48 34	108	45 08	166	42 31	205	60 53	254	29 35	327
317	21 25	039	27 38	088	49 07	109	45 16	167	42 15	206	60 19	255	29 16	328
318	21 47	040	28 14	088	49 40	110	45 24	169	41 59	208	59 45	256	28 57	328
319	22 10	040	28 49	089	50 13	111	45 30	170	41 43	209	59 10	257	28 39	329
320	22 33	041	29 24	090	50 46	112	45 36	172	41 25	210	58 36	258	28 21	329
321	22 56	042	29 59	091	51 19	113	45 40	173	41 07	211	58 01	259	28 03	330
322	23 20	042	30 35	092	51 51	114	45 44	174	40 49	213	57 27	260	27 45	330
323	23 44	043	31 10	092	52 23	116	45 47	176	40 29	214	56 52	261	27 28	331
324	24 08	044	31 45	093	52 55	117	45 49	177	40 09	215	56 17	262	27 11	331
325	24 32	044	32 20	094	53 26	118	45 51	179	39 49	216	55 42	263	26 54	332
326	24 57	045	32 55	095	53 57	119	45 51	180	39 28	217	55 07	264	26 38	333
327	25 22	045	33 31	096	54 27	120	45 51	181	39 06	219	54 32	265	26 22	333
328	25 47	046	34 06	097	54 58	122	45 49	183	38 44	220	53 57	265	26 06	334
329	26 13	047	34 41	097	55 28	123	45 47	184	38 21	221	53 22	266	25 51	334
	CAPELLA		◆Hamal		Alpheratz		Enif		◆ALTAIR		VEGA		◆Alioth	
330	26 38	047	35 16	098	55 57	124	45 44	186	37 58	222	52 46	267	25 35	335
331	27 05	048	35 50	099	56 26	125	45 40	187	37 34	223	52 11	268	25 20	335
332	27 31	048	36 25	100	56 54	127	45 35	188	37 09	224	51 36	269	25 06	336
333	27 57	049	37 00	101	57 22	128	45 30	190	36 44	226	51 01	270	24 52	336
334	28 24	050	37 34	102	57 50	130	45 23	191	36 19	227	50 25	270	24 38	337
335	28 51	050	38 09	103	58 17	131	45 16	193	35 53	228	49 50	271	24 24	338
336	29 18	051	38 43	104	58 43	132	45 08	194	35 27	229	49 15	272	24 11	338
337	29 46	051	39 17	105	59 09	134	44 59	195	35 00	230	48 40	273	23 58	339
338	30 14	052	39 52	106	59 34	135	44 49	197	34 33	231	48 04	274	23 45	339
339	30 42	053	40 25	106	59 59	137	44 38	198	34 05	232	47 29	274	23 33	340
340	31 10	053	40 59	107	60 22	138	44 27	200	33 37	233	46 54	275	23 21	340
341	31 38	054	41 33	108	60 45	140	44 15	201	33 09	234	46 19	276	23 09	341
342	32 07	054	42 06	109	61 08	142	44 02	202	32 40	235	45 44	277	22 58	342
343	32 35	055	42 39	110	61 29	143	43 48	204	32 11	236	45 09	277	22 47	342
344	33 04	056	43 12	111	61 50	145	43 34	205	31 41	237	44 34	278	22 36	343
	◆CAPELLA		ALDEBARAN		◆Diphda		Enif		ALTAIR		◆VEGA		Alioth	
345	33 34	056	16 50	085	14 39	155	43 19	206	31 12	238	43 59	279	22 26	343
346	34 03	057	17 25	086	14 54	156	43 03	207	30 41	239	43 24	279	22 16	344
347	34 33	057	18 00	087	15 08	157	42 46	209	30 11	240	42 50	280	22 06	344
348	35 03	058	18 35	087	15 22	157	42 29	210	29 40	241	42 15	281	21 57	345
349	35 33	059	19 11	088	15 35	158	42 11	211	29 09	242	41 40	282	21 48	346
350	36 03	059	19 46	089	15 48	159	41 52	213	28 38	243	41 06	282	21 39	346
351	36 33	060	20 21	090	16 00	160	41 33	214	28 06	244	40 31	283	21 31	347
352	37 04	060	20 57	091	16 12	161	41 13	215	27 34	245	39 57	284	21 23	347
353	37 35	061	21 32	091	16 22	162	40 52	216	27 02	246	39 23	284	21 15	348
354	38 05	062	22 07	092	16 33	163	40 31	217	26 30	247	38 49	285	21 08	348
355	38 37	062	22 42	093	16 43	164	40 09	219	25 58	248	38 15	286	21 01	349
356	39 08	063	23 17	094	16 52	165	39 47	220	25 25	249	37 41	286	20 55	350
357	39 39	063	23 53	095	17 01	166	39 24	221	24 52	250	37 07	287	20 49	350
358	40 11	064	24 28	095	17 09	167	39 01	222	24 19	250	36 34	288	20 43	351
359	40 43	065	25 03	096	17 16	168	38 37	223	23 45	251	36 00	288	20 37	351

LAT 53°N

LHA ♈	Hc	Zn	Hc	Zn	Hc	Zn	Hc	Zn	Hc	Zn	Hc	Zn	Hc	Zn
	♦CAPELLA		ALDEBARAN		Hamal		♦Alpheratz		ALTAIR		♦VEGA		Alioth	
0	40 49	064	25 45	097	51 56	129	66 01	176	23 30	253	35 07	290	19 33	352
1	41 22	065	26 21	098	52 24	130	66 03	178	22 55	254	34 33	290	19 28	353
2	41 54	065	26 57	098	52 52	131	66 04	180	22 21	254	33 59	291	19 24	353
3	42 27	066	27 32	099	53 19	132	66 03	182	21 46	255	33 25	292	19 20	354
4	43 00	067	28 08	100	53 45	134	66 01	184	21 11	256	32 52	292	19 16	354
5	43 34	067	28 43	101	54 11	135	65 58	186	20 36	257	32 19	293	19 13	355
6	44 07	068	29 19	102	54 36	137	65 53	189	20 01	258	31 45	294	19 10	356
7	44 40	068	29 54	103	55 00	138	65 47	191	19 25	259	31 12	294	19 07	356
8	45 14	069	30 29	104	55 24	139	65 40	193	18 50	259	30 40	295	19 05	357
9	45 48	070	31 04	104	55 47	141	65 31	195	18 14	260	30 07	296	19 03	357
10	46 22	070	31 39	105	56 10	142	65 21	197	17 39	261	29 35	296	19 01	358
11	46 56	071	32 14	106	56 31	144	65 10	199	17 03	262	29 02	297	19 00	359
12	47 30	071	32 49	107	56 52	145	64 58	201	16 27	263	28 30	298	19 00	359
13	48 04	072	33 23	108	57 12	147	64 44	203	15 51	264	27 58	298	18 59	000
14	48 39	072	33 57	109	57 32	149	64 30	205	15 15	264	27 27	299	18 59	000
	Dubhe		♦CAPELLA		ALDEBARAN		Hamal		♦Alpheratz		DENEB		♦VEGA	
15	27 04	015	49 13	073	34 31	110	57 50	150	64 14	207	48 27	286	26 55	300
16	27 14	015	49 48	074	35 05	111	58 07	152	63 57	209	47 53	287	26 24	300
17	27 24	016	50 22	074	35 39	112	58 24	154	63 39	211	47 18	288	25 53	301
18	27 34	016	50 57	075	36 12	113	58 39	155	63 20	212	46 44	288	25 22	302
19	27 44	017	51 32	075	36 45	114	58 54	157	63 01	214	46 10	289	24 51	302
20	27 55	017	52 07	076	37 18	115	59 08	159	62 40	216	45 35	289	24 21	303
21	28 06	018	52 42	077	37 51	116	59 20	160	62 18	218	45 01	290	23 50	303
22	28 17	018	53 17	077	38 23	117	59 32	162	61 56	219	44 28	291	23 20	304
23	28 29	019	53 53	078	38 56	118	59 42	164	61 32	221	43 54	291	22 51	305
24	28 41	019	54 28	079	39 27	119	59 52	166	61 08	223	43 20	292	22 21	305
25	28 53	020	55 03	079	39 59	120	60 00	168	60 44	224	42 47	292	21 52	306
26	29 05	020	55 39	080	40 30	121	60 07	169	60 18	226	42 13	293	21 23	307
27	29 18	021	56 14	080	41 01	122	60 13	171	59 52	227	41 40	294	20 54	307
28	29 31	021	56 50	081	41 31	123	60 18	173	59 25	229	41 07	294	20 25	308
29	29 44	022	57 26	082	42 01	124	60 22	175	58 57	230	40 34	295	19 57	309
	Dubhe		♦POLLUX		BETELGEUSE		RIGEL		♦Hamal		Alpheratz		♦DENEB	
30	29 58	022	24 13	075	24 23	111	16 16	129	60 25	177	58 29	232	40 02	295
31	30 12	023	24 48	076	24 56	112	16 44	130	60 26	179	58 01	233	39 29	296
32	30 26	023	25 23	076	25 30	113	17 11	131	60 26	181	57 32	234	38 57	296
33	30 40	024	25 58	077	26 03	114	17 38	132	60 25	182	57 02	236	38 24	297
34	30 55	024	26 33	078	26 35	115	18 05	133	60 23	184	56 32	237	37 52	298
35	31 10	025	27 09	079	27 08	116	18 31	134	60 20	186	56 02	238	37 20	298
36	31 25	025	27 44	079	27 40	117	18 57	135	60 16	188	55 31	240	36 49	299
37	31 41	026	28 19	080	28 12	118	19 22	136	60 10	190	54 59	241	36 17	299
38	31 56	026	28 55	081	28 44	119	19 47	137	60 03	192	54 28	242	35 46	300
39	32 12	027	29 31	082	29 16	120	20 11	138	59 56	193	53 55	243	35 14	301
40	32 29	027	30 07	082	29 47	121	20 35	139	59 47	195	53 23	244	34 43	301
41	32 45	027	30 42	083	30 18	122	20 59	140	59 37	197	52 50	246	34 13	302
42	33 02	028	31 18	084	30 48	123	21 22	141	59 26	199	52 17	247	33 42	302
43	33 19	028	31 54	085	31 18	124	21 45	142	59 13	201	51 44	248	33 12	303
44	33 36	029	32 30	085	31 48	125	22 07	143	59 00	202	51 11	249	32 41	303
	Dubhe		♦POLLUX		BETELGEUSE		RIGEL		♦Hamal		Alpheratz		♦DENEB	
45	33 54	029	33 06	086	32 18	126	22 28	144	58 46	204	50 37	250	32 11	304
46	34 12	030	33 42	087	32 47	127	22 50	145	58 31	206	50 03	251	31 41	305
47	34 30	030	34 18	088	33 15	128	23 10	146	58 15	207	49 29	252	31 12	305
48	34 48	031	34 54	088	33 44	129	23 30	147	57 58	209	48 54	253	30 42	306
49	35 07	031	35 30	089	34 11	130	23 50	148	57 40	211	48 19	254	30 13	306
50	35 25	031	36 07	090	34 39	131	24 09	149	57 21	212	47 45	255	29 44	307
51	35 44	032	36 43	091	35 06	132	24 27	150	57 01	214	47 10	256	29 16	308
52	36 03	032	37 19	092	35 32	133	24 45	151	56 41	215	46 35	257	28 47	308
53	36 23	033	37 55	092	35 59	134	25 03	152	56 19	217	45 59	258	28 19	309
54	36 43	033	38 31	093	36 24	135	25 19	153	55 57	218	45 24	259	27 51	309
55	37 02	034	39 07	094	36 49	137	25 35	154	55 35	220	44 49	260	27 23	310
56	37 23	034	39 43	095	37 14	138	25 51	155	55 11	221	44 13	260	26 55	311
57	37 43	034	40 19	096	37 38	139	26 06	156	54 47	223	43 37	261	26 28	311
58	38 03	035	40 55	097	38 01	140	26 20	157	54 22	224	43 02	262	26 01	312
59	38 24	035	41 31	098	38 24	141	26 34	158	53 57	226	42 26	263	25 34	312
	♦Dubhe		POLLUX		SIRIUS		♦RIGEL		Hamal		♦Alpheratz		DENEB	
60	38 45	036	42 06	098	11 46	140	26 47	159	53 30	227	41 50	264	25 08	313
61	39 06	036	42 42	099	12 09	141	26 59	160	53 04	228	41 14	265	24 41	314
62	39 28	036	43 18	100	12 31	142	27 11	161	52 37	230	40 38	266	24 15	314
63	39 49	037	43 53	101	12 54	143	27 22	163	52 09	231	40 02	267	23 49	315
64	40 11	037	44 28	102	13 15	143	27 33	164	51 41	232	39 26	267	23 24	315
65	40 33	038	45 04	103	13 37	144	27 43	165	51 12	233	38 50	268	22 59	316
66	40 55	038	45 39	104	13 57	145	27 52	166	50 43	235	38 14	269	22 34	317
67	41 17	038	46 14	105	14 18	146	28 00	167	50 13	236	37 38	270	22 09	317
68	41 40	039	46 49	106	14 38	147	28 08	168	49 43	237	37 02	271	21 45	318
69	42 03	039	47 23	107	14 57	148	28 15	169	49 12	238	36 25	271	21 21	318
70	42 26	040	47 58	108	15 16	149	28 22	170	48 41	239	35 49	272	20 57	319
71	42 49	040	48 32	109	15 34	150	28 27	171	48 10	241	35 13	273	20 33	320
72	43 12	040	49 06	110	15 52	151	28 32	173	47 38	242	34 37	274	20 10	320
73	43 36	041	49 40	111	16 09	152	28 37	174	47 06	243	34 01	274	19 47	321
74	43 59	041	50 14	112	16 26	153	28 40	175	46 34	244	33 25	275	19 25	322
	Dubhe		♦REGULUS		PROCYON		SIRIUS		♦RIGEL		Hamal		♦DENEB	
75	44 23	041	17 21	093	32 15	131	16 42	154	28 43	176	46 02	245	19 02	322
76	44 47	042	17 57	094	32 42	132	16 58	155	28 45	177	45 29	246	18 41	323
77	45 11	042	18 33	095	33 08	133	17 13	156	28 47	178	44 56	247	18 19	324
78	45 35	042	19 09	095	33 35	134	17 28	157	28 47	179	44 22	248	17 58	324
79	46 00	043	19 45	096	34 00	135	17 42	158	28 48	180	43 49	249	17 37	325
80	46 24	043	20 20	097	34 25	137	17 55	159	28 47	182	43 15	250	17 16	325
81	46 49	043	20 56	098	34 50	138	18 08	160	28 45	183	42 41	251	16 56	326
82	47 14	044	21 32	099	35 14	139	18 20	161	28 43	184	42 07	252	16 36	327
83	47 39	044	22 08	100	35 37	140	18 32	162	28 41	185	41 32	253	16 16	327
84	48 04	044	22 43	100	36 00	141	18 43	163	28 37	186	40 57	254	15 57	328
85	48 30	045	23 19	101	36 23	142	18 54	164	28 33	187	40 23	255	15 38	329
86	48 55	045	23 54	102	36 45	143	19 04	165	28 28	188	39 48	256	15 19	329
87	49 21	045	24 29	103	37 06	144	19 13	166	28 22	189	39 13	257	15 01	330
88	49 47	046	25 04	104	37 27	146	19 22	167	28 16	191	38 37	258	14 43	331
89	50 13	046	25 39	105	37 47	147	19 30	168	28 09	192	38 02	259	14 26	331

LHA ♈	Hc	Zn	Hc	Zn	Hc	Zn	Hc	Zn	Hc	Zn	Hc	Zn	Hc	Zn
	♦Dubhe		REGULUS		PROCYON		♦SIRIUS		RIGEL		♦Hamal		Schedar	
90	50 39	046	26 14	106	38 06	148	19 37	169	28 01	193	37 27	259	46 23	308
91	51 05	047	26 49	106	38 25	149	19 44	170	27 53	194	36 51	260	45 55	308
92	51 31	047	27 23	107	38 43	150	19 50	171	27 44	195	36 16	261	45 27	309
93	51 58	047	27 58	108	39 01	152	19 56	172	27 34	196	35 40	262	44 59	309
94	52 24	047	28 32	109	39 18	153	20 01	173	27 24	197	35 04	263	44 31	310
95	52 51	048	29 06	110	39 34	154	20 05	174	27 13	198	34 28	264	44 03	310
96	53 18	048	29 40	111	39 49	155	20 09	175	27 01	199	33 52	265	43 36	311
97	53 44	048	30 13	112	40 04	157	20 12	176	26 49	201	33 16	265	43 08	311
98	54 11	048	30 47	113	40 18	158	20 14	177	26 36	202	32 40	266	42 41	311
99	54 39	049	31 20	114	40 31	159	20 16	178	26 22	203	32 04	267	42 14	312
100	55 06	049	31 53	115	40 44	160	20 17	179	26 08	204	31 28	268	41 47	312
101	55 33	049	32 25	116	40 55	162	20 17	180	25 53	205	30 52	269	41 21	313
102	56 00	049	32 58	117	41 06	163	20 17	181	25 38	206	30 16	269	40 54	313
103	56 28	050	33 30	118	41 16	164	20 16	182	25 22	207	29 40	270	40 28	314
104	56 55	050	34 02	119	41 26	166	20 15	183	25 05	208	29 04	271	40 02	314
	♦Dubhe		Denebola		REGULUS		♦SIRIUS		RIGEL		ALDEBARAN		♦Mirfak	
105	57 23	050	22 18	095	34 33	120	20 13	184	24 48	209	43 53	232	56 59	287
106	57 51	050	22 53	096	35 05	121	20 10	185	24 30	210	43 24	233	56 24	287
107	58 18	050	23 29	097	35 35	122	20 07	186	24 12	211	42 55	234	55 50	288
108	58 46	051	24 05	098	36 06	123	20 03	187	23 53	212	42 26	235	55 15	288
109	59 14	051	24 41	099	36 36	124	19 58	188	23 33	213	41 56	236	54 41	289
110	59 42	051	25 17	099	37 06	125	19 53	189	23 13	214	41 26	237	54 07	289
111	60 10	051	25 52	100	37 36	126	19 47	190	22 53	215	40 55	238	53 33	290
112	60 38	051	26 28	101	38 05	127	19 40	191	22 32	216	40 24	239	52 59	290
113	61 07	051	27 03	102	38 33	128	19 33	192	22 10	217	39 53	240	52 25	291
114	61 35	051	27 38	103	39 01	129	19 25	193	21 48	218	39 22	241	51 52	291
115	62 03	051	28 13	104	39 29	130	19 17	194	21 25	219	38 50	242	51 18	292
116	62 31	052	28 48	105	39 57	131	19 08	195	21 02	220	38 18	243	50 45	293
117	63 00	052	29 23	105	40 23	133	18 58	196	20 39	221	37 45	244	50 11	293
118	63 28	052	29 58	106	40 50	134	18 48	197	20 15	222	37 12	245	49 38	294
119	63 56	052	30 33	107	41 16	135	18 37	198	19 51	223	36 39	246	49 05	294
	Kochab		♦ARCTURUS		REGULUS		♦SIRIUS		BETELGEUSE		CAPELLA		♦Schedar	
120	47 05	023	12 58	075	41 41	136	18 26	199	37 48	221	62 55	272	33 33	322
121	47 19	023	13 33	076	42 06	137	18 14	200	37 25	222	62 19	272	33 11	322
122	47 33	023	14 09	077	42 30	138	18 01	201	37 00	223	61 43	273	32 49	322
123	47 48	024	14 44	078	42 54	140	17 48	202	36 35	224	61 07	274	32 27	323
124	48 02	024	15 19	078	43 17	141	17 34	203	36 10	225	60 31	275	32 05	323
125	48 17	024	15 54	079	43 39	142	17 20	204	35 44	226	59 55	275	31 44	324
126	48 32	024	16 30	080	44 01	143	17 05	205	35 18	227	59 19	276	31 23	324
127	48 47	024	17 05	081	44 22	145	16 50	206	34 51	228	58 43	277	31 02	325
128	49 01	024	17 41	081	44 43	146	16 34	207	34 24	230	58 07	277	30 42	325
129	49 16	025	18 17	082	45 03	147	16 17	208	33 56	231	57 31	278	30 21	326
130	49 32	025	18 53	083	45 22	149	16 00	209	33 28	232	56 56	279	30 01	326
131	49 47	025	19 29	084	45 41	150	15 42	210	33 00	233	56 20	279	29 41	327
132	50 02	025	20 04	084	45 58	151	15 24	211	32 31	234	55 45	280	29 22	328
133	50 17	025	20 40	085	46 15	153	15 06	211	32 01	235	55 09	281	29 03	328
134	50 33	025	21 16	086	46 32	154	14 47	212	31 32	236	54 34	281	28 44	329
	Kochab		♦ARCTURUS		REGULUS		♦POLLUX		BETELGEUSE		CAPELLA		♦Schedar	
135	50 48	026	21 52	087	46 47	155	61 27	216	31 02	237	53 58	282	28 25	329
136	51 04	026	22 29	088	47 02	157	61 05	218	30 31	238	53 23	283	28 06	330
137	51 20	026	23 05	088	47 16	158	60 43	220	30 01	239	52 48	283	27 48	330
138	51 35	026	23 41	089	47 29	159	60 19	221	29 30	240	52 13	284	27 30	331
139	51 51	026	24 17	090	47 41	161	59 55	223	28 58	241	51 38	284	27 13	331
140	52 07	026	24 53	091	47 53	162	59 30	224	28 27	242	51 03	285	26 56	332
141	52 23	026	25 29	092	48 03	164	59 04	226	27 55	243	50 28	286	26 39	332
142	52 39	026	26 05	092	48 13	165	58 38	227	27 23	244	49 53	286	26 22	333
143	52 55	026	26 41	093	48 22	167	58 11	229	26 50	244	49 19	287	26 05	333
144	53 11	027	27 17	094	48 29	168	57 44	230	26 17	245	48 44	287	25 49	334
145	53 27	027	27 53	095	48 36	170	57 16	232	25 44	246	48 10	288	25 34	334
146	53 43	027	28 29	096	48 43	171	56 47	233	25 11	247	47 35	289	25 18	335
147	54 00	027	29 05	096	48 48	173	56 18	234	24 38	248	47 01	289	25 03	335
148	54 16	027	29 41	097	48 52	174	55 48	236	24 04	249	46 27	290	24 48	336
149	54 32	027	30 17	098	48 55	175	55 18	237	23 30	250	45 53	290	24 34	337
	♦VEGA		ARCTURUS		SPICA		♦REGULUS		POLLUX		♦CAPELLA		Schedar	
150	11 45	038	30 52	099	12 27	128	48 58	177	54 48	238	45 19	291	24 19	337
151	12 08	039	31 28	100	12 55	129	48 59	178	54 17	240	44 46	292	24 05	338
152	12 31	040	32 04	101	13 23	130	49 00	180	53 45	241	44 12	292	23 52	338
153	12 54	040	32 39	102	13 50	131	48 59	181	53 14	242	43 39	293	23 39	339
154	13 17	041	33 14	102	14 17	132	48 58	183	52 42	243	43 06	293	23 26	339
155	13 41	042	33 50	103	14 44	133	48 55	184	52 09	244	42 33	294	23 13	340
156	14 05	042	34 25	104	15 10	134	48 52	186	51 37	245	42 00	294	23 01	340
157	14 30	043	34 59	105	15 36	135	48 48	187	51 04	246	41 27	295	22 49	341
158	14 54	044	35 34	106	16 01	136	48 43	189	50 31	248	40 54	296	22 37	342
159	15 19	044	36 09	107	16 26	137	48 37	190	49 57	249	40 22	296	22 26	342
160	15 45	045	36 43	108	16 51	137	48 30	192	49 23	250	39 49	297	22 15	343
161	16 10	046	37 18	109	17 15	138	48 22	193	48 49	251	39 17	297	22 04	343
162	16 36	046	37 52	110	17 39	139	48 13	195	48 15	252	38 45	298	21 54	344
163	17 03	047	38 26	111	18 02	140	48 04	196	47 41	253	38 13	298	21 44	344
164	17 29	048	38 59	112	18 25	141	47 53	198	47 06	254	37 42	299	21 35	345
	DENEB		♦VEGA		ARCTURUS		♦SPICA		REGULUS		POLLUX		♦CAPELLA	
165	12 39	024	17 56	048	39 33	113	18 47	142	47 42	199	46 31	255	37 10	300
166	12 54	025	18 23	049	40 06	114	19 09	143	47 30	200	45 57	256	36 39	300
167	13 09	026	18 51	050	40 39	115	19 31	144	47 17	202	45 21	257	36 08	301
168	13 25	026	19 18	050	41 11	116	19 51	145	47 03	203	44 46	258	35 37	301
169	13 41	027	19 46	051	41 44	117	20 12	146	46 48	205	44 11	258	35 06	302
170	13 58	028	20 14	052	42 16	118	20 32	147	46 33	206	43 36	259	34 35	302
171	14 15	028	20 43	052	42 48	119	20 51	148	46 16	207	43 00	260	34 05	303
172	14 32	029	21 12	053	43 19	120	21 10	149	45 59	209	42 24	261	33 35	304
173	14 50	030	21 41	054	43 50	121	21 28	150	45 42	210	41 49	262	33 05	304
174	15 08	030	22 10	054	44 21	122	21 46	151	45 23	211	41 13	263	32 35	305
175	15 26	031	22 39	055	44 51	123	22 03	152	45 04	213	40 37	264	32 06	305
176	15 45	032	23 09	056	45 21	124	22 20	153	44 44	214	40 01	265	31 36	306
177	16 04	032	23 39	056	45 51	126	22 36	154	44 24	215	39 25	265	31 07	307
178	16 23	033	24 09	057	46 20	127	22 51	155	44 02	217	38 49	266	30 38	307
179	16 43	034	24 39	058	46 49	128	23 06	156	43 41	218	38 13	267	30 10	308

 LAT 53°N

LHA ♈	Hc	Zn	Hc	Zn	Hc	Zn	Hc	Zn	Hc	Zn	Hc	Zn	Hc	Zn
	DENEB		◆VEGA		ARCTURUS		◆SPICA		REGULUS		POLLUX		◆CAPELLA	
180	17 03	034	25 10	058	47 17	129	23 20	157	43 18	219	37 37	268	29 41	308
181	17 24	035	25 41	059	47 45	130	23 34	158	42 55	220	37 01	269	29 13	309
182	17 44	035	26 12	060	48 12	132	23 47	159	42 31	222	36 25	270	28 45	309
183	18 05	036	26 43	060	48 39	133	23 59	160	42 07	223	35 49	270	28 17	310
184	18 27	037	27 14	061	49 05	134	24 11	161	41 42	224	35 13	271	27 50	311
185	18 49	037	27 46	061	49 30	135	24 23	162	41 17	225	34 36	272	27 22	311
186	19 11	038	28 18	062	49 55	137	24 33	164	40 51	226	34 00	273	26 55	312
187	19 33	039	28 50	063	50 20	138	24 43	165	40 25	227	33 24	273	26 28	312
188	19 56	039	29 22	063	50 44	139	24 52	166	39 58	229	32 48	274	26 02	313
189	20 19	040	29 55	064	51 07	141	25 01	167	39 31	230	32 12	275	25 36	314
190	20 42	041	30 27	065	51 29	142	25 09	168	39 03	231	31 36	276	25 10	314
191	21 06	041	31 00	065	51 51	144	25 16	169	38 35	232	31 00	277	24 44	315
192	21 30	042	31 33	066	52 12	145	25 23	170	38 06	233	30 25	277	24 18	315
193	21 54	042	32 06	067	52 33	146	25 29	171	37 37	234	29 49	278	23 53	316
194	22 18	043	32 39	067	52 52	148	25 34	172	37 08	235	29 13	279	23 28	317
	◆DENEB		VEGA		Rasalhague		◆ARCTURUS		REGULUS		◆POLLUX		CAPELLA	
195	22 43	044	33 12	068	22 47	100	53 11	149	36 38	236	28 37	280	23 03	317
196	23 08	044	33 46	069	23 23	100	53 29	151	36 08	237	28 02	280	22 39	318
197	23 33	045	34 20	069	23 58	101	53 46	152	35 37	238	27 26	281	22 15	318
198	23 59	045	34 54	070	24 34	102	54 02	154	35 06	239	26 51	282	21 51	319
199	24 25	046	35 28	071	25 09	103	54 18	155	34 35	240	26 16	282	21 28	320
200	24 51	047	36 02	071	25 44	104	54 32	157	34 04	241	25 40	283	21 04	320
201	25 17	047	36 36	072	26 19	105	54 46	159	33 32	242	25 05	284	20 42	321
202	25 44	048	37 10	073	26 54	106	54 59	160	33 00	243	24 30	285	20 19	322
203	26 11	048	37 45	073	27 29	106	55 10	162	32 27	244	23 56	285	19 57	322
204	26 38	049	38 19	074	28 03	107	55 21	163	31 55	245	23 21	286	19 35	323
205	27 06	050	38 54	075	28 38	108	55 31	165	31 22	246	22 46	287	19 13	323
206	27 33	050	39 29	075	29 12	109	55 40	167	30 49	247	22 12	288	18 52	324
207	28 01	051	40 04	076	29 46	110	55 47	168	30 15	248	21 37	288	18 31	325
208	28 29	051	40 39	077	30 20	111	55 54	170	29 42	249	21 03	289	18 10	325
209	28 58	052	41 14	077	30 53	112	56 00	172	29 08	250	20 29	290	17 50	326
	◆DENEB		VEGA		Rasalhague		◆ARCTURUS		REGULUS		◆POLLUX		CAPELLA	
210	29 26	053	41 49	078	31 27	113	56 05	173	28 34	251	19 55	290	17 29	327
211	29 55	053	42 25	079	32 00	114	56 08	175	28 00	252	19 21	291	17 10	327
212	30 24	054	43 00	079	32 33	115	56 11	177	27 25	253	18 48	292	16 50	328
213	30 53	054	43 36	080	33 05	116	56 12	179	26 51	253	18 14	293	16 31	329
214	31 23	055	44 11	081	33 38	117	56 12	180	26 16	254	17 41	293	16 13	329
215	31 52	056	44 47	081	34 10	118	56 12	182	25 41	255	17 08	294	15 54	330
216	32 22	056	45 23	082	34 42	119	56 10	184	25 06	256	16 35	295	15 36	330
217	32 52	057	45 59	083	35 13	120	56 07	185	24 31	257	16 03	296	15 19	331
218	33 23	057	46 34	084	35 45	121	56 03	187	23 56	258	15 30	296	15 02	332
219	33 53	058	47 10	084	36 15	122	55 58	189	23 20	259	14 58	297	14 45	332
220	34 24	059	47 46	085	36 46	123	55 52	190	22 45	260	14 26	298	14 28	333
221	34 55	059	48 22	086	37 16	124	55 45	192	22 09	260	13 54	298	14 12	334
222	35 26	060	48 58	087	37 46	125	55 37	194	21 34	261	13 22	299	13 56	334
223	35 57	060	49 34	087	38 16	126	55 28	195	20 58	262	12 51	300	13 41	335
224	36 29	061	50 10	088	38 45	127	55 18	197	20 22	263	12 20	301	13 26	336
	Schedar		◆DENEB		VEGA		Rasalhague		◆ARCTURUS		Denebola		◆Dubhe	
225	23 12	020	37 00	061	50 47	089	39 13	128	55 07	199	36 20	243	58 10	310
226	23 24	021	37 32	062	51 23	090	39 41	129	54 55	200	35 48	244	57 43	310
227	23 37	021	38 04	063	51 59	090	40 09	130	54 42	202	35 15	245	57 15	310
228	23 50	022	38 36	063	52 35	091	40 37	131	54 28	203	34 43	246	56 47	310
229	24 04	022	39 09	064	53 11	092	41 03	133	54 14	205	34 09	247	56 20	310
230	24 18	023	39 41	064	53 47	093	41 30	134	53 58	207	33 36	248	55 52	311
231	24 32	023	40 14	065	54 23	094	41 56	135	53 41	208	33 03	249	55 25	311
232	24 46	024	40 46	066	54 59	095	42 21	136	53 24	210	32 29	250	54 58	311
233	25 01	025	41 19	066	55 35	095	42 46	137	53 06	211	31 55	251	54 31	311
234	25 16	025	41 53	067	56 11	096	43 10	139	52 47	213	31 21	251	54 04	312
235	25 32	026	42 26	067	56 47	097	43 33	140	52 27	214	30 46	252	53 37	312
236	25 48	026	42 59	068	57 23	098	43 56	141	52 06	215	30 12	253	53 10	312
237	26 04	027	43 33	068	57 58	099	44 19	142	51 45	217	29 37	254	52 43	312
238	26 20	027	44 06	069	58 34	100	44 41	144	51 23	218	29 02	255	52 16	313
239	26 37	028	44 40	070	59 09	101	45 02	145	51 00	220	28 28	256	51 50	313
	◆Schedar		DENEB		VEGA		◆Rasalhague		ARCTURUS		◆Denebola		Dubhe	
240	26 54	028	45 14	070	59 45	102	45 22	146	50 37	221	27 52	257	51 24	313
241	27 11	029	45 48	071	60 20	103	45 42	147	50 13	222	27 17	258	50 57	313
242	27 28	029	46 22	071	60 55	104	46 01	149	49 48	224	26 42	259	50 31	314
243	27 46	030	46 57	072	61 30	105	46 20	150	49 23	225	26 06	259	50 05	314
244	28 04	030	47 31	073	62 05	106	46 37	151	48 57	226	25 31	260	49 39	314
245	28 23	031	48 05	073	62 39	107	46 54	153	48 31	228	24 55	261	49 14	315
246	28 41	031	48 40	074	63 14	109	47 10	154	48 04	229	24 20	262	48 48	315
247	29 00	032	49 15	074	63 48	110	47 26	156	47 37	230	23 44	263	48 22	315
248	29 20	032	49 50	075	64 22	111	47 40	157	47 09	231	23 08	264	47 57	316
249	29 39	033	50 25	076	64 55	112	47 54	158	46 41	232	22 32	264	47 32	316
250	29 59	033	51 00	076	65 28	114	48 07	160	46 12	234	21 56	265	47 07	316
251	30 19	034	51 35	077	66 01	115	48 19	161	45 42	235	21 20	266	46 42	317
252	30 39	034	52 10	078	66 34	116	48 30	163	45 13	236	20 44	267	46 17	317
253	31 00	035	52 45	078	67 06	118	48 40	164	44 43	237	20 08	268	45 53	317
254	31 21	035	53 21	079	67 38	119	48 50	166	44 12	238	19 32	268	45 28	318
	◆Schedar		DENEB		ALTAIR		◆Rasalhague		ARCTURUS		◆Alkaid		Kochab	
255	31 42	036	53 58	079	34 05	126	48 58	167	43 41	239	60 09	283	65 06	340
256	32 03	036	54 32	080	34 34	127	49 06	169	43 10	240	59 34	283	64 53	339
257	32 25	037	55 07	081	35 03	128	49 13	170	42 39	241	58 59	284	64 41	339
258	32 47	037	55 43	081	35 31	129	49 19	171	42 07	242	58 24	284	64 27	339
259	33 09	038	56 19	082	35 59	130	49 23	173	41 35	243	57 49	285	64 14	338
260	33 31	038	56 54	083	36 26	131	49 27	174	41 02	245	57 14	285	64 01	338
261	33 54	039	57 30	083	36 53	133	49 30	176	40 29	246	56 39	286	63 47	337
262	34 16	039	58 06	084	37 19	134	49 32	177	39 56	247	56 05	287	63 33	337
263	34 39	040	58 42	085	37 45	135	49 34	179	39 23	248	55 30	287	63 19	337
264	35 03	040	59 18	086	38 10	136	49 34	180	38 50	249	54 56	288	63 05	337
265	35 26	041	59 54	086	38 35	137	49 33	182	38 16	249	54 21	288	62 50	336
266	35 50	041	60 30	087	39 00	138	49 31	183	37 42	250	53 47	289	62 36	336
267	36 14	042	61 06	088	39 23	139	49 29	185	37 08	251	53 13	289	62 21	336
268	36 38	042	61 42	089	39 47	141	49 25	186	36 34	252	52 39	290	62 06	336
269	37 02	043	62 18	089	40 09	142	49 20	188	35 59	253	52 05	290	61 51	335

LHA ♈	Hc	Zn	Hc	Zn	Hc	Zn	Hc	Zn	Hc	Zn	Hc	Zn	Hc	Zn
	◆Mirfak		Alpheratz		◆ALTAIR		Rasalhague		◆ARCTURUS		Alkaid		Kochab	
270	17 59	025	21 40	070	40 31	143	49 15	189	35 24	254	51 31	291	61 36	335
271	18 15	026	22 14	071	40 53	144	49 09	191	34 50	255	50 58	291	61 20	335
272	18 31	027	22 49	071	41 14	145	49 01	192	34 15	256	50 24	292	61 05	335
273	18 47	027	23 23	072	41 34	147	48 53	194	33 40	257	49 51	292	60 50	334
274	19 04	028	23 57	073	41 53	148	48 44	195	33 04	258	49 17	293	60 34	334
275	19 21	028	24 32	074	42 12	149	48 34	197	32 29	259	48 44	293	60 18	334
276	19 38	029	25 07	074	42 30	150	48 23	198	31 54	260	48 11	294	60 03	334
277	19 56	030	25 41	075	42 48	152	48 11	200	31 18	260	47 38	295	59 47	334
278	20 14	030	26 16	076	43 05	153	47 59	201	30 42	261	47 05	295	59 31	334
279	20 32	031	26 51	077	43 21	154	47 45	203	30 07	262	46 33	296	59 15	334
280	20 51	031	27 27	077	43 36	156	47 31	204	29 31	263	46 00	296	58 59	334
281	21 10	032	28 02	078	43 51	157	47 16	205	28 55	264	45 28	297	58 43	333
282	21 29	033	28 37	079	44 04	158	47 00	207	28 19	265	44 56	297	58 26	333
283	21 48	033	29 13	079	44 17	160	46 43	208	27 43	265	44 24	298	58 10	333
284	22 08	034	29 48	080	44 30	161	46 26	209	27 07	266	43 52	298	57 54	333
	◆Mirfak		Alpheratz		◆ALTAIR		Rasalhague		◆ARCTURUS		Alkaid		Kochab	
285	22 29	034	30 24	081	44 41	162	46 08	211	26 31	267	43 20	299	57 38	333
286	22 49	035	31 00	082	44 52	164	45 49	212	25 55	268	42 48	299	57 21	333
287	23 10	036	31 35	082	45 01	165	45 29	213	25 19	269	42 17	300	57 05	333
288	23 31	036	32 11	083	45 10	166	45 09	215	24 43	269	41 46	300	56 49	333
289	23 53	037	32 47	084	45 18	168	44 48	216	24 07	270	41 15	301	56 32	333
290	24 14	037	33 23	085	45 25	169	44 27	217	23 30	271	40 44	301	56 16	333
291	24 36	038	33 59	085	45 32	171	44 04	219	22 54	272	40 13	302	56 00	333
292	24 59	038	34 35	086	45 37	172	43 42	220	22 18	273	39 42	302	55 43	333
293	25 21	039	35 11	087	45 42	173	43 18	221	21 42	273	39 12	303	55 27	333
294	25 44	040	35 47	088	45 45	175	42 54	222	21 06	274	38 42	304	55 11	333
295	26 07	040	36 23	089	45 48	176	42 30	223	20 30	275	38 12	304	54 54	333
296	26 31	041	36 59	089	45 50	178	42 04	225	19 54	276	37 42	305	54 38	333
297	26 55	041	37 35	090	45 51	179	41 39	226	19 18	277	37 12	305	54 22	333
298	27 19	042	38 11	091	45 51	181	41 13	227	18 43	277	36 43	306	54 05	333
299	27 43	042	38 48	092	45 50	182	40 46	228	18 07	278	36 14	306	53 49	333
	CAPELLA		◆Alpheratz		Enif		ALTAIR		◆VEGA		Alphecca		◆Alioth	
300	14 59	028	39 24	093	42 02	144	45 49	183	69 51	233	35 04	269	34 29	319
301	15 16	029	40 00	093	42 23	146	45 46	185	69 21	235	34 28	270	34 05	320
302	15 34	029	40 36	094	42 43	147	45 43	186	68 51	237	33 52	271	33 42	320
303	15 52	030	41 12	095	43 02	148	45 38	188	68 21	239	33 16	271	33 19	321
304	16 10	031	41 48	096	43 21	150	45 33	189	67 50	240	32 40	272	32 57	321
305	16 28	031	42 24	097	43 39	151	45 27	190	67 18	242	32 04	273	32 34	322
306	16 47	032	42 59	098	43 56	152	45 20	192	66 46	243	31 28	274	32 12	322
307	17 07	033	43 35	099	44 13	154	45 12	193	66 14	244	30 52	274	31 50	323
308	17 26	033	44 11	099	44 29	155	45 04	195	65 41	246	30 16	275	31 29	323
309	17 46	034	44 46	100	44 44	156	44 54	196	65 08	247	29 40	276	31 07	324
310	18 07	035	45 22	101	44 58	157	44 44	197	64 35	248	29 04	277	30 46	324
311	18 27	035	45 57	102	45 11	159	44 33	199	64 01	250	28 28	278	30 25	325
312	18 48	036	46 32	103	45 24	160	44 21	200	63 27	251	27 52	278	30 04	325
313	19 10	036	47 08	104	45 36	162	44 08	201	62 53	252	27 17	279	29 44	326
314	19 31	037	47 42	105	45 47	163	43 54	203	62 18	253	26 41	280	29 24	326
	CAPELLA		◆Hamal		Alpheratz		Enif		◆ALTAIR		VEGA		◆Alioth	
315	19 53	038	26 23	085	48 17	106	45 57	164	43 40	204	61 43	254	29 04	327
316	20 15	038	26 59	086	48 52	107	46 06	166	43 25	205	61 09	255	28 45	327
317	20 38	039	27 35	087	49 26	108	46 15	167	43 09	207	60 34	256	28 25	328
318	21 01	040	28 12	088	50 01	109	46 22	169	42 53	208	59 58	258	28 06	329
319	21 24	040	28 48	089	50 35	110	46 29	170	42 35	209	59 23	259	27 48	329
320	21 47	041	29 24	089	51 08	111	46 35	171	42 17	211	58 48	260	27 29	330
321	22 11	041	30 00	090	51 42	112	46 40	173	41 59	212	58 12	260	27 11	330
322	22 35	042	30 36	091	52 15	113	46 44	174	41 39	213	57 36	261	26 53	331
323	23 00	043	31 12	092	52 48	114	46 47	176	41 19	214	57 01	262	26 36	331
324	23 24	043	31 48	093	53 21	116	46 49	177	40 58	216	56 25	263	26 18	332
325	23 49	044	32 24	093	53 53	117	46 51	179	40 37	217	55 49	264	26 01	332
326	24 14	045	33 00	094	54 25	118	46 51	180	40 15	218	55 13	265	25 45	333
327	24 40	045	33 36	095	54 57	119	46 51	181	39 53	219	54 37	266	25 28	333
328	25 05	046	34 12	096	55 29	120	46 49	183	39 30	220	54 01	267	25 12	334
329	25 31	046	34 48	097	56 00	122	46 47	184	39 06	222	53 25	268	24 56	334
	CAPELLA		◆Hamal		Alpheratz		Enif		◆ALTAIR		VEGA		◆Alioth	
330	25 58	047	35 24	098	56 30	123	46 44	186	38 42	223	52 49	268	24 41	335
331	26 24	048	36 00	098	57 00	124	46 40	187	38 17	224	52 13	269	24 26	335
332	26 51	048	36 35	099	57 30	125	46 35	189	37 52	225	51 37	270	24 11	336
333	27 18	049	37 11	100	57 59	127	46 29	190	37 26	226	51 00	271	23 57	337
334	27 45	049	37 46	101	58 28	128	46 22	191	37 00	227	50 24	272	23 42	337
335	28 13	050	38 22	102	58 56	130	46 15	193	36 33	228	49 48	272	23 29	338
336	28 40	050	38 57	103	59 23	131	46 06	194	36 06	229	49 12	273	23 15	338
337	29 08	051	39 32	104	59 50	133	45 57	196	35 38	230	48 36	274	23 02	339
338	29 37	052	40 07	105	60 16	134	45 46	197	35 10	232	48 00	275	22 49	339
339	30 05	052	40 42	106	60 42	136	45 35	198	34 42	233	47 24	275	22 36	340
340	30 34	053	41 17	107	61 07	137	45 24	200	34 13	234	46 48	276	22 24	341
341	31 02	053	41 51	108	61 31	139	45 11	201	33 44	235	46 12	277	22 12	341
342	31 32	054	42 26	109	61 54	141	44 57	203	33 14	236	45 37	278	22 01	342
343	32 01	055	43 00	109	62 17	142	44 43	204	32 44	237	45 01	278	21 50	342
344	32 30	055	43 34	110	62 39	144	44 28	205	32 14	238	44 25	279	21 39	343
	◆CAPELLA		ALDEBARAN		◆Diphda		Enif		ALTAIR		◆VEGA		Alioth	
345	33 00	056	16 44	085	15 33	155	44 12	207	31 43	239	43 50	280	21 28	343
346	33 30	056	17 20	085	15 48	155	43 56	208	31 12	240	43 14	280	21 18	344
347	34 00	057	17 56	086	16 03	156	43 39	209	30 41	241	42 38	281	21 08	345
348	34 31	057	18 33	087	16 17	157	43 21	211	30 09	242	42 03	282	20 59	345
349	35 01	058	19 09	088	16 31	158	43 02	212	29 37	243	41 28	282	20 50	346
350	35 32	059	19 45	089	16 44	159	42 42	213	29 05	244	40 53	283	20 41	346
351	36 03	059	20 21	089	16 56	160	42 22	214	28 32	244	40 17	284	20 32	347
352	36 34	060	20 57	090	17 08	161	42 02	216	28 00	245	39 42	285	20 24	347
353	37 05	060	21 33	091	17 20	162	41 40	217	27 27	246	39 08	285	20 17	348
354	37 37	061	22 09	092	17 30	163	41 18	218	26 54	247	38 33	286	20 09	349
355	38 08	061	22 45	093	17 41	164	40 56	219	26 20	248	37 58	287	20 02	349
356	38 40	062	23 21	093	17 50	165	40 33	220	25 46	249	37 24	287	19 56	350
357	39 12	063	23 57	094	17 59	166	40 09	222	25 13	250	36 49	288	19 49	350
358	39 44	063	24 33	095	18 07	167	39 45	223	24 39	251	36 15	289	19 44	351
359	40 16	064	25 09	096	18 15	168	39 20	224	24 04	252	35 41	289	19 38	351

LAT 52°N

LHA	Hc	Zn	Hc	Zn	Hc	Zn	Hc	Zn	Hc	Zn	Hc	Zn	Hc	Zn
♈	◆CAPELLA		ALDEBARAN		Hamal		◆Alpheratz		ALTAIR		◆VEGA		Alioth	
0	40 23	064	25 52	096	52 33	127	67 01	175	23 48	253	34 46	290	18 33	352
1	40 56	064	26 29	097	53 02	129	67 03	178	23 12	254	34 11	291	18 29	353
2	41 29	065	27 05	098	53 31	130	67 04	180	22 37	255	33 37	292	18 24	353
3	42 03	065	27 42	099	53 59	131	67 03	182	22 01	256	33 03	292	18 20	354
4	42 36	066	28 18	100	54 26	133	67 01	184	21 25	256	32 29	293	18 16	354
5	43 10	066	28 55	100	54 53	134	66 57	187	20 49	257	31 55	294	18 13	355
6	43 44	067	29 31	101	55 19	136	66 52	189	20 13	258	31 21	294	18 10	356
7	44 18	067	30 07	102	55 45	137	66 46	191	19 37	259	30 47	295	18 07	356
8	44 52	068	30 43	103	56 09	138	66 38	193	19 01	260	30 14	296	18 05	357
9	45 26	069	31 19	104	56 34	140	66 29	195	18 24	261	29 41	296	18 03	357
10	46 01	069	31 55	105	56 57	141	66 19	198	17 48	261	29 08	297	18 01	358
11	46 35	070	32 30	106	57 20	143	66 07	200	17 11	262	28 35	297	18 00	359
12	47 10	070	33 06	106	57 41	145	65 54	202	16 34	263	28 02	298	18 00	359
13	47 45	071	33 41	107	58 02	146	65 39	204	15 58	264	27 30	299	17 59	000
14	48 20	071	34 16	108	58 23	148	65 24	206	15 21	265	26 57	299	17 59	000
	Dubhe		◆CAPELLA		ALDEBARAN		Hamal		◆Alpheratz		DENEB		◆VEGA	
15	26 06	015	48 55	072	34 51	109	58 42	149	65 07	208	48 10	287	26 25	300
16	26 16	015	49 30	073	35 26	110	59 00	151	64 50	210	47 35	288	25 53	301
17	26 26	016	50 05	073	36 01	111	59 17	153	64 31	212	47 00	289	25 22	301
18	26 36	016	50 41	074	36 35	112	59 34	155	64 11	214	46 25	289	24 50	302
19	26 47	017	51 16	074	37 09	113	59 49	156	63 50	215	45 50	290	24 19	303
20	26 58	017	51 52	075	37 43	114	60 03	158	63 28	217	45 15	290	23 48	303
21	27 09	018	52 28	075	38 17	115	60 17	160	63 05	219	44 41	291	23 17	304
22	27 20	018	53 03	076	38 50	116	60 29	162	62 42	221	44 06	291	22 47	304
23	27 32	019	53 39	077	39 23	117	60 40	164	62 17	222	43 32	292	22 16	305
24	27 44	019	54 15	077	39 56	118	60 50	165	61 52	224	42 58	293	21 46	306
25	27 56	020	54 51	078	40 28	119	60 59	167	61 26	226	42 24	293	21 16	306
26	28 09	020	55 28	078	41 01	120	61 06	169	60 59	227	41 50	294	20 47	307
27	28 22	021	56 04	079	41 32	121	61 13	171	60 32	229	41 16	294	20 17	308
28	28 35	021	56 40	080	42 04	122	61 18	173	60 04	230	40 42	295	19 48	308
29	28 49	022	57 16	080	42 35	123	61 22	175	59 35	232	40 09	295	19 19	309
	Dubhe		◆POLLUX		BETELGEUSE		RIGEL		◆Hamal		Alpheratz		◆DENEB	
30	29 02	022	23 57	075	24 45	111	16 54	129	61 25	177	59 06	233	39 36	296
31	29 16	023	24 33	075	25 19	112	17 22	130	61 26	179	58 36	234	39 03	297
32	29 31	023	25 08	076	25 53	113	17 51	131	61 26	181	58 06	236	38 30	297
33	29 45	024	25 44	077	26 27	114	18 18	132	61 25	182	57 35	237	37 57	298
34	30 00	024	26 20	077	27 01	115	18 46	133	61 23	184	57 04	238	37 24	298
35	30 15	024	26 56	078	27 34	116	19 12	134	61 20	186	56 32	240	36 52	299
36	30 31	025	27 33	079	28 07	117	19 39	135	61 15	188	56 00	241	36 19	299
37	30 47	025	28 09	080	28 40	117	20 05	136	61 09	190	55 28	242	35 47	300
38	31 03	026	28 45	080	29 13	118	20 31	137	61 02	192	54 55	243	35 15	301
39	31 19	026	29 22	081	29 45	119	20 56	138	60 54	194	54 22	244	34 44	301
40	31 35	027	29 58	082	30 17	120	21 20	139	60 44	196	53 48	246	34 12	302
41	31 52	027	30 35	083	30 49	121	21 45	139	60 34	198	53 15	247	33 41	302
42	32 09	028	31 11	083	31 20	122	22 08	140	60 22	199	52 41	248	33 10	303
43	32 26	028	31 48	084	31 51	123	22 32	141	60 09	201	52 06	249	32 39	303
44	32 44	029	32 25	085	32 22	124	22 54	142	59 56	203	51 32	250	32 08	304
	Dubhe		◆POLLUX		BETELGEUSE		RIGEL		◆Hamal		Alpheratz		◆DENEB	
45	33 02	029	33 02	085	32 52	125	23 17	143	59 41	205	50 57	251	31 38	305
46	33 20	029	33 39	086	33 22	126	23 38	144	59 25	206	50 22	252	31 07	305
47	33 38	030	34 15	087	33 52	127	24 00	145	59 08	208	49 47	253	30 37	306
48	33 56	030	34 52	088	34 21	128	24 20	146	58 50	210	49 11	254	30 07	306
49	34 15	031	35 29	089	34 50	129	24 41	147	58 31	211	48 36	255	29 37	307
50	34 34	031	36 06	089	35 18	131	25 00	149	58 11	213	48 00	256	29 08	307
51	34 53	032	36 43	090	35 46	132	25 19	150	57 51	215	47 24	257	28 39	308
52	35 13	032	37 20	091	36 13	133	25 38	151	57 29	216	46 48	258	28 10	309
53	35 32	032	37 57	092	36 40	134	25 55	152	57 07	218	46 12	259	27 41	309
54	35 52	033	38 34	093	37 07	135	26 13	153	56 44	219	45 35	260	27 13	310
55	36 12	033	39 11	093	37 33	136	26 29	154	56 20	221	44 59	261	26 44	310
56	36 33	034	39 48	094	37 58	137	26 45	155	55 56	222	44 22	261	26 16	311
57	36 53	034	40 25	095	38 23	138	27 01	156	55 31	224	43 46	262	25 48	312
58	37 14	034	41 01	096	38 47	139	27 15	157	55 05	225	43 09	263	25 21	312
59	37 35	035	41 38	097	39 11	141	27 30	158	54 38	227	42 33	264	24 54	313
	◆Dubhe		POLLUX		SIRIUS		◆RIGEL		Hamal		◆Alpheratz		DENEB	
60	37 56	035	42 15	098	12 32	140	27 43	159	54 11	228	41 56	265	24 27	313
61	38 18	036	42 51	098	12 55	141	27 56	160	53 43	229	41 19	266	24 00	314
62	38 39	036	43 28	099	13 18	142	28 08	161	53 15	231	40 42	267	23 33	314
63	39 01	036	44 04	100	13 41	142	28 20	162	52 46	232	40 05	267	23 07	315
64	39 23	037	44 40	101	14 04	143	28 30	164	52 17	233	39 28	268	22 41	316
65	39 45	037	45 17	102	14 25	144	28 41	165	51 47	234	38 51	269	22 15	316
66	40 08	038	45 53	103	14 47	145	28 50	166	51 17	236	38 14	270	21 50	317
67	40 30	038	46 29	104	15 08	146	28 59	167	50 46	237	37 38	271	21 25	318
68	40 53	038	47 04	105	15 28	147	29 07	168	50 15	238	37 01	271	21 00	318
69	41 16	039	47 40	106	15 48	148	29 14	169	49 43	239	36 24	272	20 36	319
70	41 39	039	48 16	107	16 07	149	29 21	170	49 11	240	35 47	273	20 11	319
71	42 03	039	48 51	108	16 26	150	29 27	171	48 39	242	35 10	274	19 48	320
72	42 26	040	49 26	109	16 44	151	29 32	173	48 06	243	34 33	274	19 24	321
73	42 50	040	50 01	110	17 02	152	29 36	174	47 33	244	33 56	275	19 01	321
74	43 14	040	50 35	111	17 19	153	29 40	175	47 00	245	33 19	276	18 38	322
	Dubhe		◆REGULUS		PROCYON		SIRIUS		◆RIGEL		Hamal		◆DENEB	
75	43 38	041	17 24	093	32 54	131	17 36	154	29 43	176	46 27	246	18 15	322
76	44 02	041	18 00	094	33 22	132	17 52	155	29 45	177	45 53	247	17 53	323
77	44 26	041	18 37	094	33 49	133	18 08	156	29 47	178	45 19	248	17 31	324
78	44 51	042	19 14	095	34 16	134	18 23	157	29 47	179	44 44	249	17 09	324
79	45 16	042	19 51	096	34 43	135	18 37	158	29 48	180	44 10	250	16 48	325
80	45 40	042	20 28	097	35 09	136	18 51	158	29 47	182	43 35	251	16 26	326
81	46 05	043	21 04	098	35 34	137	19 04	159	29 45	183	43 00	252	16 06	326
82	46 31	043	21 41	098	35 59	138	19 17	160	29 43	184	42 25	253	15 45	327
83	46 56	043	22 17	099	36 23	139	19 29	161	29 40	185	41 49	254	15 25	328
84	47 21	044	22 54	100	36 47	141	19 40	162	29 37	186	41 14	255	15 06	328
85	47 47	044	23 30	101	37 10	142	19 51	163	29 32	187	40 38	256	14 47	329
86	48 13	044	24 06	102	37 33	143	20 01	164	29 27	188	40 02	257	14 28	329
87	48 39	045	24 42	103	37 55	144	20 11	165	29 22	190	39 26	258	14 09	330
88	49 05	045	25 18	103	38 16	145	20 20	166	29 15	191	38 50	258	13 51	331
89	49 31	045	25 54	104	38 37	146	20 28	167	29 08	192	38 14	259	13 33	331

LHA	Hc	Zn	Hc	Zn	Hc	Zn	Hc	Zn	Hc	Zn	Hc	Zn	Hc	Zn
♈	◆Dubhe		REGULUS		PROCYON		◆SIRIUS		RIGEL		◆Hamal		Schedar	
90	49 57	045	26 30	105	38 57	148	20 36	169	29 00	193	37 37	260	45 46	309
91	50 23	046	27 06	106	39 17	149	20 43	170	28 51	194	37 01	261	45 17	309
92	50 50	046	27 41	107	39 35	150	20 49	171	28 42	195	36 24	262	44 49	310
93	51 16	046	28 16	108	39 54	151	20 55	172	28 32	196	35 48	263	44 20	310
94	51 43	047	28 51	109	40 11	152	21 00	173	28 21	197	35 11	264	43 52	310
95	52 10	047	29 26	110	40 28	154	21 05	174	28 10	199	34 34	264	43 24	311
96	52 37	047	30 01	110	40 44	155	21 08	175	27 58	200	33 58	265	42 56	311
97	53 04	047	30 36	111	40 59	156	21 12	176	27 45	201	33 21	266	42 29	312
98	53 31	047	31 10	112	41 13	158	21 14	177	27 32	202	32 44	267	42 01	312
99	53 59	048	31 44	113	41 27	159	21 16	178	27 18	203	32 07	268	41 34	312
100	54 26	048	32 18	114	41 40	160	21 17	179	27 03	204	31 30	268	41 07	313
101	54 53	048	32 51	115	41 52	161	21 17	180	26 48	205	30 53	269	40 40	313
102	55 21	048	33 25	116	42 04	163	21 17	181	26 32	206	30 16	270	40 13	314
103	55 48	048	33 58	117	42 14	164	21 16	182	26 15	207	29 39	271	39 46	314
104	56 16	049	34 30	118	42 24	165	21 15	183	25 58	208	29 02	272	39 20	315
	◆Dubhe		Denebola		REGULUS		◆SIRIUS		RIGEL		ALDEBARAN		◆Mirfak	
105	56 44	049	22 23	095	35 03	119	21 13	184	25 40	209	44 30	232	56 41	288
106	57 12	049	23 00	096	35 35	120	21 10	185	25 22	210	44 00	234	56 06	289
107	57 40	049	23 36	096	36 07	121	21 06	186	25 03	211	43 30	235	55 31	289
108	58 08	049	24 13	097	36 38	122	21 02	187	24 43	212	43 00	236	54 56	290
109	58 36	049	24 50	098	37 09	123	20 58	188	24 23	213	42 29	237	54 21	290
110	59 04	050	25 26	099	37 40	124	20 52	189	24 03	214	41 58	238	53 46	291
111	59 32	050	26 03	100	38 10	125	20 46	190	23 42	215	41 26	239	53 12	291
112	60 00	050	26 39	101	38 40	126	20 39	191	23 20	216	40 55	240	52 38	292
113	60 28	050	27 15	101	39 10	127	20 32	192	22 58	217	40 22	241	52 03	292
114	60 57	050	27 51	102	39 39	128	20 24	193	22 35	218	39 50	242	51 29	293
115	61 25	050	28 27	103	40 08	130	20 15	194	22 12	219	39 17	243	50 55	293
116	61 53	050	29 03	104	40 36	131	20 06	195	21 48	220	38 44	244	50 21	294
117	62 22	050	29 39	105	41 04	132	19 56	196	21 24	221	38 11	245	49 47	294
118	62 50	050	30 15	106	41 31	133	19 45	197	20 59	222	37 37	246	49 14	295
119	63 18	050	30 50	107	41 58	134	19 34	198	20 34	223	37 03	247	48 40	295
	Kochab		◆ARCTURUS		REGULUS		◆SIRIUS		BETELGEUSE		CAPELLA		◆Schedar	
120	46 10	023	12 43	075	42 24	135	19 23	199	38 34	221	62 52	274	32 46	322
121	46 24	023	13 19	076	42 50	137	19 10	200	38 09	222	62 15	274	32 24	322
122	46 38	023	13 55	076	43 15	138	18 57	201	37 44	223	61 39	275	32 01	323
123	46 53	023	14 31	077	43 39	139	18 44	202	37 18	225	61 02	276	31 39	323
124	47 07	023	15 07	078	44 03	140	18 29	203	36 52	226	60 25	276	31 17	324
125	47 22	024	15 43	079	44 27	141	18 15	204	36 26	227	59 48	277	30 55	324
126	47 37	024	16 19	080	44 49	143	17 59	205	35 58	228	59 12	278	30 34	325
127	47 52	024	16 56	080	45 11	144	17 44	206	35 31	229	58 35	278	30 13	325
128	48 07	024	17 32	081	45 33	145	17 27	207	35 03	230	57 59	279	29 52	326
129	48 22	024	18 08	082	45 53	147	17 10	208	34 34	231	57 22	280	29 31	326
130	48 37	024	18 45	083	46 13	148	16 53	209	34 05	232	56 46	280	29 11	327
131	48 52	024	19 22	083	46 32	149	16 35	210	33 36	233	56 10	281	28 51	327
132	49 08	025	19 58	084	46 51	151	16 16	211	33 06	234	55 33	281	28 31	328
133	49 23	025	20 35	085	47 09	152	15 57	212	32 36	235	54 57	282	28 12	328
134	49 39	025	21 12	086	47 25	153	15 37	213	32 05	236	54 21	283	27 52	329
	Kochab		◆ARCTURUS		REGULUS		◆POLLUX		BETELGEUSE		CAPELLA		◆Schedar	
135	49 54	025	21 49	086	47 42	155	62 15	218	31 34	237	53 45	283	27 33	329
136	50 10	025	22 26	087	47 57	156	61 52	219	31 03	238	53 09	284	27 15	330
137	50 26	025	23 03	088	48 11	158	61 28	221	30 32	239	52 33	284	26 56	330
138	50 41	025	23 40	089	48 25	159	61 04	223	30 00	240	51 58	285	26 38	331
139	50 57	025	24 17	089	48 38	161	60 38	224	29 27	241	51 22	286	26 20	331
140	51 13	026	24 53	090	48 50	162	60 12	226	28 55	242	50 47	286	26 03	332
141	51 29	026	25 30	091	49 01	163	59 46	227	28 22	243	50 11	287	25 45	332
142	51 45	026	26 07	092	49 11	165	59 18	229	27 49	244	49 36	287	25 28	333
143	52 01	026	26 44	093	49 20	166	58 50	230	27 16	245	49 01	288	25 12	333
144	52 17	026	27 21	093	49 28	168	58 21	232	26 42	246	48 26	289	24 56	334
145	52 33	026	27 58	094	49 35	169	57 52	233	26 08	247	47 51	289	24 39	335
146	52 50	026	28 35	095	49 42	171	57 22	234	25 34	248	47 16	290	24 24	335
147	53 06	026	29 12	096	49 47	172	56 52	236	25 00	249	46 41	290	24 08	336
148	53 22	026	29 48	097	49 52	174	56 21	237	24 26	249	46 06	291	23 53	336
149	53 39	026	30 25	098	49 55	175	55 50	238	23 51	250	45 32	291	23 38	337
	◆Kochab		ARCTURUS		◆SPICA		REGULUS		◆POLLUX		CAPELLA		Schedar	
150	53 55	026	31 02	098	13 04	128	49 57	177	55 19	239	44 58	292	23 24	337
151	54 11	026	31 38	099	13 33	129	49 59	178	54 47	241	44 23	292	23 10	338
152	54 27	026	32 14	100	14 01	130	50 00	180	54 14	242	43 49	293	22 56	338
153	54 44	026	32 51	101	14 29	131	49 59	181	53 41	243	43 15	294	22 43	339
154	55 00	026	33 27	102	14 57	132	49 58	183	53 08	244	42 42	294	22 29	339
155	55 17	026	34 03	103	15 24	133	49 55	185	52 35	245	42 08	295	22 17	340
156	55 33	026	34 39	104	15 51	134	49 52	186	52 01	247	41 34	295	22 04	341
157	55 49	026	35 15	104	16 18	135	49 47	188	51 27	248	41 01	296	21 52	341
158	56 06	026	35 51	105	16 44	135	49 42	189	50 53	249	40 28	296	21 40	342
159	56 22	026	36 26	106	17 10	136	49 36	191	50 18	250	39 55	297	21 29	342
160	56 38	026	37 01	107	17 35	137	49 29	192	49 44	251	39 22	297	21 18	343
161	56 54	026	37 37	108	18 00	138	49 20	194	49 09	252	38 49	298	21 07	343
162	57 11	026	38 12	109	18 24	139	49 11	195	48 33	253	38 17	299	20 56	344
163	57 27	026	38 46	110	18 48	140	49 01	196	47 58	254	37 44	299	20 46	344
164	57 43	026	39 21	111	19 12	141	48 50	198	47 22	255	37 12	300	20 37	345
	DENEB		◆VEGA		ARCTURUS		◆SPICA		REGULUS		POLLUX		◆CAPELLA	
165	11 44	024	17 16	048	39 55	112	19 35	142	48 38	199	46 47	256	36 40	300
166	12 00	025	17 44	049	40 30	113	19 57	143	48 26	201	46 11	257	36 08	301
167	12 15	025	18 12	049	41 03	114	20 19	144	48 12	202	45 35	258	35 37	301
168	12 31	026	18 40	050	41 37	115	20 41	145	47 58	204	44 59	258	35 05	302
169	12 48	027	19 08	051	42 10	116	21 02	146	47 42	205	44 23	259	34 34	302
170	13 05	027	19 37	051	42 43	117	21 22	147	47 26	207	43 46	260	34 03	303
171	13 22	028	20 06	052	43 16	118	21 42	148	47 09	208	43 10	261	33 32	304
172	13 39	029	20 35	053	43 49	119	22 01	149	46 52	209	42 33	262	33 01	304
173	13 57	029	21 05	053	44 21	120	22 20	150	46 33	211	41 57	263	32 31	305
174	14 16	030	21 35	054	44 52	121	22 38	151	46 14	212	41 20	264	32 01	305
175	14 35	031	22 05	055	45 24	122	22 56	152	45 54	213	40 43	265	31 31	306
176	14 54	031	22 35	055	45 55	124	23 13	153	45 34	215	40 06	265	31 01	306
177	15 13	032	23 05	056	46 25	125	23 30	154	45 12	216	39 29	266	30 31	307
178	15 33	033	23 36	057	46 55	126	23 46	155	44 50	217	38 53	267	30 02	308
179	15 53	033	24 07	057	47 25	127	24 01	156	44 28	218	38 16	268	29 33	308

LAT 52°N

LHA ♈	Hc	Zn	Hc	Zn	Hc	Zn	Hc	Zn	Hc	Zn	Hc	Zn	Hc	Zn
	DENEB		◆VEGA		ARCTURUS		◆SPICA		REGULUS		POLLUX		◆CAPELLA	
180	16 13	034	24 38	058	47 54	128	24 16	157	44 04	220	37 39	269	29 04	309
181	16 34	035	25 10	058	48 23	130	24 30	158	43 41	221	37 02	269	28 35	309
182	16 55	035	25 41	059	48 51	131	24 43	159	43 16	222	36 25	270	28 07	310
183	17 17	036	26 13	060	49 19	132	24 56	160	42 51	223	35 48	271	27 38	310
184	17 39	037	26 45	060	49 46	133	25 08	161	42 25	225	35 11	272	27 10	311
185	18 01	037	27 17	061	50 13	135	25 20	162	41 59	226	34 34	273	26 43	312
186	18 23	038	27 50	062	50 39	136	25 31	163	41 32	227	33 57	273	26 15	312
187	18 46	038	28 22	062	51 04	137	25 41	164	41 05	228	33 20	274	25 48	313
188	19 09	039	28 55	063	51 29	139	25 50	166	40 37	229	32 44	275	25 21	313
189	19 33	040	29 28	064	51 53	140	25 59	167	40 09	230	32 07	276	24 54	314
190	19 56	040	30 01	064	52 17	141	26 07	168	39 41	231	31 30	276	24 28	315
191	20 20	041	30 35	065	52 39	143	26 15	169	39 12	233	30 53	277	24 01	315
192	20 45	041	31 08	065	53 01	144	26 22	170	38 42	234	30 17	278	23 35	316
193	21 09	042	31 42	066	53 22	146	26 28	171	38 12	235	29 40	279	23 10	316
194	21 34	043	32 16	067	53 43	147	26 33	172	37 42	236	29 04	279	22 44	317
	◆DENEB		VEGA		Rasalhague		◆ARCTURUS		REGULUS		◆POLLUX		CAPELLA	
195	21 59	043	32 50	067	22 57	099	54 02	149	37 11	237	28 27	280	22 19	318
196	22 25	044	33 24	068	23 33	100	54 21	150	36 40	238	27 51	281	21 55	318
197	22 51	045	33 58	069	24 10	101	54 39	152	36 08	239	27 15	282	21 30	319
198	23 17	045	34 33	069	24 46	102	54 56	153	35 37	240	26 39	282	21 06	319
199	23 43	046	35 07	070	25 22	102	55 12	155	35 04	241	26 02	283	20 42	320
200	24 10	046	35 42	071	25 58	103	55 28	156	34 32	242	25 27	284	20 18	321
201	24 37	047	36 17	071	26 34	104	55 42	158	33 59	243	24 51	284	19 55	321
202	25 04	048	36 52	072	27 10	105	55 55	160	33 26	244	24 15	285	19 32	322
203	25 31	048	37 27	072	27 45	106	56 07	161	32 53	245	23 39	286	19 09	322
204	25 59	049	38 02	073	28 21	107	56 19	163	32 19	246	23 04	287	18 47	323
205	26 27	049	38 38	074	28 56	108	56 29	165	31 46	247	22 29	287	18 25	324
206	26 55	050	39 13	074	29 31	109	56 38	166	31 12	248	21 53	288	18 03	324
207	27 23	050	39 49	075	30 06	110	56 46	168	30 37	249	21 18	289	17 42	325
208	27 52	051	40 25	076	30 41	110	56 53	170	30 03	249	20 43	289	17 21	326
209	28 20	052	41 01	076	31 15	111	56 59	172	29 28	250	20 09	290	17 00	326
	◆DENEB		VEGA		Rasalhague		◆ARCTURUS		REGULUS		◆POLLUX		CAPELLA	
210	28 50	052	41 37	077	31 50	112	57 04	173	28 53	251	19 34	291	16 39	327
211	29 19	053	42 13	078	32 24	113	57 08	175	28 18	252	19 00	292	16 19	327
212	29 48	053	42 49	078	32 58	114	57 11	177	27 43	253	18 25	292	16 00	328
213	30 18	054	43 25	079	33 31	115	57 12	179	27 08	254	17 51	293	15 40	329
214	30 48	055	44 01	080	34 04	116	57 12	180	26 32	255	17 17	294	15 21	329
215	31 18	055	44 38	080	34 38	117	57 12	182	25 56	256	16 43	294	15 02	330
216	31 49	056	45 14	081	35 10	118	57 10	184	25 20	257	16 10	295	14 44	331
217	32 19	056	45 51	082	35 43	119	57 07	185	24 44	257	15 37	296	14 26	331
218	32 50	057	46 27	083	36 15	120	57 03	187	24 08	258	15 03	297	14 09	332
219	33 21	057	47 04	083	36 47	121	56 58	189	23 32	259	14 30	297	13 51	333
220	33 52	058	47 41	084	37 18	122	56 51	191	22 56	260	13 58	298	13 35	333
221	34 24	059	48 17	085	37 49	123	56 44	192	22 19	261	13 25	299	13 18	334
222	34 55	059	48 54	085	38 20	124	56 36	194	21 43	262	12 53	299	13 02	335
223	35 27	060	49 31	086	38 50	125	56 26	196	21 06	262	12 21	300	12 46	335
224	35 59	060	50 08	087	39 20	126	56 16	197	20 30	263	11 49	301	12 31	336
	Schedar		◆DENEB		VEGA		Rasalhague		◆ARCTURUS		Denebola		◆Dubhe	
225	22 15	020	36 31	061	50 45	088	39 50	127	56 04	199	36 47	244	57 32	311
226	22 28	021	37 04	061	51 22	088	40 19	129	55 51	201	36 14	245	57 04	311
227	22 41	021	37 36	062	51 59	089	40 48	130	55 38	202	35 41	245	56 36	311
228	22 55	022	38 09	062	52 36	090	41 16	131	55 23	204	35 07	246	56 08	311
229	23 08	022	38 42	063	53 12	091	41 44	132	55 08	206	34 33	247	55 41	312
230	23 22	023	39 15	064	53 49	092	42 11	133	54 51	207	33 59	248	55 13	312
231	23 37	023	39 48	064	54 26	092	42 38	134	54 34	209	33 24	249	54 45	312
232	23 52	024	40 21	065	55 03	093	43 04	135	54 16	210	32 49	250	54 18	312
233	24 07	024	40 55	065	55 40	094	43 29	137	53 57	212	32 15	251	53 51	312
234	24 22	025	41 28	066	56 17	095	43 55	138	53 37	213	31 40	252	53 23	313
235	24 38	025	42 02	066	56 54	096	44 19	139	53 16	215	31 04	253	52 56	313
236	24 54	026	42 36	067	57 30	097	44 43	140	52 55	216	30 29	254	52 29	313
237	25 10	026	43 10	068	58 07	098	45 06	142	52 33	218	29 53	255	52 02	313
238	25 27	027	43 44	068	58 44	098	45 29	143	52 10	219	29 18	256	51 35	314
239	25 44	028	44 19	069	59 20	099	45 51	144	51 46	220	28 42	256	51 09	314
	◆Schedar		DENEB		VEGA		◆Rasalhague		ARCTURUS		◆Denebola		Dubhe	
240	26 01	028	44 53	069	59 57	100	46 12	146	51 22	222	28 06	257	50 42	314
241	26 18	029	45 28	070	60 33	101	46 33	147	50 57	223	27 30	258	50 16	314
242	26 36	029	46 03	070	61 09	102	46 52	148	50 32	224	26 54	259	49 49	315
243	26 54	030	46 38	071	61 45	103	47 11	150	50 05	226	26 17	260	49 23	315
244	27 13	030	47 12	072	62 21	105	47 30	151	49 39	227	25 41	261	48 57	315
245	27 31	031	47 48	072	62 56	106	47 47	152	49 11	228	25 04	262	48 31	315
246	27 50	031	48 23	073	63 32	107	48 04	154	48 43	230	24 28	262	48 05	316
247	28 09	032	48 58	073	64 07	108	48 20	155	48 15	231	23 51	263	47 40	316
248	28 29	032	49 34	074	64 42	109	48 35	157	47 46	232	23 14	264	47 14	316
249	28 49	033	50 09	075	65 17	110	48 50	158	47 17	233	22 38	265	46 49	317
250	29 09	033	50 45	075	65 51	112	49 03	159	46 47	234	22 01	266	46 23	317
251	29 29	034	51 21	076	66 26	113	49 16	161	46 17	236	21 24	266	45 58	317
252	29 50	034	51 56	076	67 00	114	49 27	162	45 46	237	20 47	267	45 33	318
253	30 11	035	52 32	077	67 33	116	49 38	164	45 15	238	20 10	268	45 08	318
254	30 32	035	53 08	078	68 06	117	49 48	165	44 43	239	19 33	269	44 44	318
	◆Schedar		DENEB		ALTAIR		◆Rasalhague		ARCTURUS		◆Alkaid		Kochab	
255	30 53	036	53 44	078	34 40	126	49 57	167	44 12	240	59 55	284	64 10	340
256	31 15	036	54 21	079	35 10	127	50 05	168	43 39	241	59 20	285	63 57	340
257	31 37	037	54 57	079	35 39	128	50 12	170	43 07	242	58 44	285	63 44	340
258	31 59	037	55 33	080	36 09	129	50 18	171	42 34	243	58 08	286	63 31	339
259	32 21	038	56 10	081	36 37	130	50 23	173	42 01	244	57 33	286	63 18	339
260	32 44	038	56 46	081	37 05	131	50 27	174	41 28	245	56 57	287	63 05	339
261	33 07	039	57 23	082	37 33	132	50 30	176	40 54	246	56 22	287	62 51	338
262	33 30	039	57 59	083	38 00	133	50 32	177	40 20	247	55 47	288	62 38	338
263	33 53	039	58 36	083	38 27	134	50 34	179	39 46	248	55 12	289	62 24	338
264	34 17	040	59 13	084	38 53	135	50 34	180	39 11	249	54 37	289	62 09	337
265	34 41	040	59 49	085	39 19	136	50 33	182	38 37	250	54 02	290	61 55	337
266	35 05	041	60 26	085	39 44	138	50 31	184	38 02	251	53 27	290	61 41	337
267	35 29	041	61 03	086	40 09	139	50 28	185	37 27	252	52 53	291	61 26	337
268	35 53	042	61 40	087	40 33	140	50 25	187	36 51	253	52 18	291	61 11	336
269	36 18	042	62 17	087	40 56	141	50 20	188	36 16	254	51 44	292	60 56	336

LHA ♈	Hc	Zn	Hc	Zn	Hc	Zn	Hc	Zn	Hc	Zn	Hc	Zn	Hc	Zn
	◆Mirfak		Alpheratz		◆ALTAIR		Rasalhague		◆ARCTURUS		Alkaid		Kochab	
270	17 05	025	21 20	070	41 19	142	50 14	190	35 40	255	51 09	292	60 41	336
271	17 21	026	21 54	070	41 41	144	50 07	191	35 05	256	50 35	293	60 26	336
272	17 37	026	22 29	071	42 03	145	50 00	193	34 29	257	50 01	293	60 11	335
273	17 54	027	23 04	072	42 24	146	49 51	194	33 53	258	49 27	294	59 55	335
274	18 11	028	23 39	073	42 44	147	49 42	196	33 17	258	48 53	294	59 40	335
275	18 28	028	24 15	073	43 04	149	49 31	197	32 40	259	48 20	295	59 24	335
276	18 45	029	24 50	074	43 22	150	49 20	199	32 04	260	47 46	295	59 08	335
277	19 03	029	25 26	075	43 41	151	49 08	200	31 28	261	47 13	296	58 53	335
278	19 22	030	26 01	075	43 58	153	48 54	202	30 51	262	46 40	296	58 37	334
279	19 40	031	26 37	076	44 15	154	48 40	203	30 15	263	46 06	297	58 21	334
280	19 59	031	27 13	077	44 31	155	48 26	204	29 38	263	45 33	297	58 05	334
281	20 19	032	27 49	077	44 46	157	48 10	206	29 01	264	45 01	298	57 49	334
282	20 38	032	28 25	078	45 00	158	47 53	207	28 24	265	44 28	298	57 33	334
283	20 58	033	29 01	079	45 14	159	47 36	209	27 48	266	43 55	299	57 16	334
284	21 18	034	29 38	080	45 26	161	47 18	210	27 11	267	43 23	299	57 00	334
	◆Mirfak		Alpheratz		◆ALTAIR		Rasalhague		◆ARCTURUS		Alkaid		Kochab	
285	21 39	034	30 14	080	45 38	162	46 59	211	26 34	268	42 51	300	56 44	334
286	22 00	035	30 51	081	45 49	163	46 40	213	25 57	268	42 19	300	56 28	334
287	22 21	035	31 27	082	45 59	165	46 19	214	25 20	269	41 47	301	56 11	334
288	22 43	036	32 04	083	46 08	166	45 58	215	24 43	270	41 15	301	55 55	334
289	23 04	036	32 40	083	46 17	168	45 37	217	24 06	271	40 44	302	55 39	334
290	23 27	037	33 17	084	46 24	169	45 14	218	23 29	272	40 12	302	55 22	334
291	23 49	038	33 54	085	46 31	170	44 51	219	22 52	272	39 41	303	55 06	334
292	24 12	038	34 31	086	46 37	172	44 27	220	22 15	273	39 10	303	54 50	334
293	24 35	039	35 08	086	46 41	173	44 03	222	21 38	274	38 39	304	54 33	334
294	24 58	039	35 44	087	46 45	175	43 38	223	21 02	275	38 09	304	54 17	334
295	25 22	040	36 21	088	46 48	176	43 13	224	20 25	275	37 38	305	54 01	334
296	25 45	040	36 58	089	46 50	178	42 47	225	19 48	276	37 08	305	53 44	334
297	26 09	041	37 35	089	46 51	179	42 20	227	19 11	277	36 38	306	53 28	334
298	26 34	042	38 12	090	46 51	181	41 53	228	18 35	278	36 08	306	53 12	334
299	26 58	042	38 49	091	46 50	182	41 26	229	17 58	278	35 38	307	52 56	334
	◆CAPELLA		Alpheratz		◆Enif		ALTAIR		Rasalhague		◆Alphecca		Alioth	
300	14 06	028	39 26	092	42 51	144	46 49	183	40 58	230	35 05	270	33 43	320
301	14 23	029	40 03	093	43 12	145	46 46	185	40 29	231	34 28	270	33 19	320
302	14 41	029	40 40	093	43 33	146	46 42	186	40 00	232	33 51	271	32 56	321
303	15 00	030	41 17	094	43 53	148	46 38	188	39 31	233	33 14	272	32 33	321
304	15 18	031	41 53	095	44 13	149	46 32	189	39 01	234	32 37	273	32 10	322
305	15 37	031	42 30	096	44 31	150	46 26	191	38 31	235	32 00	274	31 47	322
306	15 56	032	43 07	097	44 49	152	46 19	192	38 00	237	31 24	274	31 25	323
307	16 16	032	43 44	098	45 06	153	46 11	193	37 29	238	30 47	275	31 02	323
308	16 36	033	44 20	098	45 23	154	46 02	195	36 58	239	30 10	276	30 40	324
309	16 56	034	44 57	099	45 38	156	45 52	196	36 26	240	29 33	277	30 19	324
310	17 17	034	45 33	100	45 53	157	45 41	198	35 54	241	28 57	277	29 57	325
311	17 38	035	46 09	101	46 07	158	45 30	199	35 22	242	28 20	278	29 36	325
312	18 00	036	46 46	102	46 20	160	45 17	200	34 49	243	27 43	279	29 15	326
313	18 21	036	47 22	103	46 33	161	45 04	202	34 16	244	27 07	280	28 54	326
314	18 43	037	47 57	104	46 44	163	44 50	203	33 43	245	26 31	280	28 34	327
	CAPELLA		◆Hamal		Alpheratz		Enif		◆ALTAIR		VEGA		◆Alioth	
315	19 06	038	26 18	085	48 33	105	46 55	164	44 35	204	61 59	256	28 14	327
316	19 28	038	26 55	086	49 09	106	47 04	165	44 19	206	61 23	257	27 54	328
317	19 51	039	27 32	086	49 44	107	47 13	167	44 03	207	60 47	258	27 34	328
318	20 14	039	28 09	087	50 20	108	47 21	168	43 45	208	60 10	259	27 15	329
319	20 38	040	28 46	088	50 55	109	47 28	170	43 27	210	59 34	260	26 56	329
320	21 02	041	29 23	089	51 29	110	47 34	171	43 09	211	58 58	261	26 37	330
321	21 26	041	30 00	090	52 04	111	47 39	173	42 49	212	58 21	262	26 19	330
322	21 51	042	30 37	090	52 38	112	47 44	174	42 29	214	57 45	263	26 01	331
323	22 15	042	31 14	091	53 12	113	47 47	176	42 09	215	57 08	264	25 43	331
324	22 40	043	31 51	092	53 46	114	47 49	177	41 47	216	56 31	265	25 25	332
325	23 06	044	32 27	093	54 20	115	47 51	179	41 25	217	55 54	266	25 08	332
326	23 31	044	33 04	094	54 53	117	47 51	180	41 02	218	55 17	266	24 51	333
327	23 57	045	33 41	094	55 26	118	47 51	181	40 39	220	54 41	267	24 35	334
328	24 23	045	34 18	095	55 58	119	47 49	183	40 15	221	54 04	268	24 18	334
329	24 50	046	34 55	096	56 30	120	47 47	184	39 51	222	53 27	269	24 02	335
	CAPELLA		◆Hamal		Alpheratz		Enif		◆ALTAIR		VEGA		◆Alioth	
330	25 17	047	35 31	097	57 02	122	47 43	186	39 26	223	52 50	270	23 47	335
331	25 43	047	36 08	098	57 33	123	47 39	187	39 00	224	52 13	271	23 31	336
332	26 11	048	36 45	099	58 04	124	47 34	189	38 34	226	51 36	271	23 16	336
333	26 38	048	37 21	099	58 34	126	47 28	190	38 07	227	50 59	272	23 01	337
334	27 06	049	37 58	100	59 04	127	47 21	192	37 40	228	50 22	273	22 47	337
335	27 34	049	38 34	101	59 33	128	47 13	193	37 13	229	49 45	274	22 33	338
336	28 02	050	39 10	102	60 02	130	47 04	195	36 45	230	49 08	274	22 19	338
337	28 30	051	39 46	103	60 30	131	46 54	196	36 16	231	48 31	275	22 06	339
338	28 59	051	40 22	104	60 58	133	46 44	197	35 47	232	47 55	276	21 53	340
339	29 28	052	40 58	105	61 24	134	46 32	199	35 18	233	47 18	276	21 40	340
340	29 57	052	41 33	106	61 50	136	46 20	200	34 48	234	46 41	277	21 28	341
341	30 27	053	42 09	107	62 16	138	46 07	202	34 18	235	46 05	278	21 16	341
342	30 56	053	42 44	108	62 40	139	45 53	203	33 48	236	45 28	279	21 04	342
343	31 26	054	43 19	109	63 04	141	45 38	204	33 17	237	44 52	279	20 52	342
344	31 56	055	43 54	110	63 27	143	45 22	206	32 45	238	44 15	280	20 41	343
	◆CAPELLA		ALDEBARAN		◆Diphda		Enif		ALTAIR		◆VEGA		Alioth	
345	32 26	055	16 39	084	16 27	154	45 06	207	32 14	239	43 39	281	20 31	343
346	32 57	056	17 15	085	16 43	155	44 49	208	31 42	240	43 03	281	20 20	344
347	33 27	056	17 52	086	16 58	156	44 31	210	31 10	241	42 26	282	20 10	345
348	33 58	057	18 29	087	17 13	157	44 12	211	30 37	242	41 50	283	20 01	345
349	34 29	057	19 06	087	17 27	158	43 53	212	30 04	243	41 14	283	19 51	346
350	35 00	058	19 43	088	17 40	159	43 33	214	29 31	244	40 38	284	19 43	346
351	35 32	059	20 20	089	17 53	160	43 12	215	28 58	245	40 03	285	19 34	347
352	36 03	059	20 57	090	18 05	161	42 50	216	28 24	246	39 27	285	19 26	347
353	36 35	060	21 34	091	18 17	162	42 28	217	27 51	247	38 51	286	19 18	348
354	37 07	060	22 11	091	18 28	163	42 06	219	27 16	248	38 16	287	19 11	349
355	37 39	061	22 48	092	18 38	164	41 42	220	26 42	249	37 41	287	19 03	349
356	38 12	061	23 25	093	18 48	165	41 18	221	26 08	250	37 05	288	18 57	350
357	38 44	062	24 01	094	18 57	166	40 54	222	25 33	250	36 30	289	18 50	350
358	39 17	062	24 38	095	19 06	167	40 29	223	24 58	251	35 55	289	18 44	351
359	39 50	063	25 15	095	19 14	168	40 03	225	24 23	252	35 21	290	18 39	352

LAT 51°N

LHA ♈	Hc	Zn	Hc	Zn	Hc	Zn	Hc	Zn	Hc	Zn	Hc	Zn	Hc	Zn
	◆CAPELLA		ALDEBARAN		Hamal		◆Alpheratz		ALTAIR		◆VEGA		Alioth	
0	39 56	063	25 58	096	53 09	126	68 01	175	24 05	253	34 25	291	17 34	352
1	40 29	063	26 36	097	53 39	128	68 03	178	23 29	254	33 49	292	17 29	353
2	41 03	064	27 13	097	54 09	129	68 04	180	22 52	255	33 14	292	17 24	353
3	41 37	064	27 50	098	54 38	130	68 03	182	22 16	256	32 40	293	17 20	354
4	42 11	065	28 28	099	55 07	132	68 01	185	21 39	257	32 05	294	17 16	354
5	42 45	065	29 05	100	55 34	133	67 57	187	21 02	258	31 30	294	17 13	355
6	43 20	066	29 42	101	56 02	135	67 52	189	20 25	259	30 56	295	17 10	356
7	43 54	067	30 19	102	56 28	136	67 45	192	19 48	259	30 22	295	17 07	356
8	44 29	067	30 56	102	56 54	137	67 37	194	19 11	260	29 48	296	17 05	357
9	45 04	068	31 33	103	57 19	139	67 27	196	18 34	261	29 14	297	17 03	357
10	45 39	068	32 10	104	57 44	141	67 16	198	17 56	262	28 40	297	17 02	358
11	46 14	069	32 46	105	58 07	142	67 03	200	17 19	263	28 07	298	17 00	359
12	46 49	069	33 23	106	58 30	144	66 49	203	16 41	263	27 34	299	17 00	359
13	47 25	070	33 59	107	58 52	145	66 34	205	16 04	264	27 01	299	16 59	000
14	48 00	070	34 35	108	59 13	147	66 18	207	15 26	265	26 28	300	16 59	000
	Dubhe		◆CAPELLA		ALDEBARAN		Hamal		◆Alpheratz		DENEB		◆VEGA	
15	25 08	015	48 36	071	35 11	109	59 33	149	66 00	209	47 51	289	25 55	300
16	25 18	015	49 12	071	35 47	109	59 52	150	65 41	211	47 16	289	25 23	301
17	25 28	016	49 47	072	36 22	110	60 11	152	65 21	213	46 40	290	24 50	302
18	25 39	016	50 23	073	36 57	111	60 28	154	65 01	215	46 04	290	24 18	302
19	25 49	017	50 59	073	37 32	112	60 44	156	64 39	217	45 29	291	23 47	303
20	26 00	017	51 36	074	38 07	113	60 59	157	64 16	218	44 54	291	23 15	304
21	26 12	018	52 12	074	38 42	114	61 13	159	63 52	220	44 19	292	22 44	304
22	26 23	018	52 48	075	39 16	115	61 26	161	63 27	222	43 44	292	22 13	305
23	26 35	019	53 25	075	39 50	116	61 37	163	63 01	224	43 09	293	21 42	305
24	26 47	019	54 01	076	40 24	117	61 48	165	62 35	225	42 34	293	21 11	306
25	27 00	020	54 38	076	40 57	118	61 57	167	62 08	227	42 00	294	20 41	307
26	27 13	020	55 15	077	41 30	119	62 05	169	61 40	228	41 25	295	20 10	307
27	27 26	021	55 52	078	42 03	120	62 12	171	61 11	230	40 51	295	19 41	308
28	27 39	021	56 28	078	42 35	121	62 17	173	60 42	231	40 17	296	19 11	309
29	27 53	021	57 05	079	43 07	123	62 22	175	60 12	233	39 43	296	18 42	309
	Dubhe		◆POLLUX		BETELGEUSE		RIGEL		◆Hamal		Alpheratz		◆DENEB	
30	28 07	022	23 41	074	25 06	111	17 32	129	62 24	177	59 42	234	39 09	297
31	28 21	022	24 17	075	25 41	112	18 01	130	62 26	179	59 11	236	38 35	297
32	28 36	023	24 54	075	26 16	112	18 30	131	62 26	181	58 39	237	38 02	298
33	28 50	023	25 30	076	26 51	113	18 58	132	62 25	183	58 07	238	37 29	298
34	29 05	024	26 07	077	27 25	114	19 26	133	62 23	185	57 35	240	36 55	299
35	29 21	024	26 44	078	28 00	115	19 54	133	62 19	186	57 02	241	36 23	299
36	29 36	025	27 21	078	28 34	116	20 21	134	62 14	188	56 29	242	35 50	300
37	29 52	025	27 58	079	29 08	117	20 48	135	62 08	190	55 55	243	35 17	301
38	30 08	026	28 35	080	29 41	118	21 14	136	62 01	192	55 21	245	34 45	301
39	30 25	026	29 12	080	30 14	119	21 40	137	61 52	194	54 47	246	34 12	302
40	30 42	026	29 49	081	30 47	120	22 05	138	61 42	196	54 13	247	33 40	302
41	30 59	027	30 27	082	31 20	121	22 30	139	61 31	198	53 38	248	33 09	303
42	31 16	027	31 04	083	31 52	122	22 55	140	61 19	200	53 03	249	32 37	303
43	31 33	028	31 42	083	32 24	123	23 19	141	61 05	202	52 27	250	32 06	304
44	31 51	028	32 19	084	32 56	124	23 42	142	60 51	204	51 52	251	31 34	304
	Dubhe		◆POLLUX		BETELGEUSE		RIGEL		◆Hamal		Alpheratz		◆DENEB	
45	32 09	029	32 57	085	33 27	125	24 05	143	60 35	205	51 16	252	31 03	305
46	32 27	029	33 34	086	33 58	126	24 27	144	60 18	207	50 40	253	30 32	306
47	32 46	030	34 12	086	34 28	127	24 49	145	60 01	209	50 03	254	30 02	306
48	33 04	030	34 50	087	34 58	128	25 10	146	59 42	211	49 27	255	29 31	307
49	33 23	030	35 27	088	35 28	129	25 31	147	59 22	212	48 50	256	29 01	307
50	33 43	031	36 05	089	35 57	130	25 51	148	59 01	214	48 14	257	28 31	308
51	34 02	031	36 43	089	36 26	131	26 11	149	58 40	216	47 37	258	28 02	308
52	34 22	032	37 21	090	36 54	132	26 30	150	58 18	217	47 00	259	27 32	309
53	34 42	032	37 58	091	37 22	133	26 48	151	57 54	219	46 23	260	27 03	310
54	35 02	032	38 36	092	37 49	134	27 06	152	57 30	220	45 46	261	26 34	310
55	35 22	033	39 14	093	38 16	135	27 23	154	57 05	222	45 08	262	26 05	311
56	35 43	033	39 52	093	38 42	137	27 40	155	56 40	223	44 31	262	25 37	311
57	36 03	034	40 29	094	39 07	138	27 55	156	56 14	225	43 53	263	25 09	312
58	36 24	034	41 07	095	39 33	139	28 11	157	55 47	226	43 16	264	24 41	312
59	36 46	034	41 45	096	39 57	140	28 25	158	55 19	228	42 38	265	24 13	313
	◆Dubhe		POLLUX		SIRIUS		◆RIGEL		Hamal		◆Alpheratz		DENEB	
60	37 07	035	42 22	097	13 17	140	28 39	159	54 51	229	42 01	266	23 45	314
61	37 29	035	43 00	097	13 42	140	28 52	160	54 22	230	41 23	267	23 18	314
62	37 51	036	43 37	098	14 05	141	29 05	161	53 53	232	40 45	267	22 51	315
63	38 13	036	44 14	099	14 29	142	29 17	162	53 23	233	40 08	268	22 25	315
64	38 35	036	44 51	100	14 52	143	29 28	163	52 52	234	39 30	269	21 58	316
65	38 57	037	45 29	101	15 14	144	29 38	164	52 21	236	38 52	270	21 32	317
66	39 20	037	46 06	102	15 36	145	29 48	166	51 50	237	38 14	271	21 06	317
67	39 43	037	46 42	103	15 57	146	29 57	167	51 18	238	37 37	271	20 41	318
68	40 06	038	47 19	104	16 18	147	30 05	168	50 46	239	36 59	272	20 15	318
69	40 29	038	47 56	105	16 39	148	30 13	169	50 14	240	36 21	273	19 51	319
70	40 52	038	48 32	106	16 59	149	30 20	170	49 41	241	35 43	274	19 26	320
71	41 16	039	49 09	107	17 18	150	30 26	171	49 07	243	35 06	274	19 02	320
72	41 40	039	49 45	108	17 37	151	30 31	172	48 34	244	34 28	275	18 38	321
73	42 04	040	50 21	109	17 55	152	30 36	174	48 00	245	33 50	276	18 14	321
74	42 28	040	50 56	110	18 13	153	30 40	175	47 25	246	33 13	277	17 50	322
	Dubhe		◆REGULUS		PROCYON		SIRIUS		◆RIGEL		Hamal		◆DENEB	
75	42 52	040	17 26	092	33 33	130	18 30	153	30 43	176	46 51	247	17 27	323
76	43 17	041	18 04	093	34 02	131	18 46	154	30 45	177	46 16	248	17 05	323
77	43 41	041	18 42	094	34 30	132	19 02	155	30 47	178	45 41	249	16 42	324
78	44 06	041	19 19	095	34 58	133	19 18	156	30 47	179	45 05	250	16 20	325
79	44 31	041	19 57	096	35 25	134	19 33	157	30 48	180	44 30	251	15 58	325
80	44 56	042	20 34	096	35 52	136	19 47	158	30 47	182	43 54	252	15 37	326
81	45 21	042	21 12	097	36 18	137	20 00	159	30 45	183	43 18	253	15 16	326
82	45 47	042	21 49	098	36 43	138	20 13	160	30 43	184	42 42	254	14 55	327
83	46 12	043	22 27	099	37 09	139	20 26	161	30 40	185	42 05	255	14 35	328
84	46 38	043	23 04	100	37 33	140	20 38	162	30 36	186	41 29	256	14 15	328
85	47 04	043	23 41	100	37 57	141	20 49	163	30 32	187	40 52	257	13 55	329
86	47 29	044	24 18	101	38 20	142	20 59	164	30 27	189	40 16	257	13 36	330
87	47 56	044	24 55	102	38 43	144	21 09	165	30 21	190	39 39	258	13 17	330
88	48 22	044	25 32	103	39 05	145	21 18	166	30 14	191	39 02	259	12 58	331
89	48 48	044	26 09	104	39 27	146	21 27	167	30 07	192	38 24	260	12 40	332

LHA ♈	Hc	Zn	Hc	Zn	Hc	Zn	Hc	Zn	Hc	Zn	Hc	Zn	Hc	Zn
	◆Dubhe		REGULUS		PROCYON		◆SIRIUS		RIGEL		◆Hamal		Schedar	
90	49 15	045	26 45	105	39 48	147	21 35	168	29 58	193	37 47	261	45 08	310
91	49 41	045	27 22	106	40 08	148	21 42	169	29 49	194	37 10	262	44 39	310
92	50 08	045	27 58	106	40 27	150	21 49	170	29 40	195	36 32	263	44 10	310
93	50 35	045	28 34	107	40 46	151	21 54	172	29 30	196	35 55	263	43 42	311
94	51 02	046	29 10	108	41 04	152	22 00	173	29 18	198	35 17	264	43 13	311
95	51 29	046	29 46	109	41 21	153	22 04	174	29 07	199	34 40	265	42 45	312
96	51 56	046	30 22	110	41 38	155	22 08	175	28 54	200	34 02	266	42 16	312
97	52 23	046	30 57	111	41 54	156	22 11	176	28 41	201	33 25	267	41 48	312
98	52 50	046	31 32	112	42 09	157	22 14	177	28 27	202	32 47	268	41 21	313
99	53 18	047	32 07	113	42 23	159	22 16	178	28 13	203	32 09	268	40 53	313
100	53 45	047	32 42	114	42 36	160	22 17	179	27 58	204	31 31	269	40 26	314
101	54 13	047	33 16	115	42 49	161	22 17	180	27 42	205	30 54	270	39 58	314
102	54 41	047	33 51	116	43 01	162	22 17	181	27 26	206	30 16	271	39 31	314
103	55 08	047	34 25	116	43 12	164	22 16	182	27 08	207	29 38	271	39 04	315
104	55 36	048	34 58	117	43 22	165	22 15	183	26 51	208	29 00	272	38 38	315
	◆Dubhe		Denebola		REGULUS		◆SIRIUS		RIGEL		ALDEBARAN		◆Mirfak	
105	56 04	048	22 28	094	35 32	118	22 13	184	26 33	209	45 06	233	56 21	290
106	56 32	048	23 05	095	36 05	119	22 10	185	26 14	211	44 35	234	55 46	290
107	57 00	048	23 43	096	36 37	120	22 06	186	25 54	212	44 05	235	55 10	291
108	57 28	048	24 20	097	37 10	121	22 02	187	25 34	213	43 33	237	54 35	291
109	57 56	048	24 58	098	37 42	122	21 57	188	25 13	214	43 02	238	54 00	292
110	58 24	048	25 35	098	38 13	124	21 51	189	24 52	215	42 29	239	53 25	292
111	58 53	048	26 13	099	38 45	125	21 45	190	24 30	216	41 57	240	52 50	292
112	59 21	049	26 50	100	39 16	126	21 38	191	24 08	217	41 24	241	52 15	293
113	59 49	049	27 27	101	39 46	127	21 31	192	23 45	218	40 51	242	51 40	293
114	60 18	049	28 04	102	40 16	128	21 22	193	23 22	219	40 18	243	51 05	294
115	60 46	049	28 41	103	40 46	129	21 13	194	22 58	220	39 44	244	50 31	294
116	61 14	049	29 18	104	41 15	130	21 04	195	22 34	221	39 10	245	49 57	295
117	61 43	049	29 54	104	41 44	131	20 54	196	22 09	222	38 35	246	49 22	295
118	62 11	049	30 31	105	42 12	132	20 43	197	21 44	223	38 01	247	48 48	296
119	62 39	049	31 07	106	42 39	134	20 31	198	21 18	224	37 26	248	48 14	296
	Kochab		◆ARCTURUS		REGULUS		◆SIRIUS		BETELGEUSE		CAPELLA		◆Schedar	
120	45 14	022	12 27	075	43 06	135	20 19	199	39 19	222	62 47	276	31 59	322
121	45 29	022	13 04	076	43 33	136	20 06	200	38 53	223	62 10	276	31 36	323
122	45 43	023	13 40	076	43 59	137	19 53	201	38 27	224	61 32	277	31 13	323
123	45 58	023	14 17	077	44 24	138	19 39	202	38 01	225	60 55	278	30 51	324
124	46 12	023	14 54	078	44 49	140	19 25	203	37 34	226	60 17	278	30 29	324
125	46 27	023	15 31	078	45 13	141	19 10	204	37 06	227	59 40	279	30 07	325
126	46 42	023	16 08	079	45 37	142	18 54	205	36 38	228	59 03	279	29 45	325
127	46 57	023	16 45	080	46 00	143	18 38	206	36 10	230	58 26	280	29 23	326
128	47 12	024	17 22	081	46 22	145	18 21	207	35 41	231	57 48	281	29 02	326
129	47 27	024	18 00	081	46 43	146	18 03	208	35 12	232	57 11	281	28 41	327
130	47 42	024	18 37	082	47 04	147	17 45	209	34 42	233	56 34	282	28 21	327
131	47 58	024	19 15	083	47 24	149	17 27	210	34 11	234	55 57	282	28 00	328
132	48 13	024	19 52	084	47 43	150	17 08	211	33 41	235	55 21	283	27 40	328
133	48 28	024	20 30	084	48 01	152	16 48	212	33 10	236	54 44	283	27 20	329
134	48 44	024	21 07	085	48 19	153	16 28	213	32 38	237	54 07	284	27 01	329
	Kochab		◆ARCTURUS		REGULUS		◆POLLUX		BETELGEUSE		CAPELLA		◆Schedar	
135	49 00	025	21 45	086	48 36	154	63 02	219	32 07	238	53 31	285	26 42	330
136	49 15	025	22 23	087	48 52	156	62 38	220	31 34	239	52 54	285	26 23	330
137	49 31	025	23 00	087	49 07	157	62 13	222	31 02	240	52 18	286	26 04	331
138	49 47	025	23 38	088	49 21	159	61 47	224	30 29	241	51 41	286	25 46	331
139	50 03	025	24 16	089	49 34	160	61 21	225	29 56	242	51 05	287	25 27	332
140	50 19	025	24 53	090	49 47	162	60 54	227	29 23	243	50 29	287	25 10	332
141	50 35	025	25 31	091	49 58	163	60 26	228	28 49	244	49 53	288	24 52	333
142	50 51	025	26 09	091	50 09	165	59 57	230	28 15	244	49 17	288	24 35	333
143	51 07	025	26 47	092	50 18	166	59 28	231	27 41	245	48 42	289	24 18	334
144	51 23	025	27 24	093	50 27	168	58 58	233	27 07	246	48 06	290	24 02	334
145	51 39	025	28 02	094	50 34	169	58 28	234	26 32	247	47 30	290	23 45	335
146	51 56	025	28 40	095	50 41	171	57 57	236	25 57	248	46 55	291	23 29	335
147	52 12	026	29 17	095	50 47	172	57 25	237	25 22	249	46 20	291	23 14	336
148	52 28	026	29 55	096	50 51	174	56 54	238	24 46	250	45 45	292	22 58	336
149	52 45	026	30 33	097	50 55	175	56 21	240	24 11	251	45 10	292	22 43	337
	◆Kochab		ARCTURUS		◆SPICA		REGULUS		◆POLLUX		CAPELLA		Schedar	
150	53 01	026	31 10	098	13 41	128	50 57	177	55 49	241	44 35	293	22 29	337
151	53 17	026	31 47	099	14 11	129	50 59	178	55 15	242	44 00	293	22 14	338
152	53 34	026	32 25	099	14 40	130	51 00	180	54 42	243	43 25	294	22 00	338
153	53 50	026	33 02	100	15 09	131	50 59	182	54 08	244	42 51	294	21 47	339
154	54 06	026	33 39	101	15 37	132	50 58	183	53 34	245	42 17	295	21 33	340
155	54 23	026	34 16	102	16 05	132	50 55	185	52 59	247	41 42	295	21 20	340
156	54 39	026	34 53	103	16 33	133	50 51	186	52 25	248	41 08	296	21 08	341
157	54 55	026	35 30	104	17 00	134	50 47	188	51 49	249	40 35	297	20 55	341
158	55 12	026	36 06	105	17 27	135	50 41	189	51 14	250	40 01	297	20 43	342
159	55 28	026	36 43	106	17 53	136	50 35	191	50 39	251	39 27	298	20 32	342
160	55 44	025	37 19	106	18 19	137	50 27	192	50 03	252	38 54	298	20 20	343
161	56 00	025	37 55	107	18 45	138	50 19	194	49 27	253	38 21	299	20 09	343
162	56 17	025	38 31	108	19 10	139	50 09	195	48 51	254	37 48	299	19 59	344
163	56 33	025	39 07	109	19 34	140	49 59	197	48 14	255	37 15	300	19 48	345
164	56 49	025	39 42	110	19 58	141	49 47	198	47 38	256	36 42	300	19 39	345
	DENEB		◆VEGA		ARCTURUS		◆SPICA		REGULUS		POLLUX		◆CAPELLA	
165	10 49	024	16 36	048	40 17	111	20 22	142	49 35	200	47 01	257	36 10	301
166	11 05	025	17 04	049	40 53	112	20 45	143	49 22	201	46 24	258	35 37	301
167	11 21	025	17 33	049	41 27	113	21 08	144	49 08	203	45 47	259	35 05	302
168	11 37	026	18 01	050	42 02	114	21 30	145	48 53	204	45 10	259	34 33	302
169	11 54	027	18 30	050	42 36	115	21 51	146	48 37	206	44 33	260	34 02	303
170	12 11	027	19 00	051	43 10	116	22 12	147	48 20	207	43 56	261	33 30	304
171	12 29	028	19 29	052	43 44	117	22 33	148	48 02	208	43 18	262	32 59	304
172	12 47	029	19 59	052	44 17	118	22 53	149	47 44	210	42 41	263	32 28	305
173	13 05	029	20 29	053	44 51	119	23 12	150	47 25	211	42 03	264	31 57	305
174	13 24	030	20 59	054	45 23	120	23 31	151	47 05	213	41 26	265	31 26	306
175	13 43	031	21 30	054	45 56	122	23 49	152	46 44	214	40 48	265	30 55	306
176	14 02	031	22 01	055	46 28	123	24 07	153	46 23	215	40 11	266	30 25	307
177	14 22	032	22 32	056	46 59	124	24 24	154	46 01	216	39 33	267	29 55	307
178	14 42	033	23 03	056	47 30	125	24 40	155	45 38	218	38 55	268	29 25	308
179	15 03	033	23 34	057	48 01	126	24 56	156	45 15	219	38 17	269	28 55	309

LHA ♈	Hc	Zn	Hc	Zn	Hc	Zn	Hc	Zn	Hc	Zn	Hc	Zn	Hc	Zn
	DENEB		◆VEGA		ARCTURUS		◆SPICA		REGULUS		POLLUX		◆CAPELLA	
180	15 24	034	24 06	057	48 31	127	25 11	157	44 50	220	37 40	269	28 26	309
181	15 45	034	24 38	058	49 01	129	25 25	158	44 26	222	37 02	270	27 57	310
182	16 06	035	25 10	059	49 30	130	25 39	159	44 00	223	36 24	271	27 28	310
183	16 28	036	25 43	059	49 59	131	25 52	160	43 34	224	35 46	272	26 59	311
184	16 50	036	26 15	060	50 27	132	26 05	161	43 08	225	35 09	273	26 31	311
185	17 13	037	26 48	061	50 55	134	26 17	162	42 41	226	34 31	273	26 03	312
186	17 36	038	27 21	061	51 22	135	26 28	163	42 13	228	33 53	274	25 35	313
187	17 59	038	27 54	062	51 48	136	26 39	164	41 45	229	33 16	275	25 07	313
188	18 23	039	28 28	062	52 14	138	26 48	165	41 16	230	32 38	276	24 39	314
189	18 46	039	29 01	063	52 39	139	26 58	167	40 47	231	32 01	276	24 12	314
190	19 10	040	29 35	064	53 03	141	27 06	168	40 18	232	31 23	277	23 45	315
191	19 35	041	30 09	064	53 27	142	27 14	169	39 48	233	30 46	278	23 19	315
192	20 00	041	30 43	065	53 50	143	27 21	170	39 17	234	30 08	278	22 52	316
193	20 25	042	31 17	066	54 12	145	27 27	171	38 46	235	29 31	279	22 26	317
194	20 50	042	31 52	066	54 33	146	27 33	172	38 15	236	28 54	280	22 01	317
	◆DENEB		VEGA		Rasalhague		◆ARCTURUS		REGULUS		◆POLLUX		CAPELLA	
195	21 16	043	32 26	067	23 06	099	54 54	148	37 44	237	28 16	281	21 35	318
196	21 42	044	33 01	067	23 44	100	55 13	149	37 11	238	27 39	281	21 10	318
197	22 08	044	33 36	068	24 21	100	55 32	151	36 39	240	27 02	282	20 45	319
198	22 34	045	34 11	069	24 58	101	55 50	153	36 06	241	26 26	283	20 20	320
199	23 01	045	34 46	069	25 35	102	56 07	154	35 33	242	25 49	283	19 56	320
200	23 28	046	35 22	070	26 12	103	56 22	156	35 00	243	25 12	284	19 32	321
201	23 55	047	35 57	071	26 48	104	56 37	158	34 26	243	24 36	285	19 08	321
202	24 23	047	36 33	071	27 25	105	56 51	159	33 52	244	23 59	286	18 45	322
203	24 51	048	37 09	072	28 02	105	57 04	161	33 18	245	23 23	286	18 22	323
204	25 19	048	37 45	072	28 38	106	57 16	163	32 44	246	22 47	287	17 59	323
205	25 47	049	38 21	073	29 14	107	57 27	164	32 09	247	22 11	288	17 36	324
206	26 16	050	38 57	074	29 50	108	57 36	166	31 34	248	21 35	288	17 14	324
207	26 45	050	39 33	074	30 26	109	57 45	168	30 59	249	20 59	289	16 52	325
208	27 14	051	40 10	075	31 01	110	57 52	170	30 24	250	20 23	290	16 31	326
209	27 43	051	40 46	076	31 37	111	57 59	171	29 48	251	19 48	290	16 10	326
	◆DENEB		VEGA		Rasalhague		◆ARCTURUS		REGULUS		◆POLLUX		CAPELLA	
210	28 13	052	41 23	076	32 12	112	58 04	173	29 12	252	19 13	291	15 49	327
211	28 42	052	41 59	077	32 47	113	58 08	175	28 36	253	18 37	292	15 29	328
212	29 12	053	42 36	078	33 22	114	58 10	177	28 00	254	18 02	293	15 09	328
213	29 43	053	43 13	078	33 56	114	58 12	178	27 24	254	17 28	293	14 49	329
214	30 13	054	43 50	079	34 31	115	58 12	180	26 47	255	16 53	294	14 30	329
215	30 44	055	44 27	079	35 05	116	58 12	182	26 11	256	16 19	295	14 11	330
216	31 15	055	45 04	080	35 38	117	58 10	184	25 34	257	15 44	295	13 52	331
217	31 46	056	45 42	081	36 12	118	58 07	186	24 57	258	15 10	296	13 34	331
218	32 17	056	46 19	081	36 45	119	58 02	187	24 20	259	14 37	297	13 16	332
219	32 49	057	46 56	082	37 17	120	57 57	189	23 43	260	14 03	297	12 58	333
220	33 20	057	47 34	083	37 50	121	57 50	191	23 06	260	13 30	298	12 41	333
221	33 52	058	48 11	084	38 22	122	57 43	193	22 29	261	12 56	299	12 24	334
222	34 24	059	48 49	084	38 53	124	57 34	194	21 51	262	12 23	300	12 08	335
223	34 57	059	49 26	085	39 25	125	57 24	196	21 14	263	11 51	300	11 52	335
224	35 29	060	50 04	086	39 56	126	57 13	198	20 36	264	11 18	301	11 36	336
	Schedar		◆DENEB		VEGA		Rasalhague		◆ARCTURUS		Denebola		◆Dubhe	
225	21 19	020	36 02	060	50 42	086	40 26	127	57 01	200	37 14	244	56 52	312
226	21 32	020	36 35	061	51 19	087	40 56	128	56 47	201	36 40	245	56 24	312
227	21 45	021	37 08	061	51 57	088	41 26	129	56 33	203	36 05	246	55 56	312
228	21 59	021	37 41	062	52 35	089	41 55	130	56 18	205	35 30	247	55 28	312
229	22 13	022	38 14	062	53 13	089	42 24	131	56 02	206	34 56	248	55 00	313
230	22 27	023	38 48	063	53 50	090	42 52	132	55 45	208	34 20	249	54 33	313
231	22 42	023	39 21	063	54 28	091	43 19	134	55 27	209	33 45	250	54 05	313
232	22 57	024	39 55	064	55 06	092	43 46	135	55 08	211	33 09	251	53 37	313
233	23 12	024	40 29	065	55 44	093	44 13	136	54 48	212	32 34	252	53 10	313
234	23 28	025	41 04	065	56 21	093	44 39	137	54 27	214	31 58	253	52 42	314
235	23 43	025	41 38	066	56 59	094	45 04	138	54 06	216	31 22	254	52 15	314
236	24 00	026	42 12	066	57 37	095	45 29	140	53 43	217	30 45	254	51 48	314
237	24 16	026	42 47	067	58 14	096	45 53	141	53 20	218	30 09	255	51 21	314
238	24 33	027	43 22	067	58 52	097	46 16	142	52 56	220	29 32	256	50 54	314
239	24 50	027	43 57	068	59 29	098	46 39	144	52 32	221	28 56	257	50 27	315
	◆Schedar		DENEB		VEGA		◆Rasalhague		ARCTURUS		◆Denebola		Dubhe	
240	25 08	028	44 32	068	60 06	099	47 01	145	52 07	223	28 19	258	50 00	315
241	25 26	028	45 07	069	60 44	100	47 23	146	51 41	224	27 42	259	49 33	315
242	25 44	029	45 42	069	61 21	101	47 43	148	51 14	225	27 05	260	49 07	315
243	26 02	029	46 17	070	61 58	102	48 03	149	50 47	227	26 28	260	48 40	316
244	26 21	030	46 53	071	62 35	103	48 22	150	50 19	228	25 50	261	48 14	316
245	26 40	030	47 29	071	63 12	104	48 40	152	49 51	229	25 13	262	47 48	316
246	26 59	031	48 04	072	63 48	105	48 58	153	49 22	230	24 35	263	47 22	317
247	27 18	031	48 40	072	64 25	106	49 14	155	48 53	232	23 58	264	46 56	317
248	27 38	032	49 16	073	65 01	107	49 30	156	48 23	233	23 20	264	46 30	317
249	27 58	032	49 53	073	65 37	108	49 45	157	47 52	234	22 43	265	46 05	317
250	28 18	033	50 29	074	66 13	110	49 59	159	47 22	235	22 05	266	45 39	318
251	28 39	033	51 05	074	66 48	111	50 12	160	46 50	236	21 27	267	45 14	318
252	29 00	034	51 42	075	67 23	112	50 24	162	46 19	238	20 50	268	44 49	318
253	29 21	034	52 18	076	67 58	113	50 36	163	45 46	239	20 12	268	44 24	319
254	29 43	035	52 55	076	68 32	115	50 46	165	45 14	240	19 34	269	43 59	319
	◆Schedar		DENEB		ALTAIR		◆Rasalhague		ARCTURUS		◆Alkaid		Kochab	
255	30 04	035	53 31	077	35 15	125	50 55	166	44 41	241	59 40	286	63 13	341
256	30 26	036	54 08	077	35 46	126	51 04	168	44 08	242	59 03	287	63 01	341
257	30 48	036	54 45	078	36 16	127	51 11	170	43 35	243	58 27	287	62 48	340
258	31 11	037	55 22	079	36 46	128	51 17	171	43 01	244	57 51	287	62 35	340
259	31 34	037	55 59	079	37 15	129	51 23	173	42 27	245	57 15	288	62 22	340
260	31 56	038	56 36	080	37 44	130	51 27	174	41 52	246	56 39	288	62 09	339
261	32 20	038	57 13	080	38 13	131	51 30	176	41 18	247	56 03	289	61 55	339
262	32 43	039	57 51	081	38 41	132	51 32	177	40 43	248	55 28	289	61 42	339
263	33 07	039	58 28	082	39 09	134	51 34	179	40 08	249	54 52	290	61 28	338
264	33 31	039	59 05	082	39 36	135	51 34	181	39 32	250	54 17	290	61 14	338
265	33 55	040	59 43	083	40 02	136	51 33	182	38 57	251	53 41	291	61 00	338
266	34 19	040	60 20	084	40 28	137	51 31	184	38 21	252	53 06	291	60 45	337
267	34 44	041	60 58	084	40 54	138	51 28	185	37 45	253	52 31	292	60 31	337
268	35 08	041	61 35	085	41 18	139	51 24	187	37 09	254	51 56	292	60 16	337
269	35 34	042	62 13	086	41 43	141	51 19	188	36 32	255	51 21	293	60 01	337

LHA ♈	Hc	Zn	Hc	Zn	Hc	Zn	Hc	Zn	Hc	Zn	Hc	Zn	Hc	Zn
	◆Mirfak		Alpheratz		◆ALTAIR		Rasalhague		◆ARCTURUS		Alkaid		Kochab	
270	16 11	025	20 59	069	42 06	142	51 13	190	35 56	256	50 46	293	59 46	337
271	16 27	026	21 34	070	42 29	143	51 06	191	35 19	256	50 12	294	59 31	336
272	16 43	026	22 10	071	42 52	144	50 58	193	34 42	257	49 37	294	59 16	336
273	17 00	027	22 45	071	43 13	146	50 49	194	34 05	258	49 03	295	59 01	336
274	17 17	027	23 21	072	43 34	147	50 39	196	33 28	259	48 28	295	58 45	336
275	17 35	028	23 57	073	43 55	148	50 28	198	32 51	260	47 54	296	58 30	336
276	17 53	029	24 33	073	44 14	149	50 17	199	32 14	261	47 20	296	58 14	335
277	18 11	029	25 10	074	44 33	151	50 04	201	31 37	262	46 46	296	57 58	335
278	18 30	030	25 46	075	44 51	152	49 50	202	30 59	262	46 13	297	57 42	335
279	18 49	030	26 23	076	45 08	153	49 36	203	30 22	263	45 39	297	57 27	335
280	19 08	031	26 59	076	45 25	155	49 20	205	29 44	264	45 06	298	57 11	335
281	19 28	032	27 36	077	45 41	156	49 04	206	29 07	265	44 32	298	56 55	335
282	19 48	032	28 13	078	45 56	157	48 47	208	28 29	266	43 59	299	56 39	335
283	20 08	033	28 50	078	46 10	159	48 29	209	27 51	266	43 26	299	56 22	335
284	20 28	033	29 27	079	46 23	160	48 10	211	27 14	267	42 54	300	56 06	335
	◆Mirfak		Alpheratz		◆ALTAIR		Rasalhague		◆ARCTURUS		Alkaid		Kochab	
285	20 49	034	30 04	080	46 35	162	47 50	212	26 36	268	42 21	300	55 50	335
286	21 11	035	30 41	081	46 47	163	47 30	213	25 58	269	41 48	301	55 34	334
287	21 32	035	31 18	081	46 57	165	47 09	215	25 21	270	41 16	301	55 17	334
288	21 54	036	31 56	082	47 07	166	46 47	216	24 43	270	40 44	302	55 01	334
289	22 16	036	32 33	083	47 15	167	46 24	217	24 05	271	40 12	302	54 45	334
290	22 39	037	33 11	083	47 23	169	46 01	219	23 27	272	39 40	303	54 28	334
291	23 01	037	33 48	084	47 30	170	45 37	220	22 50	273	39 08	303	54 12	334
292	23 24	038	34 26	085	47 36	172	45 13	221	22 12	273	38 37	304	53 56	334
293	23 48	038	35 03	086	47 41	173	44 48	222	21 34	274	38 06	304	53 39	334
294	24 11	039	35 41	086	47 45	175	44 22	224	20 57	275	37 35	305	53 23	334
295	24 35	040	36 19	087	47 48	176	43 56	225	20 19	276	37 04	305	53 07	334
296	25 00	040	36 56	088	47 50	178	43 29	226	19 41	277	36 33	306	52 50	334
297	25 24	041	37 34	089	47 51	179	43 01	227	19 04	277	36 02	306	52 34	334
298	25 49	041	38 12	089	47 51	181	42 33	228	18 27	278	35 32	307	52 18	334
299	26 14	042	38 50	090	47 50	182	42 05	230	17 49	279	35 02	307	52 01	334
	◆CAPELLA		Alpheratz		◆Enif		ALTAIR		Rasalhague		◆Alphecca		Alioth	
300	13 13	028	39 27	091	43 39	143	47 49	183	41 36	231	35 05	270	32 57	320
301	13 31	029	40 05	092	44 02	145	47 46	185	41 07	232	34 27	271	32 33	321
302	13 49	029	40 43	093	44 23	146	47 42	186	40 37	233	33 49	272	32 09	321
303	14 08	030	41 21	093	44 44	147	47 37	188	40 06	234	33 12	273	31 46	322
304	14 26	030	41 58	094	45 04	149	47 32	189	39 36	235	32 34	273	31 22	322
305	14 46	031	42 36	095	45 23	150	47 25	191	39 05	236	31 56	274	30 59	323
306	15 05	032	43 13	096	45 42	151	47 18	192	38 33	237	31 19	275	30 37	323
307	15 25	032	43 51	097	46 00	153	47 09	194	38 01	238	30 41	276	30 14	324
308	15 46	033	44 28	098	46 17	154	47 00	195	37 29	239	30 04	276	29 52	324
309	16 07	034	45 06	098	46 33	155	46 49	197	36 56	240	29 26	277	29 30	325
310	16 28	034	45 43	099	46 48	157	46 38	198	36 23	241	28 49	278	29 08	325
311	16 49	035	46 20	100	47 03	158	46 26	199	35 50	242	28 11	279	28 47	326
312	17 11	035	46 58	101	47 17	159	46 13	201	35 16	243	27 34	279	28 25	326
313	17 33	036	47 35	102	47 29	161	45 59	202	34 42	244	26 57	280	28 04	327
314	17 55	037	48 11	103	47 41	162	45 45	204	34 08	245	26 20	281	27 44	327
	CAPELLA		◆Hamal		Alpheratz		Enif		◆ALTAIR		VEGA		◆Alioth	
315	18 18	037	26 13	084	48 48	104	47 52	164	45 29	205	62 12	258	27 23	328
316	18 41	038	26 50	085	49 25	105	48 02	165	45 13	206	61 35	259	27 03	328
317	19 04	039	27 28	086	50 01	106	48 12	167	44 56	208	60 58	260	26 43	329
318	19 28	039	28 06	087	50 37	107	48 20	168	44 38	209	60 21	261	26 24	329
319	19 52	040	28 44	087	51 13	108	48 27	170	44 19	210	59 44	262	26 04	330
320	20 16	040	29 21	088	51 49	109	48 34	171	44 00	212	59 06	263	25 45	330
321	20 41	041	29 59	089	52 25	110	48 39	173	43 40	213	58 29	264	25 27	331
322	21 06	042	30 37	090	53 00	111	48 43	174	43 19	214	57 51	265	25 08	331
323	21 31	042	31 15	091	53 36	112	48 47	176	42 58	215	57 13	265	24 50	332
324	21 56	043	31 52	091	54 10	113	48 49	177	42 35	217	56 36	266	24 32	332
325	22 22	043	32 30	092	54 45	114	48 51	179	42 13	218	55 58	267	24 15	333
326	22 48	044	33 08	093	55 19	115	48 51	180	41 49	219	55 20	268	23 58	333
327	23 15	044	33 45	094	55 53	117	48 51	182	41 25	220	54 43	269	23 41	334
328	23 41	045	34 23	095	56 27	118	48 49	183	41 00	221	54 05	270	23 24	334
329	24 08	046	35 01	095	57 00	119	48 47	185	40 35	223	53 27	270	23 08	335
	CAPELLA		◆Hamal		Alpheratz		Enif		◆ALTAIR		VEGA		◆Alioth	
330	24 35	046	35 38	096	57 33	120	48 43	186	40 09	224	52 49	271	22 52	335
331	25 03	047	36 16	097	58 05	122	48 39	187	39 43	225	52 12	272	22 37	336
332	25 30	047	36 53	098	58 37	123	48 33	189	39 16	226	51 34	273	22 21	336
333	25 58	048	37 31	099	59 09	124	48 27	190	38 48	227	50 56	273	22 06	337
334	26 26	049	38 08	100	59 40	126	48 20	192	38 21	228	50 18	274	21 52	337
335	26 55	049	38 45	100	60 10	127	48 11	193	37 52	229	49 41	275	21 37	338
336	27 23	050	39 22	101	60 40	128	48 02	195	37 23	231	49 03	275	21 23	339
337	27 52	050	39 59	102	61 09	130	47 52	196	36 54	232	48 26	276	21 10	339
338	28 21	051	40 36	103	61 38	131	47 41	198	36 24	233	47 48	277	20 57	340
339	28 51	051	41 13	104	62 06	133	47 29	199	35 54	234	47 11	278	20 44	340
340	29 20	052	41 49	105	62 33	135	47 16	201	35 23	235	46 33	278	20 31	341
341	29 50	052	42 26	106	63 00	136	47 02	202	34 52	236	45 56	279	20 19	341
342	30 20	053	43 02	107	63 25	138	46 48	203	34 21	237	45 19	280	20 07	342
343	30 50	054	43 38	108	63 50	140	46 32	205	33 49	238	44 41	280	19 55	342
344	31 21	054	44 14	109	64 14	142	46 16	206	33 17	239	44 04	281	19 44	343
	◆CAPELLA		ALDEBARAN		◆Diphda		Enif		ALTAIR		◆VEGA		Alioth	
345	31 52	055	16 33	084	17 21	154	45 59	208	32 44	240	43 27	282	19 33	344
346	32 23	055	17 10	085	17 38	155	45 41	209	32 11	241	42 50	282	19 23	344
347	32 54	056	17 48	086	17 53	156	45 23	210	31 38	242	42 14	283	19 13	345
348	33 25	056	18 25	086	18 08	157	45 03	212	31 05	243	41 37	284	19 03	345
349	33 57	057	19 03	087	18 22	158	44 43	213	30 31	244	41 00	284	18 53	346
350	34 28	057	19 41	088	18 36	159	44 22	214	29 57	245	40 24	285	18 44	346
351	35 00	058	20 19	089	18 49	160	44 01	215	29 23	246	39 47	285	18 36	347
352	35 32	059	20 56	089	19 02	161	43 39	217	28 49	246	39 11	286	18 27	348
353	36 05	059	21 34	090	19 14	162	43 16	218	28 14	247	38 35	287	18 19	348
354	36 37	060	22 12	091	19 25	163	42 52	219	27 39	248	37 58	287	18 12	349
355	37 10	060	22 50	092	19 36	164	42 28	220	27 04	249	37 22	288	18 04	349
356	37 43	061	23 27	093	19 46	165	42 03	222	26 28	250	36 47	289	17 58	350
357	38 16	061	24 05	093	19 55	166	41 38	223	25 53	251	36 11	289	17 51	350
358	38 49	062	24 43	094	20 04	167	41 12	224	25 17	252	35 35	290	17 45	351
359	39 22	062	25 20	095	20 12	168	40 46	225	24 41	253	35 00	291	17 39	352

 LAT 50°N

LHA ♈	Hc	Zn	Hc	Zn	Hc	Zn	Hc	Zn	Hc	Zn	Hc	Zn	Hc	Zn
	◆CAPELLA		ALDEBARAN		Hamal		◆Alpheratz		ALTAIR		◆VEGA		Kochab	
0	39 28	062	26 04	095	53 44	125	69 00	175	24 22	254	34 03	292	37 27	347
1	40 02	063	26 42	096	54 16	127	69 03	177	23 45	255	33 27	292	37 18	347
2	40 36	063	27 20	097	54 46	128	69 04	180	23 07	256	32 51	293	37 10	347
3	41 11	064	27 59	098	55 16	129	69 03	182	22 30	256	32 16	294	37 01	347
4	41 45	064	28 37	098	55 46	131	69 01	185	21 52	257	31 41	294	36 53	348
5	42 20	065	29 15	099	56 15	132	68 57	187	21 15	258	31 06	295	36 45	348
6	42 55	065	29 53	100	56 43	133	68 51	190	20 37	259	30 31	295	36 37	348
7	43 30	066	30 31	101	57 11	135	68 44	192	19 59	260	29 56	296	36 29	349
8	44 05	066	31 09	102	57 38	136	68 35	194	19 21	261	29 21	297	36 22	349
9	44 41	067	31 46	103	58 04	138	68 24	197	18 43	261	28 47	297	36 14	349
10	45 16	067	32 24	103	58 30	139	68 13	199	18 05	262	28 13	298	36 07	349
11	45 52	068	33 01	104	58 54	141	67 59	201	17 27	263	27 39	298	36 00	350
12	46 28	068	33 39	105	59 18	143	67 45	204	16 48	264	27 05	299	35 54	350
13	47 04	069	34 16	106	59 41	144	67 28	206	16 10	264	26 31	300	35 47	350
14	47 40	069	34 53	107	60 03	146	67 11	208	15 31	265	25 58	300	35 41	351
	◆CAPELLA		ALDEBARAN		Hamal		◆Diphda		Alpheratz		◆DENEB		Kochab	
15	48 16	070	35 30	108	60 24	148	21 53	184	66 52	210	47 32	290	35 35	351
16	48 52	070	36 06	109	60 44	149	21 50	185	66 33	212	46 55	290	35 29	351
17	49 28	071	36 43	110	61 03	151	21 46	186	66 12	214	46 19	291	35 23	352
18	50 05	071	37 19	111	61 21	153	21 41	187	65 49	216	45 43	291	35 18	352
19	50 41	072	37 55	112	61 38	155	21 36	188	65 26	218	45 07	292	35 12	352
20	51 18	072	38 31	113	61 54	157	21 30	189	65 02	220	44 32	292	35 07	353
21	51 55	073	39 06	114	62 09	159	21 24	190	64 37	222	43 56	293	35 02	353
22	52 32	073	39 41	115	62 22	160	21 16	191	64 11	223	43 20	293	34 58	353
23	53 09	074	40 16	116	62 35	162	21 08	192	63 44	225	42 45	294	34 53	354
24	53 46	075	40 51	117	62 46	164	21 00	193	63 17	227	42 10	294	34 49	354
25	54 23	075	41 25	118	62 55	166	20 51	194	62 48	228	41 35	295	34 45	354
26	55 01	076	41 59	119	63 04	168	20 41	195	62 19	230	41 00	295	34 41	355
27	55 38	076	42 33	120	63 11	170	20 30	196	61 49	231	40 25	296	34 38	355
28	56 15	077	43 06	121	63 17	172	20 19	197	61 19	233	39 50	296	34 34	355
29	56 53	077	43 39	122	63 21	174	20 07	198	60 48	234	39 16	297	34 31	356
	CAPELLA		◆BETELGEUSE		RIGEL		Hamal		◆Alpheratz		DENEB		◆Kochab	
30	57 31	078	25 27	110	18 09	129	63 24	176	60 16	236	38 42	297	34 28	356
31	58 08	078	26 03	111	18 39	130	63 26	179	59 44	237	38 08	298	34 26	356
32	58 46	079	26 39	112	19 09	130	63 26	181	59 11	238	37 34	299	34 23	356
33	59 24	079	27 14	113	19 38	131	63 25	183	58 38	240	37 00	299	34 21	357
34	60 02	080	27 50	114	20 07	132	63 23	185	58 05	241	36 26	300	34 19	357
35	60 40	081	28 25	115	20 35	133	63 19	187	57 31	242	35 53	300	34 17	357
36	61 18	081	29 00	116	21 03	134	63 14	189	56 56	244	35 19	301	34 16	358
37	61 56	082	29 35	116	21 30	135	63 07	191	56 22	245	34 46	301	34 14	358
38	62 34	082	30 09	117	21 57	136	62 59	193	55 47	246	34 13	302	34 13	358
39	63 13	083	30 43	118	22 24	137	62 50	195	55 11	247	33 41	302	34 12	359
40	63 51	084	31 17	119	22 50	138	62 40	197	54 36	248	33 08	303	34 11	359
41	64 29	084	31 50	120	23 15	139	62 28	199	54 00	249	32 36	303	34 11	359
42	65 08	085	32 23	121	23 41	140	62 15	201	53 23	250	32 04	304	34 11	000
43	65 46	085	32 56	122	24 05	141	62 01	202	52 47	251	31 32	304	34 11	000
44	66 25	086	33 29	123	24 29	142	61 46	204	52 10	252	31 00	305	34 11	000
	◆Dubhe		POLLUX		BETELGEUSE		◆RIGEL		Hamal		◆Alpheratz		DENEB	
45	31 16	028	32 51	084	34 01	124	24 53	143	61 29	206	51 34	253	30 29	306
46	31 35	029	33 29	085	34 33	125	25 16	144	61 12	208	50 56	254	29 57	306
47	31 53	029	34 08	086	35 04	126	25 38	145	60 53	210	50 19	255	29 26	307
48	32 12	030	34 46	086	35 35	127	26 00	146	60 33	212	49 42	256	28 55	307
49	32 31	030	35 25	087	36 05	128	26 21	147	60 13	213	49 04	257	28 25	308
50	32 51	030	36 03	088	36 35	129	26 42	148	59 51	215	48 27	258	27 54	308
51	33 11	031	36 42	089	37 05	130	27 02	149	59 28	217	47 49	259	27 24	309
52	33 30	031	37 20	089	37 34	132	27 22	150	59 05	218	47 11	260	26 54	309
53	33 51	032	37 59	090	38 02	133	27 41	151	58 41	220	46 33	261	26 25	310
54	34 11	032	38 38	091	38 31	134	27 59	152	58 16	221	45 55	262	25 55	311
55	34 31	032	39 16	092	38 58	135	28 17	153	57 50	223	45 17	263	25 26	311
56	34 52	033	39 55	092	39 25	136	28 34	154	57 23	224	44 38	263	24 57	312
57	35 13	033	40 33	093	39 52	137	28 50	155	56 56	226	44 00	264	24 28	312
58	35 35	034	41 12	094	40 18	138	29 06	157	56 28	227	43 22	265	24 00	313
59	35 56	034	41 50	095	40 43	140	29 21	158	55 59	229	42 43	266	23 32	313
	◆Dubhe		POLLUX		SIRIUS		◆RIGEL		Hamal		◆Alpheratz		DENEB	
60	36 18	034	42 29	096	14 03	139	29 35	159	55 30	230	42 05	267	23 04	314
61	36 40	035	43 07	097	14 28	140	29 49	160	55 00	231	41 26	267	22 36	314
62	37 02	035	43 45	097	14 52	141	30 02	161	54 29	233	40 48	268	22 09	315
63	37 24	035	44 23	098	15 16	142	30 14	162	53 58	234	40 09	269	21 42	316
64	37 46	036	45 01	099	15 40	143	30 25	163	53 27	235	39 30	270	21 15	316
65	38 09	036	45 40	100	16 03	144	30 36	164	52 55	237	38 52	271	20 48	317
66	38 32	037	46 17	101	16 25	145	30 46	165	52 23	238	38 13	271	20 22	317
67	38 55	037	46 55	102	16 47	146	30 55	167	51 50	239	37 35	272	19 56	318
68	39 18	037	47 33	103	17 08	147	31 04	168	51 16	240	36 56	273	19 31	319
69	39 42	038	48 11	104	17 29	148	31 12	169	50 43	241	36 18	274	19 05	319
70	40 05	038	48 48	105	17 50	149	31 19	170	50 09	242	35 39	274	18 40	320
71	40 29	038	49 25	105	18 10	150	31 25	171	49 34	244	35 01	275	18 15	320
72	40 53	039	50 02	106	18 29	150	31 31	172	49 00	245	34 22	276	17 51	321
73	41 17	039	50 39	107	18 48	151	31 35	174	48 25	246	33 44	276	17 27	322
74	41 42	039	51 16	108	19 06	152	31 39	175	47 49	247	33 06	277	17 03	322
	◆Dubhe		REGULUS		PROCYON		◆SIRIUS		RIGEL		◆Hamal		DENEB	
75	42 06	040	17 29	092	34 12	130	19 23	153	31 43	176	47 14	248	16 40	323
76	42 31	040	18 07	093	34 41	131	19 40	154	31 45	177	46 38	249	16 16	323
77	42 56	040	18 46	094	35 10	132	19 57	155	31 47	178	46 02	250	15 54	324
78	43 21	041	19 24	094	35 39	133	20 13	156	31 47	179	45 25	251	15 31	325
79	43 46	041	20 03	095	36 07	134	20 28	157	31 47	180	44 49	252	15 09	325
80	44 11	041	20 41	096	36 34	135	20 43	158	31 47	182	44 12	253	14 47	326
81	44 36	041	21 19	097	37 01	136	20 56	159	31 45	183	43 35	254	14 26	327
82	45 02	042	21 58	098	37 28	137	21 10	160	31 43	184	42 58	255	14 05	327
83	45 28	042	22 36	098	37 54	138	21 23	161	31 40	185	42 21	256	13 44	328
84	45 54	042	23 14	099	38 19	140	21 35	162	31 36	186	41 43	256	13 24	328
85	46 20	043	23 52	100	38 44	141	21 46	163	31 31	187	41 06	257	13 04	329
86	46 46	043	24 30	101	39 08	142	21 57	164	31 26	189	40 28	258	12 44	330
87	47 12	043	25 08	102	39 31	143	22 07	165	31 20	190	39 50	259	12 25	330
88	47 38	043	25 45	102	39 54	144	22 17	166	31 13	191	39 12	260	12 06	331
89	48 05	044	26 23	103	40 16	145	22 25	167	31 05	192	38 34	261	11 47	332

LHA ♈	Hc	Zn	Hc	Zn	Hc	Zn	Hc	Zn	Hc	Zn	Hc	Zn	Hc	Zn
	◆Dubhe		REGULUS		PROCYON		◆SIRIUS		RIGEL		ALDEBARAN		◆Mirfak	
90	48 32	044	27 00	104	40 38	147	22 33	168	30 57	193	52 26	214	65 10	285
91	48 58	044	27 38	105	40 59	148	22 41	169	30 48	194	52 03	216	64 33	285
92	49 25	044	28 15	106	41 19	149	22 48	170	30 38	195	51 40	217	63 56	286
93	49 52	045	28 52	107	41 38	150	22 54	171	30 27	197	51 17	219	63 19	286
94	50 19	045	29 29	108	41 57	152	22 59	172	30 16	198	50 52	220	62 42	286
95	50 46	045	30 05	108	42 15	153	23 04	174	30 04	199	50 27	221	62 05	287
96	51 14	045	30 42	109	42 32	154	23 08	175	29 51	200	50 01	223	61 28	287
97	51 41	045	31 18	110	42 48	156	23 11	176	29 37	201	49 35	224	60 51	288
98	52 09	046	31 54	111	43 04	157	23 14	177	29 23	202	49 07	225	60 15	288
99	52 36	046	32 30	112	43 19	158	23 16	178	29 08	203	48 40	227	59 38	289
100	53 04	046	33 06	113	43 33	160	23 17	179	28 52	204	48 11	228	59 01	289
101	53 32	046	33 41	114	43 46	161	23 17	180	28 36	205	47 42	229	58 25	289
102	53 59	046	34 16	115	43 58	162	23 17	181	28 19	207	47 13	230	57 49	290
103	54 27	046	34 51	116	44 09	164	23 16	182	28 02	208	46 43	232	57 12	290
104	54 55	047	35 26	117	44 20	165	23 15	183	27 44	209	46 12	233	56 36	291
	◆Dubhe		Denebola		REGULUS		◆SIRIUS		RIGEL		ALDEBARAN		◆Mirfak	
105	55 23	047	22 32	094	36 00	118	23 12	184	27 25	210	45 42	234	56 00	291
106	55 51	047	23 11	095	36 34	119	23 09	185	27 05	211	45 10	235	55 24	291
107	56 19	047	23 49	096	37 08	120	23 06	186	26 45	212	44 38	236	54 48	292
108	56 48	047	24 27	096	37 41	121	23 01	187	26 25	213	44 06	237	54 13	292
109	57 16	047	25 06	097	38 14	122	22 56	188	26 03	214	43 33	238	53 37	293
110	57 44	047	25 44	098	38 46	123	22 51	189	25 42	215	43 00	240	53 02	293
111	58 12	047	26 22	099	39 19	124	22 44	190	25 19	216	42 27	241	52 26	294
112	58 41	047	27 00	100	39 50	125	22 37	191	24 56	217	41 53	242	51 51	294
113	59 09	047	27 38	100	40 22	126	22 29	192	24 33	218	41 19	243	51 16	294
114	59 37	047	28 16	101	40 53	127	22 21	193	24 09	219	40 45	244	50 41	295
115	60 06	047	28 54	102	41 23	128	22 12	194	23 44	220	40 10	245	50 06	295
116	60 34	047	29 31	103	41 53	129	22 02	195	23 19	221	39 35	246	49 31	296
117	61 03	047	30 09	104	42 23	131	21 51	196	22 54	222	39 00	247	48 56	296
118	61 31	047	30 46	105	42 52	132	21 40	197	22 28	223	38 24	248	48 22	297
119	61 59	047	31 23	106	43 20	133	21 28	198	22 01	224	37 48	249	47 47	297
	◆Kochab		Denebola		◆REGULUS		SIRIUS		RIGEL		◆ALDEBARAN		CAPELLA	
120	44 19	022	32 01	106	43 48	134	21 16	199	21 34	225	37 12	250	62 41	278
121	44 33	022	32 37	107	44 16	135	21 03	200	21 07	226	36 36	250	62 02	278
122	44 48	022	33 14	108	44 43	136	20 49	201	20 39	227	36 00	251	61 24	279
123	45 02	022	33 51	109	45 09	138	20 35	202	20 11	228	35 23	252	60 46	279
124	45 17	023	34 27	110	45 35	139	20 20	203	19 42	228	34 46	253	60 08	280
125	45 32	023	35 03	111	46 00	140	20 04	204	19 13	229	34 09	254	59 30	280
126	45 47	023	35 39	112	46 24	142	19 48	205	18 44	230	33 32	255	58 52	281
127	46 02	023	36 15	113	46 48	143	19 31	206	18 14	231	32 55	256	58 14	282
128	46 17	023	36 50	114	47 11	144	19 14	207	17 44	232	32 17	257	57 37	282
129	46 32	023	37 25	115	47 33	145	18 56	208	17 13	233	31 40	258	56 59	283
130	46 47	023	38 00	116	47 54	147	18 38	209	16 42	234	31 02	258	56 21	283
131	47 03	024	38 35	117	48 15	148	18 19	210	16 11	235	30 24	259	55 44	284
132	47 18	024	39 09	118	48 35	150	17 59	211	15 39	236	29 46	260	55 06	284
133	47 34	024	39 43	119	48 54	151	17 39	212	15 07	236	29 08	261	54 29	285
134	47 49	024	40 17	120	49 12	152	17 18	213	14 35	237	28 30	262	53 52	285
	◆Kochab		ARCTURUS		Denebola		◆REGULUS		SIRIUS		BETELGEUSE		◆CAPELLA	
135	48 05	024	21 40	086	40 50	121	49 30	154	16 57	214	32 38	238	53 15	286
136	48 21	024	22 19	086	41 23	122	49 46	155	16 35	215	32 05	239	52 38	286
137	48 37	024	22 57	087	41 56	123	50 02	157	16 13	216	31 32	240	52 01	287
138	48 52	024	23 36	088	42 28	124	50 17	158	15 50	217	30 58	241	51 24	287
139	49 08	024	24 14	089	43 00	125	50 31	160	15 27	217	30 24	242	50 47	288
140	49 24	025	24 53	089	43 31	126	50 44	161	15 03	218	29 50	243	50 11	289
141	49 40	025	25 32	090	44 02	127	50 55	163	14 39	219	29 16	244	49 34	289
142	49 57	025	26 10	091	44 32	128	51 06	164	14 15	220	28 41	245	48 58	290
143	50 13	025	26 49	092	45 02	130	51 16	166	13 49	221	28 06	246	48 21	290
144	50 29	025	27 27	092	45 32	131	51 25	167	13 24	222	27 30	247	47 45	291
145	50 45	025	28 06	093	46 01	132	51 33	169	12 58	223	26 55	248	47 09	291
146	51 01	025	28 44	094	46 29	133	51 40	170	12 31	224	26 19	249	46 33	292
147	51 18	025	29 23	095	46 57	135	51 46	172	12 05	225	25 43	249	45 58	292
148	51 34	025	30 01	096	47 24	136	51 51	174	11 37	225	25 07	250	45 22	293
149	51 50	025	30 40	096	47 50	137	51 55	175	11 10	226	24 30	251	44 46	293
	◆Kochab		ARCTURUS		◆SPICA		REGULUS		PROCYON		◆BETELGEUSE		CAPELLA	
150	52 07	025	31 18	097	14 18	128	51 57	177	36 21	226	23 54	252	44 11	294
151	52 23	025	31 56	098	14 48	129	51 59	178	35 53	227	23 17	253	43 36	294
152	52 39	025	32 34	099	15 18	130	52 00	180	35 25	228	22 40	254	43 01	295
153	52 56	025	33 12	100	15 48	130	51 59	182	34 56	229	22 03	255	42 26	295
154	53 12	025	33 50	101	16 17	131	51 57	183	34 27	230	21 26	255	41 51	296
155	53 28	025	34 28	101	16 45	132	51 55	185	33 57	231	20 48	256	41 16	296
156	53 45	025	35 06	102	17 14	133	51 51	186	33 27	232	20 11	257	40 42	297
157	54 01	025	35 43	103	17 42	134	51 46	188	32 57	233	19 33	258	40 07	297
158	54 17	025	36 21	104	18 09	135	51 41	189	32 26	234	18 55	259	39 33	298
159	54 34	025	36 58	105	18 36	136	51 34	191	31 54	235	18 18	259	38 59	298
160	54 50	025	37 36	106	19 03	137	51 26	193	31 23	236	17 40	260	38 25	299
161	55 06	025	38 13	107	19 29	138	51 17	194	30 50	237	17 02	261	37 52	299
162	55 22	025	38 49	108	19 55	139	51 07	196	30 18	238	16 23	262	37 18	300
163	55 38	025	39 26	109	20 20	140	50 56	197	29 45	239	15 45	263	36 45	300
164	55 54	025	40 03	109	20 45	141	50 44	199	29 12	240	15 07	263	36 12	301
	◆VEGA		ARCTURUS		◆SPICA		REGULUS		PROCYON		◆POLLUX		CAPELLA	
165	15 56	048	40 39	110	21 09	142	50 31	200	28 38	241	47 14	258	35 39	301
166	16 24	048	41 15	111	21 33	143	50 18	202	28 05	242	46 37	259	35 06	302
167	16 53	049	41 51	112	21 56	143	50 03	203	27 31	243	45 59	260	34 33	303
168	17 23	050	42 26	113	22 19	144	49 47	205	26 56	243	45 21	260	34 01	303
169	17 52	050	43 01	114	22 41	145	49 31	206	26 22	244	44 43	261	33 29	304
170	18 22	051	43 36	115	23 02	146	49 13	208	25 47	245	44 04	262	32 57	304
171	18 52	052	44 11	116	23 23	147	48 55	209	25 12	246	43 26	263	32 25	305
172	19 22	052	44 46	117	23 44	148	48 36	210	24 36	247	42 48	264	31 53	305
173	19 53	053	45 20	119	24 04	149	48 16	212	24 00	248	42 10	265	31 22	306
174	20 24	053	45 53	120	24 23	150	47 55	213	23 25	249	41 31	266	30 51	306
175	20 55	054	46 27	121	24 42	151	47 34	214	22 49	250	40 53	266	30 20	307
176	21 26	055	47 00	122	25 00	153	47 12	216	22 12	251	40 14	267	29 49	307
177	21 58	055	47 32	123	25 17	154	46 49	217	21 36	251	39 36	268	29 18	308
178	22 29	056	48 04	124	25 34	155	46 25	218	20 59	252	38 57	269	28 48	308
179	23 01	056	48 36	125	25 50	156	46 01	220	20 22	253	38 18	269	28 18	309

LAT 50°N

LAT 50°N

LHA ♈	Hc	Zn	Hc	Zn	Hc	Zn	Hc	Zn	Hc	Zn	Hc	Zn	Hc	Zn
	◆DENEB		VEGA		ARCTURUS		◆SPICA		REGULUS		◆POLLUX		CAPELLA	
180	14 34	034	23 34	057	49 07	126	26 06	157	45 36	221	37 40	270	27 48	310
181	14 55	034	24 06	058	49 38	128	26 21	158	45 10	222	37 01	271	27 18	310
182	15 17	035	24 39	058	50 08	129	26 35	159	44 44	223	36 23	272	26 49	311
183	15 39	036	25 12	059	50 38	130	26 49	160	44 17	225	35 44	272	26 20	311
184	16 02	036	25 45	060	51 07	131	27 02	161	43 50	226	35 06	273	25 51	312
185	16 25	037	26 18	060	51 36	133	27 14	162	43 22	227	34 27	274	25 22	312
186	16 48	037	26 52	061	52 04	134	27 25	163	42 53	228	33 49	275	24 54	313
187	17 12	038	27 26	061	52 31	135	27 36	164	42 24	229	33 10	275	24 26	313
188	17 36	039	28 00	062	52 58	137	27 46	165	41 55	231	32 32	276	23 58	314
189	18 00	039	28 34	063	53 24	138	27 56	166	41 25	232	31 54	277	23 30	315
190	18 24	040	29 08	063	53 49	140	28 05	167	40 54	233	31 15	278	23 03	315
191	18 49	040	29 43	064	54 14	141	28 13	169	40 23	234	30 37	278	22 36	316
192	19 14	041	30 17	064	54 38	143	28 20	170	39 52	235	29 59	279	22 09	316
193	19 40	042	30 52	065	55 01	144	28 26	171	39 20	236	29 21	280	21 43	317
194	20 06	042	31 27	066	55 23	146	28 32	172	38 48	237	28 43	280	21 16	317
	◆DENEB		VEGA		Rasalhague		◆ARCTURUS		REGULUS		◆POLLUX		CAPELLA	
195	20 32	043	32 02	066	23 15	098	55 44	147	38 15	238	28 05	281	20 50	318
196	20 58	043	32 38	067	23 53	099	56 05	149	37 43	239	27 27	282	20 25	319
197	21 25	044	33 13	067	24 31	100	56 24	150	37 09	240	26 50	283	19 59	319
198	21 52	045	33 49	068	25 09	101	56 43	152	36 36	241	26 12	283	19 34	320
199	22 19	045	34 25	069	25 47	102	57 00	154	36 02	242	25 35	284	19 10	320
200	22 46	046	35 01	069	26 25	102	57 17	155	35 27	243	24 57	285	18 45	321
201	23 14	046	35 37	070	27 02	103	57 33	157	34 53	244	24 20	285	18 21	322
202	23 42	047	36 13	070	27 40	104	57 47	159	34 18	245	23 43	286	17 57	322
203	24 10	047	36 50	071	28 17	105	58 01	160	33 43	246	23 06	287	17 34	323
204	24 39	048	37 26	072	28 54	106	58 13	162	33 08	247	22 29	287	17 11	323
205	25 08	049	38 03	072	29 31	107	58 24	164	32 32	248	21 52	288	16 48	324
206	25 37	049	38 40	073	30 08	107	58 35	166	31 56	249	21 16	289	16 25	325
207	26 06	050	39 16	073	30 45	108	58 44	167	31 20	250	20 39	289	16 03	325
208	26 36	050	39 53	074	31 22	109	58 51	169	30 44	251	20 03	290	15 41	326
209	27 05	051	40 31	075	31 58	110	58 58	171	30 07	251	19 27	291	15 20	326
	◆DENEB		VEGA		Rasalhague		◆ARCTURUS		REGULUS		◆POLLUX		CAPELLA	
210	27 35	051	41 08	075	32 34	111	59 03	173	29 31	252	18 51	291	14 59	327
211	28 06	052	41 45	076	33 10	112	59 07	175	28 54	253	18 15	292	14 38	328
212	28 36	052	42 23	077	33 45	113	59 10	177	28 17	254	17 39	293	14 18	328
213	29 07	053	43 00	077	34 21	114	59 12	178	27 40	255	17 04	294	13 58	329
214	29 38	054	43 38	078	34 56	115	59 12	180	27 02	256	16 29	294	13 38	330
215	30 09	054	44 16	079	35 31	116	59 12	182	26 25	257	15 53	295	13 18	330
216	30 40	055	44 54	079	36 05	117	59 10	184	25 47	257	15 19	296	12 59	331
217	31 12	055	45 31	080	36 40	118	59 06	186	25 10	258	14 44	296	12 41	331
218	31 44	056	46 09	080	37 14	119	59 02	188	24 32	259	14 09	297	12 23	332
219	32 16	056	46 48	081	37 47	120	58 56	189	23 54	260	13 35	298	12 05	333
220	32 48	057	47 26	082	38 21	121	58 49	191	23 16	261	13 01	298	11 47	333
221	33 20	057	48 04	082	38 54	122	58 41	193	22 38	262	12 27	299	11 30	334
222	33 53	058	48 42	083	39 26	123	58 32	195	22 00	262	11 54	300	11 14	335
223	34 26	058	49 20	084	39 59	124	58 21	197	21 21	263	11 20	300	10 57	335
224	34 59	059	49 59	084	40 30	125	58 10	198	20 43	264	10 47	301	10 42	336
	DENEB		◆VEGA		Rasalhague		ANTARES		◆ARCTURUS		Denebola		◆Dubhe	
225	35 32	060	50 37	085	41 02	126	11 04	160	57 57	200	37 39	245	56 11	313
226	36 05	060	51 16	086	41 33	127	11 17	161	57 43	202	37 04	246	55 43	313
227	36 39	061	51 54	087	42 03	128	11 29	162	57 28	204	36 29	247	55 15	313
228	37 12	061	52 33	087	42 33	129	11 41	162	57 12	205	35 53	248	54 47	314
229	37 46	062	53 11	088	43 03	131	11 52	163	56 56	207	35 18	249	54 19	314
230	38 20	062	53 50	089	43 32	132	12 03	164	56 38	209	34 42	250	53 51	314
231	38 54	063	54 28	090	44 00	133	12 13	165	56 19	210	34 05	251	53 24	314
232	39 29	063	55 07	090	44 28	134	12 23	166	55 59	212	33 29	251	52 56	314
233	40 03	064	55 45	091	44 56	135	12 32	167	55 38	213	32 52	252	52 28	314
234	40 38	064	56 24	092	45 23	137	12 40	168	55 17	215	32 15	253	52 01	315
235	41 13	065	57 03	093	45 49	138	12 48	169	54 54	216	31 38	254	51 33	315
236	41 48	065	57 41	094	46 14	139	12 55	170	54 31	218	31 01	255	51 06	315
237	42 23	066	58 20	094	46 39	140	13 02	171	54 07	219	30 24	256	50 39	315
238	42 58	066	58 58	095	47 04	142	13 08	171	53 42	221	29 46	257	50 11	315
239	43 33	067	59 36	096	47 27	143	13 14	172	53 17	222	29 09	258	49 44	316
	◆DENEB		ALTAIR		Rasalhague		◆ANTARES		ARCTURUS		Denebola		◆Dubhe	
240	44 09	067	27 15	110	47 50	144	13 18	173	52 50	224	28 31	258	49 17	316
241	44 45	068	27 51	111	48 12	146	13 23	174	52 23	225	27 53	259	48 51	316
242	45 21	069	28 27	112	48 34	147	13 26	175	51 56	226	27 15	260	48 24	316
243	45 57	069	29 03	113	48 54	148	13 29	176	51 28	228	26 37	261	47 57	316
244	46 33	070	29 38	114	49 14	150	13 31	177	50 59	229	25 59	262	47 31	317
245	47 09	070	30 13	115	49 33	151	13 33	178	50 30	230	25 21	262	47 04	317
246	47 45	071	30 48	116	49 51	153	13 34	179	50 00	231	24 43	263	46 38	317
247	48 22	071	31 23	117	50 09	154	13 35	180	49 29	233	24 04	264	46 12	318
248	48 58	072	31 57	117	50 25	156	13 35	181	48 58	234	23 26	265	45 46	318
249	49 35	072	32 31	118	50 40	157	13 34	182	48 27	235	22 48	266	45 20	318
250	50 12	073	33 05	119	50 55	159	13 32	183	47 55	236	22 09	266	44 55	318
251	50 48	073	33 39	120	51 09	160	13 30	183	47 23	237	21 31	267	44 29	319
252	51 25	074	34 12	121	51 21	162	13 28	184	46 50	238	20 52	268	44 04	319
253	52 03	074	34 44	122	51 33	163	13 25	185	46 17	240	20 13	269	43 38	319
254	52 40	075	35 17	123	51 44	165	13 21	186	45 44	241	19 35	270	43 13	320
	◆Schedar		DENEB		ALTAIR		◆Rasalhague		ARCTURUS		◆Alkaid		Kochab	
255	29 15	035	53 17	075	35 49	124	51 54	166	45 10	242	59 22	288	62 16	342
256	29 37	035	53 54	076	36 21	125	52 02	168	44 36	243	58 45	288	62 04	341
257	30 00	036	54 32	077	36 52	126	52 10	169	44 01	244	58 09	289	61 51	341
258	30 23	036	55 09	077	37 23	128	52 16	171	43 27	245	57 32	289	61 39	341
259	30 46	037	55 47	078	37 53	129	52 22	173	42 51	246	56 56	289	61 26	340
260	31 09	037	56 25	078	38 23	130	52 27	174	42 16	247	56 20	290	61 13	340
261	31 32	038	57 03	079	38 52	131	52 30	176	41 41	248	55 43	290	60 59	340
262	31 56	038	57 41	079	39 21	132	52 32	177	41 05	249	55 07	291	60 46	339
263	32 20	039	58 18	080	39 50	133	52 34	179	40 29	250	54 31	291	60 32	339
264	32 44	039	58 56	081	40 18	134	52 34	181	39 52	251	53 55	292	60 18	339
265	33 09	040	59 35	081	40 45	135	52 33	182	39 16	252	53 19	292	60 04	338
266	33 33	040	60 13	082	41 12	136	52 31	184	38 39	253	52 44	292	59 50	338
267	33 58	040	60 51	082	41 38	138	52 28	185	38 02	254	52 08	293	59 35	338
268	34 23	041	61 29	083	42 04	139	52 24	187	37 25	254	51 33	293	59 21	338
269	34 49	041	62 07	084	42 29	140	52 19	189	36 48	255	50 57	294	59 06	337

LHA ♈	Hc	Zn	Hc	Zn	Hc	Zn	Hc	Zn	Hc	Zn	Hc	Zn	Hc	Zn
	◆Mirfak		Alpheratz		◆ALTAIR		Rasalhague		◆ARCTURUS		Alkaid		Kochab	
270	15 16	025	20 37	069	42 53	141	52 12	190	36 10	256	50 22	294	58 51	337
271	15 33	025	21 13	070	43 17	143	52 05	192	35 33	257	49 47	295	58 36	337
272	15 49	026	21 50	070	43 40	144	51 57	193	34 55	258	49 12	295	58 21	337
273	16 07	027	22 26	071	44 03	145	51 47	195	34 17	259	48 37	296	58 06	337
274	16 24	027	23 03	072	44 25	146	51 37	196	33 39	260	48 03	296	57 50	336
275	16 42	028	23 39	072	44 46	148	51 26	198	33 01	261	47 28	297	57 35	336
276	17 00	028	24 16	073	45 06	149	51 13	199	32 23	261	46 54	297	57 19	336
277	17 19	029	24 53	074	45 25	150	51 00	201	31 45	262	46 19	297	57 04	336
278	17 38	030	25 30	074	45 44	152	50 46	202	31 07	263	45 45	298	56 48	336
279	17 57	030	26 07	075	46 02	153	50 31	204	30 29	264	45 11	298	56 32	336
280	18 17	031	26 45	076	46 19	154	50 14	205	29 50	265	44 37	299	56 16	336
281	18 36	031	27 22	076	46 35	156	49 57	207	29 12	265	44 04	299	56 00	335
282	18 57	032	28 00	077	46 51	157	49 40	208	28 33	266	43 30	300	55 44	335
283	19 17	033	28 37	078	47 05	158	49 21	210	27 55	267	42 57	300	55 28	335
284	19 38	033	29 15	079	47 19	160	49 01	211	27 16	268	42 23	301	55 12	335
	◆Mirfak		Alpheratz		◆ALTAIR		Rasalhague		◆ARCTURUS		Alkaid		Kochab	
285	20 00	034	29 53	079	47 32	161	48 41	213	26 38	269	41 50	301	54 56	335
286	20 21	034	30 31	080	47 44	163	48 20	214	25 59	269	41 17	302	54 39	335
287	20 43	035	31 09	081	47 55	164	47 58	215	25 21	270	40 44	302	54 23	335
288	21 05	035	31 47	081	48 05	166	47 35	217	24 42	271	40 12	303	54 07	335
289	21 28	036	32 25	082	48 14	167	47 12	218	24 04	272	39 39	303	53 51	335
290	21 51	037	33 03	083	48 22	169	46 48	219	23 25	272	39 07	303	53 34	335
291	22 14	037	33 42	083	48 29	170	46 23	221	22 47	273	38 35	304	53 18	335
292	22 37	038	34 20	084	48 35	172	45 58	222	22 08	274	38 03	304	53 02	335
293	23 01	038	34 58	085	48 40	173	45 32	223	21 30	275	37 32	305	52 45	335
294	23 25	039	35 37	086	48 45	175	45 05	224	20 51	275	37 00	305	52 29	335
295	23 49	039	36 15	086	48 48	176	44 38	225	20 13	276	36 29	306	52 12	335
296	24 14	040	36 54	087	48 50	178	44 10	227	19 34	277	35 58	306	51 56	335
297	24 38	040	37 32	088	48 51	179	43 42	228	18 56	278	35 27	307	51 40	335
298	25 04	041	38 11	089	48 51	181	43 13	229	18 18	278	34 56	307	51 24	335
299	25 29	041	38 49	089	48 50	182	42 44	230	17 40	279	34 25	308	51 07	335
	CAPELLA		◆Alpheratz		Enif		ALTAIR		◆Rasalhague		Alphecca		◆Kochab	
300	12 20	028	39 28	090	44 27	143	48 48	184	42 14	231	35 04	271	50 51	335
301	12 38	028	40 07	091	44 50	144	48 46	185	41 43	232	34 26	272	50 35	335
302	12 57	029	40 45	092	45 13	145	48 42	187	41 13	234	33 47	273	50 19	335
303	13 15	030	41 24	092	45 34	147	48 37	188	40 41	235	33 09	273	50 02	335
304	13 35	030	42 02	093	45 55	148	48 31	190	40 10	236	32 30	274	49 46	335
305	13 54	031	42 41	094	46 15	149	48 24	191	39 38	237	31 52	275	49 30	335
306	14 14	032	43 19	095	46 34	151	48 16	192	39 05	238	31 13	276	49 14	335
307	14 35	032	43 57	096	46 53	152	48 07	194	38 32	239	30 35	276	48 58	336
308	14 55	033	44 36	097	47 11	153	47 58	195	37 59	240	29 57	277	48 42	336
309	15 17	033	45 14	097	47 27	155	47 47	197	37 26	241	29 18	278	48 26	336
310	15 38	034	45 52	098	47 43	156	47 35	198	36 52	242	28 40	278	48 11	336
311	16 00	035	46 30	099	47 58	158	47 23	200	36 18	243	28 02	279	47 55	336
312	16 22	035	47 08	100	48 13	159	47 09	201	35 43	244	27 24	280	47 39	336
313	16 44	036	47 46	101	48 26	161	46 55	203	35 08	245	26 46	281	47 24	336
314	17 07	037	48 24	102	48 38	162	46 40	204	34 33	246	26 08	281	47 08	336
	CAPELLA		◆Hamal		Alpheratz		Enif		◆ALTAIR		VEGA		◆Kochab	
315	17 30	037	26 07	084	49 02	103	48 50	163	46 24	205	62 24	260	46 53	337
316	17 54	038	26 45	085	49 39	104	49 00	165	46 07	207	61 46	261	46 37	337
317	18 17	038	27 24	085	50 17	105	49 10	166	45 49	208	61 08	262	46 22	337
318	18 41	039	28 02	086	50 54	106	49 19	168	45 30	209	60 29	263	46 07	337
319	19 06	040	28 41	087	51 31	107	49 26	169	45 11	211	59 51	264	45 52	337
320	19 30	040	29 19	088	52 08	108	49 33	171	44 51	212	59 13	264	45 37	337
321	19 55	041	29 58	088	52 45	109	49 38	172	44 30	213	58 34	265	45 22	337
322	20 21	041	30 36	089	53 21	110	49 43	174	44 09	215	57 56	266	45 07	338
323	20 46	042	31 15	090	53 57	111	49 47	175	43 46	216	57 17	267	44 53	338
324	21 12	042	31 53	091	54 33	112	49 49	177	43 23	217	56 39	268	44 38	338
325	21 38	043	32 32	091	55 09	113	49 51	178	43 00	218	56 00	269	44 24	338
326	22 05	044	33 10	092	55 44	114	49 51	180	42 36	220	55 22	269	44 09	338
327	22 32	044	33 49	093	56 19	115	49 51	182	42 11	221	54 43	270	43 55	338
328	22 59	045	34 27	094	56 54	116	49 49	183	41 45	222	54 05	271	43 41	339
329	23 26	045	35 06	095	57 29	118	49 46	185	41 19	223	53 26	272	43 27	339
	CAPELLA		◆Hamal		Diphda		Enif		◆ALTAIR		VEGA		◆Kochab	
330	23 53	046	35 44	095	13 02	140	49 43	186	40 52	224	52 48	272	43 13	339
331	24 21	046	36 23	096	13 27	141	49 38	188	40 25	226	52 09	273	43 00	339
332	24 49	047	37 01	097	13 51	142	49 33	189	39 57	227	51 30	274	42 46	339
333	25 18	048	37 39	098	14 14	143	49 26	191	39 29	228	50 52	275	42 33	340
334	25 46	048	38 17	099	14 37	144	49 18	192	39 00	229	50 14	275	42 19	340
335	26 15	049	38 55	100	14 59	145	49 10	194	38 31	230	49 35	276	42 06	340
336	26 44	049	39 33	100	15 21	146	49 00	195	38 01	231	48 57	277	41 53	340
337	27 14	050	40 11	101	15 43	147	48 50	197	37 31	232	48 19	277	41 40	341
338	27 43	050	40 49	102	16 04	148	48 38	198	37 00	233	47 40	278	41 27	341
339	28 13	051	41 27	103	16 24	148	48 26	200	36 29	234	47 02	279	41 15	341
340	28 43	051	42 04	104	16 44	149	48 12	201	35 57	235	46 24	279	41 02	341
341	29 13	052	42 42	105	17 04	150	47 58	202	35 26	236	45 46	280	40 50	342
342	29 44	053	43 19	106	17 22	151	47 43	204	34 53	237	45 08	281	40 38	342
343	30 15	053	43 56	107	17 41	152	47 27	205	34 21	238	44 30	281	40 26	342
344	30 46	054	44 33	108	17 58	153	47 10	207	33 48	239	43 52	282	40 14	342
	◆CAPELLA		ALDEBARAN		◆Diphda		Enif		ALTAIR		◆VEGA		Kochab	
345	31 17	054	16 26	084	18 15	154	46 52	208	33 14	240	43 15	283	40 02	342
346	31 48	055	17 05	084	18 32	155	46 34	209	32 41	241	42 37	283	39 51	343
347	32 20	055	17 43	085	18 48	156	46 14	211	32 07	242	42 00	284	39 39	343
348	32 52	056	18 21	086	19 03	157	45 54	212	31 32	243	41 22	284	39 28	343
349	33 24	056	19 00	087	19 18	158	45 34	213	30 58	244	40 45	285	39 17	344
350	33 56	057	19 38	087	19 32	159	45 12	215	30 23	245	40 08	286	39 06	344
351	34 28	057	20 17	088	19 46	160	44 50	216	29 48	246	39 31	286	38 56	344
352	35 01	058	20 56	089	19 59	161	44 27	217	29 12	247	38 54	287	38 45	344
353	35 33	058	21 34	090	20 11	162	44 03	219	28 37	248	38 17	288	38 35	345
354	36 06	059	22 13	091	20 23	163	43 39	220	28 01	249	37 40	288	38 25	345
355	36 40	059	22 51	091	20 34	164	43 14	221	27 25	250	37 04	289	38 15	345
356	37 13	060	23 30	092	20 44	165	42 48	222	26 49	250	36 27	289	38 05	345
357	37 46	061	24 08	093	20 54	166	42 22	223	26 12	251	35 51	290	37 55	346
358	38 20	061	24 47	094	21 03	167	41 55	225	25 36	252	35 15	291	37 46	346
359	38 54	062	25 25	094	21 11	168	41 28	226	24 59	253	34 39	291	37 36	346

LHA ♈	Hc	Zn	Hc	Zn	Hc	Zn	Hc	Zn	Hc	Zn	Hc	Zn	Hc	Zn
	◆CAPELLA		ALDEBARAN		Hamal		◆Diphda		ALTAIR		◆VEGA		Kochab	
0	38 59	061	26 09	095	54 19	124	22 18	169	24 38	254	33 40	292	36 29	347
1	39 34	062	26 48	096	54 51	125	22 25	170	24 00	255	33 04	293	36 20	347
2	40 09	062	27 27	096	55 23	127	22 32	171	23 22	256	32 28	294	36 11	347
3	40 44	063	28 06	097	55 54	128	22 37	172	22 44	257	31 52	294	36 03	348
4	41 19	063	28 45	098	56 25	130	22 43	173	22 05	258	31 16	295	35 54	348
5	41 54	064	29 24	099	56 55	131	22 47	174	21 27	258	30 40	295	35 46	348
6	42 30	064	30 03	100	57 24	132	22 51	175	20 48	259	30 05	296	35 38	348
7	43 05	065	30 42	100	57 53	134	22 54	176	20 10	260	29 29	296	35 30	349
8	43 41	065	31 21	101	58 21	135	22 56	177	19 31	261	28 54	297	35 23	349
9	44 17	066	31 59	102	58 48	137	22 58	178	18 52	262	28 19	298	35 15	349
10	44 53	066	32 38	103	59 15	138	22 59	179	18 13	262	27 44	298	35 08	350
11	45 29	067	33 16	104	59 41	140	22 59	180	17 34	263	27 10	299	35 01	350
12	46 05	067	33 54	105	60 05	142	22 59	181	16 55	264	26 36	299	34 54	350
13	46 41	068	34 32	105	60 29	143	22 57	182	16 15	265	26 01	300	34 48	351
14	47 18	068	35 10	106	60 52	145	22 56	183	15 36	266	25 27	301	34 41	351
	◆CAPELLA		ALDEBARAN		Hamal		◆Diphda		Alpheratz		◆DENEB		Kochab	
15	47 54	069	35 48	107	61 15	147	22 53	184	67 44	211	47 11	291	34 35	351
16	48 31	069	36 25	108	61 36	148	22 50	185	67 23	213	46 34	291	34 29	351
17	49 08	070	37 03	109	61 56	150	22 46	186	67 01	215	45 58	292	34 24	352
18	49 45	070	37 40	110	62 15	152	22 41	187	66 38	217	45 21	292	34 18	352
19	50 22	071	38 17	111	62 33	154	22 36	188	66 13	219	44 45	293	34 13	352
20	50 59	071	38 53	112	62 49	156	22 29	189	65 48	221	44 08	293	34 08	353
21	51 37	072	39 30	113	63 05	158	22 23	190	65 21	223	43 32	294	34 03	353
22	52 14	072	40 06	114	63 19	160	22 15	191	64 54	225	42 56	294	33 58	353
23	52 52	073	40 42	115	63 32	162	22 07	192	64 26	226	42 20	295	33 54	354
24	53 29	073	41 17	116	63 43	164	21 58	194	63 57	228	41 45	295	33 49	354
25	54 07	074	41 53	117	63 54	166	21 49	195	63 27	230	41 09	296	33 45	354
26	54 45	074	42 28	118	64 03	168	21 38	196	62 57	231	40 34	296	33 42	355
27	55 23	075	43 02	119	64 10	170	21 28	197	62 26	233	39 58	297	33 38	355
28	56 01	075	43 37	120	64 16	172	21 16	198	61 54	234	39 23	297	33 35	355
29	56 39	076	44 11	121	64 21	174	21 04	199	61 22	236	38 48	298	33 31	356
	CAPELLA		◆BETELGEUSE		RIGEL		Hamal		◆Alpheratz		DENEB		◆Kochab	
30	57 17	076	25 47	110	18 47	128	64 24	176	60 49	237	38 14	298	33 29	356
31	57 55	077	26 24	111	19 17	129	64 26	178	60 16	239	37 39	299	33 26	356
32	58 34	077	27 01	111	19 48	130	64 26	181	59 42	240	37 05	299	33 23	357
33	59 12	078	27 37	112	20 17	131	64 25	183	59 08	241	36 30	300	33 21	357
34	59 51	078	28 14	113	20 47	132	64 23	185	58 33	242	35 56	300	33 19	357
35	60 29	079	28 50	114	21 16	133	64 19	187	57 58	244	35 22	301	33 17	358
36	61 08	079	29 26	115	21 45	134	64 13	189	57 22	245	34 49	301	33 16	358
37	61 47	080	30 01	116	22 13	135	64 06	191	56 47	246	34 15	302	33 14	358
38	62 25	080	30 36	117	22 40	136	63 58	193	56 10	247	33 42	302	33 13	358
39	63 04	081	31 11	118	23 08	137	63 48	195	55 34	248	33 08	303	33 12	359
40	63 43	082	31 46	119	23 34	138	63 37	197	54 57	249	32 35	303	33 11	359
41	64 22	082	32 20	120	24 01	139	63 25	199	54 20	251	32 03	304	33 11	359
42	65 01	083	32 54	121	24 26	140	63 11	201	53 43	252	31 30	304	33 11	000
43	65 40	083	33 28	122	24 52	141	62 56	203	53 06	253	30 58	305	33 11	000
44	66 19	084	34 01	123	25 16	142	62 40	205	52 28	254	30 26	305	33 11	000
	◆Dubhe		POLLUX		BETELGEUSE		◆RIGEL		Hamal		◆Alpheratz		DENEB	
45	30 23	028	32 45	084	34 34	124	25 41	143	62 23	207	51 50	255	29 54	306
46	30 42	028	33 24	084	35 07	125	26 04	144	62 04	209	51 12	256	29 22	307
47	31 01	029	34 03	085	35 39	126	26 27	145	61 45	211	50 34	257	28 50	307
48	31 20	029	34 42	086	36 11	127	26 50	146	61 24	212	49 55	257	28 19	308
49	31 39	030	35 21	086	36 42	128	27 12	147	61 02	214	49 17	258	27 48	308
50	31 59	030	36 01	087	37 13	129	27 33	148	60 40	216	48 38	259	27 17	309
51	32 19	031	36 40	088	37 43	130	27 54	149	60 16	218	48 00	260	26 46	309
52	32 39	031	37 19	089	38 13	131	28 14	150	59 52	219	47 21	261	26 16	310
53	32 59	031	37 59	089	38 43	132	28 33	151	59 26	221	46 42	262	25 46	310
54	33 20	032	38 38	090	39 12	133	28 52	152	59 00	222	46 03	263	25 16	311
55	33 41	032	39 17	091	39 40	134	29 10	153	58 33	224	45 24	264	24 46	311
56	34 02	032	39 57	092	40 08	135	29 28	154	58 06	226	44 45	264	24 17	312
57	34 23	033	40 36	092	40 35	137	29 45	155	57 37	227	44 05	265	23 48	312
58	34 44	033	41 15	093	41 02	138	30 01	156	57 08	228	43 26	266	23 19	313
59	35 06	034	41 55	094	41 28	139	30 16	157	56 38	230	42 47	267	22 50	314
	◆Dubhe		POLLUX		SIRIUS		◆RIGEL		Hamal		◆Alpheratz		DENEB	
60	35 28	034	42 34	095	14 48	139	30 31	159	56 08	231	42 08	268	22 22	314
61	35 50	034	43 13	096	15 14	140	30 45	160	55 37	233	41 28	268	21 54	315
62	36 12	035	43 52	096	15 39	141	30 58	161	55 05	234	40 49	269	21 26	315
63	36 35	035	44 31	097	16 03	142	31 11	162	54 33	235	40 10	270	20 59	316
64	36 58	035	45 10	098	16 27	143	31 23	163	54 01	236	39 30	271	20 31	317
65	37 20	036	45 49	099	16 51	144	31 34	164	53 28	238	38 51	271	20 04	317
66	37 44	036	46 28	100	17 14	145	31 44	165	52 54	239	38 11	272	19 38	318
67	38 07	036	47 07	101	17 37	146	31 54	166	52 20	240	37 32	273	19 11	318
68	38 30	037	47 46	102	17 59	147	32 03	168	51 46	241	36 53	274	18 45	319
69	38 54	037	48 24	103	18 20	147	32 11	169	51 11	242	36 14	274	18 20	319
70	39 18	037	49 02	103	18 41	148	32 18	170	50 36	244	35 34	275	17 54	320
71	39 42	038	49 41	104	19 01	149	32 24	171	50 01	245	34 55	276	17 29	321
72	40 06	038	50 19	105	19 21	150	32 30	172	49 25	246	34 16	276	17 04	321
73	40 30	038	50 57	106	19 40	151	32 35	173	48 49	247	33 37	277	16 40	322
74	40 55	039	51 34	107	19 59	152	32 39	175	48 12	248	32 58	278	16 16	322
	◆Dubhe		REGULUS		PROCYON		◆SIRIUS		RIGEL		◆Hamal		DENEB	
75	41 20	039	17 31	092	34 50	129	20 17	153	32 42	176	47 36	249	15 52	323
76	41 44	039	18 10	093	35 20	130	20 34	154	32 45	177	46 59	250	15 28	324
77	42 10	040	18 49	093	35 50	131	20 51	155	32 47	178	46 22	251	15 05	324
78	42 35	040	19 29	094	36 19	132	21 08	156	32 47	179	45 45	252	14 42	325
79	43 00	040	20 08	095	36 48	133	21 23	157	32 47	181	45 07	253	14 20	325
80	43 26	040	20 47	096	37 16	134	21 38	158	32 47	182	44 29	254	13 58	326
81	43 51	041	21 26	096	37 44	136	21 53	159	32 45	183	43 52	255	13 36	327
82	44 17	041	22 05	097	38 12	137	22 06	160	32 43	184	43 14	256	13 14	327
83	44 43	041	22 44	098	38 38	138	22 19	161	32 40	185	42 35	256	12 53	328
84	45 09	042	23 23	099	39 04	139	22 32	162	32 36	186	41 57	257	12 33	329
85	45 35	042	24 02	100	39 30	140	22 44	163	32 31	188	41 19	258	12 12	329
86	46 01	042	24 41	100	39 55	141	22 55	164	32 25	189	40 40	259	11 52	330
87	46 28	042	25 19	101	40 19	143	23 05	165	32 19	190	40 01	260	11 33	330
88	46 54	043	25 58	102	40 43	144	23 15	166	32 12	191	39 22	261	11 13	331
89	47 21	043	26 36	103	41 06	145	23 24	167	32 04	192	38 44	262	10 55	332

LAT 49°N

LHA ♈	Hc	Zn	Hc	Zn	Hc	Zn	Hc	Zn	Hc	Zn	Hc	Zn	Hc	Zn
	◆Dubhe		REGULUS		PROCYON		◆SIRIUS		RIGEL		ALDEBARAN		◆Mirfak	
90	47 48	043	27 15	104	41 28	146	23 32	168	31 55	193	53 15	215	64 54	287
91	48 15	043	27 53	104	41 49	147	23 40	169	31 46	195	52 52	217	64 16	287
92	48 42	043	28 31	105	42 10	149	23 47	170	31 35	196	52 28	218	63 39	288
93	49 09	044	29 09	106	42 30	150	23 53	171	31 24	197	52 03	220	63 01	288
94	49 36	044	29 47	107	42 50	151	23 59	172	31 13	198	51 38	221	62 24	288
95	50 04	044	30 24	108	43 08	153	24 03	173	31 00	199	51 12	222	61 47	289
96	50 31	044	31 02	109	43 26	154	24 08	175	30 47	200	50 45	224	61 09	289
97	50 59	044	31 39	110	43 43	155	24 11	176	30 33	201	50 17	225	60 32	289
98	51 26	045	32 16	111	43 59	156	24 14	177	30 18	202	49 49	226	59 55	290
99	51 54	045	32 52	111	44 14	158	24 16	178	30 03	204	49 20	228	59 18	290
100	52 22	045	33 29	112	44 29	159	24 17	179	29 47	205	48 51	229	58 41	291
101	52 50	045	34 05	113	44 42	161	24 17	180	29 30	206	48 21	230	58 04	291
102	53 18	045	34 41	114	44 55	162	24 17	181	29 13	207	47 51	231	57 28	291
103	53 46	045	35 17	115	45 07	163	24 16	182	28 55	208	47 20	232	56 51	292
104	54 14	046	35 52	116	45 18	165	24 15	183	28 36	209	46 48	234	56 14	292
	◆Dubhe		Denebola		REGULUS		◆SIRIUS		RIGEL		ALDEBARAN		◆Mirfak	
105	54 42	046	22 36	094	36 28	117	24 12	184	28 17	210	46 16	235	55 38	292
106	55 10	046	23 15	094	37 02	118	24 09	185	27 57	211	45 44	236	55 02	293
107	55 38	046	23 55	095	37 37	119	24 05	186	27 36	212	45 11	237	54 25	293
108	56 06	046	24 34	096	38 11	120	24 01	187	27 15	213	44 38	238	53 49	294
109	56 35	046	25 13	097	38 45	121	23 56	188	26 53	214	44 04	239	53 13	294
110	57 03	046	25 52	098	39 19	122	23 50	189	26 31	215	43 30	240	52 37	294
111	57 31	046	26 31	098	39 52	123	23 43	190	26 08	216	42 56	241	52 02	295
112	58 00	046	27 10	099	40 24	124	23 36	191	25 44	217	42 21	242	51 26	295
113	58 28	046	27 49	100	40 57	125	23 28	192	25 20	218	41 46	243	50 50	296
114	58 56	046	28 27	101	41 29	126	23 19	193	24 55	219	41 11	244	50 15	296
115	59 25	046	29 06	102	42 00	128	23 10	194	24 30	220	40 35	245	49 40	296
116	59 53	046	29 44	102	42 31	129	23 00	195	24 04	221	39 59	246	49 04	297
117	60 21	046	30 23	103	43 02	130	22 49	196	23 38	222	39 23	247	48 29	297
118	60 50	046	31 01	104	43 31	131	22 37	197	23 12	223	38 47	248	47 54	298
119	61 18	046	31 39	105	44 01	132	22 25	198	22 44	224	38 10	249	47 20	298
	◆Kochab		Denebola		◆REGULUS		SIRIUS		RIGEL		◆ALDEBARAN		CAPELLA	
120	43 23	021	32 17	106	44 30	133	22 12	199	22 17	225	37 33	250	62 32	279
121	43 37	022	32 55	107	44 58	135	21 59	200	21 49	226	36 56	251	61 53	280
122	43 52	022	33 33	108	45 26	136	21 45	201	21 20	227	36 18	252	61 14	281
123	44 07	022	34 10	108	45 53	137	21 30	202	20 51	228	35 41	253	60 35	281
124	44 21	022	34 47	109	46 20	138	21 15	203	20 22	229	35 03	254	59 57	282
125	44 36	022	35 24	110	46 45	140	20 59	204	19 52	230	34 25	255	59 18	282
126	44 51	022	36 01	111	47 11	141	20 42	205	19 22	231	33 47	256	58 40	283
127	45 06	023	36 38	112	47 35	142	20 25	206	18 51	231	33 09	256	58 02	283
128	45 22	023	37 14	113	47 59	143	20 07	207	18 20	232	32 31	257	57 23	284
129	45 37	023	37 50	114	48 22	145	19 49	208	17 49	233	31 52	258	56 45	284
130	45 52	023	38 26	115	48 44	146	19 30	209	17 17	234	31 14	259	56 07	285
131	46 08	023	39 01	116	49 06	148	19 10	210	16 45	235	30 35	260	55 29	285
132	46 23	023	39 37	117	49 27	149	18 50	211	16 13	236	29 56	261	54 51	286
133	46 39	023	40 12	118	49 46	150	18 30	212	15 40	237	29 17	261	54 13	286
134	46 54	024	40 46	119	50 05	152	18 09	213	15 07	238	28 38	262	53 35	287
	◆Kochab		ARCTURUS		Denebola		◆REGULUS		SIRIUS		BETELGEUSE		◆CAPELLA	
135	47 10	024	21 36	085	41 21	120	50 24	153	17 47	214	33 10	239	52 58	287
136	47 26	024	22 15	086	41 54	121	50 41	155	17 25	215	32 36	240	52 20	288
137	47 42	024	22 54	087	42 28	122	50 57	156	17 02	216	32 02	241	51 43	288
138	47 58	024	23 33	087	43 01	123	51 12	158	16 38	217	31 27	242	51 05	289
139	48 14	024	24 13	088	43 34	124	51 27	159	16 15	218	30 52	243	50 28	289
140	48 30	024	24 52	089	44 06	125	51 40	161	15 50	219	30 17	244	49 51	290
141	48 46	024	25 31	090	44 38	127	51 53	162	15 26	219	29 42	245	49 14	290
142	49 02	024	26 11	090	45 09	128	52 04	164	15 00	220	29 06	245	48 37	291
143	49 18	024	26 50	091	45 40	129	52 15	165	14 35	221	28 30	246	48 00	291
144	49 34	024	27 30	092	46 11	130	52 24	167	14 08	222	27 54	247	47 24	292
145	49 51	024	28 09	093	46 40	131	52 32	169	13 42	223	27 17	248	46 47	292
146	50 07	024	28 48	093	47 10	132	52 39	170	13 15	224	26 41	249	46 11	293
147	50 23	024	29 27	094	47 38	134	52 45	172	12 47	225	26 04	250	45 34	293
148	50 39	025	30 07	095	48 07	135	52 50	173	12 19	226	25 27	251	44 58	294
149	50 56	025	30 46	096	48 34	136	52 54	175	11 51	226	24 50	252	44 22	294
	◆Kochab		ARCTURUS		◆SPICA		REGULUS		PROCYON		◆BETELGEUSE		CAPELLA	
150	51 12	025	31 25	097	14 55	128	52 57	177	37 03	226	24 12	252	43 46	295
151	51 29	025	32 04	097	15 26	129	52 59	178	36 34	227	23 35	253	43 11	295
152	51 45	025	32 43	098	15 56	129	53 00	180	36 05	228	22 57	254	42 35	296
153	52 01	025	33 22	099	16 26	130	52 59	182	35 36	229	22 19	255	42 00	296
154	52 18	025	34 01	100	16 56	131	52 57	183	35 06	230	21 41	256	41 24	297
155	52 34	025	34 40	101	17 26	132	52 55	185	34 35	231	21 03	257	40 49	297
156	52 50	024	35 18	102	17 55	133	52 51	186	34 04	232	20 24	257	40 14	298
157	53 07	024	35 57	102	18 23	134	52 46	188	33 33	233	19 46	258	39 40	298
158	53 23	024	36 35	103	18 52	135	52 40	190	33 01	234	19 07	259	39 05	299
159	53 39	024	37 13	104	19 19	136	52 33	191	32 29	235	18 28	260	38 30	299
160	53 55	024	37 51	105	19 47	137	52 24	193	31 56	236	17 50	261	37 56	300
161	54 12	024	38 29	106	20 13	138	52 15	194	31 23	237	17 11	261	37 22	300
162	54 28	024	39 07	107	20 40	138	52 05	196	30 50	238	16 32	262	36 48	301
163	54 44	024	39 45	108	21 06	139	51 53	198	30 16	239	15 53	263	36 14	301
164	55 00	024	40 22	109	21 31	140	51 41	199	29 42	240	15 14	264	35 41	302
	◆VEGA		ARCTURUS		◆SPICA		REGULUS		PROCYON		◆POLLUX		CAPELLA	
165	15 15	047	40 59	110	21 56	141	51 28	201	29 08	241	47 26	259	35 07	302
166	15 44	048	41 36	111	22 20	142	51 13	202	28 33	242	46 48	260	34 34	303
167	16 14	049	42 13	112	22 44	143	50 58	204	27 58	243	46 09	261	34 01	303
168	16 44	049	42 50	112	23 07	144	50 42	205	27 23	244	45 30	261	33 28	304
169	17 14	050	43 26	113	23 30	145	50 24	207	26 47	245	44 51	262	32 55	304
170	17 44	051	44 02	114	23 52	146	50 06	208	26 12	246	44 12	263	32 23	305
171	18 14	051	44 37	116	24 14	147	49 47	210	25 36	247	43 33	264	31 50	305
172	18 45	052	45 13	117	24 35	148	49 28	211	24 59	248	42 54	265	31 18	306
173	19 16	052	45 48	118	24 55	149	49 07	212	24 23	248	42 15	266	30 47	306
174	19 48	053	46 23	119	25 15	150	48 45	214	23 46	249	41 35	266	30 15	307
175	20 19	054	46 57	120	25 34	151	48 23	215	23 09	250	40 56	267	29 43	307
176	20 51	054	47 31	121	25 53	152	48 00	216	22 32	251	40 17	268	29 12	308
177	21 23	055	48 04	122	26 11	153	47 36	218	21 55	252	39 37	269	28 41	308
178	21 56	056	48 38	123	26 28	154	47 12	219	21 17	253	38 58	269	28 10	309
179	22 28	056	49 10	124	26 45	155	46 47	220	20 40	253	38 19	270	27 40	309

LHA ♈	Hc	Zn	Hc	Zn	Hc	Zn	Hc	Zn	Hc	Zn	Hc	Zn	Hc	Zn
	◆DENEB		VEGA		ARCTURUS		◆SPICA		REGULUS		◆POLLUX		CAPELLA	
180	13 44	034	23 01	057	49 43	126	27 01	156	46 21	222	37 39	271	27 10	310
181	14 06	034	23 34	057	50 14	127	27 16	158	45 55	223	37 00	272	26 40	310
182	14 28	035	24 07	058	50 46	128	27 31	159	45 27	224	36 21	272	26 10	311
183	14 51	035	24 41	059	51 16	129	27 45	160	45 00	225	35 41	273	25 40	312
184	15 14	036	25 14	059	51 47	131	27 58	161	44 31	227	35 02	274	25 11	312
185	15 37	037	25 48	060	52 16	132	28 11	162	44 02	228	34 23	275	24 42	313
186	16 00	037	26 22	060	52 45	133	28 23	163	43 33	229	33 44	275	24 13	313
187	16 24	038	26 57	061	53 14	135	28 34	164	43 03	230	33 04	276	23 44	314
188	16 49	038	27 31	062	53 41	136	28 45	165	42 33	231	32 25	277	23 16	314
189	17 13	039	28 06	062	54 08	137	28 54	166	42 02	232	31 46	277	22 48	315
190	17 38	040	28 41	063	54 35	139	29 03	167	41 30	233	31 07	278	22 20	315
191	18 04	040	29 16	063	55 00	140	29 11	168	40 58	235	30 28	279	21 53	316
192	18 29	041	29 51	064	55 25	142	29 19	170	40 26	236	29 49	280	21 26	317
193	18 55	041	30 26	064	55 49	143	29 26	171	39 53	237	29 11	280	20 59	317
194	19 21	042	31 02	065	56 12	145	29 32	172	39 20	238	28 32	281	20 32	318
	◆DENEB		VEGA		Rasalhague		◆ARCTURUS		REGULUS		◆POLLUX		CAPELLA	
195	19 48	043	31 38	066	23 24	098	56 34	146	38 47	239	27 53	282	20 06	318
196	20 14	043	32 14	066	24 03	099	56 56	148	38 13	240	27 15	282	19 40	319
197	20 41	044	32 50	067	24 41	099	57 16	150	37 39	241	26 36	283	19 14	319
198	21 09	044	33 26	067	25 20	100	57 36	151	37 04	242	25 58	284	18 49	320
199	21 36	045	34 03	068	25 59	101	57 54	153	36 29	243	25 20	284	18 23	321
200	22 04	045	34 39	069	26 37	102	58 11	155	35 54	244	24 42	285	17 59	321
201	22 33	046	35 16	069	27 16	103	58 28	156	35 19	245	24 04	286	17 34	322
202	23 01	047	35 53	070	27 54	104	58 43	158	34 43	246	23 26	286	17 10	322
203	23 30	047	36 30	070	28 32	104	58 57	160	34 07	247	22 48	287	16 46	323
204	23 59	048	37 07	071	29 10	105	59 10	162	33 31	248	22 11	288	16 22	324
205	24 28	048	37 44	072	29 48	106	59 22	163	32 54	248	21 33	288	15 59	324
206	24 57	049	38 21	072	30 26	107	59 33	165	32 18	249	20 56	289	15 36	325
207	25 27	049	38 59	073	31 04	108	59 42	167	31 41	250	20 19	290	15 14	325
208	25 57	050	39 37	073	31 41	109	59 50	169	31 04	251	19 42	290	14 52	326
209	26 27	050	40 14	074	32 18	110	59 57	171	30 26	252	19 05	291	14 30	327
	DENEB		VEGA		Rasalhague		◆ARCTURUS		Denebola		REGULUS		◆Dubhe	
210	26 58	051	40 52	075	32 55	110	60 03	173	46 23	229	29 49	253	62 35	314
211	27 28	052	41 30	075	33 32	111	60 07	175	45 52	231	29 11	254	62 06	314
212	27 59	052	42 08	076	34 09	112	60 10	176	45 22	232	28 33	255	61 38	314
213	28 31	053	42 47	076	34 45	113	60 12	178	44 51	233	27 55	255	61 10	314
214	29 02	053	43 25	077	35 21	114	60 12	180	44 19	234	27 17	256	60 42	314
215	29 34	054	44 03	078	35 57	115	60 12	182	43 47	235	26 39	257	60 13	314
216	30 05	054	44 42	078	36 32	116	60 10	184	43 14	236	26 00	258	59 45	314
217	30 37	055	45 20	079	37 07	117	60 06	186	42 41	237	25 22	259	59 16	314
218	31 10	055	45 59	079	37 42	118	60 01	188	42 08	238	24 43	260	58 48	314
219	31 42	056	46 38	080	38 17	119	59 55	190	41 34	240	24 04	260	58 20	314
220	32 15	056	47 17	081	38 51	120	59 48	192	41 00	241	23 25	261	57 51	314
221	32 48	057	47 55	081	39 25	121	59 40	193	40 26	242	22 46	262	57 23	314
222	33 21	057	48 34	082	39 59	122	59 30	195	39 51	243	22 07	263	56 55	314
223	33 54	058	49 13	083	40 32	123	59 19	197	39 16	244	21 28	264	56 26	314
224	34 27	058	49 52	083	41 04	124	59 07	199	38 40	245	20 49	264	55 58	314
	DENEB		◆VEGA		Rasalhague		ANTARES		◆ARCTURUS		Denebola		◆Dubhe	
225	35 01	059	50 32	084	41 37	125	12 00	160	58 53	201	38 05	246	55 30	314
226	35 35	059	51 11	085	42 09	126	12 13	161	58 39	202	37 29	247	55 02	314
227	36 09	060	51 50	085	42 40	128	12 26	161	58 23	204	36 52	248	54 34	314
228	36 43	060	52 29	086	43 11	129	12 38	162	58 07	206	36 16	248	54 05	315
229	37 17	061	53 08	087	43 42	130	12 50	163	57 49	208	35 39	249	53 37	315
230	37 52	062	53 48	087	44 11	131	13 01	164	57 30	209	35 02	250	53 09	315
231	38 26	062	54 27	088	44 41	132	13 11	165	57 10	211	34 25	251	52 42	315
232	39 01	063	55 06	089	45 10	133	13 21	166	56 50	213	33 48	252	52 14	315
233	39 36	063	55 46	090	45 38	135	13 30	167	56 28	214	33 10	253	51 46	315
234	40 11	064	56 25	090	46 06	136	13 39	168	56 06	216	32 32	254	51 18	315
235	40 47	064	57 05	091	46 33	137	13 47	169	55 42	217	31 54	255	50 51	316
236	41 22	065	57 44	092	46 59	138	13 55	170	55 18	219	31 16	256	50 23	316
237	41 58	065	58 23	093	47 25	140	14 01	171	54 53	220	30 38	256	49 56	316
238	42 34	066	59 03	094	47 50	141	14 07	171	54 27	222	30 00	257	49 28	316
239	43 10	066	59 42	094	48 15	142	14 13	172	54 01	223	29 21	258	49 01	316
	◆DENEB		ALTAIR		Rasalhague		◆ANTARES		ARCTURUS		Denebola		◆Dubhe	
240	43 46	067	27 36	110	48 39	144	14 18	173	53 34	224	28 43	259	48 34	317
241	44 22	067	28 13	111	49 02	145	14 22	174	53 06	226	28 04	260	48 07	317
242	44 58	068	28 49	111	49 24	146	14 26	175	52 37	227	27 25	261	47 40	317
243	45 35	068	29 26	112	49 45	148	14 29	176	52 08	229	26 47	261	47 13	317
244	46 11	069	30 02	113	50 06	149	14 31	177	51 38	230	26 08	262	46 47	317
245	46 48	069	30 38	114	50 26	151	14 33	178	51 08	231	25 29	263	46 20	318
246	47 25	070	31 14	115	50 44	152	14 34	179	50 37	232	24 49	264	45 54	318
247	48 02	070	31 50	116	51 02	154	14 35	180	50 05	234	24 10	265	45 28	318
248	48 39	071	32 25	117	51 19	155	14 35	181	49 33	235	23 31	265	45 01	318
249	49 16	071	33 00	118	51 36	157	14 34	182	49 01	236	22 52	266	44 35	318
250	49 53	072	33 34	119	51 51	158	14 32	183	48 28	237	22 13	267	44 09	319
251	50 31	072	34 09	120	52 05	160	14 30	183	47 55	238	21 33	268	43 44	319
252	51 08	073	34 43	121	52 18	161	14 28	184	47 21	239	20 54	268	43 18	320
253	51 46	073	35 16	122	52 30	163	14 24	185	46 47	241	20 15	269	42 53	320
254	52 24	074	35 50	123	52 42	164	14 20	186	46 13	242	19 35	270	42 27	320
	◆Schedar		DENEB		ALTAIR		◆Rasalhague		ARCTURUS		◆Alkaid		Kochab	
255	28 26	035	53 01	074	36 23	124	52 52	166	45 38	243	59 03	289	61 19	342
256	28 48	035	53 39	075	36 55	125	53 01	167	45 03	244	58 26	290	61 07	342
257	29 11	036	54 17	075	37 27	126	53 09	169	44 27	245	57 49	290	60 55	342
258	29 34	036	54 55	076	37 59	127	53 16	171	43 52	246	57 12	290	60 42	341
259	29 57	036	55 34	076	38 30	128	53 22	172	43 16	247	56 35	291	60 29	341
260	30 21	037	56 12	077	39 01	129	53 26	174	42 39	248	55 58	291	60 16	341
261	30 45	037	56 50	077	39 31	130	53 30	176	42 03	249	55 22	292	60 03	340
262	31 09	038	57 29	078	40 01	131	53 32	177	41 26	250	54 45	292	59 50	340
263	31 33	038	58 07	078	40 30	132	53 34	179	40 49	251	54 09	292	59 36	340
264	31 57	039	58 46	079	40 59	134	53 34	181	40 12	252	53 33	293	59 22	339
265	32 22	039	59 25	080	41 27	135	53 33	182	39 34	253	52 56	293	59 08	339
266	32 47	040	60 03	080	41 55	136	53 31	184	38 56	253	52 20	294	58 54	339
267	33 12	040	60 42	081	42 22	137	53 28	185	38 19	254	51 44	294	58 40	339
268	33 38	040	61 21	081	42 49	138	53 23	187	37 41	255	51 08	295	58 25	338
269	34 03	041	62 00	082	43 15	139	53 18	189	37 03	256	50 33	295	58 10	338

LAT 49°N

LHA ♈	Hc	Zn	Hc	Zn	Hc	Zn	Hc	Zn	Hc	Zn	Hc	Zn	Hc	Zn
	◆Mirfak		Alpheratz		◆ALTAIR		Rasalhague		◆ARCTURUS		Alkaid		Kochab	
270	14 22	025	20 16	069	43 40	141	53 11	190	36 24	257	49 57	295	57 56	338
271	14 38	025	20 52	069	44 05	142	53 04	192	35 46	258	49 21	296	57 41	338
272	14 56	026	21 29	070	44 29	143	52 55	194	35 07	259	48 46	296	57 26	337
273	15 13	027	22 06	071	44 52	144	52 45	195	34 29	260	48 11	297	57 11	337
274	15 31	027	22 43	071	45 14	146	52 34	197	33 50	260	47 36	297	56 55	337
275	15 49	028	23 21	072	45 36	147	52 23	198	33 11	261	47 01	297	56 40	337
276	16 07	028	23 58	073	45 57	148	52 10	200	32 32	262	46 26	298	56 24	337
277	16 26	029	24 36	073	46 17	150	51 56	201	31 53	263	45 51	298	56 09	337
278	16 45	030	25 14	074	46 37	151	51 41	203	31 14	264	45 17	299	55 53	336
279	17 05	030	25 52	075	46 55	152	51 25	204	30 35	264	44 42	299	55 37	336
280	17 25	031	26 30	075	47 13	154	51 08	206	29 56	265	44 08	300	55 21	336
281	17 45	031	27 08	076	47 30	155	50 51	207	29 16	266	43 34	300	55 05	336
282	18 06	032	27 46	077	47 46	157	50 32	209	28 37	267	43 00	301	54 49	336
283	18 27	032	28 24	077	48 01	158	50 13	210	27 58	268	42 26	301	54 33	336
284	18 48	033	29 03	078	48 15	160	49 53	212	27 18	268	41 52	301	54 17	336
	◆Mirfak		Alpheratz		◆ALTAIR		Rasalhague		◆ARCTURUS		Alkaid		Kochab	
285	19 10	034	29 41	079	48 29	161	49 31	213	26 39	269	41 19	302	54 01	336
286	19 31	034	30 20	079	48 41	162	49 09	215	26 00	270	40 45	302	53 45	336
287	19 54	035	30 59	080	48 53	164	48 47	216	25 20	271	40 12	303	53 29	336
288	20 16	035	31 38	081	49 03	165	48 23	217	24 41	271	39 39	303	53 12	336
289	20 39	036	32 16	081	49 12	167	47 59	219	24 02	272	39 06	304	52 56	336
290	21 02	036	32 55	082	49 21	168	47 34	220	23 22	273	38 34	304	52 40	335
291	21 26	037	33 34	083	49 28	170	47 09	221	22 43	274	38 01	305	52 23	335
292	21 49	037	34 14	084	49 35	171	46 42	222	22 04	274	37 29	305	52 07	335
293	22 14	038	34 53	084	49 40	173	46 15	224	21 24	275	36 57	306	51 51	335
294	22 38	039	35 32	085	49 44	174	45 48	225	20 45	276	36 25	306	51 34	335
295	23 03	039	36 11	086	49 48	176	45 20	226	20 06	276	35 53	307	51 18	335
296	23 27	040	36 50	086	49 50	177	44 51	227	19 27	277	35 22	307	51 02	335
297	23 53	040	37 30	087	49 51	179	44 22	229	18 48	278	34 50	307	50 45	335
298	24 18	041	38 09	088	49 51	181	43 52	230	18 09	279	34 19	308	50 29	336
299	24 44	041	38 48	089	49 50	182	43 22	231	17 30	279	33 48	308	50 13	336
	CAPELLA		◆Alpheratz		Enif		ALTAIR		◆Rasalhague		Alphecca		◆Kochab	
300	11 27	028	39 28	089	45 15	142	49 48	184	42 51	232	35 03	272	49 56	336
301	11 45	028	40 07	090	45 39	143	49 45	185	42 20	233	34 23	272	49 40	336
302	12 04	029	40 46	091	46 02	145	49 41	187	41 48	234	33 44	273	49 24	336
303	12 23	030	41 26	092	46 24	146	49 36	188	41 16	235	33 05	274	49 08	336
304	12 43	030	42 05	092	46 46	147	49 30	190	40 43	236	32 26	275	48 52	336
305	13 03	031	42 44	093	47 07	149	49 23	191	40 10	238	31 46	275	48 35	336
306	13 23	031	43 24	094	47 27	150	49 15	193	39 37	239	31 07	276	48 19	336
307	13 44	032	44 03	095	47 46	152	49 06	194	39 03	240	30 28	277	48 03	336
308	14 05	033	44 42	096	48 04	153	48 55	196	38 29	241	29 49	278	47 47	336
309	14 26	033	45 21	096	48 22	154	48 44	197	37 54	242	29 10	278	47 32	336
310	14 48	034	46 00	097	48 38	156	48 32	199	37 20	243	28 31	279	47 16	336
311	15 10	035	46 39	098	48 54	157	48 19	200	36 45	244	27 52	280	47 00	336
312	15 33	035	47 18	099	49 09	159	48 05	202	36 09	245	27 13	280	46 44	337
313	15 56	036	47 57	100	49 23	160	47 50	203	35 33	245	26 35	281	46 29	337
314	16 19	036	48 36	101	49 35	162	47 34	204	34 58	246	25 56	282	46 13	337
	CAPELLA		◆Hamal		Alpheratz		Enif		◆ALTAIR		VEGA		◆Kochab	
315	16 42	037	26 00	083	49 15	102	49 47	163	47 18	206	62 33	262	45 58	337
316	17 06	038	26 39	084	49 53	102	49 58	165	47 00	207	61 54	263	45 42	337
317	17 30	038	27 19	085	50 31	103	50 08	166	46 42	209	61 15	264	45 27	337
318	17 55	039	27 58	086	51 10	104	50 17	168	46 23	210	60 36	264	45 12	337
319	18 19	039	28 37	086	51 48	105	50 25	169	46 03	211	59 57	265	44 57	337
320	18 45	040	29 16	087	52 26	106	50 32	171	45 42	213	59 18	266	44 42	338
321	19 10	040	29 56	088	53 03	107	50 38	172	45 20	214	58 38	267	44 27	338
322	19 36	041	30 35	089	53 41	108	50 43	174	44 58	215	57 59	268	44 12	338
323	20 02	042	31 14	089	54 18	109	50 46	175	44 35	216	57 20	269	43 57	338
324	20 28	042	31 54	090	54 55	110	50 49	177	44 11	218	56 40	269	43 43	338
325	20 54	043	32 33	091	55 32	112	50 51	178	43 47	219	56 01	270	43 28	338
326	21 21	043	33 13	092	56 08	113	50 51	180	43 22	220	55 22	271	43 14	339
327	21 48	044	33 52	092	56 44	114	50 51	182	42 56	221	54 42	272	42 59	339
328	22 16	044	34 31	093	57 20	115	50 49	183	42 30	223	54 03	272	42 45	339
329	22 44	045	35 10	094	57 56	116	50 46	185	42 03	224	53 24	273	42 31	339
	CAPELLA		◆Hamal		Diphda		Enif		◆ALTAIR		VEGA		◆Kochab	
330	23 12	046	35 50	095	13 48	140	50 42	186	41 35	225	52 44	274	42 17	339
331	23 40	046	36 29	096	14 13	141	50 38	188	41 07	226	52 05	274	42 03	340
332	24 08	047	37 08	096	14 38	142	50 32	189	40 38	227	51 26	275	41 50	340
333	24 37	047	37 47	097	15 02	143	50 25	191	40 09	228	50 47	276	41 36	340
334	25 06	048	38 26	098	15 25	144	50 17	192	39 39	230	50 07	276	41 23	340
335	25 35	048	39 05	099	15 48	145	50 08	194	39 09	231	49 28	277	41 10	340
336	26 05	049	39 44	100	16 11	146	49 58	195	38 38	232	48 49	278	40 56	341
337	26 35	049	40 23	101	16 33	146	49 47	197	38 07	233	48 10	278	40 44	341
338	27 05	050	41 01	101	16 54	147	49 35	198	37 36	234	47 31	279	40 31	341
339	27 35	051	41 40	102	17 15	148	49 22	200	37 04	235	46 53	280	40 18	341
340	28 06	051	42 18	103	17 36	149	49 08	201	36 31	236	46 14	280	40 05	342
341	28 36	052	42 57	104	17 56	150	48 53	203	35 59	237	45 35	281	39 53	342
342	29 07	052	43 35	105	18 15	151	48 38	204	35 25	238	44 57	282	39 41	342
343	29 38	053	44 13	106	18 34	152	48 21	206	34 52	239	44 18	282	39 29	342
344	30 10	053	44 50	107	18 52	153	48 04	207	34 18	240	43 40	283	39 17	342
	◆CAPELLA		ALDEBARAN		◆Diphda		Enif		ALTAIR		◆VEGA		Kochab	
345	30 41	054	16 20	083	19 09	154	47 45	209	33 44	241	43 01	283	39 05	343
346	31 13	054	16 59	084	19 26	155	47 26	210	33 09	242	42 23	284	38 53	343
347	31 45	055	17 38	085	19 43	156	47 06	211	32 34	243	41 45	285	38 42	343
348	32 18	055	18 17	086	19 59	157	46 45	213	31 59	244	41 07	285	38 31	343
349	32 50	056	18 56	086	20 14	158	46 23	214	31 24	245	40 29	286	38 20	344
350	33 23	056	19 36	087	20 28	159	46 01	215	30 48	246	39 51	286	38 09	344
351	33 55	057	20 15	088	20 42	160	45 38	217	30 12	247	39 13	287	37 58	344
352	34 29	057	20 54	089	20 55	161	45 14	218	29 36	247	38 36	288	37 47	345
353	35 02	058	21 34	089	21 08	162	44 50	219	28 59	248	37 58	288	37 37	345
354	35 35	058	22 13	090	21 20	163	44 25	220	28 22	249	37 21	289	37 27	345
355	36 09	059	22 52	091	21 31	164	43 59	222	27 46	250	36 44	289	37 17	345
356	36 43	059	23 32	092	21 42	165	43 32	223	27 08	251	36 07	290	37 07	346
357	37 16	060	24 11	092	21 52	166	43 05	224	26 31	252	35 30	291	36 57	346
358	37 51	060	24 50	093	22 01	167	42 38	225	25 54	253	34 53	291	36 47	346
359	38 25	061	25 30	094	22 10	168	42 09	226	25 16	254	34 17	292	36 38	346

LAT 48°N

LHA	Hc	Zn	Hc	Zn	Hc	Zn	Hc	Zn	Hc	Zn	Hc	Zn	Hc	Zn
♈	◆CAPELLA		ALDEBARAN		Hamal		◆Diphda		ALTAIR		◆VEGA		Kochab	
0	38 30	061	26 14	094	54 52	123	23 17	169	24 54	255	33 17	293	35 31	347
1	39 05	061	26 54	095	55 25	124	23 24	170	24 15	256	32 40	294	35 22	347
2	39 41	062	27 34	096	55 58	126	23 31	171	23 36	256	32 03	294	35 13	347
3	40 16	062	28 14	097	56 31	127	23 37	172	22 57	257	31 27	295	35 04	348
4	40 52	063	28 53	097	57 02	128	23 42	173	22 18	258	30 50	295	34 56	348
5	41 27	063	29 33	098	57 34	130	23 47	174	21 39	259	30 14	296	34 47	348
6	42 03	064	30 13	099	58 04	131	23 51	175	20 59	260	29 38	296	34 39	349
7	42 39	064	30 53	100	58 34	133	23 54	176	20 20	260	29 02	297	34 32	349
8	43 15	064	31 32	101	59 03	134	23 56	177	19 40	261	28 27	298	34 24	349
9	43 52	065	32 11	101	59 32	136	23 58	178	19 00	262	27 51	298	34 16	349
10	44 28	065	32 51	102	59 59	137	23 59	179	18 21	263	27 16	299	34 09	350
11	45 05	066	33 30	103	60 26	139	23 59	180	17 41	264	26 41	299	34 02	350
12	45 41	066	34 09	104	60 52	141	23 59	181	17 01	264	26 06	300	33 55	350
13	46 18	067	34 48	105	61 17	142	23 57	182	16 21	265	25 31	300	33 49	351
14	46 55	067	35 27	106	61 41	144	23 56	183	15 41	266	24 57	301	33 42	351
	◆CAPELLA		ALDEBARAN		Hamal		◆Diphda		Alpheratz		◆DENEB		Kochab	
15	47 32	068	36 05	107	62 04	146	23 53	184	68 35	212	46 50	292	33 36	351
16	48 09	068	36 44	107	62 27	147	23 49	185	68 13	215	46 12	292	33 30	352
17	48 47	069	37 22	108	62 48	149	23 45	186	67 49	217	45 35	293	33 24	352
18	49 24	069	38 00	109	63 07	151	23 40	187	67 25	219	44 58	293	33 19	352
19	50 02	070	38 38	110	63 26	153	23 35	188	66 59	221	44 21	294	33 13	352
20	50 40	070	39 15	111	63 44	155	23 29	190	66 33	223	43 44	294	33 08	353
21	51 17	071	39 53	112	64 00	157	23 22	191	66 05	224	43 08	294	33 03	353
22	51 55	071	40 30	113	64 15	159	23 14	192	65 36	226	42 31	295	32 59	353
23	52 33	071	41 06	114	64 29	161	23 06	193	65 07	228	41 55	295	32 54	354
24	53 11	072	41 43	115	64 41	163	22 56	194	64 37	230	41 19	296	32 50	354
25	53 50	072	42 19	116	64 52	165	22 47	195	64 06	231	40 43	296	32 46	354
26	54 28	073	42 55	117	65 01	167	22 36	196	63 34	233	40 07	297	32 42	355
27	55 06	073	43 31	118	65 09	170	22 25	197	63 02	234	39 31	297	32 38	355
28	55 45	074	44 06	119	65 16	172	22 13	198	62 29	236	38 56	298	32 35	355
29	56 23	074	44 41	120	65 21	174	22 01	199	61 55	237	38 20	298	32 32	356
	CAPELLA		◆BETELGEUSE		RIGEL		Hamal		◆Alpheratz		DENEB		◆Kochab	
30	57 02	075	26 07	109	19 24	128	65 24	176	61 21	239	37 45	299	32 29	356
31	57 41	075	26 45	110	19 55	129	65 26	178	60 46	240	37 10	299	32 26	356
32	58 20	076	27 23	111	20 26	130	65 26	181	60 11	241	36 35	300	32 23	357
33	58 59	076	28 00	112	20 57	131	65 25	183	59 36	243	36 00	300	32 21	357
34	59 38	077	28 37	113	21 27	132	65 22	185	59 00	244	35 26	301	32 19	357
35	60 17	077	29 14	114	21 57	133	65 18	187	58 24	245	34 51	301	32 17	358
36	60 56	078	29 51	115	22 26	134	65 12	189	57 47	246	34 17	302	32 16	358
37	61 35	078	30 27	115	22 55	135	65 05	192	57 10	248	33 43	302	32 14	358
38	62 15	079	31 03	116	23 23	135	64 56	194	56 33	249	33 09	303	32 13	358
39	62 54	079	31 39	117	23 51	136	64 46	196	55 55	250	32 36	303	32 12	359
40	63 33	080	32 15	118	24 19	137	64 34	198	55 18	251	32 02	304	32 11	359
41	64 13	080	32 50	119	24 46	138	64 21	200	54 40	252	31 29	304	32 11	359
42	64 53	081	33 25	120	25 12	139	64 07	202	54 01	253	30 56	305	32 11	000
43	65 32	081	33 59	121	25 38	140	63 51	204	53 23	254	30 23	305	32 11	000
44	66 12	082	34 33	122	26 03	141	63 34	206	52 44	255	29 51	306	32 11	000
	◆Dubhe		POLLUX		BETELGEUSE		◆RIGEL		Hamal		◆Alpheratz		DENEB	
45	29 30	028	32 38	083	35 07	123	26 28	142	63 16	208	52 05	256	29 18	306
46	29 49	028	33 17	084	35 41	124	26 52	143	62 57	210	51 26	257	28 46	307
47	30 08	029	33 57	084	36 14	125	27 16	144	62 36	212	50 47	258	28 14	307
48	30 28	029	34 37	085	36 46	126	27 39	145	62 14	214	50 08	259	27 42	308
49	30 47	029	35 17	086	37 19	127	28 02	146	61 52	215	49 28	260	27 11	309
50	31 07	030	35 57	086	37 50	128	28 24	147	61 28	217	48 49	260	26 39	309
51	31 27	030	36 37	087	38 22	129	28 45	148	61 03	219	48 09	261	26 08	310
52	31 47	031	37 18	088	38 53	130	29 06	150	60 38	220	47 30	262	25 38	310
53	32 08	031	37 58	089	39 23	131	29 26	151	60 11	222	46 50	263	25 07	311
54	32 29	031	38 38	089	39 53	133	29 45	152	59 44	224	46 10	264	24 37	311
55	32 50	032	39 18	090	40 22	134	30 04	153	59 16	225	45 30	265	24 07	312
56	33 11	032	39 58	091	40 51	135	30 22	154	58 47	227	44 50	265	23 37	312
57	33 32	032	40 38	092	41 19	136	30 39	155	58 18	228	44 10	266	23 07	313
58	33 54	033	41 18	092	41 46	137	30 56	156	57 47	230	43 30	267	22 38	313
59	34 16	033	41 58	093	42 13	138	31 12	157	57 16	231	42 50	268	22 09	314
	◆Dubhe		POLLUX		SIRIUS		◆RIGEL		Hamal		◆Alpheratz		DENEB	
60	34 38	034	42 39	094	15 34	139	31 27	158	56 45	232	42 10	268	21 40	315
61	35 00	034	43 19	095	16 00	140	31 41	159	56 13	234	41 30	269	21 12	315
62	35 23	034	43 59	095	16 25	141	31 55	161	55 40	235	40 49	270	20 43	316
63	35 46	035	44 39	096	16 51	142	32 08	162	55 07	236	40 09	271	20 15	316
64	36 09	035	45 18	097	17 15	143	32 20	163	54 33	238	39 29	271	19 48	317
65	36 32	035	45 58	098	17 39	144	32 32	164	53 59	239	38 49	272	19 20	317
66	36 55	036	46 38	099	18 03	144	32 42	165	53 24	240	38 09	273	18 53	318
67	37 18	036	47 18	100	18 26	145	32 52	166	52 49	241	37 29	274	18 27	318
68	37 42	036	47 57	101	18 49	146	33 01	167	52 14	242	36 49	274	18 00	319
69	38 06	037	48 36	101	19 10	147	33 10	169	51 38	244	36 09	275	17 34	320
70	38 30	037	49 16	102	19 32	148	33 17	170	51 02	245	35 29	276	17 08	320
71	38 54	037	49 55	103	19 53	149	33 24	171	50 26	246	34 49	276	16 43	321
72	39 19	038	50 34	104	20 13	150	33 30	172	49 49	247	34 09	277	16 17	321
73	39 43	038	51 13	105	20 33	151	33 35	173	49 12	248	33 29	278	15 53	322
74	40 08	038	51 51	106	20 52	152	33 39	175	48 35	249	32 49	278	15 28	323
	◆Dubhe		REGULUS		PROCYON		◆SIRIUS		RIGEL		◆Hamal		Schedar	
75	40 33	038	17 32	091	35 27	129	21 10	153	33 42	176	47 57	250	50 54	308
76	40 58	039	18 12	092	35 59	130	21 28	154	33 45	177	47 19	251	50 23	308
77	41 23	039	18 53	093	36 29	131	21 46	155	33 47	178	46 41	252	49 51	308
78	41 49	039	19 33	094	36 59	132	22 02	156	33 47	179	46 03	253	49 19	308
79	42 14	040	20 13	094	37 29	133	22 18	157	33 47	181	45 24	254	48 48	309
80	42 40	040	20 53	095	37 58	134	22 34	158	33 47	182	44 46	255	48 17	309
81	43 06	040	21 33	096	38 27	135	22 49	159	33 45	183	44 07	256	47 45	309
82	43 31	040	22 13	097	38 55	136	23 03	160	33 43	184	43 28	256	47 14	309
83	43 58	041	22 52	098	39 23	137	23 16	161	33 39	185	42 49	257	46 43	310
84	44 24	041	23 32	098	39 49	138	23 29	162	33 35	186	42 10	258	46 13	310
85	44 50	041	24 12	099	40 16	140	23 41	163	33 30	188	41 30	259	45 42	310
86	45 17	041	24 51	100	40 41	141	23 52	164	33 25	189	40 51	260	45 11	311
87	45 43	042	25 31	101	41 07	142	24 03	165	33 18	190	40 11	261	44 41	311
88	46 10	042	26 10	102	41 31	143	24 13	166	33 11	191	39 32	262	44 11	311
89	46 37	042	26 50	102	41 55	144	24 22	167	33 03	192	38 52	262	43 41	312

LHA	Hc	Zn	Hc	Zn	Hc	Zn	Hc	Zn	Hc	Zn	Hc	Zn	Hc	Zn
♈	◆Dubhe		REGULUS		PROCYON		◆SIRIUS		RIGEL		ALDEBARAN		◆Mirfak	
90	47 04	042	27 29	103	42 18	146	24 31	168	32 54	194	54 04	216	64 36	289
91	47 31	043	28 08	104	42 40	147	24 39	169	32 44	195	53 40	217	63 58	289
92	47 58	043	28 47	105	43 01	148	24 46	170	32 33	196	53 15	219	63 20	289
93	48 25	043	29 25	106	43 22	149	24 52	171	32 22	197	52 49	220	62 42	290
94	48 53	043	30 04	107	43 42	151	24 58	172	32 10	198	52 23	222	62 04	290
95	49 20	043	30 42	107	44 01	152	25 03	173	31 57	199	51 56	223	61 26	290
96	49 48	043	31 21	108	44 20	153	25 07	174	31 43	200	51 28	225	60 49	291
97	50 16	044	31 59	109	44 37	155	25 11	176	31 29	202	51 00	226	60 11	291
98	50 43	044	32 36	110	44 54	156	25 14	177	31 14	203	50 30	227	59 34	291
99	51 11	044	33 14	111	45 10	157	25 16	178	30 58	204	50 01	228	58 57	292
100	51 39	044	33 51	112	45 25	159	25 17	179	30 42	205	49 30	230	58 19	292
101	52 07	044	34 29	113	45 39	160	25 17	180	30 24	206	48 59	231	57 42	292
102	52 35	044	35 06	114	45 52	162	25 17	181	30 06	207	48 28	232	57 05	293
103	53 03	044	35 42	115	46 04	163	25 16	182	29 48	208	47 56	233	56 28	293
104	53 31	045	36 19	116	46 16	164	25 15	183	29 29	209	47 24	235	55 51	293
	◆Dubhe		Denebola		REGULUS		◆SIRIUS		RIGEL		ALDEBARAN		◆Mirfak	
105	53 59	045	22 40	093	36 55	116	25 12	184	29 09	210	46 51	236	55 14	294
106	54 28	045	23 20	094	37 30	117	25 09	185	28 48	211	46 17	237	54 38	294
107	54 56	045	24 00	095	38 06	118	25 05	186	28 27	212	45 43	238	54 01	294
108	55 24	045	24 40	095	38 41	119	25 00	187	28 05	213	45 09	239	53 25	295
109	55 52	045	25 20	096	39 16	120	24 55	188	27 43	214	44 35	240	52 48	295
110	56 21	045	25 59	097	39 50	121	24 49	189	27 20	216	44 00	241	52 12	296
111	56 49	045	26 39	098	40 24	123	24 42	190	26 56	217	43 24	242	51 36	296
112	57 18	045	27 19	099	40 58	124	24 35	191	26 32	218	42 49	243	51 00	296
113	57 46	045	27 59	099	41 31	125	24 26	192	26 07	219	42 13	244	50 24	297
114	58 14	045	28 38	100	42 04	126	24 17	193	25 42	220	41 36	245	49 48	297
115	58 43	045	29 18	101	42 36	127	24 08	194	25 16	221	41 00	246	49 12	297
116	59 11	045	29 57	102	43 08	128	23 57	195	24 50	222	40 23	247	48 37	298
117	59 39	045	30 36	103	43 40	129	23 46	197	24 23	222	39 46	248	48 01	298
118	60 08	045	31 15	103	44 11	130	23 35	198	23 55	223	39 08	249	47 26	299
119	60 36	045	31 54	104	44 41	131	23 22	199	23 27	224	38 31	250	46 51	299
	◆Kochab		Denebola		◆REGULUS		SIRIUS		RIGEL		◆ALDEBARAN		CAPELLA	
120	42 27	021	32 33	105	45 11	133	23 09	200	22 59	225	37 53	251	62 21	281
121	42 41	021	33 12	106	45 40	134	22 55	201	22 30	226	37 15	252	61 42	282
122	42 56	021	33 50	107	46 09	135	22 41	202	22 01	227	36 37	253	61 02	282
123	43 11	022	34 29	108	46 37	136	22 26	203	21 31	228	35 58	254	60 23	283
124	43 26	022	35 07	109	47 04	138	22 10	204	21 01	229	35 19	255	59 44	283
125	43 41	022	35 45	110	47 31	139	21 54	205	20 31	230	34 41	255	59 05	284
126	43 56	022	36 22	110	47 57	140	21 37	206	20 00	231	34 02	256	58 26	284
127	44 11	022	37 00	111	48 22	141	21 19	207	19 28	232	33 23	257	57 47	285
128	44 26	022	37 37	112	48 47	143	21 01	208	18 57	233	32 44	258	57 08	285
129	44 41	023	38 14	113	49 11	144	20 42	208	18 25	234	32 04	259	56 30	286
130	44 57	023	38 51	114	49 34	146	20 22	209	17 52	234	31 25	260	55 51	286
131	45 12	023	39 27	115	49 56	147	20 02	210	17 19	235	30 45	260	55 12	287
132	45 28	023	40 04	116	50 18	148	19 42	211	16 46	236	30 06	261	54 34	287
133	45 44	023	40 39	117	50 38	150	19 21	212	16 13	237	29 26	262	53 56	288
134	45 59	023	41 15	118	50 58	151	18 59	213	15 39	238	28 46	263	53 17	288
	◆Kochab		ARCTURUS		Denebola		◆REGULUS		SIRIUS		BETELGEUSE		◆CAPELLA	
135	46 15	023	21 30	085	41 50	119	51 17	153	18 36	214	33 40	239	52 39	288
136	46 31	023	22 10	086	42 25	120	51 35	154	18 14	215	33 06	240	52 01	289
137	46 47	023	22 50	086	42 59	121	51 52	156	17 50	216	32 31	241	51 23	289
138	47 03	023	23 30	087	43 34	122	52 08	157	17 26	217	31 55	242	50 46	290
139	47 19	024	24 11	088	44 07	124	52 23	159	17 02	218	31 19	243	50 08	290
140	47 35	024	24 51	088	44 41	125	52 37	160	16 37	219	30 43	244	49 30	291
141	47 51	024	25 31	089	45 13	126	52 50	162	16 12	220	30 07	245	48 53	291
142	48 07	024	26 11	090	45 46	127	53 02	164	15 46	221	29 31	246	48 15	292
143	48 23	024	26 51	091	46 18	128	53 13	165	15 20	221	28 54	247	47 38	292
144	48 40	024	27 31	091	46 49	129	53 22	167	14 53	222	28 17	248	47 01	293
145	48 56	024	28 11	092	47 20	130	53 31	168	14 26	223	27 40	249	46 24	293
146	49 12	024	28 52	093	47 50	132	53 38	170	13 58	224	27 02	249	45 47	294
147	49 28	024	29 32	094	48 20	133	53 45	172	13 30	225	26 24	250	45 10	294
148	49 45	024	30 12	094	48 49	134	53 50	173	13 01	226	25 46	251	44 34	295
149	50 01	024	30 52	095	49 17	135	53 54	175	12 32	227	25 08	252	43 57	295
	◆Kochab		ARCTURUS		◆SPICA		REGULUS		PROCYON		◆BETELGEUSE		CAPELLA	
150	50 17	024	31 32	096	15 31	127	53 57	177	37 44	227	24 30	253	43 21	295
151	50 34	024	32 12	097	16 03	128	53 59	178	37 15	228	23 52	254	42 45	296
152	50 50	024	32 51	098	16 34	129	54 00	180	36 45	229	23 13	255	42 09	296
153	51 07	024	33 31	098	17 05	130	53 59	182	36 15	230	22 34	255	41 33	297
154	51 23	024	34 11	099	17 36	131	53 57	183	35 44	231	21 55	256	40 57	297
155	51 39	024	34 50	100	18 06	132	53 54	185	35 12	232	21 16	257	40 22	298
156	51 56	024	35 30	101	18 36	133	53 50	187	34 40	233	20 37	258	39 46	298
157	52 12	024	36 09	102	19 05	134	53 45	188	34 08	234	19 58	259	39 11	299
158	52 28	024	36 48	103	19 34	135	53 39	190	33 35	235	19 18	259	38 36	299
159	52 44	024	37 28	103	20 02	135	53 31	192	33 02	236	18 39	260	38 01	300
160	53 01	024	38 07	104	20 30	136	53 23	193	32 29	237	17 59	261	37 26	300
161	53 17	024	38 45	105	20 58	137	53 13	195	31 55	238	17 20	262	36 52	301
162	53 33	024	39 24	106	21 25	138	53 02	196	31 21	239	16 40	262	36 17	301
163	53 49	023	40 03	107	21 51	139	52 50	198	30 46	240	16 00	263	35 43	302
164	54 05	023	40 41	108	22 17	140	52 38	200	30 11	241	15 20	264	35 09	302
	◆VEGA		ARCTURUS		◆SPICA		REGULUS		PROCYON		◆POLLUX		CAPELLA	
165	14 35	047	41 19	109	22 43	141	52 24	201	29 36	242	47 38	260	34 35	303
166	15 04	048	41 57	110	23 08	142	52 09	203	29 01	243	46 58	261	34 01	303
167	15 34	049	42 35	111	23 32	143	51 53	204	28 25	244	46 18	262	33 28	304
168	16 04	049	43 12	112	23 56	144	51 36	206	27 49	244	45 39	262	32 54	304
169	16 35	050	43 49	113	24 19	145	51 18	207	27 13	245	44 59	263	32 21	305
170	17 06	050	44 26	114	24 42	146	50 59	209	26 36	246	44 19	264	31 48	305
171	17 37	051	45 03	115	25 04	147	50 39	210	25 59	247	43 39	265	31 16	306
172	18 08	052	45 39	116	25 26	148	50 19	212	25 22	248	42 59	266	30 43	306
173	18 40	052	46 15	117	25 47	149	49 57	213	24 45	249	42 19	267	30 11	307
174	19 12	053	46 51	118	26 07	150	49 35	214	24 07	250	41 39	267	29 39	307
175	19 44	053	47 26	119	26 27	151	49 12	216	23 29	250	40 59	268	29 07	308
176	20 16	054	48 01	120	26 46	152	48 48	217	22 51	251	40 18	269	28 35	308
177	20 49	055	48 36	121	27 05	153	48 24	218	22 13	252	39 38	270	28 04	309
178	21 21	055	49 10	122	27 22	154	47 58	220	21 35	253	38 58	270	27 33	309
179	21 55	056	49 44	123	27 40	155	47 32	221	20 57	254	38 18	271	27 02	310

LHA ♈	Hc	Zn	Hc	Zn	Hc	Zn	Hc	Zn	Hc	Zn	Hc	Zn	Hc	Zn
	◆DENEB		VEGA		ARCTURUS		◆SPICA		REGULUS		◆POLLUX		CAPELLA	
180	12 54	033	22 28	056	50 17	125	27 56	156	47 06	222	37 38	272	26 31	310
181	13 16	034	23 01	057	50 50	126	28 12	157	46 38	224	36 58	272	26 00	311
182	13 39	035	23 35	058	51 22	127	28 27	158	46 10	225	36 18	273	25 30	311
183	14 02	035	24 09	058	51 54	128	28 41	160	45 42	226	35 38	274	25 00	312
184	14 25	036	24 43	059	52 25	130	28 55	161	45 12	227	34 57	275	24 30	312
185	14 49	036	25 18	059	52 56	131	29 08	162	44 43	229	34 18	275	24 01	313
186	15 13	037	25 52	060	53 26	132	29 20	163	44 12	230	33 38	276	23 32	314
187	15 37	038	26 27	060	53 55	134	29 32	164	43 41	231	32 58	277	23 03	314
188	16 02	038	27 02	061	54 24	135	29 42	165	43 10	232	32 18	277	22 34	315
189	16 27	039	27 38	062	54 52	136	29 52	166	42 38	233	31 38	278	22 06	315
190	16 52	039	28 13	062	55 19	138	30 02	167	42 06	234	30 58	279	21 37	316
191	17 18	040	28 49	063	55 46	139	30 10	168	41 33	235	30 19	279	21 10	316
192	17 44	041	29 24	063	56 12	141	30 18	169	41 00	236	29 39	280	20 42	317
193	18 10	041	30 00	064	56 37	142	30 25	171	40 26	237	29 00	281	20 15	317
194	18 36	042	30 36	064	57 01	144	30 31	172	39 52	238	28 20	281	19 48	318
	◆DENEB		VEGA		Rasalhague		◆ARCTURUS		REGULUS		◆POLLUX		CAPELLA	
195	19 03	042	31 13	065	23 32	097	57 24	145	39 18	239	27 41	282	19 21	319
196	19 31	043	31 49	066	24 11	098	57 46	147	38 43	241	27 02	283	18 54	319
197	19 58	043	32 26	066	24 51	099	58 08	149	38 08	242	26 23	283	18 28	320
198	20 26	044	33 03	067	25 31	100	58 28	150	37 32	242	25 44	284	18 02	320
199	20 54	045	33 40	067	26 10	101	58 47	152	36 57	243	25 05	285	17 37	321
200	21 22	045	34 17	068	26 50	101	59 05	154	36 20	244	24 26	285	17 12	321
201	21 51	046	34 54	069	27 29	102	59 23	156	35 44	245	23 47	286	16 47	322
202	22 20	046	35 32	069	28 08	103	59 39	157	35 07	246	23 09	287	16 22	323
203	22 49	047	36 09	070	28 47	104	59 53	159	34 31	247	22 31	287	15 58	323
204	23 18	047	36 47	070	29 26	105	60 07	161	33 53	248	21 52	288	15 34	324
205	23 48	048	37 25	071	30 05	106	60 19	163	33 16	249	21 14	289	15 11	324
206	24 18	048	38 03	071	30 43	106	60 31	165	32 38	250	20 36	289	14 47	325
207	24 48	049	38 41	072	31 22	107	60 40	167	32 01	251	19 59	290	14 24	326
208	25 18	050	39 19	073	32 00	108	60 49	169	31 23	252	19 21	291	14 02	326
209	25 49	050	39 57	073	32 38	109	60 56	171	30 44	253	18 43	291	13 40	327
	DENEB		◆VEGA		Rasalhague		◆ARCTURUS		Denebola		REGULUS		◆Dubhe	
210	26 20	051	40 36	074	33 16	110	61 02	172	47 01	230	30 06	253	61 52	316
211	26 51	051	41 14	074	33 54	111	61 07	174	46 30	231	29 27	254	61 24	316
212	27 22	052	41 53	075	34 31	112	61 10	176	45 59	233	28 49	255	60 56	315
213	27 54	052	42 32	075	35 08	113	61 12	178	45 26	234	28 10	256	60 28	315
214	28 26	053	43 11	076	35 45	114	61 12	180	44 54	235	27 31	257	59 59	315
215	28 58	053	43 50	077	36 22	114	61 12	182	44 21	236	26 52	258	59 31	315
216	29 30	054	44 29	077	36 58	115	61 09	184	43 47	237	26 12	258	59 03	315
217	30 03	054	45 08	078	37 34	116	61 06	186	43 13	238	25 33	259	58 34	315
218	30 35	055	45 47	078	38 10	117	61 01	188	42 39	239	24 53	260	58 06	315
219	31 08	055	46 27	079	38 46	118	60 54	190	42 04	240	24 14	261	57 38	315
220	31 41	056	47 06	080	39 21	119	60 47	192	41 29	241	23 34	262	57 09	315
221	32 15	056	47 46	080	39 56	120	60 38	194	40 54	242	22 54	262	56 41	315
222	32 48	057	48 25	081	40 30	121	60 28	196	40 18	243	22 15	263	56 13	315
223	33 22	057	49 05	082	41 04	122	60 16	198	39 42	244	21 35	264	55 44	315
224	33 56	058	49 45	082	41 38	124	60 03	199	39 06	245	20 55	265	55 16	315
	DENEB		◆VEGA		Rasalhague		ANTARES		◆ARCTURUS		Denebola		◆Dubhe	
225	34 30	058	50 25	083	42 11	125	12 56	160	59 49	201	38 29	246	54 48	315
226	35 04	059	51 05	083	42 44	126	13 10	161	59 34	203	37 52	247	54 19	315
227	35 38	059	51 44	084	43 16	127	13 23	161	59 18	205	37 15	248	53 51	315
228	36 13	060	52 24	085	43 48	128	13 35	162	59 00	207	36 38	249	53 23	315
229	36 48	060	53 04	085	44 20	129	13 47	163	58 42	208	36 00	250	52 55	316
230	37 23	061	53 44	086	44 51	130	13 59	164	58 22	210	35 22	251	52 27	316
231	37 58	061	54 24	087	45 21	131	14 09	165	58 02	212	34 44	252	51 59	316
232	38 33	062	55 05	087	45 51	133	14 19	166	57 40	213	34 06	253	51 31	316
233	39 09	062	55 45	088	46 20	134	14 29	167	57 18	215	33 27	254	51 03	316
234	39 44	063	56 25	089	46 49	135	14 38	168	56 54	217	32 49	254	50 35	316
235	40 20	063	57 05	090	47 17	136	14 46	169	56 30	218	32 10	255	50 08	316
236	40 56	064	57 45	090	47 44	138	14 54	170	56 05	220	31 31	256	49 40	317
237	41 32	064	58 25	091	48 11	139	15 00	170	55 39	221	30 52	257	49 12	317
238	42 08	065	59 05	092	48 37	140	15 07	171	55 12	223	30 13	258	48 45	317
239	42 45	065	59 46	093	49 02	142	15 12	172	54 44	224	29 34	259	48 18	317
	◆DENEB		ALTAIR		Rasalhague		◆ANTARES		ARCTURUS		Denebola		◆Dubhe	
240	43 21	066	27 56	109	49 27	143	15 18	173	54 16	225	28 54	259	47 50	317
241	43 58	066	28 33	110	49 51	144	15 22	174	53 47	227	28 15	260	47 23	318
242	44 35	067	29 11	111	50 14	146	15 26	175	53 17	228	27 35	261	46 56	318
243	45 12	067	29 48	112	50 36	147	15 29	176	52 47	229	26 55	262	46 29	318
244	45 49	068	30 26	113	50 57	149	15 31	177	52 16	231	26 16	263	46 02	318
245	46 26	068	31 02	114	51 18	150	15 33	178	51 45	232	25 36	263	45 36	318
246	47 03	069	31 39	114	51 37	151	15 34	179	51 13	233	24 56	264	45 09	319
247	47 41	069	32 16	115	51 56	153	15 35	180	50 41	235	24 16	265	44 43	319
248	48 18	070	32 52	116	52 14	154	15 35	181	50 08	236	23 36	266	44 16	319
249	48 56	070	33 27	117	52 31	156	15 34	182	49 34	237	22 56	267	43 50	319
250	49 34	071	34 03	118	52 46	158	15 32	183	49 00	238	22 16	267	43 24	320
251	50 12	071	34 38	119	53 01	159	15 30	183	48 26	239	21 36	268	42 58	320
252	50 50	071	35 13	120	53 15	161	15 27	184	47 51	240	20 55	269	42 32	320
253	51 28	072	35 48	121	53 28	162	15 24	185	47 16	241	20 15	270	42 07	320
254	52 06	072	36 22	122	53 39	164	15 20	186	46 41	243	19 35	270	41 41	321
	◆Schedar		DENEB		ALTAIR		◆Rasalhague		ARCTURUS		◆Alkaid		Kochab	
255	27 36	034	52 44	073	36 56	123	53 50	166	46 05	244	58 43	291	60 22	343
256	27 59	035	53 23	073	37 29	124	53 59	167	45 29	245	58 05	291	60 10	342
257	28 22	035	54 01	074	38 02	125	54 08	169	44 52	246	57 28	292	59 58	342
258	28 46	036	54 40	074	38 35	126	54 15	170	44 16	247	56 50	292	59 45	342
259	29 09	036	55 19	075	39 07	127	54 21	172	43 39	248	56 13	292	59 32	341
260	29 33	037	55 57	075	39 38	128	54 26	174	43 01	249	55 36	293	59 20	341
261	29 57	037	56 36	076	40 10	130	54 30	176	42 24	250	54 59	293	59 06	341
262	30 21	037	57 15	076	40 40	131	54 32	177	41 46	251	54 22	293	58 53	340
263	30 46	038	57 54	077	41 11	132	54 34	179	41 08	252	53 45	294	58 40	340
264	31 11	038	58 34	077	41 40	133	54 34	181	40 30	252	53 09	294	58 26	340
265	31 36	039	59 13	078	42 09	134	54 33	182	39 52	253	52 32	294	58 12	340
266	32 01	039	59 52	078	42 38	135	54 31	184	39 13	254	51 56	295	57 58	339
267	32 26	040	60 31	079	43 06	136	54 27	186	38 34	255	51 19	295	57 44	339
268	32 52	040	61 11	079	43 33	138	54 23	187	37 56	256	50 43	296	57 29	339
269	33 18	040	61 50	080	44 00	139	54 17	189	37 17	257	50 07	296	57 15	339

LAT 48°N

LHA ♈	Hc	Zn	Hc	Zn	Hc	Zn	Hc	Zn	Hc	Zn	Hc	Zn	Hc	Zn
	◆Mirfak		Alpheratz		◆ALTAIR		Rasalhague		◆ARCTURUS		Alkaid		Kochab	
270	13 27	025	19 54	068	44 26	140	54 10	191	36 37	258	49 31	296	57 00	338
271	13 44	025	20 31	069	44 52	141	54 02	192	35 58	259	48 55	297	56 45	338
272	14 02	026	21 08	070	45 16	143	53 53	194	35 19	259	48 19	297	56 30	338
273	14 19	026	21 46	070	45 40	144	53 43	196	34 39	260	47 43	298	56 15	338
274	14 37	027	22 24	071	46 04	145	53 32	197	34 00	261	47 08	298	56 00	338
275	14 56	028	23 02	072	46 26	147	53 20	199	33 20	262	46 33	298	55 45	337
276	15 15	028	23 40	072	46 48	148	53 06	200	32 40	263	45 57	299	55 29	337
277	15 34	029	24 18	073	47 09	149	52 52	202	32 00	263	45 22	299	55 14	337
278	15 53	029	24 57	074	47 29	151	52 36	203	31 20	264	44 47	300	54 58	337
279	16 13	030	25 35	074	47 48	152	52 20	205	30 40	265	44 13	300	54 42	337
280	16 33	031	26 14	075	48 07	153	52 02	206	30 00	266	43 38	300	54 26	337
281	16 54	031	26 53	075	48 24	155	51 44	208	29 20	267	43 03	301	54 10	337
282	17 15	032	27 32	076	48 41	156	51 25	209	28 40	267	42 29	301	53 55	337
283	17 36	032	28 11	077	48 57	158	51 04	211	28 00	268	41 55	302	53 39	336
284	17 58	033	28 50	077	49 12	159	50 43	212	27 20	269	41 21	302	53 22	336
	◆Mirfak		Alpheratz		◆ALTAIR		Rasalhague		◆ARCTURUS		Alkaid		Kochab	
285	18 19	033	29 29	078	49 25	161	50 21	214	26 40	270	40 47	303	53 06	336
286	18 42	034	30 09	079	49 38	162	49 59	215	26 00	270	40 13	303	52 50	336
287	19 04	034	30 48	079	49 50	164	49 35	217	25 19	271	39 39	303	52 34	336
288	19 27	035	31 28	080	50 01	165	49 11	218	24 39	272	39 06	304	52 18	336
289	19 50	036	32 07	081	50 11	167	48 46	219	23 59	273	38 33	304	52 01	336
290	20 14	036	32 47	081	50 20	168	48 20	221	23 19	273	38 00	305	51 45	336
291	20 38	037	33 27	082	50 27	170	47 54	222	22 39	274	37 27	305	51 29	336
292	21 02	037	34 06	083	50 34	171	47 26	223	21 59	275	36 54	306	51 12	336
293	21 26	038	34 46	084	50 40	173	46 59	224	21 19	275	36 22	306	50 56	336
294	21 51	038	35 26	084	50 44	174	46 30	226	20 39	276	35 49	307	50 40	336
295	22 16	039	36 06	085	50 47	176	46 01	227	19 59	277	35 17	307	50 23	336
296	22 41	039	36 46	086	50 50	177	45 31	228	19 19	278	34 45	308	50 07	336
297	23 07	040	37 26	086	50 51	179	45 01	229	18 40	278	34 14	308	49 51	336
298	23 33	040	38 06	087	50 51	181	44 31	231	18 00	279	33 42	308	49 34	336
299	23 59	041	38 46	088	50 50	182	43 59	232	17 20	280	33 11	309	49 18	336
	Mirfak		◆Alpheratz		Enif		ALTAIR		◆Rasalhague		Alphecca		◆Kochab	
300	24 25	041	39 26	088	46 02	142	50 48	184	43 28	233	35 01	272	49 02	336
301	24 52	042	40 07	089	46 27	143	50 45	185	42 55	234	34 20	273	48 45	336
302	25 19	042	40 47	090	46 51	144	50 41	187	42 23	235	33 40	274	48 29	336
303	25 46	043	41 27	091	47 14	145	50 36	188	41 50	236	33 00	275	48 13	336
304	26 13	043	42 07	092	47 36	147	50 29	190	41 16	237	32 20	275	47 57	336
305	26 41	044	42 47	092	47 58	148	50 22	191	40 42	238	31 40	276	47 41	336
306	27 09	044	43 27	093	48 18	150	50 13	193	40 08	239	31 00	277	47 25	336
307	27 37	045	44 07	094	48 38	151	50 04	195	39 33	240	30 21	277	47 09	336
308	28 06	045	44 47	095	48 57	152	49 53	196	38 58	241	29 41	278	46 53	337
309	28 35	046	45 27	095	49 16	154	49 42	198	38 23	242	29 01	279	46 37	337
310	29 04	046	46 07	096	49 33	155	49 29	199	37 47	243	28 22	279	46 21	337
311	29 33	047	46 47	097	49 49	157	49 15	200	37 11	244	27 42	280	46 05	337
312	30 02	047	47 27	098	50 04	158	49 01	202	36 35	245	27 02	281	45 49	337
313	30 32	048	48 07	099	50 19	160	48 45	203	35 58	246	26 23	282	45 33	337
314	31 02	048	48 46	100	50 32	161	48 29	205	35 21	247	25 44	282	45 18	337
	CAPELLA		◆Hamal		Alpheratz		Enif		◆ALTAIR		VEGA		◆Kochab	
315	15 54	037	25 53	083	49 26	100	50 45	163	48 12	206	62 41	264	45 02	337
316	16 18	037	26 33	084	50 05	101	50 56	164	47 53	208	62 01	265	44 47	337
317	16 43	038	27 13	084	50 45	102	51 06	166	47 34	209	61 21	265	44 32	338
318	17 08	039	27 53	085	51 24	103	51 16	167	47 14	210	60 41	266	44 16	338
319	17 33	039	28 33	086	52 03	104	51 24	169	46 54	212	60 01	267	44 01	338
320	17 58	040	29 13	087	52 42	105	51 31	170	46 32	213	59 21	268	43 46	338
321	18 24	040	29 53	087	53 20	106	51 37	172	46 10	214	58 41	269	43 31	338
322	18 50	041	30 33	088	53 59	107	51 42	174	45 47	216	58 01	269	43 16	338
323	19 17	041	31 13	089	54 37	108	51 46	175	45 23	217	57 20	270	43 01	338
324	19 43	042	31 54	089	55 15	109	51 49	177	44 58	218	56 40	271	42 47	339
325	20 10	043	32 34	090	55 53	110	51 51	178	44 33	220	56 00	272	42 32	339
326	20 38	043	33 14	091	56 31	111	51 51	180	44 07	221	55 20	272	42 18	339
327	21 05	044	33 54	092	57 08	112	51 51	182	43 41	222	54 40	273	42 03	339
328	21 33	044	34 34	092	57 45	114	51 49	183	43 13	223	54 00	274	41 49	339
329	22 01	045	35 14	093	58 21	115	51 46	185	42 46	224	53 20	274	41 35	340
	CAPELLA		◆Hamal		Diphda		Enif		◆ALTAIR		VEGA		◆Kochab	
330	22 30	045	35 54	094	14 34	140	51 42	186	42 17	226	52 40	275	41 21	340
331	22 58	046	36 34	095	15 00	141	51 37	188	41 48	227	52 00	276	41 07	340
332	23 27	046	37 14	096	15 25	142	51 31	190	41 19	228	51 20	276	40 53	340
333	23 56	047	37 54	096	15 50	143	51 24	191	40 49	229	50 40	277	40 40	340
334	24 26	047	38 34	097	16 14	144	51 16	193	40 18	230	50 00	278	40 26	341
335	24 55	048	39 14	098	16 37	144	51 06	194	39 47	231	49 20	278	40 13	341
336	25 25	049	39 54	099	17 00	145	50 56	196	39 15	232	48 41	279	40 00	341
337	25 56	049	40 33	100	17 23	146	50 44	197	38 43	233	48 01	280	39 47	341
338	26 26	050	41 13	101	17 45	147	50 32	199	38 11	234	47 21	280	39 34	341
339	26 57	050	41 52	101	18 06	148	50 18	200	37 38	236	46 42	281	39 21	342
340	27 28	051	42 31	102	18 27	149	50 04	202	37 05	237	46 03	281	39 08	342
341	27 59	051	43 11	103	18 48	150	49 49	203	36 31	238	45 23	282	38 56	342
342	28 30	052	43 50	104	19 07	151	49 32	205	35 57	239	44 44	283	38 44	342
343	29 02	052	44 29	105	19 27	152	49 15	206	35 22	240	44 05	283	38 32	342
344	29 34	053	45 07	106	19 45	153	48 57	208	34 48	241	43 26	284	38 20	343
	◆CAPELLA		ALDEBARAN		◆Diphda		Enif		ALTAIR		◆VEGA		Kochab	
345	30 06	053	16 13	083	20 03	154	48 38	209	34 13	242	42 47	284	38 08	343
346	30 38	054	16 52	084	20 21	155	48 18	210	33 37	242	42 08	285	37 56	343
347	31 10	054	17 32	085	20 37	156	47 57	212	33 01	243	41 29	286	37 45	343
348	31 43	055	18 12	085	20 54	157	47 35	213	32 25	244	40 51	286	37 33	344
349	32 16	055	18 52	086	21 09	158	47 13	215	31 49	245	40 12	287	37 22	344
350	32 49	056	19 32	087	21 24	159	46 50	216	31 12	246	39 34	287	37 11	344
351	33 22	056	20 13	088	21 38	160	46 26	217	30 35	247	38 55	288	37 00	344
352	33 56	057	20 53	088	21 52	161	46 01	218	29 58	248	38 17	288	36 49	345
353	34 30	057	21 33	089	22 05	162	45 36	220	29 21	249	37 39	289	36 39	345
354	35 03	058	22 13	090	22 17	163	45 10	221	28 44	250	37 01	290	36 29	345
355	35 37	058	22 53	090	22 29	164	44 43	222	28 06	251	36 24	290	36 19	346
356	36 12	059	23 33	091	22 40	165	44 16	223	27 28	251	35 46	291	36 09	346
357	36 46	059	24 13	092	22 50	166	43 48	225	26 50	252	35 09	291	35 59	346
358	37 21	060	24 54	093	23 00	167	43 20	226	26 11	253	34 31	292	35 49	346
359	37 55	060	25 34	093	23 08	168	42 50	227	25 33	254	33 54	292	35 40	347

LHA ♈	Hc	Zn	Hc	Zn	Hc	Zn	Hc	Zn	Hc	Zn	Hc	Zn	Hc	Zn
	◆CAPELLA		ALDEBARAN		Hamal		◆Diphda		ALTAIR		◆VEGA		Kochab	
0	38 01	060	26 18	094	55 24	122	24 15	169	25 10	255	32 53	294	34 32	347
1	38 36	060	26 59	095	55 59	123	24 23	170	24 30	256	32 16	294	34 23	347
2	39 12	061	27 39	095	56 33	124	24 30	171	23 50	257	31 39	295	34 14	348
3	39 48	061	28 20	096	57 06	126	24 36	172	23 10	258	31 01	295	34 05	348
4	40 24	062	29 01	097	57 39	127	24 42	173	22 30	258	30 25	296	33 57	348
5	41 00	062	29 41	098	58 11	128	24 46	174	21 50	259	29 48	296	33 49	348
6	41 36	063	30 22	098	58 43	130	24 50	175	21 10	260	29 11	297	33 41	349
7	42 12	063	31 02	099	59 14	131	24 54	176	20 29	261	28 35	297	33 33	349
8	42 49	064	31 43	100	59 45	133	24 56	177	19 49	262	27 59	298	33 25	349
9	43 26	064	32 23	101	60 14	134	24 58	178	19 08	262	27 23	299	33 17	350
10	44 03	065	33 03	102	60 43	136	24 59	179	18 28	263	26 47	299	33 10	350
11	44 40	065	33 43	102	61 11	138	24 59	180	17 47	264	26 11	300	33 03	350
12	45 17	065	34 23	103	61 38	139	24 59	181	17 07	265	25 36	300	32 56	350
13	45 54	066	35 03	104	62 04	141	24 57	182	16 26	265	25 01	301	32 49	351
14	46 32	066	35 42	105	62 29	143	24 55	183	15 45	266	24 25	301	32 43	351
	◆CAPELLA		ALDEBARAN		Hamal		◆Diphda		Alpheratz		◆DENEB		Kochab	
15	47 09	067	36 22	106	62 54	145	24 53	184	69 25	214	46 27	293	32 37	351
16	47 47	067	37 01	107	63 17	146	24 49	185	69 02	216	45 49	293	32 31	352
17	48 24	068	37 40	108	63 39	148	24 45	186	68 37	218	45 12	293	32 25	352
18	49 02	068	38 19	108	64 00	150	24 40	188	68 11	220	44 34	294	32 19	352
19	49 40	069	38 58	109	64 20	152	24 34	189	67 44	222	43 57	294	32 14	353
20	50 19	069	39 36	110	64 38	154	24 28	190	67 16	224	43 20	295	32 09	353
21	50 57	069	40 15	111	64 55	156	24 21	191	66 47	226	42 43	295	32 04	353
22	51 35	070	40 53	112	65 11	158	24 13	192	66 17	228	42 06	296	31 59	353
23	52 14	070	41 30	113	65 25	160	24 04	193	65 46	230	41 29	296	31 54	354
24	52 52	071	42 08	114	65 38	163	23 55	194	65 15	231	40 52	297	31 50	354
25	53 31	071	42 45	115	65 50	165	23 45	195	64 42	233	40 16	297	31 46	354
26	54 10	072	43 22	116	66 00	167	23 34	196	64 09	235	39 39	298	31 42	355
27	54 48	072	43 59	117	66 08	169	23 23	197	63 36	236	39 03	298	31 38	355
28	55 27	072	44 35	118	66 15	171	23 10	198	63 02	238	38 27	299	31 35	355
29	56 06	073	45 11	119	66 20	174	22 58	199	62 27	239	37 51	299	31 32	356
	CAPELLA		◆BETELGEUSE		RIGEL		Hamal		◆Alpheratz		DENEB		◆Kochab	
30	56 46	073	26 27	109	20 01	128	66 24	176	61 51	240	37 16	300	31 29	356
31	57 25	074	27 05	110	20 33	129	66 26	178	61 16	242	36 40	300	31 26	356
32	58 04	074	27 44	111	21 05	130	66 26	181	60 39	243	36 05	300	31 24	357
33	58 44	075	28 22	111	21 36	131	66 25	183	60 03	244	35 30	301	31 21	357
34	59 23	075	29 00	112	22 07	131	66 22	185	59 26	245	34 55	301	31 19	357
35	60 03	075	29 38	113	22 37	132	66 18	187	58 48	247	34 20	302	31 17	358
36	60 42	076	30 15	114	23 07	133	66 11	190	58 11	248	33 45	302	31 16	358
37	61 22	076	30 53	115	23 37	134	66 04	192	57 32	249	33 11	303	31 14	358
38	62 02	077	31 30	116	24 06	135	65 54	194	56 54	250	32 37	303	31 13	359
39	62 42	077	32 06	117	24 35	136	65 44	196	56 16	251	32 02	304	31 12	359
40	63 22	078	32 43	118	25 03	137	65 31	199	55 37	252	31 29	304	31 11	359
41	64 02	078	33 19	119	25 30	138	65 18	201	54 58	253	30 55	305	31 11	359
42	64 42	079	33 55	120	25 57	139	65 02	203	54 18	254	30 21	305	31 11	000
43	65 22	079	34 30	121	26 24	140	64 46	205	53 39	255	29 48	306	31 11	000
44	66 02	079	35 05	121	26 50	141	64 28	207	52 59	256	29 15	306	31 11	000
	◆Dubhe		POLLUX		BETELGEUSE		◆RIGEL		Hamal		◆Alpheratz		DENEB	
45	28 37	028	32 30	082	35 40	122	27 15	142	64 09	209	52 19	257	28 42	307
46	28 56	028	33 10	083	36 14	123	27 40	143	63 48	211	51 39	258	28 10	307
47	29 16	028	33 51	084	36 48	124	28 05	144	63 27	213	50 59	259	27 37	308
48	29 35	029	34 32	084	37 22	126	28 28	145	63 04	215	50 19	260	27 05	308
49	29 55	029	35 12	085	37 55	127	28 52	146	62 40	216	49 39	261	26 33	309
50	30 15	030	35 53	086	38 27	128	29 14	147	62 16	218	48 58	262	26 01	309
51	30 35	030	36 34	086	38 59	129	29 36	148	61 50	220	48 18	262	25 30	310
52	30 56	030	37 15	087	39 31	130	29 57	149	61 23	222	47 37	263	24 59	311
53	31 16	031	37 56	088	40 02	131	30 18	150	60 55	223	46 57	264	24 28	311
54	31 37	031	38 37	089	40 33	132	30 38	151	60 27	225	46 16	265	23 57	312
55	31 59	031	39 18	089	41 03	133	30 57	153	59 58	226	45 35	266	23 27	312
56	32 20	032	39 59	090	41 33	134	31 15	154	59 28	228	44 54	266	22 56	313
57	32 42	032	40 39	091	42 02	135	31 33	155	58 57	229	44 13	267	22 26	313
58	33 04	032	41 20	091	42 30	137	31 50	156	58 26	231	43 33	268	21 57	314
59	33 26	033	42 01	092	42 58	138	32 07	157	57 54	232	42 52	269	21 27	314
	◆Dubhe		POLLUX		SIRIUS		◆RIGEL		Hamal		◆Alpheratz		DENEB	
60	33 48	033	42 42	093	16 19	139	32 22	158	57 21	234	42 11	269	20 58	315
61	34 11	034	43 23	094	16 46	140	32 37	159	56 48	235	41 30	270	20 29	315
62	34 33	034	44 04	095	17 12	141	32 52	160	56 14	236	40 49	271	20 00	316
63	34 56	034	44 45	095	17 38	142	33 05	162	55 40	238	40 08	272	19 32	316
64	35 19	035	45 25	096	18 03	142	33 17	163	55 05	239	39 27	272	19 04	317
65	35 43	035	46 06	097	18 28	143	33 29	164	54 30	240	38 46	273	18 36	318
66	36 06	035	46 47	098	18 52	144	33 40	165	53 54	241	38 05	274	18 09	318
67	36 30	036	47 27	099	19 15	145	33 50	166	53 18	242	37 25	274	17 42	319
68	36 54	036	48 07	099	19 38	146	34 00	167	52 41	244	36 44	275	17 15	319
69	37 18	036	48 48	100	20 01	147	34 08	169	52 04	245	36 03	276	16 48	320
70	37 42	036	49 28	101	20 23	148	34 16	170	51 27	246	35 22	276	16 22	320
71	38 06	037	50 08	102	20 44	149	34 23	171	50 50	247	34 42	277	15 56	321
72	38 31	037	50 48	103	21 05	150	34 29	172	50 12	248	34 01	278	15 30	322
73	38 56	037	51 28	104	21 25	151	34 34	173	49 34	249	33 21	278	15 05	322
74	39 21	038	52 07	105	21 45	152	34 39	174	48 56	250	32 40	279	14 40	323
	◆Dubhe		REGULUS		PROCYON		◆SIRIUS		RIGEL		◆Hamal		Schedar	
75	39 46	038	17 34	091	36 05	128	22 04	153	34 42	176	48 17	251	50 17	309
76	40 11	038	18 15	092	36 37	129	22 22	154	34 45	177	47 38	252	49 46	309
77	40 36	039	18 55	093	37 08	130	22 40	155	34 47	178	46 59	253	49 14	309
78	41 02	039	19 36	093	37 39	131	22 57	156	34 47	179	46 20	254	48 42	309
79	41 28	039	20 17	094	38 10	132	23 14	157	34 47	181	45 41	255	48 10	309
80	41 53	039	20 58	095	38 40	133	23 29	158	34 47	182	45 01	256	47 39	310
81	42 19	040	21 39	096	39 09	134	23 45	159	34 45	183	44 21	257	47 07	310
82	42 46	040	22 19	096	39 38	136	23 59	160	34 43	184	43 42	257	46 36	310
83	43 12	040	23 00	097	40 06	137	24 13	161	34 39	185	43 02	258	46 05	310
84	43 38	040	23 41	098	40 34	138	24 26	162	34 35	187	42 21	259	45 34	311
85	44 05	041	24 21	099	41 01	139	24 38	163	34 30	188	41 41	260	45 03	311
86	44 31	041	25 01	099	41 28	140	24 50	164	34 24	189	41 01	261	44 32	311
87	44 58	041	25 42	100	41 54	141	25 01	165	34 17	190	40 20	262	44 01	312
88	45 25	041	26 22	101	42 19	143	25 11	166	34 10	191	39 40	262	43 31	312
89	45 52	041	27 02	102	42 43	144	25 21	167	34 01	192	38 59	263	43 00	312

LAT 47°N

LHA ♈	Hc	Zn	Hc	Zn	Hc	Zn	Hc	Zn	Hc	Zn	Hc	Zn	Hc	Zn
	◆Dubhe		REGULUS		PROCYON		◆SIRIUS		RIGEL		ALDEBARAN		◆Mirfak	
90	46 19	042	27 42	103	43 07	145	25 30	168	33 52	194	54 52	217	64 15	291
91	46 46	042	28 22	103	43 30	146	25 38	169	33 42	195	54 27	218	63 37	291
92	47 14	042	29 02	104	43 52	148	25 45	170	33 31	196	54 01	220	62 59	291
93	47 41	042	29 41	105	44 14	149	25 52	171	33 19	197	53 35	221	62 21	292
94	48 09	042	30 21	106	44 34	150	25 58	172	33 07	198	53 07	223	61 43	292
95	48 36	043	31 00	107	44 54	152	26 03	173	32 54	199	52 39	224	61 05	292
96	49 04	043	31 39	108	45 13	153	26 07	174	32 40	201	52 10	225	60 27	292
97	49 32	043	32 18	109	45 31	154	26 11	175	32 25	202	51 41	227	59 49	293
98	50 00	043	32 57	109	45 49	156	26 13	177	32 09	203	51 11	228	59 11	293
99	50 28	043	33 35	110	46 05	157	26 16	178	31 53	204	50 40	229	58 34	293
100	50 56	043	34 13	111	46 21	158	26 17	179	31 36	205	50 09	231	57 56	294
101	51 24	043	34 51	112	46 35	160	26 17	180	31 18	206	49 37	232	57 19	294
102	51 52	043	35 29	113	46 49	161	26 17	181	31 00	207	49 04	233	56 41	294
103	52 20	044	36 07	114	47 02	163	26 16	182	30 41	208	48 31	234	56 04	294
104	52 48	044	36 44	115	47 13	164	26 15	183	30 21	209	47 58	235	55 27	295
	◆Dubhe		Denebola		REGULUS		◆SIRIUS		RIGEL		ALDEBARAN		◆Mirfak	
105	53 16	044	22 43	093	37 21	116	26 12	184	30 00	211	47 24	237	54 50	295
106	53 45	044	23 24	094	37 58	117	26 09	185	29 39	212	46 50	238	54 13	295
107	54 13	044	24 04	094	38 34	118	26 05	186	29 17	213	46 15	239	53 36	296
108	54 41	044	24 45	095	39 10	119	26 00	187	28 55	214	45 40	240	52 59	296
109	55 10	044	25 26	096	39 46	120	25 55	188	28 32	215	45 04	241	52 22	296
110	55 38	044	26 07	097	40 21	121	25 48	189	28 08	216	44 28	242	51 46	297
111	56 06	044	26 47	097	40 56	122	25 41	190	27 44	217	43 52	243	51 09	297
112	56 35	044	27 28	098	41 31	123	25 33	191	27 19	218	43 15	244	50 33	297
113	57 03	044	28 08	099	42 05	124	25 25	192	26 54	219	42 38	245	49 56	298
114	57 31	044	28 49	100	42 39	125	25 16	194	26 28	220	42 01	246	49 20	298
115	58 00	044	29 29	100	43 12	126	25 06	195	26 01	221	41 24	247	48 44	298
116	58 28	044	30 09	101	43 45	127	24 55	196	25 34	222	40 46	248	48 08	299
117	58 56	044	30 49	102	44 17	128	24 44	197	25 07	223	40 08	249	47 33	299
118	59 25	044	31 29	103	44 49	129	24 32	198	24 39	224	39 29	250	46 57	300
119	59 53	043	32 09	104	45 20	131	24 19	199	24 10	225	38 51	251	46 21	300
	◆Kochab		Denebola		◆REGULUS		SIRIUS		RIGEL		◆ALDEBARAN		CAPELLA	
120	41 31	021	32 49	105	45 51	132	24 06	200	23 41	226	38 12	252	62 08	283
121	41 46	021	33 28	105	46 21	133	23 51	201	23 12	227	37 33	253	61 28	284
122	42 00	021	34 07	106	46 51	134	23 37	202	22 42	228	36 54	253	60 49	284
123	42 15	021	34 47	107	47 20	136	23 21	203	22 11	228	36 15	254	60 09	284
124	42 30	021	35 26	108	47 48	137	23 05	204	21 41	229	35 35	255	59 29	285
125	42 45	022	36 05	109	48 16	138	22 48	205	21 09	230	34 56	256	58 50	285
126	43 00	022	36 43	110	48 43	139	22 31	206	20 38	231	34 16	257	58 10	286
127	43 15	022	37 22	111	49 09	141	22 13	207	20 06	232	33 36	258	57 31	286
128	43 31	022	38 00	112	49 35	142	21 54	208	19 33	233	32 56	259	56 52	287
129	43 46	022	38 38	113	49 59	143	21 34	209	19 00	234	32 16	259	56 13	287
130	44 01	022	39 15	114	50 23	145	21 15	210	18 27	235	31 35	260	55 34	288
131	44 17	022	39 53	114	50 46	146	20 54	211	17 53	236	30 55	261	54 55	288
132	44 33	022	40 30	115	51 09	148	20 33	212	17 20	236	30 14	262	54 16	288
133	44 48	023	41 07	116	51 30	149	20 11	212	16 45	237	29 34	263	53 37	289
134	45 04	023	41 43	117	51 51	151	19 49	213	16 11	238	28 53	263	52 58	289
	◆Kochab		ARCTURUS		Denebola		◆REGULUS		SIRIUS		BETELGEUSE		◆CAPELLA	
135	45 20	023	21 25	084	42 19	118	52 10	152	19 26	214	34 11	240	52 20	290
136	45 36	023	22 05	085	42 55	119	52 29	154	19 03	215	33 35	241	51 41	290
137	45 52	023	22 46	086	43 30	121	52 46	155	18 39	216	32 59	242	51 03	291
138	46 08	023	23 27	087	44 05	122	53 03	157	18 14	217	32 23	243	50 25	291
139	46 24	023	24 08	087	44 40	123	53 19	158	17 49	218	31 46	244	49 46	291
140	46 40	023	24 49	088	45 14	124	53 33	160	17 24	219	31 09	245	49 08	292
141	46 56	023	25 30	089	45 48	125	53 47	162	16 58	220	30 32	246	48 31	292
142	47 12	023	26 11	089	46 21	126	53 59	163	16 32	221	29 55	246	47 53	293
143	47 28	023	26 52	090	46 54	127	54 10	165	16 05	222	29 17	247	47 15	293
144	47 45	023	27 32	091	47 26	128	54 21	166	15 37	222	28 39	248	46 37	294
145	48 01	023	28 13	092	47 58	130	54 30	168	15 09	223	28 01	249	46 00	294
146	48 17	024	28 54	092	48 30	131	54 38	170	14 41	224	27 23	250	45 23	295
147	48 34	024	29 35	093	49 00	132	54 44	171	14 12	225	26 44	251	44 46	295
148	48 50	024	30 16	094	49 30	133	54 50	173	13 43	226	26 06	252	44 09	295
149	49 06	024	30 57	095	50 00	135	54 54	175	13 13	227	25 27	252	43 32	296
	◆Kochab		ARCTURUS		◆SPICA		REGULUS		PROCYON		◆BETELGEUSE		CAPELLA	
150	49 23	024	31 38	095	16 08	127	54 57	177	38 25	227	24 47	253	42 55	296
151	49 39	024	32 18	096	16 40	128	54 59	178	37 55	228	24 08	254	42 18	297
152	49 55	024	32 59	097	17 12	129	55 00	180	37 24	229	23 29	255	41 42	297
153	50 12	024	33 40	098	17 44	130	54 59	182	36 53	230	22 49	256	41 06	298
154	50 28	024	34 20	099	18 15	131	54 57	183	36 21	231	22 09	257	40 29	298
155	50 44	023	35 00	099	18 46	132	54 54	185	35 49	232	21 30	257	39 53	299
156	51 01	023	35 41	100	19 16	132	54 50	187	35 16	234	20 50	258	39 17	299
157	51 17	023	36 21	101	19 46	133	54 45	188	34 43	235	20 10	259	38 42	299
158	51 33	023	37 01	102	20 16	134	54 38	190	34 10	236	19 29	260	38 06	300
159	51 49	023	37 41	103	20 45	135	54 30	192	33 36	237	18 49	260	37 31	300
160	52 05	023	38 21	103	21 13	136	54 21	193	33 01	237	18 09	261	36 56	301
161	52 22	023	39 01	104	21 42	137	54 11	195	32 27	238	17 28	262	36 21	301
162	52 38	023	39 40	105	22 09	138	54 00	197	31 52	239	16 48	263	35 46	302
163	52 54	023	40 20	106	22 36	139	53 47	198	31 16	240	16 07	264	35 11	302
164	53 10	023	40 59	107	23 03	140	53 34	200	30 41	241	15 26	264	34 37	303
	◆VEGA		ARCTURUS		◆SPICA		REGULUS		PROCYON		◆POLLUX		CAPELLA	
165	13 54	047	41 38	108	23 29	141	53 20	202	30 05	242	47 47	261	34 02	303
166	14 24	048	42 17	109	23 55	142	53 04	203	29 28	243	47 07	262	33 28	304
167	14 54	048	42 55	110	24 20	143	52 47	205	28 52	244	46 26	263	32 54	304
168	15 25	049	43 34	111	24 44	144	52 30	206	28 15	245	45 46	263	32 20	305
169	15 56	050	44 12	112	25 08	145	52 11	208	27 37	246	45 05	264	31 47	305
170	16 27	050	44 50	113	25 32	146	51 52	209	27 00	247	44 24	265	31 14	306
171	16 59	051	45 27	114	25 54	147	51 31	211	26 22	248	43 44	266	30 40	306
172	17 31	051	46 05	115	26 17	148	51 10	212	25 44	248	43 03	267	30 08	307
173	18 03	052	46 42	116	26 38	149	50 48	214	25 06	249	42 22	267	29 35	307
174	18 35	053	47 18	117	26 59	150	50 25	215	24 28	250	41 41	268	29 02	308
175	19 08	053	47 55	118	27 19	151	50 01	216	23 49	251	41 00	269	28 30	308
176	19 41	054	48 31	119	27 39	152	49 36	218	23 10	252	40 19	270	27 58	309
177	20 14	054	49 06	120	27 58	153	49 10	219	22 32	253	39 38	270	27 26	309
178	20 47	055	49 42	121	28 16	154	48 44	221	21 52	253	38 57	271	26 54	310
179	21 21	055	50 16	122	28 34	155	48 17	222	21 13	254	38 16	272	26 23	310

LHA	Hc	Zn	Hc	Zn	Hc	Zn	Hc	Zn	Hc	Zn	Hc	Zn	Hc	Zn
♈	◆DENEB		VEGA		ARCTURUS		◆SPICA		REGULUS		◆POLLUX		CAPELLA	
180	12 04	033	21 55	056	50 51	124	28 51	156	47 50	223	37 36	273	25 52	311
181	12 26	034	22 29	057	51 25	125	29 07	157	47 21	224	36 55	273	25 21	311
182	12 49	034	23 03	057	51 58	126	29 23	158	46 52	226	36 14	274	24 50	312
183	13 13	035	23 37	058	52 31	127	29 37	159	46 23	227	35 33	275	24 20	312
184	13 36	036	24 12	058	53 03	129	29 52	160	45 53	228	34 52	275	23 50	313
185	14 00	036	24 47	059	53 35	130	30 05	162	45 22	229	34 12	276	23 20	313
186	14 25	037	25 22	059	54 06	131	30 18	163	44 51	230	33 31	277	22 50	314
187	14 49	037	25 58	060	54 36	133	30 29	164	44 19	232	32 50	277	22 21	314
188	15 15	038	26 33	061	55 06	134	30 40	165	43 47	233	32 10	278	21 52	315
189	15 40	039	27 09	061	55 35	135	30 51	166	43 14	234	31 29	279	21 23	315
190	16 06	039	27 45	062	56 04	137	31 00	167	42 40	235	30 49	279	20 54	316
191	16 32	040	28 21	062	56 31	138	31 09	168	42 07	236	30 09	280	20 26	317
192	16 58	040	28 57	063	56 58	140	31 17	169	41 33	237	29 28	281	19 58	317
193	17 25	041	29 34	063	57 24	141	31 24	171	40 58	238	28 48	281	19 30	318
194	17 52	042	30 10	064	57 49	143	31 30	172	40 23	239	28 08	282	19 03	318
	◆DENEB		VEGA		Rasalhague		◆ARCTURUS		REGULUS		◆POLLUX		CAPELLA	
195	18 19	042	30 47	065	23 39	097	58 13	145	39 48	240	27 28	283	18 36	319
196	18 46	043	31 24	065	24 20	098	58 36	146	39 12	241	26 48	283	18 09	319
197	19 14	043	32 01	066	25 00	099	58 59	148	38 36	242	26 08	284	17 43	320
198	19 43	044	32 39	066	25 41	099	59 20	150	38 00	243	25 29	285	17 16	320
199	20 11	044	33 16	067	26 21	100	59 40	151	37 23	244	24 49	285	16 50	321
200	20 40	045	33 54	067	27 01	101	59 59	153	36 46	245	24 10	286	16 25	322
201	21 09	045	34 32	068	27 41	102	60 17	155	36 09	246	23 31	287	16 00	322
202	21 38	046	35 10	068	28 21	102	60 34	157	35 31	247	22 51	287	15 35	323
203	22 08	047	35 48	069	29 01	103	60 49	159	34 53	248	22 12	288	15 10	323
204	22 37	047	36 26	070	29 41	104	61 04	160	34 15	249	21 33	289	14 46	324
205	23 07	048	37 05	070	30 20	105	61 17	162	33 37	250	20 55	289	14 22	324
206	23 38	048	37 43	071	31 00	106	61 28	164	32 59	251	20 16	290	13 58	325
207	24 08	049	38 22	071	31 39	107	61 39	166	32 20	251	19 38	290	13 35	326
208	24 39	049	39 01	072	32 18	108	61 48	168	31 41	252	18 59	291	13 12	326
209	25 10	050	39 40	072	32 57	108	61 56	170	31 02	253	18 21	292	12 50	327
	DENEB		◆VEGA		Rasalhague		◆ARCTURUS		Denebola		REGULUS		◆Dubhe	
210	25 42	050	40 19	073	33 36	109	62 02	172	47 39	231	30 23	254	61 09	317
211	26 13	051	40 58	073	34 15	110	62 07	174	47 07	232	29 43	255	60 41	317
212	26 45	051	41 37	074	34 53	111	62 10	176	46 35	233	29 04	256	60 13	317
213	27 17	052	42 16	075	35 31	112	62 12	178	46 01	235	28 24	256	59 45	317
214	27 49	052	42 56	075	36 09	113	62 12	180	45 28	236	27 44	257	59 16	316
215	28 22	053	43 36	076	36 46	114	62 12	182	44 54	237	27 04	258	58 48	316
216	28 54	053	44 15	076	37 24	115	62 09	184	44 19	238	26 24	259	58 20	316
217	29 27	054	44 55	077	38 01	116	62 05	186	43 45	239	25 44	260	57 52	316
218	30 00	054	45 35	077	38 37	117	62 00	188	43 09	240	25 04	260	57 23	316
219	30 34	055	46 15	078	39 14	118	61 53	190	42 34	241	24 23	261	56 55	316
220	31 07	055	46 55	079	39 50	119	61 45	192	41 58	242	23 43	262	56 26	316
221	31 41	056	47 35	079	40 26	120	61 36	194	41 21	243	23 02	263	55 58	316
222	32 15	056	48 15	080	41 01	121	61 25	196	40 45	244	22 22	264	55 30	316
223	32 49	057	48 56	080	41 36	122	61 13	198	40 08	245	21 41	264	55 01	316
224	33 23	057	49 36	081	42 11	123	61 00	200	39 30	246	21 00	265	54 33	316
	DENEB		◆VEGA		Rasalhague		ANTARES		◆ARCTURUS		Denebola		◆Dubhe	
225	33 58	058	50 16	082	42 45	124	13 52	160	60 45	202	38 53	247	54 05	316
226	34 33	058	50 57	082	43 19	125	14 06	160	60 29	204	38 15	248	53 36	316
227	35 08	059	51 38	083	43 52	126	14 20	161	60 12	206	37 37	249	53 08	316
228	35 43	059	52 18	083	44 25	127	14 33	162	59 54	207	36 59	250	52 40	316
229	36 18	060	52 59	084	44 57	128	14 45	163	59 34	209	36 20	251	52 12	317
230	36 53	060	53 40	085	45 29	129	14 56	164	59 14	211	35 41	252	51 44	317
231	37 29	061	54 20	085	46 00	131	15 07	165	58 52	213	35 02	253	51 15	317
232	38 05	061	55 01	086	46 31	132	15 18	166	58 30	214	34 23	253	50 47	317
233	38 41	062	55 42	087	47 01	133	15 27	167	58 06	216	33 44	254	50 19	317
234	39 17	062	56 23	087	47 31	134	15 36	168	57 42	217	33 05	255	49 52	317
235	39 53	063	57 04	088	48 00	136	15 45	169	57 17	219	32 25	256	49 24	317
236	40 29	063	57 45	089	48 28	137	15 53	170	56 50	221	31 45	257	48 56	317
237	41 06	063	58 26	089	48 56	138	16 00	170	56 23	222	31 05	258	48 28	318
238	41 42	064	59 06	090	49 23	139	16 06	171	55 56	224	30 25	258	48 01	318
239	42 19	064	59 47	091	49 49	141	16 12	172	55 27	225	29 45	259	47 33	318
	◆DENEB		ALTAIR		Rasalhague		◆ANTARES		ARCTURUS		Denebola		◆Dubhe	
240	42 56	065	28 15	109	50 14	142	16 17	173	54 58	226	29 05	260	47 06	318
241	43 33	065	28 54	110	50 39	144	16 22	174	54 28	228	28 24	261	46 39	318
242	44 11	066	29 32	110	51 03	145	16 25	175	53 57	229	27 44	262	46 11	318
243	44 48	066	30 10	111	51 26	146	16 29	176	53 26	231	27 04	262	45 44	319
244	45 26	067	30 48	112	51 48	148	16 31	177	52 54	232	26 23	263	45 17	319
245	46 03	067	31 26	113	52 09	149	16 33	178	52 21	233	25 42	264	44 51	319
246	46 41	068	32 04	114	52 30	151	16 34	179	51 48	234	25 02	265	44 24	319
247	47 19	068	32 41	115	52 49	152	16 35	180	51 15	236	24 21	265	43 57	320
248	47 57	069	33 18	116	53 08	154	16 35	181	50 41	237	23 40	266	43 31	320
249	48 35	069	33 55	117	53 25	155	16 34	182	50 07	238	22 59	267	43 04	320
250	49 13	069	34 31	118	53 42	157	16 32	183	49 32	239	22 18	268	42 38	320
251	49 52	070	35 07	119	53 57	159	16 30	183	48 56	240	21 37	268	42 12	321
252	50 30	070	35 43	120	54 12	160	16 27	184	48 21	241	20 56	269	41 46	321
253	51 09	071	36 18	121	54 25	162	16 24	185	47 44	242	20 16	270	41 20	321
254	51 47	071	36 53	122	54 37	164	16 20	186	47 08	243	19 35	271	40 55	321
	◆Schedar		DENEB		ALTAIR		◆Rasalhague		ARCTURUS		◆Alkaid		Kochab	
255	26 47	034	52 26	072	37 28	123	54 48	165	46 31	245	58 20	292	59 24	343
256	27 10	034	53 05	072	38 02	124	54 58	167	45 54	246	57 43	293	59 13	343
257	27 33	035	53 44	073	38 36	125	55 07	169	45 17	247	57 05	293	59 00	343
258	27 57	035	54 23	073	39 10	126	55 14	170	44 39	248	56 27	293	58 48	342
259	28 21	036	55 02	074	39 43	127	55 20	172	44 01	249	55 50	294	58 36	342
260	28 45	036	55 42	074	40 15	128	55 25	174	43 23	250	55 12	294	58 23	342
261	29 09	037	56 21	074	40 48	129	55 29	175	42 44	250	54 35	294	58 10	341
262	29 33	037	57 00	075	41 19	130	55 32	177	42 06	251	53 58	295	57 56	341
263	29 58	038	57 40	075	41 50	131	55 34	179	41 27	252	53 21	295	57 43	341
264	30 23	038	58 20	076	42 21	132	55 34	181	40 48	253	52 44	295	57 29	340
265	30 49	038	58 59	076	42 51	133	55 33	182	40 09	254	52 07	296	57 16	340
266	31 14	039	59 39	077	43 20	135	55 30	184	39 29	255	51 30	296	57 02	340
267	31 40	039	60 19	077	43 49	136	55 27	186	38 49	256	50 53	296	56 48	340
268	32 06	040	60 59	078	44 17	137	55 22	187	38 10	257	50 16	297	56 33	339
269	32 32	040	61 39	078	44 45	138	55 16	189	37 30	258	49 40	297	56 19	339

LAT 47°N

LHA	Hc	Zn	Hc	Zn	Hc	Zn	Hc	Zn	Hc	Zn	Hc	Zn	Hc	Zn
♈	◆Mirfak		Alpheratz		◆ALTAIR		Rasalhague		◆ARCTURUS		Alkaid		Kochab	
270	12 33	025	19 31	068	45 12	139	55 09	191	36 50	258	49 04	297	56 04	339
271	12 50	025	20 09	069	45 38	141	55 01	193	36 10	259	48 27	298	55 49	339
272	13 08	026	20 47	069	46 04	142	54 52	194	35 29	260	47 51	298	55 34	339
273	13 26	026	21 26	070	46 29	143	54 41	196	34 49	261	47 15	299	55 19	338
274	13 44	027	22 04	071	46 53	145	54 29	198	34 09	262	46 39	299	55 04	338
275	14 03	028	22 43	071	47 16	146	54 16	199	33 28	262	46 04	299	54 49	338
276	14 22	028	23 22	072	47 39	147	54 02	201	32 47	263	45 28	300	54 34	338
277	14 41	029	24 01	072	48 00	149	53 47	202	32 07	264	44 53	300	54 18	338
278	15 01	029	24 40	073	48 21	150	53 31	204	31 26	265	44 17	301	54 03	338
279	15 21	030	25 19	074	48 41	151	53 14	206	30 45	266	43 42	301	53 47	337
280	15 42	030	25 58	074	49 00	153	52 56	207	30 04	266	43 07	301	53 31	337
281	16 02	031	26 38	075	49 19	154	52 37	209	29 24	267	42 32	302	53 15	337
282	16 24	031	27 17	076	49 36	156	52 17	210	28 43	268	41 57	302	52 59	337
283	16 45	032	27 57	076	49 52	157	51 56	212	28 02	269	41 23	303	52 43	337
284	17 07	033	28 37	077	50 08	159	51 34	213	27 21	269	40 48	303	52 27	337
	◆Mirfak		Alpheratz		◆ALTAIR		Rasalhague		◆ARCTURUS		Alkaid		Kochab	
285	17 29	033	29 17	078	50 22	160	51 11	214	26 40	270	40 14	303	52 11	337
286	17 52	034	29 57	078	50 35	162	50 48	216	25 59	271	39 40	304	51 55	337
287	18 15	034	30 37	079	50 48	163	50 23	217	25 18	272	39 06	304	51 39	337
288	18 38	035	31 17	080	50 59	165	49 58	219	24 37	272	38 32	305	51 23	337
289	19 01	035	31 57	080	51 09	166	49 32	220	23 56	273	37 59	305	51 06	337
290	19 25	036	32 38	081	51 18	168	49 05	221	23 15	274	37 25	305	50 50	337
291	19 49	036	33 18	082	51 26	169	48 38	223	22 35	274	36 52	306	50 34	336
292	20 14	037	33 59	082	51 33	171	48 10	224	21 54	275	36 19	306	50 18	336
293	20 39	037	34 39	083	51 39	173	47 41	225	21 13	276	35 46	307	50 01	336
294	21 04	038	35 20	084	51 44	174	47 12	226	20 32	277	35 13	307	49 45	336
295	21 29	039	36 00	084	51 47	176	46 42	228	19 52	277	34 41	308	49 28	336
296	21 55	039	36 41	085	51 50	177	46 11	229	19 11	278	34 09	308	49 12	336
297	22 21	040	37 22	086	51 51	179	45 40	230	18 31	279	33 36	309	48 56	336
298	22 47	040	38 03	086	51 51	181	45 08	231	17 50	279	33 05	309	48 39	336
299	23 13	041	38 44	087	51 50	182	44 36	232	17 10	280	32 33	309	48 23	336
	Mirfak		◆Alpheratz		Enif		ALTAIR		◆Rasalhague		Alphecca		◆Kochab	
300	23 40	041	39 24	088	46 49	141	51 48	184	44 04	234	34 58	273	48 07	337
301	24 07	042	40 05	088	47 14	142	51 45	185	43 30	235	34 17	274	47 50	337
302	24 34	042	40 46	089	47 39	144	51 40	187	42 57	236	33 36	275	47 34	337
303	25 02	043	41 27	090	48 03	145	51 35	189	42 23	237	32 55	275	47 18	337
304	25 30	043	42 08	091	48 26	146	51 28	190	41 48	238	32 14	276	47 02	337
305	25 58	044	42 49	091	48 48	148	51 21	192	41 13	239	31 34	277	46 46	337
306	26 26	044	43 30	092	49 10	149	51 12	193	40 38	240	30 53	277	46 29	337
307	26 55	045	44 11	093	49 31	150	51 02	195	40 03	241	30 13	278	46 13	337
308	27 24	045	44 52	094	49 50	152	50 51	196	39 27	242	29 32	279	45 57	337
309	27 53	046	45 32	094	50 09	153	50 39	198	38 50	243	28 52	279	45 41	337
310	28 22	046	46 13	095	50 27	155	50 26	199	38 14	244	28 11	280	45 25	337
311	28 52	046	46 54	096	50 44	156	50 12	201	37 37	245	27 31	281	45 10	337
312	29 21	047	47 35	097	51 00	158	49 56	202	36 59	246	26 51	281	44 54	337
313	29 51	047	48 15	098	51 15	159	49 40	204	36 22	247	26 11	282	44 38	337
314	30 22	048	48 56	098	51 29	161	49 23	205	35 44	248	25 31	283	44 22	338
	CAPELLA		◆Hamal		Alpheratz		Enif		◆ALTAIR		VEGA		◆Kochab	
315	15 06	037	25 46	083	49 36	099	51 42	162	49 05	207	62 47	266	44 07	338
316	15 31	037	26 26	083	50 17	100	51 54	164	48 46	208	62 06	266	43 51	338
317	15 56	038	27 07	084	50 57	101	52 05	166	48 27	210	61 25	267	43 36	338
318	16 21	038	27 48	085	51 37	102	52 14	167	48 06	211	60 44	268	43 21	338
319	16 46	039	28 28	085	52 17	103	52 23	169	47 45	212	60 03	269	43 05	338
320	17 12	039	29 09	086	52 57	104	52 30	170	47 22	214	59 22	270	42 50	338
321	17 38	040	29 50	087	53 36	105	52 37	172	46 59	215	58 41	270	42 35	339
322	18 05	041	30 31	087	54 16	106	52 42	174	46 35	216	58 00	271	42 20	339
323	18 32	041	31 12	088	54 55	107	52 46	175	46 11	218	57 19	272	42 06	339
324	18 59	042	31 53	089	55 34	108	52 49	177	45 45	219	56 39	272	41 51	339
325	19 26	042	32 34	090	56 13	109	52 51	178	45 19	220	55 58	273	41 36	339
326	19 54	043	33 15	090	56 52	110	52 51	180	44 52	221	55 17	274	41 22	339
327	20 22	043	33 55	091	57 30	111	52 51	182	44 25	223	54 36	274	41 07	339
328	20 50	044	34 36	092	58 08	112	52 49	183	43 57	224	53 55	275	40 53	340
329	21 18	044	35 17	093	58 46	113	52 46	185	43 28	225	53 15	276	40 39	340
	CAPELLA		◆Hamal		Diphda		Enif		◆ALTAIR		VEGA		◆Kochab	
330	21 47	045	35 58	093	15 20	140	52 42	187	42 59	226	52 34	276	40 25	340
331	22 16	046	36 39	094	15 47	141	52 37	188	42 29	227	51 53	277	40 11	340
332	22 46	046	37 20	095	16 12	142	52 30	190	41 59	229	51 13	278	39 57	340
333	23 15	047	38 00	096	16 37	143	52 23	191	41 28	230	50 32	278	39 43	341
334	23 45	047	38 41	096	17 02	143	52 14	193	40 56	231	49 52	279	39 30	341
335	24 15	048	39 22	097	17 26	144	52 04	195	40 24	232	49 11	279	39 16	341
336	24 46	048	40 02	098	17 50	145	51 53	196	39 52	233	48 31	280	39 03	341
337	25 16	049	40 43	099	18 13	146	51 42	198	39 19	234	47 51	281	38 50	341
338	25 47	049	41 23	100	18 35	147	51 29	199	38 45	235	47 10	281	38 37	342
339	26 18	050	42 04	100	18 57	148	51 15	201	38 12	236	46 30	282	38 24	342
340	26 49	050	42 44	101	19 19	149	51 00	202	37 37	237	45 50	282	38 11	342
341	27 21	051	43 24	102	19 40	150	50 44	204	37 03	238	45 10	283	37 59	342
342	27 53	051	44 04	103	20 00	151	50 27	205	36 28	239	44 31	284	37 47	342
343	28 25	052	44 43	104	20 20	152	50 09	207	35 53	240	43 51	284	37 34	343
344	28 57	052	45 23	105	20 39	153	49 50	208	35 17	241	43 11	285	37 22	343
	◆CAPELLA		ALDEBARAN		◆Diphda		Enif		ALTAIR		◆VEGA		Kochab	
345	29 30	053	16 05	083	20 57	154	49 30	210	34 41	242	42 32	285	37 10	343
346	30 02	053	16 46	084	21 15	155	49 09	211	34 05	243	41 52	286	36 59	343
347	30 35	054	17 27	084	21 32	156	48 48	212	33 28	244	41 13	286	36 47	344
348	31 08	054	18 07	085	21 49	157	48 25	214	32 51	245	40 34	287	36 36	344
349	31 42	055	18 48	086	22 05	158	48 02	215	32 14	246	39 55	287	36 24	344
350	32 15	055	19 29	086	22 20	159	47 38	217	31 36	247	39 16	288	36 13	344
351	32 49	056	20 10	087	22 35	160	47 14	218	30 59	248	38 37	289	36 02	345
352	33 23	056	20 51	088	22 49	161	46 48	219	30 21	249	37 58	289	35 52	345
353	33 57	057	21 32	089	23 02	162	46 22	220	29 42	249	37 19	290	35 41	345
354	34 31	057	22 12	089	23 14	163	45 55	222	29 04	250	36 41	290	35 31	345
355	35 06	058	22 53	090	23 26	164	45 28	223	28 25	251	36 03	291	35 20	346
356	35 40	058	23 34	091	23 38	165	44 59	224	27 47	252	35 24	291	35 10	346
357	36 15	059	24 15	092	23 48	166	44 30	225	27 08	253	34 46	292	35 01	346
358	36 50	059	24 56	092	23 58	167	44 01	227	26 28	253	34 09	292	34 51	346
359	37 25	060	25 37	093	24 07	168	43 31	228	25 49	254	33 31	293	34 41	347

LAT 46°N

LHA	Hc	Zn	Hc	Zn	Hc	Zn	Hc	Zn	Hc	Zn	Hc	Zn	Hc	Zn
♈	◆CAPELLA		ALDEBARAN		Hamal		◆Diphda		ALTAIR		◆VEGA		Kochab	
0	37 30	059	26 22	093	55 55	121	25 14	169	25 25	256	32 29	294	33 34	347
1	38 06	060	27 03	094	56 31	122	25 22	170	24 44	257	31 51	295	33 25	347
2	38 42	060	27 45	095	57 06	123	25 29	171	24 04	257	31 13	295	33 16	348
3	39 18	061	28 26	095	57 41	124	25 36	172	23 23	258	30 36	296	33 07	348
4	39 55	061	29 08	096	58 15	126	25 41	173	22 42	259	29 58	296	32 58	348
5	40 31	062	29 49	097	58 48	127	25 46	174	22 01	260	29 21	297	32 50	349
6	41 08	062	30 30	098	59 21	129	25 50	175	21 20	260	28 44	297	32 42	349
7	41 45	062	31 12	099	59 53	130	25 54	176	20 39	261	28 07	298	32 34	349
8	42 22	063	31 53	099	60 25	132	25 56	177	19 58	262	27 30	299	32 26	349
9	42 59	063	32 34	100	60 56	133	25 58	178	19 16	263	26 54	299	32 18	350
10	43 36	064	33 15	101	61 26	135	25 59	179	18 35	263	26 17	300	32 11	350
11	44 14	064	33 56	102	61 55	136	25 59	180	17 54	264	25 41	300	32 04	350
12	44 51	065	34 36	103	62 23	138	25 59	181	17 12	265	25 05	301	31 57	351
13	45 29	065	35 17	103	62 50	140	25 57	182	16 30	266	24 30	301	31 50	351
14	46 07	065	35 58	104	63 17	142	25 55	183	15 49	266	23 54	302	31 44	351
	◆CAPELLA		ALDEBARAN		Hamal		◆Diphda		Enif		◆DENEB		Kochab	
15	46 45	066	36 38	105	63 42	143	25 52	184	34 53	245	46 03	294	31 37	351
16	47 23	066	37 18	106	64 07	145	25 49	185	34 15	246	45 25	294	31 31	352
17	48 01	067	37 58	107	64 30	147	25 45	187	33 37	247	44 47	294	31 25	352
18	48 39	067	38 38	108	64 52	149	25 39	188	32 59	248	44 09	295	31 20	352
19	49 18	067	39 17	109	65 12	151	25 34	189	32 20	249	43 32	295	31 14	353
20	49 56	068	39 57	110	65 32	153	25 27	190	31 41	250	42 54	296	31 09	353
21	50 35	068	40 36	110	65 50	155	25 20	191	31 02	250	42 17	296	31 04	353
22	51 14	069	41 15	111	66 07	158	25 12	192	30 22	251	41 39	297	30 59	354
23	51 53	069	41 54	112	66 22	160	25 03	193	29 43	252	41 02	297	30 55	354
24	52 32	069	42 32	113	66 35	162	24 53	194	29 03	253	40 25	297	30 50	354
25	53 11	070	43 10	114	66 48	164	24 43	195	28 23	254	39 48	298	30 46	354
26	53 50	070	43 48	115	66 58	166	24 32	196	27 43	255	39 11	298	30 42	355
27	54 29	071	44 26	116	67 07	169	24 20	197	27 03	255	38 35	299	30 39	355
28	55 09	071	45 03	117	67 14	171	24 07	198	26 22	256	37 58	299	30 35	355
29	55 48	071	45 40	118	67 20	174	23 54	199	25 42	257	37 22	300	30 32	356
	CAPELLA		◆BETELGEUSE		RIGEL		Hamal		◆Alpheratz		DENEB		◆Kochab	
30	56 28	072	26 46	108	20 37	128	67 24	176	62 20	242	36 46	300	30 29	356
31	57 07	072	27 25	109	21 10	128	67 26	178	61 43	243	36 10	301	30 26	356
32	57 47	073	28 05	110	21 43	129	67 26	181	61 06	245	35 34	301	30 24	357
33	58 27	073	28 44	111	22 15	130	67 25	183	60 28	246	34 59	302	30 21	357
34	59 07	073	29 23	112	22 46	131	67 22	185	59 50	247	34 23	302	30 19	357
35	59 47	074	30 01	113	23 18	132	67 17	188	59 11	248	33 48	303	30 17	358
36	60 27	074	30 40	113	23 48	133	67 11	190	58 32	249	33 13	303	30 16	358
37	61 07	075	31 18	114	24 19	134	67 02	192	57 53	250	32 38	303	30 14	358
38	61 47	075	31 55	115	24 48	135	66 52	195	57 14	252	32 03	304	30 13	359
39	62 27	075	32 33	116	25 18	136	66 41	197	56 34	253	31 29	304	30 12	359
40	63 08	076	33 10	117	25 46	137	66 28	199	55 54	254	30 54	305	30 11	359
41	63 48	076	33 47	118	26 15	138	66 13	202	55 14	255	30 20	305	30 11	359
42	64 29	076	34 24	119	26 43	139	65 57	204	54 34	256	29 47	306	30 11	000
43	65 09	077	35 00	120	27 10	140	65 40	206	53 54	257	29 13	306	30 11	000
44	65 50	077	35 36	121	27 37	141	65 21	208	53 13	257	28 39	307	30 11	000
	◆Dubhe		POLLUX		BETELGEUSE		◆RIGEL		Hamal		◆Alpheratz		DENEB	
45	27 44	027	32 21	082	36 12	122	28 03	142	65 01	210	52 32	258	28 06	307
46	28 03	028	33 03	082	36 47	123	28 28	143	64 40	212	51 51	259	27 33	308
47	28 23	028	33 44	083	37 22	124	28 53	144	64 17	214	51 10	260	27 00	308
48	28 42	028	34 25	084	37 56	125	29 18	145	63 53	216	50 29	261	26 28	309
49	29 02	029	35 07	084	38 30	126	29 41	146	63 28	218	49 48	262	25 55	309
50	29 23	029	35 48	085	39 04	127	30 04	147	63 02	219	49 07	263	25 23	310
51	29 43	030	36 30	086	39 37	128	30 27	148	62 35	221	48 25	263	24 51	310
52	30 04	030	37 11	086	40 09	129	30 49	149	62 08	223	47 44	264	24 20	311
53	30 25	030	37 53	087	40 41	130	31 10	150	61 39	225	47 02	265	23 48	311
54	30 46	031	38 35	088	41 13	131	31 30	151	61 09	226	46 21	266	23 17	312
55	31 07	031	39 16	088	41 44	132	31 50	152	60 39	228	45 39	267	22 46	312
56	31 29	031	39 58	089	42 14	134	32 09	153	60 07	229	44 58	267	22 16	313
57	31 51	032	40 40	090	42 44	135	32 28	154	59 36	231	44 16	268	21 45	313
58	32 13	032	41 21	091	43 14	136	32 45	156	59 03	232	43 34	269	21 15	314
59	32 35	032	42 03	091	43 42	137	33 02	157	58 30	234	42 53	270	20 45	315
	◆Dubhe		POLLUX		SIRIUS		◆RIGEL		Hamal		◆Alpheratz		DENEB	
60	32 58	033	42 45	092	17 04	139	33 18	158	57 56	235	42 11	270	20 16	315
61	33 20	033	43 26	093	17 31	140	33 33	159	57 21	236	41 29	271	19 46	316
62	33 43	034	44 08	094	17 58	140	33 48	160	56 47	238	40 48	272	19 17	316
63	34 06	034	44 50	094	18 25	141	34 02	161	56 11	239	40 06	272	18 49	317
64	34 30	034	45 31	095	18 50	142	34 15	162	55 35	240	39 24	273	18 20	317
65	34 53	034	46 13	096	19 16	143	34 27	164	54 59	241	38 43	274	17 52	318
66	35 17	035	46 54	097	19 40	144	34 38	165	54 22	242	38 01	274	17 24	318
67	35 41	035	47 35	097	20 05	145	34 49	166	53 45	244	37 20	275	16 57	319
68	36 05	035	48 17	098	20 28	146	34 58	167	53 07	245	36 38	276	16 29	319
69	36 29	036	48 58	099	20 51	147	35 07	168	52 30	246	35 57	276	16 02	320
70	36 54	036	49 39	100	21 14	148	35 15	170	51 51	247	35 15	277	15 36	321
71	37 18	036	50 20	101	21 36	149	35 22	171	51 13	248	34 34	278	15 09	321
72	37 43	037	51 01	102	21 57	150	35 28	172	50 34	249	33 53	278	14 43	322
73	38 08	037	51 42	103	22 18	151	35 34	173	49 55	250	33 11	279	14 18	322
74	38 33	037	52 22	104	22 38	152	35 38	174	49 16	251	32 30	280	13 52	323
	◆Dubhe		REGULUS		PROCYON		◆SIRIUS		RIGEL		◆Hamal		Schedar	
75	38 58	037	17 35	091	36 41	127	22 57	153	35 42	176	48 36	252	49 40	309
76	39 24	038	18 16	092	37 14	128	23 16	154	35 45	177	47 56	253	49 08	310
77	39 49	038	18 58	092	37 47	130	23 34	155	35 46	178	47 16	254	48 36	310
78	40 15	038	19 40	093	38 18	131	23 52	156	35 47	179	46 36	255	48 04	310
79	40 41	039	20 21	094	38 50	132	24 09	157	35 47	181	45 56	256	47 32	310
80	41 07	039	21 03	094	39 21	133	24 25	158	35 47	182	45 16	257	47 00	311
81	41 33	039	21 44	095	39 51	134	24 40	159	35 45	183	44 35	257	46 28	311
82	41 59	039	22 26	096	40 21	135	24 55	160	35 42	184	43 54	258	45 57	311
83	42 26	039	23 07	097	40 50	136	25 09	161	35 39	185	43 13	259	45 26	311
84	42 52	040	23 49	097	41 18	137	25 23	162	35 35	187	42 32	260	44 55	312
85	43 19	040	24 30	098	41 46	138	25 36	163	35 29	188	41 51	261	44 23	312
86	43 46	040	25 11	099	42 14	140	25 48	164	35 23	189	41 10	262	43 52	312
87	44 13	040	25 52	100	42 40	141	25 59	165	35 16	190	40 29	262	43 21	312
88	44 40	041	26 33	101	43 06	142	26 09	166	35 08	191	39 47	263	42 50	313
89	45 07	041	27 14	101	43 32	143	26 19	167	35 00	193	39 06	264	42 20	313

LHA	Hc	Zn	Hc	Zn	Hc	Zn	Hc	Zn	Hc	Zn	Hc	Zn	Hc	Zn
♈	◆Dubhe		REGULUS		PROCYON		◆SIRIUS		RIGEL		ALDEBARAN		◆Mirfak	
90	45 34	041	27 55	102	43 56	145	26 28	168	34 50	194	55 40	218	63 53	293
91	46 02	041	28 36	103	44 20	146	26 37	169	34 40	195	55 14	219	63 14	293
92	46 29	041	29 16	104	44 43	147	26 44	170	34 29	196	54 47	221	62 36	293
93	46 57	041	29 57	105	45 05	148	26 51	171	34 17	197	54 19	222	61 58	293
94	47 24	042	30 37	105	45 26	150	26 57	172	34 04	199	53 51	224	61 20	294
95	47 52	042	31 17	106	45 47	151	27 02	173	33 50	200	53 22	225	60 41	294
96	48 20	042	31 57	107	46 07	152	27 07	174	33 36	201	52 52	226	60 03	294
97	48 48	042	32 37	108	46 25	154	27 10	175	33 20	202	52 22	228	59 25	294
98	49 15	042	33 16	109	46 43	155	27 13	177	33 04	203	51 51	229	58 47	294
99	49 43	042	33 56	110	47 00	157	27 16	178	32 48	204	51 19	230	58 09	295
100	50 12	042	34 35	111	47 16	158	27 17	179	32 30	205	50 46	232	57 31	295
101	50 40	042	35 14	111	47 32	159	27 17	180	32 12	207	50 14	233	56 54	295
102	51 08	043	35 52	112	47 46	161	27 17	181	31 53	208	49 40	234	56 16	295
103	51 36	043	36 31	113	47 59	162	27 16	182	31 33	209	49 06	235	55 38	296
104	52 04	043	37 09	114	48 11	164	27 14	183	31 13	210	48 32	236	55 01	296
	◆Dubhe		Denebola		REGULUS		◆SIRIUS		RIGEL		ALDEBARAN		◆Mirfak	
105	52 33	043	22 45	092	37 47	115	27 12	184	30 52	211	47 57	237	54 24	296
106	53 01	043	23 27	093	38 24	116	27 09	185	30 30	212	47 21	239	53 46	297
107	53 29	043	24 09	094	39 02	117	27 04	186	30 08	213	46 46	240	53 09	297
108	53 58	043	24 50	095	39 39	118	27 00	187	29 45	214	46 09	241	52 32	297
109	54 26	043	25 32	095	40 15	119	26 54	188	29 21	215	45 33	242	51 55	298
110	54 54	043	26 13	096	40 52	120	26 47	189	28 57	216	44 56	243	51 18	298
111	55 23	043	26 55	097	41 28	121	26 40	190	28 32	217	44 19	244	50 41	298
112	55 51	043	27 36	098	42 03	122	26 32	192	28 06	218	43 41	245	50 05	298
113	56 19	043	28 17	098	42 38	123	26 24	193	27 40	219	43 03	246	49 28	299
114	56 48	043	28 58	099	43 13	124	26 14	194	27 14	220	42 25	247	48 52	299
115	57 16	043	29 40	100	43 47	125	26 04	195	26 47	221	41 47	248	48 15	299
116	57 44	043	30 21	101	44 21	126	25 53	196	26 19	222	41 08	249	47 39	300
117	58 12	043	31 01	101	44 54	128	25 41	197	25 51	223	40 29	250	47 03	300
118	58 41	042	31 42	102	45 27	129	25 29	198	25 22	224	39 50	251	46 27	301
119	59 09	042	32 23	103	45 59	130	25 16	199	24 53	225	39 10	252	45 51	301
	◆Kochab		Denebola		◆REGULUS		SIRIUS		RIGEL		◆ALDEBARAN		CAPELLA	
120	40 35	021	33 03	104	46 31	131	25 02	200	24 23	226	38 31	252	61 54	285
121	40 49	021	33 44	105	47 02	132	24 47	201	23 53	227	37 51	253	61 13	285
122	41 04	021	34 24	106	47 32	134	24 32	202	23 22	228	37 11	254	60 33	286
123	41 19	021	35 04	106	48 02	135	24 16	203	22 51	229	36 30	255	59 53	286
124	41 34	021	35 44	107	48 32	136	24 00	204	22 19	230	35 50	256	59 13	287
125	41 49	021	36 24	108	49 00	137	23 42	205	21 48	231	35 10	257	58 33	287
126	42 04	021	37 03	109	49 28	139	23 25	206	21 15	231	34 29	258	57 53	287
127	42 19	022	37 42	110	49 55	140	23 06	207	20 42	232	33 48	258	57 14	288
128	42 35	022	38 21	111	50 22	141	22 47	208	20 09	233	33 07	259	56 34	288
129	42 50	022	39 00	112	50 47	143	22 27	209	19 36	234	32 26	260	55 54	289
130	43 06	022	39 39	113	51 12	144	22 07	210	19 02	235	31 45	261	55 15	289
131	43 21	022	40 17	114	51 36	146	21 46	211	18 27	236	31 04	262	54 36	289
132	43 37	022	40 55	115	51 59	147	21 24	212	17 53	237	30 23	262	53 56	290
133	43 53	022	41 33	116	52 21	149	21 02	213	17 18	237	29 41	263	53 17	290
134	44 09	022	42 10	117	52 43	150	20 39	214	16 42	238	29 00	264	52 38	290
	◆Kochab		ARCTURUS		Denebola		◆REGULUS		SIRIUS		BETELGEUSE		◆CAPELLA	
135	44 24	022	21 19	084	42 47	118	53 03	152	20 16	215	34 40	241	51 59	291
136	44 40	023	22 00	085	43 24	119	53 22	153	19 52	216	34 04	242	51 20	291
137	44 56	023	22 42	085	44 00	120	53 41	155	19 27	216	33 27	243	50 41	292
138	45 12	023	23 23	086	44 36	121	53 58	156	19 02	217	32 50	243	50 03	292
139	45 28	023	24 05	087	45 12	122	54 14	158	18 37	218	32 12	244	49 24	293
140	45 45	023	24 46	087	45 47	123	54 30	159	18 11	219	31 35	245	48 46	293
141	46 01	023	25 28	088	46 22	124	54 44	161	17 44	220	30 57	246	48 07	293
142	46 17	023	26 10	089	46 56	125	54 57	163	17 17	221	30 18	247	47 29	294
143	46 33	023	26 51	090	47 30	126	55 08	164	16 49	222	29 40	248	46 51	294
144	46 49	023	27 33	090	48 03	128	55 19	166	16 21	223	29 01	249	46 13	295
145	47 06	023	28 15	091	48 36	129	55 28	168	15 53	224	28 22	250	45 35	295
146	47 22	023	28 56	092	49 08	130	55 37	170	15 24	224	27 43	250	44 57	295
147	47 38	023	29 38	093	49 40	131	55 44	171	14 55	225	27 04	251	44 20	296
148	47 55	023	30 20	093	50 11	132	55 49	173	14 25	226	26 24	252	43 42	296
149	48 11	023	31 01	094	50 42	134	55 54	175	13 55	227	25 44	253	43 05	297
	◆Kochab		ARCTURUS		◆SPICA		REGULUS		PROCYON		◆BETELGEUSE		CAPELLA	
150	48 28	023	31 43	095	16 44	127	55 57	176	39 06	228	25 04	254	42 28	297
151	48 44	023	32 24	096	17 17	128	55 59	178	38 35	229	24 24	255	41 51	298
152	49 00	023	33 06	096	17 50	129	56 00	180	38 03	230	23 44	255	41 14	298
153	49 17	023	33 47	097	18 22	130	55 59	182	37 31	231	23 04	256	40 37	298
154	49 33	023	34 29	098	18 54	130	55 57	183	36 58	232	22 23	257	40 01	299
155	49 49	023	35 10	099	19 25	131	55 54	185	36 25	233	21 43	258	39 24	299
156	50 05	023	35 51	099	19 57	132	55 49	187	35 52	234	21 02	259	38 48	300
157	50 22	023	36 32	100	20 27	133	55 44	189	35 18	235	20 21	259	38 12	300
158	50 38	023	37 13	101	20 57	134	55 37	190	34 43	236	19 40	260	37 36	301
159	50 54	023	37 54	102	21 27	135	55 29	192	34 09	237	18 59	261	37 00	301
160	51 10	023	38 35	103	21 56	136	55 19	194	33 33	238	18 18	262	36 25	302
161	51 26	023	39 15	104	22 25	137	55 09	195	32 58	239	17 36	262	35 49	302
162	51 42	023	39 56	104	22 54	138	54 57	197	32 22	240	16 55	263	35 14	302
163	51 58	022	40 36	105	23 21	139	54 44	199	31 46	241	16 14	264	34 39	303
164	52 14	022	41 16	106	23 49	140	54 30	200	31 09	242	15 32	265	34 04	303
	◆VEGA		ARCTURUS		◆SPICA		REGULUS		PROCYON		◆POLLUX		CAPELLA	
165	13 13	047	41 56	107	24 16	140	54 15	202	30 32	243	47 56	262	33 29	304
166	13 44	048	42 36	108	24 42	141	53 59	204	29 55	244	47 15	263	32 55	304
167	14 14	048	43 15	109	25 07	142	53 42	205	29 18	245	46 34	264	32 20	305
168	14 46	049	43 55	110	25 33	143	53 23	207	28 40	245	45 52	265	31 46	305
169	15 17	049	44 34	111	25 57	144	53 04	208	28 02	246	45 11	265	31 12	306
170	15 49	050	45 12	112	26 21	145	52 44	210	27 24	247	44 29	266	30 38	306
171	16 21	051	45 51	113	26 44	146	52 23	211	26 45	248	43 47	267	30 05	307
172	16 53	051	46 29	114	27 07	147	52 00	213	26 06	249	43 06	268	29 32	307
173	17 26	052	47 07	115	27 29	148	51 37	214	25 27	250	42 24	268	28 58	308
174	17 59	052	47 45	116	27 51	149	51 13	216	24 48	251	41 42	269	28 25	308
175	18 32	053	48 22	117	28 12	151	50 49	217	24 09	251	41 01	270	27 53	309
176	19 05	053	48 59	118	28 32	152	50 23	219	23 29	252	40 19	271	27 20	309
177	19 39	054	49 36	119	28 51	153	49 57	220	22 49	253	39 37	271	26 48	310
178	20 12	055	50 12	120	29 10	154	49 30	221	22 09	254	38 56	272	26 16	310
179	20 47	055	50 48	121	29 28	155	49 02	223	21 29	255	38 14	273	25 44	311

LAT 46°N

LHA	Hc	Zn	Hc	Zn	Hc	Zn	Hc	Zn	Hc	Zn	Hc	Zn	Hc	Zn
♈	♦DENEB		VEGA		ARCTURUS		♦SPICA		REGULUS		♦POLLUX		CAPELLA	
180	11 13	033	21 21	056	51 23	123	29 46	156	48 33	224	37 33	273	25 13	311
181	11 36	034	21 55	056	51 58	124	30 02	157	48 04	225	36 51	274	24 41	312
182	12 00	034	22 30	057	52 33	125	30 18	158	47 34	226	36 09	275	24 10	312
183	12 23	035	23 05	057	53 07	126	30 34	159	47 04	228	35 28	275	23 40	313
184	12 48	036	23 40	058	53 40	127	30 48	160	46 32	229	34 46	276	23 09	313
185	13 12	036	24 16	059	54 13	129	31 02	161	46 01	230	34 05	277	22 39	314
186	13 37	037	24 52	059	54 45	130	31 15	162	45 29	231	33 24	277	22 09	314
187	14 02	037	25 27	060	55 16	132	31 27	164	44 56	232	32 42	278	21 39	315
188	14 27	038	26 03	060	55 47	133	31 38	165	44 23	234	32 01	279	21 09	315
189	14 53	038	26 40	061	56 18	134	31 49	166	43 49	235	31 20	279	20 40	316
190	15 19	039	27 16	061	56 47	136	31 59	167	43 15	236	30 39	280	20 11	316
191	15 45	040	27 53	062	57 16	137	32 08	168	42 40	237	29 58	281	19 42	317
192	16 12	040	28 30	062	57 43	139	32 16	169	42 05	238	29 17	281	19 14	317
193	16 39	041	29 07	063	58 10	140	32 23	170	41 29	239	28 36	282	18 46	318
194	17 07	041	29 44	063	58 37	142	32 30	172	40 54	240	27 55	283	18 18	318
	♦DENEB		VEGA		Rasalhague		♦ARCTURUS		REGULUS		♦POLLUX		CAPELLA	
195	17 34	042	30 21	064	23 46	097	59 02	144	40 17	241	27 15	283	17 51	319
196	18 02	042	30 59	065	24 27	097	59 26	145	39 41	242	26 34	284	17 23	320
197	18 31	043	31 36	065	25 09	098	59 49	147	39 04	243	25 54	284	16 57	320
198	18 59	044	32 14	066	25 50	099	60 11	149	38 26	244	25 13	285	16 30	321
199	19 28	044	32 52	066	26 31	100	60 33	151	37 49	245	24 33	286	16 04	321
200	19 57	045	33 31	067	27 12	100	60 52	152	37 11	246	23 53	286	15 38	322
201	20 26	045	34 09	067	27 53	101	61 11	154	36 33	247	23 13	287	15 12	322
202	20 56	046	34 47	068	28 34	102	61 29	156	35 54	248	22 33	288	14 47	323
203	21 26	046	35 26	068	29 15	103	61 45	158	35 16	249	21 54	288	14 22	323
204	21 56	047	36 05	069	29 55	104	62 00	160	34 37	249	21 14	289	13 57	324
205	22 27	047	36 44	069	30 36	104	62 14	162	33 58	250	20 35	290	13 33	325
206	22 58	048	37 23	070	31 16	105	62 26	164	33 18	251	19 56	290	13 09	325
207	23 29	048	38 02	070	31 56	106	62 37	166	32 39	252	19 17	291	12 45	326
208	24 00	049	38 41	071	32 36	107	62 47	168	31 59	253	18 38	291	12 22	326
209	24 31	049	39 21	072	33 16	108	62 55	170	31 19	254	17 59	292	11 59	327
	DENEB		♦VEGA		Rasalhague		♦ARCTURUS		Denebola		REGULUS		♦Dubhe	
210	25 03	050	40 01	072	33 55	109	63 01	172	48 17	232	30 39	255	60 24	318
211	25 35	050	40 40	073	34 35	109	63 06	174	47 44	233	29 59	255	59 57	318
212	26 07	051	41 20	073	35 14	110	63 10	176	47 10	234	29 18	256	59 29	318
213	26 40	051	42 00	074	35 53	111	63 12	178	46 36	235	28 38	257	59 01	318
214	27 12	052	42 40	074	36 32	112	63 12	180	46 01	237	27 57	258	58 32	318
215	27 45	052	43 20	075	37 10	113	63 12	182	45 26	238	27 16	259	58 04	317
216	28 18	053	44 01	075	37 48	114	63 09	184	44 51	239	26 35	259	57 36	317
217	28 52	053	44 41	076	38 26	115	63 05	187	44 15	240	25 54	260	57 08	317
218	29 25	054	45 21	076	39 04	116	62 59	189	43 39	241	25 13	261	56 40	317
219	29 59	054	46 02	077	39 41	117	62 52	191	43 02	242	24 32	262	56 11	317
220	30 33	055	46 43	078	40 18	118	62 44	193	42 25	243	23 51	262	55 43	317
221	31 07	055	47 23	078	40 55	119	62 34	195	41 48	244	23 09	263	55 15	317
222	31 42	056	48 04	079	41 31	120	62 23	197	41 10	245	22 28	264	54 46	317
223	32 16	056	48 45	079	42 07	121	62 10	199	40 32	246	21 47	265	54 18	317
224	32 51	057	49 26	080	42 43	122	61 56	201	39 54	247	21 05	265	53 49	317
	DENEB		♦VEGA		Rasalhague		ANTARES		♦ARCTURUS		Denebola		♦Dubhe	
225	33 26	057	50 07	080	43 18	123	14 49	159	61 41	203	39 16	248	53 21	317
226	34 01	058	50 48	081	43 53	124	15 03	160	61 24	205	38 37	249	52 53	317
227	34 36	058	51 29	082	44 27	125	15 17	161	61 06	206	37 58	250	52 24	317
228	35 12	059	52 11	082	45 01	126	15 30	162	60 47	208	37 19	251	51 56	317
229	35 47	059	52 52	083	45 34	128	15 42	163	60 27	210	36 40	251	51 28	317
230	36 23	060	53 33	083	46 07	129	15 54	164	60 05	212	36 00	252	51 00	317
231	36 59	060	54 15	084	46 39	130	16 05	165	59 43	213	35 20	253	50 32	318
232	37 35	060	54 56	085	47 11	131	16 16	166	59 19	215	34 40	254	50 03	318
233	38 12	061	55 38	085	47 42	132	16 26	167	58 55	217	34 00	255	49 35	318
234	38 48	061	56 19	086	48 12	134	16 35	168	58 29	218	33 20	256	49 07	318
235	39 25	062	57 01	087	48 42	135	16 44	169	58 03	220	32 39	257	48 39	318
236	40 02	062	57 43	087	49 12	136	16 52	169	57 36	222	31 59	257	48 12	318
237	40 39	063	58 24	088	49 40	137	16 59	170	57 08	223	31 18	258	47 44	318
238	41 16	063	59 06	089	50 08	139	17 05	171	56 39	225	30 37	259	47 16	318
239	41 53	064	59 48	089	50 35	140	17 11	172	56 09	226	29 56	260	46 49	319
	♦DENEB		ALTAIR		Rasalhague		♦ANTARES		ARCTURUS		Denebola		♦Dubhe	
240	42 30	064	28 34	108	51 02	141	17 17	173	55 39	227	29 15	261	46 21	319
241	43 08	064	29 13	109	51 27	143	17 21	174	55 08	229	28 34	261	45 54	319
242	43 46	065	29 53	110	51 52	144	17 25	175	54 36	230	27 53	262	45 26	319
243	44 23	065	30 32	111	52 16	146	17 28	176	54 03	232	27 11	263	44 59	319
244	45 01	066	31 11	112	52 39	147	17 31	177	53 31	233	26 30	264	44 32	320
245	45 39	066	31 49	112	53 01	149	17 33	178	52 57	234	25 48	264	44 05	320
246	46 18	067	32 28	113	53 22	150	17 34	179	52 23	235	25 07	265	43 38	320
247	46 56	067	33 06	114	53 42	152	17 35	180	51 48	237	24 25	266	43 11	320
248	47 34	067	33 44	115	54 02	153	17 35	181	51 13	238	23 44	267	42 45	320
249	48 13	068	34 21	116	54 20	155	17 34	182	50 38	239	23 02	267	42 18	321
250	48 52	068	34 59	117	54 37	157	17 32	183	50 02	240	22 20	268	41 52	321
251	49 30	069	35 36	118	54 53	158	17 30	183	49 26	241	21 39	269	41 25	321
252	50 09	069	36 12	119	55 08	160	17 27	184	48 49	242	20 57	270	40 59	321
253	50 48	070	36 49	120	55 22	161	17 23	185	48 12	243	20 15	270	40 33	322
254	51 27	070	37 25	121	55 34	163	17 19	186	47 34	244	19 34	271	40 08	322
	♦Schedar		DENEB		ALTAIR		♦Rasalhague		ARCTURUS		♦Alkaid		Kochab	
255	25 57	034	52 07	070	38 00	122	55 46	165	46 57	246	57 57	294	58 27	344
256	26 20	034	52 46	071	38 35	123	55 56	167	46 19	247	57 19	294	58 15	343
257	26 44	035	53 25	071	39 10	124	56 05	168	45 40	248	56 41	294	58 03	343
258	27 08	035	54 05	072	39 45	125	56 13	170	45 01	249	56 03	295	57 51	343
259	27 32	035	54 45	072	40 18	126	56 20	172	44 23	249	55 25	295	57 38	342
260	27 56	036	55 24	073	40 52	127	56 25	174	43 43	250	54 47	295	57 26	342
261	28 21	036	56 04	073	41 25	128	56 29	175	43 04	251	54 10	296	57 13	342
262	28 46	037	56 44	073	41 57	129	56 32	177	42 24	252	53 32	296	57 00	342
263	29 11	037	57 24	074	42 29	130	56 34	179	41 45	253	52 55	296	56 46	341
264	29 36	038	58 04	074	43 01	132	56 34	181	41 05	254	52 17	296	56 33	341
265	30 01	038	58 44	075	43 32	133	56 33	182	40 24	255	51 40	297	56 19	341
266	30 27	038	59 24	075	44 02	134	56 30	184	39 44	256	51 03	297	56 05	340
267	30 53	039	60 05	075	44 32	135	56 27	186	39 04	257	50 26	297	55 51	340
268	31 19	039	60 45	076	45 01	136	56 22	188	38 23	258	49 49	298	55 37	340
269	31 46	040	61 26	076	45 30	138	56 16	189	37 42	258	49 12	298	55 23	340

LHA	Hc	Zn	Hc	Zn	Hc	Zn	Hc	Zn	Hc	Zn	Hc	Zn	Hc	Zn
♈	♦Mirfak		Alpheratz		♦ALTAIR		Rasalhague		♦ARCTURUS		Alkaid		Kochab	
270	11 38	024	19 08	068	45 57	139	56 08	191	37 01	259	48 35	298	55 08	339
271	11 56	025	19 47	068	46 24	140	56 00	193	36 20	260	47 59	299	54 53	339
272	12 13	026	20 26	069	46 51	141	55 50	195	35 39	261	47 22	299	54 39	339
273	12 32	026	21 05	069	47 17	143	55 39	196	34 58	262	46 46	300	54 24	339
274	12 50	027	21 44	070	47 42	144	55 26	198	34 17	262	46 10	300	54 08	339
275	13 09	027	22 23	071	48 06	145	55 13	200	33 36	263	45 34	300	53 53	339
276	13 29	028	23 03	071	48 29	147	54 58	201	32 54	264	44 58	301	53 38	338
277	13 48	029	23 42	072	48 51	148	54 43	203	32 13	265	44 22	301	53 23	338
278	14 09	029	24 22	073	49 13	149	54 26	205	31 31	265	43 46	301	53 07	338
279	14 29	030	25 02	073	49 34	151	54 08	206	30 50	266	43 11	302	52 51	338
280	14 50	030	25 42	074	49 54	152	53 49	208	30 08	267	42 35	302	52 36	338
281	15 11	031	26 22	075	50 13	154	53 29	209	29 26	268	42 00	302	52 20	338
282	15 32	031	27 02	075	50 30	155	53 08	211	28 45	268	41 25	303	52 04	338
283	15 54	032	27 42	076	50 47	157	52 47	212	28 03	269	40 50	303	51 48	337
284	16 16	032	28 23	076	51 03	158	52 24	214	27 21	270	40 15	304	51 32	337
	♦Mirfak		Alpheratz		♦ALTAIR		Rasalhague		♦ARCTURUS		Alkaid		Kochab	
285	16 39	033	29 03	077	51 18	160	52 00	215	26 40	271	39 41	304	51 16	337
286	17 02	034	29 44	078	51 32	161	51 36	217	25 58	271	39 06	304	51 00	337
287	17 25	034	30 25	078	51 45	163	51 11	218	25 16	272	38 32	305	50 44	337
288	17 49	035	31 06	079	51 57	164	50 45	219	24 35	273	37 58	305	50 28	337
289	18 12	035	31 47	080	52 07	166	50 18	221	23 53	273	37 24	306	50 11	337
290	18 37	036	32 28	080	52 17	168	49 50	222	23 11	274	36 50	306	49 55	337
291	19 01	036	33 09	081	52 25	169	49 22	223	22 30	275	36 17	306	49 39	337
292	19 26	037	33 50	082	52 32	171	48 53	225	21 48	275	35 43	307	49 22	337
293	19 51	037	34 31	082	52 39	172	48 23	226	21 07	276	35 10	307	49 06	337
294	20 16	038	35 13	083	52 43	174	47 53	227	20 25	277	34 37	308	48 50	337
295	20 42	038	35 54	083	52 47	176	47 22	229	19 44	278	34 04	308	48 33	337
296	21 08	039	36 35	084	52 50	177	46 50	230	19 03	278	33 31	309	48 17	337
297	21 34	039	37 17	085	52 51	179	46 18	231	18 22	279	32 59	309	48 01	337
298	22 01	040	37 58	085	52 51	181	45 46	232	17 41	280	32 27	309	47 44	337
299	22 28	040	38 40	086	52 50	182	45 12	233	16 59	280	31 55	310	47 28	337
	Mirfak		♦Alpheratz		Enif		ALTAIR		♦Rasalhague		Alphecca		♦Kochab	
300	22 55	041	39 22	087	47 35	140	52 48	184	44 39	234	34 54	274	47 12	337
301	23 22	041	40 03	088	48 02	142	52 45	185	44 05	235	34 12	275	46 55	337
302	23 50	042	40 45	088	48 27	143	52 40	187	43 30	237	33 31	275	46 39	337
303	24 18	042	41 27	089	48 52	144	52 34	189	42 55	238	32 49	276	46 23	337
304	24 46	043	42 08	090	49 16	146	52 27	190	42 20	239	32 08	277	46 07	337
305	25 14	043	42 50	090	49 39	147	52 19	192	41 44	240	31 27	277	45 50	337
306	25 43	044	43 32	091	50 01	148	52 10	194	41 08	241	30 45	278	45 34	337
307	26 12	044	44 13	092	50 23	150	52 00	195	40 31	242	30 04	279	45 18	337
308	26 41	045	44 55	093	50 43	151	51 48	197	39 54	243	29 23	279	45 02	337
309	27 11	045	45 37	093	51 03	153	51 36	198	39 17	244	28 42	280	44 46	337
310	27 40	046	46 18	094	51 21	154	51 22	200	38 40	245	28 01	281	44 30	338
311	28 10	046	47 00	095	51 39	156	51 08	201	38 02	246	27 20	281	44 14	338
312	28 40	047	47 41	096	51 56	157	50 52	203	37 24	247	26 39	282	43 58	338
313	29 11	047	48 23	096	52 11	159	50 35	204	36 45	248	25 58	282	43 43	338
314	29 41	047	49 04	097	52 26	160	50 17	206	36 07	248	25 18	283	43 27	338
	CAPELLA		♦Hamal		Alpheratz		Enif		♦ALTAIR		VEGA		♦Kochab	
315	14 18	036	25 38	082	49 45	098	52 39	162	49 59	207	62 50	268	43 11	338
316	14 43	037	26 19	083	50 27	099	52 51	164	49 39	209	62 09	268	42 56	338
317	15 08	038	27 00	083	51 08	100	53 03	165	49 19	210	61 27	269	42 40	338
318	15 34	038	27 42	084	51 49	101	53 13	167	48 57	212	60 45	270	42 25	338
319	16 00	039	28 23	085	52 30	102	53 22	168	48 35	213	60 04	270	42 10	339
320	16 26	039	29 05	085	53 10	102	53 29	170	48 12	214	59 22	271	41 55	339
321	16 52	040	29 46	086	53 51	103	53 36	172	47 48	216	58 40	272	41 39	339
322	17 19	040	30 28	087	54 31	104	53 41	173	47 23	217	57 59	273	41 24	339
323	17 46	041	31 10	088	55 12	105	53 46	175	46 58	218	57 17	273	41 10	339
324	18 14	041	31 51	088	55 52	106	53 49	177	46 32	220	56 35	274	40 55	339
325	18 42	042	32 33	089	56 32	107	53 51	178	46 05	221	55 54	275	40 40	339
326	19 10	043	33 15	090	57 11	108	53 51	180	45 37	222	55 12	275	40 25	340
327	19 38	043	33 56	090	57 51	109	53 50	182	45 09	223	54 31	276	40 11	340
328	20 07	044	34 38	091	58 30	111	53 49	183	44 40	225	53 49	276	39 57	340
329	20 35	044	35 20	092	59 09	112	53 46	185	44 10	226	53 08	277	39 42	340
	CAPELLA		♦Hamal		Diphda		Enif		♦ALTAIR		VEGA		♦Kochab	
330	21 05	045	36 01	093	16 06	140	53 41	187	43 40	227	52 27	278	39 28	340
331	21 34	045	36 43	093	16 33	141	53 36	188	43 09	228	51 45	278	39 14	340
332	22 04	046	37 24	094	16 59	141	53 29	190	42 38	229	51 04	279	39 00	341
333	22 34	046	38 06	095	17 25	142	53 21	192	42 06	230	50 23	279	38 47	341
334	23 04	047	38 47	096	17 50	143	53 12	193	41 34	232	49 42	280	38 33	341
335	23 35	047	39 29	096	18 15	144	53 02	195	41 01	233	49 01	281	38 20	341
336	24 05	048	40 10	097	18 39	145	52 51	196	40 28	234	48 20	281	38 06	341
337	24 36	048	40 52	098	19 03	146	52 39	198	39 54	235	47 39	282	37 53	342
338	25 08	049	41 33	099	19 26	147	52 25	200	39 19	236	46 58	282	37 40	342
339	25 39	049	42 14	100	19 48	148	52 11	201	38 45	237	46 18	283	37 27	342
340	26 11	050	42 55	100	20 10	149	51 55	203	38 10	238	45 37	283	37 14	342
341	26 43	050	43 36	101	20 31	150	51 38	204	37 34	239	44 56	284	37 02	343
342	27 15	051	44 17	102	20 52	151	51 21	206	36 58	240	44 16	284	36 49	343
343	27 48	051	44 57	103	21 12	152	51 02	207	36 22	241	43 36	285	36 37	343
344	28 20	052	45 38	104	21 32	153	50 43	209	35 46	242	42 56	286	36 25	343
	♦CAPELLA		ALDEBARAN		♦Diphda		Enif		ALTAIR		♦VEGA		Kochab	
345	28 53	052	15 58	083	21 51	153	50 22	210	35 09	243	42 15	286	36 13	343
346	29 26	053	16 39	083	22 09	154	50 01	212	34 31	244	41 35	287	36 01	344
347	30 00	053	17 20	084	22 27	155	49 38	213	33 54	245	40 56	287	35 49	344
348	30 33	054	18 02	085	22 44	156	49 15	214	33 16	246	40 16	288	35 38	344
349	31 07	054	18 43	085	23 00	157	48 51	216	32 38	246	39 36	288	35 27	344
350	31 41	055	19 25	086	23 16	158	48 26	217	32 00	247	38 57	289	35 15	345
351	32 15	055	20 07	087	23 31	159	48 01	219	31 21	248	38 17	289	35 04	345
352	32 49	056	20 48	087	23 45	160	47 34	220	30 42	249	37 38	290	34 54	345
353	33 24	056	21 30	088	23 59	161	47 07	221	30 03	250	36 59	290	34 43	345
354	33 58	057	22 12	089	24 12	162	46 40	222	29 24	251	36 20	291	34 33	346
355	34 33	057	22 53	090	24 24	163	46 11	224	28 45	252	35 41	291	34 22	346
356	35 08	057	23 35	090	24 35	164	45 42	225	28 05	252	35 02	292	34 12	346
357	35 43	058	24 17	091	24 46	165	45 12	226	27 25	253	34 24	293	34 02	346
358	36 19	058	24 58	092	24 56	167	44 42	227	26 45	254	33 45	293	33 53	347
359	36 54	059	25 40	093	25 06	168	44 11	228	26 05	255	33 07	294	33 43	347

LHA	Hc	Zn	Hc	Zn	Hc	Zn	Hc	Zn	Hc	Zn	Hc	Zn	Hc	Zn
♈	◆CAPELLA		ALDEBARAN		Hamal		◆Diphda		ALTAIR		◆VEGA		Kochab	
0	36 59	059	26 25	093	56 25	119	26 13	169	25 39	256	32 04	295	32 35	347
1	37 36	059	27 07	093	57 02	121	26 21	170	24 58	257	31 26	295	32 26	348
2	38 12	060	27 49	094	57 38	122	26 28	171	24 17	258	30 47	296	32 17	348
3	38 49	060	28 32	095	58 14	123	26 35	172	23 35	259	30 09	296	32 08	348
4	39 26	060	29 14	096	58 49	124	26 41	173	22 53	259	29 31	297	31 59	348
5	40 02	061	29 56	096	59 24	126	26 46	174	22 12	260	28 54	297	31 51	349
6	40 40	061	30 38	097	59 58	127	26 50	175	21 30	261	28 16	298	31 43	349
7	41 17	062	31 20	098	60 31	129	26 53	176	20 48	262	27 39	298	31 35	349
8	41 54	062	32 02	099	61 04	130	26 56	177	20 06	262	27 01	299	31 27	350
9	42 32	062	32 44	100	61 36	132	26 58	178	19 24	263	26 24	299	31 19	350
10	43 09	063	33 26	100	62 07	133	26 59	179	18 42	264	25 48	300	31 12	350
11	43 47	063	34 08	101	62 38	135	26 59	180	17 59	264	25 11	301	31 05	350
12	44 25	064	34 49	102	63 07	137	26 59	181	17 17	265	24 34	301	30 58	351
13	45 03	064	35 31	103	63 36	139	26 57	182	16 35	266	23 58	302	30 51	351
14	45 41	064	36 12	104	64 03	140	26 55	183	15 53	267	23 22	302	30 44	351
	◆CAPELLA		BETELGEUSE		RIGEL		◆Diphda		Enif		◆DENEB		Kochab	
15	46 20	065	16 43	096	12 09	115	26 52	184	35 18	246	45 39	294	30 38	352
16	46 58	065	17 25	097	12 48	116	26 49	186	34 39	247	45 01	295	30 32	352
17	47 37	066	18 07	098	13 26	117	26 44	187	34 00	247	44 22	295	30 26	352
18	48 16	066	18 49	099	14 03	117	26 39	188	33 21	248	43 44	296	30 20	352
19	48 54	066	19 31	099	14 41	118	26 33	189	32 41	249	43 06	296	30 15	353
20	49 33	067	20 13	100	15 18	119	26 26	190	32 02	250	42 28	297	30 10	353
21	50 12	067	20 55	101	15 55	120	26 19	191	31 22	251	41 50	297	30 04	353
22	50 51	067	21 36	102	16 32	120	26 10	192	30 41	252	41 12	297	30 00	354
23	51 31	068	22 18	102	17 08	121	26 01	193	30 01	253	40 34	298	29 55	354
24	52 10	068	22 59	103	17 45	122	25 51	194	29 21	253	39 57	298	29 51	354
25	52 49	069	23 41	104	18 20	123	25 41	195	28 40	254	39 20	299	29 47	355
26	53 29	069	24 22	105	18 56	124	25 29	196	27 59	255	38 42	299	29 43	355
27	54 09	069	25 03	105	19 31	125	25 17	197	27 18	256	38 05	300	29 39	355
28	54 48	070	25 43	106	20 06	126	25 05	198	26 37	257	37 29	300	29 35	355
29	55 28	070	26 24	107	20 40	126	24 51	199	25 55	257	36 52	300	29 32	356
	◆CAPELLA		BETELGEUSE		RIGEL		◆Diphda		Alpheratz		◆DENEB		Kochab	
30	56 08	070	27 05	108	21 14	127	24 37	200	62 48	244	36 15	301	29 29	356
31	56 48	071	27 45	109	21 47	128	24 22	201	62 09	245	35 39	301	29 26	356
32	57 28	071	28 25	109	22 21	129	24 06	202	61 31	246	35 03	302	29 24	357
33	58 09	071	29 05	110	22 53	130	23 50	203	60 52	247	34 27	302	29 21	357
34	58 49	072	29 45	111	23 26	131	23 33	204	60 12	249	33 51	303	29 19	357
35	59 29	072	30 24	112	23 58	132	23 15	205	59 33	250	33 15	303	29 17	358
36	60 10	072	31 03	113	24 29	133	22 57	206	58 53	251	32 40	304	29 16	358
37	60 50	073	31 42	114	25 00	134	22 38	207	58 13	252	32 05	304	29 14	358
38	61 31	073	32 21	115	25 31	134	22 19	208	57 32	253	31 30	304	29 13	359
39	62 11	073	32 59	116	26 01	135	21 58	209	56 51	254	30 55	305	29 12	359
40	62 52	074	33 37	116	26 30	136	21 37	210	56 11	255	30 20	305	29 11	359
41	63 33	074	34 15	117	26 59	137	21 16	211	55 29	256	29 45	306	29 11	359
42	64 14	074	34 53	118	27 28	138	20 54	212	54 48	257	29 11	306	29 11	000
43	64 54	075	35 30	119	27 55	139	20 31	213	54 07	258	28 37	307	29 11	000
44	65 35	075	36 07	120	28 23	140	20 08	214	53 25	259	28 03	307	29 11	000
	◆Dubhe		POLLUX		BETELGEUSE		◆RIGEL		Diphda		◆Alpheratz		DENEB	
45	26 51	027	32 12	081	36 43	121	28 50	141	19 44	215	52 44	260	27 30	308
46	27 10	027	32 54	082	37 19	122	29 16	142	19 20	215	52 02	261	26 56	308
47	27 30	028	33 36	082	37 55	123	29 41	143	18 55	216	51 20	261	26 23	309
48	27 50	028	34 18	083	38 30	124	30 06	144	18 30	217	50 38	262	25 50	309
49	28 10	029	35 01	084	39 05	125	30 31	145	18 04	218	49 56	263	25 17	310
50	28 30	029	35 43	084	39 39	126	30 55	147	17 37	219	49 14	264	24 45	310
51	28 51	029	36 25	085	40 13	127	31 18	148	17 10	220	48 31	265	24 12	311
52	29 12	030	37 07	086	40 47	128	31 40	149	16 43	221	47 49	265	23 40	311
53	29 33	030	37 50	086	41 20	130	32 02	150	16 15	222	47 07	266	23 08	312
54	29 54	030	38 32	087	41 52	131	32 23	151	15 46	223	46 25	267	22 37	312
55	30 16	031	39 14	088	42 24	132	32 43	152	15 17	223	45 42	268	22 06	313
56	30 38	031	39 57	088	42 56	133	33 03	153	14 48	224	45 00	268	21 35	313
57	31 00	031	40 39	089	43 26	134	33 22	154	14 18	225	44 17	269	21 04	314
58	31 22	032	41 22	090	43 57	135	33 40	155	13 48	226	43 35	270	20 33	314
59	31 44	032	42 04	090	44 26	136	33 57	156	13 17	227	42 53	271	20 03	315
	◆Dubhe		POLLUX		PROCYON		◆SIRIUS		RIGEL		◆Hamal		DENEB	
60	32 07	032	42 46	091	28 05	113	17 49	138	34 14	158	58 30	236	19 33	315
61	32 30	033	43 29	092	28 44	114	18 17	139	34 29	159	57 54	238	19 03	316
62	32 53	033	44 11	093	29 23	115	18 44	140	34 44	160	57 18	239	18 34	316
63	33 16	033	44 54	093	30 02	115	19 11	141	34 59	161	56 42	240	18 05	317
64	33 40	034	45 36	094	30 40	116	19 38	142	35 12	162	56 05	241	17 36	317
65	34 04	034	46 18	095	31 18	117	20 04	143	35 24	163	55 27	243	17 07	318
66	34 28	034	47 00	096	31 55	118	20 29	144	35 36	165	54 49	244	16 39	319
67	34 52	035	47 43	096	32 33	119	20 54	145	35 47	166	54 11	245	16 11	319
68	35 16	035	48 25	097	33 09	120	21 18	146	35 57	167	53 32	246	15 44	320
69	35 40	035	49 07	098	33 46	121	21 41	147	36 06	168	52 53	247	15 16	320
70	36 05	036	49 49	099	34 22	122	22 04	148	36 14	169	52 14	248	14 49	321
71	36 30	036	50 31	100	34 58	123	22 27	149	36 21	171	51 35	249	14 23	321
72	36 55	036	51 12	101	35 34	124	22 49	150	36 28	172	50 55	250	13 56	322
73	37 20	036	51 54	101	36 09	125	23 10	150	36 33	173	50 15	251	13 30	322
74	37 45	037	52 36	102	36 43	126	23 31	151	36 38	174	49 35	252	13 05	323
	◆Dubhe		REGULUS		PROCYON		◆SIRIUS		RIGEL		◆Hamal		Schedar	
75	38 10	037	17 35	091	37 17	127	23 50	152	36 42	176	48 54	253	49 01	310
76	38 36	037	18 18	091	37 51	128	24 10	153	36 45	177	48 13	254	48 29	311
77	39 02	037	19 00	092	38 24	129	24 28	154	36 46	178	47 33	255	47 57	311
78	39 28	038	19 43	093	38 57	130	24 46	155	36 47	179	46 52	256	47 25	311
79	39 54	038	20 25	093	39 29	131	25 04	156	36 47	181	46 10	257	46 53	311
80	40 20	038	21 07	094	40 01	132	25 20	157	36 47	182	45 29	258	46 21	311
81	40 46	038	21 50	095	40 32	133	25 36	158	36 45	183	44 47	258	45 49	312
82	41 13	039	22 32	096	41 03	134	25 52	159	36 42	184	44 06	259	45 17	312
83	41 39	039	23 14	096	41 33	136	26 06	161	36 39	185	43 24	260	44 46	312
84	42 06	039	23 56	097	42 02	137	26 20	162	36 34	187	42 42	261	44 14	312
85	42 33	039	24 38	098	42 31	138	26 33	163	36 29	188	42 00	262	43 43	313
86	43 00	040	25 20	099	42 59	139	26 45	164	36 22	189	41 18	263	43 12	313
87	43 27	040	26 02	099	43 27	140	26 57	165	36 15	190	40 36	263	42 41	313
88	43 54	040	26 44	100	43 53	142	27 08	166	36 07	192	39 54	264	42 10	313
89	44 21	040	27 26	101	44 20	143	27 18	167	35 58	193	39 12	265	41 39	314

LAT 45°N

LHA	Hc	Zn	Hc	Zn	Hc	Zn	Hc	Zn	Hc	Zn	Hc	Zn	Hc	Zn
♈	◆Dubhe		REGULUS		PROCYON		◆SIRIUS		RIGEL		ALDEBARAN		◆Mirfak	
90	44 49	040	28 07	102	44 45	144	27 27	168	35 48	194	56 27	219	63 29	295
91	45 16	040	28 49	102	45 09	145	27 36	169	35 38	195	56 00	220	62 50	295
92	45 44	041	29 30	103	45 33	147	27 43	170	35 26	196	55 32	222	62 12	295
93	46 11	041	30 11	104	45 56	148	27 50	171	35 14	198	55 04	223	61 33	295
94	46 39	041	30 53	105	46 18	149	27 56	172	35 01	199	54 34	225	60 55	295
95	47 07	041	31 33	106	46 39	151	28 02	173	34 47	200	54 04	226	60 16	295
96	47 35	041	32 14	106	47 00	152	28 06	174	34 32	201	53 33	227	59 38	296
97	48 03	041	32 55	107	47 19	153	28 10	175	34 16	202	53 02	229	59 00	296
98	48 31	041	33 35	108	47 38	155	28 13	176	34 00	203	52 30	230	58 22	296
99	48 59	041	34 15	109	47 55	156	28 15	178	33 42	205	51 57	231	57 44	296
100	49 27	042	34 55	110	48 12	158	28 17	179	33 24	206	51 23	233	57 05	296
101	49 55	042	35 35	111	48 28	159	28 17	180	33 06	207	50 49	234	56 28	297
102	50 23	042	36 15	112	48 42	161	28 17	181	32 46	208	50 15	235	55 50	297
103	50 52	042	36 54	113	48 56	162	28 16	182	32 26	209	49 40	236	55 12	297
104	51 20	042	37 33	113	49 08	163	28 14	183	32 05	210	49 05	237	54 34	297
	◆Dubhe		Denebola		REGULUS		◆SIRIUS		RIGEL		ALDEBARAN		◆Mirfak	
105	51 48	042	22 48	092	38 12	114	28 12	184	31 43	211	48 29	238	53 56	298
106	52 17	042	23 30	093	38 50	115	28 08	185	31 21	212	47 52	240	53 19	298
107	52 45	042	24 12	093	39 29	116	28 04	186	30 58	213	47 15	241	52 41	298
108	53 13	042	24 55	094	40 07	117	27 59	187	30 34	214	46 38	242	52 04	298
109	53 42	042	25 37	095	40 44	118	27 53	188	30 10	215	46 01	243	51 27	299
110	54 10	042	26 19	096	41 21	119	27 47	189	29 45	216	45 23	244	50 50	299
111	54 38	042	27 01	096	41 58	120	27 39	191	29 20	217	44 45	245	50 12	299
112	55 07	042	27 44	097	42 35	121	27 31	192	28 54	219	44 06	246	49 35	300
113	55 35	042	28 26	098	43 11	122	27 22	193	28 27	220	43 27	247	48 59	300
114	56 03	042	29 08	099	43 46	123	27 12	194	28 00	221	42 48	248	48 22	300
115	56 32	042	29 50	099	44 21	125	27 02	195	27 32	222	42 09	249	47 45	300
116	57 00	042	30 31	100	44 56	126	26 51	196	27 03	222	41 29	250	47 09	301
117	57 28	041	31 13	101	45 30	127	26 39	197	26 34	223	40 49	251	46 32	301
118	57 56	041	31 55	102	46 04	128	26 26	198	26 05	224	40 09	251	45 56	301
119	58 24	041	32 36	102	46 37	129	26 13	199	25 35	225	39 29	252	45 20	302
	◆Kochab		Denebola		◆REGULUS		SIRIUS		RIGEL		◆ALDEBARAN		CAPELLA	
120	39 39	020	33 18	103	47 10	130	25 58	200	25 05	226	38 48	253	61 37	287
121	39 53	020	33 59	104	47 42	131	25 43	201	24 34	227	38 08	254	60 56	287
122	40 08	021	34 40	105	48 13	133	25 28	202	24 02	228	37 27	255	60 16	287
123	40 23	021	35 21	106	48 44	134	25 12	203	23 30	229	36 46	256	59 36	288
124	40 38	021	36 01	107	49 15	135	24 55	204	22 58	230	36 04	257	58 55	288
125	40 53	021	36 42	108	49 44	137	24 37	205	22 25	231	35 23	257	58 15	288
126	41 08	021	37 22	108	50 13	138	24 19	206	21 52	232	34 42	258	57 35	289
127	41 24	021	38 03	109	50 41	139	24 00	207	21 19	233	34 00	259	56 55	289
128	41 39	021	38 42	110	51 08	141	23 40	208	20 45	234	33 18	260	56 15	290
129	41 54	021	39 22	111	51 35	142	23 20	209	20 11	234	32 36	261	55 35	290
130	42 10	022	40 02	112	52 01	143	22 59	210	19 36	235	31 55	261	54 55	290
131	42 26	022	40 41	113	52 25	145	22 37	211	19 01	236	31 13	262	54 15	291
132	42 41	022	41 20	114	52 49	146	22 15	212	18 26	237	30 30	263	53 35	291
133	42 57	022	41 58	115	53 12	148	21 52	213	17 50	238	29 48	264	52 56	291
134	43 13	022	42 37	116	53 35	149	21 29	214	17 14	239	29 06	264	52 16	292
	◆Kochab		ARCTURUS		Denebola		◆REGULUS		SIRIUS		BETELGEUSE		◆CAPELLA	
135	43 29	022	21 12	084	43 15	117	53 56	151	21 05	215	35 09	241	51 37	292
136	43 45	022	21 54	084	43 53	118	54 16	152	20 40	216	34 32	242	50 58	292
137	44 01	022	22 37	085	44 30	119	54 35	154	20 15	217	33 54	243	50 19	293
138	44 17	022	23 19	086	45 07	120	54 53	156	19 50	218	33 16	244	49 39	293
139	44 33	022	24 01	086	45 43	121	55 10	157	19 24	218	32 38	245	49 01	294
140	44 49	022	24 44	087	46 20	122	55 26	159	18 57	219	32 00	246	48 22	294
141	45 05	022	25 26	088	46 55	123	55 40	161	18 30	220	31 21	247	47 43	294
142	45 22	023	26 08	088	47 31	124	55 54	162	18 02	221	30 42	248	47 04	295
143	45 38	023	26 51	089	48 05	125	56 06	164	17 34	222	30 02	248	46 26	295
144	45 54	023	27 33	090	48 40	127	56 17	166	17 05	223	29 23	249	45 48	296
145	46 10	023	28 16	091	49 13	128	56 27	167	16 36	224	28 43	250	45 09	296
146	46 27	023	28 58	091	49 47	129	56 36	169	16 07	225	28 03	251	44 31	296
147	46 43	023	29 40	092	50 19	130	56 43	171	15 37	225	27 23	252	43 53	297
148	47 00	023	30 23	093	50 51	132	56 49	173	15 06	226	26 42	253	43 15	297
149	47 16	023	31 05	093	51 23	133	56 53	175	14 35	227	26 02	253	42 38	298
	◆Kochab		ARCTURUS		◆SPICA		REGULUS		PROCYON		◆POLLUX		CAPELLA	
150	47 32	023	31 48	094	17 20	127	56 57	176	39 46	228	58 21	249	42 00	298
151	47 49	023	32 30	095	17 54	128	56 59	178	39 14	229	57 42	250	41 23	298
152	48 05	023	33 12	096	18 27	128	57 00	180	38 42	231	57 02	251	40 46	299
153	48 21	023	33 54	096	19 00	129	56 59	182	38 09	232	56 21	252	40 08	299
154	48 38	023	34 36	097	19 33	130	56 57	184	37 35	233	55 41	253	39 31	300
155	48 54	023	35 18	098	20 05	131	56 54	185	37 01	234	55 00	254	38 55	300
156	49 10	023	36 00	099	20 37	132	56 49	187	36 27	235	54 19	255	38 18	300
157	49 26	022	36 42	100	21 08	133	56 43	189	35 52	236	53 38	256	37 41	301
158	49 43	022	37 24	100	21 39	134	56 36	191	35 17	237	52 57	257	37 05	301
159	49 59	022	38 06	101	22 09	135	56 27	192	34 41	238	52 15	258	36 29	302
160	50 15	022	38 47	102	22 39	136	56 18	194	34 05	239	51 34	259	35 53	302
161	50 31	022	39 29	103	23 09	136	56 07	196	33 29	240	50 52	260	35 17	303
162	50 47	022	40 10	104	23 38	137	55 54	198	32 52	241	50 10	261	34 41	303
163	51 03	022	40 51	104	24 06	138	55 41	199	32 15	241	49 28	262	34 06	303
164	51 19	022	41 32	105	24 34	139	55 26	201	31 37	242	48 46	262	33 31	304
	◆Kochab		ARCTURUS		◆SPICA		REGULUS		PROCYON		◆POLLUX		CAPELLA	
165	51 34	022	42 13	106	25 02	140	55 11	203	31 00	243	48 04	263	32 56	304
166	51 50	022	42 54	107	25 29	141	54 54	204	30 21	244	47 22	264	32 21	305
167	52 06	022	43 34	108	25 55	142	54 36	206	29 43	245	46 40	265	31 46	305
168	52 21	021	44 14	109	26 21	143	54 17	207	29 05	246	45 57	266	31 11	306
169	52 37	021	44 54	110	26 46	144	53 57	209	28 26	247	45 15	266	30 37	306
170	52 52	021	45 34	111	27 10	145	53 36	211	27 47	248	44 33	267	30 03	307
171	53 07	021	46 14	112	27 34	146	53 14	212	27 07	248	43 50	268	29 29	307
172	53 22	021	46 53	113	27 58	147	52 51	214	26 28	249	43 08	269	28 55	308
173	53 37	021	47 32	114	28 20	148	52 27	215	25 48	250	42 25	269	28 22	308
174	53 52	020	48 11	115	28 42	149	52 02	217	25 08	251	41 43	270	27 48	309
175	54 07	020	48 49	116	29 04	150	51 36	218	24 28	252	41 01	271	27 15	309
176	54 21	020	49 27	117	29 25	151	51 10	219	23 47	253	40 18	271	26 42	309
177	54 35	020	50 05	118	29 45	152	50 42	221	23 07	253	39 36	272	26 10	310
178	54 50	019	50 42	119	30 04	153	50 14	222	22 26	254	38 53	273	25 37	310
179	55 04	019	51 19	120	30 22	155	49 46	223	21 45	255	38 11	273	25 05	311

LAT 45°N

LHA	Hc	Zn	Hc	Zn	Hc	Zn	Hc	Zn	Hc	Zn	Hc	Zn	Hc	Zn
♈	Kochab		◆VEGA		ARCTURUS		◆SPICA		REGULUS		◆POLLUX		CAPELLA	
180	55 18	019	20 47	055	51 55	121	30 40	156	49 16	225	37 29	274	24 33	311
181	55 31	019	21 22	056	52 31	123	30 58	157	48 46	226	36 46	275	24 01	312
182	55 45	018	21 57	057	53 07	124	31 14	158	48 15	227	36 04	275	23 30	312
183	55 58	018	22 33	057	53 42	125	31 30	159	47 44	228	35 22	276	22 59	313
184	56 11	018	23 08	058	54 16	126	31 44	160	47 12	230	34 40	277	22 28	313
185	56 24	018	23 44	058	54 50	128	31 59	161	46 39	231	33 58	277	21 57	314
186	56 37	017	24 20	059	55 23	129	32 12	162	46 06	232	33 16	278	21 27	314
187	56 49	017	24 57	059	55 56	130	32 24	163	45 32	233	32 34	279	20 57	315
188	57 01	017	25 33	060	56 28	132	32 36	165	44 58	234	31 52	279	20 27	315
189	57 13	016	26 10	060	56 59	133	32 47	166	44 23	235	31 10	280	19 57	316
190	57 25	016	26 47	061	57 30	135	32 57	167	43 48	237	30 28	281	19 28	317
191	57 36	016	27 24	061	57 59	136	33 06	168	43 12	238	29 46	281	18 59	317
192	57 48	015	28 02	062	58 28	138	33 15	169	42 36	239	29 05	282	18 30	318
193	57 59	015	28 39	062	58 56	139	33 22	170	42 00	240	28 23	282	18 01	318
194	58 09	014	29 17	063	59 24	141	33 29	171	41 23	241	27 42	283	17 33	319
	◆VEGA		Rasalhague		ARCTURUS		◆SPICA		REGULUS		◆POLLUX		Dubhe	
195	29 55	063	23 53	096	59 50	143	33 35	173	40 46	242	27 01	284	66 13	325
196	30 33	064	24 35	097	60 15	144	33 40	174	40 09	243	26 20	284	65 49	325
197	31 11	065	25 17	098	60 39	146	33 44	175	39 31	244	25 38	285	65 24	324
198	31 49	065	25 59	098	61 02	148	33 47	176	38 53	245	24 58	286	64 59	323
199	32 28	066	26 41	099	61 25	150	33 50	177	38 14	246	24 17	286	64 34	323
200	33 07	066	27 23	100	61 45	151	33 51	179	37 35	246	23 36	287	64 08	323
201	33 45	067	28 04	101	62 05	153	33 52	180	36 56	247	22 55	287	63 42	322
202	34 24	067	28 46	101	62 23	155	33 52	181	36 17	248	22 15	288	63 16	322
203	35 04	068	29 28	102	62 41	157	33 51	182	35 37	249	21 35	289	62 49	321
204	35 43	068	30 09	103	62 56	159	33 49	183	34 58	250	20 55	289	62 23	321
205	36 22	069	30 50	104	63 11	161	33 46	184	34 18	251	20 15	290	61 56	321
206	37 02	069	31 31	105	63 24	163	33 42	186	33 37	252	19 35	290	61 29	320
207	37 42	070	32 12	105	63 35	165	33 37	187	32 57	253	18 55	291	61 02	320
208	38 22	070	32 53	106	63 45	167	33 32	188	32 16	254	18 16	292	60 34	320
209	39 02	071	33 34	107	63 54	170	33 26	189	31 36	254	17 36	292	60 07	320
	◆DENEB		VEGA		Rasalhague		ANTARES		◆ARCTURUS		REGULUS		◆Dubhe	
210	24 24	050	39 42	071	34 14	108	10 55	146	64 01	172	30 55	255	59 39	319
211	24 57	050	40 22	072	34 55	109	11 18	147	64 06	174	30 14	256	59 12	319
212	25 29	050	41 02	072	35 35	110	11 40	148	64 10	176	29 32	257	58 44	319
213	26 02	051	41 43	073	36 14	111	12 03	149	64 12	178	28 51	258	58 16	319
214	26 35	051	42 23	073	36 54	111	12 24	150	64 12	180	28 10	258	57 48	319
215	27 08	052	43 04	074	37 33	112	12 45	151	64 11	182	27 28	259	57 20	319
216	27 42	052	43 45	074	38 13	113	13 06	151	64 09	185	26 46	260	56 52	318
217	28 16	053	44 26	075	38 51	114	13 26	152	64 05	187	26 04	261	56 23	318
218	28 50	053	45 07	075	39 30	115	13 45	153	63 59	189	25 22	261	55 55	318
219	29 24	054	45 48	076	40 08	116	14 04	154	63 51	191	24 40	262	55 27	318
220	29 58	054	46 29	077	40 46	117	14 22	155	63 42	193	23 58	263	54 59	318
221	30 33	055	47 10	077	41 24	118	14 40	156	63 32	195	23 16	264	54 30	318
222	31 08	055	47 52	078	42 01	119	14 57	157	63 20	197	22 34	264	54 02	318
223	31 43	056	48 33	078	42 38	120	15 14	158	63 07	199	21 52	265	53 33	318
224	32 18	056	49 15	079	43 14	121	15 29	158	62 52	201	21 10	266	53 05	318
	DENEB		◆ALTAIR		Rasalhague		ANTARES		◆ARCTURUS		Denebola		◆Dubhe	
225	32 53	057	18 30	096	43 50	122	15 45	159	62 36	203	39 38	249	52 37	318
226	33 29	057	19 12	097	44 26	123	15 59	160	62 18	205	38 58	249	52 08	318
227	34 04	058	19 55	098	45 01	124	16 13	161	62 00	207	38 19	250	51 40	318
228	34 40	058	20 37	098	45 36	126	16 27	162	61 40	209	37 39	251	51 12	318
229	35 16	058	21 18	099	46 10	127	16 40	163	61 18	211	36 58	252	50 43	318
230	35 52	059	22 00	100	46 44	128	16 52	164	60 56	213	36 18	253	50 15	318
231	36 29	059	22 42	101	47 17	129	17 03	165	60 33	214	35 37	254	49 47	318
232	37 05	060	23 24	101	47 50	130	17 14	166	60 08	216	34 56	255	49 19	318
233	37 42	060	24 05	102	48 22	131	17 24	167	59 42	218	34 15	256	48 51	319
234	38 19	061	24 47	103	48 53	133	17 34	168	59 16	220	33 34	256	48 23	319
235	38 56	061	25 28	104	49 24	134	17 42	168	58 48	221	32 53	257	47 55	319
236	39 33	062	26 09	104	49 55	135	17 50	169	58 20	223	32 11	258	47 27	319
237	40 11	062	26 50	105	50 24	137	17 58	170	57 51	224	31 30	259	46 59	319
238	40 48	062	27 31	106	50 53	138	18 05	171	57 21	226	30 48	260	46 31	319
239	41 26	063	28 12	107	51 21	139	18 11	172	56 50	227	30 06	260	46 03	319
	◆DENEB		ALTAIR		Rasalhague		◆ANTARES		ARCTURUS		Denebola		◆Dubhe	
240	42 04	063	28 52	108	51 48	141	18 16	173	56 19	229	29 25	261	45 36	319
241	42 42	064	29 33	108	52 15	142	18 21	174	55 47	230	28 43	262	45 08	320
242	43 20	064	30 13	109	52 40	144	18 25	175	55 14	231	28 01	263	44 41	320
243	43 58	064	30 53	110	53 05	145	18 28	176	54 40	233	27 18	263	44 13	320
244	44 36	065	31 32	111	53 29	146	18 31	177	54 06	234	26 36	264	43 46	320
245	45 15	065	32 12	112	53 52	148	18 33	178	53 32	235	25 54	265	43 19	320
246	45 53	066	32 51	113	54 14	150	18 34	179	52 57	236	25 12	266	42 52	321
247	46 32	066	33 30	114	54 35	151	18 35	180	52 21	238	24 29	266	42 25	321
248	47 11	066	34 09	115	54 55	153	18 35	181	51 45	239	23 47	267	41 58	321
249	47 50	067	34 47	115	55 14	154	18 34	182	51 08	240	23 05	268	41 32	321
250	48 29	067	35 26	116	55 32	156	18 32	183	50 31	241	22 22	269	41 05	321
251	49 08	068	36 03	117	55 49	158	18 30	184	49 54	242	21 40	269	40 39	321
252	49 47	068	36 41	118	56 04	159	18 27	184	49 16	243	20 57	270	40 12	322
253	50 27	068	37 18	119	56 19	161	18 23	185	48 38	244	20 15	271	39 46	322
254	51 06	069	37 55	120	56 32	163	18 19	186	48 00	245	19 33	271	39 20	322
	◆DENEB		ALTAIR		Nunki		◆ANTARES		ARCTURUS		◆Alkaid		Kochab	
255	51 46	069	38 31	121	14 02	154	18 14	187	47 21	246	57 32	295	57 29	344
256	52 26	070	39 08	122	14 21	154	18 08	188	46 42	247	56 54	295	57 17	344
257	53 06	070	39 43	123	14 39	155	18 02	189	46 03	248	56 15	296	57 06	344
258	53 46	070	40 19	124	14 56	156	17 55	190	45 23	249	55 37	296	56 53	343
259	54 26	071	40 53	125	15 13	157	17 47	191	44 43	250	54 59	296	56 41	343
260	55 06	071	41 28	126	15 29	158	17 38	192	44 03	251	54 21	297	56 29	343
261	55 46	072	42 02	127	15 44	159	17 29	193	43 23	252	53 43	297	56 16	342
262	56 26	072	42 35	129	15 59	160	17 20	194	42 42	253	53 05	297	56 03	342
263	57 07	072	43 08	130	16 14	161	17 09	195	42 02	254	52 28	297	55 49	342
264	57 47	073	43 40	131	16 27	162	16 58	196	41 21	255	51 50	298	55 36	341
265	58 28	073	44 12	132	16 41	163	16 46	197	40 40	256	51 12	298	55 22	341
266	59 08	073	44 44	133	16 53	163	16 34	197	39 58	257	50 35	298	55 09	341
267	59 49	074	45 14	134	17 05	164	16 21	198	39 17	257	49 58	299	54 55	341
268	60 30	074	45 44	136	17 16	165	16 07	199	38 36	258	49 21	299	54 41	340
269	61 11	075	46 14	137	17 26	166	15 53	200	37 54	259	48 43	299	54 26	340

LHA	Hc	Zn	Hc	Zn	Hc	Zn	Hc	Zn	Hc	Zn	Hc	Zn	Hc	Zn
♈	◆Alpheratz		ALTAIR		Nunki		◆ANTARES		ARCTURUS		◆Alkaid		Kochab	
270	18 45	067	46 42	138	17 36	167	15 38	201	37 12	260	48 06	299	54 12	340
271	19 25	068	47 10	139	17 45	168	15 22	202	36 30	261	47 30	300	53 57	340
272	20 04	069	47 38	141	17 54	169	15 06	203	35 49	262	46 53	300	53 42	340
273	20 44	069	48 04	142	18 02	170	14 49	204	35 07	262	46 16	300	53 27	339
274	21 23	070	48 30	143	18 09	171	14 32	205	34 24	263	45 40	301	53 12	339
275	22 03	070	48 55	145	18 15	172	14 14	205	33 42	264	45 03	301	52 57	339
276	22 43	071	49 19	146	18 21	173	13 56	206	33 00	265	44 27	301	52 42	339
277	23 23	072	49 42	147	18 26	174	13 36	207	32 18	265	43 51	302	52 27	339
278	24 04	072	50 05	149	18 30	175	13 17	208	31 36	266	43 15	302	52 11	339
279	24 44	073	50 26	150	18 34	176	12 57	209	30 53	267	42 39	303	51 56	338
280	25 25	073	50 47	152	18 37	176	12 36	210	30 11	268	42 03	303	51 40	338
281	26 06	074	51 06	153	18 39	177	12 15	211	29 28	268	41 28	303	51 24	338
282	26 46	075	51 25	155	18 41	178	11 53	211	28 46	269	40 52	304	51 08	338
283	27 27	075	51 42	156	18 42	179	11 30	212	28 04	270	40 17	304	50 53	338
284	28 09	076	51 59	158	18 42	180	11 07	213	27 21	270	39 42	304	50 37	338
	◆Mirfak		Alpheratz		◆ALTAIR		Rasalhague		◆ARCTURUS		Alkaid		Kochab	
285	15 49	033	28 50	076	52 15	159	52 49	216	26 39	271	39 07	305	50 21	338
286	16 12	033	29 31	077	52 29	161	52 24	217	25 56	272	38 32	305	50 04	338
287	16 35	034	30 12	078	52 42	162	51 58	219	25 14	272	37 58	305	49 48	338
288	16 59	034	30 54	078	52 54	164	51 31	220	24 32	273	37 23	306	49 32	338
289	17 23	035	31 36	079	53 06	166	51 03	222	23 49	274	36 49	306	49 16	337
290	17 48	035	32 17	080	53 15	167	50 34	223	23 07	275	36 15	307	49 00	337
291	18 13	036	32 59	080	53 24	169	50 05	224	22 25	275	35 41	307	48 43	337
292	18 38	037	33 41	081	53 32	171	49 35	226	21 42	276	35 07	307	48 27	337
293	19 03	037	34 23	081	53 37	172	49 04	227	21 00	277	34 33	308	48 11	337
294	19 29	038	35 05	082	53 43	174	48 33	228	20 18	277	34 00	308	47 54	337
295	19 55	038	35 47	083	53 47	176	48 01	229	19 36	278	33 27	309	47 38	337
296	20 21	039	36 29	083	53 50	177	47 29	231	18 54	279	32 54	309	47 22	337
297	20 48	039	37 11	084	53 51	179	46 56	232	18 12	279	32 21	310	47 05	337
298	21 15	040	37 53	085	53 51	181	46 22	233	17 30	280	31 48	310	46 49	337
299	21 42	040	38 36	085	53 50	182	45 48	234	16 49	281	31 16	310	46 33	337
	Mirfak		◆Alpheratz		Enif		◆ALTAIR		Rasalhague		Alphecca		◆Kochab	
300	22 09	041	39 18	086	48 21	139	53 48	184	45 13	235	34 50	275	46 16	337
301	22 37	041	40 00	087	48 48	141	53 44	186	44 38	236	34 07	275	46 00	337
302	23 05	042	40 43	087	49 15	142	53 40	187	44 03	237	33 25	276	45 44	337
303	23 33	042	41 25	088	49 40	144	53 34	189	43 27	238	32 43	277	45 27	337
304	24 02	042	42 07	089	50 05	145	53 26	191	42 51	239	32 01	277	45 11	338
305	24 30	043	42 50	089	50 29	146	53 18	192	42 14	241	31 19	278	44 55	338
306	24 59	043	43 32	090	50 52	148	53 08	194	41 37	242	30 37	278	44 39	338
307	25 29	044	44 15	091	51 14	149	52 58	195	40 59	243	29 55	279	44 23	338
308	25 58	044	44 57	092	51 36	151	52 46	197	40 21	244	29 13	280	44 07	338
309	26 28	045	45 39	092	51 56	152	52 33	199	39 43	244	28 31	280	43 51	338
310	26 58	045	46 22	093	52 15	154	52 19	200	39 05	245	27 49	281	43 35	338
311	27 28	046	47 04	094	52 34	155	52 03	202	38 26	246	27 08	282	43 19	338
312	27 59	046	47 47	095	52 51	157	51 47	203	37 47	247	26 26	282	43 03	338
313	28 30	047	48 29	095	53 07	158	51 30	205	37 08	248	25 45	283	42 47	338
314	29 00	047	49 11	096	53 22	160	51 11	206	36 28	249	25 04	284	42 31	338
	CAPELLA		◆Alpheratz		Enif		◆ALTAIR		Rasalhague		VEGA		◆Kochab	
315	13 30	036	49 53	097	53 36	162	50 52	208	35 49	250	62 52	269	42 16	338
316	13 55	037	50 35	098	53 49	163	50 32	209	35 09	251	62 09	270	42 00	339
317	14 20	037	51 17	099	54 01	165	50 10	211	34 28	252	61 27	271	41 45	339
318	14 46	038	51 59	099	54 11	166	49 48	212	33 48	253	60 45	272	41 29	339
319	15 13	039	52 41	100	54 20	168	49 25	214	33 08	253	60 02	272	41 14	339
320	15 39	039	53 23	101	54 29	170	49 01	215	32 27	254	59 20	273	40 59	339
321	16 06	040	54 04	102	54 35	172	48 37	216	31 46	255	58 37	274	40 43	339
322	16 33	040	54 46	103	54 41	173	48 11	218	31 05	256	57 55	274	40 28	339
323	17 01	041	55 27	104	54 45	175	47 45	219	30 24	257	57 13	275	40 13	339
324	17 29	041	56 08	105	54 49	177	47 18	220	29 42	258	56 30	275	39 59	340
325	17 57	042	56 49	106	54 50	178	46 50	222	29 01	258	55 48	276	39 44	340
326	18 25	042	57 30	107	54 51	180	46 21	223	28 19	259	55 06	277	39 29	340
327	18 54	043	58 10	108	54 50	182	45 52	224	27 37	260	54 24	277	39 15	340
328	19 23	043	58 50	109	54 49	183	45 22	225	26 56	261	53 42	278	39 00	340
329	19 52	044	59 30	110	54 45	185	44 52	227	26 14	261	53 00	278	38 46	340
	◆CAPELLA		Hamal		Diphda		◆FOMALHAUT		ALTAIR		◆VEGA		Kochab	
330	20 22	044	36 04	092	16 52	139	14 13	167	44 21	228	52 18	279	38 32	341
331	20 52	045	36 46	093	17 19	140	14 22	168	43 49	229	51 36	279	38 18	341
332	21 22	045	37 28	093	17 46	141	14 31	169	43 17	230	50 54	280	38 04	341
333	21 52	046	38 11	094	18 12	142	14 38	170	42 44	231	50 13	281	37 50	341
334	22 23	047	38 53	095	18 38	143	14 46	171	42 11	232	49 31	281	37 36	341
335	22 54	047	39 35	096	19 03	144	14 52	172	41 37	233	48 49	282	37 23	341
336	23 25	048	40 17	096	19 28	145	14 58	172	41 03	234	48 08	282	37 09	342
337	23 56	048	40 59	097	19 52	146	15 03	173	40 28	235	47 26	283	36 56	342
338	24 28	049	41 42	098	20 16	147	15 08	174	39 53	237	46 45	283	36 43	342
339	25 00	049	42 24	099	20 39	148	15 12	175	39 17	238	46 04	284	36 30	342
340	25 32	049	43 05	100	21 01	149	15 15	176	38 41	239	45 23	284	36 17	343
341	26 04	050	43 47	100	21 23	149	15 17	177	38 05	240	44 42	285	36 04	343
342	26 37	050	44 29	101	21 44	150	15 19	178	37 28	241	44 01	285	35 52	343
343	27 10	051	45 10	102	22 05	151	15 20	179	36 51	241	43 20	286	35 40	343
344	27 43	051	45 52	103	22 25	152	15 21	180	36 14	242	42 39	286	35 27	343
	◆CAPELLA		ALDEBARAN		Diphda		◆FOMALHAUT		ALTAIR		◆VEGA		Kochab	
345	28 16	052	15 50	082	22 44	153	15 21	181	35 36	243	41 58	287	35 15	344
346	28 50	052	16 32	083	23 03	154	15 20	181	34 58	244	41 18	287	35 03	344
347	29 23	053	17 14	084	23 21	155	15 19	182	34 19	245	40 37	288	34 52	344
348	29 57	053	17 56	084	23 39	156	15 17	183	33 41	246	39 57	289	34 40	344
349	30 31	054	18 38	085	23 55	157	15 14	184	33 02	247	39 17	289	34 29	345
350	31 06	054	19 21	086	24 12	158	15 10	185	32 23	248	38 37	290	34 18	345
351	31 40	055	20 03	086	24 27	159	15 06	186	31 43	249	37 57	290	34 06	345
352	32 15	055	20 45	087	24 42	160	15 02	187	31 03	250	37 17	291	33 56	345
353	32 50	056	21 28	088	24 56	161	14 56	188	30 24	250	36 38	291	33 45	346
354	33 25	056	22 10	088	25 09	162	14 50	189	29 44	251	35 58	292	33 34	346
355	34 00	056	22 53	089	25 21	163	14 43	190	29 03	252	35 19	292	33 24	346
356	34 36	057	23 35	090	25 33	164	14 36	190	28 23	253	34 39	293	33 14	346
357	35 11	057	24 17	091	25 44	165	14 28	191	27 42	254	34 00	293	33 04	347
358	35 47	058	25 00	091	25 55	166	14 19	192	27 01	255	33 21	294	32 54	347
359	36 23	058	25 42	092	26 04	167	14 10	193	26 20	255	32 43	294	32 45	347

LAT 44°N

LHA	Hc	Zn	Hc	Zn	Hc	Zn	Hc	Zn	Hc	Zn	Hc	Zn	Hc	Zn
♈	◆CAPELLA		ALDEBARAN		◆Diphda		FOMALHAUT		ALTAIR		◆VEGA		Kochab	
0	36 28	058	26 27	092	27 12	168	14 58	194	25 53	257	31 39	295	31 37	347
1	37 05	058	27 10	093	27 20	169	14 47	195	25 11	257	31 00	296	31 27	348
2	37 41	059	27 53	094	27 28	171	14 36	196	24 29	258	30 21	296	31 18	348
3	38 18	059	28 37	094	27 34	172	14 24	197	23 47	259	29 42	297	31 09	348
4	38 56	060	29 20	095	27 40	173	14 11	198	23 04	260	29 04	297	31 01	349
5	39 33	060	30 03	096	27 45	174	13 58	198	22 22	260	28 26	298	30 52	349
6	40 10	060	30 45	097	27 50	175	13 44	199	21 39	261	27 48	298	30 44	349
7	40 48	061	31 28	097	27 53	176	13 29	200	20 56	262	27 10	299	30 36	349
8	41 26	061	32 11	098	27 56	177	13 14	201	20 14	263	26 32	299	30 28	350
9	42 04	062	32 54	099	27 58	178	12 59	202	19 31	263	25 55	300	30 20	350
10	42 42	062	33 36	100	27 59	179	12 42	203	18 48	264	25 17	300	30 13	350
11	43 20	062	34 19	100	27 59	180	12 25	204	18 05	265	24 40	301	30 06	350
12	43 58	063	35 01	101	27 59	181	12 08	204	17 22	266	24 03	301	29 59	351
13	44 37	063	35 43	102	27 57	182	11 50	205	16 39	266	23 27	302	29 52	351
14	45 15	064	36 26	103	27 55	183	11 31	206	15 56	267	22 50	303	29 45	351
	◆CAPELLA		BETELGEUSE		RIGEL		◆Diphda		Enif		◆DENEB		Kochab	
15	45 54	064	16 50	096	12 34	115	27 52	184	35 43	246	45 14	295	29 39	352
16	46 33	064	17 32	097	13 13	116	27 48	186	35 03	247	44 35	296	29 33	352
17	47 12	065	18 15	097	13 52	116	27 44	187	34 23	248	43 56	296	29 27	352
18	47 51	065	18 58	098	14 31	117	27 38	188	33 43	249	43 17	297	29 21	352
19	48 30	065	19 41	099	15 09	118	27 32	189	33 02	250	42 39	297	29 15	353
20	49 09	066	20 23	100	15 47	119	27 25	190	32 22	251	42 00	297	29 10	353
21	49 48	066	21 06	100	16 25	119	27 17	191	31 41	252	41 22	298	29 05	353
22	50 28	066	21 48	101	17 02	120	27 09	192	31 00	252	40 44	298	29 00	354
23	51 08	067	22 30	102	17 39	121	27 00	193	30 19	253	40 06	299	28 55	354
24	51 47	067	23 13	103	18 16	122	26 50	194	29 37	254	39 28	299	28 51	354
25	52 27	067	23 55	103	18 53	123	26 39	195	28 56	255	38 51	299	28 47	355
26	53 07	068	24 37	104	19 29	124	26 27	196	28 14	256	38 13	300	28 43	355
27	53 47	068	25 18	105	20 05	124	26 15	197	27 32	256	37 36	300	28 39	355
28	54 27	068	26 00	106	20 40	125	26 02	198	26 50	257	36 58	301	28 36	355
29	55 07	069	26 41	107	21 15	126	25 48	199	26 08	258	36 21	301	28 32	356
	◆CAPELLA		BETELGEUSE		RIGEL		◆Diphda		Alpheratz		◆DENEB		Kochab	
30	55 47	069	27 23	107	21 50	127	25 33	200	63 13	245	35 44	301	28 29	356
31	56 28	069	28 04	108	22 24	128	25 18	201	62 34	247	35 08	302	28 26	356
32	57 08	070	28 45	109	22 58	129	25 02	202	61 54	248	34 31	302	28 24	357
33	57 49	070	29 25	110	23 32	130	24 45	203	61 14	249	33 55	303	28 21	357
34	58 29	070	30 06	111	24 05	130	24 28	204	60 33	250	33 18	303	28 19	357
35	59 10	071	30 46	111	24 37	131	24 10	205	59 53	251	32 42	304	28 17	358
36	59 51	071	31 26	112	25 10	132	23 51	206	59 12	252	32 07	304	28 16	358
37	60 31	071	32 06	113	25 41	133	23 32	207	58 30	254	31 31	304	28 14	358
38	61 12	071	32 46	114	26 12	134	23 11	208	57 49	255	30 55	305	28 13	359
39	61 53	072	33 25	115	26 43	135	22 51	209	57 07	256	30 20	305	28 12	359
40	62 34	072	34 04	116	27 13	136	22 29	210	56 25	256	29 45	306	28 11	359
41	63 15	072	34 43	117	27 43	137	22 07	211	55 43	257	29 10	306	28 11	359
42	63 56	072	35 21	118	28 12	138	21 45	212	55 01	258	28 35	307	28 11	000
43	64 38	073	35 59	119	28 41	139	21 22	213	54 19	259	28 01	307	28 11	000
44	65 19	073	36 37	120	29 09	140	20 58	214	53 36	260	27 27	308	28 11	000
	◆Dubhe		POLLUX		BETELGEUSE		◆RIGEL		Diphda		◆Alpheratz		DENEB	
45	25 57	027	32 03	080	37 14	121	29 36	141	20 34	215	52 54	261	26 53	308
46	26 17	027	32 45	081	37 51	122	30 03	142	20 09	216	52 11	262	26 19	309
47	26 37	028	33 28	082	38 27	123	30 29	143	19 43	217	51 28	263	25 45	309
48	26 57	028	34 11	082	39 04	124	30 55	144	19 17	217	50 45	263	25 12	310
49	27 17	028	34 54	083	39 39	125	31 20	145	18 51	218	50 03	264	24 39	310
50	27 38	029	35 36	084	40 15	126	31 44	146	18 24	219	49 20	265	24 06	311
51	27 59	029	36 19	084	40 50	127	32 08	147	17 56	220	48 37	266	23 33	311
52	28 20	029	37 02	085	41 24	128	32 31	148	17 28	221	47 53	266	23 01	312
53	28 41	030	37 45	085	41 58	129	32 54	149	16 59	222	47 10	267	22 28	312
54	29 02	030	38 28	086	42 31	130	33 15	151	16 30	223	46 27	268	21 56	312
55	29 24	030	39 11	087	43 04	131	33 36	152	16 01	224	45 44	269	21 25	313
56	29 46	031	39 55	087	43 36	132	33 56	153	15 31	224	45 01	269	20 53	314
57	30 09	031	40 38	088	44 08	133	34 16	154	15 00	225	44 18	270	20 22	314
58	30 31	031	41 21	089	44 39	135	34 34	155	14 30	226	43 35	271	19 51	315
59	30 54	032	42 04	090	45 09	136	34 52	156	13 58	227	42 52	271	19 21	315
	◆Dubhe		POLLUX		PROCYON		◆SIRIUS		RIGEL		◆Hamal		DENEB	
60	31 16	032	42 47	090	28 28	112	18 34	138	35 09	157	59 02	238	18 50	316
61	31 40	032	43 30	091	29 08	113	19 03	139	35 25	158	58 26	239	18 20	316
62	32 03	033	44 13	092	29 48	114	19 31	140	35 41	160	57 49	240	17 50	317
63	32 26	033	44 57	092	30 27	115	19 58	141	35 55	161	57 11	242	17 21	317
64	32 50	033	45 40	093	31 06	116	20 25	142	36 09	162	56 33	243	16 52	318
65	33 14	034	46 23	094	31 45	117	20 51	143	36 22	163	55 54	244	16 23	318
66	33 38	034	47 06	095	32 23	118	21 17	144	36 34	164	55 15	245	15 54	319
67	34 02	034	47 49	095	33 01	118	21 43	145	36 45	166	54 36	246	15 26	319
68	34 27	035	48 32	096	33 39	119	22 07	145	36 55	167	53 56	247	14 58	320
69	34 51	035	49 15	097	34 17	120	22 31	146	37 05	168	53 16	248	14 30	320
70	35 16	035	49 57	098	34 54	121	22 55	147	37 13	169	52 36	249	14 03	321
71	35 41	035	50 40	098	35 30	122	23 18	148	37 21	171	51 55	250	13 36	321
72	36 06	036	51 23	099	36 07	123	23 40	149	37 27	172	51 15	251	13 09	322
73	36 31	036	52 05	100	36 43	124	24 02	150	37 33	173	50 34	252	12 43	323
74	36 57	036	52 48	101	37 18	125	24 23	151	37 38	174	49 52	253	12 17	323
	◆Dubhe		REGULUS		PROCYON		◆SIRIUS		RIGEL		◆Hamal		Schedar	
75	37 22	037	17 36	090	37 53	126	24 44	152	37 42	176	49 11	254	48 22	311
76	37 48	037	18 19	091	38 28	127	25 03	153	37 44	177	48 29	255	47 50	311
77	38 14	037	19 02	092	39 02	128	25 23	154	37 46	178	47 48	256	47 17	312
78	38 40	037	19 45	092	39 35	129	25 41	155	37 47	179	47 06	257	46 45	312
79	39 06	037	20 28	093	40 09	130	25 59	156	37 47	181	46 24	258	46 13	312
80	39 33	038	21 11	094	40 41	132	26 16	157	37 47	182	45 41	259	45 41	312
81	39 59	038	21 54	094	41 13	133	26 32	158	37 45	183	44 59	259	45 09	312
82	40 26	038	22 37	095	41 45	134	26 48	159	37 42	184	44 16	260	44 37	313
83	40 52	038	23 20	096	42 16	135	27 03	160	37 38	186	43 34	261	44 05	313
84	41 19	039	24 03	097	42 46	136	27 17	161	37 34	187	42 51	262	43 34	313
85	41 46	039	24 46	097	43 15	137	27 30	162	37 28	188	42 08	263	43 02	313
86	42 13	039	25 29	098	43 44	138	27 43	163	37 22	189	41 26	263	42 31	313
87	42 40	039	26 12	099	44 13	140	27 55	165	37 14	190	40 43	264	41 59	314
88	43 08	039	26 54	100	44 40	141	28 06	166	37 06	192	40 00	265	41 28	314
89	43 35	039	27 37	100	45 07	142	28 16	167	36 57	193	39 17	266	40 57	314

LHA	Hc	Zn	Hc	Zn	Hc	Zn	Hc	Zn	Hc	Zn	Hc	Zn	Hc	Zn
♈	◆Dubhe		REGULUS		PROCYON		◆SIRIUS		RIGEL		ALDEBARAN		◆Mirfak	
90	44 03	040	28 19	101	45 33	143	28 26	168	36 47	194	57 13	220	63 03	296
91	44 30	040	29 01	102	45 59	145	28 34	169	36 36	195	56 46	221	62 24	296
92	44 58	040	29 44	103	46 23	146	28 42	170	36 24	197	56 17	223	61 46	297
93	45 26	040	30 26	103	46 47	147	28 50	171	36 11	198	55 47	224	61 07	297
94	45 53	040	31 08	104	47 10	149	28 56	172	35 57	199	55 17	226	60 28	297
95	46 21	040	31 49	105	47 32	150	29 01	173	35 43	200	54 45	227	59 50	297
96	46 49	040	32 31	106	47 53	151	29 06	174	35 28	201	54 14	228	59 11	297
97	47 17	041	33 12	107	48 13	153	29 10	175	35 11	203	53 41	230	58 33	297
98	47 45	041	33 54	108	48 32	154	29 13	176	34 55	204	53 08	231	57 55	297
99	48 14	041	34 35	108	48 50	156	29 15	178	34 37	205	52 34	232	57 16	298
100	48 42	041	35 16	109	49 07	157	29 17	179	34 18	206	52 00	234	56 38	298
101	49 10	041	35 56	110	49 24	159	29 17	180	33 59	207	51 25	235	56 00	298
102	49 38	041	36 37	111	49 39	160	29 17	181	33 39	208	50 49	236	55 22	298
103	50 07	041	37 17	112	49 53	162	29 16	182	33 18	209	50 13	237	54 44	298
104	50 35	041	37 57	113	50 06	163	29 14	183	32 57	210	49 37	238	54 06	299
	Dubhe		◆Denebola		REGULUS		◆SIRIUS		RIGEL		ALDEBARAN		◆Mirfak	
105	51 03	041	22 49	092	38 36	114	29 12	184	32 35	212	49 00	239	53 28	299
106	51 32	041	23 33	092	39 16	115	29 08	185	32 12	213	48 22	241	52 50	299
107	52 00	041	24 16	093	39 55	116	29 04	186	31 48	214	47 44	242	52 13	299
108	52 28	041	24 59	094	40 34	117	28 59	187	31 24	215	47 06	243	51 35	299
109	52 57	041	25 42	094	41 12	117	28 53	189	30 59	216	46 28	244	50 57	300
110	53 25	041	26 25	095	41 50	118	28 46	190	30 33	217	45 49	245	50 20	300
111	53 53	041	27 08	096	42 28	119	28 38	191	30 07	218	45 10	246	49 43	300
112	54 22	041	27 51	097	43 05	121	28 30	192	29 40	219	44 30	247	49 05	301
113	54 50	041	28 34	097	43 42	122	28 21	193	29 13	220	43 50	248	48 28	301
114	55 18	041	29 16	098	44 19	123	28 11	194	28 45	221	43 10	249	47 51	301
115	55 46	041	29 59	099	44 55	124	28 00	195	28 17	222	42 30	250	47 14	301
116	56 15	041	30 42	100	45 31	125	27 48	196	27 47	223	41 50	250	46 38	302
117	56 43	040	31 24	100	46 06	126	27 36	197	27 18	224	41 09	251	46 01	302
118	57 11	040	32 07	101	46 41	127	27 23	198	26 48	225	40 28	252	45 24	302
119	57 38	040	32 49	102	47 15	128	27 09	199	26 17	226	39 47	253	44 48	303
	◆Kochab		Denebola		◆REGULUS		SIRIUS		RIGEL		◆ALDEBARAN		CAPELLA	
120	38 42	020	33 31	103	47 48	129	26 55	200	25 46	227	39 05	254	61 19	289
121	38 57	020	34 13	103	48 21	131	26 39	201	25 14	228	38 24	255	60 38	289
122	39 12	020	34 55	104	48 54	132	26 23	202	24 42	228	37 42	256	59 57	289
123	39 27	020	35 37	105	49 26	133	26 07	203	24 10	229	37 00	257	59 16	289
124	39 42	021	36 18	106	49 57	134	25 49	204	23 37	230	36 18	257	58 36	290
125	39 57	021	37 00	107	50 27	136	25 31	205	23 03	231	35 36	258	57 55	290
126	40 12	021	37 41	108	50 57	137	25 12	206	22 29	232	34 53	259	57 15	290
127	40 28	021	38 22	109	51 26	138	24 53	207	21 55	233	34 11	260	56 34	291
128	40 43	021	39 03	109	51 54	140	24 33	208	21 20	234	33 28	260	55 54	291
129	40 59	021	39 43	110	52 22	141	24 12	209	20 45	235	32 46	261	55 14	291
130	41 14	021	40 24	111	52 48	143	23 50	210	20 10	236	32 03	262	54 33	292
131	41 30	021	41 04	112	53 14	144	23 28	211	19 34	236	31 20	263	53 53	292
132	41 46	021	41 44	113	53 39	146	23 06	212	18 58	237	30 38	264	53 13	292
133	42 01	022	42 23	114	54 03	147	22 42	213	18 22	238	29 55	264	52 33	293
134	42 17	022	43 03	115	54 26	149	22 19	214	17 45	239	29 12	265	51 54	293
	◆Kochab		ARCTURUS		Denebola		◆REGULUS		SIRIUS		BETELGEUSE		◆CAPELLA	
135	42 33	022	21 05	083	43 41	116	54 48	150	21 54	215	35 38	242	51 14	293
136	42 49	022	21 48	084	44 20	117	55 09	152	21 29	216	35 00	243	50 34	294
137	43 05	022	22 31	085	44 58	118	55 29	153	21 03	217	34 21	244	49 55	294
138	43 21	022	23 14	085	45 36	119	55 47	155	20 37	218	33 42	245	49 15	294
139	43 37	022	23 57	086	46 14	120	56 05	157	20 11	219	33 03	246	48 36	295
140	43 54	022	24 40	087	46 51	121	56 22	158	19 43	220	32 24	246	47 57	295
141	44 10	022	25 23	087	47 28	122	56 37	160	19 16	220	31 44	247	47 18	295
142	44 26	022	26 06	088	48 04	123	56 51	162	18 47	221	31 04	248	46 39	296
143	44 42	022	26 50	089	48 40	125	57 04	164	18 19	222	30 24	249	46 00	296
144	44 59	022	27 33	089	49 15	126	57 15	165	17 49	223	29 44	250	45 21	296
145	45 15	022	28 16	090	49 50	127	57 26	167	17 20	224	29 03	251	44 43	297
146	45 31	022	28 59	091	50 24	128	57 34	169	16 49	225	28 22	251	44 04	297
147	45 48	022	29 42	091	50 58	129	57 42	171	16 19	226	27 41	252	43 26	298
148	46 04	022	30 25	092	51 31	131	57 48	173	15 48	226	27 00	253	42 48	298
149	46 20	022	31 08	093	52 03	132	57 53	174	15 16	227	26 19	254	42 10	298
	◆Kochab		ARCTURUS		◆SPICA		REGULUS		PROCYON		◆POLLUX		CAPELLA	
150	46 37	022	31 52	094	17 56	126	57 57	176	40 26	229	58 42	251	41 32	299
151	46 53	022	32 35	094	18 30	127	57 59	178	39 53	230	58 01	252	40 54	299
152	47 09	022	33 18	095	19 04	128	58 00	180	39 19	231	57 20	253	40 16	299
153	47 26	022	34 01	096	19 38	129	57 59	182	38 46	232	56 39	254	39 39	300
154	47 42	022	34 44	096	20 11	130	57 57	184	38 11	233	55 57	255	39 01	300
155	47 58	022	35 26	097	20 44	131	57 53	185	37 36	234	55 15	256	38 24	301
156	48 15	022	36 09	098	21 17	132	57 49	187	37 01	235	54 33	257	37 47	301
157	48 31	022	36 52	099	21 49	133	57 42	189	36 25	236	53 51	258	37 10	302
158	48 47	022	37 34	100	22 20	133	57 35	191	35 49	237	53 09	259	36 34	302
159	49 03	022	38 17	100	22 51	134	57 26	193	35 13	238	52 27	259	35 57	302
160	49 19	022	38 59	101	23 22	135	57 16	195	34 36	239	51 44	260	35 21	303
161	49 35	022	39 42	102	23 52	136	57 04	196	33 59	240	51 02	261	34 45	303
162	49 51	022	40 24	103	24 22	137	56 52	198	33 21	241	50 19	262	34 08	304
163	50 07	022	41 06	104	24 51	138	56 38	200	32 43	242	49 36	263	33 33	304
164	50 23	021	41 48	104	25 20	139	56 22	201	32 05	243	48 53	264	32 57	304
	◆Kochab		ARCTURUS		◆SPICA		REGULUS		PROCYON		◆POLLUX		CAPELLA	
165	50 39	021	42 29	105	25 48	140	56 06	203	31 26	244	48 10	264	32 21	305
166	50 54	021	43 11	106	26 15	141	55 48	205	30 47	245	47 28	265	31 46	305
167	51 10	021	43 52	107	26 42	142	55 30	206	30 08	246	46 44	266	31 11	306
168	51 25	021	44 33	108	27 09	143	55 10	208	29 29	246	46 01	267	30 36	306
169	51 41	021	45 14	109	27 34	144	54 49	210	28 49	247	45 18	267	30 01	307
170	51 56	021	45 55	110	28 00	145	54 27	211	28 09	248	44 35	268	29 27	307
171	52 11	020	46 36	111	28 24	146	54 04	213	27 29	249	43 52	269	28 52	308
172	52 26	020	47 16	112	28 48	147	53 40	214	26 49	250	43 09	269	28 18	308
173	52 41	020	47 56	113	29 11	148	53 16	216	26 08	251	42 26	270	27 44	308
174	52 56	020	48 35	114	29 34	149	52 50	217	25 27	251	41 43	271	27 11	309
175	53 10	020	49 15	115	29 56	150	52 23	219	24 46	252	40 59	272	26 37	309
176	53 25	019	49 54	116	30 17	151	51 56	220	24 05	253	40 16	272	26 04	310
177	53 39	019	50 32	117	30 38	152	51 28	222	23 24	254	39 33	273	25 31	310
178	53 53	019	51 11	118	30 57	153	50 59	223	22 42	255	38 50	274	24 58	311
179	54 07	019	51 49	119	31 17	154	50 29	224	22 00	255	38 07	274	24 26	311

LAT 44°N

LHA ♈	Hc	Zn	Hc	Zn	Hc	Zn	Hc	Zn	Hc	Zn	Hc	Zn	Hc	Zn
	Kochab		◆VEGA		ARCTURUS		◆SPICA		REGULUS		◆POLLUX		CAPELLA	
180	54 21	018	20 13	055	52 26	120	31 35	155	49 58	226	37 24	275	23 53	312
181	54 34	018	20 48	056	53 03	122	31 53	156	49 27	227	36 41	275	23 21	312
182	54 48	018	21 24	056	53 40	123	32 09	158	48 56	228	35 58	276	22 49	313
183	55 01	018	22 00	057	54 16	124	32 26	159	48 23	229	35 15	277	22 18	313
184	55 14	017	22 36	057	54 51	125	32 41	160	47 50	231	34 32	277	21 47	314
185	55 27	017	23 13	058	55 26	127	32 55	161	47 17	232	33 50	278	21 15	314
186	55 39	017	23 49	058	56 00	128	33 09	162	46 42	233	33 07	279	20 45	315
187	55 52	016	24 26	059	56 34	129	33 22	163	46 08	234	32 24	279	20 14	315
188	56 04	016	25 03	059	57 07	131	33 34	164	45 33	235	31 42	280	19 44	316
189	56 16	016	25 40	060	57 40	132	33 45	166	44 57	236	30 59	281	19 14	316
190	56 27	015	26 18	060	58 11	134	33 56	167	44 21	237	30 17	281	18 44	317
191	56 39	015	26 55	061	58 42	135	34 05	168	43 44	238	29 35	282	18 15	317
192	56 50	015	27 33	061	59 12	137	34 14	169	43 07	239	28 52	282	17 46	318
193	57 01	014	28 11	062	59 41	138	34 22	170	42 30	240	28 10	283	17 17	318
194	57 11	014	28 49	062	60 10	140	34 28	171	41 52	241	27 28	284	16 48	319
	◆VEGA		Rasalhague		ARCTURUS		◆SPICA		REGULUS		◆POLLUX		Dubhe	
195	29 28	063	23 59	096	60 37	142	34 34	173	41 14	242	26 46	284	65 24	326
196	30 06	063	24 42	096	61 03	143	34 40	174	40 36	243	26 04	285	64 59	326
197	30 45	064	25 25	097	61 29	145	34 44	175	39 57	244	25 23	285	64 35	325
198	31 24	064	26 07	098	61 53	147	34 47	176	39 18	245	24 41	286	64 10	325
199	32 03	065	26 50	099	62 16	149	34 50	177	38 39	246	24 00	287	63 45	324
200	32 42	065	27 33	099	62 38	151	34 51	179	37 59	247	23 19	287	63 20	324
201	33 21	066	28 15	100	62 58	152	34 52	180	37 19	248	22 37	288	62 54	323
202	34 01	067	28 58	101	63 18	154	34 52	181	36 39	249	21 56	288	62 28	323
203	34 40	067	29 40	102	63 36	156	34 50	182	35 58	250	21 15	289	62 02	322
204	35 20	068	30 22	102	63 52	158	34 48	183	35 18	251	20 35	290	61 36	322
205	36 00	068	31 04	103	64 07	161	34 45	185	34 37	252	19 54	290	61 09	322
206	36 40	069	31 46	104	64 21	163	34 42	186	33 56	252	19 14	291	60 42	321
207	37 21	069	32 28	105	64 33	165	34 37	187	33 15	253	18 33	291	60 15	321
208	38 01	069	33 10	106	64 44	167	34 31	188	32 33	254	17 53	292	59 48	321
209	38 41	070	33 51	106	64 53	169	34 25	189	31 52	255	17 13	293	59 21	321
	◆DENEB		VEGA		Rasalhague		ANTARES		◆ARCTURUS		REGULUS		◆Dubhe	
210	23 45	049	39 22	070	34 32	107	11 45	146	65 00	171	31 10	256	58 53	320
211	24 18	050	40 03	071	35 14	108	12 08	147	65 06	174	30 28	257	58 26	320
212	24 51	050	40 44	071	35 55	109	12 31	148	65 10	176	29 46	257	57 58	320
213	25 24	051	41 25	072	36 35	110	12 54	149	65 12	178	29 04	258	57 30	320
214	25 58	051	42 06	072	37 16	111	13 16	150	65 12	180	28 21	259	57 02	320
215	26 31	052	42 47	073	37 56	112	13 38	151	65 11	183	27 39	260	56 34	320
216	27 05	052	43 28	073	38 36	113	13 58	151	65 09	185	26 56	260	56 06	319
217	27 39	052	44 10	074	39 16	113	14 19	152	65 04	187	26 14	261	55 38	319
218	28 14	053	44 51	074	39 55	114	14 39	153	64 58	189	25 31	262	55 10	319
219	28 48	053	45 33	075	40 34	115	14 58	154	64 50	191	24 48	263	54 42	319
220	29 23	054	46 15	075	41 13	116	15 17	155	64 41	194	24 06	263	54 14	319
221	29 58	054	46 56	076	41 52	117	15 35	156	64 30	196	23 23	264	53 45	319
222	30 33	055	47 38	077	42 30	118	15 52	157	64 17	198	22 40	265	53 17	319
223	31 09	055	48 20	077	43 08	119	16 09	157	64 03	200	21 57	266	52 49	319
224	31 44	056	49 02	078	43 45	120	16 25	158	63 48	202	21 14	266	52 20	319
	DENEB		◆ALTAIR		Rasalhague		ANTARES		◆ARCTURUS		Denebola		◆Dubhe	
225	32 20	056	18 36	096	44 22	121	16 41	159	63 31	204	40 00	249	51 52	319
226	32 56	057	19 19	096	44 59	123	16 56	160	63 12	206	39 19	250	51 23	319
227	33 32	057	20 02	097	45 35	124	17 10	161	62 53	208	38 38	251	50 55	319
228	34 08	057	20 45	098	46 10	125	17 24	162	62 32	210	37 57	252	50 27	319
229	34 45	058	21 28	099	46 46	126	17 37	163	62 10	212	37 16	253	49 58	319
230	35 21	058	22 10	099	47 20	127	17 49	164	61 46	214	36 35	254	49 30	319
231	35 58	059	22 53	100	47 55	128	18 01	165	61 22	216	35 53	255	49 02	319
232	36 35	059	23 35	101	48 28	129	18 12	166	60 56	217	35 12	255	48 34	319
233	37 12	060	24 18	102	49 01	131	18 22	167	60 29	219	34 30	256	48 05	319
234	37 49	060	25 00	102	49 34	132	18 32	167	60 02	221	33 48	257	47 37	319
235	38 27	060	25 42	103	50 06	133	18 41	168	59 33	222	33 06	258	47 09	319
236	39 04	061	26 24	104	50 37	134	18 49	169	59 04	224	32 24	259	46 41	320
237	39 42	061	27 06	105	51 07	136	18 57	170	58 34	225	31 41	259	46 13	320
238	40 20	062	27 47	105	51 37	137	19 04	171	58 02	227	30 59	260	45 45	320
239	40 58	062	28 29	106	52 06	138	19 10	172	57 31	228	30 16	261	45 18	320
	◆DENEB		ALTAIR		Rasalhague		◆ANTARES		ARCTURUS		Denebola		◆Dubhe	
240	41 36	062	29 10	107	52 34	140	19 16	173	56 58	230	29 33	262	44 50	320
241	42 15	063	29 51	108	53 02	141	19 21	174	56 25	231	28 51	262	44 22	320
242	42 53	063	30 32	109	53 28	143	19 25	175	55 51	232	28 08	263	43 55	320
243	43 32	064	31 13	110	53 54	144	19 28	176	55 16	234	27 25	264	43 27	321
244	44 10	064	31 54	110	54 19	146	19 31	177	54 41	235	26 42	265	43 00	321
245	44 49	064	32 34	111	54 43	147	19 33	178	54 05	236	25 59	265	42 33	321
246	45 28	065	33 14	112	55 06	149	19 34	179	53 29	238	25 16	266	42 05	321
247	46 07	065	33 54	113	55 27	150	19 35	180	52 53	239	24 33	267	41 38	321
248	46 47	066	34 34	114	55 48	152	19 35	181	52 15	240	23 50	268	41 11	322
249	47 26	066	35 13	115	56 08	154	19 34	182	51 38	241	23 07	268	40 45	322
250	48 05	066	35 52	116	56 26	155	19 32	183	51 00	242	22 24	269	40 18	322
251	48 45	067	36 31	117	56 44	157	19 30	184	50 21	243	21 40	270	39 51	322
252	49 25	067	37 09	118	57 00	159	19 27	184	49 43	244	20 57	270	39 25	322
253	50 04	067	37 47	119	57 15	160	19 23	185	49 04	245	20 14	271	38 59	323
254	50 44	068	38 25	120	57 29	162	19 19	186	48 24	246	19 31	272	38 33	323
	◆DENEB		ALTAIR		Nunki		◆ANTARES		ARCTURUS		◆Alkaid		Kochab	
255	51 24	068	39 02	121	14 56	154	19 13	187	47 44	248	57 06	297	56 31	345
256	52 04	068	39 39	121	15 15	154	19 08	188	47 04	248	56 27	297	56 20	344
257	52 45	069	40 16	123	15 33	155	19 01	189	46 24	249	55 49	297	56 08	344
258	53 25	069	40 52	124	15 51	156	18 54	190	45 44	250	55 10	297	55 56	344
259	54 05	070	41 28	125	16 08	157	18 46	191	45 03	251	54 32	298	55 44	343
260	54 46	070	42 03	126	16 24	158	18 37	192	44 22	252	53 54	298	55 31	343
261	55 26	070	42 38	127	16 40	159	18 28	193	43 41	253	53 16	298	55 18	343
262	56 07	071	43 12	128	16 56	160	18 18	194	42 59	254	52 38	298	55 06	342
263	56 48	071	43 46	129	17 10	161	18 07	195	42 18	255	52 00	298	54 52	342
264	57 28	071	44 19	130	17 24	162	17 56	196	41 36	256	51 22	299	54 39	342
265	58 09	072	44 52	131	17 38	162	17 44	197	40 54	257	50 44	299	54 26	342
266	58 50	072	45 24	132	17 50	163	17 31	198	40 12	257	50 06	299	54 12	341
267	59 31	072	45 56	134	18 03	164	17 18	198	39 30	258	49 29	300	53 58	341
268	60 12	072	46 27	135	18 14	165	17 04	199	38 47	259	48 51	300	53 44	341
269	60 54	073	46 57	136	18 25	166	16 49	200	38 05	260	48 14	300	53 30	341

LHA ♈	Hc	Zn	Hc	Zn	Hc	Zn	Hc	Zn	Hc	Zn	Hc	Zn	Hc	Zn
	◆Alpheratz		ALTAIR		Nunki		◆ANTARES		ARCTURUS		◆Alkaid		Kochab	
270	18 22	067	47 27	137	18 35	167	16 34	201	37 22	261	47 36	300	53 15	340
271	19 02	068	47 56	139	18 44	168	16 18	202	36 40	261	46 59	301	53 01	340
272	19 42	068	48 24	140	18 53	169	16 01	203	35 57	262	46 22	301	52 46	340
273	20 22	069	48 51	141	19 01	170	15 44	204	35 14	263	45 45	301	52 31	340
274	21 02	069	49 18	143	19 08	171	15 27	205	34 31	264	45 08	302	52 16	340
275	21 43	070	49 44	144	19 15	172	15 08	206	33 48	264	44 32	302	52 01	339
276	22 24	071	50 08	145	19 20	173	14 49	206	33 05	265	43 55	302	51 46	339
277	23 04	071	50 33	147	19 26	174	14 30	207	32 22	266	43 19	303	51 31	339
278	23 45	072	50 56	148	19 30	175	14 10	208	31 39	267	42 43	303	51 15	339
279	24 26	072	51 18	150	19 34	175	13 49	209	30 56	267	42 06	303	51 00	339
280	25 08	073	51 39	151	19 37	176	13 28	210	30 13	268	41 30	304	50 44	339
281	25 49	074	52 00	153	19 39	177	13 06	211	29 30	269	40 55	304	50 28	339
282	26 30	074	52 19	154	19 41	178	12 44	212	28 47	270	40 19	304	50 13	339
283	27 12	075	52 37	156	19 42	179	12 21	212	28 04	270	39 43	305	49 57	338
284	27 54	075	52 54	157	19 42	180	11 58	213	27 21	271	39 08	305	49 41	338
	◆Mirfak		Alpheratz		◆ALTAIR		Rasalhague		◆ARCTURUS		Alkaid		Kochab	
285	14 58	033	28 35	076	53 11	159	53 37	217	26 37	272	38 33	305	49 25	338
286	15 22	033	29 17	077	53 26	160	53 11	218	25 54	272	37 57	306	49 09	338
287	15 45	034	29 59	077	53 39	162	52 44	220	25 11	273	37 22	306	48 53	338
288	16 10	034	30 41	078	53 52	164	52 16	221	24 28	274	36 48	306	48 37	338
289	16 34	035	31 24	078	54 04	165	51 47	222	23 45	274	36 13	307	48 20	338
290	16 59	035	32 06	079	54 14	167	51 18	224	23 02	275	35 39	307	48 04	338
291	17 24	036	32 48	080	54 23	169	50 48	225	22 19	276	35 04	308	47 48	338
292	17 49	036	33 31	080	54 31	170	50 17	226	21 36	276	34 30	308	47 32	338
293	18 15	037	34 13	081	54 37	172	49 45	228	20 53	277	33 56	308	47 15	338
294	18 41	037	34 56	081	54 43	174	49 13	229	20 10	278	33 23	309	46 59	338
295	19 08	038	35 39	082	54 47	175	48 40	230	19 28	278	32 49	309	46 43	338
296	19 34	038	36 22	083	54 50	177	48 07	231	18 45	279	32 16	310	46 26	338
297	20 01	039	37 04	083	54 51	179	47 33	233	18 02	280	31 43	310	46 10	338
298	20 28	039	37 47	084	54 51	181	46 58	234	17 20	280	31 10	310	45 54	338
299	20 56	040	38 30	085	54 50	182	46 23	235	16 37	281	30 37	311	45 37	338
	Mirfak		◆Alpheratz		Enif		◆ALTAIR		Rasalhague		Alphecca		◆Kochab	
300	21 24	040	39 13	085	49 06	139	54 48	184	45 47	236	34 44	275	45 21	338
301	21 52	041	39 56	086	49 35	140	54 44	186	45 11	237	34 01	276	45 05	338
302	22 20	041	40 39	087	50 02	141	54 39	187	44 35	238	33 19	277	44 48	338
303	22 48	042	41 22	087	50 28	143	54 33	189	43 58	239	32 36	277	44 32	338
304	23 17	042	42 06	088	50 54	144	54 25	191	43 21	240	31 53	278	44 16	338
305	23 46	043	42 49	089	51 19	146	54 17	193	42 43	241	31 10	278	43 59	338
306	24 16	043	43 32	089	51 43	147	54 07	194	42 05	242	30 28	279	43 43	338
307	24 45	044	44 15	090	52 06	149	53 55	196	41 27	243	29 45	280	43 27	338
308	25 15	044	44 58	091	52 28	150	53 43	197	40 48	244	29 02	280	43 11	338
309	25 45	044	45 41	091	52 49	152	53 29	199	40 09	245	28 20	281	42 55	338
310	26 16	045	46 25	092	53 09	153	53 15	201	39 29	246	27 38	282	42 39	338
311	26 46	045	47 08	093	53 28	155	52 59	202	38 50	247	26 55	282	42 23	338
312	27 17	046	47 51	093	53 46	156	52 42	204	38 10	248	26 13	283	42 07	338
313	27 48	046	48 34	094	54 03	158	52 24	205	37 30	249	25 31	283	41 51	339
314	28 19	047	49 17	095	54 18	159	52 05	207	36 49	250	24 49	284	41 35	339
	CAPELLA		◆Alpheratz		FOMALHAUT		◆ALTAIR		Rasalhague		VEGA		◆Kochab	
315	12 41	036	50 00	096	11 37	154	51 45	208	36 09	251	62 51	271	41 20	339
316	13 07	037	50 43	097	11 55	155	51 24	210	35 28	252	62 08	272	41 04	339
317	13 33	037	51 26	097	12 13	156	51 02	211	34 47	252	61 25	273	40 49	339
318	13 59	038	52 08	098	12 31	157	50 39	213	34 06	253	60 42	273	40 33	339
319	14 26	038	52 51	099	12 47	158	50 15	214	33 24	254	59 59	274	40 18	339
320	14 53	039	53 34	100	13 03	158	49 50	216	32 43	255	59 16	275	40 02	339
321	15 20	039	54 16	101	13 19	159	49 25	217	32 01	256	58 33	275	39 47	339
322	15 47	040	54 58	102	13 34	160	48 58	218	31 19	256	57 50	276	39 32	340
323	16 15	041	55 41	103	13 48	161	48 31	220	30 37	257	57 07	276	39 17	340
324	16 44	041	56 23	103	14 02	162	48 03	221	29 55	258	56 24	277	39 02	340
325	17 12	042	57 05	104	14 15	163	47 34	222	29 13	259	55 41	277	38 47	340
326	17 41	042	57 46	105	14 28	164	47 05	224	28 30	260	54 58	278	38 33	340
327	18 10	043	58 28	106	14 40	164	46 35	225	27 48	260	54 16	279	38 18	340
328	18 39	043	59 09	107	14 51	165	46 04	226	27 05	261	53 33	279	38 04	340
329	19 09	044	59 50	109	15 01	166	45 33	227	26 22	262	52 51	280	37 49	341
	◆CAPELLA		Hamal		Diphda		◆FOMALHAUT		ALTAIR		◆VEGA		Kochab	
330	19 39	044	36 05	091	17 37	139	15 11	167	45 01	228	52 08	280	37 35	341
331	20 09	045	36 48	092	18 05	140	15 21	168	44 28	230	51 26	281	37 21	341
332	20 40	045	37 31	093	18 33	141	15 29	169	43 55	231	50 43	281	37 07	341
333	21 11	046	38 14	093	19 00	142	15 37	170	43 21	232	50 01	282	36 53	341
334	21 42	046	38 58	094	19 26	143	15 45	171	42 47	233	49 19	282	36 39	342
335	22 13	047	39 41	095	19 52	144	15 51	172	42 13	234	48 37	283	36 26	342
336	22 44	047	40 24	095	20 17	145	15 57	172	41 37	235	47 54	283	36 12	342
337	23 16	048	41 06	096	20 42	146	16 03	173	41 02	236	47 13	284	35 59	342
338	23 48	048	41 49	097	21 06	146	16 07	174	40 26	237	46 31	284	35 46	342
339	24 20	049	42 32	098	21 29	147	16 11	175	39 49	238	45 49	285	35 33	343
340	24 53	049	43 15	099	21 52	148	16 15	176	39 12	239	45 07	285	35 20	343
341	25 26	050	43 57	099	22 15	149	16 17	177	38 35	240	44 26	286	35 07	343
342	25 59	050	44 40	100	22 37	150	16 19	178	37 57	241	43 44	286	34 55	343
343	26 32	051	45 22	101	22 58	151	16 20	179	37 19	242	43 03	287	34 42	343
344	27 05	051	46 05	102	23 18	152	16 21	180	36 41	243	42 22	287	34 30	344
	◆CAPELLA		ALDEBARAN		Diphda		◆FOMALHAUT		ALTAIR		◆VEGA		Kochab	
345	27 39	051	15 42	082	23 38	153	16 21	181	36 02	244	41 40	288	34 18	344
346	28 13	052	16 24	083	23 57	154	16 20	181	35 23	245	40 59	288	34 06	344
347	28 47	052	17 07	083	24 16	155	16 19	182	34 44	246	40 18	289	33 54	344
348	29 21	053	17 50	084	24 34	156	16 17	183	34 05	247	39 38	289	33 42	344
349	29 56	053	18 33	085	24 51	157	16 14	184	33 25	248	38 57	290	33 31	345
350	30 31	054	19 16	085	25 07	158	16 10	185	32 45	248	38 16	290	33 20	345
351	31 05	054	19 59	086	25 23	159	16 06	186	32 05	249	37 36	291	33 08	345
352	31 40	055	20 42	087	25 38	160	16 01	187	31 24	250	36 56	291	32 58	345
353	32 16	055	21 25	087	25 52	161	15 56	188	30 43	251	36 16	292	32 47	346
354	32 51	055	22 08	088	26 06	162	15 49	189	30 02	252	35 36	292	32 36	346
355	33 27	056	22 52	089	26 19	163	15 43	190	29 21	253	34 56	293	32 26	346
356	34 03	056	23 35	089	26 31	164	15 35	191	28 40	253	34 16	293	32 16	346
357	34 39	057	24 18	090	26 42	165	15 27	191	27 59	254	33 36	294	32 06	347
358	35 15	057	25 01	091	26 53	166	15 18	192	27 17	255	32 57	294	31 56	347
359	35 51	058	25 44	092	27 03	167	15 08	193	26 35	256	32 18	295	31 46	347

LHA ♈	Hc	Zn	Hc	Zn	Hc	Zn	Hc	Zn	Hc	Zn	Hc	Zn	Hc	Zn
	◆CAPELLA		ALDEBARAN		◆Diphda		FOMALHAUT		ALTAIR		◆VEGA		Kochab	
0	35 56	057	26 29	092	28 11	168	15 56	194	26 07	257	31 13	296	30 38	348
1	36 33	058	27 13	092	28 19	169	15 45	195	25 24	258	30 33	296	30 29	348
2	37 10	058	27 57	093	28 27	170	15 34	196	24 41	259	29 54	297	30 20	348
3	37 47	059	28 41	094	28 34	172	15 21	197	23 58	259	29 15	297	30 11	348
4	38 25	059	29 25	095	28 40	173	15 08	198	23 15	260	28 36	298	30 02	349
5	39 03	059	30 08	095	28 45	174	14 55	198	22 32	261	27 57	298	29 53	349
6	39 40	060	30 52	096	28 49	175	14 41	199	21 48	262	27 19	299	29 45	349
7	40 18	060	31 36	097	28 53	176	14 26	200	21 05	262	26 41	299	29 37	349
8	40 56	060	32 19	097	28 56	177	14 10	201	20 21	263	26 02	300	29 29	350
9	41 35	061	33 03	098	28 58	178	13 54	202	19 38	264	25 24	300	29 21	350
10	42 13	061	33 46	099	28 59	179	13 37	203	18 54	264	24 47	301	29 14	350
11	42 52	062	34 29	100	28 59	180	13 20	204	18 10	265	24 09	301	29 06	351
12	43 30	062	35 13	101	28 59	181	13 02	204	17 27	266	23 32	302	28 59	351
13	44 09	062	35 56	101	28 57	182	12 44	205	16 43	267	22 55	302	28 52	351
14	44 48	063	36 39	102	28 55	183	12 25	206	15 59	267	22 18	303	28 46	351
	◆CAPELLA		BETELGEUSE		RIGEL		◆Diphda		Enif		◆DENEB		Kochab	
15	45 27	063	16 56	096	12 59	115	28 52	185	36 06	247	44 48	296	28 39	352
16	46 06	063	17 39	096	13 39	115	28 48	186	35 26	248	44 08	297	28 33	352
17	46 45	064	18 23	097	14 19	116	28 43	187	34 45	249	43 29	297	28 27	352
18	47 25	064	19 06	098	14 58	117	28 38	188	34 04	250	42 50	297	28 21	353
19	48 04	064	19 50	099	15 37	118	28 32	189	33 23	250	42 11	298	28 16	353
20	48 44	065	20 33	099	16 16	118	28 24	190	32 41	251	41 33	298	28 10	353
21	49 24	065	21 16	100	16 54	119	28 16	191	32 00	252	40 54	299	28 05	353
22	50 03	065	22 00	101	17 32	120	28 08	192	31 18	253	40 15	299	28 00	354
23	50 43	066	22 43	101	18 10	121	27 58	193	30 36	254	39 37	299	27 56	354
24	51 23	066	23 26	102	18 48	122	27 48	194	29 54	255	38 59	300	27 51	354
25	52 03	066	24 08	103	19 25	122	27 37	195	29 11	255	38 21	300	27 47	355
26	52 44	067	24 51	104	20 02	123	27 25	196	28 29	256	37 43	300	27 43	355
27	53 24	067	25 34	105	20 39	124	27 12	197	27 46	257	37 05	301	27 39	355
28	54 04	067	26 16	105	21 15	125	26 59	198	27 03	258	36 28	301	27 36	356
29	54 45	067	26 58	106	21 50	126	26 44	199	26 20	258	35 50	302	27 32	356
	◆CAPELLA		BETELGEUSE		RIGEL		◆Diphda		Alpheratz		◆DENEB		Kochab	
30	55 25	068	27 40	107	22 26	127	26 29	200	63 37	247	35 13	302	27 29	356
31	56 06	068	28 22	108	23 01	127	26 14	201	62 57	249	34 36	302	27 27	356
32	56 47	068	29 04	108	23 36	128	25 57	202	62 16	250	33 59	303	27 24	357
33	57 27	068	29 46	109	24 10	129	25 40	203	61 34	251	33 22	303	27 22	357
34	58 08	069	30 27	110	24 44	130	25 22	204	60 53	252	32 45	304	27 19	357
35	58 49	069	31 08	111	25 17	131	25 04	205	60 11	253	32 09	304	27 17	358
36	59 30	069	31 49	112	25 50	132	24 45	206	59 29	254	31 33	305	27 16	358
37	60 11	069	32 29	113	26 22	133	24 25	207	58 47	255	30 57	305	27 14	358
38	60 52	070	33 10	114	26 54	134	24 04	208	58 04	256	30 21	305	27 13	359
39	61 34	070	33 50	114	27 26	135	23 43	209	57 21	257	29 45	306	27 12	359
40	62 15	070	34 30	115	27 56	136	23 21	210	56 39	258	29 10	306	27 11	359
41	62 56	070	35 09	116	28 27	137	22 59	211	55 56	259	28 34	307	27 11	359
42	63 37	071	35 48	117	28 57	138	22 36	212	55 12	260	27 59	307	27 11	000
43	64 19	071	36 27	118	29 26	139	22 12	213	54 29	261	27 25	308	27 11	000
44	65 00	071	37 06	119	29 55	140	21 48	214	53 46	261	26 50	308	27 11	000
	◆Dubhe		POLLUX		BETELGEUSE		◆RIGEL		Diphda		◆Alpheratz		DENEB	
45	25 04	027	31 52	080	37 44	120	30 23	141	21 23	215	53 02	262	26 15	309
46	25 23	027	32 36	080	38 22	121	30 50	142	20 57	216	52 19	263	25 41	309
47	25 43	027	33 19	081	38 59	122	31 17	143	20 31	217	51 35	264	25 07	309
48	26 04	028	34 02	082	39 37	123	31 44	144	20 05	218	50 52	265	24 33	310
49	26 24	028	34 46	082	40 13	124	32 09	145	19 38	219	50 08	265	24 00	310
50	26 45	028	35 29	083	40 49	125	32 34	146	19 10	219	49 24	266	23 27	311
51	27 06	029	36 13	083	41 25	126	32 59	147	18 42	220	48 40	267	22 54	311
52	27 27	029	36 56	084	42 00	127	33 22	148	18 13	221	47 57	268	22 21	312
53	27 49	030	37 40	085	42 35	128	33 45	149	17 44	222	47 13	268	21 48	312
54	28 11	030	38 24	085	43 09	129	34 07	150	17 14	223	46 29	269	21 16	313
55	28 32	030	39 08	086	43 43	130	34 29	151	16 44	224	45 45	270	20 44	313
56	28 55	031	39 51	087	44 16	132	34 49	152	16 14	225	45 01	270	20 12	314
57	29 17	031	40 35	087	44 49	133	35 09	154	15 43	225	44 17	271	19 40	314
58	29 40	031	41 19	088	45 21	134	35 28	155	15 11	226	43 33	272	19 09	315
59	30 03	032	42 03	089	45 52	135	35 47	156	14 39	227	42 50	272	18 38	315
	◆Dubhe		POLLUX		PROCYON		◆SIRIUS		RIGEL		◆Hamal		DENEB	
60	30 26	032	42 47	089	28 51	112	19 19	138	36 04	157	59 34	239	18 07	316
61	30 49	032	43 31	090	29 32	113	19 48	139	36 21	158	58 56	240	17 37	316
62	31 12	032	44 15	091	30 12	113	20 16	140	36 37	159	58 18	242	17 07	317
63	31 36	033	44 58	091	30 52	114	20 44	141	36 52	161	57 39	243	16 37	317
64	32 00	033	45 42	092	31 32	115	21 12	142	37 06	162	57 00	244	16 07	318
65	32 24	033	46 26	093	32 11	116	21 39	142	37 19	163	56 20	245	15 38	318
66	32 48	034	47 10	093	32 51	117	22 05	143	37 32	164	55 40	246	15 09	319
67	33 12	034	47 54	094	33 30	118	22 31	144	37 43	165	54 59	247	14 40	319
68	33 37	034	48 38	095	34 08	119	22 57	145	37 54	167	54 19	249	14 12	320
69	34 02	035	49 21	096	34 47	120	23 21	146	38 03	168	53 38	250	13 44	321
70	34 27	035	50 05	096	35 25	121	23 45	147	38 12	169	52 57	251	13 16	321
71	34 52	035	50 48	097	36 02	122	24 09	148	38 20	170	52 15	252	12 49	322
72	35 17	035	51 32	098	36 39	123	24 32	149	38 27	172	51 33	253	12 22	322
73	35 43	036	52 15	099	37 16	124	24 54	150	38 32	173	50 51	253	11 55	323
74	36 08	036	52 59	100	37 52	125	25 16	151	38 37	174	50 09	254	11 29	323
	◆Dubhe		REGULUS		PROCYON		◆SIRIUS		RIGEL		◆Hamal		Schedar	
75	36 34	036	17 36	090	38 28	126	25 37	152	38 41	175	49 27	255	47 42	312
76	37 00	036	18 20	091	39 04	127	25 57	153	38 44	177	48 44	256	47 10	312
77	37 26	037	19 04	091	39 39	128	26 16	154	38 46	178	48 02	257	46 37	312
78	37 52	037	19 47	092	40 13	129	26 35	155	38 47	179	47 19	258	46 05	313
79	38 19	037	20 31	093	40 47	130	26 54	156	38 47	181	46 36	259	45 32	313
80	38 45	037	21 15	093	41 21	131	27 11	157	38 47	182	45 53	260	45 00	313
81	39 12	037	21 59	094	41 54	132	27 28	158	38 45	183	45 09	260	44 28	313
82	39 38	038	22 43	095	42 26	133	27 44	159	38 42	184	44 26	261	43 56	313
83	40 05	038	23 26	095	42 58	134	27 59	160	38 38	186	43 43	262	43 24	313
84	40 32	038	24 10	096	43 29	135	28 14	161	38 33	187	42 59	263	42 52	314
85	40 59	038	24 54	097	43 59	137	28 27	162	38 28	188	42 16	264	42 21	314
86	41 26	038	25 37	098	44 29	138	28 40	163	38 21	189	41 32	264	41 49	314
87	41 54	039	26 21	098	44 58	139	28 52	164	38 13	191	40 48	265	41 18	314
88	42 21	039	27 04	099	45 27	140	29 04	165	38 05	192	40 05	266	40 46	315
89	42 49	039	27 47	100	45 54	142	29 14	167	37 55	193	39 21	266	40 15	315

LAT 43°N

LHA ♈	Hc	Zn	Hc	Zn	Hc	Zn	Hc	Zn	Hc	Zn	Hc	Zn	Hc	Zn
	◆Dubhe		REGULUS		PROCYON		◆SIRIUS		RIGEL		ALDEBARAN		◆Mirfak	
90	43 16	039	28 30	101	46 21	143	29 24	168	37 45	194	57 59	221	62 35	298
91	43 44	039	29 13	101	46 47	144	29 33	169	37 33	196	57 30	222	61 57	298
92	44 12	039	29 56	102	47 13	145	29 41	170	37 21	197	57 00	224	61 18	298
93	44 39	039	30 39	103	47 37	147	29 49	171	37 08	198	56 30	225	60 39	298
94	45 07	040	31 22	104	48 01	148	29 55	172	36 54	199	55 58	227	60 01	298
95	45 35	040	32 05	104	48 23	150	30 01	173	36 39	200	55 26	228	59 22	298
96	46 03	040	32 47	105	48 45	151	30 06	174	36 23	202	54 53	229	58 43	299
97	46 32	040	33 29	106	49 06	152	30 10	175	36 07	203	54 19	231	58 05	299
98	47 00	040	34 11	107	49 26	154	30 13	176	35 49	204	53 45	232	57 26	299
99	47 28	040	34 53	108	49 45	155	30 15	178	35 31	205	53 10	233	56 48	299
100	47 56	040	35 35	109	50 03	157	30 17	179	35 12	206	52 35	235	56 10	299
101	48 24	040	36 16	109	50 19	158	30 17	180	34 52	207	51 59	236	55 31	299
102	48 53	040	36 58	110	50 35	160	30 17	181	34 32	209	51 22	237	54 53	299
103	49 21	040	37 39	111	50 50	161	30 16	182	34 10	210	50 45	238	54 15	300
104	49 49	040	38 20	112	51 03	163	30 14	183	33 48	211	50 08	239	53 37	300
	Dubhe		◆Denebola		REGULUS		◆SIRIUS		RIGEL		ALDEBARAN		◆Mirfak	
105	50 18	040	22 51	091	39 00	113	30 11	184	33 26	212	49 30	240	52 59	300
106	50 46	040	23 35	092	39 40	114	30 08	185	33 02	213	48 51	242	52 21	300
107	51 15	040	24 18	092	40 20	115	30 03	186	32 38	214	48 13	243	51 43	300
108	51 43	040	25 02	093	41 00	116	29 58	187	32 13	215	47 33	244	51 05	301
109	52 11	040	25 46	094	41 39	117	29 52	189	31 47	216	46 54	245	50 27	301
110	52 40	040	26 30	095	42 18	118	29 45	190	31 21	217	46 14	246	49 50	301
111	53 08	040	27 14	095	42 57	119	29 37	191	30 54	218	45 34	247	49 12	301
112	53 36	040	27 57	096	43 35	120	29 29	192	30 27	219	44 54	248	48 35	302
113	54 04	040	28 41	097	44 13	121	29 19	193	29 59	220	44 13	249	47 57	302
114	54 32	040	29 24	097	44 51	122	29 09	194	29 30	221	43 32	249	47 20	302
115	55 01	040	30 08	098	45 28	123	28 58	195	29 01	222	42 51	250	46 43	302
116	55 29	040	30 51	099	46 05	124	28 46	196	28 31	223	42 09	251	46 06	303
117	55 57	039	31 35	100	46 41	125	28 33	197	28 01	224	41 28	252	45 29	303
118	56 24	039	32 18	100	47 16	126	28 20	198	27 30	225	40 46	253	44 52	303
119	56 52	039	33 01	101	47 52	127	28 06	199	26 59	226	40 04	254	44 15	303
	◆Kochab		Denebola		◆REGULUS		SIRIUS		RIGEL		◆ALDEBARAN		CAPELLA	
120	37 46	020	33 44	102	48 26	129	27 51	200	26 27	227	39 21	255	60 59	290
121	38 01	020	34 27	103	49 00	130	27 35	201	25 55	228	38 39	256	60 18	290
122	38 15	020	35 09	104	49 34	131	27 19	202	25 22	229	37 56	256	59 37	291
123	38 30	020	35 52	104	50 06	132	27 02	203	24 49	230	37 14	257	58 56	291
124	38 46	020	36 34	105	50 39	134	26 44	205	24 15	231	36 31	258	58 15	291
125	39 01	020	37 17	106	51 10	135	26 25	206	23 41	232	35 48	259	57 34	291
126	39 16	020	37 59	107	51 41	136	26 06	207	23 06	232	35 05	260	56 53	292
127	39 32	021	38 41	108	52 11	138	25 46	208	22 31	233	34 21	260	56 12	292
128	39 47	021	39 22	109	52 40	139	25 25	209	21 56	234	33 38	261	55 32	292
129	40 03	021	40 04	110	53 08	140	25 04	210	21 20	235	32 55	262	54 51	293
130	40 18	021	40 45	110	53 36	142	24 42	210	20 44	236	32 11	263	54 11	293
131	40 34	021	41 26	111	54 03	143	24 20	211	20 07	237	31 28	263	53 30	293
132	40 50	021	42 07	112	54 28	145	23 56	212	19 31	237	30 44	264	52 50	293
133	41 06	021	42 47	113	54 53	146	23 33	213	18 53	238	30 00	265	52 10	294
134	41 21	021	43 27	114	55 17	148	23 08	214	18 16	239	29 17	266	51 30	294
	◆Kochab		ARCTURUS		Denebola		◆REGULUS		SIRIUS		BETELGEUSE		◆CAPELLA	
135	41 37	021	20 58	083	44 07	115	55 40	149	22 43	215	36 06	242	50 50	294
136	41 53	021	21 42	084	44 47	116	56 02	151	22 18	216	35 27	243	50 10	295
137	42 09	022	22 25	084	45 26	117	56 22	153	21 51	217	34 48	244	49 30	295
138	42 26	022	23 09	085	46 05	118	56 42	154	21 25	218	34 08	245	48 50	295
139	42 42	022	23 53	085	46 44	119	57 00	156	20 57	219	33 28	246	48 11	296
140	42 58	022	24 36	086	47 22	120	57 17	158	20 29	220	32 48	247	47 31	296
141	43 14	022	25 20	087	47 59	121	57 33	160	20 01	221	32 07	248	46 52	296
142	43 30	022	26 04	087	48 37	122	57 48	161	19 32	222	31 26	249	46 12	297
143	43 47	022	26 48	088	49 13	124	58 01	163	19 03	222	30 45	250	45 33	297
144	44 03	022	27 32	089	49 50	125	58 13	165	18 33	223	30 04	250	44 54	297
145	44 19	022	28 16	089	50 25	126	58 24	167	18 03	224	29 23	251	44 15	298
146	44 36	022	28 59	090	51 01	127	58 33	169	17 32	225	28 41	252	43 36	298
147	44 52	022	29 43	091	51 35	128	58 41	171	17 01	226	27 59	253	42 58	298
148	45 08	022	30 27	092	52 09	130	58 48	172	16 29	227	27 17	254	42 19	299
149	45 25	022	31 11	092	52 43	131	58 53	174	15 57	228	26 35	254	41 41	299
	◆Kochab		ARCTURUS		◆SPICA		REGULUS		PROCYON		◆POLLUX		CAPELLA	
150	45 41	022	31 55	093	18 31	126	58 57	176	41 05	230	59 01	252	41 03	299
151	45 58	022	32 39	094	19 06	127	58 59	178	40 31	231	58 19	253	40 24	300
152	46 14	022	33 23	094	19 41	128	59 00	180	39 57	232	57 37	254	39 46	300
153	46 30	022	34 06	095	20 16	129	58 59	182	39 22	233	56 55	255	39 09	301
154	46 46	022	34 50	096	20 50	130	58 57	184	38 47	234	56 12	256	38 31	301
155	47 03	022	35 34	097	21 23	130	58 53	186	38 11	235	55 29	257	37 53	301
156	47 19	022	36 17	097	21 56	131	58 48	188	37 35	236	54 47	258	37 16	302
157	47 35	022	37 01	098	22 29	132	58 42	189	36 58	237	54 04	259	36 39	302
158	47 51	022	37 44	099	23 01	133	58 34	191	36 21	238	53 20	260	36 02	303
159	48 07	021	38 27	100	23 33	134	58 24	193	35 44	239	52 37	261	35 25	303
160	48 23	021	39 11	100	24 05	135	58 14	195	35 06	240	51 54	262	34 48	303
161	48 39	021	39 54	101	24 35	136	58 02	197	34 28	241	51 10	262	34 11	304
162	48 55	021	40 37	102	25 06	137	57 49	199	33 50	242	50 27	263	33 35	304
163	49 11	021	41 20	103	25 36	138	57 34	200	33 11	243	49 43	264	32 59	305
164	49 27	021	42 02	104	26 05	139	57 18	202	32 32	243	49 00	265	32 23	305
	◆Kochab		ARCTURUS		◆SPICA		REGULUS		PROCYON		◆POLLUX		CAPELLA	
165	49 43	021	42 45	104	26 34	140	57 01	204	31 52	244	48 16	265	31 47	305
166	49 58	021	43 27	105	27 02	141	56 43	205	31 13	245	47 32	266	31 11	306
167	50 14	021	44 09	106	27 29	142	56 23	207	30 33	246	46 48	267	30 36	306
168	50 29	020	44 52	107	27 56	143	56 03	209	29 53	247	46 04	268	30 00	307
169	50 44	020	45 33	108	28 23	144	55 41	210	29 12	248	45 21	268	29 25	307
170	51 00	020	46 15	109	28 48	145	55 18	212	28 31	249	44 37	269	28 50	308
171	51 15	020	46 56	110	29 14	146	54 54	214	27 50	249	43 53	270	28 16	308
172	51 30	020	47 38	111	29 38	147	54 30	215	27 09	250	43 09	270	27 41	308
173	51 44	020	48 18	112	30 02	148	54 04	217	26 28	251	42 25	271	27 07	309
174	51 59	019	48 59	113	30 25	149	53 37	218	25 46	252	41 41	272	26 33	309
175	52 14	019	49 39	114	30 48	150	53 10	220	25 04	253	40 57	272	25 59	310
176	52 28	019	50 19	115	31 10	151	52 41	221	24 22	253	40 14	273	25 25	310
177	52 42	019	50 59	116	31 31	152	52 12	222	23 40	254	39 30	274	24 52	311
178	52 56	019	51 38	117	31 51	153	51 42	224	22 58	255	38 46	274	24 19	311
179	53 10	018	52 17	118	32 11	154	51 12	225	22 16	256	38 02	275	23 46	312

LAT 43°N

LHA ♈	Hc	Zn	Hc	Zn	Hc	Zn	Hc	Zn	Hc	Zn	Hc	Zn	Hc	Zn
	Kochab		♦VEGA		ARCTURUS		♦SPICA		REGULUS		♦POLLUX		CAPELLA	
180	53 24	018	19 38	055	52 56	119	32 29	155	50 40	226	37 19	276	23 13	312
181	53 37	018	20 14	055	53 34	120	32 48	156	50 08	228	36 35	276	22 41	313
182	53 51	018	20 50	056	54 11	122	33 05	157	49 35	229	35 51	277	22 09	313
183	54 04	017	21 27	056	54 49	123	33 21	158	49 02	230	35 08	277	21 37	314
184	54 17	017	22 03	057	55 25	124	33 37	160	48 28	231	34 24	278	21 05	314
185	54 29	017	22 40	057	56 01	125	33 52	161	47 53	233	33 41	279	20 34	314
186	54 42	016	23 17	058	56 37	127	34 06	162	47 18	234	32 58	279	20 02	315
187	54 54	016	23 55	058	57 12	128	34 19	163	46 43	235	32 14	280	19 31	315
188	55 06	016	24 32	059	57 46	129	34 32	164	46 07	236	31 31	280	19 01	316
189	55 18	015	25 10	059	58 19	131	34 43	165	45 30	237	30 48	281	18 30	316
190	55 29	015	25 48	060	58 52	132	34 54	167	44 53	238	30 05	282	18 00	317
191	55 41	015	26 26	060	59 24	134	35 04	168	44 15	239	29 22	282	17 31	317
192	55 52	014	27 04	061	59 55	135	35 13	169	43 38	240	28 39	283	17 01	318
193	56 02	014	27 43	061	60 26	137	35 21	170	42 59	241	27 56	283	16 32	319
194	56 13	014	28 21	062	60 55	139	35 28	171	42 21	242	27 14	284	16 03	319
	♦VEGA		Rasalhague		ARCTURUS		♦SPICA		REGULUS		♦POLLUX		Dubhe	
195	29 00	062	24 05	095	61 24	140	35 34	172	41 42	243	26 31	285	64 33	328
196	29 39	063	24 48	096	61 51	142	35 39	174	41 02	244	25 49	285	64 09	327
197	30 18	063	25 32	097	62 18	144	35 44	175	40 23	245	25 07	286	63 45	326
198	30 58	064	26 15	097	62 43	146	35 47	176	39 43	246	24 25	286	63 21	326
199	31 37	064	26 59	098	63 07	148	35 50	177	39 02	247	23 42	287	62 56	325
200	32 17	065	27 42	099	63 30	150	35 51	179	38 22	248	23 01	288	62 31	325
201	32 57	065	28 26	100	63 51	152	35 52	180	37 41	249	22 19	288	62 06	324
202	33 37	066	29 09	100	64 12	154	35 52	181	37 00	250	21 37	289	61 40	324
203	34 17	066	29 52	101	64 31	156	35 50	182	36 19	251	20 56	289	61 14	324
204	34 57	067	30 35	102	64 48	158	35 48	183	35 37	251	20 14	290	60 48	323
205	35 37	067	31 18	103	65 04	160	35 45	185	34 55	252	19 33	291	60 22	323
206	36 18	068	32 01	103	65 18	162	35 41	186	34 14	253	18 52	291	59 55	323
207	36 59	068	32 43	104	65 31	164	35 36	187	33 32	254	18 11	292	59 28	322
208	37 40	069	33 26	105	65 42	166	35 31	188	32 49	255	17 31	292	59 01	322
209	38 20	069	34 08	106	65 51	169	35 24	189	32 07	256	16 50	293	58 34	322
	♦DENEB		VEGA		Rasalhague		ANTARES		♦ARCTURUS		REGULUS		♦Dubhe	
210	23 06	049	39 02	070	34 50	107	12 34	146	65 59	171	31 24	256	58 07	321
211	23 39	049	39 43	070	35 32	107	12 59	147	66 05	173	30 42	257	57 39	321
212	24 12	050	40 24	071	36 14	108	13 22	148	66 09	176	29 59	258	57 12	321
213	24 46	050	41 06	071	36 55	109	13 45	149	66 12	178	29 16	259	56 44	321
214	25 20	051	41 47	072	37 37	110	14 08	150	66 12	180	28 33	259	56 16	321
215	25 54	051	42 29	072	38 18	111	14 30	150	66 11	183	27 50	260	55 49	320
216	26 28	052	43 11	073	38 59	112	14 51	151	66 08	185	27 06	261	55 21	320
217	27 03	052	43 53	073	39 39	113	15 12	152	66 04	187	26 23	262	54 52	320
218	27 37	053	44 35	074	40 20	114	15 32	153	65 57	190	25 39	262	54 24	320
219	28 12	053	45 17	074	41 00	115	15 52	154	65 49	192	24 56	263	53 56	320
220	28 47	053	45 59	074	41 39	116	16 11	155	65 39	194	24 12	264	53 28	320
221	29 23	054	46 41	075	42 19	117	16 29	156	65 28	196	23 29	265	53 00	320
222	29 58	054	47 24	075	42 58	118	16 47	156	65 14	199	22 45	265	52 31	320
223	30 34	055	48 06	076	43 37	119	17 04	157	64 59	201	22 01	266	52 03	320
224	31 10	055	48 49	076	44 15	120	17 21	158	64 43	203	21 17	267	51 35	320
	DENEB		♦ALTAIR		Rasalhague		ANTARES		♦ARCTURUS		Denebola		♦Dubhe	
225	31 46	056	18 42	095	44 53	121	17 37	159	64 25	205	40 20	250	51 06	320
226	32 22	056	19 26	096	45 30	122	17 52	160	64 06	207	39 39	251	50 38	320
227	32 59	056	20 09	097	46 08	123	18 07	161	63 45	209	38 57	252	50 10	320
228	33 36	057	20 53	097	46 44	124	18 21	162	63 23	211	38 16	253	49 41	320
229	34 12	057	21 36	098	47 20	125	18 34	163	63 00	213	37 34	254	49 13	320
230	34 49	058	22 20	099	47 56	126	18 47	164	62 36	215	36 51	254	48 45	320
231	35 27	058	23 03	100	48 31	127	18 59	165	62 10	217	36 09	255	48 16	320
232	36 04	059	23 46	100	49 06	128	19 10	166	61 44	218	35 27	256	47 48	320
233	36 41	059	24 29	101	49 40	130	19 21	166	61 16	220	34 44	257	47 20	320
234	37 19	059	25 12	102	50 14	131	19 31	167	60 47	222	34 01	258	46 52	320
235	37 57	060	25 55	103	50 46	132	19 40	168	60 17	223	33 18	258	46 23	320
236	38 35	060	26 38	103	51 18	134	19 48	169	59 47	225	32 35	259	45 55	320
237	39 13	060	27 21	104	51 50	135	19 56	170	59 15	227	31 52	260	45 27	320
238	39 51	061	28 03	105	52 21	136	20 03	171	58 43	228	31 09	261	44 59	320
239	40 30	061	28 45	106	52 51	138	20 10	172	58 10	230	30 25	261	44 31	321
	♦DENEB		ALTAIR		Rasalhague		♦ANTARES		ARCTURUS		Denebola		♦Dubhe	
240	41 08	062	29 28	107	53 20	139	20 15	173	57 36	231	29 42	262	44 04	321
241	41 47	062	30 10	107	53 48	140	20 20	174	57 02	232	28 58	263	43 36	321
242	42 26	062	30 51	108	54 16	142	20 25	175	56 27	234	28 15	264	43 08	321
243	43 05	063	31 33	109	54 42	143	20 28	176	55 51	235	27 31	264	42 41	321
244	43 44	063	32 14	110	55 08	145	20 31	177	55 15	236	26 47	265	42 13	321
245	44 23	063	32 56	111	55 33	146	20 33	178	54 38	238	26 04	266	41 46	321
246	45 02	064	33 36	112	55 57	148	20 34	179	54 01	239	25 20	267	41 19	322
247	45 42	064	34 17	112	56 19	150	20 35	180	53 23	240	24 36	267	40 51	322
248	46 21	065	34 58	113	56 41	151	20 35	181	52 45	241	23 52	268	40 24	322
249	47 01	065	35 38	114	57 01	153	20 34	182	52 06	242	23 08	269	39 57	322
250	47 41	065	36 18	115	57 21	155	20 32	183	51 27	243	22 25	269	39 31	322
251	48 21	066	36 57	116	57 39	156	20 30	184	50 48	244	21 41	270	39 04	323
252	49 01	066	37 37	117	57 56	158	20 27	185	50 08	245	20 57	271	38 37	323
253	49 41	066	38 16	118	58 12	160	20 23	185	49 28	247	20 13	271	38 11	323
254	50 21	067	38 54	119	58 26	162	20 18	186	48 48	248	19 29	272	37 45	323
	♦DENEB		ALTAIR		Nunki		♦ANTARES		ARCTURUS		♦Alkaid		Kochab	
255	51 01	067	39 32	120	15 49	153	20 13	187	48 07	249	56 38	298	55 33	345
256	51 42	067	40 10	121	16 09	154	20 07	188	47 26	249	55 59	298	55 22	345
257	52 22	068	40 48	122	16 27	155	20 00	189	46 45	250	55 21	298	55 10	344
258	53 03	068	41 25	123	16 46	156	19 53	190	46 03	251	54 42	299	54 58	344
259	53 44	068	42 02	124	17 03	157	19 45	191	45 22	252	54 04	299	54 46	344
260	54 24	069	42 38	125	17 20	158	19 36	192	44 40	253	53 25	299	54 34	343
261	55 05	069	43 13	126	17 36	159	19 26	193	43 58	254	52 47	299	54 21	343
262	55 46	069	43 49	127	17 52	160	19 16	194	43 15	255	52 09	299	54 08	343
263	56 27	069	44 24	128	18 07	160	19 05	195	42 33	256	51 30	300	53 55	343
264	57 08	070	44 58	129	18 21	161	18 54	196	41 50	257	50 52	300	53 42	342
265	57 50	070	45 31	130	18 35	162	18 41	197	41 07	257	50 14	300	53 29	342
266	58 31	070	46 04	132	18 48	163	18 28	198	40 25	258	49 36	300	53 15	342
267	59 12	071	46 37	133	19 00	164	18 15	199	39 41	259	48 59	301	53 01	342
268	59 54	071	47 09	134	19 12	165	18 00	199	38 58	260	48 21	301	52 47	341
269	60 35	071	47 40	135	19 23	166	17 45	200	38 15	261	47 43	301	52 33	341

LAT 43°N

LHA ♈	Hc	Zn	Hc	Zn	Hc	Zn	Hc	Zn	Hc	Zn	Hc	Zn	Hc	Zn
	♦Alpheratz		ALTAIR		Nunki		♦ANTARES		ARCTURUS		♦Alkaid		Kochab	
270	17 59	067	48 10	137	19 33	167	17 30	201	37 32	261	47 06	301	52 19	341
271	18 39	067	48 40	138	19 43	168	17 14	202	36 48	262	46 28	302	52 04	341
272	19 20	068	49 09	139	19 52	169	16 57	203	36 05	263	45 51	302	51 50	340
273	20 00	068	49 38	141	20 00	170	16 39	204	35 21	264	45 14	302	51 35	340
274	20 41	069	50 05	142	20 07	171	16 21	205	34 38	264	44 37	302	51 20	340
275	21 22	070	50 32	143	20 14	172	16 02	206	33 54	265	44 00	303	51 05	340
276	22 03	070	50 58	145	20 20	173	15 43	207	33 10	266	43 23	303	50 50	340
277	22 45	071	51 23	146	20 25	174	15 23	207	32 26	267	42 46	303	50 35	340
278	23 26	071	51 47	148	20 30	175	15 03	208	31 43	267	42 10	304	50 19	339
279	24 08	072	52 10	149	20 34	175	14 42	209	30 59	268	41 33	304	50 04	339
280	24 50	073	52 32	151	20 37	176	14 20	210	30 15	269	40 57	304	49 48	339
281	25 32	073	52 53	152	20 39	177	13 58	211	29 31	269	40 21	305	49 33	339
282	26 14	074	53 13	154	20 41	178	13 35	212	28 47	270	39 45	305	49 17	339
283	26 56	074	53 32	155	20 42	179	13 12	212	28 03	271	39 09	305	49 01	339
284	27 38	075	53 50	157	20 42	180	12 48	213	27 19	271	38 33	306	48 45	339
	♦Mirfak		Alpheratz		♦ALTAIR		Rasalhague		♦ARCTURUS		Alkaid		Kochab	
285	14 08	033	28 21	075	54 06	158	54 25	218	26 35	272	37 57	306	48 29	339
286	14 31	033	29 03	076	54 22	160	53 58	219	25 52	273	37 22	306	48 13	339
287	14 56	034	29 46	077	54 36	162	53 30	221	25 08	273	36 47	307	47 57	338
288	15 20	034	30 28	077	54 50	163	53 01	222	24 24	274	36 12	307	47 41	338
289	15 45	035	31 11	078	55 02	165	52 31	223	23 40	275	35 37	307	47 25	338
290	16 10	035	31 54	078	55 12	167	52 01	225	22 57	275	35 02	308	47 09	338
291	16 35	036	32 37	079	55 22	168	51 30	226	22 13	276	34 27	308	46 52	338
292	17 01	036	33 20	080	55 30	170	50 58	227	21 29	277	33 53	309	46 36	338
293	17 27	037	34 04	080	55 37	172	50 25	229	20 46	277	33 19	309	46 20	338
294	17 53	037	34 47	081	55 42	174	49 52	230	20 02	278	32 45	309	46 03	338
295	18 20	038	35 30	081	55 47	175	49 18	231	19 19	279	32 11	310	45 47	338
296	18 47	038	36 14	082	55 49	177	48 44	232	18 35	279	31 37	310	45 31	338
297	19 14	039	36 57	083	55 51	179	48 09	233	17 52	280	31 04	310	45 14	338
298	19 42	039	37 41	083	55 51	181	47 33	235	17 09	281	30 30	311	44 58	338
299	20 10	040	38 24	084	55 50	182	46 57	236	16 26	281	29 57	311	44 42	338
	Mirfak		♦Alpheratz		Enif		♦ALTAIR		Rasalhague		Alphecca		♦Kochab	
300	20 38	040	39 08	084	49 51	138	55 48	184	46 21	237	34 39	276	44 25	338
301	21 06	041	39 52	085	50 20	139	55 44	186	45 44	238	33 55	277	44 09	338
302	21 35	041	40 35	086	50 49	141	55 38	188	45 06	239	33 11	277	43 53	338
303	22 04	041	41 19	086	51 16	142	55 32	189	44 28	240	32 28	278	43 36	338
304	22 33	042	42 03	087	51 42	143	55 24	191	43 50	241	31 44	278	43 20	338
305	23 02	042	42 47	088	52 08	145	55 15	193	43 11	242	31 01	279	43 04	338
306	23 32	043	43 31	088	52 33	146	55 05	195	42 32	243	30 18	280	42 48	338
307	24 02	043	44 14	089	52 57	148	54 53	196	41 53	244	29 35	280	42 31	338
308	24 32	044	44 58	090	53 20	149	54 40	198	41 14	245	28 51	281	42 15	338
309	25 02	044	45 42	090	53 41	151	54 26	200	40 34	246	28 08	281	41 59	339
310	25 33	045	46 26	091	54 02	152	54 11	201	39 53	247	27 25	282	41 43	339
311	26 04	045	47 10	092	54 22	154	53 54	203	39 13	248	26 43	283	41 27	339
312	26 35	045	47 54	092	54 41	156	53 37	204	38 32	249	26 00	283	41 11	339
313	27 07	046	48 38	093	54 58	157	53 18	206	37 51	250	25 17	284	40 55	339
314	27 38	046	49 21	094	55 14	159	52 58	208	37 10	251	24 35	284	40 40	339
	CAPELLA		♦Alpheratz		FOMALHAUT		♦ALTAIR		Rasalhague		VEGA		♦Kochab	
315	11 53	036	50 05	095	12 31	154	52 38	209	36 28	251	62 49	273	40 24	339
316	12 19	037	50 49	095	12 50	155	52 16	211	35 47	252	62 05	274	40 08	339
317	12 45	037	51 33	096	13 08	156	51 53	212	35 05	253	61 21	275	39 53	339
318	13 12	038	52 16	097	13 26	157	51 29	214	34 23	254	60 38	275	39 37	339
319	13 39	038	53 00	098	13 43	157	51 04	215	33 40	255	59 54	276	39 22	339
320	14 06	039	53 43	099	13 59	158	50 39	216	32 58	256	59 10	276	39 06	340
321	14 34	039	54 26	099	14 15	159	50 12	218	32 16	256	58 27	277	38 51	340
322	15 01	040	55 10	100	14 30	160	49 45	219	31 33	257	57 43	277	38 36	340
323	15 30	040	55 53	101	14 45	161	49 17	220	30 50	258	57 00	278	38 21	340
324	15 58	041	56 36	102	14 59	162	48 48	222	30 07	259	56 16	278	38 06	340
325	16 27	041	57 19	103	15 13	163	48 19	223	29 24	259	55 33	279	37 51	340
326	16 56	042	58 01	104	15 25	164	47 48	224	28 41	260	54 49	279	37 36	340
327	17 26	042	58 44	105	15 37	164	47 17	226	27 57	261	54 06	280	37 22	341
328	17 56	043	59 26	106	15 49	165	46 46	227	27 14	262	53 23	280	37 07	341
329	18 26	043	60 08	107	16 00	166	46 13	228	26 31	262	52 40	281	36 53	341
	♦CAPELLA		Hamal		Diphda		♦FOMALHAUT		ALTAIR		♦VEGA		Kochab	
330	18 56	044	36 06	090	18 23	139	16 10	167	45 40	229	51 57	281	36 38	341
331	19 27	044	36 50	091	18 51	140	16 19	168	45 07	230	51 14	282	36 24	341
332	19 57	045	37 34	092	19 19	141	16 28	169	44 33	231	50 31	282	36 10	341
333	20 29	045	38 17	092	19 47	142	16 36	170	43 58	233	49 48	283	35 56	342
334	21 00	046	39 01	093	20 14	143	16 44	171	43 23	234	49 05	283	35 42	342
335	21 32	046	39 45	094	20 40	143	16 51	172	42 47	235	48 23	284	35 29	342
336	22 03	047	40 29	095	21 06	144	16 57	172	42 11	236	47 40	284	35 15	342
337	22 36	047	41 13	095	21 31	145	17 02	173	41 35	237	46 58	285	35 02	342
338	23 08	048	41 56	096	21 56	146	17 07	174	40 58	238	46 15	285	34 49	343
339	23 41	048	42 40	097	22 20	147	17 11	175	40 20	239	45 33	286	34 36	343
340	24 14	049	43 23	098	22 43	148	17 14	176	39 43	240	44 51	286	34 23	343
341	24 47	049	44 07	098	23 06	149	17 17	177	39 04	241	44 09	287	34 10	343
342	25 20	050	44 50	099	23 29	150	17 19	178	38 26	242	43 27	287	33 57	343
343	25 54	050	45 33	100	23 50	151	17 20	179	37 47	243	42 45	288	33 45	344
344	26 27	051	46 17	101	24 11	152	17 21	180	37 08	244	42 03	288	33 32	344
	♦CAPELLA		ALDEBARAN		Diphda		♦FOMALHAUT		ALTAIR		♦VEGA		Kochab	
345	27 02	051	15 33	082	24 31	153	17 21	181	36 28	245	41 22	289	33 20	344
346	27 36	052	16 17	082	24 51	154	17 20	182	35 49	246	40 40	289	33 08	344
347	28 10	052	17 00	083	25 10	155	17 19	182	35 08	246	39 59	290	32 56	344
348	28 45	052	17 44	084	25 28	156	17 16	183	34 28	247	39 17	290	32 45	345
349	29 20	053	18 27	084	25 46	157	17 14	184	33 47	248	38 36	291	32 33	345
350	29 55	053	19 11	085	26 03	158	17 10	185	33 07	249	37 55	291	32 22	345
351	30 30	054	19 55	086	26 19	159	17 06	186	32 25	250	37 14	292	32 10	345
352	31 06	054	20 39	086	26 34	160	17 01	187	31 44	251	36 34	292	31 59	346
353	31 41	055	21 22	087	26 49	161	16 55	188	31 03	252	35 53	292	31 49	346
354	32 17	055	22 06	088	27 03	162	16 49	189	30 21	252	35 13	293	31 38	346
355	32 53	055	22 50	088	27 16	163	16 42	190	29 39	253	34 32	293	31 28	346
356	33 29	056	23 34	089	27 29	164	16 34	191	28 57	254	33 52	294	31 17	347
357	34 06	056	24 18	090	27 40	165	16 26	191	28 15	255	33 12	294	31 07	347
358	34 42	057	25 02	090	27 51	166	16 17	192	27 32	256	32 32	295	30 57	347
359	35 19	057	25 46	091	28 01	167	16 07	193	26 50	256	31 52	295	30 48	347

LAT 42°N

LHA ♈	Hc	Zn	Hc	Zn	Hc	Zn	Hc	Zn	Hc	Zn	Hc	Zn	Hc	Zn
	◆CAPELLA		ALDEBARAN		◆Diphda		FOMALHAUT		ALTAIR		◆VEGA		Kochab	
0	35 23	057	26 31	091	29 09	168	16 55	194	26 20	258	30 46	296	29 39	348
1	36 01	057	27 16	092	29 18	169	16 43	195	25 36	258	30 07	297	29 30	348
2	36 38	058	28 00	093	29 26	170	16 31	196	24 53	259	29 27	297	29 21	348
3	37 16	058	28 45	093	29 33	171	16 19	197	24 09	260	28 47	298	29 12	348
4	37 54	058	29 29	094	29 39	173	16 06	198	23 25	261	28 08	298	29 03	349
5	38 32	059	30 14	095	29 45	174	15 52	199	22 41	261	27 29	299	28 54	349
6	39 10	059	30 58	095	29 49	175	15 37	199	21 57	262	26 50	299	28 46	349
7	39 48	059	31 42	096	29 53	176	15 22	200	21 13	263	26 11	300	28 38	350
8	40 27	060	32 27	097	29 56	177	15 06	201	20 28	263	25 32	300	28 30	350
9	41 05	060	33 11	098	29 58	178	14 50	202	19 44	264	24 54	301	28 22	350
10	41 44	060	33 55	098	29 59	179	14 33	203	19 00	265	24 16	301	28 15	350
11	42 23	061	34 39	099	29 59	180	14 15	204	18 15	265	23 38	302	28 07	351
12	43 02	061	35 23	100	29 59	181	13 57	205	17 31	266	23 00	302	28 00	351
13	43 41	061	36 07	101	29 57	182	13 38	205	16 46	267	22 22	303	27 53	351
14	44 20	062	36 51	101	29 55	183	13 19	206	16 02	268	21 45	303	27 46	351
	◆CAPELLA		BETELGEUSE		RIGEL		◆Diphda		Enif		◆DENEB		Kochab	
15	44 59	062	17 01	095	13 24	114	29 52	185	36 29	248	44 21	297	27 40	352
16	45 39	062	17 46	096	14 05	115	29 48	186	35 48	249	43 41	298	27 34	352
17	46 18	063	18 30	097	14 45	116	29 43	187	35 06	249	43 02	298	27 28	352
18	46 58	063	19 14	098	15 25	117	29 37	188	34 25	250	42 22	298	27 22	353
19	47 38	063	19 59	098	16 05	117	29 31	189	33 43	251	41 43	299	27 16	353
20	48 18	064	20 43	099	16 44	118	29 23	190	33 00	252	41 04	299	27 11	353
21	48 58	064	21 27	100	17 23	119	29 15	191	32 18	253	40 25	299	27 06	354
22	49 38	064	22 11	100	18 02	120	29 06	192	31 35	254	39 46	300	27 01	354
23	50 18	064	22 54	101	18 41	120	28 56	193	30 52	254	39 07	300	26 56	354
24	50 58	065	23 38	102	19 19	121	28 46	194	30 09	255	38 29	300	26 52	354
25	51 39	065	24 22	103	19 57	122	28 34	195	29 26	256	37 50	301	26 47	355
26	52 19	065	25 05	103	20 35	123	28 22	196	28 43	257	37 12	301	26 43	355
27	53 00	066	25 48	104	21 12	124	28 09	197	27 59	257	36 34	301	26 40	355
28	53 40	066	26 32	105	21 49	125	27 55	199	27 16	258	35 56	302	26 36	356
29	54 21	066	27 15	106	22 25	125	27 41	200	26 32	259	35 18	302	26 33	356
	◆CAPELLA		BETELGEUSE		RIGEL		◆Diphda		Alpheratz		◆DENEB		Kochab	
30	55 02	066	27 58	106	23 02	126	27 26	201	64 00	249	34 41	303	26 30	356
31	55 43	067	28 40	107	23 37	127	27 10	202	63 18	250	34 03	303	26 27	356
32	56 24	067	29 23	108	24 13	128	26 53	203	62 36	252	33 26	303	26 24	357
33	57 05	067	30 05	109	24 48	129	26 35	204	61 53	253	32 49	304	26 22	357
34	57 46	067	30 47	110	25 22	130	26 17	205	61 11	254	32 12	304	26 19	357
35	58 27	067	31 29	110	25 56	131	25 58	206	60 28	255	31 35	305	26 18	358
36	59 08	068	32 11	111	26 30	132	25 38	207	59 45	256	30 59	305	26 16	358
37	59 49	068	32 52	112	27 03	132	25 18	208	59 01	257	30 22	305	26 14	358
38	60 31	068	33 33	113	27 35	133	24 57	209	58 18	258	29 46	306	26 13	359
39	61 12	068	34 14	114	28 08	134	24 35	210	57 34	259	29 10	306	26 12	359
40	61 54	068	34 55	115	28 39	135	24 13	211	56 50	259	28 34	307	26 11	359
41	62 35	069	35 35	116	29 10	136	23 50	211	56 06	260	27 58	307	26 11	359
42	63 17	069	36 16	116	29 41	137	23 26	212	55 22	261	27 23	308	26 11	000
43	63 58	069	36 55	117	30 11	138	23 02	213	54 38	262	26 48	308	26 11	000
44	64 40	069	37 35	118	30 40	139	22 37	214	53 54	263	26 13	308	26 11	000
	◆Dubhe		POLLUX		BETELGEUSE		◆RIGEL		Diphda		◆Alpheratz		DENEB	
45	24 10	026	31 41	079	38 14	119	31 09	140	22 12	215	53 10	264	25 38	309
46	24 30	027	32 25	080	38 53	120	31 37	141	21 46	216	52 26	264	25 03	309
47	24 50	027	33 09	080	39 31	121	32 05	142	21 19	217	51 41	265	24 29	310
48	25 10	028	33 53	081	40 09	122	32 32	143	20 52	218	50 57	266	23 55	310
49	25 31	028	34 37	082	40 46	123	32 58	144	20 25	219	50 12	267	23 21	311
50	25 52	028	35 21	082	41 23	124	33 24	145	19 56	220	49 28	267	22 47	311
51	26 13	029	36 06	083	42 00	125	33 49	147	19 28	221	48 43	268	22 14	312
52	26 35	029	36 50	083	42 36	126	34 13	148	18 58	221	47 59	269	21 41	312
53	26 57	029	37 34	084	43 12	127	34 36	149	18 29	222	47 14	269	21 08	313
54	27 18	030	38 19	085	43 47	129	34 59	150	17 58	223	46 29	270	20 35	313
55	27 41	030	39 03	085	44 22	130	35 21	151	17 28	224	45 45	271	20 03	314
56	28 03	030	39 47	086	44 56	131	35 42	152	16 56	225	45 00	271	19 30	314
57	28 26	031	40 32	086	45 29	132	36 03	153	16 25	226	44 16	272	18 58	315
58	28 48	031	41 16	087	46 02	133	36 23	154	15 53	226	43 31	273	18 27	315
59	29 11	031	42 01	088	46 34	134	36 41	156	15 20	227	42 47	273	17 55	316
	◆Dubhe		POLLUX		PROCYON		◆SIRIUS		RIGEL		◆Hamal		Schedar	
60	29 35	032	42 46	088	29 13	111	20 03	138	37 00	157	60 04	241	55 17	312
61	29 58	032	43 30	089	29 54	112	20 33	139	37 17	158	59 25	242	54 44	312
62	30 22	032	44 15	090	30 36	113	21 02	140	37 33	159	58 45	243	54 11	312
63	30 45	032	44 59	090	31 16	114	21 31	140	37 48	160	58 05	244	53 37	312
64	31 09	033	45 44	091	31 57	115	21 59	141	38 03	162	57 25	245	53 04	312
65	31 34	033	46 28	092	32 38	116	22 27	142	38 17	163	56 44	247	52 31	312
66	31 58	033	47 13	092	33 18	116	22 54	143	38 29	164	56 03	248	51 58	312
67	32 23	034	47 58	093	33 57	117	23 20	144	38 41	165	55 22	249	51 25	312
68	32 47	034	48 42	094	34 37	118	23 46	145	38 52	167	54 40	250	50 52	312
69	33 12	034	49 27	095	35 16	119	24 11	146	39 02	168	53 58	251	50 19	312
70	33 37	034	50 11	095	35 55	120	24 36	147	39 11	169	53 16	252	49 46	312
71	34 03	035	50 55	096	36 33	121	25 00	148	39 19	170	52 33	253	49 13	312
72	34 28	035	51 40	097	37 11	122	25 23	149	39 26	172	51 51	254	48 40	312
73	34 54	035	52 24	098	37 49	123	25 46	150	39 32	173	51 08	255	48 07	313
74	35 19	035	53 08	098	38 26	124	26 08	151	39 37	174	50 25	256	47 34	313
	◆Dubhe		REGULUS		PROCYON		◆SIRIUS		RIGEL		◆Hamal		Schedar	
75	35 45	036	17 35	090	39 03	125	26 30	152	39 41	175	49 41	256	47 02	313
76	36 11	036	18 20	090	39 39	126	26 50	153	39 44	177	48 58	257	46 29	313
77	36 38	036	19 05	091	40 15	127	27 10	154	39 46	178	48 14	258	45 56	313
78	37 04	036	19 49	092	40 51	128	27 30	155	39 47	179	47 31	259	45 24	313
79	37 30	037	20 34	092	41 25	129	27 48	156	39 47	181	46 47	260	44 52	313
80	37 57	037	21 18	093	42 00	130	28 06	157	39 47	182	46 03	261	44 19	314
81	38 24	037	22 03	094	42 33	131	28 23	158	39 45	183	45 19	261	43 47	314
82	38 51	037	22 47	094	43 07	132	28 40	159	39 42	184	44 35	262	43 15	314
83	39 18	037	23 32	095	43 39	134	28 55	160	39 38	186	43 51	263	42 43	314
84	39 45	038	24 16	096	44 11	135	29 10	161	39 33	187	43 06	264	42 11	314
85	40 12	038	25 00	096	44 43	136	29 24	162	39 27	188	42 22	264	41 39	315
86	40 39	038	25 45	097	45 13	137	29 38	163	39 20	190	41 37	265	41 07	315
87	41 07	038	26 29	098	45 43	138	29 50	164	39 12	191	40 53	266	40 36	315
88	41 34	038	27 13	099	46 13	140	30 02	165	39 03	192	40 09	267	40 04	315
89	42 02	038	27 57	099	46 41	141	30 13	166	38 54	193	39 24	267	39 33	315

LAT 42°N

LHA ♈	Hc	Zn	Hc	Zn	Hc	Zn	Hc	Zn	Hc	Zn	Hc	Zn	Hc	Zn
	◆Dubhe		REGULUS		PROCYON		◆SIRIUS		RIGEL		ALDEBARAN		◆Mirfak	
90	42 29	038	28 41	100	47 09	142	30 23	168	38 43	195	58 45	222	62 06	300
91	42 57	039	29 25	101	47 36	143	30 32	169	38 31	196	58 14	223	61 28	300
92	43 25	039	30 09	102	48 02	145	30 40	170	38 19	197	57 43	225	60 49	300
93	43 53	039	30 52	102	48 27	146	30 48	171	38 05	198	57 12	226	60 10	300
94	44 21	039	31 36	103	48 51	148	30 55	172	37 51	199	56 39	228	59 31	300
95	44 49	039	32 19	104	49 15	149	31 01	173	37 35	201	56 06	229	58 53	300
96	45 17	039	33 02	105	49 37	150	31 06	174	37 19	202	55 32	230	58 14	300
97	45 45	039	33 46	105	49 59	152	31 10	175	37 02	203	54 57	232	57 36	300
98	46 13	039	34 28	106	50 20	153	31 13	176	36 44	204	54 22	233	56 57	300
99	46 42	039	35 11	107	50 39	155	31 15	177	36 25	205	53 46	234	56 18	300
100	47 10	039	35 54	108	50 58	156	31 17	179	36 06	207	53 09	236	55 40	300
101	47 38	039	36 36	109	51 15	158	31 17	180	35 46	208	52 32	237	55 01	300
102	48 07	039	37 18	110	51 31	159	31 17	181	35 24	209	51 54	238	54 23	301
103	48 35	040	38 00	110	51 47	161	31 16	182	35 03	210	51 16	239	53 45	301
104	49 03	040	38 42	111	52 01	162	31 14	183	34 40	211	50 38	240	53 06	301
	Dubhe		◆Denebola		REGULUS		◆SIRIUS		RIGEL		ALDEBARAN		◆Mirfak	
105	49 32	040	22 52	091	39 23	112	31 11	184	34 16	212	49 59	241	52 28	301
106	50 00	040	23 36	091	40 04	113	31 08	185	33 52	213	49 19	243	51 50	301
107	50 28	039	24 21	092	40 45	114	31 03	186	33 28	214	48 40	244	51 12	301
108	50 57	039	25 05	093	41 26	115	30 58	188	33 02	215	48 00	245	50 34	302
109	51 25	039	25 50	093	42 06	116	30 51	189	32 36	217	47 19	246	49 56	302
110	51 53	039	26 34	094	42 46	117	30 44	190	32 09	218	46 38	247	49 18	302
111	52 22	039	27 19	095	43 26	118	30 36	191	31 41	219	45 57	248	48 40	302
112	52 50	039	28 03	095	44 05	119	30 27	192	31 13	220	45 16	249	48 03	302
113	53 18	039	28 48	096	44 44	120	30 18	193	30 45	221	44 34	249	47 25	303
114	53 46	039	29 32	097	45 22	121	30 07	194	30 15	222	43 52	250	46 48	303
115	54 14	039	30 16	098	46 00	122	29 56	195	29 45	223	43 10	251	46 10	303
116	54 42	039	31 00	098	46 38	123	29 44	196	29 15	224	42 28	252	45 33	303
117	55 10	039	31 44	099	47 15	124	29 31	197	28 44	225	41 45	253	44 56	304
118	55 38	038	32 28	100	47 52	125	29 17	198	28 12	226	41 03	254	44 19	304
119	56 05	038	33 12	101	48 28	126	29 02	200	27 40	226	40 20	255	43 42	304
	◆Kochab		Denebola		◆REGULUS		SIRIUS		RIGEL		◆ALDEBARAN		CAPELLA	
120	36 49	019	33 56	101	49 03	128	28 47	201	27 08	227	39 37	256	60 37	292
121	37 04	020	34 40	102	49 38	129	28 31	202	26 35	228	38 53	256	59 56	292
122	37 19	020	35 23	103	50 13	130	28 14	203	26 01	229	38 10	257	59 15	292
123	37 34	020	36 07	104	50 46	131	27 57	204	25 27	230	37 26	258	58 33	293
124	37 49	020	36 50	105	51 20	133	27 38	205	24 53	231	36 43	259	57 52	293
125	38 05	020	37 33	105	51 52	134	27 19	206	24 18	232	35 59	260	57 11	293
126	38 20	020	38 16	106	52 24	135	27 00	207	23 43	233	35 15	260	56 30	293
127	38 35	020	38 59	107	52 55	137	26 39	208	23 07	234	34 31	261	55 49	293
128	38 51	020	39 41	108	53 25	138	26 18	209	22 31	234	33 47	262	55 08	294
129	39 06	021	40 23	109	53 54	139	25 56	210	21 54	235	33 03	263	54 28	294
130	39 22	021	41 06	110	54 23	141	25 34	211	21 17	236	32 19	263	53 47	294
131	39 38	021	41 47	110	54 50	142	25 11	212	20 40	237	31 34	264	53 06	294
132	39 54	021	42 29	111	55 17	144	24 47	213	20 03	238	30 50	265	52 26	295
133	40 10	021	43 10	112	55 43	146	24 23	214	19 25	239	30 05	265	51 45	295
134	40 25	021	43 52	113	56 08	147	23 58	215	18 47	239	29 21	266	51 05	295
	◆Kochab		ARCTURUS		Denebola		◆REGULUS		SIRIUS		BETELGEUSE		◆CAPELLA	
135	40 41	021	20 50	082	44 32	114	56 31	149	23 32	215	36 33	243	50 24	295
136	40 58	021	21 35	083	45 13	115	56 54	150	23 06	216	35 54	244	49 44	296
137	41 14	021	22 19	084	45 53	116	57 15	152	22 39	217	35 13	245	49 04	296
138	41 30	021	23 03	084	46 33	117	57 36	154	22 12	218	34 33	246	48 24	296
139	41 46	021	23 48	085	47 12	118	57 55	155	21 44	219	33 52	247	47 44	297
140	42 02	021	24 32	086	47 51	119	58 13	157	21 15	220	33 11	248	47 04	297
141	42 18	021	25 17	086	48 30	120	58 29	159	20 47	221	32 29	248	46 25	297
142	42 35	021	26 01	087	49 08	121	58 45	161	20 17	222	31 48	249	45 45	298
143	42 51	021	26 46	088	49 46	123	58 59	163	19 47	223	31 06	250	45 06	298
144	43 07	021	27 30	088	50 23	124	59 11	165	19 17	224	30 24	251	44 26	298
145	43 24	022	28 15	089	51 00	125	59 22	166	18 46	224	29 42	252	43 47	299
146	43 40	022	28 59	090	51 37	126	59 32	168	18 14	225	28 59	253	43 08	299
147	43 56	022	29 44	090	52 12	127	59 40	170	17 42	226	28 17	253	42 29	299
148	44 13	022	30 29	091	52 47	129	59 47	172	17 10	227	27 34	254	41 50	300
149	44 29	022	31 13	092	53 22	130	59 53	174	16 37	228	26 51	255	41 11	300
	◆Kochab		ARCTURUS		◆SPICA		REGULUS		PROCYON		◆POLLUX		CAPELLA	
150	44 45	022	31 58	092	19 06	126	59 56	176	41 43	230	59 19	254	40 33	300
151	45 02	021	32 42	093	19 42	127	59 59	178	41 09	231	58 36	255	39 54	301
152	45 18	021	33 27	094	20 18	128	60 00	180	40 34	232	57 52	256	39 16	301
153	45 34	021	34 11	094	20 53	128	59 59	182	39 58	234	57 09	257	38 38	301
154	45 51	021	34 56	095	21 28	129	59 57	184	39 22	235	56 26	258	38 00	302
155	46 07	021	35 40	096	22 02	130	59 53	186	38 45	236	55 42	259	37 22	302
156	46 23	021	36 24	097	22 36	131	59 48	188	38 08	237	54 58	259	36 44	302
157	46 39	021	37 09	097	23 09	132	59 41	190	37 31	238	54 14	260	36 07	303
158	46 55	021	37 53	098	23 42	133	59 33	192	36 53	239	53 30	261	35 29	303
159	47 11	021	38 37	099	24 15	134	59 23	193	36 15	239	52 46	262	34 52	304
160	47 27	021	39 21	100	24 47	135	59 12	195	35 36	240	52 02	263	34 15	304
161	47 43	021	40 05	100	25 18	135	58 59	197	34 57	241	51 18	264	33 38	304
162	47 59	021	40 49	101	25 49	136	58 45	199	34 18	242	50 33	264	33 01	305
163	48 15	021	41 32	102	26 20	137	58 30	201	33 38	243	49 49	265	32 25	305
164	48 31	021	42 16	103	26 50	138	58 14	203	32 58	244	49 05	266	31 48	305
	◆Kochab		ARCTURUS		◆SPICA		REGULUS		PROCYON		◆POLLUX		CAPELLA	
165	48 46	020	42 59	104	27 19	139	57 56	204	32 18	245	48 20	267	31 12	306
166	49 02	020	43 43	104	27 48	140	57 37	206	31 38	246	47 36	267	30 36	306
167	49 17	020	44 26	105	28 16	141	57 16	208	30 57	247	46 51	268	30 00	307
168	49 33	020	45 09	106	28 44	142	56 55	210	30 16	247	46 06	269	29 24	307
169	49 48	020	45 51	107	29 11	143	56 33	211	29 35	248	45 22	269	28 49	308
170	50 03	020	46 34	108	29 37	144	56 09	213	28 53	249	44 37	270	28 14	308
171	50 18	020	47 16	109	30 03	145	55 44	214	28 11	250	43 53	271	27 39	308
172	50 33	019	47 58	110	30 28	146	55 19	216	27 29	251	43 08	271	27 04	309
173	50 48	019	48 40	111	30 53	147	54 52	217	26 47	252	42 24	272	26 29	309
174	51 02	019	49 22	112	31 16	148	54 24	219	26 05	252	41 39	273	25 55	310
175	51 17	019	50 03	113	31 39	149	53 56	220	25 22	253	40 54	273	25 21	310
176	51 31	019	50 44	114	32 02	150	53 26	222	24 39	254	40 10	274	24 47	311
177	51 45	018	51 25	115	32 23	152	52 56	223	23 56	255	39 25	275	24 13	311
178	51 59	018	52 05	116	32 44	153	52 25	225	23 13	255	38 41	275	23 39	311
179	52 13	018	52 45	117	33 04	154	51 54	226	22 30	256	37 57	276	23 06	312

LHA ♈	Hc	Zn	Hc	Zn	Hc	Zn	Hc	Zn	Hc	Zn	Hc	Zn	Hc	Zn
	Kochab		◆VEGA		ARCTURUS		◆SPICA		REGULUS		◆POLLUX		CAPELLA	
180	52 27	018	19 04	055	53 25	118	33 24	155	51 21	227	37 12	276	22 33	312
181	52 40	017	19 40	055	54 04	119	33 42	156	50 48	229	36 28	277	22 00	313
182	52 53	017	20 17	056	54 42	120	34 00	157	50 14	230	35 44	278	21 28	313
183	53 06	017	20 53	056	55 21	122	34 17	158	49 40	231	35 00	278	20 55	314
184	53 19	017	21 31	057	55 58	123	34 33	159	49 05	232	34 16	279	20 23	314
185	53 32	016	22 08	057	56 36	124	34 49	160	48 30	233	33 31	279	19 51	315
186	53 44	016	22 45	058	57 12	125	35 03	162	47 53	235	32 48	280	19 20	315
187	53 56	016	23 23	058	57 48	127	35 17	163	47 17	236	32 04	281	18 49	316
188	54 08	015	24 01	059	58 24	128	35 29	164	46 40	237	31 20	281	18 18	316
189	54 20	015	24 39	059	58 58	130	35 41	165	46 02	238	30 36	282	17 47	317
190	54 31	015	25 18	060	59 32	131	35 52	166	45 24	239	29 53	282	17 16	317
191	54 43	014	25 56	060	60 05	133	36 02	168	44 46	240	29 09	283	16 46	318
192	54 53	014	26 35	061	60 38	134	36 11	169	44 07	241	28 26	283	16 16	318
193	55 04	014	27 14	061	61 09	136	36 20	170	43 28	242	27 42	284	15 47	319
194	55 14	013	27 53	062	61 40	137	36 27	171	42 48	243	26 59	285	15 18	319
	◆VEGA		Rasalhague		ARCTURUS		◆SPICA		REGULUS		◆POLLUX		Dubhe	
195	28 32	062	24 10	095	62 10	139	36 33	172	42 08	244	26 16	285	63 42	329
196	29 12	062	24 54	095	62 38	141	36 39	174	41 28	245	25 33	286	63 19	328
197	29 51	063	25 39	096	63 06	143	36 43	175	40 47	246	24 50	286	62 55	328
198	30 31	063	26 23	097	63 32	145	36 47	176	40 06	247	24 07	287	62 31	327
199	31 11	064	27 07	098	63 57	147	36 50	177	39 25	248	23 25	287	62 06	326
200	31 51	064	27 51	098	64 21	148	36 51	178	38 44	249	22 42	288	61 42	326
201	32 31	065	28 35	099	64 44	151	36 52	180	38 02	250	22 00	289	61 17	325
202	33 12	065	29 19	100	65 05	153	36 52	181	37 20	250	21 18	289	60 51	325
203	33 52	066	30 03	101	65 25	155	36 50	182	36 38	251	20 36	290	60 26	325
204	34 33	066	30 47	101	65 43	157	36 48	183	35 56	252	19 54	290	60 00	324
205	35 14	067	31 31	102	66 00	159	36 45	185	35 13	253	19 12	291	59 33	324
206	35 55	067	32 14	103	66 15	161	36 41	186	34 31	254	18 30	291	59 07	324
207	36 36	068	32 58	104	66 29	164	36 36	187	33 48	255	17 49	292	58 40	323
208	37 17	068	33 41	104	66 40	166	36 30	188	33 05	255	17 08	293	58 14	323
209	37 59	069	34 24	105	66 50	168	36 23	189	32 22	256	16 27	293	57 47	323
	◆DENEB		VEGA		Rasalhague		ANTARES		◆ARCTURUS		REGULUS		◆Dubhe	
210	22 26	049	38 40	069	35 07	106	13 24	146	66 58	171	31 38	257	57 20	322
211	23 00	049	39 22	069	35 50	107	13 49	147	67 05	173	30 55	258	56 52	322
212	23 33	049	40 04	070	36 32	108	14 13	148	67 09	175	30 11	258	56 25	322
213	24 07	050	40 46	070	37 15	108	14 36	149	67 12	178	29 27	259	55 57	322
214	24 42	050	41 28	071	37 57	109	14 59	149	67 12	180	28 43	260	55 30	322
215	25 16	051	42 10	071	38 39	110	15 22	150	67 11	183	28 00	261	55 02	321
216	25 51	051	42 52	072	39 21	111	15 44	151	67 08	185	27 15	261	54 34	321
217	26 26	052	43 35	072	40 02	112	16 05	152	67 03	188	26 31	262	54 06	321
218	27 01	052	44 17	073	40 43	113	16 26	153	66 56	190	25 47	263	53 38	321
219	27 36	053	45 00	073	41 24	114	16 46	154	66 48	192	25 03	264	53 10	321
220	28 12	053	45 43	074	42 05	115	17 05	155	66 37	195	24 19	264	52 42	321
221	28 47	053	46 25	074	42 45	116	17 24	155	66 25	197	23 34	265	52 14	321
222	29 23	054	47 08	074	43 25	117	17 42	156	66 11	199	22 50	266	51 45	321
223	29 59	054	47 51	075	44 05	118	18 00	157	65 55	202	22 05	266	51 17	321
224	30 36	055	48 34	075	44 44	119	18 17	158	65 38	204	21 21	267	50 49	321
	DENEB		◆ALTAIR		Rasalhague		ANTARES		◆ARCTURUS		Denebola		◆Dubhe	
225	31 12	055	18 48	095	45 23	120	18 33	159	65 20	206	40 40	251	50 20	321
226	31 49	056	19 32	096	46 02	121	18 49	160	64 59	208	39 58	252	49 52	320
227	32 26	056	20 16	096	46 40	122	19 04	161	64 38	210	39 16	253	49 24	320
228	33 03	056	21 01	097	47 17	123	19 18	162	64 15	212	38 33	254	48 55	320
229	33 40	057	21 45	098	47 54	124	19 31	163	63 50	214	37 50	254	48 27	321
230	34 17	057	22 29	099	48 31	125	19 44	164	63 25	216	37 07	255	47 58	321
231	34 55	058	23 13	099	49 07	126	19 57	165	62 58	218	36 24	256	47 30	321
232	35 32	058	23 57	100	49 43	128	20 08	165	62 30	220	35 41	257	47 02	321
233	36 10	058	24 41	100	50 18	129	20 19	166	62 01	221	34 57	258	46 34	321
234	36 48	059	25 25	101	50 52	130	20 29	167	61 31	223	34 14	258	46 05	321
235	37 26	059	26 08	102	51 26	131	20 39	168	61 00	225	33 30	259	45 37	321
236	38 05	059	26 52	103	51 59	133	20 47	169	60 29	226	32 46	260	45 09	321
237	38 43	060	27 35	104	52 32	134	20 55	170	59 56	228	32 02	261	44 41	321
238	39 22	060	28 18	104	53 04	135	21 03	171	59 23	229	31 18	261	44 13	321
239	40 00	061	29 01	105	53 35	137	21 09	172	58 48	231	30 34	262	43 45	321
	◆DENEB		ALTAIR		Rasalhague		◆ANTARES		ARCTURUS		Denebola		◆Dubhe	
240	40 39	061	29 44	106	54 05	138	21 15	173	58 14	232	29 50	263	43 17	321
241	41 18	061	30 27	107	54 34	140	21 20	174	57 38	234	29 05	264	42 49	321
242	41 57	062	31 10	108	55 03	141	21 24	175	57 02	235	28 21	264	42 21	322
243	42 37	062	31 52	108	55 30	143	21 28	176	56 25	236	27 37	265	41 54	322
244	43 16	062	32 34	109	55 57	144	21 31	177	55 48	238	26 52	266	41 26	322
245	43 56	063	33 16	110	56 23	146	21 33	178	55 10	239	26 08	266	40 59	322
246	44 35	063	33 58	111	56 47	147	21 34	179	54 31	240	25 23	267	40 31	322
247	45 15	063	34 40	112	57 11	149	21 35	180	53 53	241	24 39	268	40 04	322
248	45 55	064	35 21	113	57 33	151	21 35	181	53 13	242	23 54	268	39 37	323
249	46 35	064	36 02	114	57 55	152	21 34	182	52 34	243	23 10	269	39 10	323
250	47 15	064	36 43	114	58 15	154	21 32	183	51 54	244	22 25	270	38 43	323
251	47 55	065	37 23	115	58 34	156	21 30	184	51 13	246	21 40	270	38 16	323
252	48 36	065	38 03	116	58 52	158	21 26	185	50 32	247	20 56	271	37 49	323
253	49 16	065	38 43	117	59 08	159	21 22	186	49 51	248	20 11	272	37 23	324
254	49 57	066	39 23	118	59 23	161	21 18	186	49 10	249	19 27	272	36 56	324
	◆DENEB		ALTAIR		Nunki		◆ANTARES		ARCTURUS		◆Alkaid		Kochab	
255	50 37	066	40 02	119	16 43	153	21 12	187	48 28	250	56 09	299	54 35	345
256	51 18	066	40 41	120	17 03	154	21 06	188	47 46	251	55 31	299	54 24	345
257	51 59	066	41 19	121	17 22	155	20 59	189	47 04	251	54 52	300	54 12	345
258	52 40	067	41 57	122	17 40	156	20 52	190	46 22	252	54 13	300	54 01	344
259	53 21	067	42 35	123	17 58	157	20 44	191	45 39	253	53 34	300	53 48	344
260	54 02	067	43 12	124	18 16	158	20 35	192	44 57	254	52 56	300	53 36	344
261	54 43	068	43 48	125	18 32	159	20 25	193	44 14	255	52 17	300	53 24	344
262	55 24	068	44 25	126	18 48	159	20 14	194	43 30	256	51 39	300	53 11	343
263	56 05	068	45 00	127	19 03	160	20 03	195	42 47	257	51 00	301	52 58	343
264	56 47	068	45 35	129	19 18	161	19 51	196	42 04	258	50 22	301	52 45	343
265	57 28	069	46 10	130	19 32	162	19 39	197	41 20	258	49 44	301	52 31	342
266	58 10	069	46 44	131	19 45	163	19 25	198	40 36	259	49 06	301	52 18	342
267	58 51	069	47 17	132	19 58	164	19 12	199	39 52	260	48 28	302	52 04	342
268	59 33	069	47 50	133	20 10	165	18 57	200	39 08	261	47 50	302	51 50	342
269	60 15	069	48 22	135	20 21	166	18 42	200	38 24	261	47 12	302	51 36	341

LAT 42°N

LHA ♈	Hc	Zn	Hc	Zn	Hc	Zn	Hc	Zn	Hc	Zn	Hc	Zn	Hc	Zn
	◆Alpheratz		ALTAIR		Nunki		◆ANTARES		ARCTURUS		◆Alkaid		Kochab	
270	17 35	066	48 54	136	20 32	167	18 26	201	37 40	262	46 34	302	51 22	341
271	18 16	067	49 25	137	20 41	168	18 09	202	36 56	263	45 56	303	51 08	341
272	18 57	068	49 54	138	20 50	169	17 52	203	36 12	264	45 19	303	50 53	341
273	19 38	068	50 24	140	20 59	170	17 34	204	35 27	264	44 41	303	50 38	341
274	20 20	069	50 52	141	21 06	171	17 16	205	34 43	265	44 04	303	50 23	340
275	21 01	069	51 20	143	21 13	172	16 56	206	33 59	266	43 27	304	50 09	340
276	21 43	070	51 46	144	21 19	173	16 37	207	33 14	267	42 50	304	49 53	340
277	22 25	070	52 12	145	21 25	174	16 16	208	32 30	267	42 13	304	49 38	340
278	23 07	071	52 37	147	21 29	174	15 55	208	31 45	268	41 36	304	49 23	340
279	23 49	072	53 01	148	21 33	175	15 34	209	31 00	269	40 59	305	49 08	340
280	24 31	072	53 24	150	21 37	176	15 12	210	30 16	269	40 23	305	48 52	340
281	25 14	073	53 46	151	21 39	177	14 49	211	29 31	270	39 46	305	48 36	339
282	25 57	073	54 06	153	21 41	178	14 26	212	28 47	271	39 10	306	48 21	339
283	26 39	074	54 26	155	21 42	179	14 02	213	28 02	271	38 34	306	48 05	339
284	27 22	074	54 45	156	21 42	180	13 38	213	27 18	272	37 58	306	47 49	339
	◆Mirfak		Alpheratz		◆ALTAIR		Rasalhague		◆ARCTURUS		Alkaid		Kochab	
285	13 17	032	28 05	075	55 02	158	55 13	218	26 33	273	37 22	307	47 33	339
286	13 41	033	28 48	075	55 18	160	54 44	220	25 48	273	36 46	307	47 17	339
287	14 06	033	29 32	076	55 33	161	54 15	221	25 04	274	36 11	307	47 01	339
288	14 30	034	30 15	077	55 47	163	53 45	223	24 19	275	35 35	308	46 45	339
289	14 55	034	30 58	077	56 00	165	53 15	224	23 35	275	35 00	308	46 29	339
290	15 21	035	31 42	078	56 11	166	52 43	226	22 51	276	34 25	308	46 13	339
291	15 46	035	32 25	078	56 21	168	52 11	227	22 06	276	33 50	309	45 56	339
292	16 13	036	33 09	079	56 29	170	51 38	228	21 22	277	33 15	309	45 40	339
293	16 39	036	33 53	079	56 36	172	51 04	230	20 38	278	32 41	309	45 24	339
294	17 06	037	34 37	080	56 42	173	50 30	231	19 54	278	32 07	310	45 08	339
295	17 32	037	35 21	081	56 46	175	49 55	232	19 10	279	31 32	310	44 51	339
296	18 00	038	36 05	081	56 49	177	49 20	233	18 26	280	30 58	311	44 35	338
297	18 27	038	36 49	082	56 51	179	48 44	234	17 42	280	30 25	311	44 19	338
298	18 55	039	37 33	082	56 51	181	48 07	236	16 58	281	29 51	311	44 02	338
299	19 23	039	38 17	083	56 50	182	47 30	237	16 14	281	29 18	312	43 46	338
	Mirfak		◆Alpheratz		Enif		◆ALTAIR		Rasalhague		Alphecca		◆Kochab	
300	19 52	040	39 02	084	50 36	137	56 47	184	46 53	238	34 32	277	43 30	338
301	20 20	040	39 46	084	51 06	139	56 43	186	46 15	239	33 48	277	43 13	339
302	20 49	041	40 30	085	51 35	140	56 38	188	45 37	240	33 04	278	42 57	339
303	21 19	041	41 15	085	52 03	141	56 31	190	44 58	241	32 19	278	42 41	339
304	21 48	042	41 59	086	52 30	143	56 23	191	44 19	242	31 35	279	42 24	339
305	22 18	042	42 44	087	52 57	144	56 14	193	43 39	243	30 51	280	42 08	339
306	22 48	043	43 28	087	53 23	146	56 03	195	42 59	244	30 07	280	41 52	339
307	23 18	043	44 13	088	53 47	147	55 51	197	42 19	245	29 24	281	41 36	339
308	23 49	043	44 57	089	54 11	149	55 37	198	41 38	246	28 40	281	41 19	339
309	24 19	044	45 42	089	54 34	150	55 23	200	40 58	247	27 56	282	41 03	339
310	24 50	044	46 27	090	54 55	152	55 07	202	40 17	248	27 13	283	40 47	339
311	25 21	045	47 11	091	55 16	153	54 50	203	39 35	249	26 29	283	40 31	339
312	25 53	045	47 56	091	55 35	155	54 31	205	38 53	249	25 46	284	40 15	339
313	26 25	045	48 40	092	55 53	157	54 12	207	38 12	250	25 03	284	39 59	339
314	26 56	046	49 25	093	56 10	158	53 51	208	37 29	251	24 19	285	39 43	339
	CAPELLA		◆Alpheratz		FOMALHAUT		◆ALTAIR		Rasalhague		VEGA		◆Kochab	
315	11 04	036	50 09	093	13 25	154	53 30	210	36 47	252	62 44	275	39 28	339
316	11 30	036	50 54	094	13 44	155	53 07	211	36 05	253	62 00	276	39 12	339
317	11 57	037	51 38	095	14 03	156	52 44	213	35 22	254	61 16	276	38 56	340
318	12 24	038	52 23	096	14 21	157	52 19	214	34 39	255	60 31	277	38 41	340
319	12 51	038	53 07	096	14 38	157	51 53	216	33 56	255	59 47	277	38 25	340
320	13 19	039	53 51	097	14 55	158	51 27	217	33 13	256	59 03	278	38 10	340
321	13 47	039	54 35	098	15 11	159	51 00	219	32 29	257	58 19	278	37 55	340
322	14 15	040	55 20	099	15 27	160	50 31	220	31 46	258	57 35	279	37 40	340
323	14 44	040	56 04	100	15 42	161	50 02	221	31 02	258	56 51	279	37 24	340
324	15 13	041	56 48	100	15 56	162	49 33	223	30 19	259	56 07	280	37 09	340
325	15 42	041	57 31	101	16 10	163	49 02	224	29 35	260	55 23	280	36 54	341
326	16 12	042	58 15	102	16 23	163	48 31	225	28 51	261	54 39	281	36 40	341
327	16 41	042	58 58	103	16 35	164	47 59	226	28 07	261	53 55	281	36 25	341
328	17 12	043	59 42	104	16 47	165	47 26	228	27 23	262	53 11	282	36 10	341
329	17 42	043	60 25	105	16 58	166	46 53	229	26 38	263	52 28	282	35 56	341
	◆CAPELLA		Hamal		Diphda		◆FOMALHAUT		ALTAIR		◆VEGA		Kochab	
330	18 13	044	36 06	090	19 08	139	17 08	167	46 19	230	51 44	283	35 42	341
331	18 44	044	36 50	090	19 37	140	17 18	168	45 45	231	51 01	283	35 27	341
332	19 15	045	37 35	091	20 06	141	17 27	169	45 10	232	50 17	284	35 13	342
333	19 46	045	38 20	092	20 34	141	17 35	170	44 34	233	49 34	284	34 59	342
334	20 18	046	39 04	092	21 01	142	17 43	171	43 58	234	48 51	285	34 45	342
335	20 50	046	39 49	093	21 28	143	17 50	171	43 22	236	48 08	285	34 32	342
336	21 22	047	40 33	094	21 55	144	17 56	172	42 45	237	47 25	285	34 18	342
337	21 55	047	41 18	095	22 20	145	18 02	173	42 07	238	46 42	286	34 05	343
338	22 28	048	42 02	095	22 46	146	18 07	174	41 29	239	45 59	286	33 51	343
339	23 01	048	42 47	096	23 10	147	18 11	175	40 51	240	45 16	287	33 38	343
340	23 34	048	43 31	097	23 34	148	18 14	176	40 12	241	44 34	287	33 25	343
341	24 07	049	44 15	097	23 58	149	18 17	177	39 33	242	43 51	288	33 12	343
342	24 41	049	44 59	098	24 20	150	18 19	178	38 54	243	43 09	288	33 00	344
343	25 15	050	45 43	099	24 43	151	18 20	179	38 14	244	42 26	289	32 47	344
344	25 49	050	46 27	100	25 04	152	18 21	180	37 34	244	41 44	289	32 35	344
	◆CAPELLA		ALDEBARAN		Diphda		◆FOMALHAUT		ALTAIR		◆VEGA		Kochab	
345	26 24	051	15 24	081	25 25	153	18 21	181	36 54	245	41 02	289	32 22	344
346	26 58	051	16 08	082	25 45	154	18 20	182	36 13	246	40 20	290	32 10	344
347	27 33	052	16 53	083	26 04	155	18 19	182	35 32	247	39 38	290	31 58	345
348	28 08	052	17 37	083	26 23	156	18 16	183	34 51	248	38 56	291	31 47	345
349	28 43	052	18 21	084	26 41	157	18 13	184	34 09	249	38 15	291	31 35	345
350	29 19	053	19 06	085	26 58	158	18 10	185	33 28	250	37 33	292	31 24	345
351	29 54	053	19 50	085	27 15	159	18 05	186	32 46	251	36 52	292	31 12	346
352	30 30	054	20 35	086	27 31	160	18 00	187	32 04	251	36 11	293	31 01	346
353	31 06	054	21 19	087	27 46	161	17 54	188	31 21	252	35 30	293	30 50	346
354	31 42	054	22 04	087	28 00	162	17 48	189	30 39	253	34 49	294	30 40	346
355	32 19	055	22 48	088	28 14	163	17 41	190	29 56	254	34 08	294	30 29	346
356	32 55	055	23 33	089	28 26	164	17 33	191	29 13	255	33 27	295	30 19	347
357	33 32	056	24 17	089	28 38	165	17 24	192	28 30	255	32 47	295	30 09	347
358	34 09	056	25 02	090	28 49	166	17 15	192	27 47	256	32 07	295	29 59	347
359	34 46	056	25 46	091	29 00	167	17 05	193	27 03	257	31 26	296	29 49	347

LAT 41°N

LHA	Hc	Zn	Hc	Zn	Hc	Zn	Hc	Zn	Hc	Zn	Hc	Zn	Hc	Zn
♈	◆CAPELLA		ALDEBARAN		◆Diphda		FOMALHAUT		ALTAIR		◆VEGA		Kochab	
0	34 50	056	26 32	091	30 08	168	17 53	194	26 33	258	30 19	297	28 41	348
1	35 28	057	27 17	091	30 17	169	17 41	195	25 48	259	29 39	297	28 31	348
2	36 06	057	28 03	092	30 25	170	17 29	196	25 04	260	28 59	298	28 22	348
3	36 44	057	28 48	093	30 32	171	17 16	197	24 19	260	28 19	298	28 13	349
4	37 22	058	29 33	093	30 39	172	17 03	198	23 35	261	27 39	299	28 04	349
5	38 00	058	30 18	094	30 44	174	16 49	199	22 50	262	27 00	299	27 56	349
6	38 39	058	31 03	095	30 49	175	16 34	200	22 05	262	26 20	300	27 47	349
7	39 17	059	31 48	096	30 53	176	16 18	200	21 20	263	25 41	300	27 39	350
8	39 56	059	32 34	096	30 56	177	16 02	201	20 35	264	25 02	301	27 31	350
9	40 35	059	33 19	097	30 58	178	15 45	202	19 50	264	24 23	301	27 23	350
10	41 14	060	34 03	098	30 59	179	15 28	203	19 05	265	23 44	302	27 15	350
11	41 53	060	34 48	098	30 59	180	15 10	204	18 20	266	23 06	302	27 08	351
12	42 32	060	35 33	099	30 59	181	14 51	205	17 35	266	22 28	303	27 01	351
13	43 12	061	36 18	100	30 57	182	14 32	206	16 49	267	21 50	303	26 54	351
14	43 51	061	37 02	101	30 55	184	14 12	206	16 04	268	21 12	304	26 47	352
	◆CAPELLA		BETELGEUSE		RIGEL		◆Diphda		Enif		◆DENEB		Kochab	
15	44 31	061	17 07	095	13 49	114	30 52	185	36 52	248	43 53	298	26 41	352
16	45 11	061	17 52	096	14 30	115	30 47	186	36 10	249	43 13	298	26 34	352
17	45 50	062	18 37	096	15 11	116	30 43	187	35 27	250	42 33	299	26 28	352
18	46 30	062	19 22	097	15 52	116	30 37	188	34 45	251	41 54	299	26 22	353
19	47 10	062	20 07	098	16 32	117	30 30	189	34 02	252	41 14	299	26 17	353
20	47 51	063	20 52	099	17 12	118	30 23	190	33 19	253	40 35	300	26 11	353
21	48 31	063	21 36	099	17 52	119	30 14	191	32 35	253	39 55	300	26 06	354
22	49 11	063	22 21	100	18 32	119	30 05	192	31 52	254	39 16	300	26 01	354
23	49 52	063	23 06	101	19 11	120	29 55	193	31 08	255	38 37	301	25 56	354
24	50 32	064	23 50	101	19 50	121	29 44	194	30 24	256	37 58	301	25 52	354
25	51 13	064	24 34	102	20 29	122	29 32	196	29 40	256	37 19	301	25 48	355
26	51 53	064	25 19	103	21 07	123	29 20	197	28 56	257	36 41	302	25 44	355
27	52 34	064	26 03	104	21 45	123	29 06	198	28 12	258	36 02	302	25 40	355
28	53 15	065	26 47	104	22 23	124	28 52	199	27 28	259	35 24	302	25 36	356
29	53 56	065	27 31	105	23 00	125	28 37	200	26 43	259	34 46	303	25 33	356
	◆CAPELLA		BETELGEUSE		RIGEL		◆Diphda		Alpheratz		◆DENEB		Kochab	
30	54 37	065	28 14	106	23 37	126	28 22	201	64 20	251	34 08	303	25 30	356
31	55 18	065	28 58	107	24 13	127	28 05	202	63 37	252	33 30	304	25 27	356
32	55 59	065	29 41	107	24 49	128	27 48	203	62 54	253	32 53	304	25 24	357
33	56 41	066	30 24	108	25 25	129	27 30	204	62 10	254	32 15	304	25 22	357
34	57 22	066	31 07	109	26 00	129	27 11	205	61 26	255	31 38	305	25 20	357
35	58 03	066	31 50	110	26 35	130	26 52	206	60 43	256	31 01	305	25 18	358
36	58 45	066	32 32	111	27 09	131	26 32	207	59 58	257	30 24	306	25 16	358
37	59 26	066	33 14	111	27 43	132	26 11	208	59 14	258	29 47	306	25 14	358
38	60 07	066	33 56	112	28 17	133	25 50	209	58 30	259	29 11	306	25 13	359
39	60 49	067	34 38	113	28 49	134	25 27	210	57 45	260	28 34	307	25 12	359
40	61 31	067	35 20	114	29 22	135	25 05	211	57 01	261	27 58	307	25 11	359
41	62 12	067	36 01	115	29 54	136	24 41	212	56 16	262	27 22	308	25 11	359
42	62 54	067	36 42	116	30 25	137	24 17	213	55 31	263	26 46	308	25 11	000
43	63 35	067	37 23	117	30 55	138	23 52	214	54 46	263	26 11	308	25 11	000
44	64 17	067	38 03	118	31 26	139	23 27	215	54 01	264	25 35	309	25 11	000
	◆Dubhe		POLLUX		BETELGEUSE		◆RIGEL		Diphda		◆Alpheratz		DENEB	
45	23 16	026	31 30	079	38 43	119	31 55	140	23 01	215	53 16	265	25 00	309
46	23 36	027	32 14	079	39 22	119	32 24	141	22 34	216	52 31	266	24 25	310
47	23 57	027	32 59	080	40 02	120	32 52	142	22 07	217	51 46	266	23 50	310
48	24 17	027	33 43	080	40 40	121	33 20	143	21 39	218	51 00	267	23 16	311
49	24 38	028	34 28	081	41 19	122	33 47	144	21 11	219	50 15	268	22 42	311
50	24 59	028	35 13	081	41 57	123	34 13	145	20 42	220	49 30	268	22 08	311
51	25 21	028	35 58	082	42 34	125	34 39	146	20 13	221	48 45	269	21 34	312
52	25 42	029	36 43	083	43 11	126	35 04	147	19 43	222	47 59	270	21 00	312
53	26 04	029	37 28	083	43 48	127	35 28	148	19 13	223	47 14	270	20 27	313
54	26 26	029	38 13	084	44 24	128	35 51	149	18 42	223	46 29	271	19 54	313
55	26 49	030	38 58	084	45 00	129	36 14	151	18 11	224	45 43	272	19 21	314
56	27 11	030	39 43	085	45 35	130	36 35	152	17 39	225	44 58	272	18 49	314
57	27 34	030	40 28	086	46 09	131	36 56	153	17 07	226	44 13	273	18 16	315
58	27 57	031	41 13	086	46 43	132	37 17	154	16 34	227	43 28	274	17 44	315
59	28 20	031	41 58	087	47 16	134	37 36	155	16 01	227	42 43	274	17 12	316
	◆Dubhe		POLLUX		PROCYON		◆SIRIUS		RIGEL		◆Hamal		Schedar	
60	28 43	031	42 43	087	29 34	111	20 48	138	37 55	156	60 33	242	54 36	313
61	29 07	032	43 29	088	30 17	112	21 18	138	38 12	158	59 53	243	54 03	313
62	29 31	032	44 14	089	30 59	112	21 48	139	38 29	159	59 12	245	53 30	313
63	29 55	032	44 59	089	31 40	113	22 17	140	38 45	160	58 31	246	52 57	313
64	30 19	032	45 44	090	32 22	114	22 46	141	39 00	161	57 49	247	52 24	313
65	30 43	033	46 30	091	33 03	115	23 14	142	39 14	163	57 07	248	51 51	313
66	31 08	033	47 15	091	33 44	116	23 41	143	39 27	164	56 25	249	51 18	313
67	31 33	033	48 00	092	34 25	117	24 09	144	39 39	165	55 43	250	50 44	313
68	31 57	034	48 45	093	35 05	118	24 35	145	39 50	166	55 00	251	50 11	313
69	32 23	034	49 31	093	35 45	118	25 01	146	40 01	168	54 17	252	49 38	313
70	32 48	034	50 16	094	36 25	119	25 26	147	40 10	169	53 34	253	49 05	313
71	33 13	034	51 01	095	37 04	120	25 51	148	40 18	170	52 50	254	48 32	313
72	33 39	035	51 46	096	37 43	121	26 15	149	40 25	171	52 07	255	47 59	313
73	34 05	035	52 31	096	38 21	122	26 38	150	40 32	173	51 23	256	47 26	313
74	34 30	035	53 16	097	38 59	123	27 00	151	40 37	174	50 39	257	46 53	313
	◆Dubhe		REGULUS		PROCYON		◆SIRIUS		RIGEL		◆Hamal		Schedar	
75	34 56	035	17 35	089	39 37	124	27 22	152	40 41	175	49 55	258	46 21	314
76	35 23	035	18 20	090	40 14	125	27 44	153	40 44	177	49 11	258	45 48	314
77	35 49	036	19 05	091	40 51	126	28 04	154	40 46	178	48 26	259	45 15	314
78	36 15	036	19 51	091	41 27	127	28 24	155	40 47	179	47 42	260	44 43	314
79	36 42	036	20 36	092	42 03	128	28 43	156	40 47	181	46 57	261	44 10	314
80	37 09	036	21 21	093	42 38	129	29 01	157	40 47	182	46 12	262	43 38	314
81	37 36	036	22 06	093	43 13	131	29 19	158	40 45	184	45 27	262	43 05	314
82	38 03	037	22 52	094	43 47	132	29 36	159	40 42	184	44 42	263	42 33	315
83	38 30	037	23 37	095	44 20	133	29 52	160	40 37	186	43 57	264	42 01	315
84	38 57	037	24 22	095	44 53	134	30 07	161	40 32	187	43 12	265	41 29	315
85	39 24	037	25 07	096	45 25	135	30 21	162	40 26	188	42 27	265	40 57	315
86	39 52	037	25 52	097	45 57	136	30 35	163	40 19	190	41 42	266	40 25	315
87	40 19	037	26 37	097	46 28	138	30 48	164	40 11	191	40 57	267	39 53	316
88	40 47	038	27 22	098	46 58	139	31 00	165	40 02	192	40 12	267	39 21	316
89	41 14	038	28 07	099	47 27	140	31 11	166	39 52	193	39 26	268	38 50	316

LHA	Hc	Zn	Hc	Zn	Hc	Zn	Hc	Zn	Hc	Zn	Hc	Zn	Hc	Zn
♈	◆Dubhe		REGULUS		PROCYON		◆SIRIUS		RIGEL		ALDEBARAN		◆Mirfak	
90	41 42	038	28 51	099	47 56	142	31 21	167	39 41	195	59 29	223	61 36	301
91	42 10	038	29 36	100	48 24	143	31 31	168	39 29	196	58 58	224	60 57	301
92	42 38	038	30 20	101	48 51	144	31 40	170	39 16	197	58 26	226	60 18	301
93	43 06	038	31 05	102	49 17	146	31 47	171	39 02	199	57 53	227	59 40	301
94	43 34	038	31 49	102	49 42	147	31 54	172	38 47	200	57 19	229	59 01	301
95	44 02	038	32 33	103	50 06	148	32 00	173	38 31	201	56 45	230	58 22	301
96	44 30	038	33 17	104	50 29	150	32 05	174	38 15	202	56 09	232	57 44	301
97	44 59	039	34 01	105	50 52	151	32 09	175	37 57	203	55 34	233	57 05	301
98	45 27	039	34 45	106	51 13	153	32 13	176	37 39	205	54 57	234	56 26	301
99	45 55	039	35 28	106	51 33	154	32 15	177	37 20	206	54 20	236	55 48	302
100	46 23	039	36 12	107	51 52	156	32 17	179	37 00	207	53 42	237	55 09	302
101	46 52	039	36 55	108	52 10	157	32 17	180	36 39	208	53 04	238	54 30	302
102	47 20	039	37 38	109	52 27	159	32 17	181	36 17	209	52 26	239	53 52	302
103	47 48	039	38 21	110	52 43	160	32 16	182	35 54	210	51 47	240	53 14	302
104	48 17	039	39 03	111	52 58	162	32 14	183	35 31	211	51 07	241	52 35	302
	Dubhe		◆Denebola		REGULUS		◆SIRIUS		RIGEL		ALDEBARAN		◆Mirfak	
105	48 45	039	22 52	090	39 45	111	32 11	184	35 07	213	50 27	243	51 57	302
106	49 14	039	23 37	091	40 27	112	32 07	185	34 42	214	49 47	244	51 18	302
107	49 42	039	24 23	092	41 09	113	32 03	187	34 17	215	49 06	245	50 40	302
108	50 10	039	25 08	092	41 51	114	31 57	188	33 51	216	48 25	246	50 02	303
109	50 39	039	25 53	093	42 32	115	31 51	189	33 24	217	47 43	247	49 24	303
110	51 07	039	26 38	094	43 13	116	31 43	190	32 56	218	47 02	248	48 46	303
111	51 35	038	27 24	094	43 53	117	31 35	191	32 28	219	46 20	249	48 08	303
112	52 03	038	28 09	095	44 33	118	31 26	192	31 59	220	45 37	249	47 30	303
113	52 31	038	28 54	096	45 13	119	31 16	193	31 30	221	44 55	250	46 52	304
114	52 59	038	29 39	096	45 53	120	31 05	194	31 00	222	44 12	251	46 15	304
115	53 27	038	30 24	097	46 32	121	30 54	195	30 29	223	43 29	252	45 37	304
116	53 55	038	31 09	098	47 10	122	30 41	196	29 58	224	42 46	253	45 00	304
117	54 23	038	31 54	098	47 48	123	30 28	198	29 27	225	42 02	254	44 22	304
118	54 50	038	32 38	099	48 26	124	30 14	199	28 54	226	41 19	255	43 45	305
119	55 18	037	33 23	100	49 03	126	29 59	200	28 21	227	40 35	256	43 08	305
	◆Kochab		Denebola		◆REGULUS		SIRIUS		RIGEL		◆ALDEBARAN		CAPELLA	
120	35 53	019	34 07	101	49 39	127	29 43	201	27 48	228	39 51	256	60 14	294
121	36 07	019	34 52	101	50 15	128	29 27	202	27 14	229	39 07	257	59 33	294
122	36 22	019	35 36	102	50 51	129	29 10	203	26 40	230	38 23	258	58 51	294
123	36 38	020	36 20	103	51 26	130	28 52	204	26 05	230	37 39	259	58 10	294
124	36 53	020	37 04	104	52 00	132	28 33	205	25 30	231	36 54	260	57 28	294
125	37 08	020	37 48	105	52 33	133	28 13	206	24 55	232	36 09	260	56 47	294
126	37 24	020	38 32	105	53 06	134	27 53	207	24 19	233	35 25	261	56 06	295
127	37 39	020	39 16	106	53 38	136	27 32	208	23 42	234	34 40	262	55 25	295
128	37 55	020	39 59	107	54 09	137	27 11	209	23 06	235	33 55	262	54 44	295
129	38 10	020	40 42	108	54 40	139	26 48	210	22 28	236	33 10	263	54 03	295
130	38 26	020	41 25	109	55 09	140	26 25	211	21 51	236	32 25	264	53 22	295
131	38 42	020	42 08	110	55 38	142	26 02	212	21 13	237	31 40	265	52 41	296
132	38 58	021	42 51	111	56 05	143	25 37	213	20 35	238	30 55	265	52 00	296
133	39 13	021	43 33	111	56 32	145	25 13	214	19 56	239	30 10	266	51 19	296
134	39 29	021	44 15	112	56 58	146	24 47	215	19 17	240	29 25	267	50 39	296
	◆Kochab		ARCTURUS		Denebola		◆REGULUS		SIRIUS		BETELGEUSE		◆CAPELLA	
135	39 45	021	20 42	082	44 57	113	57 22	148	24 21	216	37 00	244	49 58	297
136	40 01	021	21 27	083	45 38	114	57 46	150	23 54	217	36 19	245	49 18	297
137	40 18	021	22 12	083	46 19	115	58 08	151	23 27	218	35 38	246	48 37	297
138	40 34	021	22 57	084	47 00	116	58 29	153	22 59	219	34 57	246	47 57	297
139	40 50	021	23 42	085	47 40	117	58 49	155	22 30	219	34 15	247	47 17	298
140	41 06	021	24 27	085	48 20	118	59 08	157	22 01	220	33 33	248	46 37	298
141	41 22	021	25 12	086	49 00	119	59 25	158	21 32	221	32 51	249	45 57	298
142	41 39	021	25 58	086	49 39	120	59 41	160	21 02	222	32 09	250	45 17	298
143	41 55	021	26 43	087	50 18	122	59 56	162	20 31	223	31 26	251	44 37	299
144	42 11	021	27 28	088	50 56	123	60 09	164	20 00	224	30 43	251	43 57	299
145	42 28	021	28 13	088	51 34	124	60 21	166	19 28	225	30 00	252	43 18	299
146	42 44	021	28 59	089	52 12	125	60 31	168	18 56	225	29 17	253	42 39	300
147	43 00	021	29 44	090	52 48	126	60 39	170	18 24	226	28 34	254	41 59	300
148	43 17	021	30 29	090	53 24	128	60 47	172	17 51	227	27 50	255	41 20	300
149	43 33	021	31 14	091	54 00	129	60 52	174	17 17	228	27 06	255	40 41	301
	◆Kochab		ARCTURUS		◆SPICA		REGULUS		PROCYON		◆POLLUX		CAPELLA	
150	43 50	021	32 00	092	19 42	126	60 56	176	42 21	231	59 34	256	40 02	301
151	44 06	021	32 45	092	20 18	126	60 59	178	41 46	232	58 50	256	39 23	301
152	44 22	021	33 30	093	20 54	127	61 00	180	41 10	233	58 06	257	38 45	302
153	44 38	021	34 15	094	21 30	128	60 59	182	40 33	234	57 22	258	38 06	302
154	44 55	021	35 01	094	22 06	129	60 56	184	39 56	235	56 38	259	37 28	302
155	45 11	021	35 46	095	22 41	130	60 52	186	39 19	236	55 53	260	36 50	303
156	45 27	021	36 31	096	23 15	131	60 47	188	38 41	237	55 08	261	36 12	303
157	45 43	021	37 16	097	23 49	132	60 40	190	38 03	238	54 24	262	35 34	303
158	45 59	021	38 01	097	24 23	132	60 31	192	37 24	239	53 39	263	34 56	304
159	46 15	021	38 46	098	24 56	133	60 21	194	36 45	240	52 54	263	34 18	304
160	46 31	021	39 30	099	25 29	134	60 10	196	36 06	241	52 09	264	33 41	304
161	46 47	021	40 15	099	26 01	135	59 56	198	35 26	242	51 24	265	33 04	305
162	47 03	020	41 00	100	26 33	136	59 42	200	34 46	243	50 39	266	32 27	305
163	47 19	020	41 44	101	27 04	137	59 26	201	34 05	244	49 54	266	31 50	306
164	47 35	020	42 29	102	27 34	138	59 09	203	33 24	245	49 08	267	31 13	306
	◆Kochab		ARCTURUS		◆SPICA		REGULUS		PROCYON		◆POLLUX		CAPELLA	
165	47 50	020	43 13	103	28 04	139	58 50	205	32 43	245	48 23	268	30 37	306
166	48 06	020	43 57	103	28 34	140	58 30	207	32 02	246	47 38	268	30 00	307
167	48 21	020	44 41	104	29 03	141	58 09	209	31 20	247	46 53	269	29 24	307
168	48 36	020	45 25	105	29 31	142	57 47	210	30 39	248	46 07	270	28 48	308
169	48 52	020	46 08	106	29 59	143	57 24	212	29 56	249	45 22	270	28 12	308
170	49 07	019	46 52	107	30 26	144	56 59	214	29 14	250	44 37	271	27 37	308
171	49 22	019	47 35	108	30 52	145	56 33	215	28 32	250	43 51	272	27 01	309
172	49 36	019	48 18	109	31 18	146	56 07	217	27 49	251	43 06	272	26 26	309
173	49 51	019	49 01	110	31 43	147	55 39	218	27 06	252	42 21	273	25 51	310
174	50 06	019	49 43	111	32 07	148	55 11	220	26 23	253	41 36	274	25 16	310
175	50 20	018	50 26	112	32 31	149	54 41	221	25 39	254	40 51	274	24 42	310
176	50 34	018	51 08	113	32 54	150	54 11	223	24 56	254	40 05	275	24 07	311
177	50 48	018	51 49	114	33 16	151	53 40	224	24 12	255	39 20	275	23 33	311
178	51 02	018	52 31	115	33 37	152	53 08	226	23 28	256	38 35	276	22 59	312
179	51 16	018	53 12	116	33 58	153	52 35	227	22 44	257	37 50	277	22 26	312

LAT 41°N

LHA ♈	Hc	Zn	Hc	Zn	Hc	Zn	Hc	Zn	Hc	Zn	Hc	Zn	Hc	Zn
	Kochab		◆VEGA		ARCTURUS		◆SPICA		REGULUS		◆POLLUX		CAPELLA	
180	51 29	017	18 29	054	53 52	117	34 18	155	52 02	228	37 05	277	21 52	313
181	51 43	017	19 05	055	54 32	118	34 37	156	51 27	230	36 20	278	21 19	313
182	51 56	017	19 43	055	55 12	119	34 55	157	50 53	231	35 36	278	20 46	314
183	52 09	016	20 20	056	55 51	120	35 13	158	50 17	232	34 51	279	20 14	314
184	52 22	016	20 57	056	56 30	122	35 29	159	49 41	233	34 06	279	19 41	315
185	52 34	016	21 35	057	57 09	123	35 45	160	49 05	234	33 21	280	19 09	315
186	52 46	016	22 13	057	57 46	124	36 00	161	48 28	236	32 37	281	18 37	315
187	52 58	015	22 51	058	58 24	126	36 14	163	47 50	237	31 52	281	18 06	316
188	53 10	015	23 30	058	59 00	127	36 27	164	47 12	238	31 08	282	17 34	316
189	53 22	015	24 08	059	59 36	128	36 39	165	46 34	239	30 24	282	17 03	317
190	53 33	014	24 47	059	60 11	130	36 51	166	45 55	240	29 39	283	16 32	317
191	53 44	014	25 26	060	60 46	131	37 01	167	45 15	241	28 55	283	16 02	318
192	53 55	014	26 05	060	61 19	133	37 10	169	44 36	242	28 11	284	15 32	318
193	54 06	013	26 45	061	61 52	135	37 19	170	43 55	243	27 27	284	15 02	319
194	54 16	013	27 24	061	62 24	136	37 26	171	43 15	244	26 44	285	14 32	319
	◆VEGA		Rasalhague		ARCTURUS		◆SPICA		REGULUS		◆POLLUX		Dubhe	
195	28 04	062	24 14	094	62 55	138	37 33	172	42 34	245	26 00	286	62 51	330
196	28 44	062	25 00	095	63 24	140	37 38	174	41 53	246	25 16	286	62 28	329
197	29 24	062	25 45	096	63 53	142	37 43	175	41 11	247	24 33	287	62 04	329
198	30 04	063	26 30	096	64 21	143	37 47	176	40 30	248	23 50	287	61 40	328
199	30 44	063	27 15	097	64 47	145	37 49	177	39 48	249	23 07	288	61 16	327
200	31 25	064	28 00	098	65 12	147	37 51	178	39 05	249	22 24	288	60 52	327
201	32 06	064	28 44	099	65 36	149	37 52	180	38 23	250	21 41	289	60 27	327
202	32 46	065	29 29	099	65 58	152	37 52	181	37 40	251	20 58	290	60 02	326
203	33 27	065	30 14	100	66 19	154	37 50	182	36 57	252	20 15	290	59 36	326
204	34 09	066	30 58	101	66 38	156	37 48	183	36 14	253	19 33	291	59 11	325
205	34 50	066	31 43	101	66 56	158	37 45	185	35 31	254	18 50	291	58 45	325
206	35 31	066	32 27	102	67 12	161	37 41	186	34 47	254	18 08	292	58 18	325
207	36 13	067	33 11	103	67 26	163	37 36	187	34 03	255	17 26	292	57 52	324
208	36 55	067	33 55	104	67 38	165	37 29	188	33 20	256	16 45	293	57 26	324
209	37 37	068	34 39	105	67 49	168	37 22	190	32 36	257	16 03	293	56 59	324
	◆DENEB		VEGA		Rasalhague		ANTARES		◆ARCTURUS		REGULUS		◆Dubhe	
210	21 46	048	38 19	068	35 23	105	14 14	146	67 58	170	31 51	258	56 32	323
211	22 20	049	39 01	069	36 07	106	14 39	147	68 04	173	31 07	258	56 05	323
212	22 54	049	39 43	069	36 50	107	15 04	148	68 09	175	30 23	259	55 37	323
213	23 29	050	40 25	070	37 33	108	15 28	148	68 12	178	29 38	260	55 10	323
214	24 03	050	41 08	070	38 16	109	15 51	149	68 12	180	28 54	260	54 42	322
215	24 38	050	41 50	070	38 59	109	16 14	150	68 11	183	28 09	261	54 15	322
216	25 13	051	42 33	071	39 42	110	16 36	151	68 08	185	27 24	262	53 47	322
217	25 48	051	43 16	071	40 24	111	16 58	152	68 03	188	26 39	263	53 19	322
218	26 24	052	43 59	072	41 06	112	17 19	153	67 55	190	25 54	263	52 51	322
219	26 59	052	44 42	072	41 48	113	17 39	154	67 46	193	25 09	264	52 23	322
220	27 35	053	45 25	073	42 30	114	17 59	154	67 35	195	24 24	265	51 55	322
221	28 11	053	46 08	073	43 11	115	18 18	155	67 22	198	23 39	265	51 27	321
222	28 48	053	46 52	073	43 52	116	18 37	156	67 08	200	22 54	266	50 59	321
223	29 24	054	47 35	074	44 32	117	18 55	157	66 51	202	22 09	267	50 30	321
224	30 01	054	48 19	074	45 13	118	19 12	158	66 33	205	21 24	267	50 02	321
	DENEB		◆ALTAIR		Rasalhague		ANTARES		◆ARCTURUS		Denebola		◆Dubhe	
225	30 37	055	18 53	095	45 52	119	19 29	159	66 13	207	41 00	252	49 34	321
226	31 14	055	19 38	095	46 32	120	19 45	160	65 52	209	40 16	253	49 05	321
227	31 52	055	20 23	096	47 11	121	20 00	161	65 29	211	39 33	253	48 37	321
228	32 29	056	21 08	097	47 50	122	20 15	162	65 05	213	38 50	254	48 09	321
229	33 07	056	21 53	097	48 28	123	20 29	163	64 40	215	38 06	255	47 40	321
230	33 44	057	22 38	098	49 05	124	20 42	163	64 13	217	37 22	256	47 12	321
231	34 22	057	23 22	099	49 43	125	20 54	164	63 45	219	36 38	257	46 44	321
232	35 00	057	24 07	100	50 19	127	21 06	165	63 16	221	35 54	257	46 15	321
233	35 38	058	24 52	100	50 55	128	21 17	166	62 46	223	35 10	258	45 47	321
234	36 17	058	25 36	101	51 31	129	21 28	167	62 15	224	34 25	259	45 19	321
235	36 55	058	26 21	102	52 06	130	21 37	168	61 43	226	33 41	260	44 50	321
236	37 34	059	27 05	102	52 40	132	21 46	169	61 10	228	32 56	260	44 22	322
237	38 13	059	27 49	103	53 13	133	21 54	170	60 36	229	32 12	261	43 54	322
238	38 51	059	28 33	104	53 46	134	22 02	171	60 01	231	31 27	262	43 26	322
239	39 31	060	29 17	105	54 18	136	22 09	172	59 26	232	30 42	263	42 58	322
	◆DENEB		ALTAIR		Rasalhague		◆ANTARES		ARCTURUS		Denebola		◆Dubhe	
240	40 10	060	30 01	105	54 49	137	22 14	173	58 50	234	29 57	263	42 30	322
241	40 49	060	30 44	106	55 20	139	22 20	174	58 13	235	29 12	264	42 02	322
242	41 29	061	31 28	107	55 49	140	22 24	175	57 36	236	28 27	265	41 34	322
243	42 08	061	32 11	108	56 18	142	22 28	176	56 58	238	27 42	265	41 06	322
244	42 48	061	32 54	109	56 46	143	22 31	177	56 19	239	26 57	266	40 39	322
245	43 28	062	33 37	109	57 12	145	22 33	178	55 40	240	26 11	267	40 11	323
246	44 08	062	34 19	110	57 38	146	22 34	179	55 01	241	25 26	268	39 44	323
247	44 48	062	35 02	111	58 02	148	22 35	180	54 21	242	24 41	268	39 16	323
248	45 28	063	35 44	112	58 25	150	22 35	181	53 41	243	23 56	269	38 49	323
249	46 08	063	36 26	113	58 48	151	22 34	182	53 00	245	23 10	269	38 22	323
250	46 49	063	37 07	114	59 09	153	22 32	183	52 19	246	22 25	270	37 55	323
251	47 29	064	37 49	115	59 28	155	22 29	184	51 38	247	21 40	271	37 28	324
252	48 10	064	38 30	116	59 47	157	22 26	185	50 56	248	20 55	271	37 01	324
253	48 50	064	39 10	116	60 04	159	22 22	186	50 14	249	20 09	272	36 34	324
254	49 31	064	39 51	117	60 20	161	22 17	187	49 31	250	19 24	273	36 08	324
	◆DENEB		ALTAIR		Nunki		◆ANTARES		ARCTURUS		◆Alkaid		Kochab	
255	50 12	065	40 31	118	17 37	153	22 12	187	48 49	251	55 39	301	53 37	346
256	50 53	065	41 10	119	17 57	154	22 06	188	48 06	252	55 00	301	53 26	345
257	51 34	065	41 50	120	18 16	155	21 59	189	47 23	252	54 22	301	53 14	345
258	52 15	066	42 29	121	18 35	156	21 51	190	46 40	253	53 43	301	53 03	345
259	52 57	066	43 07	122	18 53	157	21 42	191	45 56	254	53 04	301	52 51	344
260	53 38	066	43 45	123	19 11	158	21 33	192	45 12	255	52 25	301	52 39	344
261	54 19	066	44 23	124	19 28	158	21 23	193	44 29	256	51 46	301	52 26	344
262	55 01	066	45 00	126	19 44	159	21 12	194	43 45	257	51 08	302	52 13	344
263	55 42	067	45 36	127	20 00	160	21 01	195	43 00	258	50 29	302	52 01	343
264	56 24	067	46 12	128	20 15	161	20 49	196	42 16	258	49 51	302	51 47	343
265	57 06	067	46 48	129	20 29	162	20 36	197	41 32	259	49 12	302	51 34	343
266	57 47	067	47 23	130	20 43	163	20 23	198	40 47	260	48 34	302	51 21	343
267	58 29	067	47 57	131	20 56	164	20 08	199	40 03	261	47 56	302	51 07	342
268	59 11	068	48 31	132	21 08	165	19 53	200	39 18	261	47 18	303	50 53	342
269	59 53	068	49 04	134	21 19	166	19 38	201	38 33	262	46 40	303	50 39	342

LHA ♈	Hc	Zn	Hc	Zn	Hc	Zn	Hc	Zn	Hc	Zn	Hc	Zn	Hc	Zn
	◆Alpheratz		ALTAIR		Nunki		◆ANTARES		ARCTURUS		◆Alkaid		Kochab	
270	17 11	066	49 37	135	21 30	167	19 22	202	37 48	263	46 02	303	50 25	342
271	17 52	067	50 08	136	21 40	168	19 05	202	37 03	264	45 24	303	50 11	341
272	18 34	067	50 39	138	21 49	169	18 47	203	36 18	264	44 46	304	49 56	341
273	19 16	068	51 09	139	21 58	170	18 29	204	35 33	265	44 08	304	49 42	341
274	19 58	068	51 39	140	22 06	171	18 10	205	34 48	266	43 31	304	49 27	341
275	20 40	069	52 07	142	22 13	172	17 50	206	34 03	266	42 53	304	49 12	341
276	21 22	069	52 35	143	22 19	173	17 30	207	33 17	267	42 16	305	48 57	341
277	22 04	070	53 01	145	22 24	173	17 10	208	32 32	268	41 39	305	48 42	340
278	22 47	071	53 27	146	22 29	174	16 48	209	31 47	269	41 02	305	48 27	340
279	23 30	071	53 52	148	22 33	175	16 26	209	31 02	269	40 25	305	48 11	340
280	24 13	072	54 16	149	22 36	176	16 04	210	30 16	270	39 48	306	47 56	340
281	24 56	072	54 38	151	22 39	177	15 41	211	29 31	271	39 11	306	47 40	340
282	25 39	073	55 00	152	22 41	178	15 17	212	28 46	271	38 35	306	47 25	340
283	26 22	073	55 20	154	22 42	179	14 53	213	28 01	272	37 58	307	47 09	340
284	27 06	074	55 39	156	22 42	180	14 28	214	27 15	272	37 22	307	46 53	340
	◆Mirfak		Alpheratz		◆ALTAIR		Rasalhague		◆ARCTURUS		Alkaid		Kochab	
285	12 26	032	27 49	074	55 58	157	55 59	219	26 30	273	36 46	307	46 37	339
286	12 51	033	28 33	075	56 14	159	55 30	221	25 45	274	36 10	308	46 21	339
287	13 15	033	29 17	075	56 30	161	55 00	222	25 00	274	35 34	308	46 05	339
288	13 40	034	30 01	076	56 44	162	54 29	224	24 15	275	34 58	308	45 49	339
289	14 06	034	30 45	077	56 57	164	53 57	225	23 29	276	34 23	309	45 33	339
290	14 32	035	31 29	077	57 09	166	53 25	227	22 44	276	33 48	309	45 17	339
291	14 58	035	32 13	078	57 19	168	52 52	228	21 59	277	33 12	309	45 01	339
292	15 24	036	32 57	078	57 28	170	52 18	229	21 14	277	32 37	310	44 44	339
293	15 51	036	33 42	079	57 36	171	51 43	230	20 30	278	32 03	310	44 28	339
294	16 18	037	34 26	079	57 42	173	51 08	232	19 45	279	31 28	310	44 12	339
295	16 45	037	35 11	080	57 46	175	50 32	233	19 00	279	30 54	311	43 55	339
296	17 12	038	35 55	080	57 49	177	49 55	234	18 15	280	30 19	311	43 39	339
297	17 40	038	36 40	081	57 51	179	49 18	235	17 31	281	29 45	311	43 23	339
298	18 08	039	37 25	082	57 51	181	48 41	236	16 46	281	29 11	312	43 06	339
299	18 37	039	38 10	082	57 50	183	48 03	238	16 02	282	28 38	312	42 50	339
	Mirfak		◆Alpheratz		Enif		◆ALTAIR		Rasalhague		Alphecca		◆Kochab	
300	19 05	040	38 54	083	51 19	136	57 47	184	47 25	239	34 25	277	42 34	339
301	19 34	040	39 39	083	51 50	138	57 43	186	46 46	240	33 40	278	42 17	339
302	20 04	040	40 24	084	52 20	139	57 37	188	46 06	241	32 55	278	42 01	339
303	20 33	041	41 09	085	52 50	140	57 30	190	45 27	242	32 10	279	41 45	339
304	21 03	041	41 55	085	53 18	142	57 22	192	44 46	243	31 26	280	41 28	339
305	21 33	042	42 40	086	53 45	143	57 12	194	44 06	244	30 41	280	41 12	339
306	22 03	042	43 25	086	54 12	145	57 01	195	43 25	245	29 57	281	40 56	339
307	22 34	043	44 10	087	54 38	146	56 48	197	42 44	246	29 12	281	40 40	339
308	23 05	043	44 55	088	55 02	148	56 34	199	42 03	247	28 28	282	40 23	339
309	23 36	044	45 41	088	55 26	150	56 19	201	41 21	248	27 43	282	40 07	339
310	24 07	044	46 26	089	55 48	151	56 02	202	40 39	249	26 59	283	39 51	339
311	24 39	044	47 11	089	56 09	153	55 44	204	39 57	249	26 15	284	39 35	339
312	25 10	045	47 56	090	56 29	154	55 26	206	39 14	250	25 31	284	39 19	339
313	25 42	045	48 42	091	56 48	156	55 05	207	38 31	251	24 47	285	39 03	339
314	26 15	046	49 27	091	57 06	158	54 44	209	37 48	252	24 04	285	38 47	340
	CAPELLA		◆Alpheratz		FOMALHAUT		◆ALTAIR		Rasalhague		VEGA		◆Kochab	
315	10 15	036	50 12	092	14 19	154	54 22	210	37 05	253	62 38	277	38 32	340
316	10 42	036	50 57	093	14 38	155	53 58	212	36 22	254	61 53	278	38 16	340
317	11 09	037	51 43	094	14 57	156	53 34	213	35 38	254	61 08	278	38 00	340
318	11 36	037	52 28	094	15 16	156	53 08	215	34 55	255	60 23	279	37 45	340
319	12 04	038	53 13	095	15 34	157	52 42	216	34 11	256	59 38	279	37 29	340
320	12 32	038	53 58	096	15 51	158	52 15	218	33 27	257	58 54	280	37 14	340
321	13 00	039	54 43	097	16 07	159	51 46	219	32 43	258	58 09	280	36 58	340
322	13 29	040	55 28	097	16 23	160	51 17	221	31 58	258	57 25	281	36 43	340
323	13 58	040	56 13	098	16 38	161	50 47	222	31 14	259	56 40	281	36 28	341
324	14 27	041	56 58	099	16 53	162	50 17	223	30 29	260	55 56	281	36 13	341
325	14 57	041	57 42	100	17 07	162	49 45	225	29 45	261	55 11	282	35 58	341
326	15 27	042	58 27	101	17 20	163	49 13	226	29 00	261	54 27	282	35 43	341
327	15 57	042	59 11	102	17 33	164	48 40	227	28 15	262	53 43	283	35 28	341
328	16 27	043	59 56	103	17 45	165	48 06	228	27 30	263	52 59	283	35 14	341
329	16 58	043	60 40	103	17 56	166	47 32	230	26 46	263	52 14	283	34 59	341
	◆CAPELLA		Hamal		Diphda		◆FOMALHAUT		ALTAIR		◆VEGA		Kochab	
330	17 29	044	36 05	089	19 53	139	18 07	167	46 58	231	51 30	284	34 45	342
331	18 01	044	36 50	090	20 23	139	18 17	168	46 22	232	50 47	284	34 30	342
332	18 32	044	37 36	090	20 52	140	18 26	169	45 46	233	50 03	285	34 16	342
333	19 04	045	38 21	091	21 21	141	18 34	170	45 10	234	49 19	285	34 02	342
334	19 36	045	39 06	092	21 49	142	18 42	171	44 33	235	48 35	286	33 48	342
335	20 08	046	39 52	092	22 16	143	18 49	171	43 55	236	47 52	286	33 35	342
336	20 41	046	40 37	093	22 43	144	18 56	172	43 17	237	47 08	286	33 21	343
337	21 14	047	41 22	094	23 10	145	19 01	173	42 39	238	46 25	287	33 07	343
338	21 47	047	42 07	094	23 35	146	19 06	174	42 00	239	45 42	287	32 54	343
339	22 20	048	42 52	095	24 00	147	19 11	175	41 21	240	44 58	288	32 41	343
340	22 54	048	43 37	096	24 25	148	19 14	176	40 41	241	44 15	288	32 28	343
341	23 28	049	44 22	096	24 49	149	19 17	177	40 01	242	43 32	289	32 15	344
342	24 02	049	45 07	097	25 12	150	19 19	178	39 21	243	42 50	289	32 02	344
343	24 36	049	45 52	098	25 35	151	19 20	179	38 41	244	42 07	289	31 49	344
344	25 11	050	46 37	099	25 57	151	19 21	180	38 00	245	41 24	290	31 37	344
	◆CAPELLA		ALDEBARAN		Diphda		◆FOMALHAUT		ALTAIR		◆VEGA		Kochab	
345	25 45	050	15 15	081	26 18	152	19 21	181	37 18	246	40 42	290	31 25	344
346	26 20	051	16 00	082	26 39	153	19 20	182	36 37	247	39 59	291	31 12	345
347	26 56	051	16 45	082	26 59	154	19 19	182	35 55	248	39 17	291	31 00	345
348	27 31	052	17 30	083	27 18	155	19 16	183	35 13	249	38 35	292	30 49	345
349	28 07	052	18 15	084	27 36	156	19 13	184	34 31	250	37 53	292	30 37	345
350	28 42	052	19 00	084	27 54	157	19 09	185	33 48	250	37 11	292	30 26	345
351	29 18	053	19 45	085	28 11	159	19 05	186	33 05	251	36 29	293	30 14	346
352	29 54	053	20 30	086	28 27	160	19 00	187	32 22	252	35 47	293	30 03	346
353	30 31	054	21 15	086	28 42	161	18 54	188	31 39	253	35 06	294	29 52	346
354	31 07	054	22 00	087	28 57	162	18 47	189	30 56	254	34 24	294	29 41	346
355	31 44	054	22 46	088	29 11	163	18 40	190	30 12	254	33 43	295	29 31	347
356	32 21	055	23 31	088	29 24	164	18 32	191	29 29	255	33 02	295	29 20	347
357	32 58	055	24 16	089	29 36	165	18 23	192	28 45	256	32 21	296	29 10	347
358	33 35	055	25 01	089	29 48	166	18 14	192	28 01	257	31 41	296	29 00	347
359	34 13	056	25 47	090	29 58	167	18 04	193	27 17	257	31 00	296	28 50	348

LHA ♈	Hc	Zn	Hc	Zn	Hc	Zn	Hc	Zn	Hc	Zn	Hc	Zn	Hc	Zn
	◆CAPELLA		ALDEBARAN		◆Diphda		FOMALHAUT		ALTAIR		◆VEGA		Kochab	
0	34 16	056	26 33	090	31 07	168	18 51	194	26 45	259	29 52	297	27 42	348
1	34 54	056	27 19	091	31 16	169	18 39	195	26 00	259	29 11	298	27 33	348
2	35 33	056	28 05	092	31 24	170	18 27	196	25 14	260	28 31	298	27 23	348
3	36 11	057	28 50	092	31 32	171	18 14	197	24 29	261	27 50	299	27 14	349
4	36 49	057	29 36	093	31 38	172	18 00	198	23 44	261	27 10	299	27 05	349
5	37 28	057	30 22	094	31 44	173	17 45	199	22 58	262	26 30	300	26 57	349
6	38 07	058	31 08	094	31 49	175	17 30	200	22 13	263	25 50	300	26 48	349
7	38 46	058	31 54	095	31 53	176	17 14	201	21 27	263	25 11	301	26 40	350
8	39 25	058	32 40	096	31 56	177	16 58	201	20 41	264	24 31	301	26 32	350
9	40 04	059	33 25	096	31 58	178	16 41	202	19 56	265	23 52	302	26 24	350
10	40 43	059	34 11	097	31 59	179	16 23	203	19 10	265	23 13	302	26 16	351
11	41 23	059	34 57	098	31 59	180	16 05	204	18 24	266	22 34	302	26 09	351
12	42 02	059	35 42	098	31 59	181	15 46	205	17 38	267	21 55	303	26 02	351
13	42 42	060	36 28	099	31 57	182	15 26	206	16 52	267	21 17	303	25 55	351
14	43 22	060	37 13	100	31 55	184	15 06	206	16 06	268	20 39	304	25 48	352
	◆CAPELLA		BETELGEUSE		RIGEL		◆Diphda		Enif		◆DENEB		Kochab	
15	44 01	060	17 12	095	14 13	114	31 51	185	37 14	249	43 24	299	25 41	352
16	44 41	061	17 58	095	14 55	115	31 47	186	36 31	250	42 44	299	25 35	352
17	45 22	061	18 44	096	15 37	115	31 42	187	35 47	251	42 04	299	25 29	352
18	46 02	061	19 29	097	16 18	116	31 36	188	35 04	252	41 24	300	25 23	353
19	46 42	061	20 15	097	16 59	117	31 29	189	34 20	252	40 44	300	25 17	353
20	47 22	062	21 00	098	17 40	118	31 22	190	33 36	253	40 05	300	25 12	353
21	48 03	062	21 46	099	18 21	118	31 13	191	32 52	254	39 25	301	25 06	354
22	48 44	062	22 31	100	19 01	119	31 04	192	32 08	255	38 45	301	25 01	354
23	49 24	062	23 17	100	19 41	120	30 53	194	31 23	256	38 06	301	24 57	354
24	50 05	063	24 02	101	20 21	121	30 42	195	30 39	256	37 27	302	24 52	354
25	50 46	063	24 47	102	21 00	121	30 30	196	29 54	257	36 48	302	24 48	355
26	51 27	063	25 32	102	21 39	122	30 17	197	29 09	258	36 09	302	24 44	355
27	52 08	063	26 17	103	22 18	123	30 04	198	28 24	258	35 30	303	24 40	355
28	52 49	063	27 01	104	22 56	124	29 49	199	27 39	259	34 52	303	24 36	356
29	53 30	064	27 46	105	23 34	125	29 34	200	26 54	260	34 13	303	24 33	356
	◆CAPELLA		BETELGEUSE		RIGEL		◆Diphda		Alpheratz		◆DENEB		Kochab	
30	54 11	064	28 30	105	24 12	126	29 18	201	64 38	253	33 35	304	24 30	356
31	54 52	064	29 15	106	24 49	126	29 01	202	63 54	254	32 57	304	24 27	357
32	55 34	064	29 59	107	25 26	127	28 43	203	63 10	255	32 19	305	24 24	357
33	56 15	064	30 42	108	26 02	128	28 25	204	62 25	256	31 41	305	24 22	357
34	56 57	064	31 26	108	26 38	129	28 06	205	61 41	257	31 03	305	24 20	357
35	57 38	065	32 10	109	27 14	130	27 46	206	60 56	258	30 26	306	24 18	358
36	58 20	065	32 53	110	27 49	131	27 25	207	60 11	259	29 49	306	24 16	358
37	59 01	065	33 36	111	28 23	132	27 04	208	59 25	260	29 12	306	24 14	358
38	59 43	065	34 19	112	28 57	133	26 42	209	58 40	261	28 35	307	24 13	359
39	60 24	065	35 02	112	29 31	134	26 19	210	57 55	262	27 58	307	24 12	359
40	61 06	065	35 44	113	30 04	135	25 56	211	57 09	262	27 22	308	24 11	359
41	61 48	065	36 26	114	30 36	135	25 32	212	56 24	263	26 45	308	24 11	359
42	62 29	065	37 08	115	31 08	136	25 07	213	55 38	264	26 09	308	24 11	000
43	63 11	065	37 49	116	31 40	137	24 42	214	54 52	265	25 33	309	24 11	000
44	63 53	065	38 30	117	32 11	138	24 16	215	54 06	266	24 57	309	24 11	000
	◆Dubhe		POLLUX		SIRIUS		◆RIGEL		Diphda		◆Alpheratz		DENEB	
45	22 22	026	31 18	078	12 53	125	32 41	139	23 50	216	53 20	266	24 22	310
46	22 42	026	32 03	079	13 31	126	33 10	140	23 23	217	52 35	267	23 47	310
47	23 03	027	32 48	079	14 08	127	33 39	141	22 55	218	51 49	268	23 12	310
48	23 24	027	33 33	080	14 44	128	34 08	143	22 27	218	51 03	268	22 37	311
49	23 45	027	34 18	080	15 20	128	34 35	144	21 58	219	50 17	269	22 02	311
50	24 06	028	35 04	081	15 56	129	35 02	145	21 28	220	49 31	270	21 28	312
51	24 28	028	35 49	081	16 32	130	35 28	146	20 58	221	48 45	270	20 54	312
52	24 50	028	36 34	082	17 07	131	35 54	147	20 28	222	47 59	271	20 20	313
53	25 12	029	37 20	082	17 42	131	36 19	148	19 57	223	47 13	272	19 46	313
54	25 34	029	38 06	083	18 16	132	36 43	149	19 25	224	46 27	272	19 13	314
55	25 56	029	38 51	084	18 50	133	37 06	150	18 54	224	45 41	273	18 39	314
56	26 19	030	39 37	084	19 23	134	37 28	151	18 21	225	44 55	273	18 07	315
57	26 42	030	40 23	085	19 56	135	37 50	153	17 48	226	44 09	274	17 34	315
58	27 05	030	41 08	085	20 28	136	38 11	154	17 15	227	43 24	275	17 02	315
59	27 28	031	41 54	086	21 00	136	38 30	155	16 41	228	42 38	275	16 29	316
	◆Dubhe		POLLUX		PROCYON		◆SIRIUS		RIGEL		◆Hamal		Schedar	
60	27 52	031	42 40	086	29 55	110	21 32	137	38 50	156	61 00	244	53 55	314
61	28 16	031	43 26	087	30 38	111	22 03	138	39 08	157	60 19	245	53 22	314
62	28 40	032	44 12	088	31 21	112	22 33	139	39 25	159	59 37	246	52 49	314
63	29 04	032	44 58	088	32 04	113	23 03	140	39 41	160	58 55	247	52 16	314
64	29 28	032	45 44	089	32 46	113	23 32	141	39 57	161	58 12	248	51 42	314
65	29 53	032	46 30	090	33 28	114	24 01	142	40 11	162	57 29	249	51 09	314
66	30 17	033	47 16	090	34 10	115	24 29	143	40 25	164	56 46	251	50 36	314
67	30 42	033	48 02	091	34 51	116	24 57	144	40 37	165	56 03	252	50 03	314
68	31 07	033	48 48	092	35 32	117	25 24	144	40 49	166	55 19	253	49 30	314
69	31 33	033	49 34	092	36 13	118	25 50	145	40 59	167	54 35	254	48 57	314
70	31 58	034	50 20	093	36 54	119	26 16	146	41 09	169	53 51	254	48 24	314
71	32 24	034	51 05	094	37 34	120	26 41	147	41 17	170	53 06	255	47 51	314
72	32 49	034	51 51	094	38 14	121	27 06	148	41 25	171	52 22	256	47 18	314
73	33 15	034	52 37	095	38 53	122	27 29	149	41 31	173	51 37	257	46 45	314
74	33 41	035	53 23	096	39 32	123	27 53	150	41 36	174	50 52	258	46 12	314
	◆Dubhe		POLLUX		PROCYON		◆SIRIUS		RIGEL		◆Hamal		Schedar	
75	34 07	035	54 09	097	40 10	124	28 15	151	41 41	175	50 07	259	45 39	314
76	34 34	035	54 54	097	40 48	125	28 37	152	41 44	177	49 22	260	45 06	314
77	35 00	035	55 40	098	41 26	126	28 58	153	41 46	178	48 37	260	44 33	315
78	35 27	035	56 25	099	42 03	127	29 18	154	41 47	179	47 51	261	44 01	315
79	35 53	036	57 11	100	42 40	128	29 38	155	41 47	181	47 06	262	43 28	315
80	36 20	036	57 56	101	43 16	129	29 56	156	41 46	182	46 20	263	42 55	315
81	36 47	036	58 41	101	43 52	130	30 14	157	41 44	183	45 35	263	42 23	315
82	37 14	036	59 26	102	44 27	131	30 32	159	41 41	185	44 49	264	41 51	315
83	37 42	036	60 11	103	45 01	132	30 48	160	41 37	186	44 03	265	41 18	315
84	38 09	037	60 55	104	45 35	133	31 04	161	41 32	187	43 17	266	40 46	316
85	38 36	037	61 40	105	46 08	135	31 19	162	41 26	188	42 32	266	40 14	316
86	39 04	037	62 24	106	46 40	136	31 32	163	41 18	190	41 46	267	39 42	316
87	39 31	037	63 08	107	47 12	137	31 46	164	41 10	191	41 00	268	39 10	316
88	39 59	037	63 52	108	47 43	138	31 58	165	41 01	192	40 14	268	38 38	316
89	40 27	037	64 35	110	48 13	139	32 09	166	40 50	194	39 28	269	38 06	317

LAT 40°N

LHA ♈	Hc	Zn	Hc	Zn	Hc	Zn	Hc	Zn	Hc	Zn	Hc	Zn	Hc	Zn
	◆Dubhe		REGULUS		PROCYON		◆SIRIUS		RIGEL		ALDEBARAN		◆Mirfak	
90	40 55	037	29 01	099	48 43	141	32 20	167	40 39	195	60 13	224	61 04	303
91	41 23	037	29 46	100	49 11	142	32 30	168	40 27	196	59 40	226	60 25	303
92	41 51	038	30 32	100	49 39	143	32 39	169	40 13	198	59 07	227	59 46	303
93	42 19	038	31 17	101	50 06	145	32 46	171	39 59	199	58 33	229	59 08	303
94	42 47	038	32 02	102	50 32	146	32 54	172	39 44	200	57 58	230	58 29	303
95	43 15	038	32 47	103	50 57	148	33 00	173	39 27	201	57 22	231	57 50	303
96	43 43	038	33 31	103	51 21	149	33 05	174	39 10	202	56 46	233	57 12	303
97	44 11	038	34 16	104	51 44	151	33 09	175	38 52	204	56 09	234	56 33	303
98	44 40	038	35 01	105	52 06	152	33 13	176	38 33	205	55 32	235	55 54	303
99	45 08	038	35 45	106	52 27	154	33 15	177	38 14	206	54 53	237	55 16	303
100	45 36	038	36 29	106	52 47	155	33 17	179	37 53	207	54 15	238	54 37	303
101	46 05	038	37 13	107	53 06	157	33 17	180	37 31	208	53 36	239	53 58	303
102	46 33	038	37 57	108	53 23	158	33 17	181	37 09	210	52 56	240	53 20	303
103	47 01	038	38 40	109	53 40	160	33 16	182	36 46	211	52 16	241	52 41	303
104	47 30	038	39 24	110	53 55	162	33 14	183	36 22	212	51 35	243	52 03	303
	◆Kochab		Denebola		◆REGULUS		SIRIUS		RIGEL		◆ALDEBARAN		Mirfak	
105	31 26	016	22 52	090	40 07	111	33 11	184	35 58	213	50 54	244	51 24	303
106	31 39	017	23 38	090	40 50	112	33 07	185	35 32	214	50 13	245	50 46	303
107	31 52	017	24 24	091	41 32	112	33 02	187	35 06	215	49 31	246	50 08	303
108	32 05	017	25 10	092	42 15	113	32 56	188	34 39	216	48 49	247	49 29	304
109	32 19	017	25 56	092	42 57	114	32 50	189	34 12	217	48 07	248	48 51	304
110	32 32	017	26 42	093	43 39	115	32 42	190	33 44	218	47 24	249	48 13	304
111	32 46	018	27 28	094	44 20	116	32 34	191	33 15	219	46 41	250	47 35	304
112	33 00	018	28 14	094	45 01	117	32 25	192	32 45	220	45 58	250	46 57	304
113	33 14	018	28 59	095	45 42	118	32 14	193	32 15	221	45 15	251	46 19	304
114	33 28	018	29 45	096	46 22	119	32 03	194	31 44	222	44 31	252	45 41	305
115	33 43	018	30 31	096	47 02	120	31 51	196	31 13	223	43 47	253	45 03	305
116	33 57	018	31 16	097	47 42	121	31 39	197	30 41	224	43 03	254	44 26	305
117	34 12	019	32 02	098	48 21	122	31 25	198	30 09	225	42 19	255	43 48	305
118	34 26	019	32 48	099	48 59	123	31 11	199	29 36	226	41 34	256	43 10	306
119	34 41	019	33 33	099	49 37	125	30 55	200	29 02	227	40 50	256	42 33	306
	◆Kochab		Denebola		◆REGULUS		SIRIUS		RIGEL		◆ALDEBARAN		CAPELLA	
120	34 56	019	34 18	100	50 15	126	30 39	201	28 28	228	40 05	257	59 49	295
121	35 11	019	35 03	101	50 52	127	30 22	202	27 54	229	39 20	258	59 08	295
122	35 26	019	35 49	102	51 28	128	30 05	203	27 19	230	38 35	259	58 26	295
123	35 41	019	36 34	102	52 04	129	29 46	204	26 44	231	37 50	260	57 45	295
124	35 56	019	37 18	103	52 39	131	29 27	205	26 08	232	37 05	260	57 03	296
125	36 12	020	38 03	104	53 14	132	29 07	206	25 31	233	36 19	261	56 22	296
126	36 27	020	38 48	105	53 48	133	28 47	207	24 55	233	35 34	262	55 40	296
127	36 43	020	39 32	105	54 21	135	28 25	208	24 17	234	34 48	262	54 59	296
128	36 58	020	40 16	106	54 53	136	28 03	209	23 40	235	34 03	263	54 18	296
129	37 14	020	41 00	107	55 24	138	27 40	210	23 02	236	33 17	264	53 37	296
130	37 30	020	41 44	108	55 55	139	27 17	211	22 24	237	32 31	265	52 55	297
131	37 45	020	42 28	109	56 24	141	26 53	212	21 45	238	31 46	265	52 14	297
132	38 01	020	43 11	110	56 53	142	26 28	213	21 06	238	31 00	266	51 33	297
133	38 17	020	43 54	111	57 21	144	26 02	214	20 27	239	30 14	267	50 52	297
134	38 33	020	44 37	111	57 47	145	25 36	215	19 47	240	29 28	267	50 12	297
	Kochab		◆ARCTURUS		SPICA		REGULUS		◆SIRIUS		BETELGEUSE		◆CAPELLA	
135	38 49	020	20 34	082	10 18	114	58 13	147	25 09	216	37 26	244	49 31	298
136	39 05	021	21 19	082	11 00	115	58 37	149	24 42	217	36 45	245	48 50	298
137	39 21	021	22 05	083	11 41	116	59 01	150	24 14	218	36 03	246	48 09	298
138	39 38	021	22 51	084	12 23	116	59 23	152	23 46	219	35 21	247	47 29	298
139	39 54	021	23 36	084	13 04	117	59 43	154	23 17	220	34 38	248	46 49	299
140	40 10	021	24 22	085	13 45	118	60 03	156	22 47	221	33 55	249	46 08	299
141	40 26	021	25 08	085	14 25	118	60 21	158	22 17	221	33 12	250	45 28	299
142	40 43	021	25 54	086	15 05	119	60 38	160	21 46	222	32 29	250	44 48	299
143	40 59	021	26 40	087	15 45	120	60 53	162	21 15	223	31 46	251	44 08	300
144	41 15	021	27 25	087	16 25	121	61 07	164	20 43	224	31 02	252	43 28	300
145	41 32	021	28 11	088	17 05	121	61 19	166	20 11	225	30 18	253	42 48	300
146	41 48	021	28 57	088	17 44	122	61 29	168	19 38	226	29 34	254	42 08	300
147	42 04	021	29 43	089	18 22	123	61 39	170	19 05	227	28 50	254	41 29	301
148	42 21	021	30 29	090	19 01	124	61 46	172	18 32	227	28 06	255	40 49	301
149	42 37	021	31 15	090	19 39	125	61 52	174	17 58	228	27 21	256	40 10	301
	◆Kochab		ARCTURUS		◆SPICA		REGULUS		PROCYON		◆POLLUX		CAPELLA	
150	42 54	021	32 01	091	20 16	125	61 56	176	42 59	232	59 49	257	39 31	302
151	43 10	021	32 47	092	20 54	126	61 59	178	42 22	233	59 04	258	38 52	302
152	43 26	021	33 33	092	21 31	127	62 00	180	41 46	234	58 19	259	38 13	302
153	43 42	021	34 19	093	22 07	128	61 59	182	41 08	235	57 33	260	37 34	303
154	43 59	021	35 05	094	22 43	129	61 56	184	40 30	236	56 48	261	36 56	303
155	44 15	021	35 51	094	23 19	130	61 52	186	39 52	237	56 03	262	36 17	303
156	44 31	021	36 36	095	23 54	130	61 46	188	39 13	238	55 17	262	35 39	304
157	44 47	020	37 22	096	24 29	131	61 39	190	38 34	239	54 32	263	35 01	304
158	45 03	020	38 08	096	25 03	132	61 30	192	37 55	240	53 46	264	34 23	304
159	45 19	020	38 54	097	25 37	133	61 19	194	37 15	241	53 00	265	33 45	305
160	45 35	020	39 39	098	26 10	134	61 07	196	36 34	242	52 14	265	33 07	305
161	45 51	020	40 25	099	26 43	135	60 54	198	35 54	243	51 29	266	32 29	305
162	46 07	020	41 10	099	27 16	136	60 38	200	35 13	243	50 43	267	31 52	306
163	46 23	020	41 55	100	27 48	137	60 22	202	34 31	244	49 57	267	31 15	306
164	46 38	020	42 41	101	28 19	138	60 04	204	33 50	245	49 11	268	30 38	306
	◆Kochab		ARCTURUS		◆SPICA		REGULUS		PROCYON		◆POLLUX		CAPELLA	
165	46 54	020	43 26	102	28 50	139	59 44	206	33 08	246	48 25	269	30 01	307
166	47 09	020	44 11	103	29 20	140	59 24	208	32 26	247	47 39	269	29 24	307
167	47 25	019	44 55	103	29 49	140	59 02	209	31 43	248	46 53	270	28 48	308
168	47 40	019	45 40	104	30 18	141	58 39	211	31 01	249	46 07	271	28 11	308
169	47 55	019	46 24	105	30 46	142	58 14	213	30 18	249	45 21	271	27 35	308
170	48 10	019	47 09	106	31 14	143	57 49	215	29 35	250	44 35	272	26 59	309
171	48 25	019	47 53	107	31 41	145	57 22	216	28 51	251	43 49	273	26 23	309
172	48 40	019	48 37	108	32 07	146	56 55	218	28 08	252	43 03	273	25 48	310
173	48 54	018	49 20	109	32 33	147	56 26	219	27 24	252	42 17	274	25 13	310
174	49 09	018	50 04	109	32 58	148	55 56	221	26 40	253	41 32	274	24 38	310
175	49 23	018	50 47	110	33 22	149	55 26	222	25 56	254	40 46	275	24 03	311
176	49 37	018	51 30	111	33 46	150	54 55	224	25 12	255	40 00	276	23 28	311
177	49 51	018	52 13	112	34 09	151	54 22	225	24 27	255	39 14	276	22 54	312
178	50 05	017	52 55	113	34 31	152	53 49	227	23 43	256	38 29	277	22 19	312
179	50 19	017	53 37	115	34 52	153	53 16	228	22 58	257	37 43	277	21 45	313

LAT 40°N

LHA ♈	Hc	Zn	Hc	Zn	Hc	Zn	Hc	Zn	Hc	Zn	Hc	Zn	Hc	Zn
	◆VEGA		Alphecca		ARCTURUS		◆SPICA		REGULUS		◆POLLUX		CAPELLA	
180	17 53	054	44 01	090	54 19	116	35 12	154	52 41	229	36 57	278	21 12	313
181	18 31	054	44 47	090	55 00	117	35 32	155	52 06	230	36 12	278	20 38	313
182	19 08	055	45 33	091	55 41	118	35 50	157	51 30	232	35 27	279	20 05	314
183	19 46	055	46 19	092	56 21	119	36 08	158	50 54	233	34 41	280	19 32	314
184	20 24	056	47 05	092	57 01	120	36 25	159	50 17	234	33 56	280	18 59	315
185	21 02	056	47 51	093	57 41	122	36 42	160	49 39	235	33 11	281	18 27	315
186	21 40	057	48 37	094	58 19	123	36 57	161	49 01	236	32 26	281	17 54	316
187	22 19	057	49 23	094	58 58	124	37 11	162	48 23	238	31 40	282	17 22	316
188	22 58	058	50 09	095	59 36	126	37 25	164	47 44	239	30 56	282	16 51	317
189	23 37	058	50 54	096	60 13	127	37 37	165	47 04	240	30 11	283	16 19	317
190	24 16	059	51 40	097	60 49	128	37 49	166	46 24	241	29 26	283	15 48	318
191	24 55	059	52 26	097	61 25	130	37 59	167	45 44	242	28 41	284	15 17	318
192	25 35	060	53 11	098	61 59	132	38 09	168	45 03	243	27 57	284	14 47	319
193	26 15	060	53 57	099	62 33	133	38 18	170	44 22	244	27 12	285	14 16	319
194	26 55	061	54 42	100	63 06	135	38 26	171	43 41	245	26 28	286	13 46	320
	DENEB		◆VEGA		ARCTURUS		◆SPICA		REGULUS		◆POLLUX		Dubhe	
195	13 04	041	27 35	061	63 39	137	38 32	172	42 59	246	25 44	286	61 59	331
196	13 34	041	28 15	061	64 10	138	38 38	173	42 17	247	25 00	287	61 36	330
197	14 04	042	28 56	062	64 40	140	38 43	175	41 35	248	24 16	287	61 13	329
198	14 35	042	29 36	062	65 08	142	38 47	176	40 52	248	23 32	288	60 49	329
199	15 06	043	30 17	063	65 36	144	38 49	177	40 09	249	22 48	288	60 25	328
200	15 38	043	30 58	063	66 02	146	38 51	178	39 26	250	22 04	289	60 01	328
201	16 09	044	31 39	064	66 27	148	38 52	180	38 43	251	21 21	289	59 37	327
202	16 41	044	32 21	064	66 51	150	38 52	181	37 59	252	20 38	290	59 12	327
203	17 13	045	33 02	065	67 13	153	38 50	182	37 15	253	19 54	290	58 46	327
204	17 46	045	33 44	065	67 33	155	38 48	183	36 31	254	19 11	291	58 21	326
205	18 19	046	34 25	065	67 51	157	38 45	185	35 47	254	18 29	292	57 55	326
206	18 52	046	35 07	066	68 08	160	38 40	186	35 03	255	17 46	292	57 29	325
207	19 25	047	35 49	066	68 23	162	38 35	187	34 18	256	17 03	293	57 03	325
208	19 58	047	36 31	067	68 36	165	38 29	188	33 34	257	16 21	293	56 37	325
209	20 32	047	37 14	067	68 48	167	38 22	190	32 49	257	15 39	294	56 10	324
	DENEB		◆VEGA		Rasalhague		ANTARES		◆SPICA		REGULUS		◆Dubhe	
210	21 06	048	37 56	068	35 39	105	15 04	146	38 13	191	32 04	258	55 43	324
211	21 40	048	38 38	068	36 23	105	15 29	147	38 04	192	31 19	259	55 16	324
212	22 15	049	39 21	068	37 07	106	15 54	147	37 54	193	30 34	260	54 49	324
213	22 50	049	40 04	069	37 51	107	16 19	148	37 43	195	29 49	260	54 22	323
214	23 25	050	40 47	069	38 35	108	16 43	149	37 31	196	29 03	261	53 55	323
215	24 00	050	41 30	070	39 19	109	17 06	150	37 18	197	28 18	262	53 27	323
216	24 35	051	42 13	070	40 02	110	17 29	151	37 04	198	27 32	262	52 59	323
217	25 11	051	42 56	070	40 45	110	17 51	152	36 49	199	26 47	263	52 32	323
218	25 46	051	43 39	071	41 28	111	18 12	153	36 33	201	26 01	264	52 04	323
219	26 22	052	44 23	071	42 11	112	18 33	153	36 16	202	25 15	264	51 36	322
220	26 59	052	45 06	072	42 53	113	18 53	154	35 59	203	24 30	265	51 08	322
221	27 35	053	45 50	072	43 36	114	19 13	155	35 41	204	23 44	266	50 40	322
222	28 12	053	46 34	072	44 17	115	19 32	156	35 21	205	22 58	267	50 11	322
223	28 48	053	47 18	073	44 59	116	19 50	157	35 01	206	22 12	267	49 43	322
224	29 25	054	48 02	073	45 40	117	20 08	158	34 40	207	21 26	268	49 15	322
	DENEB		◆ALTAIR		Rasalhague		ANTARES		◆SPICA		Denebola		◆Dubhe	
225	30 03	054	18 57	094	46 21	118	20 25	159	34 19	209	41 18	253	48 47	322
226	30 40	055	19 43	095	47 01	119	20 41	160	33 56	210	40 34	253	48 18	322
227	31 17	055	20 29	096	47 41	120	20 57	161	33 33	211	39 50	254	47 50	322
228	31 55	055	21 15	096	48 21	121	21 12	162	33 09	212	39 06	255	47 22	322
229	32 33	056	22 00	097	49 00	122	21 26	162	32 45	213	38 21	256	46 53	322
230	33 11	056	22 46	098	49 39	123	21 39	163	32 19	214	37 36	257	46 25	322
231	33 49	056	23 31	098	50 17	124	21 52	164	31 53	215	36 52	257	45 57	322
232	34 27	057	24 17	099	50 55	126	22 04	165	31 27	216	36 07	258	45 28	322
233	35 06	057	25 02	100	51 32	127	22 16	166	30 59	217	35 22	259	45 00	322
234	35 45	057	25 47	101	52 08	128	22 26	167	30 31	218	34 36	260	44 32	322
235	36 23	058	26 33	101	52 44	129	22 36	168	30 03	219	33 51	260	44 03	322
236	37 02	058	27 18	102	53 19	131	22 45	169	29 33	220	33 06	261	43 35	322
237	37 41	058	28 03	103	53 54	132	22 54	170	29 04	221	32 20	262	43 07	322
238	38 21	059	28 47	103	54 28	133	23 01	171	28 33	222	31 35	263	42 39	322
239	39 00	059	29 32	104	55 01	135	23 08	172	28 02	223	30 49	263	42 11	322
	◆DENEB		ALTAIR		Nunki		◆ANTARES		ARCTURUS		◆Alkaid		Kochab	
240	39 40	059	30 16	105	12 12	141	23 14	173	59 25	235	64 53	303	54 52	352
241	40 19	060	31 01	106	12 41	141	23 19	174	58 47	236	64 14	303	54 46	351
242	40 59	060	31 45	106	13 09	142	23 24	175	58 09	238	63 36	302	54 39	351
243	41 39	060	32 29	107	13 37	143	23 28	176	57 29	239	62 57	302	54 31	351
244	42 19	061	33 13	108	14 05	144	23 31	177	56 50	240	62 18	302	54 24	350
245	42 59	061	33 56	109	14 32	145	23 33	178	56 10	241	61 39	302	54 16	350
246	43 39	061	34 40	110	14 58	145	23 34	179	55 29	242	61 00	302	54 07	349
247	44 20	062	35 23	110	15 24	146	23 35	180	54 48	244	60 21	302	53 59	349
248	45 00	062	36 06	111	15 49	147	23 35	181	54 07	245	59 42	302	53 50	349
249	45 41	062	36 49	112	16 14	148	23 34	182	53 25	246	59 03	302	53 40	348
250	46 21	062	37 31	113	16 38	149	23 32	183	52 43	247	58 24	302	53 31	348
251	47 02	063	38 13	114	17 02	150	23 29	184	52 01	248	57 45	302	53 21	347
252	47 43	063	38 55	115	17 25	150	23 26	185	51 18	249	57 05	302	53 11	347
253	48 24	063	39 37	116	17 47	151	23 22	186	50 35	250	56 26	302	53 01	347
254	49 05	063	40 18	117	18 09	152	23 17	187	49 52	251	55 47	302	52 50	346
	◆DENEB		ALTAIR		Nunki		◆ANTARES		ARCTURUS		◆Alkaid		Kochab	
255	49 46	064	40 59	118	18 30	153	23 11	188	49 08	252	55 08	302	52 39	346
256	50 27	064	41 39	119	18 51	154	23 05	188	48 24	253	54 29	302	52 28	346
257	51 09	064	42 20	120	19 11	155	22 58	189	47 40	254	53 50	302	52 16	345
258	51 50	064	42 59	121	19 30	156	22 50	190	46 56	254	53 11	302	52 05	345
259	52 31	065	43 39	122	19 48	157	22 41	191	46 12	255	52 32	302	51 53	345
260	53 13	065	44 18	123	20 06	157	22 32	192	45 27	256	51 54	302	51 41	345
261	53 55	065	44 56	124	20 24	158	22 22	193	44 43	257	51 15	302	51 28	344
262	54 36	065	45 34	125	20 40	159	22 11	194	43 58	258	50 36	303	51 16	344
263	55 18	065	46 12	126	20 56	160	21 59	195	43 13	259	49 57	303	51 03	344
264	56 00	066	46 49	127	21 12	161	21 47	196	42 28	259	49 19	303	50 50	343
265	56 42	066	47 25	128	21 26	162	21 33	197	41 42	260	48 40	303	50 37	343
266	57 24	066	48 01	129	21 40	163	21 20	198	40 57	261	48 02	303	50 23	343
267	58 06	066	48 36	130	21 53	164	21 05	199	40 12	262	47 23	303	50 10	343
268	58 48	066	49 11	132	22 06	165	20 50	200	39 26	262	46 45	304	49 56	342
269	59 30	066	49 45	133	22 17	166	20 34	201	38 41	263	46 07	304	49 42	342

LHA ♈	Hc	Zn	Hc	Zn	Hc	Zn	Hc	Zn	Hc	Zn	Hc	Zn	Hc	Zn
	◆Alpheratz		ALTAIR		Nunki		◆ANTARES		ARCTURUS		◆Alkaid		Kochab	
270	16 46	066	50 19	134	22 28	167	20 17	202	37 55	264	45 28	304	49 28	342
271	17 28	066	50 51	135	22 39	168	20 00	203	37 09	264	44 50	304	49 14	342
272	18 10	067	51 23	137	22 48	169	19 42	203	36 24	265	44 12	304	48 59	342
273	18 53	067	51 54	138	22 57	170	19 23	204	35 38	266	43 35	305	48 45	341
274	19 35	068	52 24	140	23 05	171	19 04	205	34 52	266	42 57	305	48 30	341
275	20 18	069	52 54	141	23 12	171	18 44	206	34 06	267	42 19	305	48 15	341
276	21 01	069	53 22	142	23 18	172	18 24	207	33 20	268	41 42	305	48 00	341
277	21 44	070	53 50	144	23 24	173	18 03	208	32 34	268	41 04	306	47 45	341
278	22 27	070	54 17	145	23 29	174	17 41	209	31 48	269	40 27	306	47 30	341
279	23 10	071	54 42	147	23 33	175	17 18	210	31 02	270	39 50	306	47 15	340
280	23 54	071	55 07	148	23 36	176	16 56	210	30 16	270	39 13	306	46 59	340
281	24 37	072	55 30	150	23 39	177	16 32	211	29 30	271	38 36	307	46 44	340
282	25 21	072	55 53	152	23 41	178	16 08	212	28 44	272	37 59	307	46 28	340
283	26 05	073	56 14	153	23 42	179	15 43	213	27 58	272	37 22	307	46 12	340
284	26 49	073	56 34	155	23 42	180	15 18	214	27 12	273	36 46	308	45 57	340
	◆Alpheratz		Enif		ALTAIR		◆Rasalhague		ARCTURUS		Alkaid		◆Kochab	
285	27 33	074	42 49	118	56 53	157	56 45	220	26 27	274	36 09	308	45 41	340
286	28 17	074	43 30	119	57 10	158	56 15	222	25 41	274	35 33	308	45 25	340
287	29 01	075	44 10	120	57 27	160	55 44	223	24 55	275	34 57	308	45 09	340
288	29 46	075	44 49	121	57 41	162	55 12	225	24 09	275	34 21	309	44 53	340
289	30 30	076	45 28	122	57 55	164	54 39	226	23 23	276	33 45	309	44 37	339
290	31 15	077	46 07	123	58 07	166	54 06	228	22 38	277	33 10	309	44 21	339
291	32 00	077	46 45	124	58 18	167	53 31	229	21 52	277	32 34	310	44 04	339
292	32 45	078	47 23	126	58 27	169	52 56	230	21 06	278	31 59	310	43 48	339
293	33 30	078	48 00	127	58 35	171	52 21	231	20 21	278	31 24	310	43 32	339
294	34 15	079	48 36	128	58 41	173	51 44	233	19 36	279	30 49	311	43 16	339
295	35 00	079	49 12	129	58 46	175	51 08	234	18 50	280	30 14	311	42 59	339
296	35 45	080	49 48	130	58 49	177	50 30	235	18 05	280	29 40	311	42 43	339
297	36 30	080	50 22	132	58 51	179	49 52	236	17 20	281	29 05	312	42 27	339
298	37 16	081	50 56	133	58 51	181	49 14	237	16 35	281	28 31	312	42 10	339
299	38 01	081	51 30	134	58 50	183	48 35	239	15 50	282	27 57	313	41 54	339
	◆Mirfak		Alpheratz		◆Enif		ALTAIR		Rasalhague		◆Alphecca		Kochab	
300	18 19	039	38 47	082	52 02	135	58 47	185	47 55	240	34 17	278	41 38	339
301	18 48	040	39 32	083	52 34	137	58 43	186	47 15	241	33 31	279	41 21	339
302	19 18	040	40 18	083	53 05	138	58 37	188	46 35	242	32 46	279	41 05	339
303	19 48	041	41 03	084	53 36	140	58 29	190	45 54	243	32 01	280	40 49	339
304	20 18	041	41 49	084	54 05	141	58 21	192	45 13	244	31 15	280	40 32	339
305	20 48	042	42 35	085	54 33	143	58 10	194	44 32	245	30 30	281	40 16	339
306	21 19	042	43 21	085	55 01	144	57 58	196	43 50	246	29 45	281	40 00	339
307	21 50	042	44 06	086	55 27	146	57 45	198	43 08	247	29 00	282	39 43	339
308	22 21	043	44 52	087	55 53	147	57 31	199	42 26	248	28 15	282	39 27	339
309	22 52	043	45 38	087	56 17	149	57 15	201	41 43	248	27 30	283	39 11	339
310	23 24	044	46 24	088	56 40	150	56 58	203	41 00	249	26 46	284	38 55	340
311	23 56	044	47 10	088	57 02	152	56 39	205	40 17	250	26 01	284	38 39	340
312	24 28	044	47 56	089	57 23	154	56 19	206	39 34	251	25 16	285	38 23	340
313	25 00	045	48 42	090	57 43	156	55 59	208	38 50	252	24 32	285	38 07	340
314	25 32	045	49 28	090	58 01	157	55 36	210	38 07	253	23 48	286	37 51	340
	◆Mirfak		Hamal		Diphda		◆FOMALHAUT		ALTAIR		◆VEGA		Kochab	
315	26 05	046	24 39	079	12 08	126	15 13	154	55 13	211	62 29	279	37 35	340
316	26 38	046	25 24	080	12 45	127	15 33	155	54 49	213	61 44	280	37 19	340
317	27 11	046	26 10	080	13 22	128	15 52	155	54 24	214	60 58	280	37 04	340
318	27 45	047	26 55	081	13 58	129	16 11	156	53 57	216	60 13	280	36 48	340
319	28 18	047	27 40	082	14 34	129	16 29	157	53 30	217	59 28	281	36 33	340
320	28 52	047	28 26	082	15 09	130	16 46	158	53 02	219	58 43	281	36 17	340
321	29 26	048	29 11	083	15 44	131	17 03	159	52 32	220	57 58	282	36 02	341
322	30 00	048	29 57	083	16 18	132	17 19	160	52 02	222	57 13	282	35 46	341
323	30 34	048	30 43	084	16 52	133	17 35	161	51 32	223	56 28	282	35 31	341
324	31 09	049	31 28	085	17 26	133	17 50	161	51 00	224	55 43	283	35 16	341
325	31 43	049	32 14	085	17 59	134	18 04	162	50 27	226	54 58	283	35 01	341
326	32 18	049	33 00	086	18 32	135	18 18	163	49 54	227	54 14	284	34 46	341
327	32 53	050	33 46	086	19 04	136	18 31	164	49 20	228	53 29	284	34 31	341
328	33 28	050	34 32	087	19 36	137	18 43	165	48 46	229	52 44	284	34 17	341
329	34 04	050	35 18	088	20 07	138	18 54	166	48 11	230	52 00	285	34 02	342
	◆CAPELLA		Hamal		Diphda		◆FOMALHAUT		ALTAIR		◆VEGA		Kochab	
330	16 46	043	36 04	088	20 38	138	19 05	167	47 35	232	51 15	285	33 48	342
331	17 17	044	36 50	089	21 08	139	19 15	168	46 59	233	50 31	286	33 33	342
332	17 49	044	37 36	089	21 38	140	19 25	169	46 22	234	49 47	286	33 19	342
333	18 21	045	38 22	090	22 07	141	19 34	170	45 44	235	49 03	286	33 05	342
334	18 54	045	39 07	091	22 36	142	19 41	170	45 07	236	48 19	287	32 51	342
335	19 27	046	39 53	091	23 04	143	19 49	171	44 28	237	47 35	287	32 37	343
336	20 00	046	40 39	092	23 32	144	19 55	172	43 49	238	46 51	287	32 24	343
337	20 33	047	41 25	093	23 59	145	20 01	173	43 10	239	46 07	288	32 10	343
338	21 06	047	42 11	093	24 25	146	20 06	174	42 30	240	45 23	288	31 57	343
339	21 40	047	42 57	094	24 51	146	20 10	175	41 50	241	44 40	289	31 43	343
340	22 14	048	43 43	095	25 16	147	20 14	176	41 10	242	43 56	289	31 30	343
341	22 48	048	44 29	096	25 40	148	20 17	177	40 29	243	43 13	289	31 17	344
342	23 22	049	45 14	096	26 04	149	20 19	178	39 48	244	42 30	290	31 04	344
343	23 57	049	46 00	097	26 27	150	20 20	179	39 06	245	41 46	290	30 52	344
344	24 32	050	46 46	098	26 49	151	20 21	180	38 25	246	41 03	291	30 39	344
	◆CAPELLA		ALDEBARAN		Diphda		◆FOMALHAUT		ALTAIR		◆VEGA		Kochab	
345	25 07	050	15 06	081	27 11	152	20 21	181	37 42	247	40 20	291	30 27	344
346	25 42	050	15 51	082	27 32	153	20 20	182	37 00	248	39 38	292	30 15	345
347	26 18	051	16 37	082	27 53	154	20 18	182	36 17	248	38 55	292	30 03	345
348	26 53	051	17 22	083	28 12	155	20 16	183	35 35	249	38 12	292	29 51	345
349	27 29	052	18 08	083	28 31	156	20 13	184	34 51	250	37 30	293	29 39	345
350	28 05	052	18 54	084	28 49	157	20 09	185	34 08	251	36 47	293	29 27	346
351	28 42	052	19 40	085	29 07	158	20 05	186	33 25	252	36 05	294	29 16	346
352	29 18	053	20 25	085	29 23	159	19 59	187	32 41	253	35 23	294	29 05	346
353	29 55	053	21 11	086	29 39	160	19 53	188	31 57	253	34 41	294	28 54	346
354	30 32	053	21 57	086	29 54	161	19 47	189	31 13	254	34 00	295	28 43	346
355	31 09	054	22 43	087	30 08	163	19 39	190	30 28	255	33 18	295	28 32	347
356	31 46	054	23 29	088	30 22	164	19 31	191	29 44	256	32 36	296	28 22	347
357	32 23	055	24 15	088	30 34	165	19 22	192	28 59	256	31 55	296	28 12	347
358	33 01	055	25 01	089	30 46	166	19 12	193	28 15	257	31 14	297	28 02	347
359	33 39	055	25 47	090	30 57	167	19 02	193	27 30	258	30 33	297	27 52	348

LHA	Hc	Zn	Hc	Zn	Hc	Zn	Hc	Zn	Hc	Zn	Hc	Zn	Hc	Zn
♈	◆CAPELLA		ALDEBARAN		◆Diphda		FOMALHAUT		ALTAIR		◆VEGA		Kochab	
0	33 42	055	26 33	090	32 05	168	19 49	194	26 56	259	29 24	298	26 43	348
1	34 21	055	27 19	090	32 15	169	19 37	195	26 11	260	28 43	298	26 34	348
2	34 59	056	28 06	091	32 23	170	19 24	196	25 25	260	28 02	299	26 24	349
3	35 38	056	28 52	092	32 31	171	19 11	197	24 39	261	27 21	299	26 15	349
4	36 16	056	29 39	092	32 38	172	18 57	198	23 52	262	26 41	300	26 06	349
5	36 55	057	30 26	093	32 43	173	18 42	199	23 06	263	26 00	300	25 58	349
6	37 34	057	31 12	094	32 48	175	18 27	200	22 20	263	25 20	301	25 49	350
7	38 14	057	31 59	094	32 52	176	18 11	201	21 34	264	24 40	301	25 41	350
8	38 53	058	32 45	095	32 55	177	17 54	201	20 47	265	24 00	301	25 33	350
9	39 32	058	33 32	096	32 58	178	17 36	202	20 01	265	23 20	302	25 25	350
10	40 12	058	34 18	096	32 59	179	17 18	203	19 14	266	22 41	302	25 17	351
11	40 52	058	35 04	097	32 59	180	17 00	204	18 28	266	22 02	303	25 10	351
12	41 31	059	35 51	098	32 59	181	16 40	205	17 41	267	21 23	303	25 02	351
13	42 11	059	36 37	098	32 57	182	16 20	206	16 55	268	20 44	304	24 55	351
14	42 51	059	37 23	099	32 55	184	16 00	207	16 08	268	20 05	304	24 48	352
	◆CAPELLA		BETELGEUSE		RIGEL		◆Diphda		Enif		◆DENEB		Kochab	
15	43 31	060	17 17	094	14 37	114	32 51	185	37 35	250	42 55	300	24 42	352
16	44 12	060	18 03	095	15 20	114	32 47	186	36 51	251	42 15	300	24 35	352
17	44 52	060	18 50	096	16 02	115	32 42	187	36 07	251	41 34	300	24 29	353
18	45 32	060	19 36	096	16 45	116	32 36	188	35 22	252	40 54	301	24 23	353
19	46 13	060	20 23	097	17 26	117	32 29	189	34 38	253	40 14	301	24 18	353
20	46 54	061	21 09	098	18 08	117	32 21	190	33 53	254	39 34	301	24 12	353
21	47 34	061	21 55	098	18 49	118	32 12	191	33 08	255	38 54	301	24 07	354
22	48 15	061	22 41	099	19 30	119	32 02	193	32 23	255	38 14	302	24 02	354
23	48 56	061	23 27	100	20 11	120	31 52	194	31 38	256	37 35	302	23 57	354
24	49 37	062	24 13	101	20 51	120	31 40	195	30 53	257	36 55	302	23 52	355
25	50 18	062	24 59	101	21 31	121	31 28	196	30 07	258	36 16	303	23 48	355
26	50 59	062	25 44	102	22 11	122	31 15	197	29 22	258	35 37	303	23 44	355
27	51 40	062	26 30	103	22 51	123	31 01	198	28 36	259	34 58	303	23 40	355
28	52 21	062	27 15	103	23 30	124	30 46	199	27 50	260	34 19	304	23 36	356
29	53 03	062	28 01	104	24 08	124	30 30	200	27 04	260	33 40	304	23 33	356
	◆CAPELLA		BETELGEUSE		RIGEL		◆Diphda		Alpheratz		◆DENEB		Kochab	
30	53 44	063	28 46	105	24 47	125	30 14	201	64 55	255	33 01	304	23 30	356
31	54 26	063	29 31	106	25 25	126	29 57	202	64 10	256	32 23	305	23 27	357
32	55 07	063	30 16	106	26 02	127	29 39	203	63 24	257	31 45	305	23 24	357
33	55 48	063	31 00	107	26 39	128	29 20	204	62 39	258	31 07	305	23 22	357
34	56 30	063	31 45	108	27 16	129	29 00	205	61 53	259	30 29	306	23 20	357
35	57 12	063	32 29	109	27 52	129	28 40	206	61 07	260	29 51	306	23 18	358
36	57 53	063	33 13	109	28 28	130	28 19	207	60 21	261	29 13	306	23 16	358
37	58 35	063	33 57	110	29 03	131	27 57	208	59 35	262	28 36	307	23 14	358
38	59 17	063	34 41	111	29 38	132	27 35	209	58 49	262	27 59	307	23 13	359
39	59 58	063	35 24	112	30 12	133	27 11	210	58 03	263	27 22	308	23 12	359
40	60 40	063	36 07	113	30 46	134	26 47	211	57 16	264	26 45	308	23 11	359
41	61 22	063	36 50	114	31 19	135	26 23	212	56 30	265	26 08	308	23 11	000
42	62 03	063	37 33	114	31 52	136	25 58	213	55 43	266	25 32	309	23 11	000
43	62 45	063	38 15	115	32 24	137	25 32	214	54 57	266	24 55	309	23 11	000
44	63 27	063	38 57	116	32 55	138	25 05	215	54 10	267	24 19	310	23 11	000
	◆Dubhe		POLLUX		SIRIUS		◆RIGEL		Diphda		◆Alpheratz		DENEB	
45	21 28	026	31 05	077	13 28	125	33 26	139	24 38	216	53 24	268	23 44	310
46	21 49	026	31 51	078	14 06	126	33 57	140	24 11	217	52 37	268	23 08	310
47	22 09	027	32 36	078	14 43	127	34 26	141	23 42	218	51 50	269	22 33	311
48	22 30	027	33 22	079	15 21	127	34 55	142	23 13	219	51 04	270	21 57	311
49	22 52	027	34 08	080	15 58	128	35 23	143	22 44	220	50 17	270	21 22	312
50	23 13	028	34 54	080	16 34	129	35 51	144	22 14	220	49 31	271	20 48	312
51	23 35	028	35 40	081	17 10	130	36 18	145	21 44	221	48 44	271	20 13	313
52	23 57	028	36 26	081	17 46	130	36 44	146	21 12	222	47 57	272	19 39	313
53	24 19	029	37 12	082	18 21	131	37 09	148	20 41	223	47 11	273	19 05	313
54	24 41	029	37 58	082	18 56	132	37 34	149	20 09	224	46 24	273	18 31	314
55	25 04	029	38 44	083	19 31	133	37 58	150	19 36	225	45 38	274	17 58	314
56	25 27	030	39 30	083	20 05	134	38 21	151	19 03	226	44 51	274	17 24	315
57	25 50	030	40 17	084	20 38	134	38 43	152	18 30	226	44 05	275	16 51	315
58	26 13	030	41 03	084	21 11	135	39 04	153	17 56	227	43 18	275	16 19	316
59	26 37	030	41 50	085	21 44	136	39 25	155	17 21	228	42 32	276	15 46	316
	◆Dubhe		POLLUX		PROCYON		◆SIRIUS		RIGEL		◆Hamal		Schedar	
60	27 00	031	42 36	086	30 16	110	22 16	137	39 44	156	61 26	245	53 12	315
61	27 24	031	43 23	086	31 00	110	22 47	138	40 03	157	60 43	247	52 39	315
62	27 48	031	44 09	087	31 43	111	23 18	139	40 21	158	60 00	248	52 07	315
63	28 13	032	44 56	087	32 27	112	23 49	140	40 38	160	59 17	249	51 34	315
64	28 37	032	45 42	088	33 10	113	24 19	141	40 53	161	58 33	250	51 00	315
65	29 02	032	46 29	089	33 53	114	24 48	141	41 08	162	57 50	251	50 27	315
66	29 27	032	47 15	089	34 35	115	25 17	142	41 22	163	57 05	252	49 54	315
67	29 52	033	48 02	090	35 17	115	25 45	143	41 35	165	56 21	253	49 21	315
68	30 17	033	48 49	090	35 59	116	26 13	144	41 47	166	55 36	254	48 48	315
69	30 42	033	49 35	091	36 41	117	26 40	145	41 58	167	54 51	255	48 15	315
70	31 08	033	50 22	092	37 22	118	27 06	146	42 08	169	54 06	256	47 42	315
71	31 34	034	51 09	092	38 03	119	27 32	147	42 16	170	53 21	257	47 09	315
72	32 00	034	51 55	093	38 44	120	27 57	148	42 24	171	52 35	258	46 36	315
73	32 26	034	52 42	094	39 24	121	28 21	149	42 31	173	51 50	258	46 03	315
74	32 52	034	53 28	094	40 04	122	28 45	150	42 36	174	51 04	259	45 30	315
	◆Dubhe		POLLUX		PROCYON		◆SIRIUS		RIGEL		◆Hamal		Schedar	
75	33 18	034	54 15	095	40 43	123	29 08	151	42 41	175	50 18	260	44 57	315
76	33 44	035	55 01	096	41 22	124	29 30	152	42 44	177	49 32	261	44 24	315
77	34 11	035	55 47	097	42 01	125	29 51	153	42 46	178	48 46	262	43 51	315
78	34 38	035	56 34	097	42 39	126	30 12	154	42 47	179	48 00	262	43 18	315
79	35 05	035	57 20	098	43 16	127	30 32	155	42 47	181	47 14	263	42 45	315
80	35 32	035	58 06	099	43 53	128	30 51	156	42 46	182	46 27	264	42 13	316
81	35 59	036	58 52	100	44 30	129	31 10	157	42 44	183	45 41	264	41 40	316
82	36 26	036	59 38	101	45 06	130	31 27	158	42 41	185	44 55	265	41 08	316
83	36 53	036	60 24	102	45 41	131	31 44	159	42 37	186	44 08	266	40 35	316
84	37 21	036	61 09	103	46 16	133	32 00	160	42 31	187	43 22	267	40 03	316
85	37 48	036	61 55	103	46 50	134	32 15	162	42 25	189	42 35	267	39 31	316
86	38 16	036	62 40	104	47 23	135	32 30	163	42 17	190	41 48	268	38 59	316
87	38 43	036	63 25	105	47 56	136	32 43	164	42 09	191	41 02	268	38 26	317
88	39 11	037	64 10	107	48 27	137	32 56	165	41 59	193	40 15	269	37 55	317
89	39 39	037	64 54	108	48 59	139	33 08	166	41 49	194	39 29	270	37 23	317

LHA	Hc	Zn	Hc	Zn	Hc	Zn	Hc	Zn	Hc	Zn	Hc	Zn	Hc	Zn
♈	◆Dubhe		REGULUS		PROCYON		◆SIRIUS		RIGEL		ALDEBARAN		◆Mirfak	
90	40 07	037	29 10	098	49 29	140	33 18	167	41 37	195	60 55	225	60 30	304
91	40 35	037	29 56	099	49 58	141	33 28	168	41 24	196	60 22	227	59 52	304
92	41 03	037	30 42	100	50 27	143	33 38	169	41 10	198	59 47	228	59 13	304
93	41 31	037	31 28	100	50 55	144	33 46	171	40 56	199	59 12	230	58 35	304
94	41 59	037	32 14	101	51 22	146	33 53	172	40 40	200	58 36	231	57 56	304
95	42 27	037	32 59	102	51 48	147	33 59	173	40 23	202	57 59	233	57 18	304
96	42 56	037	33 45	103	52 12	148	34 05	174	40 06	203	57 22	234	56 39	304
97	43 24	037	34 30	103	52 36	150	34 09	175	39 47	204	56 44	235	56 00	304
98	43 52	037	35 16	104	52 59	152	34 13	176	39 28	205	56 05	237	55 21	304
99	44 21	037	36 01	105	53 21	153	34 15	177	39 07	206	55 26	238	54 43	304
100	44 49	037	36 46	106	53 41	155	34 17	179	38 46	208	54 46	239	54 04	304
101	45 17	037	37 31	107	54 01	156	34 17	180	38 24	209	54 06	240	53 25	304
102	45 46	037	38 15	107	54 19	158	34 17	181	38 01	210	53 25	241	52 47	304
103	46 14	037	39 00	108	54 36	159	34 16	182	37 38	211	52 44	243	52 08	304
104	46 42	037	39 44	109	54 52	161	34 14	183	37 13	212	52 02	244	51 30	304
	◆Kochab		Denebola		◆REGULUS		SIRIUS		RIGEL		◆ALDEBARAN		CAPELLA	
105	30 28	016	22 52	089	40 28	110	34 11	184	36 48	213	51 20	245	69 46	299
106	30 41	016	23 38	090	41 11	111	34 07	186	36 22	214	50 38	246	69 05	298
107	30 54	017	24 25	091	41 55	112	34 02	187	35 55	216	49 55	247	68 23	298
108	31 08	017	25 12	091	42 38	112	33 56	188	35 28	217	49 12	248	67 42	298
109	31 21	017	25 58	092	43 21	113	33 49	189	34 59	218	48 29	249	67 01	297
110	31 35	017	26 45	093	44 04	114	33 41	190	34 30	219	47 46	250	66 19	297
111	31 49	017	27 31	093	44 46	115	33 33	191	34 01	220	47 02	251	65 38	297
112	32 03	018	28 18	094	45 28	116	33 23	192	33 31	221	46 18	251	64 56	297
113	32 17	018	29 04	095	46 10	117	33 13	194	33 00	222	45 33	252	64 15	297
114	32 31	018	29 51	095	46 51	118	33 01	195	32 29	223	44 49	253	63 33	297
115	32 46	018	30 37	096	47 32	119	32 49	196	31 57	224	44 04	254	62 52	297
116	33 00	018	31 24	097	48 12	120	32 36	197	31 24	225	43 19	255	62 10	297
117	33 15	018	32 10	097	48 52	121	32 22	198	30 51	226	42 34	256	61 28	297
118	33 29	018	32 56	098	49 32	123	32 07	199	30 17	227	41 49	256	60 47	297
119	33 44	019	33 42	099	50 11	124	31 52	200	29 43	228	41 03	257	60 05	297
	◆Kochab		Denebola		◆REGULUS		SIRIUS		RIGEL		◆ALDEBARAN		CAPELLA	
120	33 59	019	34 28	099	50 50	125	31 35	201	29 08	229	40 18	258	59 23	297
121	34 14	019	35 14	100	51 28	126	31 18	202	28 33	229	39 32	259	58 41	297
122	34 29	019	36 00	101	52 05	127	31 00	203	27 57	230	38 46	260	58 00	297
123	34 44	019	36 46	102	52 42	128	30 41	204	27 21	231	38 00	260	57 18	297
124	35 00	019	37 32	102	53 18	130	30 21	205	26 45	232	37 14	261	56 37	297
125	35 15	019	38 17	103	53 54	131	30 01	206	26 08	233	36 28	262	55 55	297
126	35 31	019	39 02	104	54 28	132	29 40	208	25 30	234	35 42	262	55 14	297
127	35 46	020	39 48	105	55 03	134	29 18	209	24 52	235	34 56	263	54 32	297
128	36 02	020	40 33	105	55 36	135	28 55	210	24 14	236	34 10	264	53 51	297
129	36 17	020	41 17	106	56 08	137	28 32	211	23 35	236	33 23	264	53 09	298
130	36 33	020	42 02	107	56 40	138	28 08	212	22 56	237	32 37	265	52 28	298
131	36 49	020	42 47	108	57 10	140	27 43	212	22 17	238	31 50	266	51 47	298
132	37 05	020	43 31	109	57 40	141	27 18	213	21 37	239	31 04	266	51 06	298
133	37 21	020	44 15	110	58 09	143	26 52	214	20 57	240	30 17	267	50 25	298
134	37 37	020	44 59	111	58 36	144	26 25	215	20 17	240	29 31	268	49 43	298
	Kochab		◆ARCTURUS		SPICA		REGULUS		◆SIRIUS		BETELGEUSE		◆CAPELLA	
135	37 53	020	20 25	081	10 42	114	59 03	146	25 58	216	37 52	245	49 02	299
136	38 09	020	21 11	082	11 25	115	59 28	148	25 30	217	37 09	246	48 22	299
137	38 25	020	21 57	083	12 07	115	59 53	150	25 01	218	36 27	247	47 41	299
138	38 41	020	22 44	083	12 49	116	60 16	151	24 32	219	35 44	248	47 00	299
139	38 58	020	23 30	084	13 31	117	60 37	153	24 03	220	35 00	249	46 19	299
140	39 14	020	24 16	084	14 12	117	60 57	155	23 32	221	34 17	249	45 39	300
141	39 30	020	25 03	085	14 54	118	61 16	157	23 02	222	33 33	250	44 58	300
142	39 47	020	25 49	085	15 35	119	61 34	159	22 30	223	32 49	251	44 18	300
143	40 03	021	26 36	086	16 15	120	61 50	161	21 59	223	32 05	252	43 38	300
144	40 19	021	27 22	087	16 56	120	62 04	163	21 26	224	31 20	253	42 58	301
145	40 36	021	28 09	087	17 36	121	62 17	165	20 53	225	30 36	253	42 18	301
146	40 52	021	28 55	088	18 15	122	62 28	167	20 20	226	29 51	254	41 38	301
147	41 08	021	29 42	089	18 55	123	62 38	169	19 46	227	29 06	255	40 58	302
148	41 25	021	30 29	089	19 34	123	62 45	171	19 12	228	28 21	256	40 18	302
149	41 41	021	31 15	090	20 13	124	62 51	173	18 37	228	27 36	256	39 39	302
	◆Kochab		ARCTURUS		◆SPICA		REGULUS		PROCYON		◆POLLUX		CAPELLA	
150	41 57	020	32 02	090	20 51	125	62 56	176	43 36	233	60 01	259	38 59	302
151	42 14	020	32 49	091	21 29	126	62 59	178	42 58	234	59 15	260	38 20	303
152	42 30	020	33 35	092	22 07	127	63 00	180	42 21	235	58 29	261	37 41	303
153	42 46	020	34 22	092	22 44	127	62 59	182	41 42	236	57 43	261	37 02	303
154	43 02	020	35 08	093	23 21	128	62 56	184	41 04	237	56 57	262	36 23	304
155	43 19	020	35 55	094	23 57	129	62 52	186	40 24	238	56 11	263	35 44	304
156	43 35	020	36 41	094	24 33	130	62 46	189	39 45	239	55 24	264	35 05	304
157	43 51	020	37 28	095	25 08	131	62 38	191	39 05	240	54 38	264	34 27	305
158	44 07	020	38 14	096	25 43	132	62 29	193	38 24	241	53 52	265	33 48	305
159	44 23	020	39 01	096	26 18	133	62 17	195	37 44	241	53 05	266	33 10	305
160	44 39	020	39 47	097	26 52	134	62 05	197	37 02	242	52 19	267	32 32	306
161	44 55	020	40 33	098	27 26	134	61 50	199	36 21	243	51 32	267	31 54	306
162	45 10	020	41 19	099	27 59	135	61 35	201	35 39	244	50 45	268	31 17	306
163	45 26	020	42 05	099	28 31	136	61 17	203	34 57	245	49 59	269	30 39	307
164	45 42	019	42 51	100	29 03	137	60 58	205	34 15	246	49 12	269	30 02	307
	◆Kochab		ARCTURUS		◆SPICA		REGULUS		PROCYON		◆POLLUX		CAPELLA	
165	45 57	019	43 37	101	29 34	138	60 38	207	33 32	247	48 26	270	29 25	307
166	46 13	019	44 23	102	30 05	139	60 17	208	32 49	248	47 39	271	28 48	308
167	46 28	019	45 09	102	30 35	140	59 54	210	32 06	248	46 52	271	28 11	308
168	46 43	019	45 54	103	31 05	141	59 30	212	31 22	249	46 06	272	27 34	308
169	46 58	019	46 39	104	31 34	142	59 04	214	30 39	250	45 19	272	26 58	309
170	47 13	019	47 25	105	32 02	143	58 38	215	29 55	251	44 33	273	26 21	309
171	47 28	018	48 10	106	32 30	144	58 10	217	29 11	251	43 46	274	25 45	310
172	47 43	018	48 54	107	32 57	145	57 42	219	28 26	252	43 00	274	25 10	310
173	47 57	018	49 39	107	33 23	146	57 12	220	27 42	253	42 13	275	24 34	310
174	48 12	018	50 23	108	33 49	147	56 41	222	26 57	254	41 27	275	23 58	311
175	48 26	018	51 08	109	34 14	148	56 10	223	26 12	254	40 40	276	23 23	311
176	48 40	017	51 51	110	34 38	149	55 38	225	25 27	255	39 54	276	22 48	312
177	48 54	017	52 35	111	35 01	151	55 04	226	24 42	256	39 07	277	22 14	312
178	49 08	017	53 18	112	35 23	152	54 30	228	23 57	257	38 21	278	21 39	312
179	49 21	017	54 01	113	35 45	153	53 55	229	23 11	257	37 35	278	21 05	313

LAT 39°N

LHA ♈	Hc	Zn	Hc	Zn	Hc	Zn	Hc	Zn	Hc	Zn	Hc	Zn	Hc	Zn
	◆VEGA		Alphecca		ARCTURUS		◆SPICA		REGULUS		◆POLLUX		CAPELLA	
180	17 18	054	44 01	089	54 44	114	36 06	154	53 20	230	36 49	279	20 31	313
181	17 56	054	44 47	089	55 26	115	36 26	155	52 44	231	36 03	279	19 57	314
182	18 34	055	45 34	090	56 08	117	36 45	156	52 07	233	35 17	280	19 23	314
183	19 12	055	46 20	091	56 50	118	37 04	157	51 30	234	34 31	280	18 50	315
184	19 50	056	47 07	091	57 31	119	37 21	159	50 52	235	33 45	281	18 17	315
185	20 29	056	47 54	092	58 11	120	37 38	160	50 13	236	32 59	281	17 44	315
186	21 08	057	48 40	093	58 51	121	37 54	161	49 34	237	32 14	282	17 11	316
187	21 47	057	49 27	093	59 31	123	38 08	162	48 55	239	31 28	282	16 39	316
188	22 26	057	50 13	094	60 10	124	38 22	163	48 15	240	30 42	283	16 07	317
189	23 05	058	51 00	095	60 48	126	38 35	165	47 34	241	29 57	283	15 35	317
190	23 45	058	51 46	095	61 26	127	38 47	166	46 53	242	29 12	284	15 04	318
191	24 25	059	52 33	096	62 03	129	38 58	167	46 12	243	28 27	284	14 33	318
192	25 05	059	53 19	097	62 39	130	39 08	168	45 30	244	27 41	285	14 02	319
193	25 45	060	54 05	098	63 14	132	39 17	170	44 48	245	26 56	285	13 31	319
194	26 25	060	54 51	098	63 48	133	39 25	171	44 06	246	26 12	286	13 01	320
	DENEB		◆VEGA		Alphecca		◆SPICA		REGULUS		◆POLLUX		Dubhe	
195	12 18	041	27 06	061	55 38	099	39 32	172	43 23	247	25 27	287	61 06	332
196	12 49	041	27 46	061	56 24	100	39 38	173	42 40	247	24 42	287	60 44	331
197	13 20	042	28 27	061	57 09	101	39 43	175	41 57	248	23 58	288	60 21	330
198	13 51	042	29 08	062	57 55	102	39 46	176	41 14	249	23 13	288	59 58	330
199	14 22	043	29 49	062	58 41	103	39 49	177	40 30	250	22 29	289	59 34	329
200	14 54	043	30 31	063	59 26	103	39 51	178	39 46	251	21 45	289	59 10	329
201	15 26	044	31 12	063	60 11	104	39 52	180	39 02	252	21 01	290	58 46	328
202	15 58	044	31 54	064	60 57	105	39 52	181	38 17	253	20 17	290	58 21	328
203	16 31	044	32 36	064	61 41	106	39 50	182	37 33	253	19 33	291	57 56	327
204	17 04	045	33 18	064	62 26	107	39 48	184	36 48	254	18 50	291	57 31	327
205	17 37	045	34 00	065	63 10	109	39 45	185	36 03	255	18 07	292	57 05	327
206	18 10	046	34 42	065	63 54	110	39 40	186	35 18	256	17 23	292	56 40	326
207	18 44	046	35 25	066	64 38	111	39 35	187	34 33	257	16 40	293	56 14	326
208	19 17	047	36 07	066	65 21	112	39 28	189	33 47	257	15 57	293	55 47	326
209	19 52	047	36 50	066	66 04	113	39 21	190	33 02	258	15 15	294	55 21	325
	DENEB		◆VEGA		Rasalhague		ANTARES		◆SPICA		REGULUS		◆Dubhe	
210	20 26	048	37 33	067	35 53	104	15 53	146	39 12	191	32 16	259	54 54	325
211	21 00	048	38 16	067	36 39	105	16 19	146	39 03	192	31 30	259	54 28	325
212	21 35	049	38 59	068	37 24	105	16 45	147	38 52	194	30 44	260	54 01	325
213	22 10	049	39 42	068	38 08	106	17 10	148	38 41	195	29 58	261	53 34	324
214	22 46	049	40 25	068	38 53	107	17 34	149	38 28	196	29 12	262	53 06	324
215	23 21	050	41 08	069	39 38	108	17 58	150	38 15	197	28 26	262	52 39	324
216	23 57	050	41 52	069	40 22	109	18 21	151	38 01	199	27 40	263	52 11	324
217	24 33	051	42 36	069	41 06	110	18 43	152	37 45	200	26 54	264	51 44	324
218	25 09	051	43 19	070	41 50	110	19 05	152	37 29	201	26 07	264	51 16	323
219	25 45	051	44 03	070	42 33	111	19 27	153	37 12	202	25 21	265	50 48	323
220	26 22	052	44 47	071	43 17	112	19 47	154	36 54	203	24 34	266	50 20	323
221	26 58	052	45 31	071	44 00	113	20 07	155	36 35	204	23 48	266	49 52	323
222	27 35	053	46 15	071	44 42	114	20 27	156	36 16	206	23 01	267	49 24	323
223	28 12	053	46 59	072	45 25	115	20 45	157	35 55	207	22 15	268	48 56	323
224	28 50	053	47 44	072	46 07	116	21 03	158	35 34	208	21 28	268	48 27	323
	DENEB		◆ALTAIR		Rasalhague		ANTARES		◆SPICA		Denebola		◆Dubhe	
225	29 27	054	19 02	094	46 49	117	21 21	159	35 11	209	41 35	253	47 59	323
226	30 05	054	19 48	095	47 30	118	21 37	160	34 49	210	40 51	254	47 31	323
227	30 43	054	20 35	095	48 11	119	21 53	160	34 25	211	40 06	255	47 03	323
228	31 21	055	21 21	096	48 51	120	22 09	161	34 00	212	39 21	256	46 34	323
229	31 59	055	22 07	097	49 32	121	22 23	162	33 35	213	38 35	257	46 06	323
230	32 37	055	22 54	097	50 11	122	22 37	163	33 09	214	37 50	257	45 38	323
231	33 16	056	23 40	098	50 50	123	22 50	164	32 42	215	37 04	258	45 09	323
232	33 54	056	24 26	099	51 29	125	23 02	165	32 15	216	36 19	259	44 41	323
233	34 33	056	25 12	099	52 07	126	23 14	166	31 47	217	35 33	260	44 12	323
234	35 12	057	25 58	100	52 45	127	23 25	167	31 18	218	34 47	260	43 44	323
235	35 51	057	26 44	101	53 22	128	23 35	168	30 49	219	34 01	261	43 16	323
236	36 30	057	27 30	101	53 58	130	23 44	169	30 19	220	33 15	262	42 47	323
237	37 10	058	28 15	102	54 33	131	23 53	170	29 49	221	32 29	262	42 19	323
238	37 49	058	29 01	103	55 08	132	24 00	171	29 18	222	31 42	263	41 51	323
239	38 29	058	29 46	104	55 42	134	24 07	172	28 46	223	30 56	264	41 23	323
	◆DENEB		ALTAIR		Nunki		◆ANTARES		ARCTURUS		◆Alkaid		Kochab	
240	39 09	059	30 32	104	12 58	141	24 14	173	59 59	236	64 20	305	53 53	352
241	39 49	059	31 17	105	13 28	141	24 19	174	59 20	238	63 41	304	53 46	352
242	40 29	059	32 02	106	13 57	142	24 24	175	58 40	239	63 03	304	53 39	351
243	41 09	060	32 46	107	14 25	143	24 27	176	58 00	240	62 24	304	53 32	351
244	41 49	060	33 31	107	14 53	144	24 30	177	57 19	241	61 45	304	53 24	350
245	42 29	060	34 15	108	15 20	144	24 33	178	56 38	243	61 06	304	53 16	350
246	43 10	060	35 00	109	15 47	145	24 34	179	55 56	244	60 28	303	53 08	350
247	43 51	061	35 44	110	16 14	146	24 35	180	55 14	245	59 49	303	53 00	349
248	44 31	061	36 27	111	16 39	147	24 35	181	54 32	246	59 10	303	52 51	349
249	45 12	061	37 11	111	17 05	148	24 34	182	53 49	247	58 31	303	52 42	348
250	45 53	061	37 54	112	17 29	149	24 32	183	53 06	248	57 52	303	52 32	348
251	46 34	062	38 37	113	17 53	149	24 29	184	52 23	249	57 12	303	52 22	348
252	47 15	062	39 20	114	18 17	150	24 26	185	51 39	250	56 33	303	52 12	347
253	47 56	062	40 02	115	18 40	151	24 22	186	50 55	251	55 54	303	52 02	347
254	48 38	062	40 44	116	19 02	152	24 17	187	50 11	252	55 15	303	51 52	347
	◆DENEB		ALTAIR		Nunki		◆ANTARES		ARCTURUS		◆Alkaid		Kochab	
255	49 19	063	41 26	117	19 23	153	24 11	188	49 26	253	54 36	303	51 41	346
256	50 00	063	42 08	118	19 44	154	24 04	189	48 42	254	53 57	303	51 30	346
257	50 42	063	42 49	119	20 05	155	23 57	190	47 57	255	53 18	303	51 18	346
258	51 23	063	43 29	120	20 24	155	23 49	190	47 12	255	52 39	303	51 07	345
259	52 05	063	44 10	121	20 43	156	23 40	191	46 27	256	52 00	303	50 55	345
260	52 47	064	44 50	122	21 02	157	23 30	192	45 41	257	51 21	303	50 43	345
261	53 29	064	45 29	123	21 20	158	23 20	193	44 56	258	50 42	303	50 31	345
262	54 10	064	46 08	124	21 37	159	23 09	194	44 10	259	50 03	304	50 18	344
263	54 52	064	46 47	125	21 53	160	22 57	195	43 24	259	49 24	304	50 05	344
264	55 34	064	47 25	126	22 08	161	22 44	196	42 38	260	48 46	304	49 52	344
265	56 16	064	48 02	127	22 23	162	22 31	197	41 52	261	48 07	304	49 39	344
266	56 58	064	48 39	128	22 37	163	22 17	198	41 06	262	47 28	304	49 26	343
267	57 40	065	49 15	130	22 51	164	22 02	199	40 20	262	46 50	304	49 12	343
268	58 23	065	49 51	131	23 04	165	21 46	200	39 34	263	46 11	304	48 59	343
269	59 05	065	50 26	132	23 16	166	21 30	201	38 48	264	45 33	305	48 45	343

LHA ♈	Hc	Zn	Hc	Zn	Hc	Zn	Hc	Zn	Hc	Zn	Hc	Zn	Hc	Zn
	◆Alpheratz		ALTAIR		Nunki		◆ANTARES		ARCTURUS		◆Alkaid		Kochab	
270	16 21	066	51 00	133	23 27	167	21 13	202	38 01	265	44 55	305	48 31	342
271	17 04	066	51 34	135	23 37	168	20 55	203	37 15	265	44 16	305	48 17	342
272	17 47	067	52 06	136	23 47	169	20 37	204	36 28	266	43 38	305	48 02	342
273	18 30	067	52 39	137	23 56	169	20 18	204	35 42	267	43 00	305	47 48	342
274	19 13	068	53 10	139	24 04	170	19 58	205	34 55	267	42 22	306	47 33	342
275	19 56	068	53 40	140	24 11	171	19 38	206	34 09	268	41 44	306	47 18	341
276	20 39	069	54 10	142	24 18	172	19 17	207	33 22	268	41 07	306	47 04	341
277	21 23	069	54 38	143	24 24	173	18 56	208	32 35	269	40 29	306	46 49	341
278	22 06	070	55 06	145	24 29	174	18 33	209	31 49	270	39 51	307	46 33	341
279	22 50	070	55 32	146	24 33	175	18 11	210	31 02	270	39 14	307	46 18	341
280	23 34	071	55 58	148	24 36	176	17 47	211	30 15	271	38 37	307	46 03	341
281	24 18	071	56 22	149	24 39	177	17 23	211	29 29	272	38 00	307	45 47	341
282	25 03	072	56 45	151	24 41	178	16 59	212	28 42	272	37 23	308	45 32	340
283	25 47	072	57 07	153	24 42	179	16 33	213	27 56	273	36 46	308	45 16	340
284	26 31	073	57 28	154	24 42	180	16 08	214	27 09	273	36 09	308	45 00	340
	◆Alpheratz		Enif		ALTAIR		◆Rasalhague		ARCTURUS		Alkaid		◆Kochab	
285	27 16	073	43 17	117	57 48	156	57 31	221	26 23	274	35 32	308	44 44	340
286	28 01	074	43 59	118	58 06	158	56 59	223	25 36	275	34 56	309	44 29	340
287	28 46	074	44 39	119	58 23	160	56 27	224	24 50	275	34 20	309	44 13	340
288	29 31	075	45 20	120	58 38	161	55 54	226	24 03	276	33 43	309	43 57	340
289	30 16	075	46 00	121	58 53	163	55 20	227	23 17	276	33 07	310	43 41	340
290	31 01	076	46 39	122	59 05	165	54 46	229	22 31	277	32 31	310	43 24	340
291	31 46	076	47 18	124	59 16	167	54 10	230	21 44	278	31 56	310	43 08	340
292	32 32	077	47 57	125	59 26	169	53 34	231	20 58	278	31 20	311	42 52	340
293	33 17	078	48 35	126	59 34	171	52 58	233	20 12	279	30 45	311	42 36	340
294	34 03	078	49 13	127	59 41	173	52 20	234	19 26	279	30 10	311	42 20	340
295	34 48	079	49 50	128	59 46	175	51 42	235	18 40	280	29 35	312	42 03	340
296	35 34	079	50 26	129	59 49	177	51 04	236	17 54	281	29 00	312	41 47	339
297	36 20	080	51 02	131	59 51	179	50 25	237	17 08	281	28 25	312	41 31	339
298	37 06	080	51 37	132	59 51	181	49 46	238	16 23	282	27 51	313	41 14	339
299	37 52	081	52 11	133	59 50	183	49 06	240	15 37	282	27 16	313	40 58	339
	◆Mirfak		Alpheratz		◆Enif		ALTAIR		Rasalhague		◆Alphecca		Kochab	
300	17 33	039	38 38	081	52 45	135	59 47	185	48 25	241	34 08	279	40 41	339
301	18 02	040	39 24	082	53 18	136	59 42	187	47 44	242	33 22	279	40 25	339
302	18 32	040	40 10	082	53 50	137	59 36	189	47 03	243	32 36	280	40 09	339
303	19 02	040	40 56	083	54 21	139	59 28	190	46 22	244	31 50	280	39 52	339
304	19 33	041	41 43	083	54 51	140	59 19	192	45 40	245	31 04	281	39 36	340
305	20 03	041	42 29	084	55 21	142	59 08	194	44 57	246	30 19	281	39 20	340
306	20 34	042	43 15	084	55 49	143	58 56	196	44 15	247	29 33	282	39 04	340
307	21 05	042	44 02	085	56 17	145	58 42	198	43 32	247	28 47	282	38 47	340
308	21 37	043	44 48	086	56 43	146	58 27	200	42 48	248	28 02	283	38 31	340
309	22 08	043	45 35	086	57 08	148	58 11	202	42 05	249	27 16	284	38 15	340
310	22 40	043	46 21	087	57 32	150	57 53	203	41 21	250	26 31	284	37 59	340
311	23 12	044	47 08	087	57 55	151	57 34	205	40 37	251	25 46	285	37 43	340
312	23 45	044	47 54	088	58 17	153	57 13	207	39 53	252	25 01	285	37 27	340
313	24 17	044	48 41	089	58 38	155	56 51	209	39 09	253	24 16	286	37 11	340
314	24 50	045	49 28	089	58 57	157	56 29	210	38 24	253	23 31	286	36 55	340
	◆Mirfak		Hamal		Diphda		◆FOMALHAUT		ALTAIR		◆VEGA		Kochab	
315	25 23	045	24 28	079	12 44	126	16 06	154	56 04	212	62 19	281	36 39	340
316	25 56	046	25 13	079	13 21	127	16 27	155	55 39	213	61 33	281	36 23	340
317	26 30	046	25 59	080	13 59	128	16 47	155	55 13	215	60 47	282	36 07	340
318	27 03	046	26 45	081	14 35	128	17 06	156	54 46	217	60 02	282	35 52	340
319	27 37	047	27 31	081	15 12	129	17 24	157	54 17	218	59 16	282	35 36	341
320	28 11	047	28 17	082	15 47	130	17 42	158	53 48	220	58 31	283	35 21	341
321	28 45	047	29 04	082	16 23	131	17 59	159	53 18	221	57 45	283	35 05	341
322	29 20	048	29 50	083	16 58	132	18 16	160	52 47	222	57 00	283	34 50	341
323	29 54	048	30 36	083	17 33	132	18 32	161	52 15	224	56 14	284	34 35	341
324	30 29	048	31 22	084	18 07	133	18 47	161	51 43	225	55 29	284	34 19	341
325	31 04	049	32 09	085	18 41	134	19 01	162	51 09	226	54 44	285	34 04	341
326	31 39	049	32 55	085	19 14	135	19 15	163	50 35	228	53 59	285	33 49	341
327	32 14	049	33 42	086	19 47	136	19 28	164	50 00	229	53 14	285	33 35	342
328	32 50	050	34 28	086	20 19	136	19 41	165	49 25	230	52 29	286	33 20	342
329	33 25	050	35 15	087	20 51	137	19 53	166	48 49	231	51 44	286	33 05	342
	◆CAPELLA		Hamal		Diphda		◆FOMALHAUT		ALTAIR		◆VEGA		Kochab	
330	16 02	043	36 01	087	21 23	138	20 04	167	48 12	233	50 59	286	32 51	342
331	16 34	044	36 48	088	21 54	139	20 14	168	47 35	234	50 14	287	32 36	342
332	17 06	044	37 35	089	22 24	140	20 24	169	46 57	235	49 30	287	32 22	342
333	17 39	044	38 21	089	22 54	141	20 33	169	46 18	236	48 45	287	32 08	342
334	18 11	045	39 08	090	23 23	142	20 41	170	45 40	237	48 01	288	31 54	343
335	18 44	045	39 54	091	23 52	142	20 48	171	45 00	238	47 16	288	31 40	343
336	19 18	046	40 41	091	24 20	143	20 55	172	44 21	239	46 32	288	31 26	343
337	19 51	046	41 28	092	24 47	144	21 01	173	43 40	240	45 48	289	31 13	343
338	20 25	047	42 14	093	25 14	145	21 06	174	43 00	241	45 04	289	30 59	343
339	20 59	047	43 01	093	25 40	146	21 10	175	42 19	242	44 20	290	30 46	343
340	21 33	048	43 47	094	26 06	147	21 14	176	41 37	243	43 36	290	30 33	344
341	22 08	048	44 34	095	26 31	148	21 17	177	40 56	244	42 52	290	30 20	344
342	22 43	048	45 20	095	26 55	149	21 19	178	40 14	245	42 09	291	30 07	344
343	23 18	049	46 07	096	27 19	150	21 20	179	39 31	246	41 25	291	29 54	344
344	23 53	049	46 53	097	27 42	151	21 21	180	38 49	247	40 42	292	29 41	344
	◆CAPELLA		ALDEBARAN		Diphda		◆FOMALHAUT		ALTAIR		◆VEGA		Kochab	
345	24 28	050	14 56	081	28 04	152	21 21	181	38 06	247	39 58	292	29 29	345
346	25 04	050	15 42	081	28 26	153	21 20	182	37 23	248	39 15	292	29 17	345
347	25 40	050	16 29	082	28 47	154	21 18	182	36 39	249	38 32	293	29 05	345
348	26 16	051	17 15	082	29 07	155	21 16	183	35 55	250	37 49	293	28 53	345
349	26 52	051	18 01	083	29 26	156	21 13	184	35 11	251	37 06	293	28 41	345
350	27 28	052	18 47	084	29 45	157	21 09	185	34 27	252	36 24	294	28 29	346
351	28 05	052	19 34	084	30 02	158	21 04	186	33 43	252	35 41	294	28 18	346
352	28 42	052	20 20	085	30 19	159	20 59	187	32 58	253	34 59	295	28 07	346
353	29 19	053	21 07	085	30 36	160	20 53	188	32 14	254	34 16	295	27 56	346
354	29 56	053	21 53	086	30 51	161	20 46	189	31 29	255	33 34	295	27 45	347
355	30 33	053	22 40	087	31 05	162	20 38	190	30 44	255	32 52	296	27 34	347
356	31 11	054	23 26	087	31 19	163	20 30	191	29 59	256	32 10	296	27 24	347
357	31 48	054	24 13	088	31 32	165	20 21	192	29 13	257	31 28	297	27 13	347
358	32 26	054	24 59	089	31 44	166	20 11	193	28 28	258	30 47	297	27 03	348
359	33 04	055	25 46	089	31 55	167	20 00	194	27 42	258	30 05	298	26 53	348

LAT 38°N

LHA ♈	Hc	Zn	Hc	Zn	Hc	Zn	Hc	Zn	Hc	Zn	Hc	Zn	Hc	Zn
	◆CAPELLA		ALDEBARAN		◆Diphda		FOMALHAUT		ALTAIR		◆VEGA		Kochab	
0	33 08	055	26 32	089	33 04	168	20 47	195	27 07	260	28 56	298	25 45	348
1	33 46	055	27 19	090	33 14	169	20 35	195	26 21	260	28 14	299	25 35	348
2	34 25	055	28 07	091	33 22	170	20 22	196	25 34	261	27 33	299	25 26	349
3	35 04	055	28 54	091	33 30	171	20 08	197	24 48	262	26 52	300	25 16	349
4	35 43	056	29 41	092	33 37	172	19 54	198	24 01	262	26 11	300	25 07	349
5	36 22	056	30 28	092	33 43	173	19 39	199	23 14	263	25 30	301	24 59	349
6	37 01	056	31 16	093	33 48	174	19 23	200	22 27	264	24 49	301	24 50	350
7	37 41	057	32 03	094	33 52	176	19 07	201	21 40	264	24 09	301	24 42	350
8	38 20	057	32 50	094	33 55	177	18 50	202	20 53	265	23 29	302	24 34	350
9	39 00	057	33 37	095	33 58	178	18 32	202	20 06	266	22 49	302	24 26	350
10	39 40	057	34 24	096	33 59	179	18 14	203	19 19	266	22 09	303	24 18	351
11	40 20	058	35 11	096	33 59	180	17 54	204	18 31	267	21 29	303	24 10	351
12	41 00	058	35 58	097	33 59	181	17 35	205	17 44	267	20 49	304	24 03	351
13	41 40	058	36 45	098	33 57	182	17 14	206	16 57	268	20 10	304	23 56	352
14	42 20	058	37 32	098	33 54	184	16 54	207	16 10	269	19 31	304	23 49	352
	◆CAPELLA		BETELGEUSE		RIGEL		◆Diphda		Enif		◆DENEB		Kochab	
15	43 01	059	17 22	094	15 01	113	33 51	185	37 55	251	42 25	300	23 42	352
16	43 41	059	18 09	095	15 45	114	33 47	186	37 10	251	41 44	301	23 36	352
17	44 22	059	18 56	095	16 28	115	33 41	187	36 26	252	41 04	301	23 30	353
18	45 02	059	19 43	096	17 11	116	33 35	188	35 40	253	40 23	301	23 24	353
19	45 43	060	20 30	097	17 53	116	33 28	189	34 55	254	39 43	302	23 18	353
20	46 24	060	21 17	097	18 35	117	33 20	190	34 10	254	39 03	302	23 12	353
21	47 05	060	22 04	098	19 17	118	33 11	192	33 24	255	38 22	302	23 07	354
22	47 46	060	22 50	099	19 59	118	33 01	193	32 38	256	37 42	302	23 02	354
23	48 27	060	23 37	099	20 40	119	32 50	194	31 52	257	37 03	303	22 57	354
24	49 08	061	24 24	100	21 22	120	32 38	195	31 06	257	36 23	303	22 53	355
25	49 49	061	25 10	101	22 02	121	32 26	196	30 20	258	35 43	303	22 48	355
26	50 30	061	25 57	101	22 43	122	32 12	197	29 34	259	35 04	304	22 44	355
27	51 12	061	26 43	102	23 23	122	31 58	198	28 47	260	34 24	304	22 40	355
28	51 53	061	27 29	103	24 03	123	31 43	199	28 01	260	33 45	304	22 37	356
29	52 34	061	28 15	104	24 42	124	31 27	200	27 14	261	33 06	305	22 33	356
	◆CAPELLA		BETELGEUSE		RIGEL		◆Diphda		Alpheratz		◆DENEB		Kochab	
30	53 16	061	29 01	104	25 21	125	31 10	201	65 09	257	32 27	305	22 30	356
31	53 57	061	29 47	105	26 00	126	30 52	202	64 23	258	31 49	305	22 27	357
32	54 39	062	30 32	106	26 38	126	30 34	203	63 36	259	31 10	306	22 24	357
33	55 21	062	31 18	106	27 16	127	30 14	205	62 50	260	30 32	306	22 22	357
34	56 02	062	32 03	107	27 53	128	29 54	206	62 03	261	29 53	306	22 20	357
35	56 44	062	32 48	108	28 30	129	29 34	207	61 17	262	29 15	307	22 18	358
36	57 26	062	33 33	109	29 06	130	29 12	208	60 30	263	28 37	307	22 16	358
37	58 07	062	34 18	110	29 42	131	28 50	209	59 43	263	28 00	307	22 14	358
38	58 49	062	35 02	110	30 18	132	28 27	210	58 56	264	27 22	308	22 13	359
39	59 31	062	35 46	111	30 53	133	28 03	211	58 09	265	26 45	308	22 12	359
40	60 12	062	36 30	112	31 27	134	27 39	212	57 22	266	26 08	308	22 11	359
41	60 54	062	37 14	113	32 01	135	27 14	213	56 34	266	25 31	309	22 11	000
42	61 36	062	37 57	114	32 35	136	26 48	213	55 47	267	24 54	309	22 11	000
43	62 17	062	38 40	115	33 08	137	26 21	214	55 00	268	24 17	310	22 11	000
44	62 59	062	39 23	115	33 40	138	25 54	215	54 13	268	23 41	310	22 11	000
	◆Dubhe		POLLUX		SIRIUS		◆RIGEL		Diphda		◆Alpheratz		DENEB	
45	20 34	026	30 52	077	14 02	125	34 11	139	25 27	216	53 26	269	23 05	310
46	20 55	026	31 38	077	14 41	126	34 42	140	24 59	217	52 38	270	22 29	311
47	21 16	026	32 24	078	15 19	126	35 13	141	24 30	218	51 51	270	21 53	311
48	21 37	027	33 10	078	15 57	127	35 42	142	24 00	219	51 04	271	21 18	312
49	21 58	027	33 56	079	16 34	128	36 11	143	23 30	220	50 16	271	20 42	312
50	22 20	027	34 43	079	17 12	129	36 40	144	23 00	221	49 29	272	20 07	312
51	22 42	028	35 29	080	17 48	129	37 07	145	22 28	222	48 42	273	19 33	313
52	23 04	028	36 16	080	18 25	130	37 34	146	21 57	222	47 55	273	18 58	313
53	23 26	028	37 03	081	19 01	131	38 00	147	21 25	223	47 07	274	18 24	314
54	23 49	029	37 49	081	19 36	132	38 25	148	20 52	224	46 20	274	17 49	314
55	24 12	029	38 36	082	20 11	133	38 50	149	20 19	225	45 33	275	17 16	315
56	24 35	029	39 23	083	20 46	133	39 13	151	19 45	226	44 46	275	16 42	315
57	24 58	030	40 10	083	21 20	134	39 36	152	19 11	227	43 59	276	16 09	315
58	25 21	030	40 57	084	21 54	135	39 58	153	18 37	227	43 12	276	15 36	316
59	25 45	030	41 44	084	22 27	136	40 19	154	18 02	228	42 25	277	15 03	316
	◆Dubhe		POLLUX		PROCYON		◆SIRIUS		RIGEL		◆Hamal		Schedar	
60	26 09	030	42 31	085	30 36	109	22 59	137	40 39	155	61 50	247	52 30	316
61	26 33	031	43 18	085	31 20	110	23 32	138	40 58	157	61 06	248	51 57	316
62	26 57	031	44 05	086	32 05	111	24 03	138	41 16	158	60 22	249	51 24	316
63	27 22	031	44 52	086	32 49	111	24 34	139	41 34	159	59 38	250	50 51	316
64	27 46	032	45 40	087	33 33	112	25 05	140	41 50	160	58 53	251	50 18	316
65	28 11	032	46 27	087	34 16	113	25 35	141	42 05	162	58 08	252	49 45	316
66	28 36	032	47 14	088	35 00	114	26 04	142	42 20	163	57 23	253	49 12	316
67	29 01	032	48 01	089	35 43	115	26 33	143	42 33	164	56 38	254	48 39	316
68	29 27	033	48 49	089	36 26	116	27 01	144	42 45	166	55 52	255	48 06	316
69	29 52	033	49 36	090	37 08	117	27 29	145	42 56	167	55 06	256	47 32	316
70	30 18	033	50 23	090	37 50	117	27 56	146	43 06	168	54 20	257	46 59	316
71	30 44	033	51 10	091	38 32	118	28 22	147	43 15	170	53 34	258	46 26	316
72	31 10	033	51 58	092	39 13	119	28 47	148	43 23	171	52 48	259	45 53	316
73	31 36	034	52 45	092	39 54	120	29 12	149	43 30	172	52 01	260	45 20	316
74	32 02	034	53 32	093	40 35	121	29 37	150	43 36	174	51 15	260	44 47	316
	◆Dubhe		POLLUX		PROCYON		◆SIRIUS		RIGEL		◆Hamal		Schedar	
75	32 28	034	54 19	094	41 15	122	30 00	151	43 40	175	50 28	261	44 14	316
76	32 55	034	55 06	094	41 55	123	30 23	152	43 44	176	49 41	262	43 41	316
77	33 22	034	55 54	095	42 35	124	30 45	153	43 46	178	48 54	263	43 08	316
78	33 49	035	56 41	096	43 14	125	31 06	154	43 47	179	48 07	263	42 35	316
79	34 15	035	57 28	097	43 52	126	31 26	155	43 47	181	47 20	264	42 02	316
80	34 43	035	58 15	097	44 30	127	31 46	156	43 46	182	46 33	265	41 30	316
81	35 10	035	59 01	098	45 07	128	32 05	157	43 44	183	45 46	265	40 57	316
82	35 37	035	59 48	099	45 44	129	32 23	158	43 41	185	44 59	266	40 24	316
83	36 04	035	60 35	100	46 20	131	32 40	159	43 37	186	44 12	267	39 52	317
84	36 32	036	61 21	101	46 56	132	32 57	160	43 31	187	43 25	267	39 19	317
85	36 59	036	62 08	102	47 31	133	33 12	161	43 24	189	42 37	268	38 47	317
86	37 27	036	62 54	103	48 05	134	33 27	162	43 17	190	41 50	269	38 15	317
87	37 55	036	63 40	104	48 39	135	33 41	164	43 08	191	41 03	269	37 43	317
88	38 23	036	64 26	105	49 11	137	33 54	165	42 58	193	40 16	270	37 11	317
89	38 51	036	65 12	106	49 43	138	34 06	166	42 47	194	39 28	271	36 39	318

LHA ♈	Hc	Zn	Hc	Zn	Hc	Zn	Hc	Zn	Hc	Zn	Hc	Zn	Hc	Zn
	◆Dubhe		REGULUS		PROCYON		◆SIRIUS		RIGEL		ALDEBARAN		◆Mirfak	
90	39 19	036	29 18	098	50 15	139	34 17	167	42 35	195	61 37	227	59 56	306
91	39 47	036	30 05	099	50 45	141	34 27	168	42 22	197	61 02	228	59 17	306
92	40 15	037	30 52	099	51 15	142	34 36	169	42 07	198	60 27	230	58 39	306
93	40 43	037	31 39	100	51 43	143	34 45	170	41 52	199	59 50	231	58 01	305
94	41 11	037	32 25	101	52 11	145	34 52	172	41 36	201	59 13	233	57 22	305
95	41 39	037	33 11	101	52 38	146	34 59	173	41 19	202	58 35	234	56 43	305
96	42 08	037	33 58	102	53 03	148	35 04	174	41 01	203	57 57	235	56 05	305
97	42 36	037	34 44	103	53 28	149	35 09	175	40 42	204	57 17	237	55 26	305
98	43 04	037	35 30	103	53 52	151	35 12	176	40 22	206	56 38	238	54 47	305
99	43 33	037	36 16	104	54 14	152	35 15	177	40 01	207	55 57	239	54 09	305
100	44 01	037	37 02	105	54 35	154	35 17	179	39 39	208	55 16	240	53 30	305
101	44 30	037	37 47	106	54 56	156	35 17	180	39 17	209	54 35	242	52 51	305
102	44 58	037	38 33	107	55 14	157	35 17	181	38 53	210	53 53	243	52 13	305
103	45 26	037	39 18	107	55 32	159	35 16	182	38 29	212	53 11	244	51 34	305
104	45 55	037	40 03	108	55 48	161	35 14	183	38 04	213	52 28	245	50 55	305
	◆Kochab		Denebola		◆REGULUS		SIRIUS		RIGEL		◆ALDEBARAN		CAPELLA	
105	29 30	016	22 51	089	40 48	109	35 11	184	37 38	214	51 45	246	69 16	301
106	29 44	016	23 38	090	41 32	110	35 06	186	37 11	215	51 02	247	68 35	301
107	29 57	016	24 25	090	42 17	111	35 01	187	36 44	216	50 19	248	67 54	300
108	30 10	017	25 13	091	43 01	112	34 55	188	36 16	217	49 35	249	67 13	300
109	30 24	017	26 00	091	43 45	113	34 48	189	35 47	218	48 50	250	66 32	300
110	30 38	017	26 47	092	44 28	113	34 40	190	35 17	219	48 06	251	65 51	299
111	30 52	017	27 34	093	45 11	114	34 32	191	34 47	220	47 21	252	65 10	299
112	31 06	017	28 22	093	45 54	115	34 22	193	34 16	221	46 36	252	64 28	299
113	31 20	017	29 09	094	46 37	116	34 11	194	33 45	222	45 51	253	63 47	299
114	31 34	018	29 56	095	47 19	117	33 59	195	33 12	223	45 06	254	63 05	299
115	31 48	018	30 43	095	48 01	118	33 47	196	32 40	224	44 20	255	62 24	298
116	32 03	018	31 30	096	48 42	119	33 34	197	32 06	225	43 34	256	61 42	298
117	32 18	018	32 17	097	49 23	120	33 19	198	31 33	226	42 48	257	61 01	298
118	32 32	018	33 04	097	50 04	121	33 04	199	30 58	227	42 02	257	60 19	298
119	32 47	018	33 51	098	50 44	123	32 48	200	30 23	228	41 16	258	59 37	298
	◆Kochab		Denebola		◆REGULUS		SIRIUS		RIGEL		◆ALDEBARAN		CAPELLA	
120	33 02	019	34 38	099	51 23	124	32 31	201	29 48	229	40 30	259	58 56	298
121	33 17	019	35 24	099	52 02	125	32 14	203	29 12	230	39 43	260	58 14	298
122	33 32	019	36 11	100	52 41	126	31 55	204	28 35	231	38 57	260	57 32	298
123	33 48	019	36 58	101	53 19	127	31 36	205	27 59	232	38 10	261	56 50	298
124	34 03	019	37 44	102	53 56	129	31 16	206	27 21	233	37 23	262	56 09	298
125	34 18	019	38 30	102	54 33	130	30 55	207	26 44	233	36 37	262	55 27	298
126	34 34	019	39 16	103	55 08	131	30 33	208	26 05	234	35 50	263	54 46	298
127	34 49	019	40 02	104	55 44	133	30 11	209	25 27	235	35 03	264	54 04	299
128	35 05	019	40 48	105	56 18	134	29 48	210	24 48	236	34 16	264	53 23	299
129	35 21	019	41 34	105	56 51	136	29 24	211	24 09	237	33 29	265	52 41	299
130	35 37	020	42 19	106	57 24	137	28 59	212	23 29	238	32 41	266	52 00	299
131	35 53	020	43 05	107	57 56	139	28 34	213	22 49	238	31 54	266	51 18	299
132	36 08	020	43 50	108	58 27	140	28 08	214	22 08	239	31 07	267	50 37	299
133	36 24	020	44 35	109	58 56	142	27 41	215	21 28	240	30 20	268	49 56	299
134	36 41	020	45 19	110	59 25	143	27 14	216	20 47	241	29 33	268	49 14	299
	Kochab		◆ARCTURUS		SPICA		REGULUS		◆SIRIUS		BETELGEUSE		◆CAPELLA	
135	36 57	020	20 16	081	11 07	114	59 53	145	26 46	217	38 17	246	48 33	300
136	37 13	020	21 03	082	11 50	114	60 19	147	26 18	218	37 33	247	47 52	300
137	37 29	020	21 49	082	12 33	115	60 44	149	25 49	218	36 50	248	47 11	300
138	37 45	020	22 36	083	13 15	116	61 08	151	25 19	219	36 06	248	46 30	300
139	38 01	020	23 23	083	13 58	117	61 31	152	24 49	220	35 22	249	45 49	300
140	38 18	020	24 10	084	14 40	117	61 52	154	24 18	221	34 38	250	45 09	301
141	38 34	020	24 57	084	15 22	118	62 11	156	23 46	222	33 53	251	44 28	301
142	38 50	020	25 44	085	16 04	119	62 30	158	23 14	223	33 08	252	43 48	301
143	39 07	020	26 31	086	16 45	119	62 46	160	22 42	224	32 23	252	43 07	301
144	39 23	020	27 19	086	17 26	120	63 01	162	22 09	225	31 38	253	42 27	301
145	39 39	020	28 06	087	18 07	121	63 15	165	21 36	225	30 53	254	41 46	302
146	39 56	020	28 53	087	18 47	122	63 26	167	21 02	226	30 07	255	41 06	302
147	40 12	020	29 40	088	19 27	122	63 36	169	20 27	227	29 22	255	40 26	302
148	40 28	020	30 27	089	20 07	123	63 45	171	19 52	228	28 36	256	39 46	302
149	40 45	020	31 15	089	20 46	124	63 51	173	19 17	229	27 50	257	39 06	303
	◆Kochab		ARCTURUS		◆SPICA		REGULUS		PROCYON		◆POLLUX		CAPELLA	
150	41 01	020	32 02	090	21 25	125	63 56	175	44 12	233	60 12	261	38 27	303
151	41 17	020	32 49	090	22 04	126	63 59	178	43 34	234	59 25	261	37 47	303
152	41 34	020	33 37	091	22 42	126	64 00	180	42 55	235	58 38	262	37 08	304
153	41 50	020	34 24	092	23 20	127	63 59	182	42 16	236	57 51	263	36 28	304
154	42 06	020	35 11	092	23 58	128	63 56	184	41 36	237	57 04	264	35 49	304
155	42 22	020	35 58	093	24 35	129	63 51	187	40 56	238	56 17	264	35 10	304
156	42 38	020	36 46	094	25 11	130	63 45	189	40 16	239	55 30	265	34 31	305
157	42 54	020	37 33	094	25 47	131	63 37	191	39 35	240	54 43	266	33 53	305
158	43 11	020	38 20	095	26 23	131	63 27	193	38 54	241	53 56	267	33 14	305
159	43 26	020	39 07	096	26 58	132	63 15	195	38 12	242	53 09	267	32 35	306
160	43 42	020	39 54	096	27 33	133	63 02	197	37 30	243	52 21	268	31 57	306
161	43 58	019	40 41	097	28 07	134	62 47	199	36 48	244	51 34	269	31 19	306
162	44 14	019	41 28	098	28 41	135	62 31	202	36 05	245	50 47	269	30 41	307
163	44 30	019	42 15	098	29 14	136	62 12	204	35 22	246	50 00	270	30 03	307
164	44 45	019	43 01	099	29 47	137	61 53	206	34 39	246	49 12	270	29 26	307
	◆Kochab		ARCTURUS		◆SPICA		REGULUS		PROCYON		◆POLLUX		CAPELLA	
165	45 01	019	43 48	100	30 19	138	61 32	207	33 55	247	48 25	271	28 48	308
166	45 16	019	44 35	101	30 50	139	61 09	209	33 12	248	47 38	272	28 11	308
167	45 31	019	45 21	101	31 21	140	60 45	211	32 28	249	46 51	272	27 34	308
168	45 46	019	46 07	102	31 52	141	60 20	213	31 43	250	46 03	273	26 57	309
169	46 01	018	46 53	103	32 21	142	59 54	215	30 59	250	45 16	273	26 20	309
170	46 16	018	47 39	104	32 50	143	59 27	216	30 14	251	44 29	274	25 43	310
171	46 31	018	48 25	105	33 18	144	58 58	218	29 29	252	43 42	275	25 07	310
172	46 46	018	49 11	105	33 46	145	58 28	220	28 44	253	42 55	275	24 31	310
173	47 00	018	49 56	106	34 13	146	57 57	221	27 59	254	42 08	276	23 55	311
174	47 14	018	50 42	107	34 39	147	57 26	223	27 14	254	41 21	276	23 19	311
175	47 29	017	51 27	108	35 04	148	56 53	224	26 28	255	40 34	277	22 44	311
176	47 43	017	52 12	109	35 29	149	56 20	226	25 42	256	39 47	277	22 08	312
177	47 56	017	52 56	110	35 53	150	55 45	227	24 57	256	39 00	278	21 33	312
178	48 10	017	53 40	111	36 16	151	55 10	229	24 11	257	38 13	278	20 58	313
179	48 24	016	54 24	112	36 38	152	54 35	230	23 24	258	37 26	279	20 24	313

LHA ♈	Hc	Zn	Hc	Zn	Hc	Zn	Hc	Zn	Hc	Zn	Hc	Zn	Hc	Zn
	◆VEGA		Alphecca		ARCTURUS		◆SPICA		REGULUS		◆POLLUX		CAPELLA	
180	16 42	053	43 59	088	55 08	113	37 00	154	53 58	231	36 40	279	19 49	314
181	17 20	054	44 46	088	55 51	114	37 21	155	53 21	233	35 53	280	19 15	314
182	17 59	054	45 33	089	56 34	115	37 40	156	52 43	234	35 06	280	18 41	314
183	18 37	055	46 20	090	57 17	116	37 59	157	52 05	235	34 20	281	18 08	315
184	19 16	055	47 08	090	57 59	118	38 17	158	51 26	236	33 34	281	17 34	315
185	19 55	056	47 55	091	58 41	119	38 34	159	50 46	237	32 47	282	17 01	316
186	20 34	056	48 42	091	59 22	120	38 50	161	50 06	238	32 01	282	16 28	316
187	21 14	057	49 30	092	60 03	121	39 05	162	49 25	240	31 15	283	15 55	317
188	21 53	057	50 17	093	60 43	123	39 20	163	48 44	241	30 29	283	15 23	317
189	22 33	058	51 04	093	61 22	124	39 33	164	48 03	242	29 43	284	14 51	317
190	23 13	058	51 51	094	62 01	126	39 45	166	47 21	243	28 57	284	14 19	318
191	23 53	058	52 38	095	62 39	127	39 56	167	46 39	244	28 11	285	13 48	318
192	24 34	059	53 25	095	63 17	129	40 07	168	45 56	245	27 26	285	13 16	319
193	25 14	059	54 12	096	63 53	130	40 16	169	45 14	246	26 40	286	12 46	319
194	25 55	060	54 59	097	64 29	132	40 24	171	44 30	247	25 55	286	12 15	320
	DENEB		◆VEGA		Alphecca		◆SPICA		REGULUS		◆POLLUX		Dubhe	
195	11 33	040	26 36	060	55 46	098	40 31	172	43 47	247	25 10	287	60 13	332
196	12 04	041	27 17	061	56 33	098	40 37	173	43 03	248	24 24	287	59 51	332
197	12 35	041	27 58	061	57 20	099	40 42	175	42 19	249	23 39	288	59 28	331
198	13 06	042	28 40	061	58 06	100	40 46	176	41 35	250	22 54	289	59 06	331
199	13 38	042	29 21	062	58 53	101	40 49	177	40 50	251	22 10	289	58 42	330
200	14 10	043	30 03	062	59 39	102	40 51	178	40 05	252	21 25	290	58 18	330
201	14 42	043	30 45	063	60 26	103	40 52	180	39 20	253	20 41	290	57 54	329
202	15 15	044	31 27	063	61 12	104	40 52	181	38 35	253	19 56	291	57 30	329
203	15 48	044	32 09	063	61 57	105	40 50	182	37 50	254	19 12	291	57 05	328
204	16 21	045	32 52	064	62 43	106	40 48	184	37 04	255	18 28	292	56 40	328
205	16 54	045	33 34	064	63 28	107	40 44	185	36 18	256	17 44	292	56 15	327
206	17 28	046	34 17	065	64 14	108	40 40	186	35 32	256	17 00	293	55 50	327
207	18 02	046	35 00	065	64 59	109	40 34	187	34 46	257	16 17	293	55 24	327
208	18 36	047	35 42	065	65 43	110	40 27	189	34 00	258	15 33	294	54 58	326
209	19 11	047	36 25	066	66 27	111	40 20	190	33 14	259	14 50	294	54 31	326
	DENEB		◆VEGA		Rasalhague		ANTARES		◆SPICA		REGULUS		◆Dubhe	
210	19 45	047	37 09	066	36 07	103	16 43	146	40 11	191	32 28	259	54 05	326
211	20 20	048	37 52	066	36 53	104	17 09	146	40 01	193	31 41	260	53 38	326
212	20 55	048	38 35	067	37 39	105	17 35	147	39 50	194	30 54	261	53 12	325
213	21 31	049	39 19	067	38 25	106	18 01	148	39 39	195	30 08	261	52 45	325
214	22 06	049	40 03	068	39 10	106	18 25	149	39 26	196	29 21	262	52 17	325
215	22 42	049	40 46	068	39 56	107	18 50	150	39 12	198	28 34	263	51 50	325
216	23 18	050	41 30	068	40 41	108	19 13	151	38 57	199	27 47	263	51 23	324
217	23 54	050	42 14	069	41 26	109	19 36	151	38 42	200	27 00	264	50 55	324
218	24 31	051	42 58	069	42 10	110	19 59	152	38 25	201	26 13	265	50 27	324
219	25 08	051	43 42	069	42 55	110	20 20	153	38 08	202	25 26	265	50 00	324
220	25 44	051	44 27	070	43 39	111	20 41	154	37 49	204	24 39	266	49 32	324
221	26 21	052	45 11	070	44 23	112	21 02	155	37 30	205	23 52	267	49 04	324
222	26 59	052	45 56	070	45 07	113	21 21	156	37 10	206	23 04	267	48 36	324
223	27 36	053	46 40	071	45 50	114	21 41	157	36 49	207	22 17	268	48 08	323
224	28 14	053	47 25	071	46 33	115	21 59	158	36 27	208	21 30	269	47 40	323
	DENEB		◆ALTAIR		Rasalhague		ANTARES		◆SPICA		Denebola		◆Dubhe	
225	28 51	053	19 06	094	47 15	116	22 17	158	36 04	209	41 52	254	47 11	323
226	29 29	054	19 53	094	47 58	117	22 34	159	35 40	210	41 07	255	46 43	323
227	30 08	054	20 40	095	48 40	118	22 50	160	35 16	211	40 21	256	46 15	323
228	30 46	054	21 27	096	49 21	119	23 05	161	34 51	213	39 35	257	45 46	323
229	31 24	055	22 14	096	50 02	120	23 20	162	34 25	214	38 49	257	45 18	323
230	32 03	055	23 01	097	50 43	121	23 34	163	33 59	215	38 03	258	44 50	323
231	32 42	055	23 48	098	51 23	122	23 48	164	33 31	216	37 16	259	44 21	323
232	33 21	056	24 35	098	52 03	124	24 00	165	33 03	217	36 30	260	43 53	323
233	34 00	056	25 22	099	52 42	125	24 12	166	32 35	218	35 43	260	43 25	323
234	34 39	056	26 08	100	53 20	126	24 23	167	32 05	219	34 57	261	42 56	323
235	35 18	057	26 55	100	53 58	127	24 33	168	31 35	220	34 10	262	42 28	323
236	35 58	057	27 41	101	54 36	129	24 43	169	31 05	221	33 23	262	42 00	323
237	36 37	057	28 28	102	55 12	130	24 52	170	30 34	222	32 36	263	41 31	323
238	37 17	057	29 14	102	55 48	131	25 00	171	30 02	223	31 49	264	41 03	323
239	37 57	058	30 00	103	56 23	133	25 07	172	29 29	224	31 02	264	40 35	323
	◆DENEB		ALTAIR		Nunki		◆ANTARES		ARCTURUS		◆Alkaid		Kochab	
240	38 37	058	30 46	104	13 45	140	25 13	173	60 32	238	63 45	306	52 53	352
241	39 17	058	31 32	104	14 15	141	25 19	174	59 51	239	63 07	306	52 47	352
242	39 58	059	32 18	105	14 44	142	25 23	175	59 10	240	62 28	306	52 40	351
243	40 38	059	33 03	106	15 13	143	25 27	176	58 29	242	61 50	305	52 33	351
244	41 19	059	33 49	107	15 41	143	25 30	177	57 47	243	61 11	305	52 25	351
245	41 59	059	34 34	108	16 09	144	25 33	178	57 05	244	60 33	305	52 17	350
246	42 40	060	35 19	108	16 37	145	25 34	179	56 22	245	59 54	305	52 09	350
247	43 21	060	36 04	109	17 03	146	25 35	180	55 39	246	59 15	305	52 01	349
248	44 02	060	36 48	110	17 30	147	25 35	181	54 56	247	58 36	305	51 52	349
249	44 43	060	37 32	111	17 55	148	25 34	182	54 12	248	57 57	304	51 43	349
250	45 24	061	38 17	112	18 20	148	25 32	183	53 28	249	57 18	304	51 33	348
251	46 05	061	39 00	112	18 45	149	25 29	184	52 44	250	56 39	304	51 24	348
252	46 46	061	39 44	113	19 09	150	25 26	185	51 59	251	56 00	304	51 14	348
253	47 28	061	40 27	114	19 32	151	25 21	186	51 14	252	55 21	304	51 04	347
254	48 09	061	41 10	115	19 55	152	25 16	187	50 29	253	54 42	304	50 53	347
	◆DENEB		ALTAIR		Nunki		◆ANTARES		ARCTURUS		◆Alkaid		Kochab	
255	48 51	062	41 53	116	20 17	153	25 10	188	49 44	254	54 03	304	50 42	347
256	49 32	062	42 35	117	20 38	154	25 04	189	48 58	255	53 24	304	50 31	346
257	50 14	062	43 17	118	20 59	154	24 56	190	48 12	256	52 45	304	50 20	346
258	50 56	062	43 59	119	21 19	155	24 48	191	47 26	256	52 06	304	50 09	346
259	51 38	062	44 40	120	21 38	156	24 39	192	46 40	257	51 27	304	49 57	345
260	52 20	062	45 21	121	21 57	157	24 29	193	45 54	258	50 48	304	49 45	345
261	53 02	063	46 01	122	22 15	158	24 18	194	45 08	259	50 09	304	49 33	345
262	53 44	063	46 41	123	22 33	159	24 07	194	44 21	260	49 30	305	49 20	345
263	54 26	063	47 21	124	22 49	160	23 55	195	43 35	260	48 51	305	49 08	344
264	55 08	063	47 59	125	23 05	161	23 42	196	42 48	261	48 12	305	48 55	344
265	55 50	063	48 38	126	23 20	162	23 28	197	42 01	262	47 33	305	48 42	344
266	56 32	063	49 16	127	23 35	163	23 14	198	41 15	263	46 54	305	48 28	344
267	57 14	063	49 53	129	23 48	164	22 59	199	40 28	263	46 16	305	48 15	343
268	57 56	063	50 30	130	24 01	165	22 43	200	39 41	264	45 37	305	48 01	343
269	58 38	063	51 06	131	24 14	166	22 26	201	38 54	265	44 58	305	47 48	343

LAT 38°N

LHA ♈	Hc	Zn	Hc	Zn	Hc	Zn	Hc	Zn	Hc	Zn	Hc	Zn	Hc	Zn
	◆Alpheratz		ALTAIR		Nunki		◆ANTARES		ARCTURUS		◆Alkaid		Kochab	
270	15 56	065	51 41	132	24 25	166	22 09	202	38 06	265	44 20	306	47 34	343
271	16 39	066	52 15	134	24 36	167	21 51	203	37 19	266	43 42	306	47 20	343
272	17 23	066	52 49	135	24 46	168	21 32	204	36 32	267	43 03	306	47 05	342
273	18 06	067	53 22	136	24 55	169	21 13	205	35 45	267	42 25	306	46 51	342
274	18 50	067	53 55	138	25 03	170	20 53	206	34 58	268	41 47	306	46 36	342
275	19 33	068	54 26	139	25 11	171	20 32	206	34 10	269	41 09	307	46 22	342
276	20 17	068	54 56	141	25 17	172	20 11	207	33 23	269	40 31	307	46 07	342
277	21 01	069	55 26	142	25 23	173	19 49	208	32 36	270	39 53	307	45 52	341
278	21 45	069	55 54	144	25 28	174	19 26	209	31 49	270	39 15	307	45 37	341
279	22 30	070	56 22	145	25 33	175	19 03	210	31 01	271	38 38	307	45 21	341
280	23 14	070	56 48	147	25 36	176	18 39	211	30 14	272	38 00	308	45 06	341
281	23 59	071	57 13	148	25 39	177	18 14	212	29 27	272	37 23	308	44 51	341
282	24 44	071	57 38	150	25 41	178	17 49	212	28 40	273	36 46	308	44 35	341
283	25 29	072	58 01	152	25 42	179	17 24	213	27 52	273	36 09	308	44 19	341
284	26 14	072	58 22	154	25 42	180	16 58	214	27 05	274	35 32	309	44 04	341
	Alpheratz		◆Enif		ALTAIR		◆Rasalhague		ARCTURUS		Alkaid		◆Kochab	
285	26 59	073	43 45	117	58 42	155	58 15	222	26 18	275	34 55	309	43 48	340
286	27 44	073	44 27	118	59 01	157	57 43	224	25 31	275	34 18	309	43 32	340
287	28 29	074	45 08	119	59 19	159	57 10	225	24 44	276	33 42	309	43 16	340
288	29 15	074	45 50	120	59 35	161	56 36	227	23 57	276	33 05	310	43 00	340
289	30 00	075	46 31	121	59 50	163	56 01	228	23 10	277	32 29	310	42 44	340
290	30 46	075	47 11	122	60 03	165	55 25	230	22 23	277	31 53	310	42 28	340
291	31 32	076	47 51	123	60 15	167	54 49	231	21 36	278	31 17	311	42 12	340
292	32 18	076	48 31	124	60 25	169	54 11	232	20 49	279	30 41	311	41 56	340
293	33 04	077	49 10	125	60 33	171	53 34	234	20 03	279	30 05	311	41 40	340
294	33 50	077	49 48	126	60 40	173	52 55	235	19 16	280	29 30	312	41 23	340
295	34 36	078	50 26	127	60 45	175	52 16	236	18 29	280	28 55	312	41 07	340
296	35 22	078	51 04	128	60 49	177	51 37	237	17 43	281	28 20	312	40 51	340
297	36 09	079	51 40	130	60 51	179	50 57	238	16 57	281	27 45	313	40 34	340
298	36 55	079	52 16	131	60 51	181	50 17	239	16 10	282	27 10	313	40 18	340
299	37 42	080	52 52	132	60 50	183	49 36	241	15 24	283	26 36	313	40 02	340
	◆Mirfak		Alpheratz		◆FOMALHAUT		ALTAIR		Rasalhague		◆Alphecca		Kochab	
300	16 46	039	38 28	080	10 40	142	60 47	185	48 54	242	33 59	279	39 45	340
301	17 16	039	39 15	081	11 09	143	60 42	187	48 12	243	33 12	280	39 29	340
302	17 46	040	40 02	081	11 38	143	60 35	189	47 30	244	32 26	280	39 13	340
303	18 17	040	40 48	082	12 06	144	60 27	191	46 48	245	31 39	281	38 56	340
304	18 47	041	41 35	082	12 33	145	60 18	193	46 05	246	30 53	281	38 40	340
305	19 18	041	42 22	083	13 00	146	60 06	195	45 22	247	30 06	282	38 24	340
306	19 49	041	43 09	084	13 27	146	59 54	197	44 38	247	29 20	282	38 07	340
307	20 21	042	43 56	084	13 53	147	59 39	199	43 54	248	28 34	283	37 51	340
308	20 52	042	44 43	085	14 18	148	59 24	200	43 10	249	27 48	283	37 35	340
309	21 24	043	45 30	085	14 43	149	59 06	202	42 26	250	27 02	284	37 19	340
310	21 56	043	46 17	086	15 07	149	58 48	204	41 41	251	26 16	285	37 02	340
311	22 29	043	47 04	086	15 31	150	58 28	206	40 56	252	25 31	285	36 46	340
312	23 01	044	47 52	087	15 54	151	58 06	208	40 11	253	24 45	286	36 30	340
313	23 34	044	48 39	087	16 17	152	57 44	209	39 26	253	24 00	286	36 14	340
314	24 07	045	49 26	088	16 39	153	57 20	211	38 41	254	23 14	287	35 58	340
	◆Mirfak		Hamal		Diphda		◆FOMALHAUT		ALTAIR		◆VEGA		Kochab	
315	24 41	045	24 16	078	13 19	126	17 00	154	56 55	213	62 06	283	35 42	340
316	25 14	045	25 02	079	13 57	127	17 21	154	56 29	214	61 20	283	35 27	340
317	25 48	046	25 49	079	14 35	128	17 41	155	56 02	216	60 34	283	35 11	341
318	26 22	046	26 35	080	15 12	128	18 01	156	55 34	217	59 48	284	34 55	341
319	26 56	046	27 22	081	15 49	129	18 19	157	55 04	219	59 02	284	34 39	341
320	27 30	047	28 08	081	16 26	130	18 38	158	54 34	220	58 16	284	34 24	341
321	28 04	047	28 55	082	17 02	131	18 55	159	54 03	222	57 31	285	34 08	341
322	28 39	047	29 42	082	17 38	131	19 12	160	53 31	223	56 45	285	33 53	341
323	29 14	048	30 29	083	18 13	132	19 28	160	52 58	225	55 59	285	33 38	341
324	29 49	048	31 16	083	18 48	133	19 44	161	52 25	226	55 14	286	33 23	341
325	30 24	048	32 03	084	19 22	134	19 59	162	51 50	227	54 28	286	33 08	341
326	30 59	049	32 50	084	19 56	134	20 13	163	51 15	229	53 43	286	32 53	342
327	31 35	049	33 37	085	20 30	135	20 26	164	50 39	230	52 57	287	32 38	342
328	32 11	049	34 24	086	21 03	136	20 39	165	50 03	231	52 12	287	32 23	342
329	32 47	049	35 11	086	21 35	137	20 51	166	49 26	232	51 27	287	32 08	342
	◆CAPELLA		Hamal		Diphda		◆FOMALHAUT		ALTAIR		◆VEGA		Kochab	
330	15 18	043	35 58	087	22 07	138	21 02	167	48 48	233	50 42	287	31 54	342
331	15 50	043	36 46	087	22 39	139	21 13	168	48 10	235	49 57	288	31 39	342
332	16 23	044	37 33	088	23 10	140	21 22	168	47 31	236	49 12	288	31 25	342
333	16 56	044	38 20	089	23 40	140	21 31	169	46 52	237	48 27	288	31 11	343
334	17 29	045	39 07	089	24 10	141	21 40	170	46 12	238	47 42	289	30 57	343
335	18 02	045	39 55	090	24 39	142	21 47	171	45 32	239	46 57	289	30 43	343
336	18 36	046	40 42	090	25 08	143	21 54	172	44 51	240	46 13	289	30 29	343
337	19 10	046	41 29	091	25 36	144	22 00	173	44 10	241	45 28	290	30 15	343
338	19 44	046	42 16	092	26 03	145	22 05	174	43 28	242	44 44	290	30 02	343
339	20 18	047	43 04	092	26 30	146	22 10	175	42 47	243	43 59	291	29 48	344
340	20 53	047	43 51	093	26 56	147	22 14	176	42 04	244	43 15	291	29 35	344
341	21 28	048	44 38	094	27 22	148	22 17	177	41 22	245	42 31	291	29 22	344
342	22 03	048	45 25	094	27 47	149	22 19	178	40 39	246	41 47	292	29 09	344
343	22 38	048	46 12	095	28 11	150	22 20	179	39 56	246	41 03	292	28 56	344
344	23 14	049	47 00	096	28 34	151	22 21	180	39 12	247	40 19	292	28 44	345
	◆CAPELLA		ALDEBARAN		Diphda		◆FOMALHAUT		ALTAIR		◆VEGA		Kochab	
345	23 49	049	14 47	080	28 57	152	22 21	181	38 28	248	39 36	293	28 31	345
346	24 25	050	15 33	081	29 19	153	22 20	182	37 44	249	38 52	293	28 19	345
347	25 01	050	16 20	082	29 40	154	22 18	182	37 00	250	38 09	293	28 07	345
348	25 38	050	17 07	082	30 01	155	22 16	183	36 16	251	37 25	294	27 55	345
349	26 14	051	17 54	083	30 21	156	22 13	184	35 31	251	36 42	294	27 43	346
350	26 51	051	18 41	083	30 40	157	22 09	185	34 46	252	35 59	295	27 31	346
351	27 28	051	19 28	084	30 58	158	22 04	186	34 01	253	35 16	295	27 20	346
352	28 05	052	20 15	084	31 15	159	21 58	187	33 15	254	34 33	295	27 08	346
353	28 42	052	21 02	085	31 32	160	21 52	188	32 30	255	33 51	296	26 57	347
354	29 20	053	21 49	086	31 48	161	21 45	189	31 44	255	33 08	296	26 46	347
355	29 57	053	22 36	086	32 03	162	21 37	190	30 59	256	32 26	296	26 36	347
356	30 35	053	23 23	087	32 17	163	21 29	191	30 13	257	31 43	297	26 25	347
357	31 13	054	24 10	087	32 30	164	21 19	192	29 26	257	31 01	297	26 15	347
358	31 51	054	24 58	088	32 42	165	21 09	193	28 40	258	30 19	298	26 04	348
359	32 29	054	25 45	089	32 54	167	20 59	194	27 54	259	29 37	298	25 54	348

LAT 37°N

LHA	Hc	Zn	Hc	Zn	Hc	Zn	Hc	Zn	Hc	Zn	Hc	Zn	Hc	Zn
♈	◆CAPELLA		ALDEBARAN		◆Diphda		FOMALHAUT		ALTAIR		◆VEGA		Kochab	
0	32 33	054	26 31	089	34 03	168	21 45	195	27 18	260	28 27	299	24 46	348
1	33 12	054	27 19	089	34 13	169	21 33	196	26 31	261	27 45	299	24 36	348
2	33 51	055	28 07	090	34 21	170	21 20	196	25 43	261	27 03	300	24 27	349
3	34 30	055	28 55	091	34 30	171	21 06	197	24 56	262	26 22	300	24 18	349
4	35 09	055	29 43	091	34 37	172	20 51	198	24 09	263	25 41	301	24 09	349
5	35 48	055	30 31	092	34 43	173	20 36	199	23 21	263	24 59	301	24 00	349
6	36 28	056	31 19	092	34 48	174	20 20	200	22 33	264	24 18	301	23 51	350
7	37 08	056	32 06	093	34 52	176	20 03	201	21 46	265	23 37	302	23 43	350
8	37 47	056	32 54	094	34 55	177	19 45	202	20 58	265	22 57	302	23 34	350
9	38 27	057	33 42	094	34 58	178	19 27	203	20 10	266	22 16	303	23 26	351
10	39 07	057	34 30	095	34 59	179	19 09	203	19 22	267	21 36	303	23 19	351
11	39 47	057	35 18	096	34 59	180	18 49	204	18 34	267	20 56	303	23 11	351
12	40 28	057	36 05	096	34 59	181	18 29	205	17 47	268	20 16	304	23 04	351
13	41 08	057	36 53	097	34 57	183	18 08	206	16 59	268	19 36	304	22 57	352
14	41 49	058	37 40	098	34 54	184	17 47	207	16 11	269	18 57	305	22 50	352
	◆CAPELLA		BETELGEUSE		RIGEL		◆Diphda		Enif		◆DENEB		Kochab	
15	42 29	058	17 26	094	15 25	113	34 51	185	38 15	251	41 54	301	22 43	352
16	43 10	058	18 14	094	16 09	114	34 46	186	37 29	252	41 13	301	22 36	352
17	43 50	058	19 01	095	16 53	115	34 41	187	36 44	253	40 32	302	22 30	353
18	44 31	059	19 49	096	17 36	115	34 34	188	35 58	254	39 52	302	22 24	353
19	45 12	059	20 37	096	18 19	116	34 27	189	35 12	254	39 11	302	22 18	353
20	45 53	059	21 24	097	19 02	117	34 19	191	34 25	255	38 31	302	22 13	353
21	46 34	059	22 12	098	19 45	117	34 09	192	33 39	256	37 50	303	22 08	354
22	47 15	059	22 59	098	20 27	118	33 59	193	32 52	257	37 10	303	22 02	354
23	47 57	059	23 47	099	21 10	119	33 48	194	32 06	257	36 30	303	21 58	354
24	48 38	060	24 34	100	21 51	120	33 36	195	31 19	258	35 50	304	21 53	355
25	49 19	060	25 21	100	22 33	120	33 23	196	30 32	259	35 10	304	21 49	355
26	50 01	060	26 08	101	23 14	121	33 09	197	29 45	259	34 30	304	21 44	355
27	50 42	060	26 55	102	23 55	122	32 55	198	28 58	260	33 51	304	21 40	355
28	51 24	060	27 42	102	24 35	123	32 39	199	28 11	261	33 11	305	21 37	356
29	52 05	060	28 29	103	25 15	124	32 23	201	27 23	261	32 32	305	21 33	356
	◆CAPELLA		BETELGEUSE		RIGEL		◆Diphda		Alpheratz		◆DENEB		Kochab	
30	52 47	060	29 15	104	25 55	124	32 06	202	65 21	259	31 53	305	21 30	356
31	53 28	060	30 02	104	26 34	125	31 47	203	64 34	260	31 14	306	21 27	357
32	54 10	060	30 48	105	27 13	126	31 29	204	63 47	261	30 35	306	21 24	357
33	54 52	060	31 34	106	27 52	127	31 09	205	62 59	262	29 56	306	21 22	357
34	55 33	060	32 20	107	28 30	128	30 48	206	62 12	263	29 18	307	21 20	357
35	56 15	060	33 06	107	29 08	129	30 27	207	61 24	264	28 39	307	21 18	358
36	56 57	060	33 52	108	29 45	130	30 05	208	60 37	264	28 01	307	21 16	358
37	57 38	060	34 37	109	30 22	130	29 42	209	59 49	265	27 23	308	21 14	358
38	58 20	060	35 23	110	30 58	131	29 19	210	59 01	266	26 45	308	21 13	359
39	59 02	060	36 08	110	31 34	132	28 55	211	58 13	266	26 08	308	21 12	359
40	59 43	060	36 52	111	32 09	133	28 30	212	57 25	267	25 30	309	21 11	359
41	60 25	060	37 37	112	32 43	134	28 04	213	56 38	268	24 53	309	21 11	000
42	61 06	060	38 21	113	33 17	135	27 38	214	55 50	268	24 16	309	21 11	000
43	61 48	060	39 05	114	33 51	136	27 11	215	55 02	269	23 39	310	21 11	000
44	62 29	060	39 49	115	34 24	137	26 43	216	54 14	270	23 02	310	21 11	000
	◆Dubhe		POLLUX		SIRIUS		◆RIGEL		Diphda		◆Alpheratz		DENEB	
45	19 40	026	30 38	076	14 36	125	34 56	138	26 15	217	53 26	270	22 26	311
46	20 01	026	31 24	077	15 15	125	35 28	139	25 46	217	52 38	271	21 50	311
47	20 22	026	32 11	077	15 54	126	35 59	140	25 17	218	51 50	271	21 14	311
48	20 43	027	32 58	078	16 33	127	36 29	141	24 47	219	51 02	272	20 38	312
49	21 05	027	33 45	078	17 11	128	36 59	142	24 16	220	50 14	273	20 02	312
50	21 27	027	34 31	079	17 49	128	37 28	143	23 45	221	49 26	273	19 27	313
51	21 49	028	35 19	079	18 26	129	37 56	144	23 13	222	48 39	274	18 52	313
52	22 11	028	36 06	080	19 03	130	38 24	146	22 41	223	47 51	274	18 17	313
53	22 33	028	36 53	080	19 40	131	38 50	147	22 08	224	47 03	275	17 42	314
54	22 56	028	37 40	081	20 16	131	39 16	148	21 35	224	46 15	275	17 08	314
55	23 19	029	38 27	081	20 52	132	39 41	149	21 01	225	45 28	276	16 34	315
56	23 42	029	39 15	082	21 27	133	40 05	150	20 27	226	44 40	276	16 00	315
57	24 06	029	40 02	082	22 02	134	40 29	151	19 52	227	43 52	277	15 26	316
58	24 29	030	40 50	083	22 36	135	40 51	153	19 17	228	43 05	277	14 53	316
59	24 53	030	41 37	083	23 10	136	41 13	154	18 41	228	42 17	278	14 19	317
	◆Dubhe		POLLUX		PROCYON		◆SIRIUS		RIGEL		◆Hamal		Schedar	
60	25 17	030	42 25	084	30 55	109	23 43	136	41 33	155	62 13	249	51 46	317
61	25 41	030	43 13	084	31 40	109	24 16	137	41 53	156	61 28	250	51 13	317
62	26 06	031	44 00	085	32 26	110	24 48	138	42 12	158	60 43	251	50 40	317
63	26 30	031	44 48	085	33 11	111	25 20	139	42 30	159	59 57	252	50 08	317
64	26 55	031	45 36	086	33 55	112	25 51	140	42 46	160	59 12	253	49 35	317
65	27 20	032	46 24	086	34 40	112	26 21	141	43 02	161	58 26	254	49 02	316
66	27 45	032	47 11	087	35 24	113	26 51	142	43 17	163	57 39	255	48 29	316
67	28 10	032	47 59	088	36 08	114	27 21	143	43 31	164	56 53	256	47 56	316
68	28 36	032	48 47	088	36 51	115	27 50	144	43 43	165	56 06	257	47 22	316
69	29 02	033	49 35	089	37 34	116	28 18	145	43 55	167	55 20	258	46 49	316
70	29 27	033	50 23	089	38 17	117	28 45	145	44 05	168	54 33	258	46 16	316
71	29 53	033	51 11	090	39 00	118	29 12	146	44 14	170	53 46	259	45 43	316
72	30 19	033	51 59	090	39 42	119	29 38	147	44 22	171	52 59	260	45 10	316
73	30 46	033	52 47	091	40 24	119	30 04	148	44 30	172	52 11	261	44 37	316
74	31 12	034	53 35	092	41 06	120	30 28	149	44 35	174	51 24	262	44 04	316
	◆Dubhe		POLLUX		PROCYON		◆SIRIUS		RIGEL		◆Hamal		Schedar	
75	31 39	034	54 22	092	41 47	121	30 52	150	44 40	175	50 37	262	43 31	316
76	32 05	034	55 10	093	42 28	122	31 16	151	44 44	176	49 49	263	42 58	316
77	32 32	034	55 58	094	43 08	123	31 38	153	44 46	178	49 01	264	42 25	317
78	32 59	034	56 46	094	43 48	124	32 00	154	44 47	179	48 14	264	41 52	317
79	33 26	034	57 34	095	44 27	125	32 21	155	44 47	181	47 26	265	41 19	317
80	33 53	035	58 21	096	45 06	126	32 41	156	44 46	182	46 38	266	40 46	317
81	34 21	035	59 09	097	45 44	128	33 00	157	44 44	183	45 51	266	40 13	317
82	34 48	035	59 57	097	46 22	129	33 19	158	44 41	185	45 03	267	39 41	317
83	35 15	035	60 44	098	46 59	130	33 36	159	44 36	186	44 15	268	39 08	317
84	35 43	035	61 32	099	47 36	131	33 53	160	44 30	188	43 27	268	38 35	317
85	36 11	035	62 19	100	48 11	132	34 09	161	44 24	189	42 39	269	38 03	317
86	36 38	035	63 06	101	48 47	133	34 24	162	44 16	190	41 51	270	37 31	318
87	37 06	036	63 53	102	49 21	135	34 38	163	44 06	192	41 03	270	36 58	318
88	37 34	036	64 40	102	49 55	136	34 52	165	43 56	193	40 15	271	36 26	318
89	38 02	036	65 27	103	50 28	137	35 04	166	43 45	194	39 27	271	35 54	318

LHA	Hc	Zn	Hc	Zn	Hc	Zn	Hc	Zn	Hc	Zn	Hc	Zn	Hc	Zn
♈	◆Dubhe		REGULUS		PROCYON		◆SIRIUS		RIGEL		ALDEBARAN		◆Mirfak	
90	38 30	036	29 26	097	51 00	138	35 15	167	43 32	196	62 18	228	59 20	307
91	38 58	036	30 14	098	51 31	140	35 26	168	43 19	197	61 42	230	58 42	307
92	39 26	036	31 01	099	52 02	141	35 35	169	43 04	198	61 05	231	58 04	307
93	39 55	036	31 49	099	52 31	143	35 44	170	42 49	200	60 27	233	57 25	307
94	40 23	036	32 36	100	53 00	144	35 52	171	42 32	201	59 49	234	56 47	307
95	40 51	036	33 23	101	53 27	146	35 58	173	42 15	202	59 10	235	56 08	306
96	41 20	036	34 10	101	53 54	147	36 04	174	41 56	203	58 30	237	55 30	306
97	41 48	036	34 57	102	54 20	149	36 09	175	41 36	205	57 50	238	54 51	306
98	42 16	036	35 44	103	54 44	150	36 12	176	41 16	206	57 09	239	54 12	306
99	42 45	036	36 30	104	55 07	152	36 15	177	40 54	207	56 27	240	53 34	306
100	43 13	036	37 17	104	55 29	153	36 17	179	40 32	208	55 45	242	52 55	306
101	43 41	036	38 03	105	55 50	155	36 17	180	40 09	210	55 03	243	52 16	306
102	44 10	036	38 49	106	56 10	157	36 17	181	39 45	211	54 20	244	51 38	306
103	44 38	036	39 35	107	56 28	158	36 16	182	39 20	212	53 37	245	50 59	306
104	45 06	036	40 21	107	56 45	160	36 14	183	38 54	213	52 53	246	50 20	306
	Kochab		◆Denebola		REGULUS		◆SIRIUS		RIGEL		ALDEBARAN		◆CAPELLA	
105	28 33	016	22 50	089	41 07	108	36 10	184	38 28	214	52 09	247	68 44	303
106	28 46	016	23 37	089	41 52	109	36 06	186	38 00	215	51 25	248	68 04	303
107	28 59	016	24 25	090	42 37	110	36 01	187	37 32	216	50 41	249	67 23	302
108	29 13	016	25 13	090	43 22	111	35 55	188	37 03	218	49 56	250	66 43	302
109	29 27	017	26 01	091	44 07	112	35 48	189	36 34	219	49 11	251	66 02	301
110	29 40	017	26 49	092	44 51	113	35 39	190	36 03	220	48 25	252	65 21	301
111	29 54	017	27 37	092	45 36	113	35 30	192	35 33	221	47 40	253	64 40	301
112	30 08	017	28 25	093	46 19	114	35 20	193	35 01	222	46 54	253	63 59	301
113	30 23	017	29 13	093	47 03	115	35 09	194	34 29	223	46 08	254	63 17	300
114	30 37	017	30 01	094	47 46	116	34 57	195	33 56	224	45 22	255	62 36	300
115	30 51	018	30 48	095	48 29	117	34 45	196	33 22	225	44 35	256	61 54	300
116	31 06	018	31 36	095	49 11	118	34 31	197	32 48	226	43 49	257	61 13	300
117	31 21	018	32 24	096	49 53	119	34 16	198	32 14	227	43 02	257	60 31	300
118	31 35	018	33 11	097	50 35	120	34 01	199	31 39	228	42 15	258	59 50	300
119	31 50	018	33 59	097	51 16	122	33 44	201	31 03	229	41 28	259	59 08	300
	◆Kochab		Denebola		◆REGULUS		SIRIUS		RIGEL		◆ALDEBARAN		CAPELLA	
120	32 05	018	34 46	098	51 56	123	33 27	202	30 27	229	40 41	260	58 27	300
121	32 20	018	35 34	099	52 36	124	33 09	203	29 50	230	39 54	260	57 45	300
122	32 36	019	36 21	099	53 16	125	32 50	204	29 13	231	39 06	261	57 03	300
123	32 51	019	37 08	100	53 55	126	32 30	205	28 36	232	38 19	262	56 22	300
124	33 06	019	37 56	101	54 33	128	32 10	206	27 58	233	37 32	263	55 40	300
125	33 22	019	38 43	102	55 11	129	31 48	207	27 19	234	36 44	263	54 58	300
126	33 37	019	39 29	102	55 48	130	31 26	208	26 40	235	35 56	264	54 17	300
127	33 53	019	40 16	103	56 24	132	31 03	209	26 01	235	35 09	265	53 35	300
128	34 09	019	41 03	104	56 59	133	30 39	210	25 21	236	34 21	265	52 53	300
129	34 24	019	41 49	105	57 34	134	30 15	211	24 41	237	33 33	266	52 12	300
130	34 40	019	42 36	105	58 08	136	29 50	212	24 01	238	32 45	266	51 30	300
131	34 56	019	43 22	106	58 40	138	29 24	213	23 20	239	31 58	267	50 49	300
132	35 12	019	44 08	107	59 12	139	28 58	214	22 39	239	31 10	268	50 07	300
133	35 28	020	44 53	108	59 43	141	28 31	215	21 58	240	30 22	268	49 26	300
134	35 44	020	45 39	109	60 13	142	28 03	216	21 16	241	29 34	269	48 44	300
	Kochab		◆ARCTURUS		SPICA		REGULUS		◆SIRIUS		BETELGEUSE		◆CAPELLA	
135	36 00	020	20 06	081	11 31	114	60 42	144	27 34	217	38 41	247	48 03	301
136	36 16	020	20 54	081	12 15	114	61 09	146	27 05	218	37 57	247	47 22	301
137	36 33	020	21 41	082	12 58	115	61 35	148	26 35	219	37 12	248	46 41	301
138	36 49	020	22 28	082	13 41	116	62 00	150	26 05	220	36 28	249	46 00	301
139	37 05	020	23 16	083	14 25	116	62 24	152	25 34	221	35 43	250	45 19	301
140	37 21	020	24 04	083	15 07	117	62 46	154	25 03	221	34 58	251	44 38	301
141	37 38	020	24 51	084	15 50	118	63 06	156	24 31	222	34 12	252	43 57	302
142	37 54	020	25 39	085	16 32	118	63 25	158	23 58	223	33 27	252	43 16	302
143	38 10	020	26 27	085	17 14	119	63 43	160	23 25	224	32 41	253	42 36	302
144	38 27	020	27 14	086	17 56	120	63 58	162	22 52	225	31 55	254	41 55	302
145	38 43	020	28 02	086	18 37	121	64 12	164	22 18	226	31 09	255	41 15	302
146	38 59	020	28 50	087	19 18	121	64 25	166	21 43	227	30 23	255	40 34	303
147	39 16	020	29 38	087	19 59	122	64 35	168	21 08	227	29 36	256	39 54	303
148	39 32	020	30 26	088	20 40	123	64 44	171	20 33	228	28 50	257	39 14	303
149	39 48	020	31 14	089	21 20	124	64 51	173	19 57	229	28 03	257	38 34	303
	◆Kochab		ARCTURUS		◆SPICA		REGULUS		PROCYON		◆POLLUX		CAPELLA	
150	40 05	020	32 01	089	21 59	124	64 56	175	44 47	234	60 21	262	37 54	304
151	40 21	020	32 49	090	22 39	125	64 58	178	44 08	235	59 33	263	37 14	304
152	40 37	020	33 37	090	23 18	126	65 00	180	43 29	236	58 46	264	36 34	304
153	40 54	020	34 25	091	23 56	127	64 59	182	42 49	237	57 58	265	35 55	304
154	41 10	020	35 13	092	24 34	128	64 56	185	42 08	238	57 10	265	35 15	305
155	41 26	020	36 01	092	25 12	128	64 51	187	41 27	239	56 22	266	34 36	305
156	41 42	020	36 49	093	25 49	129	64 44	189	40 46	240	55 35	267	33 57	305
157	41 58	020	37 37	093	26 26	130	64 36	191	40 04	241	54 47	267	33 18	306
158	42 14	019	38 25	094	27 03	131	64 25	194	39 22	242	53 59	268	32 39	306
159	42 30	019	39 12	095	27 39	132	64 13	196	38 40	243	53 11	269	32 00	306
160	42 46	019	40 00	095	28 14	133	63 59	198	37 57	244	52 23	269	31 22	307
161	43 02	019	40 48	096	28 49	134	63 44	200	37 14	245	51 35	270	30 43	307
162	43 17	019	41 35	097	29 23	135	63 26	202	36 30	245	50 47	270	30 05	307
163	43 33	019	42 23	097	29 57	136	63 07	204	35 47	246	49 59	271	29 27	308
164	43 48	019	43 10	098	30 31	136	62 47	206	35 03	247	49 11	272	28 49	308
	◆Kochab		ARCTURUS		◆SPICA		REGULUS		PROCYON		◆POLLUX		CAPELLA	
165	44 04	019	43 58	099	31 03	137	62 25	208	34 18	248	48 23	272	28 11	308
166	44 19	019	44 45	100	31 35	138	62 01	210	33 34	249	47 36	273	27 34	309
167	44 34	018	45 32	100	32 07	139	61 37	212	32 49	250	46 48	273	26 56	309
168	44 49	018	46 19	101	32 38	140	61 10	214	32 04	250	46 00	274	26 19	309
169	45 04	018	47 06	102	33 08	141	60 43	216	31 19	251	45 12	274	25 42	310
170	45 19	018	47 53	103	33 38	142	60 14	217	30 33	252	44 24	275	25 05	310
171	45 34	018	48 40	103	34 07	143	59 45	219	29 48	253	43 37	275	24 28	310
172	45 48	018	49 26	104	34 35	144	59 14	221	29 02	253	42 49	276	23 52	311
173	46 03	017	50 13	105	35 02	145	58 42	222	28 16	254	42 01	277	23 16	311
174	46 17	017	50 59	106	35 29	147	58 09	224	27 30	255	41 14	277	22 40	311
175	46 31	017	51 45	107	35 55	148	57 36	225	26 43	255	40 26	278	22 04	312
176	46 45	017	52 31	108	36 21	149	57 01	227	25 57	256	39 39	278	21 28	312
177	46 59	017	53 16	109	36 45	150	56 26	228	25 10	257	38 51	279	20 53	313
178	47 13	016	54 01	110	37 09	151	55 50	230	24 24	258	38 04	279	20 18	313
179	47 26	016	54 46	111	37 32	152	55 13	231	23 37	258	37 17	280	19 43	313

LAT 37°N

LHA	Hc	Zn	Hc	Zn	Hc	Zn	Hc	Zn	Hc	Zn	Hc	Zn	Hc	Zn
♈	◆VEGA		Alphecca		ARCTURUS		◆SPICA		REGULUS		◆POLLUX		CAPELLA	
180	16 07	053	43 56	087	55 31	112	37 54	153	54 35	232	36 29	280	19 08	314
181	16 45	054	44 44	087	56 15	113	38 15	154	53 57	234	35 42	281	18 33	314
182	17 24	054	45 32	088	56 59	114	38 35	156	53 18	235	34 55	281	17 59	315
183	18 03	055	46 19	088	57 43	115	38 54	157	52 38	236	34 08	282	17 25	315
184	18 42	055	47 07	089	58 26	116	39 13	158	51 58	237	33 21	282	16 51	315
185	19 21	055	47 55	090	59 09	117	39 30	159	51 18	238	32 35	283	16 18	316
186	20 01	056	48 43	090	59 51	119	39 47	160	50 37	239	31 48	283	15 45	316
187	20 41	056	49 31	091	60 33	120	40 02	162	49 55	241	31 01	284	15 12	317
188	21 21	057	50 19	091	61 15	121	40 17	163	49 13	242	30 15	284	14 39	317
189	22 01	057	51 07	092	61 55	123	40 31	164	48 31	243	29 28	284	14 07	318
190	22 41	058	51 55	093	62 35	124	40 43	165	47 48	244	28 42	285	13 35	318
191	23 22	058	52 43	093	63 15	125	40 55	167	47 05	245	27 56	285	13 03	319
192	24 03	058	53 30	094	63 53	127	41 05	168	46 22	246	27 09	286	12 31	319
193	24 44	059	54 18	095	64 31	129	41 15	169	45 38	247	26 23	286	12 00	320
194	25 25	059	55 06	095	65 08	130	41 23	171	44 54	247	25 38	287	11 29	320
	DENEB		◆VEGA		Alphecca		◆SPICA		REGULUS		◆POLLUX		Dubhe	
195	10 47	040	26 06	060	55 54	096	41 31	172	44 09	248	24 52	287	59 20	333
196	11 18	041	26 47	060	56 41	097	41 37	173	43 25	249	24 06	288	58 58	333
197	11 50	041	27 29	061	57 29	098	41 42	174	42 40	250	23 21	288	58 36	332
198	12 22	042	28 11	061	58 16	098	41 46	176	41 55	251	22 35	289	58 13	331
199	12 54	042	28 53	061	59 04	099	41 49	177	41 09	252	21 50	289	57 50	331
200	13 26	043	29 35	062	59 51	100	41 51	178	40 24	253	21 05	290	57 26	330
201	13 59	043	30 17	062	60 38	101	41 52	180	39 38	253	20 20	290	57 03	330
202	14 32	044	31 00	062	61 25	102	41 52	181	38 52	254	19 35	291	56 39	330
203	15 05	044	31 42	063	62 12	103	41 50	182	38 06	255	18 50	291	56 14	329
204	15 38	045	32 25	063	62 58	104	41 48	184	37 19	256	18 06	292	55 49	329
205	16 12	045	33 08	064	63 45	105	41 44	185	36 33	256	17 21	292	55 24	328
206	16 46	045	33 51	064	64 31	106	41 39	186	35 46	257	16 37	293	54 59	328
207	17 20	046	34 34	064	65 17	107	41 34	188	34 59	258	15 53	293	54 33	328
208	17 55	046	35 17	065	66 03	108	41 27	189	34 12	259	15 09	294	54 07	327
209	18 30	047	36 00	065	66 48	109	41 19	190	33 25	259	14 25	294	53 41	327
	DENEB		◆VEGA		Rasalhague		ANTARES		◆SPICA		REGULUS		◆Dubhe	
210	19 05	047	36 44	065	36 21	102	17 32	145	41 10	191	32 38	260	53 15	327
211	19 40	048	37 28	066	37 07	103	17 59	146	41 00	193	31 51	261	52 49	326
212	20 15	048	38 11	066	37 54	104	18 26	147	40 49	194	31 04	261	52 22	326
213	20 51	048	38 55	066	38 40	105	18 51	148	40 37	195	30 16	262	51 55	326
214	21 27	049	39 39	067	39 27	106	19 17	149	40 23	197	29 29	263	51 28	326
215	22 03	049	40 23	067	40 13	106	19 41	149	40 09	198	28 41	263	51 01	325
216	22 39	050	41 08	067	40 59	107	20 05	150	39 54	199	27 54	264	50 34	325
217	23 16	050	41 52	068	41 44	108	20 29	151	39 38	200	27 06	265	50 06	325
218	23 53	050	42 36	068	42 30	109	20 52	152	39 21	201	26 18	265	49 39	325
219	24 30	051	43 21	068	43 15	110	21 14	153	39 03	203	25 30	266	49 11	325
220	25 07	051	44 05	069	44 00	110	21 35	154	38 44	204	24 43	267	48 43	324
221	25 44	051	44 50	069	44 45	111	21 56	155	38 24	205	23 55	267	48 15	324
222	26 22	052	45 35	069	45 29	112	22 16	156	38 03	206	23 07	268	47 47	324
223	26 59	052	46 20	070	46 14	113	22 36	156	37 42	207	22 19	268	47 19	324
224	27 37	052	47 05	070	46 58	114	22 54	157	37 19	209	21 31	269	46 51	324
	DENEB		◆ALTAIR		Rasalhague		ANTARES		◆SPICA		Denebola		◆Dubhe	
225	28 15	053	19 09	093	47 41	115	23 12	158	36 56	210	42 08	255	46 23	324
226	28 54	053	19 57	094	48 24	116	23 30	159	36 32	211	41 22	256	45 55	324
227	29 32	053	20 45	095	49 07	117	23 46	160	36 07	212	40 35	257	45 26	324
228	30 11	054	21 33	095	49 50	118	24 02	161	35 41	213	39 48	257	44 58	324
229	30 49	054	22 20	096	50 32	119	24 17	162	35 15	214	39 01	258	44 30	324
230	31 28	054	23 08	096	51 13	120	24 32	163	34 48	215	38 14	259	44 02	324
231	32 07	055	23 56	097	51 55	121	24 45	164	34 20	216	37 27	260	43 33	324
232	32 46	055	24 43	098	52 35	122	24 58	165	33 51	217	36 40	260	43 05	324
233	33 26	055	25 31	098	53 16	124	25 10	166	33 22	218	35 53	261	42 36	324
234	34 05	056	26 18	099	53 55	125	25 22	167	32 52	219	35 06	262	42 08	324
235	34 45	056	27 05	100	54 34	126	25 32	168	32 21	220	34 18	262	41 40	324
236	35 25	056	27 52	100	55 13	127	25 42	169	31 50	221	33 31	263	41 11	324
237	36 05	057	28 40	101	55 50	129	25 51	170	31 18	222	32 43	264	40 43	324
238	36 45	057	29 27	102	56 27	130	25 59	171	30 46	223	31 55	264	40 15	324
239	37 25	057	30 13	102	57 04	131	26 06	172	30 13	224	31 08	265	39 47	324
	◆DENEB		ALTAIR		Nunki		◆ANTARES		ARCTURUS		◆Alkaid		Kochab	
240	38 05	057	31 00	103	14 31	140	26 13	173	61 03	239	63 09	308	51 54	352
241	38 46	058	31 47	104	15 01	141	26 18	174	60 21	241	62 31	308	51 47	352
242	39 26	058	32 33	105	15 31	142	26 23	175	59 39	242	61 53	307	51 41	352
243	40 07	058	33 19	105	16 01	143	26 27	176	58 57	243	61 14	307	51 33	351
244	40 47	058	34 06	106	16 29	143	26 30	177	58 14	244	60 36	307	51 26	351
245	41 28	059	34 51	107	16 58	144	26 33	178	57 31	245	59 58	306	51 18	350
246	42 09	059	35 37	108	17 26	145	26 34	179	56 47	246	59 19	306	51 10	350
247	42 50	059	36 23	108	17 53	146	26 35	180	56 03	248	58 40	306	51 02	350
248	43 31	059	37 08	109	18 20	147	26 35	181	55 18	249	58 01	306	50 53	349
249	44 13	059	37 53	110	18 46	147	26 33	182	54 34	250	57 23	306	50 44	349
250	44 54	060	38 38	111	19 11	148	26 32	183	53 48	251	56 44	306	50 35	349
251	45 35	060	39 23	112	19 36	149	26 29	184	53 03	252	56 05	306	50 25	348
252	46 17	060	40 07	113	20 01	150	26 25	185	52 18	252	55 26	305	50 15	348
253	46 58	060	40 51	113	20 24	151	26 21	186	51 32	253	54 47	305	50 05	348
254	47 40	060	41 35	114	20 48	152	26 16	187	50 46	254	54 08	305	49 55	347
	◆DENEB		ALTAIR		Nunki		◆ANTARES		ARCTURUS		◆Alkaid		Kochab	
255	48 22	061	42 19	115	21 10	152	26 10	188	50 00	255	53 29	305	49 44	347
256	49 04	061	43 02	116	21 32	153	26 03	189	49 13	256	52 50	305	49 33	347
257	49 45	061	43 45	117	21 53	154	25 55	190	48 27	257	52 11	305	49 22	346
258	50 27	061	44 27	118	22 13	155	25 47	191	47 40	258	51 31	305	49 10	346
259	51 09	061	45 09	119	22 33	156	25 38	192	46 53	258	50 52	305	48 59	346
260	51 51	061	45 51	120	22 52	157	25 28	193	46 06	259	50 13	305	48 47	345
261	52 33	061	46 32	121	23 11	158	25 17	194	45 19	260	49 34	305	48 35	345
262	53 15	061	47 13	122	23 29	159	25 05	195	44 32	261	48 55	306	48 22	345
263	53 58	062	47 54	123	23 45	160	24 53	196	43 44	261	48 16	306	48 10	345
264	54 40	062	48 34	124	24 02	161	24 39	196	42 57	262	47 37	306	47 57	344
265	55 22	062	49 13	125	24 17	162	24 25	197	42 09	263	46 58	306	47 44	344
266	56 04	062	49 52	126	24 32	163	24 11	198	41 22	263	46 19	306	47 31	344
267	56 46	062	50 30	128	24 46	163	23 55	199	40 34	264	45 41	306	47 17	344
268	57 29	062	51 08	129	24 59	164	23 39	200	39 47	265	45 02	306	47 04	343
269	58 11	062	51 45	130	25 12	165	23 22	201	38 59	265	44 23	306	46 50	343

LHA	Hc	Zn	Hc	Zn	Hc	Zn	Hc	Zn	Hc	Zn	Hc	Zn	Hc	Zn
♈	◆DENEB		ALTAIR		Nunki		◆ANTARES		ARCTURUS		◆Alkaid		Kochab	
270	58 53	062	52 21	131	25 23	166	23 04	202	38 11	266	43 45	306	46 36	343
271	59 35	062	52 56	133	25 34	167	22 46	203	37 23	267	43 06	307	46 22	343
272	60 17	062	53 31	134	25 44	168	22 27	204	36 35	267	42 28	307	46 08	343
273	61 00	062	54 05	135	25 54	169	22 07	205	35 47	268	41 49	307	45 54	342
274	61 42	062	54 39	137	26 02	170	21 47	206	35 00	269	41 11	307	45 39	342
275	62 24	061	55 11	138	26 10	171	21 26	207	34 12	269	40 33	307	45 24	342
276	63 06	061	55 42	140	26 17	172	21 04	207	33 24	270	39 55	307	45 10	342
277	63 48	061	56 13	141	26 23	173	20 41	208	32 36	270	39 17	308	44 55	342
278	64 30	061	56 42	143	26 28	174	20 18	209	31 48	271	38 39	308	44 40	342
279	65 12	061	57 11	144	26 32	175	19 55	210	31 00	272	38 01	308	44 25	341
280	65 53	060	57 38	146	26 36	176	19 30	211	30 12	272	37 23	308	44 09	341
281	66 35	060	58 04	148	26 39	177	19 05	212	29 24	273	36 46	309	43 54	341
282	67 16	060	58 29	149	26 41	178	18 40	213	28 36	273	36 08	309	43 38	341
283	67 58	059	58 53	151	26 42	179	18 14	213	27 49	274	35 31	309	43 23	341
284	68 39	059	59 16	153	26 42	180	17 47	214	27 01	274	34 54	309	43 07	341
	Alpheratz		◆Enif		ALTAIR		◆Rasalhague		ARCTURUS		Alkaid		◆Kochab	
285	26 41	072	44 11	116	59 37	155	58 59	224	26 13	275	34 17	309	42 51	341
286	27 26	073	44 54	117	59 57	157	58 26	225	25 25	276	33 40	310	42 36	341
287	28 12	073	45 37	118	60 15	158	57 51	227	24 38	276	33 03	310	42 20	341
288	28 58	074	46 19	119	60 32	160	57 16	228	23 50	277	32 27	310	42 04	341
289	29 44	074	47 01	120	60 47	162	56 40	229	23 02	277	31 50	311	41 48	340
290	30 31	075	47 42	121	61 01	164	56 03	231	22 15	278	31 14	311	41 32	340
291	31 17	075	48 23	122	61 13	166	55 26	232	21 28	278	30 38	311	41 16	340
292	32 03	076	49 04	123	61 24	168	54 48	233	20 40	279	30 02	311	40 59	340
293	32 50	076	49 44	124	61 33	170	54 09	235	19 53	280	29 26	312	40 43	340
294	33 36	077	50 23	125	61 40	172	53 29	236	19 06	280	28 50	312	40 27	340
295	34 23	077	51 02	126	61 45	174	52 50	237	18 19	281	28 14	312	40 11	340
296	35 10	078	51 41	127	61 49	177	52 09	238	17 31	281	27 39	313	39 54	340
297	35 57	078	52 18	129	61 51	179	51 28	239	16 45	282	27 04	313	39 38	340
298	36 44	079	52 55	130	61 51	181	50 47	240	15 58	282	26 29	313	39 22	340
299	37 31	079	53 32	131	61 50	183	50 05	242	15 11	283	25 54	314	39 05	340
	◆Mirfak		Alpheratz		◆FOMALHAUT		ALTAIR		Rasalhague		◆Alphecca		Kochab	
300	15 59	039	38 18	080	11 28	142	61 46	185	49 22	243	33 49	280	38 49	340
301	16 30	039	39 05	080	11 57	142	61 41	187	48 40	244	33 02	280	38 33	340
302	17 00	040	39 52	081	12 26	143	61 35	189	47 56	245	32 15	281	38 16	340
303	17 31	040	40 39	081	12 54	144	61 26	191	47 13	246	31 28	281	38 00	340
304	18 02	040	41 27	082	13 22	145	61 16	193	46 29	247	30 41	282	37 43	340
305	18 33	041	42 14	082	13 50	145	61 04	195	45 45	247	29 54	282	37 27	340
306	19 04	041	43 02	083	14 17	146	60 51	197	45 01	248	29 07	283	37 11	340
307	19 36	042	43 49	083	14 43	147	60 36	199	44 16	249	28 20	284	36 55	340
308	20 08	042	44 37	084	15 09	148	60 20	201	43 31	250	27 34	284	36 38	340
309	20 40	042	45 25	084	15 34	149	60 02	203	42 46	251	26 47	285	36 22	340
310	21 13	043	46 12	085	15 59	149	59 42	205	42 00	252	26 01	285	36 06	340
311	21 45	043	47 00	085	16 23	150	59 22	207	41 15	253	25 15	286	35 50	340
312	22 18	044	47 48	086	16 47	151	58 59	208	40 29	253	24 29	286	35 34	340
313	22 51	044	48 36	086	17 10	152	58 36	210	39 43	254	23 43	287	35 18	340
314	23 24	044	49 23	087	17 32	153	58 11	212	38 57	255	22 57	287	35 02	341
	◆Mirfak		Hamal		Diphda		◆FOMALHAUT		ALTAIR		◆VEGA		Kochab	
315	23 58	045	24 04	078	13 55	126	17 54	153	57 45	214	61 52	285	34 46	341
316	24 32	045	24 50	078	14 33	127	18 15	154	57 18	215	61 06	285	34 30	341
317	25 06	045	25 37	079	15 12	127	18 35	155	56 50	217	60 19	285	34 14	341
318	25 40	046	26 24	080	15 50	128	18 55	156	56 21	218	59 33	285	33 58	341
319	26 14	046	27 12	080	16 28	129	19 15	157	55 51	220	58 47	286	33 43	341
320	26 49	046	27 59	081	17 04	130	19 33	158	55 20	221	58 01	286	33 27	341
321	27 23	047	28 46	081	17 41	130	19 51	159	54 47	223	57 15	286	33 12	341
322	27 58	047	29 34	082	18 17	131	20 08	159	54 14	224	56 29	286	32 56	341
323	28 33	047	30 21	082	18 53	132	20 25	160	53 41	226	55 43	287	32 41	341
324	29 09	048	31 09	083	19 29	133	20 41	161	53 06	227	54 57	287	32 26	342
325	29 44	048	31 56	083	20 04	133	20 56	162	52 31	228	54 11	287	32 11	342
326	30 20	048	32 44	084	20 38	134	21 10	163	51 54	230	53 25	287	31 56	342
327	30 55	048	33 31	084	21 12	135	21 24	164	51 18	231	52 40	288	31 41	342
328	31 31	049	34 19	085	21 46	136	21 37	165	50 40	232	51 54	288	31 26	342
329	32 07	049	35 07	085	22 19	137	21 49	166	50 02	233	51 09	288	31 11	342
	◆CAPELLA		Hamal		Diphda		◆FOMALHAUT		ALTAIR		◆VEGA		Kochab	
330	14 34	043	35 55	086	22 52	138	22 00	167	49 23	234	50 23	289	30 57	342
331	15 07	043	36 42	087	23 24	138	22 11	167	48 44	235	49 38	289	30 42	342
332	15 40	044	37 30	087	23 55	139	22 21	168	48 05	237	48 53	289	30 28	343
333	16 13	044	38 18	088	24 26	140	22 30	169	47 24	238	48 07	290	30 13	343
334	16 46	044	39 06	088	24 57	141	22 39	170	46 44	239	47 22	290	29 59	343
335	17 20	045	39 54	089	25 27	142	22 47	171	46 02	240	46 37	290	29 45	343
336	17 54	045	40 42	089	25 56	143	22 54	172	45 21	241	45 52	290	29 31	343
337	18 28	046	41 30	090	26 24	144	23 00	173	44 39	242	45 07	291	29 18	343
338	19 02	046	42 18	091	26 52	145	23 05	174	43 56	243	44 23	291	29 04	344
339	19 37	047	43 06	091	27 20	146	23 10	175	43 14	244	43 38	291	28 51	344
340	20 12	047	43 53	092	27 47	147	23 14	176	42 30	245	42 53	292	28 37	344
341	20 47	047	44 41	093	28 13	148	23 17	177	41 47	245	42 09	292	28 24	344
342	21 23	048	45 29	093	28 38	148	23 19	178	41 03	246	41 25	292	28 11	344
343	21 58	048	46 17	094	29 03	149	23 20	179	40 19	247	40 40	293	27 58	345
344	22 34	049	47 05	095	29 27	150	23 21	180	39 35	248	39 56	293	27 46	345
	◆CAPELLA		ALDEBARAN		Diphda		◆FOMALHAUT		ALTAIR		◆VEGA		Kochab	
345	23 10	049	14 36	080	29 50	151	23 21	181	38 50	249	39 12	293	27 33	345
346	23 46	049	15 24	081	30 12	152	23 20	182	38 05	250	38 28	294	27 21	345
347	24 23	050	16 11	081	30 34	153	23 18	183	37 20	251	37 44	294	27 09	345
348	24 59	050	16 58	082	30 55	155	23 16	183	36 35	251	37 01	294	26 57	346
349	25 36	050	17 46	082	31 15	156	23 12	184	35 50	252	36 17	295	26 45	346
350	26 13	051	18 33	083	31 35	157	23 08	185	35 04	253	35 34	295	26 33	346
351	26 50	051	19 21	084	31 54	158	23 04	186	34 18	254	34 50	296	26 21	346
352	27 28	051	20 09	084	32 11	159	22 58	187	33 32	254	34 07	296	26 10	346
353	28 05	052	20 56	085	32 28	160	22 52	188	32 46	255	33 24	296	25 59	347
354	28 43	052	21 44	085	32 44	161	22 44	189	31 59	256	32 41	297	25 48	347
355	29 21	052	22 32	086	33 00	162	22 36	190	31 13	257	31 59	297	25 37	347
356	29 59	053	23 20	086	33 14	163	22 28	191	30 26	257	31 16	297	25 27	347
357	30 37	053	24 07	087	33 28	164	22 18	192	29 39	258	30 33	298	25 16	348
358	31 15	053	24 55	088	33 40	165	22 08	193	28 52	258	29 51	298	25 06	348
359	31 54	054	25 43	088	33 52	166	21 57	194	28 05	259	29 09	299	24 56	348

LAT 36°N

LHA	Hc	Zn	Hc	Zn	Hc	Zn	Hc	Zn	Hc	Zn	Hc	Zn	Hc	Zn
♈	◆CAPELLA		ALDEBARAN		◆Diphda		FOMALHAUT		ALTAIR		◆VEGA		Kochab	
0	31 57	054	26 30	088	35 01	167	22 43	195	27 28	261	27 58	299	23 47	348
1	32 36	054	27 18	089	35 11	169	22 31	196	26 40	261	27 16	300	23 37	349
2	33 16	054	28 07	089	35 21	170	22 17	197	25 52	262	26 33	300	23 28	349
3	33 55	054	28 55	090	35 29	171	22 03	197	25 04	263	25 52	301	23 19	349
4	34 35	055	29 44	091	35 36	172	21 48	198	24 16	263	25 10	301	23 10	349
5	35 14	055	30 32	091	35 42	173	21 32	199	23 28	264	24 28	301	23 01	350
6	35 54	055	31 21	092	35 48	174	21 16	200	22 39	264	23 47	302	22 52	350
7	36 34	055	32 09	092	35 52	175	20 59	201	21 51	265	23 06	302	22 44	350
8	37 14	056	32 58	093	35 55	177	20 41	202	21 03	266	22 25	303	22 35	350
9	37 54	056	33 46	094	35 57	178	20 23	203	20 14	266	21 44	303	22 27	351
10	38 34	056	34 35	094	35 59	179	20 04	204	19 26	267	21 03	303	22 19	351
11	39 15	056	35 23	095	35 59	180	19 44	204	18 37	267	20 23	304	22 12	351
12	39 55	057	36 11	096	35 58	181	19 23	205	17 49	268	19 43	304	22 04	351
13	40 36	057	37 00	096	35 57	183	19 02	206	17 00	269	19 03	305	21 57	352
14	41 16	057	37 48	097	35 54	184	18 41	207	16 12	269	18 23	305	21 50	352
	◆CAPELLA		BETELGEUSE		RIGEL		◆Diphda		Enif		◆DENEB		Kochab	
15	41 57	057	17 30	094	15 49	113	35 51	185	38 34	252	41 23	302	21 43	352
16	42 38	057	18 18	094	16 33	114	35 46	186	37 47	253	40 42	302	21 37	352
17	43 19	058	19 06	095	17 18	114	35 40	187	37 01	254	40 01	302	21 31	353
18	44 00	058	19 55	095	18 02	115	35 34	188	36 14	254	39 20	303	21 25	353
19	44 41	058	20 43	096	18 46	116	35 26	190	35 27	255	38 39	303	21 19	353
20	45 22	058	21 31	097	19 29	116	35 18	191	34 40	256	37 58	303	21 13	354
21	46 03	058	22 20	097	20 13	117	35 08	192	33 53	257	37 18	303	21 08	354
22	46 44	058	23 08	098	20 56	118	34 58	193	33 06	257	36 37	304	21 03	354
23	47 26	058	23 56	099	21 38	119	34 46	194	32 19	258	35 57	304	20 58	354
24	48 07	059	24 44	099	22 21	119	34 34	195	31 31	259	35 16	304	20 53	355
25	48 48	059	25 32	100	23 03	120	34 21	196	30 44	259	34 36	304	20 49	355
26	49 30	059	26 19	100	23 45	121	34 07	197	29 56	260	33 56	305	20 45	355
27	50 12	059	27 07	101	24 26	122	33 52	199	29 08	261	33 17	305	20 41	355
28	50 53	059	27 55	102	25 08	122	33 36	200	28 20	261	32 37	305	20 37	356
29	51 35	059	28 42	102	25 48	123	33 19	201	27 32	262	31 57	306	20 34	356
	◆CAPELLA		BETELGEUSE		RIGEL		◆Diphda		Alpheratz		◆DENEB		Kochab	
30	52 16	059	29 29	103	26 29	124	33 01	202	65 31	262	31 18	306	20 30	356
31	52 58	059	30 17	104	27 09	125	32 43	203	64 43	262	30 39	306	20 27	357
32	53 40	059	31 04	105	27 49	126	32 23	204	63 55	263	30 00	306	20 25	357
33	54 21	059	31 51	105	28 28	126	32 03	205	63 07	264	29 21	307	20 22	357
34	55 03	059	32 37	106	29 07	127	31 42	206	62 18	265	28 42	307	20 20	357
35	55 45	059	33 24	107	29 45	128	31 21	207	61 30	265	28 03	307	20 18	358
36	56 26	059	34 10	107	30 23	129	30 58	208	60 42	266	27 25	308	20 16	358
37	57 08	059	34 56	108	31 00	130	30 35	209	59 53	267	26 46	308	20 15	358
38	57 50	059	35 42	109	31 37	131	30 11	210	59 05	267	26 08	308	20 13	359
39	58 31	059	36 28	110	32 14	132	29 46	211	58 16	268	25 30	309	20 12	359
40	59 13	059	37 14	111	32 50	133	29 21	212	57 28	269	24 53	309	20 11	359
41	59 54	059	37 59	111	33 25	134	28 54	213	56 39	269	24 15	309	20 11	000
42	60 36	059	38 44	112	34 00	135	28 28	214	55 51	270	23 38	310	20 11	000
43	61 17	058	39 29	113	34 34	136	28 00	215	55 02	271	23 00	310	20 11	000
44	61 58	058	40 13	114	35 08	137	27 32	216	54 13	271	22 23	311	20 11	000
	◆Dubhe		POLLUX		SIRIUS		◆RIGEL		Diphda		◆Alpheratz		DENEB	
45	18 46	025	30 23	076	15 10	124	35 41	138	27 03	217	53 25	272	21 47	311
46	19 07	026	31 10	076	15 50	125	36 13	139	26 34	218	52 36	272	21 10	311
47	19 28	026	31 57	077	16 30	126	36 45	140	26 04	219	51 48	273	20 34	312
48	19 50	026	32 45	077	17 09	127	37 16	141	25 33	220	50 59	273	19 58	312
49	20 11	027	33 32	078	17 48	127	37 46	142	25 02	220	50 11	274	19 22	312
50	20 33	027	34 19	078	18 26	128	38 16	143	24 30	221	49 23	274	18 46	313
51	20 55	027	35 07	078	19 04	129	38 45	144	23 58	222	48 34	275	18 11	313
52	21 18	028	35 55	079	19 42	130	39 13	145	23 25	223	47 46	275	17 35	314
53	21 40	028	36 42	079	20 19	130	39 40	146	22 52	224	46 58	276	17 00	314
54	22 03	028	37 30	080	20 56	131	40 07	147	22 18	225	46 09	276	16 26	315
55	22 26	029	38 18	080	21 32	132	40 32	149	21 43	225	45 21	277	15 51	315
56	22 50	029	39 06	081	22 08	133	40 57	150	21 08	226	44 33	277	15 17	315
57	23 13	029	39 54	081	22 43	134	41 21	151	20 33	227	43 45	278	14 43	316
58	23 37	029	40 42	082	23 18	134	41 44	152	19 57	228	42 57	278	14 09	316
59	24 01	030	41 30	082	23 52	135	42 07	153	19 21	229	42 09	279	13 36	317
	◆Dubhe		POLLUX		PROCYON		◆SIRIUS		RIGEL		◆Hamal		Schedar	
60	24 25	030	42 18	083	31 14	108	24 26	136	42 28	155	62 34	251	51 02	318
61	24 49	030	43 06	083	32 00	109	25 00	137	42 48	156	61 48	252	50 29	318
62	25 14	031	43 54	084	32 46	109	25 33	138	43 07	157	61 01	253	49 56	318
63	25 39	031	44 43	084	33 32	110	26 05	139	43 26	159	60 15	254	49 24	317
64	26 04	031	45 31	085	34 17	111	26 37	140	43 43	160	59 28	255	48 51	317
65	26 29	031	46 19	085	35 02	112	27 08	140	43 59	161	58 41	256	48 18	317
66	26 54	031	47 08	086	35 47	113	27 38	141	44 14	162	57 54	257	47 45	317
67	27 19	032	47 56	086	36 32	113	28 08	142	44 28	164	57 07	257	47 12	317
68	27 45	032	48 45	087	37 16	114	28 38	143	44 41	165	56 19	258	46 39	317
69	28 11	032	49 33	088	38 00	115	29 06	144	44 53	167	55 32	259	46 06	317
70	28 37	032	50 22	088	38 44	116	29 35	145	45 04	168	54 44	260	45 33	317
71	29 03	033	51 10	089	39 28	117	30 02	146	45 13	169	53 56	261	44 59	317
72	29 29	033	51 59	089	40 11	118	30 29	147	45 22	171	53 08	261	44 26	317
73	29 55	033	52 47	090	40 53	119	30 55	148	45 29	172	52 20	262	43 53	317
74	30 22	033	53 36	090	41 36	120	31 20	149	45 35	174	51 32	263	43 20	317
	◆Dubhe		POLLUX		PROCYON		◆SIRIUS		RIGEL		◆Hamal		Schedar	
75	30 49	033	54 24	091	42 18	121	31 44	150	45 40	175	50 44	264	42 47	317
76	31 15	034	55 13	092	42 59	122	32 08	151	45 44	176	49 56	264	42 14	317
77	31 42	034	56 01	092	43 41	123	32 31	152	45 46	178	49 07	265	41 41	317
78	32 09	034	56 50	093	44 21	124	32 53	153	45 47	179	48 19	266	41 08	317
79	32 37	034	57 38	093	45 01	125	33 15	154	45 47	181	47 31	266	40 35	317
80	33 04	034	58 27	094	45 41	126	33 35	155	45 46	182	46 42	267	40 02	317
81	33 31	034	59 15	095	46 20	127	33 55	156	45 44	183	45 54	268	39 29	317
82	33 59	035	60 03	096	46 59	128	34 14	158	45 41	185	45 05	268	38 57	318
83	34 26	035	60 52	096	47 37	129	34 32	159	45 36	186	44 17	269	38 24	318
84	34 54	035	61 40	097	48 14	130	34 50	160	45 30	188	43 28	269	37 51	318
85	35 22	035	62 28	098	48 51	131	35 06	161	45 23	189	42 40	270	37 19	318
86	35 49	035	63 16	099	49 27	133	35 21	162	45 15	190	41 51	271	36 46	318
87	36 17	035	64 04	100	50 03	134	35 36	163	45 05	192	41 02	271	36 14	318
88	36 45	035	64 52	100	50 38	135	35 49	164	44 55	193	40 14	272	35 42	318
89	37 13	035	65 39	101	51 11	136	36 02	165	44 43	195	39 25	272	35 09	319

LHA	Hc	Zn	Hc	Zn	Hc	Zn	Hc	Zn	Hc	Zn	Hc	Zn	Hc	Zn
♈	◆Dubhe		REGULUS		PROCYON		◆SIRIUS		RIGEL		ALDEBARAN		◆Mirfak	
90	37 41	035	29 34	097	51 45	138	36 14	167	44 30	196	62 57	229	58 43	309
91	38 10	035	30 22	097	52 17	139	36 25	168	44 16	197	62 20	231	58 05	308
92	38 38	036	31 10	098	52 48	140	36 34	169	44 01	199	61 42	232	57 27	308
93	39 06	036	31 58	099	53 19	142	36 43	170	43 45	200	61 03	234	56 49	308
94	39 34	036	32 46	099	53 48	143	36 51	171	43 28	201	60 24	235	56 10	308
95	40 03	036	33 34	100	54 17	145	36 58	173	43 10	203	59 43	237	55 32	308
96	40 31	036	34 21	101	54 44	146	37 04	174	42 51	204	59 02	238	54 54	308
97	40 59	036	35 09	101	55 11	148	37 08	175	42 31	205	58 21	239	54 15	307
98	41 28	036	35 57	102	55 36	149	37 12	176	42 10	206	57 39	241	53 37	307
99	41 56	036	36 44	103	56 00	151	37 15	177	41 48	208	56 56	242	52 58	307
100	42 24	036	37 31	104	56 23	153	37 17	179	41 25	209	56 13	243	52 19	307
101	42 53	036	38 18	104	56 44	154	37 17	180	41 01	210	55 30	244	51 41	307
102	43 21	036	39 05	105	57 05	156	37 17	181	40 36	211	54 46	245	51 02	307
103	43 49	036	39 52	106	57 24	158	37 16	182	40 11	212	54 02	246	50 23	307
104	44 18	036	40 39	107	57 41	160	37 14	183	39 44	214	53 17	247	49 45	307
	Kochab		◆Denebola		REGULUS		◆SIRIUS		RIGEL		ALDEBARAN		◆CAPELLA	
105	27 35	016	22 48	088	41 25	107	37 10	185	39 17	215	52 32	248	68 10	305
106	27 48	016	23 36	089	42 11	108	37 06	186	38 49	216	51 47	249	67 30	305
107	28 02	016	24 25	089	42 57	109	37 01	187	38 20	217	51 02	250	66 50	304
108	28 15	016	25 13	090	43 43	110	36 54	188	37 51	218	50 16	251	66 10	304
109	28 29	017	26 02	090	44 29	111	36 47	189	37 20	219	49 30	252	65 30	303
110	28 43	017	26 51	091	45 14	112	36 38	190	36 49	220	48 43	253	64 49	303
111	28 57	017	27 39	092	45 59	112	36 29	192	36 18	221	47 57	254	64 08	303
112	29 11	017	28 28	092	46 44	113	36 19	193	35 46	222	47 10	254	63 27	302
113	29 25	017	29 16	093	47 28	114	36 08	194	35 13	223	46 23	255	62 46	302
114	29 40	017	30 05	093	48 12	115	35 55	195	34 39	224	45 36	256	62 05	302
115	29 54	017	30 53	094	48 56	116	35 42	196	34 05	225	44 49	257	61 24	302
116	30 09	018	31 41	095	49 39	117	35 28	197	33 30	226	44 02	258	60 42	301
117	30 23	018	32 30	095	50 22	118	35 13	199	32 55	227	43 14	258	60 01	301
118	30 38	018	33 18	096	51 05	119	34 57	200	32 19	228	42 27	259	59 19	301
119	30 53	018	34 06	097	51 47	120	34 40	201	31 43	229	41 39	260	58 38	301
	◆Kochab		Denebola		◆REGULUS		SIRIUS		RIGEL		◆ALDEBARAN		CAPELLA	
120	31 08	018	34 54	097	52 28	122	34 23	202	31 06	230	40 51	261	57 56	301
121	31 23	018	35 43	098	53 09	123	34 04	203	30 28	231	40 03	261	57 15	301
122	31 39	018	36 31	099	53 50	124	33 45	204	29 51	232	39 15	262	56 33	301
123	31 54	018	37 19	099	54 30	125	33 25	205	29 12	233	38 27	263	55 51	301
124	32 09	019	38 06	100	55 09	126	33 03	206	28 34	233	37 39	263	55 10	301
125	32 25	019	38 54	101	55 48	128	32 42	207	27 54	234	36 51	264	54 28	301
126	32 40	019	39 42	101	56 26	129	32 19	208	27 15	235	36 02	265	53 46	301
127	32 56	019	40 29	102	57 03	130	31 56	209	26 35	236	35 14	265	53 05	301
128	33 12	019	41 17	103	57 40	132	31 31	210	25 54	237	34 26	266	52 23	301
129	33 28	019	42 04	104	58 16	133	31 06	211	25 14	237	33 37	266	51 41	301
130	33 43	019	42 51	104	58 50	135	30 41	212	24 33	238	32 49	267	51 00	301
131	33 59	019	43 38	105	59 24	136	30 14	213	23 51	239	32 00	268	50 18	301
132	34 15	019	44 25	106	59 57	138	29 47	214	23 09	240	31 12	268	49 37	301
133	34 31	019	45 11	107	60 29	140	29 20	215	22 27	241	30 23	269	48 55	301
134	34 48	019	45 58	108	61 00	141	28 51	216	21 45	241	29 35	269	48 14	301
	Kochab		◆ARCTURUS		SPICA		REGULUS		◆SIRIUS		BETELGEUSE		◆CAPELLA	
135	35 04	019	19 56	080	11 55	113	61 30	143	28 22	217	39 04	247	47 32	302
136	35 20	019	20 44	081	12 39	114	61 58	145	27 52	218	38 19	248	46 51	302
137	35 36	020	21 32	081	13 23	115	62 26	147	27 22	219	37 34	249	46 10	302
138	35 52	020	22 20	082	14 07	115	62 52	149	26 51	220	36 49	250	45 28	302
139	36 09	020	23 08	082	14 51	116	63 16	151	26 20	221	36 03	251	44 47	302
140	36 25	020	23 56	083	15 35	117	63 39	153	25 48	222	35 17	251	44 06	302
141	36 41	020	24 45	084	16 18	117	64 01	155	25 15	223	34 31	252	43 25	302
142	36 58	020	25 33	084	17 01	118	64 21	157	24 42	224	33 45	253	42 44	303
143	37 14	020	26 21	085	17 43	119	64 39	159	24 08	224	32 58	254	42 03	303
144	37 30	020	27 10	085	18 26	120	64 55	161	23 34	225	32 12	254	41 23	303
145	37 47	020	27 58	086	19 08	120	65 10	163	22 59	226	31 25	255	40 42	303
146	38 03	020	28 46	086	19 49	121	65 23	166	22 24	227	30 38	256	40 01	303
147	38 19	020	29 35	087	20 31	122	65 34	168	21 49	228	29 51	257	39 21	304
148	38 36	020	30 23	087	21 12	123	65 43	170	21 12	228	29 03	257	38 41	304
149	38 52	020	31 12	088	21 53	123	65 50	173	20 36	229	28 16	258	38 00	304
	◆Kochab		ARCTURUS		◆SPICA		REGULUS		PROCYON		◆POLLUX		CAPELLA	
150	39 08	020	32 00	089	22 33	124	65 55	175	45 22	235	60 28	264	37 20	304
151	39 25	020	32 49	089	23 13	125	65 58	178	44 42	236	59 39	265	36 40	305
152	39 41	020	33 37	090	23 53	126	66 00	180	44 02	237	58 51	265	36 00	305
153	39 57	019	34 26	090	24 32	126	65 59	182	43 21	238	58 03	266	35 20	305
154	40 13	019	35 14	091	25 11	127	65 56	185	42 40	239	57 14	267	34 41	305
155	40 29	019	36 03	091	25 49	128	65 51	187	41 58	240	56 26	267	34 01	306
156	40 45	019	36 51	092	26 27	129	65 44	189	41 16	241	55 37	268	33 22	306
157	41 01	019	37 40	093	27 05	130	65 35	192	40 33	242	54 49	269	32 43	306
158	41 17	019	38 28	093	27 42	131	65 24	194	39 50	243	54 00	269	32 04	306
159	41 33	019	39 17	094	28 19	131	65 11	196	39 07	244	53 12	270	31 25	307
160	41 49	019	40 05	095	28 55	132	64 56	199	38 23	244	52 23	271	30 46	307
161	42 05	019	40 54	095	29 30	133	64 40	201	37 39	245	51 35	271	30 07	307
162	42 20	019	41 42	096	30 05	134	64 22	203	36 55	246	50 46	272	29 29	308
163	42 36	019	42 30	097	30 40	135	64 02	205	36 10	247	49 58	272	28 50	308
164	42 52	019	43 18	097	31 14	136	63 40	207	35 26	248	49 09	273	28 12	308
	◆Kochab		ARCTURUS		◆SPICA		REGULUS		PROCYON		◆POLLUX		CAPELLA	
165	43 07	018	44 07	098	31 47	137	63 17	209	34 41	249	48 21	273	27 34	309
166	43 22	018	44 55	099	32 20	138	62 53	211	33 55	249	47 32	274	26 56	309
167	43 37	018	45 43	099	32 52	139	62 27	213	33 10	250	46 44	274	26 18	309
168	43 52	018	46 30	100	33 24	140	62 00	215	32 24	251	45 55	275	25 41	310
169	44 07	018	47 18	101	33 55	141	61 31	217	31 38	252	45 07	275	25 04	310
170	44 22	018	48 06	102	34 25	142	61 02	219	30 52	252	44 19	276	24 26	310
171	44 37	017	48 53	102	34 55	143	60 31	220	30 05	253	43 30	276	23 49	311
172	44 51	017	49 41	103	35 24	144	59 59	222	29 19	254	42 42	277	23 13	311
173	45 06	017	50 28	104	35 52	145	59 26	224	28 32	255	41 54	277	22 36	311
174	45 20	017	51 15	105	36 19	146	58 52	225	27 45	255	41 06	278	22 00	312
175	45 34	017	52 02	106	36 46	147	58 17	227	26 58	256	40 18	278	21 24	312
176	45 48	017	52 48	107	37 12	148	57 42	228	26 11	257	39 30	279	20 48	312
177	46 01	016	53 35	107	37 37	149	57 05	229	25 24	257	38 42	279	20 12	313
178	46 15	016	54 21	108	38 01	151	56 28	231	24 36	258	37 54	280	19 37	313
179	46 28	016	55 07	109	38 24	152	55 50	232	23 49	259	37 06	280	19 01	314

LAT 36°N

LHA ♈	Hc Zn	Hc Zn	Hc Zn	Hc Zn	Hc Zn	Hc Zn	Hc Zn
	♦VEGA	Alphecca	ARCTURUS	♦SPICA	REGULUS	♦POLLUX	Dubhe
180	15 31 053	43 52 086	55 52 110	38 47 153	55 11 233	36 19 281	62 44 345
181	16 09 053	44 40 086	56 38 111	39 09 154	54 32 235	35 31 281	62 31 344
182	16 49 054	45 29 087	57 23 112	39 30 155	53 52 236	34 43 282	62 18 344
183	17 28 054	46 17 087	58 08 113	39 49 156	53 11 237	33 56 282	62 04 343
184	18 07 055	47 06 088	58 52 115	40 08 158	52 30 238	33 08 283	61 49 342
185	18 47 055	47 54 089	59 36 116	40 26 159	51 49 239	32 21 283	61 34 341
186	19 27 056	48 43 089	60 19 117	40 43 160	51 07 241	31 34 284	61 18 340
187	20 07 056	49 31 090	61 02 118	40 59 161	50 24 242	30 47 284	61 01 339
188	20 48 056	50 20 090	61 45 120	41 14 163	49 41 243	30 00 285	60 43 339
189	21 28 057	51 08 091	62 27 121	41 28 164	48 58 244	29 13 285	60 25 338
190	22 09 057	51 57 092	63 08 122	41 41 165	48 14 245	28 26 286	60 07 337
191	22 50 058	52 46 092	63 49 124	41 53 166	47 30 246	27 39 286	59 48 336
192	23 31 058	53 34 093	64 29 125	42 04 168	46 46 247	26 53 286	59 28 336
193	24 12 059	54 23 093	65 08 127	42 14 169	46 01 247	26 06 287	59 08 335
194	24 54 059	55 11 094	65 46 129	42 22 170	45 16 248	25 20 287	58 47 335
	Kochab	♦VEGA	Alphecca	♦SPICA	REGULUS	♦POLLUX	Dubhe
195	49 32 011	25 36 059	55 59 095	42 30 172	44 31 249	24 34 288	58 26 334
196	49 42 011	26 17 060	56 48 095	42 36 173	43 46 250	23 47 288	58 05 333
197	49 51 011	26 59 060	57 36 096	42 42 174	43 00 251	23 01 289	57 43 333
198	50 00 010	27 41 060	58 24 097	42 46 176	42 14 252	22 16 289	57 20 332
199	50 08 010	28 24 061	59 12 098	42 49 177	41 28 253	21 30 290	56 57 332
200	50 16 009	29 06 061	60 00 098	42 51 178	40 41 253	20 44 290	56 34 331
201	50 24 009	29 49 062	60 48 099	42 52 180	39 55 254	19 59 291	56 11 331
202	50 32 009	30 32 062	61 36 100	42 52 181	39 08 255	19 13 291	55 47 330
203	50 39 008	31 15 062	62 24 101	42 50 182	38 21 256	18 28 292	55 22 330
204	50 46 008	31 58 063	63 12 102	42 48 184	37 34 256	17 43 292	54 58 329
205	50 52 008	32 41 063	63 59 103	42 44 185	36 46 257	16 58 293	54 33 329
206	50 58 007	33 24 063	64 46 104	42 39 186	35 59 258	16 14 293	54 08 329
207	51 04 007	34 08 064	65 33 105	42 33 188	35 12 259	15 29 294	53 42 328
208	51 10 006	34 51 064	66 20 106	42 26 189	34 24 259	14 45 294	53 17 328
209	51 15 006	35 35 064	67 07 107	42 18 190	33 36 260	14 01 295	52 51 328
	DENEB	♦VEGA	Rasalhague	ANTARES	♦SPICA	REGULUS	♦Dubhe
210	18 24 047	36 19 065	36 33 102	18 22 145	42 09 192	32 48 261	52 25 327
211	18 59 047	37 03 065	37 21 102	18 49 146	41 58 193	32 00 261	51 59 327
212	19 35 048	37 47 065	38 08 103	19 16 147	41 47 194	31 12 262	51 32 327
213	20 11 048	38 31 066	38 55 104	19 42 148	41 34 196	30 24 263	51 05 327
214	20 47 048	39 15 066	39 42 105	20 08 148	41 21 197	29 36 263	50 38 326
215	21 24 049	40 00 066	40 29 105	20 33 149	41 06 198	28 48 264	50 11 326
216	22 00 049	40 44 067	41 16 106	20 57 150	40 51 199	28 00 265	49 44 326
217	22 37 050	41 29 067	42 02 107	21 21 151	40 34 201	27 11 265	49 17 326
218	23 14 050	42 14 067	42 49 108	21 45 152	40 17 202	26 23 266	48 49 325
219	23 51 050	42 58 068	43 35 109	22 07 153	39 58 203	25 34 266	48 22 325
220	24 29 051	43 43 068	44 21 110	22 29 154	39 39 204	24 46 267	47 54 325
221	25 06 051	44 28 068	45 06 110	22 50 155	39 18 205	23 57 268	47 26 325
222	25 44 051	45 13 068	45 52 111	23 11 155	38 57 207	23 09 268	46 58 325
223	26 22 052	45 59 069	46 37 112	23 31 156	38 35 208	22 20 269	46 30 325
224	27 01 052	46 44 069	47 22 113	23 50 157	38 12 209	21 32 269	46 02 325
	DENEB	♦ALTAIR	Rasalhague	ANTARES	♦SPICA	Denebola	♦Dubhe
225	27 39 052	19 13 093	48 06 114	24 08 158	37 48 210	42 23 256	45 34 325
226	28 17 053	20 01 094	48 50 115	24 26 159	37 23 211	41 36 257	45 06 324
227	28 56 053	20 50 094	49 34 116	24 43 160	36 58 212	40 48 258	44 38 324
228	29 35 053	21 38 095	50 18 117	24 59 161	36 32 213	40 01 258	44 10 324
229	30 14 054	22 26 095	51 01 118	25 14 162	36 05 214	39 13 259	43 41 324
230	30 53 054	23 15 096	51 43 119	25 29 163	35 37 216	38 26 260	43 13 324
231	31 32 054	24 03 097	52 25 120	25 43 164	35 08 217	37 38 260	42 45 324
232	32 12 055	24 51 097	53 07 121	25 56 165	34 39 218	36 50 261	42 16 324
233	32 52 055	25 39 098	53 48 123	26 09 166	34 09 219	36 02 262	41 48 324
234	33 31 055	26 27 099	54 29 124	26 20 167	33 38 220	35 14 262	41 20 324
235	34 11 055	27 15 099	55 09 125	26 31 168	33 07 221	34 26 263	40 51 324
236	34 51 056	28 03 100	55 48 126	26 41 169	32 35 222	33 37 264	40 23 324
237	35 31 056	28 51 101	56 27 128	26 50 170	32 03 223	32 49 264	39 54 324
238	36 12 056	29 38 101	57 05 129	26 58 171	31 29 224	32 01 265	39 26 324
239	36 52 056	30 26 102	57 43 130	27 06 172	30 56 224	31 12 266	38 58 324
	♦DENEB	ALTAIR	Nunki	♦ANTARES	ARCTURUS	♦Alkaid	Kochab
240	37 32 057	31 13 103	15 17 140	27 12 173	61 33 241	62 31 309	50 54 353
241	38 13 057	32 01 103	15 48 141	27 18 174	60 50 242	61 53 309	50 48 352
242	38 54 057	32 48 104	16 18 142	27 23 175	60 07 243	61 16 309	50 41 352
243	39 35 057	33 35 105	16 48 142	27 27 176	59 23 245	60 38 308	50 34 351
244	40 16 058	34 22 105	17 17 143	27 30 177	58 39 246	60 00 308	50 27 351
245	40 57 058	35 09 106	17 46 144	27 33 178	57 55 247	59 21 308	50 19 351
246	41 38 058	35 55 107	18 15 145	27 34 179	57 10 248	58 43 308	50 11 350
247	42 19 058	36 41 108	18 42 146	27 35 180	56 25 249	58 04 307	50 03 350
248	43 00 058	37 28 108	19 10 146	27 35 181	55 40 250	57 26 307	49 54 350
249	43 42 059	38 13 109	19 36 147	27 33 182	54 54 251	56 47 307	49 45 349
250	44 23 059	38 59 110	20 02 148	27 32 183	54 08 252	56 08 307	49 36 349
251	45 05 059	39 45 111	20 28 149	27 29 184	53 22 253	55 29 307	49 26 349
252	45 47 059	40 30 112	20 53 150	27 25 185	52 35 254	54 50 307	49 17 348
253	46 28 059	41 15 113	21 17 151	27 21 186	51 48 255	54 11 307	49 06 348
254	47 10 059	42 00 113	21 40 151	27 15 187	51 01 255	53 32 306	48 56 348
	♦DENEB	ALTAIR	Nunki	♦ANTARES	ARCTURUS	♦Alkaid	Kochab
255	47 52 060	42 44 114	22 03 152	27 09 188	50 14 256	52 53 306	48 46 347
256	48 34 060	43 28 115	22 25 153	27 02 189	49 27 257	52 14 306	48 35 347
257	49 16 060	44 12 116	22 47 154	26 54 190	48 40 258	51 35 306	48 24 347
258	49 58 060	44 55 117	23 08 155	26 46 191	47 52 259	50 56 306	48 12 346
259	50 40 060	45 38 118	23 28 156	26 36 192	47 05 259	50 17 306	48 01 346
260	51 22 060	46 21 119	23 48 157	26 26 193	46 17 260	49 38 306	47 49 346
261	52 04 060	47 03 120	24 06 158	26 15 194	45 29 261	48 59 306	47 37 345
262	52 46 060	47 45 121	24 24 159	26 03 195	44 41 262	48 20 306	47 24 345
263	53 28 060	48 26 122	24 42 160	25 50 196	43 53 262	47 41 306	47 12 345
264	54 11 060	49 07 123	24 58 160	25 37 197	43 05 263	47 02 307	46 59 345
265	54 53 060	49 47 124	25 14 161	25 23 198	42 17 264	46 23 307	46 46 344
266	55 35 060	50 27 126	25 29 162	25 08 199	41 28 264	45 44 307	46 33 344
267	56 17 060	51 06 127	25 44 163	24 52 199	40 40 265	45 05 307	46 20 344
268	57 00 060	51 45 128	25 57 164	24 35 200	39 52 266	44 26 307	46 06 344
269	57 42 060	52 23 129	26 10 165	24 18 201	39 03 266	43 47 307	45 53 344

LHA ♈	Hc Zn	Hc Zn	Hc Zn	Hc Zn	Hc Zn	Hc Zn	Hc Zn
	♦DENEB	ALTAIR	Nunki	♦ANTARES	ARCTURUS	♦Alkaid	Kochab
270	58 24 060	53 00 130	26 22 166	24 00 202	38 15 267	43 09 307	45 39 343
271	59 06 060	53 37 132	26 33 167	23 41 203	37 26 267	42 30 307	45 25 343
272	59 48 060	54 13 133	26 43 168	23 22 204	36 38 268	41 51 307	45 11 343
273	60 30 060	54 48 134	26 53 169	23 02 205	35 49 269	41 13 308	44 56 343
274	61 12 060	55 22 136	27 01 170	22 41 206	35 01 269	40 35 308	44 42 343
275	61 54 060	55 55 137	27 09 171	22 19 207	34 12 270	39 56 308	44 27 342
276	62 36 060	56 28 139	27 16 172	21 57 208	33 24 270	39 18 308	44 13 342
277	63 18 059	56 59 140	27 22 173	21 34 209	32 35 271	38 40 308	43 58 342
278	64 00 059	57 30 142	27 28 174	21 11 209	31 47 272	38 02 308	43 43 342
279	64 41 059	57 59 143	27 32 175	20 47 210	30 58 272	37 24 309	43 28 342
280	65 23 059	58 28 145	27 36 176	20 22 211	30 10 273	36 46 309	43 12 342
281	66 04 058	58 55 147	27 39 177	19 56 212	29 21 273	36 08 309	42 57 342
282	66 45 058	59 21 149	27 41 178	19 30 213	28 33 274	35 31 309	42 42 341
283	67 26 057	59 46 150	27 42 179	19 04 214	27 44 274	34 53 310	42 26 341
284	68 07 057	60 09 152	27 42 180	18 37 214	26 56 275	34 16 310	42 10 341
	Alpheratz	♦Enif	ALTAIR	♦Rasalhague	ARCTURUS	Alkaid	♦Kochab
285	26 22 072	44 37 115	60 31 154	59 42 225	26 07 276	33 39 310	41 55 341
286	27 09 072	45 20 116	60 52 156	59 08 226	25 19 276	33 01 310	41 39 341
287	27 55 073	46 04 117	61 11 158	58 32 228	24 31 277	32 24 310	41 23 341
288	28 41 073	46 47 118	61 28 160	57 56 229	23 43 277	31 48 311	41 07 341
289	29 28 074	47 30 119	61 44 162	57 19 231	22 55 278	31 11 311	40 51 341
290	30 15 074	48 12 120	61 59 164	56 41 232	22 07 278	30 34 311	40 35 341
291	31 01 075	48 54 121	62 11 166	56 02 233	21 19 279	29 58 312	40 19 341
292	31 48 075	49 36 122	62 22 168	55 23 235	20 31 279	29 22 312	40 03 341
293	32 35 076	50 17 123	62 30 170	54 43 236	19 43 280	28 46 312	39 47 340
294	33 22 076	50 57 124	62 39 172	54 03 237	18 55 280	28 10 312	39 30 340
295	34 09 077	51 37 125	62 45 174	53 22 238	18 07 281	27 34 313	39 14 340
296	34 57 077	52 17 126	62 49 176	52 40 239	17 20 281	26 58 313	38 58 340
297	35 44 077	52 55 128	62 51 179	51 58 240	16 32 282	26 23 313	38 41 340
298	36 31 078	53 33 129	62 51 181	51 16 242	15 45 283	25 48 314	38 25 340
299	37 19 078	54 11 130	62 50 183	50 33 243	14 57 283	25 13 314	38 09 340
	♦Mirfak	Alpheratz	♦FOMALHAUT	ALTAIR	Rasalhague	♦Alphecca	Kochab
300	15 13 039	38 07 079	12 15 142	62 46 185	49 49 244	33 38 281	37 52 340
301	15 43 039	38 54 079	12 45 142	62 41 187	49 06 245	32 50 281	37 36 340
302	16 14 039	39 42 080	13 14 143	62 34 189	48 22 246	32 03 282	37 20 340
303	16 45 040	40 30 080	13 43 144	62 25 192	47 37 247	31 15 282	37 03 340
304	17 16 040	41 18 081	14 11 145	62 15 194	46 53 248	30 28 283	36 47 340
305	17 47 041	42 06 081	14 39 145	62 02 196	46 08 248	29 41 283	36 31 340
306	18 19 041	42 54 082	15 07 146	61 48 198	45 22 249	28 53 284	36 14 340
307	18 51 041	43 42 082	15 33 147	61 33 200	44 37 250	28 06 284	35 58 340
308	19 23 042	44 30 083	16 00 148	61 16 202	43 51 251	27 19 285	35 42 340
309	19 56 042	45 18 083	16 25 148	60 57 204	43 05 252	26 32 285	35 26 340
310	20 28 043	46 06 084	16 51 149	60 37 206	42 19 253	25 45 285	35 09 341
311	21 01 043	46 54 084	17 15 150	60 15 207	41 32 253	24 59 286	34 53 341
312	21 34 043	47 43 085	17 39 151	59 52 209	40 46 254	24 12 286	34 37 341
313	22 08 044	48 31 085	18 03 152	59 28 211	39 59 255	23 25 287	34 21 341
314	22 41 044	49 19 086	18 25 152	59 02 213	39 12 256	22 39 287	34 05 341
	♦Mirfak	Hamal	Diphda	♦FOMALHAUT	ALTAIR	♦VEGA	Kochab
315	23 15 044	23 51 078	14 30 126	18 47 153	58 35 214	61 36 286	33 49 341
316	23 49 045	24 38 078	15 09 126	19 09 154	58 07 216	60 49 287	33 33 341
317	24 23 045	25 26 079	15 48 127	19 30 155	57 38 218	60 03 287	33 17 341
318	24 58 045	26 13 079	16 26 128	19 50 156	57 08 219	59 16 287	33 02 341
319	25 32 046	27 01 080	17 05 129	20 10 157	56 36 221	58 30 287	32 46 341
320	26 07 046	27 49 080	17 42 129	20 29 158	56 04 222	57 44 287	32 30 341
321	26 42 046	28 37 081	18 20 130	20 47 158	55 31 224	56 57 288	32 15 341
322	27 17 047	29 25 081	18 57 131	21 04 159	54 57 225	56 11 288	31 59 342
323	27 52 047	30 13 082	19 33 132	21 21 160	54 22 227	55 25 288	31 44 342
324	28 28 047	31 01 082	20 09 132	21 37 161	53 47 228	54 39 288	31 29 342
325	29 04 047	31 49 083	20 45 133	21 53 162	53 10 229	53 53 289	31 14 342
326	29 39 048	32 37 083	21 20 134	22 07 163	52 33 231	53 07 289	30 59 342
327	30 15 048	33 25 084	21 55 135	22 21 164	51 55 232	52 21 289	30 44 342
328	30 51 048	34 13 084	22 29 136	22 35 165	51 17 233	51 35 289	30 29 342
329	31 28 048	35 02 085	23 03 136	22 47 166	50 38 234	50 49 290	30 14 342
	♦CAPELLA	Hamal	Diphda	♦FOMALHAUT	ALTAIR	♦VEGA	Kochab
330	13 50 043	35 50 085	23 36 137	22 59 166	49 58 235	50 03 290	29 59 343
331	14 23 043	36 38 086	24 09 138	23 10 167	49 18 236	49 18 290	29 45 343
332	14 56 043	37 27 086	24 41 139	23 20 168	48 37 238	48 32 290	29 30 343
333	15 29 044	38 15 087	25 12 140	23 29 169	47 56 239	47 47 291	29 16 343
334	16 03 044	39 04 088	25 43 141	23 38 170	47 14 240	47 01 291	29 02 343
335	16 37 045	39 52 088	26 14 142	23 46 171	46 32 241	46 16 291	28 48 343
336	17 12 045	40 41 089	26 44 143	23 53 172	45 50 242	45 31 291	28 34 343
337	17 46 046	41 29 089	27 13 143	23 59 173	45 07 243	44 46 292	28 20 344
338	18 21 046	42 18 090	27 41 144	24 05 174	44 23 244	44 01 292	28 07 344
339	18 56 046	43 06 090	28 09 145	24 10 175	43 40 245	43 16 292	27 53 344
340	19 31 047	43 55 091	28 37 146	24 13 176	42 56 245	42 31 293	27 40 344
341	20 06 047	44 44 092	29 03 147	24 17 177	42 12 246	41 46 293	27 27 344
342	20 42 047	45 32 092	29 29 148	24 19 178	41 27 247	41 01 293	27 13 344
343	21 18 048	46 21 093	29 54 149	24 20 179	40 42 248	40 17 294	27 01 345
344	21 54 048	47 09 093	30 19 150	24 21 180	39 57 249	39 32 294	26 48 345
	♦CAPELLA	ALDEBARAN	Diphda	♦FOMALHAUT	ALTAIR	♦VEGA	Kochab
345	22 30 049	14 26 080	30 43 151	24 21 181	39 11 250	38 48 294	26 35 345
346	23 07 049	15 14 080	31 06 152	24 20 182	38 26 251	38 04 294	26 23 345
347	23 44 049	16 02 081	31 28 153	24 18 183	37 40 251	37 20 295	26 11 345
348	24 21 050	16 50 082	31 49 154	24 16 183	36 54 252	36 36 295	25 58 346
349	24 58 050	17 38 082	32 10 155	24 12 184	36 08 253	35 52 295	25 47 346
350	25 35 050	18 26 083	32 30 156	24 08 185	35 21 254	35 08 296	25 35 346
351	26 12 051	19 14 083	32 49 157	24 03 186	34 34 254	34 24 296	25 23 346
352	26 50 051	20 02 084	33 07 158	23 57 187	33 48 255	33 41 296	25 12 347
353	27 28 051	20 51 084	33 25 160	23 51 188	33 01 256	32 57 297	25 01 347
354	28 06 052	21 39 085	33 41 161	23 44 189	32 14 257	32 14 297	24 50 347
355	28 44 052	22 27 085	33 57 162	23 35 190	31 26 257	31 31 298	24 39 347
356	29 22 052	23 16 086	34 11 163	23 27 191	30 39 258	30 48 298	24 28 347
357	30 01 053	24 04 087	34 25 164	23 17 192	29 51 259	30 05 298	24 17 348
358	30 40 053	24 53 087	34 38 165	23 06 193	29 04 259	29 23 299	24 07 348
359	31 18 053	25 41 088	34 50 166	22 55 194	28 16 260	28 40 299	23 57 348

LHA ♈	Hc	Zn	Hc	Zn	Hc	Zn	Hc	Zn	Hc	Zn	Hc	Zn	Hc	Zn
	◆CAPELLA		ALDEBARAN		Diphda		◆FOMALHAUT		ALTAIR		◆VEGA		Kochab	
0	31 21	053	26 27	088	36 00	167	23 41	195	27 38	261	27 28	300	22 48	348
1	32 01	053	27 17	088	36 10	168	23 28	196	26 49	262	26 46	300	22 39	349
2	32 40	054	28 06	089	36 20	170	23 15	197	26 00	262	26 03	301	22 29	349
3	33 20	054	28 55	089	36 28	171	23 00	198	25 12	263	25 21	301	22 20	349
4	34 00	054	29 44	090	36 35	172	22 45	198	24 23	264	24 39	301	22 11	349
5	34 39	054	30 33	091	36 42	173	22 29	199	23 34	264	23 57	302	22 02	350
6	35 19	055	31 22	091	36 47	174	22 12	200	22 45	265	23 15	302	21 53	350
7	35 59	055	32 11	092	36 52	175	21 55	201	21 56	265	22 34	302	21 44	350
8	36 40	055	33 01	092	36 55	177	21 37	202	21 07	266	21 52	303	21 36	350
9	37 20	055	33 50	093	36 57	178	21 18	203	20 18	267	21 11	303	21 28	351
10	38 00	055	34 39	094	36 59	179	20 59	204	19 29	267	20 30	304	21 20	351
11	38 41	056	35 28	094	36 59	180	20 38	205	18 40	268	19 49	304	21 12	351
12	39 22	056	36 17	095	36 58	181	20 18	205	17 51	268	19 09	304	21 05	351
13	40 02	056	37 06	095	36 57	183	19 56	206	17 02	269	18 28	305	20 58	352
14	40 43	056	37 55	096	36 54	184	19 34	207	16 12	270	17 48	305	20 51	352
	◆CAPELLA		RIGEL		◆Diphda		FOMALHAUT		Enif		◆DENEB		Kochab	
15	41 24	056	16 12	113	36 50	185	19 11	208	38 52	253	40 51	303	20 44	352
16	42 05	057	16 57	113	36 46	186	18 48	209	38 05	254	40 09	303	20 38	352
17	42 46	057	17 42	114	36 40	187	18 24	210	37 18	254	39 28	303	20 31	353
18	43 27	057	18 27	115	36 33	188	17 59	210	36 30	255	38 47	303	20 25	353
19	44 08	057	19 11	115	36 25	190	17 34	211	35 43	256	38 06	304	20 19	353
20	44 50	057	19 56	116	36 17	191	17 08	212	34 55	256	37 25	304	20 14	354
21	45 31	057	20 40	117	36 07	192	16 42	213	34 07	257	36 44	304	20 08	354
22	46 12	057	21 23	117	35 56	193	16 15	214	33 19	258	36 04	304	20 03	354
23	46 54	058	22 07	118	35 44	194	15 48	214	32 31	259	35 23	304	19 58	354
24	47 35	058	22 50	119	35 32	195	15 20	215	31 43	259	34 43	305	19 53	355
25	48 17	058	23 33	120	35 18	197	14 51	216	30 54	260	34 02	305	19 49	355
26	48 58	058	24 15	120	35 04	198	14 22	217	30 06	261	33 22	305	19 45	355
27	49 40	058	24 58	121	34 48	199	13 53	217	29 17	261	32 42	306	19 41	356
28	50 22	058	25 40	122	34 32	200	13 22	218	28 29	262	32 02	306	19 37	356
29	51 03	058	26 21	123	34 15	201	12 52	219	27 40	262	31 22	306	19 34	356
	◆CAPELLA		BETELGEUSE		RIGEL		◆Diphda		Alpheratz		◆DENEB		Kochab	
30	51 45	058	29 43	103	27 02	124	33 57	202	65 39	264	30 43	306	19 30	356
31	52 27	058	30 31	103	27 43	124	33 38	203	64 50	265	30 03	307	19 27	357
32	53 08	058	31 18	104	28 23	125	33 18	204	64 01	265	29 24	307	19 25	357
33	53 50	058	32 06	105	29 03	126	32 58	205	63 12	266	28 44	307	19 22	357
34	54 32	058	32 53	105	29 43	127	32 36	206	62 23	267	28 05	308	19 20	358
35	55 13	058	33 41	106	30 22	128	32 14	207	61 34	267	27 26	308	19 18	358
36	55 55	058	34 28	107	31 01	129	31 51	208	60 45	268	26 48	308	19 16	358
37	56 37	058	35 15	108	31 39	130	31 27	209	59 56	269	26 09	308	19 15	358
38	57 18	058	36 02	108	32 16	130	31 03	210	59 06	269	25 31	309	19 13	359
39	58 00	058	36 48	109	32 54	131	30 37	211	58 17	270	24 53	309	19 12	359
40	58 41	057	37 35	110	33 30	132	30 11	212	57 28	270	24 15	309	19 11	359
41	59 22	057	38 21	111	34 06	133	29 45	213	56 39	271	23 37	310	19 11	000
42	60 04	057	39 06	111	34 42	134	29 17	214	55 50	271	22 59	310	19 11	000
43	60 45	057	39 52	112	35 17	135	28 49	215	55 01	272	22 22	311	19 11	000
44	61 26	057	40 37	113	35 51	136	28 20	216	54 12	272	21 44	311	19 11	000
	◆Dubhe		POLLUX		SIRIUS		◆RIGEL		Diphda		◆Alpheratz		Schedar	
45	17 52	025	30 08	075	15 44	124	36 25	137	27 51	217	53 22	273	58 05	323
46	18 13	026	30 55	076	16 25	125	36 58	138	27 21	218	52 33	273	57 36	323
47	18 34	026	31 43	076	17 05	126	37 30	139	26 50	219	51 44	274	57 06	322
48	18 56	026	32 31	076	17 44	126	38 02	140	26 19	220	50 55	274	56 36	322
49	19 18	027	33 19	077	18 24	127	38 33	141	25 47	221	50 06	275	56 05	322
50	19 40	027	34 07	077	19 03	128	39 04	142	25 15	222	49 17	275	55 35	321
51	20 02	027	34 55	078	19 42	129	39 33	144	24 42	222	48 29	276	55 04	321
52	20 25	027	35 43	078	20 20	129	40 02	145	24 09	223	47 40	276	54 33	321
53	20 47	028	36 31	079	20 58	130	40 30	146	23 35	224	46 51	277	54 01	320
54	21 10	028	37 19	079	21 35	131	40 57	147	23 00	225	46 02	277	53 30	320
55	21 34	028	38 07	080	22 12	132	41 24	148	22 25	226	45 13	278	52 58	320
56	21 57	029	38 56	080	22 48	132	41 49	149	21 50	227	44 25	278	52 26	319
57	22 21	029	39 44	081	23 24	133	42 14	151	21 14	227	43 36	279	51 54	319
58	22 45	029	40 33	081	24 00	134	42 37	152	20 38	228	42 48	279	51 22	319
59	23 09	029	41 21	081	24 35	135	43 00	153	20 01	229	41 59	280	50 50	319
	◆Dubhe		POLLUX		PROCYON		◆SIRIUS		RIGEL		◆Hamal		Schedar	
60	23 33	030	42 10	082	31 32	107	25 10	136	43 22	154	62 53	252	50 17	319
61	23 58	030	42 59	082	32 19	108	25 44	137	43 43	156	62 06	253	49 45	318
62	24 22	030	43 47	083	33 06	109	26 17	137	44 03	157	61 18	254	49 12	318
63	24 47	031	44 36	083	33 52	110	26 50	138	44 21	158	60 31	255	48 39	318
64	25 12	031	45 25	084	34 38	110	27 22	139	44 39	160	59 43	256	48 06	318
65	25 37	031	46 14	084	35 24	111	27 54	140	44 56	161	58 55	257	47 34	318
66	26 03	031	47 03	085	36 10	112	28 25	141	45 11	162	58 07	258	47 01	318
67	26 28	031	47 52	085	36 55	113	28 56	142	45 26	164	57 19	259	46 28	318
68	26 54	032	48 41	086	37 41	114	29 26	143	45 39	165	56 31	260	45 55	318
69	27 20	032	49 30	086	38 25	114	29 55	144	45 51	166	55 42	261	45 22	318
70	27 46	032	50 19	087	39 10	115	30 24	145	46 02	168	54 54	261	44 48	318
71	28 12	032	51 08	087	39 54	116	30 52	146	46 12	169	54 05	262	44 15	318
72	28 39	033	51 57	088	40 38	117	31 19	147	46 21	171	53 17	263	43 42	318
73	29 05	033	52 46	088	41 22	118	31 45	148	46 28	172	52 28	263	43 09	318
74	29 32	033	53 35	089	42 05	119	32 11	149	46 35	173	51 39	264	42 36	318
	◆Dubhe		POLLUX		PROCYON		◆SIRIUS		RIGEL		◆Hamal		Mirfak	
75	29 58	033	54 24	090	42 48	120	32 36	150	46 40	175	50 50	265	67 04	318
76	30 25	033	55 14	090	43 30	121	33 01	151	46 43	176	50 01	265	66 30	317
77	30 52	033	56 03	091	44 12	122	33 24	152	46 46	178	49 12	266	65 56	316
78	31 19	034	56 52	091	44 54	123	33 47	153	46 47	179	48 23	267	65 22	315
79	31 47	034	57 41	092	45 35	124	34 09	154	46 47	181	47 34	267	64 47	315
80	32 14	034	58 30	093	46 16	125	34 30	155	46 46	182	46 45	268	64 12	314
81	32 42	034	59 19	093	46 56	126	34 50	156	46 44	184	45 56	269	63 37	313
82	33 09	034	60 08	094	47 35	127	35 10	157	46 40	185	45 07	269	63 01	313
83	33 37	034	60 57	095	48 14	128	35 28	158	46 35	186	44 17	270	62 25	312
84	34 04	034	61 46	095	48 53	129	35 46	160	46 29	188	43 28	270	61 48	312
85	34 32	035	62 35	096	49 31	130	36 03	161	46 22	189	42 39	271	61 12	312
86	35 00	035	63 24	097	50 08	132	36 18	162	46 14	191	41 50	271	60 35	311
87	35 28	035	64 13	097	50 44	133	36 33	163	46 04	192	41 01	272	59 58	311
88	35 56	035	65 02	098	51 20	134	36 47	164	45 53	193	40 12	273	59 20	310
89	36 24	035	65 50	099	51 55	135	37 00	165	45 41	195	39 23	273	58 43	310

LAT 35°N

LHA ♈	Hc	Zn	Hc	Zn	Hc	Zn	Hc	Zn	Hc	Zn	Hc	Zn	Hc	Zn
	◆Dubhe		REGULUS		PROCYON		◆SIRIUS		RIGEL		ALDEBARAN		◆Mirfak	
90	36 52	035	29 40	096	52 29	137	37 12	166	45 28	196	63 36	231	58 05	310
91	37 21	035	30 29	097	53 02	138	37 23	168	45 14	198	62 57	232	57 27	310
92	37 49	035	31 18	097	53 34	140	37 33	169	44 58	199	62 18	234	56 49	309
93	38 17	035	32 07	098	54 06	141	37 42	170	44 42	200	61 38	235	56 11	309
94	38 45	035	32 55	099	54 36	142	37 50	171	44 24	202	60 57	237	55 33	309
95	39 14	035	33 44	099	55 05	144	37 57	172	44 05	203	60 16	238	54 55	309
96	39 42	035	34 32	100	55 34	146	38 03	174	43 46	204	59 33	240	54 17	309
97	40 10	035	35 21	101	56 01	147	38 08	175	43 25	206	58 51	241	53 38	309
98	40 39	035	36 09	101	56 27	149	38 12	176	43 03	207	58 08	242	53 00	308
99	41 07	035	36 57	102	56 52	150	38 15	177	42 41	208	57 24	243	52 21	308
100	41 36	035	37 45	103	57 16	152	38 17	178	42 17	209	56 40	244	51 43	308
101	42 04	035	38 33	104	57 38	154	38 17	180	41 53	210	55 56	245	51 04	308
102	42 32	035	39 20	104	57 59	155	38 17	181	41 27	212	55 11	246	50 25	308
103	43 01	035	40 08	105	58 19	157	38 16	182	41 01	213	54 25	247	49 47	308
104	43 29	035	40 55	106	58 37	159	38 13	183	40 34	214	53 40	248	49 08	308
	◆Alkaid		REGULUS		Alphard		◆SIRIUS		RIGEL		ALDEBARAN		◆CAPELLA	
105	19 00	042	41 43	107	34 12	134	38 10	185	40 06	215	52 54	249	67 35	307
106	19 33	043	42 30	107	34 47	135	38 06	186	39 38	216	52 08	250	66 55	307
107	20 07	043	43 16	108	35 21	136	38 00	187	39 08	217	51 21	251	66 16	306
108	20 41	043	44 03	109	35 55	137	37 54	188	38 38	218	50 35	252	65 36	306
109	21 15	044	44 49	110	36 28	138	37 46	189	38 07	220	49 48	253	64 56	305
110	21 49	044	45 36	111	37 00	139	37 37	191	37 35	221	49 01	254	64 16	305
111	22 23	045	46 21	112	37 32	140	37 28	192	37 03	222	48 13	255	63 35	304
112	22 58	045	47 07	112	38 03	141	37 17	193	36 30	223	47 26	256	62 54	304
113	23 32	045	47 52	113	38 34	142	37 06	194	35 56	224	46 38	256	62 14	304
114	24 07	045	48 37	114	39 03	144	36 53	195	35 22	225	45 50	257	61 33	303
115	24 42	046	49 22	115	39 32	145	36 40	197	34 47	226	45 02	258	60 52	303
116	25 18	046	50 06	116	40 00	146	36 25	198	34 12	227	44 14	259	60 10	303
117	25 53	046	50 50	117	40 27	147	36 10	199	33 36	228	43 26	259	59 29	303
118	26 29	047	51 34	118	40 54	148	35 54	200	32 59	229	42 38	260	58 48	303
119	27 05	047	52 17	119	41 19	149	35 36	201	32 22	229	41 49	261	58 06	302
	◆Alkaid		REGULUS		◆Alphard		SIRIUS		RIGEL		ALDEBARAN		◆CAPELLA	
120	27 41	047	52 59	120	41 44	150	35 18	202	31 44	230	41 01	261	57 25	302
121	28 17	048	53 41	122	42 08	152	34 59	203	31 06	231	40 12	262	56 43	302
122	28 53	048	54 23	123	42 30	153	34 39	204	30 28	232	39 23	263	56 02	302
123	29 30	048	55 04	124	42 52	154	34 19	205	29 49	233	38 35	263	55 20	302
124	30 07	048	55 44	125	43 13	155	33 57	207	29 09	234	37 46	264	54 38	302
125	30 43	049	56 24	127	43 33	157	33 35	208	28 29	235	36 57	265	53 57	302
126	31 20	049	57 03	128	43 52	158	33 12	209	27 49	235	36 08	265	53 15	302
127	31 57	049	57 42	129	44 10	159	32 48	210	27 08	236	35 19	266	52 33	302
128	32 35	049	58 19	131	44 27	161	32 23	211	26 27	237	34 30	267	51 52	302
129	33 12	050	58 56	132	44 42	162	31 58	212	25 46	238	33 41	267	51 10	302
130	33 49	050	59 32	134	44 57	163	31 31	213	25 04	239	32 52	268	50 28	302
131	34 27	050	60 07	135	45 11	165	31 04	214	24 22	239	32 02	268	49 47	302
132	35 05	050	60 42	137	45 23	166	30 37	215	23 39	240	31 13	269	49 05	302
133	35 43	050	61 15	138	45 34	167	30 08	216	22 56	241	30 24	269	48 23	302
134	36 21	051	61 47	140	45 44	169	29 39	217	22 13	242	29 35	270	47 42	302
	◆Alkaid		ARCTURUS		SPICA		◆Alphard		SIRIUS		BETELGEUSE		◆CAPELLA	
135	36 59	051	19 46	080	12 18	113	45 53	170	29 10	218	39 27	248	47 00	302
136	37 37	051	20 34	080	13 03	114	46 01	172	28 39	218	38 41	249	46 19	303
137	38 15	051	21 23	081	13 48	115	46 07	173	28 09	219	37 55	250	45 38	303
138	38 53	051	22 11	081	14 33	115	46 13	175	27 37	220	37 09	251	44 56	303
139	39 32	051	23 00	082	15 17	116	46 17	176	27 05	221	36 23	251	44 15	303
140	40 10	052	23 49	083	16 01	117	46 20	177	26 32	222	35 36	252	43 34	303
141	40 49	052	24 38	083	16 45	117	46 21	179	25 59	223	34 49	253	42 53	303
142	41 27	052	25 26	084	17 29	118	46 22	180	25 25	224	34 02	254	42 12	303
143	42 06	052	26 15	084	18 12	119	46 21	182	24 51	225	33 15	254	41 31	304
144	42 45	052	27 04	085	18 55	119	46 19	183	24 16	226	32 27	255	40 50	304
145	43 24	052	27 53	085	19 38	120	46 16	185	23 41	226	31 40	256	40 09	304
146	44 03	052	28 42	086	20 20	121	46 11	186	23 05	227	30 52	256	39 28	304
147	44 42	052	29 31	086	21 02	121	46 05	187	22 29	228	30 04	257	38 47	304
148	45 20	052	30 20	087	21 44	122	45 59	189	21 52	229	29 16	258	38 07	305
149	45 59	053	31 09	087	22 26	123	45 50	190	21 15	230	28 28	258	37 26	305
	Alkaid		◆ARCTURUS		SPICA		◆Alphard		PROCYON		BETELGEUSE		◆CAPELLA	
150	46 39	053	31 58	088	23 07	124	45 41	192	45 56	236	27 40	259	36 46	305
151	47 18	053	32 48	088	23 47	124	45 31	193	45 16	237	26 52	260	36 06	305
152	47 57	053	33 37	089	24 28	125	45 19	194	44 34	238	26 03	260	35 26	305
153	48 36	053	34 26	090	25 08	126	45 06	196	43 52	239	25 15	261	34 46	306
154	49 15	053	35 15	090	25 47	127	44 52	197	43 10	240	24 26	262	34 06	306
155	49 54	053	36 04	091	26 26	128	44 37	198	42 27	241	23 38	262	33 26	306
156	50 33	053	36 53	091	27 05	129	44 21	200	41 44	242	22 49	263	32 47	306
157	51 12	053	37 42	092	27 43	129	44 04	201	41 01	243	22 00	264	32 07	307
158	51 51	053	38 32	093	28 21	130	43 46	202	40 17	243	21 11	264	31 28	307
159	52 30	053	39 21	093	28 58	131	43 27	204	39 33	244	20 22	265	30 48	307
160	53 09	052	40 10	094	29 35	132	43 06	205	38 49	245	19 33	265	30 09	307
161	53 48	052	40 59	094	30 11	133	42 45	206	38 04	246	18 44	266	29 30	308
162	54 27	052	41 48	095	30 47	134	42 23	207	37 19	247	17 55	266	28 52	308
163	55 06	052	42 37	096	31 22	135	42 00	209	36 34	248	17 06	267	28 13	308
164	55 44	052	43 25	096	31 57	136	41 36	210	35 48	248	16 17	268	27 35	309
	◆Dubhe		Alkaid		◆ARCTURUS		SPICA		REGULUS		◆PROCYON		CAPELLA	
165	63 13	001	56 23	052	44 14	097	32 31	137	64 09	210	35 02	249	26 56	309
166	63 13	000	57 02	052	45 03	098	33 04	138	63 44	212	34 16	250	26 18	309
167	63 13	359	57 40	051	45 52	098	33 37	138	63 17	214	33 30	251	25 40	310
168	63 11	358	58 18	051	46 40	099	34 10	139	62 49	216	32 43	251	25 02	310
169	63 09	357	58 56	051	47 29	100	34 41	140	62 19	218	31 57	252	24 25	310
170	63 06	356	59 34	050	48 17	100	35 12	141	61 48	220	31 10	253	23 47	311
171	63 01	355	60 12	050	49 05	101	35 42	143	61 16	221	30 23	254	23 10	311
172	62 56	354	60 50	050	49 54	102	36 12	144	60 43	223	29 35	254	22 33	311
173	62 51	353	61 27	049	50 42	103	36 41	145	60 09	225	28 48	255	21 56	312
174	62 44	352	62 05	049	51 29	104	37 09	146	59 34	226	28 00	256	21 20	312
175	62 36	351	62 42	049	52 17	104	37 36	147	58 58	228	27 13	256	20 43	312
176	62 28	350	63 18	048	53 05	105	38 03	148	58 21	229	26 25	257	20 07	313
177	62 18	349	63 55	047	53 52	106	38 28	149	57 44	231	25 37	258	19 31	313
178	62 08	348	64 31	047	54 39	107	38 53	150	57 05	232	24 49	258	18 55	313
179	61 58	347	65 06	046	55 26	108	39 17	151	56 26	233	24 01	259	18 20	314

LHA	Hc	Zn	Hc	Zn	Hc	Zn	Hc	Zn	Hc	Zn	Hc	Zn	Hc	Zn
♈	♦VEGA		Alphecca		ARCTURUS		♦SPICA		REGULUS		♦POLLUX		Dubhe	
180	14 54	053	43 47	085	56 13	109	39 40	153	55 46	235	36 07	282	61 46	346
181	15 34	053	44 36	085	56 59	110	40 03	154	55 06	236	35 19	282	61 34	345
182	16 13	054	45 25	086	57 45	111	40 24	155	54 25	237	34 31	282	61 20	344
183	16 53	054	46 14	086	58 31	112	40 44	156	53 44	238	33 43	283	61 07	343
184	17 33	055	47 03	087	59 16	113	41 04	157	53 01	239	32 55	283	60 52	342
185	18 13	055	47 52	087	60 01	114	41 22	159	52 19	241	32 07	284	60 37	342
186	18 53	055	48 41	088	60 46	115	41 40	160	51 36	242	31 20	284	60 21	341
187	19 34	056	49 30	089	61 30	117	41 56	161	50 52	243	30 32	285	60 04	340
188	20 14	056	50 20	089	62 14	118	42 12	162	50 09	244	29 44	285	59 47	339
189	20 55	057	51 09	090	62 57	119	42 26	164	49 24	245	28 57	286	59 30	339
190	21 36	057	51 58	090	63 39	121	42 39	165	48 40	246	28 10	286	59 11	338
191	22 18	057	52 47	091	64 21	122	42 51	166	47 55	247	27 23	286	58 53	337
192	22 59	058	53 36	091	65 03	124	43 03	168	47 09	248	26 35	287	58 33	336
193	23 41	058	54 25	092	65 43	125	43 13	169	46 24	248	25 49	287	58 13	336
194	24 23	059	55 14	093	66 23	127	43 22	170	45 38	249	25 02	288	57 53	335
	♦Kochab		VEGA		Rasalhague		♦ANTARES		SPICA		♦REGULUS		Dubhe	
195	48 34	011	25 05	059	24 33	092	11 10	134	43 29	172	44 52	250	57 32	335
196	48 43	011	25 47	059	25 22	092	11 45	134	43 36	173	44 06	251	57 11	334
197	48 52	010	26 29	060	26 11	093	12 20	135	43 41	174	43 19	252	56 49	333
198	49 01	010	27 12	060	27 00	093	12 55	136	43 46	176	42 32	253	56 27	333
199	49 09	010	27 54	060	27 49	094	13 29	137	43 49	177	41 45	253	56 04	332
200	49 17	009	28 37	061	28 38	095	14 02	137	43 51	178	40 58	254	55 41	332
201	49 25	009	29 20	061	29 27	095	14 35	138	43 52	180	40 11	255	55 18	331
202	49 32	009	30 03	061	30 16	096	15 08	139	43 52	181	39 23	256	54 54	331
203	49 39	008	30 46	062	31 05	096	15 40	140	43 50	182	38 35	256	54 30	331
204	49 46	008	31 30	062	31 54	097	16 12	140	43 47	184	37 47	257	54 06	330
205	49 53	007	32 13	062	32 43	098	16 43	141	43 44	185	36 59	258	53 41	330
206	49 59	007	32 57	063	33 31	098	17 13	142	43 39	186	36 11	259	53 16	329
207	50 05	007	33 41	063	34 20	099	17 44	143	43 33	188	35 23	259	52 51	329
208	50 10	006	34 25	063	35 09	100	18 13	143	43 25	189	34 35	260	52 26	329
209	50 15	006	35 09	064	35 57	100	18 42	144	43 17	190	33 46	261	52 00	328
	♦Kochab		VEGA		Rasalhague		♦ANTARES		SPICA		♦REGULUS		Dubhe	
210	50 20	005	35 53	064	36 45	101	19 11	145	43 07	192	32 58	261	51 34	328
211	50 24	005	36 37	064	37 33	102	19 39	146	42 57	193	32 09	262	51 08	328
212	50 28	005	37 21	065	38 21	102	20 06	147	42 45	194	31 20	263	50 42	327
213	50 32	004	38 06	065	39 09	103	20 33	147	42 32	196	30 32	263	50 15	327
214	50 36	004	38 51	065	39 57	104	20 59	148	42 18	197	29 43	264	49 48	327
215	50 39	003	39 35	066	40 45	105	21 25	149	42 03	198	28 54	264	49 21	327
216	50 41	003	40 20	066	41 32	105	21 49	150	41 47	200	28 05	265	48 54	326
217	50 43	002	41 05	066	42 20	106	22 14	151	41 30	201	27 16	266	48 27	326
218	50 45	002	41 50	066	43 07	107	22 37	152	41 12	202	26 27	266	48 00	326
219	50 47	002	42 35	067	43 54	108	23 00	153	40 53	203	25 38	267	47 32	326
220	50 48	001	43 20	067	44 40	109	23 23	153	40 34	205	24 49	267	47 05	326
221	50 49	001	44 06	067	45 27	109	23 44	154	40 13	206	24 00	268	46 37	326
222	50 49	000	44 51	068	46 13	110	24 05	155	39 51	207	23 11	269	46 09	325
223	50 49	000	45 36	068	46 59	111	24 26	156	39 28	208	22 21	269	45 41	325
224	50 49	359	46 22	068	47 45	112	24 45	157	39 04	209	21 32	270	45 13	325
	DENEB		VEGA		♦ALTAIR		ANTARES		♦SPICA		Denebola		♦Dubhe	
225	27 02	052	47 08	068	19 16	093	25 04	158	38 40	210	42 37	257	44 45	325
226	27 41	052	47 53	069	20 05	093	25 22	159	38 15	212	41 49	258	44 17	325
227	28 20	053	48 39	069	20 54	094	25 39	160	37 49	213	41 01	258	43 49	325
228	28 59	053	49 25	069	21 43	094	25 56	161	37 22	214	40 13	259	43 21	325
229	29 38	053	50 11	069	22 32	095	26 11	162	36 54	215	39 24	260	42 52	325
230	30 18	053	50 57	069	23 21	096	26 26	163	36 25	216	38 36	260	42 24	325
231	30 57	054	51 43	070	24 10	096	26 41	164	35 56	217	37 47	261	41 56	325
232	31 37	054	52 29	070	24 58	097	26 54	165	35 26	218	36 59	262	41 27	325
233	32 17	054	53 15	070	25 47	097	27 07	166	34 56	219	36 10	262	40 59	325
234	32 57	055	54 01	070	26 36	098	27 18	167	34 24	220	35 21	263	40 31	325
235	33 37	055	54 48	070	27 25	099	27 29	168	33 52	221	34 33	264	40 02	325
236	34 17	055	55 34	071	28 13	099	27 40	169	33 20	222	33 44	264	39 34	325
237	34 57	055	56 21	071	29 02	100	27 49	170	32 47	223	32 55	265	39 06	325
238	35 38	056	57 07	071	29 50	101	27 57	171	32 13	224	32 06	266	38 37	325
239	36 19	056	57 53	071	30 38	101	28 05	172	31 38	225	31 17	266	38 09	325
	♦DENEB		ALTAIR		Nunki		♦ANTARES		SPICA		ARCTURUS		♦Alkaid	
240	36 59	056	31 26	102	16 03	140	28 12	173	31 03	226	62 01	242	61 52	311
241	37 40	056	32 14	103	16 34	141	28 18	174	30 28	227	61 17	244	61 15	310
242	38 21	057	33 02	103	17 05	141	28 23	175	29 52	228	60 33	245	60 38	310
243	39 02	057	33 50	104	17 36	142	28 27	176	29 15	229	59 48	246	60 00	310
244	39 43	057	34 37	105	18 05	143	28 30	177	28 38	229	59 03	247	59 22	309
245	40 24	057	35 25	106	18 35	144	28 32	178	28 01	230	58 18	248	58 44	309
246	41 06	057	36 12	106	19 04	145	28 34	179	27 23	231	57 32	249	58 06	309
247	41 47	058	36 59	107	19 32	145	28 35	180	26 44	232	56 46	250	57 27	309
248	42 29	058	37 46	108	20 00	146	28 34	181	26 05	233	55 59	251	56 49	308
249	43 10	058	38 33	109	20 27	147	28 33	182	25 26	234	55 13	252	56 10	308
250	43 52	058	39 19	109	20 53	148	28 31	183	24 46	234	54 26	253	55 32	308
251	44 34	058	40 06	110	21 19	149	28 29	184	24 06	235	53 39	254	54 53	308
252	45 15	058	40 52	111	21 44	149	28 25	185	23 26	236	52 51	255	54 14	308
253	45 57	058	41 37	112	22 09	150	28 20	186	22 45	237	52 04	256	53 35	308
254	46 39	059	42 23	113	22 33	151	28 15	187	22 03	237	51 16	257	52 56	308
	♦DENEB		ALTAIR		Nunki		♦ANTARES		ARCTURUS		♦Alkaid		Kochab	
255	47 21	059	43 08	113	22 56	152	28 09	188	50 28	257	52 17	307	47 47	347
256	48 03	059	43 53	114	23 19	153	28 02	189	49 40	258	51 38	307	47 36	347
257	48 45	059	44 38	115	23 41	154	27 54	190	48 52	259	50 59	307	47 25	347
258	49 27	059	45 22	116	24 02	155	27 45	191	48 03	260	50 20	307	47 14	347
259	50 09	059	46 06	117	24 23	156	27 35	192	47 15	260	49 41	307	47 02	346
260	50 52	059	46 49	118	24 43	157	27 25	193	46 27	261	49 02	307	46 51	346
261	51 34	059	47 33	119	25 02	158	27 13	194	45 38	262	48 23	307	46 39	346
262	52 16	059	48 15	120	25 20	158	27 01	195	44 49	263	47 44	307	46 26	345
263	52 58	059	48 58	121	25 38	159	26 48	196	44 00	263	47 05	307	46 14	345
264	53 40	059	49 39	122	25 55	160	26 34	197	43 12	264	46 26	307	46 01	345
265	54 23	059	50 21	123	26 11	161	26 20	198	42 23	265	45 47	307	45 48	345
266	55 05	059	51 01	125	26 26	162	26 04	199	41 34	265	45 08	308	45 35	344
267	55 47	059	51 42	126	26 41	163	25 48	200	40 45	266	44 29	308	45 22	344
268	56 29	059	52 21	127	26 55	164	25 31	201	39 56	266	43 50	308	45 09	344
269	57 11	059	53 00	128	27 08	165	25 14	201	39 07	267	43 11	308	44 55	344

LHA	Hc	Zn	Hc	Zn	Hc	Zn	Hc	Zn	Hc	Zn	Hc	Zn	Hc	Zn
♈	♦DENEB		ALTAIR		♦Nunki		ANTARES		ARCTURUS		♦Alkaid		Kochab	
270	57 54	059	53 39	129	27 20	166	24 55	202	38 18	268	42 32	308	44 41	344
271	58 36	059	54 16	131	27 31	167	24 36	203	37 28	268	41 53	308	44 27	343
272	59 18	059	54 53	132	27 42	168	24 16	204	36 39	269	41 15	308	44 13	343
273	60 00	059	55 29	133	27 52	169	23 56	205	35 50	269	40 36	308	43 59	343
274	60 42	058	56 05	135	28 00	170	23 35	206	35 01	270	39 58	308	43 45	343
275	61 23	058	56 39	136	28 08	171	23 13	207	34 12	271	39 19	309	43 30	343
276	62 05	058	57 13	138	28 16	172	22 50	208	33 23	271	38 41	309	43 15	343
277	62 47	058	57 45	139	28 22	173	22 27	209	32 34	272	38 02	309	43 01	342
278	63 28	057	58 17	141	28 27	174	22 03	210	31 44	272	37 24	309	42 46	342
279	64 10	057	58 47	142	28 32	175	21 38	210	30 55	273	36 46	309	42 31	342
280	64 51	057	59 17	144	28 36	176	21 13	211	30 06	273	36 08	309	42 15	342
281	65 32	056	59 45	146	28 39	177	20 47	212	29 17	274	35 30	310	42 00	342
282	66 12	056	60 12	148	28 41	178	20 21	213	28 28	274	34 52	310	41 45	342
283	66 53	055	60 37	149	28 42	179	19 54	214	27 39	275	34 15	310	41 29	342
284	67 33	055	61 02	151	28 42	180	19 26	215	26 50	275	33 37	310	41 14	341
	♦Alpheratz		Enif		ALTAIR		♦ANTARES		ARCTURUS		♦Alkaid		Kochab	
285	26 04	071	45 01	114	61 25	153	18 58	215	26 01	276	33 00	311	40 58	341
286	26 50	072	45 46	115	61 46	155	18 29	216	25 13	277	32 22	311	40 42	341
287	27 37	072	46 31	116	62 06	157	18 00	217	24 24	277	31 45	311	40 26	341
288	28 24	073	47 15	117	62 24	159	17 30	218	23 35	278	31 08	311	40 10	341
289	29 11	073	47 58	118	62 41	161	17 00	219	22 46	278	30 31	311	39 54	341
290	29 58	074	48 42	119	62 56	163	16 29	219	21 58	279	29 55	312	39 38	341
291	30 45	074	49 25	120	63 10	165	15 57	220	21 09	279	29 18	312	39 22	341
292	31 33	075	50 07	121	63 21	168	15 26	221	20 21	280	28 41	312	39 06	341
293	32 20	075	50 49	122	63 31	170	14 53	222	19 32	280	28 05	313	38 50	341
294	33 08	075	51 30	123	63 39	172	14 20	222	18 44	281	27 29	313	38 34	341
295	33 55	076	52 11	124	63 45	174	13 47	223	17 56	281	26 53	313	38 18	341
296	34 43	076	52 52	125	63 49	176	13 13	224	17 08	282	26 17	313	38 01	341
297	35 31	077	53 32	127	63 51	179	12 39	224	16 19	282	25 42	314	37 45	341
298	36 19	077	54 11	128	63 51	181	12 04	225	15 32	283	25 06	314	37 29	341
299	37 07	078	54 49	129	63 50	183	11 29	226	14 44	283	24 31	314	37 12	341
	♦Mirfak		Alpheratz		♦FOMALHAUT		ALTAIR		Rasalhague		♦Alphecca		Kochab	
300	14 26	038	37 55	078	13 02	141	63 46	185	50 15	245	33 27	281	36 56	341
301	14 56	039	38 43	079	13 32	142	63 40	188	49 31	246	32 38	282	36 40	341
302	15 27	039	39 31	079	14 02	143	63 33	190	48 46	247	31 50	282	36 23	341
303	15 59	040	40 19	079	14 31	144	63 24	192	48 01	248	31 02	283	36 07	341
304	16 30	040	41 08	080	15 00	144	63 13	194	47 15	249	30 14	283	35 50	341
305	17 02	040	41 56	080	15 28	145	63 00	196	46 29	249	29 27	284	35 34	341
306	17 34	041	42 44	081	15 56	146	62 45	198	45 43	250	28 39	284	35 18	341
307	18 06	041	43 33	081	16 24	147	62 29	200	44 57	251	27 51	285	35 02	341
308	18 38	042	44 22	082	16 50	147	62 11	202	44 10	252	27 04	285	34 45	341
309	19 11	042	45 10	082	17 16	148	61 52	204	43 23	253	26 16	285	34 29	341
310	19 44	042	45 59	083	17 42	149	61 31	206	42 36	254	25 29	286	34 13	341
311	20 17	043	46 48	083	18 07	150	61 08	208	41 49	254	24 42	286	33 57	341
312	20 51	043	47 37	084	18 31	151	60 44	210	41 01	255	23 55	287	33 41	341
313	21 24	043	48 25	084	18 55	151	60 19	212	40 14	256	23 08	287	33 24	341
314	21 58	044	49 14	084	19 18	152	59 52	214	39 26	257	22 21	288	33 08	341
	♦Mirfak		Hamal		Diphda		♦FOMALHAUT		ALTAIR		♦VEGA		Kochab	
315	22 32	044	23 38	077	15 05	125	19 41	153	59 24	215	61 18	288	32 52	341
316	23 06	044	24 25	078	15 44	126	20 03	154	58 55	217	60 31	288	32 37	341
317	23 41	045	25 14	078	16 24	127	20 24	155	58 25	219	59 45	288	32 21	341
318	24 15	045	26 02	079	17 03	128	20 45	156	57 54	220	58 58	289	32 05	341
319	24 50	045	26 50	079	17 42	128	21 05	157	57 21	222	58 11	289	31 49	341
320	25 25	046	27 38	080	18 20	129	21 24	157	56 48	223	57 25	289	31 34	341
321	26 00	046	28 27	080	18 58	130	21 43	158	56 14	225	56 38	289	31 18	342
322	26 36	046	29 15	081	19 36	131	22 00	159	55 39	226	55 52	289	31 03	342
323	27 11	046	30 03	081	20 13	131	22 18	160	55 03	228	55 06	289	30 47	342
324	27 47	047	30 52	082	20 50	132	22 34	161	54 26	229	54 19	290	30 32	342
325	28 23	047	31 41	082	21 26	133	22 50	162	53 49	230	53 33	290	30 17	342
326	28 59	047	32 29	083	22 02	134	23 05	163	53 11	232	52 47	290	30 01	342
327	29 35	048	33 18	083	22 37	134	23 19	164	52 32	233	52 01	290	29 46	342
328	30 11	048	34 07	084	23 12	135	23 32	165	51 52	234	51 15	290	29 32	342
329	30 48	048	34 56	084	23 46	136	23 45	165	51 12	235	50 29	291	29 17	343
	♦CAPELLA		Hamal		Diphda		♦FOMALHAUT		ALTAIR		♦VEGA		Kochab	
330	13 06	042	35 45	085	24 20	137	23 57	166	50 32	236	49 43	291	29 02	343
331	13 39	043	36 34	085	24 53	138	24 08	167	49 51	237	48 57	291	28 47	343
332	14 12	043	37 23	086	25 26	139	24 19	168	49 09	239	48 11	291	28 33	343
333	14 46	044	38 12	086	25 58	140	24 28	169	48 27	240	47 25	292	28 19	343
334	15 20	044	39 01	087	26 30	140	24 37	170	47 44	241	46 39	292	28 04	343
335	15 55	044	39 50	087	27 01	141	24 45	171	47 01	242	45 54	292	27 50	343
336	16 29	045	40 39	088	27 31	142	24 52	172	46 18	243	45 08	292	27 36	344
337	17 04	045	41 28	088	28 01	143	24 59	173	45 34	244	44 23	293	27 23	344
338	17 39	046	42 17	089	28 30	144	25 05	174	44 50	244	43 38	293	27 09	344
339	18 14	046	43 06	089	28 59	145	25 09	175	44 05	245	42 52	293	26 55	344
340	18 50	046	43 56	090	29 26	146	25 13	176	43 20	246	42 07	293	26 42	344
341	19 25	047	44 45	091	29 54	147	25 16	177	42 35	247	41 22	294	26 29	344
342	20 01	047	45 34	091	30 20	148	25 19	178	41 50	248	40 37	294	26 16	345
343	20 38	048	46 23	092	30 46	149	25 20	179	41 04	249	39 52	294	26 03	345
344	21 14	048	47 12	092	31 11	150	25 21	180	40 18	250	39 08	295	25 50	345
	♦CAPELLA		ALDEBARAN		Diphda		♦FOMALHAUT		ALTAIR		♦VEGA		Kochab	
345	21 51	048	14 15	080	31 35	151	25 21	181	39 32	250	38 23	295	25 37	345
346	22 27	049	15 04	080	31 59	152	25 20	182	38 45	251	37 39	295	25 25	345
347	23 04	049	15 52	081	32 21	153	25 18	183	37 59	252	36 54	295	25 12	346
348	23 42	049	16 41	081	32 43	154	25 16	184	37 12	253	36 10	296	25 00	346
349	24 19	050	17 29	082	33 04	155	25 12	184	36 25	254	35 26	296	24 48	346
350	24 56	050	18 18	082	33 25	156	25 08	185	35 38	254	34 42	296	24 37	346
351	25 34	050	19 07	083	33 44	157	25 03	186	34 50	255	33 58	297	24 25	346
352	26 12	051	19 56	083	34 03	158	24 57	187	34 03	256	33 14	297	24 13	347
353	26 50	051	20 44	084	34 21	159	24 50	188	33 15	256	32 30	297	24 02	347
354	27 28	051	21 33	084	34 38	160	24 43	189	32 27	257	31 47	298	23 51	347
355	28 07	052	22 22	085	34 54	162	24 35	190	31 39	258	31 03	298	23 40	347
356	28 46	052	23 11	086	35 09	163	24 25	191	30 51	258	30 20	298	23 29	348
357	29 24	052	24 00	086	35 23	164	24 16	192	30 03	259	29 37	299	23 19	348
358	30 03	052	24 49	087	35 36	165	24 05	193	29 15	260	28 54	299	23 09	348
359	30 42	053	25 38	087	35 48	166	23 53	194	28 26	260	28 11	299	22 58	348

LAT 34°N

LHA	Hc	Zn	Hc	Zn	Hc	Zn	Hc	Zn	Hc	Zn	Hc	Zn	Hc	Zn
♈	◆CAPELLA		ALDEBARAN		Diphda		◆FOMALHAUT		ALTAIR		◆DENEB		Kochab	
0	30 45	053	26 25	087	36 58	167	24 39	195	27 47	262	50 47	302	21 50	349
1	31 25	053	27 15	088	37 09	168	24 26	196	26 57	262	50 05	302	21 40	349
2	32 04	053	28 04	088	37 19	169	24 12	197	26 08	263	49 23	302	21 30	349
3	32 44	053	28 54	089	37 27	171	23 57	198	25 19	263	48 40	302	21 21	349
4	33 24	054	29 44	089	37 35	172	23 42	199	24 29	264	47 58	302	21 12	349
5	34 04	054	30 34	090	37 41	173	23 26	200	23 40	265	47 16	302	21 03	350
6	34 44	054	31 23	091	37 47	174	23 09	200	22 50	265	46 34	302	20 54	350
7	35 25	054	32 13	091	37 51	175	22 51	201	22 01	266	45 52	302	20 45	350
8	36 05	054	33 03	092	37 55	177	22 32	202	21 11	266	45 10	302	20 37	350
9	36 46	055	33 52	092	37 57	178	22 13	203	20 21	267	44 28	303	20 29	351
10	37 26	055	34 42	093	37 59	179	21 53	204	19 32	268	43 46	303	20 21	351
11	38 07	055	35 32	093	37 59	180	21 33	205	18 42	268	43 04	303	20 13	351
12	38 48	055	36 21	094	37 58	181	21 12	206	17 52	269	42 23	303	20 06	351
13	39 28	055	37 11	095	37 57	183	20 50	206	17 02	269	41 41	303	19 58	352
14	40 09	056	38 01	095	37 54	184	20 27	207	16 13	270	40 59	303	19 51	352
	◆CAPELLA		RIGEL		◆Diphda		FOMALHAUT		Enif		◆DENEB		Kochab	
15	40 51	056	16 35	112	37 50	185	20 04	208	39 09	254	40 18	303	19 45	352
16	41 32	056	17 21	113	37 45	186	19 40	209	38 21	254	39 36	304	19 38	353
17	42 13	056	18 06	114	37 39	187	19 16	210	37 33	255	38 55	304	19 32	353
18	42 54	056	18 52	114	37 32	189	18 51	211	36 45	256	38 14	304	19 26	353
19	43 35	056	19 37	115	37 24	190	18 25	211	35 57	256	37 32	304	19 20	353
20	44 17	056	20 22	116	37 15	191	17 59	212	35 09	257	36 51	304	19 14	354
21	44 58	056	21 07	116	37 06	192	17 32	213	34 20	258	36 10	305	19 09	354
22	45 40	057	21 51	117	36 55	193	17 05	214	33 31	259	35 30	305	19 03	354
23	46 21	057	22 35	118	36 43	194	16 37	215	32 43	259	34 49	305	18 58	354
24	47 03	057	23 19	119	36 30	196	16 09	215	31 54	260	34 08	305	18 54	355
25	47 44	057	24 02	119	36 16	197	15 40	216	31 05	260	33 28	306	18 49	355
26	48 26	057	24 46	120	36 01	198	15 10	217	30 16	261	32 47	306	18 45	355
27	49 08	057	25 29	121	35 45	199	14 40	217	29 26	262	32 07	306	18 41	356
28	49 49	057	26 11	122	35 29	200	14 10	218	28 37	262	31 27	306	18 37	356
29	50 31	057	26 53	122	35 11	201	13 39	219	27 48	263	30 47	307	18 34	356
	CAPELLA		BETELGEUSE		◆RIGEL		Diphda		◆Alpheratz		DENEB		◆Kochab	
30	51 13	057	29 56	102	27 35	123	34 52	202	65 44	266	30 07	307	18 31	356
31	51 54	057	30 44	103	28 17	124	34 33	203	64 54	267	29 27	307	18 28	357
32	52 36	057	31 33	103	28 58	125	34 13	205	64 05	267	28 47	307	18 25	357
33	53 18	057	32 21	104	29 38	126	33 52	206	63 15	268	28 08	308	18 22	357
34	53 59	057	33 09	105	30 19	126	33 30	207	62 25	269	27 29	308	18 20	358
35	54 41	057	33 57	105	30 58	127	33 07	208	61 36	269	26 50	308	18 18	358
36	55 23	057	34 45	106	31 38	128	32 44	209	60 46	270	26 11	309	18 16	358
37	56 04	057	35 33	107	32 17	129	32 19	210	59 56	270	25 32	309	18 15	358
38	56 46	056	36 20	108	32 55	130	31 54	211	59 06	271	24 53	309	18 13	359
39	57 27	056	37 07	108	33 33	131	31 28	212	58 17	271	24 15	309	18 12	359
40	58 08	056	37 55	109	34 10	132	31 02	213	57 27	272	23 36	310	18 11	359
41	58 49	056	38 41	110	34 47	133	30 35	214	56 37	272	22 58	310	18 11	000
42	59 31	056	39 28	111	35 23	134	30 07	215	55 48	273	22 20	310	18 11	000
43	60 12	055	40 14	112	35 59	135	29 38	216	54 58	273	21 43	311	18 11	000
44	60 52	055	41 01	112	36 34	136	29 09	217	54 08	274	21 05	311	18 11	000
	◆Dubhe		POLLUX		SIRIUS		◆RIGEL		Diphda		◆Alpheratz		Schedar	
45	16 57	025	29 52	075	16 18	124	37 09	137	28 39	217	53 19	274	57 17	324
46	17 19	025	30 40	075	16 59	125	37 43	138	28 08	218	52 29	275	56 48	324
47	17 40	026	31 28	075	17 40	125	38 16	139	27 37	219	51 40	275	56 18	323
48	18 02	026	32 16	076	18 20	126	38 48	140	27 05	220	50 50	276	55 48	323
49	18 24	026	33 05	076	19 00	127	39 20	141	26 33	221	50 01	276	55 18	322
50	18 46	027	33 53	077	19 40	128	39 51	142	26 00	222	49 11	277	54 48	322
51	19 09	027	34 42	077	20 19	128	40 21	143	25 26	223	48 22	277	54 17	322
52	19 31	027	35 30	078	20 58	129	40 51	144	24 52	224	47 32	278	53 46	321
53	19 54	028	36 19	078	21 36	130	41 19	145	24 18	224	46 43	278	53 15	321
54	20 17	028	37 07	078	22 14	131	41 47	146	23 43	225	45 54	278	52 43	321
55	20 41	028	37 56	079	22 52	131	42 14	148	23 07	226	45 05	279	52 12	321
56	21 04	028	38 45	079	23 29	132	42 41	149	22 31	227	44 16	279	51 40	320
57	21 28	029	39 34	080	24 05	133	43 06	150	21 54	228	43 26	280	51 08	320
58	21 52	029	40 23	080	24 42	134	43 30	151	21 17	228	42 37	280	50 36	320
59	22 16	029	41 12	081	25 17	135	43 54	153	20 40	229	41 49	281	50 04	320
	◆Dubhe		POLLUX		PROCYON		◆SIRIUS		RIGEL		◆Hamal		Schedar	
60	22 41	030	42 01	081	31 50	107	25 52	135	44 16	154	63 10	254	49 32	319
61	23 06	030	42 50	081	32 37	108	26 27	136	44 37	155	62 22	255	48 59	319
62	23 30	030	43 39	082	33 25	108	27 01	137	44 58	156	61 34	256	48 27	319
63	23 55	030	44 29	082	34 12	109	27 35	138	45 17	158	60 45	257	47 54	319
64	24 20	031	45 18	083	34 59	110	28 08	139	45 35	159	59 57	258	47 21	319
65	24 46	031	46 07	083	35 45	110	28 40	140	45 52	161	59 08	259	46 49	319
66	25 11	031	46 57	084	36 32	111	29 12	141	46 08	162	58 19	260	46 16	319
67	25 37	031	47 46	084	37 18	112	29 43	142	46 23	163	57 30	260	45 43	319
68	26 03	031	48 36	085	38 04	113	30 14	143	46 37	165	56 41	261	45 10	318
69	26 29	032	49 25	085	38 50	114	30 43	144	46 50	166	55 52	262	44 37	318
70	26 55	032	50 15	086	39 35	115	31 13	145	47 01	168	55 02	263	44 04	318
71	27 21	032	51 05	086	40 20	115	31 41	145	47 11	169	54 13	263	43 31	318
72	27 48	032	51 54	087	41 05	116	32 09	146	47 20	170	53 24	264	42 58	318
73	28 14	032	52 44	087	41 50	117	32 36	147	47 28	172	52 34	265	42 25	318
74	28 41	033	53 34	088	42 34	118	33 03	148	47 34	173	51 44	265	41 51	318
	◆Dubhe		POLLUX		PROCYON		◆SIRIUS		RIGEL		◆Hamal		Mirfak	
75	29 08	033	54 23	088	43 17	119	33 28	150	47 39	175	50 55	266	66 19	319
76	29 35	033	55 13	089	44 01	120	33 53	151	47 43	176	50 05	267	65 46	318
77	30 02	033	56 03	089	44 44	121	34 17	152	47 46	178	49 16	267	65 13	318
78	30 29	033	56 53	090	45 26	122	34 40	153	47 47	179	48 26	268	64 39	317
79	30 57	033	57 42	090	46 08	123	35 03	154	47 47	181	47 36	268	64 05	316
80	31 24	034	58 32	091	46 50	124	35 24	155	47 46	182	46 46	269	63 30	316
81	31 52	034	59 22	092	47 31	125	35 45	156	47 44	184	45 57	270	62 55	315
82	32 19	034	60 11	092	48 11	126	36 05	157	47 40	185	45 07	270	62 20	314
83	32 47	034	61 01	093	48 51	127	36 24	158	47 35	187	44 17	271	61 44	314
84	33 15	034	61 51	093	49 30	128	36 42	159	47 29	188	43 27	271	61 08	313
85	33 43	034	62 40	094	50 09	130	36 59	160	47 21	189	42 38	272	60 31	313
86	34 11	034	63 30	095	50 47	131	37 15	162	47 13	191	41 48	272	59 55	312
87	34 39	034	64 20	095	51 24	132	37 31	163	47 03	192	40 58	273	59 18	312
88	35 07	034	65 09	096	52 01	133	37 45	164	46 51	194	40 09	273	58 41	312
89	35 35	034	65 58	097	52 37	135	37 58	165	46 39	195	39 19	274	58 04	311

LHA	Hc	Zn	Hc	Zn	Hc	Zn	Hc	Zn	Hc	Zn	Hc	Zn	Hc	Zn
♈	◆Dubhe		REGULUS		PROCYON		◆SIRIUS		RIGEL		ALDEBARAN		◆Mirfak	
90	36 03	035	29 46	096	53 12	136	38 10	166	46 25	197	64 13	232	57 26	311
91	36 31	035	30 36	096	53 46	137	38 22	167	46 11	198	63 33	234	56 49	311
92	37 00	035	31 25	097	54 20	139	38 32	169	45 55	199	62 53	236	56 11	311
93	37 28	035	32 15	097	54 52	140	38 41	170	45 38	201	62 11	237	55 33	310
94	37 56	035	33 04	098	55 23	142	38 49	171	45 20	202	61 29	238	54 55	310
95	38 25	035	33 53	099	55 54	143	38 57	172	45 01	203	60 46	240	54 17	310
96	38 53	035	34 42	099	56 23	145	39 03	174	44 40	205	60 03	241	53 39	310
97	39 21	035	35 31	100	56 51	146	39 08	175	44 19	206	59 19	242	53 00	310
98	39 50	035	36 20	101	57 18	148	39 12	176	43 57	207	58 35	243	52 22	309
99	40 18	035	37 09	101	57 44	150	39 15	177	43 34	208	57 51	245	51 44	309
100	40 46	035	37 58	102	58 09	151	39 17	178	43 09	210	57 05	246	51 05	309
101	41 15	035	38 46	103	58 32	153	39 17	180	42 44	211	56 20	247	50 26	309
102	41 43	035	39 35	103	58 54	155	39 17	181	42 18	212	55 34	248	49 48	309
103	42 11	035	40 23	104	59 14	157	39 16	182	41 51	213	54 48	249	49 09	309
104	42 40	035	41 11	105	59 33	158	39 13	183	41 24	215	54 01	250	48 31	309
	◆Alkaid		REGULUS		Alphard		◆SIRIUS		RIGEL		ALDEBARAN		◆CAPELLA	
105	18 16	042	41 59	106	34 54	134	39 10	185	40 55	216	53 14	251	66 58	309
106	18 49	043	42 47	106	35 29	135	39 05	186	40 26	217	52 27	252	66 19	308
107	19 23	043	43 35	107	36 04	136	39 00	187	39 56	218	51 40	252	65 40	308
108	19 57	043	44 22	108	36 39	137	38 53	188	39 25	219	50 52	253	65 00	307
109	20 31	044	45 09	109	37 12	138	38 45	190	38 53	220	50 05	254	64 21	307
110	21 06	044	45 56	110	37 46	139	38 36	191	38 21	221	49 17	255	63 41	306
111	21 40	044	46 43	111	38 18	140	38 27	192	37 47	222	48 29	256	63 01	306
112	22 15	045	47 29	111	38 50	141	38 16	193	37 14	223	47 40	257	62 20	306
113	22 50	045	48 16	112	39 21	142	38 04	194	36 39	224	46 52	257	61 40	305
114	23 25	045	49 01	113	39 51	143	37 51	196	36 04	225	46 03	258	60 59	305
115	24 01	045	49 47	114	40 21	144	37 37	197	35 29	226	45 14	259	60 18	305
116	24 36	046	50 32	115	40 49	145	37 22	198	34 53	227	44 26	260	59 37	304
117	25 12	046	51 17	116	41 17	146	37 07	199	34 16	228	43 37	260	58 56	304
118	25 48	046	52 01	117	41 44	148	36 50	200	33 39	229	42 48	261	58 15	304
119	26 24	047	52 46	118	42 11	149	36 32	201	33 01	230	41 58	262	57 33	304
	◆Alkaid		REGULUS		◆Alphard		SIRIUS		RIGEL		ALDEBARAN		◆CAPELLA	
120	27 00	047	53 29	119	42 36	150	36 14	202	32 22	231	41 09	262	56 52	304
121	27 36	047	54 12	120	43 00	151	35 54	204	31 44	232	40 20	263	56 11	304
122	28 13	047	54 55	122	43 24	153	35 34	205	31 04	233	39 30	264	55 29	303
123	28 50	048	55 37	123	43 46	154	35 13	206	30 25	233	38 41	264	54 48	303
124	29 27	048	56 18	124	44 08	155	34 51	207	29 44	234	37 51	265	54 06	303
125	30 04	048	56 59	125	44 28	156	34 28	208	29 04	235	37 02	265	53 24	303
126	30 41	048	57 40	127	44 48	158	34 04	209	28 23	236	36 12	266	52 43	303
127	31 18	049	58 19	128	45 06	159	33 40	210	27 41	237	35 23	267	52 01	303
128	31 55	049	58 58	129	45 23	160	33 14	211	27 00	238	34 33	267	51 19	303
129	32 33	049	59 36	131	45 39	162	32 48	212	26 17	238	33 43	268	50 38	303
130	33 11	049	60 13	132	45 55	163	32 22	213	25 35	239	32 54	268	49 56	303
131	33 48	049	60 49	134	46 08	164	31 54	214	24 52	240	32 04	269	49 14	303
132	34 26	050	61 25	136	46 21	166	31 26	215	24 09	241	31 14	270	48 33	303
133	35 04	050	61 59	137	46 33	167	30 57	216	23 25	241	30 24	270	47 51	303
134	35 42	050	62 32	139	46 43	169	30 27	217	22 42	242	29 35	271	47 09	303
	◆Alkaid		ARCTURUS		SPICA		◆Alphard		SIRIUS		BETELGEUSE		◆CAPELLA	
135	36 20	050	19 35	080	12 42	113	46 52	170	29 57	218	39 49	249	46 28	303
136	36 59	050	20 24	080	13 28	114	47 00	172	29 26	219	39 02	250	45 46	303
137	37 37	051	21 13	081	14 13	114	47 07	173	28 55	220	38 16	250	45 05	304
138	38 16	051	22 02	081	14 58	115	47 12	174	28 23	221	37 29	251	44 23	304
139	38 54	051	22 52	082	15 43	116	47 17	176	27 50	222	36 41	252	43 42	304
140	39 33	051	23 41	082	16 28	116	47 20	177	27 17	222	35 54	253	43 01	304
141	40 11	051	24 30	083	17 13	117	47 21	179	26 43	223	35 06	253	42 19	304
142	40 50	051	25 19	083	17 57	118	47 22	180	26 09	224	34 19	254	41 38	304
143	41 29	051	26 09	084	18 41	118	47 21	182	25 34	225	33 31	255	40 57	304
144	42 08	051	26 58	084	19 24	119	47 19	183	24 58	226	32 43	256	40 16	304
145	42 47	052	27 48	085	20 08	120	47 15	185	24 22	227	31 54	256	39 35	305
146	43 26	052	28 37	085	20 51	120	47 11	186	23 46	227	31 06	257	38 54	305
147	44 05	052	29 27	086	21 34	121	47 05	188	23 09	228	30 17	258	38 13	305
148	44 44	052	30 17	086	22 16	122	46 58	189	22 32	229	29 29	258	37 33	305
149	45 23	052	31 06	087	22 58	123	46 49	190	21 54	230	28 40	259	36 52	305
	Alkaid		◆ARCTURUS		SPICA		◆Alphard		PROCYON		BETELGEUSE		◆CAPELLA	
150	46 02	052	31 56	087	23 40	123	46 40	192	46 30	237	27 51	260	36 11	306
151	46 41	052	32 46	088	24 21	124	46 29	193	45 48	238	27 02	260	35 31	306
152	47 20	052	33 35	088	25 02	125	46 17	195	45 06	239	26 13	261	34 51	306
153	47 59	052	34 25	089	25 43	126	46 04	196	44 23	240	25 24	261	34 11	306
154	48 38	052	35 15	089	26 23	126	45 50	197	43 40	241	24 35	262	33 30	306
155	49 17	052	36 05	090	27 03	127	45 34	199	42 56	241	23 45	263	32 51	307
156	49 56	052	36 54	091	27 42	128	45 18	200	42 13	242	22 56	263	32 11	307
157	50 35	052	37 44	091	28 21	129	45 00	201	41 28	243	22 07	264	31 31	307
158	51 14	052	38 34	092	28 59	130	44 41	203	40 44	244	21 17	264	30 51	307
159	51 53	051	39 23	092	29 37	131	44 22	204	39 59	245	20 28	265	30 12	308
160	52 32	051	40 13	093	30 15	132	44 01	205	39 14	246	19 38	266	29 33	308
161	53 11	051	41 03	094	30 52	132	43 39	207	38 28	247	18 48	266	28 54	308
162	53 50	051	41 52	094	31 28	133	43 16	208	37 42	248	17 59	267	28 15	309
163	54 28	051	42 42	095	32 04	134	42 52	209	36 56	248	17 09	267	27 36	309
164	55 07	051	43 32	095	32 40	135	42 28	210	36 10	249	16 19	268	26 57	309
	◆Dubhe		Alkaid		◆ARCTURUS		SPICA		REGULUS		◆PROCYON		CAPELLA	
165	62 13	001	55 45	051	44 21	096	33 14	136	65 01	211	35 23	250	26 18	309
166	62 13	000	56 24	050	45 11	097	33 49	137	64 34	213	34 36	251	25 40	310
167	62 13	359	57 02	050	46 00	097	34 22	138	64 06	215	33 49	251	25 02	310
168	62 11	358	57 40	050	46 49	098	34 55	139	63 37	217	33 02	252	24 24	310
169	62 09	357	58 18	050	47 38	099	35 27	140	63 06	219	32 15	253	23 46	311
170	62 06	356	58 56	049	48 28	099	35 59	141	62 34	221	31 27	254	23 08	311
171	62 02	355	59 33	049	49 17	100	36 30	142	62 01	223	30 39	254	22 31	311
172	61 57	354	60 11	048	50 05	101	37 00	143	61 27	224	29 51	255	21 54	312
173	61 51	353	60 48	048	50 54	102	37 30	144	60 51	226	29 03	256	21 16	312
174	61 44	352	61 25	048	51 43	102	37 58	145	60 15	228	28 15	256	20 40	312
175	61 37	351	62 01	047	52 31	103	38 26	146	59 38	229	27 26	257	20 03	313
176	61 29	350	62 38	047	53 20	104	38 53	147	59 00	231	26 38	258	19 26	313
177	61 20	349	63 13	046	54 08	105	39 20	149	58 21	232	25 49	258	18 50	313
178	61 10	348	63 49	045	54 56	106	39 45	150	57 42	233	25 01	259	18 14	314
179	60 59	347	64 24	045	55 44	107	40 10	151	57 02	235	24 12	259	17 38	314

LHA ♈	Hc	Zn	Hc	Zn	Hc	Zn	Hc	Zn	Hc	Zn	Hc	Zn	Hc	Zn
	◆VEGA		Alphecca		ARCTURUS		◆SPICA		REGULUS		◆POLLUX		Dubhe	
180	14 18	053	43 41	084	56 31	107	40 33	152	56 21	236	35 55	282	60 48	346
181	14 58	053	44 31	084	57 19	108	40 56	153	55 39	237	35 06	283	60 36	345
182	15 37	053	45 20	085	58 06	109	41 18	155	54 57	238	34 18	283	60 23	345
183	16 17	054	46 10	085	58 52	110	41 39	156	54 15	239	33 29	284	60 09	344
184	16 58	054	46 59	086	59 39	112	41 59	157	53 31	241	32 41	284	59 55	343
185	17 38	055	47 49	086	60 25	113	42 18	158	52 48	242	31 53	284	59 40	342
186	18 19	055	48 39	087	61 11	114	42 36	160	52 04	243	31 04	285	59 24	341
187	19 00	055	49 28	087	61 56	115	42 53	161	51 19	244	30 16	285	59 08	341
188	19 41	056	50 18	088	62 41	116	43 09	162	50 35	245	29 28	286	58 51	340
189	20 22	056	51 08	088	63 25	117	43 23	163	49 49	246	28 41	286	58 34	339
190	21 04	057	51 57	089	64 09	119	43 37	165	49 04	247	27 53	287	58 16	338
191	21 45	057	52 47	089	64 52	120	43 50	166	48 18	248	27 05	287	57 57	338
192	22 27	057	53 37	090	65 35	122	44 01	167	47 32	249	26 18	287	57 38	337
193	23 09	058	54 27	091	66 17	123	44 11	169	46 45	249	25 30	288	57 18	336
194	23 51	058	55 16	091	66 58	125	44 21	170	45 59	250	24 43	288	56 58	336
	◆Kochab		VEGA		Rasalhague		◆ANTARES		SPICA		◆REGULUS		Dubhe	
195	47 35	011	24 34	059	24 35	091	11 51	134	44 29	171	45 12	251	56 38	335
196	47 44	010	25 16	059	25 24	092	12 27	134	44 35	173	44 25	252	56 17	335
197	47 53	010	25 59	059	26 14	092	13 03	135	44 41	174	43 37	253	55 55	334
198	48 01	010	26 42	060	27 04	093	13 38	136	44 46	176	42 50	254	55 33	334
199	48 10	009	27 24	060	27 53	093	14 12	136	44 49	177	42 02	254	55 11	333
200	48 18	009	28 08	060	28 43	094	14 46	137	44 51	178	41 14	255	54 48	333
201	48 25	009	28 51	061	29 33	095	15 20	138	44 52	180	40 26	256	54 25	332
202	48 33	008	29 34	061	30 22	095	15 53	139	44 52	181	39 37	257	54 02	332
203	48 40	008	30 18	061	31 12	096	16 26	139	44 50	182	38 49	257	53 38	331
204	48 47	008	31 01	062	32 01	096	16 58	140	44 47	184	38 00	258	53 14	331
205	48 53	007	31 45	062	32 50	097	17 29	141	44 43	185	37 12	259	52 49	330
206	48 59	007	32 29	062	33 40	098	18 01	142	44 38	187	36 23	259	52 25	330
207	49 05	006	33 13	063	34 29	098	18 31	142	44 32	188	35 34	260	52 00	330
208	49 10	006	33 58	063	35 18	099	19 01	143	44 24	189	34 45	261	51 34	329
209	49 15	006	34 42	063	36 07	100	19 31	144	44 16	191	33 56	261	51 09	329
	◆Kochab		VEGA		Rasalhague		◆ANTARES		SPICA		◆REGULUS		Dubhe	
210	49 20	005	35 26	063	36 56	100	20 00	145	44 06	192	33 07	262	50 43	329
211	49 25	005	36 11	064	37 45	101	20 28	146	43 55	193	32 17	263	50 17	328
212	49 29	004	36 55	064	38 34	102	20 56	146	43 43	195	31 28	263	49 51	328
213	49 32	004	37 40	064	39 23	102	21 23	147	43 30	196	30 38	264	49 25	328
214	49 36	004	38 25	065	40 11	103	21 50	148	43 16	197	29 49	264	48 58	328
215	49 39	003	39 10	065	41 00	104	22 16	149	43 00	199	28 59	265	48 31	327
216	49 41	003	39 55	065	41 48	105	22 41	150	42 44	200	28 10	266	48 04	327
217	49 44	002	40 40	065	42 36	105	23 06	151	42 26	201	27 20	266	47 37	327
218	49 45	002	41 26	066	43 24	106	23 30	151	42 08	202	26 31	267	47 10	327
219	49 47	002	42 11	066	44 11	107	23 54	152	41 48	204	25 41	267	46 42	327
220	49 48	001	42 56	066	44 59	108	24 16	153	41 28	205	24 51	268	46 15	326
221	49 49	001	43 42	066	45 46	108	24 38	154	41 07	206	24 02	268	45 47	326
222	49 49	000	44 28	067	46 33	109	25 00	155	40 44	207	23 12	269	45 20	326
223	49 49	000	45 13	067	47 20	110	25 20	156	40 21	209	22 22	270	44 52	326
224	49 49	359	45 59	067	48 07	111	25 40	157	39 57	210	21 32	270	44 24	326
	DENEB		VEGA		◆ALTAIR		ANTARES		◆SPICA		Denebola		◆Dubhe	
225	26 25	052	46 45	067	19 18	092	25 59	158	39 32	211	42 50	258	43 56	326
226	27 04	052	47 31	068	20 08	093	26 18	159	39 06	212	42 01	259	43 28	326
227	27 43	052	48 17	068	20 58	093	26 35	160	38 39	213	41 12	259	43 00	326
228	28 23	052	49 03	068	21 47	094	26 52	161	38 11	214	40 24	260	42 31	325
229	29 02	053	49 49	068	22 37	095	27 08	162	37 43	215	39 35	261	42 03	325
230	29 42	053	50 35	068	23 26	095	27 24	163	37 14	216	38 45	261	41 35	325
231	30 21	053	51 22	069	24 16	096	27 38	164	36 44	217	37 56	262	41 07	325
232	31 01	054	52 08	069	25 05	096	27 52	165	36 13	218	37 07	263	40 38	325
233	31 42	054	52 54	069	25 55	097	28 05	166	35 42	220	36 18	263	40 10	325
234	32 22	054	53 41	069	26 44	098	28 17	166	35 10	221	35 28	264	39 42	325
235	33 02	054	54 27	069	27 33	098	28 28	167	34 38	222	34 39	264	39 13	325
236	33 43	055	55 14	069	28 23	099	28 38	168	34 04	222	33 49	265	38 45	325
237	34 23	055	56 00	069	29 12	099	28 48	169	33 30	223	33 00	266	38 16	325
238	35 04	055	56 47	070	30 01	100	28 56	171	32 56	224	32 10	266	37 48	325
239	35 45	055	57 33	070	30 50	101	29 04	172	32 21	225	31 20	267	37 20	325
	◆DENEB		ALTAIR		Nunki		◆ANTARES		SPICA		ARCTURUS		◆Alkaid	
240	36 26	055	31 38	101	16 48	140	29 11	173	31 45	226	62 28	244	61 13	312
241	37 07	056	32 27	102	17 20	140	29 17	174	31 09	227	61 43	245	60 36	312
242	37 48	056	33 16	103	17 52	141	29 22	175	30 32	228	60 58	247	59 58	311
243	38 29	056	34 04	103	18 23	142	29 27	176	29 55	229	60 12	248	59 21	311
244	39 10	056	34 52	104	18 53	143	29 30	177	29 17	230	59 26	249	58 43	311
245	39 52	056	35 41	105	19 23	144	29 32	178	28 39	231	58 39	250	58 06	310
246	40 33	057	36 29	106	19 52	144	29 34	179	28 00	231	57 52	251	57 28	310
247	41 15	057	37 16	106	20 21	145	29 35	180	27 21	232	57 05	252	56 50	310
248	41 56	057	38 04	107	20 49	146	29 34	181	26 41	233	56 18	253	56 11	310
249	42 38	057	38 52	108	21 17	147	29 33	182	26 01	234	55 30	254	55 33	309
250	43 20	057	39 39	109	21 44	148	29 31	183	25 21	235	54 42	255	54 54	309
251	44 02	057	40 26	109	22 10	148	29 29	184	24 40	236	53 54	255	54 16	309
252	44 44	057	41 13	110	22 36	149	29 25	185	23 59	236	53 06	256	53 37	309
253	45 25	058	41 59	111	23 01	150	29 20	186	23 17	237	52 18	257	52 58	309
254	46 07	058	42 46	112	23 25	151	29 15	187	22 36	238	51 29	258	52 19	309
	◆DENEB		ALTAIR		Nunki		◆ANTARES		ARCTURUS		◆Alkaid		Kochab	
255	46 50	058	43 32	113	23 49	152	29 08	188	50 40	259	51 40	308	46 48	348
256	47 32	058	44 17	113	24 12	153	29 01	189	49 52	259	51 01	308	46 38	347
257	48 14	058	45 03	114	24 35	154	28 53	190	49 03	260	50 22	308	46 27	347
258	48 56	058	45 48	115	24 56	155	28 44	191	48 14	261	49 43	308	46 15	347
259	49 38	058	46 33	116	25 17	155	28 34	192	47 24	262	49 04	308	46 04	347
260	50 20	058	47 17	117	25 38	156	28 23	193	46 35	262	48 25	308	45 52	346
261	51 03	058	48 01	118	25 57	157	28 11	194	45 46	263	47 46	308	45 40	346
262	51 45	058	48 45	119	26 16	158	27 59	195	44 57	264	47 07	308	45 28	346
263	52 27	058	49 28	120	26 34	159	27 46	196	44 07	264	46 28	308	45 16	345
264	53 09	058	50 11	121	26 51	160	27 32	197	43 18	265	45 49	308	45 03	345
265	53 51	058	50 53	122	27 08	161	27 17	198	42 28	265	45 10	308	44 51	345
266	54 34	058	51 35	123	27 24	162	27 01	199	41 38	266	44 31	308	44 38	345
267	55 16	058	52 16	125	27 38	163	26 45	200	40 49	267	43 52	308	44 24	345
268	55 58	058	52 57	126	27 53	164	26 28	201	39 59	267	43 13	308	44 11	344
269	56 40	058	53 37	127	28 06	165	26 10	202	39 09	268	42 34	309	43 57	344

LAT 34°N

LHA ♈	Hc	Zn	Hc	Zn	Hc	Zn	Hc	Zn	Hc	Zn	Hc	Zn	Hc	Zn
	◆DENEB		ALTAIR		◆Nunki		ANTARES		ARCTURUS		◆Alkaid		Kochab	
270	57 22	058	54 16	128	28 18	166	25 51	203	38 20	268	41 55	309	43 44	344
271	58 04	057	54 55	130	28 30	167	25 31	204	37 30	269	41 16	309	43 30	344
272	58 46	057	55 33	131	28 41	168	25 11	204	36 40	270	40 37	309	43 16	344
273	59 28	057	56 10	132	28 51	169	24 50	205	35 50	270	39 59	309	43 02	343
274	60 10	057	56 46	134	29 00	170	24 29	206	35 01	271	39 20	309	42 47	343
275	60 51	057	57 22	135	29 08	171	24 06	207	34 11	271	38 41	309	42 33	343
276	61 33	056	57 57	137	29 15	172	23 43	208	33 21	272	38 03	309	42 18	343
277	62 14	056	58 30	138	29 22	173	23 19	209	32 32	272	37 24	309	42 03	343
278	62 55	056	59 03	140	29 27	174	22 55	210	31 42	273	36 46	310	41 48	342
279	63 36	055	59 34	141	29 32	175	22 30	211	30 52	273	36 08	310	41 33	342
280	64 17	055	60 05	143	29 36	176	22 04	211	30 03	274	35 30	310	41 18	342
281	64 58	055	60 34	145	29 38	177	21 38	212	29 13	274	34 52	310	41 03	342
282	65 38	054	61 02	147	29 40	178	21 11	213	28 23	275	34 14	310	40 48	342
283	66 18	053	61 29	148	29 42	179	20 44	214	27 34	275	33 36	311	40 32	342
284	66 58	053	61 54	150	29 42	180	20 16	215	26 44	276	32 58	311	40 17	342
	◆Alpheratz		Enif		ALTAIR		◆ANTARES		ARCTURUS		◆Alkaid		Kochab	
285	25 44	071	45 25	113	62 18	152	19 47	216	25 55	277	32 21	311	40 01	342
286	26 31	071	46 11	114	62 40	154	19 18	216	25 05	277	31 43	311	39 45	342
287	27 19	072	46 56	115	63 01	156	18 48	217	24 16	278	31 06	311	39 29	341
288	28 06	072	47 41	116	63 20	158	18 17	218	23 27	278	30 29	312	39 14	341
289	28 53	073	48 26	117	63 38	160	17 47	219	22 38	279	29 51	312	38 58	341
290	29 41	073	49 10	118	63 54	163	17 15	220	21 48	279	29 14	312	38 42	341
291	30 29	074	49 54	119	64 07	165	16 43	220	20 59	280	28 38	312	38 26	341
292	31 16	074	50 37	120	64 20	167	16 11	221	20 10	280	28 01	313	38 09	341
293	32 04	074	51 20	121	64 30	169	15 38	222	19 21	281	27 24	313	37 53	341
294	32 52	075	52 03	122	64 38	172	15 05	222	18 33	281	26 48	313	37 37	341
295	33 40	075	52 45	123	64 44	174	14 31	223	17 44	282	26 12	313	37 21	341
296	34 28	076	53 26	124	64 49	176	13 57	224	16 55	282	25 36	314	37 05	341
297	35 17	076	54 07	125	64 51	179	13 22	225	16 07	283	25 00	314	36 48	341
298	36 05	076	54 47	127	64 51	181	12 47	225	15 18	283	24 24	314	36 32	341
299	36 53	077	55 27	128	64 49	183	12 11	226	14 30	284	23 49	315	36 16	341
	◆Mirfak		Alpheratz		◆FOMALHAUT		ALTAIR		Rasalhague		◆Alphecca		Kochab	
300	13 39	038	37 42	077	13 48	141	64 46	185	50 41	246	33 15	282	35 59	341
301	14 10	039	38 30	078	14 19	142	64 40	188	49 55	247	32 26	282	35 43	341
302	14 41	039	39 19	078	14 50	143	64 32	190	49 09	248	31 37	283	35 27	341
303	15 12	039	40 08	079	15 20	143	64 23	192	48 23	249	30 49	283	35 10	341
304	15 44	040	40 57	079	15 49	144	64 11	195	47 37	250	30 01	284	34 54	341
305	16 16	040	41 45	079	16 18	145	63 58	197	46 50	250	29 12	284	34 38	341
306	16 48	041	42 34	080	16 46	146	63 42	199	46 03	251	28 24	285	34 21	341
307	17 21	041	43 23	080	17 14	146	63 25	201	45 16	252	27 36	285	34 05	341
308	17 53	041	44 12	081	17 41	147	63 07	203	44 28	253	26 48	285	33 49	341
309	18 26	042	45 02	081	18 07	148	62 46	205	43 41	254	26 00	286	33 32	341
310	19 00	042	45 51	082	18 33	149	62 24	207	42 53	254	25 12	286	33 16	341
311	19 33	042	46 40	082	18 59	150	62 01	209	42 05	255	24 25	287	33 00	341
312	20 07	043	47 29	082	19 24	150	61 36	211	41 16	256	23 37	287	32 44	341
313	20 40	043	48 19	083	19 48	151	61 09	213	40 28	257	22 50	288	32 28	341
314	21 15	043	49 08	083	20 12	152	60 42	215	39 40	257	22 02	288	32 12	341
	◆Mirfak		Hamal		Diphda		◆FOMALHAUT		ALTAIR		◆VEGA		Kochab	
315	21 49	044	23 24	077	15 39	125	20 34	153	60 13	216	60 58	290	31 56	341
316	22 23	044	24 12	077	16 20	126	20 57	154	59 43	218	60 11	290	31 40	341
317	22 58	044	25 01	078	17 00	127	21 18	155	59 12	220	59 25	290	31 24	341
318	23 33	045	25 50	078	17 40	127	21 39	155	58 39	221	58 38	290	31 08	342
319	24 08	045	26 38	079	18 19	128	22 00	156	58 06	223	57 51	290	30 52	342
320	24 43	045	27 27	079	18 58	129	22 19	157	57 31	224	57 05	290	30 37	342
321	25 18	046	28 16	080	19 36	130	22 38	158	56 56	226	56 18	290	30 21	342
322	25 54	046	29 05	080	20 15	130	22 56	159	56 20	227	55 32	291	30 06	342
323	26 30	046	29 54	080	20 52	131	23 14	160	55 43	229	54 45	291	29 50	342
324	27 06	046	30 43	081	21 30	132	23 31	161	55 05	230	53 59	291	29 35	342
325	27 42	047	31 32	081	22 07	133	23 47	162	54 27	231	53 12	291	29 19	342
326	28 18	047	32 21	082	22 43	133	24 02	163	53 48	233	52 26	291	29 04	342
327	28 54	047	33 11	082	23 19	134	24 16	163	53 08	234	51 39	291	28 49	342
328	29 31	047	34 00	083	23 54	135	24 30	164	52 27	235	50 53	292	28 34	343
329	30 07	048	34 49	083	24 29	136	24 43	165	51 46	236	50 07	292	28 20	343
	◆CAPELLA		Hamal		Diphda		◆FOMALHAUT		ALTAIR		◆VEGA		Kochab	
330	12 21	042	35 39	084	25 04	137	24 55	166	51 05	237	49 21	292	28 05	343
331	12 55	043	36 28	084	25 37	137	25 07	167	50 22	238	48 35	292	27 50	343
332	13 29	043	37 18	085	26 11	138	25 17	168	49 40	240	47 49	292	27 36	343
333	14 03	043	38 07	085	26 44	139	25 27	169	48 57	241	47 03	293	27 21	343
334	14 37	044	38 57	086	27 16	140	25 36	170	48 13	242	46 17	293	27 07	343
335	15 12	044	39 47	086	27 47	141	25 44	171	47 29	243	45 31	293	26 53	344
336	15 46	045	40 36	087	28 18	142	25 52	172	46 45	244	44 45	293	26 39	344
337	16 22	045	41 26	087	28 49	143	25 58	173	46 00	244	43 59	294	26 25	344
338	16 57	045	42 16	088	29 19	144	26 04	174	45 15	245	43 14	294	26 11	344
339	17 33	046	43 05	089	29 48	145	26 09	175	44 30	246	42 28	294	25 58	344
340	18 08	046	43 55	089	30 16	146	26 13	176	43 44	247	41 43	294	25 44	344
341	18 44	047	44 45	090	30 44	147	26 16	177	42 58	248	40 58	295	25 31	345
342	19 21	047	45 35	090	31 11	148	26 19	178	42 12	249	40 13	295	25 18	345
343	19 57	047	46 24	091	31 37	149	26 20	179	41 25	250	39 27	295	25 05	345
344	20 34	048	47 14	091	32 03	150	26 21	180	40 39	250	38 42	295	24 52	345
	◆CAPELLA		ALDEBARAN		Diphda		◆FOMALHAUT		ALTAIR		◆VEGA		Kochab	
345	21 11	048	14 04	079	32 27	151	26 21	181	39 52	251	37 57	296	24 39	345
346	21 48	048	14 53	080	32 51	152	26 20	182	39 04	252	37 13	296	24 27	346
347	22 25	049	15 42	080	33 15	153	26 18	183	38 17	253	36 28	296	24 14	346
348	23 02	049	16 31	081	33 37	154	26 15	184	37 29	254	35 43	296	24 02	346
349	23 40	049	17 21	081	33 59	155	26 12	185	36 42	254	34 59	297	23 50	346
350	24 18	050	18 10	082	34 20	156	26 08	185	35 54	255	34 15	297	23 38	346
351	24 56	050	18 59	083	34 40	157	26 02	186	35 05	256	33 30	297	23 27	347
352	25 34	050	19 48	083	34 59	158	25 56	187	34 17	256	32 46	298	23 15	347
353	26 12	051	20 38	084	35 17	159	25 50	188	33 29	257	32 02	298	23 04	347
354	26 51	051	21 27	084	35 34	160	25 42	189	32 40	258	31 18	298	22 53	347
355	27 29	051	22 17	085	35 51	161	25 34	190	31 52	258	30 35	299	22 42	347
356	28 08	051	23 06	085	36 06	162	25 24	191	31 03	259	29 51	299	22 31	348
357	28 47	052	23 56	086	36 20	164	25 14	192	30 14	260	29 08	299	22 21	348
358	29 26	052	24 46	086	36 34	165	25 03	193	29 25	260	28 24	300	22 10	348
359	30 06	052	25 35	087	36 47	166	24 52	194	28 36	261	27 41	300	22 00	348

LAT 33°N

LHA	Hc	Zn	Hc	Zn	Hc	Zn	Hc	Zn	Hc	Zn	Hc	Zn	Hc	Zn
♈	◆CAPELLA		ALDEBARAN		Diphda		◆FOMALHAUT		ALTAIR		◆DENEB		Kochab	
0	30 08	052	26 22	087	37 57	167	25 37	195	27 55	262	50 15	303	20 51	349
1	30 48	052	27 12	087	38 08	168	25 24	196	27 05	263	49 33	303	20 41	349
2	31 28	053	28 02	088	38 18	169	25 09	197	26 15	263	48 50	303	20 31	349
3	32 08	053	28 53	088	38 26	170	24 54	198	25 25	264	48 08	303	20 22	349
4	32 48	053	29 43	089	38 34	172	24 39	199	24 35	265	47 26	303	20 13	350
5	33 28	053	30 33	089	38 41	173	24 22	200	23 45	265	46 44	303	20 04	350
6	34 09	053	31 24	090	38 47	174	24 05	201	22 55	266	46 02	303	19 55	350
7	34 49	054	32 14	091	38 51	175	23 47	201	22 05	266	45 19	303	19 46	350
8	35 30	054	33 04	091	38 55	177	23 28	202	21 14	267	44 37	303	19 38	351
9	36 11	054	33 54	092	38 57	178	23 08	203	20 24	267	43 55	303	19 30	351
10	36 51	054	34 45	092	38 59	179	22 48	204	19 34	268	43 13	304	19 22	351
11	37 32	054	35 35	093	38 59	180	22 27	205	18 44	268	42 31	304	19 14	351
12	38 13	055	36 25	093	38 58	181	22 06	206	17 53	269	41 50	304	19 06	352
13	38 54	055	37 16	094	38 57	183	21 44	207	17 03	270	41 08	304	18 59	352
14	39 35	055	38 06	095	38 54	184	21 21	207	16 13	270	40 26	304	18 52	352
	◆CAPELLA		RIGEL		◆Diphda		FOMALHAUT		Enif		◆DENEB		Kochab	
15	40 16	055	16 57	112	38 50	185	20 57	208	39 26	254	39 44	304	18 45	352
16	40 58	055	17 44	113	38 45	186	20 33	209	38 37	255	39 03	304	18 39	353
17	41 39	055	18 30	113	38 39	188	20 08	210	37 49	256	38 21	304	18 32	353
18	42 20	055	19 16	114	38 32	189	19 43	211	37 00	256	37 40	305	18 26	353
19	43 02	055	20 02	115	38 24	190	19 17	212	36 11	257	36 58	305	18 20	353
20	43 43	056	20 48	115	38 14	191	18 50	212	35 22	258	36 17	305	18 14	354
21	44 25	056	21 33	116	38 04	192	18 23	213	34 32	259	35 36	305	18 09	354
22	45 06	056	22 18	117	37 53	193	17 55	214	33 43	259	34 55	305	18 04	354
23	45 48	056	23 03	117	37 41	195	17 27	215	32 53	260	34 14	306	17 59	354
24	46 29	056	23 47	118	37 27	196	16 58	215	32 04	260	33 33	306	17 54	355
25	47 11	056	24 32	119	37 13	197	16 28	216	31 14	261	32 52	306	17 49	355
26	47 53	056	25 16	120	36 58	198	15 58	217	30 24	262	32 12	306	17 45	355
27	48 34	056	25 59	120	36 42	199	15 28	218	29 35	262	31 31	307	17 41	356
28	49 16	056	26 42	121	36 25	200	14 57	218	28 45	263	30 51	307	17 37	356
29	49 58	056	27 25	122	36 07	202	14 25	219	27 55	264	30 11	307	17 34	356
	CAPELLA		BETELGEUSE		◆RIGEL		Diphda		◆Alpheratz		DENEB		◆Kochab	
30	50 40	056	30 08	101	28 08	123	35 48	203	65 47	268	29 31	307	17 31	356
31	51 21	056	30 57	102	28 50	123	35 28	204	64 57	269	28 51	308	17 28	357
32	52 03	056	31 46	103	29 32	124	35 07	205	64 07	269	28 11	308	17 25	357
33	52 44	056	32 35	103	30 13	125	34 46	206	63 16	270	27 31	308	17 22	357
34	53 26	056	33 24	104	30 54	126	34 23	207	62 26	270	26 52	308	17 20	358
35	54 08	056	34 13	105	31 35	127	34 00	208	61 36	271	26 12	309	17 18	358
36	54 49	055	35 01	106	32 15	128	33 36	209	60 45	271	25 33	309	17 16	358
37	55 30	055	35 50	106	32 54	129	33 11	210	59 55	272	24 54	309	17 15	358
38	56 12	055	36 38	107	33 33	129	32 46	211	59 05	272	24 15	310	17 13	359
39	56 53	055	37 26	108	34 12	130	32 19	212	58 14	273	23 36	310	17 12	359
40	57 34	055	38 14	108	34 50	131	31 52	213	57 24	273	22 58	310	17 11	359
41	58 15	055	39 02	109	35 28	132	31 24	214	56 34	274	22 19	310	17 11	000
42	58 56	054	39 49	110	36 05	133	30 56	215	55 44	274	21 41	311	17 11	000
43	59 37	054	40 36	111	36 41	134	30 27	216	54 54	275	21 03	311	17 11	000
44	60 17	054	41 23	112	37 17	135	29 57	217	54 04	275	20 25	311	17 11	000
	◆Dubhe		POLLUX		SIRIUS		◆RIGEL		Diphda		◆Alpheratz		Schedar	
45	16 03	025	29 36	074	16 51	124	37 52	136	29 26	218	53 13	276	56 28	325
46	16 24	025	30 24	074	17 33	124	38 27	137	28 55	219	52 23	276	55 59	325
47	16 46	026	31 13	075	18 14	125	39 01	138	28 23	220	51 33	277	55 30	324
48	17 08	026	32 01	075	18 55	126	39 34	139	27 51	221	50 43	277	55 00	324
49	17 30	026	32 50	076	19 36	127	40 06	140	27 18	221	49 53	277	54 30	323
50	17 52	027	33 39	076	20 16	127	40 38	141	26 44	222	49 04	278	54 00	323
51	18 15	027	34 28	076	20 56	128	41 09	143	26 10	223	48 14	278	53 29	323
52	18 38	027	35 17	077	21 35	129	41 39	144	25 36	224	47 24	279	52 59	322
53	19 01	027	36 06	077	22 14	129	42 09	145	25 00	225	46 34	279	52 28	322
54	19 24	028	36 55	078	22 53	130	42 37	146	24 25	226	45 45	279	51 57	322
55	19 48	028	37 44	078	23 31	131	43 05	147	23 49	226	44 55	280	51 25	321
56	20 12	028	38 33	079	24 09	132	43 32	148	23 12	227	44 05	280	50 54	321
57	20 36	029	39 23	079	24 46	133	43 58	150	22 35	228	43 16	281	50 22	321
58	21 00	029	40 12	079	25 23	133	44 23	151	21 57	229	42 26	281	49 50	321
59	21 24	029	41 02	080	25 59	134	44 47	152	21 19	230	41 37	281	49 18	320
	◆Dubhe		POLLUX		PROCYON		◆SIRIUS		RIGEL		◆Hamal		Schedar	
60	21 49	029	41 51	080	32 07	106	26 35	135	45 10	153	63 25	256	48 46	320
61	22 13	030	42 41	081	32 55	107	27 10	136	45 32	155	62 36	257	48 14	320
62	22 38	030	43 31	081	33 43	108	27 45	137	45 53	156	61 47	258	47 41	320
63	23 03	030	44 20	081	34 31	108	28 19	138	46 12	157	60 58	259	47 09	320
64	23 29	030	45 10	082	35 19	109	28 53	139	46 31	159	60 08	260	46 36	320
65	23 54	031	46 00	082	36 06	110	29 26	139	46 49	160	59 19	261	46 03	319
66	24 20	031	46 50	083	36 53	111	29 58	140	47 05	162	58 29	261	45 31	319
67	24 46	031	47 40	083	37 40	111	30 30	141	47 21	163	57 39	262	44 58	319
68	25 12	031	48 30	084	38 27	112	31 01	142	47 35	164	56 49	263	44 25	319
69	25 38	031	49 20	084	39 14	113	31 32	143	47 48	166	55 59	263	43 52	319
70	26 04	032	50 10	084	40 00	114	32 01	144	48 00	167	55 09	264	43 19	319
71	26 30	032	51 00	085	40 46	115	32 31	145	48 10	169	54 19	265	42 46	319
72	26 57	032	51 50	085	41 31	115	32 59	146	48 19	170	53 29	265	42 13	319
73	27 24	032	52 40	086	42 17	116	33 27	147	48 27	172	52 39	266	41 40	319
74	27 51	032	53 30	086	43 02	117	33 54	148	48 34	173	51 49	267	41 07	319
	◆Dubhe		POLLUX		PROCYON		◆SIRIUS		RIGEL		◆Hamal		Mirfak	
75	28 18	032	54 21	087	43 46	118	34 20	149	48 39	175	50 58	267	65 33	321
76	28 45	033	55 11	087	44 30	119	34 45	150	48 43	176	50 08	268	65 01	320
77	29 12	033	56 01	088	45 14	120	35 10	151	48 46	178	49 18	268	64 28	319
78	29 39	033	56 51	088	45 57	121	35 34	152	48 47	179	48 27	269	63 55	318
79	30 07	033	57 42	089	46 40	122	35 56	153	48 47	181	47 37	270	63 21	318
80	30 34	033	58 32	089	47 23	123	36 19	155	48 46	182	46 47	270	62 47	317
81	31 02	033	59 22	090	48 05	124	36 40	156	48 44	184	45 57	271	62 12	316
82	31 29	033	60 13	090	48 46	125	37 00	157	48 40	185	45 06	271	61 37	316
83	31 57	034	61 03	091	49 27	126	37 20	158	48 35	187	44 16	272	61 02	315
84	32 25	034	61 53	091	50 07	127	37 38	159	48 28	188	43 26	272	60 26	315
85	32 53	034	62 44	092	50 47	129	37 56	160	48 20	190	42 35	273	59 50	314
86	33 21	034	63 34	093	51 26	130	38 12	161	48 11	191	41 45	273	59 14	314
87	33 49	034	64 24	093	52 04	131	38 28	162	48 01	193	40 55	274	58 37	313
88	34 17	034	65 14	094	52 42	132	38 42	164	47 50	194	40 05	274	58 01	313
89	34 45	034	66 05	095	53 19	134	38 56	165	47 37	195	39 15	275	57 24	313

LHA	Hc	Zn	Hc	Zn	Hc	Zn	Hc	Zn	Hc	Zn	Hc	Zn	Hc	Zn
♈	◆Dubhe		REGULUS		PROCYON		◆SIRIUS		RIGEL		ALDEBARAN		◆Mirfak	
90	35 14	034	29 52	095	53 55	135	39 09	166	47 23	197	64 49	234	56 46	312
91	35 42	034	30 42	096	54 30	136	39 20	167	47 08	198	64 08	236	56 09	312
92	36 10	034	31 32	096	55 04	138	39 31	169	46 51	200	63 26	237	55 32	312
93	36 38	034	32 22	097	55 38	139	39 40	170	46 34	201	62 43	239	54 54	311
94	37 07	034	33 12	097	56 10	141	39 49	171	46 15	202	62 00	240	54 16	311
95	37 35	034	34 02	098	56 41	142	39 56	172	45 56	204	61 16	241	53 38	311
96	38 04	034	34 52	099	57 12	144	40 02	173	45 35	205	60 32	243	53 00	311
97	38 32	034	35 41	099	57 41	145	40 08	175	45 13	206	59 47	244	52 22	311
98	39 00	034	36 31	100	58 09	147	40 12	176	44 50	208	59 01	245	51 44	310
99	39 29	034	37 21	101	58 36	149	40 15	177	44 26	209	58 16	246	51 05	310
100	39 57	034	38 10	101	59 01	150	40 17	178	44 01	210	57 29	247	50 27	310
101	40 25	034	38 59	102	59 25	152	40 17	180	43 36	211	56 43	248	49 48	310
102	40 54	034	39 48	103	59 48	154	40 17	181	43 09	213	55 56	249	49 10	310
103	41 22	034	40 37	103	60 09	156	40 16	182	42 41	214	55 09	250	48 31	310
104	41 50	034	41 26	104	60 29	158	40 13	183	42 13	215	54 21	251	47 52	310
	◆Alkaid		REGULUS		Alphard		◆SIRIUS		RIGEL		ALDEBARAN		◆CAPELLA	
105	17 31	042	42 15	105	35 35	133	40 10	185	41 44	216	53 34	252	66 19	311
106	18 05	042	43 04	106	36 11	134	40 05	186	41 14	217	52 46	253	65 41	310
107	18 39	043	43 52	106	36 47	135	39 59	187	40 43	218	51 57	254	65 02	310
108	19 13	043	44 40	107	37 22	136	39 52	188	40 11	220	51 09	255	64 23	309
109	19 48	043	45 28	108	37 57	137	39 44	190	39 39	221	50 20	255	63 44	308
110	20 22	044	46 16	109	38 31	138	39 35	191	39 06	222	49 32	256	63 04	308
111	20 57	044	47 04	110	39 04	139	39 25	192	38 32	223	48 43	257	62 25	308
112	21 32	044	47 51	110	39 36	140	39 14	193	37 57	224	47 54	258	61 45	307
113	22 07	045	48 38	111	40 08	141	39 02	195	37 22	225	47 04	258	61 04	307
114	22 43	045	49 25	112	40 39	143	38 49	196	36 46	226	46 15	259	60 24	306
115	23 18	045	50 11	113	41 09	144	38 35	197	36 10	227	45 26	260	59 43	306
116	23 54	045	50 57	114	41 39	145	38 20	198	35 33	228	44 36	261	59 03	306
117	24 30	046	51 43	115	42 07	146	38 03	199	34 56	229	43 46	261	58 22	306
118	25 06	046	52 28	116	42 35	147	37 46	200	34 18	230	42 57	262	57 41	305
119	25 42	046	53 13	117	43 02	148	37 28	202	33 39	230	42 07	262	57 00	305
	◆Alkaid		REGULUS		◆Alphard		SIRIUS		RIGEL		ALDEBARAN		◆CAPELLA	
120	26 19	047	53 58	118	43 28	150	37 09	203	33 00	231	41 17	263	56 18	305
121	26 56	047	54 42	119	43 53	151	36 49	204	32 21	232	40 27	264	55 37	305
122	27 32	047	55 26	120	44 17	152	36 29	205	31 41	233	39 37	264	54 56	305
123	28 09	047	56 09	122	44 40	153	36 07	206	31 00	234	38 47	265	54 14	304
124	28 46	048	56 52	123	45 02	155	35 44	207	30 19	235	37 57	266	53 33	304
125	29 23	048	57 34	124	45 23	156	35 21	208	29 38	236	37 06	266	52 51	304
126	30 01	048	58 15	125	45 43	157	34 57	209	28 56	236	36 16	267	52 09	304
127	30 38	048	58 56	127	46 02	159	34 32	210	28 14	237	35 26	267	51 28	304
128	31 16	048	59 36	128	46 20	160	34 06	211	27 32	238	34 36	268	50 46	304
129	31 53	049	60 15	130	46 36	161	33 39	212	26 49	239	33 45	268	50 04	304
130	32 31	049	60 53	131	46 52	163	33 12	213	26 06	239	32 55	269	49 23	304
131	33 09	049	61 31	133	47 06	164	32 44	214	25 22	240	32 05	270	48 41	304
132	33 47	049	62 07	134	47 19	166	32 15	215	24 38	241	31 14	270	47 59	304
133	34 25	049	62 43	136	47 31	167	31 45	216	23 54	242	30 24	271	47 18	304
134	35 04	050	63 17	138	47 42	168	31 15	217	23 10	242	29 34	271	46 36	304
	◆Alkaid		ARCTURUS		SPICA		◆Alphard		SIRIUS		BETELGEUSE		◆CAPELLA	
135	35 42	050	19 24	079	13 05	113	47 51	170	30 44	218	40 10	250	45 55	304
136	36 20	050	20 14	080	13 52	113	48 00	171	30 13	219	39 23	250	45 13	304
137	36 59	050	21 03	080	14 38	114	48 07	173	29 41	220	38 35	251	44 31	304
138	37 37	050	21 53	081	15 24	115	48 12	174	29 08	221	37 48	252	43 50	304
139	38 16	050	22 43	081	16 09	115	48 17	176	28 35	222	37 00	253	43 08	305
140	38 55	050	23 32	082	16 54	116	48 20	177	28 01	223	36 12	253	42 27	305
141	39 33	050	24 22	082	17 40	117	48 21	179	27 26	224	35 23	254	41 46	305
142	40 12	051	25 12	083	18 24	117	48 22	180	26 51	225	34 35	255	41 04	305
143	40 51	051	26 02	083	19 09	118	48 21	182	26 16	225	33 46	256	40 23	305
144	41 30	051	26 52	084	19 53	119	48 19	183	25 40	226	32 57	256	39 42	305
145	42 09	051	27 42	084	20 37	119	48 15	185	25 03	227	32 08	257	39 01	305
146	42 48	051	28 32	085	21 21	120	48 10	186	24 26	228	31 19	258	38 20	305
147	43 27	051	29 22	085	22 04	121	48 04	188	23 49	229	30 30	258	37 39	306
148	44 06	051	30 12	086	22 47	122	47 57	189	23 11	229	29 41	259	36 58	306
149	44 45	051	31 03	086	23 30	122	47 48	191	22 32	230	28 51	259	36 17	306
	Alkaid		◆ARCTURUS		SPICA		◆Alphard		PROCYON		BETELGEUSE		◆CAPELLA	
150	45 24	051	31 53	087	24 13	123	47 39	192	47 02	237	28 02	260	35 36	306
151	46 03	051	32 43	087	24 55	124	47 28	193	46 20	238	27 12	261	34 56	306
152	46 42	051	33 33	088	25 36	125	47 15	195	45 37	239	26 22	261	34 15	307
153	47 22	051	34 24	088	26 18	125	47 02	196	44 53	240	25 33	262	33 35	307
154	48 01	051	35 14	089	26 58	126	46 47	198	44 09	241	24 43	263	32 55	307
155	48 40	051	36 04	089	27 39	127	46 31	199	43 25	242	23 53	263	32 14	307
156	49 19	051	36 54	090	28 19	128	46 14	200	42 40	243	23 03	264	31 34	307
157	49 58	051	37 45	090	28 58	129	45 56	202	41 55	244	22 13	264	30 55	308
158	50 37	051	38 35	091	29 38	129	45 37	203	41 09	245	21 23	265	30 15	308
159	51 15	051	39 25	091	30 16	130	45 16	204	40 24	246	20 33	265	29 35	308
160	51 54	050	40 16	092	30 54	131	44 55	206	39 38	247	19 42	266	28 56	308
161	52 33	050	41 06	093	31 32	132	44 32	207	38 51	247	18 52	267	28 16	309
162	53 12	050	41 56	093	32 09	133	44 09	208	38 05	248	18 02	267	27 37	309
163	53 50	050	42 46	094	32 46	134	43 45	210	37 18	249	17 12	268	26 58	309
164	54 29	050	43 37	094	33 22	135	43 19	211	36 31	250	16 21	268	26 19	309
	◆Dubhe		Alkaid		◆ARCTURUS		SPICA		REGULUS		◆PROCYON		CAPELLA	
165	61 13	001	55 07	049	44 27	095	33 57	136	65 52	212	35 43	251	25 40	310
166	61 13	000	55 45	049	45 17	096	34 32	137	65 24	215	34 56	251	25 02	310
167	61 13	359	56 23	049	46 07	096	35 06	138	64 55	217	34 08	252	24 23	310
168	61 11	358	57 01	049	46 57	097	35 40	139	64 24	219	33 20	253	23 45	311
169	61 09	357	57 39	048	47 47	098	36 13	140	63 52	220	32 32	253	23 07	311
170	61 06	356	58 16	048	48 37	098	36 45	141	63 19	222	31 44	254	22 29	311
171	61 02	355	58 53	048	49 26	099	37 17	142	62 45	224	30 55	255	21 51	312
172	60 57	354	59 30	047	50 16	100	37 48	143	62 09	226	30 07	255	21 14	312
173	60 51	353	60 07	047	51 06	100	38 18	144	61 33	227	29 18	256	20 36	312
174	60 45	352	60 44	046	51 55	101	38 47	145	60 55	229	28 29	257	19 59	313
175	60 38	351	61 20	046	52 44	102	39 16	146	60 17	230	27 40	257	19 22	313
176	60 30	350	61 56	045	53 34	103	39 44	147	59 38	232	26 51	258	18 45	313
177	60 21	349	62 31	045	54 23	103	40 11	148	58 58	233	26 01	259	18 09	314
178	60 11	348	63 06	044	55 11	104	40 37	149	58 17	235	25 12	259	17 32	314
179	60 01	348	63 41	043	56 00	105	41 02	151	57 36	236	24 22	260	16 56	314

 LAT 33°N

LHA	Hc	Zn	Hc	Zn	Hc	Zn	Hc	Zn	Hc	Zn	Hc	Zn	Hc	Zn
♈	♦VEGA		Alphecca		ARCTURUS		♦SPICA		REGULUS		♦POLLUX		Dubhe	
180	13 41	052	43 34	083	56 49	106	41 26	152	56 54	237	35 42	283	59 49	347
181	14 21	053	44 24	083	57 37	107	41 50	153	56 11	238	34 53	283	59 37	346
182	15 02	053	45 14	084	58 25	108	42 12	154	55 28	240	34 04	284	59 25	345
183	15 42	054	46 04	084	59 13	109	42 34	155	54 44	241	33 15	284	59 11	344
184	16 23	054	46 54	085	60 00	110	42 54	157	54 00	242	32 26	285	58 57	343
185	17 03	054	47 45	085	60 47	111	43 14	158	53 16	243	31 37	285	58 43	343
186	17 44	055	48 35	086	61 34	112	43 32	159	52 31	244	30 49	285	58 27	342
187	18 26	055	49 25	086	62 21	113	43 49	160	51 45	245	30 00	286	58 11	341
188	19 07	056	50 15	087	63 07	114	44 06	162	51 00	246	29 12	286	57 55	340
189	19 49	056	51 05	087	63 52	116	44 21	163	50 14	247	28 24	287	57 38	340
190	20 30	056	51 56	088	64 37	117	44 35	164	49 27	248	27 36	287	57 20	339
191	21 12	057	52 46	088	65 22	118	44 48	166	48 40	249	26 48	287	57 02	338
192	21 55	057	53 36	089	66 06	120	45 00	167	47 53	250	26 00	288	56 43	338
193	22 37	057	54 27	089	66 49	121	45 10	169	47 06	250	25 12	288	56 23	337
194	23 19	058	55 17	090	67 32	123	45 20	170	46 18	251	24 24	289	56 03	336
	♦Kochab		VEGA		Rasalhague		♦ANTARES		SPICA		♦REGULUS		Dubhe	
195	46 36	011	24 02	058	24 36	091	12 33	133	45 28	171	45 31	252	55 43	336
196	46 45	010	24 45	058	25 26	091	13 09	134	45 35	173	44 43	253	55 22	335
197	46 54	010	25 28	059	26 16	092	13 45	135	45 41	174	43 55	254	55 01	335
198	47 02	010	26 11	059	27 06	092	14 20	136	45 45	175	43 06	254	54 39	334
199	47 11	009	26 54	059	27 57	093	14 55	136	45 49	177	42 18	255	54 17	334
200	47 18	009	27 38	060	28 47	093	15 30	137	45 51	178	41 29	256	53 55	333
201	47 26	009	28 21	060	29 37	094	16 04	138	45 52	180	40 40	257	53 32	333
202	47 33	008	29 05	060	30 27	095	16 38	138	45 52	181	39 51	257	53 09	332
203	47 40	008	29 49	061	31 17	095	17 11	139	45 50	182	39 02	258	52 45	332
204	47 47	007	30 33	061	32 08	096	17 44	140	45 47	184	38 13	259	52 21	331
205	47 54	007	31 17	061	32 58	096	18 16	141	45 43	185	37 23	259	51 57	331
206	48 00	007	32 01	062	33 48	097	18 47	141	45 38	187	36 34	260	51 33	331
207	48 05	006	32 45	062	34 37	098	19 19	142	45 31	188	35 44	261	51 08	330
208	48 11	006	33 30	062	35 27	098	19 49	143	45 24	189	34 54	261	50 43	330
209	48 16	006	34 14	063	36 17	099	20 19	144	45 15	191	34 05	262	50 17	330
	♦Kochab		VEGA		Rasalhague		♦ANTARES		SPICA		♦REGULUS		Dubhe	
210	48 20	005	34 59	063	37 07	100	20 49	145	45 05	192	33 15	263	49 52	329
211	48 25	005	35 44	063	37 56	100	21 18	145	44 53	194	32 25	263	49 26	329
212	48 29	004	36 29	063	38 46	101	21 46	146	44 41	195	31 35	264	49 00	329
213	48 32	004	37 14	064	39 35	102	22 14	147	44 28	196	30 45	264	48 34	328
214	48 36	004	37 59	064	40 24	102	22 41	148	44 13	198	29 55	265	48 07	328
215	48 39	003	38 44	064	41 13	103	23 07	149	43 57	199	29 04	266	47 40	328
216	48 41	003	39 30	064	42 02	104	23 33	150	43 40	200	28 14	266	47 14	328
217	48 44	002	40 15	065	42 51	104	23 58	150	43 22	202	27 24	267	46 47	327
218	48 45	002	41 00	065	43 40	105	24 23	151	43 03	203	26 34	267	46 20	327
219	48 47	002	41 46	065	44 28	106	24 47	152	42 43	204	25 44	268	45 52	327
220	48 48	001	42 32	065	45 17	107	25 10	153	42 22	205	24 53	268	45 25	327
221	48 49	001	43 18	066	46 05	107	25 32	154	42 00	207	24 03	269	44 57	327
222	48 49	000	44 03	066	46 53	108	25 54	155	41 37	208	23 13	269	44 30	327
223	48 49	000	44 49	066	47 40	109	26 15	156	41 14	209	22 22	270	44 02	326
224	48 49	359	45 35	066	48 28	110	26 35	157	40 49	210	21 32	271	43 34	326
	DENEB		VEGA		♦ALTAIR		ANTARES		♦SPICA		Denebola		♦Dubhe	
225	25 47	051	46 21	066	19 20	092	26 55	158	40 23	211	43 02	259	43 06	326
226	26 27	051	47 07	067	20 11	092	27 14	159	39 56	212	42 13	259	42 38	326
227	27 06	052	47 54	067	21 01	093	27 32	160	39 29	214	41 23	260	42 10	326
228	27 46	052	48 40	067	21 51	094	27 49	160	39 01	215	40 34	261	41 42	326
229	28 26	052	49 26	067	22 41	094	28 05	161	38 32	216	39 44	261	41 14	326
230	29 05	053	50 13	067	23 32	095	28 21	162	38 02	217	38 54	262	40 45	326
231	29 45	053	50 59	067	24 22	095	28 36	163	37 32	218	38 04	263	40 17	326
232	30 26	053	51 45	068	25 12	096	28 50	164	37 00	219	37 14	263	39 49	326
233	31 06	053	52 32	068	26 02	096	29 03	165	36 28	220	36 24	264	39 20	326
234	31 46	054	53 19	068	26 52	097	29 15	166	35 56	221	35 34	265	38 52	326
235	32 27	054	54 05	068	27 42	098	29 27	167	35 22	222	34 44	265	38 24	326
236	33 07	054	54 52	068	28 31	098	29 37	168	34 48	223	33 54	266	37 55	326
237	33 48	054	55 38	068	29 21	099	29 47	169	34 14	224	33 04	266	37 27	326
238	34 29	054	56 25	068	30 11	100	29 56	170	33 39	225	32 13	267	36 59	326
239	35 10	055	57 12	068	31 00	100	30 04	171	33 03	226	31 23	267	36 30	326
	♦DENEB		ALTAIR		Nunki		♦ANTARES		SPICA		ARCTURUS		♦Alkaid	
240	35 51	055	31 50	101	17 34	139	30 11	172	32 26	227	62 54	246	60 32	314
241	36 32	055	32 39	101	18 07	140	30 17	173	31 49	228	62 07	247	59 55	313
242	37 14	055	33 29	102	18 39	141	30 22	175	31 12	229	61 21	248	59 18	313
243	37 55	055	34 18	103	19 10	142	30 26	176	30 34	229	60 34	249	58 41	312
244	38 37	056	35 07	103	19 41	143	30 30	177	29 56	230	59 47	250	58 04	312
245	39 18	056	35 56	104	20 11	143	30 32	178	29 17	231	58 59	251	57 26	312
246	40 00	056	36 44	105	20 41	144	30 34	179	28 37	232	58 11	252	56 49	311
247	40 41	056	37 33	106	21 10	145	30 35	180	27 58	233	57 23	253	56 11	311
248	41 23	056	38 21	106	21 39	146	30 34	181	27 17	234	56 35	254	55 33	311
249	42 05	056	39 10	107	22 07	147	30 33	182	26 37	234	55 46	255	54 54	310
250	42 47	056	39 58	108	22 34	147	30 31	183	25 55	235	54 58	256	54 16	310
251	43 29	057	40 45	109	23 01	148	30 28	184	25 14	236	54 09	257	53 38	310
252	44 11	057	41 33	109	23 27	149	30 25	185	24 32	237	53 20	258	52 59	310
253	44 53	057	42 20	110	23 53	150	30 20	186	23 50	237	52 30	258	52 20	310
254	45 35	057	43 08	111	24 18	151	30 14	187	23 07	238	51 41	259	51 42	310
	♦DENEB		ALTAIR		Nunki		♦ANTARES		ARCTURUS		♦Alkaid		Kochab	
255	46 17	057	43 54	112	24 42	152	30 08	188	50 52	260	51 03	309	45 50	348
256	46 59	057	44 41	113	25 06	153	30 00	189	50 02	261	50 24	309	45 39	348
257	47 42	057	45 27	113	25 28	153	29 52	190	49 12	261	49 45	309	45 28	347
258	48 24	057	46 13	114	25 51	154	29 43	191	48 23	262	49 06	309	45 17	347
259	49 06	057	46 59	115	26 12	155	29 32	192	47 33	263	48 27	309	45 06	347
260	49 48	057	47 44	116	26 33	156	29 22	193	46 43	263	47 48	309	44 54	347
261	50 30	057	48 29	117	26 53	157	29 10	194	45 53	264	47 09	309	44 42	346
262	51 13	057	49 14	118	27 12	158	28 57	195	45 03	265	46 30	309	44 30	346
263	51 55	057	49 58	119	27 30	159	28 44	196	44 13	265	45 51	309	44 18	346
264	52 37	057	50 42	120	27 48	160	28 29	197	43 22	266	45 11	309	44 05	345
265	53 19	057	51 25	121	28 05	161	28 14	198	42 32	266	44 32	309	43 53	345
266	54 01	057	52 08	122	28 21	162	27 58	199	41 42	267	43 53	309	43 40	345
267	54 43	057	52 50	124	28 36	163	27 41	200	40 52	268	43 14	309	43 26	345
268	55 25	057	53 32	125	28 50	164	27 24	201	40 01	268	42 35	309	43 13	345
269	56 07	056	54 13	126	29 04	165	27 05	202	39 11	269	41 56	309	43 00	344

LHA	Hc	Zn	Hc	Zn	Hc	Zn	Hc	Zn	Hc	Zn	Hc	Zn	Hc	Zn
♈	♦DENEB		ALTAIR		♦Nunki		ANTARES		ARCTURUS		♦Alkaid		Kochab	
270	56 49	056	54 53	127	29 16	166	26 46	203	38 21	269	41 17	309	42 46	344
271	57 31	056	55 33	128	29 28	167	26 26	204	37 31	270	40 38	309	42 32	344
272	58 13	056	56 12	130	29 39	168	26 06	205	36 40	270	40 00	309	42 18	344
273	58 55	056	56 50	131	29 49	169	25 44	206	35 50	271	39 21	310	42 04	344
274	59 36	055	57 28	133	29 59	170	25 22	206	35 00	271	38 42	310	41 50	343
275	60 17	055	58 04	134	30 07	171	25 00	207	34 09	272	38 03	310	41 35	343
276	60 59	055	58 40	136	30 14	172	24 36	208	33 19	272	37 25	310	41 21	343
277	61 40	055	59 15	137	30 21	173	24 12	209	32 29	273	36 46	310	41 06	343
278	62 21	054	59 48	139	30 27	174	23 47	210	31 39	273	36 08	310	40 51	343
279	63 01	054	60 21	140	30 32	175	23 22	211	30 48	274	35 29	310	40 36	343
280	63 42	053	60 53	142	30 35	176	22 55	212	29 58	274	34 51	311	40 21	342
281	64 22	053	61 23	144	30 38	177	22 29	213	29 08	275	34 13	311	40 06	342
282	65 02	052	61 52	146	30 40	178	22 01	213	28 18	276	33 35	311	39 51	342
283	65 42	052	62 20	147	30 42	179	21 33	214	27 28	276	32 57	311	39 35	342
284	66 21	051	62 46	149	30 42	180	21 05	215	26 38	277	32 19	311	39 20	342
	♦Alpheratz		Enif		ALTAIR		♦ANTARES		ARCTURUS		♦Alkaid		Kochab	
285	25 25	071	45 48	112	63 11	151	20 36	216	25 48	277	31 41	311	39 04	342
286	26 12	071	46 35	113	63 34	153	20 06	217	24 58	277	31 03	312	38 48	342
287	27 00	071	47 21	114	63 56	155	19 36	217	24 08	278	30 26	312	38 33	342
288	27 47	072	48 07	115	64 16	158	19 05	218	23 18	278	29 48	312	38 17	342
289	28 35	072	48 52	116	64 34	160	18 33	219	22 28	279	29 11	312	38 01	342
290	29 23	073	49 37	117	64 51	162	18 01	220	21 39	279	28 34	313	37 45	341
291	30 11	073	50 22	118	65 05	164	17 29	220	20 49	280	27 57	313	37 29	341
292	30 59	073	51 07	119	65 18	167	16 56	221	20 00	280	27 20	313	37 13	341
293	31 48	074	51 50	120	65 29	169	16 23	222	19 10	281	26 43	313	36 57	341
294	32 36	074	52 34	121	65 37	171	15 49	223	18 21	281	26 07	314	36 40	341
295	33 25	075	53 17	122	65 44	174	15 14	223	17 32	282	25 31	314	36 24	341
296	34 13	075	53 59	123	65 48	176	14 40	224	16 42	282	24 54	314	36 08	341
297	35 02	075	54 41	124	65 51	178	14 04	225	15 53	283	24 18	314	35 52	341
298	35 51	076	55 22	126	65 51	181	13 29	225	15 04	283	23 42	315	35 35	341
299	36 39	076	56 03	127	65 49	183	12 53	226	14 15	284	23 07	315	35 19	341
	♦Mirfak		Alpheratz		♦FOMALHAUT		ALTAIR		Rasalhague		♦Alphecca		Kochab	
300	12 51	038	37 28	077	14 35	141	65 45	186	51 05	247	33 02	283	35 03	341
301	13 23	039	38 17	077	15 06	142	65 39	188	50 18	248	32 13	283	34 46	341
302	13 54	039	39 06	077	15 37	143	65 31	190	49 31	249	31 24	283	34 30	341
303	14 26	039	39 55	078	16 08	143	65 21	193	48 44	250	30 35	284	34 14	341
304	14 58	040	40 45	078	16 37	144	65 09	195	47 57	251	29 46	284	33 57	341
305	15 30	040	41 34	079	17 07	145	64 55	197	47 09	251	28 57	285	33 41	341
306	16 03	040	42 23	079	17 35	146	64 39	200	46 22	252	28 09	285	33 24	341
307	16 35	041	43 13	079	18 04	146	64 21	202	45 34	253	27 20	286	33 08	341
308	17 08	041	44 02	080	18 31	147	64 02	204	44 45	254	26 32	286	32 52	341
309	17 42	041	44 52	080	18 58	148	63 40	206	43 57	255	25 43	286	32 36	341
310	18 15	042	45 41	081	19 25	149	63 17	208	43 08	255	24 55	287	32 19	341
311	18 49	042	46 31	081	19 51	149	62 53	210	42 20	256	24 07	287	32 03	341
312	19 22	042	47 21	081	20 16	150	62 27	212	41 31	257	23 19	288	31 47	341
313	19 57	043	48 11	082	20 41	151	62 00	214	40 42	258	22 31	288	31 31	341
314	20 31	043	49 00	082	21 05	152	61 31	216	39 52	258	21 43	289	31 15	341
	♦Mirfak		Hamal		Diphda		♦FOMALHAUT		ALTAIR		♦VEGA		Kochab	
315	21 05	043	23 10	076	16 14	125	21 28	153	61 01	217	60 37	292	30 59	341
316	21 40	044	23 59	077	16 55	126	21 51	154	60 30	219	59 50	292	30 43	342
317	22 15	044	24 48	077	17 35	126	22 13	154	59 57	221	59 03	292	30 27	342
318	22 50	044	25 37	078	18 16	127	22 34	155	59 24	222	58 17	292	30 11	342
319	23 25	045	26 26	078	18 56	128	22 55	156	58 49	224	57 30	292	29 55	342
320	24 01	045	27 15	079	19 35	128	23 15	157	58 14	226	56 43	292	29 40	342
321	24 36	045	28 05	079	20 15	129	23 34	158	57 37	227	55 56	292	29 24	342
322	25 12	045	28 54	079	20 53	130	23 52	159	57 00	229	55 10	292	29 08	342
323	25 48	046	29 44	080	21 32	131	24 10	160	56 22	230	54 23	292	28 53	342
324	26 24	046	30 33	080	22 10	132	24 27	161	55 43	231	53 36	292	28 38	342
325	27 00	046	31 23	081	22 47	132	24 44	162	55 04	232	52 50	292	28 22	342
326	27 37	046	32 13	081	23 24	133	24 59	162	54 24	234	52 03	292	28 07	342
327	28 13	047	33 02	082	24 01	134	25 14	163	53 43	235	51 17	293	27 52	343
328	28 50	047	33 52	082	24 37	135	25 28	164	53 01	236	50 30	293	27 37	343
329	29 27	047	34 42	083	25 12	135	25 41	165	52 19	237	49 44	293	27 22	343
	♦Mirfak		Hamal		Diphda		♦FOMALHAUT		ALTAIR		♦VEGA		Kochab	
330	30 04	047	35 32	083	25 47	136	25 54	166	51 36	238	48 58	293	27 07	343
331	30 41	048	36 22	084	26 22	137	26 05	167	50 53	239	48 11	293	26 53	343
332	31 18	048	37 12	084	26 56	138	26 16	168	50 10	241	47 25	293	26 38	343
333	31 55	048	38 02	085	27 29	139	26 26	169	49 26	242	46 39	294	26 24	343
334	32 33	048	38 52	085	28 02	140	26 35	170	48 41	243	45 53	294	26 09	344
335	33 10	048	39 42	086	28 34	141	26 44	171	47 56	244	45 07	294	25 55	344
336	33 48	049	40 33	086	29 06	142	26 51	172	47 11	244	44 21	294	25 41	344
337	34 26	049	41 23	087	29 36	142	26 58	173	46 26	245	43 35	294	25 27	344
338	35 04	049	42 13	087	30 07	143	27 04	174	45 40	246	42 49	295	25 14	344
339	35 42	049	43 03	088	30 36	144	27 09	175	44 54	247	42 04	295	25 00	344
340	36 20	049	43 54	088	31 05	145	27 13	176	44 07	248	41 18	295	24 46	345
341	36 58	049	44 44	089	31 34	146	27 16	177	43 20	249	40 32	295	24 33	345
342	37 36	049	45 34	089	32 01	147	27 19	178	42 33	250	39 47	296	24 20	345
343	38 14	050	46 24	090	32 28	148	27 20	179	41 46	250	39 02	296	24 07	345
344	38 53	050	47 15	090	32 54	149	27 21	180	40 58	251	38 16	296	23 54	345
	♦CAPELLA		ALDEBARAN		Diphda		♦FOMALHAUT		ALTAIR		♦VEGA		Kochab	
345	20 30	048	13 53	079	33 20	150	27 21	181	40 10	252	37 31	296	23 41	345
346	21 08	048	14 43	080	33 44	151	27 20	182	39 23	253	36 46	297	23 29	346
347	21 45	048	15 32	080	34 08	152	27 18	183	38 34	254	36 01	297	23 16	346
348	22 23	049	16 22	081	34 31	153	27 15	184	37 46	254	35 16	297	23 04	346
349	23 01	049	17 12	081	34 53	154	27 12	185	36 57	255	34 32	297	22 52	346
350	23 39	049	18 01	082	35 14	156	27 07	186	36 09	256	33 47	298	22 40	346
351	24 17	050	18 51	082	35 35	157	27 02	186	35 20	256	33 02	298	22 28	347
352	24 55	050	19 41	083	35 54	158	26 56	187	34 31	257	32 18	298	22 17	347
353	25 34	050	20 31	083	36 13	159	26 49	188	33 42	258	31 34	299	22 05	347
354	26 13	050	21 21	084	36 31	160	26 41	189	32 53	258	30 50	299	21 54	347
355	26 52	051	22 11	084	36 47	161	26 33	190	32 03	259	30 06	299	21 43	347
356	27 31	051	23 01	085	37 03	162	26 23	191	31 14	260	29 22	299	21 32	348
357	28 10	051	23 51	085	37 18	163	26 13	192	30 24	260	28 38	300	21 22	348
358	28 49	052	24 41	086	37 32	165	26 02	193	29 35	261	27 54	300	21 11	348
359	29 29	052	25 32	086	37 45	166	25 50	194	28 45	262	27 11	300	21 01	348

LAT 32°N

LHA	Hc	Zn	Hc	Zn	Hc	Zn	Hc	Zn	Hc	Zn	Hc	Zn	Hc	Zn
♈	♦CAPELLA		ALDEBARAN		Diphda		♦FOMALHAUT		ALTAIR		♦DENEB		Kochab	
0	29 31	052	26 18	086	38 55	167	26 35	195	28 03	263	49 42	304	19 52	349
1	30 11	052	27 09	087	39 06	168	26 21	196	27 12	263	49 00	304	19 42	349
2	30 51	052	28 00	087	39 16	169	26 07	197	26 22	264	48 17	304	19 32	349
3	31 32	052	28 51	088	39 26	170	25 52	198	25 31	264	47 35	304	19 23	349
4	32 12	052	29 41	088	39 34	172	25 35	199	24 41	265	46 53	304	19 14	350
5	32 52	053	30 32	089	39 40	173	25 19	200	23 50	266	46 11	304	19 05	350
6	33 33	053	31 23	089	39 46	174	25 01	201	22 59	266	45 28	304	18 56	350
7	34 13	053	32 14	090	39 51	175	24 43	202	22 08	267	44 46	304	18 47	350
8	34 54	053	33 05	090	39 55	176	24 23	203	21 18	267	44 04	304	18 39	351
9	35 35	053	33 56	091	39 57	178	24 04	203	20 27	268	43 22	304	18 30	351
10	36 16	054	34 47	092	39 59	179	23 43	204	19 36	268	42 40	304	18 22	351
11	36 57	054	35 38	092	39 59	180	23 22	205	18 45	269	41 58	304	18 15	351
12	37 38	054	36 28	093	39 58	181	23 00	206	17 54	269	41 16	304	18 07	352
13	38 19	054	37 19	093	39 57	183	22 37	207	17 03	270	40 34	305	18 00	352
14	39 00	054	38 10	094	39 54	184	22 14	208	16 12	270	39 52	305	17 53	352
	♦CAPELLA		RIGEL		♦Diphda		FOMALHAUT		Enif		♦DENEB		Kochab	
15	39 42	054	17 20	112	39 50	185	21 50	209	39 42	255	39 10	305	17 46	352
16	40 23	054	18 07	112	39 44	186	21 25	209	38 52	256	38 29	305	17 39	353
17	41 04	055	18 54	113	39 38	188	21 00	210	38 03	257	37 47	305	17 33	353
18	41 46	055	19 41	114	39 31	189	20 34	211	37 13	257	37 05	305	17 26	353
19	42 27	055	20 27	114	39 23	190	20 08	212	36 24	258	36 24	305	17 20	353
20	43 09	055	21 13	115	39 13	191	19 41	213	35 34	259	35 43	306	17 15	354
21	43 51	055	21 59	116	39 03	192	19 13	213	34 44	259	35 01	306	17 09	354
22	44 32	055	22 45	116	38 51	194	18 45	214	33 54	260	34 20	306	17 04	354
23	45 14	055	23 30	117	38 39	195	18 16	215	33 04	260	33 39	306	16 59	354
24	45 55	055	24 16	118	38 25	196	17 46	216	32 14	261	32 58	306	16 54	355
25	46 37	055	25 00	119	38 11	197	17 17	216	31 23	262	32 17	307	16 50	355
26	47 19	055	25 45	119	37 55	198	16 46	217	30 33	262	31 36	307	16 45	355
27	48 00	055	26 29	120	37 38	200	16 15	218	29 42	263	30 55	307	16 41	356
28	48 42	055	27 13	121	37 21	201	15 44	219	28 52	263	30 15	307	16 38	356
29	49 24	055	27 57	121	37 03	202	15 12	219	28 01	264	29 34	307	16 34	356
	CAPELLA		BETELGEUSE		♦RIGEL		Diphda		♦Alpheratz		DENEB		♦Kochab	
30	50 05	055	30 19	101	28 40	122	36 43	203	65 48	270	28 54	308	16 31	356
31	50 47	055	31 09	102	29 23	123	36 23	204	64 57	271	28 14	308	16 28	357
32	51 29	055	31 59	102	30 05	124	36 02	205	64 06	271	27 34	308	16 25	357
33	52 10	055	32 49	103	30 47	125	35 40	206	63 15	272	26 54	308	16 22	357
34	52 52	055	33 38	103	31 29	125	35 17	207	62 24	272	26 14	309	16 20	358
35	53 33	054	34 28	104	32 10	126	34 53	208	61 34	273	25 35	309	16 18	358
36	54 14	054	35 17	105	32 51	127	34 29	209	60 43	273	24 55	309	16 16	358
37	54 56	054	36 06	106	33 31	128	34 03	210	59 52	274	24 16	310	16 15	358
38	55 37	054	36 55	106	34 11	129	33 37	211	59 01	274	23 37	310	16 13	359
39	56 18	054	37 44	107	34 51	130	33 10	212	58 11	275	22 58	310	16 12	359
40	56 59	053	38 32	108	35 29	131	32 42	213	57 20	275	22 19	310	16 11	359
41	57 40	053	39 21	108	36 08	132	32 14	214	56 29	275	21 40	311	16 11	000
42	58 20	053	40 09	109	36 45	133	31 45	215	55 39	276	21 02	311	16 11	000
43	59 01	053	40 57	110	37 23	134	31 15	216	54 48	276	20 24	311	16 11	000
44	59 41	052	41 45	111	37 59	135	30 45	217	53 57	277	19 46	312	16 11	000
	♦Dubhe		POLLUX		SIRIUS		♦RIGEL		Diphda		♦Alpheratz		Schedar	
45	15 09	025	29 19	073	17 24	123	38 35	136	30 14	218	53 07	277	55 38	326
46	15 30	025	30 08	074	18 07	124	39 10	137	29 42	219	52 16	277	55 10	325
47	15 52	026	30 57	074	18 48	125	39 45	138	29 09	220	51 26	278	54 41	325
48	16 14	026	31 46	075	19 30	126	40 19	139	28 36	221	50 36	278	54 11	325
49	16 36	026	32 35	075	20 11	126	40 52	140	28 03	222	49 45	279	53 42	324
50	16 59	026	33 24	075	20 52	127	41 25	141	27 29	223	48 55	279	53 12	324
51	17 22	027	34 13	076	21 33	128	41 56	142	26 54	223	48 05	279	52 42	323
52	17 45	027	35 03	076	22 13	128	42 27	144	26 19	224	47 14	280	52 11	323
53	18 08	027	35 52	077	22 52	129	42 58	144	25 43	225	46 24	280	51 40	323
54	18 31	028	36 42	077	23 32	130	43 27	145	25 07	226	45 34	280	51 09	322
55	18 55	028	37 31	077	24 10	131	43 55	147	24 30	227	44 44	281	50 38	322
56	19 19	028	38 21	078	24 49	131	44 23	148	23 52	228	43 54	281	50 07	322
57	19 43	028	39 11	078	25 27	132	44 49	149	23 15	228	43 04	282	49 35	322
58	20 07	029	40 01	079	26 04	133	45 15	150	22 37	229	42 15	282	49 04	321
59	20 32	029	40 51	079	26 41	134	45 40	152	21 58	230	41 25	282	48 32	321
	♦Dubhe		POLLUX		PROCYON		♦SIRIUS		RIGEL		♦Hamal		Schedar	
60	20 56	029	41 41	079	32 23	106	27 17	135	46 03	153	63 38	258	48 00	321
61	21 21	029	42 31	080	33 12	106	27 53	136	46 26	154	62 48	259	47 27	321
62	21 46	030	43 21	080	34 01	107	28 29	136	46 47	156	61 58	260	46 55	320
63	22 11	030	44 11	080	34 50	108	29 03	137	47 08	157	61 08	261	46 23	320
64	22 37	030	45 01	081	35 38	108	29 38	138	47 27	158	60 18	261	45 50	320
65	23 02	030	45 51	081	36 26	109	30 11	139	47 45	160	59 28	262	45 18	320
66	23 28	031	46 42	082	37 14	110	30 44	140	48 02	161	58 37	263	44 45	320
67	23 54	031	47 32	082	38 02	111	31 17	141	48 18	163	57 47	264	44 12	320
68	24 20	031	48 22	082	38 49	111	31 48	142	48 33	164	56 56	264	43 39	320
69	24 46	031	49 13	083	39 37	112	32 19	143	48 46	166	56 05	265	43 06	320
70	25 13	031	50 03	083	40 24	113	32 50	144	48 58	167	55 15	266	42 33	320
71	25 39	032	50 54	084	41 10	114	33 20	145	49 09	168	54 24	266	42 00	320
72	26 06	032	51 45	084	41 57	115	33 49	146	49 18	170	53 33	267	41 27	319
73	26 33	032	52 35	085	42 43	115	34 17	147	49 27	171	52 42	267	40 54	319
74	27 00	032	53 26	085	43 29	116	34 44	148	49 33	173	51 51	268	40 21	319
	♦Dubhe		POLLUX		PROCYON		♦SIRIUS		RIGEL		♦Hamal		Mirfak	
75	27 27	032	54 17	085	44 14	117	35 11	149	49 39	175	51 01	268	64 46	322
76	27 54	032	55 07	086	44 59	118	35 37	150	49 43	176	50 10	269	64 14	321
77	28 21	032	55 58	086	45 44	119	36 02	151	49 46	178	49 19	270	63 42	320
78	28 49	033	56 49	087	46 28	120	36 27	152	49 47	179	48 28	270	63 09	320
79	29 16	033	57 40	087	47 12	121	36 50	153	49 47	181	47 37	271	62 36	319
80	29 44	033	58 30	088	47 55	122	37 13	154	49 46	182	46 46	271	62 02	318
81	30 11	033	59 21	088	48 38	123	37 34	155	49 44	184	45 55	272	61 28	318
82	30 39	033	60 12	089	49 20	124	37 55	156	49 40	185	45 04	272	60 54	317
83	31 07	033	61 03	089	50 02	125	38 15	158	49 34	187	44 14	273	60 19	316
84	31 35	033	61 54	090	50 43	126	38 34	159	49 28	188	43 23	273	59 44	316
85	32 03	033	62 45	090	51 24	128	38 52	160	49 20	190	42 32	274	59 08	315
86	32 31	034	63 36	091	52 04	129	39 09	161	49 10	191	41 41	274	58 32	315
87	32 59	034	64 27	091	52 43	130	39 25	162	49 00	193	40 51	275	57 56	314
88	33 27	034	65 17	092	53 22	131	39 40	163	48 48	194	40 00	275	57 19	314
89	33 56	034	66 08	092	54 00	133	39 54	165	48 35	196	39 09	276	56 43	314

LHA	Hc	Zn	Hc	Zn	Hc	Zn	Hc	Zn	Hc	Zn	Hc	Zn	Hc	Zn
♈	♦Dubhe		REGULUS		PROCYON		♦SIRIUS		RIGEL		ALDEBARAN		♦Mirfak	
90	34 24	034	29 57	094	54 37	134	40 07	166	48 20	197	65 23	236	56 06	313
91	34 52	034	30 48	095	55 13	135	40 19	167	48 05	199	64 41	237	55 29	313
92	35 20	034	31 38	096	55 48	137	40 30	168	47 48	200	63 58	239	54 51	313
93	35 49	034	32 29	096	56 23	138	40 39	170	47 30	201	63 14	240	54 14	312
94	36 17	034	33 19	097	56 56	140	40 48	171	47 11	203	62 29	242	53 36	312
95	36 46	034	34 10	097	57 29	141	40 56	172	46 50	204	61 44	243	52 58	312
96	37 14	034	35 00	098	58 00	143	41 02	173	46 29	205	60 59	244	52 20	312
97	37 42	034	35 51	099	58 30	144	41 07	175	46 07	207	60 13	245	51 42	312
98	38 11	034	36 41	099	58 59	146	41 12	176	45 43	208	59 26	246	51 04	311
99	38 39	034	37 31	100	59 27	148	41 15	177	45 19	209	58 39	248	50 26	311
100	39 07	034	38 21	100	59 53	150	41 17	178	44 53	211	57 52	249	49 48	311
101	39 36	034	39 11	101	60 18	151	41 17	180	44 27	212	57 04	250	49 09	311
102	40 04	034	40 01	102	60 42	153	41 17	181	43 59	213	56 17	251	48 31	311
103	40 32	034	40 51	102	61 04	155	41 16	182	43 31	214	55 29	251	47 52	311
104	41 00	034	41 40	103	61 24	157	41 13	184	43 02	216	54 40	252	47 14	311
	♦Alkaid		REGULUS		Alphard		♦SIRIUS		RIGEL		ALDEBARAN		♦CAPELLA	
105	16 47	042	42 30	104	36 16	133	41 09	185	42 32	217	53 52	253	65 39	313
106	17 21	042	43 19	105	36 53	134	41 05	186	42 01	218	53 03	254	65 01	312
107	17 55	042	44 08	105	37 29	135	40 59	187	41 30	219	52 14	255	64 23	311
108	18 29	043	44 57	106	38 05	136	40 52	189	40 57	220	51 24	256	63 45	311
109	19 04	043	45 46	107	38 40	137	40 44	190	40 24	221	50 35	257	63 06	310
110	19 39	043	46 35	108	39 15	138	40 34	191	39 50	222	49 45	257	62 27	309
111	20 14	044	47 23	109	39 49	139	40 24	192	39 16	223	48 56	258	61 47	309
112	20 49	044	48 11	109	40 22	140	40 13	194	38 40	224	48 06	259	61 08	309
113	21 25	044	48 59	110	40 55	141	40 00	195	38 05	225	47 16	259	60 28	308
114	22 00	045	49 47	111	41 26	142	39 47	196	37 28	226	46 26	260	59 48	308
115	22 36	045	50 34	112	41 57	143	39 32	197	36 51	227	45 36	261	59 07	307
116	23 12	045	51 21	113	42 27	144	39 17	198	36 13	228	44 45	261	58 27	307
117	23 48	045	52 08	114	42 57	145	39 00	200	35 35	229	43 55	262	57 46	307
118	24 24	046	52 54	115	43 25	147	38 42	201	34 56	230	43 05	263	57 05	307
119	25 01	046	53 40	116	43 53	148	38 24	202	34 17	231	42 14	263	56 25	306
	♦Alkaid		REGULUS		♦Alphard		SIRIUS		RIGEL		ALDEBARAN		♦CAPELLA	
120	25 38	046	54 26	117	44 19	149	38 05	203	33 37	232	41 24	264	55 43	306
121	26 14	046	55 11	118	44 45	150	37 44	204	32 57	233	40 33	265	55 02	306
122	26 51	047	55 56	119	45 10	152	37 23	205	32 16	234	39 42	265	54 21	306
123	27 28	047	56 40	120	45 33	153	37 01	206	31 35	234	38 51	266	53 40	306
124	28 06	047	57 23	121	45 56	154	36 38	208	30 54	235	38 01	266	52 58	305
125	28 43	047	58 07	123	46 18	156	36 14	209	30 12	236	37 10	267	52 17	305
126	29 20	048	58 49	124	46 38	157	35 49	210	29 29	237	36 19	267	51 35	305
127	29 58	048	59 31	125	46 58	158	35 23	211	28 46	238	35 28	268	50 54	305
128	30 36	048	60 12	127	47 16	160	34 57	212	28 03	238	34 37	269	50 12	305
129	31 14	048	60 52	128	47 33	161	34 30	213	27 20	239	33 47	269	49 30	305
130	31 52	048	61 32	130	47 49	162	34 02	214	26 36	240	32 56	270	48 49	305
131	32 30	049	62 11	131	48 04	164	33 33	215	25 52	241	32 05	270	48 07	305
132	33 08	049	62 49	133	48 17	165	33 04	216	25 07	241	31 14	271	47 25	305
133	33 46	049	63 25	135	48 30	167	32 34	217	24 22	242	30 23	271	46 44	305
134	34 24	049	64 01	136	48 41	168	32 03	218	23 37	243	29 32	272	46 02	305
	♦Alkaid		ARCTURUS		SPICA		♦Alphard		SIRIUS		BETELGEUSE		♦CAPELLA	
135	35 03	049	19 13	079	13 28	113	48 51	170	31 31	219	40 31	250	45 20	305
136	35 41	049	20 03	079	14 15	113	48 59	171	30 59	220	39 43	251	44 39	305
137	36 20	049	20 53	080	15 02	114	49 06	173	30 27	220	38 54	252	43 57	305
138	36 59	050	21 43	080	15 48	114	49 12	174	29 53	221	38 06	253	43 16	305
139	37 37	050	22 33	081	16 35	115	49 16	176	29 19	222	37 17	253	42 34	305
140	38 16	050	23 23	081	17 21	116	49 20	177	28 45	223	36 28	254	41 52	305
141	38 55	050	24 14	082	18 06	116	49 21	179	28 10	224	35 39	255	41 11	305
142	39 34	050	25 04	082	18 52	117	49 22	180	27 34	225	34 50	256	40 30	306
143	40 13	050	25 55	083	19 37	118	49 21	182	26 58	226	34 01	256	39 48	306
144	40 52	050	26 45	083	20 22	118	49 19	183	26 21	227	33 11	257	39 07	306
145	41 31	050	27 36	084	21 07	119	49 15	185	25 44	227	32 22	258	38 26	306
146	42 10	050	28 26	084	21 51	120	49 10	186	25 06	228	31 32	258	37 45	306
147	42 49	050	29 17	085	22 35	120	49 03	188	24 28	229	30 42	259	37 03	306
148	43 28	050	30 07	085	23 19	121	48 56	189	23 50	230	29 52	259	36 22	306
149	44 07	050	30 58	086	24 02	122	48 47	191	23 11	230	29 02	260	35 42	307
	Alkaid		♦ARCTURUS		SPICA		♦Alphard		PROCYON		BETELGEUSE		♦CAPELLA	
150	44 46	050	31 49	086	24 45	123	48 37	192	47 34	238	28 12	261	35 01	307
151	45 25	050	32 40	087	25 28	123	48 26	194	46 51	239	27 22	261	34 20	307
152	46 04	050	33 31	087	26 10	124	48 13	195	46 07	240	26 31	262	33 39	307
153	46 43	050	34 21	088	26 52	125	47 59	197	45 22	241	25 41	262	32 59	307
154	47 22	050	35 12	088	27 34	126	47 44	198	44 37	242	24 50	263	32 18	307
155	48 01	050	36 03	089	28 15	126	47 28	199	43 52	243	24 00	264	31 38	308
156	48 40	050	36 54	089	28 55	127	47 10	201	43 07	244	23 09	264	30 58	308
157	49 19	050	37 45	090	29 36	128	46 51	202	42 21	245	22 19	265	30 18	308
158	49 58	050	38 36	090	30 15	129	46 32	204	41 34	246	21 28	265	29 38	308
159	50 37	050	39 27	091	30 55	130	46 11	205	40 48	247	20 37	266	28 58	309
160	51 16	049	40 17	091	31 34	131	45 49	206	40 01	247	19 46	266	28 18	309
161	51 54	049	41 08	092	32 12	132	45 26	208	39 14	248	18 56	267	27 39	309
162	52 33	049	41 59	092	32 50	132	45 02	209	38 27	249	18 05	267	26 59	309
163	53 11	049	42 50	093	33 27	133	44 37	210	37 39	250	17 14	268	26 20	310
164	53 49	049	43 41	093	34 04	134	44 11	211	36 51	250	16 23	269	25 41	310
	Dubhe		♦Alkaid		ARCTURUS		♦SPICA		REGULUS		♦PROCYON		CAPELLA	
165	60 13	001	54 27	048	44 32	094	34 40	135	66 42	214	36 03	251	25 02	310
166	60 13	000	55 05	048	45 22	095	35 16	136	66 13	216	35 15	252	24 23	310
167	60 13	359	55 43	048	46 13	095	35 51	137	65 43	218	34 26	253	23 44	311
168	60 11	358	56 21	048	47 04	096	36 25	138	65 11	220	33 38	253	23 06	311
169	60 09	357	56 58	047	47 54	096	36 59	139	64 37	222	32 49	254	22 27	311
170	60 06	356	57 35	047	48 45	097	37 32	140	64 03	224	32 00	255	21 49	312
171	60 02	355	58 12	046	49 35	098	38 04	141	63 27	225	31 11	255	21 11	312
172	59 57	354	58 49	046	50 26	098	38 36	142	62 50	227	30 21	256	20 33	312
173	59 52	353	59 26	046	51 16	099	39 06	143	62 13	229	29 32	257	19 56	313
174	59 46	352	60 02	045	52 06	100	39 36	144	61 34	230	28 42	257	19 18	313
175	59 38	351	60 38	044	52 56	101	40 06	145	60 55	232	27 53	258	18 41	313
176	59 30	351	61 13	044	53 46	101	40 34	147	60 14	233	27 03	259	18 04	313
177	59 22	350	61 48	043	54 36	102	41 02	148	59 33	235	26 13	259	17 27	314
178	59 12	349	62 23	043	55 26	103	41 28	149	58 51	236	25 23	260	16 51	314
179	59 02	348	62 57	042	56 15	104	41 54	150	58 09	237	24 33	260	16 14	315

LAT 32°N

LHA ♈	Hc	Zn	Hc	Zn	Hc	Zn	Hc	Zn	Hc	Zn	Hc	Zn	Hc	Zn
	◆VEGA		Alphecca		ARCTURUS		◆SPICA		REGULUS		◆POLLUX		Dubhe	
180	13 05	052	43 27	082	57 04	105	42 19	151	57 26	238	35 28	284	58 51	347
181	13 45	053	44 17	082	57 54	105	42 43	152	56 42	240	34 38	284	58 39	346
182	14 26	053	45 07	083	58 43	106	43 06	154	55 58	241	33 49	284	58 27	345
183	15 06	053	45 58	083	59 31	107	43 28	155	55 13	242	33 00	285	58 14	345
184	15 47	054	46 48	084	60 20	108	43 49	156	54 28	243	32 11	285	58 00	344
185	16 28	054	47 39	084	61 08	109	44 09	158	53 43	244	31 22	286	57 45	343
186	17 10	055	48 30	085	61 56	110	44 28	159	52 57	245	30 33	286	57 30	342
187	17 51	055	49 20	085	62 43	111	44 46	160	52 10	246	29 44	286	57 14	342
188	18 33	055	50 11	085	63 30	113	45 03	161	51 24	247	28 55	287	56 58	341
189	19 15	056	51 02	086	64 17	114	45 18	163	50 37	248	28 06	287	56 41	340
190	19 57	056	51 53	086	65 04	115	45 33	164	49 49	249	27 18	288	56 24	340
191	20 39	056	52 43	087	65 49	116	45 46	166	49 02	250	26 29	288	56 06	339
192	21 22	057	53 34	087	66 35	118	45 58	167	48 14	251	25 41	288	55 47	338
193	22 04	057	54 25	088	67 19	119	46 09	168	47 26	251	24 53	289	55 28	338
194	22 47	057	55 16	088	68 03	121	46 19	170	46 37	252	24 05	289	55 08	337
	◆Kochab		VEGA		Rasalhague		◆ANTARES		SPICA		◆REGULUS		Dubhe	
195	45 37	010	23 30	058	24 36	090	13 14	133	46 27	171	45 49	253	54 48	336
196	45 46	010	24 13	058	25 27	091	13 51	134	46 34	173	45 00	254	54 28	336
197	45 55	010	24 57	058	26 18	091	14 27	135	46 40	174	44 11	255	54 07	335
198	46 03	009	25 40	059	27 09	092	15 03	135	46 45	175	43 22	255	53 45	335
199	46 11	009	26 24	059	27 59	092	15 39	136	46 49	177	42 33	256	53 23	334
200	46 19	009	27 07	059	28 50	093	16 14	137	46 51	178	41 43	257	53 01	334
201	46 27	008	27 51	060	29 41	093	16 48	137	46 52	180	40 53	257	52 38	333
202	46 34	008	28 35	060	30 32	094	17 23	138	46 52	181	40 04	258	52 15	333
203	46 41	008	29 19	060	31 23	095	17 56	139	46 50	183	39 14	259	51 52	332
204	46 48	007	30 03	061	32 13	095	18 29	140	46 47	184	38 24	259	51 28	332
205	46 54	007	30 48	061	33 04	096	19 02	140	46 43	185	37 34	260	51 04	332
206	47 00	007	31 32	061	33 55	096	19 34	141	46 37	187	36 44	261	50 40	331
207	47 06	006	32 17	061	34 45	097	20 06	142	46 31	188	35 53	261	50 15	331
208	47 11	006	33 02	062	35 36	098	20 37	143	46 23	190	35 03	262	49 51	331
209	47 16	005	33 46	062	36 26	098	21 08	144	46 14	191	34 13	263	49 25	330
	◆Kochab		VEGA		Rasalhague		◆ANTARES		SPICA		◆REGULUS		Dubhe	
210	47 21	005	34 31	062	37 16	099	21 38	144	46 03	192	33 22	263	49 00	330
211	47 25	005	35 16	062	38 07	099	22 07	145	45 52	194	32 32	264	48 34	330
212	47 29	004	36 02	063	38 57	100	22 36	146	45 39	195	31 41	264	48 08	329
213	47 33	004	36 47	063	39 47	101	23 04	147	45 25	197	30 50	265	47 42	329
214	47 36	003	37 32	063	40 37	101	23 32	148	45 10	198	30 00	266	47 16	329
215	47 39	003	38 18	063	41 26	102	23 58	148	44 54	199	29 09	266	46 49	328
216	47 41	003	39 03	064	42 16	103	24 25	149	44 36	201	28 18	267	46 23	328
217	47 44	002	39 49	064	43 06	103	24 50	150	44 18	202	27 27	267	45 56	328
218	47 45	002	40 35	064	43 55	104	25 15	151	43 59	203	26 36	268	45 29	328
219	47 47	001	41 20	064	44 44	105	25 40	152	43 38	204	25 46	268	45 02	328
220	47 48	001	42 06	064	45 33	106	26 03	153	43 16	206	24 55	269	44 34	327
221	47 49	001	42 52	065	46 22	106	26 26	154	42 54	207	24 04	269	44 07	327
222	47 49	000	43 38	065	47 11	107	26 48	155	42 30	208	23 13	270	43 39	327
223	47 49	000	44 24	065	47 59	108	27 10	156	42 06	209	22 22	270	43 12	327
224	47 49	359	45 11	065	48 48	109	27 30	156	41 41	211	21 31	271	42 44	327
	DENEB		VEGA		◆ALTAIR		ANTARES		◆SPICA		Denebola		◆Dubhe	
225	25 10	051	45 57	065	19 22	092	27 50	157	41 14	212	43 13	260	42 16	327
226	25 49	051	46 43	066	20 13	092	28 10	158	40 47	213	42 23	260	41 48	327
227	26 29	051	47 29	066	21 04	093	28 28	159	40 19	214	41 33	261	41 20	327
228	27 09	052	48 16	066	21 55	093	28 45	160	39 50	215	40 43	262	40 52	326
229	27 49	052	49 02	066	22 45	094	29 02	161	39 20	216	39 52	262	40 24	326
230	28 29	052	49 49	066	23 36	094	29 18	162	38 50	217	39 02	263	39 56	326
231	29 09	052	50 35	066	24 27	095	29 33	163	38 19	218	38 11	263	39 27	326
232	29 49	053	51 22	066	25 18	095	29 47	164	37 47	219	37 21	264	38 59	326
233	30 30	053	52 09	066	26 08	096	30 01	165	37 14	220	36 30	265	38 31	326
234	31 10	053	52 55	067	26 59	097	30 13	166	36 41	221	35 40	265	38 02	326
235	31 51	053	53 42	067	27 49	097	30 25	167	36 07	222	34 49	266	37 34	326
236	32 32	053	54 29	067	28 40	098	30 36	168	35 32	223	33 58	266	37 06	326
237	33 13	054	55 15	067	29 30	098	30 46	169	34 57	224	33 07	267	36 37	326
238	33 54	054	56 02	067	30 21	099	30 55	170	34 21	225	32 16	268	36 09	326
239	34 35	054	56 49	067	31 11	100	31 03	171	33 44	226	31 26	268	35 41	326
	◆DENEB		ALTAIR		Nunki		◆ANTARES		SPICA		ARCTURUS		◆Alkaid	
240	35 16	054	32 01	100	18 20	139	31 10	172	33 07	227	63 17	248	59 50	315
241	35 58	054	32 51	101	18 53	140	31 16	173	32 30	228	62 30	249	59 14	314
242	36 39	055	33 41	101	19 25	141	31 22	174	31 52	229	61 42	250	58 37	314
243	37 21	055	34 31	102	19 57	142	31 26	176	31 13	230	60 54	251	58 00	313
244	38 02	055	35 20	103	20 28	142	31 30	177	30 34	231	60 06	252	57 23	313
245	38 44	055	36 10	103	20 59	143	31 32	178	29 54	232	59 17	253	56 46	313
246	39 26	055	36 59	104	21 30	144	31 34	179	29 14	232	58 29	254	56 09	312
247	40 08	055	37 49	105	21 59	145	31 35	180	28 34	233	57 40	255	55 31	312
248	40 49	055	38 38	105	22 28	145	31 34	181	27 53	234	56 50	256	54 53	312
249	41 31	056	39 27	106	22 57	146	31 33	182	27 11	235	56 01	257	54 15	312
250	42 13	056	40 15	107	23 25	147	31 31	183	26 30	236	55 12	257	53 37	311
251	42 55	056	41 04	108	23 52	148	31 28	184	25 47	236	54 22	258	52 59	311
252	43 38	056	41 52	108	24 19	149	31 24	185	25 05	237	53 32	259	52 20	311
253	44 20	056	42 41	109	24 45	150	31 19	186	24 22	238	52 42	260	51 42	311
254	45 02	056	43 28	110	25 10	151	31 14	187	23 39	239	51 52	260	51 03	311
	◆DENEB		ALTAIR		Nunki		◆ANTARES		ARCTURUS		◆Alkaid		Kochab	
255	45 44	056	44 16	111	25 35	151	31 07	188	51 02	261	50 24	310	44 51	348
256	46 26	056	45 04	112	25 59	152	30 59	189	50 11	262	49 45	310	44 40	348
257	47 08	056	45 51	113	26 22	153	30 51	190	49 21	262	49 07	310	44 30	348
258	47 51	056	46 38	113	26 45	154	30 41	191	48 30	263	48 28	310	44 19	347
259	48 33	056	47 24	114	27 06	155	30 31	192	47 40	264	47 49	310	44 07	347
260	49 15	056	48 10	115	27 27	156	30 20	193	46 49	264	47 10	310	43 56	347
261	49 57	056	48 56	116	27 48	157	30 08	194	45 59	265	46 31	310	43 44	346
262	50 40	056	49 42	117	28 07	158	29 55	195	45 08	266	45 52	310	43 32	346
263	51 22	056	50 27	118	28 26	159	29 41	196	44 17	266	45 12	310	43 20	346
264	52 04	056	51 11	119	28 44	160	29 27	197	43 26	267	44 33	310	43 07	346
265	52 46	056	51 56	120	29 01	161	29 11	198	42 36	267	43 54	310	42 54	345
266	53 28	056	52 39	121	29 18	162	28 55	199	41 45	268	43 15	310	42 42	345
267	54 10	056	53 23	122	29 33	163	28 38	200	40 54	268	42 36	310	42 29	345
268	54 52	055	54 05	124	29 48	164	28 20	201	40 03	269	41 57	310	42 15	345
269	55 34	055	54 47	125	30 02	165	28 01	202	39 12	269	41 18	310	42 02	345

LHA ♈	Hc	Zn	Hc	Zn	Hc	Zn	Hc	Zn	Hc	Zn	Hc	Zn	Hc	Zn
	◆DENEB		ALTAIR		◆Nunki		ANTARES		ARCTURUS		◆Alkaid		Kochab	
270	56 16	055	55 29	126	30 15	166	27 41	203	38 21	270	40 39	310	41 48	344
271	56 57	055	56 10	127	30 27	167	27 21	204	37 30	271	40 00	310	41 35	344
272	57 39	055	56 50	129	30 38	168	27 00	205	36 40	271	39 21	310	41 21	344
273	58 20	054	57 29	130	30 48	169	26 38	206	35 49	272	38 42	310	41 07	344
274	59 02	054	58 08	131	30 58	170	26 16	207	34 58	272	38 03	310	40 52	344
275	59 43	054	58 45	133	31 06	171	25 53	208	34 07	273	37 25	310	40 38	343
276	60 24	053	59 22	134	31 14	172	25 29	208	33 16	273	36 46	311	40 23	343
277	61 04	053	59 58	136	31 21	173	25 04	209	32 25	274	36 07	311	40 09	343
278	61 45	053	60 33	138	31 26	174	24 39	210	31 35	274	35 29	311	39 54	343
279	62 25	052	61 07	139	31 31	175	24 13	211	30 44	275	34 50	311	39 39	343
280	63 05	052	61 40	141	31 35	176	23 46	212	29 53	275	34 12	311	39 24	343
281	63 45	051	62 11	143	31 38	177	23 19	213	29 02	276	33 33	311	39 09	343
282	64 25	051	62 41	145	31 40	178	22 51	214	28 12	276	32 55	311	38 53	342
283	65 04	050	63 10	146	31 42	179	22 23	214	27 21	277	32 17	312	38 38	342
284	65 43	049	63 37	148	31 42	180	21 54	215	26 31	277	31 39	312	38 23	342
	◆Alpheratz		Enif		ALTAIR		◆ANTARES		ARCTURUS		◆Alkaid		Kochab	
285	25 04	070	46 10	111	64 03	150	21 24	216	25 40	277	31 01	312	38 07	342
286	25 52	071	46 58	112	64 28	152	20 54	217	24 50	278	30 23	312	37 51	342
287	26 40	071	47 45	113	64 50	155	20 23	218	23 59	278	29 46	312	37 36	342
288	27 28	071	48 31	114	65 11	157	19 52	218	23 09	279	29 08	313	37 20	342
289	28 17	072	49 18	115	65 30	159	19 20	219	22 19	279	28 31	313	37 04	342
290	29 05	072	50 04	116	65 48	161	18 47	220	21 29	280	27 53	313	36 48	342
291	29 53	072	50 50	117	66 03	164	18 15	221	20 39	280	27 16	313	36 32	342
292	30 42	073	51 35	118	66 16	166	17 41	221	19 49	281	26 39	313	36 16	342
293	31 31	073	52 20	119	66 28	168	17 07	222	18 59	281	26 02	314	36 00	341
294	32 19	074	53 04	120	66 37	171	16 33	223	18 09	282	25 25	314	35 44	341
295	33 08	074	53 48	121	66 44	173	15 58	224	17 19	282	24 49	314	35 27	341
296	33 57	074	54 32	122	66 48	176	15 23	224	16 29	283	24 12	314	35 11	341
297	34 46	075	55 14	123	66 51	178	14 47	225	15 40	283	23 36	315	34 55	341
298	35 35	075	55 57	124	66 51	181	14 11	226	14 50	284	23 00	315	34 38	341
299	36 25	075	56 38	126	66 49	183	13 34	226	14 01	284	22 24	315	34 22	341
	◆Mirfak		Alpheratz		◆FOMALHAUT		ALTAIR		Rasalhague		◆Alphecca		Kochab	
300	12 04	038	37 14	076	15 22	141	66 45	186	51 28	248	32 48	283	34 06	341
301	12 36	038	38 03	076	15 54	142	66 39	188	50 40	249	31 59	284	33 49	341
302	13 07	039	38 53	077	16 25	142	66 30	191	49 53	250	31 10	284	33 33	341
303	13 39	039	39 42	077	16 56	143	66 20	193	49 05	251	30 20	284	33 17	341
304	14 12	040	40 32	077	17 26	144	66 07	196	48 16	252	29 31	285	33 00	341
305	14 44	040	41 22	078	17 56	145	65 52	198	47 28	252	28 42	285	32 44	341
306	15 17	040	42 11	078	18 25	145	65 35	200	46 39	253	27 53	286	32 28	341
307	15 50	041	43 01	078	18 54	146	65 17	203	45 51	254	27 04	286	32 11	341
308	16 23	041	43 51	079	19 22	147	64 56	205	45 02	255	26 15	286	31 55	341
309	16 56	041	44 41	079	19 49	148	64 34	207	44 12	256	25 26	287	31 39	341
310	17 30	042	45 31	080	20 16	148	64 10	209	43 23	256	24 38	287	31 23	341
311	18 04	042	46 21	080	20 42	149	63 45	211	42 34	257	23 49	288	31 06	341
312	18 38	042	47 11	080	21 08	150	63 18	213	41 44	258	23 01	288	30 50	341
313	19 12	043	48 01	081	21 33	151	62 49	215	40 54	258	22 12	289	30 34	342
314	19 47	043	48 52	081	21 57	152	62 19	217	40 04	259	21 24	289	30 18	342
	◆Mirfak		Hamal		Diphda		◆FOMALHAUT		ALTAIR		◆VEGA		Kochab	
315	20 22	043	22 55	076	16 48	125	22 21	153	61 48	219	60 14	293	30 02	342
316	20 57	043	23 45	076	17 30	125	22 44	153	61 16	220	59 27	293	29 46	342
317	21 32	044	24 34	076	18 11	126	23 07	154	60 42	222	58 40	293	29 30	342
318	22 07	044	25 24	077	18 52	127	23 28	155	60 08	224	57 54	293	29 14	342
319	22 42	044	26 13	078	19 32	127	23 49	156	59 32	225	57 07	293	28 58	342
320	23 18	045	27 03	078	20 13	128	24 10	157	58 55	227	56 20	293	28 43	342
321	23 54	045	27 53	078	20 52	129	24 29	158	58 18	228	55 33	293	28 27	342
322	24 30	045	28 43	079	21 32	130	24 48	159	57 40	230	54 47	293	28 11	342
323	25 06	045	29 33	079	22 11	130	25 06	160	57 00	231	54 00	293	27 56	342
324	25 42	046	30 23	080	22 49	131	25 24	160	56 20	232	53 13	293	27 40	342
325	26 19	046	31 13	080	23 27	132	25 41	161	55 40	234	52 27	294	27 25	343
326	26 55	046	32 03	081	24 05	133	25 56	162	54 59	235	51 40	294	27 10	343
327	27 32	046	32 53	081	24 42	134	26 11	163	54 17	236	50 53	294	26 55	343
328	28 09	047	33 44	082	25 19	134	26 26	164	53 34	237	50 07	294	26 40	343
329	28 46	047	34 34	082	25 55	135	26 39	165	52 51	238	49 20	294	26 25	343
	◆Mirfak		Hamal		Diphda		◆FOMALHAUT		ALTAIR		◆VEGA		Kochab	
330	29 23	047	35 24	082	26 30	136	26 52	166	52 07	239	48 34	294	26 10	343
331	30 00	047	36 15	083	27 05	137	27 04	167	51 23	241	47 47	294	25 55	343
332	30 38	047	37 05	083	27 40	138	27 15	168	50 39	242	47 01	294	25 41	343
333	31 15	048	37 56	084	28 14	139	27 25	169	49 54	243	46 15	295	25 26	344
334	31 53	048	38 47	084	28 47	139	27 34	170	49 08	244	45 28	295	25 12	344
335	32 30	048	39 37	085	29 20	140	27 43	171	48 23	245	44 42	295	24 58	344
336	33 08	048	40 28	085	29 52	141	27 51	172	47 37	245	43 56	295	24 44	344
337	33 46	048	41 19	086	30 24	142	27 57	173	46 50	246	43 10	295	24 30	344
338	34 24	048	42 09	086	30 55	143	28 03	174	46 03	247	42 24	295	24 16	344
339	35 02	049	43 00	087	31 25	144	28 09	175	45 16	248	41 38	296	24 02	344
340	35 40	049	43 51	087	31 55	145	28 13	176	44 29	249	40 52	296	23 49	345
341	36 19	049	44 42	088	32 24	146	28 16	177	43 41	250	40 06	296	23 35	345
342	36 57	049	45 33	088	32 52	147	28 19	178	42 53	251	39 21	296	23 22	345
343	37 35	049	46 24	089	33 19	148	28 20	179	42 05	251	38 35	297	23 09	345
344	38 14	049	47 14	089	33 46	149	28 21	180	41 17	252	37 50	297	22 56	345
	◆CAPELLA		ALDEBARAN		Diphda		◆FOMALHAUT		ALTAIR		◆VEGA		Kochab	
345	19 50	047	13 42	079	34 12	150	28 21	181	40 29	253	37 04	297	22 43	346
346	20 27	048	14 32	079	34 37	151	28 20	182	39 40	254	36 19	297	22 30	346
347	21 05	048	15 22	080	35 01	152	28 18	183	38 51	254	35 34	297	22 18	346
348	21 43	048	16 12	080	35 24	153	28 15	184	38 02	255	34 49	298	22 06	346
349	22 21	049	17 02	081	35 47	154	28 12	185	37 13	256	34 04	298	21 54	346
350	23 00	049	17 52	081	36 09	155	28 07	186	36 23	256	33 19	298	21 42	347
351	23 38	049	18 43	082	36 30	156	28 02	187	35 34	257	32 34	299	21 30	347
352	24 17	050	19 33	082	36 50	157	27 55	188	34 44	258	31 49	299	21 18	347
353	24 55	050	20 24	083	37 09	159	27 48	189	33 54	258	31 05	299	21 07	347
354	25 34	050	21 14	083	37 27	160	27 40	189	33 04	259	30 20	299	20 56	347
355	26 14	050	22 05	084	37 44	161	27 32	190	32 14	260	29 36	300	20 44	348
356	26 53	051	22 55	084	38 00	162	27 22	191	31 24	260	28 52	300	20 34	348
357	27 32	051	23 46	085	38 15	163	27 11	192	30 34	261	28 08	300	20 23	348
358	28 12	051	24 37	085	38 30	164	27 00	193	29 44	261	27 24	301	20 12	348
359	28 52	051	25 27	086	38 43	166	26 48	194	28 53	262	26 40	301	20 02	348

LHA ♈	Hc	Zn	Hc	Zn	Hc	Zn	Hc	Zn	Hc	Zn	Hc	Zn	Hc	Zn
	◆CAPELLA		ALDEBARAN		Diphda		◆FOMALHAUT		ALTAIR		◆DENEB		Kochab	
0	28 54	051	26 14	086	39 54	167	27 33	195	28 10	263	49 08	305	18 53	349
1	29 34	051	27 05	086	40 05	168	27 19	196	27 19	264	48 26	305	18 43	349
2	30 14	052	27 57	087	40 15	169	27 04	197	26 28	264	47 43	305	18 33	349
3	30 55	052	28 48	087	40 25	170	26 49	198	25 37	265	47 01	305	18 24	349
4	31 35	052	29 39	088	40 33	171	26 32	199	24 46	265	46 19	305	18 15	350
5	32 16	052	30 31	088	40 40	173	26 15	200	23 54	266	45 37	305	18 05	350
6	32 56	052	31 22	089	40 46	174	25 57	201	23 03	267	44 55	305	17 57	350
7	33 37	053	32 14	089	40 51	175	25 38	202	22 12	267	44 12	305	17 48	350
8	34 18	053	33 05	090	40 55	176	25 19	203	21 20	268	43 30	305	17 39	351
9	34 59	053	33 57	090	40 57	178	24 59	204	20 29	268	42 48	305	17 31	351
10	35 40	053	34 48	091	40 59	179	24 38	204	19 38	269	42 06	305	17 23	351
11	36 21	053	35 39	091	40 59	180	24 16	205	18 46	269	41 24	305	17 15	351
12	37 02	053	36 31	092	40 58	181	23 54	206	17 55	270	40 42	305	17 08	352
13	37 44	053	37 22	092	40 57	183	23 31	207	17 03	270	40 00	305	17 00	352
14	38 25	054	38 14	093	40 54	184	23 07	208	16 12	271	39 18	305	16 53	352
	◆CAPELLA		RIGEL		◆Diphda		FOMALHAUT		Enif		◆DENEB		Kochab	
15	39 06	054	17 42	112	40 49	185	22 43	209	39 57	256	38 36	306	16 46	352
16	39 48	054	18 30	112	40 44	186	22 18	210	39 07	257	37 54	306	16 40	353
17	40 29	054	19 17	113	40 38	188	21 52	210	38 16	257	37 12	306	16 33	353
18	41 11	054	20 05	113	40 30	189	21 26	211	37 26	258	36 30	306	16 27	353
19	41 52	054	20 52	114	40 22	190	20 59	212	36 36	259	35 49	306	16 21	353
20	42 34	054	21 39	115	40 12	191	20 31	213	35 45	259	35 07	306	16 15	354
21	43 16	054	22 25	115	40 01	193	20 03	214	34 55	260	34 26	306	16 10	354
22	43 57	054	23 12	116	39 50	194	19 34	214	34 04	260	33 44	307	16 04	354
23	44 39	054	23 58	117	39 37	195	19 05	215	33 13	261	33 03	307	15 59	355
24	45 21	054	24 43	117	39 23	196	18 35	216	32 23	262	32 22	307	15 54	355
25	46 02	054	25 29	118	39 08	197	18 05	217	31 32	262	31 41	307	15 50	355
26	46 44	054	26 14	119	38 52	199	17 34	217	30 41	263	31 00	307	15 46	355
27	47 26	054	26 59	120	38 35	200	17 03	218	29 50	263	30 19	307	15 42	356
28	48 07	054	27 44	120	38 17	201	16 31	219	28 58	264	29 38	308	15 38	356
29	48 49	054	28 28	121	37 58	202	15 58	219	28 07	265	28 58	308	15 34	356
	CAPELLA		BETELGEUSE		◆RIGEL		Diphda		◆Alpheratz		DENEB		◆Kochab	
30	49 31	054	30 31	100	29 12	122	37 38	203	65 46	273	28 17	308	15 31	356
31	50 12	054	31 21	101	29 55	123	37 18	204	64 55	273	27 37	308	15 28	357
32	50 54	054	32 12	102	30 39	123	36 56	205	64 04	273	26 57	309	15 25	357
33	51 35	054	33 02	102	31 21	124	36 33	207	63 12	274	26 16	309	15 22	357
34	52 16	054	33 52	103	32 04	125	36 10	208	62 21	274	25 36	309	15 20	358
35	52 58	053	34 42	103	32 46	126	35 46	209	61 30	275	24 57	309	15 18	358
36	53 39	053	35 32	104	33 27	127	35 21	210	60 39	275	24 17	310	15 16	358
37	54 20	053	36 22	105	34 08	127	34 55	211	59 47	275	23 37	310	15 15	358
38	55 01	053	37 12	105	34 49	128	34 28	212	58 56	276	22 58	310	15 13	359
39	55 42	053	38 01	106	35 29	129	34 01	213	58 05	276	22 19	310	15 12	359
40	56 23	052	38 50	107	36 08	130	33 32	214	57 14	277	21 40	311	15 11	359
41	57 03	052	39 39	108	36 47	131	33 03	215	56 23	277	21 01	311	15 11	000
42	57 44	052	40 28	108	37 26	132	32 34	216	55 32	277	20 22	311	15 11	000
43	58 24	051	41 17	109	38 04	133	32 03	217	54 41	278	19 44	312	15 11	000
44	59 04	051	42 06	110	38 41	134	31 32	218	53 50	278	19 05	312	15 11	000
	◆Dubhe		POLLUX		SIRIUS		◆RIGEL		Diphda		◆Alpheratz		Schedar	
45	14 14	025	29 02	073	17 57	123	39 18	135	31 01	219	52 59	278	54 49	327
46	14 36	025	29 51	073	18 40	124	39 54	136	30 28	219	52 08	279	54 20	326
47	14 58	025	30 40	074	19 23	125	40 29	137	29 55	220	51 17	279	53 51	326
48	15 20	026	31 30	074	20 05	125	41 04	138	29 22	221	50 26	279	53 22	325
49	15 42	026	32 19	074	20 47	126	41 38	139	28 47	222	49 36	280	52 53	325
50	16 05	026	33 09	075	21 28	127	42 11	140	28 13	223	48 45	280	52 23	325
51	16 28	027	33 58	075	22 09	127	42 44	141	27 37	224	47 54	280	51 53	324
52	16 51	027	34 48	076	22 50	128	43 15	143	27 02	225	47 04	281	51 23	324
53	17 14	027	35 38	076	23 30	129	43 46	144	26 25	225	46 13	281	50 52	323
54	17 38	027	36 28	076	24 10	130	44 16	145	25 48	226	45 23	281	50 22	323
55	18 02	028	37 18	077	24 49	130	44 45	146	25 11	227	44 32	282	49 51	323
56	18 26	028	38 08	077	25 28	131	45 13	147	24 33	228	43 42	282	49 19	323
57	18 50	028	38 58	077	26 07	132	45 41	149	23 55	229	42 52	282	48 48	322
58	19 14	028	39 48	078	26 45	133	46 07	150	23 16	229	42 02	283	48 17	322
59	19 39	029	40 39	078	27 22	134	46 32	151	22 36	230	41 12	283	47 45	322
	◆Dubhe		POLLUX		PROCYON		◆SIRIUS		RIGEL		◆Hamal		Schedar	
60	20 04	029	41 29	078	32 39	105	27 59	134	46 57	152	63 50	260	47 13	322
61	20 29	029	42 19	079	33 29	106	28 36	135	47 20	154	62 59	261	46 41	321
62	20 54	029	43 10	079	34 18	106	29 12	136	47 42	155	62 08	261	46 09	321
63	21 19	030	44 00	080	35 07	107	29 47	137	48 03	157	61 17	262	45 36	321
64	21 45	030	44 51	080	35 57	108	30 22	138	48 23	158	60 26	263	45 04	321
65	22 11	030	45 42	080	36 45	108	30 56	139	48 41	159	59 35	264	44 31	321
66	22 37	030	46 32	081	37 34	109	31 30	140	48 59	161	58 44	265	43 59	321
67	23 03	031	47 23	081	38 23	110	32 03	141	49 15	162	57 52	265	43 26	320
68	23 29	031	48 14	081	39 11	111	32 35	141	49 30	164	57 01	266	42 53	320
69	23 55	031	49 05	082	39 59	111	33 07	142	49 44	165	56 10	266	42 20	320
70	24 22	031	49 56	082	40 47	112	33 38	143	49 56	167	55 19	267	41 47	320
71	24 48	031	50 47	082	41 34	113	34 08	144	50 08	168	54 27	268	41 14	320
72	25 15	031	51 38	083	42 21	114	34 38	145	50 17	170	53 36	268	40 41	320
73	25 42	032	52 29	083	43 08	115	35 07	146	50 26	171	52 44	269	40 08	320
74	26 09	032	53 20	084	43 55	116	35 35	147	50 33	173	51 53	269	39 35	320
	◆Dubhe		POLLUX		PROCYON		◆SIRIUS		RIGEL		◆Hamal		Mirfak	
75	26 36	032	54 11	084	44 41	116	36 02	148	50 39	174	51 02	270	63 58	323
76	27 03	032	55 02	084	45 27	117	36 29	149	50 43	176	50 10	270	63 27	322
77	27 31	032	55 53	085	46 12	118	36 55	151	50 46	178	49 19	271	62 56	322
78	27 58	032	56 45	085	46 58	119	37 19	152	50 47	179	48 27	271	62 23	321
79	28 26	032	57 36	086	47 42	120	37 43	153	50 47	181	47 36	272	61 51	320
80	28 53	033	58 27	086	48 27	121	38 07	154	50 46	182	46 44	272	61 17	319
81	29 21	033	59 18	086	49 10	122	38 29	155	50 43	184	45 53	273	60 44	319
82	29 49	033	60 10	087	49 54	123	38 50	156	50 39	185	45 02	273	60 10	318
83	30 17	033	61 01	087	50 36	124	39 10	157	50 34	187	44 10	274	59 35	318
84	30 45	033	61 53	088	51 19	125	39 30	158	50 27	188	43 19	274	59 00	317
85	31 13	033	62 44	088	52 00	127	39 48	160	50 19	190	42 28	275	58 25	316
86	31 41	033	63 35	089	52 41	128	40 06	161	50 09	192	41 37	275	57 49	316
87	32 09	033	64 27	089	53 21	129	40 22	162	49 58	193	40 45	275	57 13	316
88	32 37	033	65 18	090	54 01	130	40 37	163	49 46	195	39 54	276	56 37	315
89	33 06	033	66 10	090	54 40	132	40 52	164	49 32	196	39 03	276	56 01	315

LAT 31°N

LHA ♈	Hc	Zn	Hc	Zn	Hc	Zn	Hc	Zn	Hc	Zn	Hc	Zn	Hc	Zn
	◆Dubhe		REGULUS		PROCYON		◆SIRIUS		RIGEL		ALDEBARAN		◆Mirfak	
90	33 34	033	30 01	094	55 18	133	41 05	166	49 18	197	65 56	238	55 24	314
91	34 02	033	30 52	094	55 55	134	41 17	167	49 01	199	65 12	239	54 47	314
92	34 30	033	31 44	095	56 32	136	41 28	168	48 44	200	64 28	241	54 10	314
93	34 59	033	32 35	096	57 07	137	41 38	169	48 26	202	63 43	242	53 33	313
94	35 27	033	33 26	096	57 42	139	41 47	171	48 06	203	62 57	243	52 55	313
95	35 56	033	34 17	097	58 15	140	41 55	172	47 45	205	62 11	245	52 18	313
96	36 24	033	35 08	097	58 47	142	42 02	173	47 23	206	61 24	246	51 40	313
97	36 52	033	35 59	098	59 19	143	42 07	175	47 00	207	60 37	247	51 02	312
98	37 21	033	36 50	098	59 49	145	42 11	176	46 36	209	59 49	248	50 24	312
99	37 49	033	37 41	099	60 17	147	42 15	177	46 11	210	59 01	249	49 46	312
100	38 17	033	38 32	100	60 45	149	42 17	178	45 45	211	58 13	250	49 08	312
101	38 46	033	39 22	100	61 11	151	42 17	180	45 18	212	57 25	251	48 30	312
102	39 14	033	40 13	101	61 35	152	42 17	181	44 49	214	56 36	252	47 51	312
103	39 42	033	41 03	102	61 58	154	42 16	182	44 20	215	55 47	253	47 13	312
104	40 10	033	41 54	102	62 20	156	42 13	184	43 51	216	54 58	254	46 34	311
	◆Alkaid		REGULUS		Alphard		◆SIRIUS		RIGEL		ALDEBARAN		◆CAPELLA	
105	16 02	042	42 44	103	36 56	132	42 09	185	43 20	217	54 08	255	64 58	314
106	16 36	042	43 34	104	37 34	133	42 04	186	42 48	218	53 18	255	64 21	313
107	17 10	042	44 24	104	38 11	134	41 58	187	42 16	220	52 29	256	63 43	313
108	17 45	043	45 14	105	38 48	135	41 51	189	41 43	221	51 39	257	63 05	312
109	18 20	043	46 03	106	39 24	136	41 43	190	41 09	222	50 48	258	62 27	311
110	18 55	043	46 52	107	39 59	137	41 33	191	40 34	223	49 58	258	61 48	311
111	19 30	043	47 42	107	40 34	138	41 23	193	39 59	224	49 08	259	61 09	310
112	20 06	044	48 31	108	41 08	139	41 11	194	39 23	225	48 17	260	60 30	310
113	20 41	044	49 19	109	41 41	140	40 58	195	38 47	226	47 26	261	59 50	310
114	21 17	044	50 08	110	42 14	141	40 44	196	38 09	227	46 36	261	59 10	309
115	21 53	045	50 56	111	42 45	143	40 29	197	37 32	228	45 45	262	58 30	309
116	22 29	045	51 44	112	43 16	144	40 13	199	36 53	229	44 54	262	57 50	308
117	23 06	045	52 31	113	43 46	145	39 56	200	36 14	230	44 03	263	57 10	308
118	23 42	045	53 19	114	44 15	146	39 39	201	35 35	231	43 12	264	56 29	308
119	24 19	046	54 06	115	44 43	147	39 20	202	34 55	232	42 21	264	55 49	308
	◆Alkaid		REGULUS		◆Alphard		SIRIUS		RIGEL		ALDEBARAN		◆CAPELLA	
120	24 56	046	54 52	116	45 11	149	39 00	203	34 14	232	41 29	265	55 08	307
121	25 33	046	55 38	117	45 37	150	38 39	205	33 33	233	40 38	265	54 27	307
122	26 10	046	56 24	118	46 02	151	38 17	206	32 52	234	39 47	266	53 46	307
123	26 47	047	57 09	119	46 27	152	37 54	207	32 10	235	38 55	267	53 04	307
124	27 25	047	57 54	120	46 50	154	37 31	208	31 28	236	38 04	267	52 23	307
125	28 02	047	58 38	121	47 12	155	37 06	209	30 45	237	37 13	268	51 42	306
126	28 40	047	59 22	123	47 33	156	36 41	210	30 02	237	36 21	268	51 00	306
127	29 17	047	60 05	124	47 53	158	36 15	211	29 18	238	35 30	269	50 19	306
128	29 55	048	60 47	125	48 12	159	35 48	212	28 34	239	34 38	269	49 37	306
129	30 33	048	61 29	127	48 30	161	35 20	213	27 50	240	33 47	270	48 56	306
130	31 11	048	62 10	128	48 46	162	34 52	214	27 06	240	32 56	270	48 14	306
131	31 50	048	62 50	130	49 01	164	34 22	215	26 21	241	32 04	271	47 32	306
132	32 28	048	63 29	131	49 15	165	33 52	216	25 36	242	31 13	271	46 51	306
133	33 06	048	64 07	133	49 28	166	33 22	217	24 50	243	30 21	272	46 09	306
134	33 45	048	64 44	135	49 39	168	32 50	218	24 04	243	29 30	272	45 27	306
	◆Alkaid		ARCTURUS		SPICA		◆Alphard		SIRIUS		BETELGEUSE		◆CAPELLA	
135	34 23	049	19 01	079	13 51	112	49 50	169	32 18	219	40 50	251	44 46	306
136	35 02	049	19 52	079	14 39	113	49 58	171	31 45	220	40 01	252	44 04	306
137	35 41	049	20 42	079	15 26	114	50 06	173	31 12	221	39 12	253	43 22	306
138	36 19	049	21 33	080	16 13	114	50 12	174	30 38	222	38 23	253	42 41	306
139	36 58	049	22 23	080	17 00	115	50 16	176	30 04	223	37 34	254	41 59	306
140	37 37	049	23 14	081	17 47	115	50 19	177	29 28	224	36 44	255	41 17	306
141	38 16	049	24 05	081	18 33	116	50 21	179	28 53	224	35 55	256	40 36	306
142	38 55	049	24 56	082	19 19	117	50 22	180	28 16	225	35 05	256	39 54	306
143	39 34	049	25 47	082	20 05	117	50 21	182	27 40	226	34 15	257	39 13	306
144	40 13	049	26 38	083	20 50	118	50 19	183	27 02	227	33 24	258	38 32	306
145	40 52	049	27 29	083	21 36	119	50 15	185	26 25	228	32 34	258	37 50	307
146	41 31	049	28 20	084	22 21	119	50 10	186	25 46	229	31 44	259	37 09	307
147	42 10	049	29 11	084	23 05	120	50 03	188	25 08	229	30 53	259	36 28	307
148	42 49	049	30 02	084	23 50	121	49 56	189	24 28	230	30 03	260	35 47	307
149	43 28	049	30 53	085	24 34	121	49 46	191	23 49	231	29 12	261	35 06	307
	Alkaid		◆ARCTURUS		SPICA		◆Alphard		PROCYON		BETELGEUSE		◆CAPELLA	
150	44 07	049	31 44	085	25 17	122	49 36	193	48 05	239	28 21	261	34 25	307
151	44 47	049	32 36	086	26 01	123	49 24	194	47 21	240	27 30	262	33 44	307
152	45 26	049	33 27	086	26 44	124	49 11	195	46 36	241	26 39	262	33 03	308
153	46 05	049	34 18	087	27 26	124	48 57	197	45 51	242	25 48	263	32 22	308
154	46 44	049	35 10	087	28 08	125	48 41	198	45 05	243	24 57	263	31 42	308
155	47 23	049	36 01	088	28 50	126	48 24	200	44 19	244	24 06	264	31 01	308
156	48 01	049	36 53	088	29 32	127	48 06	201	43 32	245	23 15	265	30 21	308
157	48 40	049	37 44	089	30 12	128	47 47	203	42 46	246	22 24	265	29 40	309
158	49 19	049	38 35	089	30 53	128	47 27	204	41 59	247	21 33	266	29 00	309
159	49 58	049	39 27	090	31 33	129	47 05	205	41 11	247	20 41	266	28 20	309
160	50 36	048	40 18	090	32 13	130	46 42	207	40 24	248	19 50	267	27 40	309
161	51 15	048	41 10	091	32 52	131	46 19	208	39 36	249	18 59	267	27 01	309
162	51 53	048	42 01	091	33 30	132	45 54	209	38 48	250	18 07	268	26 21	310
163	52 31	048	42 53	092	34 08	133	45 28	211	37 59	250	17 16	268	25 41	310
164	53 09	048	43 44	093	34 46	134	45 02	212	37 11	251	16 24	269	25 02	310
	Dubhe		◆Alkaid		ARCTURUS		◆SPICA		REGULUS		◆PROCYON		CAPELLA	
165	59 13	001	53 47	047	44 35	093	35 23	135	67 32	215	36 22	252	24 23	310
166	59 13	000	54 25	047	45 27	094	35 59	136	67 01	217	35 33	253	23 44	311
167	59 13	359	55 03	047	46 18	094	36 34	137	66 30	219	34 44	253	23 05	311
168	59 11	358	55 40	046	47 09	095	37 09	138	65 56	221	33 54	254	22 26	311
169	59 09	357	56 17	046	48 00	095	37 44	139	65 22	223	33 05	255	21 48	312
170	59 06	356	56 54	046	48 52	096	38 18	140	64 46	225	32 15	255	21 09	312
171	59 02	355	57 31	045	49 43	097	38 51	141	64 09	227	31 25	256	20 31	312
172	58 58	354	58 07	045	50 34	097	39 23	142	63 31	229	30 35	257	19 53	312
173	58 52	353	58 43	044	51 25	098	39 55	143	62 52	230	29 45	257	19 15	313
174	58 46	353	59 19	044	52 16	099	40 25	144	62 12	232	28 55	258	18 37	313
175	58 39	352	59 54	043	53 06	099	40 55	145	61 31	233	28 05	258	18 00	313
176	58 31	351	60 29	043	53 57	100	41 24	146	60 50	235	27 14	259	17 23	314
177	58 23	350	61 04	042	54 48	101	41 52	147	60 07	236	26 24	260	16 46	314
178	58 13	349	61 38	041	55 38	101	42 20	148	59 24	237	25 33	260	16 09	314
179	58 03	348	62 12	041	56 29	102	42 46	150	58 41	239	24 42	261	15 32	315

LAT 31°N

LHA	Hc	Zn	Hc	Zn	Hc	Zn	Hc	Zn	Hc	Zn	Hc	Zn	Hc	Zn
♈	◆VEGA		Alphecca		ARCTURUS		◆SPICA		REGULUS		◆POLLUX		Dubhe	
180	12 28	052	43 18	081	57 19	103	43 12	151	57 57	240	35 13	284	57 52	347
181	13 08	052	44 09	081	58 09	104	43 36	152	57 12	241	34 24	285	57 41	347
182	13 49	053	44 59	082	58 59	105	44 00	153	56 27	242	33 34	285	57 29	346
183	14 30	053	45 50	082	59 48	106	44 22	155	55 41	243	32 44	285	57 16	345
184	15 12	054	46 41	083	60 38	107	44 44	156	54 55	244	31 55	286	57 02	344
185	15 53	054	47 32	083	61 27	108	45 05	157	54 08	245	31 05	286	56 48	344
186	16 35	054	48 23	083	62 16	109	45 24	158	53 21	246	30 16	287	56 33	343
187	17 17	055	49 15	084	63 04	110	45 42	160	52 34	247	29 27	287	56 17	342
188	17 59	055	50 06	084	63 53	111	45 59	161	51 46	248	28 37	287	56 01	341
189	18 41	055	50 57	085	64 41	112	46 16	163	50 58	249	27 48	288	55 45	341
190	19 23	056	51 48	085	65 28	113	46 30	164	50 10	250	26 59	288	55 27	340
191	20 06	056	52 39	086	66 15	114	46 44	165	49 22	251	26 11	288	55 10	339
192	20 49	056	53 31	086	67 02	116	46 57	167	48 33	252	25 22	289	54 51	339
193	21 32	057	54 22	086	67 48	117	47 08	168	47 44	252	24 33	289	54 32	338
194	22 15	057	55 13	087	68 33	119	47 18	170	46 55	253	23 45	290	54 13	338
	◆Kochab		VEGA		Rasalhague		◆ANTARES		SPICA		◆REGULUS		Dubhe	
195	44 38	010	22 58	057	24 36	090	13 55	133	47 27	171	46 06	254	53 53	337
196	44 47	010	23 41	058	25 27	090	14 32	134	47 34	172	45 16	255	53 33	336
197	44 55	010	24 25	058	26 19	091	15 09	134	47 40	174	44 26	256	53 12	336
198	45 04	009	25 09	058	27 10	091	15 46	135	47 45	175	43 37	256	52 51	335
199	45 12	009	25 52	059	28 02	092	16 22	136	47 49	177	42 47	257	52 29	335
200	45 20	009	26 36	059	28 53	092	16 57	137	47 51	178	41 56	258	52 07	334
201	45 27	008	27 21	059	29 44	093	17 33	137	47 52	180	41 06	258	51 45	334
202	45 35	008	28 05	059	30 36	093	18 07	138	47 51	181	40 16	259	51 22	333
203	45 42	008	28 49	060	31 27	094	18 41	139	47 50	183	39 25	260	50 59	333
204	45 48	007	29 34	060	32 18	095	19 15	139	47 47	184	38 34	260	50 35	333
205	45 54	007	30 18	060	33 10	095	19 48	140	47 43	185	37 44	261	50 11	332
206	46 00	006	31 03	061	34 01	096	20 21	141	47 37	187	36 53	262	49 47	332
207	46 06	006	31 48	061	34 52	096	20 53	142	47 30	188	36 02	262	49 23	331
208	46 11	006	32 33	061	35 43	097	21 25	142	47 22	190	35 11	263	48 58	331
209	46 16	005	33 18	061	36 34	097	21 56	143	47 13	191	34 20	263	48 33	331
	◆Kochab		VEGA		Rasalhague		◆ANTARES		SPICA		◆REGULUS		Dubhe	
210	46 21	005	34 03	062	37 25	098	22 26	144	47 02	193	33 29	264	48 08	330
211	46 25	005	34 48	062	38 16	099	22 56	145	46 50	194	32 38	264	47 42	330
212	46 29	004	35 34	062	39 07	099	23 25	146	46 37	195	31 46	265	47 17	330
213	46 33	004	36 19	062	39 57	100	23 54	147	46 23	197	30 55	266	46 51	330
214	46 36	003	37 05	062	40 48	101	24 22	147	46 07	198	30 04	266	46 25	329
215	46 39	003	37 51	063	41 39	101	24 50	148	45 50	200	29 13	267	45 58	329
216	46 41	003	38 36	063	42 29	102	25 16	149	45 33	201	28 21	267	45 32	329
217	46 44	002	39 22	063	43 19	103	25 42	150	45 14	202	27 30	268	45 05	329
218	46 46	002	40 08	063	44 09	103	26 08	151	44 54	204	26 38	268	44 38	328
219	46 47	001	40 54	063	44 59	104	26 33	152	44 33	205	25 47	269	44 11	328
220	46 48	001	41 40	064	45 49	105	26 57	153	44 10	206	24 56	269	43 44	328
221	46 49	001	42 26	064	46 39	105	27 20	153	43 47	207	24 04	270	43 16	328
222	46 49	000	43 12	064	47 28	106	27 43	154	43 23	209	23 13	270	42 49	328
223	46 49	000	43 59	064	48 18	107	28 04	155	42 58	210	22 21	271	42 21	327
224	46 49	359	44 45	064	49 07	108	28 25	156	42 32	211	21 30	271	41 54	327
	DENEB		VEGA		◆ALTAIR		ANTARES		◆SPICA		Denebola		◆Dubhe	
225	24 32	050	45 31	064	19 24	091	28 46	157	42 05	212	43 24	261	41 26	327
226	25 11	051	46 18	065	20 15	092	29 05	158	41 37	213	42 33	261	40 58	327
227	25 51	051	47 04	065	21 06	092	29 24	159	41 09	214	41 42	262	40 30	327
228	26 31	051	47 51	065	21 58	093	29 42	160	40 39	216	40 51	262	40 02	327
229	27 11	051	48 37	065	22 49	093	29 59	161	40 09	217	40 00	263	39 34	327
230	27 52	052	49 24	065	23 41	094	30 15	162	39 38	218	39 09	264	39 06	327
231	28 32	052	50 11	065	24 32	094	30 31	163	39 06	219	38 18	264	38 37	327
232	29 13	052	50 57	065	25 23	095	30 45	164	38 33	220	37 27	265	38 09	327
233	29 53	052	51 44	065	26 14	096	30 59	165	38 00	221	36 35	265	37 41	327
234	30 34	053	52 31	065	27 05	096	31 12	166	37 26	222	35 44	266	37 12	327
235	31 15	053	53 18	065	27 57	097	31 24	167	36 51	223	34 53	267	36 44	327
236	31 56	053	54 04	065	28 48	097	31 35	168	36 15	224	34 01	267	36 16	327
237	32 37	053	54 51	065	29 39	098	31 45	169	35 39	225	33 10	268	35 47	327
238	33 18	053	55 38	065	30 30	098	31 54	170	35 03	226	32 19	268	35 19	327
239	34 00	054	56 25	065	31 20	099	32 02	171	34 26	227	31 27	269	34 51	327
	◆DENEB		ALTAIR		Nunki		◆ANTARES		SPICA		ARCTURUS		◆Alkaid	
240	34 41	054	32 11	100	19 05	139	32 10	172	33 48	228	63 39	250	59 07	316
241	35 23	054	33 02	100	19 39	140	32 16	173	33 10	229	62 51	251	58 31	316
242	36 04	054	33 52	101	20 11	141	32 21	174	32 31	229	62 02	252	57 55	315
243	36 46	054	34 43	101	20 44	141	32 26	175	31 51	230	61 13	253	57 19	315
244	37 28	054	35 33	102	21 16	142	32 30	177	31 12	231	60 24	254	56 42	314
245	38 09	054	36 23	103	21 47	143	32 32	178	30 31	232	59 34	255	56 05	314
246	38 51	055	37 14	103	22 18	144	32 34	179	29 51	233	58 45	256	55 28	313
247	39 33	055	38 04	104	22 48	144	32 35	180	29 09	234	57 55	256	54 50	313
248	40 15	055	38 53	105	23 18	145	32 34	181	28 28	234	57 05	257	54 13	313
249	40 57	055	39 43	105	23 47	146	32 33	182	27 46	235	56 14	258	53 35	313
250	41 39	055	40 33	106	24 15	147	32 31	183	27 03	236	55 24	259	52 57	312
251	42 21	055	41 22	107	24 43	148	32 28	184	26 21	237	54 33	259	52 19	312
252	43 04	055	42 11	108	25 10	149	32 24	185	25 37	237	53 43	260	51 40	312
253	43 46	055	43 00	108	25 37	149	32 19	186	24 54	238	52 52	261	51 02	312
254	44 28	055	43 49	109	26 02	150	32 13	187	24 10	239	52 01	262	50 24	311
	◆DENEB		ALTAIR		Nunki		◆ANTARES		ARCTURUS		◆Alkaid		Kochab	
255	45 10	055	44 37	110	26 27	151	32 06	188	51 10	262	49 45	311	43 52	348
256	45 52	055	45 25	111	26 52	152	31 59	189	50 19	263	49 06	311	43 42	348
257	46 35	055	46 13	112	27 16	153	31 50	190	49 28	264	48 28	311	43 31	348
258	47 17	055	47 01	112	27 39	154	31 40	191	48 37	264	47 49	311	43 20	347
259	47 59	055	47 48	113	28 01	155	31 30	192	47 46	265	47 10	311	43 09	347
260	48 41	055	48 35	114	28 22	156	31 18	193	46 55	265	46 31	311	42 57	347
261	49 23	055	49 22	115	28 43	157	31 06	194	46 03	266	45 52	311	42 45	347
262	50 06	055	50 09	116	29 03	158	30 53	195	45 12	267	45 13	311	42 33	346
263	50 48	055	50 55	117	29 22	159	30 39	196	44 21	267	44 34	311	42 21	346
264	51 30	055	51 40	118	29 40	160	30 24	197	43 29	268	43 55	311	42 09	346
265	52 12	055	52 25	119	29 58	161	30 08	198	42 38	268	43 16	311	41 56	346
266	52 54	055	53 10	120	30 15	162	29 51	199	41 47	269	42 36	311	41 44	345
267	53 36	054	53 54	121	30 30	163	29 34	200	40 55	269	41 57	311	41 31	345
268	54 17	054	54 38	122	30 45	164	29 16	201	40 04	270	41 18	311	41 17	345
269	54 59	054	55 21	124	30 59	165	28 57	202	39 12	270	40 39	311	41 04	345

LAT 31°N

LHA	Hc	Zn	Hc	Zn	Hc	Zn	Hc	Zn	Hc	Zn	Hc	Zn	Hc	Zn
♈	◆DENEB		ALTAIR		◆Nunki		ANTARES		ARCTURUS		◆Alkaid		Kochab	
270	55 41	054	56 04	125	31 13	166	28 37	203	38 21	271	40 00	311	40 50	345
271	56 22	054	56 45	126	31 25	167	28 16	204	37 29	271	39 21	311	40 37	344
272	57 04	053	57 27	127	31 37	168	27 55	205	36 38	272	38 42	311	40 23	344
273	57 45	053	58 07	129	31 47	169	27 32	206	35 47	272	38 03	311	40 09	344
274	58 26	053	58 47	130	31 57	170	27 10	207	34 55	273	37 24	311	39 55	344
275	59 07	052	59 26	132	32 05	171	26 46	208	34 04	273	36 45	311	39 40	344
276	59 47	052	60 04	133	32 13	172	26 22	209	33 13	274	36 07	311	39 26	344
277	60 28	052	60 41	135	32 20	173	25 56	210	32 21	274	35 28	311	39 11	343
278	61 08	051	61 17	136	32 26	174	25 31	210	31 30	275	34 49	311	38 56	343
279	61 48	051	61 52	138	32 31	175	25 04	211	30 39	275	34 11	311	38 42	343
280	62 28	050	62 26	140	32 35	176	24 37	212	29 48	276	33 32	312	38 27	343
281	63 07	050	62 58	142	32 38	177	24 10	213	28 56	276	32 54	312	38 11	343
282	63 46	049	63 30	143	32 40	178	23 41	214	28 05	277	32 15	312	37 56	343
283	64 25	048	64 00	145	32 42	179	23 12	215	27 14	277	31 37	312	37 41	343
284	65 03	048	64 28	147	32 42	180	22 43	215	26 23	277	30 59	312	37 25	342
	◆Alpheratz		Enif		ALTAIR		◆ANTARES		ARCTURUS		◆Alkaid		Kochab	
285	24 44	070	46 31	110	64 55	149	22 13	216	25 32	278	30 21	312	37 10	342
286	25 32	070	47 19	111	65 21	151	21 42	217	24 41	278	29 43	313	36 54	342
287	26 20	070	48 07	112	65 44	154	21 11	218	23 50	279	29 05	313	36 38	342
288	27 09	071	48 55	113	66 06	156	20 39	219	23 00	279	28 27	313	36 23	342
289	27 58	071	49 42	114	66 26	158	20 06	219	22 09	280	27 50	313	36 07	342
290	28 46	072	50 29	115	66 44	161	19 33	220	21 18	280	27 12	313	35 51	342
291	29 35	072	51 16	115	67 00	163	19 00	221	20 28	281	26 35	314	35 35	342
292	30 24	072	52 02	116	67 14	165	18 26	222	19 37	281	25 58	314	35 19	342
293	31 13	073	52 48	117	67 26	168	17 52	222	18 47	282	25 21	314	35 03	342
294	32 02	073	53 33	119	67 36	171	17 17	223	17 56	282	24 44	314	34 47	342
295	32 51	073	54 18	120	67 43	173	16 41	224	17 06	283	24 07	314	34 30	342
296	33 41	074	55 03	121	67 48	176	16 06	224	16 16	283	23 30	315	34 14	342
297	34 30	074	55 47	122	67 51	178	15 29	225	15 26	283	22 54	315	33 58	342
298	35 20	074	56 30	123	67 51	181	14 53	226	14 36	284	22 18	315	33 42	342
299	36 09	075	57 13	124	67 49	184	14 16	227	13 46	284	21 41	316	33 25	341
	◆Mirfak		Alpheratz		◆FOMALHAUT		ALTAIR		Rasalhague		◆Alphecca		Kochab	
300	11 17	038	36 59	075	16 08	141	67 45	186	51 49	249	32 34	284	33 09	341
301	11 49	038	37 49	075	16 41	141	67 38	189	51 01	250	31 45	284	32 53	341
302	12 21	039	38 38	076	17 12	142	67 29	191	50 13	251	30 55	285	32 36	341
303	12 53	039	39 28	076	17 44	143	67 18	194	49 24	252	30 05	285	32 20	341
304	13 25	039	40 18	076	18 14	144	67 04	196	48 35	253	29 15	285	32 04	341
305	13 58	040	41 08	077	18 45	144	66 49	199	47 46	253	28 26	286	31 47	341
306	14 31	040	41 59	077	19 14	145	66 31	201	46 56	254	27 36	286	31 31	341
307	15 04	040	42 49	078	19 43	146	66 12	203	46 07	255	26 47	287	31 15	341
308	15 38	041	43 39	078	20 12	147	65 50	206	45 17	256	25 58	287	30 58	342
309	16 11	041	44 29	078	20 40	147	65 27	208	44 27	256	25 09	287	30 42	342
310	16 45	041	45 20	079	21 07	148	65 02	210	43 37	257	24 20	288	30 26	342
311	17 19	042	46 10	079	21 34	149	64 36	212	42 47	258	23 31	288	30 09	342
312	17 54	042	47 01	079	22 00	150	64 08	214	41 56	259	22 42	289	29 53	342
313	18 28	042	47 51	080	22 25	151	63 38	216	41 06	259	21 53	289	29 37	342
314	19 03	043	48 42	080	22 50	152	63 07	218	40 15	260	21 04	289	29 21	342
	◆Mirfak		Hamal		Diphda		◆FOMALHAUT		ALTAIR		◆VEGA		Kochab	
315	19 38	043	22 40	075	17 22	124	23 14	152	62 35	220	59 50	295	29 05	342
316	20 13	043	23 30	076	18 04	125	23 38	153	62 01	222	59 03	295	28 49	342
317	20 48	043	24 20	076	18 46	126	24 01	154	61 26	223	58 16	295	28 33	342
318	21 24	044	25 10	077	19 28	127	24 23	155	60 51	225	57 29	295	28 17	342
319	21 59	044	26 00	077	20 09	127	24 44	156	60 14	226	56 43	295	28 01	342
320	22 35	044	26 50	078	20 50	128	25 05	157	59 36	228	55 56	295	27 45	342
321	23 11	045	27 41	078	21 30	129	25 25	158	58 57	229	55 09	295	27 30	342
322	23 47	045	28 31	078	22 10	129	25 44	158	58 18	231	54 22	295	27 14	342
323	24 24	045	29 21	079	22 49	130	26 03	159	57 38	232	53 36	295	26 59	342
324	25 00	045	30 12	079	23 29	131	26 20	160	56 57	234	52 49	295	26 43	343
325	25 37	046	31 02	080	24 07	132	26 37	161	56 15	235	52 02	295	26 28	343
326	26 14	046	31 53	080	24 45	132	26 54	162	55 33	236	51 15	295	26 13	343
327	26 50	046	32 44	080	25 23	133	27 09	163	54 50	237	50 29	295	25 57	343
328	27 27	046	33 35	081	26 00	134	27 23	164	54 06	238	49 42	295	25 42	343
329	28 05	046	34 25	081	26 37	135	27 37	165	53 22	240	48 55	295	25 27	343
	◆Mirfak		Hamal		Diphda		◆FOMALHAUT		ALTAIR		◆VEGA		Kochab	
330	28 42	047	35 16	082	27 13	136	27 50	166	52 37	241	48 09	295	25 13	343
331	29 19	047	36 07	082	27 49	136	28 02	167	51 52	242	47 22	295	24 58	343
332	29 57	047	36 58	083	28 24	137	28 14	168	51 07	243	46 36	295	24 43	344
333	30 34	047	37 49	083	28 59	138	28 24	169	50 21	244	45 49	295	24 29	344
334	31 12	047	38 40	083	29 33	139	28 34	170	49 35	245	45 03	296	24 14	344
335	31 50	047	39 31	084	30 06	140	28 42	171	48 48	246	44 16	296	24 00	344
336	32 28	048	40 23	084	30 39	141	28 50	172	48 01	246	43 30	296	23 46	344
337	33 06	048	41 14	085	31 11	142	28 57	173	47 14	247	42 44	296	23 32	344
338	33 44	048	42 05	085	31 43	143	29 03	174	46 26	248	41 58	296	23 18	344
339	34 22	048	42 56	086	32 14	144	29 08	175	45 38	249	41 12	296	23 04	345
340	35 00	048	43 48	086	32 44	145	29 13	176	44 50	250	40 26	297	22 51	345
341	35 39	048	44 39	087	33 13	146	29 16	177	44 02	251	39 40	297	22 37	345
342	36 17	048	45 30	087	33 42	147	29 19	178	43 13	251	38 54	297	22 24	345
343	36 56	048	46 22	088	34 10	148	29 20	179	42 24	252	38 08	297	22 11	345
344	37 34	048	47 13	088	34 37	149	29 21	180	41 35	253	37 22	297	21 58	345
	◆CAPELLA		ALDEBARAN		Diphda		◆FOMALHAUT		ALTAIR		◆VEGA		Kochab	
345	19 09	047	13 30	079	35 03	150	29 21	181	40 46	254	36 37	298	21 45	346
346	19 47	047	14 21	079	35 29	151	29 20	182	39 56	254	35 51	298	21 32	346
347	20 25	048	15 11	080	35 54	152	29 18	183	39 07	255	35 06	298	21 20	346
348	21 03	048	16 02	080	36 18	153	29 15	184	38 17	256	34 21	298	21 07	346
349	21 42	048	16 53	081	36 41	154	29 11	185	37 27	256	33 35	299	20 55	346
350	22 20	049	17 43	081	37 03	155	29 07	186	36 37	257	32 50	299	20 43	347
351	22 59	049	18 34	082	37 25	156	29 01	187	35 47	258	32 05	299	20 31	347
352	23 38	049	19 25	082	37 45	157	28 55	188	34 56	258	31 20	299	20 20	347
353	24 17	049	20 16	082	38 05	158	28 48	189	34 06	259	30 36	300	20 08	347
354	24 56	050	21 07	083	38 23	159	28 40	190	33 15	260	29 51	300	19 57	347
355	25 35	050	21 58	083	38 41	161	28 31	191	32 25	260	29 06	300	19 46	348
356	26 15	050	22 49	084	38 57	162	28 21	192	31 34	261	28 22	300	19 35	348
357	26 54	051	23 40	084	39 13	163	28 10	192	30 43	261	27 38	301	19 24	348
358	27 34	051	24 32	085	39 27	164	27 59	193	29 52	262	26 53	301	19 14	348
359	28 14	051	25 23	085	39 41	165	27 46	194	29 01	263	26 09	301	19 03	349

LAT 30°N

LHA	Hc	Zn	Hc	Zn	Hc	Zn	Hc	Zn	Hc	Zn	Hc	Zn	Hc	Zn
♈	Schedar		◆CAPELLA		ALDEBARAN		◆Diphda		FOMALHAUT		ALTAIR		◆DENEB	
0	62 34	012	28 16	051	26 09	085	40 52	166	28 31	195	28 17	264	48 33	306
1	62 44	011	28 56	051	27 01	086	41 04	168	28 17	196	27 25	264	47 51	306
2	62 53	010	29 37	051	27 53	086	41 14	169	28 01	197	26 34	265	47 09	306
3	63 02	009	30 17	051	28 45	087	41 24	170	27 46	198	25 42	265	46 27	306
4	63 09	007	30 58	052	29 37	087	41 32	171	27 29	199	24 50	266	45 44	306
5	63 15	006	31 39	052	30 29	088	41 39	173	27 11	200	23 58	266	45 02	306
6	63 20	005	32 20	052	31 21	088	41 46	174	26 53	201	23 06	267	44 20	306
7	63 24	004	33 00	052	32 13	089	41 51	175	26 34	202	22 15	267	43 38	306
8	63 27	003	33 41	052	33 04	089	41 55	176	26 14	203	21 23	268	42 55	306
9	63 29	001	34 23	052	33 56	090	41 57	178	25 54	204	20 31	268	42 13	306
10	63 29	000	35 04	052	34 48	090	41 59	179	25 32	205	19 39	269	41 31	306
11	63 29	359	35 45	053	35 40	091	41 59	180	25 10	206	18 47	270	40 49	306
12	63 27	358	36 26	053	36 32	091	41 58	181	24 47	206	17 55	270	40 07	306
13	63 25	356	37 08	053	37 24	092	41 56	183	24 24	207	17 03	271	39 25	306
14	63 21	355	37 49	053	38 16	092	41 53	184	24 00	208	16 11	271	38 43	306
	◆CAPELLA		ALDEBARAN		RIGEL		◆Diphda		FOMALHAUT		Enif		◆DENEB	
15	38 31	053	39 08	093	18 04	111	41 49	185	23 35	209	40 11	257	38 01	306
16	39 12	053	40 00	093	18 52	112	41 44	187	23 10	210	39 20	257	37 19	306
17	39 54	053	40 52	094	19 40	112	41 37	188	22 44	211	38 29	258	36 37	306
18	40 35	053	41 44	094	20 28	113	41 30	189	22 17	211	37 38	259	35 55	306
19	41 17	053	42 36	095	21 16	114	41 21	190	21 49	212	36 47	259	35 13	307
20	41 58	053	43 27	095	22 03	114	41 11	192	21 22	213	35 56	260	34 32	307
21	42 40	053	44 19	096	22 51	115	41 00	193	20 53	214	35 05	261	33 50	307
22	43 22	053	45 11	097	23 38	116	40 48	194	20 24	215	34 14	261	33 09	307
23	44 03	053	46 02	097	24 24	116	40 35	195	19 54	215	33 23	262	32 27	307
24	44 45	053	46 54	098	25 11	117	40 20	197	19 24	216	32 31	262	31 46	307
25	45 27	053	47 45	099	25 57	118	40 05	198	18 53	217	31 39	263	31 05	308
26	46 08	053	48 36	099	26 43	118	39 49	199	18 22	217	30 48	263	30 23	308
27	46 50	053	49 28	100	27 28	119	39 31	200	17 50	218	29 56	264	29 42	308
28	47 32	053	50 19	100	28 14	120	39 13	201	17 17	219	29 04	265	29 01	308
29	48 13	053	51 10	101	28 59	121	38 54	202	16 44	220	28 13	265	28 21	308
	CAPELLA		◆ALDEBARAN		RIGEL		◆Diphda		Alpheratz		DENEB		◆Kochab	
30	48 55	053	52 01	102	29 43	121	38 33	204	65 42	275	27 40	309	14 31	356
31	49 36	053	52 52	103	30 27	122	38 12	205	64 51	275	26 59	309	14 28	357
32	50 18	053	53 42	103	31 11	123	37 50	206	63 59	276	26 19	309	14 25	357
33	50 59	053	54 33	104	31 55	124	37 27	207	63 07	276	25 39	309	14 22	357
34	51 40	052	55 23	105	32 38	124	37 03	208	62 16	276	24 58	309	14 20	358
35	52 22	052	56 13	106	33 20	125	36 38	209	61 24	276	24 18	310	14 18	358
36	53 03	052	57 03	107	34 03	126	36 13	210	60 32	277	23 39	310	14 16	358
37	53 44	052	57 53	108	34 44	127	35 46	211	59 41	277	22 59	310	14 15	358
38	54 24	052	58 42	109	35 26	128	35 19	212	58 49	277	22 19	311	14 13	359
39	55 05	051	59 31	110	36 07	129	34 51	213	57 58	278	21 40	311	14 12	359
40	55 46	051	60 20	111	36 47	130	34 22	214	57 06	278	21 01	311	14 11	359
41	56 26	051	61 08	112	37 27	131	33 53	215	56 15	278	20 21	311	14 11	000
42	57 06	050	61 56	113	38 06	131	33 22	216	55 23	279	19 43	312	14 11	000
43	57 46	050	62 44	114	38 44	132	32 51	217	54 32	279	19 04	312	14 11	000
44	58 26	050	63 32	115	39 23	133	32 20	218	53 41	279	18 25	312	14 11	000
	◆CAPELLA		PROCYON		SIRIUS		◆RIGEL		Acamar		Diphda		◆Alpheratz	
45	59 05	049	20 07	096	18 30	123	40 00	134	19 40	180	31 47	219	52 50	280
46	59 45	049	20 59	096	19 13	124	40 37	135	19 40	181	31 14	220	51 58	280
47	60 24	048	21 51	097	19 57	124	41 13	136	19 38	182	30 41	221	51 07	280
48	61 02	048	22 42	097	20 39	125	41 48	138	19 36	183	30 07	222	50 16	281
49	61 40	047	23 34	098	21 22	126	42 23	139	19 33	184	29 32	222	49 25	281
50	62 18	047	24 25	098	22 04	126	42 57	140	19 30	184	28 56	223	48 34	281
51	62 56	046	25 16	099	22 45	127	43 30	141	19 25	185	28 21	224	47 43	281
52	63 33	045	26 08	100	23 27	128	44 03	142	19 20	186	27 44	225	46 52	282
53	64 10	044	26 59	100	24 08	128	44 34	143	19 14	187	27 07	226	46 01	282
54	64 46	044	27 50	101	24 48	129	45 05	144	19 08	188	26 30	227	45 10	282
55	65 22	043	28 41	101	25 28	130	45 35	146	19 00	188	25 52	227	44 20	283
56	65 57	042	29 32	102	26 08	131	46 04	147	18 52	189	25 13	228	43 29	283
57	66 31	041	30 23	102	26 47	132	46 32	148	18 44	190	24 34	229	42 39	283
58	67 05	040	31 13	103	27 25	132	46 59	149	18 34	191	23 55	230	41 48	284
59	67 38	039	32 04	104	28 04	133	47 25	151	18 24	192	23 15	230	40 58	284
	◆Dubhe		PROCYON		SIRIUS		◆RIGEL		Diphda		◆Hamal		Mirfak	
60	19 11	029	32 54	104	28 41	134	47 50	152	22 34	231	63 59	262	69 01	344
61	19 36	029	33 45	105	29 18	135	48 14	153	21 54	232	63 07	263	68 46	342
62	20 02	029	34 35	106	29 55	136	48 36	155	21 13	233	62 16	264	68 29	340
63	20 27	029	35 25	106	30 31	137	48 58	156	20 31	233	61 24	264	68 11	339
64	20 53	030	36 14	107	31 06	137	49 18	158	19 49	234	60 32	265	67 52	337
65	21 19	030	37 04	108	31 41	138	49 38	159	19 07	235	59 40	266	67 31	336
66	21 45	030	37 53	108	32 16	139	49 56	160	18 25	235	58 49	266	67 09	335
67	22 11	030	38 43	109	32 49	140	50 12	162	17 42	236	57 57	267	66 46	333
68	22 37	030	39 32	110	33 22	141	50 28	163	16 58	237	57 05	267	66 23	332
69	23 04	031	40 20	111	33 55	142	50 42	165	16 15	237	56 13	268	65 58	331
70	23 30	031	41 09	111	34 26	143	50 55	166	15 31	238	55 21	268	65 32	330
71	23 57	031	41 57	112	34 57	144	51 06	168	14 47	239	54 29	269	65 05	328
72	24 24	031	42 45	113	35 27	145	51 16	170	14 02	239	53 37	269	64 37	327
73	24 51	031	43 33	114	35 57	146	51 25	171	13 18	240	52 45	270	64 09	326
74	25 18	032	44 20	115	36 25	147	51 32	173	12 33	240	51 53	270	63 40	325
	◆Dubhe		POLLUX		PROCYON		◆SIRIUS		RIGEL		◆Hamal		Mirfak	
75	25 45	032	54 04	083	45 07	115	36 53	148	51 38	174	51 01	271	63 10	324
76	26 12	032	54 56	083	45 54	116	37 20	149	51 43	176	50 09	271	62 39	324
77	26 40	032	55 47	083	46 40	117	37 47	150	51 46	177	49 17	272	62 08	323
78	27 07	032	56 39	084	47 26	118	38 12	151	51 47	179	48 25	272	61 36	322
79	27 35	032	57 30	084	48 12	119	38 37	152	51 47	181	47 33	273	61 04	321
80	28 03	032	58 22	084	48 57	120	39 00	153	51 46	182	46 42	273	60 31	321
81	28 30	032	59 14	085	49 42	121	39 23	155	51 43	184	45 50	274	59 58	320
82	28 58	033	60 06	085	50 26	122	39 45	156	51 39	185	44 58	274	59 24	319
83	29 26	033	60 57	086	51 10	123	40 06	157	51 33	187	44 06	275	58 50	319
84	29 54	033	61 49	086	51 53	124	40 26	158	51 26	189	43 14	275	58 16	318
85	30 22	033	62 41	086	52 35	126	40 44	159	51 18	190	42 23	275	57 41	318
86	30 51	033	63 33	087	53 17	127	41 02	161	51 08	192	41 31	276	57 06	317
87	31 19	033	64 25	087	53 59	128	41 19	162	50 57	193	40 39	276	56 30	317
88	31 47	033	65 17	087	54 39	129	41 35	163	50 44	195	39 48	277	55 54	316
89	32 15	033	66 09	088	55 19	130	41 50	164	50 30	196	38 56	277	55 18	316

LHA	Hc	Zn	Hc	Zn	Hc	Zn	Hc	Zn	Hc	Zn	Hc	Zn	Hc	Zn
♈	Dubhe		◆REGULUS		PROCYON		SIRIUS		◆RIGEL		ALDEBARAN		◆Mirfak	
90	32 44	033	30 05	093	55 58	132	42 03	165	50 15	198	66 27	240	54 42	315
91	33 12	033	30 57	094	56 37	133	42 16	167	49 58	199	65 42	241	54 05	315
92	33 40	033	31 49	094	57 14	135	42 27	168	49 40	201	64 56	243	53 28	315
93	34 09	033	32 40	095	57 51	136	42 37	169	49 21	202	64 10	244	52 51	314
94	34 37	033	33 32	095	58 26	138	42 46	171	49 01	204	63 23	245	52 14	314
95	35 05	033	34 24	096	59 01	139	42 54	172	48 40	205	62 35	246	51 37	314
96	35 34	033	35 16	097	59 34	141	43 01	173	48 17	206	61 48	248	50 59	314
97	36 02	033	36 07	097	60 06	142	43 07	174	47 53	208	60 59	249	50 21	313
98	36 30	033	36 59	098	60 37	144	43 11	176	47 29	209	60 11	250	49 44	313
99	36 59	033	37 50	098	61 07	146	43 14	177	47 03	210	59 22	251	49 06	313
100	37 27	033	38 41	099	61 36	148	43 17	178	46 36	212	58 33	252	48 27	313
101	37 55	033	39 33	100	62 03	150	43 17	180	46 08	213	57 43	253	47 49	313
102	38 23	033	40 24	100	62 28	152	43 17	181	45 39	214	56 54	253	47 11	312
103	38 52	033	41 15	101	62 52	154	43 16	182	45 10	215	56 04	254	46 33	312
104	39 20	033	42 06	101	63 14	156	43 13	184	44 39	217	55 14	255	45 54	312
	◆Dubhe		REGULUS		◆Alphard		SIRIUS		RIGEL		◆ALDEBARAN		CAPELLA	
105	39 48	033	42 57	102	37 36	132	43 09	185	44 07	218	54 23	256	64 16	316
106	40 15	032	43 48	103	38 15	133	43 04	186	43 35	219	53 33	257	63 39	315
107	40 43	032	44 38	103	38 53	133	42 58	188	43 02	220	52 42	257	63 02	314
108	41 11	032	45 29	104	39 30	134	42 50	189	42 28	221	51 51	258	62 24	314
109	41 39	032	46 19	105	40 07	135	42 42	190	41 54	222	51 01	259	61 47	313
110	42 06	032	47 09	106	40 43	137	42 32	191	41 18	223	50 09	260	61 08	312
111	42 33	032	47 59	106	41 18	138	42 21	193	40 42	224	49 18	260	60 30	312
112	43 01	031	48 49	107	41 53	139	42 09	194	40 05	225	48 27	261	59 51	311
113	43 28	031	49 38	108	42 27	140	41 56	195	39 28	226	47 36	262	59 12	311
114	43 55	031	50 28	109	43 00	141	41 42	196	38 50	227	46 44	262	58 32	310
115	44 21	031	51 17	110	43 33	142	41 27	198	38 12	228	45 53	263	57 52	310
116	44 48	031	52 06	111	44 04	143	41 10	199	37 32	229	45 01	263	57 12	310
117	45 14	030	52 54	111	44 35	144	40 53	200	36 53	230	44 09	264	56 32	309
118	45 40	030	53 42	112	45 05	146	40 34	201	36 13	231	43 18	265	55 52	309
119	46 06	030	54 30	113	45 34	147	40 15	203	35 32	232	42 26	265	55 12	309
	◆Dubhe		REGULUS		◆Alphard		SIRIUS		RIGEL		◆ALDEBARAN		CAPELLA	
120	46 32	030	55 18	114	46 02	148	39 55	204	34 51	233	41 34	266	54 31	308
121	46 58	029	56 05	115	46 29	149	39 33	205	34 09	234	40 42	266	53 50	308
122	47 23	029	56 52	116	46 55	151	39 11	206	33 27	235	39 50	267	53 09	308
123	47 48	029	57 38	118	47 20	152	38 48	207	32 44	235	38 59	267	52 28	308
124	48 13	028	58 24	119	47 44	153	38 24	208	32 01	236	38 07	268	51 47	308
125	48 37	028	59 09	120	48 07	155	37 59	209	31 18	237	37 15	268	51 06	307
126	49 01	028	59 54	121	48 28	156	37 33	210	30 34	238	36 23	269	50 24	307
127	49 25	027	60 38	122	48 49	157	37 06	211	29 50	239	35 31	269	49 43	307
128	49 49	027	61 21	124	49 08	159	36 39	213	29 05	239	34 39	270	49 01	307
129	50 12	026	62 04	125	49 26	160	36 10	214	28 20	240	33 47	270	48 20	307
130	50 35	026	62 46	127	49 43	162	35 41	215	27 35	241	32 55	271	47 38	307
131	50 57	025	63 27	128	49 59	163	35 11	216	26 50	242	32 03	271	46 57	307
132	51 19	025	64 08	130	50 13	165	34 41	217	26 04	242	31 11	272	46 15	307
133	51 41	024	64 47	132	50 26	166	34 09	218	25 18	243	30 19	272	45 33	307
134	52 02	024	65 26	133	50 38	168	33 37	219	24 31	244	29 27	273	44 52	307
	Dubhe		◆ARCTURUS		SPICA		Alphard		◆SIRIUS		BETELGEUSE		◆CAPELLA	
135	52 23	023	18 49	078	14 14	112	50 48	169	33 05	219	41 09	252	44 10	307
136	52 44	023	19 40	079	15 02	113	50 57	171	32 31	220	40 20	253	43 28	307
137	53 03	022	20 31	079	15 50	113	51 05	172	31 57	221	39 30	254	42 47	307
138	53 23	022	21 22	080	16 38	114	51 11	174	31 23	222	38 40	254	42 05	307
139	53 42	021	22 13	080	17 25	115	51 16	176	30 48	223	37 50	255	41 23	307
140	54 00	021	23 04	080	18 12	115	51 19	177	30 12	224	37 00	256	40 42	307
141	54 18	020	23 56	081	18 59	116	51 21	179	29 35	225	36 09	256	40 00	307
142	54 36	019	24 47	081	19 46	116	51 22	180	28 59	226	35 19	257	39 19	307
143	54 53	019	25 38	082	20 32	117	51 21	182	28 21	226	34 28	258	38 37	307
144	55 09	018	26 30	082	21 18	118	51 18	183	27 43	227	33 37	258	37 56	307
145	55 25	017	27 21	083	22 04	118	51 15	185	27 05	228	32 46	259	37 14	307
146	55 40	017	28 13	083	22 50	119	51 09	187	26 26	229	31 55	259	36 33	307
147	55 54	016	29 04	083	23 35	120	51 03	188	25 47	230	31 04	260	35 52	307
148	56 08	015	29 56	084	24 20	120	50 55	190	25 07	230	30 13	261	35 10	308
149	56 21	014	30 48	084	25 05	121	50 45	191	24 26	231	29 22	261	34 29	308
	Dubhe		◆ARCTURUS		SPICA		◆Alphard		SIRIUS		POLLUX		◆CAPELLA	
150	56 34	014	31 39	085	25 49	122	50 34	193	23 46	232	60 32	275	33 48	308
151	56 46	013	32 31	085	26 33	123	50 22	194	23 05	233	59 40	275	33 07	308
152	56 57	012	33 23	086	27 17	123	50 09	196	22 23	233	58 48	275	32 26	308
153	57 07	011	34 15	086	28 00	124	49 54	197	21 41	234	57 57	276	31 45	308
154	57 17	010	35 07	087	28 43	125	49 38	199	20 59	235	57 05	276	31 05	308
155	57 26	010	35 59	087	29 25	126	49 21	200	20 16	235	56 13	276	30 24	309
156	57 34	009	36 50	088	30 07	126	49 02	202	19 33	236	55 22	277	29 43	309
157	57 42	008	37 42	088	30 49	127	48 42	203	18 50	237	54 30	277	29 03	309
158	57 48	007	38 34	089	31 30	128	48 21	204	18 06	237	53 39	278	28 23	309
159	57 54	006	39 26	089	32 11	129	47 59	206	17 23	238	52 47	278	27 42	309
160	57 59	005	40 18	090	32 51	130	47 36	207	16 38	239	51 56	278	27 02	310
161	58 04	004	41 10	090	33 31	131	47 12	209	15 54	239	51 04	279	26 22	310
162	58 07	003	42 02	091	34 10	131	46 46	210	15 09	240	50 13	279	25 42	310
163	58 10	003	42 54	091	34 49	132	46 20	211	14 24	241	49 22	279	25 03	310
164	58 12	002	43 46	092	35 27	133	45 53	212	13 38	241	48 30	280	24 23	311
	Dubhe		◆Alkaid		ARCTURUS		◆SPICA		Alphard		◆PROCYON		POLLUX	
165	58 13	001	53 06	046	44 38	092	36 05	134	45 24	214	36 40	253	47 39	280
166	58 13	000	53 44	046	45 30	093	36 41	135	44 55	215	35 51	253	46 48	280
167	58 13	359	54 21	046	46 22	093	37 18	136	44 25	216	35 01	254	45 57	281
168	58 11	358	54 58	045	47 14	094	37 54	137	43 54	217	34 11	255	45 06	281
169	58 09	357	55 35	045	48 05	094	38 29	138	43 22	218	33 21	255	44 15	281
170	58 06	356	56 12	045	48 57	095	39 03	139	42 49	220	32 30	256	43 24	282
171	58 03	355	56 48	044	49 49	095	39 37	140	42 16	221	31 40	257	42 33	282
172	57 58	355	57 24	044	50 41	096	40 10	141	41 42	222	30 49	257	41 42	282
173	57 53	354	58 00	043	51 32	097	40 42	142	41 07	223	29 58	258	40 51	283
174	57 47	353	58 35	043	52 24	097	41 13	143	40 31	224	29 08	258	40 01	283
175	57 40	352	59 10	042	53 15	098	41 44	144	39 55	225	28 17	259	39 10	283
176	57 32	351	59 45	042	54 07	099	42 14	146	39 18	226	27 26	260	38 20	284
177	57 23	350	60 19	041	54 58	099	42 43	147	38 40	227	26 34	260	37 29	284
178	57 14	349	60 53	040	55 49	100	43 11	148	38 02	228	25 43	261	36 39	284
179	57 04	349	61 26	040	56 41	101	43 38	149	37 23	229	24 52	261	35 48	285

LAT 30°N

LHA ♈	Hc	Zn	Hc	Zn	Hc	Zn	Hc	Zn	Hc	Zn	Hc	Zn	Hc	Zn
	◆VEGA		Alphecca		◆SPICA		Gienah		REGULUS		◆POLLUX		Dubhe	
180	11 51	052	43 08	080	44 04	150	42 20	175	58 26	241	34 58	285	56 54	348
181	12 32	052	43 59	081	44 29	152	42 24	176	57 40	242	34 08	285	56 42	347
182	13 13	053	44 50	081	44 53	153	42 27	178	56 54	243	33 18	286	56 30	346
183	13 54	053	45 42	081	45 17	154	42 29	179	56 07	245	32 28	286	56 18	345
184	14 36	053	46 33	082	45 39	155	42 29	180	55 20	246	31 38	286	56 04	345
185	15 18	054	47 25	082	46 00	157	42 28	181	54 33	247	30 48	287	55 50	344
186	16 00	054	48 16	082	46 20	158	42 27	183	53 45	248	29 59	287	55 36	343
187	16 42	054	49 08	083	46 39	159	42 23	184	52 57	249	29 09	287	55 20	343
188	17 24	055	49 59	083	46 56	161	42 19	185	52 08	249	28 19	288	55 04	342
189	18 07	055	50 51	083	47 13	162	42 14	187	51 19	250	27 30	288	54 48	341
190	18 49	055	51 42	084	47 28	164	42 07	188	50 30	251	26 41	289	54 31	341
191	19 32	056	52 34	084	47 42	165	42 00	189	49 41	252	25 51	289	54 13	340
192	20 15	056	53 26	085	47 55	166	41 51	190	48 51	253	25 02	289	53 55	339
193	20 59	056	54 18	085	48 06	168	41 41	192	48 02	254	24 13	290	53 37	339
194	21 42	057	55 09	085	48 17	169	41 30	193	47 12	254	23 24	290	53 17	338
	◆Kochab		VEGA		Rasalhague		◆ANTARES		SPICA		◆REGULUS		Dubhe	
195	43 39	010	22 25	057	24 35	089	14 36	133	48 26	171	46 22	255	52 58	338
196	43 48	010	23 09	057	25 27	090	15 14	134	48 33	172	45 31	256	52 38	337
197	43 56	009	23 53	058	26 19	090	15 51	134	48 40	174	44 41	256	52 17	336
198	44 05	009	24 37	058	27 11	091	16 28	135	48 45	175	43 50	257	51 56	336
199	44 13	009	25 21	058	28 03	091	17 05	136	48 49	177	43 00	258	51 35	335
200	44 21	008	26 05	058	28 55	092	17 41	136	48 51	178	42 09	259	51 13	335
201	44 28	008	26 50	059	29 47	092	18 17	137	48 52	180	41 18	259	50 51	334
202	44 35	008	27 34	059	30 39	093	18 52	138	48 51	181	40 27	260	50 28	334
203	44 42	007	28 19	059	31 31	093	19 26	138	48 50	183	39 35	260	50 05	334
204	44 49	007	29 04	060	32 23	094	20 01	139	48 47	184	38 44	261	49 42	333
205	44 55	007	29 48	060	33 15	094	20 34	140	48 42	186	37 53	262	49 18	333
206	45 01	006	30 33	060	34 06	095	21 07	141	48 37	187	37 01	262	48 54	332
207	45 06	006	31 18	060	34 58	096	21 40	141	48 29	189	36 10	263	48 30	332
208	45 12	006	32 04	061	35 50	096	22 12	142	48 21	190	35 18	263	48 05	332
209	45 17	005	32 49	061	36 42	097	22 44	143	48 11	191	34 27	264	47 41	331
	Kochab		◆VEGA		Rasalhague		ANTARES		◆SPICA		REGULUS		◆Dubhe	
210	45 21	005	33 34	061	37 33	097	23 15	144	48 00	193	33 35	265	47 16	331
211	45 25	005	34 20	061	38 25	098	23 45	145	47 48	194	32 43	265	46 50	331
212	45 29	004	35 05	061	39 16	098	24 15	145	47 35	196	31 51	266	46 25	330
213	45 33	004	35 51	062	40 07	099	24 44	146	47 20	197	31 00	266	45 59	330
214	45 36	003	36 37	062	40 59	100	25 13	147	47 04	199	30 08	267	45 33	330
215	45 39	003	37 23	062	41 50	100	25 40	148	46 47	200	29 16	267	45 07	330
216	45 42	003	38 09	062	42 41	101	26 08	149	46 29	201	28 24	268	44 40	329
217	45 44	002	38 55	062	43 32	102	26 34	150	46 09	203	27 32	268	44 14	329
218	45 46	002	39 41	063	44 23	102	27 00	151	45 49	204	26 40	269	43 47	329
219	45 47	001	40 27	063	45 13	103	27 25	151	45 27	205	25 48	269	43 20	329
220	45 48	001	41 13	063	46 04	104	27 50	152	45 04	207	24 56	270	42 53	328
221	45 49	001	41 59	063	46 54	104	28 14	153	44 41	208	24 04	270	42 25	328
222	45 49	000	42 46	063	47 45	105	28 37	154	44 16	209	23 12	271	41 58	328
223	45 49	000	43 32	063	48 35	106	28 59	155	43 50	210	22 20	271	41 31	328
224	45 49	359	44 19	063	49 24	107	29 20	156	43 23	212	21 28	272	41 03	328
	VEGA		◆ALTAIR		Nunki		ANTARES		◆SPICA		Denebola		◆Alkaid	
225	45 05	064	19 25	091	10 27	129	29 41	157	42 56	213	43 33	262	66 16	330
226	45 52	064	20 17	091	11 07	129	30 01	158	42 27	214	42 42	262	65 50	329
227	46 38	064	21 09	092	11 47	130	30 20	159	41 58	215	41 50	263	65 22	327
228	47 25	064	22 01	092	12 27	131	30 38	160	41 28	216	40 58	263	64 54	326
229	48 12	064	22 52	093	13 06	131	30 56	161	40 57	217	40 07	264	64 24	325
230	48 58	064	23 44	093	13 45	132	31 12	162	40 25	218	39 15	264	63 54	324
231	49 45	064	24 36	094	14 23	133	31 28	163	39 52	219	38 23	265	63 24	323
232	50 32	064	25 28	094	15 01	133	31 43	164	39 19	220	37 32	266	62 53	323
233	51 18	064	26 20	095	15 39	134	31 57	165	38 45	221	36 40	266	62 21	322
234	52 05	064	27 12	096	16 16	135	32 10	166	38 10	222	35 48	267	61 48	321
235	52 52	064	28 03	096	16 53	135	32 22	167	37 35	223	34 56	267	61 15	320
236	53 39	064	28 55	097	17 29	136	32 33	168	36 58	224	34 04	268	60 42	320
237	54 26	064	29 46	097	18 05	137	32 44	169	36 22	225	33 12	268	60 08	319
238	55 12	064	30 38	098	18 41	137	32 53	170	35 44	226	32 20	269	59 34	318
239	55 59	064	31 29	098	19 16	138	33 02	171	35 07	227	31 28	269	58 59	318
	◆VEGA		ALTAIR		Nunki		◆ANTARES		SPICA		ARCTURUS		◆Alkaid	
240	56 46	064	32 21	099	19 50	139	33 09	172	34 28	228	63 59	252	58 24	317
241	57 32	064	33 12	100	20 24	140	33 16	173	33 49	229	63 09	253	57 48	317
242	58 19	064	34 03	100	20 58	140	33 21	174	33 10	230	62 20	254	57 12	316
243	59 06	064	34 54	101	21 31	141	33 26	175	32 30	231	61 30	255	56 36	316
244	59 52	064	35 45	101	22 03	142	33 30	176	31 49	232	60 40	255	56 00	315
245	60 39	063	36 36	102	22 35	143	33 32	178	31 08	232	59 49	256	55 23	315
246	61 25	063	37 27	103	23 06	143	33 34	179	30 27	233	58 59	257	54 46	314
247	62 11	063	38 18	103	23 37	144	33 35	180	29 45	234	58 08	258	54 09	314
248	62 58	063	39 08	104	24 07	145	33 34	181	29 03	235	57 17	259	53 31	314
249	63 44	062	39 59	105	24 37	146	33 33	182	28 20	236	56 26	259	52 54	314
250	64 30	062	40 49	105	25 05	147	33 31	183	27 37	236	55 35	260	52 16	313
251	65 15	062	41 39	106	25 34	147	33 28	184	26 53	237	54 44	261	51 38	313
252	66 01	061	42 29	107	26 01	148	33 24	185	26 10	238	53 52	262	51 00	313
253	66 46	061	43 18	107	26 28	149	33 19	186	25 25	239	53 01	262	50 22	313
254	67 32	060	44 08	108	26 54	150	33 13	187	24 41	239	52 09	263	49 43	312
	◆DENEB		ALTAIR		Nunki		◆ANTARES		SPICA		ARCTURUS		◆Alkaid	
255	44 35	054	44 57	109	27 20	151	33 06	188	23 56	240	51 18	264	49 05	312
256	45 18	054	45 46	110	27 45	152	32 58	189	23 11	241	50 26	264	48 26	312
257	46 00	054	46 35	111	28 09	153	32 49	190	22 25	241	49 34	265	47 48	312
258	46 42	054	47 23	111	28 32	154	32 39	191	21 40	242	48 43	265	47 09	312
259	47 24	054	48 12	112	28 55	155	32 28	192	20 54	243	47 51	266	46 30	312
260	48 07	054	48 59	113	29 17	156	32 17	193	20 07	243	46 59	266	45 51	312
261	48 49	054	49 47	114	29 38	157	32 04	195	19 21	244	46 07	267	45 12	311
262	49 31	054	50 34	115	29 58	157	31 51	196	18 34	245	45 15	268	44 33	311
263	50 13	054	51 21	116	30 18	158	31 36	197	17 47	245	44 23	268	43 54	311
264	50 55	054	52 08	117	30 37	159	31 21	198	17 00	246	43 31	269	43 15	311
265	51 37	054	52 54	118	30 54	160	31 05	199	16 12	246	42 39	269	42 36	311
266	52 19	054	53 40	119	31 11	161	30 48	200	15 24	247	41 47	270	41 57	311
267	53 00	053	54 25	120	31 28	162	30 30	201	14 36	248	40 55	270	41 18	311
268	53 42	053	55 10	121	31 43	163	30 12	202	13 48	248	40 03	271	40 39	311
269	54 23	053	55 54	122	31 57	164	29 52	202	13 00	249	39 12	271	40 00	311

LHA ♈	Hc	Zn	Hc	Zn	Hc	Zn	Hc	Zn	Hc	Zn	Hc	Zn	Hc	Zn
	◆DENEB		ALTAIR		◆Nunki		ANTARES		◆ARCTURUS		Alkaid		Kochab	
270	55 05	053	56 37	124	32 11	165	29 32	203	38 20	272	39 21	311	39 53	345
271	55 46	052	57 20	125	32 23	166	29 11	204	37 28	272	38 42	311	39 39	345
272	56 27	052	58 03	126	32 35	167	28 49	205	36 36	273	38 03	311	39 25	344
273	57 08	052	58 44	127	32 46	169	28 26	206	35 44	273	37 24	311	39 11	344
274	57 49	051	59 25	129	32 56	170	28 03	207	34 52	273	36 45	311	38 57	344
275	58 29	051	60 05	130	33 05	171	27 39	208	34 00	274	36 06	312	38 43	344
276	59 10	051	60 44	132	33 13	172	27 14	209	33 08	274	35 27	312	38 28	344
277	59 50	050	61 22	133	33 20	173	26 49	210	32 16	275	34 48	312	38 14	344
278	60 30	050	62 00	135	33 26	174	26 22	211	31 25	275	34 09	312	37 59	343
279	61 09	049	62 36	137	33 31	175	25 56	212	30 33	276	33 31	312	37 44	343
280	61 49	049	63 11	138	33 35	176	25 28	212	29 41	276	32 52	312	37 29	343
281	62 27	048	63 45	140	33 38	177	25 00	213	28 50	277	32 14	312	37 14	343
282	63 06	048	64 17	142	33 40	178	24 31	214	27 58	277	31 35	312	36 59	343
283	63 44	047	64 49	144	33 42	179	24 02	215	27 07	278	30 57	312	36 44	343
284	64 22	046	65 18	146	33 42	180	23 31	216	26 15	278	30 18	313	36 28	343
	◆Alpheratz		Enif		ALTAIR		◆ANTARES		ARCTURUS		◆Alkaid		Kochab	
285	24 23	069	46 51	109	65 47	148	23 01	217	25 24	278	29 40	313	36 13	343
286	25 11	070	47 40	110	66 13	150	22 30	217	24 32	279	29 02	313	35 57	342
287	26 00	070	48 29	111	66 38	153	21 58	218	23 41	279	28 24	313	35 41	342
288	26 49	070	49 18	112	67 01	155	21 25	219	22 50	280	27 46	313	35 26	342
289	27 38	071	50 06	112	67 22	157	20 53	220	21 59	280	27 09	313	35 10	342
290	28 27	071	50 54	113	67 41	160	20 19	220	21 07	281	26 31	314	34 54	342
291	29 16	071	51 41	114	67 58	162	19 45	221	20 16	281	25 53	314	34 38	342
292	30 06	072	52 28	115	68 12	165	19 11	222	19 25	281	25 16	314	34 22	342
293	30 55	072	53 15	116	68 25	167	18 36	223	18 35	282	24 39	314	34 06	342
294	31 44	072	54 01	117	68 35	170	18 01	223	17 44	282	24 02	315	33 50	342
295	32 34	073	54 47	118	68 43	173	17 25	224	16 53	283	23 25	315	33 33	342
296	33 24	073	55 33	119	68 48	176	16 48	225	16 02	283	22 48	315	33 17	342
297	34 13	073	56 18	121	68 51	178	16 12	225	15 12	284	22 11	315	33 01	342
298	35 03	074	57 02	122	68 51	181	15 34	226	14 21	284	21 35	316	32 45	342
299	35 53	074	57 46	123	68 49	184	14 57	227	13 31	285	20 59	316	32 28	342
	◆Schedar		Alpheratz		◆FOMALHAUT		ALTAIR		Rasalhague		◆Alphecca		Kochab	
300	35 27	040	36 43	074	16 55	141	68 44	186	52 10	250	32 20	284	32 12	342
301	36 00	040	37 33	075	17 28	141	68 37	189	51 21	251	31 30	285	31 56	342
302	36 33	040	38 23	075	18 00	142	68 28	192	50 31	252	30 39	285	31 39	342
303	37 06	040	39 14	075	18 31	143	68 16	194	49 42	253	29 49	286	31 23	342
304	37 39	040	40 04	076	19 03	143	68 02	197	48 52	254	28 59	286	31 07	342
305	38 13	040	40 54	076	19 33	144	67 46	200	48 02	255	28 09	286	30 50	342
306	38 46	040	41 45	076	20 03	145	67 27	202	47 12	255	27 19	287	30 34	342
307	39 19	039	42 35	077	20 33	146	67 07	204	46 22	256	26 30	287	30 18	342
308	39 52	039	43 26	077	21 02	147	66 44	207	45 31	257	25 40	287	30 01	342
309	40 25	039	44 17	077	21 30	147	66 20	209	44 40	257	24 51	288	29 45	342
310	40 58	039	45 07	078	21 58	148	65 54	211	43 50	258	24 01	288	29 29	342
311	41 30	039	45 58	078	22 25	149	65 26	213	42 59	259	23 12	289	29 12	342
312	42 03	039	46 49	078	22 52	150	64 57	215	42 08	259	22 23	289	28 56	342
313	42 36	039	47 40	079	23 18	151	64 26	217	41 17	260	21 33	289	28 40	342
314	43 09	039	48 31	079	23 43	151	63 54	219	40 25	261	20 44	290	28 24	342
	◆Mirfak		Hamal		Diphda		◆FOMALHAUT		ALTAIR		◆VEGA		Kochab	
315	18 54	043	22 25	075	17 56	124	24 08	152	63 20	221	59 24	296	28 08	342
316	19 29	043	23 15	075	18 39	125	24 31	153	62 46	223	58 37	296	27 52	342
317	20 05	043	24 06	076	19 21	126	24 55	154	62 10	225	57 50	296	27 36	342
318	20 40	043	24 56	076	20 03	126	25 17	155	61 33	226	57 04	296	27 20	342
319	21 16	044	25 47	077	20 45	127	25 39	156	60 55	228	56 17	296	27 04	342
320	21 52	044	26 37	077	21 26	128	26 00	157	60 16	229	55 30	296	26 48	342
321	22 28	044	27 28	077	22 07	128	26 20	157	59 36	231	54 43	296	26 33	342
322	23 05	044	28 19	078	22 48	129	26 40	158	58 55	232	53 57	296	26 17	343
323	23 41	045	29 09	078	23 28	130	26 59	159	58 14	234	53 10	296	26 01	343
324	24 18	045	30 00	079	24 08	131	27 17	160	57 32	235	52 23	296	25 46	343
325	24 55	045	30 51	079	24 47	131	27 34	161	56 49	236	51 36	296	25 31	343
326	25 32	045	31 42	079	25 26	132	27 51	162	56 06	237	50 50	296	25 15	343
327	26 09	046	32 34	080	26 04	133	28 06	163	55 21	238	50 03	296	25 00	343
328	26 46	046	33 25	080	26 42	134	28 21	164	54 37	240	49 16	296	24 45	343
329	27 23	046	34 16	081	27 19	134	28 35	165	53 52	241	48 29	296	24 30	343
	◆Mirfak		Hamal		Diphda		◆FOMALHAUT		ALTAIR		◆VEGA		Kochab	
330	28 00	046	35 07	081	27 56	135	28 48	166	53 06	242	47 43	296	24 15	343
331	28 38	046	35 59	081	28 32	136	29 01	167	52 20	243	46 56	296	24 00	344
332	29 16	047	36 50	082	29 08	137	29 12	168	51 34	244	46 10	296	23 46	344
333	29 53	047	37 41	082	29 43	138	29 23	169	50 47	245	45 23	296	23 31	344
334	30 31	047	38 33	083	30 18	139	29 33	170	50 00	246	44 36	297	23 17	344
335	31 09	047	39 25	083	30 52	140	29 41	171	49 12	247	43 50	297	23 02	344
336	31 47	047	40 16	084	31 25	140	29 49	172	48 25	247	43 04	297	22 48	344
337	32 25	047	41 08	084	31 58	141	29 57	173	47 36	248	42 17	297	22 34	344
338	33 04	047	42 00	084	32 30	142	30 03	174	46 48	249	41 31	297	22 20	345
339	33 42	047	42 51	085	33 02	143	30 08	175	45 59	250	40 45	297	22 06	345
340	34 20	048	43 43	085	33 32	144	30 12	176	45 10	251	39 58	297	21 53	345
341	34 59	048	44 35	086	34 03	145	30 16	177	44 21	252	39 12	298	21 39	345
342	35 37	048	45 27	086	34 32	146	30 19	178	43 32	252	38 26	298	21 26	345
343	36 16	048	46 18	087	35 00	147	30 20	179	42 42	253	37 40	298	21 13	345
344	36 54	048	47 10	087	35 28	148	30 21	180	41 52	254	36 54	298	21 00	346
	◆CAPELLA		ALDEBARAN		Diphda		◆FOMALHAUT		ALTAIR		◆VEGA		Kochab	
345	18 28	047	13 18	078	35 55	149	30 21	181	41 02	255	36 09	298	20 47	346
346	19 06	047	14 09	079	36 21	150	30 20	182	40 12	255	35 23	298	20 34	346
347	19 45	048	15 00	079	36 47	151	30 18	183	39 22	256	34 37	299	20 22	346
348	20 23	048	15 51	080	37 11	152	30 15	184	38 31	257	33 52	299	20 09	346
349	21 02	048	16 43	080	37 35	153	30 11	185	37 41	257	33 06	299	19 57	346
350	21 40	048	17 34	081	37 58	155	30 06	186	36 50	258	32 21	299	19 45	347
351	22 19	049	18 25	081	38 19	156	30 01	187	35 59	258	31 36	300	19 33	347
352	22 58	049	19 16	082	38 40	157	29 54	188	35 08	259	30 51	300	19 21	347
353	23 38	049	20 08	082	39 00	158	29 47	189	34 17	260	30 06	300	19 10	347
354	24 17	049	20 59	083	39 19	159	29 39	190	33 26	260	29 21	300	18 58	347
355	24 56	050	21 51	083	39 37	160	29 30	191	32 35	261	28 36	301	18 47	348
356	25 36	050	22 43	083	39 54	161	29 20	192	31 43	261	27 51	301	18 36	348
357	26 16	050	23 34	084	40 10	163	29 09	193	30 52	262	27 07	301	18 25	348
358	26 56	050	24 26	084	40 25	164	28 57	194	30 00	263	26 22	301	18 15	348
359	27 36	051	25 18	085	40 39	165	28 44	195	29 09	263	25 38	302	18 05	349

LAT 29°N

LHA	Hc	Zn	Hc	Zn	Hc	Zn	Hc	Zn	Hc	Zn	Hc	Zn	Hc	Zn
♈	Schedar		◆CAPELLA		ALDEBARAN		◆Diphda		FOMALHAUT		ALTAIR		◆DENEB	
0	61 35	012	27 38	050	26 04	085	41 50	166	29 29	196	28 23	264	47 58	307
1	61 45	011	28 18	051	26 56	085	42 02	167	29 14	197	27 31	265	47 16	307
2	61 54	009	28 59	051	27 49	086	42 13	169	28 59	198	26 39	265	46 33	307
3	62 02	008	29 40	051	28 41	086	42 23	170	28 42	198	25 47	266	45 51	307
4	62 09	007	30 20	051	29 34	087	42 31	171	28 25	199	24 54	266	45 09	306
5	62 15	006	31 01	051	30 26	087	42 39	172	28 08	200	24 02	267	44 27	306
6	62 20	005	31 42	051	31 18	088	42 45	174	27 49	201	23 09	267	43 45	306
7	62 24	004	32 23	052	32 11	088	42 50	175	27 30	202	22 17	268	43 02	306
8	62 27	002	33 04	052	33 03	088	42 54	176	27 09	203	21 25	268	42 20	306
9	62 29	001	33 46	052	33 56	089	42 57	178	26 48	204	20 32	269	41 38	306
10	62 29	000	34 27	052	34 48	089	42 59	179	26 27	205	19 40	269	40 56	306
11	62 29	359	35 08	052	35 41	090	42 59	180	26 04	206	18 47	270	40 14	307
12	62 27	358	35 50	052	36 33	090	42 58	182	25 41	207	17 55	270	39 31	307
13	62 25	356	36 31	052	37 26	091	42 56	183	25 17	207	17 02	271	38 49	307
14	62 21	355	37 13	052	38 18	091	42 53	184	24 53	208	16 10	271	38 07	307
	◆CAPELLA		ALDEBARAN		RIGEL		◆Diphda		FOMALHAUT		Enif		◆DENEB	
15	37 54	052	39 11	092	18 25	111	42 49	185	24 28	209	40 24	258	37 25	307
16	38 36	052	40 03	092	19 14	111	42 43	187	24 02	210	39 33	258	36 43	307
17	39 17	052	40 55	093	20 03	112	42 37	188	23 35	211	38 41	259	36 01	307
18	39 59	053	41 48	093	20 52	113	42 29	189	23 08	212	37 50	259	35 19	307
19	40 41	053	42 40	094	21 40	113	42 20	191	22 40	212	36 58	260	34 37	307
20	41 22	053	43 32	095	22 28	114	42 10	192	22 12	213	36 06	261	33 56	307
21	42 04	053	44 25	095	23 16	115	41 58	193	21 43	214	35 15	261	33 14	307
22	42 46	053	45 17	096	24 03	115	41 46	194	21 13	215	34 23	262	32 32	308
23	43 27	053	46 09	096	24 51	116	41 32	196	20 43	215	33 31	262	31 51	308
24	44 09	053	47 01	097	25 38	117	41 18	197	20 12	216	32 39	263	31 09	308
25	44 51	053	47 53	097	26 25	117	41 02	198	19 41	217	31 46	263	30 28	308
26	45 32	052	48 45	098	27 11	118	40 45	199	19 09	218	30 54	264	29 46	308
27	46 14	052	49 37	099	27 57	119	40 28	200	18 37	218	30 02	265	29 05	308
28	46 55	052	50 29	099	28 43	119	40 09	202	18 04	219	29 10	265	28 24	309
29	47 37	052	51 21	100	29 29	120	39 49	203	17 31	220	28 18	266	27 43	309
	CAPELLA		◆ALDEBARAN		RIGEL		◆Diphda		Alpheratz		DENEB		◆Kochab	
30	48 18	052	52 13	101	30 14	121	39 28	204	65 36	277	27 02	309	13 31	356
31	49 00	052	53 04	101	30 59	122	39 07	205	64 44	277	26 22	309	13 28	357
32	49 41	052	53 55	102	31 44	122	38 44	206	63 52	278	25 41	309	13 25	357
33	50 22	052	54 47	103	32 28	123	38 21	207	63 00	278	25 01	310	13 23	357
34	51 03	052	55 38	104	33 11	124	37 56	208	62 08	278	24 20	310	13 20	358
35	51 44	051	56 29	104	33 55	125	37 31	209	61 16	278	23 40	310	13 18	358
36	52 25	051	57 19	105	34 38	126	37 05	210	60 24	279	23 00	310	13 16	358
37	53 06	051	58 10	106	35 20	126	36 38	212	59 32	279	22 20	311	13 15	358
38	53 47	051	59 00	107	36 02	127	36 10	213	58 41	279	21 40	311	13 13	359
39	54 27	050	59 50	108	36 44	128	35 41	214	57 49	279	21 01	311	13 12	359
40	55 07	050	60 40	109	37 25	129	35 12	215	56 57	280	20 21	311	13 11	359
41	55 48	050	61 30	110	38 05	130	34 42	216	56 05	280	19 42	312	13 11	000
42	56 27	049	62 19	111	38 45	131	34 11	217	55 14	280	19 03	312	13 11	000
43	57 07	049	63 08	112	39 25	132	33 39	217	54 22	280	18 24	312	13 11	000
44	57 47	048	63 56	113	40 04	133	33 07	218	53 30	281	17 45	312	13 11	000
	◆CAPELLA		PROCYON		SIRIUS		◆RIGEL		Acamar		Diphda		◆Alpheratz	
45	58 26	048	20 13	095	19 02	123	40 42	134	20 40	180	32 34	219	52 39	281
46	59 05	048	21 05	096	19 46	123	41 19	135	20 40	181	32 00	220	51 47	281
47	59 43	047	21 57	096	20 30	124	41 56	136	20 38	182	31 26	221	50 56	281
48	60 21	046	22 50	097	21 14	125	42 32	137	20 36	183	30 51	222	50 04	282
49	60 59	046	23 42	097	21 57	125	43 08	138	20 33	184	30 16	223	49 13	282
50	61 37	045	24 34	098	22 39	126	43 43	139	20 29	184	29 40	224	48 22	282
51	62 14	045	25 26	099	23 21	127	44 17	140	20 25	185	29 03	225	47 31	283
52	62 50	044	26 17	099	24 03	127	44 50	141	20 20	186	28 26	225	46 39	283
53	63 26	043	27 09	100	24 45	128	45 22	143	20 14	187	27 49	226	45 48	283
54	64 02	042	28 01	100	25 26	129	45 54	144	20 07	188	27 11	227	44 57	283
55	64 37	041	28 52	101	26 07	130	46 24	145	20 00	188	26 32	228	44 06	284
56	65 11	040	29 44	101	26 47	130	46 54	146	19 52	189	25 53	229	43 15	284
57	65 45	039	30 35	102	27 26	131	47 23	147	19 43	190	25 13	229	42 24	284
58	66 18	038	31 27	103	28 06	132	47 50	149	19 33	191	24 33	230	41 33	285
59	66 50	037	32 18	103	28 45	133	48 17	150	19 23	192	23 53	231	40 43	285
	◆Dubhe		PROCYON		SIRIUS		◆RIGEL		Diphda		◆Hamal		Mirfak	
60	18 19	029	33 09	104	29 23	134	48 43	151	23 12	232	64 06	264	68 04	344
61	18 44	029	34 00	104	30 01	134	49 07	153	22 31	232	63 14	265	67 49	343
62	19 09	029	34 51	105	30 38	135	49 30	154	21 49	233	62 21	266	67 33	341
63	19 35	029	35 41	106	31 14	136	49 53	156	21 07	234	61 29	266	67 15	340
64	20 01	030	36 32	106	31 51	137	50 14	157	20 25	234	60 37	267	66 56	338
65	20 27	030	37 22	107	32 26	138	50 34	159	19 42	235	59 44	267	66 36	337
66	20 53	030	38 12	108	33 01	139	50 52	160	18 59	236	58 52	268	66 15	336
67	21 19	030	39 02	108	33 35	140	51 09	162	18 15	236	57 59	268	65 53	334
68	21 45	030	39 52	109	34 09	141	51 25	163	17 31	237	57 07	269	65 29	333
69	22 12	030	40 41	110	34 42	142	51 40	165	16 47	238	56 14	269	65 05	332
70	22 39	031	41 30	111	35 14	143	51 53	166	16 03	238	55 22	270	64 40	331
71	23 05	031	42 19	111	35 46	144	52 05	168	15 18	239	54 29	270	64 14	330
72	23 32	031	43 08	112	36 16	145	52 15	169	14 33	239	53 37	271	63 46	328
73	23 59	031	43 57	113	36 46	146	52 24	171	13 48	240	52 44	271	63 19	327
74	24 27	031	44 45	114	37 16	147	52 32	173	13 02	241	51 52	272	62 50	326
	◆Dubhe		POLLUX		PROCYON		◆SIRIUS		RIGEL		◆Hamal		Mirfak	
75	24 54	031	53 56	081	45 33	115	37 44	148	52 38	174	51 00	272	62 21	326
76	25 21	032	54 47	082	46 20	115	38 12	149	52 43	176	50 07	273	61 51	325
77	25 49	032	55 39	082	47 07	116	38 39	150	52 46	177	49 15	273	61 20	324
78	26 16	032	56 31	082	47 54	117	39 05	151	52 47	179	48 22	273	60 49	323
79	26 44	032	57 23	082	48 41	118	39 30	152	52 47	181	47 30	274	60 17	322
80	27 12	032	58 15	083	49 27	119	39 54	153	52 46	182	46 38	274	59 45	322
81	27 40	032	59 08	083	50 12	120	40 17	154	52 43	184	45 45	275	59 12	321
82	28 08	032	60 00	083	50 58	121	40 40	155	52 39	186	44 53	275	58 39	320
83	28 36	032	60 52	084	51 42	122	41 01	157	52 33	187	44 01	276	58 05	320
84	29 04	032	61 44	084	52 26	123	41 21	158	52 26	189	43 09	276	57 31	319
85	29 32	032	62 36	084	53 10	124	41 41	159	52 17	190	42 16	276	56 56	319
86	30 00	033	63 28	085	53 53	126	41 59	160	52 07	192	41 24	277	56 21	318
87	30 28	033	64 21	085	54 35	127	42 16	161	51 55	194	40 32	277	55 46	318
88	30 57	033	65 13	085	55 17	128	42 32	163	51 42	195	39 40	278	55 11	317
89	31 25	033	66 05	086	55 58	129	42 47	164	51 28	197	38 48	278	54 35	317

LHA	Hc	Zn	Hc	Zn	Hc	Zn	Hc	Zn	Hc	Zn	Hc	Zn	Hc	Zn
♈	Dubhe		◆REGULUS		PROCYON		SIRIUS		◆RIGEL		ALDEBARAN		◆Mirfak	
90	31 53	033	30 08	093	56 38	131	43 01	165	51 12	198	66 57	242	53 59	316
91	32 22	033	31 00	093	57 17	132	43 14	166	50 55	200	66 10	243	53 22	316
92	32 50	033	31 53	094	57 56	133	43 26	168	50 36	201	65 23	245	52 46	316
93	33 18	033	32 45	094	58 33	135	43 36	169	50 17	203	64 35	246	52 09	315
94	33 47	033	33 38	095	59 10	136	43 46	170	49 56	204	63 47	247	51 32	315
95	34 15	033	34 30	095	59 46	138	43 54	172	49 34	206	62 59	248	50 55	315
96	34 43	033	35 22	096	60 20	140	44 01	173	49 11	207	62 10	249	50 17	314
97	35 12	033	36 14	096	60 54	141	44 07	174	48 46	208	61 21	250	49 40	314
98	35 40	033	37 06	097	61 26	143	44 11	176	48 21	210	60 31	251	49 02	314
99	36 08	033	37 58	098	61 57	145	44 14	177	47 54	211	59 41	252	48 24	314
100	36 37	033	38 50	098	62 26	147	44 17	178	47 27	212	58 51	253	47 46	314
101	37 05	032	39 42	099	62 54	149	44 17	180	46 58	214	58 01	254	47 08	313
102	37 33	032	40 34	099	63 21	151	44 17	181	46 29	215	57 10	255	46 30	313
103	38 01	032	41 26	100	63 46	153	44 16	182	45 58	216	56 19	256	45 52	313
104	38 29	032	42 17	101	64 09	155	44 13	184	45 27	217	55 28	257	45 14	313
	◆Dubhe		REGULUS		◆Alphard		SIRIUS		RIGEL		◆ALDEBARAN		CAPELLA	
105	38 57	032	43 09	101	38 16	131	44 09	185	44 55	218	54 37	257	63 32	317
106	39 25	032	44 00	102	38 55	132	44 04	186	44 22	220	53 46	258	62 56	316
107	39 52	032	44 52	103	39 34	133	43 57	188	43 48	221	52 55	259	62 20	316
108	40 20	032	45 43	103	40 12	134	43 50	189	43 13	222	52 03	259	61 43	315
109	40 48	032	46 34	104	40 49	135	43 41	190	42 38	223	51 11	260	61 05	314
110	41 15	031	47 25	105	41 26	136	43 31	192	42 02	224	50 20	261	60 27	314
111	41 42	031	48 15	105	42 02	137	43 20	193	41 25	225	49 28	261	59 49	313
112	42 09	031	49 06	106	42 38	138	43 07	194	40 47	226	48 36	262	59 11	313
113	42 36	031	49 56	107	43 13	139	42 54	195	40 09	227	47 44	263	58 32	312
114	43 03	031	50 46	108	43 47	140	42 39	197	39 31	228	46 52	263	57 53	312
115	43 30	030	51 36	108	44 20	141	42 24	198	38 51	229	46 00	264	57 13	311
116	43 56	030	52 26	109	44 52	143	42 07	199	38 11	230	45 07	264	56 34	311
117	44 22	030	53 15	110	45 24	144	41 49	200	37 31	231	44 15	265	55 54	310
118	44 48	030	54 04	111	45 54	145	41 30	202	36 50	232	43 23	266	55 14	310
119	45 14	029	54 53	112	46 24	146	41 10	203	36 08	233	42 31	266	54 34	310
	◆Dubhe		REGULUS		◆Alphard		SIRIUS		RIGEL		◆ALDEBARAN		CAPELLA	
120	45 40	029	55 42	113	46 52	147	40 50	204	35 27	233	41 38	267	53 53	309
121	46 05	029	56 30	114	47 20	149	40 28	205	34 44	234	40 46	267	53 13	309
122	46 30	028	57 18	115	47 47	150	40 05	206	34 01	235	39 53	268	52 32	309
123	46 55	028	58 05	116	48 13	151	39 41	207	33 18	236	39 01	268	51 51	309
124	47 20	028	58 52	117	48 37	153	39 16	209	32 34	237	38 08	269	51 10	309
125	47 44	027	59 38	118	49 01	154	38 51	210	31 50	238	37 16	269	50 29	308
126	48 08	027	60 24	120	49 23	156	38 24	211	31 06	238	36 23	270	49 48	308
127	48 32	027	61 09	121	49 44	157	37 57	212	30 21	239	35 31	270	49 06	308
128	48 55	026	61 54	122	50 04	158	37 29	213	29 36	240	34 39	271	48 25	308
129	49 18	026	62 38	124	50 23	160	37 00	214	28 50	241	33 46	271	47 44	308
130	49 41	025	63 21	125	50 40	161	36 30	215	28 04	241	32 54	272	47 02	308
131	50 03	025	64 04	127	50 56	163	36 00	216	27 18	242	32 01	272	46 20	308
132	50 25	024	64 46	128	51 11	164	35 29	217	26 32	243	31 09	273	45 39	308
133	50 46	024	65 26	130	51 25	166	34 57	218	25 45	243	30 16	273	44 57	307
134	51 07	023	66 06	132	51 37	167	34 24	219	24 58	244	29 24	273	44 16	307
	Dubhe		◆ARCTURUS		SPICA		Alphard		◆SIRIUS		BETELGEUSE		◆CAPELLA	
135	51 28	023	18 37	078	14 37	112	51 47	169	33 51	220	41 27	253	43 34	307
136	51 48	022	19 28	078	15 25	112	51 57	171	33 17	221	40 37	254	42 52	307
137	52 08	022	20 19	079	16 14	113	52 05	172	32 42	222	39 46	254	42 11	307
138	52 27	021	21 11	079	17 02	114	52 11	174	32 07	223	38 56	255	41 29	307
139	52 46	021	22 02	080	17 50	114	52 16	175	31 31	223	38 05	256	40 47	307
140	53 04	020	22 54	080	18 37	115	52 19	177	30 55	224	37 14	256	40 06	307
141	53 22	019	23 46	080	19 25	115	52 21	179	30 18	225	36 23	257	39 24	308
142	53 39	019	24 37	081	20 12	116	52 22	180	29 40	226	35 32	258	38 42	308
143	53 56	018	25 29	081	20 59	117	52 21	182	29 02	227	34 41	258	38 01	308
144	54 12	018	26 21	082	21 46	117	52 18	184	28 24	228	33 49	259	37 19	308
145	54 27	017	27 13	082	22 33	118	52 14	185	27 45	228	32 58	259	36 38	308
146	54 42	016	28 05	082	23 19	119	52 09	187	27 05	229	32 06	260	35 56	308
147	54 57	015	28 57	083	24 05	119	52 02	188	26 25	230	31 14	261	35 15	308
148	55 10	015	29 49	083	24 50	120	51 54	190	25 45	231	30 22	261	34 34	308
149	55 23	014	30 41	084	25 36	121	51 44	191	25 04	232	29 30	262	33 52	308
	Dubhe		◆ARCTURUS		SPICA		◆Alphard		SIRIUS		POLLUX		◆CAPELLA	
150	55 36	013	31 34	084	26 20	121	51 33	193	24 23	232	60 26	276	33 11	308
151	55 47	012	32 26	085	27 05	122	51 20	195	23 41	233	59 34	277	32 30	308
152	55 58	012	33 18	085	27 49	123	51 06	196	22 59	234	58 42	277	31 49	309
153	56 08	011	34 10	086	28 33	124	50 51	198	22 16	234	57 50	277	31 08	309
154	56 18	010	35 03	086	29 17	124	50 35	199	21 34	235	56 58	278	30 27	309
155	56 27	009	35 55	086	30 00	125	50 17	201	20 50	236	56 06	278	29 46	309
156	56 35	008	36 48	087	30 43	126	49 58	202	20 07	236	55 14	278	29 06	309
157	56 42	008	37 40	087	31 25	127	49 37	204	19 23	237	54 22	279	28 25	309
158	56 49	007	38 32	088	32 07	127	49 16	205	18 39	238	53 30	279	27 44	310
159	56 55	006	39 25	088	32 48	128	48 53	206	17 54	238	52 38	279	27 04	310
160	57 00	005	40 17	089	33 29	129	48 29	208	17 09	239	51 46	279	26 24	310
161	57 04	004	41 10	089	34 10	130	48 04	209	16 24	240	50 55	280	25 44	310
162	57 07	003	42 02	090	34 50	131	47 38	210	15 39	240	50 03	280	25 04	310
163	57 10	002	42 55	090	35 29	132	47 11	212	14 53	241	49 11	280	24 24	311
164	57 12	002	43 47	091	36 08	133	46 43	213	14 07	241	48 20	281	23 44	311
	Dubhe		◆Alkaid		ARCTURUS		◆SPICA		Alphard		◆PROCYON		POLLUX	
165	57 13	001	52 24	045	44 40	091	36 46	134	46 14	214	36 58	253	47 28	281
166	57 13	000	53 02	045	45 32	092	37 24	135	45 44	215	36 07	254	46 37	281
167	57 13	359	53 39	045	46 25	092	38 01	136	45 13	217	35 17	255	45 45	282
168	57 11	358	54 16	044	47 17	093	38 37	137	44 42	218	34 26	255	44 54	282
169	57 09	357	54 52	044	48 09	093	39 13	138	44 09	219	33 35	256	44 03	282
170	57 06	356	55 29	044	49 02	094	39 48	139	43 35	220	32 44	257	43 11	283
171	57 03	356	56 05	043	49 54	094	40 23	140	43 01	221	31 53	257	42 20	283
172	56 58	355	56 40	043	50 46	095	40 56	141	42 26	222	31 02	258	41 29	283
173	56 53	354	57 16	042	51 39	095	41 29	142	41 51	223	30 11	258	40 38	283
174	56 47	353	57 51	042	52 31	096	42 01	143	41 14	224	29 19	259	39 47	284
175	56 40	352	58 25	041	53 23	097	42 33	144	40 37	225	28 28	260	38 56	284
176	56 33	351	59 00	040	54 15	097	43 03	145	39 59	226	27 36	260	38 05	284
177	56 24	350	59 33	040	55 07	098	43 33	146	39 21	227	26 44	261	37 14	285
178	56 15	350	60 07	039	55 59	099	44 01	147	38 42	228	25 53	261	36 24	285
179	56 06	349	60 40	038	56 51	099	44 29	149	38 03	229	25 01	262	35 33	285

LHA ♈	Hc	Zn	Hc	Zn	Hc	Zn	Hc	Zn	Hc	Zn	Hc	Zn	Hc	Zn
	◆VEGA		Alphecca		◆SPICA		Glenah		REGULUS		◆POLLUX		Dubhe	
180	11 14	052	42 57	079	44 56	150	43 20	175	58 54	243	34 42	286	55 55	348
181	11 55	052	43 49	080	45 22	151	43 24	176	58 07	244	33 52	286	55 44	347
182	12 36	052	44 40	080	45 47	152	43 27	178	57 20	245	33 01	286	55 32	347
183	13 18	053	45 32	080	46 10	154	43 29	179	56 32	246	32 11	287	55 20	346
184	14 00	053	46 24	081	46 33	155	43 29	180	55 44	247	31 21	287	55 06	345
185	14 42	053	47 16	081	46 55	156	43 28	181	54 56	248	30 31	287	54 53	344
186	15 24	054	48 08	081	47 15	158	43 26	183	54 07	249	29 41	288	54 38	344
187	16 07	054	48 59	082	47 35	159	43 23	184	53 18	250	28 51	288	54 23	343
188	16 50	055	49 51	082	47 53	160	43 19	185	52 29	251	28 01	288	54 07	342
189	17 32	055	50 43	082	48 10	162	43 13	187	51 39	251	27 11	289	53 51	342
190	18 15	055	51 35	083	48 25	163	43 07	188	50 49	252	26 21	289	53 34	341
191	18 58	055	52 27	083	48 40	165	42 59	189	49 59	253	25 32	289	53 17	340
192	19 42	056	53 20	083	48 53	166	42 50	191	49 09	254	24 42	290	52 59	340
193	20 25	056	54 12	084	49 05	168	42 40	192	48 18	255	23 53	290	52 41	339
194	21 09	056	55 04	084	49 16	169	42 28	193	47 27	255	23 04	290	52 22	339
	◆Kochab		VEGA		Rasalhague		◆ANTARES		SPICA		◆REGULUS		Dubhe	
195	42 40	010	21 53	057	24 35	089	15 17	133	49 25	171	46 37	256	52 02	338
196	42 48	010	22 37	057	25 27	089	15 55	133	49 33	172	45 46	257	51 42	337
197	42 57	009	23 21	057	26 19	090	16 33	134	49 39	174	44 54	257	51 22	337
198	43 05	009	24 05	058	27 12	090	17 11	135	49 45	175	44 03	258	51 01	336
199	43 13	009	24 49	058	28 04	091	17 48	135	49 48	177	43 12	259	50 40	336
200	43 21	008	25 34	058	28 57	091	18 24	136	49 51	178	42 20	259	50 18	335
201	43 29	008	26 18	058	29 49	092	19 00	137	49 52	180	41 29	260	49 56	335
202	43 36	008	27 03	059	30 42	092	19 36	138	49 51	181	40 37	261	49 34	335
203	43 43	007	27 48	059	31 34	093	20 11	138	49 50	183	39 45	261	49 11	334
204	43 49	007	28 33	059	32 27	093	20 46	139	49 47	184	38 53	262	48 48	334
205	43 55	007	29 18	059	33 19	094	21 20	140	49 42	186	38 01	262	48 25	333
206	44 01	006	30 03	060	34 11	094	21 54	140	49 36	187	37 09	263	48 01	333
207	44 07	006	30 48	060	35 04	095	22 27	141	49 29	189	36 17	264	47 37	333
208	44 12	006	31 34	060	35 56	095	23 00	142	49 20	190	35 25	264	47 13	332
209	44 17	005	32 19	060	36 48	096	23 32	143	49 10	192	34 33	265	46 48	332
	Kochab		◆VEGA		Rasalhague		ANTARES		◆SPICA		REGULUS		◆Dubhe	
210	44 21	005	33 05	060	37 40	097	24 03	144	48 59	193	33 40	265	46 23	331
211	44 26	004	33 51	061	38 32	097	24 34	144	48 46	195	32 48	266	45 58	331
212	44 29	004	34 36	061	39 24	098	25 04	145	48 32	196	31 56	266	45 32	331
213	44 33	004	35 22	061	40 16	098	25 34	146	48 17	198	31 03	267	45 07	331
214	44 36	003	36 08	061	41 08	099	26 03	147	48 01	199	30 11	267	44 41	330
215	44 39	003	36 54	061	42 00	099	26 31	148	47 43	200	29 18	268	44 15	330
216	44 42	003	37 40	062	42 52	100	26 59	149	47 24	202	28 26	268	43 48	330
217	44 44	002	38 27	062	43 43	101	27 26	149	47 04	203	27 34	269	43 22	330
218	44 46	002	39 13	062	44 35	101	27 52	150	46 43	204	26 41	269	42 55	329
219	44 47	001	39 59	062	45 26	102	28 18	151	46 21	206	25 49	270	42 28	329
220	44 48	001	40 45	062	46 18	103	28 43	152	45 58	207	24 56	270	42 01	329
221	44 49	001	41 32	062	47 09	103	29 07	153	45 33	208	24 04	271	41 34	329
222	44 49	000	42 18	062	48 00	104	29 31	154	45 08	210	23 11	271	41 07	329
223	44 49	000	43 05	062	48 51	105	29 53	155	44 42	211	22 19	272	40 40	328
224	44 49	359	43 51	063	49 41	106	30 15	156	44 14	212	21 26	272	40 12	328
	VEGA		◆ALTAIR		Nunki		ANTARES		◆SPICA		Denebola		◆Alkaid	
225	44 38	063	19 25	091	11 04	129	30 36	157	43 46	213	43 41	262	65 24	331
226	45 25	063	20 18	091	11 45	129	30 56	158	43 17	214	42 49	263	64 58	330
227	46 11	063	21 10	091	12 26	130	31 16	159	42 47	216	41 57	264	64 31	329
228	46 58	063	22 03	092	13 06	130	31 35	160	42 16	217	41 05	264	64 03	327
229	47 45	063	22 55	092	13 45	131	31 52	161	41 44	218	40 13	265	63 35	326
230	48 31	063	23 48	093	14 25	132	32 09	162	41 12	219	39 20	265	63 05	326
231	49 18	063	24 40	093	15 04	132	32 25	163	40 38	220	38 28	266	62 35	325
232	50 05	063	25 32	094	15 42	133	32 40	164	40 04	221	37 36	266	62 05	324
233	50 52	063	26 25	095	16 21	134	32 55	165	39 29	222	36 43	267	61 33	323
234	51 39	063	27 17	095	16 58	134	33 08	166	38 54	223	35 51	267	61 01	322
235	52 25	063	28 09	096	17 36	135	33 21	167	38 18	224	34 59	268	60 29	321
236	53 12	063	29 02	096	18 12	136	33 32	168	37 41	225	34 06	268	59 56	321
237	53 59	063	29 54	097	18 49	136	33 43	169	37 04	226	33 14	269	59 22	320
238	54 45	063	30 46	097	19 25	137	33 52	170	36 26	227	32 21	269	58 48	319
239	55 32	063	31 38	098	20 00	138	34 01	171	35 47	228	31 29	270	58 14	319
	◆VEGA		ALTAIR		Nunki		◆ANTARES		SPICA		ARCTURUS		◆Alkaid	
240	56 19	063	32 30	098	20 35	139	34 08	172	35 08	229	64 17	254	57 39	318
241	57 05	063	33 22	099	21 10	139	34 15	173	34 28	230	63 27	254	57 04	318
242	57 52	062	34 13	099	21 44	140	34 21	174	33 48	230	62 36	255	56 29	317
243	58 38	062	35 05	100	22 17	141	34 26	175	33 07	231	61 45	256	55 53	317
244	59 25	062	35 57	101	22 50	142	34 29	176	32 26	232	60 54	257	55 17	316
245	60 11	062	36 48	101	23 22	142	34 32	178	31 44	233	60 03	258	54 40	316
246	60 57	062	37 40	102	23 54	143	34 34	179	31 02	234	59 11	259	54 04	315
247	61 43	061	38 31	102	24 25	144	34 35	180	30 20	235	58 20	260	53 27	315
248	62 29	061	39 22	103	24 56	145	34 34	181	29 37	235	57 28	260	52 50	315
249	63 15	061	40 13	104	25 26	146	34 33	182	28 54	236	56 36	261	52 12	314
250	64 01	060	41 04	104	25 55	146	34 31	183	28 10	237	55 44	262	51 35	314
251	64 46	060	41 55	105	26 24	147	34 28	184	27 26	238	54 52	262	50 57	314
252	65 31	059	42 45	106	26 52	148	34 24	185	26 41	238	54 00	263	50 19	314
253	66 16	059	43 36	107	27 20	149	34 18	186	25 56	239	53 08	264	49 41	313
254	67 01	058	44 26	107	27 46	150	34 12	187	25 11	240	52 16	264	49 03	313
	◆DENEB		ALTAIR		Nunki		◆ANTARES		SPICA		ARCTURUS		◆Alkaid	
255	44 00	054	45 16	108	28 12	151	34 05	188	24 26	240	51 24	265	48 24	313
256	44 42	054	46 06	109	28 38	152	33 57	189	23 40	241	50 32	265	47 46	313
257	45 25	054	46 55	110	29 02	153	33 48	190	22 54	242	49 39	266	47 07	313
258	46 07	054	47 45	110	29 26	153	33 38	192	22 08	242	48 47	266	46 29	313
259	46 49	053	48 34	111	29 49	154	33 27	193	21 21	243	47 55	267	45 50	312
260	47 31	053	49 23	112	30 11	155	33 15	194	20 34	244	47 02	268	45 11	312
261	48 13	053	50 11	113	30 33	156	33 02	195	19 47	244	46 10	268	44 32	312
262	48 55	053	50 59	114	30 54	157	32 48	196	19 00	245	45 17	269	43 54	312
263	49 37	053	51 47	115	31 14	158	32 34	197	18 12	246	44 25	269	43 15	312
264	50 19	053	52 34	116	31 33	159	32 18	198	17 24	246	43 32	270	42 36	312
265	51 01	053	53 22	117	31 51	160	32 02	199	16 36	247	42 40	270	41 57	312
266	51 42	053	54 08	118	32 08	161	31 44	200	15 48	247	41 47	271	41 17	312
267	52 24	052	54 54	119	32 25	162	31 26	201	14 59	248	40 55	271	40 38	312
268	53 05	052	55 40	120	32 40	163	31 07	202	14 10	248	40 02	271	39 59	312
269	53 47	052	56 25	121	32 55	164	30 47	203	13 22	249	39 10	272	39 20	312

LAT 29°N

LHA ♈	Hc	Zn	Hc	Zn	Hc	Zn	Hc	Zn	Hc	Zn	Hc	Zn	Hc	Zn
	◆DENEB		ALTAIR		◆Nunki		ANTARES		◆ARCTURUS		Alkaid		Kochab	
270	54 28	052	57 10	122	33 09	165	30 27	204	38 17	272	38 41	312	38 55	345
271	55 09	051	57 54	124	33 22	166	30 05	205	37 25	273	38 02	312	38 41	345
272	55 50	051	58 37	125	33 34	167	29 43	206	36 33	273	37 23	312	38 27	345
273	56 31	051	59 20	126	33 45	168	29 20	206	35 40	274	36 44	312	38 13	345
274	57 11	050	60 02	128	33 55	169	28 56	207	34 48	274	36 05	312	37 59	344
275	57 51	050	60 43	129	34 04	171	28 32	208	33 56	275	35 26	312	37 45	344
276	58 31	049	61 24	130	34 12	172	28 07	209	33 03	275	34 47	312	37 31	344
277	59 11	049	62 03	132	34 19	173	27 41	210	32 11	275	34 08	312	37 16	344
278	59 51	049	62 42	134	34 25	174	27 14	211	31 19	276	33 29	312	37 01	344
279	60 30	048	63 19	135	34 31	175	26 47	212	30 27	276	32 50	312	36 47	344
280	61 08	047	63 55	137	34 35	176	26 19	213	29 35	277	32 12	313	36 32	343
281	61 47	047	64 31	139	34 38	177	25 50	214	28 42	277	31 33	313	36 17	343
282	62 25	046	65 04	141	34 40	178	25 21	214	27 50	278	30 55	313	36 01	343
283	63 03	045	65 37	143	34 42	179	24 51	215	26 58	278	30 16	313	35 46	343
284	63 40	045	66 08	145	34 42	180	24 20	216	26 06	278	29 38	313	35 31	343
	◆Alpheratz		Enif		ALTAIR		◆ANTARES		ARCTURUS		◆Alkaid		Kochab	
285	24 01	069	47 11	108	66 37	147	23 49	217	25 15	279	28 59	313	35 15	343
286	24 50	069	48 00	109	67 05	149	23 17	218	24 23	279	28 21	313	35 00	343
287	25 39	070	48 50	110	67 31	152	22 45	218	23 31	280	27 43	314	34 44	343
288	26 29	070	49 39	111	67 55	154	22 12	219	22 39	280	27 05	314	34 28	343
289	27 18	070	50 28	111	68 17	156	21 39	220	21 48	281	26 27	314	34 13	342
290	28 07	071	51 17	112	68 37	159	21 05	221	20 56	281	25 49	314	33 57	342
291	28 57	071	52 05	113	68 55	161	20 30	221	20 05	281	25 12	314	33 41	342
292	29 46	071	52 53	114	69 10	164	19 55	222	19 13	282	24 34	314	33 25	342
293	30 36	071	53 41	115	69 23	167	19 20	223	18 22	282	23 57	315	33 09	342
294	31 26	072	54 28	116	69 34	170	18 44	224	17 31	283	23 20	315	32 53	342
295	32 16	072	55 15	117	69 42	172	18 08	224	16 40	283	22 42	315	32 36	342
296	33 06	072	56 02	118	69 48	175	17 31	225	15 49	284	22 05	315	32 20	342
297	33 56	073	56 48	119	69 51	178	16 54	226	14 58	284	21 29	316	32 04	342
298	34 46	073	57 33	120	69 51	181	16 16	226	14 07	284	20 52	316	31 48	342
299	35 36	073	58 18	122	69 49	184	15 38	227	13 16	285	20 15	316	31 31	342
	◆Schedar		Alpheratz		◆FOMALHAUT		ALTAIR		Rasalhague		◆Alphecca		Kochab	
300	34 41	039	36 27	074	17 41	140	69 44	187	52 29	252	32 05	285	31 15	342
301	35 14	039	37 17	074	18 14	141	69 37	190	51 39	253	31 14	285	30 59	342
302	35 47	039	38 08	074	18 47	142	69 27	192	50 49	253	30 23	286	30 42	342
303	36 20	039	38 58	075	19 19	143	69 14	195	49 59	254	29 33	286	30 26	342
304	36 53	039	39 49	075	19 51	143	68 59	198	49 08	255	28 43	286	30 10	342
305	37 26	039	40 39	075	20 22	144	68 42	200	48 18	256	27 52	287	29 53	342
306	37 59	039	41 30	075	20 52	145	68 23	203	47 27	256	27 02	287	29 37	342
307	38 32	039	42 21	076	21 22	146	68 01	205	46 36	257	26 12	287	29 21	342
308	39 05	039	43 12	076	21 52	146	67 38	208	45 44	258	25 22	288	29 04	342
309	39 38	039	44 03	076	22 21	147	67 12	210	44 53	258	24 32	288	28 48	342
310	40 11	039	44 54	077	22 49	148	66 45	212	44 02	259	23 42	289	28 32	342
311	40 44	039	45 45	077	23 17	149	66 16	215	43 10	260	22 53	289	28 15	342
312	41 16	039	46 36	077	23 44	149	65 45	217	42 18	260	22 03	289	27 59	342
313	41 49	038	47 27	077	24 10	150	65 13	219	41 27	261	21 13	290	27 43	342
314	42 22	038	48 18	078	24 36	151	64 40	221	40 35	262	20 24	290	27 27	342
	◆Mirfak		Hamal		Diphda		◆FOMALHAUT		ALTAIR		◆VEGA		Kochab	
315	18 10	042	22 09	075	18 30	124	25 01	152	64 05	222	58 56	298	27 11	342
316	18 45	043	23 00	075	19 13	125	25 25	153	63 29	224	58 10	298	26 55	342
317	19 21	043	23 51	075	19 56	125	25 48	154	62 52	226	57 23	298	26 39	342
318	19 57	043	24 42	076	20 39	126	26 11	155	62 14	228	56 37	297	26 23	342
319	20 33	043	25 33	076	21 21	127	26 34	155	61 34	229	55 50	297	26 07	342
320	21 09	044	26 24	077	22 03	127	26 55	156	60 54	231	55 03	297	25 51	343
321	21 45	044	27 15	077	22 44	128	27 16	157	60 13	232	54 17	297	25 35	343
322	22 22	044	28 06	077	23 26	129	27 36	158	59 32	233	53 30	297	25 20	343
323	22 58	044	28 57	078	24 06	129	27 55	159	58 49	235	52 43	297	25 04	343
324	23 35	045	29 48	078	24 47	130	28 13	160	58 06	236	51 57	297	24 49	343
325	24 12	045	30 40	078	25 26	131	28 31	161	57 22	237	51 10	297	24 33	343
326	24 49	045	31 31	079	26 06	132	28 48	162	56 37	239	50 23	297	24 18	343
327	25 26	045	32 23	079	26 45	132	29 04	163	55 52	240	49 36	297	24 03	343
328	26 04	045	33 14	080	27 23	133	29 19	164	55 07	241	48 49	297	23 48	343
329	26 41	046	34 06	080	28 01	134	29 33	165	54 21	242	48 03	297	23 33	343
	◆Mirfak		Hamal		Diphda		◆FOMALHAUT		ALTAIR		◆VEGA		Kochab	
330	27 19	046	34 58	080	28 39	135	29 46	166	53 34	243	47 16	297	23 18	344
331	27 56	046	35 49	081	29 16	136	29 59	167	52 47	244	46 29	297	23 03	344
332	28 34	046	36 41	081	29 52	137	30 11	168	52 00	245	45 43	297	22 48	344
333	29 12	046	37 33	082	30 28	137	30 22	169	51 12	246	44 56	297	22 33	344
334	29 50	046	38 25	082	31 03	138	30 32	170	50 24	247	44 09	297	22 19	344
335	30 28	047	39 17	082	31 38	139	30 41	171	49 36	248	43 23	297	22 05	344
336	31 06	047	40 09	083	32 12	140	30 49	172	48 47	249	42 36	298	21 50	344
337	31 44	047	41 01	083	32 45	141	30 56	173	47 58	249	41 50	298	21 36	344
338	32 23	047	41 53	083	33 18	142	31 02	174	47 09	250	41 03	298	21 22	345
339	33 01	047	42 45	084	33 50	143	31 08	175	46 19	251	40 17	298	21 09	345
340	33 40	047	43 38	084	34 21	144	31 12	176	45 30	252	39 31	298	20 55	345
341	34 18	047	44 30	085	34 52	145	31 16	177	44 40	253	38 44	298	20 41	345
342	34 57	047	45 22	085	35 22	146	31 19	178	43 49	253	37 58	298	20 28	345
343	35 35	047	46 14	085	35 51	147	31 20	179	42 59	254	37 12	299	20 15	345
344	36 14	047	47 07	086	36 19	148	31 21	180	42 09	255	36 26	299	20 02	346
	◆CAPELLA		ALDEBARAN		Diphda		◆FOMALHAUT		ALTAIR		◆VEGA		Kochab	
345	17 47	047	13 06	078	36 47	149	31 21	181	41 18	255	35 40	299	19 49	346
346	18 25	047	13 58	079	37 13	150	31 20	182	40 27	256	34 54	299	19 36	346
347	19 04	047	14 49	079	37 39	151	31 18	183	39 36	257	34 08	299	19 23	346
348	19 43	048	15 41	080	38 04	152	31 15	184	38 45	257	33 23	299	19 11	346
349	20 21	048	16 32	080	38 28	153	31 11	185	37 54	258	32 37	300	18 59	347
350	21 00	048	17 24	080	38 52	154	31 06	186	37 02	259	31 51	300	18 46	347
351	21 39	048	18 16	081	39 14	155	31 00	187	36 11	259	31 06	300	18 35	347
352	22 19	049	19 08	081	39 35	157	30 54	188	35 19	260	30 21	300	18 23	347
353	22 58	049	19 59	082	39 56	158	30 46	189	34 27	260	29 35	301	18 11	347
354	23 38	049	20 51	082	40 15	159	30 38	190	33 36	261	28 50	301	18 00	348
355	24 17	049	21 43	083	40 34	160	30 29	191	32 44	262	28 05	301	17 49	348
356	24 57	050	22 35	083	40 51	161	30 18	192	31 52	262	27 20	301	17 38	348
357	25 37	050	23 28	083	41 07	162	30 07	193	31 00	263	26 36	302	17 27	348
358	26 17	050	24 20	084	41 23	163	29 55	194	30 08	263	25 51	302	17 16	348
359	26 58	050	25 12	084	41 37	165	29 42	195	29 16	264	25 06	302	17 06	349

LHA ♈	Hc	Zn	Hc	Zn	Hc	Zn	Hc	Zn	Hc	Zn	Hc	Zn	Hc	Zn
	Schedar		**◆CAPELLA**		**ALDEBARAN**		**◆Diphda**		**FOMALHAUT**		**ALTAIR**		**◆DENEB**	
0	60 36	011	27 00	050	25 59	084	42 48	166	30 26	196	28 29	265	47 21	308
1	60 46	010	27 40	050	26 51	085	43 01	167	30 12	197	27 36	265	46 39	307
2	60 55	009	28 21	050	27 44	085	43 12	168	29 56	198	26 43	266	45 57	307
3	61 03	008	29 02	050	28 37	086	43 22	170	29 39	199	25 51	266	45 15	307
4	61 10	007	29 43	051	29 30	086	43 31	171	29 22	200	24 58	267	44 33	307
5	61 16	006	30 24	051	30 23	086	43 38	172	29 04	201	24 05	267	43 51	307
6	61 21	005	31 05	051	31 15	087	43 45	174	28 45	201	23 12	268	43 09	307
7	61 24	004	31 46	051	32 08	087	43 50	175	28 25	202	22 19	268	42 26	307
8	61 27	002	32 27	051	33 01	088	43 54	176	28 04	203	21 26	269	41 44	307
9	61 29	001	33 08	051	33 54	088	43 57	178	27 43	204	20 33	269	41 02	307
10	61 29	000	33 50	051	34 47	089	43 59	179	27 21	205	19 40	270	40 20	307
11	61 29	359	34 31	051	35 40	089	43 59	180	26 58	206	18 47	270	39 38	307
12	61 27	358	35 13	052	36 33	090	43 58	182	26 35	207	17 54	271	38 55	307
13	61 25	357	35 54	052	37 26	090	43 56	183	26 11	208	17 01	271	38 13	307
14	61 21	355	36 36	052	38 19	091	43 53	184	25 46	209	16 08	272	37 31	307
	◆CAPELLA		**ALDEBARAN**		**RIGEL**		**◆Diphda**		**FOMALHAUT**		**Enif**		**◆DENEB**	
15	37 17	052	39 12	091	18 47	111	43 49	185	25 20	209	40 36	258	36 49	307
16	37 59	052	40 05	092	19 36	111	43 43	187	24 54	210	39 45	259	36 07	307
17	38 40	052	40 58	092	20 25	112	43 36	188	24 27	211	38 52	260	35 25	308
18	39 22	052	41 51	093	21 15	112	43 28	189	23 59	212	38 00	260	34 43	308
19	40 04	052	42 44	093	22 03	113	43 19	191	23 31	213	37 08	261	34 01	308
20	40 46	052	43 37	094	22 52	114	43 08	192	23 02	213	36 16	261	33 19	308
21	41 27	052	44 30	094	23 41	114	42 57	193	22 32	214	35 23	262	32 37	308
22	42 09	052	45 22	095	24 29	115	42 44	195	22 02	215	34 31	263	31 55	308
23	42 51	052	46 15	095	25 17	115	42 30	196	21 32	216	33 38	263	31 14	308
24	43 32	052	47 08	096	26 04	116	42 15	197	21 01	216	32 46	264	30 32	308
25	44 14	052	48 01	096	26 52	117	41 59	198	20 29	217	31 53	264	29 51	308
26	44 55	052	48 53	097	27 39	117	41 42	199	19 57	218	31 00	265	29 09	309
27	45 37	052	49 46	097	28 26	118	41 24	201	19 24	219	30 07	265	28 28	309
28	46 18	051	50 38	098	29 12	119	41 05	202	18 50	219	29 15	266	27 47	309
29	47 00	051	51 31	099	29 59	120	40 44	203	18 17	220	28 22	266	27 06	309
	CAPELLA		**◆ALDEBARAN**		**RIGEL**		**◆Diphda**		**Alpheratz**		**DENEB**		**◆Kochab**	
30	47 41	051	52 23	099	30 45	120	40 23	204	65 28	279	26 25	309	12 31	356
31	48 22	051	53 15	100	31 30	121	40 01	205	64 35	279	25 44	310	12 28	357
32	49 04	051	54 07	101	32 15	122	39 38	206	63 43	280	25 03	310	12 25	357
33	49 45	051	54 59	101	33 00	123	39 14	208	62 51	280	24 22	310	12 23	357
34	50 26	051	55 51	102	33 45	123	38 49	209	61 59	280	23 42	310	12 20	358
35	51 07	050	56 43	103	34 29	124	38 23	210	61 07	280	23 01	310	12 18	358
36	51 47	050	57 34	104	35 12	125	37 56	211	60 14	280	22 21	311	12 16	358
37	52 28	050	58 26	105	35 56	126	37 29	212	59 22	280	21 41	311	12 15	358
38	53 08	050	59 17	105	36 38	127	37 00	213	58 30	281	21 01	311	12 13	359
39	53 48	049	60 08	106	37 21	128	36 31	214	57 38	281	20 21	311	12 12	359
40	54 28	049	60 59	107	38 02	128	36 01	215	56 46	281	19 41	312	12 11	359
41	55 08	049	61 49	108	38 44	129	35 30	216	55 54	281	19 02	312	12 11	000
42	55 48	048	62 39	109	39 24	130	34 59	217	55 02	282	18 22	312	12 11	000
43	56 27	048	63 29	110	40 04	131	34 27	218	54 10	282	17 43	312	12 11	000
44	57 06	047	64 19	111	40 44	132	33 54	219	53 19	282	17 04	313	12 11	000
	◆CAPELLA		**PROCYON**		**SIRIUS**		**◆RIGEL**		**Acamar**		**Diphda**		**◆Alpheratz**	
45	57 45	047	20 18	095	19 35	122	41 23	133	21 40	180	33 20	220	52 27	282
46	58 24	046	21 11	095	20 19	123	42 01	134	21 40	181	32 46	221	51 35	282
47	59 02	046	22 04	096	21 04	124	42 39	135	21 38	182	32 11	222	50 43	283
48	59 40	045	22 57	096	21 47	124	43 16	136	21 36	183	31 36	222	49 52	283
49	60 17	045	23 49	097	22 31	125	43 52	137	21 33	184	31 00	223	49 00	283
50	60 54	044	24 42	098	23 14	126	44 28	138	21 29	184	30 23	224	48 08	283
51	61 31	043	25 34	098	23 57	126	45 02	140	21 25	185	29 46	225	47 17	284
52	62 07	043	26 27	099	24 40	127	45 36	141	21 19	186	29 08	226	46 25	284
53	62 42	042	27 19	099	25 22	128	46 10	142	21 13	187	28 30	227	45 34	284
54	63 17	041	28 11	100	26 03	128	46 42	143	21 07	188	27 51	227	44 43	284
55	63 52	040	29 03	100	26 45	129	47 13	144	20 59	189	27 12	228	43 51	285
56	64 25	039	29 55	101	27 25	130	47 44	146	20 51	189	26 32	229	43 00	285
57	64 58	038	30 47	101	28 06	131	48 13	147	20 42	190	25 52	230	42 09	285
58	65 31	037	31 39	102	28 46	132	48 41	148	20 32	191	25 12	230	41 18	285
59	66 02	036	32 31	102	29 25	132	49 09	150	20 22	192	24 31	231	40 27	286
	◆Dubhe		**PROCYON**		**SIRIUS**		**◆RIGEL**		**Diphda**		**◆Hamal**		**Mirfak**	
60	17 26	028	33 23	103	30 04	133	49 35	151	23 49	232	64 11	266	67 06	345
61	17 51	029	34 14	104	30 42	134	50 00	152	23 07	233	63 18	267	66 52	343
62	18 17	029	35 06	104	31 20	135	50 24	154	22 25	233	62 25	267	66 36	342
63	18 43	029	35 57	105	31 58	136	50 47	155	21 42	234	61 32	268	66 19	340
64	19 08	029	36 48	106	32 34	137	51 09	157	20 59	235	60 39	268	66 00	339
65	19 34	030	37 39	106	33 10	137	51 29	158	20 16	235	59 46	269	65 41	338
66	20 01	030	38 30	107	33 46	138	51 48	160	19 32	236	58 53	269	65 20	336
67	20 27	030	39 20	108	34 21	139	52 06	161	18 48	237	58 00	270	64 58	335
68	20 53	030	40 11	108	34 55	140	52 23	163	18 04	237	57 07	270	64 36	334
69	21 20	030	41 01	109	35 29	141	52 38	164	17 19	238	56 14	271	64 12	333
70	21 47	030	41 51	110	36 01	142	52 51	166	16 34	238	55 21	271	63 47	332
71	22 14	031	42 41	110	36 34	143	53 04	167	15 49	239	54 28	272	63 22	331
72	22 41	031	43 30	111	37 05	144	53 14	169	15 03	240	53 35	272	62 55	329
73	23 08	031	44 20	112	37 36	145	53 24	171	14 18	240	52 42	273	62 28	328
74	23 35	031	45 09	113	38 06	146	53 31	172	13 31	241	51 49	273	62 00	328
	◆Dubhe		**POLLUX**		**PROCYON**		**◆SIRIUS**		**RIGEL**		**◆Hamal**		**Mirfak**	
75	24 03	031	53 46	080	45 57	114	38 35	147	53 38	174	50 57	273	61 31	327
76	24 30	031	54 38	080	46 46	114	39 03	148	53 42	176	50 04	274	61 01	326
77	24 58	031	55 30	080	47 34	115	39 30	149	53 46	177	49 11	274	60 31	325
78	25 25	032	56 22	081	48 21	116	39 57	150	53 47	179	48 18	275	60 01	324
79	25 53	032	57 15	081	49 09	117	40 23	152	53 47	181	47 25	275	59 29	323
80	26 21	032	58 07	081	49 56	118	40 47	153	53 46	182	46 33	275	58 57	323
81	26 49	032	58 59	081	50 42	119	41 11	154	53 43	184	45 40	276	58 25	322
82	27 17	032	59 52	082	51 28	120	41 34	155	53 38	186	44 47	276	57 52	321
83	27 45	032	60 44	082	52 14	121	41 56	156	53 32	187	43 54	277	57 19	321
84	28 13	032	61 37	082	52 59	122	42 17	157	53 25	189	43 02	277	56 45	320
85	28 41	032	62 29	082	53 43	123	42 36	159	53 16	191	42 09	277	56 11	320
86	29 09	032	63 22	083	54 27	124	42 55	160	53 05	192	41 17	278	55 36	319
87	29 38	032	64 14	083	55 11	126	43 13	161	52 53	194	40 24	278	55 02	319
88	30 06	032	65 07	083	55 53	127	43 29	162	52 40	196	39 32	278	54 26	318
89	30 34	032	66 00	083	56 35	128	43 45	164	52 25	197	38 39	279	53 51	318

LAT 28°N

LHA ♈	Hc	Zn	Hc	Zn	Hc	Zn	Hc	Zn	Hc	Zn	Hc	Zn	Hc	Zn
	Dubhe		**◆REGULUS**		**PROCYON**		**SIRIUS**		**◆RIGEL**		**ALDEBARAN**		**◆Mirfak**	
90	31 03	032	30 11	092	57 17	130	43 59	165	52 09	199	67 24	244	53 15	317
91	31 31	032	31 04	093	57 57	131	44 12	166	51 51	200	66 36	245	52 39	317
92	31 59	032	31 56	093	58 37	132	44 24	168	51 32	202	65 48	247	52 03	317
93	32 28	032	32 49	094	59 15	134	44 35	169	51 12	203	64 59	248	51 26	316
94	32 56	032	33 42	094	59 53	135	44 45	170	50 51	205	64 10	249	50 49	316
95	33 24	032	34 35	095	60 30	137	44 53	172	50 28	206	63 20	250	50 12	316
96	33 53	032	35 28	095	61 06	139	45 00	173	50 04	208	62 30	251	49 35	315
97	34 21	032	36 21	096	61 40	140	45 06	174	49 39	209	61 40	252	48 58	315
98	34 49	032	37 13	096	62 13	142	45 11	176	49 13	210	60 49	253	48 20	315
99	35 18	032	38 06	097	62 45	144	45 14	177	48 46	212	59 59	254	47 43	315
100	35 46	032	38 58	097	63 16	146	45 17	178	48 17	213	59 08	255	47 05	314
101	36 14	032	39 51	098	63 45	148	45 17	180	47 48	214	58 16	256	46 27	314
102	36 42	032	40 43	098	64 13	150	45 17	181	47 18	215	57 25	256	45 49	314
103	37 10	032	41 36	099	64 39	152	45 15	182	46 47	217	56 33	257	45 11	314
104	37 38	032	42 28	100	65 03	154	45 13	184	46 14	218	55 42	258	44 32	314
	◆Dubhe		**REGULUS**		**◆Alphard**		**SIRIUS**		**RIGEL**		**◆ALDEBARAN**		**CAPELLA**	
105	38 06	032	43 20	100	38 55	130	45 09	185	45 41	219	54 50	259	62 48	318
106	38 34	032	44 12	101	39 35	131	45 03	186	45 08	220	53 58	259	62 12	318
107	39 01	031	45 04	102	40 14	132	44 57	188	44 33	221	53 06	260	61 36	317
108	39 29	031	45 56	102	40 53	133	44 49	189	43 58	222	52 13	261	61 00	316
109	39 56	031	46 48	103	41 32	134	44 40	190	43 21	224	51 21	261	60 23	315
110	40 24	031	47 39	104	42 09	135	44 30	192	42 45	225	50 29	262	59 46	315
111	40 51	031	48 31	104	42 46	136	44 18	193	42 07	226	49 36	263	59 08	314
112	41 18	031	49 22	105	43 22	137	44 05	194	41 29	227	48 44	263	58 30	314
113	41 45	030	50 13	106	43 58	139	43 52	196	40 50	228	47 51	264	57 51	313
114	42 11	030	51 04	106	44 33	140	43 37	197	40 10	229	46 58	264	57 13	313
115	42 38	030	51 55	107	45 06	141	43 21	198	39 30	230	46 05	265	56 33	312
116	43 04	030	52 45	108	45 39	142	43 04	200	38 50	231	45 13	265	55 54	312
117	43 30	029	53 35	109	46 12	143	42 45	201	38 09	231	44 20	266	55 15	312
118	43 56	029	54 25	110	46 43	144	42 26	202	37 27	232	43 27	267	54 35	311
119	44 22	029	55 15	111	47 13	146	42 06	203	36 45	233	42 34	267	53 55	311
	◆Dubhe		**REGULUS**		**◆Alphard**		**SIRIUS**		**RIGEL**		**◆ALDEBARAN**		**CAPELLA**	
120	44 47	029	56 05	112	47 43	147	41 44	204	36 02	234	41 41	268	53 15	310
121	45 12	028	56 54	113	48 11	148	41 22	206	35 19	235	40 48	268	52 34	310
122	45 37	028	57 42	114	48 39	149	40 59	207	34 35	236	39 55	269	51 54	310
123	46 02	028	58 31	115	49 05	151	40 34	208	33 51	237	39 02	269	51 13	310
124	46 26	027	59 19	116	49 30	152	40 09	209	33 07	237	38 09	270	50 32	309
125	46 51	027	60 06	117	49 55	154	39 43	210	32 22	238	37 16	270	49 51	309
126	47 14	027	60 53	118	50 17	155	39 16	211	31 37	239	36 23	270	49 10	309
127	47 38	026	61 40	119	50 39	156	38 48	212	30 51	240	35 30	271	48 29	309
128	48 01	026	62 25	121	51 00	158	38 19	213	30 06	240	34 37	271	47 48	309
129	48 24	025	63 11	122	51 19	159	37 50	214	29 19	241	33 45	272	47 06	309
130	48 46	025	63 55	123	51 37	161	37 19	215	28 33	242	32 52	272	46 25	309
131	49 08	024	64 39	125	51 54	162	36 48	216	27 46	242	31 59	273	45 44	308
132	49 30	024	65 22	127	52 09	164	36 16	217	26 59	243	31 06	273	45 02	308
133	49 51	023	66 04	128	52 23	166	35 44	218	26 12	244	30 13	274	44 20	308
134	50 12	023	66 45	130	52 35	167	35 11	219	25 24	244	29 20	274	43 39	308
	Dubhe		**◆ARCTURUS**		**SPICA**		**Alphard**		**◆SIRIUS**		**BETELGEUSE**		**◆CAPELLA**	
135	50 33	022	18 24	078	14 59	112	52 46	169	34 37	220	41 44	254	42 57	308
136	50 53	022	19 16	078	15 48	112	52 56	170	34 02	221	40 53	254	42 16	308
137	51 12	021	20 07	078	16 37	113	53 04	172	33 27	222	40 02	255	41 34	308
138	51 31	021	20 59	079	17 26	113	53 11	174	32 51	223	39 11	256	40 52	308
139	51 50	020	21 51	079	18 14	114	53 16	175	32 15	224	38 19	256	40 10	308
140	52 08	020	22 43	080	19 03	115	53 19	177	31 38	225	37 28	257	39 29	308
141	52 25	019	23 35	080	19 51	115	53 21	179	31 00	226	36 36	258	38 47	308
142	52 42	018	24 28	080	20 38	116	53 22	180	30 22	226	35 44	258	38 05	308
143	52 59	018	25 20	081	21 26	116	53 21	182	29 43	227	34 52	259	37 24	308
144	53 15	017	26 12	081	22 13	117	53 18	184	29 04	228	34 00	259	36 42	308
145	53 30	016	27 05	082	23 00	118	53 14	185	28 24	229	33 08	260	36 01	308
146	53 45	016	27 57	082	23 47	118	53 09	187	27 44	230	32 16	261	35 19	308
147	53 59	015	28 50	082	24 34	119	53 01	189	27 04	230	31 24	261	34 38	309
148	54 12	014	29 42	083	25 20	120	52 53	190	26 23	231	30 31	262	33 56	309
149	54 25	014	30 35	083	26 06	120	52 43	192	25 41	232	29 39	262	33 15	309
	Dubhe		**◆ARCTURUS**		**SPICA**		**◆Alphard**		**SIRIUS**		**POLLUX**		**◆CAPELLA**	
150	54 37	013	31 27	084	26 52	121	52 31	193	24 59	233	60 18	278	32 34	309
151	54 49	012	32 20	084	27 37	122	52 18	195	24 17	233	59 26	278	31 52	309
152	54 59	011	33 13	084	28 22	122	52 04	196	23 34	234	58 34	279	31 11	309
153	55 09	011	34 05	085	29 06	123	51 48	198	22 51	235	57 41	279	30 30	309
154	55 19	010	34 58	085	29 50	124	51 31	200	22 08	235	56 49	279	29 49	309
155	55 28	009	35 51	086	30 34	125	51 13	201	21 24	236	55 57	279	29 08	310
156	55 36	008	36 44	086	31 18	125	50 53	203	20 40	237	55 04	280	28 27	310
157	55 43	007	37 37	087	32 01	126	50 32	204	19 55	237	54 12	280	27 47	310
158	55 49	007	38 30	087	32 43	127	50 10	205	19 11	238	53 20	280	27 06	310
159	55 55	006	39 22	087	33 25	128	49 47	207	18 25	239	52 28	280	26 26	310
160	56 00	005	40 15	088	34 07	129	49 22	208	17 40	239	51 36	281	25 45	310
161	56 04	004	41 08	088	34 48	129	48 57	210	16 54	240	50 44	281	25 05	311
162	56 07	003	42 01	089	35 29	130	48 30	211	16 08	240	49 52	281	24 25	311
163	56 10	002	42 54	089	36 09	131	48 02	212	15 22	241	49 00	282	23 45	311
164	56 12	002	43 47	090	36 48	132	47 33	214	14 36	242	48 08	282	23 05	311
	Dubhe		**◆Alkaid**		**ARCTURUS**		**◆SPICA**		**Alphard**		**◆PROCYON**		**POLLUX**	
165	56 13	001	51 42	045	44 40	090	37 27	133	47 04	215	37 15	254	47 16	282
166	56 13	000	52 19	044	45 33	091	38 06	134	46 33	216	36 24	255	46 24	282
167	56 13	359	52 56	044	46 26	091	38 43	135	46 01	217	35 32	255	45 33	283
168	56 12	358	53 32	043	47 19	092	39 21	136	45 29	218	34 41	256	44 41	283
169	56 09	357	54 09	043	48 12	092	39 57	137	44 55	220	33 50	257	43 49	283
170	56 07	356	54 45	043	49 05	093	40 33	138	44 21	221	32 58	257	42 58	283
171	56 03	356	55 20	042	49 58	093	41 08	139	43 46	222	32 06	258	42 06	284
172	55 59	355	55 56	042	50 51	094	41 42	140	43 10	223	31 14	258	41 15	284
173	55 53	354	56 31	041	51 44	094	42 16	141	42 34	224	30 23	259	40 23	284
174	55 47	353	57 05	041	52 37	095	42 49	142	41 57	225	29 31	260	39 32	285
175	55 41	352	57 40	040	53 29	095	43 21	143	41 19	226	28 38	260	38 41	285
176	55 33	352	58 13	039	54 22	096	43 52	145	40 41	227	27 46	261	37 50	285
177	55 25	351	58 47	039	55 15	096	44 22	146	40 02	228	26 54	261	36 59	285
178	55 16	350	59 20	038	56 07	097	44 52	147	39 22	229	26 01	262	36 08	286
179	55 07	349	59 52	037	57 00	098	45 20	148	38 42	230	25 09	262	35 17	286

LAT 28°N

LHA ♈	Hc	Zn	Hc	Zn	Hc	Zn	Hc	Zn	Hc	Zn	Hc	Zn	Hc	Zn
	♦VEGA		Alphecca		♦SPICA		Gienah		REGULUS		♦POLLUX		Dubhe	
180	10 36	052	42 46	078	45 48	149	44 20	175	59 21	244	34 26	286	54 56	348
181	11 18	052	43 37	079	46 14	151	44 24	176	58 33	245	33 35	287	54 45	348
182	12 00	052	44 29	079	46 40	152	44 27	177	57 45	246	32 44	287	54 34	347
183	12 42	053	45 21	079	47 04	153	44 29	179	56 56	247	31 54	287	54 21	346
184	13 24	053	46 14	080	47 27	155	44 29	180	56 07	248	31 03	288	54 08	345
185	14 06	053	47 06	080	47 50	156	44 28	181	55 18	249	30 13	288	53 55	345
186	14 49	054	47 58	080	48 11	157	44 26	183	54 28	250	29 22	288	53 40	344
187	15 32	054	48 50	080	48 31	159	44 23	184	53 38	251	28 32	288	53 26	343
188	16 15	054	49 42	081	48 49	160	44 19	185	52 48	252	27 42	289	53 10	343
189	16 58	055	50 35	081	49 07	161	44 13	187	51 57	253	26 52	289	52 54	342
190	17 41	055	51 27	081	49 23	163	44 06	188	51 07	253	26 02	289	52 37	341
191	18 24	055	52 19	082	49 38	164	43 58	189	50 16	254	25 12	290	52 20	341
192	19 08	056	53 12	082	49 51	166	43 49	191	49 25	255	24 22	290	52 03	340
193	19 52	056	54 04	082	50 04	167	43 38	192	48 34	256	23 32	290	51 44	340
194	20 36	056	54 57	083	50 15	169	43 27	193	47 42	256	22 43	291	51 26	339
	♦Kochab		VEGA		Rasalhague		♦ANTARES		SPICA		♦REGULUS		Dubhe	
195	41 40	010	21 20	056	24 33	088	15 57	133	50 24	170	46 51	257	51 07	338
196	41 49	009	22 04	057	25 26	089	16 36	133	50 32	172	45 59	258	50 47	338
197	41 58	009	22 48	057	26 19	089	17 15	134	50 39	173	45 07	258	50 27	337
198	42 06	009	23 33	057	27 12	090	17 53	135	50 44	175	44 15	259	50 06	337
199	42 14	009	24 17	057	28 05	090	18 30	135	50 48	177	43 23	260	49 45	336
200	42 22	008	25 02	058	28 58	091	19 07	136	50 51	178	42 31	260	49 24	336
201	42 29	008	25 47	058	29 51	091	19 44	137	50 52	180	41 38	261	49 02	335
202	42 36	008	26 32	058	30 44	092	20 20	137	50 51	181	40 46	262	48 40	335
203	42 43	007	27 17	058	31 37	092	20 56	138	50 50	183	39 54	262	48 17	335
204	42 50	007	28 02	059	32 30	093	21 31	139	50 46	184	39 01	263	47 54	334
205	42 56	006	28 47	059	33 23	093	22 06	139	50 42	186	38 09	263	47 31	334
206	43 01	006	29 33	059	34 16	094	22 40	140	50 36	187	37 16	264	47 07	333
207	43 07	006	30 18	059	35 08	094	23 14	141	50 28	189	36 23	264	46 44	333
208	43 12	005	31 04	059	36 01	095	23 47	142	50 19	190	35 31	265	46 19	333
209	43 17	005	31 49	060	36 54	095	24 19	142	50 09	192	34 38	265	45 55	332
	Kochab		♦VEGA		Rasalhague		ANTARES		♦SPICA		REGULUS		♦Dubhe	
210	43 22	005	32 35	060	37 47	096	24 51	143	49 57	193	33 45	266	45 30	332
211	43 26	004	33 21	060	38 39	096	25 23	144	49 44	195	32 52	266	45 05	332
212	43 30	004	34 07	060	39 32	097	25 53	145	49 30	196	31 59	267	44 40	331
213	43 33	004	34 53	060	40 25	097	26 24	146	49 14	198	31 06	267	44 14	331
214	43 36	003	35 39	061	41 17	098	26 53	147	48 58	199	30 13	268	43 49	331
215	43 39	003	36 25	061	42 10	099	27 22	147	48 39	201	29 20	268	43 23	331
216	43 42	003	37 11	061	43 02	099	27 50	148	48 20	202	28 27	269	42 56	330
217	43 44	002	37 58	061	43 54	100	28 18	149	48 00	203	27 35	269	42 30	330
218	43 46	002	38 44	061	44 46	100	28 44	150	47 38	205	26 42	270	42 04	330
219	43 47	001	39 31	061	45 38	101	29 10	151	47 15	206	25 49	270	41 37	330
220	43 48	001	40 17	061	46 30	102	29 36	152	46 51	207	24 56	271	41 10	329
221	43 49	001	41 04	061	47 22	102	30 00	153	46 26	209	24 03	271	40 43	329
222	43 49	000	41 50	062	48 14	103	30 24	154	46 00	210	23 10	272	40 16	329
223	43 49	000	42 37	062	49 05	104	30 47	155	45 33	211	22 17	272	39 48	329
224	43 49	000	43 23	062	49 57	104	31 10	156	45 05	213	21 24	273	39 21	329
	VEGA		♦ALTAIR		Nunki		ANTARES		♦SPICA		Denebola		♦Alkaid	
225	44 10	062	19 26	090	11 42	129	31 31	157	44 36	214	43 49	263	64 32	332
226	44 57	062	20 19	091	12 23	129	31 52	158	44 06	215	42 56	264	64 06	331
227	45 43	062	21 12	091	13 04	130	32 12	158	43 35	216	42 03	265	63 40	330
228	46 30	062	22 05	092	13 45	130	32 31	159	43 04	217	41 11	265	63 12	329
229	47 17	062	22 58	092	14 25	131	32 49	160	42 31	218	40 18	266	62 44	328
230	48 04	062	23 51	093	15 05	132	33 06	161	41 58	219	39 25	266	62 16	327
231	48 50	062	24 43	093	15 44	132	33 23	163	41 24	220	38 32	267	61 46	326
232	49 37	062	25 36	094	16 23	133	33 38	164	40 49	222	37 39	267	61 16	325
233	50 24	062	26 29	094	17 02	133	33 53	165	40 14	223	36 46	268	60 45	324
234	51 11	062	27 22	095	17 40	134	34 06	166	39 38	224	35 53	268	60 14	323
235	51 58	062	28 15	095	18 18	135	34 19	167	39 01	225	35 00	269	59 42	322
236	52 44	062	29 08	096	18 55	135	34 31	168	38 23	226	34 07	269	59 09	322
237	53 31	062	30 00	096	19 32	136	34 41	169	37 45	226	33 14	270	58 36	321
238	54 18	062	30 53	097	20 09	137	34 51	170	37 06	227	32 21	270	58 02	320
239	55 04	061	31 46	097	20 45	138	35 00	171	36 27	228	31 28	271	57 28	320
	♦VEGA		ALTAIR		Nunki		♦ANTARES		SPICA		ARCTURUS		♦Alkaid	
240	55 51	061	32 38	098	21 20	138	35 08	172	35 47	229	64 33	256	56 54	319
241	56 37	061	33 31	098	21 55	139	35 15	173	35 07	230	63 42	256	56 19	319
242	57 23	061	34 23	099	22 30	140	35 21	174	34 26	231	62 50	257	55 44	318
243	58 10	061	35 15	099	23 04	141	35 25	175	33 45	232	61 58	258	55 09	318
244	58 56	061	36 08	100	23 37	141	35 29	176	33 03	233	61 06	259	54 33	317
245	59 42	060	37 00	101	24 10	142	35 32	177	32 20	233	60 14	260	53 57	317
246	60 28	060	37 52	101	24 42	143	35 34	179	31 38	234	59 22	260	53 20	316
247	61 14	060	38 44	102	25 14	144	35 35	180	30 54	235	58 30	261	52 44	316
248	61 59	059	39 35	102	25 45	144	35 34	181	30 11	236	57 37	262	52 07	316
249	62 45	059	40 27	103	26 15	145	35 33	182	29 27	237	56 45	262	51 30	315
250	63 30	058	41 19	104	26 45	146	35 31	183	28 42	237	55 52	263	50 52	315
251	64 15	058	42 10	104	27 15	147	35 28	184	27 58	238	55 00	264	50 15	315
252	65 00	057	43 01	105	27 43	148	35 23	185	27 13	239	54 07	264	49 37	315
253	65 44	057	43 52	106	28 11	149	35 18	186	26 27	239	53 14	265	48 59	314
254	66 29	056	44 43	106	28 38	150	35 12	187	25 41	240	52 22	265	48 21	314
	♦DENEB		ALTAIR		Nunki		♦ANTARES		SPICA		ARCTURUS		♦Alkaid	
255	43 24	053	45 34	107	29 05	150	35 04	188	24 55	241	51 29	266	47 43	314
256	44 06	053	46 25	108	29 30	151	34 56	190	24 09	241	50 36	267	47 05	314
257	44 49	053	47 15	109	29 55	152	34 47	191	23 22	242	49 43	267	46 27	313
258	45 31	053	48 05	109	30 20	153	34 37	192	22 35	243	48 50	268	45 48	313
259	46 13	053	48 55	110	30 43	154	34 25	193	21 48	243	47 57	268	45 09	313
260	46 55	052	49 45	111	31 06	155	34 13	194	21 01	244	47 04	269	44 31	313
261	47 37	052	50 34	112	31 28	156	34 00	195	20 13	245	46 11	269	43 52	313
262	48 19	052	51 23	113	31 49	157	33 46	196	19 25	245	45 18	270	43 13	313
263	49 01	052	52 12	114	32 09	158	33 31	197	18 37	246	44 25	270	42 34	313
264	49 42	052	53 00	115	32 29	159	33 15	198	17 48	246	43 32	271	41 55	313
265	50 24	052	53 48	116	32 47	160	32 59	199	17 00	247	42 39	271	41 16	313
266	51 05	052	54 36	117	33 05	161	32 41	200	16 11	248	41 46	271	40 37	313
267	51 47	051	55 23	118	33 22	162	32 22	201	15 22	248	40 53	272	39 58	312
268	52 28	051	56 09	119	33 38	163	32 03	202	14 32	249	40 00	272	39 19	312
269	53 09	051	56 56	120	33 53	164	31 43	203	13 43	249	39 07	273	38 40	312

LHA ♈	Hc	Zn	Hc	Zn	Hc	Zn	Hc	Zn	Hc	Zn	Hc	Zn	Hc	Zn
	♦DENEB		ALTAIR		♦Nunki		ANTARES		♦ARCTURUS		Alkaid		Kochab	
270	53 50	051	57 41	121	34 07	165	31 22	204	38 15	273	38 01	312	37 57	345
271	54 31	050	58 27	122	34 20	166	31 00	205	37 22	274	37 22	312	37 43	345
272	55 12	050	59 11	123	34 32	167	30 37	206	36 29	274	36 43	312	37 29	345
273	55 52	050	59 55	125	34 43	168	30 14	207	35 36	274	36 04	313	37 15	345
274	56 32	049	60 38	126	34 54	169	29 50	208	34 43	275	35 25	313	37 01	345
275	57 12	049	61 20	128	35 03	170	29 25	209	33 50	275	34 46	313	36 47	344
276	57 52	048	62 02	129	35 11	172	28 59	209	32 58	276	34 07	313	36 33	344
277	58 31	048	62 43	131	35 19	173	28 32	210	32 05	276	33 28	313	36 18	344
278	59 10	047	63 23	132	35 25	174	28 05	211	31 12	277	32 49	313	36 04	344
279	59 49	047	64 01	134	35 30	175	27 37	212	30 20	277	32 10	313	35 49	344
280	60 27	046	64 39	136	35 35	176	27 09	213	29 27	277	31 31	313	35 34	344
281	61 05	045	65 15	137	35 38	177	26 40	214	28 35	278	30 52	313	35 19	343
282	61 43	045	65 50	139	35 40	178	26 10	215	27 42	278	30 14	313	35 04	343
283	62 20	044	66 24	141	35 42	179	25 40	215	26 50	279	29 35	313	34 49	343
284	62 57	043	66 56	144	35 42	180	25 09	216	25 57	279	28 57	313	34 33	343
	♦Alpheratz		Enif		Nunki		♦ANTARES		ARCTURUS		♦Alkaid		Kochab	
285	23 39	068	47 29	107	35 41	181	24 37	217	25 05	279	28 18	314	34 18	343
286	24 29	069	48 19	108	35 39	182	24 05	218	24 13	280	27 40	314	34 02	343
287	25 18	069	49 10	109	35 36	184	23 32	219	23 21	280	27 02	314	33 47	343
288	26 08	069	50 00	109	35 33	185	22 59	219	22 29	281	26 23	314	33 31	343
289	26 57	070	50 49	110	35 28	186	22 25	220	21 37	281	25 45	314	33 15	343
290	27 47	070	51 39	111	35 22	187	21 50	221	20 45	281	25 08	314	32 59	343
291	28 37	070	52 28	112	35 15	188	21 15	222	19 53	282	24 30	315	32 44	342
292	29 27	071	53 17	113	35 07	189	20 40	222	19 01	282	23 52	315	32 28	342
293	30 17	071	54 06	114	34 58	190	20 04	223	18 09	283	23 15	315	32 12	342
294	31 07	071	54 54	115	34 49	191	19 28	224	17 17	283	22 37	315	31 55	342
295	31 57	072	55 42	116	34 38	192	18 51	224	16 26	283	22 00	315	31 39	342
296	32 47	072	56 30	117	34 26	193	18 13	225	15 34	284	21 23	316	31 23	342
297	33 38	072	57 17	118	34 13	194	17 36	226	14 43	284	20 46	316	31 07	342
298	34 28	072	58 03	119	34 00	195	16 57	226	13 52	285	20 09	316	30 51	342
299	35 19	073	58 49	120	33 45	196	16 19	227	13 00	285	19 32	316	30 34	342
	Schedar		♦Alpheratz		FOMALHAUT		♦Nunki		Rasalhague		♦VEGA		Kochab	
300	33 54	039	36 09	073	18 27	140	33 30	198	52 48	253	69 38	307	30 18	342
301	34 27	039	37 00	073	19 01	141	33 13	199	51 57	254	68 55	306	30 02	342
302	35 00	039	37 51	074	19 34	142	32 56	200	51 06	255	68 12	306	29 45	342
303	35 33	039	38 42	074	20 07	142	32 38	201	50 15	255	67 29	305	29 29	342
304	36 06	039	39 33	074	20 39	143	32 19	202	49 23	256	66 45	304	29 13	342
305	36 39	039	40 24	074	21 10	144	31 59	203	48 32	257	66 01	303	28 56	342
306	37 12	039	41 15	075	21 41	145	31 38	204	47 40	257	65 17	303	28 40	342
307	37 45	038	42 06	075	22 12	145	31 17	204	46 48	258	64 32	302	28 24	342
308	38 18	038	42 57	075	22 42	146	30 54	205	45 57	259	63 47	302	28 07	342
309	38 51	038	43 48	075	23 11	147	30 31	206	45 05	259	63 02	301	27 51	342
310	39 24	038	44 39	076	23 40	148	30 07	207	44 12	260	62 17	301	27 35	342
311	39 57	038	45 31	076	24 08	148	29 42	208	43 20	261	61 31	300	27 18	342
312	40 29	038	46 22	076	24 35	149	29 17	209	42 28	261	60 45	300	27 02	342
313	41 02	038	47 14	076	25 02	150	28 51	210	41 36	262	60 00	300	26 46	342
314	41 34	038	48 05	077	25 28	151	28 24	211	40 43	262	59 14	300	26 30	342
	♦Mirfak		Hamal		Diphda		♦FOMALHAUT		ALTAIR		♦VEGA		Kochab	
315	17 25	042	21 53	074	19 03	124	25 53	152	64 49	224	58 27	299	26 14	342
316	18 01	042	22 44	075	19 47	124	26 18	153	64 12	226	57 41	299	25 58	342
317	18 37	043	23 35	075	20 30	125	26 42	153	63 33	227	56 55	299	25 42	342
318	19 13	043	24 27	075	21 14	126	27 06	154	62 54	229	56 08	299	25 26	343
319	19 49	043	25 18	076	21 57	126	27 28	155	62 13	231	55 22	299	25 10	343
320	20 25	043	26 09	076	22 39	127	27 50	156	61 32	232	54 35	298	24 54	343
321	21 02	044	27 01	076	23 21	128	28 11	157	60 50	234	53 49	298	24 38	343
322	21 39	044	27 52	077	24 03	128	28 31	158	60 07	235	53 02	298	24 22	343
323	22 15	044	28 44	077	24 44	129	28 51	159	59 23	236	52 15	298	24 07	343
324	22 52	044	29 36	078	25 25	130	29 10	160	58 39	237	51 29	298	23 51	343
325	23 30	045	30 27	078	26 06	131	29 27	161	57 54	239	50 42	298	23 36	343
326	24 07	045	31 19	078	26 46	131	29 45	162	57 08	240	49 55	298	23 20	343
327	24 44	045	32 11	079	27 25	132	30 01	163	56 22	241	49 08	298	23 05	343
328	25 22	045	33 03	079	28 04	133	30 16	164	55 35	242	48 22	298	22 50	343
329	25 59	045	33 55	079	28 43	134	30 31	165	54 48	243	47 35	298	22 35	344
	♦Mirfak		Hamal		Diphda		♦FOMALHAUT		ALTAIR		♦VEGA		Kochab	
330	26 37	045	34 47	080	29 21	134	30 45	165	54 01	244	46 48	298	22 20	344
331	27 15	046	35 39	080	29 58	135	30 57	166	53 13	245	46 01	298	22 05	344
332	27 53	046	36 32	080	30 35	136	31 09	167	52 25	246	45 15	298	21 50	344
333	28 30	046	37 24	081	31 12	137	31 20	168	51 36	247	44 28	298	21 36	344
334	29 09	046	38 16	081	31 48	138	31 31	169	50 47	248	43 41	298	21 21	344
335	29 47	046	39 08	081	32 23	139	31 40	170	49 58	249	42 55	298	21 07	344
336	30 25	046	40 01	082	32 57	140	31 48	171	49 08	250	42 08	298	20 53	344
337	31 03	046	40 53	082	33 31	141	31 56	172	48 19	250	41 21	298	20 38	345
338	31 42	046	41 46	083	34 05	141	32 02	173	47 29	251	40 35	299	20 24	345
339	32 20	047	42 38	083	34 37	142	32 08	175	46 38	252	39 48	299	20 11	345
340	32 58	047	43 31	083	35 09	143	32 12	176	45 48	253	39 02	299	19 57	345
341	33 37	047	44 24	084	35 41	144	32 16	177	44 57	253	38 16	299	19 43	345
342	34 16	047	45 16	084	36 11	145	32 18	178	44 06	254	37 29	299	19 30	345
343	34 54	047	46 09	084	36 41	146	32 20	179	43 15	255	36 43	299	19 17	346
344	35 33	047	47 02	085	37 10	147	32 21	180	42 24	256	35 57	299	19 03	346
	♦CAPELLA		ALDEBARAN		Diphda		♦FOMALHAUT		ALTAIR		♦VEGA		Kochab	
345	17 06	046	12 54	078	37 38	148	32 21	181	41 33	256	35 11	300	18 50	346
346	17 44	047	13 46	078	38 05	150	32 20	182	40 41	257	34 25	300	18 38	346
347	18 23	047	14 38	079	38 32	151	32 18	183	39 49	258	33 39	300	18 25	346
348	19 02	047	15 30	079	38 57	152	32 15	184	38 58	258	32 53	300	18 13	346
349	19 41	048	16 22	080	39 22	153	32 11	185	38 06	259	32 07	300	18 00	347
350	20 20	048	17 14	080	39 46	154	32 06	186	37 14	259	31 21	300	17 48	347
351	20 59	048	18 06	081	40 08	155	32 00	187	36 22	260	30 36	301	17 36	347
352	21 39	048	18 58	081	40 30	156	31 53	188	35 29	261	29 50	301	17 24	347
353	22 19	049	19 51	081	40 51	157	31 46	189	34 37	261	29 05	301	17 13	347
354	22 58	049	20 43	082	41 11	159	31 37	190	33 45	262	28 19	301	17 01	348
355	23 38	049	21 35	082	41 30	160	31 27	191	32 52	262	27 34	302	16 50	348
356	24 18	049	22 28	083	41 48	161	31 17	192	32 00	263	26 49	302	16 39	348
357	24 58	049	23 21	083	42 05	162	31 06	193	31 07	263	26 04	302	16 28	348
358	25 39	050	24 13	083	42 20	163	30 53	194	30 15	264	25 19	302	16 17	348
359	26 19	050	25 06	084	42 35	165	30 40	195	29 22	264	24 34	302	16 07	349

 LAT 27°N

LHA	Hc	Zn	Hc	Zn	Hc	Zn	Hc	Zn	Hc	Zn	Hc	Zn	Hc	Zn
♈	Schedar		◆CAPELLA		ALDEBARAN		◆Diphda		FOMALHAUT		ALTAIR		◆DENEB	
0	59 37	011	26 21	050	25 52	084	43 47	166	31 24	196	28 34	265	46 44	308
1	59 47	010	27 02	050	26 46	084	43 59	167	31 09	197	27 41	266	46 02	308
2	59 56	009	27 42	050	27 39	085	44 11	168	30 53	198	26 48	266	45 20	308
3	60 04	008	28 23	050	28 32	085	44 21	170	30 36	199	25 54	267	44 38	308
4	60 10	007	29 04	050	29 25	085	44 30	171	30 18	200	25 01	267	43 56	308
5	60 16	006	29 45	050	30 19	086	44 38	172	30 00	201	24 07	268	43 14	308
6	60 21	005	30 27	050	31 12	086	44 45	174	29 41	202	23 14	268	42 32	308
7	60 25	003	31 08	051	32 05	087	44 50	175	29 20	203	22 21	269	41 50	308
8	60 27	002	31 49	051	32 59	087	44 54	176	29 00	203	21 27	269	41 08	308
9	60 29	001	32 31	051	33 52	088	44 57	178	28 38	204	20 34	270	40 25	308
10	60 29	000	33 12	051	34 45	088	44 59	179	28 15	205	19 40	270	39 43	308
11	60 29	359	33 54	051	35 39	088	44 59	180	27 52	206	18 47	271	39 01	308
12	60 27	358	34 35	051	36 32	089	44 58	182	27 28	207	17 53	271	38 19	308
13	60 25	357	35 17	051	37 26	089	44 56	183	27 04	208	17 00	271	37 37	308
14	60 21	356	35 58	051	38 19	090	44 53	184	26 38	209	16 06	272	36 54	308
	◆CAPELLA		ALDEBARAN		RIGEL		◆Diphda		FOMALHAUT		Enif		◆DENEB	
15	36 40	051	39 13	090	19 08	110	44 48	186	26 12	210	40 48	259	36 12	308
16	37 21	051	40 06	091	19 58	111	44 42	187	25 45	210	39 56	260	35 30	308
17	38 03	051	41 00	091	20 48	111	44 35	188	25 18	211	39 03	260	34 48	308
18	38 45	051	41 53	092	21 37	112	44 27	190	24 50	212	38 10	261	34 06	308
19	39 27	051	42 47	092	22 27	113	44 18	191	24 21	213	37 17	262	33 24	308
20	40 08	051	43 40	093	23 16	113	44 07	192	23 52	214	36 24	262	32 42	308
21	40 50	051	44 33	093	24 05	114	43 55	193	23 22	214	35 31	263	32 00	308
22	41 32	051	45 27	094	24 54	114	43 42	195	22 52	215	34 38	263	31 18	309
23	42 13	051	46 20	094	25 42	115	43 28	196	22 20	216	33 45	264	30 36	309
24	42 55	051	47 13	095	26 31	116	43 13	197	21 49	217	32 52	264	29 55	309
25	43 36	051	48 07	095	27 19	116	42 56	199	21 17	217	31 59	265	29 13	309
26	44 18	051	49 00	096	28 06	117	42 39	200	20 44	218	31 06	265	28 32	309
27	44 59	051	49 53	096	28 54	118	42 20	201	20 11	219	30 12	266	27 50	309
28	45 41	051	50 46	097	29 41	118	42 00	202	19 37	220	29 19	266	27 09	309
29	46 22	051	51 39	097	30 28	119	41 40	203	19 02	220	28 26	267	26 27	310
	CAPELLA		◆ALDEBARAN		RIGEL		◆Diphda		Alpheratz		DENEB		◆Kochab	
30	47 03	050	52 32	098	31 15	120	41 18	205	65 17	281	25 46	310	11 31	356
31	47 44	050	53 25	099	32 01	120	40 55	206	64 25	281	25 05	310	11 28	357
32	48 25	050	54 18	099	32 47	121	40 31	207	63 32	282	24 24	310	11 25	357
33	49 06	050	55 10	100	33 32	122	40 07	208	62 40	282	23 43	310	11 23	357
34	49 47	050	56 03	101	34 17	123	39 41	209	61 48	282	23 03	311	11 20	358
35	50 28	049	56 56	101	35 02	124	39 15	210	60 55	282	22 22	311	11 18	358
36	51 08	049	57 48	102	35 47	124	38 48	211	60 03	282	21 42	311	11 16	358
37	51 49	049	58 40	103	36 30	125	38 19	212	59 11	282	21 01	311	11 15	358
38	52 29	049	59 32	104	37 14	126	37 50	213	58 18	282	20 21	311	11 13	359
39	53 09	048	60 24	105	37 57	127	37 21	214	57 26	282	19 41	312	11 12	359
40	53 49	048	61 16	105	38 39	128	36 50	215	56 34	283	19 01	312	11 11	359
41	54 28	048	62 07	106	39 21	129	36 19	216	55 42	283	18 22	312	11 11	000
42	55 07	047	62 58	107	40 03	130	35 47	217	54 50	283	17 42	312	11 11	000
43	55 46	047	63 49	108	40 44	131	35 14	218	53 58	283	17 03	313	11 11	000
44	56 25	046	64 40	109	41 24	132	34 40	219	53 06	283	16 23	313	11 11	000
	◆CAPELLA		PROCYON		SIRIUS		◆RIGEL		Acamar		Diphda		◆Alpheratz	
45	57 04	046	20 23	095	20 07	122	42 04	133	22 40	180	34 06	220	52 14	283
46	57 42	045	21 17	095	20 52	123	42 43	134	22 40	181	33 31	221	51 22	284
47	58 19	045	22 10	096	21 37	123	43 21	135	22 38	182	32 56	222	50 30	284
48	58 57	044	23 03	096	22 21	124	43 59	136	22 36	183	32 20	223	49 38	284
49	59 34	043	23 56	097	23 05	125	44 36	137	22 33	184	31 43	224	48 46	284
50	60 10	043	24 49	097	23 49	125	45 12	138	22 29	185	31 06	225	47 54	284
51	60 46	042	25 42	098	24 33	126	45 48	139	22 24	185	30 28	225	47 02	285
52	61 22	041	26 35	098	25 16	127	46 23	140	22 19	186	29 50	226	46 11	285
53	61 57	041	27 28	099	25 58	127	46 57	141	22 13	187	29 11	227	45 19	285
54	62 31	040	28 21	099	26 41	128	47 30	142	22 06	188	28 32	228	44 27	285
55	63 05	039	29 14	100	27 22	129	48 02	144	21 58	189	27 52	229	43 36	286
56	63 38	038	30 06	100	28 04	130	48 33	145	21 50	189	27 12	229	42 44	286
57	64 11	037	30 59	101	28 45	130	49 03	146	21 41	190	26 31	230	41 53	286
58	64 42	036	31 51	101	29 25	131	49 32	148	21 31	191	25 50	231	41 02	286
59	65 13	035	32 44	102	30 05	132	50 00	149	21 20	192	25 08	232	40 10	286
	◆Dubhe		PROCYON		SIRIUS		◆RIGEL		Diphda		◆Hamal		Mirfak	
60	16 33	028	33 36	102	30 45	133	50 27	150	24 26	232	64 14	268	66 08	346
61	16 59	029	34 28	103	31 24	134	50 52	152	23 44	233	63 20	269	65 54	344
62	17 24	029	35 20	104	32 02	134	51 18	153	23 01	234	62 27	269	65 39	343
63	17 50	029	36 12	104	32 40	135	51 42	155	22 18	234	61 33	270	65 22	341
64	18 16	029	37 04	105	33 18	136	52 04	156	21 34	235	60 40	270	65 04	340
65	18 42	029	37 55	105	33 54	137	52 25	158	20 50	236	59 46	271	64 45	339
66	19 08	030	38 47	106	34 31	138	52 45	159	20 06	236	58 53	271	64 25	337
67	19 35	030	39 38	107	35 06	139	53 03	161	19 21	237	57 59	272	64 04	336
68	20 02	030	40 29	107	35 41	140	53 20	162	18 36	237	57 06	272	63 42	335
69	20 28	030	41 20	108	36 15	141	53 35	164	17 51	238	56 13	272	63 18	334
70	20 55	030	42 11	109	36 49	142	53 49	166	17 06	239	55 19	273	62 54	333
71	21 22	030	43 01	110	37 21	143	54 02	167	16 20	239	54 26	273	62 29	331
72	21 49	031	43 52	110	37 54	144	54 13	169	15 34	240	53 32	274	62 03	330
73	22 16	031	44 42	111	38 25	145	54 23	170	14 47	240	52 39	274	61 36	329
74	22 44	031	45 31	112	38 55	146	54 31	172	14 01	241	51 46	274	61 09	329
	◆Dubhe		POLLUX		PROCYON		◆SIRIUS		RIGEL		◆Hamal		Mirfak	
75	23 11	031	53 35	079	46 21	113	39 25	147	54 37	174	50 52	275	60 40	328
76	23 39	031	54 27	079	47 10	114	39 54	148	54 42	176	49 59	275	60 12	327
77	24 06	031	55 19	079	47 59	114	40 22	149	54 46	177	49 06	275	59 42	326
78	24 34	031	56 12	079	48 47	115	40 49	150	54 47	179	48 13	276	59 12	325
79	25 02	031	57 05	079	49 36	116	41 15	151	54 47	181	47 19	276	58 41	324
80	25 30	032	57 57	080	50 23	117	41 41	152	54 46	182	46 26	276	58 09	324
81	25 58	032	58 50	080	51 11	118	42 05	153	54 43	184	45 33	277	57 37	323
82	26 26	032	59 42	080	51 58	119	42 28	155	54 38	186	44 40	277	57 05	322
83	26 54	032	60 35	080	52 44	120	42 51	156	54 32	188	43 47	277	56 32	322
84	27 22	032	61 28	080	53 30	121	43 12	157	54 24	189	42 54	278	55 59	321
85	27 50	032	62 20	081	54 16	122	43 32	158	54 15	191	42 01	278	55 25	321
86	28 19	032	63 13	081	55 01	123	43 52	160	54 04	193	41 08	278	54 51	320
87	28 47	032	64 06	081	55 45	124	44 10	161	53 51	194	40 15	279	54 16	320
88	29 15	032	64 59	081	56 29	126	44 27	162	53 38	196	39 23	279	53 41	319
89	29 44	032	65 52	081	57 12	127	44 42	163	53 22	197	38 30	280	53 06	319

LHA	Hc	Zn	Hc	Zn	Hc	Zn	Hc	Zn	Hc	Zn	Hc	Zn	Hc	Zn
♈	Dubhe		◆REGULUS		PROCYON		SIRIUS		◆RIGEL		ALDEBARAN		◆Mirfak	
90	30 12	032	30 12	092	57 54	128	44 57	165	53 05	199	67 49	246	52 31	318
91	30 40	032	31 06	092	58 36	130	45 11	166	52 47	201	67 00	247	51 55	318
92	31 09	032	31 59	092	59 17	131	45 23	167	52 28	202	66 11	249	51 19	317
93	31 37	032	32 53	093	59 56	133	45 34	169	52 07	204	65 21	250	50 42	317
94	32 05	032	33 46	093	60 35	134	45 44	170	51 45	205	64 30	251	50 06	317
95	32 34	032	34 39	094	61 13	136	45 52	171	51 22	207	63 40	252	49 29	316
96	33 02	032	35 33	094	61 50	137	46 00	173	50 57	208	62 49	253	48 52	316
97	33 30	032	36 26	095	62 26	139	46 06	174	50 31	209	61 58	254	48 15	316
98	33 59	032	37 19	095	63 00	141	46 11	176	50 04	211	61 06	255	47 38	316
99	34 27	032	38 13	096	63 33	143	46 14	177	49 36	212	60 14	256	47 00	315
100	34 55	032	39 06	097	64 05	145	46 16	178	49 07	214	59 23	256	46 23	315
101	35 23	032	39 59	097	64 35	146	46 17	180	48 37	215	58 30	257	45 45	315
102	35 51	032	40 52	098	65 04	149	46 17	181	48 06	216	57 38	258	45 07	315
103	36 19	031	41 45	098	65 31	151	46 15	182	47 34	217	56 46	259	44 29	315
104	36 47	031	42 38	099	65 57	153	46 13	184	47 02	219	55 53	259	43 51	314
	Dubhe		◆REGULUS		Alphard		SIRIUS		◆RIGEL		ALDEBARAN		◆CAPELLA	
105	37 15	031	43 30	099	39 34	130	46 08	185	46 28	220	55 01	260	62 02	320
106	37 42	031	44 23	100	40 14	131	46 03	187	45 53	221	54 08	261	61 28	319
107	38 10	031	45 16	101	40 55	132	45 56	188	45 18	222	53 15	261	60 52	318
108	38 37	031	46 08	101	41 34	133	45 48	189	44 42	223	52 22	262	60 16	317
109	39 05	031	47 01	102	42 13	134	45 39	191	44 05	224	51 29	262	59 40	317
110	39 32	030	47 53	102	42 52	135	45 28	192	43 27	225	50 36	263	59 03	316
111	39 59	030	48 45	103	43 29	136	45 16	193	42 49	226	49 43	264	58 25	315
112	40 26	030	49 37	104	44 06	137	45 04	195	42 10	227	48 50	264	57 48	315
113	40 53	030	50 29	105	44 43	138	44 49	196	41 30	228	47 57	265	57 10	314
114	41 19	030	51 20	105	45 18	139	44 34	197	40 50	229	47 04	265	56 31	314
115	41 46	029	52 12	106	45 53	140	44 18	199	40 09	230	46 10	266	55 53	313
116	42 12	029	53 03	107	46 27	141	44 00	200	39 28	231	45 17	266	55 14	313
117	42 38	029	53 54	108	47 00	142	43 41	201	38 46	232	44 24	267	54 34	313
118	43 04	029	54 45	108	47 32	144	43 22	202	38 03	233	43 30	267	53 55	312
119	43 29	028	55 36	109	48 03	145	43 01	204	37 20	234	42 37	268	53 15	312
	◆Dubhe		REGULUS		◆Alphard		SIRIUS		RIGEL		◆ALDEBARAN		CAPELLA	
120	43 54	028	56 26	110	48 33	146	42 39	205	36 37	235	41 43	268	52 35	311
121	44 19	028	57 16	111	49 02	148	42 16	206	35 53	236	40 50	269	51 55	311
122	44 44	027	58 06	112	49 30	149	41 52	207	35 09	236	39 56	269	51 15	311
123	45 09	027	58 55	113	49 57	150	41 27	208	34 24	237	39 03	270	50 34	311
124	45 33	027	59 44	114	50 23	152	41 01	209	33 39	238	38 09	270	49 54	310
125	45 57	026	60 32	115	50 48	153	40 35	211	32 54	239	37 16	271	49 13	310
126	46 21	026	61 21	116	51 12	155	40 07	212	32 08	239	36 23	271	48 32	310
127	46 44	026	62 08	118	51 34	156	39 39	213	31 22	240	35 29	272	47 51	310
128	47 07	025	62 55	119	51 55	157	39 09	214	30 35	241	34 36	272	47 10	310
129	47 29	025	63 42	120	52 15	159	38 39	215	29 48	242	33 42	272	46 29	309
130	47 52	024	64 28	122	52 34	161	38 08	216	29 01	242	32 49	273	45 47	309
131	48 14	024	65 13	123	52 51	162	37 36	217	28 14	243	31 55	273	45 06	309
132	48 35	023	65 57	125	53 07	164	37 04	218	27 26	244	31 02	274	44 24	309
133	48 56	023	66 40	126	53 21	165	36 31	219	26 38	244	30 09	274	43 43	309
134	49 17	022	67 23	128	53 34	167	35 57	220	25 50	245	29 15	275	43 01	309
	Dubhe		◆ARCTURUS		SPICA		Alphard		◆SIRIUS		BETELGEUSE		◆CAPELLA	
135	49 37	022	18 11	077	15 21	111	53 45	169	35 22	221	42 01	255	42 20	309
136	49 57	021	19 03	078	16 11	112	53 55	170	34 47	222	41 09	255	41 38	309
137	50 16	021	19 55	078	17 00	112	54 03	172	34 11	223	40 17	256	40 57	309
138	50 35	020	20 47	078	17 49	113	54 10	174	33 35	223	39 25	257	40 15	309
139	50 53	020	21 40	079	18 38	114	54 15	175	32 58	224	38 33	257	39 33	309
140	51 11	019	22 32	079	19 27	114	54 19	177	32 20	225	37 41	258	38 51	309
141	51 29	019	23 25	080	20 16	115	54 21	179	31 42	226	36 49	258	38 10	309
142	51 45	018	24 17	080	21 04	115	54 22	180	31 03	227	35 56	259	37 28	309
143	52 02	017	25 10	080	21 53	116	54 21	182	30 24	228	35 04	260	36 46	309
144	52 17	017	26 03	081	22 40	117	54 18	184	29 44	229	34 11	260	36 05	309
145	52 32	016	26 56	081	23 28	117	54 14	185	29 04	229	33 18	261	35 23	309
146	52 47	015	27 48	081	24 16	118	54 08	187	28 23	230	32 25	261	34 42	309
147	53 01	015	28 41	082	25 03	119	54 01	189	27 42	231	31 33	262	34 00	309
148	53 14	014	29 34	082	25 49	119	53 52	190	27 00	232	30 40	262	33 19	309
149	53 27	013	30 27	083	26 36	120	53 42	192	26 18	232	29 47	263	32 37	309
	Dubhe		◆ARCTURUS		SPICA		◆Alphard		SIRIUS		POLLUX		◆CAPELLA	
150	53 39	013	31 20	083	27 22	121	53 30	194	25 35	233	60 09	280	31 56	309
151	53 50	012	32 13	083	28 08	121	53 16	195	24 53	234	59 16	280	31 14	309
152	54 01	011	33 06	084	28 54	122	53 02	197	24 09	234	58 24	280	30 33	310
153	54 10	010	34 00	084	29 39	123	52 45	198	23 26	235	57 31	280	29 52	310
154	54 20	010	34 53	085	30 24	123	52 28	200	22 42	236	56 39	281	29 11	310
155	54 28	009	35 46	085	31 08	124	52 09	202	21 57	236	55 46	281	28 30	310
156	54 36	008	36 39	085	31 52	125	51 48	203	21 13	237	54 54	281	27 49	310
157	54 43	007	37 33	086	32 36	126	51 27	205	20 28	238	54 01	281	27 08	310
158	54 50	006	38 26	086	33 19	126	51 04	206	19 42	238	53 09	282	26 27	310
159	54 55	006	39 19	087	34 02	127	50 40	207	18 57	239	52 16	282	25 47	311
160	55 00	005	40 13	087	34 44	128	50 15	209	18 11	240	51 24	282	25 06	311
161	55 04	004	41 06	087	35 26	129	49 49	210	17 24	240	50 32	282	24 26	311
162	55 08	003	42 00	088	36 07	130	49 21	212	16 38	241	49 39	282	23 45	311
163	55 10	002	42 53	088	36 48	131	48 53	213	15 51	241	48 47	283	23 05	311
164	55 12	002	43 46	089	37 28	132	48 23	214	15 04	242	47 55	283	22 25	312
	◆Dubhe		Alkaid		ARCTURUS		◆SPICA		Alphard		◆PROCYON		POLLUX	
165	55 13	001	50 59	044	44 40	089	38 08	132	47 53	215	37 31	255	47 03	283
166	55 13	000	51 36	043	45 33	090	38 47	133	47 21	217	36 39	255	46 11	283
167	55 13	359	52 12	043	46 27	090	39 26	134	46 49	218	35 47	256	45 19	284
168	55 12	358	52 48	043	47 20	090	40 04	135	46 15	219	34 55	257	44 27	284
169	55 10	357	53 24	042	48 14	091	40 41	136	45 41	220	34 03	257	43 35	284
170	55 07	357	54 00	042	49 07	091	41 17	137	45 06	221	33 11	258	42 43	284
171	55 03	356	54 36	041	50 01	092	41 53	138	44 31	223	32 19	258	41 52	285
172	54 59	355	55 11	041	50 54	092	42 28	139	43 54	224	31 26	259	41 00	285
173	54 54	354	55 45	040	51 47	093	43 03	141	43 17	225	30 34	260	40 08	285
174	54 48	353	56 20	040	52 41	093	43 36	142	42 39	226	29 41	260	39 17	285
175	54 41	353	56 53	039	53 34	094	44 09	143	42 00	227	28 48	261	38 25	286
176	54 34	352	57 27	038	54 27	094	44 41	144	41 21	228	27 56	261	37 34	286
177	54 26	351	58 00	038	55 21	095	45 12	145	40 41	229	27 03	262	36 42	286
178	54 17	350	58 32	037	56 14	096	45 42	146	40 01	230	26 10	262	35 51	286
179	54 08	349	59 04	036	57 07	096	46 11	148	39 20	231	25 17	263	35 00	287

LHA	Hc	Zn	Hc	Zn	Hc	Zn	Hc	Zn	Hc	Zn	Hc	Zn	Hc	Zn
♈	◆Alkaid		Alphecca		◆SPICA		Gienah		REGULUS		◆POLLUX		Dubhe	
180	59 35	036	42 33	077	46 39	149	45 20	175	59 47	246	34 09	287	53 58	349
181	60 06	035	43 25	078	47 06	150	45 24	176	58 58	247	33 18	287	53 47	348
182	60 36	034	44 17	078	47 32	151	45 27	177	58 08	248	32 26	288	53 35	347
183	61 06	033	45 10	078	47 58	153	45 29	179	57 19	249	31 36	288	53 23	346
184	61 35	032	46 02	079	48 21	154	45 29	180	56 29	250	30 45	288	53 10	346
185	62 03	031	46 55	079	48 44	155	45 28	182	55 38	251	29 54	288	52 57	345
186	62 30	030	47 47	079	49 06	157	45 26	183	54 48	251	29 03	289	52 43	344
187	62 56	029	48 40	079	49 26	158	45 23	184	53 57	252	28 13	289	52 28	344
188	63 22	028	49 32	080	49 46	160	45 18	186	53 06	253	27 22	289	52 13	343
189	63 46	027	50 25	080	50 04	161	45 13	187	52 15	254	26 32	290	51 57	342
190	64 10	026	51 17	080	50 20	163	45 06	188	51 23	255	25 41	290	51 41	342
191	64 33	024	52 10	080	50 36	164	44 57	190	50 32	255	24 51	290	51 24	341
192	64 54	023	53 03	081	50 50	166	44 48	191	49 40	256	24 01	291	51 06	341
193	65 15	022	53 56	081	51 02	167	44 37	192	48 48	257	23 11	291	50 48	340
194	65 34	020	54 48	081	51 13	169	44 25	194	47 56	258	22 21	291	50 30	339
	◆Kochab		VEGA		Rasalhague		◆ANTARES		SPICA		◆REGULUS		Dubhe	
195	40 41	010	20 46	056	24 31	088	16 38	132	51 23	170	47 03	258	50 11	339
196	40 50	009	21 31	056	25 25	088	17 17	133	51 32	172	46 11	259	49 51	338
197	40 59	009	22 15	057	26 18	089	17 56	134	51 39	173	45 18	259	49 31	338
198	41 07	009	23 00	057	27 12	089	18 35	134	51 44	175	44 26	260	49 11	337
199	41 15	008	23 45	057	28 05	090	19 13	135	51 48	176	43 33	261	48 50	337
200	41 22	008	24 30	057	28 58	090	19 50	136	51 51	178	42 40	261	48 29	336
201	41 30	008	25 15	058	29 52	091	20 27	136	51 52	180	41 48	262	48 07	336
202	41 37	007	26 00	058	30 45	091	21 04	137	51 51	181	40 55	262	47 45	335
203	41 44	007	26 45	058	31 39	092	21 40	138	51 50	183	40 02	263	47 23	335
204	41 50	007	27 30	058	32 32	092	22 16	138	51 46	184	39 08	263	47 00	335
205	41 56	006	28 16	058	33 26	093	22 51	139	51 41	186	38 15	264	46 37	334
206	42 02	006	29 02	059	34 19	093	23 26	140	51 35	188	37 22	265	46 14	334
207	42 07	006	29 47	059	35 12	093	24 00	141	51 27	189	36 29	265	45 50	333
208	42 12	005	30 33	059	36 06	094	24 34	141	51 18	191	35 36	266	45 26	333
209	42 17	005	31 19	059	36 59	094	25 07	142	51 08	192	34 42	266	45 02	333
	Kochab		◆VEGA		Rasalhague		ANTARES		◆SPICA		REGULUS		◆Dubhe	
210	42 22	005	32 05	059	37 52	095	25 39	143	50 56	194	33 49	267	44 37	332
211	42 26	004	32 51	059	38 46	096	26 11	144	50 42	195	32 56	267	44 12	332
212	42 30	004	33 37	060	39 39	096	26 42	145	50 27	197	32 02	267	43 47	332
213	42 33	004	34 23	060	40 32	097	27 13	145	50 11	198	31 09	268	43 22	332
214	42 36	003	35 09	060	41 25	097	27 43	146	49 54	200	30 15	268	42 56	331
215	42 39	003	35 56	060	42 18	098	28 12	147	49 35	201	29 22	269	42 30	331
216	42 42	002	36 42	060	43 11	098	28 41	148	49 16	203	28 28	269	42 04	331
217	42 44	002	37 28	060	44 04	099	29 09	149	48 54	204	27 35	270	41 38	330
218	42 46	002	38 15	060	44 57	099	29 36	150	48 32	205	26 42	270	41 12	330
219	42 47	001	39 01	061	45 49	100	30 03	151	48 09	207	25 48	271	40 45	330
220	42 48	001	39 48	061	46 42	101	30 29	152	47 44	208	24 55	271	40 18	330
221	42 49	001	40 35	061	47 34	101	30 54	152	47 19	209	24 01	272	39 51	330
222	42 49	000	41 21	061	48 27	102	31 18	153	46 52	211	23 08	272	39 24	329
223	42 49	000	42 08	061	49 19	103	31 42	154	46 24	212	22 14	272	38 57	329
224	42 49	000	42 55	061	50 11	103	32 04	155	45 56	213	21 21	273	38 29	329
	VEGA		◆ALTAIR		Nunki		ANTARES		◆SPICA		Denebola		◆Alkaid	
225	43 41	061	19 26	090	12 19	128	32 26	156	45 26	214	43 55	264	63 38	333
226	44 28	061	20 19	090	13 01	129	32 47	157	44 55	215	43 02	265	63 14	332
227	45 15	061	21 13	091	13 42	130	33 08	158	44 24	217	42 09	265	62 48	331
228	46 02	061	22 06	091	14 23	130	33 27	159	43 51	218	41 15	266	62 21	330
229	46 48	061	23 00	092	15 04	131	33 45	160	43 18	219	40 22	266	61 53	329
230	47 35	061	23 53	092	15 44	131	34 03	161	42 44	220	39 29	267	61 25	328
231	48 22	061	24 46	093	16 24	132	34 20	162	42 10	221	38 35	267	60 56	327
232	49 09	061	25 40	093	17 04	133	34 36	163	41 34	222	37 42	268	60 27	326
233	49 55	061	26 33	094	17 43	133	34 50	164	40 58	223	36 48	268	59 56	325
234	50 42	061	27 27	094	18 22	134	35 04	165	40 21	224	35 55	269	59 25	324
235	51 29	061	28 20	094	19 00	135	35 17	167	39 43	225	35 01	269	58 54	323
236	52 15	061	29 13	095	19 38	135	35 29	168	39 05	226	34 08	270	58 22	323
237	53 02	061	30 06	095	20 15	136	35 40	169	38 26	227	33 15	270	57 49	322
238	53 48	060	31 00	096	20 52	137	35 50	170	37 47	228	32 21	271	57 16	321
239	54 35	060	31 53	097	21 29	137	35 59	171	37 07	229	31 28	271	56 42	321
	◆VEGA		ALTAIR		Nunki		◆ANTARES		SPICA		ARCTURUS		◆Alkaid	
240	55 21	060	32 46	097	22 05	138	36 07	172	36 26	230	64 47	258	56 08	320
241	56 08	060	33 39	098	22 40	139	36 14	173	35 45	231	63 55	258	55 34	320
242	56 54	060	34 32	098	23 15	139	36 20	174	35 04	232	63 02	259	54 59	319
243	57 40	059	35 25	099	23 50	140	36 25	175	34 22	232	62 10	260	54 24	319
244	58 26	059	36 17	099	24 24	141	36 29	176	33 39	233	61 17	261	53 49	318
245	59 11	059	37 10	100	24 57	142	36 32	177	32 56	234	60 24	261	53 13	318
246	59 57	058	38 03	100	25 30	143	36 34	179	32 12	235	59 31	262	52 37	317
247	60 43	058	38 55	101	26 02	143	36 35	180	31 29	236	58 38	263	52 00	317
248	61 28	058	39 48	102	26 34	144	36 34	181	30 44	236	57 45	263	51 24	317
249	62 13	057	40 40	102	27 05	145	36 33	182	30 00	237	56 52	264	50 47	316
250	62 58	057	41 32	103	27 35	146	36 31	183	29 15	238	55 59	265	50 10	316
251	63 42	056	42 24	103	28 05	147	36 28	184	28 29	238	55 06	265	49 32	316
252	64 27	056	43 16	104	28 34	148	36 23	185	27 44	239	54 12	266	48 55	315
253	65 11	055	44 08	105	29 02	148	36 18	186	26 57	240	53 19	266	48 17	315
254	65 54	054	45 00	105	29 30	149	36 11	187	26 11	241	52 26	267	47 39	315
	◆DENEB		ALTAIR		Nunki		◆ANTARES		SPICA		ARCTURUS		◆Alkaid	
255	42 48	052	45 51	106	29 57	150	36 04	189	25 24	241	51 32	267	47 01	315
256	43 30	052	46 43	107	30 23	151	35 55	190	24 37	242	50 39	268	46 23	314
257	44 12	052	47 34	108	30 48	152	35 46	191	23 50	242	49 45	268	45 45	314
258	44 54	052	48 24	108	31 13	153	35 35	192	23 03	243	48 52	269	45 07	314
259	45 36	052	49 15	109	31 37	154	35 24	193	22 15	244	47 58	269	44 28	314
260	46 18	052	50 06	110	32 00	155	35 12	194	21 27	244	47 05	270	43 49	314
261	47 00	051	50 56	111	32 23	156	34 58	195	20 38	245	46 12	270	43 11	314
262	47 42	051	51 46	112	32 44	157	34 44	196	19 50	246	45 18	271	42 32	313
263	48 23	051	52 35	112	33 05	158	34 29	197	19 01	246	44 25	271	41 53	313
264	49 05	051	53 24	113	33 25	159	34 12	198	18 12	247	43 31	271	41 14	313
265	49 46	051	54 13	114	33 44	160	33 55	199	17 23	247	42 38	272	40 35	313
266	50 28	051	55 02	115	34 02	161	33 37	200	16 34	248	41 44	272	39 56	313
267	51 09	050	55 50	116	34 19	162	33 18	201	15 44	248	40 51	273	39 17	313
268	51 50	050	56 38	117	34 35	163	32 59	202	14 54	249	39 58	273	38 38	313
269	52 31	050	57 25	118	34 50	164	32 38	203	14 04	249	39 04	274	37 59	313

LAT 27°N

LHA	Hc	Zn	Hc	Zn	Hc	Zn	Hc	Zn	Hc	Zn	Hc	Zn	Hc	Zn
♈	◆DENEB		ALTAIR		◆Nunki		ANTARES		◆ARCTURUS		Alkaid		Kochab	
270	53 12	049	58 12	120	35 05	165	32 16	204	38 11	274	37 20	313	36 59	345
271	53 52	049	58 58	121	35 18	166	31 54	205	37 18	274	36 41	313	36 45	345
272	54 33	049	59 43	122	35 31	167	31 31	206	36 24	275	36 02	313	36 31	345
273	55 13	048	60 28	123	35 42	168	31 07	207	35 31	275	35 23	313	36 18	345
274	55 53	048	61 13	125	35 53	169	30 43	208	34 38	276	34 44	313	36 04	345
275	56 32	048	61 56	126	36 02	170	30 17	209	33 45	276	34 05	313	35 49	345
276	57 11	047	62 39	128	36 11	171	29 51	210	32 51	276	33 26	313	35 35	344
277	57 50	047	63 21	129	36 18	172	29 24	211	31 58	277	32 47	313	35 21	344
278	58 29	046	64 02	131	36 25	174	28 57	212	31 05	277	32 08	313	35 06	344
279	59 07	045	64 42	132	36 30	175	28 28	212	30 12	278	31 29	313	34 51	344
280	59 45	045	65 21	134	36 35	176	27 59	213	29 19	278	30 50	313	34 37	344
281	60 23	044	65 59	136	36 38	177	27 30	214	28 26	278	30 11	314	34 22	344
282	61 00	044	66 35	138	36 40	178	26 59	215	27 33	279	29 32	314	34 06	344
283	61 36	043	67 11	140	36 42	179	26 28	216	26 41	279	28 54	314	33 51	343
284	62 12	042	67 44	142	36 42	180	25 57	217	25 48	279	28 15	314	33 36	343
	◆Alpheratz		Enif		Nunki		◆ANTARES		ARCTURUS		◆Alkaid		Kochab	
285	23 17	068	47 46	106	36 41	181	25 25	217	24 55	280	27 37	314	33 21	343
286	24 07	068	48 37	107	36 39	183	24 52	218	24 02	280	26 58	314	33 05	343
287	24 56	069	49 28	107	36 36	184	24 19	219	23 10	281	26 20	314	32 49	343
288	25 46	069	50 19	108	36 32	185	23 45	220	22 17	281	25 42	314	32 34	343
289	26 36	069	51 10	109	36 27	186	23 10	220	21 25	281	25 03	315	32 18	343
290	27 26	070	52 00	110	36 21	187	22 36	221	20 33	282	24 25	315	32 02	343
291	28 16	070	52 50	111	36 14	188	22 00	222	19 40	282	23 48	315	31 46	343
292	29 07	070	53 40	112	36 06	189	21 24	223	18 48	283	23 10	315	31 30	343
293	29 57	070	54 30	113	35 57	190	20 48	223	17 56	283	22 32	315	31 14	343
294	30 47	071	55 19	114	35 47	191	20 11	224	17 04	283	21 54	315	30 58	342
295	31 38	071	56 08	114	35 36	192	19 33	225	16 12	284	21 17	316	30 42	342
296	32 28	071	56 56	116	35 24	194	18 56	225	15 20	284	20 40	316	30 26	342
297	33 19	071	57 44	117	35 11	195	18 17	226	14 28	284	20 03	316	30 10	342
298	34 10	072	58 32	118	34 57	196	17 39	227	13 36	285	19 26	316	29 53	342
299	35 01	072	59 19	119	34 43	197	17 00	227	12 45	285	18 49	317	29 37	342
	Schedar		◆Alpheratz		FOMALHAUT		◆Nunki		Rasalhague		◆VEGA		Kochab	
300	33 07	038	35 51	072	19 13	140	34 27	198	53 05	254	69 01	309	29 21	342
301	33 40	038	36 42	073	19 47	141	34 10	199	52 13	255	68 19	308	29 05	342
302	34 13	038	37 33	073	20 21	141	33 52	200	51 21	256	67 37	307	28 48	342
303	34 46	038	38 25	073	20 54	142	33 34	201	50 29	256	66 54	307	28 32	342
304	35 19	038	39 16	073	21 27	143	33 14	202	49 37	257	66 11	306	28 16	342
305	35 52	038	40 07	074	21 59	144	32 54	203	48 45	258	65 28	305	27 59	342
306	36 25	038	40 58	074	22 30	144	32 33	204	47 53	258	64 44	305	27 43	342
307	36 58	038	41 50	074	23 01	145	32 11	205	47 00	259	63 59	304	27 26	342
308	37 31	038	42 41	074	23 31	146	31 48	206	46 08	260	63 15	303	27 10	342
309	38 04	038	43 33	074	24 01	147	31 25	207	45 15	260	62 30	303	26 54	342
310	38 37	038	44 24	075	24 30	147	31 00	208	44 22	261	61 45	302	26 37	342
311	39 09	038	45 16	075	24 59	148	30 35	208	43 30	262	61 00	302	26 21	342
312	39 42	037	46 07	075	25 27	149	30 09	209	42 37	262	60 15	302	26 05	342
313	40 14	037	46 59	075	25 54	150	29 43	210	41 44	263	59 29	301	25 49	342
314	40 47	037	47 51	076	26 20	151	29 15	211	40 51	263	58 43	301	25 33	342
	◆Mirfak		Hamal		Diphda		◆FOMALHAUT		ALTAIR		◆VEGA		Kochab	
315	16 41	042	21 37	074	19 36	123	26 46	152	65 32	225	57 57	301	25 16	342
316	17 17	042	22 28	074	20 20	124	27 11	152	64 53	227	57 11	300	25 00	343
317	17 53	043	23 20	075	21 05	125	27 36	153	64 13	229	56 25	300	24 44	343
318	18 29	043	24 11	075	21 48	125	28 00	154	63 32	231	55 39	300	24 28	343
319	19 05	043	25 03	075	22 32	126	28 23	155	62 51	232	54 53	300	24 12	343
320	19 42	043	25 55	076	23 15	127	28 45	156	62 08	234	54 06	300	23 57	343
321	20 18	043	26 46	076	23 58	127	29 06	157	61 25	235	53 20	300	23 41	343
322	20 55	044	27 38	076	24 40	128	29 27	158	60 41	236	52 33	299	23 25	343
323	21 32	044	28 30	077	25 22	129	29 47	159	59 56	238	51 47	299	23 09	343
324	22 09	044	29 22	077	26 03	129	30 06	160	59 10	239	51 00	299	22 54	343
325	22 47	044	30 14	077	26 44	130	30 24	161	58 24	240	50 13	299	22 38	343
326	23 24	044	31 07	078	27 25	131	30 42	161	57 38	241	49 27	299	22 23	343
327	24 01	045	31 59	078	28 05	132	30 58	162	56 50	242	48 40	299	22 08	343
328	24 39	045	32 51	078	28 45	132	31 14	163	56 03	243	47 53	299	21 53	344
329	25 17	045	33 44	079	29 24	133	31 29	164	55 15	244	47 06	299	21 37	344
	◆Mirfak		Hamal		Diphda		◆FOMALHAUT		ALTAIR		◆VEGA		Kochab	
330	25 55	045	34 36	079	30 03	134	31 43	165	54 26	245	46 20	299	21 22	344
331	26 32	045	35 29	079	30 41	135	31 56	166	53 38	246	45 33	299	21 08	344
332	27 10	045	36 21	080	31 18	136	32 08	167	52 48	247	44 46	299	20 53	344
333	27 49	045	37 14	080	31 55	137	32 19	168	51 59	248	43 59	299	20 38	344
334	28 27	046	38 06	080	32 32	137	32 30	169	51 09	249	43 13	299	20 24	344
335	29 05	046	38 59	081	33 08	138	32 39	170	50 19	250	42 26	299	20 09	344
336	29 43	046	39 52	081	33 43	139	32 47	171	49 29	251	41 39	299	19 55	345
337	30 22	046	40 45	081	34 18	140	32 55	172	48 38	252	40 52	299	19 41	345
338	31 00	046	41 38	082	34 52	141	33 02	173	47 47	252	40 06	299	19 27	345
339	31 39	046	42 31	082	35 25	142	33 07	174	46 56	253	39 19	299	19 13	345
340	32 17	046	43 24	082	35 57	143	33 12	175	46 05	254	38 33	300	18 59	345
341	32 56	046	44 17	083	36 29	144	33 16	177	45 14	254	37 46	300	18 45	345
342	33 34	046	45 10	083	37 00	145	33 18	178	44 22	255	37 00	300	18 32	345
343	34 13	046	46 03	083	37 31	146	33 20	179	43 30	256	36 13	300	18 18	346
344	34 52	046	46 56	084	38 00	147	33 21	180	42 38	256	35 27	300	18 05	346
	◆CAPELLA		ALDEBARAN		Diphda		◆FOMALHAUT		ALTAIR		◆VEGA		Kochab	
345	16 25	046	12 41	078	38 29	148	33 21	181	41 46	257	34 41	300	17 52	346
346	17 03	047	13 33	078	38 57	149	33 20	182	40 54	258	33 55	300	17 39	346
347	17 42	047	14 26	079	39 24	150	33 18	183	40 02	258	33 08	300	17 27	346
348	18 21	047	15 18	079	39 50	151	33 15	184	39 10	259	32 22	301	17 14	347
349	19 00	047	16 11	079	40 15	152	33 11	185	38 17	260	31 36	301	17 02	347
350	19 40	048	17 03	080	40 39	154	33 06	186	37 24	260	30 51	301	16 50	347
351	20 19	048	17 56	080	41 03	155	33 00	187	36 32	261	30 05	301	16 38	347
352	20 59	048	18 49	081	41 25	156	32 53	188	35 39	261	29 19	301	16 26	347
353	21 39	048	19 41	081	41 47	157	32 45	189	34 46	262	28 33	302	16 14	347
354	22 19	048	20 34	081	42 07	158	32 36	190	33 53	262	27 48	302	16 03	348
355	22 59	049	21 27	082	42 26	159	32 26	191	33 00	263	27 03	302	15 51	348
356	23 39	049	22 20	082	42 44	161	32 16	192	32 07	263	26 17	302	15 40	348
357	24 19	049	23 13	083	43 02	162	32 04	193	31 14	264	25 32	302	15 29	348
358	25 00	049	24 06	083	43 18	163	31 52	194	30 21	264	24 47	303	15 19	349
359	25 40	049	24 59	083	43 33	164	31 38	195	29 27	265	24 02	303	15 08	349

LAT 26°N

LHA ♈	Hc	Zn	Hc	Zn	Hc	Zn	Hc	Zn	Hc	Zn	Hc	Zn	Hc	Zn
	Schedar		◆CAPELLA		ALDEBARAN		◆Diphda		FOMALHAUT		ALTAIR		◆DENEB	
0	58 38	011	25 42	049	25 46	083	44 45	165	32 22	196	28 39	266	46 07	309
1	58 48	010	26 23	049	26 39	084	44 58	167	32 06	197	27 45	266	45 25	309
2	58 57	009	27 04	049	27 33	084	45 09	168	31 50	198	26 51	267	44 43	309
3	59 04	008	27 45	050	28 27	085	45 20	169	31 33	199	25 57	267	44 01	309
4	59 11	007	28 26	050	29 20	085	45 29	171	31 15	200	25 03	268	43 19	309
5	59 16	005	29 07	050	30 14	085	45 37	172	30 56	201	24 10	268	42 37	309
6	59 21	004	29 48	050	31 08	086	45 44	173	30 36	202	23 16	269	41 55	309
7	59 25	003	30 30	050	32 02	086	45 50	175	30 16	203	22 22	269	41 13	309
8	59 27	002	31 11	050	32 55	087	45 54	176	29 55	204	21 28	270	40 31	309
9	59 29	001	31 52	050	33 49	087	45 57	177	29 32	205	20 34	270	39 48	308
10	59 29	000	32 34	050	34 43	087	45 59	179	29 10	206	19 40	270	39 06	308
11	59 29	359	33 15	050	35 37	088	45 59	180	28 46	206	18 46	271	38 24	308
12	59 27	358	33 57	050	36 31	088	45 58	182	28 22	207	17 52	271	37 42	308
13	59 25	357	34 39	051	37 25	089	45 56	183	27 57	208	16 58	272	36 59	308
14	59 21	356	35 20	051	38 19	089	45 53	184	27 31	209	16 04	272	36 17	308
	◆CAPELLA		ALDEBARAN		RIGEL		◆Diphda		FOMALHAUT		Enif		◆DENEB	
15	36 02	051	39 13	089	19 28	110	45 48	186	27 04	210	40 59	260	35 35	309
16	36 44	051	40 07	090	20 19	110	45 42	187	26 37	211	40 06	261	34 53	309
17	37 25	051	41 00	090	21 09	111	45 35	188	26 09	211	39 12	261	34 11	309
18	38 07	051	41 54	091	21 59	112	45 26	190	25 41	212	38 19	262	33 29	309
19	38 49	051	42 48	091	22 50	112	45 17	191	25 12	213	37 26	262	32 47	309
20	39 30	051	43 42	092	23 39	113	45 06	192	24 42	214	36 32	263	32 05	309
21	40 12	051	44 36	092	24 29	113	44 53	194	24 11	215	35 39	263	31 23	309
22	40 54	050	45 30	093	25 18	114	44 40	195	23 41	215	34 45	264	30 41	309
23	41 35	050	46 24	093	26 08	115	44 26	196	23 09	216	33 51	264	29 59	309
24	42 17	050	47 18	094	26 56	115	44 10	198	22 37	217	32 58	265	29 17	309
25	42 58	050	48 11	094	27 45	116	43 53	199	22 04	218	32 04	265	28 35	309
26	43 40	050	49 05	095	28 33	117	43 35	200	21 31	218	31 10	266	27 54	309
27	44 21	050	49 59	095	29 22	117	43 16	202	20 57	219	30 16	266	27 12	310
28	45 02	050	50 53	096	30 09	118	42 56	203	20 23	220	29 23	267	26 30	310
29	45 44	050	51 46	096	30 57	119	42 35	204	19 48	220	28 29	267	25 49	310
	CAPELLA		◆ALDEBARAN		RIGEL		◆Diphda		Alpheratz		DENEB		◆Kochab	
30	46 25	050	52 40	097	31 44	119	42 12	205	65 04	283	25 08	310	10 31	357
31	47 06	049	53 33	097	32 31	120	41 49	206	64 12	283	24 27	310	10 28	357
32	47 47	049	54 27	098	33 18	121	41 25	207	63 19	283	23 46	310	10 25	357
33	48 27	049	55 20	099	34 04	121	41 00	208	62 27	284	23 05	311	10 23	357
34	49 08	049	56 13	099	34 50	122	40 34	210	61 34	284	22 24	311	10 20	358
35	49 48	048	57 07	100	35 35	123	40 07	211	60 42	284	21 43	311	10 18	358
36	50 29	048	58 00	101	36 20	124	39 39	212	59 50	284	21 02	311	10 16	358
37	51 09	048	58 53	101	37 05	125	39 10	213	58 57	284	20 22	311	10 15	358
38	51 49	048	59 45	102	37 49	125	38 40	214	58 05	284	19 41	312	10 13	359
39	52 28	047	60 38	103	38 33	126	38 10	215	57 12	284	19 01	312	10 12	359
40	53 08	047	61 31	104	39 16	127	37 39	216	56 20	284	18 21	312	10 11	359
41	53 47	046	62 23	105	39 59	128	37 07	217	55 28	284	17 41	312	10 11	000
42	54 26	046	63 15	105	40 41	129	36 34	218	54 36	284	17 02	313	10 11	000
43	55 05	046	64 07	106	41 22	130	36 01	219	53 43	284	16 22	313	10 11	000
44	55 43	045	64 59	107	42 04	131	35 27	220	52 51	285	15 42	313	10 11	000
	◆CAPELLA		PROCYON		SIRIUS		◆RIGEL		Acamar		Diphda		◆Alpheratz	
45	56 21	045	20 28	094	20 38	122	42 44	132	23 40	180	34 52	221	51 59	285
46	56 59	044	21 22	095	21 24	122	43 24	133	23 40	181	34 17	222	51 07	285
47	57 36	044	22 15	095	22 09	123	44 03	134	23 38	182	33 40	222	50 15	285
48	58 13	043	23 09	096	22 54	124	44 42	135	23 36	183	33 04	223	49 23	285
49	58 50	042	24 03	096	23 39	124	45 20	136	23 33	184	32 27	224	48 31	285
50	59 26	042	24 56	097	24 24	125	45 57	137	23 29	185	31 49	225	47 39	286
51	60 01	041	25 50	097	25 08	126	46 33	138	23 24	185	31 10	226	46 47	286
52	60 36	040	26 43	098	25 51	126	47 08	139	23 19	186	30 31	227	45 55	286
53	61 11	039	27 37	098	26 35	127	47 43	141	23 12	187	29 52	227	45 03	286
54	61 45	038	28 30	099	27 17	128	48 17	142	23 05	188	29 12	228	44 11	286
55	62 18	038	29 24	099	28 00	128	48 50	143	22 58	189	28 32	229	43 19	286
56	62 51	037	30 17	100	28 42	129	49 22	144	22 49	189	27 51	230	42 28	287
57	63 22	036	31 10	100	29 24	130	49 53	146	22 40	190	27 09	230	41 36	287
58	63 53	035	32 03	101	30 05	131	50 23	147	22 30	191	26 28	231	40 44	287
59	64 24	033	32 56	101	30 45	131	50 52	148	22 19	192	25 45	232	39 53	287
	◆Dubhe		PROCYON		SIRIUS		◆RIGEL		Diphda		◆Hamal		Mirfak	
60	15 40	028	33 49	102	31 25	132	51 19	150	25 03	233	64 14	270	65 10	346
61	16 06	028	34 41	102	32 05	133	51 46	151	24 20	233	63 20	271	64 56	345
62	16 32	029	35 34	103	32 44	134	52 11	153	23 36	234	62 26	271	64 41	343
63	16 58	029	36 27	104	33 23	135	52 36	154	22 52	235	61 32	272	64 25	342
64	17 24	029	37 19	104	34 01	136	52 59	156	22 08	235	60 39	272	64 08	341
65	17 50	029	38 11	105	34 38	137	53 20	157	21 24	236	59 45	272	63 49	339
66	18 16	029	39 03	105	35 15	137	53 41	159	20 39	237	58 51	273	63 30	338
67	18 43	030	39 55	106	35 51	138	53 59	160	19 54	237	57 57	273	63 09	337
68	19 09	030	40 47	107	36 27	139	54 17	162	19 08	238	57 03	274	62 47	336
69	19 36	030	41 38	107	37 01	140	54 33	164	18 23	238	56 09	274	62 24	335
70	20 03	030	42 30	108	37 36	141	54 48	165	17 37	239	55 15	274	62 01	333
71	20 30	030	43 21	109	38 09	142	55 01	167	16 50	240	54 22	275	61 36	332
72	20 57	030	44 12	109	38 42	143	55 12	169	16 04	240	53 28	275	61 11	331
73	21 25	030	45 03	110	39 14	144	55 22	170	15 17	241	52 34	275	60 44	330
74	21 52	031	45 53	111	39 45	145	55 30	172	14 30	241	51 41	276	60 17	329
	◆Dubhe		POLLUX		PROCYON		◆SIRIUS		RIGEL		◆Hamal		Mirfak	
75	22 20	031	53 22	077	46 43	112	40 15	146	55 37	174	50 47	276	59 50	329
76	22 47	031	54 15	077	47 33	113	40 45	147	55 42	175	49 53	276	59 21	328
77	23 15	031	55 07	078	48 23	113	41 13	148	55 46	177	49 00	277	58 52	327
78	23 43	031	56 00	078	49 12	114	41 41	150	55 47	179	48 06	277	58 22	326
79	24 11	031	56 53	078	50 01	115	42 08	151	55 47	181	47 13	277	57 52	325
80	24 39	031	57 45	078	50 50	116	42 34	152	55 46	183	46 19	277	57 21	325
81	25 07	031	58 38	078	51 38	117	42 59	153	55 43	184	45 26	278	56 49	324
82	25 35	031	59 31	078	52 26	118	43 22	154	55 38	186	44 32	278	56 17	323
83	26 03	031	60 24	078	53 14	119	43 45	155	55 31	188	43 39	278	55 45	323
84	26 31	032	61 17	079	54 01	120	44 07	157	55 23	189	42 46	279	55 12	322
85	26 59	032	62 10	079	54 47	121	44 28	158	55 14	191	41 52	279	54 38	321
86	27 28	032	63 02	079	55 33	122	44 48	159	55 02	193	40 59	279	54 04	321
87	27 56	032	63 55	079	56 19	123	45 06	161	54 50	195	40 06	280	53 30	320
88	28 24	032	64 48	079	57 03	124	45 24	162	54 35	196	39 13	280	52 56	320
89	28 53	032	65 41	079	57 47	126	45 40	163	54 19	198	38 20	280	52 21	319

LHA ♈	Hc	Zn	Hc	Zn	Hc	Zn	Hc	Zn	Hc	Zn	Hc	Zn	Hc	Zn
	Dubhe		◆REGULUS		PROCYON		SIRIUS		◆RIGEL		ALDEBARAN		◆Mirfak	
90	29 21	032	30 14	091	58 31	127	45 55	164	54 02	200	68 13	248	51 46	319
91	29 49	032	31 08	091	59 14	128	46 09	166	53 43	201	67 22	250	51 10	319
92	30 18	032	32 02	092	59 55	130	46 21	167	53 23	203	66 32	251	50 34	318
93	30 46	032	32 56	092	60 36	131	46 33	168	53 02	204	65 40	252	49 58	318
94	31 14	032	33 49	093	61 17	133	46 43	170	52 39	206	64 49	253	49 22	318
95	31 43	032	34 43	093	61 56	134	46 52	171	52 15	207	63 57	254	48 45	317
96	32 11	032	35 37	094	62 34	136	46 59	173	51 50	209	63 05	255	48 09	317
97	32 39	032	36 31	094	63 11	138	47 06	174	51 23	210	62 13	256	47 32	317
98	33 08	032	37 25	095	63 46	139	47 11	175	50 56	211	61 21	256	46 55	316
99	33 36	031	38 18	095	64 21	141	47 14	177	50 27	213	60 28	257	46 17	316
100	34 04	031	39 12	096	64 54	143	47 16	178	49 57	214	59 36	258	45 40	316
101	34 32	031	40 06	096	65 25	145	47 17	180	49 26	215	58 43	259	45 02	316
102	35 00	031	40 59	097	65 55	147	47 17	181	48 55	217	57 50	260	44 24	315
103	35 28	031	41 53	097	66 23	150	47 15	182	48 22	218	56 57	260	43 46	315
104	35 56	031	42 46	098	66 50	152	47 12	184	47 48	219	56 04	261	43 08	315
	Dubhe		◆REGULUS		Alphard		SIRIUS		◆RIGEL		ALDEBARAN		◆CAPELLA	
105	36 23	031	43 40	098	40 12	129	47 08	185	47 14	220	55 10	262	61 16	321
106	36 51	031	44 33	099	40 53	130	47 02	187	46 38	222	54 17	262	60 42	320
107	37 18	031	45 26	100	41 34	131	46 55	188	46 02	223	53 24	263	60 07	319
108	37 46	030	46 19	100	42 15	132	46 47	189	45 25	224	52 30	263	59 32	319
109	38 13	030	47 12	101	42 54	133	46 38	191	44 47	225	51 36	264	58 56	318
110	38 40	030	48 05	101	43 34	134	46 27	192	44 09	226	50 43	264	58 19	317
111	39 07	030	48 58	102	44 12	135	46 15	194	43 30	227	49 49	265	57 42	317
112	39 34	030	49 51	103	44 50	136	46 02	195	42 50	228	48 55	265	57 05	316
113	40 01	029	50 43	103	45 27	137	45 47	196	42 10	229	48 02	266	56 27	315
114	40 27	029	51 36	104	46 03	138	45 31	198	41 29	230	47 08	267	55 49	315
115	40 53	029	52 28	105	46 39	139	45 14	199	40 47	231	46 14	267	55 11	314
116	41 19	029	53 20	106	47 13	141	44 56	200	40 05	232	45 20	267	54 32	314
117	41 45	029	54 12	106	47 47	142	44 37	201	39 22	233	44 26	268	53 53	314
118	42 11	028	55 03	107	48 20	143	44 17	203	38 39	234	43 32	268	53 14	313
119	42 36	028	55 55	108	48 52	144	43 56	204	37 55	234	42 38	269	52 35	313
	◆Dubhe		REGULUS		◆Alphard		SIRIUS		RIGEL		◆ALDEBARAN		CAPELLA	
120	43 01	028	56 46	109	49 23	146	43 33	205	37 11	235	41 44	269	51 55	312
121	43 26	027	57 37	110	49 53	147	43 10	206	36 27	236	40 51	270	51 15	312
122	43 51	027	58 28	111	50 22	148	42 45	208	35 42	237	39 57	270	50 35	312
123	44 15	027	59 18	112	50 49	150	42 20	209	34 56	238	39 03	271	49 55	312
124	44 39	026	60 08	113	51 16	151	41 54	210	34 11	238	38 09	271	49 15	311
125	45 03	026	60 57	114	51 42	152	41 26	211	33 25	239	37 15	271	48 34	311
126	45 27	026	61 47	115	52 06	154	40 58	212	32 38	240	36 21	272	47 53	311
127	45 50	025	62 35	116	52 29	155	40 29	213	31 51	241	35 27	272	47 12	311
128	46 12	025	63 23	117	52 51	157	39 59	214	31 04	241	34 33	273	46 31	310
129	46 35	024	64 11	119	53 11	158	39 28	215	30 17	242	33 39	273	45 50	310
130	46 57	024	64 58	120	53 30	160	38 57	216	29 29	243	32 45	274	45 09	310
131	47 19	023	65 45	121	53 48	162	38 24	217	28 41	243	31 52	274	44 28	310
132	47 40	023	66 30	123	54 04	163	37 51	218	27 52	244	30 58	274	43 46	310
133	48 01	023	67 15	124	54 19	165	37 17	219	27 04	245	30 04	275	43 05	310
134	48 21	022	67 59	126	54 32	167	36 43	220	26 15	245	29 10	275	42 23	310
	Dubhe		◆ARCTURUS		SPICA		Alphard		◆SIRIUS		BETELGEUSE		◆CAPELLA	
135	48 41	022	17 57	077	15 43	111	54 44	168	36 08	221	42 16	256	41 42	310
136	49 01	021	18 50	077	16 33	112	54 54	170	35 32	222	41 24	256	41 00	310
137	49 20	021	19 42	078	17 23	112	55 03	172	34 55	223	40 31	257	40 19	309
138	49 39	020	20 35	078	18 13	113	55 10	173	34 18	224	39 39	257	39 37	309
139	49 57	019	21 28	078	19 02	113	55 15	175	33 40	225	38 46	258	38 55	309
140	50 14	019	22 21	079	19 52	114	55 19	177	33 02	226	37 53	259	38 14	309
141	50 32	018	23 14	079	20 41	114	55 21	179	32 23	227	37 00	259	37 32	309
142	50 48	018	24 07	079	21 30	115	55 22	180	31 44	227	36 07	260	36 50	309
143	51 04	017	25 00	080	22 19	116	55 21	182	31 04	228	35 14	260	36 09	309
144	51 20	016	25 53	080	23 07	116	55 18	184	30 24	229	34 21	261	35 27	309
145	51 35	016	26 46	081	23 55	117	55 14	185	29 43	230	33 28	261	34 45	309
146	51 49	015	27 39	081	24 43	117	55 08	187	29 01	230	32 34	262	34 04	310
147	52 03	014	28 32	081	25 31	118	55 00	189	28 19	231	31 41	262	33 22	310
148	52 16	014	29 26	082	26 19	119	54 51	191	27 37	232	30 47	263	32 41	310
149	52 28	013	30 19	082	27 06	119	54 40	192	26 55	233	29 54	263	31 59	310
	Dubhe		◆ARCTURUS		SPICA		◆Alphard		SIRIUS		POLLUX		◆CAPELLA	
150	52 40	012	31 13	082	27 52	120	54 28	194	26 11	233	59 58	282	31 18	310
151	52 51	012	32 06	083	28 39	121	54 14	196	25 28	234	59 05	282	30 36	310
152	53 02	011	33 00	083	29 25	121	53 59	197	24 44	235	58 12	282	29 55	310
153	53 11	010	33 53	083	30 11	122	53 42	199	24 00	235	57 19	282	29 14	310
154	53 21	009	34 47	084	30 56	123	53 24	200	23 15	236	56 27	282	28 32	310
155	53 29	009	35 40	084	31 42	124	53 04	202	22 30	237	55 34	282	27 51	310
156	53 37	008	36 34	085	32 26	124	52 44	204	21 45	237	54 41	282	27 10	310
157	53 44	007	37 28	085	33 11	125	52 21	205	20 59	238	53 49	283	26 29	311
158	53 50	006	38 22	085	33 54	126	51 58	207	20 14	239	52 56	283	25 48	311
159	53 56	006	39 15	086	34 38	127	51 33	208	19 27	239	52 03	283	25 07	311
160	54 00	005	40 09	086	35 21	128	51 07	209	18 41	240	51 11	283	24 27	311
161	54 04	004	41 03	087	36 03	128	50 40	211	17 54	240	50 18	283	23 46	311
162	54 08	003	41 57	087	36 45	129	50 12	212	17 07	241	49 26	284	23 06	311
163	54 10	002	42 51	087	37 27	130	49 43	214	16 20	242	48 34	284	22 25	312
164	54 12	001	43 44	088	38 08	131	49 13	215	15 32	242	47 41	284	21 45	312
	◆Dubhe		Alkaid		ARCTURUS		◆SPICA		Alphard		◆PROCYON		POLLUX	
165	54 13	001	50 15	043	44 38	088	38 48	132	48 41	216	37 46	256	46 49	284
166	54 13	000	50 52	042	45 32	089	39 28	133	48 09	217	36 54	256	45 57	284
167	54 13	359	51 28	042	46 26	089	40 07	134	47 36	219	36 01	257	45 04	285
168	54 12	358	52 04	042	47 20	089	40 46	135	47 02	220	35 09	257	44 12	285
169	54 10	357	52 40	041	48 14	090	41 24	136	46 27	221	34 16	258	43 20	285
170	54 07	357	53 15	041	49 08	090	42 01	137	45 51	222	33 23	258	42 28	285
171	54 03	356	53 50	040	50 02	091	42 38	138	45 15	223	32 30	259	41 36	285
172	53 59	355	54 25	040	50 56	091	43 14	139	44 37	224	31 37	260	40 44	286
173	53 54	354	54 59	039	51 50	092	43 49	140	43 59	225	30 44	260	39 52	286
174	53 48	353	55 33	039	52 44	092	44 23	141	43 21	226	29 51	261	39 00	286
175	53 42	353	56 06	038	53 38	093	44 57	142	42 41	227	28 58	261	38 09	286
176	53 35	352	56 39	037	54 31	093	45 29	143	42 01	228	28 05	262	37 17	287
177	53 27	351	57 12	037	55 25	094	46 01	145	41 21	229	27 11	262	36 25	287
178	53 18	350	57 44	036	56 19	094	46 32	146	40 40	230	26 18	263	35 34	287
179	53 09	350	58 15	035	57 13	095	47 02	147	39 58	231	25 24	263	34 42	287

LAT 26°N

LHA ♈	Hc	Zn	Hc	Zn	Hc	Zn	Hc	Zn	Hc	Zn	Hc	Zn	Hc	Zn
	◆Alkaid		Alphecca		◆SPICA		Gienah		REGULUS		◆POLLUX		Dubhe	
180	58 46	035	42 19	077	47 30	148	46 19	175	60 10	247	33 51	288	52 59	349
181	59 17	034	43 12	077	47 58	150	46 24	176	59 21	248	32 59	288	52 48	348
182	59 46	033	44 04	077	48 25	151	46 27	177	58 30	249	32 08	288	52 37	347
183	60 15	032	44 57	077	48 51	152	46 29	179	57 40	250	31 17	288	52 25	347
184	60 43	031	45 50	078	49 15	154	46 29	180	56 49	251	30 26	289	52 12	346
185	61 11	030	46 42	078	49 39	155	46 28	182	55 58	252	29 35	289	51 59	345
186	61 38	029	47 35	078	50 01	156	46 26	183	55 06	253	28 44	289	51 45	345
187	62 04	028	48 28	078	50 22	158	46 23	184	54 15	254	27 53	289	51 30	344
188	62 29	027	49 21	078	50 42	159	46 18	186	53 23	254	27 02	290	51 15	343
189	62 53	026	50 14	079	51 00	161	46 12	187	52 31	255	26 11	290	51 00	343
190	63 16	025	51 06	079	51 17	162	46 05	188	51 38	256	25 21	290	50 43	342
191	63 38	024	51 59	079	51 33	164	45 56	190	50 46	257	24 30	291	50 27	342
192	63 59	022	52 52	079	51 48	165	45 47	191	49 54	257	23 40	291	50 09	341
193	64 19	021	53 45	080	52 01	167	45 36	192	49 01	258	22 50	291	49 52	340
194	64 38	020	54 38	080	52 12	168	45 23	194	48 08	259	21 59	292	49 33	340
	◆Kochab		VEGA		Rasalhague		◆ANTARES		SPICA		◆REGULUS		Dubhe	
195	39 42	009	20 13	056	24 29	087	17 18	132	52 22	170	47 15	259	49 15	339
196	39 51	009	20 57	056	25 23	088	17 58	133	52 31	172	46 22	260	48 55	339
197	39 59	009	21 42	056	26 17	088	18 37	133	52 38	173	45 29	260	48 36	338
198	40 08	009	22 27	056	27 10	089	19 16	134	52 44	175	44 36	261	48 15	338
199	40 15	008	23 12	057	28 04	089	19 55	135	52 48	176	43 42	262	47 55	337
200	40 23	008	23 57	057	28 58	090	20 33	135	52 51	178	42 49	262	47 34	337
201	40 30	008	24 42	057	29 52	090	21 11	136	52 52	180	41 56	263	47 12	336
202	40 37	007	25 28	057	30 46	091	21 48	137	52 51	181	41 02	263	46 51	336
203	40 44	007	26 13	058	31 40	091	22 25	137	52 49	183	40 09	264	46 28	336
204	40 50	007	26 59	058	32 34	091	23 01	138	52 46	184	39 15	264	46 06	335
205	40 56	006	27 44	058	33 28	092	23 37	139	52 41	186	38 21	265	45 43	335
206	41 02	006	28 30	058	34 22	092	24 12	140	52 35	188	37 27	265	45 20	334
207	41 08	006	29 16	058	35 16	093	24 46	140	52 27	189	36 34	266	44 56	334
208	41 13	005	30 02	058	36 10	093	25 21	141	52 17	191	35 40	266	44 32	334
209	41 18	005	30 48	059	37 03	094	25 54	142	52 06	192	34 46	267	44 08	333
	Kochab		◆VEGA		Rasalhague		ANTARES		◆SPICA		REGULUS		◆Dubhe	
210	41 22	005	31 34	059	37 57	094	26 27	143	51 54	194	33 52	267	43 44	333
211	41 26	004	32 20	059	38 51	095	26 59	144	51 40	196	32 58	268	43 19	333
212	41 30	004	33 06	059	39 45	095	27 31	144	51 25	197	32 04	268	42 54	332
213	41 33	004	33 53	059	40 38	096	28 02	145	51 08	199	31 11	269	42 29	332
214	41 37	003	34 39	059	41 32	096	28 33	146	50 51	200	30 17	269	42 03	332
215	41 39	003	35 25	059	42 26	097	29 03	147	50 31	202	29 23	269	41 38	331
216	41 42	002	36 12	060	43 19	097	29 32	148	50 11	203	28 29	270	41 12	331
217	41 44	002	36 58	060	44 13	098	30 00	149	49 49	204	27 35	270	40 46	331
218	41 46	002	37 45	060	45 06	098	30 28	149	49 26	206	26 41	271	40 19	331
219	41 47	001	38 31	060	45 59	099	30 55	150	49 02	207	25 47	271	39 53	330
220	41 48	001	39 18	060	46 52	100	31 21	151	48 37	209	24 53	272	39 26	330
221	41 49	001	40 05	060	47 46	100	31 47	152	48 11	210	23 59	272	38 59	330
222	41 49	000	40 52	060	48 39	101	32 12	153	47 43	211	23 05	272	38 32	330
223	41 49	000	41 38	060	49 31	101	32 36	154	47 15	212	22 11	273	38 05	330
224	41 49	000	42 25	060	50 24	102	32 59	155	46 46	214	21 18	273	37 38	330
	VEGA		◆ALTAIR		Nunki		ANTARES		◆SPICA		Denebola		◆Alkaid	
225	43 12	060	19 25	089	12 56	128	33 21	156	46 15	215	44 00	265	62 45	334
226	43 59	060	20 19	090	13 38	129	33 43	157	45 44	216	43 07	266	62 20	333
227	44 45	060	21 13	090	14 20	129	34 03	158	45 12	217	42 13	266	61 55	331
228	45 32	060	22 07	091	15 02	130	34 23	159	44 39	218	41 19	267	61 29	330
229	46 19	060	23 01	091	15 43	130	34 42	160	44 05	220	40 25	267	61 02	329
230	47 06	060	23 55	092	16 24	131	35 00	161	43 30	221	39 31	268	60 34	329
231	47 52	060	24 49	092	17 04	132	35 17	162	42 55	222	38 37	268	60 06	328
232	48 39	060	25 43	093	17 44	132	35 33	163	42 18	223	37 44	269	59 37	327
233	49 26	060	26 37	093	18 24	133	35 48	164	41 41	224	36 50	269	59 07	326
234	50 12	060	27 30	093	19 03	134	36 02	165	41 04	225	35 56	270	58 36	325
235	50 59	060	28 24	094	19 42	134	36 16	166	40 25	226	35 02	270	58 05	324
236	51 45	060	29 18	094	20 20	135	36 28	167	39 46	227	34 08	270	57 34	324
237	52 32	059	30 12	095	20 58	136	36 39	169	39 07	228	33 14	271	57 01	323
238	53 18	059	31 06	095	21 36	136	36 49	170	38 27	229	32 20	271	56 29	322
239	54 05	059	31 59	096	22 13	137	36 59	171	37 46	229	31 26	272	55 55	322
	◆VEGA		ALTAIR		Nunki		◆ANTARES		SPICA		ARCTURUS		◆Alkaid	
240	54 51	059	32 53	096	22 49	138	37 07	172	37 05	230	64 59	260	55 22	321
241	55 37	059	33 46	097	23 25	138	37 14	173	36 23	231	64 06	260	54 48	321
242	56 23	058	34 40	097	24 01	139	37 20	174	35 41	232	63 12	261	54 13	320
243	57 09	058	35 33	098	24 36	140	37 25	175	34 58	233	62 19	262	53 39	320
244	57 54	058	36 27	098	25 10	141	37 29	176	34 15	234	61 26	263	53 03	319
245	58 40	057	37 20	099	25 44	141	37 32	177	33 31	234	60 32	263	52 28	319
246	59 25	057	38 13	100	26 17	142	37 34	179	32 47	235	59 39	264	51 52	318
247	60 10	057	39 06	100	26 50	143	37 35	180	32 02	236	58 45	264	51 16	318
248	60 55	056	39 59	101	27 22	144	37 34	181	31 17	237	57 51	265	50 40	317
249	61 40	056	40 52	101	27 54	145	37 33	182	30 32	238	56 58	265	50 03	317
250	62 24	055	41 45	102	28 25	146	37 31	183	29 46	238	56 04	266	49 26	317
251	63 08	055	42 38	102	28 55	146	37 27	184	29 00	239	55 10	267	48 49	316
252	63 52	054	43 30	103	29 24	147	37 23	185	28 14	240	54 16	267	48 12	316
253	64 36	053	44 23	104	29 53	148	37 17	186	27 27	240	53 22	268	47 34	316
254	65 19	053	45 15	104	30 21	149	37 11	188	26 40	241	52 28	268	46 57	316
	◆DENEB		ALTAIR		Nunki		◆ANTARES		SPICA		ARCTURUS		◆Alkaid	
255	42 10	051	46 07	105	30 49	150	37 03	189	25 53	242	51 34	269	46 19	315
256	42 53	051	46 59	106	31 15	151	36 55	190	25 06	242	50 41	269	45 41	315
257	43 35	051	47 51	106	31 41	152	36 45	191	24 18	243	49 47	269	45 03	315
258	44 17	051	48 43	107	32 07	153	36 34	192	23 30	243	48 53	270	44 25	315
259	44 59	051	49 34	108	32 31	154	36 22	193	22 41	244	47 59	270	43 46	315
260	45 40	051	50 25	109	32 55	155	36 10	194	21 53	245	47 05	271	43 08	314
261	46 22	051	51 16	110	33 17	156	35 56	195	21 04	245	46 11	271	42 29	314
262	47 04	051	52 07	110	33 39	156	35 41	196	20 15	246	45 17	272	41 50	314
263	47 45	050	52 57	111	34 00	157	35 26	197	19 25	246	44 23	272	41 12	314
264	48 27	050	53 48	112	34 21	158	35 09	198	18 36	247	43 29	272	40 33	314
265	49 08	050	54 37	113	34 40	160	34 52	199	17 46	248	42 35	273	39 54	314
266	49 49	050	55 27	114	34 58	161	34 34	200	16 56	248	41 41	273	39 15	314
267	50 30	049	56 16	115	35 16	162	34 14	201	16 06	249	40 48	274	38 36	314
268	51 11	049	57 05	116	35 32	163	33 54	202	15 16	249	39 54	274	37 57	314
269	51 52	049	57 53	117	35 48	164	33 33	203	14 25	250	39 00	274	37 18	314

LHA ♈	Hc	Zn	Hc	Zn	Hc	Zn	Hc	Zn	Hc	Zn	Hc	Zn	Hc	Zn
	◆DENEB		ALTAIR		◆Nunki		ANTARES		◆ARCTURUS		Alkaid		Kochab	
270	52 32	048	58 41	118	36 03	165	33 11	204	38 06	275	36 39	314	36 00	346
271	53 13	048	59 28	119	36 16	166	32 48	205	37 13	275	36 00	314	35 47	345
272	53 53	048	60 15	121	36 29	167	32 25	206	36 19	275	35 21	314	35 33	345
273	54 32	047	61 01	122	36 41	168	32 01	207	35 25	276	34 42	314	35 20	345
274	55 12	047	61 46	123	36 52	169	31 36	208	34 32	276	34 03	314	35 06	345
275	55 51	046	62 31	125	37 01	170	31 10	209	33 38	277	33 24	314	34 52	345
276	56 30	046	63 15	126	37 10	171	30 43	210	32 44	277	32 44	314	34 37	345
277	57 09	045	63 58	127	37 18	172	30 16	211	31 51	277	32 05	314	34 23	344
278	57 47	045	64 41	129	37 24	174	29 48	212	30 57	278	31 26	314	34 08	344
279	58 25	044	65 22	131	37 30	175	29 19	213	30 04	278	30 47	314	33 54	344
280	59 02	044	66 02	133	37 34	176	28 49	214	29 11	278	30 09	314	33 39	344
281	59 39	043	66 42	134	37 38	177	28 19	214	28 17	279	29 30	314	33 24	344
282	60 16	042	67 19	136	37 40	178	27 48	215	27 24	279	28 51	314	33 09	344
283	60 52	042	67 56	138	37 42	179	27 17	216	26 31	280	28 12	314	32 54	344
284	61 27	041	68 31	141	37 42	180	26 45	217	25 38	280	27 33	314	32 38	343
	DENEB		◆Alpheratz		Enif		◆Nunki		ANTARES		◆ARCTURUS		Kochab	
285	62 02	040	22 54	068	48 02	105	37 41	181	26 12	218	24 45	280	32 23	343
286	62 37	039	23 44	068	48 54	106	37 39	183	25 39	218	23 52	281	32 08	343
287	63 10	038	24 34	068	49 46	106	37 36	184	25 05	219	22 59	281	31 52	343
288	63 43	037	25 25	068	50 37	107	37 32	185	24 31	220	22 06	281	31 36	343
289	64 15	036	26 15	069	51 29	108	37 27	186	23 56	221	21 13	282	31 21	343
290	64 47	035	27 05	069	52 20	109	37 21	187	23 21	221	20 20	282	31 05	343
291	65 17	034	27 55	069	53 11	110	37 14	188	22 45	222	19 27	282	30 49	343
292	65 47	033	28 46	070	54 02	110	37 06	189	22 08	223	18 35	283	30 33	343
293	66 15	031	29 37	070	54 52	111	36 56	190	21 31	224	17 42	283	30 17	343
294	66 43	030	30 27	070	55 42	112	36 46	191	20 54	224	16 50	284	30 01	343
295	67 09	029	31 18	070	56 32	113	36 35	193	20 16	225	15 57	284	29 45	343
296	67 34	027	32 09	071	57 21	114	36 23	194	19 38	226	15 05	284	29 29	343
297	67 58	026	33 00	071	58 10	115	36 09	195	18 59	226	14 13	285	29 13	342
298	68 21	024	33 51	071	58 59	116	35 55	196	18 20	227	13 21	285	28 56	342
299	68 42	022	34 42	071	59 47	117	35 40	197	17 40	228	12 29	285	28 40	342
	Schedar		◆Alpheratz		FOMALHAUT		◆Nunki		Rasalhague		◆VEGA		Kochab	
300	32 20	038	35 33	072	19 59	140	35 24	198	53 20	256	68 22	311	28 24	342
301	32 53	038	36 24	072	20 34	140	35 07	199	52 28	256	67 41	310	28 07	342
302	33 26	038	37 15	072	21 08	141	34 49	200	51 35	257	66 59	309	27 51	342
303	33 59	038	38 07	072	21 41	142	34 30	201	50 43	258	66 18	308	27 35	342
304	34 32	038	38 58	073	22 14	143	34 10	202	49 50	258	65 35	308	27 18	342
305	35 05	038	39 50	073	22 47	143	33 49	203	48 57	259	64 52	307	27 02	342
306	35 38	038	40 41	073	23 19	144	33 28	204	48 04	260	64 09	306	26 46	342
307	36 11	038	41 33	073	23 50	145	33 06	205	47 11	260	63 25	306	26 29	342
308	36 44	037	42 24	073	24 21	146	32 42	206	46 18	261	62 41	305	26 13	342
309	37 16	037	43 16	074	24 51	146	32 18	207	45 25	261	61 57	305	25 57	342
310	37 49	037	44 08	074	25 21	147	31 54	208	44 31	262	61 12	304	25 40	342
311	38 22	037	45 00	074	25 50	148	31 28	209	43 38	262	60 28	304	25 24	342
312	38 54	037	45 51	074	26 18	149	31 02	210	42 44	263	59 43	303	25 08	342
313	39 27	037	46 43	074	26 46	150	30 35	211	41 51	264	58 57	303	24 52	343
314	39 59	037	47 35	074	27 13	150	30 07	211	40 57	264	58 12	302	24 35	343
	◆Mirfak		Hamal		Diphda		◆FOMALHAUT		ALTAIR		◆VEGA		Kochab	
315	15 56	042	21 20	073	20 09	123	27 39	151	66 13	227	57 26	302	24 19	343
316	16 32	042	22 12	074	20 54	124	28 04	152	65 33	229	56 40	302	24 03	343
317	17 08	042	23 03	074	21 39	124	28 29	153	64 52	230	55 54	302	23 47	343
318	17 45	043	23 55	074	22 23	125	28 53	154	64 10	232	55 08	301	23 31	343
319	18 21	043	24 47	075	23 07	126	29 17	155	63 27	234	54 22	301	23 15	343
320	18 58	043	25 39	075	23 51	126	29 39	156	62 43	235	53 36	301	22 59	343
321	19 35	043	26 32	075	24 34	127	30 01	157	61 58	237	52 50	301	22 43	343
322	20 12	043	27 24	076	25 17	128	30 22	157	61 13	238	52 03	301	22 28	343
323	20 49	044	28 16	076	25 59	128	30 43	158	60 27	239	51 17	300	22 12	343
324	21 26	044	29 09	076	26 41	129	31 02	159	59 41	240	50 30	300	21 56	343
325	22 04	044	30 01	077	27 23	130	31 21	160	58 53	242	49 44	300	21 41	343
326	22 41	044	30 54	077	28 04	131	31 38	161	58 06	243	48 57	300	21 26	343
327	23 19	044	31 46	077	28 45	131	31 55	162	57 18	244	48 10	300	21 10	344
328	23 56	044	32 39	078	29 25	132	32 11	163	56 29	245	47 24	300	20 55	344
329	24 34	045	33 31	078	30 05	133	32 26	164	55 40	246	46 37	300	20 40	344
	◆Mirfak		Hamal		Diphda		◆FOMALHAUT		ALTAIR		◆VEGA		Kochab	
330	25 12	045	34 24	078	30 44	134	32 41	165	54 51	247	45 50	300	20 25	344
331	25 50	045	35 17	079	31 23	134	32 54	166	54 01	248	45 03	300	20 10	344
332	26 28	045	36 10	079	32 01	135	33 06	167	53 11	249	44 17	300	19 55	344
333	27 06	045	37 03	079	32 39	136	33 18	168	52 21	249	43 30	300	19 40	344
334	27 45	045	37 56	080	33 16	137	33 28	169	51 30	250	42 43	300	19 26	344
335	28 23	045	38 49	080	33 52	138	33 38	170	50 39	251	41 56	300	19 11	345
336	29 01	045	39 42	080	34 28	139	33 47	171	49 48	252	41 10	300	18 57	345
337	29 40	046	40 35	081	35 04	140	33 54	172	48 57	253	40 23	300	18 43	345
338	30 18	046	41 29	081	35 38	141	34 01	173	48 05	253	39 36	300	18 29	345
339	30 57	046	42 22	081	36 12	142	34 07	174	47 13	254	38 49	300	18 15	345
340	31 35	046	43 15	081	36 45	143	34 12	175	46 21	255	38 03	300	18 01	345
341	32 14	046	44 08	082	37 18	144	34 16	176	45 29	255	37 16	300	17 47	345
342	32 53	046	45 02	082	37 49	145	34 18	178	44 37	256	36 30	300	17 34	346
343	33 31	046	45 55	082	38 20	146	34 20	179	43 45	257	35 43	301	17 20	346
344	34 10	046	46 49	083	38 50	147	34 21	180	42 52	257	34 57	301	17 07	346
	◆CAPELLA		ALDEBARAN		Diphda		◆FOMALHAUT		ALTAIR		◆VEGA		Kochab	
345	15 43	046	12 28	078	39 20	148	34 21	181	41 59	258	34 10	301	16 54	346
346	16 22	046	13 21	078	39 48	149	34 20	182	41 07	259	33 24	301	16 41	346
347	17 01	047	14 14	078	40 16	150	34 18	183	40 14	259	32 38	301	16 28	346
348	17 40	047	15 07	079	40 42	151	34 14	184	39 21	260	31 52	301	16 16	347
349	18 20	047	16 00	079	41 08	152	34 10	185	38 28	260	31 06	301	16 03	347
350	18 59	047	16 53	079	41 33	153	34 05	186	37 34	261	30 19	301	15 51	347
351	19 39	047	17 46	080	41 57	154	33 59	187	36 41	261	29 34	302	15 39	347
352	20 19	048	18 39	080	42 20	155	33 52	188	35 48	262	28 48	302	15 27	347
353	20 59	048	19 32	081	42 42	157	33 44	189	34 54	262	28 02	302	15 15	348
354	21 39	048	20 25	081	43 03	158	33 35	190	34 01	263	27 16	302	15 04	348
355	22 19	048	21 18	081	43 22	159	33 25	191	33 07	263	26 31	302	14 53	348
356	22 59	049	22 12	082	43 41	160	33 14	192	32 14	264	25 45	303	14 41	348
357	23 40	049	23 05	082	43 59	162	33 03	193	31 20	264	25 00	303	14 31	348
358	24 20	049	23 59	083	44 15	163	32 50	194	30 26	265	24 14	303	14 20	349
359	25 01	049	24 52	083	44 30	164	32 36	195	29 33	265	23 29	303	14 09	349

LHA	Hc	Zn	Hc	Zn	Hc	Zn	Hc	Zn	Hc	Zn	Hc	Zn	Hc	Zn
♈	Schedar		◆CAPELLA		ALDEBARAN		◆Diphda		FOMALHAUT		◆ALTAIR		DENEB	
0	57 39	010	25 02	049	25 39	083	45 43	165	33 19	196	28 43	266	45 28	310
1	57 49	009	25 43	049	26 32	083	45 56	166	33 04	197	27 48	267	44 47	310
2	57 57	008	26 24	049	27 27	084	46 08	168	32 47	198	26 54	267	44 05	310
3	58 05	007	27 06	049	28 21	084	46 19	169	32 30	199	26 00	268	43 23	310
4	58 11	006	27 47	049	29 15	084	46 28	171	32 11	200	25 06	268	42 41	310
5	58 17	005	28 28	049	30 09	085	46 37	172	31 52	201	24 11	269	41 59	309
6	58 21	004	29 10	050	31 03	085	46 44	173	31 32	202	23 17	269	41 17	309
7	58 25	003	29 51	050	31 57	085	46 49	175	31 11	203	22 22	270	40 35	309
8	58 27	002	30 32	050	32 51	086	46 54	176	30 49	204	21 28	270	39 53	309
9	58 29	001	31 14	050	33 46	086	46 57	177	30 27	205	20 34	270	39 11	309
10	58 29	000	31 55	050	34 40	087	46 59	179	30 04	206	19 39	271	38 29	309
11	58 29	359	32 37	050	35 34	087	46 59	180	29 40	207	18 45	271	37 46	309
12	58 28	358	33 19	050	36 29	087	46 58	182	29 15	208	17 51	272	37 04	309
13	58 25	357	34 00	050	37 23	088	46 56	183	28 49	208	16 56	272	36 22	309
14	58 22	356	34 42	050	38 17	088	46 53	184	28 23	209	16 02	272	35 40	309
	◆CAPELLA		ALDEBARAN		RIGEL		◆Diphda		FOMALHAUT		Enif		◆DENEB	
15	35 24	050	39 12	089	19 49	110	46 48	186	27 56	210	41 09	261	34 57	309
16	36 05	050	40 06	089	20 40	110	46 42	187	27 29	211	40 15	262	34 15	309
17	36 47	050	41 00	089	21 31	111	46 34	189	27 00	212	39 21	262	33 33	309
18	37 29	050	41 55	090	22 21	111	46 25	190	26 31	213	38 27	263	32 51	309
19	38 10	050	42 49	090	23 12	112	46 15	191	26 02	213	37 33	263	32 09	309
20	38 52	050	43 43	091	24 02	112	46 04	193	25 32	214	36 39	264	31 27	309
21	39 34	050	44 38	091	24 53	113	45 52	194	25 01	215	35 45	264	30 45	309
22	40 15	050	45 32	092	25 43	114	45 38	195	24 29	216	34 51	265	30 03	309
23	40 57	050	46 27	092	26 32	114	45 23	197	23 57	216	33 57	265	29 21	310
24	41 38	050	47 21	093	27 22	115	45 07	198	23 25	217	33 03	266	28 39	310
25	42 20	050	48 15	093	28 11	115	44 50	199	22 52	218	32 08	266	27 57	310
26	43 01	049	49 09	093	29 00	116	44 31	200	22 18	219	31 14	266	27 15	310
27	43 42	049	50 04	094	29 49	117	44 12	202	21 44	219	30 20	267	26 34	310
28	44 23	049	50 58	094	30 37	117	43 51	203	21 09	220	29 26	267	25 52	310
29	45 04	049	51 52	095	31 25	118	43 29	204	20 34	221	28 31	268	25 10	310
	CAPELLA		◆ALDEBARAN		RIGEL		Acamar		◆Diphda		Alpheratz		◆DENEB	
30	45 45	049	52 46	095	32 13	119	23 18	168	43 07	205	64 49	286	24 29	310
31	46 26	049	53 40	096	33 01	119	23 28	169	42 43	206	63 57	285	23 48	311
32	47 07	048	54 34	097	33 48	120	23 39	170	42 18	208	63 04	285	23 06	311
33	47 48	048	55 28	097	34 35	121	23 48	170	41 52	209	62 12	285	22 25	311
34	48 28	048	56 22	098	35 21	122	23 57	171	41 26	210	61 19	285	21 44	311
35	49 08	048	57 16	098	36 07	122	24 05	172	40 58	211	60 27	285	21 03	311
36	49 48	047	58 10	099	36 53	123	24 12	173	40 30	212	59 35	285	20 23	312
37	50 28	047	59 04	100	37 39	124	24 18	174	40 00	213	58 42	285	19 42	312
38	51 08	047	59 57	100	38 23	125	24 24	175	39 30	214	57 50	285	19 01	312
39	51 47	046	60 51	101	39 08	126	24 28	175	38 59	215	56 57	285	18 21	312
40	52 27	046	61 44	102	39 52	126	24 32	176	38 27	216	56 05	285	17 41	312
41	53 06	046	62 37	103	40 35	127	24 36	177	37 55	217	55 12	286	17 01	313
42	53 44	045	63 30	103	41 18	128	24 38	178	37 22	218	54 20	286	16 21	313
43	54 23	045	64 23	104	42 01	129	24 40	179	36 47	219	53 28	286	15 41	313
44	55 01	044	65 15	105	42 43	130	24 40	180	36 13	220	52 35	286	15 01	313
	CAPELLA		◆POLLUX		PROCYON		SIRIUS		◆Acamar		Diphda		◆Alpheratz	
45	55 38	044	27 07	070	20 32	094	21 10	121	24 40	180	35 37	221	51 43	286
46	56 16	043	27 58	070	21 26	094	21 56	122	24 40	181	35 01	222	50 51	286
47	56 53	042	28 49	070	22 21	095	22 42	123	24 38	182	34 25	223	49 59	286
48	57 29	042	29 40	071	23 15	095	23 28	123	24 36	183	33 47	224	49 06	286
49	58 05	041	30 32	071	24 09	096	24 13	124	24 33	184	33 09	225	48 14	286
50	58 41	041	31 23	071	25 03	096	24 58	125	24 29	185	32 31	225	47 22	287
51	59 16	040	32 15	071	25 57	097	25 42	125	24 24	185	31 52	226	46 30	287
52	59 50	039	33 06	072	26 51	097	26 27	126	24 18	186	31 12	227	45 38	287
53	60 24	038	33 58	072	27 45	098	27 11	127	24 12	187	30 32	228	44 46	287
54	60 57	037	34 50	072	28 39	098	27 54	127	24 05	188	29 52	229	43 54	287
55	61 30	036	35 41	072	29 33	099	28 37	128	23 57	189	29 11	229	43 02	287
56	62 02	036	36 33	073	30 26	099	29 20	129	23 48	190	28 29	230	42 10	288
57	62 33	035	37 25	073	31 20	100	30 02	129	23 39	190	27 47	231	41 18	288
58	63 04	033	38 17	073	32 14	100	30 44	130	23 29	191	27 05	232	40 26	288
59	63 33	032	39 09	073	33 07	101	31 25	131	23 18	192	26 22	232	39 35	288
	◆CAPELLA		PROCYON		◆SIRIUS		RIGEL		Diphda		◆Hamal		Mirfak	
60	64 02	031	34 01	101	32 06	132	52 11	149	25 39	233	64 13	273	64 11	347
61	64 30	030	34 54	102	32 46	133	52 38	150	24 55	234	63 18	273	63 58	345
62	64 56	029	35 47	102	33 26	133	53 05	152	24 11	234	62 24	273	63 44	344
63	65 22	027	36 40	103	34 05	134	53 29	153	23 27	235	61 30	273	63 28	343
64	65 46	026	37 33	103	34 43	135	53 53	155	22 42	236	60 35	274	63 11	341
65	66 10	025	38 26	104	35 21	136	54 15	157	21 57	236	59 41	274	62 53	340
66	66 32	023	39 19	105	35 59	137	54 36	158	21 12	237	58 47	274	62 34	339
67	66 53	022	40 11	105	36 36	138	54 56	160	20 26	237	57 53	275	62 14	338
68	67 12	020	41 04	106	37 12	139	55 14	161	19 40	238	56 59	275	61 52	336
69	67 30	019	41 56	106	37 47	140	55 30	163	18 54	239	56 04	275	61 30	335
70	67 47	017	42 48	107	38 22	141	55 45	165	18 07	239	55 10	276	61 07	334
71	68 02	015	43 40	108	38 56	142	55 59	167	17 21	240	54 16	276	60 43	333
72	68 15	013	44 31	109	39 30	143	56 11	168	16 33	240	53 22	276	60 18	332
73	68 27	012	45 23	109	40 02	144	56 21	170	15 46	241	52 28	277	59 52	331
74	68 37	010	46 14	110	40 34	145	56 30	172	14 58	241	51 34	277	59 25	330
	CAPELLA		◆Dubhe		POLLUX		PROCYON		◆SIRIUS		RIGEL		◆Hamal	
75	68 45	008	21 28	031	53 08	076	47 05	111	41 05	146	56 37	174	50 40	277
76	68 52	006	21 56	031	54 01	076	47 56	111	41 35	147	56 42	175	49 46	277
77	68 56	004	22 24	031	54 54	076	48 46	112	42 04	148	56 45	177	48 52	278
78	68 59	002	22 51	031	55 47	076	49 37	113	42 33	149	56 47	179	47 58	278
79	69 00	000	23 19	031	56 39	076	50 26	114	43 00	150	56 47	181	47 05	278
80	69 00	358	23 47	031	57 32	077	51 16	115	43 26	151	56 46	183	46 11	279
81	68 57	356	24 15	031	58 25	077	52 05	116	43 52	153	56 43	184	45 17	279
82	68 53	354	24 43	031	59 18	077	52 54	117	44 16	154	56 38	186	44 23	279
83	68 46	352	25 12	031	60 11	077	53 42	118	44 40	155	56 31	188	43 30	279
84	68 38	351	25 40	031	61 04	077	54 30	119	45 02	156	56 22	190	42 36	280
85	68 29	349	26 08	031	61 57	077	55 18	120	45 24	158	56 12	191	41 42	280
86	68 17	347	26 36	031	62 50	077	56 05	121	45 44	159	56 01	193	40 49	280
87	68 04	345	27 05	031	63 43	077	56 51	122	46 03	160	55 48	195	39 55	280
88	67 49	343	27 33	031	64 36	077	57 37	123	46 21	161	55 33	197	39 02	281
89	67 33	342	28 01	031	65 29	077	58 22	124	46 37	163	55 16	198	38 08	281

LAT 25°N

LHA	Hc	Zn	Hc	Zn	Hc	Zn	Hc	Zn	Hc	Zn	Hc	Zn	Hc	Zn
♈	Dubhe		◆REGULUS		Alphard		SIRIUS		◆RIGEL		ALDEBARAN		◆CAPELLA	
90	28 30	031	30 14	090	29 21	117	46 53	164	54 59	200	68 34	251	67 15	340
91	28 58	031	31 09	091	30 10	118	47 07	166	54 39	202	67 42	252	66 56	339
92	29 27	031	32 03	091	30 58	118	47 20	167	54 19	203	66 50	253	66 35	337
93	29 55	031	32 58	092	31 45	119	47 32	168	53 56	205	65 58	254	66 13	336
94	30 23	031	33 52	092	32 33	120	47 42	170	53 33	206	65 06	255	65 50	334
95	30 52	031	34 46	093	33 20	120	47 51	171	53 08	208	64 13	256	65 26	333
96	31 20	031	35 41	093	34 07	121	47 59	173	52 42	209	63 20	257	65 00	331
97	31 48	031	36 35	093	34 53	122	48 05	174	52 15	211	62 27	258	64 34	330
98	32 16	031	37 29	094	35 39	123	48 10	175	51 47	212	61 34	258	64 06	329
99	32 44	031	38 23	094	36 25	123	48 14	177	51 17	214	60 41	259	63 38	328
100	33 13	031	39 18	095	37 10	124	48 16	178	50 47	215	59 47	260	63 08	327
101	33 41	031	40 12	095	37 55	125	48 17	180	50 15	216	58 54	260	62 38	326
102	34 08	031	41 06	096	38 39	126	48 17	181	49 42	217	58 00	261	62 07	325
103	34 36	031	42 00	096	39 23	127	48 15	183	49 09	219	57 06	262	61 35	324
104	35 04	031	42 54	097	40 06	128	48 12	184	48 34	220	56 12	262	61 03	323
	Dubhe		◆REGULUS		Alphard		SIRIUS		◆RIGEL		ALDEBARAN		◆CAPELLA	
105	35 32	030	43 48	097	40 49	128	48 08	185	47 59	221	55 19	263	60 29	322
106	35 59	030	44 42	098	41 32	129	48 02	187	47 23	222	54 25	264	59 56	321
107	36 27	030	45 36	099	42 13	130	47 55	188	46 46	223	53 30	264	59 21	320
108	36 54	030	46 29	099	42 54	131	47 46	190	46 08	225	52 36	265	58 46	320
109	37 21	030	47 23	100	43 35	132	47 37	191	45 30	226	51 42	265	58 11	319
110	37 48	030	48 17	100	44 15	133	47 25	192	44 50	227	50 48	266	57 35	318
111	38 15	029	49 10	101	44 54	134	47 13	194	44 10	228	49 54	266	56 58	318
112	38 42	029	50 03	102	45 33	135	46 59	195	43 30	229	49 00	267	56 21	317
113	39 08	029	50 57	102	46 11	136	46 45	197	42 49	230	48 05	267	55 44	316
114	39 35	029	51 50	103	46 48	138	46 28	198	42 07	231	47 11	268	55 07	316
115	40 01	029	52 43	104	47 24	139	46 11	199	41 25	232	46 17	268	54 29	315
116	40 27	028	53 35	104	47 59	140	45 53	201	40 42	232	45 22	268	53 50	315
117	40 52	028	54 28	105	48 34	141	45 33	202	39 58	233	44 28	269	53 12	315
118	41 18	028	55 20	106	49 08	142	45 12	203	39 14	234	43 33	269	52 33	314
119	41 43	028	56 13	107	49 40	144	44 50	204	38 30	235	42 39	270	51 54	314
	◆Dubhe		REGULUS		◆Alphard		SIRIUS		RIGEL		◆ALDEBARAN		CAPELLA	
120	42 08	027	57 05	107	50 12	145	44 27	206	37 45	236	41 45	270	51 14	313
121	42 33	027	57 56	108	50 43	146	44 03	207	37 00	237	40 50	271	50 35	313
122	42 57	027	58 48	109	51 12	148	43 38	208	36 14	238	39 56	271	49 55	313
123	43 22	026	59 39	110	51 41	149	43 12	209	35 28	238	39 02	271	49 15	312
124	43 45	026	60 30	111	52 08	150	42 45	210	34 42	239	38 07	272	48 34	312
125	44 09	026	61 21	112	52 35	152	42 18	211	33 55	240	37 13	272	47 54	312
126	44 32	025	62 11	113	53 00	153	41 49	213	33 08	240	36 19	273	47 14	312
127	44 55	025	63 01	114	53 23	155	41 19	214	32 20	241	35 24	273	46 33	311
128	45 18	024	63 50	115	53 46	156	40 48	215	31 33	242	34 30	273	45 52	311
129	45 40	024	64 39	117	54 07	158	40 17	216	30 44	243	33 36	274	45 11	311
130	46 02	024	65 27	118	54 26	160	39 45	217	29 56	243	32 41	274	44 30	311
131	46 23	023	66 15	119	54 45	161	39 12	218	29 07	244	31 47	275	43 49	311
132	46 45	023	67 02	121	55 01	163	38 38	219	28 18	245	30 53	275	43 08	311
133	47 05	022	67 48	122	55 17	165	38 04	220	27 29	245	29 59	275	42 26	310
134	47 26	022	68 34	124	55 30	166	37 28	221	26 40	246	29 05	276	41 45	310
	Dubhe		◆ARCTURUS		SPICA		◆Alphard		SIRIUS		BETELGEUSE		◆CAPELLA	
135	47 45	021	17 43	077	16 04	111	55 43	168	36 53	222	42 30	256	41 03	310
136	48 05	021	18 36	077	16 55	111	55 53	170	36 16	223	41 38	257	40 22	310
137	48 24	020	19 29	077	17 45	112	56 02	171	35 39	224	40 44	258	39 40	310
138	48 42	020	20 22	078	18 36	112	56 09	173	35 01	224	39 51	258	38 59	310
139	49 00	019	21 16	078	19 26	113	56 15	175	34 23	225	38 58	259	38 17	310
140	49 18	018	22 09	078	20 16	114	56 19	177	33 44	226	38 05	259	37 35	310
141	49 35	018	23 02	079	21 06	114	56 21	179	33 04	227	37 11	260	36 54	310
142	49 51	017	23 55	079	21 55	115	56 22	180	32 24	228	36 18	260	36 12	310
143	50 07	017	24 49	079	22 44	115	56 21	182	31 44	229	35 24	261	35 30	310
144	50 22	016	25 42	080	23 33	116	56 18	184	31 03	229	34 30	261	34 49	310
145	50 37	015	26 36	080	24 22	116	56 13	186	30 21	230	33 36	262	34 07	310
146	50 51	015	27 29	080	25 11	117	56 07	187	29 39	231	32 42	263	33 25	310
147	51 04	014	28 23	081	25 59	118	55 59	189	28 57	232	31 49	263	32 44	310
148	51 17	013	29 17	081	26 47	118	55 50	191	28 14	232	30 55	263	32 02	310
149	51 30	013	30 11	081	27 35	119	55 39	193	27 31	233	30 00	264	31 21	310
	Dubhe		◆ARCTURUS		SPICA		◆Alphard		SIRIUS		POLLUX		◆CAPELLA	
150	51 41	012	31 04	082	28 22	120	55 26	194	26 47	234	59 45	283	30 39	310
151	51 52	011	31 58	082	29 09	120	55 12	196	26 03	234	58 52	283	29 58	310
152	52 03	011	32 52	082	29 56	121	54 56	198	25 19	235	57 59	283	29 16	310
153	52 12	010	33 46	083	30 43	122	54 39	199	24 34	236	57 06	284	28 35	311
154	52 21	009	34 40	083	31 29	122	54 20	201	23 49	236	56 13	284	27 53	311
155	52 30	008	35 34	084	32 15	123	54 00	203	23 03	237	55 20	284	27 12	311
156	52 37	008	36 28	084	33 00	124	53 38	204	22 17	238	54 28	284	26 31	311
157	52 44	007	37 22	084	33 45	125	53 16	206	21 31	238	53 35	284	25 50	311
158	52 50	006	38 16	085	34 29	125	52 51	207	20 45	239	52 42	284	25 09	311
159	52 56	005	39 10	085	35 14	126	52 26	209	19 58	240	51 49	284	24 28	311
160	53 01	005	40 05	085	35 57	127	51 59	210	19 11	240	50 57	284	23 47	311
161	53 05	004	40 59	086	36 40	128	51 32	211	18 24	241	50 04	285	23 06	312
162	53 08	003	41 53	086	37 23	129	51 03	213	17 36	241	49 11	285	22 26	312
163	53 10	002	42 47	086	38 05	130	50 33	214	16 48	242	48 19	285	21 45	312
164	53 12	001	43 42	087	38 47	130	50 02	215	16 00	242	47 26	285	21 05	312
	◆Dubhe		Alkaid		ARCTURUS		◆SPICA		Alphard		◆PROCYON		POLLUX	
165	53 13	001	49 31	042	44 36	087	39 28	131	49 30	217	38 01	256	46 34	285
166	53 13	000	50 07	042	45 30	088	40 09	132	48 57	218	37 08	257	45 41	285
167	53 13	359	50 43	041	46 25	088	40 49	133	48 23	219	36 15	257	44 49	286
168	53 12	358	51 19	041	47 19	088	41 28	134	47 48	220	35 22	258	43 56	286
169	53 10	358	51 54	040	48 13	089	42 07	135	47 12	222	34 28	259	43 04	286
170	53 07	357	52 29	040	49 08	089	42 45	136	46 35	223	33 35	259	42 12	286
171	53 03	356	53 04	039	50 02	089	43 22	137	45 58	224	32 42	260	41 20	286
172	52 59	355	53 38	039	50 56	090	43 59	138	45 20	225	31 48	260	40 28	286
173	52 54	354	54 12	038	51 51	090	44 34	139	44 41	226	30 54	261	39 35	287
174	52 49	354	54 46	038	52 45	091	45 10	140	44 02	227	30 01	261	38 43	287
175	52 42	353	55 19	037	53 39	091	45 44	142	43 22	228	29 07	262	37 51	287
176	52 35	352	55 51	037	54 34	092	46 17	143	42 41	229	28 13	262	36 59	287
177	52 27	351	56 24	036	55 28	092	46 50	144	42 00	230	27 19	263	36 08	288
178	52 19	351	56 55	035	56 23	093	47 21	145	41 18	231	26 25	263	35 16	288
179	52 10	350	57 26	034	57 17	093	47 52	146	40 35	232	25 31	264	34 24	288

LHA	Hc	Zn	Hc	Zn	Hc	Zn	Hc	Zn	Hc	Zn	Hc	Zn	Hc	Zn
♈	♦Alkaid		Alphecca		♦SPICA		Gienah		REGULUS		♦POLLUX		Dubhe	
180	57 57	034	42 05	076	48 21	148	47 19	175	60 33	249	33 32	288	52 00	349
181	58 26	033	42 58	076	48 50	149	47 24	176	59 42	250	32 41	288	51 49	348
182	58 56	032	43 51	076	49 17	150	47 27	177	58 51	251	31 49	289	51 38	348
183	59 24	031	44 43	076	49 44	152	47 29	179	57 59	252	30 58	289	51 26	347
184	59 52	030	45 36	077	50 09	153	47 29	180	57 07	253	30 06	289	51 14	346
185	60 19	029	46 29	077	50 33	154	47 28	182	56 15	253	29 15	289	51 01	346
186	60 45	028	47 22	077	50 56	156	47 26	183	55 23	254	28 24	290	50 47	345
187	61 10	027	48 15	077	51 17	157	47 23	184	54 31	255	27 33	290	50 33	344
188	61 35	026	49 08	077	51 38	159	47 18	186	53 38	256	26 42	290	50 18	344
189	61 59	025	50 01	078	51 57	160	47 12	187	52 45	256	25 51	291	50 02	343
190	62 21	024	50 54	078	52 14	162	47 04	189	51 52	257	25 00	291	49 46	343
191	62 43	023	51 47	078	52 31	163	46 55	190	50 59	258	24 09	291	49 30	342
192	63 03	022	52 41	078	52 46	165	46 45	191	50 06	258	23 18	291	49 13	341
193	63 23	020	53 34	078	52 59	167	46 34	193	49 13	259	22 28	292	48 55	341
194	63 41	019	54 27	078	53 11	168	46 22	194	48 19	260	21 37	292	48 37	340
	♦Kochab		VEGA		Rasalhague		♦ANTARES		SPICA		♦REGULUS		Dubhe	
195	38 43	009	19 39	055	24 26	087	17 58	132	53 21	170	47 26	260	48 18	340
196	38 52	009	20 24	056	25 20	087	18 39	132	53 30	171	46 32	261	47 59	339
197	39 00	009	21 08	056	26 15	088	19 19	133	53 38	173	45 38	261	47 40	339
198	39 08	008	21 54	056	27 09	088	19 58	134	53 44	175	44 45	262	47 20	338
199	39 16	008	22 39	056	28 03	089	20 37	134	53 48	176	43 51	263	46 59	338
200	39 24	008	23 24	057	28 58	089	21 16	135	53 51	178	42 57	263	46 39	337
201	39 31	008	24 10	057	29 52	089	21 54	136	53 52	180	42 03	264	46 17	337
202	39 38	007	24 55	057	30 46	090	22 32	136	53 51	181	41 09	264	45 56	336
203	39 44	007	25 41	057	31 41	090	23 09	137	53 49	183	40 15	265	45 34	336
204	39 51	007	26 26	057	32 35	091	23 46	138	53 46	185	39 20	265	45 11	336
205	39 57	006	27 12	057	33 30	091	24 22	139	53 41	186	38 26	266	44 49	335
206	40 02	006	27 58	058	34 24	092	24 57	139	53 34	188	37 32	266	44 26	335
207	40 08	006	28 44	058	35 18	092	25 33	140	53 26	190	36 38	267	44 02	334
208	40 13	005	29 30	058	36 13	093	26 07	141	53 16	191	35 43	267	43 39	334
209	40 18	005	30 16	058	37 07	093	26 41	142	53 05	193	34 49	267	43 15	334
	♦Kochab		VEGA		Rasalhague		♦ANTARES		SPICA		♦REGULUS		Dubhe	
210	40 22	005	31 03	058	38 01	093	27 15	142	52 52	194	33 55	268	42 50	333
211	40 26	004	31 49	058	38 55	094	27 48	143	52 38	196	33 00	268	42 26	333
212	40 30	004	32 35	059	39 50	094	28 20	144	52 22	197	32 06	269	42 01	333
213	40 33	003	33 22	059	40 44	095	28 51	145	52 05	199	31 12	269	41 36	332
214	40 37	003	34 08	059	41 38	095	29 22	146	51 47	201	30 17	270	41 10	332
215	40 39	003	34 55	059	42 32	096	29 53	147	51 27	202	29 23	270	40 45	332
216	40 42	002	35 41	059	43 26	096	30 22	147	51 06	203	28 29	270	40 19	332
217	40 44	002	36 28	059	44 20	097	30 51	148	50 44	205	27 34	271	39 53	331
218	40 46	002	37 14	059	45 14	097	31 20	149	50 20	206	26 40	271	39 27	331
219	40 47	001	38 01	059	46 08	098	31 47	150	49 56	208	25 46	272	39 01	331
220	40 48	001	38 48	059	47 02	099	32 14	151	49 30	209	24 51	272	38 34	331
221	40 49	001	39 34	059	47 56	099	32 40	152	49 03	210	23 57	273	38 07	330
222	40 49	000	40 21	059	48 49	100	33 05	153	48 35	212	23 03	273	37 40	330
223	40 49	000	41 08	059	49 43	100	33 30	154	48 06	213	22 08	273	37 13	330
224	40 49	000	41 55	059	50 36	101	33 53	155	47 35	214	21 14	274	36 46	330
	VEGA		♦ALTAIR		Nunki		ANTARES		♦SPICA		Denebola		♦Alkaid	
225	42 42	059	19 25	089	13 33	128	34 16	156	47 04	215	44 05	266	61 51	334
226	43 28	059	20 19	090	14 16	129	34 38	157	46 32	217	43 11	267	61 27	333
227	44 15	059	21 13	090	14 58	129	34 59	158	45 59	218	42 16	267	61 02	332
228	45 02	059	22 08	090	15 40	130	35 19	159	45 26	219	41 22	268	60 37	331
229	45 49	059	23 02	091	16 22	130	35 38	160	44 51	220	40 28	268	60 10	330
230	46 35	059	23 56	091	17 03	131	35 57	161	44 15	221	39 33	269	59 43	329
231	47 22	059	24 51	092	17 44	131	36 14	162	43 39	222	38 39	269	59 15	329
232	48 09	059	25 45	092	18 25	132	36 30	163	43 02	223	37 44	269	58 46	328
233	48 55	059	26 40	093	19 05	133	36 46	164	42 24	224	36 50	270	58 17	327
234	49 42	059	27 34	093	19 45	133	37 00	165	41 46	225	35 56	270	57 47	326
235	50 28	059	28 28	093	20 24	134	37 14	166	41 07	226	35 01	271	57 16	325
236	51 15	058	29 22	094	21 03	135	37 26	167	40 27	227	34 07	271	56 45	325
237	52 01	058	30 17	094	21 41	135	37 38	168	39 47	228	33 13	272	56 13	324
238	52 47	058	31 11	095	22 19	136	37 48	169	39 06	229	32 18	272	55 41	323
239	53 33	058	32 05	095	22 57	137	37 58	171	38 25	230	31 24	272	55 08	323
	♦VEGA		ALTAIR		Nunki		♦ANTARES		SPICA		ARCTURUS		♦Alkaid	
240	54 19	058	32 59	096	23 34	137	38 06	172	37 43	231	65 08	262	54 35	322
241	55 05	057	33 53	096	24 10	138	38 13	173	37 00	232	64 15	262	54 01	321
242	55 51	057	34 47	097	24 46	139	38 20	174	36 17	233	63 21	263	53 27	321
243	56 36	057	35 41	097	25 22	140	38 25	175	35 34	233	62 27	264	52 53	320
244	57 22	056	36 35	098	25 57	140	38 29	176	34 50	234	61 33	264	52 18	320
245	58 07	056	37 29	098	26 31	141	38 32	177	34 06	235	60 38	265	51 43	319
246	58 52	056	38 23	099	27 05	142	38 34	179	33 21	236	59 44	265	51 07	319
247	59 37	055	39 16	099	27 38	143	38 35	180	32 36	237	58 50	266	50 31	319
248	60 21	055	40 10	100	28 11	144	38 34	181	31 50	237	57 56	267	49 55	318
249	61 05	054	41 04	100	28 43	144	38 33	182	31 04	238	57 01	267	49 19	318
250	61 49	054	41 57	101	29 14	145	38 31	183	30 18	239	56 07	268	48 42	318
251	62 33	053	42 50	102	29 45	146	38 27	184	29 31	239	55 13	268	48 05	317
252	63 16	052	43 44	102	30 15	147	38 23	185	28 44	240	54 18	268	47 28	317
253	63 59	052	44 37	103	30 44	148	38 17	187	27 57	241	53 24	269	46 51	317
254	64 41	051	45 30	103	31 13	149	38 10	188	27 09	241	52 30	269	46 14	316
	♦DENEB		ALTAIR		♦Nunki		ANTARES		SPICA		♦ARCTURUS		Alkaid	
255	41 33	051	46 22	104	31 41	150	38 02	189	26 21	242	51 35	270	45 36	316
256	42 15	051	47 15	105	32 08	150	37 54	190	25 33	243	50 41	270	44 58	316
257	42 57	050	48 08	105	32 34	151	37 44	191	24 45	243	49 47	271	44 20	316
258	43 39	050	49 00	106	33 00	152	37 33	192	23 56	244	48 52	271	43 42	315
259	44 20	050	49 52	107	33 25	153	37 21	193	23 07	244	47 58	271	43 04	315
260	45 02	050	50 44	108	33 49	154	37 08	194	22 18	245	47 03	272	42 26	315
261	45 44	050	51 36	108	34 12	155	36 54	195	21 29	246	46 09	272	41 47	315
262	46 25	050	52 27	109	34 34	156	36 39	196	20 39	246	45 15	273	41 08	315
263	47 07	049	53 19	110	34 56	157	36 23	198	19 49	247	44 20	273	40 30	315
264	47 48	049	54 10	111	35 16	158	36 06	199	18 59	247	43 26	273	39 51	315
265	48 29	049	55 00	112	35 36	159	35 48	200	18 09	248	42 32	274	39 12	314
266	49 10	049	55 51	113	35 55	160	35 30	201	17 18	248	41 38	274	38 33	314
267	49 51	048	56 41	114	36 13	161	35 10	202	16 28	249	40 43	274	37 54	314
268	50 32	048	57 30	115	36 30	162	34 49	203	15 37	249	39 49	275	37 15	314
269	51 12	048	58 19	116	36 46	163	34 28	204	14 46	250	38 55	275	36 36	314

LAT 25°N

LHA	Hc	Zn	Hc	Zn	Hc	Zn	Hc	Zn	Hc	Zn	Hc	Zn	Hc	Zn
♈	♦DENEB		Enif		♦Nunki		ANTARES		♦ARCTURUS		Alkaid		Kochab	
270	51 52	048	34 53	095	37 01	165	34 06	205	38 01	276	35 57	314	35 02	346
271	52 32	047	35 47	096	37 15	166	33 43	206	37 07	276	35 18	314	34 49	346
272	53 12	047	36 41	096	37 28	167	33 19	207	36 13	276	34 39	314	34 35	345
273	53 51	046	37 35	097	37 40	168	32 54	208	35 19	277	34 00	314	34 22	345
274	54 31	046	38 29	097	37 51	169	32 28	209	34 25	277	33 21	314	34 08	345
275	55 09	045	39 23	098	38 00	170	32 02	209	33 31	277	32 42	314	33 54	345
276	55 48	045	40 17	098	38 09	171	31 35	210	32 37	278	32 03	314	33 39	345
277	56 26	044	41 10	099	38 17	172	31 07	211	31 43	278	31 24	314	33 25	345
278	57 04	044	42 04	100	38 24	173	30 38	212	30 49	278	30 45	314	33 11	344
279	57 42	043	42 57	100	38 30	175	30 09	213	29 55	279	30 06	314	32 56	344
280	58 19	043	43 51	101	38 34	176	29 39	214	29 02	279	29 27	314	32 41	344
281	58 55	042	44 44	101	38 38	177	29 09	215	28 08	279	28 48	314	32 26	344
282	59 31	041	45 38	102	38 40	178	28 37	216	27 14	280	28 09	314	32 11	344
283	60 07	040	46 31	103	38 42	179	28 05	216	26 21	280	27 30	314	31 56	344
284	60 42	040	47 24	103	38 42	180	27 33	217	25 27	280	26 51	315	31 41	344
	DENEB		♦Alpheratz		FOMALHAUT		♦Nunki		ANTARES		♦ARCTURUS		Kochab	
285	61 16	039	22 31	067	11 06	130	38 41	181	27 00	218	24 34	281	31 26	344
286	61 50	038	23 22	068	11 47	131	38 39	183	26 26	219	23 40	281	31 10	343
287	62 23	037	24 12	068	12 28	131	38 36	184	25 52	219	22 47	281	30 55	343
288	62 55	036	25 02	068	13 09	132	38 32	185	25 17	220	21 54	282	30 39	343
289	63 26	035	25 53	068	13 49	133	38 27	186	24 41	221	21 01	282	30 23	343
290	63 57	034	26 43	069	14 29	133	38 21	187	24 05	222	20 07	282	30 08	343
291	64 27	033	27 34	069	15 08	134	38 13	188	23 29	222	19 14	283	29 52	343
292	64 56	031	28 25	069	15 47	134	38 05	189	22 52	223	18 21	283	29 36	343
293	65 24	030	29 16	069	16 26	135	37 55	191	22 15	224	17 28	284	29 20	343
294	65 51	029	30 07	070	17 04	136	37 45	192	21 37	225	16 36	284	29 04	343
295	66 16	028	30 58	070	17 42	136	37 33	193	20 58	225	15 43	284	28 48	343
296	66 41	026	31 49	070	18 20	137	37 21	194	20 20	226	14 50	285	28 31	343
297	67 04	025	32 40	070	18 57	137	37 07	195	19 40	227	13 58	285	28 15	343
298	67 26	023	33 31	070	19 33	138	36 53	196	19 01	227	13 05	285	27 59	343
299	67 47	021	34 22	071	20 09	139	36 37	197	18 21	228	12 13	286	27 43	343
	♦DENEB		Schedar		Alpheratz		♦FOMALHAUT		Nunki		♦Rasalhague		VEGA	
300	68 06	020	31 32	037	35 14	071	20 45	139	36 21	198	53 34	257	67 41	313
301	68 23	018	32 05	037	36 05	071	21 20	140	36 03	199	52 41	258	67 01	312
302	68 39	016	32 38	037	36 57	071	21 54	141	35 45	200	51 48	258	66 21	311
303	68 54	014	33 11	037	37 48	072	22 28	142	35 26	201	50 55	259	65 39	310
304	69 06	013	33 44	037	38 40	072	23 02	142	35 06	202	50 02	259	64 58	309
305	69 17	011	34 17	037	39 31	072	23 35	143	34 45	203	49 08	260	64 15	309
306	69 26	009	34 50	037	40 23	072	24 07	144	34 23	204	48 14	261	63 33	308
307	69 34	007	35 23	037	41 15	072	24 39	145	34 00	205	47 21	261	62 50	307
308	69 39	005	35 56	037	42 07	072	25 10	145	33 36	206	46 27	262	62 06	307
309	69 43	003	36 29	037	42 59	073	25 41	146	33 12	207	45 33	262	61 22	306
310	69 44	001	37 01	037	43 50	073	26 11	147	32 46	208	44 39	263	60 38	305
311	69 44	359	37 34	037	44 42	073	26 40	148	32 20	209	43 45	263	59 54	305
312	69 42	357	38 06	037	45 34	073	27 09	148	31 54	210	42 51	264	59 09	305
313	69 37	355	38 38	036	46 27	073	27 37	149	31 26	211	41 57	264	58 24	304
314	69 31	353	39 11	036	47 19	073	28 05	150	30 58	212	41 03	265	57 39	304
	♦Alpheratz		Diphda		♦FOMALHAUT		ALTAIR		Rasalhague		♦VEGA		DENEB	
315	48 11	074	20 41	123	28 31	151	66 53	229	40 09	265	56 54	303	69 23	351
316	49 03	074	21 27	123	28 57	152	66 12	230	39 14	266	56 08	303	69 13	349
317	49 55	074	22 12	124	29 23	153	65 29	232	38 20	266	55 23	303	69 02	347
318	50 47	074	22 57	125	29 47	154	64 46	234	37 26	267	54 37	302	68 49	345
319	51 40	074	23 42	125	30 11	155	64 02	235	36 32	267	53 51	302	68 34	343
320	52 32	074	24 26	126	30 34	155	63 17	237	35 37	268	53 05	302	68 17	341
321	53 24	074	25 10	127	30 56	156	62 31	238	34 43	268	52 19	302	67 59	340
322	54 17	074	25 53	127	31 18	157	61 44	239	33 49	269	51 32	302	67 39	338
323	55 09	074	26 36	128	31 38	158	60 57	241	32 54	269	50 46	301	67 18	336
324	56 01	075	27 19	129	31 58	159	60 10	242	32 00	270	50 00	301	66 56	335
325	56 54	075	28 01	129	32 17	160	59 21	243	31 05	270	49 13	301	66 32	333
326	57 46	075	28 43	130	32 35	161	58 33	244	30 11	270	48 26	301	66 07	332
327	58 39	075	29 24	131	32 52	162	57 44	245	29 17	271	47 40	301	65 41	331
328	59 31	075	30 05	132	33 09	163	56 54	246	28 22	271	46 53	301	65 14	329
329	60 24	075	30 46	132	33 24	164	56 04	247	27 28	272	46 07	301	64 45	328
	♦Mirfak		Hamal		Diphda		♦FOMALHAUT		ALTAIR		♦VEGA		DENEB	
330	24 29	044	34 12	078	31 26	133	33 39	165	55 14	248	45 20	301	64 16	327
331	25 07	045	35 05	078	32 05	134	33 52	166	54 23	249	44 33	301	63 46	326
332	25 46	045	35 58	078	32 44	135	34 05	167	53 32	250	43 46	301	63 15	325
333	26 24	045	36 51	079	33 22	136	34 17	168	52 41	251	43 00	301	62 43	324
334	27 02	045	37 45	079	34 00	137	34 27	169	51 50	251	42 13	301	62 11	323
335	27 41	045	38 38	079	34 37	137	34 37	170	50 58	252	41 26	301	61 37	322
336	28 19	045	39 32	079	35 13	138	34 46	171	50 06	253	40 39	301	61 03	321
337	28 58	045	40 25	080	35 49	139	34 54	172	49 14	254	39 52	301	60 29	320
338	29 36	045	41 19	080	36 24	140	35 01	173	48 22	254	39 06	301	59 54	319
339	30 15	045	42 12	080	36 59	141	35 07	174	47 29	255	38 19	301	59 18	319
340	30 53	045	43 06	081	37 33	142	35 12	175	46 37	256	37 32	301	58 42	318
341	31 32	045	43 59	081	38 06	143	35 15	176	45 44	256	36 46	301	58 05	317
342	32 11	045	44 53	081	38 38	144	35 18	178	44 51	257	35 59	301	57 28	317
343	32 49	045	45 47	081	39 10	145	35 20	179	43 58	258	35 12	301	56 50	316
344	33 28	045	46 41	082	39 40	146	35 21	180	43 05	258	34 26	301	56 12	315
	Schedar		CAPELLA		♦ALDEBARAN		Diphda		♦FOMALHAUT		ALTAIR		♦DENEB	
345	53 39	023	15 01	046	12 15	077	40 10	147	35 21	181	42 11	259	55 34	315
346	54 01	023	15 40	046	13 08	078	40 39	148	35 20	182	41 18	259	54 55	314
347	54 21	022	16 20	046	14 01	078	41 07	149	35 18	183	40 25	260	54 16	314
348	54 41	021	16 59	047	14 55	078	41 35	150	35 14	184	39 31	261	53 37	314
349	55 00	020	17 39	047	15 48	079	42 01	152	35 10	185	38 37	261	52 57	313
350	55 18	019	18 18	047	16 41	079	42 26	153	35 05	186	37 44	262	52 17	313
351	55 36	019	18 58	047	17 35	080	42 51	154	34 59	187	36 50	262	51 37	312
352	55 53	018	19 38	047	18 28	080	43 14	155	34 52	188	35 56	263	50 57	312
353	56 09	017	20 18	048	19 22	080	43 37	156	34 43	189	35 02	263	50 17	312
354	56 25	016	20 59	048	20 16	081	43 58	158	34 34	190	34 08	264	49 36	311
355	56 39	015	21 39	048	21 09	081	44 18	159	34 24	191	33 14	264	48 55	311
356	56 53	014	22 19	048	22 03	081	44 37	160	34 13	192	32 20	265	48 14	311
357	57 06	013	23 00	048	22 57	082	44 56	161	34 01	193	31 25	265	47 33	311
358	57 18	012	23 41	049	23 51	082	45 12	163	33 48	194	30 31	266	46 51	310
359	57 29	011	24 22	049	24 45	083	45 28	164	33 34	195	29 37	266	46 10	310

LHA	Hc	Zn	Hc	Zn	Hc	Zn	Hc	Zn	Hc	Zn	Hc	Zn	Hc	Zn
♈	Schedar		◆CAPELLA		ALDEBARAN		◆Diphda		FOMALHAUT		◆ALTAIR		DENEB	
0	56 40	010	24 23	048	25 31	082	46 41	165	34 17	196	28 46	267	44 49	311
1	56 50	009	25 04	049	26 25	083	46 54	166	34 01	197	27 51	267	44 08	311
2	56 58	008	25 45	049	27 20	083	47 07	168	33 44	198	26 57	268	43 26	310
3	57 05	007	26 26	049	28 14	083	47 18	169	33 26	199	26 02	268	42 45	310
4	57 11	006	27 08	049	29 08	084	47 28	170	33 08	200	25 07	269	42 03	310
5	57 17	005	27 49	049	30 03	084	47 36	172	32 48	201	24 12	269	41 21	310
6	57 21	004	28 30	049	30 58	085	47 43	173	32 28	202	23 17	270	40 39	310
7	57 25	003	29 12	049	31 52	085	47 49	175	32 06	203	22 23	270	39 57	310
8	57 27	002	29 53	049	32 47	085	47 54	176	31 44	204	21 28	270	39 15	310
9	57 29	001	30 35	049	33 41	086	47 57	177	31 21	205	20 33	271	38 33	310
10	57 29	000	31 17	049	34 36	086	47 59	179	30 58	206	19 38	271	37 51	310
11	57 29	359	31 58	049	35 31	086	47 59	180	30 33	207	18 43	272	37 08	310
12	57 28	358	32 40	049	36 25	087	47 58	182	30 08	208	17 49	272	36 26	310
13	57 25	357	33 22	049	37 20	087	47 56	183	29 42	209	16 54	272	35 44	310
14	57 22	356	34 03	050	38 15	087	47 52	184	29 15	210	15 59	273	35 02	310
	◆CAPELLA		ALDEBARAN		RIGEL		◆Diphda		FOMALHAUT		Enif		◆DENEB	
15	34 45	050	39 10	088	20 08	109	47 47	186	28 48	210	41 18	262	34 19	310
16	35 27	049	40 04	088	21 00	110	47 41	187	28 20	211	40 23	262	33 37	310
17	36 08	049	40 59	089	21 52	110	47 33	189	27 51	212	39 29	263	32 55	310
18	36 50	049	41 54	089	22 43	111	47 25	190	27 22	213	38 35	263	32 13	310
19	37 32	049	42 49	089	23 34	111	47 14	191	26 52	214	37 40	264	31 31	310
20	38 13	049	43 44	090	24 25	112	47 03	193	26 21	214	36 46	264	30 48	310
21	38 55	049	44 38	090	25 16	113	46 50	194	25 50	215	35 51	265	30 06	310
22	39 36	049	45 33	091	26 06	113	46 36	196	25 18	216	34 56	265	29 24	310
23	40 18	049	46 28	091	26 57	114	46 21	197	24 45	217	34 02	266	28 42	310
24	40 59	049	47 23	091	27 47	114	46 04	198	24 12	217	33 07	266	28 00	310
25	41 40	049	48 18	092	28 37	115	45 46	199	23 39	218	32 12	267	27 18	310
26	42 22	049	49 12	092	29 26	116	45 27	201	23 05	219	31 18	267	26 37	310
27	43 03	049	50 07	093	30 15	116	45 07	202	22 30	220	30 23	267	25 55	310
28	43 44	048	51 02	093	31 05	117	44 46	203	21 55	220	29 28	268	25 13	311
29	44 25	048	51 57	094	31 53	117	44 24	205	21 19	221	28 33	268	24 31	311
	CAPELLA		◆ALDEBARAN		RIGEL		Acamar		◆Diphda		Alpheratz		◆DENEB	
30	45 06	048	52 51	094	32 42	118	24 16	168	44 01	206	64 32	288	23 50	311
31	45 46	048	53 46	095	33 30	119	24 27	169	43 37	207	63 40	287	23 08	311
32	46 27	048	54 41	095	34 18	120	24 38	170	43 11	208	62 47	287	22 27	311
33	47 07	047	55 35	096	35 05	120	24 47	170	42 45	209	61 55	287	21 46	311
34	47 48	047	56 30	096	35 53	121	24 56	171	42 18	210	61 03	287	21 05	311
35	48 28	047	57 24	097	36 39	122	25 04	172	41 49	211	60 10	287	20 24	312
36	49 07	046	58 19	097	37 26	123	25 11	173	41 20	213	59 18	287	19 43	312
37	49 47	046	59 13	098	38 12	123	25 18	174	40 50	214	58 25	287	19 02	312
38	50 26	046	60 07	099	38 57	124	25 23	175	40 20	215	57 33	287	18 21	312
39	51 06	045	61 01	099	39 43	125	25 28	175	39 48	216	56 41	287	17 41	312
40	51 45	045	61 55	100	40 27	126	25 32	176	39 16	217	55 48	287	17 00	313
41	52 23	045	62 49	101	41 11	127	25 36	177	38 42	218	54 56	287	16 20	313
42	53 01	044	63 43	102	41 55	128	25 38	178	38 08	219	54 03	287	15 40	313
43	53 39	044	64 37	102	42 38	128	25 40	179	37 34	220	53 11	287	15 00	313
44	54 17	043	65 30	103	43 21	129	25 40	180	36 58	221	52 18	287	14 20	314
	CAPELLA		◆POLLUX		PROCYON		SIRIUS		◆Acamar		Diphda		◆Alpheratz	
45	54 54	043	26 46	069	20 36	093	21 41	121	25 40	180	36 22	222	51 26	287
46	55 31	042	27 37	070	21 31	094	22 28	122	25 40	181	35 46	222	50 34	287
47	56 08	041	28 29	070	22 25	094	23 14	122	25 38	182	35 08	223	49 41	287
48	56 44	041	29 20	070	23 20	095	24 00	123	25 36	183	34 31	224	48 49	287
49	57 20	040	30 12	070	24 15	095	24 46	124	25 32	184	33 52	225	47 57	287
50	57 55	039	31 03	071	25 09	096	25 32	124	25 28	185	33 13	226	47 04	288
51	58 29	039	31 55	071	26 04	096	26 17	125	25 24	185	32 33	227	46 12	288
52	59 03	038	32 47	071	26 58	097	27 02	125	25 18	186	31 53	227	45 20	288
53	59 37	037	33 39	071	27 53	097	27 46	126	25 12	187	31 13	228	44 28	288
54	60 09	036	34 31	071	28 47	098	28 30	127	25 04	188	30 31	229	43 36	288
55	60 41	035	35 23	072	29 41	098	29 14	128	24 56	189	29 50	230	42 44	288
56	61 13	034	36 15	072	30 36	098	29 57	128	24 48	190	29 08	231	41 52	288
57	61 43	033	37 07	072	31 30	099	30 40	129	24 38	190	28 25	231	41 00	289
58	62 13	032	37 59	072	32 24	099	31 22	130	24 28	191	27 42	232	40 08	289
59	62 42	031	38 51	072	33 18	100	32 04	131	24 17	192	26 59	233	39 16	289
	◆CAPELLA		PROCYON		◆SIRIUS		RIGEL		Diphda		◆Hamal		Mirfak	
60	63 10	030	34 12	100	32 45	131	53 02	148	26 15	233	64 09	275	63 13	347
61	63 37	029	35 06	101	33 26	132	53 30	150	25 31	234	63 14	275	63 00	346
62	64 03	028	35 59	102	34 07	133	53 57	151	24 46	235	62 20	275	62 46	344
63	64 28	027	36 53	102	34 47	134	54 23	153	24 01	235	61 25	275	62 31	343
64	64 52	025	37 47	103	35 26	135	54 47	154	23 16	236	60 31	276	62 14	342
65	65 15	024	38 40	103	36 04	136	55 10	156	22 31	237	59 36	276	61 56	341
66	65 37	022	39 33	104	36 43	136	55 32	158	21 45	237	58 41	276	61 38	339
67	65 57	021	40 27	104	37 20	137	55 52	159	20 58	238	57 47	276	61 18	338
68	66 16	019	41 20	105	37 57	138	56 11	161	20 12	238	56 53	277	60 57	337
69	66 33	018	42 12	106	38 33	139	56 28	163	19 25	239	55 58	277	60 35	336
70	66 49	016	43 05	106	39 08	140	56 43	164	18 38	240	55 04	277	60 13	335
71	67 04	015	43 58	107	39 43	141	56 57	166	17 51	240	54 09	277	59 49	334
72	67 17	013	44 50	108	40 17	142	57 10	168	17 03	241	53 15	278	59 25	333
73	67 28	011	45 42	108	40 50	143	57 20	170	16 15	241	52 21	278	58 59	332
74	67 38	009	46 34	109	41 23	144	57 29	172	15 27	242	51 26	278	58 33	331
	CAPELLA		◆Dubhe		POLLUX		PROCYON		◆SIRIUS		RIGEL		◆Hamal	
75	67 46	008	20 36	030	52 53	075	47 26	110	41 54	145	57 36	173	50 32	278
76	67 52	006	21 04	030	53 46	075	48 17	110	42 25	146	57 42	175	49 38	279
77	67 57	004	21 32	031	54 39	075	49 09	111	42 55	148	57 45	177	48 44	279
78	67 59	002	22 00	031	55 32	075	50 00	112	43 24	149	57 47	179	47 50	279
79	68 00	000	22 28	031	56 25	075	50 50	113	43 52	150	57 47	181	46 55	279
80	68 00	358	22 56	031	57 17	075	51 41	114	44 19	151	57 46	183	46 01	280
81	67 57	356	23 24	031	58 10	075	52 31	115	44 45	152	57 42	184	45 07	280
82	67 53	355	23 52	031	59 03	075	53 20	116	45 10	153	57 37	186	44 13	280
83	67 47	353	24 20	031	59 56	075	54 10	116	45 34	155	57 30	188	43 19	280
84	67 39	351	24 49	031	60 49	075	54 58	117	45 57	156	57 22	190	42 25	281
85	67 30	349	25 17	031	61 42	075	55 47	118	46 19	157	57 11	192	41 32	281
86	67 19	347	25 45	031	62 35	075	56 35	120	46 40	158	56 59	194	40 38	281
87	67 06	346	26 13	031	63 28	075	57 22	121	46 59	160	56 46	195	39 44	281
88	66 52	344	26 42	031	64 21	075	58 09	122	47 17	161	56 30	197	38 50	282
89	66 36	342	27 10	031	65 14	075	58 55	123	47 35	162	56 13	199	37 57	282

LAT 24°N

LHA	Hc	Zn	Hc	Zn	Hc	Zn	Hc	Zn	Hc	Zn	Hc	Zn	Hc	Zn
♈	Dubhe		◆REGULUS		Alphard		SIRIUS		◆RIGEL		ALDEBARAN		◆CAPELLA	
90	27 39	031	30 15	090	29 48	116	47 50	164	55 55	200	68 52	253	66 19	341
91	28 07	031	31 09	090	30 37	117	48 05	165	55 35	202	68 00	254	66 00	339
92	28 35	031	32 04	091	31 26	118	48 18	167	55 14	204	67 07	255	65 40	338
93	29 04	031	32 59	091	32 14	118	48 30	168	54 51	205	66 14	256	65 19	336
94	29 32	031	33 54	091	33 02	119	48 41	169	54 27	207	65 20	257	64 56	335
95	30 00	031	34 49	092	33 50	120	48 50	171	54 01	208	64 27	258	64 32	334
96	30 28	031	35 43	092	34 37	120	48 58	172	53 34	210	63 33	259	64 08	332
97	30 57	031	36 38	093	35 25	121	49 05	174	53 07	211	62 39	259	63 42	331
98	31 25	031	37 33	093	36 11	122	49 10	175	52 37	213	61 45	260	63 15	330
99	31 53	031	38 28	094	36 58	123	49 14	177	52 07	214	60 51	261	62 47	329
100	32 21	031	39 22	094	37 43	124	49 16	178	51 36	216	59 57	261	62 18	328
101	32 49	031	40 17	095	38 29	124	49 17	180	51 03	217	59 03	262	61 48	327
102	33 17	031	41 12	095	39 14	125	49 17	181	50 30	218	58 09	263	61 18	326
103	33 45	030	42 06	095	39 59	126	49 15	183	49 55	219	57 14	263	60 47	325
104	34 12	030	43 01	096	40 43	127	49 12	184	49 20	221	56 20	264	60 15	324
	Dubhe		◆REGULUS		Alphard		SIRIUS		◆RIGEL		ALDEBARAN		◆CAPELLA	
105	34 40	030	43 55	096	41 26	128	49 08	186	48 44	222	55 25	264	59 42	323
106	35 07	030	44 50	097	42 09	129	49 02	187	48 07	223	54 31	265	59 09	322
107	35 35	030	45 44	098	42 52	130	48 54	188	47 29	224	53 36	265	58 35	321
108	36 02	030	46 38	098	43 34	131	48 45	190	46 51	225	52 41	266	58 00	321
109	36 29	030	47 33	099	44 15	132	48 35	191	46 11	226	51 47	266	57 25	320
110	36 56	029	48 27	099	44 56	133	48 24	193	45 31	227	50 52	267	56 50	319
111	37 23	029	49 21	100	45 36	134	48 11	194	44 51	228	49 57	267	56 14	319
112	37 49	029	50 15	100	46 15	135	47 57	195	44 09	229	49 02	268	55 37	318
113	38 16	029	51 09	101	46 54	136	47 42	197	43 27	230	48 08	268	55 00	317
114	38 42	028	52 02	102	47 32	137	47 26	198	42 45	231	47 13	269	54 23	317
115	39 08	028	52 56	102	48 09	138	47 08	200	42 02	232	46 18	269	53 46	316
116	39 34	028	53 49	103	48 45	139	46 49	201	41 18	233	45 23	270	53 08	316
117	39 59	028	54 43	104	49 20	140	46 29	202	40 34	234	44 28	270	52 29	315
118	40 25	027	55 36	104	49 55	142	46 07	203	39 49	235	43 34	270	51 51	315
119	40 50	027	56 29	105	50 28	143	45 45	205	39 04	236	42 39	271	51 12	315
	◆Dubhe		REGULUS		◆Alphard		SIRIUS		RIGEL		◆ALDEBARAN		CAPELLA	
120	41 15	027	57 22	106	51 01	144	45 21	206	38 19	237	41 44	271	50 33	314
121	41 39	027	58 14	107	51 32	146	44 57	207	37 33	237	40 49	272	49 53	314
122	42 04	026	59 07	108	52 03	147	44 31	208	36 46	238	39 54	272	49 14	314
123	42 28	026	59 59	108	52 32	148	44 05	210	36 00	239	39 00	272	48 34	313
124	42 51	025	60 51	109	53 00	150	43 37	211	35 12	240	38 05	273	47 54	313
125	43 15	025	61 42	110	53 27	151	43 09	212	34 25	240	37 10	273	47 14	313
126	43 38	025	62 34	111	53 53	153	42 39	213	33 37	241	36 15	273	46 33	312
127	44 01	024	63 24	112	54 18	154	42 09	214	32 49	242	35 21	274	45 53	312
128	44 23	024	64 15	114	54 41	156	41 38	215	32 01	242	34 26	274	45 12	312
129	44 45	024	65 05	115	55 02	157	41 06	216	31 12	243	33 31	274	44 31	312
130	45 07	023	65 55	116	55 23	159	40 33	217	30 23	244	32 37	275	43 50	312
131	45 28	023	66 43	117	55 41	161	39 59	218	29 33	244	31 42	275	43 09	311
132	45 49	022	67 32	119	55 59	162	39 25	219	28 44	245	30 48	276	42 28	311
133	46 10	022	68 20	120	56 14	164	38 50	220	27 54	246	29 53	276	41 47	311
134	46 30	021	69 06	122	56 29	166	38 14	221	27 04	246	28 59	276	41 06	311
	Dubhe		◆ARCTURUS		SPICA		◆Alphard		SIRIUS		BETELGEUSE		◆CAPELLA	
135	46 49	021	17 29	076	16 25	111	56 41	168	37 37	222	42 44	257	40 24	311
136	47 09	020	18 23	077	17 17	111	56 52	169	37 00	223	41 51	258	39 43	311
137	47 27	020	19 16	077	18 08	112	57 01	171	36 22	224	40 57	258	39 01	311
138	47 46	019	20 09	077	18 58	112	57 09	173	35 44	225	40 03	259	38 20	311
139	48 03	019	21 03	078	19 49	113	57 15	175	35 05	226	39 09	260	37 38	311
140	48 21	018	21 56	078	20 40	113	57 19	177	34 25	227	38 15	260	36 57	311
141	48 37	018	22 50	078	21 30	114	57 21	178	33 45	227	37 21	261	36 15	311
142	48 54	017	23 44	079	22 20	114	57 22	180	33 04	228	36 27	261	35 33	311
143	49 09	016	24 38	079	23 10	115	57 21	182	32 23	229	35 33	262	34 52	310
144	49 24	016	25 31	079	23 59	115	57 18	184	31 42	230	34 39	262	34 10	310
145	49 39	015	26 25	080	24 49	116	57 13	186	31 00	231	33 44	263	33 28	311
146	49 53	014	27 19	080	25 38	117	57 07	188	30 17	231	32 50	263	32 47	311
147	50 06	014	28 13	080	26 27	117	56 59	189	29 34	232	31 55	264	32 05	311
148	50 19	013	29 07	081	27 15	118	56 49	191	28 50	233	31 01	264	31 23	311
149	50 31	012	30 01	081	28 04	118	56 37	193	28 07	234	30 06	265	30 42	311
	Dubhe		◆ARCTURUS		SPICA		◆Alphard		SIRIUS		POLLUX		◆CAPELLA	
150	50 43	012	30 55	081	28 52	119	56 24	195	27 22	234	59 30	285	30 00	311
151	50 53	011	31 50	082	29 39	120	56 09	196	26 38	235	58 37	285	29 19	311
152	51 04	010	32 44	082	30 27	120	55 53	198	25 53	236	57 44	285	28 37	311
153	51 13	010	33 38	082	31 14	121	55 35	200	25 07	236	56 51	285	27 56	311
154	51 22	009	34 33	082	32 01	122	55 16	201	24 22	237	55 58	285	27 14	311
155	51 30	008	35 27	083	32 47	123	54 55	203	23 36	237	55 05	285	26 33	311
156	51 38	007	36 21	083	33 33	123	54 33	205	22 49	238	54 13	285	25 52	311
157	51 45	007	37 16	083	34 19	124	54 10	206	22 02	239	53 20	285	25 10	311
158	51 51	006	38 10	084	35 04	125	53 45	208	21 15	239	52 27	285	24 29	311
159	51 56	005	39 05	084	35 49	126	53 19	209	20 28	240	51 34	285	23 48	312
160	52 01	004	39 59	084	36 33	126	52 51	211	19 41	240	50 41	286	23 07	312
161	52 05	004	40 54	085	37 17	127	52 23	212	18 53	241	49 48	286	22 27	312
162	52 08	003	41 48	085	38 00	128	51 53	214	18 05	242	48 56	286	21 46	312
163	52 10	002	42 43	085	38 43	129	51 22	215	17 16	242	48 03	286	21 05	312
164	52 12	001	43 38	086	39 26	130	50 50	216	16 28	243	47 10	286	20 25	312
	◆Dubhe		Alkaid		ARCTURUS		◆SPICA		Alphard		◆PROCYON		POLLUX	
165	52 13	001	48 46	041	44 32	086	40 08	131	50 17	217	38 14	257	46 17	286
166	52 13	000	49 22	041	45 27	086	40 49	132	49 44	219	37 21	258	45 25	286
167	52 13	359	49 58	040	46 22	087	41 30	133	49 09	220	36 27	258	44 32	286
168	52 12	358	50 33	040	47 17	087	42 10	133	48 33	221	35 34	259	43 40	287
169	52 10	358	51 08	040	48 11	088	42 49	134	47 57	222	34 40	259	42 47	287
170	52 07	357	51 43	039	49 06	088	43 28	135	47 19	224	33 46	260	41 55	287
171	52 04	356	52 17	039	50 01	088	44 06	137	46 41	225	32 52	260	41 02	287
172	51 59	355	52 51	038	50 56	089	44 43	138	46 02	226	31 58	261	40 10	287
173	51 55	355	53 25	037	51 50	089	45 20	139	45 23	227	31 04	261	39 18	287
174	51 49	354	53 58	037	52 45	089	45 56	140	44 42	228	30 10	262	38 25	288
175	51 43	353	54 31	036	53 40	090	46 31	141	44 02	229	29 15	262	37 33	288
176	51 36	352	55 03	036	54 35	090	47 05	142	43 20	230	28 21	263	36 41	288
177	51 28	352	55 35	035	55 30	091	47 38	143	42 38	231	27 27	263	35 49	288
178	51 20	351	56 06	034	56 24	091	48 10	145	41 55	232	26 32	264	34 57	288
179	51 11	350	56 36	034	57 19	092	48 41	146	41 12	233	25 38	264	34 05	289

 LAT 24°N

LHA	Hc	Zn	Hc	Zn	Hc	Zn	Hc	Zn	Hc	Zn	Hc	Zn	Hc	Zn
♈	◆Alkaid		Alphecca		◆SPICA		Gienah		REGULUS		◆POLLUX		Dubhe	
180	57 06	033	41 50	075	49 12	147	48 19	174	60 54	251	33 13	289	51 01	349
181	57 36	032	42 43	075	49 41	148	48 23	176	60 02	252	32 21	289	50 50	349
182	58 05	031	43 36	075	50 09	150	48 27	177	59 10	252	31 30	289	50 39	348
183	58 33	030	44 29	075	50 36	151	48 29	179	58 17	253	30 38	290	50 28	347
184	59 00	029	45 22	076	51 02	152	48 29	180	57 25	254	29 46	290	50 15	347
185	59 26	028	46 15	076	51 27	154	48 28	182	56 32	255	28 55	290	50 02	346
186	59 52	027	47 08	076	51 50	155	48 26	183	55 39	256	28 03	290	49 49	345
187	60 17	026	48 01	076	52 13	157	48 22	184	54 46	256	27 12	290	49 35	345
188	60 41	025	48 54	076	52 34	158	48 18	186	53 52	257	26 21	291	49 20	344
189	61 04	024	49 48	076	52 53	160	48 11	187	52 59	258	25 29	291	49 05	344
190	61 26	023	50 41	076	53 11	161	48 04	189	52 05	258	24 38	291	48 49	343
191	61 47	022	51 34	077	53 28	163	47 55	190	51 11	259	23 47	291	48 33	342
192	62 07	021	52 27	077	53 43	165	47 44	192	50 18	260	22 56	292	48 16	342
193	62 26	020	53 21	077	53 57	166	47 33	193	49 24	260	22 05	292	47 58	341
194	62 44	018	54 14	077	54 10	168	47 20	194	48 30	261	21 15	292	47 40	341
	◆Kochab		VEGA		Rasalhague		◆ANTARES		SPICA		◆REGULUS		Dubhe	
195	37 44	009	19 05	055	24 23	087	18 38	132	54 20	169	47 35	261	47 22	340
196	37 52	009	19 50	055	25 17	087	19 19	132	54 30	171	46 41	262	47 03	340
197	38 01	009	20 35	056	26 12	087	19 59	133	54 37	173	45 47	262	46 44	339
198	38 09	008	21 20	056	27 07	088	20 39	134	54 43	175	44 52	263	46 24	339
199	38 17	008	22 05	056	28 02	088	21 19	134	54 48	176	43 58	263	46 04	338
200	38 24	008	22 51	056	28 56	089	21 58	135	54 51	178	43 04	264	45 43	338
201	38 31	007	23 36	056	29 51	089	22 37	135	54 52	180	42 09	264	45 22	337
202	38 38	007	24 22	057	30 46	089	23 15	136	54 51	181	41 14	265	45 01	337
203	38 45	007	25 08	057	31 41	090	23 53	137	54 49	183	40 20	265	44 39	336
204	38 51	006	25 54	057	32 36	090	24 30	138	54 46	185	39 25	266	44 17	336
205	38 57	006	26 40	057	33 30	091	25 07	138	54 40	186	38 30	266	43 54	336
206	39 03	006	27 26	057	34 25	091	25 43	139	54 33	188	37 36	267	43 31	335
207	39 08	005	28 12	057	35 20	091	26 19	140	54 25	190	36 41	267	43 08	335
208	39 13	005	28 58	058	36 15	092	26 54	141	54 15	191	35 46	268	42 45	334
209	39 18	005	29 44	058	37 10	092	27 28	141	54 03	193	34 51	268	42 21	334
	◆Kochab		VEGA		Rasalhague		◆ANTARES		SPICA		◆REGULUS		Dubhe	
210	39 22	004	30 31	058	38 04	093	28 02	142	53 50	195	33 57	269	41 57	334
211	39 26	004	31 17	058	38 59	093	28 36	143	53 35	196	33 02	269	41 32	333
212	39 30	004	32 04	058	39 54	094	29 08	144	53 19	198	32 07	269	41 07	333
213	39 34	003	32 50	058	40 49	094	29 40	145	53 02	199	31 12	270	40 43	333
214	39 37	003	33 37	058	41 43	094	30 12	145	52 43	201	30 17	270	40 17	332
215	39 39	003	34 23	058	42 38	095	30 43	146	52 23	202	29 23	271	39 52	332
216	39 42	002	35 10	058	43 32	095	31 13	147	52 01	204	28 28	271	39 26	332
217	39 44	002	35 57	058	44 27	096	31 42	148	51 38	205	27 33	271	39 00	332
218	39 46	002	36 43	058	45 21	096	32 11	149	51 14	207	26 38	272	38 34	331
219	39 47	001	37 30	058	46 16	097	32 39	150	50 49	208	25 43	272	38 08	331
220	39 48	001	38 17	059	47 10	097	33 06	151	50 22	210	24 49	273	37 42	331
221	39 49	001	39 03	059	48 05	098	33 33	152	49 54	211	23 54	273	37 15	331
222	39 49	000	39 50	059	48 59	099	33 58	153	49 26	212	22 59	273	36 48	331
223	39 49	000	40 37	059	49 53	099	34 23	154	48 56	214	22 05	274	36 21	330
224	39 49	000	41 24	059	50 47	100	34 47	154	48 25	215	21 10	274	35 54	330
	VEGA		◆ALTAIR		Nunki		ANTARES		◆SPICA		Denebola		◆Alkaid	
225	42 11	059	19 23	089	14 10	128	35 11	155	47 53	216	44 08	267	60 57	335
226	42 57	059	20 18	089	14 53	128	35 33	156	47 20	217	43 13	268	60 33	334
227	43 44	058	21 13	090	15 36	129	35 54	157	46 46	219	42 19	268	60 09	333
228	44 31	058	22 08	090	16 19	129	36 15	158	46 12	220	41 24	269	59 44	332
229	45 17	058	23 03	090	17 01	130	36 34	160	45 37	221	40 29	269	59 18	331
230	46 04	058	23 58	091	17 42	131	36 53	161	45 00	222	39 34	269	58 51	330
231	46 51	058	24 52	091	18 24	131	37 11	162	44 23	223	38 39	270	58 23	329
232	47 37	058	25 47	092	19 05	132	37 28	163	43 46	224	37 45	270	57 55	329
233	48 24	058	26 42	092	19 45	132	37 44	164	43 07	225	36 50	271	57 26	328
234	49 10	058	27 37	092	20 26	133	37 58	165	42 28	226	35 55	271	56 57	327
235	49 56	058	28 31	093	21 06	134	38 12	166	41 48	227	35 00	271	56 26	326
236	50 43	057	29 26	093	21 45	134	38 25	167	41 08	228	34 05	272	55 56	325
237	51 29	057	30 21	094	22 24	135	38 37	168	40 27	229	33 11	272	55 24	325
238	52 15	057	31 16	094	23 02	136	38 47	169	39 45	230	32 16	273	54 52	324
239	53 01	057	32 10	095	23 40	136	38 57	170	39 03	231	31 21	273	54 20	323
	◆VEGA		ALTAIR		Nunki		◆ANTARES		SPICA		ARCTURUS		◆Alkaid	
240	53 46	056	33 05	095	24 18	137	39 05	172	38 20	232	65 16	264	53 47	323
241	54 32	056	33 59	096	24 55	138	39 13	173	37 37	232	64 21	265	53 14	322
242	55 18	056	34 54	096	25 31	139	39 19	174	36 54	233	63 27	265	52 40	322
243	56 03	056	35 48	097	26 07	139	39 25	175	36 09	234	62 32	266	52 06	321
244	56 48	055	36 43	097	26 43	140	39 29	176	35 25	235	61 38	266	51 32	321
245	57 33	055	37 37	097	27 18	141	39 32	177	34 40	236	60 43	267	50 57	320
246	58 17	054	38 32	098	27 52	142	39 34	179	33 54	236	59 48	267	50 22	320
247	59 02	054	39 26	099	28 26	142	39 35	180	33 09	237	58 53	268	49 46	319
248	59 46	053	40 20	099	28 59	143	39 34	181	32 22	238	57 59	268	49 10	319
249	60 30	053	41 14	100	29 31	144	39 33	182	31 36	239	57 04	269	48 34	319
250	61 13	052	42 08	100	30 03	145	39 31	183	30 49	239	56 09	269	47 58	318
251	61 56	052	43 02	101	30 34	146	39 27	184	30 02	240	55 14	269	47 21	318
252	62 39	051	43 56	101	31 05	147	39 22	185	29 14	241	54 19	270	46 44	318
253	63 21	050	44 49	102	31 35	147	39 17	187	28 26	241	53 25	270	46 07	317
254	64 03	049	45 43	102	32 04	148	39 10	188	27 38	242	52 30	271	45 30	317
	◆DENEB		ALTAIR		◆Nunki		ANTARES		SPICA		◆ARCTURUS		Alkaid	
255	40 54	050	46 37	103	32 32	149	39 02	189	26 49	242	51 35	271	44 53	317
256	41 36	050	47 30	104	33 00	150	38 53	190	26 01	243	50 40	271	44 15	317
257	42 18	050	48 23	104	33 27	151	38 43	191	25 12	244	49 45	272	43 37	316
258	43 00	050	49 16	105	33 53	152	38 31	192	24 22	244	48 51	272	42 59	316
259	43 42	049	50 09	106	34 18	153	38 19	193	23 33	245	47 56	273	42 21	316
260	44 23	049	51 02	106	34 43	154	38 06	195	22 43	245	47 01	273	41 43	316
261	45 05	049	51 54	107	35 06	155	37 52	196	21 53	246	46 06	273	41 04	316
262	45 46	049	52 46	108	35 29	156	37 37	197	21 03	247	45 12	274	40 26	315
263	46 27	049	53 38	109	35 51	157	37 20	198	20 13	247	44 17	274	39 47	315
264	47 09	048	54 30	110	36 12	158	37 03	199	19 22	248	43 22	274	39 09	315
265	47 49	048	55 22	110	36 32	159	36 45	200	18 31	248	42 28	275	38 30	315
266	48 30	048	56 13	111	36 51	160	36 26	201	17 40	249	41 33	275	37 51	315
267	49 11	048	57 04	112	37 10	161	36 06	202	16 49	249	40 38	275	37 12	315
268	49 51	047	57 54	113	37 27	162	35 45	203	15 58	250	39 44	276	36 33	315
269	50 31	047	58 45	114	37 43	163	35 23	204	15 06	250	38 49	276	35 54	315

LHA	Hc	Zn	Hc	Zn	Hc	Zn	Hc	Zn	Hc	Zn	Hc	Zn	Hc	Zn
♈	◆DENEB		Enif		◆Nunki		ANTARES		◆ARCTURUS		Alkaid		Kochab	
270	51 11	047	34 58	095	37 58	164	35 00	205	37 55	276	35 15	315	34 04	346
271	51 51	046	35 53	095	38 13	165	34 37	206	37 00	277	34 36	315	33 51	346
272	52 30	046	36 47	096	38 26	167	34 12	207	36 06	277	33 57	315	33 37	346
273	53 10	045	37 42	096	38 38	168	33 47	208	35 11	277	33 18	315	33 24	345
274	53 48	045	38 36	097	38 49	169	33 21	209	34 17	278	32 39	315	33 10	345
275	54 27	044	39 31	097	39 00	170	32 54	210	33 23	278	32 00	315	32 56	345
276	55 05	044	40 25	098	39 09	171	32 27	211	32 29	278	31 21	315	32 42	345
277	55 43	043	41 19	098	39 17	172	31 58	212	31 34	279	30 42	315	32 27	345
278	56 20	043	42 13	099	39 24	173	31 29	212	30 40	279	30 03	315	32 13	345
279	56 57	042	43 08	099	39 29	174	30 59	213	29 46	279	29 24	315	31 58	344
280	57 34	041	44 02	100	39 34	176	30 29	214	28 52	280	28 45	315	31 43	344
281	58 10	041	44 56	100	39 38	177	29 58	215	27 58	280	28 06	315	31 29	344
282	58 46	040	45 50	101	39 40	178	29 26	216	27 04	280	27 27	315	31 14	344
283	59 21	039	46 43	102	39 41	179	28 54	217	26 10	281	26 48	315	30 59	344
284	59 55	038	47 37	102	39 42	180	28 21	217	25 16	281	26 09	315	30 43	344
	DENEB		◆Alpheratz		FOMALHAUT		◆Nunki		ANTARES		◆Alphecca		Kochab	
285	60 29	038	22 08	067	11 45	130	39 41	181	27 47	218	43 48	285	30 28	344
286	61 02	037	22 58	067	12 27	131	39 39	183	27 13	219	42 55	285	30 13	344
287	61 34	036	23 49	067	13 08	131	39 36	184	26 38	220	42 02	285	29 57	344
288	62 06	035	24 40	068	13 49	132	39 32	185	26 03	221	41 09	285	29 42	343
289	62 37	034	25 30	068	14 30	132	39 26	186	25 27	221	40 17	286	29 26	343
290	63 07	033	26 21	068	15 10	133	39 20	187	24 50	222	39 24	286	29 10	343
291	63 36	032	27 12	068	15 50	134	39 13	188	24 13	223	38 31	286	28 54	343
292	64 04	030	28 03	069	16 29	134	39 04	190	23 36	223	37 38	286	28 38	343
293	64 32	029	28 54	069	17 08	135	38 54	191	22 58	224	36 46	286	28 22	343
294	64 58	028	29 45	069	17 47	135	38 44	192	22 19	225	35 53	287	28 06	343
295	65 23	027	30 37	069	18 25	136	38 32	193	21 41	225	35 01	287	27 50	343
296	65 47	025	31 28	069	19 03	137	38 19	194	21 01	226	34 08	287	27 34	343
297	66 09	024	32 19	070	19 41	137	38 05	195	20 22	227	33 16	287	27 18	343
298	66 31	022	33 11	070	20 18	138	37 51	196	19 41	227	32 24	287	27 02	343
299	66 51	021	34 02	070	20 54	139	37 35	197	19 01	228	31 31	288	26 46	343
	◆DENEB		Schedar		Alpheratz		◆FOMALHAUT		Nunki		◆Rasalhague		VEGA	
300	67 09	019	30 44	037	34 54	070	21 30	139	37 18	198	53 47	258	67 00	315
301	67 26	017	31 18	037	35 45	070	22 06	140	37 00	199	52 54	259	66 20	314
302	67 42	016	31 51	037	36 37	071	22 41	141	36 41	201	52 00	259	65 41	313
303	67 56	014	32 24	037	37 29	071	23 15	141	36 22	202	51 06	260	65 00	312
304	68 08	012	32 57	037	38 21	071	23 49	142	36 01	203	50 12	261	64 19	311
305	68 18	010	33 29	037	39 12	071	24 23	143	35 40	204	49 18	261	63 37	310
306	68 27	008	34 02	037	40 04	071	24 56	144	35 17	205	48 24	262	62 55	309
307	68 34	006	34 35	037	40 56	071	25 28	144	34 54	206	47 29	262	62 13	309
308	68 39	004	35 08	037	41 48	072	26 00	145	34 30	207	46 35	263	61 30	308
309	68 43	003	35 40	036	42 40	072	26 31	146	34 05	208	45 41	263	60 46	307
310	68 44	001	36 13	036	43 32	072	27 01	147	33 39	208	44 46	264	60 03	307
311	68 44	359	36 45	036	44 24	072	27 31	147	33 13	209	43 52	264	59 19	306
312	68 42	357	37 18	036	45 17	072	28 00	148	32 46	210	42 57	265	58 34	306
313	68 38	355	37 50	036	46 09	072	28 29	149	32 17	211	42 02	265	57 50	305
314	68 32	353	38 22	036	47 01	072	28 57	150	31 49	212	41 08	266	57 05	305
	◆Alpheratz		Diphda		◆FOMALHAUT		ALTAIR		Rasalhague		◆VEGA		DENEB	
315	47 53	073	21 14	122	29 24	151	67 32	231	40 13	266	56 20	305	68 24	351
316	48 46	073	22 00	123	29 50	152	66 49	232	39 18	267	55 35	304	68 15	349
317	49 38	073	22 46	124	30 16	152	66 06	234	38 24	267	54 50	304	68 03	347
318	50 30	073	23 31	124	30 41	153	65 21	236	37 29	268	54 04	304	67 51	346
319	51 23	073	24 16	125	31 05	154	64 35	237	36 34	268	53 18	303	67 36	344
320	52 15	073	25 01	126	31 29	155	63 49	238	35 39	268	52 32	303	67 20	342
321	53 07	073	25 45	126	31 51	156	63 02	240	34 44	269	51 47	303	67 02	340
322	54 00	073	26 29	127	32 13	157	62 14	241	33 50	269	51 00	303	66 43	339
323	54 52	073	27 13	128	32 34	158	61 26	242	32 55	270	50 14	302	66 23	337
324	55 45	073	27 56	128	32 54	159	60 37	243	32 00	270	49 28	302	66 01	336
325	56 37	073	28 39	129	33 13	160	59 48	245	31 05	271	48 42	302	65 38	334
326	57 30	073	29 22	130	33 32	161	58 58	246	30 10	271	47 55	302	65 14	333
327	58 22	073	30 04	130	33 49	162	58 08	247	29 16	271	47 09	302	64 48	332
328	59 14	073	30 45	131	34 06	163	57 18	248	28 21	272	46 22	302	64 22	330
329	60 07	073	31 26	132	34 22	164	56 27	249	27 26	272	45 35	302	63 54	329
	◆Mirfak		Hamal		Diphda		◆FOMALHAUT		ALTAIR		◆VEGA		DENEB	
330	23 46	044	33 59	077	32 07	133	34 37	165	55 36	249	44 49	302	63 26	328
331	24 25	044	34 52	077	32 46	134	34 50	166	54 44	250	44 02	302	62 56	327
332	25 03	044	35 46	078	33 26	134	35 03	167	53 52	251	43 15	301	62 26	326
333	25 41	044	36 39	078	34 05	135	35 15	168	53 00	252	42 29	301	61 54	325
334	26 20	045	37 33	078	34 43	136	35 26	169	52 08	253	41 42	301	61 22	324
335	26 58	045	38 26	078	35 21	137	35 36	170	51 16	253	40 55	301	60 50	323
336	27 37	045	39 20	079	35 58	138	35 45	171	50 23	254	40 08	301	60 16	322
337	28 15	045	40 14	079	36 34	139	35 53	172	49 30	255	39 21	301	59 42	321
338	28 54	045	41 08	079	37 10	140	36 00	173	48 37	255	38 35	301	59 08	320
339	29 32	045	42 01	079	37 45	141	36 06	174	47 44	256	37 48	301	58 32	320
340	30 11	045	42 55	080	38 20	142	36 11	175	46 51	257	37 01	302	57 57	319
341	30 50	045	43 49	080	38 53	143	36 15	176	45 58	257	36 15	302	57 20	318
342	31 28	045	44 43	080	39 26	144	36 18	177	45 04	258	35 28	302	56 44	318
343	32 07	045	45 37	080	39 59	145	36 20	179	44 10	259	34 41	302	56 07	317
344	32 46	045	46 31	081	40 30	146	36 21	180	43 16	259	33 55	302	55 29	316
	Schedar		◆CAPELLA		ALDEBARAN		Diphda		◆FOMALHAUT		ALTAIR		◆DENEB	
345	52 44	023	14 19	046	12 02	077	41 00	147	36 21	181	42 23	260	54 51	316
346	53 05	022	14 59	046	12 55	077	41 30	148	36 20	182	41 29	260	54 13	315
347	53 25	021	15 38	046	13 49	078	41 59	149	36 17	183	40 35	261	53 34	315
348	53 45	021	16 18	046	14 43	078	42 27	150	36 14	184	39 40	261	52 55	314
349	54 04	020	16 57	047	15 36	079	42 54	151	36 10	185	38 46	262	52 16	314
350	54 22	019	17 37	047	16 30	079	43 20	152	36 05	186	37 52	262	51 36	314
351	54 39	018	18 17	047	17 24	079	43 45	153	35 58	187	36 58	263	50 57	313
352	54 56	017	18 58	047	18 18	080	44 09	155	35 51	188	36 03	263	50 17	313
353	55 12	016	19 38	047	19 12	080	44 32	156	35 43	189	35 09	264	49 36	313
354	55 27	016	20 18	048	20 06	080	44 53	157	35 33	190	34 14	264	48 56	312
355	55 41	015	20 59	048	21 00	081	45 14	158	35 23	191	33 20	265	48 15	312
356	55 55	014	21 39	048	21 54	081	45 34	160	35 12	192	32 25	265	47 34	312
357	56 08	013	22 20	048	22 48	081	45 52	161	34 59	193	31 30	266	46 53	311
358	56 19	012	23 01	048	23 42	082	46 10	162	34 46	194	30 36	266	46 12	311
359	56 30	011	23 42	048	24 37	082	46 26	164	34 32	195	29 41	267	45 31	311

LAT 23°N

LHA ♈	Hc	Zn	Hc	Zn	Hc	Zn	Hc	Zn	Hc	Zn	Hc	Zn	Hc	Zn
	Schedar		◆CAPELLA		ALDEBARAN		◆Diphda		FOMALHAUT		◆ALTAIR		DENEB	
0	55 41	010	23 43	048	25 23	082	47 38	165	35 14	197	28 49	268	44 10	312
1	55 50	009	24 24	048	26 17	082	47 53	166	34 58	198	27 54	268	43 29	311
2	55 58	008	25 05	048	27 12	083	48 05	167	34 41	199	26 59	268	42 47	311
3	56 06	007	25 47	048	28 07	083	48 17	169	34 23	200	26 03	269	42 05	311
4	56 12	006	26 28	049	29 02	083	48 27	170	34 04	201	25 08	269	41 24	311
5	56 17	005	27 10	049	29 57	084	48 35	172	33 44	202	24 13	270	40 42	311
6	56 21	004	27 51	049	30 52	084	48 43	173	33 23	203	23 18	270	40 00	311
7	56 25	003	28 33	049	31 46	084	48 49	174	33 01	204	22 23	270	39 18	311
8	56 27	002	29 14	049	32 41	085	48 54	176	32 39	204	21 27	271	38 36	310
9	56 29	001	29 56	049	33 36	085	48 57	177	32 16	205	20 32	271	37 54	310
10	56 29	000	30 37	049	34 31	085	48 59	179	31 52	206	19 37	272	37 12	310
11	56 29	359	31 19	049	35 27	086	48 59	180	31 27	207	18 42	272	36 30	310
12	56 28	358	32 01	049	36 22	086	48 58	182	31 01	208	17 46	272	35 48	310
13	56 25	357	32 42	049	37 17	086	48 56	183	30 35	209	16 51	273	35 05	310
14	56 22	356	33 24	049	38 12	087	48 52	185	30 08	210	15 56	273	34 23	310
	◆CAPELLA		ALDEBARAN		RIGEL		◆Diphda		FOMALHAUT		Enif		◆DENEB	
15	34 06	049	39 07	087	20 28	109	48 47	186	29 40	211	41 26	263	33 41	310
16	34 47	049	40 02	087	21 20	109	48 41	187	29 11	211	40 31	263	32 59	310
17	35 29	049	40 57	088	22 12	110	48 33	189	28 42	212	39 36	264	32 17	310
18	36 11	049	41 52	088	23 04	110	48 24	190	28 12	213	38 41	264	31 34	310
19	36 52	049	42 48	088	23 56	111	48 13	192	27 42	214	37 46	265	30 52	310
20	37 34	049	43 43	089	24 47	112	48 01	193	27 11	215	36 51	265	30 10	310
21	38 15	049	44 38	089	25 39	112	47 48	194	26 39	215	35 56	266	29 28	310
22	38 57	049	45 33	090	26 30	113	47 34	196	26 06	216	35 01	266	28 46	310
23	39 38	048	46 29	090	27 20	113	47 18	197	25 33	217	34 06	266	28 04	310
24	40 19	048	47 24	090	28 11	114	47 01	199	25 00	218	33 11	267	27 22	310
25	41 01	048	48 19	091	29 02	114	46 43	200	24 26	218	32 16	267	26 40	311
26	41 42	048	49 14	091	29 52	115	46 23	201	23 51	219	31 20	268	25 58	311
27	42 23	048	50 09	092	30 42	116	46 03	202	23 16	220	30 25	268	25 16	311
28	43 04	048	51 05	092	31 31	116	45 41	204	22 40	221	29 30	268	24 34	311
29	43 45	048	52 00	092	32 21	117	45 19	205	22 04	221	28 35	269	23 52	311
	CAPELLA		◆ALDEBARAN		RIGEL		Acamar		◆Diphda		Alpheratz		◆DENEB	
30	44 25	047	52 55	093	33 10	118	25 15	168	44 55	206	64 13	290	23 11	311
31	45 06	047	53 50	093	33 59	118	25 26	169	44 30	207	63 21	289	22 29	311
32	45 46	047	54 45	094	34 47	119	25 37	169	44 04	209	62 29	289	21 48	311
33	46 26	047	55 40	094	35 35	120	25 46	170	43 37	210	61 36	289	21 06	312
34	47 06	046	56 35	095	36 23	120	25 55	171	43 09	211	60 44	289	20 25	312
35	47 46	046	57 30	095	37 11	121	26 03	172	42 40	212	59 52	289	19 44	312
36	48 26	046	58 25	096	37 58	122	26 11	173	42 11	213	59 00	289	19 03	312
37	49 05	045	59 20	096	38 44	123	26 17	174	41 40	214	58 07	288	18 22	312
38	49 44	045	60 15	097	39 31	123	26 23	174	41 09	215	57 15	288	17 41	312
39	50 23	045	61 10	098	40 17	124	26 28	175	40 37	216	56 22	288	17 00	313
40	51 02	044	62 05	098	41 02	125	26 32	176	40 03	217	55 30	288	16 20	313
41	51 40	044	62 59	099	41 47	126	26 35	177	39 30	218	54 38	288	15 39	313
42	52 18	043	63 54	100	42 31	127	26 38	178	38 55	219	53 45	288	14 59	313
43	52 56	043	64 48	100	43 15	128	26 40	179	38 20	220	52 53	288	14 19	313
44	53 33	042	65 43	101	43 59	129	26 40	180	37 44	221	52 00	288	13 39	314
	CAPELLA		◆POLLUX		PROCYON		SIRIUS		◆Acamar		Diphda		◆Alpheratz	
45	54 10	042	26 25	069	20 39	093	22 11	121	26 40	180	37 07	222	51 08	288
46	54 46	041	27 16	069	21 35	094	22 59	121	26 40	181	36 30	223	50 15	288
47	55 23	040	28 08	069	22 30	094	23 46	122	26 38	182	35 52	224	49 23	288
48	55 58	040	29 00	070	23 25	094	24 33	122	26 36	183	35 13	225	48 31	288
49	56 33	039	29 51	070	24 20	095	25 19	123	26 32	184	34 34	226	47 38	289
50	57 08	038	30 43	070	25 15	095	26 05	124	26 28	185	33 55	226	46 46	289
51	57 42	038	31 35	070	26 10	096	26 51	124	26 23	186	33 14	227	45 53	289
52	58 16	037	32 27	070	27 05	096	27 36	125	26 18	186	32 34	228	45 01	289
53	58 48	036	33 19	071	28 00	097	28 21	126	26 11	187	31 52	229	44 09	289
54	59 21	035	34 11	071	28 55	097	29 06	126	26 04	188	31 11	229	43 17	289
55	59 52	034	35 04	071	29 49	097	29 50	127	25 56	189	30 28	230	42 24	289
56	60 23	033	35 56	071	30 44	098	30 34	128	25 47	190	29 46	231	41 32	289
57	60 53	032	36 48	071	31 39	098	31 17	129	25 37	191	29 02	232	40 40	289
58	61 22	031	37 40	071	32 33	099	32 00	129	25 27	191	28 19	232	39 48	289
59	61 51	030	38 33	072	33 28	099	32 43	130	25 15	192	27 35	233	38 56	290
	◆CAPELLA		PROCYON		◆SIRIUS		RIGEL		Diphda		◆Hamal		Mirfak	
60	62 18	029	34 22	100	33 25	131	53 53	148	26 51	234	64 03	277	62 14	347
61	62 45	028	35 17	100	34 06	132	54 22	149	26 06	234	63 08	277	62 02	346
62	63 10	027	36 11	101	34 47	132	54 50	151	25 21	235	62 13	277	61 48	345
63	63 35	026	37 05	101	35 28	133	55 16	152	24 35	236	61 19	277	61 33	344
64	63 58	024	37 59	102	36 08	134	55 41	154	23 50	236	60 24	277	61 17	342
65	64 20	023	38 53	102	36 47	135	56 05	155	23 03	237	59 29	278	61 00	341
66	64 41	022	39 47	103	37 26	136	56 27	157	22 17	238	58 34	278	60 41	340
67	65 01	020	40 41	104	38 04	137	56 48	159	21 30	238	57 40	278	60 22	339
68	65 19	019	41 35	104	38 41	138	57 07	160	20 43	239	56 45	278	60 02	338
69	65 36	017	42 28	105	39 18	139	57 25	162	19 56	239	55 50	278	59 40	337
70	65 51	016	43 21	105	39 54	140	57 41	164	19 08	240	54 56	278	59 18	336
71	66 06	014	44 15	106	40 30	141	57 55	166	18 20	240	54 01	279	58 55	335
72	66 18	012	45 08	107	41 04	142	58 08	168	17 32	241	53 06	279	58 31	334
73	66 29	011	46 01	107	41 38	143	58 19	169	16 44	241	52 12	279	58 06	333
74	66 38	009	46 53	108	42 11	144	58 28	171	15 55	242	51 17	279	57 40	332
	CAPELLA		◆Dubhe		POLLUX		PROCYON		◆SIRIUS		RIGEL		◆Hamal	
75	66 46	007	19 45	030	52 36	073	47 46	109	42 44	145	58 36	173	50 23	279
76	66 52	005	20 12	030	53 29	073	48 38	109	43 15	146	58 41	175	49 28	280
77	66 57	004	20 40	030	54 22	073	49 30	110	43 46	147	58 45	177	48 34	280
78	66 59	002	21 08	030	55 15	073	50 22	111	44 15	148	58 47	179	47 40	280
79	67 00	000	21 36	031	56 08	074	51 13	112	44 44	149	58 47	181	46 45	280
80	67 00	358	22 04	031	57 01	074	52 04	113	45 11	151	58 46	183	45 51	281
81	66 57	357	22 32	031	57 54	073	52 55	113	45 38	152	58 42	185	44 57	281
82	66 53	355	23 01	031	58 47	073	53 46	114	46 04	153	58 37	187	44 02	281
83	66 47	353	23 29	031	59 40	073	54 36	115	46 28	154	58 30	188	43 08	281
84	66 40	351	23 57	031	60 33	073	55 25	116	46 52	155	58 21	190	42 14	281
85	66 31	350	24 25	031	61 26	073	56 15	117	47 14	157	58 10	192	41 20	282
86	66 20	348	24 54	031	62 19	073	57 04	118	47 35	158	57 57	194	40 26	282
87	66 08	346	25 22	031	63 12	073	57 52	119	47 55	159	57 43	196	39 32	282
88	65 54	345	25 50	031	64 04	073	58 40	120	48 14	161	57 27	198	38 38	282
89	65 39	343	26 19	031	64 57	073	59 27	122	48 32	162	57 10	199	37 44	283

LHA ♈	Hc	Zn	Hc	Zn	Hc	Zn	Hc	Zn	Hc	Zn	Hc	Zn	Hc	Zn
	Dubhe		◆REGULUS		Alphard		SIRIUS		◆RIGEL		ALDEBARAN		◆CAPELLA	
90	26 47	031	30 14	089	30 15	116	48 48	164	56 51	201	69 08	256	65 22	342
91	27 16	031	31 09	090	31 04	117	49 03	165	56 30	203	68 15	257	65 04	340
92	27 44	031	32 05	090	31 53	117	49 17	166	56 08	204	67 21	258	64 44	339
93	28 12	031	33 00	090	32 42	118	49 29	168	55 45	206	66 27	258	64 23	337
94	28 40	031	33 55	091	33 31	118	49 40	169	55 20	208	65 33	259	64 02	336
95	29 09	031	34 50	091	34 20	119	49 50	171	54 54	209	64 38	260	63 38	335
96	29 37	031	35 45	092	35 08	120	49 58	172	54 26	211	63 44	261	63 14	333
97	30 05	031	36 41	092	35 55	121	50 05	174	53 58	212	62 49	261	62 49	332
98	30 33	031	37 36	092	36 43	121	50 10	175	53 28	214	61 55	262	62 22	331
99	31 01	030	38 31	093	37 30	122	50 14	177	52 56	215	61 00	263	61 55	330
100	31 29	030	39 26	093	38 16	123	50 16	178	52 24	216	60 05	263	61 27	329
101	31 57	030	40 21	094	39 02	124	50 17	180	51 51	218	59 10	264	60 58	328
102	32 25	030	41 16	094	39 48	124	50 17	181	51 17	219	58 15	264	60 28	327
103	32 53	030	42 11	095	40 33	125	50 15	183	50 42	220	57 20	265	59 57	326
104	33 20	030	43 06	095	41 18	126	50 12	184	50 05	221	56 25	265	59 26	325
	Dubhe		◆REGULUS		Alphard		SIRIUS		◆RIGEL		ALDEBARAN		◆CAPELLA	
105	33 48	030	44 01	096	42 03	127	50 07	186	49 28	223	55 30	266	58 54	324
106	34 15	030	44 56	096	42 46	128	50 01	187	48 51	224	54 35	266	58 21	323
107	34 43	029	45 51	097	43 30	129	49 54	189	48 12	225	53 40	267	57 47	322
108	35 10	029	46 46	097	44 12	130	49 45	190	47 32	226	52 45	267	57 13	322
109	35 37	029	47 41	098	44 55	131	49 34	192	46 52	227	51 50	268	56 39	321
110	36 03	029	48 36	098	45 36	132	49 23	193	46 12	228	50 55	268	56 04	320
111	36 30	029	49 30	099	46 17	133	49 09	194	45 30	229	49 59	269	55 28	320
112	36 57	029	50 25	099	46 57	134	48 55	196	44 48	230	49 04	269	54 52	319
113	37 23	028	51 19	100	47 36	135	48 39	197	44 05	231	48 09	269	54 16	318
114	37 49	028	52 14	100	48 15	136	48 22	199	43 22	232	47 14	270	53 39	318
115	38 15	028	53 08	101	48 53	137	48 04	200	42 38	233	46 18	270	53 02	317
116	38 41	028	54 02	102	49 30	138	47 45	201	41 54	234	45 23	271	52 24	317
117	39 06	027	54 56	102	50 06	140	47 24	203	41 09	235	44 28	271	51 46	316
118	39 31	027	55 50	103	50 42	141	47 02	204	40 23	236	43 33	271	51 08	316
119	39 56	027	56 44	104	51 16	142	46 39	205	39 38	236	42 38	272	50 29	316
	◆Dubhe		Denebola		◆Alphard		SIRIUS		RIGEL		◆ALDEBARAN		CAPELLA	
120	40 21	026	35 31	088	51 49	143	46 15	206	38 51	237	41 42	272	49 51	315
121	40 45	026	36 27	088	52 22	145	45 50	208	38 05	238	40 47	272	49 11	315
122	41 10	026	37 22	089	52 53	146	45 24	209	37 18	239	39 52	273	48 32	314
123	41 34	025	38 17	089	53 23	148	44 57	210	36 30	240	38 57	273	47 52	314
124	41 57	025	39 12	090	53 52	149	44 29	211	35 42	240	38 02	273	47 13	314
125	42 20	025	40 07	090	54 20	151	43 59	212	34 54	241	37 07	274	46 33	313
126	42 43	024	41 03	090	54 46	152	43 29	214	34 06	242	36 11	274	45 53	313
127	43 06	024	41 58	091	55 11	154	42 58	215	33 17	242	35 16	274	45 12	313
128	43 28	024	42 53	091	55 35	155	42 26	216	32 28	243	34 21	275	44 32	313
129	43 50	023	43 48	092	55 58	157	41 54	217	31 39	244	33 26	275	43 51	313
130	44 12	023	44 44	092	56 19	159	41 20	218	30 49	244	32 31	275	43 10	312
131	44 33	022	45 39	092	56 38	160	40 46	219	29 59	245	31 36	276	42 29	312
132	44 54	022	46 34	093	56 56	162	40 11	220	29 09	246	30 41	276	41 48	312
133	45 14	021	47 29	093	57 12	164	39 35	221	28 19	246	29 47	276	41 07	312
134	45 34	021	48 24	094	57 27	165	38 59	222	27 28	247	28 52	277	40 26	312
	Dubhe		◆ARCTURUS		SPICA		◆Alphard		SIRIUS		BETELGEUSE		◆CAPELLA	
135	45 53	020	17 15	076	16 46	110	57 40	167	38 21	223	42 57	258	39 45	312
136	46 12	020	18 09	076	17 38	111	57 51	169	37 44	224	42 03	259	39 03	311
137	46 31	019	19 02	077	18 30	111	58 01	171	37 05	225	41 08	259	38 22	311
138	46 49	019	19 56	077	19 21	112	58 08	173	36 26	225	40 14	260	37 40	311
139	47 06	018	20 50	077	20 12	112	58 15	175	35 46	226	39 20	260	36 59	311
140	47 24	018	21 44	078	21 03	113	58 19	177	35 06	227	38 25	261	36 17	311
141	47 40	017	22 38	078	21 54	113	58 21	178	34 25	228	37 31	261	35 36	311
142	47 56	017	23 32	078	22 44	114	58 22	180	33 44	229	36 36	262	34 54	311
143	48 12	016	24 26	078	23 35	114	58 21	182	33 02	230	35 41	262	34 12	311
144	48 27	015	25 20	079	24 25	115	58 18	184	32 20	230	34 46	263	33 31	311
145	48 41	015	26 14	079	25 15	116	58 13	186	31 37	231	33 52	263	32 49	311
146	48 55	014	27 08	079	26 05	116	58 06	188	30 54	232	32 57	264	32 07	311
147	49 08	014	28 03	080	26 54	117	57 58	190	30 11	233	32 02	264	31 26	311
148	49 21	013	28 57	080	27 43	117	57 48	191	29 27	233	31 07	265	30 44	311
149	49 32	012	29 52	080	28 32	118	57 36	193	28 42	234	30 12	265	30 02	311
	Dubhe		◆ARCTURUS		SPICA		◆Alphard		SIRIUS		POLLUX		◆CAPELLA	
150	49 44	012	30 46	081	29 21	119	57 22	195	27 57	235	59 14	287	29 21	311
151	49 55	011	31 41	081	30 09	119	57 07	197	27 12	235	58 21	287	28 39	311
152	50 05	010	32 35	081	30 57	120	56 50	199	26 26	236	57 28	287	27 58	311
153	50 14	009	33 30	082	31 45	121	56 32	200	25 40	237	56 35	286	27 16	311
154	50 23	009	34 24	082	32 32	121	56 12	202	24 54	237	55 42	286	26 35	311
155	50 31	008	35 19	082	33 19	122	55 50	204	24 08	238	54 49	287	25 53	311
156	50 38	007	36 14	082	34 06	123	55 28	205	23 21	238	53 56	287	25 12	312
157	50 45	007	37 08	083	34 52	123	55 03	207	22 33	239	53 03	287	24 31	312
158	50 51	006	38 03	083	35 38	124	54 38	208	21 46	240	52 10	287	23 49	312
159	50 56	005	38 58	083	36 23	125	54 11	210	20 58	240	51 17	287	23 08	312
160	51 01	004	39 53	084	37 08	126	53 43	211	20 10	241	50 24	287	22 27	312
161	51 05	004	40 48	084	37 53	127	53 13	213	19 22	241	49 32	287	21 46	312
162	51 08	003	41 43	084	38 37	127	52 43	214	18 33	242	48 39	287	21 05	312
163	51 10	002	42 38	085	39 21	128	52 11	216	17 44	242	47 46	287	20 25	313
164	51 12	001	43 33	085	40 04	129	51 38	217	16 55	243	46 53	287	19 44	313
	◆Dubhe		Alkaid		ARCTURUS		◆SPICA		Alphard		◆PROCYON		POLLUX	
165	51 13	001	48 01	041	44 28	085	40 46	130	51 05	218	38 27	258	46 00	287
166	51 13	000	48 36	040	45 23	085	41 28	131	50 30	220	37 33	258	45 08	287
167	51 13	359	49 12	040	46 18	086	42 10	132	49 55	221	36 39	259	44 15	287
168	51 12	358	49 47	039	47 13	086	42 51	133	49 18	222	35 45	259	43 22	288
169	51 10	358	50 22	039	48 08	086	43 31	134	48 41	223	34 51	260	42 29	288
170	51 07	357	50 56	038	49 03	087	44 10	135	48 03	224	33 56	260	41 37	288
171	51 04	356	51 30	038	49 58	087	44 49	136	47 24	225	33 02	261	40 44	288
172	51 00	355	52 04	037	50 54	087	45 27	137	46 44	226	32 07	261	39 52	288
173	50 55	355	52 37	037	51 49	088	46 05	138	46 04	228	31 13	262	38 59	288
174	50 49	354	53 10	036	52 44	088	46 41	139	45 22	229	30 18	262	38 07	288
175	50 43	353	53 42	035	53 39	088	47 17	140	44 41	230	29 23	263	37 15	289
176	50 36	352	54 14	035	54 34	089	47 52	141	43 58	230	28 28	263	36 22	289
177	50 29	352	54 45	034	55 30	089	48 26	143	43 16	231	27 33	264	35 30	289
178	50 20	351	55 16	033	56 25	090	48 59	144	42 32	232	26 38	264	34 38	289
179	50 11	350	55 46	033	57 20	090	49 31	145	41 48	233	25 44	265	33 46	289

 LAT 23°N

LHA	Hc	Zn	Hc	Zn	Hc	Zn	Hc	Zn	Hc	Zn	Hc	Zn	Hc	Zn
♈	♦Alkaid		Alphecca		♦SPICA		Gienah		REGULUS		♦POLLUX		Dubhe	
180	56 16	032	41 34	074	50 02	146	49 18	174	61 13	252	32 53	289	50 02	350
181	56 45	031	42 27	074	50 32	148	49 23	176	60 20	253	32 01	290	49 52	349
182	57 13	030	43 20	074	51 01	149	49 27	177	59 27	254	31 09	290	49 41	348
183	57 41	030	44 13	074	51 29	150	49 29	179	58 34	255	30 18	290	49 29	348
184	58 07	029	45 06	075	51 55	152	49 29	180	57 40	256	29 26	290	49 17	347
185	58 33	028	45 59	075	52 21	153	49 28	182	56 47	256	28 34	291	49 04	346
186	58 59	027	46 53	075	52 45	155	49 26	183	55 53	257	27 42	291	48 51	346
187	59 23	026	47 46	075	53 08	156	49 22	185	54 59	258	26 51	291	48 37	345
188	59 47	025	48 39	075	53 29	158	49 17	186	54 05	258	25 59	291	48 22	344
189	60 09	024	49 33	075	53 49	159	49 11	187	53 11	259	25 08	291	48 07	344
190	60 31	023	50 26	075	54 08	161	49 03	189	52 17	260	24 16	292	47 52	343
191	60 52	021	51 20	075	54 25	163	48 54	190	51 22	260	23 25	292	47 35	343
192	61 11	020	52 13	076	54 41	164	48 43	192	50 28	261	22 34	292	47 19	342
193	61 30	019	53 07	076	54 56	166	48 31	193	49 33	261	21 43	292	47 01	342
194	61 47	018	54 00	076	55 08	168	48 18	195	48 39	262	20 52	293	46 44	341
	♦Kochab		VEGA		Rasalhague		♦ANTARES		SPICA		♦REGULUS		Dubhe	
195	36 44	009	18 30	055	24 19	086	19 18	131	55 19	169	47 44	262	46 25	340
196	36 53	009	19 15	055	25 14	086	19 59	132	55 29	171	46 49	263	46 07	340
197	37 01	009	20 01	055	26 09	087	20 40	133	55 37	173	45 54	263	45 48	339
198	37 09	008	20 46	055	27 04	087	21 21	133	55 43	174	44 59	264	45 28	339
199	37 17	008	21 32	056	27 59	088	22 01	134	55 48	176	44 04	264	45 08	338
200	37 25	008	22 17	056	28 55	088	22 40	135	55 51	178	43 09	265	44 48	338
201	37 32	007	23 03	056	29 50	088	23 19	135	55 52	180	42 14	265	44 27	338
202	37 39	007	23 49	056	30 45	089	23 58	136	55 51	181	41 19	266	44 05	337
203	37 45	007	24 35	056	31 40	089	24 36	137	55 49	183	40 24	266	43 44	337
204	37 51	006	25 21	057	32 35	089	25 14	137	55 45	185	39 29	267	43 22	336
205	37 57	006	26 07	057	33 31	090	25 51	138	55 40	187	38 34	267	42 59	336
206	38 03	006	26 53	057	34 26	090	26 28	139	55 33	188	37 39	268	42 37	336
207	38 08	005	27 39	057	35 21	091	27 04	139	55 24	190	36 44	268	42 14	335
208	38 13	005	28 26	057	36 16	091	27 40	140	55 14	192	35 48	268	41 50	335
209	38 18	005	29 12	057	37 12	091	28 15	141	55 02	193	34 53	269	41 27	334
	♦Kochab		VEGA		Rasalhague		♦ANTARES		SPICA		♦REGULUS		Dubhe	
210	38 23	004	29 59	057	38 07	092	28 49	142	54 48	195	33 58	269	41 03	334
211	38 27	004	30 45	057	39 02	092	29 23	143	54 33	197	33 03	270	40 38	334
212	38 30	004	31 32	057	39 57	093	29 57	143	54 16	198	32 07	270	40 14	333
213	38 34	003	32 18	058	40 52	093	30 29	144	53 58	200	31 12	270	39 49	333
214	38 37	003	33 05	058	41 47	094	31 01	145	53 39	201	30 17	271	39 24	333
215	38 40	003	33 51	058	42 43	094	31 33	146	53 18	203	29 22	271	38 59	333
216	38 42	002	34 38	058	43 38	094	32 03	147	52 56	205	28 27	272	38 33	332
217	38 44	002	35 25	058	44 33	095	32 33	148	52 32	206	27 31	272	38 07	332
218	38 46	002	36 12	058	45 28	095	33 02	149	52 07	207	26 36	272	37 41	332
219	38 47	001	36 58	058	46 23	096	33 31	149	51 41	209	25 41	273	37 15	332
220	38 48	001	37 45	058	47 18	096	33 59	150	51 14	210	24 46	273	36 49	331
221	38 49	001	38 32	058	48 12	097	34 25	151	50 46	212	23 51	273	36 22	331
222	38 49	000	39 19	058	49 07	097	34 52	152	50 16	213	22 56	274	35 56	331
223	38 49	000	40 05	058	50 02	098	35 17	153	49 45	214	22 00	274	35 29	331
224	38 49	000	40 52	058	50 57	099	35 41	154	49 14	216	21 05	274	35 02	331
	VEGA		♦ALTAIR		Nunki		ANTARES		♦SPICA		Denebola		♦Alkaid	
225	41 39	058	19 22	088	14 47	128	36 05	155	48 41	217	44 11	268	60 02	336
226	42 26	058	20 17	089	15 30	128	36 28	156	48 08	218	43 15	269	59 39	335
227	43 12	058	21 12	089	16 14	129	36 50	157	47 33	219	42 20	269	59 15	334
228	43 59	058	22 08	090	16 57	129	37 11	158	46 58	220	41 25	269	58 50	333
229	44 46	058	23 03	090	17 39	130	37 31	159	46 22	221	40 30	270	58 25	332
230	45 32	057	23 58	090	18 21	130	37 50	160	45 45	223	39 34	270	57 59	331
231	46 19	057	24 53	091	19 03	131	38 08	161	45 07	224	38 39	271	57 32	330
232	47 05	057	25 49	091	19 45	132	38 25	162	44 28	225	37 44	271	57 04	329
233	47 51	057	26 44	092	20 26	132	38 41	164	43 49	226	36 49	271	56 35	329
234	48 38	057	27 39	092	21 07	133	38 56	165	43 09	227	35 54	272	56 06	328
235	49 24	057	28 34	092	21 47	133	39 10	166	42 29	228	34 58	272	55 36	327
236	50 10	056	29 29	093	22 27	134	39 23	167	41 48	229	34 03	273	55 06	326
237	50 56	056	30 24	093	23 06	135	39 35	168	41 06	230	33 08	273	54 35	326
238	51 42	056	31 20	094	23 45	135	39 46	169	40 24	230	32 13	273	54 04	325
239	52 27	056	32 15	094	24 24	136	39 56	170	39 41	231	31 18	274	53 32	324
	♦VEGA		ALTAIR		Nunki		♦ANTARES		SPICA		ARCTURUS		♦Alkaid	
240	53 13	055	33 10	094	25 02	137	40 05	171	38 57	232	65 21	266	52 59	324
241	53 58	055	34 05	095	25 39	138	40 12	173	38 14	233	64 26	267	52 26	323
242	54 43	055	35 00	095	26 16	138	40 19	174	37 29	234	63 31	267	51 53	323
243	55 28	054	35 55	096	26 53	139	40 24	175	36 44	235	62 36	268	51 19	322
244	56 13	054	36 50	096	27 29	140	40 29	176	35 59	235	61 41	268	50 45	322
245	56 58	053	37 45	097	28 04	141	40 32	177	35 13	236	60 45	268	50 10	321
246	57 42	053	38 39	097	28 39	141	40 34	178	34 27	237	59 50	269	49 35	321
247	58 26	052	39 34	098	29 13	142	40 35	180	33 41	238	58 55	269	49 00	320
248	59 09	052	40 29	098	29 47	143	40 34	181	32 54	238	58 00	270	48 25	320
249	59 53	051	41 24	099	30 20	144	40 33	182	32 07	239	57 04	270	47 49	319
250	60 36	051	42 18	099	30 52	145	40 31	183	31 19	240	56 09	271	47 13	319
251	61 18	050	43 13	100	31 24	145	40 27	184	30 31	240	55 14	271	46 36	319
252	62 00	049	44 07	100	31 55	146	40 22	186	29 43	241	54 19	271	46 00	318
253	62 42	049	45 01	101	32 25	147	40 16	187	28 55	242	53 24	272	45 23	318
254	63 23	048	45 55	101	32 55	148	40 09	188	28 06	242	52 28	272	44 46	318
	♦DENEB		ALTAIR		♦Nunki		ANTARES		SPICA		♦ARCTURUS		Alkaid	
255	40 16	049	46 50	102	33 24	149	40 01	189	27 17	243	51 33	272	44 09	317
256	40 57	049	47 44	103	33 52	150	39 52	190	26 28	244	50 38	273	43 31	317
257	41 39	049	48 37	103	34 19	151	39 41	191	25 38	244	49 43	273	42 54	317
258	42 21	049	49 31	104	34 46	152	39 30	192	24 48	245	48 48	273	42 16	317
259	43 02	049	50 25	105	35 12	153	39 18	194	23 58	245	47 53	274	41 38	316
260	43 44	049	51 18	105	35 36	154	39 04	195	23 08	246	46 57	274	41 00	316
261	44 25	048	52 11	106	36 01	155	38 50	196	22 17	246	46 02	274	40 21	316
262	45 06	048	53 04	107	36 24	156	38 34	197	21 27	247	45 07	275	39 43	316
263	45 47	048	53 57	107	36 46	157	38 17	198	20 36	247	44 12	275	39 05	316
264	46 28	048	54 50	108	37 08	158	38 00	199	19 45	248	43 17	275	38 26	316
265	47 09	047	55 42	109	37 28	159	37 41	200	18 53	248	42 22	276	37 47	316
266	47 50	047	56 34	110	37 48	160	37 22	201	18 02	249	41 27	276	37 09	315
267	48 30	047	57 26	111	38 06	161	37 01	202	17 10	249	40 32	276	36 30	315
268	49 10	046	58 17	112	38 24	162	36 40	203	16 18	250	39 37	276	35 51	315
269	49 50	046	59 08	113	38 40	163	36 18	204	15 27	250	38 43	277	35 12	315

LHA	Hc	Zn	Hc	Zn	Hc	Zn	Hc	Zn	Hc	Zn	Hc	Zn	Hc	Zn
♈	♦DENEB		Enif		♦Nunki		ANTARES		♦ARCTURUS		Alkaid		Kochab	
270	50 30	046	35 03	094	38 56	164	35 55	205	37 48	277	34 33	315	33 06	346
271	51 09	045	35 58	094	39 11	165	35 31	206	36 53	277	33 54	315	32 53	346
272	51 48	045	36 53	095	39 24	166	35 06	207	35 58	278	33 15	315	32 39	346
273	52 27	044	37 48	095	39 37	167	34 40	208	35 03	278	32 36	315	32 25	346
274	53 06	044	38 43	096	39 48	169	34 13	209	34 09	278	31 57	315	32 12	345
275	53 44	043	39 38	096	39 59	170	33 46	210	33 14	279	31 18	315	31 58	345
276	54 22	043	40 32	097	40 08	171	33 18	211	32 20	279	30 39	315	31 44	345
277	54 59	042	41 27	097	40 16	172	32 49	212	31 25	279	30 00	315	31 29	345
278	55 36	042	42 22	098	40 23	173	32 20	213	30 31	279	29 21	315	31 15	345
279	56 13	041	43 17	098	40 29	174	31 49	214	29 36	280	28 42	315	31 00	345
280	56 49	040	44 11	099	40 34	176	31 19	215	28 42	280	28 03	315	30 46	345
281	57 24	040	45 06	099	40 38	177	30 47	215	27 47	280	27 24	315	30 31	344
282	57 59	039	46 00	100	40 40	178	30 15	216	26 53	281	26 45	315	30 16	344
283	58 34	038	46 55	100	40 41	179	29 42	217	25 59	281	26 06	315	30 01	344
284	59 08	037	47 49	101	40 42	180	29 08	218	25 05	281	25 27	315	29 46	344
	DENEB		♦Alpheratz		FOMALHAUT		♦Nunki		ANTARES		♦Alphecca		Kochab	
285	59 41	037	21 44	066	12 24	130	40 41	181	28 34	219	43 32	286	29 30	344
286	60 14	036	22 35	067	13 06	131	40 39	183	27 59	219	42 39	286	29 15	344
287	60 45	035	23 26	067	13 47	131	40 36	184	27 24	220	41 46	286	29 00	344
288	61 16	034	24 17	067	14 29	132	40 31	185	26 48	221	40 53	286	28 44	344
289	61 47	033	25 08	067	15 10	132	40 26	186	26 12	222	40 00	286	28 28	343
290	62 16	032	25 59	068	15 51	133	40 20	187	25 35	222	39 07	287	28 13	343
291	62 45	031	26 50	068	16 31	133	40 12	189	24 57	223	38 14	287	27 57	343
292	63 12	029	27 41	068	17 11	134	40 03	190	24 19	224	37 21	287	27 41	343
293	63 39	028	28 32	068	17 51	135	39 53	191	23 41	224	36 29	287	27 25	343
294	64 05	027	29 24	068	18 30	135	39 42	192	23 02	225	35 36	287	27 09	343
295	64 29	026	30 15	069	19 09	136	39 30	193	22 23	226	34 43	287	26 53	343
296	64 52	024	31 06	069	19 47	136	39 17	194	21 43	226	33 50	288	26 37	343
297	65 14	023	31 58	069	20 25	137	39 03	195	21 03	227	32 58	288	26 21	343
298	65 35	021	32 50	069	21 02	138	38 48	196	20 22	228	32 05	288	26 04	343
299	65 54	020	33 41	069	21 39	138	38 32	198	19 41	228	31 13	288	25 48	343
	♦DENEB		Schedar		Alpheratz		♦FOMALHAUT		Nunki		♦Rasalhague		VEGA	
300	66 12	018	29 57	037	34 33	070	22 16	139	38 15	199	53 59	260	66 17	316
301	66 29	017	30 30	037	35 25	070	22 52	140	37 57	200	53 05	260	65 38	315
302	66 44	015	31 03	037	36 17	070	23 27	140	37 38	201	52 10	261	64 59	314
303	66 57	013	31 36	037	37 09	070	24 02	141	37 17	202	51 16	261	64 19	313
304	67 09	012	32 08	037	38 01	070	24 37	142	36 56	203	50 21	262	63 39	313
305	67 19	010	32 41	036	38 53	070	25 11	142	36 35	204	49 26	262	62 58	312
306	67 28	008	33 14	036	39 45	071	25 44	143	36 12	205	48 32	263	62 17	311
307	67 34	006	33 47	036	40 37	071	26 17	144	35 48	206	47 37	263	61 35	310
308	67 39	004	34 20	036	41 29	071	26 49	145	35 24	207	46 42	264	60 52	309
309	67 43	002	34 52	036	42 21	071	27 20	146	34 58	208	45 47	264	60 09	309
310	67 44	001	35 25	036	43 13	071	27 51	146	34 32	209	44 52	265	59 26	308
311	67 44	359	35 57	036	44 05	071	28 22	147	34 05	210	43 57	265	58 43	308
312	67 42	357	36 29	036	44 58	071	28 51	148	33 37	211	43 02	266	57 59	307
313	67 38	355	37 01	035	45 50	071	29 20	149	33 09	212	42 07	266	57 15	307
314	67 32	353	37 33	035	46 42	071	29 49	150	32 39	212	41 12	267	56 30	306
	♦Alpheratz		Diphda		♦FOMALHAUT		ALTAIR		Rasalhague		♦VEGA		DENEB	
315	47 35	071	21 46	122	30 16	150	68 10	232	40 16	267	55 46	306	67 25	351
316	48 27	072	22 32	123	30 43	151	67 25	234	39 21	268	55 01	305	67 16	350
317	49 20	072	23 19	123	31 09	152	66 40	236	38 26	268	54 16	305	67 05	348
318	50 12	072	24 05	124	31 35	153	65 54	237	37 31	268	53 30	305	66 52	346
319	51 04	072	24 50	125	31 59	154	65 07	239	36 36	269	52 45	304	66 38	344
320	51 57	072	25 36	125	32 23	155	64 19	240	35 40	269	51 59	304	66 23	343
321	52 49	072	26 21	126	32 46	156	63 31	242	34 45	270	51 13	304	66 06	341
322	53 42	072	27 05	126	33 08	157	62 42	243	33 50	270	50 28	304	65 47	340
323	54 34	072	27 50	127	33 30	158	61 53	244	32 55	270	49 42	303	65 27	338
324	55 27	072	28 33	128	33 50	159	61 03	245	32 00	271	48 55	303	65 06	337
325	56 19	072	29 17	129	34 10	160	60 13	246	31 04	271	48 09	303	64 44	335
326	57 11	072	30 00	129	34 29	161	59 22	247	30 09	272	47 23	303	64 20	334
327	58 04	072	30 42	130	34 46	162	58 31	248	29 14	272	46 37	303	63 55	333
328	58 56	071	31 24	131	35 03	163	57 40	249	28 19	272	45 50	303	63 29	331
329	59 49	071	32 06	132	35 19	164	56 48	250	27 24	273	45 04	303	63 02	330
	♦Mirfak		Hamal		Diphda		♦FOMALHAUT		ALTAIR		♦VEGA		DENEB	
330	23 03	044	33 45	076	32 47	132	35 34	165	55 56	251	44 17	302	62 34	329
331	23 41	044	34 38	077	33 28	133	35 49	166	55 04	252	43 30	302	62 05	328
332	24 20	044	35 32	077	34 08	134	36 02	167	54 11	252	42 44	302	61 36	327
333	24 58	044	36 26	077	34 47	135	36 14	168	53 18	253	41 57	302	61 05	326
334	25 37	044	37 20	077	35 26	136	36 25	169	52 25	254	41 10	302	60 34	325
335	26 15	044	38 14	078	36 05	136	36 35	170	51 32	255	40 23	302	60 02	324
336	26 54	044	39 08	078	36 42	137	36 45	171	50 39	255	39 37	302	59 29	323
337	27 32	044	40 02	078	37 19	138	36 53	172	49 45	256	38 50	302	58 55	322
338	28 11	044	40 56	078	37 56	139	37 00	173	48 52	257	38 03	302	58 21	321
339	28 50	044	41 50	078	38 32	140	37 06	174	47 58	257	37 16	302	57 46	321
340	29 28	044	42 44	079	39 07	141	37 11	175	47 04	258	36 30	302	57 11	320
341	30 07	044	43 38	079	39 41	142	37 15	176	46 10	258	35 43	302	56 35	319
342	30 46	044	44 32	079	40 15	143	37 18	177	45 16	259	34 56	302	55 59	319
343	31 24	044	45 27	079	40 47	144	37 20	179	44 22	260	34 09	302	55 22	318
344	32 03	044	46 21	080	41 19	145	37 21	180	43 27	260	33 23	302	54 45	317
	Schedar		♦CAPELLA		ALDEBARAN		Diphda		♦FOMALHAUT		ALTAIR		♦DENEB	
345	51 49	022	13 37	045	11 48	077	41 50	146	37 21	181	42 33	261	54 08	317
346	52 09	022	14 17	046	12 42	077	42 21	147	37 20	182	41 38	261	53 30	316
347	52 29	021	14 56	046	13 36	078	42 50	148	37 17	183	40 44	262	52 51	316
348	52 49	020	15 36	046	14 30	078	43 19	150	37 14	184	39 49	262	52 13	315
349	53 07	019	16 16	046	15 24	078	43 46	151	37 10	185	38 54	263	51 34	315
350	53 25	019	16 56	047	16 18	079	44 13	152	37 04	186	37 59	263	50 55	315
351	53 42	018	17 36	047	17 13	079	44 38	153	36 58	187	37 05	264	50 15	314
352	53 59	017	18 17	047	18 07	079	45 03	154	36 50	188	36 10	264	49 35	314
353	54 14	016	18 57	047	19 01	080	45 26	155	36 42	189	35 15	265	48 55	313
354	54 29	015	19 38	047	19 55	080	45 49	157	36 32	190	34 20	265	48 15	313
355	54 43	014	20 18	047	20 50	080	46 10	158	36 22	192	33 25	265	47 35	313
356	54 57	013	20 59	048	21 44	081	46 30	159	36 10	193	32 30	266	46 54	313
357	55 09	013	21 40	048	22 39	081	46 49	161	35 58	194	31 34	266	46 13	312
358	55 21	012	22 21	048	23 33	081	47 07	162	35 44	195	30 39	267	45 32	312
359	55 31	011	23 02	048	24 28	082	47 23	163	35 30	196	29 44	267	44 51	312

LHA ♈	Hc	Zn	Hc	Zn	Hc	Zn	Hc	Zn	Hc	Zn	Hc	Zn	Hc	Zn
	Schedar		◆CAPELLA		ALDEBARAN		◆Diphda		FOMALHAUT		◆ALTAIR		DENEB	
0	54 42	010	23 03	048	25 14	081	48 36	164	36 12	197	28 51	268	43 30	312
1	54 51	009	23 44	048	26 09	082	48 51	166	35 55	198	27 56	268	42 49	312
2	54 59	008	24 25	048	27 04	082	49 04	167	35 38	199	27 00	269	42 07	312
3	55 06	007	25 07	048	27 59	082	49 15	169	35 19	200	26 04	269	41 26	312
4	55 12	006	25 48	048	28 54	083	49 26	170	35 00	201	25 09	270	40 44	312
5	55 17	005	26 30	048	29 50	083	49 35	171	34 39	202	24 13	270	40 03	311
6	55 22	004	27 11	048	30 45	083	49 42	173	34 18	203	23 18	270	39 21	311
7	55 25	003	27 53	048	31 40	084	49 49	174	33 56	204	22 22	271	38 39	311
8	55 27	002	28 34	048	32 35	084	49 53	176	33 33	205	21 26	271	37 57	311
9	55 29	001	29 16	048	33 31	084	49 57	177	33 10	206	20 31	271	37 15	311
10	55 29	000	29 58	049	34 26	085	49 59	179	32 45	207	19 35	272	36 33	311
11	55 29	359	30 39	049	35 22	085	49 59	180	32 20	207	18 40	272	35 51	311
12	55 28	358	31 21	049	36 17	085	49 58	182	31 54	208	17 44	273	35 09	311
13	55 25	357	32 03	049	37 12	086	49 56	183	31 27	209	16 48	273	34 27	311
14	55 22	356	32 45	049	38 08	086	49 52	185	31 00	210	15 53	273	33 44	311
	◆CAPELLA		ALDEBARAN		RIGEL		◆Diphda		FOMALHAUT		Enif		◆DENEB	
15	33 26	048	39 03	086	20 47	109	49 47	186	30 31	211	41 33	264	33 02	311
16	34 08	048	39 59	087	21 40	109	49 40	188	30 02	212	40 38	264	32 20	311
17	34 49	048	40 54	087	22 33	110	49 32	189	29 33	213	39 42	265	31 38	311
18	35 31	048	41 50	087	23 25	110	49 23	190	29 02	213	38 47	265	30 55	311
19	36 13	048	42 46	088	24 17	111	49 12	192	28 31	214	37 51	265	30 13	311
20	36 54	048	43 41	088	25 09	111	49 00	193	28 00	215	36 56	266	29 31	311
21	37 35	048	44 37	088	26 01	112	48 46	195	27 28	216	36 00	266	28 49	311
22	38 17	048	45 32	089	26 52	112	48 31	196	26 55	217	35 05	267	28 07	311
23	38 58	048	46 28	089	27 44	113	48 15	198	26 21	217	34 09	267	27 25	311
24	39 39	048	47 24	089	28 35	113	47 58	199	25 47	218	33 14	267	26 42	311
25	40 20	048	48 19	090	29 26	114	47 39	200	25 13	219	32 18	268	26 00	311
26	41 01	047	49 15	090	30 17	114	47 19	202	24 38	219	31 23	268	25 18	311
27	41 42	047	50 10	090	31 07	115	46 58	203	24 02	220	30 27	269	24 36	311
28	42 23	047	51 06	091	31 58	116	46 36	204	23 26	221	29 31	269	23 55	311
29	43 04	047	52 02	091	32 48	116	46 13	205	22 49	222	28 36	269	23 13	311
	CAPELLA		◆ALDEBARAN		RIGEL		Acamar		◆Diphda		Alpheratz		◆DENEB	
30	43 44	047	52 57	092	33 37	117	26 13	168	45 49	207	63 52	291	22 31	311
31	44 25	046	53 53	092	34 27	118	26 25	169	45 23	208	63 00	291	21 49	312
32	45 05	046	54 49	092	35 16	118	26 36	169	44 57	209	62 08	291	21 08	312
33	45 45	046	55 44	093	36 05	119	26 45	170	44 29	210	61 16	291	20 26	312
34	46 25	046	56 40	093	36 53	120	26 55	171	44 01	211	60 24	290	19 45	312
35	47 04	045	57 35	094	37 41	120	27 03	172	43 31	212	59 32	290	19 04	312
36	47 44	045	58 31	094	38 29	121	27 10	173	43 01	214	58 40	290	18 22	312
37	48 23	045	59 26	095	39 16	122	27 17	174	42 30	215	57 47	290	17 41	312
38	49 02	044	60 22	095	40 03	123	27 23	174	41 58	216	56 55	290	17 00	313
39	49 40	044	61 17	096	40 50	124	27 28	175	41 25	217	56 03	290	16 19	313
40	50 18	043	62 12	096	41 36	124	27 32	176	40 51	218	55 10	290	15 39	313
41	50 56	043	63 08	097	42 22	125	27 35	177	40 17	219	54 18	290	14 58	313
42	51 34	042	64 03	097	43 07	126	27 38	178	39 41	220	53 26	290	14 18	313
43	52 11	042	64 58	098	43 52	127	27 40	179	39 05	221	52 33	290	13 37	314
44	52 48	041	65 53	099	44 36	128	27 40	180	38 29	222	51 41	289	12 57	314
	CAPELLA		◆POLLUX		PROCYON		SIRIUS		◆Acamar		Diphda		◆Alpheratz	
45	53 25	041	26 03	068	20 43	093	22 42	120	27 40	180	37 52	223	50 48	289
46	54 01	040	26 55	069	21 38	093	23 30	121	27 40	181	37 14	223	49 56	289
47	54 37	040	27 46	069	22 34	094	24 17	122	27 38	182	36 35	224	49 03	289
48	55 12	039	28 38	069	23 29	094	25 05	122	27 35	183	35 56	225	48 11	290
49	55 47	038	29 30	069	24 25	094	25 52	123	27 32	184	35 16	226	47 19	290
50	56 21	038	30 22	069	25 20	095	26 38	123	27 28	185	34 36	227	46 26	290
51	56 54	037	31 15	070	26 16	095	27 25	124	27 23	186	33 55	228	45 34	290
52	57 27	036	32 07	070	27 11	096	28 11	125	27 17	186	33 14	228	44 41	290
53	58 00	035	32 59	070	28 06	096	28 56	125	27 11	187	32 32	229	43 49	290
54	58 31	034	33 51	070	29 02	096	29 41	126	27 03	188	31 49	230	42 57	290
55	59 02	033	34 44	070	29 57	097	30 26	127	26 55	189	31 06	231	42 04	290
56	59 33	033	35 36	070	30 52	097	31 11	127	26 46	190	30 23	231	41 12	290
57	60 02	032	36 29	071	31 47	098	31 55	128	26 36	191	29 39	232	40 20	290
58	60 31	031	37 21	071	32 42	098	32 38	129	26 25	191	28 55	233	39 28	290
59	60 59	029	38 14	071	33 37	099	33 21	130	26 14	192	28 11	233	38 35	290
	◆CAPELLA		PROCYON		◆SIRIUS		RIGEL		Diphda		◆Hamal		Mirfak	
60	61 26	028	34 32	099	34 04	130	54 44	147	27 26	234	63 55	279	61 16	348
61	61 52	027	35 27	100	34 46	131	55 13	148	26 41	235	63 00	279	61 04	347
62	62 16	026	36 22	100	35 28	132	55 42	150	25 55	235	62 05	279	60 50	345
63	62 40	025	37 17	101	36 09	133	56 09	152	25 09	236	61 10	279	60 35	344
64	63 03	024	38 11	101	36 49	134	56 35	153	24 23	237	60 15	279	60 20	343
65	63 25	022	39 06	102	37 29	134	56 59	155	23 36	237	59 20	279	60 03	342
66	63 45	021	40 00	102	38 09	135	57 22	156	22 49	238	58 25	279	59 45	341
67	64 04	019	40 55	103	38 48	136	57 44	158	22 02	238	57 31	279	59 26	340
68	64 22	018	41 49	103	39 26	137	58 04	160	21 14	239	56 36	280	59 06	338
69	64 39	016	42 43	104	40 03	138	58 22	162	20 26	240	55 41	280	58 45	337
70	64 54	015	43 37	104	40 40	139	58 39	164	19 38	240	54 46	280	58 23	336
71	65 07	013	44 31	105	41 16	140	58 54	165	18 50	241	53 51	280	58 01	335
72	65 19	012	45 24	106	41 51	141	59 07	167	18 01	241	52 56	280	57 37	334
73	65 30	010	46 18	106	42 26	142	59 18	169	17 13	242	52 02	280	57 13	333
74	65 39	009	47 11	107	43 00	143	59 28	171	16 23	242	51 07	281	56 47	333
	CAPELLA		◆Dubhe		POLLUX		PROCYON		◆SIRIUS		RIGEL		◆Hamal	
75	65 47	007	18 53	030	52 19	072	48 04	108	43 32	144	59 35	173	50 12	281
76	65 53	005	19 20	030	53 12	072	48 57	108	44 05	145	59 41	175	49 18	281
77	65 57	004	19 48	030	54 05	072	49 50	109	44 36	146	59 45	177	48 23	281
78	65 59	002	20 16	030	54 58	072	50 42	110	45 06	148	59 47	179	47 28	281
79	66 00	000	20 44	030	55 50	072	51 35	111	45 35	149	59 47	181	46 34	281
80	66 00	358	21 13	030	56 43	072	52 27	111	46 04	150	59 46	183	45 39	282
81	65 57	357	21 41	030	57 36	072	53 18	112	46 31	151	59 42	185	44 45	282
82	65 53	355	22 09	031	58 29	072	54 10	113	46 57	152	59 36	187	43 50	282
83	65 48	353	22 37	031	59 22	072	55 01	114	47 22	154	59 29	189	42 56	282
84	65 40	352	23 05	031	60 15	072	55 51	115	47 46	155	59 20	191	42 02	282
85	65 32	350	23 34	031	61 08	072	56 42	116	48 09	156	59 09	192	41 07	283
86	65 21	348	24 02	031	62 00	071	57 31	117	48 31	158	58 56	194	40 13	283
87	65 09	347	24 30	031	62 53	071	58 21	118	48 51	159	58 41	196	39 19	283
88	64 56	345	24 59	031	63 46	071	59 10	119	49 11	160	58 25	198	38 25	283
89	64 41	344	25 27	031	64 38	071	59 58	120	49 29	162	58 07	200	37 30	283

LAT 22°N

LHA ♈	Hc	Zn	Hc	Zn	Hc	Zn	Hc	Zn	Hc	Zn	Hc	Zn	Hc	Zn
	Dubhe		◆REGULUS		Alphard		SIRIUS		◆RIGEL		ALDEBARAN		◆CAPELLA	
90	25 56	031	30 13	089	30 41	115	49 45	163	57 47	202	69 22	258	64 25	342
91	26 24	031	31 09	089	31 31	116	50 01	165	57 26	203	68 27	259	64 07	341
92	26 52	031	32 04	089	32 21	117	50 15	166	57 03	205	67 33	260	63 48	339
93	27 21	031	33 00	090	33 10	117	50 28	168	56 39	207	66 38	261	63 28	338
94	27 49	031	33 55	090	33 59	118	50 39	169	56 13	208	65 43	261	63 07	337
95	28 17	030	34 51	090	34 49	119	50 49	171	55 46	210	64 48	262	62 44	335
96	28 45	030	35 47	091	35 37	119	50 57	172	55 18	211	63 53	263	62 20	334
97	29 13	030	36 42	091	36 26	120	51 04	174	54 48	213	62 57	263	61 56	333
98	29 42	030	37 38	092	37 14	121	51 10	175	54 17	214	62 02	264	61 30	332
99	30 10	030	38 34	092	38 01	121	51 14	177	53 45	216	61 07	264	61 03	331
100	30 38	030	39 29	092	38 49	122	51 16	178	53 12	217	60 11	265	60 35	330
101	31 05	030	40 25	093	39 35	123	51 17	180	52 38	218	59 16	265	60 07	329
102	31 33	030	41 20	093	40 22	124	51 17	181	52 03	220	58 21	266	59 38	328
103	32 01	030	42 16	094	41 08	125	51 15	183	51 27	221	57 25	266	59 07	327
104	32 28	030	43 11	094	41 53	125	51 12	184	50 50	222	56 29	267	58 36	326
	Dubhe		◆REGULUS		Alphard		SIRIUS		◆RIGEL		ALDEBARAN		◆CAPELLA	
105	32 56	029	44 07	095	42 39	126	51 07	186	50 12	223	55 34	267	58 05	325
106	33 23	029	45 02	095	43 23	127	51 01	187	49 34	225	54 38	268	57 33	324
107	33 50	029	45 58	095	44 07	128	50 53	189	48 54	226	53 43	268	57 00	323
108	34 17	029	46 53	096	44 51	129	50 44	190	48 14	227	52 47	269	56 26	323
109	34 44	029	47 48	096	45 33	130	50 33	192	47 33	228	51 52	269	55 52	322
110	35 11	029	48 43	097	46 16	131	50 21	193	46 51	229	50 56	269	55 17	321
111	35 37	028	49 39	097	46 57	132	50 08	195	46 09	230	50 00	270	54 42	321
112	36 04	028	50 34	098	47 38	133	49 53	196	45 26	231	49 05	270	54 07	320
113	36 30	028	51 29	098	48 19	134	49 37	198	44 42	232	48 09	270	53 31	319
114	36 56	028	52 24	099	48 58	135	49 19	199	43 58	233	47 13	271	52 54	319
115	37 22	027	53 19	100	49 37	136	49 01	200	43 14	234	46 18	271	52 17	318
116	37 47	027	54 14	100	50 15	138	48 41	202	42 29	235	45 22	272	51 40	318
117	38 13	027	55 08	101	50 52	139	48 19	203	41 43	235	44 27	272	51 02	317
118	38 38	027	56 03	101	51 28	140	47 57	204	40 57	236	43 31	272	50 24	317
119	39 03	026	56 57	102	52 03	141	47 34	206	40 10	237	42 35	273	49 46	316
	◆Dubhe		Denebola		◆Alphard		SIRIUS		RIGEL		◆ALDEBARAN		CAPELLA	
120	39 27	026	35 29	087	52 37	143	47 09	207	39 24	238	41 40	273	49 08	316
121	39 51	026	36 25	088	53 11	144	46 43	208	38 36	239	40 44	273	48 29	316
122	40 16	025	37 20	088	53 43	145	46 16	209	37 48	239	39 49	274	47 50	315
123	40 39	025	38 16	088	54 14	147	45 48	211	37 00	240	38 53	274	47 10	315
124	41 03	025	39 11	089	54 43	148	45 20	212	36 12	241	37 58	274	46 31	315
125	41 26	024	40 07	089	55 12	150	44 50	213	35 23	242	37 02	275	45 51	314
126	41 49	024	41 03	090	55 39	151	44 19	214	34 34	242	36 07	275	45 11	314
127	42 11	024	41 58	090	56 05	153	43 47	215	33 45	243	35 11	275	44 31	314
128	42 33	023	42 54	090	56 30	155	43 15	216	32 55	244	34 16	275	43 51	313
129	42 55	023	43 49	091	56 53	156	42 42	217	32 05	244	33 21	276	43 10	313
130	43 16	022	44 45	091	57 14	158	42 07	218	31 15	245	32 25	276	42 30	313
131	43 37	022	45 41	091	57 34	160	41 32	219	30 24	245	31 30	276	41 49	313
132	43 58	021	46 36	092	57 53	161	40 57	220	29 34	246	30 35	277	41 08	313
133	44 18	021	47 32	092	58 10	163	40 20	221	28 43	247	29 39	277	40 27	312
134	44 38	021	48 28	093	58 25	165	39 43	222	27 52	247	28 44	277	39 46	312
	Dubhe		◆ARCTURUS		SPICA		◆Alphard		SIRIUS		BETELGEUSE		◆CAPELLA	
135	44 57	020	17 00	076	17 07	110	58 38	167	39 05	223	43 09	259	39 05	312
136	45 16	020	17 54	076	17 59	110	58 50	169	38 27	224	42 14	260	38 23	312
137	45 34	019	18 48	076	18 51	111	59 00	171	37 48	225	41 19	260	37 42	312
138	45 52	019	19 42	077	19 43	111	59 08	173	37 08	226	40 24	261	37 01	312
139	46 09	018	20 36	077	20 35	112	59 14	175	36 28	227	39 29	261	36 19	312
140	46 26	017	21 31	077	21 26	112	59 19	176	35 47	228	38 34	262	35 38	312
141	46 43	017	22 25	077	22 17	113	59 21	178	35 05	228	37 39	262	34 56	312
142	46 59	016	23 19	078	23 09	114	59 22	180	34 24	229	36 44	263	34 14	312
143	47 14	016	24 14	078	24 00	114	59 21	182	33 41	230	35 49	263	33 33	312
144	47 29	015	25 08	078	24 50	115	59 17	184	32 58	231	34 54	264	32 51	311
145	47 43	014	26 03	079	25 41	115	59 12	186	32 15	232	33 58	264	32 09	311
146	47 57	014	26 57	079	26 31	116	59 06	188	31 31	232	33 03	264	31 28	311
147	48 10	013	27 52	079	27 21	116	58 57	190	30 47	233	32 08	265	30 46	311
148	48 22	013	28 46	079	28 11	117	58 46	192	30 02	234	31 12	265	30 04	311
149	48 34	012	29 41	080	29 00	118	58 34	194	29 17	234	30 17	266	29 23	312
	Dubhe		◆ARCTURUS		SPICA		◆Alphard		SIRIUS		POLLUX		◆CAPELLA	
150	48 45	011	30 36	080	29 49	118	58 20	196	28 32	235	58 56	288	28 41	312
151	48 56	011	31 31	080	30 38	119	58 04	197	27 46	236	58 03	288	28 00	312
152	49 06	010	32 26	081	31 27	119	57 47	199	27 00	236	57 10	288	27 18	312
153	49 15	009	33 20	081	32 15	120	57 28	201	26 13	237	56 17	288	26 36	312
154	49 23	009	34 15	081	33 03	121	57 07	203	25 27	238	55 24	288	25 55	312
155	49 31	008	35 10	081	33 51	121	56 45	204	24 39	238	54 31	288	25 13	312
156	49 39	007	36 05	082	34 38	122	56 22	206	23 52	239	53 38	288	24 32	312
157	49 45	006	37 00	082	35 25	123	55 57	208	23 04	239	52 45	288	23 51	312
158	49 51	006	37 56	082	36 11	124	55 30	209	22 16	240	51 52	288	23 09	312
159	49 57	005	38 51	083	36 58	124	55 03	211	21 28	241	51 00	288	22 28	312
160	50 01	004	39 46	083	37 43	125	54 34	212	20 39	241	50 07	288	21 47	312
161	50 05	004	40 41	083	38 28	126	54 03	214	19 51	242	49 14	288	21 06	313
162	50 08	003	41 36	083	39 13	127	53 32	215	19 01	242	48 21	288	20 25	313
163	50 10	002	42 32	084	39 58	128	53 00	216	18 12	243	47 28	288	19 44	313
164	50 12	001	43 27	084	40 41	128	52 26	218	17 23	243	46 35	288	19 03	313
	◆Dubhe		Alkaid		ARCTURUS		◆SPICA		Alphard		◆PROCYON		POLLUX	
165	50 13	001	47 15	040	44 22	084	41 25	129	51 52	219	38 40	259	45 42	288
166	50 13	000	47 50	039	45 18	084	42 07	130	51 16	220	37 45	259	44 49	288
167	50 13	359	48 25	039	46 13	085	42 50	131	50 40	222	36 50	260	43 56	288
168	50 12	358	49 00	039	47 08	085	43 31	132	50 02	223	35 56	260	43 04	288
169	50 10	358	49 35	038	48 04	085	44 12	133	49 24	224	35 01	261	42 11	289
170	50 07	357	50 09	038	48 59	086	44 52	134	48 45	225	34 06	261	41 18	289
171	50 04	356	50 43	037	49 55	086	45 32	135	48 05	226	33 11	262	40 25	289
172	50 00	355	51 16	037	50 50	086	46 11	136	47 25	227	32 16	262	39 33	289
173	49 55	355	51 49	036	51 46	087	46 49	137	46 44	228	31 21	262	38 40	289
174	49 50	354	52 21	035	52 41	087	47 26	138	46 02	229	30 26	263	37 48	289
175	49 44	353	52 53	035	53 37	087	48 03	140	45 19	230	29 30	263	36 55	289
176	49 37	353	53 25	034	54 32	087	48 38	141	44 36	231	28 35	264	36 03	289
177	49 29	352	53 55	033	55 28	088	49 13	142	43 53	232	27 40	264	35 10	290
178	49 21	351	54 26	033	56 24	088	49 47	143	43 08	233	26 44	265	34 18	290
179	49 12	351	54 56	032	57 19	088	50 20	144	42 24	234	25 49	265	33 25	290

LHA	Hc	Zn	Hc	Zn	Hc	Zn	Hc	Zn	Hc	Zn	Hc	Zn	Hc	Zn
♈	◆Alkaid		Alphecca		◆SPICA		Gienah		REGULUS		◆POLLUX		Dubhe	
180	55 25	031	41 17	073	50 52	146	50 18	174	61 30	254	32 33	290	49 03	350
181	55 53	030	42 10	073	51 23	147	50 23	176	60 36	255	31 41	290	48 53	349
182	56 21	030	43 03	073	51 52	148	50 27	177	59 43	256	30 49	290	48 42	349
183	56 48	029	43 56	073	52 21	150	50 29	179	58 49	256	29 57	291	48 30	348
184	57 14	028	44 50	074	52 48	151	50 29	180	57 54	257	29 05	291	48 18	347
185	57 40	027	45 43	074	53 14	153	50 28	182	57 00	258	28 13	291	48 06	347
186	58 05	026	46 37	074	53 39	154	50 26	183	56 06	258	27 21	291	47 53	346
187	58 29	025	47 30	074	54 03	156	50 22	185	55 11	259	26 29	291	47 39	345
188	58 52	024	48 23	074	54 25	157	50 17	186	54 16	260	25 37	292	47 24	345
189	59 14	023	49 17	074	54 46	159	50 10	188	53 22	260	24 46	292	47 10	344
190	59 35	022	50 10	074	55 05	160	50 02	189	52 27	261	23 54	292	46 54	344
191	59 56	021	51 04	074	55 23	162	49 53	191	51 32	261	23 02	292	46 38	343
192	60 15	020	51 58	074	55 39	164	49 42	192	50 37	262	22 11	293	46 21	342
193	60 33	018	52 51	074	55 54	165	49 29	193	49 42	263	21 20	293	46 04	342
194	60 50	017	53 45	074	56 07	167	49 16	195	48 46	263	20 28	293	45 47	341
	◆Kochab		VEGA		Rasalhague		◆ANTARES		SPICA		◆REGULUS		Dubhe	
195	35 45	009	17 55	055	24 14	086	19 58	131	56 18	169	47 51	264	45 29	341
196	35 54	009	18 41	055	25 10	086	20 39	132	56 28	171	46 56	264	45 10	340
197	36 02	008	19 26	055	26 05	086	21 21	132	56 36	172	46 00	264	44 51	340
198	36 10	008	20 12	055	27 01	087	22 02	133	56 43	174	45 05	265	44 32	339
199	36 18	008	20 58	055	27 57	087	22 42	134	56 48	176	44 10	265	44 12	339
200	36 25	008	21 44	056	28 52	087	23 22	134	56 51	178	43 14	266	43 52	338
201	36 32	007	22 29	056	29 48	088	24 02	135	56 52	180	42 19	266	43 31	338
202	36 39	007	23 15	056	30 43	088	24 41	136	56 51	181	41 23	267	43 10	338
203	36 46	007	24 01	056	31 39	088	25 20	136	56 49	183	40 28	267	42 49	337
204	36 52	006	24 48	056	32 35	089	25 58	137	56 45	185	39 32	268	42 27	337
205	36 58	006	25 34	056	33 30	089	26 36	138	56 40	187	38 36	268	42 05	336
206	37 03	006	26 20	056	34 26	090	27 13	138	56 32	189	37 41	268	41 42	336
207	37 09	005	27 06	056	35 21	090	27 50	139	56 23	190	36 45	269	41 19	336
208	37 14	005	27 53	057	36 17	090	28 26	140	56 12	192	35 50	269	40 56	335
209	37 18	005	28 39	057	37 13	091	29 01	141	56 00	194	34 54	270	40 32	335
	◆Kochab		VEGA		Rasalhague		◆ANTARES		SPICA		◆REGULUS		Dubhe	
210	37 23	004	29 26	057	38 08	091	29 36	141	55 46	195	33 58	270	40 09	334
211	37 27	004	30 12	057	39 04	091	30 11	142	55 30	197	33 03	270	39 44	334
212	37 30	004	30 59	057	40 00	092	30 45	143	55 13	199	32 07	271	39 20	334
213	37 34	003	31 46	057	40 55	092	31 18	144	54 55	200	31 11	271	38 55	334
214	37 37	003	32 32	057	41 51	093	31 50	145	54 35	202	30 16	271	38 31	333
215	37 40	003	33 19	057	42 46	093	32 22	146	54 13	204	29 20	272	38 05	333
216	37 42	002	34 06	057	43 42	094	32 53	146	53 50	205	28 25	272	37 40	333
217	37 44	002	34 53	057	44 37	094	33 24	147	53 26	207	27 29	272	37 14	332
218	37 46	002	35 39	057	45 33	094	33 53	148	53 00	208	26 33	273	36 49	332
219	37 47	001	36 26	057	46 28	095	34 22	149	52 34	210	25 38	273	36 22	332
220	37 48	001	37 13	057	47 24	095	34 51	150	52 06	211	24 42	273	35 56	332
221	37 49	001	38 00	057	48 19	096	35 18	151	51 36	212	23 47	274	35 30	332
222	37 49	000	38 46	057	49 14	096	35 45	152	51 06	214	22 51	274	35 03	331
223	37 49	000	39 33	057	50 10	097	36 10	153	50 35	215	21 56	275	34 36	331
224	37 49	000	40 20	057	51 05	097	36 35	154	50 02	216	21 00	275	34 10	331
	VEGA		◆ALTAIR		Nunki		ANTARES		◆SPICA		Denebola		◆Alkaid	
225	41 07	057	19 20	088	15 23	127	36 59	155	49 29	217	44 12	269	59 07	337
226	41 53	057	20 16	088	16 07	128	37 23	156	48 55	219	43 16	270	58 44	336
227	42 40	057	21 11	089	16 51	128	37 45	157	48 19	220	42 21	270	58 21	335
228	43 26	057	22 07	089	17 34	129	38 06	158	47 43	221	41 25	270	57 57	334
229	44 13	057	23 03	090	18 17	130	38 27	159	47 06	222	40 29	271	57 32	333
230	44 59	057	23 58	090	19 00	130	38 46	160	46 29	223	39 34	271	57 06	332
231	45 46	056	24 54	090	19 43	131	39 05	161	45 50	224	38 38	271	56 39	331
232	46 32	056	25 49	091	20 25	131	39 22	162	45 11	225	37 43	272	56 12	330
233	47 18	056	26 45	091	21 06	132	39 39	163	44 31	226	36 47	272	55 44	329
234	48 04	056	27 41	091	21 47	133	39 54	164	43 50	227	35 51	272	55 15	329
235	48 50	056	28 36	092	22 28	133	40 08	166	43 09	228	34 56	273	54 46	328
236	49 36	055	29 32	092	23 08	134	40 22	167	42 27	229	34 00	273	54 16	327
237	50 22	055	30 27	093	23 48	134	40 34	168	41 45	230	33 05	274	53 45	326
238	51 08	055	31 23	093	24 28	135	40 45	169	41 02	231	32 09	274	53 14	326
239	51 53	055	32 19	093	25 07	136	40 55	170	40 18	232	31 14	274	52 43	325
	◆VEGA		ALTAIR		Nunki		◆ANTARES		SPICA		ARCTURUS		◆Alkaid	
240	52 38	054	33 14	094	25 45	137	41 04	171	39 34	233	65 24	268	52 11	324
241	53 23	054	34 10	094	26 23	137	41 12	173	38 49	234	64 28	269	51 38	324
242	54 08	054	35 05	095	27 01	138	41 19	174	38 04	234	63 33	269	51 05	323
243	54 53	053	36 01	095	27 38	139	41 24	175	37 19	235	62 37	270	50 32	323
244	55 37	053	36 56	096	28 14	139	41 28	176	36 33	236	61 42	270	49 58	322
245	56 21	052	37 51	096	28 50	140	41 32	177	35 47	237	60 46	270	49 23	322
246	57 05	052	38 47	096	29 26	141	41 34	178	35 00	238	59 50	271	48 49	321
247	57 49	051	39 42	097	30 00	142	41 35	180	34 13	238	58 55	271	48 14	321
248	58 32	051	40 37	097	30 34	143	41 34	181	33 25	239	57 59	271	47 39	320
249	59 15	050	41 32	098	31 08	143	41 33	182	32 37	240	57 03	272	47 03	320
250	59 57	049	42 27	098	31 41	144	41 30	183	31 49	240	56 08	272	46 27	320
251	60 39	049	43 22	099	32 13	145	41 27	184	31 01	241	55 12	272	45 51	319
252	61 21	048	44 17	099	32 45	146	41 22	186	30 12	242	54 17	273	45 15	319
253	62 02	047	45 12	100	33 15	147	41 16	187	29 23	242	53 21	273	44 38	319
254	62 42	046	46 07	100	33 46	148	41 09	188	28 34	243	52 26	273	44 01	318
	◆DENEB		ALTAIR		◆Nunki		ANTARES		SPICA		◆ARCTURUS		Alkaid	
255	39 36	049	47 01	101	34 15	149	41 00	189	27 44	243	51 30	274	43 24	318
256	40 18	049	47 56	102	34 44	149	40 51	190	26 54	244	50 35	274	42 47	318
257	41 00	048	48 51	102	35 11	150	40 40	192	26 04	245	49 39	274	42 09	318
258	41 41	048	49 45	103	35 39	151	40 29	193	25 14	245	48 44	274	41 32	317
259	42 23	048	50 39	103	36 05	152	40 16	194	24 23	246	47 48	275	40 54	317
260	43 04	048	51 33	104	36 30	153	40 02	195	23 32	246	46 53	275	40 16	317
261	43 45	048	52 27	105	36 55	154	39 47	196	22 41	247	45 57	275	39 38	317
262	44 26	047	53 21	105	37 18	155	39 31	197	21 50	247	45 02	276	39 00	317
263	45 07	047	54 14	106	37 41	156	39 14	198	20 59	248	44 07	276	38 21	316
264	45 48	047	55 08	107	38 03	157	38 56	199	20 07	248	43 11	276	37 43	316
265	46 28	047	56 01	108	38 24	158	38 38	200	19 15	249	42 16	276	37 04	316
266	47 09	046	56 54	108	38 44	160	38 18	201	18 23	249	41 21	277	36 26	316
267	47 49	046	57 46	109	39 03	161	37 57	203	17 31	250	40 25	277	35 47	316
268	48 29	046	58 39	110	39 21	162	37 35	204	16 39	250	39 30	277	35 08	316
269	49 08	045	59 31	111	39 38	163	37 12	205	15 46	251	38 35	278	34 29	316

LAT 22°N

LHA	Hc	Zn	Hc	Zn	Hc	Zn	Hc	Zn	Hc	Zn	Hc	Zn	Hc	Zn
♈	◆DENEB		Enif		◆Nunki		ANTARES		◆ARCTURUS		Alkaid		Kochab	
270	49 48	045	35 06	093	39 54	164	36 49	206	37 40	278	33 50	316	32 08	346
271	50 27	044	36 02	094	40 09	165	36 24	207	36 45	278	33 12	316	31 54	346
272	51 05	044	36 58	094	40 23	166	35 59	208	35 50	278	32 33	315	31 41	346
273	51 44	044	37 53	095	40 35	167	35 33	209	34 55	279	31 53	315	31 27	346
274	52 22	043	38 48	095	40 47	168	35 06	209	34 00	279	31 14	315	31 14	346
275	53 00	043	39 44	095	40 58	170	34 38	210	33 05	279	30 35	315	31 00	345
276	53 37	042	40 39	096	41 07	171	34 09	211	32 10	280	29 56	315	30 46	345
277	54 14	041	41 34	096	41 15	172	33 40	212	31 15	280	29 17	315	30 31	345
278	54 51	041	42 30	097	41 23	173	33 10	213	30 20	280	28 38	315	30 17	345
279	55 27	040	43 25	097	41 29	174	32 39	214	29 26	280	27 59	315	30 02	345
280	56 03	039	44 20	098	41 34	176	32 08	215	28 31	281	27 20	315	29 48	345
281	56 38	039	45 15	098	41 37	177	31 36	216	27 36	281	26 41	315	29 33	345
282	57 12	038	46 10	099	41 40	178	31 03	217	26 42	281	26 02	315	29 18	344
283	57 46	037	47 05	099	41 41	179	30 29	217	25 47	281	25 23	316	29 03	344
284	58 20	036	48 00	100	41 42	180	29 55	218	24 53	282	24 44	316	28 48	344
	DENEB		◆Alpheratz		FOMALHAUT		◆Nunki		ANTARES		◆Alphecca		Kochab	
285	58 52	036	21 20	066	13 02	130	41 41	182	29 21	219	43 16	287	28 33	344
286	59 24	035	22 11	066	13 45	130	41 39	183	28 45	220	42 22	287	28 17	344
287	59 56	034	23 02	067	14 27	131	41 36	184	28 10	220	41 29	287	28 02	344
288	60 26	033	23 53	067	15 09	131	41 31	185	27 33	221	40 36	287	27 46	344
289	60 56	032	24 44	067	15 50	132	41 26	186	26 56	222	39 43	287	27 31	344
290	61 25	031	25 36	067	16 31	133	41 19	187	26 19	223	38 50	287	27 15	344
291	61 53	030	26 27	067	17 12	133	41 11	189	25 41	223	37 57	287	26 59	343
292	62 20	028	27 18	068	17 53	134	41 02	190	25 02	224	37 04	288	26 43	343
293	62 46	027	28 10	068	18 33	134	40 52	191	24 24	225	36 11	288	26 28	343
294	63 11	026	29 01	068	19 12	135	40 41	192	23 44	225	35 18	288	26 12	343
295	63 35	025	29 53	068	19 51	135	40 29	193	23 04	226	34 25	288	25 55	343
296	63 57	023	30 45	068	20 30	136	40 15	194	22 24	227	33 32	288	25 39	343
297	64 19	022	31 36	068	21 09	137	40 01	196	21 43	227	32 39	288	25 23	343
298	64 39	021	32 28	069	21 46	137	39 46	197	21 02	228	31 46	289	25 07	343
299	64 58	019	33 20	069	22 24	138	39 29	198	20 21	229	30 54	289	24 51	343
	◆DENEB		Schedar		Alpheratz		◆FOMALHAUT		Nunki		◆Rasalhague		VEGA	
300	65 15	018	29 08	036	34 12	069	23 01	139	39 12	199	54 09	261	65 33	318
301	65 31	016	29 41	036	35 04	069	23 37	139	38 53	200	53 14	261	64 55	317
302	65 46	014	30 14	036	35 56	069	24 13	140	38 34	201	52 19	262	64 17	316
303	65 59	013	30 47	036	36 48	069	24 49	141	38 13	202	51 24	263	63 38	315
304	66 10	011	31 20	036	37 40	070	25 24	141	37 52	203	50 29	263	62 58	314
305	66 20	009	31 53	036	38 32	070	25 58	142	37 29	204	49 34	264	62 18	313
306	66 28	008	32 26	036	39 24	070	26 32	143	37 06	205	48 38	264	61 37	312
307	66 35	006	32 58	036	40 16	070	27 05	144	36 42	206	47 43	265	60 55	312
308	66 40	004	33 31	036	41 09	070	27 38	144	36 17	207	46 48	265	60 13	311
309	66 43	002	34 03	036	42 01	070	28 10	145	35 51	208	45 52	265	59 31	310
310	66 44	001	34 36	036	42 53	070	28 41	146	35 24	209	44 57	266	58 48	310
311	66 44	359	35 08	035	43 46	070	29 12	147	34 57	210	44 01	266	58 05	309
312	66 42	357	35 40	035	44 38	070	29 42	148	34 29	211	43 06	267	57 22	308
313	66 38	355	36 12	035	45 30	070	30 11	148	34 00	212	42 10	267	56 38	308
314	66 33	353	36 44	035	46 23	070	30 40	149	33 30	213	41 15	268	55 54	307
	◆Alpheratz		Diphda		◆FOMALHAUT		ALTAIR		Rasalhague		◆VEGA		DENEB	
315	47 15	070	22 17	122	31 08	150	68 45	234	40 19	268	55 10	307	66 25	352
316	48 08	070	23 05	122	31 36	151	68 00	236	39 23	269	54 25	307	66 17	350
317	49 00	071	23 52	123	32 02	152	67 13	238	38 28	269	53 41	306	66 06	348
318	49 52	071	24 38	124	32 28	153	66 25	239	37 32	269	52 56	306	65 54	347
319	50 45	071	25 24	124	32 53	154	65 37	241	36 37	270	52 10	306	65 40	345
320	51 37	071	26 10	125	33 17	155	64 48	242	35 41	270	51 25	305	65 25	343
321	52 30	070	26 56	125	33 41	156	63 59	243	34 45	270	50 40	305	65 09	342
322	53 22	070	27 41	126	34 03	156	63 09	244	33 50	271	49 54	305	64 51	340
323	54 15	070	28 26	127	34 25	157	62 19	246	32 54	271	49 08	304	64 32	339
324	55 07	070	29 10	127	34 46	158	61 28	247	31 58	271	48 22	304	64 11	338
325	55 59	070	29 54	128	35 06	159	60 36	248	31 03	272	47 36	304	63 49	336
326	56 52	070	30 38	129	35 25	160	59 45	249	30 07	272	46 50	304	63 26	335
327	57 44	070	31 21	130	35 43	161	58 53	250	29 12	272	46 04	304	63 02	334
328	58 36	070	32 03	130	36 01	162	58 00	251	28 16	273	45 17	304	62 36	332
329	59 29	070	32 46	131	36 17	163	57 08	251	27 21	273	44 31	303	62 10	331
	◆Mirfak		Hamal		Diphda		◆FOMALHAUT		ALTAIR		◆VEGA		DENEB	
330	22 20	044	33 30	076	33 27	132	36 32	164	56 15	252	43 44	303	61 43	330
331	22 58	044	34 24	076	34 08	133	36 47	165	55 22	253	42 58	303	61 14	329
332	23 37	044	35 18	076	34 49	133	37 00	167	54 29	254	42 11	303	60 45	328
333	24 15	044	36 12	076	35 29	134	37 13	168	53 35	254	41 25	303	60 15	327
334	24 53	044	37 06	077	36 09	135	37 24	169	52 41	255	40 38	303	59 44	326
335	25 32	044	38 00	077	36 48	136	37 34	170	51 48	256	39 51	303	59 13	325
336	26 11	044	38 55	077	37 26	137	37 44	171	50 54	256	39 04	303	58 40	324
337	26 49	044	39 49	077	38 04	138	37 52	172	49 59	257	38 18	303	58 07	323
338	27 28	044	40 43	077	38 41	139	37 59	173	49 05	258	37 31	303	57 34	322
339	28 07	044	41 37	078	39 17	140	38 06	174	48 11	258	36 44	303	57 00	322
340	28 45	044	42 32	078	39 53	141	38 11	175	47 16	259	35 57	303	56 25	321
341	29 24	044	43 26	078	40 28	142	38 15	176	46 22	259	35 11	303	55 50	320
342	30 03	044	44 21	078	41 02	143	38 18	177	45 27	260	34 24	303	55 14	320
343	30 41	044	45 15	078	41 36	144	38 20	179	44 32	261	33 37	303	54 37	319
344	31 20	044	46 10	079	42 08	145	38 21	180	43 37	261	32 50	303	54 01	318
	Schedar		◆CAPELLA		ALDEBARAN		Diphda		◆FOMALHAUT		ALTAIR		◆DENEB	
345	50 53	022	12 55	045	11 35	077	42 40	146	38 21	181	42 42	262	53 24	318
346	51 13	021	13 35	046	12 29	077	43 11	147	38 20	182	41 47	262	52 46	317
347	51 33	020	14 15	046	13 23	077	43 41	148	38 17	183	40 52	263	52 08	317
348	51 52	020	14 55	046	14 18	078	44 10	149	38 14	184	39 57	263	51 30	316
349	52 10	019	15 35	046	15 12	078	44 38	150	38 09	185	39 01	263	50 51	316
350	52 28	018	16 15	046	16 06	078	45 05	151	38 04	186	38 06	264	50 12	315
351	52 45	017	16 55	047	17 01	079	45 32	153	37 57	187	37 11	264	49 33	315
352	53 01	017	17 36	047	17 55	079	45 57	154	37 50	188	36 15	265	48 54	315
353	53 17	016	18 16	047	18 50	079	46 21	155	37 41	190	35 20	265	48 14	314
354	53 31	015	18 57	047	19 45	080	46 44	156	37 31	191	34 25	266	47 34	314
355	53 45	014	19 38	047	20 40	080	47 05	158	37 21	192	33 29	266	46 54	314
356	53 58	013	20 18	047	21 34	080	47 26	159	37 09	193	32 34	267	46 13	313
357	54 10	012	20 59	047	22 29	081	47 45	160	36 56	194	31 38	267	45 33	313
358	54 22	011	21 40	048	23 24	081	48 04	162	36 42	195	30 42	267	44 52	313
359	54 32	011	22 22	048	24 19	081	48 21	163	36 28	196	29 47	268	44 11	313

LHA ♈	Hc	Zn	Hc	Zn	Hc	Zn	Hc	Zn	Hc	Zn	Hc	Zn	Hc	Zn
	Schedar		◆CAPELLA		ALDEBARAN		◆Diphda		FOMALHAUT		◆ALTAIR		DENEB	
0	53 43	009	22 22	048	25 05	081	49 34	164	37 09	197	28 53	269	42 49	313
1	53 52	008	23 04	048	26 00	081	49 49	165	36 52	198	27 57	269	42 08	313
2	54 00	008	23 45	048	26 56	082	50 02	167	36 34	199	27 01	269	41 27	313
3	54 06	007	24 27	048	27 51	082	50 14	168	36 16	200	26 05	270	40 46	312
4	54 12	006	25 08	048	28 47	082	50 25	170	35 56	201	25 09	270	40 04	312
5	54 18	005	25 50	048	29 42	082	50 34	171	35 35	202	24 13	270	39 23	312
6	54 22	004	26 31	048	30 38	083	50 42	173	35 14	203	23 17	271	38 41	312
7	54 25	003	27 13	048	31 33	083	50 48	174	34 51	204	22 21	271	37 59	312
8	54 27	002	27 55	048	32 29	083	50 53	176	34 28	205	21 25	272	37 17	312
9	54 29	001	28 36	048	33 24	084	50 57	177	34 04	206	20 29	272	36 35	312
10	54 29	000	29 18	048	34 20	084	50 59	179	33 39	207	19 33	272	35 53	311
11	54 29	359	30 00	048	35 16	084	50 59	180	33 13	208	18 37	273	35 11	311
12	54 28	358	30 41	048	36 12	084	50 58	182	32 47	209	17 41	273	34 29	311
13	54 25	357	31 23	048	37 07	085	50 56	183	32 19	210	16 45	273	33 47	311
14	54 22	356	32 05	048	38 03	085	50 52	185	31 51	210	15 49	274	33 05	311
	◆CAPELLA		ALDEBARAN		RIGEL		◆ACHERNAR		FOMALHAUT		Enif		◆DENEB	
15	32 46	048	38 59	085	21 06	108	11 21	175	31 23	211	41 39	264	32 23	311
16	33 28	048	39 55	086	21 59	109	11 25	175	30 53	212	40 43	265	31 41	311
17	34 09	048	40 51	086	22 52	109	11 30	176	30 23	213	39 47	265	30 58	311
18	34 51	048	41 47	086	23 45	110	11 33	176	29 52	214	38 52	266	30 16	311
19	35 32	048	42 42	087	24 38	110	11 36	177	29 21	215	37 56	266	29 34	311
20	36 14	048	43 38	087	25 30	111	11 39	178	28 49	215	37 00	267	28 52	311
21	36 55	048	44 34	087	26 23	111	11 41	178	28 16	216	36 04	267	28 10	311
22	37 36	047	45 30	088	27 15	112	11 43	179	27 43	217	35 08	267	27 27	311
23	38 18	047	46 26	088	28 07	112	11 44	179	27 09	218	34 12	268	26 45	311
24	38 59	047	47 22	088	28 59	113	11 44	180	26 35	218	33 16	268	26 03	311
25	39 40	047	48 18	088	29 50	113	11 44	180	25 59	219	32 20	269	25 21	311
26	40 21	047	49 14	089	30 41	114	11 44	181	25 24	220	31 24	269	24 39	311
27	41 01	047	50 10	089	31 33	115	11 42	181	24 48	220	30 28	269	23 57	311
28	41 42	046	51 06	090	32 23	115	11 41	182	24 11	221	29 32	270	23 15	312
29	42 22	046	52 02	090	33 14	116	11 39	183	23 34	222	28 36	270	22 33	312
	◆CAPELLA		ALDEBARAN		RIGEL		◆ACHERNAR		FOMALHAUT		◆Alpheratz		DENEB	
30	43 03	046	52 58	090	34 04	116	11 36	183	22 57	222	63 29	293	21 51	312
31	43 43	046	53 54	091	34 54	117	11 32	184	22 19	223	62 38	293	21 09	312
32	44 23	045	54 50	091	35 44	118	11 29	184	21 40	224	61 46	293	20 28	312
33	45 03	045	55 46	091	36 33	118	11 24	185	21 01	224	60 54	292	19 46	312
34	45 42	045	56 42	092	37 23	119	11 19	185	20 22	225	60 02	292	19 05	312
35	46 22	044	57 38	092	38 11	120	11 14	186	19 42	226	59 10	292	18 23	312
36	47 01	044	58 34	093	39 00	121	11 08	186	19 02	226	58 18	292	17 42	313
37	47 40	044	59 30	093	39 48	121	11 02	187	18 21	227	57 26	291	17 01	313
38	48 18	043	60 26	093	40 36	122	10 55	187	17 40	227	56 34	291	16 20	313
39	48 56	043	61 22	094	41 23	123	10 47	188	16 59	228	55 42	291	15 39	313
40	49 34	042	62 18	094	42 10	124	10 39	189	16 17	228	54 50	291	14 58	313
41	50 12	042	63 14	095	42 56	124	10 30	189	15 35	229	53 57	291	14 17	313
42	50 49	042	64 09	095	43 42	125	10 21	190	14 53	229	53 05	291	13 36	314
43	51 26	041	65 05	096	44 28	126	10 12	190	14 10	230	52 13	291	12 56	314
44	52 03	040	66 01	097	45 12	127	10 02	191	13 27	230	51 20	291	12 16	314
	CAPELLA		◆POLLUX		PROCYON		SIRIUS		◆Acamar		Diphda		◆Alpheratz	
45	52 39	040	25 41	068	20 45	092	23 12	120	28 40	180	38 36	223	50 28	291
46	53 15	039	26 33	068	21 41	093	24 00	121	28 40	181	37 57	224	49 35	291
47	53 50	039	27 25	068	22 37	093	24 49	121	28 38	182	37 18	225	48 43	291
48	54 25	038	28 17	069	23 33	093	25 36	122	28 35	183	36 38	226	47 50	291
49	54 59	037	29 09	069	24 29	094	26 24	122	28 32	184	35 58	227	46 58	291
50	55 33	037	30 01	069	25 25	094	27 11	123	28 28	185	35 17	227	46 06	291
51	56 06	036	30 53	069	26 21	095	27 58	124	28 23	186	34 35	228	45 13	291
52	56 39	035	31 46	069	27 16	095	28 44	124	28 17	186	33 53	229	44 21	291
53	57 10	034	32 38	069	28 12	095	29 31	125	28 10	187	33 11	230	43 28	291
54	57 42	033	33 31	070	29 08	096	30 16	125	28 03	188	32 28	230	42 36	291
55	58 12	033	34 23	070	30 04	096	31 02	126	27 54	189	31 44	231	41 43	291
56	58 42	032	35 16	070	30 59	097	31 47	127	27 45	190	31 00	232	40 51	291
57	59 11	031	36 08	070	31 55	097	32 31	128	27 35	191	30 16	233	39 59	291
58	59 39	030	37 01	070	32 50	098	33 16	128	27 24	192	29 31	233	39 06	291
59	60 06	029	37 54	070	33 46	098	33 59	129	27 13	192	28 46	234	38 14	291
	◆CAPELLA		PROCYON		◆SIRIUS		RIGEL		Diphda		◆Hamal		Mirfak	
60	60 33	028	34 41	098	34 43	130	55 34	146	28 01	235	63 45	281	60 17	348
61	60 58	026	35 37	099	35 25	131	56 04	148	27 15	235	62 50	281	60 05	347
62	61 22	025	36 32	099	36 08	131	56 34	149	26 29	236	61 55	281	59 52	346
63	61 46	024	37 27	100	36 49	132	57 02	151	25 42	236	61 00	281	59 38	345
64	62 08	023	38 22	100	37 31	133	57 28	152	24 55	237	60 05	281	59 22	343
65	62 29	021	39 18	101	38 11	134	57 54	154	24 08	238	59 10	281	59 06	342
66	62 49	020	40 13	101	38 51	135	58 17	156	23 21	238	58 15	281	58 48	341
67	63 08	019	41 07	102	39 31	136	58 39	158	22 33	239	57 20	281	58 30	340
68	63 25	017	42 02	102	40 09	137	59 00	159	21 45	239	56 25	281	58 10	339
69	63 41	016	42 57	103	40 48	138	59 19	161	20 57	240	55 30	281	57 50	338
70	63 56	014	43 51	103	41 25	139	59 36	163	20 08	240	54 35	281	57 28	337
71	64 09	013	44 46	104	42 02	140	59 52	165	19 19	241	53 40	281	57 06	336
72	64 21	011	45 40	105	42 38	141	60 05	167	18 30	241	52 45	281	56 43	335
73	64 31	010	46 34	105	43 13	142	60 17	169	17 41	242	51 50	282	56 19	334
74	64 40	008	47 28	106	43 47	143	60 27	171	16 51	242	50 55	282	55 54	333
	CAPELLA		◆Dubhe		POLLUX		PROCYON		◆SIRIUS		RIGEL		◆Hamal	
75	64 47	007	18 01	030	52 00	071	48 22	107	44 21	144	60 35	173	50 01	282
76	64 53	005	18 28	030	52 53	071	49 15	107	44 54	145	60 41	175	49 06	282
77	64 57	003	18 56	030	53 46	071	50 09	108	45 26	146	60 45	177	48 11	282
78	64 59	002	19 24	030	54 38	071	51 02	109	45 56	147	60 47	179	47 16	282
79	65 00	000	19 53	030	55 31	071	51 55	109	46 26	148	60 47	181	46 22	282
80	65 00	358	20 21	030	56 24	071	52 48	110	46 55	149	60 46	183	45 27	283
81	64 57	357	20 49	030	57 17	070	53 40	111	47 23	151	60 42	185	44 32	283
82	64 54	355	21 17	030	58 10	070	54 32	112	47 50	152	60 36	187	43 38	283
83	64 48	354	21 45	030	59 03	070	55 24	113	48 16	153	60 28	189	42 43	283
84	64 41	352	22 14	030	59 55	070	56 16	113	48 41	155	60 19	191	41 48	283
85	64 32	350	22 42	030	60 48	070	57 07	114	49 04	156	60 07	193	40 54	283
86	64 22	349	23 10	030	61 40	070	57 58	115	49 26	157	59 54	195	39 59	284
87	64 11	347	23 39	030	62 33	069	58 48	116	49 47	159	59 39	197	39 05	284
88	63 58	346	24 07	030	63 25	069	59 38	118	50 07	160	59 22	199	38 11	284
89	63 43	344	24 36	030	64 17	069	60 28	119	50 26	161	59 03	200	37 16	284

LAT 21°N

LHA ♈	Hc	Zn	Hc	Zn	Hc	Zn	Hc	Zn	Hc	Zn	Hc	Zn	Hc	Zn
	Dubhe		◆REGULUS		Alphard		SIRIUS		◆RIGEL		ALDEBARAN		◆CAPELLA	
90	25 04	030	30 11	088	31 06	115	50 43	163	58 43	202	69 33	261	63 28	343
91	25 32	030	31 07	088	31 57	115	50 59	164	58 21	204	68 37	262	63 10	341
92	26 01	030	32 03	089	32 47	116	51 13	166	57 57	206	67 42	262	62 52	340
93	26 29	030	32 59	089	33 37	117	51 26	167	57 32	207	66 46	263	62 32	339
94	26 57	030	33 55	089	34 27	117	51 38	169	57 06	209	65 51	264	62 11	337
95	27 25	030	34 51	090	35 17	118	51 48	170	56 38	211	64 55	264	61 49	336
96	27 54	030	35 47	090	36 06	119	51 57	172	56 09	212	63 59	265	61 26	335
97	28 22	030	36 43	090	36 55	119	52 04	173	55 38	214	63 03	265	61 02	334
98	28 50	030	37 39	091	37 44	120	52 09	175	55 07	215	62 08	266	60 37	333
99	29 18	030	38 35	091	38 32	121	52 14	177	54 34	217	61 12	266	60 11	332
100	29 46	030	39 31	092	39 20	122	52 16	178	54 00	218	60 16	267	59 43	331
101	30 13	030	40 27	092	40 08	122	52 17	180	53 25	219	59 20	267	59 15	330
102	30 41	030	41 23	092	40 55	123	52 17	181	52 49	221	58 24	268	58 47	329
103	31 09	029	42 19	093	41 42	124	52 15	183	52 12	222	57 28	268	58 17	328
104	31 36	029	43 15	093	42 28	125	52 12	184	51 34	223	56 32	268	57 47	327
	Dubhe		◆REGULUS		Alphard		SIRIUS		◆RIGEL		ALDEBARAN		◆CAPELLA	
105	32 03	029	44 11	094	43 14	126	52 07	186	50 55	224	55 36	269	57 15	326
106	32 31	029	45 07	094	43 59	126	52 00	187	50 16	225	54 40	269	56 44	325
107	32 58	029	46 03	094	44 44	127	51 52	189	49 36	227	53 44	269	56 11	324
108	33 25	029	46 59	095	45 28	128	51 43	190	48 54	228	52 48	270	55 38	323
109	33 52	028	47 54	095	46 12	129	51 32	192	48 13	229	51 52	270	55 05	323
110	34 18	028	48 50	096	46 55	130	51 19	194	47 30	230	50 56	271	54 30	322
111	34 45	028	49 46	096	47 37	131	51 06	195	46 47	231	50 00	271	53 56	321
112	35 11	028	50 41	097	48 19	132	50 50	196	46 03	232	49 04	271	53 21	321
113	35 37	028	51 37	097	49 00	133	50 34	198	45 19	233	48 08	272	52 45	320
114	36 03	027	52 33	098	49 40	134	50 16	199	44 34	234	47 12	272	52 09	320
115	36 28	027	53 28	098	50 20	136	49 57	201	43 49	235	46 16	272	51 32	319
116	36 54	027	54 23	099	50 59	137	49 36	202	43 03	235	45 20	273	50 55	319
117	37 19	027	55 19	099	51 37	138	49 15	203	42 17	236	44 24	273	50 18	318
118	37 44	026	56 14	100	52 14	139	48 52	205	41 30	237	43 28	273	49 40	318
119	38 09	026	57 09	101	52 50	141	48 28	206	40 43	238	42 32	273	49 03	317
	◆Dubhe		Denebola		◆Alphard		SIRIUS		RIGEL		◆ALDEBARAN		CAPELLA	
120	38 33	026	35 26	087	53 25	142	48 02	207	39 55	239	41 36	274	48 24	317
121	38 57	025	36 22	087	53 59	143	47 36	209	39 07	239	40 40	274	47 46	316
122	39 21	025	37 18	087	54 32	145	47 08	210	38 19	240	39 45	274	47 07	316
123	39 45	025	38 14	088	55 04	146	46 40	211	37 30	241	38 49	275	46 28	316
124	40 08	024	39 10	088	55 34	148	46 10	212	36 41	242	37 53	275	45 49	315
125	40 31	024	40 06	088	56 04	149	45 40	214	35 51	242	36 57	275	45 09	315
126	40 54	024	41 02	089	56 32	151	45 09	215	35 02	243	36 01	276	44 29	315
127	41 16	023	41 58	089	56 58	152	44 36	216	34 12	244	35 06	276	43 49	314
128	41 38	023	42 54	089	57 24	154	44 03	217	33 21	244	34 10	276	43 09	314
129	41 59	022	43 50	090	57 47	156	43 29	218	32 31	245	33 14	276	42 29	314
130	42 21	022	44 46	090	58 10	157	42 54	219	31 40	245	32 19	277	41 48	314
131	42 41	022	45 42	090	58 31	159	42 19	220	30 49	246	31 23	277	41 08	313
132	43 02	021	46 38	091	58 50	161	41 42	221	29 58	247	30 27	277	40 27	313
133	43 22	021	47 34	091	59 07	163	41 05	222	29 06	247	29 32	278	39 46	313
134	43 41	020	48 30	092	59 23	165	40 27	223	28 15	248	28 36	278	39 05	313
	Dubhe		◆ARCTURUS		SPICA		◆Alphard		SIRIUS		BETELGEUSE		◆CAPELLA	
135	44 00	020	16 45	075	17 27	110	59 37	167	39 49	224	43 19	260	38 24	313
136	44 19	019	17 39	076	18 20	110	59 49	168	39 10	225	42 24	261	37 43	313
137	44 37	019	18 34	076	19 12	111	59 59	170	38 30	226	41 29	261	37 02	312
138	44 55	018	19 28	076	20 05	111	60 07	172	37 49	227	40 33	262	36 20	312
139	45 12	018	20 23	077	20 57	112	60 14	174	37 09	227	39 38	262	35 39	312
140	45 29	017	21 17	077	21 49	112	60 19	176	36 27	228	38 43	262	34 58	312
141	45 45	017	22 12	077	22 41	113	60 21	178	35 45	229	37 47	263	34 16	312
142	46 01	016	23 06	077	23 32	113	60 22	180	35 02	230	36 51	263	33 34	312
143	46 16	015	24 01	078	24 24	114	60 20	182	34 19	231	35 56	264	32 53	312
144	46 31	015	24 56	078	25 15	114	60 17	184	33 36	231	35 00	264	32 11	312
145	46 45	014	25 50	078	26 06	115	60 12	186	32 52	232	34 04	265	31 30	312
146	46 58	014	26 45	078	26 57	115	60 05	188	32 08	233	33 08	265	30 48	312
147	47 11	013	27 40	079	27 47	116	59 56	190	31 23	233	32 13	265	30 06	312
148	47 23	012	28 35	079	28 38	116	59 45	192	30 38	234	31 17	266	29 25	312
149	47 35	012	29 30	079	29 28	117	59 32	194	29 52	235	30 21	266	28 43	312
	Dubhe		◆ARCTURUS		SPICA		◆Alphard		SIRIUS		POLLUX		◆CAPELLA	
150	47 46	011	30 25	079	30 17	118	59 18	196	29 06	236	58 36	290	28 01	312
151	47 57	010	31 20	080	31 07	118	59 02	198	28 20	236	57 44	290	27 20	312
152	48 06	010	32 15	080	31 56	119	58 44	200	27 33	237	56 51	289	26 38	312
153	48 16	009	33 11	080	32 45	120	58 24	201	26 46	237	55 58	289	25 56	312
154	48 24	008	34 06	080	33 33	120	58 03	203	25 58	238	55 05	289	25 15	312
155	48 32	008	35 01	081	34 22	121	57 40	205	25 11	239	54 12	289	24 33	312
156	48 39	007	35 56	081	35 10	122	57 15	207	24 23	239	53 19	289	23 52	312
157	48 46	006	36 52	081	35 57	122	56 50	208	23 35	240	52 26	289	23 10	312
158	48 52	006	37 47	081	36 44	123	56 23	210	22 46	240	51 33	289	22 29	312
159	48 57	005	38 43	082	37 31	124	55 54	211	21 57	241	50 41	289	21 48	313
160	49 01	004	39 38	082	38 18	125	55 24	213	21 08	241	49 48	289	21 06	313
161	49 05	004	40 33	082	39 03	125	54 53	214	20 19	242	48 55	289	20 25	313
162	49 08	003	41 29	082	39 49	126	54 21	216	19 29	242	48 02	289	19 44	313
163	49 10	002	42 25	083	40 34	127	53 48	217	18 40	243	47 09	289	19 03	313
164	49 12	001	43 20	083	41 18	128	53 13	219	17 50	243	46 16	289	18 22	313
	◆Dubhe		Alkaid		ARCTURUS		◆SPICA		Alphard		◆PROCYON		POLLUX	
165	49 13	001	46 29	039	44 16	083	42 02	129	52 38	220	38 51	259	45 23	289
166	49 13	000	47 04	039	45 11	083	42 46	130	52 02	221	37 56	260	44 30	289
167	49 13	359	47 39	038	46 07	084	43 29	130	51 24	222	37 01	260	43 37	289
168	49 12	358	48 13	038	47 03	084	44 11	131	50 46	224	36 06	261	42 44	289
169	49 10	358	48 47	037	47 58	084	44 53	132	50 07	225	35 10	261	41 51	289
170	49 07	357	49 21	037	48 54	084	45 34	133	49 27	226	34 15	262	40 59	289
171	49 04	356	49 54	036	49 50	085	46 14	134	48 47	227	33 19	262	40 06	290
172	49 00	356	50 27	036	50 46	085	46 54	135	48 05	228	32 24	263	39 13	290
173	48 55	355	51 00	035	51 42	085	47 33	137	47 23	229	31 28	263	38 20	290
174	48 50	354	51 32	035	52 37	086	48 11	138	46 41	230	30 33	264	37 28	290
175	48 44	353	52 04	034	53 33	086	48 48	139	45 57	231	29 37	264	36 35	290
176	48 37	353	52 35	033	54 29	086	49 25	140	45 14	232	28 41	264	35 42	290
177	48 30	352	53 05	033	55 25	086	50 00	141	44 29	233	27 45	265	34 50	290
178	48 22	351	53 35	032	56 21	087	50 35	142	43 44	234	26 50	265	33 57	290
179	48 13	351	54 04	031	57 17	087	51 08	144	42 59	235	25 54	266	33 05	291

LHA ♈	Hc	Zn	Hc	Zn	Hc	Zn	Hc	Zn	Hc	Zn	Hc	Zn	Hc	Zn
	◆Alkaid		Alphecca		◆SPICA		Gienah		REGULUS		◆POLLUX		Dubhe	
180	54 33	030	40 59	072	51 41	145	51 18	174	61 46	256	32 12	291	48 04	350
181	55 01	030	41 52	072	52 13	146	51 23	176	60 51	257	31 20	291	47 54	349
182	55 29	029	42 45	072	52 43	148	51 26	177	59 57	257	30 28	291	47 43	349
183	55 55	028	43 39	073	53 12	149	51 29	179	59 02	258	29 35	291	47 32	348
184	56 21	027	44 32	073	53 41	151	51 29	180	58 07	259	28 43	291	47 20	347
185	56 46	026	45 26	073	54 07	152	51 28	182	57 12	259	27 51	292	47 07	347
186	57 11	025	46 19	073	54 33	154	51 26	183	56 17	260	26 59	292	46 54	346
187	57 34	024	47 13	073	54 57	155	51 22	185	55 22	261	26 07	292	46 41	346
188	57 57	023	48 06	073	55 20	157	51 17	186	54 26	261	25 15	292	46 27	345
189	58 19	022	49 00	073	55 41	158	51 10	188	53 31	262	24 23	292	46 12	344
190	58 40	021	49 53	073	56 01	160	51 01	189	52 36	262	23 31	292	45 56	344
191	58 59	020	50 47	073	56 20	162	50 52	191	51 40	263	22 39	293	45 41	343
192	59 18	019	51 41	073	56 37	163	50 40	192	50 44	263	21 48	293	45 24	343
193	59 36	018	52 34	073	56 52	165	50 28	194	49 49	264	20 56	293	45 07	342
194	59 53	017	53 28	073	57 05	167	50 14	195	48 53	264	20 05	293	44 50	342
	◆Kochab		VEGA		Rasalhague		◆ANTARES		SPICA		◆REGULUS		Dubhe	
195	34 46	009	17 21	054	24 10	085	20 37	131	57 17	169	47 57	265	44 32	341
196	34 54	009	18 06	055	25 06	086	21 19	131	57 27	170	47 02	265	44 14	341
197	35 03	008	18 52	055	26 01	086	22 01	132	57 36	172	46 06	266	43 55	340
198	35 11	008	19 38	055	26 57	086	22 42	133	57 42	174	45 10	266	43 36	340
199	35 18	008	20 23	055	27 53	087	23 23	133	57 47	176	44 14	266	43 16	339
200	35 26	007	21 09	055	28 49	087	24 04	134	57 51	178	43 18	267	42 56	339
201	35 33	007	21 55	055	29 45	087	24 44	135	57 52	180	42 22	267	42 35	338
202	35 40	007	22 42	055	30 41	088	25 24	135	57 51	181	41 26	268	42 15	338
203	35 46	006	23 28	056	31 37	088	26 03	136	57 49	183	40 30	268	41 53	337
204	35 52	006	24 14	056	32 33	088	26 42	137	57 45	185	39 34	268	41 32	337
205	35 58	006	25 00	056	33 29	089	27 20	137	57 39	187	38 38	269	41 10	337
206	36 04	006	25 47	056	34 25	089	27 58	138	57 31	189	37 42	269	40 47	336
207	36 09	005	26 33	056	35 21	089	28 35	139	57 22	191	36 46	269	40 24	336
208	36 14	005	27 20	056	36 17	090	29 12	140	57 11	192	35 50	270	40 01	336
209	36 19	005	28 06	056	37 13	090	29 48	140	56 58	194	34 54	270	39 38	335
	◆Kochab		VEGA		Rasalhague		◆ANTARES		SPICA		◆REGULUS		Dubhe	
210	36 23	004	28 53	056	38 09	090	30 23	141	56 44	196	33 58	271	39 14	335
211	36 27	004	29 39	056	39 05	091	30 58	142	56 28	198	33 02	271	38 50	335
212	36 31	004	30 26	056	40 01	091	31 32	143	56 10	199	32 06	271	38 26	334
213	36 34	003	31 13	057	40 57	091	32 06	143	55 51	201	31 10	272	38 02	334
214	36 37	003	32 00	057	41 53	092	32 39	144	55 30	203	30 14	272	37 37	334
215	36 40	003	32 46	057	42 49	092	33 11	145	55 08	204	29 18	272	37 12	333
216	36 42	002	33 33	057	43 45	093	33 43	146	54 44	206	28 22	273	36 47	333
217	36 44	002	34 20	057	44 41	093	34 14	147	54 19	207	27 26	273	36 21	333
218	36 46	002	35 07	057	45 37	093	34 44	148	53 53	209	26 30	273	35 55	333
219	36 47	001	35 53	057	46 33	094	35 14	149	53 26	210	25 34	274	35 29	332
220	36 48	001	36 40	057	47 29	094	35 42	150	52 57	212	24 39	274	35 03	332
221	36 49	001	37 27	057	48 24	095	36 10	151	52 27	213	23 43	274	34 37	332
222	36 49	000	38 14	057	49 20	095	36 37	152	51 56	214	22 47	275	34 10	332
223	36 49	000	39 00	056	50 16	096	37 04	153	51 24	216	21 51	275	33 44	331
224	36 49	000	39 47	056	51 12	096	37 29	154	50 51	217	20 55	275	33 17	331
	VEGA		◆ALTAIR		Nunki		ANTARES		◆SPICA		Denebola		◆Alkaid	
225	40 34	056	19 18	088	16 00	127	37 54	155	50 16	218	44 12	270	58 12	337
226	41 20	056	20 14	088	16 44	128	38 17	156	49 41	219	43 16	271	57 50	336
227	42 07	056	21 10	088	17 28	128	38 40	157	49 05	221	42 20	271	57 27	335
228	42 53	056	22 06	089	18 12	129	39 02	158	48 28	222	41 24	271	57 03	334
229	43 40	056	23 02	089	18 56	129	39 23	159	47 51	223	40 28	272	56 38	333
230	44 26	056	23 58	089	19 39	130	39 43	160	47 12	224	39 32	272	56 13	333
231	45 12	056	24 54	090	20 22	130	40 01	161	46 33	225	38 36	272	55 47	332
232	45 58	055	25 50	090	21 04	131	40 19	162	45 53	226	37 40	273	55 20	331
233	46 44	055	26 46	091	21 46	132	40 36	163	45 12	227	36 44	273	54 52	330
234	47 30	055	27 42	091	22 28	132	40 52	164	44 31	228	35 48	273	54 24	329
235	48 16	055	28 38	091	23 09	133	41 07	165	43 49	229	34 52	274	53 55	329
236	49 02	054	29 34	092	23 50	134	41 20	167	43 06	230	33 57	274	53 25	328
237	49 47	054	30 30	092	24 30	134	41 33	168	42 23	231	33 01	274	52 55	327
238	50 33	054	31 26	092	25 10	135	41 44	169	41 39	232	32 05	274	52 24	326
239	51 18	054	32 22	093	25 50	135	41 54	170	40 55	233	31 09	275	51 53	326
	◆VEGA		ALTAIR		Nunki		◆ANTARES		SPICA		ARCTURUS		◆Alkaid	
240	52 03	053	33 18	093	26 29	136	42 03	171	40 10	234	65 25	270	51 22	325
241	52 48	053	34 14	094	27 07	137	42 11	172	39 25	234	64 29	271	50 49	325
242	53 32	052	35 10	094	27 45	138	42 18	174	38 39	235	63 33	271	50 17	324
243	54 16	052	36 05	094	28 23	138	42 24	175	37 53	236	62 37	271	49 44	324
244	55 00	052	37 01	095	29 00	139	42 28	176	37 06	237	61 41	272	49 10	323
245	55 44	051	37 57	095	29 36	140	42 32	177	36 19	237	60 45	272	48 36	322
246	56 28	051	38 53	096	30 12	141	42 34	178	35 32	238	59 49	272	48 02	322
247	57 11	050	39 49	096	30 47	141	42 35	180	34 44	239	58 53	273	47 27	322
248	57 53	049	40 44	096	31 22	142	42 34	181	33 56	239	57 57	273	46 52	321
249	58 36	049	41 40	097	31 56	143	42 33	182	33 08	240	57 01	273	46 17	321
250	59 18	048	42 35	097	32 29	144	42 30	183	32 19	241	56 05	273	45 41	320
251	59 59	047	43 31	098	33 02	145	42 26	185	31 30	241	55 09	274	45 05	320
252	60 40	047	44 26	098	33 34	146	42 21	186	30 40	242	54 13	274	44 29	320
253	61 21	046	45 22	099	34 06	146	42 15	187	29 51	243	53 17	274	43 53	319
254	62 01	045	46 17	099	34 36	147	42 08	188	29 01	243	52 22	275	43 16	319
	◆DENEB		ALTAIR		◆Nunki		ANTARES		SPICA		◆ARCTURUS		Alkaid	
255	38 56	048	47 12	100	35 06	148	41 59	189	28 11	244	51 26	275	42 39	319
256	39 38	048	48 07	100	35 35	149	41 50	191	27 20	244	50 30	275	42 02	318
257	40 20	048	49 03	101	36 04	150	41 39	192	26 30	245	49 34	275	41 25	318
258	41 01	048	49 57	102	36 31	151	41 27	193	25 39	246	48 38	276	40 48	318
259	41 42	047	50 52	102	36 58	152	41 14	194	24 48	246	47 43	276	40 10	318
260	42 23	047	51 47	103	37 24	153	41 00	195	23 56	247	46 47	276	39 32	317
261	43 04	047	52 42	103	37 49	154	40 45	196	23 05	247	45 51	276	38 54	317
262	43 45	047	53 36	104	38 13	155	40 29	197	22 13	248	44 55	277	38 16	317
263	44 26	046	54 30	105	38 36	156	40 11	199	21 21	248	44 00	277	37 38	317
264	45 06	046	55 24	105	38 58	157	39 53	200	20 29	249	43 04	277	36 59	317
265	45 47	046	56 18	106	39 20	158	39 34	201	19 37	249	42 09	277	36 21	317
266	46 27	046	57 12	107	39 40	159	39 13	202	18 44	250	41 13	278	35 42	316
267	47 07	045	58 05	108	39 59	160	38 52	203	17 52	250	40 18	278	35 04	316
268	47 46	045	58 59	109	40 18	161	38 30	204	16 59	251	39 22	278	34 25	316
269	48 26	044	59 52	109	40 35	163	38 07	205	16 06	251	38 27	278	33 46	316

LAT 21°N

LHA ♈	Hc	Zn	Hc	Zn	Hc	Zn	Hc	Zn	Hc	Zn	Hc	Zn	Hc	Zn
	◆DENEB		Enif		◆Nunki		ANTARES		◆ARCTURUS		Alkaid		Kochab	
270	49 05	044	35 10	093	40 51	164	37 43	206	37 31	279	33 07	316	31 09	346
271	49 43	044	36 06	093	41 07	165	37 18	207	36 36	279	32 29	316	30 56	346
272	50 22	043	37 02	093	41 21	166	36 52	208	35 41	279	31 50	316	30 43	346
273	51 00	043	37 57	094	41 34	167	36 25	209	34 45	279	31 11	316	30 29	346
274	51 38	042	38 53	094	41 46	168	35 58	210	33 50	280	30 32	316	30 15	346
275	52 15	042	39 49	095	41 57	169	35 30	211	32 55	280	29 53	316	30 02	346
276	52 52	041	40 45	095	42 06	171	35 01	212	32 00	280	29 13	316	29 48	345
277	53 29	041	41 41	096	42 15	172	34 31	213	31 05	280	28 34	316	29 33	345
278	54 05	040	42 36	096	42 22	173	34 00	213	30 10	281	27 55	316	29 19	345
279	54 41	039	43 32	096	42 28	174	33 29	214	29 15	281	27 16	316	29 05	345
280	55 16	039	44 28	097	42 33	175	32 57	215	28 20	281	26 37	316	28 50	345
281	55 51	038	45 23	097	42 37	177	32 24	216	27 25	281	25 58	316	28 35	345
282	56 25	037	46 19	098	42 40	178	31 51	217	26 30	282	25 19	316	28 20	345
283	56 58	036	47 14	098	42 41	179	31 17	218	25 35	282	24 40	316	28 05	344
284	57 31	036	48 10	099	42 42	180	30 42	219	24 40	282	24 01	316	27 50	344
	DENEB		◆Alpheratz		FOMALHAUT		◆Nunki		ANTARES		◆Alphecca		Kochab	
285	58 03	035	20 56	066	13 41	130	42 41	182	30 07	219	42 58	288	27 35	344
286	58 35	034	21 47	066	14 23	130	42 39	183	29 31	220	42 05	288	27 20	344
287	59 06	033	22 38	066	15 06	131	42 35	184	28 55	221	41 11	288	27 04	344
288	59 36	032	23 29	066	15 48	131	42 31	185	28 18	222	40 18	288	26 49	344
289	60 05	031	24 21	067	16 30	132	42 25	186	27 41	222	39 25	288	26 33	344
290	60 33	030	25 12	067	17 12	132	42 19	188	27 03	223	38 31	288	26 18	344
291	61 00	029	26 04	067	17 53	133	42 11	189	26 24	224	37 38	288	26 02	344
292	61 27	028	26 55	067	18 34	133	42 01	190	25 45	224	36 45	288	25 46	344
293	61 52	026	27 47	067	19 14	134	41 51	191	25 06	225	35 52	288	25 30	343
294	62 17	025	28 39	067	19 54	135	41 40	192	24 26	226	34 59	289	25 14	343
295	62 40	024	29 30	068	20 34	135	41 27	194	23 46	226	34 06	289	24 58	343
296	63 02	023	30 22	068	21 13	136	41 14	195	23 05	227	33 13	289	24 42	343
297	63 23	021	31 14	068	21 52	136	40 59	196	22 24	228	32 20	289	24 26	343
298	63 43	020	32 06	068	22 31	137	40 43	197	21 42	228	31 27	289	24 10	343
299	64 01	018	32 58	068	23 08	138	40 26	198	21 00	229	30 34	289	23 53	343
	◆DENEB		Schedar		Alpheratz		◆FOMALHAUT		Nunki		◆Rasalhague		VEGA	
300	64 18	017	28 20	036	33 50	068	23 46	138	40 08	199	54 18	262	64 48	319
301	64 33	015	28 53	036	34 42	068	24 23	139	39 49	200	53 23	263	64 11	318
302	64 48	014	29 26	036	35 34	069	24 59	140	39 30	201	52 27	263	63 33	317
303	65 00	012	29 59	036	36 26	069	25 35	140	39 09	202	51 31	264	62 55	316
304	65 11	011	30 32	036	37 19	069	26 10	141	38 47	203	50 36	264	62 16	315
305	65 21	009	31 04	036	38 11	069	26 45	142	38 24	205	49 40	265	61 36	314
306	65 29	007	31 37	036	39 03	069	27 20	143	38 00	206	48 44	265	60 56	314
307	65 35	006	32 10	036	39 55	069	27 53	143	37 36	207	47 48	266	60 15	313
308	65 40	004	32 42	035	40 48	069	28 26	144	37 10	208	46 52	266	59 34	312
309	65 43	002	33 15	035	41 40	069	28 59	145	36 44	209	45 56	266	58 52	311
310	65 44	001	33 47	035	42 32	069	29 31	146	36 17	209	45 01	267	58 10	311
311	65 44	359	34 19	035	43 25	069	30 02	147	35 49	210	44 05	267	57 27	310
312	65 42	357	34 51	035	44 17	069	30 33	147	35 20	211	43 09	268	56 44	310
313	65 38	355	35 23	035	45 10	069	31 03	148	34 51	212	42 13	268	56 01	309
314	65 33	354	35 55	034	46 02	069	31 32	149	34 20	213	41 17	268	55 17	309
	◆Alpheratz		Diphda		◆FOMALHAUT		ALTAIR		Rasalhague		◆VEGA		DENEB	
315	46 55	069	22 49	121	32 00	150	69 19	237	40 21	269	54 33	308	65 26	352
316	47 47	069	23 37	122	32 28	151	68 32	238	39 25	269	53 49	308	65 17	350
317	48 39	069	24 24	123	32 55	152	67 44	240	38 29	270	53 05	307	65 07	349
318	49 32	069	25 11	123	33 21	153	66 55	241	37 33	270	52 20	307	64 56	347
319	50 24	069	25 58	124	33 47	153	66 06	243	36 37	270	51 35	307	64 42	346
320	51 17	069	26 44	124	34 11	154	65 16	244	35 41	271	50 50	306	64 28	344
321	52 09	069	27 30	125	34 35	155	64 25	245	34 45	271	50 05	306	64 12	343
322	53 02	069	28 16	126	34 58	156	63 34	246	33 49	271	49 19	306	63 54	341
323	53 54	069	29 01	126	35 20	157	62 43	247	32 53	272	48 34	305	63 35	340
324	54 46	069	29 46	127	35 42	158	61 51	248	31 57	272	47 48	305	63 15	338
325	55 38	069	30 31	128	36 02	159	60 58	249	31 01	272	47 02	305	62 54	337
326	56 31	069	31 15	128	36 22	160	60 06	250	30 05	273	46 16	305	62 31	336
327	57 23	069	31 59	129	36 40	161	59 13	251	29 09	273	45 30	305	62 08	334
328	58 15	068	32 42	130	36 58	162	58 20	252	28 13	273	44 44	304	61 43	333
329	59 07	068	33 25	131	37 14	163	57 26	253	27 17	274	43 58	304	61 17	332
	◆Mirfak		Hamal		Diphda		◆FOMALHAUT		ALTAIR		◆VEGA		DENEB	
330	21 36	043	33 15	075	34 07	131	37 30	164	56 33	254	43 11	304	60 50	331
331	22 15	043	34 09	075	34 49	132	37 45	165	55 39	254	42 25	304	60 23	330
332	22 53	043	35 03	075	35 30	133	37 58	166	54 45	255	41 38	304	59 54	329
333	23 32	043	35 58	076	36 11	134	38 11	167	53 51	256	40 52	304	59 25	328
334	24 10	044	36 52	076	36 51	135	38 23	169	52 56	256	40 05	304	58 54	327
335	24 49	044	37 46	076	37 31	135	38 33	170	52 02	257	39 18	304	58 23	326
336	25 27	044	38 41	076	38 10	136	38 43	171	51 07	258	38 32	303	57 52	325
337	26 06	044	39 35	076	38 48	137	38 52	172	50 12	258	37 45	303	57 19	324
338	26 45	044	40 30	077	39 26	138	38 59	173	49 17	259	36 58	303	56 46	323
339	27 23	044	41 24	077	40 03	139	39 05	174	48 22	259	36 11	303	56 12	323
340	28 02	044	42 19	077	40 39	140	39 11	175	47 27	260	35 25	303	55 38	322
341	28 41	044	43 13	077	41 15	141	39 15	176	46 32	260	34 38	303	55 03	321
342	29 19	044	44 08	077	41 50	142	39 18	177	45 37	261	33 51	303	54 28	321
343	29 58	044	45 02	077	42 24	143	39 20	178	44 41	262	33 04	303	53 52	320
344	30 37	044	45 57	078	42 57	144	39 21	180	43 46	262	32 18	303	53 16	319
	Schedar		◆CAPELLA		ALDEBARAN		Diphda		◆FOMALHAUT		ALTAIR		◆DENEB	
345	49 57	021	12 13	045	11 21	077	43 30	145	39 21	181	42 50	262	52 39	319
346	50 17	021	12 53	045	12 15	077	44 01	146	39 20	182	41 55	263	52 02	318
347	50 37	020	13 33	046	13 10	077	44 32	147	39 17	183	40 59	263	51 24	318
348	50 55	019	14 13	046	14 05	077	45 01	149	39 14	184	40 04	264	50 46	317
349	51 14	018	14 53	046	14 59	078	45 30	150	39 09	185	39 08	264	50 08	317
350	51 31	018	15 33	046	15 54	078	45 58	151	39 04	186	38 12	265	49 29	316
351	51 48	017	16 14	046	16 49	078	46 25	152	38 57	187	37 16	265	48 50	316
352	52 04	016	16 54	046	17 44	079	46 50	153	38 49	189	36 21	266	48 11	315
353	52 19	015	17 35	047	18 39	079	47 15	155	38 40	190	35 25	266	47 32	315
354	52 33	015	18 16	047	19 34	079	47 38	156	38 30	191	34 29	266	46 52	315
355	52 47	014	18 57	047	20 29	080	48 01	157	38 19	192	33 33	267	46 12	314
356	53 00	013	19 38	047	21 24	080	48 22	159	38 07	193	32 37	267	45 32	314
357	53 12	012	20 19	047	22 19	080	48 42	160	37 54	194	31 41	268	44 51	314
358	53 23	011	21 00	047	23 14	080	49 01	161	37 40	195	30 45	268	44 11	313
359	53 33	010	21 41	047	24 10	081	49 18	163	37 25	196	29 49	268	43 30	313

LAT 20°N

LHA	Hc	Zn	Hc	Zn	Hc	Zn	Hc	Zn	Hc	Zn	Hc	Zn	Hc	Zn
♈	Schedar		CAPELLA		◆ALDEBARAN		Diphda		◆FOMALHAUT		ALTAIR		◆DENEB	
0	52 44	009	21 42	047	24 55	081	50 32	164	38 07	197	28 54	269	42 08	314
1	52 52	008	22 23	047	25 51	081	50 47	165	37 49	198	27 58	270	41 27	313
2	53 00	007	23 05	047	26 47	081	51 01	167	37 31	199	27 01	270	40 46	313
3	53 07	006	23 46	047	27 42	081	51 13	168	37 12	200	26 05	270	40 05	313
4	53 13	006	24 28	048	28 38	082	51 24	170	36 52	201	25 09	271	39 24	313
5	53 18	005	25 09	048	29 34	082	51 33	171	36 31	202	24 12	271	38 42	313
6	53 22	004	25 51	048	30 30	082	51 41	173	36 09	203	23 16	271	38 01	312
7	53 25	003	26 33	048	31 26	082	51 48	174	35 46	204	22 19	272	37 19	312
8	53 27	002	27 14	048	32 21	083	51 53	176	35 22	205	21 23	272	36 37	312
9	53 29	001	27 56	048	33 17	083	51 57	177	34 58	206	20 27	272	35 55	312
10	53 29	000	28 38	048	34 13	083	51 59	179	34 32	207	19 30	273	35 14	312
11	53 29	359	29 19	048	35 09	083	51 59	180	34 06	208	18 34	273	34 32	312
12	53 28	358	30 01	048	36 05	084	51 58	182	33 39	209	17 38	273	33 50	312
13	53 25	357	30 43	048	37 01	084	51 56	183	33 11	210	16 42	274	33 07	312
14	53 22	356	31 24	048	37 58	084	51 52	185	32 43	211	15 45	274	32 25	312
	CAPELLA		◆ALDEBARAN		RIGEL		ACHERNAR		◆FOMALHAUT		Enif		◆DENEB	
15	32 06	048	38 54	085	21 25	108	12 20	175	32 14	212	41 44	265	31 43	312
16	32 47	047	39 50	085	22 18	108	12 25	175	31 44	212	40 48	266	31 01	312
17	33 29	047	40 46	085	23 12	109	12 29	176	31 13	213	39 52	266	30 19	312
18	34 10	047	41 42	085	24 05	109	12 33	176	30 42	214	38 56	267	29 37	312
19	34 52	047	42 38	086	24 58	110	12 36	177	30 10	215	37 59	267	28 54	311
20	35 33	047	43 35	086	25 51	110	12 39	178	29 38	216	37 03	267	28 12	311
21	36 14	047	44 31	086	26 44	111	12 41	178	29 05	216	36 07	268	27 30	312
22	36 56	047	45 27	086	27 37	111	12 43	179	28 31	217	35 10	268	26 48	312
23	37 37	047	46 23	087	28 29	112	12 44	179	27 56	218	34 14	268	26 05	312
24	38 18	047	47 20	087	29 22	112	12 44	180	27 21	219	33 18	269	25 23	312
25	38 59	046	48 16	087	30 14	113	12 44	180	26 46	219	32 21	269	24 41	312
26	39 39	046	49 12	088	31 05	113	12 44	181	26 10	220	31 25	269	23 59	312
27	40 20	046	50 09	088	31 57	114	12 42	181	25 33	221	30 29	270	23 17	312
28	41 00	046	51 05	088	32 49	115	12 41	182	24 56	221	29 32	270	22 35	312
29	41 41	046	52 01	089	33 40	115	12 38	183	24 19	222	28 36	271	21 53	312
	Mirfak		CAPELLA		◆ALDEBARAN		RIGEL		ACHERNAR		◆FOMALHAUT		◆DENEB	
30	55 49	024	42 21	045	52 58	089	34 31	116	12 36	183	23 41	223	21 11	312
31	56 11	023	43 01	045	53 54	089	35 21	116	12 32	184	23 02	223	20 29	312
32	56 33	022	43 41	045	54 51	090	36 12	117	12 28	184	22 23	224	19 48	312
33	56 54	021	44 20	044	55 47	090	37 02	118	12 24	185	21 44	225	19 06	312
34	57 14	020	44 59	044	56 43	090	37 51	118	12 19	185	21 04	225	18 24	312
35	57 33	019	45 39	044	57 40	091	38 41	119	12 14	186	20 24	226	17 43	313
36	57 52	018	46 17	043	58 36	091	39 30	120	12 08	186	19 43	226	17 01	313
37	58 09	017	46 56	043	59 32	091	40 19	121	12 01	187	19 02	227	16 20	313
38	58 25	016	47 34	043	60 29	092	41 07	121	11 54	187	18 21	228	15 39	313
39	58 40	015	48 12	042	61 25	092	41 55	122	11 46	188	17 39	228	14 58	313
40	58 54	014	48 50	042	62 21	092	42 43	123	11 38	189	16 57	229	14 17	313
41	59 07	013	49 27	041	63 18	093	43 30	124	11 30	189	16 15	229	13 36	314
42	59 19	011	50 04	041	64 14	093	44 16	125	11 20	190	15 32	230	12 55	314
43	59 29	010	50 41	040	65 10	094	45 03	125	11 11	190	14 49	230	12 14	314
44	59 38	009	51 17	040	66 07	094	45 48	126	11 01	191	14 05	231	11 34	314
	CAPELLA		◆POLLUX		PROCYON		SIRIUS		◆Acamar		Diphda		◆Alpheratz	
45	51 53	039	25 18	068	20 47	092	23 42	120	29 40	180	39 19	224	50 06	292
46	52 28	038	26 10	068	21 44	092	24 31	120	29 40	181	38 40	225	49 14	292
47	53 03	038	27 02	068	22 40	093	25 19	121	29 38	182	38 00	225	48 21	292
48	53 37	037	27 55	068	23 36	093	26 08	121	29 35	183	37 20	226	47 29	292
49	54 11	037	28 47	068	24 33	093	26 56	122	29 32	184	36 39	227	46 36	292
50	54 44	036	29 39	068	25 29	094	27 44	122	29 28	185	35 57	228	45 44	292
51	55 17	035	30 32	069	26 25	094	28 31	123	29 22	186	35 15	229	44 52	292
52	55 49	034	31 24	069	27 21	095	29 18	124	29 16	187	34 32	229	43 59	292
53	56 21	033	32 17	069	28 18	095	30 05	124	29 10	187	33 49	230	43 07	292
54	56 51	033	33 09	069	29 14	095	30 51	125	29 02	188	33 06	231	42 14	292
55	57 21	032	34 02	069	30 10	096	31 37	126	28 53	189	32 22	232	41 22	292
56	57 51	031	34 55	069	31 06	096	32 23	126	28 44	190	31 37	232	40 29	292
57	58 19	030	35 47	069	32 02	096	33 08	127	28 34	191	30 52	233	39 37	292
58	58 47	029	36 40	069	32 58	097	33 53	128	28 23	192	30 07	234	38 45	292
59	59 13	028	37 33	069	33 54	097	34 37	129	28 11	192	29 21	234	37 52	292
	◆CAPELLA		POLLUX		PROCYON		◆SIRIUS		CANOPUS		Diphda		◆Hamal	
60	59 39	027	38 26	070	34 50	098	35 21	129	10 54	159	28 35	235	63 33	283
61	60 04	026	39 19	070	35 46	098	36 04	130	11 14	159	27 49	236	62 38	283
62	60 28	024	40 11	070	36 42	099	36 47	131	11 33	160	27 02	236	61 43	283
63	60 51	023	41 04	070	37 37	099	37 29	132	11 53	160	26 15	237	60 48	282
64	61 12	022	41 57	070	38 33	100	38 11	132	12 11	161	25 28	237	59 53	282
65	61 33	021	42 50	070	39 28	100	38 53	133	12 30	161	24 40	238	58 58	282
66	61 52	019	43 43	070	40 24	100	39 33	134	12 47	162	23 52	239	58 03	282
67	62 11	018	44 36	070	41 19	101	40 13	135	13 05	162	23 04	239	57 08	283
68	62 28	017	45 29	070	42 15	101	40 53	136	13 21	163	22 15	240	56 13	283
69	62 43	015	46 22	070	43 10	102	41 32	137	13 38	164	21 27	240	55 18	283
70	62 57	014	47 15	070	44 05	103	42 10	138	13 53	164	20 38	241	54 23	283
71	63 10	012	48 08	070	45 00	103	42 47	139	14 08	165	19 48	241	53 28	283
72	63 22	011	49 01	070	45 55	104	43 24	140	14 23	165	18 59	242	52 33	283
73	63 32	010	49 54	070	46 49	104	44 00	141	14 37	166	18 09	242	51 38	283
74	63 40	008	50 47	070	47 44	105	44 35	142	14 51	166	17 19	243	50 43	283
	CAPELLA		◆POLLUX		PROCYON		SIRIUS		◆RIGEL		Menkar		◆Hamal	
75	63 47	006	51 39	070	48 38	105	45 09	143	61 34	173	57 09	245	49 48	283
76	63 53	005	52 32	070	49 33	106	45 43	144	61 41	175	56 18	246	48 53	283
77	63 57	003	53 25	070	50 27	107	46 15	145	61 45	177	55 26	247	47 58	283
78	63 59	002	54 18	069	51 21	107	46 47	147	61 47	179	54 34	248	47 03	283
79	64 00	000	55 11	069	52 14	108	47 17	148	61 47	181	53 42	248	46 08	283
80	64 00	359	56 04	069	53 08	109	47 47	149	61 45	183	52 49	249	45 13	284
81	63 58	357	56 56	069	54 01	110	48 15	150	61 42	185	51 57	250	44 18	284
82	63 54	355	57 49	069	54 54	110	48 43	151	61 36	187	51 04	251	43 24	284
83	63 48	354	58 41	069	55 47	111	49 09	153	61 28	189	50 10	251	42 29	284
84	63 42	352	59 34	068	56 39	112	49 35	154	61 18	191	49 17	252	41 34	284
85	63 33	351	60 26	068	57 31	113	49 59	155	61 06	193	48 23	253	40 40	284
86	63 23	349	61 19	068	58 23	114	50 22	157	60 52	195	47 29	253	39 45	284
87	63 12	348	62 11	068	59 14	115	50 43	158	60 36	197	46 35	254	38 50	285
88	63 00	346	63 03	067	60 05	116	51 04	160	60 18	199	45 41	255	37 56	285
89	62 46	345	63 55	067	60 56	117	51 23	161	59 59	201	44 46	255	37 01	285

LAT 20°N

LHA	Hc	Zn	Hc	Zn	Hc	Zn	Hc	Zn	Hc	Zn	Hc	Zn	Hc	Zn
♈	Dubhe		◆REGULUS		Alphard		SIRIUS		◆RIGEL		ALDEBARAN		◆CAPELLA	
90	24 12	030	30 09	087	31 31	114	51 40	162	59 38	203	69 41	264	62 30	343
91	24 40	030	31 05	088	32 22	115	51 56	164	59 15	205	68 45	264	62 13	342
92	25 09	030	32 01	088	33 13	115	52 11	165	58 51	206	67 49	265	61 55	341
93	25 37	030	32 58	088	34 04	116	52 25	167	58 25	208	66 53	265	61 36	339
94	26 05	030	33 54	089	34 55	117	52 37	169	57 58	210	65 56	266	61 16	338
95	26 33	030	34 51	089	35 45	117	52 47	170	57 29	211	65 00	266	60 54	337
96	27 02	030	35 47	089	36 35	118	52 56	172	56 59	213	64 04	267	60 32	336
97	27 30	030	36 43	090	37 24	119	53 03	173	56 28	214	63 07	267	60 08	335
98	27 58	030	37 40	090	38 14	119	53 09	175	55 55	216	62 11	268	59 43	333
99	28 26	030	38 36	090	39 03	120	53 14	176	55 22	217	61 15	268	59 18	332
100	28 53	030	39 32	091	39 51	121	53 16	178	54 47	219	60 18	268	58 51	331
101	29 21	029	40 29	091	40 39	122	53 17	180	54 11	220	59 22	269	58 23	330
102	29 49	029	41 25	092	41 27	122	53 17	181	53 34	221	58 26	269	57 55	329
103	30 16	029	42 22	092	42 15	123	53 15	183	52 56	223	57 29	269	57 26	328
104	30 44	029	43 18	092	43 02	124	53 11	184	52 18	224	56 33	270	56 56	328
	Dubhe		◆REGULUS		Alphard		SIRIUS		◆RIGEL		ALDEBARAN		◆CAPELLA	
105	31 11	029	44 14	093	43 48	125	53 06	186	51 38	225	55 37	270	56 26	327
106	31 38	029	45 11	093	44 34	126	53 00	188	50 58	226	54 40	271	55 54	326
107	32 05	029	46 07	093	45 20	127	52 51	189	50 16	227	53 44	271	55 22	325
108	32 32	028	47 03	094	46 05	127	52 42	191	49 35	229	52 47	271	54 50	324
109	32 59	028	47 59	094	46 49	128	52 30	192	48 52	230	51 51	271	54 17	324
110	33 25	028	48 56	095	47 33	129	52 18	194	48 09	231	50 55	272	53 43	323
111	33 52	028	49 52	095	48 17	130	52 03	195	47 25	232	49 58	272	53 09	322
112	34 18	028	50 48	096	48 59	131	51 48	197	46 40	233	49 02	272	52 34	322
113	34 44	027	51 44	096	49 41	133	51 31	198	45 55	234	48 06	273	51 58	321
114	35 09	027	52 40	096	50 22	134	51 12	200	45 10	234	47 09	273	51 23	320
115	35 35	027	53 36	097	51 03	135	50 53	201	44 23	235	46 13	273	50 47	320
116	36 00	027	54 32	097	51 42	136	50 32	203	43 37	236	45 17	274	50 10	319
117	36 25	026	55 28	098	52 21	137	50 09	204	42 50	237	44 21	274	49 33	319
118	36 50	026	56 24	099	52 59	138	49 46	205	42 02	238	43 24	274	48 56	318
119	37 15	026	57 19	099	53 36	140	49 21	207	41 14	239	42 28	274	48 18	318
	Dubhe		◆Denebola		Alphard		◆SIRIUS		RIGEL		ALDEBARAN		◆CAPELLA	
120	37 39	025	35 22	086	54 12	141	48 55	208	40 26	239	41 32	275	47 40	317
121	38 03	025	36 18	086	54 47	142	48 28	209	39 37	240	40 36	275	47 02	317
122	38 27	025	37 14	087	55 20	144	48 00	211	38 48	241	39 40	275	46 24	317
123	38 50	024	38 11	087	55 53	145	47 31	212	37 59	242	38 43	275	45 45	316
124	39 13	024	39 07	087	56 25	147	47 01	213	37 09	242	37 47	276	45 06	316
125	39 36	024	40 03	087	56 55	148	46 30	214	36 19	243	36 51	276	44 26	316
126	39 59	023	41 00	088	57 24	150	45 58	215	35 29	244	35 55	276	43 47	315
127	40 21	023	41 56	088	57 51	152	45 25	216	34 38	244	34 59	277	43 07	315
128	40 43	023	42 52	088	58 17	153	44 51	217	33 47	245	34 03	277	42 27	315
129	41 04	022	43 49	089	58 42	155	44 16	219	32 56	245	33 07	277	41 47	315
130	41 25	022	44 45	089	59 05	157	43 41	220	32 05	246	32 11	277	41 07	314
131	41 46	021	45 42	089	59 26	159	43 04	221	31 13	247	31 15	278	40 26	314
132	42 06	021	46 38	090	59 46	160	42 27	222	30 21	247	30 19	278	39 46	314
133	42 26	020	47 34	090	60 04	162	41 49	223	29 29	248	29 24	278	39 05	314
134	42 45	020	48 31	090	60 21	164	41 11	224	28 37	248	28 28	278	38 24	313
	Dubhe		◆ARCTURUS		SPICA		◆Alphard		SIRIUS		BETELGEUSE		◆CAPELLA	
135	43 04	019	16 30	075	17 47	109	60 35	166	40 32	224	43 29	261	37 43	313
136	43 22	019	17 24	075	18 41	110	60 48	168	39 52	225	42 33	261	37 02	313
137	43 40	018	18 19	076	19 33	110	60 58	170	39 12	226	41 38	262	36 21	313
138	43 58	018	19 14	076	20 26	111	61 07	172	38 31	227	40 42	262	35 40	313
139	44 15	017	20 08	076	21 19	111	61 14	174	37 49	228	39 46	263	34 58	313
140	44 32	017	21 03	076	22 11	112	61 18	176	37 07	229	38 50	263	34 17	313
141	44 48	016	21 58	077	23 04	112	61 21	178	36 24	230	37 54	264	33 36	313
142	45 03	016	22 53	077	23 56	113	61 22	180	35 41	230	36 58	264	32 54	313
143	45 18	015	23 48	077	24 48	113	61 20	182	34 57	231	36 02	265	32 13	312
144	45 33	015	24 43	077	25 39	114	61 17	184	34 13	232	35 06	265	31 31	312
145	45 47	014	25 38	078	26 31	114	61 12	187	33 29	233	34 09	265	30 49	312
146	46 00	013	26 33	078	27 22	115	61 04	189	32 44	233	33 13	266	30 08	312
147	46 13	013	27 28	078	28 13	115	60 55	191	31 58	234	32 17	266	29 26	312
148	46 25	012	28 23	078	29 04	116	60 44	193	31 12	235	31 21	266	28 44	312
149	46 36	012	29 19	079	29 55	117	60 30	195	30 26	235	30 24	267	28 03	312
	Dubhe		◆ARCTURUS		SPICA		◆Alphard		SIRIUS		BETELGEUSE		◆CAPELLA	
150	46 47	011	30 14	079	30 45	117	60 15	196	29 40	236	29 28	267	27 21	312
151	46 58	010	31 09	079	31 35	118	59 59	198	28 53	237	28 32	268	26 39	312
152	47 07	010	32 05	079	32 25	118	59 40	200	28 06	237	27 35	268	25 58	312
153	47 16	009	33 00	080	33 14	119	59 20	202	27 18	238	26 39	268	25 16	312
154	47 25	008	33 56	080	34 03	120	58 58	204	26 30	238	25 43	269	24 34	312
155	47 33	008	34 51	080	34 52	120	58 34	206	25 42	239	24 46	269	23 53	313
156	47 40	007	35 47	080	35 41	121	58 09	207	24 53	240	23 50	269	23 11	313
157	47 46	006	36 42	080	36 29	122	57 42	209	24 05	240	22 54	270	22 30	313
158	47 52	006	37 38	081	37 17	122	57 14	211	23 16	241	21 57	270	21 48	313
159	47 57	005	38 33	081	38 04	123	56 45	212	22 26	241	21 01	270	21 07	313
160	48 01	004	39 29	081	38 51	124	56 14	214	21 37	242	20 04	271	20 26	313
161	48 05	003	40 25	081	39 38	125	55 43	215	20 47	242	19 08	271	19 44	313
162	48 08	003	41 21	082	40 24	125	55 09	217	19 57	243	18 12	271	19 03	313
163	48 10	002	42 16	082	41 10	126	54 35	218	19 07	243	17 15	272	18 22	313
164	48 12	001	43 12	082	41 55	127	54 00	219	18 16	244	16 19	272	17 41	313
	◆Dubhe		Alkaid		ARCTURUS		◆SPICA		Suhail		◆PROCYON		POLLUX	
165	48 13	001	45 42	038	44 08	082	42 40	128	21 33	202	39 02	260	45 03	290
166	48 13	000	46 17	038	45 04	082	43 24	129	21 12	202	38 06	261	44 10	290
167	48 13	359	46 51	038	46 00	083	44 07	130	20 51	203	37 10	261	43 17	290
168	48 12	358	47 25	037	46 56	083	44 51	131	20 28	204	36 15	262	42 24	290
169	48 10	358	47 59	037	47 52	083	45 33	132	20 06	204	35 19	262	41 31	290
170	48 07	357	48 33	036	48 48	083	46 15	133	19 42	205	34 23	262	40 38	290
171	48 04	356	49 06	036	49 44	084	46 56	134	19 18	206	33 27	263	39 45	290
172	48 00	356	49 39	035	50 40	084	47 36	135	18 54	206	32 31	263	38 52	290
173	47 56	355	50 11	035	51 36	084	48 16	136	18 28	207	31 35	264	38 00	291
174	47 50	354	50 42	034	52 32	084	48 55	137	18 03	207	30 39	264	37 07	291
175	47 44	354	51 14	033	53 28	084	49 33	138	17 37	208	29 43	264	36 14	291
176	47 38	353	51 44	033	54 24	085	50 10	139	17 10	209	28 47	265	35 21	291
177	47 30	352	52 14	032	55 20	085	50 47	140	16 43	209	27 51	265	34 29	291
178	47 22	352	52 44	031	56 17	085	51 22	142	16 15	210	26 54	266	33 36	291
179	47 14	351	53 13	031	57 13	085	51 57	143	15 46	210	25 58	266	32 43	291

LAT 20°N

LHA ♈	Hc	Zn	Hc	Zn	Hc	Zn	Hc	Zn	Hc	Zn	Hc	Zn	Hc	Zn
	◆Alkaid		ANTARES		◆SPICA		Gienah		◆REGULUS		POLLUX		Dubhe	
180	53 41	030	09 57	123	52 30	144	52 18	174	61 59	258	31 51	291	47 05	350
181	54 09	029	10 45	123	53 02	146	52 23	175	61 04	258	30 58	291	46 55	350
182	54 36	028	11 32	124	53 34	147	52 26	177	60 09	259	30 06	292	46 44	349
183	55 02	027	12 18	124	54 04	148	52 29	179	59 13	260	29 13	292	46 33	348
184	55 28	026	13 05	125	54 33	150	52 29	180	58 18	260	28 21	292	46 21	348
185	55 52	026	13 51	125	55 00	151	52 28	182	57 22	261	27 29	292	46 09	347
186	56 16	025	14 37	126	55 27	153	52 26	183	56 27	261	26 36	292	45 56	346
187	56 40	024	15 22	126	55 51	155	52 22	185	55 31	262	25 44	292	45 43	346
188	57 02	023	16 08	127	56 15	156	52 16	186	54 35	263	24 52	292	45 29	345
189	57 23	022	16 53	127	56 37	158	52 09	188	53 39	263	24 00	293	45 14	345
190	57 44	021	17 37	128	56 58	159	52 01	189	52 43	264	23 08	293	44 59	344
191	58 03	020	18 22	128	57 17	161	51 50	191	51 47	264	22 16	293	44 43	344
192	58 21	019	19 06	129	57 34	163	51 39	193	50 51	264	21 24	293	44 27	343
193	58 39	017	19 50	129	57 50	165	51 26	194	49 55	265	20 33	293	44 10	342
194	58 55	016	20 33	130	58 04	166	51 12	195	48 59	265	19 41	294	43 53	342
	Kochab		◆VEGA		Rasalhague		ANTARES		◆SPICA		◆REGULUS		Dubhe	
195	33 47	009	16 46	054	24 04	085	21 16	131	58 16	168	48 02	266	43 35	341
196	33 55	008	17 31	054	25 01	085	21 59	131	58 26	170	47 06	266	43 17	341
197	34 03	008	18 17	054	25 57	085	22 41	132	58 35	172	46 10	267	42 59	340
198	34 11	008	19 03	055	26 53	086	23 23	132	58 42	174	45 14	267	42 39	340
199	34 19	008	19 49	055	27 49	086	24 04	133	58 47	176	44 17	267	42 20	340
200	34 26	007	20 35	055	28 46	086	24 45	134	58 50	178	43 21	268	42 00	339
201	34 33	007	21 21	055	29 42	087	25 26	134	58 52	180	42 25	268	41 40	339
202	34 40	007	22 07	055	30 38	087	26 06	135	58 51	181	41 28	268	41 19	338
203	34 46	006	22 54	055	31 34	087	26 46	136	58 49	183	40 32	269	40 58	338
204	34 53	006	23 40	055	32 31	088	27 25	136	58 45	185	39 35	269	40 36	337
205	34 58	006	24 27	055	33 27	088	28 04	137	58 39	187	38 39	270	40 14	337
206	35 04	005	25 13	056	34 23	088	28 42	138	58 31	189	37 43	270	39 52	337
207	35 09	005	26 00	056	35 20	089	29 20	138	58 21	191	36 46	270	39 30	336
208	35 14	005	26 46	056	36 16	089	29 57	139	58 10	193	35 50	271	39 07	336
209	35 19	005	27 33	056	37 13	089	30 34	140	57 56	194	34 54	271	38 44	336
	Kochab		◆VEGA		Rasalhague		◆ANTARES		SPICA		◆REGULUS		Dubhe	
210	35 23	004	28 19	056	38 09	090	31 10	141	57 41	196	33 57	271	38 20	335
211	35 27	004	29 06	056	39 05	090	31 45	141	57 25	198	33 01	272	37 56	335
212	35 31	004	29 53	056	40 02	090	32 20	142	57 07	200	32 04	272	37 32	335
213	35 34	003	30 40	056	40 58	091	32 54	143	56 47	201	31 08	272	37 08	334
214	35 37	003	31 26	056	41 54	091	33 28	144	56 25	203	30 12	273	36 43	334
215	35 40	003	32 13	056	42 51	091	34 01	145	56 03	205	29 15	273	36 18	334
216	35 42	002	33 00	056	43 47	092	34 33	146	55 38	206	28 19	273	35 53	333
217	35 44	002	33 47	056	44 44	092	35 04	147	55 13	208	27 23	273	35 28	333
218	35 46	002	34 33	056	45 40	092	35 35	147	54 46	209	26 27	274	35 02	333
219	35 47	001	35 20	056	46 36	093	36 05	148	54 17	211	25 30	274	34 36	333
220	35 48	001	36 07	056	47 32	093	36 34	149	53 48	212	24 34	274	34 10	332
221	35 49	001	36 54	056	48 29	094	37 03	150	53 17	214	23 38	275	33 44	332
222	35 49	000	37 40	056	49 25	094	37 30	151	52 45	215	22 42	275	33 18	332
223	35 49	000	38 27	056	50 21	094	37 57	152	52 12	216	21 46	275	32 51	332
224	35 49	000	39 14	056	51 17	095	38 23	153	51 38	218	20 49	276	32 24	332
	VEGA		◆ALTAIR		Nunki		ANTARES		◆SPICA		Denebola		◆Alkaid	
225	40 00	056	19 15	087	16 36	127	38 48	154	51 03	219	44 11	271	57 16	338
226	40 47	056	20 12	088	17 21	127	39 12	155	50 27	220	43 15	271	56 55	337
227	41 33	055	21 08	088	18 05	128	39 35	156	49 51	221	42 19	272	56 32	336
228	42 19	055	22 04	088	18 50	128	39 57	157	49 13	223	41 22	272	56 09	335
229	43 06	055	23 01	089	19 33	129	40 19	158	48 34	224	40 26	272	55 44	334
230	43 52	055	23 57	089	20 17	130	40 39	159	47 55	225	39 30	273	55 19	333
231	44 38	055	24 53	089	21 00	130	40 58	161	47 15	226	38 33	273	54 54	332
232	45 24	055	25 50	090	21 43	131	41 16	162	46 34	227	37 37	273	54 27	332
233	46 10	054	26 46	090	22 26	131	41 33	163	45 53	228	36 41	274	54 00	331
234	46 56	054	27 43	090	23 08	132	41 50	164	45 10	229	35 45	274	53 32	330
235	47 41	054	28 39	091	23 50	133	42 05	165	44 28	230	34 48	274	53 03	329
236	48 27	054	29 35	091	24 31	133	42 18	166	43 44	231	33 52	275	52 34	329
237	49 12	053	30 32	091	25 12	134	42 31	167	43 00	232	32 56	275	52 05	328
238	49 57	053	31 28	092	25 52	134	42 43	169	42 16	233	32 00	275	51 34	327
239	50 42	053	32 24	092	26 32	135	42 53	170	41 31	233	31 04	275	51 03	327
	◆VEGA		ALTAIR		Nunki		◆ANTARES		SPICA		◆ARCTURUS		Alkaid	
240	51 26	052	33 21	092	27 12	136	43 03	171	40 45	234	65 23	273	50 32	326
241	52 11	052	34 17	093	27 51	137	43 11	172	39 59	235	64 27	273	50 00	325
242	52 55	051	35 13	093	28 29	137	43 18	174	39 13	236	63 31	273	49 28	325
243	53 39	051	36 10	094	29 08	138	43 24	175	38 26	237	62 34	273	48 55	324
244	54 23	050	37 06	094	29 45	139	43 28	176	37 39	237	61 38	274	48 22	324
245	55 06	050	38 02	094	30 22	139	43 32	177	36 51	238	60 42	274	47 48	323
246	55 49	049	38 58	095	30 58	140	43 34	178	36 03	239	59 45	274	47 14	323
247	56 32	049	39 54	095	31 34	141	43 35	180	35 15	239	58 49	274	46 40	322
248	57 14	048	40 51	096	32 09	142	43 34	181	34 26	240	57 53	275	46 05	322
249	57 56	048	41 47	096	32 44	143	43 33	182	33 37	241	56 57	275	45 30	321
250	58 37	047	42 43	096	33 18	143	43 30	183	32 48	241	56 01	275	44 55	321
251	59 18	046	43 39	097	33 51	144	43 26	185	31 58	242	55 04	275	44 19	321
252	59 58	045	44 35	097	34 24	145	43 21	186	31 08	243	54 08	275	43 43	320
253	60 38	045	45 31	098	34 56	146	43 15	187	30 18	243	53 12	276	43 07	320
254	61 18	044	46 26	098	35 27	147	43 07	188	29 28	244	52 16	276	42 31	320
	◆DENEB		ALTAIR		◆Nunki		ANTARES		SPICA		◆ARCTURUS		Alkaid	
255	38 16	047	47 22	099	35 57	148	42 59	189	28 37	244	51 20	276	41 54	319
256	38 58	047	48 18	099	36 27	149	42 49	191	27 46	245	50 24	276	41 17	319
257	39 39	047	49 13	100	36 56	150	42 38	192	26 55	245	49 28	276	40 40	319
258	40 20	047	50 09	100	37 24	151	42 26	193	26 03	246	48 32	277	40 03	318
259	41 01	047	51 04	101	37 51	152	42 12	194	25 12	247	47 36	277	39 25	318
260	41 42	047	52 00	102	38 17	153	41 58	195	24 20	247	46 40	277	38 48	318
261	42 23	046	52 55	102	38 43	154	41 42	197	23 28	248	45 44	277	38 10	318
262	43 04	046	53 50	103	39 07	155	41 26	198	22 36	248	44 48	278	37 32	318
263	43 44	046	54 45	103	39 31	156	41 08	199	21 43	249	43 52	278	36 54	317
264	44 25	045	55 40	104	39 54	157	40 50	200	20 51	249	42 56	278	36 16	317
265	45 05	045	56 34	105	40 15	158	40 30	201	19 58	249	42 01	278	35 37	317
266	45 44	045	57 29	105	40 36	159	40 09	202	19 05	250	41 05	278	34 59	317
267	46 24	044	58 23	106	40 56	160	39 47	203	18 12	250	40 09	279	34 20	317
268	47 03	044	59 17	107	41 15	161	39 25	204	17 19	251	39 13	279	33 42	317
269	47 43	044	60 11	108	41 32	162	39 01	205	16 26	251	38 18	279	33 03	317

LAT 20°N

LHA ♈	Hc	Zn	Hc	Zn	Hc	Zn	Hc	Zn	Hc	Zn	Hc	Zn	Hc	Zn
	◆DENEB		Enif		◆Nunki		ANTARES		◆ARCTURUS		Alkaid		Kochab	
270	48 21	043	35 12	092	41 49	163	38 37	206	37 22	279	32 24	316	30 11	347
271	49 00	043	36 08	092	42 05	165	38 11	207	36 26	280	31 45	316	29 58	346
272	49 38	042	37 05	093	42 19	166	37 45	208	35 31	280	31 06	316	29 44	346
273	50 16	042	38 01	093	42 32	167	37 18	209	34 35	280	30 27	316	29 31	346
274	50 53	041	38 57	093	42 45	168	36 50	210	33 40	280	29 48	316	29 17	346
275	51 30	041	39 54	094	42 56	169	36 21	211	32 44	280	29 09	316	29 03	346
276	52 07	040	40 50	094	43 06	170	35 52	212	31 49	281	28 30	316	28 49	346
277	52 43	040	41 46	095	43 14	172	35 21	213	30 54	281	27 51	316	28 35	345
278	53 19	039	42 42	095	43 22	173	34 50	214	29 58	281	27 12	316	28 21	345
279	53 54	038	43 38	095	43 28	174	34 18	215	29 03	281	26 33	316	28 07	345
280	54 29	038	44 34	096	43 33	175	33 46	216	28 08	282	25 54	316	27 52	345
281	55 03	037	45 31	096	43 37	177	33 13	216	27 13	282	25 15	316	27 37	345
282	55 37	036	46 27	097	43 40	178	32 39	217	26 17	282	24 36	316	27 22	345
283	56 10	035	47 22	097	43 41	179	32 04	218	25 22	282	23 57	316	27 08	345
284	56 42	035	48 18	098	43 42	180	31 29	219	24 27	283	23 18	316	26 52	344
	◆DENEB		Alpheratz		◆FOMALHAUT		Nunki		ANTARES		◆Rasalhague		Kochab	
285	57 14	034	20 31	065	14 19	130	43 41	182	30 54	220	68 15	253	26 37	344
286	57 45	033	21 22	066	15 02	130	43 39	183	30 17	220	67 21	254	26 22	344
287	58 15	032	22 14	066	15 45	131	43 35	184	29 40	221	66 26	255	26 07	344
288	58 44	031	23 05	066	16 28	131	43 31	185	29 03	222	65 32	256	25 51	344
289	59 13	030	23 57	066	17 10	132	43 25	186	28 25	223	64 37	257	25 36	344
290	59 41	029	24 48	066	17 52	132	43 18	188	27 47	223	63 42	258	25 20	344
291	60 08	028	25 40	067	18 34	133	43 10	189	27 08	224	62 47	258	25 04	344
292	60 34	027	26 32	067	19 15	133	43 00	190	26 28	225	61 51	259	24 48	344
293	60 58	026	27 23	067	19 56	134	42 50	191	25 48	225	60 56	260	24 32	344
294	61 22	024	28 15	067	20 37	134	42 38	193	25 08	226	60 01	260	24 17	344
295	61 45	023	29 07	067	21 17	135	42 25	194	24 27	227	59 05	261	24 01	343
296	62 06	022	29 59	067	21 56	136	42 12	195	23 46	227	58 09	262	23 44	343
297	62 27	021	30 51	067	22 36	136	41 57	196	23 04	228	57 13	262	23 28	343
298	62 46	019	31 43	067	23 14	137	41 40	197	22 22	229	56 17	263	23 12	343
299	63 04	018	32 35	068	23 53	137	41 23	198	21 40	229	55 22	263	22 56	343
	◆DENEB		Schedar		Alpheratz		◆FOMALHAUT		Nunki		◆Rasalhague		VEGA	
300	63 20	016	27 31	036	33 28	068	24 31	138	41 05	199	54 26	264	64 02	321
301	63 35	015	28 04	036	34 20	068	25 08	139	40 46	201	53 29	264	63 25	320
302	63 49	013	28 37	036	35 12	068	25 45	139	40 25	202	52 33	265	62 49	319
303	64 01	012	29 10	036	36 04	068	26 21	140	40 04	203	51 37	265	62 11	318
304	64 12	010	29 43	035	36 57	068	26 57	141	39 42	204	50 41	265	61 33	317
305	64 22	009	30 16	035	37 49	068	27 32	142	39 19	205	49 45	266	60 54	316
306	64 29	007	30 48	035	38 41	068	28 07	142	38 54	206	48 49	266	60 14	315
307	64 35	005	31 21	035	39 34	068	28 41	143	38 29	207	47 52	267	59 34	314
308	64 40	004	31 53	035	40 26	068	29 15	144	38 03	208	46 56	267	58 53	313
309	64 43	002	32 25	035	41 18	068	29 48	145	37 37	209	46 00	267	58 12	313
310	64 44	001	32 58	035	42 11	068	30 20	145	37 09	210	45 03	268	57 30	312
311	64 44	359	33 30	035	43 03	068	30 52	146	36 40	211	44 07	268	56 48	311
312	64 42	357	34 02	034	43 56	068	31 23	147	36 11	212	43 11	269	56 05	311
313	64 38	356	34 33	034	44 48	068	31 53	148	35 41	213	42 14	269	55 23	310
314	64 33	354	35 05	034	45 41	068	32 23	149	35 10	214	41 18	269	54 39	310
	◆Alpheratz		Diphda		◆FOMALHAUT		Nunki		Rasalhague		◆VEGA		DENEB	
315	46 33	068	23 20	121	32 52	150	34 39	214	40 21	270	53 56	309	64 27	352
316	47 25	068	24 08	122	33 20	150	34 07	215	39 25	270	53 12	309	64 18	351
317	48 18	068	24 56	122	33 48	151	33 34	216	38 29	270	52 28	308	64 08	349
318	49 10	068	25 44	123	34 14	152	33 00	217	37 32	271	51 44	308	63 57	348
319	50 03	068	26 31	123	34 40	153	32 26	218	36 36	271	50 59	308	63 44	346
320	50 55	068	27 18	124	35 05	154	31 51	219	35 40	271	50 14	307	63 30	345
321	51 47	068	28 04	125	35 30	155	31 15	219	34 43	272	49 29	307	63 14	343
322	52 40	068	28 51	125	35 53	156	30 39	220	33 47	272	48 44	307	62 57	342
323	53 32	068	29 37	126	36 16	157	30 03	221	32 51	272	47 59	306	62 39	340
324	54 24	068	30 22	126	36 37	158	29 25	222	31 54	273	47 13	306	62 19	339
325	55 16	068	31 07	127	36 58	159	28 48	222	30 58	273	46 27	306	61 59	338
326	56 08	067	31 52	128	37 18	160	28 09	223	30 02	273	45 42	306	61 37	336
327	57 00	067	32 36	129	37 37	161	27 31	224	29 05	274	44 56	305	61 14	335
328	57 52	067	33 20	129	37 55	162	26 51	225	28 09	274	44 10	305	60 49	334
329	58 44	067	34 04	130	38 12	163	26 12	225	27 13	274	43 23	305	60 24	333
	◆Mirfak		Hamal		Diphda		◆FOMALHAUT		ALTAIR		◆VEGA		DENEB	
330	20 52	043	32 59	074	34 46	131	38 28	164	56 49	255	42 37	305	59 58	332
331	21 31	043	33 54	075	35 29	132	38 43	165	55 54	256	41 51	305	59 31	331
332	22 09	043	34 48	075	36 11	132	38 57	166	55 00	256	41 04	305	59 03	330
333	22 48	043	35 43	075	36 52	133	39 10	167	54 05	257	40 18	304	58 34	329
334	23 26	043	36 37	075	37 33	134	39 22	168	53 10	258	39 31	304	58 04	328
335	24 05	043	37 31	075	38 13	135	39 32	169	52 14	258	38 45	304	57 33	327
336	24 44	043	38 26	075	38 53	136	39 42	171	51 19	259	37 58	304	57 02	326
337	25 22	043	39 21	076	39 32	137	39 51	172	50 24	259	37 12	304	56 30	325
338	26 01	043	40 15	076	40 10	138	39 59	173	49 28	260	36 25	304	55 58	324
339	26 40	043	41 10	076	40 48	138	40 05	174	48 33	261	35 38	304	55 24	323
340	27 19	043	42 05	076	41 25	139	40 11	175	47 37	261	34 51	304	54 50	323
341	27 57	043	42 59	076	42 01	140	40 15	176	46 41	262	34 05	304	54 16	322
342	28 36	043	43 54	076	42 37	141	40 18	177	45 46	262	33 18	304	53 41	321
343	29 15	043	44 49	076	43 12	142	40 20	178	44 50	262	32 31	304	53 06	321
344	29 53	043	45 44	077	43 46	144	40 21	180	43 54	263	31 44	304	52 30	320
	Schedar		Mirfak		◆ALDEBARAN		Diphda		◆FOMALHAUT		ALTAIR		◆DENEB	
345	49 01	021	30 32	043	11 07	076	44 19	145	40 21	181	42 58	263	51 53	320
346	49 21	020	31 10	043	12 02	077	44 51	146	40 20	182	42 02	264	51 17	319
347	49 40	020	31 49	043	12 57	077	45 22	147	40 17	183	41 06	264	50 39	318
348	49 59	019	32 27	043	13 51	077	45 53	148	40 14	184	40 10	265	50 02	318
349	50 17	018	33 05	043	14 46	077	46 22	149	40 09	185	39 13	265	49 24	317
350	50 34	017	33 44	043	15 42	078	46 50	150	40 03	186	38 17	265	48 46	317
351	50 50	017	34 22	043	16 37	078	47 18	152	39 56	188	37 21	266	48 07	317
352	51 06	016	35 00	042	17 32	078	47 44	153	39 48	189	36 25	266	47 28	316
353	51 21	015	35 38	042	18 27	079	48 09	154	39 39	190	35 29	267	46 49	316
354	51 35	014	36 16	042	19 22	079	48 33	155	39 29	191	34 32	267	46 09	315
355	51 49	013	36 54	042	20 18	079	48 56	157	39 18	192	33 36	267	45 30	315
356	52 01	013	37 31	042	21 13	079	49 18	158	39 06	193	32 40	268	44 50	315
357	52 13	012	38 09	042	22 09	080	49 38	159	38 53	194	31 43	268	44 10	314
358	52 24	011	38 46	041	23 04	080	49 57	161	38 38	195	30 47	269	43 29	314
359	52 34	010	39 23	041	24 00	080	50 15	162	38 23	196	29 51	269	42 49	314

LAT 19°N

LAT 19°N

LHA ♈	Hc	Zn	Hc	Zn	Hc	Zn	Hc	Zn	Hc	Zn	Hc	Zn	Hc	Zn
	Schedar		CAPELLA		◆ALDEBARAN		ACHERNAR		◆FOMALHAUT		ALTAIR		◆DENEB	
0	51 45	009	21 01	047	24 45	080	11 04	167	39 04	198	28 55	270	41 26	314
1	51 53	008	21 42	047	25 41	080	11 16	167	38 46	199	27 58	270	40 46	314
2	52 00	007	22 24	047	26 37	081	11 29	168	38 28	200	27 01	270	40 05	314
3	52 07	006	23 05	047	27 33	081	11 40	168	38 08	201	26 04	271	39 24	314
4	52 13	005	23 47	047	28 29	081	11 51	169	37 47	202	25 08	271	38 43	313
5	52 18	005	24 29	047	29 25	081	12 02	169	37 26	203	24 11	271	38 01	313
6	52 22	004	25 10	047	30 21	082	12 12	170	37 04	204	23 14	272	37 20	313
7	52 25	003	25 52	047	31 17	082	12 22	170	36 40	205	22 18	272	36 38	313
8	52 28	002	26 34	047	32 14	082	12 31	171	36 16	206	21 21	272	35 57	313
9	52 29	001	27 15	047	33 10	082	12 40	172	35 51	207	20 24	273	35 15	313
10	52 29	000	27 57	047	34 06	083	12 48	172	35 26	208	19 28	273	34 33	312
11	52 29	359	28 39	047	35 02	083	12 55	173	34 59	208	18 31	273	33 51	312
12	52 28	358	29 20	047	35 59	083	13 02	173	34 32	209	17 34	274	33 09	312
13	52 26	357	30 02	047	36 55	083	13 09	174	34 03	210	16 38	274	32 27	312
14	52 22	356	30 44	047	37 51	084	13 15	174	33 35	211	15 41	274	31 45	312
	CAPELLA		◆ALDEBARAN		RIGEL		ACHERNAR		◆FOMALHAUT		Enif		◆DENEB	
15	31 25	047	38 48	084	21 43	107	13 20	175	33 05	212	41 49	266	31 03	312
16	32 07	047	39 44	084	22 37	108	13 25	175	32 35	213	40 52	267	30 21	312
17	32 48	047	40 40	084	23 31	108	13 29	176	32 03	214	39 55	267	29 39	312
18	33 30	047	41 37	084	24 25	109	13 33	176	31 32	214	38 59	267	28 57	312
19	34 11	047	42 33	085	25 18	109	13 36	177	30 59	215	38 02	268	28 14	312
20	34 52	047	43 30	085	26 12	110	13 39	178	30 26	216	37 05	268	27 32	312
21	35 33	046	44 26	085	27 05	110	13 41	178	29 53	217	36 09	268	26 50	312
22	36 14	046	45 23	085	27 58	111	13 43	179	29 18	218	35 12	269	26 08	312
23	36 55	046	46 20	086	28 51	111	13 44	179	28 44	218	34 15	269	25 26	312
24	37 36	046	47 16	086	29 44	112	13 44	180	28 08	219	33 19	269	24 43	312
25	38 17	046	48 13	086	30 37	112	13 44	180	27 32	220	32 22	270	24 01	312
26	38 58	046	49 09	087	31 29	113	13 44	181	26 56	220	31 25	270	23 19	312
27	39 38	045	50 06	087	32 21	113	13 42	181	26 19	221	30 28	270	22 37	312
28	40 18	045	51 03	087	33 13	114	13 41	182	25 41	222	29 32	271	21 55	312
29	40 58	045	51 59	087	34 05	115	13 38	183	25 03	222	28 35	271	21 13	312
	Mirfak		◆CAPELLA		ALDEBARAN		RIGEL		◆ACHERNAR		FOMALHAUT		◆Alpheratz	
30	54 54	024	41 38	045	52 56	088	34 56	115	13 36	183	24 25	223	62 38	297
31	55 16	023	42 18	044	53 53	088	35 48	116	13 32	184	23 46	224	61 47	296
32	55 38	022	42 58	044	54 49	088	36 39	116	13 28	184	23 06	224	60 56	296
33	55 58	021	43 37	044	55 46	088	37 29	117	13 24	185	22 27	225	60 05	296
34	56 18	020	44 16	043	56 43	089	38 20	118	13 19	185	21 46	226	59 14	295
35	56 37	019	44 55	043	57 39	089	39 10	118	13 13	186	21 06	226	58 23	295
36	56 55	018	45 33	043	58 36	089	39 59	119	13 07	186	20 25	227	57 31	295
37	57 11	017	46 12	042	59 33	090	40 49	120	13 01	187	19 43	227	56 40	294
38	57 27	016	46 50	042	60 30	090	41 38	121	12 53	188	19 01	228	55 48	294
39	57 42	015	47 28	041	61 26	090	42 27	121	12 46	188	18 19	228	54 56	294
40	57 56	013	48 05	041	62 23	091	43 15	122	12 38	189	17 37	229	54 04	294
41	58 08	012	48 42	040	63 20	091	44 03	123	12 29	189	16 54	229	53 12	293
42	58 20	011	49 19	040	64 17	091	44 50	124	12 20	190	16 10	230	52 20	293
43	58 30	010	49 55	039	65 13	092	45 37	125	12 10	190	15 27	230	51 28	293
44	58 39	009	50 31	039	66 10	092	46 24	125	12 00	191	14 43	231	50 36	293
	CAPELLA		◆POLLUX		PROCYON		SIRIUS		◆Acamar		Diphda		◆Alpheratz	
45	51 06	038	24 55	067	20 49	092	24 11	119	30 40	180	40 02	224	49 43	293
46	51 41	038	25 47	067	21 46	092	25 01	120	30 40	181	39 22	225	48 51	293
47	52 15	037	26 40	067	22 43	092	25 50	120	30 38	182	38 42	226	47 59	293
48	52 49	036	27 32	068	23 39	093	26 39	121	30 35	183	38 01	227	47 06	293
49	53 23	036	28 24	068	24 36	093	27 27	121	30 32	184	37 19	228	46 14	293
50	53 56	035	29 17	068	25 33	093	28 16	122	30 27	185	36 37	228	45 21	292
51	54 28	034	30 10	068	26 29	094	29 03	123	30 22	186	35 54	229	44 29	292
52	54 59	033	31 02	068	27 26	094	29 51	123	30 16	187	35 11	230	43 37	292
53	55 30	033	31 55	068	28 23	094	30 38	124	30 09	187	34 27	231	42 44	292
54	56 01	032	32 47	068	29 19	095	31 25	125	30 01	188	33 43	231	41 52	292
55	56 30	031	33 40	068	30 16	095	32 12	125	29 53	189	32 59	232	40 59	292
56	56 59	030	34 33	068	31 12	095	32 58	126	29 43	190	32 14	233	40 07	292
57	57 27	029	35 26	069	32 09	096	33 44	127	29 33	191	31 28	234	39 14	292
58	57 54	028	36 19	069	33 05	096	34 29	127	29 22	192	30 42	234	38 22	292
59	58 20	027	37 11	069	34 01	097	35 14	128	29 10	193	29 56	235	37 30	293
	CAPELLA		◆POLLUX		PROCYON		SIRIUS		◆CANOPUS		Diphda		◆Hamal	
60	58 45	026	38 04	069	34 58	097	35 59	129	11 49	159	29 10	236	63 19	285
61	59 10	025	38 57	069	35 54	097	36 43	129	12 10	159	28 23	236	62 24	284
62	59 33	024	39 50	069	36 50	098	37 26	130	12 30	160	27 35	237	61 29	284
63	59 55	022	40 43	069	37 46	098	38 09	131	12 49	160	26 48	237	60 34	284
64	60 17	021	41 36	069	38 42	099	38 52	132	13 08	161	26 00	238	59 39	284
65	60 37	020	42 29	069	39 38	099	39 34	133	13 27	161	25 12	238	58 44	284
66	60 56	019	43 22	069	40 34	100	40 15	134	13 44	162	24 23	239	57 49	284
67	61 13	018	44 15	069	41 30	100	40 56	135	14 02	162	23 35	240	56 54	284
68	61 30	016	45 08	069	42 26	101	41 36	135	14 19	163	22 46	240	55 59	284
69	61 45	015	46 01	069	43 22	101	42 15	136	14 35	164	21 56	241	55 04	284
70	61 59	013	46 54	069	44 17	102	42 54	137	14 51	164	21 07	241	54 09	284
71	62 12	012	47 47	069	45 13	102	43 32	138	15 06	165	20 17	242	53 14	284
72	62 23	011	48 40	069	46 08	103	44 10	139	15 21	165	19 27	242	52 19	284
73	62 33	009	49 32	069	47 04	103	44 46	140	15 35	166	18 37	243	51 24	284
74	62 41	008	50 25	069	47 59	104	45 22	141	15 49	166	17 46	243	50 29	284
	CAPELLA		◆POLLUX		PROCYON		SIRIUS		◆RIGEL		Menkar		◆Hamal	
75	62 48	006	51 18	069	48 54	104	45 57	142	62 34	172	57 34	246	49 34	284
76	62 53	005	52 11	068	49 49	105	46 31	144	62 40	174	56 42	247	48 39	284
77	62 57	003	53 04	068	50 43	106	47 04	145	62 45	177	55 49	248	47 44	284
78	63 00	002	53 56	068	51 38	106	47 37	146	62 47	179	54 57	249	46 49	284
79	63 00	000	54 49	068	52 32	107	48 08	147	62 47	181	54 03	250	45 54	284
80	63 00	359	55 42	068	53 27	108	48 38	148	62 45	183	53 10	250	44 59	285
81	62 58	357	56 34	068	54 21	108	49 07	150	62 41	185	52 17	251	44 04	285
82	62 54	356	57 26	067	55 14	109	49 36	151	62 35	187	51 23	252	43 09	285
83	62 49	354	58 19	067	56 08	110	50 03	152	62 27	190	50 29	253	42 14	285
84	62 42	353	59 11	067	57 01	111	50 28	154	62 16	192	49 35	253	41 19	285
85	62 34	351	60 03	067	57 54	112	50 53	155	62 04	194	48 40	254	40 24	285
86	62 25	350	60 55	066	58 47	112	51 17	156	61 50	196	47 46	254	39 30	285
87	62 14	348	61 47	066	59 39	113	51 39	158	61 33	198	46 51	255	38 35	285
88	62 01	347	62 39	065	60 31	114	52 00	159	61 15	200	45 56	256	37 40	285
89	61 48	345	63 30	065	61 22	115	52 19	161	60 55	202	45 01	256	36 46	286

LHA ♈	Hc	Zn	Hc	Zn	Hc	Zn	Hc	Zn	Hc	Zn	Hc	Zn	Hc	Zn
	Dubhe		◆REGULUS		Alphard		SIRIUS		◆RIGEL		ALDEBARAN		◆CAPELLA	
90	23 20	030	30 06	087	31 55	114	52 37	162	60 33	204	69 46	266	61 33	344
91	23 48	030	31 03	087	32 47	114	52 54	164	60 10	205	68 50	267	61 16	343
92	24 17	030	31 59	087	33 39	115	53 09	165	59 45	207	67 53	267	60 59	341
93	24 45	030	32 56	088	34 30	115	53 23	167	59 18	209	66 56	268	60 40	340
94	25 13	030	33 53	088	35 21	116	53 35	168	58 50	211	66 00	268	60 20	339
95	25 41	030	34 49	088	36 12	117	53 46	170	58 20	212	65 03	268	59 59	338
96	26 09	030	35 46	089	37 03	117	53 55	171	57 49	214	64 06	269	59 37	336
97	26 38	030	36 43	089	37 53	118	54 03	173	57 17	215	63 09	269	59 14	335
98	27 05	029	37 39	089	38 43	119	54 09	175	56 44	217	62 13	269	58 49	334
99	27 33	029	38 36	090	39 32	119	54 13	176	56 09	218	61 16	270	58 24	333
100	28 01	029	39 33	090	40 22	120	54 16	178	55 33	220	60 19	270	57 58	332
101	28 29	029	40 30	090	41 11	121	54 17	180	54 57	221	59 23	270	57 31	331
102	28 56	029	41 26	091	41 59	122	54 17	181	54 19	222	58 26	271	57 03	330
103	29 24	029	42 23	091	42 47	122	54 15	183	53 40	224	57 29	271	56 35	329
104	29 51	029	43 20	091	43 35	123	54 11	185	53 00	225	56 32	271	56 05	328
	Dubhe		◆REGULUS		Alphard		SIRIUS		◆RIGEL		ALDEBARAN		◆CAPELLA	
105	30 18	029	44 16	092	44 22	124	54 06	186	52 20	226	55 36	272	55 35	327
106	30 45	028	45 13	092	45 09	125	53 59	188	51 39	227	54 39	272	55 04	327
107	31 12	028	46 10	092	45 55	126	53 51	189	50 57	228	53 42	272	54 33	326
108	31 39	028	47 07	093	46 41	127	53 41	191	50 14	229	52 46	272	54 01	325
109	32 06	028	48 03	093	47 26	128	53 29	193	49 30	231	51 49	273	53 28	324
110	32 32	028	49 00	094	48 11	129	53 16	194	48 46	232	50 52	273	52 55	324
111	32 58	027	49 56	094	48 55	130	53 01	196	48 02	233	49 56	273	52 21	323
112	33 24	027	50 53	094	49 38	131	52 45	197	47 16	233	48 59	274	51 46	322
113	33 50	027	51 50	095	50 21	132	52 28	199	46 30	234	48 02	274	51 12	322
114	34 16	027	52 46	095	51 03	133	52 09	200	45 44	235	47 06	274	50 36	321
115	34 41	026	53 43	096	51 45	134	51 49	202	44 57	236	46 09	274	50 01	321
116	35 07	026	54 39	096	52 25	135	51 27	203	44 10	237	45 13	275	49 24	320
117	35 31	026	55 35	097	53 05	136	51 04	204	43 22	238	44 16	275	48 48	320
118	35 56	026	56 32	097	53 43	138	50 40	206	42 34	239	43 20	275	48 11	319
119	36 21	025	57 28	098	54 21	139	50 15	207	41 45	239	42 23	275	47 33	319
	Dubhe		◆Denebola		Alphard		◆SIRIUS		RIGEL		ALDEBARAN		◆CAPELLA	
120	36 45	025	35 17	085	54 58	140	49 48	209	40 56	240	41 27	276	46 56	318
121	37 09	025	36 14	086	55 34	142	49 21	210	40 07	241	40 30	276	46 18	318
122	37 32	024	37 10	086	56 09	143	48 52	211	39 17	242	39 34	276	45 40	317
123	37 56	024	38 07	086	56 42	144	48 22	212	38 27	242	38 37	276	45 01	317
124	38 19	024	39 04	086	57 15	146	47 51	214	37 37	243	37 41	277	44 22	317
125	38 41	023	40 00	087	57 46	148	47 19	215	36 46	244	36 45	277	43 43	316
126	39 03	023	40 57	087	58 16	149	46 47	216	35 55	244	35 48	277	43 04	316
127	39 25	023	41 54	087	58 44	151	46 13	217	35 04	245	34 52	277	42 24	316
128	39 47	022	42 50	087	59 11	153	45 38	218	34 13	245	33 56	277	41 45	315
129	40 08	022	43 47	088	59 36	154	45 03	219	33 21	246	32 59	278	41 05	315
130	40 29	021	44 44	088	60 00	156	44 27	220	32 29	247	32 03	278	40 25	315
131	40 50	021	45 40	088	60 22	158	43 50	221	31 37	247	31 07	278	39 44	315
132	41 10	020	46 37	089	60 43	160	43 12	222	30 44	248	30 11	278	39 04	314
133	41 29	020	47 34	089	61 01	162	42 33	223	29 52	248	29 15	279	38 23	314
134	41 49	020	48 30	089	61 18	164	41 54	224	28 59	249	28 19	279	37 43	314
	Dubhe		◆ARCTURUS		SPICA		◆Alphard		SIRIUS		BETELGEUSE		◆CAPELLA	
135	42 07	019	16 14	075	18 07	109	61 33	166	41 14	225	43 38	262	37 02	314
136	42 26	019	17 09	075	19 01	110	61 46	168	40 34	226	42 42	262	36 21	314
137	42 43	018	18 04	075	19 54	110	61 57	170	39 53	227	41 46	263	35 40	314
138	43 01	018	18 59	076	20 47	110	62 06	172	39 11	228	40 49	263	34 59	313
139	43 18	017	19 54	076	21 40	111	62 13	174	38 29	229	39 53	264	34 17	313
140	43 34	017	20 49	076	22 33	111	62 18	176	37 46	229	38 57	264	33 36	313
141	43 50	016	21 44	076	23 26	112	62 21	178	37 03	230	38 00	264	32 55	313
142	44 05	015	22 39	077	24 19	112	62 22	180	36 19	231	37 04	265	32 13	313
143	44 20	015	23 34	077	25 11	113	62 20	182	35 35	232	36 07	265	31 32	313
144	44 35	014	24 30	077	26 03	113	62 17	185	34 50	232	35 11	266	30 50	313
145	44 48	014	25 25	077	26 55	114	62 11	187	34 05	233	34 14	266	30 09	313
146	45 01	013	26 20	077	27 47	114	62 04	189	33 19	234	33 17	266	29 27	313
147	45 14	013	27 16	078	28 39	115	61 54	191	32 33	235	32 21	267	28 45	313
148	45 26	012	28 11	078	29 30	115	61 42	193	31 47	235	31 24	267	28 04	313
149	45 38	011	29 06	078	30 21	116	61 28	195	31 00	236	30 27	267	27 22	313
	Dubhe		◆ARCTURUS		SPICA		◆Alphard		SIRIUS		BETELGEUSE		◆CAPELLA	
150	45 48	011	30 02	078	31 12	117	61 13	197	30 13	236	29 31	268	26 40	313
151	45 59	010	30 58	079	32 03	117	60 55	199	29 26	237	28 34	268	25 59	313
152	46 08	009	31 53	079	32 53	118	60 36	201	28 38	238	27 37	268	25 17	313
153	46 17	009	32 49	079	33 43	118	60 15	203	27 50	238	26 41	269	24 35	313
154	46 25	008	33 45	079	34 33	119	59 52	205	27 01	239	25 44	269	23 54	313
155	46 33	007	34 40	079	35 22	120	59 28	206	26 13	239	24 47	269	23 12	313
156	46 40	007	35 36	080	36 11	120	59 02	208	25 24	240	23 50	270	22 31	313
157	46 46	006	36 32	080	37 00	121	58 35	210	24 34	241	22 54	270	21 49	313
158	46 52	005	37 28	080	37 49	122	58 06	211	23 45	241	21 57	270	21 08	313
159	46 57	005	38 24	080	38 37	122	57 36	213	22 55	242	21 00	271	20 26	313
160	47 02	004	39 20	080	39 24	123	57 04	215	22 05	242	20 03	271	19 45	313
161	47 05	003	40 15	081	40 12	124	56 31	216	21 15	243	19 07	271	19 03	313
162	47 08	003	41 11	081	40 58	125	55 57	218	20 24	243	18 10	272	18 22	313
163	47 11	002	42 07	081	41 45	126	55 22	219	19 34	244	17 13	272	17 41	314
164	47 12	001	43 03	081	42 31	126	54 46	220	18 43	244	16 17	272	17 00	314
	◆Dubhe		Alkaid		ARCTURUS		◆SPICA		Suhail		◆PROCYON		POLLUX	
165	47 13	001	44 55	038	44 00	081	43 16	127	22 29	202	39 12	261	44 42	291
166	47 13	000	45 29	037	44 56	081	44 01	128	22 08	202	38 15	261	43 49	291
167	47 13	359	46 04	037	45 52	082	44 46	129	21 46	203	37 19	262	42 56	291
168	47 12	359	46 37	037	46 48	082	45 29	130	21 23	204	36 23	262	42 03	291
169	47 10	358	47 11	036	47 44	082	46 13	131	21 00	204	35 27	263	41 10	291
170	47 07	357	47 44	036	48 40	082	46 55	132	20 37	205	34 31	263	40 17	291
171	47 04	356	48 17	035	49 36	082	47 37	133	20 12	206	33 34	264	39 24	291
172	47 00	356	48 49	034	50 33	083	48 18	134	19 47	206	32 38	264	38 31	291
173	46 56	355	49 21	034	51 29	083	48 59	135	19 22	207	31 41	264	37 38	291
174	46 51	354	49 52	033	52 25	083	49 39	136	18 56	208	30 45	265	36 45	291
175	46 45	354	50 23	033	53 22	083	50 17	137	18 29	208	29 48	265	35 52	291
176	46 38	353	50 54	032	54 18	083	50 56	138	18 02	209	28 52	265	35 00	291
177	46 31	352	51 23	031	55 14	083	51 33	140	17 35	209	27 55	266	34 07	292
178	46 23	352	51 53	031	56 11	084	52 09	141	17 07	210	26 59	266	33 14	292
179	46 15	351	52 21	030	57 07	084	52 44	142	16 38	211	26 02	267	32 21	292

 LAT 19°N

LHA ♈	Hc	Zn	Hc	Zn	Hc	Zn	Hc	Zn	Hc	Zn	Hc	Zn	Hc	Zn
	◆Alkaid		ANTARES		◆SPICA		Gienah		◆REGULUS		POLLUX		Dubhe	
180	52 49	029	10 30	123	53 19	144	53 17	174	62 11	260	31 29	292	46 05	350
181	53 16	028	11 18	123	53 52	145	53 23	175	61 15	260	30 36	292	45 56	350
182	53 43	028	12 05	124	54 24	146	53 26	177	60 19	261	29 43	292	45 45	349
183	54 09	027	12 52	124	54 55	148	53 29	179	59 23	261	28 51	292	45 34	349
184	54 34	026	13 39	125	55 24	149	53 29	180	58 27	262	27 58	292	45 23	348
185	54 58	025	14 25	125	55 53	151	53 28	182	57 31	262	27 06	292	45 10	347
186	55 22	024	15 12	126	56 20	152	53 26	183	56 35	263	26 14	293	44 58	347
187	55 44	023	15 58	126	56 46	154	53 21	185	55 38	263	25 21	293	44 44	346
188	56 06	022	16 43	127	57 10	156	53 16	187	54 42	264	24 29	293	44 30	346
189	56 27	021	17 29	127	57 32	157	53 09	188	53 46	264	23 37	293	44 16	345
190	56 47	020	18 14	128	57 54	159	53 00	190	52 49	265	22 45	293	44 01	344
191	57 06	019	18 59	128	58 13	161	52 49	191	51 53	265	21 53	293	43 46	344
192	57 24	018	19 43	129	58 31	162	52 38	193	50 56	266	21 00	294	43 29	343
193	57 41	017	20 28	129	58 47	164	52 24	194	49 59	266	20 09	294	43 13	343
194	57 58	016	21 11	130	59 02	166	52 09	196	49 03	266	19 17	294	42 56	342
	Kochab		◆VEGA		Rasalhague		ANTARES		◆SPICA		◆REGULUS		Dubhe	
195	32 47	009	16 10	054	23 59	084	21 55	130	59 15	168	48 06	267	42 38	342
196	32 56	008	16 56	054	24 55	085	22 38	131	59 26	170	47 10	267	42 20	341
197	33 04	008	17 42	054	25 52	085	23 21	131	59 35	172	46 13	268	42 02	341
198	33 12	008	18 28	054	26 48	085	24 03	132	59 42	174	45 16	268	41 43	340
199	33 19	008	19 14	054	27 45	085	24 45	133	59 47	176	44 19	268	41 24	340
200	33 27	007	20 00	055	28 41	086	25 27	133	59 50	178	43 23	269	41 04	339
201	33 34	007	20 47	055	29 38	086	26 08	134	59 52	180	42 26	269	40 44	339
202	33 40	007	21 33	055	30 35	086	26 48	135	59 51	182	41 29	269	40 23	339
203	33 47	006	22 19	055	31 31	087	27 29	135	59 49	183	40 33	270	40 02	338
204	33 53	006	23 06	055	32 28	087	28 08	136	59 44	185	39 36	270	39 41	338
205	33 59	006	23 52	055	33 25	087	28 48	137	59 38	187	38 39	270	39 19	337
206	34 04	005	24 39	055	34 21	088	29 26	137	59 30	189	37 42	271	38 57	337
207	34 09	005	25 25	055	35 18	088	30 05	138	59 20	191	36 46	271	38 35	337
208	34 14	005	26 12	055	36 15	088	30 42	139	59 08	193	35 49	271	38 12	336
209	34 19	004	26 59	055	37 11	088	31 19	140	58 54	195	34 52	272	37 49	336
	Kochab		◆VEGA		Rasalhague		◆ANTARES		SPICA		◆REGULUS		Dubhe	
210	34 23	004	27 46	055	38 08	089	31 56	140	58 39	197	33 56	272	37 25	335
211	34 27	004	28 32	055	39 05	089	32 32	141	58 22	199	32 59	272	37 02	335
212	34 31	004	29 19	056	40 01	089	33 07	142	58 03	200	32 02	273	36 38	335
213	34 34	003	30 06	056	40 58	090	33 42	143	57 43	202	31 06	273	36 14	335
214	34 37	003	30 53	056	41 55	090	34 16	144	57 20	204	30 09	273	35 49	334
215	34 40	003	31 39	056	42 52	090	34 49	144	56 57	205	29 12	273	35 24	334
216	34 42	002	32 26	056	43 48	091	35 22	145	56 32	207	28 16	274	34 59	334
217	34 44	002	33 13	056	44 45	091	35 54	146	56 06	209	27 19	274	34 34	333
218	34 46	002	34 00	055	45 42	091	36 25	147	55 38	210	26 22	274	34 09	333
219	34 47	001	34 46	055	46 39	092	36 56	148	55 09	212	25 26	275	33 43	333
220	34 48	001	35 33	055	47 35	092	37 26	149	54 38	213	24 29	275	33 17	333
221	34 49	001	36 20	055	48 32	092	37 55	150	54 07	214	23 33	275	32 51	332
222	34 49	000	37 06	055	49 29	093	38 23	151	53 34	216	22 36	275	32 25	332
223	34 49	000	37 53	055	50 25	093	38 50	152	53 00	217	21 40	276	31 58	332
224	34 49	000	38 39	055	51 22	094	39 16	153	52 25	219	20 43	276	31 31	332
	VEGA		◆ALTAIR		Nunki		ANTARES		◆SPICA		Denebola		◆Alkaid	
225	39 26	055	19 12	087	17 12	127	39 42	154	51 50	220	44 10	272	56 21	339
226	40 12	055	20 09	087	17 57	127	40 06	155	51 13	221	43 13	272	55 59	338
227	40 59	055	21 06	088	18 42	128	40 30	156	50 35	222	42 16	273	55 37	337
228	41 45	055	22 02	088	19 27	128	40 53	157	49 57	223	41 20	273	55 14	336
229	42 31	054	22 59	088	20 11	129	41 14	158	49 17	224	40 23	273	54 50	335
230	43 17	054	23 56	089	20 55	129	41 35	159	48 37	226	39 26	274	54 26	334
231	44 03	054	24 52	089	21 39	130	41 55	160	47 56	227	38 30	274	54 00	333
232	44 49	054	25 49	089	22 22	130	42 13	161	47 15	228	37 33	274	53 34	332
233	45 34	053	26 46	090	23 05	131	42 31	163	46 32	229	36 37	274	53 07	331
234	46 20	053	27 43	090	23 48	132	42 47	164	45 50	230	35 40	275	52 40	331
235	47 05	053	28 39	090	24 30	132	43 03	165	45 06	231	34 44	275	52 12	330
236	47 51	053	29 36	090	25 12	133	43 17	166	44 22	232	33 47	275	51 43	329
237	48 36	052	30 33	091	25 53	133	43 30	167	43 37	232	32 51	275	51 14	328
238	49 20	052	31 30	091	26 34	134	43 42	168	42 52	233	31 54	276	50 44	328
239	50 05	052	32 26	091	27 15	135	43 52	170	42 06	234	30 58	276	50 13	327
	◆VEGA		ALTAIR		Nunki		◆ANTARES		SPICA		◆ARCTURUS		Alkaid	
240	50 49	051	33 23	092	27 55	135	44 02	171	41 20	235	65 19	275	49 42	327
241	51 33	051	34 20	092	28 34	136	44 10	172	40 33	236	64 23	275	49 11	326
242	52 17	050	35 16	093	29 13	137	44 17	173	39 46	236	63 26	275	48 39	325
243	53 01	050	36 13	093	29 52	138	44 23	175	38 59	237	62 30	275	48 06	325
244	53 44	049	37 10	093	30 30	138	44 28	176	38 11	238	61 33	275	47 33	324
245	54 27	049	38 06	094	31 07	139	44 32	177	37 23	239	60 37	276	47 00	324
246	55 10	048	39 03	094	31 44	140	44 34	178	36 34	239	59 40	276	46 26	323
247	55 52	048	39 59	094	32 21	141	44 35	180	35 45	240	58 44	276	45 52	323
248	56 33	047	40 56	095	32 56	141	44 35	181	34 56	241	57 47	276	45 18	322
249	57 15	046	41 53	095	33 31	142	44 33	182	34 06	241	56 51	276	44 43	322
250	57 56	046	42 49	096	34 06	143	44 30	183	33 16	242	55 55	276	44 08	322
251	58 36	045	43 45	096	34 40	144	44 26	185	32 26	243	54 58	277	43 33	321
252	59 16	044	44 42	096	35 13	145	44 21	186	31 36	243	54 02	277	42 57	321
253	59 55	043	45 38	097	35 45	146	44 14	187	30 45	244	53 06	277	42 21	321
254	60 34	042	46 35	097	36 17	147	44 07	188	29 54	244	52 09	277	41 45	320
	◆DENEB		ALTAIR		◆Nunki		ANTARES		SPICA		◆ARCTURUS		Alkaid	
255	37 35	047	47 31	098	36 48	147	43 58	190	29 03	245	51 13	277	41 08	320
256	38 17	047	48 27	098	37 18	148	43 48	191	28 11	245	50 17	277	40 32	320
257	38 58	047	49 23	099	37 47	149	43 36	192	27 19	246	49 21	278	39 55	319
258	39 39	046	50 19	099	38 16	150	43 24	193	26 27	246	48 24	278	39 18	319
259	40 20	046	51 15	100	38 43	151	43 10	194	25 35	247	47 28	278	38 40	319
260	41 01	046	52 11	100	39 10	152	42 56	196	24 43	247	46 32	278	38 03	319
261	41 41	046	53 07	101	39 36	153	42 40	197	23 51	248	45 36	278	37 25	318
262	42 22	045	54 02	101	40 01	154	42 23	198	22 58	248	44 40	279	36 47	318
263	43 02	045	54 58	102	40 25	155	42 05	199	22 05	249	43 44	279	36 09	318
264	43 42	045	55 53	103	40 49	156	41 46	200	21 12	249	42 48	279	35 31	318
265	44 22	044	56 49	103	41 11	158	41 26	201	20 19	250	41 52	279	34 53	318
266	45 02	044	57 44	104	41 32	159	41 05	202	19 26	250	40 56	279	34 15	317
267	45 41	044	58 39	105	41 52	160	40 42	203	18 32	251	40 00	280	33 36	317
268	46 20	043	59 34	105	42 11	161	40 19	205	17 39	251	39 04	280	32 58	317
269	46 59	043	60 28	106	42 29	162	39 55	206	16 45	252	38 08	280	32 19	317

LHA ♈	Hc	Zn	Hc	Zn	Hc	Zn	Hc	Zn	Hc	Zn	Hc	Zn	Hc	Zn
	◆DENEB		Enif		◆Nunki		ANTARES		◆ARCTURUS		Alkaid		Kochab	
270	47 37	043	35 14	091	42 46	163	39 30	207	37 12	280	31 40	317	29 13	347
271	48 15	042	36 10	092	43 02	164	39 04	208	36 16	280	31 02	317	28 59	347
272	48 53	042	37 07	092	43 17	165	38 38	209	35 20	281	30 23	317	28 46	346
273	49 31	041	38 04	092	43 31	167	38 10	210	34 24	281	29 44	317	28 33	346
274	50 08	041	39 00	093	43 43	168	37 42	211	33 29	281	29 05	317	28 19	346
275	50 45	040	39 57	093	43 55	169	37 12	212	32 33	281	28 26	317	28 05	346
276	51 21	039	40 54	093	44 05	170	36 42	212	31 37	281	27 47	317	27 51	346
277	51 57	039	41 50	094	44 14	172	36 11	213	30 42	282	27 08	316	27 37	346
278	52 32	038	42 47	094	44 21	173	35 40	214	29 46	282	26 29	316	27 23	345
279	53 07	038	43 44	095	44 28	174	35 08	215	28 51	282	25 50	316	27 09	345
280	53 41	037	44 40	095	44 33	175	34 35	216	27 55	282	25 11	316	26 54	345
281	54 15	036	45 37	095	44 37	177	34 01	217	27 00	282	24 31	316	26 39	345
282	54 48	035	46 33	096	44 40	178	33 26	218	26 04	283	23 52	316	26 25	345
283	55 21	035	47 30	096	44 41	179	32 51	218	25 09	283	23 13	316	26 10	345
284	55 52	034	48 26	097	44 42	180	32 16	219	24 14	283	22 34	317	25 55	345
	◆DENEB		Alpheratz		◆FOMALHAUT		Nunki		ANTARES		◆Rasalhague		VEGA	
285	56 24	033	20 06	065	14 57	129	44 41	182	31 40	220	68 31	256	69 35	347
286	56 54	032	20 57	065	15 41	130	44 39	183	31 03	221	67 36	257	69 22	345
287	57 24	031	21 49	065	16 24	130	44 35	184	30 25	222	66 41	257	69 06	343
288	57 53	030	22 41	066	17 07	131	44 30	185	29 47	222	65 45	258	68 48	341
289	58 21	029	23 32	066	17 50	131	44 25	187	29 09	223	64 49	259	68 28	339
290	58 48	028	24 24	066	18 32	132	44 17	188	28 30	224	63 54	260	68 07	337
291	59 14	027	25 16	066	19 14	132	44 09	189	27 51	224	62 58	260	67 44	335
292	59 40	026	26 08	066	19 56	133	44 00	190	27 11	225	62 02	261	67 19	333
293	60 04	025	27 00	066	20 37	134	43 49	192	26 30	226	61 06	262	66 53	332
294	60 27	024	27 52	066	21 18	134	43 37	193	25 49	226	60 10	262	66 25	330
295	60 50	022	28 44	067	21 59	135	43 24	194	25 08	227	59 13	263	65 56	329
296	61 11	021	29 36	067	22 39	135	43 10	195	24 26	228	58 17	263	65 26	327
297	61 31	020	30 28	067	23 19	136	42 54	196	23 44	228	57 21	264	64 55	326
298	61 49	019	31 20	067	23 58	137	42 38	197	23 02	229	56 24	264	64 23	324
299	62 07	017	32 12	067	24 37	137	42 20	199	22 19	229	55 28	265	63 49	323
	◆DENEB		Schedar		Alpheratz		◆FOMALHAUT		Nunki		◆Rasalhague		VEGA	
300	62 23	016	26 43	035	33 05	067	25 15	138	42 01	200	54 31	265	63 15	322
301	62 37	014	27 16	035	33 57	067	25 53	138	41 42	201	53 35	265	62 39	321
302	62 51	013	27 48	035	34 49	067	26 30	139	41 21	202	52 38	266	62 03	320
303	63 03	011	28 21	035	35 41	067	27 07	140	40 59	203	51 42	266	61 26	319
304	63 13	010	28 54	035	36 34	067	27 44	141	40 37	204	50 45	267	60 48	318
305	63 22	008	29 27	035	37 26	067	28 19	141	40 13	205	49 48	267	60 10	317
306	63 30	007	29 59	035	38 19	067	28 55	142	39 48	206	48 52	267	59 31	316
307	63 36	005	30 32	035	39 11	068	29 29	143	39 23	207	47 55	268	58 51	315
308	63 40	004	31 04	035	40 03	068	30 03	143	38 56	208	46 58	268	58 11	315
309	63 43	002	31 36	035	40 56	068	30 37	144	38 29	209	46 02	269	57 31	314
310	63 44	001	32 08	034	41 48	068	31 10	145	38 01	210	45 05	269	56 49	313
311	63 44	359	32 40	034	42 41	068	31 42	146	37 32	211	44 08	269	56 08	312
312	63 42	357	33 12	034	43 33	068	32 13	147	37 02	212	43 12	270	55 26	312
313	63 39	356	33 44	034	44 26	068	32 44	147	36 32	213	42 15	270	54 43	311
314	63 34	354	34 15	034	45 18	068	33 14	148	36 00	214	41 18	270	54 01	311
	◆Alpheratz		Diphda		◆FOMALHAUT		Nunki		◆Rasalhague		VEGA		DENEB	
315	46 11	067	23 51	121	33 44	149	35 28	215	40 21	271	53 17	310	63 27	353
316	47 03	067	24 39	121	34 12	150	34 55	216	39 25	271	52 34	310	63 19	351
317	47 55	067	25 28	122	34 40	151	34 22	217	38 28	271	51 50	309	63 09	350
318	48 48	067	26 16	122	35 07	152	33 48	217	37 31	271	51 06	309	62 58	348
319	49 40	067	27 04	123	35 34	153	33 13	218	36 34	272	50 22	309	62 46	347
320	50 32	067	27 51	123	35 59	154	32 38	219	35 38	272	49 38	308	62 32	345
321	51 24	067	28 38	124	36 24	155	32 02	220	34 41	272	48 53	308	62 17	344
322	52 17	067	29 25	125	36 48	156	31 25	221	33 44	273	48 08	307	62 00	342
323	53 09	067	30 12	125	37 11	157	30 48	221	32 48	273	47 23	307	61 42	341
324	54 01	066	30 58	126	37 33	158	30 10	222	31 51	273	46 37	307	61 23	340
325	54 53	066	31 43	127	37 54	159	29 32	223	30 54	274	45 52	307	61 03	338
326	55 44	066	32 29	127	38 14	160	28 53	224	29 58	274	45 06	306	60 41	337
327	56 36	066	33 13	128	38 33	161	28 14	224	29 01	274	44 21	306	60 19	336
328	57 28	065	33 58	129	38 52	162	27 34	225	28 05	274	43 35	306	59 55	335
329	58 19	065	34 42	130	39 09	163	26 54	226	27 08	275	42 49	306	59 30	334
	◆Mirfak		Hamal		Diphda		◆FOMALHAUT		ALTAIR		◆VEGA		DENEB	
330	20 08	043	32 43	074	35 25	130	39 25	164	57 04	257	42 03	306	59 05	333
331	20 47	043	33 37	074	36 09	131	39 41	165	56 08	257	41 16	305	58 38	331
332	21 25	043	34 32	074	36 51	132	39 55	166	55 13	258	40 30	305	58 11	330
333	22 04	043	35 27	074	37 33	133	40 08	167	54 17	258	39 44	305	57 42	329
334	22 43	043	36 21	074	38 15	133	40 20	168	53 22	259	38 57	305	57 13	329
335	23 21	043	37 16	075	38 55	134	40 31	169	52 26	260	38 11	305	56 43	328
336	24 00	043	38 11	075	39 36	135	40 41	170	51 30	260	37 24	305	56 12	327
337	24 39	043	39 05	075	40 15	136	40 50	172	50 34	261	36 38	305	55 41	326
338	25 17	043	40 00	075	40 54	137	40 58	173	49 38	261	35 51	305	55 09	325
339	25 56	043	40 55	075	41 33	138	41 05	174	48 42	262	35 04	305	54 36	324
340	26 35	043	41 50	075	42 10	139	41 10	175	47 46	262	34 18	305	54 02	324
341	27 13	043	42 44	075	42 47	140	41 15	176	46 50	263	33 31	305	53 29	323
342	27 52	043	43 39	075	43 24	141	41 18	177	45 53	263	32 44	304	52 54	322
343	28 31	043	44 34	075	43 59	142	41 20	178	44 57	263	31 57	304	52 19	322
344	29 09	043	45 29	076	44 34	143	41 21	180	44 01	264	31 11	304	51 43	321
	Schedar		◆Mirfak		ALDEBARAN		Diphda		◆FOMALHAUT		ALTAIR		◆DENEB	
345	48 05	020	29 48	043	10 53	076	45 07	144	41 21	181	43 04	264	51 07	320
346	48 25	020	30 26	043	11 48	076	45 40	145	41 19	182	42 08	265	50 31	320
347	48 44	019	31 05	043	12 43	077	46 12	146	41 17	183	41 11	265	49 54	319
348	49 02	018	31 43	042	13 38	077	46 43	147	41 13	184	40 15	266	49 17	319
349	49 19	018	32 21	042	14 33	077	47 13	149	41 09	185	39 18	266	48 39	318
350	49 36	017	33 00	042	15 29	078	47 42	150	41 03	187	38 22	266	48 01	318
351	49 53	016	33 38	042	16 24	078	48 10	151	40 56	188	37 25	267	47 23	317
352	50 08	015	34 16	042	17 20	078	48 37	152	40 48	189	36 28	267	46 45	317
353	50 23	015	34 53	042	18 15	078	49 03	154	40 38	190	35 32	267	46 06	317
354	50 37	014	35 31	042	19 11	079	49 28	155	40 28	191	34 35	268	45 26	316
355	50 50	013	36 09	041	20 06	079	49 51	156	40 17	192	33 38	268	44 47	316
356	51 03	012	36 46	041	21 02	079	50 13	158	40 04	193	32 42	268	44 07	315
357	51 14	011	37 24	041	21 58	079	50 34	159	39 51	194	31 45	269	43 27	315
358	51 25	011	38 01	041	22 54	080	50 54	160	39 36	195	30 48	269	42 47	315
359	51 35	010	38 38	041	23 49	080	51 12	162	39 20	197	29 51	269	42 07	315

LAT 18°N

LHA ♈	Hc	Zn	Hc	Zn	Hc	Zn	Hc	Zn	Hc	Zn	Hc	Zn	Hc	Zn
	Schedar		CAPELLA		◆ALDEBARAN		ACHERNAR		◆FOMALHAUT		ALTAIR		◆DENEB	
0	50 45	009	20 20	047	24 35	080	12 02	167	40 01	198	28 55	270	40 44	315
1	50 54	008	21 01	047	25 31	080	12 15	167	39 43	199	27 58	271	40 04	315
2	51 01	007	21 43	047	26 27	080	12 27	168	39 24	200	27 01	271	39 23	314
3	51 08	006	22 25	047	27 23	080	12 39	168	39 04	201	26 03	271	38 42	314
4	51 13	005	23 06	047	28 19	081	12 50	169	38 43	202	25 06	272	38 01	314
5	51 18	004	23 48	047	29 16	081	13 01	169	38 21	203	24 09	272	37 20	314
6	51 22	004	24 30	047	30 12	081	13 11	170	37 59	204	23 12	272	36 39	314
7	51 25	003	25 11	047	31 09	081	13 21	170	37 35	205	22 15	272	35 57	313
8	51 28	002	25 53	047	32 05	081	13 30	171	37 10	206	21 18	273	35 16	313
9	51 29	001	26 35	047	33 01	082	13 39	172	36 45	207	20 21	273	34 34	313
10	51 29	000	27 16	047	33 58	082	13 47	172	36 19	208	19 24	273	33 53	313
11	51 29	359	27 58	047	34 54	082	13 55	173	35 52	209	18 27	274	33 11	313
12	51 28	358	28 40	047	35 51	082	14 02	173	35 24	210	17 30	274	32 29	313
13	51 26	357	29 21	047	36 47	083	14 08	174	34 55	211	16 34	274	31 47	313
14	51 23	357	30 03	047	37 44	083	14 14	174	34 26	211	15 37	274	31 05	313
	CAPELLA		◆ALDEBARAN		RIGEL		ACHERNAR		◆FOMALHAUT		Enif		◆DENEB	
15	30 44	047	38 41	083	22 01	107	14 20	175	33 56	212	41 52	267	30 23	312
16	31 26	047	39 37	083	22 55	107	14 25	175	33 25	213	40 55	268	29 41	312
17	32 07	046	40 34	083	23 50	108	14 29	176	32 53	214	39 58	268	28 59	312
18	32 48	046	41 31	084	24 44	108	14 33	176	32 21	215	39 01	268	28 16	312
19	33 30	046	42 27	084	25 38	109	14 36	177	31 48	216	38 04	269	27 34	312
20	34 11	046	43 24	084	26 32	109	14 39	178	31 15	216	37 07	269	26 52	312
21	34 52	046	44 21	084	27 26	110	14 41	178	30 41	217	36 10	269	26 10	312
22	35 33	046	45 18	084	28 19	110	14 43	179	30 06	218	35 13	269	25 28	312
23	36 14	046	46 15	085	29 13	111	14 44	179	29 31	219	34 16	270	24 45	312
24	36 54	045	47 11	085	30 06	111	14 44	180	28 55	219	33 19	270	24 03	312
25	37 35	045	48 08	085	30 59	112	14 44	180	28 18	220	32 22	270	23 21	312
26	38 15	045	49 05	085	31 52	112	14 44	181	27 41	221	31 25	271	22 39	312
27	38 56	045	50 02	086	32 45	113	14 42	181	27 04	221	30 28	271	21 57	312
28	39 36	045	50 59	086	33 37	113	14 41	182	26 26	222	29 31	271	21 14	312
29	40 16	044	51 56	086	34 30	114	14 38	183	25 47	223	28 34	272	20 32	313
	Mirfak		◆CAPELLA		ALDEBARAN		RIGEL		◆ACHERNAR		FOMALHAUT		◆Alpheratz	
30	53 59	023	40 55	044	52 53	086	35 22	114	14 35	183	25 08	223	62 10	299
31	54 21	022	41 35	044	53 50	086	36 13	115	14 32	184	24 29	224	61 20	298
32	54 42	021	42 14	043	54 47	087	37 05	116	14 28	184	23 49	225	60 29	298
33	55 02	020	42 53	043	55 44	087	37 56	116	14 24	185	23 09	225	59 39	297
34	55 21	019	43 32	043	56 41	087	38 47	117	14 19	185	22 28	226	58 48	297
35	55 40	018	44 11	042	57 38	087	39 38	118	14 13	186	21 47	226	57 57	296
36	55 57	017	44 49	042	58 35	088	40 28	118	14 07	186	21 06	227	57 06	296
37	56 14	016	45 27	042	59 32	088	41 18	119	14 00	187	20 24	227	56 14	296
38	56 29	015	46 05	041	60 29	088	42 08	120	13 53	188	19 42	228	55 23	295
39	56 44	014	46 42	041	61 26	088	42 57	121	13 45	188	18 59	229	54 31	295
40	56 57	013	47 19	040	62 23	089	43 46	121	13 37	189	18 16	229	53 39	295
41	57 10	012	47 56	040	63 20	089	44 35	122	13 28	189	17 33	230	52 48	295
42	57 21	011	48 32	039	64 17	089	45 23	123	13 19	190	16 49	230	51 56	294
43	57 31	010	49 08	039	65 14	090	46 11	124	13 09	190	16 05	231	51 04	294
44	57 40	008	49 44	038	66 11	090	46 58	125	12 59	191	15 21	231	50 12	294
	CAPELLA		◆POLLUX		PROCYON		SIRIUS		◆Acamar		Diphda		◆Alpheratz	
45	50 19	038	24 31	067	20 51	091	24 40	119	31 40	180	40 45	225	49 19	294
46	50 53	037	25 24	067	21 48	092	25 30	119	31 40	181	40 05	226	48 27	294
47	51 27	036	26 16	067	22 45	092	26 20	120	31 38	182	39 23	227	47 35	294
48	52 01	036	27 09	067	23 42	092	27 09	121	31 35	183	38 42	227	46 43	294
49	52 34	035	28 01	067	24 39	092	27 58	121	31 32	184	37 59	228	45 50	293
50	53 06	034	28 54	067	25 36	093	28 47	122	31 27	185	37 17	229	44 58	293
51	53 38	033	29 47	067	26 33	093	29 36	122	31 22	186	36 33	230	44 06	293
52	54 09	033	30 39	068	27 30	093	30 24	123	31 16	187	35 49	231	43 13	293
53	54 40	032	31 32	068	28 27	094	31 12	123	31 09	188	35 05	231	42 21	293
54	55 09	031	32 25	068	29 24	094	31 59	124	31 01	188	34 20	232	41 28	293
55	55 38	030	33 18	068	30 21	095	32 46	125	30 52	189	33 35	233	40 36	293
56	56 07	029	34 11	068	31 17	095	33 33	125	30 42	190	32 50	233	39 44	293
57	56 34	028	35 04	068	32 14	095	34 19	126	30 32	191	32 04	234	38 51	293
58	57 01	027	35 56	068	33 11	096	35 05	127	30 20	192	31 17	235	37 59	293
59	57 26	026	36 49	068	34 08	096	35 51	127	30 08	193	30 31	235	37 06	293
	CAPELLA		◆POLLUX		PROCYON		SIRIUS		◆CANOPUS		Diphda		◆Hamal	
60	57 51	025	37 42	068	35 05	096	36 36	128	12 45	159	29 43	236	63 03	286
61	58 15	024	38 35	068	36 01	097	37 21	129	13 06	159	28 56	237	62 08	286
62	58 38	023	39 28	068	36 58	097	38 05	130	13 26	160	28 08	237	61 13	286
63	59 00	022	40 21	068	37 55	098	38 48	130	13 46	160	27 20	238	60 18	286
64	59 21	021	41 14	068	38 51	098	39 31	131	14 05	161	26 32	238	59 23	286
65	59 40	020	42 07	068	39 48	098	40 14	132	14 23	161	25 43	239	58 29	286
66	59 59	018	43 00	068	40 44	099	40 56	133	14 41	162	24 54	239	57 34	286
67	60 16	017	43 53	068	41 40	099	41 38	134	14 59	162	24 05	240	56 39	285
68	60 32	016	44 46	068	42 37	100	42 18	135	15 16	163	23 15	240	55 44	285
69	60 47	014	45 39	068	43 33	100	42 59	136	15 33	163	22 26	241	54 49	285
70	61 01	013	46 32	068	44 29	101	43 38	137	15 49	164	21 36	241	53 54	285
71	61 13	012	47 25	068	45 25	101	44 17	138	16 04	165	20 45	242	52 59	285
72	61 24	010	48 17	068	46 21	102	44 55	139	16 19	165	19 55	242	52 03	285
73	61 33	009	49 10	068	47 17	102	45 32	140	16 33	166	19 04	243	51 08	285
74	61 41	007	50 03	068	48 13	103	46 09	141	16 47	166	18 13	243	50 13	285
	CAPELLA		◆POLLUX		PROCYON		SIRIUS		◆RIGEL		Menkar		◆Hamal	
75	61 48	006	50 56	067	49 08	103	46 44	142	63 33	172	57 57	248	49 18	285
76	61 53	005	51 48	067	50 04	104	47 19	143	63 40	174	57 04	249	48 23	285
77	61 57	003	52 41	067	50 59	104	47 53	144	63 45	176	56 11	249	47 28	285
78	62 00	002	53 33	067	51 54	105	48 26	145	63 47	179	55 18	250	46 33	285
79	62 00	000	54 26	067	52 49	106	48 58	146	63 47	181	54 24	251	45 38	285
80	62 00	359	55 18	066	53 44	106	49 29	148	63 45	183	53 30	252	44 43	285
81	61 58	357	56 11	066	54 39	107	49 59	149	63 41	185	52 35	252	43 48	286
82	61 54	356	57 03	066	55 33	108	50 28	150	63 35	188	51 41	253	42 53	286
83	61 49	354	57 55	066	56 27	108	50 56	152	63 26	190	50 46	254	41 58	286
84	61 43	353	58 47	065	57 21	109	51 22	153	63 15	192	49 51	254	41 03	286
85	61 35	351	59 38	065	58 15	110	51 47	154	63 02	194	48 56	255	40 08	286
86	61 25	350	60 30	065	59 09	111	52 11	156	62 47	196	48 01	256	39 14	286
87	61 15	349	61 22	064	60 02	112	52 34	157	62 30	198	47 06	256	38 19	286
88	61 03	347	62 13	064	60 55	113	52 56	159	62 11	200	46 10	257	37 24	286
89	60 49	346	63 04	063	61 47	114	53 16	160	61 51	202	45 15	257	36 29	286

LHA ♈	Hc	Zn	Hc	Zn	Hc	Zn	Hc	Zn	Hc	Zn	Hc	Zn	Hc	Zn
	Dubhe		◆REGULUS		Alphard		SIRIUS		◆RIGEL		ALDEBARAN		◆CAPELLA	
90	22 28	030	30 02	086	32 19	113	53 34	162	61 28	204	69 49	269	60 35	344
91	22 56	030	30 59	087	33 11	114	53 52	163	61 04	206	68 52	269	60 19	343
92	23 25	030	31 56	087	34 04	114	54 07	165	60 38	208	67 55	270	60 02	342
93	23 53	030	32 53	087	34 56	115	54 21	166	60 10	210	66 58	270	59 43	341
94	24 21	030	33 50	087	35 47	115	54 34	168	59 41	211	66 00	270	59 24	339
95	24 49	029	34 47	088	36 39	116	54 45	170	59 11	213	65 03	271	59 03	338
96	25 17	029	35 44	088	37 30	117	54 55	171	58 39	215	64 06	271	58 42	337
97	25 45	029	36 41	088	38 21	117	55 03	173	58 06	216	63 09	271	58 19	336
98	26 13	029	37 38	089	39 11	118	55 09	175	57 31	218	62 12	271	57 55	335
99	26 41	029	38 35	089	40 01	119	55 13	176	56 56	219	61 15	272	57 30	334
100	27 09	029	39 32	089	40 51	119	55 16	178	56 19	221	60 18	272	57 05	333
101	27 36	029	40 29	089	41 41	120	55 17	180	55 41	222	59 21	272	56 38	332
102	28 04	029	41 27	090	42 30	121	55 17	181	55 03	223	58 24	272	56 11	331
103	28 31	029	42 24	090	43 19	122	55 15	183	54 23	225	57 27	273	55 43	330
104	28 58	028	43 21	090	44 07	122	55 11	185	53 43	226	56 30	273	55 14	329
	Dubhe		◆REGULUS		Alphard		SIRIUS		◆RIGEL		ALDEBARAN		◆CAPELLA	
105	29 26	028	44 18	091	44 55	123	55 06	186	53 01	227	55 33	273	54 44	328
106	29 53	028	45 15	091	45 43	124	54 59	188	52 19	228	54 36	273	54 14	327
107	30 19	028	46 12	091	46 30	125	54 50	190	51 36	229	53 39	274	53 43	327
108	30 46	028	47 09	092	47 17	126	54 39	191	50 52	230	52 42	274	53 11	326
109	31 13	028	48 06	092	48 03	127	54 28	193	50 08	231	51 45	274	52 39	325
110	31 39	027	49 03	092	48 48	128	54 14	194	49 23	232	50 48	274	52 06	324
111	32 05	027	50 00	093	49 33	129	53 59	196	48 38	233	49 52	274	51 33	324
112	32 31	027	50 57	093	50 17	130	53 43	198	47 52	234	48 55	275	50 59	323
113	32 57	027	51 54	093	51 01	131	53 25	199	47 05	235	47 58	275	50 24	323
114	33 22	026	52 51	094	51 44	132	53 05	201	46 18	236	47 01	275	49 49	322
115	33 48	026	53 48	094	52 26	133	52 44	202	45 30	237	46 04	275	49 14	321
116	34 13	026	54 45	095	53 07	134	52 22	204	44 42	238	45 07	276	48 38	321
117	34 37	026	55 42	095	53 48	135	51 59	205	43 54	239	44 11	276	48 02	320
118	35 02	025	56 38	096	54 27	137	51 34	206	43 05	239	43 14	276	47 25	320
119	35 26	025	57 35	096	55 06	138	51 08	208	42 15	240	42 17	276	46 48	319
	Dubhe		◆Denebola		Alphard		◆SIRIUS		RIGEL		ALDEBARAN		◆CAPELLA	
120	35 50	025	35 12	085	55 44	139	50 41	209	41 26	241	41 20	276	46 11	319
121	36 14	024	36 09	085	56 21	141	50 12	210	40 36	242	40 24	277	45 33	318
122	36 38	024	37 06	085	56 56	142	49 43	212	39 45	242	39 27	277	44 55	318
123	37 01	024	38 03	085	57 31	144	49 13	213	38 55	243	38 30	277	44 17	318
124	37 24	023	38 59	086	58 04	145	48 41	214	38 04	244	37 34	277	43 38	317
125	37 46	023	39 56	086	58 36	147	48 08	215	37 13	244	36 37	278	43 00	317
126	38 08	023	40 53	086	59 07	148	47 35	217	36 21	245	35 41	278	42 21	317
127	38 30	022	41 50	086	59 36	150	47 01	218	35 29	245	34 44	278	41 41	316
128	38 51	022	42 47	087	60 04	152	46 25	219	34 37	246	33 47	278	41 02	316
129	39 13	021	43 44	087	60 30	154	45 49	220	33 45	247	32 51	278	40 22	316
130	39 33	021	44 41	087	60 55	155	45 12	221	32 53	247	31 55	279	39 42	315
131	39 54	021	45 38	087	61 18	157	44 35	222	32 00	248	30 58	279	39 02	315
132	40 13	020	46 35	088	61 39	159	43 56	223	31 07	248	30 02	279	38 22	315
133	40 33	020	47 32	088	61 58	161	43 17	224	30 14	249	29 05	279	37 41	315
134	40 52	019	48 29	088	62 16	163	42 37	225	29 21	249	28 09	279	37 01	315
	Dubhe		◆ARCTURUS		SPICA		◆Alphard		SIRIUS		BETELGEUSE		◆CAPELLA	
135	41 11	019	15 58	075	18 27	109	62 31	165	41 57	226	43 46	263	36 20	314
136	41 29	018	16 54	075	19 21	109	62 45	167	41 15	227	42 49	263	35 39	314
137	41 46	018	17 49	075	20 15	110	62 56	169	40 34	227	41 53	264	34 58	314
138	42 04	017	18 44	075	21 08	110	63 06	172	39 51	228	40 56	264	34 17	314
139	42 20	017	19 39	075	22 02	111	63 13	174	39 08	229	39 59	265	33 36	314
140	42 37	016	20 34	076	22 55	111	63 18	176	38 25	230	39 02	265	32 55	314
141	42 52	016	21 30	076	23 48	111	63 21	178	37 41	231	38 05	265	32 14	314
142	43 08	015	22 25	076	24 41	112	63 22	180	36 57	231	37 09	266	31 32	313
143	43 22	015	23 20	076	25 34	112	63 20	183	36 12	232	36 12	266	30 51	313
144	43 36	014	24 16	077	26 27	113	63 17	185	35 26	233	35 15	266	30 09	313
145	43 50	013	25 11	077	27 19	113	63 11	187	34 41	234	34 18	267	29 28	313
146	44 03	013	26 07	077	28 12	114	63 03	189	33 54	234	33 21	267	28 46	313
147	44 15	012	27 02	077	29 04	114	62 53	191	33 08	235	32 24	267	28 05	313
148	44 27	012	27 58	077	29 56	115	62 41	193	32 21	236	31 27	268	27 23	313
149	44 39	011	28 54	078	30 47	115	62 26	195	31 34	236	30 30	268	26 41	313
	Dubhe		◆ARCTURUS		SPICA		◆Alphard		SIRIUS		BETELGEUSE		◆CAPELLA	
150	44 49	010	29 50	078	31 39	116	62 10	198	30 46	237	29 33	268	26 00	313
151	44 59	010	30 45	078	32 30	117	61 52	200	29 58	238	28 36	269	25 18	313
152	45 09	009	31 41	078	33 21	117	61 32	201	29 10	238	27 39	269	24 36	313
153	45 18	009	32 37	078	34 11	118	61 10	203	28 21	239	26 42	269	23 54	313
154	45 26	008	33 33	078	35 02	118	60 47	205	27 32	239	25 44	270	23 13	313
155	45 34	007	34 29	079	35 52	119	60 22	207	26 43	240	24 47	270	22 31	313
156	45 41	007	35 25	079	36 41	120	59 55	209	25 53	240	23 50	270	21 50	313
157	45 47	006	36 21	079	37 31	120	59 27	211	25 04	241	22 53	271	21 08	313
158	45 52	005	37 17	079	38 20	121	58 57	212	24 14	241	21 56	271	20 26	313
159	45 57	005	38 13	079	39 09	122	58 26	214	23 23	242	20 59	271	19 45	313
160	46 02	004	39 09	080	39 57	122	57 53	215	22 33	242	20 02	271	19 04	313
161	46 05	003	40 05	080	40 45	123	57 20	217	21 42	243	19 05	272	18 22	314
162	46 08	003	41 01	080	41 32	124	56 45	218	20 51	243	18 08	272	17 41	314
163	46 11	002	41 58	080	42 19	125	56 09	220	20 00	244	17 11	272	17 00	314
164	46 12	001	42 54	080	43 06	126	55 32	221	19 09	244	16 14	273	16 18	314
	◆Dubhe		Alkaid		ARCTURUS		◆SPICA		Suhail		◆PROCYON		POLLUX	
165	46 13	001	44 07	037	43 50	080	43 52	126	23 25	202	39 20	262	44 20	292
166	46 13	000	44 41	037	44 46	081	44 38	127	23 03	203	38 24	262	43 27	292
167	46 13	359	45 15	036	45 43	081	45 23	128	22 41	203	37 27	263	42 34	292
168	46 12	359	45 49	036	46 39	081	46 08	129	22 18	204	36 31	263	41 41	292
169	46 10	358	46 22	035	47 35	081	46 52	130	21 55	205	35 34	263	40 48	292
170	46 07	357	46 55	035	48 32	081	47 35	131	21 31	205	34 37	264	39 55	292
171	46 04	356	47 28	034	49 28	081	48 18	132	21 06	206	33 41	264	39 02	292
172	46 01	356	48 00	034	50 24	081	49 00	133	20 41	206	32 44	265	38 09	292
173	45 56	355	48 31	033	51 21	082	49 41	134	20 15	207	31 47	265	37 16	292
174	45 51	354	49 02	033	52 17	082	50 21	135	19 49	208	30 50	265	36 23	292
175	45 45	354	49 33	032	53 14	082	51 01	136	19 22	208	29 53	266	35 30	292
176	45 39	353	50 02	031	54 10	082	51 40	138	18 55	209	28 56	266	34 37	292
177	45 31	353	50 32	031	55 07	082	52 18	139	18 27	210	27 59	266	33 45	292
178	45 24	352	51 01	030	56 03	082	52 55	140	17 59	210	27 03	267	32 52	292
179	45 15	351	51 29	029	57 00	082	53 31	141	17 30	211	26 06	267	31 59	292

LAT 18°N

LHA	Hc	Zn	Hc	Zn	Hc	Zn	Hc	Zn	Hc	Zn	Hc	Zn	Hc	Zn
♈	◆Alkaid		ANTARES		◆SPICA		Gienah		◆REGULUS		POLLUX		Dubhe	
180	51 56	028	11 02	123	54 07	143	54 17	174	62 21	261	31 06	292	45 06	351
181	52 23	028	11 50	123	54 41	144	54 22	175	61 25	262	30 13	293	44 57	350
182	52 50	027	12 38	124	55 13	145	54 26	177	60 28	263	29 21	293	44 46	349
183	53 15	026	13 26	124	55 45	147	54 29	179	59 32	263	28 28	293	44 35	349
184	53 40	025	14 13	124	56 16	148	54 29	180	58 35	264	27 35	293	44 24	348
185	54 04	024	15 00	125	56 45	150	54 28	182	57 38	264	26 43	293	44 12	348
186	54 27	023	15 46	125	57 13	152	54 26	183	56 41	264	25 50	293	43 59	347
187	54 49	023	16 33	126	57 39	153	54 21	185	55 45	265	24 58	293	43 46	346
188	55 11	022	17 19	126	58 04	155	54 15	187	54 48	265	24 05	293	43 32	346
189	55 31	021	18 05	127	58 28	157	54 08	188	53 51	266	23 13	293	43 18	345
190	55 51	020	18 50	127	58 50	158	53 59	190	52 54	266	22 21	294	43 03	345
191	56 10	019	19 36	128	59 10	160	53 48	192	51 57	267	21 28	294	42 48	344
192	56 27	018	20 21	128	59 28	162	53 36	193	51 00	267	20 36	294	42 32	344
193	56 44	017	21 05	129	59 45	164	53 22	195	50 03	267	19 44	294	42 16	343
194	57 00	015	21 50	129	60 00	166	53 07	196	49 06	268	18 52	294	41 59	343
	Kochab		◆VEGA		Rasalhague		ANTARES		◆SPICA		◆REGULUS		Dubhe	
195	31 48	009	15 35	054	23 53	084	22 34	130	60 13	168	48 09	268	41 41	342
196	31 56	008	16 21	054	24 49	084	23 17	130	60 25	170	47 12	268	41 24	342
197	32 05	008	17 07	054	25 46	084	24 00	131	60 34	172	46 15	269	41 05	341
198	32 12	008	17 53	054	26 43	085	24 43	132	60 41	174	45 18	269	40 46	341
199	32 20	007	18 39	054	27 40	085	25 26	132	60 47	176	44 21	269	40 27	340
200	32 27	007	19 25	054	28 37	085	26 08	133	60 50	178	43 24	270	40 08	340
201	32 34	007	20 12	054	29 34	085	26 49	134	60 52	180	42 27	270	39 48	339
202	32 41	007	20 58	054	30 30	086	27 30	134	60 51	182	41 30	270	39 27	339
203	32 47	006	21 45	055	31 27	086	28 11	135	60 49	184	40 32	271	39 06	338
204	32 53	006	22 31	055	32 24	086	28 51	135	60 44	186	39 35	271	38 45	338
205	32 59	006	23 18	055	33 21	087	29 31	136	60 38	188	38 38	271	38 24	338
206	33 04	005	24 04	055	34 18	087	30 10	137	60 29	190	37 41	271	38 02	337
207	33 10	005	24 51	055	35 15	087	30 49	138	60 19	191	36 44	272	37 40	337
208	33 15	005	25 38	055	36 12	087	31 27	138	60 07	193	35 47	272	37 17	336
209	33 19	004	26 25	055	37 09	088	32 05	139	59 52	195	34 50	272	36 54	336
	Kochab		◆VEGA		Rasalhague		◆ANTARES		SPICA		◆REGULUS		Dubhe	
210	33 23	004	27 11	055	38 06	088	32 42	140	59 36	197	33 53	273	36 31	336
211	33 27	004	27 58	055	39 03	088	33 18	141	59 19	199	32 56	273	36 07	335
212	33 31	003	28 45	055	40 00	089	33 54	141	58 59	201	31 59	273	35 43	335
213	33 34	003	29 32	055	40 57	089	34 30	142	58 38	203	31 02	273	35 19	335
214	33 37	003	30 18	055	41 54	089	35 04	143	58 15	204	30 05	274	34 55	335
215	33 40	003	31 05	055	42 51	089	35 38	144	57 51	206	29 08	274	34 30	334
216	33 42	002	31 52	055	43 49	090	36 11	145	57 25	208	28 11	274	34 05	334
217	33 44	002	32 39	055	44 46	090	36 44	146	56 58	209	27 15	274	33 40	334
218	33 46	002	33 25	055	45 43	090	37 16	147	56 29	211	26 18	275	33 15	333
219	33 47	001	34 12	055	46 40	091	37 47	148	56 00	212	25 21	275	32 49	333
220	33 48	001	34 59	055	47 37	091	38 17	148	55 28	214	24 24	275	32 24	333
221	33 49	001	35 45	055	48 34	091	38 46	149	54 56	215	23 27	276	31 58	333
222	33 49	000	36 32	055	49 31	092	39 15	150	54 22	217	22 30	276	31 31	333
223	33 49	000	37 18	055	50 28	092	39 43	151	53 48	218	21 34	276	31 05	332
224	33 49	000	38 05	054	51 25	092	40 09	152	53 12	219	20 37	276	30 38	332
	VEGA		◆ALTAIR		Nunki		ANTARES		◆SPICA		Denebola		◆Alkaid	
225	38 51	054	19 09	087	17 47	126	40 35	153	52 36	221	44 07	273	55 25	339
226	39 37	054	20 06	087	18 33	127	41 00	154	51 58	222	43 10	273	55 04	338
227	40 24	054	21 03	087	19 19	127	41 25	156	51 19	223	42 13	274	54 42	337
228	41 10	054	22 00	088	20 04	128	41 48	157	50 40	224	41 16	274	54 19	336
229	41 57	054	22 57	088	20 49	128	42 10	158	50 00	225	40 19	274	53 56	335
230	42 42	053	23 54	088	21 33	129	42 31	159	49 19	226	39 22	274	53 32	334
231	43 27	053	24 51	088	22 17	130	42 51	160	48 37	227	38 25	275	53 07	334
232	44 13	053	25 48	089	23 01	130	43 10	161	47 55	228	37 29	275	52 41	333
233	44 58	053	26 45	089	23 45	131	43 28	162	47 12	229	36 32	275	52 15	332
234	45 44	052	27 42	089	24 28	131	43 45	163	46 28	230	35 35	275	51 47	331
235	46 29	052	28 39	090	25 10	132	44 00	165	45 44	231	34 38	276	51 20	331
236	47 14	052	29 36	090	25 53	132	44 15	166	44 59	232	33 41	276	50 51	330
237	47 58	051	30 33	090	26 34	133	44 28	167	44 13	233	32 45	276	50 22	329
238	48 43	051	31 30	091	27 16	134	44 40	168	43 28	234	31 48	276	49 53	328
239	49 27	051	32 27	091	27 57	134	44 51	170	42 41	235	30 51	277	49 23	328
	◆VEGA		ALTAIR		Nunki		◆ANTARES		SPICA		◆ARCTURUS		Alkaid	
240	50 11	050	33 25	091	28 37	135	45 01	171	41 54	236	65 13	277	48 52	327
241	50 55	050	34 22	091	29 17	136	45 10	172	41 07	236	64 16	277	48 21	327
242	51 39	049	35 19	092	29 57	136	45 17	173	40 19	237	63 20	277	47 49	326
243	52 22	049	36 16	092	30 36	137	45 23	175	39 31	238	62 23	277	47 17	325
244	53 05	048	37 13	092	31 15	138	45 28	176	38 43	239	61 27	277	46 44	325
245	53 47	048	38 10	093	31 53	139	45 31	177	37 54	239	60 30	277	46 11	324
246	54 29	047	39 07	093	32 30	139	45 34	178	37 04	240	59 33	277	45 38	324
247	55 11	047	40 04	094	33 07	140	45 35	180	36 15	241	58 37	278	45 04	323
248	55 52	046	41 01	094	33 43	141	45 34	181	35 25	241	57 40	278	44 30	323
249	56 33	045	41 57	094	34 19	142	45 33	182	34 35	242	56 44	278	43 56	323
250	57 13	045	42 54	095	34 54	143	45 30	183	33 44	243	55 47	278	43 21	322
251	57 53	044	43 51	095	35 28	143	45 26	185	32 53	243	54 51	278	42 46	322
252	58 32	043	44 48	095	36 02	144	45 21	186	32 02	244	53 54	278	42 10	321
253	59 11	042	45 45	096	36 34	145	45 14	187	31 11	244	52 58	278	41 35	321
254	59 49	041	46 42	096	37 07	146	45 06	189	30 20	245	52 01	278	40 59	321
	◆VEGA		ALTAIR		◆Nunki		ANTARES		SPICA		◆ARCTURUS		Alkaid	
255	60 26	040	47 38	097	37 38	147	44 57	190	29 28	245	51 05	279	40 22	320
256	61 03	039	48 35	097	38 09	148	44 47	191	28 36	246	50 08	279	39 46	320
257	61 39	038	49 32	098	38 39	149	44 35	192	27 44	246	49 12	279	39 09	320
258	62 14	037	50 28	098	39 08	150	44 22	193	26 51	247	48 16	279	38 32	320
259	62 48	036	51 25	098	39 36	151	44 09	195	25 59	247	47 19	279	37 55	319
260	63 21	035	52 21	099	40 03	152	43 53	196	25 06	248	46 23	279	37 18	319
261	63 53	034	53 17	099	40 30	153	43 37	197	24 13	248	45 27	279	36 40	319
262	64 24	032	54 14	100	40 55	154	43 20	198	23 20	249	44 30	280	36 03	319
263	64 54	031	55 10	101	41 20	155	43 02	199	22 27	249	43 34	280	35 25	318
264	65 22	029	56 06	101	41 44	156	42 42	201	21 33	250	42 38	280	34 47	318
265	65 50	028	57 02	102	42 06	157	42 22	202	20 39	250	41 42	280	34 09	318
266	66 16	026	57 57	102	42 28	158	42 00	203	19 46	251	40 45	280	33 31	318
267	66 40	025	58 53	103	42 48	159	41 37	204	18 52	251	39 49	280	32 52	318
268	67 03	023	59 49	104	43 08	161	41 14	205	17 58	251	38 53	281	32 14	318
269	67 24	021	60 44	104	43 26	162	40 49	206	17 04	252	37 57	281	31 35	317

LHA	Hc	Zn	Hc	Zn	Hc	Zn	Hc	Zn	Hc	Zn	Hc	Zn	Hc	Zn
♈	VEGA		◆DENEB		Enif		◆Nunki		ANTARES		◆ARCTURUS		Alkaid	
270	67 44	019	46 53	042	35 15	091	43 44	163	40 24	207	37 01	281	30 57	317
271	68 02	017	47 31	041	36 12	091	44 00	164	39 58	208	36 05	281	30 18	317
272	68 18	015	48 08	041	37 09	091	44 15	165	39 30	209	35 09	281	29 39	317
273	68 32	013	48 45	040	38 06	091	44 29	166	39 02	210	34 13	281	29 00	317
274	68 44	011	49 22	040	39 03	092	44 42	168	38 33	211	33 17	282	28 21	317
275	68 54	009	49 58	039	40 00	092	44 53	169	38 03	212	32 21	282	27 42	317
276	69 02	007	50 34	039	40 57	093	45 04	170	37 33	213	31 25	282	27 03	317
277	69 08	005	51 10	038	41 54	093	45 13	171	37 01	214	30 30	282	26 24	317
278	69 12	003	51 45	037	42 51	093	45 21	173	36 29	215	29 34	282	25 45	317
279	69 13	000	52 19	037	43 48	094	45 27	174	35 56	216	28 38	283	25 06	317
280	69 13	358	52 53	036	44 45	094	45 33	175	35 23	216	27 42	283	24 27	317
281	69 10	356	53 26	035	45 42	094	45 37	176	34 49	217	26 47	283	23 48	317
282	69 05	354	53 59	035	46 39	095	45 40	178	34 14	218	25 51	283	23 09	317
283	68 58	352	54 31	034	47 35	095	45 41	179	33 38	219	24 56	283	22 30	317
284	68 48	350	55 02	033	48 32	096	45 42	180	33 02	220	24 00	284	21 51	317
	◆DENEB		Alpheratz		◆FOMALHAUT		Nunki		ANTARES		◆Rasalhague		VEGA	
285	55 33	032	19 40	065	15 35	129	45 41	182	32 25	220	68 44	258	68 37	348
286	56 03	031	20 32	065	16 19	130	45 39	183	31 48	221	67 48	259	68 24	345
287	56 32	030	21 24	065	17 03	130	45 35	184	31 10	222	66 52	260	68 08	343
288	57 01	029	22 16	065	17 46	131	45 30	185	30 32	223	65 56	260	67 51	342
289	57 28	028	23 07	065	18 29	131	45 24	187	29 53	223	65 00	261	67 32	340
290	57 55	027	23 59	066	19 12	132	45 17	188	29 13	224	64 03	262	67 11	338
291	58 21	026	24 51	066	19 55	132	45 08	189	28 33	225	63 07	262	66 49	336
292	58 46	025	25 43	066	20 37	133	44 59	190	27 53	225	62 10	263	66 25	334
293	59 09	024	26 35	066	21 19	133	44 48	192	27 12	226	61 14	263	66 00	333
294	59 32	023	27 27	066	22 00	134	44 35	193	26 31	227	60 17	264	65 33	331
295	59 54	022	28 20	066	22 41	134	44 22	194	25 49	227	59 20	264	65 05	330
296	60 15	021	29 12	066	23 22	135	44 07	195	25 07	228	58 23	265	64 36	328
297	60 34	019	30 04	066	24 02	136	43 52	197	24 24	229	57 27	265	64 05	327
298	60 52	018	30 56	066	24 41	136	43 35	198	23 41	229	56 30	266	63 33	326
299	61 09	017	31 49	066	25 21	137	43 17	199	22 58	230	55 33	266	63 01	324
	◆DENEB		Schedar		Alpheratz		◆FOMALHAUT		Nunki		◆Rasalhague		VEGA	
300	61 25	015	25 54	035	32 41	067	25 59	137	42 58	200	54 36	266	62 27	323
301	61 39	014	26 27	035	33 33	067	26 38	138	42 38	201	53 39	267	61 52	322
302	61 52	012	26 59	035	34 26	067	27 16	139	42 17	202	52 42	267	61 17	321
303	62 04	011	27 32	035	35 18	067	27 53	139	41 54	203	51 45	268	60 41	320
304	62 14	010	28 05	035	36 10	067	28 30	140	41 31	204	50 48	268	60 04	319
305	62 23	008	28 37	035	37 03	067	29 06	141	41 07	206	49 51	268	59 26	318
306	62 30	007	29 10	035	37 55	067	29 42	142	40 42	207	48 54	269	58 47	317
307	62 36	005	29 42	034	38 48	067	30 17	142	40 16	208	47 57	269	58 08	316
308	62 40	004	30 14	034	39 40	067	30 51	143	39 49	209	47 00	269	57 29	316
309	62 43	002	30 47	034	40 33	067	31 25	144	39 21	210	46 03	270	56 49	315
310	62 44	000	31 19	034	41 25	067	31 59	145	38 53	211	45 06	270	56 08	314
311	62 44	359	31 50	034	42 18	067	32 31	145	38 23	212	44 09	270	55 27	314
312	62 42	357	32 22	034	43 10	067	33 03	146	37 53	213	43 12	270	54 45	313
313	62 39	356	32 54	033	44 02	067	33 35	147	37 22	213	42 14	271	54 03	312
314	62 34	354	33 25	033	44 55	067	34 05	148	36 50	214	41 17	271	53 21	312
	◆Alpheratz		Diphda		◆FOMALHAUT		Nunki		◆Rasalhague		VEGA		DENEB	
315	45 47	067	24 21	120	34 35	149	36 17	215	40 20	271	52 38	311	62 28	353
316	46 39	066	25 10	121	35 04	150	35 44	216	39 23	272	51 55	311	62 20	351
317	47 32	066	25 59	121	35 33	151	35 10	217	38 26	272	51 12	310	62 10	350
318	48 24	066	26 48	122	36 00	152	34 35	218	37 29	272	50 28	310	62 00	348
319	49 16	066	27 36	122	36 27	152	34 00	219	36 32	273	49 44	309	61 48	347
320	50 08	066	28 24	123	36 53	153	33 24	219	35 35	273	49 00	309	61 34	346
321	51 00	066	29 12	124	37 18	154	32 48	220	34 38	273	48 16	309	61 19	344
322	51 52	066	29 59	124	37 42	155	32 11	221	33 41	273	47 31	308	61 03	343
323	52 44	065	30 46	125	38 06	156	31 33	222	32 44	274	46 46	308	60 46	342
324	53 36	065	31 33	126	38 28	157	30 55	222	31 47	274	46 01	308	60 27	340
325	54 28	065	32 19	126	38 50	158	30 16	223	30 50	274	45 16	307	60 07	339
326	55 19	065	33 05	127	39 10	159	29 36	224	29 54	274	44 30	307	59 46	338
327	56 11	064	33 50	128	39 30	160	28 57	225	28 57	275	43 45	307	59 24	337
328	57 02	064	34 35	128	39 49	161	28 16	225	28 00	275	42 59	307	59 01	335
329	57 54	064	35 20	129	40 06	163	27 36	226	27 03	275	42 13	306	58 37	334
	◆Mirfak		Hamal		Diphda		◆FOMALHAUT		ALTAIR		◆VEGA		DENEB	
330	19 24	042	32 26	073	36 04	130	40 23	164	57 17	258	41 27	306	58 11	333
331	20 03	043	33 21	073	36 48	130	40 39	165	56 21	259	40 41	306	57 45	332
332	20 41	043	34 15	073	37 31	131	40 53	166	55 25	259	39 55	306	57 18	331
333	21 20	043	35 10	074	38 13	132	41 07	167	54 29	260	39 09	306	56 50	330
334	21 59	043	36 05	074	38 56	133	41 19	168	53 32	260	38 23	306	56 22	329
335	22 37	043	36 59	074	39 37	134	41 30	169	52 36	261	37 36	306	55 52	328
336	23 16	043	37 54	074	40 18	135	41 41	170	51 40	261	36 50	305	55 22	328
337	23 55	043	38 49	074	40 58	135	41 50	171	50 43	262	36 03	305	54 51	327
338	24 33	043	39 44	074	41 38	136	41 58	173	49 47	262	35 17	305	54 19	326
339	25 12	043	40 39	074	42 17	137	42 04	174	48 50	263	34 30	305	53 47	325
340	25 51	043	41 34	074	42 55	138	42 10	175	47 54	263	33 43	305	53 14	324
341	26 29	043	42 29	074	43 33	139	42 15	176	46 57	264	32 57	305	52 40	324
342	27 08	043	43 24	074	44 10	140	42 18	177	46 00	264	32 10	305	52 06	323
343	27 47	043	44 19	074	44 46	141	42 20	178	45 03	264	31 23	305	51 32	322
344	28 25	043	45 14	075	45 21	142	42 21	180	44 07	265	30 36	305	50 57	322
	Schedar		◆Mirfak		ALDEBARAN		Diphda		◆FOMALHAUT		ALTAIR		◆DENEB	
345	47 09	020	29 04	042	10 38	076	45 56	143	42 21	181	43 10	265	50 21	321
346	47 28	019	29 42	042	11 33	076	46 29	145	42 19	182	42 13	266	49 45	321
347	47 47	019	30 20	042	12 29	076	47 02	146	42 17	183	41 16	266	49 09	320
348	48 05	018	30 59	042	13 24	077	47 34	147	42 13	184	40 19	266	48 32	319
349	48 22	017	31 37	042	14 20	077	48 04	148	42 08	185	39 22	267	47 54	319
350	48 39	017	32 15	042	15 16	077	48 34	149	42 02	187	38 25	267	47 17	319
351	48 55	016	32 53	042	16 11	077	49 03	151	41 55	188	37 28	267	46 39	318
352	49 10	015	33 31	042	17 07	078	49 30	152	41 47	189	36 31	268	46 00	318
353	49 25	014	34 09	041	18 03	078	49 57	153	41 38	190	35 34	268	45 22	317
354	49 39	014	34 46	041	18 59	078	50 22	154	41 27	191	34 37	268	44 43	317
355	49 52	013	35 24	041	19 55	078	50 46	156	41 15	192	33 40	269	44 04	316
356	50 04	012	36 01	041	20 50	079	51 09	157	41 03	193	32 43	269	43 24	316
357	50 16	011	36 38	041	21 46	079	51 30	159	40 49	195	31 46	269	42 45	316
358	50 26	010	37 15	040	22 42	079	51 50	160	40 34	196	30 49	270	42 05	315
359	50 36	010	37 52	040	23 39	079	52 09	161	40 18	197	29 52	270	41 25	315

LAT 17°N

LHA ♈	Hc	Zn	Hc	Zn	Hc	Zn	Hc	Zn	Hc	Zn	Hc	Zn	Hc	Zn
	Schedar		CAPELLA		◆ALDEBARAN		ACHERNAR		◆FOMALHAUT		ALTAIR		◆DENEB	
0	49 46	009	19 39	046	24 24	079	13 01	167	40 58	198	28 54	271	40 02	316
1	49 54	008	20 20	046	25 20	079	13 13	167	40 40	199	27 57	271	39 21	315
2	50 01	007	21 02	047	26 16	080	13 26	168	40 20	200	26 59	271	38 41	315
3	50 08	006	21 43	047	27 13	080	13 38	168	40 00	201	26 02	272	38 00	315
4	50 14	005	22 25	047	28 09	080	13 49	169	39 39	202	25 05	272	37 19	315
5	50 18	004	23 07	047	29 06	080	14 00	169	39 16	203	24 07	272	36 38	314
6	50 22	004	23 48	047	30 02	080	14 10	170	38 53	204	23 10	273	35 57	314
7	50 25	003	24 30	047	30 59	081	14 20	170	38 29	205	22 13	273	35 16	314
8	50 28	002	25 12	047	31 56	081	14 30	171	38 04	206	21 15	273	34 35	314
9	50 29	001	25 53	047	32 52	081	14 38	171	37 38	207	20 18	273	33 53	314
10	50 29	000	26 35	047	33 49	081	14 47	172	37 12	208	19 21	274	33 11	313
11	50 29	359	27 17	046	34 46	081	14 54	173	36 44	209	18 23	274	32 30	313
12	50 28	358	27 58	046	35 43	082	15 01	173	36 16	210	17 26	274	31 48	313
13	50 26	357	28 40	046	36 39	082	15 08	174	35 47	211	16 29	274	31 06	313
14	50 23	357	29 21	046	37 36	082	15 14	174	35 17	212	15 32	275	30 24	313
	CAPELLA		◆ALDEBARAN		RIGEL		ACHERNAR		◆FOMALHAUT		Enif		◆DENEB	
15	30 03	046	38 33	082	22 18	107	15 20	175	34 46	213	41 55	268	29 42	313
16	30 44	046	39 30	082	23 13	107	15 25	175	34 15	213	40 57	268	29 00	313
17	31 26	046	40 27	083	24 08	107	15 29	176	33 43	214	40 00	269	28 18	313
18	32 07	046	41 24	083	25 03	108	15 33	176	33 10	215	39 03	269	27 36	313
19	32 48	046	42 20	083	25 57	108	15 36	177	32 37	216	38 05	269	26 54	313
20	33 29	046	43 17	083	26 51	109	15 39	178	32 03	217	37 08	270	26 12	313
21	34 10	045	44 14	083	27 46	109	15 41	178	31 28	217	36 10	270	25 29	313
22	34 51	045	45 11	083	28 40	110	15 43	179	30 53	218	35 13	270	24 47	313
23	35 31	045	46 08	084	29 34	110	15 44	179	30 17	219	34 16	270	24 05	313
24	36 12	045	47 05	084	30 27	111	15 44	180	29 41	220	33 18	271	23 23	313
25	36 52	045	48 03	084	31 21	111	15 44	180	29 04	220	32 21	271	22 40	313
26	37 33	044	49 00	084	32 14	112	15 44	181	28 27	221	31 24	271	21 58	313
27	38 13	044	49 57	084	33 08	112	15 42	181	27 49	222	30 26	272	21 16	313
28	38 53	044	50 54	085	34 01	113	15 41	182	27 10	222	29 29	272	20 34	313
29	39 33	044	51 51	085	34 54	113	15 38	183	26 31	223	28 32	272	19 52	313
	Mirfak		◆CAPELLA		ALDEBARAN		RIGEL		◆ACHERNAR		FOMALHAUT		◆Alpheratz	
30	53 03	023	40 12	043	52 48	085	35 46	114	15 35	183	25 52	224	61 41	300
31	53 25	022	40 51	043	53 45	085	36 39	114	15 32	184	25 12	224	60 51	300
32	53 46	021	41 30	043	54 42	085	37 31	115	15 28	184	24 32	225	60 01	299
33	54 06	020	42 09	042	55 40	085	38 22	116	15 23	185	23 51	226	59 11	299
34	54 25	019	42 48	042	56 37	086	39 14	116	15 18	185	23 10	226	58 20	298
35	54 43	018	43 26	042	57 34	086	40 05	117	15 13	186	22 28	227	57 30	298
36	55 00	017	44 04	041	58 31	086	40 56	118	15 06	186	21 47	227	56 39	297
37	55 16	016	44 42	041	59 29	086	41 47	118	15 00	187	21 04	228	55 48	297
38	55 31	015	45 19	040	60 26	086	42 38	119	14 52	188	20 22	228	54 56	297
39	55 46	014	45 57	040	61 23	087	43 28	120	14 45	188	19 39	229	54 05	296
40	55 59	013	46 33	040	62 20	087	44 17	120	14 36	189	18 55	229	53 14	296
41	56 11	012	47 10	039	63 18	087	45 06	121	14 27	189	18 11	230	52 22	296
42	56 22	010	47 46	039	64 15	087	45 55	122	14 18	190	17 27	230	51 30	296
43	56 32	009	48 21	038	65 12	087	46 44	123	14 08	190	16 43	231	50 38	295
44	56 41	008	48 56	037	66 10	088	47 32	124	13 58	191	15 59	231	49 47	295
	CAPELLA		◆POLLUX		PROCYON		SIRIUS		◆Acamar		Diphda		◆Alpheratz	
45	49 31	037	24 07	066	20 52	091	25 09	118	32 40	180	41 27	225	48 55	295
46	50 05	036	25 00	066	21 49	091	26 00	119	32 40	181	40 46	226	48 03	295
47	50 39	036	25 53	067	22 47	091	26 50	119	32 38	182	40 04	227	47 10	295
48	51 12	035	26 45	067	23 44	092	27 39	120	32 35	183	39 22	228	46 18	295
49	51 44	034	27 38	067	24 41	092	28 29	121	32 31	184	38 39	229	45 26	294
50	52 16	033	28 31	067	25 39	092	29 18	121	32 27	185	37 56	230	44 34	294
51	52 48	033	29 23	067	26 36	093	30 07	122	32 22	186	37 12	230	43 41	294
52	53 18	032	30 16	067	27 33	093	30 56	122	32 15	187	36 27	231	42 49	294
53	53 48	031	31 09	067	28 31	093	31 44	123	32 08	188	35 43	232	41 57	294
54	54 18	030	32 02	067	29 28	094	32 32	123	32 00	189	34 57	233	41 04	294
55	54 46	029	32 55	067	30 25	094	33 20	124	31 51	189	34 11	233	40 12	294
56	55 14	029	33 48	067	31 22	094	34 07	125	31 41	190	33 25	234	39 20	294
57	55 41	028	34 41	067	32 19	095	34 54	125	31 31	191	32 39	235	38 27	294
58	56 07	027	35 34	067	33 17	095	35 41	126	31 19	192	31 52	235	37 35	294
59	56 33	026	36 27	067	34 14	095	36 27	127	31 07	193	31 04	236	36 42	294
	CAPELLA		◆POLLUX		PROCYON		SIRIUS		◆CANOPUS		Diphda		◆Hamal	
60	56 57	025	37 20	067	35 11	096	37 13	128	13 41	159	30 17	236	62 45	288
61	57 20	024	38 12	067	36 08	096	37 58	128	14 02	159	29 29	237	61 50	288
62	57 43	022	39 05	067	37 05	096	38 43	129	14 22	160	28 41	238	60 56	288
63	58 04	021	39 58	067	38 02	097	39 27	130	14 42	160	27 52	238	60 01	288
64	58 24	020	40 51	067	38 59	097	40 11	131	15 01	161	27 03	239	59 06	287
65	58 44	019	41 44	067	39 56	098	40 54	131	15 20	161	26 14	239	58 12	287
66	59 02	018	42 37	067	40 53	098	41 37	132	15 38	162	25 24	240	57 17	287
67	59 19	017	43 30	067	41 50	098	42 19	133	15 56	162	24 35	240	56 22	287
68	59 34	015	44 23	067	42 46	099	43 00	134	16 13	163	23 45	241	55 27	287
69	59 49	014	45 16	067	43 43	099	43 41	135	16 30	163	22 55	241	54 32	287
70	60 02	013	46 09	067	44 40	100	44 21	136	16 46	164	22 04	242	53 37	287
71	60 14	011	47 01	067	45 36	100	45 01	137	17 02	164	21 13	242	52 42	287
72	60 25	010	47 54	067	46 33	101	45 40	138	17 17	165	20 23	243	51 47	287
73	60 34	009	48 47	067	47 29	101	46 18	139	17 32	166	19 31	243	50 52	286
74	60 42	007	49 40	066	48 25	102	46 55	140	17 45	166	18 40	244	49 57	286
	CAPELLA		◆POLLUX		PROCYON		SIRIUS		◆RIGEL		Menkar		◆Hamal	
75	60 48	006	50 32	066	49 21	102	47 31	141	64 33	172	58 19	249	49 02	286
76	60 54	004	51 25	066	50 17	103	48 07	142	64 40	174	57 26	250	48 07	286
77	60 57	003	52 17	066	51 13	103	48 41	143	64 45	176	56 31	251	47 12	286
78	61 00	002	53 09	066	52 09	104	49 15	145	64 47	179	55 37	252	46 17	286
79	61 00	000	54 02	065	53 05	104	49 48	146	64 47	181	54 43	252	45 22	286
80	61 00	359	54 54	065	54 00	105	50 20	147	64 45	183	53 48	253	44 27	286
81	60 58	357	55 46	065	54 56	106	50 50	148	64 41	186	52 53	254	43 32	286
82	60 54	356	56 38	065	55 51	106	51 20	150	64 34	188	51 58	254	42 37	286
83	60 49	354	57 29	064	56 46	107	51 48	151	64 25	190	51 02	255	41 42	287
84	60 43	353	58 21	064	57 41	108	52 15	152	64 14	192	50 07	256	40 47	287
85	60 35	352	59 12	063	58 35	109	52 41	154	64 00	195	49 11	256	39 52	287
86	60 26	350	60 04	063	59 29	109	53 06	155	63 45	197	48 15	257	38 57	287
87	60 16	349	60 55	063	60 23	110	53 29	157	63 27	199	47 20	257	38 02	287
88	60 04	348	61 45	062	61 17	111	53 51	158	63 07	201	46 24	258	37 07	287
89	59 51	346	62 36	061	62 10	112	54 12	160	62 46	203	45 27	258	36 12	287

LHA ♈	Hc	Zn	Hc	Zn	Hc	Zn	Hc	Zn	Hc	Zn	Hc	Zn	Hc	Zn
	Dubhe		◆REGULUS		Alphard		SIRIUS		◆RIGEL		ALDEBARAN		◆CAPELLA	
90	21 36	030	29 58	086	32 42	113	54 31	161	62 23	205	69 48	272	59 37	345
91	22 04	030	30 55	086	33 35	113	54 49	163	61 57	207	68 51	272	59 21	344
92	22 33	029	31 53	086	34 28	114	55 05	164	61 31	209	67 54	272	59 05	342
93	23 01	029	32 50	086	35 20	114	55 20	166	61 02	211	66 56	272	58 47	341
94	23 29	029	33 47	087	36 13	115	55 33	168	60 32	212	65 59	272	58 28	340
95	23 57	029	34 44	087	37 05	115	55 44	169	60 01	214	65 02	273	58 07	339
96	24 25	029	35 42	087	37 56	116	55 54	171	59 28	216	64 04	273	57 46	338
97	24 53	029	36 39	088	38 48	117	56 02	173	58 54	217	63 07	273	57 24	337
98	25 21	029	37 36	088	39 39	117	56 08	174	58 19	219	62 10	273	57 01	336
99	25 49	029	38 34	088	40 30	118	56 13	176	57 42	220	61 13	273	56 36	335
100	26 16	029	39 31	088	41 20	119	56 16	178	57 04	222	60 15	274	56 11	334
101	26 44	029	40 28	089	42 11	119	56 17	180	56 26	223	59 18	274	55 45	333
102	27 11	029	41 26	089	43 00	120	56 17	181	55 46	224	58 21	274	55 18	332
103	27 38	028	42 23	089	43 50	121	56 15	183	55 05	226	57 24	274	54 51	331
104	28 06	028	43 21	089	44 39	122	56 11	185	54 24	227	56 26	274	54 22	330
	Dubhe		◆REGULUS		Alphard		SIRIUS		◆RIGEL		ALDEBARAN		◆CAPELLA	
105	28 33	028	44 18	090	45 28	122	56 05	186	53 42	228	55 29	275	53 53	329
106	29 00	028	45 15	090	46 16	123	55 58	188	52 59	229	54 32	275	53 23	328
107	29 26	028	46 13	090	47 04	124	55 49	190	52 15	230	53 35	275	52 53	327
108	29 53	028	47 10	091	47 51	125	55 38	192	51 30	231	52 38	275	52 21	327
109	30 19	027	48 07	091	48 38	126	55 26	193	50 45	232	51 40	275	51 50	326
110	30 46	027	49 05	091	49 24	127	55 12	195	49 59	233	50 43	275	51 17	325
111	31 12	027	50 02	092	50 10	128	54 57	196	49 13	234	49 46	276	50 44	325
112	31 37	027	50 59	092	50 55	129	54 40	198	48 26	235	48 49	276	50 10	324
113	32 03	026	51 57	092	51 40	130	54 21	200	47 39	236	47 52	276	49 36	323
114	32 28	026	52 54	093	52 23	131	54 01	201	46 51	237	46 55	276	49 02	323
115	32 54	026	53 51	093	53 06	132	53 40	203	46 03	238	45 58	276	48 27	322
116	33 19	026	54 49	093	53 49	133	53 17	204	45 14	239	45 01	277	47 51	322
117	33 43	025	55 46	094	54 30	134	52 53	206	44 25	239	44 04	277	47 15	321
118	34 08	025	56 43	094	55 11	136	52 28	207	43 35	240	43 07	277	46 39	321
119	34 32	025	57 41	094	55 50	137	52 01	208	42 45	241	42 10	277	46 02	320
	Dubhe		◆Denebola		Alphard		◆SIRIUS		RIGEL		ALDEBARAN		◆CAPELLA	
120	34 56	024	35 06	084	56 29	138	51 33	210	41 55	242	41 13	277	45 25	320
121	35 19	024	36 03	084	57 07	140	51 04	211	41 04	242	40 16	277	44 48	319
122	35 43	024	37 00	084	57 43	141	50 34	212	40 13	243	39 19	278	44 10	319
123	36 06	023	37 57	085	58 19	143	50 03	214	39 22	244	38 22	278	43 32	318
124	36 28	023	38 54	085	58 53	144	49 30	215	38 30	244	37 25	278	42 54	318
125	36 51	023	39 51	085	59 26	146	48 57	216	37 38	245	36 29	278	42 16	318
126	37 13	022	40 49	085	59 58	147	48 23	217	36 46	245	35 32	278	41 37	317
127	37 34	022	41 46	085	60 28	149	47 48	218	35 54	246	34 35	279	40 58	317
128	37 56	022	42 43	086	60 56	151	47 12	219	35 01	247	33 39	279	40 18	317
129	38 17	021	43 40	086	61 24	153	46 35	220	34 09	247	32 42	279	39 39	316
130	38 37	021	44 37	086	61 49	155	45 57	222	33 16	248	31 45	279	38 59	316
131	38 57	020	45 35	086	62 13	157	45 19	223	32 22	248	30 49	279	38 19	316
132	39 17	020	46 32	087	62 35	158	44 40	224	31 29	249	29 52	280	37 39	316
133	39 36	019	47 29	087	62 55	161	44 00	225	30 35	249	28 56	280	36 59	315
134	39 55	019	48 27	087	63 13	163	43 19	225	29 42	250	27 59	280	36 18	315
	Dubhe		◆ARCTURUS		SPICA		◆Alphard		SIRIUS		BETELGEUSE		◆CAPELLA	
135	40 14	019	15 42	074	18 46	108	63 29	165	42 38	226	43 53	264	35 38	315
136	40 32	018	16 38	074	19 40	109	63 43	167	41 56	227	42 56	264	34 57	315
137	40 49	018	17 33	075	20 35	109	63 55	169	41 14	228	41 59	265	34 16	315
138	41 06	017	18 28	075	21 29	110	64 05	171	40 31	229	41 02	265	33 35	314
139	41 23	017	19 24	075	22 23	110	64 13	174	39 47	230	40 04	265	32 54	314
140	41 39	016	20 19	075	23 16	111	64 18	176	39 03	231	39 07	266	32 13	314
141	41 55	015	21 15	076	24 10	111	64 21	178	38 19	231	38 10	266	31 32	314
142	42 10	015	22 10	076	25 03	111	64 22	180	37 34	232	37 13	266	30 51	314
143	42 24	014	23 06	076	25 57	112	64 20	183	36 48	233	36 15	267	30 09	314
144	42 38	014	24 02	076	26 50	112	64 16	185	36 02	234	35 18	267	29 28	314
145	42 52	013	24 57	076	27 43	113	64 10	187	35 16	234	34 21	267	28 46	314
146	43 05	013	25 53	076	28 36	113	64 02	189	34 29	235	33 24	268	28 05	314
147	43 17	012	26 49	077	29 28	114	63 52	192	33 42	236	32 26	268	27 23	314
148	43 29	012	27 45	077	30 21	114	63 39	194	32 55	236	31 29	268	26 42	313
149	43 40	011	28 41	077	31 13	115	63 24	196	32 07	237	30 31	269	26 00	313
	Dubhe		◆ARCTURUS		SPICA		◆Alphard		SIRIUS		BETELGEUSE		◆CAPELLA	
150	43 50	010	29 37	077	32 05	115	63 07	198	31 18	237	29 34	269	25 18	313
151	44 00	010	30 32	077	32 56	116	62 49	200	30 30	238	28 37	269	24 37	313
152	44 10	009	31 28	078	33 48	117	62 28	202	29 41	239	27 39	270	23 55	313
153	44 18	008	32 25	078	34 39	117	62 05	204	28 52	239	26 42	270	23 13	313
154	44 27	008	33 21	078	35 30	118	61 41	206	28 03	240	25 45	270	22 32	313
155	44 34	007	34 17	078	36 20	118	61 15	208	27 13	240	24 47	270	21 50	313
156	44 41	007	35 13	078	37 11	119	60 47	210	26 23	241	23 50	271	21 08	313
157	44 47	006	36 09	078	38 01	120	60 18	211	25 33	241	22 52	271	20 27	314
158	44 53	005	37 05	078	38 50	120	59 47	213	24 42	242	21 55	271	19 45	314
159	44 58	005	38 01	079	39 40	121	59 15	215	23 51	242	20 58	272	19 04	314
160	45 02	004	38 58	079	40 29	122	58 42	216	23 00	243	20 00	272	18 22	314
161	45 05	003	39 54	079	41 17	122	58 07	218	22 09	243	19 03	272	17 41	314
162	45 08	003	40 50	079	42 06	123	57 31	219	21 18	244	18 06	272	16 59	314
163	45 11	002	41 47	079	42 53	124	56 54	221	20 26	244	17 08	273	16 18	314
164	45 12	001	42 43	079	43 41	125	56 16	222	19 35	245	16 11	273	15 37	314
	◆Dubhe		Alkaid		ARCTURUS		◆SPICA		Suhail		◆PROCYON		POLLUX	
165	45 13	001	43 19	037	43 39	079	44 28	126	24 20	202	39 29	263	43 57	293
166	45 13	000	43 53	036	44 36	080	45 14	127	23 59	203	38 32	263	43 04	293
167	45 13	359	44 27	036	45 32	080	46 00	127	23 36	203	37 35	263	42 11	293
168	45 12	359	45 00	035	46 29	080	46 45	128	23 13	204	36 38	264	41 18	293
169	45 10	358	45 33	035	47 25	080	47 30	129	22 49	205	35 41	264	40 25	293
170	45 08	357	46 06	034	48 22	080	48 14	130	22 25	205	34 44	265	39 32	293
171	45 04	357	46 38	034	49 18	080	48 57	131	22 00	206	33 46	265	38 39	293
172	45 01	356	47 10	033	50 15	080	49 40	132	21 35	207	32 49	265	37 46	293
173	44 56	355	47 41	033	51 11	080	50 22	133	21 09	207	31 52	266	36 53	293
174	44 51	355	48 11	032	52 08	080	51 04	134	20 42	208	30 55	266	36 00	293
175	44 45	354	48 41	031	53 04	080	51 44	136	20 15	209	29 58	266	35 07	293
176	44 39	353	49 11	031	54 01	081	52 24	137	19 47	209	29 00	267	34 14	293
177	44 32	353	49 40	030	54 58	081	53 03	138	19 19	210	28 03	267	33 22	293
178	44 24	352	50 09	029	55 54	081	53 41	139	18 51	210	27 06	267	32 29	293
179	44 16	351	50 36	029	56 51	081	54 18	140	18 21	211	26 08	267	31 36	293

LHA ♈	Hc	Zn	Hc	Zn	Hc	Zn	Hc	Zn	Hc	Zn	Hc	Zn	Hc	Zn
	◆Alkaid		ANTARES		◆SPICA		Gienah		◆REGULUS		POLLUX		Dubhe	
180	51 04	028	11 35	123	54 54	142	55 16	173	62 29	263	30 43	293	44 07	351
181	51 30	027	12 23	123	55 29	143	55 22	175	61 32	264	29 50	293	43 58	350
182	51 56	026	13 11	123	56 03	145	55 26	177	60 35	264	28 57	293	43 47	350
183	52 21	026	13 59	124	56 35	146	55 28	179	59 38	265	28 05	293	43 37	349
184	52 45	025	14 47	124	57 07	148	55 29	180	58 41	265	27 12	293	43 25	348
185	53 09	024	15 34	125	57 37	149	55 28	182	57 44	266	26 19	293	43 13	348
186	53 32	023	16 21	125	58 05	151	55 25	184	56 46	266	25 27	294	43 01	347
187	53 54	022	17 08	126	58 33	152	55 21	185	55 49	266	24 34	294	42 48	347
188	54 15	021	17 54	126	58 58	154	55 15	187	54 52	267	23 41	294	42 34	346
189	54 35	020	18 41	126	59 23	156	55 07	189	53 55	267	22 49	294	42 20	345
190	54 54	019	19 27	127	59 45	158	54 58	190	52 57	267	21 57	294	42 05	345
191	55 13	018	20 12	127	60 06	160	54 47	192	52 00	268	21 04	294	41 50	344
192	55 30	017	20 58	128	60 25	161	54 34	193	51 03	268	20 12	294	41 34	344
193	55 46	016	21 43	129	60 43	163	54 20	195	50 05	268	19 20	294	41 18	343
194	56 02	015	22 28	129	60 58	165	54 05	197	49 08	269	18 27	295	41 01	343
	Kochab		◆VEGA		Rasalhague		ANTARES		◆SPICA		◆REGULUS		Dubhe	
195	30 49	008	14 59	053	23 46	083	23 12	130	61 12	167	48 10	269	40 44	342
196	30 57	008	15 45	054	24 43	084	23 56	130	61 24	169	47 13	269	40 27	342
197	31 05	008	16 31	054	25 40	084	24 40	131	61 33	171	46 16	270	40 08	341
198	31 13	008	17 18	054	26 37	084	25 23	131	61 41	173	45 18	270	39 50	341
199	31 20	007	18 04	054	27 34	084	26 06	132	61 47	175	44 21	270	39 31	340
200	31 28	007	18 50	054	28 31	085	26 48	133	61 50	177	43 24	271	39 11	340
201	31 35	007	19 37	054	29 29	085	27 30	133	61 52	180	42 26	271	38 52	340
202	31 41	006	20 23	054	30 26	085	28 12	134	61 51	182	41 29	271	38 31	339
203	31 47	006	21 10	054	31 23	085	28 53	134	61 49	184	40 31	271	38 11	339
204	31 54	006	21 56	054	32 20	086	29 34	135	61 44	186	39 34	272	37 50	338
205	31 59	006	22 43	054	33 17	086	30 14	136	61 37	188	38 37	272	37 28	338
206	32 05	005	23 30	054	34 15	086	30 54	136	61 28	190	37 39	272	37 06	338
207	32 10	005	24 16	055	35 12	086	31 33	137	61 18	192	36 42	272	36 44	337
208	32 15	005	25 03	055	36 09	087	32 12	138	61 05	194	35 45	273	36 22	337
209	32 19	004	25 50	055	37 06	087	32 50	139	60 50	196	34 47	273	35 59	336
	Kochab		◆VEGA		Rasalhague		◆ANTARES		SPICA		◆REGULUS		Dubhe	
210	32 23	004	26 37	055	38 04	087	33 28	139	60 34	198	33 50	273	35 36	336
211	32 27	004	27 23	055	39 01	087	34 05	140	60 15	200	32 53	274	35 13	336
212	32 31	003	28 10	055	39 58	088	34 41	141	59 55	201	31 56	274	34 49	335
213	32 34	003	28 57	055	40 56	088	35 17	142	59 33	203	30 58	274	34 25	335
214	32 37	003	29 44	055	41 53	088	35 52	143	59 10	205	30 01	274	34 01	335
215	32 40	002	30 31	055	42 50	088	36 26	144	58 45	207	29 04	275	33 36	335
216	32 42	002	31 17	055	43 48	089	37 00	144	58 18	208	28 07	275	33 11	334
217	32 44	002	32 04	054	44 45	089	37 33	145	57 50	210	27 10	275	32 46	334
218	32 46	002	32 51	054	45 42	089	38 06	146	57 21	212	26 12	275	32 21	334
219	32 47	001	33 37	054	46 40	090	38 37	147	56 50	213	25 15	276	31 56	334
220	32 48	001	34 24	054	47 37	090	39 08	148	56 18	215	24 18	276	31 30	333
221	32 49	001	35 10	054	48 35	090	39 38	149	55 45	216	23 21	276	31 04	333
222	32 49	000	35 57	054	49 32	090	40 07	150	55 10	218	22 24	276	30 38	333
223	32 49	000	36 43	054	50 29	091	40 35	151	54 35	219	21 27	277	30 12	333
224	32 49	000	37 30	054	51 27	091	41 03	152	53 58	220	20 30	277	29 45	332
	VEGA		◆ALTAIR		Nunki		ANTARES		◆SPICA		Denebola		◆Alkaid	
225	38 16	054	19 05	086	18 23	126	41 29	153	53 21	221	44 03	274	54 28	340
226	39 02	053	20 03	087	19 09	127	41 55	154	52 42	223	43 06	274	54 08	339
227	39 48	053	21 00	087	19 55	127	42 19	155	52 03	224	42 09	275	53 47	338
228	40 34	053	21 57	087	20 41	128	42 43	156	51 23	225	41 12	275	53 24	337
229	41 20	053	22 54	087	21 26	128	43 05	157	50 42	226	40 14	275	53 01	336
230	42 06	053	23 52	088	22 11	129	43 27	159	50 00	227	39 17	275	52 37	335
231	42 51	052	24 49	088	22 55	129	43 47	160	49 17	228	38 20	275	52 13	334
232	43 36	052	25 46	088	23 40	130	44 07	161	48 34	229	37 23	276	51 48	333
233	44 22	052	26 44	088	24 24	130	44 25	162	47 50	230	36 26	276	51 21	333
234	45 07	052	27 41	089	25 07	131	44 42	163	47 06	231	35 29	276	50 55	332
235	45 52	051	28 39	089	25 50	132	44 58	164	46 21	232	34 32	276	50 27	331
236	46 36	051	29 36	089	26 33	132	45 13	166	45 35	233	33 35	277	49 59	330
237	47 21	051	30 33	090	27 15	133	45 27	167	44 49	234	32 38	277	49 31	330
238	48 05	050	31 31	090	27 57	133	45 39	168	44 02	235	31 41	277	49 01	329
239	48 49	050	32 28	090	28 39	134	45 50	169	43 15	236	30 44	277	48 32	328
	◆VEGA		ALTAIR		Nunki		◆ANTARES		SPICA		◆ARCTURUS		Alkaid	
240	49 33	049	33 25	091	29 20	135	46 00	171	42 28	236	65 05	279	48 01	328
241	50 16	049	34 23	091	30 00	135	46 09	172	41 40	237	64 08	279	47 30	327
242	50 59	048	35 20	091	30 40	136	46 17	173	40 51	238	63 11	279	46 59	327
243	51 42	048	36 18	091	31 20	137	46 23	174	40 03	239	62 15	279	46 27	326
244	52 24	047	37 15	092	31 59	137	46 28	176	39 13	239	61 18	279	45 55	326
245	53 06	047	38 12	092	32 37	138	46 31	177	38 24	240	60 21	279	45 22	325
246	53 48	046	39 10	092	33 15	139	46 34	178	37 34	241	59 25	279	44 49	325
247	54 29	046	40 07	093	33 53	140	46 35	180	36 44	241	58 28	279	44 16	324
248	55 10	045	41 04	093	34 30	141	46 34	181	35 53	242	57 31	279	43 42	324
249	55 51	044	42 01	093	35 06	141	46 33	182	35 03	243	56 35	279	43 08	323
250	56 30	044	42 59	094	35 41	142	46 30	184	34 12	243	55 38	279	42 33	323
251	57 10	043	43 56	094	36 16	143	46 26	185	33 20	244	54 42	279	41 58	322
252	57 48	042	44 53	094	36 50	144	46 20	186	32 29	244	53 45	279	41 23	322
253	58 26	041	45 50	095	37 24	145	46 13	187	31 37	245	52 48	280	40 48	322
254	59 04	040	46 48	095	37 56	146	46 05	189	30 45	245	51 52	280	40 12	321
	◆VEGA		ALTAIR		◆Nunki		ANTARES		SPICA		◆ARCTURUS		Alkaid	
255	59 40	039	47 45	096	38 28	147	45 56	190	29 53	246	50 55	280	39 36	321
256	60 16	038	48 42	096	39 00	148	45 46	191	29 00	246	49 59	280	39 00	321
257	60 51	037	49 39	096	39 30	148	45 34	192	28 07	247	49 02	280	38 23	320
258	61 26	036	50 36	097	39 59	149	45 21	194	27 15	247	48 06	280	37 46	320
259	61 59	035	51 33	097	40 28	150	45 07	195	26 21	248	47 09	280	37 09	320
260	62 31	034	52 30	098	40 56	151	44 51	196	25 28	248	46 13	280	36 32	320
261	63 03	032	53 26	098	41 23	152	44 35	197	24 35	249	45 16	280	35 55	319
262	63 33	031	54 23	099	41 49	154	44 17	199	23 41	249	44 20	281	35 17	319
263	64 02	030	55 20	099	42 14	155	43 58	200	22 48	250	43 23	281	34 40	319
264	64 30	028	56 17	100	42 38	156	43 38	201	21 54	250	42 27	281	34 02	319
265	64 56	027	57 13	100	43 01	157	43 17	202	21 00	251	41 31	281	33 24	318
266	65 22	025	58 09	101	43 23	158	42 55	203	20 05	251	40 34	281	32 46	318
267	65 45	024	59 06	101	43 44	159	42 32	204	19 11	251	39 38	281	32 08	318
268	66 08	022	60 02	102	44 04	160	42 08	205	18 17	252	38 42	281	31 29	318
269	66 28	020	60 58	103	44 23	161	41 43	206	17 22	252	37 46	281	30 51	318

LAT 17°N

LHA ♈	Hc	Zn	Hc	Zn	Hc	Zn	Hc	Zn	Hc	Zn	Hc	Zn	Hc	Zn
	VEGA		◆DENEB		Enif		◆Nunki		ANTARES		◆ARCTURUS		Alkaid	
270	66 47	018	46 08	041	35 15	090	44 41	163	41 17	207	36 49	282	30 12	318
271	67 04	017	46 45	041	36 12	090	44 58	164	40 50	208	35 53	282	29 34	318
272	67 20	015	47 23	040	37 10	090	45 13	165	40 23	209	34 57	282	28 55	318
273	67 34	013	47 59	040	38 07	091	45 27	166	39 54	210	34 01	282	28 16	317
274	67 45	011	48 36	039	39 04	091	45 40	167	39 24	211	33 05	282	27 37	317
275	67 55	009	49 12	039	40 02	091	45 52	169	38 54	212	32 09	282	26 58	317
276	68 03	007	49 47	038	40 59	092	46 03	170	38 23	213	31 13	283	26 19	317
277	68 08	005	50 22	037	41 56	092	46 12	171	37 51	214	30 17	283	25 40	317
278	68 12	003	50 57	037	42 54	092	46 20	173	37 19	215	29 21	283	25 01	317
279	68 13	000	51 31	036	43 51	093	46 27	174	36 45	216	28 25	283	24 22	317
280	68 13	358	52 04	035	44 48	093	46 33	175	36 11	217	27 29	283	23 43	317
281	68 10	356	52 37	035	45 46	093	46 37	176	35 36	218	26 33	283	23 04	317
282	68 05	354	53 09	034	46 43	094	46 40	178	35 01	219	25 37	284	22 25	317
283	67 58	352	53 41	033	47 40	094	46 41	179	34 25	219	24 41	284	21 46	317
284	67 49	350	54 12	032	48 37	094	46 42	180	33 48	220	23 46	284	21 07	317
	◆DENEB		Alpheratz		◆FOMALHAUT		Nunki		ANTARES		◆Rasalhague		VEGA	
285	54 42	031	19 15	064	16 13	129	46 41	182	33 11	221	68 55	261	67 38	348
286	55 12	031	20 06	065	16 57	129	46 38	183	32 33	222	67 59	261	67 25	346
287	55 40	030	20 58	065	17 41	130	46 35	184	31 55	222	67 02	262	67 11	344
288	56 08	029	21 50	065	18 25	130	46 30	186	31 16	223	66 05	263	66 54	342
289	56 35	028	22 42	065	19 09	131	46 24	187	30 36	224	65 08	263	66 36	340
290	57 02	027	23 34	065	19 52	131	46 16	188	29 56	225	64 11	264	66 16	339
291	57 27	026	24 26	065	20 35	132	46 08	189	29 16	225	63 14	264	65 54	337
292	57 51	025	25 18	065	21 17	132	45 58	191	28 35	226	62 17	265	65 31	335
293	58 15	023	26 11	065	22 00	133	45 46	192	27 53	227	61 20	265	65 06	334
294	58 37	022	27 03	066	22 41	134	45 34	193	27 12	227	60 23	266	64 40	332
295	58 58	021	27 55	066	23 23	134	45 20	194	26 29	228	59 25	266	64 13	331
296	59 18	020	28 47	066	24 04	135	45 05	196	25 47	228	58 28	266	63 44	329
297	59 37	019	29 40	066	24 44	135	44 49	197	25 03	229	57 31	267	63 14	328
298	59 55	017	30 32	066	25 25	136	44 32	198	24 20	230	56 34	267	62 44	327
299	60 12	016	31 24	066	26 04	137	44 14	199	23 36	230	55 36	267	62 12	326
	◆DENEB		Schedar		Alpheratz		◆FOMALHAUT		Nunki		◆Rasalhague		VEGA	
300	60 27	015	25 05	035	32 17	066	26 44	137	43 54	200	54 39	268	61 39	324
301	60 41	013	25 37	035	33 09	066	27 22	138	43 34	202	53 42	268	61 05	323
302	60 54	012	26 10	035	34 02	066	28 01	138	43 13	203	52 44	268	60 30	322
303	61 05	011	26 43	035	34 54	066	28 38	139	42 49	204	51 47	269	59 54	321
304	61 15	009	27 15	035	35 46	066	29 16	140	42 26	205	50 49	269	59 18	320
305	61 23	008	27 48	034	36 39	066	29 52	141	42 01	206	49 52	269	58 41	319
306	61 30	006	28 20	034	37 31	066	30 29	141	41 36	207	48 55	270	58 03	318
307	61 36	005	28 53	034	38 24	066	31 04	142	41 09	208	47 57	270	57 25	318
308	61 40	003	29 25	034	39 16	066	31 39	143	40 42	209	47 00	270	56 45	317
309	61 43	002	29 57	034	40 09	066	32 14	144	40 13	210	46 03	271	56 06	316
310	61 44	000	30 29	034	41 01	066	32 47	144	39 44	211	45 05	271	55 26	315
311	61 44	359	31 01	034	41 53	066	33 21	145	39 14	212	44 08	271	54 45	315
312	61 42	357	31 32	033	42 46	066	33 53	146	38 43	213	43 11	271	54 04	314
313	61 39	356	32 04	033	43 38	066	34 25	147	38 12	214	42 13	272	53 23	313
314	61 34	355	32 35	033	44 30	066	34 56	148	37 39	215	41 16	272	52 41	313
	◆Alpheratz		Diphda		◆FOMALHAUT		Nunki		◆Rasalhague		VEGA		DENEB	
315	45 23	066	24 51	120	35 26	148	37 06	216	40 18	272	51 58	312	61 28	353
316	46 15	065	25 41	120	35 56	149	36 32	217	39 21	272	51 16	312	61 20	352
317	47 07	065	26 30	121	36 25	150	35 58	217	38 24	273	50 33	311	61 11	350
318	47 59	065	27 19	121	36 53	151	35 23	218	37 27	273	49 49	311	61 01	349
319	48 51	065	28 08	122	37 20	152	34 47	219	36 29	273	49 06	310	60 49	347
320	49 43	065	28 57	123	37 47	153	34 10	220	35 32	274	48 22	310	60 36	346
321	50 35	065	29 45	123	38 12	154	33 33	221	34 35	274	47 38	310	60 21	345
322	51 27	064	30 33	124	38 37	155	32 56	221	33 37	274	46 53	309	60 06	343
323	52 19	064	31 20	124	39 01	156	32 17	222	32 40	274	46 09	309	59 49	342
324	53 10	064	32 07	125	39 24	157	31 39	223	31 43	274	45 24	309	59 30	341
325	54 02	064	32 54	126	39 45	158	30 59	224	30 46	275	44 39	308	59 11	340
326	54 53	063	33 41	126	40 07	159	30 20	224	29 49	275	43 54	308	58 50	338
327	55 44	063	34 27	127	40 27	160	29 39	225	28 51	275	43 09	308	58 29	337
328	56 35	063	35 12	128	40 46	161	28 58	226	27 54	275	42 23	307	58 06	336
329	57 26	062	35 57	128	41 04	162	28 17	226	26 57	276	41 37	307	57 42	335
	◆Mirfak		Hamal		Diphda		◆FOMALHAUT		ALTAIR		◆VEGA		DENEB	
330	18 40	042	32 08	073	36 42	129	41 21	163	57 28	260	40 52	307	57 18	334
331	19 18	042	33 03	073	37 26	130	41 36	164	56 32	260	40 06	307	56 52	333
332	19 57	042	33 58	073	38 10	131	41 51	166	55 35	261	39 20	307	56 25	332
333	20 36	042	34 53	073	38 53	131	42 05	167	54 39	261	38 34	306	55 58	331
334	21 14	042	35 48	073	39 36	132	42 18	168	53 42	262	37 47	306	55 30	330
335	21 53	042	36 42	073	40 18	133	42 29	169	52 45	262	37 01	306	55 01	329
336	22 32	042	37 37	073	41 00	134	42 40	170	51 48	263	36 15	306	54 31	328
337	23 10	042	38 32	073	41 41	135	42 49	171	50 51	263	35 28	306	54 00	327
338	23 49	042	39 27	073	42 21	136	42 57	172	49 54	264	34 42	306	53 29	327
339	24 28	042	40 22	073	43 01	137	43 04	174	48 57	264	33 55	306	52 57	326
340	25 06	042	41 17	073	43 40	138	43 10	175	48 00	264	33 09	306	52 25	325
341	25 45	042	42 12	073	44 18	139	43 14	176	47 03	265	32 22	306	51 52	324
342	26 24	042	43 07	074	44 56	140	43 18	177	46 06	265	31 35	306	51 18	324
343	27 02	042	44 02	074	45 33	141	43 20	178	45 09	265	30 49	305	50 44	323
344	27 41	042	44 57	074	46 09	142	43 21	180	44 11	266	30 02	305	50 09	322
	Schedar		◆Mirfak		ALDEBARAN		Diphda		◆FOMALHAUT		ALTAIR		◆DENEB	
345	46 12	020	28 19	042	10 23	076	46 44	143	43 21	181	43 14	266	49 34	322
346	46 32	019	28 57	042	11 19	076	47 18	144	43 19	182	42 17	267	48 58	321
347	46 50	018	29 36	042	12 15	076	47 51	145	43 17	183	41 20	267	48 22	321
348	47 08	018	30 14	042	13 10	076	48 24	146	43 13	184	40 22	267	47 46	320
349	47 25	017	30 52	042	14 06	077	48 55	147	43 08	186	39 25	268	47 09	320
350	47 41	016	31 30	041	15 02	077	49 25	149	43 02	187	38 28	268	46 32	319
351	47 57	016	32 08	041	15 58	077	49 55	150	42 55	188	37 30	268	45 54	319
352	48 12	015	32 46	041	16 54	077	50 23	151	42 46	189	36 33	268	45 16	318
353	48 27	014	33 23	041	17 50	078	50 50	153	42 37	190	35 36	269	44 38	318
354	48 40	013	34 01	041	18 46	078	51 16	154	42 26	191	34 38	269	43 59	317
355	48 53	013	34 38	041	19 42	078	51 40	155	42 14	193	33 41	269	43 20	317
356	49 05	012	35 15	040	20 38	078	52 04	157	42 01	194	32 44	270	42 41	317
357	49 17	011	35 53	040	21 35	079	52 26	158	41 47	195	31 46	270	42 01	316
358	49 27	010	36 30	040	22 31	079	52 47	160	41 32	196	30 49	270	41 22	316
359	49 37	009	37 06	040	23 27	079	53 06	161	41 15	197	29 51	271	40 42	316

LHA ♈	Hc	Zn	Hc	Zn	Hc	Zn	Hc	Zn	Hc	Zn	Hc	Zn	Hc	Zn
	Schedar		CAPELLA		◆ALDEBARAN		ACHERNAR		◆FOMALHAUT		ALTAIR		◆DENEB	
0	48 47	008	18 57	046	24 12	079	13 59	167	41 55	198	28 53	271	39 19	316
1	48 55	008	19 39	046	25 09	079	14 12	167	41 36	199	27 55	272	38 39	316
2	49 02	007	20 20	046	26 05	079	14 25	168	41 17	201	26 58	272	37 58	316
3	49 08	006	21 02	046	27 02	079	14 37	168	40 56	202	26 00	272	37 18	315
4	49 14	005	21 44	046	27 59	080	14 48	169	40 34	203	25 02	272	36 37	315
5	49 19	004	22 25	046	28 55	080	14 59	169	40 12	204	24 05	273	35 56	315
6	49 22	003	23 07	046	29 52	080	15 09	170	39 48	205	23 07	273	35 15	315
7	49 25	003	23 49	046	30 49	080	15 19	170	39 23	206	22 09	273	34 34	314
8	49 28	002	24 30	046	31 46	080	15 29	171	38 58	207	21 12	273	33 53	314
9	49 29	001	25 12	046	32 43	080	15 38	171	38 32	208	20 14	274	33 12	314
10	49 29	000	25 54	046	33 40	081	15 46	172	38 04	209	19 17	274	32 30	314
11	49 29	359	26 35	046	34 36	081	15 54	173	37 37	209	18 19	274	31 48	314
12	49 28	358	27 17	046	35 33	081	16 01	173	37 08	210	17 22	274	31 07	314
13	49 26	358	27 58	046	36 30	081	16 08	174	36 38	211	16 24	275	30 25	314
14	49 23	357	28 40	046	37 27	081	16 14	174	36 08	212	15 27	275	29 43	313
	CAPELLA		◆ALDEBARAN		RIGEL		ACHERNAR		◆FOMALHAUT		Enif		◆DENEB	
15	29 21	046	38 24	081	22 35	106	16 19	175	35 37	213	41 56	269	29 01	313
16	30 02	046	39 21	082	23 30	107	16 24	175	35 05	214	40 59	269	28 19	313
17	30 44	046	40 18	082	24 26	107	16 29	176	34 32	215	40 01	270	27 37	313
18	31 25	045	41 16	082	25 21	107	16 33	176	33 59	216	39 03	270	26 55	313
19	32 06	045	42 13	082	26 16	108	16 36	177	33 25	216	38 06	270	26 13	313
20	32 47	045	43 10	082	27 11	108	16 39	178	32 51	217	37 08	270	25 31	313
21	33 28	045	44 07	082	28 05	109	16 41	178	32 16	218	36 10	271	24 49	313
22	34 08	045	45 04	082	29 00	109	16 43	179	31 40	219	35 13	271	24 06	313
23	34 49	045	46 01	083	29 54	110	16 44	179	31 04	219	34 15	271	23 24	313
24	35 29	044	46 58	083	30 48	110	16 44	180	30 27	220	33 17	271	22 42	313
25	36 10	044	47 56	083	31 42	111	16 44	180	29 50	221	32 20	272	22 00	313
26	36 50	044	48 53	083	32 36	111	16 44	181	29 12	221	31 22	272	21 17	313
27	37 30	044	49 50	083	33 30	112	16 42	181	28 33	222	30 24	272	20 35	313
28	38 09	043	50 47	083	34 24	112	16 41	182	27 55	223	29 27	272	19 53	313
29	38 49	043	51 45	083	35 17	113	16 38	183	27 15	223	28 29	273	19 11	313
	Mirfak		◆CAPELLA		ALDEBARAN		RIGEL		◆ACHERNAR		FOMALHAUT		◆Alpheratz	
30	52 08	022	39 28	043	52 42	084	36 10	113	16 35	183	26 35	224	61 10	302
31	52 29	021	40 07	043	53 39	084	37 03	114	16 32	184	25 55	225	60 21	301
32	52 50	020	40 46	042	54 37	084	37 56	114	16 28	184	25 14	225	59 31	301
33	53 09	019	41 25	042	55 34	084	38 48	115	16 23	185	24 33	226	58 41	300
34	53 28	018	42 03	041	56 31	084	39 40	116	16 18	185	23 51	226	57 51	300
35	53 46	017	42 41	041	57 29	084	40 32	116	16 12	186	23 09	227	57 01	299
36	54 03	017	43 19	041	58 26	084	41 24	117	16 06	187	22 27	228	56 10	299
37	54 19	015	43 56	040	59 24	085	42 15	117	15 59	187	21 44	228	55 20	298
38	54 33	014	44 34	040	60 21	085	43 06	118	15 52	188	21 01	229	54 29	298
39	54 47	013	45 10	039	61 19	085	43 57	119	15 44	188	20 18	229	53 38	298
40	55 00	012	45 47	039	62 16	085	44 47	120	15 36	189	19 34	230	52 47	297
41	55 12	011	46 23	038	63 13	085	45 37	120	15 27	189	18 50	230	51 55	297
42	55 23	010	46 58	038	64 11	085	46 27	121	15 17	190	18 06	231	51 04	297
43	55 33	009	47 34	037	65 08	085	47 16	122	15 07	190	17 21	231	50 12	296
44	55 41	008	48 08	037	66 06	085	48 05	123	14 56	191	16 36	232	49 21	296
	CAPELLA		◆POLLUX		PROCYON		SIRIUS		◆Acamar		Diphda		◆Alpheratz	
45	48 43	036	23 43	066	20 52	090	25 38	118	33 40	180	42 09	226	48 29	296
46	49 16	036	24 36	066	21 50	091	26 28	118	33 40	181	41 27	227	47 37	296
47	49 50	035	25 29	066	22 48	091	27 19	119	33 38	182	40 45	228	46 45	296
48	50 22	034	26 21	066	23 45	091	28 09	120	33 35	183	40 02	229	45 53	295
49	50 55	034	27 14	066	24 43	092	28 59	120	33 31	184	39 18	229	45 01	295
50	51 26	033	28 07	066	25 41	092	29 49	121	33 27	185	38 34	230	44 09	295
51	51 57	032	29 00	066	26 38	092	30 39	121	33 21	186	37 50	231	43 16	295
52	52 27	031	29 53	066	27 36	092	31 28	122	33 15	187	37 05	232	42 24	295
53	52 57	030	30 45	067	28 34	093	32 17	122	33 08	188	36 19	232	41 32	295
54	53 26	030	31 38	067	29 31	093	33 05	123	32 59	189	35 33	233	40 40	295
55	53 54	029	32 31	067	30 29	093	33 53	124	32 50	190	34 47	234	39 47	295
56	54 21	028	33 24	067	31 26	094	34 41	124	32 40	190	34 00	234	38 55	295
57	54 48	027	34 17	067	32 24	094	35 29	125	32 29	191	33 13	235	38 02	295
58	55 13	026	35 10	067	33 22	094	36 16	126	32 18	192	32 26	236	37 10	295
59	55 38	025	36 03	067	34 19	095	37 03	126	32 05	193	31 38	236	36 18	295
	CAPELLA		◆POLLUX		PROCYON		SIRIUS		◆CANOPUS		Diphda		◆Hamal	
60	56 02	024	36 56	067	35 16	095	37 49	127	14 37	158	30 50	237	62 25	290
61	56 25	023	37 49	067	36 14	095	38 35	128	14 58	159	30 01	238	61 31	290
62	56 47	022	38 42	067	37 11	096	39 20	128	15 18	159	29 13	238	60 36	290
63	57 08	021	39 35	067	38 09	096	40 05	129	15 38	160	28 23	239	59 42	289
64	57 28	020	40 28	067	39 06	096	40 50	130	15 58	161	27 34	239	58 48	289
65	57 47	018	41 21	066	40 03	097	41 34	131	16 17	161	26 44	240	57 53	289
66	58 05	017	42 14	066	41 01	097	42 17	132	16 35	162	25 54	240	56 58	289
67	58 21	016	43 06	066	41 58	097	43 00	133	16 53	162	25 04	241	56 04	288
68	58 37	015	43 59	066	42 55	098	43 42	133	17 11	163	24 14	241	55 09	288
69	58 51	014	44 52	066	43 52	098	44 23	134	17 28	163	23 23	242	54 14	288
70	59 04	012	45 45	066	44 49	099	45 04	135	17 44	164	22 32	242	53 19	288
71	59 15	011	46 37	066	45 46	099	45 45	136	18 00	164	21 41	243	52 24	288
72	59 26	010	47 30	066	46 43	100	46 24	137	18 15	165	20 50	243	51 29	288
73	59 35	008	48 23	066	47 40	100	47 03	138	18 30	166	19 58	243	50 34	288
74	59 42	007	49 15	065	48 37	100	47 41	139	18 44	166	19 07	244	49 39	288
	CAPELLA		◆POLLUX		PROCYON		SIRIUS		◆RIGEL		Menkar		◆Hamal	
75	59 49	006	50 07	065	49 33	101	48 18	140	65 32	171	58 40	251	48 44	287
76	59 54	004	51 00	065	50 30	101	48 54	142	65 40	174	57 45	252	47 49	287
77	59 57	003	51 52	065	51 26	102	49 29	143	65 45	176	56 50	252	46 54	287
78	60 00	002	52 44	064	52 23	103	50 04	144	65 47	179	55 55	253	45 59	287
79	60 00	000	53 36	064	53 19	103	50 37	145	65 47	181	55 00	254	45 04	287
80	60 00	359	54 28	064	54 15	104	51 10	146	65 45	183	54 05	254	44 09	287
81	59 58	357	55 20	064	55 11	104	51 41	148	65 40	186	53 09	255	43 14	287
82	59 54	356	56 11	063	56 07	105	52 11	149	65 33	188	52 13	256	42 19	287
83	59 50	355	57 03	063	57 03	106	52 40	150	65 24	191	51 17	256	41 24	287
84	59 44	353	57 54	062	57 58	106	53 08	152	65 12	193	50 21	257	40 29	287
85	59 36	352	58 45	062	58 53	107	53 35	153	64 58	195	49 25	257	39 34	287
86	59 27	351	59 36	061	59 48	108	54 00	155	64 42	197	48 29	258	38 39	287
87	59 17	349	60 26	061	60 43	108	54 24	156	64 24	200	47 32	258	37 44	288
88	59 06	348	61 16	060	61 38	109	54 47	158	64 03	202	46 36	259	36 49	288
89	58 53	347	62 06	060	62 32	110	55 08	159	63 41	204	45 39	259	35 54	288

LHA ♈	Hc	Zn	Hc	Zn	Hc	Zn	Hc	Zn	Hc	Zn	Hc	Zn	Hc	Zn
	Dubhe		◆REGULUS		Alphard		SIRIUS		◆RIGEL		ALDEBARAN		◆CAPELLA	
90	20 44	029	29 53	085	33 05	112	55 28	161	63 17	206	69 45	274	58 39	345
91	21 12	029	30 51	085	33 58	112	55 46	162	62 51	208	68 48	274	58 24	344
92	21 40	029	31 48	086	34 52	113	56 03	164	62 23	210	67 50	275	58 07	343
93	22 08	029	32 46	086	35 45	114	56 18	166	61 54	212	66 53	275	57 50	342
94	22 36	029	33 43	086	36 37	114	56 31	167	61 23	213	65 55	275	57 31	341
95	23 05	029	34 41	086	37 30	115	56 43	169	60 50	215	64 58	275	57 11	339
96	23 33	029	35 38	087	38 22	115	56 53	171	60 16	217	64 00	275	56 51	338
97	24 00	029	36 36	087	39 14	116	57 02	173	59 41	218	63 03	275	56 29	337
98	24 28	029	37 34	087	40 06	116	57 08	174	59 05	220	62 05	275	56 06	336
99	24 56	029	38 31	087	40 58	117	57 13	176	58 27	221	61 08	275	55 42	335
100	25 24	029	39 29	087	41 49	118	57 16	178	57 49	223	60 11	275	55 17	334
101	25 51	028	40 26	088	42 40	118	57 17	180	57 09	224	59 13	276	54 52	333
102	26 18	028	41 24	088	43 30	119	57 17	181	56 29	225	58 16	276	54 26	332
103	26 46	028	42 22	088	44 20	120	57 15	183	55 47	227	57 18	276	53 58	331
104	27 13	028	43 19	088	45 10	121	57 11	185	55 05	228	56 21	276	53 30	331
	Dubhe		◆REGULUS		Alphard		SIRIUS		◆RIGEL		ALDEBARAN		◆CAPELLA	
105	27 40	028	44 17	089	46 00	121	57 05	187	54 21	229	55 24	276	53 02	330
106	28 07	028	45 15	089	46 48	122	56 57	188	53 37	230	54 26	276	52 32	329
107	28 33	027	46 12	089	47 37	123	56 48	190	52 53	231	53 29	276	52 02	328
108	29 00	027	47 10	090	48 25	124	56 37	192	52 07	232	52 32	276	51 31	327
109	29 26	027	48 08	090	49 13	125	56 24	194	51 21	233	51 34	277	51 00	327
110	29 52	027	49 05	090	50 00	126	56 10	195	50 35	234	50 37	277	50 28	326
111	30 18	027	50 03	090	50 46	127	55 54	197	49 48	235	49 40	277	49 55	325
112	30 44	026	51 01	091	51 32	128	55 37	198	49 00	236	48 42	277	49 22	325
113	31 09	026	51 58	091	52 17	129	55 18	200	48 12	237	47 45	277	48 48	324
114	31 35	026	52 56	091	53 02	130	54 57	202	47 23	238	46 48	277	48 14	323
115	32 00	026	53 54	092	53 46	131	54 35	203	46 34	239	45 51	277	47 39	323
116	32 24	025	54 51	092	54 29	132	54 12	205	45 45	240	44 54	278	47 04	322
117	32 49	025	55 49	092	55 12	133	53 47	206	44 55	240	43 56	278	46 29	322
118	33 13	025	56 47	092	55 53	135	53 21	208	44 04	241	42 59	278	45 53	321
119	33 37	024	57 44	093	56 34	136	52 54	209	43 14	242	42 02	278	45 16	321
	Dubhe		◆Denebola		Alphard		◆SIRIUS		RIGEL		ALDEBARAN		◆CAPELLA	
120	34 01	024	34 59	083	57 13	137	52 25	210	42 23	242	41 05	278	44 40	320
121	34 25	024	35 56	083	57 52	139	51 55	212	41 31	243	40 08	278	44 02	320
122	34 48	024	36 54	084	58 30	140	51 24	213	40 40	244	39 11	278	43 25	319
123	35 11	023	37 51	084	59 06	142	50 53	214	39 48	244	38 14	279	42 47	319
124	35 33	023	38 48	084	59 41	143	50 20	215	38 56	245	37 17	279	42 09	319
125	35 55	022	39 46	084	60 15	145	49 46	217	38 03	246	36 20	279	41 31	318
126	36 17	022	40 43	084	60 48	146	49 11	218	37 11	246	35 23	279	40 52	318
127	36 39	022	41 41	085	61 19	148	48 35	219	36 18	247	34 26	279	40 14	318
128	37 00	021	42 38	085	61 49	150	47 58	220	35 25	247	33 29	279	39 35	317
129	37 21	021	43 35	085	62 17	152	47 20	221	34 32	248	32 32	280	38 55	317
130	37 41	020	44 33	085	62 43	154	46 42	222	33 38	248	31 35	280	38 16	317
131	38 01	020	45 30	085	63 08	156	46 03	223	32 44	249	30 39	280	37 36	316
132	38 21	020	46 28	085	63 31	158	45 23	224	31 50	249	29 42	280	36 56	316
133	38 40	019	47 25	086	63 51	160	44 42	225	30 56	250	28 45	280	36 16	316
134	38 58	019	48 23	086	64 10	162	44 01	226	30 02	250	27 48	281	35 36	316
	Dubhe		◆ARCTURUS		SPICA		◆Alphard		SIRIUS		BETELGEUSE		◆CAPELLA	
135	39 17	018	15 26	074	19 05	108	64 27	164	43 19	227	43 59	265	34 55	315
136	39 35	018	16 21	074	20 00	109	64 42	166	42 37	228	43 01	265	34 15	315
137	39 52	017	17 17	074	20 54	109	64 54	169	41 54	229	42 04	266	33 34	315
138	40 09	017	18 12	075	21 49	109	65 04	171	41 10	230	41 06	266	32 53	315
139	40 25	016	19 08	075	22 43	110	65 12	173	40 26	230	40 09	266	32 12	315
140	40 41	016	20 04	075	23 37	110	65 18	176	39 41	231	39 11	267	31 31	315
141	40 57	015	20 59	075	24 31	111	65 21	178	38 56	232	38 14	267	30 50	314
142	41 12	015	21 55	075	25 25	111	65 22	180	38 10	233	37 16	267	30 09	314
143	41 26	014	22 51	075	26 19	111	65 22	183	37 24	233	36 18	267	29 28	314
144	41 40	014	23 47	076	27 13	112	65 16	185	36 38	234	35 21	268	28 46	314
145	41 53	013	24 43	076	28 06	112	65 10	187	35 51	235	34 23	268	28 05	314
146	42 06	013	25 39	076	28 59	113	65 01	190	35 03	235	33 26	268	27 23	314
147	42 18	012	26 35	076	29 52	113	64 50	192	34 16	236	32 28	269	26 42	314
148	42 30	011	27 31	076	30 45	114	64 37	194	33 28	237	31 30	269	26 00	314
149	42 41	011	28 27	076	31 38	114	64 22	197	32 39	237	30 33	269	25 19	314
	Dubhe		◆ARCTURUS		SPICA		◆Alphard		SIRIUS		BETELGEUSE		◆CAPELLA	
150	42 51	010	29 23	077	32 30	115	64 04	199	31 51	238	29 35	270	24 37	314
151	43 01	010	30 19	077	33 22	115	63 45	201	31 01	239	28 37	270	23 55	314
152	43 10	009	31 15	077	34 14	116	63 23	203	30 12	239	27 40	270	23 14	314
153	43 19	008	32 11	077	35 06	117	63 00	205	29 22	240	26 42	270	22 32	314
154	43 27	008	33 08	077	35 58	117	62 35	207	28 33	240	25 44	271	21 50	314
155	43 35	007	34 04	077	36 49	118	62 08	209	27 42	241	24 47	271	21 09	314
156	43 41	006	35 00	077	37 40	118	61 39	211	26 52	241	23 49	271	20 27	314
157	43 47	006	35 57	078	38 30	119	61 09	212	26 01	242	22 51	271	19 45	314
158	43 53	005	36 53	078	39 20	120	60 37	214	25 10	242	21 54	272	19 04	314
159	43 58	004	37 49	078	40 10	120	60 04	216	24 19	243	20 56	272	18 22	314
160	44 02	004	38 46	078	41 00	121	59 30	217	23 28	243	19 58	272	17 41	314
161	44 06	003	39 42	078	41 49	122	58 54	219	22 36	244	19 01	272	16 59	314
162	44 08	003	40 38	078	42 38	122	58 17	220	21 44	244	18 03	273	16 18	314
163	44 11	002	41 35	078	43 27	123	57 39	222	20 52	245	17 05	273	15 36	314
164	44 12	001	42 31	078	44 15	124	57 00	223	20 00	245	16 08	273	14 55	314
	◆Dubhe		Alkaid		ARCTURUS		◆SPICA		Suhail		◆PROCYON		POLLUX	
165	44 13	001	42 31	036	43 28	078	45 02	125	25 16	202	39 36	263	43 33	294
166	44 13	000	43 05	036	44 24	079	45 49	126	24 54	203	38 39	264	42 40	294
167	44 13	359	43 38	035	45 21	079	46 36	127	24 31	204	37 41	264	41 47	294
168	44 12	359	44 11	035	46 18	079	47 22	127	24 08	204	36 44	265	40 55	293
169	44 10	358	44 44	034	47 14	079	48 07	128	23 44	205	35 46	265	40 02	293
170	44 08	357	45 16	034	48 11	079	48 52	129	23 19	206	34 49	265	39 09	293
171	44 05	357	45 48	033	49 07	079	49 37	130	22 54	206	33 51	266	38 16	293
172	44 01	356	46 19	033	50 04	079	50 20	131	22 28	207	32 54	266	37 23	293
173	43 56	355	46 50	032	51 01	079	51 03	132	22 02	207	31 56	266	36 30	293
174	43 51	355	47 20	031	51 57	079	51 45	134	21 35	208	30 59	266	35 37	293
175	43 46	354	47 50	031	52 54	079	52 27	135	21 08	209	30 01	267	34 44	293
176	43 39	353	48 19	030	53 50	079	53 07	136	20 40	209	29 04	267	33 51	293
177	43 32	353	48 48	029	54 47	079	53 47	137	20 11	210	28 06	267	32 58	293
178	43 25	352	49 16	029	55 44	079	54 26	138	19 42	210	27 08	268	32 05	293
179	43 17	352	49 44	028	56 40	079	55 04	140	19 13	211	26 11	268	31 12	293

LAT 16°N

LHA ♈	Hc Zn	Hc Zn	Hc Zn	Hc Zn	Hc Zn	Hc Zn	Hc Zn
	♦Alkaid	ANTARES	♦SPICA	Gienah	♦REGULUS	POLLUX	Dubhe
180	50 10 027	12 07 122	55 41 141	56 16 173	62 35 265	30 19 294	43 08 351
181	50 37 027	12 56 123	56 17 142	56 22 175	61 38 266	29 26 294	42 58 350
182	51 02 026	13 44 123	56 51 144	56 26 177	60 40 266	28 34 294	42 48 350
183	51 27 025	14 32 124	57 25 145	56 28 178	59 43 266	27 41 294	42 38 349
184	51 51 024	15 20 124	57 57 147	56 29 180	58 45 267	26 48 294	42 26 348
185	52 14 023	16 08 124	58 28 148	56 28 182	57 47 267	25 55 294	42 15 348
186	52 36 022	16 55 125	58 58 150	56 25 184	56 50 267	25 02 294	42 02 347
187	52 58 022	17 43 125	59 26 152	56 21 185	55 52 268	24 10 294	41 49 347
188	53 19 021	18 30 126	59 52 153	56 14 187	54 54 268	23 17 294	41 36 346
189	53 39 020	19 16 126	60 17 155	56 07 189	53 57 268	22 24 294	41 22 346
190	53 58 019	20 03 127	60 41 157	55 57 190	52 59 269	21 32 294	41 07 345
191	54 16 018	20 49 127	61 02 159	55 46 192	52 02 269	20 39 294	40 52 345
192	54 33 017	21 35 128	61 22 161	55 33 194	51 04 269	19 47 295	40 37 344
193	54 49 016	22 20 128	61 40 163	55 18 195	50 06 270	18 55 295	40 21 344
194	55 04 015	23 05 129	61 56 165	55 02 197	49 08 270	18 02 295	40 04 343
	Kochab	♦VEGA	Rasalhague	ANTARES	♦SPICA	♦REGULUS	Dubhe
195	29 49 008	14 23 053	23 39 083	23 50 129	62 10 167	48 11 270	39 47 343
196	29 58 008	15 09 053	24 36 083	24 35 130	62 23 169	47 13 270	39 29 342
197	30 06 008	15 56 053	25 34 083	25 19 130	62 33 171	46 15 271	39 12 342
198	30 13 008	16 42 054	26 31 084	26 02 131	62 41 173	45 18 271	38 53 341
199	30 21 007	17 28 054	27 28 084	26 46 132	62 46 175	44 20 271	38 34 341
200	30 28 007	18 15 054	28 26 084	27 29 132	62 50 177	43 22 272	38 15 340
201	30 35 007	19 01 054	29 23 084	28 11 133	62 52 180	42 25 272	37 55 340
202	30 42 006	19 48 054	30 20 085	28 53 133	62 51 182	41 27 272	37 35 339
203	30 48 006	20 35 054	31 18 085	29 35 134	62 49 184	40 30 272	37 15 339
204	30 54 006	21 21 054	32 15 085	30 16 135	62 44 186	39 32 272	36 54 339
205	31 00 006	22 08 054	33 13 085	30 57 135	62 37 188	38 34 273	36 33 338
206	31 05 005	22 55 054	34 10 085	31 37 136	62 27 190	37 37 273	36 11 338
207	31 10 005	23 41 054	35 08 086	32 17 137	62 16 192	36 39 273	35 49 337
208	31 15 005	24 28 054	36 05 086	32 56 138	62 03 194	35 42 273	35 27 337
209	31 19 004	25 15 054	37 03 086	33 35 138	61 48 196	34 44 274	35 04 337
	Kochab	♦VEGA	Rasalhague	♦ANTARES	SPICA	♦REGULUS	Dubhe
210	31 24 004	26 02 054	38 00 086	34 13 139	61 31 198	33 46 274	34 41 336
211	31 28 004	26 49 054	38 58 087	34 51 140	61 12 200	32 49 274	34 18 336
212	31 31 003	27 35 054	39 55 087	35 28 141	60 51 202	31 51 274	33 54 336
213	31 34 003	28 22 054	40 53 087	36 04 141	60 28 204	30 54 275	33 30 335
214	31 37 003	29 09 054	41 51 087	36 40 142	60 04 206	29 56 275	33 06 335
215	31 40 002	29 56 054	42 48 088	37 15 143	59 38 208	28 59 275	32 42 335
216	31 42 002	30 42 054	43 46 088	37 49 144	59 11 209	28 01 275	32 17 335
217	31 44 002	31 29 054	44 44 088	38 22 145	58 42 211	27 04 276	31 52 334
218	31 46 001	32 15 054	45 41 088	38 55 146	58 12 212	26 07 276	31 27 334
219	31 47 001	33 02 054	46 39 089	39 27 147	57 40 214	25 09 276	31 02 334
220	31 48 001	33 49 054	47 36 089	39 59 148	57 07 216	24 12 276	30 36 334
221	31 49 001	34 35 054	48 34 089	40 29 149	56 33 217	23 15 276	30 11 333
222	31 49 000	35 21 053	49 32 089	40 59 150	55 58 218	22 17 277	29 45 333
223	31 49 000	36 08 053	50 30 090	41 28 151	55 21 220	21 20 277	29 18 333
224	31 49 000	36 54 053	51 27 090	41 55 152	54 44 221	20 23 277	28 52 333
	VEGA	♦ALTAIR	Nunki	ANTARES	♦SPICA	Denebola	♦Alkaid
225	37 40 053	19 01 086	18 58 126	42 22 153	54 05 222	43 58 275	53 32 340
226	38 26 053	19 59 086	19 45 126	42 48 154	53 26 224	43 01 275	53 12 339
227	39 12 053	20 56 086	20 31 127	43 14 155	52 46 225	42 04 275	52 51 338
228	39 58 052	21 54 087	21 17 127	43 38 156	52 05 226	41 06 276	52 29 337
229	40 43 052	22 51 087	22 03 128	44 01 157	51 23 227	40 09 276	52 06 336
230	41 29 052	23 49 087	22 48 128	44 23 158	50 40 228	39 11 276	51 43 336
231	42 14 052	24 47 087	23 33 129	44 44 159	49 57 229	38 14 276	51 19 335
232	42 59 051	25 44 088	24 18 129	45 03 160	49 13 230	37 17 276	50 54 334
233	43 44 051	26 42 088	25 02 130	45 22 162	48 28 231	36 19 277	50 28 333
234	44 29 051	27 40 088	25 46 131	45 40 163	47 43 232	35 22 277	50 02 332
235	45 14 050	28 37 089	26 30 131	45 56 164	46 57 233	34 25 277	49 35 332
236	45 58 050	29 35 089	27 13 132	46 11 165	46 11 234	33 28 277	49 07 331
237	46 42 050	30 33 089	27 56 132	46 25 167	45 24 235	32 30 277	48 39 330
238	47 26 049	31 30 089	28 38 133	46 38 168	44 37 236	31 33 278	48 10 330
239	48 10 049	32 28 090	29 20 134	46 49 169	43 49 236	30 36 278	47 40 329
	♦VEGA	ALTAIR	Nunki	♦ANTARES	SPICA	♦ARCTURUS	Alkaid
240	48 53 049	33 26 090	30 02 134	47 00 170	43 01 237	64 54 281	47 10 328
241	49 36 048	34 23 090	30 43 135	47 09 172	42 12 238	63 57 281	46 40 328
242	50 19 048	35 21 090	31 23 136	47 16 173	41 23 239	63 01 281	46 09 327
243	51 01 047	36 19 091	32 04 136	47 23 174	40 34 239	62 04 281	45 37 327
244	51 43 046	37 16 091	32 43 137	47 28 176	39 44 240	61 08 281	45 05 326
245	52 25 046	38 14 091	33 22 138	47 31 177	38 54 241	60 11 281	44 33 326
246	53 06 045	39 12 092	34 01 139	47 34 178	38 03 241	59 14 281	44 00 325
247	53 47 045	40 09 092	34 38 139	47 35 180	37 12 242	58 18 281	43 27 325
248	54 27 044	41 07 092	35 16 140	47 34 181	36 21 243	57 21 281	42 54 324
249	55 07 043	42 05 092	35 52 141	47 33 182	35 30 243	56 24 281	42 20 324
250	55 46 043	43 02 093	36 28 142	47 30 184	34 38 244	55 28 281	41 45 323
251	56 25 042	44 00 093	37 04 143	47 25 185	33 47 244	54 31 281	41 11 323
252	57 03 041	44 57 093	37 39 143	47 20 186	32 54 245	53 34 281	40 36 323
253	57 41 040	45 55 094	38 13 144	47 13 188	32 02 245	52 38 281	40 00 322
254	58 17 039	46 52 094	38 46 145	47 05 189	31 10 246	51 41 281	39 25 322
	♦VEGA	ALTAIR	♦Nunki	ANTARES	SPICA	♦ARCTURUS	Alkaid
255	58 54 038	47 50 094	39 18 146	46 55 190	30 17 246	50 44 281	38 49 321
256	59 29 037	48 47 095	39 50 147	46 44 191	29 24 247	49 48 281	38 13 321
257	60 03 036	49 45 095	40 21 148	46 32 193	28 31 247	48 51 281	37 37 321
258	60 37 035	50 42 096	40 51 149	46 19 194	27 37 248	47 55 281	37 00 321
259	61 09 034	51 40 096	41 20 150	46 04 195	26 44 248	46 58 281	36 23 320
260	61 41 033	52 37 096	41 49 151	45 49 196	25 50 249	46 01 281	35 47 320
261	62 12 031	53 34 097	42 16 152	45 32 198	24 56 249	45 05 281	35 09 320
262	62 41 030	54 32 097	42 43 153	45 14 199	24 02 250	44 08 281	34 32 320
263	63 10 029	55 29 098	43 08 154	44 55 200	23 08 250	43 12 282	33 54 319
264	63 37 027	56 26 098	43 33 155	44 34 201	22 14 250	42 15 282	33 17 319
265	64 03 026	57 23 099	43 56 156	44 13 202	21 19 251	41 19 282	32 39 319
266	64 27 024	58 20 099	44 19 158	43 50 203	20 25 251	40 22 282	32 01 319
267	64 50 023	59 17 100	44 40 159	43 27 205	19 30 252	39 26 282	31 23 319
268	65 12 021	60 14 100	45 01 160	43 03 206	18 35 252	38 30 282	30 45 318
269	65 32 019	61 10 101	45 20 161	42 37 207	17 41 252	37 33 282	30 06 318

LHA ♈	Hc Zn	Hc Zn	Hc Zn	Hc Zn	Hc Zn	Hc Zn	Hc Zn
	VEGA	♦DENEB	Enif	♦Nunki	ANTARES	♦ARCTURUS	Alkaid
270	65 50 018	45 22 040	35 14 089	45 38 162	42 10 208	36 37 282	29 28 318
271	66 07 016	46 00 040	36 12 089	45 55 163	41 43 209	35 41 282	28 49 318
272	66 22 014	46 36 039	37 10 090	46 11 165	41 15 210	34 44 283	28 11 318
273	66 35 012	47 13 039	38 07 090	46 26 166	40 46 211	33 48 283	27 32 318
274	66 46 010	47 49 038	39 05 090	46 39 167	40 16 212	32 52 283	26 53 318
275	66 56 008	48 24 038	40 03 090	46 51 169	39 45 213	31 55 283	26 14 318
276	67 03 006	49 00 037	41 00 091	47 02 170	39 13 214	30 59 283	25 35 318
277	67 08 004	49 34 037	41 58 091	47 12 171	38 41 215	30 03 283	24 56 317
278	67 12 002	50 08 036	42 56 091	47 20 172	38 07 216	29 07 283	24 17 317
279	67 13 000	50 42 035	43 53 092	47 27 174	37 34 216	28 11 284	23 38 317
280	67 13 358	51 15 035	44 51 092	47 32 175	36 59 217	27 15 284	22 59 317
281	67 10 356	51 47 034	45 49 092	47 37 176	36 24 218	26 19 284	22 20 317
282	67 05 354	52 19 033	46 46 093	47 40 178	35 48 219	25 23 284	21 41 317
283	66 59 352	52 50 032	47 44 093	47 41 179	35 11 220	24 27 284	21 02 317
284	66 50 350	53 21 032	48 41 093	47 42 180	34 34 221	23 31 284	20 23 317
	♦DENEB	Alpheratz	♦FOMALHAUT	Nunki	ANTARES	♦Rasalhague	VEGA
285	53 51 031	18 49 064	16 50 129	47 41 182	33 56 221	69 04 263	66 40 349
286	54 20 030	19 41 064	17 35 129	47 38 183	33 18 222	68 06 264	66 27 347
287	54 48 029	20 33 064	18 20 130	47 35 184	32 39 223	67 09 264	66 13 345
288	55 15 028	21 25 065	19 04 130	47 30 186	31 59 224	66 12 265	65 57 343
289	55 42 027	22 17 065	19 48 131	47 23 187	31 19 224	65 14 265	65 39 341
290	56 08 026	23 09 065	20 32 131	47 16 188	30 39 225	64 17 266	65 20 340
291	56 33 025	24 01 065	21 15 132	47 07 190	29 58 226	63 19 266	64 59 338
292	56 56 024	24 53 065	21 58 132	46 56 191	29 16 226	62 22 267	64 36 336
293	57 19 023	25 45 065	22 40 133	46 45 192	28 35 227	61 24 267	64 12 335
294	57 41 022	26 38 065	23 23 133	46 32 193	27 52 228	60 26 267	63 47 333
295	58 02 021	27 30 065	24 04 134	46 18 195	27 09 228	59 29 268	63 20 332
296	58 22 019	28 22 065	24 46 134	46 03 196	26 26 229	58 31 268	62 52 330
297	58 40 018	29 15 065	25 27 135	45 47 197	25 43 229	57 33 268	62 23 329
298	58 58 017	30 07 065	26 08 136	45 29 198	24 59 230	56 36 269	61 53 328
299	59 14 016	31 00 065	26 48 136	45 10 200	24 15 230	55 38 269	61 22 327
	♦DENEB	Schedar	Alpheratz	♦FOMALHAUT	Nunki	♦Rasalhague	VEGA
300	59 29 014	24 15 035	31 52 065	27 27 137	44 50 201	54 40 269	60 49 325
301	59 42 013	24 48 035	32 44 065	28 07 137	44 29 202	53 43 270	60 16 324
302	59 55 012	25 21 034	33 37 065	28 45 138	44 07 203	52 45 270	59 42 323
303	60 06 010	25 53 034	34 29 065	29 24 139	43 44 204	51 47 270	59 07 322
304	60 16 009	26 26 034	35 22 065	30 01 139	43 20 205	50 50 270	58 31 321
305	60 24 008	26 58 034	36 14 065	30 39 140	42 55 206	49 52 271	57 55 320
306	60 31 006	27 31 034	37 07 065	31 15 141	42 29 207	48 54 271	57 18 319
307	60 36 005	28 03 034	37 59 065	31 51 142	42 02 208	47 57 271	56 40 319
308	60 40 003	28 35 034	38 51 065	32 27 142	41 34 209	46 59 271	56 01 318
309	60 43 002	29 07 034	39 44 065	33 02 143	41 05 211	46 01 272	55 22 317
310	60 44 000	29 39 033	40 36 065	33 36 144	40 35 211	45 04 272	54 43 316
311	60 44 359	30 10 033	41 29 065	34 10 145	40 05 212	44 06 272	54 03 316
312	60 42 358	30 42 033	42 21 065	34 43 146	39 34 213	43 09 272	53 22 315
313	60 39 356	31 13 033	43 13 065	35 15 146	39 01 214	42 11 273	52 41 314
314	60 34 355	31 44 033	44 05 065	35 47 147	38 28 215	41 13 273	52 00 314
	♦Alpheratz	Diphda	♦FOMALHAUT	Nunki	♦Rasalhague	VEGA	DENEB
315	44 57 065	25 21 119	36 17 148	37 55 216	40 16 273	51 18 313	60 28 353
316	45 50 065	26 11 120	36 47 149	37 20 217	39 18 273	50 35 313	60 21 352
317	46 42 064	27 01 120	37 17 150	36 45 218	38 21 274	49 53 312	60 12 351
318	47 34 064	27 50 121	37 45 151	36 10 219	37 23 274	49 10 312	60 02 349
319	48 25 064	28 40 122	38 13 152	35 33 220	36 25 274	48 27 311	59 50 348
320	49 17 064	29 29 122	38 40 153	34 56 220	35 28 274	47 43 311	59 38 346
321	50 09 064	30 17 123	39 06 154	34 19 221	34 30 274	46 59 310	59 23 345
322	51 01 063	31 06 123	39 31 155	33 41 222	33 33 275	46 15 310	59 08 344
323	51 52 063	31 54 124	39 55 156	33 02 223	32 35 275	45 31 310	58 51 343
324	52 43 063	32 42 124	40 19 157	32 22 223	31 38 275	44 46 309	58 34 341
325	53 35 062	33 29 125	40 41 158	31 43 224	30 41 275	44 02 309	58 15 340
326	54 26 062	34 16 126	41 02 159	31 02 225	29 43 276	43 17 309	57 54 339
327	55 17 062	35 02 126	41 23 160	30 21 225	28 46 276	42 32 308	57 33 338
328	56 07 061	35 49 127	41 42 161	29 40 226	27 48 276	41 46 308	57 11 337
329	56 58 061	36 34 128	42 01 162	28 58 227	26 51 276	41 01 308	56 48 336
	♦Mirfak	Hamal	Diphda	♦FOMALHAUT	ALTAIR	♦VEGA	DENEB
330	17 55 042	31 50 072	37 20 129	42 18 163	57 38 261	40 15 308	56 24 335
331	18 34 042	32 45 072	38 05 129	42 34 164	56 41 262	39 30 307	55 58 334
332	19 13 042	33 40 072	38 49 130	42 49 165	55 44 262	38 44 307	55 32 333
333	19 51 042	34 35 072	39 33 131	43 03 166	54 47 263	37 58 307	55 05 332
334	20 30 042	35 30 072	40 16 132	43 16 168	53 50 263	37 12 307	54 38 331
335	21 09 042	36 25 072	40 59 132	43 28 169	52 52 263	36 25 307	54 09 330
336	21 47 042	37 20 072	41 41 133	43 39 170	51 55 264	35 39 307	53 40 329
337	22 26 042	38 15 072	42 23 134	43 48 171	50 58 264	34 53 306	53 10 328
338	23 05 042	39 10 073	43 04 135	43 57 172	50 00 265	34 06 306	52 39 327
339	23 43 042	40 05 073	43 44 136	44 04 174	49 03 265	33 20 306	52 08 327
340	24 22 042	41 00 073	44 24 137	44 10 175	48 05 265	32 33 306	51 35 326
341	25 01 042	41 55 073	45 03 138	44 14 176	47 08 266	31 47 306	51 03 325
342	25 39 042	42 50 073	45 41 139	44 18 177	46 10 266	31 00 306	50 30 324
343	26 18 042	43 45 073	46 19 140	44 20 178	45 13 266	30 14 306	49 56 324
344	26 56 042	44 40 073	46 55 141	44 21 180	44 15 267	29 27 306	49 21 323
	Schedar	♦Mirfak	ALDEBARAN	Diphda	♦FOMALHAUT	ALTAIR	♦DENEB
345	45 16 019	27 34 042	10 09 076	47 31 142	44 21 181	43 18 267	48 47 323
346	45 35 019	28 13 042	11 04 076	48 06 143	44 19 182	42 20 267	48 11 322
347	45 53 018	28 51 041	12 00 076	48 40 144	44 17 183	41 22 268	47 36 321
348	46 11 017	29 29 041	12 56 076	49 13 146	44 13 184	40 25 268	46 59 321
349	46 28 017	30 07 041	13 52 076	49 45 147	44 08 186	39 27 268	46 23 320
350	46 44 016	30 45 041	14 49 077	50 17 148	44 02 187	38 30 269	45 46 320
351	46 59 015	31 23 041	15 45 077	50 47 149	43 54 188	37 32 269	45 09 319
352	47 14 015	32 00 041	16 41 077	51 15 151	43 45 189	36 34 269	44 31 319
353	47 28 014	32 38 040	17 37 077	51 43 152	43 36 190	35 37 270	43 53 319
354	47 42 013	33 15 040	18 33 078	52 10 153	43 25 192	34 39 270	43 14 318
355	47 55 012	33 53 040	19 30 078	52 35 155	43 12 193	33 41 270	42 36 318
356	48 07 012	34 30 040	20 26 078	52 59 156	42 59 194	32 43 270	41 57 317
357	48 18 011	35 06 040	21 23 078	53 21 158	42 45 195	31 46 271	41 18 317
358	48 28 010	35 43 039	22 19 078	53 43 159	42 29 196	30 48 271	40 38 317
359	48 38 009	36 20 039	23 16 079	54 03 161	42 13 197	29 50 271	39 59 316

LHA ♈	Hc	Zn	Hc	Zn	Hc	Zn	Hc	Zn	Hc	Zn	Hc	Zn	Hc	Zn
	CAPELLA		◆ALDEBARAN		Acamar		ACHERNAR		◆FOMALHAUT		ALTAIR		◆DENEB	
0	18 15	046	24 00	078	20 57	145	14 57	167	42 52	199	28 51	272	38 35	317
1	18 57	046	24 57	078	21 30	146	15 10	167	42 33	200	27 53	272	37 55	316
2	19 39	046	25 54	079	22 03	146	15 23	168	42 13	201	26 55	272	37 15	316
3	20 20	046	26 51	079	22 35	147	15 35	168	41 52	202	25 57	273	36 35	316
4	21 02	046	27 47	079	23 06	147	15 47	169	41 29	203	24 59	273	35 55	316
5	21 44	046	28 44	079	23 37	148	15 58	169	41 06	204	24 02	273	35 14	315
6	22 25	046	29 41	079	24 07	149	16 08	170	40 42	205	23 04	273	34 33	315
7	23 07	046	30 38	079	24 37	149	16 19	170	40 17	206	22 06	274	33 52	315
8	23 49	046	31 35	080	25 07	150	16 28	171	39 51	207	21 08	274	33 11	315
9	24 30	046	32 32	080	25 35	151	16 37	171	39 25	208	20 10	274	32 30	315
10	25 12	046	33 29	080	26 03	151	16 45	172	38 57	209	19 12	274	31 48	314
11	25 54	046	34 26	080	26 31	152	16 53	172	38 29	210	18 15	275	31 07	314
12	26 35	046	35 24	080	26 58	153	17 01	173	37 59	211	17 17	275	30 25	314
13	27 16	046	36 21	080	27 24	153	17 07	174	37 29	212	16 19	275	29 43	314
14	27 58	046	37 18	080	27 50	154	17 13	174	36 59	213	15 21	275	29 02	314
	CAPELLA		◆ALDEBARAN		RIGEL		ACHERNAR		◆FOMALHAUT		Enif		◆DENEB	
15	28 39	045	38 15	081	22 52	106	17 19	175	36 27	213	41 57	270	28 20	314
16	29 20	045	39 12	081	23 47	106	17 24	175	35 55	214	40 59	270	27 38	314
17	30 02	045	40 09	081	24 43	107	17 29	176	35 22	215	40 01	270	26 56	314
18	30 43	045	41 07	081	25 39	107	17 33	176	34 48	216	39 03	271	26 14	313
19	31 23	045	42 04	081	26 34	107	17 36	177	34 14	217	38 05	271	25 32	313
20	32 04	045	43 01	081	27 29	108	17 39	178	33 39	217	37 07	271	24 50	313
21	32 45	045	43 58	081	28 24	108	17 41	178	33 03	218	36 09	271	24 08	313
22	33 26	044	44 56	081	29 19	109	17 43	179	32 27	219	35 11	272	23 25	313
23	34 06	044	45 53	082	30 14	109	17 44	179	31 50	220	34 13	272	22 43	313
24	34 46	044	46 50	082	31 09	110	17 44	180	31 13	220	33 15	272	22 01	313
25	35 26	044	47 48	082	32 03	110	17 44	180	30 35	221	32 17	272	21 19	313
26	36 06	043	48 45	082	32 58	110	17 44	181	29 57	222	31 20	273	20 36	313
27	36 46	043	49 42	082	33 52	111	17 42	181	29 18	222	30 22	273	19 54	313
28	37 26	043	50 40	082	34 46	111	17 41	182	28 38	223	29 24	273	19 12	313
29	38 05	043	51 37	082	35 40	112	17 38	183	27 59	224	28 26	273	18 30	313
	Mirfak		◆CAPELLA		ALDEBARAN		RIGEL		◆ACHERNAR		FOMALHAUT		◆Alpheratz	
30	51 12	022	38 44	042	52 35	082	36 33	112	17 35	183	27 18	224	60 37	303
31	51 33	021	39 23	042	53 32	082	37 27	113	17 32	184	26 38	225	59 49	303
32	51 53	020	40 02	042	54 30	082	38 20	114	17 28	184	25 56	226	59 00	302
33	52 13	019	40 40	041	55 27	083	39 13	114	17 23	185	25 15	226	58 11	301
34	52 31	018	41 18	041	56 25	083	40 06	115	17 18	185	24 33	227	57 21	301
35	52 48	017	41 56	041	57 22	083	40 58	115	17 12	186	23 50	227	56 31	300
36	53 05	016	42 33	040	58 20	083	41 51	116	17 06	187	23 08	228	55 41	300
37	53 21	015	43 11	040	59 17	083	42 42	117	16 59	187	22 24	228	54 51	300
38	53 35	014	43 47	039	60 15	083	43 34	117	16 51	188	21 41	229	54 00	299
39	53 49	013	44 24	039	61 12	083	44 25	118	16 43	188	20 57	229	53 10	299
40	54 02	012	45 00	038	62 10	083	45 16	119	16 35	189	20 13	230	52 19	298
41	54 13	011	45 36	038	63 07	083	46 07	119	16 26	189	19 28	230	51 28	298
42	54 24	010	46 11	037	64 05	083	46 57	120	16 16	190	18 44	231	50 36	298
43	54 33	009	46 46	037	65 02	083	47 47	121	16 06	190	17 59	231	49 45	298
44	54 42	008	47 20	036	66 00	083	48 37	122	15 55	191	17 13	232	48 54	297
	CAPELLA		◆POLLUX		PROCYON		SIRIUS		◆ACHERNAR		Diphda		◆Alpheratz	
45	47 54	035	23 18	066	20 53	090	26 06	118	15 44	191	42 51	227	48 02	297
46	48 27	035	24 11	066	21 51	090	26 57	118	15 32	192	42 08	228	47 10	297
47	49 00	034	25 04	066	22 49	091	27 48	119	15 20	192	41 25	228	46 19	297
48	49 33	034	25 57	066	23 47	091	28 39	119	15 07	193	40 41	229	45 27	296
49	50 04	033	26 50	066	24 45	091	29 29	120	14 54	193	39 57	230	44 35	296
50	50 35	032	27 43	066	25 42	091	30 19	120	14 40	194	39 13	231	43 43	296
51	51 06	031	28 35	066	26 40	092	31 09	121	14 26	194	38 27	232	42 51	296
52	51 36	031	29 28	066	27 38	092	31 59	121	14 11	195	37 42	232	41 58	296
53	52 05	030	30 21	066	28 36	092	32 49	122	13 56	195	36 56	233	41 06	296
54	52 33	029	31 14	066	29 34	092	33 38	122	13 40	196	36 09	234	40 14	296
55	53 01	028	32 07	066	30 32	093	34 26	123	13 24	196	35 22	234	39 22	296
56	53 28	027	33 00	066	31 30	093	35 15	124	13 08	197	34 35	235	38 29	295
57	53 54	026	33 53	066	32 28	093	36 03	124	12 51	197	33 47	236	37 37	295
58	54 19	025	34 46	066	33 26	094	36 51	125	12 33	198	32 59	236	36 45	295
59	54 44	024	35 39	066	34 23	094	37 38	126	12 15	198	32 11	237	35 52	295
	CAPELLA		◆POLLUX		PROCYON		SIRIUS		◆CANOPUS		Diphda		◆Hamal	
60	55 07	023	36 32	066	35 21	094	38 25	126	15 33	158	31 22	237	62 03	292
61	55 30	022	37 25	066	36 19	095	39 11	127	15 54	159	30 33	238	61 10	292
62	55 51	021	38 18	066	37 17	095	39 57	128	16 15	159	29 44	239	60 16	291
63	56 12	020	39 11	066	38 15	095	40 43	129	16 35	160	28 54	239	59 22	291
64	56 31	019	40 04	066	39 12	096	41 28	129	16 54	160	28 05	240	58 27	291
65	56 50	018	40 56	066	40 10	096	42 13	130	17 14	161	27 14	240	57 33	290
66	57 07	017	41 49	066	41 08	096	42 57	131	17 32	161	26 24	241	56 39	290
67	57 23	016	42 42	065	42 05	097	43 40	132	17 50	162	25 33	241	55 44	290
68	57 39	014	43 35	065	43 03	097	44 23	133	18 08	163	24 43	242	54 49	290
69	57 52	013	44 27	065	44 00	097	45 05	134	18 25	163	23 52	242	53 55	289
70	58 05	012	45 20	065	44 58	098	45 47	135	18 42	164	23 00	243	53 00	289
71	58 16	011	46 12	065	45 55	098	46 28	136	18 58	164	22 09	243	52 05	289
72	58 27	009	47 05	065	46 52	099	47 08	137	19 13	165	21 17	243	51 11	289
73	58 35	008	47 57	065	47 50	099	47 47	138	19 28	165	20 25	244	50 16	289
74	58 43	007	48 50	064	48 47	099	48 26	139	19 42	166	19 33	244	49 21	289
	CAPELLA		◆POLLUX		PROCYON		SIRIUS		◆CANOPUS		Acamar		◆Hamal	
75	58 49	005	49 42	064	49 44	100	49 04	140	19 56	167	27 51	206	48 26	289
76	58 54	004	50 34	064	50 41	100	49 41	141	20 09	167	27 25	207	47 31	288
77	58 57	003	51 26	064	51 38	101	50 17	142	20 21	168	26 59	207	46 36	288
78	59 00	001	52 18	063	52 35	101	50 52	143	20 33	168	26 32	208	45 41	288
79	59 00	000	53 09	063	53 32	102	51 26	144	20 44	169	26 04	209	44 46	288
80	59 00	359	54 01	063	54 29	102	52 00	146	20 55	170	25 36	209	43 51	288
81	58 58	357	54 52	062	55 25	103	52 32	147	21 05	170	25 08	210	42 56	288
82	58 55	356	55 44	062	56 22	103	53 03	148	21 14	171	24 38	211	42 01	288
83	58 50	355	56 35	061	57 18	104	53 32	150	21 23	172	24 08	211	41 06	288
84	58 44	353	57 25	061	58 14	105	54 01	151	21 31	172	23 38	212	40 11	288
85	58 37	352	58 16	061	59 10	105	54 28	153	21 39	173	23 07	213	39 16	288
86	58 28	351	59 06	060	60 06	106	54 54	154	21 46	174	22 36	213	38 21	288
87	58 18	350	59 56	059	61 01	107	55 19	156	21 52	174	22 04	214	37 26	288
88	58 07	348	60 46	059	61 57	108	55 42	157	21 57	175	21 31	214	36 31	288
89	57 54	347	61 36	058	62 52	108	56 04	159	22 02	175	20 58	215	35 36	288

LAT 15°N

LHA ♈	Hc	Zn	Hc	Zn	Hc	Zn	Hc	Zn	Hc	Zn	Hc	Zn	Hc	Zn
	Dubhe		◆REGULUS		Suhail		SIRIUS		◆RIGEL		ALDEBARAN		◆CAPELLA	
90	19 51	029	29 48	085	17 31	146	56 25	160	64 11	207	69 39	277	57 41	346
91	20 20	029	30 46	085	18 03	147	56 43	162	63 44	209	68 42	277	57 26	345
92	20 48	029	31 43	085	18 35	147	57 00	164	63 15	211	67 44	277	57 10	343
93	21 16	029	32 41	085	19 06	148	57 16	165	62 44	213	66 47	277	56 53	342
94	21 44	029	33 39	085	19 37	148	57 30	167	62 12	214	65 49	277	56 34	341
95	22 12	029	34 37	086	20 07	149	57 42	169	61 39	216	64 52	277	56 15	340
96	22 40	029	35 34	086	20 37	149	57 52	171	61 04	218	63 54	277	55 55	339
97	23 08	029	36 32	086	21 06	150	58 01	172	60 28	219	62 57	277	55 33	338
98	23 36	029	37 30	086	21 35	151	58 08	174	59 51	221	61 59	277	55 11	337
99	24 03	028	38 28	086	22 03	151	58 13	176	59 12	222	61 02	277	54 48	336
100	24 31	028	39 26	087	22 31	152	58 16	178	58 32	224	60 04	277	54 23	335
101	24 58	028	40 24	087	22 58	152	58 17	180	57 52	225	59 07	277	53 58	334
102	25 25	028	41 22	087	23 24	153	58 17	181	57 10	227	58 09	277	53 32	333
103	25 53	028	42 19	087	23 50	154	58 15	183	56 28	228	57 12	277	53 05	332
104	26 20	028	43 17	088	24 16	154	58 10	185	55 44	229	56 14	277	52 38	331
	Dubhe		◆REGULUS		Suhail		◆SIRIUS		RIGEL		ALDEBARAN		◆CAPELLA	
105	26 47	028	44 15	088	24 41	155	58 04	187	55 00	230	55 17	277	52 10	330
106	27 13	027	45 13	088	25 05	156	57 57	189	54 15	231	54 19	278	51 41	330
107	27 40	027	46 11	088	25 28	156	57 47	190	53 30	232	53 22	278	51 11	329
108	28 06	027	47 09	088	25 51	157	57 36	192	52 44	233	52 24	278	50 40	328
109	28 32	027	48 07	089	26 14	158	57 23	194	51 57	234	51 27	278	50 09	327
110	28 59	027	49 05	089	26 35	158	57 08	196	51 09	235	50 29	278	49 38	327
111	29 24	026	50 03	089	26 56	159	56 51	197	50 21	236	49 32	278	49 05	326
112	29 50	026	51 01	089	27 17	160	56 33	199	49 33	237	48 35	278	48 33	325
113	30 15	026	51 59	090	27 36	161	56 14	201	48 44	238	47 37	278	47 59	325
114	30 41	026	52 57	090	27 55	161	55 53	202	47 55	239	46 40	278	47 26	324
115	31 05	025	53 55	090	28 13	162	55 30	204	47 05	240	45 43	278	46 51	323
116	31 30	025	54 53	090	28 31	163	55 06	205	46 15	240	44 45	279	46 16	323
117	31 55	025	55 51	091	28 48	164	54 41	207	45 24	241	43 48	279	45 41	322
118	32 19	025	56 49	091	29 04	164	54 14	208	44 33	242	42 51	279	45 06	322
119	32 43	024	57 46	091	29 19	165	53 46	210	43 42	243	41 53	279	44 30	321
	Dubhe		◆Denebola		Suhail		◆SIRIUS		RIGEL		ALDEBARAN		◆CAPELLA	
120	33 06	024	34 52	082	29 34	166	53 17	211	42 50	243	40 56	279	43 53	321
121	33 30	024	35 49	083	29 47	167	52 46	212	41 58	244	39 59	279	43 16	320
122	33 53	023	36 47	083	30 00	167	52 15	214	41 06	245	39 02	279	42 39	320
123	34 15	023	37 44	083	30 12	168	51 42	215	40 14	245	38 04	279	42 02	320
124	34 38	023	38 42	083	30 24	169	51 08	216	39 21	246	37 07	280	41 24	319
125	35 00	022	39 39	083	30 34	170	50 33	217	38 28	246	36 10	280	40 46	319
126	35 22	022	40 37	083	30 44	171	49 58	219	37 35	247	35 13	280	40 08	318
127	35 43	021	41 34	084	30 53	172	49 21	220	36 41	247	34 16	280	39 29	318
128	36 04	021	42 32	084	31 01	172	48 44	221	35 48	248	33 19	280	38 50	318
129	36 25	021	43 30	084	31 08	173	48 05	222	34 54	248	32 22	280	38 11	317
130	36 45	020	44 27	084	31 14	174	47 26	223	34 00	249	31 25	280	37 32	317
131	37 05	020	45 25	084	31 20	175	46 46	224	33 06	249	30 28	281	36 52	317
132	37 24	019	46 23	084	31 25	176	46 06	225	32 11	250	29 31	281	36 13	317
133	37 43	019	47 20	085	31 29	177	45 25	226	31 17	250	28 34	281	35 33	316
134	38 02	018	48 18	085	31 32	177	44 43	227	30 22	251	27 37	281	34 53	316
	Dubhe		◆ARCTURUS		SPICA		◆Suhail		SIRIUS		BETELGEUSE		◆CAPELLA	
135	38 20	018	15 09	074	19 23	108	31 34	178	44 00	228	44 04	266	34 12	316
136	38 37	018	16 05	074	20 18	108	31 35	179	43 17	229	43 06	266	33 32	316
137	38 55	017	17 01	074	21 13	109	31 35	180	42 33	229	42 08	266	32 51	316
138	39 11	017	17 56	074	22 08	109	31 35	181	41 49	230	41 10	267	32 11	315
139	39 28	016	18 52	074	23 03	109	31 33	182	41 04	231	40 12	267	31 30	315
140	39 43	016	19 48	075	23 58	110	31 31	183	40 19	232	39 14	267	30 49	315
141	39 59	015	20 44	075	24 52	110	31 28	183	39 33	233	38 17	268	30 08	315
142	40 14	015	21 40	075	25 47	111	31 24	184	38 46	233	37 19	268	29 27	315
143	40 28	014	22 36	075	26 41	111	31 20	185	38 00	234	36 21	268	28 46	315
144	40 42	013	23 32	075	27 35	111	31 14	186	37 13	235	35 23	268	28 04	315
145	40 55	013	24 28	075	28 29	112	31 07	187	36 25	235	34 25	269	27 23	314
146	41 07	012	25 24	076	29 22	112	31 00	188	35 37	236	33 27	269	26 42	314
147	41 19	012	26 20	076	30 16	113	30 52	188	34 49	237	32 29	269	26 00	314
148	41 31	011	27 16	076	31 09	113	30 43	189	34 00	237	31 31	270	25 19	314
149	41 42	011	28 13	076	32 02	114	30 33	190	33 11	238	30 33	270	24 37	314
	◆Dubhe		Alkaid		ARCTURUS		◆SPICA		Suhail		◆SIRIUS		POLLUX	
150	41 52	010	32 44	040	29 09	076	32 55	114	30 23	191	32 22	239	56 09	298
151	42 02	009	33 21	040	30 05	076	33 48	115	30 11	192	31 33	239	55 18	298
152	42 11	009	33 58	040	31 01	076	34 40	115	29 59	193	30 43	240	54 27	298
153	42 20	008	34 36	040	31 58	076	35 33	116	29 46	193	29 52	240	53 35	297
154	42 28	008	35 12	039	32 54	077	36 25	116	29 32	194	29 02	241	52 44	297
155	42 35	007	35 49	039	33 50	077	37 16	117	29 18	195	28 11	241	51 52	297
156	42 42	006	36 26	039	34 47	077	38 08	118	29 02	196	27 20	242	51 00	296
157	42 48	006	37 02	039	35 43	077	38 59	118	28 46	197	26 29	242	50 08	296
158	42 53	005	37 38	038	36 40	077	39 50	119	28 29	197	25 38	243	49 16	296
159	42 58	004	38 14	038	37 36	077	40 40	120	28 12	198	24 46	243	48 24	296
160	43 02	004	38 49	038	38 33	077	41 31	120	27 53	199	23 55	244	47 31	295
161	43 06	003	39 24	037	39 29	077	42 21	121	27 34	200	23 02	244	46 39	295
162	43 08	002	39 59	037	40 26	077	43 10	122	27 15	200	22 10	244	45 46	295
163	43 11	002	40 34	036	41 22	077	43 59	122	26 54	201	21 18	245	44 54	295
164	43 12	001	41 08	036	42 19	077	44 48	123	26 33	202	20 25	245	44 01	295
	◆Dubhe		ARCTURUS		◆SPICA		ACRUX		Suhail		◆PROCYON		POLLUX	
165	43 13	001	43 15	078	45 36	124	10 08	170	26 12	202	39 42	264	43 09	295
166	43 13	000	44 12	078	46 24	125	10 18	171	25 49	203	38 45	265	42 16	294
167	43 13	359	45 09	078	47 11	126	10 27	171	25 26	204	37 47	265	41 23	294
168	43 12	359	46 05	078	47 58	127	10 36	172	25 03	204	36 49	265	40 30	294
169	43 10	358	47 02	078	48 44	127	10 44	172	24 38	205	35 51	266	39 37	294
170	43 08	357	47 59	078	49 30	128	10 52	172	24 13	206	34 54	266	38 45	294
171	43 05	357	48 55	078	50 15	129	10 59	173	23 48	206	33 56	266	37 52	294
172	43 01	356	49 52	078	51 00	130	11 06	173	23 22	207	32 58	266	36 59	294
173	42 57	355	50 49	078	51 43	131	11 13	174	22 55	208	32 00	267	36 06	294
174	42 52	355	51 45	078	52 26	133	11 19	174	22 28	208	31 02	267	35 13	294
175	42 46	354	52 42	078	53 09	134	11 24	175	22 00	209	30 04	267	34 20	294
176	42 40	353	53 38	078	53 50	135	11 30	175	21 32	209	29 06	268	33 27	294
177	42 33	353	54 35	078	54 31	136	11 34	176	21 03	210	28 08	268	32 34	294
178	42 25	352	55 32	078	55 11	137	11 39	176	20 34	211	27 11	268	31 41	294
179	42 17	352	56 28	078	55 49	139	11 42	177	20 04	211	26 13	268	30 43	294

LAT 15°N

LHA	Hc	Zn	Hc	Zn	Hc	Zn	Hc	Zn	Hc	Zn	Hc	Zn	Hc	Zn
♈	Alkaid		◆Alphecca		ANTARES		◆ACRUX		REGULUS		◆POLLUX		Dubhe	
180	49 17	027	38 55	068	12 39	122	11 46	177	62 39	267	29 55	294	42 09	351
181	49 43	026	39 48	067	13 28	123	11 48	177	61 41	267	29 02	294	41 59	350
182	50 08	025	40 42	067	14 17	123	11 51	178	60 43	268	28 09	294	41 49	350
183	50 32	024	41 35	067	15 05	123	11 53	178	59 45	268	27 16	294	41 39	349
184	50 56	024	42 29	067	15 54	124	11 54	179	58 47	268	26 24	294	41 28	349
185	51 19	023	43 22	067	16 42	124	11 55	179	57 50	269	25 31	294	41 16	348
186	51 41	022	44 16	067	17 30	125	11 56	180	56 52	269	24 38	294	41 04	348
187	52 02	021	45 09	067	18 17	125	11 56	180	55 54	269	23 45	294	40 51	347
188	52 22	020	46 02	067	19 04	125	11 55	181	54 56	270	22 52	295	40 38	346
189	52 42	019	46 56	067	19 52	126	11 54	181	53 58	270	22 00	295	40 24	346
190	53 01	018	47 49	067	20 38	126	11 53	182	53 00	270	21 07	295	40 09	345
191	53 18	017	48 42	066	21 25	127	11 51	182	52 02	270	20 14	295	39 54	345
192	53 35	016	49 35	066	22 11	127	11 49	183	51 04	271	19 22	295	39 39	344
193	53 51	015	50 28	066	22 57	128	11 46	183	50 06	271	18 29	295	39 23	344
194	54 06	014	51 21	066	23 43	128	11 43	183	49 08	271	17 37	295	39 07	343
	◆VEGA		Rasalhague		ANTARES		◆RIGIL KENT.		Gienah		◆REGULUS		Dubhe	
195	13 47	053	23 31	083	24 28	129	11 38	168	55 41	199	48 10	271	38 50	343
196	14 33	053	24 29	083	25 13	129	11 50	168	55 22	201	47 12	272	38 32	342
197	15 20	053	25 26	083	25 57	130	12 01	169	55 01	202	46 14	272	38 15	342
198	16 06	053	26 24	083	26 42	131	12 12	169	54 38	204	45 16	272	37 56	341
199	16 53	053	27 22	083	27 25	131	12 23	170	54 14	205	44 18	272	37 38	341
200	17 39	053	28 19	084	28 09	132	12 33	170	53 49	207	43 20	272	37 18	341
201	18 26	054	29 17	084	28 52	132	12 42	171	53 22	208	42 22	273	36 59	340
202	19 13	054	30 14	084	29 34	133	12 51	171	52 54	209	41 25	273	36 39	340
203	19 59	054	31 12	084	30 17	134	13 00	172	52 25	211	40 27	273	36 19	339
204	20 46	054	32 10	084	30 58	134	13 08	172	51 55	212	39 29	273	35 58	339
205	21 33	054	33 07	085	31 40	135	13 16	173	51 23	213	38 31	274	35 37	338
206	22 19	054	34 05	085	32 20	136	13 23	173	50 51	215	37 33	274	35 15	338
207	23 06	054	35 03	085	33 01	136	13 30	174	50 18	216	36 35	274	34 54	338
208	23 53	054	36 01	085	33 40	137	13 36	174	49 43	217	35 38	274	34 31	337
209	24 40	054	36 58	085	34 20	138	13 41	175	49 08	218	34 40	274	34 09	337
	◆VEGA		Rasalhague		ANTARES		◆RIGIL KENT.		SPICA		◆REGULUS		Dubhe	
210	25 27	054	37 56	086	34 58	139	13 47	175	62 28	199	33 42	275	33 46	337
211	26 13	054	38 54	086	35 36	139	13 51	176	62 08	201	32 44	275	33 23	336
212	27 00	054	39 52	086	36 14	140	13 56	176	61 46	203	31 46	275	33 00	336
213	27 47	054	40 50	086	36 51	141	13 59	177	61 23	205	30 49	275	32 36	336
214	28 33	054	41 47	086	37 27	142	14 03	177	60 58	207	29 51	275	32 12	335
215	29 20	054	42 45	087	38 02	143	14 05	178	60 31	208	28 53	276	31 48	335
216	30 07	054	43 43	087	38 37	143	14 07	178	60 03	210	27 56	276	31 23	335
217	30 53	053	44 41	087	39 11	144	14 09	179	59 33	212	26 58	276	30 58	335
218	31 40	053	45 39	087	39 45	145	14 10	179	59 02	213	26 00	276	30 33	334
219	32 26	053	46 37	087	40 17	146	14 11	180	58 29	215	25 03	276	30 08	334
220	33 13	053	47 35	088	40 49	147	14 11	180	57 56	216	24 05	277	29 43	334
221	33 59	053	48 33	088	41 20	148	14 11	181	57 21	218	23 08	277	29 17	334
222	34 45	053	49 30	088	41 50	149	14 10	181	56 44	219	22 10	277	28 51	333
223	35 32	053	50 28	088	42 20	150	14 09	182	56 07	221	21 13	277	28 25	333
224	36 18	053	51 26	088	42 48	151	14 07	182	55 29	222	20 15	277	27 59	333
	◆VEGA		ALTAIR		ANTARES		◆RIGIL KENT.		SPICA		◆Denebola		Alkaid	
225	37 04	052	18 57	086	43 16	152	14 04	183	54 49	223	43 53	276	52 36	340
226	37 49	052	19 55	086	43 42	153	14 01	183	54 09	225	42 55	276	52 16	340
227	38 35	052	20 52	086	44 08	154	13 58	184	53 28	226	41 57	276	51 55	339
228	39 21	052	21 50	086	44 32	155	13 54	184	52 46	227	41 00	276	51 34	338
229	40 06	052	22 48	087	44 56	157	13 50	185	52 03	228	40 02	277	51 11	337
230	40 52	051	23 46	087	45 18	158	13 45	185	51 20	229	39 05	277	50 48	336
231	41 37	051	24 44	087	45 40	159	13 39	186	50 36	230	38 07	277	50 24	335
232	42 22	051	25 42	087	46 00	160	13 34	186	49 51	231	37 10	277	50 00	334
233	43 06	050	26 40	087	46 19	161	13 27	187	49 06	232	36 12	277	49 34	334
234	43 51	050	27 38	088	46 37	163	13 20	187	48 20	233	35 15	277	49 08	333
235	44 35	050	28 35	088	46 54	164	13 13	188	47 33	234	34 17	278	48 42	332
236	45 19	049	29 33	088	47 09	165	13 05	188	46 46	235	33 20	278	48 14	332
237	46 03	049	30 31	088	47 23	166	12 57	189	45 58	236	32 22	278	47 46	331
238	46 47	049	31 29	089	47 37	168	12 48	189	45 10	236	31 25	278	47 18	330
239	47 30	048	32 27	089	47 48	169	12 39	189	44 22	237	30 28	278	46 49	330
	◆VEGA		ALTAIR		Nunki		◆ANTARES		SPICA		◆ARCTURUS		Alkaid	
240	48 13	048	33 25	089	30 44	134	47 59	170	43 33	238	64 41	283	46 19	329
241	48 56	047	34 23	089	31 25	135	48 08	172	42 44	239	63 45	283	45 49	328
242	49 38	047	35 21	090	32 06	135	48 16	173	41 54	239	62 48	283	45 18	328
243	50 20	046	36 19	090	32 47	136	48 22	174	41 04	240	61 52	283	44 47	327
244	51 02	046	37 17	090	33 27	137	48 27	176	40 13	241	60 55	283	44 16	327
245	51 43	045	38 15	090	34 06	137	48 31	177	39 23	241	59 59	283	43 43	326
246	52 24	044	39 13	091	34 45	138	48 34	178	38 32	242	59 02	282	43 11	326
247	53 04	044	40 11	091	35 24	139	48 35	180	37 40	243	58 06	282	42 38	325
248	53 44	043	41 09	091	36 02	140	48 34	181	36 49	243	57 09	282	42 05	325
249	54 23	042	42 07	092	36 39	140	48 33	182	35 57	244	56 12	282	41 31	324
250	55 02	042	43 05	092	37 15	141	48 30	184	35 05	244	55 16	282	40 57	324
251	55 40	041	44 03	092	37 51	142	48 25	185	34 12	245	54 19	282	40 23	323
252	56 18	040	45 00	092	38 27	143	48 19	186	33 20	245	53 22	282	39 48	323
253	56 54	039	45 58	093	39 01	144	48 12	188	32 27	246	52 26	282	39 13	323
254	57 31	038	46 56	093	39 35	145	48 04	189	31 34	246	51 29	282	38 38	322
	◆VEGA		ALTAIR		◆Nunki		ANTARES		SPICA		◆ARCTURUS		Alkaid	
255	58 06	037	47 54	093	40 08	146	47 54	190	30 41	247	50 32	282	38 02	322
256	58 41	036	48 52	094	40 40	147	47 43	192	29 47	247	49 36	282	37 26	322
257	59 14	035	49 50	094	41 12	148	47 31	193	28 54	248	48 39	282	36 50	321
258	59 47	034	50 48	094	41 42	149	47 17	194	28 00	248	47 43	282	36 14	321
259	60 19	033	51 45	095	42 12	150	47 02	195	27 06	249	46 46	282	35 37	321
260	60 50	032	52 43	095	42 41	151	46 46	197	26 12	249	45 49	282	35 00	320
261	61 20	030	53 41	095	43 09	152	46 29	198	25 17	250	44 53	282	34 23	320
262	61 49	029	54 38	096	43 36	153	46 10	199	24 23	250	43 56	282	33 46	320
263	62 17	028	55 36	096	44 02	154	45 51	200	23 28	250	42 59	282	33 09	320
264	62 43	026	56 34	097	44 27	155	45 30	202	22 34	251	42 03	283	32 31	320
265	63 08	025	57 31	097	44 51	156	45 08	203	21 39	251	41 06	283	31 54	319
266	63 32	024	58 29	098	45 14	157	44 45	204	20 44	252	40 10	283	31 16	319
267	63 55	022	59 26	098	45 36	158	44 21	205	19 49	252	39 13	283	30 38	319
268	64 16	020	60 23	099	45 57	160	43 56	206	18 54	252	38 17	283	30 00	319
269	64 35	019	61 21	099	46 17	161	43 30	207	17 59	253	37 20	283	29 21	319

LHA	Hc	Zn	Hc	Zn	Hc	Zn	Hc	Zn	Hc	Zn	Hc	Zn	Hc	Zn
♈	VEGA		◆DENEB		Enif		◆Nunki		ANTARES		◆ARCTURUS		Alkaid	
270	64 53	017	44 36	040	35 13	088	46 35	162	43 03	208	36 24	283	28 43	318
271	65 09	015	45 13	039	36 11	089	46 53	163	42 35	209	35 27	283	28 04	318
272	65 24	014	45 50	039	37 09	089	47 09	164	42 07	210	34 31	283	27 26	318
273	65 36	012	46 26	038	38 07	089	47 24	166	41 37	211	33 34	283	26 47	318
274	65 47	010	47 02	038	39 05	089	47 37	167	41 06	212	32 38	283	26 09	318
275	65 56	008	47 37	037	40 03	090	47 50	168	40 35	213	31 42	284	25 30	318
276	66 03	006	48 12	037	41 01	090	48 01	170	40 03	214	30 45	284	24 51	318
277	66 09	004	48 46	036	41 59	090	48 11	171	39 30	215	29 49	284	24 12	318
278	66 12	002	49 20	035	42 57	090	48 19	172	38 56	216	28 53	284	23 33	318
279	66 13	000	49 53	035	43 55	091	48 26	174	38 22	217	27 57	284	22 54	318
280	66 13	358	50 25	034	44 52	091	48 32	175	37 47	218	27 00	284	22 15	318
281	66 10	357	50 57	033	45 50	091	48 37	176	37 11	219	26 04	284	21 36	318
282	66 06	355	51 29	032	46 48	092	48 40	178	36 34	219	25 08	285	20 57	318
283	65 59	353	51 59	032	47 46	092	48 41	179	35 57	220	24 12	285	20 18	318
284	65 51	351	52 30	031	48 44	092	48 42	180	35 19	221	23 16	285	19 39	318
	◆DENEB		Alpheratz		◆FOMALHAUT		Nunki		ANTARES		◆Rasalhague		VEGA	
285	52 59	030	18 22	064	17 27	128	48 41	182	34 41	222	69 09	266	65 41	349
286	53 27	029	19 14	064	18 13	129	48 38	183	34 02	223	68 11	266	65 29	347
287	53 55	028	20 06	064	18 58	129	48 35	184	33 23	223	67 14	267	65 15	345
288	54 22	027	20 59	064	19 42	130	48 29	186	32 43	224	66 16	267	64 59	344
289	54 48	026	21 51	064	20 27	130	48 23	187	32 02	225	65 18	267	64 42	342
290	55 14	025	22 43	064	21 11	131	48 15	188	31 21	225	64 20	268	64 23	340
291	55 38	024	23 35	064	21 55	131	48 06	190	30 40	226	63 22	268	64 03	339
292	56 02	023	24 28	064	22 38	132	47 55	191	29 58	227	62 24	268	63 41	337
293	56 24	022	25 20	065	23 21	132	47 44	192	29 15	227	61 26	269	63 18	336
294	56 45	021	26 12	065	24 04	133	47 31	194	28 33	228	60 28	269	62 53	334
295	57 06	020	27 05	065	24 46	133	47 16	195	27 49	229	59 30	269	62 27	333
296	57 25	019	27 57	065	25 28	134	47 01	196	27 06	229	58 32	270	62 00	331
297	57 43	018	28 49	065	26 09	135	46 44	197	26 22	230	57 34	270	61 32	330
298	58 00	016	29 42	065	26 50	135	46 26	199	25 37	230	56 36	270	61 02	329
299	58 16	015	30 34	065	27 31	136	46 07	200	24 53	231	55 38	270	60 31	328
	◆DENEB		Alpheratz		◆FOMALHAUT		Peacock		◆ANTARES		Rasalhague		VEGA	
300	58 31	014	31 27	065	28 11	136	18 03	176	24 08	231	54 41	271	60 00	326
301	58 44	013	32 19	065	28 51	137	18 07	177	23 22	232	53 43	271	59 27	325
302	58 56	011	33 12	065	29 30	138	18 09	178	22 36	232	52 45	271	58 54	324
303	59 07	010	34 04	065	30 09	138	18 12	178	21 50	233	51 47	271	58 19	323
304	59 16	009	34 56	065	30 47	139	18 13	179	21 04	233	50 49	272	57 44	322
305	59 24	007	35 49	065	31 25	140	18 14	179	20 17	234	49 51	272	57 08	321
306	59 31	006	36 41	065	32 02	141	18 15	180	19 30	234	48 53	272	56 32	320
307	59 37	005	37 34	065	32 38	141	18 15	180	18 43	235	47 55	272	55 55	320
308	59 40	003	38 26	065	33 14	142	18 14	181	17 56	235	46 57	272	55 17	319
309	59 43	002	39 18	064	33 50	143	18 13	182	17 08	236	45 59	273	54 38	318
310	59 44	000	40 11	064	34 24	144	18 11	182	16 20	236	45 01	273	53 59	317
311	59 44	359	41 03	064	34 59	144	18 09	183	15 32	236	44 03	273	53 19	317
312	59 42	358	41 55	064	35 32	145	18 05	183	14 43	237	43 06	273	52 39	316
313	59 39	356	42 47	064	36 05	146	18 02	184	13 55	237	42 08	274	51 59	315
314	59 35	355	43 39	064	36 37	147	17 58	184	13 06	238	41 10	274	51 18	315
	Alpheratz		◆Diphda		FOMALHAUT		◆Nunki		Rasalhague		VEGA		◆DENEB	
315	44 31	064	25 50	119	37 08	148	38 43	217	40 12	274	50 36	314	59 29	354
316	45 23	064	26 41	120	37 39	149	38 08	217	39 14	274	49 54	314	59 22	352
317	46 15	063	27 31	120	38 09	149	37 33	218	38 16	274	49 12	313	59 13	351
318	47 07	063	28 21	121	38 38	150	36 56	219	37 19	275	48 30	313	59 03	349
319	47 59	063	29 11	121	39 06	151	36 19	220	36 21	275	47 47	312	58 52	348
320	48 50	063	30 00	122	39 33	152	35 42	221	35 23	275	47 04	312	58 39	347
321	49 42	063	30 50	122	40 00	153	35 04	222	34 25	275	46 20	311	58 25	346
322	50 33	062	31 38	123	40 25	154	34 25	222	33 28	275	45 36	311	58 10	344
323	51 24	062	32 27	123	40 50	155	33 46	223	32 30	276	44 52	310	57 54	343
324	52 15	062	33 15	124	41 14	156	33 06	224	31 32	276	44 08	310	57 37	342
325	53 06	061	34 03	125	41 37	157	32 26	224	30 35	276	43 24	310	57 18	341
326	53 57	061	34 51	125	41 58	158	31 45	225	29 37	276	42 39	309	56 58	340
327	54 48	060	35 38	126	42 19	159	31 03	226	28 39	276	41 54	309	56 38	338
328	55 38	060	36 25	127	42 39	161	30 22	226	27 42	277	41 09	309	56 16	337
329	56 28	060	37 11	127	42 58	162	29 39	227	26 44	277	40 24	309	55 53	336
	◆Alpheratz		Diphda		◆FOMALHAUT		Peacock		ALTAIR		◆VEGA		DENEB	
330	57 18	059	37 57	128	43 15	163	15 35	193	57 47	263	39 38	308	55 29	335
331	58 07	059	38 42	129	43 32	164	15 21	194	56 49	263	38 53	308	55 04	334
332	58 57	058	39 27	129	43 47	165	15 07	194	55 52	264	38 07	308	54 39	333
333	59 46	057	40 12	130	44 02	166	14 53	195	54 54	264	37 21	308	54 12	332
334	60 34	057	40 56	131	44 15	167	14 38	195	53 56	264	36 35	307	53 45	331
335	61 23	056	41 39	132	44 27	169	14 22	196	52 59	265	35 49	307	53 17	331
336	62 10	055	42 22	133	44 38	170	14 06	196	52 01	265	35 03	307	52 48	330
337	62 58	054	43 05	134	44 48	171	13 50	197	51 03	266	34 17	307	52 19	329
338	63 45	054	43 46	134	44 56	172	13 33	197	50 05	266	33 31	307	51 48	328
339	64 31	053	44 27	135	45 03	173	13 15	198	49 08	266	32 44	307	51 17	327
340	65 17	052	45 08	136	45 09	175	12 57	198	48 10	267	31 58	307	50 46	327
341	66 02	051	45 47	137	45 14	176	12 39	199	47 12	267	31 11	307	50 13	326
342	66 46	049	46 26	138	45 18	177	12 20	199	46 14	267	30 25	306	49 41	325
343	67 30	048	47 05	139	45 20	178	12 01	200	45 16	267	29 38	306	49 07	324
344	68 13	047	47 42	140	45 21	180	11 42	200	44 18	268	28 51	306	48 33	324
	Schedar		◆Mirfak		Hamal		◆ACHERNAR		FOMALHAUT		◆ALTAIR		DENEB	
345	44 19	019	26 49	041	45 16	072	10 43	160	45 21	181	43 20	268	47 59	323
346	44 38	018	27 28	041	46 11	072	11 03	160	45 19	182	42 22	268	47 24	323
347	44 56	018	28 06	041	47 06	072	11 23	160	45 17	183	41 24	269	46 48	322
348	45 13	017	28 44	041	48 01	071	11 42	161	45 13	185	40 26	269	46 13	322
349	45 30	016	29 22	041	48 56	071	12 01	161	45 08	186	39 28	269	45 36	321
350	45 46	016	30 00	041	49 51	071	12 19	162	45 01	187	38 31	269	45 00	321
351	46 01	015	30 37	040	50 46	071	12 37	162	44 54	188	37 33	270	44 23	320
352	46 16	014	31 15	040	51 41	071	12 55	163	44 45	189	36 35	270	43 45	320
353	46 30	014	31 52	040	52 36	071	13 12	163	44 35	191	35 37	270	43 08	319
354	46 43	013	32 29	040	53 30	071	13 28	164	44 23	192	34 39	270	42 30	319
355	46 56	012	33 06	040	54 25	071	13 44	164	44 11	193	33 41	271	41 51	318
356	47 08	011	33 43	039	55 20	070	14 00	165	43 57	194	32 43	271	41 13	318
357	47 19	011	34 20	039	56 14	070	14 15	165	43 43	195	31 45	271	40 34	318
358	47 29	010	34 57	039	57 09	070	14 30	166	43 27	196	30 47	271	39 54	317
359	47 39	009	35 33	039	58 03	070	14 44	166	43 10	198	29 49	272	39 15	317

 LAT 14°N

LHA ♈	Hc	Zn	Hc	Zn	Hc	Zn	Hc	Zn	Hc	Zn	Hc	Zn	Hc	Zn
	CAPELLA		◆ALDEBARAN		Acamar		ACHERNAR		◆FOMALHAUT		ALTAIR		◆DENEB	
0	17 34	046	23 48	078	21 46	145	15 56	167	43 49	199	28 49	273	37 52	317
1	18 15	046	24 45	078	22 20	145	16 09	167	43 29	200	27 51	273	37 12	317
2	18 57	046	25 42	078	22 52	146	16 22	168	43 09	201	26 52	273	36 32	317
3	19 39	046	26 39	078	23 25	147	16 34	168	42 47	202	25 54	273	35 52	316
4	20 20	046	27 36	078	23 56	147	16 46	169	42 25	203	24 56	273	35 11	316
5	21 02	046	28 33	079	24 28	148	16 57	169	42 01	204	23 58	274	34 31	316
6	21 44	046	29 30	079	24 59	148	17 08	170	41 37	205	23 00	274	33 50	316
7	22 25	046	30 27	079	25 29	149	17 18	170	41 11	206	22 02	274	33 10	315
8	23 07	046	31 24	079	25 58	150	17 27	171	40 45	207	21 04	274	32 29	315
9	23 49	046	32 21	079	26 28	150	17 36	171	40 18	208	20 06	274	31 47	315
10	24 30	045	33 19	079	26 56	151	17 45	172	39 50	209	19 08	275	31 06	315
11	25 12	045	34 16	079	27 24	152	17 53	172	39 21	210	18 10	275	30 25	315
12	25 53	045	35 13	079	27 51	152	18 00	173	38 51	211	17 12	275	29 43	314
13	26 34	045	36 10	080	28 18	153	18 07	174	38 20	212	16 14	275	29 02	314
14	27 16	045	37 07	080	28 44	154	18 13	174	37 49	213	15 16	276	28 20	314
	CAPELLA		◆ALDEBARAN		RIGEL		ACHERNAR		◆FOMALHAUT		Enif		◆DENEB	
15	27 57	045	38 05	080	23 08	105	18 19	175	37 17	214	41 57	271	27 38	314
16	28 38	045	39 02	080	24 04	106	18 24	175	36 44	215	40 58	271	26 56	314
17	29 19	045	39 59	080	25 00	106	18 29	176	36 11	216	40 00	271	26 14	314
18	30 00	045	40 57	080	25 56	107	18 32	176	35 36	216	39 02	271	25 32	314
19	30 41	044	41 54	080	26 52	107	18 36	177	35 02	217	38 04	272	24 50	314
20	31 22	044	42 52	080	27 47	107	18 39	178	34 26	218	37 06	272	24 08	314
21	32 02	044	43 49	080	28 43	108	18 41	178	33 50	219	36 07	272	23 26	314
22	32 42	044	44 46	081	29 38	108	18 43	179	33 13	219	35 09	272	22 44	314
23	33 23	044	45 44	081	30 33	109	18 44	179	32 36	220	34 11	273	22 02	314
24	34 03	043	46 41	081	31 29	109	18 44	180	31 58	221	33 13	273	21 20	314
25	34 43	043	47 39	081	32 24	109	18 44	180	31 20	222	32 15	273	20 37	314
26	35 23	043	48 36	081	33 18	110	18 44	181	30 41	222	31 17	273	19 55	313
27	36 02	043	49 34	081	34 13	110	18 42	181	30 02	223	30 18	273	19 13	314
28	36 41	042	50 31	081	35 08	111	18 40	182	29 22	224	29 20	274	18 31	314
29	37 21	042	51 29	081	36 02	111	18 38	183	28 42	224	28 22	274	17 49	314
	Mirfak		◆CAPELLA		ALDEBARAN		RIGEL		◆ACHERNAR		FOMALHAUT		◆Alpheratz	
30	50 16	021	38 00	042	52 26	081	36 56	112	18 35	183	28 01	225	60 04	305
31	50 37	020	38 38	041	53 24	081	37 50	112	18 32	184	27 20	225	59 16	304
32	50 57	019	39 17	041	54 21	081	38 44	113	18 27	184	26 38	226	58 27	303
33	51 16	019	39 55	041	55 19	081	39 37	113	18 23	185	25 56	227	57 39	303
34	51 34	018	40 33	040	56 16	081	40 31	114	18 18	185	25 14	227	56 50	302
35	51 51	017	41 10	040	57 14	081	41 24	115	18 12	186	24 31	228	56 00	302
36	52 07	016	41 47	040	58 11	081	42 16	115	18 05	187	23 48	228	55 11	301
37	52 23	015	42 24	039	59 09	081	43 09	116	17 58	187	23 04	229	54 21	301
38	52 37	014	43 01	039	60 06	081	44 01	116	17 51	188	22 20	229	53 30	300
39	52 50	013	43 37	038	61 04	081	44 53	117	17 43	188	21 36	230	52 40	300
40	53 03	012	44 13	038	62 01	081	45 45	118	17 34	189	20 51	230	51 50	300
41	53 14	011	44 48	037	62 59	081	46 36	119	17 25	189	20 07	231	50 59	299
42	53 25	010	45 23	037	63 56	081	47 27	119	17 15	190	19 21	231	50 08	299
43	53 34	009	45 57	036	64 54	081	48 18	120	17 05	190	18 36	232	49 17	299
44	53 42	008	46 31	035	65 51	081	49 08	121	16 54	191	17 50	232	48 26	298
	CAPELLA		◆POLLUX		PROCYON		SIRIUS		◆ACHERNAR		Diphda		◆Alpheratz	
45	47 05	035	22 53	065	20 53	090	26 33	117	16 43	191	43 32	227	47 34	298
46	47 38	034	23 46	065	21 51	090	27 25	118	16 31	192	42 48	228	46 43	298
47	48 10	034	24 39	065	22 49	090	28 16	118	16 19	193	42 05	229	45 51	298
48	48 42	033	25 32	065	23 47	090	29 08	119	16 06	193	41 20	230	45 00	297
49	49 14	032	26 25	065	24 45	091	29 59	119	15 53	194	40 36	231	44 08	297
50	49 44	032	27 18	065	25 44	091	30 49	120	15 39	194	39 50	231	43 16	297
51	50 15	031	28 11	065	26 42	091	31 40	120	15 24	195	39 04	232	42 24	297
52	50 44	030	29 04	065	27 40	091	32 30	121	15 09	195	38 18	233	41 32	297
53	51 13	029	29 57	065	28 38	092	33 20	121	14 54	196	37 32	234	40 40	296
54	51 41	028	30 50	065	29 36	092	34 10	122	14 38	196	36 45	234	39 48	296
55	52 08	028	31 43	065	30 35	092	34 59	122	14 22	197	35 57	235	38 56	296
56	52 35	027	32 36	065	31 33	092	35 48	123	14 05	197	35 09	236	38 03	296
57	53 00	026	33 28	065	32 31	093	36 36	124	13 48	197	34 21	236	37 11	296
58	53 25	025	34 21	065	33 29	093	37 25	124	13 30	198	33 32	237	36 19	296
59	53 49	024	35 14	065	34 27	093	38 13	125	13 12	198	32 44	237	35 26	296
	CAPELLA		◆POLLUX		PROCYON		SIRIUS		◆CANOPUS		Diphda		◆Hamal	
60	54 12	023	36 07	065	35 25	094	39 00	126	16 28	158	31 54	238	61 40	294
61	54 34	022	37 00	065	36 23	094	39 47	126	16 50	159	31 05	239	60 47	293
62	54 55	021	37 53	065	37 22	094	40 34	127	17 11	159	30 15	239	59 53	293
63	55 16	020	38 46	065	38 20	094	41 20	128	17 31	160	29 25	240	58 59	292
64	55 35	019	39 39	065	39 18	095	42 06	129	17 51	160	28 35	240	58 06	292
65	55 53	018	40 31	065	40 16	095	42 51	129	18 10	161	27 44	241	57 11	292
66	56 10	016	41 24	065	41 14	095	43 36	130	18 29	161	26 53	241	56 17	291
67	56 26	015	42 17	065	42 12	096	44 20	131	18 47	162	26 02	242	55 23	291
68	56 40	014	43 09	065	43 09	096	45 03	132	19 05	163	25 11	242	54 29	291
69	56 54	013	44 02	064	44 07	096	45 46	133	19 22	163	24 19	242	53 34	291
70	57 06	012	44 54	064	45 05	097	46 29	134	19 39	164	23 28	243	52 40	290
71	57 17	010	45 47	064	46 03	097	47 10	135	19 55	164	22 36	243	51 45	290
72	57 27	009	46 39	064	47 01	097	47 51	136	20 11	165	21 44	244	50 51	290
73	57 36	008	47 31	064	47 58	098	48 31	137	20 26	165	20 51	244	49 56	290
74	57 43	007	48 23	063	48 56	098	49 11	138	20 40	166	19 59	245	49 01	290
	CAPELLA		◆POLLUX		PROCYON		SIRIUS		◆CANOPUS		Acamar		◆Hamal	
75	57 49	005	49 15	063	49 54	099	49 49	139	20 54	167	28 45	206	48 06	290
76	57 54	004	50 07	063	50 51	099	50 27	140	21 07	167	28 19	207	47 11	290
77	57 58	003	50 59	062	51 49	099	51 04	141	21 20	168	27 52	208	46 16	289
78	58 00	001	51 50	062	52 46	100	51 40	142	21 32	168	27 25	208	45 22	289
79	58 00	000	52 42	062	53 43	100	52 15	144	21 43	169	26 57	209	44 27	289
80	58 00	359	53 33	061	54 41	101	52 49	145	21 54	170	26 29	210	43 32	289
81	57 58	357	54 24	061	55 38	101	53 22	146	22 04	170	25 59	210	42 37	289
82	57 55	356	55 15	061	56 35	102	53 54	148	22 14	171	25 30	211	41 42	289
83	57 50	355	56 05	060	57 32	103	54 24	149	22 23	172	25 00	212	40 47	289
84	57 44	354	56 56	060	58 28	103	54 53	150	22 31	172	24 29	212	39 52	289
85	57 37	352	57 46	059	59 25	104	55 22	152	22 38	173	23 58	213	38 57	289
86	57 29	351	58 36	059	60 22	104	55 48	153	22 45	173	23 26	213	38 01	289
87	57 19	350	59 25	058	61 18	105	56 14	155	22 52	174	22 54	214	37 06	289
88	57 08	349	60 14	057	62 14	106	56 38	157	22 57	175	22 21	215	36 11	289
89	56 56	347	61 03	057	63 10	107	57 00	158	23 02	175	21 48	215	35 16	289

LHA ♈	Hc	Zn	Hc	Zn	Hc	Zn	Hc	Zn	Hc	Zn	Hc	Zn	Hc	Zn
	Dubhe		◆REGULUS		Suhail		SIRIUS		◆RIGEL		ALDEBARAN		◆CAPELLA	
90	18 59	029	29 42	084	18 21	146	57 21	160	65 04	208	69 30	280	56 43	346
91	19 27	029	30 40	084	18 54	147	57 40	161	64 36	210	68 33	280	56 28	345
92	19 55	029	31 38	084	19 25	147	57 58	163	64 06	212	67 36	279	56 12	344
93	20 23	029	32 36	085	19 57	148	58 14	165	63 35	214	66 38	279	55 56	343
94	20 51	029	33 34	085	20 28	148	58 28	167	63 02	215	65 41	279	55 38	342
95	21 19	029	34 32	085	20 58	149	58 41	168	62 27	217	64 43	279	55 19	340
96	21 47	029	35 30	085	21 28	149	58 52	170	61 51	219	63 46	279	54 59	339
97	22 15	028	36 28	085	21 58	150	59 00	172	61 14	221	62 48	279	54 38	338
98	22 43	028	37 26	085	22 27	150	59 08	174	60 36	222	61 51	279	54 16	337
99	23 10	028	38 24	086	22 55	151	59 13	176	59 56	224	60 53	279	53 53	336
100	23 38	028	39 22	086	23 23	152	59 16	178	59 15	225	59 56	279	53 29	335
101	24 05	028	40 20	086	23 51	152	59 17	180	58 34	226	58 58	279	53 04	334
102	24 32	028	41 18	086	24 18	153	59 17	181	57 51	228	58 01	279	52 39	334
103	25 00	028	42 16	086	24 44	153	59 14	183	57 08	229	57 03	279	52 12	333
104	25 27	028	43 14	087	25 10	154	59 10	185	56 23	230	56 06	279	51 45	332
	Dubhe		◆REGULUS		Suhail		◆SIRIUS		RIGEL		ALDEBARAN		◆CAPELLA	
105	25 53	027	44 12	087	25 35	155	59 04	187	55 38	231	55 08	279	51 17	331
106	26 20	027	45 11	087	25 59	155	58 56	189	54 52	232	54 11	279	50 49	330
107	26 46	027	46 09	087	26 23	156	58 46	191	54 06	233	53 13	279	50 19	329
108	27 13	027	47 07	087	26 47	157	58 34	193	53 19	234	52 16	279	49 49	329
109	27 39	027	48 05	088	27 09	157	58 21	194	52 31	235	51 18	279	49 19	328
110	28 05	026	49 03	088	27 31	158	58 06	196	51 43	236	50 21	279	48 47	327
111	28 31	026	50 01	088	27 52	159	57 49	198	50 54	237	49 23	279	48 16	327
112	28 56	026	50 59	088	28 13	160	57 30	199	50 05	238	48 26	279	47 43	326
113	29 21	026	51 58	088	28 33	160	57 10	201	49 15	239	47 28	279	47 10	325
114	29 46	025	52 56	089	28 52	161	56 48	203	48 25	240	46 31	279	46 37	325
115	30 11	025	53 54	089	29 11	162	56 25	204	47 35	241	45 33	279	46 03	324
116	30 36	025	54 52	089	29 28	163	56 00	206	46 44	241	44 36	280	45 28	323
117	31 00	025	55 50	089	29 45	163	55 34	207	45 53	242	43 38	280	44 54	323
118	31 24	024	56 49	089	30 01	164	55 07	209	45 01	243	42 41	280	44 18	322
119	31 48	024	57 47	090	30 17	165	54 38	210	44 09	243	41 44	280	43 43	322
	Dubhe		◆Denebola		Suhail		◆SIRIUS		RIGEL		ALDEBARAN		◆CAPELLA	
120	32 11	024	34 43	082	30 32	166	54 08	212	43 17	244	40 46	280	43 06	321
121	32 35	023	35 41	082	30 46	167	53 37	213	42 24	245	39 49	280	42 30	321
122	32 57	023	36 39	082	30 59	167	53 04	214	41 31	245	38 52	280	41 53	321
123	33 20	023	37 36	082	31 11	168	52 31	216	40 38	246	37 54	280	41 16	320
124	33 42	022	38 34	082	31 23	169	51 56	217	39 45	247	36 57	280	40 39	320
125	34 04	022	39 32	082	31 33	170	51 21	218	38 52	247	36 00	280	40 01	319
126	34 26	022	40 29	083	31 43	171	50 44	219	37 58	248	35 02	281	39 23	319
127	34 47	021	41 27	083	31 52	171	50 07	220	37 04	248	34 05	281	38 44	319
128	35 08	021	42 25	083	32 00	172	49 29	222	36 10	249	33 08	281	38 06	318
129	35 28	020	43 23	083	32 08	173	48 50	223	35 16	249	32 11	281	37 27	318
130	35 48	020	44 21	083	32 14	174	48 10	224	34 21	250	31 14	281	36 48	318
131	36 08	020	45 18	083	32 20	175	47 29	225	33 26	250	30 17	281	36 08	317
132	36 27	019	46 16	083	32 25	176	46 48	226	32 32	251	29 19	281	35 29	317
133	36 46	019	47 14	084	32 28	177	46 06	227	31 37	251	28 22	281	34 49	317
134	37 05	018	48 12	084	32 31	177	45 23	228	30 42	251	27 25	282	34 09	317
	Dubhe		◆ARCTURUS		SPICA		◆Suhail		SIRIUS		BETELGEUSE		◆CAPELLA	
135	37 23	018	14 52	073	19 42	107	32 34	178	44 40	228	44 08	267	33 29	316
136	37 40	017	15 48	074	20 37	108	32 35	179	43 56	229	43 10	267	32 49	316
137	37 57	017	16 44	074	21 32	108	32 35	180	43 12	230	42 11	267	32 09	316
138	38 14	016	17 40	074	22 28	109	32 35	181	42 27	231	41 13	268	31 28	316
139	38 30	016	18 36	074	23 23	109	32 33	182	41 41	232	40 15	268	30 47	316
140	38 46	015	19 32	074	24 18	109	32 31	183	40 55	233	39 17	268	30 07	315
141	39 01	015	20 28	074	25 13	110	32 28	183	40 09	233	38 19	268	29 26	315
142	39 15	014	21 24	075	26 07	110	32 24	184	39 22	234	37 20	269	28 45	315
143	39 30	014	22 20	075	27 02	111	32 19	185	38 35	235	36 22	269	28 03	315
144	39 43	013	23 16	075	27 56	111	32 14	186	37 47	235	35 24	269	27 22	315
145	39 56	013	24 13	075	28 51	111	32 07	187	36 59	236	34 26	269	26 41	315
146	40 09	012	25 09	075	29 45	112	32 00	188	36 10	237	33 28	270	26 00	315
147	40 21	012	26 05	075	30 39	112	31 51	189	35 22	237	32 29	270	25 18	315
148	40 32	011	27 01	075	31 33	113	31 42	189	34 32	238	31 31	270	24 37	315
149	40 43	010	27 58	075	32 26	113	31 32	190	33 43	238	30 33	270	23 55	314
	◆Dubhe		Alkaid		ARCTURUS		◆SPICA		Suhail		◆SIRIUS		POLLUX	
150	40 53	010	31 58	040	28 54	076	33 20	114	31 22	191	32 53	239	55 40	300
151	41 03	009	32 35	040	29 50	076	34 13	114	31 10	192	32 03	240	54 50	299
152	41 12	009	33 12	040	30 47	076	35 06	115	30 58	193	31 13	240	53 59	299
153	41 20	008	33 49	039	31 43	076	35 59	115	30 44	194	30 22	241	53 08	298
154	41 28	007	34 26	039	32 40	076	36 51	116	30 30	194	29 31	241	52 16	298
155	41 35	007	35 03	039	33 36	076	37 43	116	30 16	195	28 40	242	51 25	298
156	41 42	006	35 39	038	34 33	076	38 35	117	30 00	196	27 49	242	50 33	297
157	41 48	006	36 15	038	35 29	076	39 27	118	29 44	197	26 57	243	49 41	297
158	41 53	005	36 51	038	36 26	076	40 19	118	29 27	197	26 05	243	48 49	297
159	41 58	004	37 26	037	37 22	076	41 10	119	29 09	198	25 13	244	47 57	297
160	42 02	004	38 02	037	38 19	076	42 01	120	28 50	199	24 21	244	47 05	296
161	42 06	003	38 37	037	39 16	076	42 51	120	28 31	200	23 29	244	46 13	296
162	42 09	002	39 11	036	40 12	077	43 41	121	28 11	200	22 36	245	45 21	296
163	42 11	002	39 46	036	41 09	077	44 31	122	27 50	201	21 43	245	44 28	296
164	42 12	001	40 20	036	42 05	077	45 20	122	27 29	202	20 50	246	43 36	296
	◆Dubhe		ARCTURUS		◆SPICA		ACRUX		Suhail		◆PROCYON		POLLUX	
165	42 13	001	43 02	077	46 09	123	11 07	170	27 07	203	39 48	265	42 43	295
166	42 13	000	43 59	077	46 58	124	11 17	171	26 44	203	38 50	265	41 51	295
167	42 13	359	44 55	077	47 46	125	11 26	171	26 21	204	37 52	266	40 58	295
168	42 12	359	45 52	077	48 33	126	11 35	172	25 57	205	36 54	266	40 05	295
169	42 10	358	46 49	077	49 20	127	11 44	172	25 33	205	35 56	266	39 12	295
170	42 08	357	47 45	077	50 07	127	11 51	172	25 07	206	34 58	267	38 20	295
171	42 05	357	48 42	077	50 53	128	11 59	173	24 42	207	33 59	267	37 27	295
172	42 01	356	49 39	077	51 38	129	12 06	173	24 15	207	33 01	267	36 34	295
173	41 57	355	50 35	077	52 23	131	12 12	174	23 48	208	32 03	267	35 41	295
174	41 52	355	51 32	077	53 07	132	12 19	174	23 21	208	31 05	268	34 48	295
175	41 46	354	52 29	077	53 50	133	12 24	175	22 53	209	30 07	268	33 55	295
176	41 40	354	53 25	077	54 32	134	12 29	175	22 24	210	29 09	268	33 02	295
177	41 33	353	54 22	076	55 14	135	12 34	176	21 55	210	28 10	268	32 09	295
178	41 26	352	55 18	076	55 54	136	12 38	176	21 26	211	27 12	269	31 16	295
179	41 18	352	56 15	076	56 34	138	12 42	176	20 56	211	26 14	269	30 23	295

LHA	Hc	Zn	Hc	Zn	Hc	Zn	Hc	Zn	Hc	Zn	Hc	Zn	Hc	Zn
♈	Alkaid		◆Alphecca		ANTARES		◆ACRUX		REGULUS		◆POLLUX		Dubhe	
180	48 23	026	38 31	067	13 11	122	12 46	177	62 41	269	29 30	295	41 09	351
181	48 49	026	39 25	067	14 00	122	12 48	177	61 43	269	28 37	295	41 00	351
182	49 13	025	40 18	067	14 49	123	12 51	178	60 45	270	27 45	295	40 50	350
183	49 38	024	41 12	067	15 38	123	12 53	178	59 46	270	26 52	295	40 40	349
184	50 01	023	42 05	066	16 27	123	12 54	179	58 48	270	25 59	295	40 29	349
185	50 23	022	42 59	066	17 15	124	12 55	179	57 50	270	25 06	295	40 17	348
186	50 45	021	43 52	066	18 03	124	12 56	180	56 52	271	24 13	295	40 05	348
187	51 06	021	44 45	066	18 51	125	12 56	180	55 54	271	23 20	295	39 52	347
188	51 26	020	45 38	066	19 39	125	12 55	181	54 55	271	22 27	295	39 39	347
189	51 45	019	46 31	066	20 27	126	12 54	181	53 57	271	21 34	295	39 26	346
190	52 04	018	47 24	066	21 14	126	12 53	182	52 59	271	20 42	295	39 11	346
191	52 21	017	48 17	065	22 01	127	12 51	182	52 01	272	19 49	295	38 56	345
192	52 37	016	49 10	065	22 47	127	12 49	183	51 03	272	18 56	295	38 41	345
193	52 53	015	50 03	065	23 34	128	12 46	183	50 04	272	18 04	295	38 25	344
194	53 08	014	50 56	065	24 20	128	12 43	183	49 06	272	17 11	295	38 09	344
	◆VEGA		Rasalhague		ANTARES		◆RIGIL KENT.		Gienah		◆REGULUS		Dubhe	
195	13 11	053	23 24	082	25 05	129	12 36	168	56 38	200	48 08	272	37 52	343
196	13 57	053	24 21	082	25 51	129	12 48	168	56 18	201	47 10	273	37 35	343
197	14 44	053	25 19	083	26 36	130	13 00	169	55 56	203	46 12	273	37 17	342
198	15 30	053	26 17	083	27 20	130	13 11	169	55 33	204	45 14	273	36 59	342
199	16 17	053	27 14	083	28 05	131	13 22	170	55 08	206	44 15	273	36 41	341
200	17 04	053	28 12	083	28 49	131	13 32	170	54 42	207	43 17	273	36 22	341
201	17 50	053	29 10	083	29 32	132	13 41	171	54 15	209	42 19	274	36 02	340
202	18 37	053	30 08	083	30 15	133	13 51	171	53 46	210	41 21	274	35 43	340
203	19 24	053	31 06	084	30 58	133	13 59	172	53 16	211	40 23	274	35 22	340
204	20 10	053	32 04	084	31 40	134	14 07	172	52 45	213	39 25	274	35 02	339
205	20 57	053	33 01	084	32 22	135	14 15	173	52 13	214	38 27	274	34 41	339
206	21 44	053	33 59	084	33 03	135	14 22	173	51 40	215	37 29	275	34 20	338
207	22 31	053	34 57	084	33 44	136	14 29	174	51 06	217	36 31	275	33 58	338
208	23 17	053	35 55	084	34 24	137	14 35	174	50 31	218	35 33	275	33 36	338
209	24 04	053	36 53	085	35 04	137	14 41	175	49 55	219	34 35	275	33 14	337
	◆VEGA		Rasalhague		ANTARES		◆RIGIL KENT.		SPICA		◆REGULUS		Dubhe	
210	24 51	053	37 51	085	35 43	138	14 46	175	63 24	200	33 37	275	32 51	337
211	25 38	053	38 49	085	36 22	139	14 51	176	63 04	202	32 39	275	32 28	337
212	26 24	053	39 47	085	37 00	140	14 55	176	62 41	204	31 41	276	32 05	336
213	27 11	053	40 45	085	37 37	140	14 59	177	62 17	205	30 43	276	31 41	336
214	27 58	053	41 43	086	38 14	141	15 02	177	61 51	207	29 45	276	31 17	336
215	28 44	053	42 41	086	38 50	142	15 05	178	61 24	209	28 47	276	30 53	335
216	29 31	053	43 39	086	39 25	143	15 07	178	60 55	211	27 49	276	30 29	335
217	30 17	053	44 37	086	40 00	144	15 09	179	60 24	213	26 51	277	30 04	335
218	31 04	053	45 35	086	40 34	145	15 10	179	59 52	214	25 54	277	29 39	335
219	31 50	053	46 34	086	41 07	146	15 11	180	59 18	216	24 56	277	29 14	334
220	32 37	053	47 32	087	41 39	147	15 11	180	58 44	217	23 58	277	28 49	334
221	33 23	053	48 30	087	42 11	148	15 11	181	58 08	219	23 00	277	28 23	334
222	34 09	052	49 28	087	42 42	149	15 10	181	57 30	220	22 03	277	27 57	334
223	34 55	052	50 26	087	43 12	150	15 08	182	56 52	222	21 05	278	27 31	333
224	35 41	052	51 24	087	43 41	151	15 07	182	56 13	223	20 07	278	27 05	333
	◆VEGA		ALTAIR		ANTARES		◆RIGIL KENT.		SPICA		◆Denebola		Alkaid	
225	36 27	052	18 52	085	44 09	152	15 04	183	55 33	224	43 46	277	51 39	341
226	37 12	052	19 50	085	44 36	153	15 01	183	54 52	226	42 48	277	51 20	340
227	37 58	051	20 48	086	45 02	154	14 58	184	54 10	227	41 50	277	50 59	339
228	38 43	051	21 46	086	45 27	155	14 54	184	53 27	228	40 53	277	50 38	338
229	39 29	051	22 44	086	45 51	156	14 50	185	52 43	229	39 55	277	50 16	337
230	40 14	051	23 42	086	46 14	157	14 45	185	51 59	230	38 57	278	49 53	337
231	40 59	050	24 40	087	46 36	159	14 39	186	51 14	231	38 00	278	49 30	336
232	41 43	050	25 39	087	46 56	160	14 33	186	50 28	232	37 02	278	49 05	335
233	42 28	050	26 37	087	47 16	161	14 27	187	49 42	233	36 04	278	48 40	334
234	43 12	049	27 35	087	47 34	162	14 20	187	48 55	234	35 07	278	48 15	333
235	43 56	049	28 33	087	47 51	164	14 12	188	48 08	235	34 09	278	47 48	333
236	44 40	049	29 31	088	48 07	165	14 04	188	47 20	236	33 11	278	47 21	332
237	45 24	048	30 29	088	48 22	166	13 56	189	46 32	236	32 14	279	46 54	331
238	46 07	048	31 28	088	48 35	167	13 47	189	45 43	237	31 16	279	46 26	331
239	46 50	047	32 26	088	48 47	169	13 38	189	44 54	238	30 19	279	45 57	330
	◆VEGA		ALTAIR		Nunki		◆ANTARES		SPICA		◆ARCTURUS		Alkaid	
240	47 32	047	33 24	089	31 25	133	48 58	170	44 04	239	64 26	285	45 28	329
241	48 15	046	34 22	089	32 07	134	49 07	171	43 14	239	63 30	285	44 58	329
242	48 57	046	35 20	089	32 49	135	49 15	173	42 24	240	62 34	285	44 27	328
243	49 38	045	36 19	089	33 30	135	49 22	174	41 33	241	61 38	285	43 57	328
244	50 19	045	37 17	089	34 10	136	49 27	175	40 42	242	60 41	284	43 25	327
245	51 00	044	38 15	090	34 50	137	49 31	177	39 51	242	59 45	284	42 53	327
246	51 41	043	39 13	090	35 30	138	49 34	178	38 59	243	58 48	284	42 21	326
247	52 20	043	40 11	090	36 09	138	49 35	180	38 08	243	57 52	284	41 49	326
248	53 00	042	41 10	090	36 47	139	49 34	181	37 15	244	56 55	284	41 16	325
249	53 38	041	42 08	091	37 25	140	49 33	182	36 23	244	55 59	284	40 42	325
250	54 17	041	43 06	091	38 02	141	49 30	184	35 30	245	55 02	284	40 08	324
251	54 54	040	44 04	091	38 39	142	49 25	185	34 37	246	54 06	284	39 34	324
252	55 31	039	45 02	091	39 14	142	49 19	186	33 44	246	53 09	283	39 00	324
253	56 08	038	46 01	092	39 49	143	49 12	188	32 51	247	52 13	283	38 25	323
254	56 43	037	46 59	092	40 24	144	49 03	189	31 57	247	51 16	283	37 50	323
	◆VEGA		ALTAIR		◆Nunki		ANTARES		SPICA		◆ARCTURUS		Alkaid	
255	57 18	036	47 57	092	40 57	145	48 53	191	31 04	248	50 19	283	37 15	322
256	57 52	035	48 55	093	41 30	146	48 42	192	30 10	248	49 23	283	36 39	322
257	58 25	034	49 53	093	42 02	147	48 29	193	29 16	248	48 26	283	36 03	322
258	58 57	033	50 51	093	42 33	148	48 15	194	28 22	249	47 29	283	35 27	321
259	59 29	032	51 50	093	43 04	149	48 00	196	27 27	249	46 33	283	34 51	321
260	59 59	031	52 48	094	43 33	150	47 44	197	26 33	250	45 36	283	34 14	321
261	60 28	030	53 46	094	44 02	151	47 26	198	25 38	250	44 39	283	33 37	321
262	60 57	028	54 44	094	44 29	152	47 07	200	24 43	250	43 43	283	33 00	320
263	61 24	027	55 42	095	44 56	153	46 47	201	23 48	251	42 46	283	32 23	320
264	61 49	026	56 40	095	45 22	154	46 26	202	22 53	251	41 49	283	31 45	320
265	62 14	024	57 38	096	45 46	156	46 04	203	21 58	252	40 53	283	31 08	320
266	62 37	023	58 36	096	46 10	157	45 40	204	21 03	252	39 56	284	30 30	320
267	62 59	021	59 34	096	46 32	158	45 16	205	20 07	252	39 00	284	29 52	319
268	63 19	020	60 31	097	46 53	159	44 50	207	19 12	253	38 03	284	29 14	319
269	63 38	018	61 29	097	47 13	160	44 24	208	18 16	253	37 06	284	28 36	319

LAT 14°N

LHA	Hc	Zn	Hc	Zn	Hc	Zn	Hc	Zn	Hc	Zn	Hc	Zn	Hc	Zn
♈	VEGA		◆DENEB		Enif		◆Nunki		ANTARES		◆ARCTURUS		Alkaid	
270	63 55	016	43 50	039	35 11	088	47 32	162	43 56	209	36 10	284	27 58	319
271	64 11	015	44 27	039	36 09	088	47 50	163	43 28	210	35 13	284	27 20	319
272	64 25	013	45 03	038	37 07	088	48 07	164	42 58	211	34 17	284	26 41	319
273	64 37	011	45 39	038	38 05	088	48 22	165	42 28	212	33 20	284	26 02	318
274	64 48	010	46 14	037	39 04	089	48 36	167	41 57	213	32 24	284	25 24	318
275	64 57	008	46 49	036	40 02	089	48 49	168	41 25	214	31 27	284	24 45	318
276	65 04	006	47 23	036	41 00	089	49 00	169	40 52	215	30 31	284	24 06	318
277	65 09	004	47 57	035	41 58	089	49 10	171	40 19	216	29 34	284	23 27	318
278	65 12	002	48 30	035	42 57	089	49 19	172	39 44	217	28 38	284	22 48	318
279	65 13	000	49 03	034	43 55	090	49 26	173	39 09	217	27 42	285	22 09	318
280	65 13	358	49 35	033	44 53	090	49 32	175	38 34	218	26 45	285	21 30	318
281	65 10	357	50 07	033	45 51	090	49 37	176	37 57	219	25 49	285	20 51	318
282	65 06	355	50 38	032	46 49	090	49 40	178	37 20	220	24 53	285	20 12	318
283	65 00	353	51 08	031	47 48	091	49 41	179	36 43	221	23 57	285	19 33	318
284	64 52	351	51 38	030	48 46	091	49 42	180	36 04	221	23 00	285	18 54	318
	◆DENEB		Alpheratz		◆FOMALHAUT		Nunki		ANTARES		◆Rasalhague		VEGA	
285	52 07	029	17 56	064	18 05	128	49 41	182	35 26	222	69 12	269	64 42	349
286	52 35	028	18 48	064	18 50	129	49 38	183	34 46	223	68 14	269	64 30	348
287	53 02	028	19 40	064	19 36	129	49 34	185	34 06	224	67 16	269	64 17	346
288	53 29	027	20 32	064	20 21	130	49 29	186	33 26	224	66 18	269	64 02	344
289	53 55	026	21 25	064	21 05	130	49 22	187	32 45	225	65 19	270	63 45	343
290	54 19	025	22 17	064	21 50	130	49 14	189	32 03	226	64 21	270	63 27	341
291	54 43	024	23 09	064	22 34	131	49 05	190	31 21	226	63 23	270	63 07	339
292	55 06	023	24 02	064	23 18	131	48 54	191	30 39	227	62 25	270	62 46	338
293	55 28	022	24 54	064	24 01	132	48 42	193	29 56	228	61 26	271	62 23	336
294	55 49	021	25 46	064	24 44	133	48 29	194	29 13	228	60 28	271	61 59	335
295	56 09	019	26 39	064	25 27	133	48 14	195	28 29	229	59 30	271	61 34	334
296	56 28	018	27 31	064	26 09	134	47 58	197	27 45	230	58 32	271	61 07	332
297	56 46	017	28 24	064	26 51	134	47 41	198	27 00	230	57 34	271	60 39	331
298	57 03	016	29 16	064	27 33	135	47 23	199	26 16	231	56 35	272	60 10	330
299	57 18	015	30 08	064	28 14	135	47 03	200	25 30	231	55 37	272	59 41	328
	◆DENEB		Alpheratz		◆FOMALHAUT		Peacock		◆ANTARES		Rasalhague		VEGA	
300	57 32	014	31 01	064	28 54	136	19 03	176	24 45	232	54 39	272	59 10	327
301	57 45	012	31 53	064	29 35	137	19 07	177	23 59	232	53 41	272	58 38	326
302	57 57	011	32 46	064	30 14	137	19 09	177	23 13	233	52 43	272	58 05	325
303	58 08	010	33 38	064	30 53	138	19 12	178	22 26	233	51 45	273	57 31	324
304	58 17	008	34 31	064	31 32	139	19 13	179	21 40	234	50 46	273	56 57	323
305	58 25	007	35 23	064	32 10	139	19 14	179	20 53	234	49 48	273	56 21	322
306	58 31	006	36 15	064	32 48	140	19 15	180	20 05	235	48 50	273	55 45	321
307	58 37	004	37 08	064	33 25	141	19 15	180	19 18	235	47 52	273	55 09	320
308	58 41	003	38 00	064	34 01	142	19 14	181	18 30	235	46 54	274	54 31	320
309	58 43	002	38 52	064	34 37	142	19 13	182	17 42	236	45 56	274	53 53	319
310	58 44	000	39 44	064	35 13	143	19 11	182	16 53	236	44 58	274	53 15	318
311	58 44	359	40 37	064	35 47	144	19 08	183	16 05	237	44 00	274	52 36	317
312	58 42	358	41 29	063	36 21	145	19 05	183	15 16	237	43 02	274	51 56	317
313	58 39	356	42 21	063	36 54	146	19 02	184	14 27	237	42 04	274	51 16	316
314	58 35	355	43 13	063	37 27	146	18 58	184	13 38	238	41 05	275	50 35	316
	Alpheratz		◆Diphda		FOMALHAUT		◆Nunki		Rasalhague		VEGA		◆DENEB	
315	44 04	063	26 19	119	37 59	147	39 31	217	40 07	275	49 54	315	58 29	354
316	44 56	063	27 10	119	38 30	148	38 56	218	39 09	275	49 13	314	58 22	352
317	45 48	063	28 01	120	39 00	149	38 20	219	38 11	275	48 31	314	58 14	351
318	46 40	062	28 51	120	39 30	150	37 43	220	37 13	275	47 49	313	58 04	350
319	47 31	062	29 42	121	39 58	151	37 05	220	36 16	275	47 06	313	57 53	348
320	48 22	062	30 32	121	40 26	152	36 27	221	35 18	276	46 23	312	57 41	347
321	49 14	061	31 21	122	40 53	153	35 49	222	34 20	276	45 40	312	57 27	346
322	50 05	061	32 11	122	41 19	154	35 09	223	33 22	276	44 57	312	57 13	345
323	50 56	061	33 00	123	41 44	155	34 29	223	32 24	276	44 13	311	56 57	344
324	51 46	060	33 49	123	42 09	156	33 49	224	31 26	276	43 29	311	56 40	342
325	52 37	060	34 37	124	42 32	157	33 08	225	30 28	277	42 45	310	56 21	341
326	53 27	060	35 25	125	42 54	158	32 27	226	29 30	277	42 01	310	56 02	340
327	54 17	059	36 13	125	43 15	159	31 45	226	28 32	277	41 16	310	55 42	339
328	55 07	059	37 00	126	43 35	160	31 03	227	27 35	277	40 31	309	55 20	338
329	55 57	058	37 47	127	43 55	161	30 20	228	26 37	277	39 46	309	54 58	337
	◆Alpheratz		Diphda		◆FOMALHAUT		Peacock		ALTAIR		◆VEGA		DENEB	
330	56 46	058	38 34	127	44 13	163	16 33	193	57 53	264	39 01	309	54 35	336
331	57 36	057	39 20	128	44 30	164	16 19	194	56 55	265	38 16	309	54 10	335
332	58 24	057	40 05	129	44 45	165	16 05	194	55 58	265	37 30	308	53 45	334
333	59 13	056	40 50	130	45 00	166	15 51	195	54 59	265	36 44	308	53 19	333
334	60 01	055	41 35	130	45 14	167	15 35	195	54 01	266	35 59	308	52 52	332
335	60 48	055	42 19	131	45 26	168	15 20	196	53 03	266	35 13	308	52 25	331
336	61 36	054	43 03	132	45 37	170	15 04	196	52 05	266	34 27	308	51 56	330
337	62 22	053	43 46	133	45 47	171	14 47	197	51 07	267	33 41	308	51 27	330
338	63 08	052	44 28	134	45 55	172	14 30	197	50 09	267	32 54	307	50 57	329
339	63 54	051	45 10	135	46 03	173	14 12	198	49 11	267	32 08	307	50 27	328
340	64 39	050	45 51	136	46 09	175	13 54	198	48 13	268	31 22	307	49 55	327
341	65 23	049	46 31	137	46 14	176	13 36	199	47 15	268	30 35	307	49 24	327
342	66 07	048	47 11	138	46 18	177	13 17	199	46 16	268	29 49	307	48 51	326
343	66 49	046	47 50	139	46 20	178	12 58	200	45 18	268	29 02	307	48 18	325
344	67 31	045	48 28	140	46 21	180	12 38	200	44 20	269	28 16	307	47 45	325
	Schedar		◆Mirfak		Hamal		◆ACHERNAR		FOMALHAUT		◆ALTAIR		DENEB	
345	43 23	019	26 04	041	44 57	071	11 40	159	46 21	181	43 22	269	47 11	324
346	43 41	018	26 42	041	45 52	071	12 00	160	46 19	182	42 24	269	46 36	323
347	43 59	017	27 20	041	46 47	071	12 20	160	46 17	183	41 25	270	46 01	323
348	44 16	017	27 58	041	47 42	070	12 39	161	46 13	185	40 27	270	45 25	322
349	44 32	016	28 36	040	48 37	070	12 58	161	46 07	186	39 29	270	44 50	322
350	44 48	015	29 14	040	49 31	070	13 16	162	46 01	187	38 31	270	44 13	321
351	45 03	015	29 51	040	50 26	070	13 34	162	45 53	188	37 32	270	43 37	321
352	45 18	014	30 29	040	51 21	070	13 52	163	45 44	190	36 34	271	42 59	320
353	45 32	013	31 06	040	52 15	070	14 09	163	45 34	191	35 36	271	42 22	320
354	45 45	013	31 43	039	53 10	069	14 26	164	45 22	192	34 38	271	41 44	319
355	45 57	012	32 20	039	54 04	069	14 42	164	45 09	193	33 40	271	41 06	319
356	46 09	011	32 57	039	54 58	069	14 58	165	44 56	194	32 41	272	40 28	319
357	46 20	010	33 33	039	55 53	069	15 13	165	44 41	196	31 43	272	39 49	318
358	46 30	010	34 10	038	56 47	068	15 28	166	44 24	197	30 45	272	39 10	318
359	46 39	009	34 46	038	57 41	068	15 42	166	44 07	198	29 47	272	38 31	317

LHA	Hc	Zn	Hc	Zn	Hc	Zn	Hc	Zn	Hc	Zn	Hc	Zn	Hc	Zn
♈	CAPELLA		◆ALDEBARAN		Acamar		ACHERNAR		◆FOMALHAUT		ALTAIR		◆DENEB	
0	16 52	045	23 35	077	22 35	145	16 54	167	44 45	199	28 46	273	37 07	318
1	17 33	045	24 32	078	23 09	145	17 07	167	44 26	200	27 47	273	36 28	317
2	18 15	045	25 29	078	23 42	146	17 20	168	44 05	202	26 49	273	35 48	317
3	18 57	045	26 26	078	24 15	146	17 33	168	43 43	203	25 51	274	35 08	317
4	19 38	045	27 24	078	24 47	147	17 45	169	43 20	204	24 52	274	34 28	317
5	20 20	045	28 21	078	25 18	148	17 56	169	42 56	205	23 54	274	33 48	316
6	21 02	045	29 18	078	25 50	148	18 07	170	42 31	206	22 56	274	33 07	316
7	21 43	045	30 15	078	26 20	149	18 17	170	42 05	207	21 57	274	32 27	316
8	22 25	045	31 12	078	26 50	149	18 26	171	41 38	208	20 59	275	31 46	316
9	23 06	045	32 10	079	27 20	150	18 36	171	41 10	209	20 01	275	31 05	315
10	23 48	045	33 07	079	27 48	151	18 44	172	40 42	210	19 03	275	30 24	315
11	24 29	045	34 04	079	28 17	151	18 52	172	40 12	211	18 04	275	29 42	315
12	25 11	045	35 02	079	28 44	152	19 00	173	39 42	212	17 06	275	29 01	315
13	25 52	045	35 59	079	29 11	153	19 07	174	39 11	213	16 08	276	28 20	315
14	26 33	045	36 56	079	29 38	154	19 13	174	38 39	213	15 10	276	27 38	315
	CAPELLA		◆ALDEBARAN		RIGEL		ACHERNAR		◆FOMALHAUT		Enif		◆DENEB	
15	27 14	045	37 54	079	23 24	105	19 19	175	38 07	214	41 55	272	26 56	314
16	27 55	045	38 51	079	24 20	105	19 24	175	37 33	215	40 57	272	26 15	314
17	28 36	044	39 49	079	25 16	106	19 28	176	36 59	216	39 58	272	25 33	314
18	29 17	044	40 46	079	26 13	106	19 32	176	36 25	217	39 00	272	24 51	314
19	29 58	044	41 43	079	27 09	106	19 36	177	35 49	218	38 02	272	24 09	314
20	30 38	044	42 41	079	28 05	107	19 39	177	35 13	218	37 03	273	23 27	314
21	31 19	044	43 38	079	29 01	107	19 41	178	34 37	219	36 05	273	22 45	314
22	31 59	043	44 36	080	29 57	108	19 43	179	34 00	220	35 06	273	22 03	314
23	32 39	043	45 33	080	30 52	108	19 44	179	33 22	221	34 08	273	21 20	314
24	33 19	043	46 31	080	31 48	108	19 44	180	32 44	221	33 10	273	20 38	314
25	33 59	043	47 28	080	32 43	109	19 44	180	32 05	222	32 11	274	19 56	314
26	34 38	042	48 26	080	33 38	109	19 44	181	31 26	223	31 13	274	19 14	314
27	35 18	042	49 23	080	34 34	110	19 42	182	30 46	223	30 15	274	18 32	314
28	35 57	042	50 21	080	35 29	110	19 40	182	30 05	224	29 16	274	17 49	314
29	36 36	042	51 19	080	36 23	111	19 38	183	29 25	225	28 18	274	17 07	314
	Mirfak		◆CAPELLA		ALDEBARAN		RIGEL		◆ACHERNAR		FOMALHAUT		◆Alpheratz	
30	49 20	021	37 15	041	52 16	080	37 18	111	19 35	183	28 43	225	59 29	306
31	49 41	020	37 53	041	53 14	080	38 12	112	19 31	184	28 02	226	58 42	305
32	50 00	019	38 31	041	54 11	080	39 07	112	19 27	184	27 20	226	57 54	305
33	50 19	018	39 09	040	55 09	080	40 01	113	19 23	185	26 37	227	57 06	304
34	50 37	017	39 47	040	56 06	080	40 55	113	19 17	185	25 54	227	56 17	304
35	50 54	016	40 24	039	57 04	080	41 48	114	19 11	186	25 11	228	55 28	303
36	51 10	015	41 01	039	58 01	080	42 42	114	19 05	187	24 28	228	54 39	302
37	51 25	014	41 37	039	58 59	080	43 35	115	18 58	187	23 44	229	53 49	302
38	51 39	014	42 14	038	59 56	079	44 28	116	18 50	188	22 59	230	53 00	301
39	51 52	013	42 49	038	60 54	079	45 20	116	18 42	188	22 15	230	52 10	301
40	52 04	012	43 25	037	61 51	079	46 12	117	18 33	189	21 30	230	51 20	301
41	52 15	011	44 00	037	62 48	079	47 04	118	18 24	189	20 45	231	50 29	300
42	52 25	010	44 35	036	63 46	079	47 56	118	18 14	190	19 59	231	49 38	300
43	52 35	008	45 09	035	64 43	079	48 47	119	18 04	190	19 13	232	48 48	300
44	52 43	007	45 42	035	65 41	079	49 38	120	17 53	191	18 27	232	47 57	299
	CAPELLA		◆POLLUX		PROCYON		SIRIUS		◆ACHERNAR		Diphda		◆Alpheratz	
45	46 16	034	22 28	065	20 52	089	27 00	117	17 42	192	44 12	228	47 06	299
46	46 48	034	23 21	065	21 51	090	27 52	117	17 30	192	43 28	229	46 14	299
47	47 20	033	24 14	065	22 49	090	28 44	118	17 17	193	42 44	230	45 23	298
48	47 52	032	25 07	065	23 47	090	29 36	118	17 04	193	41 59	231	44 32	298
49	48 23	032	26 00	065	24 46	090	30 27	119	16 51	194	41 13	231	43 40	298
50	48 53	031	26 53	065	25 44	090	31 19	119	16 37	194	40 27	232	42 48	298
51	49 23	030	27 46	065	26 43	091	32 10	120	16 22	195	39 41	233	41 57	298
52	49 52	029	28 39	065	27 41	091	33 00	120	16 07	195	38 54	234	41 05	297
53	50 20	029	29 31	065	28 40	091	33 51	121	15 52	196	38 07	234	40 13	297
54	50 48	028	30 24	065	29 38	091	34 41	121	15 36	196	37 19	235	39 21	297
55	51 15	027	31 17	065	30 37	092	35 31	122	15 19	197	36 31	236	38 29	297
56	51 41	026	32 10	065	31 35	092	36 20	122	15 02	197	35 43	236	37 37	297
57	52 06	025	33 03	065	32 34	092	37 09	123	14 45	198	34 54	237	36 44	297
58	52 30	024	33 56	065	33 32	092	37 58	124	14 27	198	34 05	237	35 52	297
59	52 54	023	34 49	065	34 30	093	38 47	124	14 09	198	33 16	238	35 00	297
	CAPELLA		◆POLLUX		PROCYON		SIRIUS		◆CANOPUS		Diphda		◆Hamal	
60	53 17	022	35 42	065	35 29	093	39 35	125	17 24	158	32 26	238	61 15	295
61	53 38	021	36 35	065	36 27	093	40 22	126	17 46	159	31 36	239	60 22	295
62	53 59	020	37 27	064	37 25	093	41 10	126	18 07	159	30 46	240	59 29	294
63	54 19	019	38 20	064	38 24	094	41 57	127	18 27	160	29 55	240	58 36	294
64	54 38	018	39 13	064	39 22	094	42 43	128	18 47	160	29 04	241	57 42	294
65	54 55	017	40 05	064	40 20	094	43 29	129	19 07	161	28 13	241	56 49	293
66	55 12	016	40 58	064	41 19	094	44 14	130	19 26	161	27 22	242	55 55	293
67	55 28	015	41 51	064	42 17	095	44 59	130	19 45	162	26 31	242	55 01	292
68	55 42	014	42 43	064	43 15	095	45 43	131	20 02	162	25 39	242	54 07	292
69	55 55	013	43 35	063	44 14	095	46 27	132	20 20	163	24 47	243	53 13	292
70	56 07	011	44 28	063	45 12	096	47 10	133	20 37	164	23 55	243	52 18	292
71	56 18	010	45 20	063	46 10	096	47 52	134	20 53	164	23 03	244	51 24	291
72	56 28	009	46 12	063	47 08	096	48 34	135	21 09	165	22 10	244	50 29	291
73	56 36	008	47 04	063	48 06	097	49 15	136	21 24	165	21 17	244	49 35	291
74	56 44	006	47 56	062	49 04	097	49 55	137	21 38	166	20 25	245	48 40	291
	CAPELLA		◆POLLUX		PROCYON		SIRIUS		◆CANOPUS		Acamar		◆Hamal	
75	56 50	005	48 47	062	50 02	097	50 34	138	21 52	166	29 39	206	47 46	291
76	56 54	004	49 39	062	51 00	098	51 13	139	22 06	167	29 12	207	46 51	291
77	56 58	003	50 30	061	51 58	098	51 51	140	22 18	167	28 45	208	45 56	290
78	57 00	001	51 22	061	52 56	099	52 27	142	22 31	168	28 18	209	45 01	290
79	57 00	000	52 13	061	53 54	099	53 03	143	22 42	169	27 49	209	44 06	290
80	57 00	359	53 04	060	54 51	100	53 38	144	22 53	170	27 21	210	43 11	290
81	56 58	358	53 54	060	55 49	100	54 11	146	23 03	170	26 51	211	42 17	290
82	56 55	356	54 45	059	56 46	100	54 44	147	23 13	171	26 21	211	41 22	290
83	56 50	355	55 35	059	57 44	101	55 15	148	23 22	171	25 51	212	40 27	290
84	56 45	354	56 25	058	58 41	101	55 45	150	23 30	172	25 20	212	39 32	290
85	56 38	353	57 14	058	59 38	102	56 14	151	23 38	173	24 48	213	38 37	290
86	56 29	351	58 04	057	60 36	103	56 42	153	23 45	173	24 16	214	37 42	290
87	56 20	350	58 53	057	61 32	103	57 08	154	23 51	174	23 43	214	36 47	290
88	56 09	349	59 41	056	62 29	104	57 32	156	23 57	175	23 10	215	35 52	290
89	55 57	348	60 30	055	63 26	105	57 56	158	24 02	175	22 37	215	34 56	290

LAT 13°N

LHA	Hc	Zn	Hc	Zn	Hc	Zn	Hc	Zn	Hc	Zn	Hc	Zn	Hc	Zn
♈	Dubhe		◆REGULUS		Suhail		SIRIUS		◆RIGEL		ALDEBARAN		◆CAPELLA	
90	18 06	029	29 35	083	19 11	146	58 17	159	65 57	209	69 19	282	55 44	346
91	18 35	029	30 34	084	19 44	146	58 37	161	65 28	211	68 22	282	55 30	345
92	19 03	029	31 32	084	20 16	147	58 55	163	64 57	213	67 25	282	55 15	344
93	19 31	029	32 30	084	20 47	147	59 12	164	64 24	215	66 27	282	54 58	343
94	19 59	029	33 28	084	21 19	148	59 27	166	63 50	217	65 30	281	54 41	342
95	20 27	028	34 26	084	21 50	148	59 40	168	63 15	218	64 33	281	54 22	341
96	20 55	028	35 24	084	22 20	149	59 51	170	62 38	220	63 35	281	54 02	340
97	21 22	028	36 22	085	22 50	150	60 00	172	61 59	222	62 38	281	53 42	339
98	21 50	028	37 21	085	23 19	150	60 07	174	61 20	223	61 41	281	53 20	338
99	22 17	028	38 19	085	23 48	151	60 13	176	60 39	225	60 43	281	52 58	337
100	22 45	028	39 17	085	24 16	151	60 16	178	59 57	226	59 46	281	52 34	336
101	23 12	028	40 15	085	24 44	152	60 17	180	59 15	228	58 48	280	52 10	335
102	23 39	028	41 14	085	25 11	153	60 17	181	58 31	229	57 51	280	51 45	334
103	24 06	027	42 12	085	25 38	153	60 14	183	57 47	230	56 53	280	51 19	333
104	24 33	027	43 10	086	26 04	154	60 10	185	57 01	231	55 56	280	50 52	332
	Dubhe		◆REGULUS		Suhail		◆SIRIUS		RIGEL		ALDEBARAN		◆CAPELLA	
105	25 00	027	44 08	086	26 29	155	60 04	187	56 15	232	54 58	280	50 25	332
106	25 27	027	45 07	086	26 54	155	59 55	189	55 29	234	54 01	280	49 56	331
107	25 53	027	46 05	086	27 18	156	59 45	191	54 41	235	53 03	280	49 28	330
108	26 19	027	47 03	086	27 42	157	59 33	193	53 53	236	52 05	280	48 58	329
109	26 45	026	48 02	086	28 05	157	59 19	195	53 05	237	51 08	280	48 28	328
110	27 11	026	49 00	087	28 27	158	59 03	196	52 16	237	50 10	280	47 57	328
111	27 37	026	49 59	087	28 48	159	58 46	198	51 26	238	49 13	280	47 25	327
112	28 02	026	50 57	087	29 09	159	58 27	200	50 36	239	48 15	280	46 53	326
113	28 27	025	51 55	087	29 29	160	58 06	202	49 46	240	47 18	280	46 21	326
114	28 52	025	52 54	087	29 49	161	57 43	203	48 55	241	46 20	280	45 48	325
115	29 17	025	53 52	087	30 08	162	57 19	205	48 04	242	45 23	280	45 14	325
116	29 41	025	54 50	088	30 25	162	56 54	207	47 12	242	44 25	280	44 40	324
117	30 05	024	55 49	088	30 43	163	56 27	208	46 20	243	43 28	281	44 06	324
118	30 29	024	56 47	088	30 59	164	55 59	210	45 28	244	42 30	281	43 31	323
119	30 53	024	57 46	088	31 15	165	55 29	211	44 35	244	41 33	281	42 55	322
	Dubhe		◆Denebola		Suhail		◆SIRIUS		RIGEL		ALDEBARAN		◆CAPELLA	
120	31 16	023	34 34	081	31 30	166	54 59	212	43 43	245	40 36	281	42 19	322
121	31 39	023	35 32	081	31 44	166	54 27	214	42 49	246	39 38	281	41 43	322
122	32 02	023	36 30	081	31 57	167	53 54	215	41 56	246	38 41	281	41 07	321
123	32 25	022	37 28	081	32 10	168	53 19	216	41 03	247	37 43	281	40 30	321
124	32 47	022	38 26	082	32 21	169	52 44	218	40 09	247	36 46	281	39 53	320
125	33 09	022	39 23	082	32 32	170	52 08	219	39 15	248	35 48	281	39 15	320
126	33 30	021	40 21	082	32 42	171	51 31	220	38 20	248	34 51	281	38 37	319
127	33 51	021	41 19	082	32 51	171	50 52	221	37 26	249	33 54	281	37 59	319
128	34 12	021	42 17	082	33 00	172	50 13	222	36 31	249	32 56	281	37 21	319
129	34 32	020	43 15	082	33 07	173	49 34	223	35 37	250	31 59	282	36 42	318
130	34 52	020	44 13	082	33 14	174	48 53	225	34 42	250	31 02	282	36 03	318
131	35 12	019	45 11	082	33 20	175	48 12	226	33 47	251	30 05	282	35 24	318
132	35 31	019	46 09	082	33 24	176	47 30	226	32 51	251	29 07	282	34 45	318
133	35 49	018	47 07	082	33 28	177	46 47	227	31 56	252	28 10	282	34 05	317
134	36 08	018	48 05	083	33 31	177	46 04	228	31 00	252	27 13	282	33 25	317
	Dubhe		◆ARCTURUS		SPICA		◆Suhail		SIRIUS		BETELGEUSE		◆CAPELLA	
135	36 25	018	14 35	073	19 59	107	33 34	178	45 20	229	44 11	268	32 46	317
136	36 43	017	15 31	073	20 55	107	33 35	179	44 35	230	43 12	268	32 05	317
137	37 00	017	16 27	074	21 51	108	33 35	180	43 50	231	42 14	268	31 25	316
138	37 16	016	17 23	074	22 47	108	33 35	181	43 04	232	41 15	268	30 45	316
139	37 32	016	18 19	074	23 42	109	33 33	182	42 18	232	40 17	269	30 04	316
140	37 48	015	19 15	074	24 37	109	33 31	183	41 32	233	39 18	269	29 24	316
141	38 03	015	20 12	074	25 33	109	33 28	184	40 44	234	38 20	269	28 43	316
142	38 17	014	21 08	074	26 28	110	33 24	184	39 57	235	37 21	269	28 02	316
143	38 31	014	22 04	074	27 23	110	33 19	185	39 09	235	36 23	270	27 21	315
144	38 45	013	23 00	074	28 18	110	33 13	186	38 21	236	35 25	270	26 40	315
145	38 58	013	23 57	075	29 12	111	33 07	187	37 32	237	34 26	270	25 59	315
146	39 10	012	24 53	075	30 07	111	32 59	188	36 43	237	33 28	270	25 17	315
147	39 22	011	25 50	075	31 01	112	32 51	189	35 54	238	32 29	271	24 36	315
148	39 33	011	26 46	075	31 55	112	32 41	190	35 04	238	31 31	271	23 54	315
149	39 44	010	27 42	075	32 50	113	32 31	190	34 14	239	30 32	271	23 13	315
	◆Dubhe		Alkaid		ARCTURUS		◆SPICA		Suhail		◆SIRIUS		POLLUX	
150	39 54	010	31 12	040	28 39	075	33 43	113	32 20	191	33 24	240	55 10	301
151	40 04	009	31 49	039	29 35	075	34 37	114	32 09	192	32 33	240	54 20	300
152	40 13	009	32 26	039	30 32	075	35 31	114	31 56	193	31 42	241	53 29	300
153	40 21	008	33 03	039	31 28	075	36 24	115	31 43	194	30 51	241	52 38	300
154	40 29	007	33 39	039	32 25	075	37 17	115	31 28	194	30 00	242	51 47	299
155	40 36	007	34 16	038	33 21	075	38 10	116	31 13	195	29 08	242	50 56	299
156	40 42	006	34 52	038	34 18	075	39 02	116	30 58	196	28 16	243	50 05	298
157	40 48	006	35 28	038	35 15	076	39 55	117	30 41	197	27 24	243	49 13	298
158	40 54	005	36 03	037	36 11	076	40 47	117	30 24	198	26 32	244	48 22	298
159	40 58	004	36 39	037	37 08	076	41 38	118	30 06	198	25 40	244	47 30	298
160	41 02	004	37 14	037	38 04	076	42 30	119	29 47	199	24 47	244	46 38	297
161	41 06	003	37 48	036	39 01	076	43 21	119	29 27	200	23 54	245	45 46	297
162	41 09	002	38 23	036	39 58	076	44 12	120	29 07	201	23 01	245	44 54	297
163	41 11	002	38 57	035	40 54	076	45 02	121	28 46	201	22 08	246	44 02	297
164	41 12	001	39 31	035	41 51	076	45 52	122	28 25	202	21 15	246	43 09	296
	◆Dubhe		ARCTURUS		◆SPICA		ACRUX		Suhail		◆PROCYON		POLLUX	
165	41 13	001	42 48	076	46 42	122	12 07	170	28 02	203	39 53	266	42 17	296
166	41 13	000	43 44	076	47 31	123	12 16	171	27 39	203	38 54	266	41 25	296
167	41 13	359	44 41	076	48 20	124	12 26	171	27 16	204	37 56	266	40 32	296
168	41 12	359	45 38	076	49 08	125	12 34	171	26 52	205	36 57	267	39 39	296
169	41 10	358	46 34	076	49 56	126	12 43	172	26 27	205	35 59	267	38 47	296
170	41 08	357	47 31	076	50 43	127	12 51	172	26 01	206	35 01	267	37 54	296
171	41 05	357	48 28	076	51 30	127	12 58	173	25 35	207	34 02	268	37 01	295
172	41 01	356	49 24	076	52 16	128	13 05	173	25 09	207	33 04	268	36 09	295
173	40 57	356	50 21	075	53 01	130	13 12	174	24 41	208	32 05	268	35 16	295
174	40 52	355	51 17	075	53 46	131	13 18	174	24 14	209	31 07	268	34 23	295
175	40 47	354	52 14	075	54 30	132	13 24	175	23 45	209	30 09	269	33 30	295
176	40 41	354	53 11	075	55 13	133	13 29	175	23 16	210	29 10	269	32 37	295
177	40 34	353	54 07	075	55 56	134	13 34	176	22 47	210	28 12	269	31 44	295
178	40 27	352	55 04	075	56 37	135	13 38	176	22 17	211	27 13	269	30 51	295
179	40 19	352	56 00	075	57 18	137	13 42	176	21 47	211	26 15	269	29 58	295

 LAT 13°N

LHA	Hc	Zn	Hc	Zn	Hc	Zn	Hc	Zn	Hc	Zn	Hc	Zn	Hc	Zn
♈	Alkaid		◆Alphecca		ANTARES		◆ACRUX		REGULUS		◆POLLUX		Dubhe	
180	47 29	026	38 07	066	13 43	122	13 45	177	62 41	271	29 05	295	40 10	351
181	47 54	025	39 01	066	14 32	122	13 48	177	61 43	271	28 12	295	40 01	351
182	48 19	024	39 54	066	15 22	122	13 51	178	60 44	271	27 19	295	39 51	350
183	48 43	024	40 48	066	16 11	123	13 53	178	59 46	272	26 26	295	39 41	350
184	49 06	023	41 41	066	17 00	123	13 54	179	58 47	272	25 33	295	39 30	349
185	49 28	022	42 34	065	17 49	124	13 55	179	57 49	272	24 41	295	39 19	348
186	49 49	021	43 27	065	18 37	124	13 56	180	56 50	272	23 48	295	39 07	348
187	50 10	020	44 20	065	19 25	124	13 56	180	55 52	272	22 55	295	38 54	347
188	50 30	019	45 13	065	20 14	125	13 55	181	54 54	272	22 02	295	38 41	347
189	50 48	018	46 06	065	21 01	125	13 54	181	53 55	273	21 09	295	38 27	346
190	51 06	017	46 59	065	21 49	126	13 53	182	52 57	273	20 16	295	38 13	346
191	51 24	017	47 52	064	22 36	126	13 51	182	51 58	273	19 23	295	37 58	345
192	51 40	016	48 45	064	23 23	127	13 49	183	51 00	273	18 31	296	37 43	345
193	51 55	015	49 37	064	24 10	127	13 46	183	50 02	273	17 38	296	37 28	344
194	52 09	014	50 29	064	24 56	128	13 43	183	49 03	273	16 45	296	37 12	344
	◆VEGA		Rasalhague		ANTARES		◆RIGIL KENT.		Gienah		◆REGULUS		Dubhe	
195	12 35	053	23 15	082	25 43	128	13 35	168	57 34	200	48 05	274	36 55	343
196	13 21	053	24 13	082	26 28	129	13 47	168	57 14	202	47 07	274	36 38	343
197	14 08	053	25 11	082	27 14	129	13 59	169	56 51	203	46 08	274	36 20	342
198	14 54	053	26 09	082	27 59	130	14 10	169	56 27	205	45 10	274	36 02	342
199	15 41	053	27 07	082	28 44	130	14 21	170	56 02	206	44 12	274	35 44	341
200	16 28	053	28 05	083	29 28	131	14 31	170	55 35	208	43 13	274	35 25	341
201	17 14	053	29 03	083	30 12	132	14 41	171	55 07	209	42 15	274	35 06	341
202	18 01	053	30 01	083	30 56	132	14 50	171	54 38	211	41 17	275	34 46	340
203	18 48	053	30 59	083	31 39	133	14 59	172	54 07	212	40 18	275	34 26	340
204	19 34	053	31 57	083	32 22	133	15 07	172	53 36	214	39 20	275	34 06	339
205	20 21	053	32 55	083	33 04	134	15 15	173	53 03	215	38 22	275	33 45	339
206	21 08	053	33 53	083	33 46	135	15 22	173	52 29	216	37 24	275	33 24	339
207	21 55	053	34 51	084	34 27	135	15 29	174	51 54	217	36 26	275	33 02	338
208	22 42	053	35 49	084	35 08	136	15 35	174	51 18	219	35 27	276	32 40	338
209	23 28	053	36 47	084	35 48	137	15 41	175	50 41	220	34 29	276	32 18	338
	◆VEGA		Rasalhague		ANTARES		◆RIGIL KENT.		SPICA		◆REGULUS		Dubhe	
210	24 15	053	37 45	084	36 28	138	15 46	175	64 21	200	33 31	276	31 56	337
211	25 02	053	38 43	084	37 07	138	15 51	176	63 59	202	32 33	276	31 33	337
212	25 48	053	39 42	084	37 45	139	15 55	176	63 36	204	31 35	276	31 10	337
213	26 35	053	40 40	084	38 23	140	15 59	177	63 11	206	30 37	276	30 46	336
214	27 22	053	41 38	085	39 01	141	16 02	177	62 44	208	29 39	277	30 23	336
215	28 08	053	42 36	085	39 37	142	16 05	178	62 16	210	28 40	277	29 59	336
216	28 55	053	43 34	085	40 13	142	16 07	178	61 46	212	27 42	277	29 34	335
217	29 41	053	44 33	085	40 48	143	16 09	179	61 14	214	26 44	277	29 10	335
218	30 28	052	45 31	085	41 23	144	16 10	179	60 41	215	25 46	277	28 45	335
219	31 14	052	46 29	085	41 57	145	16 11	180	60 07	217	24 48	277	28 20	335
220	32 00	052	47 27	085	42 29	146	16 11	180	59 31	218	23 50	278	27 55	334
221	32 46	052	48 26	086	43 02	147	16 11	181	58 54	220	22 52	278	27 29	334
222	33 32	052	49 24	086	43 33	148	16 10	181	58 16	221	21 55	278	27 04	334
223	34 18	052	50 22	086	44 03	149	16 08	182	57 37	223	20 57	278	26 38	334
224	35 04	051	51 21	086	44 33	150	16 07	182	56 56	224	19 59	278	26 12	333
	◆VEGA		ALTAIR		ANTARES		◆RIGIL KENT.		SPICA		◆Denebola		Alkaid	
225	35 49	051	18 47	085	45 01	151	16 04	183	56 15	225	43 38	278	50 42	341
226	36 35	051	19 45	085	45 29	152	16 01	183	55 33	227	42 40	278	50 23	340
227	37 20	051	20 43	085	45 56	154	15 58	184	54 50	228	41 42	278	50 03	340
228	38 06	051	21 42	086	46 21	155	15 54	184	54 07	229	40 45	278	49 42	339
229	38 51	050	22 40	086	46 46	156	15 49	185	53 22	230	39 47	278	49 21	338
230	39 35	050	23 38	086	47 09	157	15 44	185	52 37	231	38 49	278	48 58	337
231	40 20	050	24 37	086	47 31	158	15 39	186	51 51	232	37 51	279	48 35	336
232	41 05	049	25 35	086	47 52	159	15 33	186	51 05	233	36 53	279	48 11	335
233	41 49	049	26 33	086	48 12	161	15 26	187	50 18	234	35 55	279	47 46	335
234	42 33	049	27 32	087	48 31	162	15 19	187	49 30	235	34 58	279	47 21	334
235	43 17	048	28 30	087	48 49	163	15 12	188	48 42	236	34 00	279	46 55	333
236	44 00	048	29 28	087	49 05	164	15 04	188	47 54	237	33 02	279	46 28	333
237	44 43	047	30 27	087	49 20	166	14 55	189	47 05	237	32 04	279	46 01	332
238	45 26	047	31 25	087	49 34	167	14 46	189	46 15	238	31 07	279	45 33	331
239	46 09	047	32 24	088	49 46	168	14 37	190	45 25	239	30 09	280	45 05	331
	◆VEGA		ALTAIR		Nunki		◆ANTARES		SPICA		◆ARCTURUS		Alkaid	
240	46 51	046	33 22	088	32 06	133	49 57	170	44 35	240	64 10	287	44 36	330
241	47 33	046	34 20	088	32 49	134	50 07	171	43 44	240	63 14	287	44 06	329
242	48 15	045	35 19	088	33 31	134	50 15	173	42 54	241	62 18	287	43 36	329
243	48 56	044	36 17	088	34 12	135	50 22	174	42 02	242	61 22	286	43 06	328
244	49 37	044	37 16	089	34 53	136	50 27	175	41 11	242	60 26	286	42 35	328
245	50 17	043	38 14	089	35 34	136	50 31	177	40 19	243	59 29	286	42 03	327
246	50 57	043	39 13	089	36 14	137	50 34	178	39 27	243	58 33	286	41 31	327
247	51 36	042	40 11	089	36 53	138	50 35	180	38 34	244	57 37	285	40 59	326
248	52 15	041	41 10	090	37 32	139	50 34	181	37 41	245	56 40	285	40 26	326
249	52 53	041	42 08	090	38 11	139	50 33	182	36 48	245	55 44	285	39 53	325
250	53 31	040	43 07	090	38 48	140	50 29	184	35 55	246	54 48	285	39 20	325
251	54 08	039	44 05	090	39 25	141	50 25	185	35 02	246	53 51	285	38 46	324
252	54 44	038	45 03	090	40 02	142	50 19	187	34 08	247	52 55	285	38 11	324
253	55 20	037	46 02	091	40 37	143	50 11	188	33 15	247	51 58	285	37 37	324
254	55 55	036	47 00	091	41 12	144	50 02	189	32 21	248	51 01	285	37 02	323
	◆VEGA		ALTAIR		◆Nunki		ANTARES		SPICA		◆ARCTURUS		Alkaid	
255	56 29	035	47 59	091	41 47	145	49 52	191	31 26	248	50 05	284	36 27	323
256	57 03	034	48 57	091	42 20	146	49 41	192	30 32	249	49 08	284	35 52	323
257	57 35	033	49 56	092	42 52	147	49 28	193	29 38	249	48 12	284	35 16	322
258	58 07	032	50 54	092	43 24	148	49 13	195	28 43	249	47 15	284	34 40	322
259	58 38	031	51 53	092	43 55	149	48 58	196	27 48	250	46 18	284	34 04	322
260	59 07	030	52 51	092	44 25	150	48 41	197	26 53	250	45 22	284	33 27	321
261	59 36	029	53 49	093	44 54	151	48 23	199	25 58	251	44 25	284	32 51	321
262	60 04	027	54 48	093	45 22	152	48 04	200	25 03	251	43 28	284	32 14	321
263	60 30	026	55 46	093	45 49	153	47 43	201	24 08	251	42 32	284	31 37	321
264	60 55	025	56 44	094	46 16	154	47 21	202	23 12	252	41 35	284	30 59	320
265	61 19	023	57 43	094	46 41	155	46 59	204	22 17	252	40 38	284	30 22	320
266	61 42	022	58 41	094	47 05	156	46 35	205	21 21	252	39 42	284	29 44	320
267	62 03	021	59 39	095	47 28	158	46 10	206	20 25	253	38 45	284	29 07	320
268	62 23	019	60 38	095	47 49	159	45 44	207	19 29	253	37 48	284	28 29	320
269	62 41	017	61 36	095	48 10	160	45 17	208	18 34	253	36 52	284	27 51	319

LHA	Hc	Zn	Hc	Zn	Hc	Zn	Hc	Zn	Hc	Zn	Hc	Zn	Hc	Zn
♈	VEGA		◆DENEB		Enif		◆Nunki		ANTARES		◆ARCTURUS		Alkaid	
270	62 58	016	43 03	039	35 08	087	48 29	161	44 49	209	35 55	284	27 13	319
271	63 13	014	43 40	038	36 07	087	48 47	163	44 20	210	34 59	285	26 34	319
272	63 27	013	44 15	038	37 05	087	49 04	164	43 50	211	34 02	285	25 56	319
273	63 39	011	44 51	037	38 03	088	49 20	165	43 19	212	33 05	285	25 17	319
274	63 49	009	45 26	036	39 02	088	49 34	166	42 47	213	32 09	285	24 39	319
275	63 57	007	46 00	036	40 00	088	49 47	168	42 15	214	31 12	285	24 00	319
276	64 04	006	46 34	035	40 59	088	49 59	169	41 41	215	30 16	285	23 21	318
277	64 09	004	47 08	035	41 57	088	50 09	171	41 07	216	29 19	285	22 43	318
278	64 12	002	47 41	034	42 56	089	50 18	172	40 33	217	28 23	285	22 04	318
279	64 13	000	48 13	033	43 54	089	50 26	173	39 57	218	27 26	285	21 25	318
280	64 13	359	48 45	033	44 52	089	50 32	175	39 21	219	26 30	285	20 46	318
281	64 10	357	49 16	032	45 51	089	50 36	176	38 44	220	25 34	285	20 07	318
282	64 06	355	49 47	031	46 49	089	50 40	178	38 06	220	24 37	285	19 28	318
283	64 00	353	50 17	030	47 48	090	50 41	179	37 28	221	23 41	286	18 49	318
284	63 52	351	50 46	030	48 46	090	50 42	180	36 49	222	22 44	286	18 10	318
	◆DENEB		Alpheratz		◆FOMALHAUT		Nunki		ANTARES		◆Rasalhague		VEGA	
285	51 14	029	17 29	063	18 42	128	50 41	182	36 10	223	69 12	271	63 43	350
286	51 42	028	18 21	063	19 28	128	50 38	183	35 30	223	68 14	271	63 31	348
287	52 09	027	19 13	063	20 13	129	50 34	185	34 49	224	67 15	272	63 19	346
288	52 35	026	20 06	063	20 59	129	50 29	186	34 08	225	66 17	272	63 04	345
289	53 00	025	20 58	064	21 44	130	50 22	187	33 27	226	65 19	272	62 48	343
290	53 25	024	21 50	064	22 29	130	50 14	189	32 45	226	64 20	272	62 30	342
291	53 48	023	22 43	064	23 13	131	50 04	190	32 02	227	63 22	272	62 11	340
292	54 11	022	23 35	064	23 57	131	49 53	192	31 19	228	62 23	272	61 50	339
293	54 32	021	24 28	064	24 41	132	49 41	193	30 36	228	61 25	272	61 28	337
294	54 53	020	25 20	064	25 25	132	49 27	194	29 52	229	60 26	273	61 04	336
295	55 13	019	26 12	064	26 08	133	49 12	196	29 08	229	59 28	273	60 40	334
296	55 31	018	27 05	064	26 51	133	48 56	197	28 24	230	58 30	273	60 14	333
297	55 49	017	27 57	064	27 33	134	48 38	198	27 39	230	57 31	273	59 47	332
298	56 05	016	28 50	064	28 15	134	48 19	199	26 53	231	56 33	273	59 18	331
299	56 20	014	29 42	064	28 56	135	47 59	201	26 08	232	55 35	273	58 49	329
	◆DENEB		Alpheratz		◆FOMALHAUT		Peacock		◆ANTARES		Rasalhague		VEGA	
300	56 34	013	30 35	064	29 37	136	20 03	176	25 22	232	54 36	273	58 19	328
301	56 47	012	31 27	064	30 18	136	20 07	177	24 36	233	53 38	274	57 48	327
302	56 58	011	32 19	064	30 58	137	20 09	177	23 49	233	52 40	274	57 15	326
303	57 09	010	33 12	064	31 38	138	20 12	178	23 02	234	51 41	274	56 42	325
304	57 18	008	34 04	064	32 17	138	20 13	179	22 15	234	50 43	274	56 08	324
305	57 25	007	34 56	063	32 56	139	20 14	179	21 28	234	49 45	274	55 34	323
306	57 32	006	35 49	063	33 34	140	20 15	180	20 40	235	48 46	274	54 58	322
307	57 37	004	36 41	063	34 11	140	20 15	180	19 52	235	47 48	274	54 22	321
308	57 41	003	37 33	063	34 48	141	20 14	181	19 04	236	46 50	275	53 45	321
309	57 43	002	38 25	063	35 25	142	20 13	182	18 15	236	45 51	275	53 08	320
310	57 44	000	39 17	063	36 00	143	20 11	182	17 27	237	44 53	275	52 30	319
311	57 44	359	40 09	063	36 36	143	20 08	183	16 38	237	43 55	275	51 51	318
312	57 42	358	41 01	063	37 10	144	20 05	183	15 49	237	42 57	275	51 12	318
313	57 39	356	41 53	062	37 44	145	20 02	184	14 59	238	41 58	275	50 32	317
314	57 35	355	42 45	062	38 17	146	19 57	184	14 10	238	41 00	275	49 52	316
	Alpheratz		◆Diphda		FOMALHAUT		◆Nunki		Rasalhague		VEGA		◆DENEB	
315	43 37	062	26 48	118	38 49	147	40 19	218	40 02	276	49 12	316	57 30	354
316	44 28	062	27 39	119	39 21	148	39 43	218	39 04	276	48 31	315	57 23	353
317	45 20	062	28 30	119	39 52	149	39 06	219	38 06	276	47 49	315	57 14	351
318	46 11	061	29 21	120	40 22	150	38 29	220	37 08	276	47 07	314	57 05	350
319	47 02	061	30 12	120	40 51	151	37 51	221	36 09	276	46 25	314	56 54	349
320	47 54	061	31 02	121	41 19	152	37 12	222	35 11	276	45 43	313	56 42	348
321	48 44	060	31 53	121	41 46	152	36 33	223	34 13	276	45 00	313	56 29	346
322	49 35	060	32 42	122	42 13	154	35 53	223	33 15	277	44 17	312	56 15	345
323	50 26	060	33 32	122	42 39	155	35 13	224	32 17	277	43 33	312	55 59	344
324	51 16	059	34 21	123	43 03	156	34 32	225	31 19	277	42 50	311	55 42	343
325	52 06	059	35 10	123	43 27	157	33 51	225	30 21	277	42 06	311	55 24	342
326	52 56	059	35 59	124	43 50	158	33 09	226	29 23	277	41 22	311	55 06	341
327	53 46	058	36 47	125	44 11	159	32 26	227	28 25	277	40 37	310	54 46	339
328	54 36	058	37 35	125	44 32	160	31 44	227	27 27	278	39 53	310	54 25	338
329	55 25	057	38 23	126	44 51	161	31 00	228	26 29	278	39 08	310	54 03	337
	◆Alpheratz		Diphda		◆FOMALHAUT		Peacock		ALTAIR		◆VEGA		DENEB	
330	56 14	057	39 10	127	45 10	162	17 31	193	57 59	266	38 23	310	53 40	336
331	57 02	056	39 56	127	45 27	163	17 18	194	57 00	266	37 38	309	53 16	335
332	57 51	055	40 43	128	45 43	165	17 03	194	56 02	267	36 52	309	52 51	334
333	58 39	055	41 28	129	45 58	166	16 49	195	55 04	267	36 07	309	52 26	334
334	59 26	054	42 14	130	46 12	167	16 33	195	54 05	267	35 21	309	51 59	333
335	60 13	053	42 58	130	46 25	168	16 17	196	53 07	267	34 36	308	51 32	332
336	61 00	052	43 43	131	46 36	169	16 01	196	52 08	268	33 50	308	51 04	331
337	61 46	051	44 26	132	46 46	171	15 44	197	51 10	268	33 04	308	50 35	330
338	62 31	050	45 09	133	46 55	172	15 27	197	50 11	268	32 18	308	50 06	329
339	63 16	049	45 52	134	47 02	173	15 10	198	49 13	269	31 32	308	49 36	329
340	64 00	048	46 33	135	47 09	174	14 51	198	48 15	269	30 45	308	49 05	328
341	64 43	047	47 15	136	47 14	176	14 33	199	47 16	269	29 59	308	48 33	327
342	65 26	046	47 55	137	47 17	177	14 14	199	46 18	269	29 13	307	48 01	326
343	66 07	045	48 35	138	47 20	178	13 54	200	45 19	269	28 26	307	47 29	326
344	66 48	043	49 13	139	47 21	180	13 34	200	44 21	270	27 40	307	46 56	325
	Schedar		◆Mirfak		Hamal		◆ACHERNAR		FOMALHAUT		◆ALTAIR		DENEB	
345	42 26	018	25 19	041	44 37	070	12 36	159	47 21	181	43 22	270	46 22	325
346	42 44	018	25 57	041	45 32	070	12 56	160	47 19	182	42 24	270	45 48	324
347	43 01	017	26 35	040	46 26	070	13 16	160	47 16	183	41 25	270	45 13	323
348	43 18	017	27 13	040	47 21	069	13 36	161	47 12	185	40 27	271	44 38	323
349	43 35	016	27 50	040	48 16	069	13 55	161	47 07	186	39 28	271	44 02	322
350	43 50	015	28 28	040	49 11	069	14 13	162	47 00	187	38 30	271	43 26	322
351	44 05	015	29 05	040	50 05	069	14 31	162	46 52	188	37 32	271	42 50	321
352	44 20	014	29 43	040	51 00	069	14 49	163	46 43	190	36 33	271	42 13	321
353	44 33	013	30 20	039	51 54	068	15 06	163	46 32	191	35 35	272	41 36	320
354	44 46	012	30 57	039	52 48	068	15 23	164	46 21	192	34 36	272	40 59	320
355	44 59	012	31 34	039	53 43	068	15 40	164	46 08	193	33 38	272	40 21	320
356	45 10	011	32 10	039	54 37	068	15 56	164	45 54	195	32 39	272	39 43	319
357	45 21	010	32 46	038	55 31	067	16 11	165	45 38	196	31 41	272	39 04	319
358	45 31	009	33 23	038	56 25	067	16 26	165	45 22	197	30 43	273	38 26	318
359	45 40	009	33 59	038	57 18	067	16 40	166	45 04	198	29 44	273	37 47	318

LHA ♈	Hc	Zn	Hc	Zn	Hc	Zn	Hc	Zn	Hc	Zn	Hc	Zn	Hc	Zn
	CAPELLA		◆ALDEBARAN		Acamar		ACHERNAR		◆FOMALHAUT		ALTAIR		◆DENEB	
0	16 09	045	23 22	077	23 24	144	17 52	166	45 42	200	28 42	274	36 23	318
1	16 51	045	24 19	077	23 58	145	18 06	167	45 22	201	27 44	274	35 44	318
2	17 33	045	25 16	077	24 32	146	18 19	167	45 00	202	26 45	274	35 04	318
3	18 14	045	26 13	077	25 05	146	18 31	168	44 38	203	25 47	274	34 24	317
4	18 56	045	27 11	077	25 37	147	18 43	169	44 15	204	24 48	274	33 44	317
5	19 38	045	28 08	078	26 09	147	18 55	169	43 50	205	23 50	274	33 04	317
6	20 19	045	29 05	078	26 40	148	19 06	170	43 25	206	22 51	275	32 24	316
7	21 01	045	30 03	078	27 11	149	19 16	170	42 58	207	21 53	275	31 43	316
8	21 43	045	31 00	078	27 42	149	19 26	171	42 31	208	20 54	275	31 03	316
9	22 24	045	31 57	078	28 12	150	19 35	171	42 03	209	19 56	275	30 22	316
10	23 05	045	32 55	078	28 41	150	19 44	172	41 34	210	18 57	275	29 41	316
11	23 47	045	33 52	078	29 09	151	19 52	172	41 04	211	17 59	276	29 00	315
12	24 28	045	34 50	078	29 37	152	19 59	173	40 33	212	17 00	276	28 19	315
13	25 09	045	35 47	078	30 05	153	20 06	173	40 01	213	16 02	276	27 37	315
14	25 50	044	36 45	078	30 31	153	20 12	174	39 29	214	15 04	276	26 56	315
	CAPELLA		◆ALDEBARAN		RIGEL		ACHERNAR		◆FOMALHAUT		Enif		◆DENEB	
15	26 32	044	37 42	078	23 39	105	20 18	175	38 56	215	41 53	273	26 14	315
16	27 12	044	38 39	078	24 36	105	20 24	175	38 22	216	40 55	273	25 33	315
17	27 53	044	39 37	078	25 33	105	20 28	176	37 48	216	39 56	273	24 51	315
18	28 34	044	40 34	078	26 29	106	20 32	176	37 13	217	38 57	273	24 09	314
19	29 15	044	41 32	078	27 26	106	20 36	177	36 37	218	37 59	273	23 27	314
20	29 55	043	42 29	079	28 22	106	20 39	177	36 00	219	37 00	273	22 45	314
21	30 35	043	43 27	079	29 18	107	20 41	178	35 23	220	36 01	274	22 03	314
22	31 15	043	44 25	079	30 14	107	20 43	179	34 46	220	35 03	274	21 21	314
23	31 55	043	45 22	079	31 10	107	20 44	179	34 07	221	34 04	274	20 39	314
24	32 35	043	46 20	079	32 06	108	20 44	180	33 29	222	33 06	274	19 57	314
25	33 15	042	47 17	079	33 02	108	20 44	180	32 49	222	32 07	274	19 14	314
26	33 54	042	48 15	079	33 58	109	20 44	181	32 10	223	31 09	274	18 32	314
27	34 33	042	49 12	079	34 53	109	20 42	182	31 29	224	30 10	275	17 50	314
28	35 12	041	50 10	079	35 49	109	20 40	182	30 49	224	29 12	275	17 08	314
29	35 51	041	51 07	079	36 44	110	20 38	183	30 07	225	28 13	275	16 26	314
	Mirfak		◆CAPELLA		ALDEBARAN		RIGEL		◆ACHERNAR		FOMALHAUT		◆Alpheratz	
30	48 24	020	36 29	041	52 05	078	37 39	110	20 35	183	29 26	226	58 53	308
31	48 44	020	37 07	040	53 02	078	38 34	111	20 31	184	28 44	226	58 06	307
32	49 03	019	37 45	040	54 00	078	39 29	111	20 27	184	28 01	227	57 19	306
33	49 22	018	38 23	040	54 57	078	40 23	112	20 22	185	27 18	227	56 31	305
34	49 39	017	39 00	039	55 55	078	41 18	112	20 17	186	26 35	228	55 43	305
35	49 56	016	39 37	039	56 52	078	42 12	113	20 11	186	25 51	228	54 55	304
36	50 12	015	40 14	038	57 50	078	43 06	114	20 05	187	25 07	229	54 06	304
37	50 26	014	40 50	038	58 47	078	44 00	114	19 57	187	24 23	229	53 17	303
38	50 40	013	41 26	038	59 44	078	44 53	115	19 50	188	23 38	230	52 28	303
39	50 53	012	42 02	037	60 42	078	45 46	115	19 42	188	22 53	230	51 38	302
40	51 05	011	42 37	037	61 39	077	46 39	116	19 33	189	22 08	231	50 48	302
41	51 16	010	43 12	036	62 36	077	47 32	117	19 23	189	21 22	231	49 58	301
42	51 26	009	43 46	035	63 33	077	48 24	117	19 14	190	20 36	232	49 08	301
43	51 35	008	44 20	035	64 31	077	49 16	118	19 03	191	19 50	232	48 18	301
44	51 43	007	44 53	034	65 28	076	50 08	119	18 52	191	19 04	233	47 27	300
	CAPELLA		◆POLLUX		PROCYON		SIRIUS		◆ACHERNAR		Diphda		◆Alpheratz	
45	45 26	034	22 02	064	20 51	089	27 27	116	18 41	192	44 52	229	46 36	300
46	45 58	033	22 55	064	21 50	089	28 19	117	18 28	192	44 07	230	45 45	300
47	46 30	032	23 48	064	22 49	089	29 12	117	18 16	193	43 22	231	44 54	299
48	47 01	032	24 41	064	23 47	090	30 04	118	18 03	193	42 36	231	44 03	299
49	47 32	031	25 34	064	24 46	090	30 56	118	17 49	194	41 50	232	43 12	299
50	48 02	030	26 27	064	25 45	090	31 48	119	17 35	194	41 04	233	42 20	299
51	48 31	030	27 20	064	26 43	090	32 39	119	17 20	195	40 17	234	41 28	298
52	48 59	029	28 13	064	27 42	090	33 30	120	17 05	195	39 30	234	40 37	298
53	49 27	028	29 06	064	28 41	091	34 21	120	16 50	196	38 42	235	39 45	298
54	49 55	027	29 59	064	29 39	091	35 12	121	16 33	196	37 54	236	38 53	298
55	50 21	026	30 52	064	30 38	091	36 02	121	16 17	197	37 05	236	38 01	298
56	50 47	026	31 45	064	31 37	091	36 52	122	16 00	197	36 16	237	37 09	298
57	51 12	025	32 37	064	32 35	091	37 42	122	15 42	198	35 27	237	36 17	297
58	51 36	024	33 30	064	33 34	092	38 31	123	15 24	198	34 37	238	35 25	297
59	51 59	023	34 23	064	34 33	092	39 20	124	15 06	199	33 47	238	34 33	297
	CAPELLA		◆POLLUX		PROCYON		SIRIUS		◆CANOPUS		Diphda		◆Hamal	
60	52 21	022	35 16	064	35 31	092	40 09	124	18 20	158	32 57	239	60 49	297
61	52 42	021	36 09	064	36 30	092	40 57	125	18 42	158	32 07	240	59 56	296
62	53 03	020	37 01	064	37 29	093	41 45	126	19 03	159	31 16	240	59 04	296
63	53 22	019	37 54	064	38 27	093	42 33	126	19 24	160	30 25	241	58 11	295
64	53 41	018	38 46	064	39 26	093	43 19	127	19 44	160	29 34	241	57 18	295
65	53 58	017	39 39	063	40 24	093	44 06	128	20 04	161	28 42	242	56 24	295
66	54 14	016	40 31	063	41 23	094	44 52	129	20 23	161	27 51	242	55 31	294
67	54 30	014	41 24	063	42 22	094	45 38	130	20 42	162	26 59	242	54 37	294
68	54 44	013	42 16	063	43 20	094	46 22	130	21 00	162	26 07	243	53 43	293
69	54 57	012	43 08	063	44 19	094	47 07	131	21 17	163	25 14	243	52 50	293
70	55 09	011	44 00	062	45 17	095	47 51	132	21 34	163	24 22	244	51 56	293
71	55 19	010	44 52	062	46 16	095	48 34	133	21 51	164	23 29	244	51 01	293
72	55 29	009	45 44	062	47 14	095	49 16	134	22 07	165	22 36	244	50 07	292
73	55 37	007	46 36	062	48 13	096	49 58	135	22 22	165	21 43	245	49 13	292
74	55 44	006	47 27	061	49 11	096	50 39	136	22 37	166	20 50	245	48 18	292
	CAPELLA		◆POLLUX		PROCYON		SIRIUS		◆CANOPUS		Acamar		◆Hamal	
75	55 50	005	48 19	061	50 09	096	51 19	137	22 51	166	30 32	207	47 24	292
76	55 54	004	49 10	061	51 08	097	51 58	138	23 04	167	30 06	207	46 29	292
77	55 58	003	50 01	060	52 06	097	52 37	140	23 17	168	29 38	208	45 35	291
78	56 00	001	50 52	060	53 04	097	53 14	141	23 29	168	29 10	209	44 40	291
79	56 00	000	51 43	060	54 02	098	53 51	142	23 41	169	28 42	209	43 45	291
80	56 00	359	52 33	059	55 00	098	54 26	143	23 52	169	28 13	210	42 50	291
81	55 58	358	53 24	059	55 58	099	55 01	145	24 02	170	27 43	211	41 56	291
82	55 55	356	54 14	058	56 56	099	55 34	146	24 12	171	27 12	211	41 01	291
83	55 51	355	55 03	058	57 54	099	56 06	148	24 21	171	26 42	212	40 06	291
84	55 45	354	55 53	057	58 52	100	56 37	149	24 30	172	26 10	213	39 11	291
85	55 38	353	56 42	057	59 50	100	57 07	150	24 37	173	25 38	213	38 16	291
86	55 30	352	57 31	056	60 48	101	57 35	152	24 45	173	25 06	214	37 21	290
87	55 21	350	58 19	055	61 45	101	58 02	154	24 51	174	24 33	214	36 26	290
88	55 10	349	59 07	054	62 43	102	58 27	155	24 57	175	23 59	215	35 31	290
89	54 59	348	59 55	054	63 40	103	58 51	157	25 02	175	23 25	216	34 36	290

LAT 12°N

LHA ♈	Hc	Zn	Hc	Zn	Hc	Zn	Hc	Zn	Hc	Zn	Hc	Zn	Hc	Zn
	Dubhe		◆REGULUS		Suhail		SIRIUS		◆RIGEL		ALDEBARAN		◆CAPELLA	
90	17 14	029	29 28	083	20 00	146	59 13	159	66 49	210	69 05	285	54 46	347
91	17 42	029	30 27	083	20 33	146	59 34	160	66 19	212	68 08	284	54 32	346
92	18 10	029	31 25	083	21 06	147	59 53	162	65 47	214	67 11	284	54 17	345
93	18 38	028	32 23	083	21 38	147	60 10	164	65 13	216	66 14	284	54 01	343
94	19 06	028	33 21	083	22 10	148	60 25	166	64 38	218	65 17	283	53 44	342
95	19 34	028	34 20	084	22 41	148	60 38	168	64 01	220	64 20	283	53 25	341
96	20 02	028	35 18	084	23 11	149	60 50	170	63 23	221	63 23	283	53 06	340
97	20 29	028	36 16	084	23 41	149	60 59	172	62 44	223	62 26	283	52 46	339
98	20 57	028	37 15	084	24 11	150	61 07	174	62 03	225	61 28	283	52 25	338
99	21 24	028	38 13	084	24 40	151	61 12	176	61 21	226	60 31	282	52 02	337
100	21 52	028	39 11	084	25 09	151	61 16	178	60 38	228	59 34	282	51 39	336
101	22 19	028	40 10	084	25 37	152	61 17	180	59 55	229	58 36	282	51 15	336
102	22 46	027	41 08	084	26 04	152	61 17	182	59 10	230	57 39	282	50 51	335
103	23 13	027	42 07	085	26 31	153	61 14	184	58 24	231	56 42	282	50 25	334
104	23 40	027	43 05	085	26 58	154	61 10	186	57 38	233	55 44	282	49 59	333
	Dubhe		◆REGULUS		Suhail		◆SIRIUS		RIGEL		ALDEBARAN		◆CAPELLA	
105	24 07	027	44 04	085	27 23	154	61 03	187	56 51	234	54 47	282	49 32	332
106	24 33	027	45 02	085	27 48	155	60 54	189	56 04	235	53 49	282	49 04	331
107	24 59	027	46 01	085	28 13	156	60 44	191	55 15	236	52 52	282	48 35	331
108	25 25	026	46 59	085	28 37	156	60 31	193	54 27	237	51 54	282	48 06	330
109	25 51	026	47 57	085	29 00	157	60 17	195	53 37	238	50 57	281	47 36	329
110	26 17	026	48 56	085	29 22	158	60 01	197	52 47	239	49 59	281	47 06	328
111	26 43	026	49 55	086	29 44	159	59 43	199	51 57	239	49 02	281	46 35	328
112	27 08	025	50 53	086	30 05	159	59 23	201	51 06	240	48 04	281	46 03	327
113	27 33	025	51 52	086	30 26	160	59 01	202	50 15	241	47 07	281	45 31	326
114	27 58	025	52 50	086	30 45	161	58 38	204	49 24	242	46 09	281	44 58	326
115	28 22	025	53 49	086	31 04	162	58 14	206	48 32	243	45 11	281	44 25	325
116	28 47	024	54 47	086	31 23	162	57 48	207	47 40	243	44 14	281	43 51	325
117	29 11	024	55 46	086	31 40	163	57 20	209	46 47	244	43 16	281	43 17	324
118	29 35	024	56 44	086	31 57	164	56 51	210	45 54	245	42 19	281	42 42	324
119	29 58	023	57 43	086	32 13	165	56 21	212	45 01	245	41 21	282	42 07	323
	Dubhe		◆Denebola		Suhail		◆SIRIUS		RIGEL		ALDEBARAN		◆CAPELLA	
120	30 21	023	34 25	080	32 28	165	55 49	213	44 08	246	40 24	282	41 32	323
121	30 44	023	35 23	080	32 42	166	55 16	215	43 14	246	39 26	282	40 56	322
122	31 07	023	36 21	081	32 56	167	54 42	216	42 20	247	38 29	282	40 20	322
123	31 29	022	37 18	081	33 08	168	54 07	217	41 26	248	37 31	282	39 43	321
124	31 51	022	38 16	081	33 20	169	53 31	219	40 32	248	36 34	282	39 06	321
125	32 13	021	39 14	081	33 31	170	52 54	220	39 37	249	35 37	282	38 29	320
126	32 34	021	40 12	081	33 42	170	52 16	221	38 42	249	34 39	282	37 51	320
127	32 55	021	41 10	081	33 51	171	51 37	222	37 47	250	33 42	282	37 14	320
128	33 15	020	42 08	081	33 59	172	50 57	223	36 52	250	32 44	282	36 35	319
129	33 36	020	43 06	081	34 07	173	50 17	224	35 57	250	31 47	282	35 57	319
130	33 55	020	44 04	081	34 13	174	49 35	225	35 02	251	30 50	282	35 18	319
131	34 15	019	45 02	081	34 19	175	48 53	226	34 06	251	29 52	282	34 39	318
132	34 34	019	46 00	081	34 24	176	48 11	227	33 10	252	28 55	282	34 00	318
133	34 52	018	46 58	081	34 28	177	47 27	228	32 15	252	27 58	282	33 21	318
134	35 11	018	47 56	081	34 31	177	46 43	229	31 19	253	27 00	283	32 41	318
	Dubhe		◆ARCTURUS		SPICA		◆Suhail		SIRIUS		BETELGEUSE		◆CAPELLA	
135	35 28	017	14 18	073	20 17	107	34 34	178	45 58	230	44 12	269	32 02	317
136	35 45	017	15 14	073	21 13	107	34 35	179	45 13	231	43 14	269	31 22	317
137	36 02	016	16 10	073	22 09	107	34 35	180	44 27	232	42 15	269	30 42	317
138	36 19	016	17 06	073	23 05	108	34 35	181	43 41	232	41 16	269	30 01	317
139	36 34	015	18 02	073	24 01	108	34 33	182	42 54	233	40 18	270	29 21	316
140	36 50	015	18 59	074	24 57	108	34 31	183	42 07	234	39 19	270	28 40	316
141	37 05	014	19 55	074	25 52	109	34 28	184	41 19	235	38 20	270	28 00	316
142	37 19	014	20 51	074	26 48	109	34 24	184	40 31	235	37 22	270	27 19	316
143	37 33	013	21 48	074	27 43	110	34 19	185	39 43	236	36 23	270	26 38	316
144	37 46	013	22 44	074	28 38	110	34 13	186	38 54	237	35 24	271	25 57	316
145	37 59	012	23 41	074	29 33	110	34 06	187	38 05	237	34 26	271	25 16	315
146	38 11	012	24 37	074	30 28	111	33 58	188	37 15	238	33 27	271	24 35	315
147	38 23	011	25 33	074	31 23	111	33 50	189	36 25	239	32 28	271	23 53	315
148	38 34	011	26 30	074	32 18	112	33 41	190	35 35	239	31 30	271	23 12	315
149	38 45	010	27 26	074	33 12	112	33 30	190	34 45	240	30 31	272	22 31	315
	◆Dubhe		Alkaid		ARCTURUS		◆SPICA		Suhail		◆SIRIUS		POLLUX	
150	38 55	010	30 25	039	28 23	074	34 07	112	33 19	191	33 54	240	54 39	302
151	39 04	009	31 02	039	29 20	075	35 01	113	33 07	192	33 03	241	53 49	302
152	39 13	008	31 39	039	30 16	075	35 55	113	32 55	193	32 11	241	52 59	301
153	39 21	008	32 16	038	31 13	075	36 48	114	32 41	194	31 20	242	52 08	301
154	39 29	007	32 52	038	32 09	075	37 42	114	32 27	195	30 28	242	51 18	300
155	39 36	007	33 28	038	33 06	075	38 35	115	32 11	195	29 36	243	50 27	300
156	39 43	006	34 04	038	34 03	075	39 28	116	31 55	196	28 44	243	49 36	299
157	39 49	005	34 40	037	34 59	075	40 21	116	31 39	197	27 51	244	48 45	299
158	39 54	005	35 15	037	35 56	075	41 14	117	31 21	198	26 59	244	47 53	299
159	39 58	004	35 50	037	36 53	075	42 06	117	31 03	199	26 06	244	47 02	298
160	40 02	004	36 25	036	37 49	075	42 58	118	30 44	199	25 13	245	46 10	298
161	40 06	003	37 00	036	38 46	075	43 50	119	30 24	200	24 20	245	45 18	298
162	40 09	002	37 34	035	39 43	075	44 41	119	30 03	201	23 26	246	44 27	298
163	40 11	002	38 08	035	40 39	075	45 32	120	29 42	202	22 33	246	43 35	297
164	40 12	001	38 41	035	41 36	075	46 23	121	29 20	202	21 39	246	42 42	297
	◆Dubhe		ARCTURUS		◆SPICA		ACRUX		Suhail		◆PROCYON		POLLUX	
165	40 13	001	42 32	075	47 13	121	13 06	170	28 58	203	39 56	267	41 50	297
166	40 13	000	43 29	075	48 03	122	13 16	171	28 34	204	38 58	267	40 58	297
167	40 13	359	44 26	075	48 53	123	13 25	171	28 11	204	37 59	267	40 05	297
168	40 12	359	45 22	075	49 42	124	13 34	171	27 46	205	37 00	268	39 13	297
169	40 10	358	46 19	075	50 30	125	13 42	172	27 21	206	36 02	268	38 20	296
170	40 08	357	47 16	075	51 18	126	13 50	172	26 55	206	35 03	268	37 28	296
171	40 05	357	48 12	075	52 06	126	13 58	173	26 29	207	34 05	268	36 35	296
172	40 01	356	49 09	074	52 53	127	14 05	173	26 02	208	33 06	268	35 42	296
173	39 57	356	50 05	074	53 39	128	14 12	174	25 34	208	32 07	269	34 50	296
174	39 52	355	51 02	074	54 25	130	14 18	174	25 06	209	31 09	269	33 57	296
175	39 47	354	51 58	074	55 10	131	14 24	175	24 38	210	30 10	269	33 04	296
176	39 41	354	52 55	074	55 54	132	14 29	175	24 08	210	29 11	269	32 11	296
177	39 34	353	53 51	074	56 37	133	14 34	176	23 39	211	28 12	270	31 18	296
178	39 27	353	54 47	074	57 20	134	14 38	176	23 08	211	27 14	270	30 25	296
179	39 19	352	55 44	073	58 01	135	14 42	176	22 38	212	26 15	270	29 33	296

LHA ♈	Hc Zn	Hc Zn	Hc Zn	Hc Zn	Hc Zn	Hc Zn	Hc Zn
	Alkaid	◆Alphecca	ANTARES	◆ACRUX	REGULUS	◆POLLUX	Dubhe
180	46 35 025	37 43 065	14 14 122	14 45 177	62 39 273	28 40 296	39 11 351
181	47 00 025	38 36 065	15 04 122	14 48 177	61 40 273	27 47 296	39 02 351
182	47 24 024	39 29 065	15 54 122	14 51 178	60 42 273	26 54 296	38 52 350
183	47 48 023	40 23 065	16 43 123	14 53 178	59 43 273	26 01 296	38 42 350
184	48 10 022	41 16 065	17 33 123	14 54 179	58 45 273	25 08 296	38 31 349
185	48 32 021	42 09 065	18 22 123	14 55 179	57 46 274	24 15 296	38 20 349
186	48 53 021	43 02 064	19 11 124	14 56 180	56 47 274	23 22 296	38 08 348
187	49 13 020	43 55 064	19 59 124	14 56 180	55 49 274	22 29 296	37 55 348
188	49 33 019	44 48 064	20 48 125	14 55 181	54 50 274	21 36 296	37 42 347
189	49 51 018	45 40 064	21 36 125	14 54 181	53 52 274	20 43 296	37 29 346
190	50 09 017	46 33 064	22 24 125	14 53 182	52 53 274	19 50 296	37 15 346
191	50 26 016	47 25 063	23 12 126	14 51 182	51 55 274	18 57 296	37 00 345
192	50 42 015	48 18 063	23 59 126	14 49 183	50 56 274	18 05 296	36 45 345
193	50 57 014	49 10 063	24 46 127	14 46 183	49 58 274	17 12 296	36 30 344
194	51 11 013	50 02 062	25 33 127	14 43 183	48 59 275	16 19 296	36 14 344
	◆VEGA	Rasalhague	ANTARES	◆RIGIL KENT.	Gienah	◆REGULUS	Dubhe
195	11 58 052	23 06 081	26 20 128	14 34 168	58 31 201	48 01 275	35 57 343
196	12 45 053	24 04 081	27 06 128	14 46 168	58 09 202	47 02 275	35 41 343
197	13 31 053	25 02 082	27 52 129	14 58 169	57 46 204	46 04 275	35 23 343
198	14 18 053	26 00 082	28 37 129	15 09 169	57 22 206	45 05 275	35 05 342
199	15 05 053	26 59 082	29 22 130	15 20 170	56 56 207	44 07 275	34 47 342
200	15 51 053	27 57 082	30 07 130	15 30 170	56 28 209	43 08 275	34 28 341
201	16 38 053	28 55 082	30 52 131	15 40 171	55 59 210	42 10 275	34 09 341
202	17 25 053	29 53 082	31 36 132	15 49 171	55 29 212	41 11 276	33 50 340
203	18 12 053	30 51 082	32 19 132	15 58 172	54 58 213	40 13 276	33 30 340
204	18 58 053	31 49 083	33 03 133	16 06 172	54 25 214	39 15 276	33 10 340
205	19 45 053	32 47 083	33 45 134	16 14 173	53 52 216	38 16 276	32 49 339
206	20 32 053	33 46 083	34 28 134	16 22 173	53 17 217	37 18 276	32 28 339
207	21 19 053	34 44 083	35 09 135	16 28 174	52 41 218	36 19 276	32 07 338
208	22 05 053	35 42 083	35 51 136	16 35 174	52 05 219	35 21 276	31 45 338
209	22 52 053	36 40 083	36 31 136	16 41 175	51 27 220	34 23 276	31 23 338
	◆VEGA	Rasalhague	ANTARES	◆RIGIL KENT.	SPICA	◆REGULUS	Dubhe
210	23 39 053	37 39 083	37 12 137	16 46 175	65 17 201	33 24 277	31 00 337
211	24 25 053	38 37 083	37 51 138	16 51 176	64 55 203	32 26 277	30 38 337
212	25 12 053	39 35 084	38 30 139	16 55 176	64 31 205	31 28 277	30 15 337
213	25 59 053	40 34 084	39 09 139	16 59 177	64 05 207	30 30 277	29 51 336
214	26 45 052	41 32 084	39 47 140	17 02 177	63 37 209	29 31 277	29 28 336
215	27 32 052	42 30 084	40 24 141	17 05 178	63 08 211	28 33 277	29 04 336
216	28 18 052	43 29 084	41 01 142	17 07 178	62 36 213	27 35 277	28 40 336
217	29 04 052	44 27 084	41 36 143	17 09 179	62 04 215	26 37 278	28 15 335
218	29 51 052	45 25 084	42 11 144	17 10 179	61 30 216	25 39 278	27 51 335
219	30 37 052	46 24 084	42 46 145	17 11 180	60 54 218	24 40 278	27 26 335
220	31 23 052	47 22 084	43 19 146	17 11 180	60 17 220	23 42 278	27 01 335
221	32 09 051	48 21 085	43 52 147	17 11 181	59 39 221	22 44 278	26 35 334
222	32 55 051	49 19 085	44 24 148	17 10 181	59 00 223	21 46 278	26 10 334
223	33 41 051	50 17 085	44 55 149	17 08 182	58 20 224	20 48 278	25 44 334
224	34 26 051	51 16 085	45 25 150	17 07 182	57 39 225	19 50 279	25 18 334
	◆VEGA	ALTAIR	ANTARES	◆RIGIL KENT.	SPICA	◆Denebola	Alkaid
225	35 12 051	18 41 085	45 54 151	17 04 183	56 57 226	43 29 279	49 45 342
226	35 57 050	19 40 085	46 22 152	17 01 183	56 14 228	42 32 279	49 27 341
227	36 42 050	20 38 085	46 49 153	16 58 184	55 30 229	41 34 279	49 07 340
228	37 27 050	21 37 085	47 15 154	16 54 184	54 46 230	40 36 279	48 46 339
229	38 12 050	22 35 085	47 40 155	16 49 185	54 00 231	39 38 279	48 25 338
230	38 57 049	23 34 085	48 04 157	16 44 185	53 14 232	38 40 279	48 03 337
231	39 41 049	24 32 086	48 27 158	16 39 186	52 28 233	37 42 279	47 40 337
232	40 25 049	25 31 086	48 49 159	16 32 186	51 40 234	36 44 279	47 16 336
233	41 09 048	26 29 086	49 09 160	16 26 187	50 53 235	35 46 279	46 52 335
234	41 53 048	27 28 086	49 28 162	16 19 187	50 04 236	34 48 280	46 27 334
235	42 36 048	28 26 086	49 46 163	16 11 188	49 16 237	33 50 280	46 01 334
236	43 20 047	29 25 087	50 03 164	16 03 188	48 26 237	32 52 280	45 35 333
237	44 02 047	30 24 087	50 18 166	15 55 189	47 37 238	31 55 280	45 08 332
238	44 45 046	31 22 087	50 32 167	15 46 189	46 46 239	30 57 280	44 41 332
239	45 27 046	32 21 087	50 45 168	15 36 190	45 56 240	29 59 280	44 12 331
	◆VEGA	ALTAIR	Nunki	◆ANTARES	SPICA	◆ARCTURUS	Alkaid
240	46 09 045	33 19 087	32 47 133	50 56 170	45 05 240	63 51 289	43 44 330
241	46 51 045	34 18 087	33 30 133	51 06 171	44 14 241	62 55 289	43 15 330
242	47 32 044	35 17 088	34 12 134	51 14 172	43 22 242	62 00 288	42 45 329
243	48 13 044	36 15 088	34 55 134	51 21 174	42 30 242	61 04 288	42 15 329
244	48 53 043	37 14 088	35 36 135	51 27 175	41 38 243	60 08 288	41 44 328
245	49 33 042	38 13 088	36 17 136	51 31 177	40 46 244	59 12 287	41 13 328
246	50 12 042	39 11 088	36 58 137	51 34 178	39 53 244	58 16 287	40 41 327
247	50 51 041	40 10 088	37 38 137	51 35 180	39 00 245	57 20 287	40 09 327
248	51 29 040	41 09 089	38 17 138	51 34 181	38 07 245	56 24 287	39 36 326
249	52 07 040	42 07 089	38 56 139	51 33 182	37 13 246	55 28 287	39 04 326
250	52 44 039	43 06 089	39 34 140	51 29 184	36 20 246	54 31 286	38 30 325
251	53 21 038	44 05 089	40 12 141	51 25 185	35 26 247	53 35 286	37 57 325
252	53 57 037	45 03 089	40 49 141	51 18 187	34 32 247	52 39 286	37 23 324
253	54 32 036	46 02 090	41 25 142	51 11 188	33 38 248	51 42 286	36 49 324
254	55 06 035	47 01 090	42 01 143	51 02 190	32 43 248	50 46 286	36 14 324
	◆VEGA	ALTAIR	◆Nunki	ANTARES	SPICA	◆ARCTURUS	Alkaid
255	55 40 035	47 59 090	42 35 144	50 51 191	31 49 249	49 49 286	35 39 323
256	56 13 034	48 58 090	43 09 145	50 39 192	30 54 249	48 53 286	35 04 323
257	56 45 032	49 57 090	43 42 146	50 26 194	29 59 249	47 56 285	34 28 323
258	57 16 031	50 56 091	44 15 147	50 11 195	29 04 250	47 00 285	33 53 322
259	57 46 030	51 54 091	44 46 148	49 55 196	28 09 250	46 03 285	33 17 322
260	58 15 029	52 53 091	45 17 149	49 38 198	27 13 251	45 06 285	32 40 322
261	58 43 028	53 52 091	45 46 150	49 20 199	26 18 251	44 10 285	32 04 321
262	59 10 027	54 50 092	46 15 151	49 00 200	25 22 251	43 13 285	31 27 321
263	59 36 025	55 49 092	46 43 152	48 39 202	24 27 252	42 16 285	30 50 321
264	60 01 024	56 48 092	47 09 154	48 17 203	23 31 252	41 20 285	30 13 321
265	60 24 023	57 46 092	47 35 155	47 54 204	22 35 252	40 23 285	29 36 320
266	60 46 021	58 45 093	48 00 156	47 29 205	21 39 253	39 26 285	28 58 320
267	61 07 020	59 43 093	48 23 157	47 04 206	20 43 253	38 30 285	28 21 320
268	61 26 018	60 42 093	48 45 158	46 37 207	19 47 253	37 33 285	27 43 320
269	61 44 017	61 41 094	49 06 160	46 09 209	18 51 254	36 36 285	27 05 320

LAT 12°N

LHA ♈	Hc Zn	Hc Zn	Hc Zn	Hc Zn	Hc Zn	Hc Zn	Hc Zn
	VEGA	◆DENEB	Enif	◆Nunki	ANTARES	◆ARCTURUS	Alkaid
270	62 00 015	42 16 038	35 05 086	49 26 161	45 41 210	35 40 285	26 27 320
271	62 15 014	42 52 037	36 03 086	49 45 162	45 11 211	34 43 285	25 49 319
272	62 28 012	43 28 037	37 02 087	50 02 164	44 41 212	33 47 285	25 11 319
273	62 40 011	44 03 036	38 00 087	50 18 165	44 09 213	32 50 285	24 32 319
274	62 50 009	44 37 036	38 59 087	50 33 166	43 37 214	31 53 285	23 54 319
275	62 58 007	45 12 035	39 58 087	50 46 168	43 04 215	30 57 285	23 15 319
276	63 04 006	45 45 035	40 56 087	50 58 169	42 30 216	30 00 285	22 36 319
277	63 09 004	46 18 034	41 55 087	51 08 170	41 56 217	29 04 285	21 58 319
278	63 12 002	46 51 033	42 54 088	51 18 172	41 20 218	28 07 286	21 19 319
279	63 13 000	47 23 033	43 52 088	51 25 173	40 44 218	27 10 286	20 40 318
280	63 13 359	47 54 032	44 51 088	51 31 175	40 07 219	26 14 286	20 01 318
281	63 10 357	48 25 031	45 49 088	51 36 176	39 30 220	25 17 286	19 22 318
282	63 06 355	48 55 031	46 48 088	51 40 177	38 52 221	24 21 286	18 43 318
283	63 00 353	49 25 030	47 47 089	51 41 179	38 13 222	23 25 286	18 04 318
284	62 53 352	49 53 029	48 45 089	51 42 180	37 34 222	22 28 286	17 25 318
	◆DENEB	Alpheratz	◆FOMALHAUT	Nunki	ANTARES	◆Rasalhague	VEGA
285	50 22 028	17 02 063	19 18 128	51 41 182	36 54 223	69 10 274	62 44 350
286	50 49 027	17 54 063	20 05 128	51 38 183	36 13 224	68 11 274	62 33 348
287	51 15 026	18 46 063	20 51 128	51 34 185	35 32 225	67 13 274	62 20 347
288	51 41 026	19 39 063	21 37 129	51 28 186	34 51 225	66 14 274	62 06 345
289	52 06 025	20 31 063	22 22 129	51 21 188	34 09 226	65 15 274	61 50 344
290	52 30 024	21 24 063	23 07 130	51 13 189	33 26 227	64 17 274	61 33 342
291	52 53 023	22 16 063	23 52 130	51 03 190	32 43 227	63 18 274	61 14 341
292	53 15 022	23 08 063	24 37 131	50 52 192	32 00 228	62 20 274	60 54 339
293	53 36 021	24 01 063	25 21 131	50 39 193	31 16 229	61 21 274	60 33 338
294	53 57 020	24 53 063	26 05 132	50 25 195	30 32 229	60 23 274	60 10 336
295	54 16 019	25 46 063	26 48 132	50 10 196	29 47 230	59 24 274	59 46 335
296	54 34 017	26 38 063	27 32 133	49 53 197	29 02 230	58 26 275	59 20 334
297	54 51 016	27 31 063	28 14 133	49 35 198	28 17 231	57 27 275	58 54 333
298	55 07 015	28 23 063	28 57 134	49 16 200	27 31 231	56 29 275	58 26 331
299	55 22 014	29 15 063	29 39 135	48 55 201	26 45 232	55 30 275	57 57 330
	◆DENEB	Alpheratz	◆FOMALHAUT	Peacock	◆ANTARES	Rasalhague	VEGA
300	55 36 013	30 08 063	30 20 135	21 03 176	25 59 232	54 32 275	57 28 329
301	55 48 012	31 00 063	31 01 136	21 06 177	25 12 233	53 33 275	56 57 328
302	55 59 010	31 53 063	31 42 137	21 09 177	24 25 233	52 35 275	56 25 327
303	56 09 009	32 45 063	32 22 137	21 12 178	23 38 234	51 36 275	55 53 326
304	56 18 008	33 37 063	33 02 138	21 13 179	22 50 234	50 38 275	55 19 325
305	56 26 007	34 29 063	33 41 139	21 14 179	22 02 235	49 40 275	54 45 324
306	56 32 006	35 22 063	34 19 139	21 15 180	21 14 235	48 41 275	54 10 323
307	56 37 004	36 14 063	34 57 140	21 15 180	20 26 236	47 43 276	53 35 322
308	56 41 003	37 06 062	35 35 141	21 14 181	19 37 236	46 44 276	52 59 321
309	56 43 002	37 58 062	36 12 141	21 13 182	18 49 236	45 46 276	52 22 321
310	56 44 000	38 50 062	36 48 142	21 11 182	18 00 237	44 48 276	51 44 320
311	56 44 359	39 42 062	37 24 143	21 08 183	17 10 237	43 49 276	51 06 319
312	56 42 358	40 33 062	37 59 144	21 05 183	16 21 238	42 51 276	50 27 318
313	56 40 357	41 25 062	38 33 145	21 01 184	15 31 238	41 52 276	49 48 318
314	56 35 355	42 17 061	39 06 146	20 57 185	14 42 238	40 54 276	49 08 317
	Alpheratz	◆Diphda	FOMALHAUT	◆Nunki	Rasalhague	VEGA	◆DENEB
315	43 08 061	27 16 118	39 39 146	41 06 218	39 56 276	48 28 317	56 30 354
316	44 00 061	28 08 118	40 11 147	40 30 219	38 57 277	47 48 316	56 23 353
317	44 51 061	28 59 119	40 43 148	39 52 220	37 59 277	47 07 315	56 15 352
318	45 42 060	29 51 119	41 13 149	39 14 221	37 01 277	46 25 315	56 06 350
319	46 33 060	30 42 120	41 43 150	38 36 221	36 03 277	45 43 314	55 55 349
320	47 24 060	31 33 120	42 12 151	37 57 222	35 04 277	45 01 314	55 44 348
321	48 14 059	32 23 121	42 40 152	37 17 223	34 06 277	44 19 313	55 31 347
322	49 05 059	33 14 121	43 07 153	36 37 224	33 08 277	43 36 313	55 17 346
323	49 55 059	34 04 122	43 33 154	35 56 224	32 10 277	42 53 313	55 01 344
324	50 45 058	34 54 122	43 58 155	35 14 225	31 11 278	42 10 312	54 45 343
325	51 35 058	35 43 123	44 22 156	34 33 226	30 13 278	41 26 312	54 27 342
326	52 25 057	36 32 123	44 45 157	33 50 227	29 15 278	40 42 311	54 09 341
327	53 14 057	37 21 124	45 07 158	33 07 227	28 17 278	39 58 311	53 49 340
328	54 03 056	38 09 125	45 28 160	32 24 228	27 19 278	39 14 311	53 29 339
329	54 52 056	38 58 125	45 48 161	31 40 228	26 21 278	38 29 310	53 07 338
	◆Alpheratz	Diphda	◆FOMALHAUT	Peacock	ALTAIR	◆VEGA	DENEB
330	55 40 055	39 45 126	46 07 162	18 30 193	58 02 268	37 44 310	52 45 337
331	56 28 055	40 32 127	46 25 163	18 16 194	57 03 268	36 59 310	52 21 336
332	57 16 054	41 19 127	46 41 164	18 01 194	56 05 268	36 14 310	51 57 335
333	58 03 053	42 06 128	46 56 166	17 47 195	55 06 268	35 29 309	51 32 334
334	58 50 053	42 52 129	47 10 167	17 31 195	54 07 269	34 44 309	51 06 333
335	59 36 052	43 37 130	47 23 168	17 15 196	53 09 269	33 58 309	50 39 332
336	60 22 051	44 22 131	47 35 169	16 59 196	52 10 269	33 12 309	50 11 332
337	61 08 050	45 06 131	47 45 170	16 42 197	51 11 269	32 27 309	49 43 331
338	61 52 049	45 50 132	47 54 172	16 24 197	50 13 269	31 41 308	49 14 330
339	62 36 048	46 33 133	48 02 173	16 07 198	49 14 270	30 55 308	48 44 329
340	63 19 047	47 15 134	48 08 174	15 48 198	48 15 270	30 08 308	48 14 328
341	64 02 046	47 57 135	48 14 176	15 30 199	47 17 270	29 22 308	47 43 328
342	64 43 045	48 38 136	48 17 177	15 10 199	46 18 270	28 36 308	47 11 327
343	65 24 043	49 19 137	48 20 178	14 51 200	45 19 271	27 50 308	46 39 326
344	66 04 042	49 58 138	48 21 180	14 31 200	44 21 271	27 03 308	46 06 326
	Schedar	◆Mirfak	Hamal	◆ACHERNAR	FOMALHAUT	◆ALTAIR	DENEB
345	41 29 018	24 33 040	44 16 069	13 32 159	48 21 181	43 22 271	45 33 325
346	41 47 018	25 11 040	45 10 069	13 52 160	48 19 182	42 23 271	44 59 325
347	42 04 017	25 49 040	46 05 069	14 12 160	48 16 183	41 24 271	44 25 324
348	42 21 016	26 27 040	47 00 068	14 32 161	48 12 185	40 26 271	43 50 323
349	42 37 016	27 04 040	47 54 068	14 51 161	48 07 186	39 27 272	43 15 323
350	42 52 015	27 42 040	48 49 068	15 10 162	48 00 187	38 28 272	42 39 322
351	43 07 014	28 19 039	49 43 068	15 29 162	47 52 189	37 30 272	42 03 322
352	43 21 014	28 56 039	50 37 068	15 46 162	47 42 190	36 31 272	41 26 321
353	43 35 013	29 33 039	51 31 067	16 04 163	47 31 191	35 33 272	40 50 321
354	43 48 012	30 10 039	52 26 067	16 21 163	47 19 192	34 34 273	40 12 320
355	44 00 011	30 47 038	53 19 067	16 37 164	47 06 194	33 35 273	39 35 320
356	44 11 011	31 23 038	54 13 066	16 53 164	46 52 195	32 37 273	38 57 320
357	44 22 010	31 59 038	55 07 066	17 09 165	46 36 196	31 38 273	38 19 319
358	44 32 009	32 35 038	56 01 066	17 24 165	46 19 197	30 39 273	37 40 319
359	44 41 009	33 11 037	56 54 065	17 38 166	46 01 198	29 41 273	37 02 319

LHA	Hc	Zn	Hc	Zn	Hc	Zn	Hc	Zn	Hc	Zn	Hc	Zn	Hc	Zn
♈	CAPELLA		◆ALDEBARAN		Acamar		ACHERNAR		◆FOMALHAUT		ALTAIR		◆DENEB	
0	15 27	045	23 08	077	24 13	144	18 51	166	46 38	200	28 38	274	35 38	319
1	16 09	045	24 05	077	24 47	145	19 04	167	46 18	201	27 39	274	34 59	318
2	16 50	045	25 03	077	25 21	145	19 17	167	45 56	202	26 41	274	34 20	318
3	17 32	045	26 00	077	25 54	146	19 30	168	45 33	203	25 42	275	33 40	318
4	18 14	045	26 57	077	26 27	146	19 42	168	45 09	204	24 43	275	33 00	317
5	18 55	045	27 55	077	26 59	147	19 54	169	44 44	206	23 45	275	32 20	317
6	19 37	045	28 52	077	27 31	148	20 05	170	44 18	207	22 46	275	31 40	317
7	20 19	045	29 50	077	28 03	148	20 15	170	43 52	208	21 47	275	31 00	317
8	21 00	045	30 47	077	28 33	149	20 25	171	43 24	209	20 49	275	30 19	316
9	21 41	045	31 45	077	29 03	150	20 34	171	42 55	210	19 50	276	29 39	316
10	22 23	045	32 42	077	29 33	150	20 43	172	42 25	211	18 51	276	28 58	316
11	23 04	044	33 39	077	30 02	151	20 51	172	41 55	212	17 53	276	28 17	316
12	23 45	044	34 37	077	30 30	152	20 59	173	41 24	213	16 54	276	27 36	316
13	24 26	044	35 34	077	30 58	152	21 06	173	40 52	213	15 56	276	26 55	315
14	25 08	044	36 32	078	31 25	153	21 12	174	40 19	214	14 57	276	26 13	315
	CAPELLA		◆ALDEBARAN		RIGEL		ACHERNAR		◆FOMALHAUT		Enif		◆DENEB	
15	25 48	044	37 29	078	23 54	104	21 18	175	39 45	215	41 50	273	25 32	315
16	26 29	044	38 27	078	24 51	104	21 23	175	39 11	216	40 51	274	24 50	315
17	27 10	044	39 24	078	25 48	105	21 28	176	38 36	217	39 52	274	24 09	315
18	27 51	043	40 22	078	26 45	105	21 32	176	38 00	218	38 54	274	23 27	315
19	28 31	043	41 20	078	27 42	105	21 36	177	37 24	218	37 55	274	22 45	315
20	29 11	043	42 17	078	28 39	106	21 39	177	36 47	219	36 56	274	22 03	315
21	29 51	043	43 15	078	29 35	106	21 41	178	36 09	220	35 57	274	21 21	315
22	30 31	043	44 12	078	30 32	106	21 43	179	35 31	221	34 59	274	20 39	314
23	31 11	042	45 10	078	31 28	107	21 44	179	34 53	221	34 00	275	19 57	314
24	31 51	042	46 07	078	32 24	107	21 44	180	34 13	222	33 01	275	19 15	314
25	32 30	042	47 05	078	33 21	108	21 44	180	33 34	223	32 03	275	18 33	314
26	33 09	042	48 02	077	34 17	108	21 43	181	32 53	223	31 04	275	17 50	314
27	33 48	041	49 00	077	35 13	108	21 42	182	32 13	224	30 05	275	17 08	314
28	34 27	041	49 57	077	36 09	109	21 40	182	31 31	225	29 07	275	16 26	314
29	35 05	041	50 55	077	37 04	109	21 38	183	30 50	225	28 08	275	15 44	314
	Mirfak		◆CAPELLA		ALDEBARAN		RIGEL		◆ACHERNAR		FOMALHAUT		◆Alpheratz	
30	47 28	020	35 44	040	51 52	077	38 00	110	21 35	183	30 08	226	58 16	309
31	47 48	019	36 22	040	52 50	077	38 55	110	21 31	184	29 25	227	57 30	308
32	48 06	018	36 59	040	53 47	077	39 50	111	21 27	184	28 42	227	56 43	307
33	48 25	017	37 37	039	54 44	077	40 45	111	21 22	185	27 59	228	55 56	307
34	48 42	017	38 14	039	55 42	077	41 40	112	21 17	186	27 15	228	55 09	306
35	48 58	016	38 50	038	56 39	077	42 35	112	21 11	186	26 31	229	54 21	305
36	49 14	015	39 27	038	57 36	076	43 29	113	21 04	187	25 47	229	53 33	305
37	49 28	014	40 03	037	58 34	076	44 24	113	20 57	187	25 02	230	52 44	304
38	49 42	013	40 38	037	59 31	076	45 18	114	20 49	188	24 17	230	51 55	304
39	49 55	012	41 14	036	60 28	076	46 11	114	20 41	188	23 31	231	51 06	303
40	50 06	011	41 48	036	61 25	076	47 05	115	20 32	189	22 46	231	50 16	303
41	50 17	010	42 23	035	62 22	075	47 58	116	20 23	189	22 00	232	49 27	302
42	50 27	009	42 57	035	63 19	075	48 51	116	20 13	190	21 13	232	48 37	302
43	50 36	008	43 30	034	64 16	075	49 44	117	20 02	191	20 27	232	47 47	301
44	50 44	007	44 03	034	65 13	074	50 36	118	19 51	191	19 40	233	46 56	301
	CAPELLA		◆POLLUX		PROCYON		SIRIUS		◆ACHERNAR		Diphda		◆Alpheratz	
45	44 36	033	21 36	064	20 50	089	27 53	116	19 39	192	45 31	230	46 06	301
46	45 08	033	22 29	064	21 49	089	28 46	116	19 27	192	44 46	230	45 15	300
47	45 39	032	23 22	064	22 48	089	29 39	117	19 14	193	44 00	231	44 24	300
48	46 10	031	24 15	064	23 46	089	30 31	117	19 01	193	43 14	232	43 33	300
49	46 40	030	25 08	064	24 45	089	31 24	118	18 47	194	42 27	233	42 42	300
50	47 10	030	26 01	064	25 44	089	32 16	118	18 33	194	41 40	234	41 51	299
51	47 38	029	26 54	064	26 43	090	33 08	118	18 18	195	40 52	234	41 00	299
52	48 07	028	27 47	064	27 42	090	34 00	119	18 03	195	40 04	235	40 08	299
53	48 34	027	28 40	064	28 41	090	34 51	119	17 47	196	39 16	236	39 16	299
54	49 01	027	29 33	064	29 40	090	35 42	120	17 31	196	38 27	236	38 25	299
55	49 27	026	30 25	064	30 39	090	36 33	121	17 14	197	37 38	237	37 33	298
56	49 52	025	31 18	064	31 38	091	37 23	121	16 57	197	36 49	237	36 41	298
57	50 17	024	32 11	064	32 37	091	38 14	122	16 39	198	35 59	238	35 49	298
58	50 41	023	33 04	064	33 35	091	39 04	122	16 21	198	35 09	239	34 57	298
59	51 03	022	33 57	063	34 34	091	39 53	123	16 03	199	34 18	239	34 05	298
	CAPELLA		◆POLLUX		PROCYON		SIRIUS		◆CANOPUS		Diphda		◆Hamal	
60	51 25	021	34 49	063	35 33	091	40 42	124	19 15	158	33 28	240	60 21	299
61	51 46	020	35 42	063	36 32	092	41 31	124	19 37	158	32 37	240	59 29	298
62	52 06	019	36 34	063	37 31	092	42 20	125	19 59	159	31 46	241	58 37	297
63	52 25	018	37 27	063	38 30	092	43 08	126	20 20	159	30 54	241	57 44	297
64	52 43	017	38 19	063	39 29	092	43 55	126	20 40	160	30 03	242	56 52	296
65	53 01	016	39 12	063	40 27	092	44 43	127	21 00	160	29 11	242	55 59	296
66	53 17	015	40 04	062	41 26	093	45 29	128	21 20	161	28 19	242	55 06	295
67	53 32	014	40 56	062	42 25	093	46 15	129	21 38	162	27 26	243	54 12	295
68	53 45	013	41 48	062	43 24	093	47 01	130	21 57	162	26 34	243	53 19	295
69	53 58	012	42 40	062	44 23	093	47 46	131	22 15	163	25 41	244	52 25	294
70	54 10	011	43 32	062	45 22	094	48 31	131	22 32	163	24 48	244	51 32	294
71	54 20	010	44 24	061	46 20	094	49 14	132	22 48	164	23 55	244	50 38	294
72	54 29	008	45 15	061	47 19	094	49 58	133	23 04	164	23 02	245	49 44	293
73	54 38	007	46 07	061	48 18	094	50 40	134	23 20	165	22 08	245	48 50	293
74	54 44	006	46 58	060	49 16	095	51 22	135	23 35	166	21 15	246	47 56	293
	CAPELLA		◆POLLUX		PROCYON		SIRIUS		◆CANOPUS		Acamar		◆Hamal	
75	54 50	005	47 49	060	50 15	095	52 03	137	23 49	166	31 26	207	47 01	293
76	54 55	004	48 40	060	51 14	095	52 43	138	24 03	167	30 59	208	46 07	292
77	54 58	003	49 31	059	52 12	096	53 22	139	24 16	168	30 31	208	45 12	292
78	55 00	001	50 22	059	53 11	096	54 00	140	24 28	168	30 03	209	44 18	292
79	55 00	000	51 12	059	54 10	096	54 38	141	24 40	169	29 34	210	43 23	292
80	55 00	359	52 02	058	55 08	097	55 14	143	24 51	169	29 04	210	42 29	292
81	54 58	358	52 52	058	56 07	097	55 49	144	25 02	170	28 34	211	41 34	292
82	54 55	356	53 41	057	57 05	097	56 24	145	25 11	171	28 04	212	40 39	292
83	54 51	355	54 31	056	58 03	098	56 57	147	25 21	171	27 32	212	39 44	291
84	54 45	354	55 20	056	59 02	098	57 28	148	25 29	172	27 01	213	38 49	291
85	54 39	353	56 08	055	60 00	099	57 59	150	25 37	173	26 28	214	37 55	291
86	54 31	352	56 56	055	60 58	099	58 28	151	25 44	173	25 55	214	37 00	291
87	54 22	351	57 44	054	61 56	100	58 55	153	25 51	174	25 22	215	36 05	291
88	54 11	349	58 32	053	62 54	100	59 21	155	25 56	175	24 48	215	35 10	291
89	54 00	348	59 19	052	63 52	101	59 46	156	26 02	175	24 14	216	34 15	291

LAT 11°N

LHA	Hc	Zn	Hc	Zn	Hc	Zn	Hc	Zn	Hc	Zn	Hc	Zn	Hc	Zn
♈	Dubhe		◆REGULUS		Suhail		SIRIUS		◆RIGEL		ALDEBARAN		◆CAPELLA	
90	16 21	029	29 21	082	20 50	145	60 09	158	67 41	211	68 48	287	53 47	347
91	16 49	028	30 19	082	21 23	146	60 30	160	67 09	213	67 52	287	53 34	346
92	17 17	028	31 17	083	21 56	146	60 50	162	66 36	215	66 55	286	53 19	345
93	17 45	028	32 16	083	22 28	147	61 07	164	66 01	217	65 59	286	53 03	344
94	18 13	028	33 14	083	23 00	147	61 23	165	65 25	219	65 02	286	52 46	343
95	18 41	028	34 13	083	23 32	148	61 37	167	64 47	221	64 05	285	52 28	342
96	19 09	028	35 11	083	24 03	149	61 49	169	64 07	223	63 09	285	52 09	341
97	19 36	028	36 10	083	24 33	149	61 59	171	63 27	224	62 12	285	51 50	340
98	20 04	028	37 08	083	25 03	150	62 06	173	62 45	226	61 15	284	51 29	339
99	20 31	028	38 07	083	25 32	150	62 12	175	62 02	227	60 17	284	51 07	338
100	20 59	028	39 05	083	26 01	151	62 16	177	61 18	229	59 20	284	50 44	337
101	21 26	027	40 04	084	26 30	152	62 17	180	60 34	230	58 23	284	50 21	336
102	21 53	027	41 02	084	26 57	152	62 17	182	59 48	231	57 26	283	49 56	335
103	22 20	027	42 01	084	27 25	153	62 14	184	59 01	233	56 28	283	49 31	334
104	22 46	027	42 59	084	27 51	153	62 09	186	58 14	234	55 31	283	49 05	333
	Dubhe		◆REGULUS		Suhail		◆SIRIUS		RIGEL		ALDEBARAN		◆CAPELLA	
105	23 13	027	43 58	084	28 17	154	62 02	188	57 26	235	54 34	283	48 39	333
106	23 39	027	44 56	084	28 43	155	61 54	190	56 38	236	53 36	283	48 11	332
107	24 06	026	45 55	084	29 07	155	61 43	192	55 49	237	52 39	283	47 43	331
108	24 32	026	46 53	084	29 32	156	61 30	194	54 59	238	51 42	283	47 14	330
109	24 57	026	47 52	084	29 55	157	61 15	196	54 09	239	50 44	283	46 45	330
110	25 23	026	48 51	084	30 18	158	60 58	198	53 18	240	49 47	283	46 15	329
111	25 48	025	49 49	084	30 40	158	60 39	199	52 27	241	48 49	283	45 44	328
112	26 14	025	50 48	084	31 01	159	60 19	201	51 36	241	47 52	282	45 13	328
113	26 39	025	51 46	085	31 22	160	59 57	203	50 44	242	46 54	282	44 41	327
114	27 03	025	52 45	085	31 42	161	59 33	205	49 52	243	45 57	282	44 08	326
115	27 28	024	53 44	085	32 01	161	59 08	206	48 59	244	44 59	282	43 36	326
116	27 52	024	54 42	085	32 20	162	58 41	208	48 06	244	44 02	282	43 02	325
117	28 16	024	55 41	085	32 38	163	58 12	210	47 13	245	43 04	282	42 28	325
118	28 40	024	56 40	085	32 54	164	57 43	211	46 19	246	42 07	282	41 54	324
119	29 03	023	57 38	085	33 11	164	57 11	213	45 26	246	41 09	282	41 19	324
	Dubhe		◆Denebola		Suhail		◆SIRIUS		RIGEL		ALDEBARAN		◆CAPELLA	
120	29 26	023	34 14	080	33 26	165	56 39	214	44 32	247	40 11	282	40 44	323
121	29 49	023	35 12	080	33 41	166	56 05	215	43 37	247	39 14	282	40 09	323
122	30 11	022	36 10	080	33 54	167	55 31	217	42 43	248	38 16	282	39 33	322
123	30 34	022	37 08	080	34 07	168	54 55	218	41 48	248	37 19	282	38 56	322
124	30 55	022	38 06	080	34 19	169	54 18	219	40 54	249	36 21	282	38 20	321
125	31 17	021	39 04	080	34 30	169	53 40	221	39 59	249	35 24	283	37 43	321
126	31 38	021	40 02	080	34 41	170	53 01	222	39 03	250	34 26	283	37 05	320
127	31 59	020	41 00	080	34 50	171	52 21	223	38 08	250	33 29	283	36 28	320
128	32 19	020	41 58	080	34 59	172	51 41	224	37 12	251	32 31	283	35 50	320
129	32 39	020	42 56	080	35 06	173	50 59	225	36 17	251	31 34	283	35 12	319
130	32 59	019	43 54	080	35 13	174	50 17	226	35 21	252	30 37	283	34 33	319
131	33 18	019	44 53	080	35 19	175	49 35	227	34 25	252	29 39	283	33 54	319
132	33 37	018	45 51	080	35 24	176	48 51	228	33 29	252	28 42	283	33 16	318
133	33 55	018	46 49	080	35 28	176	48 07	229	32 33	253	27 44	283	32 36	318
134	34 13	018	47 47	080	35 31	177	47 22	230	31 36	253	26 47	283	31 57	318
	Dubhe		◆ARCTURUS		SPICA		◆Suhail		SIRIUS		BETELGEUSE		◆CAPELLA	
135	34 31	017	14 00	073	20 34	106	35 34	178	46 37	231	44 13	270	31 17	318
136	34 48	017	14 56	073	21 31	107	35 35	179	45 51	232	43 14	270	30 38	317
137	35 05	016	15 52	073	22 27	107	35 35	180	45 04	232	42 16	270	29 58	317
138	35 21	016	16 49	073	23 23	107	35 35	181	44 17	233	41 17	270	29 18	317
139	35 37	015	17 45	073	24 19	108	35 33	182	43 30	234	40 18	270	28 37	317
140	35 52	015	18 42	073	25 15	108	35 31	183	42 42	235	39 19	271	27 57	317
141	36 07	014	19 38	073	26 11	108	35 28	184	41 54	235	38 20	271	27 16	316
142	36 21	014	20 34	073	27 07	109	35 24	184	41 05	236	37 21	271	26 36	316
143	36 35	013	21 31	074	28 03	109	35 19	185	40 16	237	36 22	271	25 55	316
144	36 48	013	22 27	074	28 59	109	35 13	186	39 27	237	35 23	271	25 14	316
145	37 00	012	23 24	074	29 54	110	35 06	187	38 37	238	34 24	271	24 33	316
146	37 13	012	24 20	074	30 49	110	34 58	188	37 47	239	33 26	272	23 52	316
147	37 24	011	25 17	074	31 45	111	34 49	189	36 56	239	32 27	272	23 11	316
148	37 35	011	26 14	074	32 40	111	34 40	190	36 06	240	31 28	272	22 29	315
149	37 46	010	27 10	074	33 35	111	34 29	191	35 15	240	30 29	272	21 48	315
	◆Dubhe		Alkaid		ARCTURUS		◆SPICA		Suhail		◆SIRIUS		POLLUX	
150	37 56	009	29 39	039	28 07	074	34 29	112	34 18	191	34 23	241	54 06	303
151	38 05	009	30 16	039	29 03	074	35 24	112	34 06	192	33 32	241	53 17	303
152	38 14	008	30 52	038	30 00	074	36 18	113	33 53	193	32 40	242	52 27	302
153	38 22	008	31 29	038	30 57	074	37 12	113	33 39	194	31 48	242	51 37	302
154	38 30	007	32 05	038	31 53	074	38 07	114	33 25	195	30 56	243	50 47	301
155	38 37	007	32 41	037	32 50	074	39 00	114	33 09	196	30 03	243	49 57	301
156	38 43	006	33 17	037	33 47	074	39 54	115	32 53	196	29 11	244	49 06	300
157	38 49	005	33 52	037	34 43	074	40 47	115	32 36	197	28 18	244	48 15	300
158	38 54	005	34 27	037	35 40	074	41 40	116	32 18	198	27 25	244	47 24	300
159	38 59	004	35 02	036	36 37	074	42 33	116	32 00	199	26 31	245	46 33	299
160	39 03	004	35 37	036	37 33	074	43 26	117	31 40	200	25 38	245	45 41	299
161	39 06	003	36 11	035	38 30	074	44 18	118	31 20	200	24 44	246	44 50	299
162	39 09	002	36 45	035	39 26	074	45 10	118	30 59	201	23 51	246	43 58	299
163	39 11	002	37 19	035	40 23	074	46 02	119	30 38	202	22 57	246	43 06	298
164	39 12	001	37 52	034	41 20	074	46 53	120	30 16	202	22 03	247	42 15	298
	◆Dubhe		ARCTURUS		◆SPICA		ACRUX		Suhail		◆PROCYON		POLLUX	
165	39 13	001	42 16	074	47 44	120	14 05	170	29 53	203	39 59	268	41 23	298
166	39 13	000	43 13	074	48 35	121	14 15	171	29 29	204	39 00	268	40 30	298
167	39 13	359	44 10	074	49 25	122	14 24	171	29 05	205	38 02	268	39 38	297
168	39 12	359	45 06	074	50 15	123	14 33	171	28 40	205	37 03	268	38 46	297
169	39 10	358	46 03	074	51 04	124	14 42	172	28 15	206	36 04	268	37 53	297
170	39 08	357	46 59	074	51 53	125	14 50	172	27 49	207	35 05	269	37 01	297
171	39 05	357	47 56	073	52 41	125	14 57	173	27 22	207	34 06	269	36 08	297
172	39 02	356	48 52	073	53 29	126	15 05	173	26 55	208	33 07	269	35 16	297
173	38 57	356	49 48	073	54 16	127	15 11	174	26 27	209	32 08	269	34 23	297
174	38 53	355	50 45	073	55 02	128	15 18	174	25 59	209	31 09	269	33 30	296
175	38 47	354	51 41	073	55 48	130	15 23	175	25 30	210	30 10	270	32 38	296
176	38 41	354	52 37	073	56 33	131	15 29	175	25 00	210	29 12	270	31 45	296
177	38 35	353	53 34	072	57 18	132	15 34	176	24 30	211	28 13	270	30 52	296
178	38 27	353	54 30	072	58 01	133	15 38	176	24 00	211	27 14	270	29 59	296
179	38 20	352	55 26	072	58 44	134	15 42	176	23 29	212	26 15	270	29 06	296

LAT 11°N

LHA	Hc	Zn	Hc	Zn	Hc	Zn	Hc	Zn	Hc	Zn	Hc	Zn	Hc	Zn
♈	Alkaid		◆Alphecca		ANTARES		◆ACRUX		REGULUS		◆POLLUX		Dubhe	
180	45 41	025	37 17	065	14 45	121	15 45	177	62 35	275	28 13	296	38 11	352
181	46 05	024	38 11	065	15 36	122	15 48	177	61 36	275	27 21	296	38 02	351
182	46 29	023	39 04	064	16 26	122	15 51	178	60 38	275	26 28	296	37 53	350
183	46 52	023	39 57	064	17 15	122	15 53	178	59 39	275	25 35	296	37 43	350
184	47 15	022	40 50	064	18 05	123	15 54	179	58 40	275	24 42	296	37 32	349
185	47 36	021	41 43	064	18 55	123	15 55	179	57 41	275	23 49	296	37 21	349
186	47 57	020	42 36	064	19 44	123	15 56	180	56 43	275	22 56	296	37 09	348
187	48 17	019	43 28	063	20 33	124	15 56	180	55 44	275	22 03	296	36 57	348
188	48 36	019	44 21	063	21 22	124	15 55	181	54 46	275	21 10	296	36 44	347
189	48 54	018	45 13	063	22 10	125	15 54	181	53 47	275	20 17	296	36 31	347
190	49 12	017	46 06	063	22 59	125	15 53	182	52 48	275	19 24	296	36 17	346
191	49 28	016	46 58	062	23 47	126	15 51	182	51 50	275	18 31	296	36 02	346
192	49 44	015	47 50	062	24 34	126	15 49	183	50 51	276	17 38	296	35 47	345
193	49 59	014	48 42	062	25 22	126	15 46	183	49 52	276	16 45	296	35 32	345
194	50 13	013	49 34	061	26 09	127	15 42	183	48 54	276	15 53	296	35 16	344
	◆VEGA		Rasalhague		ANTARES		◆RIGIL KENT.		Gienah		◆REGULUS		Dubhe	
195	11 21	052	22 57	081	26 56	127	15 32	168	59 27	201	47 55	276	35 00	344
196	12 08	052	23 55	081	27 43	128	15 45	168	59 05	203	46 57	276	34 43	343
197	12 55	052	24 53	081	28 29	128	15 56	169	58 41	205	45 58	276	34 26	343
198	13 41	052	25 52	081	29 15	129	16 08	169	58 16	206	44 59	276	34 08	342
199	14 28	052	26 50	081	30 01	129	16 19	170	57 49	208	44 01	276	33 50	342
200	15 15	053	27 48	081	30 46	130	16 29	170	57 21	209	43 02	276	33 32	341
201	16 02	053	28 46	082	31 31	131	16 39	171	56 51	211	42 04	276	33 13	341
202	16 48	053	29 45	082	32 15	131	16 48	171	56 20	212	41 05	276	32 53	341
203	17 35	053	30 43	082	33 00	132	16 57	172	55 48	214	40 07	276	32 33	340
204	18 22	053	31 41	082	33 43	132	17 06	172	55 15	215	39 08	277	32 13	340
205	19 09	053	32 39	082	34 26	133	17 14	172	54 40	216	38 10	277	31 53	339
206	19 55	053	33 38	082	35 09	134	17 21	173	54 05	218	37 11	277	31 32	339
207	20 42	053	34 36	082	35 52	134	17 28	173	53 28	219	36 13	277	31 11	339
208	21 29	052	35 35	082	36 33	135	17 34	174	52 51	220	35 14	277	30 49	338
209	22 16	052	36 33	082	37 15	136	17 40	174	52 12	221	34 16	277	30 27	338
	◆VEGA		Rasalhague		ANTARES		◆RIGIL KENT.		SPICA		◆REGULUS		Dubhe	
210	23 02	052	37 31	083	37 55	137	17 46	175	66 13	202	33 17	277	30 05	338
211	23 49	052	38 30	083	38 36	137	17 51	175	65 50	204	32 19	277	29 42	337
212	24 35	052	39 28	083	39 15	138	17 55	176	65 25	206	31 20	277	29 19	337
213	25 22	052	40 27	083	39 54	139	17 59	177	64 58	208	30 22	278	28 56	337
214	26 08	052	41 25	083	40 33	140	18 02	177	64 29	210	29 24	278	28 33	336
215	26 55	052	42 23	083	41 11	141	18 05	178	63 59	212	28 25	278	28 09	336
216	27 41	052	43 22	083	41 48	141	18 07	178	63 27	214	27 27	278	27 45	336
217	28 27	052	44 20	083	42 24	142	18 09	179	62 53	216	26 29	278	27 21	336
218	29 14	052	45 19	083	43 00	143	18 10	179	62 18	218	25 30	278	26 56	335
219	30 00	051	46 17	083	43 34	144	18 11	180	61 41	219	24 32	278	26 31	335
220	30 46	051	47 16	083	44 09	145	18 11	180	61 03	221	23 34	278	26 06	335
221	31 31	051	48 14	083	44 42	146	18 11	181	60 24	222	22 36	279	25 41	334
222	32 17	051	49 13	083	45 14	147	18 10	181	59 44	224	21 37	279	25 16	334
223	33 03	051	50 11	084	45 46	148	18 08	182	59 03	225	20 39	279	24 50	334
224	33 48	050	51 10	084	46 16	149	18 06	182	58 21	226	19 41	279	24 24	334
	◆VEGA		ALTAIR		ANTARES		◆RIGIL KENT.		SPICA		◆Denebola		Alkaid	
225	34 33	050	18 36	084	46 46	150	18 04	183	57 38	228	43 20	280	48 48	342
226	35 19	050	19 34	084	47 15	151	18 01	183	56 54	229	42 22	280	48 30	341
227	36 04	050	20 33	085	47 42	153	17 58	184	56 09	230	41 24	280	48 10	340
228	36 48	049	21 32	085	48 09	154	17 53	184	55 24	231	40 26	280	47 50	339
229	37 33	049	22 30	085	48 35	155	17 49	185	54 38	232	39 28	280	47 29	339
230	38 17	049	23 29	085	48 59	156	17 44	185	53 51	233	38 30	280	47 07	338
231	39 01	048	24 28	085	49 22	157	17 38	186	53 03	234	37 32	280	46 45	337
232	39 45	048	25 26	085	49 44	159	17 32	186	52 15	235	36 34	280	46 21	336
233	40 29	048	26 25	085	50 05	160	17 26	187	51 27	236	35 36	280	45 57	336
234	41 12	047	27 24	086	50 25	161	17 18	187	50 38	237	34 38	280	45 33	335
235	41 56	047	28 22	086	50 43	162	17 11	188	49 48	238	33 40	280	45 07	334
236	42 39	046	29 21	086	51 00	164	17 03	188	48 58	238	32 42	280	44 41	333
237	43 21	046	30 20	086	51 16	165	16 54	189	48 08	239	31 44	280	44 15	333
238	44 03	046	31 19	086	51 30	167	16 45	189	47 17	240	30 46	281	43 48	332
239	44 45	045	32 17	086	51 43	168	16 35	190	46 26	241	29 48	281	43 20	332
	◆VEGA		ALTAIR		Nunki		◆ANTARES		SPICA		◆ARCTURUS		Alkaid	
240	45 27	045	33 16	087	33 27	132	51 55	169	45 34	241	63 30	291	42 51	331
241	46 08	044	34 15	087	34 11	133	52 05	171	44 42	242	62 35	291	42 23	330
242	46 49	044	35 14	087	34 54	133	52 14	172	43 50	243	61 40	290	41 53	330
243	47 29	043	36 13	087	35 36	134	52 21	174	42 58	243	60 44	290	41 23	329
244	48 09	042	37 11	087	36 19	135	52 27	175	42 05	244	59 49	289	40 53	329
245	48 48	042	38 10	087	37 00	135	52 31	177	41 12	244	58 53	289	40 22	328
246	49 27	041	39 09	087	37 41	136	52 33	178	40 19	245	57 58	289	39 50	328
247	50 06	040	40 08	088	38 22	137	52 35	180	39 25	246	57 02	288	39 19	327
248	50 44	040	41 07	088	39 02	138	52 34	181	38 32	246	56 06	288	38 47	327
249	51 21	039	42 06	088	39 41	138	52 32	183	37 38	247	55 10	288	38 14	326
250	51 57	038	43 05	088	40 20	139	52 29	184	36 43	247	54 14	288	37 41	326
251	52 33	037	44 03	088	40 58	140	52 24	185	35 49	247	53 18	287	37 08	325
252	53 09	036	45 02	088	41 36	141	52 18	187	34 55	248	52 21	287	36 34	325
253	53 43	036	46 01	089	42 12	142	52 10	188	34 00	248	51 25	287	36 00	324
254	54 17	035	47 00	089	42 48	143	52 01	190	33 05	249	50 29	287	35 25	324
	◆VEGA		ALTAIR		◆Nunki		ANTARES		SPICA		◆ARCTURUS		Alkaid	
255	54 50	034	47 59	089	43 24	144	51 50	191	32 10	249	49 32	287	34 51	324
256	55 23	033	48 58	089	43 58	145	51 38	193	31 15	250	48 36	287	34 16	323
257	55 54	032	49 57	089	44 32	146	51 24	194	30 20	250	47 40	287	33 40	323
258	56 24	031	50 56	089	45 05	147	51 09	195	29 24	250	46 43	286	33 05	323
259	56 54	030	51 54	090	45 37	148	50 53	197	28 29	251	45 47	286	32 29	322
260	57 23	028	52 53	090	46 08	149	50 35	198	27 33	251	44 50	286	31 53	322
261	57 50	027	53 52	090	46 38	150	50 16	199	26 37	251	43 53	286	31 17	322
262	58 16	026	54 51	090	47 08	151	49 56	201	25 41	252	42 57	286	30 40	322
263	58 42	025	55 50	090	47 36	152	49 35	202	24 45	252	42 00	286	30 04	321
264	59 06	023	56 49	091	48 03	153	49 12	203	23 49	252	41 04	286	29 27	321
265	59 28	022	57 48	091	48 29	154	48 48	204	22 53	253	40 07	286	28 49	321
266	59 50	021	58 47	091	48 54	155	48 23	206	21 57	253	39 10	286	28 12	321
267	60 10	019	59 46	091	49 18	157	47 57	207	21 00	253	38 14	286	27 35	320
268	60 29	018	60 45	091	49 41	158	47 30	208	20 04	254	37 17	286	26 57	320
269	60 46	016	61 43	092	50 02	159	47 02	209	19 07	254	36 20	286	26 19	320

LHA	Hc	Zn	Hc	Zn	Hc	Zn	Hc	Zn	Hc	Zn	Hc	Zn	Hc	Zn
♈	VEGA		◆DENEB		Enif		◆Nunki		ANTARES		◆ARCTURUS		Alkaid	
270	61 02	015	41 29	037	35 00	086	50 23	160	46 33	210	35 24	286	25 41	320
271	61 17	013	42 04	037	35 59	086	50 42	162	46 03	211	34 27	286	25 03	320
272	61 29	012	42 40	036	36 58	086	50 59	163	45 32	212	33 30	286	24 25	320
273	61 41	010	43 14	036	37 57	086	51 16	165	45 00	213	32 34	286	23 47	319
274	61 50	009	43 49	035	38 55	086	51 31	166	44 27	214	31 37	286	23 08	319
275	61 58	007	44 22	035	39 54	086	51 44	167	43 53	215	30 41	286	22 30	319
276	62 05	005	44 56	034	40 53	086	51 57	169	43 19	216	29 44	286	21 51	319
277	62 09	004	45 28	033	41 52	087	52 08	170	42 44	217	28 47	286	21 13	319
278	62 12	002	46 01	033	42 51	087	52 17	172	42 08	218	27 51	286	20 34	319
279	62 13	000	46 32	032	43 49	087	52 25	173	41 31	219	26 54	286	19 55	319
280	62 13	359	47 03	031	44 48	087	52 31	174	40 54	220	25 57	286	19 16	319
281	62 10	357	47 34	031	45 47	087	52 36	176	40 16	221	25 01	286	18 37	319
282	62 07	355	48 03	030	46 46	087	52 40	177	39 37	221	24 04	286	17 58	319
283	62 01	354	48 32	029	47 45	087	52 41	179	38 58	222	23 08	286	17 19	318
284	61 54	352	49 01	028	48 43	088	52 42	180	38 18	223	22 11	286	16 40	318
	◆DENEB		Alpheratz		◆FOMALHAUT		Nunki		ANTARES		◆Rasalhague		VEGA	
285	49 28	028	16 34	063	19 55	127	52 41	182	37 37	224	69 04	276	61 45	350
286	49 55	027	17 27	063	20 41	128	52 38	183	36 56	224	68 06	276	61 34	349
287	50 21	026	18 19	063	21 28	128	52 34	185	36 15	225	67 07	276	61 22	347
288	50 47	025	19 12	063	22 14	129	52 28	186	35 33	226	66 09	276	61 08	346
289	51 11	024	20 04	063	23 00	129	52 21	188	34 50	227	65 10	276	60 53	344
290	51 35	023	20 56	063	23 46	130	52 12	189	34 07	227	64 12	276	60 36	343
291	51 58	022	21 49	063	24 31	130	52 02	191	33 24	228	63 13	276	60 18	341
292	52 19	021	22 41	063	25 16	130	51 51	192	32 40	228	62 14	276	59 58	340
293	52 40	020	23 34	063	26 00	131	51 38	193	31 55	229	61 16	276	59 37	338
294	53 00	019	24 26	063	26 45	131	51 23	195	31 11	230	60 17	276	59 15	337
295	53 19	018	25 19	063	27 29	132	51 07	196	30 26	230	59 19	276	58 51	336
296	53 37	017	26 11	063	28 12	133	50 50	198	29 40	231	58 20	276	58 26	334
297	53 53	016	27 03	063	28 56	133	50 32	199	28 54	231	57 22	276	58 00	333
298	54 09	015	27 56	063	29 38	134	50 12	200	28 08	232	56 23	276	57 33	332
299	54 24	014	28 48	063	30 21	134	49 51	201	27 22	232	55 25	276	57 05	331
	◆DENEB		Alpheratz		◆FOMALHAUT		Peacock		◆ANTARES		Rasalhague		VEGA	
300	54 37	013	29 41	063	31 03	135	22 03	176	26 35	233	54 26	276	56 36	330
301	54 49	011	30 33	063	31 44	135	22 06	177	25 48	233	53 27	276	56 06	329
302	55 00	010	31 25	063	32 25	136	22 09	177	25 01	234	52 29	276	55 35	328
303	55 10	009	32 17	062	33 06	137	22 12	178	24 13	234	51 30	276	55 03	327
304	55 19	008	33 10	062	33 46	137	22 13	179	23 25	235	50 32	276	54 30	326
305	55 26	007	34 02	062	34 26	138	22 14	179	22 37	235	49 33	277	53 57	325
306	55 32	005	34 54	062	35 05	139	22 15	180	21 48	236	48 35	277	53 22	324
307	55 37	004	35 46	062	35 43	140	22 15	180	21 00	236	47 36	277	52 47	323
308	55 41	003	36 38	062	36 21	140	22 14	181	20 11	236	46 38	277	52 11	322
309	55 43	002	37 30	062	36 59	141	22 13	182	19 22	237	45 39	277	51 35	321
310	55 44	000	38 21	061	37 35	142	22 11	182	18 32	237	44 41	277	50 58	321
311	55 44	359	39 13	061	38 11	143	22 08	183	17 43	237	43 42	277	50 20	320
312	55 42	358	40 05	061	38 47	143	22 05	183	16 53	238	42 44	277	49 42	319
313	55 40	357	40 56	061	39 22	144	22 01	184	16 03	238	41 45	277	49 03	319
314	55 36	355	41 48	061	39 56	145	21 57	185	15 13	238	40 47	277	48 24	318
	Alpheratz		◆Diphda		FOMALHAUT		◆Nunki		Rasalhague		VEGA		◆DENEB	
315	42 39	060	27 44	117	40 29	146	41 53	219	39 49	277	47 44	317	55 30	354
316	43 30	060	28 36	118	41 02	147	41 16	220	38 50	277	47 04	317	55 24	353
317	44 21	060	29 28	118	41 34	148	40 38	220	37 52	277	46 24	316	55 16	352
318	45 12	060	30 20	119	42 05	149	40 00	221	36 53	278	45 43	316	55 07	351
319	46 03	059	31 11	119	42 35	150	39 21	222	35 55	278	45 01	315	54 56	349
320	46 53	059	32 03	120	43 04	151	38 41	223	34 57	278	44 20	315	54 45	348
321	47 44	059	32 54	120	43 33	152	38 01	224	33 58	278	43 37	314	54 32	347
322	48 34	058	33 45	121	44 00	153	37 20	224	33 00	278	42 55	314	54 18	346
323	49 24	058	34 35	121	44 27	154	36 38	225	32 02	278	42 12	313	54 03	345
324	50 13	057	35 25	122	44 52	155	35 57	226	31 03	278	41 29	313	53 47	344
325	51 03	057	36 15	122	45 17	156	35 14	226	30 05	278	40 46	312	53 30	343
326	51 52	056	37 05	123	45 40	157	34 31	227	29 07	278	40 02	312	53 12	341
327	52 41	056	37 54	123	46 03	158	33 48	228	28 08	278	39 18	312	52 53	340
328	53 29	055	38 43	124	46 24	159	33 04	228	27 10	279	38 34	311	52 33	339
329	54 18	055	39 32	125	46 45	160	32 20	229	26 12	279	37 50	311	52 11	338
	◆Alpheratz		Diphda		◆FOMALHAUT		Peacock		ALTAIR		◆VEGA		DENEB	
330	55 06	054	40 20	125	47 04	162	19 28	194	58 04	269	37 05	311	51 49	337
331	55 53	053	41 08	126	47 22	163	19 14	194	57 05	269	36 21	310	51 26	336
332	56 40	053	41 55	127	47 39	164	19 00	195	56 06	270	35 36	310	51 02	336
333	57 27	052	42 42	127	47 54	165	18 44	195	55 07	270	34 51	310	50 38	335
334	58 13	051	43 29	128	48 09	166	18 29	196	54 08	270	34 06	310	50 12	334
335	58 59	050	44 15	129	48 22	168	18 13	196	53 09	270	33 20	309	49 46	333
336	59 44	050	45 01	130	48 34	169	17 56	197	52 10	270	32 35	309	49 18	332
337	60 28	049	45 46	131	48 44	170	17 39	197	51 11	270	31 49	309	48 50	331
338	61 12	048	46 30	131	48 54	172	17 22	198	50 13	271	31 03	309	48 22	331
339	61 55	047	47 14	132	49 02	173	17 04	198	49 14	271	30 17	309	47 52	330
340	62 38	045	47 57	133	49 08	174	16 45	199	48 15	271	29 31	309	47 22	329
341	63 19	044	48 39	134	49 13	176	16 26	199	47 16	271	28 45	308	46 52	328
342	64 00	043	49 21	135	49 17	177	16 07	199	46 17	271	27 59	308	46 21	328
343	64 40	042	50 02	136	49 20	178	15 47	200	45 18	272	27 13	308	45 49	327
344	65 19	040	50 43	137	49 21	180	15 27	200	44 19	272	26 26	308	45 16	326
	Schedar		◆Mirfak		Hamal		◆ACHERNAR		FOMALHAUT		◆ALTAIR		DENEB	
345	40 32	018	23 47	040	43 54	068	14 28	159	49 21	181	43 20	272	44 43	326
346	40 49	017	24 25	040	44 48	068	14 49	160	49 19	182	42 22	272	44 10	325
347	41 07	017	25 03	040	45 43	068	15 09	160	49 16	184	41 23	272	43 36	325
348	41 23	016	25 41	040	46 37	067	15 29	161	49 12	185	40 24	272	43 02	324
349	41 39	015	26 18	039	47 31	067	15 48	161	49 06	186	39 25	272	42 27	323
350	41 54	015	26 55	039	48 26	067	16 07	161	48 59	187	38 26	273	41 51	323
351	42 09	014	27 33	039	49 20	067	16 26	162	48 51	189	37 27	273	41 16	322
352	42 23	013	28 10	039	50 14	066	16 44	162	48 41	190	36 28	273	40 39	322
353	42 36	013	28 47	039	51 08	066	17 01	163	48 30	191	35 30	273	40 03	321
354	42 49	012	29 23	038	52 01	066	17 18	163	48 18	193	34 31	273	39 26	321
355	43 01	011	30 00	038	52 55	065	17 35	164	48 04	194	33 32	273	38 49	321
356	43 12	011	30 36	038	53 49	065	17 51	164	47 50	195	32 33	274	38 11	320
357	43 23	010	31 12	038	54 42	065	18 07	165	47 34	196	31 34	274	37 33	320
358	43 32	009	31 48	037	55 35	064	18 22	165	47 16	198	30 36	274	36 55	319
359	43 41	008	32 23	037	56 28	064	18 37	166	46 58	199	29 37	274	36 17	319

 LAT 10°N

LHA	Hc	Zn	Hc	Zn	Hc	Zn	Hc	Zn	Hc	Zn	Hc	Zn	Hc	Zn
♈	◆CAPELLA		ALDEBARAN		Acamar		◆ACHERNAR		FOMALHAUT		◆ALTAIR		DENEB	
0	14 45	045	22 54	076	25 01	144	19 49	166	47 35	200	28 34	275	34 53	319
1	15 26	045	23 51	076	25 36	144	20 03	167	47 14	202	27 35	275	34 14	319
2	16 08	045	24 49	076	26 10	145	20 16	167	46 51	203	26 36	275	33 35	318
3	16 50	045	25 46	076	26 44	146	20 29	168	46 28	204	25 37	275	32 56	318
4	17 31	045	26 44	076	27 17	146	20 41	168	46 04	205	24 38	275	32 16	318
5	18 13	045	27 41	077	27 50	147	20 53	169	45 38	206	23 39	275	31 36	318
6	18 54	045	28 39	077	28 22	147	21 04	169	45 12	207	22 40	275	30 56	317
7	19 36	045	29 36	077	28 53	148	21 14	170	44 45	208	21 42	276	30 16	317
8	20 17	044	30 34	077	29 25	149	21 24	171	44 16	209	20 43	276	29 36	317
9	20 59	044	31 31	077	29 55	149	21 33	171	43 47	210	19 44	276	28 55	317
10	21 40	044	32 29	077	30 25	150	21 42	172	43 17	211	18 45	276	28 15	316
11	22 21	044	33 26	077	30 54	151	21 51	172	42 46	212	17 47	276	27 34	316
12	23 02	044	34 24	077	31 23	151	21 58	173	42 14	213	16 48	276	26 53	316
13	23 43	044	35 21	077	31 51	152	22 05	173	41 42	214	15 49	276	26 12	316
14	24 24	044	36 19	077	32 18	153	22 12	174	41 08	215	14 50	277	25 30	316
	CAPELLA		◆ALDEBARAN		RIGEL		ACHERNAR		◆FOMALHAUT		Enif		◆DENEB	
15	25 05	044	37 16	077	24 09	104	22 18	175	40 34	216	41 46	274	24 49	315
16	25 46	043	38 14	077	25 06	104	22 23	175	39 59	217	40 47	274	24 08	315
17	26 26	043	39 11	077	26 03	104	22 28	176	39 24	217	39 48	275	23 26	315
18	27 07	043	40 09	077	27 00	105	22 32	176	38 48	218	38 49	275	22 44	315
19	27 47	043	41 06	077	27 58	105	22 36	177	38 11	219	37 50	275	22 03	315
20	28 27	043	42 04	077	28 55	105	22 38	177	37 33	220	36 51	275	21 21	315
21	29 07	042	43 01	077	29 52	106	22 41	178	36 55	220	35 53	275	20 39	315
22	29 47	042	43 59	077	30 48	106	22 43	179	36 17	221	34 54	275	19 57	315
23	30 27	042	44 56	077	31 45	106	22 44	179	35 37	222	33 55	275	19 15	315
24	31 06	042	45 54	077	32 42	107	22 44	180	34 58	223	32 56	275	18 33	315
25	31 45	041	46 51	076	33 38	107	22 44	180	34 17	223	31 57	275	17 51	315
26	32 24	041	47 49	076	34 35	107	22 43	181	33 37	224	30 58	276	17 09	314
27	33 03	041	48 46	076	35 31	108	22 42	182	32 55	225	30 00	276	16 26	314
28	33 42	041	49 43	076	36 27	108	22 40	182	32 14	225	29 01	276	15 44	314
29	34 20	040	50 41	076	37 24	108	22 38	183	31 32	226	28 02	276	15 02	314
	Mirfak		CAPELLA		◆ALDEBARAN		RIGEL		◆ACHERNAR		FOMALHAUT		◆Alpheratz	
30	46 31	020	34 58	040	51 38	076	38 20	109	22 35	183	30 49	226	57 38	310
31	46 51	019	35 35	039	52 36	076	39 15	109	22 31	184	30 06	227	56 52	309
32	47 09	018	36 13	039	53 33	076	40 11	110	22 27	184	29 23	227	56 06	308
33	47 27	017	36 50	039	54 30	076	41 07	110	22 22	185	28 39	228	55 20	308
34	47 44	016	37 27	038	55 27	075	42 02	111	22 16	186	27 55	229	54 33	307
35	48 00	015	38 03	038	56 24	075	42 57	111	22 10	186	27 10	229	53 46	306
36	48 16	015	38 39	037	57 21	075	43 52	112	22 04	187	26 26	230	52 58	306
37	48 30	014	39 15	037	58 19	075	44 47	112	21 56	187	25 41	230	52 10	305
38	48 43	013	39 50	036	59 15	074	45 41	113	21 49	188	24 55	231	51 21	305
39	48 56	012	40 25	036	60 12	074	46 36	113	21 40	188	24 09	231	50 33	304
40	49 08	011	41 00	035	61 09	074	47 30	114	21 31	189	23 23	231	49 44	304
41	49 18	010	41 34	035	62 06	074	48 24	115	21 22	190	22 37	232	48 54	303
42	49 28	009	42 07	034	63 03	073	49 17	115	21 12	190	21 50	232	48 05	303
43	49 36	008	42 41	034	63 59	073	50 11	116	21 01	191	21 03	233	47 15	302
44	49 44	007	43 13	033	64 55	072	51 04	117	20 50	191	20 16	233	46 25	302
	◆CAPELLA		PROCYON		SIRIUS		◆CANOPUS		ACHERNAR		Diphda		◆Alpheratz	
45	43 45	033	20 48	088	28 19	115	13 46	151	20 38	192	46 09	230	45 35	302
46	44 17	032	21 47	088	29 12	116	14 14	151	20 26	193	45 23	231	44 44	301
47	44 48	031	22 46	088	30 06	116	14 42	152	20 13	193	44 37	232	43 54	301
48	45 18	031	23 45	089	30 59	117	15 10	152	20 00	193	43 50	233	43 03	301
49	45 48	030	24 44	089	31 51	117	15 37	153	19 46	194	43 03	234	42 12	300
50	46 17	029	25 43	089	32 44	117	16 04	153	19 31	194	42 15	234	41 21	300
51	46 46	029	26 43	089	33 36	118	16 31	153	19 16	195	41 27	235	40 30	300
52	47 14	028	27 42	089	34 28	118	16 57	154	19 01	195	40 39	236	39 39	300
53	47 41	027	28 41	089	35 20	119	17 23	154	18 45	196	39 50	236	38 47	299
54	48 07	026	29 40	090	36 12	119	17 48	155	18 29	196	39 00	237	37 56	299
55	48 33	025	30 39	090	37 03	120	18 13	155	18 12	197	38 11	237	37 04	299
56	48 58	025	31 38	090	37 54	120	18 38	156	17 54	197	37 21	238	36 12	299
57	49 22	024	32 37	090	38 45	121	19 02	156	17 37	198	36 30	239	35 21	299
58	49 45	023	33 36	090	39 35	122	19 25	157	17 18	198	35 40	239	34 29	299
59	50 08	022	34 35	091	40 26	122	19 48	157	16 59	199	34 49	240	33 37	298
	◆CAPELLA		POLLUX		PROCYON		◆SIRIUS		CANOPUS		◆Diphda		Alpheratz	
60	50 29	021	34 22	063	35 34	091	41 15	123	20 11	158	33 58	240	32 45	298
61	50 50	020	35 15	063	36 33	091	42 05	124	20 33	158	33 06	241	31 53	298
62	51 10	019	36 07	062	37 32	091	42 54	124	20 55	159	32 15	241	31 01	298
63	51 28	018	36 59	062	38 32	091	43 43	125	21 16	159	31 23	242	30 08	298
64	51 46	017	37 52	062	39 31	091	44 31	126	21 37	160	30 31	242	29 16	298
65	52 03	016	38 44	062	40 30	092	45 19	126	21 57	160	29 39	243	28 24	298
66	52 19	015	39 36	062	41 29	092	46 06	127	22 16	161	28 46	243	27 32	298
67	52 33	014	40 28	062	42 28	092	46 53	128	22 35	161	27 53	243	26 39	298
68	52 47	013	41 20	061	43 27	092	47 39	129	22 54	162	27 00	244	25 47	298
69	52 59	012	42 12	061	44 26	092	48 25	130	23 12	163	26 07	244	24 54	298
70	53 11	011	43 03	061	45 25	093	49 10	131	23 29	163	25 14	245	24 02	297
71	53 21	009	43 55	060	46 24	093	49 55	132	23 46	164	24 21	245	23 10	297
72	53 30	008	44 46	060	47 23	093	50 38	132	24 02	164	23 27	245	22 17	297
73	53 38	007	45 37	060	48 22	093	51 22	133	24 18	165	22 33	246	21 25	297
74	53 45	006	46 28	060	49 21	094	52 04	135	24 33	166	21 39	246	20 32	297
	CAPELLA		◆POLLUX		PROCYON		SIRIUS		◆CANOPUS		Acamar		◆Hamal	
75	53 50	005	47 19	059	50 20	094	52 46	136	24 47	166	32 19	207	46 38	294
76	53 55	004	48 10	059	51 19	094	53 27	137	25 01	167	31 52	208	45 43	293
77	53 58	002	49 00	058	52 18	094	54 07	138	25 14	167	31 24	209	44 49	293
78	54 00	001	49 50	058	53 17	095	54 46	139	25 27	168	30 55	209	43 55	293
79	54 00	000	50 40	057	54 16	095	55 24	140	25 39	169	30 26	210	43 00	293
80	54 00	359	51 30	057	55 14	095	56 01	142	25 50	169	29 56	211	42 06	293
81	53 58	358	52 19	056	56 13	096	56 38	143	26 01	170	29 26	211	41 11	292
82	53 55	357	53 08	056	57 12	096	57 13	144	26 11	171	28 55	212	40 17	292
83	53 51	355	53 57	055	58 11	096	57 46	146	26 20	171	28 23	213	39 22	292
84	53 46	354	54 46	055	59 09	097	58 19	147	26 28	172	27 51	213	38 27	292
85	53 39	353	55 34	054	60 08	097	58 50	149	26 36	173	27 18	214	37 33	292
86	53 31	352	56 21	053	61 07	097	59 20	150	26 44	173	26 45	214	36 38	292
87	53 23	351	57 08	053	62 05	098	59 49	152	26 50	174	26 11	215	35 43	292
88	53 12	350	57 55	052	63 04	098	60 15	154	26 56	175	25 37	216	34 48	292
89	53 01	349	58 41	051	64 02	099	60 41	156	27 01	175	25 03	216	33 53	292

LHA	Hc	Zn	Hc	Zn	Hc	Zn	Hc	Zn	Hc	Zn	Hc	Zn	Hc	Zn
♈	◆POLLUX		REGULUS		Suhail		◆CANOPUS		RIGEL		◆ALDEBARAN		CAPELLA	
90	59 27	050	29 12	082	21 39	145	27 06	176	68 32	212	68 29	290	52 49	347
91	60 12	049	30 11	082	22 13	146	27 10	177	67 59	215	67 33	289	52 36	346
92	60 57	048	31 09	082	22 46	146	27 13	177	67 25	217	66 37	289	52 21	345
93	61 40	047	32 08	082	23 18	147	27 15	178	66 49	219	65 41	288	52 05	344
94	62 23	046	33 06	082	23 51	147	27 17	179	66 11	221	64 45	288	51 49	343
95	63 06	045	34 05	082	24 22	148	27 18	179	65 32	222	63 49	287	51 31	342
96	63 47	044	35 03	082	24 54	148	27 18	180	64 51	224	62 52	287	51 13	341
97	64 27	042	36 02	082	25 24	149	27 18	181	64 09	226	61 56	286	50 53	340
98	65 07	041	37 01	082	25 55	149	27 17	181	63 26	227	60 59	286	50 33	339
99	65 45	040	37 59	083	26 24	150	27 15	182	62 42	229	60 02	286	50 11	338
100	66 22	038	38 58	083	26 54	151	27 13	183	61 57	230	59 05	285	49 49	337
101	66 58	036	39 56	083	27 22	151	27 10	183	61 11	232	58 08	285	49 26	336
102	67 32	035	40 55	083	27 50	152	27 06	184	60 25	233	57 11	285	49 02	336
103	68 05	033	41 54	083	28 18	153	27 01	185	59 37	234	56 14	285	48 37	335
104	68 36	031	42 52	083	28 45	153	26 56	185	58 49	235	55 17	285	48 12	334
	Dubhe		◆REGULUS		Suhail		◆CANOPUS		RIGEL		ALDEBARAN		◆CAPELLA	
105	22 19	027	43 51	083	29 11	154	26 50	186	58 00	236	54 20	284	47 45	333
106	22 46	026	44 49	083	29 37	155	26 43	187	57 11	237	53 22	284	47 18	332
107	23 12	026	45 48	083	30 02	155	26 36	187	56 21	238	52 25	284	46 50	332
108	23 38	026	46 47	083	30 26	156	26 28	188	55 30	239	51 28	284	46 22	331
109	24 03	026	47 45	083	30 50	157	26 19	189	54 39	240	50 30	284	45 53	330
110	24 29	025	48 44	083	31 13	157	26 10	189	53 48	241	49 33	284	45 23	329
111	24 54	025	49 43	083	31 36	158	26 00	190	52 56	242	48 36	284	44 53	329
112	25 19	025	50 41	083	31 57	159	25 49	191	52 04	242	47 38	284	44 22	328
113	25 44	025	51 40	083	32 18	160	25 38	191	51 11	243	46 41	283	43 50	327
114	26 09	024	52 39	083	32 39	160	25 26	192	50 18	244	45 43	283	43 18	327
115	26 33	024	53 37	083	32 58	161	25 13	193	49 25	245	44 46	283	42 46	326
116	26 57	024	54 36	083	33 17	162	25 00	193	48 32	245	43 48	283	42 13	326
117	27 21	024	55 35	083	33 35	163	24 46	194	47 38	246	42 51	283	41 39	325
118	27 45	023	56 34	083	33 52	163	24 32	194	46 44	246	41 53	283	41 05	325
119	28 08	023	57 32	083	34 08	164	24 17	195	45 50	247	40 56	283	40 31	324
	Dubhe		◆Denebola		Suhail		◆SIRIUS		RIGEL		ALDEBARAN		◆CAPELLA	
120	28 31	023	34 03	079	34 24	165	57 28	215	44 55	248	39 58	283	39 56	324
121	28 53	022	35 01	079	34 39	166	56 54	216	44 00	248	39 01	283	39 21	323
122	29 16	022	35 59	079	34 53	167	56 18	218	43 05	249	38 03	283	38 45	323
123	29 38	022	36 57	079	35 06	168	55 42	219	42 10	249	37 06	283	38 09	322
124	30 00	021	37 55	079	35 18	168	55 04	220	41 15	250	36 08	283	37 33	322
125	30 21	021	38 54	079	35 29	169	54 25	222	40 19	250	35 11	283	36 56	321
126	30 42	021	39 52	079	35 40	170	53 46	223	39 24	251	34 13	283	36 19	321
127	31 03	020	40 50	079	35 49	171	53 05	224	38 28	251	33 15	283	35 41	321
128	31 23	020	41 48	079	35 58	172	52 24	225	37 32	251	32 18	283	35 04	320
129	31 43	019	42 46	079	36 06	173	51 41	226	36 36	252	31 20	283	34 26	320
130	32 02	019	43 44	079	36 13	174	50 59	227	35 40	252	30 23	283	33 48	320
131	32 21	019	44 42	079	36 19	175	50 15	228	34 43	253	29 25	283	33 09	319
132	32 40	018	45 40	079	36 24	176	49 31	229	33 47	253	28 28	283	32 30	319
133	32 58	018	46 38	079	36 28	176	48 46	230	32 50	253	27 31	283	31 52	319
134	33 16	017	47 36	079	36 31	177	48 00	231	31 54	254	26 33	284	31 12	318
	Dubhe		◆ARCTURUS		SPICA		◆Suhail		SIRIUS		BETELGEUSE		◆CAPELLA	
135	33 34	017	13 42	072	20 51	106	36 33	178	47 14	232	44 13	271	30 33	318
136	33 51	016	14 38	073	21 48	106	36 35	179	46 28	232	43 14	271	29 53	318
137	34 07	016	15 35	073	22 44	107	36 35	180	45 41	233	42 15	271	29 14	318
138	34 23	016	16 31	073	23 41	107	36 35	181	44 53	234	41 16	271	28 34	317
139	34 39	015	17 28	073	24 37	107	36 33	182	44 05	235	40 17	271	27 54	317
140	34 54	015	18 24	073	25 34	108	36 31	183	43 16	235	39 18	271	27 13	317
141	35 08	014	19 21	073	26 30	108	36 28	184	42 28	236	38 19	272	26 33	317
142	35 22	014	20 17	073	27 26	108	36 23	185	41 38	237	37 20	272	25 52	317
143	35 36	013	21 14	073	28 22	109	36 18	185	40 49	237	36 21	272	25 12	316
144	35 49	013	22 10	073	29 18	109	36 12	186	39 59	238	35 22	272	24 31	316
145	36 02	012	23 07	073	30 14	109	36 05	187	39 08	239	34 22	272	23 50	316
146	36 14	011	24 03	073	31 10	110	35 57	188	38 18	239	33 23	272	23 09	316
147	36 25	011	25 00	073	32 05	110	35 49	189	37 27	240	32 24	272	22 28	316
148	36 36	010	25 57	073	33 01	110	35 39	190	36 36	240	31 25	273	21 47	316
149	36 47	010	26 53	073	33 56	111	35 28	191	35 44	241	30 26	273	21 05	316
	Dubhe		◆ARCTURUS		SPICA		ACRUX		◆Suhail		SIRIUS		◆POLLUX	
150	36 56	009	27 50	073	34 51	111	11 44	164	35 17	192	34 52	241	53 33	304
151	37 06	009	28 47	073	35 46	112	12 00	164	35 05	192	34 00	242	52 44	304
152	37 14	008	29 43	073	36 41	112	12 16	165	34 51	193	33 08	242	51 55	303
153	37 23	008	30 40	074	37 36	113	12 31	165	34 37	194	32 16	243	51 05	303
154	37 30	007	31 37	074	38 30	113	12 46	166	34 23	195	31 23	243	50 15	302
155	37 37	006	32 33	074	39 25	113	13 01	166	34 07	196	30 30	244	49 25	302
156	37 43	006	33 30	074	40 19	114	13 15	166	33 50	197	29 37	244	48 35	301
157	37 49	005	34 26	073	41 13	115	13 29	167	33 33	197	28 44	245	47 45	301
158	37 54	005	35 23	073	42 06	115	13 42	167	33 15	198	27 50	245	46 54	301
159	37 59	004	36 20	073	43 00	116	13 55	168	32 56	199	26 57	245	46 03	300
160	38 03	004	37 16	073	43 53	116	14 08	168	32 37	200	26 03	246	45 12	300
161	38 06	003	38 13	073	44 46	117	14 20	168	32 16	200	25 09	246	44 21	300
162	38 09	002	39 10	073	45 38	117	14 31	169	31 55	201	24 15	246	43 29	299
163	38 11	002	40 06	073	46 31	118	14 43	169	31 34	202	23 21	247	42 38	299
164	38 12	001	41 03	073	47 22	119	14 54	170	31 11	203	22 26	247	41 46	299
	Dubhe		◆ARCTURUS		SPICA		◆ACRUX		Suhail		PROCYON		◆POLLUX	
165	38 13	001	41 59	073	48 14	119	15 04	170	30 48	203	40 01	268	40 54	299
166	38 13	000	42 56	073	49 05	120	15 14	171	30 24	204	39 02	269	40 02	298
167	38 13	359	43 52	073	49 56	121	15 23	171	30 00	205	38 03	269	39 10	298
168	38 12	359	44 49	073	50 47	122	15 32	171	29 35	206	37 04	269	38 18	298
169	38 10	358	45 45	073	51 37	123	15 41	172	29 09	206	36 05	269	37 26	298
170	38 08	358	46 42	073	52 26	123	15 49	172	28 42	207	35 06	269	36 33	298
171	38 05	357	47 38	072	53 15	124	15 57	173	28 15	207	34 07	270	35 41	297
172	38 02	356	48 34	072	54 04	125	16 04	173	27 48	208	33 08	270	34 49	297
173	37 58	356	49 31	072	54 52	126	16 11	174	27 20	209	32 09	270	33 56	297
174	37 53	355	50 27	072	55 39	127	16 17	174	26 51	209	31 10	270	33 03	297
175	37 47	355	51 23	072	56 26	128	16 23	175	26 22	210	30 10	270	32 11	297
176	37 42	354	52 19	071	57 12	129	16 29	175	25 52	211	29 11	270	31 18	297
177	37 35	353	53 15	071	57 57	131	16 33	175	25 22	211	28 12	271	30 25	297
178	37 28	353	54 11	071	58 41	132	16 38	176	24 51	212	27 13	271	29 33	297
179	37 20	352	55 07	071	59 25	133	16 42	176	24 19	212	26 14	271	28 40	297

LAT 10°N

LAT 10°N

LHA ♈	Hc	Zn	Hc	Zn	Hc	Zn	Hc	Zn	Hc	Zn	Hc	Zn	Hc	Zn
	◆Alphecca		ANTARES		RIGIL KENT.		◆ACRUX		Suhail		◆REGULUS		Dubhe	
180	36 51	064	15 16	121	12 33	161	16 45	177	23 48	213	62 29	277	37 12	352
181	37 44	064	16 07	121	12 52	162	16 48	177	23 15	213	61 30	277	37 03	351
182	38 37	064	16 57	122	13 10	162	16 51	178	22 43	214	60 31	277	36 54	351
183	39 30	063	17 47	122	13 28	163	16 53	178	22 10	214	59 33	277	36 44	350
184	40 23	063	18 37	122	13 45	163	16 54	179	21 36	215	58 34	277	36 33	349
185	41 16	063	19 27	123	14 03	163	16 55	179	21 02	215	57 35	277	36 22	349
186	42 09	063	20 17	123	14 19	164	16 56	180	20 28	216	56 37	277	36 10	348
187	43 01	063	21 06	124	14 36	164	16 56	180	19 53	216	55 38	277	35 58	348
188	43 53	062	21 55	124	14 52	165	16 55	181	19 18	217	54 39	277	35 45	347
189	44 46	062	22 44	124	15 07	165	16 54	181	18 42	217	53 41	277	35 32	347
190	45 38	062	23 33	125	15 22	165	16 53	182	18 06	218	52 42	277	35 18	346
191	46 30	061	24 21	125	15 37	166	16 51	182	17 30	218	51 43	277	35 04	346
192	47 22	061	25 09	126	15 51	166	16 49	183	16 53	218	50 45	277	34 49	345
193	48 13	061	25 57	126	16 05	167	16 46	183	16 16	219	49 46	277	34 34	345
194	49 05	060	26 45	127	16 18	167	16 42	184	15 39	219	48 47	277	34 18	344
	◆VEGA		Rasalhague		ANTARES		◆RIGIL KENT.		ACRUX		◆REGULUS		Dubhe	
195	10 45	052	22 47	080	27 32	127	16 31	168	16 38	184	47 49	277	34 02	344
196	11 31	052	23 46	081	28 19	127	16 43	168	16 34	184	46 50	277	33 46	343
197	12 18	052	24 44	081	29 06	128	16 55	169	16 29	185	45 51	277	33 29	343
198	13 05	052	25 42	081	29 53	128	17 07	169	16 24	185	44 53	277	33 11	343
199	13 52	052	26 41	081	30 39	129	17 18	170	16 18	186	43 54	277	32 53	342
200	14 38	052	27 39	081	31 24	130	17 28	170	16 12	186	42 55	277	32 35	342
201	15 25	052	28 37	081	32 10	130	17 38	171	16 05	187	41 57	277	32 16	341
202	16 12	052	29 36	081	32 55	131	17 48	171	15 58	187	40 58	277	31 57	341
203	16 59	052	30 34	081	33 39	131	17 57	171	15 51	188	39 59	277	31 37	340
204	17 45	052	31 32	081	34 24	132	18 05	172	15 42	188	39 01	277	31 17	340
205	18 32	052	32 31	081	35 07	133	18 13	172	15 34	189	38 02	277	30 57	340
206	19 19	052	33 29	081	35 51	133	18 21	173	15 25	189	37 04	278	30 36	339
207	20 06	052	34 28	082	36 33	134	18 28	173	15 15	189	36 05	278	30 15	339
208	20 52	052	35 26	082	37 16	135	18 34	174	15 06	190	35 07	278	29 53	339
209	21 39	052	36 25	082	37 58	135	18 40	174	14 55	190	34 08	278	29 31	338
	◆VEGA		Rasalhague		ANTARES		◆RIGIL KENT.		SPICA		◆REGULUS		Dubhe	
210	22 26	052	37 23	082	38 39	136	18 46	175	67 08	203	33 09	278	29 09	338
211	23 12	052	38 22	082	39 20	137	18 50	175	66 44	205	32 11	278	28 47	338
212	23 59	052	39 20	082	40 00	138	18 55	176	66 18	207	31 12	278	28 24	337
213	24 45	052	40 19	082	40 39	138	18 59	176	65 51	209	30 14	278	28 01	337
214	25 31	052	41 17	082	41 18	139	19 02	177	65 21	211	29 15	278	27 38	337
215	26 18	052	42 16	082	41 57	140	19 05	178	64 49	213	28 17	278	27 14	336
216	27 04	051	43 14	082	42 34	141	19 07	178	64 16	215	27 18	278	26 50	336
217	27 50	051	44 13	082	43 11	142	19 09	179	63 41	217	26 20	279	26 26	336
218	28 36	051	45 11	082	43 47	143	19 10	179	63 05	219	25 22	279	26 02	335
219	29 22	051	46 10	082	44 23	144	19 11	180	62 27	220	24 23	279	25 37	335
220	30 08	051	47 08	082	44 58	145	19 11	180	61 48	222	23 25	279	25 12	335
221	30 53	051	48 07	082	45 31	146	19 11	181	61 08	223	22 26	279	24 47	335
222	31 39	050	49 05	082	46 04	147	19 10	181	60 27	225	21 28	279	24 22	334
223	32 24	050	50 04	082	46 37	148	19 08	182	59 45	226	20 30	279	23 56	334
224	33 10	050	51 02	082	47 08	149	19 06	182	59 02	228	19 31	279	23 30	334
	VEGA		◆ALTAIR		Nunki		ANTARES		RIGIL KENT.		◆SPICA		◆Alkaid	
225	33 55	050	18 29	084	22 24	124	47 38	150	19 04	183	58 18	229	47 51	342
226	34 40	049	19 28	084	23 13	124	48 07	151	19 01	183	57 33	230	47 33	342
227	35 24	049	20 27	084	24 02	125	48 36	152	18 57	184	56 47	231	47 14	341
228	36 09	049	21 26	084	24 50	125	49 03	153	18 53	184	56 01	232	46 54	340
229	36 53	048	22 25	084	25 38	126	49 29	154	18 49	185	55 14	233	46 33	339
230	37 37	048	23 23	085	26 26	126	49 54	156	18 44	185	54 26	234	46 12	338
231	38 21	048	24 22	085	27 14	127	50 18	157	18 38	186	53 38	235	45 49	337
232	39 05	047	25 21	085	28 01	127	50 40	158	18 32	186	52 49	236	45 26	337
233	39 48	047	26 20	085	28 48	128	51 02	159	18 25	187	52 00	237	45 03	336
234	40 32	047	27 19	085	29 35	128	51 22	161	18 18	187	51 10	238	44 38	335
235	41 14	046	28 18	085	30 21	129	51 41	162	18 10	188	50 20	239	44 13	335
236	41 57	046	29 17	085	31 07	129	51 58	163	18 02	188	49 29	239	43 48	334
237	42 39	045	30 15	086	31 53	130	52 14	165	17 53	189	48 38	240	43 21	333
238	43 21	045	31 14	086	32 38	130	52 29	166	17 44	189	47 46	241	42 54	333
239	44 03	044	32 13	086	33 23	131	52 42	168	17 34	190	46 55	242	42 27	332
	◆VEGA		ALTAIR		Nunki		◆ANTARES		SPICA		◆ARCTURUS		Alkaid	
240	44 44	044	33 12	086	34 07	132	52 54	169	46 03	242	63 07	293	41 59	331
241	45 25	043	34 11	086	34 51	132	53 04	171	45 10	243	62 13	292	41 30	331
242	46 05	043	35 10	086	35 35	133	53 13	172	44 17	244	61 18	292	41 01	330
243	46 45	042	36 09	086	36 18	133	53 21	174	43 24	244	60 23	291	40 32	330
244	47 24	042	37 08	086	37 00	134	53 26	175	42 31	245	59 28	291	40 01	329
245	48 03	041	38 07	087	37 43	135	53 31	177	41 38	245	58 33	291	39 31	329
246	48 42	040	39 06	087	38 24	136	53 33	178	40 44	246	57 38	290	39 00	328
247	49 20	040	40 05	087	39 05	136	53 35	180	39 50	246	56 42	290	38 28	328
248	49 57	039	41 04	087	39 46	137	53 34	181	38 56	247	55 46	290	37 56	327
249	50 34	038	42 03	087	40 26	138	53 32	183	38 01	247	54 51	289	37 24	327
250	51 10	037	43 02	087	41 05	139	53 29	184	37 07	248	53 55	289	36 51	326
251	51 45	037	44 01	087	41 44	139	53 24	186	36 12	248	52 59	289	36 18	326
252	52 20	036	45 00	087	42 22	140	53 17	187	35 17	249	52 03	288	35 45	325
253	52 54	035	45 59	088	42 59	141	53 09	189	34 22	249	51 07	288	35 11	325
254	53 28	034	46 58	088	43 36	142	53 00	190	33 27	249	50 11	288	34 37	325
	VEGA		◆ALTAIR		Nunki		◆ANTARES		SPICA		ARCTURUS		◆Alkaid	
255	54 00	033	47 57	088	44 12	143	52 49	191	32 31	250	49 15	288	34 02	324
256	54 32	032	48 56	088	44 47	144	52 36	193	31 36	250	48 18	288	33 27	324
257	55 03	031	49 55	088	45 21	145	52 22	194	30 40	251	47 22	288	32 52	323
258	55 33	030	50 54	088	45 55	146	52 07	196	29 44	251	46 26	287	32 17	323
259	56 02	029	51 53	088	46 27	147	51 50	197	28 48	251	45 29	287	31 41	323
260	56 30	028	52 53	089	46 59	148	51 32	198	27 52	252	44 33	287	31 06	322
261	56 56	026	53 52	089	47 30	149	51 13	200	26 56	252	43 36	287	30 30	322
262	57 22	025	54 51	089	48 00	150	50 52	201	26 00	252	42 40	287	29 53	322
263	57 47	024	55 50	089	48 29	151	50 30	202	25 04	253	41 43	287	29 17	322
264	58 10	023	56 49	089	48 56	153	50 07	204	24 07	253	40 47	287	28 40	321
265	58 33	021	57 48	089	49 23	154	49 43	205	23 11	253	39 50	287	28 03	321
266	58 54	020	58 47	089	49 49	155	49 17	206	22 14	254	38 54	287	27 26	321
267	59 13	019	59 46	090	50 13	156	48 51	207	21 17	254	37 57	287	26 48	321
268	59 32	017	60 45	090	50 36	157	48 23	209	20 20	254	37 00	287	26 11	321
269	59 49	016	61 44	090	50 58	159	47 54	210	19 24	254	36 04	287	25 33	320

LHA ♈	Hc	Zn	Hc	Zn	Hc	Zn	Hc	Zn	Hc	Zn	Hc	Zn	Hc	Zn
	VEGA		◆DENEB		Enif		◆Nunki		Shaula		ANTARES		◆ARCTURUS	
270	60 04	014	40 41	037	34 55	085	51 19	160	42 29	187	47 25	211	35 07	287
271	60 18	013	41 16	036	35 54	085	51 39	161	42 21	188	46 54	212	34 10	287
272	60 31	011	41 51	036	36 53	085	51 57	163	42 12	189	46 22	213	33 14	286
273	60 42	010	42 26	035	37 52	085	52 14	164	42 02	190	45 50	214	32 17	286
274	60 51	008	42 59	035	38 51	085	52 29	166	41 51	191	45 16	215	31 20	286
275	60 59	007	43 33	034	39 50	085	52 43	167	41 38	192	44 42	216	30 24	286
276	61 05	005	44 06	034	40 49	086	52 56	168	41 25	194	44 07	217	29 27	287
277	61 09	004	44 38	033	41 48	086	53 07	170	41 11	195	43 31	218	28 30	287
278	61 12	002	45 10	032	42 47	086	53 16	171	40 55	196	42 55	219	27 34	287
279	61 13	000	45 41	032	43 46	086	53 24	173	40 39	197	42 17	220	26 37	287
280	61 13	359	46 12	031	44 44	086	53 31	174	40 22	197	41 40	220	25 41	287
281	61 11	357	46 42	030	45 43	086	53 36	176	40 04	198	41 01	221	24 44	287
282	61 07	355	47 11	029	46 42	086	53 39	177	39 44	199	40 22	222	23 47	287
283	61 01	354	47 40	029	47 41	086	53 41	179	39 24	200	39 42	223	22 51	287
284	60 54	352	48 08	028	48 40	086	53 42	180	39 03	201	39 01	224	21 54	287
	◆DENEB		Alpheratz		◆FOMALHAUT		Nunki		ANTARES		◆Rasalhague		VEGA	
285	48 35	027	16 07	062	20 31	127	53 41	182	38 20	224	68 56	279	60 45	351
286	49 02	026	16 59	062	21 18	127	53 38	183	37 39	225	67 58	279	60 35	349
287	49 27	025	17 52	063	22 05	128	53 34	185	36 57	226	66 59	279	60 23	348
288	49 52	024	18 44	063	22 51	128	53 28	186	36 14	226	66 01	278	60 10	346
289	50 16	024	19 36	063	23 38	129	53 20	188	35 31	227	65 03	278	59 55	345
290	50 40	023	20 29	063	24 24	129	53 11	189	34 48	228	64 04	278	59 38	343
291	51 02	022	21 21	063	25 09	130	53 01	191	34 04	228	63 06	278	59 21	342
292	51 23	021	22 14	063	25 55	130	52 49	192	33 19	229	62 07	278	59 01	340
293	51 44	020	23 06	063	26 40	131	52 36	194	32 35	230	61 09	278	58 41	339
294	52 03	019	23 59	063	27 24	131	52 21	195	31 49	230	60 10	278	58 19	338
295	52 22	018	24 51	062	28 09	132	52 05	197	31 04	231	59 12	278	57 56	336
296	52 39	017	25 43	062	28 53	132	51 47	198	30 18	231	58 13	278	57 32	335
297	52 56	016	26 36	062	29 36	133	51 29	199	29 32	232	57 14	278	57 07	334
298	53 11	015	27 28	062	30 20	133	51 08	201	28 45	232	56 16	278	56 40	333
299	53 25	013	28 20	062	31 02	134	50 47	202	27 58	233	55 17	278	56 12	332
	◆DENEB		Alpheratz		◆FOMALHAUT		Peacock		◆ANTARES		Rasalhague		VEGA	
300	53 38	012	29 13	062	31 45	134	23 03	176	27 11	233	54 19	278	55 44	331
301	53 50	011	30 05	062	32 27	135	23 06	177	26 24	234	53 20	278	55 14	329
302	54 01	010	30 57	062	33 08	136	23 09	177	25 36	234	52 22	278	54 44	328
303	54 11	009	31 49	062	33 49	136	23 12	178	24 48	235	51 23	278	54 13	327
304	54 19	008	32 41	062	34 30	137	23 13	179	24 00	235	50 24	278	53 40	326
305	54 27	006	33 33	062	35 10	138	23 14	179	23 11	235	49 26	278	53 07	326
306	54 33	005	34 25	062	35 50	138	23 15	180	22 22	236	48 27	278	52 34	325
307	54 37	004	35 17	061	36 29	139	23 15	180	21 33	236	47 29	278	51 59	324
308	54 41	003	36 09	061	37 07	140	23 14	181	20 44	237	46 30	278	51 24	323
309	54 43	002	37 01	061	37 45	141	23 13	182	19 54	237	45 32	278	50 48	322
310	54 44	000	37 52	061	38 22	141	23 11	182	19 05	237	44 33	278	50 11	321
311	54 44	359	38 44	061	38 59	142	23 08	183	18 15	238	43 35	278	49 34	321
312	54 42	358	39 35	060	39 35	143	23 05	183	17 25	238	42 36	278	48 56	320
313	54 40	357	40 27	060	40 10	144	23 01	184	16 35	238	41 38	278	48 18	319
314	54 36	356	41 18	060	40 45	145	22 57	185	15 44	239	40 39	278	47 39	319
	Alpheratz		◆Diphda		FOMALHAUT		◆Nunki		Rasalhague		VEGA		◆DENEB	
315	42 09	060	28 11	117	41 19	145	42 40	219	39 41	278	47 00	318	54 31	354
316	43 00	059	29 03	117	41 52	146	42 02	220	38 42	278	46 20	317	54 24	353
317	43 51	059	29 56	118	42 24	147	41 24	221	37 44	278	45 40	317	54 16	352
318	44 41	059	30 48	118	42 56	148	40 45	222	36 45	278	45 00	316	54 07	351
319	45 32	058	31 40	119	43 26	149	40 05	223	35 47	278	44 19	316	53 57	350
320	46 22	058	32 32	119	43 56	150	39 25	223	34 48	278	43 37	315	53 46	348
321	47 12	058	33 23	119	44 25	151	38 44	224	33 50	278	42 55	315	53 34	347
322	48 02	057	34 15	120	44 53	152	38 03	225	32 51	279	42 13	314	53 20	346
323	48 51	057	35 06	121	45 20	153	37 21	226	31 53	279	41 31	314	53 06	345
324	49 40	056	35 57	121	45 46	154	36 38	226	30 54	279	40 48	313	52 50	344
325	50 30	056	36 47	122	46 11	155	35 55	227	29 56	279	40 05	313	52 33	343
326	51 18	055	37 37	122	46 36	157	35 12	228	28 58	279	39 22	313	52 15	342
327	52 07	055	38 27	123	46 58	158	34 28	228	27 59	279	38 38	312	51 56	341
328	52 55	054	39 17	123	47 20	159	33 44	229	27 01	279	37 54	312	51 36	340
329	53 43	054	40 06	124	47 41	160	32 59	229	26 03	279	37 10	312	51 16	339
	◆Alpheratz		Diphda		◆FOMALHAUT		Peacock		ALTAIR		◆VEGA		DENEB	
330	54 30	053	40 55	125	48 01	161	20 26	194	58 04	271	36 26	311	50 54	338
331	55 17	052	41 43	125	48 19	162	20 12	194	57 05	271	35 42	311	50 31	337
332	56 03	052	42 31	126	48 36	164	19 58	195	56 06	271	34 57	311	50 08	336
333	56 49	051	43 19	127	48 52	165	19 42	195	55 07	271	34 12	310	49 43	335
334	57 35	050	44 06	127	49 07	166	19 27	196	54 08	271	33 27	310	49 18	334
335	58 20	049	44 52	128	49 21	167	19 10	196	53 08	271	32 42	310	48 52	333
336	59 04	048	45 39	129	49 33	169	18 54	197	52 09	272	31 56	310	48 25	333
337	59 48	047	46 24	130	49 43	170	18 37	197	51 10	272	31 11	310	47 58	332
338	60 31	046	47 09	131	49 53	171	18 19	198	50 11	272	30 25	309	47 29	331
339	61 14	045	47 54	132	50 01	173	18 01	198	49 12	272	29 40	309	47 00	330
340	61 55	044	48 38	132	50 08	174	17 42	199	48 13	272	28 54	309	46 31	330
341	62 36	043	49 21	133	50 13	175	17 23	199	47 14	272	28 08	309	46 01	329
342	63 16	042	50 04	134	50 17	177	17 03	200	46 15	272	27 22	309	45 30	328
343	63 55	040	50 45	135	50 20	178	16 43	200	45 16	273	26 35	309	44 58	328
344	64 32	039	51 27	136	50 21	180	16 23	200	44 17	273	25 49	308	44 26	327
	◆Mirfak		Hamal		Acamar		◆ACHERNAR		FOMALHAUT		◆ALTAIR		DENEB	
345	23 01	040	43 31	067	15 34	137	15 24	159	50 21	181	43 18	273	43 54	326
346	23 39	040	44 25	067	16 15	137	15 45	160	50 19	182	42 19	273	43 21	326
347	24 17	039	45 19	067	16 54	138	16 05	160	50 16	184	41 20	273	42 47	325
348	24 54	039	46 13	066	17 34	138	16 25	160	50 12	185	40 21	273	42 13	324
349	25 32	039	47 08	066	18 13	139	16 45	161	50 06	186	39 22	273	41 38	324
350	26 09	039	48 02	066	18 52	139	17 04	161	49 59	188	38 23	273	41 03	323
351	26 46	039	48 55	066	19 31	139	17 23	162	49 50	189	37 24	274	40 28	323
352	27 23	038	49 49	065	20 09	140	17 41	162	49 40	190	36 25	274	39 52	322
353	27 59	038	50 43	065	20 47	140	17 59	163	49 29	192	35 26	274	39 16	322
354	28 36	038	51 36	065	21 25	141	18 16	163	49 16	193	34 27	274	38 39	321
355	29 12	038	52 30	064	22 02	141	18 33	164	49 03	194	33 28	274	38 02	321
356	29 48	037	53 23	064	22 38	142	18 49	164	48 48	195	32 29	274	37 25	321
357	30 24	037	54 16	063	23 15	142	19 05	165	48 31	197	31 30	274	36 47	320
358	31 00	037	55 09	063	23 51	143	19 20	165	48 14	198	30 31	274	36 09	320
359	31 35	037	56 01	063	24 26	143	19 35	166	47 55	199	29 32	275	35 31	319

LHA	Hc	Zn	Hc	Zn	Hc	Zn	Hc	Zn	Hc	Zn	Hc	Zn	Hc	Zn
♈	◆CAPELLA		ALDEBARAN		Acamar		◆ACHERNAR		FOMALHAUT		◆ALTAIR		DENEB	
0	14 02	045	22 39	076	25 50	144	20 47	166	48 31	201	28 28	275	34 07	320
1	14 44	045	23 37	076	26 25	144	21 01	167	48 09	202	27 29	275	33 29	319
2	15 25	045	24 34	076	26 59	145	21 15	167	47 47	203	26 30	275	32 50	319
3	16 07	045	25 32	076	27 33	145	21 27	168	47 23	204	25 31	276	32 11	319
4	16 49	045	26 29	076	28 07	146	21 40	168	46 58	205	24 32	276	31 31	318
5	17 30	044	27 27	076	28 40	146	21 51	169	46 32	206	23 33	276	30 52	318
6	18 12	044	28 24	076	29 12	147	22 03	169	46 05	208	22 34	276	30 12	318
7	18 53	044	29 22	076	29 44	148	22 13	170	45 37	209	21 36	276	29 32	317
8	19 34	044	30 19	076	30 16	148	22 23	171	45 09	210	20 37	276	28 52	317
9	20 16	044	31 17	076	30 46	149	22 33	171	44 39	211	19 38	276	28 12	317
10	20 57	044	32 14	076	31 17	150	22 42	172	44 08	212	18 39	276	27 31	317
11	21 38	044	33 12	076	31 46	150	22 50	172	43 37	213	17 40	276	26 50	317
12	22 19	044	34 10	076	32 15	151	22 58	173	43 04	214	16 41	277	26 10	316
13	23 00	044	35 07	076	32 44	152	23 05	173	42 31	214	15 42	277	25 29	316
14	23 41	043	36 05	076	33 12	152	23 12	174	41 57	215	14 43	277	24 47	316
	CAPELLA		◆ALDEBARAN		RIGEL		ACHERNAR		◆FOMALHAUT		Enif		◆DENEB	
15	24 22	043	37 02	076	24 23	103	23 17	174	41 23	216	41 41	275	24 06	316
16	25 02	043	38 00	076	25 20	104	23 23	175	40 47	217	40 42	275	23 25	316
17	25 43	043	38 57	076	26 18	104	23 28	176	40 11	218	39 43	275	22 43	316
18	26 23	043	39 55	076	27 15	104	23 32	176	39 35	219	38 44	275	22 02	315
19	27 03	043	40 52	076	28 13	104	23 35	177	38 57	219	37 45	276	21 20	315
20	27 43	042	41 50	076	29 10	105	23 38	177	38 19	220	36 46	276	20 38	315
21	28 23	042	42 47	076	30 07	105	23 41	178	37 41	221	35 47	276	19 57	315
22	29 02	042	43 44	076	31 05	105	23 43	179	37 02	222	34 48	276	19 15	315
23	29 42	042	44 42	076	32 02	106	23 44	179	36 22	222	33 49	276	18 33	315
24	30 21	041	45 39	076	32 59	106	23 44	180	35 42	223	32 50	276	17 51	315
25	31 00	041	46 37	075	33 56	106	23 44	180	35 01	224	31 51	276	17 09	315
26	31 39	041	47 34	075	34 52	107	23 43	181	34 20	224	30 52	276	16 26	315
27	32 17	040	48 31	075	35 49	107	23 42	182	33 38	225	29 53	276	15 44	315
28	32 56	040	49 29	075	36 46	107	23 40	182	32 56	226	28 54	276	15 02	315
29	33 34	040	50 26	075	37 42	108	23 38	183	32 13	226	27 56	276	14 20	315
	Mirfak		CAPELLA		◆ALDEBARAN		RIGEL		◆ACHERNAR		FOMALHAUT		◆Alpheratz	
30	45 35	019	34 12	039	51 23	075	38 39	108	23 35	183	31 30	227	56 59	311
31	45 54	018	34 49	039	52 20	075	39 35	109	23 31	184	30 47	227	56 14	310
32	46 12	018	35 26	039	53 17	074	40 31	109	23 27	184	30 03	228	55 29	310
33	46 30	017	36 03	038	54 14	074	41 27	109	23 22	185	29 19	228	54 43	309
34	46 47	016	36 40	038	55 11	074	42 23	110	23 16	186	28 34	229	53 56	308
35	47 03	015	37 16	037	56 08	074	43 18	110	23 10	186	27 50	229	53 10	308
36	47 18	014	37 52	037	57 05	073	44 14	111	23 03	187	27 04	230	52 22	307
37	47 32	013	38 27	036	58 02	073	45 09	111	22 56	187	26 19	230	51 35	306
38	47 45	012	39 02	036	58 59	073	46 04	112	22 48	188	25 33	231	50 47	306
39	47 57	012	39 37	035	59 55	073	46 59	112	22 40	189	24 47	231	49 59	305
40	48 09	011	40 11	035	60 52	072	47 54	113	22 31	189	24 01	232	49 10	305
41	48 19	010	40 45	034	61 48	072	48 48	114	22 21	190	23 14	232	48 21	304
42	48 29	009	41 18	034	62 44	071	49 42	114	22 11	190	22 27	233	47 32	304
43	48 37	008	41 51	033	63 40	071	50 36	115	22 00	191	21 40	233	46 42	303
44	48 45	007	42 23	033	64 36	070	51 30	116	21 49	191	20 52	233	45 53	303
	◆CAPELLA		PROCYON		SIRIUS		◆CANOPUS		ACHERNAR		Diphda		◆Alpheratz	
45	42 55	032	20 46	088	28 44	115	14 38	151	21 37	192	46 47	231	45 03	303
46	43 26	031	21 45	088	29 38	115	15 07	151	21 24	192	46 01	232	44 13	302
47	43 57	031	22 44	088	30 32	116	15 35	152	21 11	193	45 14	233	43 23	302
48	44 27	030	23 44	088	31 25	116	16 03	152	20 58	193	44 26	234	42 32	301
49	44 56	029	24 43	088	32 18	116	16 30	152	20 44	194	43 38	234	41 42	301
50	45 25	029	25 42	088	33 11	117	16 58	153	20 29	194	42 50	235	40 51	301
51	45 53	028	26 41	089	34 04	117	17 24	153	20 14	195	42 01	236	40 00	301
52	46 21	027	27 41	089	34 57	118	17 51	154	19 59	195	41 12	236	39 09	300
53	46 47	026	28 40	089	35 49	118	18 17	154	19 43	196	40 23	237	38 18	300
54	47 13	026	29 39	089	36 41	119	18 42	155	19 26	196	39 33	238	37 26	300
55	47 39	025	30 38	089	37 33	119	19 08	155	19 09	197	38 43	238	36 35	300
56	48 03	024	31 38	089	38 24	120	19 32	156	18 52	197	37 52	239	35 43	299
57	48 27	023	32 37	090	39 16	120	19 56	156	18 34	198	37 01	239	34 52	299
58	48 50	022	33 36	090	40 07	121	20 20	157	18 15	198	36 10	240	34 00	299
59	49 12	021	34 35	090	40 57	122	20 44	157	17 56	199	35 19	240	33 08	299
	◆CAPELLA		POLLUX		PROCYON		◆SIRIUS		CANOPUS		◆Diphda		Alpheratz	
60	49 33	020	33 54	062	35 35	090	41 48	122	21 06	158	34 27	241	32 16	299
61	49 53	020	34 47	062	36 34	090	42 38	123	21 29	158	33 36	241	31 24	299
62	50 13	019	35 39	062	37 33	090	43 27	123	21 51	159	32 44	242	30 32	299
63	50 31	018	36 31	062	38 32	090	44 17	124	22 12	159	31 51	242	29 40	298
64	50 49	017	37 23	061	39 32	091	45 05	125	22 33	160	30 59	243	28 48	298
65	51 05	016	38 15	061	40 31	091	45 54	126	22 53	160	30 06	243	27 56	298
66	51 21	015	39 07	061	41 30	091	46 42	126	23 13	161	29 13	243	27 03	298
67	51 35	014	39 59	061	42 29	091	47 29	127	23 32	161	28 20	244	26 11	298
68	51 48	012	40 51	061	43 29	091	48 16	128	23 51	162	27 27	244	25 19	298
69	52 01	011	41 42	060	44 28	091	49 03	129	24 09	162	26 33	245	24 26	298
70	52 12	010	42 34	060	45 27	092	49 49	130	24 27	163	25 40	245	23 34	298
71	52 22	009	43 25	060	46 26	092	50 34	131	24 44	164	24 46	245	22 42	298
72	52 31	008	44 16	059	47 26	092	51 19	132	25 00	164	23 52	246	21 49	298
73	52 38	007	45 07	059	48 25	092	52 03	133	25 16	165	22 58	246	20 57	298
74	52 45	006	45 57	059	49 24	092	52 46	134	25 31	165	22 04	246	20 04	298
	CAPELLA		◆POLLUX		PROCYON		SIRIUS		◆CANOPUS		Acamar		◆Hamal	
75	52 50	005	46 48	058	50 23	093	53 28	135	25 46	166	33 13	208	46 13	295
76	52 55	004	47 38	058	51 22	093	54 10	136	25 59	167	32 45	208	45 19	294
77	52 58	002	48 28	057	52 22	093	54 51	137	26 13	167	32 16	209	44 25	294
78	53 00	001	49 18	057	53 21	093	55 31	138	26 26	168	31 47	210	43 31	294
79	53 00	000	50 07	056	54 20	094	56 10	139	26 38	169	31 18	210	42 37	294
80	53 00	359	50 57	056	55 19	094	56 48	141	26 49	169	30 48	211	41 42	293
81	52 58	358	51 46	055	56 18	094	57 25	142	27 00	170	30 17	212	40 48	293
82	52 55	357	52 34	055	57 17	094	58 01	143	27 10	171	29 45	212	39 54	293
83	52 51	355	53 23	054	58 16	095	58 36	145	27 19	171	29 13	213	38 59	293
84	52 46	354	54 10	054	59 15	095	59 09	146	27 28	172	28 41	214	38 04	293
85	52 40	353	54 58	053	60 14	095	59 41	148	27 36	173	28 08	214	37 10	293
86	52 32	352	55 45	052	61 13	096	60 12	150	27 43	173	27 34	215	36 15	293
87	52 23	351	56 32	051	62 12	096	60 41	151	27 50	174	27 00	215	35 20	292
88	52 13	350	57 18	051	63 11	096	61 09	153	27 56	175	26 26	216	34 25	292
89	52 02	349	58 03	050	64 10	097	61 35	155	28 01	175	25 51	216	33 31	292

LAT 9°N

LHA	Hc	Zn	Hc	Zn	Hc	Zn	Hc	Zn	Hc	Zn	Hc	Zn	Hc	Zn
♈	◆POLLUX		REGULUS		Suhail		◆CANOPUS		RIGEL		◆ALDEBARAN		CAPELLA	
90	58 48	049	29 03	081	22 28	145	28 06	176	69 22	214	68 08	292	51 50	348
91	59 32	048	30 02	081	23 02	145	28 10	177	68 48	216	67 12	291	51 37	347
92	60 16	047	31 01	081	23 36	146	28 13	177	68 12	218	66 17	291	51 23	346
93	60 59	046	31 59	081	24 09	146	28 15	178	67 35	220	65 22	290	51 08	345
94	61 41	045	32 58	081	24 41	147	28 17	179	66 56	222	64 26	290	50 51	344
95	62 23	044	33 56	082	25 13	148	28 18	179	66 15	224	63 30	289	50 34	343
96	63 03	042	34 55	082	25 45	148	28 18	180	65 33	226	62 34	289	50 16	342
97	63 42	041	35 54	082	26 16	149	28 18	181	64 50	227	61 38	288	49 57	341
98	64 21	040	36 52	082	26 46	149	28 17	181	64 06	229	60 41	288	49 36	340
99	64 58	038	37 51	082	27 16	150	28 15	182	63 21	230	59 45	287	49 15	339
100	65 34	037	38 50	082	27 46	150	28 13	183	62 35	232	58 48	287	48 54	338
101	66 09	035	39 48	082	28 15	151	28 09	183	61 48	233	57 52	287	48 31	337
102	66 42	033	40 47	082	28 43	152	28 05	184	61 00	234	56 55	286	48 07	336
103	67 14	032	41 46	082	29 11	152	28 01	185	60 12	235	55 58	286	47 43	335
104	67 44	030	42 44	082	29 38	153	27 55	186	59 23	237	55 01	286	47 18	334
	Dubhe		◆REGULUS		Suhail		◆CANOPUS		RIGEL		ALDEBARAN		◆CAPELLA	
105	21 26	026	43 43	082	30 05	154	27 49	186	58 33	238	54 04	286	46 52	334
106	21 52	026	44 42	082	30 31	154	27 43	187	57 43	239	53 07	286	46 25	333
107	22 18	026	45 40	082	30 56	155	27 35	188	56 52	240	52 10	285	45 57	332
108	22 44	026	46 39	082	31 21	156	27 27	188	56 01	240	51 13	285	45 29	331
109	23 09	026	47 38	082	31 45	156	27 18	189	55 09	241	50 15	285	45 01	331
110	23 35	025	48 36	082	32 09	157	27 09	190	54 17	242	49 18	285	44 31	330
111	24 00	025	49 35	082	32 31	158	26 59	190	53 24	243	48 21	285	44 01	329
112	24 25	025	50 34	082	32 53	159	26 48	191	52 31	244	47 23	285	43 31	329
113	24 50	025	51 32	082	33 15	159	26 37	191	51 38	244	46 26	285	43 00	328
114	25 14	024	52 31	082	33 35	160	26 25	192	50 44	245	45 29	284	42 28	327
115	25 38	024	53 30	082	33 55	161	26 12	193	49 50	246	44 31	284	41 56	327
116	26 02	024	54 28	082	34 14	162	25 58	193	48 56	246	43 34	284	41 23	326
117	26 26	023	55 27	082	34 32	162	25 44	194	48 02	247	42 36	284	40 50	326
118	26 49	023	56 26	082	34 50	163	25 30	195	47 07	247	41 39	284	40 16	325
119	27 13	023	57 24	082	35 06	164	25 15	195	46 12	248	40 42	284	39 42	325
	Dubhe		◆Denebola		Suhail		◆SIRIUS		RIGEL		ALDEBARAN		◆CAPELLA	
120	27 35	023	33 52	078	35 22	165	58 17	216	45 17	249	39 44	284	39 08	324
121	27 58	022	34 50	078	35 37	166	57 42	217	44 22	249	38 47	284	38 33	324
122	28 20	022	35 48	078	35 51	167	57 06	219	43 27	250	37 49	284	37 57	323
123	28 42	022	36 46	078	36 04	167	56 28	220	42 31	250	36 52	284	37 21	323
124	29 04	021	37 44	078	36 17	168	55 49	221	41 35	251	35 54	284	36 45	322
125	29 25	021	38 42	078	36 28	169	55 10	223	40 39	251	34 56	284	36 09	322
126	29 46	020	39 40	078	36 39	170	54 29	224	39 43	251	33 59	284	35 32	321
127	30 06	020	40 38	078	36 49	171	53 48	225	38 47	252	33 01	284	34 55	321
128	30 26	020	41 36	078	36 57	172	53 06	226	37 51	252	32 04	284	34 18	321
129	30 46	019	42 34	078	37 05	173	52 23	227	36 54	253	31 06	284	33 40	320
130	31 05	019	43 32	078	37 12	174	51 39	228	35 58	253	30 09	284	33 02	320
131	31 24	018	44 30	078	37 19	175	50 55	229	35 01	253	29 11	284	32 24	320
132	31 43	018	45 28	078	37 24	175	50 10	230	34 04	254	28 14	284	31 45	319
133	32 01	018	46 26	078	37 28	176	49 24	231	33 07	254	27 16	284	31 06	319
134	32 19	017	47 24	078	37 31	177	48 38	232	32 10	254	26 19	284	30 27	319
	Dubhe		◆ARCTURUS		SPICA		◆Suhail		SIRIUS		BETELGEUSE		◆CAPELLA	
135	32 36	017	13 24	072	21 07	106	37 33	178	47 51	232	44 12	272	29 48	318
136	32 53	016	14 20	072	22 04	106	37 35	179	47 04	233	43 13	272	29 09	318
137	33 09	016	15 17	072	23 01	106	37 35	180	46 16	234	42 14	272	28 29	318
138	33 25	015	16 13	072	23 58	107	37 35	181	45 28	235	41 14	272	27 49	318
139	33 41	015	17 10	073	24 55	107	37 33	182	44 39	236	40 15	272	27 09	317
140	33 56	014	18 06	073	25 52	107	37 31	183	43 50	236	39 16	272	26 29	317
141	34 10	014	19 03	073	26 48	107	37 28	184	43 01	237	38 17	272	25 49	317
142	34 24	013	19 59	073	27 45	108	37 23	185	42 11	238	37 17	272	25 09	317
143	34 38	013	20 56	073	28 41	108	37 18	186	41 21	238	36 18	273	24 28	317
144	34 51	012	21 53	073	29 37	108	37 12	186	40 30	239	35 19	273	23 47	317
145	35 03	012	22 49	073	30 34	109	37 05	187	39 39	239	34 20	273	23 07	316
146	35 15	011	23 46	073	31 30	109	36 57	188	38 48	240	33 21	273	22 26	316
147	35 26	011	24 43	073	32 26	109	36 48	189	37 57	240	32 21	273	21 45	316
148	35 37	010	25 39	073	33 21	110	36 38	190	37 05	241	31 22	273	21 03	316
149	35 47	010	26 36	073	34 17	110	36 27	191	36 13	242	30 23	273	20 22	316
	Dubhe		◆ARCTURUS		SPICA		ACRUX		◆Suhail		SIRIUS		◆POLLUX	
150	35 57	009	27 33	073	35 13	111	12 42	164	36 16	192	35 21	242	52 59	305
151	36 06	009	28 29	073	36 08	111	12 58	164	36 03	193	34 28	242	52 10	305
152	36 15	008	29 26	073	37 03	111	13 14	165	35 50	193	33 36	243	51 21	304
153	36 23	008	30 23	073	37 58	112	13 29	165	35 36	194	32 43	243	50 32	304
154	36 31	007	31 19	073	38 53	112	13 44	165	35 21	195	31 50	244	49 43	303
155	36 37	006	32 16	073	39 48	113	13 59	166	35 05	196	30 56	244	48 53	303
156	36 44	006	33 12	073	40 43	113	14 13	166	34 48	197	30 03	245	48 03	302
157	36 49	005	34 09	073	41 37	114	14 27	167	34 30	198	29 09	245	47 13	302
158	36 54	005	35 06	073	42 31	114	14 40	167	34 12	198	28 16	245	46 23	302
159	36 59	004	36 02	073	43 25	115	14 54	167	33 53	199	27 22	246	45 32	301
160	37 03	003	36 59	073	44 19	115	15 06	168	33 33	200	26 27	246	44 42	301
161	37 06	003	37 55	073	45 12	116	15 18	168	33 13	201	25 33	247	43 51	301
162	37 09	002	38 52	073	46 05	117	15 30	169	32 51	201	24 39	247	42 59	300
163	37 11	002	39 49	072	46 58	117	15 42	169	32 29	202	23 44	247	42 08	300
164	37 12	001	40 45	072	47 51	118	15 53	170	32 06	203	22 49	248	41 17	300
	Dubhe		◆ARCTURUS		SPICA		◆ACRUX		Suhail		PROCYON		◆POLLUX	
165	37 13	001	41 41	072	48 43	118	16 03	170	31 43	204	40 02	269	40 25	299
166	37 13	000	42 38	072	49 35	119	16 13	171	31 19	204	39 03	269	39 33	299
167	37 13	359	43 34	072	50 27	120	16 23	171	30 54	205	38 04	270	38 42	299
168	37 12	359	44 30	072	51 18	121	16 32	171	30 29	206	37 05	270	37 50	299
169	37 10	358	45 27	072	52 09	121	16 41	172	30 03	206	36 05	270	36 58	298
170	37 08	358	46 23	072	52 59	122	16 49	172	29 36	207	35 06	270	36 05	298
171	37 05	357	47 19	071	53 49	123	16 57	173	29 09	208	34 07	270	35 13	298
172	37 02	356	48 15	071	54 38	124	17 04	173	28 41	208	33 08	270	34 21	298
173	36 58	356	49 11	071	55 27	125	17 11	174	28 12	209	32 08	271	33 28	298
174	36 53	355	50 07	071	56 15	126	17 17	174	27 43	210	31 09	271	32 36	298
175	36 48	355	51 03	070	57 03	127	17 23	175	27 14	210	30 10	271	31 43	297
176	36 42	354	51 59	070	57 49	128	17 28	175	26 44	211	29 11	271	30 51	297
177	36 35	353	52 55	070	58 36	129	17 33	175	26 13	211	28 11	271	29 58	297
178	36 28	353	53 51	070	59 21	131	17 38	176	25 42	212	27 12	271	29 05	297
179	36 21	352	54 46	069	60 06	132	17 42	176	25 10	213	26 13	271	28 13	297

LAT 9°N

LHA	Hc	Zn	Hc	Zn	Hc	Zn	Hc	Zn	Hc	Zn	Hc	Zn	Hc	Zn
♈	◆ARCTURUS		ANTARES		RIGIL KENT.		◆ACRUX		Suhail		◆REGULUS		Dubhe	
180	55 41	069	15 47	121	13 30	161	17 45	177	24 38	213	62 21	279	36 13	352
181	56 37	069	16 38	121	13 49	162	17 48	177	24 05	214	61 22	279	36 04	351
182	57 32	068	17 29	121	14 07	162	17 51	178	23 32	214	60 23	278	35 54	351
183	58 27	068	18 19	122	14 25	162	17 53	178	22 59	215	59 25	278	35 45	350
184	59 21	067	19 09	122	14 43	163	17 54	179	22 25	215	58 26	278	35 34	350
185	60 16	067	20 00	122	15 00	163	17 55	179	21 51	216	57 28	278	35 23	349
186	61 10	066	20 49	123	15 17	164	17 56	180	21 16	216	56 29	278	35 12	349
187	62 04	066	21 39	123	15 33	164	17 56	180	20 41	217	55 30	278	35 00	348
188	62 58	065	22 29	124	15 49	165	17 55	181	20 06	217	54 32	278	34 47	347
189	63 52	064	23 18	124	16 05	165	17 54	181	19 30	217	53 33	278	34 34	347
190	64 45	064	24 07	124	16 20	165	17 53	182	18 54	218	52 34	278	34 20	346
191	65 38	063	24 56	125	16 35	166	17 51	182	18 17	218	51 36	278	34 06	346
192	66 30	062	25 44	125	16 49	166	17 48	183	17 40	219	50 37	278	33 51	345
193	67 23	061	26 33	126	17 03	167	17 46	183	17 03	219	49 38	278	33 36	345
194	68 14	060	27 21	126	17 17	167	17 42	184	16 26	219	48 39	278	33 21	345
	◆Alphecca		Rasalhague		ANTARES		◆RIGIL KENT.		ACRUX		◆REGULUS		Dubhe	
195	49 26	059	22 37	080	28 08	127	17 30	168	17 38	184	47 41	278	33 05	344
196	50 16	059	23 36	080	28 56	127	17 42	168	17 34	184	46 42	278	32 48	344
197	51 07	058	24 34	080	29 43	128	17 54	169	17 29	185	45 43	278	32 31	343
198	51 57	058	25 32	080	30 30	128	18 06	169	17 24	185	44 45	278	32 14	343
199	52 47	057	26 31	080	31 16	129	18 17	169	17 18	186	43 46	278	31 56	342
200	53 36	056	27 29	080	32 02	129	18 27	170	17 12	186	42 47	278	31 38	342
201	54 26	056	28 28	081	32 48	130	18 37	170	17 05	187	41 49	278	31 19	341
202	55 14	055	29 26	081	33 34	130	18 47	171	16 58	187	40 50	278	31 00	341
203	56 03	055	30 25	081	34 19	131	18 56	171	16 50	188	39 51	278	30 40	341
204	56 51	054	31 23	081	35 03	131	19 05	172	16 42	188	38 53	278	30 21	340
205	57 39	053	32 22	081	35 48	132	19 13	172	16 33	189	37 54	278	30 00	340
206	58 26	052	33 20	081	36 31	133	19 20	173	16 24	189	36 55	278	29 40	339
207	59 12	051	34 19	081	37 15	133	19 27	173	16 15	189	35 57	278	29 19	339
208	59 58	051	35 17	081	37 58	134	19 34	174	16 05	190	34 58	278	28 57	339
209	60 44	050	36 16	081	38 40	135	19 40	174	15 54	190	33 59	278	28 36	338
	◆VEGA		Rasalhague		ANTARES		◆RIGIL KENT.		SPICA		◆REGULUS		Dubhe	
210	21 48	052	37 14	081	39 22	135	19 45	175	68 03	204	33 01	279	28 14	338
211	22 35	052	38 13	081	40 03	136	19 50	175	67 39	206	32 02	279	27 51	338
212	23 21	052	39 11	081	40 44	137	19 55	176	67 12	208	31 04	279	27 29	337
213	24 08	051	40 10	081	41 24	138	19 59	176	66 43	210	30 05	279	27 06	337
214	24 54	051	41 08	081	42 04	139	20 02	177	66 12	212	29 07	279	26 43	337
215	25 40	051	42 07	081	42 42	139	20 05	178	65 39	214	28 08	279	26 19	336
216	26 26	051	43 05	081	43 21	140	20 07	178	65 05	216	27 09	279	25 55	336
217	27 12	051	44 04	081	43 58	141	20 09	179	64 29	218	26 11	279	25 31	336
218	27 58	051	45 03	081	44 35	142	20 10	179	63 51	220	25 12	279	25 07	336
219	28 44	050	46 01	081	45 11	143	20 11	180	63 13	222	24 14	279	24 42	335
220	29 30	050	47 00	081	45 46	144	20 11	180	62 33	223	23 15	279	24 18	335
221	30 15	050	47 58	081	46 21	145	20 11	181	61 51	225	22 17	279	23 53	335
222	31 01	050	48 57	081	46 54	146	20 10	181	61 09	226	21 18	279	23 27	335
223	31 46	050	49 55	081	47 27	147	20 08	182	60 26	228	20 20	280	23 02	334
224	32 31	049	50 54	081	47 59	148	20 06	182	59 42	229	19 22	280	22 36	334
	VEGA		◆ALTAIR		Nunki		ANTARES		RIGIL KENT.		◆SPICA		◆Alkaid	
225	33 16	049	18 23	084	22 58	124	48 30	149	20 04	183	58 57	230	46 54	343
226	34 00	049	19 22	084	23 47	124	49 00	150	20 01	183	58 11	231	46 36	342
227	34 45	049	20 21	084	24 36	124	49 28	151	19 57	184	57 24	232	46 17	341
228	35 29	048	21 20	084	25 25	125	49 56	153	19 53	184	56 37	233	45 57	340
229	36 13	048	22 19	084	26 13	125	50 23	154	19 49	185	55 49	234	45 37	339
230	36 57	048	23 18	084	27 01	126	50 48	155	19 43	185	55 01	235	45 16	339
231	37 41	047	24 17	084	27 49	126	51 13	156	19 38	186	54 12	236	44 54	338
232	38 24	047	25 15	084	28 37	127	51 36	158	19 31	186	53 22	237	44 31	337
233	39 07	046	26 14	085	29 24	127	51 58	159	19 25	187	52 32	238	44 08	336
234	39 50	046	27 13	085	30 12	128	52 18	160	19 17	187	51 41	239	43 44	336
235	40 33	046	28 12	085	30 58	128	52 38	162	19 10	188	50 50	240	43 19	335
236	41 15	045	29 11	085	31 45	129	52 55	163	19 01	188	49 59	240	42 54	334
237	41 57	045	30 11	085	32 31	129	53 12	165	18 53	189	49 07	241	42 28	334
238	42 38	044	31 10	085	33 16	130	53 27	166	18 43	189	48 15	242	42 01	333
239	43 20	044	32 09	085	34 02	130	53 41	167	18 34	190	47 23	243	41 34	332
	◆VEGA		ALTAIR		Nunki		◆ANTARES		SPICA		◆ARCTURUS		Alkaid	
240	44 00	043	33 08	085	34 47	131	53 53	169	46 30	243	62 43	295	41 06	332
241	44 41	043	34 07	085	35 31	132	54 03	170	45 37	244	61 49	294	40 38	331
242	45 21	042	35 06	085	36 15	132	54 13	172	44 44	244	60 55	294	40 09	331
243	46 00	042	36 05	086	36 59	133	54 20	173	43 50	245	60 01	293	39 40	330
244	46 39	041	37 04	086	37 42	134	54 26	175	42 56	246	59 06	293	39 10	329
245	47 18	040	38 03	086	38 25	134	54 31	176	42 02	246	58 11	292	38 39	329
246	47 56	040	39 02	086	39 07	135	54 33	178	41 08	247	57 16	292	38 09	328
247	48 33	039	40 01	086	39 49	136	54 35	180	40 14	247	56 21	291	37 37	328
248	49 10	038	41 00	086	40 30	136	54 34	181	39 19	248	55 26	291	37 06	327
249	49 46	037	41 59	086	41 10	137	54 32	183	38 24	248	54 30	291	36 34	327
250	50 22	037	42 59	086	41 50	138	54 29	184	37 29	248	53 35	290	36 01	327
251	50 57	036	43 58	086	42 29	139	54 24	186	36 34	249	52 39	290	35 28	326
252	51 31	035	44 57	086	43 08	140	54 17	187	35 38	249	51 43	290	34 55	326
253	52 05	034	45 56	087	43 46	141	54 09	189	34 43	250	50 48	289	34 22	325
254	52 38	033	46 55	087	44 23	142	53 59	190	33 47	250	49 52	289	33 48	325
	VEGA		◆ALTAIR		Nunki		◆ANTARES		SPICA		ARCTURUS		◆Alkaid	
255	53 10	032	47 54	087	45 00	142	53 48	192	32 52	250	48 56	289	33 14	325
256	53 41	031	48 54	087	45 35	143	53 35	193	31 56	251	48 00	289	32 39	324
257	54 11	030	49 53	087	46 10	144	53 20	195	31 00	251	47 03	289	32 04	324
258	54 40	029	50 52	087	46 44	145	53 05	196	30 04	251	46 07	288	31 29	323
259	55 09	028	51 51	087	47 18	146	52 48	198	29 07	252	45 11	288	30 54	323
260	55 36	027	52 50	087	47 50	147	52 29	199	28 11	252	44 15	288	30 18	323
261	56 03	026	53 49	087	48 21	149	52 09	200	27 14	252	43 18	288	29 42	323
262	56 28	025	54 49	087	48 52	150	51 48	202	26 18	253	42 22	288	29 06	322
263	56 52	023	55 48	087	49 21	151	51 26	203	25 21	253	41 25	288	28 29	322
264	57 15	022	56 47	088	49 50	152	51 02	204	24 25	253	40 29	288	27 53	322
265	57 37	021	57 46	088	50 17	153	50 37	205	23 28	254	39 33	288	27 16	321
266	57 57	020	58 45	088	50 43	154	50 11	207	22 31	254	38 36	287	26 39	321
267	58 17	018	59 45	088	51 08	156	49 44	208	21 34	254	37 39	287	26 02	321
268	58 34	017	60 44	088	51 32	157	49 16	209	20 37	254	36 43	287	25 24	321
269	58 51	015	61 43	088	51 54	158	48 46	210	19 40	255	35 46	287	24 47	321

LHA	Hc	Zn	Hc	Zn	Hc	Zn	Hc	Zn	Hc	Zn	Hc	Zn	Hc	Zn
♈	VEGA		◆DENEB		Enif		◆Nunki		Shaula		ANTARES		◆ARCTURUS	
270	59 06	014	39 53	036	34 50	084	52 15	160	43 29	187	48 16	211	34 50	287
271	59 20	013	40 28	036	35 49	084	52 35	161	43 20	188	47 45	212	33 53	287
272	59 32	011	41 02	035	36 48	084	52 54	162	43 11	190	47 12	213	32 56	287
273	59 42	010	41 36	035	37 47	084	53 11	164	43 01	191	46 39	214	32 00	287
274	59 52	008	42 10	034	38 46	085	53 27	165	42 49	192	46 05	215	31 03	287
275	59 59	007	42 43	034	39 45	085	53 41	167	42 37	193	45 31	216	30 06	287
276	60 05	005	43 16	033	40 44	085	53 54	168	42 23	194	44 55	217	29 10	287
277	60 09	003	43 48	032	41 43	085	54 06	170	42 09	195	44 19	218	28 13	287
278	60 12	002	44 19	032	42 42	085	54 16	171	41 53	196	43 41	219	27 17	287
279	60 13	000	44 50	031	43 41	085	54 24	173	41 37	197	43 04	220	26 20	287
280	60 13	359	45 20	030	44 40	085	54 31	174	41 19	198	42 25	221	25 23	287
281	60 11	357	45 50	030	45 39	085	54 36	176	41 01	199	41 46	222	24 27	287
282	60 07	356	46 19	029	46 38	085	54 39	177	40 41	200	41 06	223	23 30	287
283	60 02	354	46 47	028	47 37	085	54 41	179	40 21	201	40 26	223	22 33	287
284	59 55	353	47 15	027	48 36	085	54 42	180	39 59	202	39 45	224	21 37	287
	DENEB		◆Alpheratz		FOMALHAUT		◆Nunki		ANTARES		◆Rasalhague		VEGA	
285	47 42	027	15 39	062	21 07	127	54 41	182	39 03	225	68 46	282	59 46	351
286	48 08	026	16 31	062	21 54	127	54 38	184	38 21	226	67 47	281	59 36	349
287	48 33	025	17 24	062	22 42	128	54 33	185	37 38	226	66 49	281	59 24	348
288	48 58	024	18 16	062	23 28	128	54 27	187	36 55	227	65 51	281	59 11	347
289	49 21	023	19 09	062	24 15	128	54 20	188	36 12	228	64 53	280	58 57	345
290	49 44	022	20 01	062	25 01	129	54 11	190	35 28	228	63 55	280	58 41	344
291	50 06	021	20 54	062	25 47	129	54 00	191	34 43	229	62 56	280	58 24	342
292	50 27	020	21 46	062	26 33	130	53 48	193	33 59	229	61 58	280	58 05	341
293	50 47	019	22 38	062	27 19	130	53 34	194	33 13	230	60 59	280	57 45	340
294	51 06	018	23 31	062	28 04	131	53 19	196	32 28	231	60 01	280	57 24	338
295	51 25	017	24 23	062	28 48	131	53 02	197	31 42	231	59 03	279	57 01	337
296	51 42	016	25 15	062	29 33	132	52 44	198	30 55	232	58 04	279	56 37	336
297	51 58	015	26 08	062	30 17	132	52 25	200	30 09	232	57 06	279	56 12	335
298	52 13	014	27 00	062	31 01	133	52 05	201	29 22	233	56 07	279	55 47	333
299	52 27	013	27 52	062	31 44	133	51 43	202	28 34	233	55 09	279	55 20	332
	◆DENEB		Alpheratz		◆FOMALHAUT		Peacock		◆ANTARES		Rasalhague		VEGA	
300	52 40	012	28 45	062	32 27	134	24 03	176	27 47	234	54 10	279	54 51	331
301	52 52	011	29 37	062	33 09	135	24 06	177	26 59	234	53 12	279	54 23	330
302	53 02	010	30 29	061	33 51	135	24 09	177	26 11	235	52 13	279	53 53	329
303	53 12	009	31 21	061	34 33	136	24 11	178	25 22	235	51 14	279	53 22	328
304	53 20	007	32 13	061	35 14	136	24 13	179	24 34	235	50 16	279	52 50	327
305	53 27	006	33 05	061	35 54	137	24 14	179	23 45	236	49 17	279	52 18	326
306	53 33	005	33 57	061	36 34	138	24 15	180	22 56	236	48 19	279	51 44	325
307	53 38	004	34 48	061	37 14	139	24 15	180	22 06	237	47 20	279	51 10	325
308	53 41	003	35 40	061	37 53	139	24 14	181	21 17	237	46 22	279	50 36	324
309	53 43	002	36 31	060	38 31	140	24 13	182	20 27	237	45 23	279	50 00	323
310	53 44	000	37 23	060	39 09	141	24 11	182	19 37	238	44 25	279	49 24	322
311	53 44	359	38 14	060	39 46	142	24 08	183	18 47	238	43 26	279	48 47	321
312	53 43	358	39 05	060	40 23	142	24 05	183	17 57	238	42 27	279	48 10	321
313	53 40	357	39 57	059	40 58	143	24 01	184	17 06	239	41 29	279	47 32	320
314	53 36	356	40 47	059	41 34	144	23 57	185	16 15	239	40 30	279	46 54	319
	Alpheratz		◆Diphda		FOMALHAUT		◆Nunki		Rasalhague		VEGA		◆DENEB	
315	41 38	059	28 38	116	42 08	145	43 26	220	39 32	279	46 15	319	53 31	354
316	42 29	059	29 31	117	42 42	146	42 48	221	38 33	279	45 36	318	53 24	353
317	43 19	058	30 23	117	43 15	147	42 09	222	37 35	279	44 56	318	53 17	352
318	44 10	058	31 16	118	43 47	148	41 29	222	36 36	279	44 16	317	53 08	351
319	45 00	058	32 09	118	44 18	149	40 49	223	35 38	279	43 35	316	52 58	350
320	45 50	057	33 01	118	44 48	150	40 08	224	34 39	279	42 54	316	52 47	349
321	46 39	057	33 53	119	45 18	151	39 27	225	33 41	279	42 13	315	52 35	348
322	47 29	056	34 45	119	45 46	152	38 45	225	32 42	279	41 31	315	52 22	347
323	48 18	056	35 36	120	46 14	153	38 03	226	31 44	279	40 49	315	52 07	345
324	49 07	055	36 27	120	46 40	154	37 20	227	30 45	279	40 07	314	51 52	344
325	49 55	055	37 18	121	47 06	155	36 36	227	29 47	279	39 24	314	51 36	343
326	50 44	054	38 09	122	47 30	156	35 52	228	28 48	279	38 41	313	51 18	342
327	51 32	054	38 59	122	47 54	157	35 08	229	27 50	280	37 58	313	50 59	341
328	52 19	053	39 49	123	48 16	158	34 23	229	26 51	280	37 14	313	50 40	340
329	53 07	053	40 39	123	48 37	160	33 38	230	25 53	280	36 30	312	50 19	339
	◆Alpheratz		Diphda		◆FOMALHAUT		Peacock		ALTAIR		◆VEGA		DENEB	
330	53 53	052	41 28	124	48 57	161	21 25	194	58 02	272	35 46	312	49 58	338
331	54 40	051	42 17	125	49 16	162	21 10	194	57 03	272	35 02	312	49 36	337
332	55 26	050	43 06	125	49 34	163	20 56	195	56 04	273	34 18	311	49 13	337
333	56 11	050	43 54	126	49 50	165	20 40	195	55 05	273	33 33	311	48 49	336
334	56 56	049	44 42	127	50 05	166	20 24	196	54 05	273	32 48	311	48 24	335
335	57 40	048	45 29	127	50 19	167	20 08	196	53 06	273	32 03	310	47 58	334
336	58 24	047	46 16	128	50 32	169	19 51	197	52 07	273	31 18	310	47 32	333
337	59 07	046	47 02	129	50 43	170	19 34	197	51 08	273	30 33	310	47 05	332
338	59 49	045	47 48	130	50 52	171	19 16	198	50 09	273	29 47	310	46 37	332
339	60 31	044	48 33	131	51 01	173	18 58	198	49 09	273	29 02	310	46 08	331
340	61 12	043	49 18	132	51 08	174	18 39	199	48 10	273	28 16	309	45 39	330
341	61 52	042	50 02	133	51 13	175	18 20	199	47 11	273	27 30	309	45 09	329
342	62 31	040	50 45	134	51 17	177	18 00	200	46 12	273	26 44	309	44 39	329
343	63 08	039	51 28	135	51 20	178	17 40	200	45 13	274	25 58	309	44 08	328
344	63 45	038	52 10	136	51 21	180	17 19	201	44 14	274	25 12	309	43 36	327
	◆Mirfak		Hamal		Acamar		◆ACHERNAR		FOMALHAUT		◆ALTAIR		DENEB	
345	22 15	040	43 07	066	16 18	137	16 20	159	51 21	181	43 15	274	43 04	327
346	22 53	039	44 01	066	16 59	137	16 41	159	51 19	182	42 15	274	42 31	326
347	23 30	039	44 55	066	17 39	138	17 02	160	51 16	184	41 16	274	41 58	326
348	24 08	039	45 49	065	18 19	138	17 22	160	51 11	185	40 17	274	41 24	325
349	24 45	039	46 43	065	18 58	138	17 42	161	51 06	186	39 18	274	40 50	324
350	25 22	039	47 37	065	19 37	139	18 01	161	50 58	188	38 19	274	40 15	324
351	25 59	038	48 30	065	20 16	139	18 20	162	50 49	189	37 20	274	39 40	323
352	26 36	038	49 24	064	20 55	140	18 38	162	50 39	191	36 21	274	39 04	323
353	27 12	038	50 17	064	21 33	140	18 56	163	50 28	192	35 22	275	38 28	322
354	27 49	038	51 10	064	22 11	141	19 13	163	50 15	193	34 23	275	37 52	322
355	28 25	037	52 03	063	22 48	141	19 30	164	50 01	194	33 24	275	37 15	322
356	29 01	037	52 56	063	23 25	142	19 47	164	49 45	196	32 25	275	36 38	321
357	29 36	037	53 48	062	24 02	142	20 03	165	49 29	197	31 25	275	36 01	321
358	30 12	037	54 41	062	24 38	143	20 18	165	49 11	198	30 26	275	35 23	320
359	30 47	036	55 33	061	25 14	143	20 33	166	48 51	200	29 27	275	34 45	320

LAT 8°N

LHA ♈	Hc	Zn	Hc	Zn	Hc	Zn	Hc	Zn	Hc	Zn	Hc	Zn	Hc	Zn
	◆CAPELLA		ALDEBARAN		RIGEL		◆ACHERNAR		FOMALHAUT		◆ALTAIR		DENEB	
0	13 19	045	22 25	075	10 02	100	21 46	166	49 27	201	28 23	276	33 21	320
1	14 01	044	23 22	075	11 01	100	22 00	167	49 05	202	27 24	276	32 43	320
2	14 43	044	24 20	075	11 59	100	22 13	167	48 42	204	26 24	276	32 04	319
3	15 24	044	25 17	075	12 58	100	22 26	168	48 18	205	25 25	276	31 26	319
4	16 06	044	26 15	075	13 56	101	22 38	168	47 52	206	24 26	276	30 46	319
5	16 47	044	27 12	076	14 55	101	22 50	169	47 26	207	23 27	276	30 07	318
6	17 29	044	28 10	076	15 53	101	23 02	169	46 58	208	22 28	276	29 28	318
7	18 10	044	29 07	076	16 51	101	23 12	170	46 30	209	21 29	276	28 48	318
8	18 51	044	30 05	076	17 49	101	23 22	170	46 01	210	20 30	277	28 08	318
9	19 33	044	31 02	076	18 48	102	23 32	171	45 30	211	19 31	277	27 28	317
10	20 14	044	32 00	076	19 46	102	23 41	172	44 59	212	18 32	277	26 47	317
11	20 55	044	32 57	075	20 44	102	23 49	172	44 27	213	17 33	277	26 07	317
12	21 36	043	33 55	075	21 42	102	23 57	173	43 54	214	16 34	277	25 26	317
13	22 17	043	34 52	075	22 40	102	24 05	173	43 21	215	15 35	277	24 45	316
14	22 57	043	35 50	075	23 38	103	24 11	174	42 46	216	14 36	277	24 04	316
	CAPELLA		◆ALDEBARAN		RIGEL		ACHERNAR		◆FOMALHAUT		Enif		◆DENEB	
15	23 38	043	36 47	075	24 36	103	24 17	174	42 11	217	41 35	276	23 23	316
16	24 18	043	37 45	075	25 34	103	24 23	175	41 35	218	40 36	276	22 42	316
17	24 59	043	38 42	075	26 32	103	24 27	176	40 58	218	39 37	276	22 00	316
18	25 39	042	39 40	075	27 30	104	24 32	176	40 21	219	38 38	276	21 19	316
19	26 19	042	40 37	075	28 27	104	24 35	177	39 43	220	37 39	276	20 37	316
20	26 59	042	41 34	075	29 25	104	24 38	177	39 05	221	36 40	276	19 56	315
21	27 38	042	42 32	075	30 23	104	24 41	178	38 26	221	35 41	276	19 14	315
22	28 18	041	43 29	075	31 20	105	24 43	179	37 46	222	34 42	276	18 32	315
23	28 57	041	44 27	075	32 18	105	24 44	179	37 06	223	33 43	277	17 50	315
24	29 36	041	45 24	075	33 15	105	24 44	180	36 25	224	32 44	277	17 08	315
25	30 15	041	46 21	074	34 12	106	24 44	180	35 44	224	31 45	277	16 26	315
26	30 53	040	47 18	074	35 09	106	24 43	181	35 02	225	30 46	277	15 44	315
27	31 32	040	48 15	074	36 06	106	24 42	182	34 20	225	29 47	277	15 02	315
28	32 10	040	49 13	074	37 03	107	24 40	182	33 38	226	28 48	277	14 20	315
29	32 48	039	50 10	074	38 00	107	24 38	183	32 55	227	27 49	277	13 38	315
	Mirfak		CAPELLA		◆ALDEBARAN		RIGEL		◆ACHERNAR		FOMALHAUT		◆Alpheratz	
30	44 38	019	33 25	039	51 07	074	38 57	107	24 35	183	32 11	227	56 19	312
31	44 57	018	34 02	039	52 04	073	39 54	108	24 31	184	31 27	228	55 35	312
32	45 15	017	34 39	038	53 01	073	40 50	108	24 26	185	30 43	228	54 50	311
33	45 32	017	35 16	038	53 57	073	41 46	109	24 21	185	29 59	229	54 05	310
34	45 49	016	35 52	037	54 54	073	42 43	109	24 16	186	29 14	229	53 19	309
35	46 05	015	36 28	037	55 51	072	43 39	109	24 10	186	28 28	230	52 33	309
36	46 19	014	37 03	036	56 47	072	44 35	110	24 03	187	27 43	230	51 46	308
37	46 33	013	37 39	036	57 44	072	45 31	110	23 55	187	26 57	231	50 59	307
38	46 46	012	38 13	036	58 40	071	46 26	111	23 48	188	26 11	231	50 11	307
39	46 58	011	38 48	035	59 36	071	47 22	111	23 39	189	25 24	232	49 24	306
40	47 10	010	39 21	035	60 32	070	48 17	112	23 30	189	24 37	232	48 35	306
41	47 20	010	39 55	034	61 28	070	49 12	113	23 20	190	23 50	233	47 47	305
42	47 29	009	40 28	033	62 24	069	50 06	113	23 10	190	23 03	233	46 58	305
43	47 38	008	41 00	033	63 20	069	51 01	114	22 59	191	22 16	233	46 09	304
44	47 45	007	41 32	032	64 15	068	51 55	114	22 48	191	21 28	234	45 20	304
	◆CAPELLA		PROCYON		SIRIUS		◆CANOPUS		ACHERNAR		Diphda		◆Alpheratz	
45	42 04	032	20 43	087	29 09	114	15 30	151	22 36	192	47 24	232	44 30	303
46	42 35	031	21 43	088	30 03	115	15 59	151	22 23	192	46 37	233	43 41	303
47	43 05	030	22 42	088	30 57	115	16 28	152	22 10	193	45 50	234	42 51	303
48	43 35	030	23 42	088	31 51	115	16 56	152	21 56	194	45 02	234	42 00	302
49	44 04	029	24 41	088	32 45	116	17 24	152	21 42	194	44 13	235	41 10	302
50	44 32	028	25 40	088	33 38	116	17 51	153	21 28	195	43 24	236	40 20	302
51	45 00	028	26 40	088	34 31	117	18 18	153	21 12	195	42 35	236	39 29	301
52	45 27	027	27 39	088	35 24	117	18 45	154	20 57	196	41 45	237	38 38	301
53	45 54	026	28 38	088	36 17	118	19 11	154	20 40	196	40 55	238	37 47	301
54	46 19	025	29 38	089	37 09	118	19 37	155	20 24	197	40 05	238	36 56	301
55	46 44	024	30 37	089	38 02	119	20 02	155	20 07	197	39 14	239	36 05	300
56	47 08	024	31 37	089	38 54	119	20 27	155	19 49	198	38 23	239	35 13	300
57	47 32	023	32 36	089	39 46	120	20 51	156	19 31	198	37 32	240	34 22	300
58	47 54	022	33 35	089	40 37	120	21 15	156	19 12	198	36 40	240	33 30	300
59	48 16	021	34 35	089	41 28	121	21 39	157	18 53	199	35 49	241	32 39	300
	◆CAPELLA		POLLUX		PROCYON		◆SIRIUS		CANOPUS		◆Diphda		Alpheratz	
60	48 37	020	33 26	062	35 34	089	42 19	121	22 02	157	34 56	241	31 47	299
61	48 57	019	34 18	061	36 34	089	43 10	122	22 24	158	34 04	242	30 55	299
62	49 16	018	35 10	061	37 33	090	44 00	123	22 47	158	33 12	242	30 03	299
63	49 34	017	36 02	061	38 32	090	44 50	123	23 08	159	32 19	243	29 11	299
64	49 51	016	36 54	061	39 32	090	45 39	124	23 29	160	31 26	243	28 19	299
65	50 07	015	37 46	061	40 31	090	46 28	125	23 50	160	30 33	244	27 27	299
66	50 22	014	38 38	060	41 31	090	47 17	125	24 10	161	29 40	244	26 35	299
67	50 37	013	39 29	060	42 30	090	48 05	126	24 29	161	28 46	244	25 43	298
68	50 50	012	40 21	060	43 30	090	48 53	127	24 48	162	27 53	245	24 51	298
69	51 02	011	41 12	059	44 29	091	49 40	128	25 06	162	26 59	245	23 58	298
70	51 13	010	42 03	059	45 28	091	50 27	129	25 24	163	26 05	245	23 06	298
71	51 23	009	42 54	059	46 28	091	51 13	130	25 41	164	25 11	246	22 14	298
72	51 31	008	43 45	058	47 27	091	51 58	131	25 58	164	24 17	246	21 21	298
73	51 39	007	44 35	058	48 27	091	52 43	132	26 14	165	23 22	246	20 29	298
74	51 45	006	45 26	058	49 26	091	53 27	133	26 29	165	22 28	247	19 36	298
	CAPELLA		◆POLLUX		PROCYON		SIRIUS		◆CANOPUS		Acamar		◆Hamal	
75	51 51	005	46 16	057	50 25	091	54 10	134	26 44	166	34 06	208	45 48	296
76	51 55	003	47 06	057	51 25	092	54 53	135	26 58	167	33 38	209	44 54	295
77	51 58	002	47 55	056	52 24	092	55 35	136	27 11	167	33 09	209	44 00	295
78	52 00	001	48 45	056	53 24	092	56 15	137	27 24	168	32 40	210	43 06	295
79	52 00	000	49 34	055	54 23	092	56 55	138	27 36	168	32 10	211	42 12	295
80	52 00	359	50 23	055	55 22	092	57 34	140	27 48	169	31 39	211	41 18	294
81	51 58	358	51 11	054	56 22	093	58 12	141	27 59	170	31 08	212	40 24	294
82	51 55	357	51 59	054	57 21	093	58 49	142	28 09	170	30 36	213	39 30	294
83	51 51	356	52 47	053	58 20	093	59 25	144	28 18	171	30 04	213	38 35	294
84	51 46	354	53 34	052	59 20	093	59 59	145	28 27	172	29 31	214	37 41	294
85	51 40	353	54 21	052	60 19	093	60 32	147	28 35	172	28 58	214	36 46	293
86	51 33	352	55 08	051	61 18	094	61 04	149	28 43	173	28 24	215	35 52	293
87	51 24	351	55 54	050	62 18	094	61 34	150	28 50	174	27 49	216	34 57	293
88	51 14	350	56 39	049	63 17	094	62 02	152	28 56	174	27 14	216	34 02	293
89	51 04	349	57 24	049	64 16	095	62 29	154	29 01	175	26 39	217	33 08	293

LHA ♈	Hc	Zn	Hc	Zn	Hc	Zn	Hc	Zn	Hc	Zn	Hc	Zn	Hc	Zn
	◆POLLUX		REGULUS		Suhail		◆CANOPUS		Acamar		◆ALDEBARAN		CAPELLA	
90	58 08	048	28 54	081	23 17	145	29 06	176	26 03	217	67 44	294	50 52	348
91	58 52	047	29 53	081	23 52	145	29 10	177	25 27	218	66 50	294	50 39	347
92	59 35	046	30 51	081	24 25	146	29 13	177	24 51	218	65 55	293	50 25	346
93	60 17	045	31 50	081	24 59	146	29 15	178	24 14	219	65 00	292	50 10	345
94	60 58	044	32 49	081	25 31	147	29 17	179	23 36	219	64 05	292	49 54	344
95	61 39	042	33 47	081	26 04	147	29 18	179	22 59	220	63 09	291	49 37	343
96	62 18	041	34 46	081	26 36	148	29 18	180	22 20	220	62 14	290	49 19	342
97	62 57	040	35 45	081	27 07	148	29 18	181	21 42	221	61 18	290	49 00	341
98	63 34	038	36 43	081	27 38	149	29 17	181	21 03	221	60 22	289	48 40	340
99	64 11	037	37 42	081	28 08	150	29 15	182	20 24	221	59 26	289	48 19	339
100	64 46	035	38 41	081	28 38	150	29 13	183	19 44	222	58 30	289	47 58	338
101	65 19	034	39 39	081	29 07	151	29 09	183	19 05	222	57 33	288	47 35	337
102	65 52	032	40 38	081	29 36	151	29 05	184	18 24	223	56 37	288	47 12	337
103	66 22	030	41 37	081	30 04	152	29 01	185	17 44	223	55 40	288	46 48	336
104	66 52	028	42 35	081	30 32	153	28 55	186	17 03	223	54 44	287	46 23	335
	Dubhe		◆REGULUS		Suhail		◆CANOPUS		RIGEL		ALDEBARAN		◆CAPELLA	
105	20 32	026	43 34	081	30 59	153	28 49	186	59 05	239	53 47	287	45 58	334
106	20 58	026	44 33	081	31 25	154	28 42	187	58 13	240	52 50	287	45 31	333
107	21 24	026	45 31	081	31 51	155	28 35	188	57 22	241	51 53	287	45 04	333
108	21 50	026	46 30	081	32 16	155	28 27	188	56 30	242	50 56	286	44 37	332
109	22 15	025	47 29	081	32 40	156	28 18	189	55 37	243	49 59	286	44 08	331
110	22 40	025	48 27	081	33 04	157	28 08	190	54 44	243	49 02	286	43 39	330
111	23 05	025	49 26	081	33 27	158	27 58	190	53 51	244	48 05	286	43 10	330
112	23 30	025	50 25	081	33 49	158	27 47	191	52 57	245	47 08	286	42 39	329
113	23 55	024	51 23	081	34 11	159	27 35	192	52 03	246	46 11	286	42 09	328
114	24 19	024	52 22	081	34 31	160	27 23	192	51 09	246	45 13	285	41 37	328
115	24 43	024	53 21	081	34 52	161	27 10	193	50 15	247	44 16	285	41 05	327
116	25 07	024	54 19	081	35 11	161	26 57	193	49 20	247	43 19	285	40 33	327
117	25 31	023	55 18	080	35 29	162	26 43	194	48 25	248	42 21	285	40 00	326
118	25 54	023	56 17	080	35 47	163	26 28	195	47 30	248	41 24	285	39 27	326
119	26 17	023	57 15	080	36 04	164	26 12	195	46 34	249	40 27	285	38 53	325
	Dubhe		◆Denebola		Suhail		◆SIRIUS		RIGEL		ALDEBARAN		◆CAPELLA	
120	26 40	022	33 39	078	36 20	165	59 06	217	45 39	250	39 29	285	38 19	325
121	27 02	022	34 37	078	36 35	166	58 29	218	44 43	250	38 32	285	37 44	324
122	27 24	022	35 35	078	36 49	166	57 52	220	43 47	250	37 34	285	37 09	324
123	27 46	021	36 33	078	37 03	167	57 14	221	42 51	251	36 37	285	36 34	323
124	28 08	021	37 31	078	37 15	168	56 34	222	41 55	251	35 39	285	35 58	323
125	28 29	021	38 30	078	37 27	169	55 54	224	40 58	252	34 42	285	35 22	322
126	28 49	020	39 28	078	37 38	170	55 12	225	40 02	252	33 44	285	34 45	322
127	29 10	020	40 26	078	37 48	171	54 30	226	39 05	253	32 47	285	34 08	321
128	29 30	019	41 24	078	37 57	172	53 47	227	38 09	253	31 49	284	33 31	321
129	29 49	019	42 22	078	38 05	173	53 03	228	37 12	253	30 52	284	32 54	321
130	30 09	019	43 20	077	38 12	174	52 19	229	36 15	254	29 54	284	32 16	320
131	30 27	018	44 18	077	38 18	174	51 34	230	35 18	254	28 57	284	31 38	320
132	30 46	018	45 16	077	38 23	175	50 48	231	34 21	254	27 59	284	30 59	320
133	31 04	017	46 14	077	38 28	176	50 02	232	33 23	255	27 02	284	30 21	319
134	31 21	017	47 12	077	38 31	177	49 15	233	32 26	255	26 04	285	29 42	319
	Dubhe		◆ARCTURUS		SPICA		◆Suhail		SIRIUS		BETELGEUSE		◆CAPELLA	
135	31 39	017	13 05	072	21 23	105	38 33	178	48 27	233	44 10	273	29 03	319
136	31 55	016	14 02	072	22 21	106	38 35	179	47 39	234	43 10	273	28 24	319
137	32 12	016	14 59	072	23 18	106	38 35	180	46 51	235	42 11	273	27 45	318
138	32 27	015	15 55	072	24 15	106	38 35	181	46 02	236	41 12	273	27 05	318
139	32 43	015	16 52	072	25 12	106	38 33	182	45 13	236	40 12	273	26 25	318
140	32 58	014	17 48	072	26 09	107	38 31	183	44 23	237	39 13	273	25 45	318
141	33 12	014	18 45	072	27 06	107	38 27	184	43 33	238	38 14	273	25 05	317
142	33 26	013	19 42	072	28 03	107	38 23	185	42 43	238	37 14	273	24 25	317
143	33 39	013	20 38	072	28 59	108	38 18	186	41 52	239	36 15	273	23 44	317
144	33 52	012	21 35	072	29 56	108	38 11	187	41 01	240	35 16	273	23 04	317
145	34 04	012	22 31	072	30 53	108	38 04	187	40 10	240	34 16	274	22 23	317
146	34 16	011	23 28	072	31 49	108	37 56	188	39 18	241	33 17	274	21 42	317
147	34 27	011	24 25	072	32 45	109	37 47	189	38 26	241	32 18	274	21 01	316
148	34 38	010	25 21	072	33 41	109	37 37	190	37 34	242	31 19	274	20 20	316
149	34 48	010	26 18	072	34 38	110	37 26	191	36 41	242	30 19	274	19 39	316
	Dubhe		◆ARCTURUS		SPICA		ACRUX		◆Suhail		SIRIUS		◆POLLUX	
150	34 58	009	27 15	072	35 33	110	13 39	164	37 14	192	35 49	243	52 23	307
151	35 07	009	28 11	072	36 29	110	13 56	164	37 02	193	34 56	243	51 35	306
152	35 16	008	29 08	072	37 25	111	14 11	165	36 48	194	34 03	244	50 47	305
153	35 24	007	30 05	072	38 20	111	14 27	165	36 34	194	33 09	244	49 59	305
154	35 31	007	31 01	072	39 16	112	14 42	165	36 18	195	32 16	244	49 10	304
155	35 38	006	31 58	072	40 11	112	14 57	166	36 02	196	31 22	245	48 20	304
156	35 44	006	32 55	072	41 06	112	15 11	166	35 45	197	30 28	245	47 31	303
157	35 50	005	33 51	072	42 01	113	15 25	167	35 28	198	29 34	246	46 41	303
158	35 55	005	34 48	072	42 55	113	15 39	167	35 09	199	28 40	246	45 51	302
159	35 59	004	35 44	072	43 50	114	15 52	167	34 50	199	27 46	246	45 01	302
160	36 03	003	36 41	072	44 44	114	16 05	168	34 30	200	26 51	247	44 10	302
161	36 06	003	37 37	072	45 38	115	16 17	168	34 09	201	25 57	247	43 20	301
162	36 09	002	38 34	072	46 32	116	16 29	169	33 47	202	25 02	247	42 29	301
163	36 11	002	39 30	072	47 25	116	16 41	169	33 25	202	24 07	248	41 38	301
164	36 12	001	40 26	072	48 18	117	16 52	170	33 02	203	23 12	248	40 47	300
	Dubhe		◆ARCTURUS		SPICA		◆ACRUX		Suhail		PROCYON		◆POLLUX	
165	36 13	001	41 23	071	49 11	117	17 02	170	32 38	204	40 03	270	39 55	300
166	36 13	000	42 19	071	50 04	118	17 12	170	32 13	205	39 03	270	39 04	300
167	36 13	359	43 15	071	50 56	119	17 22	171	31 48	205	38 04	270	38 12	300
168	36 12	359	44 11	071	51 48	120	17 31	171	31 23	206	37 05	271	37 20	299
169	36 10	358	45 08	071	52 39	120	17 40	172	30 56	207	36 05	271	36 29	299
170	36 08	358	46 04	071	53 30	121	17 48	172	30 29	207	35 06	271	35 37	299
171	36 05	357	47 00	070	54 21	122	17 56	173	30 02	208	34 06	271	34 45	299
172	36 02	356	47 56	070	55 11	123	18 03	173	29 33	209	33 07	271	33 52	299
173	35 58	356	48 51	070	56 01	124	18 10	174	29 05	209	32 08	271	33 00	298
174	35 53	355	49 47	070	56 50	125	18 17	174	28 35	210	31 08	271	32 08	298
175	35 48	355	50 43	069	57 38	126	18 23	175	28 05	210	30 09	271	31 15	298
176	35 42	354	51 38	069	58 26	127	18 28	175	27 35	211	29 09	272	30 23	298
177	35 36	354	52 34	069	59 13	128	18 33	175	27 04	212	28 10	272	29 30	298
178	35 29	353	53 29	068	60 00	129	18 37	176	26 33	212	27 11	272	28 38	298
179	35 21	352	54 24	068	60 45	131	18 41	176	26 01	213	26 11	272	27 45	298

LAT 8°N

LHA ♈	Hc	Zn	Hc	Zn	Hc	Zn	Hc	Zn	Hc	Zn	Hc	Zn	Hc	Zn
	♦ARCTURUS		ANTARES		RIGIL KENT.		♦ACRUX		Suhail		♦REGULUS		Dubhe	
180	55 19	068	16 18	121	14 27	161	18 45	177	25 28	213	62 11	281	35 13	352
181	56 14	067	17 09	121	14 46	162	18 48	177	24 55	214	61 12	280	35 05	351
182	57 09	067	18 00	121	15 04	162	18 51	178	24 22	214	60 14	280	34 55	351
183	58 03	066	18 51	122	15 22	162	18 53	178	23 48	215	59 15	280	34 45	350
184	58 58	066	19 41	122	15 40	163	18 54	179	23 14	215	58 17	280	34 35	350
185	59 52	065	20 32	122	15 58	163	18 55	179	22 40	216	57 18	280	34 24	349
186	60 45	065	21 22	123	16 15	164	18 56	180	22 05	216	56 20	280	34 13	349
187	61 39	064	22 12	123	16 31	164	18 56	180	21 29	217	55 21	280	34 01	348
188	62 32	063	23 02	123	16 47	164	18 55	181	20 54	217	54 22	279	33 48	348
189	63 25	063	23 51	124	17 03	165	18 54	181	20 17	218	53 24	279	33 35	347
190	64 18	062	24 41	124	17 18	165	18 53	182	19 41	218	52 25	279	33 22	347
191	65 10	061	25 30	124	17 33	166	18 51	182	19 04	218	51 27	279	33 08	346
192	66 01	060	26 19	125	17 48	166	18 48	183	18 27	219	50 28	279	32 53	346
193	66 52	059	27 07	125	18 02	167	18 45	183	17 50	219	49 29	279	32 38	345
194	67 43	058	27 56	126	18 15	167	18 42	184	17 12	220	48 31	279	32 23	345
	♦Alphecca		Rasalhague		ANTARES		♦RIGIL KENT.		ACRUX		♦REGULUS		Dubhe	
195	48 54	058	22 27	080	28 44	126	18 28	168	18 38	184	47 32	279	32 07	344
196	49 45	058	23 25	080	29 32	127	18 41	168	18 34	184	46 33	279	31 51	344
197	50 35	057	24 23	080	30 19	127	18 53	168	18 29	185	45 35	279	31 34	343
198	51 24	056	25 22	080	31 07	128	19 04	169	18 23	185	44 36	279	31 16	343
199	52 14	056	26 20	080	31 53	128	19 16	169	18 18	186	43 37	279	30 59	342
200	53 03	055	27 19	080	32 40	129	19 26	170	18 11	186	42 38	279	30 41	342
201	53 51	055	28 17	080	33 26	129	19 36	170	18 04	187	41 40	279	30 22	342
202	54 40	054	29 16	080	34 12	130	19 46	171	17 57	187	40 41	279	30 03	341
203	55 28	053	30 14	080	34 58	130	19 55	171	17 49	188	39 42	279	29 44	341
204	56 15	053	31 13	080	35 43	131	20 04	172	17 41	188	38 44	279	29 24	340
205	57 02	052	32 12	080	36 28	131	20 12	172	17 33	189	37 45	279	29 04	340
206	57 48	051	33 10	080	37 12	132	20 20	173	17 23	189	36 46	279	28 43	340
207	58 34	050	34 09	080	37 56	133	20 27	173	17 14	190	35 48	279	28 23	339
208	59 20	049	35 07	080	38 39	133	20 33	174	17 04	190	34 49	279	28 01	339
209	60 04	048	36 06	080	39 22	134	20 40	174	16 53	190	33 50	279	27 40	339
	♦VEGA		Rasalhague		ANTARES		♦RIGIL KENT.		SPICA		♦REGULUS		Dubhe	
210	21 11	051	37 04	080	40 04	135	20 45	175	68 58	205	32 52	279	27 18	338
211	21 58	051	38 03	080	40 46	136	20 50	175	68 32	207	31 53	279	26 56	338
212	22 44	051	39 01	080	41 27	136	20 55	176	68 04	209	30 54	279	26 33	338
213	23 30	051	40 00	080	42 08	137	20 59	176	67 34	212	29 56	279	26 11	337
214	24 16	051	40 59	080	42 48	138	21 02	177	67 02	214	28 57	279	25 47	337
215	25 02	051	41 57	080	43 28	139	21 05	177	66 28	216	27 58	279	25 24	337
216	25 48	051	42 56	080	44 07	140	21 07	178	65 52	218	27 00	279	25 00	336
217	26 34	050	43 54	080	44 45	141	21 09	179	65 15	220	26 01	279	24 36	336
218	27 20	050	44 53	080	45 22	141	21 10	179	64 37	221	25 03	280	24 12	336
219	28 06	050	45 51	080	45 59	142	21 11	180	63 57	223	24 04	280	23 48	336
220	28 51	050	46 50	080	46 35	143	21 11	180	63 16	225	23 05	280	23 23	335
221	29 36	050	47 48	080	47 10	144	21 11	181	62 34	226	22 07	280	22 58	335
222	30 22	049	48 47	080	47 44	145	21 10	181	61 50	227	21 08	280	22 33	335
223	31 07	049	49 45	080	48 17	146	21 08	182	61 06	229	20 10	280	22 08	335
224	31 52	049	50 44	080	48 50	147	21 06	182	60 21	230	19 11	280	21 42	334
	VEGA		♦ALTAIR		Nunki		ANTARES		RIGIL KENT.		♦SPICA		♦Alkaid	
225	32 36	049	18 16	083	23 31	123	49 21	149	21 04	183	59 35	231	45 57	343
226	33 21	048	19 15	083	24 21	124	49 52	150	21 01	183	58 48	233	45 39	342
227	34 05	048	20 14	083	25 10	124	50 21	151	20 57	184	58 01	234	45 20	341
228	34 49	048	21 13	084	25 59	124	50 49	152	20 53	184	57 12	235	45 01	341
229	35 33	047	22 12	084	26 48	125	51 17	153	20 48	185	56 24	236	44 41	340
230	36 17	047	23 11	084	27 36	125	51 43	155	20 43	185	55 34	237	44 20	339
231	37 00	047	24 10	084	28 25	126	52 08	156	20 37	186	54 44	238	43 58	338
232	37 43	046	25 09	084	29 13	126	52 31	157	20 31	186	53 54	238	43 36	338
233	38 26	046	26 08	084	30 01	127	52 54	159	20 24	187	53 03	239	43 13	337
234	39 08	045	27 08	084	30 48	127	53 15	160	20 17	187	52 12	240	42 49	336
235	39 51	045	28 07	084	31 35	128	53 34	161	20 09	188	51 20	241	42 25	335
236	40 32	045	29 06	084	32 22	128	53 53	163	20 01	188	50 28	242	42 00	335
237	41 14	044	30 05	084	33 09	129	54 10	164	19 52	189	49 36	242	41 34	334
238	41 55	044	31 04	084	33 55	129	54 25	166	19 43	189	48 43	243	41 08	333
239	42 36	043	32 03	085	34 40	130	54 39	167	19 33	190	47 50	244	40 41	333
	♦VEGA		ALTAIR		Nunki		♦ANTARES		SPICA		♦ARCTURUS		Alkaid	
240	43 16	043	33 02	085	35 26	130	54 52	169	46 57	244	62 17	297	40 13	332
241	43 56	042	34 01	085	36 11	131	55 03	170	46 03	245	61 24	296	39 45	332
242	44 36	041	35 01	085	36 55	132	55 12	172	45 09	245	60 30	295	39 17	331
243	45 15	041	36 00	085	37 40	132	55 20	173	44 15	246	59 36	295	38 48	330
244	45 54	040	36 59	085	38 23	133	55 26	175	43 21	246	58 42	294	38 18	330
245	46 32	040	37 58	085	39 06	134	55 30	176	42 26	247	57 48	294	37 48	329
246	47 09	039	38 57	085	39 49	134	55 33	178	41 31	247	56 53	293	37 17	329
247	47 46	038	39 57	085	40 31	135	55 35	180	40 37	248	55 59	293	36 47	328
248	48 23	037	40 56	085	41 13	136	55 34	181	39 41	248	55 04	292	36 15	328
249	48 59	037	41 55	085	41 54	137	55 32	183	38 46	249	54 09	292	35 43	327
250	49 34	036	42 54	085	42 34	137	55 29	184	37 51	249	53 13	292	35 11	327
251	50 08	035	43 53	085	43 14	138	55 23	186	36 55	250	52 18	291	34 39	327
252	50 42	034	44 53	085	43 54	139	55 16	187	35 59	250	51 23	291	34 06	326
253	51 15	033	45 52	086	44 32	140	55 08	189	35 04	250	50 27	291	33 32	326
254	51 47	032	46 51	086	45 10	141	54 58	191	34 08	251	49 31	290	32 59	325
	VEGA		♦ALTAIR		Nunki		♦ANTARES		SPICA		ARCTURUS		♦Alkaid	
255	52 19	032	47 50	086	45 47	142	54 46	192	33 11	251	48 36	290	32 25	325
256	52 49	031	48 50	086	46 23	143	54 33	194	32 15	251	47 40	290	31 50	325
257	53 19	030	49 49	086	46 59	144	54 18	195	31 19	252	46 44	290	31 16	324
258	53 48	028	50 48	086	47 34	145	54 02	196	30 22	252	45 48	289	30 41	324
259	54 16	027	51 47	086	48 07	146	53 45	198	29 26	252	44 52	289	30 05	324
260	54 43	026	52 47	086	48 40	147	53 26	199	28 29	253	43 56	289	29 30	323
261	55 08	025	53 46	086	49 12	148	53 05	201	27 32	253	42 59	289	28 54	323
262	55 33	024	54 45	086	49 43	149	52 44	202	26 35	253	42 03	289	28 18	323
263	55 57	023	55 44	086	50 13	150	52 21	203	25 39	254	41 07	289	27 42	322
264	56 19	022	56 44	086	50 42	151	51 57	205	24 42	254	40 10	288	27 06	322
265	56 41	020	57 43	086	51 10	153	51 31	206	23 44	254	39 14	288	26 29	322
266	57 01	019	58 42	086	51 37	154	51 04	207	22 47	254	38 18	288	25 52	322
267	57 19	018	59 42	086	52 02	155	50 37	208	21 50	255	37 21	288	25 15	321
268	57 37	016	60 41	086	52 27	156	50 08	210	20 53	255	36 25	288	24 38	321
269	57 53	015	61 40	086	52 50	158	49 38	211	19 55	255	35 28	288	24 00	321

LHA ♈	Hc	Zn	Hc	Zn	Hc	Zn	Hc	Zn	Hc	Zn	Hc	Zn	Hc	Zn
	VEGA		♦DENEB		Enif		♦Nunki		Shaula		ANTARES		♦ARCTURUS	
270	58 08	014	39 04	036	34 43	084	53 12	159	44 28	187	49 07	212	34 32	288
271	58 21	012	39 39	035	35 42	084	53 32	161	44 20	189	48 35	213	33 35	288
272	58 33	011	40 13	035	36 41	084	53 51	162	44 10	190	48 02	214	32 38	288
273	58 43	009	40 47	034	37 41	084	54 09	163	44 00	191	47 29	215	31 42	288
274	58 52	008	41 20	034	38 40	084	54 25	165	43 48	192	46 54	216	30 45	288
275	58 59	006	41 53	033	39 39	084	54 40	166	43 35	193	46 19	217	29 49	288
276	59 05	005	42 25	033	40 38	084	54 53	168	43 22	194	45 42	218	28 52	288
277	59 09	003	42 57	032	41 37	084	55 05	169	43 07	195	45 05	219	27 55	288
278	59 12	002	43 28	031	42 36	084	55 15	171	42 51	196	44 28	220	26 59	288
279	59 13	000	43 58	031	43 35	084	55 23	173	42 34	197	43 49	221	26 02	288
280	59 13	359	44 28	030	44 34	084	55 30	174	42 16	198	43 10	222	25 05	288
281	59 11	357	44 58	029	45 33	084	55 36	176	41 57	199	42 30	222	24 09	288
282	59 07	356	45 26	028	46 32	084	55 39	177	41 37	200	41 50	223	23 12	288
283	59 02	354	45 54	028	47 31	084	55 41	179	41 17	201	41 09	224	22 15	288
284	58 55	353	46 21	027	48 30	084	55 42	180	40 55	202	40 27	225	21 19	288
	DENEB		♦Alpheratz		FOMALHAUT		♦Nunki		ANTARES		♦Rasalhague		VEGA	
285	46 48	026	15 11	062	21 43	126	55 40	182	39 45	225	68 32	284	58 47	351
286	47 14	025	16 03	062	22 31	127	55 38	184	39 03	226	67 35	284	58 37	350
287	47 39	024	16 56	062	23 18	127	55 33	185	38 20	227	66 37	283	58 26	348
288	48 03	024	17 48	062	24 05	128	55 27	187	37 36	228	65 39	283	58 13	347
289	48 26	023	18 41	062	24 52	128	55 19	188	36 52	228	64 41	283	57 59	345
290	48 49	022	19 33	062	25 39	128	55 10	190	36 08	229	63 43	282	57 43	344
291	49 10	021	20 25	062	26 25	129	54 59	191	35 23	229	62 45	282	57 26	343
292	49 31	020	21 18	062	27 11	129	54 46	193	34 37	230	61 47	282	57 08	341
293	49 51	019	22 10	062	27 57	130	54 32	194	33 52	231	60 48	281	56 48	340
294	50 09	018	23 03	062	28 43	130	54 17	196	33 06	231	59 50	281	56 28	339
295	50 27	017	23 55	062	29 28	131	54 00	197	32 19	232	58 52	281	56 06	338
296	50 44	016	24 47	062	30 13	131	53 41	199	31 32	232	57 54	281	55 42	336
297	51 00	015	25 39	062	30 57	132	53 22	200	30 45	233	56 55	281	55 18	335
298	51 15	014	26 32	061	31 41	132	53 00	202	29 58	233	55 57	281	54 53	334
299	51 28	013	27 24	061	32 25	133	52 38	203	29 10	234	54 58	280	54 26	333
	♦DENEB		Alpheratz		♦FOMALHAUT		Peacock		♦ANTARES		Rasalhague		VEGA	
300	51 41	012	28 16	061	33 08	133	25 02	176	28 22	234	54 00	280	53 59	332
301	51 53	011	29 08	061	33 51	134	25 06	177	27 34	235	53 01	280	53 30	331
302	52 03	010	30 00	061	34 34	135	25 09	177	26 45	235	52 03	280	53 01	330
303	52 12	008	30 52	061	35 16	135	25 11	178	25 57	235	51 05	280	52 31	329
304	52 20	007	31 44	061	35 57	136	25 13	179	25 08	236	50 06	280	51 59	328
305	52 27	006	32 35	061	36 38	137	25 14	179	24 18	236	49 08	280	51 27	327
306	52 33	005	33 27	060	37 19	137	25 15	180	23 29	237	48 09	280	50 55	326
307	52 38	004	34 19	060	37 59	138	25 15	180	22 39	237	47 10	280	50 21	325
308	52 41	003	35 10	060	38 38	139	25 14	181	21 49	237	46 12	280	49 47	324
309	52 43	002	36 02	060	39 17	139	25 13	182	20 59	238	45 13	280	49 12	324
310	52 44	000	36 53	059	39 55	140	25 11	182	20 09	238	44 15	280	48 37	323
311	52 44	359	37 44	059	40 33	141	25 08	183	19 19	238	43 16	280	48 00	322
312	52 43	358	38 35	059	41 10	142	25 05	183	18 28	239	42 18	280	47 24	321
313	52 40	357	39 26	059	41 46	143	25 01	184	17 37	239	41 19	280	46 46	321
314	52 36	356	40 16	058	42 22	144	24 56	185	16 46	239	40 21	280	46 08	320
	Alpheratz		♦Diphda		FOMALHAUT		♦Nunki		Rasalhague		VEGA		♦DENEB	
315	41 07	058	29 04	116	42 57	144	44 12	220	39 22	280	45 30	319	52 31	355
316	41 57	058	29 57	116	43 31	145	43 33	221	38 23	280	44 51	319	52 25	353
317	42 47	057	30 51	117	44 05	146	42 54	222	37 25	280	44 12	318	52 17	352
318	43 37	057	31 44	117	44 37	147	42 13	223	36 26	280	43 32	318	52 09	351
319	44 27	057	32 36	117	45 09	148	41 33	224	35 28	280	42 52	317	51 59	350
320	45 17	056	33 29	118	45 40	149	40 51	225	34 29	280	42 11	317	51 48	349
321	46 06	056	34 22	118	46 10	150	40 09	225	33 31	280	41 30	316	51 37	348
322	46 55	055	35 14	119	46 39	151	39 27	226	32 32	280	40 48	316	51 23	347
323	47 44	055	36 06	119	47 07	152	38 44	227	31 34	280	40 07	315	51 09	346
324	48 32	054	36 57	120	47 34	153	38 00	227	30 35	280	39 25	315	50 54	345
325	49 20	054	37 49	120	48 00	155	37 17	228	29 37	280	38 42	314	50 38	344
326	50 08	053	38 40	121	48 25	156	36 32	229	28 38	280	38 00	314	50 21	343
327	50 56	053	39 31	121	48 49	157	35 47	229	27 39	280	37 17	313	50 03	342
328	51 43	052	40 21	122	49 12	158	35 02	230	26 41	280	36 33	313	49 43	341
329	52 30	051	41 12	123	49 34	159	34 17	230	25 43	280	35 50	313	49 23	340
	♦Alpheratz		Diphda		♦FOMALHAUT		Peacock		ALTAIR		♦VEGA		DENEB	
330	53 16	051	42 02	123	49 54	160	22 23	194	57 59	274	35 06	312	49 02	339
331	54 02	050	42 51	124	50 13	162	22 09	194	57 00	274	34 22	312	48 40	338
332	54 47	049	43 40	124	50 31	163	21 54	195	56 00	274	33 38	312	48 18	337
333	55 32	049	44 29	125	50 48	164	21 38	195	55 01	274	32 53	311	47 54	336
334	56 16	048	45 17	126	51 03	166	21 22	196	54 02	274	32 09	311	47 29	335
335	57 00	047	46 05	127	51 18	167	21 06	196	53 03	274	31 24	311	47 04	334
336	57 43	046	46 53	127	51 30	168	20 49	197	52 03	274	30 39	311	46 38	334
337	58 25	045	47 40	128	51 42	170	20 31	197	51 04	274	29 54	310	46 11	333
338	59 07	044	48 26	129	51 52	171	20 13	198	50 05	274	29 09	310	45 44	332
339	59 47	043	49 12	130	52 00	172	19 55	198	49 06	274	28 23	310	45 16	331
340	60 27	042	49 57	131	52 07	174	19 36	199	48 06	274	27 38	310	44 47	331
341	61 06	040	50 42	132	52 13	175	19 16	199	47 07	274	26 52	310	44 17	330
342	61 44	039	51 26	133	52 17	177	18 56	200	46 08	274	26 06	309	43 47	329
343	62 22	038	52 10	134	52 20	178	18 36	200	45 09	275	25 20	309	43 17	329
344	62 57	037	52 52	135	52 21	180	18 15	201	44 09	275	24 34	309	42 45	328
	♦Mirfak		Hamal		Acamar		♦ACHERNAR		FOMALHAUT		♦ALTAIR		DENEB	
345	21 29	039	42 42	065	17 02	137	17 16	159	52 21	181	43 10	275	42 13	327
346	22 06	039	43 36	065	17 43	137	17 37	159	52 19	182	42 11	275	41 41	327
347	22 44	039	44 30	065	18 23	137	17 58	160	52 16	184	41 12	275	41 08	326
348	23 21	039	45 24	065	19 03	138	18 18	160	52 11	185	40 13	275	40 35	325
349	23 58	039	46 17	064	19 43	138	18 38	161	52 05	187	39 13	275	40 01	325
350	24 35	038	47 11	064	20 22	139	18 58	161	51 58	188	38 14	275	39 26	324
351	25 12	038	48 04	064	21 02	139	19 17	162	51 49	189	37 15	275	38 52	324
352	25 48	038	48 57	063	21 41	139	19 35	162	51 38	191	36 16	275	38 16	323
353	26 25	038	49 50	063	22 19	140	19 53	163	51 26	192	35 17	275	37 41	323
354	27 01	037	50 43	062	22 57	140	20 11	163	51 13	193	34 17	275	37 05	322
355	27 37	037	51 35	062	23 35	141	20 28	164	50 59	195	33 18	275	36 28	322
356	28 13	037	52 28	062	24 12	141	20 44	164	50 43	196	32 19	275	35 52	322
357	28 48	037	53 20	061	24 49	142	21 00	165	50 26	197	31 20	276	35 14	321
358	29 23	036	54 12	061	25 26	142	21 16	165	50 07	199	30 21	276	34 37	321
359	29 58	036	55 03	060	26 02	143	21 31	166	49 48	200	29 22	276	33 59	320

LHA ♈	Hc	Zn	Hc	Zn	Hc	Zn	Hc	Zn	Hc	Zn	Hc	Zn	Hc	Zn
	◆CAPELLA		ALDEBARAN		RIGEL		◆ACHERNAR		FOMALHAUT		◆ALTAIR		DENEB	
0	12 37	044	22 09	075	10 12	100	22 44	166	50 23	202	28 16	276	32 35	320
1	13 18	044	23 07	075	11 11	100	22 58	167	50 00	203	27 17	276	31 57	320
2	14 00	044	24 04	075	12 10	100	23 12	167	49 37	204	26 18	276	31 19	320
3	14 41	044	25 02	075	13 08	100	23 25	168	49 12	205	25 19	277	30 40	319
4	15 23	044	25 59	075	14 07	100	23 37	168	48 46	206	24 20	277	30 01	319
5	16 04	044	26 57	075	15 06	100	23 49	169	48 19	207	23 20	277	29 22	319
6	16 46	044	27 54	075	16 04	101	24 00	169	47 51	209	22 21	277	28 43	318
7	17 27	044	28 52	075	17 03	101	24 11	170	47 22	210	21 22	277	28 03	318
8	18 08	044	29 49	075	18 01	101	24 22	170	46 52	211	20 23	277	27 23	318
9	18 49	044	30 47	075	19 00	101	24 31	171	46 21	212	19 24	277	26 43	318
10	19 30	044	31 44	075	19 58	101	24 40	171	45 50	213	18 25	277	26 03	317
11	20 11	043	32 42	075	20 56	102	24 49	172	45 17	214	17 26	277	25 23	317
12	20 52	043	33 39	075	21 55	102	24 57	173	44 44	215	16 27	277	24 42	317
13	21 33	043	34 37	075	22 53	102	25 04	173	44 10	216	15 28	277	24 02	317
14	22 13	043	35 34	075	23 51	102	25 11	174	43 35	216	14 28	277	23 21	317
	CAPELLA		◆ALDEBARAN		RIGEL		ACHERNAR		◆FOMALHAUT		Enif		◆DENEB	
15	22 54	043	36 32	075	24 49	102	25 17	174	42 59	217	41 28	277	22 40	316
16	23 34	043	37 29	075	25 47	103	25 22	175	42 22	218	40 29	277	21 59	316
17	24 14	042	38 27	074	26 46	103	25 27	176	41 45	219	39 30	277	21 17	316
18	24 54	042	39 24	074	27 44	103	25 32	176	41 08	220	38 31	277	20 36	316
19	25 34	042	40 21	074	28 42	103	25 35	177	40 29	221	37 32	277	19 54	316
20	26 14	042	41 19	074	29 39	104	25 38	177	39 50	221	36 33	277	19 13	316
21	26 53	041	42 16	074	30 37	104	25 41	178	39 11	222	35 34	277	18 31	316
22	27 33	041	43 13	074	31 35	104	25 43	179	38 30	223	34 34	277	17 49	315
23	28 12	041	44 10	074	32 33	104	25 44	179	37 50	223	33 35	277	17 08	315
24	28 50	041	45 07	074	33 30	105	25 44	180	37 09	224	32 36	277	16 26	315
25	29 29	040	46 04	073	34 28	105	25 44	180	36 27	225	31 37	277	15 44	315
26	30 07	040	47 02	073	35 25	105	25 43	181	35 45	225	30 38	277	15 02	315
27	30 46	040	47 59	073	36 23	106	25 42	182	35 02	226	29 39	277	14 20	315
28	31 23	039	48 55	073	37 20	106	25 40	182	34 19	227	28 40	277	13 38	315
29	32 01	039	49 52	073	38 17	106	25 38	183	33 36	227	27 41	277	12 55	315
	Mirfak		CAPELLA		◆ALDEBARAN		RIGEL		◆ACHERNAR		FOMALHAUT		◆Alpheratz	
30	43 41	019	32 38	039	50 49	072	39 15	107	25 34	183	32 52	228	55 38	313
31	44 00	018	33 15	038	51 46	072	40 12	107	25 31	184	32 08	228	54 54	313
32	44 18	017	33 52	038	52 43	072	41 08	107	25 26	185	31 23	229	54 10	312
33	44 35	016	34 28	037	53 39	072	42 05	108	25 21	185	30 38	229	53 26	311
34	44 51	015	35 04	037	54 36	071	43 02	108	25 16	186	29 53	230	52 40	310
35	45 07	015	35 40	037	55 32	071	43 58	109	25 09	186	29 07	230	51 55	310
36	45 21	014	36 15	036	56 28	071	44 55	109	25 02	187	28 21	231	51 09	309
37	45 35	013	36 50	036	57 24	070	45 51	109	24 55	187	27 35	231	50 22	308
38	45 48	012	37 24	035	58 20	070	46 47	110	24 47	188	26 48	232	49 35	308
39	46 00	011	37 58	035	59 16	069	47 43	110	24 38	189	26 01	232	48 48	307
40	46 11	010	38 32	034	60 12	069	48 39	111	24 29	189	25 14	232	48 00	307
41	46 21	009	39 05	034	61 07	068	49 34	111	24 19	190	24 27	233	47 12	306
42	46 30	008	39 38	033	62 02	068	50 30	112	24 09	190	23 39	233	46 24	306
43	46 38	008	40 10	032	62 57	067	51 25	113	23 58	191	22 51	234	45 35	305
44	46 45	007	40 41	032	63 52	066	52 19	113	23 46	191	22 03	234	44 46	305
	◆CAPELLA		PROCYON		SIRIUS		◆CANOPUS		ACHERNAR		Diphda		◆Alpheratz	
45	41 12	031	20 40	087	29 34	114	16 23	151	23 34	192	48 01	233	43 57	304
46	41 43	031	21 40	087	30 28	114	16 52	151	23 22	193	47 13	234	43 08	304
47	42 13	030	22 39	087	31 22	114	17 20	151	23 08	193	46 25	234	42 18	303
48	42 42	029	23 39	087	32 17	115	17 49	152	22 55	194	45 36	235	41 28	303
49	43 11	029	24 38	087	33 10	115	18 17	152	22 40	194	44 47	236	40 38	303
50	43 39	028	25 38	088	34 04	116	18 44	153	22 26	195	43 58	237	39 48	302
51	44 07	027	26 37	088	34 58	116	19 12	153	22 10	195	43 08	237	38 57	302
52	44 33	026	27 37	088	35 51	116	19 38	153	21 54	196	42 17	238	38 07	302
53	45 00	026	28 36	088	36 44	117	20 05	154	21 38	196	41 27	238	37 16	301
54	45 25	025	29 36	088	37 37	117	20 31	154	21 21	197	40 36	239	36 25	301
55	45 49	024	30 35	088	38 30	118	20 56	155	21 04	197	39 45	240	35 34	301
56	46 13	023	31 35	088	39 23	118	21 21	155	20 46	198	38 53	240	34 43	301
57	46 36	022	32 35	088	40 15	119	21 46	156	20 28	198	38 02	241	33 52	300
58	46 59	021	33 34	088	41 07	119	22 10	156	20 09	199	37 10	241	33 00	300
59	47 20	021	34 34	088	41 59	120	22 34	157	19 50	199	36 17	242	32 09	300
	◆CAPELLA		POLLUX		PROCYON		◆SIRIUS		CANOPUS		◆Diphda		Alpheratz	
60	47 40	020	32 57	061	35 33	089	42 50	121	22 57	157	35 25	242	31 17	300
61	48 00	019	33 49	061	36 33	089	43 41	121	23 20	158	34 32	242	30 26	300
62	48 19	018	34 41	061	37 32	089	44 32	122	23 42	158	33 39	243	29 34	300
63	48 37	017	35 33	060	38 32	089	45 22	122	24 04	159	32 46	243	28 42	299
64	48 54	016	36 25	060	39 31	089	46 12	123	24 25	159	31 53	244	27 50	299
65	49 09	015	37 16	060	40 31	089	47 02	124	24 46	160	30 59	244	26 58	299
66	49 24	014	38 08	060	41 30	089	47 51	125	25 06	160	30 06	244	26 06	299
67	49 38	013	38 59	059	42 30	089	48 40	125	25 26	161	29 12	245	25 14	299
68	49 51	012	39 50	059	43 29	089	49 29	126	25 45	162	28 18	245	24 22	299
69	50 03	011	40 41	059	44 29	090	50 16	127	26 03	162	27 24	246	23 30	299
70	50 14	010	41 32	058	45 29	090	51 04	128	26 21	163	26 30	246	22 37	299
71	50 23	009	42 23	058	46 28	090	51 51	129	26 39	163	25 35	246	21 45	299
72	50 32	008	43 13	058	47 28	090	52 37	130	26 55	164	24 41	246	20 53	298
73	50 39	007	44 03	057	48 27	090	53 22	131	27 12	165	23 46	247	20 00	298
74	50 46	006	44 53	057	49 27	090	54 07	132	27 27	165	22 51	247	19 08	298
	CAPELLA		◆POLLUX		PROCYON		SIRIUS		◆CANOPUS		Acamar		◆Hamal	
75	50 51	004	45 43	056	50 26	090	54 51	133	27 42	166	34 59	208	45 21	296
76	50 55	003	46 33	056	51 26	090	55 35	134	27 56	166	34 30	209	44 28	296
77	50 58	002	47 22	056	52 25	091	56 17	135	28 10	167	34 01	210	43 34	296
78	51 00	001	48 11	055	53 25	091	56 59	136	28 23	168	33 31	210	42 41	296
79	51 00	000	48 59	055	54 24	091	57 40	137	28 35	168	33 01	211	41 47	295
80	51 00	359	49 48	054	55 24	091	58 20	139	28 47	169	32 30	212	40 53	295
81	50 58	358	50 36	053	56 24	091	58 59	140	28 58	170	31 59	212	39 59	295
82	50 56	357	51 23	053	57 23	091	59 36	141	29 08	170	31 26	213	39 05	295
83	50 52	356	52 11	052	58 23	091	60 13	143	29 18	171	30 54	214	38 11	294
84	50 47	355	52 57	051	59 22	092	60 48	144	29 27	172	30 21	214	37 17	294
85	50 40	354	53 44	051	60 22	092	61 22	146	29 35	172	29 47	215	36 22	294
86	50 33	352	54 30	050	61 21	092	61 55	148	29 42	173	29 13	215	35 28	294
87	50 25	351	55 15	049	62 21	092	62 26	149	29 49	174	28 38	216	34 33	294
88	50 15	350	56 00	048	63 20	092	62 55	151	29 55	174	28 03	216	33 39	294
89	50 05	349	56 44	047	64 20	092	63 23	153	30 01	175	27 27	217	32 44	293

LAT 7°N

LHA ♈	Hc	Zn	Hc	Zn	Hc	Zn	Hc	Zn	Hc	Zn	Hc	Zn	Hc	Zn
	POLLUX		◆REGULUS		Suhail		◆CANOPUS		Acamar		◆ALDEBARAN		CAPELLA	
90	57 27	047	28 44	080	24 06	144	30 05	176	26 51	218	67 18	297	49 53	348
91	58 10	046	29 43	080	24 41	145	30 09	177	26 15	218	66 24	296	49 40	347
92	58 52	045	30 41	080	25 15	145	30 13	177	25 38	219	65 31	295	49 27	346
93	59 34	043	31 40	080	25 48	146	30 15	178	25 00	219	64 36	294	49 12	345
94	60 14	042	32 39	080	26 21	146	30 17	179	24 23	220	63 42	293	48 56	344
95	60 54	041	33 37	080	26 54	147	30 18	179	23 45	220	62 47	293	48 39	343
96	61 33	040	34 36	080	27 26	148	30 18	180	23 06	220	61 52	292	48 22	342
97	62 10	039	35 35	080	27 58	148	30 18	181	22 27	221	60 57	292	48 03	341
98	62 47	037	36 33	080	28 29	149	30 17	181	21 48	221	60 01	291	47 44	340
99	63 22	036	37 32	080	29 00	149	30 15	182	21 09	222	59 06	291	47 23	340
100	63 56	034	38 31	080	29 30	150	30 12	183	20 29	222	58 10	290	47 02	339
101	64 29	033	39 30	080	30 00	150	30 09	184	19 49	223	57 14	290	46 40	338
102	65 01	031	40 28	080	30 29	151	30 05	184	19 09	223	56 18	289	46 17	337
103	65 30	029	41 27	080	30 57	152	30 00	185	18 28	223	55 22	289	45 53	336
104	65 59	027	42 26	080	31 25	152	29 55	186	17 47	224	54 25	289	45 29	335
	POLLUX		◆REGULUS		Suhail		◆CANOPUS		RIGEL		ALDEBARAN		◆CAPELLA	
105	66 25	026	43 24	080	31 52	153	29 49	186	59 35	240	53 29	288	45 04	335
106	66 50	024	44 23	080	32 19	154	29 42	187	58 43	241	52 32	288	44 38	334
107	67 13	021	45 22	080	32 45	154	29 34	188	57 50	242	51 36	288	44 11	333
108	67 33	019	46 20	080	33 10	155	29 26	188	56 57	243	50 39	288	43 44	332
109	67 52	017	47 19	080	33 35	156	29 17	189	56 04	244	49 42	287	43 16	332
110	68 09	015	48 17	080	33 59	157	29 07	190	55 10	245	48 45	287	42 47	331
111	68 23	013	49 16	080	34 22	157	28 57	190	54 16	245	47 48	287	42 18	330
112	68 35	010	50 15	080	34 45	158	28 46	191	53 22	246	46 51	287	41 48	330
113	68 44	008	51 13	080	35 07	159	28 34	192	52 28	247	45 54	287	41 17	329
114	68 51	006	52 12	079	35 28	160	28 22	192	51 33	247	44 57	286	40 46	328
115	68 56	003	53 10	079	35 48	160	28 09	193	50 38	248	44 00	286	40 15	328
116	68 58	001	54 09	079	36 08	161	27 55	194	49 42	248	43 03	286	39 43	327
117	68 57	358	55 07	079	36 26	162	27 41	194	48 47	249	42 05	286	39 10	327
118	68 54	356	56 06	079	36 44	163	27 26	195	47 51	250	41 08	286	38 37	326
119	68 48	353	57 04	079	37 01	164	27 10	195	46 55	250	40 11	286	38 04	326
	Dubhe		◆Denebola		Suhail		◆SIRIUS		RIGEL		ALDEBARAN		◆CAPELLA	
120	25 44	022	33 26	077	37 18	165	59 53	218	45 59	250	39 13	286	37 30	325
121	26 07	022	34 24	077	37 33	165	59 16	219	45 03	251	38 16	286	36 55	325
122	26 29	021	35 22	077	37 48	166	58 38	221	44 07	251	37 19	285	36 21	324
123	26 50	021	36 20	077	38 01	167	57 58	222	43 10	252	36 21	285	35 45	324
124	27 12	021	37 18	077	38 14	168	57 18	223	42 14	252	35 24	285	35 10	323
125	27 32	020	38 16	077	38 26	169	56 37	225	41 17	253	34 26	285	34 34	323
126	27 53	020	39 14	077	38 37	170	55 55	226	40 20	253	33 29	285	33 58	322
127	28 13	020	40 12	077	38 47	171	55 12	227	39 23	253	32 31	285	33 21	322
128	28 33	019	41 10	077	38 56	172	54 28	228	38 26	254	31 34	285	32 44	321
129	28 53	019	42 08	077	39 04	173	53 43	229	37 29	254	30 36	285	32 07	321
130	29 12	018	43 06	077	39 12	173	52 58	230	36 31	254	29 39	285	31 29	321
131	29 30	018	44 04	076	39 18	174	52 12	231	35 34	255	28 41	285	30 52	320
132	29 49	018	45 02	076	39 23	175	51 25	232	34 37	255	27 44	285	30 14	320
133	30 07	017	46 00	076	39 28	176	50 38	233	33 39	255	26 46	285	29 35	320
134	30 24	017	46 58	076	39 31	177	49 51	234	32 41	256	25 49	285	28 57	319
	Dubhe		◆ARCTURUS		SPICA		◆Suhail		SIRIUS		BETELGEUSE		◆CAPELLA	
135	30 41	016	12 47	072	21 39	105	39 33	178	49 03	234	44 07	274	28 18	319
136	30 58	016	13 43	072	22 37	105	39 35	179	48 14	235	43 07	274	27 39	319
137	31 14	015	14 40	072	23 34	105	39 35	180	47 25	236	42 08	274	27 00	319
138	31 29	015	15 37	072	24 31	106	39 35	181	46 36	237	41 08	274	26 20	318
139	31 45	015	16 33	072	25 29	106	39 33	182	45 46	237	40 09	274	25 41	318
140	31 59	014	17 30	072	26 26	106	39 31	183	44 55	238	39 09	274	25 01	318
141	32 14	014	18 27	072	27 23	106	39 27	184	44 05	239	38 10	274	24 21	318
142	32 27	013	19 23	072	28 20	107	39 23	185	43 14	239	37 11	274	23 41	318
143	32 41	013	20 20	072	29 17	107	39 17	186	42 23	240	36 11	274	23 00	317
144	32 53	012	21 17	072	30 14	107	39 11	187	41 31	240	35 12	274	22 20	317
145	33 06	012	22 13	072	31 11	108	39 04	188	40 39	241	34 12	274	21 39	317
146	33 17	011	23 10	072	32 08	108	38 55	188	39 47	241	33 13	274	20 59	317
147	33 28	011	24 06	072	33 04	108	38 46	189	38 55	242	32 14	274	20 18	317
148	33 39	010	25 03	072	34 01	109	38 36	190	38 02	242	31 14	274	19 37	317
149	33 49	009	26 00	072	34 57	109	38 25	191	37 09	243	30 15	275	18 56	316
	Dubhe		◆ARCTURUS		SPICA		ACRUX		◆Suhail		SIRIUS		◆POLLUX	
150	33 59	009	26 56	072	35 54	109	14 37	164	38 13	192	36 16	243	51 47	308
151	34 08	008	27 53	072	36 50	110	14 53	164	38 00	193	35 23	244	51 00	307
152	34 16	008	28 50	072	37 46	110	15 09	165	37 46	194	34 29	244	50 12	306
153	34 24	007	29 46	072	38 42	110	15 25	165	37 32	195	33 35	245	49 24	306
154	34 31	007	30 43	072	39 37	111	15 40	165	37 16	196	32 42	245	48 35	305
155	34 38	006	31 39	072	40 33	111	15 55	166	37 00	196	31 48	245	47 47	305
156	34 44	006	32 36	072	41 28	112	16 10	166	36 43	197	30 53	246	46 57	304
157	34 50	005	33 32	072	42 24	112	16 24	167	36 25	198	29 59	246	46 08	304
158	34 55	005	34 29	071	43 19	113	16 37	167	36 06	199	29 04	246	45 18	303
159	34 59	004	35 25	071	44 14	113	16 51	167	35 46	200	28 10	247	44 29	303
160	35 03	003	36 22	071	45 08	114	17 03	168	35 26	200	27 15	247	43 38	303
161	35 06	003	37 18	071	46 03	114	17 16	168	35 05	201	26 20	247	42 48	302
162	35 09	002	38 14	071	46 57	115	17 28	169	34 43	202	25 25	248	41 58	302
163	35 11	002	39 11	071	47 51	115	17 39	169	34 20	203	24 30	248	41 07	301
164	35 12	001	40 07	071	48 45	116	17 51	169	33 57	203	23 35	248	40 16	301
	Dubhe		◆ARCTURUS		SPICA		◆ACRUX		Suhail		PROCYON		◆POLLUX	
165	35 13	000	41 03	071	49 38	116	18 01	170	33 33	204	40 02	271	39 25	301
166	35 13	000	41 59	070	50 32	117	18 11	170	33 08	205	39 03	271	38 34	301
167	35 13	359	42 55	070	51 25	118	18 21	171	32 43	206	38 03	271	37 42	300
168	35 12	359	43 51	070	52 17	118	18 30	171	32 16	206	37 04	271	36 51	300
169	35 10	358	44 47	070	53 09	119	18 39	172	31 50	207	36 04	271	35 59	300
170	35 08	358	45 43	070	54 01	120	18 48	172	31 22	208	35 05	271	35 07	300
171	35 05	357	46 39	069	54 52	121	18 56	173	30 54	208	34 05	272	34 16	299
172	35 02	356	47 35	069	55 43	122	19 03	173	30 26	209	33 05	272	33 24	299
173	34 58	356	48 30	069	56 34	123	19 10	174	29 57	210	32 06	272	32 31	299
174	34 53	355	49 26	069	57 24	124	19 16	174	29 27	210	31 06	272	31 39	299
175	34 48	355	50 21	068	58 13	125	19 22	174	28 57	211	30 07	272	30 47	299
176	34 43	354	51 16	068	59 02	126	19 28	175	28 26	211	29 07	272	29 55	298
177	34 36	354	52 11	068	59 50	127	19 33	175	27 55	212	28 08	272	29 02	298
178	34 29	353	53 06	067	60 37	128	19 37	176	27 23	212	27 08	272	28 10	298
179	34 22	353	54 01	067	61 24	129	19 41	176	26 51	213	26 09	272	27 17	298

 LAT 7°N

LHA ♈	Hc	Zn	Hc	Zn	Hc	Zn	Hc	Zn	Hc	Zn	Hc	Zn	Hc	Zn
	◆ARCTURUS		ANTARES		RIGIL KENT.		◆ACRUX		Suhail		◆REGULUS		Dubhe	
180	54 56	066	16 48	120	15 23	161	19 45	177	26 18	214	61 59	282	34 14	352
181	55 50	066	17 40	121	15 42	161	19 48	177	25 45	214	61 01	282	34 05	351
182	56 44	065	18 31	121	16 01	162	19 50	178	25 12	215	60 02	282	33 56	351
183	57 38	065	19 22	121	16 19	162	19 53	178	24 38	215	59 04	282	33 46	350
184	58 32	064	20 13	122	16 37	163	19 54	179	24 03	216	58 06	281	33 36	350
185	59 26	064	21 03	122	16 55	163	19 55	179	23 28	216	57 07	281	33 25	349
186	60 19	063	21 54	122	17 12	163	19 56	180	22 53	217	56 09	281	33 14	349
187	61 12	062	22 44	123	17 29	164	19 56	180	22 17	217	55 10	281	33 02	348
188	62 04	062	23 34	123	17 45	164	19 55	181	21 41	217	54 12	281	32 50	348
189	62 57	061	24 24	123	18 01	165	19 54	181	21 05	218	53 13	281	32 37	347
190	63 48	060	25 14	124	18 16	165	19 53	182	20 28	218	52 15	281	32 23	347
191	64 40	059	26 03	124	18 31	166	19 51	182	19 51	219	51 16	280	32 10	346
192	65 30	058	26 53	124	18 46	166	19 48	183	19 14	219	50 18	280	31 55	346
193	66 21	057	27 42	125	19 00	167	19 45	183	18 36	219	49 19	280	31 40	345
194	67 10	056	28 31	125	19 14	167	19 42	184	17 58	220	48 21	280	31 25	345
	◆Alphecca		Rasalhague		ANTARES		◆RIGIL KENT.		ACRUX		◆REGULUS		Dubhe	
195	48 22	057	22 16	079	29 19	126	19 27	167	19 38	184	47 22	280	31 09	344
196	49 12	057	23 14	079	30 07	126	19 39	168	19 34	185	46 23	280	30 53	344
197	50 01	056	24 13	079	30 55	127	19 52	168	19 29	185	45 25	280	30 36	343
198	50 51	055	25 11	079	31 43	127	20 03	169	19 23	185	44 26	280	30 19	343
199	51 40	055	26 10	079	32 30	128	20 15	169	19 17	186	43 27	280	30 02	343
200	52 28	054	27 08	079	33 17	128	20 25	170	19 11	186	42 29	280	29 44	342
201	53 16	054	28 07	079	34 04	129	20 36	170	19 04	187	41 30	280	29 25	342
202	54 04	053	29 05	079	34 50	129	20 45	171	18 57	187	40 31	280	29 06	341
203	54 51	052	30 04	080	35 36	130	20 55	171	18 49	188	39 33	280	28 47	341
204	55 38	051	31 02	080	36 22	130	21 03	172	18 41	188	38 34	280	28 27	341
205	56 24	051	32 01	080	37 07	131	21 12	172	18 32	189	37 35	280	28 08	340
206	57 10	050	33 00	080	37 52	132	21 19	173	18 23	189	36 37	280	27 47	340
207	57 55	049	33 58	080	38 36	132	21 26	173	18 13	190	35 38	280	27 26	339
208	58 40	048	34 57	080	39 20	133	21 33	174	18 03	190	34 39	280	27 05	339
209	59 24	047	35 55	080	40 04	134	21 39	174	17 52	190	33 40	280	26 44	339
	◆VEGA		Rasalhague		ANTARES		◆RIGIL KENT.		ACRUX		◆REGULUS		Dubhe	
210	20 34	051	36 54	080	40 46	134	21 45	175	17 41	191	32 42	280	26 22	338
211	21 20	051	37 52	080	41 29	135	21 50	175	17 30	191	31 43	280	26 00	338
212	22 06	051	38 51	079	42 11	136	21 54	176	17 18	192	30 44	280	25 38	338
213	22 52	051	39 49	079	42 52	137	21 58	176	17 05	192	29 46	280	25 15	337
214	23 38	051	40 48	079	43 33	137	22 02	177	16 53	193	28 47	280	24 52	337
215	24 24	050	41 47	079	44 13	138	22 05	177	16 40	193	27 48	280	24 29	337
216	25 10	050	42 45	079	44 52	139	22 07	178	16 26	193	26 50	280	24 05	337
217	25 56	050	43 44	079	45 31	140	22 09	179	16 12	194	25 51	280	23 42	336
218	26 41	050	44 42	079	46 09	141	22 10	179	15 57	194	24 52	280	23 18	336
219	27 27	050	45 41	079	46 46	142	22 11	180	15 43	195	23 54	280	22 53	336
220	28 12	049	46 39	079	47 23	143	22 11	180	15 27	195	22 55	280	22 29	335
221	28 57	049	47 38	079	47 58	144	22 11	181	15 12	195	21 57	280	22 04	335
222	29 42	049	48 36	079	48 33	145	22 10	181	14 56	196	20 58	280	21 39	335
223	30 27	049	49 34	079	49 07	146	22 08	182	14 39	196	19 59	280	21 13	335
224	31 12	048	50 33	079	49 40	147	22 06	182	14 23	197	19 01	280	20 48	335
	VEGA		◆ALTAIR		Nunki		ANTARES		RIGIL KENT.		◆SPICA		◆Alkaid	
225	31 56	048	18 09	083	24 04	123	50 12	148	22 04	183	60 12	233	44 59	343
226	32 41	048	19 08	083	24 54	123	50 43	149	22 01	183	59 24	234	44 42	342
227	33 25	048	20 07	083	25 43	124	51 13	150	21 57	184	58 36	235	44 23	342
228	34 09	047	21 06	083	26 33	124	51 42	152	21 53	184	57 47	236	44 04	341
229	34 52	047	22 05	083	27 22	124	52 10	153	21 48	185	56 57	237	43 44	340
230	35 35	047	23 04	083	28 11	125	52 37	154	21 43	185	56 07	238	43 24	339
231	36 19	046	24 04	083	29 00	125	53 02	155	21 37	186	55 16	239	43 02	339
232	37 01	046	25 03	083	29 48	126	53 26	157	21 31	186	54 25	240	42 40	338
233	37 44	045	26 02	084	30 36	126	53 49	158	21 24	187	53 33	240	42 18	337
234	38 26	045	27 01	084	31 24	127	54 11	159	21 16	187	52 41	241	41 54	336
235	39 08	044	28 00	084	32 12	127	54 31	161	21 09	188	51 49	242	41 30	336
236	39 50	044	29 00	084	32 59	128	54 50	162	21 00	188	50 56	243	41 05	335
237	40 31	044	29 59	084	33 46	128	55 07	164	20 51	189	50 03	243	40 40	334
238	41 12	043	30 58	084	34 32	129	55 23	165	20 42	189	49 10	244	40 14	334
239	41 52	042	31 57	084	35 19	129	55 38	167	20 32	190	48 16	245	39 47	333
	VEGA		◆ALTAIR		Nunki		◆ANTARES		SPICA		◆ARCTURUS		Alkaid	
240	42 32	042	32 56	084	36 05	130	55 50	168	47 22	245	61 49	298	39 20	333
241	43 12	041	33 56	084	36 50	131	56 02	170	46 28	246	60 57	297	38 52	332
242	43 51	041	34 55	084	37 35	131	56 11	171	45 34	246	60 04	297	38 24	331
243	44 29	040	35 54	084	38 20	132	56 19	173	44 39	247	59 11	296	37 55	331
244	45 08	040	36 53	084	39 04	132	56 26	175	43 44	247	58 17	296	37 26	330
245	45 45	039	37 53	084	39 48	133	56 30	176	42 49	248	57 23	295	36 56	330
246	46 22	038	38 52	084	40 31	134	56 33	178	41 54	248	56 29	295	36 26	329
247	46 59	038	39 51	084	41 14	135	56 35	180	40 59	249	55 35	294	35 55	329
248	47 35	037	40 50	084	41 56	135	56 34	181	40 03	249	54 40	294	35 24	328
249	48 10	036	41 50	084	42 37	136	56 32	183	39 08	249	53 46	293	34 53	328
250	48 45	035	42 49	084	43 18	137	56 28	184	38 12	250	52 51	293	34 21	327
251	49 19	034	43 48	084	43 59	138	56 23	186	37 16	250	51 56	292	33 48	327
252	49 52	034	44 47	084	44 39	138	56 16	188	36 20	251	51 01	292	33 16	326
253	50 25	033	45 47	084	45 18	139	56 07	189	35 23	251	50 05	292	32 43	326
254	50 56	032	46 46	085	45 56	140	55 57	191	34 27	251	49 10	291	32 09	326
	VEGA		◆ALTAIR		Nunki		◆ANTARES		SPICA		ARCTURUS		◆Alkaid	
255	51 27	031	47 45	085	46 34	141	55 45	192	33 31	252	48 14	291	31 35	325
256	51 57	030	48 44	085	47 11	142	55 31	194	32 34	252	47 19	291	31 01	325
257	52 27	029	49 44	085	47 47	143	55 16	195	31 37	252	46 23	291	30 27	325
258	52 55	028	50 43	085	48 22	144	55 00	197	30 41	253	45 27	290	29 52	324
259	53 22	027	51 42	085	48 57	145	54 42	198	29 44	253	44 31	290	29 17	324
260	53 49	026	52 42	085	49 30	146	54 22	200	28 47	253	43 36	290	28 42	324
261	54 14	025	53 41	085	50 03	147	54 01	201	27 50	253	42 40	290	28 06	323
262	54 38	023	54 40	085	50 35	148	53 39	203	26 53	254	41 43	290	27 31	323
263	55 01	022	55 39	085	51 05	150	53 16	204	25 55	254	40 47	289	26 54	323
264	55 23	021	56 39	085	51 35	151	52 51	205	24 58	254	39 51	289	26 18	322
265	55 44	020	57 38	084	52 03	152	52 25	207	24 01	254	38 55	289	25 42	322
266	56 04	019	58 37	084	52 31	153	51 58	208	23 03	255	37 59	289	25 05	322
267	56 22	017	59 37	084	52 57	155	51 29	209	22 06	255	37 02	289	24 28	322
268	56 39	016	60 36	084	53 22	156	51 00	210	21 08	255	36 06	289	23 51	321
269	56 55	015	61 35	084	53 45	157	50 29	211	20 11	255	35 09	289	23 14	321

LHA ♈	Hc	Zn	Hc	Zn	Hc	Zn	Hc	Zn	Hc	Zn	Hc	Zn	Hc	Zn
	VEGA		◆DENEB		Enif		◆Nunki		Shaula		ANTARES		◆ARCTURUS	
270	57 09	013	38 16	035	34 36	083	54 08	159	45 28	188	49 58	213	34 13	289
271	57 22	012	38 50	035	35 35	083	54 29	160	45 19	189	49 25	214	33 16	288
272	57 34	010	39 24	034	36 34	083	54 48	162	45 09	190	48 52	215	32 20	288
273	57 44	009	39 57	034	37 34	083	55 06	163	44 59	191	48 18	216	31 23	288
274	57 53	008	40 30	033	38 33	083	55 23	165	44 47	192	47 42	217	30 27	288
275	58 00	006	41 03	033	39 32	083	55 38	166	44 34	193	47 06	218	29 30	288
276	58 05	005	41 34	032	40 31	083	55 52	168	44 20	194	46 29	219	28 34	288
277	58 10	003	42 06	031	41 30	083	56 04	169	44 05	195	45 52	220	27 37	288
278	58 12	002	42 36	031	42 29	083	56 14	171	43 49	196	45 13	221	26 40	288
279	58 13	000	43 07	030	43 28	083	56 23	172	43 31	197	44 34	221	25 44	288
280	58 13	359	43 36	029	44 27	083	56 30	174	43 13	198	43 55	222	24 47	288
281	58 11	357	44 05	029	45 26	083	56 35	176	42 54	199	43 14	223	23 50	288
282	58 07	356	44 33	028	46 26	083	56 39	177	42 34	200	42 33	224	22 54	288
283	58 02	354	45 01	027	47 25	083	56 41	179	42 13	201	41 52	225	21 57	288
284	57 56	353	45 28	026	48 24	083	56 42	180	41 51	202	41 10	225	21 00	288
	DENEB		◆Alpheratz		FOMALHAUT		◆Nunki		ANTARES		◆Rasalhague		VEGA	
285	45 54	026	14 42	062	22 18	126	56 40	182	40 27	226	68 16	286	57 48	351
286	46 19	025	15 35	062	23 06	126	56 37	184	39 44	227	67 19	286	57 38	350
287	46 44	024	16 27	062	23 54	127	56 33	185	39 00	227	66 22	285	57 27	349
288	47 08	023	17 20	062	24 42	127	56 26	187	38 16	228	65 24	285	57 15	347
289	47 31	022	18 12	062	25 29	128	56 18	189	37 32	229	64 27	285	57 01	346
290	47 53	021	19 05	062	26 16	128	56 09	190	36 47	229	63 29	284	56 45	345
291	48 14	020	19 57	062	27 03	128	55 58	192	36 01	230	62 31	284	56 29	343
292	48 34	020	20 49	062	27 49	129	55 45	193	35 16	231	61 33	284	56 11	342
293	48 54	019	21 42	061	28 35	129	55 30	195	34 30	231	60 36	283	55 52	341
294	49 12	018	22 34	061	29 21	130	55 14	196	33 43	232	59 38	283	55 32	339
295	49 30	017	23 26	061	30 07	130	54 57	198	32 56	232	58 39	283	55 10	338
296	49 46	016	24 18	061	30 52	131	54 38	199	32 09	233	57 41	282	54 47	337
297	50 02	015	25 11	061	31 37	131	54 18	201	31 22	233	56 43	282	54 23	336
298	50 16	014	26 03	061	32 21	132	53 56	202	30 34	234	55 45	282	53 59	335
299	50 30	013	26 55	061	33 06	132	53 33	203	29 46	234	54 47	282	53 33	334
	◆DENEB		Alpheratz		◆FOMALHAUT		Peacock		◆ANTARES		Rasalhague		VEGA	
300	50 42	011	27 47	061	33 49	133	26 02	176	28 57	235	53 48	282	53 06	333
301	50 54	010	28 39	061	34 33	134	26 06	177	28 09	235	52 50	282	52 38	332
302	51 04	009	29 31	060	35 16	134	26 09	177	27 20	235	51 52	281	52 09	330
303	51 13	008	30 22	060	35 58	135	26 11	178	26 31	236	50 53	281	51 39	330
304	51 21	007	31 14	060	36 40	135	26 13	179	25 41	236	49 55	281	51 08	329
305	51 28	006	32 06	060	37 22	136	26 14	179	24 52	237	48 57	281	50 37	328
306	51 33	005	32 57	060	38 03	137	26 15	180	24 02	237	47 58	281	50 05	327
307	51 38	004	33 49	060	38 43	137	26 15	180	23 12	237	47 00	281	49 32	326
308	51 41	003	34 40	059	39 23	138	26 14	181	22 22	238	46 01	281	48 58	325
309	51 43	001	35 31	059	40 02	139	26 13	182	21 31	238	45 03	281	48 24	324
310	51 44	000	36 22	059	40 41	140	26 11	182	20 41	238	44 04	281	47 49	324
311	51 44	359	37 13	059	41 19	140	26 08	183	19 50	239	43 06	281	47 13	323
312	51 43	358	38 04	058	41 57	141	26 05	183	18 59	239	42 07	281	46 36	322
313	51 40	357	38 54	058	42 34	142	26 01	184	18 08	239	41 09	281	46 00	321
314	51 36	356	39 45	058	43 10	143	25 56	185	17 17	239	40 10	281	45 22	321
	Alpheratz		◆Diphda		FOMALHAUT		◆Nunki		Rasalhague		VEGA		◆DENEB	
315	40 35	057	29 30	115	43 46	144	44 57	221	39 11	281	44 44	320	51 31	355
316	41 25	057	30 24	116	44 20	145	44 18	222	38 13	281	44 06	319	51 25	354
317	42 15	057	31 17	116	44 54	146	43 38	223	37 14	280	43 27	319	51 18	352
318	43 04	056	32 11	116	45 27	147	42 57	224	36 16	280	42 47	318	51 10	351
319	43 54	056	33 04	117	46 00	148	42 16	224	35 17	280	42 07	318	51 00	350
320	44 43	055	33 57	117	46 31	149	41 34	225	34 19	280	41 27	317	50 50	349
321	45 32	055	34 50	118	47 02	150	40 51	226	33 20	280	40 46	317	50 38	348
322	46 21	055	35 42	118	47 31	151	40 08	227	32 22	280	40 05	316	50 25	347
323	47 09	054	36 35	119	48 00	152	39 25	227	31 23	280	39 24	316	50 11	346
324	47 57	054	37 27	119	48 28	153	38 41	228	30 24	280	38 42	315	49 56	345
325	48 45	053	38 19	120	48 54	154	37 56	229	29 26	281	38 00	315	49 40	344
326	49 32	052	39 10	120	49 20	155	37 12	229	28 27	281	37 18	314	49 23	343
327	50 19	052	40 02	121	49 44	156	36 26	230	27 29	281	36 35	314	49 06	342
328	51 06	051	40 53	121	50 08	158	35 41	230	26 30	281	35 52	314	48 47	341
329	51 52	050	41 44	122	50 30	159	34 55	231	25 32	281	35 09	313	48 27	340
	◆Alpheratz		Diphda		◆FOMALHAUT		Peacock		ALTAIR		◆VEGA		DENEB	
330	52 38	050	42 34	122	50 51	160	23 21	194	57 54	276	34 25	313	48 06	339
331	53 23	049	43 24	123	51 10	161	23 07	194	56 55	275	33 42	313	47 45	338
332	54 07	048	44 14	124	51 29	163	22 52	195	55 55	275	32 58	312	47 22	337
333	54 52	047	45 03	124	51 46	164	22 36	195	54 56	275	32 13	312	46 59	337
334	55 35	047	45 52	125	52 02	165	22 20	196	53 57	275	31 29	312	46 35	336
335	56 18	046	46 41	126	52 16	167	22 03	196	52 58	275	30 44	311	46 10	335
336	57 01	045	47 29	127	52 29	168	21 46	197	51 58	275	30 00	311	45 44	334
337	57 42	044	48 17	127	52 41	169	21 28	197	50 59	275	29 15	311	45 18	333
338	58 23	043	49 04	128	52 51	171	21 10	198	50 00	275	28 30	311	44 51	333
339	59 03	042	49 50	129	53 00	172	20 52	198	49 00	275	27 44	310	44 23	332
340	59 42	041	50 36	130	53 07	174	20 33	199	48 01	275	26 59	310	43 54	331
341	60 20	039	51 22	131	53 13	175	20 13	199	47 02	275	26 13	310	43 25	330
342	60 58	038	52 07	132	53 17	177	19 53	200	46 03	276	25 28	310	42 56	330
343	61 34	037	52 51	133	53 20	178	19 32	200	45 03	276	24 42	310	42 25	329
344	62 09	035	53 34	134	53 21	179	19 12	201	44 04	276	23 56	309	41 54	328
	◆Mirfak		Hamal		Acamar		◆ACHERNAR		FOMALHAUT		◆ALTAIR		DENEB	
345	20 42	039	42 17	065	17 45	136	18 12	159	53 21	181	43 05	276	41 23	328
346	21 20	039	43 10	064	18 26	137	18 33	159	53 19	182	42 06	276	40 51	327
347	21 57	039	44 04	064	19 07	137	18 54	160	53 16	184	41 06	276	40 18	327
348	22 34	038	44 58	064	19 47	137	19 15	160	53 11	185	40 07	276	39 45	326
349	23 11	038	45 51	063	20 28	138	19 35	161	53 05	187	39 08	276	39 11	325
350	23 48	038	46 44	063	21 07	138	19 54	161	52 57	188	38 09	276	38 37	325
351	24 25	038	47 37	063	21 47	139	20 13	162	52 48	190	37 09	276	38 03	324
352	25 01	038	48 30	062	22 26	139	20 32	162	52 37	191	36 10	276	37 28	324
353	25 37	037	49 22	062	23 05	140	20 50	162	52 25	192	35 11	276	36 53	323
354	26 13	037	50 15	061	23 43	140	21 08	163	52 12	194	34 12	276	36 17	323
355	26 49	037	51 07	061	24 21	140	21 25	163	51 57	195	33 12	276	35 41	322
356	27 25	037	51 59	060	24 59	141	21 42	164	51 41	196	32 13	276	35 04	322
357	28 00	036	52 50	060	25 36	141	21 58	164	51 23	198	31 14	276	34 28	322
358	28 35	036	53 42	059	26 13	142	22 14	165	51 04	199	30 15	276	33 50	321
359	29 10	036	54 33	059	26 50	142	22 29	165	50 44	200	29 15	276	33 13	321

 LAT 6°N

LHA ♈	Hc	Zn	Hc	Zn	Hc	Zn	Hc	Zn	Hc	Zn	Hc	Zn	Hc	Zn
	◆CAPELLA		ALDEBARAN		RIGEL		◆ACHERNAR		FOMALHAUT		◆ALTAIR		DENEB	
0	11 54	044	21 53	075	10 22	100	23 42	166	51 19	202	28 09	277	31 49	321
1	12 35	044	22 51	075	11 21	100	23 56	166	50 56	203	27 10	277	31 11	320
2	13 17	044	23 49	075	12 20	100	24 10	167	50 31	204	26 11	277	30 33	320
3	13 58	044	24 46	075	13 19	100	24 23	167	50 06	206	25 12	277	29 54	320
4	14 40	044	25 44	075	14 18	100	24 36	168	49 40	207	24 12	277	29 16	319
5	15 21	044	26 41	075	15 16	100	24 48	169	49 12	208	23 13	277	28 37	319
6	16 02	044	27 39	074	16 15	100	24 59	169	48 44	209	22 14	277	27 58	319
7	16 44	044	28 36	074	17 14	101	25 10	170	48 14	210	21 15	277	27 18	319
8	17 25	044	29 34	074	18 12	101	25 21	170	47 44	211	20 16	277	26 39	318
9	18 06	043	30 31	074	19 11	101	25 31	171	47 12	212	19 16	277	25 59	318
10	18 47	043	31 28	074	20 10	101	25 40	171	46 40	213	18 17	277	25 19	318
11	19 27	043	32 26	074	21 08	101	25 48	172	46 07	214	17 18	277	24 39	318
12	20 08	043	33 23	074	22 07	101	25 56	173	45 33	215	16 19	277	23 58	317
13	20 49	043	34 21	074	23 05	102	26 04	173	44 58	216	15 20	278	23 18	317
14	21 29	043	35 18	074	24 04	102	26 10	174	44 23	217	14 21	278	22 37	317
	CAPELLA		◆ALDEBARAN		RIGEL		ACHERNAR		◆FOMALHAUT		Enif		◆DENEB	
15	22 10	042	36 15	074	25 02	102	26 17	174	43 46	218	41 20	278	21 56	317
16	22 50	042	37 13	074	26 00	102	26 22	175	43 09	219	40 21	278	21 15	317
17	23 30	042	38 10	074	26 59	102	26 27	176	42 32	220	39 22	278	20 34	316
18	24 10	042	39 07	074	27 57	103	26 31	176	41 54	220	38 23	278	19 53	316
19	24 49	042	40 05	073	28 55	103	26 35	177	41 15	221	37 24	278	19 11	316
20	25 29	041	41 02	073	29 53	103	26 38	177	40 35	222	36 25	278	18 30	316
21	26 08	041	41 59	073	30 51	103	26 41	178	39 55	223	35 26	278	17 48	316
22	26 47	041	42 56	073	31 49	104	26 42	179	39 14	223	34 27	278	17 07	316
23	27 26	041	43 53	073	32 47	104	26 44	179	38 33	224	33 28	278	16 25	316
24	28 05	040	44 50	073	33 45	104	26 44	180	37 52	225	32 28	278	15 43	315
25	28 43	040	45 47	072	34 43	104	26 44	180	37 09	225	31 29	278	15 01	315
26	29 21	040	46 44	072	35 41	105	26 43	181	36 27	226	30 30	278	14 19	315
27	29 59	039	47 41	072	36 39	105	26 42	182	35 44	227	29 31	278	13 37	315
28	30 37	039	48 37	072	37 36	105	26 40	182	35 00	227	28 32	278	12 55	315
29	31 14	039	49 34	072	38 34	106	26 38	183	34 16	228	27 33	278	12 13	315
	Mirfak		CAPELLA		◆ALDEBARAN		RIGEL		◆ACHERNAR		FOMALHAUT		◆Alpheratz	
30	42 44	018	31 51	038	50 30	071	39 31	106	26 34	183	33 32	228	54 56	314
31	43 03	018	32 28	038	51 27	071	40 29	106	26 30	184	32 47	229	54 13	314
32	43 20	017	33 04	037	52 23	071	41 26	107	26 26	185	32 02	229	53 30	313
33	43 37	016	33 40	037	53 20	070	42 23	107	26 21	185	31 17	230	52 46	312
34	43 53	015	34 16	037	54 16	070	43 20	107	26 15	186	30 31	230	52 01	311
35	44 08	014	34 51	036	55 12	070	44 17	108	26 09	186	29 45	231	51 16	311
36	44 23	013	35 26	036	56 07	069	45 14	108	26 02	187	28 59	231	50 31	310
37	44 36	013	36 01	035	57 03	069	46 10	108	25 54	188	28 12	232	49 45	309
38	44 49	012	36 35	035	57 59	068	47 07	109	25 46	188	27 25	232	48 58	309
39	45 01	011	37 09	034	58 54	068	48 03	109	25 38	189	26 38	232	48 11	308
40	45 11	010	37 42	034	59 49	067	49 00	110	25 28	189	25 51	233	47 24	307
41	45 21	009	38 15	033	60 44	067	49 56	110	25 18	190	25 03	233	46 36	307
42	45 30	008	38 47	033	61 39	066	50 51	111	25 08	190	24 15	234	45 48	306
43	45 39	007	39 19	032	62 33	065	51 47	111	24 57	191	23 27	234	45 00	306
44	45 46	006	39 50	031	63 27	065	52 43	112	24 45	192	22 38	234	44 12	305
	◆CAPELLA		PROCYON		SIRIUS		◆CANOPUS		ACHERNAR		Diphda		◆Alpheratz	
45	40 21	031	20 37	087	29 58	113	17 15	150	24 33	192	48 37	234	43 23	305
46	40 51	030	21 37	087	30 52	114	17 44	151	24 20	193	47 48	235	42 34	305
47	41 21	029	22 36	087	31 47	114	18 13	151	24 07	193	46 59	235	41 45	304
48	41 50	029	23 36	087	32 41	114	18 42	152	23 53	194	46 10	236	40 55	304
49	42 18	028	24 35	087	33 36	115	19 10	152	23 39	194	45 20	237	40 05	303
50	42 46	027	25 35	087	34 30	115	19 38	152	23 24	195	44 30	237	39 15	303
51	43 13	027	26 35	087	35 24	115	20 05	153	23 08	195	43 40	238	38 25	303
52	43 40	026	27 34	087	36 18	116	20 32	153	22 52	196	42 49	239	37 35	302
53	44 05	025	28 34	087	37 11	116	20 59	154	22 36	196	41 58	239	36 45	302
54	44 30	024	29 33	087	38 05	117	21 25	154	22 19	197	41 07	240	35 54	302
55	44 55	024	30 33	087	38 58	117	21 51	155	22 01	197	40 15	240	35 03	302
56	45 18	023	31 33	088	39 51	118	22 16	155	21 43	198	39 23	241	34 12	301
57	45 41	022	32 32	088	40 44	118	22 41	156	21 25	198	38 31	241	33 21	301
58	46 03	021	33 32	088	41 36	119	23 05	156	21 06	199	37 38	242	32 30	301
59	46 24	020	34 32	088	42 28	119	23 29	157	20 46	199	36 46	242	31 39	301
	◆CAPELLA		POLLUX		PROCYON		◆SIRIUS		CANOPUS		◆Diphda		Alpheratz	
60	46 44	019	32 28	060	35 31	088	43 20	120	23 53	157	35 53	243	30 47	300
61	47 03	018	33 20	060	36 30	088	44 12	120	24 16	158	35 00	243	29 56	300
62	47 22	018	34 11	060	37 30	088	45 03	121	24 38	158	34 06	243	29 04	300
63	47 39	017	35 03	060	38 30	088	45 54	122	25 00	159	33 13	244	28 12	300
64	47 56	016	35 55	060	39 30	088	46 45	122	25 21	159	32 19	244	27 21	300
65	48 11	015	36 46	059	40 29	088	47 35	123	25 42	160	31 25	245	26 29	300
66	48 26	014	37 37	059	41 29	088	48 25	124	26 03	160	30 31	245	25 37	299
67	48 40	013	38 28	059	42 29	088	49 14	124	26 23	161	29 37	245	24 45	299
68	48 52	012	39 19	058	43 28	088	50 04	125	26 42	161	28 43	246	23 53	299
69	49 04	011	40 10	058	44 28	089	50 52	126	27 00	162	27 49	246	23 01	299
70	49 14	010	41 00	058	45 28	089	51 40	127	27 19	163	26 54	246	22 08	299
71	49 24	009	41 51	057	46 27	089	52 28	128	27 36	163	25 59	247	21 16	299
72	49 32	008	42 41	057	47 27	089	53 15	129	27 53	164	25 05	247	20 24	299
73	49 40	007	43 31	057	48 27	089	54 01	130	28 09	164	24 10	247	19 32	299
74	49 46	005	44 20	056	49 26	089	54 47	131	28 25	165	23 15	247	18 39	299
	CAPELLA		◆POLLUX		PROCYON		SIRIUS		◆CANOPUS		Acamar		◆Hamal	
75	49 51	004	45 10	056	50 26	089	55 32	132	28 40	166	35 51	209	44 54	297
76	49 55	003	45 59	055	51 26	089	56 16	133	28 55	166	35 23	209	44 01	297
77	49 58	002	46 48	055	52 25	089	56 59	134	29 08	167	34 53	210	43 08	297
78	50 00	001	47 36	054	53 25	089	57 42	135	29 21	168	34 23	211	42 14	296
79	50 00	000	48 24	054	54 25	089	58 24	136	29 34	168	33 52	211	41 21	296
80	50 00	359	49 12	053	55 24	089	59 04	138	29 46	169	33 21	212	40 27	296
81	49 58	358	50 00	052	56 24	090	59 44	139	29 57	170	32 49	213	39 34	296
82	49 56	357	50 47	052	57 24	090	60 23	140	30 07	170	32 17	213	38 40	295
83	49 52	356	51 33	051	58 23	090	61 00	142	30 17	171	31 44	214	37 46	295
84	49 47	355	52 20	050	59 23	090	61 37	143	30 26	172	31 10	215	36 52	295
85	49 41	354	53 05	050	60 23	090	62 11	145	30 34	172	30 36	215	35 57	295
86	49 34	353	53 50	049	61 22	090	62 45	147	30 42	173	30 02	216	35 03	295
87	49 25	352	54 35	048	62 22	090	63 17	148	30 49	174	29 26	216	34 09	294
88	49 16	350	55 19	047	63 22	090	63 47	150	30 55	174	28 51	217	33 14	294
89	49 06	349	56 03	046	64 21	090	64 16	152	31 01	175	28 15	217	32 20	294

LHA ♈	Hc	Zn	Hc	Zn	Hc	Zn	Hc	Zn	Hc	Zn	Hc	Zn	Hc	Zn
	POLLUX		◆REGULUS		Suhail		◆CANOPUS		Acamar		◆ALDEBARAN		CAPELLA	
90	56 46	045	28 33	080	24 55	144	31 05	176	27 39	218	66 50	299	48 54	348
91	57 28	044	29 32	080	25 30	145	31 09	176	27 02	218	65 57	298	48 42	347
92	58 09	043	30 31	080	26 04	145	31 13	177	26 25	219	65 04	297	48 28	346
93	58 50	042	31 29	080	26 38	146	31 15	178	25 47	219	64 11	296	48 14	345
94	59 30	041	32 28	080	27 11	146	31 17	179	25 09	220	63 17	295	47 58	345
95	60 08	040	33 27	080	27 44	147	31 18	179	24 30	220	62 23	295	47 42	344
96	60 46	039	34 26	080	28 17	147	31 18	180	23 52	221	61 29	294	47 25	343
97	61 23	037	35 24	080	28 49	148	31 18	181	23 13	221	60 34	293	47 06	342
98	61 59	036	36 23	080	29 20	148	31 17	181	22 33	222	59 39	293	46 47	341
99	62 33	035	37 22	079	29 51	149	31 15	182	21 53	222	58 44	292	46 27	340
100	63 06	033	38 20	079	30 22	150	31 12	183	21 13	222	57 48	292	46 06	339
101	63 38	032	39 19	079	30 52	150	31 09	184	20 33	223	56 53	291	45 44	338
102	64 09	030	40 18	079	31 21	151	31 05	184	19 52	223	55 57	291	45 22	337
103	64 38	028	41 16	079	31 50	151	31 00	185	19 11	223	55 01	290	44 58	337
104	65 05	026	42 15	079	32 18	152	30 55	186	18 30	224	54 05	290	44 34	336
	POLLUX		◆REGULUS		Suhail		◆CANOPUS		RIGEL		ALDEBARAN		◆CAPELLA	
105	65 31	025	43 13	079	32 46	153	30 48	186	60 04	242	53 09	290	44 09	335
106	65 55	023	44 12	079	33 13	153	30 41	187	59 11	243	52 13	289	43 44	334
107	66 17	021	45 11	079	33 39	154	30 34	188	58 18	244	51 17	289	43 17	333
108	66 37	019	46 09	079	34 05	155	30 25	188	57 24	244	50 20	289	42 50	333
109	66 55	017	47 08	079	34 30	156	30 16	189	56 30	245	49 24	288	42 23	332
110	67 11	014	48 06	079	34 54	156	30 06	190	55 35	246	48 27	288	41 54	331
111	67 24	012	49 05	079	35 18	157	29 56	190	54 41	247	47 30	288	41 25	331
112	67 36	010	50 03	078	35 40	158	29 45	191	53 46	247	46 33	288	40 56	330
113	67 45	008	51 02	078	36 03	159	29 33	192	52 51	248	45 36	288	40 26	329
114	67 51	005	52 00	078	36 24	159	29 20	192	51 55	248	44 40	287	39 55	329
115	67 56	003	52 59	078	36 45	160	29 07	193	51 00	249	43 43	287	39 24	328
116	67 58	001	53 57	078	37 04	161	28 53	194	50 04	250	42 45	287	38 52	328
117	67 57	358	54 55	078	37 23	162	28 39	194	49 08	250	41 48	287	38 20	327
118	67 54	356	55 53	077	37 42	163	28 24	195	48 12	251	40 51	287	37 47	326
119	67 48	354	56 52	077	37 59	163	28 08	196	47 15	251	39 54	287	37 14	326
	Dubhe		◆Denebola		Suhail		◆SIRIUS		RIGEL		ALDEBARAN		◆CAPELLA	
120	24 49	022	33 12	076	38 16	164	60 40	219	46 19	251	38 57	286	36 40	325
121	25 11	022	34 10	076	38 31	165	60 02	220	45 22	252	38 00	286	36 06	325
122	25 33	021	35 08	076	38 46	166	59 23	222	44 25	252	37 02	286	35 32	324
123	25 54	021	36 06	076	39 00	167	58 43	223	43 29	253	36 05	286	34 57	324
124	26 15	021	37 04	076	39 13	168	58 01	224	42 32	253	35 08	286	34 22	324
125	26 36	020	38 02	076	39 25	169	57 19	226	41 34	253	34 10	286	33 46	323
126	26 57	020	39 00	076	39 36	170	56 36	227	40 37	254	33 13	286	33 10	323
127	27 17	019	39 58	076	39 46	171	55 52	228	39 40	254	32 15	286	32 34	322
128	27 37	019	40 56	076	39 56	172	55 07	229	38 42	254	31 18	286	31 57	322
129	27 56	019	41 54	076	40 04	172	54 22	230	37 45	255	30 20	286	31 20	322
130	28 15	018	42 52	076	40 11	173	53 36	231	36 47	255	29 23	286	30 43	321
131	28 33	018	43 49	075	40 18	174	52 49	232	35 50	255	28 26	286	30 05	321
132	28 52	018	44 47	075	40 23	175	52 02	233	34 52	256	27 28	285	29 27	320
133	29 09	017	45 45	075	40 27	176	51 14	234	33 54	256	26 31	285	28 49	320
134	29 27	017	46 43	075	40 31	177	50 26	234	32 56	256	25 33	285	28 11	320
	Dubhe		◆ARCTURUS		SPICA		◆Suhail		SIRIUS		BETELGEUSE		◆CAPELLA	
135	29 43	016	12 28	072	21 54	105	40 33	178	49 37	235	44 02	275	27 32	320
136	30 00	016	13 25	072	22 52	105	40 35	179	48 48	236	43 03	275	26 53	319
137	30 16	015	14 21	072	23 50	105	40 35	180	47 58	237	42 03	275	26 14	319
138	30 31	015	15 18	072	24 47	105	40 35	181	47 08	237	41 04	275	25 35	319
139	30 47	014	16 15	072	25 45	105	40 33	182	46 18	238	40 04	275	24 56	318
140	31 01	014	17 11	072	26 42	106	40 31	183	45 27	239	39 05	275	24 16	318
141	31 15	013	18 08	072	27 40	106	40 27	184	44 36	239	38 05	275	23 36	318
142	31 29	013	19 05	072	28 37	106	40 23	185	43 44	240	37 06	275	22 56	318
143	31 42	012	20 01	072	29 34	106	40 17	186	42 52	241	36 07	275	22 16	318
144	31 55	012	20 58	072	30 32	107	40 11	187	42 00	241	35 07	275	21 36	317
145	32 07	011	21 54	072	31 29	107	40 03	188	41 08	242	34 08	275	20 55	317
146	32 18	011	22 51	072	32 26	107	39 55	189	40 15	242	33 08	275	20 15	317
147	32 29	010	23 48	072	33 23	108	39 45	189	39 23	243	32 09	275	19 34	317
148	32 40	010	24 44	072	34 20	108	39 35	190	38 29	243	31 09	275	18 53	317
149	32 50	009	25 41	072	35 16	108	39 24	191	37 36	244	30 10	275	18 12	317
	Dubhe		◆ARCTURUS		SPICA		ACRUX		◆Suhail		SIRIUS		◆POLLUX	
150	32 59	009	26 38	071	36 13	109	15 34	164	39 12	192	36 43	244	51 10	309
151	33 08	008	27 34	071	37 09	109	15 51	164	38 59	193	35 49	244	50 23	308
152	33 17	008	28 31	071	38 06	109	16 07	164	38 45	194	34 55	245	49 36	307
153	33 25	007	29 27	071	39 02	110	16 23	165	38 30	195	34 01	245	48 48	307
154	33 32	007	30 24	071	39 58	110	16 38	165	38 14	196	33 07	246	48 00	306
155	33 38	006	31 20	071	40 54	110	16 53	166	37 57	197	32 12	246	47 12	306
156	33 45	006	32 17	071	41 50	111	17 08	166	37 40	197	31 18	246	46 23	305
157	33 50	005	33 13	071	42 46	111	17 22	166	37 22	198	30 23	247	45 34	305
158	33 55	004	34 10	071	43 41	112	17 36	167	37 03	199	29 28	247	44 45	304
159	33 59	004	35 06	071	44 37	112	17 49	167	36 43	200	28 33	247	43 56	304
160	34 03	003	36 02	070	45 32	113	18 02	168	36 22	201	27 38	248	43 06	303
161	34 06	003	36 58	070	46 27	113	18 15	168	36 01	201	26 43	248	42 16	303
162	34 09	002	37 55	070	47 22	114	18 27	169	35 38	202	25 48	248	41 26	303
163	34 11	002	38 51	070	48 16	114	18 38	169	35 15	203	24 52	248	40 35	302
164	34 12	001	39 47	070	49 11	115	18 50	169	34 52	204	23 57	249	39 45	302
	Dubhe		◆ARCTURUS		SPICA		◆ACRUX		Suhail		PROCYON		◆POLLUX	
165	34 13	000	40 43	070	50 05	115	19 00	170	34 27	204	40 01	272	38 54	302
166	34 13	000	41 39	070	50 58	116	19 11	170	34 02	205	39 01	272	38 03	301
167	34 13	359	42 35	069	51 52	117	19 20	171	33 37	206	38 01	272	37 12	301
168	34 12	359	43 30	069	52 45	117	19 30	171	33 10	207	37 02	272	36 21	301
169	34 10	358	44 26	069	53 38	118	19 39	172	32 43	207	36 02	272	35 29	300
170	34 08	358	45 22	069	54 30	119	19 47	172	32 15	208	35 03	272	34 38	300
171	34 05	357	46 17	068	55 23	120	19 55	173	31 47	209	34 03	272	33 46	300
172	34 02	356	47 13	068	56 14	120	20 02	173	31 18	209	33 03	272	32 54	300
173	33 58	356	48 08	068	57 05	121	20 09	173	30 49	210	32 04	272	32 02	299
174	33 54	355	49 03	067	57 56	122	20 16	174	30 19	210	31 04	273	31 10	299
175	33 49	355	49 58	067	58 46	123	20 22	174	29 48	211	30 04	273	30 18	299
176	33 43	354	50 53	067	59 36	124	20 28	175	29 17	212	29 05	273	29 26	299
177	33 37	354	51 48	066	60 25	125	20 33	175	28 46	212	28 05	273	28 34	299
178	33 30	353	52 42	066	61 13	126	20 37	176	28 14	213	27 06	273	27 41	299
179	33 22	353	53 37	065	62 01	128	20 41	176	27 41	213	26 06	273	26 49	298

LAT 6°N

LHA ♈	Hc	Zn	Hc	Zn	Hc	Zn	Hc	Zn	Hc	Zn	Hc	Zn	Hc	Zn
	◆ARCTURUS		ANTARES		RIGIL KENT.		◆ACRUX		Suhail		◆REGULUS		Dubhe	
180	54 31	065	17 19	120	16 20	161	20 45	177	27 08	214	61 45	284	33 14	352
181	55 25	065	18 10	120	16 39	161	20 48	177	26 35	214	60 47	284	33 06	352
182	56 19	064	19 02	121	16 58	162	20 50	178	26 01	215	59 49	284	32 57	351
183	57 12	063	19 53	121	17 17	162	20 52	178	25 26	215	58 51	283	32 47	350
184	58 05	063	20 44	121	17 35	163	20 54	179	24 52	216	57 53	283	32 37	350
185	58 58	062	21 35	122	17 52	163	20 55	179	24 17	216	56 55	283	32 26	349
186	59 51	061	22 26	122	18 10	163	20 56	180	23 41	217	55 57	283	32 15	349
187	60 43	061	23 16	122	18 26	164	20 56	180	23 05	217	54 58	282	32 03	348
188	61 35	060	24 07	122	18 43	164	20 55	181	22 29	218	54 00	282	31 51	348
189	62 26	059	24 57	123	18 59	165	20 54	181	21 52	218	53 02	282	31 38	347
190	63 17	058	25 47	123	19 14	165	20 53	182	21 15	219	52 03	282	31 25	347
191	64 08	057	26 37	124	19 29	166	20 51	182	20 38	219	51 05	282	31 11	346
192	64 58	056	27 26	124	19 44	166	20 48	183	20 00	219	50 06	282	30 57	346
193	65 47	055	28 16	124	19 58	166	20 45	183	19 22	220	49 08	281	30 42	345
194	66 36	054	29 05	125	20 12	167	20 42	184	18 44	220	48 09	281	30 27	345
	◆Alphecca		Rasalhague		ANTARES		◆RIGIL KENT.		ACRUX		◆REGULUS		Dubhe	
195	47 49	056	22 04	079	29 54	125	20 25	167	20 38	184	47 11	281	30 11	345
196	48 38	056	23 03	079	30 42	126	20 38	168	20 33	185	46 12	281	29 55	344
197	49 27	055	24 01	079	31 31	126	20 50	168	20 28	185	45 14	281	29 39	344
198	50 16	054	25 00	079	32 19	127	21 02	169	20 23	185	44 15	281	29 22	343
199	51 05	054	25 58	079	33 07	127	21 14	169	20 17	186	43 17	281	29 04	343
200	51 53	053	26 57	079	33 54	128	21 24	170	20 11	186	42 18	281	28 46	342
201	52 40	053	27 56	079	34 41	128	21 35	170	20 04	187	41 19	281	28 28	342
202	53 27	052	28 54	079	35 28	129	21 45	171	19 56	187	40 21	281	28 09	342
203	54 14	051	29 53	079	36 14	129	21 54	171	19 48	188	39 22	281	27 50	341
204	55 00	050	30 51	079	37 01	130	22 03	172	19 40	188	38 23	281	27 31	341
205	55 46	049	31 50	079	37 46	130	22 11	172	19 31	189	37 25	281	27 11	340
206	56 31	049	32 48	079	38 31	131	22 19	173	19 22	189	36 26	280	26 51	340
207	57 15	048	33 47	079	39 16	132	22 26	173	19 12	190	35 27	280	26 30	340
208	57 59	047	34 45	079	40 01	132	22 33	174	19 02	190	34 29	280	26 09	339
209	58 42	046	35 44	079	40 45	133	22 39	174	18 51	191	33 30	280	25 48	339
	◆VEGA		Rasalhague		ANTARES		◆RIGIL KENT.		ACRUX		◆REGULUS		Dubhe	
210	19 56	051	36 43	079	41 28	134	22 45	175	18 40	191	32 31	280	25 26	339
211	20 42	051	37 41	079	42 11	134	22 50	175	18 29	191	31 33	280	25 05	338
212	21 28	051	38 40	079	42 53	135	22 54	176	18 17	192	30 34	280	24 42	338
213	22 14	050	39 38	079	43 35	136	22 58	176	18 04	192	29 35	280	24 20	338
214	23 00	050	40 37	079	44 17	137	23 02	177	17 51	193	28 37	280	23 57	337
215	23 46	050	41 35	079	44 57	137	23 05	177	17 38	193	27 38	280	23 34	337
216	24 32	050	42 34	078	45 37	138	23 07	178	17 24	193	26 39	280	23 10	337
217	25 17	050	43 32	078	46 16	139	23 09	179	17 10	194	25 40	280	22 47	336
218	26 03	049	44 30	078	46 55	140	23 10	179	16 56	194	24 42	280	22 23	336
219	26 48	049	45 29	078	47 33	141	23 11	180	16 41	195	23 43	280	21 58	336
220	27 33	049	46 27	078	48 10	142	23 11	180	16 25	195	22 44	281	21 34	336
221	28 18	049	47 26	078	48 46	143	23 11	181	16 10	195	21 46	281	21 09	335
222	29 03	049	48 24	078	49 22	144	23 10	181	15 54	196	20 47	281	20 44	335
223	29 48	048	49 22	078	49 56	145	23 08	182	15 37	196	19 48	281	20 19	335
224	30 32	048	50 21	078	50 30	146	23 06	182	15 20	197	18 50	281	19 54	335
	VEGA		◆ALTAIR		Nunki		ANTARES		RIGIL KENT.		◆SPICA		◆Alkaid	
225	31 16	048	18 01	083	24 36	123	51 03	147	23 04	183	60 47	234	44 02	344
226	32 00	047	19 00	083	25 26	123	51 35	148	23 01	183	59 59	235	43 44	343
227	32 44	047	20 00	083	26 16	123	52 05	150	22 57	184	59 09	236	43 26	342
228	33 28	047	20 59	083	27 06	124	52 35	151	22 53	184	58 20	237	43 08	341
229	34 11	046	21 58	083	27 56	124	53 03	152	22 48	185	57 29	238	42 48	340
230	34 54	046	22 57	083	28 45	124	53 31	153	22 43	185	56 38	239	42 28	340
231	35 37	046	23 57	083	29 34	125	53 57	155	22 37	186	55 47	240	42 06	339
232	36 19	045	24 56	083	30 23	125	54 21	156	22 30	186	54 55	241	41 45	338
233	37 01	045	25 55	083	31 12	126	54 45	158	22 23	187	54 02	242	41 22	337
234	37 43	044	26 54	083	32 00	126	55 07	159	22 16	187	53 10	242	40 59	337
235	38 25	044	27 53	083	32 48	127	55 28	160	22 08	188	52 17	243	40 35	336
236	39 06	043	28 53	083	33 35	127	55 47	162	22 00	188	51 23	244	40 11	335
237	39 47	043	29 52	083	34 23	128	56 05	163	21 51	189	50 30	244	39 46	335
238	40 27	042	30 51	083	35 10	128	56 21	165	21 41	189	49 36	245	39 20	334
239	41 08	042	31 50	083	35 57	129	56 36	166	21 31	190	48 42	246	38 54	334
	VEGA		◆ALTAIR		Nunki		◆ANTARES		SPICA		◆ARCTURUS		Alkaid	
240	41 47	041	32 50	083	36 43	129	56 49	168	47 47	246	61 20	300	38 27	333
241	42 26	041	33 49	083	37 29	130	57 01	170	46 52	247	60 28	299	37 59	332
242	43 05	040	34 48	083	38 14	131	57 11	171	45 58	247	59 36	298	37 31	332
243	43 43	040	35 48	083	38 59	131	57 19	173	45 02	248	58 43	298	37 03	331
244	44 21	039	36 47	083	39 44	132	57 25	175	44 07	248	57 50	297	36 34	331
245	44 58	038	37 46	083	40 28	132	57 30	176	43 12	249	56 57	296	36 04	330
246	45 35	038	38 45	083	41 12	133	57 33	178	42 16	249	56 03	296	35 34	330
247	46 11	037	39 45	083	41 55	134	57 35	180	41 20	249	55 10	295	35 04	329
248	46 47	036	40 44	083	42 38	135	57 34	181	40 24	250	54 16	295	34 33	329
249	47 21	035	41 43	083	43 20	135	57 32	183	39 28	250	53 21	294	34 02	328
250	47 56	035	42 42	083	44 02	136	57 28	185	38 32	251	52 27	294	33 30	328
251	48 29	034	43 42	083	44 43	137	57 23	186	37 36	251	51 32	294	32 58	327
252	49 02	033	44 41	083	45 23	138	57 15	188	36 39	251	50 38	293	32 26	327
253	49 34	032	45 40	083	46 03	139	57 06	189	35 43	252	49 43	293	31 53	326
254	50 05	031	46 40	083	46 42	140	56 56	191	34 46	252	48 48	292	31 19	326
	VEGA		◆ALTAIR		Nunki		◆ANTARES		SPICA		ARCTURUS		◆Alkaid	
255	50 36	030	47 39	083	47 21	140	56 43	193	33 49	252	47 52	292	30 46	326
256	51 05	029	48 38	083	47 58	141	56 30	194	32 52	253	46 57	292	30 12	325
257	51 34	028	49 37	083	48 35	142	56 14	196	31 55	253	46 02	292	29 38	325
258	52 02	027	50 37	083	49 11	143	55 57	197	30 58	253	45 06	291	29 03	325
259	52 29	026	51 36	083	49 46	144	55 39	199	30 01	253	44 10	291	28 29	324
260	52 55	025	52 35	083	50 20	146	55 19	200	29 04	254	43 15	291	27 53	324
261	53 19	024	53 35	083	50 53	147	54 57	202	28 06	254	42 19	291	27 18	324
262	53 43	023	54 34	083	51 26	148	54 35	203	27 09	254	41 23	290	26 43	323
263	54 06	022	55 33	083	51 57	149	54 10	204	26 12	254	40 27	290	26 07	323
264	54 27	021	56 32	083	52 27	150	53 45	206	25 14	255	39 31	290	25 31	323
265	54 48	019	57 31	083	52 56	151	53 18	207	24 17	255	38 35	290	24 54	322
266	55 07	018	58 31	083	53 24	153	52 51	208	23 19	255	37 39	290	24 18	322
267	55 25	017	59 30	083	53 51	154	52 22	210	22 21	255	36 42	290	23 41	322
268	55 42	016	60 29	083	54 16	155	51 52	211	21 23	256	35 46	289	23 04	322
269	55 57	014	61 28	082	54 41	157	51 20	212	20 26	256	34 50	289	22 27	321

LHA ♈	Hc	Zn	Hc	Zn	Hc	Zn	Hc	Zn	Hc	Zn	Hc	Zn	Hc	Zn
	VEGA		◆DENEB		Enif		◆Nunki		Shaula		ANTARES		◆ARCTURUS	
270	56 11	013	37 27	035	34 28	082	55 03	158	46 27	188	50 48	213	33 54	289
271	56 24	012	38 01	034	35 28	082	55 25	160	46 18	189	50 15	214	32 57	289
272	56 35	010	38 34	034	36 27	082	55 45	161	46 09	190	49 41	215	32 01	289
273	56 45	009	39 07	033	37 26	082	56 04	163	45 58	191	49 06	216	31 04	289
274	56 53	007	39 40	033	38 25	082	56 21	164	45 46	192	48 30	217	30 08	289
275	57 00	006	40 12	032	39 24	082	56 36	166	45 32	193	47 53	218	29 11	289
276	57 06	005	40 43	032	40 23	082	56 50	167	45 18	194	47 16	219	28 15	289
277	57 10	003	41 14	031	41 22	082	57 03	169	45 03	195	46 38	220	27 18	289
278	57 12	002	41 45	030	42 21	082	57 13	170	44 46	197	45 59	221	26 22	288
279	57 13	000	42 15	030	43 20	082	57 22	172	44 29	198	45 19	222	25 25	288
280	57 13	359	42 44	029	44 20	082	57 30	174	44 10	199	44 39	223	24 28	288
281	57 11	357	43 12	028	45 19	082	57 35	175	43 51	200	43 58	224	23 32	288
282	57 07	356	43 40	027	46 18	082	57 39	177	43 30	201	43 17	224	22 35	288
283	57 02	355	44 07	027	47 17	082	57 41	179	43 09	202	42 34	225	21 38	288
284	56 56	353	44 34	026	48 16	082	57 42	180	42 46	203	41 52	226	20 42	288
	DENEB		Alpheratz		◆FOMALHAUT		Nunki		◆ANTARES		Rasalhague		◆VEGA	
285	45 00	025	14 14	061	22 54	126	57 40	182	41 09	227	67 58	289	56 48	352
286	45 25	024	15 06	061	23 42	126	57 37	184	40 25	227	67 02	288	56 39	350
287	45 49	024	15 59	061	24 30	126	57 33	185	39 41	228	66 05	288	56 28	349
288	46 12	023	16 51	061	25 18	127	57 26	187	38 56	229	65 08	287	56 16	348
289	46 35	022	17 44	061	26 05	127	57 18	189	38 11	229	64 11	287	56 02	346
290	46 57	021	18 36	061	26 53	128	57 08	190	37 26	230	63 13	286	55 48	345
291	47 18	020	19 28	061	27 40	128	56 56	192	36 40	231	62 16	286	55 31	344
292	47 38	019	20 21	061	28 27	129	56 43	194	35 54	231	61 19	285	55 14	342
293	47 57	018	21 13	061	29 13	129	56 28	195	35 07	232	60 21	285	54 55	341
294	48 15	017	22 05	061	30 00	129	56 12	197	34 20	232	59 23	285	54 35	340
295	48 32	016	22 57	061	30 45	130	55 54	198	33 33	233	58 25	284	54 14	339
296	48 49	015	23 49	061	31 31	130	55 35	200	32 45	233	57 28	284	53 52	338
297	49 04	014	24 41	061	32 16	131	55 14	201	31 57	234	56 30	284	53 29	336
298	49 18	013	25 33	061	33 01	131	54 52	203	31 09	234	55 32	283	53 04	335
299	49 31	012	26 25	060	33 46	132	54 28	204	30 21	235	54 34	283	52 39	334
	◆DENEB		Alpheratz		◆FOMALHAUT		Peacock		◆ANTARES		Rasalhague		VEGA	
300	49 43	011	27 17	060	34 30	133	27 02	176	29 32	235	53 36	283	52 12	333
301	49 55	010	28 09	060	35 14	133	27 06	177	28 43	235	52 37	283	51 45	332
302	50 05	009	29 01	060	35 57	134	27 09	177	27 54	236	51 39	283	51 16	331
303	50 13	008	29 52	060	36 40	134	27 11	178	27 04	236	50 41	283	50 47	330
304	50 21	007	30 44	060	37 23	135	27 13	179	26 14	237	49 43	282	50 17	329
305	50 28	006	31 35	059	38 05	136	27 14	179	25 25	237	48 44	282	49 46	328
306	50 34	005	32 27	059	38 46	136	27 15	180	24 34	237	47 46	282	49 14	327
307	50 38	004	33 18	059	39 27	137	27 15	180	23 44	238	46 48	282	48 42	327
308	50 41	003	34 09	059	40 08	138	27 14	181	22 54	238	45 49	282	48 09	326
309	50 43	001	35 00	058	40 47	138	27 13	182	22 03	238	44 51	282	47 35	325
310	50 44	000	35 51	058	41 27	139	27 11	182	21 12	239	43 52	282	47 00	324
311	50 44	359	36 41	058	42 06	140	27 08	183	20 21	239	42 54	282	46 25	323
312	50 43	358	37 32	058	42 44	141	27 05	183	19 30	239	41 56	282	45 49	323
313	50 40	357	38 22	057	43 21	142	27 01	184	18 38	239	40 57	281	45 12	322
314	50 36	356	39 12	057	43 58	142	26 56	185	17 47	240	39 59	281	44 35	321
	Alpheratz		◆Diphda		FOMALHAUT		◆Nunki		Rasalhague		VEGA		◆DENEB	
315	40 02	057	29 55	115	44 34	143	45 42	222	39 00	281	43 58	321	50 32	355
316	40 52	056	30 49	115	45 09	144	45 02	223	38 02	281	43 20	320	50 26	354
317	41 41	056	31 43	116	45 44	145	44 22	223	37 03	281	42 41	319	50 19	353
318	42 31	056	32 37	116	46 17	146	43 40	224	36 05	281	42 02	319	50 10	352
319	43 20	055	33 31	116	46 50	147	42 58	225	35 06	281	41 23	318	50 01	350
320	44 09	055	34 24	117	47 22	148	42 16	226	34 07	281	40 43	318	49 51	349
321	44 57	054	35 17	117	47 53	149	41 33	226	33 09	281	40 03	317	49 39	348
322	45 45	054	36 10	118	48 24	150	40 49	227	32 10	281	39 22	317	49 27	347
323	46 33	053	37 03	118	48 53	151	40 05	228	31 12	281	38 41	316	49 13	346
324	47 21	053	37 56	118	49 21	152	39 21	229	30 13	281	37 59	316	48 58	345
325	48 08	052	38 48	119	49 48	154	38 36	229	29 15	281	37 18	315	48 43	344
326	48 55	051	39 40	119	50 14	155	37 51	230	28 16	281	36 36	315	48 26	343
327	49 42	051	40 32	120	50 39	156	37 05	230	27 18	281	35 53	315	48 08	342
328	50 28	050	41 24	120	51 03	157	36 19	231	26 19	281	35 10	314	47 50	341
329	51 13	050	42 15	121	51 26	158	35 32	231	25 20	281	34 28	314	47 30	340
	◆Alpheratz		Diphda		◆FOMALHAUT		Peacock		ALTAIR		◆VEGA		DENEB	
330	51 58	049	43 06	122	51 47	160	24 20	194	57 47	277	33 44	313	47 10	340
331	52 43	048	43 57	122	52 07	161	24 05	195	56 48	277	33 01	313	46 49	339
332	53 27	047	44 47	123	52 26	162	23 50	195	55 49	277	32 17	313	46 27	338
333	54 11	046	45 37	123	52 43	164	23 34	196	54 50	277	31 33	312	46 04	337
334	54 54	046	46 26	124	53 00	165	23 18	196	53 50	277	30 49	312	45 40	336
335	55 36	045	47 15	125	53 14	166	23 01	197	52 51	277	30 05	312	45 16	335
336	56 18	044	48 04	126	53 28	168	22 43	197	51 52	277	29 20	312	44 50	335
337	56 58	043	48 53	126	53 40	169	22 26	198	50 53	277	28 35	311	44 24	334
338	57 39	042	49 40	127	53 50	171	22 07	198	49 53	277	27 50	311	43 57	333
339	58 18	041	50 28	128	53 59	172	21 48	199	48 54	277	27 05	311	43 30	332
340	58 56	039	51 14	129	54 06	174	21 29	199	47 55	277	26 20	311	43 02	332
341	59 34	038	52 00	130	54 12	175	21 09	200	46 56	277	25 35	310	42 33	331
342	60 10	037	52 46	131	54 17	177	20 49	200	45 56	277	24 49	310	42 04	330
343	60 45	036	53 31	132	54 20	178	20 29	200	44 57	277	24 04	310	41 34	329
344	61 20	034	54 15	133	54 21	179	20 08	201	43 58	277	23 18	310	41 03	329
	◆Mirfak		Hamal		Acamar		◆ACHERNAR		FOMALHAUT		◆ALTAIR		DENEB	
345	19 56	039	41 51	064	18 29	136	19 08	159	54 21	181	42 58	277	40 32	328
346	20 33	039	42 44	063	19 10	137	19 30	159	54 19	182	41 59	277	40 00	328
347	21 10	038	43 37	063	19 51	137	19 51	160	54 16	184	41 00	277	39 28	327
348	21 47	038	44 30	063	20 32	137	20 11	160	54 11	185	40 01	277	38 55	326
349	22 24	038	45 23	062	21 12	138	20 31	160	54 04	187	39 01	277	38 22	326
350	23 01	038	46 16	062	21 52	138	20 51	161	53 56	188	38 02	277	37 48	325
351	23 37	038	47 09	062	22 32	138	21 10	161	53 47	190	37 03	277	37 14	325
352	24 13	037	48 01	061	23 11	139	21 29	162	53 36	191	36 04	277	36 39	324
353	24 49	037	48 53	061	23 50	139	21 48	162	53 24	193	35 04	277	36 04	324
354	25 25	037	49 45	060	24 29	140	22 05	163	53 10	194	34 05	277	35 29	323
355	26 01	037	50 37	060	25 08	140	22 23	163	52 55	195	33 06	277	34 53	323
356	26 36	036	51 29	059	25 46	141	22 40	164	52 38	197	32 06	277	34 17	322
357	27 11	036	52 20	059	26 23	141	22 56	164	52 20	198	31 07	277	33 40	322
358	27 46	036	53 11	058	27 00	142	23 12	165	52 01	199	30 08	277	33 04	322
359	28 21	035	54 01	058	27 37	142	23 27	165	51 40	201	29 09	277	32 26	321

LAT 5°N

LHA ♈	Hc	Zn	Hc	Zn	Hc	Zn	Hc	Zn	Hc	Zn	Hc	Zn	Hc	Zn
	◆CAPELLA		ALDEBARAN		RIGEL		◆ACHERNAR		FOMALHAUT		◆ALTAIR		DENEB	
0	11 10	044	21 37	074	10 32	099	24 40	166	52 14	203	28 02	277	31 02	321
1	11 52	044	22 35	074	11 31	099	24 55	166	51 51	204	27 03	277	30 25	321
2	12 34	044	23 32	074	12 30	100	25 08	167	51 26	205	26 03	277	29 47	320
3	13 15	044	24 30	074	13 29	100	25 22	167	51 00	206	25 04	277	29 08	320
4	13 56	044	25 27	074	14 28	100	25 35	168	50 33	207	24 05	277	28 30	320
5	14 38	044	26 25	074	15 27	100	25 47	169	50 05	209	23 06	278	27 51	319
6	15 19	044	27 22	074	16 26	100	25 58	169	49 36	210	22 06	278	27 12	319
7	16 00	043	28 20	074	17 25	100	26 09	170	49 06	211	21 07	278	26 33	319
8	16 41	043	29 17	074	18 23	100	26 20	170	48 35	212	20 08	278	25 54	319
9	17 22	043	30 15	074	19 22	101	26 30	171	48 03	213	19 09	278	25 14	318
10	18 03	043	31 12	074	20 21	101	26 39	171	47 30	214	18 09	278	24 34	318
11	18 44	043	32 09	074	21 20	101	26 48	172	46 56	215	17 10	278	23 54	318
12	19 24	043	33 07	074	22 18	101	26 56	173	46 22	216	16 11	278	23 14	318
13	20 05	043	34 04	073	23 17	101	27 03	173	45 47	217	15 12	278	22 34	317
14	20 45	042	35 01	073	24 16	101	27 10	174	45 10	218	14 13	278	21 53	317
	CAPELLA		◆ALDEBARAN		RIGEL		◆ACHERNAR		FOMALHAUT		Enif		◆DENEB	
15	21 25	042	35 58	073	25 14	102	27 16	174	44 34	218	41 12	279	21 12	317
16	22 05	042	36 56	073	26 13	102	27 22	175	43 56	219	40 13	279	20 32	317
17	22 45	042	37 53	073	27 11	102	27 27	176	43 18	220	39 14	279	19 51	317
18	23 25	042	38 50	073	28 10	102	27 31	176	42 39	221	38 15	279	19 09	316
19	24 04	041	39 47	073	29 08	102	27 35	177	42 00	222	37 15	279	18 28	316
20	24 44	041	40 44	072	30 07	102	27 38	177	41 20	222	36 16	279	17 47	316
21	25 23	041	41 41	072	31 05	103	27 41	178	40 39	223	35 17	279	17 05	316
22	26 02	040	42 38	072	32 03	103	27 42	179	39 58	224	34 18	279	16 24	316
23	26 40	040	43 35	072	33 01	103	27 44	179	39 16	225	33 19	279	15 42	316
24	27 19	040	44 32	072	34 00	103	27 44	180	38 34	225	32 20	279	15 00	316
25	27 57	040	45 28	071	34 58	104	27 44	180	37 51	226	31 21	279	14 18	316
26	28 35	039	46 25	071	35 56	104	27 43	181	37 08	226	30 22	278	13 36	315
27	29 13	039	47 22	071	36 54	104	27 42	182	36 25	227	29 23	279	12 55	315
28	29 50	039	48 18	071	37 52	104	27 40	182	35 41	228	28 23	279	12 12	315
29	30 27	038	49 14	070	38 49	105	27 37	183	34 56	228	27 24	279	11 30	315
	Mirfak		CAPELLA		◆ALDEBARAN		RIGEL		◆ACHERNAR		FOMALHAUT		◆Alpheratz	
30	41 47	018	31 04	038	50 11	070	39 47	105	27 34	183	34 12	229	54 14	315
31	42 06	017	31 40	037	51 07	070	40 45	105	27 30	184	33 27	229	53 32	315
32	42 23	017	32 17	037	52 03	069	41 43	106	27 26	185	32 41	230	52 49	314
33	42 39	016	32 52	037	52 59	069	42 40	106	27 21	185	31 55	230	52 05	313
34	42 55	015	33 28	036	53 54	069	43 37	106	27 15	186	31 09	231	51 21	312
35	43 10	014	34 03	036	54 50	068	44 35	107	27 09	186	30 23	231	50 37	311
36	43 24	013	34 37	035	55 45	068	45 32	107	27 02	187	29 36	232	49 52	311
37	43 38	012	35 12	035	56 41	067	46 29	107	26 54	188	28 49	232	49 06	310
38	43 50	012	35 46	034	57 36	067	47 26	108	26 46	188	28 02	232	48 20	309
39	44 02	011	36 19	034	58 31	066	48 23	108	26 37	189	27 14	233	47 34	309
40	44 12	010	36 52	033	59 25	066	49 19	109	26 27	189	26 27	233	46 47	308
41	44 22	009	37 24	033	60 19	065	50 16	109	26 17	190	25 39	234	46 00	308
42	44 31	008	37 56	032	61 13	064	51 12	110	26 07	191	24 50	234	45 13	307
43	44 39	007	38 28	032	62 07	064	52 08	110	25 56	191	24 02	234	44 25	307
44	44 46	006	38 59	031	63 01	063	53 04	111	25 44	192	23 13	235	43 37	306
	CAPELLA		◆PROCYON		SIRIUS		CANOPUS		◆ACHERNAR		Diphda		◆Alpheratz	
45	39 29	030	20 33	086	30 21	113	18 07	150	25 32	192	49 12	235	42 48	306
46	39 59	030	21 33	086	31 16	113	18 37	151	25 19	193	48 23	236	42 00	305
47	40 28	029	22 33	086	32 11	113	19 06	151	25 05	193	47 33	236	41 11	305
48	40 57	028	23 32	086	33 06	114	19 34	151	24 51	194	46 43	237	40 22	304
49	41 25	028	24 32	087	34 01	114	20 03	152	24 37	194	45 53	238	39 32	304
50	41 53	027	25 32	087	34 55	114	20 31	152	24 22	195	45 02	238	38 42	304
51	42 19	026	26 31	087	35 49	115	20 58	153	24 06	195	44 11	239	37 53	303
52	42 46	026	27 31	087	36 44	115	21 26	153	23 50	196	43 20	239	37 03	303
53	43 11	025	28 31	087	37 38	116	21 52	154	23 33	196	42 28	240	36 12	303
54	43 36	024	29 30	087	38 31	116	22 19	154	23 16	197	41 37	240	35 22	302
55	44 00	023	30 30	087	39 25	116	22 45	154	22 58	197	40 44	241	34 31	302
56	44 23	022	31 30	087	40 18	117	23 10	155	22 40	198	39 52	241	33 41	302
57	44 45	022	32 29	087	41 12	117	23 35	155	22 22	198	38 59	242	32 50	302
58	45 07	021	33 29	087	42 05	118	24 00	156	22 03	199	38 06	242	31 59	301
59	45 27	020	34 29	087	42 57	118	24 24	156	21 43	199	37 13	243	31 08	301
	◆CAPELLA		PROCYON		◆SIRIUS		CANOPUS		ACHERNAR		◆Diphda		Alpheratz	
60	45 47	019	35 29	087	43 50	119	24 48	157	21 23	200	36 20	243	30 17	301
61	46 06	018	36 28	087	44 42	120	25 11	157	21 03	200	35 27	244	29 25	301
62	46 24	017	37 28	087	45 34	120	25 34	158	20 42	201	34 33	244	28 34	301
63	46 42	016	38 28	087	46 25	121	25 56	158	20 21	201	33 39	244	27 42	300
64	46 58	015	39 27	087	47 16	121	26 17	159	19 59	202	32 45	245	26 51	300
65	47 13	014	40 27	087	48 07	122	26 39	160	19 37	202	31 51	245	25 59	300
66	47 28	013	41 27	087	48 58	123	26 59	160	19 14	202	30 57	246	25 07	300
67	47 41	012	42 26	087	49 48	123	27 19	161	18 51	203	30 02	246	24 15	300
68	47 54	011	43 26	088	50 38	124	27 39	161	18 28	203	29 08	246	23 23	300
69	48 05	010	44 26	088	51 27	125	27 58	162	18 04	204	28 13	246	22 31	299
70	48 15	009	45 26	088	52 16	126	28 16	162	17 40	204	27 18	247	21 39	299
71	48 25	008	46 25	088	53 04	127	28 34	163	17 16	204	26 23	247	20 47	299
72	48 33	007	47 25	088	53 52	128	28 51	164	16 51	205	25 28	247	19 55	299
73	48 40	006	48 25	088	54 39	128	29 07	164	16 26	205	24 33	248	19 03	299
74	48 46	005	49 25	088	55 25	129	29 23	165	16 00	205	23 37	248	18 10	299
	CAPELLA		◆POLLUX		PROCYON		SIRIUS		◆CANOPUS		Acamar		◆Hamal	
75	48 51	004	44 35	055	50 24	088	56 11	130	29 38	166	36 44	209	44 26	298
76	48 55	003	45 24	054	51 24	088	56 56	132	29 53	166	36 15	210	43 33	298
77	48 58	002	46 12	054	52 24	088	57 40	133	30 07	167	35 45	210	42 40	298
78	49 00	001	47 00	053	53 23	088	58 24	134	30 20	167	35 15	211	41 47	297
79	49 00	000	47 48	053	54 23	088	59 07	135	30 33	168	34 44	212	40 54	297
80	49 00	359	48 36	052	55 23	088	59 48	136	30 45	169	34 12	212	40 01	297
81	48 58	358	49 23	051	56 23	088	60 29	138	30 56	169	33 40	213	39 07	296
82	48 56	357	50 09	051	57 22	088	61 09	139	31 06	170	33 07	214	38 14	296
83	48 52	356	50 55	050	58 22	088	61 47	141	31 16	171	32 33	214	37 20	296
84	48 47	355	51 41	049	59 22	088	62 24	142	31 25	172	31 59	215	36 26	296
85	48 41	354	52 26	049	60 22	088	63 00	144	31 34	172	31 25	215	35 32	295
86	48 34	353	53 11	048	61 21	088	63 35	146	31 42	173	30 50	216	34 38	295
87	48 26	352	53 55	047	62 21	088	64 08	147	31 49	174	30 15	217	33 44	295
88	48 17	351	54 38	046	63 21	088	64 39	149	31 55	174	29 39	217	32 50	295
89	48 07	350	55 21	045	64 21	088	65 09	151	32 00	175	29 03	218	31 55	295

LHA ♈	Hc	Zn	Hc	Zn	Hc	Zn	Hc	Zn	Hc	Zn	Hc	Zn	Hc	Zn
	POLLUX		◆REGULUS		Suhail		◆CANOPUS		Acamar		◆ALDEBARAN		CAPELLA	
90	56 03	044	28 22	079	25 44	144	32 05	176	28 26	218	66 20	301	47 55	349
91	56 45	043	29 21	079	26 19	144	32 09	176	27 49	219	65 29	300	47 43	348
92	57 25	042	30 20	079	26 53	145	32 13	177	27 11	219	64 36	299	47 30	347
93	58 05	041	31 18	079	27 27	145	32 15	178	26 33	220	63 44	298	47 16	346
94	58 44	040	32 17	079	28 01	146	32 17	179	25 55	220	62 51	297	47 00	345
95	59 22	039	33 16	079	28 34	146	32 18	179	25 16	221	61 57	296	46 44	344
96	59 59	038	34 14	079	29 07	147	32 18	180	24 37	221	61 04	296	46 27	343
97	60 35	036	35 13	079	29 40	148	32 18	181	23 58	221	60 09	295	46 09	342
98	61 10	035	36 12	079	30 11	148	32 17	181	23 18	222	59 15	294	45 50	341
99	61 43	034	37 10	079	30 43	149	32 15	182	22 38	222	58 20	294	45 31	340
100	62 16	032	38 09	079	31 14	149	32 12	183	21 58	223	57 26	293	45 10	339
101	62 47	031	39 07	079	31 44	150	32 09	184	21 17	223	56 31	293	44 49	339
102	63 17	029	40 06	079	32 13	151	32 05	184	20 36	223	55 35	292	44 26	338
103	63 45	027	41 05	078	32 43	151	32 00	185	19 55	224	54 40	292	44 03	337
104	64 11	025	42 03	078	33 11	152	31 54	186	19 13	224	53 44	291	43 39	336
	POLLUX		◆REGULUS		Suhail		◆CANOPUS		RIGEL		◆ALDEBARAN		CAPELLA	
105	64 36	024	43 02	078	33 39	152	31 48	186	60 31	244	52 48	291	43 15	335
106	64 59	022	44 00	078	34 06	153	31 41	187	59 37	244	51 53	291	42 50	335
107	65 20	020	44 59	078	34 33	154	31 33	188	58 43	245	50 56	290	42 24	334
108	65 40	018	45 57	078	34 59	155	31 25	189	57 49	246	50 00	290	41 57	333
109	65 57	016	46 56	078	35 24	155	31 15	189	56 54	247	49 04	290	41 30	332
110	66 12	014	47 54	078	35 49	156	31 06	190	55 59	247	48 08	289	41 02	332
111	66 26	012	48 52	077	36 13	157	30 55	191	55 04	248	47 11	289	40 33	331
112	66 37	009	49 51	077	36 36	158	30 44	191	54 09	249	46 15	289	40 04	330
113	66 45	007	50 49	077	36 58	158	30 32	192	53 13	249	45 18	288	39 34	330
114	66 52	005	51 47	077	37 20	159	30 19	193	52 17	250	44 21	288	39 04	329
115	66 56	003	52 45	077	37 41	160	30 06	193	51 21	250	43 24	288	38 33	329
116	66 58	001	53 44	077	38 01	161	29 52	194	50 24	251	42 28	288	38 01	328
117	66 57	358	54 42	076	38 20	162	29 37	194	49 28	251	41 31	288	37 30	327
118	66 54	356	55 40	076	38 39	162	29 22	195	48 31	252	40 34	287	36 57	327
119	66 49	354	56 38	076	38 57	163	29 06	196	47 34	252	39 37	287	36 24	326
	Dubhe		◆Denebola		Suhail		◆CANOPUS		SIRIUS		BETELGEUSE		◆CAPELLA	
120	23 53	022	32 58	076	39 13	164	28 49	196	61 27	220	58 49	276	35 51	326
121	24 15	021	33 56	076	39 29	165	28 32	197	60 47	222	57 50	276	35 17	325
122	24 37	021	34 54	076	39 44	166	28 14	198	60 07	223	56 50	276	34 43	325
123	24 58	021	35 52	076	39 58	167	27 56	198	59 26	224	55 51	276	34 08	324
124	25 19	020	36 50	075	40 12	168	27 37	199	58 44	226	54 52	276	33 33	324
125	25 40	020	37 48	075	40 24	169	27 18	199	58 01	227	53 52	276	32 58	323
126	26 00	020	38 45	075	40 35	170	26 58	200	57 17	228	52 53	276	32 22	323
127	26 20	019	39 43	075	40 46	170	26 37	200	56 32	229	51 53	276	31 46	323
128	26 40	019	40 41	075	40 55	171	26 16	201	55 46	230	50 54	276	31 10	322
129	26 59	019	41 39	075	41 03	172	25 54	202	55 00	231	49 54	276	30 33	322
130	27 18	018	42 36	075	41 11	173	25 32	202	54 13	232	48 55	276	29 56	322
131	27 36	018	43 34	075	41 17	174	25 09	203	53 26	233	47 55	276	29 19	321
132	27 54	017	44 32	074	41 23	175	24 46	203	52 38	234	46 56	276	28 41	321
133	28 12	017	45 29	074	41 27	176	24 22	204	51 50	235	45 56	276	28 03	320
134	28 29	016	46 27	074	41 31	177	23 58	204	51 01	235	44 57	275	27 25	320
	Dubhe		◆SPICA		ACRUX		◆CANOPUS		SIRIUS		BETELGEUSE		◆CAPELLA	
135	28 46	016	22 09	104	11 42	159	23 33	205	50 11	236	43 57	275	26 47	320
136	29 02	016	23 07	104	12 03	159	23 08	205	49 21	237	42 58	275	26 08	320
137	29 18	015	24 05	105	12 24	159	22 43	206	48 31	238	41 58	275	25 29	319
138	29 33	015	25 03	105	12 45	160	22 17	206	47 40	238	40 59	275	24 50	319
139	29 48	014	26 01	105	13 06	160	21 50	206	46 49	239	39 59	275	24 11	319
140	30 03	014	26 58	105	13 26	160	21 23	207	45 58	240	39 00	275	23 31	319
141	30 17	013	27 56	105	13 46	161	20 56	207	45 06	240	38 00	276	22 52	318
142	30 30	013	28 54	106	14 06	161	20 28	208	44 14	241	37 01	276	22 12	318
143	30 43	012	29 51	106	14 26	161	20 00	208	43 22	241	36 01	276	21 32	318
144	30 56	012	30 49	106	14 45	162	19 32	209	42 29	242	35 02	276	20 51	318
145	31 08	011	31 46	106	15 03	162	19 03	209	41 36	242	34 02	276	20 11	317
146	31 19	011	32 43	107	15 22	162	18 34	209	40 43	243	33 03	276	19 31	317
147	31 30	010	33 41	107	15 40	163	18 05	210	39 50	243	32 03	276	18 50	317
148	31 41	010	34 38	107	15 58	163	17 35	210	38 56	244	31 04	276	18 09	317
149	31 51	009	35 35	108	16 15	163	17 05	210	38 03	244	30 04	276	17 29	317
	◆Dubhe		ARCTURUS		◆SPICA		ACRUX		◆SIRIUS		PROCYON		POLLUX	
150	32 00	009	26 18	071	36 32	108	16 32	164	37 09	245	54 54	272	50 32	310
151	32 09	008	27 15	071	37 28	108	16 49	164	36 15	245	53 55	272	49 46	309
152	32 17	008	28 11	071	38 25	108	17 05	164	35 20	245	52 55	272	48 59	308
153	32 25	007	29 08	071	39 22	109	17 21	165	34 26	246	51 55	272	48 12	308
154	32 32	007	30 04	071	40 18	109	17 36	165	33 31	246	50 55	272	47 25	307
155	32 39	006	31 01	071	41 15	110	17 51	166	32 36	247	49 56	272	46 37	306
156	32 45	006	31 57	070	42 11	110	18 06	166	31 42	247	48 56	272	45 49	306
157	32 50	005	32 53	070	43 07	110	18 20	166	30 47	247	47 56	272	45 00	305
158	32 55	004	33 49	070	44 03	111	18 34	167	29 51	247	46 56	272	44 11	305
159	32 59	004	34 46	070	44 59	111	18 48	167	28 56	248	45 57	272	43 22	305
160	33 03	003	35 42	070	45 55	112	19 01	168	28 01	248	44 57	272	42 33	304
161	33 06	003	36 38	070	46 50	112	19 13	168	27 05	248	43 57	272	41 43	304
162	33 09	002	37 34	070	47 45	113	19 25	168	26 10	249	42 58	273	40 53	303
163	33 11	002	38 30	069	48 40	113	19 37	169	25 14	249	41 58	273	40 03	303
164	33 12	001	39 26	069	49 35	114	19 48	169	24 18	249	40 58	273	39 13	303
	Dubhe		◆ARCTURUS		SPICA		◆ACRUX		Suhail		PROCYON		◆POLLUX	
165	33 13	000	40 22	069	50 30	114	19 59	170	35 22	205	39 58	273	38 22	302
166	33 13	000	41 17	069	51 24	115	20 10	170	34 57	205	38 59	273	37 32	302
167	33 13	359	42 13	068	52 18	115	20 20	171	34 30	206	37 59	273	36 41	302
168	33 12	359	43 09	068	53 12	116	20 29	171	34 04	207	36 59	273	35 50	301
169	33 10	358	44 04	068	54 06	117	20 38	172	33 36	208	36 00	273	34 58	301
170	33 08	358	44 59	068	54 59	118	20 46	172	33 08	208	35 00	273	34 07	301
171	33 06	357	45 55	067	55 52	118	20 55	173	32 40	209	34 00	273	33 16	300
172	33 02	357	46 50	067	56 44	119	21 02	173	32 11	210	33 01	273	32 24	300
173	32 58	356	47 45	067	57 36	120	21 09	173	31 41	210	32 01	273	31 32	300
174	32 54	355	48 40	066	58 28	121	21 16	174	31 11	211	31 01	273	30 41	300
175	32 49	355	49 34	066	59 19	122	21 22	174	30 40	211	30 01	273	29 49	300
176	32 43	354	50 29	066	60 09	123	21 27	175	30 08	212	29 02	273	28 57	299
177	32 37	354	51 23	065	60 59	124	21 32	175	29 37	213	28 02	273	28 05	299
178	32 30	353	52 17	065	61 48	125	21 37	176	29 04	213	27 02	273	27 12	299
179	32 23	353	53 11	064	62 37	126	21 41	176	28 31	214	26 03	273	26 20	299

LAT 5°N

LHA ♈	Hc	Zn	Hc	Zn	Hc	Zn	Hc	Zn	Hc	Zn	Hc	Zn	Hc	Zn
	◆ARCTURUS		ANTARES		RIGIL KENT.		◆ACRUX		Suhail		◆REGULUS		Dubhe	
180	54 05	064	17 49	120	17 17	161	21 45	177	27 58	214	61 29	286	32 15	352
181	54 59	063	18 40	120	17 36	161	21 48	177	27 24	215	60 32	286	32 07	352
182	55 52	063	19 32	120	17 55	162	21 50	178	26 50	215	59 34	285	31 58	351
183	56 45	062	20 24	121	18 14	162	21 52	178	26 15	216	58 36	285	31 48	351
184	57 37	061	21 15	121	18 32	162	21 54	179	25 40	216	57 39	285	31 38	350
185	58 30	061	22 06	121	18 50	163	21 55	179	25 05	217	56 41	284	31 27	350
186	59 22	060	22 57	121	19 07	163	21 56	180	24 29	217	55 43	284	31 16	349
187	60 13	059	23 48	122	19 24	164	21 56	180	23 53	218	54 45	284	31 05	349
188	61 04	058	24 39	122	19 41	164	21 55	181	23 16	218	53 47	284	30 52	348
189	61 55	057	25 29	122	19 57	165	21 54	181	22 39	218	52 48	283	30 40	348
190	62 45	057	26 20	123	20 12	165	21 53	182	22 02	219	51 50	283	30 27	347
191	63 35	056	27 10	123	20 28	165	21 51	182	21 24	219	50 52	283	30 13	347
192	64 24	054	28 00	124	20 42	166	21 48	183	20 46	220	49 54	283	29 59	346
193	65 12	053	28 49	124	20 57	166	21 45	183	20 08	220	48 55	283	29 44	346
194	65 59	052	29 39	124	21 10	167	21 42	184	19 30	220	47 57	282	29 29	345
	◆Alphecca		Rasalhague		ANTARES		◆RIGIL KENT.		ACRUX		◆REGULUS		Dubhe	
195	47 15	055	21 52	078	30 28	125	21 24	167	21 38	184	46 59	282	29 14	345
196	48 04	055	22 51	078	31 17	125	21 37	168	21 33	185	46 00	282	28 58	344
197	48 53	054	23 49	078	32 06	126	21 49	168	21 28	185	45 02	282	28 41	344
198	49 41	054	24 48	078	32 54	126	22 01	169	21 23	186	44 03	282	28 24	343
199	50 29	053	25 47	078	33 43	127	22 13	169	21 17	186	43 05	282	28 07	343
200	51 16	052	26 45	078	34 30	127	22 23	170	21 10	186	42 06	282	27 49	343
201	52 03	052	27 44	078	35 18	128	22 34	170	21 03	187	41 08	282	27 31	342
202	52 50	051	28 42	078	36 05	128	22 44	171	20 56	187	40 09	281	27 12	342
203	53 36	050	29 41	078	36 52	129	22 53	171	20 48	188	39 11	281	26 54	341
204	54 21	049	30 39	078	37 39	129	23 02	172	20 39	188	38 12	281	26 34	341
205	55 06	048	31 38	078	38 25	130	23 10	172	20 30	189	37 13	281	26 14	341
206	55 51	048	32 36	078	39 11	130	23 18	173	20 21	189	36 15	281	25 54	340
207	56 35	047	33 35	078	39 56	131	23 26	173	20 11	190	35 16	281	25 34	340
208	57 18	046	34 33	078	40 41	132	23 32	174	20 01	190	34 17	281	25 13	339
209	58 00	045	35 32	078	41 25	132	23 39	174	19 50	191	33 19	281	24 52	339
	◆VEGA		Rasalhague		ANTARES		◆RIGIL KENT.		ACRUX		◆REGULUS		Dubhe	
210	19 18	051	36 30	078	42 09	133	23 44	175	19 39	191	32 20	281	24 31	339
211	20 04	050	37 29	078	42 53	134	23 50	175	19 27	191	31 21	281	24 09	338
212	20 50	050	38 27	078	43 36	134	23 54	176	19 15	192	30 23	281	23 47	338
213	21 36	050	39 26	078	44 18	135	23 58	176	19 03	192	29 24	281	23 24	338
214	22 22	050	40 24	078	45 00	136	24 02	177	18 50	193	28 25	281	23 01	337
215	23 07	050	41 23	078	45 41	137	24 05	177	18 36	193	27 27	281	22 38	337
216	23 53	050	42 21	078	46 22	138	24 07	178	18 23	194	26 28	281	22 15	337
217	24 38	049	43 19	077	47 02	139	24 09	178	18 08	194	25 29	281	21 52	337
218	25 24	049	44 18	077	47 41	139	24 10	179	17 54	194	24 31	281	21 28	336
219	26 09	049	45 16	077	48 19	140	24 11	180	17 39	195	23 32	281	21 04	336
220	26 54	049	46 14	077	48 57	141	24 11	180	17 23	195	22 33	281	20 39	336
221	27 38	048	47 12	077	49 34	142	24 11	181	17 08	196	21 35	281	20 15	336
222	28 23	048	48 11	077	50 10	143	24 10	181	16 51	196	20 36	281	19 50	335
223	29 07	048	49 09	077	50 45	144	24 08	182	16 35	196	19 37	281	19 25	335
224	29 52	048	50 07	076	51 20	146	24 06	182	16 18	197	18 39	281	18 59	335
	VEGA		ALTAIR		◆Nunki		ANTARES		RIGIL KENT.		◆SPICA		◆Alkaid	
225	30 36	047	17 53	082	25 08	122	51 53	147	24 04	183	61 22	236	43 04	344
226	31 19	047	18 53	082	25 59	123	52 26	148	24 00	183	60 32	237	42 47	343
227	32 03	047	19 52	082	26 49	123	52 57	149	23 57	184	59 42	238	42 29	342
228	32 46	046	20 51	082	27 39	123	53 27	150	23 52	184	58 51	239	42 11	341
229	33 29	046	21 50	082	28 29	124	53 56	152	23 48	185	58 00	240	41 51	341
230	34 12	046	22 50	082	29 19	124	54 24	153	23 42	185	57 08	240	41 31	340
231	34 55	045	23 49	083	30 08	124	54 51	154	23 36	186	56 16	241	41 10	339
232	35 37	045	24 48	083	30 57	125	55 16	156	23 30	186	55 23	242	40 49	339
233	36 19	044	25 47	083	31 46	125	55 40	157	23 23	187	54 30	243	40 27	338
234	37 00	044	26 47	083	32 35	126	56 03	158	23 15	187	53 37	244	40 04	337
235	37 42	043	27 46	083	33 23	126	56 24	160	23 07	188	52 43	244	39 40	336
236	38 22	043	28 45	083	34 12	127	56 44	161	22 59	188	51 49	245	39 16	336
237	39 03	042	29 45	083	34 59	127	57 02	163	22 50	189	50 55	245	38 51	335
238	39 43	042	30 44	083	35 47	128	57 19	164	22 40	189	50 01	246	38 26	335
239	40 23	041	31 43	083	36 34	128	57 34	166	22 30	190	49 06	247	38 00	334
	VEGA		◆ALTAIR		Nunki		◆ANTARES		SPICA		◆ARCTURUS		Alkaid	
240	41 02	041	32 42	083	37 21	129	57 48	168	48 11	247	60 50	301	37 33	333
241	41 41	040	33 42	083	38 07	129	58 00	169	47 16	248	59 59	301	37 06	333
242	42 19	040	34 41	083	38 53	130	58 10	171	46 20	248	59 07	300	36 38	332
243	42 57	039	35 40	083	39 39	131	58 18	173	45 25	249	58 15	299	36 10	332
244	43 34	038	36 40	083	40 24	131	58 25	174	44 29	249	57 22	298	35 41	331
245	44 11	038	37 39	083	41 09	132	58 30	176	43 33	249	56 30	298	35 12	330
246	44 47	037	38 38	083	41 53	133	58 33	178	42 37	250	55 37	297	34 42	330
247	45 23	036	39 37	083	42 37	133	58 35	180	41 41	250	54 43	297	34 12	329
248	45 58	036	40 37	083	43 20	134	58 34	181	40 45	251	53 50	296	33 42	329
249	46 32	035	41 36	083	44 03	135	58 32	183	39 48	251	52 56	296	33 11	328
250	47 06	034	42 35	083	44 45	135	58 28	185	38 52	251	52 02	295	32 39	328
251	47 39	033	43 34	083	45 27	136	58 22	186	37 55	252	51 08	295	32 07	328
252	48 11	032	44 34	082	46 08	137	58 15	188	36 58	252	50 13	294	31 35	327
253	48 43	031	45 33	082	46 48	138	58 06	190	36 01	252	49 19	294	31 03	327
254	49 14	031	46 32	082	47 28	139	57 55	191	35 04	253	48 24	294	30 30	326
	◆VEGA		ALTAIR		◆Nunki		ANTARES		◆SPICA		ARCTURUS		Alkaid	
255	49 44	030	47 31	082	48 07	140	57 42	193	34 07	253	47 29	293	29 56	326
256	50 13	029	48 31	082	48 45	141	57 28	195	33 10	253	46 34	293	29 23	326
257	50 41	028	49 30	082	49 22	142	57 12	196	32 13	253	45 39	293	28 49	325
258	51 08	027	50 29	082	49 59	143	56 54	198	31 15	254	44 44	292	28 14	325
259	51 35	026	51 28	082	50 35	144	56 35	199	30 18	254	43 48	292	27 40	324
260	52 00	025	52 28	082	51 09	145	56 15	201	29 20	254	42 53	292	27 05	324
261	52 24	024	53 27	082	51 43	146	55 53	202	28 23	254	41 57	291	26 30	324
262	52 48	022	54 26	082	52 16	147	55 30	204	27 25	255	41 02	291	25 54	324
263	53 10	021	55 25	082	52 48	148	55 05	205	26 27	255	40 06	291	25 19	323
264	53 31	020	56 24	082	53 19	150	54 39	206	25 30	255	39 10	291	24 43	323
265	53 51	019	57 23	081	53 49	151	54 12	208	24 32	255	38 14	291	24 07	323
266	54 10	018	58 22	081	54 17	152	53 43	209	23 34	256	37 18	290	23 30	322
267	54 27	016	59 21	081	54 45	153	53 14	210	22 36	256	36 22	290	22 54	322
268	54 44	015	60 20	081	55 11	155	52 43	212	21 38	256	35 26	290	22 17	322
269	54 59	014	61 19	081	55 36	156	52 11	213	20 40	256	34 30	290	21 40	322

LHA ♈	Hc	Zn	Hc	Zn	Hc	Zn	Hc	Zn	Hc	Zn	Hc	Zn	Hc	Zn
	VEGA		◆DENEB		FOMALHAUT		◆Peacock		Shaula		ANTARES		◆ARCTURUS	
270	55 12	013	36 37	035	10 59	122	21 33	160	47 26	188	51 38	214	33 34	290
271	55 25	011	37 11	034	11 50	122	21 53	160	47 18	189	51 04	215	32 37	290
272	55 36	010	37 44	033	12 41	122	22 13	160	47 08	190	50 30	216	31 41	290
273	55 45	009	38 17	033	13 31	122	22 33	161	46 56	191	49 54	217	30 45	289
274	55 54	007	38 49	032	14 22	122	22 52	161	46 44	192	49 18	218	29 48	289
275	56 00	006	39 21	032	15 12	123	23 11	162	46 31	194	48 40	219	28 52	289
276	56 06	004	39 52	031	16 03	123	23 30	162	46 16	195	48 02	220	27 55	289
277	56 10	003	40 23	030	16 53	123	23 47	163	46 00	196	47 23	221	26 59	289
278	56 12	002	40 53	030	17 43	123	24 05	163	45 44	197	46 44	222	26 02	289
279	56 13	000	41 22	029	18 33	124	24 21	164	45 26	198	46 04	223	25 06	289
280	56 13	359	41 51	028	19 22	124	24 38	164	45 07	199	45 23	224	24 09	289
281	56 11	357	42 19	028	20 12	124	24 53	165	44 47	200	44 41	224	23 13	289
282	56 08	356	42 47	027	21 01	124	25 08	166	44 26	201	43 59	225	22 16	289
283	56 03	355	43 14	026	21 51	125	25 23	166	44 04	202	43 16	226	21 19	289
284	55 56	353	43 40	026	22 40	125	25 37	167	43 41	203	42 33	227	20 23	289
	DENEB		Alpheratz		◆FOMALHAUT		Peacock		◆ANTARES		Rasalhague		◆VEGA	
285	44 05	025	13 45	061	23 28	125	25 51	167	41 50	227	67 38	291	55 49	352
286	44 30	024	14 38	061	24 17	126	26 04	168	41 05	228	66 42	290	55 40	351
287	44 54	023	15 30	061	25 05	126	26 16	168	40 21	229	65 46	290	55 29	349
288	45 17	022	16 22	061	25 54	126	26 28	169	39 36	229	64 49	289	55 17	348
289	45 39	021	17 15	061	26 42	127	26 39	169	38 50	230	63 53	289	55 04	347
290	46 01	021	18 07	061	27 29	127	26 50	170	38 04	231	62 56	288	54 50	345
291	46 21	020	18 59	061	28 17	128	27 00	171	37 18	231	61 59	287	54 34	344
292	46 41	019	19 52	061	29 04	128	27 09	171	36 31	232	61 02	287	54 17	343
293	47 00	018	20 44	061	29 51	129	27 18	172	35 44	232	60 05	287	53 58	342
294	47 18	017	21 36	061	30 37	129	27 26	172	34 57	233	59 07	286	53 39	340
295	47 35	016	22 28	061	31 24	129	27 34	173	34 09	233	58 10	286	53 18	339
296	47 51	015	23 20	060	32 10	130	27 41	174	33 21	234	57 12	286	52 56	338
297	48 06	014	24 12	060	32 55	130	27 47	174	32 33	234	56 15	285	52 34	337
298	48 20	013	25 04	060	33 41	131	27 53	175	31 44	235	55 17	285	52 10	336
299	48 33	012	25 56	060	34 26	131	27 58	175	30 55	235	54 19	285	51 45	335
	◆DENEB		Alpheratz		◆FOMALHAUT		Peacock		◆ANTARES		Rasalhague		VEGA	
300	48 45	011	26 47	060	35 10	132	28 02	176	30 06	235	53 21	284	51 19	334
301	48 55	010	27 39	060	35 55	133	28 06	177	29 17	236	52 23	284	50 52	333
302	49 05	009	28 31	060	36 38	133	28 09	177	28 27	236	51 25	284	50 24	332
303	49 14	008	29 22	059	37 22	134	28 11	178	27 37	237	50 27	284	49 55	331
304	49 22	007	30 13	059	38 05	134	28 13	179	26 47	237	49 29	284	49 25	330
305	49 28	006	31 05	059	38 47	135	28 14	179	25 57	237	48 31	283	48 55	329
306	49 34	005	31 56	059	39 29	136	28 15	180	25 07	238	47 33	283	48 24	328
307	49 38	004	32 47	058	40 11	136	28 15	180	24 16	238	46 35	283	47 52	327
308	49 41	003	33 38	058	40 52	137	28 14	181	23 25	238	45 36	283	47 19	326
309	49 43	001	34 28	058	41 32	138	28 13	182	22 34	239	44 38	283	46 45	326
310	49 44	000	35 19	058	42 12	139	28 11	182	21 43	239	43 40	283	46 11	325
311	49 44	359	36 09	057	42 51	139	28 08	183	20 52	239	42 42	282	45 36	324
312	49 43	358	36 59	057	43 30	140	28 04	184	20 00	240	41 43	282	45 01	323
313	49 40	357	37 49	057	44 08	141	28 00	184	19 09	240	40 45	282	44 25	323
314	49 37	356	38 39	056	44 45	142	27 56	185	18 17	240	39 46	282	43 48	322
	◆Alpheratz		◆FOMALHAUT		Peacock		Nunki		◆Rasalhague		VEGA		DENEB	
315	39 29	056	45 22	143	27 51	185	46 27	222	38 48	282	43 11	321	49 32	355
316	40 18	056	45 58	144	27 45	186	45 46	223	37 49	282	42 34	321	49 26	354
317	41 08	055	46 33	145	27 38	187	45 05	224	36 51	282	41 56	320	49 19	353
318	41 56	055	47 07	145	27 31	187	44 23	225	35 53	282	41 17	319	49 11	352
319	42 45	054	47 40	146	27 23	188	43 41	226	34 54	282	40 38	319	49 02	351
320	43 34	054	48 13	147	27 15	188	42 58	226	33 56	282	39 58	318	48 52	350
321	44 22	053	48 45	149	27 06	189	42 14	227	32 57	282	39 18	318	48 40	349
322	45 10	053	49 15	150	26 56	190	41 30	228	31 58	282	38 38	317	48 28	348
323	45 57	052	49 45	151	26 46	190	40 45	228	31 00	282	37 57	317	48 15	347
324	46 44	052	50 14	152	26 35	191	40 00	229	30 01	282	37 16	316	48 00	346
325	47 31	051	50 42	153	26 24	191	39 15	230	29 03	282	36 35	316	47 45	345
326	48 17	051	51 08	154	26 11	192	38 29	230	28 04	282	35 53	315	47 29	344
327	49 03	050	51 34	155	25 59	192	37 43	231	27 06	282	35 11	315	47 11	343
328	49 49	049	51 58	157	25 46	193	36 56	232	26 07	282	34 29	315	46 53	342
329	50 34	049	52 21	158	25 32	194	36 09	232	25 09	282	33 46	314	46 34	341
	◆Alpheratz		Diphda		◆FOMALHAUT		Peacock		◆ALTAIR		VEGA		DENEB	
330	51 19	048	43 37	121	52 43	159	25 18	194	57 39	279	33 03	314	46 14	340
331	52 03	047	44 28	121	53 04	161	25 03	195	56 40	279	32 20	314	45 53	339
332	52 46	046	45 19	122	53 23	162	24 48	195	55 41	278	31 36	313	45 31	338
333	53 29	045	46 10	123	53 41	163	24 32	196	54 42	278	30 53	313	45 08	337
334	54 11	045	47 00	123	53 57	165	24 15	196	53 43	278	30 09	313	44 45	337
335	54 53	044	47 49	124	54 13	166	23 58	197	52 44	278	29 24	312	44 21	336
336	55 34	043	48 39	125	54 26	167	23 41	197	51 44	278	28 40	312	43 56	335
337	56 14	042	49 28	125	54 39	169	23 23	198	50 45	278	27 56	312	43 30	334
338	56 53	041	50 16	126	54 49	170	23 04	198	49 46	278	27 11	311	43 04	333
339	57 32	040	51 04	127	54 58	172	22 45	199	48 47	278	26 26	311	42 37	333
340	58 10	038	51 52	128	55 06	173	22 26	199	47 47	278	25 41	311	42 09	332
341	58 46	037	52 38	129	55 12	175	22 06	200	46 48	278	24 56	311	41 41	331
342	59 22	036	53 25	130	55 17	176	21 46	200	45 49	278	24 10	311	41 12	331
343	59 56	035	54 10	131	55 20	178	21 25	201	44 50	278	23 25	310	40 42	330
344	60 30	033	54 55	132	55 21	179	21 04	201	43 50	277	22 39	310	40 12	329
	Alpheratz		◆Hamal		Acamar		◆ACHERNAR		FOMALHAUT		◆ALTAIR		DENEB	
345	61 02	032	41 24	063	19 12	136	20 04	159	55 21	181	42 51	277	39 41	329
346	61 33	030	42 17	063	19 53	136	20 26	159	55 19	183	41 52	277	39 09	328
347	62 03	029	43 10	062	20 35	137	20 47	159	55 15	184	40 53	277	38 37	327
348	62 31	027	44 03	062	21 16	137	21 08	160	55 10	186	39 53	277	38 05	327
349	62 57	026	44 55	062	21 56	137	21 28	160	55 04	187	38 54	277	37 32	326
350	63 23	024	45 48	061	22 37	138	21 48	161	54 56	189	37 55	277	36 59	326
351	63 46	022	46 40	061	23 17	138	22 07	161	54 46	190	36 56	277	36 25	325
352	64 08	020	47 32	060	23 56	139	22 26	162	54 35	192	35 56	277	35 51	325
353	64 28	019	48 24	060	24 36	139	22 45	162	54 22	193	34 57	277	35 16	324
354	64 46	017	49 15	059	25 15	139	23 03	163	54 08	194	33 58	277	34 41	324
355	65 02	015	50 07	059	25 53	140	23 20	163	53 52	196	32 58	277	34 05	323
356	65 16	013	50 58	058	26 32	140	23 37	164	53 35	197	31 59	277	33 29	323
357	65 28	011	51 48	058	27 10	141	23 54	164	53 17	199	31 00	277	32 53	322
358	65 38	009	52 39	057	27 47	141	24 10	165	52 57	200	30 01	277	32 16	322
359	65 46	006	53 29	056	28 25	142	24 25	165	52 36	201	29 01	277	31 39	322

LHA	Hc	Zn	Hc	Zn	Hc	Zn	Hc	Zn	Hc	Zn	Hc	Zn	Hc	Zn
♈	Alpheratz		◆Mirfak		ALDEBARAN		◆ACHERNAR		FOMALHAUT		◆ALTAIR		DENEB	
0	64 52	004	27 16	034	21 21	074	25 38	166	53 09	203	27 54	278	30 15	322
1	64 55	002	27 50	034	22 18	074	25 53	166	52 45	204	26 55	278	29 38	321
2	64 56	000	28 23	034	23 16	074	26 07	167	52 20	206	25 55	278	29 00	321
3	64 55	358	28 56	033	24 13	074	26 20	167	51 54	207	24 56	278	28 22	320
4	64 52	356	29 29	033	25 11	074	26 33	168	51 26	208	23 57	278	27 44	320
5	64 47	354	30 01	032	26 08	074	26 46	168	50 58	209	22 58	278	27 06	320
6	64 39	352	30 33	032	27 05	073	26 57	169	50 28	210	21 58	278	26 27	320
7	64 30	350	31 04	032	28 03	073	27 08	170	49 57	211	20 59	278	25 48	319
8	64 18	348	31 36	031	29 00	073	27 19	170	49 26	212	20 00	278	25 09	319
9	64 05	346	32 06	031	29 58	073	27 29	171	48 53	213	19 00	278	24 29	319
10	63 49	344	32 36	030	30 55	073	27 38	171	48 20	214	18 01	278	23 50	318
11	63 32	342	33 06	030	31 52	073	27 47	172	47 45	215	17 02	278	23 10	318
12	63 13	340	33 36	029	32 49	073	27 55	172	47 10	216	16 03	278	22 30	318
13	62 52	339	34 05	029	33 47	073	28 03	173	46 34	217	15 03	278	21 49	318
14	62 29	337	34 33	028	34 44	073	28 10	174	45 58	218	14 04	278	21 09	317
	CAPELLA		ALDEBARAN		◆RIGEL		ACHERNAR		◆FOMALHAUT		Enif		◆Alpheratz	
15	20 41	042	35 41	073	25 26	101	28 16	174	45 20	219	41 02	280	62 05	335
16	21 21	042	36 38	072	26 25	101	28 22	175	44 42	220	40 03	280	61 39	334
17	22 00	041	37 35	072	27 23	101	28 27	175	44 04	221	39 04	279	61 12	332
18	22 40	041	38 32	072	28 22	102	28 31	176	43 24	222	38 05	279	60 43	331
19	23 19	041	39 29	072	29 21	102	28 35	177	42 44	222	37 06	279	60 13	329
20	23 58	041	40 26	072	30 19	102	28 38	177	42 04	223	36 07	279	59 41	328
21	24 37	040	41 22	071	31 18	102	28 41	178	41 23	224	35 08	279	59 09	326
22	25 16	040	42 19	071	32 16	102	28 42	179	40 41	224	34 09	279	58 35	325
23	25 54	040	43 16	071	33 15	103	28 44	179	39 59	225	33 10	279	58 00	324
24	26 33	040	44 12	071	34 13	103	28 44	180	39 16	226	32 11	279	57 25	323
25	27 11	039	45 09	071	35 11	103	28 44	180	38 33	226	31 12	279	56 48	321
26	27 48	039	46 05	070	36 10	103	28 43	181	37 49	227	30 12	279	56 10	320
27	28 26	039	47 01	070	37 08	103	28 42	182	37 05	228	29 13	279	55 31	319
28	29 03	038	47 58	070	38 06	104	28 40	182	36 21	228	28 14	279	54 52	318
29	29 40	038	48 54	069	39 04	104	28 37	183	35 36	229	27 15	279	54 12	317
	Mirfak		CAPELLA		◆ALDEBARAN		RIGEL		◆ACHERNAR		FOMALHAUT		◆Alpheratz	
30	40 50	018	30 16	037	49 50	069	40 02	104	28 34	183	34 51	229	53 31	316
31	41 08	017	30 53	037	50 45	069	41 00	105	28 30	184	34 06	230	52 49	316
32	41 25	016	31 28	037	51 41	068	41 58	105	28 26	185	33 20	230	52 07	315
33	41 42	015	32 04	036	52 37	068	42 56	105	28 20	185	32 34	231	51 24	314
34	41 57	015	32 39	036	53 32	067	43 54	105	28 15	186	31 47	231	50 41	313
35	42 12	014	33 14	035	54 27	067	44 52	106	28 08	186	31 00	232	49 57	312
36	42 26	013	33 48	035	55 22	066	45 49	106	28 01	187	30 13	232	49 12	312
37	42 39	012	34 22	034	56 17	066	46 47	106	27 53	188	29 26	232	48 27	311
38	42 51	011	34 56	034	57 11	065	47 44	107	27 45	188	28 38	233	47 42	310
39	43 03	011	35 29	033	58 06	065	48 41	107	27 36	189	27 50	233	46 56	310
40	43 13	010	36 02	033	59 00	064	49 38	108	27 27	189	27 02	234	46 10	309
41	43 23	009	36 34	032	59 53	063	50 35	108	27 17	190	26 14	234	45 23	308
42	43 32	008	37 05	032	60 47	063	51 32	109	27 06	191	25 25	234	44 36	308
43	43 40	007	37 37	031	61 40	062	52 29	109	26 55	191	24 37	235	43 49	307
44	43 47	006	38 07	030	62 32	061	53 25	110	26 43	192	23 48	235	43 01	307
	CAPELLA		◆PROCYON		SIRIUS		CANOPUS		◆ACHERNAR		Diphda		◆Alpheratz	
45	38 37	030	20 29	086	30 44	112	18 59	150	26 30	192	49 46	236	42 13	306
46	39 07	029	21 29	086	31 39	112	19 29	151	26 17	193	48 56	236	41 25	306
47	39 36	029	22 29	086	32 35	113	19 58	151	26 04	193	48 06	237	40 36	306
48	40 04	028	23 28	086	33 30	113	20 27	151	25 49	194	47 15	238	39 47	305
49	40 32	027	24 28	086	34 25	113	20 56	152	25 35	194	46 25	238	38 58	305
50	40 59	027	25 28	086	35 20	114	21 24	152	25 20	195	45 33	239	38 09	304
51	41 26	026	26 28	086	36 14	114	21 52	153	25 04	196	44 42	240	37 19	304
52	41 51	025	27 27	086	37 09	114	22 19	153	24 48	196	43 50	240	36 30	304
53	42 16	024	28 27	086	38 03	115	22 46	153	24 31	197	42 58	241	35 40	303
54	42 41	024	29 27	086	38 57	115	23 13	154	24 13	197	42 06	241	34 50	303
55	43 04	023	30 27	086	39 51	116	23 39	154	23 56	198	41 13	242	33 59	303
56	43 27	022	31 26	086	40 45	116	24 05	155	23 37	198	40 20	242	33 09	302
57	43 49	021	32 26	086	41 39	117	24 30	155	23 19	199	39 27	243	32 18	302
58	44 10	020	33 26	086	42 32	117	24 55	156	22 59	199	38 34	243	31 28	302
59	44 31	020	34 25	086	43 25	118	25 19	156	22 40	199	37 40	244	30 37	302
	◆CAPELLA		PROCYON		◆SIRIUS		CANOPUS		ACHERNAR		◆Diphda		Alpheratz	
60	44 50	019	35 25	086	44 18	118	25 43	157	22 20	200	36 47	244	29 46	301
61	45 09	018	36 25	086	45 11	119	26 06	157	21 59	200	35 53	244	28 54	301
62	45 27	017	37 25	086	46 03	119	26 29	158	21 38	201	34 59	245	28 03	301
63	45 44	016	38 24	086	46 55	120	26 52	158	21 16	201	34 05	245	27 12	301
64	46 00	015	39 24	087	47 47	120	27 13	159	20 55	202	33 10	245	26 20	301
65	46 15	014	40 24	087	48 39	121	27 35	159	20 32	202	32 16	246	25 29	300
66	46 29	013	41 24	087	49 30	122	27 56	160	20 10	202	31 21	246	24 37	300
67	46 43	012	42 23	087	50 21	122	28 16	161	19 46	203	30 26	246	23 45	300
68	46 55	011	43 23	087	51 11	123	28 35	161	19 23	203	29 32	247	22 53	300
69	47 06	010	44 23	087	52 01	124	28 55	162	18 59	204	28 37	247	22 02	300
70	47 16	009	45 23	087	52 50	125	29 13	162	18 35	204	27 41	247	21 10	300
71	47 25	008	46 22	087	53 39	126	29 31	163	18 10	204	26 46	248	20 18	300
72	47 33	007	47 22	087	54 28	126	29 48	164	17 45	205	25 51	248	19 25	299
73	47 40	006	48 22	087	55 15	127	30 05	164	17 20	205	24 55	248	18 33	299
74	47 46	005	49 22	087	56 03	128	30 21	165	16 55	206	24 00	248	17 41	299
	CAPELLA		◆POLLUX		PROCYON		SIRIUS		◆CANOPUS		Acamar		◆Hamal	
75	47 51	004	44 00	054	50 21	087	56 49	129	30 36	165	37 37	209	43 57	299
76	47 55	003	44 49	054	51 21	087	57 35	130	30 51	166	37 07	210	43 05	299
77	47 58	002	45 37	053	52 21	087	58 21	131	31 05	166	36 37	211	42 12	298
78	48 00	001	46 24	052	53 21	087	59 05	133	31 19	167	36 06	211	41 20	298
79	48 00	000	47 12	052	54 20	087	59 49	134	31 31	168	35 35	212	40 27	298
80	48 00	359	47 58	051	55 20	087	60 31	135	31 43	169	35 02	213	39 34	297
81	47 58	358	48 45	051	56 20	087	61 13	137	31 55	169	34 30	213	38 40	297
82	47 56	357	49 31	050	57 20	087	61 54	138	32 05	170	33 57	214	37 47	297
83	47 52	356	50 16	049	58 19	087	62 33	139	32 15	171	33 23	215	36 53	296
84	47 47	355	51 01	048	59 19	086	63 11	141	32 25	171	32 49	215	36 00	296
85	47 41	354	51 46	048	60 19	086	63 48	143	32 33	172	32 14	216	35 06	296
86	47 35	353	52 30	047	61 19	086	64 24	144	32 41	173	31 39	216	34 12	296
87	47 27	352	53 13	046	62 18	086	64 58	146	32 48	174	31 03	217	33 18	296
88	47 18	351	53 56	045	63 18	086	65 31	148	32 55	174	30 27	217	32 24	295
89	47 08	350	54 38	044	64 18	086	66 01	150	33 00	175	29 50	218	31 30	295

LAT 4°N

LHA	Hc	Zn	Hc	Zn	Hc	Zn	Hc	Zn	Hc	Zn	Hc	Zn	Hc	Zn
♈	POLLUX		◆REGULUS		Suhail		◆CANOPUS		Acamar		◆ALDEBARAN		CAPELLA	
90	55 20	043	28 11	079	26 32	144	33 05	176	29 13	219	65 49	303	46 57	349
91	56 01	042	29 09	078	27 07	144	33 09	176	28 35	219	64 58	302	46 45	348
92	56 40	041	30 08	078	27 42	145	33 12	177	27 58	220	64 07	301	46 32	347
93	57 20	040	31 07	078	28 17	145	33 15	178	27 19	220	63 15	300	46 18	346
94	57 58	039	32 05	078	28 51	146	33 17	179	26 41	220	62 23	299	46 03	345
95	58 35	038	33 04	078	29 24	146	33 18	179	26 02	221	61 30	298	45 47	344
96	59 11	037	34 02	078	29 57	147	33 18	180	25 22	221	60 37	297	45 30	343
97	59 46	035	35 01	078	30 30	147	33 18	181	24 43	222	59 44	296	45 12	342
98	60 20	034	36 00	078	31 02	148	33 17	181	24 03	222	58 50	296	44 54	341
99	60 53	033	36 58	078	31 34	148	33 15	182	23 22	223	57 56	295	44 34	341
100	61 25	031	37 57	078	32 05	149	33 12	183	22 42	223	57 01	295	44 14	340
101	61 55	030	38 55	078	32 36	150	33 09	184	22 01	223	56 07	294	43 53	339
102	62 24	028	39 54	078	33 06	150	33 05	184	21 20	224	55 12	293	43 31	338
103	62 51	026	40 52	078	33 35	151	33 00	185	20 38	224	54 17	293	43 08	337
104	63 17	025	41 51	077	34 04	152	32 54	186	19 57	224	53 22	293	42 45	337
	POLLUX		◆REGULUS		Suhail		◆CANOPUS		RIGEL		◆ALDEBARAN		CAPELLA	
105	63 41	023	42 49	077	34 32	152	32 48	187	60 57	245	52 26	292	42 20	336
106	64 03	021	43 47	077	35 00	153	32 40	187	60 03	246	51 31	292	41 55	335
107	64 24	019	44 46	077	35 27	154	32 33	188	59 08	247	50 35	291	41 30	334
108	64 43	017	45 44	077	35 53	154	32 24	189	58 13	247	49 39	291	41 03	334
109	64 59	015	46 42	077	36 19	155	32 15	189	57 17	248	48 43	291	40 36	333
110	65 14	013	47 41	077	36 44	156	32 05	190	56 22	249	47 47	290	40 09	332
111	65 27	011	48 39	076	37 08	156	31 54	191	55 26	249	46 51	290	39 40	331
112	65 37	009	49 37	076	37 31	157	31 42	191	54 30	250	45 55	290	39 12	331
113	65 46	007	50 35	076	37 54	158	31 30	192	53 34	250	44 58	289	38 42	330
114	65 52	005	51 33	076	38 16	159	31 18	193	52 37	251	44 02	289	38 12	330
115	65 56	003	52 31	075	38 37	160	31 04	193	51 40	251	43 05	289	37 42	329
116	65 58	001	53 29	075	38 58	160	30 50	194	50 44	252	42 09	289	37 10	328
117	65 57	358	54 27	075	39 17	161	30 35	195	49 47	252	41 12	288	36 39	328
118	65 54	356	55 24	075	39 36	162	30 20	195	48 50	253	40 15	288	36 07	327
119	65 49	354	56 22	074	39 54	163	30 04	196	47 52	253	39 18	288	35 34	327
	Dubhe		◆Denebola		Suhail		◆CANOPUS		SIRIUS		BETELGEUSE		◆CAPELLA	
120	22 57	022	32 43	075	40 11	164	29 47	197	62 12	221	58 42	278	35 01	326
121	23 19	021	33 41	075	40 27	165	29 30	197	61 32	223	57 43	278	34 28	326
122	23 41	021	34 39	075	40 42	166	29 12	198	60 51	224	56 43	277	33 54	325
123	24 02	021	35 37	075	40 57	167	28 53	198	60 08	226	55 44	277	33 19	325
124	24 23	020	36 34	075	41 10	168	28 34	199	59 25	227	54 45	277	32 45	324
125	24 44	020	37 32	075	41 23	168	28 14	200	58 41	228	53 45	277	32 10	324
126	25 04	020	38 30	074	41 34	169	27 54	200	57 56	229	52 46	277	31 34	323
127	25 24	019	39 27	074	41 45	170	27 33	201	57 11	230	51 47	277	30 58	323
128	25 43	019	40 25	074	41 54	171	27 12	201	56 24	231	50 47	277	30 22	323
129	26 02	018	41 23	074	42 03	172	26 50	202	55 38	232	49 48	277	29 46	322
130	26 21	018	42 20	074	42 10	173	26 27	202	54 50	233	48 48	277	29 09	322
131	26 39	018	43 18	074	42 17	174	26 04	203	54 02	234	47 49	277	28 32	322
132	26 57	017	44 15	073	42 23	175	25 41	203	53 13	235	46 49	277	27 54	321
133	27 14	017	45 12	073	42 27	176	25 17	204	52 24	236	45 50	277	27 17	321
134	27 32	016	46 10	073	42 31	177	24 53	204	51 34	237	44 50	276	26 39	320
	Dubhe		◆SPICA		ACRUX		◆CANOPUS		SIRIUS		BETELGEUSE		◆CAPELLA	
135	27 48	016	22 24	104	12 37	159	24 28	205	50 44	237	43 51	276	26 01	320
136	28 04	015	23 22	104	12 59	159	24 03	205	49 53	238	42 51	276	25 22	320
137	28 20	015	24 20	104	13 20	159	23 37	206	49 02	239	41 52	276	24 43	320
138	28 35	015	25 18	104	13 41	160	23 11	206	48 11	239	40 52	276	24 05	319
139	28 50	014	26 16	105	14 02	160	22 44	207	47 20	240	39 53	276	23 25	319
140	29 05	014	27 14	105	14 23	160	22 17	207	46 28	241	38 53	276	22 46	319
141	29 18	013	28 12	105	14 43	160	21 49	208	45 35	241	37 54	276	22 07	319
142	29 32	013	29 10	105	15 03	161	21 22	208	44 43	242	36 54	276	21 27	318
143	29 45	012	30 07	105	15 22	161	20 53	208	43 50	242	35 55	276	20 47	318
144	29 57	012	31 05	106	15 42	161	20 25	209	42 57	243	34 56	276	20 07	318
145	30 09	011	32 03	106	16 00	162	19 56	209	42 04	243	33 56	276	19 27	318
146	30 21	011	33 00	106	16 19	162	19 26	210	41 10	244	32 57	276	18 47	318
147	30 31	010	33 58	106	16 37	162	18 57	210	40 16	244	31 57	276	18 06	317
148	30 42	010	34 55	107	16 55	163	18 27	210	39 22	245	30 58	276	17 25	317
149	30 52	009	35 52	107	17 12	163	17 56	211	38 28	245	29 58	276	16 45	317
	◆Dubhe		ARCTURUS		◆SPICA		ACRUX		◆SIRIUS		PROCYON		POLLUX	
150	31 01	009	25 59	071	36 50	107	17 30	164	37 34	245	54 51	273	49 54	310
151	31 10	008	26 55	070	37 47	107	17 46	164	36 40	246	53 52	273	49 08	310
152	31 18	008	27 51	070	38 44	108	18 03	164	35 45	246	52 52	273	48 22	309
153	31 25	007	28 48	070	39 41	108	18 19	165	34 50	246	51 52	273	47 35	308
154	31 33	007	29 44	070	40 38	108	18 34	165	33 55	247	50 52	273	46 48	308
155	31 39	006	30 40	070	41 34	109	18 50	165	33 00	247	49 53	273	46 01	307
156	31 45	005	31 37	070	42 31	109	19 04	166	32 05	247	48 53	273	45 13	307
157	31 51	005	32 33	070	43 28	109	19 19	166	31 10	248	47 53	273	44 25	306
158	31 55	004	33 29	070	44 24	110	19 33	167	30 14	248	46 53	273	43 36	306
159	32 00	004	34 25	069	45 20	110	19 46	167	29 19	248	45 54	273	42 48	305
160	32 03	003	35 21	069	46 16	111	19 59	168	28 23	249	44 54	273	41 59	305
161	32 06	003	36 17	069	47 12	111	20 12	168	27 27	249	43 54	273	41 09	304
162	32 09	002	37 13	069	48 08	112	20 24	168	26 31	249	42 54	273	40 20	304
163	32 11	002	38 08	069	49 03	112	20 36	169	25 35	249	41 55	273	39 30	304
164	32 12	001	39 04	068	49 59	113	20 47	169	24 39	250	40 55	273	38 40	303
	Dubhe		◆ARCTURUS		SPICA		◆ACRUX		Suhail		PROCYON		◆POLLUX	
165	32 13	000	40 00	068	50 54	113	20 58	170	36 16	205	39 55	273	37 50	303
166	32 13	000	40 55	068	51 49	114	21 09	170	35 51	206	38 55	274	37 00	302
167	32 13	359	41 51	068	52 44	114	21 19	171	35 24	207	37 56	274	36 09	302
168	32 12	359	42 46	067	53 38	115	21 28	171	34 57	207	36 56	274	35 18	302
169	32 10	358	43 41	067	54 32	116	21 37	172	34 30	208	35 56	274	34 27	302
170	32 08	358	44 36	067	55 26	116	21 46	172	34 01	209	34 57	274	33 36	301
171	32 06	357	45 31	066	56 19	117	21 54	172	33 32	209	33 57	274	32 45	301
172	32 02	357	46 26	066	57 13	118	22 02	173	33 03	210	32 57	274	31 54	301
173	31 58	356	47 21	066	58 05	119	22 09	173	32 33	210	31 57	274	31 02	301
174	31 54	355	48 15	065	58 58	119	22 15	174	32 02	211	30 58	274	30 11	300
175	31 49	355	49 09	065	59 50	120	22 21	174	31 31	212	29 58	274	29 19	300
176	31 43	354	50 04	065	60 41	121	22 27	175	30 59	212	28 58	274	28 27	300
177	31 37	354	50 57	064	61 32	122	22 32	175	30 27	213	27 58	274	27 35	300
178	31 31	353	51 51	064	62 22	123	22 37	176	29 54	213	26 59	274	26 43	299
179	31 23	353	52 45	063	63 12	125	22 41	176	29 21	214	25 59	274	25 51	299

 LAT 4°N

LHA ♈	Hc	Zn	Hc	Zn	Hc	Zn	Hc	Zn	Hc	Zn	Hc	Zn	Hc	Zn
	◆ARCTURUS		ANTARES		RIGIL KENT.		◆ACRUX		Suhail		◆REGULUS		Dubhe	
180	53 38	063	18 18	120	18 14	161	22 45	177	28 48	214	61 12	288	31 16	352
181	54 31	062	19 10	120	18 33	161	22 48	177	28 13	215	60 15	287	31 07	352
182	55 24	061	20 02	120	18 52	162	22 50	178	27 39	215	59 18	287	30 58	351
183	56 16	061	20 54	120	19 11	162	22 52	178	27 04	216	58 20	286	30 49	351
184	57 08	060	21 46	121	19 29	162	22 54	179	26 29	216	57 23	286	30 39	350
185	58 00	059	22 37	121	19 47	163	22 55	179	25 53	217	56 25	286	30 28	350
186	58 51	059	23 28	121	20 05	163	22 56	180	25 17	217	55 28	285	30 17	349
187	59 42	058	24 20	121	20 22	164	22 56	180	24 40	218	54 30	285	30 06	349
188	60 32	057	25 11	122	20 38	164	22 55	181	24 03	218	53 32	285	29 54	348
189	61 22	056	26 01	122	20 55	164	22 54	181	23 26	219	52 34	285	29 41	348
190	62 11	055	26 52	122	21 10	165	22 53	182	22 49	219	51 36	284	29 28	347
191	63 00	054	27 42	123	21 26	165	22 51	182	22 11	219	50 38	284	29 15	347
192	63 48	053	28 33	123	21 41	166	22 48	183	21 33	220	49 40	284	29 00	346
193	64 35	052	29 23	123	21 55	166	22 45	183	20 54	220	48 42	284	28 46	346
194	65 22	050	30 13	124	22 09	167	22 42	184	20 15	221	47 44	283	28 31	345
	◆Alphecca		Rasalhague		ANTARES		◆RIGIL KENT.		ACRUX		◆REGULUS		Dubhe	
195	46 41	054	21 40	078	31 02	124	22 22	167	22 38	184	46 45	283	28 16	345
196	47 29	054	22 39	078	31 52	125	22 35	168	22 33	185	45 47	283	28 00	344
197	48 17	053	23 37	078	32 41	125	22 48	168	22 28	185	44 49	283	27 43	344
198	49 05	053	24 36	078	33 30	126	23 00	169	22 22	186	43 50	283	27 27	344
199	49 52	052	25 34	078	34 18	126	23 11	169	22 16	186	42 52	283	27 09	343
200	50 39	051	26 33	078	35 06	126	23 22	170	22 10	187	41 54	283	26 52	343
201	51 26	051	27 31	078	35 54	127	23 33	170	22 03	187	40 55	282	26 34	342
202	52 12	050	28 30	078	36 42	127	23 43	171	21 55	187	39 57	282	26 15	342
203	52 57	049	29 28	078	37 29	128	23 53	171	21 47	188	38 58	282	25 57	341
204	53 42	048	30 27	078	38 16	129	24 01	172	21 39	188	38 00	282	25 37	341
205	54 26	047	31 25	078	39 03	129	24 10	172	21 30	189	37 01	282	25 18	341
206	55 10	046	32 24	078	39 49	130	24 18	173	21 20	189	36 03	282	24 58	340
207	55 53	046	33 22	078	40 35	130	24 25	173	21 10	190	35 04	282	24 38	340
208	56 35	045	34 21	078	41 20	131	24 32	174	21 00	190	34 06	282	24 17	340
209	57 17	043	35 19	077	42 05	132	24 38	174	20 49	191	33 07	282	23 56	339
	◆VEGA		Rasalhague		ANTARES		◆RIGIL KENT.		ACRUX		◆REGULUS		Dubhe	
210	18 40	050	36 18	077	42 50	132	24 44	175	20 38	191	32 08	282	23 35	339
211	19 26	050	37 16	077	43 34	133	24 49	175	20 26	192	31 10	282	23 13	339
212	20 11	050	38 14	077	44 17	134	24 54	176	20 14	192	30 11	282	22 51	338
213	20 57	050	39 13	077	45 00	134	24 58	176	20 01	192	29 12	282	22 29	338
214	21 43	050	40 11	077	45 43	135	25 02	177	19 48	193	28 14	281	22 06	338
215	22 28	049	41 09	077	46 25	136	25 05	177	19 35	193	27 15	281	21 43	337
216	23 14	049	42 08	077	47 06	137	25 07	178	19 21	194	26 16	281	21 20	337
217	23 59	049	43 06	077	47 46	138	25 09	178	19 07	194	25 18	281	20 56	337
218	24 44	049	44 04	076	48 26	139	25 10	179	18 52	194	24 19	281	20 33	337
219	25 29	049	45 02	076	49 05	140	25 11	180	18 37	195	23 20	281	20 09	336
220	26 14	048	46 00	076	49 44	141	25 11	180	18 21	195	22 22	281	19 45	336
221	26 58	048	46 58	076	50 21	142	25 11	181	18 05	196	21 23	281	19 20	336
222	27 43	048	47 56	076	50 58	143	25 10	181	17 49	196	20 24	281	18 55	335
223	28 27	047	48 54	075	51 34	144	25 08	182	17 32	196	19 26	281	18 30	335
224	29 11	047	49 52	075	52 09	145	25 06	182	17 15	197	18 27	281	18 05	335
	VEGA		ALTAIR		◆Nunki		ANTARES		RIGIL KENT.		◆SPICA		◆Alkaid	
225	29 55	047	17 45	082	25 40	122	52 43	146	25 04	183	61 55	237	42 06	344
226	30 38	047	18 44	082	26 31	122	53 16	147	25 00	183	61 05	238	41 50	343
227	31 22	046	19 44	082	27 22	122	53 48	148	24 57	184	60 14	239	41 32	343
228	32 05	046	20 43	082	28 12	123	54 19	150	24 52	184	59 22	240	41 14	342
229	32 47	045	21 42	082	29 02	123	54 49	151	24 47	185	58 30	241	40 55	341
230	33 30	045	22 42	082	29 52	124	55 17	152	24 42	185	57 37	242	40 35	340
231	34 12	045	23 41	082	30 42	124	55 45	154	24 36	186	56 44	243	40 14	340
232	34 54	044	24 40	082	31 31	124	56 11	155	24 30	186	55 51	243	39 53	339
233	35 36	044	25 39	082	32 21	125	56 35	156	24 23	187	54 57	244	39 31	338
234	36 17	043	26 39	082	33 10	125	56 59	158	24 15	188	54 03	245	39 08	337
235	36 58	043	27 38	082	33 59	126	57 20	159	24 07	188	53 09	245	38 45	337
236	37 38	042	28 37	082	34 47	126	57 41	161	23 58	189	52 14	246	38 21	336
237	38 18	042	29 37	082	35 35	127	58 00	162	23 49	189	51 20	247	37 57	335
238	38 58	041	30 36	082	36 23	127	58 17	164	23 39	190	50 24	247	37 32	335
239	39 37	041	31 35	082	37 11	128	58 33	166	23 29	190	49 29	248	37 06	334
	VEGA		◆ALTAIR		Nunki		◆ANTARES		SPICA		◆ARCTURUS		Alkaid	
240	40 16	040	32 34	082	37 58	128	58 46	167	48 34	248	60 18	303	36 39	334
241	40 55	040	33 34	082	38 45	129	58 59	169	47 38	249	59 27	302	36 13	333
242	41 33	039	34 33	082	39 31	129	59 09	171	46 42	249	58 36	301	35 45	332
243	42 10	038	35 32	082	40 17	130	59 18	172	45 46	250	57 45	300	35 17	332
244	42 47	038	36 31	082	41 03	131	59 25	174	44 50	250	56 53	300	34 49	331
245	43 23	037	37 31	082	41 48	131	59 30	176	43 54	250	56 01	299	34 20	331
246	43 59	036	38 30	082	42 33	132	59 33	178	42 57	251	55 09	298	33 50	330
247	44 34	036	39 29	082	43 18	133	59 35	180	42 01	251	54 16	298	33 21	330
248	45 09	035	40 29	082	44 01	133	59 34	181	41 04	251	53 23	297	32 50	329
249	45 43	034	41 28	082	44 45	134	59 32	183	40 07	252	52 30	297	32 19	329
250	46 16	033	42 27	082	45 28	135	59 28	185	39 10	252	51 36	296	31 48	328
251	46 49	033	43 26	082	46 10	136	59 22	187	38 13	252	50 42	296	31 17	328
252	47 21	032	44 25	082	46 51	136	59 14	188	37 16	253	49 48	295	30 45	327
253	47 52	031	45 25	081	47 32	137	59 05	190	36 19	253	48 54	295	30 12	327
254	48 22	030	46 24	081	48 13	138	58 53	192	35 22	253	48 00	295	29 40	327
	◆VEGA		ALTAIR		◆Nunki		ANTARES		◆SPICA		ARCTURUS		Alkaid	
255	48 51	029	47 23	081	48 52	139	58 40	193	34 24	254	47 05	294	29 07	326
256	49 20	028	48 22	081	49 31	140	58 26	195	33 27	254	46 10	294	28 33	326
257	49 48	027	49 21	081	50 09	141	58 09	197	32 29	254	45 16	293	27 59	325
258	50 15	026	50 20	081	50 46	142	57 51	198	31 32	254	44 21	293	27 25	325
259	50 41	025	51 19	081	51 23	143	57 32	200	30 34	255	43 26	293	26 51	325
260	51 05	024	52 18	081	51 58	144	57 11	201	29 36	255	42 30	293	26 16	324
261	51 29	023	53 18	081	52 33	145	56 48	203	28 39	255	41 35	292	25 41	324
262	51 52	022	54 17	080	53 06	146	56 24	204	27 41	255	40 40	292	25 06	324
263	52 14	021	55 16	080	53 39	148	55 59	206	26 43	255	39 44	292	24 31	323
264	52 35	020	56 15	080	54 11	149	55 32	207	25 45	256	38 48	292	23 55	323
265	52 54	018	57 13	080	54 41	150	55 05	208	24 47	256	37 53	291	23 19	323
266	53 13	017	58 12	080	55 10	151	54 35	210	23 49	256	36 57	291	22 43	323
267	53 30	016	59 11	079	55 38	153	54 05	211	22 51	256	36 01	291	22 06	322
268	53 46	015	60 10	079	56 05	154	53 34	212	21 53	256	35 05	291	21 30	322
269	54 00	014	61 09	079	56 30	156	53 01	213	20 54	256	34 09	291	20 53	322

LHA ♈	Hc	Zn	Hc	Zn	Hc	Zn	Hc	Zn	Hc	Zn	Hc	Zn	Hc	Zn
	VEGA		◆DENEB		FOMALHAUT		◆Peacock		Shaula		ANTARES		◆ARCTURUS	
270	54 14	012	35 48	034	11 30	121	22 29	159	48 26	188	52 28	215	33 13	290
271	54 26	011	36 21	034	12 21	122	22 50	160	48 17	189	51 53	216	32 17	290
272	54 37	010	36 54	033	13 12	122	23 10	160	48 07	190	51 18	217	31 21	290
273	54 46	008	37 27	033	14 03	122	23 30	161	47 55	192	50 42	218	30 24	290
274	54 54	007	37 59	032	14 54	122	23 49	161	47 43	193	50 05	219	29 28	290
275	55 01	006	38 30	031	15 45	122	24 08	162	47 29	194	49 27	220	28 32	290
276	55 06	004	39 01	031	16 35	123	24 27	162	47 14	195	48 48	221	27 35	290
277	55 10	003	39 31	030	17 25	123	24 45	163	46 58	196	48 08	222	26 39	290
278	55 12	002	40 01	029	18 16	123	25 02	163	46 41	197	47 28	223	25 43	289
279	55 13	000	40 30	029	19 06	123	25 19	164	46 23	198	46 47	223	24 46	289
280	55 13	359	40 58	028	19 56	124	25 35	164	46 04	199	46 06	224	23 50	289
281	55 11	358	41 26	027	20 46	124	25 51	165	45 43	200	45 24	225	22 53	289
282	55 08	356	41 53	027	21 35	124	26 07	165	45 22	201	44 41	226	21 57	289
283	55 03	355	42 20	026	22 25	124	26 21	166	45 00	202	43 58	227	21 00	289
284	54 57	353	42 46	025	23 14	125	26 36	167	44 37	203	43 14	227	20 03	289
	DENEB		Alpheratz		◆FOMALHAUT		Peacock		◆ANTARES		Rasalhague		◆VEGA	
285	43 11	024	13 16	061	24 03	125	26 49	167	42 30	228	67 15	293	54 49	352
286	43 35	024	14 09	061	24 52	125	27 02	168	41 45	229	66 20	293	54 40	351
287	43 59	023	15 01	061	25 41	126	27 15	168	41 00	229	65 24	292	54 30	349
288	44 21	022	15 53	061	26 29	126	27 27	169	40 14	230	64 29	291	54 19	348
289	44 43	021	16 46	061	27 17	126	27 38	169	39 28	231	63 33	290	54 06	347
290	45 04	020	17 38	061	28 05	127	27 49	170	38 42	231	62 36	290	53 52	346
291	45 25	019	18 30	061	28 53	127	27 59	171	37 55	232	61 40	289	53 36	344
292	45 44	018	19 22	061	29 41	128	28 08	171	37 08	232	60 43	289	53 19	343
293	46 03	018	20 14	060	30 28	128	28 17	172	36 21	233	59 47	288	53 01	342
294	46 20	017	21 06	060	31 15	128	28 26	172	35 33	233	58 50	288	52 42	341
295	46 37	016	21 58	060	32 02	129	28 33	173	34 45	234	57 53	287	52 22	340
296	46 53	015	22 50	060	32 48	129	28 40	174	33 56	234	56 56	287	52 01	339
297	47 07	014	23 42	060	33 34	130	28 47	174	33 08	235	55 58	287	51 38	337
298	47 21	013	24 34	060	34 20	130	28 52	175	32 19	235	55 01	286	51 15	336
299	47 34	012	25 25	060	35 05	131	28 57	175	31 29	236	54 03	286	50 50	335
	◆DENEB		Alpheratz		◆FOMALHAUT		Peacock		◆ANTARES		Rasalhague		VEGA	
300	47 46	011	26 17	059	35 50	131	29 02	176	30 40	236	53 06	286	50 25	334
301	47 56	010	27 08	059	36 35	132	29 06	177	29 50	236	52 08	285	49 58	333
302	48 06	009	28 00	059	37 19	133	29 09	177	29 00	237	51 10	285	49 31	332
303	48 15	008	28 51	059	38 03	133	29 11	178	28 10	237	50 13	285	49 02	331
304	48 22	007	29 42	059	38 46	134	29 13	179	27 20	237	49 15	285	48 33	330
305	48 29	006	30 33	058	39 29	134	29 14	179	26 29	238	48 17	284	48 03	329
306	48 34	005	31 24	058	40 12	135	29 15	180	25 39	238	47 19	284	47 33	329
307	48 38	004	32 15	058	40 54	136	29 15	180	24 48	238	46 21	284	47 01	328
308	48 41	002	33 06	058	41 35	136	29 14	181	23 57	239	45 23	284	46 29	327
309	48 43	001	33 56	057	42 16	137	29 13	182	23 05	239	44 24	284	45 56	326
310	48 44	000	34 46	057	42 57	138	29 10	182	22 14	239	43 26	284	45 22	325
311	48 44	359	35 37	057	43 37	139	29 08	183	21 22	240	42 28	283	44 48	325
312	48 43	358	36 26	056	44 16	140	29 04	184	20 31	240	41 30	283	44 13	324
313	48 40	357	37 16	056	44 54	140	29 00	184	19 39	240	40 32	283	43 37	323
314	48 37	356	38 06	056	45 32	141	28 56	185	18 47	240	39 33	283	43 01	323
	◆Alpheratz		◆FOMALHAUT		Peacock		Nunki		◆Rasalhague		VEGA		DENEB	
315	38 55	055	46 09	142	28 50	185	47 11	223	38 35	283	42 24	322	48 32	355
316	39 44	055	46 46	143	28 44	186	46 30	224	37 37	283	41 47	321	48 26	354
317	40 33	054	47 21	144	28 38	187	45 48	225	36 38	283	41 09	321	48 19	353
318	41 22	054	47 56	145	28 30	187	45 05	226	35 40	283	40 31	320	48 12	352
319	42 10	054	48 30	146	28 23	188	44 22	226	34 41	283	39 52	319	48 03	351
320	42 58	053	49 03	147	28 14	188	43 39	227	33 43	282	39 13	319	47 53	350
321	43 46	053	49 36	148	28 05	189	42 55	228	32 44	282	38 34	318	47 41	349
322	44 33	052	50 07	149	27 55	190	42 10	228	31 46	282	37 54	318	47 29	348
323	45 20	052	50 37	150	27 45	190	41 25	229	30 48	282	37 13	317	47 16	347
324	46 07	051	51 07	151	27 34	191	40 39	230	29 49	282	36 33	317	47 02	346
325	46 53	050	51 35	152	27 22	191	39 53	230	28 51	282	35 51	316	46 47	345
326	47 39	050	52 02	154	27 10	192	39 07	231	27 52	282	35 10	316	46 31	344
327	48 24	049	52 28	155	26 57	193	38 20	232	26 53	282	34 28	316	46 14	343
328	49 09	048	52 53	156	26 44	193	37 33	232	25 55	282	33 46	315	45 56	342
329	49 54	048	53 17	157	26 30	194	36 46	233	24 56	282	33 04	315	45 37	341
	◆Alpheratz		Diphda		◆FOMALHAUT		Peacock		◆ALTAIR		VEGA		DENEB	
330	50 38	047	44 07	120	53 39	159	26 16	194	57 29	280	32 21	314	45 17	340
331	51 21	046	44 59	121	54 00	160	26 01	195	56 30	280	31 38	314	44 57	339
332	52 04	045	45 50	121	54 20	161	25 45	195	55 31	280	30 55	314	44 35	339
333	52 47	045	46 41	122	54 38	163	25 29	196	54 32	280	30 12	313	44 13	338
334	53 28	044	47 32	122	54 55	164	25 13	196	53 33	280	29 28	313	43 50	337
335	54 09	043	48 23	123	55 11	166	24 56	197	52 34	279	28 44	313	43 26	336
336	54 49	042	49 13	124	55 25	167	24 38	197	51 35	279	28 00	312	43 02	335
337	55 29	041	50 02	124	55 37	169	24 20	198	50 36	279	27 16	312	42 36	335
338	56 08	040	50 51	125	55 48	170	24 01	198	49 37	279	26 31	312	42 10	334
339	56 45	039	51 40	126	55 58	172	23 42	199	48 38	279	25 46	312	41 43	333
340	57 22	037	52 28	127	56 06	173	23 23	199	47 39	279	25 01	311	41 16	332
341	57 58	036	53 16	128	56 12	175	23 02	200	46 40	279	24 16	311	40 48	332
342	58 33	035	54 03	129	56 17	176	22 42	200	45 41	279	23 31	311	40 19	331
343	59 07	034	54 49	130	56 20	178	22 21	201	44 41	279	22 46	311	39 50	330
344	59 39	032	55 35	131	56 21	179	22 00	201	43 42	278	22 00	310	39 20	330
	Alpheratz		◆Hamal		Acamar		◆ACHERNAR		FOMALHAUT		◆ALTAIR		DENEB	
345	60 11	031	40 56	062	19 55	136	21 00	158	56 21	181	42 43	278	38 49	329
346	60 41	030	41 49	062	20 37	136	21 22	159	56 19	183	41 44	278	38 18	328
347	61 10	028	42 41	061	21 18	136	21 43	159	56 15	184	40 44	278	37 47	328
348	61 37	026	43 34	061	21 59	137	22 04	160	56 10	186	39 45	278	37 15	327
349	62 03	025	44 26	061	22 40	137	22 24	160	56 03	187	38 46	278	36 42	327
350	62 28	023	45 18	060	23 21	137	22 44	161	55 55	189	37 47	278	36 09	326
351	62 50	021	46 10	060	24 01	138	23 04	161	55 45	190	36 47	278	35 35	326
352	63 11	020	47 02	059	24 41	138	23 23	162	55 34	192	35 48	278	35 01	325
353	63 31	018	47 53	059	25 21	139	23 42	162	55 21	193	34 49	278	34 27	325
354	63 48	016	48 44	058	26 00	139	24 00	163	55 06	195	33 50	278	33 52	324
355	64 04	014	49 35	058	26 39	140	24 18	163	54 50	196	32 50	278	33 17	324
356	64 18	012	50 26	057	27 18	140	24 35	164	54 33	198	31 51	278	32 41	323
357	64 29	010	51 16	057	27 56	140	24 51	164	54 14	199	30 52	278	32 05	323
358	64 39	008	52 05	056	28 34	141	25 08	165	53 54	200	29 53	278	31 29	322
359	64 46	006	52 55	055	29 12	141	25 23	165	53 32	202	28 53	278	30 52	322

LHA ♈	Hc	Zn	Hc	Zn	Hc	Zn	Hc	Zn	Hc	Zn	Hc	Zn	Hc	Zn
	Alpheratz		**◆Mirfak**		**ALDEBARAN**		**◆ACHERNAR**		**FOMALHAUT**		**◆ALTAIR**		**DENEB**	
0	63 52	004	26 27	034	21 04	073	26 36	166	54 05	204	27 45	278	29 28	322
1	63 55	002	27 00	034	22 01	073	26 51	166	53 40	205	26 46	278	28 51	322
2	63 56	000	27 33	033	22 59	073	27 05	167	53 14	206	25 47	278	28 14	321
3	63 55	358	28 06	033	23 56	073	27 19	167	52 47	207	24 48	278	27 36	321
4	63 52	356	28 38	032	24 54	073	27 32	168	52 19	209	23 48	278	26 58	320
5	63 47	354	29 10	032	25 51	073	27 44	168	51 50	210	22 49	278	26 20	320
6	63 40	352	29 42	032	26 48	073	27 56	169	51 20	211	21 50	278	25 41	320
7	63 31	350	30 13	031	27 45	073	28 07	169	50 48	212	20 51	278	25 02	320
8	63 20	348	30 44	031	28 43	073	28 18	170	50 16	213	19 51	278	24 23	319
9	63 07	346	31 14	030	29 40	073	28 28	171	49 43	214	18 52	278	23 44	319
10	62 52	345	31 45	030	30 37	073	28 38	171	49 09	215	17 53	278	23 05	319
11	62 35	343	32 14	029	31 34	072	28 46	172	48 34	216	16 53	278	22 25	318
12	62 16	341	32 43	029	32 31	072	28 55	172	47 58	217	15 54	278	21 45	318
13	61 56	339	33 12	028	33 28	072	29 02	173	47 22	218	14 55	278	21 05	318
14	61 34	338	33 40	028	34 25	072	29 09	174	46 45	219	13 56	278	20 25	318
	CAPELLA		**ALDEBARAN**		**◆RIGEL**		**ACHERNAR**		**◆FOMALHAUT**		**Enif**		**◆Alpheratz**	
15	19 56	042	35 22	072	25 37	101	29 16	174	46 07	220	40 52	280	61 10	336
16	20 36	041	36 19	072	26 36	101	29 21	175	45 28	221	39 53	280	60 45	334
17	21 15	041	37 16	071	27 35	101	29 27	175	44 49	221	38 54	280	60 18	333
18	21 55	041	38 13	071	28 34	101	29 31	176	44 09	222	37 55	280	59 50	331
19	22 34	041	39 10	071	29 33	101	29 35	177	43 28	223	36 56	280	59 21	330
20	23 13	040	40 06	071	30 31	101	29 38	177	42 47	224	35 57	280	58 51	329
21	23 51	040	41 03	071	31 30	102	29 41	178	42 06	224	34 58	280	58 19	327
22	24 30	040	41 59	070	32 29	102	29 42	179	41 24	225	33 59	280	57 46	326
23	25 08	040	42 56	070	33 27	102	29 44	179	40 41	226	33 00	280	57 12	325
24	25 46	039	43 52	070	34 26	102	29 44	180	39 58	226	32 01	280	56 37	324
25	26 24	039	44 48	070	35 25	102	29 44	180	39 14	227	31 02	280	56 00	322
26	27 02	039	45 44	069	36 23	102	29 43	181	38 30	228	30 03	280	55 23	321
27	27 39	038	46 40	069	37 22	103	29 42	182	37 46	228	29 04	280	54 46	320
28	28 16	038	47 36	069	38 20	103	29 40	182	37 01	229	28 05	280	54 07	319
29	28 52	037	48 32	068	39 18	103	29 37	183	36 16	229	27 05	280	53 27	318
	Mirfak		**CAPELLA**		**◆ALDEBARAN**		**RIGEL**		**◆ACHERNAR**		**FOMALHAUT**		**◆Alpheratz**	
30	39 53	018	29 29	037	49 28	068	40 17	103	29 34	183	35 30	230	52 47	317
31	40 11	017	30 05	037	50 23	067	41 15	104	29 30	184	34 44	230	52 06	316
32	40 28	016	30 40	036	51 18	067	42 13	104	29 25	185	33 58	231	51 25	316
33	40 44	015	31 15	036	52 13	067	43 11	104	29 20	185	33 11	231	50 42	315
34	40 59	014	31 50	035	53 08	066	44 09	104	29 14	186	32 25	232	49 59	314
35	41 14	014	32 25	035	54 03	066	45 07	105	29 08	187	31 37	232	49 16	313
36	41 27	013	32 59	034	54 57	065	46 05	105	29 01	187	30 50	233	48 32	312
37	41 40	012	33 33	034	55 52	065	47 03	105	28 53	188	30 02	233	47 48	312
38	41 53	011	34 06	033	56 46	064	48 01	106	28 44	188	29 14	233	47 03	311
39	42 04	010	34 39	033	57 39	063	48 58	106	28 35	189	28 26	234	46 17	310
40	42 14	010	35 11	032	58 33	063	49 56	107	28 26	190	27 38	234	45 32	310
41	42 24	009	35 43	032	59 26	062	50 53	107	28 16	190	26 49	234	44 45	309
42	42 32	008	36 14	031	60 19	061	51 50	107	28 05	191	26 00	235	43 59	309
43	42 40	007	36 45	031	61 11	060	52 48	108	27 53	191	25 11	235	43 12	308
44	42 47	006	37 15	030	62 03	059	53 45	108	27 41	192	24 22	235	42 25	308
	CAPELLA		**◆PROCYON**		**SIRIUS**		**CANOPUS**		**◆ACHERNAR**		**Diphda**		**◆Alpheratz**	
45	37 45	029	20 25	086	31 06	112	19 51	150	27 29	192	50 19	237	41 37	307
46	38 14	029	21 25	086	32 02	112	20 21	150	27 16	193	49 29	237	40 49	307
47	38 43	028	22 24	086	32 57	112	20 50	151	27 02	194	48 38	238	40 01	306
48	39 11	028	23 24	086	33 53	112	21 20	151	26 48	194	47 47	239	39 12	306
49	39 39	027	24 24	086	34 48	113	21 48	152	26 33	195	46 56	239	38 24	305
50	40 05	026	25 24	086	35 43	113	22 17	152	26 18	195	46 04	240	37 35	305
51	40 31	025	26 23	086	36 38	113	22 45	152	26 02	196	45 12	241	36 46	305
52	40 57	025	27 23	086	37 33	114	23 12	153	25 45	196	44 20	241	35 56	304
53	41 22	024	28 23	086	38 28	114	23 40	153	25 28	197	43 27	242	35 07	304
54	41 46	023	29 23	086	39 23	115	24 06	154	25 11	197	42 34	242	34 17	304
55	42 09	022	30 22	086	40 17	115	24 33	154	24 53	198	41 41	243	33 27	303
56	42 31	022	31 22	086	41 11	115	24 59	155	24 34	198	40 48	243	32 36	303
57	42 53	021	32 22	086	42 05	116	25 24	155	24 16	199	39 54	243	31 46	303
58	43 14	020	33 22	086	42 59	116	25 49	156	23 56	199	39 01	244	30 56	302
59	43 34	019	34 21	086	43 53	117	26 14	156	23 36	200	38 07	244	30 05	302
	◆CAPELLA		**PROCYON**		**◆SIRIUS**		**CANOPUS**		**ACHERNAR**		**◆Diphda**		**Alpheratz**	
60	43 54	018	35 21	086	44 46	117	26 38	157	23 16	200	37 13	245	29 14	302
61	44 12	018	36 21	086	45 39	118	27 02	157	22 55	201	36 19	245	28 23	302
62	44 30	017	37 21	086	46 32	118	27 25	158	22 34	201	35 24	245	27 32	301
63	44 46	016	38 20	086	47 25	119	27 47	158	22 12	201	34 30	246	26 41	301
64	45 02	015	39 20	086	48 17	119	28 09	159	21 50	202	33 35	246	25 50	301
65	45 17	014	40 20	086	49 09	120	28 31	159	21 28	202	32 40	246	24 58	301
66	45 31	013	41 20	086	50 01	121	28 52	160	21 05	203	31 45	247	24 07	301
67	45 44	012	42 19	086	50 52	121	29 12	160	20 42	203	30 50	247	23 15	300
68	45 56	011	43 19	086	51 43	122	29 32	161	20 18	203	29 55	247	22 23	300
69	46 07	010	44 19	086	52 34	123	29 51	162	19 54	204	29 00	247	21 32	300
70	46 17	009	45 19	086	53 24	124	30 10	162	19 30	204	28 04	248	20 40	300
71	46 26	008	46 18	086	54 14	124	30 28	163	19 05	205	27 09	248	19 48	300
72	46 34	007	47 18	086	55 03	125	30 46	163	18 40	205	26 13	248	18 56	300
73	46 41	006	48 18	085	55 51	126	31 03	164	18 14	205	25 18	248	18 04	300
74	46 47	005	49 17	085	56 39	127	31 19	165	17 49	206	24 22	249	17 12	300
	CAPELLA		**◆POLLUX**		**PROCYON**		**SIRIUS**		**◆CANOPUS**		**Acamar**		**◆Hamal**	
75	46 52	004	43 25	053	50 17	085	57 27	128	31 34	165	38 29	210	43 28	300
76	46 55	003	44 13	053	51 17	085	58 14	129	31 49	166	37 59	210	42 36	300
77	46 58	002	45 00	052	52 17	085	59 00	130	32 04	167	37 28	211	41 43	299
78	47 00	001	45 47	052	53 16	085	59 45	131	32 17	167	36 57	212	40 51	299
79	47 00	000	46 34	051	54 16	085	60 30	133	32 30	168	36 25	212	39 58	298
80	47 00	359	47 20	050	55 16	085	61 13	134	32 42	169	35 53	213	39 06	298
81	46 58	358	48 06	050	56 15	085	61 56	135	32 54	169	35 20	214	38 13	298
82	46 56	357	48 52	049	57 15	085	62 38	137	33 05	170	34 46	214	37 20	297
83	46 52	356	49 37	048	58 15	085	63 18	138	33 15	171	34 12	215	36 26	297
84	46 48	355	50 21	048	59 15	085	63 58	140	33 24	171	33 38	216	35 33	297
85	46 42	354	51 05	047	60 14	085	64 36	141	33 33	172	33 02	216	34 39	297
86	46 35	353	51 49	046	61 14	085	65 12	143	33 41	173	32 27	217	33 46	296
87	46 27	352	52 32	045	62 13	084	65 48	145	33 48	173	31 51	217	32 52	296
88	46 18	351	53 14	044	63 13	084	66 21	147	33 54	174	31 14	218	31 58	296
89	46 09	350	53 55	043	64 13	084	66 53	149	34 00	175	30 37	218	31 04	296

LAT 3°N

LHA ♈	Hc	Zn	Hc	Zn	Hc	Zn	Hc	Zn	Hc	Zn	Hc	Zn	Hc	Zn
	POLLUX		**◆REGULUS**		**Suhail**		**◆CANOPUS**		**Acamar**		**◆ALDEBARAN**		**CAPELLA**	
90	54 36	042	27 58	078	27 20	143	34 05	176	30 00	219	65 16	304	45 58	349
91	55 16	041	28 57	078	27 56	144	34 09	176	29 22	219	64 26	303	45 46	348
92	55 55	040	29 56	078	28 31	144	34 12	177	28 44	220	63 36	302	45 33	347
93	56 33	039	30 54	078	29 06	145	34 15	178	28 05	220	62 45	301	45 19	346
94	57 11	038	31 53	078	29 40	145	34 17	179	27 26	221	61 53	300	45 05	345
95	57 47	037	32 51	078	30 14	146	34 18	179	26 47	221	61 01	299	44 49	344
96	58 23	036	33 50	078	30 47	146	34 18	180	26 07	222	60 09	299	44 32	344
97	58 57	034	34 48	077	31 20	147	34 18	181	25 27	222	59 16	298	44 15	343
98	59 30	033	35 47	077	31 53	147	34 17	181	24 47	222	58 23	297	43 57	342
99	60 02	032	36 45	077	32 25	148	34 15	182	24 06	223	57 30	297	43 37	341
100	60 33	030	37 44	077	32 56	149	34 12	183	23 26	223	56 36	296	43 17	340
101	61 03	029	38 42	077	33 27	149	34 09	184	22 44	224	55 42	295	42 57	339
102	61 31	027	39 40	077	33 58	150	34 04	184	22 03	224	54 48	295	42 35	338
103	61 57	025	40 39	077	34 27	151	33 59	185	21 21	224	53 53	294	42 13	338
104	62 22	024	41 37	077	34 57	151	33 54	186	20 39	225	52 58	294	41 49	337
	POLLUX		**◆REGULUS**		**Suhail**		**◆CANOPUS**		**RIGEL**		**◆ALDEBARAN**		**CAPELLA**	
105	62 45	022	42 35	076	35 25	152	33 47	187	61 22	247	52 03	293	41 26	336
106	63 07	020	43 34	076	35 53	153	33 40	187	60 26	248	51 08	293	41 01	335
107	63 27	019	44 32	076	36 20	153	33 32	188	59 31	248	50 13	292	40 36	335
108	63 45	017	45 30	076	36 47	154	33 23	189	58 35	249	49 17	292	40 10	334
109	64 01	015	46 28	076	37 13	155	33 14	189	57 39	250	48 22	292	39 43	333
110	64 16	013	47 26	075	37 38	155	33 04	190	56 43	250	47 26	291	39 16	333
111	64 28	011	48 24	075	38 03	156	32 53	191	55 46	251	46 30	291	38 48	332
112	64 38	009	49 22	075	38 27	157	32 41	191	54 50	251	45 34	291	38 19	331
113	64 46	007	50 20	075	38 50	158	32 29	192	53 53	252	44 38	290	37 50	331
114	64 52	005	51 18	074	39 12	159	32 16	193	52 56	252	43 42	290	37 20	330
115	64 56	003	52 15	074	39 34	159	32 02	193	51 59	253	42 45	290	36 50	329
116	64 58	001	53 13	074	39 54	160	31 48	194	51 02	253	41 49	290	36 19	329
117	64 57	358	54 10	074	40 14	161	31 33	195	50 04	253	40 53	289	35 48	328
118	64 54	356	55 08	073	40 33	162	31 18	195	49 07	254	39 56	289	35 16	328
119	64 49	354	56 05	073	40 51	163	31 01	196	48 09	254	38 59	289	34 44	327
	Dubhe		**◆Denebola**		**Suhail**		**◆CANOPUS**		**SIRIUS**		**BETELGEUSE**		**◆CAPELLA**	
120	22 02	021	32 27	075	41 09	164	30 44	197	62 57	223	58 33	279	34 11	327
121	22 23	021	33 25	074	41 25	165	30 27	197	62 15	224	57 34	279	33 38	326
122	22 45	021	34 23	074	41 40	165	30 09	198	61 33	225	56 35	279	33 04	326
123	23 06	020	35 21	074	41 55	166	29 50	199	60 50	227	55 36	279	32 30	325
124	23 27	020	36 18	074	42 09	167	29 31	199	60 06	228	54 37	279	31 56	325
125	23 47	020	37 16	074	42 21	168	29 11	200	59 21	229	53 37	278	31 21	324
126	24 07	019	38 13	074	42 33	169	28 50	200	58 35	230	52 38	278	30 46	324
127	24 27	019	39 11	074	42 44	170	28 29	201	57 49	231	51 39	278	30 10	323
128	24 46	019	40 08	073	42 54	171	28 08	201	57 02	232	50 39	278	29 34	323
129	25 05	018	41 06	073	43 02	172	27 46	202	56 14	233	49 40	278	28 58	323
130	25 24	018	42 03	073	43 10	173	27 23	202	55 25	234	48 41	278	28 22	322
131	25 42	017	43 00	073	43 17	174	27 00	203	54 36	235	47 41	278	27 45	322
132	26 00	017	43 57	073	43 22	175	26 36	204	53 47	236	46 42	278	27 08	321
133	26 17	017	44 55	072	43 27	176	26 12	204	52 57	237	45 43	278	26 30	321
134	26 34	016	45 52	072	43 31	177	25 47	205	52 07	238	44 43	277	25 52	321
	Dubhe		**◆SPICA**		**ACRUX**		**◆CANOPUS**		**SIRIUS**		**BETELGEUSE**		**◆CAPELLA**	
135	26 50	016	22 38	103	13 33	159	25 22	205	51 16	238	43 44	277	25 14	320
136	27 07	015	23 36	104	13 55	159	24 57	205	50 25	239	42 44	277	24 36	320
137	27 22	015	24 34	104	14 17	159	24 31	206	49 33	240	41 45	277	23 58	320
138	27 37	014	25 33	104	14 38	159	24 04	206	48 41	240	40 45	277	23 19	320
139	27 52	014	26 31	104	14 59	160	23 37	207	47 49	241	39 46	277	22 40	319
140	28 06	014	27 29	104	15 19	160	23 10	207	46 57	241	38 46	277	22 01	319
141	28 20	013	28 27	104	15 39	160	22 43	208	46 04	242	37 47	277	21 22	319
142	28 33	013	29 25	105	15 59	161	22 15	208	45 11	243	36 48	277	20 42	319
143	28 46	012	30 23	105	16 19	161	21 46	209	44 18	243	35 48	277	20 02	318
144	28 58	012	31 21	105	16 38	161	21 17	209	43 24	244	34 49	277	19 22	318
145	29 10	011	32 19	105	16 57	162	20 48	209	42 30	244	33 49	277	18 42	318
146	29 22	011	33 16	105	17 16	162	20 19	210	41 36	244	32 50	277	18 02	318
147	29 32	010	34 14	106	17 34	162	19 49	210	40 42	245	31 50	277	17 22	318
148	29 43	010	35 12	106	17 52	163	19 19	210	39 48	245	30 51	277	16 41	317
149	29 52	009	36 09	106	18 10	163	18 48	211	38 53	246	29 51	277	16 01	317
	◆Dubhe		**ARCTURUS**		**◆SPICA**		**ACRUX**		**◆SIRIUS**		**PROCYON**		**POLLUX**	
150	30 02	009	25 38	070	37 07	106	18 27	163	37 59	246	54 47	275	49 15	311
151	30 10	008	26 35	070	38 04	107	18 44	164	37 04	246	53 47	275	48 29	311
152	30 18	008	27 31	070	39 02	107	19 00	164	36 09	247	52 48	275	47 44	310
153	30 26	007	28 27	070	39 59	107	19 17	165	35 14	247	51 48	275	46 58	309
154	30 33	006	29 23	070	40 56	108	19 32	165	34 19	247	50 48	275	46 11	309
155	30 39	006	30 20	069	41 53	108	19 48	165	33 23	248	49 49	275	45 24	308
156	30 45	005	31 16	069	42 50	108	20 02	166	32 28	248	48 49	275	44 37	308
157	30 51	005	32 12	069	43 47	109	20 17	166	31 32	248	47 49	274	43 49	307
158	30 56	004	33 08	069	44 44	109	20 31	167	30 36	249	46 49	274	43 01	307
159	31 00	004	34 04	069	45 40	109	20 45	167	29 41	249	45 50	274	42 13	306
160	31 03	003	34 59	069	46 37	110	20 58	167	28 45	249	44 50	274	41 24	306
161	31 07	003	35 55	068	47 33	110	21 11	168	27 49	249	43 50	274	40 35	305
162	31 09	002	36 51	068	48 29	111	21 23	168	26 52	250	42 50	274	39 46	305
163	31 11	002	37 46	068	49 26	111	21 35	169	25 56	250	41 51	274	38 57	304
164	31 12	001	38 42	068	50 21	111	21 46	169	25 00	250	40 51	274	38 07	304
	Dubhe		**◆ARCTURUS**		**SPICA**		**◆ACRUX**		**Suhail**		**PROCYON**		**◆POLLUX**	
165	31 13	000	39 37	067	51 17	112	21 57	170	37 11	205	39 51	274	37 17	303
166	31 13	000	40 32	067	52 12	113	22 08	170	36 45	206	38 51	274	36 27	303
167	31 13	359	41 27	067	53 08	113	22 18	171	36 18	207	37 52	274	35 37	303
168	31 12	359	42 23	067	54 03	114	22 28	171	35 50	208	36 52	274	34 46	302
169	31 10	358	43 17	066	54 57	114	22 37	171	35 22	208	35 52	274	33 56	302
170	31 08	358	44 12	066	55 52	115	22 45	172	34 54	209	34 52	274	33 05	302
171	31 06	357	45 07	066	56 46	116	22 53	172	34 25	210	33 53	274	32 14	302
172	31 02	357	46 01	065	57 40	116	23 01	173	33 55	210	32 53	274	31 23	301
173	30 59	356	46 56	065	58 33	117	23 08	173	33 24	211	31 53	274	30 31	301
174	30 54	356	47 50	064	59 27	118	23 15	174	32 53	211	30 53	274	29 40	301
175	30 49	355	48 44	064	60 19	119	23 21	174	32 22	212	29 54	274	28 49	301
176	30 44	354	49 37	063	61 11	120	23 27	175	31 50	213	28 54	274	27 57	300
177	30 38	354	50 31	063	62 03	121	23 32	175	31 17	213	27 54	274	27 05	300
178	30 31	353	51 24	062	62 54	122	23 37	176	30 44	214	26 54	274	26 13	300
179	30 24	353	52 17	062	63 45	123	23 41	176	30 11	214	25 55	274	25 21	300

LAT 3°N

LHA	Hc	Zn	Hc	Zn	Hc	Zn	Hc	Zn	Hc	Zn	Hc	Zn	Hc	Zn
♈	◆ARCTURUS		ANTARES		RIGIL KENT.		◆ACRUX		Suhail		◆REGULUS		Dubhe	
180	53 10	061	18 48	119	19 10	161	23 45	177	29 37	215	60 53	289	30 16	352
181	54 02	061	19 40	119	19 30	161	23 48	177	29 03	215	59 56	289	30 08	352
182	54 54	060	20 32	120	19 49	161	23 50	178	28 28	216	58 59	288	29 59	351
183	55 46	059	21 24	120	20 08	162	23 52	178	27 52	216	58 02	288	29 50	351
184	56 37	059	22 16	120	20 26	162	23 54	179	27 17	217	57 05	288	29 40	350
185	57 28	058	23 08	120	20 44	163	23 55	179	26 41	217	56 08	287	29 29	350
186	58 19	057	23 59	121	21 02	163	23 56	180	26 04	218	55 11	287	29 18	349
187	59 09	056	24 51	121	21 19	164	23 56	180	25 28	218	54 13	286	29 07	349
188	59 59	055	25 42	121	21 36	164	23 55	181	24 50	218	53 16	286	28 55	348
189	60 48	054	26 33	122	21 52	164	23 54	181	24 13	219	52 18	286	28 42	348
190	61 36	053	27 24	122	22 08	165	23 53	182	23 35	219	51 21	286	28 30	347
191	62 24	052	28 15	122	22 24	165	23 51	182	22 57	220	50 23	285	28 16	347
192	63 11	051	29 05	123	22 39	166	23 48	183	22 19	220	49 25	285	28 02	346
193	63 57	050	29 56	123	22 53	166	23 45	183	21 40	220	48 27	285	27 48	346
194	64 43	049	30 46	123	23 07	167	23 41	184	21 01	221	47 29	285	27 33	345
	◆Alphecca		Rasalhague		ANTARES		◆RIGIL KENT.		ACRUX		◆REGULUS		Dubhe	
195	46 05	053	21 28	078	31 36	124	23 21	167	23 37	184	46 31	284	27 18	345
196	46 53	053	22 26	078	32 25	124	23 34	168	23 33	185	45 33	284	27 02	345
197	47 41	052	23 25	078	33 15	125	23 47	168	23 28	185	44 35	284	26 46	344
198	48 28	052	24 23	078	34 04	125	23 59	169	23 22	186	43 37	284	26 29	344
199	49 15	051	25 22	078	34 53	125	24 10	169	23 16	186	42 38	284	26 12	343
200	50 01	050	26 20	077	35 42	126	24 22	170	23 09	187	41 40	283	25 55	343
201	50 47	050	27 19	077	36 30	126	24 32	170	23 02	187	40 42	283	25 37	342
202	51 32	049	28 17	077	37 18	127	24 42	171	22 55	188	39 44	283	25 18	342
203	52 17	048	29 16	077	38 06	127	24 52	171	22 47	188	38 45	283	25 00	342
204	53 02	047	30 14	077	38 53	128	25 01	172	22 38	188	37 47	283	24 41	341
205	53 45	046	31 12	077	39 41	128	25 09	172	22 29	189	36 48	283	24 21	341
206	54 28	045	32 11	077	40 27	129	25 17	173	22 20	189	35 50	283	24 01	341
207	55 11	044	33 09	077	41 14	130	25 25	173	22 10	190	34 51	283	23 41	340
208	55 52	043	34 08	077	41 59	130	25 32	174	21 59	190	33 53	282	23 21	340
209	56 33	042	35 06	077	42 45	131	25 38	174	21 48	191	32 54	282	23 00	339
	◆VEGA		Rasalhague		ANTARES		◆RIGIL KENT.		ACRUX		◆REGULUS		Dubhe	
210	18 01	050	36 04	077	43 30	132	25 44	175	21 37	191	31 56	282	22 39	339
211	18 47	050	37 02	077	44 15	132	25 49	175	21 25	192	30 57	282	22 17	339
212	19 33	050	38 01	076	44 59	133	25 54	176	21 13	192	29 59	282	21 55	338
213	20 18	050	38 59	076	45 42	134	25 58	176	21 00	192	29 00	282	21 33	338
214	21 04	049	39 57	076	46 25	135	26 02	177	20 47	193	28 02	282	21 11	338
215	21 49	049	40 55	076	47 08	135	26 05	177	20 33	193	27 03	282	20 48	338
216	22 34	049	41 53	076	47 49	136	26 07	178	20 19	194	26 04	282	20 25	337
217	23 20	049	42 51	076	48 31	137	26 09	178	20 05	194	25 06	282	20 01	337
218	24 04	048	43 49	075	49 11	138	26 10	179	19 50	195	24 07	282	19 38	337
219	24 49	048	44 47	075	49 51	139	26 11	180	19 35	195	23 08	282	19 14	336
220	25 34	048	45 45	075	50 30	140	26 11	180	19 19	195	22 10	282	18 50	336
221	26 18	048	46 43	075	51 08	141	26 11	181	19 03	196	21 11	282	18 25	336
222	27 02	047	47 41	075	51 46	142	26 10	181	18 47	196	20 12	282	18 01	336
223	27 46	047	48 39	074	52 22	143	26 08	182	18 30	196	19 14	282	17 36	335
224	28 30	047	49 36	074	52 58	144	26 06	182	18 13	197	18 15	282	17 11	335
	VEGA		ALTAIR		◆Nunki		ANTARES		RIGIL KENT.		◆SPICA		◆Alkaid	
225	29 14	046	17 37	082	26 12	121	53 33	145	26 03	183	62 27	239	41 09	344
226	29 57	046	18 36	082	27 03	122	54 06	146	26 00	183	61 36	240	40 52	344
227	30 40	046	19 35	082	27 54	122	54 39	148	25 56	184	60 44	241	40 35	343
228	31 23	045	20 35	082	28 44	122	55 11	149	25 52	184	59 51	242	40 17	342
229	32 05	045	21 34	082	29 35	123	55 41	150	25 47	185	58 58	242	39 58	341
230	32 47	045	22 33	082	30 25	123	56 10	151	25 42	185	58 05	243	39 38	341
231	33 29	044	23 32	082	31 15	123	56 38	153	25 36	186	57 11	244	39 18	340
232	34 11	044	24 32	082	32 05	124	57 05	154	25 29	187	56 17	245	38 57	339
233	34 52	043	25 31	082	32 55	124	57 30	156	25 22	187	55 23	245	38 35	338
234	35 33	043	26 30	082	33 44	125	57 54	157	25 14	188	54 28	246	38 13	338
235	36 14	042	27 29	082	34 33	125	58 17	159	25 06	188	53 33	247	37 50	337
236	36 54	042	28 29	082	35 22	126	58 37	160	24 58	189	52 38	247	37 26	336
237	37 33	041	29 28	082	36 11	126	58 57	162	24 48	189	51 43	248	37 02	336
238	38 13	041	30 27	081	36 59	127	59 15	164	24 39	190	50 47	248	36 37	335
239	38 52	040	31 27	081	37 47	127	59 31	165	24 28	190	49 51	249	36 12	335
	VEGA		◆ALTAIR		Nunki		◆ANTARES		SPICA		◆ARCTURUS		Alkaid	
240	39 30	040	32 26	081	38 35	128	59 45	167	48 56	249	59 45	304	35 46	334
241	40 08	039	33 25	081	39 22	128	59 58	169	47 59	250	58 55	303	35 19	333
242	40 46	038	34 24	081	40 09	129	60 08	170	47 03	250	58 05	303	34 52	333
243	41 23	038	35 23	081	40 56	129	60 17	172	46 07	251	57 14	302	34 24	332
244	41 59	037	36 23	081	41 42	130	60 25	174	45 10	251	56 23	301	33 56	332
245	42 35	037	37 22	081	42 28	130	60 30	176	44 13	251	55 31	300	33 27	331
246	43 11	036	38 21	081	43 13	131	60 33	178	43 17	252	54 40	300	32 58	331
247	43 45	035	39 20	081	43 58	132	60 35	179	42 20	252	53 47	299	32 29	330
248	44 20	034	40 19	081	44 42	133	60 34	181	41 23	252	52 55	298	31 59	330
249	44 53	034	41 19	081	45 26	133	60 32	183	40 26	253	52 02	298	31 28	329
250	45 26	033	42 18	081	46 10	134	60 28	185	39 28	253	51 09	297	30 57	329
251	45 58	032	43 17	081	46 52	135	60 22	187	38 31	253	50 16	297	30 26	328
252	46 29	031	44 16	081	47 35	136	60 14	189	37 34	253	49 22	296	29 54	328
253	47 00	030	45 15	080	48 16	136	60 04	190	36 36	254	48 28	296	29 22	327
254	47 30	029	46 14	080	48 57	137	59 52	192	35 39	254	47 34	296	28 49	327
	◆VEGA		ALTAIR		◆Nunki		ANTARES		◆SPICA		ARCTURUS		Alkaid	
255	47 59	029	47 13	080	49 37	138	59 39	194	34 41	254	46 40	295	28 17	327
256	48 27	028	48 12	080	50 17	139	59 24	195	33 43	254	45 46	295	27 43	326
257	48 54	027	49 11	080	50 55	140	59 07	197	32 45	255	44 51	294	27 10	326
258	49 21	026	50 10	080	51 33	141	58 48	199	31 48	255	43 57	294	26 36	325
259	49 46	025	51 09	080	52 10	142	58 28	200	30 50	255	43 02	294	26 02	325
260	50 11	024	52 08	079	52 47	143	58 07	202	29 52	255	42 07	293	25 27	325
261	50 34	023	53 07	079	53 22	144	57 44	203	28 54	256	41 12	293	24 53	324
262	50 56	021	54 06	079	53 56	146	57 19	205	27 56	256	40 17	293	24 18	324
263	51 18	020	55 05	079	54 30	147	56 53	206	26 58	256	39 21	293	23 42	324
264	51 38	019	56 03	079	55 02	148	56 26	208	26 00	256	38 26	292	23 07	323
265	51 57	018	57 02	078	55 33	149	55 57	209	25 01	256	37 30	292	22 31	323
266	52 15	017	58 01	078	56 03	151	55 27	210	24 03	256	36 35	292	21 55	323
267	52 32	016	58 59	078	56 31	152	54 56	212	23 05	257	35 39	292	21 19	323
268	52 48	014	59 58	077	56 59	153	54 24	213	22 07	257	34 43	291	20 42	322
269	53 02	013	60 56	077	57 25	155	53 51	214	21 08	257	33 48	291	20 06	322

LAT 3°N

LHA	Hc	Zn	Hc	Zn	Hc	Zn	Hc	Zn	Hc	Zn	Hc	Zn	Hc	Zn
♈	VEGA		◆DENEB		FOMALHAUT		◆Peacock		Shaula		ANTARES		◆ARCTURUS	
270	53 15	012	34 58	034	12 01	121	23 25	159	49 25	188	53 17	215	32 52	291
271	53 27	011	35 31	033	12 53	121	23 46	160	49 16	189	52 42	216	31 56	291
272	53 38	009	36 04	033	13 44	122	24 07	160	49 06	191	52 06	218	31 00	291
273	53 47	008	36 36	032	14 35	122	24 27	161	48 54	192	51 29	219	30 04	291
274	53 55	007	37 07	032	15 26	122	24 46	161	48 41	193	50 51	220	29 07	290
275	54 01	006	37 39	031	16 17	122	25 05	162	48 27	194	50 12	221	28 11	290
276	54 06	004	38 09	030	17 07	122	25 24	162	48 12	195	49 33	222	27 15	290
277	54 10	003	38 39	030	17 58	123	25 42	163	47 56	196	48 53	222	26 19	290
278	54 12	002	39 08	029	18 48	123	26 00	163	47 38	197	48 12	223	25 22	290
279	54 13	000	39 37	028	19 39	123	26 17	164	47 20	199	47 31	224	24 26	290
280	54 13	359	40 05	028	20 29	123	26 33	164	47 00	200	46 49	225	23 30	290
281	54 11	358	40 33	027	21 19	124	26 49	165	46 40	201	46 06	226	22 33	290
282	54 08	356	41 00	026	22 09	124	27 05	165	46 18	202	45 23	227	21 37	289
283	54 03	355	41 26	026	22 58	124	27 20	166	45 55	203	44 39	227	20 40	289
284	53 57	354	41 51	025	23 48	124	27 34	166	45 32	204	43 55	228	19 44	289
	DENEB		Alpheratz		◆FOMALHAUT		Peacock		◆ANTARES		Rasalhague		◆VEGA	
285	42 16	024	12 47	061	24 37	125	27 48	167	43 10	229	66 50	296	53 50	352
286	42 40	023	13 39	061	25 27	125	28 01	168	42 25	229	65 56	295	53 41	351
287	43 03	022	14 32	061	26 15	125	28 14	168	41 39	230	65 01	294	53 31	350
288	43 26	022	15 24	061	27 04	126	28 26	169	40 53	231	64 06	293	53 20	348
289	43 47	021	16 16	061	27 53	126	28 37	169	40 06	231	63 11	292	53 07	347
290	44 08	020	17 08	060	28 41	126	28 48	170	39 19	232	62 15	292	52 53	346
291	44 28	019	18 00	060	29 29	127	28 58	170	38 32	232	61 19	291	52 38	345
292	44 47	018	18 53	060	30 17	127	29 08	171	37 45	233	60 23	290	52 22	344
293	45 05	017	19 45	060	31 05	128	29 17	172	36 57	233	59 27	290	52 04	342
294	45 23	016	20 36	060	31 52	128	29 25	172	36 08	234	58 31	289	51 46	341
295	45 39	015	21 28	060	32 39	128	29 33	173	35 20	234	57 34	289	51 26	340
296	45 55	014	22 20	060	33 26	129	29 40	174	34 31	235	56 37	288	51 05	339
297	46 09	014	23 12	060	34 12	129	29 46	174	33 42	235	55 40	288	50 43	338
298	46 23	013	24 03	059	34 59	130	29 52	175	32 53	236	54 43	288	50 20	337
299	46 35	012	24 55	059	35 44	130	29 57	175	32 03	236	53 46	287	49 56	336
	◆DENEB		Alpheratz		◆FOMALHAUT		Peacock		◆ANTARES		Rasalhague		VEGA	
300	46 47	011	25 46	059	36 30	131	30 02	176	31 13	236	52 49	287	49 31	335
301	46 57	010	26 38	059	37 15	131	30 06	177	30 23	237	51 52	287	49 05	334
302	47 07	009	27 29	059	38 00	132	30 09	177	29 33	237	50 54	286	48 38	333
303	47 15	008	28 20	058	38 44	133	30 11	178	28 43	238	49 57	286	48 10	332
304	47 23	007	29 11	058	39 28	133	30 13	179	27 52	238	48 59	286	47 41	331
305	47 29	006	30 02	058	40 11	134	30 14	179	27 01	238	48 01	286	47 12	330
306	47 34	005	30 52	058	40 54	134	30 15	180	26 10	239	47 03	285	46 41	329
307	47 38	003	31 43	057	41 37	135	30 15	180	25 19	239	46 06	285	46 10	328
308	47 41	002	32 33	057	42 19	136	30 14	181	24 28	239	45 08	285	45 38	328
309	47 43	001	33 23	057	43 00	137	30 13	182	23 36	239	44 10	285	45 06	327
310	47 44	000	34 13	056	43 41	137	30 10	182	22 44	240	43 12	284	44 32	326
311	47 44	359	35 03	056	44 21	138	30 08	183	21 53	240	42 14	284	43 59	325
312	47 43	358	35 53	056	45 01	139	30 04	184	21 01	240	41 16	284	43 24	324
313	47 40	357	36 42	055	45 40	140	30 00	184	20 09	240	40 18	284	42 49	324
314	47 37	356	37 32	055	46 19	141	29 55	185	19 16	241	39 19	284	42 13	323
	◆Alpheratz		◆FOMALHAUT		Peacock		Nunki		◆Rasalhague		VEGA		DENEB	
315	38 21	055	46 56	141	29 50	185	47 54	224	38 21	284	41 37	322	47 32	355
316	39 09	054	47 33	142	29 44	186	47 13	225	37 23	284	41 00	322	47 27	354
317	39 58	054	48 10	143	29 37	187	46 30	226	36 25	283	40 23	321	47 20	353
318	40 46	053	48 45	144	29 30	187	45 47	226	35 26	283	39 45	321	47 12	352
319	41 34	053	49 20	145	29 22	188	45 03	227	34 28	283	39 07	320	47 03	351
320	42 22	052	49 53	146	29 13	189	44 19	228	33 30	283	38 28	319	46 53	350
321	43 09	052	50 26	147	29 04	189	43 35	229	32 31	283	37 49	319	46 43	349
322	43 56	051	50 58	148	28 54	190	42 49	229	31 33	283	37 09	318	46 31	348
323	44 42	051	51 29	149	28 44	190	42 04	230	30 34	283	36 29	318	46 18	347
324	45 29	050	51 59	151	28 33	191	41 18	230	29 36	283	35 49	317	46 04	346
325	46 14	050	52 28	152	28 21	192	40 31	231	28 38	283	35 08	317	45 49	345
326	47 00	049	52 56	153	28 09	192	39 45	232	27 39	283	34 27	316	45 33	344
327	47 45	048	53 22	154	27 56	193	38 57	232	26 41	283	33 45	316	45 16	343
328	48 29	048	53 48	156	27 43	193	38 10	233	25 42	283	33 03	316	44 59	342
329	49 13	047	54 12	157	27 29	194	37 22	233	24 44	282	32 21	315	44 40	342
	◆Alpheratz		Diphda		◆FOMALHAUT		Peacock		◆ALTAIR		VEGA		DENEB	
330	49 57	046	44 37	119	54 35	158	27 14	194	57 18	282	31 39	315	44 21	341
331	50 40	045	45 29	120	54 56	160	26 59	195	56 19	281	30 56	314	44 01	340
332	51 22	044	46 21	120	55 17	161	26 43	195	55 20	281	30 13	314	43 39	339
333	52 03	044	47 13	121	55 36	162	26 27	196	54 22	281	29 30	314	43 17	338
334	52 44	043	48 04	121	55 53	164	26 10	196	53 23	281	28 47	313	42 55	337
335	53 25	042	48 55	122	56 09	165	25 53	197	52 24	281	28 03	313	42 31	336
336	54 04	041	49 45	122	56 23	167	25 35	198	51 25	280	27 19	313	42 07	336
337	54 43	040	50 36	123	56 36	168	25 17	198	50 26	280	26 35	312	41 42	335
338	55 21	039	51 25	124	56 47	170	24 58	199	49 27	280	25 51	312	41 16	334
339	55 58	038	52 15	125	56 57	171	24 39	199	48 28	280	25 06	312	40 50	333
340	56 34	037	53 04	126	57 05	173	24 19	199	47 29	280	24 22	312	40 23	333
341	57 09	035	53 52	127	57 12	175	23 59	200	46 30	280	23 37	311	39 55	332
342	57 44	034	54 40	128	57 16	176	23 38	200	45 31	280	22 52	311	39 27	331
343	58 17	033	55 27	128	57 20	178	23 17	201	44 32	279	22 07	311	38 58	331
344	58 48	031	56 13	129	57 21	179	22 55	201	43 33	279	21 21	311	38 28	330
	Alpheratz		◆Hamal		Acamar		◆ACHERNAR		FOMALHAUT		◆ALTAIR		DENEB	
345	59 19	030	40 28	061	20 38	135	21 56	158	57 21	181	42 34	279	37 58	329
346	59 49	029	41 20	061	21 20	136	22 17	159	57 19	183	41 35	279	37 27	329
347	60 17	027	42 12	061	22 01	136	22 39	159	57 15	184	40 35	279	36 56	328
348	60 43	026	43 05	060	22 43	136	23 00	160	57 10	186	39 36	279	36 24	328
349	61 09	024	43 56	060	23 24	137	23 21	160	57 03	187	38 37	279	35 52	327
350	61 32	022	44 48	059	24 05	137	23 41	161	56 54	189	37 38	279	35 19	327
351	61 54	021	45 40	059	24 46	138	24 01	161	56 44	191	36 39	279	34 46	326
352	62 15	019	46 31	058	25 26	138	24 20	161	56 32	192	35 39	279	34 12	326
353	62 34	017	47 22	058	26 06	138	24 39	162	56 19	194	34 40	279	33 38	325
354	62 51	016	48 12	057	26 45	139	24 57	162	56 04	195	33 41	279	33 03	325
355	63 06	014	49 03	057	27 25	139	25 15	163	55 48	197	32 42	279	32 28	324
356	63 19	012	49 53	056	28 04	140	25 32	163	55 30	198	31 43	279	31 53	324
357	63 30	010	50 42	056	28 42	140	25 49	164	55 11	199	30 43	279	31 17	323
358	63 39	008	51 31	055	29 21	141	26 05	164	54 50	201	29 44	278	30 41	323
359	63 47	006	52 20	054	29 58	141	26 21	165	54 28	202	28 45	278	30 05	322

LHA ♈	Hc	Zn	Hc	Zn	Hc	Zn	Hc	Zn	Hc	Zn	Hc	Zn	Hc	Zn
	Alpheratz		♦Mirfak		ALDEBARAN		♦ACHERNAR		FOMALHAUT		♦ALTAIR		DENEB	
0	62 52	004	25 37	034	20 47	073	27 35	165	54 59	204	27 36	279	28 41	322
1	62 55	002	26 10	033	21 44	073	27 49	166	54 34	205	26 37	279	28 04	322
2	62 56	000	26 43	033	22 41	073	28 04	167	54 08	207	25 38	279	27 27	322
3	62 55	358	27 15	033	23 39	073	28 17	167	53 40	208	24 39	279	26 49	321
4	62 52	356	27 48	032	24 36	073	28 30	168	53 12	209	23 39	279	26 12	321
5	62 47	354	28 19	032	25 33	073	28 43	168	52 42	210	22 40	279	25 33	320
6	62 40	352	28 51	031	26 30	073	28 55	169	52 11	212	21 41	279	24 55	320
7	62 32	351	29 22	031	27 28	072	29 06	169	51 39	213	20 42	279	24 17	320
8	62 21	349	29 52	030	28 25	072	29 17	170	51 06	214	19 42	279	23 38	320
9	62 08	347	30 23	030	29 22	072	29 27	171	50 33	215	18 43	279	22 59	319
10	61 54	345	30 52	030	30 19	072	29 37	171	49 58	216	17 44	279	22 19	319
11	61 37	343	31 22	029	31 16	072	29 46	172	49 22	217	16 45	279	21 40	319
12	61 19	342	31 51	029	32 13	072	29 54	172	48 46	218	15 45	279	21 00	318
13	61 00	340	32 19	028	33 10	072	30 02	173	48 09	219	14 46	279	20 20	318
14	60 38	338	32 47	027	34 07	071	30 09	174	47 31	220	13 47	279	19 40	318
	CAPELLA		ALDEBARAN		♦RIGEL		ACHERNAR		♦FOMALHAUT		Enif		♦Alpheratz	
15	19 11	041	35 03	071	25 48	100	30 15	174	46 53	220	40 41	281	60 15	337
16	19 51	041	36 00	071	26 47	100	30 21	175	46 13	221	39 42	281	59 51	335
17	20 30	041	36 57	071	27 46	100	30 26	175	45 34	222	38 43	281	59 25	334
18	21 09	041	37 53	071	28 45	100	30 31	176	44 53	223	37 44	281	58 58	332
19	21 48	040	38 50	070	29 44	101	30 35	177	44 12	224	36 45	281	58 29	331
20	22 27	040	39 46	070	30 43	101	30 38	177	43 30	224	35 46	281	57 59	329
21	23 06	040	40 43	070	31 42	101	30 41	178	42 48	225	34 47	281	57 28	328
22	23 44	040	41 39	070	32 41	101	30 42	179	42 06	226	33 48	281	56 56	327
23	24 22	039	42 35	069	33 39	101	30 44	179	41 23	226	32 49	280	56 22	326
24	25 00	039	43 31	069	34 38	101	30 44	180	40 39	227	31 50	280	55 48	324
25	25 37	039	44 27	069	35 37	102	30 44	180	39 55	228	30 51	280	55 13	323
26	26 15	038	45 23	068	36 36	102	30 43	181	39 10	228	29 52	280	54 36	322
27	26 52	038	46 18	068	37 34	102	30 42	182	38 26	229	28 53	280	53 59	321
28	27 28	038	47 14	068	38 33	102	30 40	182	37 40	229	27 54	280	53 21	320
29	28 05	037	48 09	067	39 32	102	30 37	183	36 55	230	26 55	280	52 42	319
	Mirfak		CAPELLA		♦ALDEBARAN		RIGEL		♦ACHERNAR		FOMALHAUT		♦Alpheratz	
30	38 56	017	28 41	037	49 04	067	40 30	103	30 34	184	36 09	230	52 03	318
31	39 13	017	29 16	036	50 00	066	41 29	103	30 30	184	35 22	231	51 22	317
32	39 30	016	29 52	036	50 54	066	42 27	103	30 25	185	34 36	231	50 41	316
33	39 46	015	30 27	035	51 49	065	43 26	103	30 20	185	33 49	232	50 00	316
34	40 01	014	31 01	035	52 43	065	44 24	104	30 14	186	33 02	232	49 17	315
35	40 15	013	31 36	035	53 38	064	45 22	104	30 07	187	32 14	233	48 35	314
36	40 29	013	32 09	034	54 32	064	46 20	104	30 00	187	31 26	233	47 51	313
37	40 42	012	32 43	034	55 25	063	47 18	104	29 52	188	30 38	233	47 07	313
38	40 54	011	33 16	033	56 19	063	48 16	105	29 44	188	29 50	234	46 23	312
39	41 05	010	33 48	033	57 12	062	49 14	105	29 35	189	29 01	234	45 38	311
40	41 15	009	34 20	032	58 05	061	50 12	105	29 25	190	28 13	235	44 53	311
41	41 24	009	34 52	031	58 57	060	51 10	106	29 15	190	27 24	235	44 07	310
42	41 33	008	35 23	031	59 49	060	52 08	106	29 04	191	26 35	235	43 21	309
43	41 40	007	35 53	030	60 40	059	53 05	107	28 52	191	25 45	235	42 34	309
44	41 47	006	36 23	030	61 31	058	54 03	107	28 40	192	24 56	236	41 48	308
	CAPELLA		♦PROCYON		SIRIUS		CANOPUS		♦ACHERNAR		Diphda		♦Alpheratz	
45	36 53	029	20 20	085	31 28	111	20 43	150	28 27	193	50 52	238	41 00	308
46	37 22	028	21 20	085	32 24	111	21 13	150	28 14	193	50 01	238	40 13	307
47	37 50	028	22 20	085	33 20	112	21 43	151	28 00	194	49 09	239	39 25	307
48	38 18	027	23 19	085	34 15	112	22 12	151	27 46	194	48 18	240	38 37	306
49	38 45	027	24 19	085	35 11	112	22 41	151	27 31	195	47 26	240	37 49	306
50	39 11	026	25 19	085	36 07	112	23 10	152	27 15	195	46 34	241	37 00	306
51	39 37	025	26 19	085	37 02	113	23 38	152	26 59	196	45 41	241	36 11	305
52	40 02	024	27 18	085	37 57	113	24 06	153	26 43	196	44 48	242	35 22	305
53	40 27	024	28 18	085	38 52	113	24 33	153	26 26	197	43 55	242	34 33	304
54	40 50	023	29 18	085	39 47	114	25 00	153	26 08	197	43 02	243	33 43	304
55	41 13	022	30 17	085	40 42	114	25 27	154	25 50	198	42 08	243	32 54	304
56	41 36	021	31 17	085	41 37	115	25 53	154	25 31	198	41 15	244	32 04	303
57	41 57	021	32 17	085	42 31	115	26 19	155	25 12	199	40 21	244	31 14	303
58	42 18	020	33 17	085	43 25	115	26 44	155	24 53	199	39 27	245	30 23	303
59	42 38	019	34 16	085	44 19	116	27 09	156	24 33	200	38 33	245	29 33	303
	♦CAPELLA		PROCYON		♦SIRIUS		CANOPUS		ACHERNAR		♦Diphda		Alpheratz	
60	42 57	018	35 16	085	45 13	116	27 33	156	24 12	200	37 38	245	28 42	302
61	43 15	017	36 16	085	46 07	117	27 57	157	23 51	201	36 44	246	27 51	302
62	43 32	016	37 16	085	47 00	117	28 20	157	23 30	201	35 49	246	27 01	302
63	43 49	015	38 15	085	47 53	118	28 43	158	23 08	202	34 54	246	26 10	302
64	44 04	015	39 15	085	48 46	118	29 05	158	22 46	202	33 59	247	25 19	301
65	44 19	014	40 15	085	49 39	119	29 27	159	22 23	202	33 04	247	24 27	301
66	44 32	013	41 15	085	50 31	120	29 48	160	22 00	203	32 09	247	23 36	301
67	44 45	012	42 14	085	51 23	120	30 09	160	21 37	203	31 13	247	22 44	301
68	44 57	011	43 14	085	52 15	121	30 29	161	21 13	204	30 18	248	21 53	301
69	45 08	010	44 14	085	53 06	122	30 48	161	20 49	204	29 22	248	21 01	300
70	45 18	009	45 13	085	53 57	122	31 07	162	20 24	204	28 27	248	20 10	300
71	45 26	008	46 13	085	54 47	123	31 26	163	19 59	205	27 31	248	19 18	300
72	45 34	007	47 13	084	55 37	124	31 43	163	19 34	205	26 35	249	18 26	300
73	45 41	006	48 12	084	56 26	125	32 00	164	19 09	205	25 39	249	17 34	300
74	45 47	005	49 12	084	57 15	126	32 17	164	18 43	206	24 43	249	16 42	300
	CAPELLA		♦POLLUX		PROCYON		SIRIUS		♦CANOPUS		Acamar		♦Hamal	
75	45 52	004	42 49	052	50 12	084	58 03	127	32 32	165	39 21	210	42 57	301
76	45 55	003	43 36	052	51 11	084	58 51	128	32 47	166	38 51	211	42 06	300
77	45 58	002	44 23	051	52 11	084	59 38	129	33 02	166	38 20	211	41 14	300
78	46 00	001	45 10	051	53 11	084	60 24	130	33 16	167	37 48	212	40 22	300
79	46 00	000	45 56	050	54 10	084	61 10	131	33 29	168	37 16	213	39 30	299
80	46 00	359	46 42	050	55 10	084	61 54	133	33 41	168	36 43	214	38 37	299
81	45 58	358	47 27	049	56 10	084	62 38	134	33 53	169	36 10	214	37 44	298
82	45 56	357	48 12	048	57 09	083	63 21	135	34 04	170	35 36	215	36 52	298
83	45 52	356	48 57	047	58 09	083	64 02	137	34 14	171	35 01	215	35 59	298
84	45 48	355	49 41	047	59 08	083	64 43	138	34 23	171	34 26	216	35 06	298
85	45 42	354	50 24	046	60 08	083	65 22	140	34 32	172	33 51	217	34 12	297
86	45 35	353	51 07	045	61 07	083	66 00	142	34 40	173	33 15	217	33 19	297
87	45 28	352	51 49	044	62 07	083	66 36	144	34 47	173	32 38	218	32 25	297
88	45 19	351	52 30	043	63 06	082	67 11	146	34 54	174	32 01	218	31 32	296
89	45 09	350	53 11	042	64 06	082	67 44	148	35 00	175	31 24	219	30 38	296

LHA ♈	Hc	Zn	Hc	Zn	Hc	Zn	Hc	Zn	Hc	Zn	Hc	Zn	Hc	Zn
	POLLUX		♦REGULUS		Suhail		♦CANOPUS		Acamar		♦ALDEBARAN		CAPELLA	
90	53 51	041	27 46	077	28 08	143	35 05	176	30 46	219	64 41	306	44 59	349
91	54 31	040	28 44	077	28 44	143	35 09	176	30 08	220	63 52	305	44 47	348
92	55 09	039	29 43	077	29 20	144	35 12	177	29 30	220	63 03	304	44 34	347
93	55 47	038	30 41	077	29 55	144	35 15	178	28 51	221	62 13	303	44 21	346
94	56 23	037	31 40	077	30 29	145	35 17	179	28 12	221	61 22	302	44 06	346
95	56 59	036	32 38	077	31 04	145	35 18	179	27 32	222	60 31	301	43 51	345
96	57 34	035	33 37	077	31 37	146	35 18	180	26 52	222	59 39	300	43 35	344
97	58 07	033	34 35	077	32 11	147	35 18	181	26 12	222	58 47	299	43 18	343
98	58 40	032	35 33	077	32 43	147	35 17	182	25 31	223	57 55	299	43 00	342
99	59 11	031	36 32	077	33 16	148	35 15	182	24 50	223	57 02	298	42 41	341
100	59 41	029	37 30	076	33 47	148	35 12	183	24 09	224	56 09	297	42 21	340
101	60 10	028	38 28	076	34 19	149	35 09	184	23 28	224	55 15	297	42 00	340
102	60 37	026	39 26	076	34 49	150	35 04	184	22 46	224	54 22	296	41 39	339
103	61 03	025	40 25	076	35 20	150	34 59	185	22 04	225	53 28	295	41 17	338
104	61 27	023	41 23	076	35 49	151	34 53	186	21 22	225	52 34	295	40 54	337
	POLLUX		♦REGULUS		Suhail		♦CANOPUS		RIGEL		♦ALDEBARAN		CAPELLA	
105	61 50	021	42 21	076	36 18	152	34 47	187	61 44	248	51 39	294	40 31	336
106	62 11	020	43 19	075	36 46	152	34 40	187	60 49	249	50 44	294	40 06	336
107	62 30	018	44 17	075	37 14	153	34 31	188	59 52	250	49 49	294	39 41	335
108	62 48	016	45 15	075	37 41	154	34 23	189	58 56	250	48 54	293	39 16	334
109	63 03	014	46 13	075	38 07	154	34 13	190	57 59	251	47 59	293	38 49	334
110	63 17	012	47 10	074	38 33	155	34 03	190	57 03	252	47 04	292	38 22	333
111	63 29	010	48 08	074	38 58	156	33 52	191	56 06	252	46 08	292	37 55	332
112	63 39	008	49 06	074	39 22	157	33 40	192	55 09	253	45 13	292	37 26	332
113	63 47	007	50 03	074	39 45	157	33 28	192	54 11	253	44 17	291	36 58	331
114	63 52	005	51 01	073	40 08	158	33 15	193	53 14	253	43 21	291	36 28	330
115	63 56	003	51 58	073	40 30	159	33 01	194	52 16	254	42 25	291	35 58	330
116	63 58	001	52 56	073	40 51	160	32 46	194	51 19	254	41 29	290	35 28	329
117	63 57	358	53 53	072	41 11	161	32 31	195	50 21	255	40 32	290	34 57	329
118	63 54	356	54 50	072	41 30	162	32 15	196	49 23	255	39 36	290	34 25	328
119	63 50	354	55 47	071	41 49	163	31 59	196	48 25	255	38 40	290	33 53	328
	Dubhe		♦Denebola		Suhail		♦CANOPUS		SIRIUS		BETELGEUSE		♦CAPELLA	
120	21 06	021	32 11	074	42 06	163	31 42	197	63 40	224	58 23	281	33 21	327
121	21 27	021	33 09	074	42 23	164	31 24	197	62 58	225	57 24	281	32 48	326
122	21 49	021	34 06	074	42 39	165	31 06	198	62 15	227	56 25	280	32 15	326
123	22 10	020	35 04	074	42 53	166	30 47	199	61 31	228	55 26	280	31 41	326
124	22 30	020	36 01	073	43 07	167	30 27	199	60 46	229	54 27	280	31 07	325
125	22 51	020	36 59	073	43 20	168	30 07	200	60 00	231	53 28	280	30 32	325
126	23 11	019	37 56	073	43 32	169	29 47	200	59 13	232	52 29	280	29 57	324
127	23 30	019	38 53	073	43 43	170	29 25	201	58 26	233	51 30	279	29 22	324
128	23 49	018	39 51	073	43 53	171	29 03	202	57 38	234	50 30	279	28 46	323
129	24 08	018	40 48	072	44 02	172	28 41	202	56 49	235	49 31	279	28 10	323
130	24 27	018	41 45	072	44 10	173	28 18	203	56 00	236	48 32	279	27 34	323
131	24 45	017	42 42	072	44 16	174	27 55	203	55 10	236	47 33	279	26 57	322
132	25 02	017	43 39	072	44 22	175	27 31	204	54 20	237	46 33	279	26 21	322
133	25 20	016	44 36	071	44 27	176	27 07	204	53 29	238	45 34	279	25 43	321
134	25 36	016	45 33	071	44 31	177	26 42	205	52 38	239	44 35	278	25 06	321
	Dubhe		♦SPICA		ACRUX		♦CANOPUS		SIRIUS		BETELGEUSE		♦CAPELLA	
135	25 53	016	22 52	103	14 29	159	26 17	205	51 47	239	43 36	278	24 28	321
136	26 09	015	23 50	103	14 51	159	25 51	206	50 55	240	42 36	278	23 50	320
137	26 24	015	24 48	103	15 13	159	25 25	206	50 03	241	41 37	278	23 12	320
138	26 39	014	25 47	103	15 34	159	24 58	207	49 11	241	40 37	278	22 33	320
139	26 54	014	26 45	104	15 55	160	24 31	207	48 18	242	39 38	278	21 54	320
140	27 08	013	27 43	104	16 16	160	24 03	207	47 25	242	38 39	278	21 15	319
141	27 22	013	28 42	104	16 36	160	23 36	208	46 32	243	37 39	278	20 36	319
142	27 35	012	29 40	104	16 56	161	23 07	208	45 38	243	36 40	278	19 57	319
143	27 47	012	30 38	104	17 16	161	22 39	209	44 44	244	35 40	278	19 17	319
144	28 00	012	31 36	104	17 35	161	22 10	209	43 50	244	34 41	278	18 38	318
145	28 11	011	32 34	105	17 54	162	21 40	210	42 56	245	33 42	278	17 58	318
146	28 23	011	33 32	105	18 13	162	21 11	210	42 02	245	32 42	278	17 18	318
147	28 33	010	34 30	105	18 32	162	20 41	210	41 07	246	31 43	277	16 37	318
148	28 43	010	35 28	105	18 50	163	20 10	211	40 13	246	30 43	277	15 57	318
149	28 53	009	36 26	105	19 07	163	19 40	211	39 18	246	29 44	277	15 17	317
	♦Dubhe		ARCTURUS		♦SPICA		ACRUX		♦SIRIUS		PROCYON		POLLUX	
150	29 02	008	25 18	070	37 24	106	19 25	163	38 23	247	54 41	276	48 35	312
151	29 11	008	26 14	070	38 21	106	19 42	164	37 28	247	53 42	276	47 50	311
152	29 19	007	27 10	069	39 19	106	19 58	164	36 32	247	52 42	276	47 05	311
153	29 26	007	28 06	069	40 16	106	20 14	165	35 37	248	51 42	276	46 19	310
154	29 33	006	29 02	069	41 14	107	20 30	165	34 41	248	50 43	276	45 33	310
155	29 40	006	29 58	069	42 11	107	20 46	165	33 46	248	49 43	276	44 47	309
156	29 46	005	30 54	069	43 09	107	21 01	166	32 50	249	48 43	276	44 00	308
157	29 51	005	31 50	069	44 06	108	21 15	166	31 54	249	47 44	276	43 13	308
158	29 56	004	32 46	068	45 03	108	21 29	167	30 58	249	46 44	276	42 25	307
159	30 00	004	33 41	068	46 00	108	21 43	167	30 02	249	45 44	275	41 37	307
160	30 04	003	34 37	068	46 57	109	21 56	167	29 06	250	44 45	275	40 49	306
161	30 07	003	35 33	068	47 53	109	22 09	168	28 10	250	43 45	275	40 00	306
162	30 09	002	36 28	067	48 50	109	22 22	168	27 13	250	42 45	275	39 12	305
163	30 11	002	37 23	067	49 46	110	22 34	169	26 17	250	41 46	275	38 23	305
164	30 12	001	38 19	067	50 43	110	22 45	169	25 20	250	40 46	275	37 33	305
	Dubhe		♦ARCTURUS		SPICA		♦ACRUX		Suhail		PROCYON		♦POLLUX	
165	30 13	000	39 14	067	51 39	111	22 56	170	38 05	206	39 46	275	36 44	304
166	30 13	000	40 09	066	52 35	111	23 07	170	37 38	206	38 46	275	35 54	304
167	30 13	359	41 03	066	53 31	112	23 17	170	37 11	207	37 47	275	35 04	303
168	30 12	359	41 58	066	54 26	112	23 27	171	36 44	208	36 47	275	34 14	303
169	30 10	358	42 53	065	55 21	113	23 36	171	36 15	209	35 47	275	33 24	303
170	30 08	358	43 47	065	56 17	114	23 45	172	35 46	209	34 48	275	32 33	302
171	30 06	357	44 41	065	57 11	114	23 53	172	35 17	210	33 48	275	31 42	302
172	30 03	357	45 36	064	58 06	115	24 01	173	34 47	211	32 48	275	30 51	302
173	29 59	356	46 29	064	59 00	116	24 08	173	34 16	211	31 48	275	30 00	302
174	29 54	356	47 23	063	59 54	116	24 15	174	33 45	212	30 49	275	29 09	301
175	29 49	355	48 17	063	60 47	117	24 21	174	33 13	212	29 49	275	28 18	301
176	29 44	354	49 10	062	61 41	118	24 27	175	32 40	213	28 49	275	27 26	301
177	29 38	354	50 03	062	62 33	119	24 32	175	32 08	213	27 49	275	26 35	301
178	29 31	353	50 56	061	63 25	120	24 37	176	31 34	214	26 50	275	25 43	300
179	29 24	353	51 48	061	64 17	121	24 41	176	31 00	215	25 50	275	24 51	300

LAT 2°N

LHA ♈	Hc	Zn	Hc	Zn	Hc	Zn	Hc	Zn	Hc	Zn	Hc	Zn	Hc	Zn
	◆ARCTURUS		ANTARES		RIGIL KENT.		◆ACRUX		Suhail		◆REGULUS		Dubhe	
180	52 40	060	19 17	119	20 07	161	24 44	177	30 26	215	60 32	291	29 17	352
181	53 32	060	20 09	119	20 27	161	24 48	177	29 51	216	59 36	291	29 08	352
182	54 24	059	21 02	119	20 46	161	24 50	178	29 16	216	58 40	290	29 00	351
183	55 15	058	21 54	120	21 05	162	24 52	178	28 41	217	57 43	290	28 50	351
184	56 06	057	22 46	120	21 23	162	24 54	179	28 05	217	56 47	289	28 41	350
185	56 56	057	23 38	120	21 42	163	24 55	179	27 28	218	55 50	289	28 30	350
186	57 46	056	24 30	120	21 59	163	24 56	180	26 52	218	54 53	288	28 19	349
187	58 35	055	25 21	121	22 17	163	24 56	180	26 15	218	53 56	288	28 08	349
188	59 24	054	26 13	121	22 34	164	24 55	181	25 37	219	52 59	287	27 56	348
189	60 12	053	27 04	121	22 50	164	24 54	181	25 00	219	52 01	287	27 44	348
190	61 00	052	27 55	121	23 06	165	24 53	182	24 22	220	51 04	287	27 31	347
191	61 47	051	28 47	122	23 22	165	24 51	182	23 43	220	50 07	286	27 18	347
192	62 33	050	29 37	122	23 37	166	24 48	183	23 04	220	49 09	286	27 04	346
193	63 18	048	30 28	122	23 51	166	24 45	183	22 26	221	48 11	286	26 50	346
194	64 03	047	31 19	123	24 06	167	24 41	184	21 46	221	47 14	286	26 35	346
	◆Alphecca		Rasalhague		ANTARES		◆RIGIL KENT.		ACRUX		◆REGULUS		Dubhe	
195	45 29	053	21 15	077	32 09	123	24 19	167	24 37	184	46 16	285	26 20	345
196	46 17	052	22 13	077	32 59	124	24 33	167	24 33	185	45 18	285	26 04	345
197	47 04	051	23 12	077	33 49	124	24 45	168	24 27	185	44 20	285	25 48	344
198	47 50	051	24 10	077	34 38	124	24 58	168	24 22	186	43 22	285	25 32	344
199	48 37	050	25 08	077	35 28	125	25 09	169	24 16	186	42 24	284	25 15	343
200	49 22	049	26 07	077	36 17	125	25 20	169	24 09	187	41 26	284	24 57	343
201	50 08	049	27 05	077	37 06	126	25 31	170	24 02	187	40 28	284	24 39	343
202	50 53	048	28 04	077	37 54	126	25 41	170	23 54	188	39 30	284	24 21	342
203	51 37	047	29 02	077	38 42	127	25 51	171	23 46	188	38 31	284	24 03	342
204	52 20	046	30 00	077	39 30	127	26 00	171	23 37	189	37 33	284	23 44	341
205	53 03	045	30 59	077	40 18	128	26 09	172	23 28	189	36 35	283	23 24	341
206	53 46	044	31 57	076	41 05	128	26 17	173	23 19	189	35 36	283	23 05	341
207	54 27	044	32 55	076	41 52	129	26 24	173	23 09	190	34 38	283	22 45	340
208	55 08	043	33 54	076	42 38	130	26 31	174	22 58	190	33 40	283	22 24	340
209	55 48	041	34 52	076	43 24	130	26 38	174	22 47	191	32 41	283	22 04	340
	◆VEGA		Rasalhague		ANTARES		◆RIGIL KENT.		ACRUX		◆REGULUS		Dubhe	
210	17 22	050	35 50	076	44 10	131	26 44	175	22 36	191	31 43	283	21 43	339
211	18 08	050	36 48	076	44 55	132	26 49	175	22 24	192	30 44	283	21 21	339
212	18 54	049	37 46	076	45 39	132	26 54	176	22 11	192	29 46	283	20 59	339
213	19 39	049	38 44	075	46 23	133	26 58	176	21 59	193	28 47	283	20 37	338
214	20 25	049	39 42	075	47 07	134	27 01	177	21 45	193	27 49	283	20 15	338
215	21 10	049	40 40	075	47 50	135	27 04	177	21 32	193	26 50	282	19 52	338
216	21 55	049	41 38	075	48 32	135	27 07	178	21 18	194	25 52	282	19 29	337
217	22 40	048	42 36	075	49 14	136	27 09	178	21 03	194	24 53	282	19 06	337
218	23 24	048	43 34	075	49 55	137	27 10	179	20 48	195	23 55	282	18 43	337
219	24 09	048	44 32	074	50 36	138	27 11	180	20 33	195	22 56	282	18 19	337
220	24 53	048	45 29	074	51 15	139	27 11	180	20 17	195	21 57	282	17 55	336
221	25 38	047	46 27	074	51 54	140	27 11	181	20 01	196	20 59	282	17 31	336
222	26 22	047	47 24	074	52 32	141	27 10	181	19 44	196	20 00	282	17 06	336
223	27 05	047	48 22	073	53 10	142	27 08	182	19 27	197	19 01	282	16 41	335
224	27 49	046	49 19	073	53 46	143	27 06	182	19 10	197	18 03	282	16 16	335
	VEGA		◆ALTAIR		Nunki		ANTARES		◆RIGIL KENT.		SPICA		◆ARCTURUS	
225	28 32	046	17 28	081	26 43	121	54 22	144	27 03	183	62 57	240	69 38	328
226	29 15	046	18 27	081	27 34	121	54 56	146	27 00	183	62 05	241	69 05	326
227	29 58	045	19 26	081	28 25	122	55 29	147	26 56	184	61 12	242	68 31	324
228	30 40	045	20 26	081	29 16	122	56 02	148	26 52	184	60 19	243	67 55	322
229	31 23	045	21 25	081	30 07	122	56 33	149	26 47	185	59 25	244	67 17	320
230	32 04	044	22 24	081	30 58	123	57 03	151	26 41	186	58 31	245	66 38	319
231	32 46	044	23 23	081	31 48	123	57 31	152	26 35	186	57 37	245	65 58	317
232	33 27	043	24 23	081	32 38	123	57 59	154	26 29	187	56 42	246	65 16	315
233	34 08	043	25 22	081	33 28	124	58 25	155	26 22	187	55 47	247	64 34	314
234	34 49	042	26 21	081	34 18	124	58 49	157	26 14	188	54 52	247	63 50	312
235	35 29	042	27 20	081	35 08	125	59 12	158	26 06	188	53 56	248	63 05	311
236	36 09	041	28 20	081	35 57	125	59 34	160	25 57	189	53 01	248	62 20	310
237	36 48	041	29 19	081	36 46	125	59 54	161	25 48	189	52 05	249	61 33	309
238	37 27	040	30 18	081	37 35	126	60 12	163	25 38	190	51 09	249	60 46	308
239	38 06	040	31 17	081	38 23	126	60 29	165	25 27	190	50 13	250	59 59	307
	VEGA		◆ALTAIR		Nunki		◆ANTARES		SPICA		◆ARCTURUS		Alkaid	
240	38 44	039	32 16	081	39 11	127	60 43	167	49 16	250	59 10	306	34 52	334
241	39 22	039	33 16	081	39 59	127	60 56	168	48 20	251	58 21	305	34 25	334
242	39 59	038	34 15	081	40 46	128	61 08	170	47 23	251	57 32	304	33 58	333
243	40 35	037	35 14	081	41 33	129	61 17	172	46 26	252	56 42	303	33 31	333
244	41 11	037	36 13	080	42 20	129	61 24	174	45 29	252	55 51	302	33 03	332
245	41 47	036	37 12	080	43 06	130	61 30	176	44 32	252	55 00	302	32 35	331
246	42 22	035	38 11	080	43 52	130	61 33	178	43 35	253	54 09	301	32 06	331
247	42 56	035	39 11	080	44 38	131	61 35	179	42 38	253	53 18	300	31 37	330
248	43 30	034	40 10	080	45 23	132	61 34	181	41 41	253	52 26	300	31 07	330
249	44 03	033	41 09	080	46 07	133	61 32	183	40 43	253	51 33	299	30 36	329
250	44 35	032	42 08	080	46 51	133	61 27	185	39 46	254	50 41	298	30 06	329
251	45 07	031	43 07	080	47 34	134	61 21	187	38 48	254	49 48	298	29 35	329
252	45 38	031	44 06	080	48 17	135	61 13	189	37 50	254	48 55	297	29 03	328
253	46 08	030	45 05	079	48 59	136	61 03	191	36 53	254	48 02	297	28 31	328
254	46 37	029	46 04	079	49 41	137	60 51	192	35 55	255	47 08	297	27 59	327
	◆VEGA		ALTAIR		◆Nunki		ANTARES		◆SPICA		ARCTURUS		Alkaid	
255	47 06	028	47 02	079	50 22	137	60 37	194	34 57	255	46 14	296	27 26	327
256	47 34	027	48 01	079	51 02	138	60 21	196	33 59	255	45 20	296	26 53	326
257	48 00	026	49 00	079	51 41	139	60 04	198	33 01	255	44 26	295	26 20	326
258	48 26	025	49 59	079	52 20	140	59 45	199	32 03	256	43 32	295	25 46	326
259	48 51	024	50 58	078	52 58	141	59 24	201	31 05	256	42 37	295	25 13	325
260	49 15	023	51 56	078	53 35	143	59 02	203	30 07	256	41 43	294	24 38	325
261	49 38	022	52 55	078	54 11	144	58 38	204	29 09	256	40 48	294	24 04	325
262	50 01	021	53 54	078	54 46	145	58 13	206	28 10	256	39 53	294	23 29	324
263	50 21	020	54 52	077	55 20	146	57 47	207	27 12	256	38 58	293	22 54	324
264	50 41	019	55 51	077	55 52	147	57 19	209	26 14	257	38 03	293	22 18	324
265	51 00	018	56 49	077	56 24	149	56 49	210	25 15	257	37 08	293	21 43	323
266	51 18	017	57 48	076	56 55	150	56 19	211	24 17	257	36 12	292	21 07	323
267	51 34	015	58 46	076	57 24	151	55 47	213	23 19	257	35 17	292	20 31	323
268	51 50	014	59 44	076	57 52	153	55 14	214	22 20	257	34 21	292	19 55	323
269	52 04	013	60 42	075	58 19	154	54 41	215	21 22	257	33 26	292	19 18	322

LHA ♈	Hc	Zn	Hc	Zn	Hc	Zn	Hc	Zn	Hc	Zn	Hc	Zn	Hc	Zn
	VEGA		◆DENEB		FOMALHAUT		◆Peacock		Shaula		ANTARES		◆ARCTURUS	
270	52 16	012	34 08	033	12 32	121	24 21	159	50 25	188	54 06	216	32 30	292
271	52 28	011	34 41	033	13 24	121	24 42	160	50 15	190	53 30	217	31 34	291
272	52 38	009	35 13	032	14 15	121	25 03	160	50 05	191	52 53	218	30 38	291
273	52 47	008	35 45	032	15 06	121	25 23	161	49 53	192	52 15	219	29 42	291
274	52 55	007	36 16	031	15 57	122	25 43	161	49 40	193	51 37	220	28 46	291
275	53 01	005	36 47	031	16 48	122	26 02	162	49 25	194	50 58	221	27 50	291
276	53 06	004	37 17	030	17 39	122	26 21	162	49 10	196	50 18	222	26 54	291
277	53 10	003	37 47	029	18 30	122	26 39	163	48 53	197	49 37	223	25 58	290
278	53 12	002	38 16	029	19 21	122	26 57	163	48 35	198	48 55	224	25 02	290
279	53 13	000	38 44	028	20 11	123	27 14	164	48 17	199	48 13	225	24 06	290
280	53 13	359	39 12	027	21 02	123	27 31	164	47 57	200	47 31	226	23 09	290
281	53 11	358	39 39	027	21 52	123	27 47	165	47 36	201	46 47	227	22 13	290
282	53 08	356	40 06	026	22 42	123	28 03	165	47 14	202	46 04	227	21 17	290
283	53 03	355	40 32	025	23 32	124	28 18	166	46 51	203	45 19	228	20 20	290
284	52 58	354	40 57	024	24 22	124	28 32	166	46 26	204	44 34	229	19 24	290
	DENEB		Alpheratz		◆FOMALHAUT		Peacock		◆ANTARES		Rasalhague		◆VEGA	
285	41 21	024	12 18	061	25 11	124	28 46	167	43 49	229	66 23	298	52 50	352
286	41 45	023	13 10	061	26 01	125	28 59	167	43 03	230	65 30	297	52 42	351
287	42 08	022	14 02	061	26 50	125	29 12	168	42 17	231	64 36	296	52 32	350
288	42 30	021	14 54	060	27 39	125	29 24	169	41 31	231	63 42	295	52 21	349
289	42 51	020	15 47	060	28 28	126	29 36	169	40 44	232	62 47	294	52 09	348
290	43 12	020	16 39	060	29 17	126	29 47	170	39 56	232	61 52	293	51 55	346
291	43 31	019	17 31	060	30 05	126	29 57	170	39 09	233	60 57	293	51 40	345
292	43 50	018	18 23	060	30 53	127	30 07	171	38 21	233	60 02	292	51 24	344
293	44 08	017	19 14	060	31 41	127	30 16	172	37 32	234	59 06	291	51 07	343
294	44 25	016	20 06	060	32 29	128	30 24	172	36 44	234	58 10	291	50 49	342
295	44 41	015	20 58	060	33 16	128	30 32	173	35 55	235	57 14	290	50 29	341
296	44 56	014	21 50	059	34 03	128	30 39	173	35 05	235	56 18	290	50 09	339
297	45 11	013	22 41	059	34 50	129	30 46	174	34 16	236	55 21	289	49 47	338
298	45 24	012	23 33	059	35 37	129	30 52	175	33 26	236	54 24	289	49 24	337
299	45 36	011	24 24	059	36 23	130	30 57	175	32 36	237	53 28	289	49 01	336
	◆DENEB		Alpheratz		◆FOMALHAUT		Peacock		◆ANTARES		Rasalhague		VEGA	
300	45 48	010	25 15	059	37 09	130	31 02	176	31 46	237	52 31	288	48 36	335
301	45 58	009	26 06	058	37 54	131	31 05	177	30 56	237	51 34	288	48 11	334
302	46 07	008	26 57	058	38 40	131	31 09	177	30 05	238	50 37	287	47 44	333
303	46 16	007	27 48	058	39 24	132	31 11	178	29 15	238	49 39	287	47 17	332
304	46 23	006	28 39	058	40 09	133	31 13	179	28 24	238	48 42	287	46 48	331
305	46 29	005	29 30	057	40 53	133	31 14	179	27 32	239	47 45	287	46 19	331
306	46 34	004	30 20	057	41 36	134	31 15	180	26 41	239	46 47	286	45 50	330
307	46 38	003	31 10	057	42 19	135	31 15	180	25 50	239	45 50	286	45 19	329
308	46 41	002	32 00	057	43 02	135	31 14	181	24 58	240	44 52	286	44 48	328
309	46 43	001	32 50	056	43 44	136	31 12	182	24 06	240	43 54	286	44 15	327
310	46 44	000	33 40	056	44 25	137	31 10	182	23 15	240	42 56	285	43 43	326
311	46 44	359	34 30	056	45 06	137	31 08	183	22 23	240	41 58	285	43 09	326
312	46 43	358	35 19	055	45 46	138	31 04	184	21 30	241	41 01	285	42 35	325
313	46 40	357	36 08	055	46 26	139	31 00	184	20 38	241	40 03	285	42 00	324
314	46 37	356	36 57	054	47 05	140	30 55	185	19 46	241	39 05	285	41 25	324
	◆Alpheratz		◆FOMALHAUT		Peacock		Nunki		◆Rasalhague		VEGA		DENEB	
315	37 46	054	47 43	141	30 50	186	48 37	225	38 07	284	40 49	323	46 32	355
316	38 34	054	48 21	142	30 44	186	47 55	226	37 08	284	40 13	322	46 27	354
317	39 22	053	48 58	143	30 37	187	47 12	226	36 10	284	39 36	322	46 20	353
318	40 10	053	49 34	144	30 29	187	46 28	227	35 12	284	38 58	321	46 13	352
319	40 57	052	50 09	145	30 21	188	45 44	228	34 14	284	38 21	320	46 04	351
320	41 45	052	50 43	146	30 13	189	44 59	229	33 16	284	37 42	320	45 54	350
321	42 32	051	51 17	147	30 03	189	44 14	229	32 17	284	37 03	319	45 44	349
322	43 18	051	51 49	148	29 53	190	43 28	230	31 19	284	36 24	319	45 32	348
323	44 04	050	52 21	149	29 43	190	42 42	231	30 21	283	35 44	318	45 19	347
324	44 50	049	52 51	150	29 32	191	41 56	231	29 23	283	35 04	318	45 06	346
325	45 35	049	53 21	151	29 20	192	41 09	232	28 24	283	34 24	317	44 51	345
326	46 20	048	53 49	152	29 08	192	40 22	232	27 26	283	33 43	317	44 35	345
327	47 05	047	54 16	154	28 55	193	39 34	233	26 27	283	33 02	316	44 19	344
328	47 48	047	54 42	155	28 41	193	38 46	233	25 29	283	32 20	316	44 02	343
329	48 32	046	55 07	156	28 27	194	37 58	234	24 31	283	31 39	316	43 43	342
	◆Alpheratz		Diphda		◆FOMALHAUT		Peacock		◆ALTAIR		VEGA		DENEB	
330	49 15	045	45 06	118	55 30	158	28 12	194	57 05	283	30 57	315	43 24	341
331	49 57	044	45 58	119	55 53	159	27 57	195	56 07	283	30 14	315	43 04	340
332	50 39	044	46 51	119	56 13	160	27 41	196	55 08	283	29 32	314	42 43	339
333	51 20	043	47 43	120	56 33	162	27 25	196	54 10	282	28 49	314	42 22	338
334	52 00	042	48 35	120	56 51	163	27 08	197	53 11	282	28 05	314	41 59	338
335	52 40	041	49 26	121	57 07	165	26 50	197	52 12	282	27 22	313	41 36	337
336	53 19	040	50 17	122	57 22	166	26 32	198	51 14	282	26 38	313	41 12	336
337	53 57	039	51 08	122	57 35	168	26 14	198	50 15	281	25 54	313	40 47	335
338	54 34	038	51 59	123	57 47	170	25 55	199	49 16	281	25 10	313	40 22	335
339	55 10	037	52 49	124	57 56	171	25 36	199	48 17	281	24 26	312	39 56	334
340	55 46	036	53 38	125	58 05	173	25 16	200	47 18	281	23 42	312	39 29	333
341	56 20	034	54 27	126	58 11	174	24 55	200	46 19	281	22 57	312	39 02	332
342	56 54	033	55 16	126	58 16	176	24 34	201	45 21	281	22 12	312	38 34	332
343	57 26	032	56 04	127	58 20	178	24 13	201	44 22	280	21 27	311	38 05	331
344	57 57	031	56 51	128	58 21	179	23 51	201	43 23	280	20 42	311	37 36	330
	Alpheratz		◆Hamal		Acamar		◆ACHERNAR		FOMALHAUT		◆ALTAIR		DENEB	
345	58 27	029	39 59	061	21 20	135	22 51	158	58 21	181	42 24	280	37 06	330
346	58 56	028	40 51	060	22 03	135	23 13	159	58 19	183	41 25	280	36 36	329
347	59 23	026	41 43	060	22 45	136	23 35	159	58 15	184	40 26	280	36 05	329
348	59 49	025	42 34	059	23 26	136	23 56	159	58 10	186	39 26	280	35 33	328
349	60 14	023	43 26	059	24 08	136	24 17	160	58 02	188	38 27	280	35 01	328
350	60 37	022	44 17	059	24 49	137	24 38	160	57 54	189	37 28	280	34 29	327
351	60 58	020	45 08	058	25 30	137	24 57	161	57 43	191	36 29	280	33 56	326
352	61 18	018	45 59	058	26 10	138	25 17	161	57 31	192	35 30	279	33 23	326
353	61 36	017	46 49	057	26 51	138	25 36	162	57 17	194	34 31	279	32 49	325
354	61 53	015	47 40	056	27 30	138	25 54	162	57 02	196	33 32	279	32 14	325
355	62 07	013	48 29	056	28 10	139	26 12	163	56 45	197	32 32	279	31 40	324
356	62 20	011	49 19	055	28 49	139	26 30	163	56 27	199	31 33	279	31 05	324
357	62 31	010	50 08	055	29 28	140	26 47	164	56 07	200	30 34	279	30 29	324
358	62 40	008	50 57	054	30 07	140	27 03	164	55 46	201	29 35	279	29 53	323
359	62 47	006	51 45	053	30 45	141	27 19	165	55 23	203	28 36	279	29 17	323

LAT 1°N

LHA ♈	Hc	Zn	Hc	Zn	Hc	Zn	Hc	Zn	Hc	Zn	Hc	Zn	Hc	Zn
	Alpheratz		◆Mirfak		ALDEBARAN		◆ACHERNAR		FOMALHAUT		◆ALTAIR		DENEB	
0	61 52	004	24 47	033	20 29	073	28 33	165	55 54	205	27 27	279	27 53	323
1	61 55	002	25 20	033	21 26	073	28 48	166	55 28	206	26 28	279	27 17	322
2	61 56	000	25 53	033	22 24	073	29 02	166	55 01	207	25 28	279	26 40	322
3	61 55	358	26 25	032	23 21	072	29 16	167	54 33	209	24 29	279	26 02	321
4	61 52	356	26 57	032	24 18	072	29 29	168	54 04	210	23 30	279	25 25	321
5	61 48	355	27 28	032	25 15	072	29 42	168	53 33	211	22 31	279	24 47	321
6	61 41	353	27 59	031	26 12	072	29 54	169	53 02	212	21 32	279	24 09	320
7	61 32	351	28 30	031	27 09	072	30 05	169	52 29	213	20 32	279	23 31	320
8	61 22	349	29 01	030	28 06	072	30 16	170	51 56	214	19 33	279	22 52	320
9	61 10	347	29 31	030	29 03	072	30 27	170	51 22	215	18 34	279	22 13	319
10	60 56	346	30 00	029	30 00	071	30 36	171	50 46	217	17 35	279	21 34	319
11	60 40	344	30 29	029	30 57	071	30 45	172	50 10	217	16 35	279	20 55	319
12	60 22	342	30 58	028	31 54	071	30 54	172	49 33	218	15 36	279	20 15	319
13	60 03	341	31 26	028	32 50	071	31 01	173	48 56	219	14 37	279	19 36	318
14	59 42	339	31 54	027	33 47	071	31 09	173	48 17	220	13 38	279	18 56	318
	CAPELLA		ALDEBARAN		◆RIGEL		ACHERNAR		◆FOMALHAUT		Enif		◆Alpheratz	
15	18 26	041	34 44	071	25 58	100	31 15	174	47 38	221	40 28	282	59 20	337
16	19 05	041	35 40	070	26 57	100	31 21	175	46 58	222	39 30	282	58 56	336
17	19 45	041	36 37	070	27 57	100	31 26	175	46 18	223	38 31	282	58 31	334
18	20 24	040	37 33	070	28 56	100	31 31	176	45 37	224	37 32	282	58 04	333
19	21 02	040	38 29	070	29 55	100	31 35	177	44 55	224	36 33	282	57 36	332
20	21 41	040	39 25	069	30 54	100	31 38	177	44 13	225	35 35	281	57 07	330
21	22 19	040	40 22	069	31 53	100	31 40	178	43 30	226	34 36	281	56 37	329
22	22 58	039	41 18	069	32 52	100	31 42	178	42 47	226	33 37	281	56 05	328
23	23 35	039	42 13	068	33 51	101	31 44	179	42 04	227	32 38	281	55 33	326
24	24 13	039	43 09	068	34 50	101	31 44	180	41 20	228	31 39	281	54 59	325
25	24 50	038	44 05	068	35 49	101	31 44	180	40 35	228	30 40	281	54 24	324
26	25 27	038	45 00	067	36 48	101	31 43	181	39 50	229	29 41	281	53 49	323
27	26 04	038	45 55	067	37 46	101	31 42	182	39 05	229	28 42	281	53 12	322
28	26 41	037	46 51	067	38 45	101	31 40	182	38 19	230	27 44	281	52 35	321
29	27 17	037	47 46	066	39 44	102	31 37	183	37 33	230	26 45	281	51 57	320
	Mirfak		CAPELLA		◆ALDEBARAN		RIGEL		◆ACHERNAR		FOMALHAUT		◆Alpheratz	
30	37 59	017	27 53	036	48 40	066	40 43	102	31 34	184	36 47	231	51 18	319
31	38 16	016	28 28	036	49 35	065	41 42	102	31 30	184	36 00	231	50 38	318
32	38 32	016	29 03	036	50 29	065	42 40	102	31 25	185	35 13	232	49 58	317
33	38 48	015	29 38	035	51 24	064	43 39	102	31 20	185	34 26	232	49 17	316
34	39 03	014	30 12	035	52 18	064	44 37	103	31 14	186	33 38	233	48 35	316
35	39 17	013	30 46	034	53 11	063	45 36	103	31 07	187	32 50	233	47 53	315
36	39 30	012	31 20	034	54 05	063	46 34	103	31 00	187	32 02	234	47 10	314
37	39 43	012	31 53	033	54 58	062	47 33	103	30 52	188	31 14	234	46 26	313
38	39 55	011	32 25	033	55 51	061	48 31	104	30 43	189	30 25	234	45 43	313
39	40 06	010	32 58	032	56 43	061	49 29	104	30 34	189	29 36	235	44 58	312
40	40 16	009	33 29	032	57 35	060	50 28	104	30 24	190	28 47	235	44 13	311
41	40 25	008	34 01	031	58 27	059	51 26	105	30 14	190	27 58	235	43 28	311
42	40 33	008	34 31	031	59 18	058	52 24	105	30 03	191	27 09	236	42 43	310
43	40 41	007	35 02	030	60 09	057	53 22	105	29 51	191	26 19	236	41 57	310
44	40 48	006	35 31	029	60 59	056	54 20	106	29 39	192	25 29	236	41 10	309
	CAPELLA		◆PROCYON		SIRIUS		CANOPUS		◆ACHERNAR		Diphda		◆Alpheratz	
45	36 00	029	20 15	085	31 49	110	21 35	150	29 26	193	51 23	239	40 23	309
46	36 29	028	21 14	085	32 45	111	22 05	150	29 13	193	50 32	239	39 36	308
47	36 57	027	22 14	085	33 42	111	22 35	150	28 59	194	49 40	240	38 49	308
48	37 24	027	23 14	085	34 37	111	23 04	151	28 44	194	48 48	241	38 01	307
49	37 51	026	24 14	085	35 33	111	23 34	151	28 29	195	47 55	241	37 13	307
50	38 17	025	25 13	085	36 29	112	24 02	152	28 13	195	47 02	242	36 25	306
51	38 43	025	26 13	085	37 25	112	24 31	152	27 57	196	46 09	242	35 36	306
52	39 08	024	27 13	085	38 20	112	24 59	152	27 40	196	45 16	243	34 48	305
53	39 32	023	28 13	085	39 16	113	25 27	153	27 23	197	44 23	243	33 59	305
54	39 55	023	29 12	085	40 11	113	25 54	153	27 05	197	43 29	244	33 09	305
55	40 18	022	30 12	085	41 06	113	26 21	154	26 47	198	42 35	244	32 20	304
56	40 40	021	31 12	084	42 01	114	26 47	154	26 28	198	41 41	245	31 30	304
57	41 01	020	32 11	084	42 56	114	27 13	155	26 09	199	40 47	245	30 40	304
58	41 21	019	33 11	084	43 51	115	27 38	155	25 49	199	39 52	245	29 50	303
59	41 41	019	34 11	084	44 45	115	28 03	156	25 29	200	38 58	246	29 00	303
	◆CAPELLA		PROCYON		◆SIRIUS		CANOPUS		ACHERNAR		◆Diphda		Alpheratz	
60	42 00	018	35 11	084	45 39	115	28 28	156	25 09	200	38 03	246	28 10	303
61	42 17	017	36 10	084	46 33	116	28 52	157	24 47	201	37 08	246	27 19	303
62	42 34	016	37 10	084	47 27	116	29 15	157	24 26	201	36 13	247	26 29	302
63	42 51	015	38 10	084	48 21	117	29 39	158	24 04	202	35 18	247	25 38	302
64	43 06	014	39 09	084	49 14	117	30 01	158	23 42	202	34 23	247	24 47	302
65	43 20	013	40 09	084	50 07	118	30 23	159	23 19	203	33 27	248	23 56	302
66	43 34	013	41 09	084	51 00	119	30 44	159	22 56	203	32 32	248	23 05	301
67	43 46	012	42 08	084	51 53	119	31 05	160	22 32	203	31 36	248	22 14	301
68	43 58	011	43 08	084	52 45	120	31 25	161	22 08	204	30 40	248	21 22	301
69	44 09	010	44 08	084	53 37	121	31 45	161	21 44	204	29 45	249	20 31	301
70	44 18	009	45 07	084	54 28	121	32 04	162	21 19	205	28 49	249	19 39	301
71	44 27	008	46 07	083	55 19	122	32 23	162	20 54	205	27 53	249	18 48	300
72	44 35	007	47 06	083	56 10	123	32 41	163	20 29	205	26 57	249	17 56	300
73	44 41	006	48 06	083	57 00	124	32 58	164	20 03	206	26 01	249	17 04	300
74	44 47	005	49 06	083	57 50	125	33 14	164	19 37	206	25 05	250	16 12	300
	CAPELLA		◆POLLUX		PROCYON		SIRIUS		◆CANOPUS		Acamar		◆Hamal	
75	44 52	004	42 12	052	50 05	083	58 39	126	33 30	165	40 13	210	42 26	302
76	44 56	003	42 59	051	51 05	083	59 27	127	33 46	166	39 42	211	41 35	301
77	44 58	002	43 45	051	52 04	083	60 15	128	34 00	166	39 11	212	40 44	301
78	45 00	001	44 31	050	53 04	083	61 02	129	34 14	167	38 39	213	39 52	300
79	45 00	000	45 17	049	54 03	082	61 49	130	34 27	168	38 06	213	39 00	300
80	45 00	359	46 03	049	55 03	082	62 34	131	34 40	168	37 33	214	38 08	299
81	44 59	358	46 47	048	56 02	082	63 19	132	34 52	169	36 59	215	37 16	299
82	44 56	357	47 32	047	57 01	082	64 03	134	35 03	170	36 25	215	36 23	299
83	44 53	356	48 16	047	58 01	082	64 46	135	35 13	170	35 50	216	35 30	298
84	44 48	355	48 59	046	59 00	081	65 27	137	35 23	171	35 15	216	34 38	298
85	44 42	354	49 42	045	59 59	081	66 08	139	35 32	172	34 39	217	33 45	298
86	44 36	353	50 24	044	60 59	081	66 47	140	35 40	173	34 02	218	32 51	298
87	44 28	352	51 06	043	61 58	081	67 24	142	35 47	173	33 26	218	31 58	297
88	44 20	351	51 46	042	62 57	080	68 00	144	35 54	174	32 48	219	31 05	297
89	44 10	350	52 27	042	63 56	080	68 34	146	35 59	175	32 11	219	30 11	297

LHA ♈	Hc	Zn	Hc	Zn	Hc	Zn	Hc	Zn	Hc	Zn	Hc	Zn	Hc	Zn
	POLLUX		◆REGULUS		Suhail		◆CANOPUS		Acamar		◆ALDEBARAN		CAPELLA	
90	53 06	041	27 32	077	28 56	143	36 04	176	31 33	220	64 05	308	44 00	349
91	53 45	040	28 31	077	29 32	143	36 09	176	30 54	220	63 17	307	43 48	349
92	54 22	038	29 29	077	30 08	144	36 12	177	30 15	221	62 29	306	43 36	348
93	54 59	037	30 28	077	30 43	144	36 15	178	29 36	221	61 39	304	43 23	347
94	55 35	036	31 26	077	31 18	145	36 17	179	28 57	221	60 50	303	43 08	346
95	56 10	035	32 24	076	31 53	145	36 18	179	28 17	222	59 59	303	42 53	345
96	56 44	034	33 23	076	32 27	146	36 18	180	27 36	222	59 09	302	42 37	344
97	57 17	033	34 21	076	33 01	146	36 18	181	26 56	223	58 17	301	42 20	343
98	57 49	031	35 19	076	33 34	147	36 17	182	26 15	223	57 26	300	42 02	342
99	58 19	030	36 17	076	34 06	147	36 15	182	25 34	223	56 33	299	41 44	342
100	58 48	028	37 15	076	34 38	148	36 12	183	24 53	224	55 41	299	41 24	341
101	59 16	027	38 14	075	35 10	149	36 08	184	24 11	224	54 48	298	41 04	340
102	59 43	026	39 12	075	35 41	149	36 04	185	23 29	225	53 55	297	40 43	339
103	60 08	024	40 10	075	36 11	150	35 59	185	22 47	225	53 01	297	40 21	338
104	60 32	022	41 08	075	36 41	150	35 53	186	22 04	225	52 08	296	39 59	338
	POLLUX		◆REGULUS		Suhail		◆CANOPUS		RIGEL		◆ALDEBARAN		CAPELLA	
105	60 54	021	42 05	075	37 11	151	35 46	187	62 06	250	51 14	296	39 35	337
106	61 14	019	43 03	074	37 39	152	35 39	187	61 09	251	50 19	295	39 11	336
107	61 33	017	44 01	074	38 07	153	35 31	188	60 12	251	49 25	295	38 47	335
108	61 50	016	44 59	074	38 35	153	35 22	189	59 15	252	48 30	294	38 21	335
109	62 05	014	45 56	074	39 01	154	35 12	190	58 18	253	47 36	294	37 55	334
110	62 18	012	46 54	073	39 27	155	35 02	190	57 21	253	46 40	293	37 29	333
111	62 30	010	47 51	073	39 52	156	34 51	191	56 23	253	45 45	293	37 01	333
112	62 39	008	48 49	073	40 17	156	34 39	192	55 26	254	44 50	293	36 34	332
113	62 47	006	49 46	072	40 41	157	34 26	192	54 28	254	43 54	292	36 05	331
114	62 53	004	50 43	072	41 03	158	34 13	193	53 30	255	42 59	292	35 36	331
115	62 56	002	51 40	072	41 26	159	33 59	194	52 32	255	42 03	292	35 06	330
116	62 58	000	52 37	071	41 47	160	33 44	194	51 34	255	41 07	291	34 36	330
117	62 57	359	53 34	071	42 07	160	33 29	195	50 36	256	40 11	291	34 06	329
118	62 55	357	54 30	071	42 27	161	33 13	196	49 38	256	39 15	291	33 34	328
119	62 50	355	55 27	070	42 46	162	32 56	196	48 40	256	38 19	290	33 03	328
	Dubhe		◆Denebola		Suhail		◆CANOPUS		SIRIUS		BETELGEUSE		◆CAPELLA	
120	20 10	021	31 54	073	43 04	163	32 39	197	64 23	225	58 11	282	32 31	327
121	20 31	021	32 52	073	43 21	164	32 21	198	63 40	227	57 12	282	31 58	327
122	20 52	021	33 49	073	43 37	165	32 03	198	62 55	228	56 13	282	31 25	326
123	21 13	020	34 47	073	43 52	166	31 44	199	62 10	230	55 15	282	30 51	326
124	21 34	020	35 44	073	44 06	167	31 24	199	61 24	231	54 16	281	30 18	325
125	21 54	019	36 41	072	44 19	168	31 04	200	60 37	232	53 17	281	29 43	325
126	22 14	019	37 38	072	44 31	169	30 43	201	59 50	233	52 18	281	29 09	324
127	22 33	019	38 35	072	44 42	170	30 21	201	59 01	234	51 19	281	28 34	324
128	22 52	018	39 32	072	44 52	171	29 59	202	58 13	235	50 20	280	27 58	324
129	23 11	018	40 29	072	45 01	172	29 37	202	57 23	236	49 21	280	27 22	323
130	23 29	018	41 26	071	45 09	173	29 14	203	56 33	237	48 22	280	26 46	323
131	23 47	017	42 23	071	45 16	174	28 50	203	55 43	238	47 23	280	26 10	322
132	24 05	017	43 20	071	45 22	175	28 26	204	54 52	238	46 24	280	25 33	322
133	24 22	016	44 16	070	45 27	176	28 01	204	54 01	239	45 25	280	24 56	322
134	24 39	016	45 13	070	45 31	177	27 36	205	53 09	240	44 26	279	24 19	321
	Dubhe		◆SPICA		ACRUX		◆CANOPUS		SIRIUS		BETELGEUSE		◆CAPELLA	
135	24 55	016	23 05	103	15 25	158	27 11	205	52 17	240	43 26	279	23 41	321
136	25 11	015	24 03	103	15 47	159	26 45	206	51 25	241	42 27	279	23 04	321
137	25 26	015	25 02	103	16 09	159	26 18	206	50 32	242	41 28	279	22 25	320
138	25 41	014	26 00	103	16 30	159	25 52	207	49 39	242	40 29	279	21 47	320
139	25 56	014	26 59	103	16 51	160	25 24	207	48 46	243	39 29	279	21 09	320
140	26 10	013	27 57	103	17 12	160	24 57	208	47 52	243	38 30	279	20 30	320
141	26 23	013	28 56	103	17 32	160	24 29	208	46 58	244	37 31	279	19 51	319
142	26 36	012	29 54	103	17 53	160	24 00	209	46 05	244	36 31	278	19 12	319
143	26 49	012	30 52	104	18 12	161	23 31	209	45 10	245	35 32	278	18 32	319
144	27 01	011	31 51	104	18 32	161	23 02	209	44 16	245	34 33	278	17 53	319
145	27 12	011	32 49	104	18 51	161	22 33	210	43 21	246	33 33	278	17 13	318
146	27 24	010	33 47	104	19 10	162	22 03	210	42 27	246	32 34	278	16 33	318
147	27 34	010	34 45	104	19 29	162	21 32	210	41 32	246	31 35	278	15 53	318
148	27 44	009	35 43	105	19 47	163	21 02	211	40 37	247	30 35	278	15 13	318
149	27 54	009	36 41	105	20 05	163	20 31	211	39 41	247	29 36	278	14 32	318
	◆Dubhe		ARCTURUS		◆SPICA		ACRUX		◆SIRIUS		PROCYON		POLLUX	
150	28 03	008	24 57	069	37 39	105	20 22	163	38 46	247	54 34	278	47 54	313
151	28 11	008	25 53	069	38 37	105	20 39	164	37 51	248	53 35	277	47 10	312
152	28 19	007	26 49	069	39 35	105	20 56	164	36 55	248	52 35	277	46 25	312
153	28 27	007	27 45	069	40 33	106	21 12	164	35 59	248	51 36	277	45 40	311
154	28 34	006	28 41	069	41 31	106	21 28	165	35 03	249	50 36	277	44 55	310
155	28 40	006	29 36	068	42 28	106	21 44	165	34 07	249	49 37	277	44 09	310
156	28 46	005	30 32	068	43 26	106	21 59	166	33 11	249	48 37	277	43 22	309
157	28 51	005	31 28	068	44 23	107	22 13	166	32 15	249	47 37	277	42 36	309
158	28 56	004	32 23	068	45 21	107	22 28	166	31 19	250	46 38	277	41 48	308
159	29 00	004	33 19	068	46 18	107	22 42	167	30 23	250	45 38	276	41 01	307
160	29 04	003	34 14	067	47 15	108	22 55	167	29 26	250	44 39	276	40 13	307
161	29 07	003	35 09	067	48 13	108	23 08	168	28 30	250	43 39	276	39 25	306
162	29 09	002	36 05	067	49 09	108	23 21	168	27 33	251	42 39	276	38 37	306
163	29 11	002	37 00	066	50 06	109	23 33	169	26 37	251	41 40	276	37 48	306
164	29 12	001	37 55	066	51 03	109	23 44	169	25 40	251	40 40	276	36 59	305
	Dubhe		◆ARCTURUS		SPICA		◆ACRUX		Suhail		PROCYON		◆POLLUX	
165	29 13	000	38 50	066	52 00	110	23 55	170	38 59	206	39 40	276	36 10	305
166	29 13	000	39 44	066	52 56	110	24 06	170	38 32	207	38 41	276	35 20	304
167	29 13	359	40 39	065	53 52	111	24 16	170	38 05	208	37 41	276	34 31	304
168	29 12	359	41 33	065	54 48	111	24 26	171	37 37	208	36 41	276	33 41	304
169	29 11	358	42 27	065	55 44	112	24 35	171	37 08	209	35 42	276	32 51	303
170	29 08	358	43 21	064	56 40	112	24 44	172	36 39	210	34 42	276	32 01	303
171	29 06	357	44 15	064	57 35	113	24 52	172	36 09	210	33 42	276	31 10	303
172	29 03	357	45 09	063	58 31	113	25 00	173	35 38	211	32 43	276	30 20	302
173	28 59	356	46 03	063	59 25	114	25 07	173	35 07	212	31 43	276	29 29	302
174	28 55	356	46 56	062	60 20	115	25 14	174	34 35	212	30 43	275	28 38	302
175	28 50	355	47 49	062	61 14	116	25 21	174	34 03	213	29 43	275	27 47	301
176	28 44	355	48 42	061	62 08	117	25 26	175	33 31	213	28 44	275	26 55	301
177	28 38	354	49 34	061	63 02	117	25 32	175	32 57	214	27 44	275	26 04	301
178	28 32	353	50 26	060	63 55	118	25 36	176	32 24	214	26 44	275	25 13	301
179	28 25	353	51 18	060	64 47	119	25 41	176	31 50	215	25 45	275	24 21	301

LAT 1°N

LHA ♈	Hc	Zn	Hc	Zn	Hc	Zn	Hc	Zn	Hc	Zn	Hc	Zn	Hc	Zn
	◆ARCTURUS		ANTARES		RIGIL KENT.		◆ACRUX		Suhail		◆REGULUS		Dubhe	
180	52 10	059	19 46	119	21 03	160	25 44	177	31 15	215	60 09	293	28 17	352
181	53 01	058	20 38	119	21 23	161	25 48	177	30 40	216	59 14	292	28 09	352
182	53 52	058	21 31	119	21 43	161	25 50	178	30 05	216	58 18	292	28 00	351
183	54 43	057	22 23	119	22 02	162	25 52	178	29 29	217	57 22	291	27 51	351
184	55 33	056	23 16	119	22 21	162	25 54	179	28 53	217	56 26	290	27 41	350
185	56 22	055	24 08	120	22 39	162	25 55	179	28 16	218	55 30	290	27 31	350
186	57 12	055	25 00	120	22 57	163	25 56	180	27 39	218	54 33	290	27 20	349
187	58 00	054	25 52	120	23 14	163	25 56	180	27 02	219	53 37	289	27 09	349
188	58 48	053	26 44	120	23 31	164	25 55	181	26 24	219	52 40	289	26 57	348
189	59 36	052	27 35	121	23 48	164	25 54	181	25 46	219	51 43	288	26 45	348
190	60 22	051	28 27	121	24 04	165	25 53	182	25 08	220	50 46	288	26 32	348
191	61 08	049	29 18	121	24 20	165	25 51	182	24 29	220	49 49	288	26 19	347
192	61 53	048	30 09	122	24 35	166	25 48	183	23 50	221	48 52	287	26 06	347
193	62 38	047	31 00	122	24 50	166	25 45	183	23 11	221	47 54	287	25 51	346
194	63 21	046	31 51	122	25 04	166	25 41	184	22 32	221	46 57	287	25 37	346
	◆Alphecca		Rasalhague		ANTARES		◆RIGIL KENT.		ACRUX		◆REGULUS		Dubhe	
195	44 52	052	21 01	077	32 41	123	25 18	167	25 37	184	45 59	286	25 22	345
196	45 39	051	22 00	077	33 32	123	25 31	167	25 32	185	45 02	286	25 06	345
197	46 26	051	22 58	077	34 22	123	25 44	168	25 27	185	44 04	286	24 50	344
198	47 12	050	23 56	077	35 12	124	25 56	168	25 22	186	43 06	286	24 34	344
199	47 58	049	24 55	077	36 02	124	26 08	169	25 15	186	42 09	285	24 17	344
200	48 43	049	25 53	077	36 51	125	26 19	169	25 09	187	41 11	285	24 00	343
201	49 28	048	26 51	076	37 40	125	26 30	170	25 01	187	40 13	285	23 42	343
202	50 12	047	27 50	076	38 29	126	26 41	170	24 54	188	39 15	285	23 24	342
203	50 56	046	28 48	076	39 18	126	26 50	171	24 45	188	38 17	285	23 06	342
204	51 39	045	29 46	076	40 06	127	27 00	171	24 37	189	37 19	284	22 47	342
205	52 21	044	30 44	076	40 54	127	27 08	172	24 28	189	36 20	284	22 28	341
206	53 03	044	31 43	076	41 42	128	27 16	172	24 18	190	35 22	284	22 08	341
207	53 44	043	32 41	076	42 29	128	27 24	173	24 08	190	34 24	284	21 48	340
208	54 24	042	33 39	076	43 16	129	27 31	174	23 57	190	33 26	284	21 28	340
209	55 03	041	34 37	075	44 02	129	27 37	174	23 46	191	32 27	284	21 07	340
	◆VEGA		Rasalhague		ANTARES		◆RIGIL KENT.		ACRUX		◆REGULUS		Dubhe	
210	16 44	050	35 35	075	44 49	130	27 43	175	23 34	191	31 29	283	20 46	339
211	17 29	049	36 33	075	45 34	131	27 49	175	23 22	192	30 31	283	20 25	339
212	18 15	049	37 31	075	46 19	132	27 53	176	23 10	192	29 32	283	20 03	339
213	19 00	049	38 29	075	47 04	132	27 58	176	22 57	193	28 34	283	19 42	338
214	19 45	049	39 27	074	47 48	133	28 01	177	22 44	193	27 36	283	19 19	338
215	20 30	049	40 24	074	48 32	134	28 04	177	22 30	193	26 37	283	18 57	338
216	21 15	048	41 22	074	49 15	135	28 07	178	22 16	194	25 39	283	18 34	337
217	22 00	048	42 20	074	49 57	135	28 09	178	22 01	194	24 40	283	18 11	337
218	22 44	048	43 17	074	50 39	136	28 10	179	21 46	195	23 42	283	17 47	337
219	23 29	048	44 15	073	51 20	137	28 11	180	21 31	195	22 43	283	17 24	337
220	24 13	047	45 12	073	52 00	138	28 11	180	21 15	196	21 44	283	17 00	336
221	24 57	047	46 10	073	52 40	139	28 11	181	20 59	196	20 46	282	16 36	336
222	25 40	047	47 07	072	53 19	140	28 10	181	20 42	196	19 47	282	16 11	336
223	26 24	046	48 04	072	53 57	141	28 08	182	20 25	197	18 49	282	15 47	336
224	27 07	046	49 01	072	54 34	142	28 06	182	20 07	197	17 50	282	15 22	335
	VEGA		◆ALTAIR		Nunki		ANTARES		◆RIGIL KENT.		SPICA		◆ARCTURUS	
225	27 50	046	17 19	081	27 13	120	55 10	144	28 03	183	63 26	242	68 46	330
226	28 33	045	18 18	081	28 05	121	55 45	145	28 00	183	62 33	243	68 15	328
227	29 15	045	19 17	081	28 56	121	56 19	146	27 56	184	61 39	244	67 42	326
228	29 58	045	20 16	081	29 48	121	56 52	147	27 52	185	60 45	245	67 07	324
229	30 40	044	21 16	081	30 39	122	57 24	149	27 47	185	59 51	246	66 31	322
230	31 21	044	22 15	081	31 30	122	57 55	150	27 41	186	58 56	246	65 53	320
231	32 03	043	23 14	081	32 20	122	58 24	151	27 35	186	58 01	247	65 14	318
232	32 43	043	24 13	081	33 11	123	58 52	153	27 28	187	57 06	248	64 33	317
233	33 24	042	25 12	081	34 01	123	59 19	154	27 21	187	56 10	248	63 52	315
234	34 04	042	26 12	081	34 51	124	59 44	156	27 13	188	55 14	249	63 09	314
235	34 44	041	27 11	081	35 41	124	60 08	158	27 05	188	54 18	249	62 25	313
236	35 24	041	28 10	080	36 31	124	60 30	159	26 56	189	53 22	250	61 41	311
237	36 03	040	29 09	080	37 20	125	60 51	161	26 47	189	52 26	250	60 55	310
238	36 41	040	30 08	080	38 10	125	61 09	163	26 37	190	51 29	251	60 09	309
239	37 20	039	31 07	080	38 58	126	61 26	164	26 26	190	50 33	251	59 22	308
	VEGA		◆ALTAIR		Nunki		◆ANTARES		SPICA		◆ARCTURUS		Alkaid	
240	37 57	039	32 07	080	39 47	126	61 42	166	49 36	251	58 35	307	33 58	335
241	38 34	038	33 06	080	40 35	127	61 55	168	48 39	252	57 46	306	33 32	334
242	39 11	037	34 05	080	41 23	127	62 07	170	47 42	252	56 58	305	33 05	333
243	39 47	037	35 04	080	42 11	128	62 16	172	46 45	253	56 08	304	32 38	333
244	40 23	036	36 03	080	42 58	128	62 24	174	45 48	253	55 19	304	32 10	332
245	40 58	035	37 02	080	43 44	129	62 29	176	44 50	253	54 28	303	31 42	332
246	41 33	035	38 01	080	44 31	130	62 33	178	43 53	253	53 38	302	31 13	331
247	42 07	034	39 00	079	45 17	130	62 35	179	42 55	254	52 47	301	30 44	331
248	42 40	033	39 59	079	46 02	131	62 34	181	41 57	254	51 55	301	30 15	330
249	43 12	033	40 58	079	46 47	132	62 32	183	41 00	254	51 04	300	29 45	330
250	43 44	032	41 57	079	47 32	132	62 27	185	40 02	255	50 12	300	29 14	329
251	44 16	031	42 56	079	48 16	133	62 21	187	39 04	255	49 19	299	28 43	329
252	44 46	030	43 54	079	48 59	134	62 12	189	38 06	255	48 27	298	28 12	328
253	45 16	029	44 53	078	49 42	135	62 02	191	37 08	255	47 34	298	27 41	328
254	45 45	028	45 52	078	50 24	136	61 49	193	36 10	255	46 41	297	27 08	328
	◆VEGA		ALTAIR		◆Nunki		ANTARES		◆SPICA		ARCTURUS		Alkaid	
255	46 13	027	46 51	078	51 06	137	61 35	195	35 12	256	45 47	297	26 36	327
256	46 40	027	47 49	078	51 46	138	61 19	196	34 14	256	44 54	297	26 03	327
257	47 06	026	48 48	078	52 27	139	61 01	198	33 16	256	44 00	296	25 30	326
258	47 32	025	49 47	077	53 06	140	60 42	200	32 18	256	43 06	296	24 57	326
259	47 57	024	50 45	077	53 44	141	60 20	202	31 19	256	42 12	295	24 23	326
260	48 20	023	51 44	077	54 22	142	59 57	203	30 21	256	41 18	295	23 49	325
261	48 43	022	52 42	077	54 59	143	59 33	205	29 23	257	40 23	295	23 15	325
262	49 04	021	53 40	076	55 34	144	59 07	206	28 24	257	39 29	294	22 40	325
263	49 25	020	54 38	076	56 09	145	58 40	208	27 26	257	38 34	294	22 05	324
264	49 45	018	55 37	076	56 43	147	58 11	209	26 27	257	37 39	294	21 30	324
265	50 03	017	56 35	075	57 15	148	57 41	211	25 29	257	36 44	293	20 55	324
266	50 20	016	57 33	075	57 47	149	57 10	212	24 31	257	35 49	293	20 19	323
267	50 36	015	58 31	075	58 17	151	56 38	213	23 32	257	34 54	293	19 43	323
268	50 51	014	59 28	074	58 45	152	56 04	215	22 33	258	33 58	293	19 07	323
269	51 05	013	60 26	074	59 13	154	55 30	216	21 35	258	33 03	292	18 31	323

LHA ♈	Hc	Zn	Hc	Zn	Hc	Zn	Hc	Zn	Hc	Zn	Hc	Zn	Hc	Zn
	VEGA		◆DENEB		FOMALHAUT		◆Peacock		Shaula		ANTARES		◆ARCTURUS	
270	51 18	011	33 18	033	13 03	121	25 17	159	51 24	189	54 54	217	32 08	292
271	51 29	010	33 50	032	13 55	121	25 38	159	51 14	190	54 17	218	31 12	292
272	51 39	009	34 22	032	14 46	121	25 59	160	51 04	191	53 40	219	30 16	292
273	51 48	008	34 54	031	15 37	121	26 20	160	50 51	192	53 02	220	29 21	292
274	51 55	007	35 25	031	16 29	121	26 40	161	50 38	193	52 22	221	28 25	291
275	52 02	005	35 55	030	17 20	122	26 59	161	50 23	195	51 42	222	27 29	291
276	52 06	004	36 25	030	18 11	122	27 18	162	50 08	196	51 02	223	26 33	291
277	52 10	003	36 54	029	19 02	122	27 36	162	49 51	197	50 20	224	25 37	291
278	52 12	002	37 23	028	19 53	122	27 54	163	49 33	198	49 38	225	24 41	291
279	52 13	000	37 51	028	20 43	122	28 12	163	49 13	199	48 56	226	23 45	291
280	52 13	359	38 19	027	21 34	123	28 29	164	48 53	200	48 12	227	22 48	290
281	52 11	358	38 45	026	22 24	123	28 45	164	48 32	201	47 28	227	21 52	290
282	52 08	356	39 12	026	23 15	123	29 01	165	48 09	203	46 44	228	20 56	290
283	52 04	355	39 37	025	24 05	123	29 16	166	47 46	204	45 59	229	20 00	290
284	51 58	354	40 02	024	24 55	124	29 30	166	47 21	205	45 14	229	19 03	290
	DENEB		Alpheratz		◆FOMALHAUT		Peacock		◆ANTARES		Rasalhague		◆VEGA	
285	40 26	023	11 48	060	25 45	124	29 45	167	44 28	230	65 54	300	51 51	353
286	40 49	023	12 41	060	26 35	124	29 58	167	43 42	231	65 02	299	51 43	351
287	41 12	022	13 33	060	27 24	124	30 11	168	42 55	231	64 09	298	51 33	350
288	41 34	021	14 25	060	28 14	125	30 23	168	42 08	232	63 16	297	51 22	349
289	41 55	020	15 17	060	29 03	125	30 35	169	41 20	233	62 22	296	51 10	348
290	42 15	019	16 09	060	29 52	125	30 46	170	40 33	233	61 28	295	50 57	347
291	42 34	018	17 01	060	30 40	126	30 56	170	39 44	234	60 33	294	50 42	345
292	42 53	018	17 52	060	31 29	126	31 06	171	38 56	234	59 38	294	50 27	344
293	43 11	017	18 44	060	32 17	127	31 15	172	38 07	235	58 43	293	50 10	343
294	43 27	016	19 36	059	33 05	127	31 24	172	37 18	235	57 48	292	49 52	342
295	43 43	015	20 27	059	33 53	127	31 32	173	36 29	236	56 52	292	49 33	341
296	43 58	014	21 19	059	34 40	128	31 39	173	35 39	236	55 57	291	49 12	340
297	44 12	013	22 10	059	35 28	128	31 46	174	34 50	236	55 01	291	48 51	339
298	44 25	012	23 02	059	36 15	129	31 52	175	33 59	237	54 04	290	48 29	338
299	44 38	011	23 53	058	37 01	129	31 57	175	33 09	237	53 08	290	48 06	337
	◆DENEB		Alpheratz		◆FOMALHAUT		Peacock		◆ANTARES		Rasalhague		VEGA	
300	44 49	010	24 44	058	37 48	130	32 01	176	32 19	237	52 11	289	47 42	336
301	44 59	009	25 35	058	38 33	130	32 05	177	31 28	238	51 15	289	47 16	335
302	45 08	008	26 26	058	39 19	131	32 09	177	30 37	238	50 18	289	46 50	334
303	45 16	007	27 16	057	40 04	131	32 11	178	29 46	238	49 21	288	46 23	333
304	45 23	006	28 07	057	40 49	132	32 13	179	28 55	239	48 24	288	45 56	332
305	45 29	005	28 57	057	41 33	133	32 14	179	28 03	239	47 27	288	45 27	331
306	45 34	004	29 47	057	42 17	133	32 15	180	27 12	239	46 30	287	44 58	330
307	45 38	003	30 37	056	43 01	134	32 15	180	26 20	240	45 32	287	44 27	329
308	45 41	002	31 27	056	43 44	135	32 14	181	25 28	240	44 35	287	43 56	329
309	45 43	001	32 17	056	44 26	135	32 12	182	24 36	240	43 38	286	43 25	328
310	45 44	000	33 06	055	45 08	136	32 10	182	23 44	240	42 40	286	42 52	327
311	45 44	359	33 55	055	45 50	137	32 07	183	22 52	241	41 42	286	42 19	326
312	45 43	358	34 44	055	46 31	138	32 04	184	22 00	241	40 45	286	41 46	326
313	45 40	357	35 33	054	47 11	138	32 00	184	21 07	241	39 47	286	41 12	325
314	45 37	356	36 22	054	47 50	139	31 55	185	20 15	241	38 49	285	40 37	324
	◆Alpheratz		◆FOMALHAUT		Peacock		Nunki		◆Rasalhague		VEGA		DENEB	
315	37 10	053	48 29	140	31 49	186	49 20	226	37 51	285	40 01	323	45 33	355
316	37 58	053	49 07	141	31 43	186	48 37	226	36 53	285	39 25	323	45 27	354
317	38 46	052	49 45	142	31 36	187	47 53	227	35 55	285	38 49	322	45 21	353
318	39 33	052	50 22	143	31 29	187	47 09	228	34 57	285	38 12	322	45 13	352
319	40 20	051	50 57	144	31 21	188	46 24	229	33 59	285	37 34	321	45 05	351
320	41 07	051	51 32	145	31 12	189	45 39	229	33 01	284	36 56	320	44 55	350
321	41 54	050	52 07	146	31 03	189	44 53	230	32 03	284	36 18	320	44 45	349
322	42 40	050	52 40	147	30 53	190	44 07	231	31 05	284	35 39	319	44 33	348
323	43 25	049	53 12	148	30 42	191	43 20	231	30 07	284	34 59	319	44 21	348
324	44 11	049	53 43	149	30 31	191	42 33	232	29 08	284	34 20	318	44 07	347
325	44 55	048	54 13	151	30 19	192	41 46	232	28 10	284	33 40	318	43 53	346
326	45 40	047	54 42	152	30 06	192	40 58	233	27 12	284	32 59	317	43 38	345
327	46 24	047	55 10	153	29 53	193	40 10	234	26 14	284	32 18	317	43 21	344
328	47 07	046	55 36	154	29 39	193	39 22	234	25 15	283	31 37	316	43 04	343
329	47 50	045	56 02	156	29 25	194	38 33	235	24 17	283	30 56	316	42 46	342
	◆Alpheratz		Diphda		◆FOMALHAUT		Peacock		◆ALTAIR		VEGA		DENEB	
330	48 32	044	45 34	117	56 26	157	29 10	195	56 50	285	30 14	316	42 27	341
331	49 14	044	46 27	118	56 49	159	28 55	195	55 52	284	29 32	315	42 08	340
332	49 55	043	47 20	118	57 10	160	28 39	196	54 54	284	28 49	315	41 47	340
333	50 35	042	48 12	119	57 30	161	28 22	196	53 56	284	28 07	314	41 26	339
334	51 15	041	49 05	119	57 48	163	28 05	197	52 58	283	27 24	314	41 04	338
335	51 54	040	49 57	120	58 05	165	27 48	197	51 59	283	26 41	314	40 41	337
336	52 32	039	50 49	121	58 20	166	27 30	198	51 01	283	25 57	313	40 17	336
337	53 10	038	51 40	121	58 34	168	27 11	198	50 02	283	25 14	313	39 53	336
338	53 46	037	52 31	122	58 45	169	26 52	199	49 04	282	24 30	313	39 28	335
339	54 22	036	53 22	123	58 56	171	26 32	199	48 05	282	23 46	313	39 02	334
340	54 57	035	54 12	124	59 04	173	26 12	200	47 06	282	23 01	312	38 36	334
341	55 31	034	55 02	124	59 11	174	25 52	200	46 08	282	22 17	312	38 09	333
342	56 03	032	55 51	125	59 16	176	25 31	201	45 09	282	21 32	312	37 41	332
343	56 35	031	56 40	126	59 19	178	25 09	201	44 10	281	20 47	312	37 13	331
344	57 05	030	57 28	127	59 21	179	24 47	202	43 11	281	20 03	311	36 44	331
	Alpheratz		◆Hamal		Acamar		◆ACHERNAR		FOMALHAUT		◆ALTAIR		DENEB	
345	57 35	029	39 29	060	22 03	135	23 47	158	59 21	181	42 13	281	36 14	330
346	58 03	027	40 21	060	22 45	135	24 09	158	59 19	183	41 14	281	35 44	330
347	58 29	026	41 12	059	23 27	135	24 31	159	59 15	185	40 15	281	35 13	329
348	58 55	024	42 04	059	24 09	136	24 52	159	59 09	186	39 16	281	34 42	328
349	59 18	023	42 55	058	24 51	136	25 13	160	59 02	188	38 17	281	34 11	328
350	59 41	021	43 46	058	25 32	137	25 34	160	58 53	190	37 18	280	33 38	327
351	60 02	020	44 36	057	26 14	137	25 54	161	58 42	191	36 19	280	33 06	327
352	60 21	018	45 26	057	26 54	137	26 14	161	58 29	193	35 20	280	32 33	326
353	60 39	016	46 16	056	27 35	138	26 33	162	58 15	194	34 21	280	31 59	326
354	60 55	015	47 06	056	28 15	138	26 51	162	58 00	196	33 22	280	31 25	325
355	61 09	013	47 55	055	28 55	139	27 10	163	57 42	198	32 23	280	30 51	325
356	61 21	011	48 44	054	29 35	139	27 27	163	57 24	199	31 23	280	30 16	324
357	61 32	009	49 33	054	30 14	139	27 44	164	57 03	200	30 24	280	29 41	324
358	61 40	007	50 21	053	30 53	140	28 01	164	56 42	202	29 25	280	29 05	323
359	61 47	006	51 09	052	31 31	140	28 17	165	56 18	203	28 26	280	28 29	323

LHA ♈	Hc Zn	Hc Zn	Hc Zn	Hc Zn	Hc Zn	Hc Zn	Hc Zn
	Alpheratz	◆Mirfak	ALDEBARAN	◆ACHERNAR	FOMALHAUT	◆ALTAIR	DENEB
0	60 52 004	23 57 033	20 11 072	29 31 165	56 48 205	27 17 280	27 05 323
1	60 55 002	24 30 033	21 08 072	29 46 166	56 22 207	26 18 280	26 29 323
2	60 56 000	25 02 032	22 05 072	30 00 166	55 54 208	25 18 280	25 52 322
3	60 55 358	25 34 032	23 02 072	30 14 167	55 26 209	24 19 280	25 15 322
4	60 53 356	26 06 032	24 00 072	30 28 167	54 56 211	23 20 280	24 38 321
5	60 48 355	26 37 031	24 57 072	30 40 168	54 25 212	22 21 280	24 01 321
6	60 41 353	27 08 031	25 53 072	30 53 169	53 52 213	21 22 280	23 23 321
7	60 33 351	27 39 030	26 50 071	31 04 169	53 19 214	20 23 279	22 45 320
8	60 23 349	28 09 030	27 47 071	31 15 170	52 45 215	19 23 279	22 06 320
9	60 11 348	28 38 029	28 44 071	31 26 170	52 10 216	18 24 279	21 28 320
10	59 57 346	29 08 029	29 41 071	31 35 171	51 34 217	17 25 279	20 49 319
11	59 42 344	29 37 028	30 37 071	31 45 172	50 58 218	16 26 279	20 10 319
12	59 25 343	30 05 028	31 34 071	31 53 172	50 20 219	15 27 279	19 30 319
13	59 06 341	30 33 027	32 31 070	32 01 173	49 42 220	14 27 279	18 51 319
14	58 46 340	31 00 027	33 27 070	32 08 173	49 03 221	13 28 279	18 11 318
	CAPELLA	ALDEBARAN	◆RIGEL	ACHERNAR	◆FOMALHAUT	Enif	◆Alpheratz
15	17 41 041	34 23 070	26 08 099	32 15 174	48 23 222	40 15 283	58 24 338
16	18 20 041	35 20 070	27 07 099	32 21 175	47 43 223	39 17 283	58 01 337
17	18 59 040	36 16 069	28 07 099	32 26 175	47 02 223	38 18 283	57 37 335
18	19 38 040	37 12 069	29 06 099	32 31 176	46 20 224	37 20 282	57 11 334
19	20 17 040	38 08 069	30 05 099	32 35 177	45 38 225	36 21 282	56 43 332
20	20 55 040	39 04 069	31 04 100	32 38 177	44 55 226	35 22 282	56 15 331
21	21 33 039	40 00 068	32 03 100	32 40 178	44 12 226	34 24 282	55 45 330
22	22 11 039	40 55 068	33 02 100	32 42 178	43 29 227	33 25 282	55 14 328
23	22 49 039	41 51 068	34 01 100	32 44 179	42 44 228	32 26 282	54 42 327
24	23 26 038	42 46 067	35 01 100	32 44 180	42 00 228	31 28 282	54 09 326
25	24 03 038	43 42 067	36 00 100	32 44 180	41 15 229	30 29 281	53 35 325
26	24 40 038	44 37 066	36 59 100	32 43 181	40 29 229	29 30 281	53 00 324
27	25 16 037	45 32 066	37 58 100	32 42 182	39 44 230	28 31 281	52 25 323
28	25 53 037	46 26 066	38 57 101	32 40 182	38 58 230	27 32 281	51 48 322
29	26 29 037	47 21 065	39 56 101	32 37 183	38 11 231	26 33 281	51 10 321
	Mirfak	CAPELLA	◆ALDEBARAN	RIGEL	◆ACHERNAR	FOMALHAUT	◆Alpheratz
30	37 01 017	27 04 036	48 15 065	40 55 101	32 34 184	37 24 231	50 32 320
31	37 18 016	27 39 036	49 09 064	41 53 101	32 30 184	36 37 232	49 53 319
32	37 34 015	28 14 035	50 03 064	42 52 101	32 25 185	35 50 232	49 13 318
33	37 50 015	28 49 035	50 57 063	43 51 101	32 19 186	35 02 233	48 33 317
34	38 05 014	29 23 034	51 50 063	44 50 102	32 13 186	34 14 233	47 52 316
35	38 19 013	29 56 034	52 44 062	45 49 102	32 07 187	33 26 234	47 10 316
36	38 32 012	30 30 033	53 36 061	46 47 102	31 59 187	32 38 234	46 28 315
37	38 44 012	31 02 033	54 29 061	47 46 102	31 51 188	31 49 234	45 45 314
38	38 56 011	31 35 032	55 21 060	48 45 103	31 42 189	31 00 235	45 02 313
39	39 07 010	32 07 032	56 13 059	49 43 103	31 33 189	30 11 235	44 18 313
40	39 17 009	32 38 031	57 04 059	50 42 103	31 23 190	29 22 235	43 33 312
41	39 26 008	33 09 031	57 55 058	51 40 103	31 13 190	28 32 236	42 49 311
42	39 34 008	33 40 030	58 46 057	52 39 104	31 02 191	27 42 236	42 04 311
43	39 41 007	34 10 030	59 36 056	53 37 104	30 50 192	26 53 236	41 18 310
44	39 48 006	34 39 029	60 25 055	54 35 104	30 37 192	26 03 237	40 32 310
	CAPELLA	◆PROCYON	SIRIUS	CANOPUS	◆ACHERNAR	Diphda	◆Hamal
45	35 08 028	20 09 084	32 10 110	22 27 149	30 25 193	51 54 240	63 15 332
46	35 36 028	21 09 084	33 06 110	22 57 150	30 11 193	51 02 241	62 46 330
47	36 04 027	22 09 084	34 03 110	23 27 150	29 57 194	50 09 241	62 16 329
48	36 31 027	23 08 084	34 59 111	23 57 150	29 42 194	49 16 242	61 44 327
49	36 57 026	24 08 084	35 55 111	24 26 151	29 27 195	48 23 242	61 10 326
50	37 23 025	25 08 084	36 51 111	24 55 151	29 11 196	47 30 243	60 36 324
51	37 48 024	26 07 084	37 47 111	25 24 152	28 55 196	46 37 243	60 00 323
52	38 13 024	27 07 084	38 43 112	25 52 152	28 38 197	45 43 244	59 23 321
53	38 37 023	28 07 084	39 39 112	26 20 153	28 21 197	44 49 244	58 45 320
54	39 00 022	29 06 084	40 34 112	26 47 153	28 03 198	43 55 245	58 06 319
55	39 22 022	30 06 084	41 30 113	27 14 153	27 44 198	43 01 245	57 26 318
56	39 44 021	31 06 084	42 25 113	27 41 154	27 25 199	42 06 245	56 45 317
57	40 05 020	32 05 084	43 20 113	28 07 154	27 06 199	41 12 246	56 04 315
58	40 25 019	33 05 084	44 15 114	28 33 155	26 46 200	40 17 246	55 21 314
59	40 44 018	34 05 084	45 10 114	28 58 155	26 26 200	39 22 246	54 38 313
	CAPELLA	◆PROCYON	SIRIUS	CANOPUS	◆ACHERNAR	Diphda	◆Hamal
60	41 02 018	35 04 084	46 05 114	29 23 156	26 05 201	38 27 247	53 54 312
61	41 20 017	36 04 084	46 59 115	29 47 156	25 44 201	37 32 247	53 09 312
62	41 37 016	37 03 083	47 54 115	30 11 157	25 22 201	36 36 247	52 24 311
63	41 53 015	38 03 083	48 48 116	30 34 157	25 00 202	35 41 248	51 38 310
64	42 08 014	39 03 083	49 42 116	30 57 158	24 37 202	34 46 248	50 52 309
65	42 22 013	40 02 083	50 35 117	31 19 159	24 14 203	33 50 248	50 05 308
66	42 35 012	41 02 083	51 28 117	31 40 159	23 51 203	32 54 248	49 18 308
67	42 48 011	42 01 083	52 22 118	32 02 160	23 27 204	31 58 249	48 30 307
68	42 59 011	43 01 083	53 14 119	32 22 160	23 03 204	31 02 249	47 42 306
69	43 10 010	44 00 083	54 07 119	32 42 161	22 38 204	30 06 249	46 53 306
70	43 19 009	45 00 083	54 59 120	33 01 162	22 14 205	29 10 249	46 04 305
71	43 28 008	45 59 082	55 51 121	33 20 162	21 48 205	28 14 249	45 15 304
72	43 35 007	46 59 082	56 42 122	33 38 163	21 23 205	27 18 250	44 25 304
73	43 42 006	47 58 082	57 33 122	33 55 163	20 57 206	26 22 250	43 35 303
74	43 47 005	48 58 082	58 23 123	34 12 164	20 31 206	25 25 250	42 45 303
	CAPELLA	◆POLLUX	PROCYON	SIRIUS	◆CANOPUS	Acamar	◆Hamal
75	43 52 004	41 34 051	49 57 082	59 13 124	34 28 165	41 04 211	41 55 302
76	43 56 003	42 21 051	50 57 082	60 03 125	34 44 165	40 33 212	41 04 302
77	43 58 002	43 07 050	51 56 081	60 51 126	34 58 166	40 02 212	40 13 301
78	44 00 001	43 53 049	52 55 081	61 39 127	35 13 167	39 29 213	39 21 301
79	44 00 000	44 38 049	53 55 081	62 27 128	35 26 167	38 56 214	38 30 301
80	44 00 359	45 23 048	54 54 081	63 13 130	35 39 168	38 23 214	37 38 300
81	43 59 358	46 07 047	55 53 081	63 59 131	35 50 169	37 48 215	36 46 300
82	43 56 357	46 51 047	56 52 080	64 44 132	36 02 170	37 14 216	35 54 299
83	43 53 356	47 34 046	57 51 080	65 28 134	36 12 170	36 38 216	35 01 299
84	43 48 355	48 17 045	58 50 080	66 11 135	36 22 171	36 03 217	34 09 299
85	43 43 354	48 59 044	59 49 080	66 52 137	36 31 172	35 27 217	33 16 298
86	43 36 353	49 41 043	60 48 079	67 32 139	36 39 172	34 50 218	32 23 298
87	43 29 352	50 22 043	61 47 079	68 11 141	36 47 173	34 13 219	31 30 298
88	43 20 351	51 02 042	62 46 078	68 48 143	36 53 174	33 35 219	30 37 298
89	43 11 351	51 41 041	63 45 078	69 24 145	36 59 175	32 57 220	29 44 297

LAT 0°

LHA ♈	Hc Zn	Hc Zn	Hc Zn	Hc Zn	Hc Zn	Hc Zn	Hc Zn
	POLLUX	◆REGULUS	Suhail	◆CANOPUS	Acamar	◆ALDEBARAN	CAPELLA
90	52 20 040	27 19 076	29 44 142	37 04 175	32 19 220	63 27 309	43 01 350
91	52 58 039	28 17 076	30 20 143	37 09 176	31 40 221	62 41 308	42 49 349
92	53 35 038	29 15 076	30 56 143	37 12 177	31 01 221	61 53 307	42 37 348
93	54 11 037	30 14 076	31 32 144	37 15 178	30 21 221	61 05 306	42 24 347
94	54 47 035	31 12 076	32 07 144	37 17 179	29 41 222	60 16 305	42 10 346
95	55 21 034	32 10 076	32 42 145	37 18 179	29 01 222	59 26 304	41 55 345
96	55 54 033	33 08 076	33 16 145	37 18 180	28 21 223	58 36 303	41 39 344
97	56 26 032	34 06 075	33 50 146	37 18 181	27 40 223	57 46 302	41 23 343
98	56 57 030	35 04 075	34 24 146	37 17 182	26 59 223	56 55 301	41 05 343
99	57 27 029	36 02 075	34 57 147	37 15 182	26 17 224	56 03 301	40 47 342
100	57 56 028	37 00 075	35 29 148	37 12 183	25 36 224	55 12 300	40 28 341
101	58 23 026	37 58 075	36 01 148	37 08 184	24 54 224	54 19 299	40 08 340
102	58 49 025	38 56 075	36 32 149	37 04 185	24 12 225	53 27 298	39 47 339
103	59 13 023	39 54 074	37 03 149	36 59 185	23 29 225	52 34 298	39 26 339
104	59 36 022	40 52 074	37 33 150	36 53 186	22 47 225	51 41 297	39 03 338
	POLLUX	◆REGULUS	Suhail	◆CANOPUS	RIGEL	◆ALDEBARAN	CAPELLA
105	59 58 020	41 49 074	38 03 151	36 46 187	62 25 252	50 47 297	38 40 337
106	60 17 018	42 47 074	38 32 151	36 38 188	61 28 253	49 53 296	38 17 336
107	60 36 017	43 44 073	39 00 152	36 30 188	60 31 253	49 00 296	37 52 336
108	60 52 015	44 42 073	39 28 153	36 21 189	59 33 254	48 05 295	37 27 335
109	61 07 013	45 39 073	39 55 154	36 11 190	58 35 254	47 11 295	37 01 334
110	61 20 012	46 36 072	40 21 154	36 01 190	57 38 255	46 16 294	36 35 334
111	61 31 010	47 33 072	40 47 155	35 50 191	56 40 255	45 22 294	36 08 333
112	61 40 008	48 30 072	41 12 156	35 38 192	55 42 255	44 27 293	35 40 332
113	61 47 006	49 27 071	41 36 157	35 25 193	54 44 256	43 31 293	35 12 332
114	61 53 004	50 24 071	41 59 158	35 11 193	53 45 256	42 36 293	34 44 331
115	61 56 002	51 21 071	42 21 158	34 57 194	52 47 256	41 41 292	34 14 330
116	61 58 000	52 17 070	42 43 159	34 42 195	51 49 257	40 45 292	33 44 330
117	61 57 359	53 14 070	43 04 160	34 27 195	50 50 257	39 49 292	33 14 329
118	61 55 357	54 10 069	43 24 161	34 11 196	49 52 257	38 54 291	32 43 329
119	61 50 355	55 06 069	43 43 162	33 54 197	48 53 257	37 58 291	32 12 328
	Dubhe	◆REGULUS	Suhail	◆CANOPUS	SIRIUS	BETELGEUSE	◆CAPELLA
120	19 14 021	56 02 068	44 01 163	33 37 197	65 04 227	57 57 284	31 40 328
121	19 35 021	56 57 068	44 18 164	33 18 198	64 20 228	56 59 284	31 08 327
122	19 56 020	57 53 067	44 34 165	33 00 198	63 35 230	56 00 283	30 35 327
123	20 17 020	58 48 066	44 50 166	32 40 199	62 48 231	55 02 283	30 02 326
124	20 37 020	59 42 066	45 04 167	32 20 200	62 01 232	54 03 283	29 28 326
125	20 57 019	60 37 065	45 17 168	32 00 200	61 14 233	53 05 282	28 54 325
126	21 17 019	61 31 064	45 30 169	31 39 201	60 25 234	52 06 282	28 20 325
127	21 36 019	62 25 063	45 41 170	31 17 201	59 36 235	51 07 282	27 45 324
128	21 55 018	63 18 062	45 51 171	30 55 202	58 47 236	50 09 282	27 10 324
129	22 14 018	64 11 061	46 00 172	30 32 203	57 56 237	49 10 281	26 34 324
130	22 32 017	65 04 060	46 09 173	30 09 203	57 06 238	48 11 281	25 58 323
131	22 50 017	65 56 059	46 16 174	29 45 204	56 15 239	47 12 281	25 22 323
132	23 07 017	66 47 058	46 22 175	29 21 204	55 23 240	46 13 281	24 46 322
133	23 24 016	67 38 057	46 27 176	28 56 205	54 31 240	45 14 281	24 09 322
134	23 41 016	68 28 056	46 30 177	28 31 205	53 39 241	44 15 280	23 32 322
	Dubhe	◆SPICA	ACRUX	◆CANOPUS	SIRIUS	BETELGEUSE	◆POLLUX
135	23 57 015	23 18 102	16 21 158	28 05 206	52 46 242	43 16 280	56 42 329
136	24 13 015	24 16 102	16 43 159	27 39 206	51 53 242	42 17 280	56 10 328
137	24 28 015	25 15 102	17 05 159	27 12 207	51 00 243	41 18 280	55 38 326
138	24 43 014	26 13 102	17 26 159	26 45 207	50 06 243	40 19 280	55 04 325
139	24 57 014	27 12 103	17 47 159	26 18 207	49 13 244	39 20 280	54 29 324
140	25 11 013	28 11 103	18 08 160	25 50 208	48 19 244	38 21 279	53 53 323
141	25 25 013	29 09 103	18 29 160	25 21 208	47 24 245	37 21 279	53 17 322
142	25 38 012	30 08 103	18 49 160	24 53 209	46 30 245	36 22 279	52 39 321
143	25 50 012	31 06 103	19 09 161	24 24 209	45 35 246	35 23 279	52 01 320
144	26 02 011	32 05 103	19 29 161	23 54 210	44 41 246	34 24 279	51 22 319
145	26 14 011	33 03 103	19 48 161	23 25 210	43 46 247	33 24 279	50 42 318
146	26 25 010	34 01 103	20 07 162	22 55 210	42 51 247	32 25 279	50 01 317
147	26 35 010	35 00 104	20 26 162	22 24 211	41 55 247	31 26 279	49 20 316
148	26 45 009	35 58 104	20 44 162	21 53 211	41 00 248	30 26 279	48 38 315
149	26 55 009	36 56 104	21 02 163	21 22 211	40 04 248	29 27 279	47 56 315
	◆Dubhe	ARCTURUS	◆SPICA	ACRUX	◆SIRIUS	PROCYON	POLLUX
150	27 04 008	24 35 069	37 54 104	21 20 163	39 09 248	54 25 279	47 13 314
151	27 12 008	25 31 069	38 53 104	21 37 164	38 13 249	53 26 279	46 29 313
152	27 20 007	26 27 068	39 51 105	21 54 164	37 17 249	52 27 279	45 45 312
153	27 27 007	27 23 068	40 49 105	22 10 164	36 21 249	51 27 278	45 01 312
154	27 34 006	28 18 068	41 47 105	22 26 165	35 25 249	50 28 278	44 16 311
155	27 40 006	29 14 068	42 45 105	22 42 165	34 29 250	49 29 278	43 30 310
156	27 46 005	30 10 068	43 42 105	22 57 166	33 32 250	48 29 278	42 44 310
157	27 51 005	31 05 067	44 40 106	23 12 166	32 36 250	47 30 278	41 58 309
158	27 56 004	32 00 067	45 38 106	23 26 166	31 40 250	46 30 278	41 11 309
159	28 00 004	32 56 067	46 36 106	23 40 167	30 43 250	45 31 277	40 24 308
160	28 04 003	33 51 067	47 33 107	23 54 167	29 47 251	44 31 277	39 37 308
161	28 07 003	34 46 066	48 31 107	24 07 168	28 50 251	43 32 277	38 49 307
162	28 09 002	35 41 066	49 28 107	24 19 168	27 53 251	42 32 277	38 01 307
163	28 11 002	36 35 066	50 25 108	24 31 169	26 56 251	41 33 277	37 13 306
164	28 12 001	37 30 065	51 22 108	24 43 169	26 00 251	40 33 277	36 24 306
	Dubhe	◆ARCTURUS	SPICA	◆ACRUX	Suhail	PROCYON	◆POLLUX
165	28 13 000	38 25 065	52 19 108	24 54 169	39 53 206	39 34 277	35 35 305
166	28 13 000	39 19 065	53 16 109	25 05 170	39 25 207	38 34 277	34 46 305
167	28 13 359	40 13 064	54 13 109	25 15 170	38 58 208	37 34 277	33 57 305
168	28 12 359	41 07 064	55 09 110	25 25 171	38 29 209	36 35 277	33 07 304
169	28 11 358	42 01 064	56 06 110	25 35 171	38 00 209	35 35 276	32 18 304
170	28 08 358	42 55 063	57 02 111	25 44 172	37 31 210	34 36 276	31 28 303
171	28 06 357	43 48 063	57 58 111	25 52 172	37 00 211	33 36 276	30 38 303
172	28 03 357	44 42 062	58 54 112	26 00 173	36 30 211	32 36 276	29 47 303
173	27 59 356	45 35 062	59 49 113	26 07 173	35 58 212	31 37 276	28 57 302
174	27 55 356	46 28 061	60 44 113	26 14 174	35 26 213	30 37 276	28 06 302
175	27 50 355	47 20 061	61 39 114	26 20 174	34 54 213	29 37 276	27 15 302
176	27 45 355	48 13 060	62 34 115	26 26 175	34 21 214	28 38 276	26 24 302
177	27 39 354	49 05 060	63 28 116	26 31 175	33 47 214	27 38 276	25 33 301
178	27 32 354	49 56 059	64 22 117	26 36 176	33 13 215	26 38 276	24 42 301
179	27 25 353	50 48 059	65 16 117	26 40 176	32 39 215	25 39 276	23 50 301

LAT 0°

LHA ♈	Hc	Zn	Hc	Zn	Hc	Zn	Hc	Zn	Hc	Zn	Hc	Zn	Hc	Zn
	◆ARCTURUS		ANTARES		RIGIL KENT.		◆ACRUX		Suhail		◆REGULUS		Dubhe	
180	51 39	058	20 14	118	22 00	160	26 44	177	32 04	216	59 45	294	27 18	353
181	52 29	057	21 07	118	22 20	161	26 47	177	31 29	216	58 51	294	27 10	352
182	53 20	057	22 00	119	22 39	161	26 50	178	30 53	217	57 55	293	27 01	352
183	54 10	056	22 52	119	22 59	161	26 52	178	30 17	217	57 00	292	26 52	351
184	54 59	055	23 45	119	23 18	162	26 54	179	29 40	218	56 05	292	26 42	351
185	55 48	054	24 37	119	23 36	162	26 55	179	29 03	218	55 09	291	26 32	350
186	56 36	053	25 30	120	23 54	163	26 56	180	28 26	219	54 13	291	26 21	350
187	57 24	052	26 22	120	24 12	163	26 56	180	27 48	219	53 17	290	26 10	349
188	58 11	051	27 14	120	24 29	164	26 55	181	27 10	219	52 20	290	25 59	349
189	58 58	050	28 06	120	24 46	164	26 54	181	26 32	220	51 24	289	25 46	348
190	59 44	049	28 57	121	25 02	164	26 53	182	25 54	220	50 27	289	25 34	348
191	60 29	048	29 49	121	25 18	165	26 51	182	25 15	221	49 30	289	25 21	347
192	61 13	047	30 40	121	25 33	165	26 48	183	24 36	221	48 33	288	25 07	347
193	61 56	046	31 32	121	25 48	166	26 45	183	23 56	221	47 36	288	24 53	346
194	62 39	044	32 23	122	26 02	166	26 41	184	23 16	222	46 39	288	24 39	346
	◆Alphecca		Rasalhague		ANTARES		◆RIGIL KENT.		ACRUX		◆REGULUS		Dubhe	
195	44 15	051	20 47	077	33 14	122	26 16	167	26 37	184	45 42	287	24 24	345
196	45 02	050	21 46	076	34 04	122	26 30	167	26 32	185	44 45	287	24 08	345
197	45 48	050	22 44	076	34 55	123	26 43	168	26 27	185	43 47	287	23 52	344
198	46 33	049	23 42	076	35 45	123	26 55	168	26 21	186	42 50	286	23 36	344
199	47 18	048	24 41	076	36 35	124	27 07	169	26 15	186	41 52	286	23 20	344
200	48 03	048	25 39	076	37 25	124	27 18	169	26 08	187	40 55	286	23 02	343
201	48 47	047	26 37	076	38 15	125	27 29	170	26 01	187	39 57	286	22 45	343
202	49 31	046	27 35	076	39 04	125	27 40	170	25 53	188	38 59	286	22 27	342
203	50 14	045	28 33	076	39 53	125	27 50	171	25 45	188	38 01	285	22 09	342
204	50 56	044	29 32	076	40 42	126	27 59	171	25 36	189	37 03	285	21 50	342
205	51 38	044	30 30	075	41 30	126	28 08	172	25 27	189	36 05	285	21 31	341
206	52 19	043	31 28	075	42 18	127	28 16	172	25 17	190	35 07	285	21 11	341
207	52 59	042	32 26	075	43 06	128	28 23	173	25 07	190	34 09	285	20 52	341
208	53 38	041	33 24	075	43 53	128	28 31	173	24 56	191	33 11	284	20 32	340
209	54 17	040	34 21	075	44 40	129	28 37	174	24 45	191	32 13	284	20 11	340
	◆VEGA		Rasalhague		ANTARES		◆RIGIL KENT.		ACRUX		◆REGULUS		Dubhe	
210	16 05	049	35 19	075	45 27	129	28 43	175	24 33	191	31 15	284	19 50	340
211	16 50	049	36 17	074	46 13	130	28 48	175	24 21	192	30 17	284	19 29	339
212	17 35	049	37 15	074	46 59	131	28 53	176	24 09	192	29 18	284	19 08	339
213	18 21	049	38 13	074	47 44	131	28 58	176	23 56	193	28 20	284	18 46	339
214	19 06	048	39 10	074	48 29	132	29 01	177	23 42	193	27 22	284	18 24	338
215	19 50	048	40 08	073	49 13	133	29 04	177	23 28	194	26 23	283	18 01	338
216	20 35	048	41 05	073	49 57	134	29 07	178	23 14	194	25 25	283	17 38	338
217	21 20	048	42 03	073	50 40	135	29 09	178	22 59	194	24 27	283	17 15	337
218	22 04	047	43 00	073	51 22	135	29 10	179	22 44	195	23 28	283	16 52	337
219	22 48	047	43 57	072	52 04	136	29 11	180	22 29	195	22 30	283	16 29	337
220	23 32	047	44 54	072	52 45	137	29 11	180	22 13	196	21 31	283	16 05	336
221	24 16	047	45 51	072	53 25	138	29 11	181	21 56	196	20 33	283	15 41	336
222	24 59	046	46 48	071	54 05	139	29 10	181	21 39	196	19 34	283	15 16	336
223	25 42	046	47 45	071	54 43	140	29 08	182	21 22	197	18 36	283	14 52	336
224	26 25	046	48 42	071	55 21	142	29 06	182	21 05	197	17 37	283	14 27	335
	VEGA		◆ALTAIR		Nunki		ANTARES		◆RIGIL KENT.		SPICA		◆ARCTURUS	
225	27 08	045	17 09	081	27 43	120	55 58	143	29 03	183	63 53	244	67 54	331
226	27 51	045	18 08	081	28 35	120	56 34	144	29 00	183	62 59	245	67 24	329
227	28 33	045	19 08	081	29 27	121	57 09	145	28 56	184	62 05	246	66 52	327
228	29 15	044	20 07	081	30 19	121	57 43	146	28 52	185	61 10	246	66 18	325
229	29 56	044	21 06	081	31 10	121	58 15	148	28 47	185	60 15	247	65 43	323
230	30 38	043	22 05	080	32 01	122	58 47	149	28 41	186	59 20	248	65 06	321
231	31 19	043	23 04	080	32 52	122	59 17	151	28 35	186	58 24	248	64 28	320
232	31 59	042	24 03	080	33 43	122	59 46	152	28 28	187	57 28	249	63 49	318
233	32 40	042	25 03	080	34 34	123	60 13	154	28 21	187	56 32	250	63 08	317
234	33 19	041	26 02	080	35 24	123	60 39	155	28 13	188	55 36	250	62 27	315
235	33 59	041	27 01	080	36 15	123	61 03	157	28 04	188	54 39	250	61 44	314
236	34 38	040	28 00	080	37 05	124	61 26	159	27 55	189	53 42	251	61 00	313
237	35 17	040	28 59	080	37 54	124	61 47	160	27 46	189	52 46	251	60 16	312
238	35 55	039	29 58	080	38 44	125	62 07	162	27 36	190	51 49	252	59 31	310
239	36 33	039	30 57	080	39 33	125	62 24	164	27 25	190	50 52	252	58 45	309
	VEGA		◆ALTAIR		Nunki		◆ANTARES		SPICA		◆ARCTURUS		Alkaid	
240	37 10	038	31 56	080	40 22	126	62 40	166	49 54	253	57 58	308	33 03	335
241	37 47	038	32 55	079	41 11	126	62 54	168	48 57	253	57 11	307	32 38	334
242	38 23	037	33 54	079	41 59	127	63 06	170	48 00	253	56 23	306	32 11	334
243	38 59	036	34 53	079	42 47	127	63 16	171	47 02	254	55 34	306	31 44	333
244	39 34	036	35 52	079	43 35	128	63 23	173	46 05	254	54 45	305	31 17	333
245	40 09	035	36 51	079	44 22	128	63 29	175	45 07	254	53 55	304	30 49	332
246	40 43	034	37 50	079	45 09	129	63 33	177	44 09	254	53 05	303	30 21	332
247	41 17	034	38 48	079	45 55	130	63 35	179	43 11	255	52 15	303	29 52	331
248	41 50	033	39 47	078	46 41	130	63 34	181	42 14	255	51 24	302	29 23	331
249	42 22	032	40 46	078	47 27	131	63 32	183	41 16	255	50 33	301	28 53	330
250	42 53	031	41 45	078	48 12	132	63 27	185	40 18	255	49 42	301	28 23	330
251	43 24	030	42 43	078	48 57	132	63 20	187	39 20	256	48 50	300	27 52	329
252	43 54	030	43 42	078	49 41	133	63 11	189	38 21	256	47 58	299	27 21	329
253	44 23	029	44 41	077	50 24	134	63 01	191	37 23	256	47 05	299	26 50	328
254	44 52	028	45 39	077	51 07	135	62 48	193	36 25	256	46 13	298	26 18	328
	◆VEGA		ALTAIR		◆Nunki		ANTARES		◆SPICA		ARCTURUS		Alkaid	
255	45 19	027	46 38	077	51 49	136	62 33	195	35 27	256	45 20	298	25 46	327
256	45 46	026	47 36	077	52 30	137	62 16	197	34 28	256	44 27	297	25 13	327
257	46 12	025	48 35	077	53 11	138	61 58	199	33 30	257	43 33	297	24 40	327
258	46 37	024	49 33	076	53 51	139	61 38	201	32 32	257	42 40	297	24 07	326
259	47 02	023	50 31	076	54 30	140	61 16	202	31 33	257	41 46	296	23 34	326
260	47 25	022	51 29	076	55 09	141	60 52	204	30 35	257	40 52	296	23 00	325
261	47 47	021	52 27	075	55 46	142	60 27	206	29 36	257	39 58	295	22 26	325
262	48 08	020	53 25	075	56 23	143	60 01	207	28 38	257	39 04	295	21 51	325
263	48 28	019	54 23	075	56 58	144	59 33	209	27 39	257	38 09	295	21 16	324
264	48 48	018	55 21	074	57 33	146	59 03	210	26 41	258	37 15	294	20 41	324
265	49 06	017	56 19	074	58 06	147	58 33	211	25 42	258	36 20	294	20 06	324
266	49 23	016	57 16	073	58 38	148	58 01	213	24 44	258	35 25	294	19 31	324
267	49 38	015	58 14	073	59 09	150	57 27	214	23 45	258	34 30	294	18 55	323
268	49 53	014	59 11	073	59 38	151	56 53	215	22 46	258	33 35	293	18 19	323
269	50 07	012	60 08	072	60 06	153	56 18	217	21 48	258	32 40	293	17 43	323

LHA ♈	Hc	Zn	Hc	Zn	Hc	Zn	Hc	Zn	Hc	Zn	Hc	Zn	Hc	Zn
	VEGA		◆DENEB		FOMALHAUT		◆Peacock		Shaula		ANTARES		◆ARCTURUS	
270	50 19	011	32 28	033	13 34	121	26 13	159	52 23	189	55 42	218	31 45	293
271	50 30	010	33 00	032	14 25	121	26 35	159	52 13	190	55 04	219	30 49	293
272	50 40	009	33 31	032	15 17	121	26 56	160	52 02	191	54 26	220	29 54	292
273	50 48	008	34 02	031	16 08	121	27 16	160	51 50	193	53 47	221	28 58	292
274	50 56	006	34 33	030	17 00	121	27 36	161	51 36	194	53 07	222	28 03	292
275	51 02	005	35 03	030	17 51	121	27 56	161	51 21	195	52 26	223	27 07	292
276	51 07	004	35 33	029	18 42	121	28 15	162	51 05	196	51 45	224	26 11	292
277	51 10	003	36 02	029	19 34	122	28 34	162	50 48	197	51 03	225	25 15	291
278	51 12	001	36 30	028	20 25	122	28 52	163	50 30	199	50 20	226	24 19	291
279	51 13	000	36 58	027	21 15	122	29 09	163	50 10	200	49 37	227	23 23	291
280	51 13	359	37 25	027	22 06	122	29 26	164	49 49	201	48 53	227	22 27	291
281	51 11	358	37 52	026	22 57	122	29 43	164	49 27	202	48 09	228	21 31	291
282	51 08	357	38 17	025	23 47	123	29 59	165	49 04	203	47 24	229	20 35	291
283	51 04	355	38 43	024	24 38	123	30 14	165	48 41	204	46 38	230	19 39	290
284	50 58	354	39 07	024	25 28	123	30 29	166	48 16	205	45 52	230	18 43	290
	DENEB		Alpheratz		◆FOMALHAUT		Peacock		◆ANTARES		Rasalhague		◆VEGA	
285	39 31	023	11 19	060	26 18	123	30 43	167	45 06	231	65 24	302	50 51	353
286	39 54	022	12 11	060	27 08	124	30 57	167	44 19	232	64 33	300	50 43	352
287	40 16	021	13 03	060	27 58	124	31 10	168	43 32	232	63 40	299	50 34	350
288	40 38	021	13 55	060	28 48	124	31 22	168	42 44	233	62 48	298	50 23	349
289	40 58	020	14 47	060	29 37	125	31 34	169	41 57	233	61 55	298	50 11	348
290	41 18	019	15 39	060	30 26	125	31 45	170	41 08	234	61 02	297	49 58	347
291	41 37	018	16 30	060	31 15	125	31 55	170	40 20	234	60 08	296	49 44	346
292	41 56	017	17 22	059	32 04	126	32 05	171	39 31	235	59 14	295	49 29	345
293	42 13	016	18 14	059	32 53	126	32 15	171	38 42	235	58 19	294	49 12	343
294	42 30	016	19 05	059	33 41	126	32 23	172	37 52	236	57 24	294	48 55	342
295	42 45	015	19 57	059	34 29	127	32 31	173	37 03	236	56 29	293	48 36	341
296	43 00	014	20 48	059	35 17	127	32 39	173	36 13	237	55 34	293	48 16	340
297	43 14	013	21 39	058	36 05	128	32 45	174	35 23	237	54 39	292	47 55	339
298	43 27	012	22 30	058	36 52	128	32 51	175	34 32	237	53 43	292	47 33	338
299	43 39	011	23 21	058	37 39	129	32 57	175	33 42	238	52 47	291	47 11	337
	◆DENEB		Alpheratz		◆FOMALHAUT		Peacock		◆ANTARES		Rasalhague		VEGA	
300	43 50	010	24 12	058	38 26	129	33 01	176	32 51	238	51 51	291	46 47	336
301	44 00	009	25 03	058	39 12	130	33 05	177	32 00	238	50 55	290	46 22	335
302	44 09	008	25 53	057	39 58	130	33 09	177	31 09	239	49 58	290	45 56	334
303	44 17	007	26 44	057	40 44	131	33 11	178	30 17	239	49 02	289	45 30	333
304	44 24	006	27 34	057	41 29	131	33 13	178	29 26	239	48 05	289	45 03	332
305	44 30	005	28 24	056	42 14	132	33 14	179	28 34	240	47 08	289	44 34	332
306	44 35	004	29 14	056	42 58	133	33 15	180	27 42	240	46 11	288	44 05	331
307	44 39	003	30 04	056	43 42	133	33 15	180	26 50	240	45 14	288	43 36	330
308	44 41	002	30 53	056	44 26	134	33 14	181	25 58	240	44 17	288	43 05	329
309	44 43	001	31 43	055	45 09	135	33 12	182	25 06	241	43 20	287	42 34	328
310	44 44	000	32 32	055	45 51	135	33 10	182	24 14	241	42 23	287	42 02	327
311	44 44	359	33 21	054	46 33	136	33 07	183	23 21	241	41 25	287	41 29	327
312	44 43	358	34 09	054	47 15	137	33 04	184	22 29	241	40 28	287	40 56	326
313	44 41	357	34 58	054	47 55	138	33 00	184	21 36	241	39 30	286	40 22	325
314	44 37	356	35 46	053	48 36	138	32 55	185	20 43	242	38 33	286	39 48	325
	◆Alpheratz		◆FOMALHAUT		Peacock		Nunki		◆Rasalhague		VEGA		DENEB	
315	36 34	053	49 15	139	32 49	186	50 01	226	37 35	286	39 13	324	44 33	355
316	37 22	052	49 54	140	32 43	186	49 18	227	36 37	286	38 37	323	44 28	354
317	38 09	052	50 32	141	32 36	187	48 33	228	35 40	286	38 01	323	44 21	353
318	38 56	051	51 09	142	32 28	188	47 49	229	34 42	285	37 24	322	44 14	352
319	39 43	051	51 46	143	32 20	188	47 03	229	33 44	285	36 47	321	44 05	351
320	40 29	050	52 21	144	32 11	189	46 17	230	32 46	285	36 10	321	43 56	351
321	41 15	050	52 56	145	32 02	189	45 31	231	31 48	285	35 32	320	43 46	350
322	42 01	049	53 30	146	31 52	190	44 44	231	30 50	285	34 53	320	43 34	349
323	42 46	049	54 03	147	31 41	191	43 57	232	29 52	285	34 14	319	43 22	348
324	43 31	048	54 34	149	31 29	191	43 10	233	28 54	284	33 35	319	43 09	347
325	44 15	047	55 05	150	31 17	192	42 22	233	27 56	284	32 55	318	42 55	346
326	44 59	047	55 35	151	31 05	192	41 34	234	26 58	284	32 15	318	42 40	345
327	45 42	046	56 03	152	30 52	193	40 45	234	25 59	284	31 34	317	42 24	344
328	46 25	045	56 30	154	30 38	194	39 57	235	25 01	284	30 53	317	42 07	343
329	47 07	044	56 56	155	30 23	194	39 07	235	24 03	284	30 12	316	41 49	342
	◆Alpheratz		Diphda		◆FOMALHAUT		Peacock		◆ALTAIR		VEGA		DENEB	
330	47 49	044	46 01	116	57 21	156	30 08	195	56 34	286	29 31	316	41 30	342
331	48 30	043	46 54	117	57 44	158	29 53	195	55 37	286	28 49	316	41 11	341
332	49 11	042	47 48	117	58 06	159	29 37	196	54 39	285	28 07	315	40 51	340
333	49 50	041	48 41	118	58 26	161	29 20	196	53 41	285	27 25	315	40 30	339
334	50 30	040	49 34	118	58 45	162	29 03	197	52 43	285	26 42	315	40 08	338
335	51 08	039	50 26	119	59 03	164	28 45	197	51 45	284	25 59	314	39 45	338
336	51 46	038	51 19	120	59 18	166	28 27	198	50 47	284	25 16	314	39 22	337
337	52 22	037	52 11	120	59 32	167	28 08	198	49 49	284	24 32	314	38 58	336
338	52 58	036	53 02	121	59 44	169	27 49	199	48 50	284	23 49	313	38 33	335
339	53 33	035	53 54	122	59 55	171	27 29	199	47 52	283	23 05	313	38 08	335
340	54 07	034	54 45	122	60 04	172	27 09	200	46 53	283	22 21	313	37 42	334
341	54 40	033	55 35	123	60 11	174	26 48	200	45 55	283	21 37	312	37 15	333
342	55 12	032	56 25	124	60 16	176	26 27	201	44 56	283	20 52	312	36 48	333
343	55 43	030	57 14	125	60 19	178	26 05	201	43 58	282	20 08	312	36 20	332
344	56 13	029	58 03	126	60 21	179	25 43	202	42 59	282	19 23	312	35 51	331
	Alpheratz		◆Hamal		Acamar		◆ACHERNAR		FOMALHAUT		◆ALTAIR		DENEB	
345	56 42	028	38 58	059	22 45	135	24 43	158	60 21	181	42 01	282	35 22	331
346	57 09	026	39 50	059	23 28	135	25 05	158	60 19	183	41 02	282	34 52	330
347	57 35	025	40 41	058	24 10	135	25 27	159	60 15	185	40 03	282	34 22	329
348	58 00	024	41 32	058	24 52	136	25 49	159	60 09	186	39 04	281	33 51	329
349	58 23	022	42 23	057	25 34	136	26 10	160	60 01	188	38 05	281	33 20	328
350	58 45	021	43 13	057	26 16	136	26 30	160	59 52	190	37 07	281	32 48	328
351	59 05	019	44 03	056	26 57	137	26 51	161	59 41	192	36 08	281	32 16	327
352	59 24	017	44 53	056	27 38	137	27 10	161	59 28	193	35 09	281	31 43	327
353	59 41	016	45 43	055	28 19	137	27 30	161	59 13	195	34 10	281	31 10	326
354	59 56	014	46 32	055	29 00	138	27 49	162	58 57	196	33 11	281	30 36	326
355	60 10	012	47 20	054	29 40	138	28 07	162	58 40	198	32 12	280	30 02	325
356	60 22	011	48 09	053	30 20	139	28 25	163	58 20	200	31 13	280	29 27	325
357	60 33	009	48 57	053	30 59	139	28 42	164	57 59	201	30 14	280	28 52	324
358	60 41	007	49 44	052	31 39	139	28 59	164	57 37	203	29 15	280	28 17	324
359	60 48	005	50 31	051	32 17	140	29 15	165	57 13	204	28 16	280	27 41	323

LHA	Hc	Zn	Hc	Zn	Hc	Zn	Hc	Zn	Hc	Zn	Hc	Zn	Hc	Zn
♈	Alpheratz		◆ALDEBARAN		RIGEL		ACHERNAR		◆FOMALHAUT		ALTAIR		◆DENEB	
0	59 52	004	19 53	072	11 27	098	30 29	165	57 42	206	27 06	280	26 17	323
1	59 55	002	20 50	072	12 26	098	30 44	166	57 15	207	26 07	280	25 41	323
2	59 56	000	21 47	072	13 26	098	30 58	166	56 47	209	25 08	280	25 05	322
3	59 55	358	22 44	072	14 25	098	31 13	167	56 18	210	24 09	280	24 28	322
4	59 53	357	23 41	071	15 25	098	31 26	167	55 47	211	23 10	280	23 51	322
5	59 48	355	24 38	071	16 24	098	31 39	168	55 15	213	22 11	280	23 14	321
6	59 42	353	25 34	071	17 23	098	31 51	168	54 43	214	21 12	280	22 36	321
7	59 34	351	26 31	071	18 23	098	32 03	169	54 09	215	20 13	280	21 58	321
8	59 24	350	27 28	071	19 22	098	32 14	170	53 34	216	19 14	280	21 20	320
9	59 12	348	28 24	071	20 21	098	32 25	170	52 58	217	18 14	280	20 42	320
10	58 59	346	29 21	070	21 21	098	32 35	171	52 22	218	17 15	280	20 03	320
11	58 44	345	30 17	070	22 20	098	32 44	171	51 44	219	16 16	280	19 24	319
12	58 28	343	31 14	070	23 19	099	32 53	172	51 06	220	15 17	279	18 45	319
13	58 10	342	32 10	070	24 19	099	33 01	173	50 27	221	14 18	279	18 06	319
14	57 50	340	33 06	069	25 18	099	33 08	173	49 48	222	13 19	279	17 26	319
	CAPELLA		ALDEBARAN		◆RIGEL		ACHERNAR		◆FOMALHAUT		Enif		◆Alpheratz	
15	16 55	041	34 02	069	26 17	099	33 14	174	49 07	223	40 01	284	57 29	339
16	17 34	041	34 58	069	27 17	099	33 20	175	48 27	223	39 03	284	57 06	337
17	18 13	040	35 54	069	28 16	099	33 26	175	47 45	224	38 05	283	56 42	336
18	18 52	040	36 50	068	29 15	099	33 30	176	47 03	225	37 06	283	56 17	334
19	19 30	040	37 46	068	30 15	099	33 34	177	46 20	226	36 08	283	55 50	333
20	20 09	039	38 42	068	31 14	099	33 38	177	45 37	226	35 09	283	55 22	332
21	20 47	039	39 37	067	32 13	099	33 40	178	44 53	227	34 11	283	54 53	330
22	21 24	039	40 32	067	33 12	099	33 42	178	44 09	228	33 12	282	54 23	329
23	22 02	038	41 28	067	34 11	099	33 44	179	43 25	228	32 14	282	53 52	328
24	22 39	038	42 23	066	35 11	099	33 44	180	42 39	229	31 15	282	53 19	327
25	23 16	038	43 18	066	36 10	099	33 44	180	41 54	230	30 17	282	52 46	326
26	23 52	037	44 12	066	37 09	100	33 43	181	41 08	230	29 18	282	52 12	325
27	24 29	037	45 07	065	38 08	100	33 42	182	40 22	231	28 19	282	51 37	324
28	25 05	037	46 01	065	39 07	100	33 40	182	39 35	231	27 20	282	51 01	323
29	25 40	036	46 55	064	40 06	100	33 37	183	38 49	232	26 22	282	50 24	322
	CAPELLA		◆BETELGEUSE		SIRIUS		CANOPUS		◆ACHERNAR		FOMALHAUT		◆Alpheratz	
30	26 16	036	30 50	081	18 15	107	15 07	145	33 34	184	38 01	232	49 46	321
31	26 51	035	31 49	081	19 12	107	15 41	145	33 29	184	37 14	233	49 07	320
32	27 25	035	32 48	081	20 09	107	16 15	145	33 25	185	36 26	233	48 28	319
33	27 59	035	33 47	080	21 07	108	16 49	146	33 19	186	35 38	233	47 48	318
34	28 33	034	34 46	080	22 04	108	17 23	146	33 13	186	34 50	234	47 08	317
35	29 06	034	35 46	080	23 01	108	17 56	146	33 06	187	34 01	234	46 27	316
36	29 39	033	36 45	080	23 58	108	18 30	146	32 59	187	33 13	235	45 45	316
37	30 12	033	37 44	080	24 55	108	19 03	147	32 51	188	32 24	235	45 03	315
38	30 44	032	38 43	080	25 52	108	19 36	147	32 42	189	31 34	235	44 20	314
39	31 16	032	39 42	080	26 49	108	20 08	147	32 32	189	30 45	236	43 37	313
40	31 47	031	40 41	079	27 46	108	20 41	148	32 22	190	29 55	236	42 53	313
41	32 18	030	41 40	079	28 43	109	21 13	148	32 12	191	29 06	236	42 09	312
42	32 48	030	42 39	079	29 40	109	21 44	148	32 00	191	28 16	236	41 24	312
43	33 17	029	43 37	079	30 37	109	22 16	148	31 49	192	27 26	237	40 39	311
44	33 46	029	44 36	079	31 33	109	22 47	149	31 36	192	26 35	237	39 53	310
	CAPELLA		◆BETELGEUSE		SIRIUS		CANOPUS		◆ACHERNAR		Diphda		◆Hamal	
45	34 15	028	45 35	078	32 30	109	23 18	149	31 23	193	52 23	241	62 22	333
46	34 43	027	46 34	078	33 27	109	23 49	150	31 09	193	51 31	242	61 54	331
47	35 10	027	47 32	078	34 23	110	24 19	150	30 55	194	50 38	242	61 24	330
48	35 37	026	48 31	078	35 20	110	24 49	150	30 40	195	49 44	243	60 53	328
49	36 03	026	49 30	077	36 16	110	25 18	151	30 25	195	48 51	243	60 21	327
50	36 29	025	50 28	077	37 12	110	25 48	151	30 09	196	47 57	244	59 47	325
51	36 54	024	51 27	077	38 08	111	26 17	151	29 52	196	47 03	244	59 12	324
52	37 18	023	52 25	076	39 05	111	26 45	152	29 35	197	46 09	245	58 36	322
53	37 41	023	53 23	076	40 01	111	27 13	152	29 18	197	45 15	245	57 59	321
54	38 04	022	54 21	076	40 56	111	27 41	153	29 00	198	44 20	245	57 21	320
55	38 26	021	55 20	075	41 52	112	28 08	153	28 41	198	43 26	246	56 42	319
56	38 47	020	56 18	075	42 48	112	28 35	154	28 22	199	42 31	246	56 01	318
57	39 08	020	57 15	075	43 43	112	29 01	154	28 03	199	41 36	247	55 21	316
58	39 28	019	58 13	074	44 39	113	29 27	155	27 42	200	40 41	247	54 39	315
59	39 47	018	59 11	074	45 34	113	29 52	155	27 22	200	39 46	247	53 56	314
	◆CAPELLA		POLLUX		◆SIRIUS		CANOPUS		ACHERNAR		◆Diphda		Hamal	
60	40 05	017	28 49	057	46 29	114	30 17	156	27 01	201	38 50	247	53 13	313
61	40 22	016	29 39	057	47 24	114	30 42	156	26 40	201	37 55	248	52 29	313
62	40 39	016	30 29	056	48 19	114	31 06	157	26 18	202	36 59	248	51 45	312
63	40 55	015	31 19	056	49 13	115	31 29	157	25 55	202	36 04	248	51 00	311
64	41 10	014	32 09	056	50 08	115	31 52	158	25 33	202	35 08	249	50 14	310
65	41 24	013	32 58	055	51 02	116	32 15	158	25 09	203	34 12	249	49 28	309
66	41 37	012	33 47	055	51 56	116	32 37	159	24 46	203	33 16	249	48 41	308
67	41 49	011	34 36	054	52 49	117	32 58	160	24 22	204	32 20	249	47 54	308
68	42 00	010	35 25	054	53 43	118	33 18	160	23 58	204	31 24	249	47 06	307
69	42 10	009	36 13	053	54 36	118	33 39	161	23 33	204	30 28	250	46 18	306
70	42 20	009	37 01	053	55 28	119	33 58	161	23 08	205	29 31	250	45 30	306
71	42 28	008	37 49	053	56 21	120	34 17	162	22 43	205	28 35	250	44 41	305
72	42 36	007	38 36	052	57 13	120	34 35	163	22 17	206	27 39	250	43 52	305
73	42 42	006	39 23	051	58 04	121	34 53	163	21 51	206	26 42	250	43 02	304
74	42 48	005	40 10	051	58 56	122	35 10	164	21 24	206	25 46	250	42 12	304
	CAPELLA		◆POLLUX		REGULUS		Suhail		◆CANOPUS		ACHERNAR		◆Hamal	
75	42 52	004	40 56	050	12 28	077	20 42	137	35 26	165	20 58	207	41 22	303
76	42 56	003	41 42	050	13 27	077	21 23	137	35 42	165	20 31	207	40 32	303
77	42 58	002	42 28	049	14 25	077	22 04	137	35 57	166	20 03	207	39 41	302
78	43 00	001	43 13	049	15 24	077	22 45	138	36 11	167	19 36	208	38 50	302
79	43 00	000	43 58	048	16 22	077	23 25	138	36 24	167	19 08	208	37 59	301
80	43 00	359	44 42	047	17 21	077	24 05	138	36 37	168	18 40	208	37 08	301
81	42 59	358	45 26	047	18 19	077	24 45	138	36 49	169	18 11	208	36 16	300
82	42 56	357	46 09	046	19 18	077	25 25	139	37 01	169	17 43	209	35 24	300
83	42 53	356	46 52	045	20 16	077	26 04	139	37 11	170	17 14	209	34 32	300
84	42 48	355	47 34	044	21 15	077	26 43	140	37 21	171	16 45	209	33 40	299
85	42 43	354	48 16	043	22 13	077	27 22	140	37 30	172	16 16	209	32 47	299
86	42 37	353	48 57	043	23 11	076	28 00	140	37 39	172	15 46	210	31 55	299
87	42 29	353	49 37	042	24 10	076	28 38	141	37 46	173	15 16	210	31 02	298
88	42 21	352	50 17	041	25 08	076	29 16	141	37 53	174	14 46	210	30 09	298
89	42 12	351	50 56	040	26 06	076	29 54	142	37 59	175	14 16	210	29 16	298

LAT 1°S

LHA	Hc	Zn	Hc	Zn	Hc	Zn	Hc	Zn	Hc	Zn	Hc	Zn	Hc	Zn
♈	POLLUX		◆REGULUS		Suhail		◆CANOPUS		Acamar		◆ALDEBARAN		CAPELLA	
90	51 34	039	27 04	076	30 31	142	38 04	175	33 05	220	62 49	311	42 02	350
91	52 11	038	28 03	076	31 08	142	38 08	176	32 25	221	62 03	310	41 51	349
92	52 47	037	29 01	076	31 44	143	38 12	177	31 46	221	61 16	309	41 39	348
93	53 23	036	29 59	076	32 20	143	38 15	178	31 06	222	60 29	307	41 26	347
94	53 57	035	30 57	075	32 56	144	38 17	178	30 26	222	59 41	306	41 12	346
95	54 31	033	31 55	075	33 31	144	38 18	179	29 46	223	58 52	305	40 57	345
96	55 03	032	32 53	075	34 06	145	38 18	180	29 05	223	58 03	304	40 42	345
97	55 35	031	33 51	075	34 40	145	38 18	181	28 24	223	57 13	303	40 25	344
98	56 05	030	34 49	075	35 14	146	38 17	182	27 42	224	56 23	303	40 08	343
99	56 34	028	35 47	074	35 47	147	38 15	182	27 01	224	55 32	302	39 50	342
100	57 02	027	36 44	074	36 20	147	38 12	183	26 19	224	54 41	301	39 31	341
101	57 29	026	37 42	074	36 52	148	38 08	184	25 37	225	53 50	300	39 11	340
102	57 54	024	38 40	074	37 24	148	38 04	185	24 54	225	52 58	300	38 51	340
103	58 18	023	39 37	073	37 55	149	37 58	185	24 11	225	52 05	299	38 30	339
104	58 40	021	40 35	073	38 25	150	37 52	186	23 29	226	51 13	298	38 08	338
	POLLUX		◆REGULUS		Suhail		◆CANOPUS		RIGEL		◆ALDEBARAN		CAPELLA	
105	59 01	020	41 32	073	38 55	150	37 46	187	62 43	254	50 20	298	37 45	337
106	59 20	018	42 29	073	39 25	151	37 38	188	61 45	254	49 27	297	37 22	337
107	59 38	016	43 27	072	39 53	152	37 30	188	60 47	255	48 33	297	36 57	336
108	59 54	015	44 24	072	40 21	153	37 20	189	59 49	255	47 39	296	36 33	335
109	60 08	013	45 21	072	40 49	153	37 10	190	58 51	256	46 45	296	36 07	335
110	60 21	011	46 18	071	41 15	154	37 00	191	57 53	256	45 51	295	35 41	334
111	60 32	009	47 14	071	41 41	155	36 48	191	56 55	256	44 57	295	35 15	333
112	60 41	008	48 11	071	42 06	156	36 36	192	55 56	257	44 02	294	34 47	333
113	60 48	006	49 08	070	42 31	156	36 23	193	54 58	257	43 07	294	34 19	332
114	60 53	004	50 04	070	42 54	157	36 10	193	53 59	257	42 13	294	33 51	331
115	60 56	002	51 00	069	43 17	158	35 55	194	53 01	258	41 17	293	33 22	331
116	60 58	000	51 56	069	43 39	159	35 41	195	52 02	258	40 22	293	32 52	330
117	60 57	359	52 52	068	44 00	160	35 25	195	51 03	258	39 27	292	32 22	330
118	60 55	357	53 48	068	44 20	161	35 08	196	50 05	258	38 31	292	31 52	329
119	60 51	355	54 43	067	44 40	162	34 51	197	49 06	259	37 36	292	31 21	329
	◆REGULUS		Gienah		◆ACRUX		CANOPUS		SIRIUS		◆BETELGEUSE		POLLUX	
120	55 39	067	25 09	109	11 17	155	34 34	197	65 45	229	57 41	286	60 44	353
121	56 34	066	26 05	109	11 42	155	34 15	198	64 59	230	56 44	285	60 36	351
122	57 28	066	27 02	109	12 07	155	33 57	199	64 13	231	55 46	285	60 27	350
123	58 23	065	27 59	109	12 32	155	33 37	199	63 26	233	54 48	284	60 15	348
124	59 17	064	28 55	110	12 57	156	33 17	200	62 38	234	53 49	284	60 01	346
125	60 11	063	29 52	110	13 22	156	32 56	201	61 49	235	52 51	284	59 46	345
126	61 04	063	30 48	110	13 46	156	32 35	201	61 00	236	51 53	283	59 29	343
127	61 57	062	31 45	110	14 10	156	32 13	202	60 10	237	50 54	283	59 11	341
128	62 50	061	32 41	110	14 34	156	31 50	202	59 19	238	49 56	283	58 51	340
129	63 42	060	33 37	110	14 58	157	31 28	203	58 28	239	48 57	282	58 29	338
130	64 33	059	34 33	111	15 22	157	31 04	203	57 37	239	47 59	282	58 06	337
131	65 24	057	35 29	111	15 45	157	30 40	204	56 45	240	47 00	282	57 42	335
132	66 15	056	36 25	111	16 08	157	30 15	204	55 53	241	46 01	282	57 16	334
133	67 04	055	37 21	111	16 31	158	29 50	205	55 00	242	45 03	282	56 49	332
134	67 53	054	38 17	112	16 54	158	29 25	205	54 07	242	44 04	281	56 20	331
	Alioth		◆SPICA		ACRUX		◆CANOPUS		SIRIUS		BETELGEUSE		◆POLLUX	
135	16 09	030	23 30	102	17 16	158	28 59	206	53 14	243	43 05	281	55 50	330
136	16 39	029	24 29	102	17 39	158	28 33	206	52 21	243	42 06	281	55 20	328
137	17 08	029	25 27	102	18 01	159	28 06	207	51 27	244	41 07	281	54 48	327
138	17 37	029	26 26	102	18 22	159	27 39	207	50 33	244	40 08	281	54 14	326
139	18 06	029	27 25	102	18 43	159	27 11	208	49 39	245	39 09	280	53 40	325
140	18 35	028	28 24	102	19 04	160	26 43	208	48 44	245	38 10	280	53 05	324
141	19 03	028	29 22	102	19 25	160	26 14	209	47 49	246	37 11	280	52 29	323
142	19 31	028	30 21	102	19 46	160	25 45	209	46 55	246	36 12	280	51 53	322
143	19 59	027	31 19	102	20 06	161	25 16	209	46 00	247	35 13	280	51 15	321
144	20 26	027	32 18	103	20 26	161	24 47	210	45 04	247	34 14	280	50 36	320
145	20 53	027	33 16	103	20 45	161	24 17	210	44 09	247	33 15	280	49 57	319
146	21 20	026	34 15	103	21 04	162	23 46	211	43 14	248	32 16	279	49 17	318
147	21 46	026	35 13	103	21 23	162	23 16	211	42 18	248	31 16	279	48 37	317
148	22 12	026	36 12	103	21 41	162	22 45	211	41 22	248	30 17	279	47 55	316
149	22 38	025	37 10	103	21 59	163	22 13	212	40 26	249	29 18	279	47 14	315
	Alioth		◆ARCTURUS		SPICA		◆ACRUX		CANOPUS		SIRIUS		◆POLLUX	
150	23 03	025	24 13	068	38 09	103	22 17	163	21 42	212	39 31	249	46 31	315
151	23 28	024	25 09	068	39 07	104	22 34	163	21 10	212	38 34	249	45 48	314
152	23 52	024	26 05	068	40 05	104	22 51	164	20 38	212	37 38	250	45 04	313
153	24 17	023	27 00	068	41 04	104	23 08	164	20 06	213	36 42	250	44 20	312
154	24 40	023	27 56	068	42 02	104	23 24	165	19 33	213	35 46	250	43 36	312
155	25 03	023	28 51	067	43 00	104	23 40	165	19 00	213	34 49	250	42 51	311
156	25 26	022	29 46	067	43 58	105	23 55	165	18 27	214	33 53	250	42 06	310
157	25 49	022	30 42	067	44 56	105	24 10	166	17 54	214	32 56	251	41 20	310
158	26 11	021	31 37	067	45 54	105	24 24	166	17 20	214	32 00	251	40 33	309
159	26 32	021	32 32	066	46 52	105	24 38	167	16 47	214	31 03	251	39 47	309
160	26 53	020	33 27	066	47 50	106	24 52	167	16 13	215	30 06	251	39 00	308
161	27 14	020	34 21	066	48 47	106	25 05	168	15 38	215	29 09	251	38 13	308
162	27 34	019	35 16	065	49 45	106	25 18	168	15 04	215	28 12	252	37 25	307
163	27 53	019	36 11	065	50 43	106	25 30	168	14 30	215	27 16	252	36 37	307
164	28 12	018	37 05	065	51 40	107	25 42	169	13 55	215	26 19	252	35 49	306
	Alioth		◆ARCTURUS		SPICA		◆ACRUX		Suhail		SIRIUS		◆POLLUX	
165	28 30	018	37 59	064	52 38	107	25 53	169	40 46	207	25 22	252	35 01	306
166	28 48	017	38 53	064	53 35	108	26 04	170	40 19	208	24 25	252	34 12	305
167	29 06	017	39 47	064	54 32	108	26 15	170	39 51	208	23 27	252	33 23	305
168	29 23	016	40 41	063	55 29	108	26 25	171	39 22	209	22 30	252	32 34	305
169	29 39	015	41 34	063	56 26	109	26 34	171	38 53	210	21 33	252	31 44	304
170	29 55	015	42 28	062	57 23	109	26 43	172	38 23	210	20 36	253	30 54	304
171	30 10	014	43 21	062	58 19	110	26 51	172	37 52	211	19 39	253	30 05	304
172	30 24	014	44 14	062	59 15	110	26 59	173	37 21	212	18 41	253	29 15	303
173	30 38	013	45 06	061	60 11	111	27 07	173	36 49	212	17 44	253	28 24	303
174	30 51	013	45 59	061	61 07	112	27 14	174	36 17	213	16 47	253	27 34	303
175	31 04	012	46 51	060	62 03	112	27 20	174	35 44	213	15 50	253	26 43	302
176	31 16	011	47 42	059	62 58	113	27 26	175	35 10	214	14 52	253	25 53	302
177	31 28	011	48 34	059	63 53	114	27 31	175	34 37	215	13 55	253	25 02	302
178	31 39	010	49 25	058	64 48	115	27 36	176	34 02	215	12 57	253	24 11	302
179	31 49	009	50 16	058	65 43	116	27 40	176	33 28	216	12 00	253	23 19	301

LHA	Hc	Zn	Hc	Zn	Hc	Zn	Hc	Zn	Hc	Zn	Hc	Zn	Hc	Zn
♈	Alioth		◆ARCTURUS		ANTARES		◆ACRUX		Suhail		Alphard		◆REGULUS	
180	31 58	009	51 06	057	20 43	118	27 44	177	32 52	216	51 15	257	59 20	296
181	32 07	008	51 57	056	21 36	118	27 47	177	32 17	217	50 16	258	58 26	295
182	32 15	008	52 46	055	22 29	118	27 50	178	31 41	217	49 17	258	57 31	294
183	32 23	007	53 35	055	23 21	119	27 52	178	31 04	218	48 19	258	56 37	294
184	32 30	006	54 24	054	24 14	119	27 54	179	30 27	218	47 20	258	55 42	293
185	32 36	006	55 12	053	25 07	119	27 55	179	29 50	219	46 21	259	54 46	293
186	32 41	005	56 00	052	25 59	119	27 56	180	29 13	219	45 23	259	53 51	292
187	32 46	004	56 47	051	26 51	119	27 56	180	28 35	219	44 24	259	52 55	292
188	32 50	004	57 33	050	27 44	120	27 55	181	27 57	220	43 25	259	51 59	291
189	32 54	003	58 19	049	28 36	120	27 54	181	27 18	220	42 26	259	51 03	291
190	32 57	002	59 04	048	29 28	120	27 53	182	26 39	221	41 27	259	50 07	290
191	32 59	002	59 48	047	30 20	120	27 50	182	26 00	221	40 28	259	49 11	290
192	33 00	001	60 31	046	31 11	121	27 48	183	25 21	221	39 29	260	48 14	289
193	33 01	000	61 14	044	32 03	121	27 45	183	24 41	222	38 30	260	47 17	289
194	33 01	000	61 55	043	32 54	121	27 41	184	24 01	222	37 31	260	46 21	289
	ARCTURUS		◆Rasalhague		ANTARES		◆ACRUX		Alphard		◆REGULUS		Alioth	
195	62 36	042	20 33	076	33 45	122	27 37	184	36 32	260	45 24	288	33 00	359
196	63 15	040	21 31	076	34 36	122	27 32	185	35 33	260	44 27	288	32 59	358
197	63 53	038	22 30	076	35 27	122	27 27	185	34 34	260	43 30	288	32 56	358
198	64 29	037	23 28	076	36 18	123	27 21	186	33 35	260	42 32	287	32 54	357
199	65 05	035	24 26	076	37 08	123	27 15	186	32 35	260	41 35	287	32 50	356
200	65 38	033	25 24	076	37 58	123	27 08	187	31 36	260	40 38	287	32 46	356
201	66 10	031	26 22	075	38 48	124	27 00	187	30 37	261	39 40	287	32 41	355
202	66 41	029	27 20	075	39 38	124	26 53	188	29 38	261	38 43	286	32 35	354
203	67 09	027	28 18	075	40 27	125	26 44	188	28 39	261	37 45	286	32 29	354
204	67 36	025	29 16	075	41 17	125	26 35	189	27 39	261	36 47	286	32 22	353
205	68 00	023	30 14	075	42 05	126	26 26	189	26 40	261	35 50	286	32 15	352
206	68 22	021	31 12	075	42 54	126	26 16	190	25 41	261	34 52	285	32 06	352
207	68 42	018	32 10	074	43 42	127	26 06	190	24 42	261	33 54	285	31 57	351
208	68 59	016	33 08	074	44 30	127	25 55	191	23 43	261	32 56	285	31 48	350
209	69 14	013	34 05	074	45 18	128	25 44	191	22 43	261	31 58	285	31 38	350
	ARCTURUS		◆VEGA		ANTARES		◆RIGIL KENT.		ACRUX		Gienah		◆REGULUS	
210	69 26	010	15 25	049	46 05	129	29 43	175	25 32	192	59 29	236	31 00	285
211	69 36	008	16 11	049	46 51	129	29 48	175	25 20	192	58 39	237	30 02	284
212	69 43	005	16 56	049	47 38	130	29 53	176	25 07	192	57 48	238	29 04	284
213	69 46	002	17 41	048	48 23	131	29 57	176	24 54	193	56 58	238	28 06	284
214	69 47	000	18 26	048	49 09	131	30 01	177	24 41	193	56 06	239	27 07	284
215	69 46	357	19 10	048	49 54	132	30 04	177	24 27	194	55 15	240	26 09	284
216	69 41	354	19 55	048	50 38	133	30 07	178	24 12	194	54 23	241	25 11	284
217	69 33	351	20 39	047	51 21	134	30 09	178	23 57	195	53 30	241	24 13	284
218	69 23	349	21 23	047	52 05	135	30 10	179	23 42	195	52 38	242	23 14	284
219	69 10	346	22 07	047	52 47	135	30 11	180	23 26	195	51 45	242	22 16	283
220	68 55	344	22 51	047	53 29	136	30 11	180	23 10	196	50 51	243	21 18	283
221	68 37	341	23 34	046	54 10	137	30 11	181	22 54	196	49 58	243	20 19	283
222	68 16	339	24 18	046	54 50	138	30 10	181	22 37	197	49 04	244	19 21	283
223	67 53	337	25 01	046	55 29	139	30 08	182	22 20	197	48 10	244	18 22	283
224	67 28	334	25 43	045	56 08	141	30 06	182	22 02	197	47 16	245	17 24	283
	Alphecca		VEGA		◆ALTAIR		ANTARES		◆RIGIL KENT.		SPICA		◆ARCTURUS	
225	61 03	016	26 26	045	16 59	080	56 45	142	30 03	183	64 19	246	67 01	332
226	61 19	014	27 08	045	17 58	080	57 22	143	30 00	183	63 24	247	66 32	330
227	61 33	012	27 50	044	18 58	080	57 58	144	29 56	184	62 29	247	66 01	328
228	61 45	011	28 32	044	19 57	080	58 32	146	29 51	185	61 33	248	65 29	326
229	61 55	009	29 13	043	20 56	080	59 06	147	29 46	185	60 37	249	64 55	324
230	62 03	007	29 54	043	21 55	080	59 38	148	29 40	186	59 41	249	64 19	323
231	62 09	005	30 35	042	22 54	080	60 09	150	29 34	186	58 45	250	63 42	321
232	62 13	003	31 15	042	23 53	080	60 38	151	29 28	187	57 49	250	63 04	320
233	62 16	001	31 55	042	24 52	080	61 06	153	29 20	187	56 52	251	62 24	318
234	62 16	359	32 34	041	25 51	080	61 33	154	29 12	188	55 55	251	61 44	317
235	62 14	357	33 14	041	26 50	080	61 58	156	29 04	188	54 58	252	61 02	315
236	62 10	355	33 52	040	27 49	079	62 22	158	28 55	189	54 01	252	60 19	314
237	62 05	354	34 31	039	28 48	079	62 43	160	28 45	189	53 04	253	59 36	313
238	61 57	352	35 09	039	29 47	079	63 04	161	28 35	190	52 07	253	58 51	312
239	61 47	350	35 46	038	30 46	079	63 22	163	28 24	190	51 09	253	58 06	311
	VEGA		◆ALTAIR		Nunki		◆RIGIL KENT.		SPICA		◆ARCTURUS		Alphecca	
240	36 23	038	31 45	079	40 57	125	28 13	191	50 12	254	57 20	310	61 36	348
241	36 59	037	32 44	079	41 46	125	28 01	192	49 14	254	56 34	309	61 22	346
242	37 35	036	33 43	079	42 35	126	27 49	192	48 17	254	55 46	308	61 07	344
243	38 11	036	34 41	079	43 23	126	27 36	193	47 19	255	54 59	307	60 50	343
244	38 46	035	35 40	078	44 11	127	27 23	193	46 21	255	54 10	306	60 31	341
245	39 20	034	36 39	078	44 59	128	27 09	193	45 23	255	53 21	305	60 11	339
246	39 54	034	37 38	078	45 46	128	26 55	194	44 25	255	52 32	304	59 49	338
247	40 27	033	38 36	078	46 33	129	26 41	194	43 27	256	51 42	304	59 25	336
248	40 59	032	39 35	078	47 20	129	26 25	195	42 29	256	50 52	303	59 00	334
249	41 31	032	40 33	077	48 06	130	26 10	195	41 31	256	50 02	302	58 33	333
250	42 02	031	41 32	077	48 52	131	25 54	196	40 32	256	49 11	302	58 05	332
251	42 32	030	42 30	077	49 37	132	25 37	196	39 34	256	48 19	301	57 36	330
252	43 02	029	43 29	077	50 21	132	25 20	197	38 36	257	47 28	300	57 06	329
253	43 31	028	44 27	077	51 05	133	25 02	197	37 37	257	46 36	300	56 34	327
254	43 59	027	45 25	076	51 49	134	24 45	198	36 39	257	45 44	299	56 01	326
	◆VEGA		ALTAIR		◆Nunki		Shaula		ANTARES		◆SPICA		ARCTURUS	
255	44 26	027	46 24	076	52 32	135	53 05	169	63 31	196	35 41	257	44 51	299
256	44 52	026	47 22	076	53 14	136	53 16	170	63 14	198	34 42	257	43 59	298
257	45 18	025	48 20	075	53 55	137	53 26	172	62 55	199	33 44	257	43 06	298
258	45 43	024	49 18	075	54 36	138	53 34	173	62 34	201	32 45	257	42 12	297
259	46 06	023	50 16	075	55 16	139	53 41	174	62 11	203	31 47	258	41 19	297
260	46 29	022	51 14	074	55 55	140	53 46	176	61 47	205	30 48	258	40 26	297
261	46 51	021	52 12	074	56 33	141	53 50	177	61 21	206	29 49	258	39 32	296
262	47 12	020	53 09	074	57 10	142	53 53	178	60 54	208	28 51	258	38 38	296
263	47 32	019	54 07	073	57 47	143	53 54	180	60 25	209	27 52	258	37 44	295
264	47 50	018	55 04	073	58 22	145	53 54	181	59 55	211	26 53	258	36 50	295
265	48 08	017	56 01	072	58 56	146	53 52	182	59 24	212	25 55	258	35 55	295
266	48 25	016	56 59	072	59 29	147	53 49	184	58 51	214	24 56	258	35 01	294
267	48 40	014	57 56	071	60 00	149	53 44	185	58 17	215	23 57	258	34 06	294
268	48 55	013	58 52	071	60 31	150	53 38	186	57 42	216	22 59	258	33 11	294
269	49 08	012	59 49	070	61 00	152	53 31	188	57 06	218	22 00	258	32 16	294

LAT 1°S

LHA	Hc	Zn	Hc	Zn	Hc	Zn	Hc	Zn	Hc	Zn	Hc	Zn	Hc	Zn
♈	◆VEGA		DENEB		Enif		◆FOMALHAUT		Peacock		◆ANTARES		ARCTURUS	
270	49 20	011	31 37	032	33 14	078	14 04	120	27 09	159	56 29	219	31 21	293
271	49 31	010	32 09	032	34 13	077	14 56	120	27 31	159	55 51	220	30 26	293
272	49 41	009	32 40	031	35 11	077	15 48	121	27 52	160	55 12	221	29 31	293
273	49 49	007	33 11	031	36 10	077	16 39	121	28 13	160	54 32	222	28 35	293
274	49 56	006	33 41	030	37 08	077	17 31	121	28 33	161	53 51	223	27 40	292
275	50 02	005	34 11	029	38 07	077	18 22	121	28 53	161	53 10	224	26 44	292
276	50 07	004	34 40	029	39 05	076	19 14	121	29 12	162	52 28	225	25 49	292
277	50 10	003	35 09	028	40 03	076	20 05	121	29 31	162	51 45	226	24 53	292
278	50 12	001	35 37	028	41 01	076	20 56	122	29 49	163	51 02	227	23 57	292
279	50 13	000	36 04	027	42 00	076	21 47	122	30 07	163	50 18	227	23 02	291
280	50 13	359	36 31	026	42 58	076	22 38	122	30 24	164	49 33	228	22 06	291
281	50 11	358	36 57	026	43 56	075	23 29	122	30 40	164	48 48	229	21 10	291
282	50 08	357	37 23	025	44 54	075	24 20	122	30 56	165	48 03	230	20 14	291
283	50 04	355	37 48	024	45 52	075	25 10	123	31 12	165	47 17	230	19 18	291
284	49 59	354	38 12	023	46 49	074	26 01	123	31 27	166	46 30	231	18 22	291
	DENEB		Enif		◆FOMALHAUT		Peacock		◆ANTARES		Rasalhague		◆VEGA	
285	38 36	023	47 47	074	26 51	123	31 41	166	45 43	232	64 52	303	49 52	353
286	38 58	022	48 45	074	27 41	123	31 55	167	44 56	232	64 01	302	49 44	352
287	39 20	021	49 42	073	28 31	124	32 08	168	44 09	233	63 10	301	49 35	351
288	39 41	020	50 40	073	29 21	124	32 21	168	43 21	233	62 19	300	49 24	349
289	40 02	020	51 37	073	30 11	124	32 33	169	42 32	234	61 26	299	49 13	348
290	40 22	019	52 34	072	31 00	125	32 44	169	41 43	234	60 34	298	49 00	347
291	40 40	018	53 31	072	31 50	125	32 55	170	40 54	235	59 41	297	48 46	346
292	40 58	017	54 28	071	32 39	125	33 05	171	40 05	235	58 47	297	48 31	345
293	41 16	016	55 25	071	33 28	126	33 14	171	39 16	236	57 54	296	48 15	344
294	41 32	015	56 22	070	34 17	126	33 23	172	38 26	236	56 59	295	47 57	343
295	41 47	014	57 18	070	35 05	126	33 31	173	37 36	237	56 05	295	47 39	342
296	42 02	014	58 14	069	35 53	127	33 38	173	36 46	237	55 10	294	47 20	341
297	42 15	013	59 10	069	36 41	127	33 45	174	35 55	238	54 15	293	46 59	340
298	42 28	012	60 06	068	37 29	128	33 51	175	35 04	238	53 20	293	46 38	339
299	42 40	011	61 02	067	38 16	128	33 56	175	34 13	238	52 25	292	46 15	338
	◆DENEB		Alpheratz		◆FOMALHAUT		Peacock		◆ANTARES		Rasalhague		VEGA	
300	42 51	010	23 40	057	39 03	129	34 01	176	33 22	239	51 29	292	45 52	337
301	43 00	009	24 30	057	39 50	129	34 05	176	32 31	239	50 33	291	45 27	336
302	43 09	008	25 21	057	40 36	130	34 08	177	31 40	239	49 37	291	45 02	335
303	43 17	007	26 11	057	41 23	130	34 11	178	30 48	239	48 41	290	44 36	334
304	43 24	006	27 01	056	42 08	131	34 13	178	29 56	240	47 45	290	44 09	333
305	43 30	005	27 51	056	42 54	131	34 14	179	29 04	240	46 49	290	43 42	332
306	43 35	004	28 40	056	43 38	132	34 15	180	28 12	240	45 52	289	43 13	331
307	43 39	003	29 30	055	44 23	132	34 15	180	27 20	241	44 55	289	42 44	330
308	43 42	002	30 19	055	45 07	133	34 14	181	26 28	241	43 59	289	42 14	330
309	43 43	001	31 08	055	45 51	134	34 12	182	25 35	241	43 02	288	41 43	329
310	43 44	000	31 57	054	46 34	135	34 10	182	24 43	241	42 05	288	41 11	328
311	43 44	359	32 46	054	47 16	135	34 07	183	23 50	241	41 08	288	40 39	327
312	43 43	358	33 34	054	47 58	136	34 04	184	22 58	242	40 10	287	40 06	326
313	43 41	357	34 22	053	48 39	137	33 59	184	22 05	242	39 13	287	39 33	326
314	43 37	356	35 10	053	49 20	138	33 54	185	21 12	242	38 16	287	38 59	325
	◆Alpheratz		◆FOMALHAUT		Peacock		Nunki		◆Rasalhague		VEGA		DENEB	
315	35 58	052	50 00	139	33 49	186	50 42	227	37 18	287	38 24	324	43 33	355
316	36 45	052	50 40	139	33 43	186	49 58	228	36 21	286	37 49	324	43 28	354
317	37 32	051	51 18	140	33 36	187	49 13	229	35 23	286	37 13	323	43 22	354
318	38 18	051	51 56	141	33 28	188	48 28	230	34 26	286	36 37	323	43 14	353
319	39 05	050	52 33	142	33 20	188	47 42	230	33 28	286	36 00	322	43 06	352
320	39 51	050	53 10	143	33 11	189	46 56	231	32 30	286	35 23	321	42 57	351
321	40 36	049	53 45	144	33 01	190	46 09	232	31 32	285	34 45	321	42 47	350
322	41 21	049	54 19	145	32 51	190	45 22	232	30 34	285	34 07	320	42 36	349
323	42 06	048	54 53	147	32 40	191	44 34	233	29 37	285	33 29	320	42 23	348
324	42 50	047	55 25	148	32 28	191	43 46	233	28 39	285	32 49	319	42 10	347
325	43 34	047	55 57	149	32 16	192	42 58	234	27 41	285	32 10	319	41 56	346
326	44 17	046	56 27	150	32 03	193	42 09	234	26 43	285	31 30	318	41 42	345
327	45 00	045	56 56	152	31 50	193	41 20	235	25 45	284	30 50	318	41 26	344
328	45 42	044	57 24	153	31 36	194	40 31	235	24 46	284	30 10	317	41 09	344
329	46 24	044	57 51	154	31 21	194	39 41	236	23 48	284	29 29	317	40 52	343
	◆Alpheratz		Hamal		Diphda		◆FOMALHAUT		Nunki		◆ALTAIR		DENEB	
330	47 05	043	25 19	063	46 27	116	58 16	156	38 52	236	56 17	288	40 34	342
331	47 46	042	26 13	063	47 21	116	58 40	157	38 02	237	55 20	287	40 14	341
332	48 26	041	27 06	063	48 15	116	59 02	159	37 11	237	54 22	287	39 54	340
333	49 05	040	27 59	063	49 09	117	59 23	160	36 21	237	53 25	286	39 34	339
334	49 43	039	28 53	062	50 02	117	59 42	162	35 30	238	52 27	286	39 12	339
335	50 21	039	29 46	062	50 55	118	60 00	164	34 39	238	51 29	286	38 50	338
336	50 58	038	30 39	062	51 48	119	60 16	165	33 48	239	50 32	285	38 27	337
337	51 34	037	31 31	061	52 40	119	60 31	167	32 57	239	49 34	285	38 03	336
338	52 10	035	32 24	061	53 33	120	60 43	169	32 05	239	48 36	285	37 39	336
339	52 44	034	33 17	061	54 25	120	60 54	170	31 14	239	47 38	284	37 14	335
340	53 17	033	34 09	060	55 16	121	61 03	172	30 22	240	46 39	284	36 48	334
341	53 50	032	35 01	060	56 07	122	61 10	174	29 30	240	45 41	284	36 21	334
342	54 21	031	35 53	060	56 58	123	61 16	176	28 38	240	44 43	284	35 54	333
343	54 51	030	36 45	059	57 48	124	61 19	178	27 46	241	43 45	283	35 27	332
344	55 20	028	37 36	059	58 38	124	61 21	179	26 54	241	42 46	283	34 58	332
	Alpheratz		◆Hamal		◆ACHERNAR		FOMALHAUT		Nunki		◆ALTAIR		DENEB	
345	55 48	027	38 27	059	25 38	158	61 21	181	26 01	241	41 48	283	34 30	331
346	56 15	026	39 18	058	26 01	158	61 18	183	25 09	241	40 49	283	34 00	330
347	56 41	024	40 09	058	26 23	159	61 14	185	24 16	241	39 51	282	33 30	330
348	57 05	023	41 00	057	26 45	159	61 08	187	23 23	242	38 52	282	33 00	329
349	57 27	021	41 50	057	27 06	159	61 01	188	22 31	242	37 53	282	32 29	329
350	57 49	020	42 40	056	27 27	160	60 51	190	21 38	242	36 55	282	31 57	328
351	58 08	018	43 30	056	27 47	160	60 40	192	20 45	242	35 56	282	31 25	327
352	58 27	017	44 19	055	28 07	161	60 26	194	19 52	242	34 57	282	30 53	327
353	58 43	015	45 08	054	28 27	161	60 11	195	18 58	242	33 58	281	30 20	326
354	58 58	014	45 57	054	28 46	162	59 55	197	18 05	243	33 00	281	29 46	326
355	59 12	012	46 45	053	29 04	162	59 37	199	17 12	243	32 01	281	29 12	325
356	59 23	010	47 33	053	29 22	163	59 17	200	16 19	243	31 02	281	28 38	325
357	59 33	009	48 20	052	29 39	163	58 55	202	15 25	243	30 03	281	28 04	325
358	59 41	007	49 07	051	29 56	164	58 32	203	14 32	243	29 04	281	27 29	324
359	59 48	005	49 53	050	30 13	164	58 08	205	13 38	243	28 05	281	26 53	324

 LAT 2°S

LHA	Hc	Zn	Hc	Zn	Hc	Zn	Hc	Zn	Hc	Zn	Hc	Zn	Hc	Zn
ϒ	Alpheratz		◆ALDEBARAN		RIGEL		ACHERNAR		◆FOMALHAUT		ALTAIR		◆DENEB	
0	58 53	003	19 34	072	11 35	098	31 26	165	58 36	207	26 55	281	25 29	324
1	58 55	002	20 31	072	12 35	098	31 42	165	58 09	208	25 56	281	24 53	323
2	58 56	000	21 28	071	13 34	098	31 57	166	57 40	210	24 57	281	24 17	323
3	58 55	358	22 25	071	14 34	098	32 11	167	57 09	211	23 58	281	23 41	322
4	58 53	357	23 21	071	15 33	098	32 25	167	56 38	212	22 59	280	23 04	322
5	58 48	355	24 18	071	16 32	098	32 38	168	56 06	213	22 00	280	22 27	322
6	58 42	353	25 15	071	17 32	098	32 50	168	55 32	215	21 01	280	21 49	321
7	58 34	352	26 11	070	18 31	098	33 02	169	54 58	216	20 02	280	21 12	321
8	58 25	350	27 08	070	19 31	098	33 13	169	54 22	217	19 03	280	20 34	321
9	58 14	348	28 04	070	20 30	098	33 24	170	53 46	218	18 04	280	19 56	320
10	58 01	347	29 00	070	21 29	098	33 34	171	53 09	219	17 05	280	19 17	320
11	57 46	345	29 57	070	22 29	098	33 43	171	52 31	220	16 06	280	18 38	320
12	57 30	344	30 53	069	23 28	098	33 52	172	51 52	221	15 07	280	17 59	319
13	57 13	342	31 49	069	24 27	098	34 00	173	51 12	222	14 08	280	17 20	319
14	56 53	341	32 45	069	25 27	098	34 07	173	50 32	223	13 09	280	16 41	319
	CAPELLA		ALDEBARAN		◆RIGEL		ACHERNAR		◆FOMALHAUT		Enif		◆Alpheratz	
15	16 10	041	33 41	069	26 26	098	34 14	174	49 51	223	39 47	285	56 33	339
16	16 49	040	34 37	068	27 25	098	34 20	175	49 10	224	38 49	284	56 11	338
17	17 27	040	35 32	068	28 25	098	34 26	175	48 28	225	37 51	284	55 47	336
18	18 06	040	36 28	068	29 24	098	34 30	176	47 45	226	36 52	284	55 23	335
19	18 44	039	37 23	067	30 24	098	34 34	176	47 02	226	35 54	284	54 57	334
20	19 22	039	38 19	067	31 23	098	34 38	177	46 18	227	34 56	283	54 29	332
21	20 00	039	39 14	067	32 22	098	34 40	178	45 34	228	33 58	283	54 01	331
22	20 37	039	40 09	066	33 21	099	34 42	178	44 49	228	32 59	283	53 31	330
23	21 15	038	41 04	066	34 21	099	34 44	179	44 04	229	32 01	283	53 01	329
24	21 52	038	41 58	066	35 20	099	34 44	180	43 19	230	31 02	283	52 29	328
25	22 28	038	42 53	065	36 19	099	34 44	180	42 33	230	30 04	283	51 56	326
26	23 05	037	43 47	065	37 19	099	34 43	181	41 46	231	29 05	282	51 23	325
27	23 41	037	44 41	064	38 18	099	34 42	182	41 00	231	28 07	282	50 48	324
28	24 16	036	45 35	064	39 17	099	34 40	182	40 13	232	27 08	282	50 13	323
29	24 52	036	46 29	063	40 16	099	34 37	183	39 26	232	26 09	282	49 36	322
	CAPELLA		◆BETELGEUSE		SIRIUS		CANOPUS		◆ACHERNAR		FOMALHAUT		◆Alpheratz	
30	25 27	036	30 40	080	18 33	107	15 56	145	34 33	184	38 38	233	48 59	321
31	26 02	035	31 39	080	19 30	107	16 30	145	34 29	184	37 50	233	48 21	320
32	26 36	035	32 38	080	20 27	107	17 05	145	34 24	185	37 02	234	47 43	320
33	27 10	034	33 37	080	21 24	107	17 39	145	34 19	186	36 14	234	47 04	319
34	27 43	034	34 36	080	22 22	107	18 13	146	34 13	186	35 25	234	46 24	318
35	28 16	033	35 35	079	23 19	107	18 46	146	34 06	187	34 36	235	45 43	317
36	28 49	033	36 34	079	24 16	107	19 20	146	33 58	188	33 47	235	45 02	316
37	29 21	032	37 33	079	25 13	108	19 53	146	33 50	188	32 58	235	44 20	316
38	29 53	032	38 32	079	26 11	108	20 26	147	33 41	189	32 08	236	43 38	315
39	30 24	031	39 30	079	27 08	108	20 59	147	33 32	189	31 19	236	42 55	314
40	30 55	031	40 29	078	28 05	108	21 31	147	33 21	190	30 29	236	42 12	313
41	31 26	030	41 28	078	29 02	108	22 03	148	33 11	191	29 39	237	41 28	313
42	31 56	030	42 27	078	29 59	108	22 35	148	32 59	191	28 49	237	40 44	312
43	32 25	029	43 25	078	30 56	108	23 07	148	32 47	192	27 58	237	39 59	312
44	32 54	028	44 24	078	31 53	108	23 38	149	32 35	192	27 08	237	39 14	311
	CAPELLA		◆BETELGEUSE		SIRIUS		CANOPUS		◆ACHERNAR		Diphda		◆Hamal	
45	33 22	028	45 22	077	32 49	109	24 10	149	32 21	193	52 52	242	61 28	334
46	33 50	027	46 21	077	33 46	109	24 40	149	32 08	194	51 58	243	61 01	332
47	34 17	027	47 19	077	34 43	109	25 11	150	31 53	194	51 05	243	60 32	331
48	34 43	026	48 18	077	35 40	109	25 41	150	31 38	195	50 11	244	60 02	329
49	35 09	025	49 16	076	36 36	109	26 11	150	31 23	195	49 17	244	59 30	328
50	35 34	025	50 14	076	37 33	110	26 40	151	31 07	196	48 23	245	58 57	326
51	35 59	024	51 12	076	38 29	110	27 09	151	30 50	196	47 29	245	58 23	325
52	36 23	023	52 10	075	39 26	110	27 38	152	30 33	197	46 34	246	57 48	323
53	36 46	022	53 08	075	40 22	110	28 06	152	30 15	197	45 40	246	57 12	322
54	37 08	022	54 06	074	41 18	111	28 34	153	29 57	198	44 45	246	56 34	321
55	37 30	021	55 04	074	42 14	111	29 01	153	29 38	198	43 50	247	55 56	320
56	37 51	020	56 01	074	43 10	111	29 28	153	29 19	199	42 55	247	55 17	319
57	38 12	019	56 59	073	44 06	112	29 55	154	28 59	199	41 59	247	54 37	317
58	38 31	019	57 56	073	45 02	112	30 21	154	28 39	200	41 04	248	53 56	316
59	38 50	018	58 53	072	45 57	112	30 47	155	28 18	200	40 09	248	53 14	315
	◆CAPELLA		POLLUX		◆SIRIUS		CANOPUS		ACHERNAR		◆Diphda		Hamal	
60	39 08	017	28 16	056	46 53	113	31 12	155	27 57	201	39 13	248	52 32	314
61	39 25	016	29 06	056	47 48	113	31 37	156	27 35	201	38 17	248	51 48	313
62	39 41	015	29 56	056	48 43	113	32 01	156	27 13	202	37 21	249	51 05	313
63	39 57	015	30 45	055	49 38	114	32 25	157	26 51	202	36 25	249	50 20	312
64	40 11	014	31 34	055	50 33	114	32 48	158	26 28	203	35 29	249	49 35	311
65	40 25	013	32 23	055	51 27	115	33 10	158	26 05	203	34 33	249	48 49	310
66	40 38	012	33 12	054	52 22	115	33 32	159	25 41	203	33 37	250	48 03	309
67	40 50	011	34 01	054	53 16	116	33 54	159	25 17	204	32 41	250	47 17	309
68	41 01	010	34 49	053	54 10	116	34 15	160	24 52	204	31 45	250	46 30	308
69	41 11	009	35 37	053	55 03	117	34 35	161	24 28	205	30 48	250	45 42	307
70	41 20	008	36 25	052	55 57	118	34 55	161	24 02	205	29 52	250	44 54	307
71	41 29	007	37 12	052	56 50	118	35 14	162	23 37	205	28 55	250	44 06	306
72	41 36	007	37 59	051	57 42	119	35 32	162	23 11	206	27 59	251	43 17	305
73	41 42	006	38 46	051	58 35	120	35 50	163	22 45	206	27 02	251	42 28	305
74	41 48	005	39 32	050	59 27	120	36 07	164	22 18	206	26 06	251	41 39	304
	CAPELLA		◆POLLUX		REGULUS		Suhail		◆CANOPUS		ACHERNAR		◆Hamal	
75	41 52	004	40 18	050	12 15	077	21 26	136	36 24	164	21 51	207	40 49	304
76	41 56	003	41 03	049	13 14	077	22 07	137	36 40	165	21 24	207	39 59	303
77	41 58	002	41 49	049	14 12	077	22 48	137	36 55	166	20 57	207	39 09	303
78	42 00	001	42 33	048	15 11	077	23 29	137	37 09	166	20 29	208	38 18	302
79	42 00	000	43 18	047	16 09	077	24 09	138	37 23	167	20 01	208	37 28	302
80	42 00	359	44 01	047	17 07	077	24 50	138	37 36	168	19 33	208	36 37	301
81	41 59	358	44 45	046	18 06	077	25 30	138	37 48	169	19 04	209	35 45	301
82	41 56	357	45 27	045	19 04	077	26 10	139	38 00	169	18 35	209	34 54	301
83	41 53	356	46 09	044	20 02	076	26 49	139	38 10	170	18 06	209	34 02	300
84	41 49	355	46 51	044	21 01	076	27 29	139	38 20	171	17 37	209	33 10	300
85	41 43	354	47 32	043	21 59	076	28 08	140	38 30	172	17 08	210	32 18	300
86	41 37	354	48 12	042	22 57	076	28 46	140	38 38	172	16 38	210	31 26	299
87	41 30	353	48 52	041	23 55	076	29 25	140	38 46	173	16 08	210	30 33	299
88	41 22	352	49 31	040	24 53	076	30 03	141	38 53	174	15 38	210	29 41	299
89	41 13	351	50 09	039	25 52	076	30 41	141	38 59	175	15 08	210	28 48	298

LHA	Hc	Zn	Hc	Zn	Hc	Zn	Hc	Zn	Hc	Zn	Hc	Zn	Hc	Zn
ϒ	POLLUX		◆REGULUS		Suhail		◆CANOPUS		Acamar		◆ALDEBARAN		CAPELLA	
90	50 47	038	26 50	075	31 18	142	39 04	175	33 50	221	62 09	312	41 03	350
91	51 23	037	27 48	075	31 55	142	39 08	176	33 11	221	61 24	311	40 52	349
92	51 59	036	28 46	075	32 32	143	39 12	177	32 31	222	60 38	310	40 40	348
93	52 34	035	29 44	075	33 08	143	39 15	178	31 51	222	59 52	309	40 27	347
94	53 08	034	30 41	075	33 44	143	39 17	178	31 10	223	59 05	308	40 14	346
95	53 41	033	31 39	075	34 20	144	39 18	179	30 30	223	58 17	307	39 59	346
96	54 13	031	32 37	074	34 55	145	39 18	180	29 48	223	57 29	306	39 44	345
97	54 43	030	33 35	074	35 29	145	39 18	181	29 07	224	56 40	305	39 28	344
98	55 13	029	34 32	074	36 03	146	39 17	182	28 25	224	55 50	304	39 11	343
99	55 41	028	35 30	074	36 37	146	39 15	182	27 44	225	55 00	303	38 53	342
100	56 09	026	36 28	073	37 10	147	39 12	183	27 01	225	54 10	302	38 34	342
101	56 35	025	37 25	073	37 43	147	39 08	184	26 19	225	53 19	302	38 15	341
102	56 59	024	38 22	073	38 15	148	39 04	185	25 36	225	52 27	301	37 54	340
103	57 22	022	39 20	073	38 46	149	38 58	185	24 53	226	51 36	300	37 34	339
104	57 44	021	40 17	072	39 17	149	38 52	186	24 10	226	50 44	299	37 12	338
	POLLUX		◆REGULUS		Suhail		◆CANOPUS		RIGEL		◆ALDEBARAN		CAPELLA	
105	58 04	019	41 14	072	39 47	150	38 45	187	62 58	256	49 51	299	36 49	338
106	58 23	017	42 11	072	40 17	151	38 37	188	62 00	256	48 59	298	36 26	337
107	58 40	016	43 08	071	40 46	151	38 29	189	61 02	257	48 06	298	36 03	336
108	58 56	014	44 05	071	41 14	152	38 20	189	60 03	257	47 12	297	35 38	336
109	59 10	013	45 01	071	41 42	153	38 10	190	59 05	257	46 19	297	35 13	335
110	59 22	011	45 58	070	42 09	154	37 59	191	58 06	258	45 25	296	34 47	334
111	59 32	009	46 54	070	42 35	154	37 47	191	57 08	258	44 31	296	34 21	334
112	59 41	007	47 51	070	43 01	155	37 35	192	56 09	258	43 37	295	33 54	333
113	59 48	006	48 47	069	43 26	156	37 22	193	55 10	258	42 43	295	33 26	332
114	59 53	004	49 43	069	43 50	157	37 08	194	54 12	259	41 48	294	32 58	332
115	59 56	002	50 39	068	44 13	158	36 54	194	53 13	259	40 54	294	32 30	331
116	59 58	000	51 34	068	44 35	159	36 38	195	52 14	259	39 59	294	32 00	331
117	59 57	359	52 30	067	44 57	159	36 23	196	51 15	259	39 04	293	31 31	330
118	59 55	357	53 25	067	45 17	160	36 06	196	50 16	260	38 08	293	31 00	329
119	59 51	355	54 20	066	45 37	161	35 49	197	49 17	260	37 13	293	30 29	329
	◆REGULUS		Gienah		◆ACRUX		CANOPUS		SIRIUS		◆BETELGEUSE		POLLUX	
120	55 14	065	25 28	108	12 11	155	35 31	198	66 24	230	57 25	287	59 45	353
121	56 09	065	26 25	109	12 36	155	35 12	198	65 37	232	56 27	287	59 37	352
122	57 03	064	27 22	109	13 02	155	34 53	199	64 50	233	55 30	286	59 27	350
123	57 57	063	28 18	109	13 27	155	34 34	200	64 01	234	54 32	286	59 16	348
124	58 50	063	29 15	109	13 52	156	34 13	200	63 12	235	53 34	285	59 03	347
125	59 43	062	30 12	109	14 16	156	33 52	201	62 23	236	52 36	285	58 48	345
126	60 36	061	31 08	109	14 41	156	33 31	201	61 33	237	51 38	285	58 32	343
127	61 28	060	32 05	109	15 05	156	33 09	202	60 42	238	50 40	284	58 14	342
128	62 20	059	33 01	110	15 29	156	32 46	202	59 51	239	49 42	284	57 55	340
129	63 11	058	33 58	110	15 53	157	32 23	203	58 59	240	48 44	284	57 34	339
130	64 01	057	34 54	110	16 17	157	31 59	204	58 07	241	47 46	283	57 11	337
131	64 51	056	35 51	110	16 41	157	31 35	204	57 14	241	46 47	283	56 47	336
132	65 40	054	36 47	110	17 04	157	31 10	205	56 21	242	45 49	283	56 22	334
133	66 29	053	37 43	111	17 27	158	30 45	205	55 28	243	44 50	283	55 55	333
134	67 16	052	38 39	111	17 50	158	30 19	206	54 35	243	43 52	282	55 27	332
	Alioth		◆SPICA		ACRUX		◆CANOPUS		SIRIUS		BETELGEUSE		◆POLLUX	
135	15 17	030	23 42	101	18 12	158	29 53	206	53 41	244	42 53	282	54 58	330
136	15 46	029	24 41	101	18 34	158	29 26	207	52 47	245	41 54	282	54 28	329
137	16 16	029	25 40	101	18 56	159	28 59	207	51 53	245	40 56	282	53 57	328
138	16 45	029	26 38	101	19 18	159	28 32	208	50 58	246	39 57	281	53 25	327
139	17 13	028	27 37	102	19 40	159	28 04	208	50 03	246	38 58	281	52 51	326
140	17 42	028	28 36	102	20 01	160	27 36	208	49 09	246	37 59	281	52 17	324
141	18 10	028	29 35	102	20 22	160	27 07	209	48 14	247	37 00	281	51 41	323
142	18 38	027	30 33	102	20 42	160	26 38	209	47 18	247	36 01	281	51 05	322
143	19 05	027	31 32	102	21 02	160	26 08	210	46 23	248	35 03	280	50 28	321
144	19 32	027	32 31	102	21 22	161	25 39	210	45 27	248	34 04	280	49 50	320
145	19 59	026	33 29	102	21 42	161	25 08	210	44 32	248	33 05	280	49 12	319
146	20 26	026	34 28	102	22 01	161	24 38	211	43 36	249	32 05	280	48 33	319
147	20 52	026	35 27	102	22 20	162	24 07	211	42 40	249	31 06	280	47 53	318
148	21 18	025	36 25	102	22 38	162	23 36	211	41 44	249	30 07	280	47 12	317
149	21 43	025	37 24	103	22 57	163	23 05	212	40 48	250	29 08	280	46 31	316
	Alioth		ARCTURUS		◆SPICA		ACRUX		◆CANOPUS		SIRIUS		◆POLLUX	
150	22 09	025	23 51	068	38 22	103	23 14	163	22 33	212	39 52	250	45 49	315
151	22 33	024	24 46	068	39 21	103	23 32	163	22 01	212	38 55	250	45 06	315
152	22 58	024	25 42	068	40 19	103	23 49	164	21 29	213	37 59	250	44 23	314
153	23 21	023	26 37	067	41 18	103	24 05	164	20 56	213	37 02	250	43 40	313
154	23 45	023	27 33	067	42 16	103	24 22	164	20 23	213	36 06	251	42 56	312
155	24 08	022	28 28	067	43 14	103	24 38	165	19 50	214	35 09	251	42 11	312
156	24 31	022	29 23	067	44 13	104	24 53	165	19 17	214	34 13	251	41 26	311
157	24 53	021	30 18	066	45 11	104	25 08	166	18 44	214	33 16	251	40 41	311
158	25 15	021	31 13	066	46 09	104	25 23	166	18 10	214	32 19	251	39 55	310
159	25 36	021	32 07	066	47 07	104	25 37	167	17 36	215	31 22	252	39 09	309
160	25 57	020	33 02	065	48 05	105	25 51	167	17 02	215	30 25	252	38 23	309
161	26 17	020	33 56	065	49 03	105	26 04	167	16 28	215	29 28	252	37 36	308
162	26 37	019	34 51	065	50 01	105	26 17	168	15 53	215	28 31	252	36 48	308
163	26 56	019	35 45	064	50 59	105	26 29	168	15 19	215	27 34	252	36 01	307
164	27 15	018	36 39	064	51 57	106	26 41	169	14 44	216	26 37	252	35 13	307
	Alioth		◆ARCTURUS		SPICA		◆ACRUX		Suhail		SIRIUS		◆POLLUX	
165	27 33	017	37 33	064	52 55	106	26 52	169	41 40	207	25 40	252	34 25	306
166	27 51	017	38 27	063	53 52	106	27 03	170	41 12	208	24 43	253	33 37	306
167	28 08	016	39 20	063	54 50	107	27 14	170	40 43	209	23 46	253	32 48	306
168	28 25	016	40 13	063	55 47	107	27 24	171	40 14	209	22 48	253	31 59	305
169	28 41	015	41 07	062	56 44	107	27 33	171	39 45	210	21 51	253	31 10	305
170	28 57	015	41 59	062	57 42	108	27 42	172	39 14	211	20 54	253	30 21	304
171	29 12	014	42 52	061	58 39	108	27 51	172	38 43	211	19 56	253	29 31	304
172	29 26	014	43 45	061	59 35	109	27 59	173	38 12	212	18 59	253	28 41	304
173	29 40	013	44 37	060	60 32	109	28 06	173	37 40	213	18 02	253	27 51	303
174	29 53	012	45 29	060	61 29	110	28 13	174	37 07	213	17 04	253	27 01	303
175	30 05	012	46 20	059	62 25	111	28 20	174	36 34	214	16 07	253	26 11	303
176	30 17	011	47 12	059	63 21	111	28 26	175	36 00	214	15 10	253	25 20	302
177	30 29	011	48 03	058	64 17	112	28 31	175	35 26	215	14 12	253	24 30	302
178	30 39	010	48 53	057	65 12	113	28 36	176	34 51	216	13 15	253	23 39	302
179	30 50	009	49 43	057	66 07	113	28 40	176	34 16	216	12 17	253	22 48	302

LHA	Hc	Zn	Hc	Zn	Hc	Zn	Hc	Zn	Hc	Zn	Hc	Zn	Hc	Zn
♈	Alioth		◆ARCTURUS		ANTARES		◆ACRUX		Suhail		Alphard		◆REGULUS	
180	30 59	009	50 33	056	21 11	118	28 44	177	33 41	217	51 27	259	58 53	297
181	31 08	008	51 23	055	22 04	118	28 47	177	33 05	217	50 28	259	58 00	297
182	31 16	007	52 12	054	22 57	118	28 50	178	32 28	218	49 29	259	57 06	296
183	31 23	007	53 00	054	23 50	118	28 52	178	31 52	218	48 31	259	56 12	295
184	31 30	006	53 48	053	24 43	118	28 54	179	31 15	218	47 32	259	55 17	295
185	31 36	006	54 36	052	25 35	118	28 55	179	30 37	219	46 33	260	54 23	294
186	31 42	005	55 23	051	26 28	119	28 56	180	29 59	219	45 34	260	53 28	293
187	31 46	004	56 09	050	27 21	119	28 56	180	29 21	220	44 35	260	52 33	293
188	31 51	004	56 54	049	28 13	119	28 55	181	28 43	220	43 36	260	51 37	292
189	31 54	003	57 39	048	29 05	119	28 54	181	28 04	220	42 37	260	50 42	292
190	31 57	002	58 23	047	29 58	120	28 53	182	27 25	221	41 38	260	49 46	291
191	31 59	002	59 07	046	30 50	120	28 50	182	26 45	221	40 38	260	48 50	291
192	32 00	001	59 49	044	31 42	120	28 48	183	26 06	222	39 39	260	47 54	290
193	32 01	000	60 30	043	32 33	120	28 45	183	25 26	222	38 40	261	46 57	290
194	32 01	000	61 11	042	33 25	121	28 41	184	24 46	222	37 41	261	46 01	290
	ARCTURUS		◆Rasalhague		ANTARES		◆ACRUX		Alphard		◆REGULUS		Alioth	
195	61 50	040	20 19	076	34 16	121	28 37	184	36 42	261	45 04	289	32 00	359
196	62 28	039	21 17	076	35 08	121	28 32	185	35 43	261	44 08	289	31 59	358
197	63 05	037	22 15	076	35 59	122	28 26	185	34 44	261	43 11	289	31 56	358
198	63 41	036	23 13	075	36 50	122	28 21	186	33 44	261	42 14	288	31 54	357
199	64 15	034	24 11	075	37 41	122	28 14	186	32 45	261	41 17	288	31 50	356
200	64 48	032	25 09	075	38 31	123	28 07	187	31 46	261	40 20	288	31 46	356
201	65 19	030	26 07	075	39 21	123	28 00	187	30 47	261	39 23	287	31 41	355
202	65 48	028	27 05	075	40 11	124	27 52	188	29 47	261	38 25	287	31 36	354
203	66 16	026	28 03	075	41 01	124	27 44	188	28 48	261	37 28	287	31 29	354
204	66 41	024	29 00	074	41 51	125	27 35	189	27 49	261	36 31	287	31 23	353
205	67 04	022	29 58	074	42 40	125	27 25	189	26 50	261	35 33	286	31 15	352
206	67 26	020	30 56	074	43 29	125	27 15	190	25 50	261	34 35	286	31 07	352
207	67 45	017	31 53	074	44 18	126	27 05	190	24 51	261	33 38	286	30 58	351
208	68 01	015	32 51	074	45 06	127	26 54	191	23 52	261	32 40	286	30 49	351
209	68 16	012	33 49	073	45 54	127	26 43	191	22 52	261	31 42	285	30 38	350
	ARCTURUS		◆VEGA		ANTARES		◆RIGIL KENT.		ACRUX		Gienah		◆REGULUS	
210	68 27	010	14 46	049	46 42	128	30 43	174	26 31	192	60 02	237	30 45	285
211	68 36	007	15 31	049	47 29	128	30 48	175	26 19	192	59 11	238	29 47	285
212	68 43	005	16 16	048	48 16	129	30 53	176	26 06	193	58 20	239	28 49	285
213	68 46	002	17 01	048	49 02	130	30 57	176	25 53	193	57 28	240	27 51	285
214	68 47	000	17 46	048	49 48	130	31 01	177	25 39	193	56 37	240	26 53	285
215	68 46	357	18 30	048	50 33	131	31 04	177	25 25	194	55 44	241	25 55	284
216	68 41	354	19 14	047	51 18	132	31 07	178	25 10	194	54 52	242	24 57	284
217	68 34	352	19 58	047	52 03	133	31 09	178	24 55	195	53 59	242	23 58	284
218	68 24	349	20 42	047	52 46	134	31 10	179	24 40	195	53 05	243	23 00	284
219	68 12	347	21 26	047	53 29	135	31 11	180	24 24	195	52 12	243	22 02	284
220	67 57	344	22 09	046	54 12	135	31 11	180	24 08	196	51 18	244	21 04	284
221	67 40	342	22 53	046	54 53	136	31 11	181	23 51	196	50 24	245	20 05	284
222	67 20	340	23 36	046	55 34	137	31 10	181	23 35	197	49 30	245	19 07	283
223	66 58	337	24 18	045	56 15	139	31 08	182	23 17	197	48 35	245	18 09	283
224	66 34	335	25 01	045	56 54	140	31 06	182	22 59	197	47 41	246	17 10	283
	Alphecca		VEGA		◆ALTAIR		ANTARES		◆RIGIL KENT.		SPICA		◆ARCTURUS	
225	60 05	016	25 43	045	16 49	080	57 32	141	31 03	183	64 43	248	66 08	333
226	60 21	014	26 25	044	17 48	080	58 10	142	31 00	184	63 47	249	65 40	331
227	60 34	012	27 07	044	18 47	080	58 46	143	30 56	184	62 51	249	65 10	329
228	60 46	010	27 48	043	19 46	080	59 21	145	30 51	185	61 55	250	64 39	327
229	60 56	009	28 29	043	20 45	080	59 56	146	30 46	185	60 58	250	64 06	326
230	61 03	007	29 10	043	21 44	080	60 29	147	30 40	186	60 02	251	63 31	324
231	61 09	005	29 50	042	22 43	080	61 00	149	30 34	186	59 05	251	62 55	322
232	61 14	003	30 30	042	23 42	079	61 31	150	30 27	187	58 08	252	62 18	321
233	61 16	001	31 10	041	24 41	079	62 00	152	30 20	187	57 11	252	61 39	319
234	61 16	359	31 49	041	25 40	079	62 27	154	30 12	188	56 14	253	61 00	318
235	61 14	357	32 28	040	26 39	079	62 53	155	30 03	188	55 16	253	60 19	317
236	61 11	356	33 06	040	27 38	079	63 17	157	29 54	189	54 19	254	59 37	315
237	61 05	354	33 44	039	28 37	079	63 40	159	29 44	190	53 21	254	58 54	314
238	60 58	352	34 22	038	29 35	079	64 00	161	29 34	190	52 24	254	58 11	313
239	60 48	350	34 59	038	30 34	078	64 19	163	29 23	191	51 26	255	57 27	312
	VEGA		◆ALTAIR		Nunki		◆RIGIL KENT.		SPICA		◆ARCTURUS		Alphecca	
240	35 35	037	31 33	078	41 31	124	29 12	191	50 28	255	56 42	311	60 37	348
241	36 11	037	32 32	078	42 20	125	29 00	192	49 30	255	55 56	310	60 24	347
242	36 47	036	33 30	078	43 09	125	28 48	192	48 32	255	55 09	309	60 09	345
243	37 22	035	34 29	078	43 58	126	28 35	193	47 34	256	54 22	308	59 53	343
244	37 56	035	35 28	078	44 47	126	28 22	193	46 36	256	53 35	307	59 34	341
245	38 30	034	36 26	077	45 35	127	28 08	194	45 38	256	52 47	306	59 15	340
246	39 04	033	37 25	077	46 23	127	27 53	194	44 40	256	51 58	305	58 53	338
247	39 36	033	38 23	077	47 10	128	27 39	195	43 41	256	51 09	305	58 30	337
248	40 08	032	39 22	077	47 58	129	27 23	195	42 43	257	50 19	304	58 06	335
249	40 39	031	40 20	077	48 44	129	27 07	195	41 45	257	49 29	303	57 40	334
250	41 10	030	41 18	076	49 30	130	26 51	196	40 46	257	48 39	303	57 13	332
251	41 40	030	42 16	076	50 16	131	26 35	196	39 48	257	47 48	302	56 44	331
252	42 09	029	43 15	076	51 01	131	26 17	197	38 49	257	46 57	301	56 14	330
253	42 38	028	44 13	076	51 46	132	26 00	197	37 51	257	46 06	301	55 43	328
254	43 05	027	45 11	075	52 30	133	25 42	198	36 52	258	45 14	300	55 11	327
	◆VEGA		ALTAIR		◆Nunki		Shaula		ANTARES		◆SPICA		ARCTURUS	
255	43 32	026	46 09	075	53 14	134	54 04	169	64 29	196	35 54	258	44 22	300
256	43 58	025	47 07	075	53 57	135	54 15	170	64 11	198	34 55	258	43 30	299
257	44 23	024	48 04	074	54 39	136	54 25	171	63 51	200	33 57	258	42 37	299
258	44 48	023	49 02	074	55 20	137	54 34	173	63 30	202	32 58	258	41 45	298
259	45 11	022	50 00	074	56 01	138	54 40	174	63 06	204	31 59	258	40 52	298
260	45 33	021	50 57	073	56 41	139	54 46	175	62 41	205	31 01	258	39 58	297
261	45 55	020	51 55	073	57 19	140	54 50	177	62 15	207	30 02	258	39 05	297
262	46 15	019	52 52	072	57 58	141	54 53	178	61 47	209	29 03	258	38 11	297
263	46 35	018	53 49	072	58 35	142	54 54	180	61 17	210	28 04	258	37 18	296
264	46 53	017	54 46	072	59 11	144	54 54	181	60 46	212	27 06	259	36 24	296
265	47 11	016	55 43	071	59 45	145	54 52	182	60 14	213	26 07	259	35 30	295
266	47 27	015	56 39	071	60 19	147	54 49	184	59 40	215	25 08	259	34 35	295
267	47 42	014	57 36	070	60 51	148	54 44	185	59 06	216	24 09	259	33 41	295
268	47 56	013	58 32	069	61 23	149	54 38	186	58 30	217	23 10	259	32 47	294
269	48 09	012	59 28	069	61 52	151	54 31	188	57 53	219	22 12	259	31 52	294

LAT 2°S

LHA	Hc	Zn	Hc	Zn	Hc	Zn	Hc	Zn	Hc	Zn	Hc	Zn	Hc	Zn
♈	◆VEGA		DENEB		Enif		◆FOMALHAUT		Peacock		◆ANTARES		ARCTURUS	
270	48 21	011	30 46	032	33 01	077	14 35	120	28 05	158	57 15	220	30 57	294
271	48 32	010	31 18	031	33 59	077	15 26	120	28 27	159	56 36	221	30 02	294
272	48 41	009	31 49	031	34 58	077	16 18	120	28 48	159	55 57	222	29 07	293
273	48 49	007	32 19	030	35 56	076	17 10	120	29 09	160	55 16	223	28 12	293
274	48 56	006	32 49	030	36 54	076	18 01	121	29 29	160	54 35	224	27 17	293
275	49 02	005	33 19	029	37 52	076	18 53	121	29 49	161	53 53	225	26 22	293
276	49 07	004	33 48	029	38 50	076	19 45	121	30 09	161	53 10	226	25 26	292
277	49 10	003	34 16	028	39 48	075	20 36	121	30 28	162	52 27	227	24 31	292
278	49 12	001	34 44	027	40 46	075	21 27	121	30 46	162	51 43	228	23 35	292
279	49 13	000	35 11	027	41 44	075	22 19	121	31 04	163	50 58	228	22 40	292
280	49 13	359	35 37	026	42 42	075	23 10	122	31 21	163	50 13	229	21 44	292
281	49 11	358	36 03	025	43 40	074	24 01	122	31 38	164	49 27	230	20 48	291
282	49 08	357	36 29	025	44 38	074	24 52	122	31 54	165	48 41	231	19 52	291
283	49 04	355	36 53	024	45 35	074	25 42	122	32 10	165	47 55	231	18 56	291
284	48 59	354	37 17	023	46 33	073	26 33	122	32 25	166	47 08	232	18 00	291
	DENEB		Enif		◆FOMALHAUT		Peacock		◆ANTARES		Rasalhague		◆VEGA	
285	37 40	022	47 30	073	27 24	123	32 40	166	46 20	233	64 18	305	48 52	353
286	38 03	022	48 28	073	28 14	123	32 53	167	45 33	233	63 29	304	48 45	352
287	38 24	021	49 25	072	29 04	123	33 07	167	44 44	234	62 39	303	48 35	351
288	38 45	020	50 22	072	29 55	123	33 19	168	43 56	234	61 48	302	48 25	350
289	39 05	019	51 19	071	30 44	124	33 31	169	43 07	235	60 57	301	48 14	349
290	39 25	018	52 16	071	31 34	124	33 43	169	42 18	235	60 05	300	48 01	347
291	39 43	018	53 12	071	32 24	124	33 54	170	41 29	236	59 12	299	47 48	346
292	40 01	017	54 09	070	33 13	125	34 04	171	40 39	236	58 20	298	47 33	345
293	40 18	016	55 05	070	34 03	125	34 13	171	39 49	237	57 27	297	47 17	344
294	40 34	015	56 01	069	34 52	125	34 22	172	38 59	237	56 33	297	47 00	343
295	40 49	014	56 57	068	35 40	126	34 30	173	38 08	237	55 39	296	46 42	342
296	41 03	013	57 53	068	36 29	126	34 38	173	37 18	238	54 45	295	46 23	341
297	41 17	012	58 48	067	37 17	127	34 45	174	36 27	238	53 51	295	46 03	340
298	41 29	012	59 43	066	38 05	127	34 51	174	35 36	238	52 56	294	45 42	339
299	41 41	011	60 38	066	38 53	127	34 56	175	34 45	239	52 01	293	45 20	338
	◆DENEB		Alpheratz		◆FOMALHAUT		Peacock		◆ANTARES		Rasalhague		VEGA	
300	41 51	010	23 07	057	39 40	128	35 01	176	33 53	239	51 06	293	44 57	337
301	42 01	009	23 58	057	40 28	128	35 05	176	33 02	239	50 11	292	44 33	336
302	42 10	008	24 48	057	41 14	129	35 08	177	32 10	240	49 16	292	44 08	335
303	42 18	007	25 38	056	42 01	129	35 11	178	31 18	240	48 20	292	43 42	334
304	42 24	006	26 27	056	42 47	130	35 13	178	30 26	240	47 24	291	43 16	333
305	42 30	005	27 17	056	43 33	131	35 14	179	29 34	241	46 28	291	42 48	332
306	42 35	004	28 06	055	44 18	131	35 15	180	28 42	241	45 32	290	42 20	332
307	42 39	003	28 56	055	45 03	132	35 15	180	27 49	241	44 35	290	41 51	331
308	42 42	002	29 45	055	45 48	132	35 14	181	26 57	241	43 39	290	41 22	330
309	42 43	001	30 33	054	46 32	133	35 12	182	26 04	241	42 42	289	40 51	329
310	42 44	000	31 22	054	47 15	134	35 10	182	25 12	242	41 46	289	40 20	328
311	42 44	359	32 10	053	47 58	134	35 07	183	24 19	242	40 49	289	39 49	328
312	42 43	358	32 58	053	48 41	135	35 04	184	23 26	242	39 52	288	39 16	327
313	42 41	357	33 46	053	49 23	136	34 59	184	22 33	242	38 55	288	38 43	326
314	42 37	356	34 33	052	50 04	137	34 54	185	21 40	242	37 58	288	38 09	326
	◆Alpheratz		◆FOMALHAUT		Peacock		Nunki		◆Rasalhague		VEGA		DENEB	
315	35 21	052	50 45	138	34 49	186	51 23	228	37 01	287	37 35	325	42 33	356
316	36 07	051	51 25	139	34 42	186	50 38	229	36 03	287	37 00	324	42 28	355
317	36 54	051	52 04	140	34 35	187	49 52	230	35 06	287	36 25	324	42 22	354
318	37 40	050	52 43	140	34 27	188	49 06	230	34 09	287	35 49	323	42 15	353
319	38 26	050	53 21	141	34 19	188	48 20	231	33 11	286	35 13	322	42 07	352
320	39 11	049	53 57	142	34 10	189	47 33	232	32 14	286	34 36	322	41 58	351
321	39 56	048	54 34	144	34 00	190	46 46	232	31 16	286	33 59	321	41 48	350
322	40 41	048	55 09	145	33 50	190	45 58	233	30 18	286	33 21	321	41 37	349
323	41 25	047	55 43	146	33 39	191	45 10	234	29 21	286	32 43	320	41 25	348
324	42 09	047	56 16	147	33 27	192	44 22	234	28 23	285	32 04	320	41 12	347
325	42 53	046	56 48	148	33 15	192	43 33	235	27 25	285	31 25	319	40 58	346
326	43 35	045	57 19	150	33 02	193	42 44	235	26 27	285	30 45	319	40 44	345
327	44 18	045	57 49	151	32 48	193	41 54	236	25 29	285	30 06	318	40 28	345
328	44 59	044	58 17	152	32 34	194	41 05	236	24 31	285	29 25	318	40 12	344
329	45 41	043	58 45	154	32 20	194	40 15	237	23 33	285	28 45	317	39 55	343
	◆Alpheratz		Hamal		Diphda		◆FOMALHAUT		Nunki		◆ALTAIR		DENEB	
330	46 21	042	24 52	063	46 53	115	59 10	155	39 25	237	55 58	289	39 36	342
331	47 01	041	25 45	063	47 47	115	59 35	157	38 34	237	55 01	289	39 18	341
332	47 40	041	26 39	062	48 41	115	59 58	158	37 44	238	54 04	288	38 58	340
333	48 19	040	27 32	062	49 35	116	60 19	160	36 53	238	53 07	288	38 38	340
334	48 57	039	28 25	062	50 29	116	60 39	161	36 02	238	52 10	287	38 16	339
335	49 34	038	29 17	062	51 23	117	60 58	163	35 11	239	51 13	287	37 54	338
336	50 10	037	30 10	061	52 16	117	61 14	165	34 19	239	50 15	286	37 32	337
337	50 46	036	31 03	061	53 09	118	61 29	167	33 28	239	49 18	286	37 08	337
338	51 20	035	31 55	061	54 02	119	61 42	168	32 36	240	48 20	286	36 44	336
339	51 54	034	32 47	060	54 54	119	61 53	170	31 44	240	47 22	285	36 19	335
340	52 27	033	33 39	060	55 47	120	62 03	172	30 52	240	46 24	285	35 54	335
341	52 59	031	34 31	060	56 38	121	62 10	174	30 00	241	45 26	285	35 28	334
342	53 29	030	35 22	059	57 30	121	62 16	176	29 08	241	44 28	284	35 01	333
343	53 59	029	36 14	059	58 21	122	62 19	177	28 15	241	43 30	284	34 34	333
344	54 28	028	37 05	058	59 11	123	62 21	179	27 23	241	42 32	284	34 06	332
	Alpheratz		◆Hamal		◆ACHERNAR		FOMALHAUT		Nunki		◆ALTAIR		DENEB	
345	54 55	026	37 56	058	26 33	157	62 21	181	26 30	241	41 34	284	33 37	331
346	55 21	025	38 46	057	26 56	158	62 18	183	25 37	242	40 36	283	33 08	331
347	55 46	024	39 37	057	27 19	158	62 14	185	24 45	242	39 37	283	32 38	330
348	56 09	022	40 27	056	27 41	159	62 08	187	23 52	242	38 39	283	32 08	329
349	56 31	021	41 17	056	28 02	159	62 00	189	22 59	242	37 40	283	31 37	329
350	56 52	019	42 06	055	28 23	160	61 50	190	22 06	242	36 42	283	31 06	328
351	57 11	018	42 55	055	28 44	160	61 38	192	21 13	242	35 43	282	30 34	328
352	57 29	016	43 44	054	29 04	161	61 25	194	20 19	243	34 45	282	30 02	327
353	57 45	015	44 33	054	29 23	161	61 09	196	19 26	243	33 46	282	29 30	327
354	58 00	013	45 21	053	29 43	162	60 52	197	18 33	243	32 48	282	28 56	326
355	58 13	012	46 09	052	30 01	162	60 33	199	17 39	243	31 49	282	28 23	326
356	58 24	010	46 56	052	30 19	163	60 13	201	16 46	243	30 50	282	27 49	325
357	58 34	008	47 43	051	30 37	163	59 51	202	15 52	243	29 51	281	27 15	325
358	58 42	007	48 29	050	30 54	164	59 27	204	14 59	243	28 53	281	26 40	324
359	58 48	005	49 15	049	31 10	164	59 03	205	14 05	243	27 54	281	26 05	324

LHA ♈	Hc	Zn	Hc	Zn	Hc	Zn	Hc	Zn	Hc	Zn	Hc	Zn	Hc	Zn
	Alpheratz		**◆ALDEBARAN**		**RIGEL**		**ACHERNAR**		**◆FOMALHAUT**		**ALTAIR**		**◆DENEB**	
0	57 53	003	19 15	071	11 44	098	32 24	165	59 30	208	26 43	281	24 41	324
1	57 55	002	20 12	071	12 43	098	32 40	165	59 01	209	25 44	281	24 05	323
2	57 56	000	21 08	071	13 42	098	32 55	166	58 32	210	24 46	281	23 29	323
3	57 55	358	22 05	071	14 42	098	33 09	166	58 01	212	23 47	281	22 53	323
4	57 53	357	23 02	071	15 41	098	33 23	167	57 29	213	22 48	281	22 17	322
5	57 49	355	23 58	070	16 41	098	33 36	168	56 56	214	21 49	281	21 40	322
6	57 43	353	24 55	070	17 40	098	33 49	168	56 21	215	20 50	281	21 03	321
7	57 35	352	25 51	070	18 39	098	34 01	169	55 46	217	19 51	281	20 25	321
8	57 26	350	26 47	070	19 39	098	34 12	169	55 10	218	18 53	280	19 47	321
9	57 15	349	27 43	070	20 38	098	34 23	170	54 33	219	17 54	280	19 09	320
10	57 02	347	28 40	069	21 37	098	34 33	171	53 55	220	16 55	280	18 31	320
11	56 48	346	29 36	069	22 37	098	34 43	171	53 17	221	15 56	280	17 53	320
12	56 33	344	30 32	069	23 36	098	34 51	172	52 37	222	14 57	280	17 14	320
13	56 15	343	31 27	069	24 36	098	35 00	173	51 57	223	13 58	280	16 35	319
14	55 57	341	32 23	068	25 35	098	35 07	173	51 16	223	12 59	280	15 56	319
	CAPELLA		**ALDEBARAN**		**◆RIGEL**		**ACHERNAR**		**◆FOMALHAUT**		**Enif**		**◆Alpheratz**	
15	15 24	040	33 19	068	26 34	098	35 14	174	50 35	224	39 31	285	55 37	340
16	16 03	040	34 14	068	27 34	098	35 20	174	49 53	225	38 33	285	55 15	338
17	16 41	040	35 10	067	28 33	098	35 25	175	49 10	226	37 36	285	54 52	337
18	17 20	040	36 05	067	29 33	098	35 30	176	48 27	227	36 38	285	54 28	336
19	17 58	039	37 00	067	30 32	098	35 34	176	47 43	227	35 40	284	54 03	334
20	18 36	039	37 55	066	31 31	098	35 38	177	46 59	228	34 42	284	53 36	333
21	19 13	039	38 50	066	32 31	098	35 40	178	46 14	229	33 43	284	53 08	332
22	19 50	038	39 44	066	33 30	098	35 42	178	45 29	229	32 45	284	52 39	331
23	20 27	038	40 39	065	34 29	098	35 44	179	44 43	230	31 47	284	52 09	329
24	21 04	038	41 33	065	35 29	098	35 44	180	43 57	230	30 49	284	51 38	328
25	21 41	037	42 27	064	36 28	098	35 44	180	43 11	231	29 50	283	51 06	327
26	22 17	037	43 21	064	37 27	098	35 43	181	42 24	231	28 52	283	50 33	326
27	22 53	037	44 15	063	38 27	098	35 42	182	41 37	232	27 54	283	49 59	325
28	23 28	036	45 08	063	39 26	098	35 40	182	40 50	232	26 55	283	49 24	324
29	24 03	036	46 01	062	40 25	098	35 37	183	40 02	233	25 57	282	48 49	323
	CAPELLA		**◆BETELGEUSE**		**SIRIUS**		**CANOPUS**		**◆ACHERNAR**		**FOMALHAUT**		**◆Alpheratz**	
30	24 38	035	30 29	080	18 50	107	16 45	145	35 33	184	39 14	233	48 12	322
31	25 12	035	31 28	079	19 47	107	17 20	145	35 29	184	38 26	234	47 35	321
32	25 46	034	32 27	079	20 45	107	17 54	145	35 24	185	37 37	234	46 57	320
33	26 20	034	33 26	079	21 42	107	18 28	145	35 19	186	36 49	235	46 18	319
34	26 53	033	34 25	079	22 39	107	19 02	146	35 12	186	36 00	235	45 39	318
35	27 26	033	35 24	079	23 37	107	19 36	146	35 05	187	35 11	235	44 59	318
36	27 59	033	36 22	079	24 34	107	20 10	146	34 58	188	34 21	236	44 18	317
37	28 31	032	37 21	078	25 31	107	20 43	146	34 49	188	33 32	236	43 37	316
38	29 02	031	38 20	078	26 29	107	21 16	147	34 40	189	32 42	236	42 56	315
39	29 33	031	39 18	078	27 26	107	21 49	147	34 31	190	31 52	237	42 13	315
40	30 04	030	40 17	078	28 23	107	22 22	147	34 21	190	31 02	237	41 30	314
41	30 34	030	41 15	077	29 20	108	22 54	147	34 10	191	30 12	237	40 47	313
42	31 03	029	42 14	077	30 17	108	23 26	148	33 58	191	29 21	237	40 03	313
43	31 32	029	43 12	077	31 14	108	23 58	148	33 46	192	28 31	238	39 19	312
44	32 01	028	44 10	077	32 11	108	24 30	148	33 33	193	27 40	238	38 35	312
	CAPELLA		**◆BETELGEUSE**		**SIRIUS**		**CANOPUS**		**◆ACHERNAR**		**Diphda**		**◆Hamal**	
45	32 29	027	45 09	076	33 08	108	25 01	149	33 20	193	53 19	243	60 34	335
46	32 56	027	46 07	076	34 05	108	25 32	149	33 06	194	52 25	244	60 07	333
47	33 23	026	47 05	076	35 02	108	26 03	149	32 51	194	51 31	244	59 40	331
48	33 49	026	48 03	075	35 59	109	26 33	150	32 36	195	50 37	245	59 10	330
49	34 15	025	49 01	075	36 56	109	27 03	150	32 21	195	49 43	245	58 39	328
50	34 40	024	49 59	075	37 53	109	27 32	151	32 04	196	48 48	246	58 07	327
51	35 04	024	50 57	074	38 49	109	28 02	151	31 48	197	47 54	246	57 34	326
52	35 27	023	51 54	074	39 46	109	28 31	151	31 30	197	46 59	247	57 00	324
53	35 50	022	52 52	074	40 42	110	28 59	152	31 12	198	46 04	247	56 24	323
54	36 13	021	53 49	073	41 39	110	29 27	152	30 54	198	45 09	247	55 48	322
55	36 34	021	54 47	073	42 35	110	29 55	153	30 35	199	44 13	248	55 10	321
56	36 55	020	55 44	072	43 31	110	30 22	153	30 16	199	43 18	248	54 32	319
57	37 15	019	56 41	072	44 27	111	30 49	154	29 56	200	42 22	248	53 52	318
58	37 34	018	57 37	071	45 23	111	31 15	154	29 35	200	41 26	248	53 12	317
59	37 53	018	58 34	071	46 19	111	31 41	155	29 14	201	40 31	249	52 31	316
	◆CAPELLA		**POLLUX**		**◆SIRIUS**		**CANOPUS**		**ACHERNAR**		**◆Diphda**		**Hamal**	
60	38 10	017	27 43	056	47 15	112	32 07	155	28 53	201	39 35	249	51 49	315
61	38 27	016	28 32	056	48 11	112	32 31	156	28 31	202	38 39	249	51 07	314
62	38 43	015	29 22	055	49 06	112	32 56	156	28 09	202	37 43	249	50 24	313
63	38 59	014	30 11	055	50 02	113	33 20	157	27 46	202	36 47	250	49 40	313
64	39 13	013	31 00	055	50 57	113	33 43	157	27 23	203	35 50	250	48 55	312
65	39 27	013	31 48	054	51 52	114	34 06	158	27 00	203	34 54	250	48 10	311
66	39 39	012	32 37	054	52 47	114	34 28	158	26 36	204	33 58	250	47 25	310
67	39 51	011	33 25	053	53 41	115	34 50	159	26 12	204	33 01	250	46 39	309
68	40 02	010	34 13	053	54 36	115	35 11	160	25 47	204	32 05	251	45 52	309
69	40 12	009	35 00	052	55 30	116	35 32	160	25 22	205	31 08	251	45 05	308
70	40 21	008	35 48	052	56 24	116	35 52	161	24 57	205	30 12	251	44 18	307
71	40 29	007	36 35	051	57 17	117	36 11	162	24 31	206	29 15	251	43 30	307
72	40 36	006	37 21	051	58 11	117	36 30	162	24 05	206	28 18	251	42 42	306
73	40 43	006	38 07	050	59 04	118	36 48	163	23 39	206	27 22	251	41 54	306
74	40 48	005	38 53	050	59 56	119	37 05	163	23 12	207	26 25	251	41 05	305
	CAPELLA		**◆POLLUX**		**REGULUS**		**Suhail**		**◆CANOPUS**		**ACHERNAR**		**◆Hamal**	
75	40 52	004	39 39	049	12 02	077	22 09	136	37 22	164	22 45	207	40 16	304
76	40 56	003	40 24	048	13 00	077	22 50	136	37 38	165	22 18	207	39 26	304
77	40 58	002	41 09	048	13 59	077	23 32	137	37 53	166	21 50	208	38 36	303
78	41 00	001	41 53	047	14 57	077	24 13	137	38 08	166	21 22	208	37 46	303
79	41 00	000	42 37	047	15 55	077	24 53	137	38 21	167	20 54	208	36 56	303
80	41 00	359	43 20	046	16 53	077	25 34	138	38 35	168	20 26	208	36 05	302
81	40 59	358	44 03	045	17 52	076	26 14	138	38 47	168	19 57	209	35 14	302
82	40 56	357	44 45	044	18 50	076	26 54	138	38 59	169	19 28	209	34 23	301
83	40 53	356	45 26	044	19 48	076	27 34	139	39 10	170	18 59	209	33 32	301
84	40 49	355	46 07	043	20 46	076	28 14	139	39 20	171	18 29	209	32 40	300
85	40 44	355	46 48	042	21 44	076	28 53	139	39 29	171	18 00	210	31 48	300
86	40 37	354	47 27	041	22 42	076	29 32	140	39 38	172	17 30	210	30 56	300
87	40 30	353	48 07	040	23 40	076	30 11	140	39 45	173	17 00	210	30 04	299
88	40 22	352	48 45	039	24 38	075	30 49	140	39 52	174	16 30	210	29 12	299
89	40 13	351	49 22	038	25 36	075	31 27	141	39 58	175	16 00	211	28 19	299

LAT 3°S

LHA ♈	Hc	Zn	Hc	Zn	Hc	Zn	Hc	Zn	Hc	Zn	Hc	Zn	Hc	Zn
	POLLUX		**◆REGULUS**		**Suhail**		**◆CANOPUS**		**Acamar**		**◆ALDEBARAN**		**CAPELLA**	
90	49 59	037	26 34	075	32 05	141	40 04	175	34 35	221	61 28	314	40 04	350
91	50 35	036	27 32	075	32 42	142	40 08	176	33 56	222	60 44	312	39 53	349
92	51 10	035	28 30	075	33 19	142	40 12	177	33 15	222	59 59	311	39 41	348
93	51 44	034	29 28	074	33 56	143	40 15	178	32 35	223	59 14	310	39 29	347
94	52 18	033	30 25	074	34 32	143	40 17	178	31 54	223	58 28	309	39 15	347
95	52 50	032	31 23	074	35 08	144	40 18	179	31 13	223	57 41	308	39 01	346
96	53 21	031	32 21	074	35 43	144	40 18	180	30 32	224	56 53	307	38 46	345
97	53 51	030	33 18	074	36 18	145	40 18	181	29 50	224	56 05	306	38 30	344
98	54 20	028	34 16	073	36 53	145	40 17	182	29 08	225	55 16	305	38 13	343
99	54 48	027	35 13	073	37 27	146	40 15	182	28 26	225	54 27	304	37 56	343
100	55 15	026	36 10	073	38 00	146	40 12	183	27 44	225	53 37	303	37 37	342
101	55 40	024	37 07	073	38 33	147	40 08	184	27 01	226	52 47	303	37 18	341
102	56 04	023	38 04	072	39 05	148	40 03	185	26 18	226	51 56	302	36 58	340
103	56 27	021	39 01	072	39 37	148	39 58	186	25 35	226	51 05	301	36 37	339
104	56 48	020	39 58	072	40 09	149	39 52	186	24 52	226	50 14	300	36 16	339
	POLLUX		**◆REGULUS**		**Suhail**		**◆CANOPUS**		**RIGEL**		**◆ALDEBARAN**		**CAPELLA**	
105	57 08	019	40 55	071	40 39	150	39 45	187	63 12	258	49 22	300	35 54	338
106	57 26	017	41 52	071	41 09	150	39 37	188	62 14	258	48 30	299	35 31	337
107	57 43	015	42 48	071	41 39	151	39 28	189	61 15	258	47 37	299	35 08	337
108	57 58	014	43 45	070	42 07	152	39 19	189	60 16	259	46 45	298	34 43	336
109	58 11	012	44 41	070	42 35	152	39 09	190	59 17	259	45 52	298	34 19	335
110	58 23	011	45 37	069	43 03	153	38 58	191	58 19	259	44 58	297	33 53	335
111	58 33	009	46 33	069	43 29	154	38 46	192	57 20	259	44 05	297	33 27	334
112	58 42	007	47 29	069	43 55	155	38 34	192	56 21	260	43 11	296	33 00	333
113	58 48	006	48 25	068	44 20	156	38 20	193	55 22	260	42 17	296	32 33	333
114	58 53	004	49 20	068	44 45	156	38 06	194	54 23	260	41 23	295	32 05	332
115	58 56	002	50 16	067	45 08	157	37 52	195	53 24	260	40 29	295	31 37	331
116	58 58	000	51 11	067	45 31	158	37 36	195	52 25	260	39 34	294	31 08	331
117	58 57	359	52 06	066	45 53	159	37 20	196	51 26	261	38 40	294	30 38	330
118	58 55	357	53 00	065	46 14	160	37 04	197	50 26	261	37 45	294	30 09	330
119	58 51	355	53 55	065	46 34	161	36 46	197	49 27	261	36 50	293	29 38	329
	◆REGULUS		**Gienah**		**◆ACRUX**		**CANOPUS**		**SIRIUS**		**◆BETELGEUSE**		**POLLUX**	
120	54 49	064	25 47	108	13 05	155	36 28	198	67 01	232	57 06	289	58 45	354
121	55 43	064	26 44	108	13 31	155	36 09	199	66 14	233	56 09	288	58 38	352
122	56 36	063	27 41	108	13 56	155	35 50	199	65 25	235	55 12	288	58 28	350
123	57 29	062	28 38	108	14 21	155	35 30	200	64 36	236	54 15	287	58 17	349
124	58 22	061	29 34	108	14 46	155	35 10	200	63 46	237	53 18	287	58 05	347
125	59 14	060	30 31	109	15 11	156	34 48	201	62 55	238	52 20	286	57 50	345
126	60 06	059	31 28	109	15 36	156	34 27	202	62 04	239	51 23	286	57 35	344
127	60 57	058	32 25	109	16 00	156	34 04	202	61 13	240	50 25	285	57 17	342
128	61 48	057	33 21	109	16 24	156	33 41	203	60 21	241	49 27	285	56 58	341
129	62 38	056	34 18	109	16 48	156	33 18	203	59 28	241	48 29	285	56 38	339
130	63 28	055	35 15	109	17 12	157	32 54	204	58 35	242	47 31	284	56 16	338
131	64 17	054	36 11	110	17 36	157	32 30	204	57 42	243	46 33	284	55 52	336
132	65 05	053	37 07	110	17 59	157	32 05	205	56 49	244	45 35	284	55 28	335
133	65 52	051	38 04	110	18 22	157	31 39	205	55 55	244	44 37	283	55 02	334
134	66 38	050	39 00	110	18 45	158	31 13	206	55 01	245	43 38	283	54 34	332
	Alioth		**◆SPICA**		**ACRUX**		**◆CANOPUS**		**SIRIUS**		**BETELGEUSE**		**◆POLLUX**	
135	14 25	029	23 53	101	19 08	158	30 47	206	54 07	245	42 40	283	54 06	331
136	14 54	029	24 52	101	19 30	158	30 20	207	53 12	246	41 42	283	53 37	330
137	15 23	029	25 51	101	19 52	158	29 53	207	52 17	246	40 43	282	53 06	329
138	15 52	029	26 50	101	20 14	159	29 25	208	51 22	247	39 45	282	52 34	327
139	16 21	028	27 49	101	20 36	159	28 57	208	50 27	247	38 46	282	52 01	326
140	16 49	028	28 48	101	20 57	159	28 28	209	49 32	248	37 47	282	51 28	325
141	17 17	028	29 46	101	21 18	160	27 59	209	48 37	248	36 49	282	50 53	324
142	17 45	027	30 45	101	21 38	160	27 30	209	47 41	248	35 50	281	50 18	323
143	18 12	027	31 44	101	21 59	160	27 00	210	46 45	249	34 51	281	49 41	322
144	18 39	027	32 43	101	22 19	161	26 30	210	45 49	249	33 52	281	49 04	321
145	19 06	026	33 42	101	22 39	161	26 00	211	44 53	249	32 54	281	48 26	320
146	19 32	026	34 40	101	22 58	161	25 29	211	43 57	250	31 55	281	47 47	319
147	19 58	026	35 39	102	23 17	162	24 58	211	43 01	250	30 56	280	47 08	318
148	20 24	025	36 38	102	23 35	162	24 27	212	42 05	250	29 57	280	46 28	318
149	20 49	025	37 36	102	23 54	162	23 56	212	41 08	250	28 58	280	45 47	317
	Alioth		**ARCTURUS**		**◆SPICA**		**ACRUX**		**◆CANOPUS**		**SIRIUS**		**◆POLLUX**	
150	21 14	024	23 28	068	38 35	102	24 12	163	23 24	212	40 12	251	45 06	316
151	21 38	024	24 24	067	39 34	102	24 29	163	22 51	213	39 15	251	44 24	315
152	22 03	024	25 19	067	40 32	102	24 46	164	22 19	213	38 19	251	43 41	315
153	22 26	023	26 14	067	41 31	102	25 03	164	21 46	213	37 22	251	42 58	314
154	22 50	023	27 09	067	42 29	102	25 20	164	21 13	213	36 25	251	42 15	313
155	23 13	022	28 04	066	43 28	103	25 35	165	20 40	214	35 29	252	41 31	312
156	23 35	022	28 59	066	44 26	103	25 51	165	20 07	214	34 32	252	40 47	312
157	23 57	021	29 53	066	45 25	103	26 06	166	19 33	214	33 35	252	40 02	311
158	24 19	021	30 48	065	46 23	103	26 21	166	18 59	214	32 38	252	39 16	311
159	24 40	020	31 42	065	47 21	103	26 35	166	18 25	215	31 41	252	38 31	310
160	25 00	020	32 37	065	48 20	103	26 49	167	17 51	215	30 44	252	37 45	309
161	25 20	019	33 31	065	49 18	104	27 02	167	17 17	215	29 47	252	36 58	309
162	25 40	019	34 25	064	50 16	104	27 15	168	16 42	215	28 49	253	36 11	308
163	25 59	018	35 19	064	51 14	104	27 28	168	16 08	216	27 52	253	35 24	308
164	26 18	018	36 13	063	52 12	104	27 40	169	15 33	216	26 55	253	34 37	307
	Alioth		**◆ARCTURUS**		**SPICA**		**◆ACRUX**		**Suhail**		**SIRIUS**		**◆POLLUX**	
165	26 36	017	37 06	063	53 10	105	27 51	169	42 33	208	25 58	253	33 49	307
166	26 54	017	37 59	063	54 08	105	28 02	170	42 05	208	25 01	253	33 01	307
167	27 11	016	38 53	062	55 06	105	28 13	170	41 36	209	24 03	253	32 13	306
168	27 27	016	39 45	062	56 04	106	28 23	171	41 06	210	23 06	253	31 24	306
169	27 43	015	40 38	061	57 02	106	28 33	171	40 36	211	22 09	253	30 36	305
170	27 59	015	41 31	061	57 59	106	28 42	172	40 06	211	21 11	253	29 47	305
171	28 13	014	42 23	060	58 57	107	28 50	172	39 34	212	20 14	253	28 57	305
172	28 28	013	43 15	060	59 54	107	28 58	173	39 02	212	19 16	253	28 08	304
173	28 41	013	44 07	059	60 51	108	29 06	173	38 30	213	18 19	253	27 18	304
174	28 54	012	44 58	059	61 48	108	29 13	174	37 57	214	17 22	253	26 28	304
175	29 07	012	45 49	058	62 45	109	29 19	174	37 23	214	16 24	253	25 38	303
176	29 19	011	46 40	058	63 42	109	29 25	175	36 49	215	15 27	253	24 48	303
177	29 30	010	47 30	057	64 38	110	29 31	175	36 15	215	14 29	254	23 58	303
178	29 40	010	48 20	056	65 34	111	29 36	176	35 40	216	13 32	254	23 07	302
179	29 50	009	49 10	056	66 30	111	29 40	176	35 05	216	12 34	254	22 16	302

LAT 3°S

LHA	Hc	Zn	Hc	Zn	Hc	Zn	Hc	Zn	Hc	Zn	Hc	Zn	Hc	Zn
♈	Alioth		◆ARCTURUS		ANTARES		◆ACRUX		Suhail		Alphard		◆REGULUS	
180	30 00	009	49 59	055	21 38	117	29 44	177	34 29	217	51 38	260	58 25	299
181	30 08	008	50 48	054	22 32	117	29 47	177	33 53	217	50 39	260	57 32	298
182	30 16	007	51 36	053	23 25	118	29 50	178	33 16	218	49 40	260	56 39	297
183	30 24	007	52 24	053	24 18	118	29 52	178	32 39	218	48 41	260	55 45	297
184	30 30	006	53 11	052	25 11	118	29 54	179	32 01	219	47 42	260	54 52	296
185	30 36	005	53 58	051	26 04	118	29 55	179	31 24	219	46 43	261	53 58	295
186	30 42	005	54 44	050	26 57	118	29 56	180	30 46	220	45 44	261	53 03	295
187	30 47	004	55 30	049	27 49	118	29 56	180	30 07	220	44 45	261	52 09	294
188	30 51	004	56 15	048	28 42	119	29 55	181	29 28	220	43 46	261	51 14	293
189	30 54	003	56 59	047	29 34	119	29 54	181	28 49	221	42 47	261	50 19	293
190	30 57	002	57 42	046	30 27	119	29 53	182	28 10	221	41 47	261	49 23	292
191	30 59	002	58 24	044	31 19	119	29 50	182	27 31	222	40 48	261	48 28	292
192	31 00	001	59 06	043	32 11	120	29 48	183	26 51	222	39 49	261	47 32	291
193	31 01	000	59 46	042	33 03	120	29 44	183	26 11	222	38 50	261	46 36	291
194	31 01	000	60 26	041	33 55	120	29 41	184	25 30	222	37 50	261	45 40	291
	ARCTURUS		◆Rasalhague		ANTARES		◆ACRUX		Alphard		◆REGULUS		Alioth	
195	61 04	039	20 04	075	34 47	120	29 36	184	36 51	261	44 44	290	31 00	359
196	61 41	038	21 02	075	35 39	121	29 32	185	35 52	262	43 48	290	30 59	358
197	62 17	036	22 00	075	36 30	121	29 26	185	34 53	262	42 51	289	30 56	358
198	62 52	034	22 58	075	37 21	121	29 20	186	33 53	262	41 55	289	30 54	357
199	63 25	033	23 56	075	38 12	122	29 14	186	32 54	262	40 58	289	30 50	356
200	63 57	031	24 53	075	39 03	122	29 07	187	31 55	262	40 01	288	30 46	356
201	64 27	029	25 51	074	39 54	122	28 59	187	30 56	262	39 04	288	30 41	355
202	64 55	027	26 49	074	40 44	123	28 51	188	29 56	262	38 07	288	30 36	354
203	65 21	025	27 46	074	41 35	123	28 43	188	28 57	262	37 10	288	30 30	354
204	65 46	023	28 44	074	42 25	124	28 34	189	27 58	262	36 13	287	30 23	353
205	66 09	021	29 42	074	43 14	124	28 24	189	26 58	262	35 16	287	30 16	353
206	66 29	019	30 39	073	44 04	125	28 14	190	25 59	262	34 19	287	30 08	352
207	66 47	017	31 36	073	44 53	125	28 04	190	25 00	262	33 21	286	29 59	351
208	67 03	014	32 34	073	45 42	126	27 53	191	24 00	262	32 24	286	29 49	351
209	67 17	012	33 31	073	46 30	126	27 42	191	23 01	262	31 26	286	29 39	350
	ARCTURUS		◆VEGA		ANTARES		◆RIGIL KENT.		ACRUX		Gienah		◆REGULUS	
210	67 28	010	14 07	049	47 18	127	31 42	174	27 30	192	60 34	239	30 28	286
211	67 37	007	14 51	048	48 06	127	31 48	175	27 17	192	59 42	239	29 31	286
212	67 43	005	15 36	048	48 53	128	31 53	176	27 04	193	58 50	240	28 33	285
213	67 47	002	16 21	048	49 40	129	31 57	176	26 51	193	57 58	241	27 35	285
214	67 47	000	17 05	048	50 27	130	32 01	177	26 37	193	57 06	242	26 37	285
215	67 46	357	17 50	047	51 13	130	32 04	177	26 23	194	56 13	242	25 39	285
216	67 42	355	18 34	047	51 58	131	32 07	178	26 09	194	55 19	243	24 42	285
217	67 35	352	19 18	047	52 43	132	32 09	178	25 53	195	54 26	244	23 44	284
218	67 25	350	20 01	047	53 27	133	32 10	179	25 38	195	53 32	244	22 46	284
219	67 13	347	20 45	046	54 11	134	32 11	180	25 22	196	52 38	245	21 47	284
220	66 59	345	21 28	046	54 54	134	32 11	180	25 06	196	51 44	245	20 49	284
221	66 42	343	22 11	046	55 37	135	32 11	181	24 49	196	50 49	246	19 51	284
222	66 24	340	22 54	045	56 18	136	32 10	181	24 32	197	49 55	246	18 53	284
223	66 02	338	23 36	045	56 59	137	32 08	182	24 14	197	49 00	246	17 55	284
224	65 39	336	24 18	045	57 39	139	32 06	182	23 57	198	48 05	247	16 57	284
	Alphecca		VEGA		◆ALTAIR		ANTARES		◆RIGIL KENT.		SPICA		◆ARCTURUS	
225	59 07	015	25 00	044	16 39	080	58 18	140	32 03	183	65 04	250	65 14	334
226	59 22	013	25 42	044	17 38	080	58 57	141	32 00	184	64 08	250	64 47	332
227	59 35	012	26 23	043	18 37	080	59 34	142	31 56	184	63 11	251	64 18	330
228	59 47	010	27 04	043	19 36	079	60 10	144	31 51	185	62 15	252	63 48	328
229	59 56	008	27 45	043	20 34	079	60 45	145	31 46	185	61 18	252	63 16	327
230	60 04	006	28 25	042	21 33	079	61 19	146	31 40	186	60 21	253	62 42	325
231	60 10	005	29 05	042	22 32	079	61 51	148	31 34	186	59 23	253	62 07	323
232	60 14	003	29 45	041	23 31	079	62 23	149	31 27	187	58 26	253	61 31	322
233	60 16	001	30 24	041	24 30	079	62 52	151	31 19	187	57 28	254	60 53	320
234	60 16	359	31 03	040	25 29	079	63 21	153	31 11	188	56 31	254	60 15	319
235	60 14	358	31 42	040	26 27	079	63 47	154	31 02	189	55 33	255	59 35	318
236	60 11	356	32 20	039	27 26	078	64 12	156	30 53	189	54 35	255	58 54	316
237	60 05	354	32 57	039	28 25	078	64 35	158	30 43	190	53 37	255	58 12	315
238	59 58	352	33 35	038	29 23	078	64 57	160	30 33	190	52 39	256	57 30	314
239	59 49	350	34 11	037	30 22	078	65 16	162	30 22	191	51 41	256	56 46	313
	VEGA		◆ALTAIR		Nunki		◆RIGIL KENT.		SPICA		◆ARCTURUS		Alphecca	
240	34 47	037	31 21	078	42 04	123	30 11	191	50 43	256	56 02	312	59 38	349
241	35 23	036	32 19	078	42 54	124	29 59	192	49 45	256	55 17	311	59 26	347
242	35 58	036	33 18	077	43 44	124	29 47	192	48 47	256	54 31	310	59 11	345
243	36 33	035	34 16	077	44 33	125	29 34	193	47 49	257	53 45	309	58 55	344
244	37 07	034	35 14	077	45 22	125	29 20	193	46 50	257	52 58	308	58 37	342
245	37 40	034	36 13	077	46 11	126	29 06	194	45 52	257	52 11	307	58 18	340
246	38 13	033	37 11	077	46 59	126	28 52	194	44 53	257	51 23	306	57 57	339
247	38 46	032	38 09	076	47 47	127	28 37	195	43 55	257	50 34	306	57 35	337
248	39 17	031	39 07	076	48 35	128	28 21	195	42 56	258	49 45	305	57 11	336
249	39 48	031	40 06	076	49 22	128	28 05	196	41 58	258	48 56	304	56 46	334
250	40 18	030	41 04	076	50 09	129	27 49	196	40 59	258	48 06	303	56 19	333
251	40 48	029	42 02	075	50 55	130	27 32	197	40 01	258	47 16	303	55 51	332
252	41 17	028	42 59	075	51 41	130	27 15	197	39 02	258	46 25	302	55 22	330
253	41 45	027	43 57	075	52 26	131	26 57	197	38 04	258	45 35	302	54 52	329
254	42 12	027	44 55	074	53 11	132	26 39	198	37 05	258	44 43	301	54 21	328
	◆VEGA		ALTAIR		◆Nunki		Shaula		ANTARES		◆SPICA		ARCTURUS	
255	42 38	026	45 53	074	53 55	133	55 03	168	65 26	197	36 06	258	43 52	300
256	43 04	025	46 50	074	54 38	134	55 15	170	65 08	199	35 08	258	43 00	300
257	43 29	024	47 48	073	55 21	135	55 24	171	64 47	201	34 09	259	42 08	299
258	43 52	023	48 45	073	56 03	136	55 33	172	64 25	203	33 10	259	41 16	299
259	44 15	022	49 42	073	56 45	137	55 40	174	64 01	205	32 11	259	40 23	298
260	44 37	021	50 39	072	57 25	138	55 46	175	63 35	206	31 13	259	39 30	298
261	44 59	020	51 36	072	58 05	139	55 50	177	63 08	208	30 14	259	38 38	298
262	45 19	019	52 33	071	58 44	140	55 53	178	62 39	210	29 15	259	37 44	297
263	45 38	018	53 30	071	59 22	141	55 54	180	62 09	211	28 16	259	36 51	297
264	45 56	017	54 26	070	59 59	143	55 54	181	61 37	213	27 17	259	35 57	296
265	46 13	016	55 23	070	60 34	144	55 52	182	61 04	214	26 19	259	35 04	296
266	46 29	015	56 19	069	61 09	146	55 49	184	60 29	216	25 20	259	34 10	296
267	46 44	014	57 14	069	61 42	147	55 44	185	59 54	217	24 21	259	33 16	295
268	46 58	013	58 10	068	62 14	149	55 38	187	59 17	218	23 22	259	32 21	295
269	47 11	012	59 05	067	62 45	150	55 30	188	58 40	220	22 23	259	31 27	295

LHA	Hc	Zn	Hc	Zn	Hc	Zn	Hc	Zn	Hc	Zn	Hc	Zn	Hc	Zn
♈	◆VEGA		DENEB		Enif		◆FOMALHAUT		Peacock		◆ANTARES		ARCTURUS	
270	47 22	011	29 55	032	32 47	076	15 05	120	29 01	158	58 01	221	30 33	294
271	47 33	009	30 26	031	33 45	076	15 56	120	29 23	159	57 21	222	29 38	294
272	47 42	008	30 57	031	34 43	076	16 48	120	29 44	159	56 41	223	28 43	294
273	47 50	007	31 27	030	35 41	076	17 40	120	30 05	160	56 00	224	27 48	294
274	47 57	006	31 57	029	36 39	075	18 32	120	30 26	160	55 18	225	26 53	293
275	48 03	005	32 26	029	37 37	075	19 24	120	30 46	161	54 35	226	25 58	293
276	48 07	004	32 55	028	38 35	075	20 15	121	31 06	161	53 51	227	25 03	293
277	48 10	003	33 23	028	39 33	075	21 07	121	31 25	162	53 07	228	24 08	293
278	48 12	001	33 50	027	40 31	074	21 58	121	31 43	162	52 23	229	23 13	292
279	48 13	000	34 17	026	41 28	074	22 50	121	32 01	163	51 38	229	22 17	292
280	48 13	359	34 43	026	42 26	074	23 41	121	32 19	163	50 52	230	21 22	292
281	48 11	358	35 09	025	43 23	073	24 32	121	32 36	164	50 06	231	20 26	292
282	48 08	357	35 34	024	44 21	073	25 23	122	32 52	164	49 19	231	19 30	292
283	48 04	356	35 58	024	45 18	073	26 14	122	33 08	165	48 32	232	18 35	291
284	47 59	354	36 22	023	46 15	072	27 05	122	33 23	166	47 44	233	17 39	291
	DENEB		Enif		◆FOMALHAUT		Peacock		◆ANTARES		Rasalhague		◆VEGA	
285	36 45	022	47 12	072	27 56	122	33 38	166	46 56	233	63 43	307	47 53	353
286	37 07	021	48 09	072	28 47	122	33 52	167	46 08	234	62 54	306	47 45	352
287	37 28	021	49 06	071	29 37	123	34 05	167	45 20	234	62 05	304	47 36	351
288	37 49	020	50 03	071	30 27	123	34 18	168	44 31	235	61 16	303	47 26	350
289	38 09	019	50 59	070	31 18	123	34 30	169	43 41	236	60 25	302	47 15	349
290	38 28	018	51 55	070	32 08	124	34 42	169	42 52	236	59 34	301	47 03	348
291	38 46	017	52 52	069	32 57	124	34 53	170	42 02	236	58 43	300	46 49	347
292	39 04	017	53 48	069	33 47	124	35 03	170	41 12	237	57 51	300	46 35	345
293	39 20	016	54 43	068	34 37	124	35 13	171	40 22	237	56 59	299	46 19	344
294	39 36	015	55 39	068	35 26	125	35 22	172	39 31	238	56 06	298	46 03	343
295	39 51	014	56 34	067	36 15	125	35 30	172	38 40	238	55 13	297	45 45	342
296	40 05	013	57 29	066	37 04	126	35 37	173	37 49	238	54 19	297	45 26	341
297	40 18	012	58 24	066	37 53	126	35 44	174	36 58	239	53 25	296	45 06	340
298	40 30	011	59 18	065	38 41	126	35 50	174	36 07	239	52 31	295	44 46	339
299	40 42	010	60 12	064	39 29	127	35 56	175	35 15	239	51 37	295	44 24	338
	◆DENEB		Alpheratz		◆FOMALHAUT		Peacock		◆ANTARES		Rasalhague		VEGA	
300	40 52	010	22 35	057	40 17	127	36 01	176	34 24	240	50 42	294	44 01	337
301	41 02	009	23 25	056	41 05	128	36 05	176	33 32	240	49 48	294	43 38	336
302	41 10	008	24 15	056	41 52	128	36 08	177	32 40	240	48 53	293	43 13	335
303	41 18	007	25 04	056	42 39	129	36 11	178	31 48	241	47 57	293	42 48	335
304	41 25	006	25 54	056	43 25	129	36 13	178	30 56	241	47 02	292	42 22	334
305	41 30	005	26 43	055	44 12	130	36 14	179	30 03	241	46 06	292	41 55	333
306	41 35	004	27 32	055	44 57	130	36 15	180	29 11	241	45 11	291	41 27	332
307	41 39	003	28 21	054	45 43	131	36 15	180	28 18	241	44 15	291	40 59	331
308	41 42	002	29 10	054	46 28	132	36 14	181	27 26	242	43 19	290	40 30	330
309	41 43	001	29 58	054	47 12	132	36 12	182	26 33	242	42 22	290	40 00	330
310	41 44	000	30 46	053	47 57	133	36 10	183	25 40	242	41 26	290	39 29	329
311	41 44	359	31 34	053	48 40	134	36 07	183	24 47	242	40 29	289	38 58	328
312	41 43	358	32 22	052	49 23	134	36 03	184	23 54	242	39 33	289	38 26	327
313	41 41	357	33 09	052	50 06	135	35 59	185	23 01	242	38 36	289	37 53	327
314	41 38	357	33 56	052	50 48	136	35 54	185	22 08	243	37 39	288	37 20	326
	◆Alpheratz		◆FOMALHAUT		Peacock		Nunki		◆Rasalhague		VEGA		DENEB	
315	34 43	051	51 29	137	35 48	186	52 02	229	36 42	288	36 46	325	41 33	356
316	35 30	051	52 10	138	35 42	187	51 17	230	35 45	288	36 12	325	41 28	355
317	36 16	050	52 50	139	35 35	187	50 31	231	34 48	288	35 37	324	41 22	354
318	37 01	050	53 29	140	35 27	188	49 44	231	33 51	287	35 01	323	41 15	353
319	37 47	049	54 07	141	35 18	188	48 57	232	32 54	287	34 25	323	41 07	352
320	38 32	048	54 45	142	35 09	189	48 10	233	31 57	287	33 49	322	40 58	351
321	39 16	048	55 22	143	34 59	190	47 22	233	30 59	287	33 12	322	40 49	350
322	40 01	047	55 57	144	34 49	190	46 34	234	30 02	286	32 34	321	40 38	349
323	40 44	047	56 32	145	34 38	191	45 45	234	29 04	286	31 56	321	40 26	348
324	41 28	046	57 06	146	34 26	192	44 56	235	28 07	286	31 18	320	40 13	347
325	42 11	045	57 39	147	34 13	192	44 07	235	27 09	286	30 39	319	40 00	347
326	42 53	045	58 11	149	34 00	193	43 18	236	26 11	286	30 00	319	39 45	346
327	43 35	044	58 41	150	33 47	193	42 28	236	25 14	285	29 21	319	39 30	345
328	44 16	043	59 10	151	33 32	194	41 38	237	24 16	285	28 41	318	39 14	344
329	44 56	042	59 38	153	33 18	195	40 48	237	23 18	285	28 01	318	38 57	343
	◆Alpheratz		Hamal		Diphda		◆FOMALHAUT		Nunki		◆ALTAIR		DENEB	
330	45 36	042	24 24	063	47 17	114	60 05	154	39 57	238	55 38	290	38 39	342
331	46 16	041	25 18	062	48 12	114	60 30	156	39 06	238	54 41	290	38 21	342
332	46 55	040	26 11	062	49 06	114	60 54	158	38 15	238	53 45	289	38 01	341
333	47 33	039	27 03	062	50 01	115	61 16	159	37 24	239	52 48	289	37 41	340
334	48 10	038	27 56	061	50 55	115	61 36	161	36 33	239	51 52	288	37 20	339
335	48 46	037	28 49	061	51 49	116	61 55	163	35 41	239	50 55	288	36 59	338
336	49 22	036	29 41	061	52 43	116	62 12	164	34 50	240	49 58	288	36 36	338
337	49 57	035	30 33	060	53 37	117	62 27	166	33 58	240	49 00	287	36 13	337
338	50 31	034	31 25	060	54 30	117	62 41	168	33 06	240	48 03	287	35 49	336
339	51 04	033	32 17	060	55 23	118	62 52	170	32 14	241	47 06	286	35 25	336
340	51 36	032	33 09	059	56 16	119	63 02	172	31 22	241	46 08	286	35 00	335
341	52 07	031	34 00	059	57 08	119	63 10	174	30 29	241	45 11	286	34 34	334
342	52 37	030	34 51	059	58 00	120	63 16	175	29 37	241	44 13	285	34 07	333
343	53 06	028	35 42	058	58 52	121	63 19	177	28 44	241	43 15	285	33 40	333
344	53 34	027	36 33	058	59 43	122	63 21	179	27 51	242	42 17	285	33 13	332
	Alpheratz		◆Hamal		◆ACHERNAR		FOMALHAUT		Nunki		◆ALTAIR		DENEB	
345	54 01	026	37 24	057	27 29	157	63 21	181	26 59	242	41 19	285	32 44	332
346	54 26	025	38 14	057	27 52	158	63 18	183	26 06	242	40 21	284	32 16	331
347	54 51	023	39 04	056	28 14	158	63 14	185	25 13	242	39 23	284	31 46	330
348	55 14	022	39 53	056	28 36	159	63 08	187	24 20	242	38 25	284	31 16	330
349	55 35	020	40 43	055	28 58	159	62 59	189	23 27	243	37 27	284	30 46	329
350	55 55	019	41 32	055	29 19	159	62 49	191	22 33	243	36 28	283	30 15	329
351	56 14	018	42 21	054	29 40	160	62 37	193	21 40	243	35 30	283	29 43	328
352	56 31	016	43 09	053	30 00	160	62 23	194	20 47	243	34 32	283	29 12	328
353	56 47	015	43 57	053	30 20	161	62 07	196	19 53	243	33 33	283	28 39	327
354	57 01	013	44 44	052	30 39	161	61 49	198	19 00	243	32 35	282	28 06	327
355	57 14	011	45 32	052	30 58	162	61 30	200	18 06	243	31 36	282	27 33	326
356	57 25	010	46 18	051	31 17	162	61 09	201	17 13	243	30 38	282	27 00	326
357	57 35	008	47 05	050	31 34	163	60 46	203	16 19	243	29 39	282	26 25	325
358	57 42	007	47 50	049	31 52	164	60 22	205	15 26	244	28 41	282	25 51	325
359	57 48	005	48 35	049	32 08	164	59 57	206	14 32	244	27 42	282	25 16	324

LHA ♈	Hc	Zn	Hc	Zn	Hc	Zn	Hc	Zn	Hc	Zn	Hc	Zn	Hc	Zn
	Alpheratz		◆ALDEBARAN		RIGEL		ACHERNAR		◆FOMALHAUT		ALTAIR		◆DENEB	
0	56 53	003	18 56	071	11 52	098	33 22	164	60 23	208	26 31	282	23 52	324
1	56 55	002	19 52	071	12 51	098	33 38	165	59 54	210	25 32	282	23 17	324
2	56 56	000	20 49	071	13 50	097	33 53	166	59 23	211	24 34	282	22 42	323
3	56 55	358	21 45	070	14 50	097	34 08	166	58 52	213	23 35	281	22 06	323
4	56 53	357	22 42	070	15 49	097	34 22	167	58 19	214	22 36	281	21 29	322
5	56 49	355	23 38	070	16 48	097	34 35	167	57 45	215	21 38	281	20 53	322
6	56 43	354	24 34	070	17 48	097	34 48	168	57 10	216	20 39	281	20 16	322
7	56 36	352	25 30	070	18 47	097	35 00	169	56 34	217	19 40	281	19 38	321
8	56 27	351	26 26	069	19 47	097	35 11	169	55 57	219	18 41	281	19 01	321
9	56 16	349	27 22	069	20 46	097	35 22	170	55 20	220	17 43	281	18 23	321
10	56 04	347	28 18	069	21 45	097	35 32	170	54 41	221	16 44	280	17 45	320
11	55 50	346	29 14	069	22 45	097	35 42	171	54 02	222	15 45	280	17 07	320
12	55 35	344	30 10	068	23 44	097	35 51	172	53 22	223	14 46	280	16 28	320
13	55 18	343	31 05	068	24 43	097	35 59	172	52 41	223	13 47	280	15 49	319
14	55 00	342	32 01	068	25 43	097	36 07	173	51 59	224	12 48	280	15 10	319
	CAPELLA		ALDEBARAN		◆RIGEL		ACHERNAR		◆FOMALHAUT		Enif		◆Alpheratz	
15	14 38	040	32 56	067	26 42	097	36 13	174	51 17	225	39 15	286	54 40	340
16	15 17	040	33 51	067	27 42	097	36 20	174	50 35	226	38 17	286	54 19	339
17	15 55	040	34 46	067	28 41	097	36 25	175	49 51	227	37 20	286	53 57	337
18	16 33	039	35 41	066	29 40	097	36 30	176	49 08	227	36 22	285	53 33	336
19	17 11	039	36 36	066	30 40	097	36 34	176	48 23	228	35 24	285	53 08	335
20	17 49	039	37 30	066	31 39	097	36 38	177	47 38	229	34 26	285	52 42	334
21	18 26	038	38 25	065	32 38	097	36 40	178	46 53	229	33 29	285	52 15	332
22	19 03	038	39 19	065	33 38	097	36 42	178	46 08	230	32 31	284	51 47	331
23	19 40	038	40 13	064	34 37	097	36 44	179	45 22	231	31 33	284	51 17	330
24	20 17	037	41 07	064	35 37	097	36 44	180	44 35	231	30 35	284	50 47	329
25	20 53	037	42 01	064	36 36	097	36 44	180	43 48	232	29 36	284	50 16	328
26	21 29	037	42 54	063	37 35	097	36 43	181	43 01	232	28 38	284	49 43	327
27	22 04	036	43 47	063	38 35	097	36 42	182	42 14	233	27 40	283	49 10	326
28	22 39	036	44 40	062	39 34	097	36 40	182	41 26	233	26 42	283	48 36	325
29	23 14	035	45 33	061	40 33	097	36 37	183	40 38	234	25 43	283	48 01	324
	CAPELLA		◆BETELGEUSE		SIRIUS		CANOPUS		◆ACHERNAR		FOMALHAUT		◆Alpheratz	
30	23 49	035	30 18	079	19 07	106	17 34	145	36 33	184	39 50	234	47 25	323
31	24 23	035	31 17	079	20 04	106	18 09	145	36 29	184	39 01	234	46 48	322
32	24 57	034	32 16	079	21 02	106	18 43	145	36 24	185	38 12	235	46 11	321
33	25 30	034	33 14	078	21 59	106	19 17	145	36 18	186	37 23	235	45 33	320
34	26 03	033	34 13	078	22 57	106	19 52	145	36 12	186	36 34	236	44 54	319
35	26 36	033	35 11	078	23 54	107	20 25	146	36 05	187	35 45	236	44 14	318
36	27 08	032	36 10	078	24 51	107	20 59	146	35 57	188	34 55	236	43 34	318
37	27 40	032	37 08	078	25 49	107	21 33	146	35 49	188	34 05	237	42 54	317
38	28 11	031	38 07	077	26 46	107	22 06	146	35 40	189	33 15	237	42 13	316
39	28 42	031	39 05	077	27 43	107	22 39	147	35 30	190	32 25	237	41 31	315
40	29 12	030	40 04	077	28 41	107	23 12	147	35 20	190	31 34	237	40 48	315
41	29 42	030	41 02	077	29 38	107	23 45	147	35 09	191	30 44	238	40 06	314
42	30 11	029	42 00	076	30 35	107	24 17	147	34 57	192	29 53	238	39 22	313
43	30 40	028	42 58	076	31 33	107	24 49	148	34 45	192	29 03	238	38 39	313
44	31 08	028	43 56	076	32 30	107	25 21	148	34 32	193	28 12	238	37 55	312
	CAPELLA		◆BETELGEUSE		SIRIUS		CANOPUS		◆ACHERNAR		Diphda		◆Hamal	
45	31 35	027	44 54	075	33 27	107	25 52	148	34 18	193	53 45	245	59 40	335
46	32 03	027	45 52	075	34 24	108	26 23	149	34 04	194	52 51	245	59 14	334
47	32 29	026	46 50	075	35 21	108	26 54	149	33 50	194	51 57	246	58 47	332
48	32 55	025	47 47	074	36 18	108	27 25	150	33 34	195	51 02	246	58 18	331
49	33 20	025	48 45	074	37 15	108	27 55	150	33 18	196	50 07	246	57 48	329
50	33 45	024	49 43	074	38 12	108	28 25	150	33 02	196	49 13	247	57 17	328
51	34 09	023	50 40	073	39 08	108	28 54	151	32 45	197	48 17	247	56 44	326
52	34 32	023	51 37	073	40 05	109	29 23	151	32 28	197	47 22	248	56 11	325
53	34 55	022	52 34	072	41 02	109	29 52	152	32 09	198	46 27	248	55 36	324
54	35 17	021	53 31	072	41 59	109	30 20	152	31 51	198	45 31	248	55 00	323
55	35 38	020	54 28	071	42 55	109	30 48	152	31 32	199	44 36	249	54 23	321
56	35 58	020	55 25	071	43 52	109	31 16	153	31 12	199	43 40	249	53 46	320
57	36 18	019	56 21	070	44 48	110	31 43	153	30 52	200	42 44	249	53 07	319
58	36 37	018	57 17	070	45 44	110	32 09	154	30 32	200	41 48	249	52 28	318
59	36 55	017	58 13	069	46 41	110	32 35	154	30 11	201	40 52	250	51 47	317
	◆CAPELLA		POLLUX		◆SIRIUS		CANOPUS		ACHERNAR		◆Diphda		Hamal	
60	37 13	017	27 09	056	47 37	111	33 01	155	29 49	201	39 56	250	51 06	316
61	37 30	016	27 58	055	48 33	111	33 26	155	29 27	202	39 00	250	50 24	315
62	37 45	015	28 47	055	49 29	111	33 51	156	29 05	202	38 03	250	49 42	314
63	38 00	014	29 36	054	50 24	112	34 15	156	28 42	203	37 07	250	48 59	313
64	38 15	013	30 25	054	51 20	112	34 39	157	28 19	203	36 11	251	48 15	313
65	38 28	012	31 13	054	52 15	112	35 02	158	27 55	203	35 14	251	47 31	312
66	38 40	012	32 01	053	53 11	113	35 24	158	27 31	204	34 18	251	46 46	311
67	38 52	011	32 49	053	54 06	113	35 46	159	27 06	204	33 21	251	46 00	310
68	39 03	010	33 36	052	55 01	114	36 07	159	26 42	205	32 25	251	45 15	310
69	39 13	009	34 24	052	55 55	114	36 28	160	26 17	205	31 28	251	44 28	309
70	39 22	008	35 10	051	56 50	115	36 48	161	25 51	205	30 31	251	43 41	308
71	39 30	007	35 57	051	57 44	115	37 08	161	25 25	206	29 34	252	42 54	308
72	39 37	006	36 43	050	58 38	116	37 27	162	24 59	206	28 38	252	42 06	307
73	39 43	005	37 29	050	59 31	117	37 45	163	24 32	206	27 41	252	41 18	306
74	39 48	005	38 14	049	60 25	117	38 02	163	24 05	207	26 44	252	40 30	306
	CAPELLA		◆POLLUX		REGULUS		Suhail		◆CANOPUS		ACHERNAR		◆Hamal	
75	39 53	004	38 59	048	11 48	077	22 52	136	38 19	164	23 38	207	39 41	305
76	39 56	003	39 44	048	12 47	077	23 34	136	38 36	165	23 11	207	38 52	305
77	39 58	002	40 28	047	13 45	077	24 15	136	38 51	165	22 43	208	38 03	304
78	40 00	001	41 12	047	14 43	076	24 56	137	39 06	166	22 15	208	37 13	304
79	40 00	000	41 55	046	15 41	076	25 37	137	39 20	167	21 47	208	36 23	303
80	40 00	359	42 38	045	16 39	076	26 18	137	39 33	168	21 18	209	35 33	303
81	39 59	358	43 20	044	17 37	076	26 59	138	39 46	168	20 49	209	34 42	302
82	39 56	357	44 02	044	18 36	076	27 39	138	39 58	169	20 20	209	33 52	302
83	39 53	356	44 43	043	19 34	076	28 19	138	40 09	170	19 51	209	33 01	301
84	39 49	356	45 23	042	20 32	076	28 59	138	40 19	171	19 22	210	32 09	301
85	39 44	355	46 03	041	21 29	075	29 38	139	40 28	171	18 52	210	31 18	301
86	39 38	354	46 42	040	22 27	075	30 18	139	40 37	172	18 22	210	30 26	300
87	39 31	353	47 20	040	23 25	075	30 57	140	40 45	173	17 52	210	29 35	300
88	39 23	352	47 58	039	24 23	075	31 35	140	40 52	174	17 22	211	28 43	300
89	39 14	351	48 35	038	25 21	075	32 14	140	40 58	174	16 51	211	27 50	299

LAT 4°S

LHA ♈	Hc	Zn	Hc	Zn	Hc	Zn	Hc	Zn	Hc	Zn	Hc	Zn	Hc	Zn
	POLLUX		◆REGULUS		Suhail		◆CANOPUS		Acamar		◆ALDEBARAN		CAPELLA	
90	49 11	037	26 19	075	32 52	141	41 04	175	35 20	222	60 46	315	39 04	350
91	49 47	036	27 16	074	33 29	141	41 08	176	34 40	222	60 03	314	38 54	349
92	50 21	035	28 14	074	34 07	142	41 12	177	34 00	223	59 19	313	38 42	349
93	50 55	034	29 11	074	34 43	142	41 15	178	33 19	223	58 35	311	38 30	348
94	51 27	032	30 09	074	35 20	143	41 17	178	32 38	223	57 49	310	38 17	347
95	51 59	031	31 06	073	35 56	143	41 18	179	31 57	224	57 03	309	38 03	346
96	52 29	030	32 04	073	36 32	144	41 18	180	31 15	224	56 17	308	37 48	345
97	52 59	029	33 01	073	37 07	144	41 18	181	30 33	225	55 29	307	37 32	344
98	53 27	028	33 58	073	37 42	145	41 17	182	29 51	225	54 41	306	37 16	344
99	53 54	026	34 55	072	38 16	145	41 15	182	29 09	225	53 53	305	36 58	343
100	54 20	025	35 52	072	38 50	146	41 12	183	28 26	226	53 04	305	36 40	342
101	54 45	024	36 49	072	39 23	147	41 08	184	27 43	226	52 14	304	36 21	341
102	55 09	022	37 46	071	39 56	147	41 03	185	27 00	226	51 24	303	36 02	340
103	55 31	021	38 42	071	40 28	148	40 58	186	26 17	226	50 34	302	35 41	340
104	55 51	019	39 39	071	41 00	148	40 51	186	25 33	227	49 43	302	35 20	339
	POLLUX		◆REGULUS		Suhail		◆CANOPUS		RIGEL		◆ALDEBARAN		CAPELLA	
105	56 11	018	40 36	070	41 31	149	40 44	187	63 24	260	48 52	301	34 58	338
106	56 28	017	41 32	070	42 01	150	40 36	188	62 25	260	48 00	300	34 36	338
107	56 45	015	42 28	070	42 31	151	40 28	189	61 26	260	47 08	300	34 13	337
108	56 59	013	43 24	069	43 00	151	40 18	190	60 27	260	46 16	299	33 49	336
109	57 13	012	44 20	069	43 29	152	40 08	190	59 28	261	45 23	298	33 24	335
110	57 24	010	45 16	069	43 56	153	39 57	191	58 29	261	44 31	298	32 59	335
111	57 34	009	46 11	068	44 23	154	39 45	192	57 30	261	43 38	297	32 33	334
112	57 42	007	47 07	068	44 50	154	39 32	193	56 31	261	42 44	297	32 07	334
113	57 49	005	48 02	067	45 15	155	39 19	193	55 32	261	41 51	296	31 40	333
114	57 53	004	48 57	067	45 40	156	39 05	194	54 32	261	40 57	296	31 12	332
115	57 56	002	49 52	066	46 04	157	38 50	195	53 33	262	40 03	296	30 44	332
116	57 58	000	50 47	065	46 27	158	38 34	195	52 34	262	39 09	295	30 16	331
117	57 57	359	51 41	065	46 49	159	38 18	196	51 35	262	38 15	295	29 46	331
118	57 55	357	52 35	064	47 10	160	38 01	197	50 35	262	37 20	294	29 17	330
119	57 51	355	53 29	064	47 30	161	37 43	197	49 36	262	36 26	294	28 46	329
	◆REGULUS		Gienah		◆ACRUX		CANOPUS		SIRIUS		◆BETELGEUSE		POLLUX	
120	54 22	063	26 05	108	13 59	155	37 25	198	67 37	234	56 46	290	57 46	354
121	55 15	062	27 02	108	14 25	155	37 06	199	66 49	235	55 50	289	57 38	352
122	56 08	061	27 59	108	14 50	155	36 47	199	65 59	237	54 53	289	57 29	351
123	57 00	061	28 56	108	15 16	155	36 27	200	65 09	238	53 57	288	57 19	349
124	57 52	060	29 53	108	15 41	155	36 06	201	64 18	239	53 00	288	57 06	347
125	58 44	059	30 50	108	16 06	156	35 44	201	63 26	240	52 03	287	56 52	346
126	59 35	058	31 47	108	16 30	156	35 22	202	62 34	241	51 06	287	56 37	344
127	60 25	057	32 44	108	16 55	156	35 00	202	61 42	241	50 08	287	56 20	343
128	61 15	056	33 41	108	17 19	156	34 37	203	60 49	242	49 11	286	56 01	341
129	62 04	055	34 37	109	17 43	156	34 13	204	59 56	243	48 13	286	55 41	340
130	62 53	054	35 34	109	18 07	157	33 49	204	59 03	244	47 16	285	55 20	338
131	63 41	052	36 31	109	18 31	157	33 24	205	58 09	244	46 18	285	54 57	337
132	64 28	051	37 27	109	18 54	157	32 59	205	57 15	245	45 20	285	54 33	336
133	65 14	050	38 24	109	19 18	157	32 33	206	56 21	245	44 22	284	54 08	334
134	65 59	048	39 20	109	19 41	158	32 07	206	55 26	246	43 24	284	53 41	333
	Alioth		◆SPICA		ACRUX		◆CANOPUS		SIRIUS		BETELGEUSE		◆POLLUX	
135	13 32	029	24 05	100	20 03	158	31 40	207	54 31	247	42 26	284	53 13	332
136	14 02	029	25 03	100	20 26	158	31 13	207	53 36	247	41 28	284	52 44	330
137	14 31	029	26 02	100	20 48	158	30 46	208	52 41	247	40 30	283	52 14	329
138	14 59	028	27 01	100	21 10	159	30 18	208	51 46	248	39 32	283	51 43	328
139	15 28	028	28 00	100	21 32	159	29 50	208	50 50	248	38 33	283	51 11	327
140	15 56	028	28 59	101	21 53	159	29 21	209	49 54	249	37 35	283	50 38	326
141	16 24	028	29 58	101	22 14	160	28 52	209	48 59	249	36 37	282	50 04	325
142	16 51	027	30 57	101	22 35	160	28 22	210	48 03	249	35 38	282	49 29	324
143	17 18	027	31 55	101	22 55	160	27 52	210	47 07	250	34 39	282	48 54	323
144	17 45	027	32 54	101	23 15	161	27 22	210	46 10	250	33 41	282	48 17	322
145	18 12	026	33 53	101	23 35	161	26 52	211	45 14	250	32 42	281	47 40	321
146	18 38	026	34 52	101	23 55	161	26 21	211	44 18	250	31 43	281	47 02	320
147	19 04	025	35 51	101	24 14	162	25 50	212	43 21	251	30 45	281	46 23	319
148	19 29	025	36 49	101	24 33	162	25 18	212	42 25	251	29 46	281	45 43	318
149	19 54	025	37 48	101	24 51	162	24 46	212	41 28	251	28 47	281	45 03	317
	Alioth		ARCTURUS		◆SPICA		ACRUX		◆CANOPUS		SIRIUS		◆POLLUX	
150	20 19	024	23 05	067	38 47	101	25 09	163	24 14	213	40 32	251	44 22	317
151	20 44	024	24 00	067	39 46	101	25 27	163	23 42	213	39 35	252	43 41	316
152	21 08	023	24 55	067	40 44	101	25 44	163	23 09	213	38 38	252	42 59	315
153	21 31	023	25 50	066	41 43	101	26 01	164	22 36	213	37 41	252	42 17	314
154	21 54	023	26 45	066	42 42	101	26 17	164	22 03	214	36 44	252	41 34	314
155	22 17	022	27 40	066	43 40	102	26 33	165	21 30	214	35 47	252	40 50	313
156	22 39	022	28 34	066	44 39	102	26 49	165	20 57	214	34 50	252	40 06	312
157	23 01	021	29 29	065	45 38	102	27 04	165	20 23	214	33 53	253	39 22	312
158	23 22	021	30 23	065	46 36	102	27 19	166	19 49	215	32 56	253	38 37	311
159	23 43	020	31 17	065	47 35	102	27 33	166	19 15	215	31 59	253	37 52	311
160	24 04	020	32 11	064	48 33	102	27 47	167	18 40	215	31 02	253	37 06	310
161	24 24	019	33 05	064	49 32	102	28 01	167	18 06	215	30 04	253	36 20	310
162	24 43	019	33 59	064	50 30	103	28 14	168	17 31	216	29 07	253	35 34	309
163	25 02	018	34 52	063	51 28	103	28 26	168	16 56	216	28 10	253	34 47	308
164	25 21	018	35 45	063	52 27	103	28 39	169	16 21	216	27 13	253	34 00	308
	Alioth		◆ARCTURUS		SPICA		◆ACRUX		Suhail		SIRIUS		◆POLLUX	
165	25 39	017	36 39	062	53 25	103	28 50	169	43 26	208	26 15	253	33 13	308
166	25 56	017	37 32	062	54 23	104	29 01	170	42 57	209	25 18	253	32 25	307
167	26 13	016	38 24	062	55 21	104	29 12	170	42 28	210	24 21	253	31 37	307
168	26 29	016	39 17	061	56 19	104	29 22	170	41 58	210	23 23	254	30 49	306
169	26 45	015	40 09	061	57 17	104	29 32	171	41 28	211	22 26	254	30 01	306
170	27 00	014	41 01	060	58 15	105	29 41	171	40 57	212	21 28	254	29 12	305
171	27 15	014	41 53	060	59 13	105	29 50	172	40 25	212	20 31	254	28 23	305
172	27 29	013	42 44	059	60 11	106	29 58	172	39 53	213	19 34	254	27 34	305
173	27 43	013	43 36	059	61 09	106	30 05	173	39 20	214	18 36	254	26 45	304
174	27 56	012	44 26	058	62 06	106	30 12	173	38 47	214	17 39	254	25 55	304
175	28 08	012	45 17	057	63 03	107	30 19	174	38 13	215	16 41	254	25 05	304
176	28 20	011	46 07	057	64 01	107	30 25	174	37 39	215	15 44	254	24 15	303
177	28 31	010	46 57	056	64 58	108	30 30	175	37 04	216	14 46	254	23 25	303
178	28 41	010	47 47	055	65 54	109	30 35	175	36 28	216	13 49	254	22 35	303
179	28 51	009	48 36	055	66 51	109	30 40	176	35 53	217	12 51	254	21 44	302

LAT 4°S

LHA ♈	Hc	Zn	Hc	Zn	Hc	Zn	Hc	Zn	Hc	Zn	Hc	Zn	Hc	Zn
	Alioth		◆ARCTURUS		ANTARES		◆ACRUX		Suhail		Alphard		◆REGULUS	
180	29 00	009	49 24	054	22 06	117	30 44	177	35 17	217	51 48	261	57 55	300
181	29 09	008	50 12	053	22 59	117	30 47	177	34 40	218	50 49	261	57 03	299
182	29 17	007	51 00	052	23 52	117	30 50	178	34 03	218	49 50	261	56 11	299
183	29 24	007	51 47	052	24 46	117	30 52	178	33 26	219	48 51	261	55 18	298
184	29 31	006	52 34	051	25 39	117	30 54	179	32 48	219	47 52	262	54 25	297
185	29 37	005	53 20	050	26 32	118	30 55	179	32 10	220	46 52	262	53 32	296
186	29 42	005	54 05	049	27 25	118	30 56	180	31 32	220	45 53	262	52 38	296
187	29 47	004	54 50	048	28 18	118	30 56	180	30 53	220	44 54	262	51 44	295
188	29 51	004	55 34	047	29 10	118	30 55	181	30 14	221	43 55	262	50 49	295
189	29 54	003	56 17	046	30 03	118	30 54	181	29 35	221	42 55	262	49 55	294
190	29 57	002	56 59	045	30 56	119	30 52	182	28 55	222	41 56	262	49 00	293
191	29 59	002	57 41	043	31 48	119	30 50	182	28 15	222	40 57	262	48 05	293
192	30 00	001	58 22	042	32 41	119	30 48	183	27 35	222	39 58	262	47 10	292
193	30 01	000	59 01	041	33 33	119	30 44	183	26 55	223	38 58	262	46 14	292
194	30 01	000	59 40	039	34 25	120	30 41	184	26 14	223	37 59	262	45 19	292
	ARCTURUS		◆Rasalhague		ANTARES		◆ACRUX		Alphard		◆REGULUS		Alioth	
195	60 17	038	19 49	075	35 17	120	30 36	184	37 00	262	44 23	291	30 00	359
196	60 53	037	20 46	075	36 09	120	30 31	185	36 00	262	43 27	291	29 59	358
197	61 28	035	21 44	075	37 01	120	30 26	185	35 01	262	42 31	290	29 57	358
198	62 02	033	22 42	075	37 52	121	30 20	186	34 02	262	41 35	290	29 54	357
199	62 34	032	23 40	074	38 44	121	30 13	186	33 02	262	40 39	290	29 50	356
200	63 05	030	24 37	074	39 35	121	30 06	187	32 03	262	39 42	289	29 46	356
201	63 34	028	25 35	074	40 26	122	29 59	187	31 04	262	38 45	289	29 42	355
202	64 01	026	26 32	074	41 17	122	29 51	188	30 05	262	37 49	289	29 36	355
203	64 27	024	27 30	074	42 07	123	29 42	188	29 05	262	36 52	288	29 30	354
204	64 51	022	28 27	073	42 58	123	29 33	189	28 06	262	35 55	288	29 23	353
205	65 13	020	29 24	073	43 48	123	29 24	189	27 07	262	34 58	288	29 16	353
206	65 32	018	30 22	073	44 37	124	29 14	190	26 07	262	34 01	287	29 08	352
207	65 50	016	31 19	073	45 27	124	29 03	190	25 08	262	33 04	287	28 59	351
208	66 05	014	32 16	072	46 16	125	28 52	191	24 09	262	32 07	287	28 50	351
209	66 18	011	33 13	072	47 05	125	28 40	191	23 09	262	31 09	287	28 40	350
	ARCTURUS		◆VEGA		ANTARES		◆RIGIL KENT.		ACRUX		Gienah		◆REGULUS	
210	66 29	009	13 27	049	47 54	126	32 42	174	28 28	192	61 04	240	30 12	286
211	66 37	007	14 12	048	48 42	127	32 48	175	28 16	192	60 12	241	29 14	286
212	66 43	004	14 56	048	49 30	127	32 53	175	28 03	193	59 20	242	28 17	286
213	66 47	002	15 41	048	50 17	128	32 57	176	27 50	193	58 27	242	27 19	286
214	66 47	000	16 25	048	51 04	129	33 01	177	27 36	194	57 33	243	26 22	285
215	66 46	357	17 09	047	51 51	129	33 04	177	27 21	194	56 40	244	25 24	285
216	66 42	355	17 53	047	52 37	130	33 07	178	27 07	194	55 46	244	24 26	285
217	66 35	353	18 37	047	53 23	131	33 09	178	26 51	195	54 52	245	23 28	285
218	66 26	350	19 20	046	54 08	132	33 10	179	26 36	195	53 58	245	22 30	285
219	66 15	348	20 03	046	54 52	133	33 11	180	26 20	196	53 03	246	21 33	285
220	66 01	346	20 46	046	55 36	133	33 11	180	26 03	196	52 08	246	20 35	284
221	65 45	343	21 29	045	56 19	134	33 11	181	25 47	197	51 14	247	19 37	284
222	65 27	341	22 11	045	57 01	135	33 10	181	25 29	197	50 18	247	18 39	284
223	65 07	339	22 54	045	57 43	136	33 08	182	25 12	197	49 23	248	17 41	284
224	64 44	337	23 36	044	58 24	138	33 06	182	24 54	198	48 28	248	16 42	284
	Alphecca		VEGA		◆ALTAIR		ANTARES		◆RIGIL KENT.		SPICA		◆ARCTURUS	
225	58 10	015	24 17	044	16 28	080	59 04	139	33 03	183	65 24	252	64 20	335
226	58 24	013	24 59	044	17 27	079	59 43	140	32 59	184	64 27	252	63 54	333
227	58 37	011	25 40	043	18 26	079	60 21	141	32 55	184	63 30	253	63 26	331
228	58 48	010	26 20	043	19 24	079	60 58	142	32 51	185	62 33	253	62 56	330
229	58 57	008	27 01	042	20 23	079	61 34	144	32 46	185	61 35	254	62 25	328
230	59 04	006	27 41	042	21 22	079	62 08	145	32 40	186	60 38	254	61 53	326
231	59 10	005	28 21	041	22 21	079	62 42	147	32 33	186	59 40	255	61 19	325
232	59 14	003	29 00	041	23 19	079	63 14	148	32 26	187	58 42	255	60 43	323
233	59 16	001	29 39	040	24 18	078	63 45	150	32 19	188	57 44	255	60 07	322
234	59 16	359	30 17	040	25 17	078	64 14	152	32 11	188	56 46	256	59 29	320
235	59 14	358	30 55	039	26 15	078	64 41	154	32 02	189	55 48	256	58 50	319
236	59 11	356	31 33	039	27 14	078	65 07	155	31 52	189	54 50	256	58 10	318
237	59 06	354	32 10	038	28 12	078	65 31	157	31 43	190	53 52	257	57 29	316
238	58 59	352	32 47	038	29 11	078	65 53	159	31 32	190	52 54	257	56 48	315
239	58 50	351	33 23	037	30 09	077	66 13	161	31 21	191	51 55	257	56 05	314
	VEGA		◆ALTAIR		Nunki		◆RIGIL KENT.		SPICA		◆ARCTURUS		Alphecca	
240	33 59	036	31 07	077	42 37	123	31 10	191	50 57	257	55 21	313	58 39	349
241	34 35	036	32 06	077	43 27	123	30 58	192	49 59	257	54 37	312	58 27	347
242	35 09	035	33 04	077	44 17	124	30 45	192	49 00	258	53 52	311	58 13	346
243	35 44	035	34 02	076	45 07	124	30 32	193	48 02	258	53 07	310	57 57	344
244	36 17	034	35 01	076	45 56	125	30 18	193	47 03	258	52 21	309	57 40	342
245	36 50	033	35 59	076	46 45	125	30 04	194	46 05	258	51 34	308	57 21	341
246	37 23	032	36 57	076	47 34	126	29 50	194	45 06	258	50 47	307	57 01	339
247	37 55	032	37 55	076	48 23	126	29 35	195	44 08	258	49 59	307	56 39	338
248	38 26	031	38 53	075	49 11	127	29 19	195	43 09	258	49 11	306	56 16	336
249	38 56	030	39 50	075	49 59	127	29 03	196	42 10	259	48 22	305	55 52	335
250	39 26	029	40 48	075	50 46	128	28 47	196	41 12	259	47 33	304	55 26	334
251	39 55	029	41 46	074	51 33	129	28 30	197	40 13	259	46 43	304	54 58	332
252	40 24	028	42 43	074	52 19	130	28 12	197	39 14	259	45 53	303	54 30	331
253	40 51	027	43 41	074	53 05	130	27 54	198	38 15	259	45 03	302	54 00	330
254	41 18	026	44 38	073	53 51	131	27 36	198	37 17	259	44 12	302	53 30	328
	◆VEGA		ALTAIR		◆Nunki		Shaula		ANTARES		◆SPICA		ARCTURUS	
255	41 44	025	45 36	073	54 35	132	56 02	168	66 23	197	36 18	259	43 21	301
256	42 09	024	46 33	073	55 20	133	56 14	169	66 04	200	35 19	259	42 30	301
257	42 34	024	47 30	072	56 03	134	56 24	171	65 43	202	34 20	259	41 38	300
258	42 57	023	48 27	072	56 46	135	56 33	172	65 20	204	33 22	259	40 46	300
259	43 20	022	49 24	071	57 28	136	56 40	174	64 56	205	32 23	259	39 54	299
260	43 41	021	50 20	071	58 10	137	56 46	175	64 29	207	31 24	259	39 02	299
261	44 02	020	51 17	070	58 50	138	56 50	177	64 01	209	30 25	259	38 09	298
262	44 22	019	52 13	070	59 30	139	56 53	178	63 31	211	29 26	259	37 17	298
263	44 41	018	53 09	069	60 08	140	56 54	180	63 00	212	28 27	259	36 24	297
264	44 59	017	54 05	069	60 46	142	56 54	181	62 27	214	27 29	260	35 30	297
265	45 15	016	55 01	068	61 23	143	56 52	182	61 53	215	26 30	260	34 37	297
266	45 31	015	55 57	068	61 58	144	56 48	184	61 18	217	25 31	260	33 43	296
267	45 46	014	56 52	067	62 32	146	56 44	185	60 41	218	24 32	260	32 50	296
268	45 59	013	57 47	066	63 05	148	56 37	187	60 04	219	23 33	260	31 56	296
269	46 12	011	58 41	066	63 36	149	56 29	188	59 25	221	22 34	260	31 02	295

LAT 4°S

LHA ♈	Hc	Zn	Hc	Zn	Hc	Zn	Hc	Zn	Hc	Zn	Hc	Zn	Hc	Zn
	◆VEGA		DENEB		Enif		◆FOMALHAUT		Peacock		◆ANTARES		ARCTURUS	
270	46 23	010	29 04	031	32 32	076	15 34	120	29 56	158	58 46	222	30 08	295
271	46 33	009	29 35	031	33 30	075	16 26	120	30 18	158	58 06	223	29 13	295
272	46 42	008	30 05	030	34 28	075	17 18	120	30 40	159	57 24	224	28 19	294
273	46 50	007	30 35	030	35 26	075	18 10	120	31 01	159	56 42	225	27 24	294
274	46 57	006	31 05	029	36 24	075	19 02	120	31 22	160	56 00	226	26 29	294
275	47 03	005	31 34	029	37 21	074	19 54	120	31 43	160	55 16	227	25 35	294
276	47 07	004	32 02	028	38 19	074	20 46	120	32 02	161	54 32	228	24 40	293
277	47 10	003	32 30	027	39 17	074	21 37	120	32 22	161	53 47	229	23 45	293
278	47 12	001	32 57	027	40 14	074	22 29	121	32 40	162	53 02	230	22 49	293
279	47 13	000	33 23	026	41 11	073	23 20	121	32 59	163	52 16	230	21 54	293
280	47 13	359	33 49	025	42 09	073	24 12	121	33 16	163	51 30	231	20 59	292
281	47 11	358	34 14	025	43 06	073	25 03	121	33 33	164	50 43	232	20 04	292
282	47 09	357	34 39	024	44 03	072	25 54	121	33 50	164	49 56	232	19 08	292
283	47 05	356	35 03	023	45 00	072	26 46	121	34 06	165	49 08	233	18 12	292
284	46 59	355	35 26	023	45 57	071	27 37	122	34 21	165	48 20	234	17 17	292
	DENEB		Enif		◆FOMALHAUT		Peacock		◆ANTARES		Rasalhague		◆VEGA	
285	35 49	022	46 53	071	28 28	122	34 36	166	47 32	234	63 06	308	46 53	353
286	36 11	021	47 50	071	29 18	122	34 50	167	46 43	235	62 19	307	46 46	352
287	36 32	020	48 46	070	30 09	122	35 04	167	45 54	235	61 31	306	46 37	351
288	36 52	020	49 42	070	31 00	122	35 17	168	45 05	236	60 42	305	46 27	350
289	37 12	019	50 38	069	31 50	123	35 29	168	44 15	236	59 53	304	46 16	349
290	37 31	018	51 34	069	32 41	123	35 41	169	43 25	237	59 02	303	46 04	348
291	37 49	017	52 30	068	33 31	123	35 52	170	42 35	237	58 12	302	45 51	347
292	38 06	016	53 25	068	34 21	124	36 02	170	41 44	238	57 21	301	45 37	346
293	38 22	015	54 20	067	35 10	124	36 12	171	40 54	238	56 29	300	45 21	345
294	38 38	015	55 15	066	36 00	124	36 21	172	40 03	238	55 37	299	45 05	344
295	38 53	014	56 10	066	36 49	125	36 29	172	39 12	239	54 45	298	44 48	343
296	39 07	013	57 04	065	37 39	125	36 37	173	38 21	239	53 52	298	44 29	342
297	39 20	012	57 58	064	38 28	125	36 44	174	37 29	239	52 59	297	44 10	341
298	39 32	011	58 52	063	39 16	126	36 50	174	36 38	240	52 05	296	43 50	340
299	39 43	010	59 46	063	40 05	126	36 56	175	35 46	240	51 11	296	43 28	339
	◆DENEB		Alpheratz		◆FOMALHAUT		Peacock		◆ANTARES		Rasalhague		VEGA	
300	39 53	009	22 02	056	40 53	127	37 01	176	34 54	240	50 17	295	43 06	338
301	40 02	009	22 51	056	41 41	127	37 05	176	34 02	241	49 23	295	42 43	337
302	40 11	008	23 41	056	42 29	127	37 08	177	33 10	241	48 29	294	42 19	336
303	40 18	007	24 30	055	43 16	128	37 11	178	32 17	241	47 34	294	41 54	335
304	40 25	006	25 20	055	44 03	128	37 13	178	31 25	241	46 39	293	41 28	334
305	40 31	005	26 09	055	44 50	129	37 14	179	30 32	242	45 44	293	41 02	333
306	40 35	004	26 57	054	45 36	130	37 15	180	29 39	242	44 48	292	40 34	332
307	40 39	003	27 46	054	46 22	130	37 15	180	28 47	242	43 53	292	40 06	332
308	40 42	002	28 34	054	47 07	131	37 14	181	27 54	242	42 57	291	39 37	331
309	40 43	001	29 22	053	47 52	131	37 12	182	27 01	242	42 01	291	39 08	330
310	40 44	000	30 10	053	48 37	132	37 10	183	26 08	242	41 05	291	38 38	329
311	40 44	359	30 58	052	49 21	133	37 07	183	25 15	243	40 09	290	38 07	329
312	40 43	358	31 45	052	50 05	134	37 03	184	24 22	243	39 13	290	37 35	328
313	40 41	358	32 32	052	50 48	134	36 59	185	23 28	243	38 17	289	37 03	327
314	40 38	357	33 19	051	51 31	135	36 54	185	22 35	243	37 20	289	36 30	326
	◆Alpheratz		◆FOMALHAUT		Peacock		Nunki		◆Rasalhague		VEGA		DENEB	
315	34 05	051	52 12	136	36 48	186	52 41	230	36 23	289	35 57	326	40 34	356
316	34 51	050	52 54	137	36 41	187	51 55	231	35 27	289	35 23	325	40 29	355
317	35 37	050	53 34	138	36 34	187	51 08	232	34 30	288	34 48	324	40 23	354
318	36 22	049	54 14	139	36 26	188	50 21	232	33 33	288	34 13	324	40 16	353
319	37 07	048	54 53	140	36 18	189	49 34	233	32 36	288	33 37	323	40 08	352
320	37 52	048	55 32	141	36 08	189	48 46	234	31 39	287	33 01	323	39 59	351
321	38 36	047	56 09	142	35 58	190	47 57	234	30 42	287	32 25	322	39 49	350
322	39 20	047	56 46	143	35 48	191	47 09	235	29 45	287	31 48	321	39 39	349
323	40 03	046	57 21	144	35 37	191	46 20	235	28 47	287	31 10	321	39 27	348
324	40 46	045	57 56	145	35 25	192	45 30	236	27 50	286	30 32	320	39 15	348
325	41 28	045	58 29	147	35 12	192	44 41	236	26 53	286	29 54	320	39 02	347
326	42 10	044	59 02	148	34 59	193	43 51	237	25 55	286	29 15	319	38 47	346
327	42 51	043	59 33	149	34 45	194	43 01	237	24 57	286	28 36	319	38 32	345
328	43 32	042	60 03	151	34 31	194	42 10	238	24 00	286	27 56	318	38 16	344
329	44 12	042	60 31	152	34 16	195	41 20	238	23 02	285	27 16	318	38 00	343
	◆Alpheratz		Hamal		Diphda		◆FOMALHAUT		Nunki		◆ALTAIR		DENEB	
330	44 51	041	23 57	062	47 40	113	60 59	154	40 29	238	55 16	292	37 42	343
331	45 30	040	24 49	062	48 36	113	61 25	155	39 38	239	54 20	291	37 24	342
332	46 08	039	25 42	062	49 31	113	61 49	157	38 47	239	53 24	291	37 05	341
333	46 46	038	26 35	061	50 26	114	62 12	159	37 55	239	52 28	290	36 45	340
334	47 22	037	27 27	061	51 20	114	62 33	160	37 03	240	51 32	290	36 24	339
335	47 58	036	28 19	061	52 15	115	62 52	162	36 12	240	50 36	289	36 03	339
336	48 33	035	29 11	060	53 09	115	63 10	164	35 20	240	49 39	289	35 41	338
337	49 08	034	30 03	060	54 03	116	63 26	166	34 28	241	48 42	288	35 18	337
338	49 41	033	30 55	060	54 57	116	63 39	168	33 35	241	47 45	288	34 54	336
339	50 14	032	31 47	059	55 51	117	63 51	169	32 43	241	46 48	287	34 30	336
340	50 45	031	32 38	059	56 44	117	64 01	171	31 51	241	45 51	287	34 05	335
341	51 16	030	33 29	058	57 37	118	64 09	173	30 58	242	44 54	287	33 40	334
342	51 45	029	34 20	058	58 30	119	64 15	175	30 05	242	43 56	286	33 14	334
343	52 13	028	35 10	058	59 22	119	64 19	177	29 13	242	42 59	286	32 47	333
344	52 41	027	36 01	057	60 14	120	64 21	179	28 20	242	42 01	286	32 19	332
	Alpheratz		◆Hamal		◆ACHERNAR		FOMALHAUT		Nunki		◆ALTAIR		DENEB	
345	53 07	025	36 51	057	28 24	157	64 21	181	27 27	242	41 04	285	31 52	332
346	53 32	024	37 41	056	28 47	157	64 18	183	26 34	243	40 06	285	31 23	331
347	53 55	023	38 30	056	29 10	158	64 14	185	25 41	243	39 08	285	30 54	331
348	54 18	021	39 19	055	29 32	158	64 07	187	24 47	243	38 10	285	30 24	330
349	54 39	020	40 08	055	29 54	159	63 59	189	23 54	243	37 12	284	29 54	330
350	54 59	019	40 57	054	30 16	159	63 48	191	23 01	243	36 14	284	29 24	329
351	55 17	017	41 45	053	30 36	160	63 35	193	22 07	243	35 16	284	28 52	328
352	55 34	016	42 33	053	30 57	160	63 21	195	21 14	243	34 18	284	28 21	328
353	55 49	014	43 20	052	31 17	161	63 04	197	20 20	243	33 20	283	27 49	327
354	56 03	013	44 07	051	31 36	161	62 46	199	19 27	243	32 22	283	27 16	327
355	56 15	011	44 54	051	31 55	162	62 26	200	18 33	244	31 23	283	26 43	326
356	56 26	010	45 40	050	32 14	162	62 05	202	17 40	244	30 25	283	26 10	326
357	56 35	008	46 26	049	32 32	163	61 41	204	16 46	244	29 26	283	25 36	325
358	56 43	006	47 11	049	32 49	163	61 17	205	15 52	244	28 28	282	25 02	325
359	56 49	005	47 55	048	33 06	164	60 50	207	14 59	244	27 30	282	24 27	324

 LAT 5°S

LHA	Hc	Zn	Hc	Zn	Hc	Zn	Hc	Zn	Hc	Zn	Hc	Zn	Hc	Zn
♈	Alpheratz		◆ALDEBARAN		RIGEL		ACHERNAR		◆FOMALHAUT		ALTAIR		◆DENEB	
0	55 53	003	18 36	071	11 59	097	34 20	164	61 15	209	26 18	282	23 04	324
1	55 55	002	19 32	071	12 59	097	34 36	165	60 45	211	25 20	282	22 29	324
2	55 56	000	20 29	070	13 58	097	34 51	165	60 14	212	24 21	282	21 53	323
3	55 55	358	21 25	070	14 57	097	35 06	166	59 42	213	23 23	282	21 18	323
4	55 53	357	22 21	070	15 57	097	35 20	167	59 08	215	22 24	282	20 42	323
5	55 49	355	23 17	070	16 56	097	35 33	167	58 34	216	21 26	282	20 05	322
6	55 43	354	24 13	069	17 55	097	35 46	168	57 58	217	20 27	281	19 28	322
7	55 36	352	25 09	069	18 55	097	35 59	168	57 22	218	19 29	281	18 51	322
8	55 27	351	26 05	069	19 54	097	36 10	169	56 44	219	18 30	281	18 14	321
9	55 17	349	27 01	069	20 53	097	36 21	170	56 06	221	17 31	281	17 37	321
10	55 05	348	27 56	068	21 53	097	36 31	170	55 26	222	16 33	281	16 59	321
11	54 52	346	28 52	068	22 52	097	36 41	171	54 46	223	15 34	281	16 21	320
12	54 37	345	29 47	068	23 51	097	36 50	172	54 06	223	14 35	281	15 42	320
13	54 21	343	30 42	067	24 51	097	36 58	172	53 24	224	13 36	280	15 04	320
14	54 03	342	31 38	067	25 50	097	37 06	173	52 42	225	12 38	280	14 25	319
	CAPELLA		ALDEBARAN		◆RIGEL		ACHERNAR		◆FOMALHAUT		Enif		◆Alpheratz	
15	13 53	040	32 33	067	26 49	097	37 13	174	51 59	226	38 58	287	53 44	341
16	14 31	040	33 27	066	27 49	097	37 19	174	51 16	227	38 01	287	53 23	339
17	15 09	039	34 22	066	28 48	097	37 25	175	50 32	228	37 03	286	53 01	338
18	15 47	039	35 17	066	29 48	097	37 30	176	49 48	228	36 06	286	52 38	337
19	16 25	039	36 11	065	30 47	097	37 34	176	49 03	229	35 08	286	52 14	335
20	17 02	039	37 05	065	31 46	097	37 37	177	48 18	230	34 11	285	51 49	334
21	17 39	038	37 59	065	32 46	097	37 40	178	47 32	230	33 13	285	51 22	333
22	18 16	038	38 53	064	33 45	097	37 42	178	46 46	231	32 15	285	50 54	332
23	18 53	038	39 47	064	34 44	097	37 44	179	45 59	231	31 18	285	50 25	331
24	19 29	037	40 40	063	35 44	097	37 44	180	45 12	232	30 20	284	49 55	329
25	20 05	037	41 34	063	36 43	097	37 44	180	44 25	232	29 22	284	49 25	328
26	20 40	036	42 27	062	37 43	097	37 43	181	43 38	233	28 24	284	48 53	327
27	21 16	036	43 19	062	38 42	097	37 42	182	42 50	233	27 26	284	48 20	326
28	21 51	036	44 12	061	39 41	097	37 40	182	42 02	234	26 28	284	47 46	325
29	22 25	035	45 04	061	40 41	097	37 37	183	41 13	234	25 30	283	47 12	324
	CAPELLA		◆BETELGEUSE		SIRIUS		CANOPUS		◆ACHERNAR		FOMALHAUT		◆Alpheratz	
30	23 00	035	30 06	078	19 24	106	18 23	144	37 33	184	40 25	235	46 37	323
31	23 34	034	31 05	078	20 21	106	18 57	145	37 29	185	39 36	235	46 01	322
32	24 07	034	32 03	078	21 18	106	19 32	145	37 24	185	38 47	235	45 24	322
33	24 40	033	33 02	078	22 16	106	20 07	145	37 18	186	37 57	236	44 46	321
34	25 13	033	34 00	078	23 13	106	20 41	145	37 11	187	37 08	236	44 08	320
35	25 45	032	34 59	077	24 11	106	21 15	145	37 04	187	36 18	237	43 29	319
36	26 17	032	35 57	077	25 08	106	21 49	146	36 57	188	35 28	237	42 50	318
37	26 48	031	36 55	077	26 06	106	22 22	146	36 48	188	34 38	237	42 10	317
38	27 19	031	37 53	077	27 03	106	22 56	146	36 39	189	33 48	237	41 29	317
39	27 50	030	38 51	076	28 00	106	23 29	146	36 29	190	32 57	238	40 48	316
40	28 20	030	39 49	076	28 58	106	24 02	147	36 19	190	32 07	238	40 06	315
41	28 49	029	40 47	076	29 55	106	24 35	147	36 08	191	31 16	238	39 24	315
42	29 18	029	41 45	075	30 52	106	25 07	147	35 56	192	30 25	238	38 41	314
43	29 47	028	42 43	075	31 50	107	25 40	148	35 43	192	29 34	239	37 58	313
44	30 15	028	43 41	075	32 47	107	26 12	148	35 30	193	28 43	239	37 14	313
	CAPELLA		◆BETELGEUSE		SIRIUS		CANOPUS		◆ACHERNAR		Diphda		◆Hamal	
45	30 42	027	44 39	074	33 44	107	26 43	148	35 17	193	54 11	246	58 45	336
46	31 09	026	45 36	074	34 41	107	27 15	149	35 02	194	53 16	246	58 20	334
47	31 35	026	46 33	074	35 39	107	27 46	149	34 48	195	52 21	247	57 53	333
48	32 01	025	47 31	073	36 36	107	28 16	149	34 32	195	51 26	247	57 26	331
49	32 26	024	48 28	073	37 33	107	28 47	150	34 16	196	50 31	248	56 56	330
50	32 50	024	49 25	072	38 30	107	29 17	150	34 00	196	49 36	248	56 26	329
51	33 14	023	50 22	072	39 27	108	29 46	150	33 42	197	48 40	248	55 54	327
52	33 37	022	51 19	072	40 24	108	30 16	151	33 25	197	47 45	249	55 21	326
53	33 59	022	52 15	071	41 21	108	30 45	151	33 07	198	46 49	249	54 47	325
54	34 21	021	53 12	071	42 18	108	31 13	152	32 48	199	45 53	249	54 12	324
55	34 42	020	54 08	070	43 14	108	31 41	152	32 29	199	44 57	249	53 36	322
56	35 02	019	55 04	069	44 11	109	32 09	153	32 09	200	44 01	250	52 59	321
57	35 21	019	56 00	069	45 08	109	32 36	153	31 49	200	43 05	250	52 21	320
58	35 40	018	56 56	068	46 04	109	33 03	154	31 28	201	42 09	250	51 43	319
59	35 58	017	57 51	068	47 01	109	33 29	154	31 07	201	41 13	250	51 03	318
	◆CAPELLA		POLLUX		◆SIRIUS		CANOPUS		ACHERNAR		◆Diphda		Hamal	
60	36 15	016	26 35	055	47 57	110	33 55	155	30 45	201	40 16	251	50 23	317
61	36 32	016	27 24	055	48 54	110	34 21	155	30 23	202	39 20	251	49 41	316
62	36 47	015	28 13	054	49 50	110	34 45	156	30 00	202	38 23	251	49 00	315
63	37 02	014	29 01	054	50 46	110	35 10	156	29 37	203	37 27	251	48 17	314
64	37 16	013	29 49	054	51 42	111	35 34	157	29 14	203	36 30	251	47 34	313
65	37 29	012	30 37	053	52 38	111	35 57	157	28 50	204	35 34	251	46 50	313
66	37 42	011	31 25	053	53 33	112	36 20	158	28 26	204	34 37	252	46 06	312
67	37 53	011	32 12	052	54 29	112	36 42	159	28 01	205	33 40	252	45 21	311
68	38 04	010	32 59	052	55 24	112	37 03	159	27 36	205	32 44	252	44 36	310
69	38 13	009	33 46	051	56 19	113	37 24	160	27 11	205	31 47	252	43 50	310
70	38 22	008	34 33	051	57 14	113	37 45	160	26 45	206	30 50	252	43 04	309
71	38 30	007	35 19	050	58 09	114	38 05	161	26 19	206	29 53	252	42 17	308
72	38 37	006	36 04	050	59 03	115	38 24	162	25 53	206	28 56	252	41 30	308
73	38 43	005	36 50	049	59 58	115	38 42	162	25 26	207	27 59	252	40 43	307
74	38 48	005	37 35	048	60 52	116	39 00	163	24 59	207	27 02	252	39 55	306
	CAPELLA		◆POLLUX		REGULUS		Suhail		◆CANOPUS		ACHERNAR		◆Hamal	
75	38 53	004	38 19	048	11 35	077	23 35	136	39 17	164	24 32	207	39 06	306
76	38 56	003	39 03	047	12 33	077	24 17	136	39 33	164	24 04	208	38 18	305
77	38 58	002	39 47	047	13 31	076	24 59	136	39 49	165	23 36	208	37 29	305
78	39 00	001	40 30	046	14 29	076	25 40	136	40 04	166	23 08	208	36 40	304
79	39 00	000	41 13	045	15 27	076	26 21	137	40 18	167	22 40	209	35 50	304
80	39 00	359	41 55	045	16 25	076	27 02	137	40 32	167	22 11	209	35 00	303
81	38 59	358	42 37	044	17 23	076	27 43	137	40 44	168	21 42	209	34 10	303
82	38 56	357	43 18	043	18 21	076	28 23	137	40 56	169	21 13	209	33 20	302
83	38 53	357	43 58	042	19 19	075	29 04	138	41 08	170	20 43	210	32 29	302
84	38 49	356	44 38	041	20 16	075	29 44	138	41 18	170	20 14	210	31 38	302
85	38 44	355	45 18	041	21 14	075	30 23	138	41 28	171	19 44	210	30 47	301
86	38 38	354	45 56	040	22 12	075	31 03	139	41 36	172	19 14	210	29 56	301
87	38 31	353	46 34	039	23 10	075	31 42	139	41 44	173	18 44	210	29 04	300
88	38 24	352	47 11	038	24 07	074	32 21	140	41 51	174	18 13	211	28 13	300
89	38 15	351	47 47	037	25 05	074	33 00	140	41 58	174	17 43	211	27 21	300

LHA	Hc	Zn	Hc	Zn	Hc	Zn	Hc	Zn	Hc	Zn	Hc	Zn	Hc	Zn
♈	POLLUX		◆REGULUS		Suhail		◆CANOPUS		Acamar		◆ALDEBARAN		CAPELLA	
90	48 23	036	26 02	074	33 38	140	42 03	175	36 05	222	60 03	316	38 05	350
91	48 58	035	27 00	074	34 16	141	42 08	176	35 24	223	59 21	315	37 55	350
92	49 32	034	27 57	074	34 53	141	42 12	177	34 44	223	58 38	314	37 44	349
93	50 04	033	28 54	073	35 31	142	42 15	178	34 03	224	57 55	313	37 31	348
94	50 36	032	29 52	073	36 08	142	42 17	178	33 21	224	57 10	311	37 18	347
95	51 07	031	30 49	073	36 44	143	42 18	179	32 40	224	56 25	310	37 05	346
96	51 37	029	31 46	073	37 20	143	42 18	180	31 58	225	55 39	309	36 50	345
97	52 06	028	32 43	072	37 56	144	42 18	181	31 16	225	54 52	308	36 34	345
98	52 34	027	33 40	072	38 31	144	42 17	182	30 33	225	54 05	307	36 18	344
99	53 01	026	34 37	072	39 05	145	42 14	182	29 51	226	53 17	306	36 01	343
100	53 26	024	35 33	071	39 39	145	42 11	183	29 08	226	52 29	306	35 43	342
101	53 50	023	36 30	071	40 13	146	42 08	184	28 25	226	51 40	305	35 24	341
102	54 13	022	37 26	071	40 46	147	42 03	185	27 41	227	50 51	304	35 05	341
103	54 35	020	38 23	070	41 19	147	41 57	186	26 58	227	50 01	303	34 45	340
104	54 55	019	39 19	070	41 51	148	41 51	187	26 14	227	49 11	303	34 24	339
	POLLUX		◆REGULUS		Suhail		◆CANOPUS		RIGEL		◆ALDEBARAN		CAPELLA	
105	55 14	018	40 15	070	42 22	149	41 44	187	63 34	262	48 20	302	34 02	339
106	55 31	016	41 11	069	42 53	149	41 36	188	62 35	262	47 29	301	33 40	338
107	55 47	015	42 07	069	43 23	150	41 27	189	61 35	262	46 38	301	33 17	337
108	56 01	013	43 03	068	43 53	151	41 17	190	60 36	262	45 46	300	32 54	336
109	56 14	012	43 58	068	44 21	152	41 07	190	59 37	262	44 54	299	32 29	336
110	56 25	010	44 53	068	44 50	152	40 55	191	58 38	262	44 02	299	32 05	335
111	56 35	008	45 49	067	45 17	153	40 43	192	57 38	263	43 10	298	31 39	334
112	56 42	007	46 44	067	45 44	154	40 31	193	56 39	263	42 17	298	31 13	334
113	56 49	005	47 38	066	46 09	155	40 17	193	55 40	263	41 24	297	30 46	333
114	56 53	004	48 33	066	46 34	156	40 03	194	54 41	263	40 31	297	30 19	333
115	56 56	002	49 27	065	46 59	157	39 48	195	53 41	263	39 37	296	29 51	332
116	56 58	000	50 21	064	47 22	157	39 32	196	52 42	263	38 43	296	29 23	331
117	56 57	359	51 15	064	47 45	158	39 16	196	51 43	263	37 49	295	28 54	331
118	56 55	357	52 08	063	48 06	159	38 59	197	50 43	263	36 55	295	28 25	330
119	56 51	356	53 01	062	48 27	160	38 41	198	49 44	263	36 01	295	27 55	330
	◆REGULUS		Gienah		◆ACRUX		CANOPUS		SIRIUS		◆BETELGEUSE		POLLUX	
120	53 54	062	26 23	107	14 54	155	38 22	198	68 12	236	56 25	291	56 46	354
121	54 47	061	27 20	107	15 19	155	38 03	199	67 22	237	55 29	291	56 39	352
122	55 39	060	28 17	107	15 45	155	37 43	200	66 31	238	54 33	290	56 30	351
123	56 30	059	29 14	107	16 10	155	37 23	200	65 40	240	53 37	290	56 20	349
124	57 22	058	30 11	107	16 35	155	37 02	201	64 48	241	52 41	289	56 08	348
125	58 12	058	31 08	107	17 00	155	36 40	202	63 56	241	51 44	289	55 54	346
126	59 02	057	32 05	108	17 25	156	36 18	202	63 03	242	50 48	288	55 39	345
127	59 52	056	33 02	108	17 50	156	35 55	203	62 10	243	49 51	288	55 22	343
128	60 41	054	33 59	108	18 14	156	35 32	203	61 17	244	48 54	287	55 04	342
129	61 29	053	34 56	108	18 38	156	35 08	204	60 23	245	47 57	287	54 45	340
130	62 17	052	35 53	108	19 02	156	34 44	204	59 29	245	46 59	286	54 24	339
131	63 03	051	36 50	108	19 26	157	34 19	205	58 34	246	46 02	286	54 02	338
132	63 49	049	37 47	108	19 50	157	33 53	205	57 40	246	45 04	286	53 38	336
133	64 34	048	38 43	108	20 13	157	33 27	206	56 45	247	44 07	285	53 14	335
134	65 18	046	39 40	109	20 36	157	33 01	206	55 50	247	43 09	285	52 48	334
	Alioth		◆SPICA		ACRUX		◆CANOPUS		SIRIUS		BETELGEUSE		◆POLLUX	
135	12 40	029	24 15	100	20 59	158	32 34	207	54 54	248	42 11	285	52 20	332
136	13 09	029	25 14	100	21 21	158	32 07	207	53 59	248	41 14	284	51 52	331
137	13 38	029	26 13	100	21 44	158	31 39	208	53 03	249	40 16	284	51 23	330
138	14 07	028	27 12	100	22 06	158	31 11	208	52 08	249	39 18	284	50 52	329
139	14 35	028	28 11	100	22 28	159	30 42	209	51 12	249	38 20	284	50 21	328
140	15 03	028	29 10	100	22 49	159	30 13	209	50 16	250	37 21	283	49 48	327
141	15 30	027	30 08	100	23 10	159	29 44	210	49 20	250	36 23	283	49 15	326
142	15 58	027	31 07	100	23 31	160	29 14	210	48 23	250	35 25	283	48 41	325
143	16 25	027	32 06	100	23 52	160	28 44	210	47 27	251	34 27	283	48 06	324
144	16 52	026	33 05	100	24 12	160	28 14	211	46 31	251	33 28	282	47 30	323
145	17 18	026	34 04	100	24 32	161	27 43	211	45 34	251	32 30	282	46 53	322
146	17 44	026	35 03	100	24 51	161	27 12	211	44 37	251	31 31	282	46 15	321
147	18 10	025	36 02	100	25 11	161	26 41	212	43 41	252	30 33	282	45 37	320
148	18 35	025	37 00	100	25 30	162	26 09	212	42 44	252	29 34	281	44 58	319
149	19 00	024	37 59	100	25 48	162	25 37	212	41 47	252	28 36	281	44 19	318
	Alioth		ARCTURUS		◆SPICA		ACRUX		◆CANOPUS		SIRIUS		◆POLLUX	
150	19 24	024	22 42	067	38 58	100	26 06	163	25 05	213	40 50	252	43 38	317
151	19 49	024	23 37	067	39 57	100	26 24	163	24 32	213	39 53	252	42 58	317
152	20 12	023	24 31	066	40 56	100	26 41	163	24 00	213	38 56	253	42 16	316
153	20 36	023	25 26	066	41 54	100	26 58	164	23 26	214	37 59	253	41 34	315
154	20 59	022	26 20	066	42 53	101	27 15	164	22 53	214	37 02	253	40 52	314
155	21 21	022	27 15	065	43 52	101	27 31	164	22 20	214	36 05	253	40 09	314
156	21 43	021	28 09	065	44 51	101	27 47	165	21 46	214	35 08	253	39 26	313
157	22 05	021	29 03	065	45 49	101	28 02	165	21 12	215	34 11	253	38 42	312
158	22 26	021	29 57	064	46 48	101	28 17	166	20 38	215	33 14	253	37 57	312
159	22 47	020	30 51	064	47 47	101	28 32	166	20 04	215	32 16	253	37 13	311
160	23 07	020	31 45	064	48 45	101	28 46	167	19 29	215	31 19	253	36 27	311
161	23 27	019	32 38	063	49 44	101	28 59	167	18 55	216	30 22	254	35 42	310
162	23 46	019	33 32	063	50 43	101	29 12	168	18 20	216	29 24	254	34 56	310
163	24 05	018	34 25	063	51 41	102	29 25	168	17 45	216	28 27	254	34 10	309
164	24 24	018	35 18	062	52 40	102	29 37	168	17 10	216	27 30	254	33 23	309
	Alioth		◆ARCTURUS		SPICA		◆ACRUX		Suhail		SIRIUS		◆POLLUX	
165	24 41	017	36 11	062	53 38	102	29 49	169	44 19	209	26 32	254	32 36	308
166	24 59	017	37 03	061	54 37	102	30 00	169	43 50	209	25 35	254	31 49	308
167	25 15	016	37 55	061	55 35	102	30 11	170	43 20	210	24 37	254	31 01	307
168	25 32	015	38 47	060	56 33	103	30 21	170	42 50	211	23 40	254	30 14	307
169	25 47	015	39 39	060	57 32	103	30 31	171	42 19	211	22 43	254	29 25	306
170	26 02	014	40 31	059	58 30	103	30 40	171	41 48	212	21 45	254	28 37	306
171	26 17	014	41 22	059	59 28	103	30 49	172	41 16	213	20 48	254	27 49	305
172	26 31	013	42 13	058	60 26	104	30 57	172	40 43	213	19 50	254	27 00	305
173	26 44	013	43 04	058	61 24	104	31 05	173	40 10	214	18 53	254	26 11	305
174	26 57	012	43 54	057	62 22	105	31 12	173	39 36	215	17 55	254	25 21	304
175	27 09	011	44 44	057	63 20	105	31 19	174	39 02	215	16 58	254	24 32	304
176	27 21	011	45 34	056	64 17	105	31 25	174	38 27	216	16 00	254	23 42	304
177	27 32	010	46 23	055	65 15	106	31 30	175	37 52	216	15 03	254	22 52	303
178	27 42	010	47 12	055	66 12	106	31 35	175	37 17	217	14 05	254	22 02	303
179	27 52	009	48 01	054	67 10	107	31 40	176	36 41	217	13 08	254	21 12	303

LAT 5°S

LHA ♈	Hc	Zn	Hc	Zn	Hc	Zn	Hc	Zn	Hc	Zn	Hc	Zn	Hc	Zn
	Alioth		◆ARCTURUS		ANTARES		◆ACRUX		Suhail		Alphard		◆REGULUS	
180	28 01	008	48 49	053	22 33	117	31 44	177	36 04	218	51 57	262	57 24	302
181	28 09	008	49 36	052	23 26	117	31 47	177	35 27	218	50 58	262	56 33	301
182	28 17	007	50 23	051	24 20	117	31 50	178	34 50	219	49 58	263	55 42	300
183	28 25	007	51 10	051	25 13	117	31 52	178	34 12	219	48 59	263	54 49	299
184	28 31	006	51 55	050	26 06	117	31 54	179	33 34	220	48 00	263	53 57	298
185	28 37	005	52 41	049	26 59	117	31 55	179	32 56	220	47 01	263	53 04	298
186	28 42	005	53 25	048	27 53	117	31 56	180	32 17	220	46 01	263	52 11	297
187	28 47	004	54 09	047	28 46	117	31 56	180	31 38	221	45 02	263	51 18	296
188	28 51	003	54 52	046	29 39	118	31 55	181	30 59	221	44 03	263	50 24	296
189	28 54	003	55 35	045	30 31	118	31 54	181	30 20	222	43 03	263	49 30	295
190	28 57	002	56 16	043	31 24	118	31 52	182	29 40	222	42 04	263	48 36	294
191	28 59	002	56 57	042	32 17	118	31 50	182	29 00	222	41 05	263	47 41	294
192	29 00	001	57 37	041	33 10	118	31 48	183	28 20	223	40 05	263	46 46	293
193	29 01	000	58 15	040	34 02	119	31 44	183	27 39	223	39 06	263	45 51	293
194	29 01	000	58 53	038	34 54	119	31 40	184	26 58	223	38 07	263	44 56	292
	ARCTURUS		◆Rasalhague		ANTARES		◆ACRUX		Alphard		◆REGULUS		Alioth	
195	59 30	037	19 33	075	35 47	119	31 36	184	37 07	263	44 01	292	29 00	359
196	60 05	035	20 31	075	36 39	119	31 31	185	36 08	263	43 06	292	28 59	358
197	60 39	034	21 28	074	37 31	120	31 26	186	35 09	263	42 10	291	28 57	358
198	61 12	032	22 26	074	38 23	120	31 20	186	34 10	263	41 14	291	28 54	357
199	61 43	031	23 23	074	39 14	120	31 13	187	33 10	263	40 18	290	28 51	356
200	62 13	029	24 21	074	40 06	121	31 06	187	32 11	263	39 22	290	28 47	356
201	62 41	027	25 18	074	40 57	121	30 58	188	31 12	263	38 26	290	28 42	355
202	63 07	025	26 15	073	41 48	121	30 50	188	30 12	263	37 29	289	28 37	355
203	63 32	024	27 13	073	42 39	122	30 42	189	29 13	263	36 33	289	28 31	354
204	63 55	022	28 10	073	43 30	122	30 33	189	28 14	263	35 36	289	28 24	353
205	64 16	020	29 07	073	44 20	123	30 23	190	27 14	263	34 40	288	28 17	353
206	64 35	018	30 04	072	45 11	123	30 13	190	26 15	263	33 43	288	28 09	352
207	64 52	015	31 01	072	46 00	124	30 02	191	25 16	263	32 46	288	28 00	351
208	65 07	013	31 58	072	46 50	124	29 51	191	24 16	263	31 49	287	27 51	351
209	65 19	011	32 54	072	47 40	125	29 39	191	23 17	263	30 52	287	27 41	350
	ARCTURUS		◆VEGA		ANTARES		◆RIGIL KENT.		ACRUX		Gienah		◆REGULUS	
210	65 30	009	12 47	048	48 29	125	33 42	174	29 27	192	61 33	242	29 55	287
211	65 38	007	13 32	048	49 17	126	33 47	175	29 15	192	60 40	243	28 57	287
212	65 43	004	14 16	048	50 06	126	33 52	175	29 01	193	59 47	243	28 00	286
213	65 47	002	15 00	048	50 54	127	33 57	176	28 48	193	58 54	244	27 03	286
214	65 47	000	15 44	047	51 41	128	34 01	177	28 34	194	58 00	245	26 05	286
215	65 46	357	16 28	047	52 28	128	34 04	177	28 20	194	57 06	245	25 08	286
216	65 42	355	17 12	047	53 15	129	34 07	178	28 05	195	56 11	246	24 10	286
217	65 36	353	17 55	046	54 01	130	34 09	178	27 49	195	55 17	246	23 13	285
218	65 27	351	18 38	046	54 47	131	34 10	179	27 34	195	54 22	247	22 15	285
219	65 16	348	19 21	046	55 32	131	34 11	180	27 18	196	53 27	247	21 17	285
220	65 03	346	20 04	045	56 17	132	34 11	180	27 01	196	52 32	248	20 20	285
221	64 47	344	20 47	045	57 00	133	34 11	181	26 44	197	51 37	248	19 22	285
222	64 30	342	21 29	045	57 44	134	34 10	181	26 27	197	50 41	248	18 24	284
223	64 10	340	22 11	044	58 26	135	34 08	182	26 09	197	49 46	249	17 26	284
224	63 49	338	22 52	044	59 08	136	34 06	182	25 51	198	48 50	249	16 28	284
	Alphecca		VEGA		◆ALTAIR		ANTARES		◆RIGIL KENT.		SPICA		◆ARCTURUS	
225	57 11	014	23 34	044	16 17	079	59 48	138	34 03	183	65 42	254	63 25	336
226	57 25	013	24 15	043	17 16	079	60 28	139	33 59	184	64 44	254	63 00	334
227	57 38	011	24 56	043	18 14	079	61 07	140	33 55	184	63 47	255	62 33	332
228	57 48	009	25 36	042	19 13	079	61 45	141	33 51	185	62 49	255	62 04	330
229	57 57	008	26 16	042	20 12	079	62 22	143	33 45	185	61 51	256	61 34	329
230	58 05	006	26 56	041	21 10	078	62 58	144	33 39	186	60 53	256	61 03	327
231	58 10	004	27 35	041	22 09	078	63 32	146	33 33	187	59 55	256	60 29	326
232	58 14	003	28 14	040	23 07	078	64 05	147	33 26	187	58 57	257	59 55	324
233	58 16	001	28 53	040	24 06	078	64 36	149	33 18	188	57 59	257	59 19	323
234	58 16	359	29 31	039	25 04	078	65 06	151	33 10	188	57 00	257	58 42	321
235	58 14	358	30 09	039	26 03	078	65 35	153	33 01	189	56 02	257	58 04	320
236	58 11	356	30 46	038	27 01	077	66 01	154	32 52	189	55 04	258	57 25	319
237	58 06	354	31 23	038	27 59	077	66 26	156	32 42	190	54 05	258	56 45	317
238	57 59	353	32 00	037	28 57	077	66 49	158	32 31	190	53 07	258	56 05	316
239	57 51	351	32 35	037	29 56	077	67 10	161	32 20	191	52 08	258	55 23	315
	VEGA		◆ALTAIR		Nunki		◆RIGIL KENT.		SPICA		◆ARCTURUS		Alphecca	
240	33 11	036	30 54	077	43 09	122	32 09	191	51 10	258	54 40	314	57 40	349
241	33 46	035	31 52	076	43 59	122	31 56	192	50 11	259	53 57	313	57 28	348
242	34 20	035	32 50	076	44 50	123	31 44	193	49 13	259	53 13	312	57 15	346
243	34 54	034	33 48	076	45 40	123	31 31	193	48 14	259	52 28	311	57 00	344
244	35 27	033	34 46	076	46 30	124	31 17	194	47 15	259	51 42	310	56 43	343
245	36 00	033	35 44	075	47 19	124	31 03	194	46 17	259	50 56	309	56 25	341
246	36 32	032	36 42	075	48 09	125	30 48	195	45 18	259	50 10	308	56 05	340
247	37 03	031	37 39	075	48 58	125	30 33	195	44 19	259	49 23	308	55 44	338
248	37 34	031	38 37	074	49 46	126	30 17	195	43 20	259	48 35	307	55 21	337
249	38 04	030	39 34	074	50 35	126	30 01	196	42 22	259	47 47	306	54 57	336
250	38 34	029	40 32	074	51 23	127	29 44	196	41 23	260	46 58	305	54 32	334
251	39 02	028	41 29	074	52 10	128	29 27	197	40 24	260	46 09	305	54 05	333
252	39 30	027	42 27	073	52 57	128	29 09	197	39 25	260	45 20	304	53 37	332
253	39 58	027	43 24	073	53 44	129	28 51	198	38 26	260	44 30	303	53 08	330
254	40 24	026	44 21	072	54 30	130	28 33	198	37 28	260	43 40	303	52 38	329
	◆VEGA		ALTAIR		◆Nunki		Shaula		ANTARES		◆SPICA		ARCTURUS	
255	40 50	025	45 18	072	55 15	131	57 01	168	67 21	198	36 29	260	42 50	302
256	41 15	024	46 14	072	56 00	132	57 13	169	67 01	200	35 30	260	41 59	302
257	41 39	023	47 11	071	56 44	133	57 23	171	66 39	202	34 31	260	41 08	301
258	42 02	022	48 08	071	57 28	134	57 32	172	66 15	204	33 32	260	40 16	300
259	42 24	021	49 04	070	58 11	135	57 39	174	65 50	206	32 33	260	39 25	300
260	42 45	020	50 00	070	58 53	136	57 45	175	65 22	208	31 35	260	38 33	299
261	43 06	019	50 56	069	59 34	137	57 50	177	64 53	210	30 36	260	37 41	299
262	43 25	018	51 52	069	60 15	138	57 53	178	64 22	212	29 37	260	36 48	299
263	43 44	018	52 48	068	60 54	139	57 54	180	63 50	213	28 38	260	35 56	298
264	44 01	017	53 43	068	61 33	141	57 54	181	63 17	215	27 39	260	35 03	298
265	44 18	015	54 38	067	62 10	142	57 52	183	62 42	216	26 40	260	34 10	297
266	44 33	014	55 33	066	62 46	143	57 48	184	62 06	218	25 41	260	33 17	297
267	44 47	013	56 28	066	63 22	145	57 43	186	61 28	219	24 43	260	32 23	297
268	45 01	012	57 22	065	63 55	146	57 37	187	60 50	221	23 44	260	31 30	296
269	45 13	011	58 16	064	64 28	148	57 29	188	60 11	222	22 45	260	30 36	296

LHA ♈	Hc	Zn	Hc	Zn	Hc	Zn	Hc	Zn	Hc	Zn	Hc	Zn	Hc	Zn
	◆VEGA		DENEB		Enif		◆FOMALHAUT		Peacock		◆ANTARES		ARCTURUS	
270	45 24	010	28 13	031	32 17	075	16 04	119	30 52	158	59 30	223	29 42	295
271	45 34	009	28 43	031	33 15	075	16 56	120	31 14	158	58 49	224	28 48	295
272	45 43	008	29 14	030	34 13	075	17 48	120	31 36	159	58 07	225	27 54	295
273	45 51	007	29 43	029	35 10	074	18 40	120	31 58	159	57 24	226	26 59	295
274	45 57	006	30 12	029	36 08	074	19 32	120	32 19	160	56 41	227	26 05	294
275	46 03	005	30 41	028	37 05	074	20 24	120	32 39	160	55 57	228	25 10	294
276	46 07	004	31 09	028	38 02	073	21 16	120	32 59	161	55 12	229	24 16	294
277	46 10	002	31 36	027	38 59	073	22 07	120	33 18	161	54 27	230	23 21	293
278	46 12	001	32 03	026	39 57	073	22 59	120	33 37	162	53 41	231	22 26	293
279	46 13	000	32 29	026	40 54	072	23 51	120	33 56	162	52 54	231	21 31	293
280	46 13	359	32 55	025	41 51	072	24 42	120	34 14	163	52 07	232	20 36	293
281	46 11	358	33 20	024	42 47	072	25 34	121	34 31	163	51 20	233	19 41	292
282	46 09	357	33 44	024	43 44	071	26 25	121	34 48	164	50 32	233	18 45	292
283	46 05	356	34 08	023	44 41	071	27 17	121	35 04	165	49 44	234	17 50	292
284	46 00	355	34 31	022	45 37	070	28 08	121	35 19	165	48 55	235	16 55	292
	DENEB		Enif		◆FOMALHAUT		Peacock		◆ANTARES		Rasalhague		◆VEGA	
285	34 53	022	46 33	070	28 59	121	35 34	166	48 07	235	62 29	310	45 54	353
286	35 15	021	47 29	070	29 50	121	35 49	166	47 17	236	61 42	309	45 46	352
287	35 36	020	48 25	069	30 41	122	36 02	167	46 28	236	60 55	307	45 38	351
288	35 56	019	49 21	069	31 32	122	36 15	168	45 38	237	60 07	306	45 28	350
289	36 15	018	50 16	068	32 22	122	36 28	168	44 48	237	59 19	305	45 17	349
290	36 34	018	51 12	067	33 13	122	36 40	169	43 58	238	58 29	304	45 06	348
291	36 51	017	52 07	067	34 03	123	36 51	170	43 07	238	57 40	303	44 53	347
292	37 08	016	53 02	066	34 54	123	37 01	170	42 16	238	56 49	302	44 39	346
293	37 25	015	53 56	066	35 44	123	37 11	171	41 25	239	55 58	301	44 24	345
294	37 40	014	54 51	065	36 33	124	37 20	172	40 34	239	55 07	300	44 07	344
295	37 54	014	55 45	064	37 23	124	37 29	172	39 43	239	54 15	300	43 50	343
296	38 08	013	56 38	064	38 13	124	37 36	173	38 51	240	53 23	299	43 32	342
297	38 21	012	57 32	063	39 02	125	37 44	174	37 59	240	52 31	298	43 13	341
298	38 33	011	58 25	062	39 51	125	37 50	174	37 07	240	51 38	298	42 53	340
299	38 44	010	59 17	061	40 40	125	37 56	175	36 15	241	50 45	297	42 32	339
	◆DENEB		Alpheratz		◆FOMALHAUT		Peacock		◆ANTARES		Rasalhague		VEGA	
300	38 54	009	21 28	056	41 28	126	38 00	176	35 23	241	49 51	296	42 10	338
301	39 03	008	22 18	056	42 17	126	38 05	176	34 31	241	48 58	296	41 48	337
302	39 11	008	23 07	055	43 05	127	38 08	177	33 38	241	48 04	295	41 24	336
303	39 19	007	23 56	055	43 53	127	38 11	178	32 46	242	47 09	295	40 59	335
304	39 25	006	24 45	055	44 40	128	38 13	178	31 53	242	46 15	294	40 34	334
305	39 31	005	25 34	054	45 27	128	38 14	179	31 01	242	45 20	294	40 08	334
306	39 35	004	26 22	054	46 14	129	38 15	180	30 08	242	44 25	293	39 41	333
307	39 39	003	27 11	054	47 00	129	38 15	180	29 15	242	43 30	293	39 13	332
308	39 42	002	27 59	053	47 46	130	38 14	181	28 22	243	42 35	292	38 45	331
309	39 43	001	28 46	053	48 32	131	38 12	182	27 29	243	41 39	292	38 16	330
310	39 44	000	29 34	052	49 17	131	38 10	183	26 35	243	40 44	291	37 46	330
311	39 44	359	30 21	052	50 02	132	38 07	183	25 42	243	39 48	291	37 15	329
312	39 43	358	31 08	052	50 46	133	38 03	184	24 49	243	38 52	291	36 44	328
313	39 41	358	31 55	051	51 30	133	37 59	185	23 56	243	37 56	290	36 12	327
314	39 38	357	32 41	051	52 13	134	37 53	185	23 02	243	37 00	290	35 40	327
	◆Alpheratz		◆FOMALHAUT		Peacock		Nunki		◆Rasalhague		VEGA		DENEB	
315	33 27	050	52 55	135	37 48	186	53 19	231	36 04	290	35 07	326	39 34	356
316	34 12	050	53 37	136	37 41	187	52 33	232	35 07	289	34 33	325	39 29	355
317	34 58	049	54 18	137	37 34	187	51 45	233	34 11	289	33 59	325	39 23	354
318	35 43	048	54 59	138	37 26	188	50 58	233	33 14	289	33 24	324	39 16	353
319	36 27	048	55 39	139	37 17	189	50 10	234	32 18	288	32 49	324	39 08	352
320	37 11	047	56 18	140	37 08	189	49 21	234	31 21	288	32 13	323	39 00	351
321	37 55	047	56 56	141	36 57	190	48 32	235	30 24	288	31 37	322	38 50	350
322	38 38	046	57 33	142	36 47	191	47 43	236	29 27	287	31 00	322	38 40	349
323	39 21	045	58 09	143	36 35	191	46 54	236	28 30	287	30 23	321	38 28	349
324	40 03	045	58 45	144	36 23	192	46 04	237	27 33	287	29 46	321	38 16	348
325	40 45	044	59 19	146	36 11	193	45 14	237	26 35	287	29 08	320	38 03	347
326	41 26	043	59 52	147	35 57	193	44 23	238	25 38	287	28 29	320	37 49	346
327	42 07	043	60 24	148	35 43	194	43 33	238	24 41	286	27 50	319	37 34	345
328	42 47	042	60 55	150	35 29	194	42 42	238	23 43	286	27 11	319	37 19	344
329	43 27	041	61 24	151	35 14	195	41 51	239	22 46	286	26 32	318	37 02	344
	◆Alpheratz		Hamal		Diphda		◆FOMALHAUT		Nunki		◆ALTAIR		DENEB	
330	44 06	040	23 28	062	48 03	112	61 52	153	41 00	239	54 53	293	36 45	343
331	44 44	039	24 21	061	48 59	112	62 19	154	40 09	239	53 58	293	36 27	342
332	45 22	039	25 13	061	49 54	112	62 44	156	39 17	240	53 03	292	36 08	341
333	45 58	038	26 06	061	50 49	113	63 07	158	38 25	240	52 07	291	35 48	340
334	46 34	037	26 58	060	51 44	113	63 29	160	37 33	240	51 11	291	35 28	340
335	47 10	036	27 50	060	52 39	113	63 49	161	36 41	241	50 15	290	35 07	339
336	47 44	035	28 41	060	53 34	114	64 07	163	35 49	241	49 19	290	34 45	338
337	48 18	034	29 33	059	54 29	114	64 24	165	34 57	241	48 23	289	34 22	337
338	48 51	033	30 24	059	55 23	115	64 38	167	34 04	241	47 26	289	33 59	337
339	49 23	032	31 16	059	56 17	115	64 50	169	33 12	242	46 30	288	33 35	336
340	49 54	031	32 07	058	57 11	116	65 01	171	32 19	242	45 33	288	33 11	335
341	50 23	029	32 57	058	58 05	117	65 09	173	31 26	242	44 36	288	32 46	335
342	50 52	028	33 48	057	58 58	117	65 15	175	30 33	242	43 39	287	32 20	334
343	51 20	027	34 38	057	59 51	118	65 19	177	29 41	242	42 42	287	31 53	333
344	51 47	026	35 28	056	60 44	119	65 21	179	28 47	243	41 45	287	31 26	333
	Alpheratz		◆Hamal		◆ACHERNAR		FOMALHAUT		Nunki		◆ALTAIR		DENEB	
345	52 12	025	36 18	056	29 19	157	65 21	181	27 54	243	40 47	286	30 59	332
346	52 37	023	37 07	055	29 43	157	65 18	183	27 01	243	39 50	286	30 30	332
347	53 00	022	37 56	055	30 06	158	65 13	186	26 08	243	38 52	286	30 02	331
348	53 22	021	38 45	054	30 28	158	65 07	188	25 15	243	37 55	285	29 32	330
349	53 42	019	39 33	054	30 50	159	64 58	190	24 21	243	36 57	285	29 02	330
350	54 02	018	40 21	053	31 12	159	64 47	192	23 28	243	35 59	285	28 32	329
351	54 19	017	41 09	053	31 33	160	64 34	194	22 34	244	35 02	284	28 01	329
352	54 36	015	41 56	052	31 53	160	64 19	195	21 41	244	34 04	284	27 30	328
353	54 51	014	42 43	051	32 13	161	64 02	197	20 47	244	33 06	284	26 58	328
354	55 04	012	43 30	051	32 33	161	63 43	199	19 53	244	32 08	284	26 26	327
355	55 16	011	44 16	050	32 52	162	63 22	201	19 00	244	31 10	284	25 53	327
356	55 27	009	45 01	049	33 11	162	63 00	203	18 06	244	30 11	283	25 20	326
357	55 36	008	45 46	049	33 29	163	62 36	204	17 12	244	29 13	283	24 47	326
358	55 43	006	46 31	048	33 47	163	62 11	206	16 19	244	28 15	283	24 13	325
359	55 49	005	47 15	047	34 04	164	61 44	208	15 25	244	27 17	283	23 38	325

LAT 6°S

LHA	Hc	Zn	Hc	Zn	Hc	Zn	Hc	Zn	Hc	Zn	Hc	Zn	Hc	Zn
♈	◆Alpheratz		ALDEBARAN		◆RIGEL		ACHERNAR		◆FOMALHAUT		ALTAIR		DENEB	
0	54 53	003	18 16	070	12 07	097	35 18	164	62 07	210	26 05	283	22 15	325
1	54 55	002	19 12	070	13 06	097	35 34	165	61 37	212	25 07	283	21 40	324
2	54 56	000	20 08	070	14 05	097	35 49	165	61 05	213	24 09	283	21 05	324
3	54 55	359	21 04	070	15 05	097	36 04	166	60 32	214	23 10	282	20 30	323
4	54 53	357	22 00	069	16 04	097	36 18	166	59 57	216	22 12	282	19 54	323
5	54 49	355	22 56	069	17 03	097	36 32	167	59 22	217	21 14	282	19 18	323
6	54 44	354	23 52	069	18 02	097	36 45	168	58 46	218	20 15	282	18 41	322
7	54 37	352	24 48	069	19 02	097	36 57	168	58 08	219	19 17	282	18 04	322
8	54 28	351	25 43	068	20 01	097	37 09	169	57 30	220	18 18	281	17 27	321
9	54 18	350	26 39	068	21 00	097	37 20	170	56 51	222	17 20	281	16 50	321
10	54 07	348	27 34	068	22 00	096	37 31	170	56 11	223	16 21	281	16 12	321
11	53 53	347	28 29	068	22 59	096	37 40	171	55 30	224	15 23	281	15 34	320
12	53 39	345	29 24	067	23 58	096	37 49	172	54 49	224	14 24	281	14 56	320
13	53 23	344	30 19	067	24 57	096	37 58	172	54 07	225	13 26	281	14 18	320
14	53 06	342	31 14	067	25 57	096	38 06	173	53 24	226	12 27	280	13 39	320
	Hamal		ALDEBARAN		◆RIGEL		ACHERNAR		◆FOMALHAUT		Enif		◆Alpheratz	
15	56 20	028	32 09	066	26 56	096	38 13	174	52 41	227	38 40	288	52 47	341
16	56 48	027	33 03	066	27 55	096	38 19	174	51 57	228	37 43	287	52 27	340
17	57 14	026	33 58	066	28 55	096	38 25	175	51 12	228	36 46	287	52 06	338
18	57 39	024	34 52	065	29 54	096	38 30	176	50 27	229	35 49	287	51 43	337
19	58 03	022	35 46	065	30 53	096	38 34	176	49 42	230	34 52	286	51 19	336
20	58 25	021	36 40	064	31 53	096	38 37	177	48 56	230	33 54	286	50 54	335
21	58 45	019	37 33	064	32 52	096	38 40	178	48 10	231	32 57	286	50 28	333
22	59 04	018	38 27	063	33 52	096	38 42	178	47 23	232	32 00	286	50 01	332
23	59 21	016	39 20	063	34 51	096	38 44	179	46 36	232	31 02	285	49 33	331
24	59 37	014	40 13	062	35 50	096	38 44	180	45 49	233	30 05	285	49 04	330
25	59 50	012	41 06	062	36 50	096	38 44	180	45 01	233	29 07	285	48 33	329
26	60 02	011	41 58	061	37 49	096	38 43	181	44 13	234	28 09	285	48 02	328
27	60 12	009	42 50	061	38 48	096	38 42	182	43 25	234	27 11	284	47 30	327
28	60 20	007	43 42	060	39 48	096	38 40	183	42 37	235	26 14	284	46 57	326
29	60 26	005	44 34	060	40 47	096	38 37	183	41 48	235	25 16	284	46 23	325
	CAPELLA		◆BETELGEUSE		SIRIUS		CANOPUS		◆ACHERNAR		FOMALHAUT		◆Alpheratz	
30	22 10	035	29 54	078	19 40	106	19 11	144	38 33	184	40 59	235	45 48	324
31	22 44	034	30 52	078	20 37	106	19 46	144	38 29	185	40 10	236	45 13	323
32	23 17	034	31 51	077	21 35	106	20 21	144	38 23	185	39 20	236	44 37	322
33	23 50	033	32 49	077	22 32	106	20 56	145	38 18	186	38 31	236	44 00	321
34	24 22	033	33 47	077	23 30	106	21 30	145	38 11	187	37 41	237	43 22	320
35	24 54	032	34 45	077	24 27	106	22 04	145	38 04	187	36 51	237	42 44	320
36	25 26	032	35 43	076	25 25	106	22 38	146	37 56	188	36 01	237	42 05	319
37	25 57	031	36 41	076	26 22	106	23 12	146	37 47	189	35 10	238	41 25	318
38	26 28	031	37 39	076	27 20	106	23 46	146	37 38	189	34 20	238	40 45	317
39	26 58	030	38 37	076	28 17	106	24 19	146	37 28	190	33 29	238	40 04	317
40	27 28	030	39 35	075	29 14	106	24 52	146	37 18	191	32 38	238	39 23	316
41	27 57	029	40 32	075	30 12	106	25 25	147	37 06	191	31 47	239	38 41	315
42	28 26	028	41 30	075	31 09	106	25 58	147	36 54	192	30 56	239	37 59	315
43	28 54	028	42 27	074	32 07	106	26 30	147	36 42	192	30 05	239	37 16	314
44	29 21	027	43 25	074	33 04	106	27 02	148	36 29	193	29 14	239	36 33	313
	CAPELLA		◆BETELGEUSE		SIRIUS		CANOPUS		◆ACHERNAR		FOMALHAUT		◆Alpheratz	
45	29 48	027	44 22	073	34 01	106	27 34	148	36 15	194	28 23	239	35 50	313
46	30 15	026	45 19	073	34 59	106	28 06	148	36 01	194	27 31	240	35 05	312
47	30 41	025	46 16	073	35 56	106	28 37	149	35 46	195	26 40	240	34 21	312
48	31 06	025	47 13	072	36 53	106	29 08	149	35 30	195	25 48	240	33 36	311
49	31 31	024	48 10	072	37 50	107	29 38	149	35 14	196	24 56	240	32 51	310
50	31 55	023	49 06	071	38 48	107	30 09	150	34 57	197	24 05	240	32 05	310
51	32 18	023	50 03	071	39 45	107	30 38	150	34 40	197	23 13	240	31 20	309
52	32 41	022	50 59	070	40 42	107	31 08	151	34 22	198	22 21	240	30 33	309
53	33 03	021	51 55	070	41 39	107	31 37	151	34 04	198	21 29	240	29 47	309
54	33 25	021	52 51	069	42 36	107	32 06	151	33 45	199	20 37	241	29 00	308
55	33 45	020	53 47	069	43 33	107	32 34	152	33 25	199	19 45	241	28 13	308
56	34 05	019	54 42	068	44 30	108	33 02	152	33 05	200	18 53	241	27 25	307
57	34 24	018	55 38	067	45 27	108	33 30	153	32 45	200	18 01	241	26 38	307
58	34 43	018	56 33	067	46 23	108	33 57	153	32 24	201	17 09	241	25 50	306
59	35 01	017	57 27	066	47 20	108	34 23	154	32 03	201	16 17	241	25 02	306
	◆CAPELLA		POLLUX		◆SIRIUS		CANOPUS		ACHERNAR		◆Diphda		Hamal	
60	35 18	016	26 01	055	48 17	108	34 49	154	31 41	202	40 36	251	49 38	318
61	35 34	015	26 49	054	49 13	109	35 15	155	31 18	202	39 39	252	48 58	317
62	35 49	015	27 38	054	50 10	109	35 40	155	30 56	203	38 43	252	48 17	316
63	36 04	014	28 26	054	51 06	109	36 05	156	30 32	203	37 46	252	47 35	315
64	36 18	013	29 13	053	52 02	110	36 29	156	30 09	204	36 49	252	46 53	314
65	36 31	012	30 01	053	52 59	110	36 52	157	29 45	204	35 53	252	46 10	313
66	36 43	011	30 48	052	53 55	110	37 15	158	29 20	204	34 56	252	45 26	313
67	36 54	010	31 35	052	54 51	111	37 38	158	28 56	205	33 59	252	44 42	312
68	37 05	010	32 22	051	55 46	111	38 00	159	28 31	205	33 02	252	43 57	311
69	37 14	009	33 08	051	56 42	112	38 21	159	28 05	206	32 05	252	43 12	310
70	37 23	008	33 54	050	57 37	112	38 41	160	27 39	206	31 08	253	42 26	310
71	37 31	007	34 40	050	58 33	112	39 01	161	27 13	206	30 11	253	41 40	309
72	37 37	006	35 25	049	59 28	113	39 21	161	26 46	207	29 14	253	40 53	308
73	37 43	005	36 10	049	60 22	114	39 39	162	26 20	207	28 17	253	40 06	308
74	37 49	004	36 55	048	61 17	114	39 57	163	25 52	207	27 20	253	39 19	307
	CAPELLA		POLLUX		◆REGULUS		CANOPUS		◆ACHERNAR		Diphda		◆Hamal	
75	37 53	004	37 39	047	11 21	076	40 15	163	25 25	208	26 23	253	38 31	307
76	37 56	003	38 22	047	12 19	076	40 31	164	24 57	208	25 26	253	37 43	306
77	37 58	002	39 06	046	13 17	076	40 47	165	24 29	208	24 29	253	36 54	305
78	38 00	001	39 48	045	14 14	076	41 02	166	24 01	208	23 32	253	36 06	305
79	38 00	000	40 31	045	15 12	076	41 17	166	23 32	209	22 35	253	35 16	304
80	38 00	359	41 12	044	16 10	076	41 30	167	23 03	209	21 38	253	34 27	304
81	37 59	358	41 53	043	17 08	075	41 43	168	22 34	209	20 41	253	33 37	303
82	37 56	357	42 34	042	18 06	075	41 55	169	22 05	210	19 44	253	32 47	303
83	37 53	357	43 14	042	19 03	075	42 07	169	21 36	210	18 47	253	31 57	302
84	37 49	356	43 53	041	20 01	075	42 17	170	21 06	210	17 50	253	31 07	302
85	37 44	355	44 32	040	20 59	075	42 27	171	20 36	210	16 53	253	30 16	302
86	37 38	354	45 10	039	21 56	074	42 36	172	20 06	210	15 56	253	29 25	301
87	37 32	353	45 47	038	22 54	074	42 44	173	19 35	211	14 59	253	28 34	301
88	37 24	352	46 24	037	23 51	074	42 51	173	19 05	211	14 02	253	27 43	300
89	37 16	351	46 59	036	24 48	074	42 58	174	18 34	211	13 05	253	26 51	300

LHA	Hc	Zn	Hc	Zn	Hc	Zn	Hc	Zn	Hc	Zn	Hc	Zn	Hc	Zn
♈	POLLUX		◆REGULUS		Suhail		◆CANOPUS		Acamar		◆ALDEBARAN		CAPELLA	
90	47 34	035	25 46	074	34 24	140	43 03	175	36 49	223	59 19	317	37 06	351
91	48 08	034	26 43	073	35 02	140	43 08	176	36 08	223	58 38	316	36 56	350
92	48 42	033	27 40	073	35 40	141	43 12	177	35 27	224	57 56	315	36 45	349
93	49 14	032	28 37	073	36 18	141	43 15	178	34 46	224	57 14	314	36 33	348
94	49 45	031	29 34	073	36 55	142	43 17	178	34 04	224	56 30	313	36 20	347
95	50 16	030	30 31	072	37 32	142	43 18	179	33 22	225	55 46	311	36 06	346
96	50 45	029	31 28	072	38 08	143	43 18	180	32 40	225	55 01	310	35 52	346
97	51 13	028	32 24	072	38 44	143	43 18	181	31 58	225	54 15	309	35 36	345
98	51 40	026	33 21	071	39 19	144	43 17	182	31 15	226	53 28	308	35 20	344
99	52 06	025	34 17	071	39 54	144	43 14	183	30 32	226	52 41	308	35 04	343
100	52 31	024	35 14	071	40 29	145	43 11	183	29 49	226	51 54	307	34 46	342
101	52 55	023	36 10	070	41 03	146	43 07	184	29 06	227	51 06	306	34 27	342
102	53 17	021	37 06	070	41 36	146	43 03	185	28 22	227	50 17	305	34 08	341
103	53 38	020	38 02	070	42 09	147	42 57	186	27 39	227	49 28	304	33 48	340
104	53 58	019	38 58	069	42 42	147	42 51	187	26 55	227	48 38	303	33 28	339
	POLLUX		◆REGULUS		Suhail		◆CANOPUS		RIGEL		◆ALDEBARAN		CAPELLA	
105	54 16	017	39 54	069	43 13	148	42 43	187	63 41	264	47 48	303	33 07	339
106	54 33	016	40 49	068	43 44	149	42 35	188	62 42	264	46 58	302	32 45	338
107	54 49	014	41 45	068	44 15	150	42 26	189	61 43	264	46 07	301	32 22	337
108	55 03	013	42 40	068	44 45	150	42 16	190	60 43	264	45 16	301	31 59	337
109	55 15	011	43 35	067	45 14	151	42 06	191	59 44	264	44 25	300	31 35	336
110	55 26	010	44 30	067	45 43	152	41 54	191	58 45	264	43 33	300	31 10	335
111	55 35	008	45 25	066	46 10	153	41 42	192	57 45	264	42 41	299	30 45	335
112	55 43	007	46 19	066	46 37	153	41 29	193	56 46	264	41 49	298	30 19	334
113	55 49	005	47 13	065	47 04	154	41 15	194	55 47	264	40 56	298	29 53	333
114	55 54	004	48 07	065	47 29	155	41 01	194	54 47	264	40 03	297	29 26	333
115	55 56	002	49 01	064	47 54	156	40 46	195	53 48	264	39 10	297	28 58	332
116	55 58	000	49 55	063	48 17	157	40 30	196	52 49	264	38 17	297	28 30	332
117	55 57	359	50 48	063	48 40	158	40 13	197	51 49	264	37 23	296	28 02	331
118	55 55	357	51 41	062	49 02	159	39 56	197	50 50	264	36 30	296	27 32	331
119	55 51	356	52 33	061	49 23	160	39 38	198	49 50	264	35 36	295	27 03	330
	◆REGULUS		Gienah		◆ACRUX		CANOPUS		RIGEL		◆BETELGEUSE		CAPELLA	
120	53 25	061	26 40	107	15 48	154	39 19	199	48 51	264	56 02	293	26 33	329
121	54 17	060	27 38	107	16 13	155	39 00	199	47 52	264	55 07	292	26 02	329
122	55 08	059	28 35	107	16 39	155	38 40	200	46 52	264	54 12	292	25 31	328
123	55 59	058	29 32	107	17 04	155	38 19	201	45 53	264	53 16	291	25 00	328
124	56 50	057	30 29	107	17 30	155	37 58	201	44 53	264	52 21	290	24 28	327
125	57 39	056	31 26	107	17 55	155	37 36	202	43 54	264	51 24	290	23 55	327
126	58 29	055	32 23	107	18 20	155	37 13	202	42 55	264	50 28	289	23 22	326
127	59 17	054	33 20	107	18 44	156	36 50	203	41 55	264	49 32	289	22 49	326
128	60 05	053	34 17	107	19 09	156	36 27	204	40 56	264	48 35	288	22 16	326
129	60 53	052	35 14	107	19 33	156	36 03	204	39 57	264	47 39	288	21 42	325
130	61 39	051	36 11	107	19 57	156	35 38	205	38 57	264	46 42	287	21 07	325
131	62 25	049	37 08	107	20 21	157	35 13	205	37 58	264	45 45	287	20 33	324
132	63 10	048	38 05	108	20 45	157	34 47	206	36 58	264	44 48	287	19 58	324
133	63 53	046	39 02	108	21 08	157	34 21	206	35 59	264	43 50	286	19 22	323
134	64 36	045	39 59	108	21 31	157	33 55	207	35 00	264	42 53	286	18 47	323
	Dubhe		Denebola		◆SPICA		ACRUX		◆CANOPUS		SIRIUS		◆POLLUX	
135	18 09	015	43 22	063	24 25	100	21 54	158	33 27	207	55 16	249	51 27	333
136	18 24	014	44 15	063	25 24	100	22 17	158	33 00	208	54 21	250	50 59	332
137	18 39	014	45 08	062	26 23	099	22 39	158	32 32	208	53 25	250	50 31	331
138	18 53	014	46 01	062	27 22	099	23 02	158	32 04	209	52 29	250	50 01	329
139	19 07	013	46 53	061	28 21	099	23 24	159	31 35	209	51 32	251	49 30	328
140	19 20	013	47 45	060	29 20	099	23 45	159	31 06	209	50 36	251	48 58	327
141	19 33	012	48 37	060	30 19	099	24 06	159	30 36	210	49 40	251	48 25	326
142	19 45	012	49 28	059	31 17	099	24 27	160	30 06	210	48 43	251	47 52	325
143	19 57	011	50 19	058	32 16	099	24 48	160	29 36	211	47 46	252	47 17	324
144	20 09	011	51 10	058	33 15	099	25 08	160	29 05	211	46 50	252	46 42	323
145	20 20	010	52 00	057	34 14	099	25 28	161	28 34	211	45 53	252	46 06	322
146	20 30	010	52 50	056	35 13	099	25 48	161	28 03	212	44 56	252	45 29	321
147	20 40	009	53 39	055	36 12	099	26 08	161	27 32	212	43 59	253	44 51	320
148	20 50	009	54 28	054	37 11	099	26 27	162	27 00	212	43 02	253	44 13	320
149	20 59	008	55 16	053	38 09	099	26 45	162	26 28	213	42 05	253	43 34	319
	Dubhe		ARCTURUS		◆SPICA		ACRUX		◆CANOPUS		SIRIUS		◆POLLUX	
150	21 07	008	22 18	066	39 08	099	27 03	162	25 55	213	41 08	253	42 54	318
151	21 15	007	23 13	066	40 07	100	27 21	163	25 22	213	40 11	253	42 14	317
152	21 23	007	24 07	066	41 06	100	27 39	163	24 50	214	39 14	253	41 33	316
153	21 30	006	25 01	066	42 05	100	27 56	164	24 16	214	38 17	253	40 52	316
154	21 36	006	25 56	065	43 04	100	28 13	164	23 43	214	37 20	254	40 10	315
155	21 42	006	26 50	065	44 03	100	28 29	164	23 09	214	36 22	254	39 27	314
156	21 48	005	27 44	065	45 01	100	28 45	165	22 36	215	35 25	254	38 44	314
157	21 53	004	28 37	064	46 00	100	29 00	165	22 01	215	34 28	254	38 01	313
158	21 57	004	29 31	064	46 59	100	29 15	166	21 27	215	33 31	254	37 17	312
159	22 01	003	30 25	064	47 58	100	29 30	166	20 53	215	32 33	254	36 33	312
160	22 04	003	31 18	063	48 56	100	29 44	167	20 18	216	31 36	254	35 48	311
161	22 07	002	32 11	063	49 55	100	29 58	167	19 43	216	30 38	254	35 03	311
162	22 09	002	33 04	062	50 54	100	30 11	167	19 09	216	29 41	254	34 18	310
163	22 11	001	33 57	062	51 53	100	30 24	168	18 34	216	28 44	254	33 32	310
164	22 12	001	34 49	062	52 51	100	30 36	168	17 58	216	27 46	254	32 45	309
	Dubhe		◆ARCTURUS		SPICA		◆ACRUX		Suhail		SIRIUS		◆POLLUX	
165	22 13	000	35 42	061	53 50	101	30 48	169	45 11	209	26 49	254	31 59	309
166	22 13	000	36 34	061	54 49	101	30 59	169	44 42	210	25 51	254	31 12	308
167	22 13	359	37 26	060	55 47	101	31 10	170	44 12	210	24 54	254	30 25	308
168	22 12	359	38 18	060	56 46	101	31 20	170	43 42	211	23 56	254	29 38	307
169	22 11	358	39 09	059	57 44	101	31 30	171	43 10	212	22 59	254	28 50	307
170	22 09	358	40 00	059	58 43	102	31 40	171	42 38	213	22 01	254	28 02	306
171	22 06	357	40 51	058	59 41	102	31 48	172	42 06	213	21 04	254	27 14	306
172	22 03	357	41 41	058	60 40	102	31 57	172	41 33	214	20 07	254	26 25	305
173	22 00	356	42 32	057	61 38	102	32 04	173	41 00	214	19 09	254	25 36	305
174	21 56	356	43 21	056	62 36	103	32 12	173	40 26	215	18 12	254	24 47	305
175	21 51	355	44 11	056	63 34	103	32 18	174	39 51	216	17 14	254	23 58	304
176	21 46	355	45 00	055	64 32	103	32 24	174	39 16	216	16 17	254	23 09	304
177	21 40	354	45 49	054	65 30	104	32 30	175	38 40	217	15 19	254	22 19	304
178	21 34	354	46 37	054	66 28	104	32 35	175	38 04	217	14 22	254	21 30	303
179	21 28	353	47 25	053	67 26	105	32 40	176	37 28	218	13 24	254	20 40	303

LHA ♈	Hc	Zn	Hc	Zn	Hc	Zn	Hc	Zn	Hc	Zn	Hc	Zn	Hc	Zn
	ARCTURUS		♦ANTARES		ACRUX		♦Suhail		Alphard		REGULUS		♦Dubhe	
180	48 12	052	22 59	116	32 44	176	36 51	218	52 04	264	56 52	303	21 21	353
181	48 59	051	23 53	116	32 47	177	36 14	219	51 05	264	56 02	302	21 13	352
182	49 45	050	24 46	116	32 50	178	35 37	219	50 06	264	55 11	301	21 05	352
183	50 31	050	25 40	116	32 52	178	34 59	220	49 06	264	54 20	300	20 56	351
184	51 16	049	26 33	117	32 54	179	34 20	220	48 07	264	53 28	300	20 47	351
185	52 01	048	27 27	117	32 55	179	33 42	220	47 08	264	52 36	299	20 37	350
186	52 45	047	28 20	117	32 56	180	33 03	221	46 08	264	51 43	298	20 27	350
187	53 28	046	29 13	117	32 56	180	32 24	221	45 09	264	50 51	297	20 17	350
188	54 10	045	30 06	117	32 55	181	31 44	222	44 10	264	49 57	297	20 05	349
189	54 52	044	30 59	117	32 54	181	31 04	222	43 10	264	49 04	296	19 54	349
190	55 32	042	31 52	117	32 52	182	30 24	222	42 11	264	48 10	296	19 42	348
191	56 12	041	32 45	118	32 50	182	29 44	223	41 12	264	47 16	295	19 29	348
192	56 51	040	33 38	118	32 47	183	29 04	223	40 12	264	46 22	294	19 16	347
193	57 29	039	34 31	118	32 44	183	28 23	223	39 13	264	45 28	294	19 03	347
194	58 06	037	35 23	118	32 40	184	27 42	224	38 14	264	44 33	293	18 49	346
	ARCTURUS		♦Rasalhague		ANTARES		♦ACRUX		Suhail		♦REGULUS		Dubhe	
195	58 41	036	19 17	074	36 16	119	32 36	185	27 01	224	43 38	293	18 35	346
196	59 16	035	20 15	074	37 08	119	32 31	185	26 19	224	42 43	292	18 20	346
197	59 49	033	21 12	074	38 00	119	32 25	186	25 38	224	41 48	292	18 05	345
198	60 21	031	22 09	074	38 52	119	32 19	186	24 56	225	40 52	292	17 49	345
199	60 51	030	23 07	074	39 44	120	32 13	187	24 14	225	39 57	291	17 33	344
200	61 20	028	24 04	073	40 36	120	32 06	187	23 32	225	39 01	291	17 17	344
201	61 47	026	25 01	073	41 28	120	31 58	188	22 50	225	38 05	290	17 00	343
202	62 13	025	25 58	073	42 19	121	31 50	188	22 07	225	37 09	290	16 43	343
203	62 37	023	26 55	073	43 10	121	31 41	189	21 25	225	36 13	290	16 26	343
204	62 59	021	27 52	072	44 01	121	31 32	189	20 42	226	35 17	289	16 08	342
205	63 19	019	28 49	072	44 52	122	31 22	190	20 00	226	34 20	289	15 49	342
206	63 38	017	29 45	072	45 43	122	31 12	190	19 17	226	33 24	289	15 31	342
207	63 54	015	30 42	072	46 33	123	31 01	191	18 34	226	32 27	288	15 12	341
208	64 08	013	31 39	071	47 23	123	30 50	191	17 51	226	31 31	288	14 52	341
209	64 20	011	32 35	071	48 13	124	30 38	192	17 08	226	30 34	288	14 32	340
	ARCTURUS		♦Rasalhague		ANTARES		♦RIGIL KENT.		ACRUX		Gienah		♦REGULUS	
210	64 30	008	33 31	071	49 03	124	34 41	174	30 26	192	62 01	243	29 37	287
211	64 38	006	34 28	070	49 52	125	34 47	175	30 13	193	61 07	244	28 40	287
212	64 44	004	35 24	070	50 41	125	34 52	175	30 00	193	60 13	245	27 43	287
213	64 47	002	36 20	070	51 29	126	34 57	176	29 46	193	59 19	245	26 46	287
214	64 47	000	37 16	069	52 17	127	35 01	177	29 32	194	58 25	246	25 49	286
215	64 46	357	38 11	069	53 05	127	35 04	177	29 18	194	57 30	247	24 51	286
216	64 42	355	39 07	068	53 52	128	35 07	178	29 03	195	56 36	247	23 54	286
217	64 36	353	40 02	068	54 39	129	35 09	178	28 47	195	55 40	247	22 57	286
218	64 28	351	40 58	068	55 26	129	35 10	179	28 32	196	54 45	248	21 59	286
219	64 17	349	41 53	067	56 11	130	35 11	180	28 15	196	53 50	248	21 02	285
220	64 05	347	42 48	067	56 57	131	35 11	180	27 59	196	52 54	249	20 04	285
221	63 50	345	43 42	066	57 41	132	35 11	181	27 42	197	51 59	249	19 06	285
222	63 33	343	44 37	066	58 25	133	35 10	181	27 24	197	51 03	249	18 09	285
223	63 14	341	45 31	065	59 08	134	35 08	182	27 06	198	50 07	250	17 11	285
224	62 53	339	46 25	065	59 51	135	35 06	182	26 48	198	49 11	250	16 13	284
	Alphecca		♦VEGA		ALTAIR		♦Shaula		RIGIL KENT.		♦SPICA		ARCTURUS	
225	56 13	014	22 50	043	16 06	079	43 16	137	35 03	183	65 57	256	62 30	337
226	56 27	012	23 31	043	17 04	079	43 56	138	34 59	184	64 59	256	62 06	335
227	56 39	011	24 12	042	18 03	079	44 36	138	34 55	184	64 01	257	61 40	333
228	56 49	009	24 52	042	19 01	078	45 15	139	34 50	185	63 03	257	61 12	331
229	56 58	008	25 31	042	20 00	078	45 54	140	34 45	185	62 05	258	60 43	330
230	57 05	006	26 11	041	20 58	078	46 33	140	34 39	186	61 07	258	60 12	328
231	57 10	004	26 50	041	21 56	078	47 10	141	34 32	187	60 08	258	59 40	327
232	57 14	003	27 29	040	22 55	078	47 47	142	34 25	187	59 10	258	59 06	325
233	57 16	001	28 07	040	23 53	078	48 24	143	34 18	188	58 11	259	58 31	324
234	57 16	359	28 45	039	24 51	077	49 00	143	34 09	188	57 13	259	57 55	322
235	57 14	358	29 22	039	25 49	077	49 35	144	34 00	189	56 14	259	57 18	321
236	57 11	356	29 59	038	26 48	077	50 09	145	33 51	189	55 16	259	56 40	320
237	57 06	354	30 36	037	27 46	077	50 43	146	33 41	190	54 17	259	56 01	318
238	57 00	353	31 12	037	28 44	076	51 16	147	33 30	191	53 19	259	55 21	317
239	56 51	351	31 47	036	29 42	076	51 48	148	33 19	191	52 20	260	54 40	316
	VEGA		♦ALTAIR		Shaula		♦RIGIL KENT.		SPICA		♦ARCTURUS		Alphecca	
240	32 22	036	30 40	076	52 19	149	33 07	192	51 21	260	53 58	315	56 41	350
241	32 57	035	31 37	076	52 50	150	32 55	192	50 22	260	53 16	314	56 30	348
242	33 31	034	32 35	075	53 19	151	32 42	193	49 24	260	52 32	313	56 17	346
243	34 04	034	33 33	075	53 48	152	32 29	193	48 25	260	51 48	312	56 02	345
244	34 37	033	34 31	075	54 15	153	32 15	194	47 26	260	51 03	311	55 46	343
245	35 09	032	35 28	075	54 42	154	32 01	194	46 27	260	50 18	310	55 28	342
246	35 41	032	36 26	074	55 07	155	31 46	195	45 29	260	49 32	309	55 08	340
247	36 12	031	37 23	074	55 31	157	31 31	195	44 30	260	48 46	308	54 48	339
248	36 43	030	38 20	074	55 54	158	31 15	196	43 31	260	47 59	308	54 26	338
249	37 12	029	39 18	073	56 16	159	30 58	196	42 32	260	47 11	307	54 02	336
250	37 41	029	40 15	073	56 37	160	30 42	197	41 33	260	46 23	306	53 38	335
251	38 10	028	41 12	073	56 56	162	30 24	197	40 35	260	45 35	305	53 12	334
252	38 37	027	42 09	072	57 14	163	30 07	197	39 36	260	44 46	305	52 44	332
253	39 04	026	43 06	072	57 30	165	29 49	198	38 37	261	43 57	304	52 16	331
254	39 30	025	44 02	072	57 45	166	29 30	198	37 38	261	43 07	303	51 47	330
	VEGA		♦ALTAIR		Peacock		♦RIGIL KENT.		SPICA		♦ARCTURUS		Alphecca	
255	39 55	025	44 59	071	25 21	152	29 11	199	36 39	261	42 17	303	51 16	329
256	40 20	024	45 55	071	25 49	152	28 52	199	35 40	261	41 27	302	50 45	327
257	40 43	023	46 51	070	26 17	152	28 32	200	34 41	261	40 36	302	50 12	326
258	41 06	022	47 47	070	26 45	153	28 12	200	33 42	261	39 46	301	49 38	325
259	41 28	021	48 43	069	27 12	153	27 51	200	32 44	261	38 54	301	49 04	324
260	41 49	020	49 39	069	27 39	153	27 30	201	31 45	261	38 03	300	48 29	323
261	42 09	019	50 35	068	28 05	154	27 09	201	30 46	261	37 11	300	47 53	322
262	42 28	018	51 30	068	28 32	154	26 47	202	29 47	261	36 19	299	47 16	321
263	42 46	017	52 25	067	28 57	155	26 25	202	28 48	261	35 27	299	46 38	320
264	43 03	016	53 20	066	29 23	155	26 02	202	27 49	261	34 35	298	46 00	319
265	43 20	015	54 14	066	29 48	155	25 40	203	26 50	261	33 42	298	45 20	319
266	43 35	014	55 08	065	30 13	156	25 17	203	25 52	261	32 49	297	44 41	318
267	43 49	013	56 02	064	30 37	156	24 53	203	24 53	260	31 56	297	44 00	317
268	44 02	012	56 56	064	31 01	157	24 30	204	23 54	260	31 03	297	43 19	316
269	44 14	011	57 49	063	31 24	157	24 06	204	22 55	260	30 09	296	42 37	315

LAT 6°S

LHA ♈	Hc	Zn	Hc	Zn	Hc	Zn	Hc	Zn	Hc	Zn	Hc	Zn	Hc	Zn
	♦VEGA		DENEB		Enif		♦FOMALHAUT		Peacock		♦ANTARES		ARCTURUS	
270	44 25	010	27 21	031	32 01	074	16 33	119	31 47	158	60 14	224	29 16	296
271	44 35	009	27 52	030	32 59	074	17 25	119	32 10	158	59 32	225	28 22	296
272	44 44	008	28 21	030	33 56	074	18 18	119	32 32	158	58 49	226	27 28	295
273	44 51	007	28 51	029	34 53	074	19 10	119	32 54	159	58 05	227	26 34	295
274	44 58	006	29 20	029	35 51	073	20 02	119	33 15	159	57 21	228	25 40	295
275	45 03	005	29 48	028	36 48	073	20 54	119	33 35	160	56 36	229	24 46	294
276	45 07	004	30 16	027	37 45	073	21 45	120	33 56	161	55 51	230	23 51	294
277	45 11	002	30 43	027	38 42	072	22 37	120	34 15	161	55 05	231	22 57	294
278	45 12	001	31 09	026	39 38	072	23 29	120	34 34	162	54 18	232	22 02	294
279	45 13	000	31 35	025	40 35	072	24 21	120	34 53	162	53 31	232	21 08	293
280	45 13	359	32 00	025	41 32	071	25 13	120	35 11	163	52 44	233	20 13	293
281	45 11	358	32 25	024	42 28	071	26 04	120	35 28	163	51 56	234	19 18	293
282	45 09	357	32 49	023	43 24	070	26 56	120	35 45	164	51 08	234	18 23	293
283	45 05	356	33 13	023	44 20	070	27 47	120	36 02	164	50 19	235	17 27	292
284	45 00	355	33 35	022	45 16	069	28 39	121	36 17	165	49 30	236	16 32	292
	DENEB		Enif		♦FOMALHAUT		Peacock		♦ANTARES		Rasalhague		♦VEGA	
285	33 57	021	46 12	069	29 30	121	36 32	166	48 40	236	61 50	311	44 54	354
286	34 19	021	47 08	069	30 21	121	36 47	166	47 51	237	61 04	310	44 47	353
287	34 39	020	48 03	068	31 12	121	37 01	167	47 01	237	60 18	309	44 38	351
288	34 59	019	48 58	067	32 03	121	37 14	168	46 11	238	59 31	308	44 29	350
289	35 18	018	49 53	067	32 54	122	37 27	168	45 20	238	58 44	306	44 18	349
290	35 36	017	50 48	066	33 45	122	37 39	169	44 29	238	57 55	305	44 07	348
291	35 54	017	51 43	066	34 35	122	37 50	169	43 39	239	57 06	304	43 54	347
292	36 11	016	52 37	065	35 26	122	38 00	170	42 47	239	56 17	303	43 40	346
293	36 27	015	53 31	064	36 16	123	38 10	171	41 56	240	55 27	303	43 26	345
294	36 42	014	54 25	064	37 06	123	38 20	171	41 05	240	54 36	302	43 10	344
295	36 56	013	55 18	063	37 56	123	38 28	172	40 13	240	53 45	301	42 53	343
296	37 10	013	56 11	062	38 46	124	38 36	173	39 21	241	52 54	300	42 35	342
297	37 22	012	57 04	061	39 36	124	38 43	173	38 29	241	52 02	299	42 16	341
298	37 34	011	57 56	061	40 25	124	38 50	174	37 37	241	51 10	299	41 57	340
299	37 45	010	58 47	060	41 14	125	38 55	175	36 44	241	50 17	298	41 36	339
	♦DENEB		Alpheratz		♦FOMALHAUT		Peacock		♦ANTARES		Rasalhague		VEGA	
300	37 55	009	20 55	056	42 03	125	39 00	176	35 52	242	49 24	297	41 15	338
301	38 04	008	21 44	055	42 52	126	39 05	176	35 00	242	48 31	297	40 52	337
302	38 12	007	22 33	055	43 40	126	39 08	177	34 07	242	47 38	296	40 29	337
303	38 19	007	23 22	055	44 29	126	39 11	178	33 14	242	46 44	296	40 05	336
304	38 26	006	24 10	054	45 16	127	39 13	178	32 21	242	45 50	295	39 40	335
305	38 31	005	24 59	054	46 04	127	39 14	179	31 28	243	44 56	294	39 14	334
306	38 36	004	25 47	054	46 51	128	39 15	180	30 35	243	44 01	294	38 48	333
307	38 39	003	26 35	053	47 38	129	39 15	180	29 42	243	43 07	293	38 20	332
308	38 42	002	27 22	053	48 24	129	39 14	181	28 49	243	42 12	293	37 52	332
309	38 43	001	28 10	052	49 11	130	39 12	182	27 56	243	41 17	293	37 23	331
310	38 44	000	28 57	052	49 56	130	39 10	183	27 03	243	40 22	292	36 54	330
311	38 44	359	29 44	052	50 41	131	39 07	183	26 09	243	39 26	292	36 24	329
312	38 43	358	30 30	051	51 26	132	39 03	184	25 16	244	38 31	291	35 53	329
313	38 41	358	31 17	051	52 10	133	38 58	185	24 22	244	37 35	291	35 22	328
314	38 38	357	32 03	050	52 54	133	38 53	185	23 29	244	36 39	291	34 50	327
	Alpheratz		♦Diphda		ACHERNAR		Peacock		♦Nunki		VEGA		♦DENEB	
315	32 48	050	34 18	108	16 06	148	38 47	186	53 57	232	34 17	326	38 34	356
316	33 33	049	35 15	108	16 37	148	38 41	187	53 09	233	33 44	326	38 29	355
317	34 18	048	36 11	108	17 09	149	38 33	187	52 21	234	33 10	325	38 23	354
318	35 03	048	37 08	108	17 40	149	38 25	188	51 33	234	32 36	325	38 17	353
319	35 47	047	38 05	108	18 11	149	38 16	189	50 44	235	32 01	324	38 09	352
320	36 30	047	39 01	108	18 41	149	38 07	189	49 56	235	31 25	323	38 01	351
321	37 14	046	39 58	109	19 12	149	37 57	190	49 06	236	30 50	323	37 51	350
322	37 56	045	40 55	109	19 43	149	37 46	191	48 17	237	30 13	322	37 41	350
323	38 39	045	41 51	109	20 13	150	37 34	191	47 27	237	29 36	322	37 30	349
324	39 20	044	42 47	109	20 43	150	37 22	192	46 36	237	28 59	321	37 18	348
325	40 02	043	43 44	109	21 13	150	37 09	193	45 46	238	28 21	321	37 05	347
326	40 43	043	44 40	109	21 43	150	36 56	193	44 55	238	27 43	320	36 51	346
327	41 23	042	45 36	110	22 12	151	36 42	194	44 04	239	27 05	320	36 36	345
328	42 02	041	46 33	110	22 41	151	36 27	195	43 13	239	26 26	319	36 21	345
329	42 41	040	47 29	110	23 10	151	36 12	195	42 22	240	25 47	319	36 05	344
	♦Alpheratz		Diphda		♦ACHERNAR		Peacock		Nunki		♦ALTAIR		DENEB	
330	43 20	040	48 25	110	23 39	151	35 56	196	41 30	240	54 29	294	35 48	343
331	43 57	039	49 20	111	24 08	152	35 39	196	40 39	240	53 34	294	35 30	342
332	44 34	038	50 16	111	24 36	152	35 22	197	39 47	241	52 40	293	35 11	341
333	45 11	037	51 12	111	25 04	152	35 04	198	38 55	241	51 45	293	34 52	341
334	45 46	036	52 07	112	25 31	153	34 46	198	38 03	241	50 49	292	34 32	340
335	46 21	035	53 02	112	25 59	153	34 27	199	37 10	241	49 54	291	34 11	339
336	46 55	034	53 58	113	26 26	153	34 08	199	36 18	242	48 58	291	33 49	338
337	47 28	033	54 53	113	26 53	154	33 48	200	35 25	242	48 02	290	33 27	338
338	48 00	032	55 48	113	27 19	154	33 28	200	34 33	242	47 06	290	33 04	337
339	48 31	031	56 42	114	27 45	154	33 07	201	33 40	242	46 10	289	32 40	336
340	49 02	030	57 37	115	28 11	155	32 46	201	32 47	242	45 14	289	32 16	336
341	49 31	029	58 31	115	28 36	155	32 24	202	31 54	243	44 17	289	31 51	335
342	49 59	028	59 25	116	29 02	155	32 02	202	31 01	243	43 21	288	31 26	334
343	50 27	027	60 18	116	29 26	156	31 39	203	30 08	243	42 24	288	30 59	334
344	50 53	025	61 12	117	29 51	156	31 16	203	29 15	243	41 27	287	30 33	333
	Alpheratz		♦Hamal		♦ACHERNAR		Peacock		Nunki		♦ALTAIR		DENEB	
345	51 18	024	35 44	055	30 14	157	30 52	204	28 22	243	40 30	287	30 05	332
346	51 42	023	36 33	055	30 38	157	30 28	204	27 28	243	39 33	287	29 38	332
347	52 04	022	37 21	054	31 01	157	30 04	204	26 35	244	38 36	286	29 09	331
348	52 25	020	38 10	054	31 24	158	29 39	205	25 41	244	37 39	286	28 40	331
349	52 46	019	38 58	053	31 46	158	29 13	205	24 48	244	36 41	286	28 11	330
350	53 05	018	39 45	053	32 08	159	28 48	206	23 54	244	35 44	285	27 40	330
351	53 22	016	40 32	052	32 29	159	28 22	206	23 01	244	34 46	285	27 10	329
352	53 38	015	41 19	051	32 50	160	27 56	206	22 07	244	33 49	285	26 39	328
353	53 53	013	42 05	051	33 10	160	27 29	207	21 13	244	32 51	285	26 07	328
354	54 06	012	42 51	050	33 30	161	27 02	207	20 20	244	31 53	284	25 36	327
355	54 17	011	43 37	049	33 49	161	26 35	207	19 26	244	30 55	284	25 03	327
356	54 28	009	44 22	049	34 08	162	26 07	208	18 32	244	29 57	284	24 30	326
357	54 36	008	45 06	048	34 26	162	25 39	208	17 39	244	28 59	284	23 57	326
358	54 43	006	45 50	047	34 44	163	25 11	208	16 45	244	28 01	283	23 23	325
359	54 49	005	46 33	046	35 01	164	24 42	209	15 51	244	27 03	283	22 49	325

LHA ♈	Hc	Zn	Hc	Zn	Hc	Zn	Hc	Zn	Hc	Zn	Hc	Zn	Hc	Zn
	♦Alpheratz		ALDEBARAN		♦RIGEL		ACHERNAR		♦FOMALHAUT		ALTAIR		DENEB	
0	53 53	003	17 56	070	12 14	097	36 15	164	62 59	211	25 51	283	21 26	325
1	53 55	002	18 52	070	13 14	097	36 32	165	62 28	213	24 53	283	20 52	324
2	53 56	000	19 48	070	14 13	097	36 47	165	61 55	214	23 55	283	20 17	324
3	53 56	359	20 43	069	15 12	097	37 02	166	61 21	215	22 57	283	19 41	324
4	53 53	357	21 39	069	16 11	097	37 17	166	60 46	217	21 59	283	19 06	323
5	53 49	356	22 35	069	17 10	096	37 30	167	60 10	218	21 01	282	18 30	323
6	53 44	354	23 30	069	18 09	096	37 44	168	59 32	219	20 03	282	17 54	322
7	53 37	353	24 26	068	19 08	096	37 56	168	58 54	220	19 05	282	17 17	322
8	53 29	351	25 21	068	20 08	096	38 08	169	58 15	222	18 06	282	16 40	322
9	53 19	350	26 16	068	21 07	096	38 19	169	57 35	223	17 08	282	16 03	321
10	53 08	348	27 11	067	22 06	096	38 30	170	56 55	224	16 10	281	15 26	321
11	52 55	347	28 06	067	23 05	096	38 40	171	56 13	225	15 11	281	14 48	321
12	52 41	346	29 01	067	24 05	096	38 49	171	55 31	225	14 13	281	14 10	320
13	52 25	344	29 55	066	25 04	096	38 57	172	54 48	226	13 14	281	13 32	320
14	52 08	343	30 50	066	26 03	096	39 05	173	54 05	227	12 16	281	12 54	320
	Hamal		ALDEBARAN		♦RIGEL		ACHERNAR		♦FOMALHAUT		Enif		♦Alpheratz	
15	55 27	028	31 44	066	27 02	096	39 12	173	53 21	228	38 21	288	51 50	341
16	55 54	026	32 38	065	28 02	096	39 19	174	52 37	229	37 25	288	51 31	340
17	56 20	025	33 32	065	29 01	096	39 24	175	51 52	229	36 28	288	51 10	339
18	56 44	023	34 26	064	30 00	095	39 29	176	51 06	230	35 31	287	50 48	338
19	57 07	022	35 20	064	30 59	095	39 34	176	50 20	231	34 34	287	50 24	336
20	57 29	020	36 13	064	31 59	095	39 37	177	49 34	231	33 37	287	50 00	335
21	57 48	019	37 06	063	32 58	095	39 40	178	48 47	232	32 40	286	49 35	334
22	58 07	017	37 59	063	33 57	095	39 42	178	48 00	233	31 43	286	49 08	333
23	58 23	015	38 52	062	34 57	095	39 44	179	47 13	233	30 46	286	48 40	332
24	58 38	014	39 45	062	35 56	095	39 44	180	46 25	234	29 49	286	48 12	331
25	58 52	012	40 37	061	36 55	095	39 44	180	45 37	234	28 51	285	47 42	330
26	59 03	010	41 29	061	37 55	095	39 43	181	44 49	235	27 54	285	47 11	329
27	59 13	008	42 21	060	38 54	095	39 42	182	44 00	235	26 56	285	46 40	327
28	59 21	007	43 12	059	39 53	095	39 39	183	43 11	235	25 59	285	46 07	326
29	59 27	005	44 04	059	40 53	095	39 36	183	42 22	236	25 01	284	45 34	326
	CAPELLA		♦BETELGEUSE		SIRIUS		CANOPUS		♦ACHERNAR		FOMALHAUT		♦Alpheratz	
30	21 21	034	29 41	077	19 56	105	20 00	144	39 33	184	41 33	236	45 00	325
31	21 54	034	30 39	077	20 53	105	20 35	144	39 28	185	40 43	237	44 25	324
32	22 27	033	31 37	077	21 51	105	21 10	144	39 23	185	39 53	237	43 49	323
33	23 00	033	32 35	077	22 48	105	21 44	144	39 17	186	39 03	237	43 13	322
34	23 32	032	33 33	076	23 46	105	22 19	145	39 11	187	38 13	237	42 36	321
35	24 04	032	34 31	076	24 43	105	22 53	145	39 03	187	37 23	238	41 58	320
36	24 35	031	35 29	076	25 41	105	23 27	145	38 55	188	36 33	238	41 19	319
37	25 06	031	36 26	075	26 38	105	24 01	145	38 47	189	35 42	238	40 40	319
38	25 36	030	37 24	075	27 36	105	24 35	146	38 37	189	34 51	239	40 01	318
39	26 06	030	38 21	075	28 33	105	25 09	146	38 27	190	34 00	239	39 21	317
40	26 36	029	39 19	074	29 30	105	25 42	146	38 17	191	33 09	239	38 40	316
41	27 04	029	40 16	074	30 28	105	26 15	146	38 05	191	32 18	239	37 59	316
42	27 33	028	41 13	074	31 25	105	26 48	147	37 53	192	31 27	239	37 17	315
43	28 01	028	42 11	073	32 23	105	27 21	147	37 41	193	30 36	240	36 35	314
44	28 28	027	43 08	073	33 20	105	27 53	147	37 27	193	29 44	240	35 52	314
	CAPELLA		♦BETELGEUSE		SIRIUS		CANOPUS		♦ACHERNAR		FOMALHAUT		♦Alpheratz	
45	28 55	026	44 04	073	34 18	105	28 25	148	37 13	194	28 53	240	35 09	313
46	29 21	026	45 01	072	35 15	106	28 57	148	36 59	194	28 01	240	34 25	313
47	29 47	025	45 58	072	36 12	106	29 28	148	36 44	195	27 10	240	33 41	312
48	30 12	025	46 54	071	37 10	106	29 59	149	36 28	196	26 18	240	32 57	312
49	30 36	024	47 51	071	38 07	106	30 30	149	36 12	196	25 26	240	32 12	311
50	31 00	023	48 47	070	39 04	106	31 00	149	35 55	197	24 34	241	31 27	310
51	31 23	023	49 43	070	40 02	106	31 30	150	35 37	197	23 43	241	30 41	310
52	31 45	022	50 38	069	40 59	106	32 00	150	35 19	198	22 51	241	29 55	309
53	32 07	021	51 34	069	41 56	106	32 30	151	35 01	198	21 59	241	29 09	309
54	32 28	020	52 29	068	42 53	106	32 58	151	34 41	199	21 07	241	28 23	309
55	32 49	020	53 25	067	43 50	106	33 27	152	34 22	199	20 15	241	27 36	308
56	33 09	019	54 19	067	44 47	107	33 55	152	34 02	200	19 23	241	26 49	308
57	33 28	018	55 14	066	45 44	107	34 23	152	33 41	201	18 30	241	26 02	307
58	33 46	017	56 08	065	46 41	107	34 50	153	33 20	201	17 38	241	25 14	307
59	34 03	017	57 02	065	47 38	107	35 17	153	32 58	201	16 46	241	24 26	306
	♦CAPELLA		POLLUX		♦SIRIUS		CANOPUS		ACHERNAR		♦Diphda		Hamal	
60	34 20	016	25 26	054	48 35	107	35 43	154	32 36	202	40 55	252	48 54	319
61	34 36	015	26 14	054	49 32	108	36 09	155	32 14	202	39 58	252	48 14	318
62	34 51	014	27 02	054	50 29	108	36 35	155	31 51	203	39 01	252	47 33	317
63	35 06	014	27 50	053	51 25	108	36 59	156	31 28	203	38 04	253	46 52	316
64	35 19	013	28 37	053	52 22	108	37 24	156	31 04	204	37 08	253	46 10	315
65	35 32	012	29 24	052	53 18	109	37 48	157	30 40	204	36 11	253	45 28	314
66	35 44	011	30 11	052	54 15	109	38 11	157	30 15	205	35 14	253	44 45	313
67	35 55	010	30 58	051	55 11	109	38 33	158	29 50	205	34 17	253	44 01	313
68	36 05	010	31 44	051	56 07	110	38 55	159	29 25	205	33 20	253	43 17	312
69	36 15	009	32 30	050	57 03	110	39 17	159	28 59	206	32 23	253	42 33	311
70	36 23	008	33 16	050	57 59	110	39 38	160	28 33	206	31 26	253	41 47	310
71	36 31	007	34 01	049	58 55	111	39 58	161	28 07	206	30 29	253	41 02	310
72	36 38	006	34 46	049	59 50	111	40 17	161	27 40	207	29 32	253	40 16	309
73	36 44	005	35 30	048	60 46	112	40 36	162	27 13	207	28 35	253	39 29	308
74	36 49	004	36 14	047	61 41	113	40 55	163	26 46	207	27 38	253	38 42	308
	CAPELLA		POLLUX		♦REGULUS		CANOPUS		♦ACHERNAR		Diphda		♦Hamal	
75	36 53	004	36 58	047	11 06	076	41 12	163	26 18	208	26 41	253	37 55	307
76	36 56	003	37 41	046	12 04	076	41 29	164	25 50	208	25 44	253	37 07	307
77	36 58	002	38 24	045	13 02	076	41 45	165	25 22	208	24 47	253	36 19	306
78	37 00	001	39 06	045	14 00	076	42 00	165	24 53	209	23 50	253	35 31	305
79	37 00	000	39 48	044	14 57	076	42 15	166	24 25	209	22 53	253	34 42	305
80	37 00	359	40 29	043	15 55	075	42 29	167	23 56	209	21 56	253	33 53	304
81	36 59	358	41 09	043	16 53	075	42 42	168	23 27	209	20 58	253	33 04	304
82	36 57	357	41 49	042	17 50	075	42 54	168	22 57	210	20 01	253	32 15	303
83	36 53	357	42 29	041	18 48	075	43 06	169	22 28	210	19 04	253	31 25	303
84	36 49	356	43 08	040	19 45	075	43 16	170	21 58	210	18 07	253	30 35	303
85	36 45	355	43 46	039	20 43	074	43 26	171	21 28	210	17 10	253	29 44	302
86	36 39	354	44 23	038	21 40	074	43 35	172	20 57	211	16 13	253	28 54	302
87	36 32	353	45 00	038	22 37	074	43 43	172	20 27	211	15 16	253	28 03	301
88	36 25	352	45 36	037	23 34	074	43 51	173	19 56	211	14 19	253	27 12	301
89	36 16	351	46 11	036	24 31	073	43 57	174	19 26	211	13 22	253	26 21	301

LHA ♈	Hc	Zn	Hc	Zn	Hc	Zn	Hc	Zn	Hc	Zn	Hc	Zn	Hc	Zn
	POLLUX		♦REGULUS		Suhail		♦CANOPUS		Acamar		♦ALDEBARAN		CAPELLA	
90	46 45	035	25 28	073	35 10	140	44 03	175	37 33	223	58 34	319	36 07	351
91	47 19	034	26 25	073	35 48	140	44 08	176	36 52	224	57 54	317	35 57	350
92	47 51	033	27 22	073	36 27	140	44 12	177	36 11	224	57 14	316	35 46	349
93	48 23	032	28 19	072	37 04	141	44 15	177	35 29	225	56 32	315	35 34	348
94	48 54	031	29 16	072	37 42	141	44 17	178	34 47	225	55 49	314	35 21	347
95	49 24	029	30 12	072	38 19	142	44 18	179	34 05	225	55 06	313	35 08	347
96	49 52	028	31 09	071	38 55	142	44 18	180	33 22	226	54 21	311	34 54	346
97	50 20	027	32 05	071	39 32	143	44 18	181	32 40	226	53 36	310	34 39	345
98	50 47	026	33 01	071	40 08	143	44 17	182	31 57	226	52 51	309	34 23	344
99	51 12	025	33 58	070	40 43	144	44 14	183	31 14	227	52 04	309	34 06	343
100	51 36	023	34 54	070	41 18	144	44 11	183	30 30	227	51 18	308	33 49	343
101	51 59	022	35 50	070	41 52	145	44 07	184	29 47	227	50 30	307	33 30	342
102	52 21	021	36 45	069	42 26	146	44 02	185	29 03	227	49 42	306	33 12	341
103	52 42	020	37 41	069	42 59	146	43 57	186	28 19	228	48 54	305	32 52	340
104	53 01	018	38 37	069	43 32	147	43 50	187	27 35	228	48 05	304	32 32	340
	POLLUX		♦REGULUS		Suhail		♦CANOPUS		RIGEL		♦ALDEBARAN		CAPELLA	
105	53 19	017	39 32	068	44 04	148	43 43	188	63 47	266	47 15	304	32 11	339
106	53 35	015	40 27	068	44 36	148	43 34	188	62 48	266	46 26	303	31 49	338
107	53 50	014	41 22	067	45 07	149	43 25	189	61 48	266	45 36	302	31 27	338
108	54 04	012	42 17	067	45 37	150	43 15	190	60 49	266	44 45	302	31 04	337
109	54 16	011	43 11	066	46 06	151	43 05	191	59 49	266	43 54	301	30 40	336
110	54 27	010	44 06	066	46 35	151	42 53	192	58 50	266	43 03	300	30 16	336
111	54 36	008	45 00	065	47 04	152	42 41	192	57 51	266	42 11	300	29 51	335
112	54 43	007	45 54	065	47 31	153	42 28	193	56 51	266	41 20	299	29 25	334
113	54 49	005	46 48	064	47 58	154	42 14	194	55 52	266	40 28	299	28 59	334
114	54 54	003	47 41	064	48 23	155	41 59	195	54 53	266	39 35	298	28 32	333
115	54 56	002	48 34	063	48 48	156	41 44	195	53 53	266	38 43	298	28 05	332
116	54 58	000	49 27	062	49 12	157	41 28	196	52 54	266	37 50	297	27 37	332
117	54 57	359	50 20	062	49 36	158	41 11	197	51 54	266	36 57	297	27 09	331
118	54 55	357	51 12	061	49 58	158	40 53	198	50 55	266	36 03	296	26 40	331
119	54 52	356	52 04	060	50 19	159	40 35	198	49 56	266	35 10	296	26 11	330
	♦REGULUS		Gienah		♦ACRUX		CANOPUS		RIGEL		♦BETELGEUSE		CAPELLA	
120	52 55	059	26 57	106	16 42	154	40 16	199	48 56	266	55 39	294	25 41	330
121	53 46	059	27 54	106	17 08	154	39 56	200	47 57	265	54 44	293	25 11	329
122	54 37	058	28 52	106	17 33	155	39 36	200	46 58	265	53 49	293	24 40	329
123	55 27	057	29 49	106	17 59	155	39 15	201	45 58	265	52 54	292	24 09	328
124	56 16	056	30 46	106	18 24	155	38 54	201	44 59	265	51 59	292	23 37	328
125	57 06	055	31 43	106	18 49	155	38 32	202	43 59	265	51 04	291	23 05	327
126	57 54	054	32 40	106	19 14	155	38 09	203	43 00	265	50 08	290	22 32	327
127	58 42	053	33 38	106	19 39	156	37 46	203	42 01	265	49 12	290	21 59	326
128	59 29	052	34 35	106	20 04	156	37 22	204	41 01	265	48 16	289	21 26	326
129	60 15	050	35 32	107	20 28	156	36 57	204	40 02	265	47 20	289	20 52	325
130	61 01	049	36 29	107	20 52	156	36 33	205	39 03	265	46 23	288	20 18	325
131	61 45	048	37 26	107	21 16	156	36 07	206	38 03	265	45 27	288	19 44	324
132	62 29	046	38 23	107	21 40	157	35 41	206	37 04	265	44 30	288	19 09	324
133	63 12	045	39 20	107	22 03	157	35 15	207	36 05	265	43 33	287	18 34	324
134	63 53	043	40 17	107	22 27	157	34 48	207	35 05	265	42 36	287	17 59	323
	Dubhe		Denebola		♦SPICA		ACRUX		♦CANOPUS		SIRIUS		♦POLLUX	
135	17 11	015	42 54	063	24 35	099	22 50	157	34 21	208	55 37	251	50 34	333
136	17 26	014	43 47	062	25 34	099	23 13	158	33 53	208	54 41	251	50 06	332
137	17 41	014	44 40	061	26 33	099	23 35	158	33 25	208	53 45	251	49 38	331
138	17 55	013	45 32	061	27 32	099	23 57	158	32 56	209	52 48	251	49 09	330
139	18 08	013	46 24	060	28 30	099	24 19	158	32 27	209	51 52	252	48 39	329
140	18 22	013	47 15	060	29 29	099	24 41	159	31 58	210	50 55	252	48 07	328
141	18 34	012	48 06	059	30 28	099	25 02	159	31 28	210	49 58	252	47 35	327
142	18 47	012	48 57	058	31 27	099	25 24	159	30 58	211	49 02	253	47 02	326
143	18 58	011	49 47	057	32 26	099	25 44	160	30 27	211	48 05	253	46 28	325
144	19 10	011	50 37	057	33 25	099	26 05	160	29 57	211	47 08	253	45 54	324
145	19 21	010	51 27	056	34 23	099	26 25	160	29 26	212	46 11	253	45 18	323
146	19 31	010	52 16	055	35 22	099	26 45	161	28 54	212	45 14	253	44 42	322
147	19 41	009	53 04	054	36 21	099	27 04	161	28 22	212	44 17	253	44 05	321
148	19 50	009	53 52	053	37 20	099	27 23	161	27 50	213	43 20	254	43 27	320
149	19 59	008	54 40	052	38 19	099	27 42	162	27 18	213	42 23	254	42 48	319
	Dubhe		ARCTURUS		♦SPICA		ACRUX		♦CANOPUS		SIRIUS		♦POLLUX	
150	20 08	008	21 54	066	39 18	099	28 01	162	26 45	213	41 25	254	42 09	319
151	20 16	007	22 48	066	40 17	099	28 19	163	26 13	214	40 28	254	41 30	318
152	20 23	007	23 42	065	41 16	099	28 36	163	25 39	214	39 31	254	40 49	317
153	20 30	006	24 36	065	42 14	099	28 53	163	25 06	214	38 34	254	40 08	316
154	20 36	006	25 30	065	43 13	099	29 10	164	24 33	214	37 36	254	39 27	316
155	20 42	005	26 24	064	44 12	099	29 27	164	23 59	215	36 39	254	38 45	315
156	20 48	005	27 18	064	45 11	099	29 43	165	23 25	215	35 42	254	38 03	314
157	20 53	004	28 11	064	46 10	099	29 58	165	22 51	215	34 44	254	37 20	314
158	20 57	004	29 04	063	47 09	099	30 13	165	22 16	215	33 47	255	36 36	313
159	21 01	003	29 58	063	48 08	099	30 28	166	21 42	216	32 49	255	35 53	312
160	21 04	003	30 51	063	49 06	099	30 42	166	21 07	216	31 52	255	35 08	312
161	21 07	002	31 43	062	50 05	099	30 56	167	20 32	216	30 55	255	34 24	311
162	21 09	002	32 36	062	51 04	099	31 10	167	19 57	216	29 57	255	33 39	311
163	21 11	001	33 28	061	52 03	099	31 22	168	19 22	216	29 00	255	32 53	310
164	21 12	001	34 21	061	53 02	099	31 35	168	18 47	216	28 02	255	32 08	310
	Dubhe		♦ARCTURUS		SPICA		♦ACRUX		Suhail		SIRIUS		♦POLLUX	
165	21 13	000	35 13	061	54 00	099	31 47	169	46 04	209	27 05	255	31 21	309
166	21 13	000	36 04	060	54 59	099	31 58	169	45 34	210	26 07	255	30 35	309
167	21 13	359	36 56	060	55 58	100	32 09	170	45 04	211	25 10	255	29 48	308
168	21 12	359	37 47	059	56 57	100	32 20	170	44 33	212	24 12	255	29 01	308
169	21 11	358	38 38	059	57 55	100	32 29	171	44 01	212	23 15	255	28 14	307
170	21 09	358	39 29	058	58 54	100	32 39	171	43 29	213	22 17	255	27 26	307
171	21 06	357	40 19	057	59 53	100	32 48	172	42 56	214	21 20	255	26 38	306
172	21 03	357	41 09	057	60 51	100	32 56	172	42 23	214	20 23	255	25 50	306
173	21 00	356	41 58	056	61 50	101	33 04	173	41 49	215	19 25	255	25 02	305
174	20 56	356	42 48	056	62 48	101	33 11	173	41 14	216	18 28	255	24 13	305
175	20 51	355	43 37	055	63 47	101	33 18	174	40 40	216	17 30	255	23 24	305
176	20 46	355	44 25	054	64 45	101	33 24	174	40 04	217	16 33	255	22 35	304
177	20 41	354	45 13	054	65 44	102	33 30	175	39 28	217	15 35	255	21 46	304
178	20 35	354	46 01	053	66 42	102	33 35	175	38 52	218	14 38	254	20 56	304
179	20 28	353	46 48	052	67 40	102	33 39	176	38 15	218	13 41	254	20 07	303

LAT 7°S

LHA	Hc	Zn	Hc	Zn	Hc	Zn	Hc	Zn	Hc	Zn	Hc	Zn	Hc	Zn
♈	ARCTURUS		◆ANTARES		ACRUX		◆Suhail		Alphard		REGULUS		◆Dubhe	
180	47 35	051	23 26	116	33 43	176	37 38	219	52 10	265	56 19	304	20 21	353
181	48 21	050	24 19	116	33 47	177	37 01	219	51 11	265	55 30	303	20 13	352
182	49 07	050	25 13	116	33 50	178	36 23	220	50 12	265	54 39	302	20 05	352
183	49 52	049	26 06	116	33 52	178	35 45	220	49 12	265	53 49	301	19 57	351
184	50 36	048	27 00	116	33 54	179	35 06	221	48 13	265	52 58	301	19 48	351
185	51 20	047	27 53	116	33 55	179	34 27	221	47 14	265	52 07	300	19 38	351
186	52 03	046	28 47	116	33 56	180	33 48	221	46 14	265	51 15	299	19 28	350
187	52 45	045	29 40	116	33 56	180	33 09	222	45 15	265	50 23	298	19 18	350
188	53 27	044	30 33	117	33 55	181	32 29	222	44 16	265	49 30	298	19 07	349
189	54 08	043	31 27	117	33 54	181	31 49	222	43 16	265	48 37	297	18 55	349
190	54 48	041	32 20	117	33 52	182	31 09	223	42 17	265	47 44	297	18 43	348
191	55 27	040	33 13	117	33 50	182	30 28	223	41 18	265	46 51	296	18 31	348
192	56 05	039	34 06	117	33 47	183	29 47	223	40 19	265	45 57	295	18 18	347
193	56 42	038	34 59	117	33 44	183	29 06	224	39 19	265	45 03	295	18 05	347
194	57 18	036	35 51	118	33 40	184	28 25	224	38 20	265	44 09	294	17 51	346
	ARCTURUS		◆Rasalhague		ANTARES		◆ACRUX		Suhail		◆REGULUS		Dubhe	
195	57 52	035	19 01	074	36 44	118	33 36	185	27 44	224	43 14	294	17 37	346
196	58 26	034	19 58	074	37 37	118	33 31	185	27 02	224	42 20	293	17 22	346
197	58 58	032	20 55	074	38 29	118	33 25	186	26 21	225	41 25	293	17 07	345
198	59 29	031	21 52	073	39 21	119	33 19	186	25 39	225	40 30	292	16 52	345
199	59 59	029	22 49	073	40 14	119	33 12	187	24 57	225	39 35	292	16 36	344
200	60 27	027	23 46	073	41 06	119	33 05	187	24 14	225	38 39	291	16 19	344
201	60 53	026	24 43	073	41 58	120	32 57	188	23 32	225	37 44	291	16 03	344
202	61 18	024	25 40	072	42 49	120	32 49	188	22 50	226	36 48	291	15 46	343
203	61 42	022	26 37	072	43 41	120	32 40	189	22 07	226	35 53	290	15 28	343
204	62 03	020	27 33	072	44 32	121	32 31	189	21 24	226	34 57	290	15 10	342
205	62 23	018	28 30	072	45 24	121	32 21	190	20 41	226	34 01	290	14 52	342
206	62 40	016	29 26	071	46 14	121	32 11	190	19 58	226	33 04	289	14 34	342
207	62 56	014	30 23	071	47 05	122	32 00	191	19 15	226	32 08	289	14 15	341
208	63 10	012	31 19	071	47 56	122	31 49	191	18 32	226	31 12	289	13 55	341
209	63 21	010	32 15	070	48 46	123	31 37	192	17 49	227	30 15	288	13 36	341
	ARCTURUS		◆Rasalhague		ANTARES		◆RIGIL KENT.		ACRUX		Gienah		◆REGULUS	
210	63 31	008	33 11	070	49 36	123	35 41	174	31 24	192	62 27	245	29 19	288
211	63 38	006	34 07	070	50 26	124	35 47	175	31 12	193	61 33	246	28 22	288
212	63 44	004	35 03	069	51 15	124	35 52	175	30 58	193	60 38	246	27 25	287
213	63 47	002	35 58	069	52 04	125	35 57	176	30 45	194	59 44	247	26 28	287
214	63 47	000	36 54	069	52 53	125	36 01	176	30 30	194	58 49	248	25 31	287
215	63 46	358	37 49	068	53 41	126	36 04	177	30 16	194	57 54	248	24 34	287
216	63 42	355	38 44	068	54 29	127	36 07	178	30 01	195	56 58	248	23 37	286
217	63 37	353	39 39	067	55 16	128	36 09	178	29 45	195	56 03	249	22 40	286
218	63 29	351	40 34	067	56 03	128	36 10	179	29 29	196	55 07	249	21 43	286
219	63 18	349	41 29	066	56 50	129	36 11	180	29 13	196	54 11	250	20 46	286
220	63 06	347	42 23	066	57 36	130	36 11	180	28 56	197	53 15	250	19 48	285
221	62 52	345	43 18	065	58 21	131	36 11	181	28 39	197	52 19	250	18 51	285
222	62 35	343	44 12	065	59 06	132	36 10	181	28 21	197	51 23	251	17 53	285
223	62 17	341	45 05	064	59 50	133	36 08	182	28 03	198	50 27	251	16 56	285
224	61 57	339	45 59	064	60 33	134	36 06	183	27 45	198	49 31	251	15 58	285
	Alphecca		◆VEGA		ALTAIR		◆Shaula		RIGIL KENT.		◆SPICA		ARCTURUS	
225	55 15	014	22 07	043	15 54	079	44 00	137	36 03	183	66 11	258	61 35	337
226	55 28	012	22 47	043	16 52	079	44 41	137	35 59	184	65 12	259	61 11	336
227	55 40	011	23 27	042	17 51	078	45 21	138	35 55	184	64 14	259	60 46	334
228	55 50	009	24 07	042	18 49	078	46 01	138	35 50	185	63 15	259	60 19	332
229	55 58	007	24 46	041	19 47	078	46 40	139	35 45	186	62 17	259	59 51	331
230	56 05	006	25 26	041	20 45	078	47 19	140	35 39	186	61 18	260	59 21	329
231	56 10	004	26 04	040	21 44	078	47 57	140	35 32	187	60 20	260	58 49	327
232	56 14	003	26 43	040	22 42	077	48 34	141	35 25	187	59 21	260	58 17	326
233	56 16	001	27 21	039	23 40	077	49 11	142	35 17	188	58 23	260	57 43	325
234	56 16	359	27 58	039	24 38	077	49 48	143	35 09	188	57 24	260	57 08	323
235	56 15	358	28 35	038	25 36	077	50 23	144	35 00	189	56 25	260	56 31	322
236	56 11	356	29 12	038	26 34	076	50 58	144	34 50	190	55 26	261	55 54	321
237	56 07	355	29 48	037	27 32	076	51 33	145	34 40	190	54 28	261	55 16	319
238	56 00	353	30 24	037	28 29	076	52 06	146	34 29	191	53 29	261	54 37	318
239	55 52	351	30 59	036	29 27	076	52 39	147	34 18	191	52 30	261	53 56	317
	VEGA		◆ALTAIR		Shaula		◆RIGIL KENT.		SPICA		◆ARCTURUS		Alphecca	
240	31 33	035	30 25	075	53 11	148	34 06	192	51 31	261	53 15	316	55 42	350
241	32 08	035	31 22	075	53 42	149	33 54	192	50 33	261	52 34	315	55 31	348
242	32 41	034	32 20	075	54 12	150	33 41	193	49 34	261	51 51	314	55 18	347
243	33 14	033	33 17	075	54 41	151	33 27	193	48 35	261	51 08	313	55 04	345
244	33 47	033	34 15	074	55 09	153	33 13	194	47 36	261	50 24	312	54 48	344
245	34 19	032	35 12	074	55 36	154	32 59	194	46 37	261	49 39	311	54 31	342
246	34 50	031	36 09	074	56 01	155	32 44	195	45 38	261	48 54	310	54 12	341
247	35 21	031	37 06	073	56 26	156	32 28	195	44 39	261	48 08	309	53 52	339
248	35 51	030	38 03	073	56 50	157	32 13	196	43 41	261	47 22	309	53 30	338
249	36 20	029	39 00	073	57 12	159	31 56	196	42 42	261	46 35	308	53 07	337
250	36 49	028	39 57	072	57 33	160	31 39	197	41 43	261	45 48	307	52 43	335
251	37 16	028	40 54	072	57 53	161	31 22	197	40 44	261	45 00	306	52 18	334
252	37 44	027	41 50	071	58 11	163	31 04	198	39 45	261	44 12	306	51 51	333
253	38 10	026	42 46	071	58 28	164	30 46	198	38 46	261	43 23	305	51 23	332
254	38 36	025	43 43	071	58 44	166	30 27	199	37 47	261	42 34	304	50 55	330
	VEGA		◆ALTAIR		Peacock		◆RIGIL KENT.		SPICA		◆ARCTURUS		Alphecca	
255	39 01	024	44 39	070	26 14	152	30 08	199	36 49	261	41 44	304	50 25	329
256	39 25	023	45 35	070	26 42	152	29 48	199	35 50	261	40 55	303	49 54	328
257	39 48	023	46 31	069	27 10	152	29 28	200	34 51	261	40 05	302	49 22	327
258	40 10	022	47 26	069	27 38	152	29 08	200	33 52	261	39 14	302	48 49	326
259	40 32	021	48 22	068	28 05	153	28 47	201	32 53	261	38 23	301	48 15	325
260	40 53	020	49 17	068	28 32	153	28 26	201	31 54	261	37 32	301	47 40	324
261	41 12	019	50 12	067	28 59	154	28 05	201	30 55	261	36 41	300	47 05	323
262	41 31	018	51 06	066	29 26	154	27 43	202	29 57	261	35 50	300	46 29	322
263	41 49	017	52 01	066	29 52	154	27 21	202	28 58	261	34 58	299	45 52	321
264	42 06	016	52 55	065	30 17	155	26 58	202	27 59	261	34 06	299	45 14	320
265	42 22	015	53 49	064	30 42	155	26 35	203	27 00	261	33 14	298	44 35	319
266	42 37	014	54 43	064	31 07	156	26 12	203	26 01	261	32 21	298	43 56	318
267	42 51	013	55 36	063	31 32	156	25 48	203	25 02	261	31 28	298	43 16	318
268	43 03	012	56 29	062	31 56	156	25 24	204	24 04	261	30 36	297	42 36	317
269	43 15	011	57 21	061	32 20	157	25 00	204	23 05	261	29 43	297	41 54	316

LHA	Hc	Zn	Hc	Zn	Hc	Zn	Hc	Zn	Hc	Zn	Hc	Zn	Hc	Zn
♈	◆VEGA		DENEB		Enif		◆FOMALHAUT		Peacock		◆ANTARES		ARCTURUS	
270	43 26	010	26 30	031	31 45	074	17 03	119	32 43	157	60 56	225	28 49	296
271	43 36	009	27 00	030	32 42	074	17 55	119	33 05	158	60 13	227	27 56	296
272	43 44	008	27 29	029	33 39	073	18 47	119	33 28	158	59 30	228	27 02	296
273	43 52	007	27 58	029	34 36	073	19 39	119	33 50	159	58 46	229	26 09	295
274	43 58	006	28 27	028	35 33	073	20 31	119	34 11	159	58 01	229	25 15	295
275	44 03	005	28 55	028	36 30	072	21 23	119	34 32	160	57 15	230	24 21	295
276	44 08	003	29 22	027	37 26	072	22 15	119	34 52	160	56 29	231	23 27	294
277	44 11	002	29 49	026	38 23	072	23 07	119	35 12	161	55 42	232	22 33	294
278	44 12	001	30 15	026	39 19	071	23 59	119	35 31	161	54 55	233	21 38	294
279	44 13	000	30 41	025	40 16	071	24 51	119	35 50	162	54 07	233	20 44	294
280	44 13	359	31 06	025	41 12	070	25 42	120	36 08	162	53 19	234	19 49	293
281	44 11	358	31 30	024	42 08	070	26 34	120	36 26	163	52 31	235	18 54	293
282	44 09	357	31 54	023	43 04	069	27 26	120	36 43	164	51 42	235	17 59	293
283	44 05	356	32 17	022	43 59	069	28 17	120	36 59	164	50 53	236	17 05	293
284	44 00	355	32 40	022	44 55	069	29 09	120	37 15	165	50 03	237	16 10	292
	DENEB		Enif		◆FOMALHAUT		Peacock		◆ANTARES		Rasalhague		◆VEGA	
285	33 01	021	45 50	068	30 00	120	37 31	165	49 14	237	61 09	313	43 54	354
286	33 22	020	46 45	068	30 52	120	37 45	166	48 23	238	60 25	311	43 47	353
287	33 43	020	47 40	067	31 43	121	37 59	167	47 33	238	59 40	310	43 39	352
288	34 02	019	48 35	066	32 34	121	38 13	167	46 42	238	58 54	309	43 30	351
289	34 21	018	49 29	066	33 25	121	38 25	168	45 52	239	58 07	308	43 19	349
290	34 39	017	50 24	065	34 16	121	38 37	169	45 00	239	57 20	307	43 08	348
291	34 56	016	51 18	065	35 07	122	38 49	169	44 09	240	56 32	306	42 56	347
292	35 13	016	52 11	064	35 58	122	39 00	170	43 18	240	55 43	305	42 42	346
293	35 29	015	53 05	063	36 48	122	39 10	171	42 26	240	54 54	304	42 28	345
294	35 44	014	53 58	063	37 39	122	39 19	171	41 34	241	54 04	303	42 12	344
295	35 58	013	54 50	062	38 29	123	39 28	172	40 42	241	53 14	302	41 56	343
296	36 11	012	55 42	061	39 19	123	39 35	173	39 50	241	52 23	301	41 38	342
297	36 23	012	56 34	060	40 09	123	39 43	173	38 58	241	51 32	300	41 20	341
298	36 35	011	57 26	059	40 59	124	39 49	174	38 06	242	50 41	300	41 00	341
299	36 46	010	58 17	058	41 48	124	39 55	175	37 13	242	49 49	299	40 40	340
	◆DENEB		Alpheratz		◆FOMALHAUT		Peacock		◆ANTARES		Rasalhague		VEGA	
300	36 55	009	20 21	055	42 37	124	40 00	175	36 20	242	48 56	298	40 19	339
301	37 04	008	21 10	055	43 26	125	40 04	176	35 28	242	48 04	298	39 57	338
302	37 12	007	21 58	055	44 15	125	40 08	177	34 35	243	47 11	297	39 34	337
303	37 20	006	22 47	054	45 04	126	40 11	178	33 42	243	46 18	296	39 10	336
304	37 26	006	23 35	054	45 52	126	40 13	178	32 49	243	45 24	296	38 45	335
305	37 31	005	24 23	054	46 40	127	40 14	179	31 56	243	44 31	295	38 20	334
306	37 36	004	25 11	053	47 28	127	40 15	180	31 03	243	43 37	295	37 54	334
307	37 39	003	25 59	053	48 15	128	40 15	180	30 09	243	42 42	294	37 27	333
308	37 42	002	26 46	052	49 02	128	40 14	181	29 16	244	41 48	294	36 59	332
309	37 43	001	27 33	052	49 49	129	40 12	182	28 23	244	40 54	293	36 31	331
310	37 44	000	28 20	052	50 35	129	40 10	183	27 29	244	39 59	293	36 02	330
311	37 44	359	29 06	051	51 21	130	40 07	183	26 36	244	39 04	292	35 32	330
312	37 43	359	29 52	051	52 06	131	40 03	184	25 42	244	38 09	292	35 02	329
313	37 41	358	30 38	050	52 51	132	39 58	185	24 49	244	37 13	292	34 31	328
314	37 38	357	31 24	050	53 35	132	39 53	185	23 55	244	36 18	291	33 59	328
	Alpheratz		◆Diphda		ACHERNAR		Peacock		◆Nunki		VEGA		◆DENEB	
315	32 09	049	34 36	107	16 57	148	39 47	186	54 33	233	33 27	327	37 34	356
316	32 54	049	35 33	107	17 28	148	39 40	187	53 45	234	32 54	326	37 29	355
317	33 38	048	36 29	107	18 00	148	39 33	188	52 56	235	32 21	326	37 24	354
318	34 22	047	37 26	107	18 31	148	39 24	188	52 08	235	31 47	325	37 17	353
319	35 06	047	38 23	107	19 02	149	39 16	189	51 19	236	31 12	324	37 10	352
320	35 49	046	39 20	108	19 33	149	39 06	190	50 29	236	30 37	324	37 01	351
321	36 32	046	40 17	108	20 04	149	38 56	190	49 39	237	30 02	323	36 52	351
322	37 14	045	41 13	108	20 34	149	38 45	191	48 49	237	29 26	323	36 42	350
323	37 56	044	42 10	108	21 05	149	38 33	192	47 59	238	28 49	322	36 31	349
324	38 37	044	43 07	108	21 35	150	38 21	192	47 08	238	28 12	321	36 19	348
325	39 18	043	44 03	108	22 05	150	38 08	193	46 17	239	27 35	321	36 06	347
326	39 58	042	45 00	109	22 35	150	37 54	194	45 26	239	26 57	320	35 53	346
327	40 38	041	45 56	109	23 04	150	37 40	194	44 35	240	26 19	320	35 38	346
328	41 17	041	46 52	109	23 34	151	37 25	195	43 44	240	25 40	319	35 23	345
329	41 55	040	47 49	109	24 03	151	37 09	195	42 52	240	25 02	319	35 07	344
	◆Alpheratz		Diphda		◆ACHERNAR		Peacock		Nunki		◆ALTAIR		DENEB	
330	42 33	039	48 45	109	24 32	151	36 53	196	42 00	241	54 03	296	34 50	343
331	43 10	038	49 41	110	25 00	151	36 37	197	41 08	241	53 10	295	34 33	342
332	43 47	037	50 37	110	25 29	152	36 19	197	40 16	241	52 16	294	34 14	342
333	44 23	036	51 33	110	25 57	152	36 01	198	39 24	242	51 21	294	33 55	341
334	44 57	035	52 29	111	26 25	152	35 43	198	38 31	242	50 26	293	33 35	340
335	45 32	035	53 25	111	26 52	153	35 24	199	37 39	242	49 32	293	33 15	339
336	46 05	034	54 20	111	27 19	153	35 05	199	36 46	242	48 36	292	32 53	339
337	46 38	033	55 16	112	27 46	153	34 45	200	35 53	242	47 41	291	32 31	338
338	47 09	032	56 11	112	28 13	154	34 24	200	35 01	243	46 46	291	32 09	337
339	47 40	031	57 06	113	28 39	154	34 03	201	34 08	243	45 50	290	31 45	337
340	48 10	029	58 01	113	29 05	154	33 42	201	33 15	243	44 54	290	31 21	336
341	48 38	028	58 55	114	29 31	155	33 20	202	32 21	243	43 58	290	30 57	335
342	49 06	027	59 50	114	29 56	155	32 57	202	31 28	243	43 02	289	30 32	335
343	49 33	026	60 44	115	30 21	156	32 34	203	30 35	244	42 05	289	30 06	334
344	49 58	025	61 38	115	30 45	156	32 11	203	29 42	244	41 09	288	29 39	333
	Alpheratz		◆Hamal		◆ACHERNAR		Peacock		Nunki		◆ALTAIR		DENEB	
345	50 23	024	35 09	055	31 09	156	31 47	204	28 48	244	40 12	288	29 12	333
346	50 46	022	35 58	054	31 33	157	31 23	204	27 55	244	39 15	287	28 45	332
347	51 08	021	36 46	054	31 56	157	30 58	205	27 01	244	38 19	287	28 16	331
348	51 29	020	37 34	053	32 19	158	30 33	205	26 08	244	37 22	287	27 48	331
349	51 49	019	38 21	053	32 42	158	30 08	205	25 14	244	36 25	286	27 18	330
350	52 07	017	39 08	052	33 04	159	29 42	206	24 21	244	35 27	286	26 49	330
351	52 24	016	39 55	051	33 25	159	29 16	206	23 27	244	34 30	286	26 18	329
352	52 40	015	40 41	051	33 46	160	28 49	207	22 33	244	33 33	286	25 48	329
353	52 54	013	41 27	050	34 06	160	28 22	207	21 40	244	32 35	285	25 17	328
354	53 07	012	42 12	049	34 26	161	27 55	207	20 46	244	31 38	285	24 45	328
355	53 18	010	42 57	049	34 46	161	27 28	208	19 52	245	30 40	285	24 13	327
356	53 28	009	43 42	048	35 05	162	27 00	208	18 58	245	29 43	284	23 40	327
357	53 37	007	44 26	047	35 23	162	26 32	208	18 05	245	28 45	284	23 07	326
358	53 44	006	45 09	046	35 41	163	26 04	209	17 11	245	27 47	284	22 34	326
359	53 49	004	45 52	045	35 59	163	25 35	209	16 17	245	26 49	284	22 00	325

 LAT 8°S

LHA ♈	Hc	Zn	Hc	Zn	Hc	Zn	Hc	Zn	Hc	Zn	Hc	Zn	Hc	Zn
	◆Alpheratz		ALDEBARAN		◆RIGEL		ACHERNAR		◆FOMALHAUT		ALTAIR		DENEB	
0	52 53	003	17 35	070	12 22	097	37 13	164	63 50	212	25 37	284	20 37	325
1	52 55	001	18 31	070	13 21	097	37 29	164	63 18	214	24 40	284	20 03	325
2	52 56	000	19 27	069	14 20	096	37 45	165	62 44	215	23 42	283	19 28	324
3	52 56	359	20 22	069	15 19	096	38 00	166	62 10	217	22 44	283	18 53	324
4	52 53	357	21 18	069	16 18	096	38 15	166	61 34	218	21 46	283	18 18	323
5	52 50	356	22 13	068	17 17	096	38 29	167	60 57	219	20 48	283	17 42	323
6	52 44	354	23 08	068	18 16	096	38 42	167	60 19	220	19 50	282	17 06	323
7	52 38	353	24 03	068	19 15	096	38 55	168	59 40	222	18 52	282	16 30	322
8	52 30	351	24 58	068	20 14	096	39 07	169	59 00	223	17 54	282	15 53	322
9	52 20	350	25 53	067	21 13	096	39 18	169	58 19	224	16 56	282	15 16	321
10	52 09	349	26 48	067	22 12	096	39 29	170	57 38	225	15 58	282	14 39	321
11	51 57	347	27 42	067	23 11	096	39 39	171	56 56	226	15 00	281	14 02	321
12	51 43	346	28 37	066	24 11	095	39 48	171	56 13	227	14 01	281	13 24	320
13	51 28	345	29 31	066	25 10	095	39 57	172	55 30	227	13 03	281	12 46	320
14	51 11	343	30 25	065	26 09	095	40 05	173	54 46	228	12 05	281	12 08	320
	Hamal		ALDEBARAN		◆RIGEL		ACHERNAR		◆FOMALHAUT		Enif		◆Alpheratz	
15	54 34	027	31 19	065	27 08	095	40 12	173	54 01	229	38 02	289	50 53	342
16	55 00	026	32 13	065	28 07	095	40 18	174	53 16	230	37 06	289	50 34	341
17	55 25	024	33 07	064	29 06	095	40 24	175	52 30	230	36 09	288	50 14	339
18	55 49	023	34 00	064	30 06	095	40 29	175	51 44	231	35 13	288	49 52	338
19	56 11	021	34 53	063	31 05	095	40 34	176	50 58	232	34 16	288	49 29	337
20	56 32	020	35 46	063	32 04	095	40 37	177	50 11	232	33 20	287	49 06	336
21	56 52	018	36 39	062	33 03	095	40 40	178	49 24	233	32 23	287	48 40	335
22	57 09	017	37 32	062	34 02	095	40 42	178	48 36	233	31 26	287	48 14	333
23	57 26	015	38 24	061	35 02	094	40 44	179	47 49	234	30 29	286	47 47	332
24	57 40	013	39 16	061	36 01	094	40 44	180	47 00	234	29 32	286	47 19	331
25	57 53	012	40 08	060	37 00	094	40 44	180	46 12	235	28 35	286	46 50	330
26	58 04	010	40 59	060	37 59	094	40 43	181	45 23	235	27 38	286	46 20	329
27	58 13	008	41 51	059	38 59	094	40 42	182	44 34	236	26 41	285	45 49	328
28	58 21	007	42 42	059	39 58	094	40 39	183	43 45	236	25 43	285	45 17	327
29	58 27	005	43 32	058	40 57	094	40 36	183	42 55	237	24 46	285	44 44	326
	CAPELLA		◆BETELGEUSE		SIRIUS		CANOPUS		◆ACHERNAR		FOMALHAUT		◆Alpheratz	
30	20 31	034	29 28	077	20 12	105	20 48	144	40 33	184	42 06	237	44 11	325
31	21 04	034	30 26	076	21 09	105	21 23	144	40 28	185	41 16	237	43 36	324
32	21 37	033	31 23	076	22 06	105	21 58	144	40 23	185	40 26	238	43 01	323
33	22 09	033	32 21	076	23 04	105	22 33	144	40 17	186	39 36	238	42 25	322
34	22 41	032	33 19	076	24 01	105	23 08	144	40 10	187	38 45	238	41 49	322
35	23 13	032	34 16	075	24 59	105	23 42	145	40 03	187	37 55	238	41 12	321
36	23 44	031	35 14	075	25 56	105	24 17	145	39 55	188	37 04	239	40 34	320
37	24 14	031	36 11	075	26 54	105	24 51	145	39 46	189	36 13	239	39 55	319
38	24 44	030	37 08	074	27 51	105	25 25	145	39 37	190	35 22	239	39 16	318
39	25 14	030	38 05	074	28 49	105	25 58	146	39 26	190	34 31	239	38 36	318
40	25 43	029	39 02	074	29 46	105	26 32	146	39 16	191	33 40	240	37 56	317
41	26 12	029	39 59	073	30 43	105	27 05	146	39 04	191	32 49	240	37 15	316
42	26 40	028	40 56	073	31 41	105	27 38	146	38 52	192	31 57	240	36 34	316
43	27 07	027	41 53	072	32 38	105	28 11	147	38 39	193	31 06	240	35 52	315
44	27 34	027	42 50	072	33 36	105	28 43	147	38 26	193	30 14	240	35 10	314
	CAPELLA		◆BETELGEUSE		SIRIUS		CANOPUS		◆ACHERNAR		FOMALHAUT		◆Alpheratz	
45	28 01	026	43 46	072	34 33	105	29 15	147	38 12	194	29 23	240	34 27	314
46	28 27	026	44 42	071	35 31	105	29 47	148	37 57	195	28 31	241	33 44	313
47	28 52	025	45 38	071	36 28	105	30 19	148	37 42	195	27 39	241	33 01	313
48	29 17	024	46 34	070	37 26	105	30 50	148	37 26	196	26 48	241	32 17	312
49	29 41	024	47 30	070	38 23	105	31 21	149	37 09	196	25 56	241	31 32	311
50	30 05	023	48 26	069	39 20	105	31 52	149	36 52	197	25 04	241	30 48	311
51	30 27	022	49 21	069	40 18	105	32 22	150	36 34	198	24 12	241	30 03	310
52	30 50	022	50 17	068	41 15	105	32 52	150	36 16	198	23 20	241	29 17	310
53	31 11	021	51 12	068	42 12	105	33 22	150	35 57	199	22 28	241	28 31	309
54	31 32	020	52 06	067	43 10	105	33 51	151	35 38	199	21 36	241	27 45	309
55	31 52	019	53 01	066	44 07	106	34 20	151	35 18	200	20 44	241	26 59	308
56	32 12	019	53 55	066	45 04	106	34 48	152	34 58	200	19 52	241	26 12	308
57	32 31	018	54 49	065	46 01	106	35 16	152	34 37	201	18 59	241	25 25	308
58	32 49	017	55 43	064	46 58	106	35 44	153	34 16	201	18 07	241	24 38	307
59	33 06	017	56 36	063	47 56	106	36 11	153	33 54	202	17 15	241	23 51	307
	◆CAPELLA		POLLUX		◆SIRIUS		CANOPUS		ACHERNAR		◆Diphda		Hamal	
60	33 22	016	24 51	054	48 53	106	36 37	154	33 32	202	41 13	253	48 08	319
61	33 38	015	25 39	054	49 50	106	37 03	154	33 09	203	40 16	253	47 29	318
62	33 53	014	26 26	053	50 47	107	37 29	155	32 46	203	39 19	253	46 49	317
63	34 07	013	27 14	053	51 43	107	37 54	155	32 23	204	38 22	253	46 09	317
64	34 21	013	28 01	052	52 40	107	38 19	156	31 59	204	37 25	253	45 28	316
65	34 33	012	28 48	052	53 37	107	38 43	156	31 34	204	36 28	253	44 46	315
66	34 45	011	29 34	051	54 34	108	39 06	157	31 10	205	35 31	254	44 04	314
67	34 56	010	30 20	051	55 30	108	39 29	158	30 44	205	34 34	254	43 21	313
68	35 06	009	31 06	050	56 27	108	39 51	158	30 19	206	33 37	254	42 37	312
69	35 15	009	31 52	050	57 23	109	40 13	159	29 53	206	32 40	254	41 53	312
70	35 24	008	32 37	049	58 19	109	40 34	160	29 27	206	31 43	254	41 08	311
71	35 31	007	33 22	049	59 15	109	40 54	160	29 00	207	30 46	254	40 23	310
72	35 38	006	34 06	048	60 11	110	41 14	161	28 34	207	29 49	254	39 38	310
73	35 44	005	34 50	047	61 07	110	41 33	162	28 06	207	28 52	254	38 52	309
74	35 49	004	35 34	047	62 03	111	41 52	162	27 39	208	27 55	254	38 05	308
	CAPELLA		POLLUX		◆REGULUS		CANOPUS		◆ACHERNAR		Diphda		◆Hamal	
75	35 53	003	36 17	046	10 52	076	42 09	163	27 11	208	26 58	254	37 19	308
76	35 56	003	36 59	046	11 50	076	42 26	164	26 43	208	26 01	254	36 31	307
77	35 58	002	37 41	045	12 47	076	42 43	164	26 15	209	25 04	254	35 44	307
78	36 00	001	38 23	044	13 45	075	42 58	165	25 46	209	24 07	254	34 56	306
79	36 00	000	39 04	043	14 42	075	43 13	166	25 17	209	23 10	254	34 08	305
80	36 00	359	39 45	043	15 40	075	43 27	167	24 48	209	22 13	254	33 19	305
81	35 59	358	40 25	042	16 37	075	43 40	168	24 19	210	21 16	254	32 30	304
82	35 57	357	41 04	041	17 35	075	43 53	168	23 49	210	20 18	254	31 41	304
83	35 54	357	41 43	040	18 32	074	44 04	169	23 19	210	19 21	254	30 52	303
84	35 50	356	42 21	040	19 29	074	44 15	170	22 50	210	18 24	254	30 02	303
85	35 45	355	42 59	039	20 26	074	44 25	171	22 19	211	17 27	254	29 12	303
86	35 39	354	43 36	038	21 23	074	44 34	172	21 49	211	16 30	253	28 22	302
87	35 33	353	44 12	037	22 20	073	44 43	172	21 18	211	15 34	253	27 32	302
88	35 25	352	44 47	036	23 17	073	44 50	173	20 48	211	14 37	253	26 41	301
89	35 17	352	45 22	035	24 14	073	44 57	174	20 17	211	13 40	253	25 50	301

LHA ♈	Hc	Zn	Hc	Zn	Hc	Zn	Hc	Zn	Hc	Zn	Hc	Zn	Hc	Zn
	POLLUX		◆REGULUS		Suhail		◆CANOPUS		Acamar		◆ALDEBARAN		CAPELLA	
90	45 56	034	25 11	073	35 55	139	45 03	175	38 16	224	57 49	320	35 08	351
91	46 29	033	26 07	072	36 34	139	45 07	176	37 35	224	57 10	318	34 58	350
92	47 01	032	27 04	072	37 13	140	45 11	177	36 53	225	56 30	317	34 47	349
93	47 32	031	28 00	072	37 51	140	45 15	177	36 12	225	55 49	316	34 35	348
94	48 02	030	28 57	071	38 28	141	45 17	178	35 29	225	55 07	315	34 23	347
95	48 31	029	29 53	071	39 06	141	45 18	179	34 47	226	54 25	314	34 10	347
96	48 59	028	30 49	071	39 43	142	45 18	180	34 04	226	53 41	312	33 55	346
97	49 26	027	31 45	071	40 19	142	45 18	181	33 21	226	52 57	311	33 41	345
98	49 53	025	32 41	070	40 55	143	45 17	182	32 38	227	52 12	310	33 25	344
99	50 17	024	33 37	070	41 31	143	45 14	183	31 55	227	51 27	310	33 09	344
100	50 41	023	34 33	069	42 06	144	45 11	183	31 11	227	50 40	309	32 51	343
101	51 04	022	35 28	069	42 41	145	45 07	184	30 28	228	49 54	308	32 33	342
102	51 25	020	36 24	069	43 15	145	45 02	185	29 44	228	49 07	307	32 15	341
103	51 45	019	37 19	068	43 49	146	44 56	186	29 00	228	48 19	306	31 55	341
104	52 04	018	38 14	068	44 22	146	44 50	187	28 15	228	47 31	305	31 35	340
	POLLUX		◆REGULUS		Suhail		◆CANOPUS		RIGEL		◆ALDEBARAN		CAPELLA	
105	52 21	016	39 09	067	44 55	147	44 42	188	63 51	268	46 42	305	31 15	339
106	52 37	015	40 04	067	45 27	148	44 34	189	62 51	268	45 53	304	30 53	338
107	52 52	014	40 58	066	45 58	149	44 25	189	61 52	268	45 03	303	30 31	338
108	53 05	012	41 53	066	46 29	149	44 14	190	60 52	267	44 13	302	30 08	337
109	53 17	011	42 47	065	46 59	150	44 04	191	59 53	267	43 23	302	29 45	336
110	53 27	009	43 41	065	47 28	151	43 52	192	58 54	267	42 32	301	29 21	336
111	53 36	008	44 35	064	47 56	152	43 39	193	57 54	267	41 41	301	28 56	335
112	53 44	006	45 28	064	48 24	153	43 26	193	56 55	267	40 50	300	28 31	335
113	53 49	005	46 21	063	48 51	153	43 12	194	55 56	267	39 58	299	28 05	334
114	53 54	003	47 14	063	49 18	154	42 57	195	54 56	267	39 07	299	27 39	333
115	53 56	002	48 07	062	49 43	155	42 41	196	53 57	267	38 14	298	27 12	333
116	53 58	000	48 59	061	50 07	156	42 25	196	52 58	267	37 22	298	26 44	332
117	53 57	359	49 51	061	50 31	157	42 08	197	51 58	267	36 29	297	26 16	332
118	53 55	357	50 42	060	50 54	158	41 50	198	50 59	267	35 37	297	25 48	331
119	53 52	356	51 33	059	51 16	159	41 32	198	50 00	267	34 43	297	25 19	330
	◆REGULUS		SPICA		◆ACRUX		CANOPUS		RIGEL		◆BETELGEUSE		POLLUX	
120	52 24	058	10 05	100	17 36	154	41 13	199	49 00	267	55 13	296	53 47	354
121	53 14	057	11 03	100	18 02	154	40 53	200	48 01	267	54 20	295	53 40	353
122	54 04	057	12 02	100	18 27	154	40 32	200	47 02	267	53 25	294	53 32	351
123	54 54	056	13 00	100	18 53	155	40 11	201	46 02	266	52 31	293	53 23	350
124	55 42	055	13 59	100	19 18	155	39 49	202	45 03	266	51 36	293	53 11	349
125	56 30	054	14 57	099	19 44	155	39 27	202	44 04	266	50 41	292	52 59	347
126	57 18	053	15 56	099	20 09	155	39 04	203	43 05	266	49 46	292	52 45	346
127	58 05	052	16 55	099	20 34	155	38 41	204	42 05	266	48 51	291	52 30	344
128	58 51	050	17 53	099	20 58	156	38 17	204	41 06	266	47 55	290	52 13	343
129	59 36	049	18 52	099	21 23	156	37 52	205	40 07	266	47 00	290	51 55	342
130	60 21	048	19 51	099	21 47	156	37 27	205	39 07	266	46 04	289	51 35	340
131	61 04	047	20 49	099	22 11	156	37 01	206	38 08	266	45 08	289	51 15	339
132	61 47	045	21 48	099	22 35	156	36 35	206	37 09	266	44 11	289	50 53	338
133	62 29	044	22 47	099	22 59	157	36 08	207	36 10	266	43 15	288	50 30	336
134	63 09	042	23 46	099	23 22	157	35 41	207	35 10	266	42 18	288	50 05	335
	Dubhe		Denebola		◆SPICA		ACRUX		◆CANOPUS		SIRIUS		◆POLLUX	
135	16 13	015	42 26	062	24 44	099	23 45	157	35 14	208	55 56	252	49 40	334
136	16 28	014	43 19	061	25 43	099	24 08	157	34 46	208	55 00	252	49 13	333
137	16 43	014	44 10	061	26 42	098	24 31	158	34 17	209	54 03	252	48 46	332
138	16 56	013	45 02	060	27 41	098	24 53	158	33 49	209	53 07	253	48 17	331
139	17 10	013	45 53	059	28 39	098	25 15	158	33 19	210	52 10	253	47 47	329
140	17 23	012	46 44	059	29 38	098	25 37	159	32 50	210	51 13	253	47 16	328
141	17 36	012	47 35	058	30 37	098	25 59	159	32 20	211	50 16	253	46 45	327
142	17 48	012	48 25	057	31 36	098	26 20	159	31 49	211	49 19	254	46 12	326
143	18 00	011	49 15	056	32 35	098	26 41	160	31 19	211	48 22	254	45 39	325
144	18 11	011	50 04	056	33 33	098	27 01	160	30 48	212	47 25	254	45 05	324
145	18 22	010	50 53	055	34 32	098	27 22	160	30 16	212	46 28	254	44 30	323
146	18 32	010	51 41	054	35 31	098	27 41	161	29 45	212	45 31	254	43 54	323
147	18 42	009	52 29	053	36 30	098	28 01	161	29 13	213	44 33	254	43 18	322
148	18 51	009	53 16	052	37 29	098	28 20	161	28 41	213	43 36	255	42 41	321
149	19 00	008	54 03	051	38 28	098	28 39	162	28 08	213	42 39	255	42 03	320
	Dubhe		ARCTURUS		◆SPICA		ACRUX		◆CANOPUS		SIRIUS		◆POLLUX	
150	19 08	008	21 29	066	39 26	098	28 58	162	27 35	214	41 42	255	41 24	319
151	19 16	007	22 23	065	40 25	098	29 16	162	27 02	214	40 44	255	40 45	318
152	19 24	007	23 17	065	41 24	098	29 34	163	26 29	214	39 47	255	40 05	318
153	19 30	006	24 11	065	42 23	098	29 51	163	25 56	214	38 50	255	39 25	317
154	19 37	006	25 04	064	43 22	098	30 08	164	25 22	215	37 52	255	38 44	316
155	19 43	005	25 58	064	44 21	098	30 24	164	24 48	215	36 55	255	38 03	315
156	19 48	005	26 51	064	45 20	098	30 41	164	24 14	215	35 57	255	37 21	315
157	19 53	004	27 44	063	46 18	098	30 56	165	23 40	215	35 00	255	36 38	314
158	19 57	004	28 37	063	47 17	098	31 12	165	23 05	216	34 02	255	35 55	313
159	20 01	003	29 30	063	48 16	098	31 26	166	22 30	216	33 05	255	35 12	313
160	20 04	003	30 23	062	49 15	098	31 41	166	21 56	216	32 08	255	34 28	312
161	20 07	002	31 15	062	50 14	098	31 55	167	21 21	216	31 10	255	33 44	312
162	20 09	002	32 07	061	51 13	098	32 08	167	20 46	216	30 13	255	32 59	311
163	20 11	001	32 59	061	52 12	098	32 21	168	20 10	216	29 15	255	32 14	311
164	20 12	001	33 51	060	53 11	098	32 34	168	19 35	217	28 18	255	31 29	310
	Dubhe		◆ARCTURUS		SPICA		◆ACRUX		Suhail		SIRIUS		◆POLLUX	
165	20 13	000	34 43	060	54 09	098	32 46	169	46 56	210	27 20	255	30 43	309
166	20 13	000	35 34	059	55 08	098	32 57	169	46 26	211	26 23	255	29 57	309
167	20 13	359	36 25	059	56 07	098	33 08	170	45 55	212	25 25	255	29 11	308
168	20 12	359	37 16	058	57 06	098	33 19	170	45 24	212	24 28	255	28 24	308
169	20 11	358	38 06	058	58 05	098	33 29	171	44 52	213	23 30	255	27 37	308
170	20 09	358	38 56	057	59 03	098	33 38	171	44 19	214	22 33	255	26 50	307
171	20 06	357	39 46	057	60 02	098	33 47	172	43 46	214	21 36	255	26 03	307
172	20 03	357	40 36	056	61 01	099	33 56	172	43 12	215	20 38	255	25 15	306
173	20 00	356	41 25	055	62 00	099	34 03	173	42 38	216	19 41	255	24 27	306
174	19 56	356	42 14	055	62 58	099	34 11	173	42 03	216	18 43	255	23 38	305
175	19 52	355	43 02	054	63 57	099	34 18	174	41 28	217	17 46	255	22 50	305
176	19 47	355	43 50	053	64 56	099	34 24	174	40 52	217	16 49	255	22 01	305
177	19 41	354	44 37	053	65 54	099	34 30	175	40 16	218	15 51	255	21 12	304
178	19 35	354	45 24	052	66 53	100	34 35	175	39 39	218	14 54	255	20 23	304
179	19 28	353	46 11	051	67 52	100	34 39	176	39 02	219	13 57	255	19 34	304

LAT 8°S

LHA ♈	Hc	Zn	Hc	Zn	Hc	Zn	Hc	Zn	Hc	Zn	Hc	Zn	Hc	Zn
	ARCTURUS		◆ANTARES		ACRUX		◆Suhail		Alphard		REGULUS		◆Dubhe	
180	46 57	050	23 52	115	34 43	176	38 25	219	52 15	266	55 45	305	19 21	353
181	47 43	050	24 45	115	34 47	177	37 47	220	51 16	266	54 56	304	19 14	352
182	48 28	049	25 39	116	34 50	177	37 09	220	50 16	266	54 07	304	19 06	352
183	49 12	048	26 33	116	34 52	178	36 30	221	49 17	266	53 17	303	18 57	352
184	49 56	047	27 26	116	34 54	179	35 52	221	48 18	266	52 27	302	18 48	351
185	50 39	046	28 20	116	34 55	179	35 13	221	47 18	266	51 36	301	18 39	351
186	51 21	045	29 13	116	34 56	180	34 33	222	46 19	266	50 45	300	18 29	350
187	52 03	044	30 07	116	34 56	180	33 53	222	45 20	266	49 53	300	18 19	350
188	52 43	043	31 00	116	34 55	181	33 13	222	44 21	266	49 02	299	18 08	349
189	53 23	042	31 53	116	34 54	181	32 33	223	43 21	266	48 09	298	17 56	349
190	54 02	041	32 47	116	34 52	182	31 53	223	42 22	266	47 17	297	17 44	348
191	54 41	039	33 40	117	34 50	182	31 12	223	41 23	266	46 24	297	17 32	348
192	55 18	038	34 33	117	34 47	183	30 31	224	40 24	265	45 31	296	17 19	347
193	55 54	037	35 26	117	34 44	184	29 50	224	39 24	265	44 37	296	17 06	347
194	56 29	036	36 19	117	34 40	184	29 08	224	38 25	265	43 44	295	16 53	347
	ARCTURUS		◆Rasalhague		ANTARES		◆ACRUX		Suhail		◆REGULUS		Dubhe	
195	57 03	034	18 44	074	37 12	117	34 35	185	28 27	225	42 50	295	16 38	346
196	57 36	033	19 41	074	38 05	117	34 30	185	27 45	225	41 56	294	16 24	346
197	58 07	031	20 38	073	38 57	118	34 25	186	27 03	225	41 01	294	16 09	345
198	58 37	030	21 35	073	39 50	118	34 19	186	26 21	225	40 07	293	15 54	345
199	59 06	028	22 32	073	40 42	118	34 12	187	25 39	225	39 12	293	15 38	344
200	59 33	027	23 29	073	41 35	118	34 05	187	24 57	226	38 17	292	15 22	344
201	59 59	025	24 25	072	42 27	119	33 57	188	24 14	226	37 22	292	15 05	344
202	60 23	023	25 22	072	43 19	119	33 48	188	23 31	226	36 27	291	14 48	343
203	60 46	021	26 18	072	44 11	119	33 40	189	22 49	226	35 31	291	14 31	343
204	61 07	020	27 14	071	45 02	120	33 30	189	22 06	226	34 36	291	14 13	342
205	61 26	018	28 11	071	45 54	120	33 20	190	21 23	226	33 40	290	13 55	342
206	61 43	016	29 07	071	46 45	120	33 10	190	20 40	226	32 44	290	13 37	342
207	61 58	014	30 03	070	47 36	121	32 59	191	19 57	227	31 48	290	13 18	341
208	62 11	012	30 59	070	48 27	121	32 47	191	19 14	227	30 52	289	12 59	341
209	62 22	010	31 55	070	49 18	122	32 36	192	18 30	227	29 56	289	12 39	341
	ARCTURUS		◆Rasalhague		ANTARES		◆RIGIL KENT.		ACRUX		Gienah		◆REGULUS	
210	62 32	008	32 50	069	50 08	122	36 41	174	32 23	192	62 51	247	29 00	289
211	62 39	006	33 46	069	50 59	123	36 47	175	32 10	193	61 56	248	28 03	288
212	62 44	004	34 41	069	51 48	123	36 52	175	31 57	193	61 01	248	27 07	288
213	62 47	002	35 37	068	52 38	124	36 56	176	31 43	194	60 06	249	26 10	288
214	62 47	000	36 32	068	53 27	124	37 00	176	31 29	194	59 11	249	25 14	287
215	62 46	358	37 27	067	54 16	125	37 04	177	31 14	195	58 15	249	24 17	287
216	62 43	356	38 21	067	55 04	126	37 06	178	30 59	195	57 20	250	23 20	287
217	62 37	354	39 16	067	55 52	126	37 09	178	30 43	195	56 24	250	22 23	287
218	62 29	352	40 10	066	56 40	127	37 10	179	30 27	196	55 28	251	21 26	286
219	62 19	350	41 04	066	57 27	128	37 11	180	30 11	196	54 32	251	20 29	286
220	62 08	348	41 58	065	58 14	129	37 11	180	29 54	197	53 35	251	19 32	286
221	61 54	346	42 52	065	59 00	130	37 11	181	29 36	197	52 39	252	18 35	286
222	61 38	344	43 46	064	59 45	131	37 10	181	29 19	198	51 43	252	17 38	285
223	61 20	342	44 39	063	60 30	132	37 08	182	29 01	198	50 46	252	16 40	285
224	61 01	340	45 32	063	61 14	133	37 06	183	28 42	198	49 50	252	15 43	285
	Alphecca		◆VEGA		ALTAIR		◆Shaula		RIGIL KENT.		◆SPICA		ARCTURUS	
225	54 17	013	21 23	043	15 42	078	44 43	136	37 03	183	66 22	260	60 40	338
226	54 29	012	22 03	042	16 40	078	45 24	136	36 59	184	65 23	261	60 17	336
227	54 41	010	22 43	042	17 38	078	46 05	137	36 55	184	64 24	261	59 52	335
228	54 51	009	23 22	041	18 36	078	46 45	138	36 50	185	63 26	261	59 26	333
229	54 59	007	24 01	041	19 35	078	47 25	138	36 44	186	62 27	261	58 58	331
230	55 06	006	24 40	040	20 33	077	48 04	139	36 38	186	61 28	261	58 29	330
231	55 11	004	25 18	040	21 30	077	48 43	140	36 32	187	60 30	262	57 58	328
232	55 14	003	25 56	039	22 28	077	49 21	140	36 24	187	59 31	262	57 27	327
233	55 16	001	26 34	039	23 26	077	49 58	141	36 16	188	58 32	262	56 54	325
234	55 16	359	27 11	038	24 24	076	50 35	142	36 08	189	57 33	262	56 19	324
235	55 15	358	27 48	038	25 22	076	51 12	143	35 59	189	56 34	262	55 44	323
236	55 12	356	28 24	037	26 19	076	51 47	144	35 49	190	55 36	262	55 07	322
237	55 07	355	29 00	037	27 17	076	52 22	145	35 39	190	54 37	262	54 30	320
238	55 01	353	29 35	036	28 15	075	52 56	146	35 28	191	53 38	262	53 51	319
239	54 53	352	30 10	036	29 12	075	53 29	146	35 17	191	52 39	262	53 12	318
	VEGA		◆ALTAIR		Shaula		◆RIGIL KENT.		SPICA		◆ARCTURUS		Alphecca	
240	30 44	035	30 09	075	54 01	147	35 05	192	51 40	262	52 32	317	54 43	350
241	31 18	034	31 07	075	54 33	149	34 52	192	50 41	262	51 51	316	54 32	349
242	31 51	034	32 04	074	55 03	150	34 39	193	49 42	262	51 09	315	54 20	347
243	32 24	033	33 01	074	55 33	151	34 26	193	48 44	262	50 27	314	54 06	346
244	32 56	032	33 58	074	56 02	152	34 12	194	47 45	262	49 43	313	53 50	344
245	33 28	032	34 55	073	56 29	153	33 57	194	46 46	262	48 59	312	53 33	343
246	33 59	031	35 52	073	56 56	154	33 42	195	45 47	262	48 15	311	53 15	341
247	34 29	030	36 49	073	57 21	155	33 26	195	44 48	262	47 30	310	52 55	340
248	34 58	030	37 45	072	57 45	157	33 10	196	43 49	262	46 44	309	52 34	339
249	35 27	029	38 42	072	58 08	158	32 54	196	42 50	262	45 58	309	52 12	337
250	35 56	028	39 38	071	58 29	159	32 37	197	41 51	262	45 11	308	51 48	336
251	36 23	027	40 34	071	58 49	161	32 19	197	40 53	262	44 24	307	51 24	335
252	36 50	026	41 31	071	59 08	162	32 01	198	39 54	262	43 36	306	50 58	333
253	37 16	026	42 27	070	59 26	164	31 43	198	38 55	262	42 48	306	50 30	332
254	37 41	025	43 22	070	59 42	165	31 24	199	37 56	262	42 00	305	50 02	331
	VEGA		◆ALTAIR		Peacock		◆RIGIL KENT.		SPICA		◆ARCTURUS		Alphecca	
255	38 06	024	44 18	069	27 07	151	31 05	199	36 57	262	41 11	304	49 33	330
256	38 30	023	45 13	069	27 35	152	30 45	200	35 58	262	40 22	304	49 03	329
257	38 53	022	46 09	068	28 03	152	30 25	200	35 00	262	39 32	303	48 31	328
258	39 15	021	47 04	068	28 31	152	30 04	200	34 01	262	38 42	303	47 59	327
259	39 36	020	47 59	067	28 59	153	29 43	201	33 02	262	37 52	302	47 26	326
260	39 56	020	48 53	067	29 26	153	29 22	201	32 03	262	37 01	301	46 52	325
261	40 15	019	49 48	066	29 53	153	29 00	202	31 04	262	36 11	301	46 17	324
262	40 34	018	50 42	065	30 19	154	28 38	202	30 05	262	35 20	300	45 41	323
263	40 52	017	51 36	065	30 46	154	28 16	202	29 07	262	34 28	300	45 05	322
264	41 08	016	52 29	064	31 11	154	27 53	203	28 08	262	33 37	299	44 27	321
265	41 24	015	53 23	063	31 37	155	27 30	203	27 09	262	32 45	299	43 50	320
266	41 38	014	54 15	063	32 02	155	27 07	203	26 10	261	31 53	299	43 11	319
267	41 52	013	55 08	062	32 27	156	26 43	204	25 12	261	31 00	298	42 32	318
268	42 05	012	56 00	061	32 51	156	26 19	204	24 13	261	30 08	298	41 52	317
269	42 16	011	56 52	060	33 15	157	25 55	204	23 14	261	29 15	297	41 11	317

LHA ♈	Hc	Zn	Hc	Zn	Hc	Zn	Hc	Zn	Hc	Zn	Hc	Zn	Hc	Zn
	◆VEGA		DENEB		Enif		◆FOMALHAUT		Peacock		◆ANTARES		ARCTURUS	
270	42 27	010	25 38	030	31 28	073	17 31	119	33 38	157	61 38	227	28 22	297
271	42 36	009	26 08	030	32 25	073	18 24	119	34 01	158	60 54	228	27 29	297
272	42 45	008	26 37	029	33 22	073	19 16	119	34 23	158	60 10	229	26 36	296
273	42 52	007	27 06	029	34 18	072	20 08	119	34 45	158	59 25	230	25 43	296
274	42 58	006	27 34	028	35 15	072	21 00	119	35 07	159	58 39	231	24 49	296
275	43 04	004	28 02	027	36 11	072	21 52	119	35 28	160	57 53	232	23 56	295
276	43 08	003	28 29	027	37 07	071	22 44	119	35 49	160	57 06	232	23 02	295
277	43 11	002	28 55	026	38 04	071	23 36	119	36 09	161	56 19	233	22 08	295
278	43 12	001	29 21	026	39 00	070	24 28	119	36 28	161	55 31	234	21 14	294
279	43 13	000	29 47	025	39 55	070	25 20	119	36 47	162	54 43	235	20 19	294
280	43 13	359	30 11	024	40 51	070	26 12	119	37 05	162	53 54	235	19 25	294
281	43 11	358	30 35	024	41 47	069	27 04	119	37 23	163	53 05	236	18 31	293
282	43 09	357	30 59	023	42 42	069	27 55	119	37 40	163	52 16	236	17 36	293
283	43 05	356	31 22	022	43 37	068	28 47	120	37 57	164	51 26	237	16 41	293
284	43 00	355	31 44	022	44 32	068	29 39	120	38 13	165	50 36	238	15 47	293
	DENEB		Enif		◆FOMALHAUT		Peacock		◆ANTARES		Rasalhague		◆VEGA	
285	32 05	021	45 27	067	30 30	120	38 29	165	49 46	238	60 28	314	42 55	354
286	32 26	020	46 22	067	31 22	120	38 43	166	48 55	238	59 45	313	42 48	353
287	32 46	019	47 16	066	32 13	120	38 58	167	48 04	239	59 01	311	42 40	352
288	33 05	019	48 10	065	33 05	120	39 11	167	47 13	239	58 16	310	42 31	351
289	33 24	018	49 04	065	33 56	120	39 24	168	46 22	240	57 30	309	42 20	350
290	33 42	017	49 58	064	34 47	121	39 36	168	45 31	240	56 44	308	42 09	349
291	33 59	016	50 51	063	35 38	121	39 48	169	44 39	240	55 56	307	41 57	348
292	34 15	016	51 44	063	36 29	121	39 59	170	43 47	241	55 08	306	41 44	347
293	34 31	015	52 37	062	37 20	121	40 09	170	42 55	241	54 20	305	41 29	346
294	34 45	014	53 29	061	38 11	122	40 18	171	42 03	241	53 31	304	41 14	345
295	34 59	013	54 21	060	39 01	122	40 27	172	41 11	242	52 42	303	40 58	344
296	35 12	012	55 13	060	39 51	122	40 35	173	40 19	242	51 52	302	40 41	343
297	35 25	011	56 04	059	40 42	123	40 42	173	39 26	242	51 01	302	40 23	342
298	35 36	011	56 54	058	41 32	123	40 49	174	38 34	242	50 10	301	40 04	341
299	35 46	010	57 44	057	42 21	123	40 55	175	37 41	243	49 19	300	39 44	340
	◆DENEB		Alpheratz		◆FOMALHAUT		Peacock		◆ANTARES		Rasalhague		VEGA	
300	35 56	009	19 47	055	43 11	124	41 00	175	36 48	243	48 27	299	39 23	339
301	36 05	008	20 35	055	44 00	124	41 04	176	35 55	243	47 35	299	39 01	338
302	36 13	007	21 24	054	44 49	124	41 08	177	35 02	243	46 43	298	38 39	337
303	36 20	006	22 12	054	45 38	125	41 11	178	34 09	243	45 51	297	38 15	336
304	36 26	006	23 00	054	46 27	125	41 13	178	33 16	244	44 58	297	37 51	336
305	36 31	005	23 48	053	47 15	126	41 14	179	32 23	244	44 04	296	37 26	335
306	36 36	004	24 35	053	48 04	126	41 15	180	31 29	244	43 11	296	37 00	334
307	36 39	003	25 22	052	48 51	127	41 15	181	30 36	244	42 17	295	36 34	333
308	36 42	002	26 09	052	49 39	127	41 14	181	29 43	244	41 23	295	36 06	332
309	36 43	001	26 56	052	50 26	128	41 12	182	28 49	244	40 29	294	35 38	332
310	36 44	000	27 42	051	51 12	129	41 10	183	27 56	244	39 35	294	35 10	331
311	36 44	359	28 28	051	51 59	129	41 07	183	27 02	244	38 41	293	34 40	330
312	36 43	359	29 14	050	52 45	130	41 03	184	26 08	244	37 46	293	34 10	329
313	36 41	358	30 00	050	53 30	131	40 58	185	25 15	244	36 51	292	33 40	329
314	36 38	357	30 45	049	54 15	131	40 53	186	24 21	245	35 56	292	33 08	328
	Alpheratz		◆Diphda		ACHERNAR		Peacock		◆Nunki		VEGA		◆DENEB	
315	31 29	049	34 53	106	17 48	148	40 47	186	55 08	234	32 37	327	36 34	356
316	32 14	048	35 50	106	18 19	148	40 40	187	54 20	235	32 04	327	36 30	355
317	32 58	047	36 47	107	18 51	148	40 32	188	53 31	236	31 31	326	36 24	354
318	33 41	047	37 44	107	19 22	148	40 24	188	52 41	236	30 57	325	36 17	353
319	34 25	046	38 41	107	19 53	148	40 15	189	51 52	237	30 23	325	36 10	352
320	35 07	046	39 38	107	20 24	149	40 05	190	51 02	237	29 49	324	36 02	352
321	35 50	045	40 35	107	20 55	149	39 55	190	50 12	238	29 14	323	35 53	351
322	36 31	044	41 31	107	21 26	149	39 44	191	49 21	238	28 38	323	35 43	350
323	37 13	044	42 28	107	21 56	149	39 32	192	48 30	239	28 02	322	35 32	349
324	37 54	043	43 25	107	22 26	149	39 19	192	47 39	239	27 25	322	35 20	348
325	38 34	042	44 22	107	22 57	150	39 06	193	46 48	240	26 48	321	35 08	347
326	39 14	042	45 18	108	23 27	150	38 52	194	45 57	240	26 11	321	34 54	347
327	39 53	041	46 15	108	23 56	150	38 38	194	45 05	240	25 33	320	34 40	346
328	40 31	040	47 11	108	24 26	150	38 23	195	44 13	241	24 55	320	34 25	345
329	41 09	039	48 08	108	24 55	151	38 07	196	43 21	241	24 16	319	34 09	344
	◆Alpheratz		Diphda		◆ACHERNAR		Peacock		Nunki		◆ALTAIR		DENEB	
330	41 46	038	49 04	108	25 24	151	37 51	196	42 29	241	53 37	297	33 53	343
331	42 23	038	50 01	109	25 53	151	37 34	197	41 37	242	52 44	296	33 35	343
332	42 59	037	50 57	109	26 21	151	37 17	197	40 45	242	51 50	295	33 17	342
333	43 34	036	51 53	109	26 50	152	36 59	198	39 52	242	50 56	295	32 58	341
334	44 08	035	52 49	109	27 18	152	36 40	199	38 59	242	50 02	294	32 39	340
335	44 42	034	53 45	110	27 45	152	36 21	199	38 07	243	49 08	294	32 18	340
336	45 15	033	54 41	110	28 13	153	36 01	200	37 14	243	48 13	293	31 57	339
337	45 47	032	55 37	110	28 40	153	35 41	200	36 21	243	47 19	292	31 36	338
338	46 18	031	56 33	111	29 07	153	35 20	201	35 28	243	46 24	292	31 13	338
339	46 48	030	57 28	111	29 33	154	34 59	201	34 35	243	45 28	291	30 50	337
340	47 17	029	58 24	112	29 59	154	34 37	202	33 42	244	44 33	291	30 27	336
341	47 45	028	59 19	112	30 25	154	34 15	202	32 48	244	43 37	290	30 02	335
342	48 13	027	60 14	113	30 50	155	33 52	203	31 55	244	42 42	290	29 37	335
343	48 39	026	61 08	113	31 15	155	33 29	203	31 02	244	41 46	290	29 12	334
344	49 04	024	62 03	114	31 40	156	33 06	204	30 08	244	40 50	289	28 46	334
	Alpheratz		◆Hamal		◆ACHERNAR		Peacock		Nunki		◆ALTAIR		DENEB	
345	49 28	023	34 34	054	32 04	156	32 42	204	29 15	244	39 53	289	28 19	333
346	49 51	022	35 23	054	32 28	157	32 17	204	28 21	244	38 57	288	27 51	332
347	50 12	021	36 10	053	32 52	157	31 53	205	27 27	244	38 01	288	27 24	332
348	50 33	019	36 58	053	33 15	157	31 27	205	26 34	245	37 04	288	26 55	331
349	50 52	018	37 45	052	33 37	158	31 02	206	25 40	245	36 07	287	26 26	331
350	51 10	017	38 31	051	33 59	158	30 36	206	24 46	245	35 10	287	25 57	330
351	51 27	016	39 17	051	34 21	159	30 10	206	23 53	245	34 13	286	25 27	329
352	51 42	014	40 03	050	34 42	159	29 43	207	22 59	245	33 16	286	24 56	329
353	51 56	013	40 48	049	35 03	160	29 16	207	22 05	245	32 19	286	24 26	328
354	52 08	011	41 33	049	35 23	160	28 49	208	21 12	245	31 22	286	23 54	328
355	52 19	010	42 17	048	35 43	161	28 21	208	20 18	245	30 25	285	23 22	327
356	52 29	009	43 01	047	36 02	161	27 53	208	19 24	245	29 27	285	22 50	327
357	52 37	007	43 44	046	36 20	162	27 25	209	18 30	245	28 30	285	22 17	326
358	52 44	006	44 27	046	36 39	163	26 56	209	17 36	245	27 33	284	21 44	326
359	52 49	004	45 09	045	36 56	163	26 27	209	16 43	245	26 35	284	21 11	325

LAT 9°S

LHA ♈	Hc	Zn	Hc	Zn	Hc	Zn	Hc	Zn	Hc	Zn	Hc	Zn	Hc	Zn
	♦Alpheratz		ALDEBARAN		♦RIGEL		ACHERNAR		♦FOMALHAUT		ALTAIR		DENEB	
0	51 53	003	17 14	069	12 28	096	38 11	164	64 41	213	25 23	284	19 48	325
1	51 55	001	18 10	069	13 27	096	38 27	164	64 07	215	24 25	284	19 14	325
2	51 56	000	19 05	069	14 26	096	38 43	165	63 33	216	23 28	284	18 39	324
3	51 56	359	20 00	069	15 25	096	38 58	165	62 57	218	22 30	284	18 05	324
4	51 53	357	20 56	068	16 24	096	39 13	166	62 21	219	21 32	283	17 30	324
5	51 50	356	21 51	068	17 23	096	39 27	167	61 43	220	20 35	283	16 54	323
6	51 45	354	22 46	068	18 22	096	39 41	167	61 04	222	19 37	283	16 18	323
7	51 38	353	23 40	067	19 21	096	39 53	168	60 24	223	18 39	283	15 42	322
8	51 30	352	24 35	067	20 20	095	40 06	168	59 44	224	17 41	282	15 06	322
9	51 21	350	25 30	067	21 19	095	40 17	169	59 02	225	16 43	282	14 29	322
10	51 10	349	26 24	066	22 18	095	40 28	170	58 20	226	15 45	282	13 52	321
11	50 58	348	27 18	066	23 17	095	40 38	171	57 37	227	14 47	282	13 15	321
12	50 45	346	28 12	066	24 16	095	40 47	171	56 54	228	13 49	282	12 38	321
13	50 30	345	29 06	065	25 15	095	40 56	172	56 10	228	12 51	281	12 00	320
14	50 14	344	30 00	065	26 14	095	41 04	173	55 25	229	11 53	281	11 22	320
	Hamal		ALDEBARAN		♦RIGEL		ACHERNAR		♦FOMALHAUT		Enif		♦Alpheratz	
15	53 40	026	30 54	065	27 13	095	41 11	173	54 40	230	37 42	290	49 56	342
16	54 06	025	31 47	064	28 12	095	41 18	174	53 54	231	36 46	290	49 37	341
17	54 31	024	32 40	064	29 11	094	41 24	175	53 08	231	35 50	289	49 18	340
18	54 54	022	33 33	063	30 10	094	41 29	175	52 22	232	34 54	289	48 56	339
19	55 15	021	34 26	063	31 09	094	41 33	176	51 35	233	33 58	288	48 34	337
20	55 36	019	35 19	062	32 09	094	41 37	177	50 47	233	33 02	288	48 11	336
21	55 54	018	36 11	062	33 08	094	41 40	178	50 00	234	32 05	288	47 46	335
22	56 12	016	37 03	061	34 07	094	41 42	178	49 12	234	31 09	287	47 21	334
23	56 27	015	37 55	061	35 06	094	41 44	179	48 23	235	30 12	287	46 54	333
24	56 42	013	38 47	060	36 05	094	41 44	180	47 35	235	29 15	287	46 26	332
25	56 54	011	39 38	060	37 04	094	41 44	180	46 46	236	28 18	286	45 58	331
26	57 05	010	40 29	059	38 03	093	41 43	181	45 57	236	27 22	286	45 28	330
27	57 14	008	41 20	059	39 03	093	41 42	182	45 08	237	26 25	286	44 58	329
28	57 21	006	42 10	058	40 02	093	41 39	183	44 18	237	25 28	286	44 26	328
29	57 27	005	43 00	057	41 01	093	41 36	183	43 28	237	24 30	285	43 54	327
	CAPELLA		♦BETELGEUSE		SIRIUS		CANOPUS		♦ACHERNAR		FOMALHAUT		♦Alpheratz	
30	19 41	034	29 14	076	20 27	105	21 37	143	41 32	184	42 38	238	43 21	326
31	20 14	033	30 11	076	21 24	105	22 12	144	41 28	185	41 48	238	42 47	325
32	20 47	033	31 09	076	22 22	104	22 47	144	41 23	185	40 58	238	42 13	324
33	21 19	032	32 06	075	23 19	104	23 22	144	41 17	186	40 07	239	41 38	323
34	21 50	032	33 03	075	24 16	104	23 57	144	41 10	187	39 17	239	41 02	322
35	22 22	032	34 01	075	25 14	104	24 31	144	41 02	188	38 26	239	40 25	321
36	22 52	031	34 58	074	26 11	104	25 06	145	40 54	188	37 35	239	39 48	321
37	23 23	030	35 55	074	27 09	104	25 40	145	40 45	189	36 44	240	39 10	320
38	23 52	030	36 52	074	28 06	104	26 14	145	40 36	190	35 53	240	38 31	319
39	24 22	029	37 48	073	29 04	104	26 48	145	40 25	190	35 02	240	37 52	318
40	24 51	029	38 45	073	30 01	104	27 21	146	40 14	191	34 10	240	37 12	318
41	25 19	028	39 42	072	30 58	104	27 55	146	40 03	192	33 19	240	36 32	317
42	25 47	028	40 38	072	31 56	104	28 28	146	39 51	192	32 27	240	35 51	316
43	26 14	027	41 34	072	32 53	104	29 01	146	39 38	193	31 36	241	35 10	315
44	26 41	027	42 31	071	33 51	104	29 33	147	39 24	194	30 44	241	34 28	315
	CAPELLA		♦BETELGEUSE		SIRIUS		CANOPUS		♦ACHERNAR		FOMALHAUT		♦Alpheratz	
45	27 07	026	43 27	071	34 48	104	30 06	147	39 10	194	29 52	241	33 46	314
46	27 33	025	44 23	070	35 46	104	30 38	147	38 55	195	29 00	241	33 03	314
47	27 58	025	45 18	070	36 43	104	31 10	148	38 39	195	28 09	241	32 20	313
48	28 22	024	46 14	069	37 41	104	31 41	148	38 23	196	27 17	241	31 36	312
49	28 46	023	47 09	069	38 38	104	32 12	148	38 07	197	26 25	241	30 52	312
50	29 09	023	48 04	068	39 36	104	32 43	149	37 49	197	25 33	241	30 08	311
51	29 32	022	48 59	068	40 33	104	33 14	149	37 32	198	24 41	241	29 23	311
52	29 54	021	49 54	067	41 30	104	33 44	150	37 13	198	23 49	241	28 38	310
53	30 15	021	50 48	066	42 28	104	34 14	150	36 54	199	22 57	242	27 53	310
54	30 36	020	51 42	066	43 25	105	34 43	150	36 35	199	22 04	242	27 07	309
55	30 56	019	52 36	065	44 23	105	35 12	151	36 15	200	21 12	242	26 21	309
56	31 15	019	53 30	064	45 20	105	35 41	151	35 54	200	20 20	242	25 35	308
57	31 33	018	54 23	064	46 17	105	36 09	152	35 33	201	19 28	242	24 49	308
58	31 51	017	55 16	063	47 14	105	36 37	152	35 12	201	18 36	242	24 02	308
59	32 08	016	56 08	062	48 12	105	37 04	153	34 50	202	17 44	242	23 15	307
	♦CAPELLA		POLLUX		♦SIRIUS		CANOPUS		ACHERNAR		♦Diphda		Hamal	
60	32 25	016	24 15	054	49 09	105	37 31	153	34 27	202	41 30	254	47 23	320
61	32 40	015	25 03	053	50 06	105	37 57	154	34 05	203	40 33	254	46 44	319
62	32 55	014	25 50	053	51 03	105	38 23	154	33 41	203	39 36	254	46 05	318
63	33 09	013	26 37	052	52 00	106	38 48	155	33 18	204	38 39	254	45 25	317
64	33 22	012	27 24	052	52 57	106	39 13	156	32 53	204	37 42	254	44 45	316
65	33 35	012	28 10	051	53 54	106	39 38	156	32 29	205	36 45	254	44 03	316
66	33 46	011	28 56	051	54 51	106	40 01	157	32 04	205	35 48	254	43 22	315
67	33 57	010	29 42	050	55 48	107	40 24	157	31 39	205	34 51	254	42 39	314
68	34 07	009	30 28	050	56 45	107	40 47	158	31 13	206	33 54	254	41 56	313
69	34 16	008	31 13	049	57 41	107	41 09	159	30 47	206	32 57	254	41 13	312
70	34 24	008	31 57	049	58 38	107	41 30	159	30 21	207	32 00	254	40 29	312
71	34 32	007	32 42	048	59 35	108	41 51	160	29 54	207	31 03	254	39 44	311
72	34 38	006	33 26	048	60 31	108	42 11	161	29 27	207	30 05	254	38 59	310
73	34 44	005	34 09	047	61 27	109	42 30	161	29 00	208	29 08	254	38 14	310
74	34 49	004	34 52	046	62 23	109	42 49	162	28 32	208	28 11	254	37 28	309
	CAPELLA		POLLUX		♦REGULUS		CANOPUS		♦ACHERNAR		Diphda		♦Hamal	
75	34 53	003	35 35	046	10 38	076	43 07	163	28 04	208	27 14	254	36 42	308
76	34 56	003	36 17	045	11 35	076	43 24	163	27 36	209	26 17	254	35 55	308
77	34 59	002	36 59	044	12 32	075	43 41	164	27 07	209	25 20	254	35 08	307
78	35 00	001	37 40	044	13 30	075	43 56	165	26 39	209	24 23	254	34 20	307
79	35 00	000	38 21	043	14 27	075	44 11	166	26 10	209	23 26	254	33 33	306
80	35 00	359	39 01	042	15 24	075	44 26	167	25 40	210	22 29	254	32 45	305
81	34 59	358	39 40	041	16 21	075	44 39	167	25 11	210	21 32	254	31 56	305
82	34 57	358	40 19	041	17 19	074	44 52	168	24 41	210	20 35	254	31 07	304
83	34 54	357	40 57	040	18 16	074	45 03	169	24 11	210	19 38	254	30 18	304
84	34 50	356	41 35	039	19 12	074	45 14	170	23 41	211	18 41	254	29 29	304
85	34 45	355	42 12	038	20 09	074	45 25	171	23 11	211	17 44	254	28 40	303
86	34 39	354	42 48	037	21 06	073	45 34	171	22 40	211	16 47	254	27 50	303
87	34 33	353	43 24	036	22 03	073	45 42	172	22 10	211	15 50	254	27 00	302
88	34 26	352	43 59	035	23 00	073	45 50	173	21 39	211	14 54	254	26 10	302
89	34 18	352	44 33	035	23 56	073	45 57	174	21 08	212	13 57	254	25 19	301

LHA ♈	Hc	Zn	Hc	Zn	Hc	Zn	Hc	Zn	Hc	Zn	Hc	Zn	Hc	Zn
	POLLUX		♦REGULUS		Suhail		♦CANOPUS		Acamar		♦ALDEBARAN		CAPELLA	
90	45 06	034	24 53	072	36 40	139	46 02	175	38 59	224	57 03	321	34 09	351
91	45 38	033	25 49	072	37 20	139	46 07	176	38 18	225	56 25	319	33 59	350
92	46 10	032	26 45	072	37 58	139	46 11	177	37 36	225	55 46	318	33 48	349
93	46 40	031	27 41	071	38 37	140	46 14	177	36 54	226	55 06	317	33 37	348
94	47 10	029	28 38	071	39 15	140	46 17	178	36 11	226	54 25	316	33 24	348
95	47 38	028	29 34	071	39 52	141	46 18	179	35 29	226	53 43	315	33 11	347
96	48 06	027	30 29	070	40 30	141	46 18	180	34 46	227	53 00	313	32 57	346
97	48 33	026	31 25	070	41 07	142	46 18	181	34 03	227	52 17	312	32 43	345
98	48 58	025	32 21	070	41 43	142	46 17	182	33 19	227	51 33	311	32 27	345
99	49 23	024	33 16	069	42 19	143	46 14	183	32 36	227	50 48	310	32 11	344
100	49 46	022	34 11	069	42 55	143	46 11	184	31 52	228	50 03	310	31 54	343
101	50 08	021	35 07	068	43 30	144	46 07	184	31 08	228	49 17	309	31 36	342
102	50 29	020	36 02	068	44 04	145	46 02	185	30 24	228	48 30	308	31 18	342
103	50 48	019	36 56	068	44 38	145	45 56	186	29 40	228	47 43	307	30 59	341
104	51 07	017	37 51	067	45 12	146	45 49	187	28 55	229	46 55	306	30 39	340
	POLLUX		♦REGULUS		Suhail		♦CANOPUS		RIGEL		♦ALDEBARAN		CAPELLA	
105	51 24	016	38 46	067	45 45	147	45 42	188	63 52	270	46 07	305	30 18	339
106	51 39	015	39 40	066	46 17	147	45 33	189	62 53	270	45 19	305	29 57	339
107	51 54	013	40 34	066	46 49	148	45 24	190	61 53	269	44 30	304	29 35	338
108	52 07	012	41 28	065	47 20	149	45 14	190	60 54	269	43 41	303	29 13	337
109	52 18	010	42 22	065	47 50	150	45 02	191	59 55	269	42 51	303	28 50	337
110	52 28	009	43 15	064	48 20	150	44 51	192	58 56	269	42 01	302	28 26	336
111	52 37	008	44 08	063	48 49	151	44 38	193	57 56	269	41 10	301	28 02	335
112	52 44	006	45 01	063	49 17	152	44 24	194	56 57	269	40 20	301	27 37	335
113	52 50	005	45 54	062	49 45	153	44 10	194	55 58	269	39 29	300	27 11	334
114	52 54	003	46 46	062	50 11	154	43 55	195	54 59	269	38 37	300	26 45	334
115	52 56	002	47 38	061	50 37	155	43 39	196	53 59	268	37 46	299	26 18	333
116	52 58	000	48 29	060	51 02	156	43 23	197	53 00	268	36 54	299	25 51	332
117	52 57	359	49 21	060	51 26	157	43 05	197	52 01	268	36 01	298	25 23	332
118	52 55	357	50 12	059	51 49	158	42 47	198	51 02	268	35 09	298	24 55	331
119	52 52	356	51 02	058	52 11	159	42 29	199	50 02	268	34 16	297	24 26	331
	♦REGULUS		SPICA		♦ACRUX		CANOPUS		RIGEL		♦BETELGEUSE		POLLUX	
120	51 52	057	10 15	100	18 30	154	42 09	199	49 03	268	54 47	297	52 47	355
121	52 42	056	11 13	100	18 56	154	41 49	200	48 04	268	53 54	296	52 41	353
122	53 31	055	12 12	100	19 22	154	41 28	201	47 05	268	53 00	295	52 33	352
123	54 19	054	13 10	099	19 47	154	41 07	201	46 06	267	52 07	295	52 23	350
124	55 07	053	14 09	099	20 13	155	40 45	202	45 06	267	51 13	294	52 13	349
125	55 54	052	15 07	099	20 38	155	40 23	203	44 07	267	50 18	293	52 00	347
126	56 41	051	16 06	099	21 03	155	39 59	203	43 08	267	49 24	293	51 47	346
127	57 27	050	17 04	099	21 28	155	39 36	204	42 09	267	48 29	292	51 32	345
128	58 12	049	18 03	099	21 53	155	39 11	205	41 10	267	47 34	292	51 15	343
129	58 57	048	19 01	099	22 17	156	38 46	205	40 10	267	46 39	291	50 58	342
130	59 40	047	20 00	099	22 42	156	38 21	206	39 11	267	45 43	290	50 39	341
131	60 23	045	20 59	099	23 06	156	37 55	206	38 12	267	44 48	290	50 19	339
132	61 04	044	21 57	098	23 30	156	37 29	207	37 13	266	43 52	289	49 57	338
133	61 45	042	22 56	098	23 54	157	37 02	207	36 14	266	42 56	289	49 34	337
134	62 24	041	23 54	098	24 17	157	36 35	208	35 15	266	42 00	289	49 11	336
	Dubhe		Denebola		♦SPICA		ACRUX		♦CANOPUS		SIRIUS		♦POLLUX	
135	15 15	015	41 58	061	24 53	098	24 40	157	36 07	208	56 14	253	48 46	335
136	15 30	014	42 49	060	25 52	098	25 03	157	35 39	209	55 18	254	48 20	333
137	15 44	014	43 41	060	26 50	098	25 26	158	35 10	209	54 21	254	47 53	332
138	15 58	013	44 32	059	27 49	098	25 49	158	34 41	210	53 24	254	47 24	331
139	16 12	013	45 22	058	28 48	098	26 11	158	34 11	210	52 27	254	46 55	330
140	16 24	012	46 13	058	29 47	098	26 33	158	33 42	210	51 30	254	46 25	329
141	16 37	012	47 02	057	30 45	098	26 54	159	33 11	211	50 33	255	45 54	328
142	16 49	012	47 52	056	31 44	098	27 16	159	32 41	211	49 35	255	45 22	327
143	17 01	011	48 41	055	32 43	098	27 37	159	32 10	212	48 38	255	44 50	326
144	17 12	011	49 30	055	33 41	097	27 58	160	31 39	212	47 41	255	44 16	325
145	17 23	010	50 18	054	34 40	097	28 18	160	31 07	212	46 44	255	43 42	324
146	17 33	010	51 05	053	35 39	097	28 38	160	30 35	213	45 46	255	43 06	323
147	17 42	009	51 52	052	36 38	097	28 58	161	30 03	213	44 49	255	42 30	322
148	17 52	009	52 39	051	37 37	097	29 17	161	29 31	213	43 52	255	41 54	321
149	18 01	008	53 24	050	38 35	097	29 36	161	28 58	214	42 54	256	41 17	321
	Dubhe		ARCTURUS		♦SPICA		ACRUX		♦CANOPUS		SIRIUS		♦POLLUX	
150	18 09	008	21 04	065	39 34	097	29 55	162	28 25	214	41 57	256	40 39	320
151	18 17	007	21 58	065	40 33	097	30 13	162	27 52	214	41 00	256	40 00	319
152	18 24	007	22 51	065	41 32	097	30 31	163	27 19	214	40 02	256	39 21	318
153	18 31	006	23 45	064	42 31	097	30 48	163	26 45	215	39 05	256	38 41	317
154	18 37	006	24 38	064	43 30	097	31 05	163	26 11	215	38 07	256	38 01	317
155	18 43	005	25 31	064	44 28	097	31 22	164	25 37	215	37 10	256	37 20	316
156	18 48	005	26 24	063	45 27	097	31 38	164	25 03	215	36 12	256	36 38	315
157	18 53	004	27 17	063	46 26	097	31 54	165	24 28	216	35 15	256	35 56	315
158	18 57	004	28 10	062	47 25	097	32 10	165	23 54	216	34 17	256	35 14	314
159	19 01	003	29 02	062	48 24	097	32 25	166	23 19	216	33 20	256	34 31	313
160	19 04	003	29 55	062	49 23	097	32 39	166	22 44	216	32 23	256	33 48	313
161	19 07	002	30 47	061	50 21	097	32 53	167	22 09	216	31 25	256	33 04	312
162	19 09	002	31 38	061	51 20	097	33 07	167	21 34	217	30 28	256	32 20	312
163	19 11	001	32 30	060	52 19	097	33 20	167	20 58	217	29 30	256	31 35	311
164	19 12	001	33 21	060	53 18	097	33 32	168	20 23	217	28 33	256	30 50	310
	Dubhe		♦ARCTURUS		SPICA		♦ACRUX		Suhail		SIRIUS		♦POLLUX	
165	19 13	000	34 12	059	54 17	097	33 44	168	47 48	211	27 35	256	30 05	310
166	19 13	000	35 03	059	55 16	097	33 56	169	47 17	211	26 38	256	29 20	309
167	19 13	359	35 54	058	56 15	097	34 07	169	46 46	212	25 40	256	28 34	309
168	19 12	359	36 44	058	57 14	097	34 18	170	46 14	213	24 43	256	27 47	308
169	19 11	358	37 34	057	58 12	097	34 28	170	45 42	213	23 46	256	27 01	308
170	19 09	358	38 24	057	59 11	097	34 37	171	45 09	214	22 48	256	26 14	307
171	19 06	357	39 13	056	60 10	097	34 46	171	44 35	215	21 51	256	25 27	307
172	19 04	357	40 02	055	61 09	097	34 55	172	44 01	215	20 53	255	24 39	307
173	19 00	356	40 50	055	62 08	097	35 03	173	43 27	216	19 56	255	23 52	306
174	18 56	356	41 39	054	63 07	097	35 10	173	42 51	217	18 59	255	23 04	306
175	18 52	355	42 26	053	64 06	097	35 17	174	42 16	217	18 01	255	22 15	305
176	18 47	355	43 14	053	65 04	097	35 24	174	41 40	218	17 04	255	21 27	305
177	18 41	354	44 01	052	66 03	097	35 29	175	41 03	218	16 07	255	20 38	305
178	18 35	354	44 47	051	67 02	097	35 35	175	40 26	219	15 09	255	19 49	304
179	18 29	353	45 33	050	68 01	097	35 39	176	39 49	219	14 12	255	19 00	304

LAT 9°S

LHA ♈	Hc Zn	Hc Zn	Hc Zn	Hc Zn	Hc Zn	Hc Zn	Hc Zn
	ARCTURUS	♦ANTARES	ACRUX	♦Suhail	Alphard	REGULUS	♦Dubhe
180	46 18 050	24 17 115	35 43 176	39 11 220	52 18 267	55 10 307	18 22 353
181	47 03 049	25 11 115	35 47 177	38 33 220	51 19 267	54 22 306	18 14 353
182	47 48 048	26 05 115	35 50 177	37 55 221	50 20 267	53 33 305	18 06 352
183	48 31 047	26 58 115	35 52 178	37 16 221	49 21 267	52 44 304	17 58 352
184	49 14 046	27 52 115	35 54 179	36 37 221	48 21 267	51 55 303	17 49 351
185	49 57 045	28 45 115	35 55 179	35 57 222	47 22 267	51 05 302	17 40 351
186	50 38 044	29 39 115	35 56 180	35 18 222	46 23 267	50 14 301	17 30 350
187	51 19 043	30 33 115	35 56 180	34 38 223	45 24 267	49 23 301	17 19 350
188	51 59 042	31 26 116	35 55 181	33 57 223	44 25 267	48 32 300	17 09 349
189	52 38 041	32 20 116	35 54 181	33 17 223	43 25 267	47 41 299	16 57 349
190	53 17 040	33 13 116	35 52 182	32 36 224	42 26 266	46 49 298	16 46 348
191	53 54 039	34 06 116	35 50 182	31 55 224	41 27 266	45 56 298	16 33 348
192	54 30 037	35 00 116	35 47 183	31 14 224	40 28 266	45 04 297	16 21 347
193	55 06 036	35 53 116	35 44 184	30 33 224	39 29 266	44 11 297	16 08 347
194	55 40 035	36 46 116	35 40 184	29 51 225	38 30 266	43 18 296	15 54 347
	ARCTURUS	♦Rasalhague	ANTARES	♦ACRUX	Suhail	♦REGULUS	Dubhe
195	56 13 033	18 27 073	37 39 117	35 35 185	29 09 225	42 24 295	15 40 346
196	56 45 032	19 24 073	38 32 117	35 30 185	28 28 225	41 31 295	15 26 346
197	57 16 030	20 21 073	39 25 117	35 24 186	27 45 225	40 37 294	15 11 345
198	57 45 029	21 17 073	40 18 117	35 18 186	27 03 226	39 43 294	14 56 345
199	58 13 027	22 14 072	41 10 117	35 11 187	26 21 226	38 49 293	14 40 344
200	58 40 026	23 10 072	42 03 118	35 04 187	25 38 226	37 54 293	14 24 344
201	59 05 024	24 07 072	42 55 118	34 56 188	24 56 226	36 59 292	14 08 344
202	59 28 022	25 03 072	43 48 118	34 48 188	24 13 226	36 05 292	13 51 343
203	59 50 021	25 59 071	44 40 118	34 39 189	23 30 226	35 10 292	13 34 343
204	60 10 019	26 55 071	45 32 119	34 29 189	22 47 227	34 14 291	13 16 343
205	60 28 017	27 51 071	46 24 119	34 19 190	22 04 227	33 19 291	12 58 342
206	60 45 015	28 47 070	47 15 119	34 09 190	21 21 227	32 24 290	12 40 342
207	61 00 013	29 43 070	48 07 120	33 58 191	20 38 227	31 28 290	12 21 341
208	61 12 012	30 38 070	48 58 120	33 46 191	19 55 227	30 32 290	12 02 341
209	61 23 010	31 34 069	49 49 121	33 33 192	19 11 227	29 36 289	11 43 341
	ARCTURUS	♦Rasalhague	ANTARES	♦RIGIL KENT.	ACRUX	Gienah	♦REGULUS
210	61 32 008	32 29 069	50 40 121	37 40 174	33 22 192	63 14 249	28 41 289
211	61 39 006	33 24 068	51 30 122	37 46 175	33 09 193	62 19 249	27 44 289
212	61 44 004	34 19 068	52 21 122	37 52 175	32 55 193	61 23 250	26 48 288
213	61 47 002	35 14 068	53 11 123	37 56 176	32 41 194	60 27 250	25 52 288
214	61 47 000	36 09 067	54 01 123	38 00 176	32 27 194	59 32 251	24 56 288
215	61 46 358	37 03 067	54 50 124	38 04 177	32 12 195	58 36 251	23 59 288
216	61 43 356	37 57 066	55 39 125	38 06 178	31 57 195	57 39 251	23 03 287
217	61 37 354	38 52 066	56 28 125	38 09 178	31 41 196	56 43 252	22 06 287
218	61 30 352	39 46 065	57 16 126	38 10 179	31 25 196	55 47 252	21 09 287
219	61 20 350	40 39 065	58 04 127	38 11 179	31 08 196	54 51 252	20 12 286
220	61 09 348	41 33 064	58 51 128	38 11 180	30 51 197	53 54 253	19 16 286
221	60 56 346	42 26 064	59 38 128	38 11 181	30 34 197	52 57 253	18 19 286
222	60 40 344	43 19 063	60 24 129	38 10 181	30 16 198	52 01 253	17 22 286
223	60 23 342	44 12 063	61 09 130	38 08 182	29 58 198	51 04 253	16 24 285
224	60 04 341	45 04 062	61 54 131	38 05 183	29 39 198	50 07 253	15 27 285
	Alphecca	♦VEGA	ALTAIR	♦Shaula	RIGIL KENT.	♦SPICA	ARCTURUS
225	53 18 013	20 38 042	15 30 078	45 26 135	38 02 183	66 30 263	59 44 339
226	53 31 011	21 18 042	16 28 078	46 08 136	37 59 184	65 32 263	59 21 337
227	53 42 010	21 58 042	17 26 078	46 49 136	37 55 184	64 33 263	58 58 335
228	53 51 009	22 37 041	18 24 077	47 29 137	37 50 185	63 34 263	58 32 334
229	53 59 007	23 16 041	19 21 077	48 10 138	37 44 186	62 35 263	58 05 332
230	54 06 006	23 54 040	20 19 077	48 49 138	37 38 186	61 36 263	57 37 331
231	54 11 004	24 32 040	21 17 077	49 28 139	37 31 187	60 38 263	57 07 329
232	54 14 002	25 10 039	22 15 077	50 07 140	37 24 187	59 39 263	56 36 328
233	54 16 001	25 47 039	23 12 076	50 45 140	37 16 188	58 40 263	56 04 326
234	54 16 359	26 24 038	24 10 076	51 22 141	37 07 189	57 41 263	55 30 325
235	54 15 358	27 00 038	25 07 076	51 59 142	36 58 189	56 42 263	54 56 324
236	54 12 356	27 36 037	26 05 075	52 35 143	36 48 190	55 43 263	54 20 322
237	54 07 355	28 12 036	27 02 075	53 11 144	36 38 190	54 44 263	53 43 321
238	54 01 353	28 47 036	27 59 075	53 45 145	36 27 191	53 45 263	53 06 320
239	53 53 352	29 21 035	28 56 075	54 19 146	36 16 191	52 47 263	52 27 319
	VEGA	♦ALTAIR	Shaula	♦RIGIL KENT.	SPICA	♦ARCTURUS	Alphecca
240	29 55 035	29 53 074	54 52 147	36 04 192	51 48 263	51 48 318	53 44 350
241	30 29 034	30 50 074	55 24 148	35 51 193	50 49 263	51 08 317	53 33 349
242	31 01 033	31 47 074	55 55 149	35 38 193	49 50 263	50 26 316	53 21 347
243	31 34 033	32 44 073	56 25 150	35 24 194	48 51 263	49 45 315	53 08 346
244	32 05 032	33 41 073	56 54 151	35 10 194	47 52 263	49 02 314	52 53 345
245	32 37 031	34 37 073	57 22 152	34 55 195	46 53 263	48 19 313	52 36 343
246	33 07 031	35 34 072	57 49 154	34 40 195	45 55 263	47 35 312	52 18 342
247	33 37 030	36 30 072	58 15 155	34 24 196	44 56 263	46 51 311	51 59 340
248	34 06 029	37 27 071	58 40 156	34 08 196	43 57 263	46 06 310	51 38 339
249	34 35 028	38 23 071	59 03 157	33 51 197	42 58 263	45 20 309	51 17 338
250	35 03 028	39 19 071	59 25 159	33 34 197	41 59 263	44 34 309	50 54 336
251	35 30 027	40 15 070	59 46 160	33 16 198	41 00 263	43 47 308	50 29 335
252	35 56 026	41 10 070	60 05 162	32 58 198	40 02 263	43 00 307	50 04 334
253	36 22 025	42 06 069	60 23 163	32 40 198	39 03 263	42 13 306	49 37 333
254	36 47 024	43 01 069	60 40 165	32 21 199	38 04 263	41 25 306	49 10 332
	VEGA	♦ALTAIR	Peacock	♦RIGIL KENT.	SPICA	♦ARCTURUS	Alphecca
255	37 11 024	43 56 068	27 59 151	32 01 199	37 05 263	40 37 305	48 41 330
256	37 34 023	44 51 068	28 28 151	31 41 200	36 06 263	39 48 304	48 11 329
257	37 57 022	45 46 067	28 56 152	31 21 200	35 08 263	38 59 304	47 40 328
258	38 19 021	46 41 067	29 24 152	31 00 201	34 09 263	38 10 303	47 09 327
259	38 39 020	47 35 066	29 52 152	30 39 201	33 10 263	37 20 303	46 36 326
260	38 59 019	48 29 066	30 19 153	30 18 201	32 11 262	36 30 302	46 03 325
261	39 19 018	49 23 065	30 46 153	29 56 202	31 13 262	35 39 302	45 28 324
262	39 37 017	50 16 064	31 13 153	29 34 202	30 14 262	34 49 301	44 53 323
263	39 54 016	51 10 064	31 39 154	29 12 202	29 15 262	33 58 301	44 17 322
264	40 10 016	52 02 063	32 05 154	28 49 203	28 16 262	33 07 300	43 41 321
265	40 26 015	52 55 062	32 31 155	28 26 203	27 18 262	32 15 300	43 03 320
266	40 40 014	53 47 061	32 56 155	28 02 203	26 19 262	31 24 299	42 25 320
267	40 53 013	54 39 060	33 21 155	27 38 204	25 20 262	30 32 299	41 47 319
268	41 06 012	55 30 060	33 46 156	27 14 204	24 22 262	29 40 298	41 07 318
269	41 17 011	56 21 059	34 10 156	26 50 204	23 23 262	28 47 298	40 27 317

LHA ♈	Hc Zn	Hc Zn	Hc Zn	Hc Zn	Hc Zn	Hc Zn	Hc Zn
	♦VEGA	DENEB	Enif	♦FOMALHAUT	Peacock	♦ANTARES	ARCTURUS
270	41 28 010	24 46 030	31 10 073	18 00 118	34 33 157	62 19 228	27 55 297
271	41 37 009	25 16 030	32 07 072	18 52 118	34 56 157	61 34 229	27 02 297
272	41 45 008	25 45 029	33 03 072	19 44 118	35 19 158	60 49 230	26 09 297
273	41 52 006	26 13 028	34 00 072	20 36 118	35 41 158	60 03 231	25 16 296
274	41 59 005	26 41 028	34 56 071	21 29 118	36 03 159	59 17 232	24 23 296
275	42 04 004	27 08 027	35 52 071	22 21 118	36 24 159	58 30 233	23 30 296
276	42 08 003	27 35 027	36 48 070	23 13 118	36 45 160	57 42 234	22 36 295
277	42 11 002	28 01 026	37 43 070	24 05 119	37 05 160	56 54 234	21 43 295
278	42 12 001	28 27 025	38 39 070	24 57 119	37 25 161	56 06 235	20 49 295
279	42 13 000	28 52 025	39 34 069	25 49 119	37 44 161	55 17 236	19 55 294
280	42 13 359	29 17 024	40 30 069	26 41 119	38 03 162	54 28 236	19 01 294
281	42 11 358	29 40 023	41 25 068	27 33 119	38 21 163	53 38 237	18 07 294
282	42 09 357	30 04 023	42 20 068	28 25 119	38 38 163	52 48 238	17 12 293
283	42 05 356	30 26 022	43 15 067	29 17 119	38 55 164	51 58 238	16 18 293
284	42 01 355	30 48 021	44 09 067	30 08 119	39 11 164	51 08 239	15 23 293
	DENEB	Enif	♦FOMALHAUT	Peacock	♦ANTARES	Rasalhague	♦VEGA
285	31 09 021	45 03 066	31 00 119	39 27 165	50 17 239	59 46 315	41 55 354
286	31 30 020	45 58 066	31 52 119	39 42 166	49 26 239	59 04 314	41 48 353
287	31 49 019	46 51 065	32 43 120	39 56 166	48 35 240	58 21 313	41 40 352
288	32 08 018	47 45 064	33 35 120	40 10 167	47 44 240	57 37 311	41 31 351
289	32 27 018	48 38 064	34 26 120	40 23 168	46 52 241	56 52 310	41 21 350
290	32 44 017	49 31 063	35 18 120	40 35 168	46 00 241	56 06 309	41 10 349
291	33 01 016	50 24 062	36 09 120	40 47 169	45 08 241	55 20 308	40 58 348
292	33 17 015	51 16 062	37 00 121	40 58 170	44 16 242	54 33 307	40 45 347
293	33 33 015	52 08 061	37 51 121	41 08 170	43 24 242	53 45 306	40 31 346
294	33 47 014	53 00 060	38 42 121	41 17 171	42 32 242	52 57 305	40 16 345
295	34 01 013	53 51 059	39 32 121	41 26 172	41 39 242	52 08 304	40 00 344
296	34 14 012	54 42 058	40 23 122	41 34 172	40 47 243	51 19 303	39 43 343
297	34 26 011	55 32 058	41 13 122	41 42 173	39 54 243	50 29 303	39 26 342
298	34 37 011	56 22 057	42 04 122	41 49 174	39 01 243	49 39 302	39 07 341
299	34 47 010	57 11 056	42 54 122	41 55 175	38 08 243	48 49 301	38 47 340
	♦DENEB	Alpheratz	♦FOMALHAUT	Peacock	♦ANTARES	Rasalhague	VEGA
300	34 57 009	19 12 055	43 44 123	42 00 175	37 15 244	47 58 300	38 27 339
301	35 06 008	20 00 054	44 34 123	42 04 176	36 22 244	47 06 300	38 05 338
302	35 13 007	20 49 054	45 23 124	42 08 177	35 29 244	46 15 299	37 43 338
303	35 20 006	21 36 054	46 12 124	42 11 178	34 36 244	45 23 298	37 20 337
304	35 26 005	22 24 053	47 01 124	42 13 178	33 42 244	44 30 298	36 56 336
305	35 32 005	23 11 053	47 50 125	42 14 179	32 49 244	43 38 297	36 32 335
306	35 36 004	23 59 053	48 39 125	42 15 180	31 55 244	42 45 297	36 06 334
307	35 39 003	24 46 052	49 27 126	42 15 181	31 02 245	41 52 296	35 40 333
308	35 42 002	25 32 052	50 15 126	42 14 181	30 09 245	40 58 295	35 13 333
309	35 43 001	26 18 051	51 02 127	42 12 182	29 15 245	40 04 295	34 46 332
310	35 44 000	27 04 051	51 49 128	42 10 183	28 21 245	39 11 294	34 17 331
311	35 44 359	27 50 050	52 36 128	42 07 183	27 28 245	38 17 294	33 48 330
312	35 43 359	28 36 050	53 23 129	42 03 184	26 34 245	37 22 293	33 19 330
313	35 41 358	29 21 049	54 09 129	41 58 185	25 40 245	36 28 293	32 48 329
314	35 38 357	30 05 049	54 54 130	41 52 186	24 47 245	35 33 293	32 17 328
	Alpheratz	♦Diphda	ACHERNAR	Peacock	♦Nunki	VEGA	♦DENEB
315	30 50 048	35 10 106	18 39 148	41 46 186	55 43 236	31 46 328	35 34 356
316	31 34 048	36 07 106	19 10 148	41 39 187	54 53 236	31 14 327	35 30 355
317	32 17 047	37 04 106	19 42 148	41 32 188	54 04 237	30 41 326	35 24 354
318	33 00 046	38 01 106	20 13 148	41 23 189	53 14 237	30 08 326	35 18 353
319	33 43 046	38 58 106	20 44 148	41 14 189	52 24 238	29 34 325	35 11 353
320	34 25 045	39 55 106	21 15 148	41 04 190	51 34 239	29 00 324	35 02 352
321	35 07 045	40 52 106	21 46 149	40 54 191	50 43 239	28 25 324	34 53 351
322	35 48 044	41 48 106	22 17 149	40 42 191	49 52 239	27 50 323	34 44 350
323	36 29 043	42 45 106	22 48 149	40 30 192	49 01 240	27 14 323	34 33 349
324	37 10 043	43 42 106	23 18 149	40 18 193	48 10 240	26 38 322	34 21 348
325	37 49 042	44 39 106	23 48 149	40 05 193	47 18 241	26 01 322	34 09 348
326	38 28 041	45 36 107	24 18 150	39 51 194	46 26 241	25 24 321	33 56 347
327	39 07 040	46 33 107	24 48 150	39 36 195	45 34 241	24 47 320	33 42 346
328	39 45 040	47 29 107	25 18 150	39 21 195	44 42 242	24 09 320	33 27 345
329	40 23 039	48 26 107	25 47 150	39 05 196	43 50 242	23 31 320	33 11 344
	♦Alpheratz	Diphda	♦ACHERNAR	Peacock	Nunki	♦ALTAIR	DENEB
330	40 59 038	49 23 107	26 16 151	38 49 196	42 58 242	53 09 298	32 55 344
331	41 35 037	50 19 107	26 45 151	38 32 197	42 05 243	52 17 297	32 38 343
332	42 11 036	51 16 108	27 14 151	38 14 198	41 12 243	51 24 297	32 20 342
333	42 45 035	52 12 108	27 42 151	37 56 198	40 20 243	50 31 296	32 02 341
334	43 19 034	53 09 108	28 11 152	37 37 199	39 27 243	49 37 295	31 42 341
335	43 52 033	54 05 108	28 38 152	37 18 199	38 34 243	48 44 295	31 22 340
336	44 24 032	55 01 109	29 06 152	36 58 200	37 41 244	47 50 294	31 01 339
337	44 56 031	55 57 109	29 33 153	36 37 200	36 48 244	46 55 293	30 40 338
338	45 26 030	56 53 109	30 00 153	36 16 201	35 54 244	46 01 293	30 18 338
339	45 56 029	57 49 110	30 27 153	35 55 201	35 01 244	45 06 292	29 55 337
340	46 25 028	58 45 110	30 53 154	35 33 202	34 08 244	44 11 292	29 32 336
341	46 52 027	59 40 110	31 19 154	35 11 202	33 14 244	43 16 291	29 08 336
342	47 19 026	60 36 111	31 45 155	34 48 203	32 21 244	42 21 291	28 43 335
343	47 45 025	61 31 111	32 10 155	34 24 203	31 28 245	41 25 290	28 18 334
344	48 09 024	62 26 112	32 35 155	34 01 203	30 34 245	40 30 290	27 52 334
	Alpheratz	♦Hamal	♦ACHERNAR	Peacock	Nunki	♦ALTAIR	DENEB
345	48 33 023	33 59 054	32 59 156	33 37 204	29 40 245	39 34 289	27 25 333
346	48 55 022	34 47 053	33 23 156	33 12 205	28 47 245	38 38 289	26 58 333
347	49 16 020	35 34 053	33 47 157	32 47 205	27 53 245	37 42 289	26 31 332
348	49 36 019	36 21 052	34 10 157	32 22 206	26 59 245	36 46 288	26 03 331
349	49 55 018	37 07 051	34 33 158	31 56 206	26 06 245	35 49 288	25 34 331
350	50 12 017	37 53 051	34 55 158	31 30 206	25 12 245	34 53 287	25 05 330
351	50 29 015	38 39 050	35 17 159	31 03 207	24 18 245	33 56 287	24 35 330
352	50 44 014	39 24 049	35 38 159	30 36 207	23 24 245	32 59 287	24 05 329
353	50 57 013	40 09 049	35 59 160	30 09 207	22 31 245	32 03 286	23 34 329
354	51 09 011	40 53 048	36 20 160	29 42 208	21 37 245	31 06 286	23 03 328
355	51 20 010	41 37 047	36 39 161	29 14 208	20 43 245	30 09 286	22 32 328
356	51 30 008	42 20 046	36 59 161	28 46 208	19 49 245	29 12 285	22 00 327
357	51 38 007	43 03 046	37 18 162	28 17 209	18 56 245	28 15 285	21 27 327
358	51 44 006	43 45 045	37 36 162	27 49 209	18 02 245	27 17 285	20 55 326
359	51 49 004	44 26 044	37 53 163	27 20 209	17 08 245	26 20 285	20 21 326

LAT 10°S

LHA ♈	Hc	Zn	Hc	Zn	Hc	Zn	Hc	Zn	Hc	Zn	Hc	Zn	Hc	Zn
	◆Alpheratz		Hamal		ALDEBARAN		◆RIGEL		ACHERNAR		◆FOMALHAUT		Enif	
0	50 53	003	44 23	042	16 53	069	12 35	096	39 08	163	65 30	214	50 46	299
1	50 55	001	45 03	042	17 48	069	13 34	096	39 25	164	64 56	216	49 54	299
2	50 56	000	45 42	041	18 43	069	14 33	096	39 41	164	64 21	218	49 02	298
3	50 56	359	46 20	040	19 38	068	15 31	096	39 56	165	63 44	219	48 10	297
4	50 53	357	46 57	039	20 33	068	16 30	096	40 11	166	63 07	220	47 17	297
5	50 50	356	47 34	038	21 28	068	17 29	096	40 26	166	62 28	222	46 24	296
6	50 45	355	48 09	037	22 23	067	18 28	095	40 39	167	61 48	223	45 31	295
7	50 39	353	48 44	036	23 17	067	19 27	095	40 52	168	61 08	224	44 37	295
8	50 31	352	49 18	034	24 11	067	20 25	095	41 04	168	60 26	225	43 43	294
9	50 22	350	49 51	033	25 06	066	21 24	095	41 16	169	59 44	226	42 49	294
10	50 11	349	50 23	032	26 00	066	22 23	095	41 27	170	59 01	227	41 55	293
11	49 59	348	50 54	031	26 54	066	23 22	095	41 37	170	58 18	228	41 01	293
12	49 46	346	51 24	030	27 47	065	24 21	095	41 47	171	57 34	229	40 06	292
13	49 32	345	51 52	028	28 41	065	25 20	094	41 56	172	56 49	230	39 11	292
14	49 16	344	52 20	027	29 34	064	26 19	094	42 04	172	56 04	230	38 16	291
	Hamal		ALDEBARAN		◆RIGEL		ACHERNAR		◆FOMALHAUT		Enif		◆Alpheratz	
15	52 46	026	30 28	064	27 18	094	42 11	173	55 18	231	37 21	291	48 59	343
16	53 12	025	31 21	064	28 17	094	42 18	174	54 32	232	36 26	290	48 41	341
17	53 36	023	32 13	063	29 16	094	42 24	175	53 45	232	35 30	290	48 21	340
18	53 58	022	33 06	063	30 15	094	42 29	175	52 58	233	34 34	289	48 00	339
19	54 19	020	33 58	062	31 14	094	42 33	176	52 11	234	33 39	289	47 39	338
20	54 39	019	34 50	062	32 13	093	42 37	177	51 23	234	32 43	289	47 16	337
21	54 57	017	35 42	061	33 11	093	42 40	178	50 35	235	31 47	288	46 52	335
22	55 14	016	36 34	061	34 10	093	42 42	178	49 46	235	30 50	288	46 27	334
23	55 29	014	37 25	060	35 09	093	42 44	179	48 58	236	29 54	288	46 01	333
24	55 43	013	38 17	060	36 08	093	42 44	180	48 09	236	28 58	287	45 33	332
25	55 55	011	39 07	059	37 08	093	42 44	180	47 19	237	28 01	287	45 05	331
26	56 06	009	39 58	058	38 07	093	42 43	181	46 30	237	27 05	287	44 36	330
27	56 15	008	40 48	058	39 06	092	42 42	182	45 40	237	26 08	286	44 06	329
28	56 22	006	41 38	057	40 05	092	42 39	183	44 50	238	25 11	286	43 36	328
29	56 27	005	42 27	056	41 04	092	42 36	183	44 00	238	24 14	286	43 04	327
	CAPELLA		◆BETELGEUSE		SIRIUS		CANOPUS		◆ACHERNAR		FOMALHAUT		◆Alpheratz	
30	18 52	034	28 59	076	20 42	104	22 25	143	42 32	184	43 10	238	42 32	326
31	19 24	033	29 56	075	21 39	104	23 00	143	42 28	185	42 20	239	41 58	325
32	19 56	033	30 54	075	22 36	104	23 35	144	42 22	186	41 29	239	41 24	324
33	20 28	032	31 51	075	23 34	104	24 10	144	42 16	186	40 38	239	40 50	324
34	20 59	032	32 48	074	24 31	104	24 45	144	42 09	187	39 47	240	40 14	323
35	21 30	031	33 44	074	25 28	104	25 20	144	42 02	188	38 56	240	39 38	322
36	22 01	031	34 41	074	26 26	104	25 54	144	41 54	188	38 05	240	39 01	321
37	22 31	030	35 38	073	27 23	104	26 29	145	41 45	189	37 14	240	38 24	320
38	23 00	030	36 34	073	28 21	104	27 03	145	41 35	190	36 23	240	37 46	320
39	23 29	029	37 31	073	29 18	104	27 37	145	41 24	190	35 31	241	37 07	319
40	23 58	029	38 27	072	30 15	104	28 11	145	41 13	191	34 40	241	36 28	318
41	24 26	028	39 23	072	31 13	104	28 44	146	41 02	192	33 48	241	35 48	317
42	24 54	028	40 19	071	32 10	104	29 18	146	40 49	192	32 57	241	35 08	317
43	25 21	027	41 15	071	33 08	104	29 51	146	40 36	193	32 05	241	34 27	316
44	25 47	026	42 11	070	34 05	103	30 24	146	40 22	194	31 13	241	33 45	315
	CAPELLA		◆BETELGEUSE		SIRIUS		CANOPUS		◆ACHERNAR		FOMALHAUT		◆Alpheratz	
45	26 13	026	43 06	070	35 03	103	30 56	147	40 08	194	30 21	241	33 04	315
46	26 38	025	44 02	069	36 00	103	31 28	147	39 53	195	29 29	241	32 21	314
47	27 03	024	44 57	069	36 58	103	32 00	147	39 37	196	28 37	242	31 39	313
48	27 27	024	45 52	068	37 55	103	32 32	148	39 21	196	27 45	242	30 56	313
49	27 51	023	46 47	068	38 52	103	33 04	148	39 04	197	26 53	242	30 12	312
50	28 14	023	47 41	067	39 50	103	33 35	148	38 47	197	26 01	242	29 28	312
51	28 36	022	48 36	067	40 47	104	34 05	149	38 29	198	25 09	242	28 44	311
52	28 58	021	49 30	066	41 45	104	34 36	149	38 10	199	24 17	242	27 59	311
53	29 19	021	50 24	065	42 42	104	35 06	150	37 51	199	23 25	242	27 15	310
54	29 39	020	51 17	065	43 40	104	35 35	150	37 31	200	22 33	242	26 29	310
55	29 59	019	52 10	064	44 37	104	36 05	151	37 11	200	21 41	242	25 44	309
56	30 18	018	53 03	063	45 35	104	36 33	151	36 50	201	20 49	242	24 58	309
57	30 36	018	53 56	062	46 32	104	37 02	151	36 29	201	19 56	242	24 11	308
58	30 54	017	54 48	062	47 29	104	37 30	152	36 08	202	19 04	242	23 25	308
59	31 11	016	55 40	061	48 27	104	37 57	152	35 45	202	18 12	242	22 38	307
	◆CAPELLA		POLLUX		◆SIRIUS		CANOPUS		ACHERNAR		◆Diphda		Hamal	
60	31 27	015	23 39	053	49 24	104	38 24	153	35 23	203	41 46	255	46 36	321
61	31 42	015	24 27	053	50 21	104	38 51	154	35 00	203	40 49	255	45 59	320
62	31 57	014	25 14	052	51 19	104	39 17	154	34 36	204	39 52	255	45 20	319
63	32 11	013	26 00	052	52 16	104	39 43	155	34 12	204	38 55	255	44 41	318
64	32 24	012	26 47	051	53 13	105	40 08	155	33 48	205	37 58	255	44 01	317
65	32 36	012	27 33	051	54 10	105	40 32	156	33 23	205	37 01	255	43 20	316
66	32 47	011	28 18	050	55 07	105	40 56	156	32 58	205	36 04	255	42 39	315
67	32 58	010	29 04	050	56 04	105	41 20	157	32 33	206	35 07	255	41 57	315
68	33 08	009	29 49	049	57 01	105	41 42	158	32 07	206	34 10	255	41 15	314
69	33 17	008	30 33	049	57 58	106	42 05	158	31 41	207	33 13	255	40 32	313
70	33 25	008	31 18	048	58 55	106	42 26	159	31 14	207	32 15	255	39 49	312
71	33 32	007	32 02	048	59 52	106	42 47	160	30 47	207	31 18	255	39 05	312
72	33 39	006	32 45	047	60 49	106	43 07	160	30 20	208	30 21	255	38 20	311
73	33 44	005	33 28	046	61 45	107	43 27	161	29 53	208	29 24	255	37 35	310
74	33 49	004	34 11	046	62 42	107	43 46	162	29 25	208	28 27	255	36 50	310
	CAPELLA		POLLUX		◆REGULUS		CANOPUS		◆ACHERNAR		Diphda		◆Hamal	
75	33 53	003	34 53	045	10 23	076	44 04	162	28 57	209	27 30	255	36 04	309
76	33 56	003	35 34	045	11 20	075	44 22	163	28 28	209	26 33	255	35 18	308
77	33 59	002	36 16	044	12 17	075	44 38	164	28 00	209	25 36	255	34 31	308
78	34 00	001	36 56	043	13 14	075	44 54	165	27 31	209	24 39	255	33 44	307
79	34 00	000	37 36	042	14 11	075	45 09	165	27 02	210	23 42	255	32 57	307
80	34 00	359	38 16	042	15 08	075	45 24	166	26 32	210	22 45	255	32 09	306
81	33 59	358	38 55	041	16 05	074	45 37	167	26 03	210	21 48	254	31 22	305
82	33 57	358	39 33	040	17 02	074	45 50	168	25 33	210	20 51	254	30 33	305
83	33 54	357	40 11	039	17 59	074	46 02	169	25 03	211	19 55	254	29 45	304
84	33 50	356	40 48	038	18 56	074	46 13	170	24 33	211	18 58	254	28 56	304
85	33 45	355	41 25	038	19 52	073	46 24	170	24 02	211	18 01	254	28 07	304
86	33 40	354	42 00	037	20 49	073	46 33	171	23 32	211	17 04	254	27 17	303
87	33 33	353	42 35	036	21 45	073	46 42	172	23 01	211	16 07	254	26 28	303
88	33 26	353	43 10	035	22 42	072	46 49	173	22 30	212	15 10	254	25 38	302
89	33 18	352	43 43	034	23 38	072	46 56	174	21 59	212	14 14	254	24 48	302

LHA ♈	Hc	Zn	Hc	Zn	Hc	Zn	Hc	Zn	Hc	Zn	Hc	Zn	Hc	Zn
	POLLUX		◆REGULUS		Suhail		◆CANOPUS		Acamar		◆ALDEBARAN		CAPELLA	
90	44 16	033	24 34	072	37 25	138	47 02	175	39 42	225	56 16	322	33 09	351
91	44 47	032	25 30	071	38 05	138	47 07	176	39 00	225	55 39	320	33 00	350
92	45 18	031	26 26	071	38 44	139	47 11	176	38 18	226	55 01	319	32 49	349
93	45 48	030	27 22	071	39 22	139	47 14	177	37 36	226	54 22	318	32 38	349
94	46 17	029	28 18	070	40 01	140	47 17	178	36 53	226	53 41	317	32 26	348
95	46 46	028	29 13	070	40 39	140	47 18	179	36 10	227	53 00	316	32 13	347
96	47 13	027	30 09	070	41 16	141	47 18	180	35 27	227	52 19	314	31 59	346
97	47 39	026	31 04	069	41 54	141	47 18	181	34 43	227	51 36	313	31 45	345
98	48 04	024	31 59	069	42 30	142	47 17	182	34 00	228	50 53	312	31 29	345
99	48 28	023	32 55	069	43 07	142	47 14	183	33 16	228	50 09	311	31 13	344
100	48 50	022	33 49	068	43 43	143	47 11	184	32 32	228	49 24	310	30 57	343
101	49 12	021	34 44	068	44 18	143	47 07	184	31 48	228	48 39	310	30 39	342
102	49 32	020	35 39	067	44 53	144	47 02	185	31 04	229	47 53	309	30 21	342
103	49 51	018	36 33	067	45 28	145	46 56	186	30 19	229	47 07	308	30 02	341
104	50 09	017	37 27	066	46 01	145	46 49	187	29 35	229	46 20	307	29 43	340
	POLLUX		◆REGULUS		Suhail		◆CANOPUS		RIGEL		◆ALDEBARAN		CAPELLA	
105	50 26	016	38 21	066	46 35	146	46 41	188	63 51	272	45 32	306	29 22	340
106	50 41	014	39 15	065	47 08	147	46 32	189	62 52	272	44 44	306	29 01	339
107	50 55	013	40 09	065	47 40	147	46 23	190	61 53	271	43 56	305	28 40	338
108	51 08	012	41 02	064	48 11	148	46 13	191	60 54	271	43 07	304	28 18	338
109	51 19	010	41 55	064	48 42	149	46 01	191	59 55	271	42 18	303	27 55	337
110	51 29	009	42 48	063	49 12	150	45 49	192	58 56	271	41 29	303	27 31	336
111	51 37	007	43 41	063	49 42	151	45 36	193	57 57	270	40 39	302	27 07	336
112	51 44	006	44 33	062	50 10	151	45 23	194	56 58	270	39 49	301	26 42	335
113	51 50	005	45 25	061	50 38	152	45 08	195	55 59	270	38 58	301	26 17	334
114	51 54	003	46 17	061	51 05	153	44 53	195	54 59	270	38 07	300	25 51	334
115	51 57	002	47 08	060	51 31	154	44 37	196	54 00	270	37 16	300	25 25	333
116	51 58	000	47 59	059	51 57	155	44 20	197	53 01	270	36 25	299	24 58	333
117	51 57	359	48 50	059	52 21	156	44 03	198	52 02	269	35 33	299	24 31	332
118	51 55	357	49 40	058	52 45	157	43 44	198	51 03	269	34 41	298	24 03	331
119	51 52	356	50 30	057	53 07	158	43 25	199	50 04	269	33 49	298	23 34	331
	◆REGULUS		SPICA		◆ACRUX		CANOPUS		RIGEL		◆BETELGEUSE		POLLUX	
120	51 19	056	10 25	100	19 24	154	43 06	200	49 05	269	54 19	298	51 47	355
121	52 08	055	11 23	099	19 50	154	42 45	200	48 06	269	53 27	297	51 41	353
122	52 56	054	12 22	099	20 16	154	42 25	201	47 07	269	52 34	296	51 33	352
123	53 44	053	13 20	099	20 41	154	42 03	202	46 08	269	51 41	296	51 24	350
124	54 31	052	14 18	099	21 07	154	41 41	202	45 09	268	50 48	295	51 14	349
125	55 17	051	15 17	099	21 32	155	41 18	203	44 10	268	49 54	294	51 02	348
126	56 03	050	16 15	099	21 57	155	40 54	204	43 11	268	49 00	294	50 49	346
127	56 48	049	17 13	099	22 22	155	40 30	204	42 11	268	48 06	293	50 34	345
128	57 32	048	18 12	099	22 47	155	40 06	205	41 12	268	47 11	293	50 18	344
129	58 16	047	19 10	098	23 12	155	39 41	205	40 13	268	46 17	292	50 01	342
130	58 58	045	20 09	098	23 36	156	39 15	206	39 14	268	45 22	291	49 42	341
131	59 40	044	21 07	098	24 01	156	38 49	207	38 15	267	44 27	291	49 22	340
132	60 20	043	22 06	098	24 25	156	38 22	207	37 16	267	43 31	290	49 01	339
133	61 00	041	23 04	098	24 49	156	37 55	208	36 17	267	42 36	290	48 39	337
134	61 38	040	24 03	098	25 12	157	37 28	208	35 18	267	41 40	289	48 16	336
	Dubhe		Denebola		◆SPICA		ACRUX		◆CANOPUS		SIRIUS		◆POLLUX	
135	14 17	014	41 28	060	25 01	098	25 36	157	37 00	209	56 31	255	47 51	335
136	14 32	014	42 19	060	26 00	098	25 59	157	36 31	209	55 34	255	47 26	334
137	14 46	014	43 10	059	26 58	097	26 22	157	36 02	209	54 37	255	46 59	333
138	15 00	013	44 00	058	27 57	097	26 44	158	35 33	210	53 40	255	46 32	332
139	15 13	013	44 50	058	28 56	097	27 07	158	35 03	210	52 42	256	46 03	331
140	15 26	012	45 40	057	29 54	097	27 29	158	34 33	211	51 45	256	45 34	329
141	15 38	012	46 29	056	30 53	097	27 50	159	34 03	211	50 48	256	45 03	328
142	15 50	011	47 18	055	31 52	097	28 12	159	33 32	212	49 51	256	44 32	327
143	16 02	011	48 07	055	32 50	097	28 33	159	33 01	212	48 53	256	44 00	326
144	16 13	011	48 55	054	33 49	097	28 54	160	32 30	212	47 56	256	43 27	326
145	16 23	010	49 42	053	34 48	097	29 14	160	31 58	213	46 59	256	42 53	325
146	16 34	010	50 29	052	35 46	097	29 34	160	31 26	213	46 01	256	42 18	324
147	16 43	009	51 15	051	36 45	096	29 54	161	30 54	213	45 04	256	41 43	323
148	16 52	009	52 01	050	37 44	096	30 14	161	30 21	214	44 06	256	41 07	322
149	17 01	008	52 46	049	38 42	096	30 33	161	29 48	214	43 09	256	40 30	321
	Dubhe		ARCTURUS		◆SPICA		ACRUX		◆CANOPUS		SIRIUS		◆POLLUX	
150	17 09	008	20 39	065	39 41	096	30 52	162	29 15	214	42 12	256	39 53	320
151	17 17	007	21 32	065	40 40	096	31 10	162	28 42	214	41 14	256	39 14	319
152	17 24	007	22 26	064	41 39	096	31 28	162	28 08	215	40 17	257	38 36	319
153	17 31	006	23 19	064	42 37	096	31 46	163	27 34	215	39 19	257	37 57	318
154	17 37	006	24 12	064	43 36	096	32 03	163	27 00	215	38 22	257	37 17	317
155	17 43	005	25 05	063	44 35	096	32 20	164	26 26	215	37 24	257	36 36	317
156	17 48	005	25 57	063	45 34	096	32 36	164	25 52	216	36 27	257	35 55	316
157	17 53	004	26 50	062	46 33	096	32 52	165	25 17	216	35 29	257	35 14	315
158	17 58	004	27 42	062	47 31	096	33 08	165	24 42	216	34 32	257	34 32	314
159	18 01	003	28 34	062	48 30	096	33 23	165	24 07	216	33 34	257	33 50	314
160	18 05	003	29 26	061	49 29	095	33 37	166	23 32	216	32 37	256	33 07	313
161	18 07	002	30 17	061	50 28	095	33 51	166	22 57	217	31 39	256	32 24	313
162	18 10	002	31 09	060	51 27	095	34 05	167	22 22	217	30 42	256	31 40	312
163	18 11	001	32 00	060	52 25	095	34 18	167	21 46	217	29 45	256	30 56	311
164	18 12	001	32 51	059	53 24	095	34 31	168	21 11	217	28 47	256	30 11	311
	Dubhe		◆ARCTURUS		SPICA		◆ACRUX		Suhail		SIRIUS		◆POLLUX	
165	18 13	000	33 42	059	54 23	095	34 43	168	48 39	211	27 50	256	29 26	310
166	18 13	000	34 32	058	55 22	095	34 55	169	48 08	212	26 52	256	28 41	310
167	18 13	359	35 22	058	56 21	095	35 06	169	47 37	213	25 55	256	27 56	309
168	18 12	359	36 12	057	57 20	095	35 17	170	47 05	213	24 58	256	27 10	309
169	18 11	358	37 01	057	58 19	095	35 27	170	46 32	214	24 00	256	26 24	308
170	18 09	358	37 50	056	59 17	095	35 37	171	45 58	215	23 03	256	25 37	308
171	18 07	357	38 39	055	60 16	095	35 46	171	45 24	215	22 06	256	24 50	307
172	18 04	357	39 28	055	61 15	095	35 54	172	44 50	216	21 08	256	24 03	307
173	18 00	356	40 16	054	62 14	095	36 02	172	44 15	217	20 11	256	23 16	307
174	17 56	356	41 03	053	63 13	095	36 10	173	43 39	217	19 14	256	22 28	306
175	17 52	355	41 50	053	64 12	095	36 17	174	43 03	218	18 17	256	21 40	306
176	17 47	355	42 37	052	65 11	095	36 23	174	42 27	218	17 19	255	20 52	305
177	17 42	354	43 23	051	66 09	095	36 29	175	41 50	219	16 22	255	20 04	305
178	17 36	354	44 09	050	67 08	095	36 34	175	41 13	219	15 25	255	19 16	305
179	17 29	354	44 55	050	68 07	095	36 39	176	40 35	220	14 28	255	18 27	304

LHA	Hc	Zn	Hc	Zn	Hc	Zn	Hc	Zn	Hc	Zn	Hc	Zn	Hc	Zn
♈	ARCTURUS		◆ANTARES		ACRUX		◆Suhail		Alphard		REGULUS		◆Dubhe	
180	45 39	049	24 42	115	36 43	176	39 57	220	52 20	269	54 33	308	17 22	353
181	46 23	048	25 36	115	36 47	177	39 19	221	51 21	269	53 46	307	17 15	353
182	47 07	047	26 30	115	36 50	177	38 40	221	50 22	269	52 59	306	17 07	352
183	47 50	046	27 24	115	36 52	178	38 01	222	49 23	268	52 10	305	16 59	352
184	48 32	045	28 17	115	36 54	179	37 22	222	48 24	268	51 22	304	16 50	351
185	49 14	044	29 11	115	36 55	179	36 42	222	47 25	268	50 32	303	16 41	351
186	49 55	043	30 05	115	36 56	180	36 02	223	46 26	268	49 43	302	16 31	350
187	50 35	042	30 58	115	36 56	180	35 22	223	45 27	268	48 53	302	16 20	350
188	51 14	041	31 52	115	36 55	181	34 41	223	44 28	268	48 02	301	16 10	349
189	51 53	040	32 45	115	36 54	181	34 01	224	43 29	268	47 11	300	15 59	349
190	52 30	039	33 39	115	36 52	182	33 20	224	42 30	267	46 20	299	15 47	348
191	53 07	038	34 32	115	36 50	182	32 38	224	41 30	267	45 28	299	15 35	348
192	53 42	036	35 26	115	36 47	183	31 57	225	40 31	267	44 36	298	15 22	348
193	54 17	035	36 19	116	36 44	184	31 15	225	39 32	267	43 44	297	15 09	347
194	54 50	034	37 12	116	36 40	184	30 34	225	38 33	267	42 51	297	14 56	347
	ARCTURUS		◆Rasalhague		ANTARES		◆ACRUX		Suhail		◆REGULUS		Denebola	
195	55 23	032	18 10	073	38 05	116	36 35	185	29 52	225	41 58	296	59 44	324
196	55 54	031	19 06	073	38 59	116	36 30	185	29 10	226	41 05	296	59 08	323
197	56 24	030	20 03	073	39 52	116	36 24	186	28 28	226	40 12	295	58 32	321
198	56 52	028	20 59	072	40 45	116	36 18	186	27 45	226	39 18	295	57 54	320
199	57 20	027	21 55	072	41 37	117	36 11	187	27 03	226	38 24	294	57 15	318
200	57 45	025	22 52	072	42 30	117	36 04	187	26 20	226	37 30	294	56 36	317
201	58 10	023	23 48	071	43 23	117	35 56	188	25 37	226	36 36	293	55 55	316
202	58 32	022	24 44	071	44 15	117	35 47	189	24 54	227	35 42	293	55 13	315
203	58 54	020	25 39	071	45 08	118	35 38	189	24 11	227	34 47	292	54 31	313
204	59 13	018	26 35	070	46 00	118	35 29	190	23 28	227	33 52	292	53 48	312
205	59 31	017	27 31	070	46 52	118	35 18	190	22 45	227	32 57	291	53 04	311
206	59 47	015	28 26	070	47 44	119	35 08	191	22 02	227	32 02	291	52 19	310
207	60 01	013	29 22	069	48 36	119	34 57	191	21 19	227	31 07	291	51 33	309
208	60 14	011	30 17	069	49 28	119	34 45	192	20 35	227	30 12	290	50 47	308
209	60 24	009	31 12	069	50 19	120	34 33	192	19 52	227	29 16	290	50 01	307
	ARCTURUS		◆Rasalhague		ANTARES		◆RIGIL KENT.		ACRUX		Gienah		◆REGULUS	
210	60 33	007	32 07	068	51 10	120	38 40	174	34 20	193	63 35	251	28 21	290
211	60 39	006	33 02	068	52 01	121	38 46	175	34 07	193	62 39	251	27 25	289
212	60 44	004	33 56	067	52 52	121	38 51	175	33 54	194	61 43	252	26 29	289
213	60 47	002	34 51	067	53 43	122	38 56	176	33 39	194	60 47	252	25 33	289
214	60 47	000	35 45	066	54 33	122	39 00	176	33 25	194	59 51	252	24 37	288
215	60 46	358	36 39	066	55 23	123	39 04	177	33 10	195	58 54	253	23 41	288
216	60 43	356	37 33	066	56 12	123	39 06	178	32 55	195	57 58	253	22 45	288
217	60 38	354	38 27	065	57 02	124	39 09	178	32 39	196	57 01	253	21 48	287
218	60 30	352	39 20	065	57 50	125	39 10	179	32 22	196	56 05	253	20 52	287
219	60 21	350	40 13	064	58 39	125	39 11	179	32 06	197	55 08	254	19 55	287
220	60 10	348	41 06	063	59 27	126	39 11	180	31 48	197	54 11	254	18 59	286
221	59 57	346	41 59	063	60 14	127	39 11	181	31 31	197	53 15	254	18 02	286
222	59 43	345	42 52	062	61 01	128	39 10	181	31 13	198	52 18	254	17 05	286
223	59 26	343	43 44	062	61 47	129	39 08	182	30 55	198	51 21	254	16 08	286
224	59 08	341	44 36	061	62 33	130	39 05	183	30 36	199	50 24	255	15 11	285
	Alphecca		◆VEGA		ALTAIR		◆Shaula		RIGIL KENT.		◆SPICA		ARCTURUS	
225	52 20	013	19 54	042	15 17	078	46 08	134	39 02	183	66 37	265	58 48	339
226	52 32	011	20 34	042	16 15	078	46 50	135	38 59	184	65 38	265	58 26	338
227	52 43	010	21 13	041	17 13	077	47 32	136	38 54	184	64 39	265	58 03	336
228	52 52	008	21 52	041	18 10	077	48 13	136	38 49	185	63 40	265	57 38	334
229	53 00	007	22 30	040	19 08	077	48 54	137	38 44	186	62 41	265	57 12	333
230	53 06	005	23 08	040	20 06	077	49 34	138	38 38	186	61 42	265	56 44	331
231	53 11	004	23 46	039	21 03	076	50 13	138	38 31	187	60 44	265	56 15	330
232	53 14	002	24 23	039	22 00	076	50 52	139	38 23	188	59 45	265	55 45	329
233	53 16	001	25 00	038	22 58	076	51 31	140	38 15	188	58 46	265	55 14	327
234	53 16	359	25 37	038	23 55	076	52 09	140	38 07	189	57 47	265	54 41	326
235	53 15	358	26 13	037	24 52	075	52 46	141	37 57	189	56 48	265	54 07	324
236	53 12	356	26 48	037	25 49	075	53 23	142	37 47	190	55 49	265	53 32	323
237	53 07	355	27 23	036	26 46	075	53 59	143	37 37	190	54 50	265	52 56	322
238	53 01	353	27 58	036	27 43	074	54 34	144	37 26	191	53 52	265	52 20	321
239	52 54	352	28 32	035	28 40	074	55 08	145	37 14	192	52 53	265	51 42	320
	VEGA		◆ALTAIR		Shaula		◆RIGIL KENT.		SPICA		◆ARCTURUS		Alphecca	
240	29 06	034	29 37	074	55 42	146	37 02	192	51 54	265	51 03	319	52 45	351
241	29 39	034	30 34	073	56 14	147	36 49	193	50 55	265	50 24	317	52 35	349
242	30 11	033	31 30	073	56 46	148	36 36	193	49 56	265	49 43	316	52 23	348
243	30 43	032	32 27	073	57 17	149	36 22	194	48 57	264	49 02	315	52 09	346
244	31 14	032	33 23	072	57 47	150	36 08	194	47 59	264	48 20	314	51 55	345
245	31 45	031	34 19	072	58 15	152	35 53	195	47 00	264	47 38	314	51 39	344
246	32 15	030	35 15	072	58 43	153	35 38	195	46 01	264	46 55	313	51 21	342
247	32 45	030	36 11	071	59 09	154	35 22	196	45 02	264	46 11	312	51 02	341
248	33 14	029	37 07	071	59 35	155	35 05	196	44 03	264	45 27	311	50 42	339
249	33 42	028	38 03	070	59 59	157	34 49	197	43 05	264	44 42	310	50 21	338
250	34 09	027	38 58	070	60 21	158	34 31	197	42 06	264	43 56	309	49 58	337
251	34 36	027	39 54	069	60 42	160	34 13	198	41 07	264	43 10	309	49 35	336
252	35 02	026	40 49	069	61 02	161	33 55	198	40 08	264	42 24	308	49 10	334
253	35 28	025	41 44	068	61 21	163	33 36	199	39 10	264	41 37	307	48 44	333
254	35 52	024	42 39	068	61 37	164	33 17	199	38 11	264	40 50	306	48 17	332
	VEGA		◆ALTAIR		Peacock		◆RIGIL KENT.		SPICA		◆ARCTURUS		Alphecca	
255	36 16	023	43 34	067	28 52	151	32 58	200	37 12	264	40 02	306	47 49	331
256	36 39	022	44 28	067	29 21	151	32 38	200	36 14	263	39 14	305	47 19	330
257	37 01	022	45 22	066	29 49	151	32 17	200	35 15	263	38 25	304	46 49	329
258	37 23	021	46 16	066	30 17	152	31 57	201	34 16	263	37 36	304	46 18	328
259	37 43	020	47 10	065	30 45	152	31 35	201	33 17	263	36 47	303	45 46	327
260	38 03	019	48 04	065	31 13	152	31 14	202	32 19	263	35 58	303	45 13	326
261	38 22	018	48 57	064	31 40	153	30 52	202	31 20	263	35 08	302	44 40	325
262	38 39	017	49 50	063	32 07	153	30 30	202	30 22	263	34 18	302	44 05	324
263	38 56	016	50 42	062	32 33	153	30 07	203	29 23	263	33 27	301	43 30	323
264	39 12	015	51 35	062	32 59	154	29 44	203	28 24	263	32 36	301	42 54	322
265	39 28	014	52 26	061	33 25	154	29 21	203	27 26	263	31 45	300	42 17	321
266	39 42	013	53 18	060	33 51	155	28 57	204	26 27	262	30 54	300	41 39	320
267	39 55	012	54 09	059	34 16	155	28 33	204	25 29	262	30 03	299	41 01	319
268	40 07	011	54 59	058	34 40	156	28 09	204	24 30	262	29 11	299	40 23	319
269	40 18	010	55 49	057	35 05	156	27 44	205	23 31	262	28 19	298	39 43	318

LAT 10°S

LHA	Hc	Zn	Hc	Zn	Hc	Zn	Hc	Zn	Hc	Zn	Hc	Zn	Hc	Zn
♈	◆VEGA		DENEB		Enif		◆FOMALHAUT		Peacock		◆ANTARES		ARCTURUS	
270	40 28	009	23 54	030	30 52	072	18 28	118	35 28	157	62 58	230	27 27	298
271	40 38	008	24 23	029	31 48	072	19 21	118	35 52	157	62 13	231	26 35	298
272	40 46	007	24 52	029	32 44	071	20 13	118	36 15	157	61 27	232	25 42	297
273	40 53	006	25 20	028	33 40	071	21 05	118	36 37	158	60 40	233	24 50	297
274	40 59	005	25 48	028	34 36	071	21 57	118	36 59	158	59 53	233	23 57	296
275	41 04	004	26 15	027	35 32	070	22 49	118	37 20	159	59 05	234	23 04	296
276	41 08	003	26 41	026	36 27	070	23 41	118	37 41	160	58 17	235	22 10	296
277	41 11	002	27 07	026	37 23	069	24 33	118	38 02	160	57 29	236	21 17	295
278	41 13	001	27 33	025	38 18	069	25 25	118	38 21	161	56 40	236	20 24	295
279	41 13	000	27 58	024	39 13	068	26 18	118	38 41	161	55 50	237	19 30	295
280	41 13	359	28 22	024	40 08	068	27 10	118	39 00	162	55 00	238	18 36	294
281	41 12	358	28 45	023	41 02	067	28 02	118	39 18	162	54 10	238	17 42	294
282	41 09	357	29 08	022	41 57	067	28 54	118	39 35	163	53 20	239	16 48	294
283	41 06	356	29 30	022	42 51	066	29 45	119	39 52	164	52 29	239	15 54	293
284	41 01	355	29 52	021	43 45	066	30 37	119	40 09	164	51 39	240	15 00	293
	DENEB		Enif		◆FOMALHAUT		Peacock		◆ANTARES		Rasalhague		◆VEGA	
285	30 13	020	44 39	065	31 29	119	40 25	165	50 47	240	59 03	316	40 55	354
286	30 33	020	45 32	065	32 21	119	40 40	165	49 56	241	58 22	315	40 49	353
287	30 53	019	46 26	064	33 13	119	40 54	166	49 05	241	57 40	314	40 41	352
288	31 12	018	47 19	063	34 04	119	41 08	167	48 13	241	56 57	313	40 32	351
289	31 30	017	48 11	063	34 56	119	41 21	167	47 21	242	56 13	311	40 22	350
290	31 47	017	49 04	062	35 47	119	41 34	168	46 29	242	55 28	310	40 12	349
291	32 04	016	49 56	061	36 39	120	41 46	169	45 37	242	54 42	309	40 00	348
292	32 19	015	50 47	061	37 30	120	41 57	169	44 44	243	53 56	308	39 47	347
293	32 35	014	51 39	060	38 21	120	42 07	170	43 52	243	53 09	307	39 33	346
294	32 49	014	52 30	059	39 12	120	42 17	171	42 59	243	52 22	306	39 18	345
295	33 02	013	53 20	058	40 03	121	42 26	172	42 07	243	51 34	305	39 03	344
296	33 15	012	54 10	057	40 54	121	42 34	172	41 14	243	50 46	304	38 46	343
297	33 27	011	54 59	056	41 45	121	42 41	173	40 21	244	49 57	304	38 29	342
298	33 38	010	55 48	055	42 35	121	42 48	174	39 28	244	49 07	303	38 10	341
299	33 48	010	56 37	054	43 26	122	42 54	175	38 35	244	48 17	302	37 51	340
	◆DENEB		Alpheratz		◆FOMALHAUT		Peacock		◆ANTARES		Rasalhague		VEGA	
300	33 58	009	18 37	055	44 16	122	42 59	175	37 41	244	47 27	301	37 31	340
301	34 06	008	19 25	054	45 06	122	43 04	176	36 48	244	46 36	301	37 10	339
302	34 14	007	20 13	054	45 56	123	43 08	177	35 55	245	45 45	300	36 48	338
303	34 21	006	21 01	053	46 45	123	43 11	178	35 02	245	44 54	299	36 25	337
304	34 27	005	21 48	053	47 35	123	43 13	178	34 08	245	44 02	299	36 01	336
305	34 32	005	22 35	053	48 24	124	43 14	179	33 15	245	43 10	298	35 37	335
306	34 36	004	23 22	052	49 13	124	43 15	180	32 21	245	42 18	297	35 12	335
307	34 39	003	24 08	052	50 02	125	43 15	181	31 28	245	41 25	297	34 46	334
308	34 42	002	24 55	051	50 50	125	43 14	181	30 34	245	40 32	296	34 20	333
309	34 43	001	25 41	051	51 38	126	43 12	182	29 40	245	39 39	296	33 53	332
310	34 44	000	26 26	050	52 26	126	43 10	183	28 47	245	38 45	295	33 25	331
311	34 44	359	27 12	050	53 13	127	43 06	184	27 53	245	37 52	295	32 56	331
312	34 43	359	27 57	049	54 00	128	43 02	184	26 59	245	36 58	294	32 27	330
313	34 41	358	28 41	049	54 46	128	42 58	185	26 06	245	36 04	294	31 57	329
314	34 38	357	29 25	048	55 32	129	42 52	186	25 12	245	35 10	293	31 26	329
	Alpheratz		◆Diphda		ACHERNAR		Peacock		◆Nunki		VEGA		◆DENEB	
315	30 09	048	35 26	105	19 29	148	42 46	186	56 16	237	30 55	328	34 35	356
316	30 53	047	36 23	105	20 01	148	42 39	187	55 26	238	30 24	327	34 30	355
317	31 36	047	37 20	105	20 32	148	42 31	188	54 36	238	29 51	327	34 25	354
318	32 19	046	38 17	105	21 04	148	42 22	189	53 46	239	29 18	326	34 18	353
319	33 01	045	39 14	105	21 35	148	42 13	189	52 55	239	28 45	325	34 11	353
320	33 43	045	40 11	105	22 06	148	42 03	190	52 04	240	28 11	325	34 03	352
321	34 24	044	41 08	105	22 37	148	41 53	191	51 13	240	27 37	324	33 54	351
322	35 05	043	42 05	105	23 08	149	41 41	191	50 22	240	27 02	324	33 45	350
323	35 45	043	43 02	105	23 39	149	41 29	192	49 31	241	26 26	323	33 34	349
324	36 25	042	43 59	105	24 10	149	41 16	193	48 39	241	25 51	322	33 23	348
325	37 04	041	44 56	106	24 40	149	41 03	193	47 47	242	25 14	322	33 10	348
326	37 43	041	45 53	106	25 10	149	40 49	194	46 55	242	24 38	321	32 57	347
327	38 21	040	46 49	106	25 40	150	40 34	195	46 03	242	24 00	321	32 44	346
328	38 59	039	47 46	106	26 10	150	40 19	195	45 10	243	23 23	320	32 29	345
329	39 36	038	48 43	106	26 39	150	40 03	196	44 18	243	22 45	320	32 14	345
	◆Alpheratz		Diphda		◆ACHERNAR		Peacock		Nunki		◆ALTAIR		DENEB	
330	40 12	037	49 40	106	27 09	150	39 46	197	43 25	243	52 40	299	31 58	344
331	40 47	037	50 37	106	27 38	151	39 29	197	42 32	243	51 49	298	31 41	343
332	41 22	036	51 33	106	28 07	151	39 11	198	41 40	244	50 56	298	31 23	342
333	41 56	035	52 30	107	28 35	151	38 53	198	40 47	244	50 04	297	31 05	342
334	42 29	034	53 27	107	29 03	152	38 34	199	39 54	244	49 11	296	30 46	341
335	43 02	033	54 23	107	29 31	152	38 14	200	39 00	244	48 18	296	30 26	340
336	43 34	032	55 20	107	29 59	152	37 54	200	38 07	244	47 25	295	30 05	339
337	44 05	031	56 16	108	30 27	153	37 33	201	37 14	244	46 31	294	29 44	339
338	44 34	030	57 12	108	30 54	153	37 12	201	36 21	245	45 37	294	29 22	338
339	45 04	029	58 08	108	31 20	153	36 51	202	35 27	245	44 43	293	29 00	337
340	45 32	028	59 05	108	31 47	154	36 29	202	34 34	245	43 48	293	28 37	337
341	45 59	027	60 01	109	32 13	154	36 06	203	33 40	245	42 54	292	28 13	336
342	46 25	026	60 56	109	32 39	154	35 43	203	32 47	245	41 59	292	27 48	335
343	46 50	025	61 52	110	33 04	155	35 19	204	31 53	245	41 04	291	27 23	335
344	47 14	023	62 48	110	33 29	155	34 56	204	30 59	245	40 09	291	26 58	334
	Alpheratz		◆Hamal		◆ACHERNAR		Peacock		Nunki		◆ALTAIR		DENEB	
345	47 37	022	33 23	053	33 54	156	34 31	205	30 06	245	39 13	290	26 32	333
346	47 59	021	34 11	053	34 18	156	34 06	205	29 12	245	38 18	290	26 05	333
347	48 20	020	34 57	052	34 42	156	33 41	205	28 18	245	37 22	289	25 38	332
348	48 39	019	35 44	051	35 05	157	33 16	206	27 25	245	36 26	289	25 10	332
349	48 58	017	36 30	051	35 28	157	32 50	206	26 31	245	35 30	289	24 42	331
350	49 15	016	37 15	050	35 51	158	32 23	207	25 37	246	34 34	288	24 13	330
351	49 31	015	38 00	049	36 13	158	31 57	207	24 43	246	33 38	288	23 43	330
352	49 45	014	38 45	049	36 34	159	31 30	207	23 49	246	32 42	287	23 13	329
353	49 59	012	39 29	048	36 55	159	31 02	208	22 56	246	31 45	287	22 43	329
354	50 11	011	40 13	047	37 16	160	30 35	208	22 02	246	30 49	287	22 12	328
355	50 21	010	40 56	047	37 36	160	30 07	208	21 08	246	29 52	286	21 41	328
356	50 30	008	41 39	046	37 55	161	29 38	209	20 14	246	28 55	286	21 09	327
357	50 38	007	42 21	045	38 14	162	29 10	209	19 21	245	27 59	286	20 37	327
358	50 45	006	43 02	044	38 33	162	28 41	209	18 27	245	27 02	285	20 05	326
359	50 50	004	43 43	043	38 51	163	28 12	210	17 33	245	26 05	285	19 32	326

LHA ♈	Hc	Zn	Hc	Zn	Hc	Zn	Hc	Zn	Hc	Zn	Hc	Zn	Hc	Zn
	◆Alpheratz		Hamal		ALDEBARAN		◆RIGEL		ACHERNAR		◆FOMALHAUT		Enif	
0	49 53	003	43 39	042	16 32	069	12 41	096	40 06	163	66 20	216	50 16	300
1	49 55	001	44 18	041	17 27	069	13 40	096	40 22	164	65 44	217	49 25	300
2	49 56	000	44 56	040	18 21	068	14 39	096	40 39	164	65 08	219	48 34	299
3	49 56	359	45 33	039	19 16	068	15 37	096	40 54	165	64 31	220	47 42	298
4	49 54	357	46 10	038	20 11	068	16 36	095	41 09	166	63 52	222	46 50	298
5	49 50	356	46 46	037	21 05	067	17 35	095	41 24	166	63 13	223	45 57	297
6	49 45	355	47 21	036	21 59	067	18 33	095	41 38	167	62 32	224	45 05	296
7	49 39	353	47 55	035	22 54	067	19 32	095	41 51	167	61 51	225	44 12	296
8	49 31	352	48 28	034	23 48	066	20 31	095	42 03	168	61 08	226	43 18	295
9	49 23	351	49 01	033	24 41	066	21 29	095	42 15	169	60 26	227	42 25	294
10	49 12	349	49 32	032	25 35	066	22 28	094	42 26	170	59 42	228	41 31	294
11	49 01	348	50 02	030	26 29	065	23 27	094	42 36	170	58 58	229	40 37	293
12	48 48	347	50 31	029	27 22	065	24 25	094	42 46	171	58 13	230	39 43	293
13	48 34	345	51 00	028	28 15	064	25 24	094	42 55	172	57 28	231	38 49	292
14	48 18	344	51 27	027	29 08	064	26 23	094	43 03	172	56 42	232	37 54	292
	Hamal		◆ALDEBARAN		RIGEL		◆ACHERNAR		FOMALHAUT		◆Enif		Alpheratz	
15	51 52	025	30 01	063	27 22	094	43 11	173	55 55	232	36 59	291	48 02	343
16	52 17	024	30 54	063	28 21	093	43 17	174	55 08	233	36 04	291	47 44	342
17	52 40	023	31 46	063	29 19	093	43 23	175	54 21	234	35 09	290	47 25	341
18	53 02	021	32 38	062	30 18	093	43 29	175	53 34	234	34 14	290	47 04	339
19	53 23	020	33 30	062	31 17	093	43 33	176	52 46	235	33 19	290	46 43	338
20	53 42	018	34 22	061	32 16	093	43 37	177	51 58	235	32 23	289	46 21	337
21	54 00	017	35 13	061	33 15	093	43 40	177	51 09	236	31 28	289	45 57	336
22	54 16	015	36 04	060	34 13	093	43 42	178	50 20	236	30 32	288	45 32	335
23	54 31	014	36 55	059	35 12	092	43 44	179	49 31	237	29 36	288	45 07	334
24	54 45	012	37 46	059	36 11	092	43 44	180	48 42	237	28 40	288	44 40	333
25	54 56	011	38 36	058	37 10	092	43 44	180	47 52	238	27 44	287	44 13	332
26	55 07	009	39 26	058	38 09	092	43 43	181	47 02	238	26 47	287	43 44	331
27	55 15	008	40 16	057	39 08	092	43 42	182	46 12	238	25 51	287	43 15	330
28	55 22	006	41 05	056	40 07	092	43 39	183	45 22	239	24 55	286	42 45	329
29	55 27	004	41 54	056	41 06	091	43 36	183	44 32	239	23 58	286	42 13	328
	CAPELLA		◆BETELGEUSE		SIRIUS		CANOPUS		◆ACHERNAR		FOMALHAUT		◆Alpheratz	
30	18 02	034	28 44	075	20 56	104	23 13	143	43 32	184	43 41	239	41 42	327
31	18 34	033	29 41	075	21 53	104	23 48	143	43 27	185	42 50	240	41 09	326
32	19 06	033	30 38	074	22 51	104	24 23	143	43 22	186	42 00	240	40 35	325
33	19 37	032	31 34	074	23 48	104	24 59	143	43 16	186	41 09	240	40 01	324
34	20 08	032	32 31	074	24 45	103	25 34	144	43 09	187	40 17	240	39 26	323
35	20 39	031	33 28	073	25 42	103	26 08	144	43 01	188	39 26	240	38 51	322
36	21 09	031	34 24	073	26 40	103	26 43	144	42 53	189	38 35	241	38 14	322
37	21 39	030	35 20	073	27 37	103	27 18	144	42 44	189	37 44	241	37 37	321
38	22 08	030	36 16	072	28 34	103	27 52	144	42 34	190	36 52	241	37 00	320
39	22 37	029	37 12	072	29 32	103	28 26	145	42 23	191	36 00	241	36 22	319
40	23 05	028	38 08	071	30 29	103	29 00	145	42 12	191	35 09	241	35 43	319
41	23 33	028	39 04	071	31 27	103	29 34	145	42 00	192	34 17	241	35 04	318
42	24 00	027	40 00	070	32 24	103	30 07	145	41 48	193	33 25	242	34 24	317
43	24 27	027	40 55	070	33 21	103	30 40	146	41 34	193	32 33	242	33 43	316
44	24 53	026	41 50	069	34 19	103	31 13	146	41 20	194	31 42	242	33 03	316
	◆CAPELLA		BETELGEUSE		SIRIUS		◆CANOPUS		ACHERNAR		◆FOMALHAUT		Alpheratz	
45	25 19	026	42 45	069	35 16	103	31 46	146	41 06	195	30 50	242	32 21	315
46	25 44	025	43 40	068	36 14	103	32 19	147	40 51	195	29 58	242	31 40	315
47	26 09	024	44 35	068	37 11	103	32 51	147	40 35	196	29 06	242	30 57	314
48	26 32	024	45 29	067	38 09	103	33 23	147	40 18	197	28 14	242	30 15	313
49	26 56	023	46 24	067	39 06	103	33 54	148	40 01	197	27 22	242	29 32	313
50	27 18	022	47 18	066	40 04	103	34 26	148	39 44	198	26 29	242	28 48	312
51	27 41	022	48 11	066	41 01	103	34 57	149	39 26	198	25 37	242	28 04	312
52	28 02	021	49 05	065	41 58	103	35 27	149	39 07	199	24 45	242	27 20	311
53	28 23	020	49 58	064	42 56	103	35 57	149	38 48	199	23 53	242	26 36	311
54	28 43	020	50 51	063	43 53	103	36 27	150	38 28	200	23 01	242	25 51	310
55	29 02	019	51 43	063	44 51	103	36 57	150	38 07	200	22 09	242	25 06	310
56	29 21	018	52 36	062	45 48	103	37 26	151	37 46	201	21 17	242	24 20	309
57	29 39	017	53 27	061	46 46	103	37 54	151	37 25	202	20 24	242	23 34	309
58	29 56	017	54 19	060	47 43	103	38 23	152	37 03	202	19 32	242	22 48	308
59	30 13	016	55 10	059	48 41	103	38 50	152	36 41	203	18 40	242	22 02	308
	CAPELLA		POLLUX		◆SIRIUS		CANOPUS		◆ACHERNAR		Diphda		◆Hamal	
60	30 29	015	23 03	053	49 38	103	39 18	153	36 18	203	42 01	256	45 50	321
61	30 44	015	23 50	052	50 35	103	39 45	153	35 55	203	41 04	256	45 13	320
62	30 58	014	24 37	052	51 33	103	40 11	154	35 31	204	40 07	256	44 35	320
63	31 12	013	25 23	052	52 30	103	40 37	154	35 07	204	39 10	256	43 56	319
64	31 25	012	26 09	051	53 28	103	41 02	155	34 43	205	38 13	256	43 17	318
65	31 37	011	26 55	051	54 25	103	41 27	155	34 18	205	37 16	256	42 37	317
66	31 48	011	27 40	050	55 22	104	41 51	156	33 52	206	36 19	256	41 56	316
67	31 59	010	28 25	049	56 19	104	42 15	157	33 27	206	35 22	256	41 15	315
68	32 09	009	29 09	049	57 17	104	42 38	157	33 01	206	34 25	256	40 33	314
69	32 17	008	29 54	048	58 14	104	43 00	158	32 34	207	33 28	256	39 51	314
70	32 25	007	30 37	048	59 11	104	43 22	159	32 08	207	32 31	256	39 08	313
71	32 33	007	31 21	047	60 08	104	43 43	159	31 41	208	31 34	256	38 25	312
72	32 39	006	32 04	047	61 05	105	44 04	160	31 13	208	30 37	255	37 41	311
73	32 45	005	32 47	046	62 02	105	44 24	161	30 46	208	29 40	255	36 56	311
74	32 49	004	33 29	045	62 59	105	44 43	161	30 18	208	28 43	255	36 11	310
	CAPELLA		POLLUX		◆PROCYON		Suhail		CANOPUS		◆ACHERNAR		◆Hamal	
75	32 53	003	34 10	045	47 14	070	27 48	134	45 01	162	29 49	209	35 26	309
76	32 56	003	34 52	044	48 09	069	28 31	134	45 19	163	29 21	209	34 41	309
77	32 59	002	35 32	043	49 04	069	29 13	134	45 36	164	28 52	209	33 54	308
78	33 00	001	36 12	043	49 58	068	29 56	134	45 52	164	28 23	210	33 08	308
79	33 00	000	36 52	042	50 53	067	30 38	134	46 07	165	27 54	210	32 21	307
80	33 00	359	37 31	041	51 47	067	31 20	135	46 22	166	27 24	210	31 34	307
81	32 59	358	38 09	040	52 41	066	32 02	135	46 36	167	26 55	210	30 47	306
82	32 57	358	38 47	040	53 34	065	32 44	135	46 49	168	26 25	211	29 59	305
83	32 54	357	39 25	039	54 28	064	33 25	135	47 01	168	25 55	211	29 11	305
84	32 50	356	40 01	038	55 21	064	34 06	136	47 12	169	25 24	211	28 22	304
85	32 46	355	40 37	037	56 13	063	34 48	136	47 23	170	24 54	211	27 33	304
86	32 40	354	41 12	036	57 05	062	35 28	136	47 32	171	24 23	211	26 44	303
87	32 34	353	41 47	035	57 57	061	36 09	136	47 41	172	23 52	212	25 55	303
88	32 27	353	42 20	034	58 48	060	36 50	137	47 49	173	23 21	212	25 06	303
89	32 19	352	42 53	033	59 39	059	37 30	137	47 56	174	22 50	212	24 16	302

LAT 11°S

LHA ♈	Hc	Zn	Hc	Zn	Hc	Zn	Hc	Zn	Hc	Zn	Hc	Zn	Hc	Zn
	POLLUX		◆REGULUS		Suhail		◆CANOPUS		ACHERNAR		◆ALDEBARAN		CAPELLA	
90	43 25	033	24 15	071	38 10	138	48 02	175	22 19	212	55 29	322	32 10	351
91	43 56	032	25 11	071	38 49	138	48 07	175	21 47	212	54 53	321	32 00	350
92	44 27	031	26 07	071	39 29	138	48 11	176	21 16	212	54 15	320	31 50	349
93	44 56	029	27 02	070	40 08	139	48 14	177	20 44	213	53 37	319	31 39	349
94	45 25	028	27 57	070	40 46	139	48 17	178	20 12	213	52 57	318	31 27	348
95	45 52	027	28 53	070	41 25	140	48 18	179	19 40	213	52 17	316	31 14	347
96	46 19	026	29 48	069	42 03	140	48 18	180	19 09	213	51 36	315	31 01	346
97	46 44	025	30 43	069	42 40	141	48 18	181	18 36	213	50 54	314	30 46	346
98	47 09	024	31 38	068	43 17	141	48 16	182	18 04	213	50 12	313	30 31	345
99	47 32	023	32 32	068	43 54	142	48 14	183	17 32	213	49 29	312	30 16	344
100	47 55	022	33 27	068	44 30	142	48 11	184	17 00	213	48 45	311	29 59	343
101	48 16	020	34 21	067	45 06	143	48 07	185	16 28	213	48 00	310	29 42	343
102	48 36	019	35 15	067	45 41	143	48 01	185	15 55	213	47 15	310	29 24	342
103	48 54	018	36 09	066	46 16	144	47 55	186	15 23	213	46 29	309	29 05	341
104	49 12	017	37 03	066	46 51	145	47 48	187	14 51	213	45 43	308	28 46	340
	POLLUX		◆REGULUS		ACRUX		◆CANOPUS		RIGEL		◆ALDEBARAN		CAPELLA	
105	49 28	015	37 57	065	13 36	153	47 40	188	63 48	274	44 56	307	28 26	340
106	49 43	014	38 50	065	14 03	153	47 32	189	62 50	273	44 09	306	28 05	339
107	49 57	013	39 43	064	14 31	153	47 22	190	61 51	273	43 22	306	27 44	338
108	50 09	011	40 36	064	14 58	153	47 12	191	60 52	273	42 33	305	27 22	338
109	50 20	010	41 29	063	15 25	153	47 00	192	59 53	273	41 45	304	26 59	337
110	50 30	009	42 21	062	15 52	153	46 48	192	58 54	272	40 56	303	26 36	336
111	50 38	007	43 13	062	16 19	153	46 35	193	57 55	272	40 07	303	26 12	336
112	50 45	006	44 05	061	16 45	153	46 21	194	56 56	272	39 17	302	25 48	335
113	50 50	005	44 56	060	17 12	153	46 06	195	55 58	272	38 27	302	25 23	335
114	50 54	003	45 47	060	17 39	153	45 51	196	54 59	271	37 37	301	24 57	334
115	50 57	002	46 38	059	18 06	153	45 35	196	54 00	271	36 46	300	24 31	333
116	50 58	000	47 28	058	18 32	153	45 18	197	53 01	271	35 55	300	24 05	333
117	50 57	359	48 18	058	18 59	153	45 00	198	52 02	271	35 04	299	23 37	332
118	50 55	358	49 08	057	19 25	153	44 41	199	51 03	271	34 12	299	23 10	332
119	50 52	356	49 57	056	19 51	154	44 22	199	50 04	270	33 21	298	22 42	331
	◆REGULUS		SPICA		◆ACRUX		CANOPUS		◆RIGEL		BETELGEUSE		POLLUX	
120	50 45	055	10 35	099	20 18	154	44 02	200	49 05	270	53 50	299	50 48	355
121	51 33	054	11 33	099	20 44	154	43 42	201	48 06	270	52 59	298	50 42	353
122	52 21	053	12 31	099	21 10	154	43 20	201	47 08	270	52 07	298	50 34	352
123	53 08	052	13 29	099	21 35	154	42 59	202	46 09	270	51 15	297	50 25	351
124	53 54	051	14 28	099	22 01	154	42 36	203	45 10	269	50 22	296	50 15	349
125	54 39	050	15 26	099	22 26	154	42 13	203	44 11	269	49 29	295	50 03	348
126	55 24	049	16 24	099	22 52	155	41 49	204	43 12	269	48 35	295	49 50	347
127	56 08	048	17 22	098	23 17	155	41 25	205	42 13	269	47 42	294	49 36	345
128	56 52	047	18 21	098	23 42	155	41 00	205	41 14	269	46 48	294	49 20	344
129	57 34	046	19 19	098	24 06	155	40 35	206	40 15	269	45 54	293	49 03	343
130	58 16	044	20 17	098	24 31	155	40 09	206	39 16	268	44 59	292	48 45	341
131	58 56	043	21 16	098	24 55	156	39 43	207	38 18	268	44 05	292	48 26	340
132	59 36	041	22 14	098	25 20	156	39 16	207	37 19	268	43 10	291	48 05	339
133	60 14	040	23 12	098	25 44	156	38 48	208	36 20	268	42 15	291	47 44	338
134	60 52	038	24 11	097	26 07	156	38 20	208	35 21	268	41 20	290	47 21	337
	REGULUS		◆SPICA		ACRUX		◆CANOPUS		SIRIUS		BETELGEUSE		◆POLLUX	
135	61 28	037	25 09	097	26 31	157	37 52	209	56 46	256	40 25	290	46 57	335
136	62 02	035	26 08	097	26 54	157	37 24	209	55 49	256	39 29	289	46 32	334
137	62 35	033	27 06	097	27 17	157	36 54	210	54 51	257	38 33	289	46 06	333
138	63 07	032	28 05	097	27 40	157	36 25	210	53 54	257	37 38	288	45 39	332
139	63 37	030	29 03	097	28 02	158	35 55	211	52 57	257	36 42	288	45 11	331
140	64 05	028	30 01	097	28 24	158	35 25	211	51 59	257	35 46	288	44 42	330
141	64 32	026	31 00	096	28 46	158	34 54	212	51 02	257	34 49	287	44 12	329
142	64 57	024	31 59	096	29 08	159	34 23	212	50 05	257	33 53	287	43 41	328
143	65 19	022	32 57	096	29 29	159	33 52	212	49 07	257	32 57	286	43 09	327
144	65 40	019	33 56	096	29 50	159	33 20	213	48 10	257	32 00	286	42 37	326
145	65 58	017	34 54	096	30 11	160	32 48	213	47 12	257	31 03	286	42 04	325
146	66 15	015	35 53	096	30 31	160	32 16	213	46 15	257	30 07	285	41 30	324
147	66 28	012	36 51	096	30 51	160	31 44	214	45 17	257	29 10	285	40 55	323
148	66 40	010	37 50	096	31 10	161	31 11	214	44 20	257	28 13	285	40 19	322
149	66 49	008	38 49	096	31 30	161	30 38	214	43 23	257	27 16	284	39 43	322
	REGULUS		◆ARCTURUS		SPICA		◆ACRUX		CANOPUS		SIRIUS		◆PROCYON	
150	66 55	005	20 13	065	39 47	095	31 49	161	30 05	215	42 25	257	51 21	293
151	66 59	003	21 06	064	40 46	095	32 07	162	29 31	215	41 28	257	50 27	292
152	67 00	000	21 59	064	41 45	095	32 25	162	28 57	215	40 30	257	49 32	292
153	66 59	358	22 52	064	42 43	095	32 43	163	28 23	215	39 33	257	48 37	291
154	66 55	355	23 45	063	43 42	095	33 00	163	27 49	216	38 35	257	47 42	291
155	66 49	353	24 37	063	44 41	095	33 17	164	27 15	216	37 38	257	46 47	290
156	66 40	350	25 30	062	45 39	095	33 34	164	26 40	216	36 40	257	45 52	290
157	66 29	348	26 22	062	46 38	095	33 50	164	26 06	216	35 43	257	44 56	289
158	66 15	345	27 13	062	47 37	095	34 05	165	25 31	216	34 45	257	44 00	288
159	65 59	343	28 05	061	48 35	094	34 21	165	24 56	217	33 48	257	43 04	288
160	65 41	341	28 57	061	49 34	094	34 35	166	24 21	217	32 51	257	42 08	288
161	65 20	339	29 48	060	50 33	094	34 50	166	23 45	217	31 53	257	41 12	287
162	64 58	336	30 39	060	51 32	094	35 03	167	23 10	217	30 56	257	40 16	287
163	64 33	334	31 29	059	52 30	094	35 17	167	22 34	217	29 58	257	39 19	286
164	64 07	332	32 20	059	53 29	094	35 30	168	21 59	217	29 01	257	38 22	286
	Denebola		◆ARCTURUS		SPICA		◆ACRUX		SIRIUS		◆PROCYON		REGULUS	
165	61 41	026	33 10	058	54 28	094	35 42	168	28 04	257	37 26	285	63 39	330
166	62 06	024	34 00	058	55 27	094	35 54	169	27 06	257	36 29	285	63 09	328
167	62 29	022	34 50	057	56 25	094	36 05	169	26 09	257	35 32	285	62 37	327
168	62 49	020	35 39	057	57 24	094	36 16	170	25 12	257	34 35	284	62 04	325
169	63 08	018	36 28	056	58 23	093	36 26	170	24 14	256	33 38	284	61 29	323
170	63 25	016	37 16	055	59 22	093	36 36	171	23 17	256	32 41	284	60 54	322
171	63 40	014	38 05	055	60 21	093	36 45	171	22 20	256	31 43	283	60 16	320
172	63 53	011	38 53	054	61 19	093	36 54	172	21 23	256	30 46	283	59 38	319
173	64 04	009	39 40	053	62 18	093	37 02	172	20 26	256	29 49	283	58 59	317
174	64 12	007	40 27	053	63 17	093	37 09	173	19 28	256	28 51	282	58 18	316
175	64 18	005	41 14	052	64 16	093	37 16	173	18 31	256	27 53	282	57 36	315
176	64 22	003	42 00	051	65 15	093	37 23	174	17 34	256	26 56	282	56 54	313
177	64 24	000	42 46	050	66 13	093	37 29	175	16 37	256	25 58	281	56 11	312
178	64 23	358	43 31	050	67 12	093	37 34	175	15 40	256	25 00	281	55 27	311
179	64 20	356	44 15	049	68 11	093	37 39	176	14 43	255	24 03	281	54 42	310

LAT 11°S

LHA	Hc	Zn	Hc	Zn	Hc	Zn	Hc	Zn	Hc	Zn	Hc	Zn	Hc	Zn
♈	ARCTURUS		◆ANTARES		ACRUX		◆Suhail		Alphard		REGULUS		◆Denebola	
180	44 59	048	25 07	114	37 43	176	40 43	221	52 21	270	53 56	309	64 15	354
181	45 43	047	26 01	114	37 47	177	40 04	221	51 22	270	53 10	308	64 07	352
182	46 26	046	26 55	114	37 50	177	39 25	222	50 23	270	52 23	307	63 58	349
183	47 08	045	27 48	114	37 52	178	38 46	222	49 24	270	51 36	306	63 46	347
184	47 50	044	28 42	114	37 54	179	38 06	223	48 25	269	50 48	305	63 32	345
185	48 31	043	29 36	114	37 55	179	37 26	223	47 26	269	49 59	304	63 16	343
186	49 11	042	30 29	114	37 56	180	36 46	223	46 27	269	49 10	303	62 57	341
187	49 50	041	31 23	114	37 56	180	36 05	224	45 28	269	48 21	302	62 37	339
188	50 29	040	32 17	114	37 55	181	35 25	224	44 30	269	47 31	302	62 15	337
189	51 06	039	33 10	114	37 54	181	34 44	224	43 31	268	46 41	301	61 51	335
190	51 43	038	34 04	115	37 52	182	34 03	224	42 32	268	45 50	300	61 26	333
191	52 19	037	34 58	115	37 50	183	33 21	225	41 33	268	44 59	300	60 59	332
192	52 54	036	35 51	115	37 47	183	32 40	225	40 34	268	44 07	299	60 30	330
193	53 28	034	36 44	115	37 44	184	31 58	225	39 35	268	43 16	298	59 59	328
194	54 00	033	37 38	115	37 40	184	31 16	225	38 36	268	42 24	298	59 28	327
	ARCTURUS		◆Rasalhague		ANTARES		◆ACRUX		Suhail		◆REGULUS		Denebola	
195	54 32	032	17 52	073	38 31	115	37 35	185	30 34	226	41 31	297	58 55	325
196	55 02	030	18 49	073	39 25	115	37 30	185	29 52	226	40 39	296	58 20	324
197	55 31	029	19 45	072	40 18	115	37 24	186	29 09	226	39 46	296	57 45	322
198	55 59	027	20 41	072	41 11	116	37 18	186	28 27	226	38 53	295	57 08	321
199	56 26	026	21 37	072	42 04	116	37 11	187	27 44	226	37 59	295	56 30	319
200	56 51	024	22 33	071	42 57	116	37 03	188	27 01	227	37 06	294	55 51	318
201	57 14	023	23 28	071	43 50	116	36 55	188	26 18	227	36 12	294	55 11	317
202	57 37	021	24 24	071	44 43	116	36 47	189	25 36	227	35 18	293	54 31	316
203	57 57	020	25 19	070	45 35	117	36 37	189	24 52	227	34 24	293	53 49	314
204	58 16	018	26 15	070	46 28	117	36 28	190	24 09	227	33 30	292	53 07	313
205	58 33	016	27 10	070	47 20	117	36 18	190	23 26	227	32 35	292	52 24	312
206	58 49	014	28 05	069	48 13	118	36 07	191	22 43	227	31 41	292	51 40	311
207	59 03	013	29 00	069	49 05	118	35 56	191	21 59	227	30 46	291	50 55	310
208	59 15	011	29 55	068	49 57	118	35 44	192	21 16	227	29 51	291	50 10	309
209	59 25	009	30 50	068	50 48	119	35 32	192	20 33	228	28 56	290	49 24	308
	ARCTURUS		◆Rasalhague		ANTARES		◆RIGIL KENT.		ACRUX		Gienah		◆REGULUS	
210	59 33	007	31 44	068	51 40	119	39 40	174	35 19	193	63 54	253	28 00	290
211	59 40	005	32 39	067	52 31	119	39 46	174	35 06	193	62 57	253	27 05	290
212	59 44	003	33 33	067	53 23	120	39 51	175	34 52	194	62 01	253	26 09	289
213	59 47	002	34 27	066	54 14	120	39 56	176	34 38	194	61 05	254	25 14	289
214	59 47	000	35 21	066	55 04	121	40 00	176	34 23	195	60 08	254	24 18	289
215	59 46	358	36 14	065	55 55	121	40 04	177	34 08	195	59 11	254	23 22	288
216	59 43	356	37 08	065	56 45	122	40 06	178	33 52	196	58 15	254	22 26	288
217	59 38	354	38 01	064	57 35	123	40 09	178	33 36	196	57 18	255	21 30	288
218	59 31	352	38 54	064	58 24	123	40 10	179	33 20	196	56 21	255	20 34	287
219	59 22	350	39 47	063	59 13	124	40 11	179	33 03	197	55 24	255	19 38	287
220	59 11	349	40 39	063	60 02	125	40 11	180	32 46	197	54 27	255	18 41	287
221	58 59	347	41 31	062	60 50	126	40 11	181	32 28	198	53 30	255	17 45	287
222	58 45	345	42 23	061	61 37	126	40 10	181	32 10	198	52 33	255	16 49	286
223	58 29	343	43 15	061	62 24	127	40 08	182	31 52	198	51 36	256	15 52	286
224	58 11	342	44 06	060	63 11	128	40 05	183	31 33	199	50 39	256	14 55	286
	Alphecca		◆VEGA		ALTAIR		◆Shaula		RIGIL KENT.		◆SPICA		ARCTURUS	
225	51 21	012	19 10	042	15 05	078	46 50	134	40 02	183	66 41	267	57 51	340
226	51 33	011	19 49	042	16 02	077	47 32	134	39 59	184	65 42	267	57 30	338
227	51 43	010	20 28	041	17 00	077	48 14	135	39 54	185	64 43	267	57 08	337
228	51 53	008	21 06	041	17 57	077	48 56	135	39 49	185	63 44	267	56 44	335
229	52 00	007	21 44	040	18 54	077	49 37	136	39 44	186	62 46	267	56 18	334
230	52 06	005	22 22	040	19 51	076	50 18	137	39 37	186	61 47	267	55 52	332
231	52 11	004	23 00	039	20 49	076	50 58	137	39 30	187	60 48	267	55 23	331
232	52 14	002	23 37	039	21 46	076	51 37	138	39 23	188	59 49	267	54 54	329
233	52 16	001	24 13	038	22 43	075	52 16	139	39 15	188	58 50	267	54 23	328
234	52 16	359	24 49	038	23 40	075	52 55	140	39 06	189	57 51	267	53 51	327
235	52 15	358	25 25	037	24 37	075	53 33	140	38 57	189	56 53	266	53 18	325
236	52 12	357	26 00	036	25 34	075	54 10	141	38 47	190	55 54	266	52 44	324
237	52 08	355	26 35	036	26 30	074	54 46	142	38 36	191	54 55	266	52 09	323
238	52 02	354	27 09	035	27 27	074	55 22	143	38 25	191	53 56	266	51 33	322
239	51 55	352	27 43	035	28 23	074	55 57	144	38 13	192	52 58	266	50 56	320
	VEGA		◆ALTAIR		Shaula		◆RIGIL KENT.		SPICA		◆ARCTURUS		Alphecca	
240	28 16	034	29 20	073	56 31	145	38 01	192	51 59	266	50 18	319	51 46	351
241	28 49	033	30 16	073	57 04	146	37 48	193	51 00	266	49 39	318	51 36	349
242	29 21	033	31 12	072	57 37	147	37 35	193	50 01	266	49 00	317	51 24	348
243	29 52	032	32 08	072	58 08	148	37 21	194	49 03	266	48 19	316	51 11	347
244	30 23	031	33 04	072	58 39	150	37 06	195	48 04	266	47 38	315	50 57	345
245	30 54	031	34 00	071	59 08	151	36 51	195	47 05	265	46 56	314	50 41	344
246	31 23	030	34 56	071	59 36	152	36 36	196	46 07	265	46 14	313	50 24	343
247	31 53	029	35 52	070	60 03	153	36 20	196	45 08	265	45 31	313	50 06	341
248	32 21	029	36 47	070	60 29	155	36 03	197	44 09	265	44 47	312	49 46	340
249	32 49	028	37 42	070	60 54	156	35 46	197	43 10	265	44 03	311	49 25	339
250	33 16	027	38 37	069	61 17	158	35 29	198	42 12	265	43 18	310	49 03	337
251	33 42	026	39 32	069	61 39	159	35 11	198	41 13	265	42 33	309	48 40	336
252	34 08	025	40 27	068	61 59	161	34 52	198	40 14	265	41 47	309	48 16	335
253	34 33	025	41 22	068	62 18	162	34 33	199	39 16	265	41 01	308	47 50	334
254	34 57	024	42 16	067	62 35	164	34 14	199	38 17	264	40 14	307	47 24	333
	◆VEGA		ALTAIR		◆Peacock		RIGIL KENT.		◆SPICA		ARCTURUS		Alphecca	
255	35 21	023	43 10	067	29 44	150	33 54	200	37 19	264	39 27	306	46 56	332
256	35 44	022	44 04	066	30 13	151	33 34	200	36 20	264	38 39	306	46 27	330
257	36 05	021	44 58	065	30 42	151	33 14	201	35 21	264	37 51	305	45 58	329
258	36 26	021	45 51	065	31 10	151	32 53	201	34 23	264	37 03	305	45 27	328
259	36 47	020	46 45	064	31 38	152	32 31	201	33 24	264	36 14	304	44 56	327
260	37 06	019	47 37	064	32 06	152	32 10	202	32 26	264	35 25	303	44 24	326
261	37 24	018	48 30	063	32 33	152	31 47	202	31 27	264	34 36	303	43 51	325
262	37 42	017	49 22	062	33 00	153	31 25	203	30 29	263	33 46	302	43 17	324
263	37 59	016	50 14	061	33 27	153	31 02	203	29 30	263	32 56	302	42 42	323
264	38 15	015	51 06	061	33 53	154	30 39	203	28 32	263	32 06	301	42 06	322
265	38 29	014	51 57	060	34 19	154	30 16	204	27 33	263	31 15	301	41 30	322
266	38 43	013	52 47	059	34 45	154	29 52	204	26 35	263	30 24	300	40 53	321
267	38 56	012	53 38	058	35 10	155	29 28	204	25 36	263	29 33	300	40 16	320
268	39 08	011	54 27	057	35 35	155	29 04	205	24 38	263	28 42	299	39 37	319
269	39 19	010	55 17	056	35 59	156	28 39	205	23 39	263	27 51	299	38 58	318

LHA	Hc	Zn	Hc	Zn	Hc	Zn	Hc	Zn	Hc	Zn	Hc	Zn	Hc	Zn
♈	◆VEGA		ALTAIR		◆FOMALHAUT		Peacock		RIGIL KENT.		◆ANTARES		ARCTURUS	
270	39 29	009	56 05	055	18 57	118	36 23	156	28 14	205	63 36	231	26 59	298
271	39 38	008	56 53	054	19 49	118	36 47	157	27 49	205	62 50	232	26 07	298
272	39 46	007	57 41	053	20 41	118	37 10	157	27 23	206	62 03	233	25 15	298
273	39 53	006	58 28	052	21 33	118	37 32	158	26 58	206	61 16	234	24 22	297
274	39 59	005	59 14	051	22 25	118	37 55	158	26 32	206	60 28	235	23 30	297
275	40 04	004	59 59	049	23 17	118	38 16	159	26 06	206	59 40	236	22 37	296
276	40 08	003	60 43	048	24 09	118	38 37	159	25 39	207	58 51	236	21 44	296
277	40 11	002	61 26	047	25 01	118	38 58	160	25 13	207	58 02	237	20 51	296
278	40 13	001	62 09	045	25 54	118	39 18	160	24 46	207	57 12	238	19 58	295
279	40 13	000	62 50	044	26 46	118	39 38	161	24 19	207	56 22	238	19 05	295
280	40 13	359	63 30	042	27 38	118	39 56	162	23 52	208	55 32	239	18 11	295
281	40 12	358	64 09	040	28 30	118	40 15	162	23 25	208	54 42	239	17 18	294
282	40 09	357	64 47	039	29 22	118	40 33	163	22 57	208	53 51	240	16 24	294
283	40 06	356	65 23	037	30 14	118	40 50	163	22 29	208	53 00	240	15 30	294
284	40 01	355	65 57	035	31 06	118	41 06	164	22 02	208	52 08	241	14 36	293
	◆DENEB		◆FOMALHAUT		Peacock		RIGIL KENT.		◆ANTARES		Rasalhague		VEGA	
285	29 17	020	31 58	118	41 22	165	21 34	208	51 17	241	58 19	317	39 56	354
286	29 37	019	32 50	118	41 38	165	21 05	209	50 25	242	57 39	316	39 49	353
287	29 56	019	33 41	118	41 52	166	20 37	209	49 33	242	56 58	315	39 41	352
288	30 14	018	34 33	119	42 06	167	20 09	209	48 41	242	56 16	314	39 33	351
289	30 32	017	35 25	119	42 20	167	19 40	209	47 49	243	55 33	312	39 23	350
290	30 49	017	36 17	119	42 32	168	19 12	209	46 57	243	54 49	311	39 13	349
291	31 06	016	37 08	119	42 44	169	18 43	209	46 04	243	54 04	310	39 01	348
292	31 22	015	38 00	119	42 56	169	18 14	209	45 12	243	53 19	309	38 48	347
293	31 36	014	38 51	119	43 06	170	17 46	209	44 19	244	52 33	308	38 35	346
294	31 50	013	39 42	120	43 16	171	17 17	209	43 26	244	51 46	307	38 20	345
295	32 04	013	40 33	120	43 25	171	16 48	210	42 33	244	50 59	306	38 05	344
296	32 16	012	41 24	120	43 33	172	16 19	210	41 40	244	50 11	305	37 49	343
297	32 28	011	42 15	120	43 41	173	15 50	210	40 47	244	49 23	305	37 31	343
298	32 39	010	43 06	121	43 48	174	15 21	210	39 54	245	48 34	304	37 13	342
299	32 49	009	43 57	121	43 54	174	14 51	210	39 01	245	47 45	303	36 54	341
	◆DENEB		Alpheratz		◆FOMALHAUT		Peacock		◆ANTARES		Rasalhague		VEGA	
300	32 58	009	18 02	054	44 47	121	43 59	175	38 07	245	46 55	302	36 34	340
301	33 07	008	18 50	054	45 38	121	44 04	176	37 14	245	46 05	301	36 14	339
302	33 14	007	19 38	054	46 28	122	44 08	177	36 20	245	45 15	301	35 52	338
303	33 21	006	20 25	053	47 18	122	44 11	177	35 27	245	44 24	300	35 30	337
304	33 27	005	21 12	053	48 08	123	44 13	178	34 33	245	43 33	299	35 07	336
305	33 32	004	21 59	052	48 57	123	44 14	179	33 40	245	42 41	299	34 43	336
306	33 36	004	22 45	052	49 46	123	44 15	180	32 46	246	41 50	298	34 18	335
307	33 39	003	23 31	051	50 35	124	44 15	181	31 53	246	40 58	298	33 52	334
308	33 42	002	24 17	051	51 24	124	44 14	181	30 59	246	40 05	297	33 26	333
309	33 44	001	25 03	050	52 13	125	44 12	182	30 05	246	39 13	296	32 59	332
310	33 44	000	25 48	050	53 01	125	44 10	183	29 12	246	38 20	296	32 32	332
311	33 44	359	26 33	049	53 49	126	44 06	184	28 18	246	37 27	295	32 04	331
312	33 43	359	27 17	049	54 36	127	44 02	184	27 24	246	36 33	295	31 35	330
313	33 41	358	28 02	048	55 23	127	43 57	185	26 30	246	35 40	294	31 05	330
314	33 38	357	28 45	048	56 10	128	43 52	186	25 37	246	34 46	294	30 35	329
	Alpheratz		◆Diphda		ACHERNAR		Peacock		◆ANTARES		VEGA		◆DENEB	
315	29 29	047	35 41	104	20 20	147	43 45	187	24 43	246	30 04	328	33 35	356
316	30 12	047	36 38	104	20 52	147	43 38	187	23 49	246	29 33	328	33 30	355
317	30 55	046	37 35	104	21 23	148	43 30	188	22 56	246	29 01	327	33 25	354
318	31 37	046	38 32	104	21 55	148	43 22	189	22 02	246	28 29	326	33 19	354
319	32 19	045	39 29	104	22 26	148	43 12	190	21 08	246	27 56	326	33 12	353
320	33 00	044	40 26	104	22 57	148	43 02	190	20 14	246	27 22	325	33 04	352
321	33 41	044	41 23	104	23 28	148	42 51	191	19 21	246	26 48	324	32 55	351
322	34 21	043	42 20	104	23 59	148	42 40	192	18 27	246	26 14	324	32 45	350
323	35 01	042	43 17	104	24 30	149	42 28	192	17 33	246	25 39	323	32 35	349
324	35 40	042	44 14	105	25 01	149	42 15	193	16 40	245	25 03	323	32 24	349
325	36 19	041	45 11	105	25 31	149	42 01	194	15 46	245	24 27	322	32 12	348
326	36 57	040	46 08	105	26 02	149	41 47	194	14 53	245	23 51	322	31 59	347
327	37 35	039	47 05	105	26 32	149	41 32	195	13 59	245	23 14	321	31 45	346
328	38 12	038	48 02	105	27 02	150	41 17	196	13 06	245	22 37	321	31 31	345
329	38 48	038	48 59	105	27 31	150	41 00	196	12 12	245	21 59	320	31 16	345
	◆Alpheratz		Diphda		◆ACHERNAR		Peacock		◆Nunki		ALTAIR		DENEB	
330	39 24	037	49 56	105	28 01	150	40 44	197	43 52	244	52 10	300	31 00	344
331	39 59	036	50 53	105	28 30	150	40 26	198	42 59	244	51 19	300	30 43	343
332	40 33	035	51 50	105	28 59	151	40 08	198	42 06	244	50 28	299	30 26	342
333	41 07	034	52 47	105	29 28	151	39 49	199	41 13	245	49 36	298	30 08	342
334	41 39	033	53 43	106	29 56	151	39 30	199	40 19	245	48 44	297	29 49	341
335	42 11	032	54 40	106	30 24	152	39 11	200	39 26	245	47 52	297	29 29	340
336	42 43	031	55 37	106	30 52	152	38 50	200	38 33	245	46 59	296	29 09	340
337	43 13	030	56 33	106	31 20	152	38 30	201	37 39	245	46 06	295	28 48	339
338	43 42	030	57 30	106	31 47	153	38 08	201	36 46	245	45 12	295	28 27	338
339	44 11	028	58 26	107	32 14	153	37 46	202	35 52	245	44 19	294	28 04	337
340	44 39	027	59 23	107	32 41	153	37 24	203	34 59	245	43 25	294	27 42	337
341	45 05	026	60 19	107	33 07	154	37 01	203	34 05	246	42 31	293	27 18	336
342	45 31	025	61 15	108	33 33	154	36 38	203	33 12	246	41 36	292	26 54	336
343	45 56	024	62 11	108	33 58	154	36 14	204	32 18	246	40 42	292	26 29	335
344	46 19	023	63 07	108	34 24	155	35 50	204	31 24	246	39 47	291	26 04	334
	Alpheratz		◆Hamal		◆ACHERNAR		Peacock		Nunki		◆ALTAIR		DENEB	
345	46 42	022	32 47	053	34 48	155	35 26	205	30 31	246	38 52	291	25 38	334
346	47 03	021	33 34	052	35 13	156	35 01	205	29 37	246	37 57	291	25 12	333
347	47 23	020	34 20	051	35 37	156	34 35	206	28 43	246	37 02	290	24 45	332
348	47 43	018	35 06	051	36 00	157	34 10	206	27 49	246	36 07	290	24 17	332
349	48 00	017	35 51	050	36 24	157	33 43	207	26 55	246	35 11	289	23 49	331
350	48 17	016	36 36	050	36 46	158	33 17	207	26 02	246	34 15	289	23 20	331
351	48 33	015	37 21	049	37 08	158	32 50	207	25 08	246	33 20	288	22 51	330
352	48 47	013	38 05	048	37 30	159	32 23	208	24 14	246	32 24	288	22 22	330
353	49 00	012	38 49	047	37 51	159	31 55	208	23 20	246	31 28	288	21 52	329
354	49 12	011	39 32	047	38 12	160	31 28	208	22 27	246	30 31	287	21 21	329
355	49 22	009	40 14	046	38 32	160	30 59	209	21 33	246	29 35	287	20 50	328
356	49 31	008	40 56	045	38 52	161	30 31	209	20 39	246	28 39	287	20 19	328
357	49 39	007	41 38	044	39 11	161	30 02	209	19 45	246	27 42	286	19 47	327
358	49 45	005	42 19	044	39 30	162	29 33	210	18 52	246	26 46	286	19 15	327
359	49 50	004	42 59	043	39 48	162	29 04	210	17 58	246	25 49	286	18 42	326

LHA	Hc	Zn	Hc	Zn	Hc	Zn	Hc	Zn	Hc	Zn	Hc	Zn	Hc	Zn
♈	◆Alpheratz		Hamal		ALDEBARAN		◆RIGEL		ACHERNAR		◆FOMALHAUT		Enif	
0	48 53	003	42 54	041	16 10	069	12 48	096	41 03	163	67 08	217	49 45	301
1	48 56	001	43 32	040	17 05	068	13 46	096	41 20	163	66 32	219	48 55	301
2	48 56	000	44 10	039	17 59	068	14 45	095	41 36	164	65 54	220	48 04	300
3	48 56	359	44 46	038	18 54	068	15 43	095	41 52	165	65 16	222	47 13	299
4	48 54	357	45 22	037	19 48	067	16 41	095	42 07	165	64 36	223	46 22	298
5	48 50	356	45 58	036	20 42	067	17 40	095	42 22	166	63 56	224	45 30	298
6	48 46	355	46 32	035	21 36	067	18 38	095	42 36	167	63 15	225	44 38	297
7	48 39	353	47 06	034	22 30	066	19 37	095	42 49	167	62 32	227	43 45	296
8	48 32	352	47 38	033	23 23	066	20 35	094	43 02	168	61 49	228	42 53	296
9	48 23	351	48 10	032	24 17	066	21 34	094	43 14	169	61 06	229	42 00	295
10	48 13	350	48 41	031	25 10	065	22 32	094	43 25	169	60 21	230	41 07	295
11	48 02	348	49 10	030	26 03	065	23 31	094	43 35	170	59 36	230	40 13	294
12	47 49	347	49 39	029	26 56	064	24 30	094	43 45	171	58 51	231	39 19	294
13	47 36	346	50 06	027	27 49	064	25 28	093	43 54	171	58 05	232	38 26	293
14	47 20	344	50 33	026	28 42	063	26 27	093	44 03	172	57 18	233	37 31	293
	Hamal		◆ALDEBARAN		RIGEL		◆ACHERNAR		FOMALHAUT		◆Enif		Alpheratz	
15	50 58	025	29 34	063	27 25	093	44 10	173	56 32	233	36 37	292	47 04	343
16	51 22	023	30 26	063	28 24	093	44 17	174	55 44	234	35 43	292	46 47	342
17	51 45	022	31 18	062	29 23	093	44 23	174	54 56	235	34 48	291	46 28	341
18	52 06	021	32 10	062	30 21	093	44 28	175	54 08	235	33 53	291	46 08	340
19	52 26	019	33 01	061	31 20	092	44 33	176	53 20	236	32 58	290	45 47	339
20	52 45	018	33 52	061	32 18	092	44 37	177	52 31	236	32 03	290	45 25	337
21	53 02	016	34 43	060	33 17	092	44 40	177	51 42	237	31 08	289	45 02	336
22	53 18	015	35 34	059	34 16	092	44 42	178	50 53	237	30 12	289	44 38	335
23	53 33	014	36 24	059	35 14	092	44 44	179	50 03	238	29 17	289	44 13	334
24	53 46	012	37 15	058	36 13	091	44 44	180	49 14	238	28 21	288	43 47	333
25	53 57	011	38 04	058	37 12	091	44 44	180	48 24	239	27 25	288	43 20	332
26	54 07	009	38 54	057	38 10	091	44 43	181	47 34	239	26 30	288	42 52	331
27	54 16	007	39 43	056	39 09	091	44 42	182	46 43	239	25 33	287	42 23	330
28	54 22	006	40 31	056	40 08	091	44 39	183	45 53	240	24 37	287	41 53	329
29	54 28	004	41 20	055	41 06	090	44 36	184	45 02	240	23 41	287	41 23	328
	CAPELLA		◆BETELGEUSE		SIRIUS		CANOPUS		◆ACHERNAR		FOMALHAUT		◆Alpheratz	
30	17 12	033	28 28	075	21 10	103	24 01	143	44 32	184	44 11	240	40 51	327
31	17 44	033	29 25	074	22 08	103	24 36	143	44 27	185	43 20	240	40 19	326
32	18 15	032	30 21	074	23 05	103	25 11	143	44 22	186	42 29	241	39 46	325
33	18 46	032	31 18	074	24 02	103	25 47	143	44 15	186	41 38	241	39 12	325
34	19 17	031	32 14	073	24 59	103	26 22	143	44 08	187	40 47	241	38 38	324
35	19 48	031	33 10	073	25 56	103	26 57	144	44 01	188	39 55	241	38 03	323
36	20 18	030	34 06	072	26 53	103	27 32	144	43 52	189	39 04	241	37 27	322
37	20 47	030	35 02	072	27 51	103	28 06	144	43 43	189	38 12	242	36 51	321
38	21 16	029	35 58	072	28 48	103	28 41	144	43 33	190	37 21	242	36 14	320
39	21 44	029	36 53	071	29 45	103	29 15	144	43 22	191	36 29	242	35 36	320
40	22 13	028	37 49	071	30 42	102	29 49	145	43 11	192	35 37	242	34 58	319
41	22 40	028	38 44	070	31 40	102	30 23	145	42 59	192	34 45	242	34 19	318
42	23 07	027	39 39	070	32 37	102	30 57	145	42 46	193	33 54	242	33 40	318
43	23 33	027	40 34	069	33 34	102	31 30	145	42 33	194	33 02	242	33 00	317
44	23 59	026	41 29	069	34 32	102	32 03	146	42 19	194	32 10	242	32 19	316
	◆CAPELLA		BETELGEUSE		SIRIUS		◆CANOPUS		ACHERNAR		◆FOMALHAUT		Alpheratz	
45	24 25	025	42 23	068	35 29	102	32 36	146	42 04	195	31 18	242	31 39	316
46	24 50	025	43 18	068	36 27	102	33 09	146	41 49	196	30 26	243	30 57	315
47	25 14	024	44 12	067	37 24	102	33 41	147	41 33	196	29 34	243	30 16	314
48	25 37	023	45 06	066	38 21	102	34 13	147	41 16	197	28 41	243	29 33	314
49	26 00	023	45 59	066	39 19	102	34 45	147	40 59	197	27 49	243	28 51	313
50	26 23	022	46 53	065	40 16	102	35 16	148	40 41	198	26 57	243	28 08	313
51	26 45	021	47 46	065	41 14	102	35 48	148	40 23	199	26 05	243	27 24	312
52	27 06	021	48 39	064	42 11	102	36 18	149	40 04	199	25 13	243	26 41	311
53	27 26	020	49 31	063	43 09	102	36 49	149	39 44	200	24 21	243	25 56	311
54	27 46	019	50 24	062	44 06	102	37 19	149	39 24	200	23 29	243	25 12	310
55	28 05	019	51 15	062	45 04	102	37 49	150	39 04	201	22 36	243	24 27	310
56	28 24	018	52 07	061	46 01	102	38 18	150	38 42	201	21 44	243	23 42	309
57	28 42	017	52 58	060	46 58	102	38 47	151	38 21	202	20 52	243	22 57	309
58	28 59	017	53 48	059	47 56	102	39 15	151	37 59	202	20 00	243	22 11	309
59	29 15	016	54 38	058	48 53	102	39 43	152	37 36	203	19 08	243	21 25	308
	CAPELLA		POLLUX		◆SIRIUS		CANOPUS		◆ACHERNAR		Diphda		◆Hamal	
60	29 31	015	22 27	053	49 51	102	40 11	152	37 13	203	42 16	256	45 03	322
61	29 46	014	23 13	052	50 48	102	40 38	153	36 50	204	41 19	256	44 26	321
62	30 00	014	24 00	052	51 46	102	41 05	153	36 26	204	40 22	256	43 49	320
63	30 14	013	24 46	051	52 43	102	41 31	154	36 02	205	39 25	256	43 11	319
64	30 26	012	25 31	051	53 41	102	41 56	154	35 37	205	38 28	256	42 32	318
65	30 38	011	26 16	050	54 38	102	42 21	155	35 12	206	37 31	256	41 53	317
66	30 49	011	27 01	050	55 35	102	42 46	156	34 46	206	36 33	256	41 13	317
67	31 00	010	27 46	049	56 33	102	43 10	156	34 21	206	35 36	256	40 32	316
68	31 09	009	28 30	049	57 30	102	43 33	157	33 54	207	34 39	256	39 51	315
69	31 18	008	29 14	048	58 27	102	43 56	158	33 28	207	33 42	256	39 09	314
70	31 26	007	29 57	047	59 25	103	44 18	158	33 01	207	32 45	256	38 27	313
71	31 33	007	30 40	047	60 22	103	44 39	159	32 34	208	31 48	256	37 44	313
72	31 39	006	31 23	046	61 19	103	45 00	160	32 06	208	30 51	256	37 01	312
73	31 45	005	32 05	046	62 16	103	45 20	160	31 38	208	29 55	256	36 17	311
74	31 50	004	32 46	045	63 14	103	45 40	161	31 10	209	28 58	256	35 33	311
	CAPELLA		POLLUX		◆PROCYON		Suhail		CANOPUS		◆ACHERNAR		◆Hamal	
75	31 53	003	33 28	044	46 52	069	28 29	133	45 58	162	30 42	209	34 48	310
76	31 56	003	34 08	044	47 47	068	29 12	133	46 16	163	30 13	209	34 03	309
77	31 59	002	34 48	043	48 41	067	29 55	133	46 33	163	29 44	210	33 17	309
78	32 00	001	35 28	042	49 35	067	30 37	134	46 50	164	29 15	210	32 31	308
79	32 00	000	36 07	041	50 29	066	31 20	134	47 05	165	28 46	210	31 45	308
80	32 00	359	36 46	041	51 23	065	32 02	134	47 20	166	28 16	210	30 58	307
81	31 59	358	37 24	040	52 16	065	32 44	134	47 34	167	27 46	211	30 11	306
82	31 57	358	38 01	039	53 09	064	33 26	135	47 47	167	27 16	211	29 24	306
83	31 54	357	38 38	038	54 01	063	34 08	135	48 00	168	26 46	211	28 36	305
84	31 50	356	39 14	037	54 53	062	34 49	135	48 11	169	26 16	211	27 48	305
85	31 46	355	39 49	037	55 45	061	35 30	135	48 22	170	25 45	212	27 00	304
86	31 40	354	40 24	036	56 36	061	36 10	136	48 32	171	25 14	212	26 11	304
87	31 34	354	40 58	035	57 27	060	36 52	136	48 41	172	24 43	212	25 22	303
88	31 27	353	41 31	034	58 18	059	37 33	136	48 48	173	24 12	212	24 33	303
89	31 19	352	42 03	033	59 07	057	38 14	137	48 55	174	23 41	212	23 44	303

LHA	Hc	Zn	Hc	Zn	Hc	Zn	Hc	Zn	Hc	Zn	Hc	Zn	Hc	Zn
♈	POLLUX		◆REGULUS		Suhail		◆CANOPUS		ACHERNAR		◆ALDEBARAN		CAPELLA	
90	42 35	032	23 56	071	38 54	137	49 02	174	23 09	212	54 41	323	31 11	351
91	43 05	031	24 51	071	39 34	137	49 07	175	22 38	213	54 06	322	31 01	350
92	43 35	030	25 46	070	40 13	138	49 11	176	22 06	213	53 29	321	30 51	350
93	44 04	029	26 42	070	40 53	138	49 14	177	21 35	213	52 51	320	30 40	349
94	44 32	028	27 37	069	41 32	139	49 17	178	21 03	213	52 13	318	30 28	348
95	44 59	027	28 32	069	42 10	139	49 18	179	20 31	213	51 34	317	30 16	347
96	45 25	026	29 26	069	42 48	140	49 18	180	19 59	213	50 53	316	30 02	346
97	45 50	025	30 21	068	43 26	140	49 18	181	19 27	213	50 12	315	29 48	346
98	46 14	024	31 15	068	44 04	141	49 16	182	18 55	213	49 31	314	29 33	345
99	46 37	022	32 10	067	44 41	141	49 14	183	18 22	213	48 48	313	29 18	344
100	46 59	021	33 04	067	45 18	142	49 11	184	17 50	213	48 05	312	29 02	344
101	47 19	020	33 58	066	45 54	142	49 06	185	17 18	213	47 21	311	28 45	343
102	47 39	019	34 51	066	46 29	143	49 01	186	16 45	213	46 37	310	28 27	342
103	47 57	018	35 45	065	47 05	143	48 55	186	16 13	214	45 52	309	28 08	341
104	48 14	016	36 38	065	47 39	144	48 48	187	15 41	214	45 06	309	27 49	341
	POLLUX		◆REGULUS		ACRUX		◆CANOPUS		RIGEL		◆ALDEBARAN		CAPELLA	
105	48 30	015	37 31	064	14 30	152	48 40	188	63 43	276	44 20	308	27 30	340
106	48 45	014	38 24	064	14 57	152	48 31	189	62 45	275	43 33	307	27 09	339
107	48 58	012	39 17	063	15 24	152	48 21	190	61 46	275	42 46	306	26 48	339
108	49 10	011	40 09	063	15 51	153	48 10	191	60 48	275	41 59	306	26 26	338
109	49 21	010	41 01	062	16 18	153	47 59	192	59 49	274	41 11	305	26 04	337
110	49 30	009	41 53	062	16 45	153	47 46	193	58 51	274	40 22	304	25 41	337
111	49 38	007	42 44	061	17 12	153	47 33	193	57 52	274	39 34	304	25 18	336
112	49 45	006	43 35	060	17 39	153	47 19	194	56 54	273	38 45	303	24 53	335
113	49 50	004	44 26	060	18 06	153	47 04	195	55 55	273	37 55	302	24 29	335
114	49 54	003	45 17	059	18 32	153	46 49	196	54 57	273	37 05	302	24 03	334
115	49 57	002	46 07	058	18 59	153	46 32	197	53 58	273	36 15	301	23 38	334
116	49 58	000	46 56	057	19 26	153	46 15	197	52 59	272	35 25	300	23 11	333
117	49 57	359	47 45	057	19 52	153	45 57	198	52 01	272	34 34	300	22 44	332
118	49 56	358	48 34	056	20 19	153	45 38	199	51 02	272	33 43	299	22 17	332
119	49 52	356	49 23	055	20 45	153	45 19	200	50 03	272	32 52	299	21 49	331
	◆REGULUS		SPICA		◆ACRUX		CANOPUS		◆RIGEL		BETELGEUSE		POLLUX	
120	50 10	054	10 45	099	21 11	154	44 58	200	49 05	271	53 21	300	49 48	355
121	50 58	053	11 43	099	21 37	154	44 38	201	48 06	271	52 30	300	49 42	354
122	51 44	052	12 41	099	22 03	154	44 16	202	47 07	271	51 39	299	49 35	352
123	52 30	051	13 39	099	22 29	154	43 54	202	46 09	271	50 47	298	49 26	351
124	53 16	050	14 37	099	22 55	154	43 31	203	45 10	270	49 55	297	49 16	349
125	54 01	049	15 35	098	23 20	154	43 08	204	44 11	270	49 03	296	49 05	348
126	54 45	048	16 33	098	23 46	154	42 44	204	43 13	270	48 10	296	48 52	347
127	55 28	047	17 31	098	24 11	155	42 19	205	42 14	270	47 17	295	48 38	346
128	56 10	046	18 29	098	24 36	155	41 54	206	41 15	270	46 24	294	48 23	344
129	56 52	044	19 27	098	25 01	155	41 29	206	40 16	269	45 30	294	48 06	343
130	57 32	043	20 25	098	25 26	155	41 03	207	39 18	269	44 36	293	47 48	342
131	58 12	042	21 24	097	25 50	155	40 36	207	38 19	269	43 42	293	47 29	341
132	58 51	040	22 22	097	26 14	156	40 09	208	37 20	269	42 48	292	47 09	339
133	59 28	039	23 20	097	26 38	156	39 41	208	36 22	269	41 53	292	46 48	338
134	60 04	037	24 18	097	27 02	156	39 13	209	35 23	268	40 59	291	46 26	337
	REGULUS		◆SPICA		ACRUX		◆CANOPUS		SIRIUS		BETELGEUSE		◆POLLUX	
135	60 39	036	25 16	097	27 26	156	38 45	209	56 59	258	40 04	291	46 02	336
136	61 13	034	26 15	097	27 49	157	38 16	210	56 02	258	39 09	290	45 38	335
137	61 45	032	27 13	096	28 12	157	37 46	210	55 05	258	38 14	290	45 12	334
138	62 16	031	28 11	096	28 35	157	37 17	211	54 07	258	37 18	289	44 46	333
139	62 45	029	29 10	096	28 58	158	36 46	211	53 10	258	36 23	289	44 18	332
140	63 12	027	30 08	096	29 20	158	36 16	212	52 12	258	35 27	288	43 50	330
141	63 38	025	31 06	096	29 42	158	35 45	212	51 15	258	34 31	288	43 20	329
142	64 02	023	32 05	096	30 04	158	35 14	212	50 17	258	33 35	287	42 50	328
143	64 23	021	33 03	096	30 25	159	34 42	213	49 20	258	32 39	287	42 19	328
144	64 43	019	34 02	095	30 46	159	34 11	213	48 23	258	31 43	287	41 47	327
145	65 01	016	35 00	095	31 07	159	33 39	213	47 25	258	30 47	286	41 14	326
146	65 16	014	35 59	095	31 27	160	33 06	214	46 28	258	29 50	286	40 41	325
147	65 30	012	36 57	095	31 47	160	32 33	214	45 30	258	28 54	286	40 07	324
148	65 41	010	37 55	095	32 07	161	32 00	214	44 33	258	27 57	285	39 32	323
149	65 49	007	38 54	095	32 26	161	31 27	215	43 35	258	27 01	285	38 56	322
	REGULUS		◆ARCTURUS		SPICA		◆ACRUX		CANOPUS		SIRIUS		◆PROCYON	
150	65 55	005	19 47	064	39 52	095	32 46	161	30 54	215	42 38	258	50 57	294
151	65 59	002	20 40	064	40 51	094	33 04	162	30 20	215	41 40	258	50 03	294
152	66 00	000	21 33	064	41 49	094	33 22	162	29 46	215	40 43	258	49 09	293
153	65 59	358	22 25	063	42 48	094	33 40	162	29 12	216	39 45	258	48 15	292
154	65 56	355	23 18	063	43 47	094	33 58	163	28 38	216	38 48	258	47 21	292
155	65 50	353	24 10	062	44 45	094	34 15	163	28 03	216	37 51	258	46 26	291
156	65 41	351	25 02	062	45 44	094	34 31	164	27 29	216	36 53	258	45 31	290
157	65 30	348	25 53	062	46 42	094	34 48	164	26 54	216	35 56	258	44 36	290
158	65 17	346	26 45	061	47 41	093	35 03	165	26 19	217	34 58	258	43 41	289
159	65 02	344	27 36	061	48 39	093	35 19	165	25 44	217	34 01	258	42 45	289
160	64 44	341	28 27	060	49 38	093	35 34	166	25 09	217	33 04	258	41 50	288
161	64 24	339	29 18	060	50 37	093	35 48	166	24 33	217	32 06	258	40 54	288
162	64 03	337	30 08	059	51 35	093	36 02	167	23 58	217	31 09	258	39 58	287
163	63 39	335	30 59	059	52 34	093	36 15	167	23 22	217	30 12	258	39 02	287
164	63 14	333	31 49	058	53 32	093	36 28	168	22 46	218	29 14	257	38 06	287
	Denebola		◆ARCTURUS		SPICA		◆ACRUX		SIRIUS		◆PROCYON		REGULUS	
165	60 47	025	32 38	058	54 31	092	36 41	168	28 17	257	37 09	286	62 46	331
166	61 11	023	33 28	057	55 30	092	36 53	169	27 20	257	36 13	286	62 17	329
167	61 33	021	34 17	056	56 28	092	37 04	169	26 23	257	35 16	285	61 47	328
168	61 53	019	35 06	056	57 27	092	37 15	170	25 25	257	34 20	285	61 15	326
169	62 11	017	35 54	055	58 26	092	37 25	170	24 28	257	33 23	285	60 41	324
170	62 28	015	36 42	055	59 24	092	37 35	171	23 31	257	32 26	284	60 06	323
171	62 42	013	37 30	054	60 23	091	37 44	171	22 34	257	31 29	284	59 30	321
172	62 54	011	38 17	053	61 22	091	37 53	172	21 37	257	30 32	284	58 53	320
173	63 05	009	39 04	053	62 20	091	38 01	172	20 40	256	29 35	283	58 14	318
174	63 13	007	39 50	052	63 19	091	38 09	173	19 43	256	28 38	283	57 35	317
175	63 19	005	40 36	051	64 18	091	38 16	173	18 46	256	27 41	283	56 54	316
176	63 22	003	41 22	051	65 16	091	38 23	174	17 49	256	26 43	282	56 12	314
177	63 24	000	42 07	050	66 15	090	38 29	174	16 52	256	25 46	282	55 30	313
178	63 23	358	42 52	049	67 14	090	38 34	175	15 55	256	24 49	282	54 47	312
179	63 20	356	43 36	048	68 12	090	38 39	176	14 58	256	23 51	281	54 03	311

LAT 12°S

LHA	Hc	Zn	Hc	Zn	Hc	Zn	Hc	Zn	Hc	Zn	Hc	Zn	Hc	Zn
♈	ARCTURUS		◆ANTARES		ACRUX		◆Suhail		Alphard		REGULUS		◆Denebola	
180	44 19	047	25 32	114	38 43	176	41 28	221	52 20	271	53 18	310	63 15	354
181	45 02	046	26 25	114	38 46	177	40 49	222	51 21	271	52 33	309	63 08	352
182	45 44	046	27 19	114	38 49	177	40 10	222	50 23	271	51 47	308	62 59	350
183	46 26	045	28 13	114	38 52	178	39 30	223	49 24	271	51 00	307	62 47	348
184	47 07	044	29 06	114	38 54	178	38 50	223	48 25	270	50 13	306	62 34	346
185	47 47	043	30 00	114	38 55	179	38 10	223	47 26	270	49 25	305	62 18	344
186	48 26	042	30 54	114	38 56	180	37 29	224	46 28	270	48 37	304	62 00	342
187	49 05	041	31 48	114	38 56	180	36 49	224	45 29	270	47 48	303	61 41	340
188	49 43	040	32 41	114	38 55	181	36 08	224	44 30	270	46 59	303	61 20	338
189	50 20	038	33 35	114	38 54	181	35 27	225	43 32	269	46 09	302	60 57	336
190	50 56	037	34 29	114	38 52	182	34 45	225	42 33	269	45 19	301	60 32	334
191	51 31	036	35 22	114	38 50	183	34 04	225	41 34	269	44 29	300	60 06	332
192	52 05	035	36 16	114	38 47	183	33 22	225	40 36	269	43 38	300	59 38	331
193	52 38	034	37 09	114	38 43	184	32 40	226	39 37	269	42 47	299	59 08	329
194	53 10	032	38 03	114	38 39	184	31 58	226	38 38	268	41 55	298	58 37	328
	ARCTURUS		◆Rasalhague		ANTARES		◆ACRUX		Suhail		◆REGULUS		Denebola	
195	53 41	031	17 35	073	38 56	114	38 35	185	31 16	226	41 04	298	58 05	326
196	54 10	030	18 30	072	39 50	115	38 29	185	30 33	226	40 12	297	57 32	324
197	54 39	028	19 26	072	40 43	115	38 24	186	29 51	227	39 19	297	56 57	323
198	55 06	027	20 22	072	41 36	115	38 17	187	29 08	227	38 27	296	56 21	322
199	55 32	025	21 18	071	42 30	115	38 10	187	28 25	227	37 34	296	55 44	320
200	55 56	024	22 13	071	43 23	115	38 03	188	27 42	227	36 41	295	55 06	319
201	56 19	022	23 09	071	44 16	115	37 54	188	26 59	227	35 48	295	54 27	318
202	56 41	021	24 04	070	45 09	116	37 46	189	26 16	227	34 54	294	53 47	317
203	57 01	019	24 59	070	46 02	116	37 37	189	25 33	227	34 00	294	53 07	315
204	57 19	017	25 54	069	46 55	116	37 27	190	24 50	227	33 06	293	52 25	314
205	57 36	016	26 49	069	47 47	116	37 17	190	24 07	228	32 12	293	51 43	313
206	57 51	014	27 44	069	48 40	117	37 06	191	23 23	228	31 18	292	51 00	312
207	58 04	012	28 38	068	49 32	117	36 54	191	22 40	228	30 24	292	50 16	311
208	58 16	011	29 33	068	50 25	117	36 43	192	21 57	228	29 29	291	49 31	310
209	58 26	009	30 27	067	51 17	118	36 30	192	21 13	228	28 34	291	48 46	309
	ARCTURUS		◆Rasalhague		ANTARES		◆RIGIL KENT.		ACRUX		Gienah		◆REGULUS	
210	58 34	007	31 21	067	52 09	118	40 39	174	36 17	193	64 11	255	27 40	291
211	58 40	005	32 15	067	53 00	118	40 45	174	36 04	193	63 14	255	26 45	290
212	58 44	003	33 09	066	53 52	119	40 51	175	35 50	194	62 17	255	25 49	290
213	58 47	002	34 03	066	54 43	119	40 56	176	35 36	194	61 21	255	24 54	289
214	58 47	000	34 56	065	55 35	120	41 00	176	35 21	195	60 24	256	23 59	289
215	58 46	358	35 49	065	56 25	120	41 03	177	35 06	195	59 27	256	23 03	289
216	58 43	356	36 42	064	57 16	121	41 06	178	34 50	196	58 30	256	22 08	288
217	58 38	354	37 35	064	58 06	121	41 08	178	34 34	196	57 33	256	21 12	288
218	58 32	352	38 27	063	58 56	122	41 10	179	34 17	197	56 36	256	20 16	288
219	58 23	351	39 19	063	59 46	123	41 11	179	34 00	197	55 39	256	19 20	287
220	58 13	349	40 11	062	60 35	123	41 11	180	33 43	197	54 42	257	18 24	287
221	58 00	347	41 03	061	61 24	124	41 11	181	33 25	198	53 45	257	17 28	287
222	57 47	345	41 54	061	62 12	125	41 10	181	33 07	198	52 48	257	16 32	287
223	57 31	344	42 45	060	63 00	126	41 08	182	32 48	199	51 51	257	15 35	286
224	57 14	342	43 36	059	63 48	127	41 05	183	32 30	199	50 54	257	14 39	286
	Alphecca		◆VEGA		ALTAIR		◆Shaula		RIGIL KENT.		◆SPICA		ARCTURUS	
225	50 22	012	18 25	042	14 52	077	47 31	133	41 02	183	66 42	270	56 55	340
226	50 34	011	19 04	041	15 49	077	48 14	133	40 58	184	65 44	269	56 35	339
227	50 44	009	19 42	041	16 46	077	48 56	134	40 54	185	64 45	269	56 13	337
228	50 53	008	20 21	040	17 43	077	49 38	135	40 49	185	63 46	269	55 49	336
229	51 01	007	20 59	040	18 40	076	50 20	135	40 43	186	62 48	269	55 25	334
230	51 07	005	21 36	039	19 37	076	51 01	136	40 37	187	61 49	269	54 58	333
231	51 11	004	22 13	039	20 34	076	51 42	137	40 30	187	60 50	269	54 31	331
232	51 14	002	22 50	038	21 31	075	52 22	137	40 22	188	59 52	268	54 02	330
233	51 16	001	23 26	038	22 28	075	53 01	138	40 14	188	58 53	268	53 32	329
234	51 16	359	24 02	037	23 24	075	53 40	139	40 05	189	57 54	268	53 01	327
235	51 15	358	24 37	037	24 21	074	54 19	140	39 56	190	56 56	268	52 29	326
236	51 12	357	25 12	036	25 17	074	54 56	140	39 46	190	55 57	268	51 55	325
237	51 08	355	25 46	036	26 14	074	55 34	141	39 35	191	54 58	268	51 21	324
238	51 02	354	26 20	035	27 10	073	56 10	142	39 24	191	54 00	268	50 46	322
239	50 55	352	26 53	034	28 06	073	56 45	143	39 12	192	53 01	267	50 09	321
	VEGA		◆ALTAIR		Shaula		◆RIGIL KENT.		SPICA		◆ARCTURUS		Alphecca	
240	27 26	034	29 02	073	57 20	144	38 59	193	52 02	267	49 32	320	50 47	351
241	27 59	033	29 58	072	57 54	145	38 46	193	51 04	267	48 54	319	50 37	350
242	28 30	032	30 54	072	58 27	146	38 33	194	50 05	267	48 15	318	50 25	348
243	29 01	032	31 50	072	58 59	147	38 19	194	49 07	267	47 36	317	50 13	347
244	29 32	031	32 45	071	59 30	149	38 04	195	48 08	267	46 55	316	49 59	346
245	30 02	030	33 41	071	60 00	150	37 49	195	47 09	266	46 14	315	49 43	344
246	30 31	030	34 36	070	60 29	151	37 33	196	46 11	266	45 32	314	49 27	343
247	31 00	029	35 31	070	60 57	153	37 17	196	45 12	266	44 50	313	49 09	342
248	31 28	028	36 26	069	61 23	154	37 01	197	44 14	266	44 07	312	48 50	340
249	31 56	028	37 21	069	61 48	155	36 43	197	43 15	266	43 23	312	48 29	339
250	32 23	027	38 16	068	62 12	157	36 26	198	42 17	266	42 39	311	48 08	338
251	32 49	026	39 10	068	62 34	158	36 08	198	41 18	266	41 54	310	47 45	337
252	33 14	025	40 05	067	62 55	160	35 49	199	40 20	265	41 09	309	47 21	335
253	33 39	024	40 59	067	63 15	161	35 30	199	39 21	265	40 23	308	46 56	334
254	34 02	024	41 52	066	63 33	163	35 11	200	38 23	265	39 37	308	46 30	333
	◆VEGA		ALTAIR		◆Peacock		RIGIL KENT.		◆SPICA		ARCTURUS		Alphecca	
255	34 26	023	42 46	066	30 36	150	34 51	200	37 24	265	38 51	307	46 03	332
256	34 48	022	43 39	065	31 05	150	34 30	200	36 26	265	38 04	306	45 35	331
257	35 09	021	44 33	065	31 34	151	34 10	201	35 27	265	37 16	306	45 06	330
258	35 30	020	45 25	064	32 03	151	33 49	201	34 29	265	36 28	305	44 36	329
259	35 50	019	46 18	063	32 31	151	33 27	202	33 30	264	35 40	305	44 05	328
260	36 09	019	47 10	063	32 59	152	33 05	202	32 32	264	34 52	304	43 34	327
261	36 27	018	48 02	062	33 26	152	32 43	202	31 34	264	34 03	303	43 01	326
262	36 45	017	48 54	061	33 53	153	32 20	203	30 35	264	33 14	303	42 28	325
263	37 01	016	49 45	060	34 20	153	31 57	203	29 37	264	32 24	302	41 53	324
264	37 17	015	50 36	060	34 47	153	31 34	204	28 38	264	31 34	302	41 19	323
265	37 31	014	51 26	059	35 13	154	31 11	204	27 40	264	30 44	301	40 43	322
266	37 45	013	52 16	058	35 39	154	30 47	204	26 42	263	29 54	301	40 07	321
267	37 58	012	53 05	057	36 04	155	30 23	204	25 44	263	29 03	300	39 30	320
268	38 09	011	53 54	056	36 29	155	29 58	205	24 45	263	28 13	300	38 52	320
269	38 20	010	54 43	055	36 54	155	29 33	205	23 47	263	27 21	299	38 13	319

LHA	Hc	Zn	Hc	Zn	Hc	Zn	Hc	Zn	Hc	Zn	Hc	Zn	Hc	Zn
♈	◆VEGA		ALTAIR		◆FOMALHAUT		Peacock		RIGIL KENT.		◆ANTARES		ARCTURUS	
270	38 30	009	55 31	054	19 24	117	37 18	156	29 08	205	64 13	233	26 30	299
271	38 39	008	56 18	053	20 16	117	37 42	156	28 43	206	63 26	234	25 39	298
272	38 47	007	57 04	052	21 09	117	38 05	157	28 17	206	62 39	235	24 47	298
273	38 54	006	57 50	051	22 01	117	38 28	157	27 52	206	61 51	235	23 55	298
274	38 59	005	58 35	049	22 53	117	38 50	158	27 26	206	61 02	236	23 03	297
275	39 04	004	59 19	048	23 45	117	39 12	158	26 59	207	60 13	237	22 10	297
276	39 08	003	60 03	047	24 37	117	39 33	159	26 33	207	59 24	238	21 18	296
277	39 11	002	60 45	045	25 29	117	39 54	160	26 06	207	58 34	238	20 25	296
278	39 13	001	61 26	044	26 21	117	40 14	160	25 39	207	57 44	239	19 32	296
279	39 13	000	62 06	042	27 13	117	40 34	161	25 12	208	56 53	240	18 39	295
280	39 13	359	62 45	041	28 06	117	40 53	161	24 45	208	56 03	240	17 46	295
281	39 12	358	63 23	039	28 58	117	41 12	162	24 18	208	55 12	241	16 53	295
282	39 09	357	63 59	037	29 50	117	41 30	162	23 50	208	54 20	241	16 00	294
283	39 06	356	64 34	036	30 42	118	41 47	163	23 22	208	53 29	241	15 06	294
284	39 01	355	65 08	034	31 34	118	42 04	164	22 54	208	52 37	242	14 12	294
	◆DENEB		◆FOMALHAUT		Peacock		RIGIL KENT.		◆ANTARES		Rasalhague		VEGA	
285	28 20	020	32 26	118	42 20	164	22 26	209	51 45	242	57 35	319	38 56	354
286	28 40	019	33 18	118	42 36	165	21 58	209	50 53	243	56 55	317	38 49	353
287	28 59	019	34 10	118	42 51	166	21 30	209	50 01	243	56 15	316	38 42	352
288	29 17	018	35 02	118	43 05	166	21 01	209	49 09	243	55 34	315	38 34	351
289	29 35	017	35 53	118	43 18	167	20 33	209	48 16	244	54 52	314	38 24	350
290	29 52	016	36 45	118	43 31	168	20 04	209	47 24	244	54 09	312	38 14	349
291	30 08	016	37 37	118	43 43	168	19 36	209	46 31	244	53 25	311	38 02	348
292	30 24	015	38 29	118	43 55	169	19 07	209	45 38	244	52 40	310	37 50	347
293	30 38	014	39 20	119	44 05	170	18 38	210	44 45	245	51 55	309	37 37	346
294	30 52	013	40 12	119	44 15	171	18 09	210	43 52	245	51 09	308	37 22	345
295	31 05	013	41 03	119	44 24	171	17 40	210	42 59	245	50 23	307	37 07	345
296	31 18	012	41 54	119	44 33	172	17 11	210	42 06	245	49 36	306	36 51	344
297	31 29	011	42 45	119	44 41	173	16 42	210	41 12	245	48 49	305	36 34	343
298	31 40	010	43 36	120	44 48	174	16 13	210	40 19	245	48 01	305	36 16	342
299	31 50	009	44 27	120	44 54	174	15 43	210	39 26	246	47 12	304	35 58	341
	◆DENEB		Alpheratz		◆FOMALHAUT		Peacock		◆ANTARES		Rasalhague		VEGA	
300	31 59	009	17 27	054	45 18	120	44 59	175	38 32	246	46 23	303	35 38	340
301	32 07	008	18 15	054	46 09	121	45 04	176	37 39	246	45 34	302	35 18	339
302	32 15	007	19 02	053	46 59	121	45 08	177	36 45	246	44 44	302	34 56	338
303	32 21	006	19 49	053	47 49	121	45 11	177	35 52	246	43 54	301	34 34	338
304	32 27	005	20 35	052	48 39	122	45 13	178	34 58	246	43 03	300	34 11	337
305	32 32	004	21 22	052	49 29	122	45 14	179	34 04	246	42 12	300	33 48	336
306	32 36	004	22 08	051	50 19	122	45 15	180	33 11	246	41 21	299	33 24	335
307	32 40	003	22 54	051	51 08	123	45 15	181	32 17	246	40 29	298	32 58	334
308	32 42	002	23 39	051	51 58	123	45 14	181	31 23	246	39 38	298	32 33	334
309	32 44	001	24 24	050	52 46	124	45 12	182	30 30	246	38 45	297	32 06	333
310	32 44	000	25 09	050	53 35	124	45 09	183	29 36	246	37 53	297	31 39	332
311	32 44	359	25 54	049	54 23	125	45 06	184	28 42	246	37 01	296	31 11	331
312	32 43	359	26 38	049	55 11	125	45 02	184	27 49	246	36 08	295	30 43	331
313	32 41	358	27 22	048	55 59	126	44 57	185	26 55	246	35 15	295	30 13	330
314	32 38	357	28 05	047	56 46	127	44 51	186	26 01	246	34 21	294	29 44	329
	Alpheratz		◆Diphda		ACHERNAR		Peacock		◆ANTARES		VEGA		◆DENEB	
315	28 48	047	35 55	104	21 10	147	44 45	187	25 07	246	29 13	328	32 35	356
316	29 31	046	36 52	104	21 42	147	44 38	187	24 14	246	28 42	328	32 30	355
317	30 13	046	37 49	104	22 14	147	44 30	188	23 20	246	28 11	327	32 25	354
318	30 55	045	38 46	104	22 45	147	44 21	189	22 26	246	27 39	327	32 19	354
319	31 36	044	39 43	104	23 17	148	44 12	190	21 33	246	27 06	326	32 12	353
320	32 17	044	40 41	104	23 48	148	44 01	190	20 39	246	26 33	325	32 04	352
321	32 57	043	41 38	104	24 19	148	43 50	191	19 45	246	25 59	325	31 56	351
322	33 37	042	42 35	104	24 50	148	43 39	192	18 52	246	25 25	324	31 46	350
323	34 16	042	43 32	104	25 21	148	43 26	193	17 58	246	24 50	324	31 36	350
324	34 55	041	44 29	104	25 52	149	43 13	193	17 05	246	24 15	323	31 25	349
325	35 34	040	45 26	104	26 23	149	43 00	194	16 11	246	23 40	322	31 13	348
326	36 11	040	46 23	104	26 53	149	42 45	195	15 18	246	23 04	322	31 00	347
327	36 48	039	47 20	104	27 23	149	42 30	195	14 24	245	22 27	321	30 47	346
328	37 25	038	48 17	104	27 53	149	42 14	196	13 31	245	21 50	321	30 33	346
329	38 01	037	49 14	104	28 23	150	41 58	197	12 38	245	21 13	320	30 18	345
	◆Alpheratz		Diphda		◆ACHERNAR		Peacock		◆Nunki		ALTAIR		DENEB	
330	38 36	036	50 11	104	28 53	150	41 41	197	44 18	245	51 40	301	30 02	344
331	39 10	036	51 08	104	29 22	150	41 23	198	43 25	245	50 49	301	29 46	343
332	39 44	035	52 05	104	29 51	150	41 05	198	42 31	245	49 59	300	29 29	343
333	40 17	034	53 02	104	30 20	151	40 46	199	41 38	245	49 08	299	29 11	342
334	40 49	033	53 59	104	30 49	151	40 27	200	40 45	245	48 16	298	28 52	341
335	41 21	032	54 56	104	31 17	151	40 07	200	39 51	246	47 24	298	28 33	340
336	41 51	031	55 52	104	31 45	152	39 47	201	38 58	246	46 32	297	28 13	340
337	42 21	030	56 49	105	32 13	152	39 26	201	38 04	246	45 40	296	27 52	339
338	42 50	029	57 46	105	32 40	152	39 04	202	37 11	246	44 47	296	27 31	338
339	43 18	028	58 43	105	33 07	153	38 42	202	36 17	246	43 54	295	27 09	338
340	43 45	027	59 39	105	33 34	153	38 19	203	35 23	246	43 00	294	26 46	337
341	44 11	026	60 36	105	34 01	153	37 56	203	34 30	246	42 07	294	26 23	336
342	44 37	025	61 33	106	34 27	154	37 33	204	33 36	246	41 13	293	25 59	336
343	45 01	024	62 29	106	34 53	154	37 09	204	32 42	246	40 19	293	25 35	335
344	45 24	023	63 25	106	35 18	155	36 45	205	31 49	246	39 25	292	25 10	334
	Alpheratz		◆Hamal		◆ACHERNAR		Peacock		Nunki		◆ALTAIR		DENEB	
345	45 46	022	32 11	052	35 43	155	36 20	205	30 55	246	38 30	292	24 44	334
346	46 07	020	32 57	052	36 07	155	35 55	206	30 01	246	37 36	291	24 18	333
347	46 27	019	33 42	051	36 32	156	35 29	206	29 07	246	36 41	291	23 51	333
348	46 46	018	34 28	050	36 55	156	35 03	206	28 13	246	35 46	290	23 24	332
349	47 03	017	35 13	050	37 19	157	34 37	207	27 20	246	34 51	290	22 56	331
350	47 19	016	35 57	049	37 42	157	34 10	207	26 26	246	33 56	289	22 28	331
351	47 35	014	36 41	048	38 04	158	33 43	208	25 32	246	33 00	289	21 59	330
352	47 49	013	37 25	048	38 26	158	33 16	208	24 38	246	32 05	289	21 30	330
353	48 01	012	38 08	047	38 47	159	32 48	208	23 45	246	31 09	288	21 00	329
354	48 13	011	38 50	046	39 08	159	32 20	209	22 51	246	30 13	288	20 30	329
355	48 23	009	39 32	045	39 29	160	31 52	209	21 57	246	29 17	287	19 59	328
356	48 32	008	40 14	045	39 49	160	31 23	209	21 03	246	28 21	287	19 28	328
357	48 39	007	40 55	044	40 08	161	30 54	210	20 10	246	27 25	287	18 57	327
358	48 45	005	41 35	043	40 27	162	30 25	210	19 16	246	26 29	286	18 25	327
359	48 50	004	42 15	042	40 45	162	29 56	210	18 22	246	25 32	286	17 52	326

 LAT 13°S

LHA ♈	Hc	Zn	Hc	Zn	Hc	Zn	Hc	Zn	Hc	Zn	Hc	Zn	Hc	Zn
	◆Alpheratz		Hamal		ALDEBARAN		◆RIGEL		ACHERNAR		◆FOMALHAUT		Enif	
0	47 53	003	42 08	041	15 48	068	12 54	096	42 00	163	67 55	219	49 13	302
1	47 56	001	42 46	040	16 42	068	13 52	095	42 17	163	67 18	220	48 24	302
2	47 56	000	43 23	039	17 37	068	14 50	095	42 34	164	66 40	222	47 34	301
3	47 56	359	43 59	038	18 31	067	15 48	095	42 50	164	66 00	223	46 43	300
4	47 54	357	44 35	037	19 24	067	16 47	095	43 05	165	65 20	224	45 53	299
5	47 50	356	45 09	036	20 18	067	17 45	095	43 20	166	64 38	226	45 01	299
6	47 46	355	45 43	035	21 12	066	18 43	094	43 34	166	63 56	227	44 10	298
7	47 40	354	46 16	034	22 05	066	19 41	094	43 48	167	63 13	228	43 18	297
8	47 33	352	46 48	033	22 59	066	20 40	094	44 00	168	62 29	229	42 26	297
9	47 24	351	47 19	032	23 52	065	21 38	094	44 13	168	61 45	230	41 34	296
10	47 14	350	47 49	030	24 45	065	22 36	094	44 24	169	61 00	231	40 41	295
11	47 03	348	48 18	029	25 37	064	23 35	093	44 35	170	60 14	232	39 48	295
12	46 51	347	48 46	028	26 30	064	24 33	093	44 44	171	59 28	233	38 55	294
13	46 37	346	49 13	027	27 22	063	25 31	093	44 54	171	58 41	233	38 02	294
14	46 23	345	49 39	026	28 15	063	26 30	093	45 02	172	57 54	234	37 08	293
	Hamal		◆ALDEBARAN		RIGEL		◆ACHERNAR		FOMALHAUT		◆Enif		Alpheratz	
15	50 03	024	29 06	062	27 28	093	45 10	173	57 07	235	36 14	293	46 07	344
16	50 27	023	29 58	062	28 27	092	45 17	174	56 19	235	35 20	292	45 50	342
17	50 49	022	30 50	061	29 25	092	45 23	174	55 31	236	34 26	292	45 31	341
18	51 10	020	31 41	061	30 24	092	45 28	175	54 42	236	33 32	291	45 12	340
19	51 30	019	32 32	060	31 22	092	45 33	176	53 53	237	32 37	291	44 51	339
20	51 48	018	33 23	060	32 20	092	45 37	177	53 04	237	31 42	290	44 30	338
21	52 05	016	34 13	059	33 19	091	45 40	177	52 15	238	30 48	290	44 07	337
22	52 20	015	35 03	059	34 17	091	45 42	178	51 25	238	29 53	290	43 44	336
23	52 35	013	35 53	058	35 16	091	45 44	179	50 35	239	28 57	289	43 19	335
24	52 47	012	36 43	058	36 14	091	45 44	180	49 45	239	28 02	289	42 53	333
25	52 58	010	37 32	057	37 13	090	45 44	180	48 55	239	27 07	288	42 27	332
26	53 08	009	38 21	056	38 11	090	45 43	181	48 04	240	26 11	288	41 59	331
27	53 16	007	39 09	056	39 10	090	45 42	182	47 14	240	25 16	288	41 31	330
28	53 23	006	39 57	055	40 08	090	45 39	183	46 23	240	24 20	287	41 02	330
29	53 28	004	40 45	054	41 06	090	45 36	184	45 32	241	23 24	287	40 32	329
	CAPELLA		◆BETELGEUSE		SIRIUS		CANOPUS		◆ACHERNAR		FOMALHAUT		◆Alpheratz	
30	16 21	033	28 12	074	21 24	103	24 48	142	45 32	184	44 41	241	40 01	328
31	16 53	033	29 08	074	22 21	103	25 24	143	45 27	185	43 50	241	39 29	327
32	17 25	032	30 05	073	23 18	103	25 59	143	45 21	186	42 58	241	38 57	326
33	17 55	032	31 00	073	24 15	103	26 35	143	45 15	187	42 07	242	38 23	325
34	18 26	031	31 56	073	25 12	103	27 10	143	45 08	187	41 16	242	37 49	324
35	18 56	031	32 52	072	26 09	102	27 45	143	45 00	188	40 24	242	37 15	323
36	19 26	030	33 48	072	27 06	102	28 20	143	44 52	189	39 32	242	36 40	322
37	19 55	030	34 43	071	28 04	102	28 55	144	44 42	190	38 41	242	36 04	322
38	20 24	029	35 38	071	29 01	102	29 29	144	44 32	190	37 49	242	35 27	321
39	20 52	029	36 33	070	29 58	102	30 04	144	44 21	191	36 57	243	34 50	320
40	21 20	028	37 28	070	30 55	102	30 38	144	44 10	192	36 05	243	34 12	319
41	21 47	027	38 23	069	31 52	102	31 12	145	43 58	192	35 13	243	33 34	319
42	22 14	027	39 18	069	32 50	102	31 46	145	43 45	193	34 21	243	32 55	318
43	22 40	026	40 12	068	33 47	102	32 19	145	43 31	194	33 29	243	32 16	317
44	23 05	026	41 07	068	34 44	101	32 53	145	43 17	194	32 37	243	31 36	317
	◆CAPELLA		BETELGEUSE		SIRIUS		◆CANOPUS		ACHERNAR		◆FOMALHAUT		Alpheratz	
45	23 30	025	42 01	067	35 41	101	33 26	146	43 02	195	31 45	243	30 56	316
46	23 55	025	42 54	067	36 39	101	33 59	146	42 46	196	30 53	243	30 15	315
47	24 19	024	43 48	066	37 36	101	34 31	146	42 30	196	30 01	243	29 33	315
48	24 42	023	44 41	066	38 33	101	35 03	147	42 13	197	29 09	243	28 52	314
49	25 05	023	45 34	065	39 31	101	35 35	147	41 56	198	28 17	243	28 10	314
50	25 27	022	46 27	064	40 28	101	36 07	147	41 38	198	27 25	243	27 27	313
51	25 49	021	47 20	064	41 26	101	36 38	148	41 19	199	26 32	243	26 44	312
52	26 10	021	48 12	063	42 23	101	37 09	148	41 00	199	25 40	243	26 01	312
53	26 30	020	49 04	062	43 20	101	37 40	149	40 41	200	24 48	243	25 17	311
54	26 50	019	49 55	061	44 18	101	38 11	149	40 20	201	23 56	243	24 33	311
55	27 09	019	50 46	061	45 15	101	38 40	149	40 00	201	23 04	243	23 48	310
56	27 27	018	51 37	060	46 13	101	39 10	150	39 38	202	22 12	243	23 04	310
57	27 45	017	52 27	059	47 10	101	39 39	150	39 17	202	21 20	243	22 19	309
58	28 01	016	53 17	058	48 08	101	40 08	151	38 54	203	20 28	243	21 33	309
59	28 18	016	54 06	057	49 05	101	40 36	151	38 32	203	19 36	243	20 48	308
	CAPELLA		POLLUX		◆SIRIUS		CANOPUS		◆ACHERNAR		Diphda		◆Hamal	
60	28 33	015	21 50	052	50 02	101	41 04	152	38 08	204	42 29	257	44 15	323
61	28 48	014	22 36	052	51 00	101	41 31	152	37 45	204	41 32	257	43 39	322
62	29 02	013	23 22	051	51 57	101	41 58	153	37 21	205	40 35	257	43 03	321
63	29 15	013	24 08	051	52 55	101	42 25	153	36 56	205	39 38	257	42 25	320
64	29 28	012	24 53	050	53 52	101	42 50	154	36 31	205	38 41	257	41 47	319
65	29 39	011	25 38	050	54 50	101	43 16	155	36 06	206	37 44	257	41 08	318
66	29 50	010	26 22	049	55 47	101	43 40	155	35 40	206	36 47	257	40 29	317
67	30 01	010	27 06	049	56 45	101	44 05	156	35 14	207	35 50	257	39 49	316
68	30 10	009	27 50	048	57 42	101	44 28	157	34 48	207	34 53	257	39 08	316
69	30 19	008	28 33	048	58 40	101	44 51	157	34 21	207	33 56	257	38 27	315
70	30 26	007	29 16	047	59 37	101	45 14	158	33 54	208	32 59	257	37 45	314
71	30 34	007	29 59	046	60 34	101	45 35	159	33 27	208	32 03	257	37 03	313
72	30 40	006	30 41	046	61 32	101	45 56	159	32 59	208	31 06	257	36 20	313
73	30 45	005	31 23	045	62 29	101	46 17	160	32 31	209	30 09	257	35 37	312
74	30 50	004	32 04	044	63 26	101	46 36	161	32 03	209	29 12	256	34 53	311
	CAPELLA		POLLUX		◆PROCYON		Suhail		CANOPUS		◆ACHERNAR		◆Hamal	
75	30 54	003	32 44	044	46 30	068	29 10	133	46 55	161	31 34	209	34 09	311
76	30 56	002	33 25	043	47 24	067	29 53	133	47 13	162	31 05	210	33 24	310
77	30 59	002	34 04	042	48 18	066	30 36	133	47 31	163	30 36	210	32 39	309
78	31 00	001	34 43	042	49 11	066	31 18	133	47 47	164	30 07	210	31 54	309
79	31 00	000	35 22	041	50 04	065	32 01	133	48 03	165	29 38	210	31 08	308
80	31 00	359	36 00	040	50 57	064	32 43	134	48 18	165	29 08	211	30 22	307
81	30 59	358	36 37	039	51 50	064	33 26	134	48 33	166	28 38	211	29 35	307
82	30 57	358	37 14	039	52 42	063	34 08	134	48 46	167	28 08	211	28 48	306
83	30 54	357	37 50	038	53 34	062	34 50	134	48 59	168	27 37	211	28 01	306
84	30 50	356	38 26	037	54 25	061	35 31	135	49 10	169	27 07	212	27 14	305
85	30 46	355	39 01	036	55 16	060	36 13	135	49 21	170	26 36	212	26 26	305
86	30 41	354	39 35	035	56 06	059	36 54	135	49 31	171	26 05	212	25 37	304
87	30 35	354	40 08	034	56 56	058	37 35	135	49 40	172	25 34	212	24 49	304
88	30 28	353	40 41	033	57 46	057	38 16	136	49 48	173	25 03	212	24 00	303
89	30 20	352	41 13	032	58 35	056	38 57	136	49 55	173	24 32	212	23 11	303

LHA ♈	Hc	Zn	Hc	Zn	Hc	Zn	Hc	Zn	Hc	Zn	Hc	Zn	Hc	Zn
	POLLUX		◆REGULUS		Suhail		◆CANOPUS		ACHERNAR		◆ALDEBARAN		CAPELLA	
90	41 44	032	23 36	071	39 37	136	50 01	174	24 00	213	53 53	324	30 11	351
91	42 14	031	24 31	070	40 18	137	50 07	175	23 28	213	53 18	323	30 02	350
92	42 43	030	25 26	070	40 57	137	50 11	176	22 57	213	52 42	322	29 52	350
93	43 11	029	26 21	069	41 37	138	50 14	177	22 25	213	52 05	320	29 41	349
94	43 39	028	27 15	069	42 16	138	50 17	178	21 53	213	51 28	319	29 30	348
95	44 05	026	28 10	069	42 55	138	50 18	179	21 21	213	50 49	318	29 17	347
96	44 31	025	29 04	068	43 34	139	50 18	180	20 49	213	50 10	317	29 04	347
97	44 55	024	29 58	068	44 12	139	50 18	181	20 17	213	49 29	316	28 50	346
98	45 19	023	30 52	067	44 50	140	50 16	182	19 45	213	48 48	315	28 35	345
99	45 41	022	31 46	067	45 27	140	50 14	183	19 12	214	48 07	314	28 20	344
100	46 03	021	32 40	066	46 04	141	50 11	184	18 40	214	47 24	313	28 04	344
101	46 23	020	33 33	066	46 41	142	50 06	185	18 08	214	46 41	312	27 47	343
102	46 42	018	34 27	065	47 17	142	50 01	186	17 35	214	45 57	311	27 30	342
103	47 00	017	35 20	065	47 53	143	49 55	187	17 03	214	45 13	310	27 12	342
104	47 17	016	36 12	064	48 28	143	49 47	188	16 31	214	44 28	309	26 53	341
	POLLUX		◆REGULUS		ACRUX		◆CANOPUS		RIGEL		◆ALDEBARAN		CAPELLA	
105	47 32	015	37 05	064	15 23	152	49 39	188	63 36	278	43 43	309	26 33	340
106	47 47	014	37 57	063	15 50	152	49 30	189	62 38	277	42 57	308	26 13	339
107	48 00	012	38 49	063	16 17	152	49 20	190	61 40	277	42 10	307	25 52	339
108	48 11	011	39 41	062	16 44	152	49 09	191	60 42	276	41 24	306	25 31	338
109	48 22	010	40 33	061	17 11	152	48 58	192	59 44	276	40 36	306	25 09	337
110	48 31	008	41 24	061	17 38	152	48 45	193	58 46	276	39 48	305	24 46	337
111	48 39	007	42 15	060	18 05	153	48 32	194	57 48	275	39 00	304	24 23	336
112	48 45	006	43 05	059	18 32	153	48 17	195	56 49	275	38 12	304	23 59	336
113	48 50	004	43 55	059	18 59	153	48 02	195	55 51	275	37 23	303	23 34	335
114	48 54	003	44 45	058	19 26	153	47 46	196	54 53	274	36 34	302	23 09	334
115	48 57	002	45 35	057	19 53	153	47 29	197	53 55	274	35 44	302	22 44	334
116	48 58	000	46 24	057	20 19	153	47 12	198	52 56	274	34 54	301	22 18	333
117	48 57	359	47 12	056	20 46	153	46 54	199	51 58	273	34 04	301	21 51	333
118	48 56	358	48 00	055	21 12	153	46 35	199	51 00	273	33 13	300	21 24	332
119	48 53	356	48 48	054	21 39	153	46 15	200	50 01	273	32 23	299	20 56	331
	◆REGULUS		SPICA		◆ACRUX		CANOPUS		◆RIGEL		BETELGEUSE		POLLUX	
120	49 35	053	10 54	099	22 05	153	45 55	201	49 03	272	52 50	302	48 48	355
121	50 21	052	11 52	099	22 31	153	45 34	201	48 04	272	52 00	301	48 42	354
122	51 07	051	12 50	099	22 57	154	45 12	202	47 06	272	51 09	300	48 35	352
123	51 52	050	13 48	098	23 23	154	44 49	203	46 07	272	50 18	299	48 27	351
124	52 37	049	14 45	098	23 49	154	44 26	204	45 09	271	49 27	298	48 17	350
125	53 21	048	15 43	098	24 14	154	44 03	204	44 11	271	48 35	298	48 06	348
126	54 04	047	16 41	098	24 40	154	43 39	205	43 12	271	47 43	297	47 53	347
127	54 46	046	17 39	098	25 05	154	43 14	205	42 14	271	46 51	296	47 40	346
128	55 28	045	18 37	098	25 30	155	42 48	206	41 15	270	45 58	295	47 25	345
129	56 09	043	19 35	097	25 55	155	42 23	207	40 17	270	45 05	295	47 09	343
130	56 48	042	20 33	097	26 20	155	41 56	207	39 18	270	44 12	294	46 51	342
131	57 27	041	21 31	097	26 45	155	41 29	208	38 20	270	43 19	294	46 33	341
132	58 04	039	22 29	097	27 09	155	41 02	208	37 21	270	42 25	293	46 13	340
133	58 41	038	23 27	097	27 33	156	40 34	209	36 23	269	41 31	292	45 52	339
134	59 16	036	24 25	096	27 57	156	40 06	209	35 24	269	40 37	292	45 30	337
	REGULUS		◆SPICA		ACRUX		◆CANOPUS		SIRIUS		BETELGEUSE		◆POLLUX	
135	59 50	035	25 23	096	28 21	156	39 37	210	57 11	259	39 42	291	45 07	336
136	60 23	033	26 21	096	28 44	156	39 08	210	56 14	259	38 48	291	44 43	335
137	60 54	031	27 20	096	29 07	157	38 38	211	55 16	259	37 53	290	44 18	334
138	61 24	030	28 18	096	29 30	157	38 08	211	54 19	259	36 58	290	43 52	333
139	61 52	028	29 16	096	29 53	157	37 38	211	53 22	259	36 03	289	43 25	332
140	62 18	026	30 14	095	30 15	158	37 07	212	52 24	259	35 08	289	42 57	331
141	62 43	024	31 12	095	30 38	158	36 36	212	51 27	259	34 13	288	42 29	330
142	63 06	022	32 11	095	30 59	158	36 05	213	50 29	259	33 17	288	41 59	329
143	63 27	020	33 09	095	31 21	159	35 33	213	49 32	259	32 21	288	41 28	328
144	63 46	018	34 07	095	31 42	159	35 01	213	48 34	259	31 26	287	40 57	327
145	64 03	016	35 05	095	32 03	159	34 29	214	47 37	259	30 30	287	40 25	326
146	64 18	014	36 04	094	32 24	160	33 56	214	46 39	259	29 34	286	39 52	325
147	64 31	011	37 02	094	32 44	160	33 23	214	45 42	259	28 38	286	39 18	324
148	64 41	009	38 00	094	33 04	160	32 50	215	44 44	259	27 41	286	38 44	323
149	64 50	007	38 58	094	33 23	161	32 17	215	43 47	259	26 45	285	38 08	323
	REGULUS		◆ARCTURUS		SPICA		◆ACRUX		CANOPUS		SIRIUS		◆PROCYON	
150	64 56	005	19 21	064	39 57	094	33 42	161	31 43	215	42 50	259	50 32	295
151	64 59	002	20 14	064	40 55	094	34 01	161	31 09	215	41 52	259	49 39	295
152	65 00	000	21 06	063	41 53	093	34 19	162	30 35	216	40 55	259	48 45	294
153	64 59	358	21 58	063	42 52	093	34 37	162	30 01	216	39 57	259	47 52	293
154	64 56	355	22 50	062	43 50	093	34 55	163	29 26	216	39 00	259	46 58	293
155	64 50	353	23 42	062	44 49	093	35 12	163	28 52	216	38 03	259	46 04	292
156	64 42	351	24 33	062	45 47	093	35 29	164	28 17	217	37 05	259	45 10	291
157	64 32	349	25 24	061	46 45	093	35 45	164	27 42	217	36 08	259	44 15	291
158	64 19	346	26 15	061	47 44	092	36 01	164	27 07	217	35 11	259	43 20	290
159	64 04	344	27 06	060	48 42	092	36 17	165	26 32	217	34 13	258	42 25	290
160	63 47	342	27 57	060	49 41	092	36 32	165	25 56	217	33 16	258	41 30	289
161	63 28	340	28 47	059	50 39	092	36 46	166	25 21	217	32 19	258	40 35	289
162	63 07	338	29 37	059	51 38	092	37 00	166	24 45	218	31 22	258	39 40	288
163	62 44	336	30 27	058	52 36	091	37 14	167	24 10	218	30 24	258	38 44	288
164	62 20	334	31 17	058	53 34	091	37 27	167	23 34	218	29 27	258	37 48	287
	Denebola		◆ARCTURUS		SPICA		◆ACRUX		SIRIUS		◆PROCYON		REGULUS	
165	59 52	024	32 06	057	54 33	091	37 39	168	28 30	258	36 52	287	61 53	332
166	60 15	022	32 55	057	55 31	091	37 51	168	27 33	258	35 56	286	61 25	330
167	60 37	020	33 43	056	56 30	091	38 03	169	26 36	258	35 00	286	60 56	329
168	60 56	019	34 32	055	57 28	090	38 14	169	25 39	257	34 04	286	60 25	327
169	61 14	017	35 20	055	58 27	090	38 24	170	24 42	257	33 08	285	59 52	325
170	61 30	015	36 07	054	59 25	090	38 34	170	23 45	257	32 11	285	59 18	324
171	61 44	013	36 54	053	60 24	090	38 44	171	22 48	257	31 15	284	58 43	322
172	61 55	011	37 41	053	61 22	089	38 53	172	21 51	257	30 18	284	58 06	321
173	62 05	009	38 27	052	62 21	089	39 01	172	20 54	257	29 21	284	57 29	319
174	62 13	007	39 13	051	63 19	089	39 09	173	19 57	257	28 24	283	56 50	318
175	62 19	005	39 59	051	64 17	089	39 16	173	19 00	257	27 27	283	56 11	317
176	62 22	003	40 44	050	65 16	088	39 22	174	18 03	256	26 30	283	55 30	315
177	62 24	000	41 28	049	66 14	088	39 28	174	17 06	256	25 33	282	54 49	314
178	62 23	358	42 12	048	67 13	088	39 33	175	16 09	256	24 36	282	54 06	313
179	62 21	356	42 55	047	68 11	088	39 38	176	15 13	256	23 39	282	53 23	312

LAT 13°S

LHA	Hc	Zn	Hc	Zn	Hc	Zn	Hc	Zn	Hc	Zn	Hc	Zn	Hc	Zn
♈	ARCTURUS		♦ANTARES		ACRUX		♦Suhail		Alphard		REGULUS		♦Denebola	
180	43 38	047	25 55	113	39 43	176	42 13	222	52 18	273	52 39	311	62 16	354
181	44 20	046	26 49	113	39 46	177	41 33	222	51 19	272	51 55	310	62 09	352
182	45 02	045	27 43	113	39 49	177	40 54	223	50 21	272	51 10	309	62 00	350
183	45 43	044	28 37	113	39 52	178	40 14	223	49 23	272	50 24	308	61 49	348
184	46 23	043	29 30	113	39 54	178	39 34	224	48 24	272	49 37	307	61 35	346
185	47 03	042	30 24	113	39 55	179	38 53	224	47 26	271	48 50	306	61 20	344
186	47 41	041	31 18	113	39 56	180	38 12	224	46 27	271	48 03	305	61 03	342
187	48 19	040	32 12	113	39 56	180	37 32	225	45 29	271	47 15	304	60 45	340
188	48 56	039	33 05	113	39 55	181	36 50	225	44 30	271	46 26	304	60 24	338
189	49 32	038	33 59	113	39 54	181	36 09	225	43 32	270	45 37	303	60 02	337
190	50 08	037	34 53	113	39 52	182	35 27	225	42 33	270	44 48	302	59 38	335
191	50 42	035	35 46	113	39 50	183	34 46	226	41 35	270	43 58	301	59 12	333
192	51 15	034	36 40	113	39 47	183	34 04	226	40 36	270	43 08	301	58 45	332
193	51 48	033	37 34	113	39 43	184	33 22	226	39 38	269	42 17	300	58 17	330
194	52 19	032	38 27	114	39 39	184	32 39	226	38 40	269	41 26	299	57 47	328
	ARCTURUS		♦Rasalhague		ANTARES		♦ACRUX		Suhail		♦REGULUS		Denebola	
195	52 49	030	17 16	072	39 21	114	39 34	185	31 57	227	40 35	299	57 15	327
196	53 18	029	18 12	072	40 14	114	39 29	186	31 14	227	39 44	298	56 43	325
197	53 46	028	19 07	072	41 08	114	39 23	186	30 32	227	38 52	297	56 09	324
198	54 12	026	20 03	071	42 01	114	39 17	187	29 49	227	38 00	297	55 34	323
199	54 37	025	20 58	071	42 55	114	39 10	187	29 06	227	37 08	296	54 58	321
200	55 01	023	21 53	071	43 48	114	39 02	188	28 23	227	36 15	296	54 21	320
201	55 23	022	22 48	070	44 41	114	38 54	188	27 40	228	35 22	295	53 43	319
202	55 44	020	23 43	070	45 34	115	38 45	189	26 57	228	34 29	295	53 04	318
203	56 04	019	24 38	069	46 27	115	38 36	189	26 14	228	33 36	294	52 24	316
204	56 22	017	25 33	069	47 20	115	38 26	190	25 30	228	32 43	294	51 43	315
205	56 38	015	26 27	069	48 13	115	38 16	191	24 47	228	31 49	293	51 01	314
206	56 52	014	27 22	068	49 06	116	38 05	191	24 04	228	30 55	293	50 19	313
207	57 05	012	28 16	068	49 59	116	37 53	192	23 20	228	30 01	292	49 36	312
208	57 17	010	29 10	067	50 51	116	37 41	192	22 37	228	29 07	292	48 52	311
209	57 26	009	30 04	067	51 44	116	37 29	193	21 53	228	28 13	291	48 08	310
	ARCTURUS		♦Rasalhague		ANTARES		♦RIGIL KENT.		ACRUX		Gienah		♦REGULUS	
210	57 34	007	30 58	066	52 36	117	41 39	174	37 16	193	64 26	257	27 18	291
211	57 40	005	31 51	066	53 28	117	41 45	174	37 02	194	63 29	257	26 24	291
212	57 44	003	32 44	066	54 20	117	41 51	175	36 48	194	62 32	257	25 29	290
213	57 47	002	33 37	065	55 12	118	41 56	176	36 34	195	61 35	257	24 34	290
214	57 47	000	34 30	065	56 04	118	42 00	176	36 19	195	60 38	257	23 39	290
215	57 46	358	35 23	064	56 55	119	42 03	177	36 04	195	59 41	257	22 44	289
216	57 43	356	36 16	064	57 46	119	42 06	177	35 48	196	58 44	258	21 48	289
217	57 39	354	37 08	063	58 37	120	42 08	178	35 32	196	57 47	258	20 53	288
218	57 32	353	38 00	062	59 27	120	42 10	179	35 15	197	56 50	258	19 57	288
219	57 24	351	38 51	062	60 18	121	42 11	179	34 58	197	55 52	258	19 02	288
220	57 14	349	39 43	061	61 07	122	42 11	180	34 40	198	54 55	258	18 06	287
221	57 02	348	40 34	061	61 57	123	42 11	181	34 22	198	53 58	258	17 10	287
222	56 48	346	41 24	060	62 46	123	42 09	181	34 04	198	53 01	258	16 14	287
223	56 33	344	42 15	059	63 35	124	42 08	182	33 45	199	52 04	258	15 18	287
224	56 17	343	43 05	059	64 23	125	42 05	183	33 26	199	51 06	258	14 22	286
	Alphecca		♦VEGA		ALTAIR		♦Shaula		RIGIL KENT.		♦SPICA		ARCTURUS	
225	49 24	012	17 40	042	14 38	077	48 12	132	42 02	183	66 41	272	55 58	341
226	49 35	011	18 19	041	15 35	077	48 55	133	41 58	184	65 43	272	55 39	339
227	49 45	009	18 57	041	16 32	077	49 38	133	41 54	185	64 45	271	55 17	338
228	49 54	008	19 35	040	17 29	076	50 20	134	41 49	185	63 46	271	54 54	336
229	50 01	006	20 12	040	18 26	076	51 02	134	41 43	186	62 48	271	54 30	335
230	50 07	005	20 50	039	19 22	076	51 44	135	41 37	187	61 49	271	54 05	333
231	50 11	004	21 26	039	20 19	075	52 25	136	41 29	187	60 51	270	53 38	332
232	50 14	002	22 02	038	21 15	075	53 06	136	41 22	188	59 52	270	53 10	331
233	50 16	001	22 38	038	22 12	075	53 46	137	41 13	189	58 54	270	52 41	329
234	50 16	359	23 14	037	23 08	074	54 25	138	41 04	189	57 55	270	52 10	328
235	50 15	358	23 49	036	24 04	074	55 04	139	40 55	190	56 57	270	51 39	327
236	50 12	357	24 23	036	25 01	074	55 42	139	40 45	190	55 58	269	51 06	325
237	50 08	355	24 57	035	25 57	073	56 20	140	40 34	191	55 00	269	50 32	324
238	50 02	354	25 31	035	26 53	073	56 57	141	40 23	192	54 02	269	49 58	323
239	49 56	353	26 04	034	27 48	073	57 33	142	40 11	192	53 03	269	49 22	322
	VEGA		♦ALTAIR		Shaula		♦RIGIL KENT.		SPICA		♦ARCTURUS		Alphecca	
240	26 36	033	28 44	072	58 09	143	39 58	193	52 05	269	48 46	321	49 47	351
241	27 08	033	29 40	072	58 43	144	39 45	193	51 06	268	48 08	320	49 38	350
242	27 40	032	30 35	071	59 17	145	39 31	194	50 08	268	47 30	319	49 27	348
243	28 10	031	31 30	071	59 49	147	39 17	194	49 09	268	46 51	318	49 14	347
244	28 41	031	32 26	070	60 21	148	39 02	195	48 11	268	46 12	317	49 01	346
245	29 10	030	33 21	070	60 52	149	38 47	195	47 13	268	45 31	316	48 46	344
246	29 39	029	34 16	070	61 21	150	38 31	196	46 14	267	44 50	315	48 29	343
247	30 08	029	35 10	069	61 50	152	38 15	196	45 16	267	44 08	314	48 12	342
248	30 35	028	36 05	069	62 17	153	37 58	197	44 17	267	43 26	313	47 53	341
249	31 02	027	36 59	068	62 43	155	37 41	197	43 19	267	42 43	312	47 33	339
250	31 29	026	37 53	068	63 07	156	37 23	198	42 21	267	42 00	311	47 12	338
251	31 55	026	38 47	067	63 30	158	37 05	198	41 22	267	41 16	311	46 50	337
252	32 20	025	39 41	067	63 52	159	36 46	199	40 24	266	40 31	310	46 26	336
253	32 44	024	40 35	066	64 12	161	36 27	199	39 26	266	39 46	309	46 02	335
254	33 07	023	41 28	065	64 30	163	36 07	200	38 27	266	39 00	308	45 36	334
	♦VEGA		ALTAIR		♦Peacock		RIGIL KENT.		♦SPICA		ARCTURUS		Alphecca	
255	33 30	023	42 21	065	31 28	150	35 47	200	37 29	266	38 14	308	45 10	332
256	33 52	022	43 14	064	31 57	150	35 27	201	36 31	266	37 28	307	44 42	331
257	34 13	021	44 06	064	32 26	150	35 06	201	35 32	265	36 41	306	44 14	330
258	34 34	020	44 59	063	32 55	151	34 44	202	34 34	265	35 54	306	43 45	329
259	34 53	019	45 51	062	33 23	151	34 23	202	33 36	265	35 06	305	43 14	328
260	35 12	018	46 42	062	33 51	151	34 01	202	32 38	265	34 18	304	42 43	327
261	35 30	017	47 33	061	34 19	152	33 38	203	31 39	265	33 30	304	42 11	326
262	35 47	017	48 24	060	34 47	152	33 16	203	30 41	265	32 41	303	41 38	325
263	36 03	016	49 15	059	35 14	153	32 53	203	29 43	264	31 52	303	41 05	324
264	36 19	015	50 05	059	35 40	153	32 29	204	28 45	264	31 03	302	40 30	324
265	36 33	014	50 54	058	36 07	153	32 05	204	27 47	264	30 13	302	39 55	323
266	36 46	013	51 44	057	36 33	154	31 41	204	26 48	264	29 23	301	39 20	322
267	36 59	012	52 32	056	36 58	154	31 17	205	25 50	264	28 33	301	38 43	321
268	37 11	011	53 20	055	37 24	155	30 53	205	24 52	264	27 43	300	38 06	320
269	37 21	010	54 08	054	37 48	155	30 28	205	23 54	263	26 52	300	37 28	319

LHA	Hc	Zn	Hc	Zn	Hc	Zn	Hc	Zn	Hc	Zn	Hc	Zn	Hc	Zn
♈	♦VEGA		ALTAIR		♦FOMALHAUT		Peacock		RIGIL KENT.		♦ANTARES		ARCTURUS	
270	37 31	009	54 55	053	19 52	117	38 13	156	30 02	206	64 49	234	26 01	299
271	37 39	008	55 41	052	20 44	117	38 37	156	29 37	206	64 01	235	25 10	299
272	37 47	007	56 27	051	21 36	117	39 00	157	29 11	206	63 13	236	24 19	298
273	37 54	006	57 12	049	22 28	117	39 23	157	28 45	206	62 24	237	23 27	298
274	38 00	005	57 56	048	23 20	117	39 46	158	28 19	207	61 35	238	22 35	297
275	38 04	004	58 39	047	24 12	117	40 08	158	27 53	207	60 45	239	21 43	297
276	38 08	003	59 21	046	25 04	117	40 29	159	27 26	207	59 55	239	20 51	297
277	38 11	002	60 02	044	25 57	117	40 50	159	27 00	207	59 05	240	19 59	296
278	38 13	001	60 43	043	26 49	117	41 11	160	26 33	208	58 14	240	19 06	296
279	38 13	000	61 22	041	27 41	117	41 31	160	26 05	208	57 23	241	18 14	296
280	38 13	359	62 00	040	28 33	117	41 50	161	25 38	208	56 32	241	17 21	295
281	38 12	358	62 36	038	29 25	117	42 09	162	25 11	208	55 40	242	16 28	295
282	38 09	357	63 11	036	30 17	117	42 27	162	24 43	208	54 49	242	15 35	295
283	38 06	356	63 45	034	31 09	117	42 45	163	24 15	209	53 57	243	14 41	294
284	38 02	355	64 17	032	32 01	117	43 02	163	23 47	209	53 05	243	13 48	294
	♦DENEB		♦FOMALHAUT		Peacock		RIGIL KENT.		♦ANTARES		Rasalhague		VEGA	
285	27 24	020	32 53	117	43 18	164	23 19	209	52 13	243	56 49	320	37 56	354
286	27 43	019	33 46	117	43 34	165	22 51	209	51 20	244	56 11	318	37 50	353
287	28 02	018	34 38	117	43 49	165	22 22	209	50 28	244	55 32	317	37 43	352
288	28 20	018	35 29	117	44 03	166	21 54	209	49 35	244	54 51	316	37 34	351
289	28 38	017	36 21	117	44 17	167	21 25	209	48 42	245	54 10	315	37 25	350
290	28 54	016	37 13	118	44 30	168	20 57	209	47 50	245	53 28	313	37 15	349
291	29 10	015	38 05	118	44 42	168	20 28	210	46 57	245	52 45	312	37 03	348
292	29 26	015	38 57	118	44 53	169	19 59	210	46 04	245	52 01	311	36 51	348
293	29 40	014	39 49	118	45 04	170	19 30	210	45 10	245	51 17	310	36 38	347
294	29 54	013	40 40	118	45 14	170	19 01	210	44 17	246	50 32	309	36 24	346
295	30 07	012	41 32	118	45 24	171	18 32	210	43 24	246	49 46	308	36 09	345
296	30 19	012	42 23	118	45 32	172	18 03	210	42 31	246	49 00	307	35 53	344
297	30 30	011	43 14	119	45 40	173	17 34	210	41 37	246	48 13	306	35 37	343
298	30 41	010	44 06	119	45 47	173	17 05	210	40 44	246	47 26	306	35 19	342
299	30 51	009	44 57	119	45 53	174	16 35	210	39 50	246	46 38	305	35 01	341
	♦DENEB		Alpheratz		♦FOMALHAUT		Peacock		♦ANTARES		Rasalhague		VEGA	
300	31 00	008	16 52	054	45 48	119	45 59	175	38 57	246	45 50	304	34 41	340
301	31 08	008	17 39	053	46 39	120	46 04	176	38 03	246	45 01	303	34 21	339
302	31 15	007	18 26	053	47 29	120	46 07	177	37 10	247	44 12	302	34 00	339
303	31 22	006	19 12	053	48 20	120	46 10	177	36 16	247	43 22	302	33 39	338
304	31 27	005	19 59	052	49 10	121	46 13	178	35 22	247	42 33	301	33 16	337
305	31 32	004	20 45	052	50 01	121	46 14	179	34 29	247	41 42	300	32 53	336
306	31 36	004	21 30	051	50 51	121	46 15	180	33 35	247	40 52	300	32 29	335
307	31 40	003	22 16	051	51 40	122	46 15	181	32 41	247	40 01	299	32 04	335
308	31 42	002	23 01	050	52 30	122	46 14	181	31 47	247	39 09	298	31 39	334
309	31 44	001	23 46	050	53 19	123	46 12	182	30 54	247	38 18	298	31 13	333
310	31 44	000	24 30	049	54 08	123	46 09	183	30 00	247	37 26	297	30 46	332
311	31 44	359	25 14	049	54 57	124	46 06	184	29 06	247	36 34	297	30 18	332
312	31 43	359	25 58	048	55 46	124	46 02	184	28 12	247	35 42	296	29 50	331
313	31 41	358	26 41	048	56 34	125	45 57	185	27 19	247	34 49	296	29 21	330
314	31 39	357	27 24	047	57 22	126	45 51	186	26 25	247	33 56	295	28 52	329
	Alpheratz		♦Diphda		ACHERNAR		Peacock		♦ANTARES		VEGA		♦DENEB	
315	28 07	046	36 09	103	22 01	147	45 45	187	25 31	247	28 22	329	31 35	356
316	28 49	046	37 06	103	22 32	147	45 37	188	24 38	247	27 51	328	31 31	355
317	29 31	045	38 03	103	23 04	147	45 29	188	23 44	247	27 20	327	31 25	354
318	30 12	045	39 00	103	23 36	147	45 20	189	22 50	247	26 49	327	31 19	354
319	30 53	044	39 57	103	24 07	147	45 11	190	21 57	246	26 16	326	31 13	353
320	31 33	043	40 54	103	24 39	148	45 00	191	21 03	246	25 43	326	31 05	352
321	32 13	043	41 51	103	25 10	148	44 49	191	20 10	246	25 10	325	30 56	351
322	32 53	042	42 48	103	25 41	148	44 37	192	19 16	246	24 36	324	30 47	350
323	33 32	041	43 45	103	26 12	148	44 25	193	18 23	246	24 02	324	30 37	350
324	34 10	041	44 42	103	26 43	148	44 12	193	17 29	246	23 27	323	30 26	349
325	34 48	040	45 39	103	27 14	148	43 58	194	16 36	246	22 52	323	30 14	348
326	35 25	039	46 36	103	27 44	149	43 43	195	15 42	246	22 16	322	30 02	347
327	36 01	038	47 34	103	28 15	149	43 28	195	14 49	246	21 40	322	29 49	347
328	36 37	038	48 31	103	28 45	149	43 12	196	13 56	246	21 04	321	29 35	346
329	37 13	037	49 28	103	29 15	149	42 55	197	13 03	245	20 27	321	29 20	345
	♦Alpheratz		Diphda		♦ACHERNAR		Peacock		♦Nunki		ALTAIR		DENEB	
330	37 47	036	50 25	103	29 45	150	42 38	197	44 43	246	51 08	303	29 04	344
331	38 21	035	51 22	103	30 14	150	42 20	198	43 50	246	50 18	302	28 48	344
332	38 54	034	52 19	103	30 43	150	42 02	199	42 56	246	49 28	301	28 31	343
333	39 27	033	53 16	103	31 12	150	41 43	199	42 03	246	48 38	300	28 14	342
334	39 59	032	54 13	103	31 41	151	41 23	200	41 09	246	47 47	299	27 55	341
335	40 30	031	55 10	103	32 09	151	41 03	200	40 16	246	46 56	299	27 36	341
336	41 00	031	56 07	103	32 38	151	40 43	201	39 22	246	46 04	298	27 17	340
337	41 29	030	57 04	103	33 06	152	40 21	202	38 28	247	45 13	297	26 56	339
338	41 57	029	58 01	103	33 33	152	40 00	202	37 35	247	44 20	297	26 35	339
339	42 25	028	58 57	103	34 01	152	39 37	203	36 41	247	43 28	296	26 13	338
340	42 52	027	59 54	104	34 28	153	39 15	203	35 47	247	42 35	295	25 51	337
341	43 17	026	60 51	104	34 54	153	38 52	204	34 54	247	41 42	295	25 28	337
342	43 42	024	61 48	104	35 20	153	38 28	204	34 00	247	40 49	294	25 05	336
343	44 06	023	62 45	104	35 46	154	38 04	205	33 06	247	39 56	294	24 40	335
344	44 28	022	63 41	104	36 12	154	37 39	205	32 12	247	39 02	293	24 16	335
	Alpheratz		♦Hamal		♦ACHERNAR		Peacock		Nunki		♦ALTAIR		DENEB	
345	44 50	021	31 33	052	36 37	155	37 14	205	31 19	247	38 08	292	23 50	334
346	45 11	020	32 19	051	37 02	155	36 49	206	30 25	247	37 14	292	23 24	333
347	45 30	019	33 04	050	37 26	156	36 23	206	29 31	247	36 19	291	22 58	333
348	45 48	018	33 49	050	37 50	156	35 57	207	28 37	247	35 25	291	22 31	332
349	46 06	017	34 34	049	38 14	157	35 30	207	27 43	247	34 30	291	22 04	332
350	46 22	015	35 18	048	38 37	157	35 04	208	26 50	247	33 35	290	21 36	331
351	46 36	014	36 01	048	39 00	157	34 36	208	25 56	247	32 40	290	21 07	331
352	46 50	013	36 44	047	39 22	158	34 09	208	25 02	247	31 45	289	20 38	330
353	47 03	012	37 27	046	39 43	159	33 41	209	24 08	247	30 50	289	20 09	329
354	47 14	010	38 09	046	40 05	159	33 13	209	23 15	247	29 55	288	19 39	329
355	47 24	009	38 50	045	40 25	160	32 44	209	22 21	247	28 59	288	19 08	328
356	47 32	008	39 31	044	40 45	160	32 16	210	21 27	247	28 03	288	18 37	328
357	47 39	007	40 11	043	41 05	161	31 47	210	20 34	246	27 08	287	18 06	327
358	47 45	005	40 51	042	41 24	161	31 17	210	19 40	246	26 12	287	17 34	327
359	47 50	004	41 30	041	41 42	162	30 48	211	18 47	246	25 16	286	17 02	326

 LAT 14°S

LHA ♈	Hc Zn	Hc Zn	Hc Zn	Hc Zn	Hc Zn	Hc Zn	Hc Zn
	◆Alpheratz	Hamal	ALDEBARAN	◆RIGEL	ACHERNAR	◆FOMALHAUT	Enif
0	46 54 003	41 23 040	15 26 068	12 59 095	42 57 162	68 42 220	48 41 303
1	46 56 001	42 00 039	16 20 068	13 57 095	43 15 163	68 03 222	47 52 303
2	46 56 000	42 36 038	17 14 067	14 55 095	43 32 163	67 24 223	47 03 302
3	46 56 359	43 12 037	18 07 067	15 53 095	43 48 164	66 44 225	46 13 301
4	46 54 357	43 46 036	19 01 067	16 51 094	44 03 165	66 02 226	45 23 300
5	46 51 356	44 20 035	19 54 066	17 49 094	44 18 165	65 20 227	44 32 300
6	46 46 355	44 54 034	20 48 066	18 48 094	44 33 166	64 37 228	43 41 299
7	46 40 354	45 26 033	21 41 066	19 46 094	44 46 167	63 53 229	42 50 298
8	46 33 352	45 57 032	22 34 065	20 44 094	44 59 168	63 08 231	41 59 297
9	46 25 351	46 28 031	23 26 065	21 42 093	45 11 168	62 23 231	41 07 297
10	46 15 350	46 57 030	24 19 064	22 40 093	45 23 169	61 37 232	40 15 296
11	46 04 349	47 26 029	25 11 064	23 38 093	45 34 170	60 51 233	39 23 296
12	45 52 347	47 53 027	26 03 063	24 36 093	45 44 170	60 04 234	38 30 295
13	45 39 346	48 19 026	26 55 063	25 34 093	45 53 171	59 17 235	37 37 295
14	45 25 345	48 45 025	27 47 062	26 33 092	46 01 172	58 29 235	36 44 294
	Hamal	◆ALDEBARAN	RIGEL	◆ACHERNAR	FOMALHAUT	◆Enif	Alpheratz
15	49 09 024	28 39 062	27 31 092	46 09 173	57 41 236	35 51 293	45 09 344
16	49 32 023	29 30 061	28 29 092	46 16 173	56 52 237	34 57 293	44 52 343
17	49 53 021	30 21 061	29 27 092	46 23 174	56 04 237	34 03 292	44 34 342
18	50 14 020	31 12 060	30 25 091	46 28 175	55 15 238	33 10 292	44 15 340
19	50 33 019	32 02 060	31 23 091	46 33 176	54 25 238	32 15 291	43 55 339
20	50 51 017	32 52 059	32 22 091	46 37 177	53 36 239	31 21 291	43 34 338
21	51 07 016	33 42 059	33 20 091	46 40 177	52 46 239	30 27 291	43 12 337
22	51 22 014	34 32 058	34 18 090	46 42 178	51 56 239	29 32 290	42 49 336
23	51 36 013	35 21 058	35 16 090	46 44 179	51 06 240	28 37 290	42 25 335
24	51 48 012	36 10 057	36 15 090	46 44 180	50 15 240	27 43 289	42 00 334
25	51 59 010	36 59 056	37 13 090	46 44 180	49 25 240	26 48 289	41 33 333
26	52 09 009	37 47 056	38 11 089	46 43 181	48 34 241	25 52 288	41 06 332
27	52 17 007	38 35 055	39 09 089	46 42 182	47 43 241	24 57 288	40 39 331
28	52 23 006	39 23 054	40 07 089	46 39 183	46 52 241	24 02 288	40 10 330
29	52 28 004	40 10 054	41 06 089	46 36 184	46 01 242	23 06 287	39 40 329
	CAPELLA	◆BETELGEUSE	SIRIUS	CANOPUS	◆ACHERNAR	FOMALHAUT	◆Alpheratz
30	15 31 033	27 56 074	21 38 103	25 36 142	46 32 184	45 10 242	39 10 328
31	16 03 033	28 51 073	22 34 103	26 11 142	46 27 185	44 18 242	38 39 327
32	16 34 032	29 47 073	23 31 102	26 47 142	46 21 186	43 27 242	38 07 326
33	17 04 032	30 43 072	24 28 102	27 22 143	46 15 187	42 35 242	37 34 325
34	17 35 031	31 38 072	25 25 102	27 58 143	46 07 187	41 44 243	37 01 325
35	18 04 031	32 33 072	26 22 102	28 33 143	46 00 188	40 52 243	36 27 324
36	18 34 030	33 28 071	27 19 102	29 08 143	45 51 189	40 00 243	35 52 323
37	19 03 030	34 23 071	28 16 102	29 43 143	45 41 190	39 08 243	35 16 322
38	19 31 029	35 18 070	29 13 102	30 18 143	45 31 190	38 16 243	34 40 321
39	19 59 028	36 13 070	30 10 101	30 52 144	45 20 191	37 24 243	34 04 321
40	20 27 028	37 07 069	31 07 101	31 26 144	45 09 192	36 32 243	33 27 320
41	20 54 027	38 02 069	32 04 101	32 01 144	44 56 193	35 40 243	32 49 319
42	21 20 027	38 56 068	33 01 101	32 35 144	44 43 193	34 48 243	32 10 318
43	21 46 026	39 50 068	33 58 101	33 08 145	44 29 194	33 56 243	31 32 318
44	22 11 026	40 44 067	34 56 101	33 42 145	44 15 195	33 04 244	30 52 317
	◆CAPELLA	BETELGEUSE	SIRIUS	◆CANOPUS	ACHERNAR	◆FOMALHAUT	Alpheratz
45	22 36 025	41 37 066	35 53 101	34 15 145	44 00 195	32 12 244	30 12 316
46	23 00 024	42 30 066	36 50 101	34 48 146	43 44 196	31 20 244	29 32 316
47	23 24 024	43 23 065	37 47 100	35 21 146	43 28 197	30 28 244	28 51 315
48	23 47 023	44 16 065	38 45 100	35 53 146	43 11 197	29 36 244	28 10 315
49	24 10 022	45 09 064	39 42 100	36 26 147	42 53 198	28 44 244	27 28 314
50	24 32 022	46 01 063	40 39 100	36 58 147	42 35 199	27 51 244	26 46 313
51	24 53 021	46 53 063	41 36 100	37 29 147	42 16 199	26 59 244	26 03 313
52	25 14 020	47 44 062	42 34 100	38 00 148	41 57 200	26 07 244	25 20 312
53	25 34 020	48 35 061	43 31 100	38 31 148	41 37 200	25 15 244	24 37 312
54	25 53 019	49 26 060	44 29 100	39 02 149	41 16 201	24 23 243	23 54 311
55	26 12 018	50 16 059	45 26 100	39 32 149	40 55 201	23 31 243	23 10 311
56	26 30 018	51 06 059	46 23 100	40 02 149	40 34 202	22 39 243	22 25 310
57	26 47 017	51 56 058	47 21 100	40 31 150	40 12 202	21 47 243	21 40 310
58	27 04 016	52 45 057	48 18 100	41 00 150	39 50 203	20 55 243	20 55 309
59	27 20 016	53 33 056	49 15 099	41 29 151	39 27 203	20 03 243	20 10 309
	CAPELLA	POLLUX	◆SIRIUS	CANOPUS	◆ACHERNAR	Diphda	◆Hamal
60	27 35 015	21 14 052	50 13 099	41 57 151	39 03 204	42 42 258	43 27 323
61	27 50 014	21 59 051	51 10 099	42 24 152	38 39 204	41 45 258	42 52 322
62	28 03 013	22 45 051	52 08 099	42 52 152	38 15 205	40 48 258	42 16 321
63	28 17 013	23 30 050	53 05 099	43 18 153	37 50 205	39 51 258	41 39 320
64	28 29 012	24 14 050	54 03 099	43 44 154	37 25 206	38 54 258	41 02 319
65	28 41 011	24 59 049	55 00 099	44 10 154	37 00 206	37 57 258	40 24 319
66	28 51 010	25 43 049	55 58 099	44 35 155	36 34 207	37 00 258	39 45 318
67	29 01 010	26 26 048	56 55 099	44 59 155	36 08 207	36 03 258	39 05 317
68	29 11 009	27 10 048	57 53 099	45 23 156	35 41 207	35 07 258	38 25 316
69	29 19 008	27 53 047	58 50 099	45 46 157	35 14 208	34 10 258	37 45 315
70	29 27 007	28 35 047	59 47 099	46 09 157	34 47 208	33 13 257	37 03 315
71	29 34 006	29 17 046	60 45 099	46 31 158	34 20 208	32 16 257	36 22 314
72	29 40 006	29 59 045	61 42 099	46 52 159	33 52 209	31 19 257	35 39 313
73	29 45 005	30 40 045	62 40 099	47 13 160	33 24 209	30 23 257	34 57 312
74	29 50 004	31 21 044	63 37 099	47 33 160	32 55 209	29 26 257	34 14 312
	CAPELLA	POLLUX	◆PROCYON	Suhail	CANOPUS	◆ACHERNAR	◆Hamal
75	29 54 003	32 01 043	46 07 067	29 51 132	47 52 161	32 26 210	33 30 311
76	29 57 002	32 41 043	47 00 066	30 34 132	48 10 162	31 57 210	32 46 310
77	29 59 002	33 20 042	47 53 065	31 17 133	48 28 163	31 28 210	32 01 310
78	30 00 001	33 58 041	48 46 065	31 59 133	48 45 164	30 59 211	31 16 309
79	30 00 000	34 37 040	49 38 064	32 42 133	49 01 164	30 29 211	30 31 309
80	30 00 359	35 14 040	50 31 063	33 25 133	49 16 165	29 59 211	29 45 308
81	29 59 358	35 51 039	51 22 062	34 07 133	49 31 166	29 29 211	28 59 307
82	29 57 358	36 27 038	52 14 062	34 49 134	49 45 167	28 59 211	28 13 307
83	29 54 357	37 03 037	53 05 061	35 31 134	49 57 168	28 28 212	27 26 306
84	29 51 356	37 38 036	53 55 060	36 13 134	50 09 169	27 58 212	26 39 306
85	29 46 355	38 12 036	54 45 059	36 55 134	50 20 170	27 27 212	25 51 305
86	29 41 354	38 46 035	55 35 058	37 37 135	50 30 171	26 56 212	25 04 305
87	29 35 354	39 18 034	56 24 057	38 18 135	50 39 171	26 25 212	24 15 304
88	29 28 353	39 51 033	57 13 056	38 59 135	50 47 172	25 53 213	23 27 304
89	29 21 352	40 22 032	58 01 055	39 40 136	50 55 173	25 22 213	22 39 303

LHA ♈	Hc Zn	Hc Zn	Hc Zn	Hc Zn	Hc Zn	Hc Zn	Hc Zn
	POLLUX	◆REGULUS	Suhail	◆CANOPUS	ACHERNAR	◆ALDEBARAN	CAPELLA
90	40 52 031	23 16 070	40 21 136	51 01 174	24 51 213	53 04 325	29 12 351
91	41 22 030	24 10 070	41 01 136	51 06 175	24 19 213	52 30 324	29 03 351
92	41 51 029	25 05 069	41 41 137	51 11 176	23 47 213	51 55 322	28 53 350
93	42 19 028	25 59 069	42 21 137	51 14 177	23 15 213	51 19 321	28 42 349
94	42 46 027	26 54 068	43 01 137	51 17 178	22 43 213	50 42 320	28 31 348
95	43 12 026	27 48 068	43 40 138	51 18 179	22 11 213	50 04 319	28 19 347
96	43 37 025	28 42 068	44 19 138	51 18 180	21 39 214	49 25 318	28 06 347
97	44 01 024	29 35 067	44 57 139	51 18 181	21 07 214	48 46 317	27 52 346
98	44 24 023	30 29 067	45 36 139	51 16 182	20 35 214	48 06 316	27 37 345
99	44 46 022	31 22 066	46 13 140	51 14 183	20 02 214	47 25 315	27 22 345
100	45 07 020	32 15 066	46 51 140	51 10 184	19 30 214	46 43 314	27 06 344
101	45 26 019	33 08 065	47 28 141	51 06 185	18 58 214	46 01 313	26 50 343
102	45 45 018	34 01 065	48 04 141	51 01 186	18 25 214	45 18 312	26 33 342
103	46 03 017	34 54 064	48 40 142	50 54 187	17 53 214	44 34 311	26 15 342
104	46 19 016	35 46 064	49 16 143	50 47 188	17 20 214	43 50 310	25 56 341
	POLLUX	◆REGULUS	ACRUX	◆CANOPUS	RIGEL	◆ALDEBARAN	CAPELLA
105	46 34 015	36 38 063	16 16 152	50 39 189	63 27 280	43 05 309	25 37 340
106	46 48 013	37 30 063	16 43 152	50 29 190	62 30 279	42 20 309	25 17 340
107	47 01 012	38 21 062	17 10 152	50 19 190	61 32 279	41 34 308	24 56 339
108	47 12 011	39 13 061	17 37 152	50 08 191	60 35 278	40 48 307	24 35 338
109	47 23 009	40 04 061	18 04 152	49 56 192	59 37 278	40 01 306	24 13 338
110	47 32 008	40 54 060	18 31 152	49 43 193	58 39 277	39 14 306	23 51 337
111	47 39 007	41 44 059	18 58 152	49 30 194	57 41 277	38 26 305	23 28 336
112	47 46 006	42 34 059	19 25 152	49 15 195	56 44 276	37 38 304	23 04 336
113	47 51 004	43 24 058	19 52 152	49 00 196	55 46 276	36 50 304	22 40 335
114	47 54 003	44 13 057	20 19 153	48 44 197	54 48 276	36 01 303	22 15 335
115	47 57 002	45 02 056	20 46 153	48 27 197	53 50 275	35 12 302	21 50 334
116	47 58 000	45 50 056	21 13 153	48 09 198	52 52 275	34 23 302	21 24 333
117	47 57 359	46 38 055	21 39 153	47 51 199	51 54 275	33 33 301	20 58 333
118	47 56 358	47 25 054	22 06 153	47 31 200	50 56 274	32 43 301	20 31 332
119	47 53 356	48 12 053	22 32 153	47 11 200	49 58 274	31 53 300	20 04 332
	◆REGULUS	SPICA	◆ACRUX	CANOPUS	◆RIGEL	BETELGEUSE	POLLUX
120	48 58 052	11 04 099	22 59 153	46 51 201	49 00 274	52 18 303	47 48 355
121	49 44 051	12 01 099	23 25 153	46 29 202	48 01 273	51 29 302	47 43 354
122	50 29 050	12 59 098	23 51 153	46 07 203	47 03 273	50 39 301	47 36 352
123	51 14 049	13 56 098	24 17 154	45 45 203	46 05 273	49 49 300	47 27 351
124	51 57 048	14 54 098	24 43 154	45 21 204	45 07 272	48 58 299	47 18 350
125	52 41 047	15 52 098	25 08 154	44 57 205	44 09 272	48 07 298	47 07 349
126	53 23 046	16 49 098	25 34 154	44 33 205	43 11 272	47 16 298	46 55 347
127	54 04 045	17 47 097	25 59 154	44 08 206	42 12 272	46 24 297	46 41 346
128	54 45 044	18 45 097	26 25 154	43 42 206	41 14 271	45 32 296	46 27 345
129	55 25 042	19 43 097	26 50 155	43 16 207	40 16 271	44 40 296	46 11 344
130	56 03 041	20 40 097	27 14 155	42 49 208	39 18 271	43 47 295	45 54 342
131	56 41 040	21 38 097	27 39 155	42 22 208	38 20 271	42 54 294	45 36 341
132	57 18 038	22 36 096	28 04 155	41 55 209	37 21 270	42 01 294	45 17 340
133	57 53 037	23 34 096	28 28 156	41 26 209	36 23 270	41 08 293	44 56 339
134	58 28 035	24 32 096	28 52 156	40 58 210	35 25 270	40 14 293	44 35 338
	REGULUS	◆SPICA	ACRUX	◆CANOPUS	SIRIUS	BETELGEUSE	◆POLLUX
135	59 01 034	25 30 096	29 16 156	40 29 210	57 22 261	39 20 292	44 12 337
136	59 32 032	26 28 096	29 39 156	39 59 211	56 24 261	38 26 292	43 49 336
137	60 03 031	27 26 095	30 02 157	39 30 211	55 27 261	37 32 291	43 24 335
138	60 31 029	28 23 095	30 25 157	38 59 211	54 29 261	36 38 291	42 59 333
139	60 59 027	29 21 095	30 48 157	38 29 212	53 32 261	35 43 290	42 32 332
140	61 24 025	30 19 095	31 11 157	37 58 212	52 34 261	34 48 290	42 05 331
141	61 48 023	31 17 095	31 33 158	37 27 213	51 37 261	33 53 289	41 37 330
142	62 10 021	32 16 094	31 55 158	36 55 213	50 40 261	32 58 289	41 07 329
143	62 31 019	33 14 094	32 17 158	36 23 213	49 42 261	32 03 288	40 37 328
144	62 49 017	34 12 094	32 38 159	35 51 214	48 45 260	31 08 288	40 06 328
145	63 05 015	35 10 094	32 59 159	35 18 214	47 47 260	30 12 287	39 35 327
146	63 20 013	36 08 094	33 20 159	34 46 214	46 50 260	29 16 287	39 02 326
147	63 32 011	37 06 093	33 40 160	34 12 215	45 52 260	28 21 287	38 29 325
148	63 42 009	38 04 093	34 00 160	33 39 215	44 55 260	27 25 286	37 55 324
149	63 50 007	39 02 093	34 20 160	33 06 215	43 58 260	26 29 286	37 21 323
	REGULUS	◆ARCTURUS	SPICA	◆ACRUX	CANOPUS	SIRIUS	◆PROCYON
150	63 56 005	18 55 064	40 00 093	34 39 161	32 32 216	43 00 260	50 06 296
151	63 59 002	19 47 063	40 58 093	34 58 161	31 58 216	42 03 260	49 13 296
152	64 00 000	20 39 063	41 57 092	35 16 162	31 24 216	41 06 260	48 21 295
153	63 59 358	21 30 062	42 55 092	35 35 162	30 49 216	40 08 260	47 28 294
154	63 56 356	22 22 062	43 53 092	35 52 162	30 15 217	39 11 260	46 34 294
155	63 50 353	23 13 062	44 51 092	36 10 163	29 40 217	38 14 260	45 41 293
156	63 43 351	24 04 061	45 49 092	36 27 163	29 05 217	37 17 259	44 47 292
157	63 33 349	24 55 061	46 47 091	36 43 164	28 30 217	36 19 259	43 53 292
158	63 21 347	25 46 060	47 46 091	36 59 164	27 55 217	35 22 259	42 59 291
159	63 06 345	26 36 060	48 44 091	37 15 165	27 20 217	34 25 259	42 05 291
160	62 50 343	27 26 059	49 42 091	37 30 165	26 44 218	33 28 259	41 10 290
161	62 32 341	28 16 059	50 40 091	37 44 166	26 08 218	32 31 259	40 15 290
162	62 12 339	29 06 058	51 39 090	37 58 166	25 33 218	31 34 259	39 20 289
163	61 49 337	29 55 058	52 37 090	38 12 167	24 57 218	30 36 259	38 25 289
164	61 26 335	30 44 057	53 35 090	38 25 167	24 21 218	29 39 259	37 30 288
	Denebola	◆ARCTURUS	SPICA	◆ACRUX	SIRIUS	◆PROCYON	REGULUS
165	58 57 023	31 33 057	54 33 090	38 38 168	28 42 258	36 35 288	61 00 333
166	59 20 022	32 21 056	55 31 089	38 50 168	27 45 258	35 39 287	60 33 331
167	59 40 020	33 10 055	56 30 089	39 02 169	26 48 258	34 43 287	60 04 330
168	59 59 018	33 57 055	57 28 089	39 13 169	25 51 258	33 48 286	59 34 328
169	60 16 016	34 45 054	58 26 089	39 23 170	24 55 258	32 52 286	59 02 326
170	60 32 014	35 32 054	59 24 088	39 33 170	23 58 258	31 56 285	58 29 325
171	60 45 012	36 18 053	60 22 088	39 43 171	23 01 258	30 59 285	57 55 323
172	60 56 010	37 04 052	61 21 088	39 52 171	22 04 257	30 03 285	57 20 322
173	61 06 008	37 50 051	62 19 087	40 00 172	21 07 257	29 07 284	56 43 320
174	61 14 006	38 36 051	63 17 087	40 08 173	20 10 257	28 10 284	56 05 319
175	61 19 004	39 20 050	64 15 087	40 15 173	19 14 257	27 14 284	55 27 318
176	61 23 002	40 05 049	65 13 086	40 22 174	18 17 257	26 17 283	54 47 316
177	61 24 000	40 49 048	66 11 086	40 28 174	17 20 257	25 20 283	54 06 315
178	61 23 358	41 32 048	67 09 085	40 33 175	16 24 256	24 23 282	53 25 314
179	61 21 356	42 15 047	68 07 085	40 38 175	15 27 256	23 27 282	52 43 313

LAT 14°S

LHA ♈	Hc Zn	Hc Zn	Hc Zn	Hc Zn	Hc Zn	Hc Zn	Hc Zn
	ARCTURUS	◆ANTARES	ACRUX	◆Suhail	Alphard	REGULUS	◆Denebola
180	42 57 046	26 19 113	40 43 176	42 57 223	52 14 274	52 00 312	61 16 354
181	43 38 045	27 13 113	40 46 177	42 17 223	51 16 274	51 16 311	61 09 352
182	44 19 044	28 06 113	40 49 177	41 38 223	50 18 273	50 32 310	61 01 350
183	44 59 043	29 00 113	40 52 178	40 57 224	49 20 273	49 47 309	60 50 348
184	45 39 042	29 54 113	40 54 178	40 17 224	48 22 273	49 01 308	60 37 346
185	46 18 041	30 47 113	40 55 179	39 36 225	47 24 272	48 15 307	60 23 345
186	46 56 040	31 41 113	40 56 180	38 55 225	46 25 272	47 28 306	60 06 343
187	47 33 039	32 35 113	40 56 180	38 14 225	45 27 272	46 40 305	59 48 341
188	48 09 038	33 29 113	40 55 181	37 33 225	44 29 272	45 53 304	59 28 339
189	48 45 037	34 22 113	40 54 181	36 51 226	43 31 271	45 04 304	59 07 337
190	49 19 036	35 16 113	40 52 182	36 09 226	42 33 271	44 16 303	58 43 336
191	49 53 035	36 10 113	40 50 183	35 27 226	41 35 271	43 27 302	58 18 334
192	50 26 033	37 04 113	40 47 183	34 45 226	40 36 271	42 37 301	57 52 332
193	50 57 032	37 57 113	40 43 184	34 03 227	39 38 270	41 47 301	57 24 331
194	51 28 031	38 51 113	40 39 184	33 21 227	38 40 270	40 57 300	56 55 329
	ARCTURUS	◆Rasalhague	ANTARES	◆ACRUX	Suhail	REGULUS	◆Denebola
195	51 57 030	16 58 072	39 45 113	40 34 185	32 38 227	40 06 299	56 25 328
196	52 25 028	17 53 072	40 38 113	40 29 186	31 55 227	39 15 299	55 53 326
197	52 52 027	18 48 071	41 32 113	40 23 186	31 13 227	38 24 298	55 20 325
198	53 18 026	19 43 071	42 25 113	40 16 187	30 30 228	37 33 297	54 46 323
199	53 43 024	20 38 071	43 19 113	40 09 187	29 47 228	36 41 297	54 11 322
200	54 06 023	21 33 070	44 12 113	40 01 188	29 04 228	35 49 296	53 34 321
201	54 28 021	22 28 070	45 06 114	39 53 188	28 20 228	34 57 296	52 57 320
202	54 48 020	23 22 069	45 59 114	39 44 189	27 37 228	34 04 295	52 19 318
203	55 07 018	24 17 069	46 52 114	39 35 190	26 54 228	33 11 295	51 40 317
204	55 24 017	25 11 069	47 45 114	39 25 190	26 11 228	32 18 294	51 00 316
205	55 40 015	26 05 068	48 39 114	39 15 191	25 27 228	31 25 294	50 19 315
206	55 54 013	26 59 068	49 32 114	39 04 191	24 44 228	30 32 293	49 38 314
207	56 07 012	27 53 067	50 25 115	38 52 192	24 00 228	29 38 293	48 55 313
208	56 18 010	28 47 067	51 17 115	38 40 192	23 17 228	28 44 292	48 12 312
209	56 27 008	29 40 066	52 10 115	38 27 192	22 33 228	27 51 292	47 29 311
	ARCTURUS	◆Rasalhague	ANTARES	◆RIGIL KENT.	ACRUX	Gienah	◆REGULUS
210	56 35 007	30 33 066	53 03 116	42 39 174	38 14 193	64 39 259	26 56 292
211	56 40 005	31 26 065	53 55 116	42 45 174	38 01 194	63 42 259	26 02 291
212	56 44 003	32 19 065	54 47 116	42 50 175	37 47 194	62 45 259	25 08 291
213	56 47 001	33 12 064	55 40 117	42 55 175	37 32 195	61 47 259	24 13 290
214	56 47 000	34 04 064	56 31 117	43 00 176	37 17 195	60 50 259	23 19 290
215	56 46 358	34 57 063	57 23 117	43 03 177	37 01 196	59 53 259	22 24 290
216	56 43 356	35 48 063	58 15 118	43 06 177	36 46 196	58 56 259	21 29 289
217	56 39 355	36 40 062	59 06 118	43 08 178	36 29 197	57 59 259	20 34 289
218	56 33 353	37 32 062	59 57 119	43 10 179	36 12 197	57 01 259	19 39 288
219	56 24 351	38 23 061	60 48 120	43 11 179	35 55 197	56 04 259	18 43 288
220	56 15 350	39 13 060	61 38 120	43 11 180	35 37 198	55 07 259	17 48 288
221	56 03 348	40 04 060	62 29 121	43 11 181	35 19 198	54 10 259	16 52 287
222	55 50 346	40 54 059	63 18 122	43 09 181	35 01 199	53 13 259	15 57 287
223	55 36 345	41 44 058	64 08 122	43 08 182	34 42 199	52 15 259	15 01 287
224	55 19 343	42 33 058	64 56 123	43 05 183	34 23 199	51 18 259	14 05 286
	Alphecca	◆VEGA	ALTAIR	◆Shaula	RIGIL KENT.	◆SPICA	ARCTURUS
225	48 25 012	16 55 041	14 25 077	48 52 131	43 02 183	66 38 274	55 02 341
226	48 36 010	17 33 041	15 21 077	49 35 132	42 58 184	65 40 274	54 42 340
227	48 46 009	18 11 040	16 18 076	50 18 132	42 54 185	64 42 274	54 22 338
228	48 54 008	18 49 040	17 14 076	51 01 133	42 48 185	63 44 273	53 59 337
229	49 01 006	19 26 039	18 11 076	51 44 133	42 43 186	62 46 273	53 36 335
230	49 07 005	20 03 039	19 07 075	52 26 134	42 36 187	61 48 273	53 11 334
231	49 11 004	20 39 038	20 04 075	53 07 135	42 29 187	60 49 272	52 45 333
232	49 14 002	21 15 038	21 00 075	53 49 135	42 21 188	59 51 272	52 18 331
233	49 16 001	21 51 037	21 56 074	54 29 136	42 13 189	58 53 272	51 49 330
234	49 16 359	22 26 037	22 52 074	55 09 137	42 04 189	57 55 271	51 19 329
235	49 15 358	23 00 036	23 48 074	55 49 138	41 54 190	56 57 271	50 48 327
236	49 12 357	23 35 036	24 43 073	56 28 139	41 44 190	55 58 271	50 16 326
237	49 08 355	24 08 035	25 39 073	57 06 139	41 33 191	55 00 271	49 44 325
238	49 03 354	24 41 034	26 35 072	57 43 140	41 21 192	54 02 270	49 10 324
239	48 56 353	25 14 034	27 30 072	58 20 141	41 09 192	53 04 270	48 35 323
	VEGA	◆ALTAIR	Shaula	◆RIGIL KENT.	SPICA	◆ARCTURUS	Alphecca
240	25 46 033	28 25 072	58 56 142	40 57 193	52 06 270	47 59 322	48 48 351
241	26 18 033	29 21 071	59 31 143	40 43 193	51 07 270	47 22 320	48 39 350
242	26 49 032	30 16 071	60 06 144	40 29 194	50 09 269	46 45 319	48 28 349
243	27 19 031	31 11 070	60 39 146	40 15 195	49 11 269	46 07 318	48 16 347
244	27 49 031	32 05 070	61 12 147	40 00 195	48 13 269	45 28 317	48 02 346
245	28 18 030	33 00 069	61 43 148	39 45 196	47 15 269	44 48 317	47 48 345
246	28 47 029	33 54 069	62 13 149	39 29 196	46 16 268	44 08 316	47 32 344
247	29 15 028	34 49 068	62 42 151	39 12 197	45 18 268	43 27 315	47 15 342
248	29 42 028	35 43 068	63 10 152	38 55 197	44 20 268	42 45 314	46 56 341
249	30 09 027	36 37 067	63 37 154	38 38 198	43 22 268	42 03 313	46 37 340
250	30 35 026	37 30 067	64 02 155	38 20 198	42 24 268	41 20 312	46 16 339
251	31 00 025	38 24 066	64 25 157	38 01 199	41 25 267	40 36 311	45 54 337
252	31 25 025	39 17 066	64 48 158	37 43 199	40 27 267	39 52 311	45 32 336
253	31 49 024	40 10 065	65 08 160	37 23 200	39 29 267	39 08 310	45 08 335
254	32 12 023	41 03 065	65 27 162	37 03 200	38 31 267	38 23 309	44 43 334
	◆VEGA	ALTAIR	◆Peacock	RIGIL KENT.	◆SPICA	ARCTURUS	Alphecca
255	32 35 022	41 55 064	32 20 150	36 43 201	37 33 267	37 37 308	44 17 333
256	32 56 021	42 47 063	32 49 150	36 23 201	36 35 266	36 51 308	43 50 332
257	33 17 021	43 39 063	33 18 150	36 02 201	35 37 266	36 05 307	43 22 331
258	33 37 020	44 31 062	33 47 150	35 40 202	34 39 266	35 18 306	42 53 330
259	33 57 019	45 22 061	34 16 151	35 18 202	33 41 266	34 31 306	42 23 329
260	34 15 018	46 13 061	34 44 151	34 56 203	32 42 266	33 44 305	41 53 328
261	34 33 017	47 04 060	35 12 152	34 34 203	31 44 265	32 56 304	41 21 327
262	34 50 016	47 54 059	35 40 152	34 11 203	30 46 265	32 08 304	40 49 326
263	35 06 015	48 44 058	36 07 152	33 48 204	29 48 265	31 19 303	40 16 325
264	35 21 015	49 33 058	36 34 153	33 24 204	28 50 265	30 30 303	39 42 324
265	35 35 014	50 22 057	37 00 153	33 00 204	27 52 265	29 41 302	39 08 323
266	35 48 013	51 10 056	37 27 153	32 36 205	26 54 264	28 52 302	38 32 322
267	36 00 012	51 58 055	37 52 154	32 12 205	25 57 264	28 02 301	37 56 321
268	36 12 011	52 45 054	38 18 154	31 47 205	24 59 264	27 12 301	37 20 321
269	36 22 010	53 32 053	38 43 155	31 22 206	24 01 264	26 22 300	36 42 320

LHA ♈	Hc Zn	Hc Zn	Hc Zn	Hc Zn	Hc Zn	Hc Zn	Hc Zn
	◆VEGA	ALTAIR	◆FOMALHAUT	Peacock	RIGIL KENT.	◆ANTARES	ARCTURUS
270	36 32 009	54 18 052	20 19 117	39 07 155	30 56 206	65 23 236	25 31 300
271	36 40 008	55 04 051	21 11 117	39 32 156	30 31 206	64 35 237	24 41 299
272	36 48 007	55 48 050	22 03 117	39 55 156	30 05 206	63 45 238	23 50 299
273	36 54 006	56 32 048	22 55 117	40 18 157	29 39 207	62 56 239	22 59 298
274	37 00 005	57 15 047	23 47 117	40 41 157	29 13 207	62 06 239	22 07 298
275	37 05 004	57 57 046	24 39 117	41 03 158	28 46 207	61 16 240	21 16 297
276	37 08 003	58 39 044	25 31 116	41 25 158	28 20 207	60 25 241	20 24 297
277	37 11 002	59 19 043	26 23 116	41 46 159	27 53 208	59 34 241	19 32 297
278	37 13 001	59 58 042	27 16 116	42 07 159	27 26 208	58 43 242	18 40 296
279	37 13 000	60 36 040	28 08 116	42 27 160	26 58 208	57 52 242	17 48 296
280	37 13 359	61 13 038	29 00 116	42 47 161	26 31 208	57 00 243	16 55 295
281	37 12 358	61 48 037	29 52 116	43 06 161	26 03 208	56 08 243	16 03 295
282	37 09 357	62 23 035	30 44 116	43 24 162	25 36 209	55 16 244	15 10 295
283	37 06 356	62 55 033	31 36 116	43 42 163	25 08 209	54 24 244	14 17 294
284	37 02 355	63 26 031	32 28 116	43 59 163	24 40 209	53 32 244	13 24 294
	◆DENEB	◆FOMALHAUT	Peacock	RIGIL KENT.	◆ANTARES	Rasalhague	VEGA
285	26 27 020	33 21 117	44 16 164	24 11 209	52 39 245	56 03 321	36 57 354
286	26 47 019	34 13 117	44 31 165	23 43 209	51 46 245	55 26 319	36 50 353
287	27 05 018	35 05 117	44 47 165	23 15 209	50 54 245	54 47 318	36 43 352
288	27 23 018	35 57 117	45 01 166	22 46 209	50 01 245	54 08 317	36 35 351
289	27 40 017	36 49 117	45 15 167	22 18 210	49 08 246	53 28 315	36 26 351
290	27 57 016	37 41 117	45 28 167	21 49 210	48 15 246	52 46 314	36 16 350
291	28 12 015	38 33 117	45 41 168	21 20 210	47 21 246	52 04 313	36 05 349
292	28 27 015	39 24 117	45 52 169	20 51 210	46 28 246	51 21 312	35 53 348
293	28 42 014	40 16 117	46 03 170	20 22 210	45 35 246	50 38 311	35 40 347
294	28 55 013	41 08 117	46 14 170	19 53 210	44 42 247	49 54 310	35 26 346
295	29 08 012	42 00 117	46 23 171	19 24 210	43 48 247	49 09 309	35 11 345
296	29 20 012	42 51 118	46 32 172	18 55 210	42 55 247	48 23 308	34 56 344
297	29 31 011	43 43 118	46 40 173	18 26 210	42 01 247	47 37 307	34 39 343
298	29 42 010	44 34 118	46 47 173	17 57 210	41 08 247	46 51 306	34 22 342
299	29 51 009	45 26 118	46 53 174	17 27 210	40 14 247	46 04 306	34 04 341
	◆DENEB	Alpheratz	◆FOMALHAUT	Peacock	◆ANTARES	Rasalhague	VEGA
300	30 00 008	16 16 054	46 17 118	46 59 175	39 20 247	45 16 305	33 45 341
301	30 08 008	17 03 053	47 08 119	47 03 176	38 27 247	44 28 304	33 25 340
302	30 16 007	17 50 053	47 59 119	47 07 177	37 33 247	43 40 303	33 05 339
303	30 22 006	18 36 052	48 50 119	47 10 177	36 39 247	42 51 302	32 43 338
304	30 28 005	19 22 052	49 40 120	47 13 178	35 46 247	42 01 302	32 21 337
305	30 33 004	20 07 051	50 31 120	47 14 179	34 52 247	41 12 301	31 58 336
306	30 37 004	20 53 051	51 21 120	47 15 180	33 58 247	40 22 300	31 34 336
307	30 40 003	21 38 050	52 12 121	47 15 181	33 04 247	39 31 300	31 10 335
308	30 42 002	22 22 050	53 02 121	47 14 181	32 11 247	38 40 299	30 45 334
309	30 44 001	23 07 049	53 51 122	47 12 182	31 17 247	37 49 299	30 19 333
310	30 44 000	23 51 049	54 41 122	47 09 183	30 23 247	36 58 298	29 53 333
311	30 44 359	24 34 048	55 30 123	47 06 184	29 30 247	36 07 297	29 26 332
312	30 43 359	25 18 048	56 19 123	47 02 185	28 36 247	35 15 297	28 58 331
313	30 41 358	26 01 047	57 08 124	46 57 185	27 42 247	34 23 296	28 29 330
314	30 39 357	26 43 047	57 56 124	46 51 186	26 48 247	33 30 296	28 00 330
	Alpheratz	◆Diphda	ACHERNAR	Peacock	◆ANTARES	VEGA	◆DENEB
315	27 25 046	36 22 102	22 51 147	46 44 187	25 55 247	27 31 329	30 35 356
316	28 07 046	37 19 102	23 23 147	46 37 188	25 01 247	27 00 328	30 31 355
317	28 48 045	38 16 102	23 55 147	46 29 188	24 08 247	26 30 328	30 26 355
318	29 29 044	39 13 102	24 26 147	46 20 189	23 14 247	25 58 327	30 20 354
319	30 10 044	40 10 102	24 58 147	46 10 190	22 21 247	25 26 326	30 13 353
320	30 50 043	41 07 102	25 29 147	45 59 191	21 27 247	24 54 326	30 05 352
321	31 29 042	42 04 102	26 01 147	45 48 192	20 34 247	24 21 325	29 57 351
322	32 08 042	43 01 102	26 32 148	45 36 192	19 40 247	23 48 325	29 48 351
323	32 46 041	43 58 102	27 03 148	45 23 193	18 47 246	23 14 324	29 38 350
324	33 24 040	44 55 102	27 34 148	45 10 194	17 53 246	22 39 323	29 27 349
325	34 01 039	45 52 102	28 05 148	44 56 194	17 00 246	22 04 323	29 16 348
326	34 38 039	46 49 102	28 36 148	44 41 195	16 07 246	21 29 322	29 03 347
327	35 14 038	47 46 102	29 06 149	44 26 196	15 14 246	20 53 322	28 50 347
328	35 50 037	48 43 101	29 36 149	44 10 196	14 20 246	20 17 321	28 37 346
329	36 24 036	49 40 101	30 06 149	43 53 197	13 27 246	19 40 321	28 22 345
	◆Alpheratz	Diphda	◆ACHERNAR	Peacock	◆Nunki	ALTAIR	DENEB
330	36 59 035	50 37 101	30 36 149	43 35 198	45 07 247	50 35 304	28 07 344
331	37 32 035	51 34 101	31 06 150	43 17 198	44 14 247	49 46 303	27 51 344
332	38 05 034	52 31 101	31 35 150	42 59 199	43 20 247	48 57 302	27 34 343
333	38 37 033	53 28 101	32 04 150	42 40 200	42 27 247	48 07 301	27 17 342
334	39 08 032	54 25 101	32 33 150	42 20 200	41 33 247	47 17 300	26 58 342
335	39 38 031	55 22 102	33 02 151	41 59 201	40 39 247	46 27 300	26 40 341
336	40 08 030	56 20 102	33 30 151	41 39 201	39 46 247	45 36 299	26 20 340
337	40 37 029	57 17 102	33 58 151	41 17 202	38 52 247	44 45 298	26 00 339
338	41 05 028	58 14 102	34 26 152	40 55 202	37 58 247	43 53 297	25 39 339
339	41 32 027	59 11 102	34 54 152	40 33 203	37 05 247	43 01 297	25 18 338
340	41 58 026	60 08 102	35 21 152	40 10 203	36 11 247	42 09 296	24 56 337
341	42 23 025	61 04 102	35 48 153	39 46 204	35 17 247	41 17 295	24 33 337
342	42 47 024	62 01 102	36 14 153	39 23 204	34 23 247	40 24 295	24 10 336
343	43 11 023	62 58 102	36 40 154	38 58 205	33 29 248	39 31 294	23 46 335
344	43 33 022	63 55 102	37 06 154	38 34 205	32 36 248	38 38 294	23 21 335
	Alpheratz	◆Hamal	◆ACHERNAR	Peacock	Nunki	◆ALTAIR	DENEB
345	43 54 021	30 56 051	37 31 154	38 08 206	31 42 247	37 45 293	22 56 334
346	44 14 020	31 41 051	37 56 155	37 43 206	30 48 247	36 51 293	22 31 334
347	44 33 019	32 26 050	38 21 155	37 17 207	29 54 247	35 57 292	22 05 333
348	44 51 017	33 10 049	38 45 156	36 50 207	29 01 247	35 03 292	21 38 332
349	45 08 016	33 54 049	39 09 156	36 24 208	28 07 247	34 09 291	21 11 332
350	45 24 015	34 38 048	39 32 157	35 57 208	27 13 247	33 15 291	20 43 331
351	45 38 014	35 21 047	39 55 157	35 29 208	26 19 247	32 20 290	20 15 331
352	45 52 013	36 03 046	40 17 158	35 02 209	25 26 247	31 25 290	19 46 330
353	46 04 011	36 45 046	40 39 158	34 34 209	24 32 247	30 30 289	19 17 330
354	46 15 010	37 26 045	41 00 159	34 05 209	23 38 247	29 35 289	18 47 329
355	46 24 009	38 07 044	41 21 159	33 37 210	22 45 247	28 40 288	18 17 329
356	46 33 008	38 48 043	41 42 160	33 08 210	21 51 247	27 45 288	17 46 328
357	46 40 006	39 27 043	42 01 160	32 38 210	20 58 247	26 50 288	17 15 328
358	46 46 005	40 06 042	42 21 161	32 09 211	20 04 247	25 54 287	16 44 327
359	46 50 004	40 45 041	42 39 162	31 39 211	19 11 247	24 58 287	16 12 327

LAT 15°S

LHA ♈	Hc	Zn	Hc	Zn	Hc	Zn	Hc	Zn	Hc	Zn	Hc	Zn	Hc	Zn
	◆Alpheratz		Hamal		ALDEBARAN		◆RIGEL		ACHERNAR		◆FOMALHAUT		Enif	
0	45 54	003	40 36	039	15 03	068	13 05	095	43 54	162	69 27	222	48 07	304
1	45 56	001	41 13	039	15 57	068	14 03	095	44 12	163	68 48	224	47 19	303
2	45 56	000	41 49	038	16 51	067	15 00	095	44 29	163	68 07	225	46 31	303
3	45 56	359	42 24	037	17 44	067	15 58	094	44 46	164	67 26	226	45 42	302
4	45 54	358	42 58	036	18 37	066	16 56	094	45 01	165	66 43	228	44 52	301
5	45 51	356	43 31	035	19 30	066	17 54	094	45 16	165	66 00	229	44 02	300
6	45 46	355	44 04	034	20 23	066	18 52	094	45 31	166	65 16	230	43 12	300
7	45 41	354	44 35	033	21 16	065	19 49	093	45 45	167	64 31	231	42 22	299
8	45 34	353	45 06	032	22 08	065	20 47	093	45 58	167	63 46	232	41 31	298
9	45 26	351	45 36	030	23 00	064	21 45	093	46 10	168	63 00	233	40 40	298
10	45 16	350	46 05	029	23 53	064	22 43	093	46 22	169	62 13	234	39 48	297
11	45 06	349	46 33	028	24 45	063	23 41	093	46 33	170	61 26	235	38 56	296
12	44 54	348	47 00	027	25 36	063	24 39	092	46 43	170	60 39	235	38 04	296
13	44 41	346	47 25	026	26 28	062	25 37	092	46 52	171	59 51	236	37 12	295
14	44 27	345	47 50	025	27 19	062	26 35	092	47 01	172	59 02	237	36 19	295
	Hamal		◆ALDEBARAN		RIGEL		◆ACHERNAR		FOMALHAUT		◆Enif		Alpheratz	
15	48 14	023	28 10	062	27 33	092	47 09	173	58 14	237	35 27	294	44 11	344
16	48 36	022	29 01	061	28 31	091	47 16	173	57 25	238	34 34	294	43 55	343
17	48 57	021	29 51	060	29 28	091	47 22	174	56 36	238	33 40	293	43 37	342
18	49 17	019	30 42	060	30 26	091	47 28	175	55 46	239	32 47	293	43 19	341
19	49 36	018	31 32	059	31 24	091	47 33	176	54 56	239	31 53	292	42 59	340
20	49 53	017	32 22	059	32 22	090	47 36	176	54 06	240	30 59	292	42 38	338
21	50 09	015	33 11	058	33 20	090	47 40	177	53 16	240	30 05	291	42 17	337
22	50 24	014	34 00	058	34 18	090	47 42	178	52 26	241	29 11	291	41 54	336
23	50 38	013	34 49	057	35 16	090	47 44	179	51 35	241	28 17	290	41 30	335
24	50 50	011	35 37	056	36 14	089	47 44	180	50 45	241	27 23	290	41 06	334
25	51 00	010	36 25	056	37 12	089	47 44	180	49 54	242	26 28	289	40 40	333
26	51 09	008	37 13	055	38 10	089	47 43	181	49 03	242	25 33	289	40 13	332
27	51 17	007	38 00	054	39 08	088	47 41	182	48 12	242	24 38	289	39 46	331
28	51 23	005	38 47	054	40 06	088	47 39	183	47 20	242	23 43	288	39 18	330
29	51 28	004	39 34	053	41 04	088	47 36	184	46 29	243	22 48	288	38 49	329
	CAPELLA		◆BETELGEUSE		SIRIUS		CANOPUS		◆ACHERNAR		FOMALHAUT		◆Alpheratz	
30	14 41	033	27 38	073	21 51	102	26 23	142	47 31	184	45 38	243	38 19	329
31	15 12	032	28 34	073	22 47	102	26 59	142	47 27	185	44 46	243	37 48	328
32	15 43	032	29 29	072	23 44	102	27 34	142	47 21	186	43 54	243	37 17	327
33	16 13	031	30 24	072	24 41	102	28 10	142	47 14	187	43 03	243	36 45	326
34	16 43	031	31 19	071	25 37	102	28 45	142	47 07	188	42 11	243	36 12	325
35	17 13	030	32 14	071	26 34	101	29 21	143	46 59	188	41 19	243	35 38	324
36	17 42	030	33 09	070	27 31	101	29 56	143	46 50	189	40 27	244	35 04	323
37	18 10	029	34 03	070	28 28	101	30 31	143	46 40	190	39 35	244	34 29	323
38	18 39	029	34 58	070	29 25	101	31 06	143	46 30	191	38 43	244	33 53	322
39	19 06	028	35 52	069	30 22	101	31 40	143	46 19	191	37 51	244	33 17	321
40	19 34	028	36 46	068	31 19	101	32 15	144	46 07	192	36 59	244	32 41	320
41	20 00	027	37 40	068	32 16	101	32 49	144	45 55	193	36 07	244	32 03	320
42	20 26	027	38 33	067	33 12	100	33 23	144	45 41	194	35 15	244	31 25	319
43	20 52	026	39 27	067	34 10	100	33 57	144	45 28	194	34 23	244	30 47	318
44	21 17	025	40 20	066	35 07	100	34 31	145	45 13	195	33 31	244	30 08	317
	◆CAPELLA		BETELGEUSE		SIRIUS		◆CANOPUS		ACHERNAR		◆FOMALHAUT		Alpheratz	
45	21 42	025	41 13	066	36 04	100	35 04	145	44 58	196	32 39	244	29 29	317
46	22 06	024	42 05	065	37 01	100	35 38	145	44 42	196	31 46	244	28 49	316
47	22 29	024	42 58	064	37 58	100	36 10	146	44 25	197	30 54	244	28 08	316
48	22 52	023	43 50	064	38 55	100	36 43	146	44 08	198	30 02	244	27 28	315
49	23 14	022	44 42	063	39 52	099	37 16	146	43 50	198	29 10	244	26 46	314
50	23 36	022	45 33	062	40 49	099	37 48	147	43 32	199	28 18	244	26 05	314
51	23 57	021	46 24	062	41 47	099	38 19	147	43 13	199	27 26	244	25 23	313
52	24 17	020	47 15	061	42 44	099	38 51	147	42 53	200	26 34	244	24 40	313
53	24 37	020	48 06	060	43 41	099	39 22	148	42 33	201	25 42	244	23 57	312
54	24 56	019	48 56	059	44 38	099	39 53	148	42 12	201	24 50	244	23 14	311
55	25 15	018	49 45	058	45 36	099	40 23	149	41 51	202	23 58	244	22 30	311
56	25 33	018	50 35	058	46 33	099	40 53	149	41 30	202	23 06	244	21 46	310
57	25 50	017	51 23	057	47 30	099	41 23	149	41 07	203	22 14	244	21 02	310
58	26 06	016	52 11	056	48 27	098	41 52	150	40 45	203	21 22	244	20 17	309
59	26 22	015	52 59	055	49 25	098	42 21	150	40 22	204	20 30	243	19 33	309
	CAPELLA		POLLUX		◆SIRIUS		CANOPUS		◆ACHERNAR		Diphda		◆Hamal	
60	26 37	015	20 36	052	50 22	098	42 49	151	39 58	204	42 54	259	42 39	324
61	26 51	014	21 22	051	51 19	098	43 17	152	39 34	205	41 57	259	42 04	323
62	27 05	013	22 07	051	52 17	098	43 45	152	39 10	205	41 00	259	41 29	322
63	27 18	013	22 51	050	53 14	098	44 12	153	38 45	206	40 03	259	40 53	321
64	27 30	012	23 36	050	54 12	098	44 38	153	38 19	206	39 06	259	40 16	320
65	27 42	011	24 20	049	55 09	098	45 04	154	37 54	206	38 09	259	39 38	319
66	27 52	010	25 03	049	56 06	098	45 29	154	37 28	207	37 13	259	39 00	318
67	28 02	009	25 46	048	57 04	098	45 54	155	37 01	207	36 16	258	38 21	317
68	28 11	009	26 29	047	58 01	098	46 18	156	36 35	208	35 19	258	37 42	317
69	28 20	008	27 12	047	58 59	098	46 42	156	36 07	208	34 22	258	37 02	316
70	28 27	007	27 54	046	59 56	098	47 04	157	35 40	208	33 26	258	36 21	315
71	28 34	006	28 35	046	60 54	098	47 27	158	35 12	209	32 29	258	35 40	314
72	28 40	006	29 16	045	61 51	097	47 48	159	34 44	209	31 32	258	34 58	314
73	28 46	005	29 57	044	62 49	097	48 09	159	34 16	209	30 36	258	34 16	313
74	28 50	004	30 37	044	63 46	097	48 29	160	33 47	210	29 39	258	33 33	312
	CAPELLA		POLLUX		◆PROCYON		Suhail		◆CANOPUS		ACHERNAR		◆Hamal	
75	28 54	003	31 17	043	45 43	066	30 31	132	48 49	161	33 19	210	32 50	312
76	28 57	002	31 56	042	46 35	065	31 14	132	49 07	162	32 49	210	32 07	311
77	28 59	002	32 35	042	47 28	064	31 57	132	49 25	162	32 20	211	31 23	310
78	29 00	001	33 13	041	48 20	064	32 40	132	49 43	163	31 50	211	30 38	310
79	29 00	000	33 51	040	49 12	063	33 23	132	49 59	164	31 21	211	29 53	309
80	29 00	359	34 28	039	50 03	062	34 05	133	50 14	165	30 51	211	29 08	308
81	28 59	358	35 04	038	50 54	061	34 48	133	50 29	166	30 20	212	28 22	308
82	28 57	358	35 40	038	51 45	060	35 30	133	50 43	167	29 50	212	27 36	307
83	28 54	357	36 15	037	52 35	060	36 13	133	50 56	168	29 19	212	26 50	307
84	28 51	356	36 49	036	53 25	059	36 55	134	51 08	168	28 49	212	26 03	306
85	28 46	355	37 23	035	54 14	058	37 37	134	51 19	169	28 18	212	25 16	306
86	28 41	355	37 56	034	55 03	057	38 19	134	51 29	170	27 47	213	24 29	305
87	28 35	354	38 28	033	55 51	056	39 00	134	51 39	171	27 15	213	23 42	305
88	28 29	353	39 00	033	56 38	055	39 41	135	51 47	172	26 44	213	22 54	304
89	28 21	352	39 31	032	57 25	053	40 23	135	51 54	173	26 12	213	22 06	304

LHA ♈	Hc	Zn	Hc	Zn	Hc	Zn	Hc	Zn	Hc	Zn	Hc	Zn	Hc	Zn
	POLLUX		◆REGULUS		Suhail		◆CANOPUS		ACHERNAR		◆ALDEBARAN		CAPELLA	
90	40 01	031	22 55	070	41 04	135	52 01	174	25 41	213	52 14	326	28 13	351
91	40 30	030	23 49	069	41 44	136	52 06	175	25 09	213	51 41	324	28 04	351
92	40 58	029	24 44	069	42 25	136	52 11	176	24 37	213	51 07	323	27 54	350
93	41 26	028	25 38	068	43 05	136	52 14	177	24 05	213	50 32	322	27 43	349
94	41 52	027	26 31	068	43 45	137	52 17	178	23 33	214	49 56	321	27 32	348
95	42 18	026	27 25	068	44 24	137	52 18	179	23 01	214	49 19	320	27 20	348
96	42 42	025	28 18	067	45 03	138	52 18	180	22 29	214	48 41	319	27 07	347
97	43 06	023	29 12	067	45 42	138	52 18	181	21 57	214	48 02	318	26 54	346
98	43 28	022	30 05	066	46 21	139	52 16	182	21 25	214	47 23	316	26 39	345
99	43 50	021	30 58	066	46 59	139	52 14	183	20 52	214	46 42	315	26 24	345
100	44 10	020	31 51	065	47 37	140	52 10	184	20 20	214	46 01	315	26 09	344
101	44 30	019	32 43	065	48 14	140	52 06	185	19 48	214	45 20	314	25 52	343
102	44 48	018	33 35	064	48 51	141	52 00	186	19 15	214	44 37	313	25 35	343
103	45 05	017	34 27	064	49 27	141	51 54	187	18 43	214	43 54	312	25 18	342
104	45 21	015	35 19	063	50 03	142	51 46	188	18 10	214	43 11	311	24 59	341
	POLLUX		◆REGULUS		ACRUX		◆CANOPUS		RIGEL		◆ALDEBARAN		CAPELLA	
105	45 36	014	36 11	062	17 09	152	51 38	189	63 16	282	42 27	310	24 40	340
106	45 50	013	37 02	062	17 36	152	51 28	190	62 19	281	41 42	309	24 21	340
107	46 02	012	37 53	061	18 03	152	51 18	191	61 22	281	40 57	308	24 00	339
108	46 14	011	38 44	061	18 30	152	51 07	192	60 25	280	40 11	308	23 39	338
109	46 24	009	39 34	060	18 57	152	50 55	193	59 28	279	39 25	307	23 18	338
110	46 32	008	40 24	059	19 24	152	50 42	193	58 31	279	38 39	306	22 56	337
111	46 40	007	41 14	059	19 52	152	50 28	194	57 33	278	37 52	306	22 33	337
112	46 46	005	42 03	058	20 19	152	50 13	195	56 36	278	37 04	305	22 09	336
113	46 51	004	42 52	057	20 45	152	49 58	196	55 39	277	36 17	304	21 46	335
114	46 54	003	43 40	056	21 12	152	49 41	197	54 41	277	35 28	304	21 21	335
115	46 57	002	44 28	056	21 39	152	49 24	198	53 44	277	34 40	303	20 56	334
116	46 58	000	45 16	055	22 06	153	49 06	198	52 46	276	33 51	302	20 30	334
117	46 57	359	46 03	054	22 33	153	48 47	199	51 48	276	33 02	302	20 04	333
118	46 56	358	46 50	053	22 59	153	48 28	200	50 51	275	32 12	301	19 38	332
119	46 53	356	47 36	052	23 26	153	48 07	201	49 53	275	31 23	301	19 11	332
	◆REGULUS		SPICA		◆ACRUX		CANOPUS		◆RIGEL		BETELGEUSE		POLLUX	
120	48 21	051	11 13	099	23 52	153	47 47	202	48 55	275	51 45	304	46 49	355
121	49 06	050	12 10	098	24 18	153	47 25	202	47 57	274	50 57	303	46 43	354
122	49 51	049	13 07	098	24 45	153	47 03	203	47 00	274	50 08	302	46 36	353
123	50 34	048	14 05	098	25 11	153	46 40	204	46 02	274	49 18	301	46 28	351
124	51 17	047	15 02	098	25 36	154	46 16	204	45 04	273	48 28	300	46 19	350
125	51 59	046	16 00	098	26 02	154	45 52	205	44 06	273	47 38	299	46 08	349
126	52 41	045	16 57	097	26 28	154	45 27	206	43 08	273	46 47	299	45 56	348
127	53 21	044	17 55	097	26 53	154	45 02	206	42 10	272	45 56	298	45 43	346
128	54 01	043	18 52	097	27 19	154	44 36	207	41 12	272	45 05	297	45 29	345
129	54 40	041	19 50	097	27 44	154	44 09	207	40 15	272	44 13	297	45 14	344
130	55 18	040	20 47	096	28 09	155	43 42	208	39 17	272	43 21	296	44 57	343
131	55 55	039	21 45	096	28 33	155	43 15	209	38 19	271	42 29	295	44 39	342
132	56 30	037	22 42	096	28 58	155	42 47	209	37 21	271	41 36	295	44 20	340
133	57 05	036	23 40	096	29 22	155	42 19	210	36 23	271	40 44	294	44 00	339
134	57 38	034	24 38	096	29 46	156	41 50	210	35 25	271	39 51	293	43 39	338
	REGULUS		◆SPICA		ACRUX		◆CANOPUS		SIRIUS		BETELGEUSE		◆POLLUX	
135	58 10	033	25 35	095	30 10	156	41 21	211	57 31	262	38 57	293	43 17	337
136	58 41	031	26 33	095	30 34	156	40 51	211	56 33	262	38 04	292	42 54	336
137	59 11	030	27 31	095	30 57	156	40 21	211	55 36	262	37 10	292	42 30	335
138	59 39	028	28 29	095	31 21	157	39 50	212	54 38	262	36 16	291	42 05	334
139	60 05	026	29 26	094	31 44	157	39 20	212	53 41	262	35 22	291	41 39	333
140	60 30	024	30 24	094	32 06	157	38 48	213	52 43	262	34 28	290	41 12	332
141	60 53	023	31 22	094	32 29	157	38 17	213	51 46	262	33 33	290	40 44	331
142	61 14	021	32 20	094	32 51	158	37 45	213	50 49	262	32 39	289	40 16	330
143	61 34	019	33 18	094	33 12	158	37 13	214	49 51	262	31 44	289	39 46	329
144	61 52	017	34 16	093	33 34	158	36 41	214	48 54	262	30 49	288	39 16	328
145	62 08	015	35 13	093	33 55	159	36 08	215	47 57	262	29 54	288	38 45	327
146	62 21	013	36 11	093	34 16	159	35 35	215	46 59	261	28 59	288	38 13	326
147	62 33	011	37 09	093	34 36	159	35 02	215	46 02	261	28 03	287	37 40	325
148	62 43	009	38 07	093	34 56	160	34 28	215	45 05	261	27 08	287	37 07	324
149	62 51	006	39 05	092	35 16	160	33 54	216	44 08	261	26 12	286	36 32	324
	REGULUS		◆ARCTURUS		SPICA		◆ACRUX		CANOPUS		SIRIUS		◆PROCYON	
150	62 56	004	18 28	063	40 03	092	35 36	161	33 21	216	43 10	261	49 38	297
151	62 59	002	19 20	063	41 01	092	35 55	161	32 46	216	42 13	261	48 47	297
152	63 00	000	20 11	063	41 59	092	36 13	161	32 12	216	41 16	261	47 55	296
153	62 59	358	21 02	062	42 57	091	36 32	162	31 38	217	40 19	261	47 03	295
154	62 56	356	21 54	062	43 55	091	36 49	162	31 03	217	39 21	260	46 10	295
155	62 51	354	22 44	061	44 53	091	37 07	163	30 28	217	38 24	260	45 17	294
156	62 43	351	23 35	061	45 50	091	37 24	163	29 53	217	37 27	260	44 24	293
157	62 34	349	24 26	060	46 48	090	37 41	164	29 18	217	36 30	260	43 31	293
158	62 22	347	25 16	060	47 46	090	37 57	164	28 42	218	35 33	260	42 37	292
159	62 08	345	26 06	059	48 44	090	38 12	165	28 07	218	34 36	260	41 43	291
160	61 53	343	26 55	059	49 42	090	38 28	165	27 32	218	33 39	260	40 49	291
161	61 35	341	27 45	058	50 40	089	38 42	165	26 56	218	32 42	260	39 55	290
162	61 15	339	28 34	058	51 38	089	38 57	166	26 20	218	31 45	259	39 00	290
163	60 54	337	29 23	057	52 36	089	39 10	166	25 44	218	30 48	259	38 06	289
164	60 31	336	30 11	057	53 34	088	39 24	167	25 08	218	29 51	259	37 11	289
	Denebola		◆ARCTURUS		SPICA		◆ACRUX		SIRIUS		◆PROCYON		REGULUS	
165	58 02	023	31 00	056	54 32	088	39 37	168	28 54	259	36 16	288	60 06	334
166	58 24	021	31 48	055	55 30	088	39 49	168	27 57	259	35 21	288	59 40	332
167	58 44	019	32 35	055	56 28	088	40 01	169	27 00	259	34 26	287	59 12	330
168	59 02	018	33 22	054	57 26	087	40 12	169	26 04	258	33 30	287	58 43	329
169	59 19	016	34 09	054	58 24	087	40 22	170	25 07	258	32 35	286	58 12	327
170	59 33	014	34 56	053	59 22	087	40 33	170	24 10	258	31 39	286	57 40	326
171	59 46	012	35 42	052	60 19	086	40 42	171	23 14	258	30 43	286	57 07	324
172	59 57	010	36 27	052	61 17	086	40 51	171	22 17	258	29 48	285	56 32	323
173	60 07	008	37 13	051	62 15	085	41 00	172	21 20	258	28 52	285	55 57	321
174	60 14	006	37 57	050	63 13	085	41 07	172	20 24	257	27 56	284	55 20	320
175	60 19	004	38 42	049	64 10	085	41 15	173	19 27	257	26 59	284	54 42	319
176	60 23	002	39 25	049	65 08	084	41 22	174	18 31	257	26 03	284	54 03	317
177	60 24	000	40 09	048	66 06	084	41 28	174	17 34	257	25 07	283	53 24	316
178	60 23	358	40 51	047	67 03	083	41 33	175	16 38	257	24 10	283	52 43	315
179	60 21	356	41 33	046	68 01	083	41 38	175	15 41	256	23 14	283	52 02	314

LAT 15°S

LHA	Hc	Zn	Hc	Zn	Hc	Zn	Hc	Zn	Hc	Zn	Hc	Zn	Hc	Zn
♈	ARCTURUS		◆ANTARES		ACRUX		◆Suhail		Alphard		REGULUS		◆Denebola	
180	42 15	045	26 42	112	41 42	176	43 41	223	52 09	275	51 19	313	60 16	355
181	42 56	044	27 36	112	41 46	177	43 01	224	51 12	275	50 36	312	60 10	353
182	43 36	043	28 29	112	41 49	177	42 21	224	50 14	275	49 53	311	60 01	351
183	44 15	043	29 23	112	41 52	178	41 40	224	49 16	274	49 09	310	59 51	349
184	44 54	042	30 17	112	41 54	178	41 00	225	48 18	274	48 24	309	59 39	347
185	45 32	041	31 10	112	41 55	179	40 19	225	47 21	274	47 38	308	59 25	345
186	46 10	040	32 04	112	41 56	180	39 38	225	46 23	273	46 52	307	59 09	343
187	46 46	039	32 58	112	41 56	180	38 56	226	45 25	273	46 05	306	58 51	341
188	47 22	037	33 51	112	41 55	181	38 15	226	44 27	273	45 18	305	58 32	340
189	47 57	036	34 45	112	41 54	181	37 33	226	43 29	272	44 31	304	58 11	338
190	48 31	035	35 39	112	41 52	182	36 51	227	42 31	272	43 43	304	57 48	336
191	49 03	034	36 33	112	41 50	183	36 09	227	41 33	272	42 54	303	57 24	335
192	49 35	033	37 26	112	41 47	183	35 26	227	40 35	271	42 05	302	56 59	333
193	50 06	032	38 20	112	41 43	184	34 44	227	39 37	271	41 16	301	56 32	331
194	50 36	030	39 14	112	41 39	184	34 01	227	38 39	271	40 27	301	56 03	330
	ARCTURUS		◆Rasalhague		ANTARES		◆ACRUX		Suhail		◆REGULUS		Denebola	
195	51 05	029	16 39	072	40 07	112	41 34	185	33 19	228	39 37	300	55 34	328
196	51 32	028	17 34	071	41 01	112	41 29	186	32 36	228	38 46	299	55 03	327
197	51 59	026	18 29	071	41 55	112	41 23	186	31 53	228	37 56	299	54 31	326
198	52 24	025	19 24	071	42 48	112	41 16	187	31 10	228	37 05	298	53 57	324
199	52 48	024	20 18	070	43 42	112	41 09	187	30 27	228	36 13	298	53 23	323
200	53 10	022	21 13	070	44 36	113	41 01	188	29 44	228	35 22	297	52 48	322
201	53 32	021	22 07	069	45 29	113	40 53	189	29 01	228	34 30	296	52 11	320
202	53 51	019	23 01	069	46 23	113	40 44	189	28 17	228	33 38	296	51 34	319
203	54 10	018	23 55	069	47 16	113	40 34	190	27 34	228	32 46	295	50 56	318
204	54 27	016	24 49	068	48 09	113	40 24	190	26 50	229	31 53	295	50 16	317
205	54 42	015	25 43	068	49 03	113	40 13	191	26 07	229	31 01	294	49 36	316
206	54 56	013	26 36	067	49 56	113	40 02	191	25 23	229	30 08	294	48 56	315
207	55 08	011	27 30	067	50 49	114	39 51	192	24 40	229	29 15	293	48 14	314
208	55 19	010	28 23	066	51 42	114	39 39	192	23 56	229	28 21	293	47 32	313
209	55 28	008	29 16	066	52 35	114	39 26	193	23 13	229	27 28	292	46 49	312
	ARCTURUS		◆Rasalhague		ANTARES		◆RIGIL KENT.		ACRUX		Gienah		◆REGULUS	
210	55 35	006	30 09	065	53 28	114	43 38	173	39 13	193	64 50	261	26 34	292
211	55 41	005	31 01	065	54 21	115	43 45	174	38 59	194	63 52	261	25 40	292
212	55 45	003	31 54	064	55 13	115	43 50	175	38 45	194	62 55	261	24 46	291
213	55 47	001	32 46	064	56 06	115	43 55	175	38 30	195	61 58	261	23 52	291
214	55 47	000	33 38	063	56 58	116	44 00	176	38 15	195	61 01	261	22 58	290
215	55 46	358	34 29	063	57 50	116	44 03	177	37 59	196	60 03	261	22 04	290
216	55 44	356	35 21	062	58 42	116	44 06	177	37 43	196	59 06	261	21 09	290
217	55 39	355	36 12	062	59 34	117	44 08	178	37 27	197	58 09	261	20 14	289
218	55 33	353	37 03	061	60 26	117	44 10	179	37 10	197	57 12	261	19 20	289
219	55 25	351	37 53	060	61 17	118	44 11	179	36 52	198	56 15	261	18 25	288
220	55 16	350	38 44	060	62 08	119	44 11	180	36 34	198	55 17	261	17 30	288
221	55 05	348	39 33	059	62 59	119	44 11	181	36 16	199	54 20	261	16 34	288
222	54 52	347	40 23	058	63 49	120	44 09	181	35 58	199	53 23	261	15 39	287
223	54 38	345	41 12	058	64 39	121	44 08	182	35 39	199	52 26	261	14 44	287
224	54 22	343	42 01	057	65 29	122	44 05	183	35 19	200	51 29	261	13 48	287
	Alphecca		◆VEGA		ALTAIR		◆Shaula		RIGIL KENT.		◆SPICA		ARCTURUS	
225	47 26	011	16 10	041	14 11	077	49 31	130	44 02	184	66 32	277	54 05	342
226	47 37	010	16 48	041	15 07	076	50 15	131	43 58	184	65 35	276	53 46	340
227	47 47	009	17 26	040	16 04	076	50 59	131	43 53	185	64 37	276	53 26	339
228	47 55	007	18 03	040	17 00	076	51 42	132	43 48	186	63 40	275	53 04	337
229	48 02	006	18 40	039	17 56	075	52 25	133	43 42	186	62 42	275	52 41	336
230	48 07	005	19 16	039	18 52	075	53 07	133	43 36	187	61 44	274	52 17	335
231	48 11	004	19 52	038	19 48	075	53 49	134	43 28	187	60 46	274	51 51	333
232	48 14	002	20 28	038	20 44	074	54 31	134	43 21	188	59 48	274	51 25	332
233	48 16	001	21 03	037	21 39	074	55 12	135	43 12	189	58 51	273	50 57	331
234	48 16	359	21 38	037	22 35	073	55 53	136	43 03	189	57 53	273	50 28	329
235	48 15	358	22 12	036	23 30	073	56 33	137	42 53	190	56 55	273	49 58	328
236	48 12	357	22 46	035	24 26	073	57 12	137	42 43	191	55 57	272	49 26	327
237	48 08	355	23 19	035	25 21	072	57 51	138	42 32	191	54 59	272	48 54	326
238	48 03	354	23 52	034	26 16	072	58 29	139	42 20	192	54 01	272	48 21	324
239	47 57	353	24 24	034	27 11	072	59 07	140	42 08	192	53 03	271	47 47	323
	VEGA		◆ALTAIR		Shaula		◆RIGIL KENT.		SPICA		◆ARCTURUS		Alphecca	
240	24 56	033	28 06	071	59 43	141	41 55	193	52 05	271	47 12	322	47 49	352
241	25 27	032	29 01	071	60 19	142	41 42	194	51 07	271	46 36	321	47 39	350
242	25 58	032	29 56	070	60 54	143	41 28	194	50 09	271	45 59	320	47 29	349
243	26 28	031	30 50	070	61 28	145	41 13	195	49 11	270	45 22	319	47 17	348
244	26 57	030	31 44	069	62 02	146	40 58	195	48 13	270	44 43	318	47 04	346
245	27 26	030	32 39	069	62 34	147	40 42	196	47 15	270	44 04	317	46 50	345
246	27 54	029	33 32	068	63 05	148	40 26	196	46 17	269	43 25	316	46 34	344
247	28 22	028	34 26	068	63 34	150	40 10	197	45 19	269	42 44	315	46 17	343
248	28 49	027	35 20	067	64 03	151	39 53	197	44 22	269	42 03	314	46 00	341
249	29 16	027	36 13	067	64 30	153	39 35	198	43 24	269	41 21	314	45 40	340
250	29 41	026	37 06	066	64 56	154	39 17	198	42 26	269	40 39	313	45 20	339
251	30 06	025	37 59	066	65 20	156	38 58	199	41 28	268	39 56	312	44 59	338
252	30 31	024	38 52	065	65 43	158	38 39	199	40 30	268	39 13	311	44 37	337
253	30 54	024	39 45	065	66 04	159	38 20	200	39 32	268	38 29	310	44 13	336
254	31 17	023	40 37	064	66 24	161	38 00	200	38 34	268	37 45	310	43 49	334
	◆VEGA		ALTAIR		◆Peacock		RIGIL KENT.		◆SPICA		ARCTURUS		Alphecca	
255	31 39	022	41 29	063	33 12	149	37 39	201	37 36	267	37 00	309	43 23	333
256	32 01	021	42 20	063	33 41	150	37 19	201	36 38	267	36 15	308	42 57	332
257	32 21	020	43 11	062	34 10	150	36 57	202	35 40	267	35 29	308	42 29	331
258	32 41	020	44 02	061	34 39	150	36 36	202	34 42	267	34 43	307	42 01	330
259	33 00	019	44 53	061	35 08	150	36 14	202	33 45	266	33 56	306	41 32	329
260	33 18	018	45 43	060	35 37	151	35 52	203	32 47	266	33 09	306	41 02	328
261	33 35	017	46 33	059	36 05	151	35 29	203	31 49	266	32 22	305	40 31	327
262	33 52	016	47 23	058	36 32	152	35 06	204	30 51	266	31 34	304	39 59	326
263	34 08	015	48 12	057	37 00	152	34 42	204	29 53	266	30 46	304	39 27	325
264	34 22	014	49 00	057	37 27	152	34 19	204	28 56	265	29 58	303	38 53	325
265	34 36	013	49 48	056	37 54	153	33 55	205	27 58	265	29 09	303	38 19	324
266	34 49	013	50 36	055	38 20	153	33 30	205	27 00	265	28 20	302	37 45	323
267	35 01	012	51 23	054	38 46	154	33 06	205	26 02	265	27 31	302	37 09	322
268	35 13	011	52 10	053	39 12	154	32 41	206	25 05	265	26 41	301	36 33	321
269	35 23	010	52 55	052	39 37	154	32 16	206	24 07	264	25 52	301	35 56	320

LHA	Hc	Zn	Hc	Zn	Hc	Zn	Hc	Zn	Hc	Zn	Hc	Zn	Hc	Zn
♈	◆VEGA		ALTAIR		◆FOMALHAUT		Peacock		RIGIL KENT.		◆ANTARES		ARCTURUS	
270	35 32	009	53 41	051	20 46	116	40 02	155	31 50	206	65 56	238	25 02	300
271	35 41	008	54 25	050	21 38	116	40 26	155	31 25	206	65 06	239	24 11	300
272	35 48	007	55 09	048	22 30	116	40 50	156	30 59	207	64 17	240	23 21	299
273	35 55	006	55 52	047	23 22	116	41 13	156	30 33	207	63 26	240	22 30	299
274	36 00	005	56 34	046	24 14	116	41 36	157	30 06	207	62 36	241	21 39	298
275	36 05	004	57 15	045	25 06	116	41 59	157	29 40	207	61 45	242	20 48	298
276	36 08	003	57 55	043	25 58	116	42 21	158	29 13	208	60 54	242	19 57	297
277	36 11	002	58 35	042	26 50	116	42 42	159	28 46	208	60 02	243	19 05	297
278	36 13	001	59 13	040	27 42	116	43 03	159	28 19	208	59 11	243	18 13	297
279	36 13	000	59 50	039	28 34	116	43 24	160	27 51	208	58 19	244	17 21	296
280	36 13	359	60 26	037	29 26	116	43 43	160	27 24	208	57 27	244	16 29	296
281	36 12	358	61 00	036	30 18	116	44 03	161	26 56	209	56 35	245	15 37	295
282	36 09	357	61 33	034	31 11	116	44 21	162	26 28	209	55 42	245	14 44	295
283	36 06	356	62 05	032	32 03	116	44 39	162	26 00	209	54 50	245	13 52	295
284	36 02	355	62 35	030	32 55	116	44 56	163	25 32	209	53 57	246	12 59	294
	◆DENEB		◆FOMALHAUT		Peacock		RIGIL KENT.		◆ANTARES		Rasalhague		VEGA	
285	25 31	019	33 47	116	45 13	164	25 04	209	53 04	246	55 17	321	35 57	354
286	25 50	019	34 39	116	45 29	164	24 35	209	52 11	246	54 40	320	35 51	353
287	26 08	018	35 31	116	45 45	165	24 07	210	51 18	246	54 03	319	35 44	353
288	26 26	017	36 23	116	45 59	166	23 38	210	50 25	246	53 24	318	35 36	352
289	26 43	017	37 15	116	46 13	166	23 10	210	49 32	247	52 44	316	35 27	351
290	26 59	016	38 07	116	46 27	167	22 41	210	48 39	247	52 04	315	35 17	350
291	27 15	015	38 59	116	46 39	168	22 12	210	47 45	247	51 23	314	35 06	349
292	27 29	014	39 51	116	46 51	169	21 43	210	46 52	247	50 41	313	34 54	348
293	27 43	014	40 43	116	47 02	169	21 14	210	45 59	247	49 58	312	34 41	347
294	27 57	013	41 35	117	47 13	170	20 45	210	45 05	247	49 15	311	34 28	346
295	28 09	012	42 27	117	47 22	171	20 16	210	44 12	248	48 31	310	34 13	345
296	28 21	011	43 19	117	47 31	172	19 47	210	43 18	248	47 46	309	33 58	344
297	28 32	011	44 10	117	47 39	172	19 18	210	42 24	248	47 01	308	33 42	343
298	28 43	010	45 02	117	47 46	173	18 48	210	41 31	248	46 15	307	33 25	342
299	28 52	009	45 54	117	47 53	174	18 19	210	40 37	248	45 28	306	33 07	342
	◆DENEB		Alpheratz		◆FOMALHAUT		Peacock		◆ANTARES		Rasalhague		VEGA	
300	29 01	008	15 41	053	46 45	118	47 58	175	39 43	248	44 41	306	32 48	341
301	29 09	007	16 27	053	47 36	118	48 03	176	38 50	248	43 54	305	32 29	340
302	29 16	007	17 13	052	48 28	118	48 07	176	37 56	248	43 06	304	32 09	339
303	29 22	006	17 59	052	49 19	118	48 10	177	37 02	248	42 18	303	31 47	338
304	29 28	005	18 44	052	50 10	119	48 13	178	36 08	248	41 29	303	31 26	337
305	29 33	004	19 30	051	51 00	119	48 14	179	35 15	248	40 40	302	31 03	337
306	29 37	003	20 15	051	51 51	119	48 15	180	34 21	248	39 51	301	30 40	336
307	29 40	003	20 59	050	52 42	120	48 15	181	33 27	248	39 01	300	30 16	335
308	29 42	002	21 44	050	53 32	120	48 14	181	32 34	248	38 11	300	29 51	334
309	29 44	001	22 27	049	54 22	120	48 12	182	31 40	248	37 21	299	29 25	334
310	29 44	000	23 11	049	55 12	121	48 09	183	30 46	248	36 30	299	28 59	333
311	29 44	359	23 54	048	56 02	121	48 06	184	29 52	248	35 39	298	28 33	332
312	29 43	359	24 37	047	56 51	122	48 01	185	28 59	248	34 47	297	28 05	331
313	29 41	358	25 20	047	57 40	122	47 56	185	28 05	248	33 56	297	27 37	331
314	29 39	357	26 02	046	58 28	123	47 50	186	27 12	248	33 04	296	27 08	330
	Alpheratz		◆Diphda		ACHERNAR		Peacock		◆ANTARES		VEGA		◆DENEB	
315	26 44	046	36 35	102	23 41	146	47 44	187	26 18	248	26 39	329	29 35	356
316	27 25	045	37 31	101	24 13	147	47 36	188	25 24	247	26 09	329	29 31	355
317	28 06	045	38 28	101	24 45	147	47 28	189	24 31	247	25 39	328	29 26	355
318	28 46	044	39 25	101	25 17	147	47 19	189	23 37	247	25 08	327	29 20	354
319	29 26	043	40 22	101	25 48	147	47 09	190	22 44	247	24 36	327	29 13	353
320	30 06	043	41 19	101	26 20	147	46 58	191	21 51	247	24 04	326	29 06	352
321	30 44	042	42 16	101	26 51	147	46 47	192	20 57	247	23 32	325	28 58	351
322	31 23	041	43 13	101	27 23	147	46 35	192	20 04	247	22 59	325	28 49	351
323	32 01	040	44 10	101	27 54	148	46 22	193	19 11	247	22 25	324	28 39	350
324	32 38	040	45 07	101	28 25	148	46 08	194	18 17	247	21 51	324	28 28	349
325	33 15	039	46 03	101	28 56	148	45 54	195	17 24	246	21 16	323	28 17	348
326	33 51	038	47 00	100	29 27	148	45 39	195	16 31	246	20 41	323	28 05	348
327	34 27	037	47 57	100	29 57	148	45 23	196	15 38	246	20 06	322	27 52	347
328	35 02	037	48 54	100	30 28	149	45 07	197	14 45	246	19 30	322	27 38	346
329	35 36	036	49 51	100	30 58	149	44 50	197	13 52	246	18 54	321	27 24	345
	◆Alpheratz		Diphda		◆ACHERNAR		Peacock		◆Nunki		ALTAIR		DENEB	
330	36 10	035	50 48	100	31 28	149	44 32	198	45 31	248	50 01	305	27 09	345
331	36 42	034	51 46	100	31 57	149	44 14	199	44 37	248	49 13	304	26 53	344
332	37 15	033	52 43	100	32 27	149	43 55	199	43 43	248	48 25	303	26 37	343
333	37 46	032	53 40	100	32 56	150	43 36	200	42 50	248	47 36	302	26 19	342
334	38 17	032	54 37	100	33 25	150	43 16	200	41 56	248	46 47	301	26 02	342
335	38 47	031	55 34	100	33 54	150	42 55	201	41 02	248	45 57	300	25 43	341
336	39 16	030	56 31	100	34 23	151	42 34	202	40 09	248	45 07	300	25 24	340
337	39 44	029	57 28	100	34 51	151	42 13	202	39 15	248	44 16	299	25 04	340
338	40 12	028	58 25	100	35 19	151	41 51	203	38 21	248	43 25	298	24 43	339
339	40 38	027	59 22	100	35 47	152	41 28	203	37 27	248	42 34	298	24 22	338
340	41 04	026	60 19	100	36 14	152	41 05	204	36 33	248	41 43	297	24 00	338
341	41 29	025	61 16	100	36 41	152	40 41	204	35 40	248	40 51	296	23 38	337
342	41 52	024	62 13	100	37 08	153	40 17	205	34 46	248	39 59	296	23 15	336
343	42 15	023	63 10	100	37 34	153	39 53	205	33 52	248	39 06	295	22 51	336
344	42 37	022	64 07	100	38 00	154	39 28	206	32 58	248	38 14	294	22 27	335
	Alpheratz		◆Hamal		◆ACHERNAR		Peacock		Nunki		◆ALTAIR		DENEB	
345	42 58	020	30 18	051	38 25	154	39 02	206	32 05	248	37 21	294	22 02	334
346	43 18	019	31 03	050	38 51	154	38 37	207	31 11	248	36 28	293	21 37	334
347	43 36	018	31 47	049	39 15	155	38 10	207	30 17	248	35 34	293	21 11	333
348	43 54	017	32 31	049	39 40	155	37 44	207	29 23	248	34 41	292	20 45	333
349	44 10	016	33 14	048	40 04	156	37 17	208	28 30	248	33 47	292	20 18	332
350	44 26	015	33 57	047	40 27	156	36 50	208	27 36	248	32 53	291	19 50	331
351	44 40	014	34 40	047	40 50	157	36 22	209	26 42	248	31 59	291	19 22	331
352	44 53	012	35 21	046	41 13	157	35 54	209	25 49	248	31 05	290	18 54	330
353	45 05	011	36 03	045	41 35	158	35 26	209	24 55	248	30 10	290	18 25	330
354	45 16	010	36 44	044	41 56	158	34 57	210	24 02	247	29 16	289	17 56	329
355	45 25	009	37 24	044	42 17	159	34 29	210	23 08	247	28 21	289	17 26	329
356	45 33	008	38 04	043	42 38	160	34 00	210	22 15	247	27 26	289	16 55	328
357	45 40	006	38 43	042	42 58	160	33 30	211	21 21	247	26 31	288	16 25	328
358	45 46	005	39 21	041	43 17	161	33 01	211	20 28	247	25 36	288	15 54	327
359	45 50	004	39 59	040	43 36	161	32 31	211	19 34	247	24 41	287	15 22	327

LHA ♈	Hc	Zn	Hc	Zn	Hc	Zn	Hc	Zn	Hc	Zn	Hc	Zn	Hc	Zn
	◆Alpheratz		Hamal		ALDEBARAN		◆RIGEL		ACHERNAR		◆Peacock		Enif	
0	44 54	003	39 50	039	14 41	068	13 10	095	44 51	162	32 52	212	47 33	305
1	44 56	001	40 26	038	15 34	067	14 08	095	45 09	162	32 21	212	46 46	304
2	44 56	000	41 01	037	16 27	067	15 05	094	45 27	163	31 51	212	45 58	304
3	44 56	359	41 35	036	17 20	066	16 03	094	45 43	164	31 20	212	45 10	303
4	44 54	358	42 09	035	18 13	066	17 00	094	45 59	164	30 49	213	44 21	302
5	44 51	356	42 42	034	19 05	066	17 58	094	46 14	165	30 18	213	43 32	301
6	44 46	355	43 14	033	19 58	065	18 55	093	46 29	166	29 46	213	42 42	300
7	44 41	354	43 45	032	20 50	065	19 53	093	46 43	166	29 15	213	41 52	300
8	44 34	353	44 15	031	21 42	064	20 50	093	46 56	167	28 43	213	41 02	299
9	44 26	351	44 44	030	22 34	064	21 48	093	47 09	168	28 11	214	40 11	298
10	44 17	350	45 12	029	23 26	063	22 46	092	47 21	169	27 39	214	39 21	298
11	44 07	349	45 40	028	24 18	063	23 43	092	47 32	169	27 07	214	38 29	297
12	43 55	348	46 06	027	25 09	063	24 41	092	47 42	170	26 35	214	37 38	296
13	43 42	347	46 31	025	26 00	062	25 39	092	47 51	171	26 03	214	36 46	296
14	43 29	346	46 55	024	26 51	062	26 36	091	48 00	172	25 30	214	35 54	295
	◆Hamal		ALDEBARAN		RIGEL		◆CANOPUS		ACHERNAR		◆FOMALHAUT		Alpheratz	
15	47 18	023	27 41	061	27 34	091	18 06	141	48 08	172	58 46	239	43 14	344
16	47 40	022	28 32	061	28 32	091	18 43	141	48 15	173	57 56	239	42 58	343
17	48 01	020	29 22	060	29 29	090	19 19	141	48 22	174	57 07	240	42 40	342
18	48 21	019	30 11	059	30 27	090	19 55	141	48 28	175	56 17	240	42 22	341
19	48 39	018	31 01	059	31 25	090	20 32	141	48 32	176	55 26	241	42 03	340
20	48 56	016	31 50	058	32 22	090	21 08	141	48 36	176	54 36	241	41 43	339
21	49 12	015	32 39	058	33 20	089	21 45	141	48 40	177	53 46	241	41 21	338
22	49 26	014	33 28	057	34 18	089	22 21	141	48 42	178	52 55	242	40 59	337
23	49 39	012	34 16	056	35 15	089	22 57	141	48 43	179	52 04	242	40 36	336
24	49 51	011	35 04	056	36 13	089	23 33	141	48 44	180	51 13	242	40 11	335
25	50 01	010	35 51	055	37 11	088	24 10	141	48 44	181	50 22	243	39 46	334
26	50 10	008	36 38	054	38 08	088	24 46	141	48 43	181	49 31	243	39 20	333
27	50 18	007	37 25	054	39 06	088	25 22	141	48 41	182	48 39	243	38 53	332
28	50 24	005	38 11	053	40 04	087	25 58	141	48 39	183	47 48	243	38 26	331
29	50 28	004	38 57	052	41 01	087	26 34	141	48 35	184	46 56	243	37 57	330
	◆Hamal		ALDEBARAN		◆SIRIUS		CANOPUS		ACHERNAR		◆FOMALHAUT		Alpheratz	
30	50 32	002	39 43	052	22 03	102	27 10	142	48 31	185	46 05	244	37 28	329
31	50 33	001	40 28	051	23 00	102	27 46	142	48 26	185	45 13	244	36 57	328
32	50 34	000	41 12	050	23 56	102	28 22	142	48 20	186	44 21	244	36 26	327
33	50 32	358	41 56	049	24 53	101	28 57	142	48 14	187	43 29	244	35 55	326
34	50 30	357	42 39	048	25 49	101	29 33	142	48 06	188	42 37	244	35 22	325
35	50 26	355	43 22	047	26 46	101	30 08	142	47 58	189	41 45	244	34 49	325
36	50 20	354	44 04	046	27 42	101	30 43	142	47 49	189	40 53	244	34 16	324
37	50 14	352	44 46	046	28 39	101	31 19	143	47 40	190	40 01	244	33 41	323
38	50 05	351	45 26	045	29 36	100	31 54	143	47 29	191	39 09	245	33 06	322
39	49 55	350	46 06	044	30 33	100	32 28	143	47 18	192	38 17	245	32 30	321
40	49 44	348	46 46	043	31 29	100	33 03	143	47 06	192	37 25	245	31 54	321
41	49 32	347	47 25	042	32 26	100	33 37	143	46 53	193	36 33	245	31 17	320
42	49 18	345	48 02	040	33 23	100	34 12	144	46 40	194	35 41	245	30 40	319
43	49 03	344	48 39	039	34 20	100	34 46	144	46 26	195	34 49	245	30 02	319
44	48 47	343	49 16	038	35 17	099	35 20	144	46 11	195	33 57	245	29 24	318
	◆CAPELLA		BETELGEUSE		SIRIUS		◆CANOPUS		ACHERNAR		◆FOMALHAUT		Hamal	
45	20 47	025	40 48	065	36 14	099	35 53	144	45 55	196	33 05	245	48 29	341
46	21 11	024	41 40	064	37 11	099	36 27	145	45 39	197	32 12	245	48 10	340
47	21 34	023	42 32	064	38 08	099	37 00	145	45 23	197	31 20	245	47 50	339
48	21 57	023	43 23	063	39 05	099	37 33	145	45 05	198	30 28	245	47 28	338
49	22 19	022	44 14	062	40 02	099	38 05	146	44 47	199	29 36	245	47 06	336
50	22 40	021	45 05	061	40 59	098	38 38	146	44 29	199	28 44	245	46 42	335
51	23 01	021	45 56	061	41 56	098	39 10	146	44 09	200	27 52	244	46 17	334
52	23 21	020	46 46	060	42 53	098	39 41	147	43 50	200	27 00	244	45 52	333
53	23 41	019	47 35	059	43 50	098	40 13	147	43 29	201	26 08	244	45 25	332
54	24 00	019	48 25	058	44 47	098	40 44	148	43 08	201	25 16	244	44 57	331
55	24 18	018	49 14	057	45 44	098	41 14	148	42 47	202	24 24	244	44 28	329
56	24 35	017	50 02	057	46 41	098	41 45	149	42 25	203	23 32	244	43 58	328
57	24 52	017	50 50	056	47 38	097	42 15	149	42 03	203	22 40	244	43 28	327
58	25 09	016	51 37	055	48 36	097	42 44	149	41 40	204	21 48	244	42 56	326
59	25 24	015	52 24	054	49 33	097	43 13	150	41 16	204	20 56	244	42 24	325
	CAPELLA		BETELGEUSE		◆SIRIUS		CANOPUS		◆ACHERNAR		Diphda		◆Hamal	
60	25 39	015	53 10	053	50 30	097	43 42	151	40 53	205	43 05	260	41 50	324
61	25 53	014	53 56	052	51 27	097	44 10	151	40 28	205	42 08	260	41 16	323
62	26 07	013	54 40	050	52 25	097	44 38	152	40 04	206	41 11	260	40 42	322
63	26 19	012	55 25	049	53 22	097	45 05	152	39 39	206	40 14	260	40 06	321
64	26 31	012	56 08	048	54 19	097	45 31	153	39 13	206	39 18	260	39 30	321
65	26 43	011	56 50	047	55 16	096	45 58	153	38 47	207	38 21	259	38 53	320
66	26 53	010	57 32	046	56 14	096	46 23	154	38 21	207	37 24	259	38 15	319
67	27 03	009	58 13	044	57 11	096	46 48	155	37 55	208	36 27	259	37 37	318
68	27 12	009	58 52	043	58 09	096	47 13	155	37 28	208	35 31	259	36 58	317
69	27 20	008	59 31	041	59 06	096	47 36	156	37 00	208	34 34	259	36 19	316
70	27 28	007	60 08	040	60 03	096	48 00	157	36 33	209	33 38	259	35 38	316
71	27 35	006	60 45	038	61 01	096	48 22	157	36 05	209	32 41	259	34 58	315
72	27 41	006	61 20	036	61 58	096	48 44	158	35 37	209	31 45	258	34 17	314
73	27 46	005	61 53	035	62 55	096	49 05	159	35 08	210	30 48	258	33 35	313
74	27 50	004	62 25	033	63 53	095	49 26	160	34 39	210	29 52	258	32 53	313
	CAPELLA		POLLUX		◆PROCYON		Suhail		◆CANOPUS		ACHERNAR		◆Hamal	
75	27 54	003	30 33	043	45 18	065	31 11	131	49 45	160	34 10	210	32 10	312
76	27 57	002	31 12	042	46 10	064	31 54	132	50 04	161	33 41	211	31 27	311
77	27 59	002	31 50	041	47 01	063	32 37	132	50 23	162	33 12	211	30 44	311
78	28 00	001	32 28	040	47 53	063	33 20	132	50 40	163	32 42	211	30 00	310
79	28 00	000	33 05	040	48 44	062	34 03	132	50 57	164	32 12	211	29 15	309
80	28 00	359	33 41	039	49 35	061	34 46	132	51 12	165	31 42	212	28 31	309
81	27 59	358	34 17	038	50 25	060	35 29	132	51 27	165	31 11	212	27 45	308
82	27 57	358	34 52	037	51 15	059	36 11	133	51 41	166	30 41	212	27 00	308
83	27 54	357	35 27	036	52 04	059	36 54	133	51 54	167	30 10	212	26 14	307
84	27 51	356	36 01	036	52 53	058	37 36	133	52 07	168	29 39	212	25 28	307
85	27 47	355	36 34	035	53 41	057	38 18	133	52 18	169	29 08	213	24 41	306
86	27 41	355	37 07	034	54 29	056	39 00	133	52 28	170	28 37	213	23 55	305
87	27 36	354	37 38	033	55 17	055	39 42	134	52 38	171	28 06	213	23 07	305
88	27 29	353	38 09	032	56 03	053	40 23	134	52 46	172	27 34	213	22 20	304
89	27 22	352	38 40	031	56 49	052	41 05	134	52 54	173	27 03	213	21 32	304

LAT 16°S

LHA ♈	Hc	Zn	Hc	Zn	Hc	Zn	Hc	Zn	Hc	Zn	Hc	Zn	Hc	Zn
	POLLUX		◆REGULUS		Suhail		◆CANOPUS		ACHERNAR		◆ALDEBARAN		CAPELLA	
90	39 09	030	22 34	069	41 46	135	53 00	174	26 31	213	51 25	326	27 13	351
91	39 38	029	23 28	069	42 27	135	53 06	175	25 59	214	50 52	325	27 05	351
92	40 05	028	24 22	068	43 08	135	53 10	176	25 27	214	50 19	324	26 55	350
93	40 32	027	25 15	068	43 48	136	53 14	177	24 55	214	49 44	323	26 44	349
94	40 58	026	26 09	068	44 28	136	53 16	178	24 23	214	49 09	322	26 33	348
95	41 23	025	27 02	067	45 08	136	53 18	179	23 51	214	48 33	320	26 21	348
96	41 47	024	27 55	067	45 48	137	53 18	180	23 19	214	47 55	319	26 09	347
97	42 11	023	28 48	066	46 27	137	53 18	181	22 47	214	47 18	318	25 55	346
98	42 33	022	29 40	066	47 06	138	53 16	182	22 14	214	46 39	317	25 41	346
99	42 54	021	30 33	065	47 44	138	53 14	183	21 42	214	45 59	316	25 27	345
100	43 14	020	31 25	065	48 22	139	53 10	184	21 10	214	45 19	315	25 11	344
101	43 33	019	32 17	064	49 00	139	53 06	185	20 37	214	44 38	314	24 55	343
102	43 51	018	33 09	064	49 37	140	53 00	186	20 05	214	43 56	313	24 38	343
103	44 08	016	34 00	063	50 14	141	52 53	187	19 32	214	43 14	312	24 21	342
104	44 23	015	34 52	062	50 50	141	52 46	188	19 00	214	42 31	312	24 02	341
	POLLUX		◆REGULUS		◆ACRUX		CANOPUS		◆RIGEL		ALDEBARAN		CAPELLA	
105	44 38	014	35 43	062	18 02	152	52 37	189	63 03	284	41 48	311	23 44	341
106	44 51	013	36 33	061	18 29	152	52 28	190	62 07	283	41 04	310	23 24	340
107	45 04	012	37 24	061	18 56	152	52 17	191	61 10	282	40 19	309	23 04	339
108	45 15	010	38 14	060	19 23	152	52 06	192	60 14	282	39 34	308	22 43	339
109	45 24	009	39 04	059	19 50	152	51 53	193	59 17	281	38 49	308	22 22	338
110	45 33	008	39 53	059	20 17	152	51 40	194	58 21	280	38 03	307	22 00	337
111	45 40	007	40 42	058	20 45	152	51 26	195	57 24	280	37 17	306	21 38	337
112	45 46	005	41 31	057	21 12	152	51 11	195	56 27	279	36 30	305	21 15	336
113	45 51	004	42 19	056	21 39	152	50 55	196	55 30	279	35 43	305	20 51	335
114	45 54	003	43 07	056	22 05	152	50 39	197	54 33	278	34 55	304	20 27	335
115	45 57	002	43 54	055	22 32	152	50 21	198	53 36	278	34 07	303	20 02	334
116	45 58	000	44 41	054	22 59	152	50 03	199	52 39	278	33 19	303	19 37	334
117	45 57	359	45 27	053	23 26	152	49 44	200	51 42	277	32 30	302	19 11	333
118	45 56	358	46 13	052	23 52	153	49 24	200	50 44	277	31 41	302	18 45	333
119	45 53	357	46 59	051	24 19	153	49 04	201	49 47	276	30 52	301	18 18	332
	◆REGULUS		SPICA		◆ACRUX		CANOPUS		◆RIGEL		BETELGEUSE		POLLUX	
120	47 43	050	11 22	098	24 45	153	48 42	202	48 50	276	51 11	305	45 49	355
121	48 28	049	12 19	098	25 12	153	48 20	203	47 52	275	50 24	304	45 43	354
122	49 11	048	13 16	098	25 38	153	47 58	203	46 55	275	49 35	303	45 37	353
123	49 54	047	14 13	098	26 04	153	47 35	204	45 57	275	48 47	302	45 29	351
124	50 36	046	15 10	097	26 30	153	47 11	205	45 00	274	47 58	301	45 20	350
125	51 17	045	16 07	097	26 56	153	46 46	205	44 02	274	47 08	300	45 09	349
126	51 58	044	17 05	097	27 22	154	46 21	206	43 05	274	46 18	300	44 58	348
127	52 38	043	18 02	097	27 47	154	45 55	207	42 07	273	45 28	299	44 45	347
128	53 17	042	18 59	097	28 13	154	45 29	207	41 10	273	44 37	298	44 31	345
129	53 55	041	19 56	096	28 38	154	45 03	208	40 12	273	43 46	297	44 16	344
130	54 32	039	20 54	096	29 03	154	44 35	208	39 14	272	42 55	297	44 00	343
131	55 08	038	21 51	096	29 28	155	44 08	209	38 17	272	42 03	296	43 42	342
132	55 42	036	22 49	096	29 52	155	43 39	210	37 19	272	41 11	295	43 24	341
133	56 16	035	23 46	095	30 17	155	43 11	210	36 22	272	40 19	295	43 04	340
134	56 49	034	24 43	095	30 41	155	42 42	211	35 24	271	39 26	294	42 44	339
	REGULUS		◆SPICA		ACRUX		◆CANOPUS		SIRIUS		BETELGEUSE		◆POLLUX	
135	57 20	032	25 41	095	31 05	156	42 12	211	57 38	264	38 34	294	42 22	337
136	57 50	030	26 38	095	31 29	156	41 42	211	56 40	264	37 41	293	41 59	336
137	58 18	029	27 36	094	31 52	156	41 12	212	55 43	264	36 47	292	41 36	335
138	58 45	027	28 33	094	32 16	156	40 41	212	54 46	264	35 54	292	41 11	334
139	59 11	025	29 31	094	32 39	157	40 10	213	53 48	263	35 00	291	40 46	333
140	59 35	024	30 28	094	33 01	157	39 39	213	52 51	263	34 07	291	40 19	332
141	59 57	022	31 26	093	33 24	157	39 07	214	51 54	263	33 13	290	39 52	331
142	60 18	020	32 24	093	33 46	158	38 35	214	50 57	263	32 19	290	39 24	330
143	60 37	018	33 21	093	34 08	158	38 03	214	49 59	263	31 24	289	38 55	329
144	60 54	016	34 19	093	34 30	158	37 30	215	49 02	263	30 30	289	38 25	328
145	61 09	014	35 16	092	34 51	159	36 57	215	48 05	263	29 35	288	37 54	327
146	61 23	012	36 14	092	35 12	159	36 24	215	47 08	262	28 40	288	37 23	327
147	61 34	010	37 12	092	35 32	159	35 51	216	46 11	262	27 45	288	36 51	326
148	61 44	008	38 09	092	35 53	160	35 17	216	45 13	262	26 50	287	36 18	325
149	61 51	006	39 07	091	36 13	160	34 43	216	44 16	262	25 55	287	35 44	324
	REGULUS		◆ARCTURUS		SPICA		◆ACRUX		CANOPUS		SIRIUS		◆PROCYON	
150	61 56	004	18 01	063	40 05	091	36 32	160	34 09	216	43 19	262	49 10	298
151	61 59	002	18 52	063	41 02	091	36 51	161	33 35	217	42 22	262	48 19	298
152	62 00	000	19 43	062	42 00	091	37 10	161	33 00	217	41 25	262	47 28	297
153	61 59	358	20 34	062	42 58	090	37 29	162	32 26	217	40 28	261	46 37	296
154	61 56	356	21 25	061	43 55	090	37 47	162	31 51	217	39 31	261	45 45	295
155	61 51	354	22 15	061	44 53	090	38 04	162	31 16	217	38 34	261	44 52	295
156	61 44	352	23 06	060	45 51	090	38 21	163	30 41	218	37 37	261	44 00	294
157	61 35	350	23 56	060	46 48	089	38 38	163	30 05	218	36 40	261	43 07	293
158	61 23	348	24 45	059	47 46	089	38 54	164	29 30	218	35 43	261	42 14	293
159	61 10	346	25 35	059	48 44	089	39 10	164	28 54	218	34 46	261	41 21	292
160	60 55	344	26 24	058	49 41	088	39 26	165	28 19	218	33 49	260	40 27	292
161	60 38	342	27 13	058	50 39	088	39 40	165	27 43	218	32 53	260	39 34	291
162	60 19	340	28 02	057	51 37	088	39 55	166	27 07	218	31 56	260	38 40	291
163	59 59	338	28 50	057	52 34	087	40 09	166	26 31	219	30 59	260	37 46	290
164	59 36	336	29 38	056	53 32	087	40 22	167	25 55	219	30 02	260	36 51	289
	Denebola		◆ARCTURUS		SPICA		◆ACRUX		SIRIUS		◆PROCYON		REGULUS	
165	57 07	022	30 26	056	54 29	087	40 35	167	29 05	259	35 57	289	59 12	335
166	57 28	020	31 13	055	55 27	086	40 47	168	28 09	259	35 02	288	58 47	333
167	57 47	019	32 00	054	56 25	086	40 59	168	27 12	259	34 08	288	58 20	331
168	58 05	017	32 47	054	57 22	086	41 11	169	26 15	259	33 13	288	57 51	330
169	58 21	015	33 33	053	58 20	085	41 21	170	25 19	259	32 18	287	57 21	328
170	58 35	013	34 19	052	59 17	085	41 32	170	24 22	259	31 22	287	56 50	327
171	58 48	012	35 05	052	60 14	084	41 41	171	23 26	258	30 27	286	56 18	325
172	58 58	010	35 50	051	61 12	084	41 50	171	22 29	258	29 32	286	55 44	324
173	59 07	008	36 34	050	62 09	084	41 59	172	21 33	258	28 36	285	55 09	322
174	59 14	006	37 19	050	63 06	083	42 07	172	20 37	258	27 40	285	54 34	321
175	59 19	004	38 02	049	64 04	083	42 14	173	19 40	258	26 45	285	53 57	320
176	59 23	002	38 45	048	65 01	082	42 21	174	18 44	257	25 49	284	53 19	318
177	59 24	000	39 28	047	65 58	081	42 27	174	17 48	257	24 53	284	52 40	317
178	59 23	358	40 10	046	66 55	081	42 33	175	16 51	257	23 57	283	52 00	316
179	59 21	357	40 51	046	67 52	080	42 38	175	15 55	257	23 00	283	51 20	315

LHA ♈	Hc	Zn	Hc	Zn	Hc	Zn	Hc	Zn	Hc	Zn	Hc	Zn	Hc	Zn
	ARCTURUS		◆ANTARES		ACRUX		◆Suhail		Alphard		REGULUS		◆Denebola	
180	41 32	045	27 05	112	42 42	176	44 24	224	52 03	277	50 38	314	59 17	355
181	42 13	044	27 58	112	42 46	177	43 44	224	51 06	276	49 56	313	59 10	353
182	42 52	043	28 52	112	42 49	177	43 04	225	50 09	276	49 13	312	59 02	351
183	43 31	042	29 45	112	42 52	178	42 23	225	49 11	275	48 30	311	58 52	349
184	44 09	041	30 39	112	42 54	178	41 42	225	48 14	275	47 46	310	58 40	347
185	44 47	040	31 33	112	42 55	179	41 01	226	47 16	275	47 01	309	58 27	345
186	45 23	039	32 26	111	42 56	180	40 19	226	46 19	274	46 16	308	58 11	344
187	45 59	038	33 20	111	42 56	180	39 38	226	45 21	274	45 30	307	57 54	342
188	46 34	037	34 14	111	42 55	181	38 56	227	44 24	274	44 43	306	57 36	340
189	47 08	036	35 07	111	42 54	181	38 14	227	43 26	273	43 57	305	57 15	339
190	47 41	035	36 01	111	42 52	182	37 32	227	42 29	273	43 09	304	56 53	337
191	48 14	033	36 55	111	42 50	183	36 50	227	41 31	273	42 21	304	56 30	335
192	48 45	032	37 49	111	42 47	183	36 07	227	40 33	272	41 33	303	56 05	334
193	49 15	031	38 42	111	42 43	184	35 25	228	39 36	272	40 45	302	55 39	332
194	49 44	030	39 36	111	42 39	185	34 42	228	38 38	272	39 56	301	55 11	331
	ARCTURUS		◆Rasalhague		ANTARES		◆ACRUX		Suhail		◆REGULUS		Denebola	
195	50 12	028	16 20	071	40 30	111	42 34	185	33 59	228	39 06	301	54 42	329
196	50 39	027	17 15	071	41 23	111	42 28	186	33 16	228	38 16	300	54 12	328
197	51 05	026	18 09	071	42 17	111	42 22	186	32 33	228	37 26	299	53 41	326
198	51 29	024	19 03	070	43 11	111	42 15	187	31 50	228	36 36	299	53 09	325
199	51 53	023	19 58	070	44 04	112	42 08	188	31 07	229	35 45	298	52 35	324
200	52 15	022	20 52	069	44 58	112	42 00	188	30 24	229	34 54	298	52 00	322
201	52 35	020	21 46	069	45 52	112	41 52	189	29 40	229	34 03	297	51 25	321
202	52 55	019	22 39	069	46 45	112	41 43	189	28 57	229	33 12	296	50 48	320
203	53 12	017	23 33	068	47 39	112	41 33	190	28 13	229	32 20	296	50 11	319
204	53 29	016	24 26	068	48 32	112	41 23	190	27 30	229	31 28	295	49 32	318
205	53 44	014	25 20	067	49 26	112	41 12	191	26 46	229	30 36	295	48 53	317
206	53 57	013	26 13	067	50 19	112	41 01	192	26 03	229	29 43	294	48 13	316
207	54 09	011	27 06	066	51 13	112	40 49	192	25 19	229	28 51	294	47 32	315
208	54 19	010	27 59	066	52 06	113	40 37	193	24 36	229	27 58	293	46 51	314
209	54 28	008	28 51	065	52 59	113	40 24	193	23 52	229	27 05	293	46 09	313
	◆ARCTURUS		Rasalhague		◆ANTARES		RIGIL KENT.		ACRUX		◆Suhail		REGULUS	
210	54 35	006	29 43	065	53 52	113	44 38	173	40 11	194	23 09	229	26 11	292
211	54 41	005	30 36	064	54 45	113	44 44	174	39 57	194	22 25	229	25 18	292
212	54 45	003	31 27	064	55 38	114	44 50	175	39 43	195	21 41	229	24 25	292
213	54 47	001	32 19	063	56 31	114	44 55	175	39 28	195	20 58	229	23 31	291
214	54 47	000	33 10	063	57 23	114	44 59	176	39 13	196	20 14	229	22 37	291
215	54 46	358	34 02	062	58 16	115	45 03	177	38 57	196	19 31	229	21 43	290
216	54 44	356	34 53	062	59 08	115	45 06	177	38 41	197	18 47	229	20 49	290
217	54 39	355	35 43	061	60 00	115	45 08	178	38 24	197	18 04	229	19 54	289
218	54 33	353	36 33	060	60 52	116	45 10	179	38 07	197	17 21	229	19 00	289
219	54 26	352	37 23	060	61 44	116	45 11	179	37 49	198	16 37	229	18 05	289
220	54 17	350	38 13	059	62 36	117	45 11	180	37 31	198	15 54	229	17 11	288
221	54 06	348	39 02	058	63 27	118	45 11	181	37 13	199	15 11	228	16 16	288
222	53 54	347	39 51	058	64 18	118	45 09	182	36 54	199	14 28	228	15 21	288
223	53 40	345	40 40	057	65 09	119	45 08	182	36 35	200	13 45	228	14 26	287
224	53 24	344	41 28	056	65 59	120	45 05	183	36 16	200	13 02	228	13 31	287
	Alphecca		◆VEGA		ALTAIR		◆Peacock		RIGIL KENT.		◆SPICA		ARCTURUS	
225	46 27	011	15 25	041	13 57	076	18 04	145	45 02	184	66 24	279	53 08	342
226	46 38	010	16 02	041	14 53	076	18 37	145	44 58	184	65 27	278	52 49	341
227	46 47	009	16 40	040	15 49	076	19 10	145	44 53	185	64 30	278	52 30	339
228	46 55	007	17 17	040	16 45	075	19 43	145	44 48	186	63 33	277	52 09	338
229	47 02	006	17 53	039	17 40	075	20 16	145	44 42	186	62 36	277	51 46	337
230	47 07	005	18 29	038	18 36	075	20 48	145	44 35	187	61 38	276	51 23	335
231	47 12	003	19 05	038	19 32	074	21 21	145	44 28	188	60 41	276	50 58	334
232	47 14	002	19 40	037	20 27	074	21 54	145	44 20	188	59 44	275	50 32	332
233	47 16	001	20 15	037	21 22	073	22 27	145	44 11	189	58 46	275	50 04	331
234	47 16	359	20 49	036	22 18	073	23 00	145	44 02	190	57 49	274	49 36	330
235	47 15	358	21 23	036	23 13	073	23 32	145	43 52	190	56 51	274	49 07	329
236	47 12	357	21 57	035	24 08	072	24 05	146	43 42	191	55 54	274	48 36	327
237	47 09	356	22 30	035	25 03	072	24 38	146	43 30	191	54 56	273	48 04	326
238	47 03	354	23 02	034	25 57	071	25 10	146	43 19	192	53 59	273	47 32	325
239	46 57	353	23 34	033	26 52	071	25 42	146	43 06	193	53 01	273	46 59	324
	◆VEGA		ALTAIR		◆Peacock		RIGIL KENT.		◆SPICA		ARCTURUS		Alphecca	
240	24 05	033	27 47	071	26 15	146	42 53	193	52 03	272	46 24	323	46 49	352
241	24 36	032	28 41	070	26 47	146	42 40	194	51 06	272	45 49	322	46 40	350
242	25 06	031	29 35	070	27 19	146	42 26	194	50 08	272	45 13	321	46 30	349
243	25 36	031	30 29	069	27 51	146	42 11	195	49 10	271	44 36	320	46 18	348
244	26 05	030	31 23	069	28 23	147	41 56	196	48 13	271	43 58	319	46 06	347
245	26 34	029	32 17	068	28 55	147	41 40	196	47 15	271	43 20	318	45 52	345
246	27 02	029	33 10	068	29 26	147	41 24	197	46 17	271	42 41	317	45 37	344
247	27 29	028	34 03	067	29 58	147	41 07	197	45 20	270	42 01	316	45 20	343
248	27 56	027	34 56	067	30 29	147	40 50	198	44 22	270	41 21	315	45 03	342
249	28 22	026	35 49	066	31 00	147	40 32	198	43 24	270	40 40	314	44 44	341
250	28 47	026	36 42	066	31 31	148	40 14	199	42 27	269	39 58	313	44 24	339
251	29 12	025	37 34	065	32 02	148	39 55	199	41 29	269	39 16	313	44 03	338
252	29 36	024	38 26	064	32 33	148	39 36	200	40 31	269	38 33	312	43 41	337
253	29 59	023	39 18	064	33 03	148	39 16	200	39 34	269	37 50	311	43 18	336
254	30 22	023	40 10	063	33 33	149	38 56	201	38 36	268	37 06	310	42 54	335
	◆VEGA		ALTAIR		◆Peacock		RIGIL KENT.		◆SPICA		ARCTURUS		Alphecca	
255	30 44	022	41 01	063	34 03	149	38 35	201	37 38	268	36 22	310	42 29	334
256	31 05	021	41 52	062	34 33	149	38 14	202	36 41	268	35 37	309	42 03	333
257	31 25	020	42 43	061	35 02	149	37 53	202	35 43	268	34 52	308	41 37	332
258	31 44	019	43 33	060	35 31	150	37 31	202	34 46	267	34 06	307	41 09	331
259	32 03	019	44 23	060	36 00	150	37 09	203	33 48	267	33 20	307	40 40	330
260	32 21	018	45 13	059	36 29	150	36 47	203	32 50	267	32 34	306	40 11	329
261	32 38	017	46 02	058	36 57	151	36 24	204	31 53	267	31 47	305	39 40	328
262	32 54	016	46 51	057	37 25	151	36 01	204	30 55	266	31 00	305	39 09	327
263	33 10	015	47 39	056	37 53	152	35 37	204	29 58	266	30 13	304	38 37	326
264	33 24	014	48 27	056	38 20	152	35 13	205	29 00	266	29 25	304	38 04	325
265	33 38	013	49 14	055	38 47	152	34 49	205	28 03	266	28 37	303	37 31	324
266	33 51	012	50 01	054	39 14	153	34 25	205	27 05	265	27 48	303	36 57	323
267	34 03	011	50 47	053	39 40	153	34 00	206	26 08	265	26 59	302	36 22	322
268	34 14	011	51 33	052	40 06	154	33 35	206	25 10	265	26 10	301	35 46	322
269	34 24	010	52 18	051	40 31	154	33 10	206	24 13	265	25 21	301	35 10	321

LAT 16°S

LHA ♈	Hc	Zn	Hc	Zn	Hc	Zn	Hc	Zn	Hc	Zn	Hc	Zn	Hc	Zn
	◆VEGA		ALTAIR		◆FOMALHAUT		Peacock		RIGIL KENT.		◆ANTARES		Alphecca	
270	34 33	009	53 02	050	21 13	116	40 56	155	32 44	206	66 27	240	34 33	320
271	34 41	008	53 46	049	22 04	116	41 21	155	32 18	207	65 37	241	33 56	319
272	34 49	007	54 29	047	22 56	116	41 45	156	31 52	207	64 46	241	33 18	318
273	34 55	006	55 11	046	23 48	116	42 08	156	31 26	207	63 55	242	32 39	318
274	35 00	005	55 52	045	24 40	116	42 32	157	31 00	207	63 04	243	32 00	317
275	35 05	004	56 32	044	25 32	116	42 54	157	30 33	208	62 13	243	31 21	316
276	35 08	003	57 11	042	26 24	116	43 16	158	30 06	208	61 21	244	30 41	316
277	35 11	002	57 50	041	27 16	116	43 38	158	29 39	208	60 29	244	30 00	315
278	35 13	001	58 27	039	28 08	115	43 59	159	29 12	208	59 37	245	29 19	314
279	35 13	000	59 03	038	29 00	115	44 20	159	28 44	209	58 45	245	28 37	314
280	35 13	359	59 38	036	29 52	115	44 40	160	28 16	209	57 52	246	27 55	313
281	35 12	358	60 11	035	30 44	115	44 59	161	27 49	209	57 00	246	27 13	312
282	35 09	357	60 43	033	31 37	115	45 18	161	27 21	209	56 07	246	26 30	312
283	35 06	356	61 14	031	32 29	115	45 36	162	26 53	209	55 14	247	25 47	311
284	35 02	355	61 43	029	33 21	115	45 54	163	26 24	209	54 21	247	25 03	311
	DENEB		◆FOMALHAUT		Peacock		RIGIL KENT.		◆ANTARES		Rasalhague		◆VEGA	
285	24 34	019	34 13	115	46 11	163	25 56	209	53 28	247	54 30	322	34 57	354
286	24 53	019	35 05	115	46 27	164	25 28	210	52 35	247	53 54	321	34 51	354
287	25 11	018	35 57	115	46 43	165	24 59	210	51 42	247	53 17	320	34 44	353
288	25 29	017	36 49	115	46 57	165	24 31	210	50 49	248	52 39	318	34 36	352
289	25 45	017	37 41	115	47 12	166	24 02	210	49 55	248	52 01	317	34 27	351
290	26 01	016	38 34	115	47 25	167	23 33	210	49 02	248	51 21	316	34 18	350
291	26 17	015	39 26	115	47 38	168	23 04	210	48 08	248	50 41	315	34 07	349
292	26 31	014	40 18	116	47 50	168	22 35	210	47 15	248	50 00	314	33 55	348
293	26 45	014	41 10	116	48 01	169	22 06	210	46 21	248	49 18	313	33 43	347
294	26 58	013	42 02	116	48 12	170	21 37	210	45 28	248	48 35	312	33 30	346
295	27 11	012	42 54	116	48 21	171	21 08	210	44 34	248	47 52	311	33 15	345
296	27 22	011	43 45	116	48 30	171	20 39	210	43 40	249	47 08	310	33 00	344
297	27 33	011	44 37	116	48 39	172	20 09	210	42 47	249	46 23	309	32 44	344
298	27 44	010	45 29	116	48 46	173	19 40	210	41 53	249	45 38	308	32 28	343
299	27 53	009	46 21	116	48 52	174	19 11	210	40 59	249	44 52	307	32 10	342
	◆DENEB		Alpheratz		◆FOMALHAUT		Peacock		◆ANTARES		Rasalhague		VEGA	
300	28 02	008	15 05	053	47 12	117	48 58	175	40 06	249	44 06	306	31 52	341
301	28 09	007	15 51	053	48 04	117	49 03	176	39 12	249	43 20	306	31 32	340
302	28 16	007	16 36	052	48 55	117	49 07	176	38 18	249	42 32	305	31 12	339
303	28 23	006	17 22	052	49 47	117	49 10	177	37 24	249	41 45	304	30 52	338
304	28 28	005	18 07	051	50 38	118	49 13	178	36 31	249	40 57	303	30 30	338
305	28 33	004	18 52	051	51 29	118	49 14	179	35 37	249	40 08	303	30 08	337
306	28 37	003	19 36	050	52 20	118	49 15	180	34 43	249	39 20	302	29 45	336
307	28 40	003	20 21	050	53 11	118	49 15	181	33 49	249	38 30	301	29 21	335
308	28 42	002	21 05	049	54 01	119	49 14	181	32 56	249	37 41	300	28 57	335
309	28 44	001	21 48	049	54 52	119	49 12	182	32 02	248	36 51	300	28 32	334
310	28 44	000	22 31	048	55 42	120	49 09	183	31 08	248	36 01	299	28 06	333
311	28 44	359	23 14	048	56 32	120	49 06	184	30 15	248	35 10	299	27 39	332
312	28 43	359	23 57	047	57 22	120	49 01	185	29 21	248	34 20	298	27 12	332
313	28 41	358	24 39	047	58 12	121	48 56	186	28 28	248	33 28	297	26 45	331
314	28 39	357	25 20	046	59 01	121	48 50	186	27 34	248	32 37	297	26 16	330
	Alpheratz		◆Diphda		ACHERNAR		Peacock		◆ANTARES		VEGA		◆DENEB	
315	26 02	045	36 46	101	24 31	146	48 43	187	26 41	248	25 47	330	28 35	356
316	26 42	045	37 43	101	25 03	146	48 36	188	25 47	248	25 18	329	28 31	355
317	27 23	044	38 40	101	25 35	146	48 27	189	24 54	248	24 48	328	28 26	355
318	28 03	044	39 36	100	26 07	147	48 18	190	24 00	248	24 17	328	28 20	354
319	28 42	043	40 33	100	26 38	147	48 08	190	23 07	248	23 46	327	28 14	353
320	29 21	042	41 30	100	27 10	147	47 57	191	22 14	247	23 14	326	28 07	352
321	30 00	041	42 27	100	27 42	147	47 46	192	21 20	247	22 42	326	27 58	352
322	30 38	041	43 23	100	28 13	147	47 33	193	20 27	247	22 09	325	27 50	351
323	31 15	040	44 20	100	28 44	147	47 20	193	19 34	247	21 36	325	27 40	350
324	31 52	039	45 17	100	29 16	147	47 06	194	18 41	247	21 02	324	27 29	349
325	32 28	039	46 14	100	29 47	148	46 52	195	17 48	247	20 28	323	27 18	348
326	33 04	038	47 11	099	30 17	148	46 37	196	16 55	247	19 54	323	27 06	348
327	33 39	037	48 08	099	30 48	148	46 21	196	16 02	246	19 19	322	26 54	347
328	34 13	036	49 05	099	31 19	148	46 04	197	15 09	246	18 43	322	26 40	346
329	34 47	035	50 02	099	31 49	148	45 47	198	14 16	246	18 07	321	26 26	345
	◆Alpheratz		Diphda		◆ACHERNAR		Peacock		◆Nunki		ALTAIR		DENEB	
330	35 20	035	50 59	099	32 19	149	45 30	198	45 53	249	49 27	306	26 11	345
331	35 53	034	51 56	099	32 49	149	45 11	199	44 59	249	48 40	305	25 55	344
332	36 24	033	52 52	099	33 19	149	44 52	200	44 06	249	47 52	304	25 39	343
333	36 55	032	53 49	099	33 48	149	44 32	200	43 12	249	47 04	303	25 22	343
334	37 26	031	54 46	099	34 17	150	44 12	201	42 18	249	46 15	302	25 05	342
335	37 55	030	55 44	099	34 46	150	43 51	201	41 24	249	45 26	301	24 46	341
336	38 24	029	56 41	099	35 15	150	43 30	202	40 31	249	44 37	301	24 27	340
337	38 52	028	57 38	099	35 43	151	43 08	203	39 37	249	43 47	300	24 08	340
338	39 19	027	58 35	098	36 11	151	42 46	203	38 43	249	42 57	299	23 47	339
339	39 45	026	59 32	098	36 39	151	42 23	204	37 49	249	42 06	298	23 26	338
340	40 10	025	60 29	098	37 07	152	42 00	204	36 55	249	41 15	298	23 05	338
341	40 34	024	61 26	098	37 34	152	41 36	205	36 02	249	40 24	297	22 43	337
342	40 57	023	62 23	098	38 01	152	41 12	205	35 08	249	39 32	296	22 20	336
343	41 20	022	63 20	098	38 27	153	40 47	206	34 14	249	38 40	296	21 57	336
344	41 41	021	64 17	098	38 53	153	40 22	206	33 20	249	37 48	295	21 33	335
	Alpheratz		◆Hamal		◆ACHERNAR		Peacock		Nunki		◆ALTAIR		DENEB	
345	42 02	020	29 40	050	39 19	154	39 56	207	32 27	249	36 56	295	21 08	335
346	42 21	019	30 24	050	39 45	154	39 30	207	31 33	249	36 03	294	20 43	334
347	42 39	018	31 08	049	40 10	155	39 04	207	30 39	249	35 11	293	20 17	333
348	42 57	017	31 51	048	40 34	155	38 37	208	29 46	248	34 18	293	19 51	333
349	43 13	016	32 34	048	40 58	155	38 10	208	28 52	248	33 24	292	19 25	332
350	43 28	015	33 16	047	41 22	156	37 42	209	27 58	248	32 31	292	18 58	332
351	43 42	013	33 58	046	41 45	156	37 15	209	27 05	248	31 37	291	18 30	331
352	43 54	012	34 40	045	42 08	157	36 47	209	26 11	248	30 44	291	18 02	331
353	44 06	011	35 20	045	42 30	158	36 18	210	25 18	248	29 50	290	17 33	330
354	44 16	010	36 01	044	42 52	158	35 49	210	24 24	248	28 56	290	17 04	329
355	44 26	009	36 40	043	43 13	159	35 21	210	23 31	248	28 01	290	16 34	329
356	44 34	007	37 20	042	43 34	159	34 51	211	22 37	248	27 07	289	16 04	328
357	44 41	006	37 58	042	43 54	160	34 22	211	21 44	248	26 12	289	15 34	328
358	44 46	005	38 36	041	44 14	160	33 52	211	20 51	247	25 17	288	15 03	327
359	44 51	004	39 13	040	44 33	161	33 22	211	19 58	247	24 23	288	14 32	327

LAT 17°S

LHA	Hc	Zn	Hc	Zn	Hc	Zn	Hc	Zn	Hc	Zn	Hc	Zn	Hc	Zn
♈	♦Alpheratz		Hamal		ALDEBARAN		♦RIGEL		ACHERNAR		♦Peacock		Enif	
0	43 54	002	39 03	038	14 18	067	13 15	095	45 48	161	33 43	212	46 58	306
1	43 56	001	39 38	037	15 11	067	14 12	094	46 06	162	33 12	212	46 11	305
2	43 56	000	40 13	037	16 03	067	15 10	094	46 24	163	32 41	213	45 24	304
3	43 56	359	40 47	036	16 56	066	16 07	094	46 41	163	32 10	213	44 37	304
4	43 54	358	41 20	035	17 48	066	17 04	094	46 57	164	31 39	213	43 49	303
5	43 51	356	41 52	034	18 41	065	18 01	093	47 12	165	31 08	213	43 00	302
6	43 47	355	42 23	033	19 33	065	18 59	093	47 27	165	30 36	213	42 11	301
7	43 41	354	42 54	032	20 25	064	19 56	093	47 41	166	30 05	214	41 22	301
8	43 35	353	43 23	031	21 16	064	20 53	092	47 55	167	29 33	214	40 33	300
9	43 27	352	43 52	029	22 08	064	21 51	092	48 07	168	29 01	214	39 43	299
10	43 18	350	44 20	028	22 59	063	22 48	092	48 19	168	28 29	214	38 52	298
11	43 08	349	44 47	027	23 50	063	23 45	092	48 31	169	27 57	214	38 02	298
12	42 56	348	45 12	026	24 41	062	24 43	091	48 41	170	27 25	214	37 11	297
13	42 44	347	45 37	025	25 32	062	25 40	091	48 51	171	26 52	214	36 20	297
14	42 31	346	46 01	024	26 22	061	26 37	091	49 00	171	26 20	214	35 28	296
	♦Hamal		ALDEBARAN		RIGEL		♦CANOPUS		ACHERNAR		♦FOMALHAUT		Alpheratz	
15	46 23	023	27 12	061	27 35	090	18 53	141	49 08	172	59 16	240	42 16	345
16	46 44	021	28 02	060	28 32	090	19 29	141	49 15	173	58 26	241	42 00	344
17	47 05	020	28 51	059	29 29	090	20 06	141	49 22	174	57 36	241	41 43	342
18	47 24	019	29 41	059	30 27	090	20 42	141	49 27	175	56 46	241	41 25	341
19	47 42	017	30 30	058	31 24	089	21 18	141	49 32	176	55 55	242	41 06	340
20	47 58	016	31 18	058	32 22	089	21 55	141	49 36	176	55 05	242	40 47	339
21	48 14	015	32 07	057	33 19	089	22 31	141	49 39	177	54 14	243	40 26	338
22	48 28	014	32 55	057	34 16	088	23 07	141	49 42	178	53 23	243	40 04	337
23	48 40	012	33 43	056	35 14	088	23 44	141	49 43	179	52 32	243	39 41	336
24	48 52	011	34 30	055	36 11	088	24 20	141	49 44	180	51 40	243	39 17	335
25	49 02	009	35 17	055	37 08	087	24 56	141	49 44	181	50 49	244	38 52	334
26	49 11	008	36 03	054	38 06	087	25 32	141	49 43	181	49 58	244	38 27	333
27	49 18	007	36 49	053	39 03	087	26 09	141	49 41	182	49 06	244	38 00	332
28	49 24	005	37 35	052	40 00	086	26 45	141	49 39	183	48 14	244	37 33	331
29	49 28	004	38 20	052	40 58	086	27 21	141	49 35	184	47 23	244	37 05	330
	♦Hamal		ALDEBARAN		♦SIRIUS		CANOPUS		ACHERNAR		♦FOMALHAUT		Alpheratz	
30	49 32	002	39 05	051	22 15	102	27 57	141	49 31	185	46 31	245	36 36	329
31	49 33	001	39 49	050	23 12	101	28 33	141	49 26	185	45 39	245	36 06	328
32	49 34	000	40 33	049	24 08	101	29 09	141	49 20	186	44 47	245	35 36	328
33	49 33	358	41 16	048	25 04	101	29 44	142	49 13	187	43 55	245	35 05	327
34	49 30	357	41 59	048	26 01	101	30 20	142	49 06	188	43 03	245	34 33	326
35	49 26	355	42 41	047	26 57	100	30 55	142	48 58	189	42 11	245	34 00	325
36	49 21	354	43 22	046	27 53	100	31 31	142	48 48	190	41 19	245	33 27	324
37	49 14	353	44 03	045	28 50	100	32 06	142	48 39	190	40 27	245	32 53	323
38	49 06	351	44 43	044	29 46	100	32 41	142	48 28	191	39 35	245	32 19	323
39	48 56	350	45 23	043	30 43	100	33 16	143	48 17	192	38 43	245	31 43	322
40	48 46	348	46 01	042	31 40	099	33 51	143	48 04	193	37 51	245	31 08	321
41	48 33	347	46 39	041	32 36	099	34 26	143	47 52	193	36 58	245	30 31	320
42	48 20	346	47 16	040	33 33	099	35 00	143	47 38	194	36 06	245	29 54	320
43	48 05	344	47 53	039	34 29	099	35 34	144	47 24	195	35 14	245	29 17	319
44	47 49	343	48 28	038	35 26	099	36 08	144	47 09	195	34 22	245	28 39	318
	♦CAPELLA		BETELGEUSE		SIRIUS		♦CANOPUS		ACHERNAR		♦FOMALHAUT		Hamal	
45	19 53	024	40 22	064	36 23	099	36 42	144	46 53	196	33 30	245	47 32	342
46	20 16	024	41 13	063	37 20	098	37 16	144	46 37	197	32 38	245	47 13	341
47	20 39	023	42 04	063	38 16	098	37 49	145	46 20	198	31 46	245	46 54	339
48	21 01	023	42 55	062	39 13	098	38 22	145	46 02	198	30 54	245	46 33	338
49	21 23	022	43 46	061	40 10	098	38 55	145	45 44	199	30 02	245	46 11	337
50	21 44	021	44 36	061	41 07	098	39 27	146	45 25	199	29 09	245	45 48	336
51	22 05	021	45 26	060	42 04	097	40 00	146	45 06	200	28 17	245	45 23	334
52	22 25	020	46 15	059	43 01	097	40 31	146	44 46	201	27 25	245	44 58	333
53	22 44	019	47 04	058	43 58	097	41 03	147	44 25	201	26 34	245	44 32	332
54	23 03	019	47 53	057	44 55	097	41 34	147	44 04	202	25 42	245	44 05	331
55	23 21	018	48 41	057	45 52	097	42 05	148	43 43	202	24 50	245	43 36	330
56	23 38	017	49 28	056	46 49	097	42 36	148	43 20	203	23 58	245	43 07	329
57	23 55	017	50 16	055	47 46	096	43 06	149	42 58	203	23 06	244	42 37	328
58	24 11	016	51 02	054	48 43	096	43 36	149	42 35	204	22 14	244	42 06	327
59	24 26	015	51 48	053	49 40	096	44 05	150	42 11	204	21 23	244	41 34	326
	CAPELLA		BETELGEUSE		♦SIRIUS		CANOPUS		♦ACHERNAR		Diphda		♦Hamal	
60	24 41	014	52 33	052	50 37	096	44 34	150	41 47	205	43 14	261	41 02	325
61	24 55	014	53 18	051	51 34	096	45 02	151	41 23	205	42 18	261	40 28	324
62	25 08	013	54 02	049	52 31	095	45 30	151	40 58	206	41 21	261	39 54	323
63	25 21	012	54 45	048	53 28	095	45 58	152	40 33	206	40 25	261	39 19	322
64	25 33	012	55 27	047	54 25	095	46 25	152	40 07	207	39 28	260	38 43	321
65	25 44	011	56 09	046	55 22	095	46 51	153	39 41	207	38 31	260	38 07	320
66	25 54	010	56 50	044	56 20	095	47 17	154	39 14	208	37 35	260	37 30	319
67	26 04	009	57 29	043	57 17	095	47 42	154	38 48	208	36 38	260	36 52	319
68	26 13	009	58 08	042	58 14	094	48 07	155	38 20	208	35 42	260	36 14	318
69	26 21	008	58 46	040	59 11	094	48 31	155	37 53	209	34 45	260	35 35	317
70	26 28	007	59 22	039	60 08	094	48 55	156	37 25	209	33 49	259	34 55	316
71	26 35	006	59 57	037	61 06	094	49 17	157	36 57	209	32 53	259	34 15	315
72	26 41	005	60 31	035	62 03	094	49 40	158	36 29	210	31 56	259	33 35	315
73	26 46	005	61 04	034	63 00	094	50 01	158	36 00	210	31 00	259	32 54	314
74	26 50	004	61 35	032	63 57	093	50 22	159	35 31	210	30 04	259	32 12	313
	CAPELLA		POLLUX		♦PROCYON		Suhail		♦CANOPUS		ACHERNAR		♦Hamal	
75	26 54	003	29 49	042	44 52	064	31 50	131	50 42	160	35 02	211	31 30	312
76	26 57	002	30 27	041	45 43	063	32 34	131	51 01	161	34 33	211	30 47	312
77	26 59	002	31 05	041	46 34	062	33 17	131	51 20	162	34 03	211	30 04	311
78	27 00	001	31 42	040	47 25	062	34 00	131	51 37	162	33 33	211	29 21	310
79	27 00	000	32 18	039	48 15	061	34 43	131	51 54	163	33 03	212	28 37	310
80	27 00	359	32 54	038	49 05	060	35 26	132	52 10	164	32 33	212	27 53	309
81	26 59	359	33 30	038	49 55	059	36 09	132	52 25	165	32 02	212	27 08	309
82	26 57	358	34 04	037	50 44	058	36 52	132	52 40	166	31 32	212	26 23	308
83	26 54	357	34 38	036	51 32	057	37 34	132	52 53	167	31 01	213	25 38	307
84	26 51	356	35 12	035	52 20	056	38 17	132	53 05	168	30 30	213	24 52	307
85	26 47	355	35 45	034	53 08	055	38 59	133	53 17	169	29 59	213	24 06	306
86	26 42	355	36 17	033	53 55	054	39 41	133	53 27	170	29 27	213	23 20	306
87	26 36	354	36 48	033	54 41	053	40 23	133	53 37	171	28 56	213	22 33	305
88	26 29	353	37 18	032	55 27	052	41 05	133	53 46	172	28 24	213	21 46	305
89	26 22	352	37 48	031	56 12	051	41 46	134	53 53	173	27 53	214	20 59	304

LHA	Hc	Zn	Hc	Zn	Hc	Zn	Hc	Zn	Hc	Zn	Hc	Zn	Hc	Zn
♈	POLLUX		♦REGULUS		Suhail		♦CANOPUS		ACHERNAR		♦ALDEBARAN		CAPELLA	
90	38 17	030	22 13	069	42 28	134	54 00	174	27 21	214	50 35	327	26 14	352
91	38 45	029	23 06	068	43 09	134	54 06	175	26 49	214	50 03	326	26 05	351
92	39 12	028	23 59	068	43 50	135	54 10	176	26 17	214	49 30	325	25 56	350
93	39 39	027	24 53	068	44 31	135	54 14	177	25 45	214	48 56	323	25 46	349
94	40 04	026	25 46	067	45 11	135	54 16	178	25 13	214	48 22	322	25 35	349
95	40 29	025	26 38	067	45 51	136	54 18	179	24 41	214	47 46	321	25 23	348
96	40 53	024	27 31	066	46 31	136	54 18	180	24 09	214	47 10	320	25 10	347
97	41 15	023	28 23	066	47 11	137	54 18	181	23 36	214	46 32	319	24 57	346
98	41 37	022	29 15	065	47 50	137	54 16	182	23 04	214	45 54	318	24 43	346
99	41 58	021	30 07	065	48 29	138	54 14	183	22 32	214	45 16	317	24 29	345
100	42 17	020	30 59	064	49 07	138	54 10	184	21 59	214	44 36	316	24 13	344
101	42 36	018	31 51	064	49 45	139	54 05	185	21 27	214	43 56	315	23 57	344
102	42 54	017	32 42	063	50 23	139	54 00	186	20 54	214	43 15	314	23 41	343
103	43 10	016	33 33	062	51 00	140	53 53	187	20 22	214	42 33	313	23 24	342
104	43 25	015	34 24	062	51 37	140	53 45	188	19 49	214	41 51	312	23 06	341
	POLLUX		♦REGULUS		♦ACRUX		CANOPUS		♦RIGEL		ALDEBARAN		CAPELLA	
105	43 40	014	35 14	061	18 55	152	53 36	189	62 47	286	41 08	311	22 47	341
106	43 53	013	36 04	061	19 22	152	53 27	190	61 52	285	40 25	311	22 28	340
107	44 05	011	36 54	060	19 49	152	53 16	191	60 57	284	39 41	310	22 08	339
108	44 15	010	37 43	059	20 16	152	53 04	192	60 01	283	38 57	309	21 48	339
109	44 25	009	38 33	059	20 43	152	52 52	193	59 05	283	38 12	308	21 26	338
110	44 33	008	39 21	058	21 10	152	52 38	194	58 09	282	37 27	307	21 05	338
111	44 41	007	40 10	057	21 37	152	52 24	195	57 13	281	36 41	307	20 43	337
112	44 46	005	40 58	056	22 05	152	52 09	196	56 16	281	35 55	306	20 20	336
113	44 51	004	41 45	056	22 32	152	51 53	197	55 20	280	35 08	305	19 56	336
114	44 55	003	42 32	055	22 59	152	51 36	198	54 24	280	34 21	305	19 32	335
115	44 57	002	43 19	054	23 25	152	51 18	198	53 27	279	33 34	304	19 08	334
116	44 58	000	44 05	053	23 52	152	51 00	199	52 30	279	32 46	303	18 43	334
117	44 57	359	44 51	052	24 19	152	50 40	200	51 34	278	31 58	303	18 17	333
118	44 56	358	45 36	051	24 46	152	50 20	201	50 37	278	31 10	302	17 51	333
119	44 53	357	46 21	051	25 12	152	49 59	202	49 40	277	30 21	302	17 25	332
	♦REGULUS		SPICA		♦ACRUX		CANOPUS		♦RIGEL		BETELGEUSE		POLLUX	
120	47 05	050	11 30	098	25 39	153	49 38	202	48 43	277	50 37	306	44 49	355
121	47 48	049	12 27	098	26 05	153	49 16	203	47 46	277	49 50	305	44 44	354
122	48 31	048	13 24	098	26 31	153	48 53	204	46 49	276	49 02	304	44 37	353
123	49 13	047	14 21	097	26 58	153	48 29	205	45 52	276	48 14	303	44 29	352
124	49 54	046	15 18	097	27 24	153	48 05	205	44 55	275	47 26	302	44 20	350
125	50 35	044	16 15	097	27 50	153	47 40	206	43 58	275	46 37	301	44 10	349
126	51 15	043	17 12	097	28 15	153	47 15	207	43 00	275	45 48	301	43 59	348
127	51 54	042	18 09	096	28 41	154	46 49	207	42 03	274	44 58	300	43 46	347
128	52 32	041	19 06	096	29 06	154	46 23	208	41 06	274	44 08	299	43 33	346
129	53 09	040	20 03	096	29 32	154	45 56	208	40 09	274	43 18	298	43 18	344
130	53 45	038	21 00	096	29 57	154	45 28	209	39 11	273	42 27	298	43 02	343
131	54 20	037	21 57	095	30 22	154	45 00	209	38 14	273	41 36	297	42 45	342
132	54 54	036	22 54	095	30 47	155	44 32	210	37 17	273	40 45	296	42 27	341
133	55 27	034	23 51	095	31 11	155	44 03	211	36 20	272	39 53	296	42 08	340
134	55 58	033	24 48	095	31 35	155	43 33	211	35 22	272	39 01	295	41 48	339
	REGULUS		♦SPICA		ACRUX		♦CANOPUS		SIRIUS		BETELGEUSE		♦POLLUX	
135	56 29	031	25 46	094	32 00	155	43 03	211	57 43	265	38 09	294	41 26	338
136	56 58	030	26 43	094	32 23	156	42 33	212	56 46	265	37 17	294	41 04	337
137	57 26	028	27 40	094	32 47	156	42 03	212	55 49	265	36 24	293	40 41	336
138	57 52	026	28 37	094	33 10	156	41 32	213	54 52	265	35 31	293	40 17	335
139	58 17	025	29 35	093	33 34	156	41 01	213	53 55	265	34 38	292	39 52	334
140	58 40	023	30 32	093	33 57	157	40 29	214	52 57	265	33 45	291	39 26	333
141	59 02	021	31 29	093	34 19	157	39 57	214	52 00	264	32 52	291	38 59	332
142	59 22	020	32 27	093	34 41	157	39 25	214	51 03	264	31 58	290	38 31	331
143	59 40	018	33 24	092	35 04	158	38 52	215	50 06	264	31 04	290	38 03	330
144	59 57	016	34 21	092	35 25	158	38 19	215	49 09	264	30 10	289	37 34	329
145	60 11	014	35 19	092	35 47	158	37 46	215	48 12	264	29 16	289	37 03	328
146	60 24	012	36 16	091	36 08	159	37 13	216	47 15	264	28 22	289	36 32	327
147	60 35	010	37 13	091	36 29	159	36 39	216	46 18	263	27 27	288	36 01	326
148	60 44	008	38 11	091	36 49	159	36 05	216	45 21	263	26 32	288	35 28	325
149	60 51	006	39 08	091	37 09	160	35 31	217	44 24	263	25 38	287	34 55	324
	REGULUS		♦ARCTURUS		SPICA		♦ACRUX		CANOPUS		SIRIUS		♦PROCYON	
150	60 56	004	17 34	063	40 05	090	37 29	160	34 57	217	43 27	263	48 41	300
151	60 59	002	18 24	062	41 03	090	37 48	161	34 23	217	42 30	263	47 51	299
152	61 00	000	19 15	062	42 00	090	38 07	161	33 48	217	41 33	262	47 00	298
153	61 00	358	20 06	061	42 58	089	38 25	161	33 13	217	40 36	262	46 10	297
154	60 57	356	20 56	061	43 55	089	38 44	162	32 38	218	39 40	262	45 18	296
155	60 52	354	21 46	060	44 52	089	39 01	162	32 03	218	38 43	262	44 27	296
156	60 45	352	22 36	060	45 50	089	39 19	163	31 28	218	37 46	262	43 35	295
157	60 36	350	23 25	060	46 47	088	39 36	163	30 53	218	36 49	262	42 43	294
158	60 25	348	24 15	059	47 44	088	39 52	164	30 17	218	35 52	261	41 50	294
159	60 12	346	25 04	058	48 42	088	40 08	164	29 41	218	34 56	261	40 58	293
160	59 57	344	25 53	058	49 39	087	40 23	165	29 06	219	33 59	261	40 05	292
161	59 41	342	26 41	057	50 36	087	40 38	165	28 30	219	33 02	261	39 12	292
162	59 23	341	27 29	057	51 34	087	40 53	166	27 54	219	32 06	261	38 18	291
163	59 03	339	28 17	056	52 31	086	41 07	166	27 18	219	31 09	260	37 25	291
164	58 41	337	29 05	056	53 28	086	41 21	167	26 42	219	30 13	260	36 31	290
	Denebola		♦ARCTURUS		SPICA		♦ACRUX		SIRIUS		♦PROCYON		REGULUS	
165	56 11	022	29 52	055	54 25	085	41 34	167	29 16	260	35 37	290	58 18	335
166	56 31	020	30 39	054	55 22	085	41 46	168	28 20	260	34 43	289	57 53	334
167	56 50	018	31 25	054	56 20	085	41 58	168	27 23	260	33 49	289	57 27	332
168	57 07	017	32 11	053	57 17	084	42 10	169	26 27	259	32 54	288	56 59	330
169	57 23	015	32 57	053	58 14	084	42 20	169	25 30	259	32 00	288	56 30	329
170	57 37	013	33 43	052	59 11	083	42 31	170	24 34	259	31 05	287	56 00	327
171	57 49	011	34 27	051	60 08	083	42 41	170	23 38	259	30 10	287	55 28	326
172	57 59	010	35 12	050	61 05	082	42 50	171	22 41	259	29 15	286	54 56	324
173	58 08	008	35 56	050	62 01	082	42 58	172	21 45	258	28 20	286	54 22	323
174	58 15	006	36 39	049	62 58	081	43 06	172	20 49	258	27 25	285	53 47	322
175	58 20	004	37 22	048	63 55	080	43 14	173	19 53	258	26 29	285	53 11	320
176	58 23	002	38 05	047	64 51	080	43 21	173	18 57	258	25 34	285	52 34	319
177	58 24	000	38 47	047	65 48	079	43 27	174	18 01	257	24 38	284	51 56	318
178	58 23	359	39 28	046	66 44	078	43 33	175	17 05	257	23 43	284	51 17	317
179	58 21	357	40 09	045	67 40	078	43 38	175	16 09	257	22 47	283	50 37	316

LHA ♈	ARCTURUS Hc	Zn	◆ANTARES Hc	Zn	ACRUX Hc	Zn	◆Suhail Hc	Zn	Alphard Hc	Zn	REGULUS Hc	Zn	◆Denebola Hc	Zn
180	40 49	044	27 27	111	43 42	176	45 07	225	51 56	278	49 57	315	58 17	355
181	41 29	043	28 20	111	43 46	177	44 27	225	50 59	277	49 15	313	58 11	353
182	42 08	042	29 14	111	43 49	177	43 46	225	50 02	277	48 33	312	58 03	351
183	42 46	041	30 07	111	43 52	178	43 05	226	49 05	276	47 51	311	57 53	349
184	43 24	040	31 01	111	43 54	178	42 24	226	48 08	276	47 07	310	57 42	348
185	44 00	039	31 54	111	43 55	179	41 43	226	47 11	276	46 23	309	57 29	346
186	44 36	038	32 48	111	43 56	180	41 01	227	46 14	275	45 39	309	57 14	344
187	45 11	037	33 42	111	43 56	180	40 19	227	45 17	275	44 53	308	56 57	342
188	45 46	036	34 35	111	43 55	181	39 37	227	44 19	275	44 08	307	56 39	341
189	46 19	035	35 29	111	43 54	182	38 55	227	43 22	274	43 22	306	56 19	339
190	46 52	034	36 23	111	43 52	182	38 13	228	42 25	274	42 35	305	55 58	337
191	47 23	033	37 16	111	43 50	183	37 30	228	41 28	273	41 48	304	55 35	336
192	47 54	032	38 10	111	43 46	183	36 48	228	40 30	273	41 00	304	55 11	334
193	48 23	030	39 04	111	43 43	184	36 05	228	39 33	273	40 12	303	54 46	333
194	48 52	029	39 57	111	43 38	185	35 22	228	38 36	272	39 24	302	54 19	331

LHA ♈	ARCTURUS Hc	Zn	◆Rasalhague Hc	Zn	ANTARES Hc	Zn	◆ACRUX Hc	Zn	Suhail Hc	Zn	◆REGULUS Hc	Zn	Denebola Hc	Zn
195	49 19	028	16 01	071	40 51	111	43 33	185	34 39	229	38 35	301	53 51	330
196	49 46	027	16 55	071	41 45	111	43 28	186	33 56	229	37 46	301	53 21	329
197	50 11	025	17 49	070	42 39	111	43 22	186	33 13	229	36 57	300	52 51	327
198	50 35	024	18 43	070	43 32	111	43 15	187	32 30	229	36 07	299	52 19	326
199	50 57	023	19 37	069	44 26	111	43 08	188	31 46	229	35 17	299	51 46	324
200	51 19	021	20 30	069	45 20	111	43 00	188	31 03	229	34 26	298	51 12	323
201	51 39	020	21 24	069	46 13	111	42 51	189	30 20	229	33 36	298	50 38	322
202	51 58	018	22 17	068	47 07	111	42 42	189	29 36	229	32 45	297	50 02	321
203	52 15	017	23 11	068	48 01	111	42 32	190	28 53	229	31 53	296	49 25	320
204	52 31	015	24 04	067	48 54	111	42 22	191	28 09	229	31 02	296	48 48	319
205	52 46	014	24 56	067	49 48	111	42 11	191	27 26	229	30 10	295	48 09	317
206	52 59	012	25 49	066	50 41	111	42 00	192	26 42	229	29 18	295	47 30	316
207	53 10	011	26 42	066	51 35	111	41 48	192	25 59	229	28 26	294	46 50	315
208	53 20	009	27 34	065	52 28	111	41 36	193	25 15	229	27 34	294	46 09	314
209	53 29	008	28 26	065	53 22	112	41 23	193	24 31	229	26 41	293	45 28	313

LHA ♈	◆ARCTURUS Hc	Zn	Rasalhague Hc	Zn	◆ANTARES Hc	Zn	RIGIL KENT. Hc	Zn	ACRUX Hc	Zn	◆Suhail Hc	Zn	REGULUS Hc	Zn
210	53 36	006	29 18	064	54 15	112	45 37	173	41 09	194	23 48	229	25 48	293
211	53 41	005	30 09	064	55 08	112	45 44	174	40 55	194	23 04	229	24 55	292
212	53 45	003	31 01	063	56 01	112	45 50	175	40 41	195	22 21	229	24 02	292
213	53 47	001	31 52	063	56 54	113	45 55	175	40 26	195	21 37	229	23 09	292
214	53 47	000	32 43	062	57 47	113	45 59	176	40 10	196	20 54	229	22 16	291
215	53 46	358	33 33	062	58 40	113	46 03	177	39 55	196	20 10	229	21 22	291
216	53 44	357	34 24	061	59 33	113	46 06	177	39 38	197	19 27	229	20 28	290
217	53 40	355	35 14	060	60 25	114	46 08	178	39 21	197	18 43	229	19 34	290
218	53 34	353	36 03	060	61 18	114	46 10	179	39 04	198	18 00	229	18 40	289
219	53 26	352	36 53	059	62 10	115	46 11	179	38 46	198	17 17	229	17 46	289
220	53 18	350	37 42	058	63 02	115	46 11	180	38 28	199	16 34	229	16 52	289
221	53 07	349	38 31	058	63 54	116	46 11	181	38 10	199	15 50	229	15 57	288
222	52 55	347	39 19	057	64 46	116	46 09	182	37 51	199	15 07	229	15 03	288
223	52 42	346	40 07	056	65 37	117	46 08	182	37 32	200	14 24	228	14 08	288
224	52 27	344	40 54	056	66 28	118	46 05	183	37 12	200	13 42	228	13 13	287

LHA ♈	Alphecca Hc	Zn	◆VEGA Hc	Zn	ALTAIR Hc	Zn	◆Peacock Hc	Zn	RIGIL KENT. Hc	Zn	◆SPICA Hc	Zn	ARCTURUS Hc	Zn
225	45 28	011	14 39	041	13 43	076	18 53	145	46 02	184	66 14	281	52 10	343
226	45 39	010	15 17	040	14 38	076	19 26	145	45 58	184	65 18	280	51 53	341
227	45 48	008	15 54	040	15 34	075	19 59	145	45 53	185	64 21	280	51 33	340
228	45 56	007	16 30	039	16 29	075	20 32	145	45 48	186	63 24	279	51 13	338
229	46 02	006	17 06	039	17 25	075	21 05	145	45 42	186	62 28	279	50 51	337
230	46 08	005	17 42	038	18 20	074	21 38	145	45 35	187	61 31	278	50 28	336
231	46 12	003	18 18	038	19 15	074	22 11	145	45 27	188	60 34	278	50 04	334
232	46 14	002	18 52	037	20 10	074	22 43	145	45 19	188	59 37	277	49 38	333
233	46 16	001	19 27	037	21 05	073	23 16	145	45 11	189	58 40	277	49 12	332
234	46 16	000	20 01	036	22 00	073	23 49	145	45 01	190	57 43	276	48 44	330
235	46 15	358	20 34	035	22 55	072	24 22	145	44 51	190	56 46	276	48 15	329
236	46 12	357	21 07	035	23 49	072	24 54	145	44 41	191	55 49	275	47 45	328
237	46 09	356	21 40	034	24 44	071	25 27	145	44 29	192	54 52	275	47 14	327
238	46 04	354	22 12	034	25 38	071	26 00	145	44 17	192	53 55	274	46 43	326
239	45 57	353	22 44	033	26 32	071	26 32	146	44 05	193	52 57	274	46 10	325

LHA ♈	◆VEGA Hc	Zn	ALTAIR Hc	Zn	◆Peacock Hc	Zn	RIGIL KENT. Hc	Zn	◆SPICA Hc	Zn	ARCTURUS Hc	Zn	Alphecca Hc	Zn
240	23 15	032	27 26	070	27 04	146	43 52	193	52 00	274	45 36	324	45 50	352
241	23 45	032	28 20	070	27 37	146	43 38	194	51 03	273	45 02	322	45 41	351
242	24 15	031	29 14	069	28 09	146	43 24	195	50 06	273	44 26	321	45 31	349
243	24 45	030	30 08	069	28 41	146	43 09	195	49 08	273	43 50	320	45 20	348
244	25 13	030	31 01	068	29 13	146	42 54	196	48 11	272	43 13	319	45 07	347
245	25 42	029	31 54	068	29 45	146	42 38	196	47 14	272	42 35	318	44 54	346
246	26 09	028	32 47	067	30 17	147	42 21	197	46 16	272	41 57	318	44 39	344
247	26 36	028	33 40	067	30 48	147	42 04	197	45 19	271	41 18	317	44 23	343
248	27 02	027	34 32	066	31 20	147	41 47	198	44 22	271	40 38	316	44 06	342
249	27 28	026	35 25	065	31 51	147	41 29	199	43 24	271	39 58	315	43 47	341
250	27 53	025	36 17	065	32 22	147	41 10	199	42 27	270	39 17	314	43 28	340
251	28 17	025	37 09	064	32 53	148	40 51	200	41 29	270	38 35	313	43 07	339
252	28 41	024	38 00	064	33 24	148	40 32	200	40 32	270	37 53	312	42 46	337
253	29 04	023	38 51	063	33 54	148	40 12	200	39 35	269	37 10	312	42 23	336
254	29 26	022	39 42	062	34 24	148	39 52	201	38 37	269	36 27	311	42 00	335

LHA ♈	◆VEGA Hc	Zn	ALTAIR Hc	Zn	◆Peacock Hc	Zn	RIGIL KENT. Hc	Zn	◆SPICA Hc	Zn	ARCTURUS Hc	Zn	Alphecca Hc	Zn
255	29 48	022	40 33	062	34 54	149	39 31	201	37 40	269	35 43	310	41 35	334
256	30 09	021	41 24	061	35 24	149	39 10	202	36 43	269	34 59	309	41 10	333
257	30 29	020	42 14	060	35 54	149	38 49	202	35 45	268	34 15	309	40 44	332
258	30 48	019	43 03	060	36 23	149	38 27	203	34 48	268	33 30	308	40 16	331
259	31 06	018	43 53	059	36 52	150	38 05	203	33 51	268	32 44	307	39 48	330
260	31 24	017	44 41	058	37 21	150	37 42	203	32 53	268	31 58	307	39 19	329
261	31 41	017	45 30	057	37 49	150	37 19	204	31 56	267	31 12	306	38 49	328
262	31 57	016	46 18	056	38 18	151	36 56	204	30 59	267	30 26	305	38 19	327
263	32 12	015	47 06	056	38 45	151	36 32	205	30 01	267	29 39	305	37 47	326
264	32 26	014	47 53	055	39 13	152	36 08	205	29 04	266	28 51	304	37 15	325
265	32 40	013	48 39	054	39 40	152	35 44	205	28 07	266	28 04	304	36 42	325
266	32 52	012	49 25	053	40 07	152	35 19	206	27 10	266	27 16	303	36 08	324
267	33 04	011	50 11	052	40 33	153	34 54	206	26 12	266	26 27	303	35 34	323
268	33 15	010	50 55	051	40 59	153	34 29	206	25 15	265	25 39	302	34 59	322
269	33 25	010	51 40	050	41 25	154	34 04	206	24 18	265	24 50	301	34 24	321

LHA ♈	◆VEGA Hc	Zn	ALTAIR Hc	Zn	◆FOMALHAUT Hc	Zn	Peacock Hc	Zn	RIGIL KENT. Hc	Zn	◆ANTARES Hc	Zn	Alphecca Hc	Zn
270	33 34	009	52 23	049	21 39	116	41 50	154	33 38	207	66 56	242	33 47	320
271	33 42	008	53 06	048	22 31	116	42 15	155	33 12	207	66 05	243	33 10	320
272	33 49	007	53 48	046	23 22	116	42 39	155	32 46	207	65 14	243	32 33	319
273	33 55	006	54 29	045	24 14	115	43 03	156	32 19	208	64 22	244	31 55	318
274	34 01	005	55 09	044	25 06	115	43 27	156	31 53	208	63 31	245	31 16	317
275	34 05	004	55 48	043	25 58	115	43 49	157	31 26	208	62 39	245	30 37	317
276	34 08	003	56 27	041	26 50	115	44 12	157	30 59	208	61 47	246	29 58	316
277	34 11	002	57 04	040	27 42	115	44 34	158	30 32	208	60 54	246	29 17	315
278	34 13	001	57 40	038	28 34	115	44 55	158	30 04	209	60 02	246	28 37	315
279	34 13	000	58 15	037	29 26	115	45 16	159	29 37	209	59 09	247	27 56	314
280	34 13	359	58 49	035	30 18	115	45 36	160	29 09	209	58 16	247	27 14	313
281	34 12	358	59 21	034	31 10	115	45 56	160	28 41	209	57 24	247	26 32	313
282	34 10	357	59 53	032	32 02	115	46 15	161	28 13	209	56 31	248	25 50	312
283	34 06	356	60 22	030	32 54	115	46 33	162	27 45	209	55 37	248	25 07	312
284	34 02	355	60 50	029	33 46	115	46 51	162	27 17	210	54 44	248	24 24	311

LHA ♈	DENEB Hc	Zn	◆FOMALHAUT Hc	Zn	Peacock Hc	Zn	RIGIL KENT. Hc	Zn	◆ANTARES Hc	Zn	Rasalhague Hc	Zn	◆VEGA Hc	Zn
285	23 38	019	34 38	115	47 08	163	26 48	210	53 51	248	53 42	323	33 57	355
286	23 56	018	35 30	115	47 25	164	26 20	210	52 58	248	53 07	322	33 51	354
287	24 14	018	36 23	115	47 40	164	25 51	210	52 04	249	52 31	321	33 45	353
288	24 31	017	37 15	115	47 56	165	25 22	210	51 11	249	51 54	319	33 37	352
289	24 48	016	38 07	115	48 10	166	24 54	210	50 17	249	51 16	318	33 28	351
290	25 04	016	38 59	115	48 24	167	24 25	210	49 24	249	50 38	317	33 19	350
291	25 19	015	39 51	115	48 37	167	23 56	210	48 30	249	49 58	316	33 08	349
292	25 33	014	40 43	115	48 49	168	23 27	210	47 37	249	49 18	315	32 57	348
293	25 47	013	41 35	115	49 00	169	22 58	210	46 43	249	48 36	314	32 44	347
294	26 00	013	42 27	115	49 11	170	22 29	211	45 49	249	47 55	313	32 31	346
295	26 12	012	43 19	115	49 21	170	22 00	211	44 56	249	47 12	312	32 17	345
296	26 24	011	44 11	115	49 30	171	21 30	211	44 02	249	46 29	311	32 03	345
297	26 34	010	45 03	115	49 38	172	21 01	211	43 08	249	45 45	310	31 47	344
298	26 44	010	45 55	115	49 45	173	20 32	211	42 14	249	45 01	309	31 30	343
299	26 54	009	46 47	115	49 52	174	20 03	211	41 21	249	44 16	308	31 13	342

LHA ♈	◆DENEB Hc	Zn	Alpheratz Hc	Zn	◆FOMALHAUT Hc	Zn	Peacock Hc	Zn	◆ANTARES Hc	Zn	Rasalhague Hc	Zn	VEGA Hc	Zn
300	27 02	008	14 29	053	47 39	116	49 58	175	40 27	249	43 30	307	30 55	341
301	27 10	007	15 14	052	48 30	116	50 03	175	39 33	249	42 44	306	30 36	340
302	27 17	007	16 00	052	49 22	116	50 07	176	38 40	249	41 58	306	30 16	340
303	27 23	006	16 45	052	50 14	116	50 10	177	37 46	249	41 11	305	29 56	339
304	27 28	005	17 29	051	51 05	116	50 13	178	36 52	249	40 24	304	29 35	338
305	27 33	004	18 14	051	51 56	117	50 14	179	35 58	249	39 36	303	29 13	337
306	27 37	003	18 58	050	52 48	117	50 15	180	35 05	249	38 48	303	28 50	336
307	27 40	003	19 42	050	53 39	117	50 15	181	34 11	249	37 59	302	28 27	336
308	27 42	002	20 25	049	54 30	118	50 14	181	33 17	249	37 10	301	28 03	335
309	27 44	001	21 08	048	55 20	118	50 12	182	32 24	249	36 21	300	27 38	334
310	27 44	000	21 51	048	56 11	118	50 09	183	31 30	249	35 31	300	27 12	333
311	27 44	359	22 34	047	57 02	119	50 06	184	30 37	249	34 41	299	26 46	333
312	27 43	359	23 16	047	57 52	119	50 01	185	29 43	249	33 51	299	26 20	332
313	27 41	358	23 57	046	58 42	120	49 56	186	28 50	249	33 01	298	25 52	331
314	27 39	357	24 39	046	59 32	120	49 50	187	27 56	249	32 10	297	25 24	330

LHA ♈	Alpheratz Hc	Zn	◆Diphda Hc	Zn	ACHERNAR Hc	Zn	Peacock Hc	Zn	◆ANTARES Hc	Zn	VEGA Hc	Zn	◆DENEB Hc	Zn
315	25 19	045	36 57	100	25 21	146	49 43	187	27 03	249	24 56	330	27 36	356
316	26 00	044	37 54	100	25 53	146	49 35	188	26 09	248	24 27	329	27 31	355
317	26 40	044	38 50	100	26 25	146	49 26	189	25 16	248	23 57	328	27 27	355
318	27 19	043	39 47	100	26 57	146	49 17	190	24 23	248	23 27	328	27 21	354
319	27 58	042	40 43	099	27 29	146	49 07	191	23 30	248	22 56	327	27 14	353
320	28 37	042	41 40	099	28 00	146	48 56	191	22 36	248	22 24	327	27 07	352
321	29 15	041	42 37	099	28 32	147	48 44	192	21 43	248	21 52	326	26 59	352
322	29 52	040	43 33	099	29 03	147	48 32	193	20 50	248	21 20	325	26 50	351
323	30 29	040	44 30	099	29 35	147	48 19	194	19 57	247	20 47	325	26 41	350
324	31 05	039	45 27	099	30 06	147	48 05	194	19 04	247	20 14	324	26 30	349
325	31 41	038	46 23	099	30 37	147	47 50	195	18 11	247	19 40	324	26 19	349
326	32 16	037	47 20	098	31 08	147	47 35	196	17 19	247	19 06	323	26 08	348
327	32 51	037	48 17	098	31 39	148	47 19	197	16 26	247	18 31	322	25 55	347
328	33 25	036	49 14	098	32 09	148	47 02	197	15 33	247	17 56	322	25 42	346
329	33 58	035	50 11	098	32 40	148	46 44	198	14 41	246	17 20	321	25 28	346

LHA ♈	◆Alpheratz Hc	Zn	Diphda Hc	Zn	◆ACHERNAR Hc	Zn	Peacock Hc	Zn	◆Nunki Hc	Zn	ALTAIR Hc	Zn	DENEB Hc	Zn
330	34 31	034	51 07	098	33 10	148	46 26	199	46 15	249	48 52	306	25 13	345
331	35 03	033	52 04	098	33 40	149	46 08	199	45 21	250	48 05	306	24 58	344
332	35 34	033	53 01	098	34 10	149	45 49	200	44 27	250	47 18	305	24 42	343
333	36 04	032	53 58	097	34 40	149	45 29	201	43 33	250	46 31	304	24 25	343
334	36 34	031	54 55	097	35 09	149	45 08	201	42 39	250	45 43	303	24 07	342
335	37 03	030	55 52	097	35 38	150	44 47	202	41 46	250	44 55	302	23 49	341
336	37 31	029	56 49	097	36 07	150	44 26	202	40 52	250	44 06	301	23 31	341
337	37 59	028	57 46	097	36 35	150	44 04	203	39 58	250	43 17	301	23 11	340
338	38 25	027	58 43	097	37 04	151	43 41	203	39 04	250	42 27	300	22 51	339
339	38 51	026	59 40	097	37 32	151	43 18	204	38 11	250	41 37	299	22 30	339
340	39 16	025	60 37	097	37 59	151	42 54	205	37 17	250	40 47	298	22 09	338
341	39 39	024	61 34	097	38 27	152	42 30	205	36 23	249	39 56	298	21 47	337
342	40 02	023	62 31	096	38 54	152	42 06	206	35 29	249	39 05	297	21 25	337
343	40 24	022	63 28	096	39 21	152	41 41	206	34 36	249	38 14	296	21 02	336
344	40 45	021	64 25	096	39 47	153	41 15	206	33 42	249	37 23	296	20 38	335

LHA ♈	Alpheratz Hc	Zn	◆Hamal Hc	Zn	◆ACHERNAR Hc	Zn	Peacock Hc	Zn	Nunki Hc	Zn	◆ALTAIR Hc	Zn	DENEB Hc	Zn
345	41 05	020	29 01	050	40 13	153	40 50	207	32 48	249	36 31	295	20 14	335
346	41 24	019	29 45	049	40 39	154	40 23	207	31 55	249	35 39	295	19 49	334
347	41 42	018	30 28	049	41 04	154	39 57	208	31 01	249	34 46	294	19 24	334
348	41 59	017	31 11	048	41 28	155	39 30	208	30 07	249	33 54	294	18 58	333
349	42 15	015	31 53	047	41 53	155	39 03	209	29 14	249	33 01	293	18 32	332
350	42 30	014	32 35	046	42 17	156	38 35	209	28 20	249	32 08	292	18 05	332
351	42 43	013	33 17	046	42 40	156	38 07	209	27 27	249	31 15	292	17 37	331
352	42 56	012	33 57	045	43 03	157	37 39	210	26 33	249	30 22	291	17 09	331
353	43 07	011	34 38	044	43 26	157	37 10	210	25 40	248	29 28	291	16 41	330
354	43 17	010	35 17	043	43 48	158	36 41	210	24 47	248	28 35	290	16 12	330
355	43 26	008	35 57	043	44 09	158	36 12	211	23 53	248	27 41	290	15 43	329
356	43 34	007	36 35	042	44 30	159	35 43	211	23 00	248	26 47	290	15 13	329
357	43 41	006	37 13	041	44 51	159	35 13	211	22 07	248	25 53	289	14 43	328
358	43 46	005	37 50	040	45 10	160	34 43	212	21 14	248	24 59	289	14 13	328
359	43 51	004	38 27	039	45 30	161	34 13	212	20 21	248	24 04	288	13 42	327

LAT 18°S

LHA ♈	Hc	Zn	Hc	Zn	Hc	Zn	Hc	Zn	Hc	Zn	Hc	Zn	Hc	Zn
	◆Alpheratz		Hamal		ALDEBARAN		◆RIGEL		ACHERNAR		◆Peacock		Enif	
0	42 54	002	38 16	038	13 55	067	13 20	094	46 45	161	34 33	212	46 22	307
1	42 56	001	38 51	037	14 47	067	14 17	094	47 03	162	34 03	213	45 36	306
2	42 56	000	39 25	036	15 39	066	15 14	094	47 21	162	33 32	213	44 50	305
3	42 56	359	39 58	035	16 32	066	16 11	094	47 38	163	33 01	213	44 03	304
4	42 54	358	40 30	034	17 24	065	17 08	093	47 54	164	32 29	213	43 16	304
5	42 51	356	41 02	033	18 15	065	18 05	093	48 10	164	31 58	214	42 28	303
6	42 47	355	41 33	032	19 07	065	19 02	093	48 25	165	31 26	214	41 40	302
7	42 42	354	42 03	031	19 59	064	19 59	092	48 39	166	30 55	214	40 51	301
8	42 35	353	42 32	030	20 50	064	20 56	092	48 53	167	30 23	214	40 02	301
9	42 28	352	43 00	029	21 41	063	21 53	092	49 06	167	29 51	214	39 13	300
10	42 19	351	43 27	028	22 32	063	22 50	091	49 18	168	29 19	214	38 23	299
11	42 09	349	43 53	027	23 22	062	23 47	091	49 29	169	28 47	214	37 33	298
12	41 58	348	44 18	026	24 13	062	24 44	091	49 40	170	28 14	215	36 43	298
13	41 46	347	44 42	024	25 03	061	25 41	091	49 50	170	27 42	215	35 52	297
14	41 32	346	45 06	023	25 53	061	26 38	090	49 59	171	27 09	215	35 02	297
	◆Hamal		ALDEBARAN		RIGEL		◆CANOPUS		ACHERNAR		◆FOMALHAUT		Alpheratz	
15	45 28	022	26 42	060	27 35	090	19 39	141	50 07	172	59 45	242	41 18	345
16	45 49	021	27 32	060	28 32	090	20 16	140	50 15	173	58 55	242	41 03	344
17	46 08	020	28 21	059	29 29	089	20 52	140	50 21	174	58 05	242	40 46	343
18	46 27	018	29 10	058	30 26	089	21 28	140	50 27	175	57 14	243	40 29	342
19	46 44	017	29 58	058	31 23	089	22 05	140	50 32	175	56 23	243	40 10	340
20	47 01	016	30 46	057	32 20	088	22 41	140	50 36	176	55 32	244	39 50	339
21	47 16	015	31 34	057	33 17	088	23 17	140	50 39	177	54 41	244	39 30	338
22	47 29	013	32 22	056	34 14	088	23 54	140	50 42	178	53 50	244	39 08	337
23	47 42	012	33 09	055	35 11	087	24 30	140	50 43	179	52 58	244	38 46	336
24	47 53	011	33 55	055	36 08	087	25 06	140	50 44	180	52 07	245	38 23	335
25	48 03	009	34 42	054	37 05	087	25 43	141	50 44	181	51 15	245	37 58	334
26	48 11	008	35 28	053	38 02	086	26 19	141	50 43	181	50 23	245	37 33	333
27	48 18	007	36 13	053	38 59	086	26 55	141	50 41	182	49 32	245	37 07	332
28	48 24	005	36 58	052	39 56	086	27 31	141	50 39	183	48 40	245	36 40	331
29	48 29	004	37 43	051	40 53	085	28 07	141	50 35	184	47 48	245	36 13	331
	◆Hamal		ALDEBARAN		◆SIRIUS		CANOPUS		ACHERNAR		◆FOMALHAUT		Alpheratz	
30	48 32	002	38 27	050	22 27	101	28 44	141	50 31	185	46 56	246	35 44	330
31	48 33	001	39 11	049	23 23	101	29 19	141	50 26	186	46 04	246	35 15	329
32	48 34	000	39 54	049	24 19	101	29 55	141	50 20	186	45 12	246	34 45	328
33	48 33	358	40 36	048	25 15	100	30 31	141	50 13	187	44 20	246	34 15	327
34	48 30	357	41 18	047	26 11	100	31 07	141	50 05	188	43 28	246	33 43	326
35	48 26	355	42 00	046	27 08	100	31 43	141	49 57	189	42 36	246	33 11	325
36	48 21	354	42 40	045	28 04	100	32 18	142	49 48	190	41 44	246	32 38	325
37	48 15	353	43 20	044	29 00	100	32 53	142	49 38	191	40 52	246	32 05	324
38	48 07	351	44 00	043	29 56	099	33 29	142	49 27	191	40 00	246	31 31	323
39	47 57	350	44 39	042	30 53	099	34 04	142	49 15	192	39 07	246	30 56	322
40	47 47	349	45 17	041	31 49	099	34 39	142	49 03	193	38 15	246	30 21	321
41	47 35	347	45 54	040	32 46	099	35 13	143	48 50	194	37 23	246	29 45	321
42	47 22	346	46 30	039	33 42	098	35 48	143	48 36	194	36 31	246	29 09	320
43	47 07	345	47 06	038	34 38	098	36 22	143	48 22	195	35 39	246	28 32	319
44	46 52	343	47 40	037	35 35	098	36 56	143	48 06	196	34 47	246	27 54	319
	◆CAPELLA		BETELGEUSE		SIRIUS		◆CANOPUS		ACHERNAR		◆FOMALHAUT		Hamal	
45	18 58	024	39 55	063	36 31	098	37 30	144	47 51	196	33 55	246	46 35	342
46	19 21	024	40 46	063	37 28	098	38 04	144	47 34	197	33 03	246	46 17	341
47	19 44	023	41 37	062	38 25	097	38 38	144	47 17	198	32 11	246	45 57	340
48	20 06	022	42 27	061	39 21	097	39 11	144	46 59	199	31 19	246	45 37	338
49	20 27	022	43 17	061	40 18	097	39 44	145	46 41	199	30 27	246	45 16	337
50	20 48	021	44 06	060	41 14	097	40 17	145	46 22	200	29 35	246	44 53	336
51	21 09	021	44 55	059	42 11	097	40 49	146	46 02	200	28 43	245	44 29	335
52	21 28	020	45 44	058	43 08	096	41 21	146	45 42	201	27 51	245	44 04	334
53	21 47	019	46 32	057	44 05	096	41 53	146	45 21	202	26 59	245	43 39	333
54	22 06	019	47 20	056	45 01	096	42 25	147	45 00	202	26 07	245	43 12	332
55	22 24	018	48 07	056	45 58	096	42 56	147	44 38	203	25 15	245	42 44	330
56	22 41	017	48 54	055	46 55	095	43 26	148	44 16	203	24 24	245	42 16	329
57	22 57	016	49 40	054	47 52	095	43 57	148	43 53	204	23 32	245	41 46	328
58	23 13	016	50 26	053	48 49	095	44 27	149	43 29	204	22 40	245	41 16	327
59	23 28	015	51 11	052	49 45	095	44 56	149	43 06	205	21 49	245	40 44	326
	CAPELLA		BETELGEUSE		◆SIRIUS		CANOPUS		◆ACHERNAR		Diphda		◆Hamal	
60	23 43	014	51 56	051	50 42	095	45 26	150	42 41	205	43 23	262	40 12	325
61	23 57	014	52 39	050	51 39	094	45 54	150	42 17	206	42 27	262	39 39	324
62	24 10	013	53 22	048	52 36	094	46 23	151	41 52	206	41 30	262	39 06	323
63	24 22	012	54 05	047	53 33	094	46 50	151	41 26	207	40 34	261	38 31	322
64	24 34	011	54 46	046	54 30	094	47 18	152	41 00	207	39 38	261	37 56	322
65	24 45	011	55 27	045	55 27	094	47 44	152	40 34	208	38 41	261	37 21	321
66	24 55	010	56 06	043	56 24	093	48 11	153	40 07	208	37 45	261	36 44	320
67	25 05	009	56 45	042	57 21	093	48 36	154	39 40	208	36 49	261	36 07	319
68	25 13	008	57 23	041	58 18	093	49 01	154	39 13	209	35 52	260	35 29	318
69	25 21	008	57 59	039	59 15	093	49 26	155	38 45	209	34 56	260	34 51	317
70	25 29	007	58 35	038	60 12	092	49 49	156	38 18	210	34 00	260	34 12	317
71	25 35	006	59 09	036	61 09	092	50 13	156	37 49	210	33 04	260	33 32	316
72	25 41	005	59 42	034	62 06	092	50 35	157	37 21	210	32 07	260	32 52	315
73	25 46	005	60 14	033	63 03	092	50 57	158	36 52	210	31 11	259	32 12	314
74	25 50	004	60 44	031	64 00	091	51 18	159	36 23	211	30 15	259	31 31	314
	CAPELLA		POLLUX		◆PROCYON		Suhail		◆CANOPUS		ACHERNAR		◆Hamal	
75	25 54	003	29 04	042	44 25	063	32 30	131	51 38	160	35 54	211	30 49	313
76	25 57	002	29 42	041	45 16	062	33 13	131	51 58	160	35 25	211	30 07	312
77	25 59	002	30 19	040	46 06	062	33 56	131	52 16	161	34 54	212	29 25	312
78	26 00	001	30 56	040	46 56	061	34 40	131	52 34	162	34 24	212	28 42	311
79	26 00	000	31 32	039	47 46	060	35 23	131	52 52	163	33 54	212	27 58	310
80	26 00	359	32 07	038	48 35	059	36 06	131	53 08	164	33 24	212	27 15	310
81	25 59	359	32 42	037	49 23	058	36 49	131	53 23	165	32 53	213	26 31	309
82	25 57	358	33 16	036	50 12	057	37 31	131	53 38	166	32 22	213	25 46	308
83	25 54	357	33 50	036	51 00	056	38 14	132	53 51	167	31 51	213	25 01	308
84	25 51	356	34 23	035	51 47	055	38 57	132	54 04	168	31 20	213	24 16	307
85	25 47	355	34 55	034	52 33	054	39 39	132	54 16	169	30 49	213	23 30	307
86	25 42	355	35 26	033	53 20	053	40 22	132	54 27	170	30 18	213	22 44	306
87	25 36	354	35 57	032	54 05	052	41 04	132	54 36	171	29 46	214	21 58	306
88	25 30	353	36 27	031	54 50	051	41 46	133	54 45	172	29 14	214	21 12	305
89	25 23	352	36 56	030	55 34	050	42 28	133	54 53	173	28 43	214	20 25	305

LHA ♈	Hc	Zn	Hc	Zn	Hc	Zn	Hc	Zn	Hc	Zn	Hc	Zn	Hc	Zn
	POLLUX		◆REGULUS		Suhail		◆CANOPUS		ACHERNAR		◆ALDEBARAN		CAPELLA	
90	37 25	029	21 51	069	43 09	133	55 00	174	28 11	214	49 44	328	25 15	352
91	37 53	028	22 44	068	43 51	134	55 05	175	27 39	214	49 13	326	25 06	351
92	38 19	028	23 37	068	44 32	134	55 10	176	27 07	214	48 41	325	24 57	350
93	38 45	027	24 30	067	45 13	134	55 14	177	26 35	214	48 08	324	24 47	349
94	39 10	026	25 22	067	45 54	135	55 16	178	26 03	214	47 34	323	24 36	349
95	39 34	025	26 14	066	46 34	135	55 18	179	25 30	214	46 59	322	24 24	348
96	39 58	023	27 06	066	47 14	135	55 18	180	24 58	214	46 24	321	24 12	347
97	40 20	022	27 58	065	47 54	136	55 18	181	24 26	215	45 47	320	23 59	346
98	40 41	021	28 50	065	48 34	136	55 16	182	23 53	215	45 10	319	23 45	346
99	41 02	020	29 41	064	49 13	137	55 14	183	23 21	215	44 32	318	23 31	345
100	41 21	019	30 33	064	49 52	137	55 10	184	22 49	215	43 53	317	23 16	344
101	41 39	018	31 24	063	50 30	138	55 05	185	22 16	215	43 13	316	23 00	344
102	41 56	017	32 14	062	51 08	138	54 59	186	21 44	215	42 33	315	22 44	343
103	42 12	016	33 05	062	51 46	139	54 52	187	21 11	215	41 52	314	22 26	342
104	42 27	015	33 55	061	52 23	140	54 44	188	20 39	215	41 11	313	22 09	342
	POLLUX		◆REGULUS		◆ACRUX		CANOPUS		◆RIGEL		ALDEBARAN		CAPELLA	
105	42 41	014	34 45	061	19 47	152	54 36	189	62 30	287	40 28	312	21 50	341
106	42 54	012	35 34	060	20 15	152	54 26	190	61 36	287	39 46	311	21 31	340
107	43 06	011	36 24	059	20 42	152	54 15	191	60 41	286	39 03	310	21 12	340
108	43 16	010	37 12	059	21 09	152	54 03	192	59 46	285	38 19	310	20 52	339
109	43 26	009	38 01	058	21 36	152	53 50	193	58 51	284	37 35	309	20 31	338
110	43 34	008	38 49	057	22 03	152	53 37	194	57 56	284	36 50	308	20 09	338
111	43 41	006	39 37	056	22 30	152	53 22	195	57 00	283	36 05	307	19 47	337
112	43 47	005	40 24	056	22 57	152	53 07	196	56 04	282	35 19	307	19 25	336
113	43 51	004	41 11	055	23 24	152	52 50	197	55 08	282	34 33	306	19 02	336
114	43 55	003	41 58	054	23 51	152	52 33	198	54 13	281	33 47	305	18 38	335
115	43 57	002	42 44	053	24 18	152	52 15	199	53 17	281	33 00	305	18 14	335
116	43 58	000	43 29	052	24 45	152	51 56	200	52 20	280	32 13	304	17 49	334
117	43 57	359	44 14	052	25 12	152	51 37	201	51 24	280	31 25	303	17 24	333
118	43 56	358	44 59	051	25 39	152	51 16	201	50 28	279	30 37	303	16 58	333
119	43 53	357	45 42	050	26 05	152	50 55	202	49 31	279	29 49	302	16 32	332
	◆REGULUS		SPICA		◆ACRUX		CANOPUS		◆RIGEL		BETELGEUSE		POLLUX	
120	46 26	049	11 39	098	26 32	152	50 33	203	48 35	278	50 01	307	43 49	355
121	47 08	048	12 35	098	26 58	152	50 11	204	47 39	278	49 15	306	43 44	354
122	47 50	047	13 32	098	27 25	153	49 48	204	46 42	277	48 28	305	43 38	353
123	48 32	046	14 29	097	27 51	153	49 24	205	45 45	277	47 41	304	43 30	352
124	49 12	045	15 25	097	28 17	153	48 59	206	44 49	276	46 54	303	43 21	351
125	49 52	044	16 22	097	28 43	153	48 34	206	43 52	276	46 06	302	43 11	349
126	50 31	042	17 19	096	29 09	153	48 08	207	42 55	276	45 17	301	43 00	348
127	51 09	041	18 15	096	29 35	153	47 42	208	41 58	275	44 28	301	42 48	347
128	51 46	040	19 12	096	30 00	153	47 15	208	41 01	275	43 39	300	42 35	346
129	52 22	039	20 09	096	30 26	154	46 48	209	40 05	274	42 49	299	42 20	345
130	52 57	038	21 06	095	30 51	154	46 20	209	39 08	274	41 59	298	42 05	344
131	53 32	036	22 02	095	31 16	154	45 52	210	38 11	274	41 09	298	41 48	342
132	54 05	035	22 59	095	31 41	154	45 23	211	37 14	273	40 18	297	41 30	341
133	54 37	033	23 56	094	32 05	155	44 54	211	36 17	273	39 27	296	41 11	340
134	55 08	032	24 53	094	32 30	155	44 25	212	35 20	273	38 36	296	40 52	339
	REGULUS		◆SPICA		ACRUX		◆CANOPUS		SIRIUS		BETELGEUSE		◆POLLUX	
135	55 37	030	25 50	094	32 54	155	43 54	212	57 47	267	37 44	295	40 31	338
136	56 06	029	26 47	094	33 18	155	43 24	212	56 50	267	36 52	294	40 09	337
137	56 32	027	27 44	093	33 42	156	42 53	213	55 53	267	36 00	294	39 46	336
138	56 58	026	28 41	093	34 05	156	42 22	213	54 56	266	35 08	293	39 23	335
139	57 22	024	29 38	093	34 29	156	41 51	214	53 59	266	34 15	293	38 58	334
140	57 45	022	30 35	093	34 52	156	41 19	214	53 02	266	33 23	292	38 33	333
141	58 06	021	31 32	092	35 14	157	40 47	214	52 06	266	32 30	292	38 06	332
142	58 25	019	32 29	092	35 37	157	40 14	215	51 09	265	31 37	291	37 39	331
143	58 43	017	33 26	092	35 59	157	39 41	215	50 12	265	30 43	291	37 11	330
144	58 59	015	34 23	091	36 21	158	39 08	216	49 15	265	29 50	290	36 42	329
145	59 13	014	35 20	091	36 42	158	38 35	216	48 18	265	28 56	290	36 12	328
146	59 25	012	36 17	091	37 04	158	38 01	216	47 21	265	28 02	289	35 42	327
147	59 36	010	37 14	090	37 24	159	37 28	216	46 24	264	27 08	289	35 11	326
148	59 45	008	38 11	090	37 45	159	36 54	217	45 28	264	26 14	288	34 39	326
149	59 52	006	39 08	090	38 05	159	36 20	217	44 31	264	25 20	288	34 06	325
	REGULUS		◆ARCTURUS		SPICA		◆ACRUX		CANOPUS		SIRIUS		◆PROCYON	
150	59 56	004	17 06	062	40 05	090	38 25	160	35 45	217	43 34	264	48 11	300
151	59 59	002	17 56	062	41 02	089	38 45	160	35 11	217	42 37	264	47 22	300
152	60 00	000	18 47	062	41 59	089	39 04	161	34 36	218	41 41	263	46 32	299
153	60 00	358	19 37	061	42 57	089	39 22	161	34 01	218	40 44	263	45 42	298
154	59 57	356	20 27	061	43 54	088	39 41	162	33 26	218	39 47	263	44 51	297
155	59 52	354	21 16	060	44 51	088	39 58	162	32 51	218	38 51	263	44 00	297
156	59 45	352	22 06	060	45 48	088	40 16	162	32 15	218	37 54	263	43 09	296
157	59 37	350	22 55	059	46 45	087	40 33	163	31 40	219	36 58	262	42 18	295
158	59 26	348	23 44	059	47 42	087	40 49	163	31 04	219	36 01	262	41 26	295
159	59 14	347	24 32	058	48 39	086	41 06	164	30 28	219	35 05	262	40 34	294
160	59 00	345	25 21	058	49 36	086	41 21	164	29 53	219	34 08	262	39 42	293
161	58 44	343	26 09	057	50 32	086	41 36	165	29 17	219	33 12	261	38 49	293
162	58 26	341	26 56	056	51 29	085	41 51	165	28 41	219	32 15	261	37 56	292
163	58 07	339	27 44	056	52 26	085	42 05	166	28 05	219	31 19	261	37 03	291
164	57 46	338	28 31	055	53 23	084	42 19	166	27 28	219	30 23	261	36 10	291
	Denebola		◆ARCTURUS		SPICA		◆ACRUX		SIRIUS		◆PROCYON		REGULUS	
165	55 15	021	29 17	055	54 20	084	42 32	167	29 26	261	35 17	290	57 23	336
166	55 35	019	30 04	054	55 17	084	42 45	167	28 30	260	34 23	290	56 59	334
167	55 53	018	30 50	053	56 13	083	42 57	168	27 34	260	33 29	289	56 34	333
168	56 10	016	31 35	053	57 10	083	43 08	169	26 38	260	32 35	289	56 07	331
169	56 25	014	32 20	052	58 06	082	43 19	169	25 41	260	31 41	288	55 39	330
170	56 38	013	33 05	051	59 03	082	43 30	170	24 45	259	30 47	288	55 09	328
171	56 50	011	33 50	051	59 59	081	43 40	170	23 49	259	29 52	287	54 39	327
172	57 00	009	34 33	050	60 56	080	43 49	171	22 53	259	28 58	287	54 07	325
173	57 08	008	35 17	049	61 52	080	43 58	172	21 57	259	28 03	286	53 33	324
174	57 15	006	36 00	048	62 48	079	44 06	172	21 01	259	27 08	286	52 59	323
175	57 20	004	36 42	048	63 44	078	44 13	173	20 05	258	26 13	285	52 24	321
176	57 23	002	37 24	047	64 40	078	44 20	173	19 09	258	25 18	285	51 48	320
177	57 24	000	38 06	046	65 35	077	44 27	174	18 14	258	24 23	285	51 11	319
178	57 23	359	38 46	045	66 31	076	44 32	175	17 18	258	23 28	284	50 33	318
179	57 21	357	39 27	044	67 26	075	44 37	175	16 22	257	22 33	284	49 54	316

LAT 18°S

LHA	Hc	Zn	Hc	Zn	Hc	Zn	Hc	Zn	Hc	Zn	Hc	Zn	Hc	Zn
♈	ARCTURUS		♦ANTARES		ACRUX		♦Suhail		Alphard		REGULUS		♦Denebola	
180	40 06	043	27 49	111	44 42	176	45 50	225	51 47	279	49 14	315	57 17	355
181	40 45	043	28 42	111	44 46	176	45 09	226	50 51	279	48 34	314	57 11	353
182	41 23	042	29 35	111	44 49	177	44 28	226	49 54	278	47 52	313	57 04	351
183	42 01	041	30 29	111	44 52	178	43 47	226	48 58	278	47 11	312	56 54	350
184	42 38	040	31 22	110	44 54	178	43 05	227	48 01	277	46 28	311	56 43	348
185	43 14	039	32 16	110	44 55	179	42 24	227	47 04	277	45 45	310	56 30	346
186	43 49	038	33 09	110	44 56	180	41 42	227	46 08	276	45 01	309	56 16	345
187	44 24	037	34 03	110	44 56	180	41 00	228	45 11	276	44 16	308	56 00	343
188	44 57	036	34 56	110	44 55	181	40 18	228	44 14	276	43 32	308	55 42	341
189	45 30	034	35 50	110	44 54	182	39 35	228	43 17	275	42 46	307	55 23	340
190	46 02	033	36 43	110	44 52	182	38 53	228	42 21	275	42 00	306	55 03	338
191	46 33	032	37 37	110	44 50	183	38 10	228	41 24	274	41 14	305	54 41	336
192	47 03	031	38 31	110	44 46	183	37 27	229	40 27	274	40 27	304	54 17	335
193	47 32	030	39 24	110	44 43	184	36 45	229	39 30	274	39 39	304	53 52	333
194	47 59	029	40 18	110	44 38	185	36 02	229	38 33	273	38 52	303	53 26	332
	ARCTURUS		♦Rasalhague		ANTARES		♦ACRUX		Suhail		♦REGULUS		Denebola	
195	48 26	027	15 41	071	41 12	110	44 33	185	35 19	229	38 04	302	52 59	331
196	48 52	026	16 35	070	42 06	110	44 28	186	34 35	229	37 15	301	52 30	329
197	49 16	025	17 29	070	42 59	110	44 21	187	33 52	229	36 26	301	52 00	328
198	49 40	023	18 22	070	43 53	110	44 15	187	33 09	229	35 37	300	51 29	327
199	50 02	022	19 16	069	44 47	110	44 07	188	32 26	229	34 47	299	50 57	325
200	50 23	021	20 09	069	45 40	110	43 59	188	31 42	230	33 58	299	50 24	324
201	50 42	019	21 02	068	46 34	110	43 50	189	30 59	230	33 07	298	49 50	323
202	51 01	018	21 55	068	47 28	110	43 41	190	30 15	230	32 17	298	49 15	322
203	51 18	017	22 48	067	48 22	110	43 31	190	29 32	230	31 26	297	48 39	320
204	51 33	015	23 40	067	49 15	110	43 21	191	28 48	230	30 35	296	48 02	319
205	51 47	014	24 33	066	50 09	110	43 10	191	28 05	230	29 44	296	47 25	318
206	52 00	012	25 25	066	51 03	110	42 59	192	27 21	230	28 53	295	46 46	317
207	52 11	011	26 17	065	51 56	110	42 47	192	26 37	230	28 01	295	46 07	316
208	52 21	009	27 09	065	52 50	110	42 34	193	25 54	230	27 09	294	45 27	315
209	52 29	008	28 00	064	53 43	110	42 21	194	25 10	230	26 17	294	44 46	314
	♦ARCTURUS		Rasalhague		♦ANTARES		RIGIL KENT.		ACRUX		♦Suhail		REGULUS	
210	52 36	006	28 52	064	54 37	111	46 37	173	42 07	194	24 27	230	25 25	293
211	52 41	004	29 43	063	55 30	111	46 44	174	41 53	195	23 43	230	24 32	293
212	52 45	003	30 34	063	56 23	111	46 49	174	41 39	195	23 00	230	23 40	292
213	52 47	001	31 24	062	57 17	111	46 55	175	41 24	196	22 16	230	22 47	292
214	52 47	000	32 14	062	58 10	111	46 59	176	41 08	196	21 33	230	21 54	291
215	52 46	358	33 05	061	59 03	112	47 03	177	40 52	197	20 49	229	21 01	291
216	52 44	357	33 54	060	59 56	112	47 06	177	40 36	197	20 06	229	20 07	291
217	52 40	355	34 44	060	60 49	112	47 08	178	40 19	198	19 23	229	19 14	290
218	52 34	354	35 33	059	61 42	113	47 10	179	40 01	198	18 39	229	18 20	290
219	52 27	352	36 22	058	62 34	113	47 11	179	39 43	198	17 56	229	17 26	289
220	52 18	350	37 10	058	63 27	113	47 11	180	39 25	199	17 13	229	16 32	289
221	52 08	349	37 58	057	64 19	114	47 11	181	39 07	199	16 30	229	15 38	289
222	51 57	347	38 46	056	65 11	114	47 09	182	38 48	200	15 47	229	14 44	288
223	51 43	346	39 33	056	66 03	115	47 07	182	38 28	200	15 04	229	13 50	288
224	51 29	345	40 20	055	66 55	116	47 05	183	38 08	200	14 21	228	12 56	287
	Alphecca		♦VEGA		ALTAIR		♦Peacock		RIGIL KENT.		♦SPICA		ARCTURUS	
225	44 30	011	13 54	041	13 28	076	19 42	145	47 02	184	66 01	283	51 13	343
226	44 40	010	14 31	040	14 23	075	20 15	145	46 57	184	65 06	283	50 56	342
227	44 49	008	15 08	040	15 19	075	20 48	145	46 53	185	64 10	282	50 37	340
228	44 56	007	15 44	039	16 14	075	21 21	145	46 47	186	63 14	281	50 17	339
229	45 03	006	16 20	039	17 09	074	21 54	145	46 41	187	62 18	280	49 56	337
230	45 08	005	16 55	038	18 04	074	22 27	145	46 34	187	61 22	280	49 33	336
231	45 12	003	17 30	038	18 58	074	23 00	145	46 27	188	60 25	279	49 10	335
232	45 14	002	18 05	037	19 53	073	23 32	145	46 19	189	59 29	279	48 45	334
233	45 16	001	18 39	036	20 48	073	24 05	145	46 10	189	58 33	278	48 19	332
234	45 16	000	19 12	036	21 42	072	24 38	145	46 00	190	57 36	278	47 52	331
235	45 15	358	19 46	035	22 36	072	25 11	145	45 50	191	56 39	277	47 23	330
236	45 13	357	20 18	035	23 31	071	25 44	145	45 39	191	55 43	277	46 54	329
237	45 09	356	20 50	034	24 25	071	26 16	145	45 28	192	54 46	276	46 24	327
238	45 04	354	21 22	033	25 18	071	26 49	145	45 16	192	53 49	276	45 53	326
239	44 58	353	21 53	033	26 12	070	27 21	145	45 03	193	52 53	275	45 21	325
	♦VEGA		ALTAIR		♦Peacock		RIGIL KENT.		♦SPICA		ARCTURUS		Alphecca	
240	22 24	032	27 06	070	27 54	145	44 50	194	51 56	275	44 48	324	44 51	352
241	22 54	032	27 59	069	28 26	145	44 36	194	50 59	274	44 14	323	44 42	351
242	23 24	031	28 52	069	28 59	146	44 22	195	50 02	274	43 39	322	44 32	349
243	23 53	030	29 45	068	29 31	146	44 07	196	49 05	274	43 04	321	44 21	348
244	24 21	030	30 38	068	30 03	146	43 51	196	48 08	273	42 27	320	44 09	347
245	24 49	029	31 31	067	30 35	146	43 35	197	47 11	273	41 50	319	43 55	346
246	25 16	028	32 23	067	31 07	146	43 19	197	46 14	273	41 12	318	43 41	345
247	25 43	027	33 16	066	31 38	146	43 02	198	45 17	272	40 34	317	43 25	343
248	26 09	027	34 08	065	32 10	147	42 44	198	44 20	272	39 55	316	43 08	342
249	26 34	026	34 59	065	32 41	147	42 26	199	43 23	272	39 15	315	42 51	341
250	26 59	025	35 51	064	33 12	147	42 07	199	42 26	271	38 35	315	42 32	340
251	27 23	025	36 42	064	33 43	147	41 48	200	41 29	271	37 54	314	42 12	339
252	27 46	024	37 33	063	34 14	147	41 28	200	40 32	271	37 12	313	41 51	338
253	28 09	023	38 24	062	34 45	148	41 08	201	39 35	270	36 30	312	41 28	337
254	28 31	022	39 14	062	35 15	148	40 48	201	38 38	270	35 48	311	41 05	336
	♦VEGA		ALTAIR		♦Peacock		RIGIL KENT.		♦SPICA		ARCTURUS		Alphecca	
255	28 52	021	40 04	061	35 45	148	40 27	202	37 41	270	35 05	311	40 41	335
256	29 12	021	40 54	060	36 15	148	40 06	202	36 44	269	34 21	310	40 16	334
257	29 32	020	41 44	060	36 45	149	39 44	203	35 47	269	33 37	309	39 50	333
258	29 51	019	42 33	059	37 15	149	39 22	203	34 50	269	32 53	308	39 24	332
259	30 09	018	43 21	058	37 44	149	39 00	203	33 52	268	32 08	308	38 56	331
260	30 27	017	44 09	057	38 13	150	38 37	204	32 55	268	31 22	307	38 28	330
261	30 43	016	44 57	056	38 41	150	38 14	204	31 58	268	30 37	306	37 58	329
262	30 59	016	45 44	056	39 10	150	37 50	204	31 01	268	29 51	306	37 28	328
263	31 14	015	46 31	055	39 38	151	37 26	205	30 04	267	29 04	305	36 57	327
264	31 28	014	47 18	054	40 06	151	37 02	205	29 07	267	28 17	305	36 26	326
265	31 41	013	48 03	053	40 33	152	36 38	206	28 10	267	27 30	304	35 53	325
266	31 53	012	48 49	052	41 00	152	36 13	206	27 13	266	26 43	303	35 20	324
267	32 05	011	49 33	051	41 27	152	35 48	206	26 16	266	25 55	303	34 46	323
268	32 16	010	50 17	050	41 53	153	35 23	206	25 20	266	25 07	302	34 12	322
269	32 25	009	51 00	049	42 19	153	34 57	207	24 23	266	24 18	302	33 37	322

LHA	Hc	Zn	Hc	Zn	Hc	Zn	Hc	Zn	Hc	Zn	Hc	Zn	Hc	Zn
♈	♦VEGA		ALTAIR		♦FOMALHAUT		Peacock		RIGIL KENT.		♦ANTARES		Alphecca	
270	32 34	008	51 43	048	22 05	115	42 44	154	34 31	207	67 23	244	33 01	321
271	32 42	008	52 25	047	22 56	115	43 09	154	34 05	207	66 32	245	32 25	320
272	32 49	007	53 06	045	23 48	115	43 34	155	33 39	208	65 40	245	31 48	319
273	32 56	006	53 46	044	24 40	115	43 58	155	33 12	208	64 48	246	31 10	319
274	33 01	005	54 25	043	25 31	115	44 21	156	32 46	208	63 56	246	30 32	318
275	33 05	004	55 04	042	26 23	115	44 45	156	32 19	208	63 03	247	29 53	317
276	33 09	003	55 41	040	27 15	115	45 07	157	31 52	208	62 11	247	29 14	316
277	33 11	002	56 17	039	28 07	115	45 29	157	31 24	209	61 18	248	28 35	316
278	33 13	001	56 53	037	28 59	114	45 51	158	30 57	209	60 25	248	27 55	315
279	33 13	000	57 27	036	29 51	114	46 12	159	30 29	209	59 32	248	27 14	314
280	33 13	359	58 00	034	30 43	114	46 32	159	30 01	209	58 39	249	26 33	314
281	33 12	358	58 31	033	31 35	114	46 52	160	29 33	209	57 46	249	25 52	313
282	33 10	357	59 01	031	32 27	114	47 11	161	29 05	210	56 53	249	25 10	312
283	33 07	356	59 30	029	33 19	114	47 30	161	28 37	210	55 59	249	24 27	312
284	33 03	356	59 58	028	34 11	114	47 48	162	28 09	210	55 06	249	23 45	311
	DENEB		♦FOMALHAUT		Peacock		RIGIL KENT.		♦ANTARES		Rasalhague		♦VEGA	
285	22 41	019	35 03	114	48 05	163	27 40	210	54 13	250	52 54	324	32 58	355
286	22 59	018	35 55	114	48 22	163	27 12	210	53 19	250	52 19	323	32 52	354
287	23 17	018	36 47	114	48 38	164	26 43	210	52 26	250	51 44	321	32 45	353
288	23 34	017	37 39	114	48 53	165	26 14	210	51 32	250	51 08	320	32 37	352
289	23 50	016	38 32	114	49 08	166	25 46	210	50 38	250	50 31	319	32 29	351
290	24 06	016	39 24	114	49 22	166	25 17	210	49 45	250	49 53	318	32 20	350
291	24 21	015	40 16	114	49 35	167	24 48	211	48 51	250	49 15	317	32 09	349
292	24 35	014	41 08	114	49 47	168	24 19	211	47 57	250	48 35	316	31 58	348
293	24 48	013	42 00	114	49 59	169	23 49	211	47 04	250	47 55	315	31 46	347
294	25 01	013	42 52	114	50 10	169	23 20	211	46 10	250	47 14	313	31 33	346
295	25 13	012	43 44	114	50 20	170	22 51	211	45 16	250	46 32	312	31 19	346
296	25 25	011	44 36	114	50 29	171	22 22	211	44 23	250	45 49	312	31 05	345
297	25 35	010	45 28	114	50 37	172	21 53	211	43 29	250	45 06	311	30 49	344
298	25 45	010	46 20	114	50 45	173	21 23	211	42 35	250	44 23	310	30 33	343
299	25 54	009	47 12	114	50 52	174	20 54	211	41 41	250	43 39	309	30 16	342
	♦DENEB		Alpheratz		♦FOMALHAUT		Peacock		♦ANTARES		Rasalhague		VEGA	
300	26 03	008	13 52	053	48 04	115	50 58	175	40 48	250	42 54	308	29 58	341
301	26 10	007	14 38	052	48 56	115	51 03	175	39 54	250	42 09	307	29 39	341
302	26 17	007	15 23	052	49 48	115	51 07	176	39 00	250	41 23	306	29 20	340
303	26 23	006	16 07	051	50 40	115	51 10	177	38 07	250	40 36	305	29 00	339
304	26 29	005	16 52	051	51 31	115	51 13	178	37 13	250	39 50	305	28 39	338
305	26 33	004	17 36	050	52 23	115	51 14	179	36 19	250	39 03	304	28 17	337
306	26 37	003	18 19	050	53 14	116	51 15	180	35 26	250	38 15	303	27 55	337
307	26 40	003	19 03	049	54 06	116	51 15	181	34 32	250	37 27	302	27 32	336
308	26 42	002	19 46	049	54 57	116	51 14	181	33 38	250	36 39	302	27 08	335
309	26 44	001	20 29	048	55 48	117	51 12	182	32 45	250	35 50	301	26 44	334
310	26 44	000	21 11	048	56 39	117	51 09	183	31 51	250	35 01	300	26 19	334
311	26 44	359	21 53	047	57 30	117	51 05	184	30 58	249	34 12	300	25 53	333
312	26 43	359	22 34	047	58 20	118	51 01	185	30 05	249	33 22	299	25 27	332
313	26 41	358	23 16	046	59 11	118	50 56	186	29 11	249	32 32	299	25 00	331
314	26 39	357	23 56	045	60 01	119	50 49	187	28 18	249	31 42	298	24 32	331
	Alpheratz		♦Diphda		ACHERNAR		Peacock		♦ANTARES		VEGA		♦DENEB	
315	24 37	045	37 07	099	26 10	146	50 42	188	27 25	249	24 04	330	26 36	356
316	25 17	044	38 03	099	26 42	146	50 34	188	26 31	249	23 35	329	26 32	356
317	25 56	043	39 00	099	27 14	146	50 26	189	25 38	249	23 06	329	26 27	355
318	26 35	043	39 56	099	27 46	146	50 16	190	24 45	249	22 36	328	26 21	354
319	27 14	042	40 53	099	28 18	146	50 06	191	23 52	248	22 05	327	26 15	353
320	27 52	041	41 49	098	28 50	146	49 55	192	22 59	248	21 34	327	26 08	352
321	28 29	041	42 46	098	29 22	146	49 43	192	22 06	248	21 03	326	26 00	352
322	29 06	040	43 42	098	29 53	146	49 30	193	21 13	248	20 31	326	25 51	351
323	29 43	039	44 39	098	30 25	147	49 17	194	20 20	248	19 58	325	25 42	350
324	30 19	039	45 35	098	30 56	147	49 03	194	19 27	248	19 25	324	25 31	349
325	30 54	038	46 32	097	31 28	147	48 48	195	18 35	247	18 52	324	25 21	349
326	31 29	037	47 28	097	31 59	147	48 32	196	17 42	247	18 18	323	25 09	348
327	32 03	036	48 25	097	32 29	147	48 16	197	16 49	247	17 43	323	24 57	347
328	32 36	035	49 22	097	33 00	148	47 59	198	15 57	247	17 09	322	24 44	346
329	33 09	035	50 18	097	33 31	148	47 41	198	15 05	247	16 33	322	24 30	346
	♦Alpheratz		Diphda		♦ACHERNAR		Peacock		♦Nunki		ALTAIR		DENEB	
330	33 41	034	51 15	097	34 01	148	47 23	199	46 35	250	48 16	307	24 15	345
331	34 13	033	52 12	096	34 31	148	47 04	200	45 41	250	47 30	306	24 00	344
332	34 43	032	53 08	096	35 01	148	46 45	200	44 47	250	46 44	306	23 44	343
333	35 13	031	54 05	096	35 31	149	46 25	201	43 54	250	45 57	305	23 28	343
334	35 43	030	55 02	096	36 00	149	46 04	202	43 00	250	45 10	304	23 10	342
335	36 11	029	55 59	096	36 30	149	45 43	202	42 06	250	44 22	303	22 53	341
336	36 39	029	56 55	096	36 59	150	45 21	203	41 12	250	43 34	302	22 34	341
337	37 06	028	57 52	095	37 28	150	44 59	203	40 19	250	42 46	301	22 15	340
338	37 32	027	58 49	095	37 56	150	44 36	204	39 25	250	41 57	301	21 55	339
339	37 57	026	59 46	095	38 24	151	44 13	204	38 31	250	41 08	300	21 35	339
340	38 21	025	60 43	095	38 52	151	43 49	205	37 37	250	40 18	299	21 14	338
341	38 45	024	61 39	095	39 20	151	43 25	205	36 44	250	39 28	298	20 52	337
342	39 07	023	62 36	095	39 47	152	43 00	206	35 50	250	38 38	298	20 30	337
343	39 29	022	63 33	094	40 14	152	42 35	206	34 56	250	37 47	297	20 07	336
344	39 49	021	64 30	094	40 40	152	42 09	207	34 03	250	36 56	297	19 44	335
	Alpheratz		♦Hamal		♦ACHERNAR		Peacock		Nunki		♦ALTAIR		DENEB	
345	40 09	020	28 23	049	41 06	153	41 43	207	33 09	250	36 05	296	19 20	335
346	40 27	018	29 06	049	41 32	153	41 17	208	32 16	250	35 13	295	18 55	334
347	40 45	017	29 48	048	41 58	154	40 50	208	31 22	250	34 22	295	18 30	334
348	41 02	016	30 31	047	42 23	154	40 23	209	30 29	250	33 30	294	18 04	333
349	41 17	015	31 12	047	42 47	155	39 55	209	29 35	249	32 38	294	17 38	333
350	41 31	014	31 54	046	43 11	155	39 27	209	28 42	249	31 45	293	17 12	332
351	41 45	013	32 34	045	43 35	156	38 59	210	27 48	249	30 53	293	16 45	331
352	41 57	012	33 15	045	43 58	156	38 31	210	26 55	249	30 00	292	16 17	331
353	42 08	011	33 54	044	44 21	157	38 02	210	26 02	249	29 07	291	15 49	330
354	42 18	010	34 34	043	44 43	157	37 33	211	25 09	249	28 14	291	15 21	330
355	42 27	008	35 12	042	45 05	158	37 04	211	24 15	249	27 20	290	14 52	329
356	42 35	007	35 50	041	45 26	159	36 34	211	23 22	248	26 27	290	14 22	329
357	42 41	006	36 28	041	45 47	159	36 04	212	22 29	248	25 33	290	13 52	328
358	42 47	005	37 04	040	46 07	160	35 34	212	21 36	248	24 39	289	13 22	328
359	42 51	004	37 40	039	46 26	160	35 04	212	20 43	248	23 45	289	12 51	327

 LAT 19°S

LHA	Hc	Zn	Hc	Zn	Hc	Zn	Hc	Zn	Hc	Zn	Hc	Zn	Hc	Zn
♈	◆Alpheratz		Hamal		ALDEBARAN		◆RIGEL		ACHERNAR		◆Peacock		Enif	
0	41 54	002	37 28	037	13 31	067	13 24	094	47 42	161	35 24	213	45 46	308
1	41 56	001	38 03	037	14 23	067	14 21	094	48 00	161	34 53	213	45 01	307
2	41 56	000	38 36	036	15 15	066	15 18	094	48 18	162	34 22	213	44 15	306
3	41 56	359	39 09	035	16 07	066	16 14	093	48 35	163	33 51	213	43 29	305
4	41 54	358	39 40	034	16 59	065	17 11	093	48 52	163	33 19	214	42 42	304
5	41 51	357	40 11	033	17 50	065	18 08	093	49 08	164	32 48	214	41 55	304
6	41 47	355	40 42	032	18 41	064	19 04	092	49 23	165	32 16	214	41 08	303
7	41 42	354	41 11	031	19 32	064	20 01	092	49 38	166	31 44	214	40 20	302
8	41 36	353	41 40	030	20 23	063	20 58	092	49 51	166	31 12	214	39 32	301
9	41 28	352	42 07	029	21 14	063	21 54	091	50 04	167	30 40	214	38 43	301
10	41 20	351	42 34	027	22 04	062	22 51	091	50 17	168	30 08	215	37 54	300
11	41 10	350	42 59	026	22 54	062	23 48	091	50 28	169	29 36	215	37 04	299
12	40 59	348	43 24	025	23 44	061	24 44	090	50 39	169	29 04	215	36 15	298
13	40 47	347	43 48	024	24 34	061	25 41	090	50 49	170	28 31	215	35 25	298
14	40 34	346	44 10	023	25 23	060	26 38	090	50 58	171	27 59	215	34 34	297
	◆Hamal		ALDEBARAN		RIGEL		◆CANOPUS		ACHERNAR		◆FOMALHAUT		Alpheratz	
15	44 32	022	26 12	060	27 35	089	20 26	140	51 07	172	60 13	243	40 20	345
16	44 52	021	27 01	059	28 31	089	21 02	140	51 14	173	59 22	244	40 05	344
17	45 12	019	27 50	059	29 28	089	21 38	140	51 21	174	58 32	244	39 49	343
18	45 30	018	28 38	058	30 25	088	22 14	140	51 27	174	57 41	244	39 32	342
19	45 47	017	29 26	057	31 22	088	22 51	140	51 32	175	56 49	245	39 13	341
20	46 03	016	30 14	057	32 18	088	23 27	140	51 36	176	55 58	245	38 54	340
21	46 17	014	31 01	056	33 15	087	24 03	140	51 39	177	55 07	245	38 34	339
22	46 31	013	31 48	055	34 12	087	24 40	140	51 42	178	54 15	245	38 13	338
23	46 43	012	32 34	055	35 08	087	25 16	140	51 43	179	53 24	246	37 51	337
24	46 54	010	33 20	054	36 05	086	25 53	140	51 44	180	52 32	246	37 28	336
25	47 03	009	34 06	053	37 01	086	26 29	140	51 44	181	51 40	246	37 04	335
26	47 12	008	34 52	053	37 58	086	27 05	140	51 43	181	50 48	246	36 40	334
27	47 19	006	35 36	052	38 55	085	27 41	140	51 41	182	49 56	246	36 14	333
28	47 24	005	36 21	051	39 51	085	28 18	140	51 39	183	49 04	246	35 48	332
29	47 29	004	37 05	050	40 48	084	28 54	140	51 35	184	48 13	246	35 20	331
	◆Hamal		ALDEBARAN		◆SIRIUS		CANOPUS		ACHERNAR		◆FOMALHAUT		Alpheratz	
30	47 32	002	37 48	050	22 39	101	29 30	141	51 31	185	47 21	247	34 53	330
31	47 33	001	38 31	049	23 34	100	30 06	141	51 25	186	46 28	247	34 24	329
32	47 34	000	39 14	048	24 30	100	30 42	141	51 19	187	45 36	247	33 54	328
33	47 33	358	39 56	047	25 26	100	31 18	141	51 12	187	44 44	247	33 24	327
34	47 30	357	40 37	046	26 22	100	31 54	141	51 05	188	43 52	247	32 53	327
35	47 26	356	41 18	045	27 18	099	32 29	141	50 56	189	43 00	247	32 22	326
36	47 21	354	41 58	044	28 14	099	33 05	141	50 47	190	42 08	247	31 49	325
37	47 15	353	42 37	044	29 10	099	33 40	141	50 37	191	41 16	247	31 16	324
38	47 07	352	43 16	043	30 06	099	34 16	142	50 26	192	40 24	247	30 43	323
39	46 58	350	43 54	042	31 02	098	34 51	142	50 14	192	39 31	247	30 09	323
40	46 48	349	44 31	041	31 58	098	35 26	142	50 01	193	38 39	247	29 34	322
41	46 36	348	45 08	040	32 54	098	36 01	142	49 48	194	37 47	247	28 58	321
42	46 24	346	45 43	038	33 50	098	36 36	142	49 34	195	36 55	247	28 23	320
43	46 09	345	46 18	037	34 47	098	37 10	143	49 20	195	36 03	247	27 46	320
44	45 54	344	46 52	036	35 43	097	37 44	143	49 04	196	35 11	247	27 09	319
	◆CAPELLA		BETELGEUSE		SIRIUS		◆CANOPUS		ACHERNAR		◆FOMALHAUT		Hamal	
45	18 03	024	39 28	063	36 39	097	38 19	143	48 48	197	34 19	247	45 38	342
46	18 26	024	40 18	062	37 35	097	38 52	143	48 31	198	33 27	246	45 20	341
47	18 49	023	41 08	061	38 32	097	39 26	144	48 14	198	32 35	246	45 01	340
48	19 10	022	41 58	060	39 28	096	40 00	144	47 56	199	31 43	246	44 41	339
49	19 32	022	42 47	060	40 25	096	40 33	144	47 37	200	30 51	246	44 20	338
50	19 52	021	43 36	059	41 21	096	41 06	145	47 18	200	29 59	246	43 58	336
51	20 12	020	44 24	058	42 17	096	41 38	145	46 58	201	29 07	246	43 35	335
52	20 32	020	45 12	057	43 14	095	42 11	145	46 38	201	28 15	246	43 11	334
53	20 51	019	45 59	056	44 10	095	42 43	146	46 17	202	27 24	246	42 45	333
54	21 09	018	46 46	056	45 07	095	43 15	146	45 55	203	26 32	246	42 19	332
55	21 27	018	47 33	055	46 03	095	43 46	147	45 33	203	25 40	245	41 52	331
56	21 43	017	48 19	054	47 00	094	44 17	147	45 11	204	24 49	245	41 24	330
57	22 00	016	49 04	053	47 57	094	44 48	148	44 48	204	23 57	245	40 55	329
58	22 15	016	49 49	052	48 53	094	45 18	148	44 24	205	23 06	245	40 25	328
59	22 30	015	50 34	051	49 50	094	45 48	148	44 00	205	22 14	245	39 54	327
	CAPELLA		BETELGEUSE		◆SIRIUS		CANOPUS		◆ACHERNAR		Diphda		◆Hamal	
60	22 45	014	51 17	050	50 46	093	46 17	149	43 36	206	43 31	263	39 23	326
61	22 58	014	52 00	049	51 43	093	46 46	150	43 11	206	42 35	263	38 51	325
62	23 11	013	52 42	047	52 40	093	47 15	150	42 45	207	41 39	262	38 18	324
63	23 24	012	53 23	046	53 36	093	47 43	151	42 20	207	40 43	262	37 44	323
64	23 35	011	54 04	045	54 33	092	48 10	151	41 54	208	39 46	262	37 09	322
65	23 46	011	54 44	044	55 30	092	48 37	152	41 27	208	38 50	262	36 34	321
66	23 56	010	55 22	042	56 26	092	49 04	152	41 00	208	37 54	262	35 58	320
67	24 05	009	56 00	041	57 23	091	49 30	153	40 33	209	36 58	261	35 22	319
68	24 14	008	56 37	040	58 20	091	49 55	154	40 06	209	36 02	261	34 44	319
69	24 22	008	57 12	038	59 17	091	50 20	155	39 38	210	35 06	261	34 07	318
70	24 29	007	57 47	037	60 13	091	50 44	155	39 10	210	34 10	261	33 28	317
71	24 36	006	58 20	035	61 10	090	51 07	156	38 41	210	33 14	260	32 49	316
72	24 41	005	58 52	034	62 07	090	51 30	157	38 13	211	32 18	260	32 10	316
73	24 46	005	59 23	032	63 04	090	51 52	157	37 44	211	31 22	260	31 30	315
74	24 51	004	59 52	030	64 00	089	52 14	158	37 14	211	30 26	260	30 49	314
	CAPELLA		POLLUX		◆PROCYON		Suhail		◆CANOPUS		ACHERNAR		◆Hamal	
75	24 54	003	28 19	041	43 57	062	33 08	130	52 34	159	36 45	211	30 08	313
76	24 57	002	28 56	041	44 47	061	33 52	130	52 54	160	36 15	212	29 27	313
77	24 59	002	29 33	040	45 37	061	34 35	130	53 13	161	35 45	212	28 45	312
78	25 00	001	30 09	039	46 26	060	35 19	130	53 31	162	35 15	212	28 02	311
79	25 00	000	30 45	038	47 15	059	36 02	130	53 49	163	34 45	212	27 20	311
80	25 00	359	31 20	038	48 03	058	36 45	131	54 05	163	34 14	213	26 36	310
81	24 59	359	31 54	037	48 51	057	37 28	131	54 21	164	33 43	213	25 53	309
82	24 57	358	32 28	036	49 39	056	38 11	131	54 36	165	33 13	213	25 09	309
83	24 55	357	33 01	035	50 26	055	38 54	131	54 50	166	32 42	213	24 24	308
84	24 51	356	33 33	034	51 12	054	39 37	131	55 03	167	32 10	213	23 39	308
85	24 47	355	34 05	034	51 58	053	40 19	131	55 15	168	31 39	214	22 54	307
86	24 42	355	34 36	033	52 43	052	41 02	132	55 26	169	31 08	214	22 09	306
87	24 37	354	35 06	032	53 28	051	41 44	132	55 35	170	30 36	214	21 23	306
88	24 30	353	35 36	031	54 12	050	42 26	132	55 44	171	30 04	214	20 37	305
89	24 23	352	36 05	030	54 55	049	43 08	132	55 52	172	29 32	214	19 51	305

LHA	Hc	Zn	Hc	Zn	Hc	Zn	Hc	Zn	Hc	Zn	Hc	Zn	Hc	Zn
♈	POLLUX		◆REGULUS		Suhail		◆CANOPUS		ACHERNAR		◆ALDEBARAN		CAPELLA	
90	36 33	029	21 29	068	43 50	133	55 59	174	29 01	214	48 53	328	24 15	352
91	37 00	028	22 21	068	44 32	133	56 05	175	28 29	214	48 23	327	24 07	351
92	37 26	027	23 14	067	45 13	133	56 10	176	27 56	214	47 51	326	23 58	350
93	37 52	026	24 06	067	45 54	134	56 14	177	27 24	215	47 19	325	23 48	349
94	38 16	025	24 58	066	46 35	134	56 16	178	26 52	215	46 46	324	23 37	349
95	38 40	024	25 50	066	47 16	134	56 18	179	26 20	215	46 12	322	23 25	348
96	39 03	023	26 42	065	47 57	135	56 18	180	25 48	215	45 37	321	23 13	347
97	39 24	022	27 33	065	48 37	135	56 18	181	25 15	215	45 01	320	23 00	347
98	39 45	021	28 24	064	49 17	136	56 16	182	24 43	215	44 24	319	22 47	346
99	40 05	020	29 15	064	49 56	136	56 13	183	24 10	215	43 47	318	22 33	345
100	40 24	019	30 06	063	50 35	137	56 10	184	23 38	215	43 09	317	22 18	344
101	40 42	018	30 56	062	51 14	137	56 05	185	23 05	215	42 30	316	22 02	344
102	40 59	017	31 46	062	51 53	138	55 59	187	22 33	215	41 50	315	21 46	343
103	41 15	016	32 36	061	52 31	138	55 52	188	22 01	215	41 10	314	21 29	342
104	41 29	014	33 26	061	53 08	139	55 44	189	21 28	215	40 29	314	21 12	342
	POLLUX		◆REGULUS		◆ACRUX		CANOPUS		◆RIGEL		ALDEBARAN		CAPELLA	
105	41 43	013	34 15	060	20 40	151	55 35	190	62 12	289	39 48	313	20 54	341
106	41 56	012	35 04	059	21 07	151	55 25	191	61 18	288	39 06	312	20 35	340
107	42 07	011	35 53	059	21 35	151	55 14	192	60 24	287	38 24	311	20 16	340
108	42 17	010	36 41	058	22 02	151	55 02	193	59 30	287	37 40	310	19 56	339
109	42 26	009	37 29	057	22 29	151	54 49	194	58 35	286	36 57	309	19 35	338
110	42 34	008	38 16	057	22 56	151	54 35	195	57 41	285	36 13	309	19 14	338
111	42 41	006	39 03	056	23 23	151	54 20	196	56 46	284	35 28	308	18 52	337
112	42 47	005	39 50	055	23 50	151	54 04	197	55 51	284	34 43	307	18 30	337
113	42 51	004	40 36	054	24 17	152	53 47	197	54 56	283	33 58	306	18 07	336
114	42 55	003	41 22	053	24 44	152	53 30	198	54 00	283	33 12	306	17 44	335
115	42 57	002	42 07	053	25 11	152	53 12	199	53 05	282	32 26	305	17 20	335
116	42 58	000	42 52	052	25 38	152	52 53	200	52 09	281	31 39	304	16 55	334
117	42 57	359	43 37	051	26 05	152	52 33	201	51 14	281	30 52	304	16 30	334
118	42 56	358	44 20	050	26 32	152	52 12	202	50 18	280	30 05	303	16 05	333
119	42 53	357	45 03	049	26 58	152	51 51	203	49 22	280	29 17	303	15 39	332
	◆REGULUS		SPICA		◆ACRUX		CANOPUS		◆RIGEL		BETELGEUSE		POLLUX	
120	45 46	048	11 47	098	27 25	152	51 28	203	48 26	279	49 25	308	42 49	355
121	46 28	047	12 43	098	27 52	152	51 06	204	47 30	279	48 40	307	42 44	354
122	47 09	046	13 40	097	28 18	152	50 42	205	46 34	278	47 54	306	42 38	353
123	47 49	045	14 36	097	28 44	152	50 18	206	45 38	278	47 07	305	42 31	352
124	48 29	044	15 32	097	29 10	153	49 53	206	44 41	277	46 21	304	42 22	351
125	49 08	043	16 29	096	29 37	153	49 28	207	43 45	277	45 33	303	42 12	350
126	49 46	042	17 25	096	30 02	153	49 02	208	42 49	276	44 46	302	42 02	348
127	50 23	040	18 22	096	30 28	153	48 35	208	41 52	276	43 57	301	41 50	347
128	51 00	039	19 18	095	30 54	153	48 08	209	40 56	276	43 09	301	41 36	346
129	51 35	038	20 14	095	31 19	153	47 41	209	39 59	275	42 20	300	41 22	345
130	52 10	037	21 11	095	31 45	154	47 12	210	39 03	275	41 30	299	41 07	344
131	52 43	035	22 08	095	32 10	154	46 44	211	38 06	274	40 41	298	40 51	343
132	53 15	034	23 04	094	32 35	154	46 15	211	37 10	274	39 51	298	40 33	342
133	53 47	033	24 01	094	32 59	154	45 45	212	36 13	274	39 00	297	40 15	341
134	54 17	031	24 57	094	33 24	154	45 16	212	35 17	273	38 10	296	39 56	339
	REGULUS		◆SPICA		ACRUX		◆CANOPUS		SIRIUS		BETELGEUSE		◆POLLUX	
135	54 45	030	25 54	093	33 48	155	44 45	213	57 49	269	37 19	296	39 35	338
136	55 13	028	26 51	093	34 12	155	44 15	213	56 53	268	36 27	295	39 14	337
137	55 39	027	27 47	093	34 36	155	43 43	213	55 56	268	35 36	294	38 51	336
138	56 04	025	28 44	093	35 00	156	43 12	214	54 59	268	34 44	294	38 28	335
139	56 27	024	29 41	092	35 23	156	42 40	214	54 03	268	33 52	293	38 04	334
140	56 49	022	30 37	092	35 46	156	42 08	215	53 06	267	33 00	293	37 39	333
141	57 09	020	31 34	092	36 09	156	41 36	215	52 09	267	32 07	292	37 13	332
142	57 28	018	32 31	091	36 32	157	41 03	215	51 13	267	31 15	292	36 46	331
143	57 45	017	33 27	091	36 54	157	40 30	216	50 16	266	30 22	291	36 19	330
144	58 01	015	34 24	091	37 16	157	39 57	216	49 19	266	29 29	291	35 50	330
145	58 15	013	35 21	090	37 38	158	39 24	216	48 23	266	28 36	290	35 21	329
146	58 27	011	36 18	090	37 59	158	38 50	217	47 26	266	27 42	290	34 51	328
147	58 37	009	37 14	090	38 20	158	38 16	217	46 30	265	26 49	289	34 21	327
148	58 45	008	38 11	089	38 41	159	37 42	217	45 33	265	25 55	289	33 49	326
149	58 52	006	39 08	089	39 01	159	37 07	217	44 37	265	25 01	288	33 17	325
	REGULUS		◆ARCTURUS		SPICA		◆ACRUX		CANOPUS		SIRIUS		◆PROCYON	
150	58 57	004	16 38	062	40 04	089	39 21	160	36 33	218	43 40	265	47 40	301
151	59 00	002	17 28	062	41 01	088	39 41	160	35 58	218	42 44	264	46 52	301
152	59 00	000	18 18	061	41 58	088	40 00	160	35 23	218	41 47	264	46 03	300
153	59 00	358	19 08	061	42 55	088	40 19	161	34 48	218	40 51	264	45 13	299
154	58 57	356	19 57	060	43 51	087	40 37	161	34 13	218	39 54	264	44 23	298
155	58 52	354	20 46	060	44 48	087	40 55	162	33 38	219	38 58	264	43 33	297
156	58 46	352	21 35	059	45 45	087	41 13	162	33 02	219	38 02	263	42 43	297
157	58 37	351	22 24	059	46 41	086	41 30	163	32 26	219	37 05	263	41 52	296
158	58 27	349	23 12	058	47 38	086	41 47	163	31 51	219	36 09	263	41 01	295
159	58 15	347	24 00	058	48 34	085	42 03	164	31 15	219	35 13	263	40 09	295
160	58 02	345	24 48	057	49 31	085	42 19	164	30 39	219	34 16	262	39 18	294
161	57 46	343	25 36	057	50 27	084	42 34	165	30 03	219	33 20	262	38 26	293
162	57 29	342	26 23	056	51 24	084	42 49	165	29 27	219	32 24	262	37 33	293
163	57 10	340	27 10	055	52 20	084	43 03	166	28 51	220	31 28	262	36 41	292
164	56 50	338	27 56	055	53 17	083	43 17	166	28 15	220	30 32	261	35 48	292
	Denebola		◆ARCTURUS		SPICA		◆ACRUX		SIRIUS		◆PROCYON		REGULUS	
165	54 19	021	28 42	054	54 13	083	43 31	167	29 36	261	34 55	291	56 28	337
166	54 38	019	29 28	054	55 09	082	43 43	167	28 40	261	34 02	290	56 05	335
167	54 56	017	30 14	053	56 05	082	43 55	168	27 44	261	33 09	290	55 40	333
168	55 12	016	30 59	052	57 01	081	44 07	168	26 48	260	32 16	289	55 14	332
169	55 27	014	31 43	052	57 57	080	44 18	169	25 52	260	31 22	289	54 47	330
170	55 40	012	32 28	051	58 53	080	44 29	170	24 56	260	30 28	288	54 18	329
171	55 51	011	33 11	050	59 49	079	44 39	170	24 00	260	29 34	288	53 48	327
172	56 01	009	33 55	049	60 45	079	44 48	171	23 04	259	28 40	287	53 17	326
173	56 09	007	34 37	049	61 40	078	44 57	171	22 09	259	27 46	287	52 45	325
174	56 15	006	35 20	048	62 36	077	45 05	172	21 13	259	26 52	286	52 11	323
175	56 20	004	36 02	047	63 31	077	45 13	173	20 17	259	25 57	286	51 37	322
176	56 23	002	36 43	046	64 26	076	45 20	173	19 22	258	25 03	286	51 02	321
177	56 24	000	37 24	045	65 21	075	45 26	174	18 26	258	24 08	285	50 25	320
178	56 23	359	38 04	045	66 15	074	45 32	174	17 31	258	23 13	285	49 48	318
179	56 21	357	38 43	044	67 10	073	45 37	175	16 35	258	22 18	284	49 10	317

LAT 19°S

LHA ♈	Hc	Zn	Hc	Zn	Hc	Zn	Hc	Zn	Hc	Zn	Hc	Zn	Hc	Zn
	ARCTURUS		◆ANTARES		ACRUX		◆Suhail		Alphard		REGULUS		◆Denebola	
180	39 22	043	28 10	110	45 42	176	46 32	226	51 37	280	48 31	316	56 17	355
181	40 01	042	29 03	110	45 46	176	45 51	226	50 41	280	47 51	315	56 12	353
182	40 38	041	29 56	110	45 49	177	45 09	227	49 45	279	47 11	314	56 04	352
183	41 15	040	30 49	110	45 52	178	44 28	227	48 49	279	46 30	313	55 55	350
184	41 51	039	31 43	110	45 54	178	43 46	227	47 53	278	45 48	312	55 44	348
185	42 27	038	32 36	110	45 55	179	43 04	228	46 57	278	45 06	311	55 32	347
186	43 01	037	33 30	110	45 56	180	42 22	228	46 01	277	44 23	310	55 18	345
187	43 35	036	34 23	110	45 56	180	41 40	228	45 04	277	43 39	309	55 03	343
188	44 08	035	35 16	109	45 55	181	40 58	228	44 08	276	42 55	308	54 45	342
189	44 40	034	36 10	109	45 54	182	40 15	229	43 12	276	42 10	307	54 27	340
190	45 12	033	37 04	109	45 52	182	39 33	229	42 15	276	41 25	307	54 07	339
191	45 42	032	37 57	109	45 49	183	38 50	229	41 19	275	40 39	306	53 45	337
192	46 11	031	38 51	109	45 46	184	38 07	229	40 22	275	39 53	305	53 23	336
193	46 39	029	39 44	109	45 42	184	37 24	229	39 26	274	39 06	304	52 58	334
194	47 07	028	40 38	109	45 38	185	36 41	229	38 29	274	38 19	304	52 33	333
	ARCTURUS		◆Rasalhague		ANTARES		◆ACRUX		Suhail		◆REGULUS		Denebola	
195	47 33	027	15 21	071	41 32	109	45 33	185	35 58	230	37 31	303	52 06	331
196	47 58	026	16 15	070	42 25	109	45 27	186	35 14	230	36 43	302	51 38	330
197	48 22	024	17 08	070	43 19	109	45 21	187	34 31	230	35 55	301	51 09	328
198	48 45	023	18 01	069	44 13	109	45 14	187	33 48	230	35 07	301	50 39	327
199	49 06	022	18 54	069	45 06	109	45 07	188	33 04	230	34 18	300	50 08	326
200	49 27	020	19 47	068	46 00	109	44 58	189	32 21	230	33 28	299	49 35	325
201	49 46	019	20 40	068	46 54	109	44 50	189	31 37	230	32 39	299	49 02	323
202	50 04	018	21 32	067	47 48	109	44 40	190	30 54	230	31 49	298	48 28	322
203	50 20	016	22 24	067	48 41	109	44 30	190	30 10	230	30 59	298	47 53	321
204	50 35	015	23 16	067	49 35	109	44 20	191	29 27	230	30 08	297	47 17	320
205	50 49	013	24 08	066	50 29	109	44 09	192	28 43	230	29 18	296	46 40	319
206	51 01	012	25 00	066	51 23	109	43 57	192	28 00	230	28 27	296	46 02	318
207	51 12	010	25 52	065	52 16	109	43 45	193	27 16	230	27 36	295	45 24	317
208	51 22	009	26 43	064	53 10	109	43 33	193	26 32	230	26 44	295	44 44	316
209	51 30	007	27 34	064	54 03	109	43 19	194	25 49	230	25 53	294	44 04	315
	◆ARCTURUS		Rasalhague		◆ANTARES		RIGIL KENT.		ACRUX		◆Suhail		REGULUS	
210	51 36	006	28 25	063	54 57	109	47 37	173	43 06	194	25 05	230	25 01	294
211	51 41	004	29 16	063	55 51	109	47 43	174	42 51	195	24 22	230	24 09	293
212	51 45	003	30 06	062	56 44	109	47 49	174	42 37	195	23 38	230	23 17	293
213	51 47	001	30 56	062	57 38	110	47 54	175	42 21	196	22 55	230	22 24	292
214	51 47	000	31 46	061	58 31	110	47 59	176	42 06	196	22 11	230	21 32	292
215	51 47	358	32 35	060	59 24	110	48 03	176	41 50	197	21 28	230	20 39	291
216	51 44	357	33 24	060	60 18	110	48 06	177	41 33	197	20 45	230	19 46	291
217	51 40	355	34 13	059	61 11	111	48 08	178	41 16	198	20 02	230	18 53	290
218	51 35	354	35 02	059	62 04	111	48 10	179	40 58	198	19 18	229	18 00	290
219	51 28	352	35 50	058	62 57	111	48 11	179	40 40	199	18 35	229	17 06	290
220	51 19	351	36 38	057	63 50	111	48 11	180	40 22	199	17 52	229	16 13	289
221	51 09	349	37 25	056	64 42	112	48 11	181	40 03	200	17 09	229	15 19	289
222	50 58	348	38 12	056	65 35	112	48 09	182	39 44	200	16 27	229	14 25	288
223	50 45	346	38 59	055	66 27	113	48 07	182	39 24	200	15 44	229	13 32	288
224	50 31	345	39 45	054	67 20	113	48 05	183	39 05	201	15 01	229	12 37	288
	Alphecca		◆VEGA		ALTAIR		◆Peacock		RIGIL KENT.		◆SPICA		ARCTURUS	
225	43 31	011	13 08	040	13 13	076	20 31	145	48 01	184	65 46	286	50 16	343
226	43 41	009	13 45	040	14 08	075	21 04	145	47 57	185	64 51	285	49 59	342
227	43 49	008	14 21	039	15 03	075	21 37	145	47 52	185	63 56	284	49 40	341
228	43 57	007	14 57	039	15 58	074	22 10	145	47 47	186	63 01	283	49 21	339
229	44 03	006	15 33	038	16 52	074	22 43	145	47 41	187	62 06	282	49 00	338
230	44 08	005	16 08	038	17 47	074	23 16	145	47 34	187	61 10	282	48 38	337
231	44 12	003	16 42	037	18 41	073	23 49	145	47 26	188	60 15	281	48 15	335
232	44 15	002	17 17	037	19 35	073	24 21	145	47 18	189	59 19	280	47 51	334
233	44 16	001	17 50	036	20 30	072	24 54	145	47 09	189	58 23	280	47 26	333
234	44 16	000	18 24	036	21 24	072	25 27	145	46 59	190	57 27	279	46 59	332
235	44 15	358	18 56	035	22 17	072	26 00	145	46 49	191	56 31	279	46 31	330
236	44 13	357	19 29	034	23 11	071	26 33	145	46 38	191	55 35	278	46 03	329
237	44 09	356	20 01	034	24 05	071	27 05	145	46 27	192	54 39	278	45 33	328
238	44 04	355	20 32	033	24 58	070	27 38	145	46 15	193	53 43	277	45 03	327
239	43 58	353	21 03	033	25 52	070	28 11	145	46 02	193	52 46	277	44 31	326
	◆VEGA		ALTAIR		◆Peacock		RIGIL KENT.		◆SPICA		ARCTURUS		Alphecca	
240	21 33	032	26 45	069	28 43	145	45 48	194	51 50	276	43 59	325	43 51	352
241	22 03	031	27 38	069	29 16	145	45 34	195	50 54	276	43 26	324	43 43	351
242	22 32	031	28 30	068	29 48	145	45 20	195	49 57	275	42 52	323	43 33	350
243	23 01	030	29 23	068	30 20	145	45 05	196	49 01	275	42 17	322	43 22	348
244	23 29	029	30 15	067	30 52	146	44 49	196	48 04	274	41 41	321	43 10	347
245	23 56	029	31 07	067	31 24	146	44 33	197	47 07	274	41 05	320	42 57	346
246	24 23	028	31 59	066	31 56	146	44 16	197	46 11	274	40 28	319	42 43	345
247	24 50	027	32 51	065	32 28	146	43 59	198	45 14	273	39 50	318	42 28	344
248	25 15	027	33 42	065	33 00	146	43 41	199	44 18	273	39 11	317	42 11	343
249	25 40	026	34 34	064	33 31	146	43 22	199	43 21	273	38 32	316	41 54	341
250	26 05	025	35 25	064	34 03	147	43 04	200	42 24	272	37 53	315	41 35	340
251	26 28	024	36 15	063	34 34	147	42 44	200	41 28	272	37 12	314	41 16	339
252	26 51	024	37 06	062	35 05	147	42 25	201	40 31	271	36 31	313	40 55	338
253	27 14	023	37 56	062	35 35	147	42 04	201	39 34	271	35 50	313	40 33	337
254	27 35	022	38 46	061	36 06	148	41 44	202	38 37	271	35 08	312	40 11	336
	◆VEGA		ALTAIR		◆Peacock		RIGIL KENT.		◆SPICA		ARCTURUS		Alphecca	
255	27 56	021	39 35	060	36 36	148	41 23	202	37 41	270	34 25	311	39 47	335
256	28 16	020	40 24	060	37 06	148	41 01	202	36 44	270	33 42	310	39 23	334
257	28 36	020	41 13	059	37 36	148	40 40	203	35 47	270	32 59	310	38 57	333
258	28 54	019	42 01	058	38 06	149	40 17	203	34 50	269	32 15	309	38 31	332
259	29 12	018	42 49	057	38 35	149	39 55	204	33 54	269	31 31	308	38 04	331
260	29 29	017	43 37	056	39 05	149	39 32	204	32 57	269	30 46	308	37 36	330
261	29 46	016	44 24	056	39 33	150	39 08	204	32 00	269	30 01	307	37 07	329
262	30 01	015	45 10	055	40 02	150	38 45	205	31 04	268	29 15	306	36 37	328
263	30 16	015	45 56	054	40 30	150	38 21	205	30 07	268	28 29	306	36 07	327
264	30 30	014	46 42	053	40 58	151	37 56	206	29 10	268	27 43	305	35 36	326
265	30 43	013	47 27	052	41 26	151	37 32	206	28 14	267	26 56	304	35 04	325
266	30 55	012	48 11	051	41 53	152	37 07	206	27 17	267	26 10	304	34 31	325
267	31 06	011	48 55	050	42 20	152	36 42	206	26 20	267	25 22	303	33 58	324
268	31 17	010	49 38	049	42 46	152	36 16	207	25 24	266	24 35	303	33 24	323
269	31 26	009	50 21	048	43 12	153	35 51	207	24 27	266	23 47	302	32 49	322

LHA ♈	Hc	Zn	Hc	Zn	Hc	Zn	Hc	Zn	Hc	Zn	Hc	Zn	Hc	Zn
	◆VEGA		ALTAIR		◆FOMALHAUT		Peacock		RIGIL KENT.		◆ANTARES		Alphecca	
270	31 35	008	51 02	047	22 30	115	43 38	153	35 25	207	67 48	246	32 14	321
271	31 43	008	51 43	046	23 22	115	44 03	154	34 59	208	66 56	247	31 38	320
272	31 50	007	52 23	044	24 13	115	44 28	154	34 32	208	66 04	247	31 02	320
273	31 56	006	53 03	043	25 05	115	44 52	155	34 05	208	65 11	248	30 25	319
274	32 01	005	53 41	042	25 57	114	45 16	155	33 39	208	64 19	248	29 47	318
275	32 05	004	54 19	041	26 48	114	45 39	156	33 12	209	63 26	249	29 09	317
276	32 09	003	54 55	039	27 40	114	46 02	157	32 44	209	62 33	249	28 31	317
277	32 11	002	55 30	038	28 32	114	46 25	157	32 17	209	61 40	249	27 52	316
278	32 13	001	56 05	037	29 23	114	46 46	158	31 49	209	60 47	250	27 12	315
279	32 13	000	56 38	035	30 15	114	47 08	158	31 22	209	59 54	250	26 32	315
280	32 13	359	57 10	034	31 07	114	47 28	159	30 54	210	59 00	250	25 51	314
281	32 12	358	57 41	032	31 59	114	47 48	160	30 26	210	58 07	250	25 10	313
282	32 10	357	58 10	030	32 51	114	48 08	160	29 58	210	57 14	250	24 29	313
283	32 07	356	58 38	029	33 43	114	48 27	161	29 29	210	56 20	251	23 47	312
284	32 03	356	59 04	027	34 35	113	48 45	162	29 01	210	55 27	251	23 05	312
	DENEB		◆FOMALHAUT		Peacock		RIGIL KENT.		◆ANTARES		Rasalhague		◆VEGA	
285	21 44	019	35 27	113	49 03	162	28 32	210	54 33	251	52 05	325	31 58	355
286	22 02	018	36 19	113	49 20	163	28 04	210	53 39	251	51 32	323	31 52	354
287	22 20	018	37 11	113	49 36	164	27 35	210	52 46	251	50 57	322	31 46	353
288	22 36	017	38 03	113	49 51	165	27 06	211	51 52	251	50 22	321	31 38	352
289	22 53	016	38 56	113	50 06	165	26 37	211	50 58	251	49 46	320	31 30	351
290	23 08	015	39 48	113	50 20	166	26 08	211	50 05	251	49 09	319	31 20	350
291	23 23	015	40 40	113	50 33	167	25 39	211	49 11	251	48 31	317	31 10	349
292	23 37	014	41 32	113	50 46	168	25 10	211	48 17	251	47 52	316	30 59	348
293	23 50	013	42 24	113	50 58	168	24 41	211	47 23	251	47 12	315	30 47	347
294	24 03	013	43 16	113	51 09	169	24 12	211	46 30	251	46 32	314	30 35	347
295	24 15	012	44 08	113	51 19	170	23 43	211	45 36	251	45 51	313	30 21	346
296	24 26	011	45 01	113	51 28	171	23 13	211	44 42	251	45 09	312	30 07	345
297	24 36	010	45 53	113	51 37	172	22 44	211	43 49	251	44 27	311	29 52	344
298	24 46	010	46 45	113	51 44	173	22 15	211	42 55	251	43 44	310	29 36	343
299	24 55	009	47 37	114	51 51	174	21 46	211	42 01	251	43 01	309	29 19	342
	◆DENEB		Alpheratz		◆FOMALHAUT		Peacock		◆ANTARES		Rasalhague		VEGA	
300	25 03	008	13 16	052	48 29	114	51 57	174	41 07	251	42 17	309	29 01	342
301	25 11	007	14 01	052	49 21	114	52 02	175	40 14	251	41 32	308	28 43	341
302	25 18	006	14 45	052	50 13	114	52 07	176	39 20	251	40 47	307	28 24	340
303	25 24	006	15 30	051	51 05	114	52 10	177	38 27	251	40 01	306	28 04	339
304	25 29	005	16 14	051	51 56	114	52 13	178	37 33	251	39 15	305	27 43	338
305	25 33	004	16 57	050	52 48	114	52 14	179	36 39	251	38 29	305	27 22	338
306	25 37	003	17 41	050	53 40	115	52 15	180	35 46	251	37 42	304	27 00	337
307	25 40	003	18 24	049	54 31	115	52 15	181	34 52	251	36 55	303	26 37	336
308	25 42	002	19 06	048	55 23	115	52 14	182	33 59	250	36 07	302	26 14	335
309	25 44	001	19 48	048	56 14	115	52 12	182	33 05	250	35 19	302	25 50	334
310	25 44	000	20 30	047	57 05	116	52 09	183	32 12	250	34 30	301	25 25	334
311	25 44	359	21 12	047	57 57	116	52 05	184	31 19	250	33 42	300	25 00	333
312	25 43	359	21 53	046	58 48	116	52 01	185	30 25	250	32 53	300	24 33	332
313	25 41	358	22 34	046	59 38	117	51 55	186	29 32	250	32 03	299	24 07	332
314	25 39	357	23 14	045	60 29	117	51 49	187	28 39	250	31 13	299	23 40	331
	Alpheratz		◆Diphda		ACHERNAR		Peacock		◆ANTARES		VEGA		◆DENEB	
315	23 54	044	37 16	099	27 00	145	51 42	188	27 46	249	23 12	330	25 36	356
316	24 34	044	38 13	098	27 32	145	51 34	189	26 53	249	22 43	330	25 32	356
317	25 13	043	39 09	098	28 04	146	51 25	189	26 00	249	22 14	329	25 27	355
318	25 51	042	40 05	098	28 36	146	51 15	190	25 07	249	21 45	328	25 21	354
319	26 29	042	41 01	098	29 08	146	51 05	191	24 14	249	21 15	328	25 15	353
320	27 07	041	41 57	097	29 40	146	50 53	192	23 21	249	20 44	327	25 08	352
321	27 44	040	42 54	097	30 12	146	50 41	193	22 28	248	20 13	326	25 00	352
322	28 20	040	43 50	097	30 43	146	50 29	193	21 35	248	19 41	326	24 52	351
323	28 56	039	44 46	097	31 15	146	50 15	194	20 43	248	19 09	325	24 43	350
324	29 32	038	45 43	097	31 46	146	50 01	195	19 50	248	18 36	325	24 33	349
325	30 06	037	46 39	096	32 18	147	49 46	196	18 58	248	18 03	324	24 22	349
326	30 41	037	47 35	096	32 49	147	49 30	197	18 05	248	17 30	323	24 10	348
327	31 14	036	48 32	096	33 20	147	49 13	197	17 13	247	16 56	323	23 58	347
328	31 47	035	49 28	096	33 51	147	48 56	198	16 20	247	16 21	322	23 45	346
329	32 20	034	50 25	096	34 21	147	48 38	199	15 28	247	15 46	322	23 32	346
	◆Alpheratz		Diphda		◆ACHERNAR		Peacock		◆Nunki		ALTAIR		DENEB	
330	32 51	033	51 21	095	34 52	148	48 20	199	46 55	251	47 39	307	23 17	345
331	33 22	033	52 18	095	35 22	148	48 01	200	46 01	251	46 54	307	23 02	344
332	33 52	032	53 14	095	35 52	148	47 41	201	45 07	251	46 09	306	22 47	344
333	34 22	031	54 11	095	36 22	148	47 21	201	44 13	251	45 23	305	22 30	343
334	34 51	030	55 07	094	36 52	149	47 00	202	43 20	251	44 36	305	22 13	342
335	35 19	029	56 04	094	37 21	149	46 38	203	42 26	251	43 49	304	21 56	342
336	35 46	028	57 00	094	37 50	149	46 16	203	41 32	251	43 02	303	21 37	341
337	36 12	027	57 57	094	38 19	150	45 54	204	40 38	251	42 14	302	21 18	340
338	36 38	026	58 54	094	38 48	150	45 31	204	39 45	251	41 26	301	20 59	339
339	37 03	025	59 50	093	39 16	150	45 07	205	38 51	251	40 37	301	20 39	339
340	37 27	024	60 47	093	39 44	151	44 43	205	37 57	251	39 48	300	20 18	338
341	37 50	023	61 43	093	40 12	151	44 19	206	37 04	251	38 59	299	19 57	338
342	38 12	022	62 40	093	40 40	151	43 54	206	36 10	251	38 09	299	19 35	337
343	38 33	021	63 37	092	41 07	152	43 28	207	35 17	251	37 19	298	19 12	336
344	38 53	020	64 33	092	41 33	152	43 02	207	34 23	251	36 29	297	18 49	336
	Alpheratz		◆Hamal		◆ACHERNAR		Peacock		Nunki		◆ALTAIR		DENEB	
345	39 12	019	27 43	049	42 00	153	42 36	208	33 30	250	35 38	297	18 25	335
346	39 30	018	28 26	048	42 26	153	42 10	208	32 36	250	34 47	296	18 01	334
347	39 48	017	29 08	048	42 51	153	41 43	209	31 43	250	33 56	295	17 36	334
348	40 04	016	29 50	047	43 17	154	41 15	209	30 49	250	33 05	295	17 11	333
349	40 19	015	30 31	046	43 41	154	40 48	209	29 56	250	32 13	294	16 45	333
350	40 33	014	31 12	046	44 06	155	40 20	210	29 03	250	31 21	294	16 19	332
351	40 46	013	31 52	045	44 30	155	39 51	210	28 09	250	30 29	293	15 52	332
352	40 58	012	32 32	044	44 53	156	39 23	210	27 16	250	29 37	293	15 25	331
353	41 09	011	33 11	043	45 16	156	38 54	211	26 23	249	28 44	292	14 57	330
354	41 19	009	33 50	043	45 38	157	38 24	211	25 30	249	27 52	291	14 29	330
355	41 28	008	34 28	042	46 00	158	37 55	211	24 37	249	26 59	291	14 00	329
356	41 35	007	35 05	041	46 22	158	37 25	212	23 44	249	26 06	290	13 31	329
357	41 42	006	35 42	040	46 43	159	36 55	212	22 51	249	25 13	290	13 01	328
358	41 47	005	36 18	039	47 03	159	36 25	212	21 58	249	24 19	290	12 31	328
359	41 51	004	36 54	038	47 23	160	35 55	213	21 06	248	23 26	289	12 01	327

LHA ♈	Hc	Zn	Hc	Zn	Hc	Zn	Hc	Zn	Hc	Zn	Hc	Zn	Hc	Zn
	♦Alpheratz		Hamal		ALDEBARAN		♦RIGEL		ACHERNAR		♦Peacock		Enif	
0	40 54	002	36 41	037	13 08	067	13 29	094	48 38	160	36 14	213	45 09	309
1	40 56	001	37 14	036	13 59	066	14 25	094	48 57	161	35 43	213	44 24	308
2	40 56	000	37 47	035	14 51	066	15 21	093	49 15	162	35 12	214	43 39	307
3	40 56	359	38 19	034	15 42	065	16 17	093	49 33	162	34 41	214	42 54	306
4	40 54	358	38 50	033	16 33	065	17 14	093	49 49	163	34 09	214	42 08	305
5	40 51	357	39 21	032	17 24	064	18 10	092	50 06	164	33 38	214	41 22	304
6	40 47	355	39 51	031	18 15	064	19 06	092	50 21	164	33 06	214	40 35	303
7	40 42	354	40 19	030	19 06	064	20 03	092	50 36	165	32 34	215	39 48	303
8	40 36	353	40 47	029	19 56	063	20 59	091	50 50	166	32 02	215	39 00	302
9	40 29	352	41 14	028	20 46	063	21 56	091	51 03	167	31 30	215	38 12	301
10	40 20	351	41 40	027	21 36	062	22 52	091	51 15	168	30 57	215	37 24	300
11	40 11	350	42 06	026	22 26	061	23 48	090	51 27	168	30 25	215	36 35	300
12	40 00	349	42 30	025	23 15	061	24 45	090	51 38	169	29 53	215	35 46	299
13	39 49	347	42 53	024	24 04	060	25 41	090	51 48	170	29 20	215	34 56	298
14	39 36	346	43 15	023	24 53	060	26 37	089	51 57	171	28 48	215	34 07	298
	♦Hamal		ALDEBARAN		RIGEL		♦CANOPUS		ACHERNAR		♦FOMALHAUT		Alpheratz	
15	43 36	021	25 42	059	27 34	089	21 12	140	52 06	172	60 40	245	39 22	345
16	43 56	020	26 30	059	28 30	089	21 48	140	52 14	173	59 48	245	39 07	344
17	44 15	019	27 18	058	29 27	088	22 24	140	52 20	173	58 57	245	38 51	343
18	44 33	018	28 06	058	30 23	088	23 00	140	52 26	174	58 06	246	38 35	342
19	44 49	017	28 53	057	31 19	087	23 37	140	52 32	175	57 15	246	38 17	341
20	45 05	015	29 40	056	32 16	087	24 13	140	52 36	176	56 23	246	37 58	340
21	45 19	014	30 27	056	33 12	087	24 49	140	52 39	177	55 31	246	37 38	339
22	45 32	013	31 13	055	34 08	086	25 26	140	52 42	178	54 40	247	37 17	338
23	45 44	011	31 59	054	35 04	086	26 02	140	52 43	179	53 48	247	36 56	337
24	45 55	010	32 45	054	36 01	086	26 39	140	52 44	180	52 56	247	36 33	336
25	46 04	009	33 30	053	36 57	085	27 15	140	52 44	181	52 04	247	36 10	335
26	46 12	008	34 15	052	37 53	085	27 51	140	52 43	181	51 12	247	35 46	334
27	46 19	006	34 59	051	38 49	084	28 27	140	52 41	182	50 20	247	35 21	333
28	46 25	005	35 43	051	39 45	084	29 04	140	52 39	183	49 28	247	34 55	332
29	46 29	004	36 26	050	40 41	084	29 40	140	52 35	184	48 36	247	34 28	331
	♦Hamal		ALDEBARAN		♦SIRIUS		CANOPUS		ACHERNAR		♦FOMALHAUT		Alpheratz	
30	46 32	002	37 09	049	22 49	100	30 16	140	52 30	185	47 44	248	34 00	330
31	46 33	001	37 52	048	23 45	100	30 52	140	52 25	186	46 52	248	33 32	329
32	46 34	000	38 33	047	24 40	100	31 28	140	52 19	187	46 00	248	33 03	329
33	46 33	358	39 15	047	25 36	099	32 04	140	52 12	188	45 08	248	32 33	328
34	46 30	357	39 55	046	26 32	099	32 40	141	52 04	188	44 15	248	32 03	327
35	46 27	356	40 35	045	27 27	099	33 16	141	51 55	189	43 23	248	31 32	326
36	46 22	354	41 15	044	28 23	099	33 52	141	51 46	190	42 31	248	31 00	325
37	46 15	353	41 53	043	29 19	098	34 27	141	51 35	191	41 39	248	30 28	324
38	46 08	352	42 31	042	30 15	098	35 03	141	51 24	192	40 47	248	29 55	324
39	45 59	350	43 09	041	31 10	098	35 38	141	51 12	193	39 55	248	29 21	323
40	45 49	349	43 45	040	32 06	098	36 13	141	51 00	193	39 03	248	28 47	322
41	45 38	348	44 21	039	33 02	097	36 48	142	50 46	194	38 11	247	28 12	321
42	45 25	347	44 56	038	33 58	097	37 23	142	50 32	195	37 19	247	27 36	321
43	45 11	345	45 30	037	34 54	097	37 58	142	50 17	196	36 26	247	27 00	320
44	44 57	344	46 04	036	35 50	097	38 32	142	50 02	196	35 34	247	26 24	319
	♦CAPELLA		BETELGEUSE		SIRIUS		♦CANOPUS		ACHERNAR		♦FOMALHAUT		Hamal	
45	17 08	024	39 00	062	36 46	096	39 06	143	49 45	197	34 42	247	44 40	343
46	17 31	023	39 50	061	37 42	096	39 41	143	49 29	198	33 51	247	44 23	342
47	17 53	023	40 39	060	38 38	096	40 14	143	49 11	199	32 59	247	44 05	340
48	18 15	022	41 28	060	39 34	096	40 48	144	48 53	199	32 07	247	43 45	339
49	18 36	022	42 16	059	40 31	095	41 21	144	48 34	200	31 15	247	43 25	338
50	18 56	021	43 04	058	41 27	095	41 55	144	48 14	201	30 23	247	43 03	337
51	19 16	020	43 52	057	42 23	095	42 27	145	47 54	201	29 31	246	42 40	336
52	19 35	020	44 39	056	43 19	094	43 00	145	47 34	202	28 40	246	42 16	335
53	19 54	019	45 26	056	44 15	094	43 32	145	47 12	202	27 48	246	41 52	333
54	20 12	018	46 12	055	45 12	094	44 04	146	46 51	203	26 57	246	41 26	332
55	20 29	018	46 58	054	46 08	094	44 36	146	46 28	204	26 05	246	40 59	331
56	20 46	017	47 43	053	47 04	093	45 07	147	46 05	204	25 14	246	40 32	330
57	21 02	016	48 28	052	48 00	093	45 38	147	45 42	205	24 22	246	40 03	329
58	21 18	016	49 12	051	48 57	093	46 09	147	45 18	205	23 31	245	39 34	328
59	21 32	015	49 55	050	49 53	092	46 39	148	44 54	206	22 40	245	39 04	327
	CAPELLA		BETELGEUSE		♦SIRIUS		CANOPUS		♦ACHERNAR		Diphda		♦Hamal	
60	21 47	014	50 38	049	50 49	092	47 09	148	44 29	206	43 38	264	38 33	326
61	22 00	013	51 20	048	51 46	092	47 38	149	44 04	207	42 42	264	38 01	325
62	22 13	013	52 01	046	52 42	092	48 07	150	43 39	207	41 46	263	37 29	324
63	22 25	012	52 42	045	53 38	091	48 35	150	43 13	208	40 50	263	36 56	323
64	22 36	011	53 21	044	54 35	091	49 03	151	42 47	208	39 54	263	36 22	323
65	22 47	011	54 00	043	55 31	091	49 30	151	42 20	208	38 58	263	35 47	322
66	22 57	010	54 38	041	56 28	090	49 57	152	41 53	209	38 02	262	35 12	321
67	23 06	009	55 15	040	57 24	090	50 23	153	41 26	209	37 07	262	34 36	320
68	23 15	008	55 50	039	58 20	090	50 49	153	40 58	210	36 11	262	33 59	319
69	23 23	008	56 25	037	59 17	089	51 14	154	40 30	210	35 15	262	33 22	318
70	23 30	007	56 59	036	60 13	089	51 38	155	40 02	210	34 19	261	32 44	317
71	23 36	006	57 31	034	61 09	088	52 02	155	39 33	211	33 23	261	32 06	317
72	23 42	005	58 02	033	62 06	088	52 25	156	39 04	211	32 28	261	31 27	316
73	23 47	005	58 32	031	63 02	088	52 48	157	38 35	211	31 32	261	30 47	315
74	23 51	004	59 00	029	63 58	087	53 09	158	38 06	212	30 37	260	30 07	314
	CAPELLA		POLLUX		♦PROCYON		Suhail		♦CANOPUS		ACHERNAR		♦Hamal	
75	23 54	003	27 34	041	43 29	061	33 47	130	53 30	159	37 36	212	29 27	314
76	23 57	002	28 11	040	44 18	061	34 30	130	53 50	159	37 06	212	28 46	313
77	23 59	002	28 47	040	45 07	060	35 14	130	54 10	160	36 36	213	28 05	312
78	24 00	001	29 23	039	45 56	059	35 57	130	54 28	161	36 06	213	27 23	312
79	24 00	000	29 58	038	46 44	058	36 40	130	54 46	162	35 35	213	26 40	311
80	24 00	359	30 32	037	47 31	057	37 24	130	55 03	163	35 05	213	25 58	310
81	23 59	359	31 06	037	48 19	056	38 07	130	55 19	164	34 34	213	25 14	310
82	23 57	358	31 39	036	49 05	055	38 50	130	55 34	165	34 03	213	24 31	309
83	23 55	357	32 12	035	49 51	054	39 33	130	55 48	166	33 32	214	23 47	309
84	23 51	356	32 44	034	50 37	053	40 16	131	56 01	167	33 00	214	23 03	308
85	23 47	356	33 15	033	51 22	052	40 59	131	56 13	168	32 29	214	22 18	307
86	23 43	355	33 45	032	52 06	051	41 41	131	56 24	169	31 57	214	21 33	307
87	23 37	354	34 15	031	52 50	050	42 24	131	56 35	170	31 26	214	20 48	306
88	23 31	353	34 44	031	53 33	049	43 06	131	56 44	171	30 54	214	20 02	306
89	23 24	353	35 13	030	54 15	048	43 48	132	56 52	172	30 22	214	19 16	305

LAT 20°S

LHA ♈	Hc	Zn	Hc	Zn	Hc	Zn	Hc	Zn	Hc	Zn	Hc	Zn	Hc	Zn
	POLLUX		♦REGULUS		Suhail		♦CANOPUS		ACHERNAR		♦ALDEBARAN		CAPELLA	
90	35 40	029	21 06	068	44 30	132	56 59	173	29 50	215	48 02	329	23 16	352
91	36 07	028	21 58	067	45 12	132	57 05	174	29 18	215	47 32	328	23 08	351
92	36 33	027	22 50	067	45 54	132	57 10	176	28 46	215	47 02	327	22 58	350
93	36 58	026	23 42	066	46 35	133	57 14	177	28 14	215	46 30	325	22 49	350
94	37 22	025	24 34	066	47 17	133	57 16	178	27 41	215	45 58	324	22 38	349
95	37 45	024	25 25	065	47 58	134	57 18	179	27 09	215	45 24	323	22 27	348
96	38 07	023	26 16	065	48 39	134	57 18	180	26 37	215	44 50	322	22 15	347
97	38 29	022	27 07	064	49 19	134	57 18	181	26 04	215	44 15	321	22 02	347
98	38 49	021	27 58	064	49 59	135	57 16	182	25 32	215	43 39	320	21 49	346
99	39 09	020	28 48	063	50 39	135	57 13	183	25 00	215	43 02	319	21 35	345
100	39 27	019	29 38	063	51 19	136	57 09	185	24 27	215	42 25	318	21 20	345
101	39 45	018	30 28	062	51 58	136	57 04	186	23 55	215	41 46	317	21 05	344
102	40 01	016	31 18	061	52 37	137	56 58	187	23 22	215	41 08	316	20 49	343
103	40 17	015	32 07	061	53 15	137	56 51	188	22 50	215	40 28	315	20 32	342
104	40 31	014	32 56	060	53 53	138	56 43	189	22 17	215	39 48	314	20 15	342
	POLLUX		♦REGULUS		♦ACRUX		CANOPUS		♦RIGEL		ALDEBARAN		CAPELLA	
105	40 45	013	33 45	059	21 33	151	56 34	190	61 51	291	39 07	313	19 57	341
106	40 57	012	34 33	059	22 00	151	56 24	191	60 58	290	38 26	312	19 38	340
107	41 08	011	35 21	058	22 27	151	56 12	192	60 05	289	37 44	312	19 19	340
108	41 18	010	36 09	057	22 54	151	56 00	193	59 12	288	37 01	311	18 59	339
109	41 27	009	36 56	057	23 21	151	55 47	194	58 18	287	36 19	310	18 39	339
110	41 35	007	37 43	056	23 49	151	55 33	195	57 24	287	35 35	309	18 18	338
111	41 42	006	38 29	055	24 16	151	55 18	196	56 30	286	34 51	308	17 57	337
112	41 47	005	39 15	054	24 43	151	55 02	197	55 36	285	34 07	308	17 35	337
113	41 52	004	40 01	054	25 10	151	54 45	198	54 41	285	33 22	307	17 12	336
114	41 55	003	40 46	053	25 37	151	54 27	199	53 47	284	32 37	306	16 49	335
115	41 57	001	41 31	052	26 04	151	54 08	200	52 52	283	31 51	306	16 25	335
116	41 58	000	42 15	051	26 31	151	53 49	201	51 57	283	31 05	305	16 01	334
117	41 57	359	42 58	050	26 58	152	53 29	201	51 02	282	30 19	304	15 36	334
118	41 56	358	43 41	049	27 25	152	53 08	202	50 06	281	29 32	304	15 11	333
119	41 53	357	44 24	048	27 51	152	52 46	203	49 11	281	28 45	303	14 45	333
	♦REGULUS		SPICA		♦ACRUX		CANOPUS		♦RIGEL		BETELGEUSE		POLLUX	
120	45 05	047	11 55	098	28 18	152	52 23	204	48 16	280	48 48	309	41 49	356
121	45 47	046	12 51	097	28 45	152	52 00	205	47 20	280	48 03	308	41 45	354
122	46 27	045	13 47	097	29 11	152	51 36	205	46 25	279	47 18	307	41 38	353
123	47 07	044	14 43	097	29 37	152	51 12	206	45 29	279	46 33	306	41 31	352
124	47 46	043	15 39	096	30 04	152	50 47	207	44 33	278	45 47	305	41 23	351
125	48 24	042	16 35	096	30 30	152	50 21	207	43 37	278	45 00	304	41 13	350
126	49 01	041	17 31	096	30 56	153	49 55	208	42 42	277	44 13	303	41 03	349
127	49 38	040	18 27	095	31 22	153	49 28	209	41 46	277	43 26	302	40 51	347
128	50 13	038	19 24	095	31 47	153	49 01	209	40 50	277	42 38	301	40 38	346
129	50 48	037	20 20	095	32 13	153	48 33	210	39 54	276	41 50	301	40 24	345
130	51 21	036	21 16	095	32 38	153	48 04	211	38 57	276	41 01	300	40 09	344
131	51 54	035	22 12	094	33 04	154	47 35	211	38 01	275	40 12	299	39 53	343
132	52 25	033	23 08	094	33 29	154	47 06	212	37 05	275	39 22	298	39 36	342
133	52 56	032	24 05	094	33 53	154	46 36	212	36 09	274	38 33	298	39 18	341
134	53 25	031	25 01	093	34 18	154	46 06	213	35 13	274	37 43	297	38 59	340
	REGULUS		♦SPICA		ACRUX		♦CANOPUS		SIRIUS		BETELGEUSE		♦POLLUX	
135	53 53	029	25 57	093	34 42	154	45 36	213	57 50	270	36 52	296	38 39	339
136	54 20	028	26 54	093	35 07	155	45 05	214	56 54	270	36 02	296	38 18	338
137	54 45	026	27 50	092	35 31	155	44 33	214	55 57	270	35 11	295	37 56	337
138	55 09	025	28 46	092	35 54	155	44 02	214	55 01	269	34 19	294	37 34	336
139	55 32	023	29 43	092	36 18	155	43 30	215	54 05	269	33 28	294	37 10	335
140	55 53	021	30 39	091	36 41	156	42 57	215	53 08	269	32 36	293	36 45	334
141	56 13	020	31 35	091	37 04	156	42 25	216	52 12	268	31 44	293	36 20	333
142	56 31	018	32 32	091	37 27	156	41 52	216	51 15	268	30 52	292	35 54	332
143	56 48	016	33 28	090	37 49	157	41 19	216	50 19	268	30 00	292	35 27	331
144	57 03	015	34 24	090	38 12	157	40 45	217	49 23	267	29 08	291	34 59	330
145	57 16	013	35 21	090	38 33	157	40 12	217	48 26	267	28 15	291	34 30	329
146	57 28	011	36 17	089	38 55	158	39 38	217	47 30	267	27 22	290	34 01	328
147	57 38	009	37 14	089	39 16	158	39 04	217	46 34	267	26 29	290	33 30	327
148	57 46	007	38 10	089	39 37	158	38 29	218	45 38	266	25 36	289	33 00	326
149	57 52	006	39 06	088	39 57	159	37 55	218	44 41	266	24 42	289	32 28	326
	REGULUS		♦ARCTURUS		SPICA		♦ACRUX		CANOPUS		SIRIUS		♦PROCYON	
150	57 57	004	16 10	062	40 03	088	40 18	159	37 20	218	43 45	266	47 09	302
151	58 00	002	17 00	061	40 59	087	40 37	160	36 45	218	42 49	265	46 21	301
152	58 00	000	17 49	061	41 55	087	40 57	160	36 10	219	41 53	265	45 32	301
153	58 00	358	18 38	061	42 52	087	41 16	161	35 35	219	40 57	265	44 44	300
154	57 57	356	19 27	060	43 48	086	41 34	161	35 00	219	40 00	265	43 55	299
155	57 52	355	20 16	059	44 44	086	41 52	161	34 24	219	39 04	264	43 05	298
156	57 46	353	21 04	059	45 40	085	42 10	162	33 49	219	38 08	264	42 15	298
157	57 38	351	21 53	058	46 37	085	42 27	162	33 13	219	37 12	264	41 25	297
158	57 28	349	22 41	058	47 33	085	42 44	163	32 37	219	36 16	264	40 35	296
159	57 17	347	23 28	057	48 29	084	43 01	163	32 01	220	35 20	263	39 44	295
160	57 04	346	24 15	057	49 25	084	43 17	164	31 25	220	34 24	263	38 53	295
161	56 49	344	25 02	056	50 21	083	43 32	164	30 49	220	33 28	263	38 01	294
162	56 32	342	25 49	056	51 17	083	43 47	165	30 13	220	32 32	262	37 10	293
163	56 14	340	26 35	055	52 13	082	44 01	165	29 37	220	31 36	262	36 18	293
164	55 54	339	27 21	054	53 09	082	44 15	166	29 01	220	30 41	262	35 26	292
	Denebola		♦ARCTURUS		SPICA		♦ACRUX		SIRIUS		♦PROCYON		REGULUS	
165	53 23	020	28 07	054	54 04	081	44 29	166	29 45	262	34 34	292	55 33	337
166	53 41	019	28 52	053	55 00	081	44 42	167	28 49	261	33 41	291	55 11	336
167	53 59	017	29 37	052	55 56	080	44 54	168	27 53	261	32 48	291	54 47	334
168	54 14	015	30 22	052	56 51	080	45 06	168	26 58	261	31 55	290	54 21	332
169	54 28	014	31 06	051	57 47	079	45 17	169	26 02	261	31 02	289	53 55	331
170	54 41	012	31 49	050	58 42	078	45 28	169	25 06	260	30 09	289	53 27	330
171	54 52	010	32 33	050	59 37	078	45 38	170	24 11	260	29 16	288	52 57	328
172	55 01	009	33 15	049	60 32	077	45 47	171	23 15	260	28 22	288	52 27	327
173	55 09	007	33 58	048	61 27	076	45 56	171	22 20	260	27 28	287	51 56	325
174	55 15	005	34 39	047	62 21	075	46 05	172	21 24	259	26 34	287	51 23	324
175	55 20	004	35 21	047	63 16	075	46 12	172	20 29	259	25 40	286	50 49	323
176	55 23	002	36 01	046	64 10	074	46 20	173	19 34	259	24 46	286	50 15	322
177	55 24	000	36 41	045	65 04	073	46 26	174	18 38	258	23 52	286	49 39	320
178	55 23	359	37 21	044	65 58	072	46 32	174	17 43	258	22 58	285	49 03	319
179	55 21	357	38 00	043	66 51	071	46 37	175	16 48	258	22 03	285	48 26	318

LHA ♈	Hc	Zn	Hc	Zn	Hc	Zn	Hc	Zn	Hc	Zn	Hc	Zn	Hc	Zn
	ARCTURUS		**◆ANTARES**		**ACRUX**		**◆Suhail**		**Alphard**		**REGULUS**		**◆Denebola**	
180	38 38	042	28 30	110	46 42	176	47 13	227	51 26	282	47 48	317	55 17	355
181	39 16	041	29 24	110	46 46	176	46 32	227	50 30	281	47 09	316	55 12	354
182	39 53	041	30 17	110	46 49	177	45 50	228	49 35	280	46 29	315	55 05	352
183	40 29	040	31 10	109	46 52	178	45 08	228	48 39	280	45 49	314	54 56	350
184	41 05	039	32 03	109	46 54	178	44 27	228	47 44	279	45 08	313	54 46	349
185	41 39	038	32 56	109	46 55	179	43 44	228	46 48	279	44 26	312	54 34	347
186	42 13	037	33 49	109	46 56	180	43 02	229	45 52	278	43 44	311	54 20	345
187	42 47	036	34 43	109	46 56	180	42 20	229	44 57	278	43 01	310	54 05	344
188	43 19	034	35 36	109	46 55	181	41 37	229	44 01	277	42 17	309	53 48	342
189	43 50	033	36 29	109	46 54	182	40 55	229	43 05	277	41 33	308	53 30	341
190	44 21	032	37 23	109	46 52	182	40 12	229	42 09	277	40 48	307	53 11	339
191	44 51	031	38 16	108	46 49	183	39 29	230	41 13	276	40 03	307	52 50	338
192	45 19	030	39 10	108	46 46	184	38 46	230	40 17	276	39 18	306	52 28	336
193	45 47	029	40 03	108	46 42	184	38 03	230	39 21	275	38 32	305	52 04	335
194	46 14	028	40 57	108	46 38	185	37 20	230	38 24	275	37 45	304	51 40	333
	ARCTURUS		**◆Rasalhague**		**ANTARES**		**◆ACRUX**		**Suhail**		**◆REGULUS**		**Denebola**	
195	46 39	026	15 01	070	41 51	108	46 33	186	36 36	230	36 59	303	51 13	332
196	47 04	025	15 54	070	42 44	108	46 27	186	35 53	230	36 11	303	50 46	330
197	47 27	024	16 47	069	43 38	108	46 21	187	35 10	230	35 24	302	50 18	329
198	47 49	023	17 40	069	44 32	108	46 14	187	34 26	230	34 36	301	49 48	328
199	48 10	021	18 32	069	45 25	108	46 06	188	33 43	230	33 47	301	49 18	327
200	48 30	020	19 25	068	46 19	108	45 58	189	32 59	231	32 59	300	48 46	325
201	48 49	019	20 17	068	47 13	108	45 49	189	32 16	231	32 10	299	48 14	324
202	49 06	017	21 09	067	48 06	108	45 39	190	31 32	231	31 20	299	47 40	323
203	49 22	016	22 01	067	49 00	108	45 29	191	30 49	231	30 31	298	47 06	322
204	49 37	014	22 52	066	49 54	108	45 19	191	30 05	231	29 41	298	46 30	321
205	49 51	013	23 44	066	50 48	108	45 08	192	29 21	231	28 51	297	45 54	320
206	50 03	012	24 35	065	51 41	108	44 56	192	28 38	231	28 00	296	45 17	318
207	50 13	010	25 26	065	52 35	108	44 44	193	27 54	231	27 10	296	44 40	317
208	50 22	009	26 17	064	53 29	108	44 31	193	27 11	231	26 19	295	44 01	316
209	50 30	007	27 07	063	54 22	108	44 18	194	26 27	231	25 28	295	43 22	315
	◆ARCTURUS		**Rasalhague**		**◆ANTARES**		**RIGIL KENT.**		**ACRUX**		**◆Suhail**		**REGULUS**	
210	50 37	006	27 58	063	55 16	108	48 36	173	44 04	195	25 44	230	24 36	294
211	50 42	004	28 48	062	56 10	108	48 43	173	43 49	195	25 00	230	23 45	294
212	50 45	003	29 38	062	57 03	108	48 49	174	43 34	196	24 17	230	22 53	293
213	50 47	001	30 27	061	57 57	108	48 54	175	43 19	196	23 33	230	22 01	293
214	50 47	000	31 16	061	58 51	108	48 59	176	43 03	197	22 50	230	21 09	292
215	50 47	358	32 05	060	59 44	108	49 03	176	42 47	197	22 07	230	20 17	292
216	50 44	357	32 54	059	60 38	109	49 06	177	42 30	198	21 24	230	19 24	291
217	50 40	355	33 42	059	61 31	109	49 08	178	42 13	198	20 40	230	18 32	291
218	50 35	354	34 30	058	62 24	109	49 10	179	41 55	198	19 57	230	17 39	290
219	50 28	352	35 18	057	63 18	109	49 11	179	41 37	199	19 14	230	16 46	290
220	50 20	351	36 05	057	64 11	110	49 11	180	41 19	199	18 31	230	15 53	289
221	50 10	349	36 52	056	65 04	110	49 11	181	41 00	200	17 49	229	15 00	289
222	49 59	348	37 38	055	65 57	110	49 09	182	40 40	200	17 06	229	14 06	289
223	49 47	347	38 24	054	66 50	111	49 07	182	40 21	201	16 23	229	13 13	288
224	49 33	345	39 10	053	67 42	111	49 05	183	40 01	201	15 41	229	12 19	288
	Alphecca		**◆VEGA**		**ALTAIR**		**◆Peacock**		**RIGIL KENT.**		**◆SPICA**		**ARCTURUS**	
225	42 32	010	12 23	040	12 58	075	21 20	144	49 01	184	65 29	288	49 18	344
226	42 41	009	12 59	040	13 53	075	21 53	144	48 57	185	64 35	287	49 02	342
227	42 50	008	13 35	039	14 47	075	22 26	144	48 52	185	63 41	286	48 44	341
228	42 57	007	14 11	039	15 42	074	22 59	144	48 47	186	62 47	285	48 25	340
229	43 03	006	14 46	038	16 36	074	23 32	144	48 40	187	61 52	284	48 05	338
230	43 08	004	15 20	038	17 30	073	24 05	144	48 33	187	60 57	283	47 43	337
231	43 12	003	15 55	037	18 24	073	24 37	144	48 26	188	60 03	283	47 21	336
232	43 15	002	16 28	037	19 18	072	25 10	144	48 17	189	59 07	282	46 57	334
233	43 16	001	17 02	036	20 11	072	25 43	144	48 08	190	58 12	281	46 32	333
234	43 16	000	17 35	035	21 05	072	26 16	144	47 58	190	57 17	281	46 06	332
235	43 15	358	18 07	035	21 58	071	26 49	144	47 48	191	56 21	280	45 39	331
236	43 13	357	18 39	034	22 52	071	27 22	144	47 37	192	55 26	280	45 11	330
237	43 09	356	19 11	034	23 45	070	27 54	145	47 25	192	54 30	279	44 42	329
238	43 05	355	19 42	033	24 38	070	28 27	145	47 13	193	53 35	278	44 12	327
239	42 59	353	20 12	032	25 30	069	29 00	145	47 00	194	52 39	278	43 42	326
	◆VEGA		**ALTAIR**		**◆Peacock**		**RIGIL KENT.**		**◆SPICA**		**ARCTURUS**		**Alphecca**	
240	20 42	032	26 23	069	29 32	145	46 47	194	51 43	277	43 10	325	42 52	352
241	21 12	031	27 15	068	30 05	145	46 32	195	50 47	277	42 37	324	42 43	351
242	21 41	030	28 08	068	30 37	145	46 18	195	49 51	276	42 04	323	42 34	350
243	22 09	030	29 00	067	31 10	145	46 02	196	48 55	276	41 30	322	42 24	349
244	22 37	029	29 52	067	31 42	145	45 46	197	47 59	276	40 55	321	42 12	347
245	23 04	028	30 43	066	32 14	145	45 30	197	47 03	275	40 19	320	41 59	346
246	23 30	028	31 35	065	32 46	145	45 13	198	46 06	275	39 42	319	41 45	345
247	23 56	027	32 26	065	33 18	146	44 56	198	45 10	274	39 05	318	41 30	344
248	24 21	026	33 17	064	33 49	146	44 38	199	44 14	274	38 27	317	41 14	343
249	24 46	026	34 07	064	34 21	146	44 19	199	43 18	273	37 49	317	40 57	342
250	25 10	025	34 58	063	34 53	146	44 00	200	42 21	273	37 10	316	40 39	341
251	25 34	024	35 48	062	35 24	146	43 41	200	41 25	273	36 30	315	40 19	340
252	25 56	023	36 38	062	35 55	147	43 21	201	40 29	272	35 50	314	39 59	338
253	26 18	023	37 27	061	36 26	147	43 00	201	39 32	272	35 09	313	39 38	337
254	26 39	022	38 16	060	36 56	147	42 40	202	38 36	272	34 28	312	39 16	336
	◆VEGA		**ALTAIR**		**◆Peacock**		**RIGIL KENT.**		**◆SPICA**		**ARCTURUS**		**Alphecca**	
255	27 00	021	39 05	060	37 27	147	42 18	202	37 40	271	33 46	312	38 53	335
256	27 20	020	39 53	059	37 57	148	41 57	203	36 43	271	33 03	311	38 29	334
257	27 39	019	40 41	058	38 27	148	41 35	203	35 47	271	32 20	310	38 04	333
258	27 57	019	41 29	057	38 57	148	41 12	204	34 51	270	31 37	309	37 38	332
259	28 15	018	42 16	056	39 27	149	40 50	204	33 54	270	30 53	309	37 11	331
260	28 32	017	43 03	056	39 56	149	40 26	204	32 58	269	30 09	308	36 44	330
261	28 48	016	43 49	055	40 25	149	40 03	205	32 01	269	29 25	307	36 15	329
262	29 03	015	44 35	054	40 54	150	39 39	205	31 05	269	28 40	307	35 46	328
263	29 18	014	45 20	053	41 22	150	39 15	206	30 09	268	27 54	306	35 16	328
264	29 31	014	46 05	052	41 50	150	38 51	206	29 12	268	27 08	305	34 46	327
265	29 44	013	46 49	051	42 18	151	38 26	206	28 16	268	26 22	305	34 14	326
266	29 56	012	47 33	050	42 45	151	38 01	207	27 20	268	25 36	304	33 42	325
267	30 07	011	48 16	049	43 12	152	37 35	207	26 23	267	24 49	304	33 10	324
268	30 18	010	48 58	048	43 39	152	37 10	207	25 27	267	24 02	303	32 36	323
269	30 27	009	49 40	047	44 05	152	36 44	207	24 31	267	23 15	303	32 02	322

LAT 20°S

LHA ♈	Hc	Zn	Hc	Zn	Hc	Zn	Hc	Zn	Hc	Zn	Hc	Zn	Hc	Zn
	◆VEGA		**ALTAIR**		**◆FOMALHAUT**		**Peacock**		**RIGIL KENT.**		**◆ANTARES**		**Alphecca**	
270	30 36	008	50 21	046	22 56	115	44 31	153	36 18	208	68 11	249	31 27	322
271	30 43	007	51 01	045	23 47	115	44 57	153	35 52	208	67 19	249	30 52	321
272	30 50	007	51 40	044	24 38	114	45 22	154	35 25	208	66 26	250	30 16	320
273	30 56	006	52 19	042	25 30	114	45 46	154	34 58	208	65 33	250	29 40	319
274	31 01	005	52 56	041	26 21	114	46 10	155	34 31	209	64 40	250	29 03	319
275	31 05	004	53 33	040	27 13	114	46 34	156	34 04	209	63 47	251	28 25	318
276	31 09	003	54 08	038	28 04	114	46 57	156	33 37	209	62 54	251	27 47	317
277	31 11	002	54 43	037	28 56	114	47 20	157	33 09	209	62 00	251	27 08	316
278	31 13	001	55 16	036	29 48	113	47 42	157	32 42	209	61 07	251	26 29	316
279	31 13	000	55 49	034	30 39	113	48 03	158	32 14	210	60 14	251	25 50	315
280	31 13	359	56 20	033	31 31	113	48 24	159	31 46	210	59 20	252	25 10	314
281	31 12	358	56 49	031	32 23	113	48 45	159	31 18	210	58 26	252	24 29	314
282	31 10	357	57 18	030	33 15	113	49 04	160	30 49	210	57 33	252	23 48	313
283	31 07	357	57 45	028	34 07	113	49 23	161	30 21	210	56 39	252	23 07	313
284	31 03	356	58 11	026	34 59	113	49 42	161	29 53	210	55 46	252	22 25	312
	DENEB		**◆FOMALHAUT**		**Peacock**		**RIGIL KENT.**		**◆ANTARES**		**Rasalhague**		**◆VEGA**	
285	20 47	019	35 51	113	50 00	162	29 24	211	54 52	252	51 16	325	30 58	355
286	21 05	018	36 43	113	50 17	163	28 55	211	53 58	252	50 43	324	30 53	354
287	21 22	017	37 35	113	50 33	163	28 27	211	53 05	252	50 10	323	30 46	353
288	21 39	017	38 27	113	50 49	164	27 58	211	52 11	252	49 35	322	30 39	352
289	21 55	016	39 19	112	51 04	165	27 29	211	51 17	252	49 00	320	30 30	351
290	22 10	015	40 11	112	51 18	166	27 00	211	50 23	252	48 23	319	30 21	350
291	22 25	015	41 03	112	51 32	167	26 31	211	49 30	252	47 46	318	30 11	349
292	22 38	014	41 55	112	51 45	167	26 02	211	48 36	252	47 08	317	30 00	348
293	22 52	013	42 47	112	51 57	168	25 32	211	47 42	252	46 29	316	29 49	348
294	23 04	012	43 40	112	52 08	169	25 03	211	46 48	252	45 50	315	29 36	347
295	23 16	012	44 32	112	52 18	170	24 34	211	45 55	252	45 10	314	29 23	346
296	23 27	011	45 24	112	52 28	171	24 05	211	45 01	252	44 29	313	29 09	345
297	23 37	010	46 16	112	52 36	172	23 36	211	44 07	252	43 47	312	28 54	344
298	23 47	009	47 08	112	52 44	172	23 06	211	43 14	252	43 05	311	28 38	343
299	23 56	009	48 00	112	52 51	173	22 37	211	42 20	252	42 22	310	28 22	343
	◆DENEB		**Alpheratz**		**◆FOMALHAUT**		**Peacock**		**◆ANTARES**		**Rasalhague**		**VEGA**	
300	24 04	008	12 39	052	48 52	113	52 57	174	41 27	252	41 39	309	28 04	342
301	24 11	007	13 24	052	49 44	113	53 02	175	40 33	252	40 55	308	27 46	341
302	24 18	006	14 08	051	50 36	113	53 07	176	39 39	252	40 11	308	27 27	340
303	24 24	006	14 52	051	51 28	113	53 10	177	38 46	252	39 26	307	27 08	339
304	24 29	005	15 35	050	52 20	113	53 12	178	37 52	252	38 40	306	26 47	339
305	24 34	004	16 19	050	53 12	113	53 14	179	36 59	251	37 54	305	26 26	338
306	24 37	003	17 01	049	54 04	113	53 15	180	36 05	251	37 08	304	26 05	337
307	24 40	003	17 44	049	54 56	113	53 15	181	35 12	251	36 22	304	25 42	336
308	24 42	002	18 26	048	55 47	114	53 14	182	34 19	251	35 34	303	25 19	335
309	24 44	001	19 08	048	56 39	114	53 12	182	33 25	251	34 47	302	24 55	335
310	24 44	000	19 50	047	57 31	114	53 09	183	32 32	251	33 59	302	24 31	334
311	24 44	359	20 31	047	58 22	114	53 05	184	31 39	251	33 11	301	24 06	333
312	24 43	359	21 11	046	59 13	115	53 00	185	30 46	250	32 22	300	23 40	333
313	24 42	358	21 52	045	60 04	115	52 55	186	29 53	250	31 34	300	23 14	332
314	24 39	357	22 32	045	60 55	115	52 48	187	29 00	250	30 45	299	22 47	331
	Alpheratz		**◆Diphda**		**ACHERNAR**		**Peacock**		**◆ANTARES**		**VEGA**		**◆DENEB**	
315	23 11	044	37 25	098	27 49	145	52 41	188	28 07	250	22 20	330	24 36	356
316	23 50	043	38 21	098	28 21	145	52 33	189	27 14	250	21 51	330	24 32	356
317	24 29	043	39 17	097	28 53	145	52 24	190	26 21	250	21 23	329	24 27	355
318	25 07	042	40 13	097	29 26	145	52 14	190	25 28	249	20 54	328	24 22	354
319	25 44	041	41 09	097	29 58	145	52 04	191	24 35	249	20 24	328	24 16	353
320	26 21	041	42 05	097	30 30	146	51 52	192	23 43	249	19 54	327	24 09	353
321	26 58	040	43 01	096	31 01	146	51 40	193	22 50	249	19 23	327	24 01	352
322	27 34	039	43 57	096	31 33	146	51 27	194	21 57	249	18 51	326	23 53	351
323	28 09	039	44 53	096	32 05	146	51 13	195	21 05	248	18 20	325	23 43	350
324	28 44	038	45 49	096	32 36	146	50 59	195	20 12	248	17 47	325	23 34	350
325	29 19	037	46 45	095	33 08	146	50 43	196	19 20	248	17 15	324	23 23	349
326	29 52	036	47 41	095	33 39	146	50 27	197	18 28	248	16 41	324	23 12	348
327	30 26	036	48 37	095	34 10	147	50 11	198	17 36	248	16 08	323	23 00	347
328	30 58	035	49 34	095	34 41	147	49 53	198	16 44	247	15 34	322	22 47	347
329	31 30	034	50 30	094	35 12	147	49 35	199	15 52	247	14 59	322	22 33	346
	◆Alpheratz		**Diphda**		**◆ACHERNAR**		**Peacock**		**◆Nunki**		**ALTAIR**		**DENEB**	
330	32 01	033	51 26	094	35 42	147	49 16	200	47 13	252	47 01	309	22 19	345
331	32 32	032	52 22	094	36 13	147	48 57	200	46 19	252	46 17	308	22 05	344
332	33 01	031	53 19	094	36 43	148	48 37	201	45 26	252	45 33	307	21 49	344
333	33 30	031	54 15	093	37 13	148	48 17	202	44 32	252	44 47	306	21 33	343
334	33 59	030	55 11	093	37 43	148	47 55	202	43 38	252	44 02	305	21 16	342
335	34 26	029	56 07	093	38 13	149	47 34	203	42 45	252	43 16	305	20 59	342
336	34 53	028	57 04	092	38 42	149	47 11	204	41 51	252	42 29	304	20 41	341
337	35 19	027	58 00	092	39 11	149	46 49	204	40 57	252	41 42	303	20 22	340
338	35 44	026	58 56	092	39 40	149	46 25	205	40 04	252	40 54	302	20 03	340
339	36 08	025	59 53	092	40 08	150	46 02	205	39 10	252	40 06	301	19 43	339
340	36 32	024	60 49	091	40 37	150	45 37	206	38 16	252	39 18	301	19 22	338
341	36 54	023	61 46	091	41 04	150	45 13	206	37 23	252	38 29	300	19 01	338
342	37 16	022	62 42	091	41 32	151	44 47	207	36 29	251	37 40	299	18 39	337
343	37 37	021	63 38	090	41 59	151	44 22	207	35 36	251	36 51	299	18 17	336
344	37 57	020	64 35	090	42 26	152	43 56	208	34 43	251	36 01	298	17 54	336
	Alpheratz		**◆Hamal**		**◆ACHERNAR**		**Peacock**		**Nunki**		**◆ALTAIR**		**DENEB**	
345	38 15	019	27 04	049	42 53	152	43 29	208	33 49	251	35 11	297	17 31	335
346	38 33	018	27 46	048	43 19	153	43 02	209	32 56	251	34 21	297	17 07	335
347	38 50	017	28 28	047	43 45	153	42 35	209	32 03	251	33 30	296	16 42	334
348	39 06	016	29 09	047	44 10	153	42 08	209	31 09	251	32 40	295	16 17	333
349	39 21	015	29 49	046	44 35	154	41 40	210	30 16	251	31 48	295	15 52	333
350	39 35	014	30 30	045	45 00	154	41 12	210	29 23	250	30 57	294	15 26	332
351	39 48	013	31 09	044	45 24	155	40 43	211	28 30	250	30 06	294	14 59	332
352	39 59	011	31 48	044	45 48	155	40 14	211	27 37	250	29 14	293	14 32	331
353	40 10	010	32 27	043	46 11	156	39 45	211	26 44	250	28 22	292	14 05	331
354	40 20	009	33 05	042	46 34	157	39 16	212	25 51	250	27 30	292	13 37	330
355	40 28	008	33 43	041	46 56	157	38 46	212	24 58	249	26 37	291	13 08	329
356	40 36	007	34 20	040	47 17	158	38 16	212	24 06	249	25 45	291	12 39	329
357	40 42	006	34 56	040	47 38	158	37 46	212	23 13	249	24 52	290	12 10	328
358	40 47	005	35 31	039	47 59	159	37 16	213	22 20	249	23 59	290	11 40	328
359	40 51	004	36 06	038	48 19	160	36 45	213	21 28	249	23 06	289	11 10	327

LHA	Hc	Zn	Hc	Zn	Hc	Zn	Hc	Zn	Hc	Zn	Hc	Zn	Hc	Zn
♈	♦Alpheratz		Hamal		♦RIGEL		CANOPUS		ACHERNAR		♦Peacock		Enif	
0	39 54	002	35 53	037	13 33	094	13 05	142	49 35	160	37 04	214	44 31	309
1	39 56	001	36 26	036	14 29	093	13 39	142	49 54	161	36 33	214	43 47	308
2	39 56	000	36 58	035	15 25	093	14 14	141	50 12	161	36 02	214	43 03	308
3	39 56	359	37 29	034	16 20	093	14 49	141	50 30	162	35 31	214	42 19	307
4	39 54	358	38 00	033	17 16	092	15 24	141	50 47	163	34 59	214	41 33	306
5	39 51	357	38 30	032	18 12	092	16 00	141	51 03	163	34 27	215	40 48	305
6	39 47	355	38 59	031	19 08	092	16 35	141	51 19	164	33 55	215	40 02	304
7	39 43	354	39 27	030	20 04	091	17 11	141	51 34	165	33 23	215	39 15	303
8	39 36	353	39 55	029	21 00	091	17 46	141	51 48	166	32 51	215	38 28	303
9	39 29	352	40 21	028	21 56	091	18 22	140	52 01	167	32 19	215	37 41	302
10	39 21	351	40 47	027	22 52	090	18 58	140	52 14	167	31 47	215	36 53	301
11	39 12	350	41 12	026	23 48	090	19 33	140	52 26	168	31 14	215	36 05	300
12	39 01	349	41 35	024	24 44	090	20 09	140	52 37	169	30 42	216	35 16	300
13	38 50	348	41 58	023	25 40	089	20 45	140	52 47	170	30 09	216	34 28	299
14	38 37	347	42 20	022	26 36	089	21 21	140	52 57	171	29 36	216	33 38	298
	♦Hamal		ALDEBARAN		RIGEL		♦CANOPUS		ACHERNAR		♦FOMALHAUT		Alpheratz	
15	42 40	021	25 11	059	27 32	088	21 58	140	53 05	172	61 04	246	38 24	346
16	43 00	020	25 59	058	28 28	088	22 34	140	53 13	172	60 13	247	38 09	344
17	43 18	019	26 46	058	29 24	088	23 10	140	53 20	173	59 22	247	37 54	343
18	43 36	017	27 34	057	30 20	087	23 46	140	53 26	174	58 30	247	37 37	342
19	43 52	016	28 20	056	31 16	087	24 23	140	53 31	175	57 38	247	37 20	341
20	44 07	015	29 07	056	32 12	086	24 59	140	53 36	176	56 46	248	37 01	340
21	44 21	014	29 53	055	33 08	086	25 35	140	53 39	177	55 55	248	36 42	339
22	44 34	013	30 39	054	34 04	086	26 12	140	53 42	178	55 03	248	36 22	338
23	44 45	011	31 24	054	35 00	085	26 48	140	53 43	179	54 11	248	36 01	337
24	44 56	010	32 09	053	35 56	085	27 24	140	53 44	180	53 19	248	35 38	336
25	45 05	009	32 54	052	36 51	084	28 01	140	53 44	181	52 27	248	35 15	335
26	45 13	007	33 38	052	37 47	084	28 37	140	53 43	181	51 35	248	34 52	334
27	45 19	006	34 22	051	38 43	084	29 13	140	53 41	182	50 43	248	34 27	333
28	45 25	005	35 05	050	39 38	083	29 50	140	53 38	183	49 51	248	34 02	332
29	45 29	004	35 48	049	40 34	083	30 26	140	53 35	184	48 58	249	33 35	332
	♦Hamal		ALDEBARAN		♦SIRIUS		CANOPUS		ACHERNAR		♦FOMALHAUT		Alpheratz	
30	45 32	002	36 30	049	23 00	100	31 02	140	53 30	185	48 06	249	33 08	331
31	45 33	001	37 11	048	23 55	100	31 38	140	53 25	186	47 14	249	32 40	330
32	45 34	000	37 53	047	24 50	099	32 14	140	53 19	187	46 22	249	32 12	329
33	45 33	358	38 33	046	25 46	099	32 50	140	53 11	188	45 30	249	31 43	328
34	45 30	357	39 13	045	26 41	099	33 26	140	53 03	189	44 38	249	31 13	327
35	45 27	356	39 53	044	27 36	098	34 02	140	52 55	190	43 46	249	30 42	326
36	45 22	354	40 31	043	28 32	098	34 38	140	52 45	190	42 54	248	30 11	326
37	45 16	353	41 09	042	29 27	098	35 14	141	52 34	191	42 01	248	29 39	325
38	45 09	352	41 47	041	30 23	098	35 49	141	52 23	192	41 09	248	29 06	324
39	45 00	351	42 23	040	31 18	097	36 25	141	52 11	193	40 17	248	28 33	323
40	44 50	349	42 59	039	32 14	097	37 00	141	51 58	194	39 25	248	27 59	322
41	44 39	348	43 34	038	33 10	097	37 35	141	51 45	194	38 33	248	27 25	322
42	44 27	347	44 09	037	34 05	096	38 10	141	51 30	195	37 41	248	26 50	321
43	44 13	346	44 42	036	35 01	096	38 45	142	51 15	196	36 49	248	26 14	320
44	43 59	344	45 15	035	35 57	096	39 19	142	50 59	197	35 57	248	25 38	320
	♦CAPELLA		BETELGEUSE		SIRIUS		♦CANOPUS		ACHERNAR		♦FOMALHAUT		Hamal	
45	16 14	024	38 31	061	36 52	096	39 54	142	50 43	198	35 05	248	43 43	343
46	16 36	023	39 20	060	37 48	095	40 28	142	50 26	198	34 14	248	43 26	342
47	16 58	023	40 09	060	38 44	095	41 02	143	50 08	199	33 22	248	43 08	341
48	17 19	022	40 57	059	39 40	095	41 36	143	49 49	200	32 30	247	42 49	339
49	17 40	021	41 45	058	40 36	094	42 10	143	49 30	200	31 38	247	42 29	338
50	18 00	021	42 32	057	41 31	094	42 43	144	49 10	201	30 47	247	42 08	337
51	18 20	020	43 19	057	42 27	094	43 16	144	48 50	202	29 55	247	41 45	336
52	18 39	019	44 06	056	43 23	093	43 49	144	48 29	202	29 04	247	41 22	335
53	18 57	019	44 52	055	44 19	093	44 21	145	48 08	203	28 12	247	40 58	334
54	19 15	018	45 37	054	45 15	093	44 54	145	47 46	203	27 21	247	40 33	333
55	19 32	018	46 22	053	46 11	093	45 26	146	47 23	204	26 29	246	40 07	332
56	19 49	017	47 07	052	47 07	092	45 57	146	47 00	205	25 38	246	39 40	331
57	20 05	016	47 50	051	48 03	092	46 28	146	46 37	205	24 47	246	39 12	330
58	20 20	015	48 34	050	48 59	092	46 59	147	46 13	206	23 56	246	38 43	329
59	20 34	015	49 16	049	49 55	091	47 29	147	45 48	206	23 05	246	38 13	328
	CAPELLA		BETELGEUSE		♦SIRIUS		CANOPUS		♦ACHERNAR		Diphda		♦Hamal	
60	20 48	014	49 58	048	50 51	091	47 59	148	45 23	207	43 44	265	37 43	327
61	21 02	013	50 39	047	51 47	091	48 29	148	44 58	207	42 48	264	37 12	326
62	21 14	013	51 19	046	52 43	090	48 58	149	44 32	208	41 53	264	36 40	325
63	21 26	012	51 59	044	53 39	090	49 27	150	44 06	208	40 57	264	36 07	324
64	21 37	011	52 38	043	54 35	090	49 55	150	43 40	208	40 01	264	35 34	323
65	21 48	010	53 16	042	55 31	089	50 23	151	43 13	209	39 06	263	35 00	322
66	21 58	010	53 52	041	56 27	089	50 50	151	42 45	209	38 10	263	34 25	321
67	22 07	009	54 28	039	57 23	088	51 16	152	42 18	210	37 14	263	33 50	320
68	22 15	008	55 03	038	58 19	088	51 42	153	41 50	210	36 19	263	33 14	320
69	22 23	008	55 37	036	59 15	088	52 08	153	41 22	210	35 23	262	32 37	319
70	22 30	007	56 10	035	60 11	087	52 32	154	40 53	211	34 28	262	32 00	318
71	22 36	006	56 41	033	61 07	087	52 57	155	40 24	211	33 32	262	31 22	317
72	22 42	005	57 11	032	62 03	086	53 20	156	39 55	211	32 37	262	30 44	316
73	22 47	005	57 40	030	62 59	086	53 43	156	39 26	212	31 42	261	30 05	316
74	22 51	004	58 07	029	63 54	085	54 05	157	38 57	212	30 46	261	29 25	315
	♦CAPELLA		POLLUX		PROCYON		♦Suhail		CANOPUS		♦ACHERNAR		Hamal	
75	22 54	003	26 49	041	43 00	061	34 25	129	54 26	158	38 27	212	28 45	314
76	22 57	002	27 25	040	43 48	060	35 08	129	54 46	159	37 57	213	28 05	313
77	22 59	002	28 01	039	44 37	059	35 52	129	55 06	160	37 27	213	27 24	313
78	23 00	001	28 36	038	45 24	058	36 35	129	55 25	161	36 56	213	26 43	312
79	23 00	000	29 10	038	46 12	057	37 19	129	55 43	162	36 26	213	26 01	311
80	23 00	359	29 44	037	46 59	056	38 02	129	56 00	163	35 55	213	25 19	311
81	22 59	359	30 18	036	47 45	055	38 45	129	56 16	164	35 24	214	24 36	310
82	22 57	358	30 50	035	48 31	054	39 28	130	56 32	165	34 53	214	23 53	309
83	22 55	357	31 22	035	49 16	053	40 12	130	56 46	166	34 21	214	23 09	309
84	22 51	356	31 54	034	50 01	052	40 55	130	57 00	167	33 50	214	22 26	308
85	22 47	356	32 25	033	50 45	051	41 38	130	57 12	168	33 19	214	21 42	308
86	22 43	355	32 55	032	51 28	050	42 20	130	57 23	169	32 47	214	20 57	307
87	22 37	354	33 24	031	52 11	049	43 03	130	57 34	170	32 15	215	20 12	307
88	22 31	353	33 53	030	52 53	048	43 46	131	57 43	171	31 43	215	19 27	306
89	22 24	353	34 20	029	53 34	047	44 28	131	57 51	172	31 11	215	18 42	305

LHA	Hc	Zn	Hc	Zn	Hc	Zn	Hc	Zn	Hc	Zn	Hc	Zn	Hc	Zn
♈	POLLUX		♦REGULUS		ACRUX		CANOPUS		♦ACHERNAR		♦ALDEBARAN		CAPELLA	
90	34 47	028	20 44	067	15 44	152	57 59	173	30 39	215	47 10	329	22 17	352
91	35 14	027	21 35	067	16 10	152	58 05	174	30 07	215	46 41	328	22 08	351
92	35 39	026	22 27	066	16 36	152	58 10	175	29 35	215	46 11	327	21 59	350
93	36 04	026	23 18	066	17 03	152	58 13	177	29 03	215	45 40	326	21 50	350
94	36 27	025	24 09	065	17 29	152	58 16	178	28 30	215	45 09	325	21 39	349
95	36 50	024	25 00	065	17 56	152	58 18	179	27 58	215	44 36	324	21 28	348
96	37 12	023	25 50	064	18 23	151	58 18	180	27 26	215	44 02	323	21 16	347
97	37 33	022	26 41	064	18 49	151	58 18	181	26 53	215	43 28	322	21 04	347
98	37 53	021	27 31	063	19 16	151	58 16	182	26 21	215	42 53	320	20 51	346
99	38 12	019	28 21	063	19 43	151	58 13	183	25 49	215	42 17	319	20 37	345
100	38 30	018	29 10	062	20 10	151	58 09	185	25 16	215	41 40	318	20 22	345
101	38 48	017	30 00	061	20 37	151	58 04	186	24 44	215	41 02	318	20 07	344
102	39 04	016	30 49	061	21 04	151	57 58	187	24 11	215	40 24	317	19 51	343
103	39 19	015	31 37	060	21 31	151	57 51	188	23 39	215	39 45	316	19 35	343
104	39 33	014	32 26	059	21 58	151	57 42	189	23 06	215	39 06	315	19 18	342
	POLLUX		♦REGULUS		Gienah		♦ACRUX		CANOPUS		♦RIGEL		BETELGEUSE	
105	39 46	013	33 14	059	16 14	103	22 25	151	57 33	190	61 29	293	57 24	329
106	39 58	012	34 02	058	17 08	103	22 52	151	57 22	191	60 37	292	56 55	327
107	40 09	011	34 49	057	18 03	102	23 20	151	57 11	192	59 45	291	56 24	326
108	40 19	010	35 36	057	18 58	102	23 47	151	56 58	193	58 52	290	55 52	324
109	40 28	008	36 23	056	19 53	102	24 14	151	56 45	194	57 59	289	55 19	323
110	40 35	007	37 09	055	20 47	101	24 41	151	56 31	195	57 06	288	54 44	321
111	40 42	006	37 55	054	21 42	101	25 08	151	56 15	196	56 13	287	54 09	320
112	40 47	005	38 40	054	22 37	101	25 35	151	55 59	197	55 19	287	53 32	319
113	40 52	004	39 25	053	23 32	101	26 03	151	55 42	198	54 25	286	52 55	317
114	40 55	003	40 09	052	24 27	100	26 30	151	55 24	199	53 31	285	52 17	316
115	40 57	001	40 53	051	25 23	100	26 57	151	55 05	200	52 37	285	51 37	315
116	40 58	000	41 37	050	26 18	100	27 24	151	54 45	201	51 43	284	50 57	314
117	40 57	359	42 20	049	27 13	100	27 51	151	54 24	202	50 49	283	50 17	313
118	40 56	358	43 02	048	28 08	099	28 17	151	54 03	203	49 54	283	49 35	312
119	40 53	357	43 43	048	29 03	099	28 44	151	53 41	204	48 59	282	48 53	311
	♦REGULUS		SPICA		♦ACRUX		CANOPUS		♦RIGEL		BETELGEUSE		POLLUX	
120	44 24	047	12 03	097	29 11	152	53 18	204	48 04	281	48 10	309	40 50	356
121	45 05	046	12 59	097	29 37	152	52 55	205	47 09	281	47 26	308	40 45	354
122	45 44	045	13 54	097	30 04	152	52 30	206	46 14	280	46 42	307	40 39	353
123	46 23	043	14 50	096	30 30	152	52 06	207	45 19	280	45 57	307	40 32	352
124	47 02	042	15 46	096	30 57	152	51 40	207	44 24	279	45 12	306	40 24	351
125	47 39	041	16 41	096	31 23	152	51 14	208	43 29	279	44 26	305	40 14	350
126	48 15	040	17 37	095	31 49	152	50 48	209	42 33	278	43 40	304	40 04	349
127	48 51	039	18 33	095	32 15	152	50 20	209	41 38	278	42 53	303	39 52	348
128	49 26	038	19 29	095	32 41	153	49 53	210	40 42	277	42 06	302	39 40	347
129	50 00	037	20 25	094	33 06	153	49 25	211	39 47	277	41 19	301	39 26	345
130	50 33	035	21 20	094	33 32	153	48 56	211	38 51	276	40 31	301	39 12	344
131	51 04	034	22 16	094	33 57	153	48 27	212	37 55	276	39 42	300	38 56	343
132	51 35	033	23 12	093	34 22	153	47 57	212	37 00	276	38 53	299	38 39	342
133	52 05	031	24 08	093	34 47	154	47 27	213	36 04	275	38 04	298	38 22	341
134	52 33	030	25 04	093	35 12	154	46 57	213	35 08	275	37 15	298	38 03	340
	REGULUS		♦SPICA		ACRUX		♦CANOPUS		SIRIUS		BETELGEUSE		♦POLLUX	
135	53 00	028	26 00	092	35 37	154	46 26	214	57 49	272	36 25	297	37 43	339
136	53 27	027	26 56	092	36 01	154	45 55	214	56 53	271	35 35	296	37 23	338
137	53 51	025	27 52	092	36 25	155	45 23	215	55 57	271	34 45	296	37 01	337
138	54 15	024	28 48	091	36 49	155	44 51	215	55 01	271	33 54	295	36 39	336
139	54 37	022	29 44	091	37 13	155	44 19	215	54 05	270	33 03	294	36 16	335
140	54 57	021	30 40	091	37 36	155	43 46	216	53 09	270	32 12	294	35 51	334
141	55 16	019	31 36	090	37 59	156	43 14	216	52 13	270	31 21	293	35 27	333
142	55 34	018	32 32	090	38 22	156	42 40	216	51 17	269	30 29	293	35 01	332
143	55 50	016	33 28	090	38 44	156	42 07	217	50 21	269	29 38	292	34 34	331
144	56 05	014	34 24	089	39 07	157	41 33	217	49 25	269	28 46	292	34 07	330
145	56 18	012	35 20	089	39 29	157	41 00	217	48 29	268	27 54	291	33 38	329
146	56 29	011	36 16	089	39 50	157	40 26	218	47 33	268	27 01	291	33 10	328
147	56 38	009	37 12	088	40 12	158	39 51	218	46 37	268	26 09	290	32 40	328
148	56 46	007	38 08	088	40 33	158	39 17	218	45 41	267	25 16	290	32 09	327
149	56 52	005	39 04	087	40 53	159	38 42	218	44 45	267	24 23	289	31 38	326
	REGULUS		♦ARCTURUS		SPICA		♦ACRUX		CANOPUS		SIRIUS		♦PROCYON	
150	56 57	004	15 42	062	40 00	087	41 14	159	38 07	219	43 49	267	46 36	303
151	57 00	002	16 31	061	40 56	087	41 33	159	37 32	219	42 53	266	45 49	302
152	57 00	000	17 20	061	41 52	086	41 53	160	36 57	219	41 57	266	45 01	302
153	57 00	358	18 09	060	42 48	086	42 12	160	36 22	219	41 01	266	44 14	301
154	56 57	356	18 57	060	43 44	085	42 31	161	35 46	219	40 06	265	43 25	300
155	56 53	355	19 45	059	44 39	085	42 49	161	35 11	219	39 10	265	42 36	299
156	56 47	353	20 33	059	45 35	084	43 07	162	34 35	220	38 14	265	41 47	298
157	56 39	351	21 21	058	46 31	084	43 25	162	33 59	220	37 18	265	40 58	298
158	56 29	349	22 08	058	47 27	084	43 42	163	33 23	220	36 22	264	40 08	297
159	56 18	348	22 56	057	48 22	083	43 58	163	32 47	220	35 27	264	39 18	296
160	56 05	346	23 42	056	49 18	083	44 14	164	32 11	220	34 31	264	38 27	295
161	55 51	344	24 29	056	50 13	082	44 30	164	31 35	220	33 35	263	37 37	295
162	55 35	343	25 15	055	51 09	082	44 45	165	30 59	220	32 40	263	36 46	294
163	55 17	341	26 01	055	52 04	081	45 00	165	30 23	220	31 44	263	35 54	293
164	54 58	339	26 46	054	52 59	080	45 14	166	29 47	220	30 49	263	35 03	293
	♦ARCTURUS		ANTARES		♦ACRUX		CANOPUS		SIRIUS		♦PROCYON		REGULUS	
165	27 31	053	15 46	113	45 27	166	29 10	220	29 53	262	34 11	292	54 38	338
166	28 16	053	16 38	113	45 40	167	28 34	220	28 58	262	33 19	292	54 16	336
167	29 01	052	17 30	112	45 53	167	27 58	220	28 02	262	32 27	291	53 53	335
168	29 44	051	18 22	112	46 05	168	27 21	220	27 07	261	31 35	291	53 28	333
169	30 28	051	19 14	112	46 16	169	26 45	220	26 11	261	30 42	290	53 02	332
170	31 11	050	20 06	112	46 27	169	26 09	220	25 16	261	29 49	289	52 35	330
171	31 54	049	20 58	111	46 37	170	25 32	220	24 21	261	28 57	289	52 06	329
172	32 36	048	21 50	111	46 47	170	24 56	220	23 26	260	28 03	288	51 37	327
173	33 17	048	22 43	111	46 56	171	24 20	220	22 30	260	27 10	288	51 06	326
174	33 58	047	23 35	111	47 04	172	23 43	220	21 35	260	26 17	287	50 34	325
175	34 39	046	24 27	110	47 12	172	23 07	220	20 40	259	25 23	287	50 01	324
176	35 19	045	25 20	110	47 19	173	22 31	220	19 45	259	24 30	286	49 28	322
177	35 59	044	26 13	110	47 26	174	21 55	220	18 50	259	23 36	286	48 53	321
178	36 38	044	27 05	110	47 32	174	21 19	220	17 55	258	22 42	285	48 17	320
179	37 16	043	27 58	110	47 37	175	20 43	220	17 00	258	21 48	285	47 41	319

LHA ♈	Hc	Zn	Hc	Zn	Hc	Zn	Hc	Zn	Hc	Zn	Hc	Zn	Hc	Zn
	ARCTURUS		◆ANTARES		ACRUX		◆Suhail		Alphard		REGULUS		◆Denebola	
180	37 54	042	28 51	109	47 41	176	47 54	228	51 13	283	47 04	318	54 18	355
181	38 31	041	29 44	109	47 45	176	47 12	228	50 18	282	46 25	317	54 12	354
182	39 07	040	30 36	109	47 49	177	46 30	228	49 23	282	45 47	316	54 05	352
183	39 43	039	31 29	109	47 51	178	45 48	229	48 28	281	45 07	314	53 57	350
184	40 17	038	32 22	109	47 54	178	45 06	229	47 33	280	44 27	313	53 47	349
185	40 52	037	33 16	109	47 55	179	44 24	229	46 38	280	43 46	313	53 35	347
186	41 25	036	34 09	108	47 56	180	43 42	229	45 43	279	43 04	312	53 22	346
187	41 58	035	35 02	108	47 56	180	42 59	230	44 48	279	42 22	311	53 07	344
188	42 29	034	35 55	108	47 55	181	42 16	230	43 52	278	41 39	310	52 51	343
189	43 00	033	36 48	108	47 54	182	41 33	230	42 57	278	40 56	309	52 34	341
190	43 30	032	37 42	108	47 52	182	40 51	230	42 01	277	40 12	308	52 15	340
191	43 59	031	38 35	108	47 49	183	40 08	230	41 06	277	39 27	307	51 55	338
192	44 27	030	39 28	108	47 46	184	39 24	230	40 10	277	38 43	306	51 33	337
193	44 54	028	40 22	107	47 42	184	38 41	231	39 15	276	37 57	306	51 10	335
194	45 20	027	41 15	107	47 38	185	37 58	231	38 19	276	37 11	305	50 46	334
	ARCTURUS		◆Rasalhague		ANTARES		◆ACRUX		Suhail		◆REGULUS		Denebola	
195	45 45	026	14 41	070	42 09	107	47 32	186	37 15	231	36 25	304	50 20	332
196	46 09	025	15 33	070	43 02	107	47 27	186	36 31	231	35 39	303	49 54	331
197	46 32	023	16 26	069	43 56	107	47 20	187	35 48	231	34 52	303	49 26	330
198	46 54	022	17 18	069	44 49	107	47 13	188	35 04	231	34 04	302	48 57	328
199	47 15	021	18 10	068	45 43	107	47 05	188	34 21	231	33 16	301	48 28	327
200	47 34	020	19 02	068	46 37	107	46 57	189	33 37	231	32 28	301	47 57	326
201	47 52	018	19 54	067	47 30	107	46 48	189	32 54	231	31 40	300	47 25	325
202	48 09	017	20 45	067	48 24	107	46 39	190	32 10	231	30 51	299	46 52	324
203	48 25	016	21 37	066	49 18	107	46 28	191	31 26	231	30 02	299	46 18	322
204	48 39	014	22 28	066	50 11	106	46 18	191	30 43	231	29 13	298	45 44	321
205	48 52	013	23 19	065	51 05	106	46 06	192	29 59	231	28 23	297	45 08	320
206	49 04	011	24 10	065	51 59	106	45 55	192	29 16	231	27 34	297	44 32	319
207	49 14	010	25 00	064	52 53	106	45 42	193	28 32	231	26 43	296	43 55	318
208	49 23	009	25 50	064	53 46	106	45 29	194	27 49	231	25 53	296	43 17	317
209	49 31	007	26 40	063	54 40	106	45 16	194	27 05	231	25 02	295	42 39	316
	ARCTURUS		◆Rasalhague		ANTARES		◆RIGIL KENT.		ACRUX		Gienah		◆Denebola	
210	49 37	006	27 30	062	55 34	106	49 36	173	45 02	195	65 08	274	42 00	315
211	49 42	004	28 20	062	56 28	106	49 42	173	44 47	195	64 12	273	41 20	314
212	49 45	003	29 09	061	57 21	107	49 48	174	44 32	196	63 16	273	40 39	313
213	49 47	001	29 58	061	58 15	107	49 54	175	44 17	196	62 20	272	39 58	312
214	49 47	000	30 47	060	59 09	107	49 59	176	44 01	197	61 24	272	39 17	312
215	49 47	358	31 35	059	60 02	107	50 02	176	43 44	197	60 28	271	38 34	311
216	49 44	357	32 23	059	60 56	107	50 06	177	43 27	198	59 32	271	37 52	310
217	49 40	355	33 11	058	61 49	107	50 08	178	43 10	198	58 36	271	37 08	309
218	49 35	354	33 58	057	62 43	107	50 10	179	42 52	199	57 40	270	36 25	308
219	49 29	353	34 45	057	63 36	107	50 11	179	42 34	199	56 44	270	35 40	307
220	49 21	351	35 32	056	64 30	108	50 11	180	42 15	200	55 48	270	34 56	307
221	49 11	350	36 18	055	65 23	108	50 11	181	41 56	200	54 52	269	34 11	306
222	49 01	348	37 04	054	66 17	108	50 09	182	41 37	201	53 56	269	33 25	305
223	48 48	347	37 49	054	67 10	108	50 07	182	41 17	201	53 00	268	32 39	305
224	48 35	345	38 34	053	68 03	109	50 05	183	40 57	201	52 04	268	31 53	304
	Alphecca		Rasalhague		◆ALTAIR		Peacock		◆RIGIL KENT.		◆SPICA		ARCTURUS	
225	41 33	010	39 18	052	12 43	075	22 09	144	50 01	184	65 10	290	48 20	344
226	41 42	009	40 02	051	13 37	075	22 42	144	49 57	185	64 17	289	48 04	343
227	41 50	008	40 46	050	14 31	074	23 15	144	49 52	185	63 24	288	47 47	341
228	41 58	007	41 28	049	15 25	074	23 47	144	49 46	186	62 30	287	47 29	340
229	42 04	006	42 11	049	16 19	073	24 20	144	49 40	187	61 37	286	47 09	339
230	42 08	004	42 52	048	17 12	073	24 53	144	49 33	188	60 43	285	46 48	337
231	42 12	003	43 33	047	18 06	073	25 26	144	49 25	188	59 48	284	46 26	336
232	42 15	002	44 14	046	18 59	072	25 59	144	49 17	189	58 54	284	46 03	335
233	42 16	001	44 53	045	19 53	072	26 32	144	49 07	190	58 00	283	45 38	334
234	42 16	000	45 33	044	20 46	071	27 05	144	48 58	190	57 05	282	45 13	333
235	42 15	358	46 11	043	21 39	071	27 38	144	48 47	191	56 10	282	44 47	331
236	42 13	357	46 48	041	22 31	070	28 10	144	48 36	192	55 15	281	44 19	330
237	42 09	356	47 25	040	23 24	070	28 43	144	48 24	193	54 20	280	43 51	329
238	42 05	355	48 01	039	24 17	069	29 16	144	48 11	193	53 25	280	43 22	328
239	41 59	354	48 36	038	25 09	069	29 49	144	47 58	194	52 30	279	42 51	327
	VEGA		◆ALTAIR		Peacock		◆RIGIL KENT.		SPICA		◆ARCTURUS		Alphecca	
240	19 51	032	26 01	068	30 21	144	47 45	194	51 34	279	42 20	326	41 52	352
241	20 20	031	26 53	068	30 54	145	47 30	195	50 39	278	41 48	325	41 44	351
242	20 49	030	27 45	067	31 26	145	47 15	196	49 44	278	41 16	324	41 35	350
243	21 17	030	28 36	067	31 59	145	47 00	196	48 48	277	40 42	323	41 25	349
244	21 44	029	29 27	066	32 31	145	46 44	197	47 52	277	40 08	322	41 13	348
245	22 11	028	30 19	065	33 03	145	46 27	198	46 57	276	39 33	321	41 01	346
246	22 37	028	31 09	065	33 35	145	46 10	198	46 01	276	38 57	320	40 47	345
247	23 03	027	32 00	064	34 07	145	45 53	199	45 05	275	38 20	319	40 32	344
248	23 28	026	32 50	064	34 39	145	45 34	199	44 09	275	37 43	318	40 17	343
249	23 52	025	33 40	063	35 11	146	45 16	200	43 14	274	37 05	317	40 00	342
250	24 16	025	34 30	062	35 42	146	44 56	200	42 18	274	36 27	316	39 42	341
251	24 39	024	35 20	062	36 14	146	44 37	201	41 22	274	35 48	315	39 23	340
252	25 01	023	36 09	061	36 45	146	44 17	201	40 26	273	35 08	315	39 03	339
253	25 23	022	36 58	060	37 16	146	43 56	202	39 30	273	34 28	314	38 43	338
254	25 44	022	37 46	060	37 47	147	43 35	202	38 34	272	33 47	313	38 21	337
	◆VEGA		ALTAIR		◆Peacock		RIGIL KENT.		◆SPICA		ARCTURUS		Alphecca	
255	26 04	021	38 34	059	38 17	147	43 14	203	37 38	272	33 06	312	37 58	336
256	26 24	020	39 22	058	38 48	147	42 52	203	36 42	272	32 24	311	37 35	335
257	26 42	019	40 09	057	39 18	148	42 30	204	35 46	271	31 42	311	37 10	334
258	27 01	018	40 56	057	39 48	148	42 07	204	34 50	271	30 59	310	36 45	333
259	27 18	018	41 43	056	40 18	148	41 44	204	33 54	270	30 16	309	36 18	332
260	27 34	017	42 29	055	40 47	148	41 21	205	32 58	270	29 32	309	35 51	331
261	27 50	016	43 14	054	41 16	149	40 57	205	32 02	270	28 48	308	35 24	330
262	28 05	015	44 00	053	41 45	149	40 33	206	31 06	269	28 04	307	34 55	329
263	28 19	014	44 44	052	42 14	149	40 09	206	30 10	269	27 19	307	34 26	328
264	28 33	013	45 28	051	42 42	150	39 44	206	29 14	269	26 34	306	33 56	327
265	28 46	013	46 11	050	43 10	150	39 20	207	28 18	268	25 48	305	33 25	326
266	28 57	012	46 54	049	43 38	151	38 54	207	27 22	268	25 02	305	32 53	325
267	29 08	011	47 36	048	44 05	151	38 29	207	26 26	268	24 16	304	32 21	324
268	29 18	010	48 18	047	44 32	152	38 03	207	25 30	267	23 29	303	31 48	324
269	29 28	009	48 59	046	44 58	152	37 37	208	24 34	267	22 42	303	31 14	323

LAT 21°S

LHA ♈	Hc	Zn	Hc	Zn	Hc	Zn	Hc	Zn	Hc	Zn	Hc	Zn	Hc	Zn
	◆VEGA		ALTAIR		◆FOMALHAUT		Peacock		RIGIL KENT.		◆ANTARES		Alphecca	
270	29 36	008	49 39	045	23 21	114	45 25	152	37 11	208	68 32	251	30 40	322
271	29 44	007	50 18	044	24 12	114	45 50	153	36 45	208	67 39	251	30 05	321
272	29 51	006	50 56	043	25 03	114	46 16	153	36 18	209	66 46	252	29 30	320
273	29 56	006	51 34	041	25 54	114	46 40	154	35 51	209	65 53	252	28 54	320
274	30 01	005	52 11	040	26 45	114	47 05	155	35 24	209	64 59	252	28 17	319
275	30 06	004	52 46	039	27 37	113	47 29	155	34 57	209	64 06	253	27 40	318
276	30 09	003	53 21	038	28 28	113	47 52	156	34 29	209	63 12	253	27 03	317
277	30 11	002	53 55	036	29 20	113	48 15	156	34 02	210	62 19	253	26 25	317
278	30 13	001	54 27	035	30 11	113	48 37	157	33 34	210	61 25	253	25 46	316
279	30 13	000	54 59	033	31 03	113	48 59	157	33 06	210	60 32	253	25 07	315
280	30 13	359	55 29	032	31 55	113	49 20	158	32 38	210	59 38	253	24 27	315
281	30 12	358	55 58	030	32 46	113	49 41	159	32 10	210	58 44	253	23 47	314
282	30 10	357	56 25	029	33 38	112	50 01	159	31 41	210	57 51	253	23 07	313
283	30 07	357	56 52	027	34 30	112	50 20	160	31 13	211	56 57	254	22 26	313
284	30 03	356	57 17	026	35 22	112	50 39	161	30 44	211	56 03	254	21 45	312
	◆DENEB		FOMALHAUT		◆ACHERNAR		RIGIL KENT.		◆ANTARES		Rasalhague		VEGA	
285	19 51	019	36 14	112	12 40	147	30 16	211	55 10	254	50 26	326	29 58	355
286	20 08	018	37 05	112	13 10	147	29 47	211	54 16	254	49 54	325	29 53	354
287	20 25	017	37 57	112	13 41	146	29 18	211	53 22	254	49 22	324	29 46	353
288	20 41	017	38 49	112	14 12	146	28 49	211	52 28	254	48 48	322	29 39	352
289	20 57	016	39 41	112	14 43	146	28 20	211	51 35	254	48 13	321	29 31	351
290	21 12	015	40 34	112	15 14	146	27 51	211	50 41	254	47 38	320	29 22	350
291	21 27	014	41 26	112	15 46	146	27 22	211	49 47	253	47 01	319	29 12	349
292	21 40	014	42 18	112	16 17	146	26 53	211	48 54	253	46 24	318	29 02	349
293	21 53	013	43 10	111	16 49	146	26 24	211	48 00	253	45 46	317	28 50	348
294	22 05	012	44 02	111	17 21	145	25 55	211	47 06	253	45 07	316	28 38	347
295	22 17	012	44 54	111	17 52	145	25 25	211	46 13	253	44 28	315	28 25	346
296	22 28	011	45 46	111	18 24	145	24 56	211	45 19	253	43 48	314	28 11	345
297	22 38	010	46 38	111	18 56	145	24 27	211	44 25	253	43 07	313	27 56	344
298	22 48	009	47 31	111	19 28	145	23 58	211	43 32	253	42 25	312	27 41	344
299	22 56	009	48 23	111	20 00	145	23 28	211	42 38	253	41 43	311	27 24	343
	DENEB		Enif		◆FOMALHAUT		ACHERNAR		◆RIGIL KENT.		ANTARES		◆VEGA	
300	23 04	008	49 57	042	49 15	111	20 32	145	22 59	211	41 45	253	27 07	342
301	23 12	007	50 34	041	50 07	112	21 05	145	22 30	211	40 51	253	26 49	341
302	23 18	006	51 11	040	50 59	112	21 37	145	22 01	211	39 58	253	26 31	340
303	23 24	006	51 46	038	51 51	112	22 09	145	21 31	211	39 04	252	26 12	339
304	23 29	005	52 20	037	52 43	112	22 42	145	21 02	211	38 11	252	25 52	339
305	23 34	004	52 53	036	53 35	112	23 14	145	20 33	211	37 18	252	25 31	338
306	23 37	003	53 26	034	54 27	112	23 46	145	20 04	211	36 24	252	25 09	337
307	23 40	003	53 57	033	55 19	112	24 19	145	19 35	211	35 31	252	24 47	336
308	23 42	002	54 27	032	56 11	112	24 51	145	19 06	211	34 38	252	24 25	336
309	23 44	001	54 55	030	57 03	112	25 24	145	18 37	211	33 45	252	24 01	335
310	23 44	000	55 23	029	57 54	113	25 56	145	18 09	211	32 52	251	23 37	334
311	23 44	359	55 49	027	58 46	113	26 29	145	17 40	211	31 59	251	23 12	333
312	23 43	359	56 14	025	59 38	113	27 01	145	17 11	211	31 06	251	22 47	333
313	23 42	358	56 37	024	60 29	113	27 33	145	16 43	211	30 13	251	22 21	332
314	23 39	357	56 59	022	61 20	114	28 06	145	16 14	210	29 20	251	21 54	331
	Alpheratz		◆Diphda		ACHERNAR		◆Peacock		ANTARES		◆ALTAIR		DENEB	
315	22 28	044	37 33	097	28 38	145	53 41	188	28 27	250	55 37	329	23 36	356
316	23 06	043	38 28	097	29 10	145	53 32	189	27 34	250	55 07	327	23 32	356
317	23 45	043	39 24	097	29 43	145	53 23	190	26 41	250	54 36	326	23 28	355
318	24 22	042	40 20	096	30 15	145	53 13	191	25 49	250	54 04	324	23 22	354
319	24 59	041	41 15	096	30 47	145	53 02	192	24 56	250	53 30	323	23 16	353
320	25 36	040	42 11	096	31 19	145	52 51	192	24 04	249	52 56	321	23 09	353
321	26 12	040	43 07	095	31 51	145	52 38	193	23 11	249	52 20	320	23 02	352
322	26 47	039	44 03	095	32 23	145	52 25	194	22 19	249	51 44	319	22 53	351
323	27 22	038	44 58	095	32 54	146	52 11	195	21 27	249	51 07	318	22 44	350
324	27 57	038	45 54	095	33 26	146	51 56	196	20 35	249	50 28	316	22 35	350
325	28 31	037	46 50	094	33 57	146	51 41	196	19 42	248	49 49	315	22 24	349
326	29 04	036	47 46	094	34 29	146	51 25	197	18 50	248	49 10	314	22 13	348
327	29 37	035	48 42	094	35 00	146	51 08	198	17 58	248	48 29	313	22 01	347
328	30 09	034	49 38	093	35 31	146	50 50	199	17 07	248	47 48	312	21 48	347
329	30 40	034	50 34	093	36 02	147	50 32	199	16 15	247	47 06	311	21 35	346
	◆Alpheratz		Diphda		◆ACHERNAR		Peacock		◆Nunki		ALTAIR		DENEB	
330	31 11	033	51 30	093	36 33	147	50 13	200	47 31	254	46 23	310	21 21	345
331	31 41	032	52 26	093	37 03	147	49 53	201	46 37	253	45 40	309	21 07	345
332	32 10	031	53 22	092	37 34	147	49 33	201	45 43	253	44 56	308	20 51	344
333	32 39	030	54 18	092	38 04	148	49 12	202	44 50	253	44 12	307	20 36	343
334	33 06	029	55 14	092	38 34	148	48 51	203	43 56	253	43 27	306	20 19	342
335	33 34	028	56 10	091	39 04	148	48 29	203	43 02	253	42 41	305	20 02	342
336	34 00	028	57 06	091	39 33	148	48 06	204	42 09	253	41 55	304	19 44	341
337	34 25	027	58 02	091	40 02	149	47 43	205	41 15	253	41 09	304	19 25	340
338	34 50	026	58 58	090	40 31	149	47 20	205	40 22	253	40 22	303	19 06	340
339	35 14	025	59 54	090	41 00	149	46 56	206	39 28	253	39 35	302	18 47	339
340	35 37	024	60 50	090	41 28	150	46 31	206	38 35	252	38 47	301	18 26	338
341	35 59	023	61 46	089	41 57	150	46 06	207	37 42	252	37 59	301	18 06	338
342	36 20	022	62 42	089	42 24	150	45 41	207	36 48	252	37 11	300	17 44	337
343	36 41	021	63 38	088	42 52	151	45 15	208	35 55	252	36 22	299	17 22	337
344	37 00	020	64 34	088	43 19	151	44 49	208	35 02	252	35 33	298	16 59	336
	Alpheratz		◆Hamal		◆ACHERNAR		Peacock		Nunki		◆ALTAIR		Enif	
345	37 19	019	26 24	048	43 46	152	44 22	209	34 08	252	34 43	298	53 56	327
346	37 36	018	27 06	048	44 12	152	43 55	209	33 15	252	33 54	297	53 25	326
347	37 53	017	27 47	047	44 38	153	43 28	209	32 22	251	33 04	297	52 52	324
348	38 08	016	28 27	046	45 04	153	43 00	210	31 29	251	32 14	296	52 19	323
349	38 23	015	29 07	045	45 29	153	42 32	210	30 36	251	31 23	295	51 45	322
350	38 37	013	29 47	045	45 54	154	42 03	211	29 43	251	30 32	295	51 09	320
351	38 49	012	30 26	044	46 18	154	41 35	211	28 50	251	29 41	294	50 33	319
352	39 01	011	31 05	043	46 42	155	41 06	211	27 57	251	28 50	294	49 56	318
353	39 11	010	31 43	042	47 06	156	40 36	212	27 05	250	27 59	293	49 18	317
354	39 20	009	32 20	042	47 29	156	40 07	212	26 12	250	27 07	292	48 39	316
355	39 29	008	32 57	041	47 51	157	39 37	212	25 19	250	26 15	292	48 00	314
356	39 36	007	33 34	040	48 13	157	39 07	213	24 27	250	25 23	291	47 19	313
357	39 42	006	34 09	039	48 34	158	38 37	213	23 34	249	24 31	291	46 38	312
358	39 47	005	34 44	038	48 55	159	38 06	213	22 42	249	23 38	290	45 56	311
359	39 51	003	35 19	037	49 15	159	37 35	213	21 49	249	22 46	290	45 14	310

LHA ♈	Hc	Zn	Hc	Zn	Hc	Zn	Hc	Zn	Hc	Zn	Hc	Zn	Hc	Zn
	◆Alpheratz		Hamal		◆RIGEL		CANOPUS		ACHERNAR		◆Peacock		Enif	
0	38 54	002	35 04	036	13 36	093	13 52	142	50 31	159	37 54	214	43 53	310
1	38 56	001	35 37	035	14 32	093	14 26	141	50 50	160	37 23	214	43 10	309
2	38 56	000	36 08	034	15 28	093	15 01	141	51 09	161	36 52	214	42 26	308
3	38 56	359	36 39	033	16 23	092	15 36	141	51 27	162	36 20	215	41 42	307
4	38 54	358	37 10	032	17 19	092	16 11	141	51 44	162	35 48	215	40 58	307
5	38 51	357	37 39	031	18 14	092	16 46	141	52 01	163	35 16	215	40 13	306
6	38 48	356	38 07	030	19 10	091	17 22	141	52 16	164	34 44	215	39 28	305
7	38 43	354	38 35	029	20 06	091	17 57	140	52 32	165	34 12	215	38 42	304
8	38 37	353	39 02	028	21 01	091	18 32	140	52 46	165	33 40	215	37 55	303
9	38 30	352	39 28	027	21 57	090	19 08	140	53 00	166	33 08	216	37 09	303
10	38 22	351	39 53	026	22 52	090	19 44	140	53 12	167	32 35	216	36 22	302
11	38 13	350	40 17	025	23 48	089	20 19	140	53 25	168	32 03	216	35 34	301
12	38 02	349	40 40	024	24 44	089	20 55	140	53 36	169	31 30	216	34 46	300
13	37 51	348	41 03	023	25 39	089	21 31	140	53 46	170	30 58	216	33 58	300
14	37 39	347	41 24	022	26 35	088	22 07	140	53 56	170	30 25	216	33 10	299
	◆Hamal		ALDEBARAN		RIGEL		◆CANOPUS		ACHERNAR		◆FOMALHAUT		Alpheratz	
15	41 44	021	24 40	058	27 31	088	22 43	140	54 05	171	61 28	248	37 26	346
16	42 03	020	25 27	058	28 26	087	23 19	139	54 13	172	60 36	248	37 12	345
17	42 21	018	26 14	057	29 22	087	23 56	139	54 20	173	59 44	249	36 56	344
18	42 38	017	27 01	057	30 17	087	24 32	139	54 26	174	58 52	249	36 40	343
19	42 54	016	27 47	056	31 13	086	25 08	139	54 31	175	58 01	249	36 23	342
20	43 09	015	28 33	055	32 08	086	25 44	139	54 36	176	57 09	249	36 05	341
21	43 23	014	29 19	055	33 04	085	26 21	139	54 39	177	56 17	249	35 46	340
22	43 35	012	30 04	054	33 59	085	26 57	139	54 42	178	55 25	249	35 26	339
23	43 47	011	30 49	053	34 55	085	27 33	139	54 43	179	54 33	249	35 05	338
24	43 57	010	31 33	053	35 50	084	28 10	139	54 44	180	53 41	249	34 43	337
25	44 06	009	32 17	052	36 45	084	28 46	139	54 44	181	52 48	250	34 21	336
26	44 13	007	33 00	051	37 41	083	29 23	139	54 43	182	51 56	250	33 57	335
27	44 20	006	33 44	050	38 36	083	29 59	139	54 41	182	51 04	250	33 33	334
28	44 25	005	34 26	050	39 31	082	30 35	139	54 38	183	50 12	250	33 08	333
29	44 29	004	35 08	049	40 26	082	31 11	139	54 35	184	49 20	250	32 42	332
	Hamal		◆ALDEBARAN		SIRIUS		◆CANOPUS		ACHERNAR		◆FOMALHAUT		Alpheratz	
30	44 32	002	35 50	048	23 10	099	31 48	139	54 30	185	48 28	250	32 16	331
31	44 33	001	36 31	047	24 05	099	32 24	139	54 24	186	47 36	250	31 48	330
32	44 34	000	37 11	046	25 00	099	33 00	140	54 18	187	46 44	250	31 20	329
33	44 33	358	37 51	045	25 55	099	33 36	140	54 11	188	45 51	250	30 52	328
34	44 30	357	38 31	045	26 50	098	34 12	140	54 03	189	44 59	249	30 22	328
35	44 27	356	39 09	044	27 45	098	34 48	140	53 54	190	44 07	249	29 52	327
36	44 22	355	39 47	043	28 40	098	35 24	140	53 44	191	43 15	249	29 21	326
37	44 16	353	40 25	042	29 35	097	36 00	140	53 33	191	42 23	249	28 50	325
38	44 09	352	41 01	041	30 30	097	36 35	140	53 22	192	41 31	249	28 17	324
39	44 01	351	41 37	040	31 26	097	37 11	140	53 09	193	40 39	249	27 45	324
40	43 51	349	42 13	039	32 21	096	37 46	141	52 56	194	39 47	249	27 11	323
41	43 40	348	42 47	038	33 16	096	38 22	141	52 43	195	38 55	249	26 37	322
42	43 28	347	43 21	037	34 12	096	38 57	141	52 28	196	38 03	249	26 03	321
43	43 15	346	43 53	036	35 07	095	39 32	141	52 13	196	37 11	249	25 28	321
44	43 01	345	44 25	034	36 02	095	40 07	141	51 57	197	36 20	249	24 52	320
	◆CAPELLA		BETELGEUSE		SIRIUS		◆CANOPUS		ACHERNAR		◆FOMALHAUT		Hamal	
45	15 19	024	38 02	060	36 58	095	40 41	142	51 40	198	35 28	248	42 46	343
46	15 41	023	38 50	060	37 53	095	41 16	142	51 23	199	34 36	248	42 29	342
47	16 03	023	39 38	059	38 49	094	41 50	142	51 04	199	33 44	248	42 11	341
48	16 24	022	40 26	058	39 44	094	42 24	142	50 46	200	32 53	248	41 53	340
49	16 44	021	41 13	057	40 40	094	42 58	143	50 26	201	32 01	248	41 33	339
50	17 04	021	42 00	057	41 35	093	43 31	143	50 06	201	31 10	248	41 12	338
51	17 23	020	42 46	056	42 31	093	44 04	143	49 46	202	30 18	248	40 51	336
52	17 42	019	43 31	055	43 26	093	44 38	144	49 25	203	29 27	247	40 28	335
53	18 00	019	44 17	054	44 22	092	45 10	144	49 03	203	28 36	247	40 04	334
54	18 18	018	45 01	053	45 18	092	45 43	145	48 41	204	27 44	247	39 39	333
55	18 35	017	45 46	052	46 13	092	46 15	145	48 18	204	26 53	247	39 14	332
56	18 51	017	46 29	051	47 09	091	46 47	145	47 55	205	26 02	247	38 47	331
57	19 07	016	47 12	050	48 04	091	47 18	146	47 31	206	25 11	246	38 20	330
58	19 22	015	47 55	049	49 00	090	47 49	146	47 07	206	24 20	246	37 52	329
59	19 36	015	48 36	048	49 56	090	48 20	147	46 42	207	23 29	246	37 23	328
	CAPELLA		BETELGEUSE		◆SIRIUS		CANOPUS		◆ACHERNAR		Diphda		◆Hamal	
60	19 50	014	49 17	047	50 51	090	48 50	147	46 17	207	43 49	266	36 53	327
61	20 03	013	49 58	046	51 47	089	49 20	148	45 51	208	42 54	265	36 22	326
62	20 16	013	50 37	045	52 43	089	49 49	148	45 25	208	41 58	265	35 51	325
63	20 27	012	51 16	043	53 38	089	50 18	149	44 59	209	41 03	265	35 19	324
64	20 39	011	51 54	042	54 34	088	50 47	150	44 32	209	40 07	265	34 46	323
65	20 49	010	52 31	041	55 29	088	51 15	150	44 05	209	39 12	264	34 12	323
66	20 59	010	53 07	040	56 25	087	51 42	151	43 38	210	38 17	264	33 38	322
67	21 08	009	53 42	038	57 20	087	52 09	151	43 10	210	37 21	264	33 03	321
68	21 16	008	54 16	037	58 16	086	52 35	152	42 42	211	36 26	263	32 28	320
69	21 24	008	54 48	036	59 12	086	53 01	153	42 13	211	35 31	263	31 52	319
70	21 31	007	55 20	034	60 07	085	53 26	154	41 45	211	34 36	263	31 15	318
71	21 37	006	55 51	033	61 02	085	53 51	154	41 16	212	33 41	262	30 38	318
72	21 42	005	56 20	031	61 58	084	54 15	155	40 47	212	32 45	262	30 00	317
73	21 47	005	56 48	029	62 53	084	54 38	156	40 17	212	31 50	262	29 22	316
74	21 51	004	57 15	028	63 48	083	55 00	157	39 47	212	30 55	262	28 43	315
	◆CAPELLA		POLLUX		PROCYON		◆Suhail		CANOPUS		◆ACHERNAR		Hamal	
75	21 54	003	26 03	040	42 30	060	35 02	128	55 22	158	39 17	213	28 03	315
76	21 57	002	26 39	040	43 18	059	35 46	128	55 42	158	38 47	213	27 23	314
77	21 59	002	27 14	039	44 05	058	36 29	129	56 02	159	38 17	213	26 43	313
78	22 00	001	27 49	038	44 52	057	37 13	129	56 22	160	37 46	213	26 02	312
79	22 00	000	28 23	037	45 39	056	37 56	129	56 40	161	37 16	214	25 21	312
80	22 00	359	28 56	037	46 25	055	38 40	129	56 57	162	36 45	214	24 39	311
81	21 59	359	29 29	036	47 10	054	39 23	129	57 14	163	36 14	214	23 57	310
82	21 57	358	30 01	035	47 55	054	40 06	129	57 30	164	35 42	214	23 15	310
83	21 55	357	30 33	034	48 40	053	40 50	129	57 44	165	35 11	214	22 32	309
84	21 52	356	31 04	033	49 24	051	41 33	129	57 58	166	34 40	215	21 48	309
85	21 48	356	31 34	033	50 07	050	42 16	129	58 11	167	34 08	215	21 05	308
86	21 43	355	32 04	032	50 49	049	42 59	130	58 22	168	33 36	215	20 21	307
87	21 38	354	32 31	031	51 31	048	43 42	130	58 33	170	33 04	215	19 36	307
88	21 32	353	33 01	030	52 12	047	44 24	130	58 42	171	32 32	215	18 52	306
89	21 25	353	33 28	029	52 53	046	45 07	130	58 51	172	32 00	215	18 07	306

LAT 22°S

LHA ♈	Hc	Zn	Hc	Zn	Hc	Zn	Hc	Zn	Hc	Zn	Hc	Zn	Hc	Zn
	POLLUX		◆REGULUS		ACRUX		CANOPUS		◆ACHERNAR		◆ALDEBARAN		CAPELLA	
90	33 54	028	20 20	067	16 37	152	58 58	173	31 28	215	46 18	330	21 17	352
91	34 20	027	21 12	067	17 03	152	59 04	174	30 56	215	45 50	329	21 09	351
92	34 45	026	22 02	066	17 29	152	59 09	175	30 24	215	45 21	328	21 00	350
93	35 09	025	22 53	066	17 56	152	59 13	176	29 52	215	44 51	326	20 51	350
94	35 33	024	23 44	065	18 22	152	59 16	178	29 19	216	44 19	325	20 40	349
95	35 55	023	24 34	064	18 49	151	59 18	179	28 47	216	43 47	324	20 29	348
96	36 17	022	25 24	064	19 15	151	59 18	180	28 15	216	43 14	323	20 18	348
97	36 37	021	26 14	063	19 42	151	59 18	181	27 42	216	42 41	322	20 05	347
98	36 57	020	27 04	063	20 09	151	59 16	182	27 10	216	42 06	321	19 52	346
99	37 16	019	27 53	062	20 36	151	59 13	184	26 37	216	41 31	320	19 39	345
100	37 33	018	28 42	062	21 02	151	59 09	185	26 05	216	40 55	319	19 24	345
101	37 50	017	29 31	061	21 29	151	59 04	186	25 32	216	40 18	318	19 09	344
102	38 06	016	30 19	060	21 56	151	58 58	187	25 00	216	39 40	317	18 54	343
103	38 21	015	31 07	060	22 24	151	58 50	188	24 28	216	39 02	316	18 38	343
104	38 35	014	31 55	059	22 51	151	58 42	189	23 55	216	38 23	315	18 21	342
	POLLUX		◆REGULUS		Gienah		◆ACRUX		CANOPUS		◆RIGEL		BETELGEUSE	
105	38 48	013	32 43	058	16 27	103	23 18	151	58 32	191	61 05	294	56 33	330
106	38 59	012	33 30	058	17 21	102	23 45	151	58 21	192	60 14	293	56 04	328
107	39 10	011	34 17	057	18 16	102	24 12	151	58 10	193	59 23	292	55 34	327
108	39 20	009	35 03	056	19 10	102	24 39	151	57 57	194	58 31	291	55 03	325
109	39 28	008	35 49	055	20 05	101	25 06	151	57 43	195	57 39	290	54 30	324
110	39 36	007	36 35	055	20 59	101	25 34	151	57 28	196	56 47	290	53 57	322
111	39 42	006	37 20	054	21 54	101	26 01	151	57 13	197	55 54	289	53 22	321
112	39 48	005	38 04	053	22 48	100	26 28	151	56 56	198	55 01	288	52 47	320
113	39 52	004	38 49	052	23 43	100	26 55	151	56 38	199	54 08	287	52 10	318
114	39 55	003	39 32	051	24 38	100	27 22	151	56 20	200	53 15	286	51 33	317
115	39 57	001	40 16	051	25 33	100	27 49	151	56 01	201	52 22	286	50 55	316
116	39 58	000	40 58	050	26 28	099	28 16	151	55 41	202	51 28	285	50 15	315
117	39 57	359	41 40	049	27 23	099	28 43	151	55 20	202	50 34	284	49 36	314
118	39 56	358	42 22	048	28 18	099	29 10	151	54 58	203	49 40	284	48 55	312
119	39 53	357	43 03	047	29 13	098	29 37	151	54 36	204	48 46	283	48 13	311
	◆REGULUS		SPICA		◆ACRUX		CANOPUS		◆RIGEL		BETELGEUSE		POLLUX	
120	43 43	046	12 11	097	30 04	151	54 13	205	47 52	283	47 31	310	39 50	356
121	44 23	045	13 06	097	30 30	151	53 49	206	46 58	282	46 49	309	39 45	355
122	45 01	044	14 01	097	30 57	152	53 24	207	46 03	281	46 05	308	39 39	353
123	45 40	043	14 57	096	31 23	152	52 59	207	45 08	281	45 21	307	39 32	352
124	46 17	042	15 52	096	31 50	152	52 33	208	44 14	280	44 37	306	39 24	351
125	46 54	041	16 47	095	32 16	152	52 07	209	43 19	280	43 52	306	39 15	350
126	47 29	039	17 43	095	32 42	152	51 40	209	42 24	279	43 06	305	39 05	349
127	48 04	038	18 38	095	33 08	152	51 13	210	41 29	279	42 20	304	38 54	348
128	48 38	037	19 34	094	33 34	152	50 45	211	40 34	278	41 34	303	38 42	347
129	49 11	036	20 29	094	34 00	153	50 16	211	39 39	278	40 47	302	38 28	346
130	49 43	035	21 25	094	34 25	153	49 47	212	38 44	277	40 00	301	38 14	345
131	50 14	033	22 20	093	34 51	153	49 18	212	37 49	277	39 12	301	37 58	343
132	50 44	032	23 16	093	35 16	153	48 48	213	36 53	276	38 24	300	37 42	342
133	51 13	031	24 11	093	35 41	153	48 17	213	35 58	276	37 36	299	37 25	341
134	51 41	029	25 07	092	36 06	154	47 47	214	35 03	275	36 47	298	37 07	340
	REGULUS		◆SPICA		ACRUX		◆CANOPUS		SIRIUS		BETELGEUSE		◆POLLUX	
135	52 08	028	26 02	092	36 30	154	47 16	214	57 46	273	35 58	298	36 47	339
136	52 33	026	26 58	092	36 55	154	46 44	215	56 51	273	35 08	297	36 27	338
137	52 57	025	27 54	091	37 19	154	46 12	215	55 55	273	34 19	296	36 06	337
138	53 20	023	28 49	091	37 43	155	45 40	216	55 00	272	33 29	296	35 44	336
139	53 41	022	29 45	091	38 07	155	45 08	216	54 04	272	32 38	295	35 21	335
140	54 01	020	30 40	090	38 30	155	44 35	216	53 08	271	31 48	294	34 57	334
141	54 20	019	31 36	090	38 54	155	44 02	217	52 13	271	30 57	294	34 33	333
142	54 37	017	32 32	089	39 17	156	43 29	217	51 17	270	30 06	293	34 08	332
143	54 52	015	33 27	089	39 39	156	42 55	217	50 21	270	29 15	293	33 41	331
144	55 06	014	34 23	089	40 02	156	42 21	218	49 26	270	28 23	292	33 14	331
145	55 19	012	35 19	088	40 24	157	41 47	218	48 30	269	27 32	292	32 47	330
146	55 30	010	36 14	088	40 46	157	41 13	218	47 35	269	26 40	291	32 18	329
147	55 39	009	37 10	087	41 07	157	40 38	218	46 39	269	25 48	290	31 49	328
148	55 47	007	38 05	087	41 28	158	40 04	219	45 43	268	24 56	290	31 19	327
149	55 53	005	39 01	087	41 49	158	39 29	219	44 48	268	24 03	289	30 49	326
	REGULUS		◆ARCTURUS		SPICA		◆ACRUX		CANOPUS		SIRIUS		◆PROCYON	
150	55 57	004	15 13	061	39 56	086	42 10	159	38 54	219	43 52	268	46 03	304
151	56 00	002	16 02	061	40 52	086	42 30	159	38 19	219	42 57	267	45 17	303
152	56 00	000	16 50	060	41 47	085	42 49	159	37 44	219	42 01	267	44 30	302
153	56 00	358	17 39	060	42 43	085	43 09	160	37 08	220	41 05	267	43 43	301
154	55 57	357	18 27	059	43 38	084	43 27	160	36 33	220	40 10	266	42 55	301
155	55 53	355	19 15	059	44 34	084	43 46	161	35 57	220	39 14	266	42 07	300
156	55 47	353	20 02	058	45 29	083	44 04	161	35 21	220	38 19	266	41 18	299
157	55 40	351	20 49	058	46 24	083	44 22	162	34 45	220	37 24	265	40 30	298
158	55 30	350	21 36	057	47 19	082	44 39	162	34 09	220	36 28	265	39 40	298
159	55 20	348	22 23	057	48 14	082	44 55	163	33 33	220	35 33	265	38 51	297
160	55 07	346	23 09	056	49 09	081	45 12	163	32 57	220	34 37	264	38 01	296
161	54 53	345	23 55	055	50 04	081	45 27	164	32 21	221	33 42	264	37 11	295
162	54 38	343	24 41	055	50 59	080	45 43	164	31 45	221	32 47	264	36 21	295
163	54 21	341	25 26	054	51 54	080	45 57	165	31 09	221	31 51	263	35 30	294
164	54 02	340	26 11	054	52 49	079	46 12	165	30 32	221	30 56	263	34 39	294
	◆ARCTURUS		ANTARES		◆ACRUX		CANOPUS		SIRIUS		◆PROCYON		REGULUS	
165	26 55	053	16 10	112	46 25	166	29 56	221	30 01	263	33 48	293	53 42	338
166	27 40	052	17 01	112	46 39	167	29 20	221	29 06	263	32 57	292	53 21	337
167	28 23	052	17 53	112	46 51	167	28 43	221	28 11	262	32 05	292	52 58	335
168	29 07	051	18 44	112	47 03	168	28 07	221	27 15	262	31 13	291	52 34	334
169	29 50	050	19 36	111	47 15	168	27 31	221	26 20	262	30 21	291	52 09	332
170	30 32	049	20 28	111	47 26	169	26 54	221	25 25	261	29 29	290	51 42	331
171	31 14	049	21 20	111	47 36	170	26 18	221	24 30	261	28 37	289	51 15	329
172	31 56	048	22 12	111	47 46	170	25 42	221	23 35	261	27 44	289	50 46	328
173	32 37	047	23 04	110	47 55	171	25 05	221	22 41	260	26 52	288	50 16	327
174	33 17	046	23 56	110	48 03	172	24 29	221	21 46	260	25 59	288	49 45	325
175	33 57	046	24 48	110	48 11	172	23 53	221	20 51	260	25 06	287	49 13	324
176	34 37	045	25 40	110	48 19	173	23 17	221	19 56	259	24 12	287	48 40	323
177	35 16	044	26 33	110	48 25	173	22 40	220	19 02	259	23 19	286	48 06	322
178	35 54	043	27 25	109	48 31	174	22 04	220	18 07	259	22 26	286	47 31	321
179	36 32	042	28 18	109	48 37	175	21 28	220	17 13	258	21 32	285	46 55	319

LAT 22°S

LHA ♈	Hc	Zn	Hc	Zn	Hc	Zn	Hc	Zn	Hc	Zn	Hc	Zn	Hc	Zn
	ARCTURUS		◆ANTARES		ACRUX		◆Suhail		Alphard		REGULUS		◆Denebola	
180	37 09	041	29 10	109	48 41	175	48 34	229	50 59	284	46 19	318	53 18	355
181	37 45	040	30 03	109	48 45	176	47 52	229	50 05	283	45 42	317	53 13	354
182	38 21	039	30 56	109	48 49	177	47 10	229	49 11	283	45 03	316	53 06	352
183	38 56	039	31 49	108	48 51	178	46 28	229	48 16	282	44 25	315	52 58	351
184	39 30	038	32 41	108	48 53	178	45 45	230	47 22	282	43 45	314	52 48	349
185	40 04	037	33 34	108	48 55	179	45 03	230	46 27	281	43 05	313	52 37	348
186	40 36	036	34 27	108	48 56	180	44 20	230	45 33	280	42 24	312	52 24	346
187	41 08	035	35 20	108	48 56	180	43 38	230	44 38	280	41 42	311	52 10	344
188	41 39	033	36 13	107	48 55	181	42 55	230	43 43	279	41 00	310	51 54	343
189	42 10	032	37 06	107	48 54	182	42 12	231	42 48	279	40 18	310	51 37	341
190	42 39	031	38 00	107	48 52	182	41 29	231	41 53	278	39 35	309	51 19	340
191	43 07	030	38 53	107	48 49	183	40 46	231	40 58	278	38 51	308	50 59	338
192	43 35	029	39 46	107	48 46	184	40 02	231	40 03	277	38 07	307	50 38	337
193	44 01	028	40 39	107	48 42	184	39 19	231	39 08	277	37 22	306	50 15	336
194	44 27	027	41 33	106	48 37	185	38 36	231	38 13	276	36 37	305	49 52	334
	ARCTURUS		◆Rasalhague		ANTARES		◆ACRUX		Suhail		◆REGULUS		Denebola	
195	44 51	026	14 20	070	42 26	106	48 32	186	37 52	231	35 51	305	49 27	333
196	45 15	024	15 12	069	43 19	106	48 26	186	37 09	231	35 05	304	49 01	332
197	45 37	023	16 04	069	44 13	106	48 20	187	36 25	231	34 19	303	48 34	330
198	45 58	022	16 56	068	45 06	106	48 13	188	35 42	232	33 32	302	48 06	329
199	46 18	021	17 48	068	46 00	106	48 05	188	34 58	232	32 45	302	47 37	328
200	46 37	019	18 39	067	46 53	106	47 56	189	34 15	232	31 58	301	47 07	327
201	46 55	018	19 30	067	47 47	106	47 47	190	33 31	232	31 10	300	46 36	325
202	47 12	017	20 22	066	48 41	106	47 38	190	32 48	232	30 22	300	46 04	324
203	47 27	015	21 12	066	49 34	105	47 27	191	32 04	232	29 33	299	45 31	323
204	47 41	014	22 03	065	50 28	105	47 17	192	31 20	232	28 45	299	44 57	322
205	47 53	013	22 54	065	51 22	105	47 05	192	30 37	232	27 56	298	44 22	321
206	48 05	011	23 44	064	52 15	105	46 53	193	29 53	231	27 06	297	43 47	320
207	48 15	010	24 34	064	53 09	105	46 41	193	29 10	231	26 17	297	43 10	319
208	48 24	008	25 24	063	54 03	105	46 28	194	28 26	231	25 27	296	42 33	318
209	48 31	007	26 13	063	54 56	105	46 14	194	27 43	231	24 37	296	41 55	317
	ARCTURUS		◆Rasalhague		ANTARES		◆RIGIL KENT.		ACRUX		Gienah		◆Denebola	
210	48 37	006	27 02	062	55 50	105	50 35	172	46 00	195	65 03	276	41 17	316
211	48 42	004	27 51	061	56 44	105	50 42	173	45 45	196	64 08	275	40 38	315
212	48 45	003	28 40	061	57 38	105	50 48	174	45 30	196	63 12	275	39 58	314
213	48 47	001	29 28	060	58 31	105	50 54	175	45 14	197	62 17	274	39 18	313
214	48 47	000	30 16	060	59 25	105	50 58	176	44 58	197	61 21	274	38 37	312
215	48 47	358	31 04	059	60 19	105	51 02	176	44 41	198	60 26	273	37 55	311
216	48 44	357	31 52	058	61 12	105	51 06	177	44 24	198	59 30	273	37 13	310
217	48 41	356	32 39	057	62 06	105	51 08	178	44 07	199	58 35	272	36 30	310
218	48 36	354	33 26	057	63 00	105	51 10	179	43 49	199	57 39	272	35 47	309
219	48 29	353	34 12	056	63 53	105	51 11	179	43 30	200	56 43	271	35 04	308
220	48 21	351	34 58	055	64 47	106	51 11	180	43 12	200	55 48	271	34 20	307
221	48 12	350	35 43	055	65 41	106	51 11	181	42 52	200	54 52	271	33 35	306
222	48 02	348	36 28	054	66 34	106	51 09	182	42 33	201	53 57	270	32 50	306
223	47 50	347	37 13	053	67 28	106	51 07	182	42 13	201	53 01	270	32 05	305
224	47 37	346	37 57	052	68 21	106	51 04	183	41 52	202	52 05	269	31 19	304
	Alphecca		Rasalhague		◆ALTAIR		Peacock		◆RIGIL KENT.		◆SPICA		ARCTURUS	
225	40 34	010	38 41	051	12 28	075	22 57	144	51 01	184	64 49	292	47 23	344
226	40 43	009	39 24	051	13 21	075	23 30	144	50 57	185	63 57	291	47 07	343
227	40 51	008	40 07	050	14 15	074	24 03	144	50 52	186	63 05	290	46 50	342
228	40 58	007	40 49	049	15 08	074	24 36	144	50 46	186	62 12	289	46 32	340
229	41 04	005	41 31	048	16 02	073	25 09	144	50 39	187	61 19	288	46 13	339
230	41 09	004	42 12	047	16 55	073	25 42	144	50 32	188	60 26	287	45 52	338
231	41 12	003	42 52	046	17 48	072	26 15	144	50 24	189	59 33	286	45 31	337
232	41 15	002	43 32	045	18 41	072	26 47	144	50 16	189	58 39	285	45 08	335
233	41 16	001	44 11	044	19 34	071	27 20	144	50 06	190	57 45	284	44 44	334
234	41 16	000	44 49	043	20 26	071	27 53	144	49 56	191	56 51	284	44 20	333
235	41 15	358	45 26	042	21 19	070	28 26	144	49 46	191	55 57	283	43 54	332
236	41 13	357	46 03	041	22 11	070	28 59	144	49 34	192	55 03	282	43 27	331
237	41 10	356	46 39	040	23 03	069	29 32	144	49 22	193	54 09	282	42 59	330
238	41 05	355	47 14	039	23 55	069	30 04	144	49 10	193	53 14	281	42 31	328
239	40 59	354	47 48	037	24 47	068	30 37	144	48 57	194	52 19	281	42 01	327
	VEGA		◆ALTAIR		Peacock		◆RIGIL KENT.		SPICA		◆ARCTURUS		Alphecca	
240	19 00	031	25 39	068	31 10	144	48 43	195	51 25	280	41 31	326	40 53	352
241	19 29	031	26 30	067	31 43	144	48 28	195	50 30	279	40 59	325	40 45	351
242	19 57	030	27 21	067	32 15	144	48 13	196	49 35	279	40 27	324	40 36	350
243	20 25	029	28 12	066	32 48	144	47 57	197	48 40	278	39 54	323	40 26	349
244	20 52	029	29 03	066	33 20	144	47 41	197	47 45	278	39 20	322	40 15	348
245	21 18	028	29 53	065	33 52	145	47 24	198	46 50	277	38 46	321	40 02	347
246	21 44	027	30 44	064	34 24	145	47 07	198	45 54	277	38 11	320	39 49	346
247	22 09	027	31 34	064	34 56	145	46 49	199	44 59	276	37 35	319	39 35	344
248	22 34	026	32 23	063	35 28	145	46 31	200	44 04	276	36 58	318	39 19	343
249	22 58	025	33 13	062	36 00	145	46 12	200	43 09	275	36 21	318	39 03	342
250	23 21	024	34 02	062	36 32	145	45 53	201	42 13	275	35 43	317	38 45	341
251	23 44	024	34 51	061	37 03	146	45 33	201	41 18	274	35 05	316	38 27	340
252	24 06	023	35 39	060	37 35	146	45 13	202	40 22	274	34 26	315	38 07	339
253	24 27	022	36 28	060	38 06	146	44 52	202	39 27	274	33 46	314	37 47	338
254	24 48	021	37 15	059	38 37	146	44 31	203	38 31	273	33 06	313	37 26	337
	◆VEGA		ALTAIR		◆Peacock		RIGIL KENT.		◆SPICA		ARCTURUS		Alphecca	
255	25 08	021	38 03	058	39 08	147	44 09	203	37 36	273	32 25	313	37 03	336
256	25 27	020	38 50	057	39 38	147	43 47	204	36 40	272	31 44	312	36 40	335
257	25 46	019	39 37	057	40 09	147	43 25	204	35 44	272	31 02	311	36 16	334
258	26 04	018	40 23	056	40 39	147	43 02	204	34 49	272	30 20	310	35 51	333
259	26 21	017	41 09	055	41 09	148	42 39	205	33 53	271	29 38	310	35 26	332
260	26 37	017	41 54	054	41 38	148	42 15	205	32 58	271	28 55	309	34 59	331
261	26 53	016	42 39	053	42 08	148	41 52	206	32 02	270	28 11	308	34 32	330
262	27 07	015	43 23	052	42 37	149	41 27	206	31 06	270	27 27	308	34 04	329
263	27 21	014	44 07	051	43 05	149	41 03	206	30 11	270	26 43	307	33 35	328
264	27 35	013	44 50	051	43 34	149	40 38	207	29 15	269	25 58	306	33 05	327
265	27 47	012	45 33	050	44 02	150	40 13	207	28 19	269	25 13	306	32 35	327
266	27 59	012	46 15	049	44 30	150	39 48	207	27 24	269	24 28	305	32 04	326
267	28 09	011	46 56	048	44 57	151	39 22	208	26 28	268	23 42	304	31 32	325
268	28 19	010	47 37	046	45 25	151	38 56	208	25 33	268	22 56	304	31 00	324
269	28 29	009	48 17	045	45 51	151	38 30	208	24 37	267	22 10	303	30 26	323

LHA ♈	Hc	Zn	Hc	Zn	Hc	Zn	Hc	Zn	Hc	Zn	Hc	Zn	Hc	Zn
	◆VEGA		ALTAIR		◆FOMALHAUT		Peacock		RIGIL KENT.		◆ANTARES		Alphecca	
270	28 37	008	48 56	044	23 45	114	46 18	152	38 04	208	68 50	253	29 53	322
271	28 44	007	49 34	043	24 36	114	46 44	152	37 37	209	67 57	254	29 19	322
272	28 51	006	50 12	042	25 27	113	47 09	153	37 10	209	67 04	254	28 44	321
273	28 57	006	50 49	041	26 18	113	47 34	153	36 43	209	66 10	254	28 08	320
274	29 02	005	51 25	039	27 09	113	47 59	154	36 16	209	65 16	254	27 32	319
275	29 06	004	51 59	038	28 00	113	48 23	155	35 49	210	64 23	254	26 56	319
276	29 09	003	52 33	037	28 52	113	48 47	155	35 21	210	63 29	255	26 18	318
277	29 11	002	53 06	035	29 43	113	49 10	156	34 54	210	62 36	255	25 41	317
278	29 13	001	53 38	034	30 34	112	49 32	156	34 26	210	61 42	255	25 03	316
279	29 13	000	54 08	033	31 26	112	49 54	157	33 58	210	60 48	255	24 24	316
280	29 13	359	54 38	031	32 17	112	50 16	158	33 30	210	59 55	255	23 45	315
281	29 12	358	55 06	030	33 09	112	50 36	158	33 01	211	59 01	255	23 06	314
282	29 10	357	55 33	028	34 01	112	50 57	159	32 33	211	58 07	255	22 26	314
283	29 07	357	55 58	026	34 52	112	51 16	160	32 04	211	57 13	255	21 45	313
284	29 03	356	56 22	025	35 44	112	51 35	160	31 36	211	56 20	255	21 04	312
	◆DENEB		FOMALHAUT		◆ACHERNAR		RIGIL KENT.		◆ANTARES		Rasalhague		VEGA	
285	18 54	019	36 36	111	13 30	147	31 07	211	55 26	255	49 36	327	28 59	355
286	19 11	018	37 28	111	14 00	147	30 38	211	54 32	255	49 05	326	28 53	354
287	19 28	017	38 19	111	14 31	146	30 09	211	53 38	255	48 33	324	28 47	353
288	19 44	016	39 11	111	15 02	146	29 40	211	52 45	255	48 00	323	28 40	352
289	19 59	016	40 03	111	15 33	146	29 11	211	51 51	255	47 26	322	28 32	351
290	20 14	015	40 55	111	16 04	146	28 42	212	50 57	255	46 51	321	28 23	350
291	20 28	014	41 47	111	16 35	146	28 13	212	50 04	255	46 16	320	28 13	350
292	20 42	014	42 39	111	17 07	146	27 44	212	49 10	255	45 39	319	28 03	349
293	20 55	013	43 31	111	17 38	145	27 15	212	48 17	254	45 02	317	27 52	348
294	21 07	012	44 23	111	18 10	145	26 46	212	47 23	254	44 24	316	27 39	347
295	21 18	012	45 16	111	18 42	145	26 16	212	46 29	254	43 45	315	27 27	346
296	21 29	011	46 08	110	19 13	145	25 47	212	45 36	254	43 06	314	27 13	345
297	21 39	010	47 00	110	19 45	145	25 18	212	44 42	254	42 26	313	26 58	344
298	21 48	009	47 52	110	20 17	145	24 49	212	43 49	254	41 45	312	26 43	344
299	21 57	009	48 44	110	20 49	145	24 19	212	42 55	254	41 04	312	26 27	343
	DENEB		Enif		◆FOMALHAUT		ACHERNAR		◆RIGIL KENT.		ANTARES		◆VEGA	
300	22 05	008	49 12	041	49 36	110	21 21	145	23 50	212	42 02	254	26 10	342
301	22 12	007	49 49	040	50 28	110	21 54	145	23 21	212	41 09	253	25 53	341
302	22 19	006	50 24	039	51 20	110	22 26	145	22 52	212	40 15	253	25 34	340
303	22 24	006	50 59	038	52 13	110	22 58	144	22 23	212	39 22	253	25 15	340
304	22 30	005	51 32	036	53 05	111	23 30	144	21 53	212	38 29	253	24 56	339
305	22 34	004	52 05	035	53 57	111	24 03	144	21 24	211	37 36	253	24 35	338
306	22 37	003	52 36	034	54 49	111	24 35	144	20 55	211	36 43	253	24 14	337
307	22 40	003	53 06	032	55 41	111	25 08	144	20 26	211	35 49	253	23 52	337
308	22 42	002	53 35	031	56 33	111	25 40	144	19 58	211	34 56	252	23 30	336
309	22 44	001	54 03	029	57 25	111	26 13	144	19 29	211	34 03	252	23 07	335
310	22 44	000	54 30	028	58 17	111	26 45	144	19 00	211	33 11	252	22 43	334
311	22 44	359	54 55	026	59 09	111	27 17	144	18 31	211	32 18	252	22 19	334
312	22 43	359	55 19	025	60 00	112	27 50	144	18 03	211	31 25	252	21 54	333
313	22 42	358	55 42	023	60 52	112	28 22	144	17 34	211	30 32	251	21 28	332
314	22 39	357	56 03	021	61 44	112	28 55	144	17 06	211	29 39	251	21 02	332
	Alpheratz		◆Diphda		ACHERNAR		◆Peacock		ANTARES		◆ALTAIR		DENEB	
315	21 44	044	37 40	096	29 27	144	54 40	188	28 47	251	54 45	329	22 36	356
316	22 23	043	38 35	096	29 59	145	54 32	189	27 54	251	54 16	328	22 32	356
317	23 00	042	39 30	096	30 32	145	54 22	190	27 02	251	53 46	326	22 28	355
318	23 37	042	40 26	095	31 04	145	54 12	191	26 09	250	53 15	325	22 22	354
319	24 14	041	41 21	095	31 36	145	54 01	192	25 17	250	52 42	324	22 16	353
320	24 50	040	42 17	095	32 08	145	53 49	193	24 25	250	52 09	322	22 10	353
321	25 26	039	43 12	094	32 40	145	53 37	194	23 32	250	51 34	321	22 02	352
322	26 01	039	44 08	094	33 12	145	53 23	194	22 40	249	50 59	320	21 54	351
323	26 35	038	45 03	094	33 44	145	53 09	195	21 48	249	50 22	318	21 45	350
324	27 09	037	45 59	094	34 15	145	52 54	196	20 56	249	49 45	317	21 35	350
325	27 43	036	46 54	093	34 47	145	52 38	197	20 04	249	49 07	316	21 25	349
326	28 15	036	47 50	093	35 18	146	52 22	198	19 13	248	48 28	315	21 14	348
327	28 48	035	48 45	093	35 50	146	52 05	198	18 21	248	47 48	314	21 02	347
328	29 19	034	49 41	092	36 21	146	51 47	199	17 29	248	47 07	313	20 50	347
329	29 50	033	50 36	092	36 52	146	51 28	200	16 38	248	46 26	312	20 37	346
	◆Alpheratz		Diphda		◆ACHERNAR		Peacock		◆Nunki		ALTAIR		DENEB	
330	30 20	032	51 32	092	37 23	146	51 09	201	47 47	255	45 44	311	20 23	345
331	30 50	032	52 28	091	37 54	147	50 49	201	46 54	254	45 02	310	20 09	345
332	31 19	031	53 23	091	38 24	147	50 29	202	46 00	254	44 19	309	19 54	344
333	31 47	030	54 19	091	38 54	147	50 08	203	45 06	254	43 35	308	19 38	343
334	32 14	029	55 14	090	39 24	147	49 46	203	44 13	254	42 51	307	19 22	343
335	32 41	028	56 10	090	39 54	148	49 24	204	43 19	254	42 06	306	19 05	342
336	33 07	027	57 06	089	40 24	148	49 01	204	42 26	254	41 21	305	18 47	341
337	33 32	026	58 01	089	40 53	148	48 38	205	41 33	254	40 35	304	18 29	341
338	33 56	025	58 57	089	41 23	149	48 14	206	40 39	254	39 49	304	18 10	340
339	34 19	024	59 53	088	41 51	149	47 50	206	39 46	253	39 03	303	17 51	339
340	34 42	024	60 48	088	42 20	149	47 25	207	38 53	253	38 16	302	17 31	339
341	35 04	023	61 44	087	42 48	150	47 00	207	37 59	253	37 28	301	17 10	338
342	35 25	022	62 39	087	43 16	150	46 34	208	37 06	253	36 41	301	16 49	337
343	35 45	021	63 35	086	43 44	150	46 08	208	36 13	253	35 52	300	16 27	337
344	36 04	020	64 30	086	44 11	151	45 41	209	35 20	253	35 04	299	16 05	336
	Alpheratz		◆Hamal		◆ACHERNAR		Peacock		Nunki		◆ALTAIR		Enif	
345	36 22	019	25 44	048	44 38	151	45 15	209	34 27	252	34 15	298	53 05	328
346	36 39	018	26 25	047	45 05	152	44 47	210	33 34	252	33 26	298	52 35	326
347	36 55	016	27 06	046	45 31	152	44 20	210	32 41	252	32 37	297	52 04	325
348	37 11	015	27 46	046	45 57	153	43 52	210	31 48	252	31 47	296	51 31	324
349	37 25	014	28 25	045	46 23	153	43 23	211	30 55	252	30 57	296	50 58	322
350	37 38	013	29 04	044	46 48	154	42 55	211	30 02	251	30 07	295	50 23	321
351	37 51	012	29 43	044	47 12	154	42 26	211	29 10	251	29 17	295	49 48	320
352	38 02	011	30 21	043	47 36	155	41 57	212	28 17	251	28 26	294	49 11	319
353	38 12	010	30 59	042	48 00	155	41 27	212	27 24	251	27 35	293	48 34	317
354	38 21	009	31 36	041	48 23	156	40 57	212	26 32	251	26 44	293	47 56	316
355	38 29	008	32 12	040	48 46	156	40 27	213	25 40	250	25 52	292	47 17	315
356	38 36	007	32 48	040	49 08	157	39 57	213	24 47	250	25 01	292	46 38	314
357	38 42	006	33 23	039	49 30	157	39 27	213	23 55	250	24 09	291	45 57	313
358	38 47	005	33 57	038	49 51	158	38 56	214	23 03	250	23 17	291	45 16	312
359	38 51	003	34 31	037	50 11	159	38 25	214	22 11	249	22 25	290	44 35	311

 LAT 23°S

LHA ♈	Hc	Zn	Hc	Zn	Hc	Zn	Hc	Zn	Hc	Zn	Hc	Zn	Hc	Zn
	◆Alpheratz		Hamal		◆RIGEL		CANOPUS		ACHERNAR		◆Peacock		Enif	
0	37 54	002	34 16	036	13 40	093	14 39	141	51 27	159	38 44	214	43 14	311
1	37 56	001	34 48	035	14 35	093	15 13	141	51 47	160	38 12	215	42 32	310
2	37 56	000	35 19	034	15 30	092	15 48	141	52 05	160	37 41	215	41 49	309
3	37 56	359	35 49	033	16 25	092	16 23	141	52 24	161	37 09	215	41 06	308
4	37 54	358	36 19	032	17 21	092	16 58	141	52 41	162	36 37	215	40 22	307
5	37 52	357	36 48	031	18 16	091	17 33	141	52 58	163	36 05	215	39 38	306
6	37 48	356	37 16	030	19 11	091	18 08	140	53 14	163	35 33	216	38 53	306
7	37 43	355	37 43	029	20 06	091	18 43	140	53 29	164	35 01	216	38 08	305
8	37 37	353	38 09	028	21 02	090	19 18	140	53 44	165	34 29	216	37 22	304
9	37 30	352	38 35	027	21 57	090	19 54	140	53 58	166	33 57	216	36 36	303
10	37 22	351	38 59	026	22 52	089	20 30	140	54 11	167	33 24	216	35 50	302
11	37 14	350	39 23	025	23 47	089	21 05	140	54 23	168	32 51	216	35 03	302
12	37 04	349	39 46	024	24 42	089	21 41	140	54 35	168	32 19	216	34 16	301
13	36 53	348	40 07	023	25 38	088	22 17	139	54 45	169	31 46	216	33 28	300
14	36 41	347	40 28	022	26 33	088	22 53	139	54 55	170	31 13	216	32 40	300
	◆Hamal		ALDEBARAN		RIGEL		◆CANOPUS		ACHERNAR		◆FOMALHAUT		Alpheratz	
15	40 48	020	24 08	058	27 28	087	23 29	139	55 04	171	61 49	250	36 28	346
16	41 07	019	24 55	057	28 23	087	24 05	139	55 12	172	60 57	250	36 14	345
17	41 24	018	25 41	057	29 18	087	24 41	139	55 19	173	60 05	250	35 59	344
18	41 41	017	26 28	056	30 13	086	25 17	139	55 26	174	59 13	250	35 43	343
19	41 57	016	27 13	056	31 09	086	25 53	139	55 31	175	58 21	250	35 26	342
20	42 11	015	27 59	055	32 04	085	26 30	139	55 35	176	57 29	250	35 08	341
21	42 24	013	28 44	054	32 59	085	27 06	139	55 39	177	56 37	251	34 50	340
22	42 37	012	29 28	054	33 54	084	27 42	139	55 42	178	55 45	251	34 30	339
23	42 48	011	30 13	053	34 49	084	28 19	139	55 43	179	54 53	251	34 10	338
24	42 58	010	30 56	052	35 43	083	28 55	139	55 44	180	54 01	251	33 48	337
25	43 06	008	31 40	051	36 38	083	29 31	139	55 44	181	53 09	251	33 26	336
26	43 14	007	32 23	051	37 33	082	30 08	139	55 43	182	52 17	251	33 03	335
27	43 20	006	33 05	050	38 28	082	30 44	139	55 41	183	51 25	251	32 39	334
28	43 25	005	33 47	049	39 22	082	31 21	139	55 38	183	50 32	251	32 15	333
29	43 29	003	34 29	048	40 17	081	31 57	139	55 34	184	49 40	251	31 49	332
	Hamal		◆ALDEBARAN		SIRIUS		◆CANOPUS		ACHERNAR		◆FOMALHAUT		Alpheratz	
30	43 32	002	35 09	047	23 20	099	32 33	139	55 30	185	48 48	251	31 23	331
31	43 33	001	35 50	047	24 14	099	33 09	139	55 24	186	47 56	251	30 56	330
32	43 34	000	36 30	046	25 09	098	33 46	139	55 18	187	47 04	251	30 29	330
33	43 33	358	37 09	045	26 04	098	34 22	139	55 10	188	46 12	251	30 00	329
34	43 31	357	37 48	044	26 58	098	34 58	139	55 02	189	45 20	250	29 31	328
35	43 27	356	38 26	043	27 53	097	35 34	139	54 53	190	44 28	250	29 02	327
36	43 23	355	39 03	042	28 48	097	36 10	139	54 43	191	43 36	250	28 31	326
37	43 17	353	39 40	041	29 43	097	36 46	140	54 32	192	42 44	250	28 00	325
38	43 10	352	40 16	040	30 37	096	37 21	140	54 20	193	41 52	250	27 29	325
39	43 02	351	40 51	039	31 32	096	37 57	140	54 08	193	41 00	250	26 56	324
40	42 52	350	41 26	038	32 27	096	38 33	140	53 55	194	40 08	250	26 24	323
41	42 42	348	41 59	037	33 22	095	39 08	140	53 41	195	39 16	250	25 50	322
42	42 30	347	42 32	036	34 17	095	39 43	140	53 26	196	38 25	250	25 16	322
43	42 17	346	43 04	035	35 12	095	40 18	141	53 10	197	37 33	249	24 41	321
44	42 03	345	43 36	034	36 07	094	40 53	141	52 54	198	36 41	249	24 06	320
	◆CAPELLA		BETELGEUSE		SIRIUS		◆CANOPUS		ACHERNAR		◆FOMALHAUT		Hamal	
45	14 24	024	37 32	060	37 02	094	41 28	141	52 37	198	35 50	249	41 48	344
46	14 46	023	38 20	059	37 58	094	42 03	141	52 19	199	34 58	249	41 32	342
47	15 07	022	39 07	058	38 53	093	42 37	142	52 01	200	34 06	249	41 15	341
48	15 28	022	39 54	057	39 48	093	43 11	142	51 42	200	33 15	249	40 56	340
49	15 48	021	40 40	057	40 43	093	43 45	142	51 22	201	32 24	248	40 37	339
50	16 08	021	41 26	056	41 38	092	44 19	142	51 02	202	31 32	248	40 17	338
51	16 27	020	42 12	055	42 33	092	44 52	143	50 41	202	30 41	248	39 55	337
52	16 46	019	42 57	054	43 29	092	45 26	143	50 20	203	29 50	248	39 33	336
53	17 04	019	43 41	053	44 24	091	45 59	144	49 58	204	28 59	248	39 10	335
54	17 21	018	44 25	052	45 19	091	46 31	144	49 35	204	28 08	248	38 46	334
55	17 38	017	45 08	051	46 14	090	47 04	144	49 12	205	27 17	247	38 21	332
56	17 54	017	45 51	050	47 09	090	47 36	145	48 49	206	26 26	247	37 55	331
57	18 09	016	46 33	049	48 05	090	48 08	145	48 25	206	25 35	247	37 28	330
58	18 24	015	47 15	048	49 00	089	48 39	146	48 00	207	24 44	247	37 00	329
59	18 38	015	47 56	047	49 55	089	49 10	146	47 35	207	23 53	246	36 32	328
	CAPELLA		BETELGEUSE		◆SIRIUS		CANOPUS		◆ACHERNAR		Diphda		◆Hamal	
60	18 52	014	48 36	046	50 50	088	49 40	147	47 10	208	43 53	267	36 02	327
61	19 05	013	49 15	045	51 46	088	50 11	147	46 44	208	42 58	266	35 32	327
62	19 17	012	49 54	044	52 41	088	50 40	148	46 18	209	42 03	266	35 01	326
63	19 29	012	50 32	043	53 36	087	51 10	148	45 51	209	41 08	266	34 30	325
64	19 40	011	51 09	041	54 31	087	51 38	149	45 25	209	40 13	265	33 58	324
65	19 50	010	51 45	040	55 26	086	52 07	149	44 57	210	39 18	265	33 25	323
66	20 00	010	52 20	039	56 21	086	52 34	150	44 30	210	38 23	265	32 51	322
67	20 08	009	52 54	037	57 16	085	53 02	151	44 02	211	37 28	264	32 17	321
68	20 17	008	53 27	036	58 11	085	53 28	151	43 33	211	36 33	264	31 42	320
69	20 24	007	53 59	035	59 06	084	53 54	152	43 05	211	35 38	264	31 06	320
70	20 31	007	54 30	033	60 01	084	54 20	153	42 36	212	34 43	263	30 30	319
71	20 37	006	55 00	032	60 56	083	54 45	154	42 07	212	33 48	263	29 53	318
72	20 42	005	55 28	030	61 51	082	55 09	154	41 37	212	32 53	263	29 16	317
73	20 47	005	55 55	029	62 46	082	55 32	155	41 08	213	31 59	262	28 38	316
74	20 51	004	56 21	027	63 40	081	55 55	156	40 38	213	31 04	262	28 00	316
	◆CAPELLA		POLLUX		PROCYON		◆Suhail		CANOPUS		◆ACHERNAR		Hamal	
75	20 54	003	25 17	040	41 59	059	35 39	128	56 17	157	40 08	213	27 21	315
76	20 57	002	25 52	039	42 46	058	36 23	128	56 38	158	39 37	213	26 42	314
77	20 59	002	26 27	039	43 33	057	37 07	128	56 58	159	39 07	214	26 02	313
78	21 00	001	27 01	038	44 19	056	37 50	128	57 18	160	38 36	214	25 22	313
79	21 00	000	27 35	037	45 05	056	38 34	128	57 37	161	38 05	214	24 41	312
80	21 00	359	28 08	036	45 50	055	39 17	128	57 54	162	37 34	214	24 00	311
81	20 59	359	28 40	035	46 35	054	40 01	128	58 11	163	37 03	214	23 18	311
82	20 57	358	29 12	035	47 19	053	40 44	128	58 27	164	36 32	215	22 36	310
83	20 55	357	29 43	034	48 03	052	41 27	128	58 42	165	36 00	215	21 54	310
84	20 52	356	30 14	033	48 46	051	42 10	128	58 56	166	35 29	215	21 11	309
85	20 48	356	30 43	032	49 28	050	42 54	129	59 09	167	34 57	215	20 28	308
86	20 43	355	31 12	031	50 10	048	43 37	129	59 21	168	34 25	215	19 44	308
87	20 38	354	31 41	030	50 51	047	44 20	129	59 32	169	33 53	215	19 00	307
88	20 32	353	32 08	030	51 31	046	45 03	129	59 41	170	33 21	215	18 16	307
89	20 25	353	32 35	029	52 11	045	45 45	129	59 50	172	32 49	216	17 31	306

LHA ♈	Hc	Zn	Hc	Zn	Hc	Zn	Hc	Zn	Hc	Zn	Hc	Zn	Hc	Zn
	POLLUX		◆REGULUS		ACRUX		CANOPUS		◆ACHERNAR		◆ALDEBARAN		CAPELLA	
90	33 01	028	19 57	067	17 30	152	59 58	173	32 17	216	45 26	331	20 18	352
91	33 27	027	20 48	066	17 56	152	60 04	174	31 45	216	44 59	329	20 10	351
92	33 51	026	21 38	066	18 22	152	60 09	175	31 13	216	44 30	328	20 01	350
93	34 15	025	22 28	065	18 48	151	60 13	176	30 40	216	44 00	327	19 51	350
94	34 38	024	23 18	065	19 15	151	60 16	178	30 08	216	43 30	326	19 41	349
95	35 00	023	24 08	064	19 41	151	60 18	179	29 36	216	42 59	325	19 31	348
96	35 21	022	24 58	063	20 08	151	60 18	180	29 03	216	42 26	324	19 19	348
97	35 41	021	25 47	063	20 34	151	60 18	181	28 31	216	41 53	323	19 07	347
98	36 01	020	26 36	062	21 01	151	60 16	182	27 58	216	41 19	322	18 54	346
99	36 19	019	27 25	062	21 28	151	60 13	184	27 26	216	40 45	321	18 41	346
100	36 36	018	28 13	061	21 55	151	60 09	185	26 54	216	40 09	320	18 26	345
101	36 53	017	29 01	060	22 22	151	60 04	186	26 21	216	39 33	319	18 12	344
102	37 09	016	29 49	060	22 49	151	59 57	187	25 49	216	38 56	318	17 56	343
103	37 23	015	30 37	059	23 16	151	59 49	188	25 16	216	38 19	317	17 40	343
104	37 37	014	31 24	058	23 43	151	59 41	190	24 44	216	37 41	316	17 24	342
	POLLUX		◆REGULUS		Gienah		◆ACRUX		CANOPUS		◆RIGEL		BETELGEUSE	
105	37 49	013	32 11	058	16 40	102	24 10	151	59 31	191	60 39	296	55 41	330
106	38 01	012	32 57	057	17 34	102	24 37	151	59 20	192	59 49	295	55 13	329
107	38 11	010	33 44	056	18 28	102	25 04	151	59 08	193	58 59	294	54 44	327
108	38 21	009	34 29	056	19 22	101	25 32	151	58 55	194	58 08	293	54 13	326
109	38 29	008	35 15	055	20 16	101	25 59	151	58 41	195	57 17	292	53 42	325
110	38 36	007	36 00	054	21 10	101	26 26	151	58 26	196	56 26	291	53 09	323
111	38 43	006	36 44	053	22 05	100	26 53	151	58 10	197	55 34	290	52 36	322
112	38 48	005	37 28	052	22 59	100	27 20	151	57 53	198	54 42	289	52 01	320
113	38 52	004	38 12	052	23 54	100	27 47	151	57 35	199	53 50	289	51 25	319
114	38 55	003	38 55	051	24 48	099	28 14	151	57 16	200	52 57	288	50 49	318
115	38 57	001	39 37	050	25 42	099	28 41	151	56 57	201	52 05	287	50 11	317
116	38 58	000	40 19	049	26 37	099	29 09	151	56 36	202	51 12	286	49 33	316
117	38 57	359	41 00	048	27 32	098	29 36	151	56 15	203	50 19	286	48 54	314
118	38 56	358	41 41	047	28 26	098	30 02	151	55 53	204	49 25	285	48 14	313
119	38 54	357	42 21	046	29 21	098	30 29	151	55 30	205	48 32	284	47 33	312
	◆REGULUS		SPICA		◆ACRUX		CANOPUS		◆RIGEL		BETELGEUSE		POLLUX	
120	43 01	045	12 18	097	30 56	151	55 07	206	47 38	284	46 52	311	38 50	356
121	43 40	044	13 13	097	31 23	151	54 43	206	46 45	283	46 10	310	38 45	355
122	44 18	043	14 08	096	31 49	151	54 18	207	45 51	282	45 28	309	38 40	354
123	44 55	042	15 03	096	32 16	151	53 52	208	44 57	282	44 45	308	38 33	352
124	45 32	041	15 58	096	32 42	151	53 26	209	44 03	281	44 01	307	38 25	351
125	46 08	040	16 53	095	33 09	152	53 00	209	43 08	281	43 17	306	38 16	350
126	46 43	039	17 48	095	33 35	152	52 32	210	42 14	280	42 32	305	38 06	349
127	47 17	038	18 43	094	34 01	152	52 04	211	41 20	280	41 47	305	37 55	348
128	47 50	036	19 38	094	34 27	152	51 36	211	40 25	279	41 01	304	37 43	347
129	48 22	035	20 33	094	34 53	152	51 07	212	39 31	279	40 15	303	37 30	346
130	48 54	034	21 28	093	35 18	152	50 38	212	38 36	278	39 28	302	37 16	345
131	49 24	033	22 23	093	35 44	153	50 08	213	37 41	278	38 41	301	37 01	344
132	49 53	031	23 19	093	36 09	153	49 38	213	36 46	277	37 54	301	36 45	343
133	50 21	030	24 14	092	36 34	153	49 07	214	35 52	277	37 06	300	36 28	342
134	50 48	029	25 09	092	36 59	153	48 36	214	34 57	276	36 18	299	36 10	341
	REGULUS		◆SPICA		ACRUX		◆CANOPUS		SIRIUS		BETELGEUSE		◆POLLUX	
135	51 14	027	26 04	091	37 24	153	48 05	215	57 42	275	35 29	298	35 51	340
136	51 39	026	26 59	091	37 49	154	47 33	215	56 47	274	34 41	298	35 31	339
137	52 02	024	27 55	091	38 13	154	47 01	216	55 52	274	33 52	297	35 11	338
138	52 25	023	28 50	090	38 37	154	46 29	216	54 57	274	33 02	296	34 49	337
139	52 45	021	29 45	090	39 01	155	45 56	216	54 01	273	32 13	296	34 27	336
140	53 05	020	30 40	090	39 25	155	45 23	217	53 06	273	31 23	295	34 03	335
141	53 23	018	31 36	089	39 48	155	44 50	217	52 11	272	30 33	294	33 39	334
142	53 39	017	32 31	089	40 11	155	44 16	218	51 16	272	29 42	294	33 14	333
143	53 55	015	33 26	088	40 34	156	43 43	218	50 21	271	28 51	293	32 49	332
144	54 08	013	34 21	088	40 57	156	43 09	218	49 26	271	28 01	293	32 22	331
145	54 20	012	35 16	088	41 19	156	42 34	218	48 30	270	27 09	292	31 55	330
146	54 31	010	36 12	087	41 41	157	42 00	219	47 35	270	26 18	291	31 27	329
147	54 40	009	37 07	087	42 03	157	41 25	219	46 40	270	25 27	291	30 58	328
148	54 47	007	38 02	086	42 24	158	40 50	219	45 45	269	24 35	290	30 29	327
149	54 53	005	38 57	086	42 45	158	40 15	219	44 49	269	23 43	290	29 59	327
	REGULUS		◆ARCTURUS		SPICA		◆ACRUX		CANOPUS		SIRIUS		◆PROCYON	
150	54 57	003	14 44	061	39 52	085	43 05	158	39 40	220	43 54	269	45 29	305
151	55 00	002	15 33	061	40 47	085	43 26	159	39 05	220	42 59	268	44 43	304
152	55 00	000	16 21	060	41 42	084	43 45	159	38 30	220	42 04	268	43 57	303
153	55 00	358	17 09	060	42 37	084	44 05	160	37 54	220	41 09	267	43 11	302
154	54 57	357	17 56	059	43 32	083	44 24	160	37 18	220	40 13	267	42 24	301
155	54 53	355	18 43	059	44 27	083	44 43	161	36 43	220	39 18	267	41 37	301
156	54 48	353	19 30	058	45 21	082	45 01	161	36 07	221	38 23	266	40 49	300
157	54 40	352	20 17	057	46 16	082	45 19	161	35 31	221	37 28	266	40 01	299
158	54 31	350	21 03	057	47 11	081	45 36	162	34 55	221	36 33	266	39 12	298
159	54 21	348	21 50	056	48 05	081	45 53	162	34 19	221	35 38	265	38 24	298
160	54 09	347	22 35	056	49 00	080	46 09	163	33 43	221	34 43	265	37 34	297
161	53 55	345	23 21	055	49 54	080	46 25	164	33 06	221	33 48	265	36 45	296
162	53 40	343	24 06	054	50 49	079	46 40	164	32 30	221	32 53	264	35 55	295
163	53 24	342	24 51	054	51 43	079	46 55	165	31 54	221	31 58	264	35 05	295
164	53 06	340	25 35	053	52 37	078	47 10	165	31 18	221	31 03	264	34 15	294
	◆ARCTURUS		ANTARES		◆ACRUX		CANOPUS		SIRIUS		◆PROCYON		REGULUS	
165	26 19	053	16 32	112	47 24	166	30 41	221	30 08	263	33 24	294	52 46	339
166	27 03	052	17 24	112	47 37	166	30 05	221	29 13	263	32 34	293	52 26	337
167	27 46	051	18 15	112	47 50	167	29 29	221	28 18	263	31 43	292	52 04	336
168	28 29	050	19 06	111	48 02	168	28 52	221	27 24	262	30 52	292	51 40	334
169	29 11	050	19 58	111	48 14	168	28 16	221	26 29	262	30 00	291	51 16	333
170	29 53	049	20 49	111	48 25	169	27 39	221	25 34	262	29 08	290	50 50	331
171	30 34	048	21 41	111	48 35	169	27 03	221	24 40	261	28 17	290	50 23	330
172	31 15	047	22 33	110	48 45	170	26 27	221	23 45	261	27 25	289	49 55	329
173	31 56	047	23 24	110	48 54	171	25 51	221	22 50	261	26 32	289	49 26	327
174	32 36	046	24 16	110	49 03	171	25 14	221	21 56	260	25 40	288	48 56	326
175	33 15	045	25 08	110	49 11	172	24 38	221	21 02	260	24 48	288	48 24	325
176	33 54	044	26 01	109	49 18	173	24 02	221	20 07	260	23 55	287	47 52	324
177	34 32	043	26 53	109	49 25	173	23 26	221	19 13	259	23 02	287	47 19	322
178	35 10	043	27 45	109	49 31	174	22 50	221	18 19	259	22 09	286	46 45	321
179	35 47	042	28 37	109	49 36	175	22 14	221	17 24	259	21 16	286	46 10	320

LHA ♈	Hc	Zn	Hc	Zn	Hc	Zn	Hc	Zn	Hc	Zn	Hc	Zn	Hc	Zn
	ARCTURUS		♦ANTARES		ACRUX		♦Suhail		Alphard		REGULUS		♦Denebola	
180	36 23	041	29 30	108	49 41	175	49 13	229	50 44	285	45 34	319	52 18	356
181	36 59	040	30 22	108	49 45	176	48 31	230	49 51	284	44 57	318	52 13	354
182	37 34	039	31 15	108	49 49	177	47 49	230	48 57	284	44 20	317	52 06	352
183	38 09	038	32 07	108	49 51	177	47 06	230	48 03	283	43 42	316	51 58	351
184	38 42	037	33 00	108	49 53	178	46 24	230	47 09	283	43 03	315	51 49	349
185	39 15	036	33 52	107	49 55	179	45 41	231	46 16	282	42 24	314	51 38	348
186	39 47	035	34 45	107	49 56	180	44 59	231	45 21	281	41 43	313	51 26	346
187	40 19	034	35 38	107	49 56	180	44 16	231	44 27	281	41 03	312	51 12	345
188	40 49	033	36 31	107	49 55	181	43 33	231	43 33	280	40 21	311	50 57	343
189	41 19	032	37 24	106	49 54	182	42 50	231	42 39	280	39 39	310	50 40	342
190	41 48	031	38 17	106	49 52	182	42 06	231	41 44	279	38 57	309	50 22	340
191	42 15	030	39 10	106	49 49	183	41 23	232	40 50	279	38 14	308	50 03	339
192	42 42	029	40 03	106	49 46	184	40 40	232	39 55	278	37 30	308	49 42	338
193	43 08	027	40 56	106	49 42	184	39 56	232	39 00	278	36 46	307	49 21	336
194	43 33	026	41 49	106	49 37	185	39 13	232	38 05	277	36 02	306	48 58	335
	ARCTURUS		♦Rasalhague		ANTARES		♦ACRUX		Suhail		♦REGULUS		Denebola	
195	43 57	025	13 59	070	42 42	105	49 32	186	38 30	232	35 17	305	48 34	333
196	44 20	024	14 51	069	43 36	105	49 26	187	37 46	232	34 32	304	48 08	332
197	44 42	023	15 43	069	44 29	105	49 19	187	37 02	232	33 46	304	47 42	331
198	45 03	021	16 34	068	45 22	105	49 12	188	36 19	232	33 00	303	47 15	330
199	45 22	020	17 25	068	46 16	105	49 04	189	35 35	232	32 13	302	46 46	328
200	45 41	019	18 16	067	47 09	105	48 56	189	34 52	232	31 26	302	46 17	327
201	45 58	018	19 07	067	48 03	105	48 46	190	34 08	232	30 39	301	45 46	326
202	46 14	016	19 57	066	48 56	104	48 37	190	33 25	232	29 52	300	45 15	325
203	46 29	015	20 48	066	49 50	104	48 26	191	32 41	232	29 04	300	44 43	324
204	46 42	014	21 38	065	50 43	104	48 15	192	31 57	232	28 16	299	44 09	323
205	46 55	012	22 28	064	51 37	104	48 04	192	31 14	232	27 27	298	43 35	321
206	47 06	011	23 18	064	52 30	104	47 52	193	30 30	232	26 38	298	43 01	320
207	47 16	010	24 07	063	53 24	104	47 39	194	29 47	232	25 49	297	42 25	319
208	47 24	008	24 56	063	54 18	104	47 26	194	29 04	232	25 00	297	41 49	318
209	47 32	007	25 45	062	55 11	104	47 12	195	28 20	232	24 11	296	41 12	317
	ARCTURUS		♦Rasalhague		ANTARES		♦RIGIL KENT.		ACRUX		Gienah		♦Denebola	
210	47 37	005	26 34	062	56 05	104	51 35	172	46 58	195	64 56	278	40 34	316
211	47 42	004	27 22	061	56 59	104	51 42	173	46 43	196	64 01	277	39 55	315
212	47 45	003	28 10	060	57 52	103	51 48	174	46 28	196	63 06	277	39 16	314
213	47 47	001	28 58	060	58 46	103	51 53	175	46 12	197	62 12	276	38 37	314
214	47 47	000	29 46	059	59 40	103	51 58	175	45 55	197	61 17	275	37 56	313
215	47 47	358	30 33	058	60 33	103	52 02	176	45 39	198	60 22	275	37 15	312
216	47 44	357	31 20	058	61 27	103	52 06	177	45 21	198	59 27	274	36 34	311
217	47 41	356	32 06	057	62 21	103	52 08	178	45 04	199	58 31	274	35 52	310
218	47 36	354	32 52	056	63 15	103	52 10	179	44 45	199	57 36	273	35 09	309
219	47 30	353	33 38	056	64 08	103	52 11	179	44 27	200	56 41	273	34 27	309
220	47 22	351	34 23	055	65 02	104	52 11	180	44 08	200	55 46	272	33 43	308
221	47 13	350	35 08	054	65 56	104	52 11	181	43 49	201	54 51	272	32 59	307
222	47 03	349	35 53	053	66 49	104	52 09	182	43 29	201	53 56	272	32 15	306
223	46 52	347	36 37	052	67 43	104	52 07	183	43 09	202	53 00	271	31 30	306
224	46 39	346	37 20	052	68 37	104	52 04	183	42 48	202	52 05	271	30 45	305
	Alphecca		Rasalhague		♦ALTAIR		Peacock		♦RIGIL KENT.		♦SPICA		ARCTURUS	
225	39 35	010	38 03	051	12 12	075	23 46	144	52 01	184	64 26	294	46 25	345
226	39 44	009	38 46	050	13 05	074	24 19	144	51 56	185	63 35	293	46 10	343
227	39 52	008	39 28	049	13 58	074	24 51	144	51 51	186	62 44	291	45 53	342
228	39 58	007	40 09	048	14 51	073	25 24	144	51 46	186	61 52	290	45 35	341
229	40 04	005	40 50	047	15 44	073	25 57	143	51 39	187	61 00	289	45 17	339
230	40 09	004	41 30	046	16 37	072	26 30	143	51 32	188	60 08	289	44 57	338
231	40 12	003	42 10	045	17 29	072	27 03	143	51 24	189	59 15	288	44 36	337
232	40 15	002	42 49	044	18 22	072	27 36	143	51 15	189	58 23	287	44 14	336
233	40 16	001	43 27	043	19 14	071	28 09	143	51 06	190	57 30	286	43 50	335
234	40 16	000	44 05	042	20 06	071	28 41	143	50 55	191	56 36	285	43 26	333
235	40 15	358	44 41	041	20 58	070	29 14	143	50 45	192	55 43	284	43 01	332
236	40 13	357	45 17	040	21 50	069	29 47	144	50 33	192	54 49	284	42 35	331
237	40 10	356	45 53	039	22 42	069	30 20	144	50 21	193	53 56	283	42 08	330
238	40 05	355	46 27	038	23 33	068	30 53	144	50 08	194	53 02	282	41 39	329
239	40 00	354	47 00	037	24 25	068	31 26	144	49 55	194	52 08	282	41 10	328
	VEGA		♦ALTAIR		Peacock		♦RIGIL KENT.		SPICA		♦ARCTURUS		Alphecca	
240	18 09	031	25 16	067	31 58	144	49 41	195	51 14	281	40 41	327	39 53	353
241	18 37	031	26 07	067	32 31	144	49 26	196	50 20	281	40 10	326	39 46	351
242	19 05	030	26 57	066	33 04	144	49 11	196	49 25	280	39 38	325	39 37	350
243	19 32	029	27 48	066	33 36	144	48 55	197	48 31	279	39 06	324	39 27	349
244	19 59	029	28 38	065	34 09	144	48 39	198	47 36	279	38 33	323	39 16	348
245	20 25	028	29 28	064	34 41	144	48 22	198	46 42	278	37 59	322	39 04	347
246	20 51	027	30 17	064	35 13	144	48 04	199	45 47	278	37 24	321	38 51	346
247	21 15	026	31 07	063	35 45	144	47 46	199	44 52	277	36 49	320	38 37	345
248	21 40	026	31 56	063	36 17	145	47 27	200	43 57	277	36 13	319	38 22	344
249	22 03	025	32 45	062	36 49	145	47 08	200	43 02	276	35 37	318	38 06	342
250	22 26	024	33 33	061	37 21	145	46 49	201	42 08	276	34 59	317	37 48	341
251	22 49	024	34 22	060	37 53	145	46 29	202	41 13	275	34 22	316	37 30	340
252	23 11	023	35 09	060	38 24	145	46 08	202	40 18	275	33 43	315	37 11	339
253	23 32	022	35 57	059	38 55	146	45 47	203	39 22	274	33 04	315	36 51	338
254	23 52	021	36 44	058	39 27	146	45 26	203	38 27	274	32 25	314	36 30	337
	♦VEGA		ALTAIR		♦Peacock		RIGIL KENT.		♦SPICA		ARCTURUS		Alphecca	
255	24 12	021	37 31	058	39 57	146	45 04	203	37 32	274	31 44	313	36 09	336
256	24 31	020	38 17	057	40 28	146	44 42	204	36 37	273	31 04	312	35 46	335
257	24 49	019	39 03	056	40 59	146	44 20	204	35 42	273	30 23	312	35 22	334
258	25 07	018	39 49	055	41 29	147	43 57	205	34 47	272	29 41	311	34 58	333
259	25 23	017	40 34	054	41 59	147	43 33	205	33 52	272	28 59	310	34 33	332
260	25 39	017	41 19	053	42 29	147	43 10	206	32 56	271	28 17	309	34 06	331
261	25 55	016	42 03	053	42 58	148	42 46	206	32 01	271	27 34	309	33 40	330
262	26 09	015	42 46	052	43 28	148	42 21	206	31 06	271	26 50	308	33 12	330
263	26 23	014	43 29	051	43 57	149	41 57	207	30 11	270	26 07	307	32 44	329
264	26 36	013	44 12	050	44 25	149	41 32	207	29 16	270	25 23	307	32 14	328
265	26 48	012	44 54	049	44 54	149	41 07	207	28 20	269	24 38	306	31 45	327
266	27 00	012	45 35	048	45 22	150	40 41	208	27 25	269	23 53	305	31 14	326
267	27 10	011	46 15	047	45 50	150	40 15	208	26 30	269	23 08	305	30 43	325
268	27 20	010	46 55	046	46 17	151	39 49	208	25 35	268	22 22	304	30 11	324
269	27 29	009	47 34	045	46 44	151	39 23	209	24 39	268	21 37	304	29 38	323

LAT 23°S

LHA ♈	Hc	Zn	Hc	Zn	Hc	Zn	Hc	Zn	Hc	Zn	Hc	Zn	Hc	Zn
	♦VEGA		ALTAIR		♦FOMALHAUT		Peacock		RIGIL KENT.		♦ANTARES		Alphecca	
270	27 37	008	48 13	043	24 09	113	47 11	151	38 57	209	69 06	256	29 05	323
271	27 45	007	48 50	042	25 00	113	47 37	152	38 30	209	68 13	256	28 31	322
272	27 51	006	49 27	041	25 51	113	48 02	152	38 03	209	67 19	256	27 57	321
273	27 57	005	50 03	040	26 42	113	48 28	153	37 36	210	66 25	256	27 22	320
274	28 02	005	50 38	039	27 32	113	48 53	154	37 08	210	65 32	256	26 47	320
275	28 06	004	51 12	037	28 23	112	49 17	154	36 41	210	64 38	257	26 10	319
276	28 09	003	51 45	036	29 15	112	49 41	155	36 13	210	63 44	257	25 34	318
277	28 11	002	52 17	035	30 06	112	50 04	155	35 46	210	62 50	257	24 57	317
278	28 13	001	52 48	033	30 57	112	50 27	156	35 18	211	61 57	257	24 19	317
279	28 13	000	53 18	032	31 48	112	50 49	157	34 49	211	61 03	257	23 41	316
280	28 13	359	53 46	030	32 40	111	51 11	157	34 21	211	60 09	257	23 02	315
281	28 12	358	54 14	029	33 31	111	51 32	158	33 53	211	59 16	257	22 23	315
282	28 10	358	54 40	027	34 23	111	51 53	159	33 24	211	58 22	257	21 44	314
283	28 07	357	55 04	026	35 14	111	52 13	159	32 56	211	57 28	257	21 04	313
284	28 03	356	55 28	024	36 06	111	52 32	160	32 27	211	56 34	256	20 24	313
	♦DENEB		FOMALHAUT		♦ACHERNAR		RIGIL KENT.		♦ANTARES		Rasalhague		VEGA	
285	17 57	018	36 57	111	14 20	147	31 58	211	55 41	256	48 46	327	27 59	355
286	18 14	018	37 49	111	14 50	146	31 29	212	54 47	256	48 15	326	27 54	354
287	18 31	017	38 41	110	15 21	146	31 01	212	53 53	256	47 44	325	27 47	353
288	18 46	016	39 33	110	15 52	146	30 32	212	53 00	256	47 12	324	27 40	352
289	19 02	016	40 24	110	16 23	146	30 02	212	52 06	256	46 39	323	27 32	351
290	19 16	015	41 16	110	16 54	146	29 33	212	51 13	256	46 05	321	27 24	351
291	19 30	014	42 08	110	17 25	146	29 04	212	50 19	256	45 30	320	27 14	350
292	19 44	014	43 00	110	17 56	145	28 35	212	49 26	256	44 54	319	27 04	349
293	19 56	013	43 52	110	18 28	145	28 06	212	48 32	256	44 18	318	26 53	348
294	20 08	012	44 44	110	18 59	145	27 37	212	47 39	255	43 40	317	26 41	347
295	20 20	011	45 36	110	19 31	145	27 07	212	46 45	255	43 02	316	26 28	346
296	20 30	011	46 28	109	20 03	145	26 38	212	45 52	255	42 24	315	26 15	345
297	20 40	010	47 20	109	20 34	145	26 09	212	44 58	255	41 44	314	26 00	345
298	20 49	009	48 12	109	21 06	145	25 40	212	44 05	255	41 04	313	25 45	344
299	20 58	009	49 04	109	21 38	145	25 10	212	43 12	255	40 24	312	25 30	343
	DENEB		Enif		♦FOMALHAUT		ACHERNAR		♦RIGIL KENT.		ANTARES		♦VEGA	
300	21 06	008	48 27	041	49 57	109	22 10	144	24 41	212	42 19	254	25 13	342
301	21 13	007	49 03	039	50 49	109	22 42	144	24 12	212	41 25	254	24 56	341
302	21 19	006	49 37	038	51 41	109	23 15	144	23 43	212	40 32	254	24 38	341
303	21 25	006	50 11	037	52 33	109	23 47	144	23 14	212	39 39	254	24 19	340
304	21 30	005	50 44	036	53 25	109	24 19	144	22 45	212	38 46	254	24 00	339
305	21 34	004	51 15	034	54 17	109	24 52	144	22 16	212	37 53	254	23 40	338
306	21 37	003	51 46	033	55 09	109	25 24	144	21 47	212	37 00	253	23 19	337
307	21 40	003	52 15	032	56 02	109	25 56	144	21 18	212	36 07	253	22 57	337
308	21 42	002	52 44	030	56 54	109	26 29	144	20 49	211	35 14	253	22 35	336
309	21 44	001	53 11	029	57 46	110	27 01	144	20 20	211	34 21	253	22 12	335
310	21 44	000	53 37	027	58 38	110	27 34	144	19 51	211	33 29	253	21 49	335
311	21 44	359	54 01	026	59 30	110	28 06	144	19 23	211	32 36	252	21 25	334
312	21 43	359	54 25	024	60 22	110	28 39	144	18 54	211	31 43	252	21 00	333
313	21 42	358	54 46	023	61 14	110	29 11	144	18 26	211	30 51	252	20 35	332
314	21 39	357	55 07	021	62 05	110	29 43	144	17 58	211	29 58	252	20 09	332
	Alpheratz		♦Diphda		ACHERNAR		♦Peacock		ANTARES		♦ALTAIR		DENEB	
315	21 01	043	37 46	096	30 16	144	55 39	188	29 06	252	53 54	330	21 36	356
316	21 38	043	38 41	095	30 48	144	55 31	189	28 14	251	53 25	329	21 32	356
317	22 16	042	39 36	095	31 20	144	55 21	190	27 21	251	52 56	327	21 28	355
318	22 52	041	40 31	095	31 53	144	55 11	191	26 29	251	52 25	326	21 23	354
319	23 28	041	41 26	094	32 25	144	55 00	192	25 37	251	51 54	324	21 17	353
320	24 04	040	42 21	094	32 57	144	54 48	193	24 45	250	51 21	323	21 10	353
321	24 39	039	43 16	094	33 29	145	54 35	194	23 53	250	50 47	322	21 03	352
322	25 14	038	44 11	093	34 01	145	54 21	195	23 01	250	50 13	320	20 55	351
323	25 48	038	45 07	093	34 33	145	54 07	196	22 09	250	49 37	319	20 46	350
324	26 21	037	46 02	093	35 05	145	53 52	196	21 18	249	49 00	318	20 36	350
325	26 54	036	46 57	092	35 36	145	53 36	197	20 26	249	48 23	317	20 26	349
326	27 27	035	47 52	092	36 08	145	53 19	198	19 34	249	47 45	316	20 15	348
327	27 58	035	48 47	091	36 39	145	53 02	199	18 43	248	47 06	315	20 04	348
328	28 29	034	49 43	091	37 11	146	52 43	200	17 52	248	46 26	314	19 52	347
329	29 00	033	50 38	091	37 42	146	52 25	200	17 00	248	45 46	313	19 39	346
	♦Alpheratz		Diphda		♦ACHERNAR		Peacock		♦Nunki		ALTAIR		DENEB	
330	29 30	032	51 33	090	38 13	146	52 05	201	48 03	256	45 05	311	19 25	345
331	29 59	031	52 28	090	38 44	146	51 45	202	47 09	256	44 23	311	19 11	345
332	30 27	030	53 23	090	39 14	146	51 24	202	46 16	255	43 41	310	18 56	344
333	30 55	030	54 19	089	39 45	147	51 03	203	45 22	255	42 58	309	18 41	343
334	31 22	029	55 14	089	40 15	147	50 41	204	44 29	255	42 15	308	18 24	343
335	31 48	028	56 09	088	40 45	147	50 19	204	43 36	255	41 31	307	18 08	342
336	32 13	027	57 04	088	41 15	147	49 56	205	42 42	255	40 46	306	17 50	341
337	32 38	026	57 59	087	41 44	148	49 32	206	41 49	255	40 01	305	17 32	341
338	33 02	025	58 55	087	42 14	148	49 08	206	40 56	254	39 16	304	17 14	340
339	33 25	024	59 50	086	42 43	148	48 43	207	40 03	254	38 30	303	16 55	339
340	33 47	023	60 45	086	43 12	149	48 18	207	39 09	254	37 44	303	16 35	339
341	34 08	022	61 40	085	43 40	149	47 53	208	38 16	254	36 57	302	16 14	338
342	34 29	021	62 35	085	44 08	149	47 27	208	37 23	254	36 10	301	15 53	337
343	34 48	020	63 30	084	44 36	150	47 01	209	36 30	253	35 22	300	15 32	337
344	35 07	019	64 25	084	45 04	150	46 34	209	35 38	253	34 35	300	15 10	336
	Alpheratz		♦Hamal		♦ACHERNAR		Peacock		Nunki		♦ALTAIR		Enif	
345	35 25	018	25 04	048	45 31	151	46 07	210	34 45	253	33 46	299	52 14	328
346	35 42	017	25 44	047	45 58	151	45 39	210	33 52	253	32 58	298	51 45	327
347	35 58	016	26 24	046	46 24	152	45 11	210	32 59	253	32 09	298	51 14	326
348	36 13	015	27 04	045	46 50	152	44 43	211	32 06	252	31 20	297	50 43	324
349	36 27	014	27 43	045	47 16	153	44 15	211	31 14	252	30 31	296	50 10	323
350	36 40	013	28 21	044	47 41	153	43 46	212	30 21	252	29 41	296	49 36	322
351	36 52	012	28 59	043	48 06	154	43 17	212	29 29	252	28 51	295	49 02	321
352	37 03	011	29 37	042	48 31	154	42 48	212	28 36	252	28 01	295	48 26	319
353	37 13	010	30 14	042	48 54	155	42 18	213	27 44	251	27 11	294	47 50	318
354	37 22	009	30 50	041	49 18	155	41 48	213	26 52	251	26 20	293	47 12	317
355	37 30	008	31 26	040	49 41	156	41 18	213	26 00	251	25 29	293	46 34	316
356	37 37	007	32 01	039	50 03	156	40 47	214	25 07	251	24 38	292	45 56	315
357	37 43	006	32 36	038	50 25	157	40 17	214	24 15	250	23 47	292	45 16	314
358	37 48	004	33 10	037	50 46	158	39 46	214	23 23	250	22 56	291	44 36	313
359	37 51	003	33 43	037	51 07	158	39 15	214	22 31	250	22 04	291	43 55	312

LHA ♈	Hc	Zn	Hc	Zn	Hc	Zn	Hc	Zn	Hc	Zn	Hc	Zn	Hc	Zn
	◆Alpheratz		Hamal		◆RIGEL		CANOPUS		ACHERNAR		◆Peacock		Enif	
0	36 54	002	33 27	035	13 43	093	15 26	141	52 23	159	39 33	215	42 34	312
1	36 56	001	33 58	034	14 38	093	16 00	141	52 43	159	39 02	215	41 53	311
2	36 56	000	34 29	033	15 33	092	16 35	141	53 02	160	38 30	215	41 11	310
3	36 56	359	34 59	033	16 28	092	17 09	141	53 20	161	37 58	216	40 28	309
4	36 54	358	35 28	032	17 22	091	17 44	141	53 38	161	37 26	216	39 45	308
5	36 52	357	35 56	031	18 17	091	18 19	140	53 55	162	36 54	216	39 02	307
6	36 48	356	36 24	030	19 12	091	18 54	140	54 11	163	36 22	216	38 18	306
7	36 43	355	36 50	029	20 07	090	19 29	140	54 27	164	35 50	216	37 33	305
8	36 38	353	37 16	028	21 02	090	20 04	140	54 42	165	35 17	216	36 48	305
9	36 31	352	37 41	027	21 56	089	20 40	140	54 56	166	34 45	216	36 03	304
10	36 23	351	38 05	026	22 51	089	21 15	140	55 09	166	34 12	216	35 17	303
11	36 14	350	38 28	024	23 46	089	21 51	139	55 22	167	33 40	217	34 31	302
12	36 05	349	38 51	023	24 41	088	22 27	139	55 33	168	33 07	217	33 45	302
13	35 54	348	39 12	022	25 36	088	23 02	139	55 44	169	32 34	217	32 58	301
14	35 42	347	39 32	021	26 30	087	23 38	139	55 54	170	32 02	217	32 10	300
	◆Hamal		ALDEBARAN		RIGEL		◆CANOPUS		ACHERNAR		◆FOMALHAUT		Alpheratz	
15	39 52	020	23 36	058	27 25	087	24 14	139	56 03	171	62 09	252	35 29	346
16	40 10	019	24 23	057	28 20	086	24 50	139	56 11	172	61 17	252	35 16	345
17	40 27	018	25 09	056	29 14	086	25 26	139	56 19	173	60 25	252	35 01	344
18	40 44	017	25 54	056	30 09	086	26 02	139	56 25	174	59 33	252	34 46	343
19	40 59	016	26 39	055	31 04	085	26 39	139	56 31	175	58 41	252	34 29	342
20	41 13	014	27 24	054	31 58	085	27 15	139	56 35	176	57 49	252	34 12	341
21	41 26	013	28 08	054	32 53	084	27 51	139	56 39	177	56 57	252	33 53	340
22	41 38	012	28 52	053	33 47	084	28 27	139	56 42	178	56 04	252	33 34	339
23	41 49	011	29 36	052	34 42	083	29 04	138	56 43	179	55 12	252	33 14	338
24	41 58	010	30 19	052	35 36	083	29 40	138	56 44	180	54 20	252	32 53	337
25	42 07	008	31 02	051	36 31	082	30 16	138	56 44	181	53 28	252	32 31	336
26	42 14	007	31 44	050	37 25	082	30 53	138	56 43	182	52 36	252	32 09	335
27	42 20	006	32 26	049	38 19	081	31 29	138	56 41	183	51 44	252	31 45	334
28	42 25	005	33 08	049	39 13	081	32 06	138	56 38	184	50 52	252	31 21	333
29	42 29	003	33 48	048	40 07	080	32 42	138	56 34	185	50 00	252	30 56	332
	Hamal		◆ALDEBARAN		SIRIUS		◆CANOPUS		ACHERNAR		◆FOMALHAUT		Alpheratz	
30	42 32	002	34 29	047	23 29	099	33 18	139	56 29	186	49 07	252	30 31	332
31	42 33	001	35 08	046	24 23	098	33 54	139	56 24	186	48 15	252	30 04	331
32	42 34	000	35 48	045	25 17	098	34 31	139	56 17	187	47 23	252	29 37	330
33	42 33	358	36 26	044	26 12	098	35 07	139	56 10	188	46 31	251	29 09	329
34	42 31	357	37 04	043	27 06	097	35 43	139	56 01	189	45 39	251	28 40	328
35	42 27	356	37 42	043	28 00	097	36 19	139	55 52	190	44 48	251	28 11	327
36	42 23	355	38 18	042	28 55	097	36 55	139	55 42	191	43 56	251	27 41	327
37	42 17	353	38 54	041	29 49	096	37 31	139	55 31	192	43 04	251	27 11	326
38	42 10	352	39 30	040	30 44	096	38 07	139	55 19	193	42 12	251	26 40	325
39	42 02	351	40 04	039	31 38	095	38 43	139	55 06	194	41 20	251	26 08	324
40	41 53	350	40 38	038	32 33	095	39 18	140	54 53	195	40 28	251	25 35	323
41	41 43	349	41 11	037	33 28	095	39 54	140	54 38	196	39 37	250	25 02	323
42	41 31	347	41 44	036	34 22	094	40 29	140	54 23	196	38 45	250	24 29	322
43	41 19	346	42 15	034	35 17	094	41 04	140	54 08	197	37 54	250	23 55	321
44	41 05	345	42 46	033	36 12	094	41 40	140	53 51	198	37 02	250	23 20	320
	◆CAPELLA		BETELGEUSE		SIRIUS		◆CANOPUS		ACHERNAR		◆FOMALHAUT		Hamal	
45	13 29	024	37 02	059	37 06	093	42 15	141	53 34	199	36 11	250	40 51	344
46	13 51	023	37 49	058	38 01	093	42 49	141	53 16	199	35 19	250	40 35	343
47	14 12	022	38 35	058	38 56	093	43 24	141	52 57	200	34 28	249	40 18	342
48	14 32	022	39 21	057	39 51	092	43 58	141	52 38	201	33 37	249	40 00	340
49	14 52	021	40 07	056	40 45	092	44 32	142	52 18	202	32 45	249	39 41	339
50	15 12	020	40 52	055	41 40	091	45 06	142	51 58	202	31 54	249	39 21	338
51	15 31	020	41 37	054	42 35	091	45 40	142	51 37	203	31 03	249	39 00	337
52	15 49	019	42 21	053	43 30	091	46 14	143	51 15	204	30 12	248	38 38	336
53	16 07	019	43 05	052	44 25	090	46 47	143	50 53	204	29 21	248	38 16	335
54	16 24	018	43 48	052	45 19	090	47 20	143	50 30	205	28 30	248	37 52	334
55	16 40	017	44 31	051	46 14	089	47 52	144	50 07	205	27 39	248	37 27	333
56	16 56	017	45 13	050	47 09	089	48 25	144	49 43	206	26 49	248	37 02	332
57	17 12	016	45 54	049	48 04	089	48 57	145	49 19	207	25 58	247	36 36	331
58	17 26	015	46 35	047	48 59	088	49 28	145	48 54	207	25 08	247	36 08	330
59	17 40	014	47 15	046	49 53	088	49 59	145	48 29	208	24 17	247	35 40	329
	CAPELLA		BETELGEUSE		◆SIRIUS		CANOPUS		◆ACHERNAR		Diphda		◆Hamal	
60	17 54	014	47 54	045	50 48	087	50 30	146	48 03	208	43 56	268	35 12	328
61	18 06	013	48 33	044	51 43	087	51 01	146	47 37	209	43 01	267	34 42	327
62	18 19	012	49 11	043	52 38	086	51 31	147	47 11	209	42 07	267	34 12	326
63	18 30	012	49 48	042	53 32	086	52 00	148	46 44	210	41 12	267	33 41	325
64	18 41	011	50 24	041	54 27	085	52 30	148	46 17	210	40 17	266	33 09	324
65	18 51	010	50 59	039	55 21	085	52 58	149	45 49	210	39 23	266	32 37	323
66	19 00	010	51 33	038	56 16	084	53 26	149	45 21	211	38 28	265	32 04	322
67	19 09	009	52 06	037	57 11	084	53 54	150	44 53	211	37 33	265	31 30	322
68	19 17	008	52 39	035	58 05	083	54 21	151	44 25	211	36 39	265	30 55	321
69	19 25	007	53 10	034	58 59	083	54 47	152	43 56	212	35 44	264	30 20	320
70	19 31	007	53 40	032	59 54	082	55 13	152	43 27	212	34 50	264	29 45	319
71	19 37	006	54 09	031	60 48	081	55 38	153	42 58	212	33 55	264	29 09	318
72	19 43	005	54 36	030	61 42	081	56 03	154	42 28	213	33 01	263	28 32	318
73	19 47	004	55 03	028	62 36	080	56 27	155	41 58	213	32 06	263	27 55	317
74	19 51	004	55 28	026	63 30	079	56 50	156	41 28	213	31 12	263	27 17	316
	◆CAPELLA		POLLUX		PROCYON		◆Suhail		CANOPUS		◆ACHERNAR		Hamal	
75	19 55	003	24 31	040	41 28	058	36 16	127	57 12	156	40 58	214	26 39	315
76	19 57	002	25 06	039	42 14	057	37 00	127	57 34	157	40 27	214	26 00	315
77	19 59	002	25 40	038	43 00	056	37 43	127	57 54	158	39 57	214	25 20	314
78	20 00	001	26 14	038	43 46	056	38 27	127	58 14	159	39 26	214	24 41	313
79	20 00	000	26 47	037	44 31	055	39 10	127	58 33	160	38 55	215	24 00	312
80	20 00	359	27 20	036	45 15	054	39 54	127	58 51	161	38 24	215	23 20	312
81	19 59	359	27 51	035	45 59	053	40 37	128	59 09	162	37 53	215	22 39	311
82	19 57	358	28 23	034	46 43	052	41 21	128	59 25	163	37 21	215	21 57	310
83	19 55	357	28 53	034	47 25	051	42 04	128	59 40	164	36 50	215	21 15	310
84	19 52	356	29 23	033	48 08	050	42 48	128	59 54	165	36 18	215	20 33	309
85	19 48	356	29 53	032	48 49	049	43 31	128	60 07	167	35 46	216	19 50	309
86	19 43	355	30 21	031	49 30	048	44 14	128	60 20	168	35 14	216	19 07	308
87	19 38	354	30 49	030	50 10	046	44 57	128	60 31	169	34 42	216	18 24	307
88	19 32	353	31 16	029	50 49	045	45 40	128	60 41	170	34 10	216	17 40	307
89	19 26	353	31 43	028	51 28	044	46 23	129	60 49	171	33 38	216	16 56	306

LAT 24°S

LHA ♈	Hc	Zn	Hc	Zn	Hc	Zn	Hc	Zn	Hc	Zn	Hc	Zn	Hc	Zn
	POLLUX		◆REGULUS		ACRUX		CANOPUS		◆ACHERNAR		◆ALDEBARAN		CAPELLA	
90	32 08	027	19 33	066	18 22	152	60 57	173	33 06	216	44 34	331	19 18	352
91	32 33	027	20 23	066	18 49	152	61 04	174	32 34	216	44 07	330	19 10	351
92	32 57	026	21 13	065	19 15	151	61 09	175	32 01	216	43 39	329	19 02	351
93	33 21	025	22 03	065	19 41	151	61 13	176	31 29	216	43 10	328	18 52	350
94	33 43	024	22 52	064	20 07	151	61 16	178	30 57	216	42 40	326	18 42	349
95	34 05	023	23 42	064	20 34	151	61 18	179	30 24	216	42 09	325	18 32	348
96	34 25	022	24 31	063	21 00	151	61 18	180	29 52	216	41 38	324	18 20	348
97	34 45	021	25 19	062	21 27	151	61 18	181	29 19	216	41 05	323	18 08	347
98	35 04	020	26 08	062	21 54	151	61 16	183	28 47	216	40 32	322	17 56	346
99	35 22	019	26 56	061	22 20	151	61 13	184	28 14	216	39 58	321	17 42	346
100	35 39	018	27 44	061	22 47	151	61 09	185	27 42	216	39 23	320	17 29	345
101	35 56	017	28 31	060	23 14	151	61 03	186	27 10	216	38 48	319	17 14	344
102	36 11	016	29 19	059	23 41	151	60 57	188	26 37	216	38 12	318	16 59	344
103	36 25	015	30 06	059	24 08	150	60 49	189	26 05	216	37 35	317	16 43	343
104	36 38	013	30 52	058	24 35	150	60 40	190	25 32	216	36 57	316	16 26	342
	POLLUX		◆REGULUS		Gienah		◆ACRUX		CANOPUS		◆RIGEL		BETELGEUSE	
105	36 51	012	31 39	057	16 53	102	25 02	150	60 30	191	60 12	298	54 48	331
106	37 02	011	32 24	056	17 46	102	25 29	150	60 19	192	59 23	296	54 21	330
107	37 12	010	33 10	056	18 40	101	25 57	150	60 06	193	58 34	295	53 53	328
108	37 21	009	33 55	055	19 34	101	26 24	150	59 53	195	57 44	294	53 23	327
109	37 30	008	34 40	054	20 28	101	26 51	150	59 39	196	56 54	293	52 53	325
110	37 37	007	35 24	053	21 21	100	27 18	150	59 23	197	56 04	292	52 21	324
111	37 43	006	36 08	053	22 15	100	27 45	150	59 07	198	55 13	292	51 48	323
112	37 48	005	36 51	052	23 09	100	28 12	150	58 50	199	54 22	291	51 14	321
113	37 52	004	37 34	051	24 03	099	28 40	150	58 32	200	53 30	290	50 40	320
114	37 55	003	38 16	050	24 58	099	29 07	150	58 13	201	52 39	289	50 04	319
115	37 57	001	38 58	049	25 52	099	29 34	150	57 53	202	51 47	288	49 27	318
116	37 58	000	39 39	048	26 46	098	30 01	150	57 32	203	50 54	287	48 50	316
117	37 57	359	40 20	047	27 40	098	30 28	151	57 10	204	50 02	287	48 12	315
118	37 56	358	41 00	046	28 35	098	30 55	151	56 48	205	49 09	286	47 33	314
119	37 54	357	41 40	046	29 29	097	31 22	151	56 25	205	48 17	285	46 53	313
	◆REGULUS		SPICA		◆ACRUX		CANOPUS		◆RIGEL		BETELGEUSE		POLLUX	
120	42 18	045	12 26	097	31 49	151	56 01	206	47 24	285	46 12	312	37 50	356
121	42 56	044	13 20	096	32 15	151	55 36	207	46 31	284	45 31	311	37 46	355
122	43 34	042	14 15	096	32 42	151	55 11	208	45 37	283	44 50	310	37 40	354
123	44 11	041	15 09	096	33 09	151	54 45	209	44 44	283	44 07	309	37 33	352
124	44 46	040	16 04	095	33 35	151	54 19	209	43 50	282	43 24	308	37 26	351
125	45 22	039	16 58	095	34 01	151	53 52	210	42 57	282	42 41	307	37 17	350
126	45 56	038	17 53	095	34 28	151	53 24	211	42 03	281	41 57	306	37 07	349
127	46 29	037	18 47	094	34 54	152	52 56	211	41 09	280	41 12	305	36 56	348
128	47 02	036	19 42	094	35 20	152	52 27	212	40 15	280	40 27	304	36 45	347
129	47 33	035	20 37	093	35 46	152	51 58	212	39 21	279	39 42	304	36 32	346
130	48 04	033	21 32	093	36 12	152	51 28	213	38 27	279	38 56	303	36 18	345
131	48 33	032	22 26	093	36 37	152	50 58	214	37 33	278	38 10	302	36 03	344
132	49 02	031	23 21	092	37 03	152	50 28	214	36 39	278	37 23	301	35 48	343
133	49 29	029	24 16	092	37 28	153	49 57	215	35 44	277	36 36	300	35 31	342
134	49 56	028	25 11	091	37 53	153	49 26	215	34 50	277	35 48	300	35 13	341
	REGULUS		◆SPICA		ACRUX		◆CANOPUS		SIRIUS		BETELGEUSE		◆POLLUX	
135	50 21	027	26 05	091	38 18	153	48 54	216	57 36	277	35 01	299	34 55	340
136	50 45	025	27 00	091	38 42	153	48 22	216	56 41	276	34 13	298	34 35	339
137	51 08	024	27 55	090	39 07	154	47 50	216	55 47	275	33 24	298	34 15	338
138	51 29	022	28 50	090	39 31	154	47 17	217	54 52	275	32 35	297	33 54	337
139	51 49	021	29 45	089	39 55	154	46 44	217	53 58	274	31 46	296	33 32	336
140	52 08	019	30 40	089	40 19	154	46 11	217	53 03	274	30 57	296	33 09	335
141	52 26	018	31 34	089	40 43	155	45 37	218	52 08	273	30 07	295	32 45	334
142	52 42	016	32 29	088	41 06	155	45 04	218	51 13	273	29 18	294	32 21	333
143	52 57	015	33 24	088	41 29	155	44 30	218	50 19	273	28 28	294	31 56	332
144	53 10	013	34 19	087	41 51	156	43 56	219	49 24	272	27 37	293	31 30	331
145	53 22	012	35 13	087	42 14	156	43 21	219	48 29	272	26 47	293	31 03	330
146	53 32	010	36 08	086	42 36	156	42 47	219	47 34	271	25 56	292	30 35	329
147	53 40	008	37 03	086	42 58	157	42 12	220	46 40	271	25 05	291	30 07	329
148	53 48	007	37 57	085	43 19	157	41 37	220	45 45	270	24 14	291	29 38	328
149	53 53	005	38 52	085	43 40	158	41 02	220	44 50	270	23 23	290	29 09	327
	REGULUS		◆ARCTURUS		SPICA		◆ACRUX		CANOPUS		SIRIUS		◆PROCYON	
150	53 57	003	14 15	061	39 47	085	44 01	158	40 26	220	43 55	270	44 54	306
151	54 00	002	15 03	060	40 41	084	44 21	158	39 51	220	43 00	269	44 09	305
152	54 00	000	15 51	060	41 36	084	44 41	159	39 15	220	42 06	269	43 24	304
153	54 00	358	16 38	059	42 30	083	45 01	159	38 40	221	41 11	268	42 38	303
154	53 57	357	17 25	059	43 25	083	45 20	160	38 04	221	40 16	268	41 52	302
155	53 53	355	18 12	058	44 19	082	45 39	160	37 28	221	39 21	268	41 06	301
156	53 48	353	18 58	058	45 13	081	45 57	161	36 52	221	38 26	267	40 19	301
157	53 41	352	19 45	057	46 07	081	46 15	161	36 16	221	37 32	267	39 31	300
158	53 32	350	20 31	057	47 01	080	46 33	162	35 40	221	36 37	266	38 44	299
159	53 22	349	21 16	056	47 55	080	46 50	162	35 04	221	35 42	266	37 55	298
160	53 10	347	22 01	055	48 49	079	47 06	163	34 28	221	34 48	266	37 07	298
161	52 57	345	22 46	055	49 43	079	47 23	163	33 52	221	33 53	265	36 18	297
162	52 43	344	23 31	054	50 37	078	47 38	164	33 15	221	32 58	265	35 29	296
163	52 27	342	24 15	053	51 30	077	47 53	164	32 39	222	32 04	265	34 40	295
164	52 09	341	24 59	053	52 24	077	48 08	165	32 03	222	31 09	264	33 50	295
	◆ARCTURUS		ANTARES		◆ACRUX		CANOPUS		SIRIUS		◆PROCYON		REGULUS	
165	25 42	052	16 55	112	48 22	165	31 26	222	30 15	264	33 00	294	51 50	339
166	26 25	051	17 46	112	48 35	166	30 50	222	29 20	264	32 10	293	51 30	338
167	27 08	051	18 37	111	48 48	167	30 14	222	28 26	263	31 20	293	51 09	336
168	27 50	050	19 28	111	49 00	167	29 37	222	27 31	263	30 29	292	50 46	335
169	28 32	049	20 19	111	49 12	168	29 01	222	26 37	263	29 38	292	50 22	333
170	29 13	049	21 10	110	49 23	169	28 25	221	25 43	262	28 47	291	49 57	332
171	29 54	048	22 02	110	49 34	169	27 48	221	24 48	262	27 56	290	49 31	331
172	30 35	047	22 53	110	49 44	170	27 12	221	23 54	262	27 04	290	49 03	329
173	31 15	046	23 45	110	49 53	171	26 36	221	23 00	261	26 13	289	48 35	328
174	31 54	045	24 37	109	50 02	171	26 00	221	22 06	261	25 21	289	48 05	327
175	32 33	045	25 28	109	50 10	172	25 23	221	21 12	261	24 29	288	47 35	326
176	33 11	044	26 20	109	50 18	173	24 47	221	20 18	260	23 37	288	47 03	324
177	33 49	043	27 12	109	50 24	173	24 11	221	19 24	260	22 45	287	46 31	323
178	34 26	042	28 04	108	50 31	174	23 35	221	18 30	259	21 52	287	45 58	322
179	35 02	041	28 56	108	50 36	175	23 00	221	17 36	259	21 00	286	45 23	321

LHA	Hc	Zn	Hc	Zn	Hc	Zn	Hc	Zn	Hc	Zn	Hc	Zn	Hc	Zn
♈	ARCTURUS		◆ANTARES		ACRUX		◆Suhail		Alphard		REGULUS		◆Denebola	
180	35 38	040	29 48	108	50 41	175	49 52	230	50 28	286	44 48	320	51 18	356
181	36 13	039	30 40	108	50 45	176	49 10	231	49 35	286	44 12	319	51 13	354
182	36 48	038	31 33	107	50 49	177	48 27	231	48 42	285	43 36	318	51 07	353
183	37 21	038	32 25	107	50 51	177	47 45	231	47 49	284	42 58	317	50 59	351
184	37 54	037	33 18	107	50 53	178	47 02	231	46 56	284	42 20	316	50 50	350
185	38 27	036	34 10	107	50 55	179	46 19	231	46 03	283	41 42	315	50 39	348
186	38 58	035	35 03	106	50 56	180	45 36	232	45 09	282	41 02	314	50 27	347
187	39 29	034	35 55	106	50 56	180	44 53	232	44 15	282	40 22	313	50 14	345
188	39 59	033	36 48	106	50 55	181	44 10	232	43 22	281	39 42	312	49 59	344
189	40 28	031	37 41	106	50 54	182	43 27	232	42 28	281	39 00	311	49 43	342
190	40 56	030	38 33	106	50 52	182	42 43	232	41 34	280	38 19	310	49 26	341
191	41 23	029	39 26	105	50 49	183	42 00	232	40 40	280	37 36	309	49 07	339
192	41 49	028	40 19	105	50 46	184	41 17	232	39 46	279	36 53	308	48 47	338
193	42 15	027	41 12	105	50 42	185	40 33	232	38 52	278	36 10	307	48 26	337
194	42 39	026	42 05	105	50 37	185	39 50	233	37 58	278	35 26	307	48 03	335
	ARCTURUS		◆Rasalhague		ANTARES		◆ACRUX		Suhail		◆REGULUS		Denebola	
195	43 03	025	13 38	069	42 58	105	50 31	186	39 06	233	34 42	306	47 40	334
196	43 25	023	14 29	069	43 51	104	50 25	187	38 23	233	33 57	305	47 15	333
197	43 46	022	15 20	068	44 44	104	50 19	187	37 39	233	33 12	304	46 49	331
198	44 07	021	16 11	068	45 37	104	50 11	188	36 56	233	32 27	304	46 23	330
199	44 26	020	17 02	067	46 31	104	50 03	189	36 12	233	31 41	303	45 55	329
200	44 44	019	17 53	067	47 24	104	49 55	189	35 28	233	30 55	302	45 26	328
201	45 01	017	18 43	066	48 17	103	49 45	190	34 45	233	30 08	301	44 56	327
202	45 16	016	19 33	066	49 10	103	49 36	191	34 01	233	29 21	301	44 26	325
203	45 31	015	20 23	065	50 04	103	49 25	191	33 18	233	28 34	300	43 54	324
204	45 44	013	21 12	065	50 57	103	49 14	192	32 34	233	27 46	299	43 22	323
205	45 56	012	22 02	064	51 51	103	49 02	193	31 51	232	26 59	299	42 48	322
206	46 07	011	22 51	064	52 44	103	48 50	193	31 07	232	26 10	298	42 14	321
207	46 17	009	23 40	063	53 38	103	48 37	194	30 24	232	25 22	298	41 39	320
208	46 25	008	24 29	062	54 31	102	48 24	194	29 40	232	24 33	297	41 04	319
209	46 32	007	25 17	062	55 25	102	48 10	195	28 57	232	23 44	296	40 27	318
	ARCTURUS		◆Rasalhague		ANTARES		◆RIGIL KENT.		ACRUX		Gienah		◆Denebola	
210	46 38	005	26 05	061	56 18	102	52 34	172	47 56	196	64 47	280	39 50	317
211	46 42	004	26 53	060	57 12	102	52 41	173	47 41	196	63 53	279	39 12	316
212	46 45	003	27 41	060	58 05	102	52 47	174	47 25	197	62 59	279	38 34	315
213	46 47	001	28 28	059	58 59	102	52 53	174	47 09	197	62 04	278	37 55	314
214	46 47	000	29 15	059	59 53	102	52 58	175	46 53	198	61 10	277	37 15	313
215	46 47	358	30 01	058	60 46	102	53 02	176	46 36	198	60 16	277	36 35	312
216	46 44	357	30 47	057	61 40	102	53 05	177	46 18	199	59 21	276	35 54	312
217	46 41	356	31 33	056	62 34	102	53 08	178	46 00	199	58 27	275	35 13	311
218	46 36	354	32 19	056	63 28	101	53 10	179	45 42	200	57 32	275	34 31	310
219	46 30	353	33 04	055	64 21	101	53 11	179	45 23	200	56 37	274	33 49	309
220	46 23	352	33 49	054	65 15	101	53 11	180	45 04	201	55 43	274	33 06	308
221	46 14	350	34 33	053	66 09	101	53 11	181	44 45	201	54 48	273	32 23	308
222	46 04	349	35 17	053	67 02	101	53 09	182	44 25	202	53 53	273	31 39	307
223	45 53	348	36 00	052	67 56	101	53 07	183	44 04	202	52 58	272	30 55	306
224	45 41	346	36 43	051	68 50	101	53 04	183	43 44	202	52 04	272	30 11	305
	Alphecca		Rasalhague		◆ALTAIR		Peacock		◆RIGIL KENT.		◆SPICA		ARCTURUS	
225	38 35	010	37 25	050	11 56	075	24 34	143	53 01	184	64 01	296	45 27	345
226	38 44	009	38 07	049	12 49	074	25 07	143	52 56	185	63 11	294	45 12	344
227	38 52	008	38 48	048	13 42	074	25 40	143	52 51	186	62 21	293	44 56	342
228	38 59	006	39 29	048	14 34	073	26 12	143	52 45	187	61 30	292	44 39	341
229	39 04	005	40 09	047	15 26	073	26 45	143	52 39	187	60 39	291	44 20	340
230	39 09	004	40 49	046	16 19	072	27 18	143	52 31	188	59 48	290	44 01	339
231	39 12	003	41 28	045	17 11	072	27 51	143	52 23	189	58 56	289	43 40	337
232	39 15	002	42 06	044	18 03	071	28 24	143	52 14	190	58 04	288	43 19	336
233	39 16	001	42 43	043	18 55	071	28 57	143	52 05	190	57 12	287	42 56	335
234	39 16	000	43 20	042	19 46	070	29 30	143	51 54	191	56 20	287	42 32	334
235	39 15	358	43 56	041	20 38	070	30 02	143	51 43	192	55 27	286	42 08	333
236	39 13	357	44 31	039	21 29	069	30 35	143	51 32	193	54 34	285	41 42	332
237	39 10	356	45 06	038	22 20	069	31 08	143	51 19	193	53 41	284	41 15	330
238	39 06	355	45 39	037	23 11	068	31 41	143	51 06	194	52 48	284	40 48	329
239	39 00	354	46 12	036	24 02	067	32 14	143	50 53	195	51 55	283	40 20	328
	VEGA		◆ALTAIR		Peacock		◆RIGIL KENT.		SPICA		◆ARCTURUS		Alphecca	
240	17 17	031	24 52	067	32 47	143	50 39	195	51 02	282	39 50	327	38 54	353
241	17 45	030	25 43	066	33 19	143	50 24	196	50 08	282	39 20	326	38 46	352
242	18 13	030	26 33	066	33 52	143	50 08	197	49 14	281	38 49	325	38 38	350
243	18 40	029	27 23	065	34 25	144	49 52	197	48 20	281	38 18	324	38 28	349
244	19 06	028	28 12	065	34 57	144	49 36	198	47 26	280	37 45	323	38 17	348
245	19 32	028	29 02	064	35 29	144	49 18	199	46 32	279	37 12	322	38 05	347
246	19 57	027	29 51	063	36 02	144	49 01	199	45 38	279	36 38	321	37 53	346
247	20 22	026	30 39	063	36 34	144	48 43	200	44 44	278	36 03	320	37 39	345
248	20 46	026	31 28	062	37 06	144	48 24	200	43 50	278	35 28	319	37 24	344
249	21 09	025	32 16	061	37 38	144	48 04	201	42 55	277	34 52	319	37 08	343
250	21 32	024	33 04	061	38 10	145	47 45	201	42 01	277	34 15	318	36 51	342
251	21 54	023	33 52	060	38 42	145	47 25	202	41 07	276	33 38	317	36 34	341
252	22 15	023	34 39	059	39 13	145	47 04	202	40 12	276	33 00	316	36 15	340
253	22 36	022	35 26	058	39 45	145	46 43	203	39 17	275	32 22	315	35 55	339
254	22 56	021	36 12	058	40 16	145	46 21	203	38 23	275	31 43	314	35 35	338
	◆VEGA		ALTAIR		◆Peacock		RIGIL KENT.		◆SPICA		ARCTURUS		Alphecca	
255	23 16	020	36 58	057	40 47	146	45 59	204	37 28	274	31 03	314	35 14	337
256	23 34	020	37 44	056	41 18	146	45 37	204	36 34	274	30 23	313	34 51	336
257	23 52	019	38 29	055	41 49	146	45 14	205	35 39	273	29 43	312	34 28	335
258	24 10	018	39 14	054	42 19	146	44 51	205	34 44	273	29 02	311	34 04	334
259	24 26	017	39 59	054	42 49	147	44 27	206	33 49	273	28 20	310	33 39	333
260	24 42	016	40 43	053	43 19	147	44 04	206	32 55	272	27 38	310	33 14	332
261	24 57	016	41 26	052	43 49	147	43 39	206	32 00	272	26 56	309	32 47	331
262	25 11	015	42 09	051	44 19	148	43 15	207	31 05	271	26 13	308	32 20	330
263	25 25	014	42 51	050	44 48	148	42 50	207	30 10	271	25 30	308	31 52	329
264	25 38	013	43 33	049	45 17	148	42 25	207	29 15	270	24 47	307	31 24	328
265	25 50	012	44 14	048	45 45	149	42 00	208	28 21	270	24 03	306	30 54	327
266	26 01	011	44 54	047	46 14	149	41 34	208	27 26	270	23 18	306	30 24	326
267	26 11	011	45 34	046	46 41	150	41 08	208	26 31	269	22 34	305	29 53	325
268	26 21	010	46 13	045	47 09	150	40 42	209	25 36	269	21 49	304	29 22	325
269	26 30	009	46 51	044	47 36	150	40 16	209	24 41	268	21 03	304	28 50	324

LAT 24°S

LHA	Hc	Zn	Hc	Zn	Hc	Zn	Hc	Zn	Hc	Zn	Hc	Zn	Hc	Zn
♈	◆VEGA		ALTAIR		◆FOMALHAUT		Peacock		RIGIL KENT.		◆ANTARES		Alphecca	
270	26 38	008	47 29	043	24 33	113	48 03	151	39 49	209	69 19	258	28 17	323
271	26 45	007	48 06	042	25 23	113	48 30	151	39 22	209	68 26	259	27 44	322
272	26 52	006	48 42	040	26 14	113	48 56	152	38 55	210	67 32	259	27 10	321
273	26 57	005	49 17	039	27 05	112	49 21	152	38 28	210	66 38	259	26 36	321
274	27 02	005	49 51	038	27 55	112	49 46	153	38 00	210	65 45	259	26 01	320
275	27 06	004	50 24	037	28 46	112	50 11	154	37 33	210	64 51	259	25 25	319
276	27 09	003	50 56	035	29 37	112	50 35	154	37 05	211	63 57	259	24 49	318
277	27 11	002	51 27	034	30 28	111	50 59	155	36 37	211	63 03	259	24 12	318
278	27 13	001	51 57	033	31 19	111	51 22	155	36 09	211	62 10	258	23 35	317
279	27 13	000	52 26	031	32 10	111	51 44	156	35 41	211	61 16	258	22 58	316
280	27 13	359	52 54	030	33 01	111	52 06	157	35 13	211	60 22	258	22 20	316
281	27 12	358	53 21	028	33 53	111	52 28	157	34 44	211	59 29	258	21 41	315
282	27 10	358	53 46	027	34 44	111	52 48	158	34 16	211	58 35	258	21 02	314
283	27 07	357	54 10	025	35 35	110	53 09	159	33 47	212	57 41	258	20 23	314
284	27 04	356	54 33	024	36 27	110	53 28	160	33 18	212	56 48	258	19 43	313
	◆DENEB		FOMALHAUT		◆ACHERNAR		RIGIL KENT.		◆ANTARES		Rasalhague		VEGA	
285	17 00	018	37 18	110	15 10	146	32 49	212	55 54	258	47 55	328	26 59	355
286	17 17	018	38 10	110	15 40	146	32 21	212	55 01	258	47 25	327	26 54	354
287	17 33	017	39 01	110	16 11	146	31 52	212	54 07	258	46 55	326	26 48	353
288	17 49	016	39 53	109	16 41	146	31 22	212	53 14	257	46 23	324	26 41	352
289	18 04	016	40 45	109	17 12	146	30 53	212	52 20	257	45 51	323	26 33	351
290	18 18	015	41 36	109	17 43	146	30 24	212	51 27	257	45 18	322	26 25	351
291	18 32	014	42 28	109	18 14	145	29 55	212	50 33	257	44 44	321	26 15	350
292	18 45	014	43 20	109	18 46	145	29 26	212	49 40	257	44 09	320	26 05	349
293	18 58	013	44 12	109	19 17	145	28 57	212	48 46	257	43 33	319	25 54	348
294	19 10	012	45 04	109	19 48	145	28 27	212	47 53	256	42 56	318	25 42	347
295	19 21	011	45 56	109	20 20	145	27 58	212	47 00	256	42 19	317	25 30	346
296	19 31	011	46 48	108	20 52	145	27 29	212	46 07	256	41 41	316	25 17	346
297	19 41	010	47 40	108	21 23	145	27 00	212	45 13	256	41 02	315	25 03	345
298	19 50	009	48 32	108	21 55	144	26 30	212	44 20	256	40 23	314	24 48	344
299	19 58	008	49 24	108	22 27	144	26 01	212	43 27	256	39 43	313	24 32	343
	DENEB		Enif		◆FOMALHAUT		ACHERNAR		◆RIGIL KENT.		ANTARES		◆VEGA	
300	20 06	008	47 41	040	50 16	108	22 59	144	25 32	212	42 34	255	24 16	342
301	20 13	007	48 16	039	51 08	108	23 31	144	25 03	212	41 41	255	23 59	342
302	20 19	006	48 50	037	52 00	108	24 03	144	24 34	212	40 48	255	23 41	341
303	20 25	005	49 23	036	52 52	108	24 36	144	24 05	212	39 55	255	23 23	340
304	20 30	005	49 55	035	53 44	108	25 08	144	23 36	212	39 02	255	23 04	339
305	20 34	004	50 25	034	54 36	108	25 40	144	23 07	212	38 10	254	22 44	338
306	20 38	003	50 55	032	55 29	108	26 12	144	22 38	212	37 17	254	22 23	338
307	20 40	002	51 24	031	56 21	108	26 45	144	22 09	212	36 24	254	22 02	337
308	20 42	002	51 52	030	57 13	108	27 17	144	21 40	212	35 31	254	21 40	336
309	20 44	001	52 18	028	58 05	108	27 50	144	21 11	212	34 39	253	21 18	335
310	20 44	000	52 43	027	58 57	108	28 22	144	20 43	211	33 46	253	20 55	335
311	20 44	359	53 07	025	59 49	108	28 55	144	20 14	211	32 54	253	20 31	334
312	20 43	359	53 30	024	60 41	108	29 27	144	19 46	211	32 02	253	20 07	333
313	20 42	358	53 51	022	61 33	108	29 59	144	19 17	211	31 09	253	19 42	333
314	20 39	357	54 11	020	62 25	108	30 32	144	18 49	211	30 17	252	19 16	332
	Alpheratz		◆Diphda		ACHERNAR		◆Peacock		ANTARES		◆ALTAIR		DENEB	
315	20 17	043	37 51	095	31 04	144	56 39	189	29 25	252	53 02	331	20 36	356
316	20 54	042	38 46	094	31 37	144	56 30	190	28 33	252	52 34	329	20 33	356
317	21 31	042	39 41	094	32 09	144	56 20	191	27 41	252	52 05	328	20 28	355
318	22 07	041	40 35	094	32 41	144	56 10	192	26 49	251	51 36	326	20 23	354
319	22 43	040	41 30	093	33 14	144	55 59	192	25 57	251	51 05	325	20 17	353
320	23 18	040	42 25	093	33 46	144	55 46	193	25 05	251	50 33	324	20 11	353
321	23 53	039	43 19	093	34 18	144	55 33	194	24 13	251	50 00	322	20 03	352
322	24 27	038	44 14	092	34 50	144	55 19	195	23 22	250	49 26	321	19 55	351
323	25 00	037	45 09	092	35 22	144	55 05	196	22 30	250	48 51	320	19 47	351
324	25 33	037	46 04	091	35 54	145	54 49	197	21 39	250	48 16	319	19 37	350
325	26 06	036	46 59	091	36 25	145	54 33	198	20 47	249	47 39	318	19 27	349
326	26 38	035	47 53	091	36 57	145	54 16	198	19 56	249	47 02	316	19 17	348
327	27 09	034	48 48	090	37 29	145	53 58	199	19 05	249	46 23	315	19 05	348
328	27 39	034	49 43	090	38 00	145	53 40	200	18 14	249	45 45	314	18 53	347
329	28 09	033	50 38	089	38 31	145	53 21	201	17 23	248	45 05	313	18 41	346
	◆Alpheratz		Diphda		◆ACHERNAR		Peacock		◆Nunki		ALTAIR		DENEB	
330	28 39	032	51 33	089	39 02	146	53 01	201	48 17	257	44 25	312	18 27	346
331	29 07	031	52 27	089	39 33	146	52 41	202	47 24	257	43 44	311	18 13	345
332	29 35	030	53 22	088	40 04	146	52 20	203	46 30	256	43 02	310	17 58	344
333	30 02	029	54 17	088	40 35	146	51 58	204	45 37	256	42 20	309	17 43	343
334	30 29	028	55 12	087	41 05	146	51 36	204	44 44	256	41 38	308	17 27	343
335	30 55	028	56 07	087	41 35	147	51 13	205	43 51	256	40 54	307	17 11	342
336	31 20	027	57 01	086	42 05	147	50 50	205	42 58	256	40 11	307	16 53	341
337	31 44	026	57 56	086	42 35	147	50 26	206	42 05	255	39 26	306	16 36	341
338	32 07	025	58 51	085	43 04	148	50 02	207	41 12	255	38 42	305	16 17	340
339	32 30	024	59 45	085	43 34	148	49 37	207	40 19	255	37 56	304	15 58	339
340	32 52	023	60 40	084	44 03	148	49 12	208	39 26	255	37 11	303	15 39	339
341	33 13	022	61 34	084	44 31	149	48 46	208	38 33	255	36 25	303	15 19	338
342	33 33	021	62 29	083	45 00	149	48 20	209	37 40	254	35 38	302	14 58	338
343	33 52	020	63 23	082	45 28	149	47 53	209	36 47	254	34 52	301	14 37	337
344	34 10	019	64 17	082	45 56	150	47 26	210	35 54	254	34 05	300	14 15	336
	Alpheratz		◆Hamal		◆ACHERNAR		Peacock		Nunki		◆ALTAIR		Enif	
345	34 28	018	24 23	047	46 23	150	46 59	210	35 02	254	33 17	300	51 23	329
346	34 45	017	25 03	046	46 50	151	46 31	211	34 09	253	32 29	299	50 54	328
347	35 00	016	25 42	046	47 17	151	46 03	211	33 17	253	31 41	298	50 25	326
348	35 15	015	26 21	045	47 43	152	45 35	211	32 24	253	30 53	298	49 54	325
349	35 29	014	27 00	044	48 09	152	45 06	212	31 32	253	30 04	297	49 22	324
350	35 41	013	27 38	044	48 35	153	44 37	212	30 40	253	29 15	296	48 49	322
351	35 53	012	28 15	043	49 00	153	44 08	212	29 47	252	28 26	296	48 15	321
352	36 04	011	28 52	042	49 24	154	43 38	213	28 55	252	27 36	295	47 40	320
353	36 14	010	29 29	041	49 49	154	43 08	213	28 03	252	26 46	294	47 05	319
354	36 23	009	30 05	040	50 12	155	42 38	213	27 11	252	25 56	294	46 28	318
355	36 30	008	30 40	040	50 35	155	42 08	214	26 19	251	25 06	293	45 51	317
356	36 37	007	31 15	039	50 58	156	41 37	214	25 27	251	24 15	293	45 13	316
357	36 43	005	31 49	038	51 20	157	41 07	214	24 35	251	23 25	292	44 34	315
358	36 48	004	32 22	037	51 42	157	40 36	215	23 44	250	22 34	292	43 55	314
359	36 51	003	32 55	036	52 03	158	40 04	215	22 52	250	21 43	291	43 15	313

LHA ♈	Hc	Zn	Hc	Zn	Hc	Zn	Hc	Zn	Hc	Zn	Hc	Zn	Hc	Zn
	◆Alpheratz		Hamal		◆RIGEL		CANOPUS		ACHERNAR		◆Peacock		Enif	
0	35 54	002	32 38	035	13 46	093	16 12	141	53 19	158	40 22	215	41 54	312
1	35 56	001	33 09	034	14 41	092	16 47	141	53 39	159	39 51	216	41 13	311
2	35 56	000	33 39	033	15 35	092	17 21	141	53 58	160	39 19	216	40 32	310
3	35 56	359	34 08	032	16 29	092	17 56	140	54 17	160	38 47	216	39 51	309
4	35 54	358	34 37	031	17 24	091	18 30	140	54 35	161	38 15	216	39 08	309
5	35 52	357	35 04	030	18 18	091	19 05	140	54 52	162	37 43	216	38 26	308
6	35 48	356	35 31	029	19 12	090	19 40	140	55 09	163	37 10	216	37 42	307
7	35 44	355	35 58	028	20 07	090	20 15	140	55 25	163	36 38	217	36 58	306
8	35 38	354	36 23	027	21 01	089	20 50	140	55 40	164	36 06	217	36 14	305
9	35 31	353	36 47	026	21 55	089	21 25	139	55 54	165	35 33	217	35 30	304
10	35 24	351	37 11	025	22 50	089	22 01	139	56 07	166	35 01	217	34 44	304
11	35 15	350	37 34	024	23 44	088	22 36	139	56 20	167	34 28	217	33 59	303
12	35 06	349	37 56	023	24 39	088	23 12	139	56 32	168	33 55	217	33 13	302
13	34 55	348	38 16	022	25 33	087	23 48	139	56 43	169	33 22	217	32 27	301
14	34 44	347	38 36	021	26 27	087	24 24	139	56 53	170	32 50	217	31 40	301
	◆Hamal		ALDEBARAN		RIGEL		◆CANOPUS		ACHERNAR		◆FOMALHAUT		Alpheratz	
15	38 55	020	23 04	057	27 21	086	24 59	139	57 02	171	62 27	253	34 31	346
16	39 13	019	23 50	057	28 16	086	25 35	139	57 11	172	61 35	253	34 18	345
17	39 30	018	24 35	056	29 10	085	26 11	138	57 18	173	60 43	253	34 03	344
18	39 46	016	25 20	055	30 04	085	26 47	138	57 25	174	59 51	254	33 48	343
19	40 01	015	26 05	055	30 58	084	27 24	138	57 30	175	58 59	254	33 32	342
20	40 15	014	26 49	054	31 52	084	28 00	138	57 35	176	58 07	254	33 15	341
21	40 28	013	27 33	053	32 46	084	28 36	138	57 39	177	57 14	254	32 57	340
22	40 39	012	28 16	053	33 40	083	29 12	138	57 42	178	56 22	253	32 38	339
23	40 50	011	28 59	052	34 34	083	29 49	138	57 43	179	55 30	253	32 18	338
24	40 59	009	29 42	051	35 28	082	30 25	138	57 44	180	54 38	253	31 58	337
25	41 07	008	30 24	050	36 22	082	31 01	138	57 44	181	53 46	253	31 36	336
26	41 15	007	31 06	050	37 16	081	31 38	138	57 43	182	52 54	253	31 14	335
27	41 21	006	31 47	049	38 09	080	32 14	138	57 41	183	52 02	253	30 51	335
28	41 26	005	32 28	048	39 03	080	32 50	138	57 38	184	51 10	253	30 28	334
29	41 29	003	33 08	047	39 57	079	33 27	138	57 34	185	50 18	253	30 03	333
	Hamal		◆ALDEBARAN		SIRIUS		◆CANOPUS		ACHERNAR		◆FOMALHAUT		Alpheratz	
30	41 32	002	33 48	046	23 38	098	34 03	138	57 29	186	49 26	253	29 38	332
31	41 33	001	34 27	046	24 32	098	34 39	138	57 23	187	48 34	253	29 12	331
32	41 34	000	35 05	045	25 25	097	35 16	138	57 17	188	47 42	253	28 45	330
33	41 33	358	35 43	044	26 19	097	35 52	138	57 09	189	46 50	253	28 18	329
34	41 31	357	36 21	043	27 13	097	36 28	138	57 00	190	45 58	252	27 49	328
35	41 27	356	36 57	042	28 07	096	37 04	138	56 51	190	45 06	252	27 21	328
36	41 23	355	37 33	041	29 01	096	37 40	138	56 41	191	44 15	252	26 51	327
37	41 18	354	38 09	040	29 56	096	38 16	139	56 29	192	43 23	252	26 21	326
38	41 11	352	38 43	039	30 50	095	38 52	139	56 17	193	42 31	252	25 50	325
39	41 03	351	39 17	038	31 44	095	39 28	139	56 04	194	41 40	252	25 19	324
40	40 54	350	39 51	037	32 38	094	40 04	139	55 51	195	40 48	251	24 47	324
41	40 44	349	40 23	036	33 32	094	40 39	139	55 36	196	39 56	251	24 15	323
42	40 33	348	40 55	035	34 26	094	41 15	139	55 21	197	39 05	251	23 42	322
43	40 21	346	41 25	034	35 21	093	41 50	140	55 05	198	38 14	251	23 08	321
44	40 07	345	41 55	033	36 15	093	42 26	140	54 48	198	37 22	251	22 34	321
	◆CAPELLA		BETELGEUSE		SIRIUS		◆CANOPUS		ACHERNAR		◆FOMALHAUT		Hamal	
45	12 34	024	36 31	059	37 09	093	43 01	140	54 31	199	36 31	250	39 53	344
46	12 55	023	37 17	058	38 04	092	43 36	140	54 12	200	35 40	250	39 37	343
47	13 16	022	38 03	057	38 58	092	44 10	140	53 54	201	34 49	250	39 21	342
48	13 37	022	38 48	056	39 52	091	44 45	141	53 34	201	33 57	250	39 03	341
49	13 56	021	39 33	055	40 47	091	45 19	141	53 14	202	33 06	250	38 45	340
50	14 16	020	40 18	054	41 41	091	45 53	141	52 53	203	32 16	249	38 25	338
51	14 34	020	41 02	054	42 36	090	46 27	142	52 32	203	31 25	249	38 05	337
52	14 52	019	41 45	053	43 30	090	47 01	142	52 10	204	30 34	249	37 44	336
53	15 10	018	42 28	052	44 24	089	47 34	142	51 47	205	29 43	249	37 21	335
54	15 27	018	43 10	051	45 19	089	48 08	143	51 24	205	28 52	249	36 58	334
55	15 43	017	43 52	050	46 13	088	48 40	143	51 01	206	28 02	248	36 34	333
56	15 59	016	44 33	049	47 07	088	49 13	143	50 37	207	27 11	248	36 09	332
57	16 14	016	45 14	048	48 02	087	49 45	144	50 12	207	26 21	248	35 43	331
58	16 28	015	45 54	047	48 56	087	50 17	144	49 47	208	25 31	248	35 16	330
59	16 42	014	46 33	046	49 50	087	50 49	145	49 22	208	24 41	247	34 49	329
	CAPELLA		BETELGEUSE		◆SIRIUS		CANOPUS		◆ACHERNAR		Diphda		◆Hamal	
60	16 55	014	47 12	045	50 45	086	51 20	145	48 56	209	43 58	269	34 21	328
61	17 08	013	47 49	043	51 39	086	51 51	146	48 30	209	43 03	268	33 52	327
62	17 20	012	48 26	042	52 33	085	52 21	146	48 03	210	42 09	268	33 22	326
63	17 31	012	49 03	041	53 27	084	52 51	147	47 36	210	41 15	267	32 52	325
64	17 42	011	49 38	040	54 21	084	53 20	147	47 09	210	40 21	267	32 20	325
65	17 52	010	50 12	039	55 15	083	53 49	148	46 41	211	39 26	267	31 48	324
66	18 01	010	50 46	037	56 09	083	54 18	149	46 13	211	38 32	266	31 16	323
67	18 10	009	51 18	036	57 03	082	54 46	149	45 44	212	37 38	266	30 43	322
68	18 18	008	51 49	035	57 57	082	55 13	150	45 16	212	36 44	266	30 09	321
69	18 25	007	52 20	033	58 51	081	55 40	151	44 47	212	35 49	265	29 34	320
70	18 32	007	52 49	032	59 44	080	56 06	152	44 17	213	34 55	265	28 59	319
71	18 38	006	53 17	030	60 38	080	56 32	152	43 48	213	34 01	264	28 24	319
72	18 43	005	53 44	029	61 31	079	56 57	153	43 18	213	33 07	264	27 48	318
73	18 48	004	54 09	027	62 25	078	57 21	154	42 48	214	32 13	264	27 11	317
74	18 51	004	54 34	026	63 18	077	57 44	155	42 18	214	31 19	263	26 34	316
	◆CAPELLA		POLLUX		PROCYON		◆Suhail		CANOPUS		◆ACHERNAR		Hamal	
75	18 55	003	23 45	039	40 56	057	36 52	127	58 07	156	41 48	214	25 56	316
76	18 57	002	24 19	039	41 42	057	37 36	127	58 29	157	41 17	214	25 18	315
77	18 59	002	24 53	038	42 27	056	38 19	127	58 50	158	40 46	215	24 39	314
78	19 00	001	25 26	037	43 12	055	39 03	127	59 10	159	40 15	215	24 00	313
79	19 00	000	25 59	036	43 56	054	39 47	127	59 30	160	39 44	215	23 20	313
80	19 00	359	26 31	036	44 40	053	40 30	127	59 48	161	39 13	215	22 40	312
81	18 59	359	27 02	035	45 23	052	41 14	127	60 06	162	38 42	215	21 59	311
82	18 57	358	27 33	034	46 05	051	41 57	127	60 22	163	38 10	216	21 18	311
83	18 55	357	28 03	033	46 47	050	42 41	127	60 38	164	37 38	216	20 37	310
84	18 52	356	28 33	032	47 28	049	43 24	127	60 52	165	37 07	216	19 55	309
85	18 48	356	29 02	032	48 09	048	44 07	127	61 06	166	36 35	216	19 13	309
86	18 44	355	29 30	031	48 49	047	44 51	127	61 18	167	36 03	216	18 30	308
87	18 39	354	29 57	030	49 28	046	45 34	127	61 30	169	35 31	216	17 47	308
88	18 33	353	30 24	029	50 07	044	46 17	128	61 40	170	34 59	216	17 04	307
89	18 26	353	30 50	028	50 44	043	47 00	128	61 49	171	34 27	216	16 21	306

LAT 25°S

LHA ♈	Hc	Zn	Hc	Zn	Hc	Zn	Hc	Zn	Hc	Zn	Hc	Zn	Hc	Zn
	POLLUX		◆REGULUS		ACRUX		CANOPUS		◆ACHERNAR		◆ALDEBARAN		CAPELLA	
90	31 15	027	19 09	066	19 15	152	61 57	172	33 54	216	43 41	331	18 19	352
91	31 39	026	19 59	066	19 41	151	62 03	174	33 22	216	43 15	330	18 11	351
92	32 03	025	20 48	065	20 07	151	62 09	175	32 50	217	42 48	329	18 03	351
93	32 26	024	21 37	064	20 34	151	62 13	176	32 17	217	42 19	328	17 53	350
94	32 48	023	22 26	064	21 00	151	62 16	177	31 45	217	41 50	327	17 44	349
95	33 09	022	23 15	063	21 26	151	62 18	179	31 12	217	41 20	326	17 33	348
96	33 30	021	24 03	063	21 53	151	62 18	180	30 40	217	40 49	325	17 22	348
97	33 49	021	24 51	062	22 19	151	62 18	181	30 08	217	40 17	324	17 10	347
98	34 08	020	25 39	061	22 46	151	62 16	183	29 35	217	39 45	323	16 57	346
99	34 25	019	26 27	061	23 13	151	62 13	184	29 03	217	39 11	322	16 44	346
100	34 42	017	27 14	060	23 40	150	62 08	185	28 30	217	38 37	321	16 31	345
101	34 58	016	28 01	059	24 06	150	62 03	187	27 58	217	38 02	320	16 16	344
102	35 13	015	28 48	059	24 33	150	61 56	188	27 25	217	37 27	319	16 01	344
103	35 27	014	29 34	058	25 00	150	61 48	189	26 53	216	36 51	318	15 46	343
104	35 40	013	30 20	057	25 27	150	61 39	190	26 21	216	36 14	317	15 29	342
	POLLUX		◆REGULUS		Gienah		◆ACRUX		CANOPUS		◆RIGEL		BETELGEUSE	
105	35 52	012	31 06	057	17 05	102	25 54	150	61 29	192	59 44	299	53 56	332
106	36 03	011	31 51	056	17 58	101	26 21	150	61 17	193	58 56	298	53 29	330
107	36 13	010	32 36	055	18 52	101	26 49	150	61 05	194	58 08	297	53 02	329
108	36 22	009	33 20	054	19 45	101	27 16	150	60 51	195	57 19	296	52 33	327
109	36 30	008	34 05	054	20 39	100	27 43	150	60 37	196	56 30	295	52 03	326
110	36 37	007	34 48	053	21 32	100	28 10	150	60 21	197	55 40	294	51 32	325
111	36 43	006	35 31	052	22 26	100	28 37	150	60 04	198	54 50	293	51 00	323
112	36 48	005	36 14	051	23 19	099	29 04	150	59 47	199	54 00	292	50 27	322
113	36 52	004	36 56	050	24 13	099	29 32	150	59 28	200	53 09	291	49 53	321
114	36 55	002	37 38	050	25 07	099	29 59	150	59 09	201	52 18	290	49 19	320
115	36 57	001	38 19	049	26 01	098	30 26	150	58 48	202	51 27	289	48 43	318
116	36 58	000	38 59	048	26 54	098	30 53	150	58 27	203	50 36	289	48 06	317
117	36 57	359	39 39	047	27 48	097	31 20	150	58 05	204	49 44	288	47 29	316
118	36 56	358	40 19	046	28 42	097	31 47	150	57 42	205	48 52	287	46 51	315
119	36 54	357	40 57	045	29 36	097	32 14	150	57 19	206	48 00	286	46 12	314
	◆REGULUS		SPICA		◆ACRUX		CANOPUS		◆RIGEL		BETELGEUSE		POLLUX	
120	41 35	044	12 33	097	32 41	150	56 55	207	47 08	286	45 32	313	36 50	356
121	42 13	043	13 27	096	33 08	151	56 30	208	46 16	285	44 52	312	36 46	355
122	42 49	042	14 21	096	33 34	151	56 04	208	45 23	284	44 11	311	36 40	354
123	43 25	041	15 15	095	34 01	151	55 38	209	44 30	284	43 29	310	36 34	353
124	44 00	040	16 09	095	34 28	151	55 11	210	43 37	283	42 47	309	36 26	351
125	44 35	039	17 03	095	34 54	151	54 44	211	42 44	282	42 04	308	36 18	350
126	45 08	037	17 57	094	35 20	151	54 16	211	41 51	282	41 21	307	36 08	349
127	45 41	036	18 52	094	35 47	151	53 47	212	40 58	281	40 37	306	35 58	348
128	46 13	035	19 46	093	36 13	151	53 18	213	40 04	281	39 53	305	35 46	347
129	46 44	034	20 40	093	36 39	152	52 49	213	39 11	280	39 08	304	35 34	346
130	47 13	033	21 34	093	37 05	152	52 19	214	38 17	280	38 23	303	35 20	345
131	47 42	031	22 29	092	37 30	152	51 48	214	37 24	279	37 38	303	35 06	344
132	48 10	030	23 23	092	37 56	152	51 17	215	36 30	279	36 52	302	34 50	343
133	48 37	029	24 18	091	38 21	152	50 46	215	35 36	278	36 05	301	34 34	342
134	49 03	028	25 12	091	38 46	153	50 15	216	34 42	278	35 18	300	34 17	341
	REGULUS		◆SPICA		ACRUX		◆CANOPUS		SIRIUS		BETELGEUSE		◆POLLUX	
135	49 27	026	26 06	091	39 11	153	49 43	216	57 28	278	34 31	300	33 59	340
136	49 51	025	27 01	090	39 36	153	49 10	217	56 34	278	33 44	299	33 39	339
137	50 13	023	27 55	090	40 01	153	48 38	217	55 40	277	32 56	298	33 20	338
138	50 34	022	28 49	089	40 25	154	48 05	217	54 46	276	32 08	297	32 59	337
139	50 53	020	29 44	089	40 49	154	47 32	218	53 52	276	31 20	297	32 37	336
140	51 12	019	30 38	088	41 13	154	46 58	218	52 58	275	30 31	296	32 15	335
141	51 29	017	31 32	088	41 37	154	46 25	218	52 04	275	29 42	295	31 51	334
142	51 44	016	32 27	087	42 00	155	45 51	219	51 10	274	28 53	295	31 27	333
143	51 59	014	33 21	087	42 23	155	45 17	219	50 15	274	28 03	294	31 03	332
144	52 11	013	34 15	087	42 46	155	44 42	219	49 21	273	27 13	294	30 37	331
145	52 23	011	35 10	086	43 09	156	44 08	220	48 27	273	26 24	293	30 11	331
146	52 33	010	36 04	086	43 31	156	43 33	220	47 33	272	25 33	292	29 44	330
147	52 41	008	36 58	085	43 53	156	42 58	220	46 38	272	24 43	292	29 16	329
148	52 48	007	37 52	085	44 14	157	42 23	220	45 44	271	23 52	291	28 47	328
149	52 53	005	38 46	084	44 36	157	41 47	221	44 50	271	23 02	291	28 18	327
	REGULUS		◆ARCTURUS		SPICA		◆ACRUX		CANOPUS		SIRIUS		◆PROCYON	
150	52 57	003	13 46	061	39 41	084	44 57	158	41 12	221	43 55	270	44 19	307
151	53 00	002	14 34	060	40 35	083	45 17	158	40 37	221	43 01	270	43 35	306
152	53 00	000	15 21	060	41 28	083	45 37	158	40 01	221	42 06	270	42 50	305
153	53 00	358	16 07	059	42 22	082	45 57	159	39 25	221	41 12	269	42 05	304
154	52 57	357	16 54	059	43 16	082	46 16	159	38 49	221	40 18	269	41 20	303
155	52 54	355	17 40	058	44 10	081	46 35	160	38 13	221	39 23	268	40 34	302
156	52 48	354	18 26	057	45 04	080	46 54	160	37 37	222	38 29	268	39 48	301
157	52 41	352	19 12	057	45 57	080	47 12	161	37 01	222	37 35	268	39 01	300
158	52 33	350	19 57	056	46 51	079	47 30	161	36 25	222	36 40	267	38 14	300
159	52 23	349	20 42	056	47 44	079	47 47	162	35 49	222	35 46	267	37 27	299
160	52 12	347	21 27	055	48 37	078	48 04	162	35 13	222	34 52	266	36 39	298
161	51 59	346	22 12	054	49 30	077	48 20	163	34 36	222	33 57	266	35 51	297
162	51 45	344	22 56	054	50 23	077	48 36	163	34 00	222	33 03	265	35 02	297
163	51 29	343	23 39	053	51 16	076	48 51	164	33 24	222	32 09	265	34 14	296
164	51 13	341	24 23	052	52 09	075	49 06	165	32 47	222	31 15	265	33 25	295
	◆ARCTURUS		ANTARES		◆ACRUX		CANOPUS		SIRIUS		◆PROCYON		REGULUS	
165	25 05	052	17 17	112	49 20	165	32 11	222	30 21	265	32 35	295	50 54	340
166	25 48	051	18 08	111	49 33	166	31 35	222	29 27	264	31 46	294	50 35	338
167	26 30	050	18 58	111	49 46	166	30 58	222	28 32	264	30 56	293	50 14	337
168	27 12	050	19 49	111	49 59	167	30 22	222	27 38	263	30 06	293	49 52	335
169	27 53	049	20 40	110	50 11	168	29 46	222	26 44	263	29 16	292	49 28	334
170	28 34	048	21 31	110	50 22	168	29 09	222	25 50	263	28 25	292	49 04	333
171	29 14	047	22 22	110	50 33	169	28 33	222	24 56	262	27 35	291	48 38	331
172	29 54	047	23 14	110	50 43	170	27 57	222	24 03	262	26 44	290	48 12	330
173	30 33	046	24 05	109	50 52	170	27 21	222	23 09	262	25 53	290	47 44	329
174	31 12	045	24 56	109	51 01	171	26 45	222	22 15	261	25 01	289	47 15	327
175	31 50	044	25 48	109	51 10	172	26 08	222	21 21	261	24 10	289	46 45	326
176	32 27	043	26 39	108	51 17	172	25 32	221	20 28	261	23 18	288	46 14	325
177	33 05	043	27 31	108	51 24	173	24 57	221	19 34	260	22 27	288	45 43	324
178	33 41	042	28 23	108	51 30	174	24 21	221	18 40	260	21 35	287	45 10	323
179	34 17	041	29 14	108	51 36	174	23 45	221	17 47	259	20 43	286	44 37	321

LHA	Hc	Zn	Hc	Zn	Hc	Zn	Hc	Zn	Hc	Zn	Hc	Zn	Hc	Zn
♈	ARCTURUS		◆ANTARES		ACRUX		◆Suhail		Alphard		REGULUS		◆Denebola	
180	34 52	040	30 06	107	51 41	175	50 30	231	50 10	287	44 02	320	50 18	356
181	35 27	039	30 58	107	51 45	176	49 47	231	49 18	287	43 27	319	50 14	354
182	36 00	038	31 50	107	51 48	177	49 05	232	48 26	286	42 51	318	50 07	353
183	36 34	037	32 42	107	51 51	177	48 22	232	47 34	285	42 15	317	50 00	351
184	37 06	036	33 35	106	51 53	178	47 39	232	46 41	285	41 37	316	49 51	350
185	37 38	035	34 27	106	51 55	179	46 56	232	45 49	284	40 59	315	49 41	348
186	38 09	034	35 19	106	51 56	180	46 13	232	44 56	283	40 21	314	49 29	347
187	38 39	033	36 12	105	51 56	180	45 30	233	44 03	283	39 41	313	49 16	345
188	39 08	032	37 04	105	51 55	181	44 47	233	43 10	282	39 01	312	49 01	344
189	39 36	031	37 56	105	51 54	182	44 03	233	42 16	282	38 21	311	48 46	343
190	40 04	030	38 49	105	51 52	183	43 20	233	41 23	281	37 40	311	48 29	341
191	40 31	029	39 42	105	51 49	183	42 36	233	40 30	280	36 58	310	48 11	340
192	40 56	028	40 34	104	51 46	184	41 53	233	39 36	280	36 16	309	47 51	338
193	41 21	027	41 27	104	51 41	185	41 10	233	38 42	279	35 34	308	47 31	337
194	41 45	025	42 20	104	51 37	185	40 26	233	37 49	279	34 50	307	47 09	336
	ARCTURUS		◆Rasalhague		ANTARES		◆ACRUX		Suhail		◆REGULUS		Denebola	
195	42 08	024	13 17	069	43 13	104	51 31	186	39 42	233	34 07	306	46 46	334
196	42 30	023	14 08	069	44 05	103	51 25	187	38 59	233	33 23	306	46 22	333
197	42 51	022	14 58	068	44 58	103	51 18	188	38 15	233	32 38	305	45 57	332
198	43 11	021	15 49	068	45 51	103	51 11	188	37 32	233	31 54	304	45 31	331
199	43 29	019	16 39	067	46 44	103	51 03	189	36 48	233	31 08	303	45 03	329
200	43 47	018	17 29	067	47 37	103	50 54	190	36 05	233	30 23	303	44 35	328
201	44 03	017	18 19	066	48 31	102	50 45	190	35 21	233	29 37	302	44 06	327
202	44 19	016	19 08	065	49 24	102	50 35	191	34 37	233	28 50	301	43 36	326
203	44 33	014	19 57	065	50 17	102	50 24	192	33 54	233	28 04	301	43 05	325
204	44 46	013	20 47	064	51 10	102	50 13	192	33 10	233	27 17	300	42 33	324
205	44 58	012	21 35	064	52 03	102	50 01	193	32 27	233	26 29	299	42 01	323
206	45 08	011	22 24	063	52 57	101	49 48	193	31 44	233	25 42	299	41 27	322
207	45 17	009	23 12	063	53 50	101	49 36	194	31 00	233	24 54	298	40 53	320
208	45 26	008	24 01	062	54 43	101	49 22	195	30 17	233	24 06	297	40 18	319
209	45 32	007	24 48	061	55 37	101	49 08	195	29 34	233	23 17	297	39 42	318
	ARCTURUS		◆Rasalhague		ANTARES		◆RIGIL KENT.		ACRUX		Gienah		◆Denebola	
210	45 38	005	25 36	061	56 30	101	53 33	172	48 53	196	64 35	282	39 06	318
211	45 42	004	26 23	060	57 24	100	53 41	173	48 38	196	63 42	281	38 29	317
212	45 45	003	27 10	059	58 17	100	53 47	174	48 22	197	62 49	280	37 51	316
213	45 47	001	27 57	059	59 11	100	53 53	174	48 06	198	61 55	280	37 13	315
214	45 47	000	28 43	058	60 04	100	53 58	175	47 50	198	61 01	279	36 34	314
215	45 47	358	29 29	057	60 58	100	54 02	176	47 33	199	60 08	278	35 54	313
216	45 45	357	30 15	057	61 51	100	54 05	177	47 15	199	59 14	278	35 14	312
217	45 41	356	31 00	056	62 45	100	54 08	178	46 57	200	58 20	277	34 34	311
218	45 36	354	31 45	055	63 38	099	54 10	178	46 38	200	57 26	276	33 53	310
219	45 31	353	32 29	054	64 32	099	54 11	179	46 19	201	56 32	276	33 11	310
220	45 23	352	33 13	054	65 26	099	54 11	180	46 00	201	55 38	275	32 29	309
221	45 15	350	33 57	053	66 19	099	54 11	181	45 40	201	54 44	275	31 46	308
222	45 05	349	34 40	052	67 13	099	54 09	182	45 20	202	53 49	274	31 03	307
223	44 54	348	35 23	051	68 07	099	54 07	183	45 00	202	52 55	274	30 20	307
224	44 42	346	36 05	050	69 01	099	54 04	183	44 39	203	52 01	273	29 36	306
	Alphecca		Rasalhague		◆ALTAIR		Peacock		◆RIGIL KENT.		◆SPICA		ARCTURUS	
225	37 36	010	36 47	050	11 40	074	25 22	143	54 00	184	63 34	297	44 29	345
226	37 45	009	37 28	049	12 32	074	25 55	143	53 56	185	62 45	296	44 14	344
227	37 53	007	38 08	048	13 24	073	26 28	143	53 51	186	61 56	295	43 59	343
228	37 59	006	38 48	047	14 17	073	27 00	143	53 45	187	61 07	294	43 42	341
229	38 05	005	39 28	046	15 08	072	27 33	143	53 38	188	60 17	293	43 24	340
230	38 09	004	40 06	045	16 00	072	28 06	143	53 31	188	59 26	292	43 05	339
231	38 12	003	40 45	044	16 52	071	28 39	143	53 22	189	58 36	291	42 45	338
232	38 15	002	41 22	043	17 43	071	29 12	143	53 13	190	57 45	290	42 24	337
233	38 16	001	41 59	042	18 35	070	29 45	143	53 04	191	56 54	289	42 02	335
234	38 16	000	42 35	041	19 26	070	30 17	143	52 53	191	56 02	288	41 38	334
235	38 15	358	43 10	040	20 17	069	30 50	143	52 42	192	55 10	287	41 14	333
236	38 13	357	43 45	039	21 07	069	31 23	143	52 30	193	54 18	286	40 49	332
237	38 10	356	44 19	038	21 58	068	31 56	143	52 18	194	53 26	286	40 23	331
238	38 06	355	44 51	037	22 48	068	32 29	143	52 05	194	52 33	285	39 56	330
239	38 01	354	45 23	035	23 39	067	33 02	143	51 51	195	51 41	284	39 28	329
	VEGA		◆ALTAIR		Peacock		◆RIGIL KENT.		SPICA		◆ARCTURUS		Alphecca	
240	16 26	031	24 29	066	33 35	143	51 36	196	50 48	284	39 00	328	37 54	353
241	16 54	030	25 18	066	34 07	143	51 21	196	49 55	283	38 30	327	37 47	352
242	17 21	030	26 08	065	34 40	143	51 06	197	49 02	282	38 00	326	37 38	351
243	17 47	029	26 57	065	35 13	143	50 49	198	48 09	282	37 29	325	37 29	349
244	18 13	028	27 46	064	35 45	143	50 33	198	47 16	281	36 57	324	37 18	348
245	18 39	028	28 35	063	36 18	143	50 15	199	46 22	280	36 24	323	37 07	347
246	19 04	027	29 23	063	36 50	143	49 57	200	45 29	280	35 51	322	36 54	346
247	19 28	026	30 12	062	37 22	144	49 39	200	44 35	279	35 17	321	36 41	345
248	19 52	025	31 00	061	37 55	144	49 20	201	43 41	279	34 42	320	36 26	344
249	20 15	025	31 47	061	38 27	144	49 00	201	42 47	278	34 07	319	36 11	343
250	20 37	024	32 34	060	38 59	144	48 41	202	41 54	278	33 31	318	35 55	342
251	20 59	023	33 21	059	39 31	144	48 20	202	41 00	277	32 54	317	35 37	341
252	21 20	022	34 08	059	40 02	144	47 59	203	40 06	277	32 17	316	35 19	340
253	21 40	022	34 54	058	40 34	145	47 38	203	39 12	276	31 39	316	35 00	339
254	22 00	021	35 40	057	41 05	145	47 16	204	38 17	276	31 01	315	34 39	338
	◆VEGA		ALTAIR		◆Peacock		RIGIL KENT.		◆SPICA		ARCTURUS		Alphecca	
255	22 19	020	36 25	056	41 36	145	46 54	204	37 23	275	30 22	314	34 18	337
256	22 38	019	37 10	055	42 08	145	46 31	205	36 29	275	29 42	313	33 57	336
257	22 55	019	37 55	055	42 38	146	46 08	205	35 35	274	29 02	312	33 34	335
258	23 12	018	38 39	054	43 09	146	45 45	206	34 41	274	28 22	312	33 10	334
259	23 29	017	39 23	053	43 39	146	45 22	206	33 46	273	27 41	311	32 46	333
260	23 44	016	40 06	052	44 10	146	44 57	206	32 52	273	27 00	310	32 21	332
261	23 59	015	40 49	051	44 39	147	44 33	207	31 58	272	26 18	309	31 55	331
262	24 13	015	41 31	050	45 09	147	44 08	207	31 03	272	25 36	309	31 28	330
263	24 27	014	42 12	049	45 39	147	43 44	208	30 09	271	24 53	308	31 01	329
264	24 39	013	42 53	048	46 08	148	43 18	208	29 15	271	24 10	307	30 33	328
265	24 51	012	43 33	047	46 36	148	42 53	208	28 20	271	23 27	307	30 04	327
266	25 02	011	44 13	046	47 05	149	42 27	208	27 26	270	22 43	306	29 34	327
267	25 12	010	44 52	045	47 33	149	42 01	209	26 32	270	21 59	305	29 04	326
268	25 22	010	45 30	044	48 01	149	41 35	209	25 37	269	21 14	305	28 33	325
269	25 31	009	46 08	043	48 28	150	41 08	209	24 43	269	20 30	304	28 02	324

LAT 25°S

LHA	Hc	Zn	Hc	Zn	Hc	Zn	Hc	Zn	Hc	Zn	Hc	Zn	Hc	Zn
♈	VEGA		◆ALTAIR		FOMALHAUT		◆Peacock		RIGIL KENT.		◆ANTARES		Alphecca	
270	25 39	008	46 45	042	24 56	113	48 55	150	40 41	210	69 30	261	27 29	323
271	25 46	007	47 21	041	25 46	112	49 22	151	40 14	210	68 36	261	26 57	323
272	25 52	006	47 56	040	26 37	112	49 48	151	39 47	210	67 43	261	26 23	322
273	25 57	005	48 30	038	27 27	112	50 14	152	39 20	210	66 49	261	25 49	321
274	26 02	005	49 03	037	28 18	112	50 40	152	38 52	211	65 55	261	25 15	320
275	26 06	004	49 36	036	29 08	111	51 04	153	38 25	211	65 02	261	24 40	319
276	26 09	003	50 07	035	29 59	111	51 29	154	37 57	211	64 08	261	24 04	319
277	26 11	002	50 37	033	30 50	111	51 53	154	37 29	211	63 14	260	23 28	318
278	26 13	001	51 07	032	31 41	111	52 16	155	37 01	211	62 21	260	22 51	317
279	26 13	000	51 35	031	32 31	110	52 39	156	36 32	211	61 27	260	22 14	317
280	26 13	359	52 02	029	33 22	110	53 01	156	36 04	212	60 34	260	21 37	316
281	26 12	358	52 28	028	34 13	110	53 23	157	35 35	212	59 40	260	20 59	315
282	26 10	358	52 52	026	35 05	110	53 44	158	35 07	212	58 46	260	20 20	315
283	26 07	357	53 16	025	35 56	110	54 04	158	34 38	212	57 53	260	19 41	314
284	26 04	356	53 38	023	36 47	109	54 24	159	34 09	212	57 00	259	19 02	313
	◆DENEB		FOMALHAUT		◆ACHERNAR		RIGIL KENT.		◆ANTARES		Rasalhague		VEGA	
285	16 03	018	37 38	109	16 00	146	33 40	212	56 06	259	47 04	329	25 59	355
286	16 20	018	38 30	109	16 30	146	33 11	212	55 13	259	46 35	327	25 54	354
287	16 36	017	39 21	109	17 01	146	32 42	212	54 19	259	46 05	326	25 48	353
288	16 51	016	40 13	109	17 31	146	32 13	212	53 26	259	45 34	325	25 41	352
289	17 06	016	41 04	109	18 02	146	31 44	212	52 33	259	45 03	324	25 34	352
290	17 20	015	41 56	108	18 33	145	31 15	212	51 39	258	44 30	323	25 25	351
291	17 34	014	42 47	108	19 04	145	30 46	212	50 46	258	43 57	322	25 16	350
292	17 47	013	43 39	108	19 35	145	30 17	213	49 53	258	43 22	320	25 06	349
293	17 59	013	44 31	108	20 06	145	29 47	213	49 00	258	42 47	319	24 55	348
294	18 11	012	45 23	108	20 37	145	29 18	213	48 07	258	42 12	318	24 44	347
295	18 22	011	46 14	108	21 09	145	28 49	213	47 14	257	41 35	317	24 32	347
296	18 32	011	47 06	107	21 40	144	28 20	213	46 21	257	40 58	316	24 19	346
297	18 42	010	47 58	107	22 12	144	27 50	213	45 28	257	40 20	315	24 05	345
298	18 51	009	48 50	107	22 44	144	27 21	213	44 35	257	39 41	314	23 50	344
299	18 59	008	49 42	107	23 16	144	26 52	212	43 42	256	39 02	313	23 35	343
	DENEB		Enif		◆FOMALHAUT		ACHERNAR		◆RIGIL KENT.		ANTARES		◆VEGA	
300	19 07	008	46 55	039	50 34	107	23 48	144	26 23	212	42 49	256	23 19	342
301	19 14	007	47 29	038	51 26	107	24 20	144	25 54	212	41 56	256	23 02	342
302	19 20	006	48 02	037	52 18	107	24 52	144	25 24	212	41 03	256	22 45	341
303	19 25	005	48 34	036	53 10	107	25 24	144	24 55	212	40 11	256	22 26	340
304	19 30	005	49 05	034	54 02	107	25 56	144	24 26	212	39 18	255	22 08	339
305	19 34	004	49 35	033	54 54	107	26 28	144	23 57	212	38 25	255	21 48	339
306	19 38	003	50 04	032	55 46	107	27 01	144	23 29	212	37 33	255	21 28	338
307	19 40	002	50 32	030	56 39	107	27 33	143	23 00	212	36 40	255	21 07	337
308	19 42	002	50 59	029	57 31	106	28 06	143	22 31	212	35 48	254	20 45	336
309	19 44	001	51 25	027	58 23	106	28 38	143	22 02	212	34 56	254	20 23	336
310	19 44	000	51 49	026	59 15	106	29 10	143	21 34	212	34 03	254	20 00	335
311	19 44	359	52 13	025	60 07	107	29 43	143	21 05	212	33 11	254	19 37	334
312	19 43	359	52 35	023	60 59	107	30 15	143	20 37	211	32 19	253	19 13	333
313	19 42	358	52 55	022	61 51	107	30 48	143	20 09	211	31 27	253	18 48	333
314	19 39	357	53 15	020	62 43	107	31 20	143	19 41	211	30 35	253	18 23	332
	Alpheratz		◆Diphda		ACHERNAR		◆Peacock		ANTARES		◆ALTAIR		DENEB	
315	19 33	043	37 56	094	31 53	143	57 38	189	29 43	253	52 09	331	19 36	357
316	20 10	042	38 50	094	32 25	143	57 29	190	28 51	252	51 42	330	19 33	356
317	20 46	041	39 44	093	32 57	143	57 19	191	27 59	252	51 14	328	19 28	355
318	21 22	041	40 39	093	33 30	144	57 09	192	27 08	252	50 45	327	19 23	354
319	21 57	040	41 33	092	34 02	144	56 57	193	26 16	251	50 15	326	19 18	354
320	22 32	039	42 27	092	34 34	144	56 45	194	25 25	251	49 44	324	19 11	353
321	23 06	039	43 22	092	35 06	144	56 31	195	24 33	251	49 12	323	19 04	352
322	23 39	038	44 16	091	35 38	144	56 17	195	23 42	251	48 39	322	18 56	351
323	24 13	037	45 10	091	36 10	144	56 02	196	22 50	250	48 05	321	18 48	351
324	24 45	036	46 05	090	36 42	144	55 47	197	21 59	250	47 30	319	18 38	350
325	25 17	036	46 59	090	37 14	144	55 30	198	21 08	250	46 54	318	18 28	349
326	25 48	035	47 54	090	37 46	144	55 13	199	20 17	249	46 18	317	18 18	348
327	26 19	034	48 48	089	38 18	145	54 55	200	19 26	249	45 41	316	18 07	348
328	26 49	033	49 42	089	38 49	145	54 36	200	18 36	249	45 02	315	17 55	347
329	27 19	032	50 37	088	39 20	145	54 17	201	17 45	249	44 24	314	17 42	346
	◆Alpheratz		Diphda		◆ACHERNAR		Peacock		◆Nunki		ALTAIR		DENEB	
330	27 48	032	51 31	088	39 52	145	53 57	202	48 30	258	43 44	313	17 29	346
331	28 16	031	52 25	087	40 23	145	53 36	203	47 37	258	43 04	312	17 15	345
332	28 43	030	53 20	087	40 54	145	53 15	203	46 44	257	42 23	311	17 01	344
333	29 10	029	54 14	086	41 24	146	52 53	204	45 51	257	41 42	310	16 46	344
334	29 36	028	55 08	086	41 55	146	52 31	205	44 58	257	41 00	309	16 30	343
335	30 01	027	56 02	085	42 25	146	52 07	205	44 05	257	40 18	308	16 14	342
336	30 26	026	56 57	085	42 55	146	51 44	206	43 12	257	39 34	307	15 57	342
337	30 50	026	57 51	084	43 25	147	51 20	207	42 19	256	38 51	306	15 39	341
338	31 13	025	58 45	084	43 55	147	50 55	207	41 26	256	38 07	306	15 21	340
339	31 35	024	59 39	083	44 24	147	50 30	208	40 34	256	37 23	305	15 02	340
340	31 56	023	60 33	082	44 54	148	50 05	208	39 41	256	36 38	304	14 43	339
341	32 17	022	61 27	082	45 22	148	49 39	209	38 48	255	35 52	303	14 23	338
342	32 37	021	62 20	081	45 51	148	49 12	209	37 56	255	35 07	302	14 03	338
343	32 56	020	63 14	080	46 19	149	48 45	210	37 03	255	34 20	302	13 42	337
344	33 14	019	64 08	080	46 47	149	48 18	210	36 11	255	33 34	301	13 20	336
	Alpheratz		◆Hamal		◆ACHERNAR		Peacock		Nunki		◆ALTAIR		Enif	
345	33 31	018	23 42	047	47 15	150	47 51	211	35 18	254	32 47	300	50 32	330
346	33 47	017	24 21	046	47 42	150	47 23	211	34 26	254	32 00	299	50 04	328
347	34 02	016	25 00	045	48 09	151	46 54	212	33 34	254	31 12	299	49 34	327
348	34 17	015	25 39	045	48 36	151	46 26	212	32 41	254	30 25	298	49 04	326
349	34 30	014	26 17	044	49 02	151	45 57	212	31 49	253	29 36	297	48 33	324
350	34 43	013	26 54	043	49 28	152	45 28	213	30 57	253	28 48	297	48 01	323
351	34 54	012	27 31	042	49 53	152	44 58	213	30 05	253	27 59	296	47 28	322
352	35 05	011	28 08	042	50 18	153	44 28	213	29 13	253	27 10	295	46 54	321
353	35 15	010	28 44	041	50 42	154	43 58	214	28 21	252	26 21	295	46 19	320
354	35 23	009	29 19	040	51 06	154	43 28	214	27 30	252	25 32	294	45 44	318
355	35 31	008	29 54	039	51 30	155	42 58	214	26 38	252	24 42	294	45 07	317
356	35 38	006	30 28	038	51 53	155	42 27	215	25 46	251	23 52	293	44 30	316
357	35 43	005	31 01	038	52 15	156	41 56	215	24 55	251	23 02	293	43 52	315
358	35 48	004	31 34	037	52 37	157	41 25	215	24 03	251	22 12	292	43 13	314
359	35 52	003	32 06	036	52 58	157	40 54	215	23 12	251	21 21	291	42 34	313

LAT 26°S

LHA	Hc	Zn	Hc	Zn	Hc	Zn	Hc	Zn	Hc	Zn	Hc	Zn	Hc	Zn
♈	◆Alpheratz		Hamal		◆RIGEL		CANOPUS		ACHERNAR		◆Peacock		Enif	
0	34 54	002	31 48	035	13 49	092	16 59	141	54 14	158	41 11	216	41 13	313
1	34 56	001	32 19	034	14 43	092	17 33	141	54 35	158	40 39	216	40 34	312
2	34 56	000	32 48	033	15 37	092	18 07	140	54 54	159	40 07	216	39 53	311
3	34 56	359	33 17	032	16 31	091	18 42	140	55 13	160	39 35	216	39 12	310
4	34 54	358	33 45	031	17 25	091	19 16	140	55 31	161	39 03	217	38 31	309
5	34 52	357	34 12	030	18 18	090	19 51	140	55 49	161	38 31	217	37 49	308
6	34 48	356	34 39	029	19 12	090	20 26	140	56 06	162	37 59	217	37 06	307
7	34 44	355	35 05	028	20 06	089	21 01	140	56 22	163	37 26	217	36 23	307
8	34 38	354	35 29	027	21 00	089	21 36	139	56 37	164	36 54	217	35 39	306
9	34 32	353	35 54	026	21 54	089	22 11	139	56 52	165	36 21	217	34 55	305
10	34 25	352	36 17	025	22 48	088	22 46	139	57 06	166	35 48	217	34 11	304
11	34 16	351	36 39	024	23 42	088	23 22	139	57 19	167	35 16	217	33 26	303
12	34 07	349	37 00	023	24 36	087	23 57	139	57 31	168	34 43	217	32 41	303
13	33 56	348	37 21	022	25 30	087	24 33	139	57 42	168	34 10	217	31 55	302
14	33 45	347	37 40	021	26 24	086	25 09	138	57 52	169	33 37	218	31 09	301
	◆Hamal		ALDEBARAN		RIGEL		◆CANOPUS		ACHERNAR		◆FOMALHAUT		Alpheratz	
15	37 59	020	22 32	057	27 17	086	25 44	138	58 02	170	62 43	255	33 33	346
16	38 16	018	23 17	056	28 11	085	26 20	138	58 10	171	61 51	255	33 20	345
17	38 33	017	24 02	056	29 05	085	26 56	138	58 18	172	60 59	255	33 06	344
18	38 49	016	24 46	055	29 59	084	27 32	138	58 24	173	60 07	255	32 51	343
19	39 03	015	25 30	054	30 52	084	28 08	138	58 30	174	59 15	255	32 35	342
20	39 17	014	26 14	054	31 46	083	28 44	138	58 35	175	58 23	255	32 18	341
21	39 29	013	26 57	053	32 39	083	29 21	138	58 39	176	57 31	255	32 00	340
22	39 40	012	27 40	052	33 33	082	29 57	138	58 41	178	56 39	255	31 42	339
23	39 51	010	28 22	052	34 26	082	30 33	138	58 43	179	55 47	255	31 23	339
24	40 00	009	29 04	051	35 20	081	31 09	138	58 44	180	54 54	255	31 02	338
25	40 08	008	29 46	050	36 13	081	31 46	138	58 44	181	54 02	255	30 41	337
26	40 15	007	30 27	049	37 06	080	32 22	138	58 43	182	53 10	255	30 20	336
27	40 21	006	31 07	048	37 59	080	32 58	138	58 41	183	52 19	254	29 57	335
28	40 26	004	31 47	048	38 52	079	33 35	138	58 38	184	51 27	254	29 34	334
29	40 29	003	32 27	047	39 45	079	34 11	138	58 34	185	50 35	254	29 10	333
	Hamal		ALDEBARAN		◆SIRIUS		CANOPUS		◆ACHERNAR		FOMALHAUT		◆Alpheratz	
30	40 32	002	33 06	046	23 46	098	34 47	138	58 29	186	49 43	254	28 45	332
31	40 33	001	33 44	045	24 39	097	35 24	138	58 23	187	48 51	254	28 19	331
32	40 34	000	34 22	044	25 33	097	36 00	138	58 16	188	47 59	254	27 53	330
33	40 33	358	35 00	043	26 26	097	36 36	138	58 08	189	47 08	254	27 26	330
34	40 31	357	35 36	042	27 20	096	37 13	138	57 59	190	46 16	253	26 58	329
35	40 28	356	36 12	042	28 14	096	37 49	138	57 50	191	45 24	253	26 30	328
36	40 23	355	36 48	041	29 07	095	38 25	138	57 39	192	44 33	253	26 01	327
37	40 18	354	37 23	040	30 01	095	39 01	138	57 28	193	43 41	253	25 31	326
38	40 11	352	37 57	039	30 55	095	39 37	138	57 16	194	42 50	253	25 01	325
39	40 04	351	38 30	038	31 49	094	40 13	138	57 03	195	41 58	252	24 30	325
40	39 55	350	39 03	037	32 42	094	40 49	138	56 49	195	41 07	252	23 59	324
41	39 45	349	39 34	036	33 36	093	41 25	139	56 34	196	40 15	252	23 27	323
42	39 34	348	40 05	035	34 30	093	42 00	139	56 18	197	39 24	252	22 54	322
43	39 22	347	40 36	033	35 24	093	42 36	139	56 02	198	38 33	252	22 21	322
44	39 09	345	41 05	032	36 18	092	43 11	139	55 45	199	37 42	251	21 47	321
	◆ALDEBARAN		BETELGEUSE		SIRIUS		◆CANOPUS		ACHERNAR		◆FOMALHAUT		Hamal	
45	41 33	031	35 59	058	37 12	092	43 46	139	55 27	200	36 51	251	38 55	344
46	42 01	030	36 45	057	38 06	091	44 21	140	55 09	200	36 00	251	38 40	343
47	42 27	029	37 30	056	38 59	091	44 56	140	54 50	201	35 09	251	38 24	342
48	42 53	028	38 14	055	39 53	091	45 31	140	54 30	202	34 18	251	38 07	341
49	43 18	027	38 59	055	40 47	090	46 06	140	54 09	203	33 27	250	37 49	340
50	43 42	025	39 42	054	41 41	090	46 40	141	53 48	203	32 36	250	37 30	339
51	44 04	024	40 26	053	42 35	089	47 14	141	53 27	204	31 46	250	37 10	338
52	44 26	023	41 08	052	43 29	089	47 48	141	53 04	205	30 55	250	36 49	337
53	44 46	022	41 50	051	44 23	088	48 22	142	52 42	205	30 05	249	36 27	336
54	45 06	020	42 32	050	45 17	088	48 55	142	52 18	206	29 14	249	36 04	335
55	45 24	019	43 13	049	46 11	087	49 28	142	51 55	207	28 24	249	35 40	333
56	45 41	018	43 54	048	47 05	087	50 01	143	51 30	207	27 34	249	35 16	332
57	45 57	017	44 33	047	47 58	086	50 33	143	51 05	208	26 44	248	34 50	331
58	46 12	015	45 12	046	48 52	086	51 06	144	50 40	208	25 53	248	34 24	331
59	46 25	014	45 51	045	49 46	085	51 37	144	50 15	209	25 04	248	33 57	330
	ALDEBARAN		BETELGEUSE		◆SIRIUS		CANOPUS		◆ACHERNAR		FOMALHAUT		◆Hamal	
60	46 38	012	46 29	044	50 40	085	52 09	145	49 48	209	24 14	247	33 30	329
61	46 49	011	47 06	043	51 33	084	52 40	145	49 22	210	23 24	247	33 01	328
62	46 58	010	47 42	041	52 27	084	53 11	146	48 55	210	22 34	247	32 32	327
63	47 07	008	48 17	040	53 21	083	53 41	146	48 28	211	21 45	247	32 02	326
64	47 14	007	48 51	039	54 14	083	54 11	147	48 00	211	20 55	246	31 31	325
65	47 20	006	49 25	038	55 08	082	54 40	147	47 32	211	20 06	246	31 00	324
66	47 24	004	49 58	037	56 01	081	55 09	148	47 04	212	19 17	246	30 28	323
67	47 28	003	50 29	035	56 54	081	55 37	149	46 35	212	18 28	245	29 55	322
68	47 29	001	51 00	034	57 47	080	56 05	149	46 06	213	17 39	245	29 22	321
69	47 30	000	51 29	033	58 40	079	56 32	150	45 37	213	16 50	245	28 48	321
70	47 29	358	51 58	031	59 33	079	56 59	151	45 08	213	16 01	244	28 14	320
71	47 27	357	52 25	030	60 26	078	57 25	152	44 38	214	15 13	244	27 39	319
72	47 24	356	52 51	028	61 19	077	57 50	152	44 08	214	14 24	244	27 03	318
73	47 19	354	53 16	027	62 11	076	58 14	153	43 38	214	13 36	243	26 27	317
74	47 13	353	53 40	025	63 03	075	58 38	154	43 08	214	12 48	243	25 50	317
	BETELGEUSE		PROCYON		◆Suhail		CANOPUS		◆ACHERNAR		Hamal		◆ALDEBARAN	
75	54 02	024	40 24	057	37 28	126	59 01	155	42 37	215	25 13	316	47 05	351
76	54 23	022	41 08	056	38 11	126	59 24	156	42 07	215	24 35	315	46 57	350
77	54 42	020	41 53	055	38 55	126	59 45	157	41 36	215	23 57	314	46 47	349
78	55 00	019	42 37	054	39 39	126	60 06	158	41 05	215	23 18	314	46 35	347
79	55 17	017	43 20	053	40 22	126	60 26	159	40 33	215	22 39	313	46 23	346
80	55 32	015	44 03	052	41 06	126	60 45	160	40 02	216	21 59	312	46 09	345
81	55 46	014	44 45	051	41 49	126	61 02	161	39 30	216	21 19	312	45 54	343
82	55 58	012	45 27	050	42 33	126	61 19	162	38 59	216	20 39	311	45 38	342
83	56 08	010	46 08	049	43 16	126	61 35	163	38 27	216	19 58	310	45 21	341
84	56 17	008	46 49	048	44 00	126	61 50	165	37 55	216	19 17	310	45 02	339
85	56 24	007	47 28	047	44 43	126	62 04	166	37 23	216	18 35	309	44 43	338
86	56 29	005	48 08	046	45 27	126	62 17	167	36 51	216	17 53	308	44 22	337
87	56 33	003	48 46	045	46 10	127	62 28	168	36 19	217	17 11	308	44 00	336
88	56 35	001	49 23	044	46 53	127	62 39	169	35 47	217	16 28	307	43 37	334
89	56 36	000	50 00	042	47 36	127	62 48	171	35 15	217	15 45	307	43 13	333

LHA	Hc	Zn	Hc	Zn	Hc	Zn	Hc	Zn	Hc	Zn	Hc	Zn	Hc	Zn
♈	POLLUX		◆REGULUS		ACRUX		◆CANOPUS		ACHERNAR		◆RIGEL		BETELGEUSE	
90	30 22	027	18 45	066	20 08	151	62 56	172	34 42	217	69 09	327	56 34	358
91	30 46	026	19 34	065	20 34	151	63 03	173	34 10	217	68 39	324	56 31	356
92	31 09	025	20 22	065	21 00	151	63 09	175	33 38	217	68 06	322	56 27	354
93	31 31	024	21 11	064	21 26	151	63 13	176	33 05	217	67 32	320	56 20	352
94	31 53	023	21 59	063	21 52	151	63 16	177	32 33	217	66 57	318	56 12	351
95	32 14	022	22 48	063	22 19	151	63 18	179	32 01	217	66 20	316	56 03	349
96	32 34	021	23 35	062	22 45	151	63 18	180	31 28	217	65 41	314	55 51	347
97	32 53	020	24 23	062	23 12	150	63 18	181	30 56	217	65 02	312	55 38	345
98	33 11	019	25 10	061	23 38	150	63 16	183	30 23	217	64 22	310	55 24	344
99	33 28	018	25 57	060	24 05	150	63 13	184	29 51	217	63 40	309	55 08	342
100	33 45	017	26 44	060	24 32	150	63 08	185	29 18	217	62 58	307	54 51	340
101	34 00	016	27 31	059	24 58	150	63 02	187	28 46	217	62 14	306	54 32	339
102	34 15	015	28 17	058	25 25	150	62 55	188	28 14	217	61 30	304	54 11	337
103	34 29	014	29 02	058	25 52	150	62 47	189	27 41	217	60 45	303	53 50	336
104	34 42	013	29 48	057	26 19	150	62 38	191	27 09	217	60 00	302	53 27	334
	POLLUX		◆REGULUS		Gienah		◆ACRUX		CANOPUS		◆RIGEL		BETELGEUSE	
105	34 53	012	30 33	056	17 17	101	26 46	150	62 27	192	59 14	301	53 02	332
106	35 04	011	31 17	055	18 10	101	27 13	150	62 16	193	58 27	299	52 37	331
107	35 14	010	32 02	055	19 03	101	27 41	150	62 03	194	57 40	298	52 10	330
108	35 23	009	32 45	054	19 56	100	28 08	150	61 49	196	56 52	297	51 42	328
109	35 31	008	33 29	053	20 49	100	28 35	150	61 34	197	56 04	296	51 13	327
110	35 38	007	34 12	052	21 42	100	29 02	150	61 18	198	55 15	295	50 43	325
111	35 44	006	34 54	052	22 35	099	29 29	150	61 01	199	54 26	294	50 12	324
112	35 48	005	35 36	051	23 29	099	29 56	150	60 43	200	53 37	293	49 40	323
113	35 52	004	36 18	050	24 22	098	30 23	150	60 24	201	52 47	292	49 07	321
114	35 55	002	36 59	049	25 15	098	30 51	150	60 04	202	51 57	291	48 33	320
115	35 57	001	37 39	048	26 09	098	31 18	150	59 44	203	51 07	291	47 58	319
116	35 58	000	38 19	047	27 02	097	31 45	150	59 22	204	50 16	290	47 22	318
117	35 57	359	38 58	046	27 56	097	32 12	150	59 00	205	49 25	289	46 45	317
118	35 56	358	39 37	045	28 49	097	32 39	150	58 36	206	48 34	288	46 08	316
119	35 54	357	40 15	044	29 43	096	33 06	150	58 13	207	47 43	287	45 30	314
	◆REGULUS		SPICA		◆ACRUX		CANOPUS		◆RIGEL		BETELGEUSE		POLLUX	
120	40 52	043	12 39	096	33 33	150	57 48	208	46 51	287	44 51	313	35 50	356
121	41 29	042	13 33	096	34 00	150	57 23	208	45 59	286	44 12	312	35 46	355
122	42 05	041	14 27	096	34 27	150	56 57	209	45 08	285	43 31	311	35 41	354
123	42 40	040	15 20	095	34 53	150	56 30	210	44 15	285	42 51	310	35 34	353
124	43 14	039	16 14	095	35 20	150	56 03	211	43 23	284	42 09	309	35 27	352
125	43 48	038	17 08	094	35 46	151	55 35	211	42 31	283	41 27	308	35 19	351
126	44 21	037	18 02	094	36 13	151	55 07	212	41 38	283	40 45	308	35 09	349
127	44 52	036	18 55	093	36 39	151	54 38	213	40 46	282	40 02	307	34 59	348
128	45 24	035	19 49	093	37 05	151	54 08	213	39 53	282	39 18	306	34 48	347
129	45 54	033	20 43	093	37 31	151	53 39	214	39 00	281	38 34	305	34 35	346
130	46 23	032	21 37	092	37 57	151	53 08	214	38 07	280	37 50	304	34 22	345
131	46 51	031	22 31	092	38 23	152	52 38	215	37 14	280	37 05	303	34 08	344
132	47 18	030	23 25	091	38 49	152	52 06	215	36 21	279	36 20	302	33 53	343
133	47 44	028	24 19	091	39 14	152	51 35	216	35 27	279	35 34	302	33 37	342
134	48 09	027	25 13	090	39 39	152	51 03	216	34 34	278	34 48	301	33 20	341
	REGULUS		◆SPICA		ACRUX		◆CANOPUS		RIGEL		BETELGEUSE		◆POLLUX	
135	48 33	026	26 07	090	40 05	152	50 31	217	33 41	278	34 02	300	33 02	340
136	48 56	024	27 00	090	40 29	153	49 58	217	32 47	277	33 15	299	32 43	339
137	49 17	023	27 54	089	40 54	153	49 26	218	31 54	277	32 28	299	32 24	338
138	49 38	021	28 48	089	41 19	153	48 52	218	31 00	276	31 40	298	32 03	337
139	49 57	020	29 42	088	41 43	153	48 19	218	30 07	276	30 52	297	31 42	336
140	50 15	019	30 36	088	42 07	154	47 45	219	29 13	275	30 04	297	31 20	335
141	50 31	017	31 30	087	42 31	154	47 11	219	28 19	275	29 16	296	30 57	334
142	50 47	016	32 24	087	42 54	154	46 37	219	27 25	274	28 27	295	30 34	334
143	51 00	014	33 18	086	43 17	155	46 03	220	26 32	274	27 38	295	30 09	333
144	51 13	013	34 11	086	43 40	155	45 28	220	25 38	273	26 49	294	29 44	332
145	51 24	011	35 05	085	44 03	155	44 54	220	24 44	273	26 00	293	29 18	331
146	51 33	010	35 59	085	44 26	156	44 19	220	23 50	272	25 10	293	28 52	330
147	51 42	008	36 53	084	44 48	156	43 44	221	22 56	272	24 20	292	28 24	329
148	51 48	006	37 46	084	45 09	156	43 08	221	22 02	271	23 30	292	27 56	328
149	51 54	005	38 40	083	45 31	157	42 33	221	21 08	271	22 40	291	27 28	327
	REGULUS		◆Denebola		SPICA		◆ACRUX		CANOPUS		◆SIRIUS		PROCYON	
150	51 57	003	41 32	036	39 33	083	45 52	157	41 57	221	43 54	271	43 43	307
151	52 00	002	42 04	035	40 27	082	46 13	158	41 22	221	43 00	271	42 59	306
152	52 00	000	42 34	034	41 20	082	46 33	158	40 46	222	42 06	271	42 16	305
153	52 00	358	43 04	033	42 14	081	46 53	159	40 10	222	41 12	270	41 32	305
154	51 58	357	43 33	032	43 07	081	47 13	159	39 34	222	40 18	270	40 47	304
155	51 54	355	44 01	031	44 00	080	47 32	159	38 58	222	39 24	269	40 02	303
156	51 49	354	44 28	029	44 53	079	47 50	160	38 22	222	38 31	269	39 16	302
157	51 42	352	44 53	028	45 46	079	48 09	160	37 46	222	37 37	268	38 30	301
158	51 34	351	45 18	027	46 39	078	48 27	161	37 10	222	36 43	268	37 44	300
159	51 24	349	45 42	026	47 32	078	48 44	161	36 34	222	35 49	268	36 57	300
160	51 13	347	46 05	024	48 24	077	49 01	162	35 57	222	34 55	267	36 10	299
161	51 01	346	46 27	023	49 17	076	49 17	163	35 21	222	34 01	267	35 23	298
162	50 47	344	46 47	022	50 09	076	49 33	163	34 45	222	33 07	266	34 35	297
163	50 32	343	47 07	020	51 01	075	49 49	164	34 08	222	32 14	266	33 47	297
164	50 16	341	47 25	019	51 53	074	50 03	164	33 32	222	31 20	266	32 59	296
	ARCTURUS		◆ANTARES		ACRUX		◆CANOPUS		SIRIUS		PROCYON		◆REGULUS	
165	24 28	051	17 39	111	50 18	165	32 56	222	30 26	265	32 10	295	49 58	340
166	25 10	051	18 29	111	50 32	165	32 20	222	29 32	265	31 21	295	49 39	339
167	25 52	050	19 20	111	50 45	166	31 43	222	28 39	264	30 32	294	49 19	337
168	26 33	049	20 10	110	50 57	167	31 06	222	27 45	264	29 43	293	48 57	336
169	27 13	049	21 01	110	51 09	167	30 30	222	26 51	264	28 53	293	48 34	334
170	27 53	048	21 52	110	51 21	168	29 54	222	25 58	263	28 03	292	48 11	333
171	28 33	047	22 42	109	51 32	169	29 18	222	25 04	263	27 13	291	47 46	332
172	29 12	046	23 33	109	51 42	169	28 42	222	24 11	262	26 23	291	47 20	330
173	29 51	045	24 24	109	51 52	170	28 05	222	23 17	262	25 32	290	46 52	329
174	30 29	045	25 15	108	52 01	171	27 29	222	22 24	262	24 42	290	46 24	328
175	31 07	044	26 07	108	52 09	171	26 53	222	21 31	261	23 51	289	45 55	327
176	31 44	043	26 58	108	52 17	172	26 17	222	20 37	261	23 00	288	45 25	326
177	32 20	042	27 49	108	52 24	173	25 41	222	19 44	261	22 08	288	44 54	324
178	32 56	041	28 41	107	52 30	174	25 06	222	18 51	260	21 17	287	44 22	323
179	33 31	040	29 32	107	52 36	174	24 30	221	17 58	260	20 25	287	43 50	322

LHA ♈	Hc	Zn	Hc	Zn	Hc	Zn	Hc	Zn	Hc	Zn	Hc	Zn	Hc	Zn
	◆ARCTURUS		ANTARES		◆RIGIL KENT.		ACRUX		CANOPUS		◆Alphard		REGULUS	
180	34 06	039	30 24	107	46 00	153	52 40	175	23 54	221	49 52	289	43 16	321
181	34 40	039	31 16	106	46 24	154	52 45	176	23 19	221	49 00	288	42 42	320
182	35 13	038	32 07	106	46 48	154	52 48	177	22 43	221	48 09	287	42 06	319
183	35 46	037	32 59	106	47 11	155	52 51	177	22 08	221	47 17	286	41 31	318
184	36 17	036	33 51	106	47 34	155	52 53	178	21 33	221	46 26	286	40 54	317
185	36 48	035	34 43	105	47 57	155	52 55	179	20 58	220	45 34	285	40 17	316
186	37 19	034	35 35	105	48 19	156	52 56	180	20 23	220	44 41	284	39 39	315
187	37 48	033	36 27	105	48 41	156	52 56	180	19 48	220	43 49	284	39 00	314
188	38 17	032	37 19	105	49 02	157	52 55	181	19 14	220	42 57	283	38 21	313
189	38 45	031	38 12	104	49 23	157	52 54	182	18 39	220	42 04	282	37 41	312
190	39 12	030	39 04	104	49 44	158	52 52	183	18 05	219	41 11	282	37 01	311
191	39 38	028	39 56	104	50 04	159	52 49	183	17 30	219	40 18	281	36 20	310
192	40 03	027	40 49	103	50 23	159	52 45	184	16 56	219	39 25	281	35 38	309
193	40 28	026	41 41	103	50 42	160	52 41	185	16 22	219	38 32	280	34 56	309
194	40 51	025	42 34	103	51 01	160	52 36	186	15 49	219	37 39	280	34 14	308
	ARCTURUS		◆Rasalhague		ANTARES		RIGIL KENT.		◆ACRUX		Suhail		◆REGULUS	
195	41 13	024	12 55	069	43 26	103	51 19	161	52 31	186	40 18	234	33 31	307
196	41 35	023	13 46	068	44 19	102	51 36	162	52 25	187	39 34	234	32 48	306
197	41 55	022	14 36	068	45 12	102	51 53	162	52 18	188	38 51	234	32 04	305
198	42 14	020	15 26	067	46 04	102	52 09	163	52 10	188	38 07	234	31 20	305
199	42 33	019	16 15	067	46 57	102	52 25	163	52 02	189	37 24	234	30 35	304
200	42 50	018	17 05	066	47 50	101	52 40	164	51 53	190	36 40	234	29 50	303
201	43 06	017	17 54	066	48 43	101	52 54	165	51 44	190	35 57	234	29 05	302
202	43 21	015	18 43	065	49 36	101	53 08	166	51 33	191	35 13	234	28 19	302
203	43 35	014	19 32	065	50 29	101	53 21	166	51 23	192	34 30	234	27 33	301
204	43 47	013	20 20	064	51 22	101	53 33	167	51 11	192	33 46	234	26 47	300
205	43 59	012	21 09	063	52 15	100	53 45	168	50 59	193	33 03	233	26 00	300
206	44 09	010	21 57	063	53 08	100	53 56	169	50 47	194	32 20	233	25 13	299
207	44 18	009	22 45	062	54 01	100	54 06	169	50 34	194	31 36	233	24 26	298
208	44 26	008	23 32	062	54 54	100	54 16	170	50 20	195	30 53	233	23 38	298
209	44 33	006	24 19	061	55 47	099	54 25	171	50 06	196	30 10	233	22 50	297
	◆ARCTURUS		Rasalhague		◆Nunki		RIGIL KENT.		ACRUX		◆Suhail		Denebola	
210	44 38	005	25 06	060	24 50	109	54 33	172	49 51	196	29 27	233	38 22	318
211	44 42	004	25 53	060	25 41	108	54 40	173	49 36	197	28 44	233	37 45	317
212	44 45	002	26 39	059	26 32	108	54 47	173	49 20	197	28 01	233	37 08	316
213	44 47	001	27 25	058	27 23	108	54 53	174	49 03	198	27 18	233	36 31	315
214	44 47	000	28 11	058	28 15	107	54 58	175	48 47	198	26 35	232	35 52	314
215	44 47	358	28 57	057	29 06	107	55 02	176	48 29	199	25 53	232	35 13	313
216	44 45	357	29 42	056	29 58	107	55 05	177	48 12	199	25 10	232	34 34	313
217	44 41	356	30 26	055	30 50	106	55 08	178	47 53	200	24 28	232	33 54	312
218	44 37	355	31 10	055	31 41	106	55 10	178	47 35	200	23 45	232	33 13	311
219	44 31	353	31 54	054	32 33	106	55 11	179	47 16	201	23 03	231	32 32	310
220	44 24	352	32 38	053	33 25	106	55 11	180	46 56	201	22 21	231	31 51	309
221	44 16	351	33 20	052	34 17	105	55 11	181	46 36	202	21 39	231	31 09	308
222	44 06	349	34 03	052	35 09	105	55 09	182	46 16	202	20 57	231	30 27	308
223	43 56	348	34 45	051	36 01	105	55 07	183	45 55	203	20 16	231	29 44	307
224	43 44	347	35 26	050	36 53	104	55 04	184	45 34	203	19 34	230	29 00	306
	Alphecca		Rasalhague		◆Nunki		Peacock		◆RIGIL KENT.		SPICA		◆ARCTURUS	
225	36 37	010	36 07	049	37 46	104	26 10	143	55 00	184	63 06	299	43 31	345
226	36 46	008	36 48	048	38 38	104	26 43	143	54 56	185	62 18	298	43 17	344
227	36 53	007	37 28	047	39 30	104	27 15	143	54 50	186	61 30	297	43 01	343
228	36 59	006	38 07	046	40 23	103	27 48	143	54 44	187	60 42	295	42 45	342
229	37 05	005	38 46	045	41 15	103	28 21	143	54 37	188	59 53	294	42 28	340
230	37 09	004	39 24	044	42 08	103	28 54	143	54 30	189	59 04	293	42 09	339
231	37 12	003	40 01	043	43 00	103	29 27	142	54 21	189	58 14	292	41 49	338
232	37 15	002	40 38	042	43 53	102	29 59	142	54 12	190	57 24	291	41 29	337
233	37 16	001	41 14	041	44 46	102	30 32	142	54 02	191	56 33	290	41 07	336
234	37 16	000	41 50	040	45 38	102	31 05	142	53 52	192	55 43	290	40 44	335
235	37 15	358	42 24	039	46 31	102	31 38	142	53 41	192	54 52	289	40 21	333
236	37 13	357	42 58	038	47 24	101	32 11	142	53 29	193	54 00	288	39 56	332
237	37 10	356	43 31	037	48 17	101	32 44	142	53 16	194	53 09	287	39 31	331
238	37 06	355	44 03	036	49 10	101	33 17	142	53 03	195	52 17	286	39 04	330
239	37 01	354	44 34	035	50 03	101	33 50	142	52 49	195	51 25	285	38 37	329
	◆VEGA		ALTAIR		◆Peacock		RIGIL KENT.		◆SPICA		ARCTURUS		Alphecca	
240	15 34	031	24 04	066	34 22	143	52 34	196	50 33	285	38 09	328	36 55	353
241	16 02	030	24 54	065	34 55	143	52 19	197	49 41	284	37 40	327	36 47	352
242	16 29	029	25 43	065	35 28	143	52 03	197	48 49	283	37 10	326	36 39	351
243	16 55	029	26 31	064	36 01	143	51 47	198	47 56	283	36 40	325	36 30	350
244	17 20	028	27 20	064	36 33	143	51 30	199	47 04	282	36 08	324	36 20	348
245	17 46	027	28 08	063	37 06	143	51 12	199	46 11	281	35 36	323	36 08	347
246	18 10	027	28 56	062	37 38	143	50 54	200	45 18	281	35 04	322	35 56	346
247	18 34	026	29 43	062	38 11	143	50 35	201	44 25	280	34 30	321	35 43	345
248	18 57	025	30 31	061	38 43	143	50 16	201	43 32	280	33 56	320	35 29	344
249	19 20	025	31 18	060	39 15	143	49 56	202	42 38	279	33 21	319	35 14	343
250	19 42	024	32 04	059	39 47	144	49 36	202	41 45	278	32 46	319	34 57	342
251	20 04	023	32 50	059	40 19	144	49 16	203	40 52	278	32 10	318	34 40	341
252	20 24	022	33 36	058	40 51	144	48 54	203	39 58	277	31 33	317	34 22	340
253	20 45	022	34 22	057	41 23	144	48 33	204	39 05	277	30 56	316	34 04	339
254	21 04	021	35 07	056	41 54	144	48 11	204	38 11	276	30 18	315	33 44	338
	◆VEGA		ALTAIR		FOMALHAUT		◆Peacock		RIGIL KENT.		◆SPICA		ARCTURUS	
255	21 23	020	35 52	056	13 03	117	42 26	145	47 49	205	37 18	276	29 40	314
256	21 41	019	36 36	055	13 51	117	42 57	145	47 26	205	36 24	275	29 01	314
257	21 59	019	37 20	054	14 39	116	43 28	145	47 03	206	35 30	275	28 22	313
258	22 15	018	38 03	053	15 28	116	43 58	145	46 39	206	34 36	274	27 42	312
259	22 31	017	38 46	052	16 17	116	44 29	146	46 15	206	33 43	274	27 02	311
260	22 47	016	39 29	051	17 05	115	44 59	146	45 51	207	32 49	273	26 21	311
261	23 01	015	40 11	051	17 54	115	45 29	146	45 27	207	31 55	273	25 40	310
262	23 15	015	40 52	050	18 43	115	45 59	147	45 02	208	31 01	272	24 58	309
263	23 28	014	41 33	049	19 32	114	46 29	147	44 37	208	30 07	272	24 16	308
264	23 41	013	42 13	048	20 21	114	46 58	147	44 11	208	29 13	272	23 34	308
265	23 52	012	42 52	047	21 11	114	47 27	148	43 46	209	28 19	271	22 51	307
266	24 03	011	43 31	046	22 00	113	47 56	148	43 20	209	27 26	271	22 08	306
267	24 13	010	44 10	045	22 50	113	48 24	148	42 53	209	26 32	270	21 24	306
268	24 23	010	44 47	044	23 39	113	48 52	149	42 27	210	25 38	270	20 40	305
269	24 31	009	45 24	042	24 29	112	49 20	149	42 00	210	24 44	269	19 56	304

LHA ♈	Hc	Zn	Hc	Zn	Hc	Zn	Hc	Zn	Hc	Zn	Hc	Zn	Hc	Zn
	VEGA		◆ALTAIR		FOMALHAUT		◆Peacock		RIGIL KENT.		◆ANTARES		Alphecca	
270	24 39	008	46 00	041	25 19	112	49 47	150	41 33	210	69 38	264	26 41	324
271	24 46	007	46 35	040	26 09	112	50 14	150	41 06	210	68 44	264	26 09	323
272	24 52	006	47 09	039	26 59	112	50 41	151	40 39	211	67 51	263	25 36	322
273	24 58	005	47 43	038	27 49	111	51 07	151	40 11	211	66 57	263	25 03	321
274	25 02	004	48 15	037	28 40	111	51 33	152	39 44	211	66 04	263	24 29	321
275	25 06	004	48 47	035	29 30	111	51 58	152	39 16	211	65 10	263	23 54	320
276	25 09	003	49 17	034	30 20	111	52 22	153	38 48	211	64 17	263	23 19	319
277	25 11	002	49 47	033	31 11	110	52 47	154	38 20	211	63 23	262	22 43	318
278	25 13	001	50 16	031	32 01	110	53 10	154	37 52	212	62 30	262	22 07	318
279	25 13	000	50 43	030	32 52	110	53 33	155	37 23	212	61 36	262	21 31	317
280	25 13	359	51 09	029	33 43	110	53 56	156	36 55	212	60 43	262	20 53	316
281	25 12	358	51 35	027	34 34	109	54 18	156	36 26	212	59 50	262	20 16	316
282	25 10	358	51 59	026	35 25	109	54 39	157	35 58	212	58 56	261	19 38	315
283	25 07	357	52 21	024	36 16	109	55 00	158	35 29	212	58 03	261	18 59	314
284	25 04	356	52 43	023	37 07	109	55 20	159	35 00	212	57 10	261	18 21	314
	◆DENEB		FOMALHAUT		◆ACHERNAR		RIGIL KENT.		◆ANTARES		Rasalhague		VEGA	
285	15 06	018	37 58	109	16 50	146	34 31	212	56 16	261	46 12	329	25 00	355
286	15 22	017	38 49	108	17 20	146	34 02	213	55 23	261	45 44	328	24 55	354
287	15 38	017	39 40	108	17 50	146	33 33	213	54 30	260	45 15	327	24 49	353
288	15 54	016	40 31	108	18 21	146	33 04	213	53 37	260	44 45	326	24 42	352
289	16 08	015	41 23	108	18 51	145	32 35	213	52 44	260	44 14	324	24 34	352
290	16 22	015	42 14	107	19 22	145	32 06	213	51 51	260	43 42	323	24 26	351
291	16 36	014	43 06	107	19 53	145	31 36	213	50 58	259	43 10	322	24 17	350
292	16 49	013	43 57	107	20 24	145	31 07	213	50 05	259	42 36	321	24 07	349
293	17 01	013	44 49	107	20 55	145	30 38	213	49 12	259	42 02	320	23 57	348
294	17 12	012	45 40	107	21 26	144	30 09	213	48 19	259	41 27	319	23 45	347
295	17 23	011	46 32	107	21 58	144	29 39	213	47 26	258	40 51	318	23 33	347
296	17 33	010	47 24	106	22 29	144	29 10	213	46 33	258	40 14	317	23 20	346
297	17 43	010	48 15	106	23 01	144	28 41	213	45 41	258	39 37	316	23 07	345
298	17 52	009	49 07	106	23 32	144	28 12	213	44 48	258	38 59	315	22 52	344
299	18 00	008	49 59	106	24 04	144	27 42	213	43 55	257	38 21	314	22 37	343
	DENEB		Enif		◆FOMALHAUT		ACHERNAR		◆RIGIL KENT.		ANTARES		◆VEGA	
300	18 07	008	46 08	039	50 51	106	24 36	144	27 13	213	43 03	257	22 22	343
301	18 14	007	46 41	037	51 43	106	25 08	144	26 44	213	42 10	257	22 05	342
302	18 20	006	47 14	036	52 35	106	25 40	144	26 15	213	41 18	257	21 48	341
303	18 26	005	47 45	035	53 27	105	26 12	143	25 46	213	40 25	256	21 30	340
304	18 30	005	48 15	034	54 19	105	26 44	143	25 17	212	39 33	256	21 11	339
305	18 34	004	48 45	032	55 11	105	27 17	143	24 48	212	38 40	256	20 52	339
306	18 38	003	49 13	031	56 03	105	27 49	143	24 19	212	37 48	256	20 32	338
307	18 40	002	49 40	030	56 55	105	28 21	143	23 51	212	36 56	255	20 12	337
308	18 42	002	50 07	028	57 47	105	28 54	143	23 22	212	36 04	255	19 50	336
309	18 44	001	50 32	027	58 39	105	29 26	143	22 53	212	35 12	255	19 29	336
310	18 44	000	50 55	025	59 31	105	29 58	143	22 25	212	34 20	255	19 06	335
311	18 44	359	51 18	024	60 23	105	30 31	143	21 56	212	33 28	254	18 43	334
312	18 43	359	51 39	023	61 15	105	31 03	143	21 28	212	32 36	254	18 19	334
313	18 42	358	51 59	021	62 08	105	31 36	143	21 00	211	31 44	254	17 55	333
314	18 40	357	52 18	020	63 00	105	32 08	143	20 32	211	30 52	253	17 30	332
	Alpheratz		◆Diphda		ACHERNAR		◆RIGIL KENT.		ANTARES		◆ALTAIR		DENEB	
315	18 49	043	38 00	093	32 41	143	20 04	211	30 01	253	51 16	332	18 37	357
316	19 25	042	38 53	093	33 13	143	19 36	211	29 09	253	50 50	330	18 33	356
317	20 01	041	39 47	092	33 45	143	19 08	211	28 18	253	50 23	329	18 29	355
318	20 36	040	40 41	092	34 18	143	18 41	211	27 26	252	49 55	328	18 24	354
319	21 11	040	41 35	092	34 50	143	18 14	210	26 35	252	49 26	326	18 18	354
320	21 45	039	42 29	091	35 22	143	17 46	210	25 44	252	48 55	325	18 12	353
321	22 19	038	43 23	091	35 55	143	17 19	210	24 52	251	48 24	324	18 05	352
322	22 52	038	44 17	090	36 27	143	16 52	210	24 01	251	47 52	323	17 57	351
323	23 25	037	45 11	090	36 59	144	16 25	210	23 10	251	47 18	321	17 48	351
324	23 57	036	46 05	089	37 31	144	15 59	209	22 20	250	46 44	320	17 39	350
325	24 28	035	46 59	089	38 03	144	15 32	209	21 29	250	46 09	319	17 30	349
326	24 59	035	47 53	088	38 35	144	15 06	209	20 38	250	45 34	318	17 19	349
327	25 29	034	48 46	088	39 06	144	14 40	209	19 48	249	44 57	317	17 08	348
328	25 59	033	49 40	088	39 38	144	14 14	209	18 57	249	44 20	316	16 56	347
329	26 28	032	50 34	087	40 09	144	13 49	208	18 07	249	43 42	315	16 44	346
	◆Alpheratz		Diphda		◆ACHERNAR		Peacock		◆Nunki		ALTAIR		Enif	
330	26 56	031	51 28	087	40 41	145	54 52	203	48 42	259	43 03	314	53 56	353
331	27 24	031	52 22	086	41 12	145	54 31	203	47 49	259	42 24	313	53 49	352
332	27 51	030	53 16	086	41 43	145	54 10	204	46 56	258	41 44	312	53 40	350
333	28 18	029	54 09	085	42 14	145	53 48	205	46 04	258	41 03	311	53 30	348
334	28 43	028	55 03	084	42 44	145	53 25	205	45 11	258	40 22	310	53 19	347
335	29 08	027	55 57	084	43 15	146	53 02	206	44 18	258	39 40	309	53 05	345
336	29 32	026	56 50	083	43 45	146	52 38	207	43 26	257	38 58	308	52 51	344
337	29 56	025	57 44	083	44 15	146	52 13	207	42 33	257	38 15	307	52 35	342
338	30 18	024	58 37	082	44 45	147	51 48	208	41 40	257	37 32	306	52 18	340
339	30 40	023	59 31	081	45 15	147	51 23	208	40 48	257	36 48	305	51 59	339
340	31 01	022	60 24	081	45 44	147	50 57	209	39 55	256	36 04	305	51 39	337
341	31 21	022	61 17	080	46 13	148	50 31	209	39 03	256	35 19	304	51 17	336
342	31 41	021	62 10	079	46 42	148	50 04	210	38 11	256	34 35	303	50 55	334
343	31 59	020	63 03	078	47 11	148	49 37	210	37 18	256	33 49	302	50 31	333
344	32 17	019	63 56	078	47 39	149	49 10	211	36 26	255	33 03	301	50 06	332
	Alpheratz		◆Hamal		◆ACHERNAR		Peacock		Nunki		◆ALTAIR		Enif	
345	32 34	018	23 01	047	48 07	149	48 42	211	35 34	255	32 17	301	49 40	330
346	32 50	017	23 40	046	48 34	150	48 14	212	34 42	255	31 30	300	49 12	329
347	33 05	016	24 18	045	49 01	150	47 45	212	33 50	255	30 43	299	48 44	328
348	33 19	015	24 56	044	49 28	150	47 16	213	32 58	254	29 56	299	48 15	326
349	33 32	014	25 34	044	49 55	151	46 47	213	32 06	254	29 09	298	47 44	325
350	33 44	013	26 10	043	50 21	151	46 18	213	31 14	254	28 21	297	47 13	324
351	33 56	012	26 47	042	50 46	152	45 48	214	30 23	253	27 33	297	46 40	323
352	34 06	011	27 23	041	51 11	152	45 18	214	29 31	253	26 44	296	46 07	321
353	34 16	010	27 58	041	51 36	153	44 48	214	28 39	253	25 56	295	45 33	320
354	34 24	009	28 33	040	52 00	154	44 18	215	27 48	253	25 07	295	44 58	319
355	34 32	007	29 07	039	52 24	154	43 47	215	26 57	252	24 18	294	44 23	318
356	34 38	006	29 41	038	52 47	155	43 16	215	26 05	252	23 28	293	43 46	317
357	34 44	005	30 14	037	53 10	155	42 45	215	25 14	252	22 39	293	43 09	316
358	34 48	004	30 46	036	53 32	156	42 14	216	24 23	251	21 49	292	42 31	315
359	34 52	003	31 17	035	53 53	157	41 42	216	23 32	251	20 59	292	41 53	314

LAT 27°S

LHA	Hc	Zn	Hc	Zn	Hc	Zn	Hc	Zn	Hc	Zn	Hc	Zn	Hc	Zn
♈	◆Alpheratz		Hamal		◆RIGEL		CANOPUS		ACHERNAR		◆Peacock		Enif	
0	33 54	002	30 59	034	13 51	092	17 46	141	55 10	157	41 59	216	40 32	314
1	33 56	001	31 29	033	14 45	092	18 20	141	55 30	158	41 27	217	39 53	313
2	33 56	000	31 58	032	15 38	091	18 54	140	55 50	158	40 55	217	39 14	312
3	33 56	359	32 26	031	16 32	091	19 28	140	56 09	159	40 23	217	38 33	311
4	33 54	358	32 54	031	17 25	090	20 02	140	56 28	160	39 51	217	37 53	310
5	33 52	357	33 20	030	18 19	090	20 37	140	56 46	161	39 19	217	37 11	309
6	33 48	356	33 46	029	19 12	090	21 11	139	57 03	162	38 46	217	36 29	308
7	33 44	355	34 12	028	20 06	089	21 46	139	57 19	163	38 14	217	35 47	307
8	33 39	354	34 36	027	20 59	089	22 21	139	57 35	163	37 41	218	35 04	306
9	33 32	353	34 59	026	21 52	088	22 56	139	57 50	164	37 09	218	34 21	306
10	33 25	352	35 22	025	22 46	088	23 31	139	58 04	165	36 36	218	33 37	305
11	33 17	351	35 44	024	23 39	087	24 07	139	58 17	166	36 03	218	32 53	304
12	33 08	350	36 05	023	24 33	087	24 42	138	58 29	167	35 30	218	32 08	303
13	32 58	349	36 25	021	25 26	086	25 18	138	58 41	168	34 58	218	31 23	302
14	32 47	348	36 44	020	26 19	086	25 53	138	58 51	169	34 25	218	30 38	302
	◆Hamal		ALDEBARAN		RIGEL		◆CANOPUS		ACHERNAR		◆FOMALHAUT		Alpheratz	
15	37 02	019	21 59	057	27 13	085	26 29	138	59 01	170	62 58	257	32 35	347
16	37 19	018	22 43	056	28 06	085	27 05	138	59 09	171	62 06	257	32 22	346
17	37 36	017	23 28	055	28 59	084	27 41	138	59 17	172	61 14	257	32 08	345
18	37 51	016	24 11	055	29 52	084	28 17	138	59 24	173	60 21	257	31 53	344
19	38 05	015	24 55	054	30 45	083	28 53	138	59 30	174	59 29	257	31 38	343
20	38 18	014	25 38	053	31 39	083	29 29	138	59 35	175	58 37	257	31 21	342
21	38 30	013	26 21	053	32 32	082	30 05	137	59 39	176	57 45	257	31 04	341
22	38 42	011	27 03	052	33 24	082	30 41	137	59 41	177	56 53	256	30 46	340
23	38 52	010	27 45	051	34 17	081	31 17	137	59 43	179	56 01	256	30 27	339
24	39 01	009	28 26	050	35 10	081	31 54	137	59 44	180	55 10	256	30 07	338
25	39 09	008	29 07	050	36 03	080	32 30	137	59 44	181	54 18	256	29 46	337
26	39 16	007	29 47	049	36 55	079	33 06	137	59 43	182	53 26	256	29 25	336
27	39 21	006	30 27	048	37 48	079	33 43	137	59 41	183	52 34	256	29 03	335
28	39 26	004	31 07	047	38 40	078	34 19	137	59 38	184	51 42	255	28 40	334
29	39 30	003	31 46	046	39 33	078	34 55	137	59 34	185	50 51	255	28 16	333
	Hamal		ALDEBARAN		◆SIRIUS		CANOPUS		◆ACHERNAR		FOMALHAUT		◆Alpheratz	
30	39 32	002	32 24	046	23 54	097	35 32	137	59 29	186	49 59	255	27 52	332
31	39 33	001	33 02	045	24 47	097	36 08	137	59 22	187	49 07	255	27 27	332
32	39 34	000	33 39	044	25 40	096	36 44	137	59 15	188	48 16	255	27 01	331
33	39 33	358	34 16	043	26 33	096	37 21	137	59 08	189	47 24	255	26 34	330
34	39 31	357	34 52	042	27 26	096	37 57	137	58 59	190	46 33	254	26 07	329
35	39 28	356	35 27	041	28 20	095	38 33	137	58 49	191	45 41	254	25 39	328
36	39 24	355	36 02	040	29 13	095	39 09	137	58 38	192	44 50	254	25 10	327
37	39 18	354	36 36	039	30 06	094	39 46	138	58 26	193	43 58	254	24 41	327
38	39 12	353	37 10	038	30 59	094	40 22	138	58 14	194	43 07	254	24 12	326
39	39 04	351	37 42	037	31 53	094	40 58	138	58 01	195	42 16	253	23 41	325
40	38 56	350	38 14	036	32 46	093	41 34	138	57 46	196	41 25	253	23 10	324
41	38 46	349	38 45	035	33 39	093	42 09	138	57 31	197	40 33	253	22 39	323
42	38 36	348	39 16	034	34 33	092	42 45	138	57 16	198	39 42	253	22 06	323
43	38 24	347	39 45	033	35 26	092	43 21	138	56 59	198	38 51	252	21 34	322
44	38 11	346	40 14	032	36 20	091	43 56	138	56 42	199	38 00	252	21 01	321
	◆ALDEBARAN		BETELGEUSE		SIRIUS		◆CANOPUS		ACHERNAR		◆FOMALHAUT		Hamal	
45	40 42	031	35 27	057	37 13	091	44 32	139	56 24	200	37 10	252	37 57	345
46	41 09	030	36 12	056	38 07	091	45 07	139	56 05	201	36 19	252	37 43	343
47	41 35	029	36 56	056	39 00	090	45 42	139	55 45	202	35 28	251	37 27	342
48	42 00	027	37 40	055	39 54	090	46 17	139	55 25	202	34 38	251	37 10	341
49	42 24	026	38 24	054	40 47	089	46 52	140	55 05	203	33 47	251	36 52	340
50	42 47	025	39 07	053	41 40	089	47 26	140	54 43	204	32 56	251	36 34	339
51	43 09	024	39 49	052	42 34	088	48 00	140	54 21	205	32 06	250	36 14	338
52	43 30	023	40 31	051	43 27	088	48 35	141	53 59	205	31 16	250	35 53	337
53	43 50	021	41 12	050	44 21	087	49 08	141	53 36	206	30 26	250	35 32	336
54	44 09	020	41 53	049	45 14	087	49 42	141	53 12	207	29 35	250	35 10	335
55	44 27	019	42 34	048	46 07	086	50 15	142	52 48	207	28 45	249	34 47	334
56	44 44	018	43 13	047	47 01	086	50 48	142	52 24	208	27 55	249	34 22	333
57	44 59	016	43 52	046	47 54	085	51 21	142	51 58	208	27 06	249	33 58	332
58	45 14	015	44 31	045	48 47	085	51 54	143	51 33	209	26 16	248	33 32	331
59	45 27	014	45 08	044	49 41	084	52 26	143	51 07	209	25 26	248	33 06	330
	ALDEBARAN		BETELGEUSE		◆SIRIUS		CANOPUS		◆ACHERNAR		FOMALHAUT		◆Hamal	
60	45 39	012	45 45	043	50 34	084	52 58	144	50 41	210	24 37	248	32 38	329
61	45 50	011	46 21	042	51 27	083	53 29	144	50 14	210	23 47	248	32 10	328
62	45 59	010	46 57	041	52 20	082	54 00	145	49 47	211	22 58	247	31 42	327
63	46 07	008	47 31	040	53 13	082	54 31	145	49 19	211	22 08	247	31 12	326
64	46 14	007	48 05	038	54 06	081	55 01	146	48 51	212	21 19	247	30 42	325
65	46 20	005	48 37	037	54 58	081	55 30	147	48 23	212	20 30	246	30 11	324
66	46 25	004	49 09	036	55 51	080	56 00	147	47 55	212	19 41	246	29 40	324
67	46 28	003	49 40	035	56 44	079	56 28	148	47 26	213	18 53	246	29 08	323
68	46 29	001	50 10	033	57 36	078	56 56	149	46 57	213	18 04	245	28 35	322
69	46 30	000	50 39	032	58 28	078	57 24	149	46 27	213	17 16	245	28 02	321
70	46 29	358	51 06	030	59 21	077	57 51	150	45 58	214	16 27	245	27 28	320
71	46 27	357	51 33	029	60 13	076	58 17	151	45 28	214	15 39	244	26 53	319
72	46 24	356	51 58	028	61 04	075	58 43	152	44 58	214	14 51	244	26 18	319
73	46 19	354	52 22	026	61 56	074	59 08	153	44 28	215	14 03	244	25 42	318
74	46 13	353	52 45	025	62 47	073	59 32	153	43 57	215	13 15	243	25 06	317
	BETELGEUSE		PROCYON		◆Suhail		CANOPUS		◆ACHERNAR		Hamal		◆ALDEBARAN	
75	53 07	023	39 50	056	38 03	125	59 56	154	43 26	215	24 30	316	46 06	352
76	53 27	022	40 34	055	38 46	125	60 18	155	42 56	216	23 52	316	45 58	350
77	53 46	020	41 18	054	39 30	125	60 40	156	42 25	216	23 15	315	45 48	349
78	54 03	018	42 01	053	40 14	125	61 01	157	41 53	216	22 37	314	45 37	347
79	54 20	017	42 44	052	40 57	125	61 22	158	41 22	216	21 58	313	45 25	346
80	54 34	015	43 26	051	41 41	125	61 41	159	40 51	216	21 19	313	45 11	345
81	54 47	013	44 07	050	42 24	125	61 59	161	40 19	216	20 39	312	44 57	344
82	54 59	012	44 48	049	43 08	125	62 16	162	39 47	216	19 59	311	44 41	342
83	55 09	010	45 29	048	43 51	125	62 33	163	39 15	217	19 19	311	44 24	341
84	55 17	008	46 08	047	44 35	126	62 48	164	38 43	217	18 38	310	44 06	340
85	55 24	007	46 47	046	45 19	126	63 02	165	38 11	217	17 57	309	43 47	338
86	55 30	005	47 25	045	46 02	126	63 15	167	37 39	217	17 16	309	43 27	337
87	55 33	003	48 03	044	46 45	126	63 27	168	37 07	217	16 34	308	43 05	336
88	55 35	001	48 40	043	47 29	126	63 38	169	36 35	217	15 52	308	42 43	335
89	55 36	000	49 16	042	48 12	126	63 47	170	36 03	217	15 09	307	42 20	334

LHA	Hc	Zn	Hc	Zn	Hc	Zn	Hc	Zn	Hc	Zn	Hc	Zn	Hc	Zn
♈	POLLUX		◆REGULUS		ACRUX		◆CANOPUS		ACHERNAR		◆RIGEL		BETELGEUSE	
90	29 28	027	18 20	066	21 01	151	63 55	172	35 30	217	68 19	328	55 34	358
91	29 52	026	19 08	065	21 26	151	64 02	173	34 58	217	67 49	326	55 31	356
92	30 14	025	19 57	064	21 52	151	64 08	174	34 26	217	67 18	323	55 27	354
93	30 36	024	20 45	064	22 18	151	64 13	176	33 53	217	66 46	321	55 21	353
94	30 58	023	21 33	063	22 45	151	64 16	177	33 21	217	66 12	319	55 13	351
95	31 18	022	22 20	063	23 11	150	64 18	179	32 48	217	65 36	317	55 04	349
96	31 38	021	23 07	062	23 37	150	64 18	180	32 16	217	64 59	315	54 53	347
97	31 56	020	23 54	061	24 04	150	64 18	181	31 43	217	64 21	314	54 40	346
98	32 14	019	24 41	061	24 30	150	64 16	183	31 11	217	63 42	312	54 26	344
99	32 31	018	25 28	060	24 57	150	64 12	184	30 39	217	63 02	310	54 11	342
100	32 48	017	26 14	059	25 24	150	64 08	186	30 06	217	62 21	309	53 54	341
101	33 03	016	26 59	059	25 50	150	64 02	187	29 34	217	61 39	307	53 36	339
102	33 17	015	27 45	058	26 17	150	63 55	188	29 02	217	60 56	306	53 16	338
103	33 31	014	28 30	057	26 44	150	63 46	190	28 29	217	60 12	305	52 55	336
104	33 43	013	29 15	056	27 11	150	63 37	191	27 57	217	59 28	303	52 33	335
	POLLUX		◆REGULUS		Gienah		◆ACRUX		CANOPUS		◆RIGEL		BETELGEUSE	
105	33 55	012	29 59	056	17 29	101	27 38	150	63 26	192	58 43	302	52 09	333
106	34 05	011	30 43	055	18 21	101	28 05	150	63 14	194	57 57	301	51 44	332
107	34 15	010	31 27	054	19 14	100	28 32	150	63 01	195	57 11	300	51 18	330
108	34 24	009	32 10	053	20 07	100	28 59	150	62 47	196	56 24	299	50 51	329
109	34 31	008	32 53	053	20 59	100	29 27	149	62 31	197	55 37	297	50 23	327
110	34 38	007	33 35	052	21 52	099	29 54	149	62 15	198	54 49	296	49 54	326
111	34 44	006	34 17	051	22 45	099	30 21	149	61 58	200	54 01	295	49 23	325
112	34 49	005	34 58	050	23 38	098	30 48	149	61 39	201	53 13	294	48 52	323
113	34 52	003	35 39	049	24 31	098	31 15	149	61 20	202	52 24	294	48 20	322
114	34 55	002	36 19	048	25 24	098	31 42	149	61 00	203	51 35	293	47 46	321
115	34 57	001	36 59	048	26 17	097	32 10	149	60 39	204	50 45	292	47 12	320
116	34 58	000	37 38	047	27 10	097	32 37	150	60 17	205	49 55	291	46 37	319
117	34 57	359	38 16	046	28 03	096	33 04	150	59 54	206	49 05	290	46 01	317
118	34 56	358	38 54	045	28 56	096	33 31	150	59 30	207	48 15	289	45 25	316
119	34 54	357	39 31	044	29 49	096	33 58	150	59 06	208	47 24	289	44 48	315
	◆REGULUS		SPICA		◆ACRUX		CANOPUS		◆RIGEL		BETELGEUSE		POLLUX	
120	40 08	043	12 46	096	34 25	150	58 41	208	46 33	288	44 10	314	34 51	356
121	40 44	042	13 39	096	34 52	150	58 15	209	45 42	287	43 31	313	34 46	355
122	41 19	041	14 32	095	35 19	150	57 49	210	44 51	286	42 51	312	34 41	354
123	41 54	040	15 25	095	35 45	150	57 22	211	44 00	286	42 12	311	34 35	353
124	42 27	039	16 19	094	36 12	150	56 54	211	43 08	285	41 31	310	34 28	352
125	43 00	037	17 12	094	36 39	150	56 26	212	42 16	284	40 50	309	34 19	351
126	43 32	036	18 05	094	37 05	150	55 57	213	41 25	284	40 08	308	34 10	350
127	44 04	035	18 59	093	37 31	151	55 28	213	40 33	283	39 26	307	34 00	349
128	44 34	034	19 52	093	37 58	151	54 58	214	39 40	282	38 43	306	33 49	348
129	45 03	033	20 46	092	38 24	151	54 28	215	38 48	282	38 00	306	33 37	346
130	45 32	032	21 39	092	38 50	151	53 58	215	37 56	281	37 16	305	33 24	345
131	45 59	030	22 32	091	39 16	151	53 27	216	37 03	281	36 32	304	33 10	344
132	46 26	029	23 26	091	39 41	151	52 55	216	36 11	280	35 47	303	32 55	343
133	46 51	028	24 19	090	40 07	152	52 23	217	35 18	279	35 02	302	32 40	342
134	47 16	026	25 13	090	40 32	152	51 51	217	34 25	279	34 17	301	32 23	341
	REGULUS		◆SPICA		ACRUX		◆CANOPUS		RIGEL		BETELGEUSE		◆POLLUX	
135	47 39	025	26 06	090	40 58	152	51 19	218	33 32	278	33 31	301	32 06	340
136	48 01	024	27 00	089	41 23	152	50 46	218	32 39	278	32 45	300	31 47	339
137	48 22	022	27 53	089	41 47	152	50 13	218	31 46	277	31 58	299	31 28	338
138	48 42	021	28 47	088	42 12	153	49 39	219	30 53	277	31 12	299	31 08	338
139	49 01	020	29 40	088	42 36	153	49 06	219	30 00	276	30 25	298	30 47	337
140	49 18	018	30 33	087	43 01	153	48 32	220	29 07	276	29 37	297	30 26	336
141	49 34	017	31 27	087	43 25	154	47 58	220	28 14	275	28 49	296	30 03	335
142	49 49	015	32 20	086	43 48	154	47 23	220	27 21	275	28 01	296	29 40	334
143	50 02	014	33 14	086	44 12	154	46 49	220	26 27	274	27 13	295	29 16	333
144	50 14	012	34 07	085	44 35	155	46 14	221	25 34	274	26 25	294	28 51	332
145	50 25	011	35 00	085	44 58	155	45 39	221	24 41	273	25 36	294	28 26	331
146	50 34	009	35 53	084	45 20	155	45 04	221	23 47	273	24 47	293	28 00	330
147	50 42	008	36 46	084	45 42	156	44 29	221	22 54	272	23 58	293	27 33	329
148	50 49	006	37 40	083	46 04	156	43 53	222	22 00	272	23 08	292	27 05	329
149	50 54	005	38 33	083	46 26	156	43 18	222	21 07	271	22 18	291	26 37	328
	REGULUS		◆Denebola		SPICA		◆ACRUX		CANOPUS		◆SIRIUS		PROCYON	
150	50 58	003	40 44	036	39 26	082	46 47	157	42 42	222	43 52	272	43 06	308
151	51 00	002	41 14	035	40 19	081	47 08	157	42 07	222	42 59	272	42 23	307
152	51 00	000	41 44	034	41 11	081	47 29	158	41 31	222	42 05	271	41 41	306
153	51 00	358	42 13	032	42 04	080	47 49	158	40 55	222	41 12	271	40 57	305
154	50 58	357	42 42	031	42 57	080	48 08	159	40 19	222	40 18	271	40 13	304
155	50 54	355	43 09	030	43 49	079	48 28	159	39 43	223	39 25	270	39 29	304
156	50 49	354	43 35	029	44 42	079	48 47	160	39 07	223	38 31	270	38 44	303
157	50 43	352	44 00	028	45 34	078	49 05	160	38 30	223	37 38	269	37 59	302
158	50 35	351	44 25	026	46 26	077	49 23	161	37 54	223	36 44	269	37 13	301
159	50 25	349	44 48	025	47 18	077	49 41	161	37 18	223	35 51	268	36 27	300
160	50 15	348	45 10	024	48 10	076	49 58	162	36 41	223	34 58	268	35 41	299
161	50 03	346	45 32	023	49 02	075	50 14	162	36 05	223	34 04	267	34 54	299
162	49 49	345	45 52	021	49 54	074	50 31	163	35 29	223	33 11	267	34 07	298
163	49 35	343	46 10	020	50 45	074	50 46	163	34 52	223	32 17	267	33 20	297
164	49 19	342	46 28	019	51 36	073	51 01	164	34 16	223	31 24	266	32 32	297
	ARCTURUS		◆ANTARES		ACRUX		◆CANOPUS		SIRIUS		PROCYON		◆REGULUS	
165	23 51	051	18 01	111	51 16	165	33 40	223	30 31	266	31 44	296	49 01	340
166	24 32	050	18 51	111	51 30	165	33 03	223	29 37	265	30 56	295	48 43	339
167	25 13	050	19 41	110	51 43	166	32 27	223	28 44	265	30 07	294	48 23	338
168	25 53	049	20 31	110	51 56	166	31 51	223	27 51	265	29 19	294	48 02	336
169	26 33	048	21 21	110	52 08	167	31 14	223	26 58	264	28 30	293	47 40	335
170	27 13	047	22 12	109	52 20	168	30 38	223	26 05	264	27 40	293	47 17	334
171	27 52	047	23 02	109	52 31	168	30 02	223	25 11	263	26 51	292	46 53	332
172	28 31	046	23 53	109	52 41	169	29 26	222	24 18	263	26 01	291	46 27	331
173	29 09	045	24 44	108	52 51	170	28 50	222	23 25	262	25 11	291	46 01	330
174	29 46	044	25 34	108	53 00	171	28 14	222	22 32	262	24 21	290	45 33	328
175	30 23	043	26 25	108	53 08	171	27 38	222	21 39	262	23 31	289	45 05	327
176	31 00	043	27 16	107	53 16	172	27 02	222	20 47	261	22 40	289	44 36	326
177	31 35	042	28 07	107	53 23	173	26 26	222	19 54	261	21 50	288	44 05	325
178	32 11	041	28 58	107	53 29	173	25 51	222	19 01	260	20 59	288	43 34	324
179	32 45	040	29 50	106	53 35	174	25 15	222	18 08	260	20 08	287	43 02	323

LHA ♈	Hc	Zn	Hc	Zn	Hc	Zn	Hc	Zn	Hc	Zn	Hc	Zn	Hc	Zn
	◆ARCTURUS		ANTARES		◆RIGIL KENT.		ACRUX		CANOPUS		◆Alphard		REGULUS	
180	33 19	039	30 41	106	46 54	153	53 40	175	24 39	222	49 32	290	42 29	322
181	33 53	038	31 32	106	47 18	153	53 45	176	24 04	221	48 42	289	41 55	320
182	34 25	037	32 24	106	47 42	154	53 48	176	23 29	221	47 51	288	41 21	319
183	34 57	036	33 15	105	48 05	154	53 51	177	22 54	221	47 00	287	40 46	318
184	35 29	035	34 07	105	48 29	155	53 53	178	22 18	221	46 09	287	40 10	317
185	35 59	034	34 59	105	48 51	155	53 55	179	21 44	221	45 18	286	39 33	316
186	36 29	033	35 50	104	49 14	155	53 56	180	21 09	221	44 26	285	38 56	315
187	36 58	032	36 42	104	49 36	156	53 56	180	20 34	220	43 34	285	38 18	314
188	37 26	031	37 34	104	49 57	156	53 55	181	19 59	220	42 43	284	37 40	313
189	37 53	030	38 26	103	50 19	157	53 54	182	19 25	220	41 51	283	37 01	313
190	38 20	029	39 18	103	50 39	158	53 52	183	18 51	220	40 59	283	36 21	312
191	38 45	028	40 10	103	50 59	158	53 49	183	18 17	219	40 06	282	35 41	311
192	39 10	027	41 02	103	51 19	159	53 45	184	17 43	219	39 14	281	35 00	310
193	39 34	026	41 54	102	51 38	159	53 41	185	17 09	219	38 21	281	34 19	309
194	39 56	025	42 47	102	51 57	160	53 36	186	16 36	219	37 29	280	33 37	308
	ARCTURUS		◆Rasalhague		ANTARES		RIGIL KENT.		◆ACRUX		Suhail		◆REGULUS	
195	40 18	024	12 34	069	43 39	102	52 15	160	53 30	186	40 53	235	32 55	307
196	40 39	022	13 23	068	44 31	101	52 33	161	53 24	187	40 10	235	32 12	307
197	40 59	021	14 13	068	45 24	101	52 50	162	53 17	188	39 26	235	31 29	306
198	41 18	020	15 02	067	46 16	101	53 06	162	53 09	189	38 42	235	30 45	305
199	41 36	019	15 51	067	47 09	101	53 22	163	53 01	189	37 59	234	30 01	304
200	41 53	018	16 40	066	48 01	100	53 37	164	52 52	190	37 15	234	29 17	304
201	42 08	016	17 29	065	48 54	100	53 52	165	52 43	191	36 32	234	28 32	303
202	42 23	015	18 18	065	49 47	100	54 06	165	52 32	191	35 48	234	27 47	302
203	42 36	014	19 06	064	50 39	100	54 19	166	52 21	192	35 05	234	27 02	301
204	42 49	013	19 54	064	51 32	099	54 32	167	52 10	193	34 22	234	26 16	301
205	43 00	011	20 42	063	52 25	099	54 44	168	51 58	193	33 38	234	25 30	300
206	43 10	010	21 29	062	53 18	099	54 55	168	51 45	194	32 55	234	24 44	299
207	43 19	009	22 16	062	54 10	098	55 05	169	51 32	195	32 12	234	23 57	299
208	43 27	008	23 03	061	55 03	098	55 15	170	51 18	195	31 29	234	23 10	298
209	43 33	006	23 50	061	55 56	098	55 24	171	51 03	196	30 46	234	22 23	298
	◆ARCTURUS		Rasalhague		◆Nunki		RIGIL KENT.		ACRUX		◆Suhail		Denebola	
210	43 38	005	24 36	060	25 08	108	55 32	172	50 49	197	30 03	233	37 37	319
211	43 43	004	25 23	059	25 59	108	55 40	172	50 33	197	29 20	233	37 01	318
212	43 45	002	26 08	059	26 50	107	55 46	173	50 17	198	28 37	233	36 25	317
213	43 47	001	26 54	058	27 41	107	55 52	174	50 00	198	27 55	233	35 48	316
214	43 47	000	27 39	057	28 32	107	55 57	175	49 43	199	27 12	233	35 10	315
215	43 47	359	28 24	056	29 24	106	56 02	176	49 26	199	26 29	233	34 32	314
216	43 45	357	29 08	056	30 15	106	56 05	177	49 08	200	25 47	232	33 53	313
217	43 41	356	29 52	055	31 06	106	56 08	178	48 50	200	25 05	232	33 14	312
218	43 37	355	30 35	054	31 58	106	56 10	178	48 31	201	24 23	232	32 34	311
219	43 31	353	31 19	053	32 49	105	56 11	179	48 12	201	23 41	232	31 54	311
220	43 25	352	32 01	053	33 41	105	56 11	180	47 52	202	22 59	232	31 13	310
221	43 17	351	32 44	052	34 33	105	56 11	181	47 32	202	22 17	231	30 31	309
222	43 07	349	33 25	051	35 24	104	56 09	182	47 11	203	21 35	231	29 50	308
223	42 57	348	34 07	050	36 16	104	56 07	183	46 51	203	20 54	231	29 07	307
224	42 45	347	34 48	049	37 08	104	56 04	184	46 29	204	20 12	231	28 25	307
	Alphecca		Rasalhague		◆Nunki		Peacock		◆RIGIL KENT.		SPICA		◆ARCTURUS	
225	35 38	009	35 28	048	38 00	103	26 58	143	56 00	185	62 36	301	42 33	346
226	35 46	008	36 08	048	38 52	103	27 30	142	55 56	185	61 49	299	42 19	344
227	35 54	007	36 47	047	39 44	103	28 03	142	55 50	186	61 03	298	42 04	343
228	36 00	006	37 25	046	40 36	103	28 36	142	55 44	187	60 15	297	41 48	342
229	36 05	005	38 03	045	41 29	102	29 08	142	55 37	188	59 27	296	41 31	341
230	36 09	004	38 41	044	42 21	102	29 41	142	55 29	189	58 39	295	41 13	340
231	36 13	003	39 18	043	43 13	102	30 14	142	55 21	190	57 50	294	40 54	338
232	36 15	002	39 54	042	44 05	101	30 47	142	55 11	190	57 01	293	40 33	337
233	36 16	001	40 29	041	44 58	101	31 20	142	55 01	191	56 12	292	40 12	336
234	36 16	000	41 04	040	45 50	101	31 53	142	54 51	192	55 22	291	39 50	335
235	36 15	358	41 38	039	46 43	101	32 25	142	54 39	193	54 32	290	39 27	334
236	36 13	357	42 11	038	47 35	100	32 58	142	54 27	194	53 41	289	39 03	333
237	36 10	356	42 43	037	48 28	100	33 31	142	54 14	194	52 51	288	38 38	332
238	36 06	355	43 14	035	49 21	100	34 04	142	54 01	195	52 00	287	38 12	331
239	36 01	354	43 45	034	50 13	100	34 37	142	53 47	196	51 09	287	37 45	330
	◆VEGA		ALTAIR		◆Peacock		RIGIL KENT.		◆SPICA		ARCTURUS		Alphecca	
240	14 43	031	23 40	066	35 10	142	53 32	196	50 18	286	37 18	328	35 55	353
241	15 10	030	24 29	065	35 43	142	53 16	197	49 26	285	36 50	327	35 48	352
242	15 36	029	25 17	064	36 15	142	53 00	198	48 34	284	36 20	326	35 40	351
243	16 02	029	26 05	064	36 48	142	52 44	198	47 42	284	35 50	325	35 31	350
244	16 27	028	26 53	063	37 21	142	52 26	199	46 50	283	35 20	324	35 21	349
245	16 52	027	27 40	062	37 53	142	52 09	200	45 58	282	34 48	324	35 10	348
246	17 16	027	28 28	062	38 26	143	51 50	200	45 06	282	34 16	323	34 58	346
247	17 40	026	29 15	061	38 58	143	51 31	201	44 14	281	33 43	322	34 45	345
248	18 03	025	30 01	060	39 31	143	51 12	202	43 21	281	33 10	321	34 31	344
249	18 25	024	30 48	060	40 03	143	50 52	202	42 29	280	32 36	320	34 16	343
250	18 47	024	31 34	059	40 35	143	50 32	203	41 36	279	32 01	319	34 00	342
251	19 08	023	32 19	058	41 07	143	50 11	203	40 43	279	31 25	318	33 44	341
252	19 29	022	33 04	057	41 39	143	49 49	204	39 50	278	30 49	317	33 26	340
253	19 49	021	33 49	057	42 11	144	49 28	204	38 57	278	30 13	316	33 08	339
254	20 08	021	34 34	056	42 43	144	49 06	205	38 04	277	29 36	316	32 48	338
	◆VEGA		ALTAIR		FOMALHAUT		◆Peacock		RIGIL KENT.		◆SPICA		ARCTURUS	
255	20 27	020	35 18	055	13 30	117	43 14	144	48 43	205	37 11	277	28 58	315
256	20 44	019	36 01	054	14 18	116	43 46	144	48 20	206	36 18	276	28 20	314
257	21 02	018	36 45	053	15 06	116	44 17	145	47 57	206	35 25	276	27 41	313
258	21 18	018	37 27	053	15 54	116	44 48	145	47 33	207	34 32	275	27 02	312
259	21 34	017	38 09	052	16 42	115	45 18	145	47 09	207	33 38	275	26 22	312
260	21 49	016	38 51	051	17 31	115	45 49	145	46 45	207	32 45	274	25 42	311
261	22 03	015	39 32	050	18 19	115	46 19	146	46 20	208	31 52	274	25 01	310
262	22 17	014	40 13	049	19 08	114	46 49	146	45 55	208	30 58	273	24 20	309
263	22 30	014	40 53	048	19 57	114	47 19	146	45 30	208	30 05	273	23 39	309
264	22 42	013	41 32	047	20 46	114	47 49	147	45 04	209	29 11	272	22 57	308
265	22 54	012	42 11	046	21 35	113	48 18	147	44 38	209	28 18	272	22 15	307
266	23 04	011	42 49	045	22 24	113	48 47	147	44 12	209	27 25	271	21 32	307
267	23 14	010	43 27	044	23 13	113	49 15	148	43 46	210	26 31	271	20 49	306
268	23 24	009	44 03	043	24 03	112	49 44	148	43 19	210	25 38	270	20 05	305
269	23 32	009	44 39	042	24 52	112	50 12	149	42 52	210	24 44	270	19 22	305

LAT 27°S

LHA ♈	Hc	Zn	Hc	Zn	Hc	Zn	Hc	Zn	Hc	Zn	Hc	Zn	Hc	Zn
	VEGA		◆ALTAIR		FOMALHAUT		◆Peacock		RIGIL KENT.		◆ANTARES		Alphecca	
270	23 40	008	45 14	041	25 42	112	50 39	149	42 25	210	69 43	266	25 53	324
271	23 47	007	45 49	039	26 31	111	51 06	150	41 58	211	68 50	266	25 21	323
272	23 53	006	46 22	038	27 21	111	51 33	150	41 31	211	67 56	266	24 49	322
273	23 58	005	46 55	037	28 11	111	51 59	151	41 03	211	67 03	266	24 16	322
274	24 03	004	47 27	036	29 01	111	52 25	151	40 35	211	66 10	265	23 42	321
275	24 06	004	47 58	035	29 51	110	52 51	152	40 07	212	65 17	265	23 08	320
276	24 09	003	48 28	033	30 41	110	53 16	152	39 39	212	64 23	265	22 33	319
277	24 11	002	48 56	032	31 31	110	53 40	153	39 11	212	63 30	264	21 58	319
278	24 13	001	49 24	031	32 22	110	54 04	154	38 43	212	62 37	264	21 23	318
279	24 13	000	49 51	029	33 12	109	54 28	154	38 14	212	61 44	264	20 47	317
280	24 13	359	50 17	028	34 03	109	54 51	155	37 46	212	60 51	264	20 10	316
281	24 12	358	50 41	027	34 53	109	55 13	156	37 17	212	59 58	263	19 33	316
282	24 10	358	51 04	025	35 44	109	55 34	156	36 48	213	59 04	263	18 55	315
283	24 07	357	51 26	024	36 35	108	55 55	157	36 19	213	58 11	263	18 18	314
284	24 04	356	51 47	022	37 26	108	56 16	158	35 51	213	57 18	262	17 39	314
	◆DENEB		FOMALHAUT		◆ACHERNAR		RIGIL KENT.		◆ANTARES		Rasalhague		VEGA	
285	14 09	018	38 16	108	17 39	146	35 22	213	56 25	262	45 21	330	24 00	355
286	14 25	017	39 07	108	18 09	146	34 52	213	55 32	262	44 53	328	23 55	354
287	14 41	017	39 58	107	18 40	146	34 23	213	54 40	262	44 25	327	23 49	353
288	14 56	016	40 49	107	19 10	145	33 54	213	53 47	261	43 56	326	23 42	353
289	15 10	015	41 41	107	19 40	145	33 25	213	52 54	261	43 25	325	23 35	352
290	15 24	015	42 32	107	20 11	145	32 56	213	52 01	261	42 54	324	23 27	351
291	15 38	014	43 23	106	20 42	145	32 27	213	51 08	261	42 22	323	23 18	350
292	15 50	013	44 14	106	21 13	145	31 57	213	50 16	260	41 49	322	23 08	349
293	16 02	013	45 06	106	21 44	144	31 28	213	49 23	260	41 16	321	22 58	348
294	16 13	012	45 57	106	22 15	144	30 59	213	48 30	260	40 41	319	22 47	348
295	16 24	011	46 49	106	22 46	144	30 30	213	47 38	259	40 06	318	22 35	347
296	16 34	010	47 40	105	23 18	144	30 00	213	46 45	259	39 30	317	22 22	346
297	16 44	010	48 32	105	23 49	144	29 31	213	45 53	259	38 54	316	22 09	345
298	16 52	009	49 23	105	24 21	144	29 02	213	45 00	259	38 17	315	21 55	344
299	17 00	008	50 15	105	24 53	144	28 33	213	44 08	258	37 39	315	21 40	343
	DENEB		Enif		◆FOMALHAUT		ACHERNAR		◆RIGIL KENT.		ANTARES		◆VEGA	
300	17 08	008	45 21	038	51 07	105	25 24	143	28 04	213	43 15	258	21 24	343
301	17 14	007	45 54	037	51 58	104	25 56	143	27 35	213	42 23	258	21 08	342
302	17 20	006	46 25	036	52 50	104	26 28	143	27 06	213	41 31	258	20 51	341
303	17 26	005	46 56	034	53 42	104	27 00	143	26 37	213	40 39	257	20 33	340
304	17 31	005	47 25	033	54 34	104	27 32	143	26 08	213	39 47	257	20 15	340
305	17 35	004	47 54	032	55 26	104	28 05	143	25 39	213	38 55	257	19 56	339
306	17 38	003	48 22	030	56 18	104	28 37	143	25 10	213	38 03	256	19 37	338
307	17 40	002	48 48	029	57 10	104	29 09	143	24 41	212	37 11	256	19 16	337
308	17 42	002	49 14	028	58 02	103	29 41	143	24 13	212	36 19	256	18 55	337
309	17 44	001	49 38	026	58 54	103	30 14	143	23 44	212	35 27	256	18 34	336
310	17 44	000	50 01	025	59 46	103	30 46	143	23 16	212	34 35	255	18 12	335
311	17 44	359	50 23	024	60 38	103	31 19	143	22 47	212	33 44	255	17 49	334
312	17 43	359	50 44	022	61 30	103	31 51	143	22 19	212	32 52	255	17 26	334
313	17 42	358	51 03	021	62 22	103	32 24	143	21 51	212	32 01	254	17 02	333
314	17 40	357	51 21	019	63 14	103	32 56	143	21 23	211	31 09	254	16 37	332
	Alpheratz		◆Diphda		ACHERNAR		◆RIGIL KENT.		ANTARES		◆ALTAIR		DENEB	
315	18 05	042	38 03	092	33 28	143	20 55	211	30 18	254	50 23	332	17 37	357
316	18 40	042	38 56	092	34 01	143	20 28	211	29 27	253	49 58	331	17 33	356
317	19 16	041	39 49	092	34 33	143	20 00	211	28 35	253	49 32	330	17 29	355
318	19 50	040	40 43	091	35 06	143	19 32	211	27 44	253	49 04	328	17 24	354
319	20 25	040	41 36	091	35 38	143	19 05	211	26 53	252	48 35	327	17 18	354
320	20 58	039	42 30	090	36 10	143	18 38	210	26 02	252	48 06	326	17 12	353
321	21 32	038	43 23	090	36 43	143	18 11	210	25 11	252	47 35	324	17 05	352
322	22 04	037	44 17	089	37 15	143	17 44	210	24 21	251	47 04	323	16 57	351
323	22 37	037	45 10	089	37 47	143	17 18	210	23 30	251	46 31	322	16 49	351
324	23 08	036	46 04	088	38 19	143	16 51	210	22 40	251	45 58	321	16 40	350
325	23 39	035	46 57	088	38 51	143	16 25	209	21 49	250	45 24	320	16 31	349
326	24 10	034	47 50	087	39 23	143	15 59	209	20 59	250	44 49	319	16 20	349
327	24 39	034	48 44	087	39 55	144	15 33	209	20 08	250	44 13	317	16 09	348
328	25 09	033	49 37	086	40 26	144	15 07	209	19 18	249	43 37	316	15 58	347
329	25 37	032	50 31	086	40 58	144	14 41	208	18 28	249	42 59	315	15 46	346
	◆Alpheratz		Diphda		◆ACHERNAR		Peacock		◆Nunki		ALTAIR		Enif	
330	26 05	031	51 24	085	41 29	144	55 48	203	48 53	260	42 21	314	52 57	353
331	26 32	030	52 17	085	42 01	144	55 26	204	48 01	260	41 43	313	52 50	352
332	26 59	029	53 10	084	42 32	144	55 05	205	47 08	260	41 04	312	52 41	350
333	27 25	029	54 03	084	43 03	145	54 42	205	46 15	259	40 24	311	52 31	349
334	27 50	028	54 57	083	43 34	145	54 19	206	45 23	259	39 43	310	52 20	347
335	28 15	027	55 50	082	44 04	145	53 55	207	44 30	259	39 02	309	52 07	345
336	28 38	026	56 43	082	44 35	145	53 31	207	43 38	258	38 21	309	51 53	344
337	29 01	025	57 35	081	45 05	146	53 07	208	42 46	258	37 39	308	51 38	342
338	29 23	024	58 28	080	45 35	146	52 41	208	41 53	258	36 56	307	51 21	341
339	29 45	023	59 21	080	46 05	146	52 16	209	41 01	258	36 13	306	51 03	339
340	30 06	022	60 13	079	46 34	147	51 50	209	40 09	257	35 30	305	50 43	338
341	30 25	021	61 06	078	47 04	147	51 23	210	39 17	257	34 46	304	50 22	336
342	30 44	020	61 58	077	47 33	147	50 56	210	38 25	257	34 01	304	50 00	335
343	31 03	019	62 50	077	48 01	148	50 29	211	37 33	256	33 17	303	49 37	334
344	31 20	018	63 42	076	48 30	148	50 01	211	36 41	256	32 31	302	49 13	332
	Alpheratz		◆Hamal		◆ACHERNAR		Peacock		Nunki		◆ALTAIR		Enif	
345	31 37	017	22 19	046	48 58	148	49 33	212	35 49	256	31 46	301	48 47	331
346	31 52	017	22 58	046	49 26	149	49 05	212	34 57	255	31 00	301	48 21	330
347	32 07	016	23 36	045	49 53	149	48 36	213	34 06	255	30 14	300	47 53	328
348	32 21	015	24 13	044	50 20	150	48 07	213	33 14	255	29 27	299	47 24	327
349	32 34	014	24 50	043	50 47	150	47 38	213	32 22	255	28 40	298	46 55	326
350	32 46	013	25 26	043	51 13	151	47 08	214	31 31	254	27 53	298	46 24	324
351	32 57	011	26 02	042	51 39	151	46 38	214	30 39	254	27 06	297	45 53	323
352	33 07	010	26 38	041	52 04	152	46 08	214	29 48	254	26 18	296	45 20	322
353	33 16	009	27 12	040	52 29	152	45 38	215	28 57	253	25 30	296	44 47	321
354	33 25	008	27 47	039	52 54	153	45 07	215	28 06	253	24 42	295	44 13	320
355	33 32	007	28 20	039	53 18	154	44 36	215	27 15	253	23 53	294	43 38	319
356	33 38	006	28 53	038	53 41	154	44 05	216	26 24	252	23 04	294	43 02	318
357	33 44	005	29 26	037	54 04	155	43 34	216	25 33	252	22 15	293	42 26	317
358	33 48	004	29 57	036	54 27	156	43 03	216	24 42	252	21 26	293	41 49	316
359	33 52	003	30 29	035	54 48	156	42 31	216	23 51	251	20 37	292	41 11	314

LAT 28°S

LHA	Hc	Zn	Hc	Zn	Hc	Zn	Hc	Zn	Hc	Zn	Hc	Zn	Hc	Zn
♈	♦Alpheratz		Hamal		♦RIGEL		CANOPUS		ACHERNAR		♦Peacock		Enif	
0	32 54	002	30 09	034	13 54	092	18 32	141	56 05	156	42 47	217	39 51	314
1	32 56	001	30 38	033	14 47	092	19 06	140	56 26	157	42 15	217	39 13	313
2	32 56	000	31 07	032	15 40	091	19 40	140	56 46	158	41 43	217	38 34	312
3	32 56	359	31 35	031	16 33	091	20 14	140	57 05	159	41 11	218	37 54	311
4	32 54	358	32 02	030	17 26	090	20 48	140	57 24	160	40 39	218	37 14	310
5	32 52	357	32 28	029	18 19	090	21 22	139	57 42	160	40 06	218	36 33	309
6	32 49	356	32 54	028	19 11	089	21 57	139	58 00	161	39 34	218	35 52	309
7	32 44	355	33 18	027	20 04	089	22 32	139	58 17	162	39 01	218	35 10	308
8	32 39	354	33 42	026	20 57	088	23 06	139	58 32	163	38 29	218	34 28	307
9	32 33	353	34 05	025	21 50	088	23 41	139	58 47	164	37 56	218	33 46	306
10	32 26	352	34 28	024	22 43	087	24 17	138	59 02	165	37 23	218	33 03	305
11	32 18	351	34 49	023	23 36	087	24 52	138	59 15	166	36 50	218	32 19	304
12	32 09	350	35 09	022	24 29	086	25 27	138	59 28	167	36 18	218	31 35	304
13	31 59	349	35 29	021	25 22	086	26 02	138	59 39	168	35 45	218	30 51	303
14	31 48	348	35 48	020	26 15	085	26 38	138	59 50	169	35 12	218	30 06	302

LHA	Hc	Zn	Hc	Zn	Hc	Zn	Hc	Zn	Hc	Zn	Hc	Zn	Hc	Zn
	♦Hamal		ALDEBARAN		RIGEL		♦CANOPUS		ACHERNAR		♦FOMALHAUT		Alpheratz	
15	36 06	019	21 26	056	27 08	085	27 14	138	60 00	170	63 10	259	31 36	347
16	36 22	018	22 10	056	28 00	084	27 49	138	60 09	171	62 18	259	31 24	346
17	36 38	017	22 53	055	28 53	084	28 25	137	60 17	172	61 26	259	31 10	345
18	36 53	016	23 37	054	29 46	083	29 01	137	60 24	173	60 34	259	30 56	344
19	37 07	015	24 19	054	30 38	083	29 37	137	60 30	174	59 42	258	30 40	343
20	37 20	014	25 02	053	31 31	082	30 13	137	60 34	175	58 50	258	30 24	342
21	37 32	012	25 44	052	32 23	082	30 49	137	60 38	176	57 59	258	30 07	341
22	37 43	011	26 26	051	33 15	081	31 25	137	60 41	177	57 07	258	29 49	340
23	37 53	010	27 07	051	34 08	080	32 01	137	60 43	178	56 15	258	29 31	339
24	38 01	009	27 47	050	35 00	080	32 38	137	60 44	180	55 23	258	29 11	338
25	38 09	008	28 28	049	35 52	079	33 14	137	60 44	181	54 32	257	28 51	337
26	38 16	007	29 08	048	36 44	079	33 50	137	60 43	182	53 40	257	28 30	336
27	38 22	006	29 47	048	37 36	078	34 26	137	60 41	183	52 48	257	28 08	335
28	38 26	004	30 26	047	38 28	078	35 03	137	60 38	184	51 57	257	27 46	334
29	38 30	003	31 04	046	39 19	077	35 39	137	60 33	185	51 05	257	27 22	333

LHA	Hc	Zn	Hc	Zn	Hc	Zn	Hc	Zn	Hc	Zn	Hc	Zn	Hc	Zn
	Hamal		ALDEBARAN		♦SIRIUS		CANOPUS		♦ACHERNAR		FOMALHAUT		♦Alpheratz	
30	38 32	002	31 42	045	24 01	097	36 15	137	60 28	186	50 14	256	26 58	333
31	38 33	001	32 19	044	24 54	096	36 52	137	60 22	187	49 22	256	26 34	332
32	38 34	000	32 56	043	25 47	096	37 28	137	60 15	188	48 31	256	26 08	331
33	38 33	359	33 32	042	26 39	096	38 05	137	60 07	189	47 39	256	25 42	330
34	38 31	357	34 07	042	27 32	095	38 41	137	59 58	190	46 48	255	25 15	329
35	38 28	356	34 42	041	28 25	095	39 17	137	59 48	191	45 57	255	24 48	328
36	38 24	355	35 16	040	29 18	094	39 53	137	59 37	192	45 06	255	24 20	328
37	38 19	354	35 50	039	30 10	094	40 30	137	59 25	193	44 15	255	23 51	327
38	38 12	353	36 22	038	31 03	093	41 06	137	59 12	194	43 24	254	23 22	326
39	38 05	352	36 54	037	31 56	093	41 42	137	58 58	195	42 33	254	22 52	325
40	37 57	350	37 26	036	32 49	093	42 18	137	58 44	196	41 42	254	22 21	324
41	37 47	349	37 56	035	33 42	092	42 54	137	58 29	197	40 51	254	21 50	324
42	37 37	348	38 26	034	34 35	092	43 30	138	58 13	198	40 00	253	21 19	323
43	37 25	347	38 55	033	35 28	091	44 05	138	57 56	199	39 09	253	20 46	322
44	37 13	346	39 23	031	36 21	091	44 41	138	57 38	200	38 19	253	20 14	321

LHA	Hc	Zn	Hc	Zn	Hc	Zn	Hc	Zn	Hc	Zn	Hc	Zn	Hc	Zn
	♦ALDEBARAN		BETELGEUSE		SIRIUS		♦CANOPUS		ACHERNAR		♦FOMALHAUT		Hamal	
45	39 50	030	34 54	057	37 14	090	45 16	138	57 20	201	37 28	253	36 59	345
46	40 17	029	35 38	056	38 07	090	45 52	138	57 01	201	36 37	252	36 45	344
47	40 42	028	36 22	055	39 00	089	46 27	138	56 41	202	35 47	252	36 30	342
48	41 07	027	37 05	054	39 53	089	47 02	139	56 21	203	34 57	252	36 13	341
49	41 30	026	37 48	053	40 46	088	47 37	139	56 00	204	34 06	252	35 56	340
50	41 53	025	38 30	052	41 39	088	48 12	139	55 38	204	33 16	251	35 38	339
51	42 14	023	39 12	052	42 32	087	48 46	139	55 16	205	32 26	251	35 18	338
52	42 35	022	39 53	051	43 25	087	49 21	140	54 53	206	31 36	251	34 58	337
53	42 55	021	40 34	050	44 17	086	49 55	140	54 30	206	30 46	250	34 37	336
54	43 13	020	41 14	049	45 10	086	50 29	140	54 06	207	29 56	250	34 15	335
55	43 30	019	41 53	048	46 03	085	51 02	141	53 41	208	29 06	250	33 53	334
56	43 47	017	42 32	047	46 56	085	51 35	141	53 17	208	28 17	250	33 29	333
57	44 02	016	43 11	046	47 49	084	52 09	142	52 51	209	27 27	249	33 05	332
58	44 16	015	43 48	045	48 41	084	52 41	142	52 25	209	26 38	249	32 40	331
59	44 29	013	44 25	044	49 34	083	53 14	143	51 59	210	25 48	249	32 14	330

LHA	Hc	Zn	Hc	Zn	Hc	Zn	Hc	Zn	Hc	Zn	Hc	Zn	Hc	Zn
	ALDEBARAN		BETELGEUSE		♦SIRIUS		CANOPUS		♦ACHERNAR		FOMALHAUT		♦Hamal	
60	44 40	012	45 01	042	50 26	082	53 46	143	51 33	210	24 59	248	31 47	329
61	44 51	011	45 36	041	51 19	082	54 17	143	51 05	211	24 10	248	31 19	328
62	45 00	009	46 11	040	52 11	081	54 49	144	50 38	211	23 21	248	30 51	327
63	45 08	008	46 45	039	53 04	080	55 20	145	50 10	212	22 32	247	30 22	326
64	45 15	007	47 17	038	53 56	080	55 50	145	49 42	212	21 43	247	29 53	326
65	45 20	005	47 49	036	54 48	079	56 20	146	49 14	213	20 54	247	29 22	325
66	45 25	004	48 20	035	55 40	078	56 50	146	48 45	213	20 06	246	28 52	324
67	45 28	003	48 50	034	56 32	078	57 19	147	48 16	213	19 17	246	28 20	323
68	45 30	001	49 19	033	57 23	077	57 47	148	47 47	214	18 29	246	27 48	322
69	45 30	000	49 47	031	58 15	076	58 15	149	47 17	214	17 41	245	27 15	321
70	45 29	359	50 14	030	59 06	075	58 43	149	46 48	214	16 53	245	26 42	320
71	45 27	357	50 40	028	59 57	074	59 09	150	46 18	215	16 05	245	26 08	320
72	45 24	356	51 05	027	60 48	074	59 36	151	45 47	215	15 17	244	25 33	319
73	45 19	354	51 28	026	61 39	073	60 01	152	45 17	215	14 29	244	24 58	318
74	45 14	353	51 50	024	62 29	072	60 26	153	44 46	215	13 42	243	24 22	317

LHA	Hc	Zn	Hc	Zn	Hc	Zn	Hc	Zn	Hc	Zn	Hc	Zn	Hc	Zn
	BETELGEUSE		PROCYON		♦Suhail		CANOPUS		♦ACHERNAR		Hamal		♦ALDEBARAN	
75	52 11	023	39 17	055	38 38	125	60 50	154	44 15	216	23 46	317	45 07	352
76	52 31	021	40 00	054	39 21	125	61 13	155	43 44	216	23 09	316	44 58	350
77	52 49	019	40 43	054	40 05	125	61 35	156	43 13	216	22 32	315	44 49	349
78	53 06	018	41 25	053	40 48	125	61 57	157	42 42	216	21 55	314	44 38	348
79	53 22	016	42 07	052	41 32	125	62 17	158	42 10	217	21 17	314	44 26	346
80	53 36	015	42 48	051	42 15	125	62 37	159	41 39	217	20 38	313	44 13	345
81	53 49	013	43 29	050	42 59	125	62 56	160	41 07	217	19 59	312	43 59	344
82	54 00	011	44 09	049	43 42	125	63 13	161	40 35	217	19 20	312	43 44	343
83	54 10	010	44 49	048	44 26	125	63 30	162	40 03	217	18 40	311	43 27	341
84	54 18	008	45 27	047	45 10	125	63 46	163	39 31	217	18 00	310	43 10	340
85	54 25	006	46 05	045	45 53	125	64 00	165	38 59	217	17 19	310	42 51	339
86	54 30	005	46 43	044	46 37	125	64 13	166	38 27	217	16 38	309	42 31	338
87	54 33	003	47 19	043	47 20	125	64 26	167	37 55	217	15 57	308	42 11	336
88	54 35	001	47 55	042	48 04	125	64 37	169	37 23	218	15 15	308	41 49	335
89	54 36	000	48 30	041	48 47	125	64 46	170	36 50	218	14 33	307	41 26	334

LHA	Hc	Zn	Hc	Zn	Hc	Zn	Hc	Zn	Hc	Zn	Hc	Zn	Hc	Zn
♈	POLLUX		♦REGULUS		ACRUX		♦CANOPUS		ACHERNAR		♦RIGEL		BETELGEUSE	
90	28 34	026	17 55	065	21 53	151	64 55	171	36 18	218	67 28	329	54 34	358
91	28 57	025	18 43	065	22 19	151	65 02	173	35 46	218	67 00	327	54 32	356
92	29 20	025	19 31	064	22 45	151	65 08	174	35 13	218	66 30	325	54 27	354
93	29 42	024	20 18	063	23 11	151	65 13	176	34 41	218	65 59	323	54 21	353
94	30 02	023	21 05	063	23 37	150	65 16	177	34 08	218	65 26	321	54 14	351
95	30 22	022	21 52	062	24 03	150	65 18	179	33 36	218	64 51	319	54 05	349
96	30 42	021	22 39	062	24 29	150	65 18	180	33 03	218	64 16	317	53 54	348
97	31 00	020	23 25	061	24 56	150	65 18	182	32 31	218	63 39	315	53 42	346
98	31 18	019	24 12	060	25 22	150	65 16	183	31 59	218	63 01	313	53 29	344
99	31 34	018	24 57	060	25 49	150	65 12	184	31 26	218	62 22	312	53 14	343
100	31 50	017	25 43	059	26 16	150	65 08	186	30 54	218	61 42	310	52 57	341
101	32 05	016	26 28	058	26 42	150	65 01	187	30 22	218	61 02	309	52 40	340
102	32 19	015	27 13	057	27 09	150	64 54	189	29 49	218	60 20	307	52 20	338
103	32 32	014	27 57	057	27 36	149	64 46	190	29 17	217	59 37	306	52 00	337
104	32 45	013	28 41	056	28 03	149	64 36	191	28 45	217	58 54	305	51 38	335

LHA	Hc	Zn	Hc	Zn	Hc	Zn	Hc	Zn	Hc	Zn	Hc	Zn	Hc	Zn
	POLLUX		♦REGULUS		Gienah		♦ACRUX		CANOPUS		♦RIGEL		BETELGEUSE	
105	32 56	012	29 25	055	17 40	101	28 30	149	64 25	193	58 10	303	51 16	334
106	33 06	011	30 08	054	18 32	100	28 57	149	64 12	194	57 26	302	50 51	332
107	33 16	010	30 51	054	19 25	100	29 24	149	63 59	195	56 40	301	50 26	331
108	33 24	009	31 34	053	20 17	100	29 51	149	63 44	197	55 55	300	50 00	329
109	33 32	008	32 16	052	21 09	099	30 18	149	63 29	198	55 09	299	49 32	328
110	33 39	007	32 57	051	22 01	099	30 45	149	63 12	199	54 22	298	49 04	327
111	33 44	006	33 39	050	22 54	098	31 13	149	62 54	200	53 35	297	48 34	325
112	33 49	005	34 19	050	23 46	098	31 40	149	62 35	201	52 47	296	48 03	324
113	33 52	003	34 59	049	24 39	098	32 07	149	62 16	202	51 59	295	47 32	323
114	33 55	002	35 39	048	25 31	097	32 34	149	61 55	203	51 11	294	46 59	322
115	33 57	001	36 18	047	26 24	097	33 01	149	61 33	205	50 22	293	46 26	320
116	33 58	000	36 56	046	27 16	096	33 28	149	61 11	206	49 33	292	45 52	319
117	33 57	359	37 34	045	28 09	096	33 55	149	60 48	206	48 44	291	45 17	318
118	33 56	358	38 11	044	29 02	095	34 22	149	60 24	207	47 54	290	44 41	317
119	33 54	357	38 48	043	29 55	095	34 50	149	59 59	208	47 05	290	44 05	316

LHA	Hc	Zn	Hc	Zn	Hc	Zn	Hc	Zn	Hc	Zn	Hc	Zn	Hc	Zn
	♦REGULUS		SPICA		♦ACRUX		CANOPUS		♦RIGEL		BETELGEUSE		POLLUX	
120	39 24	042	12 52	096	35 17	149	59 33	209	46 15	289	43 28	315	33 51	356
121	39 59	041	13 45	095	35 43	149	59 07	210	45 24	288	42 50	314	33 47	355
122	40 34	040	14 38	095	36 10	150	58 41	211	44 34	287	42 11	313	33 41	354
123	41 07	039	15 30	095	36 37	150	58 13	212	43 43	287	41 32	312	33 35	353
124	41 40	038	16 23	094	37 04	150	57 45	212	42 52	286	40 52	311	33 28	352
125	42 12	037	17 16	094	37 31	150	57 17	213	42 01	285	40 12	310	33 20	351
126	42 44	036	18 09	093	37 57	150	56 47	214	41 10	284	39 31	309	33 11	350
127	43 14	035	19 02	093	38 24	150	56 18	214	40 19	284	38 49	308	33 01	349
128	43 44	033	19 55	092	38 50	150	55 48	215	39 27	283	38 07	307	32 50	348
129	44 13	032	20 48	092	39 16	150	55 17	215	38 35	283	37 25	306	32 39	347
130	44 41	031	21 41	091	39 42	151	54 46	216	37 44	282	36 42	305	32 26	346
131	45 07	030	22 34	091	40 08	151	54 15	217	36 52	281	35 58	304	32 12	345
132	45 33	029	23 27	090	40 34	151	53 43	217	36 00	281	35 14	304	31 58	344
133	45 58	027	24 20	090	41 00	151	53 11	218	35 08	280	34 30	303	31 42	343
134	46 22	026	25 13	090	41 25	151	52 39	218	34 16	280	33 45	302	31 26	342

LHA	Hc	Zn	Hc	Zn	Hc	Zn	Hc	Zn	Hc	Zn	Hc	Zn	Hc	Zn
	REGULUS		♦SPICA		ACRUX		♦CANOPUS		RIGEL		BETELGEUSE		♦POLLUX	
135	46 45	025	26 06	089	41 50	152	52 06	218	33 23	279	33 00	301	31 09	341
136	47 06	023	26 59	089	42 16	152	51 33	219	32 31	278	32 15	301	30 51	340
137	47 27	022	27 51	088	42 41	152	50 59	219	31 38	278	31 29	300	30 32	339
138	47 46	021	28 44	088	43 05	152	50 26	220	30 46	277	30 43	299	30 13	338
139	48 04	019	29 37	087	43 30	153	49 52	220	29 53	277	29 56	298	29 52	337
140	48 21	018	30 30	087	43 54	153	49 18	220	29 01	276	29 10	298	29 31	336
141	48 36	016	31 23	086	44 18	153	48 44	221	28 08	276	28 22	297	29 09	335
142	48 51	015	32 16	086	44 42	153	48 09	221	27 15	275	27 35	296	28 46	334
143	49 04	014	33 09	085	45 06	154	47 34	221	26 23	275	26 47	296	28 23	333
144	49 16	012	34 01	085	45 29	154	46 59	221	25 30	274	26 00	295	27 58	332
145	49 26	011	34 54	084	45 52	154	46 24	222	24 37	274	25 11	294	27 33	331
146	49 35	009	35 47	083	46 15	155	45 49	222	23 44	273	24 23	294	27 08	331
147	49 43	008	36 39	083	46 37	155	45 14	222	22 51	273	23 34	293	26 41	330
148	49 49	006	37 32	082	46 59	156	44 38	222	21 58	272	22 45	292	26 14	329
149	49 54	005	38 24	082	47 21	156	44 02	222	21 05	272	21 56	292	25 46	328

LHA	Hc	Zn	Hc	Zn	Hc	Zn	Hc	Zn	Hc	Zn	Hc	Zn	Hc	Zn
	REGULUS		♦Denebola		SPICA		♦ACRUX		CANOPUS		♦SIRIUS		PROCYON	
150	49 58	003	39 55	035	39 17	081	47 42	156	43 27	223	43 49	273	42 29	309
151	50 00	002	40 25	034	40 09	081	48 03	157	42 51	223	42 56	273	41 47	308
152	50 00	000	40 54	033	41 01	080	48 24	157	42 15	223	42 03	272	41 05	307
153	50 00	359	41 23	032	41 54	079	48 44	158	41 39	223	41 10	272	40 22	306
154	49 58	357	41 50	031	42 46	079	49 04	158	41 03	223	40 17	271	39 39	305
155	49 54	355	42 17	030	43 38	078	49 24	159	40 27	223	39 24	271	38 56	304
156	49 49	354	42 42	028	44 29	078	49 43	159	39 50	223	38 31	270	38 12	303
157	49 43	352	43 07	027	45 21	077	50 01	160	39 14	223	37 38	270	37 27	303
158	49 36	351	43 31	026	46 12	076	50 20	160	38 38	223	36 45	269	36 42	302
159	49 27	349	43 54	025	47 04	076	50 37	161	38 02	223	35 52	269	35 57	301
160	49 16	348	44 15	024	47 55	075	50 55	161	37 25	223	35 00	269	35 11	300
161	49 04	347	44 36	022	48 46	074	51 11	162	36 49	223	34 07	268	34 25	299
162	48 51	345	44 56	021	49 37	073	51 28	162	36 13	223	33 14	268	33 39	299
163	48 37	344	45 14	020	50 28	073	51 44	163	35 36	223	32 21	267	32 52	298
164	48 22	342	45 31	018	51 18	072	51 59	164	35 00	223	31 28	267	32 05	297

LHA	Hc	Zn	Hc	Zn	Hc	Zn	Hc	Zn	Hc	Zn	Hc	Zn	Hc	Zn
	ARCTURUS		♦ANTARES		ACRUX		♦CANOPUS		SIRIUS		PROCYON		♦REGULUS	
165	23 13	051	18 22	111	52 13	164	34 24	223	30 35	266	31 18	296	48 05	341
166	23 53	050	19 12	110	52 28	165	33 47	223	29 42	266	30 30	296	47 47	339
167	24 34	049	20 02	110	52 41	166	33 11	223	28 49	265	29 42	295	47 28	338
168	25 14	049	20 51	110	52 54	166	32 35	223	27 56	265	28 54	294	47 07	337
169	25 53	048	21 41	109	53 06	167	31 58	223	27 04	265	28 06	294	46 46	335
170	26 32	047	22 31	109	53 18	168	31 22	223	26 11	264	27 17	293	46 23	334
171	27 11	046	23 22	109	53 29	168	30 46	223	25 18	264	26 28	292	45 59	333
172	27 49	045	24 12	108	53 40	169	30 10	223	24 26	263	25 39	292	45 35	331
173	28 26	045	25 02	108	53 50	170	29 34	223	23 33	263	24 50	291	45 09	330
174	29 03	044	25 53	108	53 59	170	28 58	223	22 40	262	24 00	290	44 42	329
175	29 39	043	26 43	107	54 08	171	28 22	223	21 48	262	23 11	290	44 14	328
176	30 15	042	27 34	107	54 15	172	27 46	222	20 56	262	22 21	289	43 46	327
177	30 51	041	28 25	107	54 23	173	27 11	222	20 03	261	21 31	289	43 16	325
178	31 25	040	29 15	106	54 29	173	26 35	222	19 11	261	20 40	288	42 45	324
179	31 59	040	30 06	106	54 35	174	26 00	222	18 19	260	19 50	288	42 14	323

LAT 28°S

LHA ♈	Hc	Zn	Hc	Zn	Hc	Zn	Hc	Zn	Hc	Zn	Hc	Zn	Hc	Zn
	◆ARCTURUS		ANTARES		◆RIGIL KENT.		ACRUX		CANOPUS		◆Alphard		REGULUS	
180	32 33	039	30 57	106	47 47	152	54 40	175	25 24	222	49 11	291	41 42	322
181	33 05	038	31 48	105	48 11	153	54 44	176	24 49	222	48 22	290	41 09	321
182	33 37	037	32 40	105	48 35	153	54 48	176	24 14	222	47 32	289	40 35	320
183	34 09	036	33 31	105	48 59	154	54 51	177	23 39	221	46 41	288	40 01	319
184	34 39	035	34 22	104	49 23	154	54 53	178	23 04	221	45 51	288	39 26	318
185	35 09	034	35 13	104	49 46	154	54 55	179	22 29	221	45 01	287	38 50	317
186	35 38	033	36 05	104	50 08	155	54 56	180	21 54	221	44 10	286	38 13	316
187	36 07	032	36 56	103	50 31	155	54 56	180	21 20	221	43 19	286	37 36	315
188	36 34	031	37 48	103	50 52	156	54 55	181	20 45	220	42 28	285	36 58	314
189	37 01	030	38 40	103	51 14	156	54 54	182	20 11	220	41 36	284	36 20	313
190	37 27	029	39 31	102	51 35	157	54 51	183	19 37	220	40 45	284	35 41	312
191	37 52	028	40 23	102	51 55	158	54 49	183	19 03	220	39 53	283	35 01	311
192	38 16	027	41 15	102	52 15	158	54 45	184	18 29	219	39 02	282	34 21	310
193	38 40	026	42 07	101	52 34	159	54 41	185	17 56	219	38 10	282	33 41	310
194	39 02	024	42 59	101	52 53	159	54 36	186	17 22	219	37 18	281	33 00	309
	ARCTURUS		◆Rasalhague		ANTARES		RIGIL KENT.		◆ACRUX		Suhail		◆REGULUS	
195	39 23	023	12 12	068	43 51	101	53 12	160	54 30	187	41 28	235	32 18	308
196	39 44	022	13 01	068	44 43	101	53 29	161	54 24	187	40 44	235	31 36	307
197	40 03	021	13 50	067	45 35	100	53 47	161	54 17	188	40 00	235	30 54	306
198	40 22	020	14 39	067	46 27	100	54 03	162	54 09	189	39 17	235	30 11	306
199	40 39	019	15 27	066	47 19	100	54 19	163	54 00	190	38 33	235	29 27	305
200	40 55	017	16 16	066	48 12	099	54 35	163	53 51	190	37 50	235	28 44	304
201	41 11	016	17 04	065	49 04	099	54 50	164	53 41	191	37 07	235	28 00	303
202	41 25	015	17 52	065	49 56	099	55 04	165	53 31	192	36 23	235	27 15	303
203	41 38	014	18 40	064	50 49	098	55 17	166	53 20	192	35 40	235	26 30	302
204	41 50	013	19 27	063	51 41	098	55 30	166	53 08	193	34 57	235	25 45	301
205	42 01	011	20 14	063	52 34	098	55 42	167	52 56	194	34 13	235	25 00	300
206	42 11	010	21 01	062	53 26	097	55 53	168	52 43	194	33 30	234	24 14	300
207	42 20	009	21 48	061	54 19	097	56 04	169	52 30	195	32 47	234	23 28	299
208	42 27	008	22 34	061	55 11	097	56 14	170	52 16	196	32 04	234	22 41	299
209	42 33	006	23 20	060	56 04	096	56 23	170	52 01	196	31 21	234	21 55	298
	◆ARCTURUS		Rasalhague		◆Nunki		RIGIL KENT.		ACRUX		◆Suhail		Denebola	
210	42 39	005	24 06	059	25 27	108	56 31	171	51 46	197	30 38	234	36 52	319
211	42 43	004	24 52	059	26 17	107	56 39	172	51 30	198	29 56	234	36 17	318
212	42 45	002	25 37	058	27 08	107	56 46	173	51 14	198	29 13	234	35 41	317
213	42 47	001	26 22	057	27 59	107	56 52	174	50 57	199	28 31	233	35 05	316
214	42 47	000	27 06	057	28 50	106	56 57	175	50 40	199	27 48	233	34 28	315
215	42 47	359	27 50	056	29 40	106	57 01	176	50 23	200	27 06	233	33 50	314
216	42 45	357	28 34	055	30 31	106	57 05	177	50 04	200	26 24	233	33 12	314
217	42 42	356	29 17	055	31 22	105	57 08	177	49 46	201	25 41	233	32 33	313
218	42 37	355	30 00	054	32 14	105	57 10	178	49 27	201	24 59	232	31 54	312
219	42 32	353	30 43	053	33 05	105	57 11	179	49 07	202	24 18	232	31 14	311
220	42 25	352	31 25	052	33 56	104	57 11	180	48 47	202	23 36	232	30 34	310
221	42 17	351	32 06	051	34 47	104	57 11	181	48 27	203	22 54	232	29 54	309
222	42 08	350	32 47	051	35 39	104	57 09	182	48 07	203	22 13	231	29 12	309
223	41 58	348	33 28	050	36 30	103	57 07	183	47 46	204	21 31	231	28 31	308
224	41 47	347	34 08	049	37 22	103	57 04	184	47 24	204	20 50	231	27 49	307
	Alphecca		Rasalhague		◆Nunki		Peacock		◆RIGIL KENT.		SPICA		◆ARCTURUS	
225	34 39	009	34 48	048	38 14	103	27 45	142	57 00	185	62 04	302	41 35	346
226	34 47	008	35 27	047	39 05	102	28 18	142	56 55	186	61 19	301	41 21	345
227	34 54	007	36 05	046	39 57	102	28 50	142	56 50	186	60 33	300	41 07	343
228	35 00	006	36 43	045	40 49	102	29 23	142	56 43	187	59 47	299	40 51	342
229	35 05	005	37 21	044	41 41	101	29 56	142	56 36	188	59 00	297	40 34	341
230	35 10	004	37 57	043	42 33	101	30 28	142	56 28	189	58 13	296	40 17	340
231	35 13	003	38 33	042	43 25	101	31 01	142	56 20	190	57 25	295	39 58	339
232	35 15	002	39 09	041	44 17	101	31 34	142	56 10	191	56 37	294	39 38	338
233	35 16	001	39 43	040	45 09	100	32 07	142	56 00	191	55 49	293	39 17	336
234	35 16	000	40 17	039	46 01	100	32 40	142	55 49	192	55 00	292	38 56	335
235	35 15	358	40 51	038	46 53	100	33 13	142	55 38	193	54 11	291	38 33	334
236	35 13	357	41 23	037	47 46	099	33 46	142	55 25	194	53 21	290	38 10	333
237	35 10	356	41 55	036	48 38	099	34 18	142	55 12	195	52 31	290	37 45	332
238	35 06	355	42 25	035	49 30	099	34 51	142	54 59	195	51 41	289	37 20	331
239	35 02	354	42 55	034	50 23	098	35 24	142	54 44	196	50 51	288	36 54	330
	◆VEGA		ALTAIR		◆Peacock		RIGIL KENT.		◆SPICA		ARCTURUS		Alphecca	
240	13 51	030	23 15	065	35 57	142	54 29	197	50 00	287	36 27	329	34 56	353
241	14 18	030	24 03	065	36 30	142	54 14	198	49 10	286	35 59	328	34 49	352
242	14 44	029	24 51	064	37 03	142	53 57	198	48 19	286	35 30	327	34 41	351
243	15 09	028	25 38	063	37 35	142	53 40	199	47 28	285	35 01	326	34 32	350
244	15 34	028	26 25	063	38 08	142	53 23	200	46 36	284	34 31	325	34 22	349
245	15 59	027	27 12	062	38 41	142	53 05	200	45 45	283	34 00	324	34 11	348
246	16 23	026	27 59	061	39 14	142	52 46	201	44 53	283	33 28	323	33 59	347
247	16 46	026	28 45	061	39 46	142	52 27	201	44 02	282	32 56	322	33 47	346
248	17 09	025	29 31	060	40 18	142	52 08	202	43 10	281	32 23	321	33 33	345
249	17 31	024	30 17	059	40 51	142	51 47	203	42 18	281	31 50	320	33 19	344
250	17 52	024	31 02	058	41 23	143	51 27	203	41 26	280	31 15	319	33 03	343
251	18 13	023	31 47	058	41 55	143	51 06	204	40 33	280	30 41	318	32 47	341
252	18 33	022	32 32	057	42 27	143	50 44	204	39 41	279	30 05	318	32 30	340
253	18 53	021	33 16	056	42 59	143	50 22	205	38 49	278	29 29	317	32 11	339
254	19 12	021	34 00	055	43 31	143	50 00	205	37 56	278	28 53	316	31 52	339
	◆VEGA		ALTAIR		FOMALHAUT		◆Peacock		RIGIL KENT.		◆SPICA		ARCTURUS	
255	19 30	020	34 43	055	13 57	116	44 03	143	49 37	206	37 04	277	28 16	315
256	19 48	019	35 26	054	14 44	116	44 34	144	49 14	206	36 11	277	27 38	314
257	20 05	018	36 09	053	15 32	116	45 05	144	48 50	207	35 19	276	27 00	314
258	20 21	017	36 51	052	16 20	115	45 36	144	48 27	207	34 26	276	26 21	313
259	20 36	017	37 32	051	17 08	115	46 07	144	48 02	207	33 33	275	25 42	312
260	20 51	016	38 13	050	17 56	115	46 38	145	47 38	208	32 40	275	25 02	311
261	21 06	015	38 53	049	18 44	114	47 09	145	47 13	208	31 48	274	24 22	311
262	21 19	014	39 33	048	19 32	114	47 39	145	46 48	209	30 55	274	23 42	310
263	21 32	014	40 12	047	20 21	114	48 09	146	46 22	209	30 02	273	23 01	309
264	21 44	013	40 51	046	21 10	113	48 39	146	45 56	209	29 09	273	22 20	308
265	21 55	012	41 29	045	21 58	113	49 08	146	45 30	210	28 16	272	21 38	308
266	22 06	011	42 06	044	22 47	113	49 37	147	45 04	210	27 23	272	20 56	307
267	22 15	010	42 43	043	23 36	112	50 06	147	44 38	210	26 30	271	20 13	306
268	22 24	009	43 19	042	24 25	112	50 34	148	44 11	210	25 37	271	19 31	306
269	22 33	009	43 54	041	25 14	112	51 03	148	43 44	211	24 44	270	18 47	305

LHA ♈	Hc	Zn	Hc	Zn	Hc	Zn	Hc	Zn	Hc	Zn	Hc	Zn	Hc	Zn
	VEGA		◆ALTAIR		FOMALHAUT		◆Peacock		RIGIL KENT.		◆ANTARES		Alphecca	
270	22 40	008	44 29	040	26 04	111	51 30	149	43 17	211	69 45	269	25 04	324
271	22 47	007	45 02	039	26 53	111	51 58	149	42 49	211	68 52	269	24 33	323
272	22 53	006	45 35	038	27 43	111	52 25	150	42 22	211	68 00	268	24 01	323
273	22 58	005	46 07	036	28 32	110	52 52	150	41 54	212	67 07	268	23 29	322
274	23 03	004	46 38	035	29 22	110	53 18	151	41 26	212	66 14	268	22 56	321
275	23 06	004	47 08	034	30 12	110	53 44	151	40 58	212	65 21	267	22 22	320
276	23 09	003	47 37	033	31 02	110	54 09	152	40 30	212	64 28	267	21 48	320
277	23 11	002	48 05	031	31 52	109	54 34	152	40 02	212	63 35	266	21 13	319
278	23 13	001	48 32	030	32 42	109	54 58	153	39 33	213	62 42	266	20 38	318
279	23 13	000	48 58	029	33 32	109	55 22	154	39 05	213	61 49	266	20 03	317
280	23 13	359	49 23	027	34 22	108	55 45	154	38 36	213	60 56	265	19 26	317
281	23 12	358	49 47	026	35 12	108	56 07	155	38 08	213	60 04	265	18 50	316
282	23 10	358	50 10	025	36 03	108	56 29	156	37 39	213	59 11	265	18 13	315
283	23 08	357	50 31	023	36 53	108	56 51	157	37 10	213	58 18	264	17 35	315
284	23 04	356	50 51	022	37 44	107	57 11	157	36 41	213	57 25	264	16 58	314
	◆DENEB		FOMALHAUT		◆ACHERNAR		RIGIL KENT.		◆ANTARES		Rasalhague		VEGA	
285	13 12	018	38 34	107	18 29	146	36 12	213	56 33	264	44 29	330	23 00	355
286	13 28	017	39 25	107	18 59	146	35 43	213	55 40	263	44 02	329	22 55	354
287	13 43	017	40 16	106	19 29	145	35 14	213	54 47	263	43 34	328	22 49	353
288	13 58	016	41 07	106	19 59	145	34 44	213	53 55	263	43 06	327	22 43	353
289	14 13	015	41 58	106	20 30	145	34 15	213	53 02	262	42 36	325	22 36	352
290	14 26	015	42 48	106	21 00	145	33 46	213	52 10	262	42 05	324	22 28	351
291	14 39	014	43 39	105	21 31	145	33 17	214	51 17	262	41 34	323	22 19	350
292	14 52	013	44 31	105	22 02	144	32 48	214	50 25	262	41 02	322	22 09	349
293	15 04	013	45 22	105	22 33	144	32 18	214	49 33	261	40 29	321	21 59	348
294	15 15	012	46 13	105	23 04	144	31 49	214	48 40	261	39 55	320	21 48	348
295	15 25	011	47 04	105	23 35	144	31 20	213	47 48	261	39 21	319	21 36	347
296	15 35	010	47 55	104	24 06	144	30 51	213	46 56	260	38 46	318	21 24	346
297	15 44	010	48 47	104	24 38	144	30 21	213	46 04	260	38 10	317	21 11	345
298	15 53	009	49 38	104	25 09	143	29 52	213	45 11	260	37 34	316	20 57	344
299	16 01	008	50 30	104	25 41	143	29 23	213	44 19	259	36 57	315	20 42	344
	DENEB		Enif		◆FOMALHAUT		ACHERNAR		◆RIGIL KENT.		ANTARES		◆VEGA	
300	16 08	008	44 34	037	51 21	103	26 13	143	28 54	213	43 27	259	20 27	343
301	16 15	007	45 05	036	52 13	103	26 44	143	28 25	213	42 35	259	20 11	342
302	16 21	006	45 36	035	53 04	103	27 16	143	27 56	213	41 43	258	19 54	341
303	16 26	005	46 06	034	53 56	103	27 48	143	27 27	213	40 52	258	19 37	340
304	16 31	005	46 35	032	54 48	103	28 20	143	26 58	213	40 00	258	19 19	340
305	16 35	004	47 03	031	55 39	102	28 52	143	26 29	213	39 08	257	19 00	339
306	16 38	003	47 30	030	56 31	102	29 25	143	26 01	213	38 16	257	18 41	338
307	16 41	002	47 56	029	57 23	102	29 57	142	25 32	213	37 25	257	18 21	337
308	16 42	002	48 20	027	58 15	102	30 29	142	25 03	213	36 33	257	18 00	337
309	16 44	001	48 44	026	59 07	102	31 01	142	24 35	212	35 42	256	17 39	336
310	16 44	000	49 07	024	59 59	102	31 34	142	24 07	212	34 50	256	17 17	335
311	16 44	359	49 28	023	60 50	101	32 06	142	23 38	212	33 59	256	16 55	335
312	16 43	359	49 48	022	61 42	101	32 39	142	23 10	212	33 08	255	16 32	334
313	16 42	358	50 07	020	62 34	101	33 11	142	22 42	212	32 16	255	16 08	333
314	16 40	357	50 25	019	63 26	101	33 44	142	22 14	212	31 25	255	15 44	332
	Alpheratz		◆Diphda		ACHERNAR		◆RIGIL KENT.		ANTARES		◆ALTAIR		DENEB	
315	17 20	042	38 05	092	34 16	142	21 46	212	30 34	254	49 30	333	16 37	357
316	17 55	041	38 58	091	34 48	142	21 19	211	29 43	254	49 05	332	16 33	356
317	18 30	041	39 51	091	35 21	142	20 51	211	28 53	254	48 40	330	16 29	355
318	19 05	040	40 44	090	35 53	142	20 24	211	28 02	253	48 13	329	16 24	354
319	19 38	039	41 37	090	36 26	142	19 57	211	27 11	253	47 45	328	16 19	354
320	20 12	039	42 30	089	36 58	142	19 30	211	26 20	253	47 16	326	16 13	353
321	20 44	038	43 22	089	37 30	142	19 03	210	25 30	252	46 46	325	16 06	352
322	21 17	037	44 15	088	38 03	143	18 36	210	24 40	252	46 15	324	15 58	351
323	21 48	036	45 08	088	38 35	143	18 10	210	23 49	252	45 44	323	15 50	351
324	22 19	036	46 01	087	39 07	143	17 43	210	22 59	251	45 11	321	15 41	350
325	22 50	035	46 54	087	39 39	143	17 17	210	22 09	251	44 38	320	15 32	349
326	23 20	034	47 47	086	40 11	143	16 51	209	21 19	251	44 04	319	15 21	349
327	23 49	033	48 40	086	40 43	143	16 25	209	20 29	250	43 29	318	15 11	348
328	24 18	032	49 33	085	41 15	143	15 59	209	19 39	250	42 53	317	14 59	347
329	24 46	032	50 26	085	41 46	143	15 34	209	18 50	249	42 16	316	14 47	347
	◆Alpheratz		Diphda		◆ACHERNAR		Peacock		◆Nunki		ALTAIR		Enif	
330	25 14	031	51 18	084	42 18	144	56 43	204	49 03	261	41 39	315	51 57	354
331	25 40	030	52 11	083	42 49	144	56 21	204	48 11	261	41 01	314	51 50	352
332	26 07	029	53 04	083	43 21	144	55 59	205	47 18	261	40 23	313	51 42	350
333	26 32	028	53 56	082	43 52	144	55 36	206	46 26	260	39 44	312	51 33	349
334	26 57	027	54 48	082	44 23	144	55 13	207	45 34	260	39 04	311	51 22	347
335	27 21	027	55 41	081	44 53	145	54 49	207	44 42	260	38 24	310	51 09	346
336	27 44	026	56 33	080	45 24	145	54 24	208	43 50	259	37 43	309	50 56	344
337	28 07	025	57 25	080	45 54	145	53 59	208	42 58	259	37 02	308	50 41	343
338	28 29	024	58 17	079	46 25	145	53 34	209	42 06	259	36 20	307	50 24	341
339	28 50	023	59 09	078	46 55	146	53 08	210	41 14	258	35 38	307	50 06	340
340	29 10	022	60 01	077	47 24	146	52 42	210	40 22	258	34 55	306	49 48	338
341	29 30	021	60 53	076	47 54	146	52 15	211	39 30	258	34 12	305	49 27	337
342	29 48	020	61 44	076	48 23	147	51 48	211	38 38	257	33 28	304	49 06	335
343	30 06	019	62 35	075	48 52	147	51 20	212	37 47	257	32 44	303	48 43	334
344	30 23	018	63 26	074	49 21	147	50 52	212	36 55	257	31 59	303	48 20	333
	Alpheratz		◆Hamal		◆ACHERNAR		Peacock		Nunki		◆ALTAIR		Enif	
345	30 39	017	21 38	046	49 49	148	50 24	213	36 04	256	31 14	302	47 55	331
346	30 55	016	22 16	045	50 17	148	49 55	213	35 12	256	30 29	301	47 29	330
347	31 09	015	22 53	044	50 45	149	49 26	213	34 21	256	29 44	300	47 02	329
348	31 23	014	23 30	044	51 12	149	48 57	214	33 29	256	28 58	300	46 34	328
349	31 35	013	24 06	043	51 39	150	48 27	214	32 38	255	28 12	299	46 05	326
350	31 47	012	24 42	042	52 05	150	47 58	214	31 47	255	27 25	298	45 35	325
351	31 58	011	25 17	041	52 31	151	47 28	215	30 56	255	26 38	297	45 04	324
352	32 08	010	25 52	041	52 57	151	46 57	215	30 05	254	25 51	297	44 33	323
353	32 17	009	26 26	040	53 22	152	46 27	215	29 14	254	25 04	296	44 00	322
354	32 25	008	27 00	039	53 47	152	45 56	216	28 23	254	24 16	296	43 27	320
355	32 33	007	27 33	038	54 11	153	45 25	216	27 32	253	23 28	295	42 53	319
356	32 39	006	28 06	037	54 35	154	44 54	216	26 42	253	22 40	294	42 18	318
357	32 44	005	28 38	037	54 58	154	44 22	216	25 51	253	21 51	294	41 42	317
358	32 48	004	29 09	036	55 21	155	43 51	217	25 00	252	21 03	293	41 06	316
359	32 52	003	29 39	035	55 43	156	43 19	217	24 10	252	20 14	292	40 29	315

LAT 29°S

LHA	Hc	Zn	Hc	Zn	Hc	Zn	Hc	Zn	Hc	Zn	Hc	Zn	Hc	Zn
♈	◆Alpheratz		Hamal		◆RIGEL		CANOPUS		ACHERNAR		◆Peacock		Enif	
0	31 54	002	29 19	034	13 56	092	19 18	140	57 00	156	43 35	218	39 09	315
1	31 56	001	29 48	033	14 48	091	19 52	140	57 21	157	43 03	218	38 31	314
2	31 56	000	30 16	032	15 41	091	20 26	140	57 41	157	42 31	218	37 53	313
3	31 56	359	30 43	031	16 33	090	21 00	140	58 01	158	41 59	218	37 14	312
4	31 54	358	31 10	030	17 26	090	21 34	139	58 20	159	41 26	218	36 35	311
5	31 52	357	31 36	029	18 18	089	22 08	139	58 39	160	40 54	218	35 55	310
6	31 49	356	32 01	028	19 10	089	22 42	139	58 57	161	40 21	218	35 14	309
7	31 45	355	32 25	027	20 03	088	23 17	139	59 14	162	39 48	218	34 34	308
8	31 39	354	32 48	026	20 55	088	23 52	139	59 30	163	39 16	219	33 52	307
9	31 33	353	33 11	025	21 48	087	24 26	138	59 45	163	38 43	219	33 10	307
10	31 26	352	33 33	024	22 40	087	25 01	138	60 00	164	38 10	219	32 28	306
11	31 19	351	33 54	023	23 33	086	25 36	138	60 13	165	37 37	219	31 45	305
12	31 10	350	34 14	022	24 25	086	26 12	138	60 26	166	37 05	219	31 02	304
13	31 00	349	34 33	021	25 17	085	26 47	138	60 38	167	36 32	219	30 18	303
14	30 49	348	34 51	020	26 10	085	27 22	137	60 49	168	35 59	219	29 34	303
	◆Hamal		ALDEBARAN		RIGEL		◆CANOPUS		ACHERNAR		◆FOMALHAUT		Alpheratz	
15	35 09	019	20 52	056	27 02	084	27 58	137	60 59	170	63 21	261	30 38	347
16	35 25	018	21 36	055	27 54	084	28 33	137	61 08	171	62 29	261	30 25	346
17	35 41	017	22 19	055	28 46	083	29 09	137	61 16	172	61 37	261	30 12	345
18	35 55	016	23 01	054	29 38	083	29 45	137	61 23	173	60 45	260	29 58	344
19	36 09	014	23 44	053	30 30	082	30 21	137	61 29	174	59 53	260	29 43	343
20	36 22	013	24 25	053	31 22	082	30 57	137	61 34	175	59 02	260	29 27	342
21	36 33	012	25 07	052	32 14	081	31 33	137	61 38	176	58 10	260	29 10	341
22	36 44	011	25 48	051	33 06	080	32 09	137	61 41	177	57 19	259	28 53	340
23	36 54	010	26 29	050	33 58	080	32 45	136	61 43	178	56 27	259	28 35	339
24	37 02	009	27 09	050	34 49	079	33 21	136	61 44	180	55 35	259	28 16	338
25	37 10	008	27 48	049	35 41	079	33 57	136	61 44	181	54 44	259	27 56	337
26	37 16	007	28 28	048	36 32	078	34 34	136	61 43	182	53 53	258	27 35	336
27	37 22	005	29 06	047	37 23	077	35 10	136	61 41	183	53 01	258	27 14	335
28	37 26	004	29 45	046	38 14	077	35 46	136	61 37	184	52 10	258	26 52	335
29	37 30	003	30 22	045	39 05	076	36 23	136	61 33	185	51 18	258	26 29	334
	Hamal		ALDEBARAN		◆SIRIUS		CANOPUS		◆ACHERNAR		FOMALHAUT		◆Alpheratz	
30	37 32	002	30 59	045	24 08	096	36 59	136	61 28	186	50 27	257	26 05	333
31	37 33	001	31 36	044	25 00	096	37 35	136	61 22	187	49 36	257	25 41	332
32	37 34	000	32 12	043	25 53	096	38 12	136	61 14	189	48 45	257	25 16	331
33	37 33	359	32 47	042	26 45	095	38 48	136	61 06	190	47 54	257	24 50	330
34	37 31	357	33 22	041	27 37	095	39 24	136	60 57	191	47 03	256	24 24	329
35	37 28	356	33 56	040	28 29	094	40 01	136	60 46	192	46 12	256	23 57	329
36	37 24	355	34 30	039	29 22	094	40 37	136	60 35	193	45 21	256	23 29	328
37	37 19	354	35 03	038	30 14	093	41 13	136	60 23	194	44 30	256	23 01	327
38	37 13	353	35 35	037	31 07	093	41 49	136	60 10	195	43 39	255	22 32	326
39	37 06	352	36 06	036	31 59	092	42 26	137	59 56	196	42 48	255	22 03	325
40	36 58	350	36 37	035	32 51	092	43 02	137	59 42	197	41 58	255	21 33	325
41	36 48	349	37 07	034	33 44	091	43 38	137	59 26	198	41 07	255	21 02	324
42	36 38	348	37 36	033	34 36	091	44 14	137	59 10	199	40 17	254	20 31	323
43	36 27	347	38 04	032	35 29	090	44 49	137	58 52	200	39 26	254	19 59	322
44	36 15	346	38 32	031	36 21	090	45 25	137	58 35	200	38 36	254	19 27	322
	◆ALDEBARAN		BETELGEUSE		SIRIUS		◆CANOPUS		ACHERNAR		◆FOMALHAUT		Hamal	
45	38 58	030	34 21	056	37 14	090	46 01	137	58 16	201	37 45	253	36 02	345
46	39 24	029	35 05	055	38 06	089	46 36	138	57 57	202	36 55	253	35 47	344
47	39 49	028	35 47	054	38 59	089	47 12	138	57 37	203	36 05	253	35 32	343
48	40 13	027	36 30	054	39 51	088	47 47	138	57 16	204	35 15	252	35 16	342
49	40 36	025	37 12	053	40 44	088	48 22	138	56 55	204	34 25	252	34 59	341
50	40 58	024	37 53	052	41 36	087	48 57	138	56 33	205	33 35	252	34 41	340
51	41 19	023	38 34	051	42 28	086	49 32	139	56 10	206	32 45	252	34 23	338
52	41 39	022	39 15	050	43 21	086	50 06	139	55 47	206	31 55	251	34 03	337
53	41 58	021	39 55	049	44 13	085	50 40	139	55 23	207	31 06	251	33 42	336
54	42 16	019	40 34	048	45 05	085	51 15	140	54 59	208	30 16	251	33 21	335
55	42 33	018	41 13	047	45 58	084	51 48	140	54 34	208	29 27	250	32 59	334
56	42 49	017	41 51	046	46 50	084	52 22	140	54 09	209	28 37	250	32 35	333
57	43 04	016	42 28	045	47 42	083	52 55	141	53 44	210	27 48	250	32 12	332
58	43 18	014	43 05	044	48 34	082	53 28	141	53 17	210	26 59	249	31 47	331
59	43 30	013	43 41	043	49 26	082	54 01	142	52 51	211	26 10	249	31 21	331
	ALDEBARAN		BETELGEUSE		◆SIRIUS		CANOPUS		◆ACHERNAR		FOMALHAUT		◆Hamal	
60	43 42	012	44 16	042	50 18	081	54 33	142	52 24	211	25 21	249	30 55	330
61	43 52	011	44 51	041	51 10	081	55 05	143	51 57	212	24 32	248	30 28	329
62	44 01	009	45 25	039	52 01	080	55 37	143	51 29	212	23 43	248	30 01	328
63	44 09	008	45 58	038	52 53	079	56 08	144	51 01	212	22 55	248	29 32	327
64	44 15	007	46 30	037	53 44	078	56 39	144	50 33	213	22 06	247	29 03	326
65	44 21	005	47 01	036	54 36	078	57 10	145	50 04	213	21 18	247	28 33	325
66	44 25	004	47 31	035	55 27	077	57 40	146	49 35	214	20 30	247	28 03	324
67	44 28	003	48 00	033	56 18	076	58 09	146	49 06	214	19 42	246	27 32	323
68	44 30	001	48 29	032	57 09	075	58 38	147	48 37	214	18 54	246	27 00	322
69	44 30	000	48 56	031	58 00	075	59 06	148	48 07	215	18 06	246	26 28	322
70	44 29	359	49 22	029	58 50	074	59 34	148	47 37	215	17 18	245	25 55	321
71	44 27	357	49 47	028	59 40	073	60 01	149	47 07	215	16 31	245	25 22	320
72	44 24	356	50 11	026	60 30	072	60 28	150	46 36	216	15 43	244	24 48	319
73	44 20	355	50 34	025	61 20	071	60 54	151	46 06	216	14 56	244	24 13	318
74	44 14	353	50 56	024	62 09	070	61 19	152	45 35	216	14 09	244	23 38	318
	BETELGEUSE		PROCYON		◆Suhail		CANOPUS		◆ACHERNAR		Hamal		◆ALDEBARAN	
75	51 16	022	38 42	055	39 12	124	61 43	153	45 04	216	23 02	317	44 07	352
76	51 35	021	39 25	054	39 55	124	62 07	154	44 33	217	22 26	316	43 59	351
77	51 53	019	40 07	053	40 38	124	62 30	155	44 01	217	21 50	315	43 50	349
78	52 09	018	40 48	052	41 22	124	62 52	156	43 30	217	21 13	315	43 40	348
79	52 24	016	41 29	051	42 05	124	63 13	157	42 58	217	20 35	314	43 28	347
80	52 38	014	42 10	050	42 49	124	63 33	158	42 27	217	19 57	313	43 15	345
81	52 50	013	42 50	049	43 33	124	63 52	159	41 55	217	19 19	313	43 01	344
82	53 01	011	43 29	048	44 16	124	64 10	160	41 23	217	18 40	312	42 46	343
83	53 11	010	44 08	047	45 00	124	64 27	162	40 51	218	18 00	311	42 30	342
84	53 19	008	44 46	046	45 43	124	64 43	163	40 19	218	17 21	311	42 13	340
85	53 25	006	45 23	045	46 27	124	64 58	164	39 47	218	16 41	310	41 55	339
86	53 30	005	46 00	044	47 10	124	65 12	166	39 15	218	16 00	309	41 36	338
87	53 33	003	46 35	042	47 54	124	65 24	167	38 42	218	15 19	309	41 15	337
88	53 35	001	47 10	041	48 38	124	65 35	168	38 10	218	14 38	308	40 54	335
89	53 36	000	47 45	040	49 21	124	65 45	170	37 38	218	13 57	307	40 32	334

LHA	Hc	Zn	Hc	Zn	Hc	Zn	Hc	Zn	Hc	Zn	Hc	Zn	Hc	Zn
♈	POLLUX		◆REGULUS		ACRUX		◆CANOPUS		ACHERNAR		◆RIGEL		BETELGEUSE	
90	27 40	026	17 30	065	22 46	151	65 54	171	37 05	218	66 36	330	53 34	358
91	28 03	025	18 17	064	23 11	151	66 02	173	36 33	218	66 09	328	53 32	356
92	28 25	024	19 04	064	23 37	150	66 08	174	36 01	218	65 40	326	53 27	355
93	28 46	023	19 51	063	24 03	150	66 12	176	35 28	218	65 10	324	53 22	353
94	29 07	022	20 38	062	24 29	150	66 16	177	34 56	218	64 39	322	53 14	351
95	29 27	022	21 24	062	24 55	150	66 18	179	34 23	218	64 06	320	53 06	350
96	29 46	021	22 10	061	25 21	150	66 18	180	33 51	218	63 32	318	52 55	348
97	30 04	020	22 56	060	25 48	150	66 18	182	33 18	218	62 56	317	52 44	346
98	30 21	019	23 42	060	26 14	150	66 16	183	32 46	218	62 19	315	52 31	345
99	30 37	018	24 27	059	26 41	150	66 12	185	32 14	218	61 42	313	52 16	343
100	30 53	017	25 12	058	27 07	149	66 07	186	31 41	218	61 03	312	52 00	342
101	31 07	016	25 56	058	27 34	149	66 01	188	31 09	218	60 23	310	51 43	340
102	31 21	015	26 40	057	28 01	149	65 53	189	30 37	218	59 43	309	51 25	339
103	31 34	014	27 24	056	28 28	149	65 45	190	30 05	218	59 01	307	51 05	337
104	31 46	013	28 08	056	28 54	149	65 34	192	29 32	218	58 19	306	50 44	336
	POLLUX		◆REGULUS		Gienah		◆ACRUX		CANOPUS		◆RIGEL		BETELGEUSE	
105	31 57	012	28 51	055	17 51	101	29 21	149	65 23	193	57 37	305	50 22	334
106	32 07	011	29 33	054	18 43	100	29 48	149	65 11	195	56 53	303	49 58	333
107	32 17	010	30 16	053	19 35	100	30 15	149	64 57	196	56 09	302	49 34	331
108	32 25	009	30 58	052	20 27	099	30 43	149	64 42	197	55 24	301	49 08	330
109	32 32	008	31 39	052	21 18	099	31 10	149	64 26	198	54 39	300	48 41	329
110	32 39	007	32 20	051	22 10	098	31 37	149	64 09	200	53 53	299	48 13	327
111	32 44	006	33 00	050	23 02	098	32 04	149	63 50	201	53 07	298	47 44	326
112	32 49	004	33 40	049	23 54	098	32 31	149	63 31	202	52 21	297	47 15	325
113	32 53	003	34 20	048	24 46	097	32 58	149	63 11	203	51 34	296	46 44	324
114	32 55	002	34 58	047	25 38	097	33 25	149	62 50	204	50 46	295	46 12	322
115	32 57	001	35 37	046	26 31	096	33 53	149	62 28	205	49 58	294	45 40	321
116	32 58	000	36 14	046	27 23	096	34 20	149	62 05	206	49 10	293	45 06	320
117	32 57	359	36 52	045	28 15	095	34 47	149	61 41	207	48 22	292	44 32	319
118	32 56	358	37 28	044	29 07	095	35 14	149	61 17	208	47 33	291	43 57	318
119	32 54	357	38 04	043	30 00	094	35 41	149	60 52	209	46 44	291	43 21	317
	◆REGULUS		SPICA		◆ACRUX		CANOPUS		◆RIGEL		BETELGEUSE		POLLUX	
120	38 39	042	12 58	096	36 08	149	60 26	210	45 55	290	42 45	315	32 51	356
121	39 14	041	13 50	095	36 35	149	59 59	211	45 05	289	42 08	314	32 47	355
122	39 47	040	14 43	095	37 02	149	59 32	212	44 16	288	41 30	313	32 42	354
123	40 21	039	15 35	094	37 29	149	59 04	212	43 26	287	40 52	312	32 36	353
124	40 53	037	16 27	094	37 56	149	58 36	213	42 35	287	40 13	311	32 29	352
125	41 24	036	17 20	093	38 22	149	58 07	214	41 45	286	39 33	310	32 21	351
126	41 55	035	18 12	093	38 49	150	57 37	215	40 55	285	38 53	310	32 12	350
127	42 25	034	19 05	092	39 15	150	57 07	215	40 04	285	38 12	309	32 02	349
128	42 54	033	19 57	092	39 42	150	56 37	216	39 13	284	37 31	308	31 52	348
129	43 22	032	20 49	091	40 08	150	56 06	216	38 22	283	36 49	307	31 40	347
130	43 49	031	21 42	091	40 34	150	55 35	217	37 31	283	36 07	306	31 28	346
131	44 15	029	22 34	090	41 00	150	55 03	217	36 40	282	35 24	305	31 14	345
132	44 40	028	23 27	090	41 26	151	54 31	218	35 48	281	34 41	304	31 00	344
133	45 05	027	24 19	090	41 52	151	53 58	218	34 57	281	33 57	303	30 45	343
134	45 28	026	25 12	089	42 18	151	53 26	219	34 05	280	33 13	303	30 29	342
	REGULUS		◆SPICA		ACRUX		◆CANOPUS		RIGEL		BETELGEUSE		◆POLLUX	
135	45 50	024	26 04	089	42 43	151	52 53	219	33 14	280	32 29	302	30 12	341
136	46 11	023	26 57	088	43 08	151	52 19	220	32 22	279	31 44	301	29 55	340
137	46 31	022	27 49	088	43 33	152	51 46	220	31 30	279	30 59	300	29 36	339
138	46 50	020	28 42	087	43 58	152	51 12	220	30 38	278	30 13	300	29 17	338
139	47 07	019	29 34	087	44 23	152	50 38	221	29 46	277	29 28	299	28 57	337
140	47 24	018	30 26	086	44 47	152	50 03	221	28 54	277	28 41	298	28 36	336
141	47 39	016	31 19	085	45 12	153	49 29	221	28 02	276	27 55	297	28 14	335
142	47 53	015	32 11	085	45 36	153	48 54	222	27 10	276	27 08	297	27 52	334
143	48 05	013	33 03	084	45 59	153	48 19	222	26 17	275	26 21	296	27 29	333
144	48 17	012	33 55	084	46 23	154	47 44	222	25 25	275	25 34	295	27 05	333
145	48 27	010	34 48	083	46 46	154	47 09	222	24 33	274	24 46	295	26 40	332
146	48 36	009	35 40	083	47 09	154	46 33	223	23 40	274	23 59	294	26 15	331
147	48 43	007	36 32	082	47 31	155	45 58	223	22 48	273	23 11	293	25 49	330
148	48 49	006	37 24	082	47 54	155	45 22	223	21 56	273	22 22	293	25 23	329
149	48 54	005	38 16	081	48 16	155	44 47	223	21 03	272	21 34	292	24 55	328
	REGULUS		◆Denebola		SPICA		◆ACRUX		CANOPUS		◆SIRIUS		PROCYON	
150	48 58	003	39 06	035	39 07	080	48 37	156	44 11	223	43 45	274	41 51	310
151	49 00	002	39 35	034	39 59	080	48 58	156	43 35	223	42 53	274	41 10	309
152	49 00	000	40 04	033	40 51	079	49 19	157	42 59	223	42 00	273	40 29	308
153	49 00	359	40 32	031	41 42	079	49 40	157	42 23	224	41 08	273	39 47	307
154	48 58	357	40 58	030	42 33	078	50 00	158	41 47	224	40 15	272	39 04	306
155	48 54	356	41 25	029	43 25	077	50 20	158	41 10	224	39 23	272	38 22	305
156	48 50	354	41 50	028	44 16	077	50 39	159	40 34	224	38 31	271	37 38	304
157	48 44	353	42 14	027	45 07	076	50 58	159	39 58	224	37 38	271	36 55	303
158	48 36	351	42 37	026	45 58	075	51 16	160	39 21	224	36 46	270	36 10	302
159	48 28	350	42 59	024	46 48	074	51 34	160	38 45	224	35 53	270	35 26	301
160	48 17	348	43 20	023	47 39	074	51 51	161	38 09	224	35 01	269	34 41	301
161	48 06	347	43 40	022	48 29	073	52 08	161	37 32	224	34 08	269	33 56	300
162	47 53	345	44 00	021	49 19	072	52 25	162	36 56	224	33 16	268	33 10	299
163	47 40	344	44 17	019	50 09	071	52 41	163	36 20	224	32 23	268	32 24	298
164	47 24	343	44 34	018	50 59	071	52 56	163	35 43	224	31 31	267	31 38	298
	ARCTURUS		◆ANTARES		ACRUX		◆CANOPUS		SIRIUS		PROCYON		◆REGULUS	
165	22 35	050	18 43	110	53 11	164	35 07	224	30 38	267	30 51	297	47 08	341
166	23 15	050	19 33	110	53 25	165	34 31	224	29 46	266	30 04	296	46 51	340
167	23 54	049	20 22	110	53 39	165	33 54	224	28 54	266	29 17	295	46 32	338
168	24 34	048	21 11	109	53 52	166	33 18	224	28 01	266	28 29	295	46 12	337
169	25 13	047	22 01	109	54 05	167	32 42	224	27 09	265	27 42	294	45 51	336
170	25 51	047	22 51	109	54 17	167	32 06	223	26 17	265	26 54	293	45 29	334
171	26 29	046	23 41	108	54 28	168	31 30	223	25 25	264	26 05	293	45 06	333
172	27 06	045	24 30	108	54 39	169	30 54	223	24 32	264	25 17	292	44 42	332
173	27 43	044	25 20	107	54 49	169	30 18	223	23 40	263	24 28	292	44 17	331
174	28 20	043	26 11	107	54 58	170	29 42	223	22 48	263	23 39	291	43 50	330
175	28 55	043	27 01	107	55 07	171	29 06	223	21 56	262	22 50	290	43 23	328
176	29 31	042	27 51	106	55 15	172	28 31	223	21 04	262	22 01	290	42 55	327
177	30 05	041	28 41	106	55 22	172	27 55	223	20 12	262	21 11	289	42 26	326
178	30 39	040	29 32	106	55 29	173	27 19	223	19 20	261	20 22	288	41 57	325
179	31 13	039	30 22	105	55 35	174	26 44	222	18 28	261	19 32	288	41 26	324

LHA	Hc	Zn	Hc	Zn	Hc	Zn	Hc	Zn	Hc	Zn	Hc	Zn	Hc	Zn
♈	◆ARCTURUS		ANTARES		◆RIGIL KENT.		ACRUX		CANOPUS		◆Alphard		REGULUS	
180	31 46	038	31 13	105	48 40	152	55 40	175	26 09	222	48 49	292	40 54	323
181	32 18	037	32 04	105	49 05	152	55 44	176	25 34	222	48 00	291	40 22	322
182	32 49	036	32 55	104	49 29	153	55 48	176	24 59	222	47 11	290	39 49	320
183	33 20	035	33 46	104	49 53	153	55 51	177	24 24	222	46 22	289	39 15	319
184	33 50	034	34 37	104	50 16	153	55 53	178	23 49	221	45 32	289	38 41	318
185	34 19	033	35 28	103	50 40	154	55 55	179	23 14	221	44 43	288	38 06	317
186	34 48	033	36 19	103	51 03	154	55 56	180	22 40	221	43 53	287	37 30	316
187	35 16	032	37 10	103	51 25	155	55 56	180	22 05	221	43 02	286	36 54	315
188	35 43	030	38 01	102	51 47	155	55 55	181	21 31	221	42 12	286	36 16	315
189	36 09	029	38 52	102	52 09	156	55 54	182	20 57	220	41 21	285	35 39	314
190	36 34	028	39 44	102	52 30	157	55 51	183	20 23	220	40 30	284	35 00	313
191	36 59	027	40 35	101	52 50	157	55 48	184	19 49	220	39 40	284	34 22	312
192	37 23	026	41 27	101	53 11	158	55 45	184	19 16	220	38 49	283	33 42	311
193	37 45	025	42 18	101	53 30	158	55 41	185	18 42	219	37 57	282	33 02	310
194	38 07	024	43 10	100	53 49	159	55 35	186	18 09	219	37 06	282	32 22	309
	ARCTURUS		◆Rasalhague		ANTARES		RIGIL KENT.		◆ACRUX		Suhail		◆REGULUS	
195	38 28	023	11 50	068	44 02	100	54 08	160	55 30	187	42 01	236	31 41	308
196	38 48	022	12 38	068	44 53	100	54 26	160	55 23	187	41 18	236	31 00	308
197	39 07	021	13 27	067	45 45	099	54 44	161	55 16	188	40 34	236	30 18	307
198	39 25	020	14 15	067	46 37	099	55 00	162	55 08	189	39 51	236	29 36	306
199	39 42	018	15 03	066	47 29	099	55 17	162	55 00	190	39 07	236	28 53	305
200	39 58	017	15 51	065	48 21	098	55 32	163	54 50	191	38 24	236	28 10	304
201	40 13	016	16 39	065	49 13	098	55 47	164	54 40	191	37 41	236	27 27	304
202	40 27	015	17 26	064	50 05	097	56 02	165	54 30	192	36 57	236	26 43	303
203	40 40	014	18 13	064	50 57	097	56 15	165	54 19	193	36 14	235	25 59	302
204	40 52	012	19 00	063	51 49	097	56 28	166	54 07	193	35 31	235	25 14	302
205	41 02	011	19 47	062	52 41	096	56 41	167	53 54	194	34 48	235	24 29	301
206	41 12	010	20 33	062	53 33	096	56 52	168	53 41	195	34 05	235	23 44	300
207	41 20	009	21 19	061	54 25	096	57 03	169	53 28	195	33 22	235	22 58	300
208	41 28	007	22 05	060	55 17	095	57 13	169	53 13	196	32 39	235	22 13	299
209	41 34	006	22 50	060	56 10	095	57 22	170	52 59	197	31 56	235	21 27	298
	◆ARCTURUS		Rasalhague		◆Nunki		RIGIL KENT.		ACRUX		◆Suhail		Denebola	
210	41 39	005	23 36	059	25 45	107	57 31	171	52 43	197	31 14	234	36 06	320
211	41 43	004	24 20	058	26 35	107	57 39	172	52 27	198	30 31	234	35 32	319
212	41 45	002	25 05	058	27 25	106	57 45	173	52 11	198	29 49	234	34 57	318
213	41 47	001	25 49	057	28 16	106	57 52	174	51 54	199	29 06	234	34 21	317
214	41 47	000	26 33	056	29 06	106	57 57	175	51 37	200	28 24	234	33 45	316
215	41 47	359	27 16	056	29 57	105	58 01	176	51 19	200	27 42	233	33 08	315
216	41 45	357	28 00	055	30 47	105	58 05	176	51 01	201	27 00	233	32 30	314
217	41 42	356	28 42	054	31 38	105	58 08	177	50 42	201	26 18	233	31 52	313
218	41 38	355	29 24	053	32 29	104	58 10	178	50 23	202	25 36	233	31 14	312
219	41 32	354	30 06	052	33 20	104	58 11	179	50 03	202	24 54	232	30 35	311
220	41 26	352	30 48	052	34 11	104	58 11	180	49 43	203	24 13	232	29 55	311
221	41 18	351	31 29	051	35 02	103	58 11	181	49 22	203	23 31	232	29 15	310
222	41 09	350	32 09	050	35 53	103	58 09	182	49 02	204	22 50	232	28 35	309
223	40 59	349	32 49	049	36 44	103	58 07	183	48 40	204	22 09	231	27 54	308
224	40 48	347	33 29	048	37 35	102	58 04	184	48 19	204	21 28	231	27 12	308
	Alphecca		Rasalhague		◆Nunki		Peacock		◆RIGIL KENT.		SPICA		◆ARCTURUS	
225	33 40	009	34 08	047	38 26	102	28 33	142	58 00	185	61 31	304	40 36	346
226	33 48	008	34 46	047	39 18	102	29 05	142	57 55	186	60 47	303	40 23	345
227	33 55	007	35 24	046	40 09	101	29 38	142	57 49	187	60 03	301	40 09	344
228	34 01	006	36 01	045	41 01	101	30 10	142	57 43	187	59 18	300	39 54	342
229	34 06	005	36 38	044	41 52	101	30 43	142	57 36	188	58 32	299	39 37	341
230	34 10	004	37 14	043	42 44	100	31 16	141	57 28	189	57 46	298	39 20	340
231	34 13	003	37 49	042	43 36	100	31 48	141	57 19	190	56 59	297	39 02	339
232	34 15	002	38 24	041	44 27	100	32 21	141	57 09	191	56 12	296	38 43	338
233	34 16	001	38 58	040	45 19	099	32 54	141	56 59	192	55 25	295	38 22	337
234	34 16	000	39 31	039	46 11	099	33 27	141	56 48	193	54 37	294	38 01	336
235	34 15	359	40 03	038	47 03	099	34 00	141	56 36	193	53 48	293	37 39	335
236	34 13	357	40 35	037	47 55	098	34 32	141	56 24	194	53 00	292	37 16	333
237	34 10	356	41 06	036	48 47	098	35 05	141	56 10	195	52 11	291	36 52	332
238	34 07	355	41 36	034	49 39	097	35 38	141	55 56	196	51 22	290	36 27	331
239	34 02	354	42 05	033	50 31	097	36 11	141	55 42	197	50 32	289	36 02	330
	◆VEGA		ALTAIR		◆Peacock		RIGIL KENT.		◆SPICA		ARCTURUS		Alphecca	
240	12 59	030	22 50	065	36 44	141	55 27	197	49 42	288	35 35	329	33 56	353
241	13 26	030	23 37	064	37 17	141	55 11	198	48 52	287	35 08	328	33 49	352
242	13 51	029	24 24	064	37 50	141	54 54	199	48 02	287	34 40	327	33 42	351
243	14 17	028	25 11	063	38 23	141	54 37	199	47 12	286	34 11	326	33 33	350
244	14 41	028	25 58	062	38 55	141	54 19	200	46 21	285	33 42	325	33 23	349
245	15 05	027	26 44	062	39 28	141	54 01	201	45 30	284	33 11	324	33 13	348
246	15 29	026	27 30	061	40 01	142	53 42	201	44 40	284	32 40	323	33 01	347
247	15 52	026	28 16	060	40 33	142	53 23	202	43 49	283	32 09	322	32 49	346
248	16 14	025	29 01	059	41 06	142	53 03	203	42 57	282	31 36	322	32 35	345
249	16 36	024	29 46	059	41 38	142	52 43	203	42 06	282	31 03	321	32 21	344
250	16 57	023	30 31	058	42 11	142	52 22	204	41 15	281	30 30	320	32 06	343
251	17 18	023	31 15	057	42 43	142	52 01	204	40 23	280	29 56	319	31 50	342
252	17 38	022	31 59	056	43 15	142	51 39	205	39 31	280	29 21	318	31 33	341
253	17 57	021	32 42	056	43 47	142	51 17	205	38 40	279	28 45	317	31 15	340
254	18 16	020	33 25	055	44 19	143	50 54	206	37 48	279	28 09	316	30 57	339
	◆VEGA		ALTAIR		FOMALHAUT		◆Peacock		RIGIL KENT.		◆SPICA		ARCTURUS	
255	18 34	020	34 08	054	14 23	116	44 51	143	50 31	206	36 56	278	27 33	316
256	18 51	019	34 50	053	15 11	116	45 22	143	50 08	207	36 04	277	26 56	315
257	19 08	018	35 32	052	15 58	115	45 54	143	49 44	207	35 12	277	26 18	314
258	19 24	017	36 13	051	16 45	115	46 25	144	49 20	208	34 20	276	25 40	313
259	19 39	017	36 54	050	17 33	115	46 56	144	48 55	208	33 27	276	25 02	312
260	19 54	016	37 34	050	18 21	114	47 27	144	48 31	208	32 35	275	24 23	312
261	20 08	015	38 14	049	19 09	114	47 58	144	48 06	209	31 43	275	23 43	311
262	20 21	014	38 53	048	19 57	114	48 28	145	47 40	209	30 51	274	23 03	310
263	20 33	013	39 32	047	20 45	113	48 58	145	47 15	209	29 58	274	22 23	309
264	20 45	013	40 09	046	21 33	113	49 28	145	46 49	210	29 06	273	21 42	309
265	20 56	012	40 47	045	22 21	113	49 58	146	46 22	210	28 14	273	21 01	308
266	21 07	011	41 23	044	23 10	112	50 27	146	45 56	210	27 21	272	20 20	307
267	21 16	010	41 59	043	23 59	112	50 56	147	45 29	211	26 29	272	19 38	307
268	21 25	009	42 34	042	24 47	112	51 25	147	45 02	211	25 36	271	18 55	306
269	21 33	009	43 09	040	25 36	111	51 53	147	44 35	211	24 44	271	18 13	305

LAT 29°S

LHA	Hc	Zn	Hc	Zn	Hc	Zn	Hc	Zn	Hc	Zn	Hc	Zn	Hc	Zn
♈	VEGA		◆ALTAIR		FOMALHAUT		◆Peacock		RIGIL KENT.		◆ANTARES		Alphecca	
270	21 41	008	43 42	039	26 25	111	52 21	148	44 08	211	69 45	272	24 15	324
271	21 47	007	44 15	038	27 14	111	52 49	148	43 41	212	68 52	271	23 45	324
272	21 53	006	44 47	037	28 04	110	53 17	149	43 13	212	68 00	271	23 13	323
273	21 59	005	45 19	036	28 53	110	53 43	149	42 45	212	67 08	270	22 41	322
274	22 03	004	45 49	035	29 42	110	54 10	150	42 17	212	66 15	270	22 09	321
275	22 06	004	46 18	033	30 32	109	54 36	151	41 49	212	65 23	269	21 36	321
276	22 09	003	46 47	032	31 21	109	55 02	151	41 21	213	64 30	269	21 02	320
277	22 11	002	47 14	031	32 11	109	55 27	152	40 52	213	63 38	268	20 28	319
278	22 13	001	47 40	030	33 01	108	55 51	152	40 24	213	62 45	268	19 53	318
279	22 13	000	48 06	028	33 51	108	56 15	153	39 55	213	61 53	268	19 18	318
280	22 13	359	48 30	027	34 41	108	56 39	154	39 27	213	61 00	267	18 43	317
281	22 12	358	48 53	025	35 31	107	57 02	154	38 58	213	60 08	267	18 07	316
282	22 10	358	49 15	024	36 21	107	57 24	155	38 29	213	59 16	266	17 30	316
283	22 08	357	49 36	023	37 11	107	57 46	156	38 00	214	58 23	266	16 53	315
284	22 04	356	49 56	021	38 01	107	58 07	157	37 31	214	57 31	266	16 16	314
	◆DENEB		FOMALHAUT		◆ACHERNAR		RIGIL KENT.		◆ANTARES		Rasalhague		VEGA	
285	12 15	018	38 51	106	19 19	146	37 02	214	56 39	265	43 37	331	22 00	355
286	12 31	017	39 42	106	19 48	145	36 33	214	55 46	265	43 11	329	21 55	354
287	12 46	017	40 32	106	20 18	145	36 04	214	54 54	264	42 43	328	21 50	353
288	13 01	016	41 23	105	20 48	145	35 34	214	54 02	264	42 15	327	21 43	353
289	13 15	015	42 14	105	21 19	145	35 05	214	53 10	264	41 46	326	21 36	352
290	13 28	015	43 04	105	21 49	144	34 36	214	52 17	263	41 17	325	21 28	351
291	13 41	014	43 55	105	22 20	144	34 07	214	51 25	263	40 46	324	21 20	350
292	13 53	013	44 46	104	22 50	144	33 37	214	50 33	263	40 15	323	21 11	349
293	14 05	012	45 37	104	23 21	144	33 08	214	49 41	262	39 42	322	21 00	349
294	14 16	012	46 28	104	23 52	144	32 39	214	48 49	262	39 09	321	20 50	348
295	14 26	011	47 19	103	24 23	144	32 10	214	47 57	262	38 36	320	20 38	347
296	14 36	010	48 10	103	24 54	143	31 40	214	47 05	261	38 01	319	20 26	346
297	14 45	010	49 01	103	25 26	143	31 11	214	46 14	261	37 26	318	20 13	345
298	14 54	009	49 52	103	25 57	143	30 42	214	45 22	261	36 50	317	19 59	344
299	15 02	008	50 43	102	26 29	143	30 13	214	44 30	260	36 14	316	19 45	344
	DENEB		Enif		◆FOMALHAUT		ACHERNAR		◆RIGIL KENT.		ANTARES		◆VEGA	
300	15 09	008	43 46	037	51 35	102	27 00	143	29 44	214	43 38	260	19 30	343
301	15 15	007	44 17	036	52 26	102	27 32	143	29 15	214	42 47	260	19 14	342
302	15 21	006	44 47	034	53 17	102	28 04	143	28 46	213	41 55	259	18 57	341
303	15 26	005	45 16	033	54 09	101	28 36	142	28 17	213	41 04	259	18 40	341
304	15 31	005	45 44	032	55 00	101	29 08	142	27 48	213	40 12	259	18 23	340
305	15 35	004	46 11	031	55 52	101	29 40	142	27 20	213	39 21	258	18 04	339
306	15 38	003	46 38	029	56 43	101	30 12	142	26 51	213	38 29	258	17 45	338
307	15 41	002	47 03	028	57 35	100	30 44	142	26 22	213	37 38	258	17 25	338
308	15 42	002	47 27	027	58 26	100	31 17	142	25 54	213	36 47	257	17 05	337
309	15 44	001	47 50	025	59 18	100	31 49	142	25 25	213	35 56	257	16 44	336
310	15 44	000	48 12	024	60 10	100	32 21	142	24 57	213	35 05	257	16 23	335
311	15 44	359	48 33	023	61 01	100	32 54	142	24 29	212	34 14	256	16 01	335
312	15 43	359	48 52	021	61 53	099	33 26	142	24 01	212	33 23	256	15 38	334
313	15 42	358	49 11	020	62 45	099	33 58	142	23 33	212	32 32	256	15 15	333
314	15 40	357	49 28	018	63 37	099	34 31	142	23 05	212	31 41	255	14 51	333
	Alpheratz		◆Diphda		ACHERNAR		◆RIGIL KENT.		ANTARES		◆ALTAIR		DENEB	
315	16 36	042	38 06	091	35 03	142	22 38	212	30 50	255	48 36	334	15 37	357
316	17 10	041	38 58	090	35 36	142	22 10	212	30 00	254	48 12	332	15 33	356
317	17 45	040	39 51	090	36 08	142	21 43	211	29 09	254	47 47	331	15 29	355
318	18 18	040	40 43	089	36 41	142	21 15	211	28 19	254	47 21	329	15 25	354
319	18 52	039	41 36	089	37 13	142	20 48	211	27 28	253	46 54	328	15 19	354
320	19 25	038	42 28	088	37 45	142	20 21	211	26 38	253	46 26	327	15 13	353
321	19 57	038	43 21	088	38 18	142	19 55	211	25 48	253	45 57	326	15 06	352
322	20 29	037	44 13	087	38 50	142	19 28	210	24 58	252	45 27	324	14 59	352
323	21 00	036	45 06	087	39 22	142	19 01	210	24 08	252	44 56	323	14 51	351
324	21 31	035	45 58	086	39 54	142	18 35	210	23 18	252	44 24	322	14 42	350
325	22 01	035	46 50	086	40 27	142	18 09	210	22 28	251	43 51	321	14 33	349
326	22 30	034	47 43	085	40 59	142	17 43	209	21 39	251	43 18	320	14 23	349
327	22 59	033	48 35	085	41 31	143	17 18	209	20 49	251	42 44	319	14 12	348
328	23 27	032	49 27	084	42 03	143	16 52	209	20 00	250	42 09	318	14 01	347
329	23 55	031	50 19	083	42 34	143	16 27	209	19 11	250	41 33	317	13 49	347
	◆Alpheratz		Diphda		◆ACHERNAR		Peacock		◆Nunki		ALTAIR		Enif	
330	24 22	031	51 11	083	43 06	143	57 38	204	49 12	262	40 57	316	50 57	354
331	24 48	030	52 03	082	43 37	143	57 16	205	48 20	262	40 20	315	50 51	352
332	25 14	029	52 55	082	44 09	143	56 53	206	47 28	262	39 42	314	50 43	351
333	25 39	028	53 47	081	44 40	144	56 30	206	46 36	261	39 03	313	50 34	349
334	26 04	027	54 39	080	45 11	144	56 06	207	45 44	261	38 25	312	50 23	348
335	26 27	026	55 31	079	45 42	144	55 42	208	44 52	261	37 45	311	50 11	346
336	26 50	025	56 22	079	46 13	144	55 17	208	44 00	260	37 05	310	49 58	345
337	27 12	025	57 14	078	46 43	145	54 52	209	43 09	260	36 24	309	49 43	343
338	27 34	024	58 05	077	47 14	145	54 26	210	42 17	260	35 43	308	49 27	342
339	27 54	023	58 56	076	47 44	145	54 00	210	41 25	259	35 02	307	49 10	340
340	28 14	022	59 47	076	48 14	145	53 33	211	40 34	259	34 20	306	48 52	339
341	28 34	021	60 38	075	48 44	146	53 06	211	39 42	259	33 37	305	48 32	337
342	28 52	020	61 28	074	49 13	146	52 39	212	38 51	258	32 54	305	48 11	336
343	29 09	019	62 18	073	49 42	146	52 11	212	38 00	258	32 11	304	47 49	335
344	29 26	018	63 08	072	50 11	147	51 43	213	37 08	258	31 27	303	47 26	333
	Alpheratz		◆Hamal		◆ACHERNAR		Peacock		Nunki		◆ALTAIR		Enif	
345	29 42	017	20 56	046	50 40	147	51 14	213	36 17	257	30 43	302	47 02	332
346	29 57	016	21 33	045	51 08	148	50 45	214	35 26	257	29 58	302	46 37	331
347	30 11	015	22 10	044	51 36	148	50 16	214	34 35	257	29 13	301	46 10	329
348	30 25	014	22 46	043	52 03	149	49 47	214	33 44	256	28 28	300	45 43	328
349	30 37	013	23 22	043	52 30	149	49 17	215	32 53	256	27 42	299	45 15	327
350	30 49	012	23 58	042	52 57	150	48 47	215	32 02	255	26 56	299	44 46	326
351	30 59	011	24 32	041	53 24	150	48 17	215	31 11	255	26 10	298	44 16	324
352	31 09	010	25 07	040	53 50	151	47 46	216	30 21	255	25 24	297	43 45	323
353	31 18	009	25 40	040	54 15	151	47 15	216	29 30	254	24 37	297	43 13	322
354	31 26	008	26 13	039	54 40	152	46 44	216	28 40	254	23 50	296	42 40	321
355	31 33	007	26 46	038	55 05	152	46 13	217	27 49	254	23 03	295	42 07	320
356	31 39	006	27 18	037	55 29	153	45 42	217	26 59	253	22 15	295	41 33	319
357	31 44	005	27 49	036	55 52	154	45 10	217	26 09	253	21 27	294	40 58	318
358	31 49	004	28 20	035	56 15	154	44 39	217	25 19	253	20 39	293	40 22	317
359	31 52	003	28 50	034	56 38	155	44 07	217	24 29	252	19 51	293	39 46	316

 LAT 30°S

LHA	Hc	Zn	Hc	Zn	Hc	Zn	Hc	Zn	Hc	Zn	Hc	Zn	Hc	Zn
♈	♦Alpheratz		Hamal		♦RIGEL		CANOPUS		ACHERNAR		♦Peacock		Enif	
0	30 54	002	28 29	033	13 57	092	20 04	140	57 54	155	44 22	218	38 26	315
1	30 56	001	28 57	032	14 49	091	20 38	140	58 16	156	43 50	218	37 50	314
2	30 56	000	29 25	031	15 41	091	21 11	140	58 37	157	43 18	218	37 12	313
3	30 56	359	29 52	031	16 33	090	21 45	139	58 57	158	42 46	219	36 34	312
4	30 54	358	30 18	030	17 25	090	22 19	139	59 16	158	42 13	219	35 55	311
5	30 52	357	30 43	029	18 17	089	22 53	139	59 35	159	41 41	219	35 16	311
6	30 49	356	31 08	028	19 09	089	23 27	139	59 53	160	41 08	219	34 36	310
7	30 45	355	31 31	027	20 01	088	24 02	138	60 10	161	40 35	219	33 56	309
8	30 40	354	31 54	026	20 53	087	24 36	138	60 27	162	40 03	219	33 15	308
9	30 34	353	32 17	025	21 45	087	25 11	138	60 43	163	39 30	219	32 34	307
10	30 27	352	32 38	024	22 37	086	25 46	138	60 57	164	38 57	219	31 53	306
11	30 19	351	32 59	023	23 29	086	26 21	138	61 11	165	38 24	219	31 10	305
12	30 11	350	33 18	022	24 20	085	26 56	137	61 24	166	37 51	219	30 28	305
13	30 01	349	33 37	021	25 12	085	27 31	137	61 36	167	37 18	219	29 45	304
14	29 51	348	33 55	020	26 04	084	28 06	137	61 48	168	36 45	219	29 02	303
	♦Hamal		ALDEBARAN		RIGEL		♦CANOPUS		ACHERNAR		♦FOMALHAUT		Alpheratz	
15	34 12	019	20 19	056	26 56	084	28 42	137	61 58	169	63 29	263	29 39	347
16	34 28	018	21 02	055	27 47	083	29 17	137	62 07	170	62 37	263	29 27	346
17	34 43	016	21 44	054	28 39	083	29 53	137	62 15	171	61 46	262	29 14	345
18	34 58	015	22 26	054	29 30	082	30 29	137	62 23	173	60 54	262	29 00	344
19	35 11	014	23 08	053	30 22	082	31 04	136	62 29	174	60 03	262	28 46	343
20	35 23	013	23 49	052	31 13	081	31 40	136	62 34	175	59 11	262	28 30	342
21	35 35	012	24 30	051	32 04	080	32 16	136	62 38	176	58 20	261	28 14	341
22	35 45	011	25 10	051	32 56	080	32 52	136	62 41	177	57 29	261	27 57	340
23	35 54	010	25 50	050	33 47	079	33 28	136	62 43	178	56 37	261	27 39	339
24	36 03	009	26 30	049	34 38	079	34 04	136	62 44	180	55 46	260	27 20	338
25	36 10	008	27 09	048	35 28	078	34 41	136	62 44	181	54 55	260	27 00	337
26	36 17	007	27 47	048	36 19	077	35 17	136	62 43	182	54 04	260	26 40	337
27	36 22	005	28 25	047	37 10	077	35 53	136	62 41	183	53 13	260	26 19	336
28	36 26	004	29 03	046	38 00	076	36 29	136	62 37	184	52 22	259	25 57	335
29	36 30	003	29 40	045	38 51	075	37 06	136	62 33	185	51 31	259	25 35	334
	Hamal		ALDEBARAN		♦SIRIUS		CANOPUS		♦ACHERNAR		FOMALHAUT		♦Alpheratz	
30	36 32	002	30 16	044	24 15	096	37 42	136	62 27	187	50 40	259	25 12	333
31	36 33	001	30 52	043	25 06	096	38 18	136	62 21	188	49 49	258	24 48	332
32	36 34	000	31 28	042	25 58	095	38 55	136	62 14	189	48 58	258	24 23	331
33	36 33	359	32 03	042	26 50	095	39 31	136	62 05	190	48 07	258	23 58	331
34	36 31	357	32 37	041	27 42	094	40 07	136	61 56	191	47 16	257	23 32	330
35	36 28	356	33 10	040	28 34	094	40 44	136	61 45	192	46 26	257	23 06	329
36	36 24	355	33 43	039	29 25	093	41 20	136	61 34	193	45 35	257	22 38	328
37	36 19	354	34 15	038	30 17	093	41 56	136	61 21	194	44 44	257	22 11	327
38	36 13	353	34 47	037	31 09	092	42 33	136	61 08	195	43 54	256	21 42	326
39	36 06	352	35 18	036	32 01	092	43 09	136	60 54	196	43 03	256	21 13	326
40	35 58	351	35 48	035	32 53	091	43 45	136	60 39	197	42 13	256	20 44	325
41	35 49	349	36 17	034	33 45	091	44 21	136	60 23	198	41 23	255	20 13	324
42	35 39	348	36 46	033	34 37	090	44 57	136	60 06	199	40 33	255	19 43	323
43	35 28	347	37 13	032	35 29	090	45 33	136	59 49	200	39 42	255	19 11	323
44	35 17	346	37 40	031	36 21	089	46 09	136	59 31	201	38 52	254	18 40	322
	♦ALDEBARAN		BETELGEUSE		SIRIUS		♦CANOPUS		ACHERNAR		♦FOMALHAUT		Hamal	
45	38 06	030	33 47	056	37 13	089	46 45	137	59 12	202	38 02	254	35 04	345
46	38 32	028	34 30	055	38 05	088	47 20	137	58 52	203	37 12	254	34 50	344
47	38 56	027	35 12	054	38 57	088	47 56	137	58 32	203	36 22	253	34 35	343
48	39 19	026	35 54	053	39 49	087	48 31	137	58 11	204	35 33	253	34 19	342
49	39 42	025	36 35	052	40 41	087	49 06	137	57 49	205	34 43	253	34 03	341
50	40 03	024	37 16	051	41 32	086	49 42	138	57 27	206	33 53	253	33 45	340
51	40 24	023	37 56	050	42 24	086	50 16	138	57 04	206	33 04	252	33 27	339
52	40 44	022	38 36	049	43 16	085	50 51	138	56 40	207	32 14	252	33 07	338
53	41 02	020	39 15	048	44 08	084	51 26	139	56 16	208	31 25	252	32 47	337
54	41 20	019	39 54	047	44 59	084	52 00	139	55 52	208	30 36	251	32 26	336
55	41 36	018	40 32	046	45 51	083	52 34	139	55 27	209	29 47	251	32 04	335
56	41 52	017	41 09	045	46 43	083	53 08	140	55 01	210	28 58	251	31 42	334
57	42 06	015	41 46	044	47 34	082	53 42	140	54 36	210	28 09	250	31 18	333
58	42 20	014	42 22	043	48 26	081	54 15	140	54 09	211	27 20	250	30 54	332
59	42 32	013	42 57	042	49 17	081	54 48	141	53 42	211	26 31	249	30 29	331
	ALDEBARAN		BETELGEUSE		♦SIRIUS		CANOPUS		♦ACHERNAR		FOMALHAUT		♦Hamal	
60	42 43	012	43 31	041	50 08	080	55 20	141	53 15	212	25 43	249	30 03	330
61	42 53	010	44 05	040	50 59	079	55 53	142	52 48	212	24 54	249	29 37	329
62	43 02	009	44 38	039	51 50	079	56 25	142	52 20	213	24 06	248	29 10	328
63	43 09	008	45 10	038	52 41	078	56 56	143	51 52	213	23 17	248	28 42	327
64	43 16	006	45 42	036	53 32	077	57 28	143	51 23	214	22 29	248	28 13	326
65	43 21	005	46 12	035	54 22	076	57 58	144	50 54	214	21 41	247	27 44	325
66	43 25	004	46 42	034	55 13	076	58 29	145	50 25	214	20 53	247	27 14	324
67	43 28	003	47 10	033	56 03	075	58 59	145	49 56	215	20 06	247	26 44	324
68	43 30	001	47 38	031	56 53	074	59 28	146	49 26	215	19 18	246	26 13	323
69	43 30	000	48 04	030	57 43	073	59 57	147	48 56	215	18 30	246	25 41	322
70	43 29	359	48 30	029	58 32	072	60 25	147	48 26	216	17 43	245	25 09	321
71	43 27	357	48 54	027	59 22	071	60 53	148	47 55	216	16 56	245	24 36	320
72	43 24	356	49 17	026	60 11	070	61 20	149	47 25	216	16 09	245	24 02	319
73	43 20	355	49 40	025	60 59	069	61 46	150	46 54	216	15 22	244	23 28	319
74	43 15	353	50 01	023	61 48	068	62 11	151	46 23	217	14 35	244	22 54	318
	BETELGEUSE		PROCYON		♦Suhail		CANOPUS		♦ACHERNAR		Hamal		♦ALDEBARAN	
75	50 20	022	38 07	054	39 45	124	62 36	152	45 52	217	22 19	317	43 08	352
76	50 39	020	38 49	053	40 28	123	63 00	153	45 21	217	21 43	316	43 00	351
77	50 56	019	39 30	052	41 12	123	63 24	154	44 49	217	21 07	316	42 51	349
78	51 12	017	40 11	051	41 55	123	63 46	155	44 18	217	20 30	315	42 41	348
79	51 27	016	40 51	050	42 39	123	64 08	156	43 46	218	19 53	314	42 30	347
80	51 40	014	41 31	049	43 22	123	64 28	157	43 14	218	19 16	313	42 17	346
81	51 52	012	42 10	048	44 06	123	64 48	158	42 43	218	18 38	313	42 04	344
82	52 02	011	42 49	047	44 49	123	65 06	160	42 11	218	18 00	312	41 49	343
83	52 11	009	43 27	046	45 33	123	65 24	161	41 38	218	17 21	311	41 33	342
84	52 19	008	44 04	045	46 16	123	65 40	162	41 06	218	16 42	311	41 17	341
85	52 25	006	44 40	044	47 00	123	65 55	164	40 34	218	16 02	310	40 59	339
86	52 30	004	45 16	043	47 44	123	66 10	165	40 02	218	15 22	309	40 40	338
87	52 33	003	45 51	042	48 27	123	66 22	166	39 30	218	14 42	309	40 20	337
88	52 35	001	46 25	041	49 11	123	66 34	168	38 57	219	14 01	308	40 00	336
89	52 36	000	46 59	039	49 54	123	66 44	169	38 25	219	13 20	308	39 38	335

LHA	Hc	Zn	Hc	Zn	Hc	Zn	Hc	Zn	Hc	Zn	Hc	Zn	Hc	Zn
♈	POLLUX		♦REGULUS		ACRUX		♦CANOPUS		ACHERNAR		♦RIGEL		BETELGEUSE	
90	26 47	026	17 04	065	23 38	151	66 53	171	37 52	219	65 43	332	52 34	358
91	27 09	025	17 51	064	24 03	150	67 01	172	37 20	219	65 18	329	52 32	356
92	27 31	024	18 37	063	24 29	150	67 07	174	36 48	219	64 50	327	52 28	355
93	27 51	023	19 24	063	24 55	150	67 12	175	36 15	219	64 21	325	52 22	353
94	28 11	022	20 10	062	25 21	150	67 16	177	35 43	219	63 51	323	52 15	351
95	28 31	021	20 56	061	25 47	150	67 18	178	35 10	219	63 19	321	52 07	350
96	28 49	020	21 41	061	26 13	150	67 18	180	34 38	219	62 46	320	51 57	348
97	29 07	019	22 26	060	26 39	150	67 18	182	34 05	219	62 12	318	51 45	347
98	29 24	019	23 11	059	27 06	149	67 15	183	33 33	219	61 37	316	51 33	345
99	29 40	018	23 56	059	27 32	149	67 12	185	33 01	218	61 00	315	51 19	344
100	29 55	017	24 40	058	27 59	149	67 07	186	32 28	218	60 23	313	51 03	342
101	30 10	016	25 24	057	28 26	149	67 00	188	31 56	218	59 44	311	50 47	341
102	30 23	015	26 08	057	28 52	149	66 53	189	31 24	218	59 05	310	50 29	339
103	30 36	014	26 51	056	29 19	149	66 44	191	30 52	218	58 25	309	50 10	338
104	30 48	013	27 34	055	29 46	149	66 33	192	30 20	218	57 44	307	49 49	336
	POLLUX		♦REGULUS		Gienah		♦ACRUX		CANOPUS		♦RIGEL		BETELGEUSE	
105	30 58	012	28 16	054	18 02	100	30 13	149	66 21	194	57 02	306	49 27	335
106	31 08	011	28 58	054	18 53	100	30 40	149	66 09	195	56 19	305	49 05	333
107	31 18	010	29 40	053	19 45	099	31 07	149	65 54	197	55 36	303	48 41	332
108	31 26	009	30 21	052	20 36	099	31 34	149	65 39	198	54 53	302	48 16	331
109	31 33	008	31 01	051	21 27	098	32 01	149	65 22	199	54 09	301	47 50	329
110	31 39	006	31 42	050	22 19	098	32 28	149	65 05	200	53 24	300	47 23	328
111	31 45	005	32 21	049	23 10	098	32 55	148	64 46	202	52 39	299	46 54	327
112	31 49	004	33 01	049	24 02	097	33 22	148	64 27	203	51 53	298	46 25	325
113	31 53	003	33 39	048	24 53	097	33 50	148	64 06	204	51 07	297	45 55	324
114	31 55	002	34 18	047	25 45	096	34 17	148	63 44	205	50 20	296	45 25	323
115	31 57	001	34 55	046	26 37	096	34 44	148	63 22	206	49 34	295	44 53	322
116	31 58	000	35 32	045	27 28	095	35 11	148	62 58	207	48 46	294	44 20	321
117	31 57	359	36 09	044	28 20	095	35 38	149	62 34	208	47 59	293	43 47	319
118	31 56	358	36 44	043	29 12	094	36 05	149	62 09	209	47 11	292	43 13	318
119	31 54	357	37 20	042	30 04	094	36 32	149	61 44	210	46 23	292	42 38	317
	♦REGULUS		SPICA		♦ACRUX		CANOPUS		♦RIGEL		BETELGEUSE		POLLUX	
120	37 54	041	13 04	095	36 59	149	61 17	211	45 34	291	42 02	316	31 51	356
121	38 28	040	13 56	095	37 26	149	60 50	212	44 45	290	41 26	315	31 47	355
122	39 01	039	14 48	094	37 53	149	60 23	213	43 56	289	40 49	314	31 42	354
123	39 33	038	15 39	094	38 20	149	59 54	213	43 07	288	40 11	313	31 36	353
124	40 05	037	16 31	094	38 47	149	59 26	214	42 18	288	39 33	312	31 29	352
125	40 36	036	17 23	093	39 14	149	58 56	215	41 28	287	38 54	311	31 22	351
126	41 06	035	18 15	093	39 41	149	58 26	215	40 38	286	38 14	310	31 13	350
127	41 35	034	19 07	092	40 07	149	57 56	216	39 48	285	37 34	309	31 04	349
128	42 03	032	19 59	092	40 34	149	57 25	217	38 58	285	36 54	308	30 53	348
129	42 31	031	20 51	091	41 00	150	56 54	217	38 08	284	36 13	307	30 42	347
130	42 57	030	21 43	091	41 26	150	56 22	218	37 17	283	35 31	307	30 30	346
131	43 23	029	22 35	090	41 52	150	55 50	218	36 27	283	34 49	306	30 17	345
132	43 47	028	23 27	090	42 18	150	55 18	219	35 36	282	34 07	305	30 03	344
133	44 11	026	24 19	089	42 44	150	54 45	219	34 45	282	33 24	304	29 48	343
134	44 34	025	25 11	089	43 10	150	54 12	220	33 54	281	32 41	303	29 32	342
	REGULUS		♦SPICA		ACRUX		♦CANOPUS		RIGEL		BETELGEUSE		♦POLLUX	
135	44 55	024	26 03	088	43 36	151	53 39	220	33 03	280	31 57	302	29 16	341
136	45 16	023	26 54	088	44 01	151	53 05	220	32 12	280	31 13	302	28 58	340
137	45 35	021	27 46	087	44 26	151	52 31	221	31 21	279	30 28	301	28 40	339
138	45 53	020	28 38	086	44 51	151	51 57	221	30 29	279	29 44	300	28 21	338
139	46 10	019	29 30	086	45 16	152	51 23	222	29 38	278	28 58	299	28 02	337
140	46 26	017	30 22	085	45 40	152	50 48	222	28 46	277	28 13	299	27 41	336
141	46 41	016	31 14	085	46 05	152	50 14	222	27 55	277	27 27	298	27 20	335
142	46 55	014	32 05	084	46 29	153	49 39	222	27 03	276	26 41	297	26 58	335
143	47 07	013	32 57	084	46 53	153	49 04	223	26 12	276	25 55	296	26 35	334
144	47 18	012	33 49	083	47 16	153	48 28	223	25 20	275	25 08	296	26 12	333
145	47 28	010	34 40	083	47 40	153	47 53	223	24 28	275	24 21	295	25 48	332
146	47 37	009	35 32	082	48 03	154	47 17	223	23 36	274	23 34	294	25 23	331
147	47 44	007	36 23	081	48 25	154	46 42	223	22 44	274	22 47	294	24 57	330
148	47 50	006	37 15	081	48 48	155	46 06	224	21 53	273	21 59	293	24 31	329
149	47 54	004	38 06	080	49 10	155	45 30	224	21 01	273	21 11	293	24 04	328
	REGULUS		♦Denebola		SPICA		♦ACRUX		CANOPUS		♦SIRIUS		PROCYON	
150	47 58	003	38 16	034	38 57	080	49 32	155	44 54	224	43 40	275	41 12	310
151	48 00	002	38 45	033	39 48	079	49 53	156	44 18	224	42 48	275	40 32	309
152	48 00	000	39 13	032	40 39	078	50 14	156	43 42	224	41 56	274	39 52	308
153	48 00	359	39 40	031	41 30	078	50 35	157	43 06	224	41 05	274	39 11	307
154	47 58	357	40 07	030	42 20	077	50 55	157	42 30	224	40 13	273	38 29	306
155	47 55	356	40 32	029	43 11	076	51 15	158	41 53	224	39 21	273	37 47	306
156	47 50	354	40 57	028	44 01	076	51 35	158	41 17	224	38 29	272	37 04	305
157	47 44	353	41 20	026	44 52	075	51 54	159	40 41	224	37 37	271	36 21	304
158	47 37	351	41 43	025	45 42	074	52 12	159	40 05	224	36 45	271	35 38	303
159	47 28	350	42 04	024	46 32	073	52 30	160	39 28	224	35 53	270	34 54	302
160	47 19	348	42 25	023	47 21	073	52 48	160	38 52	224	35 01	270	34 10	301
161	47 08	347	42 45	022	48 11	072	53 05	161	38 15	224	34 09	269	33 25	300
162	46 55	346	43 03	020	49 00	071	53 22	162	37 39	224	33 17	269	32 40	300
163	46 42	344	43 21	019	49 49	070	53 38	162	37 03	224	32 25	268	31 55	299
164	46 27	343	43 37	018	50 38	069	53 54	163	36 26	224	31 33	268	31 10	298
	ARCTURUS		♦ANTARES		ACRUX		♦CANOPUS		SIRIUS		PROCYON		♦REGULUS	
165	21 56	050	19 04	110	54 09	163	35 50	224	30 41	268	30 24	297	46 11	342
166	22 36	049	19 53	110	54 23	164	35 14	224	29 49	267	29 37	297	45 54	340
167	23 15	049	20 42	109	54 37	165	34 38	224	28 58	267	28 51	296	45 36	339
168	23 54	048	21 31	109	54 50	165	34 01	224	28 06	266	28 04	295	45 17	338
169	24 32	047	22 20	109	55 03	166	33 25	224	27 14	266	27 17	295	44 56	336
170	25 10	046	23 10	108	55 15	167	32 49	224	26 22	265	26 29	294	44 35	335
171	25 47	046	23 59	108	55 27	168	32 13	224	25 30	265	25 42	293	44 12	334
172	26 24	045	24 49	107	55 38	168	31 37	224	24 39	264	24 54	293	43 49	332
173	27 00	044	25 38	107	55 48	169	31 02	224	23 47	264	24 06	292	43 24	331
174	27 36	043	26 28	107	55 57	170	30 26	223	22 55	263	23 18	291	42 59	330
175	28 11	042	27 18	106	56 06	171	29 50	223	22 04	263	22 29	291	42 32	329
176	28 46	041	28 08	106	56 14	171	29 14	223	21 12	262	21 40	290	42 05	328
177	29 20	041	28 58	106	56 22	172	28 39	223	20 21	262	20 51	289	41 37	326
178	29 53	040	29 48	105	56 28	173	28 04	223	19 29	261	20 02	289	41 07	325
179	30 26	039	30 38	105	56 34	174	27 28	223	18 38	261	19 13	288	40 37	324

 LAT 30°S

LHA	Hc	Zn	Hc	Zn	Hc	Zn	Hc	Zn	Hc	Zn	Hc	Zn	Hc	Zn
♈	◆ARCTURUS		ANTARES		◆RIGIL KENT.		ACRUX		CANOPUS		◆Alphard		REGULUS	
180	30 58	038	31 28	104	49 33	151	56 40	175	26 53	223	48 26	293	40 07	323
181	31 30	037	32 19	104	49 57	152	56 44	175	26 18	222	47 38	292	39 35	322
182	32 01	036	33 09	104	50 22	152	56 48	176	25 43	222	46 50	291	39 03	321
183	32 31	035	34 00	103	50 46	152	56 51	177	25 08	222	46 02	290	38 30	320
184	33 00	034	34 50	103	51 10	153	56 53	178	24 34	222	45 13	290	37 56	319
185	33 29	033	35 41	103	51 33	153	56 55	179	23 59	222	44 24	289	37 21	318
186	33 57	032	36 32	102	51 57	154	56 56	180	23 25	221	43 34	288	36 46	317
187	34 25	031	37 23	102	52 19	154	56 56	180	22 50	221	42 45	287	36 11	316
188	34 51	030	38 13	101	52 41	155	56 55	181	22 16	221	41 55	287	35 34	315
189	35 17	029	39 04	101	53 03	155	56 54	182	21 42	221	41 05	286	34 57	314
190	35 42	028	39 55	101	53 25	156	56 51	183	21 09	220	40 15	285	34 20	313
191	36 06	027	40 46	100	53 46	157	56 48	184	20 35	220	39 25	284	33 41	312
192	36 29	026	41 38	100	54 06	157	56 45	184	20 02	220	38 35	284	33 03	311
193	36 51	025	42 29	100	54 26	158	56 40	185	19 28	220	37 44	283	32 24	311
194	37 12	024	43 20	099	54 45	158	56 35	186	18 55	219	36 53	283	31 44	310
	ARCTURUS		◆Rasalhague		ANTARES		RIGIL KENT.		◆ACRUX		Suhail		◆REGULUS	
195	37 33	023	11 27	068	44 11	099	55 04	159	56 29	187	42 35	237	31 04	309
196	37 52	022	12 16	068	45 03	099	55 22	160	56 23	188	41 51	237	30 23	308
197	38 11	020	13 03	067	45 54	098	55 40	160	56 15	188	41 08	237	29 42	307
198	38 29	019	13 51	066	46 46	098	55 57	161	56 07	189	40 24	237	29 00	306
199	38 45	018	14 39	066	47 37	097	56 14	162	55 59	190	39 41	236	28 18	306
200	39 01	017	15 26	065	48 29	097	56 30	163	55 49	191	38 58	236	27 36	305
201	39 15	016	16 13	065	49 20	097	56 45	163	55 39	192	38 14	236	26 53	304
202	39 29	015	17 00	064	50 12	096	56 59	164	55 28	192	37 31	236	26 10	303
203	39 42	013	17 46	063	51 04	096	57 13	165	55 17	193	36 48	236	25 26	303
204	39 53	012	18 33	063	51 55	096	57 26	166	55 05	194	36 05	236	24 42	302
205	40 03	011	19 19	062	52 47	095	57 39	167	54 52	194	35 22	236	23 58	301
206	40 13	010	20 04	061	53 39	095	57 51	167	54 39	195	34 39	236	23 14	301
207	40 21	009	20 50	061	54 31	094	58 02	168	54 25	196	33 56	235	22 29	300
208	40 28	007	21 35	060	55 22	094	58 12	169	54 11	196	33 14	235	21 43	299
209	40 34	006	22 20	059	56 14	093	58 21	170	53 56	197	32 31	235	20 58	299
	◆ARCTURUS		Rasalhague		◆Nunki		RIGIL KENT.		ACRUX		◆Suhail		Denebola	
210	40 39	005	23 05	059	26 02	107	58 30	171	53 41	198	31 48	235	35 20	320
211	40 43	004	23 49	058	26 52	106	58 38	172	53 24	198	31 06	235	34 47	319
212	40 46	002	24 33	057	27 42	106	58 45	173	53 08	199	30 24	234	34 12	318
213	40 47	001	25 16	057	28 32	106	58 51	174	52 51	199	29 41	234	33 37	317
214	40 47	000	25 59	056	29 22	105	58 57	175	52 33	200	28 59	234	33 02	316
215	40 47	359	26 42	055	30 12	105	59 01	175	52 15	201	28 17	234	32 25	315
216	40 45	357	27 25	054	31 03	104	59 05	176	51 57	201	27 35	234	31 49	314
217	40 42	356	28 07	054	31 53	104	59 08	177	51 38	202	26 54	233	31 11	314
218	40 38	355	28 48	053	32 43	104	59 10	178	51 18	202	26 12	233	30 33	313
219	40 33	354	29 30	052	33 34	103	59 11	179	50 58	203	25 31	233	29 55	312
220	40 26	352	30 10	051	34 24	103	59 11	180	50 38	203	24 49	233	29 16	311
221	40 19	351	30 51	050	35 15	103	59 10	181	50 18	204	24 08	232	28 37	310
222	40 10	350	31 30	050	36 06	102	59 09	182	49 57	204	23 27	232	27 57	309
223	40 01	349	32 10	049	36 57	102	59 07	183	49 35	205	22 46	232	27 16	309
224	39 50	347	32 49	048	37 48	102	59 04	184	49 13	205	22 05	232	26 36	308
	Alphecca		Rasalhague		◆Nunki		Peacock		◆RIGIL KENT.		SPICA		◆ARCTURUS	
225	32 40	009	33 27	047	38 38	101	29 20	142	59 00	185	60 57	305	39 38	346
226	32 48	008	34 04	046	39 29	101	29 52	141	58 55	186	60 14	304	39 25	345
227	32 55	007	34 42	045	40 21	100	30 25	141	58 49	187	59 31	303	39 11	344
228	33 01	006	35 18	044	41 12	100	30 57	141	58 42	188	58 47	301	38 56	343
229	33 06	005	35 54	043	42 03	100	31 30	141	58 35	189	58 03	300	38 41	342
230	33 10	004	36 29	042	42 54	099	32 02	141	58 27	189	57 17	299	38 24	340
231	33 13	003	37 04	041	43 45	099	32 35	141	58 18	190	56 32	298	38 06	339
232	33 15	002	37 38	040	44 37	099	33 08	141	58 08	191	55 46	297	37 47	338
233	33 16	001	38 11	039	45 28	098	33 41	141	57 58	192	54 59	296	37 27	337
234	33 16	000	38 44	038	46 20	098	34 13	141	57 46	193	54 12	295	37 06	336
235	33 15	359	39 16	037	47 11	097	34 46	141	57 34	194	53 25	294	36 45	335
236	33 13	357	39 47	036	48 03	097	35 19	141	57 22	195	52 37	293	36 22	334
237	33 11	356	40 17	035	48 54	097	35 52	141	57 08	195	51 49	292	35 59	333
238	33 07	355	40 46	034	49 46	096	36 25	141	56 54	196	51 01	291	35 35	332
239	33 02	354	41 15	033	50 38	096	36 58	141	56 39	197	50 12	290	35 09	331
	◆VEGA		ALTAIR		◆Peacock		RIGIL KENT.		◆SPICA		ARCTURUS		Alphecca	
240	12 08	030	22 24	065	37 31	141	56 24	198	49 23	289	34 44	330	32 56	353
241	12 34	030	23 11	064	38 03	141	56 08	198	48 34	288	34 17	329	32 50	352
242	12 59	029	23 57	063	38 36	141	55 51	199	47 44	288	33 49	328	32 42	351
243	13 24	028	24 44	063	39 09	141	55 34	200	46 55	287	33 21	327	32 34	350
244	13 48	028	25 30	062	39 42	141	55 16	201	46 05	286	32 52	326	32 24	349
245	14 12	027	26 15	061	40 15	141	54 57	201	45 15	285	32 22	325	32 14	348
246	14 35	026	27 01	060	40 47	141	54 38	202	44 25	285	31 52	324	32 03	347
247	14 58	025	27 46	060	41 20	141	54 19	202	43 34	284	31 21	323	31 50	346
248	15 20	025	28 30	059	41 53	141	53 58	203	42 44	283	30 49	322	31 37	345
249	15 41	024	29 15	058	42 25	141	53 38	204	41 53	283	30 17	321	31 23	344
250	16 02	023	29 59	057	42 58	141	53 17	204	41 03	282	29 44	320	31 09	343
251	16 22	023	30 42	057	43 30	142	52 55	205	40 12	281	29 10	319	30 53	342
252	16 42	022	31 25	056	44 02	142	52 33	205	39 21	281	28 36	318	30 36	341
253	17 01	021	32 08	055	44 34	142	52 11	206	38 30	280	28 01	318	30 19	340
254	17 19	020	32 51	054	45 06	142	51 48	206	37 38	279	27 26	317	30 01	339
	◆VEGA		ALTAIR		FOMALHAUT		◆Peacock		RIGIL KENT.		◆SPICA		ARCTURUS	
255	17 37	020	33 33	053	14 50	116	45 38	142	51 25	207	36 47	279	26 50	316
256	17 54	019	34 14	053	15 37	116	46 10	142	51 01	207	35 56	278	26 13	315
257	18 11	018	34 55	052	16 24	115	46 42	143	50 37	208	35 04	278	25 37	314
258	18 26	017	35 36	051	17 11	115	47 13	143	50 13	208	34 13	277	24 59	313
259	18 42	017	36 16	050	17 58	114	47 44	143	49 48	209	33 21	276	24 21	313
260	18 56	016	36 55	049	18 45	114	48 15	143	49 23	209	32 29	276	23 43	312
261	19 10	015	37 34	048	19 33	114	48 46	144	48 58	209	31 38	275	23 04	311
262	19 23	014	38 12	047	20 20	113	49 17	144	48 32	210	30 46	275	22 25	310
263	19 35	013	38 50	046	21 08	113	49 47	144	48 07	210	29 54	274	21 45	310
264	19 47	013	39 27	045	21 56	113	50 17	145	47 41	210	29 02	274	21 05	309
265	19 58	012	40 04	044	22 44	112	50 47	145	47 14	211	28 10	273	20 24	308
266	20 08	011	40 40	043	23 32	112	51 17	145	46 48	211	27 19	273	19 43	308
267	20 17	010	41 15	042	24 21	111	51 46	146	46 21	211	26 27	272	19 02	307
268	20 26	009	41 49	041	25 09	111	52 15	146	45 54	211	25 35	272	18 20	306
269	20 34	008	42 23	040	25 58	111	52 44	147	45 27	212	24 43	271	17 38	306

LHA	Hc	Zn	Hc	Zn	Hc	Zn	Hc	Zn	Hc	Zn	Hc	Zn	Hc	Zn
♈	VEGA		◆ALTAIR		FOMALHAUT		◆Peacock		RIGIL KENT.		◆ANTARES		Alphecca	
270	20 41	008	42 56	039	26 46	110	53 12	147	44 59	212	69 42	275	23 26	325
271	20 48	007	43 28	038	27 35	110	53 40	148	44 31	212	68 50	274	22 56	324
272	20 54	006	43 59	036	28 24	110	54 08	148	44 04	212	67 58	273	22 25	323
273	20 59	005	44 30	035	29 13	109	54 35	149	43 36	213	67 06	273	21 54	322
274	21 03	004	44 59	034	30 02	109	55 02	149	43 08	213	66 14	272	21 22	322
275	21 07	004	45 28	033	30 51	109	55 28	150	42 39	213	65 22	272	20 49	321
276	21 09	003	45 56	032	31 40	108	55 54	150	42 11	213	64 30	271	20 16	320
277	21 11	002	46 22	030	32 30	108	56 19	151	41 43	213	63 38	270	19 43	319
278	21 13	001	46 48	029	33 19	108	56 44	152	41 14	213	62 46	270	19 08	319
279	21 13	000	47 13	028	34 09	107	57 09	152	40 45	214	61 54	269	18 34	318
280	21 13	359	47 36	026	34 58	107	57 32	153	40 17	214	61 02	269	17 59	317
281	21 12	358	47 59	025	35 48	107	57 56	154	39 48	214	60 10	269	17 23	316
282	21 10	358	48 20	024	36 38	106	58 18	155	39 19	214	59 18	268	16 47	316
283	21 08	357	48 40	022	37 28	106	58 40	155	38 50	214	58 27	268	16 11	315
284	21 05	356	49 00	021	38 18	106	59 02	156	38 21	214	57 35	267	15 34	314
	◆DENEB		FOMALHAUT		◆ACHERNAR		RIGIL KENT.		◆ANTARES		Rasalhague		VEGA	
285	11 18	018	39 08	105	20 08	145	37 52	214	56 43	267	42 44	331	21 01	355
286	11 33	017	39 58	105	20 38	145	37 23	214	55 51	266	42 19	330	20 56	354
287	11 48	017	40 48	105	21 07	145	36 53	214	54 59	266	41 52	329	20 50	353
288	12 03	016	41 38	105	21 37	145	36 24	214	54 07	265	41 25	328	20 44	353
289	12 17	015	42 29	104	22 08	144	35 55	214	53 15	265	40 57	326	20 37	352
290	12 30	014	43 19	104	22 38	144	35 26	214	52 24	265	40 27	325	20 29	351
291	12 43	014	44 10	104	23 08	144	34 56	214	51 32	264	39 57	324	20 21	350
292	12 55	013	45 00	103	23 39	144	34 27	214	50 40	264	39 27	323	20 12	349
293	13 06	012	45 51	103	24 10	144	33 58	214	49 49	264	38 55	322	20 02	349
294	13 17	012	46 41	103	24 40	143	33 29	214	48 57	263	38 23	321	19 51	348
295	13 28	011	47 32	102	25 11	143	32 59	214	48 05	263	37 50	320	19 40	347
296	13 37	010	48 23	102	25 43	143	32 30	214	47 14	262	37 16	319	19 28	346
297	13 46	010	49 14	102	26 14	143	32 01	214	46 22	262	36 42	318	19 15	345
298	13 54	009	50 05	102	26 45	143	31 32	214	45 31	262	36 07	317	19 01	345
299	14 02	008	50 56	101	27 17	143	31 03	214	44 40	261	35 31	316	18 47	344
	DENEB		Enif		◆FOMALHAUT		ACHERNAR		◆RIGIL KENT.		ANTARES		◆VEGA	
300	14 09	007	42 57	036	51 47	101	27 48	143	30 34	214	43 48	261	18 32	343
301	14 16	007	43 28	035	52 38	101	28 20	142	30 05	214	42 57	261	18 17	342
302	14 22	006	43 57	034	53 29	100	28 52	142	29 36	214	42 06	260	18 01	341
303	14 27	005	44 25	033	54 20	100	29 23	142	29 07	214	41 15	260	17 44	341
304	14 31	005	44 53	031	55 11	100	29 55	142	28 38	214	40 23	259	17 26	340
305	14 35	004	45 20	030	56 02	100	30 27	142	28 10	213	39 32	259	17 08	339
306	14 38	003	45 45	029	56 53	099	30 59	142	27 41	213	38 41	259	16 49	338
307	14 41	002	46 10	028	57 45	099	31 32	142	27 13	213	37 51	258	16 30	338
308	14 42	002	46 33	026	58 36	099	32 04	142	26 44	213	37 00	258	16 10	337
309	14 44	001	46 56	025	59 28	098	32 36	142	26 16	213	36 09	258	15 49	336
310	14 44	000	47 17	024	60 19	098	33 08	142	25 48	213	35 18	257	15 28	336
311	14 44	000	47 37	022	61 10	098	33 41	141	25 20	213	34 28	257	15 06	335
312	14 43	359	47 56	021	62 02	098	34 13	141	24 52	212	33 37	257	14 44	334
313	14 42	358	48 14	019	62 53	097	34 45	141	24 24	212	32 46	256	14 21	333
314	14 40	357	48 31	018	63 45	097	35 18	141	23 56	212	31 56	256	13 57	333
	Alpheratz		◆Diphda		ACHERNAR		◆RIGIL KENT.		ANTARES		◆ALTAIR		DENEB	
315	15 51	042	38 06	090	35 50	141	23 29	212	31 06	255	47 43	334	14 37	357
316	16 25	041	38 58	090	36 23	141	23 01	212	30 15	255	47 19	333	14 34	356
317	16 59	040	39 50	089	36 55	141	22 34	212	29 25	255	46 55	331	14 30	355
318	17 32	040	40 42	089	37 28	141	22 07	211	28 35	254	46 29	330	14 25	354
319	18 05	039	41 34	088	38 00	141	21 40	211	27 45	254	46 03	329	14 19	354
320	18 38	038	42 26	087	38 32	141	21 13	211	26 55	254	45 35	328	14 13	353
321	19 09	037	43 18	087	39 05	141	20 46	211	26 06	253	45 07	326	14 07	352
322	19 41	037	44 10	086	39 37	142	20 20	211	25 16	253	44 38	325	13 59	352
323	20 11	036	45 02	086	40 09	142	19 53	210	24 26	252	44 08	324	13 51	351
324	20 42	035	45 54	085	40 42	142	19 27	210	23 37	252	43 37	323	13 43	350
325	21 11	034	46 45	085	41 14	142	19 01	210	22 47	252	43 05	322	13 34	349
326	21 40	034	47 37	084	41 46	142	18 35	210	21 58	251	42 32	320	13 24	349
327	22 09	033	48 29	083	42 18	142	18 10	209	21 09	251	41 58	319	13 13	348
328	22 37	032	49 20	083	42 50	142	17 44	209	20 20	250	41 24	318	13 02	347
329	23 04	031	50 12	082	43 22	142	17 19	209	19 31	250	40 49	317	12 51	347
	◆Alpheratz		Diphda		◆ACHERNAR		Peacock		◆Nunki		ALTAIR		Enif	
330	23 30	030	51 03	082	43 54	142	58 32	205	49 19	263	40 14	316	49 58	354
331	23 56	030	51 55	081	44 25	143	58 10	206	48 27	263	39 37	315	49 51	352
332	24 22	029	52 46	080	44 57	143	57 47	206	47 36	263	39 00	314	49 44	351
333	24 46	028	53 37	080	45 28	143	57 24	207	46 44	262	38 23	313	49 35	349
334	25 10	027	54 28	079	45 59	143	57 00	208	45 53	262	37 44	312	49 25	348
335	25 33	026	55 19	078	46 30	143	56 35	209	45 01	262	37 06	311	49 13	346
336	25 56	025	56 10	077	47 01	144	56 10	209	44 10	261	36 26	310	49 00	345
337	26 18	024	57 00	076	47 32	144	55 44	210	43 19	261	35 46	309	48 46	343
338	26 39	023	57 51	076	48 03	144	55 18	210	42 27	260	35 06	309	48 30	342
339	26 59	023	58 41	075	48 33	144	54 52	211	41 36	260	34 25	308	48 14	341
340	27 19	022	59 31	074	49 03	145	54 25	212	40 45	260	33 44	307	47 56	339
341	27 37	021	60 21	073	49 33	145	53 57	212	39 54	259	33 02	306	47 37	338
342	27 55	020	61 10	072	50 03	145	53 30	213	39 03	259	32 20	305	47 16	336
343	28 13	019	62 00	071	50 32	146	53 02	213	38 12	259	31 37	304	46 55	335
344	28 29	018	62 49	070	51 01	146	52 33	213	37 21	258	30 54	304	46 32	334
	Alpheratz		◆Hamal		◆ACHERNAR		Peacock		Nunki		◆ALTAIR		Enif	
345	28 45	017	20 14	045	51 30	147	52 04	214	36 30	258	30 10	303	46 09	332
346	28 59	016	20 51	045	51 58	147	51 35	214	35 39	258	29 27	302	45 44	331
347	29 13	015	21 27	044	52 26	147	51 06	215	34 49	257	28 42	301	45 19	330
348	29 26	014	22 03	043	52 54	148	50 36	215	33 58	257	27 58	301	44 52	329
349	29 39	013	22 38	042	53 22	148	50 06	215	33 07	256	27 13	300	44 25	327
350	29 50	012	23 13	042	53 49	149	49 36	216	32 17	256	26 28	299	43 56	326
351	30 00	011	23 47	041	54 15	149	49 05	216	31 27	256	25 42	298	43 27	325
352	30 10	010	24 21	040	54 42	150	48 35	216	30 36	255	24 56	298	42 56	324
353	30 19	009	24 54	039	55 08	150	48 04	217	29 46	255	24 10	297	42 25	323
354	30 27	008	25 27	038	55 33	151	47 33	217	28 56	255	23 23	296	41 53	322
355	30 33	007	25 59	038	55 58	152	47 01	217	28 06	254	22 37	296	41 21	320
356	30 39	006	26 30	037	56 22	152	46 30	217	27 16	254	21 50	295	40 47	319
357	30 45	005	27 01	036	56 46	153	45 58	218	26 26	253	21 03	294	40 13	318
358	30 49	004	27 31	035	57 09	154	45 26	218	25 36	253	20 15	294	39 38	317
359	30 52	003	28 00	034	57 32	154	44 54	218	24 47	253	19 27	293	39 03	316

LHA ♈	◆Alpheratz Hc	Zn	Hamal Hc	Zn	◆RIGEL Hc	Zn	CANOPUS Hc	Zn	ACHERNAR Hc	Zn	◆Peacock Hc	Zn	Enif Hc	Zn
0	29 54	002	27 39	033	13 59	091	20 50	140	58 48	154	45 09	219	37 44	316
1	29 56	001	28 07	032	14 50	091	21 23	140	59 10	155	44 37	219	37 07	315
2	29 56	000	28 34	031	15 42	090	21 57	139	59 32	156	44 05	219	36 31	314
3	29 56	359	29 00	030	16 33	090	22 31	139	59 52	157	43 32	219	35 53	313
4	29 54	358	29 26	029	17 25	089	23 04	139	60 12	158	43 00	219	35 15	312
5	29 52	357	29 50	028	18 16	089	23 38	139	60 31	159	42 27	219	34 37	311
6	29 49	356	30 14	027	19 07	088	24 12	138	60 49	160	41 54	219	33 58	310
7	29 45	355	30 38	026	19 59	088	24 47	138	61 07	160	41 22	220	33 18	309
8	29 40	354	31 00	026	20 50	087	25 21	138	61 24	161	40 49	220	32 38	308
9	29 34	353	31 22	025	21 41	087	25 56	138	61 40	162	40 16	220	31 58	308
10	29 28	352	31 43	024	22 33	086	26 30	138	61 55	163	39 43	220	31 17	307
11	29 20	351	32 03	023	23 24	085	27 05	137	62 09	164	39 10	220	30 35	306
12	29 12	350	32 22	022	24 15	085	27 40	137	62 22	166	38 38	220	29 54	305
13	29 02	349	32 41	020	25 07	084	28 15	137	62 35	167	38 05	220	29 11	304
14	28 52	348	32 58	019	25 58	084	28 50	137	62 46	168	37 32	220	28 29	304
	◆Hamal		ALDEBARAN		RIGEL		◆CANOPUS		ACHERNAR		◆FOMALHAUT		Alpheratz	
15	33 15	018	19 45	055	26 49	083	29 26	137	62 57	169	63 35	265	28 41	347
16	33 31	017	20 27	055	27 40	083	30 01	136	63 06	170	62 44	265	28 29	346
17	33 46	016	21 09	054	28 31	082	30 36	136	63 15	171	61 53	264	28 16	345
18	34 00	015	21 50	053	29 22	082	31 12	136	63 22	172	61 02	264	28 03	344
19	34 13	014	22 31	053	30 13	081	31 48	136	63 28	173	60 11	264	27 48	343
20	34 25	013	23 12	052	31 03	080	32 23	136	63 34	175	59 19	263	27 33	342
21	34 36	012	23 52	051	31 54	080	32 59	136	63 38	176	58 28	263	27 17	341
22	34 46	011	24 32	050	32 45	079	33 35	136	63 41	177	57 37	263	27 00	340
23	34 55	010	25 11	050	33 35	079	34 11	136	63 43	178	56 46	262	26 42	340
24	35 04	009	25 50	049	34 25	078	34 47	135	63 44	180	55 55	262	26 24	339
25	35 11	008	26 29	048	35 16	077	35 23	135	63 44	181	55 05	262	26 05	338
26	35 17	006	27 07	047	36 06	077	36 00	135	63 43	182	54 14	261	25 45	337
27	35 22	005	27 44	046	36 56	076	36 36	135	63 41	183	53 23	261	25 24	336
28	35 27	004	28 21	046	37 45	075	37 12	135	63 37	184	52 32	261	25 03	335
29	35 30	003	28 57	045	38 35	075	37 48	135	63 33	186	51 41	260	24 41	334
	Hamal		ALDEBARAN		◆SIRIUS		CANOPUS		◆ACHERNAR		FOMALHAUT		◆Alpheratz	
30	35 32	002	29 33	044	24 21	096	38 25	135	63 27	187	50 51	260	24 18	333
31	35 33	001	30 09	043	25 12	095	39 01	135	63 20	188	50 00	260	23 55	332
32	35 34	000	30 43	042	26 03	095	39 37	135	63 13	189	49 10	259	23 30	332
33	35 33	359	31 18	041	26 54	094	40 14	135	63 04	190	48 19	259	23 06	331
34	35 31	357	31 51	040	27 46	094	40 50	135	62 54	191	47 29	259	22 40	330
35	35 28	356	32 24	039	28 37	093	41 26	135	62 44	193	46 38	258	22 14	329
36	35 24	355	32 56	038	29 28	093	42 03	135	62 32	194	45 48	258	21 47	328
37	35 20	354	33 28	037	30 20	092	42 39	135	62 19	195	44 58	258	21 20	327
38	35 14	353	33 59	036	31 11	092	43 15	135	62 06	196	44 08	257	20 52	327
39	35 07	352	34 29	035	32 03	091	43 52	135	61 51	197	43 18	257	20 23	326
40	34 59	351	34 58	034	32 54	091	44 28	135	61 36	198	42 27	257	19 54	325
41	34 50	350	35 27	033	33 45	090	45 04	135	61 20	199	41 37	256	19 25	324
42	34 41	349	35 55	032	34 37	090	45 40	135	61 03	200	40 48	256	18 54	324
43	34 30	347	36 22	031	35 28	089	46 16	136	60 45	201	39 58	256	18 24	323
44	34 18	346	36 49	030	36 20	089	46 52	136	60 27	202	39 08	255	17 52	322
	◆ALDEBARAN		BETELGEUSE		SIRIUS		◆CANOPUS		ACHERNAR		◆FOMALHAUT		Hamal	
45	37 14	029	33 13	055	37 11	088	47 28	136	60 07	202	38 18	255	34 06	345
46	37 39	028	33 55	054	38 03	087	48 04	136	59 47	203	37 29	255	33 52	344
47	38 03	027	34 37	053	38 54	087	48 39	136	59 27	204	36 39	254	33 38	343
48	38 25	026	35 18	052	39 45	086	49 15	136	59 05	205	35 50	254	33 22	342
49	38 47	025	35 58	052	40 37	086	49 50	137	58 43	206	35 00	254	33 06	341
50	39 09	024	36 38	051	41 28	085	50 26	137	58 21	206	34 11	253	32 49	340
51	39 29	022	37 18	050	42 19	085	51 01	137	57 57	207	33 22	253	32 31	339
52	39 48	021	37 57	049	43 10	084	51 36	137	57 34	208	32 33	252	32 12	338
53	40 06	020	38 35	048	44 01	083	52 10	138	57 09	209	31 44	252	31 52	337
54	40 23	019	39 13	047	44 52	083	52 45	138	56 45	209	30 55	252	31 31	336
55	40 39	018	39 50	046	45 43	082	53 19	138	56 19	210	30 06	251	31 10	335
56	40 54	016	40 27	045	46 34	082	53 53	139	55 53	210	29 17	251	30 48	334
57	41 08	015	41 03	044	47 25	081	54 27	139	55 27	211	28 29	251	30 25	333
58	41 21	014	41 38	043	48 16	080	55 01	139	55 01	211	27 40	250	30 01	332
59	41 33	013	42 12	042	49 07	080	55 34	140	54 34	212	26 52	250	29 37	331
	ALDEBARAN		◆BETELGEUSE		SIRIUS		◆CANOPUS		ACHERNAR		◆FOMALHAUT		Hamal	
60	41 44	011	42 46	040	49 57	079	56 07	140	54 06	212	26 04	250	29 11	330
61	41 54	010	43 19	039	50 47	078	56 40	141	53 38	213	25 16	249	28 45	329
62	42 02	009	43 51	038	51 38	077	57 12	141	53 10	213	24 28	249	28 19	328
63	42 10	008	44 23	037	52 28	077	57 44	142	52 42	214	23 40	248	27 51	327
64	42 16	006	44 53	036	53 18	076	58 15	142	52 13	214	22 52	248	27 23	327
65	42 21	005	45 23	035	54 08	075	58 47	143	51 44	215	22 04	248	26 55	326
66	42 25	004	45 52	033	54 57	074	59 17	144	51 14	215	21 17	247	26 25	325
67	42 28	002	46 19	032	55 46	073	59 48	144	50 45	215	20 29	247	25 55	324
68	42 30	001	46 46	031	56 36	072	60 17	145	50 15	216	19 42	247	25 25	323
69	42 30	000	47 12	030	57 25	072	60 47	146	49 45	216	18 55	246	24 54	322
70	42 29	359	47 37	028	58 13	071	61 15	147	49 14	216	18 08	246	24 22	321
71	42 27	357	48 01	027	59 02	070	61 43	147	48 44	217	17 21	245	23 49	321
72	42 24	356	48 23	025	59 50	069	62 11	148	48 13	217	16 34	245	23 16	320
73	42 20	355	48 45	024	60 37	067	62 38	149	47 42	217	15 48	245	22 43	319
74	42 15	353	49 05	023	61 25	066	63 04	150	47 11	217	15 02	244	22 09	318
	BETELGEUSE		◆PROCYON		Suhail		◆ACRUX		ACHERNAR		◆Diphda		ALDEBARAN	
75	49 24	021	37 32	053	40 18	123	18 27	154	46 40	218	30 58	267	42 08	352
76	49 42	020	38 13	053	41 01	123	18 50	153	46 08	218	30 07	266	42 01	351
77	49 59	018	38 53	052	41 44	123	19 13	153	45 37	218	29 15	266	41 52	350
78	50 15	017	39 33	051	42 28	122	19 36	153	45 05	218	28 24	265	41 42	348
79	50 29	015	40 13	050	43 11	122	20 00	153	44 34	218	27 33	265	41 31	347
80	50 42	014	40 52	049	43 55	122	20 23	152	44 02	218	26 42	264	41 19	346
81	50 53	012	41 30	048	44 38	122	20 47	152	43 30	218	25 50	264	41 06	345
82	51 03	011	42 08	047	45 22	122	21 11	152	42 58	219	24 59	263	40 52	343
83	51 12	009	42 45	046	46 05	122	21 36	152	42 25	219	24 08	263	40 36	342
84	51 20	008	43 21	044	46 49	122	22 00	152	41 53	219	23 17	262	40 20	341
85	51 26	006	43 57	043	47 32	122	22 25	151	41 21	219	22 26	262	40 03	340
86	51 30	004	44 32	042	48 16	122	22 50	151	40 49	219	21 35	261	39 44	339
87	51 33	003	45 06	041	48 59	122	23 14	151	40 16	219	20 45	261	39 25	337
88	51 35	001	45 39	040	49 43	122	23 40	151	39 44	219	19 54	260	39 05	336
89	51 36	000	46 12	039	50 27	122	24 05	151	39 12	219	19 03	260	38 43	335

LAT 31°S

LHA ♈	PROCYON Hc	Zn	REGULUS Hc	Zn	◆Suhail Hc	Zn	ACRUX Hc	Zn	◆ACHERNAR Hc	Zn	RIGEL Hc	Zn	◆BETELGEUSE Hc	Zn
90	46 44	037	16 38	064	51 10	122	24 30	150	38 39	219	64 50	333	51 34	358
91	47 14	036	17 24	064	51 54	122	24 56	150	38 07	219	64 26	330	51 32	356
92	47 44	035	18 10	063	52 37	122	25 21	150	37 34	219	63 59	328	51 28	355
93	48 13	034	18 56	062	53 21	122	25 47	150	37 02	219	63 32	326	51 23	353
94	48 41	032	19 42	062	54 04	122	26 13	150	36 29	219	63 03	324	51 16	352
95	49 08	031	20 27	061	54 47	123	26 39	150	35 57	219	62 32	323	51 08	350
96	49 34	030	21 12	060	55 31	123	27 05	149	35 25	219	62 00	321	50 58	348
97	49 59	028	21 56	060	56 14	123	27 31	149	34 52	219	61 27	319	50 47	347
98	50 23	027	22 41	059	56 57	123	27 57	149	34 20	219	60 53	317	50 35	345
99	50 45	025	23 24	058	57 40	123	28 24	149	33 48	219	60 18	316	50 21	344
100	51 07	024	24 08	058	58 23	124	28 50	149	33 15	219	59 41	314	50 06	342
101	51 27	022	24 51	057	59 06	124	29 17	149	32 43	219	59 04	313	49 50	341
102	51 46	021	25 34	056	59 48	124	29 44	149	32 11	219	58 26	311	49 33	339
103	52 04	019	26 17	055	60 31	124	30 10	149	31 39	219	57 47	310	49 14	338
104	52 20	018	26 59	055	61 13	125	30 37	149	31 07	218	57 07	309	48 54	337
	PROCYON		REGULUS		◆Gienah		ACRUX		◆ACHERNAR		RIGEL		◆BETELGEUSE	
105	52 35	016	27 41	054	18 13	100	31 04	148	30 35	218	56 26	307	48 33	335
106	52 48	015	28 22	053	19 03	099	31 31	148	30 03	218	55 45	306	48 11	334
107	53 01	013	29 03	052	19 54	099	31 58	148	29 31	218	55 03	305	47 48	332
108	53 11	011	29 44	052	20 45	098	32 25	148	29 00	218	54 20	303	47 23	331
109	53 21	010	30 24	051	21 36	098	32 52	148	28 28	218	53 37	302	46 58	330
110	53 29	008	31 03	050	22 27	098	33 19	148	27 57	218	52 53	301	46 32	328
111	53 35	006	31 42	049	23 18	097	33 46	148	27 25	217	52 09	300	46 04	327
112	53 40	005	32 21	048	24 09	097	34 13	148	26 54	217	51 24	299	45 36	326
113	53 43	003	32 59	047	25 00	096	34 41	148	26 23	217	50 39	298	45 07	325
114	53 45	001	33 36	046	25 51	096	35 08	148	25 52	217	49 54	297	44 36	324
115	53 46	000	34 13	045	26 42	095	35 35	148	25 21	217	49 08	296	44 05	322
116	53 45	358	34 50	044	27 34	095	36 02	148	24 51	217	48 21	295	43 34	321
117	53 42	356	35 25	044	28 25	094	36 29	148	24 20	216	47 35	294	43 01	320
118	53 38	355	36 00	043	29 16	094	36 56	148	23 50	216	46 47	293	42 28	319
119	53 32	353	36 35	042	30 08	093	37 23	148	23 19	216	46 00	292	41 53	318
	REGULUS		◆SPICA		ACRUX		◆CANOPUS		RIGEL		BETELGEUSE		◆PROCYON	
120	37 09	041	13 10	095	37 51	148	62 09	212	45 12	292	41 18	317	53 25	351
121	37 42	040	14 01	095	38 18	148	61 41	213	44 24	291	40 43	316	53 16	350
122	38 14	039	14 52	094	38 45	148	61 13	214	43 36	290	40 07	315	53 06	348
123	38 46	037	15 43	094	39 12	148	60 44	214	42 48	289	39 30	314	52 55	346
124	39 17	036	16 35	093	39 38	149	60 15	215	41 59	288	38 52	313	52 42	345
125	39 47	035	17 26	093	40 05	149	59 45	216	41 10	288	38 14	312	52 28	343
126	40 16	034	18 17	092	40 32	149	59 15	216	40 21	287	37 35	311	52 12	342
127	40 45	033	19 09	092	40 59	149	58 44	217	39 32	286	36 56	310	51 55	340
128	41 13	032	20 00	091	41 25	149	58 13	218	38 42	286	36 16	309	51 37	338
129	41 39	031	20 52	091	41 52	149	57 41	218	37 53	285	35 36	308	51 18	337
130	42 05	030	21 43	090	42 18	149	57 09	219	37 03	284	34 55	307	50 57	335
131	42 30	028	22 35	090	42 44	149	56 37	219	36 13	284	34 14	306	50 35	334
132	42 54	027	23 26	089	43 10	150	56 04	220	35 23	283	33 32	305	50 12	333
133	43 17	026	24 17	089	43 36	150	55 31	220	34 33	282	32 50	305	49 47	331
134	43 39	025	25 09	088	44 02	150	54 58	221	33 42	282	32 08	304	49 22	330
	◆REGULUS		SPICA		◆ACRUX		CANOPUS		◆RIGEL		BETELGEUSE		PROCYON	
135	44 00	023	26 00	088	44 28	150	54 24	221	32 52	281	31 25	303	48 56	328
136	44 20	022	26 52	087	44 53	150	53 51	221	32 02	280	30 41	302	48 28	327
137	44 39	021	27 43	086	45 18	151	53 16	222	31 11	280	29 57	301	48 00	326
138	44 57	020	28 34	086	45 44	151	52 42	222	30 20	279	29 13	301	47 30	324
139	45 13	018	29 26	085	46 09	151	52 08	222	29 29	279	28 29	300	47 00	323
140	45 29	017	30 17	085	46 33	151	51 33	223	28 38	278	27 44	299	46 29	322
141	45 43	016	31 08	084	46 58	152	50 58	223	27 47	277	26 59	298	45 56	321
142	45 57	014	31 59	084	47 22	152	50 23	223	26 56	277	26 14	298	45 23	320
143	46 09	013	32 50	083	47 46	152	49 48	223	26 05	276	25 28	297	44 50	318
144	46 19	011	33 41	083	48 10	153	49 12	224	25 14	276	24 42	296	44 15	317
145	46 29	010	34 32	082	48 33	153	48 37	224	24 23	275	23 56	296	43 40	316
146	46 37	009	35 23	081	48 56	153	48 01	224	23 32	275	23 09	295	43 04	315
147	46 44	007	36 14	081	49 19	154	47 25	224	22 40	274	22 22	294	42 27	314
148	46 50	006	37 05	080	49 42	154	46 49	224	21 49	274	21 35	294	41 50	313
149	46 55	004	37 55	079	50 04	155	46 13	224	20 58	273	20 48	293	41 12	312
	REGULUS		◆Denebola		SPICA		◆ACRUX		CANOPUS		◆SIRIUS		PROCYON	
150	46 58	003	37 26	034	38 46	079	50 26	155	45 37	225	43 34	276	40 33	311
151	47 00	001	37 55	033	39 36	078	50 48	155	45 01	225	42 43	276	39 54	310
152	47 00	000	38 22	032	40 26	078	51 09	156	44 25	225	41 52	275	39 14	309
153	47 00	359	38 49	031	41 17	077	51 30	156	43 49	225	41 00	274	38 34	308
154	46 58	357	39 14	029	42 07	076	51 50	157	43 13	225	40 09	274	37 53	307
155	46 55	356	39 39	028	42 56	075	52 11	157	42 36	225	39 18	273	37 12	306
156	46 50	354	40 03	027	43 46	075	52 30	158	42 00	225	38 26	273	36 30	305
157	46 45	353	40 26	026	44 36	074	52 50	158	41 24	225	37 35	272	35 48	304
158	46 38	351	40 48	025	45 25	073	53 08	159	40 47	225	36 44	272	35 05	304
159	46 29	350	41 10	024	46 14	072	53 27	159	40 11	225	35 52	271	34 22	303
160	46 20	349	41 30	022	47 03	072	53 45	160	39 34	225	35 01	271	33 39	302
161	46 09	347	41 49	021	47 52	071	54 02	161	38 58	225	34 09	270	32 55	301
162	45 57	346	42 07	020	48 40	070	54 19	161	38 22	225	33 18	270	32 10	300
163	45 44	345	42 24	019	49 28	069	54 35	162	37 45	225	32 26	269	31 26	299
164	45 30	343	42 40	017	50 16	068	54 51	162	37 09	225	31 35	269	30 41	299
	ARCTURUS		◆ANTARES		ACRUX		◆CANOPUS		SIRIUS		PROCYON		◆REGULUS	
165	21 17	050	19 25	110	55 06	163	36 33	225	30 44	268	29 56	298	45 14	342
166	21 57	049	20 13	109	55 21	164	35 57	225	29 52	268	29 10	297	44 58	340
167	22 35	048	21 02	109	55 35	164	35 21	225	29 01	267	28 24	296	44 40	339
168	23 13	048	21 50	109	55 48	165	34 44	225	28 10	267	27 38	296	44 21	338
169	23 51	047	22 39	108	56 01	166	34 08	224	27 18	266	26 52	295	44 01	337
170	24 28	046	23 28	108	56 14	167	33 32	224	26 27	266	26 05	294	43 40	335
171	25 05	045	24 17	107	56 25	167	32 56	224	25 36	265	25 18	294	43 18	334
172	25 41	044	25 06	107	56 36	168	32 21	224	24 44	265	24 31	293	42 55	333
173	26 17	044	25 56	107	56 47	169	31 45	224	23 53	264	23 43	292	42 31	332
174	26 52	043	26 45	106	56 56	170	31 09	224	23 02	264	22 56	292	42 07	330
175	27 27	042	27 34	106	57 05	170	30 34	224	22 11	263	22 08	291	41 41	329
176	28 01	041	28 24	105	57 13	171	29 58	224	21 20	263	21 20	290	41 14	328
177	28 34	040	29 14	105	57 21	172	29 23	223	20 29	262	20 31	290	40 46	327
178	29 07	039	30 03	105	57 28	173	28 47	223	19 38	262	19 43	289	40 18	326
179	29 39	038	30 53	104	57 34	174	28 12	223	18 47	261	18 54	289	39 49	325

LHA ♈	Hc	Zn	Hc	Zn	Hc	Zn	Hc	Zn	Hc	Zn	Hc	Zn	Hc	Zn
	◆ARCTURUS		ANTARES		◆RIGIL KENT.		ACRUX		CANOPUS		◆Alphard		REGULUS	
180	30 11	037	31 43	104	50 25	151	57 39	174	27 37	223	48 03	294	39 18	324
181	30 42	037	32 33	103	50 50	151	57 44	175	27 02	223	47 15	293	38 48	323
182	31 12	036	33 23	103	51 15	151	57 48	176	26 27	222	46 28	292	38 16	322
183	31 42	035	34 13	103	51 39	152	57 51	177	25 53	222	45 40	291	37 44	320
184	32 11	034	35 03	102	52 03	152	57 53	178	25 18	222	44 52	291	37 10	319
185	32 39	033	35 54	102	52 27	153	57 55	179	24 44	222	44 04	290	36 37	318
186	33 06	032	36 44	101	52 50	153	57 56	180	24 10	222	43 15	289	36 02	317
187	33 33	031	37 34	101	53 13	154	57 56	180	23 36	221	42 27	288	35 27	317
188	33 59	030	38 25	101	53 36	154	57 55	181	23 02	221	41 38	287	34 51	316
189	34 24	029	39 16	100	53 58	155	57 53	182	22 28	221	40 48	287	34 15	315
190	34 48	028	40 06	100	54 19	155	57 51	183	21 54	221	39 59	286	33 38	314
191	35 12	027	40 57	100	54 40	156	57 48	184	21 21	220	39 10	285	33 01	313
192	35 35	026	41 48	099	55 01	157	57 45	185	20 48	220	38 20	285	32 23	312
193	35 56	025	42 38	099	55 21	157	57 40	185	20 15	220	37 30	284	31 44	311
194	36 17	023	43 29	098	55 41	158	57 35	186	19 42	220	36 40	283	31 05	310
	ARCTURUS		◆ANTARES		RIGIL KENT.		◆ACRUX		Suhail		REGULUS		◆Denebola	
195	36 37	022	44 20	098	56 00	159	57 29	187	43 07	238	30 26	309	41 18	337
196	36 56	021	45 11	098	56 19	159	57 22	188	42 24	238	29 46	309	40 57	336
197	37 15	020	46 02	097	56 37	160	57 15	189	41 40	237	29 05	308	40 35	334
198	37 32	019	46 53	097	56 54	161	57 07	190	40 57	237	28 24	307	40 13	333
199	37 48	018	47 44	096	57 11	161	56 58	190	40 14	237	27 43	306	39 49	332
200	38 03	017	48 35	096	57 27	162	56 48	191	39 30	237	27 01	305	39 25	331
201	38 18	016	49 27	096	57 42	163	56 38	192	38 47	237	26 19	305	38 59	330
202	38 31	014	50 18	095	57 57	164	56 27	193	38 04	237	25 37	304	38 33	329
203	38 43	013	51 09	095	58 11	164	56 16	193	37 21	237	24 54	303	38 06	328
204	38 54	012	52 00	094	58 25	165	56 03	194	36 38	236	24 10	302	37 38	327
205	39 05	011	52 52	094	58 37	166	55 51	195	35 56	236	23 27	302	37 09	326
206	39 14	010	53 43	093	58 49	167	55 37	195	35 13	236	22 43	301	36 40	324
207	39 22	008	54 34	093	59 00	168	55 23	196	34 30	236	21 59	300	36 09	323
208	39 29	007	55 26	092	59 11	169	55 09	197	33 48	236	21 14	300	35 38	322
209	39 35	006	56 17	092	59 20	170	54 53	197	33 05	236	20 29	299	35 07	321
	◆ARCTURUS		Rasalhague		◆Nunki		RIGIL KENT.		ACRUX		◆Suhail		SPICA	
210	39 39	005	22 33	058	26 19	106	59 29	171	54 38	198	32 23	235	68 32	336
211	39 43	004	23 17	058	27 09	106	59 37	172	54 21	199	31 40	235	68 10	333
212	39 46	002	24 00	057	27 58	105	59 44	172	54 05	199	30 58	235	67 46	331
213	39 47	001	24 43	056	28 48	105	59 51	173	53 47	200	30 16	235	67 21	329
214	39 47	000	25 26	055	29 38	105	59 56	174	53 30	201	29 34	234	66 53	326
215	39 47	359	26 08	055	30 27	104	60 01	175	53 11	201	28 53	234	66 24	324
216	39 45	357	26 50	054	31 17	104	60 05	176	52 53	202	28 11	234	65 53	322
217	39 42	356	27 31	053	32 07	103	60 08	177	52 33	202	27 29	234	65 21	320
218	39 38	355	28 12	052	32 57	103	60 10	178	52 14	203	26 48	234	64 47	318
219	39 33	354	28 53	052	33 47	103	60 11	179	51 54	203	26 07	233	64 12	317
220	39 27	352	29 33	051	34 38	102	60 11	180	51 33	204	25 26	233	63 37	315
221	39 19	351	30 12	050	35 28	102	60 10	181	51 12	204	24 45	233	62 59	313
222	39 11	350	30 51	049	36 18	102	60 09	182	50 51	205	24 04	232	62 21	311
223	39 02	349	31 30	048	37 09	101	60 07	183	50 30	205	23 23	232	61 42	310
224	38 51	348	32 08	047	37 59	101	60 03	184	50 08	205	22 42	232	61 02	308
	Alphecca		Rasalhague		◆Nunki		Peacock		◆RIGIL KENT.		SPICA		◆ARCTURUS	
225	31 41	009	32 46	046	38 50	100	30 07	141	59 59	185	60 22	307	38 40	346
226	31 49	008	33 23	046	39 40	100	30 39	141	59 54	186	59 40	306	38 27	345
227	31 55	007	33 59	045	40 31	100	31 11	141	59 49	187	58 58	304	38 14	344
228	32 01	006	34 35	044	41 22	099	31 44	141	59 42	188	58 15	303	37 59	343
229	32 06	005	35 10	043	42 13	099	32 16	141	59 34	189	57 32	302	37 44	342
230	32 10	004	35 45	042	43 03	098	32 49	141	59 26	190	56 48	300	37 27	341
231	32 13	003	36 19	041	43 54	098	33 22	141	59 17	191	56 03	299	37 10	340
232	32 15	002	36 52	040	44 45	098	33 54	141	59 07	192	55 18	298	36 51	338
233	32 16	001	37 25	039	45 36	097	34 27	140	58 56	192	54 32	297	36 32	337
234	32 16	000	37 57	038	46 27	097	35 00	140	58 45	193	53 46	296	36 12	336
235	32 15	359	38 28	037	47 18	096	35 33	140	58 33	194	53 00	295	35 50	335
236	32 13	357	38 58	036	48 10	096	36 05	140	58 20	195	52 13	294	35 28	334
237	32 11	356	39 28	035	49 01	096	36 38	140	58 06	196	51 26	293	35 05	333
238	32 07	355	39 56	033	49 52	095	37 11	140	57 52	197	50 38	292	34 42	332
239	32 02	354	40 24	032	50 43	095	37 44	140	57 37	197	49 51	291	34 17	331
	Rasalhague		ALTAIR		◆Nunki		Peacock		◆RIGIL KENT.		SPICA		◆ARCTURUS	
240	40 52	031	21 58	064	51 34	094	38 17	140	57 21	198	49 03	290	33 52	330
241	41 18	030	22 44	063	52 26	094	38 50	140	57 05	199	48 14	290	33 26	329
242	41 43	029	23 30	063	53 17	093	39 23	140	56 48	200	47 26	289	32 59	328
243	42 07	028	24 16	062	54 08	093	39 56	140	56 30	200	46 37	288	32 31	327
244	42 31	026	25 01	061	55 00	092	40 28	140	56 12	201	45 48	287	32 03	326
245	42 53	025	25 46	061	55 51	092	41 01	140	55 53	202	44 59	286	31 33	325
246	43 15	024	26 31	060	56 43	092	41 34	140	55 34	202	44 09	286	31 04	324
247	43 35	023	27 15	059	57 34	091	42 07	141	55 14	203	43 20	285	30 33	323
248	43 54	021	27 59	058	58 25	091	42 39	141	54 54	204	42 30	284	30 02	322
249	44 13	020	28 43	058	59 17	090	43 12	141	54 33	204	41 40	283	29 30	321
250	44 30	019	29 26	057	60 08	090	43 44	141	54 11	205	40 50	283	28 58	320
251	44 46	018	30 09	056	61 00	089	44 17	141	53 50	205	40 00	282	28 25	320
252	45 01	016	30 52	055	61 51	088	44 49	141	53 27	206	39 09	281	27 51	319
253	45 15	015	31 34	055	62 42	088	45 21	141	53 05	206	38 19	281	27 17	318
254	45 27	014	32 15	054	63 34	087	45 54	141	52 42	207	37 28	280	26 42	317
	VEGA		◆ALTAIR		FOMALHAUT		◆Peacock		RIGIL KENT.		◆SPICA		ARCTURUS	
255	16 41	019	32 57	053	15 16	116	46 26	142	52 18	207	36 37	280	26 07	316
256	16 57	019	33 37	052	16 02	115	46 57	142	51 54	208	35 47	279	25 31	315
257	17 14	018	34 18	051	16 49	115	47 29	142	51 30	208	34 56	278	24 55	315
258	17 29	017	34 57	050	17 36	115	48 01	142	51 06	209	34 05	278	24 18	314
259	17 44	016	35 37	049	18 23	114	48 32	143	50 41	209	33 14	277	23 40	313
260	17 58	016	36 16	048	19 10	114	49 03	143	50 16	209	32 23	277	23 02	312
261	18 12	015	36 54	047	19 57	113	49 34	143	49 50	210	31 32	276	22 24	312
262	18 25	014	37 31	047	20 44	113	50 05	143	49 24	210	30 41	275	21 45	311
263	18 37	013	38 08	046	21 31	113	50 36	144	48 58	211	29 49	275	21 06	310
264	18 48	012	38 45	045	22 19	112	51 06	144	48 32	211	28 58	274	20 27	309
265	18 59	012	39 21	044	23 07	112	51 36	144	48 06	211	28 07	274	19 47	309
266	19 09	011	39 56	043	23 55	111	52 06	145	47 39	211	27 15	273	19 06	308
267	19 18	010	40 30	041	24 42	111	52 36	145	47 12	212	26 24	273	18 26	307
268	19 27	009	41 04	040	25 31	111	53 05	145	46 45	212	25 33	272	17 44	307
269	19 35	008	41 37	039	26 19	110	53 34	146	46 17	212	24 41	272	17 03	306

LAT 31°S

LHA ♈	Hc	Zn	Hc	Zn	Hc	Zn	Hc	Zn	Hc	Zn	Hc	Zn	Hc	Zn
	VEGA		ALTAIR		◆FOMALHAUT		ACHERNAR		◆RIGIL KENT.		ANTARES		◆Rasalhague	
270	19 42	008	42 09	038	27 07	110	14 00	149	45 50	213	69 35	277	46 01	351
271	19 48	007	42 40	037	27 55	110	14 26	149	45 22	213	68 44	276	45 52	350
272	19 54	006	43 11	036	28 44	109	14 53	149	44 54	213	67 53	276	45 43	348
273	19 59	005	43 41	035	29 33	109	15 19	149	44 26	213	67 02	275	45 32	347
274	20 03	004	44 09	034	30 21	108	15 46	148	43 58	213	66 11	274	45 19	346
275	20 07	003	44 37	032	31 10	108	16 13	148	43 30	213	65 19	274	45 06	344
276	20 09	003	45 04	031	31 59	108	16 41	148	43 01	214	64 28	273	44 51	343
277	20 11	002	45 30	030	32 48	107	17 09	147	42 33	214	63 37	272	44 36	342
278	20 13	001	45 55	029	33 37	107	17 36	147	42 04	214	62 45	272	44 19	340
279	20 13	000	46 19	027	34 26	107	18 05	147	41 35	214	61 54	271	44 01	339
280	20 13	359	46 42	026	35 16	106	18 33	146	41 06	214	61 02	271	43 42	338
281	20 12	358	47 04	025	36 05	106	19 01	146	40 38	214	60 11	270	43 22	336
282	20 10	358	47 25	023	36 55	106	19 30	146	40 09	214	59 20	270	43 01	335
283	20 08	357	47 45	022	37 44	105	19 59	146	39 39	214	58 28	269	42 39	334
284	20 05	356	48 03	020	38 34	105	20 28	145	39 10	214	57 37	269	42 16	333
	ALTAIR		Enif		◆FOMALHAUT		ACHERNAR		◆RIGIL KENT.		ANTARES		◆Rasalhague	
285	48 21	019	33 20	051	39 23	105	20 57	145	38 41	215	56 45	268	41 52	332
286	48 37	018	33 59	050	40 13	104	21 27	145	38 12	215	55 54	268	41 27	330
287	48 52	016	34 38	049	41 03	104	21 57	145	37 43	215	55 03	267	41 01	329
288	49 05	015	35 17	048	41 53	104	22 26	144	37 14	215	54 11	267	40 34	328
289	49 18	013	35 54	047	42 43	103	22 56	144	36 44	215	53 20	266	40 06	327
290	49 29	012	36 32	046	43 33	103	23 26	144	36 15	215	52 29	266	39 38	326
291	49 38	010	37 09	045	44 23	103	23 57	144	35 46	215	51 37	266	39 09	325
292	49 47	009	37 45	044	45 14	102	24 27	144	35 17	215	50 46	265	38 38	324
293	49 54	007	38 20	043	46 04	102	24 58	143	34 47	215	49 55	265	38 08	323
294	50 00	006	38 55	042	46 54	102	25 29	143	34 18	215	49 04	264	37 36	322
295	50 04	004	39 29	041	47 45	101	25 59	143	33 49	215	48 12	264	37 04	321
296	50 07	003	40 03	040	48 35	101	26 30	143	33 20	215	47 21	263	36 31	320
297	50 08	001	40 35	039	49 25	101	27 02	143	32 51	214	46 30	263	35 57	319
298	50 09	359	41 07	038	50 16	100	27 33	143	32 22	214	45 39	263	35 22	318
299	50 07	358	41 38	037	51 07	100	28 04	142	31 53	214	44 48	262	34 47	317
	Enif		◆FOMALHAUT		ACHERNAR		◆RIGIL KENT.		ANTARES		Rasalhague		◆ALTAIR	
300	42 09	036	51 57	100	28 36	142	31 24	214	43 57	262	34 12	316	50 05	356
301	42 38	034	52 48	099	29 07	142	30 55	214	43 06	261	33 36	315	50 01	355
302	43 07	033	53 39	099	29 39	142	30 26	214	42 16	261	32 59	314	49 56	353
303	43 35	032	54 30	099	30 11	142	29 57	214	41 25	261	32 21	313	49 49	352
304	44 02	031	55 20	098	30 43	142	29 28	214	40 34	260	31 43	312	49 41	350
305	44 27	030	56 11	098	31 14	142	29 00	214	39 43	260	31 05	311	49 31	349
306	44 52	028	57 02	098	31 46	141	28 31	214	38 53	259	30 26	310	49 21	347
307	45 16	027	57 53	097	32 19	141	28 03	213	38 02	259	29 47	310	49 09	346
308	45 39	026	58 44	097	32 51	141	27 34	213	37 12	259	29 07	309	48 55	344
309	46 01	024	59 35	097	33 23	141	27 06	213	36 21	258	28 26	308	48 41	343
310	46 22	023	60 26	096	33 55	141	26 38	213	35 31	258	27 45	307	48 25	341
311	46 42	022	61 18	096	34 27	141	26 10	213	34 41	258	27 04	306	48 08	340
312	47 00	020	62 09	096	35 00	141	25 42	213	33 51	257	26 23	306	47 50	339
313	47 17	019	63 00	095	35 32	141	25 14	213	33 01	257	25 41	305	47 31	337
314	47 34	018	63 51	095	36 05	141	24 47	212	32 10	256	24 58	304	47 10	336
	Alpheratz		◆Diphda		ACHERNAR		◆RIGIL KENT.		ANTARES		Rasalhague		◆ALTAIR	
315	15 06	041	38 06	089	36 37	141	24 19	212	31 21	256	24 15	303	46 48	334
316	15 40	041	38 57	089	37 09	141	23 52	212	30 31	256	23 32	303	46 26	333
317	16 13	040	39 49	088	37 42	141	23 25	212	29 41	255	22 49	302	46 02	332
318	16 46	039	40 40	088	38 14	141	22 58	212	28 51	255	22 05	301	45 37	331
319	17 18	039	41 32	087	38 47	141	22 31	211	28 02	254	21 21	300	45 11	329
320	17 50	038	42 23	087	39 19	141	22 04	211	27 12	254	20 36	300	44 45	328
321	18 22	037	43 14	086	39 52	141	21 38	211	26 23	254	19 51	299	44 17	327
322	18 52	036	44 06	085	40 24	141	21 11	211	25 33	253	19 06	298	43 48	326
323	19 23	036	44 57	085	40 56	141	20 45	211	24 44	253	18 21	298	43 19	324
324	19 52	035	45 48	084	41 29	141	20 19	210	23 55	252	17 35	297	42 49	323
325	20 22	034	46 39	084	42 01	141	19 53	210	23 06	252	16 49	297	42 17	322
326	20 50	033	47 30	083	42 33	141	19 28	210	22 17	252	16 03	296	41 45	321
327	21 18	033	48 21	082	43 05	141	19 02	210	21 29	251	15 17	295	41 13	320
328	21 46	032	49 12	082	43 37	142	18 37	209	20 40	251	14 30	295	40 39	319
329	22 12	031	50 03	081	44 09	142	18 12	209	19 51	250	13 43	294	40 05	318
	◆Alpheratz		Diphda		◆ACHERNAR		Peacock		◆Nunki		ALTAIR		Enif	
330	22 39	030	50 54	080	44 41	142	59 26	206	49 25	265	39 30	317	48 58	354
331	23 04	029	51 44	080	45 13	142	59 04	206	48 34	264	38 55	316	48 52	352
332	23 29	028	52 35	079	45 44	142	58 41	207	47 43	264	38 18	315	48 45	351
333	23 53	028	53 25	078	46 16	142	58 17	208	46 52	263	37 41	314	48 36	350
334	24 17	027	54 16	077	46 47	143	57 52	209	46 01	263	37 04	313	48 26	348
335	24 39	026	55 06	077	47 18	143	57 28	209	45 10	263	36 26	312	48 15	347
336	25 02	025	55 56	076	47 50	143	57 02	210	44 19	262	35 47	311	48 02	345
337	25 23	024	56 46	075	48 20	143	56 36	211	43 28	262	35 08	310	47 48	344
338	25 44	023	57 35	074	48 51	143	56 10	211	42 37	261	34 28	309	47 33	342
339	26 04	022	58 25	073	49 22	144	55 43	212	41 46	261	33 48	308	47 17	341
340	26 23	021	59 14	072	49 52	144	55 16	212	40 55	261	33 08	307	47 00	340
341	26 41	021	60 03	071	50 22	144	54 48	213	40 05	260	32 27	306	46 41	338
342	26 59	020	60 51	070	50 52	145	54 20	213	39 14	260	31 45	306	46 21	337
343	27 16	019	61 39	069	51 21	145	53 52	214	38 23	259	31 03	305	46 00	335
344	27 32	018	62 27	068	51 51	145	53 23	214	37 33	259	30 21	304	45 39	334
	Alpheratz		◆Hamal		◆ACHERNAR		Peacock		◆Nunki		ALTAIR		Enif	
345	27 47	017	19 32	045	52 20	146	52 54	215	36 42	259	29 38	303	45 16	333
346	28 02	016	20 08	044	52 48	146	52 25	215	35 52	258	28 55	302	44 52	332
347	28 15	015	20 44	044	53 17	147	51 55	215	35 02	258	28 11	302	44 27	330
348	28 28	014	21 19	043	53 45	147	51 25	216	34 11	257	27 27	301	44 01	329
349	28 40	013	21 54	042	54 13	148	50 55	216	33 21	257	26 43	300	43 34	328
350	28 51	012	22 28	041	54 40	148	50 24	216	32 31	257	25 58	299	43 06	327
351	29 02	011	23 02	041	55 07	149	49 54	217	31 41	256	25 13	299	42 37	326
352	29 11	010	23 35	040	55 33	149	49 23	217	30 51	256	24 28	298	42 08	324
353	29 19	009	24 07	039	56 00	150	48 52	217	30 01	256	23 42	297	41 38	323
354	29 27	008	24 39	038	56 25	150	48 20	218	29 12	255	22 57	297	41 06	322
355	29 34	007	25 11	037	56 50	151	47 49	218	28 22	255	22 11	296	40 34	321
356	29 40	006	25 42	036	57 15	152	47 17	218	27 32	254	21 24	295	40 02	320
357	29 45	005	26 13	036	57 39	152	46 46	218	26 43	254	20 38	295	39 28	319
358	29 49	004	26 42	035	58 03	153	46 14	218	25 53	254	19 51	294	38 54	318
359	29 52	003	27 11	034	58 26	154	45 42	219	25 04	253	19 04	293	38 19	317

LHA	Hc	Zn	Hc	Zn	Hc	Zn	Hc	Zn	Hc	Zn	Hc	Zn	Hc	Zn
♈	◆Alpheratz		Hamal		◆RIGEL		CANOPUS		ACHERNAR		◆Peacock		Enif	
0	28 54	002	26 49	033	14 00	091	21 36	140	59 42	154	45 56	219	37 00	316
1	28 56	001	27 16	032	14 51	090	22 09	139	60 05	155	45 24	220	36 25	315
2	28 56	000	27 42	031	15 42	090	22 42	139	60 26	155	44 51	220	35 49	314
3	28 56	359	28 08	030	16 33	089	23 16	139	60 47	156	44 19	220	35 12	313
4	28 55	358	28 33	029	17 24	089	23 49	139	61 07	157	43 46	220	34 35	313
5	28 52	357	28 58	028	18 14	088	24 23	138	61 27	158	43 13	220	33 57	312
6	28 49	356	29 21	027	19 05	088	24 57	138	61 46	159	42 41	220	33 19	311
7	28 45	355	29 44	026	19 56	087	25 31	138	62 03	160	42 08	220	32 40	310
8	28 40	354	30 06	025	20 47	087	26 05	138	62 21	161	41 35	220	32 01	309
9	28 35	353	30 27	024	21 38	086	26 40	137	62 37	162	41 02	220	31 21	308
10	28 28	352	30 48	023	22 28	086	27 14	137	62 52	163	40 29	220	30 41	307
11	28 21	351	31 08	022	23 19	085	27 49	137	63 07	164	39 56	220	30 00	306
12	28 12	350	31 27	021	24 10	084	28 24	137	63 20	165	39 23	220	29 19	306
13	28 03	349	31 45	020	25 00	084	28 59	137	63 33	166	38 51	220	28 37	305
14	27 53	348	32 02	019	25 51	083	29 34	136	63 45	167	38 18	220	27 55	304
	◆Hamal		ALDEBARAN		RIGEL		◆CANOPUS		ACHERNAR		◆FOMALHAUT		Alpheratz	
15	32 18	018	19 11	055	26 42	083	30 09	136	63 56	168	63 39	267	27 42	347
16	32 34	017	19 52	054	27 32	082	30 44	136	64 05	170	62 49	267	27 31	346
17	32 48	016	20 33	054	28 22	082	31 20	136	64 14	171	61 58	266	27 18	345
18	33 02	015	21 14	053	29 13	081	31 55	136	64 22	172	61 07	266	27 05	344
19	33 15	014	21 55	052	30 03	080	32 31	136	64 28	173	60 16	265	26 51	343
20	33 26	013	22 35	051	30 53	080	33 06	135	64 34	174	59 26	265	26 36	342
21	33 37	012	23 14	051	31 43	079	33 42	135	64 38	176	58 35	264	26 20	342
22	33 47	011	23 53	050	32 33	079	34 18	135	64 41	177	57 44	264	26 04	341
23	33 56	010	24 32	049	33 23	078	34 54	135	64 43	178	56 54	264	25 46	340
24	34 04	009	25 11	048	34 12	077	35 30	135	64 44	180	56 03	263	25 28	339
25	34 11	007	25 48	048	35 02	077	36 06	135	64 44	181	55 13	263	25 09	338
26	34 18	006	26 26	047	35 51	076	36 42	135	64 43	182	54 22	263	24 50	337
27	34 23	005	27 03	046	36 41	075	37 18	135	64 40	183	53 32	262	24 30	336
28	34 27	004	27 39	045	37 30	074	37 54	135	64 37	185	52 41	262	24 09	335
29	34 30	003	28 15	044	38 19	074	38 31	135	64 32	186	51 51	261	23 47	334
	Hamal		ALDEBARAN		◆SIRIUS		CANOPUS		◆ACHERNAR		FOMALHAUT		◆Alpheratz	
30	34 32	002	28 50	043	24 26	095	39 07	134	64 27	187	51 01	261	23 24	333
31	34 33	001	29 25	043	25 17	095	39 43	134	64 20	188	50 11	261	23 01	333
32	34 34	000	29 59	042	26 08	094	40 20	134	64 12	189	49 20	260	22 38	332
33	34 33	359	30 32	041	26 58	094	40 56	134	64 03	191	48 30	260	22 13	331
34	34 31	357	31 05	040	27 49	093	41 32	134	63 53	192	47 40	260	21 48	330
35	34 28	356	31 37	039	28 40	093	42 09	134	63 42	193	46 50	259	21 23	329
36	34 25	355	32 09	038	29 31	092	42 45	134	63 30	194	46 00	259	20 56	328
37	34 20	354	32 40	037	30 22	092	43 21	134	63 17	195	45 10	259	20 29	328
38	34 14	353	33 10	036	31 13	091	43 58	134	63 04	196	44 20	258	20 02	327
39	34 08	352	33 40	035	32 03	090	44 34	135	62 49	197	43 31	258	19 34	326
40	34 00	351	34 09	034	32 54	090	45 10	135	62 33	198	42 41	257	19 05	325
41	33 51	350	34 37	033	33 45	089	45 46	135	62 17	199	41 51	257	18 36	325
42	33 42	349	35 04	032	34 36	089	46 23	135	61 59	200	41 02	257	18 06	324
43	33 31	348	35 31	031	35 27	088	46 59	135	61 41	201	40 12	256	17 36	323
44	33 20	346	35 57	030	36 18	088	47 35	135	61 22	202	39 23	256	17 05	322
	◆ALDEBARAN		BETELGEUSE		SIRIUS		◆CANOPUS		ACHERNAR		◆FOMALHAUT		Hamal	
45	36 22	029	32 39	054	37 09	087	48 11	135	61 03	203	38 34	256	33 08	345
46	36 46	028	33 20	054	37 59	087	48 47	135	60 42	204	37 44	255	32 54	344
47	37 09	027	34 01	053	38 50	086	49 22	135	60 21	205	36 55	255	32 40	343
48	37 31	026	34 41	052	39 41	086	49 58	136	59 59	206	36 06	255	32 25	342
49	37 53	024	35 21	051	40 32	085	50 34	136	59 37	206	35 17	254	32 09	341
50	38 13	023	36 00	050	41 22	084	51 09	136	59 14	207	34 28	254	31 52	340
51	38 33	022	36 39	049	42 13	084	51 44	136	58 51	208	33 39	253	31 35	339
52	38 52	021	37 17	048	43 04	083	52 19	136	58 27	209	32 51	253	31 16	338
53	39 10	020	37 55	047	43 54	082	52 54	137	58 02	209	32 02	253	30 57	337
54	39 26	019	38 32	046	44 44	082	53 29	137	57 37	210	31 13	252	30 37	336
55	39 42	017	39 08	045	45 35	081	54 04	137	57 11	211	30 25	252	30 16	335
56	39 57	016	39 44	044	46 25	081	54 38	138	56 45	211	29 37	252	29 54	334
57	40 11	015	40 19	043	47 15	080	55 12	138	56 18	212	28 48	251	29 31	333
58	40 23	014	40 53	042	48 05	079	55 46	138	55 52	212	28 00	251	29 08	332
59	40 35	013	41 27	041	48 55	078	56 20	139	55 24	213	27 12	250	28 44	331
	ALDEBARAN		◆BETELGEUSE		SIRIUS		◆CANOPUS		ACHERNAR		◆FOMALHAUT		Hamal	
60	40 45	011	42 00	040	49 45	078	56 53	139	54 56	213	26 24	250	28 19	330
61	40 55	010	42 32	039	50 34	077	57 26	140	54 28	214	25 37	250	27 54	329
62	41 03	009	43 04	038	51 24	076	57 58	140	54 00	214	24 49	249	27 28	329
63	41 10	008	43 35	036	52 13	075	58 31	141	53 31	215	24 01	249	27 01	328
64	41 16	006	44 04	035	53 02	075	59 03	141	53 02	215	23 14	248	26 33	327
65	41 21	005	44 33	034	53 51	074	59 34	142	52 33	215	22 27	248	26 05	326
66	41 25	004	45 01	033	54 40	073	60 05	143	52 03	216	21 40	248	25 36	325
67	41 28	002	45 28	032	55 29	072	60 36	143	51 33	216	20 53	247	25 07	324
68	41 30	001	45 55	030	56 17	071	61 06	144	51 03	216	20 06	247	24 37	323
69	41 30	000	46 20	029	57 05	070	61 36	145	50 33	217	19 19	246	24 06	323
70	41 29	359	46 44	028	57 52	069	62 05	145	50 03	217	18 32	246	23 35	322
71	41 28	357	47 07	026	58 40	068	62 34	146	49 32	217	17 46	246	23 03	321
72	41 25	356	47 29	025	59 27	067	63 01	147	49 01	218	17 00	245	22 31	320
73	41 21	355	47 50	024	60 13	066	63 29	148	48 30	218	16 14	245	21 58	319
74	41 15	354	48 10	022	61 00	065	63 55	149	47 59	218	15 28	244	21 24	318
	BETELGEUSE		◆PROCYON		Suhail		◆ACRUX		ACHERNAR		◆Diphda		ALDEBARAN	
75	48 28	021	36 56	053	40 50	122	19 21	153	47 27	218	31 01	267	41 09	352
76	48 46	019	37 36	052	41 33	122	19 43	153	46 56	218	30 10	267	41 02	351
77	49 02	018	38 16	051	42 16	122	20 06	153	46 24	219	29 20	266	40 53	350
78	49 17	016	38 55	050	43 00	122	20 30	153	45 52	219	28 29	266	40 43	348
79	49 31	015	39 34	049	43 43	122	20 53	152	45 20	219	27 38	265	40 33	347
80	49 43	013	40 12	048	44 26	121	21 17	152	44 48	219	26 47	265	40 21	346
81	49 55	012	40 49	047	45 10	121	21 40	152	44 16	219	25 57	264	40 08	345
82	50 04	010	41 26	046	45 53	121	22 04	152	43 44	219	25 06	264	39 54	344
83	50 13	009	42 03	045	46 37	121	22 29	152	43 12	219	24 16	263	39 39	342
84	50 20	007	42 38	044	47 20	121	22 53	151	42 40	219	23 25	263	39 23	341
85	50 26	006	43 13	043	48 04	121	23 17	151	42 08	219	22 35	262	39 06	340
86	50 30	004	43 47	042	48 47	121	23 42	151	41 35	220	21 44	262	38 48	339
87	50 34	003	44 20	040	49 31	121	24 07	151	41 03	220	20 54	261	38 30	338
88	50 35	001	44 53	039	50 14	121	24 32	151	40 30	220	20 04	261	38 10	336
89	50 36	000	45 25	038	50 58	121	24 57	150	39 58	220	19 13	260	37 49	335

LHA	Hc	Zn	Hc	Zn	Hc	Zn	Hc	Zn	Hc	Zn	Hc	Zn	Hc	Zn
♈	PROCYON		REGULUS		◆Suhail		ACRUX		◆ACHERNAR		RIGEL		◆BETELGEUSE	
90	45 56	037	16 12	064	51 42	121	25 22	150	39 26	220	63 57	333	50 34	358
91	46 26	036	16 58	063	52 25	121	25 48	150	38 53	220	63 33	331	50 32	356
92	46 55	034	17 43	063	53 09	121	26 13	150	38 21	220	63 08	329	50 28	355
93	47 23	033	18 28	062	53 52	121	26 39	150	37 48	220	62 41	327	50 23	353
94	47 51	032	19 13	061	54 36	121	27 05	149	37 16	220	62 13	326	50 16	352
95	48 17	030	19 58	061	55 19	121	27 30	149	36 43	220	61 44	324	50 08	350
96	48 42	029	20 42	060	56 03	122	27 57	149	36 11	219	61 13	322	49 59	349
97	49 06	028	21 26	059	56 46	122	28 23	149	35 39	219	60 41	320	49 49	347
98	49 29	026	22 10	059	57 29	122	28 49	149	35 06	219	60 08	319	49 37	346
99	49 51	025	22 53	058	58 12	122	29 15	149	34 34	219	59 34	317	49 23	344
100	50 12	023	23 36	057	58 55	122	29 42	149	34 02	219	58 59	315	49 09	343
101	50 31	022	24 19	057	59 38	122	30 08	149	33 30	219	58 23	314	48 53	341
102	50 50	020	25 01	056	60 21	123	30 35	148	32 58	219	57 46	312	48 36	340
103	51 07	019	25 43	055	61 04	123	31 01	148	32 26	219	57 08	311	48 18	338
104	51 23	017	26 24	054	61 47	123	31 28	148	31 54	219	56 29	310	47 59	337
	PROCYON		REGULUS		◆Gienah		ACRUX		◆ACHERNAR		RIGEL		◆BETELGEUSE	
105	51 37	016	27 05	054	18 23	100	31 55	148	31 22	219	55 49	308	47 39	336
106	51 50	014	27 46	053	19 13	099	32 22	148	30 50	219	55 09	307	47 17	334
107	52 02	013	28 26	052	20 03	099	32 49	148	30 19	218	54 28	306	46 54	333
108	52 13	011	29 06	051	20 54	098	33 16	148	29 47	218	53 47	305	46 31	332
109	52 22	009	29 45	050	21 44	098	33 43	148	29 16	218	53 04	303	46 06	330
110	52 29	008	30 24	049	22 35	097	34 10	148	28 44	218	52 22	302	45 40	329
111	52 35	006	31 03	049	23 25	097	34 37	148	28 13	218	51 38	301	45 14	328
112	52 40	005	31 41	048	24 16	096	35 04	148	27 42	218	50 55	300	44 46	327
113	52 43	003	32 18	047	25 06	096	35 31	148	27 11	217	50 11	299	44 17	325
114	52 45	001	32 55	046	25 57	095	35 59	148	26 40	217	49 26	298	43 48	324
115	52 46	000	33 31	045	26 48	095	36 26	148	26 09	217	48 41	297	43 18	323
116	52 45	358	34 07	044	27 38	094	36 53	148	25 39	217	47 55	296	42 47	322
117	52 42	356	34 42	043	28 29	094	37 20	148	25 08	217	47 09	295	42 15	321
118	52 38	355	35 16	042	29 20	093	37 47	148	24 38	216	46 23	294	41 42	319
119	52 33	353	35 50	041	30 11	093	38 14	148	24 08	216	45 37	293	41 09	318
	REGULUS		◆SPICA		ACRUX		◆CANOPUS		RIGEL		BETELGEUSE		◆PROCYON	
120	36 23	040	13 15	095	38 41	148	62 59	213	44 50	293	40 35	317	52 26	351
121	36 55	039	14 06	094	39 09	148	62 31	214	44 03	292	40 00	316	52 17	350
122	37 27	038	14 56	094	39 36	148	62 03	215	43 15	291	39 24	315	52 08	348
123	37 58	037	15 47	093	40 03	148	61 34	215	42 28	290	38 48	314	51 57	347
124	38 28	036	16 38	093	40 29	148	61 04	216	41 40	289	38 11	313	51 44	345
125	38 58	035	17 29	092	40 56	148	60 34	217	40 52	289	37 34	312	51 30	343
126	39 27	034	18 20	092	41 23	148	60 03	217	40 03	288	36 56	311	51 15	342
127	39 54	033	19 10	091	41 50	148	59 32	218	39 15	287	36 17	310	50 59	340
128	40 21	032	20 01	091	42 16	149	59 00	219	38 26	286	35 38	309	50 41	339
129	40 48	030	20 52	090	42 43	149	58 28	219	37 37	286	34 59	309	50 22	337
130	41 13	029	21 43	090	43 09	149	57 56	220	36 48	285	34 19	308	50 02	336
131	41 37	028	22 34	089	43 36	149	57 23	220	35 59	284	33 38	307	49 41	335
132	42 01	027	23 25	089	44 02	149	56 50	221	35 09	284	32 57	306	49 18	333
133	42 23	026	24 16	088	44 28	149	56 17	221	34 20	283	32 16	305	48 55	332
134	42 45	024	25 07	088	44 54	149	55 43	222	33 30	282	31 34	304	48 30	330
	◆REGULUS		SPICA		◆ACRUX		CANOPUS		◆RIGEL		BETELGEUSE		PROCYON	
135	43 05	023	25 57	087	45 20	150	55 09	222	32 40	282	30 52	303	48 04	329
136	43 25	022	26 48	087	45 45	150	54 35	222	31 50	281	30 09	303	47 38	328
137	43 43	021	27 39	086	46 11	150	54 01	223	31 00	280	29 26	302	47 10	326
138	44 00	019	28 30	085	46 36	150	53 26	223	30 10	280	28 43	301	46 41	325
139	44 16	018	29 20	085	47 01	151	52 52	223	29 20	279	27 59	300	46 12	324
140	44 32	017	30 11	084	47 26	151	52 17	224	28 30	278	27 15	300	45 41	323
141	44 46	015	31 02	084	47 50	151	51 42	224	27 39	278	26 30	299	45 10	321
142	44 58	014	31 52	083	48 15	151	51 06	224	26 49	277	25 45	298	44 38	320
143	45 10	013	32 43	082	48 39	152	50 31	224	25 59	277	25 00	297	44 05	319
144	45 21	011	33 33	082	49 03	152	49 55	224	25 08	276	24 15	297	43 31	318
145	45 30	010	34 23	081	49 27	152	49 20	225	24 17	276	23 29	296	42 56	317
146	45 38	008	35 14	081	49 50	153	48 44	225	23 27	275	22 44	295	42 21	316
147	45 45	007	36 04	080	50 13	153	48 08	225	22 36	274	21 57	295	41 45	315
148	45 50	006	36 54	079	50 36	154	47 32	225	21 45	274	21 11	294	41 09	314
149	45 55	004	37 44	079	50 58	154	46 56	225	20 54	273	20 24	293	40 31	313
	REGULUS		◆Denebola		SPICA		◆ACRUX		CANOPUS		◆SIRIUS		PROCYON	
150	45 58	003	36 36	033	38 34	078	51 20	154	46 20	225	43 27	277	39 54	312
151	46 00	001	37 04	032	39 23	077	51 42	155	45 44	225	42 36	277	39 15	311
152	46 00	000	37 31	031	40 13	077	52 04	155	45 07	225	41 46	276	38 36	310
153	46 00	359	37 57	030	41 02	076	52 25	156	44 31	225	40 55	275	37 57	309
154	45 58	357	38 22	029	41 52	075	52 46	156	43 55	226	40 05	275	37 17	308
155	45 55	356	38 46	028	42 41	075	53 06	157	43 18	226	39 14	274	36 36	307
156	45 51	354	39 10	027	43 30	074	53 26	157	42 42	226	38 23	274	35 55	306
157	45 45	353	39 32	026	44 19	073	53 45	158	42 06	226	37 32	273	35 14	305
158	45 38	352	39 54	025	45 07	072	54 04	158	41 29	226	36 41	272	34 32	304
159	45 30	350	40 15	023	45 56	071	54 23	159	40 53	226	35 51	272	33 49	303
160	45 21	349	40 34	022	46 44	071	54 41	159	40 17	226	35 00	271	33 07	302
161	45 11	347	40 53	021	47 32	070	54 58	160	39 40	226	34 09	271	32 24	302
162	44 59	346	41 11	020	48 19	069	55 15	161	39 04	226	33 18	270	31 40	301
163	44 46	345	41 27	018	49 07	068	55 32	161	38 28	225	32 27	270	30 56	300
164	44 32	343	41 43	017	49 54	067	55 48	162	37 51	225	31 36	269	30 12	299
	ARCTURUS		◆ANTARES		ACRUX		◆CANOPUS		SIRIUS		PROCYON		◆REGULUS	
165	20 39	049	19 45	109	56 03	163	37 15	225	30 45	269	29 27	298	44 17	342
166	21 17	049	20 33	109	56 18	163	36 39	225	29 54	268	28 42	298	44 01	341
167	21 55	048	21 21	109	56 33	164	36 03	225	29 04	268	27 57	297	43 44	340
168	22 33	047	22 09	108	56 46	165	35 27	225	28 13	267	27 12	296	43 26	338
169	23 10	046	22 58	108	56 59	165	34 51	225	27 22	267	26 26	296	43 06	337
170	23 46	046	23 46	107	57 12	166	34 15	225	26 31	266	25 40	295	42 46	336
171	24 22	045	24 35	107	57 24	167	33 39	225	25 40	266	24 54	294	42 24	334
172	24 58	044	25 24	107	57 35	168	33 04	225	24 50	265	24 07	293	42 02	333
173	25 33	043	26 12	106	57 45	169	32 28	224	23 59	265	23 20	293	41 39	332
174	26 08	042	27 01	106	57 55	169	31 52	224	23 08	264	22 33	292	41 14	331
175	26 42	042	27 50	105	58 04	170	31 17	224	22 18	264	21 46	291	40 49	330
176	27 15	041	28 40	105	58 13	171	30 41	224	21 27	263	20 58	291	40 23	329
177	27 48	040	29 29	104	58 20	172	30 06	224	20 37	263	20 11	290	39 56	327
178	28 20	039	30 18	104	58 27	173	29 31	224	19 46	262	19 23	290	39 28	326
179	28 52	038	31 08	104	58 33	173	28 56	223	18 56	262	18 35	289	38 59	325

LAT 32°S

LAT 32°S

LHA ♈	Hc	Zn	Hc	Zn	Hc	Zn	Hc	Zn	Hc	Zn	Hc	Zn	Hc	Zn
	◆ARCTURUS		ANTARES		◆RIGIL KENT.		ACRUX		CANOPUS		◆Alphard		REGULUS	
180	29 23	037	31 57	103	51 17	150	58 39	174	28 21	223	47 38	295	38 30	324
181	29 54	036	32 47	103	51 42	150	58 44	175	27 46	223	46 51	294	38 00	323
182	30 23	035	33 36	102	52 07	151	58 48	176	27 12	223	46 05	293	37 29	322
183	30 52	034	34 26	102	52 32	151	58 51	177	26 37	223	45 18	292	36 57	321
184	31 21	033	35 16	102	52 56	152	58 53	178	26 03	222	44 31	291	36 25	320
185	31 48	032	36 06	101	53 20	152	58 55	179	25 28	222	43 43	291	35 52	319
186	32 15	031	36 56	101	53 44	153	58 56	179	24 54	222	42 55	290	35 18	318
187	32 41	030	37 46	100	54 07	153	58 56	180	24 20	222	42 07	289	34 44	317
188	33 07	029	38 36	100	54 30	154	58 55	181	23 47	221	41 19	288	34 08	316
189	33 31	028	39 26	100	54 52	154	58 53	182	23 13	221	40 31	288	33 33	315
190	33 55	027	40 16	099	55 14	155	58 51	183	22 40	221	39 42	287	32 57	314
191	34 18	026	41 06	099	55 35	155	58 48	184	22 06	221	38 53	286	32 20	313
192	34 40	025	41 57	098	55 56	156	58 44	185	21 33	220	38 04	285	31 43	312
193	35 02	024	42 47	098	56 17	157	58 40	186	21 00	220	37 15	285	31 05	312
194	35 22	023	43 38	097	56 36	157	58 34	186	20 28	220	36 26	284	30 26	311
	ARCTURUS		◆ANTARES		RIGIL KENT.		◆ACRUX		Suhail		REGULUS		◆Denebola	
195	35 42	022	44 28	097	56 56	158	58 28	187	43 39	238	29 47	310	40 23	337
196	36 01	021	45 19	097	57 15	159	58 21	188	42 55	238	29 08	309	40 02	336
197	36 18	020	46 09	096	57 33	159	58 14	189	42 12	238	28 28	308	39 41	335
198	36 35	019	47 00	096	57 50	160	58 06	190	41 29	238	27 48	307	39 19	334
199	36 51	018	47 50	095	58 07	161	57 57	191	40 46	238	27 07	307	38 56	332
200	37 06	016	48 41	095	58 24	162	57 47	191	40 03	238	26 26	306	38 32	331
201	37 20	015	49 32	094	58 40	162	57 37	192	39 20	238	25 45	305	38 07	330
202	37 33	014	50 23	094	58 55	163	57 26	193	38 37	237	25 03	304	37 41	329
203	37 45	013	51 13	093	59 09	164	57 14	194	37 54	237	24 21	303	37 15	328
204	37 56	012	52 04	093	59 23	165	57 02	194	37 11	237	23 38	303	36 48	327
205	38 06	011	52 55	092	59 35	166	56 49	195	36 29	237	22 55	302	36 20	326
206	38 14	009	53 46	092	59 48	167	56 35	196	35 46	237	22 12	301	35 51	325
207	38 22	008	54 37	092	59 59	168	56 21	197	35 03	237	21 28	301	35 21	324
208	38 29	007	55 27	091	60 10	168	56 06	197	34 21	236	20 44	300	34 51	323
209	38 35	006	56 18	090	60 19	169	55 51	198	33 39	236	20 00	299	34 20	322
	◆ARCTURUS		Rasalhague		◆Nunki		RIGIL KENT.		ACRUX		◆Suhail		SPICA	
210	38 40	005	22 02	058	26 36	106	60 28	170	55 35	199	32 57	236	67 37	337
211	38 43	003	22 45	057	27 25	105	60 37	171	55 18	199	32 15	236	67 16	334
212	38 46	002	23 27	057	28 14	105	60 44	172	55 01	200	31 33	235	66 54	332
213	38 47	001	24 10	056	29 03	105	60 50	173	54 44	200	30 51	235	66 29	330
214	38 47	000	24 52	055	29 52	104	60 56	174	54 26	201	30 09	235	66 03	328
215	38 47	359	25 33	054	30 42	104	61 01	175	54 07	202	29 27	235	65 35	326
216	38 45	357	26 14	054	31 31	103	61 05	176	53 48	202	28 46	234	65 05	324
217	38 42	356	26 55	053	32 21	103	61 08	177	53 29	203	28 05	234	64 34	322
218	38 38	355	27 35	052	33 11	102	61 10	178	53 09	203	27 23	234	64 02	320
219	38 33	354	28 15	051	34 00	102	61 11	179	52 49	204	26 42	234	63 28	318
220	38 27	353	28 54	050	34 50	102	61 11	180	52 28	204	26 01	233	62 54	316
221	38 20	351	29 33	050	35 40	101	61 10	181	52 07	205	25 21	233	62 18	314
222	38 12	350	30 12	049	36 30	101	61 09	182	51 46	205	24 40	233	61 41	313
223	38 03	349	30 50	048	37 20	100	61 07	183	51 24	206	24 00	232	61 03	311
224	37 53	348	31 27	047	38 10	100	61 03	184	51 02	206	23 19	232	60 25	310
	Alphecca		Rasalhague		◆Nunki		Peacock		◆RIGIL KENT.		SPICA		◆ARCTURUS	
225	30 42	009	32 04	046	39 00	100	30 53	141	60 59	185	59 45	308	37 41	347
226	30 49	008	32 41	045	39 50	099	31 26	141	60 54	186	59 05	307	37 29	346
227	30 56	007	33 16	044	40 41	099	31 58	141	60 48	187	58 24	306	37 16	344
228	31 02	006	33 51	043	41 31	098	32 30	140	60 41	188	57 42	304	37 02	343
229	31 06	005	34 26	042	42 21	098	33 03	140	60 34	189	57 00	303	36 47	342
230	31 10	004	35 00	041	43 12	097	33 35	140	60 25	190	56 17	302	36 30	341
231	31 13	003	35 33	040	44 02	097	34 08	140	60 16	191	55 33	301	36 13	340
232	31 15	002	36 06	039	44 53	097	34 40	140	60 06	192	54 49	299	35 55	339
233	31 16	001	36 38	038	45 43	096	35 13	140	59 55	193	54 04	298	35 36	338
234	31 16	000	37 09	037	46 34	096	35 46	140	59 43	194	53 19	297	35 17	337
235	31 15	359	37 40	036	47 24	095	36 19	140	59 31	195	52 34	296	34 56	335
236	31 13	358	38 09	035	48 15	095	36 51	140	59 18	195	51 48	295	34 34	334
237	31 11	356	38 38	034	49 06	094	37 24	140	59 04	196	51 02	294	34 12	333
238	31 07	355	39 06	033	49 57	094	37 57	140	58 49	197	50 15	293	33 49	332
239	31 03	354	39 34	032	50 47	093	38 30	140	58 34	198	49 28	292	33 25	331
	Rasalhague		ALTAIR		◆Nunki		Peacock		◆RIGIL KENT.		SPICA		◆ARCTURUS	
240	40 00	031	21 32	064	51 38	093	39 03	140	58 18	199	48 41	291	33 00	330
241	40 26	030	22 17	063	52 29	093	39 36	140	58 01	199	47 54	291	32 34	329
242	40 50	028	23 03	062	53 20	092	40 09	140	57 44	200	47 06	290	32 08	328
243	41 14	027	23 48	062	54 11	092	40 42	140	57 26	201	46 18	289	31 41	327
244	41 37	026	24 32	061	55 02	091	41 14	140	57 08	202	45 30	288	31 13	326
245	41 59	025	25 17	060	55 52	091	41 47	140	56 49	202	44 41	287	30 44	325
246	42 20	024	26 01	060	56 43	090	42 20	140	56 29	203	43 53	287	30 15	324
247	42 40	022	26 44	059	57 34	089	42 53	140	56 09	204	43 04	286	29 45	324
248	42 59	021	27 28	058	58 25	089	43 25	140	55 48	204	42 15	285	29 14	323
249	43 16	020	28 11	057	59 16	088	43 58	140	55 27	205	41 25	284	28 43	322
250	43 33	019	28 53	056	60 07	088	44 31	140	55 06	205	40 36	284	28 11	321
251	43 49	017	29 35	056	60 58	087	45 03	140	54 44	206	39 46	283	27 39	320
252	44 03	016	30 17	055	61 48	087	45 36	140	54 21	206	38 57	282	27 06	319
253	44 17	015	30 59	054	62 39	086	46 08	141	53 58	207	38 07	282	26 32	318
254	44 29	013	31 40	053	63 30	085	46 40	141	53 35	207	37 17	281	25 58	317
	VEGA		◆ALTAIR		FOMALHAUT		◆Peacock		RIGIL KENT.		◆SPICA		ARCTURUS	
255	15 44	019	32 20	052	15 42	115	47 12	141	53 11	208	36 27	280	25 23	317
256	16 01	019	33 00	052	16 28	115	47 44	141	52 47	208	35 37	280	24 48	316
257	16 16	018	33 40	051	17 14	115	48 16	141	52 23	209	34 47	279	24 12	315
258	16 32	017	34 19	050	18 00	114	48 48	142	51 58	209	33 57	278	23 36	314
259	16 46	016	34 57	049	18 47	114	49 19	142	51 33	210	33 06	278	22 59	313
260	17 00	016	35 36	048	19 34	113	49 51	142	51 08	210	32 16	277	22 22	313
261	17 14	015	36 13	047	20 20	113	50 22	142	50 42	210	31 25	277	21 44	312
262	17 26	014	36 50	046	21 07	113	50 53	143	50 16	211	30 35	276	21 06	311
263	17 38	013	37 26	045	21 54	112	51 24	143	49 50	211	29 44	275	20 28	310
264	17 50	012	38 02	044	22 41	112	51 54	143	49 24	211	28 53	275	19 49	310
265	18 00	012	38 37	043	23 29	111	52 25	144	48 57	212	28 03	274	19 09	309
266	18 10	011	39 11	042	24 16	111	52 55	144	48 30	212	27 12	274	18 29	308
267	18 19	010	39 45	041	25 04	111	53 25	144	48 03	212	26 21	273	17 49	307
268	18 28	009	40 18	040	25 51	110	53 54	145	47 36	213	25 30	273	17 09	307
269	18 35	008	40 50	039	26 39	110	54 23	145	47 08	213	24 39	272	16 28	306

LHA ♈	Hc	Zn	Hc	Zn	Hc	Zn	Hc	Zn	Hc	Zn	Hc	Zn	Hc	Zn
	VEGA		ALTAIR		◆FOMALHAUT		ACHERNAR		◆RIGIL KENT.		ANTARES		◆Rasalhague	
270	18 42	008	41 22	038	27 27	109	14 51	149	46 40	213	69 26	280	45 02	351
271	18 49	007	41 52	037	28 15	109	15 18	149	46 12	213	68 36	279	44 53	350
272	18 54	006	42 22	035	29 03	109	15 44	149	45 44	214	67 46	278	44 44	349
273	18 59	005	42 51	034	29 52	108	16 10	148	45 16	214	66 55	277	44 33	347
274	19 03	004	43 19	033	30 40	108	16 37	148	44 48	214	66 05	277	44 21	346
275	19 07	003	43 46	032	31 29	108	17 04	148	44 20	214	65 14	276	44 08	345
276	19 10	003	44 13	031	32 17	107	17 31	147	43 51	214	64 24	275	43 54	343
277	19 12	002	44 38	029	33 06	107	17 59	147	43 22	214	63 33	274	43 39	342
278	19 13	001	45 03	028	33 55	106	18 27	147	42 54	214	62 42	274	43 22	341
279	19 13	000	45 26	027	34 43	106	18 55	147	42 25	215	61 52	273	43 05	339
280	19 13	359	45 48	025	35 32	106	19 23	146	41 56	215	61 01	273	42 46	338
281	19 12	359	46 10	024	36 21	105	19 51	146	41 27	215	60 10	272	42 27	337
282	19 10	358	46 30	023	37 10	105	20 20	146	40 58	215	59 19	271	42 06	336
283	19 08	357	46 49	021	38 00	105	20 48	145	40 29	215	58 28	271	41 45	334
284	19 05	356	47 07	020	38 49	104	21 17	145	40 00	215	57 37	270	41 22	333
	ALTAIR		Enif		◆FOMALHAUT		ACHERNAR		◆RIGIL KENT.		ANTARES		◆Rasalhague	
285	47 24	019	32 41	050	39 38	104	21 47	145	39 31	215	56 46	270	40 59	332
286	47 40	017	33 20	049	40 28	104	22 16	145	39 01	215	55 55	269	40 34	331
287	47 54	016	33 58	048	41 17	103	22 45	144	38 32	215	55 05	269	40 09	330
288	48 07	014	34 36	047	42 07	103	23 15	144	38 03	215	54 14	268	39 43	328
289	48 19	013	35 13	047	42 56	102	23 45	144	37 34	215	53 23	268	39 16	327
290	48 30	011	35 50	046	43 46	102	24 15	144	37 04	215	52 32	267	38 48	326
291	48 39	010	36 26	045	44 36	102	24 45	144	36 35	215	51 41	267	38 19	325
292	48 48	008	37 01	044	45 26	101	25 15	143	36 06	215	50 50	266	37 50	324
293	48 54	007	37 36	043	46 16	101	25 46	143	35 37	215	50 00	266	37 20	323
294	49 00	005	38 10	042	47 06	101	26 17	143	35 07	215	49 09	265	36 49	322
295	49 04	004	38 44	041	47 56	100	26 47	143	34 38	215	48 18	265	36 17	321
296	49 07	002	39 16	040	48 46	100	27 18	143	34 09	215	47 28	265	35 45	320
297	49 08	001	39 48	038	49 36	100	27 49	142	33 40	215	46 37	264	35 12	319
298	49 09	359	40 20	037	50 26	099	28 20	142	33 11	215	45 46	264	34 38	318
299	49 07	358	40 50	036	51 16	099	28 52	142	32 42	215	44 56	263	34 04	317
	Enif		◆FOMALHAUT		ACHERNAR		◆RIGIL KENT.		ANTARES		Rasalhague		◆ALTAIR	
300	41 20	035	52 07	098	29 23	142	32 13	215	44 05	263	33 29	316	49 05	356
301	41 49	034	52 57	098	29 54	142	31 44	215	43 15	262	32 53	315	49 01	355
302	42 17	033	53 48	098	30 26	142	31 15	214	42 24	262	32 17	314	48 56	353
303	42 44	032	54 38	097	30 58	141	30 47	214	41 34	262	31 40	313	48 49	352
304	43 10	030	55 28	097	31 29	141	30 18	214	40 44	261	31 03	313	48 42	350
305	43 35	029	56 19	097	32 01	141	29 49	214	39 54	261	30 25	312	48 33	349
306	44 00	028	57 10	096	32 33	141	29 21	214	39 03	260	29 47	311	48 22	348
307	44 23	027	58 00	096	33 05	141	28 53	214	38 13	260	29 08	310	48 11	346
308	44 45	025	58 51	095	33 37	141	28 24	214	37 23	259	28 29	309	47 58	345
309	45 06	024	59 42	095	34 10	141	27 56	213	36 33	259	27 49	308	47 44	343
310	45 27	023	60 32	095	34 42	141	27 28	213	35 43	259	27 09	307	47 28	342
311	45 46	021	61 23	094	35 14	141	27 00	213	34 53	258	26 29	307	47 12	340
312	46 04	020	62 14	094	35 46	141	26 33	213	34 04	258	25 48	306	46 54	339
313	46 21	019	63 04	093	36 19	141	26 05	213	33 14	257	25 06	305	46 35	338
314	46 36	017	63 55	093	36 51	140	25 37	213	32 24	257	24 24	304	46 15	336
	Alpheratz		◆Diphda		ACHERNAR		◆RIGIL KENT.		ANTARES		Rasalhague		◆ALTAIR	
315	14 21	041	38 05	088	37 23	140	25 10	212	31 35	257	23 42	304	45 54	335
316	14 54	041	38 56	088	37 56	140	24 43	212	30 45	256	23 00	303	45 32	334
317	15 27	040	39 47	087	38 28	140	24 16	212	29 56	256	22 17	302	45 09	332
318	16 00	039	40 37	087	39 01	140	23 49	212	29 07	255	21 34	302	44 45	331
319	16 31	038	41 28	086	39 33	140	23 22	212	28 17	255	20 50	301	44 20	330
320	17 03	038	42 19	086	40 06	140	22 55	211	27 28	255	20 06	300	43 54	329
321	17 34	037	43 10	085	40 38	140	22 29	211	26 39	254	19 22	299	43 27	327
322	18 04	036	44 00	084	41 10	140	22 03	211	25 50	254	18 38	299	42 59	326
323	18 34	035	44 51	084	41 43	141	21 37	211	25 02	253	17 53	298	42 30	325
324	19 03	035	45 42	083	42 15	141	21 11	210	24 13	253	17 08	297	42 00	324
325	19 32	034	46 32	083	42 47	141	20 45	210	23 24	252	16 23	297	41 30	323
326	20 00	033	47 22	082	43 20	141	20 20	210	22 36	252	15 37	296	40 59	322
327	20 28	032	48 13	081	43 52	141	19 54	210	21 48	252	14 51	296	40 27	320
328	20 55	032	49 03	081	44 24	141	19 29	209	20 59	251	14 05	295	39 54	319
329	21 21	031	49 53	080	44 56	141	19 04	209	20 11	251	13 19	294	39 20	318
	◆Alpheratz		Diphda		◆ACHERNAR		Peacock		◆Nunki		ALTAIR		Enif	
330	21 47	030	50 43	079	45 28	141	60 20	206	49 30	266	38 46	317	47 58	354
331	22 12	029	51 33	078	46 00	141	59 57	207	48 39	265	38 11	316	47 52	353
332	22 36	028	52 23	078	46 32	141	59 34	208	47 49	265	37 36	315	47 45	351
333	23 00	027	53 13	077	47 03	142	59 10	209	46 58	264	37 00	314	47 37	350
334	23 23	027	54 02	076	47 35	142	58 45	209	46 07	264	36 23	313	47 27	348
335	23 45	026	54 51	075	48 06	142	58 20	210	45 17	264	35 46	312	47 16	347
336	24 07	025	55 40	074	48 37	142	57 54	211	44 26	263	35 08	311	47 04	345
337	24 28	024	56 29	074	49 08	143	57 28	211	43 36	263	34 29	311	46 51	344
338	24 49	023	57 18	073	49 39	143	57 01	212	42 45	262	33 50	310	46 36	343
339	25 08	022	58 06	072	50 10	143	56 34	213	41 55	262	33 11	309	46 20	341
340	25 27	021	58 55	071	50 40	143	56 06	213	41 05	261	32 31	308	46 03	340
341	25 45	020	59 42	070	51 11	144	55 38	214	40 14	261	31 51	307	45 45	339
342	26 02	019	60 30	069	51 41	144	55 10	214	39 24	261	31 10	306	45 26	337
343	26 19	019	61 17	068	52 10	144	54 41	215	38 34	260	30 28	305	45 06	336
344	26 35	018	62 04	066	52 40	145	54 12	215	37 44	260	29 47	305	44 44	335
	Alpheratz		◆Hamal		◆ACHERNAR		Peacock		◆Nunki		ALTAIR		Enif	
345	26 50	017	18 49	045	53 09	145	53 43	215	36 54	259	29 05	304	44 22	333
346	27 04	016	19 25	044	53 38	146	53 13	216	36 04	259	28 22	303	43 59	332
347	27 17	015	20 00	043	54 07	146	52 44	216	35 14	259	27 39	302	43 34	331
348	27 30	014	20 35	043	54 35	146	52 13	217	34 24	258	26 56	301	43 09	330
349	27 42	013	21 09	042	55 03	147	51 43	217	33 34	258	26 12	301	42 43	328
350	27 53	012	21 43	041	55 31	147	51 12	217	32 45	257	25 28	300	42 16	327
351	28 03	011	22 16	040	55 58	148	50 41	217	31 55	257	24 44	299	41 48	326
352	28 12	010	22 49	040	56 25	148	50 10	218	31 05	257	24 00	298	41 19	325
353	28 20	009	23 21	039	56 51	149	49 39	218	30 16	256	23 15	298	40 49	324
354	28 28	008	23 52	038	57 17	150	49 08	218	29 27	256	22 30	297	40 19	323
355	28 34	007	24 23	037	57 43	150	48 36	219	28 37	255	21 44	296	39 48	322
356	28 40	006	24 53	036	58 08	151	48 04	219	27 48	255	20 58	296	39 16	320
357	28 45	005	25 23	035	58 32	152	47 32	219	26 59	254	20 12	295	38 43	319
358	28 49	004	25 52	034	58 56	152	47 00	219	26 10	254	19 26	294	38 09	318
359	28 52	003	26 21	034	59 20	153	46 28	219	25 21	254	18 40	294	37 35	317

LAT 33°S

LHA	Hc	Zn	Hc	Zn	Hc	Zn	Hc	Zn	Hc	Zn	Hc	Zn	Hc	Zn
♈	◆Alpheratz		Hamal		◆RIGEL		CANOPUS		ACHERNAR		◆Peacock		Enif	
0	27 54	002	25 58	032	14 01	091	22 22	139	60 36	153	46 42	220	36 17	317
1	27 56	001	26 25	032	14 51	090	22 54	139	60 59	154	46 10	220	35 42	316
2	27 56	000	26 51	031	15 42	090	23 28	139	61 21	155	45 37	220	35 07	315
3	27 56	359	27 16	030	16 32	089	24 01	138	61 42	155	45 04	220	34 31	314
4	27 55	358	27 41	029	17 22	089	24 34	138	62 02	156	44 32	221	33 54	313
5	27 52	357	28 05	028	18 12	088	25 08	138	62 22	157	43 59	221	33 17	312
6	27 49	356	28 28	027	19 03	087	25 42	138	62 41	158	43 26	221	32 40	311
7	27 46	355	28 50	026	19 53	087	26 16	137	63 00	159	42 53	221	32 01	310
8	27 41	354	29 12	025	20 43	086	26 50	137	63 17	160	42 21	221	31 23	309
9	27 35	353	29 33	024	21 33	086	27 24	137	63 34	161	41 48	221	30 44	309
10	27 29	352	29 53	023	22 24	085	27 58	137	63 50	162	41 15	221	30 04	308
11	27 21	351	30 12	022	23 14	085	28 33	137	64 04	163	40 42	221	29 24	307
12	27 13	350	30 31	021	24 04	084	29 07	136	64 18	164	40 09	221	28 44	306
13	27 04	349	30 48	020	24 54	083	29 42	136	64 31	166	39 36	221	28 03	305
14	26 55	348	31 05	019	25 44	083	30 17	136	64 43	167	39 03	221	27 22	304
	◆Hamal		ALDEBARAN		RIGEL		◆CANOPUS		ACHERNAR		◆FOMALHAUT		Alpheratz	
15	31 21	018	18 36	055	26 34	082	30 52	136	64 54	168	63 42	269	26 44	347
16	31 36	017	19 17	054	27 24	082	31 27	136	65 04	169	62 51	268	26 32	346
17	31 50	016	19 58	053	28 13	081	32 03	135	65 13	170	62 01	268	26 20	345
18	32 04	015	20 38	053	29 03	080	32 38	135	65 21	172	61 11	268	26 07	344
19	32 16	014	21 18	052	29 53	080	33 13	135	65 28	173	60 20	267	25 53	344
20	32 28	013	21 57	051	30 42	079	33 49	135	65 33	174	59 30	267	25 39	343
21	32 39	012	22 36	050	31 31	079	34 25	135	65 38	176	58 40	266	25 23	342
22	32 48	011	23 15	050	32 21	078	35 00	135	65 41	177	57 50	266	25 07	341
23	32 57	010	23 53	049	33 10	077	35 36	135	65 43	178	57 00	265	24 50	340
24	33 05	008	24 31	048	33 59	077	36 12	134	65 44	179	56 09	265	24 32	339
25	33 12	007	25 08	047	34 48	076	36 48	134	65 44	181	55 19	264	24 14	338
26	33 18	006	25 44	046	35 36	075	37 24	134	65 43	182	54 29	264	23 55	337
27	33 23	005	26 21	046	36 25	074	38 00	134	65 40	183	53 39	264	23 35	336
28	33 27	004	26 56	045	37 13	074	38 36	134	65 37	185	52 49	263	23 14	335
29	33 30	003	27 32	044	38 02	073	39 12	134	65 32	186	51 59	263	22 53	335
	Hamal		ALDEBARAN		◆SIRIUS		CANOPUS		◆ACHERNAR		FOMALHAUT		◆Alpheratz	
30	33 32	002	28 06	043	24 31	095	39 49	134	65 26	187	51 09	262	22 31	334
31	33 33	001	28 40	042	25 21	094	40 25	134	65 19	189	50 20	262	22 08	333
32	33 34	000	29 14	041	26 12	094	41 01	134	65 11	190	49 30	261	21 45	332
33	33 33	359	29 47	040	27 02	093	41 38	134	65 02	191	48 40	261	21 21	331
34	33 31	357	30 19	039	27 52	093	42 14	134	64 52	192	47 50	261	20 56	330
35	33 28	356	30 51	039	28 42	092	42 50	134	64 41	193	47 01	260	20 31	329
36	33 25	355	31 22	038	29 33	091	43 27	134	64 28	195	46 11	260	20 05	329
37	33 20	354	31 52	037	30 23	091	44 03	134	64 15	196	45 22	260	19 39	328
38	33 15	353	32 22	036	31 13	090	44 39	134	64 01	197	44 32	259	19 12	327
39	33 08	352	32 51	035	32 04	090	45 16	134	63 46	198	43 43	259	18 44	326
40	33 01	351	33 19	034	32 54	089	45 52	134	63 30	199	42 54	258	18 16	326
41	32 52	350	33 46	033	33 44	089	46 28	134	63 13	200	42 04	258	17 47	325
42	32 43	349	34 13	032	34 35	088	47 05	134	62 55	201	41 15	258	17 18	324
43	32 33	348	34 39	031	35 25	088	47 41	134	62 37	202	40 26	257	16 48	323
44	32 22	347	35 05	030	36 15	087	48 17	134	62 18	203	39 37	257	16 17	322
	◆ALDEBARAN		BETELGEUSE		SIRIUS		◆CANOPUS		ACHERNAR		◆FOMALHAUT		Hamal	
45	35 29	029	32 04	054	37 05	086	48 53	134	61 58	204	38 48	256	32 09	346
46	35 53	027	32 44	053	37 56	086	49 29	134	61 37	205	37 59	256	31 57	345
47	36 15	026	33 24	052	38 46	085	50 05	135	61 15	206	37 10	256	31 43	343
48	36 37	025	34 04	051	39 36	085	50 41	135	60 53	206	36 22	255	31 28	342
49	36 58	024	34 43	050	40 26	084	51 16	135	60 31	207	35 33	255	31 12	341
50	37 18	023	35 21	050	41 16	083	51 52	135	60 07	208	34 44	254	30 56	340
51	37 38	022	35 59	049	42 06	083	52 27	135	59 43	209	33 56	254	30 39	339
52	37 56	021	36 37	048	42 56	082	53 03	136	59 19	209	33 08	254	30 20	338
53	38 13	020	37 14	047	43 46	082	53 38	136	58 54	210	32 19	253	30 01	337
54	38 29	018	37 50	046	44 35	081	54 13	136	58 29	211	31 31	253	29 42	336
55	38 45	017	38 26	045	45 25	080	54 48	136	58 03	211	30 43	253	29 21	335
56	38 59	016	39 01	044	46 15	079	55 22	137	57 36	212	29 55	252	29 00	334
57	39 13	015	39 35	043	47 04	079	55 56	137	57 09	213	29 07	252	28 38	333
58	39 25	014	40 09	042	47 53	078	56 31	137	56 42	213	28 20	251	28 15	333
59	39 36	012	40 42	040	48 42	077	57 04	138	56 14	214	27 32	251	27 51	332
	ALDEBARAN		◆BETELGEUSE		SIRIUS		◆CANOPUS		ACHERNAR		◆FOMALHAUT		Hamal	
60	39 46	011	41 14	039	49 31	077	57 38	138	55 46	214	26 45	251	27 27	331
61	39 56	010	41 45	038	50 20	076	58 11	139	55 18	215	25 57	250	27 02	330
62	40 04	009	42 16	037	51 09	075	58 44	139	54 49	215	25 10	250	26 36	329
63	40 11	007	42 46	036	51 57	074	59 17	140	54 20	215	24 23	249	26 10	328
64	40 17	006	43 15	035	52 46	073	59 49	140	53 51	216	23 36	249	25 43	327
65	40 22	005	43 43	034	53 34	072	60 21	141	53 22	216	22 49	248	25 15	326
66	40 25	004	44 11	032	54 22	071	60 53	142	52 52	217	22 02	248	24 47	325
67	40 28	002	44 37	031	55 09	071	61 24	142	52 22	217	21 16	248	24 18	324
68	40 30	001	45 03	030	55 57	070	61 54	143	51 51	217	20 29	247	23 48	324
69	40 30	000	45 27	028	56 44	069	62 24	144	51 21	217	19 43	247	23 18	323
70	40 29	359	45 51	027	57 30	068	62 54	144	50 50	218	18 57	246	22 48	322
71	40 28	357	46 13	026	58 17	067	63 23	145	50 19	218	18 11	246	22 16	321
72	40 25	356	46 35	025	59 03	065	63 52	146	49 48	218	17 25	246	21 44	320
73	40 21	355	46 55	023	59 48	064	64 19	147	49 17	218	16 39	245	21 12	320
74	40 16	354	47 14	022	60 33	063	64 46	148	48 46	219	15 53	245	20 39	319
	BETELGEUSE		◆PROCYON		Suhail		◆ACRUX		ACHERNAR		◆Diphda		ALDEBARAN	
75	47 32	020	36 19	052	41 21	121	20 14	153	48 14	219	31 04	268	40 10	352
76	47 49	019	36 59	051	42 04	121	20 37	153	47 43	219	30 14	267	40 02	351
77	48 05	018	37 38	050	42 48	121	21 00	153	47 11	219	29 23	267	39 54	350
78	48 20	016	38 16	049	43 31	121	21 23	153	46 39	219	28 33	266	39 45	349
79	48 33	015	38 54	048	44 14	121	21 46	152	46 07	219	27 43	266	39 34	347
80	48 45	013	39 31	047	44 57	121	22 10	152	45 35	220	26 53	265	39 23	346
81	48 56	012	40 08	046	45 41	121	22 33	152	45 03	220	26 03	265	39 10	345
82	49 05	010	40 44	045	46 24	120	22 57	152	44 31	220	25 13	264	38 57	344
83	49 14	009	41 20	044	47 07	120	23 21	151	43 58	220	24 22	264	38 42	343
84	49 21	007	41 55	043	47 51	120	23 45	151	43 26	220	23 32	263	38 27	341
85	49 26	006	42 29	042	48 34	120	24 10	151	42 54	220	22 43	263	38 10	340
86	49 31	004	43 02	041	49 18	120	24 34	151	42 21	220	21 53	262	37 52	339
87	49 34	003	43 35	040	50 01	120	24 59	150	41 49	220	21 03	262	37 34	338
88	49 35	001	44 06	039	50 45	120	25 24	150	41 16	220	20 13	261	37 15	337
89	49 36	000	44 37	037	51 28	120	25 49	150	40 44	220	19 23	261	36 54	336

LHA	Hc	Zn	Hc	Zn	Hc	Zn	Hc	Zn	Hc	Zn	Hc	Zn	Hc	Zn
♈	PROCYON		REGULUS		◆Suhail		ACRUX		◆ACHERNAR		RIGEL		◆BETELGEUSE	
90	45 08	036	15 46	064	52 12	120	26 14	150	40 12	220	63 03	334	49 35	358
91	45 37	035	16 31	063	52 56	120	26 39	150	39 39	220	62 40	332	49 32	357
92	46 05	034	17 16	063	53 39	120	27 05	150	39 07	220	62 16	330	49 28	355
93	46 33	032	18 00	062	54 23	120	27 30	149	38 34	220	61 51	328	49 23	353
94	46 59	031	18 44	061	55 06	120	27 56	149	38 02	220	61 24	327	49 17	352
95	47 25	030	19 28	061	55 50	120	28 22	149	37 29	220	60 55	325	49 09	350
96	47 49	028	20 12	060	56 33	120	28 48	149	36 57	220	60 26	323	49 00	349
97	48 13	027	20 55	059	57 17	120	29 14	149	36 25	220	59 55	321	48 50	347
98	48 35	026	21 38	058	58 00	120	29 40	149	35 53	220	59 23	320	48 39	346
99	48 57	024	22 21	058	58 44	121	30 06	148	35 20	220	58 50	318	48 26	345
100	49 17	023	23 03	057	59 27	121	30 33	148	34 48	220	58 16	317	48 12	343
101	49 36	021	23 45	056	60 10	121	30 59	148	34 16	220	57 41	315	47 56	342
102	49 53	020	24 27	055	60 53	121	31 26	148	33 44	219	57 05	314	47 40	340
103	50 10	018	25 08	055	61 36	121	31 52	148	33 12	219	56 28	312	47 22	339
104	50 25	017	25 49	054	62 19	122	32 19	148	32 40	219	55 50	311	47 04	337
	PROCYON		REGULUS		◆Gienah		ACRUX		◆ACHERNAR		RIGEL		◆BETELGEUSE	
105	50 39	015	26 29	053	18 33	099	32 46	148	32 09	219	55 12	309	46 44	336
106	50 52	014	27 09	052	19 22	099	33 13	148	31 37	219	54 32	308	46 23	335
107	51 04	012	27 49	051	20 12	098	33 40	148	31 05	219	53 53	307	46 01	333
108	51 14	011	28 28	051	21 02	098	34 07	148	30 34	219	53 12	306	45 38	332
109	51 22	009	29 07	050	21 52	097	34 34	148	30 03	218	52 31	305	45 14	331
110	51 30	008	29 45	049	22 42	097	35 01	147	29 31	218	51 49	303	44 49	330
111	51 36	006	30 23	048	23 32	096	35 28	147	29 00	218	51 07	302	44 23	328
112	51 40	004	31 00	047	24 22	096	35 55	147	28 29	218	50 24	301	43 56	327
113	51 44	003	31 37	046	25 12	095	36 22	147	27 58	218	49 41	300	43 28	326
114	51 45	001	32 13	045	26 02	095	36 49	147	27 28	218	48 57	299	42 59	325
115	51 46	000	32 48	044	26 52	094	37 16	147	26 57	217	48 13	298	42 30	323
116	51 45	358	33 23	044	27 43	094	37 44	147	26 27	217	47 28	297	41 59	322
117	51 42	356	33 58	043	28 33	093	38 11	147	25 56	217	46 43	296	41 28	321
118	51 38	355	34 31	042	29 23	093	38 38	147	25 26	217	45 58	295	40 56	320
119	51 33	353	35 05	041	30 13	092	39 05	147	24 56	217	45 12	294	40 24	319
	REGULUS		◆SPICA		ACRUX		◆CANOPUS		RIGEL		BETELGEUSE		◆PROCYON	
120	35 37	040	13 20	095	39 32	147	63 49	214	44 26	293	39 50	318	51 26	352
121	36 09	039	14 10	094	39 59	147	63 21	215	43 40	293	39 16	317	51 18	350
122	36 40	038	15 00	094	40 26	147	62 52	216	42 53	292	38 41	316	51 09	348
123	37 10	037	15 51	093	40 53	148	62 22	216	42 07	291	38 06	315	50 58	347
124	37 40	035	16 41	093	41 20	148	61 52	217	41 19	290	37 30	314	50 46	345
125	38 09	034	17 31	092	41 47	148	61 21	218	40 32	289	36 53	313	50 33	344
126	38 37	033	18 21	092	42 14	148	60 50	219	39 45	289	36 16	312	50 18	342
127	39 04	032	19 12	091	42 41	148	60 19	219	38 57	288	35 38	311	50 02	341
128	39 30	031	20 02	090	43 07	148	59 47	220	38 09	287	35 00	310	49 45	339
129	39 56	030	20 52	090	43 34	148	59 14	220	37 21	286	34 21	309	49 27	338
130	40 20	029	21 43	089	44 01	148	58 42	221	36 32	286	33 42	308	49 07	336
131	40 44	028	22 33	089	44 27	148	58 09	221	35 44	285	33 02	307	48 46	335
132	41 07	026	23 23	088	44 53	149	57 35	222	34 55	284	32 22	306	48 25	334
133	41 29	025	24 14	088	45 19	149	57 02	222	34 06	284	31 41	306	48 02	332
134	41 50	024	25 04	087	45 45	149	56 28	223	33 17	283	31 00	305	47 38	331
	◆REGULUS		SPICA		◆ACRUX		CANOPUS		◆RIGEL		BETELGEUSE		PROCYON	
135	42 10	023	25 54	087	46 11	149	55 54	223	32 28	282	30 18	304	47 13	330
136	42 29	021	26 44	086	46 37	149	55 19	223	31 39	282	29 37	303	46 47	328
137	42 47	020	27 35	085	47 03	150	54 45	224	30 49	281	28 54	302	46 20	327
138	43 04	019	28 25	085	47 28	150	54 10	224	30 00	280	28 11	301	45 52	326
139	43 19	018	29 15	084	47 53	150	53 35	224	29 10	280	27 28	301	45 23	324
140	43 34	016	30 05	084	48 18	150	53 00	224	28 21	279	26 45	300	44 53	323
141	43 48	015	30 55	083	48 43	151	52 25	225	27 31	278	26 01	299	44 23	322
142	44 00	014	31 45	082	49 07	151	51 49	225	26 41	278	25 17	298	43 51	321
143	44 11	012	32 35	082	49 32	151	51 14	225	25 51	277	24 33	298	43 19	320
144	44 22	011	33 24	081	49 56	152	50 38	225	25 01	277	23 48	297	42 46	319
145	44 31	010	34 14	081	50 20	152	50 02	225	24 11	276	23 03	296	42 12	317
146	44 39	008	35 04	080	50 43	152	49 26	226	23 21	275	22 18	296	41 38	316
147	44 45	007	35 53	079	51 06	153	48 50	226	22 31	275	21 32	295	41 03	315
148	44 51	006	36 42	079	51 29	153	48 14	226	21 41	274	20 47	294	40 27	314
149	44 55	004	37 32	078	51 52	153	47 38	226	20 51	274	20 01	294	39 51	313
	REGULUS		◆Denebola		SPICA		◆ACRUX		CANOPUS		◆SIRIUS		PROCYON	
150	44 58	003	35 46	033	38 21	077	52 14	154	47 02	226	43 19	278	39 14	312
151	45 00	001	36 13	032	39 10	077	52 36	154	46 25	226	42 29	277	38 36	311
152	45 00	000	36 40	031	39 59	076	52 58	155	45 49	226	41 39	277	37 58	310
153	45 00	359	37 05	030	40 48	075	53 19	155	45 13	226	40 49	276	37 19	309
154	44 58	357	37 30	029	41 36	074	53 40	156	44 37	226	39 59	276	36 40	308
155	44 55	356	37 53	028	42 24	074	54 01	156	44 00	226	39 09	275	36 00	307
156	44 51	355	38 16	026	43 13	073	54 21	157	43 24	226	38 19	274	35 20	306
157	44 46	353	38 38	025	44 01	072	54 41	157	42 47	226	37 29	274	34 39	306
158	44 39	352	38 59	024	44 48	071	55 00	158	42 11	226	36 38	273	33 58	305
159	44 31	350	39 19	023	45 36	070	55 19	158	41 35	226	35 48	273	33 16	304
160	44 22	349	39 39	022	46 23	070	55 37	159	40 58	226	34 58	272	32 34	303
161	44 12	348	39 57	021	47 10	069	55 55	160	40 22	226	34 08	271	31 52	302
162	44 01	346	40 14	019	47 57	068	56 12	160	39 46	226	33 17	271	31 09	301
163	43 48	345	40 30	018	48 44	067	56 29	161	39 10	226	32 27	270	30 26	301
164	43 35	344	40 45	017	49 30	066	56 45	162	38 33	226	31 37	270	29 42	300
	ARCTURUS		◆ANTARES		ACRUX		◆CANOPUS		SIRIUS		PROCYON		◆REGULUS	
165	19 59	049	20 05	109	57 01	162	37 57	226	30 46	269	28 59	299	43 20	342
166	20 37	048	20 52	109	57 16	163	37 21	226	29 56	269	28 14	298	43 04	341
167	21 15	048	21 40	108	57 30	164	36 45	226	29 06	268	27 30	297	42 48	340
168	21 52	047	22 28	108	57 44	164	36 09	226	28 15	268	26 45	297	42 30	339
169	22 28	046	23 16	107	57 58	165	35 33	225	27 25	267	26 00	296	42 11	337
170	23 04	045	24 04	107	58 10	166	34 57	225	26 35	267	25 15	295	41 51	336
171	23 40	045	24 52	106	58 22	167	34 22	225	25 45	266	24 29	295	41 30	335
172	24 15	044	25 40	106	58 34	167	33 46	225	24 55	266	23 43	294	41 08	334
173	24 49	043	26 29	106	58 44	168	33 11	225	24 04	265	22 57	293	40 45	332
174	25 23	042	27 17	105	58 54	169	32 35	225	23 14	265	22 10	292	40 22	331
175	25 57	041	28 06	105	59 03	170	32 00	225	22 24	264	21 24	292	39 57	330
176	26 30	040	28 55	104	59 12	171	31 24	224	21 34	264	20 37	291	39 32	329
177	27 02	039	29 43	104	59 20	172	30 49	224	20 44	263	19 50	290	39 05	328
178	27 34	039	30 32	103	59 27	172	30 14	224	19 54	263	19 03	290	38 38	327
179	28 05	038	31 21	103	59 33	173	29 39	224	19 04	262	18 15	289	38 10	326

LHA ♈	Hc	Zn	Hc	Zn	Hc	Zn	Hc	Zn	Hc	Zn	Hc	Zn	Hc	Zn
	◆ARCTURUS		ANTARES		◆RIGIL KENT.		ACRUX		CANOPUS		◆Alphard		REGULUS	
180	28 35	037	32 10	103	52 09	149	59 39	174	29 05	224	47 12	296	37 41	325
181	29 05	036	33 00	102	52 35	150	59 43	175	28 30	223	46 26	295	37 12	323
182	29 34	035	33 49	102	53 00	150	59 47	176	27 55	223	45 41	294	36 41	322
183	30 03	034	34 38	101	53 25	151	59 51	177	27 21	223	44 55	293	36 10	321
184	30 31	033	35 27	101	53 49	151	59 53	178	26 47	223	44 08	292	35 39	320
185	30 58	032	36 17	100	54 13	152	59 55	179	26 13	222	43 22	292	35 06	319
186	31 24	031	37 06	100	54 37	152	59 56	179	25 39	222	42 35	291	34 33	318
187	31 50	030	37 56	100	55 00	153	59 56	180	25 05	222	41 47	290	33 59	317
188	32 15	029	38 46	099	55 23	153	59 55	181	24 32	222	41 00	289	33 25	316
189	32 39	028	39 35	099	55 46	154	59 53	182	23 58	221	40 12	288	32 50	316
190	33 02	027	40 25	098	56 08	154	59 51	183	23 25	221	39 24	288	32 15	315
191	33 24	026	41 15	098	56 30	155	59 48	184	22 52	221	38 36	287	31 39	314
192	33 46	025	42 05	097	56 51	155	59 44	185	22 19	221	37 48	286	31 02	313
193	34 07	024	42 55	097	57 11	156	59 39	186	21 46	220	37 00	285	30 25	312
194	34 27	023	43 45	096	57 32	157	59 34	187	21 14	220	36 11	285	29 47	311
	ARCTURUS		◆ANTARES		RIGIL KENT.		◆ACRUX		Suhail		REGULUS		◆Denebola	
195	34 46	022	44 35	096	57 51	157	59 28	188	44 10	239	29 09	310	39 27	337
196	35 04	021	45 25	096	58 10	158	59 21	188	43 27	239	28 30	309	39 07	336
197	35 22	020	46 15	095	58 29	159	59 13	189	42 44	239	27 51	309	38 47	335
198	35 38	019	47 05	095	58 47	160	59 05	190	42 00	239	27 12	308	38 25	334
199	35 54	017	47 55	094	59 04	160	58 56	191	41 17	239	26 32	307	38 03	333
200	36 08	016	48 45	094	59 21	161	58 46	192	40 34	238	25 51	306	37 39	332
201	36 22	015	49 36	093	59 37	162	58 35	192	39 52	238	25 10	305	37 15	331
202	36 35	014	50 26	093	59 52	163	58 24	193	39 09	238	24 29	305	36 50	330
203	36 46	013	51 16	092	60 07	164	58 12	194	38 26	238	23 47	304	36 24	328
204	36 57	012	52 07	092	60 20	164	58 00	195	37 44	238	23 06	303	35 57	327
205	37 07	011	52 57	091	60 34	165	57 46	196	37 01	238	22 23	302	35 30	326
206	37 15	009	53 47	091	60 46	166	57 33	196	36 19	237	21 41	302	35 01	325
207	37 23	008	54 37	090	60 58	167	57 18	197	35 36	237	20 58	301	34 32	324
208	37 30	007	55 28	090	61 08	168	57 03	198	34 54	237	20 14	300	34 03	323
209	37 35	006	56 18	089	61 18	169	56 48	198	34 12	237	19 31	300	33 32	322
	◆ARCTURUS		Rasalhague		◆Nunki		RIGIL KENT.		ACRUX		◆Suhail		SPICA	
210	37 40	005	21 30	058	26 52	105	61 28	170	56 31	199	33 30	236	66 42	338
211	37 43	003	22 12	057	27 40	105	61 36	171	56 15	200	32 48	236	66 22	335
212	37 46	002	22 54	056	28 29	104	61 43	172	55 57	200	32 06	236	66 00	333
213	37 47	001	23 36	055	29 18	104	61 50	173	55 40	201	31 25	236	65 37	331
214	37 47	000	24 17	055	30 07	104	61 56	174	55 22	202	30 43	235	65 12	329
215	37 47	359	24 58	054	30 56	103	62 01	175	55 03	202	30 02	235	64 45	327
216	37 45	357	25 38	053	31 45	103	62 04	176	54 44	203	29 21	235	64 16	325
217	37 42	356	26 18	052	32 34	102	62 07	177	54 24	203	28 40	235	63 47	323
218	37 39	355	26 58	052	33 23	102	62 10	178	54 04	204	27 59	234	63 16	321
219	37 34	354	27 37	051	34 12	101	62 11	179	53 44	204	27 18	234	62 43	319
220	37 28	353	28 16	050	35 02	101	62 11	180	53 23	205	26 37	234	62 10	317
221	37 21	352	28 54	049	35 51	101	62 10	181	53 02	205	25 57	233	61 35	316
222	37 13	350	29 32	048	36 41	100	62 09	182	52 40	206	25 16	233	61 00	314
223	37 04	349	30 09	047	37 30	100	62 06	183	52 18	206	24 36	233	60 23	313
224	36 54	348	30 46	046	38 20	099	62 03	184	51 56	207	23 56	233	59 46	311
	Alphecca		Rasalhague		◆Nunki		Peacock		◆RIGIL KENT.		SPICA		◆ARCTURUS	
225	29 43	009	31 22	046	39 10	099	31 40	140	61 59	185	59 07	310	36 43	347
226	29 50	008	31 58	045	39 59	098	32 12	140	61 54	186	58 28	308	36 31	346
227	29 56	007	32 33	044	40 49	098	32 44	140	61 48	187	57 48	307	36 18	345
228	30 02	006	33 08	043	41 39	097	33 16	140	61 41	188	57 08	305	36 04	343
229	30 06	005	33 42	042	42 29	097	33 49	140	61 33	189	56 26	304	35 49	342
230	30 10	004	34 15	041	43 19	097	34 21	140	61 24	190	55 44	303	35 34	341
231	30 13	003	34 47	040	44 09	096	34 54	140	61 15	191	55 02	302	35 17	340
232	30 15	002	35 19	039	44 59	096	35 26	140	61 04	192	54 19	301	34 59	339
233	30 16	001	35 51	038	45 49	095	35 59	140	60 53	193	53 35	300	34 41	338
234	30 16	000	36 21	037	46 39	095	36 32	139	60 41	194	52 51	298	34 21	337
235	30 15	359	36 51	036	47 29	094	37 04	139	60 29	195	52 07	297	34 01	336
236	30 14	358	37 20	035	48 20	094	37 37	139	60 15	196	51 22	296	33 40	335
237	30 11	357	37 48	034	49 10	093	38 10	139	60 01	197	50 37	295	33 18	334
238	30 07	355	38 16	033	50 00	093	38 43	139	59 46	198	49 51	294	32 55	333
239	30 03	354	38 43	031	50 50	092	39 16	139	59 31	198	49 05	293	32 32	332
	Rasalhague		ALTAIR		◆Nunki		Peacock		◆RIGIL KENT.		SPICA		◆ARCTURUS	
240	39 08	030	21 05	063	51 41	092	39 48	139	59 15	199	48 19	293	32 08	331
241	39 33	029	21 50	063	52 31	091	40 21	139	58 58	200	47 32	292	31 42	330
242	39 58	028	22 35	062	53 21	091	40 54	139	58 40	201	46 45	291	31 17	329
243	40 21	027	23 19	061	54 12	090	41 27	139	58 22	201	45 58	290	30 50	328
244	40 43	026	24 03	061	55 02	090	42 00	139	58 03	202	45 11	289	30 23	327
245	41 04	024	24 47	060	55 52	089	42 33	139	57 44	203	44 23	288	29 55	326
246	41 25	023	25 30	059	56 43	088	43 06	139	57 24	204	43 35	287	29 26	325
247	41 44	022	26 13	058	57 33	088	43 38	139	57 04	204	42 47	287	28 57	324
248	42 03	021	26 56	058	58 23	087	44 11	139	56 43	205	41 59	286	28 27	323
249	42 20	020	27 38	057	59 13	087	44 44	139	56 22	205	41 10	285	27 56	322
250	42 36	018	28 20	056	60 04	086	45 17	140	56 00	206	40 21	284	27 25	321
251	42 51	017	29 01	055	60 54	085	45 49	140	55 37	207	39 33	284	26 53	320
252	43 06	016	29 43	054	61 44	085	46 22	140	55 15	207	38 44	283	26 20	319
253	43 19	014	30 23	054	62 34	084	46 54	140	54 52	208	37 55	282	25 47	319
254	43 31	013	31 04	053	63 24	083	47 26	140	54 28	208	37 05	282	25 14	318
	VEGA		◆ALTAIR		FOMALHAUT		◆Peacock		RIGIL KENT.		◆SPICA		ARCTURUS	
255	14 47	019	31 43	052	16 08	115	47 59	140	54 04	209	36 16	281	24 40	317
256	15 04	019	32 23	051	16 53	115	48 31	140	53 40	209	35 27	280	24 05	316
257	15 19	018	33 02	050	17 39	114	49 03	141	53 15	210	34 37	280	23 30	315
258	15 34	017	33 40	049	18 25	114	49 35	141	52 50	210	33 47	279	22 54	314
259	15 49	016	34 18	048	19 11	113	50 06	141	52 25	210	32 58	278	22 18	314
260	16 03	015	34 55	047	19 57	113	50 38	141	52 00	211	32 08	278	21 41	313
261	16 16	015	35 32	046	20 44	113	51 09	142	51 34	211	31 18	277	21 04	312
262	16 28	014	36 08	045	21 30	112	51 40	142	51 08	211	30 28	277	20 27	311
263	16 40	013	36 44	044	22 17	112	52 11	142	50 41	212	29 38	276	19 49	311
264	16 51	012	37 19	043	23 04	111	52 42	142	50 15	212	28 48	275	19 10	310
265	17 01	012	37 53	042	23 50	111	53 13	143	49 48	212	27 58	275	18 31	309
266	17 11	011	38 26	041	24 38	111	53 43	143	49 21	213	27 08	274	17 52	308
267	17 20	010	38 59	040	25 25	110	54 13	144	48 53	213	26 17	274	17 13	308
268	17 28	009	39 32	039	26 12	110	54 43	144	48 26	213	25 27	273	16 33	307
269	17 36	008	40 03	038	26 59	109	55 12	144	47 58	213	24 37	273	15 52	306

LAT 33°S

LHA ♈	Hc	Zn	Hc	Zn	Hc	Zn	Hc	Zn	Hc	Zn	Hc	Zn	Hc	Zn
	VEGA		ALTAIR		◆FOMALHAUT		ACHERNAR		◆RIGIL KENT.		ANTARES		◆Rasalhague	
270	17 43	008	40 34	037	27 47	109	15 43	149	47 30	214	69 15	283	44 02	351
271	17 49	007	41 04	036	28 35	109	16 09	149	47 02	214	68 26	282	43 54	350
272	17 55	006	41 33	035	29 22	108	16 35	149	46 34	214	67 36	281	43 45	349
273	17 59	005	42 01	034	30 10	108	17 01	148	46 06	214	66 47	280	43 34	347
274	18 04	004	42 29	032	30 58	107	17 28	148	45 38	214	65 57	279	43 23	346
275	18 07	003	42 55	031	31 46	107	17 55	148	45 09	215	65 07	278	43 10	345
276	18 10	003	43 21	030	32 34	107	18 22	147	44 40	215	64 17	277	42 57	343
277	18 12	002	43 46	029	33 23	106	18 49	147	44 12	215	63 27	276	42 42	342
278	18 13	001	44 10	028	34 11	106	19 17	147	43 43	215	62 37	276	42 26	341
279	18 13	000	44 32	026	35 00	105	19 45	146	43 14	215	61 47	275	42 09	340
280	18 13	359	44 54	025	35 48	105	20 13	146	42 45	215	60 57	274	41 51	338
281	18 12	359	45 15	024	36 37	105	20 41	146	42 16	215	60 07	274	41 32	337
282	18 10	358	45 35	022	37 26	104	21 09	146	41 47	215	59 17	273	41 12	336
283	18 08	357	45 53	021	38 14	104	21 38	145	41 18	215	58 26	272	40 51	335
284	18 05	356	46 11	020	39 03	103	22 07	145	40 49	215	57 36	272	40 29	334
	ALTAIR		Enif		◆FOMALHAUT		ACHERNAR		◆RIGIL KENT.		ANTARES		◆Rasalhague	
285	46 27	018	32 03	050	39 52	103	22 36	145	40 20	215	56 46	271	40 06	332
286	46 42	017	32 41	049	40 41	103	23 05	144	39 50	216	55 55	271	39 42	331
287	46 56	015	33 18	048	41 30	102	23 34	144	39 21	216	55 05	270	39 17	330
288	47 09	014	33 55	047	42 20	102	24 04	144	38 52	216	54 15	270	38 52	329
289	47 21	013	34 32	046	43 09	102	24 33	144	38 23	216	53 25	269	38 25	328
290	47 31	011	35 08	045	43 58	101	25 03	143	37 53	216	52 34	269	37 58	327
291	47 40	010	35 43	044	44 48	101	25 33	143	37 24	216	51 44	268	37 30	326
292	47 48	008	36 18	043	45 37	100	26 03	143	36 55	215	50 54	268	37 01	325
293	47 55	007	36 52	042	46 27	100	26 34	143	36 26	215	50 03	267	36 32	323
294	48 00	005	37 25	041	47 16	100	27 04	143	35 56	215	49 13	267	36 01	322
295	48 04	004	37 58	040	48 06	099	27 35	142	35 27	215	48 23	266	35 30	321
296	48 07	002	38 30	039	48 56	099	28 06	142	34 58	215	47 33	266	34 59	320
297	48 08	001	39 01	038	49 45	098	28 37	142	34 29	215	46 43	265	34 26	319
298	48 09	359	39 32	037	50 35	098	29 08	142	34 00	215	45 52	265	33 53	318
299	48 07	358	40 02	036	51 25	098	29 39	142	33 31	215	45 02	264	33 20	318
	Enif		◆FOMALHAUT		ACHERNAR		◆RIGIL KENT.		ANTARES		Rasalhague		◆ALTAIR	
300	40 31	035	52 15	097	30 10	142	33 02	215	44 12	264	32 45	317	48 05	357
301	40 59	033	53 05	097	30 41	141	32 34	215	43 22	263	32 10	316	48 01	355
302	41 26	032	53 55	096	31 13	141	32 05	215	42 32	263	31 35	315	47 56	354
303	41 53	031	54 45	096	31 44	141	31 36	215	41 42	262	30 59	314	47 50	352
304	42 18	030	55 35	095	32 16	141	31 08	215	40 53	262	30 22	313	47 42	351
305	42 43	029	56 25	095	32 48	141	30 39	214	40 03	262	29 45	312	47 34	349
306	43 06	027	57 15	095	33 20	141	30 11	214	39 13	261	29 08	311	47 24	348
307	43 29	026	58 05	094	33 52	141	29 42	214	38 23	261	28 30	310	47 12	346
308	43 51	025	58 56	094	34 24	140	29 14	214	37 34	260	27 51	310	47 00	345
309	44 12	024	59 46	093	34 56	140	28 46	214	36 44	260	27 12	309	46 46	343
310	44 31	022	60 36	093	35 28	140	28 18	214	35 55	259	26 32	308	46 31	342
311	44 50	021	61 26	092	36 00	140	27 51	213	35 05	259	25 52	307	46 15	341
312	45 07	020	62 17	092	36 32	140	27 23	213	34 16	258	25 12	306	45 58	339
313	45 24	018	63 07	091	37 05	140	26 55	213	33 27	258	24 31	306	45 40	338
314	45 39	017	63 57	091	37 37	140	26 28	213	32 37	258	23 50	305	45 20	337
	Alpheratz		◆Diphda		ACHERNAR		◆RIGIL KENT.		ANTARES		Rasalhague		◆ALTAIR	
315	13 36	041	38 03	088	38 09	140	26 01	213	31 48	257	23 09	304	45 00	335
316	14 09	040	38 53	087	38 42	140	25 33	213	30 59	257	22 27	303	44 38	334
317	14 41	040	39 43	087	39 14	140	25 07	212	30 10	256	21 45	303	44 16	333
318	15 13	039	40 34	086	39 47	140	24 40	212	29 22	256	21 02	302	43 52	332
319	15 44	038	41 24	085	40 19	140	24 13	212	28 33	255	20 19	301	43 28	330
320	16 15	038	42 14	085	40 52	140	23 47	212	27 44	255	19 36	300	43 02	329
321	16 46	037	43 04	084	41 24	140	23 20	211	26 56	255	18 52	300	42 36	328
322	17 16	036	43 54	083	41 57	140	22 54	211	26 07	254	18 09	299	42 09	327
323	17 45	035	44 44	083	42 29	140	22 28	211	25 19	254	17 24	298	41 41	325
324	18 14	035	45 34	082	43 01	140	22 02	211	24 30	253	16 40	298	41 12	324
325	18 42	034	46 24	082	43 34	140	21 37	210	23 42	253	15 55	297	40 42	323
326	19 10	033	47 13	081	44 06	140	21 12	210	22 54	252	15 10	296	40 11	322
327	19 37	032	48 03	080	44 38	140	20 46	210	22 06	252	14 25	296	39 40	321
328	20 03	031	48 53	079	45 10	140	20 21	210	21 19	252	13 40	295	39 08	320
329	20 29	031	49 42	079	45 42	140	19 57	209	20 31	251	12 54	295	38 35	319
	◆Alpheratz		Diphda		◆ACHERNAR		Peacock		◆Nunki		ALTAIR		Enif	
330	20 55	030	50 31	078	46 14	141	61 14	207	49 34	267	38 02	318	46 59	354
331	21 19	029	51 20	077	46 46	141	60 51	208	48 44	266	37 28	317	46 53	353
332	21 43	028	52 09	076	47 18	141	60 27	209	47 53	266	36 53	316	46 46	351
333	22 07	027	52 58	076	47 50	141	60 02	210	47 03	266	36 18	315	46 38	350
334	22 29	026	53 47	075	48 22	141	59 37	210	46 13	265	35 42	314	46 28	348
335	22 51	026	54 35	074	48 53	141	59 11	211	45 23	265	35 05	313	46 18	347
336	23 13	025	55 24	073	49 24	142	58 45	212	44 33	264	34 28	312	46 06	346
337	23 33	024	56 12	072	49 56	142	58 19	212	43 43	264	33 50	311	45 53	344
338	23 53	023	56 59	071	50 27	142	57 52	213	42 53	263	33 12	310	45 39	343
339	24 13	022	57 47	070	50 57	142	57 24	213	42 03	263	32 33	309	45 23	342
340	24 31	021	58 34	069	51 28	143	56 56	214	41 13	262	31 54	308	45 07	340
341	24 49	020	59 21	068	51 59	143	56 28	214	40 23	262	31 14	307	44 49	339
342	25 06	019	60 07	067	52 29	143	56 00	215	39 33	261	30 34	307	44 31	338
343	25 22	018	60 54	066	52 59	144	55 31	215	38 44	261	29 54	306	44 11	336
344	25 38	017	61 39	065	53 29	144	55 01	216	37 54	261	29 13	305	43 50	335
	Alpheratz		◆Hamal		◆ACHERNAR		Peacock		◆Nunki		ALTAIR		Enif	
345	25 52	017	18 07	045	53 58	144	54 32	216	37 04	260	28 31	304	43 28	334
346	26 06	016	18 42	044	54 27	145	54 02	217	36 15	260	27 49	303	43 06	332
347	26 19	015	19 16	043	54 56	145	53 32	217	35 25	259	27 07	303	42 42	331
348	26 32	014	19 51	042	55 25	146	53 01	217	34 36	259	26 25	302	42 17	330
349	26 43	013	20 24	042	55 53	146	52 31	218	33 47	258	25 42	301	41 52	329
350	26 54	012	20 57	041	56 21	147	52 00	218	32 57	258	24 58	300	41 25	328
351	27 04	011	21 30	040	56 49	147	51 29	218	32 08	258	24 15	300	40 58	327
352	27 13	010	22 02	039	57 16	148	50 58	219	31 19	257	23 31	299	40 30	325
353	27 21	009	22 34	038	57 42	148	50 26	219	30 30	257	22 47	298	40 01	324
354	27 28	008	23 05	038	58 09	149	49 55	219	29 41	256	22 02	297	39 31	323
355	27 35	007	23 35	037	58 35	149	49 23	219	28 52	256	21 17	297	39 00	322
356	27 40	006	24 05	036	59 00	150	48 51	219	28 04	255	20 32	296	38 29	321
357	27 45	005	24 34	035	59 25	151	48 19	220	27 15	255	19 47	295	37 57	320
358	27 49	004	25 03	034	59 49	151	47 47	220	26 26	255	19 01	295	37 24	319
359	27 52	003	25 31	033	60 13	152	47 14	220	25 38	254	18 15	294	36 51	318

LHA ♈	Hc	Zn	Hc	Zn	Hc	Zn	Hc	Zn	Hc	Zn	Hc	Zn	Hc	Zn
	◆Alpheratz		Hamal		◆RIGEL		CANOPUS		ACHERNAR		◆Peacock		Enif	
0	26 54	002	25 07	032	14 02	091	23 07	139	61 29	152	47 28	221	35 33	317
1	26 56	001	25 34	031	14 51	090	23 40	139	61 52	153	46 55	221	34 59	316
2	26 56	000	25 59	030	15 41	089	24 13	138	62 15	154	46 23	221	34 24	315
3	26 56	359	26 24	029	16 31	089	24 46	138	62 36	155	45 50	221	33 49	314
4	26 55	358	26 48	029	17 21	088	25 19	138	62 57	156	45 17	221	33 13	314
5	26 52	357	27 11	028	18 10	088	25 52	138	63 17	156	44 44	221	32 37	313
6	26 50	356	27 34	027	19 00	087	26 26	137	63 37	157	44 12	221	32 00	312
7	26 46	355	27 56	026	19 50	087	27 00	137	63 56	158	43 39	221	31 23	311
8	26 41	354	28 17	025	20 39	086	27 34	137	64 13	159	43 06	221	30 45	310
9	26 36	353	28 38	024	21 29	085	28 08	137	64 31	161	42 33	221	30 06	309
10	26 29	352	28 58	023	22 18	085	28 42	136	64 47	162	42 00	221	29 27	308
11	26 22	351	29 16	022	23 08	084	29 16	136	65 02	163	41 27	221	28 48	307
12	26 14	350	29 35	021	23 57	084	29 51	136	65 16	164	40 54	221	28 08	306
13	26 05	349	29 52	020	24 47	083	30 25	136	65 29	165	40 21	221	27 28	306
14	25 56	348	30 08	019	25 36	082	31 00	136	65 42	166	39 49	221	26 47	305
	◆Hamal		ALDEBARAN		RIGEL		◆CANOPUS		ACHERNAR		◆FOMALHAUT		Alpheratz	
15	30 24	018	18 02	055	26 25	082	31 35	135	65 53	168	63 42	271	25 45	347
16	30 39	017	18 42	054	27 15	081	32 10	135	66 03	169	62 52	270	25 34	346
17	30 53	016	19 22	053	28 04	081	32 45	135	66 12	170	62 02	270	25 22	346
18	31 06	015	20 01	052	28 53	080	33 20	135	66 20	171	61 12	269	25 09	345
19	31 18	014	20 41	052	29 42	079	33 56	135	66 27	173	60 23	269	24 56	344
20	31 29	013	21 19	051	30 30	079	34 31	134	66 33	174	59 33	268	24 41	343
21	31 40	012	21 58	050	31 19	078	35 07	134	66 37	175	58 43	268	24 26	342
22	31 49	011	22 36	049	32 08	077	35 42	134	66 41	177	57 53	267	24 10	341
23	31 58	009	23 13	049	32 56	077	36 18	134	66 43	178	57 04	267	23 54	340
24	32 06	008	23 50	048	33 45	076	36 54	134	66 44	179	56 14	266	23 36	339
25	32 12	007	24 27	047	34 33	075	37 30	134	66 44	181	55 24	266	23 18	338
26	32 18	006	25 03	046	35 21	074	38 06	134	66 43	182	54 35	265	22 59	337
27	32 23	005	25 39	045	36 09	074	38 42	134	66 40	184	53 45	265	22 40	336
28	32 27	004	26 14	044	36 56	073	39 18	133	66 37	185	52 56	264	22 19	336
29	32 30	003	26 48	044	37 44	072	39 54	133	66 32	186	52 06	264	21 59	335
	Hamal		ALDEBARAN		◆SIRIUS		CANOPUS		◆ACHERNAR		FOMALHAUT		◆Alpheratz	
30	32 32	002	27 22	043	24 36	094	40 30	133	66 26	188	51 17	264	21 37	334
31	32 33	001	27 56	042	25 25	094	41 06	133	66 18	189	50 27	263	21 15	333
32	32 34	000	28 29	041	26 15	093	41 43	133	66 10	190	49 38	263	20 52	332
33	32 33	359	29 01	040	27 05	093	42 19	133	66 01	191	48 49	262	20 28	331
34	32 31	358	29 33	039	27 55	092	42 55	133	65 50	193	48 00	262	20 04	330
35	32 29	356	30 04	038	28 44	091	43 32	133	65 39	194	47 10	261	19 39	330
36	32 25	355	30 34	037	29 34	091	44 08	133	65 26	195	46 21	261	19 14	329
37	32 21	354	31 04	036	30 24	090	44 44	133	65 13	196	45 32	261	18 48	328
38	32 15	353	31 33	035	31 13	090	45 21	133	64 58	198	44 43	260	18 21	327
39	32 09	352	32 01	034	32 03	089	45 57	133	64 43	199	43 54	260	17 54	326
40	32 01	351	32 29	033	32 53	089	46 33	133	64 27	200	43 05	259	17 26	326
41	31 53	350	32 56	032	33 43	088	47 10	133	64 09	201	42 16	259	16 58	325
42	31 44	349	33 22	031	34 32	088	47 46	133	63 51	202	41 28	258	16 29	324
43	31 34	348	33 48	030	35 22	087	48 22	133	63 32	203	40 39	258	16 00	323
44	31 23	347	34 12	029	36 12	086	48 58	133	63 13	204	39 50	258	15 30	323
	◆ALDEBARAN		BETELGEUSE		SIRIUS		◆CANOPUS		ACHERNAR		◆FOMALHAUT		Hamal	
45	34 36	028	31 28	053	37 01	086	49 35	133	62 52	205	39 02	257	31 11	346
46	34 59	027	32 08	053	37 51	085	50 11	134	62 31	206	38 13	257	30 59	345
47	35 21	026	32 47	052	38 40	085	50 47	134	62 09	206	37 25	256	30 45	344
48	35 43	025	33 26	051	39 30	084	51 23	134	61 47	207	36 37	256	30 31	343
49	36 03	024	34 04	050	40 19	083	51 58	134	61 24	208	35 48	256	30 15	342
50	36 23	023	34 42	049	41 09	083	52 34	134	61 00	209	35 00	255	29 59	341
51	36 42	022	35 19	048	41 58	082	53 10	134	60 36	210	34 12	255	29 42	340
52	37 00	020	35 56	047	42 47	081	53 45	135	60 11	210	33 24	254	29 25	339
53	37 16	019	36 32	046	43 36	081	54 20	135	59 46	211	32 36	254	29 06	338
54	37 32	018	37 08	045	44 25	080	54 56	135	59 20	212	31 49	254	28 47	337
55	37 47	017	37 43	044	45 14	079	55 31	135	58 54	212	31 01	253	28 27	336
56	38 01	016	38 17	043	46 03	078	56 05	136	58 27	213	30 13	253	28 06	335
57	38 14	015	38 51	042	46 52	078	56 40	136	58 00	213	29 26	252	27 44	334
58	38 27	013	39 24	041	47 40	077	57 14	136	57 32	214	28 39	252	27 22	333
59	38 38	012	39 56	040	48 29	076	57 49	137	57 04	214	27 51	251	26 59	332
	ALDEBARAN		◆BETELGEUSE		SIRIUS		◆CANOPUS		ACHERNAR		◆FOMALHAUT		Hamal	
60	38 48	011	40 27	039	49 17	075	58 22	137	56 36	215	27 04	251	26 35	331
61	38 56	010	40 58	038	50 05	075	58 56	138	56 07	215	26 17	251	26 10	330
62	39 04	009	41 28	037	50 53	074	59 29	138	55 38	216	25 31	250	25 45	329
63	39 11	007	41 57	035	51 40	073	60 02	139	55 09	216	24 44	250	25 19	328
64	39 17	006	42 26	034	52 28	072	60 35	139	54 40	217	23 57	249	24 53	327
65	39 22	005	42 53	033	53 15	071	61 07	140	54 10	217	23 11	249	24 25	326
66	39 25	004	43 20	032	54 02	070	61 39	140	53 40	217	22 24	248	23 58	326
67	39 28	002	43 46	031	54 49	069	62 11	141	53 09	218	21 38	248	23 29	325
68	39 30	001	44 10	029	55 35	068	62 42	142	52 39	218	20 52	248	23 00	324
69	39 30	000	44 34	028	56 21	067	63 12	142	52 08	218	20 06	247	22 30	323
70	39 29	359	44 57	027	57 07	066	63 42	143	51 37	219	19 20	247	22 00	322
71	39 28	357	45 19	025	57 52	065	64 12	144	51 06	219	18 35	246	21 29	321
72	39 25	356	45 40	024	58 37	064	64 41	145	50 35	219	17 49	246	20 58	321
73	39 21	355	46 00	023	59 21	063	65 09	146	50 04	219	17 04	245	20 26	320
74	39 16	354	46 18	021	60 05	062	65 37	147	49 32	219	16 19	245	19 54	319
	BETELGEUSE		◆PROCYON		Suhail		◆ACRUX		ACHERNAR		◆Diphda		ALDEBARAN	
75	46 36	020	35 42	052	41 52	121	21 08	153	49 01	220	31 06	268	39 10	352
76	46 52	019	36 21	051	42 35	120	21 30	153	48 29	220	30 16	268	39 03	351
77	47 08	017	36 59	050	43 18	120	21 53	153	47 57	220	29 26	267	38 55	350
78	47 22	016	37 37	049	44 01	120	22 16	152	47 25	220	28 37	267	38 46	349
79	47 35	014	38 14	048	44 44	120	22 39	152	46 53	220	27 47	266	38 36	348
80	47 47	013	38 51	047	45 27	120	23 03	152	46 21	220	26 57	266	38 24	346
81	47 57	011	39 27	046	46 11	120	23 26	152	45 49	220	26 08	265	38 12	345
82	48 06	010	40 02	045	46 54	120	23 50	151	45 16	220	25 18	265	37 59	344
83	48 14	009	40 37	044	47 37	119	24 14	151	44 44	221	24 29	264	37 45	343
84	48 21	007	41 11	043	48 21	119	24 38	151	44 12	221	23 39	264	37 30	342
85	48 27	006	41 44	041	49 04	119	25 02	151	43 39	221	22 50	263	37 13	341
86	48 31	004	42 16	040	49 47	119	25 27	150	43 07	221	22 01	263	36 56	339
87	48 34	003	42 48	039	50 31	119	25 51	150	42 35	221	21 11	262	36 38	338
88	48 35	001	43 19	038	51 14	119	26 16	150	42 02	221	20 22	262	36 19	337
89	48 36	000	43 50	037	51 58	119	26 41	150	41 30	221	19 33	261	36 00	336

LAT 34°S

LHA ♈	Hc	Zn	Hc	Zn	Hc	Zn	Hc	Zn	Hc	Zn	Hc	Zn	Hc	Zn
	PROCYON		REGULUS		◆Suhail		ACRUX		◆ACHERNAR		RIGEL		◆BETELGEUSE	
90	44 19	036	15 19	064	52 42	119	27 06	150	40 57	221	62 09	335	48 35	358
91	44 48	034	16 04	063	53 25	119	27 31	149	40 25	221	61 47	333	48 32	357
92	45 15	033	16 48	062	54 09	119	27 57	149	39 52	221	61 24	331	48 29	355
93	45 42	032	17 32	062	54 52	119	28 22	149	39 20	221	60 59	329	48 24	354
94	46 08	031	18 15	061	55 36	119	28 48	149	38 48	221	60 33	328	48 18	352
95	46 33	029	18 59	060	56 19	119	29 13	149	38 15	221	60 06	326	48 10	351
96	46 57	028	19 42	059	57 03	119	29 39	149	37 43	220	59 37	324	48 01	349
97	47 19	027	20 24	059	57 46	119	30 05	148	37 11	220	59 08	322	47 51	348
98	47 41	025	21 07	058	58 30	119	30 31	148	36 39	220	58 37	321	47 40	346
99	48 02	024	21 49	057	59 13	119	30 58	148	36 06	220	58 05	319	47 28	345
100	48 21	022	22 30	057	59 57	119	31 24	148	35 34	220	57 32	318	47 14	343
101	48 40	021	23 12	056	60 40	119	31 50	148	35 02	220	56 58	316	46 59	342
102	48 57	020	23 53	055	61 23	120	32 17	148	34 30	220	56 23	315	46 43	341
103	49 13	018	24 33	054	62 07	120	32 43	148	33 59	220	55 47	313	46 26	339
104	49 28	017	25 13	054	62 50	120	33 10	148	33 27	220	55 10	312	46 08	338
	PROCYON		REGULUS		◆Gienah		ACRUX		◆ACHERNAR		RIGEL		◆BETELGEUSE	
105	49 42	015	25 53	053	18 42	099	33 37	147	32 55	219	54 33	311	45 49	337
106	49 54	014	26 33	052	19 31	098	34 03	147	32 24	219	53 55	309	45 29	335
107	50 05	012	27 12	051	20 21	098	34 30	147	31 52	219	53 16	308	45 07	334
108	50 15	011	27 50	050	21 10	097	34 57	147	31 21	219	52 37	307	44 45	333
109	50 23	009	28 28	049	21 59	097	35 24	147	30 50	219	51 56	306	44 21	331
110	50 30	007	29 06	049	22 49	096	35 51	147	30 18	219	51 16	304	43 57	330
111	50 36	006	29 43	048	23 38	096	36 18	147	29 47	218	50 34	303	43 32	329
112	50 41	004	30 19	047	24 28	095	36 45	147	29 16	218	49 53	302	43 05	328
113	50 44	003	30 55	046	25 17	095	37 12	147	28 46	218	49 10	301	42 38	326
114	50 45	001	31 30	045	26 07	094	37 40	147	28 15	218	48 27	300	42 10	325
115	50 46	000	32 05	044	26 56	094	38 07	147	27 45	218	47 44	299	41 41	324
116	50 45	358	32 40	043	27 46	093	38 34	147	27 14	217	47 01	298	41 12	323
117	50 42	356	33 13	042	28 36	093	39 01	147	26 44	217	46 16	297	40 41	322
118	50 39	355	33 46	041	29 25	092	39 28	147	26 14	217	45 32	296	40 10	321
119	50 33	353	34 19	040	30 15	092	39 55	147	25 44	217	44 47	295	39 38	320
	REGULUS		◆SPICA		ACRUX		◆CANOPUS		RIGEL		BETELGEUSE		◆PROCYON	
120	34 51	039	13 25	095	40 23	147	64 39	215	44 02	294	39 06	318	50 27	352
121	35 22	038	14 14	094	40 50	147	64 10	216	43 17	293	38 32	317	50 19	350
122	35 52	037	15 04	093	41 17	147	63 40	217	42 31	293	37 58	316	50 10	349
123	36 22	036	15 54	093	41 44	147	63 10	218	41 45	292	37 24	315	50 00	347
124	36 51	035	16 43	092	42 11	147	62 39	218	40 58	291	36 48	314	49 48	346
125	37 19	034	17 33	092	42 38	147	62 08	219	40 12	290	36 12	313	49 35	344
126	37 46	033	18 23	091	43 05	147	61 37	220	39 25	289	35 36	312	49 21	343
127	38 13	032	19 13	091	43 31	147	61 05	220	38 38	289	34 59	311	49 05	341
128	38 39	031	20 02	090	43 58	148	60 33	221	37 51	288	34 21	310	48 49	340
129	39 04	030	20 52	090	44 25	148	60 00	221	37 03	287	33 43	310	48 31	338
130	39 28	028	21 42	089	44 51	148	59 27	222	36 16	286	33 05	309	48 12	337
131	39 51	027	22 32	088	45 18	148	58 53	222	35 28	286	32 26	308	47 52	335
132	40 13	026	23 21	088	45 44	148	58 20	223	34 40	285	31 46	307	47 31	334
133	40 35	025	24 11	087	46 11	148	57 46	223	33 52	284	31 06	306	47 09	333
134	40 55	024	25 01	087	46 37	148	57 12	224	33 03	284	30 26	305	46 45	331
	◆REGULUS		SPICA		◆ACRUX		CANOPUS		◆RIGEL		BETELGEUSE		PROCYON	
135	41 14	022	25 50	086	47 03	149	56 37	224	32 15	283	29 45	304	46 21	330
136	41 33	021	26 40	086	47 28	149	56 03	224	31 26	282	29 04	304	45 56	329
137	41 50	020	27 29	085	47 54	149	55 28	225	30 38	281	28 22	303	45 29	327
138	42 07	019	28 19	084	48 20	149	54 53	225	29 49	281	27 40	302	45 02	326
139	42 22	017	29 08	084	48 45	150	54 18	225	29 00	280	26 57	301	44 34	325
140	42 36	016	29 58	083	49 10	150	53 42	225	28 11	280	26 15	300	44 05	324
141	42 50	015	30 47	082	49 35	150	53 07	226	27 22	279	25 32	300	43 35	323
142	43 02	013	31 37	082	50 00	150	52 31	226	26 33	278	24 48	299	43 05	321
143	43 13	012	32 26	081	50 24	151	51 56	226	25 43	278	24 05	298	42 33	320
144	43 23	011	33 15	081	50 48	151	51 20	226	24 54	277	23 21	297	42 01	319
145	43 32	009	34 04	080	51 12	151	50 44	226	24 05	276	22 36	297	41 28	318
146	43 39	008	34 53	079	51 36	152	50 08	226	23 15	276	21 52	296	40 54	317
147	43 46	007	35 42	079	52 00	152	49 32	227	22 26	275	21 07	295	40 20	316
148	43 51	005	36 30	078	52 23	152	48 56	227	21 36	275	20 22	295	39 45	315
149	43 55	004	37 19	077	52 46	153	48 19	227	20 47	274	19 36	294	39 09	314
	REGULUS		◆Denebola		SPICA		◆ACRUX		CANOPUS		◆SIRIUS		PROCYON	
150	43 58	003	34 56	033	38 07	077	53 08	153	47 43	227	43 10	279	38 33	313
151	44 00	001	35 22	032	38 56	076	53 30	154	47 07	227	42 21	278	37 56	312
152	44 00	000	35 48	031	39 44	075	53 52	154	46 30	227	41 32	278	37 19	311
153	44 00	359	36 13	029	40 32	074	54 14	155	45 54	227	40 42	277	36 41	310
154	43 58	357	36 37	028	41 20	074	54 35	155	45 18	227	39 53	276	36 02	309
155	43 55	356	37 00	027	42 07	073	54 56	156	44 41	227	39 03	276	35 23	308
156	43 51	355	37 22	026	42 55	072	55 16	156	44 05	227	38 14	275	34 44	307
157	43 46	353	37 44	025	43 42	071	55 36	157	43 29	227	37 24	275	34 04	306
158	43 40	352	38 04	024	44 29	070	55 55	157	42 52	227	36 35	274	33 24	305
159	43 32	351	38 24	023	45 15	070	56 14	158	42 16	227	35 45	273	32 43	304
160	43 23	349	38 43	022	46 02	069	56 33	158	41 40	227	34 55	273	32 01	303
161	43 13	348	39 01	020	46 48	068	56 51	159	41 03	227	34 06	272	31 20	303
162	43 02	347	39 17	019	47 34	067	57 08	160	40 27	227	33 16	272	30 38	302
163	42 50	345	39 33	018	48 20	066	57 25	160	39 51	227	32 26	271	29 55	301
164	42 37	344	39 48	017	49 05	065	57 42	161	39 15	227	31 37	270	29 12	300
	ARCTURUS		◆ANTARES		ACRUX		◆CANOPUS		SIRIUS		PROCYON		◆REGULUS	
165	19 20	049	20 24	109	57 58	162	38 39	226	30 47	270	28 29	299	42 23	343
166	19 57	048	21 11	108	58 13	162	38 03	226	29 57	269	27 46	299	42 08	341
167	20 34	047	21 59	108	58 28	163	37 27	226	29 07	269	27 02	298	41 51	340
168	21 11	047	22 46	107	58 42	164	36 51	226	28 18	268	26 18	297	41 34	339
169	21 47	046	23 34	107	58 55	165	36 15	226	27 28	268	25 33	296	41 15	338
170	22 22	045	24 21	106	59 08	165	35 39	226	26 38	267	24 49	296	40 56	336
171	22 57	044	25 09	106	59 20	166	35 04	226	25 49	267	24 04	295	40 36	335
172	23 31	043	25 57	106	59 32	167	34 28	226	24 59	266	23 19	294	40 14	334
173	24 05	043	26 45	105	59 43	168	33 53	225	24 09	266	22 33	294	39 52	333
174	24 39	042	27 33	105	59 53	169	33 18	225	23 20	265	21 47	293	39 29	332
175	25 12	041	28 21	104	60 02	170	32 42	225	22 30	264	21 01	292	39 05	331
176	25 44	040	29 09	104	60 11	170	32 07	225	21 41	264	20 15	291	38 40	329
177	26 16	039	29 58	103	60 19	171	31 32	225	20 51	263	19 29	291	38 14	328
178	26 47	038	30 46	103	60 26	172	30 57	224	20 02	263	18 42	290	37 48	327
179	27 17	037	31 35	102	60 33	173	30 23	224	19 13	262	17 55	290	37 20	326

LHA ♈	Hc	Zn	Hc	Zn	Hc	Zn	Hc	Zn	Hc	Zn	Hc	Zn	Hc	Zn
	◆ARCTURUS		ANTARES		◆RIGIL KENT.		ACRUX		CANOPUS		◆Alphard		REGULUS	
180	27 47	036	32 23	102	53 01	149	60 38	174	29 48	224	46 45	297	36 52	325
181	28 16	036	33 12	102	53 26	149	60 43	175	29 13	224	46 01	296	36 23	324
182	28 45	035	34 01	101	53 52	150	60 47	176	28 39	224	45 16	295	35 54	323
183	29 13	034	34 50	101	54 17	150	60 51	177	28 05	223	44 31	294	35 23	322
184	29 40	033	35 38	100	54 41	150	60 53	178	27 31	223	43 45	293	34 52	321
185	30 07	032	36 27	100	55 06	151	60 55	179	26 57	223	42 59	292	34 21	320
186	30 33	031	37 17	099	55 30	151	60 56	179	26 23	223	42 13	292	33 48	319
187	30 58	030	38 06	099	55 53	152	60 56	180	25 50	222	41 27	291	33 15	318
188	31 22	029	38 55	098	56 17	152	60 55	181	25 16	222	40 40	290	32 41	317
189	31 46	028	39 44	098	56 39	153	60 53	182	24 43	222	39 53	289	32 07	316
190	32 08	027	40 33	097	57 02	154	60 51	183	24 10	222	39 06	288	31 32	315
191	32 31	026	41 23	097	57 24	154	60 48	184	23 37	221	38 19	288	30 57	314
192	32 52	025	42 12	096	57 45	155	60 44	185	23 04	221	37 31	287	30 21	313
193	33 12	024	43 02	096	58 06	155	60 39	186	22 32	221	36 43	286	29 44	312
194	33 32	023	43 51	096	58 27	156	60 34	187	22 00	220	35 56	285	29 08	312
	ARCTURUS		◆ANTARES		RIGIL KENT.		◆ACRUX		Suhail		REGULUS		◆Denebola	
195	33 50	022	44 41	095	58 47	157	60 27	188	44 40	240	28 30	311	38 32	338
196	34 08	020	45 30	095	59 06	157	60 20	189	43 57	240	27 52	310	38 12	337
197	34 25	019	46 20	094	59 25	158	60 12	189	43 14	240	27 14	309	37 52	335
198	34 41	018	47 09	094	59 43	159	60 04	190	42 31	240	26 35	308	37 31	334
199	34 56	017	47 59	093	60 00	160	59 55	191	41 48	239	25 55	307	37 09	333
200	35 11	016	48 49	093	60 17	161	59 44	192	41 06	239	25 16	307	36 46	332
201	35 24	015	49 38	092	60 34	161	59 34	193	40 23	239	24 35	306	36 23	331
202	35 36	014	50 28	091	60 49	162	59 22	194	39 40	239	23 55	305	35 58	330
203	35 48	013	51 18	091	61 04	163	59 10	194	38 58	239	23 14	304	35 33	329
204	35 58	012	52 08	090	61 18	164	58 58	195	38 15	238	22 33	303	35 07	328
205	36 08	010	52 57	090	61 32	165	58 44	196	37 33	238	21 51	303	34 40	327
206	36 16	009	53 47	089	61 44	166	58 30	197	36 51	238	21 09	302	34 12	326
207	36 24	008	54 37	089	61 56	167	58 15	197	36 09	238	20 27	301	33 44	325
208	36 30	007	55 27	088	62 07	168	58 00	198	35 27	237	19 44	301	33 14	324
209	36 35	006	56 16	087	62 17	169	57 44	199	34 45	237	19 01	300	32 45	323
	◆ARCTURUS		Rasalhague		◆Nunki		RIGIL KENT.		ACRUX		◆Suhail		SPICA	
210	36 40	005	20 58	057	27 07	105	62 27	170	57 28	200	34 03	237	65 46	339
211	36 43	003	21 39	057	27 56	104	62 35	171	57 11	200	33 21	237	65 27	336
212	36 46	002	22 21	056	28 44	104	62 43	172	56 54	201	32 40	236	65 06	334
213	36 47	001	23 02	055	29 32	103	62 49	173	56 36	201	31 58	236	64 44	332
214	36 47	000	23 42	054	30 21	103	62 55	174	56 17	202	31 17	236	64 20	330
215	36 47	359	24 22	054	31 09	103	63 00	175	55 58	203	30 36	236	63 54	328
216	36 45	357	25 02	053	31 58	102	63 04	176	55 39	203	29 55	235	63 27	326
217	36 42	356	25 42	052	32 46	102	63 07	177	55 19	204	29 14	235	62 59	324
218	36 39	355	26 21	051	33 35	101	63 10	178	54 59	204	28 33	235	62 29	322
219	36 34	354	26 59	050	34 24	101	63 11	179	54 38	205	27 53	234	61 58	320
220	36 28	353	27 37	050	35 13	100	63 11	180	54 17	205	27 12	234	61 25	319
221	36 22	352	28 15	049	36 02	100	63 10	181	53 56	206	26 32	234	60 52	317
222	36 14	350	28 52	048	36 51	099	63 09	182	53 34	206	25 52	234	60 18	315
223	36 05	349	29 29	047	37 40	099	63 06	183	53 12	207	25 12	233	59 42	314
224	35 55	348	30 05	046	38 29	098	63 03	185	52 49	207	24 32	233	59 06	312
	Alphecca		Rasalhague		◆Nunki		Peacock		◆RIGIL KENT.		SPICA		◆ARCTURUS	
225	28 43	009	30 40	045	39 18	098	32 26	140	62 59	186	58 29	311	35 45	347
226	28 50	008	31 15	044	40 08	097	32 58	140	62 53	187	57 51	309	35 33	346
227	28 57	007	31 50	043	40 57	097	33 30	140	62 47	188	57 12	308	35 20	345
228	29 02	006	32 23	042	41 46	097	34 02	140	62 40	189	56 32	307	35 07	344
229	29 07	005	32 57	041	42 36	096	34 34	139	62 32	190	55 52	305	34 52	343
230	29 10	004	33 29	040	43 25	096	35 07	139	62 23	191	55 11	304	34 37	341
231	29 13	003	34 01	039	44 15	095	35 39	139	62 14	192	54 30	303	34 21	340
232	29 15	002	34 33	038	45 04	095	36 12	139	62 03	193	53 48	302	34 03	339
233	29 16	001	35 03	037	45 54	094	36 44	139	61 52	194	53 05	301	33 45	338
234	29 16	000	35 33	036	46 44	094	37 17	139	61 40	195	52 22	300	33 26	337
235	29 15	359	36 02	035	47 33	093	37 50	139	61 27	195	51 39	298	33 06	336
236	29 14	358	36 31	034	48 23	093	38 22	139	61 13	196	50 55	297	32 46	335
237	29 11	357	36 58	033	49 13	092	38 55	139	60 59	197	50 11	296	32 24	334
238	29 08	356	37 25	032	50 02	092	39 28	139	60 43	198	49 26	295	32 02	333
239	29 03	355	37 51	031	50 52	091	40 01	139	60 28	199	48 41	294	31 39	332
	Rasalhague		ALTAIR		◆Nunki		Peacock		◆RIGIL KENT.		SPICA		◆ARCTURUS	
240	38 17	030	20 38	063	51 42	090	40 34	139	60 11	200	47 55	294	31 15	331
241	38 41	029	21 22	062	52 32	090	41 07	139	59 54	201	47 09	293	30 51	330
242	39 04	028	22 06	062	53 21	089	41 39	139	59 36	201	46 23	292	30 25	329
243	39 27	027	22 50	061	54 11	089	42 12	139	59 18	202	45 37	291	29 59	328
244	39 49	025	23 33	060	55 01	088	42 45	139	58 59	203	44 50	290	29 32	327
245	40 10	024	24 16	059	55 51	088	43 18	139	58 39	204	44 04	289	29 05	326
246	40 30	023	24 59	059	56 40	087	43 51	139	58 19	204	43 17	288	28 37	325
247	40 48	022	25 41	058	57 30	086	44 24	139	57 58	205	42 29	288	28 08	324
248	41 06	020	26 23	057	58 19	086	44 57	139	57 37	205	41 42	287	27 39	323
249	41 23	019	27 05	056	59 09	085	45 29	139	57 16	206	40 54	286	27 09	322
250	41 39	018	27 46	056	59 59	084	46 02	139	56 54	207	40 06	285	26 38	321
251	41 54	017	28 27	055	60 48	084	46 35	139	56 31	207	39 18	285	26 07	321
252	42 08	015	29 07	054	61 37	083	47 07	139	56 08	208	38 30	284	25 35	320
253	42 20	014	29 47	053	62 27	082	47 40	139	55 45	208	37 41	283	25 02	319
254	42 32	013	30 27	052	63 16	081	48 12	139	55 21	209	36 53	282	24 29	318
	VEGA		◆ALTAIR		FOMALHAUT		◆Peacock		RIGIL KENT.		◆SPICA		ARCTURUS	
255	13 51	019	31 06	051	16 33	115	48 45	140	54 57	209	36 04	282	23 56	317
256	14 07	018	31 45	051	17 18	115	49 17	140	54 32	210	35 15	281	23 22	316
257	14 22	018	32 23	050	18 04	114	49 49	140	54 07	210	34 27	280	22 47	316
258	14 37	017	33 01	049	18 49	114	50 21	140	53 42	211	33 38	280	22 12	315
259	14 51	016	33 38	048	19 35	113	50 53	140	53 17	211	32 49	279	21 36	314
260	15 05	015	34 14	047	20 21	113	51 24	141	52 51	211	31 59	278	21 00	313
261	15 18	015	34 50	046	21 07	112	51 56	141	52 25	212	31 10	278	20 24	312
262	15 30	014	35 26	045	21 53	112	52 27	141	51 59	212	30 21	277	19 47	312
263	15 41	013	36 01	044	22 39	111	52 59	141	51 32	212	29 31	277	19 09	311
264	15 52	012	36 35	043	23 25	111	53 30	142	51 05	213	28 42	276	18 32	310
265	16 02	011	37 08	042	24 12	111	54 00	142	50 38	213	27 52	275	17 53	309
266	16 12	011	37 41	041	24 58	110	54 31	142	50 11	213	27 03	275	17 15	309
267	16 21	010	38 13	040	25 45	110	55 01	143	49 43	214	26 13	274	16 36	308
268	16 29	009	38 45	039	26 32	109	55 31	143	49 16	214	25 24	274	15 56	307
269	16 37	008	39 16	038	27 19	109	56 01	144	48 48	214	24 34	273	15 17	307

LAT 34°S

LHA ♈	Hc	Zn	Hc	Zn	Hc	Zn	Hc	Zn	Hc	Zn	Hc	Zn	Hc	Zn
	VEGA		ALTAIR		◆FOMALHAUT		ACHERNAR		◆RIGIL KENT.		ANTARES		◆Rasalhague	
270	16 43	007	39 46	037	28 06	108	16 35	149	48 20	214	69 00	285	43 03	352
271	16 50	007	40 15	035	28 53	108	17 00	149	47 52	214	68 12	284	42 55	350
272	16 55	006	40 44	034	29 41	108	17 26	148	47 24	215	67 24	283	42 46	349
273	17 00	005	41 11	033	30 28	107	17 52	148	46 55	215	66 35	282	42 36	348
274	17 04	004	41 38	032	31 16	107	18 19	148	46 27	215	65 47	281	42 25	346
275	17 07	003	42 04	031	32 04	106	18 45	147	45 58	215	64 58	280	42 12	345
276	17 10	003	42 29	030	32 51	106	19 12	147	45 30	215	64 09	279	41 59	344
277	17 12	002	42 53	028	33 39	106	19 40	147	45 01	215	63 20	278	41 44	342
278	17 13	001	43 16	027	34 27	105	20 07	146	44 32	216	62 30	278	41 29	341
279	17 13	000	43 38	026	35 15	105	20 35	146	44 03	216	61 41	277	41 12	340
280	17 13	359	44 00	025	36 03	104	21 02	146	43 34	216	60 52	276	40 55	339
281	17 12	359	44 20	023	36 52	104	21 30	146	43 05	216	60 02	275	40 36	337
282	17 10	358	44 39	022	37 40	103	21 59	145	42 36	216	59 12	275	40 17	336
283	17 08	357	44 57	021	38 28	103	22 27	145	42 07	216	58 23	274	39 56	335
284	17 05	356	45 14	019	39 17	103	22 56	145	41 37	216	57 33	273	39 35	334
	ALTAIR		Enif		◆FOMALHAUT		ACHERNAR		◆RIGIL KENT.		ANTARES		◆Rasalhague	
285	45 30	018	31 24	049	40 05	102	23 24	144	41 08	216	56 44	273	39 13	333
286	45 45	017	32 01	048	40 54	102	23 54	144	40 39	216	55 54	272	38 49	332
287	45 58	015	32 38	047	41 43	101	24 23	144	40 10	216	55 04	272	38 25	330
288	46 11	014	33 14	046	42 32	101	24 52	144	39 40	216	54 14	271	38 00	329
289	46 22	012	33 50	046	43 20	101	25 22	143	39 11	216	53 25	270	37 34	328
290	46 32	011	34 25	045	44 09	100	25 51	143	38 42	216	52 35	270	37 08	327
291	46 41	010	35 00	044	44 58	100	26 21	143	38 13	216	51 45	269	36 40	326
292	46 49	008	35 34	043	45 47	099	26 51	143	37 44	216	50 56	269	36 12	325
293	46 55	007	36 07	042	46 37	099	27 21	143	37 14	216	50 06	268	35 43	324
294	47 00	005	36 40	041	47 26	098	27 52	142	36 45	216	49 16	268	35 14	323
295	47 04	004	37 12	040	48 15	098	28 22	142	36 16	216	48 26	267	34 43	322
296	47 07	002	37 43	038	49 04	098	28 53	142	35 47	216	47 37	267	34 12	321
297	47 08	001	38 14	037	49 54	097	29 24	142	35 18	216	46 47	266	33 41	320
298	47 09	359	38 44	036	50 43	097	29 55	141	34 49	216	45 57	266	33 08	319
299	47 08	358	39 13	035	51 32	096	30 26	141	34 20	215	45 08	265	32 35	318
	Enif		◆FOMALHAUT		ACHERNAR		◆RIGIL KENT.		ANTARES		Rasalhague		◆ALTAIR	
300	39 41	034	52 22	096	30 57	141	33 51	215	44 18	265	32 02	317	47 05	357
301	40 09	033	53 11	095	31 28	141	33 23	215	43 29	264	31 27	316	47 02	355
302	40 35	032	54 01	095	32 00	141	32 54	215	42 39	264	30 53	315	46 57	354
303	41 01	031	54 50	095	32 31	141	32 25	215	41 50	263	30 17	314	46 51	352
304	41 26	029	55 40	094	33 03	141	31 57	215	41 01	263	29 41	313	46 43	351
305	41 50	028	56 30	094	33 34	140	31 29	215	40 11	262	29 05	312	46 35	349
306	42 13	027	57 19	093	34 06	140	31 00	215	39 22	262	28 28	312	46 25	348
307	42 35	026	58 09	093	34 38	140	30 32	214	38 33	261	27 51	311	46 14	347
308	42 56	025	58 59	092	35 10	140	30 04	214	37 44	261	27 13	310	46 02	345
309	43 17	023	59 48	092	35 42	140	29 36	214	36 54	261	26 34	309	45 49	344
310	43 36	022	60 38	091	36 14	140	29 08	214	36 05	260	25 55	308	45 34	342
311	43 54	021	61 28	090	36 46	140	28 40	214	35 16	260	25 16	307	45 18	341
312	44 11	019	62 18	090	37 18	140	28 13	214	34 28	259	24 36	307	45 02	340
313	44 27	018	63 07	089	37 51	140	27 45	213	33 39	259	23 56	306	44 44	338
314	44 42	017	63 57	089	38 23	139	27 18	213	32 50	258	23 16	305	44 25	337
	Alpheratz		◆Diphda		ACHERNAR		◆RIGIL KENT.		ANTARES		Rasalhague		◆ALTAIR	
315	12 51	041	38 00	087	38 55	139	26 51	213	32 01	258	22 35	304	44 05	336
316	13 23	040	38 50	086	39 28	139	26 24	213	31 13	257	21 54	304	43 44	334
317	13 55	040	39 39	086	40 00	139	25 57	213	30 24	257	21 12	303	43 22	333
318	14 26	039	40 29	085	40 32	139	25 30	212	29 36	256	20 30	302	42 59	332
319	14 57	038	41 18	084	41 05	139	25 04	212	28 48	256	19 48	301	42 36	331
320	15 28	037	42 08	084	41 37	139	24 38	212	27 59	256	19 05	301	42 11	330
321	15 58	037	42 57	083	42 10	139	24 11	212	27 11	255	18 22	300	41 45	328
322	16 27	036	43 47	083	42 42	139	23 45	211	26 23	255	17 39	299	41 19	327
323	16 56	035	44 36	082	43 15	139	23 20	211	25 35	254	16 56	299	40 51	326
324	17 24	034	45 25	081	43 47	139	22 54	211	24 47	254	16 12	298	40 23	325
325	17 52	034	46 14	081	44 19	139	22 29	211	24 00	253	15 28	297	39 54	324
326	18 19	033	47 03	080	44 52	139	22 03	210	23 12	253	14 44	297	39 24	323
327	18 46	032	47 52	079	45 24	140	21 38	210	22 25	252	13 59	296	38 53	322
328	19 12	031	48 41	078	45 56	140	21 13	210	21 37	252	13 14	295	38 22	320
329	19 38	030	49 30	078	46 28	140	20 49	210	20 50	251	12 29	295	37 50	319
	◆Alpheratz		Diphda		◆ACHERNAR		Peacock		◆Nunki		ALTAIR		Enif	
330	20 02	030	50 18	077	47 01	140	62 07	208	49 36	268	37 17	318	45 59	354
331	20 27	029	51 07	076	47 33	140	61 43	209	48 47	268	36 44	317	45 53	353
332	20 50	028	51 55	075	48 05	140	61 19	210	47 57	267	36 10	316	45 47	351
333	21 13	027	52 43	074	48 36	140	60 54	210	47 07	267	35 35	315	45 39	350
334	21 36	026	53 30	073	49 08	140	60 29	211	46 18	266	35 00	314	45 30	349
335	21 57	025	54 18	073	49 40	141	60 03	212	45 28	266	34 24	313	45 19	347
336	22 18	024	55 05	072	50 11	141	59 36	212	44 39	265	33 48	312	45 08	346
337	22 38	024	55 52	071	50 43	141	59 09	213	43 49	265	33 11	312	44 55	345
338	22 58	023	56 39	070	51 14	141	58 42	214	42 59	264	32 33	311	44 41	343
339	23 17	022	57 26	069	51 45	142	58 14	214	42 10	264	31 55	310	44 26	342
340	23 35	021	58 12	068	52 16	142	57 46	215	41 21	263	31 17	309	44 10	341
341	23 52	020	58 58	067	52 46	142	57 17	215	40 31	263	30 38	308	43 53	339
342	24 09	019	59 43	065	53 17	142	56 48	216	39 42	262	29 58	307	43 35	338
343	24 25	018	60 28	064	53 47	143	56 19	216	38 53	262	29 18	306	43 16	337
344	24 40	017	61 13	063	54 17	143	55 50	217	38 03	261	28 38	305	42 56	335
	Alpheratz		◆Hamal		◆ACHERNAR		Peacock		◆Nunki		ALTAIR		Enif	
345	24 55	016	17 24	044	54 47	143	55 20	217	37 14	261	27 57	305	42 35	334
346	25 08	015	17 58	044	55 16	144	54 50	217	36 25	260	27 16	304	42 12	333
347	25 21	015	18 33	043	55 45	144	54 19	218	35 36	260	26 35	303	41 49	332
348	25 33	014	19 06	042	56 14	145	53 49	218	34 47	259	25 53	302	41 25	330
349	25 45	013	19 39	041	56 43	145	53 18	218	33 58	259	25 10	301	41 00	329
350	25 55	012	20 12	041	57 11	146	52 47	219	33 10	259	24 28	301	40 34	328
351	26 05	011	20 44	040	57 39	146	52 16	219	32 21	258	23 45	300	40 08	327
352	26 14	010	21 16	039	58 06	147	51 44	219	31 32	258	23 02	299	39 40	326
353	26 22	009	21 47	038	58 33	147	51 13	220	30 44	257	22 18	299	39 12	325
354	26 29	008	22 17	037	59 00	148	50 41	220	29 55	257	21 34	298	38 43	324
355	26 35	007	22 47	037	59 26	149	50 09	220	29 07	256	20 50	297	38 13	323
356	26 41	006	23 16	036	59 52	149	49 37	220	28 19	256	20 06	296	37 42	321
357	26 45	005	23 45	035	60 17	150	49 05	220	27 30	255	19 21	296	37 11	320
358	26 49	004	24 13	034	60 42	151	48 33	221	26 42	255	18 36	295	36 39	319
359	26 52	003	24 41	033	61 06	151	48 00	221	25 54	255	17 51	294	36 06	318

LHA ♈	Hc	Zn	Hc	Zn	Hc	Zn	Hc	Zn	Hc	Zn	Hc	Zn	Hc	Zn
	◆Alpheratz		Hamal		◆RIGEL		CANOPUS		ACHERNAR		◆Peacock		Enif	
0	25 55	002	24 17	032	14 02	090	23 52	139	62 22	151	48 13	222	34 49	318
1	25 56	001	24 42	031	14 51	090	24 25	138	62 45	152	47 40	222	34 15	317
2	25 56	000	25 07	030	15 40	089	24 57	138	63 08	153	47 08	222	33 41	316
3	25 56	359	25 32	029	16 29	089	25 30	138	63 30	154	46 35	222	33 07	315
4	25 55	358	25 55	028	17 19	088	26 03	138	63 52	155	46 02	222	32 32	314
5	25 53	357	26 18	027	18 08	087	26 37	137	64 12	156	45 29	222	31 56	313
6	25 50	356	26 41	026	18 57	087	27 10	137	64 32	157	44 56	222	31 20	312
7	25 46	355	27 02	026	19 46	086	27 44	137	64 51	158	44 24	222	30 43	311
8	25 41	354	27 23	025	20 35	086	28 17	137	65 10	159	43 51	222	30 06	310
9	25 36	353	27 43	024	21 24	085	28 51	136	65 27	160	43 18	222	29 28	309
10	25 30	352	28 02	023	22 13	084	29 25	136	65 43	161	42 45	222	28 50	309
11	25 23	351	28 21	022	23 02	084	29 59	136	65 59	162	42 12	222	28 11	308
12	25 15	350	28 38	021	23 50	083	30 34	136	66 14	163	41 39	222	27 32	307
13	25 06	349	28 55	020	24 39	083	31 08	135	66 27	165	41 06	222	26 53	306
14	24 57	348	29 12	019	25 28	082	31 43	135	66 40	166	40 33	222	26 13	305
	◆Hamal		ALDEBARAN		RIGEL		◆CANOPUS		ACHERNAR		◆FOMALHAUT		Alpheratz	
15	29 27	018	17 27	054	26 17	081	32 17	135	66 51	167	63 39	273	24 47	348
16	29 41	017	18 06	054	27 05	081	32 52	135	67 02	168	62 50	272	24 36	347
17	29 55	016	18 46	053	27 54	080	33 27	135	67 11	170	62 01	272	24 24	346
18	30 08	015	19 25	052	28 42	079	34 02	134	67 20	171	61 12	271	24 11	345
19	30 20	014	20 03	051	29 30	079	34 38	134	67 27	172	60 23	271	23 58	344
20	30 31	012	20 41	051	30 18	078	35 13	134	67 33	174	59 34	270	23 44	343
21	30 41	011	21 19	050	31 06	077	35 48	134	67 37	175	58 45	269	23 29	342
22	30 50	010	21 57	049	31 54	077	36 24	134	67 41	177	57 55	269	23 13	341
23	30 59	009	22 33	048	32 42	076	37 00	133	67 43	178	57 06	268	22 57	340
24	31 06	008	23 10	047	33 30	075	37 35	133	67 44	179	56 17	268	22 40	339
25	31 13	007	23 46	047	34 17	075	38 11	133	67 44	181	55 28	267	22 22	338
26	31 19	006	24 21	046	35 04	074	38 47	133	67 43	182	54 39	267	22 04	337
27	31 23	005	24 56	045	35 51	073	39 23	133	67 40	184	53 50	266	21 45	337
28	31 27	004	25 31	044	36 38	072	39 59	133	67 36	185	53 01	266	21 25	336
29	31 30	003	26 05	043	37 25	072	40 35	133	67 31	187	52 12	265	21 04	335
	Hamal		ALDEBARAN		◆SIRIUS		CANOPUS		◆ACHERNAR		FOMALHAUT		◆Alpheratz	
30	31 32	002	26 38	042	24 40	094	41 11	133	67 25	188	51 23	265	20 43	334
31	31 33	001	27 11	041	25 29	093	41 47	133	67 18	189	50 34	264	20 21	333
32	31 34	000	27 43	041	26 18	093	42 23	132	67 09	191	49 45	264	19 59	332
33	31 33	359	28 15	040	27 07	092	43 00	132	67 00	192	48 56	263	19 36	331
34	31 31	358	28 46	039	27 56	091	43 36	132	66 49	193	48 08	263	19 12	331
35	31 29	356	29 16	038	28 46	091	44 12	132	66 37	195	47 19	262	18 47	330
36	31 25	355	29 46	037	29 35	090	44 49	132	66 24	196	46 30	262	18 22	329
37	31 21	354	30 15	036	30 24	090	45 25	132	66 10	197	45 41	262	17 57	328
38	31 16	353	30 44	035	31 13	089	46 01	132	65 55	198	44 53	261	17 31	327
39	31 09	352	31 12	034	32 02	089	46 38	132	65 40	199	44 04	261	17 04	327
40	31 02	351	31 39	033	32 51	088	47 14	132	65 23	200	43 16	260	16 37	326
41	30 54	350	32 05	032	33 40	087	47 50	132	65 05	202	42 27	260	16 09	325
42	30 45	349	32 31	031	34 29	087	48 27	132	64 47	203	41 39	259	15 40	324
43	30 35	348	32 56	030	35 18	086	49 03	132	64 28	204	40 51	259	15 11	324
44	30 25	347	33 20	029	36 07	086	49 39	132	64 07	205	40 03	258	14 42	323
	◆ALDEBARAN		BETELGEUSE		SIRIUS		◆CANOPUS		ACHERNAR		◆FOMALHAUT		Hamal	
45	33 43	028	30 52	053	36 56	085	50 15	133	63 47	206	39 15	258	30 13	346
46	34 06	027	31 31	052	37 45	084	50 52	133	63 25	206	38 27	258	30 01	345
47	34 27	026	32 10	051	38 34	084	51 28	133	63 03	207	37 39	257	29 48	344
48	34 48	025	32 48	050	39 23	083	52 04	133	62 40	208	36 51	257	29 33	343
49	35 08	024	33 25	049	40 12	082	52 40	133	62 17	209	36 03	256	29 18	342
50	35 28	022	34 03	048	41 01	082	53 16	133	61 52	210	35 15	256	29 03	341
51	35 46	021	34 39	048	41 49	081	53 51	133	61 28	210	34 28	255	28 46	340
52	36 03	020	35 15	047	42 38	080	54 27	134	61 03	211	33 40	255	28 29	339
53	36 20	019	35 50	046	43 26	080	55 02	134	60 37	212	32 53	255	28 11	338
54	36 35	018	36 25	045	44 14	079	55 38	134	60 11	213	32 05	254	27 52	337
55	36 50	017	36 59	044	45 03	078	56 13	134	59 44	213	31 18	254	27 32	336
56	37 04	016	37 33	043	45 51	077	56 48	135	59 17	214	30 31	253	27 11	335
57	37 16	014	38 06	042	46 38	077	57 23	135	58 50	214	29 44	253	26 50	334
58	37 28	013	38 38	040	47 26	076	57 58	135	58 22	215	28 57	252	26 28	333
59	37 39	012	39 10	039	48 14	075	58 32	136	57 53	215	28 10	252	26 06	332
	ALDEBARAN		◆BETELGEUSE		SIRIUS		◆CANOPUS		ACHERNAR		◆FOMALHAUT		Hamal	
60	37 49	011	39 40	038	49 01	074	59 06	136	57 25	216	27 24	251	25 42	331
61	37 57	010	40 10	037	49 48	073	59 40	137	56 56	216	26 37	251	25 18	330
62	38 05	008	40 40	036	50 35	073	60 14	137	56 27	217	25 51	251	24 53	329
63	38 12	007	41 08	035	51 22	072	60 47	138	55 57	217	25 04	250	24 28	328
64	38 17	006	41 36	034	52 09	071	61 20	138	55 27	217	24 18	250	24 02	328
65	38 22	005	42 03	032	52 55	070	61 53	139	54 57	218	23 32	249	23 35	327
66	38 26	004	42 29	031	53 41	069	62 25	139	54 27	218	22 46	249	23 08	326
67	38 28	002	42 54	030	54 27	068	62 57	140	53 57	218	22 00	248	22 40	325
68	38 30	001	43 18	029	55 12	067	63 29	140	53 26	219	21 15	248	22 12	324
69	38 30	000	43 41	028	55 57	066	64 00	141	52 55	219	20 29	247	21 42	323
70	38 29	359	44 03	026	56 42	065	64 30	142	52 24	219	19 44	247	21 13	322
71	38 28	357	44 25	025	57 26	064	65 00	143	51 53	220	18 59	247	20 43	322
72	38 25	356	44 45	024	58 10	062	65 30	144	51 22	220	18 14	246	20 12	321
73	38 21	355	45 04	022	58 53	061	65 58	144	50 50	220	17 29	246	19 40	320
74	38 16	354	45 22	021	59 36	060	66 27	145	50 18	220	16 44	245	19 09	319
	BETELGEUSE		◆PROCYON		Suhail		◆ACRUX		ACHERNAR		◆Diphda		ALDEBARAN	
75	45 39	020	35 05	051	42 22	120	22 01	153	49 47	220	31 07	269	38 11	353
76	45 53	018	35 43	050	43 05	120	22 24	153	49 15	220	30 18	268	38 04	351
77	46 10	017	36 20	049	43 48	119	22 46	152	48 43	221	29 29	268	37 56	350
78	46 24	016	36 57	048	44 31	119	23 09	152	48 11	221	28 40	267	37 47	349
79	46 37	014	37 34	047	45 14	119	23 32	152	47 39	221	27 51	267	37 37	348
80	46 48	013	38 09	046	45 57	119	23 55	152	47 06	221	27 02	266	37 26	347
81	46 58	011	38 45	045	46 40	119	24 19	151	46 34	221	26 13	266	37 14	345
82	47 07	010	39 19	044	47 23	119	24 43	151	46 02	221	25 24	265	37 01	344
83	47 15	008	39 53	043	48 06	118	25 06	151	45 30	221	24 35	265	36 47	343
84	47 22	007	40 26	042	48 49	118	25 30	151	44 57	221	23 46	264	36 33	342
85	47 27	005	40 59	041	49 33	118	25 54	150	44 25	221	22 57	263	36 17	341
86	47 31	004	41 31	040	50 16	118	26 19	150	43 52	221	22 08	263	36 00	340
87	47 34	003	42 02	039	50 59	118	26 43	150	43 20	221	21 19	262	35 43	338
88	47 35	001	42 32	037	51 43	118	27 08	150	42 47	221	20 31	262	35 24	337
89	47 36	000	43 01	036	52 26	118	27 33	150	42 15	221	19 42	261	35 05	336

LAT 35°S

LHA ♈	Hc	Zn	Hc	Zn	Hc	Zn	Hc	Zn	Hc	Zn	Hc	Zn	Hc	Zn
	PROCYON		REGULUS		◆Suhail		ACRUX		◆ACHERNAR		RIGEL		◆BETELGEUSE	
90	43 30	035	14 52	063	53 10	118	27 58	149	41 43	221	61 14	336	47 35	358
91	43 58	034	15 36	063	53 53	118	28 23	149	41 10	221	60 53	334	47 32	357
92	44 25	033	16 20	062	54 37	118	28 48	149	40 38	221	60 31	332	47 29	355
93	44 51	031	17 03	061	55 21	118	29 13	149	40 05	221	60 07	330	47 24	354
94	45 16	030	17 46	061	56 04	118	29 39	149	39 33	221	59 42	329	47 18	352
95	45 40	029	18 29	060	56 48	118	30 05	148	39 01	221	59 16	327	47 11	351
96	46 03	027	19 11	059	57 31	118	30 30	148	38 28	221	58 48	325	47 02	349
97	46 26	026	19 53	058	58 15	118	30 56	148	37 56	221	58 20	323	46 53	348
98	46 47	025	20 35	058	58 58	118	31 22	148	37 24	221	57 50	322	46 42	347
99	47 07	023	21 16	057	59 42	118	31 48	148	36 52	221	57 19	320	46 30	345
100	47 26	022	21 57	056	60 26	118	32 15	148	36 20	221	56 47	319	46 17	344
101	47 44	021	22 38	055	61 09	118	32 41	148	35 48	220	56 14	317	46 02	342
102	48 00	019	23 18	055	61 52	118	33 07	147	35 16	220	55 40	316	45 47	341
103	48 16	018	23 58	054	62 36	118	33 34	147	34 45	220	55 06	314	45 30	340
104	48 30	016	24 38	053	63 19	118	34 00	147	34 13	220	54 30	313	45 13	338
	PROCYON		REGULUS		◆Gienah		ACRUX		◆ACHERNAR		RIGEL		◆BETELGEUSE	
105	48 44	015	25 17	052	18 51	099	34 27	147	33 41	220	53 54	312	44 54	337
106	48 56	013	25 55	052	19 40	098	34 54	147	33 10	220	53 17	310	44 34	336
107	49 06	012	26 34	051	20 29	098	35 21	147	32 39	220	52 39	309	44 13	334
108	49 16	010	27 11	050	21 17	097	35 48	147	32 07	219	52 00	308	43 51	333
109	49 24	009	27 49	049	22 06	096	36 14	147	31 36	219	51 21	307	43 29	332
110	49 31	007	28 26	048	22 55	096	36 41	147	31 05	219	50 41	305	43 05	330
111	49 36	006	29 02	047	23 44	095	37 08	147	30 34	219	50 01	304	42 40	329
112	49 41	004	29 38	046	24 33	095	37 35	147	30 03	219	49 20	303	42 15	328
113	49 44	003	30 13	045	25 22	094	38 03	147	29 33	218	48 39	302	41 48	327
114	49 45	001	30 48	045	26 11	094	38 30	146	29 02	218	47 57	301	41 21	326
115	49 46	000	31 22	044	27 00	093	38 57	146	28 32	218	47 15	300	40 53	325
116	49 45	358	31 56	043	27 49	093	39 24	146	28 02	218	46 32	299	40 24	323
117	49 42	357	32 29	042	28 38	092	39 51	146	27 32	218	45 49	298	39 54	322
118	49 39	355	33 01	041	29 27	092	40 18	146	27 02	217	45 05	297	39 24	321
119	49 34	353	33 33	040	30 16	091	40 46	146	26 32	217	44 21	296	38 52	320
	REGULUS		◆SPICA		ACRUX		◆CANOPUS		RIGEL		BETELGEUSE		◆PROCYON	
120	34 04	039	13 29	094	41 13	146	65 27	216	43 37	295	38 20	319	49 28	352
121	34 34	038	14 18	094	41 40	146	64 58	217	42 52	294	37 48	318	49 20	350
122	35 04	037	15 08	093	42 07	147	64 28	218	42 07	293	37 15	317	49 11	349
123	35 33	036	15 57	093	42 34	147	63 57	219	41 22	293	36 41	316	49 01	347
124	36 01	035	16 46	092	43 01	147	63 26	220	40 37	292	36 06	315	48 50	346
125	36 29	034	17 35	091	43 28	147	62 55	220	39 51	291	35 31	314	48 37	344
126	36 56	032	18 24	091	43 55	147	62 23	221	39 05	290	34 55	313	48 24	343
127	37 22	031	19 13	090	44 22	147	61 50	221	38 19	289	34 19	312	48 09	342
128	37 47	030	20 02	090	44 49	147	61 17	222	37 32	289	33 42	311	47 53	340
129	38 11	029	20 51	089	45 15	147	60 44	223	36 45	288	33 05	310	47 35	339
130	38 35	028	21 41	089	45 42	147	60 11	223	35 58	287	32 27	309	47 17	337
131	38 57	027	22 30	088	46 09	147	59 37	223	35 11	286	31 49	308	46 57	336
132	39 19	026	23 19	087	46 35	148	59 03	224	34 24	286	31 10	307	46 37	335
133	39 40	024	24 08	087	47 01	148	58 29	224	33 37	285	30 31	307	46 15	333
134	40 00	023	24 57	086	47 28	148	57 55	224	32 49	284	29 51	306	45 52	332
	◆REGULUS		SPICA		◆ACRUX		CANOPUS		◆RIGEL		BETELGEUSE		PROCYON	
135	40 19	022	25 46	086	47 54	148	57 20	225	32 01	283	29 11	305	45 29	331
136	40 37	021	26 35	085	48 20	148	56 45	225	31 13	283	28 30	304	45 04	329
137	40 54	020	27 24	084	48 45	148	56 10	226	30 25	282	27 49	303	44 39	328
138	41 10	018	28 13	084	49 11	149	55 35	226	29 37	281	27 08	302	44 12	327
139	41 25	017	29 02	083	49 37	149	55 00	226	28 49	281	26 26	302	43 45	326
140	41 39	016	29 50	083	50 02	149	54 24	226	28 01	280	25 44	301	43 16	324
141	41 52	015	30 39	082	50 27	149	53 48	227	27 12	279	25 02	300	42 47	323
142	42 03	013	31 28	081	50 52	150	53 13	227	26 24	279	24 19	299	42 17	322
143	42 14	012	32 16	081	51 16	150	52 37	227	25 35	278	23 36	299	41 47	321
144	42 24	011	33 05	080	51 41	150	52 01	227	24 46	278	22 53	298	41 15	320
145	42 32	009	33 53	079	52 05	151	51 25	227	23 58	277	22 09	297	40 43	319
146	42 40	008	34 41	079	52 29	151	50 49	227	23 09	276	21 25	296	40 10	317
147	42 46	007	35 29	078	52 52	151	50 13	227	22 20	276	20 41	296	39 37	316
148	42 51	005	36 17	077	53 16	152	49 36	228	21 31	275	19 57	295	39 03	315
149	42 55	004	37 05	076	53 39	152	49 00	228	20 42	274	19 12	294	38 28	314
	REGULUS		◆Denebola		SPICA		◆ACRUX		CANOPUS		◆SIRIUS		PROCYON	
150	42 58	003	34 05	032	37 53	076	54 02	153	48 24	228	43 00	280	37 52	313
151	43 00	001	34 31	031	38 40	075	54 24	153	47 48	228	42 12	279	37 16	312
152	43 00	000	34 56	030	39 28	074	54 46	154	47 11	228	41 23	279	36 40	311
153	43 00	359	35 20	029	40 15	074	55 08	154	46 35	228	40 34	278	36 02	310
154	42 58	357	35 44	028	41 02	073	55 29	155	45 58	228	39 46	277	35 25	309
155	42 55	356	36 07	027	41 49	072	55 50	155	45 22	228	38 57	277	34 46	308
156	42 51	355	36 28	026	42 36	071	56 11	156	44 46	228	38 08	276	34 08	308
157	42 46	353	36 49	025	43 22	070	56 31	156	44 09	228	37 19	275	33 28	307
158	42 40	352	37 10	024	44 08	069	56 50	157	43 33	228	36 30	275	32 49	306
159	42 33	351	37 29	022	44 54	069	57 10	157	42 57	228	35 41	274	32 09	305
160	42 24	349	37 47	021	45 40	068	57 28	158	42 21	227	34 52	273	31 28	304
161	42 15	348	38 04	020	46 25	067	57 47	158	41 44	227	34 03	273	30 47	303
162	42 04	347	38 21	019	47 10	066	58 05	159	41 08	227	33 14	272	30 06	302
163	41 52	346	38 36	018	47 55	065	58 22	160	40 32	227	32 25	272	29 24	301
164	41 39	344	38 51	016	48 39	064	58 38	160	39 56	227	31 36	271	28 42	301
	ARCTURUS		◆ANTARES		ACRUX		◆CANOPUS		SIRIUS		PROCYON		◆REGULUS	
165	18 41	049	20 43	108	58 55	161	39 20	227	30 47	270	28 00	300	41 26	343
166	19 17	048	21 30	108	59 10	162	38 44	227	29 57	270	27 17	299	41 11	342
167	19 54	047	22 17	107	59 25	163	38 08	227	29 08	269	26 34	298	40 55	340
168	20 29	046	23 04	107	59 39	163	37 33	227	28 19	269	25 50	298	40 38	339
169	21 05	046	23 51	107	59 53	164	36 57	226	27 30	268	25 07	297	40 20	338
170	21 39	045	24 38	106	60 06	165	36 21	226	26 41	268	24 23	296	40 01	337
171	22 14	044	25 25	106	60 19	166	35 46	226	25 52	267	23 38	295	39 41	336
172	22 48	043	26 13	105	60 30	167	35 10	226	25 03	267	22 54	295	39 20	334
173	23 21	042	27 00	105	60 41	167	34 35	226	24 14	266	22 09	294	38 59	333
174	23 54	041	27 48	104	60 52	168	34 00	226	23 25	265	21 24	293	38 36	332
175	24 26	041	28 35	104	61 01	169	33 25	225	22 36	265	20 39	293	38 13	331
176	24 58	040	29 23	103	61 10	170	32 50	225	21 47	264	19 53	292	37 48	330
177	25 29	039	30 11	103	61 18	171	32 15	225	20 58	264	19 07	291	37 23	329
178	25 59	038	30 59	102	61 26	172	31 40	225	20 09	263	18 21	290	36 57	328
179	26 29	037	31 47	102	61 32	173	31 05	225	19 20	263	17 35	290	36 30	326

LHA ♈	Hc	Zn	Hc	Zn	Hc	Zn	Hc	Zn	Hc	Zn	Hc	Zn	Hc	Zn
	◆ARCTURUS		ANTARES		◆RIGIL KENT.		ACRUX		CANOPUS		◆Alphard		REGULUS	
180	26 59	036	32 35	101	53 52	148	61 38	174	30 31	224	46 18	298	36 03	325
181	27 27	035	33 24	101	54 18	148	61 43	175	29 57	224	45 34	297	35 35	324
182	27 55	034	34 12	100	54 43	149	61 47	176	29 22	224	44 50	296	35 06	323
183	28 23	033	35 00	100	55 08	149	61 50	177	28 48	224	44 06	295	34 36	322
184	28 50	032	35 49	099	55 33	150	61 53	178	28 15	223	43 21	294	34 06	321
185	29 16	031	36 37	099	55 58	150	61 55	178	27 41	223	42 36	293	33 35	320
186	29 41	031	37 26	099	56 22	151	61 56	179	27 07	223	41 51	292	33 03	319
187	30 06	030	38 14	098	56 46	151	61 56	180	26 34	223	41 05	292	32 30	318
188	30 29	029	39 03	098	57 10	152	61 55	181	26 01	222	40 19	291	31 57	317
189	30 53	028	39 52	097	57 33	152	61 53	182	25 28	222	39 33	290	31 24	316
190	31 15	027	40 41	097	57 55	153	61 51	183	24 55	222	38 47	289	30 50	316
191	31 36	026	41 30	096	58 18	153	61 48	184	24 22	222	38 00	288	30 15	315
192	31 57	024	42 18	096	58 39	154	61 44	185	23 50	221	37 13	288	29 40	314
193	32 17	023	43 07	095	59 01	155	61 39	186	23 17	221	36 26	287	29 04	313
194	32 36	022	43 56	095	59 21	155	61 33	187	22 45	221	35 39	286	28 28	312
	ARCTURUS		◆ANTARES		RIGIL KENT.		◆ACRUX		Suhail		REGULUS		◆Denebola	
195	32 55	021	44 45	094	59 42	156	61 27	188	45 10	241	27 51	311	37 36	338
196	33 12	020	45 34	094	60 01	157	61 20	189	44 27	241	27 13	310	37 17	337
197	33 29	019	46 23	093	60 20	158	61 12	190	43 44	241	26 36	309	36 58	336
198	33 44	018	47 13	092	60 39	158	61 03	191	43 01	240	25 57	309	36 37	335
199	33 59	017	48 02	092	60 57	159	60 53	192	42 18	240	25 19	308	36 16	334
200	34 13	016	48 51	091	61 14	160	60 43	192	41 36	240	24 40	307	35 53	332
201	34 26	015	49 40	091	61 30	161	60 32	193	40 53	240	24 00	306	35 30	331
202	34 38	014	50 29	090	61 46	162	60 21	194	40 11	240	23 20	305	35 06	330
203	34 49	013	51 18	090	62 01	163	60 08	195	39 29	239	22 40	305	34 41	329
204	34 59	011	52 07	089	62 16	163	59 55	196	38 46	239	21 59	304	34 16	328
205	35 09	010	52 57	089	62 29	164	59 42	196	38 04	239	21 18	303	33 49	327
206	35 17	009	53 46	088	62 42	165	59 27	197	37 22	239	20 37	302	33 22	326
207	35 24	008	54 35	087	62 54	166	59 13	198	36 40	238	19 55	302	32 55	325
208	35 30	007	55 24	087	63 06	167	58 57	199	35 59	238	19 13	301	32 26	324
209	35 36	006	56 13	086	63 16	168	58 41	199	35 17	238	18 31	300	31 57	323
	◆ARCTURUS		Rasalhague		◆Nunki		RIGIL KENT.		ACRUX		◆Suhail		SPICA	
210	35 40	004	20 25	057	27 22	104	63 26	169	58 24	200	34 35	238	64 50	339
211	35 43	003	21 06	056	28 10	104	63 34	170	58 07	201	33 54	237	64 32	337
212	35 46	002	21 47	056	28 58	103	63 42	171	57 50	201	33 13	237	64 12	335
213	35 47	001	22 27	055	29 46	103	63 49	172	57 31	202	32 32	237	63 51	333
214	35 47	000	23 07	054	30 34	102	63 55	174	57 13	203	31 50	236	63 28	331
215	35 47	359	23 47	053	31 22	102	64 00	175	56 54	203	31 10	236	63 03	329
216	35 45	358	24 26	052	32 10	101	64 04	176	56 34	204	30 29	236	62 37	327
217	35 43	356	25 05	052	32 58	101	64 07	177	56 14	204	29 48	236	62 10	325
218	35 39	355	25 43	051	33 46	101	64 10	178	55 53	205	29 08	235	61 41	323
219	35 34	354	26 21	050	34 35	100	64 11	179	55 33	205	28 27	235	61 11	322
220	35 29	353	26 58	049	35 23	100	64 11	180	55 11	206	27 47	235	60 40	320
221	35 22	352	27 35	048	36 12	099	64 10	181	54 50	206	27 07	234	60 08	318
222	35 15	351	28 12	047	37 00	099	64 09	182	54 28	207	26 28	234	59 34	317
223	35 06	349	28 47	047	37 49	098	64 06	184	54 05	207	25 48	234	59 00	315
224	34 57	348	29 23	046	38 38	098	64 03	185	53 42	208	25 08	233	58 25	314
	Alphecca		Rasalhague		◆Nunki		Peacock		◆RIGIL KENT.		SPICA		◆ARCTURUS	
225	27 44	009	29 58	045	39 26	097	33 12	140	63 58	186	57 49	312	34 46	347
226	27 51	008	30 32	044	40 15	097	33 44	139	63 53	187	57 12	311	34 35	346
227	27 57	007	31 06	043	41 04	096	34 16	139	63 47	188	56 34	309	34 22	345
228	28 02	006	31 39	042	41 53	096	34 48	139	63 39	189	55 56	308	34 09	344
229	28 07	005	32 12	041	42 42	095	35 20	139	63 31	190	55 17	307	33 55	343
230	28 10	004	32 43	040	43 31	095	35 52	139	63 22	191	54 37	305	33 40	342
231	28 13	003	33 15	039	44 20	094	36 25	139	63 12	192	53 57	304	33 24	341
232	28 15	002	33 45	038	45 09	094	36 57	139	63 01	193	53 16	303	33 07	339
233	28 16	001	34 15	037	45 58	093	37 29	139	62 50	194	52 34	302	32 49	338
234	28 16	000	34 45	036	46 47	093	38 02	138	62 38	195	51 52	301	32 31	337
235	28 15	359	35 13	035	47 36	092	38 35	138	62 24	196	51 10	300	32 12	336
236	28 14	358	35 41	034	48 25	091	39 07	138	62 10	197	50 27	299	31 51	335
237	28 11	357	36 08	033	49 14	091	39 40	138	61 56	198	49 43	297	31 30	334
238	28 08	356	36 34	032	50 03	090	40 13	138	61 40	199	49 00	296	31 09	333
239	28 04	355	37 00	031	50 53	090	40 46	138	61 24	200	48 15	296	30 46	332
	Rasalhague		ALTAIR		◆Nunki		Peacock		◆RIGIL KENT.		SPICA		◆ARCTURUS	
240	37 24	030	20 11	063	51 42	089	41 18	138	61 07	200	47 31	295	30 23	331
241	37 48	028	20 54	062	52 31	089	41 51	138	60 50	201	46 46	294	29 59	330
242	38 11	027	21 38	061	53 20	088	42 24	138	60 32	202	46 01	293	29 34	329
243	38 33	026	22 21	061	54 09	087	42 57	138	60 13	203	45 15	292	29 08	328
244	38 55	025	23 03	060	54 58	087	43 30	138	59 54	203	44 29	291	28 42	327
245	39 15	024	23 46	059	55 47	086	44 03	138	59 34	204	43 43	290	28 15	326
246	39 34	023	24 28	058	56 36	085	44 36	138	59 14	205	42 57	289	27 48	325
247	39 53	021	25 09	058	57 25	085	45 09	138	58 53	206	42 11	288	27 19	324
248	40 10	020	25 51	057	58 14	084	45 41	138	58 31	206	41 24	288	26 50	324
249	40 27	019	26 32	056	59 03	083	46 14	138	58 09	207	40 37	287	26 21	323
250	40 42	018	27 12	055	59 52	083	46 47	138	57 47	207	39 50	286	25 51	322
251	40 56	016	27 52	054	60 40	082	47 20	138	57 24	208	39 03	285	25 20	321
252	41 10	015	28 32	054	61 29	081	47 52	138	57 01	208	38 15	285	24 49	320
253	41 22	014	29 11	053	62 18	080	48 25	139	56 37	209	37 27	284	24 17	319
254	41 34	013	29 50	052	63 06	079	48 58	139	56 13	210	36 40	283	23 45	318
	VEGA		◆ALTAIR		FOMALHAUT		◆Peacock		RIGIL KENT.		◆SPICA		ARCTURUS	
255	12 54	019	30 29	051	16 58	115	49 30	139	55 49	210	35 52	282	23 12	317
256	13 10	018	31 06	050	17 43	114	50 02	139	55 24	210	35 04	282	22 38	317
257	13 25	018	31 44	049	18 28	114	50 35	139	54 59	211	34 15	281	22 04	316
258	13 40	017	32 21	048	19 13	113	51 07	139	54 34	211	33 27	280	21 30	315
259	13 54	016	32 57	047	19 58	113	51 39	139	54 08	212	32 39	280	20 55	314
260	14 07	015	33 33	046	20 44	112	52 11	140	53 42	212	31 50	279	20 19	313
261	14 20	015	34 08	045	21 29	112	52 42	140	53 16	212	31 02	278	19 43	313
262	14 32	014	34 43	044	22 15	111	53 14	140	52 49	213	30 13	278	19 07	312
263	14 43	013	35 17	043	23 01	111	53 45	140	52 22	213	29 24	277	18 30	311
264	14 54	012	35 51	042	23 47	111	54 16	141	51 55	213	28 35	276	17 53	310
265	15 04	011	36 24	041	24 33	110	54 47	141	51 28	214	27 47	276	17 15	310
266	15 13	011	36 56	040	25 19	110	55 18	141	51 01	214	26 58	275	16 37	309
267	15 22	010	37 27	039	26 05	109	55 48	142	50 33	214	26 09	275	15 59	308
268	15 30	009	37 58	038	26 52	109	56 19	142	50 06	215	25 20	274	15 20	307
269	15 37	008	38 28	037	27 38	108	56 49	143	49 38	215	24 31	273	14 41	307

LAT 35°S

LHA ♈	Hc	Zn	Hc	Zn	Hc	Zn	Hc	Zn	Hc	Zn	Hc	Zn	Hc	Zn
	VEGA		ALTAIR		◆FOMALHAUT		ACHERNAR		◆RIGIL KENT.		ANTARES		◆Rasalhague	
270	15 44	007	38 58	036	28 25	108	17 26	149	49 10	215	68 44	288	42 04	352
271	15 50	007	39 26	035	29 12	107	17 51	149	48 41	215	67 57	286	41 56	350
272	15 55	006	39 54	034	29 59	107	18 17	148	48 13	215	67 09	285	41 47	349
273	16 00	005	40 21	033	30 46	107	18 43	148	47 44	215	66 22	284	41 37	348
274	16 04	004	40 47	032	31 33	106	19 09	148	47 16	216	65 34	283	41 26	347
275	16 07	003	41 12	030	32 20	106	19 36	147	46 47	216	64 46	282	41 14	345
276	16 10	003	41 37	029	33 07	105	20 03	147	46 18	216	63 58	281	41 01	344
277	16 12	002	42 00	028	33 55	105	20 30	147	45 50	216	63 10	280	40 47	343
278	16 13	001	42 23	027	34 42	104	20 57	146	45 21	216	62 21	280	40 32	341
279	16 13	000	42 44	025	35 30	104	21 24	146	44 52	216	61 33	279	40 16	340
280	16 13	359	43 05	024	36 18	104	21 52	146	44 23	216	60 44	278	39 59	339
281	16 12	359	43 25	023	37 06	103	22 20	145	43 53	216	59 55	277	39 41	338
282	16 11	358	43 43	022	37 54	103	22 48	145	43 24	216	59 07	276	39 22	337
283	16 08	357	44 01	020	38 42	102	23 16	145	42 55	216	58 18	276	39 02	335
284	16 05	356	44 17	019	39 30	102	23 45	144	42 26	216	57 29	275	38 41	334
	ALTAIR		Enif		◆FOMALHAUT		ACHERNAR		◆RIGIL KENT.		ANTARES		◆Rasalhague	
285	44 33	018	30 44	049	40 18	101	24 13	144	41 57	217	56 40	274	38 19	333
286	44 47	016	31 21	048	41 06	101	24 42	144	41 27	217	55 51	274	37 56	332
287	45 00	015	31 57	047	41 54	101	25 11	144	40 58	217	55 02	273	37 33	331
288	45 13	014	32 33	046	42 43	100	25 40	143	40 29	217	54 13	272	37 09	330
289	45 23	012	33 08	045	43 31	100	26 10	143	40 00	217	53 24	272	36 43	329
290	45 33	011	33 42	044	44 19	099	26 39	143	39 30	216	52 34	271	36 17	328
291	45 42	009	34 16	043	45 08	099	27 09	143	39 01	216	51 45	271	35 51	326
292	45 49	008	34 49	042	45 57	098	27 39	142	38 32	216	50 56	270	35 23	325
293	45 56	007	35 22	041	46 45	098	28 09	142	38 03	216	50 07	269	34 55	324
294	46 01	005	35 54	040	47 34	097	28 39	142	37 34	216	49 18	269	34 26	323
295	46 04	004	36 25	039	48 23	097	29 10	142	37 05	216	48 29	268	33 56	322
296	46 07	002	36 56	038	49 12	096	29 40	142	36 36	216	47 40	268	33 26	321
297	46 08	001	37 26	037	50 00	096	30 11	141	36 07	216	46 50	267	32 55	320
298	46 09	359	37 55	036	50 49	096	30 42	141	35 38	216	46 01	267	32 23	319
299	46 08	358	38 24	035	51 38	095	31 12	141	35 09	216	45 12	266	31 50	318
	Enif		◆FOMALHAUT		ACHERNAR		◆RIGIL KENT.		ANTARES		Rasalhague		◆ALTAIR	
300	38 51	034	52 27	095	31 43	141	34 40	216	44 23	266	31 18	317	46 05	357
301	39 18	033	53 16	094	32 15	141	34 12	216	43 34	265	30 44	316	46 02	355
302	39 44	031	54 05	094	32 46	140	33 43	216	42 45	265	30 10	316	45 57	354
303	40 09	030	54 54	093	33 17	140	33 14	215	41 56	264	29 35	315	45 51	352
304	40 34	029	55 43	093	33 49	140	32 46	215	41 08	264	29 00	314	45 44	351
305	40 57	028	56 33	092	34 20	140	32 18	215	40 19	263	28 24	313	45 36	350
306	41 20	027	57 22	092	34 52	140	31 50	215	39 30	263	27 48	312	45 26	348
307	41 41	025	58 11	091	35 24	140	31 21	215	38 41	262	27 11	311	45 16	347
308	42 02	024	59 00	090	35 56	140	30 53	215	37 53	262	26 34	310	45 04	345
309	42 21	023	59 49	090	36 28	139	30 26	214	37 04	261	25 56	309	44 51	344
310	42 40	022	60 38	089	37 00	139	29 58	214	36 15	261	25 18	309	44 37	343
311	42 58	020	61 27	089	37 32	139	29 30	214	35 27	260	24 39	308	44 22	341
312	43 14	019	62 17	088	38 04	139	29 03	214	34 38	260	24 00	307	44 05	340
313	43 30	018	63 06	087	38 36	139	28 35	214	33 50	259	23 21	306	43 48	339
314	43 44	016	63 55	087	39 08	139	28 08	213	33 02	259	22 41	305	43 30	337
	Alpheratz		◆Diphda		ACHERNAR		◆RIGIL KENT.		ANTARES		Rasalhague		◆ALTAIR	
315	12 05	041	37 56	086	39 41	139	27 41	213	32 14	258	22 01	305	43 10	336
316	12 37	040	38 45	086	40 13	139	27 14	213	31 26	258	21 20	304	42 50	335
317	13 09	039	39 34	085	40 45	139	26 48	213	30 38	257	20 39	303	42 29	334
318	13 40	039	40 23	084	41 18	139	26 21	213	29 50	257	19 58	302	42 06	332
319	14 10	038	41 12	084	41 50	139	25 55	212	29 02	257	19 17	302	41 43	331
320	14 40	037	42 01	083	42 23	139	25 29	212	28 14	256	18 35	301	41 19	330
321	15 10	036	42 50	082	42 55	139	25 02	212	27 26	256	17 52	300	40 54	329
322	15 38	036	43 38	082	43 27	139	24 37	212	26 39	255	17 10	300	40 28	328
323	16 07	035	44 27	081	44 00	139	24 11	211	25 51	255	16 27	299	40 01	326
324	16 35	034	45 16	080	44 32	139	23 45	211	25 04	254	15 44	298	39 34	325
325	17 02	033	46 04	079	45 05	139	23 20	211	24 17	254	15 00	298	39 05	324
326	17 29	033	46 52	079	45 37	139	22 55	211	23 30	253	14 17	297	38 36	323
327	17 55	032	47 40	078	46 09	139	22 30	210	22 43	253	13 33	296	38 06	322
328	18 21	031	48 28	077	46 42	139	22 06	210	21 56	252	12 48	296	37 36	321
329	18 46	030	49 16	076	47 14	139	21 41	210	21 09	252	12 04	295	37 04	320
	◆Alpheratz		Diphda		◆ACHERNAR		Peacock		◆Nunki		ALTAIR		Enif	
330	19 10	029	50 04	076	47 46	139	63 00	209	49 38	269	36 32	319	44 59	354
331	19 34	029	50 51	075	48 18	139	62 36	210	48 49	269	36 00	318	44 54	353
332	19 57	028	51 39	074	48 50	139	62 11	211	47 59	268	35 26	317	44 47	352
333	20 20	027	52 26	073	49 22	140	61 45	211	47 10	268	34 52	316	44 40	350
334	20 42	026	53 13	072	49 54	140	61 20	212	46 21	267	34 18	315	44 31	349
335	21 03	025	53 59	071	50 26	140	60 53	213	45 32	267	33 43	314	44 21	348
336	21 24	024	54 46	070	50 57	140	60 26	213	44 43	266	33 07	313	44 09	346
337	21 43	023	55 32	069	51 29	140	59 59	214	43 54	266	32 31	312	43 57	345
338	22 03	023	56 18	068	52 00	141	59 32	215	43 05	265	31 54	311	43 44	344
339	22 21	022	57 03	067	52 31	141	59 03	215	42 16	265	31 17	310	43 29	342
340	22 39	021	57 48	066	53 02	141	58 35	216	41 27	264	30 39	309	43 14	341
341	22 56	020	58 33	065	53 33	141	58 06	216	40 38	264	30 01	308	42 57	340
342	23 12	019	59 18	064	54 04	142	57 37	217	39 50	263	29 22	308	42 39	338
343	23 28	018	60 02	063	54 34	142	57 07	217	39 01	263	28 43	307	42 21	337
344	23 43	017	60 45	061	55 05	142	56 38	218	38 12	262	28 03	306	42 01	336
	Alpheratz		◆Hamal		◆ACHERNAR		Peacock		◆Nunki		ALTAIR		Enif	
345	23 57	016	16 41	044	55 35	143	56 07	218	37 23	262	27 23	305	41 40	335
346	24 11	015	17 15	043	56 04	143	55 37	218	36 35	261	26 43	304	41 19	333
347	24 23	014	17 49	043	56 34	143	55 07	219	35 46	261	26 02	303	40 56	332
348	24 35	013	18 22	042	57 03	144	54 36	219	34 58	260	25 20	303	40 33	331
349	24 46	013	18 54	041	57 32	144	54 05	219	34 09	260	24 39	302	40 09	330
350	24 56	012	19 26	040	58 00	145	53 33	220	33 21	259	23 57	301	39 43	329
351	25 06	011	19 58	040	58 28	145	53 02	220	32 33	259	23 15	300	39 17	327
352	25 15	010	20 29	039	58 56	146	52 30	220	31 45	258	22 32	300	38 50	326
353	25 22	009	20 59	038	59 23	146	51 59	220	30 57	258	21 49	299	38 23	325
354	25 29	008	21 29	037	59 50	147	51 27	221	30 09	257	21 06	298	37 54	324
355	25 36	007	21 59	036	60 17	148	50 55	221	29 21	257	20 23	297	37 25	323
356	25 41	006	22 27	035	60 43	148	50 23	221	28 33	256	19 39	297	36 55	322
357	25 46	005	22 56	035	61 09	149	49 50	221	27 45	256	18 55	296	36 25	321
358	25 49	004	23 23	034	61 34	150	49 18	221	26 58	255	18 11	295	35 53	320
359	25 52	003	23 50	033	61 58	150	48 45	221	26 10	255	17 26	295	35 21	319

LHA ♈	Hc	Zn	Hc	Zn	Hc	Zn	Hc	Zn	Hc	Zn	Hc	Zn	Hc	Zn
	Alpheratz		◆Hamal		RIGEL		◆CANOPUS		Peacock		◆Nunki		Enif	
0	24 55	002	23 26	032	14 02	090	24 37	138	48 58	222	25 38	255	34 04	318
1	24 56	001	23 51	031	14 51	089	25 10	138	48 25	222	24 52	255	33 31	317
2	24 56	000	24 15	030	15 39	089	25 42	138	47 52	222	24 05	254	32 58	316
3	24 56	359	24 39	029	16 28	088	26 15	138	47 19	223	23 18	254	32 24	315
4	24 55	358	25 02	028	17 16	088	26 48	137	46 47	223	22 32	253	31 50	314
5	24 53	357	25 25	027	18 05	087	27 21	137	46 14	223	21 45	253	31 15	313
6	24 50	356	25 47	026	18 53	086	27 54	137	45 41	223	20 59	252	30 39	313
7	24 46	355	26 08	025	19 42	086	28 27	136	45 08	223	20 13	252	30 03	312
8	24 42	354	26 28	024	20 30	085	29 01	136	44 35	223	19 27	251	29 27	311
9	24 36	353	26 48	023	21 18	085	29 34	136	44 02	223	18 41	251	28 50	310
10	24 30	352	27 07	022	22 07	084	30 08	136	43 29	223	17 55	250	28 12	309
11	24 24	351	27 25	021	22 55	083	30 42	135	42 56	223	17 10	250	27 34	308
12	24 16	350	27 42	020	23 43	083	31 16	135	42 23	223	16 24	249	26 56	307
13	24 07	350	27 59	019	24 31	082	31 51	135	41 51	223	15 39	249	26 17	306
14	23 58	349	28 15	018	25 19	081	32 25	135	41 18	222	14 54	248	25 38	306
	◆Hamal		ALDEBARAN		RIGEL		◆CANOPUS		Peacock		◆FOMALHAUT		Alpheratz	
15	28 30	017	16 52	054	26 07	081	33 00	134	40 45	222	63 35	275	23 48	348
16	28 44	016	17 31	053	26 55	080	33 34	134	40 12	222	62 47	274	23 37	347
17	28 57	015	18 09	053	27 43	079	34 09	134	39 40	222	61 58	274	23 26	346
18	29 10	014	18 48	052	28 31	079	34 44	134	39 07	222	61 10	273	23 13	345
19	29 21	013	19 26	051	29 18	078	35 19	134	38 35	222	60 21	272	23 00	344
20	29 32	012	20 03	050	30 06	077	35 54	133	38 02	222	59 33	272	22 47	343
21	29 42	011	20 40	050	30 53	077	36 30	133	37 30	222	58 44	271	22 32	342
22	29 51	010	21 17	049	31 40	076	37 05	133	36 57	222	57 56	270	22 17	341
23	29 59	009	21 53	048	32 27	075	37 41	133	36 25	222	57 07	270	22 01	340
24	30 07	008	22 29	047	33 14	075	38 16	133	35 53	221	56 19	269	21 44	339
25	30 13	007	23 04	046	34 01	074	38 52	133	35 21	221	55 30	269	21 27	339
26	30 19	006	23 39	045	34 47	073	39 28	132	34 49	221	54 42	268	21 08	338
27	30 24	005	24 14	045	35 34	072	40 03	132	34 17	221	53 53	268	20 50	337
28	30 27	004	24 47	044	36 20	072	40 39	132	33 46	221	53 05	267	20 30	336
29	30 30	003	25 21	043	37 06	071	41 15	132	33 14	221	52 16	267	20 10	335
	Hamal		◆ALDEBARAN		RIGEL		SIRIUS		◆CANOPUS		Peacock		◆FOMALHAUT	
30	30 32	002	25 54	042	37 51	070	24 44	093	41 51	132	32 42	220	51 28	266
31	30 33	001	26 26	041	38 37	069	25 32	093	42 28	132	32 11	220	50 39	266
32	30 34	000	26 57	040	39 22	068	26 21	092	43 04	132	31 40	220	49 51	265
33	30 33	359	27 28	039	40 07	068	27 09	092	43 40	132	31 09	220	49 03	265
34	30 31	358	27 59	038	40 52	067	27 58	091	44 16	132	30 38	220	48 14	264
35	30 29	357	28 29	037	41 36	066	28 46	090	44 52	132	30 07	219	47 26	264
36	30 25	355	28 58	037	42 21	065	29 35	090	45 29	132	29 36	219	46 38	263
37	30 21	354	29 27	036	43 04	064	30 23	089	46 05	132	29 06	219	45 50	263
38	30 16	353	29 54	035	43 48	063	31 12	089	46 41	131	28 35	219	45 02	262
39	30 10	352	30 22	034	44 31	062	32 00	088	47 18	131	28 05	218	44 14	262
40	30 03	351	30 48	033	45 14	061	32 49	087	47 54	131	27 35	218	43 26	261
41	29 55	350	31 14	032	45 56	060	33 37	087	48 30	131	27 05	218	42 38	261
42	29 46	349	31 39	031	46 38	059	34 26	086	49 07	132	26 35	218	41 50	260
43	29 37	348	32 04	030	47 19	058	35 14	086	49 43	132	26 06	217	41 02	260
44	29 26	347	32 27	029	48 00	057	36 03	085	50 19	132	25 37	217	40 14	259
	◆ALDEBARAN		RIGEL		SIRIUS		◆CANOPUS		ACHERNAR		◆FOMALHAUT		Hamal	
45	32 50	028	48 41	056	36 51	084	50 56	132	64 41	206	39 27	259	29 15	346
46	33 12	026	49 21	055	37 39	084	51 32	132	64 19	207	38 39	258	29 03	345
47	33 33	025	50 00	054	38 27	083	52 08	132	63 56	208	37 52	258	28 50	344
48	33 54	024	50 39	053	39 15	082	52 44	132	63 33	209	37 04	257	28 36	343
49	34 13	023	51 17	051	40 04	082	53 20	132	63 09	210	36 17	257	28 21	342
50	34 32	022	51 55	050	40 51	081	53 56	132	62 44	211	35 30	257	28 06	341
51	34 50	021	52 32	049	41 39	080	54 32	132	62 19	211	34 42	256	27 50	340
52	35 07	020	53 08	048	42 27	079	55 08	133	61 54	212	33 55	256	27 33	339
53	35 23	019	53 44	046	43 15	079	55 44	133	61 28	213	33 08	255	27 15	338
54	35 38	018	54 18	045	44 02	078	56 19	133	61 01	213	32 21	255	26 56	337
55	35 53	017	54 52	043	44 50	077	56 55	133	60 34	214	31 35	254	26 37	336
56	36 06	015	55 25	042	45 37	076	57 30	134	60 07	215	30 48	254	26 17	335
57	36 18	014	55 57	041	46 24	076	58 05	134	59 39	215	30 01	253	25 56	334
58	36 30	013	56 28	039	47 11	075	58 40	134	59 11	216	29 15	253	25 35	333
59	36 40	012	56 58	037	47 58	074	59 14	135	58 42	216	28 29	252	25 13	332
	ALDEBARAN		◆BETELGEUSE		SIRIUS		◆ACRUX		ACHERNAR		◆FOMALHAUT		Hamal	
60	36 50	011	38 53	038	48 44	073	17 48	158	58 13	217	27 42	252	24 50	331
61	36 58	010	39 22	037	49 31	072	18 07	157	57 44	217	26 56	252	24 26	330
62	37 06	008	39 51	036	50 17	071	18 26	157	57 15	218	26 10	251	24 02	330
63	37 12	007	40 19	034	51 03	071	18 45	157	56 45	218	25 24	251	23 37	329
64	37 18	006	40 46	033	51 48	070	19 04	156	56 15	218	24 39	250	23 11	328
65	37 22	005	41 12	032	52 34	069	19 24	156	55 45	219	23 53	250	22 45	327
66	37 26	004	41 37	031	53 19	068	19 44	156	55 14	219	23 08	249	22 18	326
67	37 28	002	42 02	030	54 04	067	20 04	155	54 43	219	22 22	249	21 51	325
68	37 30	001	42 25	028	54 48	066	20 25	155	54 13	220	21 37	248	21 23	324
69	37 30	000	42 48	027	55 32	065	20 45	155	53 42	220	20 52	248	20 54	323
70	37 29	359	43 10	026	56 16	063	21 06	154	53 10	220	20 07	247	20 25	323
71	37 28	357	43 30	025	56 59	062	21 28	154	52 39	220	19 23	247	19 55	322
72	37 25	356	43 50	023	57 41	061	21 49	154	52 07	221	18 38	246	19 25	321
73	37 21	355	44 09	022	58 24	060	22 11	153	51 36	221	17 54	246	18 54	320
74	37 17	354	44 26	021	59 05	059	22 33	153	51 04	221	17 09	245	18 23	319
	BETELGEUSE		◆SIRIUS		Suhail		◆ACRUX		ACHERNAR		◆Diphda		ALDEBARAN	
75	44 43	019	59 46	057	42 52	119	22 55	153	50 32	221	31 08	270	37 11	353
76	44 58	018	60 27	056	43 34	119	23 17	153	50 00	221	30 20	269	37 04	351
77	45 13	017	61 07	054	44 17	119	23 39	152	49 28	221	29 31	268	36 57	350
78	45 26	015	61 46	053	45 00	118	24 02	152	48 56	222	28 42	268	36 48	349
79	45 38	014	62 24	051	45 42	118	24 25	152	48 24	222	27 54	267	36 38	348
80	45 49	012	63 02	050	46 25	118	24 48	151	47 51	222	27 05	267	36 28	347
81	45 59	011	63 39	048	47 08	118	25 12	151	47 19	222	26 17	266	36 16	346
82	46 08	010	64 14	047	47 51	118	25 35	151	46 47	222	25 29	266	36 03	344
83	46 16	008	64 49	045	48 34	117	25 59	151	46 14	222	24 40	265	35 50	343
84	46 22	007	65 23	043	49 17	117	26 23	150	45 42	222	23 52	264	35 36	342
85	46 27	005	65 55	041	50 01	117	26 47	150	45 10	222	23 04	264	35 20	341
86	46 31	004	66 26	039	50 44	117	27 11	150	44 37	222	22 15	263	35 04	340
87	46 34	002	66 56	037	51 27	117	27 35	150	44 05	222	21 27	263	34 47	339
88	46 35	001	67 25	035	52 10	117	28 00	150	43 32	222	20 39	262	34 29	338
89	46 36	000	67 52	033	52 54	117	28 24	149	43 00	222	19 51	262	34 10	337

LAT 36°S

LHA ♈	Hc	Zn	Hc	Zn	Hc	Zn	Hc	Zn	Hc	Zn	Hc	Zn	Hc	Zn
	PROCYON		REGULUS		◆Suhail		ACRUX		◆ACHERNAR		RIGEL		◆BETELGEUSE	
90	42 41	035	14 25	063	53 37	116	28 49	149	42 27	222	60 19	337	46 35	358
91	43 08	033	15 09	062	54 21	116	29 14	149	41 55	222	59 59	335	46 32	357
92	43 34	032	15 51	062	55 04	116	29 39	149	41 23	222	59 38	333	46 29	355
93	44 00	031	16 34	061	55 48	116	30 05	149	40 50	222	59 15	331	46 24	354
94	44 24	030	17 16	060	56 31	116	30 30	148	40 18	222	58 51	329	46 19	352
95	44 48	028	17 58	060	57 15	116	30 56	148	39 46	222	58 26	328	46 12	351
96	45 10	027	18 40	059	57 58	116	31 21	148	39 14	221	57 59	326	46 03	350
97	45 32	026	19 22	058	58 42	116	31 47	148	38 41	221	57 31	324	45 54	348
98	45 52	024	20 03	057	59 26	116	32 13	148	38 09	221	57 02	323	45 44	347
99	46 12	023	20 43	057	60 09	116	32 39	147	37 37	221	56 33	321	45 32	345
100	46 30	022	21 24	056	60 53	116	33 05	147	37 05	221	56 02	320	45 19	344
101	46 47	020	22 04	055	61 36	116	33 31	147	36 34	221	55 30	318	45 05	343
102	47 04	019	22 43	054	62 20	116	33 58	147	36 02	221	54 57	317	44 50	341
103	47 19	017	23 23	054	63 03	116	34 24	147	35 30	221	54 23	315	44 34	340
104	47 33	016	24 01	053	63 47	117	34 51	147	34 59	220	53 49	314	44 17	339
	PROCYON		REGULUS		◆Gienah		ACRUX		◆ACHERNAR		RIGEL		◆BETELGEUSE	
105	47 46	015	24 40	052	19 00	098	35 17	147	34 27	220	53 13	313	43 59	337
106	47 57	013	25 18	051	19 48	098	35 44	147	33 56	220	52 37	311	43 39	336
107	48 07	012	25 55	050	20 36	097	36 11	147	33 25	220	52 00	310	43 19	335
108	48 17	010	26 33	049	21 24	097	36 38	146	32 53	220	51 23	309	42 58	333
109	48 25	009	27 09	049	22 13	096	37 04	146	32 22	220	50 45	308	42 36	332
110	48 31	007	27 45	048	23 01	095	37 31	146	31 52	219	50 06	306	42 12	331
111	48 37	006	28 21	047	23 49	095	37 58	146	31 21	219	49 27	305	41 48	330
112	48 41	004	28 56	046	24 38	094	38 25	146	30 50	219	48 47	304	41 24	329
113	48 44	003	29 31	045	25 26	094	38 53	146	30 20	219	48 06	303	40 58	327
114	48 45	001	30 05	044	26 15	093	39 20	146	29 49	219	47 26	302	40 31	326
115	48 46	000	30 38	043	27 03	093	39 47	146	29 19	218	46 44	301	40 04	325
116	48 45	358	31 11	042	27 52	092	40 14	146	28 49	218	46 02	300	39 35	324
117	48 43	357	31 44	041	28 40	092	40 41	146	28 19	218	45 20	299	39 06	323
118	48 39	355	32 16	040	29 29	091	41 08	146	27 50	218	44 37	298	38 37	322
119	48 34	354	32 47	039	30 17	090	41 35	146	27 20	217	43 54	297	38 06	321
	REGULUS		◆SPICA		ACRUX		◆CANOPUS		RIGEL		BETELGEUSE		◆PROCYON	
120	33 17	038	13 34	094	42 03	146	66 15	218	43 11	296	37 35	319	48 28	352
121	33 47	037	14 22	093	42 30	146	65 45	219	42 27	295	37 03	318	48 21	351
122	34 16	036	15 11	093	42 57	146	65 15	219	41 43	294	36 31	317	48 12	349
123	34 44	035	15 59	092	43 24	146	64 43	220	40 59	293	35 57	316	48 03	348
124	35 12	034	16 48	092	43 51	146	64 12	221	40 14	293	35 24	315	47 52	346
125	35 39	033	17 36	091	44 18	146	63 40	222	39 29	292	34 49	314	47 40	345
126	36 05	032	18 25	091	44 45	146	63 08	222	38 44	291	34 14	313	47 26	343
127	36 30	031	19 13	090	45 12	146	62 35	223	37 58	290	33 39	312	47 12	342
128	36 55	030	20 02	089	45 39	146	62 02	223	37 13	289	33 03	312	46 56	340
129	37 19	029	20 50	089	46 06	147	61 28	224	36 27	288	32 26	311	46 39	339
130	37 42	028	21 39	088	46 32	147	60 54	224	35 41	288	31 49	310	46 21	338
131	38 04	026	22 27	088	46 59	147	60 20	225	34 54	287	31 11	309	46 02	336
132	38 25	025	23 16	087	47 26	147	59 46	225	34 08	286	30 33	308	45 42	335
133	38 45	024	24 04	086	47 52	147	59 12	225	33 21	285	29 55	307	45 21	334
134	39 05	023	24 53	086	48 18	147	58 37	226	32 34	285	29 16	306	44 59	332
	◆REGULUS		SPICA		◆ACRUX		CANOPUS		◆RIGEL		BETELGEUSE		PROCYON	
135	39 23	022	25 41	085	48 44	147	58 02	226	31 47	284	28 36	305	44 36	331
136	39 41	021	26 29	085	49 11	148	57 27	226	31 00	283	27 56	304	44 12	330
137	39 57	019	27 18	084	49 36	148	56 52	227	30 13	283	27 16	304	43 48	329
138	40 13	018	28 06	083	50 02	148	56 16	227	29 25	282	26 36	303	43 22	327
139	40 27	017	28 54	083	50 28	148	55 41	227	28 38	281	25 55	302	42 55	326
140	40 41	016	29 42	082	50 53	149	55 05	227	27 50	281	25 13	301	42 28	325
141	40 53	014	30 30	081	51 18	149	54 29	228	27 02	280	24 32	300	41 59	324
142	41 05	013	31 18	081	51 43	149	53 53	228	26 14	279	23 50	300	41 30	323
143	41 15	012	32 06	080	52 08	149	53 17	228	25 26	279	23 07	299	41 00	321
144	41 25	010	32 54	079	52 33	150	52 41	228	24 38	278	22 25	298	40 29	320
145	41 33	009	33 42	079	52 57	150	52 05	228	23 50	277	21 42	297	39 58	319
146	41 40	008	34 29	078	53 21	150	51 29	228	23 02	277	20 58	297	39 26	318
147	41 46	007	35 16	077	53 45	151	50 53	228	22 14	276	20 15	296	38 53	317
148	41 51	005	36 04	076	54 09	151	50 17	228	21 26	275	19 31	295	38 20	316
149	41 55	004	36 51	076	54 32	152	49 40	228	20 37	275	18 47	295	37 46	315
	REGULUS		◆SPICA		ACRUX		◆Miaplacidus		CANOPUS		SIRIUS		◆PROCYON	
150	41 58	003	37 38	075	54 55	152	55 42	187	49 04	228	42 49	281	37 11	314
151	42 00	001	38 25	074	55 17	152	55 36	188	48 28	229	42 02	280	36 36	313
152	42 00	000	39 11	073	55 40	153	55 29	188	47 51	229	41 14	279	36 00	312
153	42 00	359	39 58	073	56 02	153	55 22	189	47 15	229	40 26	279	35 23	311
154	41 58	357	40 44	072	56 23	154	55 14	189	46 38	229	39 38	278	34 46	310
155	41 56	356	41 30	071	56 44	154	55 06	190	46 02	228	38 50	277	34 09	309
156	41 52	355	42 16	070	57 05	155	54 57	191	45 26	228	38 01	277	33 31	308
157	41 47	353	43 01	069	57 26	155	54 48	191	44 50	228	37 13	276	32 52	307
158	41 41	352	43 47	069	57 45	156	54 38	192	44 13	228	36 25	275	32 13	306
159	41 34	351	44 32	068	58 05	157	54 28	192	43 37	228	35 37	275	31 34	305
160	41 25	350	45 16	067	58 24	157	54 18	193	43 01	228	34 48	274	30 54	305
161	41 16	348	46 01	066	58 42	158	54 07	193	42 25	228	34 00	274	30 14	304
162	41 06	347	46 45	065	59 00	159	53 56	194	41 49	228	33 11	273	29 34	303
163	40 54	346	47 29	064	59 18	159	53 44	194	41 12	228	32 23	272	28 53	302
164	40 42	344	48 12	063	59 35	160	53 32	195	40 37	228	31 34	272	28 11	301
	ARCTURUS		◆ANTARES		ACRUX		◆CANOPUS		SIRIUS		PROCYON		◆REGULUS	
165	18 01	048	21 02	108	59 51	161	40 01	228	30 46	271	27 29	300	40 28	343
166	18 37	048	21 48	108	60 07	161	39 25	228	29 57	270	26 47	300	40 14	342
167	19 13	047	22 35	107	60 22	162	38 49	227	29 09	270	26 05	299	39 58	341
168	19 48	046	23 21	107	60 37	163	38 13	227	28 20	269	25 22	298	39 42	340
169	20 23	045	24 08	106	60 51	164	37 38	227	27 32	269	24 39	297	39 24	338
170	20 57	044	24 54	106	61 04	165	37 02	227	26 43	268	23 56	296	39 06	337
171	21 31	044	25 41	105	61 17	165	36 27	227	25 55	268	23 12	296	38 46	336
172	22 04	043	26 28	105	61 29	166	35 52	227	25 06	267	22 29	295	38 26	335
173	22 37	042	27 15	104	61 40	167	35 16	226	24 18	266	21 44	294	38 05	334
174	23 09	041	28 02	104	61 50	168	34 41	226	23 29	266	21 00	294	37 43	332
175	23 40	040	28 49	103	62 00	169	34 06	226	22 41	265	20 15	293	37 20	331
176	24 12	039	29 37	103	62 09	170	33 32	226	21 52	265	19 31	292	36 56	330
177	24 42	039	30 24	102	62 18	171	32 57	226	21 04	264	18 46	291	36 32	329
178	25 12	038	31 12	102	62 25	172	32 22	225	20 16	264	18 00	291	36 06	328
179	25 41	037	31 59	101	62 32	173	31 48	225	19 28	263	17 15	290	35 40	327

LAT 36°S

LHA	Hc	Zn	Hc	Zn	Hc	Zn	Hc	Zn	Hc	Zn	Hc	Zn	Hc	Zn
♈	ARCTURUS		◆ANTARES		RIGIL KENT.		ACRUX		◆CANOPUS		Alphard		◆REGULUS	
180	26 10	036	32 47	101	54 42	147	62 38	174	31 14	225	45 49	299	35 13	326
181	26 38	035	33 35	100	55 08	148	62 43	174	30 40	225	45 06	298	34 46	325
182	27 06	034	34 22	100	55 34	148	62 47	175	30 06	224	44 23	297	34 17	324
183	27 33	033	35 10	099	56 00	149	62 50	176	29 32	224	43 40	296	33 48	323
184	27 59	032	35 58	099	56 25	149	62 53	177	28 58	224	42 56	295	33 19	322
185	28 24	031	36 46	098	56 50	149	62 55	178	28 24	224	42 12	294	32 48	321
186	28 49	030	37 34	098	57 14	150	62 56	179	27 51	223	41 27	293	32 17	320
187	29 13	029	38 22	097	57 39	150	62 56	180	27 18	223	40 43	292	31 45	319
188	29 37	028	39 11	097	58 02	151	62 55	181	26 45	223	39 58	292	31 13	318
189	29 59	027	39 59	096	58 26	152	62 53	182	26 12	222	39 12	291	30 40	317
190	30 21	026	40 47	096	58 49	152	62 51	183	25 39	222	38 27	290	30 07	316
191	30 42	025	41 35	095	59 11	153	62 48	184	25 07	222	37 41	289	29 33	315
192	31 03	024	42 24	095	59 33	153	62 43	185	24 35	222	36 55	288	28 58	314
193	31 22	023	43 12	094	59 55	154	62 38	186	24 03	221	36 09	288	28 23	313
194	31 41	022	44 01	094	60 16	155	62 33	187	23 31	221	35 22	287	27 47	312
	ARCTURUS		◆ANTARES		RIGIL KENT.		◆ACRUX		Suhail		◆REGULUS		Denebola	
195	31 59	021	44 49	093	60 36	155	62 26	188	45 38	242	27 11	311	36 40	338
196	32 16	020	45 38	093	60 56	156	62 19	189	44 56	242	26 35	311	36 22	337
197	32 32	019	46 26	092	61 16	157	62 11	190	44 13	241	25 57	310	36 03	336
198	32 47	018	47 15	091	61 34	158	62 02	191	43 30	241	25 20	309	35 43	335
199	33 02	017	48 03	091	61 53	158	61 52	192	42 48	241	24 42	308	35 22	334
200	33 15	016	48 52	090	62 10	159	61 42	193	42 06	241	24 04	307	35 00	333
201	33 28	015	49 40	090	62 27	160	61 31	194	41 23	241	23 25	306	34 37	332
202	33 40	013	50 29	089	62 43	161	61 19	195	40 41	240	22 45	306	34 14	331
203	33 51	012	51 17	088	62 58	162	61 06	195	39 59	240	22 06	305	33 50	330
204	34 01	011	52 06	088	63 13	163	60 53	196	39 17	240	21 26	304	33 25	328
205	34 10	010	52 54	087	63 27	164	60 39	197	38 35	240	20 45	303	32 59	327
206	34 18	009	53 43	087	63 40	165	60 25	198	37 53	239	20 05	303	32 32	326
207	34 25	008	54 31	086	63 53	166	60 10	199	37 12	239	19 24	302	32 05	325
208	34 31	007	55 20	085	64 04	167	59 54	199	36 30	239	18 42	301	31 37	324
209	34 36	006	56 08	085	64 15	168	59 38	200	35 49	238	18 01	300	31 09	323
	ARCTURUS		◆ANTARES		Peacock		RIGIL KENT.		◆ACRUX		Suhail		◆SPICA	
210	34 40	004	56 56	084	26 17	143	64 24	169	59 21	201	35 07	238	63 54	340
211	34 44	003	57 44	083	26 47	142	64 33	170	59 03	201	34 26	238	63 37	338
212	34 46	002	58 33	082	27 16	142	64 41	171	58 45	202	33 45	238	63 18	336
213	34 47	001	59 21	082	27 46	142	64 48	172	58 27	203	33 04	237	62 57	334
214	34 47	000	60 09	081	28 16	142	64 55	173	58 08	203	32 23	237	62 35	332
215	34 47	359	60 56	080	28 47	141	65 00	174	57 49	204	31 43	237	62 11	330
216	34 45	358	61 44	079	29 17	141	65 04	176	57 29	204	31 02	236	61 46	328
217	34 43	356	62 32	078	29 48	141	65 07	177	57 08	205	30 22	236	61 20	326
218	34 39	355	63 19	077	30 18	141	65 09	178	56 48	206	29 42	236	60 52	324
219	34 35	354	64 06	076	30 49	140	65 11	179	56 27	206	29 02	235	60 24	323
220	34 29	353	64 53	075	31 20	140	65 11	180	56 05	207	28 22	235	59 54	321
221	34 23	352	65 40	074	31 51	140	65 10	181	55 43	207	27 42	235	59 23	319
222	34 15	351	66 27	073	32 23	140	65 09	183	55 21	207	27 03	234	58 50	318
223	34 07	350	67 13	072	32 54	140	65 06	184	54 58	208	26 23	234	58 17	316
224	33 58	348	67 59	071	33 26	139	65 03	185	54 35	208	25 44	234	57 43	315
	Alphecca		Rasalhague		◆Nunki		Peacock		◆RIGIL KENT.		◆SPICA		ARCTURUS	
225	26 45	009	29 15	044	39 33	096	33 57	139	64 58	186	57 08	313	33 48	347
226	26 51	008	29 49	043	40 22	096	34 29	139	64 52	187	56 32	312	33 36	346
227	26 58	007	30 22	043	41 10	095	35 01	139	64 46	188	55 56	310	33 24	345
228	27 03	006	30 54	042	41 58	095	35 33	139	64 39	189	55 19	309	33 12	344
229	27 07	005	31 26	041	42 47	094	36 05	138	64 30	190	54 41	308	32 58	343
230	27 11	004	31 57	040	43 35	094	36 37	138	64 21	191	54 02	306	32 43	342
231	27 13	003	32 28	039	44 24	093	37 10	138	64 11	193	53 23	305	32 27	341
232	27 15	002	32 58	038	45 12	093	37 42	138	64 00	194	52 43	304	32 11	340
233	27 16	001	33 27	037	46 00	092	38 14	138	63 48	195	52 02	303	31 54	339
234	27 16	000	33 56	036	46 49	092	38 47	138	63 35	196	51 21	302	31 36	338
235	27 15	359	34 24	035	47 38	091	39 19	138	63 22	197	50 40	301	31 17	337
236	27 14	358	34 51	034	48 26	090	39 52	138	63 08	198	49 58	300	30 57	335
237	27 11	357	35 18	032	49 15	090	40 25	138	62 53	198	49 15	299	30 36	334
238	27 08	356	35 43	031	50 03	089	40 57	138	62 37	199	48 32	298	30 15	333
239	27 04	355	36 08	030	50 52	089	41 30	138	62 21	200	47 49	297	29 53	332
	Rasalhague		ALTAIR		◆Nunki		Peacock		◆RIGIL KENT.		SPICA		◆ARCTURUS	
240	36 32	029	19 43	062	51 40	088	42 03	137	62 04	201	47 05	296	29 30	331
241	36 55	028	20 26	062	52 29	087	42 36	137	61 46	202	46 21	295	29 07	330
242	37 18	027	21 09	061	53 17	087	43 09	137	61 27	203	45 37	294	28 42	329
243	37 39	026	21 51	060	54 06	086	43 41	137	61 08	203	44 53	293	28 17	329
244	38 00	025	22 33	059	54 54	085	44 14	137	60 49	204	44 08	292	27 52	328
245	38 20	023	23 15	059	55 42	085	44 47	137	60 29	205	43 22	291	27 25	327
246	38 39	022	23 56	058	56 31	084	45 20	137	60 08	206	42 37	290	26 58	326
247	38 57	021	24 37	057	57 19	083	45 53	137	59 47	206	41 51	289	26 30	325
248	39 14	020	25 18	056	58 07	082	46 26	137	59 25	207	41 05	288	26 02	324
249	39 30	019	25 58	056	58 55	082	46 59	137	59 03	208	40 19	288	25 33	323
250	39 45	017	26 38	055	59 43	081	47 32	137	58 40	208	39 33	287	25 04	322
251	39 59	016	27 17	054	60 31	080	48 04	138	58 17	209	38 46	286	24 33	321
252	40 12	015	27 56	053	61 19	079	48 37	138	57 53	209	38 00	285	24 03	320
253	40 24	014	28 35	052	62 06	078	49 10	138	57 30	210	37 13	285	23 31	319
254	40 35	012	29 13	051	62 54	077	49 42	138	57 05	210	36 26	284	23 00	319
	Rasalhague		◆ALTAIR		FOMALHAUT		◆ACHERNAR		RIGIL KENT.		◆SPICA		ARCTURUS	
255	40 45	011	29 51	050	17 23	114	12 31	155	56 41	211	35 38	283	22 27	318
256	40 54	010	30 28	050	18 07	114	12 52	154	56 16	211	34 51	282	21 54	317
257	41 02	009	31 04	049	18 52	113	13 14	154	55 50	212	34 04	282	21 21	316
258	41 08	007	31 41	048	19 37	113	13 35	153	55 25	212	33 16	281	20 47	315
259	41 14	006	32 16	047	20 21	112	13 57	153	54 59	212	32 28	280	20 13	314
260	41 19	005	32 51	046	21 06	112	14 19	153	54 33	213	31 41	280	19 38	314
261	41 22	003	33 26	045	21 51	112	14 42	152	54 06	213	30 53	279	19 03	313
262	41 25	002	34 00	044	22 37	111	15 05	152	53 39	214	30 05	278	18 27	312
263	41 26	001	34 33	043	23 22	111	15 28	151	53 12	214	29 17	278	17 51	311
264	41 26	000	35 06	042	24 07	110	15 51	151	52 45	214	28 28	277	17 14	311
265	41 25	358	35 38	041	24 53	110	16 15	151	52 18	214	27 40	276	16 37	310
266	41 23	357	36 10	040	25 39	109	16 39	150	51 50	215	26 52	276	15 59	309
267	41 20	356	36 41	039	26 25	109	17 03	150	51 23	215	26 04	275	15 22	308
268	41 16	354	37 11	038	27 11	108	17 27	149	50 55	215	25 15	275	14 43	308
269	41 11	353	37 40	037	27 57	108	17 52	149	50 27	215	24 27	274	14 05	307

LAT 36°S

LHA	Hc	Zn	Hc	Zn	Hc	Zn	Hc	Zn	Hc	Zn	Hc	Zn	Hc	Zn
♈	ALTAIR		Enif		◆FOMALHAUT		ACHERNAR		◆RIGIL KENT.		ANTARES		◆Rasalhague	
270	38 09	036	20 12	060	28 43	107	18 17	149	49 58	216	68 24	290	41 04	352
271	38 37	035	20 54	060	29 29	107	18 43	148	49 30	216	67 39	289	40 57	351
272	39 04	033	21 36	059	30 16	106	19 08	148	49 02	216	66 52	288	40 48	349
273	39 30	032	22 17	058	31 03	106	19 34	148	48 33	216	66 06	286	40 39	348
274	39 56	031	22 58	057	31 49	106	20 00	147	48 04	216	65 19	285	40 28	347
275	40 20	030	23 39	057	32 36	105	20 26	147	47 36	216	64 32	284	40 16	345
276	40 44	029	24 20	056	33 23	105	20 53	147	47 07	217	63 45	283	40 04	344
277	41 07	028	25 00	055	34 10	104	21 20	146	46 38	217	62 58	282	39 50	343
278	41 29	026	25 39	054	34 57	104	21 47	146	46 09	217	62 10	281	39 35	342
279	41 50	025	26 18	053	35 44	103	22 14	146	45 40	217	61 23	281	39 19	341
280	42 10	024	26 57	053	36 32	103	22 41	145	45 11	217	60 35	280	39 03	339
281	42 29	023	27 36	052	37 19	102	23 09	145	44 42	217	59 47	279	38 45	338
282	42 47	021	28 14	051	38 06	102	23 37	145	44 12	217	58 59	278	38 27	337
283	43 05	020	28 51	050	38 54	102	24 05	144	43 43	217	58 11	277	38 07	336
284	43 21	019	29 28	049	39 41	101	24 33	144	43 14	217	57 23	277	37 47	335
	ALTAIR		Enif		◆FOMALHAUT		ACHERNAR		◆RIGIL KENT.		ANTARES		◆Rasalhague	
285	43 36	017	30 05	048	40 29	101	25 02	144	42 45	217	56 34	276	37 26	333
286	43 50	016	30 40	047	41 17	100	25 30	144	42 15	217	55 46	275	37 03	332
287	44 02	015	31 16	046	42 05	100	25 59	143	41 46	217	54 58	274	36 40	331
288	44 14	013	31 51	046	42 53	099	26 28	143	41 17	217	54 09	274	36 17	330
289	44 25	012	32 25	045	43 41	099	26 58	143	40 48	217	53 21	273	35 52	329
290	44 34	011	32 59	044	44 29	098	27 27	143	40 18	217	52 32	273	35 27	328
291	44 43	009	33 32	043	45 17	098	27 57	142	39 49	217	51 44	272	35 00	327
292	44 50	008	34 05	042	46 05	097	28 26	142	39 20	217	50 55	271	34 34	326
293	44 56	006	34 37	041	46 53	097	28 56	142	38 51	217	50 07	271	34 06	325
294	45 01	005	35 08	040	47 41	096	29 26	142	38 22	217	49 18	270	33 37	324
295	45 05	004	35 39	039	48 29	096	29 57	141	37 53	217	48 30	269	33 08	323
296	45 07	002	36 09	038	49 18	095	30 27	141	37 24	217	47 41	269	32 39	322
297	45 08	001	36 38	037	50 06	095	30 57	141	36 55	216	46 53	268	32 08	321
298	45 09	359	37 06	035	50 54	094	31 28	141	36 26	216	46 04	268	31 37	320
299	45 08	358	37 34	034	51 43	094	31 59	141	35 57	216	45 16	267	31 06	319
	Enif		◆FOMALHAUT		ACHERNAR		◆RIGIL KENT.		ANTARES		◆Rasalhague		ALTAIR	
300	38 01	033	52 31	093	32 30	140	35 29	216	44 27	267	30 33	318	45 05	357
301	38 27	032	53 20	093	33 01	140	35 00	216	43 39	266	30 00	317	45 02	355
302	38 53	031	54 08	092	33 32	140	34 32	216	42 50	266	29 27	316	44 57	354
303	39 17	030	54 57	092	34 03	140	34 03	216	42 02	265	28 53	315	44 52	353
304	39 41	029	55 45	091	34 35	140	33 35	216	41 14	265	28 18	314	44 45	351
305	40 04	027	56 34	091	35 06	139	33 07	215	40 25	264	27 43	313	44 37	350
306	40 26	026	57 22	090	35 38	139	32 39	215	39 37	264	27 08	312	44 28	348
307	40 47	025	58 11	089	36 09	139	32 11	215	38 49	263	26 32	312	44 17	347
308	41 07	024	59 00	089	36 41	139	31 43	215	38 01	263	25 55	311	44 06	346
309	41 26	023	59 48	088	37 13	139	31 15	215	37 13	262	25 18	310	43 53	344
310	41 44	021	60 37	087	37 45	139	30 47	215	36 25	262	24 40	309	43 39	343
311	42 01	020	61 25	087	38 17	139	30 20	214	35 37	261	24 03	308	43 25	342
312	42 17	019	62 13	086	38 49	139	29 53	214	34 49	261	23 24	307	43 09	340
313	42 33	017	63 02	085	39 21	138	29 25	214	34 01	260	22 45	307	42 52	339
314	42 47	016	63 50	085	39 53	138	28 58	214	33 13	260	22 06	306	42 34	338
	Enif		◆Diphda		ACHERNAR		◆RIGIL KENT.		ANTARES		Nunki		◆ALTAIR	
315	43 00	015	37 52	085	40 26	138	28 31	214	32 25	259	61 42	281	42 15	337
316	43 12	014	38 40	085	40 58	138	28 05	213	31 38	259	60 54	280	41 56	335
317	43 22	012	39 29	084	41 30	138	27 38	213	30 50	258	60 06	279	41 35	334
318	43 32	011	40 17	083	42 03	138	27 12	213	30 03	258	59 18	279	41 13	333
319	43 41	010	41 05	083	42 35	138	26 45	213	29 15	257	58 30	278	40 50	332
320	43 48	008	41 53	082	43 07	138	26 19	212	28 28	257	57 42	277	40 27	330
321	43 54	007	42 41	081	43 40	138	25 53	212	27 41	256	56 54	276	40 02	329
322	44 00	005	43 29	081	44 12	138	25 28	212	26 54	256	56 05	276	39 37	328
323	44 04	004	44 17	080	44 45	138	25 02	212	26 07	255	55 17	275	39 11	327
324	44 07	003	45 05	079	45 17	138	24 37	211	25 20	255	54 29	274	38 44	326
325	44 08	001	45 52	078	45 50	138	24 12	211	24 33	254	53 40	274	38 17	325
326	44 09	000	46 40	078	46 22	138	23 47	211	23 47	254	52 52	273	37 48	324
327	44 08	359	47 27	077	46 54	138	23 22	210	23 00	253	52 03	272	37 19	322
328	44 07	357	48 14	076	47 27	138	22 57	210	22 14	253	51 15	272	36 49	321
329	44 04	356	49 02	075	47 59	138	22 33	210	21 27	252	50 26	271	36 18	320
	◆Alpheratz		Diphda		◆ACHERNAR		RIGIL KENT.		◆Nunki		ALTAIR		Enif	
330	18 18	029	49 48	074	48 31	138	22 09	210	49 38	271	35 47	319	44 00	354
331	18 41	028	50 35	074	49 04	139	21 45	209	48 49	270	35 15	318	43 54	353
332	19 04	028	51 22	073	49 36	139	21 21	209	48 01	269	34 42	317	43 48	352
333	19 26	027	52 08	072	50 08	139	20 58	209	47 12	269	34 09	316	43 40	350
334	19 48	026	52 54	071	50 40	139	20 35	208	46 24	268	33 35	315	43 32	349
335	20 09	025	53 40	070	51 11	139	20 12	208	45 35	268	33 01	314	43 22	348
336	20 29	024	54 25	069	51 43	139	19 49	208	44 47	267	32 26	313	43 11	346
337	20 48	023	55 10	068	52 15	139	19 27	207	43 58	267	31 50	312	42 59	345
338	21 07	022	55 55	067	52 46	140	19 05	207	43 10	266	31 14	312	42 46	344
339	21 25	022	56 39	066	53 18	140	18 43	207	42 21	265	30 38	311	42 32	342
340	21 43	021	57 24	065	53 49	140	18 21	206	41 33	265	30 01	310	42 17	341
341	22 00	020	58 07	064	54 20	140	18 00	206	40 45	264	29 23	309	42 01	340
342	22 16	019	58 51	062	54 51	141	17 38	206	39 56	264	28 45	308	41 44	339
343	22 31	018	59 33	061	55 21	141	17 17	205	39 08	263	28 07	307	41 25	337
344	22 46	017	60 16	060	55 52	141	16 57	205	38 20	263	27 28	306	41 06	336
	◆Alpheratz		Diphda		◆CANOPUS		RIGIL KENT.		◆Nunki		ALTAIR		Enif	
345	23 00	016	60 57	059	16 59	144	16 37	205	37 32	262	26 48	305	40 46	335
346	23 13	015	61 38	057	17 28	143	16 16	204	36 44	262	26 09	305	40 25	334
347	23 25	014	62 19	056	17 57	143	15 57	204	35 56	261	25 28	304	40 03	333
348	23 37	013	62 59	054	18 26	143	15 37	203	35 08	261	24 48	303	39 40	331
349	23 47	012	63 38	053	18 56	142	15 18	203	34 20	260	24 07	302	39 17	330
350	23 58	012	64 16	051	19 26	142	14 59	203	33 32	260	23 26	301	38 52	329
351	24 07	011	64 53	049	19 56	141	14 41	202	32 44	259	22 44	301	38 27	328
352	24 15	010	65 30	048	20 26	141	14 22	202	31 57	259	22 02	300	38 00	327
353	24 23	009	66 05	046	20 57	141	14 04	202	31 09	258	21 20	299	37 33	326
354	24 30	008	66 40	044	21 28	140	13 47	201	30 21	258	20 38	298	37 06	325
355	24 36	007	67 13	042	21 59	140	13 30	201	29 34	257	19 55	298	36 37	323
356	24 41	006	67 45	040	22 30	140	13 13	200	28 47	257	19 12	297	36 08	322
357	24 46	005	68 15	038	23 02	139	12 56	200	27 59	256	18 28	296	35 38	321
358	24 50	004	68 44	036	23 33	139	12 40	199	27 12	256	17 45	296	35 07	320
359	24 52	003	69 12	033	24 05	139	12 23	199	26 25	255	17 01	295	34 36	319

LAT 37°S

LHA ♈	Hc	Zn	Hc	Zn	Hc	Zn	Hc	Zn	Hc	Zn	Hc	Zn	Hc	Zn
	Alpheratz		◆Hamal		RIGEL		◆CANOPUS		Peacock		◆Nunki		Enif	
0	23 55	002	22 34	031	14 02	090	25 22	138	49 42	223	25 54	255	33 19	319
1	23 56	001	22 59	031	14 50	089	25 54	138	49 09	223	25 07	255	32 47	318
2	23 56	000	23 23	030	15 38	089	26 26	138	48 36	223	24 21	254	32 14	317
3	23 56	359	23 47	029	16 26	088	26 59	137	48 03	223	23 35	254	31 41	316
4	23 55	358	24 09	028	17 14	087	27 31	137	47 30	223	22 49	253	31 08	315
5	23 53	357	24 31	027	18 01	087	28 04	137	46 58	223	22 03	253	30 33	314
6	23 50	356	24 53	026	18 49	086	28 37	136	46 25	223	21 17	252	29 59	313
7	23 46	355	25 14	025	19 37	085	29 10	136	45 52	223	20 32	252	29 23	312
8	23 42	354	25 34	024	20 25	085	29 44	136	45 19	223	19 46	251	28 48	311
9	23 37	353	25 53	023	21 13	084	30 17	135	44 46	223	19 01	251	28 11	310
10	23 31	352	26 11	022	22 00	084	30 51	135	44 13	223	18 16	250	27 34	309
11	23 24	351	26 29	021	22 48	083	31 25	135	43 40	223	17 31	250	26 57	309
12	23 17	351	26 46	020	23 35	082	31 59	135	43 07	223	16 46	249	26 20	308
13	23 08	350	27 02	019	24 23	082	32 33	134	42 35	223	16 01	249	25 41	307
14	22 59	349	27 18	018	25 10	081	33 07	134	42 02	223	15 16	248	25 03	306
	◆Hamal		ALDEBARAN		RIGEL		◆CANOPUS		Peacock		◆FOMALHAUT		Alpheratz	
15	27 32	017	16 16	054	25 57	080	33 42	134	41 29	223	63 29	277	22 50	348
16	27 46	016	16 55	053	26 45	080	34 16	134	40 57	223	62 41	276	22 39	347
17	27 59	015	17 33	052	27 32	079	34 51	134	40 24	223	61 54	276	22 28	346
18	28 12	014	18 11	052	28 19	078	35 26	133	39 51	223	61 06	275	22 16	345
19	28 23	013	18 48	051	29 06	078	36 00	133	39 19	223	60 18	274	22 03	344
20	28 34	012	19 25	050	29 52	077	36 36	133	38 47	222	59 30	273	21 49	343
21	28 43	011	20 01	049	30 39	076	37 11	133	38 14	222	58 42	273	21 35	342
22	28 52	010	20 37	048	31 25	075	37 46	133	37 42	222	57 54	272	21 20	341
23	29 00	009	21 13	048	32 12	075	38 21	132	37 10	222	57 07	271	21 04	340
24	29 07	008	21 48	047	32 58	074	38 57	132	36 38	222	56 19	271	20 48	340
25	29 14	007	22 23	046	33 44	073	39 32	132	36 06	222	55 31	270	20 31	339
26	29 19	006	22 57	045	34 30	072	40 08	132	35 34	222	54 43	270	20 13	338
27	29 24	005	23 31	044	35 15	072	40 44	132	35 02	221	53 55	269	19 54	337
28	29 28	004	24 04	043	36 01	071	41 19	132	34 31	221	53 07	268	19 35	336
29	29 30	003	24 37	043	36 46	070	41 55	131	33 59	221	52 19	268	19 15	335
	Hamal		◆ALDEBARAN		RIGEL		SIRIUS		◆CANOPUS		Peacock		◆FOMALHAUT	
30	29 32	002	25 09	042	37 31	069	24 47	093	42 31	131	33 28	221	51 31	267
31	29 33	001	25 40	041	38 15	068	25 35	092	43 07	131	32 57	221	50 43	267
32	29 34	000	26 11	040	39 00	068	26 23	092	43 43	131	32 26	220	49 56	266
33	29 33	359	26 42	039	39 44	067	27 10	091	44 20	131	31 55	220	49 08	266
34	29 31	358	27 12	038	40 28	066	27 58	090	44 56	131	31 24	220	48 20	265
35	29 29	357	27 41	037	41 12	065	28 46	090	45 32	131	30 53	220	47 32	265
36	29 26	355	28 10	036	41 55	064	29 34	089	46 08	131	30 23	219	46 45	264
37	29 21	354	28 38	035	42 38	063	30 22	089	46 44	131	29 52	219	45 57	264
38	29 16	353	29 05	034	43 20	062	31 10	088	47 21	131	29 22	219	45 09	263
39	29 10	352	29 32	033	44 03	061	31 58	087	47 57	131	28 52	219	44 22	263
40	29 04	351	29 58	032	44 44	060	32 46	087	48 34	131	28 22	218	43 34	262
41	28 56	350	30 23	031	45 26	059	33 34	086	49 10	131	27 52	218	42 47	262
42	28 47	349	30 48	030	46 07	058	34 21	085	49 46	131	27 23	218	42 00	261
43	28 38	348	31 11	029	46 47	057	35 09	085	50 23	131	26 54	218	41 12	261
44	28 28	347	31 34	028	47 27	056	35 57	084	50 59	131	26 24	217	40 25	260
	◆ALDEBARAN		RIGEL		SIRIUS		◆CANOPUS		ACHERNAR		◆FOMALHAUT		Hamal	
45	31 57	027	48 07	055	36 44	083	51 35	131	65 34	207	39 38	260	28 17	346
46	32 18	026	48 46	054	37 32	083	52 12	131	65 12	208	38 51	259	28 05	345
47	32 39	025	49 24	053	38 20	082	52 48	131	64 49	209	38 04	259	27 52	344
48	32 59	024	50 02	052	39 07	081	53 24	131	64 25	210	37 17	258	27 39	343
49	33 18	023	50 40	050	39 54	081	54 00	131	64 01	211	36 30	258	27 24	342
50	33 36	022	51 16	049	40 42	080	54 36	131	63 36	212	35 43	257	27 09	341
51	33 54	021	51 52	048	41 29	079	55 12	131	63 10	212	34 56	257	26 53	340
52	34 11	020	52 27	047	42 16	079	55 48	132	62 44	213	34 10	256	26 37	339
53	34 26	019	53 02	045	43 03	078	56 24	132	62 18	214	33 23	256	26 19	338
54	34 41	017	53 35	044	43 49	077	57 00	132	61 51	215	32 37	255	26 01	337
55	34 55	016	54 08	043	44 36	076	57 35	132	61 23	215	31 51	255	25 42	336
56	35 08	015	54 40	041	45 22	075	58 11	132	60 56	216	31 04	254	25 23	335
57	35 20	014	55 11	040	46 09	075	58 46	133	60 28	216	30 18	254	25 02	334
58	35 31	013	55 41	038	46 55	074	59 21	133	59 59	217	29 32	253	24 41	333
59	35 41	012	56 10	037	47 41	073	59 56	133	59 30	217	28 46	253	24 19	333
	ALDEBARAN		◆BETELGEUSE		SIRIUS		◆ACRUX		ACHERNAR		◆FOMALHAUT		Hamal	
60	35 51	011	38 06	037	48 27	072	18 43	157	59 01	218	28 01	252	23 57	332
61	35 59	009	38 34	036	49 12	071	19 02	157	58 32	218	27 15	252	23 34	331
62	36 06	008	39 02	035	49 57	070	19 21	157	58 02	219	26 30	252	23 10	330
63	36 13	007	39 29	034	50 42	069	19 40	156	57 32	219	25 44	251	22 46	329
64	36 18	006	39 56	033	51 27	068	19 59	156	57 02	219	24 59	251	22 20	328
65	36 22	005	40 21	032	52 11	067	20 19	156	56 31	220	24 14	250	21 55	327
66	36 26	003	40 46	030	52 55	066	20 39	155	56 00	220	23 29	250	21 28	326
67	36 28	002	41 09	029	53 39	065	20 59	155	55 30	220	22 44	249	21 02	325
68	36 30	001	41 32	028	54 23	064	21 19	155	54 58	221	21 59	249	20 34	325
69	36 30	000	41 54	027	55 05	063	21 40	154	54 27	221	21 15	248	20 06	324
70	36 29	359	42 15	025	55 48	062	22 00	154	53 56	221	20 30	248	19 37	323
71	36 28	358	42 36	024	56 30	061	22 22	154	53 24	221	19 46	247	19 08	322
72	36 25	356	42 55	023	57 12	060	22 43	153	52 53	221	19 02	247	18 38	321
73	36 22	355	43 13	022	57 53	058	23 04	153	52 21	222	18 18	246	18 08	320
74	36 17	354	43 30	020	58 33	057	23 26	153	51 49	222	17 34	246	17 37	320
	BETELGEUSE		◆SIRIUS		Suhail		◆ACRUX		ACHERNAR		◆Diphda		ALDEBARAN	
75	43 46	019	59 13	056	43 21	118	23 48	153	51 17	222	31 08	270	36 12	353
76	44 01	018	59 53	054	44 03	118	24 10	152	50 45	222	30 20	270	36 05	352
77	44 15	016	60 31	053	44 45	118	24 33	152	50 13	222	29 32	269	35 58	350
78	44 28	015	61 09	052	45 28	117	24 55	152	49 41	222	28 44	268	35 49	349
79	44 40	014	61 46	050	46 10	117	25 18	151	49 08	222	27 57	268	35 40	348
80	44 51	012	62 23	048	46 53	117	25 41	151	48 36	222	27 09	267	35 29	347
81	45 00	011	62 58	047	47 36	117	26 04	151	48 04	223	26 21	267	35 18	346
82	45 09	009	63 32	045	48 19	117	26 27	151	47 31	223	25 33	266	35 06	345
83	45 16	008	64 06	043	49 01	116	26 51	150	46 59	223	24 45	265	34 52	343
84	45 22	007	64 38	041	49 44	116	27 15	150	46 26	223	23 57	265	34 38	342
85	45 27	005	65 09	040	50 27	116	27 39	150	45 54	223	23 10	264	34 23	341
86	45 31	004	65 39	038	51 10	116	28 03	150	45 22	223	22 22	264	34 08	340
87	45 34	002	66 08	036	51 54	116	28 27	149	44 49	223	21 35	263	33 51	339
88	45 35	001	66 35	033	52 37	116	28 51	149	44 17	223	20 47	263	33 33	338
89	45 36	000	67 01	031	53 20	115	29 16	149	43 44	223	19 59	262	33 15	337

LHA ♈	Hc	Zn	Hc	Zn	Hc	Zn	Hc	Zn	Hc	Zn	Hc	Zn	Hc	Zn
	PROCYON		REGULUS		◆Suhail		ACRUX		◆ACHERNAR		RIGEL		◆BETELGEUSE	
90	41 51	034	13 58	063	54 03	115	29 41	149	43 12	223	59 24	337	45 35	358
91	42 18	033	14 41	062	54 47	115	30 06	149	42 39	222	59 05	336	45 33	357
92	42 43	032	15 23	062	55 30	115	30 31	148	42 07	222	58 44	334	45 29	355
93	43 08	030	16 05	061	56 14	115	30 56	148	41 35	222	58 22	332	45 25	354
94	43 32	029	16 47	060	56 57	115	31 21	148	41 03	222	57 59	330	45 19	353
95	43 55	028	17 28	059	57 41	115	31 47	148	40 30	222	57 35	329	45 12	351
96	44 17	027	18 09	059	58 24	115	32 12	148	39 58	222	57 09	327	45 04	350
97	44 37	025	18 50	058	59 08	115	32 38	147	39 26	222	56 42	325	44 55	348
98	44 57	024	19 30	057	59 51	115	33 04	147	38 54	222	56 14	324	44 45	347
99	45 16	023	20 10	056	60 35	115	33 30	147	38 22	222	55 45	322	44 34	346
100	45 34	021	20 50	056	61 18	115	33 56	147	37 51	222	55 16	321	44 21	344
101	45 51	020	21 29	055	62 02	115	34 22	147	37 19	221	54 45	319	44 08	343
102	46 07	018	22 08	054	62 46	115	34 48	147	36 47	221	54 13	318	43 53	342
103	46 21	017	22 47	053	63 29	115	35 14	147	36 16	221	53 40	316	43 38	340
104	46 35	016	23 25	052	64 13	115	35 41	146	35 44	221	53 07	315	43 21	339
	PROCYON		REGULUS		◆Gienah		ACRUX		◆ACHERNAR		RIGEL		◆BETELGEUSE	
105	46 47	014	24 03	052	19 09	098	36 07	146	35 13	221	52 32	314	43 03	338
106	46 59	013	24 40	051	19 56	097	36 34	146	34 42	221	51 57	312	42 44	336
107	47 09	011	25 17	050	20 44	097	37 01	146	34 10	220	51 21	311	42 25	335
108	47 18	010	25 53	049	21 31	096	37 27	146	33 39	220	50 45	310	42 04	334
109	47 25	008	26 29	048	22 19	096	37 54	146	33 09	220	50 08	309	41 42	333
110	47 32	007	27 05	047	23 07	095	38 21	146	32 38	220	49 30	307	41 20	331
111	47 37	006	27 40	046	23 54	094	38 48	146	32 07	220	48 52	306	40 57	330
112	47 41	004	28 14	046	24 42	094	39 15	146	31 37	219	48 13	305	40 32	329
113	47 44	003	28 48	045	25 30	093	39 42	146	31 06	219	47 33	304	40 07	328
114	47 45	001	29 22	044	26 18	093	40 09	146	30 36	219	46 53	303	39 41	327
115	47 46	000	29 55	043	27 06	092	40 36	146	30 06	219	46 13	302	39 14	325
116	47 45	358	30 27	042	27 54	092	41 03	145	29 36	218	45 32	301	38 47	324
117	47 43	357	30 59	041	28 41	091	41 31	145	29 07	218	44 51	300	38 19	323
118	47 39	355	31 30	040	29 29	090	41 58	145	28 37	218	44 09	299	37 49	322
119	47 35	354	32 00	039	30 17	090	42 25	145	28 08	218	43 27	298	37 20	321
	REGULUS		◆SPICA		ACRUX		◆CANOPUS		RIGEL		BETELGEUSE		◆PROCYON	
120	32 30	038	13 38	094	42 52	145	67 02	219	42 44	297	36 49	320	47 29	352
121	32 59	037	14 26	093	43 19	145	66 31	220	42 01	296	36 18	319	47 22	351
122	33 27	036	15 14	093	43 46	145	66 00	221	41 18	295	35 46	318	47 13	349
123	33 55	035	16 01	092	44 14	146	65 29	222	40 34	294	35 14	317	47 04	348
124	34 22	034	16 49	091	44 41	146	64 57	223	39 51	293	34 41	316	46 53	346
125	34 49	033	17 37	091	45 08	146	64 24	223	39 06	292	34 07	315	46 42	345
126	35 14	032	18 25	090	45 35	146	63 52	224	38 22	292	33 33	314	46 29	344
127	35 39	031	19 13	090	46 02	146	63 18	224	37 37	291	32 58	313	46 15	342
128	36 03	030	20 01	089	46 29	146	62 45	225	36 52	290	32 23	312	45 59	341
129	36 26	028	20 49	088	46 56	146	62 11	225	36 07	289	31 47	311	45 43	339
130	36 49	027	21 37	088	47 22	146	61 37	226	35 22	288	31 10	310	45 26	338
131	37 10	026	22 25	087	47 49	146	61 03	226	34 36	288	30 34	309	45 07	337
132	37 31	025	23 13	087	48 16	146	60 28	226	33 51	287	29 56	308	44 48	335
133	37 51	024	24 00	086	48 42	146	59 53	227	33 05	286	29 18	307	44 28	334
134	38 09	023	24 48	085	49 09	147	59 18	227	32 18	285	28 40	307	44 06	333
	◆REGULUS		SPICA		◆ACRUX		CANOPUS		◆RIGEL		BETELGEUSE		PROCYON	
135	38 27	021	25 36	085	49 35	147	58 43	227	31 32	285	28 01	306	43 44	332
136	38 44	020	26 24	084	50 01	147	58 08	228	30 46	284	27 22	305	43 20	330
137	39 01	019	27 11	083	50 27	147	57 32	228	29 59	283	26 43	304	42 56	329
138	39 16	018	27 59	083	50 53	147	56 57	228	29 12	283	26 03	303	42 31	328
139	39 30	017	28 46	082	51 19	148	56 21	228	28 26	282	25 23	302	42 05	327
140	39 43	015	29 34	081	51 44	148	55 45	228	27 39	281	24 42	302	41 38	325
141	39 55	014	30 21	081	52 10	148	55 09	229	26 52	280	24 01	301	41 11	324
142	40 07	013	31 08	080	52 35	148	54 33	229	26 04	280	23 20	300	40 42	323
143	40 17	012	31 55	079	53 00	149	53 57	229	25 17	279	22 38	299	40 13	322
144	40 26	010	32 42	079	53 24	149	53 21	229	24 30	278	21 56	299	39 43	321
145	40 34	009	33 29	078	53 49	149	52 45	229	23 42	278	21 14	298	39 12	320
146	40 41	008	34 16	077	54 13	150	52 09	229	22 55	277	20 31	297	38 41	319
147	40 47	006	35 03	077	54 37	150	51 32	229	22 07	277	19 49	296	38 09	318
148	40 52	005	35 49	076	55 01	150	50 56	229	21 20	276	19 05	296	37 36	316
149	40 56	004	36 36	075	55 24	151	50 20	229	20 32	275	18 22	295	37 03	315
	REGULUS		◆SPICA		ACRUX		◆Miaplacidus		CANOPUS		SIRIUS		◆PROCYON	
150	40 58	003	37 22	074	55 48	151	56 42	187	49 43	229	42 37	282	36 29	314
151	41 00	001	38 08	073	56 10	152	56 35	188	49 07	229	41 50	281	35 55	313
152	41 00	000	38 54	073	56 33	152	56 28	189	48 31	229	41 03	280	35 19	312
153	41 00	359	39 39	072	56 55	153	56 21	189	47 54	229	40 16	280	34 44	311
154	40 58	357	40 25	071	57 17	153	56 13	190	47 18	229	39 29	279	34 08	310
155	40 56	356	41 10	070	57 38	154	56 05	190	46 42	229	38 42	278	33 31	310
156	40 52	355	41 55	069	57 59	154	55 56	191	46 05	229	37 54	277	32 54	309
157	40 47	354	42 40	069	58 20	155	55 47	191	45 29	229	37 07	277	32 16	308
158	40 41	352	43 24	068	58 40	155	55 37	192	44 53	229	36 19	276	31 38	307
159	40 34	351	44 08	067	59 00	156	55 27	192	44 17	229	35 31	275	30 59	306
160	40 26	350	44 52	066	59 19	157	55 16	193	43 41	229	34 44	275	30 20	305
161	40 17	348	45 36	065	59 38	157	55 05	194	43 04	229	33 56	274	29 41	304
162	40 07	347	46 19	064	59 56	158	54 54	194	42 28	229	33 08	274	29 01	303
163	39 56	346	47 02	063	60 14	159	54 42	195	41 52	229	32 20	273	28 21	302
164	39 44	345	47 45	062	60 31	159	54 30	195	41 17	228	31 32	272	27 40	302
	ARCTURUS		◆ANTARES		ACRUX		◆CANOPUS		SIRIUS		PROCYON		◆REGULUS	
165	17 21	048	21 20	108	60 48	160	40 41	228	30 44	272	26 59	301	39 31	343
166	17 56	047	22 06	107	61 04	161	40 05	228	29 56	271	26 18	300	39 17	342
167	18 31	047	22 52	107	61 19	162	39 29	228	29 09	270	25 36	299	39 01	341
168	19 06	046	23 38	106	61 34	162	38 54	228	28 21	270	24 54	298	38 45	340
169	19 40	045	24 24	106	61 48	163	38 18	228	27 33	269	24 12	298	38 28	339
170	20 14	044	25 10	105	62 02	164	37 43	227	26 45	269	23 29	297	38 10	337
171	20 47	043	25 57	105	62 15	165	37 08	227	25 57	268	22 46	296	37 52	336
172	21 20	043	26 43	104	62 27	166	36 33	227	25 09	267	22 03	295	37 32	335
173	21 52	042	27 30	104	62 38	167	35 58	227	24 21	267	21 20	295	37 11	334
174	22 23	041	28 16	103	62 49	168	35 23	227	23 33	266	20 36	294	36 50	333
175	22 55	040	29 03	103	62 59	168	34 48	226	22 45	266	19 52	293	36 27	332
176	23 25	039	29 50	102	63 08	169	34 13	226	21 58	265	19 08	292	36 04	331
177	23 55	038	30 37	102	63 17	170	33 39	226	21 10	265	18 23	292	35 40	329
178	24 24	037	31 24	101	63 24	171	33 04	226	20 22	264	17 39	291	35 15	328
179	24 53	037	32 11	101	63 31	172	32 30	226	19 35	263	16 54	290	34 50	327

LAT 37°S

LHA	Hc	Zn	Hc	Zn	Hc	Zn	Hc	Zn	Hc	Zn	Hc	Zn	Hc	Zn
♈	ARCTURUS		♦ANTARES		RIGIL KENT.		ACRUX		♦CANOPUS		Alphard		♦REGULUS	
180	25 22	036	32 58	100	55 33	147	63 37	173	31 56	225	45 20	300	34 24	326
181	25 49	035	33 45	100	55 59	147	63 42	174	31 22	225	44 38	299	33 57	325
182	26 16	034	34 32	099	56 25	147	63 47	175	30 48	225	43 56	298	33 29	324
183	26 42	033	35 20	099	56 51	148	63 50	176	30 15	224	43 13	297	33 01	323
184	27 08	032	36 07	098	57 16	148	63 53	177	29 41	224	42 30	296	32 31	322
185	27 33	031	36 54	098	57 41	149	63 55	178	29 08	224	41 47	295	32 02	321
186	27 57	030	37 42	097	58 06	149	63 56	179	28 35	224	41 03	294	31 31	320
187	28 21	029	38 30	096	58 30	150	63 56	180	28 02	223	40 19	293	31 00	319
188	28 44	028	39 17	096	58 55	150	63 55	181	27 29	223	39 35	292	30 29	318
189	29 06	027	40 05	095	59 18	151	63 53	182	26 56	223	38 51	291	29 56	317
190	29 27	026	40 53	095	59 41	151	63 51	184	26 24	222	38 06	291	29 24	316
191	29 48	025	41 40	094	60 04	152	63 47	185	25 52	222	37 21	290	28 50	315
192	30 08	024	42 28	094	60 27	153	63 43	186	25 19	222	36 36	289	28 16	314
193	30 27	023	43 16	093	60 48	153	63 38	187	24 48	221	35 50	288	27 42	314
194	30 45	022	44 04	093	61 10	154	63 32	188	24 16	221	35 05	287	27 07	313
	ARCTURUS		♦ANTARES		RIGIL KENT.		♦ACRUX		Suhail		♦REGULUS		Denebola	
195	31 03	021	44 52	092	61 31	155	63 25	189	46 06	243	26 31	312	35 45	339
196	31 19	020	45 40	091	61 51	155	63 18	189	45 24	243	25 55	311	35 27	337
197	31 35	019	46 28	091	62 11	156	63 10	190	44 41	242	25 19	310	35 08	336
198	31 50	018	47 15	090	62 30	157	63 01	191	43 59	242	24 42	309	34 48	335
199	32 04	017	48 03	090	62 48	158	62 51	192	43 17	242	24 05	308	34 28	334
200	32 17	016	48 51	089	63 06	159	62 40	193	42 34	242	23 27	308	34 07	333
201	32 30	014	49 39	088	63 23	159	62 29	194	41 52	241	22 49	307	33 44	332
202	32 41	013	50 27	088	63 40	160	62 17	195	41 10	241	22 10	306	33 22	331
203	32 52	012	51 15	087	63 55	161	62 04	196	40 29	241	21 31	305	32 58	330
204	33 02	011	52 03	087	64 10	162	61 51	197	39 47	241	20 52	304	32 33	329
205	33 10	010	52 51	086	64 25	163	61 36	198	39 05	240	20 12	304	32 08	328
206	33 18	009	53 38	085	64 38	164	61 22	198	38 24	240	19 32	303	31 42	327
207	33 25	008	54 26	084	64 51	165	61 06	199	37 42	240	18 52	302	31 16	326
208	33 31	007	55 14	084	65 02	166	60 50	200	37 01	239	18 11	301	30 48	325
209	33 36	006	56 01	083	65 13	167	60 34	201	36 20	239	17 30	301	30 20	324
	ARCTURUS		♦ANTARES		Peacock		RIGIL KENT.		♦ACRUX		Suhail		♦SPICA	
210	33 40	004	56 49	082	27 04	142	65 23	168	60 17	201	35 39	239	62 57	341
211	33 44	003	57 36	082	27 34	142	65 32	170	59 59	202	34 58	238	62 41	339
212	33 46	002	58 24	081	28 03	142	65 41	171	59 41	203	34 17	238	62 23	337
213	33 47	001	59 11	080	28 33	141	65 48	172	59 22	203	33 36	238	62 03	335
214	33 47	000	59 58	079	29 03	141	65 54	173	59 03	204	32 56	238	61 42	333
215	33 47	359	60 45	078	29 33	141	65 59	174	58 43	205	32 16	237	61 19	331
216	33 45	358	61 32	077	30 04	141	66 04	175	58 23	205	31 35	237	60 55	329
217	33 43	356	62 18	076	30 34	140	66 07	177	58 03	206	30 55	237	60 30	327
218	33 39	355	63 05	075	31 05	140	66 09	178	57 42	206	30 15	236	60 03	325
219	33 35	354	63 51	074	31 35	140	66 11	179	57 20	207	29 36	236	59 36	324
220	33 30	353	64 37	073	32 06	140	66 11	180	56 59	207	28 56	235	59 07	322
221	33 23	352	65 23	072	32 37	140	66 10	181	56 36	208	28 17	235	58 37	320
222	33 16	351	66 08	071	33 08	139	66 09	183	56 14	208	27 37	235	58 06	319
223	33 08	350	66 54	070	33 40	139	66 06	184	55 51	209	26 58	234	57 34	317
224	32 59	349	67 38	068	34 11	139	66 02	185	55 28	209	26 20	234	57 01	316
	Alphecca		Rasalhague		♦Nunki		Peacock		♦RIGIL KENT.		♦SPICA		ARCTURUS	
225	25 45	009	28 32	044	39 40	096	34 43	139	65 58	186	56 27	314	32 49	347
226	25 52	008	29 05	043	40 27	095	35 14	139	65 52	187	55 52	313	32 38	346
227	25 58	007	29 37	042	41 15	094	35 46	138	65 45	189	55 17	312	32 26	345
228	26 03	006	30 09	041	42 03	094	36 18	138	65 38	190	54 40	310	32 14	344
229	26 07	005	30 40	040	42 51	093	36 50	138	65 29	191	54 03	309	32 00	343
230	26 11	004	31 11	039	43 38	093	37 22	138	65 20	192	53 26	308	31 46	342
231	26 13	003	31 41	038	44 26	092	37 54	138	65 09	193	52 47	306	31 31	341
232	26 15	002	32 10	037	45 14	092	38 26	138	64 58	194	52 09	305	31 15	340
233	26 16	001	32 39	036	46 02	091	38 59	138	64 46	195	51 29	304	30 58	339
234	26 16	000	33 07	035	46 50	090	39 31	137	64 33	196	50 49	303	30 40	338
235	26 15	359	33 34	034	47 38	090	40 04	137	64 19	197	50 09	302	30 22	337
236	26 14	358	34 01	033	48 26	089	40 36	137	64 05	198	49 27	301	30 02	336
237	26 11	357	34 27	032	49 14	089	41 09	137	63 50	199	48 46	300	29 42	335
238	26 08	356	34 52	031	50 02	088	41 41	137	63 34	200	48 04	298	29 21	334
239	26 04	355	35 16	030	50 50	087	42 14	137	63 17	201	47 22	297	29 00	333
	Rasalhague		ALTAIR		♦Nunki		Peacock		♦RIGIL KENT.		SPICA		♦ARCTURUS	
240	35 40	029	19 15	062	51 37	087	42 47	137	62 59	202	46 39	296	28 37	332
241	36 02	028	19 58	061	52 25	086	43 20	137	62 41	203	45 56	296	28 14	331
242	36 24	027	20 39	061	53 13	085	43 53	137	62 23	203	45 13	295	27 50	330
243	36 45	025	21 21	060	54 01	085	44 25	137	62 03	204	44 29	294	27 26	329
244	37 06	024	22 02	059	54 48	084	44 58	137	61 43	205	43 45	293	27 01	328
245	37 25	023	22 43	058	55 36	083	45 31	137	61 23	206	43 00	292	26 35	327
246	37 43	022	23 24	058	56 24	082	46 04	137	61 02	206	42 16	291	26 08	326
247	38 01	021	24 04	057	57 11	082	46 37	137	60 40	207	41 31	290	25 41	325
248	38 17	020	24 44	056	57 58	081	47 10	137	60 18	208	40 46	289	25 14	324
249	38 33	018	25 24	055	58 46	080	47 43	137	59 56	208	40 01	288	24 45	323
250	38 48	017	26 03	054	59 33	079	48 16	137	59 33	209	39 15	288	24 16	322
251	39 01	016	26 42	054	60 20	078	48 48	137	59 09	210	38 29	287	23 47	321
252	39 14	015	27 20	053	61 07	077	49 21	137	58 46	210	37 43	286	23 17	321
253	39 26	014	27 58	052	61 53	077	49 54	137	58 21	211	36 57	285	22 46	320
254	39 36	012	28 35	051	62 40	076	50 27	137	57 57	211	36 11	285	22 15	319
	Rasalhague		♦ALTAIR		FOMALHAUT		♦ACHERNAR		RIGIL KENT.		♦SPICA		ARCTURUS	
255	39 46	011	29 12	050	17 48	114	13 26	155	57 32	212	35 24	284	21 43	318
256	39 55	010	29 49	049	18 32	114	13 46	154	57 07	212	34 38	283	21 10	317
257	40 02	009	30 25	048	19 16	113	14 07	154	56 41	212	33 51	282	20 38	316
258	40 09	007	31 00	047	20 00	113	14 29	153	56 15	213	33 04	282	20 04	316
259	40 14	006	31 35	046	20 44	112	14 50	153	55 49	213	32 17	281	19 31	315
260	40 19	005	32 10	045	21 29	112	15 12	152	55 23	214	31 30	280	18 56	314
261	40 22	003	32 43	044	22 13	111	15 35	152	54 56	214	30 43	280	18 22	313
262	40 25	002	33 17	044	22 58	111	15 57	152	54 29	214	29 56	279	17 46	312
263	40 26	001	33 49	043	23 43	110	16 20	151	54 02	215	29 08	278	17 11	312
264	40 26	000	34 22	042	24 28	110	16 43	151	53 35	215	28 21	278	16 35	311
265	40 25	358	34 53	041	25 13	109	17 07	150	53 07	215	27 33	277	15 58	310
266	40 23	357	35 24	039	25 58	109	17 31	150	52 39	215	26 46	276	15 21	309
267	40 20	356	35 54	038	26 44	108	17 55	150	52 12	216	25 58	276	14 44	309
268	40 16	354	36 23	037	27 29	108	18 19	149	51 44	216	25 10	275	14 07	308
269	40 11	353	36 52	036	28 15	107	18 44	149	51 15	216	24 23	274	13 29	307

LHA	Hc	Zn	Hc	Zn	Hc	Zn	Hc	Zn	Hc	Zn	Hc	Zn	Hc	Zn
♈	ALTAIR		Enif		♦FOMALHAUT		ACHERNAR		♦RIGIL KENT.		ANTARES		♦Rasalhague	
270	37 20	035	19 42	060	29 01	107	19 09	149	50 47	216	68 03	292	40 05	352
271	37 47	034	20 24	059	29 47	106	19 34	148	50 19	217	67 18	291	39 58	351
272	38 14	033	21 05	059	30 33	106	19 59	148	49 50	217	66 33	290	39 49	349
273	38 39	032	21 46	058	31 19	105	20 25	147	49 21	217	65 48	289	39 40	348
274	39 04	031	22 26	057	32 05	105	20 51	147	48 53	217	65 02	287	39 30	347
275	39 28	030	23 06	056	32 51	105	21 17	147	48 24	217	64 17	286	39 18	346
276	39 51	028	23 46	056	33 38	104	21 43	146	47 55	217	63 30	285	39 06	344
277	40 14	027	24 25	055	34 24	104	22 10	146	47 26	217	62 44	284	38 52	343
278	40 35	026	25 04	054	35 11	103	22 36	146	46 57	217	61 58	283	38 38	342
279	40 56	025	25 43	053	35 58	103	23 03	145	46 28	217	61 11	282	38 23	341
280	41 15	023	26 21	052	36 45	102	23 31	145	45 59	218	60 24	281	38 07	340
281	41 34	022	26 58	051	37 31	102	23 58	145	45 29	218	59 37	281	37 49	338
282	41 51	021	27 36	051	38 18	101	24 26	145	45 00	218	58 50	280	37 31	337
283	42 08	020	28 12	050	39 05	101	24 54	144	44 31	218	58 02	279	37 12	336
284	42 24	018	28 49	049	39 53	100	25 22	144	44 02	218	57 15	278	36 53	335
	ALTAIR		Enif		♦FOMALHAUT		ACHERNAR		♦RIGIL KENT.		ANTARES		♦Rasalhague	
285	42 38	017	29 24	048	40 40	100	25 50	144	43 32	218	56 28	277	36 32	334
286	42 52	016	30 00	047	41 27	099	26 19	143	43 03	218	55 40	277	36 10	333
287	43 04	014	30 34	046	42 14	099	26 47	143	42 34	218	54 52	276	35 48	332
288	43 16	013	31 09	045	43 02	098	27 16	143	42 05	218	54 05	275	35 25	330
289	43 26	012	31 42	044	43 49	098	27 45	143	41 35	218	53 17	274	35 00	329
290	43 35	010	32 15	043	44 37	097	28 15	142	41 06	218	52 29	274	34 36	328
291	43 43	009	32 48	042	45 24	097	28 44	142	40 37	217	51 41	273	34 10	327
292	43 50	008	33 20	041	46 12	096	29 14	142	40 08	217	50 53	273	33 44	326
293	43 56	006	33 51	040	47 00	096	29 43	141	39 39	217	50 06	272	33 17	325
294	44 01	005	34 22	039	47 47	095	30 13	141	39 10	217	49 18	271	32 49	324
295	44 05	004	34 52	038	48 35	095	30 43	141	38 41	217	48 30	271	32 21	323
296	44 07	002	35 21	037	49 23	094	31 14	141	38 12	217	47 42	270	31 51	322
297	44 08	001	35 50	036	50 11	094	31 44	141	37 43	217	46 54	269	31 22	321
298	44 09	359	36 17	035	50 58	093	32 14	140	37 14	217	46 06	269	30 51	320
299	44 08	358	36 45	034	51 46	093	32 45	140	36 46	217	45 18	268	30 20	319
	Enif		♦FOMALHAUT		ACHERNAR		♦RIGIL KENT.		ANTARES		♦Rasalhague		ALTAIR	
300	37 11	033	52 34	092	33 16	140	36 17	217	44 30	268	29 49	318	44 05	357
301	37 36	032	53 22	091	33 47	140	35 49	216	43 42	267	29 16	317	44 02	355
302	38 01	031	54 10	091	34 18	140	35 20	216	42 54	267	28 44	316	43 58	354
303	38 25	029	54 58	090	34 49	139	34 52	216	42 07	266	28 10	315	43 52	353
304	38 48	028	55 46	090	35 20	139	34 24	216	41 19	265	27 36	315	43 45	351
305	39 11	027	56 34	089	35 52	139	33 55	216	40 31	265	27 02	314	43 38	350
306	39 32	026	57 22	088	36 23	139	33 27	216	39 43	264	26 27	313	43 29	349
307	39 52	025	58 09	088	36 55	139	33 00	215	38 56	264	25 52	312	43 19	347
308	40 12	023	58 57	087	37 26	139	32 32	215	38 08	263	25 16	311	43 08	346
309	40 31	022	59 45	086	37 58	138	32 04	215	37 21	263	24 39	310	42 55	345
310	40 48	021	60 33	086	38 30	138	31 37	215	36 33	262	24 03	309	42 42	343
311	41 05	020	61 21	085	39 02	138	31 09	215	35 46	262	23 25	309	42 28	342
312	41 21	018	62 08	084	39 34	138	30 42	215	34 58	261	22 48	308	42 12	341
313	41 35	017	62 56	083	40 06	138	30 15	214	34 11	261	22 09	307	41 56	339
314	41 49	016	63 44	083	40 38	138	29 48	214	33 24	260	21 31	306	41 39	338
	Enif		♦Diphda		ACHERNAR		♦RIGIL KENT.		ANTARES		Nunki		♦ALTAIR	
315	42 02	015	37 47	085	41 10	138	29 21	214	32 36	260	61 29	283	41 20	337
316	42 13	013	38 34	084	41 42	138	28 55	214	31 49	259	60 42	282	41 01	336
317	42 24	012	39 22	083	42 15	138	28 28	213	31 02	259	59 55	281	40 41	334
318	42 33	011	40 10	083	42 47	138	28 02	213	30 15	258	59 08	280	40 20	333
319	42 41	009	40 57	082	43 19	137	27 36	213	29 29	258	58 21	280	39 58	332
320	42 49	008	41 44	081	43 52	137	27 10	213	28 42	257	57 34	279	39 35	331
321	42 55	007	42 32	080	44 24	137	26 44	212	27 55	257	56 46	278	39 11	330
322	43 00	005	43 19	080	44 57	137	26 19	212	27 09	256	55 59	277	38 46	328
323	43 04	004	44 06	079	45 29	137	25 53	212	26 22	256	55 11	276	38 21	327
324	43 07	003	44 53	078	46 02	137	25 28	212	25 36	255	54 24	276	37 54	326
325	43 08	001	45 40	077	46 34	137	25 03	211	24 49	255	53 36	275	37 27	325
326	43 09	000	46 27	077	47 07	137	24 38	211	24 03	254	52 48	274	37 00	324
327	43 08	359	47 13	076	47 39	137	24 14	211	23 17	254	52 00	274	36 31	323
328	43 07	357	48 00	075	48 11	138	23 49	210	22 31	253	51 12	273	36 02	322
329	43 04	356	48 46	074	48 44	138	23 25	210	21 46	253	50 25	272	35 32	321
	♦Alpheratz		Diphda		♦ACHERNAR		RIGIL KENT.		♦Nunki		ALTAIR		Enif	
330	17 25	029	49 32	073	49 16	138	23 01	210	49 37	272	35 01	320	43 00	355
331	17 48	028	50 18	072	49 48	138	22 37	210	48 49	271	34 30	319	42 55	353
332	18 11	027	51 03	072	50 20	138	22 14	209	48 01	270	33 58	318	42 49	352
333	18 33	027	51 48	071	50 53	138	21 51	209	47 13	270	33 26	317	42 41	351
334	18 54	026	52 34	070	51 25	138	21 28	209	46 25	269	32 52	316	42 33	349
335	19 14	025	53 18	069	51 57	138	21 05	208	45 37	269	32 19	315	42 23	348
336	19 34	024	54 03	068	52 28	138	20 42	208	44 49	268	31 44	314	42 13	347
337	19 53	023	54 47	067	53 00	139	20 20	208	44 01	268	31 10	313	42 01	345
338	20 12	022	55 31	066	53 32	139	19 58	207	43 13	267	30 34	312	41 49	344
339	20 30	021	56 14	065	54 03	139	19 36	207	42 26	266	29 58	311	41 35	343
340	20 47	021	56 57	063	54 35	139	19 15	207	41 38	266	29 22	310	41 20	341
341	21 03	020	57 40	062	55 06	140	18 53	206	40 50	265	28 45	309	41 04	340
342	21 19	019	58 22	061	55 37	140	18 32	206	40 02	265	28 08	308	40 48	339
343	21 34	018	59 04	060	56 08	140	18 12	205	39 15	264	27 30	308	40 30	338
344	21 48	017	59 45	058	56 38	140	17 51	205	38 27	264	26 52	307	40 11	336
	♦Alpheratz		Diphda		♦CANOPUS		RIGIL KENT.		♦Nunki		ALTAIR		Enif	
345	22 02	016	60 25	057	17 47	144	17 31	205	37 39	263	26 13	306	39 52	335
346	22 15	015	61 05	056	18 16	143	17 11	204	36 52	263	25 34	305	39 31	334
347	22 27	014	61 45	054	18 45	143	16 52	204	36 04	262	24 55	304	39 10	333
348	22 38	013	62 23	053	19 14	142	16 32	204	35 17	262	24 15	303	38 48	332
349	22 49	012	63 01	051	19 43	142	16 13	203	34 30	261	23 35	303	38 24	331
350	22 59	011	63 38	050	20 13	142	15 55	203	33 42	261	22 54	302	38 00	329
351	23 08	010	64 14	048	20 43	141	15 36	202	32 55	260	22 13	301	37 36	328
352	23 16	010	64 49	046	21 13	141	15 18	202	32 08	260	21 32	300	37 10	327
353	23 24	009	65 23	044	21 43	141	15 00	202	31 21	259	20 51	300	36 44	326
354	23 31	008	65 56	042	22 14	140	14 43	201	30 34	258	20 09	299	36 17	325
355	23 37	007	66 28	041	22 45	140	14 26	201	29 47	258	19 27	298	35 49	324
356	23 42	006	66 58	038	23 16	139	14 09	200	29 00	257	18 44	297	35 20	323
357	23 46	005	67 27	036	23 47	139	13 52	200	28 13	257	18 02	297	34 51	322
358	23 50	004	67 55	034	24 19	139	13 36	200	27 27	256	17 19	296	34 21	321
359	23 53	003	68 21	032	24 50	138	13 20	199	26 40	256	16 35	295	33 50	320

LHA ♈	Hc	Zn	Hc	Zn	Hc	Zn	Hc	Zn	Hc	Zn	Hc	Zn	Hc	Zn
	Alpheratz		◆Hamal		RIGEL		◆CANOPUS		Peacock		◆Nunki		Enif	
0	22 55	002	21 43	031	14 02	090	26 07	138	50 25	224	26 08	256	32 34	319
1	22 56	001	22 07	030	14 49	089	26 38	138	49 52	224	25 23	255	32 02	318
2	22 56	000	22 31	030	15 36	088	27 11	137	49 20	224	24 37	255	31 31	317
3	22 56	359	22 54	029	16 23	088	27 43	137	48 47	224	23 51	254	30 58	316
4	22 55	358	23 16	028	17 11	087	28 15	137	48 14	224	23 06	254	30 25	315
5	22 53	357	23 38	027	17 58	086	28 48	136	47 41	224	22 21	253	29 52	314
6	22 50	356	23 59	026	18 45	086	29 21	136	47 08	224	21 35	253	29 18	313
7	22 47	355	24 19	025	19 32	085	29 54	136	46 35	224	20 50	252	28 43	313
8	22 42	354	24 39	024	20 19	084	30 27	135	46 02	224	20 05	252	28 08	312
9	22 37	353	24 58	023	21 06	084	31 00	135	45 29	224	19 20	251	27 32	311
10	22 31	352	25 16	022	21 53	083	31 33	135	44 57	224	18 36	251	26 56	310
11	22 25	352	25 33	021	22 40	083	32 07	135	44 24	224	17 51	250	26 20	309
12	22 18	351	25 50	020	23 27	082	32 41	134	43 51	224	17 07	250	25 43	308
13	22 09	350	26 06	019	24 14	081	33 15	134	43 18	224	16 23	249	25 05	307
14	22 01	349	26 21	018	25 00	081	33 49	134	42 46	224	15 38	249	24 27	306
	◆Hamal		ALDEBARAN		RIGEL		◆CANOPUS		Peacock		◆FOMALHAUT		Alpheratz	
15	26 35	017	15 41	054	25 47	080	34 23	134	42 13	224	63 20	279	21 51	348
16	26 49	016	16 19	053	26 34	079	34 57	133	41 40	223	62 34	278	21 41	347
17	27 01	015	16 56	052	27 20	078	35 32	133	41 08	223	61 47	277	21 29	346
18	27 13	014	17 33	051	28 06	078	36 07	133	40 35	223	61 00	277	21 18	345
19	27 25	013	18 10	051	28 52	077	36 41	133	40 03	223	60 13	276	21 05	344
20	27 35	012	18 46	050	29 38	076	37 16	132	39 31	223	59 26	275	20 52	343
21	27 44	011	19 22	049	30 24	076	37 51	132	38 59	223	58 39	274	20 38	342
22	27 53	010	19 57	048	31 10	075	38 26	132	38 26	223	57 52	274	20 23	341
23	28 01	009	20 32	047	31 56	074	39 01	132	37 54	223	57 04	273	20 08	341
24	28 08	008	21 07	047	32 41	073	39 37	132	37 22	222	56 17	272	19 52	340
25	28 14	007	21 41	046	33 26	073	40 12	131	36 51	222	55 30	272	19 35	339
26	28 20	006	22 15	045	34 11	072	40 48	131	36 19	222	54 43	271	19 17	338
27	28 24	005	22 48	044	34 56	071	41 23	131	35 47	222	53 55	270	18 59	337
28	28 28	004	23 20	043	35 41	070	41 59	131	35 16	222	53 08	270	18 40	336
29	28 30	003	23 52	042	36 25	069	42 35	131	34 44	222	52 21	269	18 21	335
	Hamal		◆ALDEBARAN		RIGEL		SIRIUS		◆CANOPUS		Peacock		◆FOMALHAUT	
30	28 32	002	24 24	041	37 09	069	24 49	092	43 11	131	34 13	221	51 33	269
31	28 33	001	24 55	041	37 53	068	25 37	092	43 47	130	33 42	221	50 46	268
32	28 34	000	25 25	040	38 37	067	26 24	091	44 23	130	33 11	221	49 59	267
33	28 33	359	25 55	039	39 20	066	27 11	091	44 59	130	32 40	221	49 12	267
34	28 31	358	26 24	038	40 03	065	27 59	090	45 35	130	32 10	220	48 25	266
35	28 29	357	26 53	037	40 46	064	28 46	089	46 11	130	31 39	220	47 37	266
36	28 26	356	27 21	036	41 28	063	29 33	089	46 47	130	31 09	220	46 50	265
37	28 22	354	27 49	035	42 10	062	30 20	088	47 23	130	30 39	220	46 03	265
38	28 17	353	28 15	034	42 52	061	31 08	087	48 00	130	30 09	219	45 16	264
39	28 11	352	28 41	033	43 33	060	31 55	087	48 36	130	29 39	219	44 29	264
40	28 04	351	29 07	032	44 14	059	32 42	086	49 12	130	29 09	219	43 42	263
41	27 57	350	29 32	031	44 55	058	33 29	085	49 49	130	28 39	219	42 55	262
42	27 48	349	29 56	030	45 35	057	34 16	085	50 25	130	28 10	218	42 08	262
43	27 39	348	30 19	029	46 14	056	35 03	084	51 01	130	27 41	218	41 22	261
44	27 29	347	30 42	028	46 54	055	35 50	083	51 38	130	27 12	218	40 35	261
	◆ALDEBARAN		RIGEL		SIRIUS		◆CANOPUS		ACHERNAR		◆FOMALHAUT		Hamal	
45	31 03	027	47 32	054	36 37	083	52 14	130	66 27	208	39 48	260	27 18	346
46	31 24	026	48 10	053	37 24	082	52 50	130	66 04	209	39 02	260	27 07	345
47	31 45	025	48 48	052	38 11	081	53 27	130	65 41	210	38 15	259	26 54	344
48	32 04	024	49 25	051	38 58	081	54 03	130	65 16	211	37 29	259	26 41	343
49	32 23	023	50 01	049	39 44	080	54 39	130	64 52	212	36 42	258	26 27	342
50	32 41	022	50 37	048	40 31	079	55 15	130	64 26	213	35 56	258	26 12	341
51	32 58	021	51 11	047	41 17	078	55 51	130	64 00	214	35 10	257	25 57	340
52	33 14	019	51 46	046	42 03	078	56 28	130	63 34	214	34 24	257	25 41	339
53	33 29	018	52 19	044	42 49	077	57 03	131	63 07	215	33 38	256	25 24	338
54	33 44	017	52 52	043	43 35	076	57 39	131	62 40	216	32 52	256	25 06	337
55	33 57	016	53 24	042	44 21	075	58 15	131	62 12	216	32 06	255	24 47	336
56	34 10	015	53 54	040	45 07	074	58 51	131	61 44	217	31 20	255	24 28	336
57	34 22	014	54 25	039	45 52	074	59 26	131	61 16	217	30 35	254	24 08	335
58	34 33	013	54 54	037	46 38	073	60 02	132	60 47	218	29 49	254	23 47	334
59	34 43	012	55 22	036	47 23	072	60 37	132	60 18	218	29 04	254	23 26	333
	ALDEBARAN		◆BETELGEUSE		SIRIUS		◆ACRUX		ACHERNAR		◆FOMALHAUT		Hamal	
60	34 52	010	37 18	037	48 08	071	19 39	157	59 48	219	28 19	253	23 04	332
61	35 00	009	37 46	036	48 52	070	19 57	157	59 18	219	27 33	253	22 41	331
62	35 07	008	38 13	035	49 36	069	20 16	157	58 48	220	26 48	252	22 18	330
63	35 13	007	38 39	033	50 21	068	20 35	156	58 18	220	26 03	252	21 54	329
64	35 18	006	39 05	032	51 04	067	20 54	156	57 48	220	25 19	251	21 30	328
65	35 23	005	39 30	031	51 48	066	21 13	156	57 17	221	24 34	251	21 04	327
66	35 26	003	39 54	030	52 31	065	21 33	155	56 46	221	23 49	250	20 39	326
67	35 28	002	40 17	029	53 14	064	21 53	155	56 15	221	23 05	250	20 12	326
68	35 30	001	40 39	028	53 56	063	22 13	155	55 44	221	22 21	249	19 45	325
69	35 30	000	41 01	026	54 38	062	22 34	154	55 12	221	21 37	249	19 18	324
70	35 29	359	41 21	025	55 19	061	22 54	154	54 41	222	20 53	248	18 49	323
71	35 28	358	41 41	024	56 00	060	23 15	154	54 09	222	20 09	248	18 21	322
72	35 25	356	41 59	023	56 41	058	23 36	153	53 37	222	19 25	247	17 52	321
73	35 22	355	42 17	021	57 21	057	23 58	153	53 05	223	18 42	247	17 22	321
74	35 17	354	42 34	020	58 00	056	24 19	153	52 33	223	17 59	246	16 52	320
	BETELGEUSE		◆SIRIUS		Suhail		◆ACRUX		ACHERNAR		◆Diphda		ALDEBARAN	
75	42 49	019	58 39	054	43 49	117	24 41	152	52 01	223	31 08	271	35 12	353
76	43 04	017	59 17	053	44 31	117	25 03	152	51 29	223	30 20	270	35 06	352
77	43 18	016	59 55	052	45 13	117	25 25	152	50 57	223	29 33	270	34 58	351
78	43 30	015	60 31	050	45 55	117	25 48	152	50 25	223	28 46	269	34 50	349
79	43 42	013	61 07	049	46 37	116	26 11	151	49 52	223	27 59	268	34 41	348
80	43 52	012	61 42	047	47 20	116	26 33	151	49 20	223	27 11	268	34 31	347
81	44 02	011	62 16	045	48 02	116	26 56	151	48 48	223	26 24	267	34 20	346
82	44 10	009	62 50	044	48 45	116	27 20	150	48 15	223	25 37	267	34 08	345
83	44 17	008	63 22	042	49 28	115	27 43	150	47 43	223	24 50	266	33 55	344
84	44 23	007	63 53	040	50 10	115	28 07	150	47 10	223	24 03	265	33 41	343
85	44 28	005	64 23	038	50 53	115	28 30	150	46 38	223	23 15	265	33 27	341
86	44 31	004	64 51	036	51 36	115	28 54	149	46 05	223	22 28	264	33 11	340
87	44 34	002	65 19	034	52 19	115	29 19	149	45 33	223	21 41	264	32 55	339
88	44 35	001	65 45	032	53 02	114	29 43	149	45 01	223	20 54	263	32 38	338
89	44 36	000	66 09	030	53 45	114	30 07	149	44 28	223	20 08	262	32 20	337

LAT 38°S

LHA ♈	Hc	Zn	Hc	Zn	Hc	Zn	Hc	Zn	Hc	Zn	Hc	Zn	Hc	Zn
	PROCYON		REGULUS		◆Suhail		ACRUX		◆ACHERNAR		RIGEL		◆BETELGEUSE	
90	41 01	034	13 31	063	54 28	114	30 32	149	43 56	223	58 28	338	44 35	358
91	41 27	032	14 13	062	55 12	114	30 57	148	43 24	223	58 10	336	44 33	357
92	41 52	031	14 54	061	55 55	114	31 22	148	42 51	223	57 50	334	44 29	355
93	42 16	030	15 35	061	56 38	114	31 47	148	42 19	223	57 29	333	44 25	354
94	42 39	029	16 17	060	57 22	113	32 12	148	41 47	223	57 07	331	44 20	353
95	43 01	027	16 57	059	58 05	113	32 37	147	41 15	223	56 43	329	44 13	351
96	43 23	026	17 38	058	58 48	113	33 03	147	40 43	223	56 18	328	44 05	350
97	43 43	025	18 18	058	59 32	113	33 28	147	40 11	222	55 53	326	43 57	349
98	44 02	024	18 57	057	60 15	113	33 54	147	39 39	222	55 26	325	43 47	347
99	44 21	022	19 37	056	60 59	113	34 20	147	39 07	222	54 58	323	43 36	346
100	44 38	021	20 16	055	61 43	113	34 46	147	38 35	222	54 29	322	43 24	345
101	44 55	020	20 55	054	62 26	113	35 12	146	38 04	222	53 59	320	43 10	343
102	45 10	018	21 33	054	63 10	113	35 38	146	37 32	222	53 28	319	42 56	342
103	45 24	017	22 11	053	63 53	113	36 04	146	37 01	222	52 56	317	42 41	341
104	45 37	015	22 48	052	64 37	113	36 31	146	36 29	221	52 24	316	42 25	339
	PROCYON		REGULUS		◆Gienah		ACRUX		◆ACHERNAR		RIGEL		◆BETELGEUSE	
105	45 49	014	23 25	051	19 17	098	36 57	146	35 58	221	51 51	315	42 08	338
106	46 00	013	24 02	050	20 03	097	37 24	146	35 27	221	51 17	313	41 49	337
107	46 10	011	24 38	050	20 50	096	37 50	146	34 56	221	50 42	312	41 30	335
108	46 18	010	25 14	049	21 37	096	38 17	146	34 25	221	50 06	311	41 10	334
109	46 26	008	25 49	048	22 25	095	38 44	145	33 54	220	49 30	309	40 49	333
110	46 32	007	26 24	047	23 12	095	39 11	145	33 24	220	48 53	308	40 27	332
111	46 37	005	26 58	046	23 59	094	39 38	145	32 53	220	48 16	307	40 04	331
112	46 41	004	27 32	045	24 46	093	40 05	145	32 23	220	47 38	306	39 41	329
113	46 44	003	28 06	044	25 33	093	40 32	145	31 53	220	46 59	305	39 16	328
114	46 45	001	28 38	043	26 20	092	40 59	145	31 23	219	46 20	304	38 51	327
115	46 46	000	29 10	042	27 08	092	41 26	145	30 53	219	45 41	303	38 25	326
116	46 45	358	29 42	041	27 55	091	41 53	145	30 23	219	45 01	302	37 58	325
117	46 43	357	30 13	041	28 42	090	42 20	145	29 54	219	44 20	301	37 30	324
118	46 39	355	30 43	040	29 29	090	42 47	145	29 24	218	43 39	300	37 02	323
119	46 35	354	31 13	039	30 17	089	43 14	145	28 55	218	42 58	299	36 33	322
	REGULUS		◆SPICA		ACRUX		◆CANOPUS		RIGEL		BETELGEUSE		◆PROCYON	
120	31 42	038	13 42	094	43 41	145	67 48	221	42 17	298	36 03	320	46 29	352
121	32 11	037	14 29	093	44 09	145	67 17	222	41 35	297	35 33	319	46 22	351
122	32 39	036	15 16	092	44 36	145	66 45	222	40 52	296	35 02	318	46 14	350
123	33 06	035	16 03	092	45 03	145	66 13	223	40 09	295	34 30	317	46 05	348
124	33 32	033	16 51	091	45 30	145	65 41	224	39 26	294	33 58	316	45 55	347
125	33 58	032	17 38	091	45 57	145	65 08	224	38 43	293	33 25	315	45 44	345
126	34 23	031	18 25	090	46 24	145	64 34	225	38 00	292	32 51	314	45 31	344
127	34 47	030	19 12	089	46 51	145	64 01	226	37 16	292	32 17	313	45 17	343
128	35 11	029	20 00	089	47 18	145	63 27	226	36 32	291	31 42	312	45 03	341
129	35 33	028	20 47	088	47 45	145	62 53	226	35 47	290	31 07	312	44 47	340
130	35 55	027	21 34	087	48 12	145	62 19	227	35 03	289	30 32	311	44 30	338
131	36 16	026	22 21	087	48 39	146	61 44	227	34 18	288	29 55	310	44 12	337
132	36 36	025	23 09	086	49 05	146	61 09	228	33 33	287	29 19	309	43 53	336
133	36 56	024	23 56	085	49 32	146	60 34	228	32 48	287	28 42	308	43 33	335
134	37 14	022	24 43	085	49 59	146	59 59	228	32 02	286	28 04	307	43 13	333
	◆REGULUS		SPICA		◆ACRUX		CANOPUS		◆RIGEL		BETELGEUSE		PROCYON	
135	37 32	021	25 30	084	50 25	146	59 23	229	31 17	285	27 26	306	42 51	332
136	37 48	020	26 17	084	50 51	146	58 48	229	30 31	284	26 48	305	42 28	331
137	38 04	019	27 04	083	51 17	147	58 12	229	29 45	284	26 09	304	42 05	329
138	38 19	018	27 51	082	51 43	147	57 36	229	28 59	283	25 30	304	41 40	328
139	38 32	016	28 38	082	52 09	147	57 01	229	28 13	282	24 50	303	41 15	327
140	38 45	015	29 24	081	52 35	147	56 25	230	27 27	282	24 10	302	40 49	326
141	38 57	014	30 11	080	53 00	147	55 49	230	26 40	281	23 30	301	40 22	325
142	39 08	013	30 58	079	53 26	148	55 12	230	25 54	280	22 50	300	39 54	324
143	39 18	011	31 44	079	53 51	148	54 36	230	25 07	280	22 09	300	39 26	322
144	39 27	010	32 30	078	54 16	148	54 00	230	24 21	279	21 27	299	38 56	321
145	39 35	009	33 16	077	54 40	149	53 24	230	23 34	278	20 46	298	38 27	320
146	39 41	008	34 03	077	55 05	149	52 47	230	22 47	278	20 04	297	37 56	319
147	39 47	006	34 48	076	55 29	149	52 11	230	22 00	277	19 22	297	37 25	318
148	39 52	005	35 34	075	55 53	150	51 35	230	21 13	276	18 39	296	36 53	317
149	39 56	004	36 20	074	56 17	150	50 58	230	20 26	276	17 57	295	36 20	316
	REGULUS		◆SPICA		ACRUX		◆Miaplacidus		CANOPUS		SIRIUS		◆PROCYON	
150	39 58	003	37 05	074	56 40	151	57 41	188	50 22	230	42 25	283	35 47	315
151	40 00	001	37 51	073	57 03	151	57 35	188	49 46	230	41 39	282	35 13	314
152	40 00	000	38 36	072	57 26	151	57 28	189	49 09	230	40 52	281	34 39	313
153	40 00	359	39 20	071	57 48	152	57 20	189	48 33	230	40 06	280	34 04	312
154	39 58	357	40 05	070	58 10	152	57 12	190	47 57	230	39 19	280	33 28	311
155	39 56	356	40 49	069	58 32	153	57 04	191	47 20	230	38 33	279	32 52	310
156	39 52	355	41 34	069	58 53	154	56 55	191	46 44	230	37 46	278	32 16	309
157	39 47	354	42 17	068	59 14	154	56 46	192	46 08	230	36 59	278	31 39	308
158	39 42	352	43 01	067	59 35	155	56 36	192	45 32	230	36 12	277	31 02	307
159	39 35	351	43 44	066	59 55	155	56 26	193	44 56	230	35 25	276	30 24	306
160	39 27	350	44 27	065	60 14	156	56 15	193	44 20	230	34 38	276	29 46	305
161	39 18	349	45 10	064	60 33	157	56 04	194	43 44	229	33 51	275	29 07	305
162	39 09	347	45 52	063	60 52	157	55 52	194	43 08	229	33 04	274	28 28	304
163	38 58	346	46 34	062	61 10	158	55 40	195	42 32	229	32 17	274	27 48	303
164	38 46	345	47 16	061	61 27	159	55 28	195	41 56	229	31 29	273	27 08	302
	ARCTURUS		◆ANTARES		ACRUX		◆CANOPUS		SIRIUS		PROCYON		◆REGULUS	
165	16 41	048	21 39	107	61 44	159	41 20	229	30 42	272	26 28	301	38 33	344
166	17 16	047	22 24	107	62 00	160	40 45	229	29 55	272	25 47	300	38 19	342
167	17 50	046	23 09	106	62 16	161	40 09	229	29 08	271	25 06	300	38 05	341
168	18 24	046	23 55	106	62 31	162	39 34	228	28 20	270	24 25	299	37 49	340
169	18 58	045	24 40	105	62 46	163	38 59	228	27 33	270	23 44	298	37 32	339
170	19 31	044	25 26	105	63 00	163	38 23	228	26 46	269	23 02	297	37 15	338
171	20 03	043	26 12	104	63 13	164	37 48	228	25 59	269	22 20	296	36 57	337
172	20 35	042	26 57	104	63 25	165	37 13	228	25 11	268	21 37	296	36 37	335
173	21 07	041	27 43	103	63 37	166	36 39	227	24 24	267	20 54	295	36 17	334
174	21 38	041	28 30	103	63 48	167	36 04	227	23 37	267	20 11	294	35 56	333
175	22 09	040	29 16	102	63 58	168	35 29	227	22 50	266	19 28	294	35 34	332
176	22 39	039	30 02	102	64 07	169	34 55	227	22 03	266	18 45	293	35 12	331
177	23 08	038	30 48	101	64 16	170	34 20	226	21 15	265	18 01	292	34 48	330
178	23 37	037	31 35	101	64 24	171	33 46	226	20 28	264	17 17	291	34 24	329
179	24 05	036	32 21	100	64 31	172	33 12	226	19 41	264	16 33	291	33 59	328

LAT 38°S

LHA	Hc	Zn	Hc	Zn	Hc	Zn	Hc	Zn	Hc	Zn	Hc	Zn	Hc	Zn
♈	ARCTURUS		◆ANTARES		RIGIL KENT.		ACRUX		◆CANOPUS		Alphard		◆REGULUS	
180	24 33	035	33 08	099	56 22	146	64 37	173	32 38	226	44 50	301	33 34	327
181	25 00	034	33 55	099	56 49	146	64 42	174	32 04	225	44 09	300	33 07	326
182	25 26	033	34 41	098	57 15	146	64 47	175	31 31	225	43 27	299	32 40	325
183	25 52	033	35 28	098	57 41	147	64 50	176	30 57	225	42 46	298	32 12	323
184	26 17	032	36 15	097	58 07	147	64 53	177	30 24	225	42 04	297	31 44	322
185	26 41	031	37 02	097	58 32	148	64 55	178	29 51	224	41 21	296	31 15	321
186	27 05	030	37 49	096	58 57	148	64 56	179	29 18	224	40 38	295	30 45	321
187	27 28	029	38 36	096	59 22	149	64 56	180	28 45	224	39 55	294	30 15	320
188	27 51	028	39 23	095	59 46	149	64 55	182	28 13	223	39 12	293	29 44	319
189	28 12	027	40 10	095	60 10	150	64 53	183	27 40	223	38 28	292	29 12	318
190	28 33	026	40 57	094	60 34	150	64 51	184	27 08	223	37 44	291	28 40	317
191	28 53	025	41 44	093	60 57	151	64 47	185	26 36	222	37 00	291	28 07	316
192	29 13	024	42 32	093	61 20	152	64 43	186	26 04	222	36 16	290	27 34	315
193	29 32	023	43 19	092	61 42	152	64 38	187	25 32	222	35 31	289	27 00	314
194	29 49	022	44 06	092	62 03	153	64 32	188	25 01	221	34 46	288	26 26	313
	ARCTURUS		◆ANTARES		RIGIL KENT.		◆ACRUX		Suhail		◆REGULUS		Denebola	
195	30 07	021	44 53	091	62 25	154	64 25	189	46 33	244	25 51	312	34 49	339
196	30 23	020	45 41	090	62 45	155	64 17	190	45 51	243	25 16	311	34 31	338
197	30 38	019	46 28	090	63 05	155	64 09	191	45 09	243	24 40	310	34 13	337
198	30 53	017	47 15	089	63 25	156	63 59	192	44 27	243	24 04	310	33 54	336
199	31 07	016	48 03	089	63 44	157	63 49	193	43 45	243	23 27	309	33 34	334
200	31 20	015	48 50	088	64 02	158	63 39	194	43 03	242	22 50	308	33 13	333
201	31 32	014	49 37	087	64 19	159	63 27	195	42 21	242	22 13	307	32 51	332
202	31 43	013	50 24	087	64 36	160	63 15	196	41 39	242	21 35	306	32 29	331
203	31 53	012	51 11	086	64 52	161	63 02	196	40 57	242	20 57	306	32 06	330
204	32 03	011	51 59	085	65 07	162	62 48	197	40 16	241	20 18	305	31 42	329
205	32 11	010	52 46	085	65 22	163	62 34	198	39 35	241	19 39	304	31 17	328
206	32 19	009	53 33	084	65 36	164	62 19	199	38 53	241	19 00	303	30 52	327
207	32 26	008	54 20	083	65 49	165	62 03	200	38 12	240	18 20	302	30 26	326
208	32 32	007	55 07	082	66 01	166	61 47	200	37 31	240	17 40	302	29 59	325
209	32 37	005	55 53	082	66 12	167	61 30	201	36 50	240	16 59	301	29 32	324
	ARCTURUS		◆ANTARES		Peacock		RIGIL KENT.		◆ACRUX		Suhail		◆SPICA	
210	32 41	004	56 40	081	27 52	142	66 22	168	61 13	202	36 10	239	62 01	341
211	32 44	003	57 27	080	28 21	142	66 31	169	60 55	203	35 29	239	61 45	339
212	32 46	002	58 13	079	28 50	141	66 40	170	60 36	203	34 48	239	61 27	337
213	32 47	001	59 00	078	29 20	141	66 47	172	60 17	204	34 08	238	61 09	336
214	32 47	000	59 46	077	29 50	141	66 54	173	59 58	205	33 28	238	60 48	334
215	32 47	359	60 32	076	30 20	141	66 59	174	59 38	205	32 48	238	60 27	332
216	32 45	358	61 18	075	30 50	140	67 04	175	59 17	206	32 08	237	60 04	330
217	32 43	356	62 03	074	31 20	140	67 07	176	58 57	206	31 28	237	59 39	328
218	32 40	355	62 49	073	31 51	140	67 09	178	58 35	207	30 49	237	59 14	326
219	32 35	354	63 34	072	32 21	140	67 11	179	58 14	207	30 09	236	58 47	325
220	32 30	353	64 19	071	32 52	139	67 11	180	57 52	208	29 30	236	58 19	323
221	32 24	352	65 03	070	33 23	139	67 10	181	57 29	208	28 51	236	57 50	321
222	32 17	351	65 48	069	33 54	139	67 09	183	57 07	209	28 12	235	57 20	320
223	32 09	350	66 32	068	34 25	139	67 06	184	56 44	209	27 33	235	56 49	318
224	32 00	349	67 15	066	34 56	138	67 02	185	56 20	210	26 55	234	56 17	317
	Alphecca		Rasalhague		◆Nunki		Peacock		◆RIGIL KENT.		◆SPICA		ARCTURUS	
225	24 46	008	27 49	044	39 45	095	35 28	138	66 57	186	55 44	315	31 50	348
226	24 53	007	28 21	043	40 32	094	35 59	138	66 51	188	55 11	314	31 40	347
227	24 58	007	28 53	042	41 19	094	36 31	138	66 45	189	54 36	313	31 28	345
228	25 03	006	29 24	041	42 06	093	37 02	138	66 37	190	54 01	311	31 16	344
229	25 07	005	29 55	040	42 54	092	37 34	138	66 28	191	53 25	310	31 03	343
230	25 11	004	30 24	039	43 41	092	38 06	137	66 18	192	52 49	309	30 49	342
231	25 13	003	30 54	038	44 28	091	38 38	137	66 08	194	52 11	307	30 34	341
232	25 15	002	31 23	037	45 15	091	39 10	137	65 56	195	51 34	306	30 18	340
233	25 16	001	31 51	036	46 03	090	39 43	137	65 44	196	50 55	305	30 02	339
234	25 16	000	32 18	035	46 50	089	40 15	137	65 31	197	50 16	304	29 44	338
235	25 15	359	32 45	034	47 37	089	40 47	137	65 17	198	49 37	303	29 26	337
236	25 14	358	33 11	033	48 24	088	41 20	137	65 02	199	48 57	302	29 07	336
237	25 11	357	33 36	032	49 12	087	41 52	136	64 46	200	48 16	301	28 48	335
238	25 08	356	34 00	031	49 59	087	42 25	136	64 30	201	47 35	299	28 27	334
239	25 04	355	34 24	030	50 46	086	42 58	136	64 13	202	46 54	298	28 06	333
	Rasalhague		ALTAIR		◆Nunki		Peacock		◆RIGIL KENT.		SPICA		◆ARCTURUS	
240	34 47	029	18 47	062	51 33	085	43 30	136	63 55	203	46 12	297	27 44	332
241	35 09	027	19 29	061	52 20	085	44 03	136	63 37	203	45 30	296	27 22	331
242	35 31	026	20 10	060	53 07	084	44 36	136	63 18	204	44 47	295	26 59	330
243	35 51	025	20 51	060	53 54	083	45 09	136	62 58	205	44 04	295	26 35	329
244	36 11	024	21 31	059	54 41	083	45 42	136	62 38	206	43 21	294	26 10	328
245	36 30	023	22 12	058	55 28	082	46 15	136	62 17	206	42 38	293	25 45	327
246	36 48	022	22 52	057	56 15	081	46 47	136	61 55	207	41 54	292	25 19	326
247	37 05	021	23 31	056	57 02	080	47 20	136	61 34	208	41 10	291	24 52	325
248	37 21	019	24 10	056	57 48	079	47 53	136	61 11	209	40 26	290	24 25	324
249	37 36	018	24 49	055	58 34	078	48 26	136	60 48	209	39 41	289	23 57	324
250	37 50	017	25 28	054	59 21	078	48 59	136	60 25	210	38 56	288	23 29	323
251	38 04	016	26 06	053	60 07	077	49 32	136	60 01	210	38 11	288	23 00	322
252	38 16	015	26 43	052	60 53	076	50 05	136	59 37	211	37 26	287	22 30	321
253	38 27	013	27 20	051	61 38	075	50 37	136	59 13	211	36 41	286	22 00	320
254	38 38	012	27 57	051	62 24	074	51 10	136	58 48	212	35 55	285	21 29	319
	Rasalhague		◆ALTAIR		FOMALHAUT		◆ACHERNAR		RIGIL KENT.		◆SPICA		ARCTURUS	
255	38 47	011	28 33	050	18 12	114	14 20	154	58 23	212	35 10	285	20 58	318
256	38 56	010	29 09	049	18 56	113	14 40	154	57 57	213	34 24	284	20 26	317
257	39 03	008	29 45	048	19 39	113	15 01	154	57 32	213	33 38	283	19 54	317
258	39 09	007	30 19	047	20 23	112	15 22	153	57 05	214	32 52	282	19 21	316
259	39 15	006	30 54	046	21 07	112	15 44	153	56 39	214	32 06	282	18 48	315
260	39 19	005	31 27	045	21 51	111	16 06	152	56 12	214	31 19	281	18 15	314
261	39 22	003	32 00	044	22 35	111	16 28	152	55 46	215	30 33	280	17 40	313
262	39 25	002	32 33	043	23 19	110	16 50	152	55 18	215	29 46	279	17 06	313
263	39 26	001	33 05	042	24 03	110	17 13	151	54 51	215	28 59	279	16 31	312
264	39 26	000	33 36	041	24 48	109	17 36	151	54 24	216	28 13	278	15 55	311
265	39 25	358	34 07	040	25 33	109	17 59	150	53 56	216	27 26	277	15 20	310
266	39 23	357	34 37	039	26 17	108	18 23	150	53 28	216	26 39	277	14 43	310
267	39 20	356	35 07	038	27 02	108	18 47	150	53 00	216	25 52	276	14 07	309
268	39 16	355	35 35	037	27 47	107	19 11	149	52 32	217	25 05	275	13 30	308
269	39 11	353	36 04	036	28 33	107	19 35	149	52 04	217	24 18	275	12 52	307

LHA	Hc	Zn	Hc	Zn	Hc	Zn	Hc	Zn	Hc	Zn	Hc	Zn	Hc	Zn
♈	ALTAIR		Enif		◆FOMALHAUT		ACHERNAR		◆RIGIL KENT.		ANTARES		◆Rasalhague	
270	36 31	035	19 12	060	29 18	106	20 00	148	51 35	217	67 39	295	39 05	352
271	36 57	034	19 53	059	30 03	106	20 25	148	51 07	217	66 56	293	38 58	351
272	37 23	033	20 34	058	30 49	105	20 50	148	50 38	217	66 12	292	38 50	350
273	37 48	031	21 14	058	31 35	105	21 15	147	50 09	218	65 28	291	38 41	348
274	38 13	030	21 53	057	32 20	104	21 41	147	49 40	218	64 43	289	38 31	347
275	38 36	029	22 33	056	33 06	104	22 07	147	49 11	218	63 59	288	38 20	346
276	38 59	028	23 12	055	33 52	103	22 33	146	48 42	218	63 14	287	38 08	345
277	39 20	027	23 50	054	34 38	103	22 59	146	48 13	218	62 28	286	37 55	343
278	39 41	026	24 29	054	35 24	102	23 26	146	47 44	218	61 43	285	37 41	342
279	40 01	024	25 06	053	36 10	102	23 53	145	47 15	218	60 57	284	37 26	341
280	40 20	023	25 44	052	36 57	101	24 20	145	46 46	218	60 11	283	37 10	340
281	40 38	022	26 21	051	37 43	101	24 47	145	46 17	218	59 25	282	36 54	339
282	40 55	021	26 57	050	38 30	100	25 15	144	45 47	218	58 39	281	36 36	338
283	41 12	019	27 33	049	39 16	100	25 42	144	45 18	218	57 52	281	36 17	336
284	41 27	018	28 09	048	40 03	099	26 10	144	44 49	218	57 06	280	35 58	335
	ALTAIR		Enif		◆FOMALHAUT		ACHERNAR		◆RIGIL KENT.		ANTARES		◆Rasalhague	
285	41 41	017	28 44	047	40 49	099	26 38	143	44 20	218	56 19	279	35 38	334
286	41 54	016	29 19	047	41 36	098	27 07	143	43 50	218	55 32	278	35 17	333
287	42 06	014	29 53	046	42 23	098	27 35	143	43 21	218	54 45	277	34 55	332
288	42 17	013	30 26	045	43 10	097	28 04	142	42 52	218	53 59	277	34 32	331
289	42 27	012	30 59	044	43 57	097	28 33	142	42 23	218	53 12	276	34 09	330
290	42 36	010	31 32	043	44 44	096	29 02	142	41 54	218	52 24	275	33 45	329
291	42 44	009	32 03	042	45 31	096	29 31	142	41 25	218	51 37	274	33 20	328
292	42 51	008	32 35	041	46 18	095	30 01	141	40 55	218	50 50	274	32 54	327
293	42 57	006	33 05	040	47 05	095	30 30	141	40 26	218	50 03	273	32 27	325
294	43 01	005	33 35	039	47 52	094	31 00	141	39 57	218	49 16	272	32 00	324
295	43 05	004	34 04	038	48 39	094	31 30	141	39 29	218	48 29	272	31 32	323
296	43 07	002	34 33	037	49 26	093	32 00	140	39 00	218	47 41	271	31 04	322
297	43 08	001	35 01	036	50 14	092	32 30	140	38 31	217	46 54	270	30 35	321
298	43 09	000	35 28	035	51 01	092	33 00	140	38 02	217	46 07	270	30 05	321
299	43 08	358	35 55	034	51 48	091	33 31	140	37 34	217	45 19	269	29 35	320
	Enif		◆FOMALHAUT		ACHERNAR		◆RIGIL KENT.		ANTARES		◆Rasalhague		ALTAIR	
300	36 20	032	52 35	091	34 02	140	37 05	217	44 32	269	29 04	319	43 06	357
301	36 45	031	53 23	090	34 32	139	36 37	217	43 45	268	28 32	318	43 02	355
302	37 09	030	54 10	089	35 03	139	36 08	217	42 58	267	28 00	317	42 58	354
303	37 33	029	54 57	089	35 34	139	35 40	217	42 10	267	27 27	316	42 53	353
304	37 55	028	55 45	088	36 05	139	35 12	216	41 23	266	26 54	315	42 46	351
305	38 17	027	56 32	088	36 37	139	34 44	216	40 36	266	26 20	314	42 39	350
306	38 38	026	57 19	087	37 08	138	34 16	216	39 49	265	25 46	313	42 30	349
307	38 58	024	58 06	086	37 40	138	33 48	216	39 02	265	25 11	312	42 20	347
308	39 17	023	58 53	085	38 11	138	33 21	216	38 15	264	24 36	311	42 09	346
309	39 35	022	59 41	085	38 43	138	32 53	215	37 28	264	24 00	311	41 57	345
310	39 52	021	60 28	084	39 15	138	32 26	215	36 41	263	23 24	310	41 45	344
311	40 08	019	61 15	083	39 46	138	31 58	215	35 54	262	22 48	309	41 31	342
312	40 24	018	62 01	082	40 18	137	31 31	215	35 07	262	22 11	308	41 16	341
313	40 38	017	62 48	082	40 50	137	31 04	215	34 20	261	21 33	307	41 00	340
314	40 51	016	63 35	081	41 22	137	30 38	214	33 34	261	20 55	306	40 43	338
	Enif		◆Diphda		ACHERNAR		◆RIGIL KENT.		ANTARES		Nunki		◆ALTAIR	
315	41 03	014	37 41	084	41 54	137	30 11	214	32 47	260	61 15	285	40 25	337
316	41 15	013	38 28	083	42 27	137	29 45	214	32 00	260	60 29	284	40 06	336
317	41 25	012	39 15	082	42 59	137	29 18	214	31 14	259	59 43	283	39 47	335
318	41 34	011	40 01	082	43 31	137	28 52	213	30 27	259	58 57	282	39 26	334
319	41 42	009	40 48	081	44 03	137	28 26	213	29 41	258	58 10	281	39 05	332
320	41 49	008	41 35	080	44 36	137	28 00	213	28 55	258	57 24	280	38 42	331
321	41 55	007	42 21	080	45 08	137	27 35	213	28 09	257	56 37	279	38 19	330
322	42 00	005	43 08	079	45 41	137	27 09	212	27 23	257	55 51	279	37 55	329
323	42 04	004	43 54	078	46 13	137	26 44	212	26 37	256	55 04	278	37 30	328
324	42 07	003	44 40	077	46 46	137	26 19	212	25 51	256	54 17	277	37 04	327
325	42 08	001	45 26	076	47 18	137	25 54	212	25 05	255	53 30	276	36 38	326
326	42 09	000	46 12	076	47 50	137	25 30	211	24 20	255	52 43	276	36 11	324
327	42 08	359	46 58	075	48 23	137	25 05	211	23 34	254	51 56	275	35 43	323
328	42 07	357	47 44	074	48 55	137	24 41	211	22 49	254	51 09	274	35 15	322
329	42 04	356	48 29	073	49 28	137	24 17	210	22 03	253	50 21	274	34 45	321
	◆Alpheratz		Diphda		◆ACHERNAR		RIGIL KENT.		◆Nunki		ALTAIR		Enif	
330	16 33	029	49 14	072	50 00	137	23 53	210	49 34	273	34 15	320	42 00	355
331	16 56	028	49 59	071	50 32	137	23 30	210	48 47	272	33 45	319	41 55	353
332	17 18	027	50 44	070	51 05	137	23 06	209	48 00	272	33 14	318	41 49	352
333	17 39	026	51 28	069	51 37	137	22 43	209	47 13	271	32 42	317	41 42	351
334	18 00	026	52 12	068	52 09	137	22 20	209	46 25	270	32 09	316	41 34	349
335	18 20	025	52 56	067	52 41	137	21 58	208	45 38	270	31 36	315	41 25	348
336	18 39	024	53 39	066	53 13	138	21 35	208	44 51	269	31 03	314	41 14	347
337	18 58	023	54 23	065	53 45	138	21 13	208	44 03	268	30 29	313	41 03	346
338	19 16	022	55 05	064	54 17	138	20 51	207	43 16	268	29 54	312	40 51	344
339	19 34	021	55 48	063	54 48	138	20 30	207	42 29	267	29 19	312	40 37	343
340	19 50	020	56 30	062	55 20	138	20 08	207	41 42	267	28 43	311	40 23	342
341	20 07	020	57 11	061	55 51	139	19 47	206	40 55	266	28 07	310	40 08	340
342	20 22	019	57 52	060	56 22	139	19 26	206	40 07	266	27 30	309	39 52	339
343	20 37	018	58 33	058	56 53	139	19 06	206	39 20	265	26 53	308	39 34	338
344	20 51	017	59 13	057	57 24	139	18 46	205	38 33	264	26 16	307	39 16	337
	◆Alpheratz		Diphda		◆CANOPUS		RIGIL KENT.		◆Nunki		ALTAIR		Enif	
345	21 04	016	59 52	056	18 36	143	18 26	205	37 46	264	25 38	306	38 57	336
346	21 17	015	60 31	054	19 04	143	18 06	204	36 59	263	25 00	305	38 37	334
347	21 29	014	61 09	053	19 33	143	17 46	204	36 12	263	24 21	305	38 16	333
348	21 40	013	61 46	051	20 01	142	17 27	204	35 25	262	23 42	304	37 55	332
349	21 50	012	62 23	050	20 31	142	17 08	203	34 38	262	23 02	303	37 32	331
350	22 00	011	62 58	048	21 00	141	16 50	203	33 52	261	22 23	302	37 09	330
351	22 09	010	63 33	046	21 30	141	16 32	203	33 05	261	21 42	301	36 45	329
352	22 17	009	64 07	045	22 00	141	16 14	202	32 18	260	21 02	301	36 20	328
353	22 24	009	64 39	043	22 30	140	15 56	202	31 32	260	20 21	300	35 54	326
354	22 31	008	65 11	041	23 00	140	15 39	201	30 45	259	19 40	299	35 27	325
355	22 37	007	65 41	039	23 31	140	15 22	201	29 59	259	18 58	298	35 00	324
356	22 42	006	66 11	037	24 01	139	15 05	200	29 13	258	18 17	298	34 32	323
357	22 46	005	66 38	035	24 32	139	14 49	200	28 27	258	17 35	297	34 04	322
358	22 50	004	67 05	033	25 04	139	14 33	200	27 40	257	16 52	296	33 34	321
359	22 53	003	67 30	031	25 35	138	14 17	199	26 54	256	16 10	295	33 04	320

 LAT 39°S

LHA	Hc	Zn	Hc	Zn	Hc	Zn	Hc	Zn	Hc	Zn	Hc	Zn	Hc	Zn
♈	Alpheratz		◆Hamal		RIGEL		◆CANOPUS		Peacock		◆Nunki		Enif	
0	21 55	002	20 52	031	14 01	089	26 51	137	51 08	225	26 23	256	31 48	320
1	21 56	001	21 16	030	14 48	089	27 23	137	50 35	225	25 38	256	31 18	319
2	21 56	000	21 39	029	15 34	088	27 54	137	50 02	225	24 52	255	30 46	318
3	21 56	359	22 01	028	16 21	087	28 26	137	49 30	225	24 07	255	30 15	317
4	21 55	358	22 23	027	17 07	087	28 59	136	48 57	225	23 22	254	29 42	316
5	21 53	357	22 44	027	17 54	086	29 31	136	48 24	225	22 38	254	29 10	315
6	21 50	356	23 05	026	18 40	085	30 04	136	47 51	225	21 53	253	28 36	314
7	21 47	355	23 25	025	19 27	085	30 36	135	47 18	225	21 08	253	28 02	313
8	21 43	354	23 44	024	20 13	084	31 09	135	46 45	225	20 24	252	27 28	312
9	21 38	353	24 02	023	21 00	083	31 42	135	46 12	225	19 40	252	26 53	311
10	21 32	353	24 20	022	21 46	083	32 16	134	45 39	225	18 55	251	26 18	310
11	21 26	352	24 37	021	22 32	082	32 49	134	45 07	225	18 11	250	25 42	309
12	21 18	351	24 53	020	23 18	081	33 23	134	44 34	225	17 27	250	25 06	308
13	21 10	350	25 09	019	24 04	081	33 56	134	44 01	224	16 44	249	24 29	308
14	21 02	349	25 24	018	24 50	080	34 30	133	43 29	224	16 00	249	23 52	307
	◆Hamal		ALDEBARAN		RIGEL		◆CANOPUS		Peacock		◆FOMALHAUT		Alpheratz	
15	25 38	017	15 05	053	25 36	079	35 04	133	42 56	224	63 10	281	20 52	348
16	25 51	016	15 42	053	26 22	079	35 38	133	42 24	224	62 24	280	20 42	347
17	26 03	015	16 19	052	27 08	078	36 13	133	41 51	224	61 38	279	20 31	346
18	26 15	014	16 56	051	27 53	077	36 47	132	41 19	224	60 52	278	20 20	345
19	26 26	013	17 32	050	28 39	077	37 22	132	40 47	224	60 06	278	20 07	344
20	26 36	012	18 07	049	29 24	076	37 56	132	40 14	224	59 20	277	19 54	343
21	26 46	011	18 43	049	30 09	075	38 31	132	39 42	223	58 33	276	19 41	342
22	26 54	010	19 17	048	30 54	074	39 06	131	39 10	223	57 47	275	19 26	342
23	27 02	009	19 52	047	31 39	074	39 41	131	38 38	223	57 00	274	19 11	341
24	27 09	008	20 26	046	32 23	073	40 16	131	38 07	223	56 14	274	18 55	340
25	27 15	007	20 59	045	33 08	072	40 52	131	37 35	223	55 27	273	18 39	339
26	27 20	006	21 32	045	33 52	071	41 27	131	37 03	223	54 41	272	18 22	338
27	27 24	005	22 04	044	34 36	070	42 02	130	36 32	222	53 54	272	18 04	337
28	27 28	004	22 36	043	35 20	070	42 38	130	36 01	222	53 08	271	17 46	336
29	27 31	003	23 08	042	36 04	069	43 14	130	35 29	222	52 21	270	17 26	335
	Hamal		◆ALDEBARAN		RIGEL		SIRIUS		◆CANOPUS		Peacock		◆FOMALHAUT	
30	27 32	002	23 39	041	36 47	068	24 52	092	43 49	130	34 58	222	51 34	270
31	27 33	001	24 09	040	37 30	067	25 38	091	44 25	130	34 27	221	50 48	269
32	27 34	000	24 39	039	38 13	066	26 25	091	45 01	130	33 57	221	50 01	269
33	27 33	359	25 08	038	38 55	065	27 12	090	45 37	129	33 26	221	49 14	268
34	27 31	358	25 37	037	39 38	064	27 58	089	46 13	129	32 55	221	48 28	267
35	27 29	357	26 05	037	40 19	063	28 45	089	46 49	129	32 25	220	47 41	267
36	27 26	356	26 32	036	41 01	063	29 31	088	47 25	129	31 55	220	46 55	266
37	27 22	355	26 59	035	41 42	062	30 18	087	48 01	129	31 25	220	46 08	266
38	27 17	354	27 26	034	42 23	061	31 05	087	48 38	129	30 55	220	45 22	265
39	27 11	352	27 51	033	43 03	060	31 51	086	49 14	129	30 25	219	44 35	265
40	27 05	351	28 16	032	43 43	059	32 38	085	49 50	129	29 56	219	43 49	264
41	26 58	350	28 40	031	44 23	058	33 24	085	50 27	129	29 26	219	43 03	263
42	26 49	349	29 04	030	45 02	057	34 10	084	51 03	129	28 57	219	42 16	263
43	26 40	348	29 26	029	45 41	056	34 57	083	51 39	129	28 28	218	41 30	262
44	26 31	347	29 49	028	46 19	054	35 43	083	52 16	129	27 59	218	40 44	262
	◆ALDEBARAN		RIGEL		SIRIUS		◆CANOPUS		ACHERNAR		◆FOMALHAUT		Hamal	
45	30 10	027	46 57	053	36 29	082	52 52	129	67 20	210	39 58	261	26 20	346
46	30 30	026	47 34	052	37 15	081	53 28	129	66 56	211	39 12	261	26 09	345
47	30 50	025	48 10	051	38 01	081	54 05	129	66 32	211	38 26	260	25 57	344
48	31 09	024	48 46	050	38 47	080	54 41	129	66 07	212	37 40	260	25 44	343
49	31 28	023	49 22	049	39 33	079	55 17	129	65 42	213	36 54	259	25 30	342
50	31 45	021	49 56	047	40 19	078	55 54	129	65 16	214	36 08	259	25 16	341
51	32 02	020	50 30	046	41 05	078	56 30	129	64 50	215	35 23	258	25 00	340
52	32 17	019	51 03	045	41 50	077	57 06	129	64 23	215	34 37	258	24 44	340
53	32 32	018	51 36	043	42 35	076	57 42	129	63 56	216	33 51	257	24 28	339
54	32 46	017	52 08	042	43 21	075	58 18	130	63 28	217	33 06	257	24 10	338
55	33 00	016	52 38	041	44 06	074	58 54	130	63 00	217	32 21	256	23 52	337
56	33 12	015	53 08	039	44 50	074	59 30	130	62 32	218	31 36	256	23 33	336
57	33 24	014	53 37	038	45 35	073	60 05	130	62 03	218	30 50	255	23 14	335
58	33 34	013	54 06	036	46 19	072	60 41	130	61 34	219	30 05	255	22 54	334
59	33 44	011	54 33	035	47 04	071	61 16	131	61 04	219	29 21	254	22 33	333
	ALDEBARAN		◆BETELGEUSE		SIRIUS		◆ACRUX		ACHERNAR		◆FOMALHAUT		Hamal	
60	33 53	010	36 29	036	47 48	070	20 34	157	60 35	220	28 36	254	22 11	332
61	34 01	009	36 57	035	48 31	069	20 52	157	60 05	220	27 51	253	21 49	331
62	34 08	008	37 23	034	49 15	068	21 11	156	59 34	221	27 07	253	21 26	330
63	34 14	007	37 49	033	49 58	067	21 30	156	59 04	221	26 22	252	21 03	329
64	34 19	006	38 14	032	50 41	066	21 49	156	58 33	221	25 38	251	20 38	328
65	34 23	005	38 38	031	51 23	065	22 08	155	58 02	222	24 54	251	20 14	328
66	34 26	003	39 02	030	52 05	064	22 28	155	57 31	222	24 10	250	19 48	327
67	34 28	002	39 24	028	52 47	063	22 47	155	57 00	222	23 26	250	19 23	326
68	34 30	001	39 46	027	53 28	062	23 07	154	56 28	222	22 42	249	18 56	325
69	34 30	000	40 07	026	54 09	061	23 28	154	55 57	223	21 59	249	18 29	324
70	34 29	359	40 27	025	54 50	060	23 48	154	55 25	223	21 15	248	18 01	323
71	34 28	358	40 46	023	55 29	058	24 09	153	54 53	223	20 32	248	17 33	322
72	34 26	356	41 04	022	56 09	057	24 30	153	54 21	223	19 49	247	17 05	322
73	34 22	355	41 21	021	56 48	056	24 51	153	53 49	223	19 06	247	16 35	321
74	34 18	354	41 37	020	57 26	055	25 13	152	53 17	224	18 23	246	16 06	320
	BETELGEUSE		◆SIRIUS		Suhail		◆ACRUX		ACHERNAR		◆Diphda		ALDEBARAN	
75	41 53	018	58 04	053	44 16	116	25 34	152	52 45	224	31 07	271	34 13	353
76	42 07	017	58 41	052	44 58	116	25 56	152	52 13	224	30 20	271	34 06	352
77	42 20	016	59 17	050	45 39	116	26 18	152	51 41	224	29 33	270	33 59	351
78	42 32	015	59 52	049	46 21	116	26 41	151	51 08	224	28 47	269	33 51	349
79	42 43	013	60 27	047	47 04	115	27 03	151	50 36	224	28 00	269	33 42	348
80	42 54	012	61 01	046	47 46	115	27 26	151	50 03	224	27 13	268	33 32	347
81	43 03	011	61 34	044	48 28	115	27 49	150	49 31	224	26 27	268	33 22	346
82	43 11	009	62 06	042	49 10	115	28 12	150	48 59	224	25 40	267	33 10	345
83	43 17	008	62 37	041	49 53	114	28 35	150	48 26	224	24 54	266	32 57	344
84	43 23	006	63 06	039	50 35	114	28 59	150	47 54	224	24 07	266	32 44	343
85	43 28	005	63 35	037	51 18	114	29 22	149	47 21	224	23 21	265	32 30	342
86	43 31	004	64 03	035	52 01	114	29 46	149	46 49	224	22 34	265	32 15	341
87	43 34	002	64 29	033	52 43	113	30 10	149	46 16	224	21 48	264	31 59	339
88	43 35	001	64 54	031	53 26	113	30 34	149	45 44	224	21 02	263	31 42	338
89	43 36	000	65 17	029	54 09	113	30 58	148	45 12	224	20 15	263	31 24	337

LHA	Hc	Zn	Hc	Zn	Hc	Zn	Hc	Zn	Hc	Zn	Hc	Zn	Hc	Zn
♈	PROCYON		REGULUS		◆Suhail		ACRUX		◆ACHERNAR		RIGEL		◆BETELGEUSE	
90	40 11	033	13 03	062	54 52	113	31 23	148	44 39	224	57 32	339	43 35	358
91	40 36	032	13 44	062	55 35	112	31 48	148	44 07	224	57 15	337	43 33	357
92	41 01	031	14 25	061	56 18	112	32 12	148	43 35	224	56 56	335	43 30	356
93	41 24	029	15 06	060	57 02	112	32 37	148	43 03	224	56 36	333	43 25	354
94	41 46	028	15 46	060	57 45	112	33 02	147	42 31	223	56 14	332	43 20	353
95	42 08	027	16 26	059	58 28	112	33 28	147	41 59	223	55 51	330	43 14	351
96	42 29	026	17 06	058	59 11	112	33 53	147	41 27	223	55 28	328	43 06	350
97	42 49	024	17 45	057	59 55	111	34 19	147	40 55	223	55 03	327	42 58	349
98	43 07	023	18 25	057	60 38	111	34 44	147	40 23	223	54 37	325	42 48	347
99	43 25	022	19 03	056	61 22	111	35 10	146	39 51	223	54 10	324	42 37	346
100	43 42	021	19 42	055	62 05	111	35 36	146	39 20	223	53 42	322	42 26	345
101	43 58	019	20 20	054	62 49	111	36 02	146	38 48	222	53 13	321	42 13	343
102	44 13	018	20 57	053	63 32	111	36 28	146	38 17	222	52 43	319	41 59	342
103	44 27	017	21 34	053	64 16	111	36 54	146	37 45	222	52 12	318	41 44	341
104	44 39	015	22 11	052	64 59	111	37 20	146	37 14	222	51 41	317	41 29	340
	PROCYON		REGULUS		◆Gienah		ACRUX		◆ACHERNAR		RIGEL		◆BETELGEUSE	
105	44 51	014	22 48	051	19 24	097	37 47	145	36 43	222	51 08	315	41 12	338
106	45 02	012	23 24	050	20 11	097	38 13	145	36 12	222	50 35	314	40 54	337
107	45 11	011	23 59	049	20 57	096	38 40	145	35 41	221	50 01	313	40 36	336
108	45 19	010	24 34	048	21 43	095	39 06	145	35 10	221	49 27	312	40 16	335
109	45 27	008	25 09	048	22 30	095	39 33	145	34 40	221	48 52	310	39 56	333
110	45 33	007	25 43	047	23 16	094	40 00	145	34 09	221	48 16	309	39 34	332
111	45 38	005	26 17	046	24 03	094	40 27	145	33 39	220	47 39	308	39 12	331
112	45 41	004	26 50	045	24 49	093	40 54	145	33 09	220	47 02	307	38 49	330
113	45 44	003	27 22	044	25 36	092	41 21	145	32 39	220	46 25	306	38 25	329
114	45 45	001	27 55	043	26 22	092	41 48	145	32 09	220	45 47	305	38 00	327
115	45 46	000	28 26	042	27 09	091	42 15	145	31 39	219	45 08	304	37 35	326
116	45 45	358	28 57	041	27 56	091	42 42	144	31 10	219	44 29	303	37 09	325
117	45 43	357	29 27	040	28 42	090	43 09	144	30 40	219	43 49	302	36 42	324
118	45 40	355	29 57	039	29 29	089	43 36	144	30 11	219	43 09	300	36 14	323
119	45 35	354	30 26	038	30 16	089	44 03	144	29 42	218	42 29	300	35 46	322
	REGULUS		◆SPICA		ACRUX		◆CANOPUS		RIGEL		BETELGEUSE		◆PROCYON	
120	30 55	037	13 45	093	44 30	144	68 33	222	41 48	299	35 17	321	45 30	353
121	31 23	036	14 32	093	44 58	144	68 01	223	41 07	298	34 47	320	45 23	351
122	31 50	035	15 18	092	45 25	144	67 29	224	40 26	297	34 17	319	45 15	350
123	32 16	034	16 05	091	45 52	144	66 56	225	39 44	296	33 46	318	45 07	348
124	32 42	033	16 52	091	46 19	144	66 23	225	39 02	295	33 14	317	44 57	347
125	33 07	032	17 38	090	46 46	144	65 50	226	38 19	294	32 42	316	44 46	346
126	33 32	031	18 25	090	47 13	144	65 16	226	37 36	293	32 09	315	44 33	344
127	33 55	030	19 12	089	47 40	145	64 42	227	36 53	292	31 36	314	44 20	343
128	34 18	029	19 58	088	48 07	145	64 08	227	36 10	291	31 02	313	44 06	341
129	34 40	028	20 45	088	48 34	145	63 34	228	35 26	291	30 27	312	43 51	340
130	35 02	027	21 31	087	49 01	145	62 59	228	34 43	290	29 52	311	43 34	339
131	35 22	025	22 18	086	49 28	145	62 24	229	33 59	289	29 17	310	43 17	338
132	35 42	024	23 04	086	49 55	145	61 49	229	33 14	288	28 41	309	42 58	336
133	36 00	023	23 51	085	50 21	145	61 14	229	32 30	287	28 05	308	42 39	335
134	36 18	022	24 37	084	50 48	145	60 38	230	31 45	287	27 28	307	42 19	334
	◆REGULUS		SPICA		◆ACRUX		CANOPUS		◆RIGEL		BETELGEUSE		PROCYON	
135	36 36	021	25 24	084	51 15	145	60 03	230	31 01	286	26 51	307	41 58	332
136	36 52	020	26 10	083	51 41	146	59 27	230	30 16	285	26 13	306	41 36	331
137	37 07	019	26 56	082	52 07	146	58 51	230	29 31	284	25 35	305	41 13	330
138	37 21	017	27 42	082	52 33	146	58 15	230	28 45	284	24 56	304	40 49	329
139	37 35	016	28 29	081	52 59	146	57 39	231	28 00	283	24 18	303	40 24	328
140	37 47	015	29 15	080	53 25	147	57 03	231	27 14	282	23 38	302	39 59	326
141	37 59	014	30 00	080	53 51	147	56 27	231	26 29	281	22 59	302	39 33	325
142	38 09	013	30 46	079	54 16	147	55 51	231	25 43	281	22 19	301	39 06	324
143	38 19	011	31 32	078	54 41	147	55 14	231	24 57	280	21 39	300	38 38	323
144	38 28	010	32 18	077	55 07	148	54 38	231	24 11	279	20 58	299	38 10	322
145	38 35	009	33 03	077	55 31	148	54 02	231	23 25	279	20 17	298	37 40	321
146	38 42	008	33 48	076	55 56	148	53 25	231	22 39	278	19 36	298	37 10	320
147	38 48	006	34 33	075	56 20	149	52 49	231	21 53	277	18 55	297	36 40	319
148	38 52	005	35 18	074	56 45	149	52 13	231	21 07	277	18 13	296	36 09	317
149	38 56	004	36 03	074	57 08	149	51 36	231	20 20	276	17 31	295	35 37	316
	REGULUS		◆SPICA		ACRUX		◆Miaplacidus		CANOPUS		SIRIUS		◆PROCYON	
150	38 58	003	36 48	073	57 32	150	58 41	188	51 00	231	42 11	284	35 04	315
151	39 00	001	37 32	072	57 55	150	58 34	188	50 24	231	41 26	283	34 31	314
152	39 00	000	38 17	071	58 18	151	58 27	189	49 47	231	40 40	282	33 58	313
153	39 00	359	39 01	070	58 41	151	58 19	190	49 11	231	39 55	281	33 24	312
154	38 58	358	39 44	070	59 03	152	58 11	190	48 35	231	39 09	280	32 49	311
155	38 56	356	40 28	069	59 25	152	58 03	191	47 59	231	38 23	280	32 14	310
156	38 52	355	41 11	068	59 47	153	57 54	191	47 22	231	37 37	279	31 38	310
157	38 48	354	41 54	067	60 08	153	57 44	192	46 46	231	36 51	278	31 02	309
158	38 42	353	42 37	066	60 29	154	57 34	193	46 10	231	36 05	278	30 25	308
159	38 36	351	43 19	065	60 49	155	57 24	193	45 34	231	35 18	277	29 48	307
160	38 28	350	44 02	064	61 09	155	57 13	194	44 58	230	34 32	276	29 11	306
161	38 20	349	44 43	063	61 28	156	57 02	194	44 22	230	33 46	276	28 33	305
162	38 10	348	45 25	062	61 47	157	56 50	195	43 47	230	32 59	275	27 54	304
163	37 59	346	46 06	061	62 05	157	56 38	195	43 11	230	32 13	274	27 15	303
164	37 48	345	46 46	060	62 23	158	56 26	196	42 35	230	31 26	274	26 36	302
	ARCTURUS		◆ANTARES		ACRUX		◆CANOPUS		SIRIUS		PROCYON		◆REGULUS	
165	16 01	048	21 56	107	62 40	159	42 00	230	30 40	273	25 57	302	37 35	344
166	16 35	047	22 41	106	62 57	160	41 24	229	29 53	272	25 17	301	37 22	343
167	17 09	046	23 26	106	63 13	160	40 49	229	29 06	272	24 37	300	37 08	342
168	17 42	045	24 11	105	63 28	161	40 14	229	28 20	271	23 56	299	36 53	340
169	18 15	045	24 56	105	63 43	162	39 38	229	27 33	270	23 15	298	36 36	339
170	18 47	044	25 41	104	63 57	163	39 03	229	26 47	270	22 34	298	36 19	338
171	19 19	043	26 26	104	64 10	164	38 28	228	26 00	269	21 53	297	36 01	337
172	19 51	042	27 11	103	64 23	165	37 54	228	25 13	268	21 11	296	35 43	336
173	20 22	041	27 57	103	64 35	166	37 19	228	24 27	268	20 29	295	35 23	335
174	20 52	040	28 42	102	64 46	167	36 44	228	23 40	267	19 47	295	35 03	333
175	21 22	040	29 28	102	64 56	168	36 10	227	22 54	267	19 04	294	34 41	332
176	21 52	039	30 14	101	65 06	169	35 36	227	22 07	266	18 21	293	34 19	331
177	22 21	038	31 00	100	65 15	170	35 01	227	21 20	265	17 38	292	33 57	330
178	22 49	037	31 45	100	65 23	171	34 27	227	20 34	265	16 55	292	33 33	329
179	23 17	036	32 31	099	65 30	172	33 54	226	19 48	264	16 12	291	33 09	328

LAT 39°S

LHA ♈	Hc	Zn	Hc	Zn	Hc	Zn	Hc	Zn	Hc	Zn	Hc	Zn	Hc	Zn
	ARCTURUS		◆ANTARES		RIGIL KENT.		ACRUX		◆CANOPUS		Alphard		◆REGULUS	
180	23 44	035	33 17	099	57 12	145	65 36	173	33 20	226	44 19	301	32 44	327
181	24 10	034	34 04	098	57 38	145	65 42	174	32 46	226	43 39	300	32 18	326
182	24 36	033	34 50	098	58 05	146	65 46	175	32 13	226	42 58	299	31 51	325
183	25 01	032	35 36	097	58 31	146	65 50	176	31 40	225	42 18	298	31 24	324
184	25 26	031	36 22	097	58 57	146	65 53	177	31 07	225	41 36	297	30 56	323
185	25 50	030	37 09	096	59 23	147	65 55	178	30 34	225	40 55	296	30 28	322
186	26 13	029	37 55	095	59 48	147	65 56	179	30 01	224	40 13	296	29 59	321
187	26 36	028	38 41	095	60 13	148	65 56	180	29 28	224	39 31	295	29 29	320
188	26 58	028	39 28	094	60 38	148	65 55	182	28 56	224	38 48	294	28 59	319
189	27 19	027	40 14	094	61 02	149	65 53	183	28 24	223	38 05	293	28 28	318
190	27 39	026	41 01	093	61 26	150	65 50	184	27 52	223	37 22	292	27 56	317
191	27 59	025	41 48	093	61 49	150	65 47	185	27 20	223	36 39	291	27 24	316
192	28 18	024	42 34	092	62 12	151	65 43	186	26 48	222	35 55	290	26 52	315
193	28 36	023	43 21	091	62 35	151	65 37	187	26 17	222	35 12	290	26 18	314
194	28 54	021	44 07	091	62 57	152	65 31	188	25 46	222	34 27	289	25 45	313
	ARCTURUS		◆ANTARES		RIGIL KENT.		◆ACRUX		Suhail		◆REGULUS		Denebola	
195	29 10	020	44 54	090	63 18	153	65 24	189	47 00	245	25 11	313	33 53	339
196	29 26	019	45 41	089	63 39	154	65 16	190	46 17	244	24 36	312	33 36	338
197	29 41	018	46 27	089	64 00	154	65 08	191	45 35	244	24 01	311	33 18	337
198	29 56	017	47 14	088	64 19	155	64 58	192	44 54	244	23 26	310	32 59	336
199	30 09	016	48 00	087	64 39	156	64 48	193	44 12	244	22 50	309	32 40	335
200	30 22	015	48 47	087	64 57	157	64 37	194	43 30	243	22 13	308	32 19	334
201	30 34	014	49 34	086	65 15	158	64 25	195	42 49	243	21 36	307	31 58	333
202	30 45	013	50 20	085	65 32	159	64 12	196	42 07	243	20 59	307	31 36	332
203	30 55	012	51 07	085	65 49	160	63 59	197	41 26	242	20 22	306	31 14	330
204	31 04	011	51 53	084	66 04	161	63 45	198	40 45	242	19 44	305	30 50	329
205	31 12	010	52 39	083	66 19	162	63 31	199	40 03	242	19 05	304	30 26	328
206	31 20	009	53 26	083	66 33	163	63 15	200	39 22	241	18 27	303	30 02	327
207	31 26	008	54 12	082	66 46	164	62 59	200	38 42	241	17 48	303	29 36	326
208	31 32	006	54 58	081	66 59	165	62 43	201	38 01	241	17 08	302	29 10	325
209	31 37	005	55 44	080	67 10	166	62 26	202	37 20	240	16 28	301	28 43	324
	ARCTURUS		◆ANTARES		Peacock		RIGIL KENT.		◆ACRUX		Suhail		◆SPICA	
210	31 41	004	56 30	079	28 39	142	67 21	168	62 08	203	36 40	240	61 04	342
211	31 44	003	57 15	078	29 08	141	67 30	169	61 50	203	35 59	240	60 49	340
212	31 46	002	58 01	078	29 37	141	67 39	170	61 31	204	35 19	239	60 32	338
213	31 47	001	58 46	077	30 07	141	67 47	171	61 12	205	34 39	239	60 14	336
214	31 47	000	59 32	076	30 36	140	67 53	172	60 52	205	33 59	239	59 54	334
215	31 47	359	60 17	075	31 06	140	67 59	174	60 32	206	33 20	238	59 34	333
216	31 45	358	61 02	074	31 36	140	68 03	175	60 11	207	32 40	238	59 11	331
217	31 43	357	61 46	073	32 06	140	68 07	176	59 50	207	32 01	238	58 48	329
218	31 40	355	62 31	072	32 36	139	68 09	178	59 29	208	31 21	237	58 23	327
219	31 36	354	63 15	070	33 07	139	68 11	179	59 07	208	30 42	237	57 58	326
220	31 30	353	63 59	069	33 37	139	68 11	180	58 45	209	30 03	236	57 31	324
221	31 25	352	64 42	068	34 08	139	68 10	182	58 22	209	29 25	236	57 03	322
222	31 18	351	65 25	067	34 39	138	68 09	183	57 59	210	28 46	236	56 34	321
223	31 10	350	66 08	065	35 10	138	68 06	184	57 36	210	28 08	235	56 04	319
224	31 01	349	66 50	064	35 41	138	68 02	185	57 12	211	27 29	235	55 33	318
	Alphecca		Rasalhague		◆Nunki		Peacock		◆RIGIL KENT.		◆SPICA		ARCTURUS	
225	23 47	008	27 05	043	39 49	094	36 12	138	67 57	187	55 01	316	30 52	348
226	23 53	007	27 37	042	40 36	093	36 44	138	67 51	188	54 29	315	30 41	347
227	23 59	006	28 08	041	41 22	093	37 15	137	67 44	189	53 55	314	30 30	346
228	24 04	005	28 38	040	42 09	092	37 47	137	67 36	190	53 21	312	30 18	345
229	24 08	005	29 08	039	42 56	091	38 18	137	67 27	192	52 46	311	30 05	343
230	24 11	004	29 38	039	43 42	091	38 50	137	67 17	193	52 11	310	29 52	342
231	24 13	003	30 06	038	44 29	090	39 22	137	67 06	194	51 35	308	29 37	341
232	24 15	002	30 34	037	45 15	090	39 54	137	66 54	195	50 58	307	29 22	340
233	24 16	001	31 02	036	46 02	089	40 26	136	66 42	196	50 20	306	29 06	339
234	24 16	000	31 29	035	46 49	088	40 59	136	66 28	197	49 42	305	28 49	338
235	24 15	359	31 55	033	47 35	088	41 31	136	66 14	198	49 04	304	28 31	337
236	24 14	358	32 20	032	48 22	087	42 03	136	65 58	200	48 25	303	28 13	336
237	24 12	357	32 45	031	49 08	086	42 36	136	65 43	201	47 45	301	27 53	335
238	24 08	356	33 09	030	49 55	086	43 08	136	65 26	201	47 05	300	27 33	334
239	24 05	355	33 32	029	50 41	085	43 41	136	65 08	202	46 25	299	27 13	333
	Rasalhague		ALTAIR		◆Nunki		Peacock		◆RIGIL KENT.		SPICA		◆ARCTURUS	
240	33 54	028	18 19	062	51 28	084	44 13	136	64 50	203	45 44	298	26 51	332
241	34 16	027	19 00	061	52 14	083	44 46	135	64 32	204	45 03	297	26 29	331
242	34 37	026	19 40	060	53 00	083	45 19	135	64 12	205	44 21	296	26 07	330
243	34 57	025	20 20	059	53 47	082	45 52	135	63 52	206	43 39	295	25 43	329
244	35 16	024	21 00	058	54 33	081	46 25	135	63 31	207	42 57	294	25 19	328
245	35 34	023	21 40	058	55 19	080	46 57	135	63 10	207	42 14	294	24 54	327
246	35 52	021	22 19	057	56 05	079	47 30	135	62 49	208	41 31	293	24 29	326
247	36 08	020	22 58	056	56 51	079	48 03	135	62 26	209	40 48	292	24 03	326
248	36 24	019	23 36	055	57 36	078	48 36	135	62 04	209	40 05	291	23 36	325
249	36 39	018	24 15	054	58 22	077	49 09	135	61 41	210	39 21	290	23 09	324
250	36 53	017	24 52	054	59 07	076	49 42	135	61 17	211	38 37	289	22 41	323
251	37 06	016	25 30	053	59 52	075	50 15	135	60 53	211	37 53	288	22 12	322
252	37 18	014	26 06	052	60 37	074	50 48	135	60 29	212	37 09	288	21 43	321
253	37 29	013	26 43	051	61 22	073	51 20	135	60 04	212	36 24	287	21 14	320
254	37 39	012	27 19	050	62 06	072	51 53	135	59 39	213	35 39	286	20 44	319
	Rasalhague		◆ALTAIR		FOMALHAUT		◆ACHERNAR		RIGIL KENT.		◆SPICA		ARCTURUS	
255	37 48	011	27 54	049	18 36	114	15 14	154	59 13	213	34 54	285	20 13	319
256	37 56	010	28 30	048	19 19	113	15 34	154	58 47	214	34 09	284	19 42	318
257	38 04	008	29 04	047	20 02	112	15 55	153	58 21	214	33 24	284	19 10	317
258	38 10	007	29 38	046	20 45	112	16 16	153	57 55	215	32 39	283	18 38	316
259	38 15	006	30 12	046	21 29	111	16 37	153	57 29	215	31 53	282	18 06	315
260	38 19	005	30 45	045	22 12	111	16 59	152	57 02	215	31 08	281	17 33	314
261	38 23	003	31 17	044	22 56	110	17 21	152	56 35	216	30 22	281	16 59	314
262	38 25	002	31 49	043	23 40	110	17 43	151	56 07	216	29 36	280	16 25	313
263	38 26	001	32 20	042	24 23	109	18 05	151	55 40	216	28 50	279	15 51	312
264	38 26	000	32 51	041	25 08	109	18 28	151	55 12	217	28 04	279	15 16	311
265	38 25	358	33 21	040	25 52	108	18 51	150	54 44	217	27 18	278	14 41	311
266	38 23	357	33 51	039	26 36	108	19 15	150	54 16	217	26 32	277	14 05	310
267	38 21	356	34 19	038	27 20	107	19 38	149	53 48	217	25 45	277	13 29	309
268	38 17	355	34 47	037	28 05	107	20 02	149	53 20	217	24 59	276	12 53	308
269	38 12	353	35 15	035	28 50	106	20 26	149	52 51	218	24 12	275	12 16	308

LHA ♈	Hc	Zn	Hc	Zn	Hc	Zn	Hc	Zn	Hc	Zn	Hc	Zn	Hc	Zn
	ALTAIR		Enif		◆FOMALHAUT		ACHERNAR		◆RIGIL KENT.		ANTARES		◆Rasalhague	
270	35 41	034	18 42	060	29 35	106	20 51	148	52 23	218	67 13	297	38 06	352
271	36 07	033	19 22	059	30 19	105	21 15	148	51 54	218	66 31	295	37 59	351
272	36 33	032	20 02	058	31 05	105	21 40	147	51 25	218	65 49	294	37 51	350
273	36 57	031	20 41	057	31 50	104	22 06	147	50 57	218	65 06	293	37 42	348
274	37 21	030	21 20	056	32 35	104	22 31	147	50 28	218	64 23	291	37 33	347
275	37 43	029	21 59	056	33 20	103	22 57	146	49 59	219	63 39	290	37 22	346
276	38 05	028	22 37	055	34 06	103	23 23	146	49 30	219	62 55	289	37 10	345
277	38 27	026	23 15	054	34 51	102	23 49	146	49 00	219	62 11	288	36 57	344
278	38 47	025	23 53	053	35 37	102	24 15	145	48 31	219	61 26	287	36 44	342
279	39 06	024	24 30	052	36 23	101	24 42	145	48 02	219	60 42	286	36 29	341
280	39 25	023	25 07	051	37 08	101	25 09	145	47 33	219	59 57	285	36 14	340
281	39 42	022	25 43	051	37 54	100	25 36	144	47 04	219	59 11	284	35 58	339
282	39 59	020	26 19	050	38 40	100	26 03	144	46 34	219	58 26	283	35 40	338
283	40 15	019	26 54	049	39 26	099	26 31	144	46 05	219	57 41	282	35 22	337
284	40 30	018	27 29	048	40 12	099	26 59	143	45 36	219	56 55	281	35 04	336
	ALTAIR		Enif		◆FOMALHAUT		ACHERNAR		◆RIGIL KENT.		ANTARES		◆Rasalhague	
285	40 43	017	28 03	047	40 58	098	27 26	143	45 07	219	56 09	280	34 44	334
286	40 56	015	28 37	046	41 45	098	27 55	143	44 37	219	55 23	280	34 23	333
287	41 08	014	29 11	045	42 31	097	28 23	142	44 08	219	54 37	279	34 02	332
288	41 19	013	29 43	044	43 17	096	28 51	142	43 39	219	53 51	278	33 40	331
289	41 29	011	30 16	043	44 03	096	29 20	142	43 10	219	53 05	277	33 17	330
290	41 37	010	30 47	042	44 50	095	29 49	142	42 41	219	52 18	276	32 53	329
291	41 45	009	31 19	041	45 36	095	30 18	141	42 12	219	51 32	276	32 29	328
292	41 52	007	31 49	040	46 23	094	30 47	141	41 43	218	50 46	275	32 04	327
293	41 57	006	32 19	039	47 09	094	31 17	141	41 14	218	49 59	274	31 38	326
294	42 02	005	32 48	038	47 56	093	31 46	140	40 45	218	49 13	274	31 11	325
295	42 05	004	33 17	037	48 42	092	32 16	140	40 16	218	48 26	273	30 44	324
296	42 07	002	33 45	036	49 29	092	32 46	140	39 47	218	47 39	272	30 16	323
297	42 08	001	34 12	035	50 16	091	33 16	140	39 18	218	46 53	272	29 48	322
298	42 09	000	34 39	034	51 02	091	33 46	140	38 50	218	46 06	271	29 19	321
299	42 08	358	35 04	033	51 49	090	34 17	139	38 21	218	45 20	270	28 49	320
	Enif		◆FOMALHAUT		ACHERNAR		◆RIGIL KENT.		ANTARES		◆Rasalhague		ALTAIR	
300	35 30	032	52 36	089	34 47	139	37 53	218	44 33	270	28 19	319	42 06	357
301	35 54	031	53 22	089	35 18	139	37 25	217	43 46	269	27 48	318	42 03	356
302	36 17	030	54 09	088	35 48	139	36 56	217	43 00	268	27 16	317	41 58	354
303	36 40	029	54 55	087	36 19	138	36 28	217	42 13	268	26 44	316	41 53	353
304	37 02	028	55 42	087	36 50	138	36 00	217	41 27	267	26 12	315	41 47	352
305	37 23	026	56 28	086	37 21	138	35 32	217	40 40	267	25 39	314	41 39	350
306	37 44	025	57 15	085	37 53	138	35 04	216	39 53	266	25 05	313	41 31	349
307	38 03	024	58 01	085	38 24	138	34 37	216	39 07	265	24 31	313	41 22	348
308	38 22	023	58 48	084	38 56	137	34 09	216	38 21	265	23 56	312	41 11	346
309	38 39	022	59 34	083	39 27	137	33 42	216	37 34	264	23 21	311	41 00	345
310	38 56	020	60 20	082	39 59	137	33 15	216	36 48	264	22 46	310	40 47	344
311	39 12	019	61 06	081	40 30	137	32 47	215	36 01	263	22 10	309	40 33	343
312	39 27	018	61 53	081	41 02	137	32 21	215	35 15	263	21 34	308	40 19	341
313	39 40	017	62 38	080	41 34	137	31 54	215	34 29	262	20 57	308	40 03	340
314	39 53	015	63 24	079	42 06	137	31 27	215	33 43	261	20 20	307	39 47	339
	Enif		◆Diphda		ACHERNAR		◆RIGIL KENT.		ANTARES		Nunki		◆ALTAIR	
315	40 05	014	37 34	083	42 38	137	31 01	215	32 57	261	60 58	287	39 30	338
316	40 16	013	38 20	082	43 10	136	30 34	214	32 11	260	60 14	285	39 11	336
317	40 26	012	39 06	082	43 42	136	30 08	214	31 25	260	59 29	285	38 52	335
318	40 35	010	39 52	081	44 15	136	29 42	214	30 39	259	58 43	284	38 32	334
319	40 43	009	40 38	080	44 47	136	29 16	213	29 53	259	57 58	283	38 11	333
320	40 50	008	41 24	079	45 19	136	28 51	213	29 07	258	57 12	282	37 49	332
321	40 56	007	42 10	079	45 52	136	28 25	213	28 22	258	56 27	281	37 27	330
322	41 00	005	42 56	078	46 24	136	28 00	213	27 36	257	55 41	280	37 03	329
323	41 04	004	43 41	077	46 56	136	27 35	212	26 51	257	54 55	279	36 39	328
324	41 07	003	44 27	076	47 29	136	27 10	212	26 06	256	54 09	278	36 14	327
325	41 08	001	45 12	075	48 01	136	26 45	212	25 20	256	53 23	278	35 49	326
326	41 09	000	45 57	075	48 34	136	26 21	212	24 35	255	52 36	277	35 22	325
327	41 08	359	46 42	074	49 06	136	25 56	211	23 50	254	51 50	276	34 55	324
328	41 07	357	47 26	073	49 39	136	25 32	211	23 05	254	51 04	275	34 27	323
329	41 04	356	48 11	072	50 11	136	25 09	211	22 21	253	50 17	275	33 58	322
	◆Alpheratz		Diphda		◆ACHERNAR		RIGIL KENT.		◆Nunki		ALTAIR		Enif	
330	15 40	029	48 55	071	50 43	136	24 45	210	49 31	274	33 29	321	41 00	355
331	16 03	028	49 39	070	51 16	136	24 22	210	48 44	273	32 59	320	40 56	353
332	16 24	027	50 23	069	51 48	136	23 58	210	47 58	273	32 29	319	40 50	352
333	16 45	026	51 06	068	52 20	136	23 36	209	47 11	272	31 58	318	40 43	351
334	17 05	025	51 50	067	52 53	136	23 13	209	46 24	271	31 26	317	40 35	350
335	17 25	025	52 32	066	53 25	136	22 50	209	45 38	271	30 54	316	40 26	348
336	17 44	024	53 15	065	53 57	137	22 28	208	44 51	270	30 21	315	40 16	347
337	18 03	023	53 57	064	54 29	137	22 06	208	44 04	269	29 47	314	40 05	346
338	18 21	022	54 39	063	55 01	137	21 45	208	43 18	269	29 13	313	39 53	344
339	18 38	021	55 20	062	55 33	137	21 23	207	42 31	268	28 39	312	39 40	343
340	18 54	020	56 01	061	56 04	137	21 02	207	41 45	268	28 04	311	39 26	342
341	19 10	019	56 41	059	56 36	138	20 41	206	40 58	267	27 29	310	39 11	341
342	19 25	019	57 21	058	57 07	138	20 20	206	40 12	266	26 53	309	38 55	340
343	19 40	018	58 01	057	57 38	138	20 00	206	39 25	266	26 16	308	38 39	338
344	19 53	017	58 40	056	58 09	138	19 40	205	38 39	265	25 40	307	38 21	337
	◆Alpheratz		Diphda		◆CANOPUS		RIGIL KENT.		◆Nunki		ALTAIR		Enif	
345	20 06	016	59 18	054	19 24	143	19 20	205	37 52	265	25 02	307	38 02	336
346	20 19	015	59 55	053	19 52	143	19 00	205	37 06	264	24 25	306	37 43	335
347	20 31	014	60 32	051	20 20	142	18 41	204	36 19	264	23 47	305	37 23	334
348	20 41	013	61 08	050	20 49	142	18 22	204	35 33	263	23 08	304	37 02	332
349	20 52	012	61 43	048	21 18	142	18 03	203	34 47	262	22 30	303	36 40	331
350	21 01	011	62 18	047	21 47	141	17 45	203	34 01	262	21 50	303	36 17	330
351	21 10	010	62 51	045	22 16	141	17 27	203	33 14	261	21 11	302	35 53	329
352	21 18	009	63 24	043	22 46	140	17 09	202	32 28	261	20 31	301	35 29	328
353	21 25	008	63 55	041	23 16	140	16 52	202	31 42	260	19 51	300	35 04	327
354	21 32	008	64 25	040	23 46	140	16 35	201	30 56	260	19 11	299	34 38	326
355	21 37	007	64 54	038	24 16	139	16 18	201	30 11	259	18 30	299	34 11	325
356	21 42	006	65 22	036	24 47	139	16 01	201	29 25	259	17 49	298	33 44	324
357	21 47	005	65 49	034	25 17	139	15 45	200	28 39	258	17 07	297	33 16	323
358	21 50	004	66 14	032	25 48	138	15 29	200	27 54	257	16 26	296	32 47	322
359	21 53	003	66 38	029	26 20	138	15 14	199	27 08	257	15 44	296	32 18	321

LAT 40°S

LAT 40°S

LHA ♈	Hc	Zn	Hc	Zn	Hc	Zn	Hc	Zn	Hc	Zn	Hc	Zn	Hc	Zn
	Alpheratz		◆Hamal		RIGEL		◆CANOPUS		Peacock		◆Nunki		Enif	
0	20 55	002	20 00	031	14 00	089	27 35	137	51 50	226	26 37	257	31 02	320
1	20 56	001	20 24	030	14 46	088	28 06	137	51 18	226	25 52	256	30 32	319
2	20 56	000	20 46	029	15 32	088	28 38	136	50 45	226	25 07	256	30 02	318
3	20 56	359	21 08	028	16 18	087	29 10	136	50 12	226	24 23	255	29 31	317
4	20 55	358	21 30	027	17 04	086	29 42	136	49 39	226	23 38	255	28 59	316
5	20 53	357	21 51	026	17 50	086	30 14	135	49 06	226	22 54	254	28 27	315
6	20 50	356	22 11	025	18 35	085	30 46	135	48 33	226	22 10	254	27 55	314
7	20 47	355	22 30	025	19 21	084	31 19	135	48 00	226	21 26	253	27 21	313
8	20 43	354	22 49	024	20 07	084	31 51	135	47 27	226	20 42	252	26 48	312
9	20 38	353	23 07	023	20 53	083	32 24	134	46 55	225	19 58	252	26 13	311
10	20 32	353	23 24	022	21 38	082	32 57	134	46 22	225	19 15	251	25 39	311
11	20 26	352	23 41	021	22 24	082	33 31	134	45 49	225	18 31	251	25 04	310
12	20 19	351	23 57	020	23 09	081	34 04	133	45 16	225	17 48	250	24 28	309
13	20 11	350	24 12	019	23 55	080	34 37	133	44 44	225	17 05	250	23 52	308
14	20 03	349	24 27	018	24 40	080	35 11	133	44 11	225	16 22	249	23 16	307
	◆Hamal		ALDEBARAN		RIGEL		◆CANOPUS		Peacock		◆FOMALHAUT		Alpheratz	
15	24 40	017	14 29	053	25 25	079	35 45	132	43 39	225	62 58	283	19 54	348
16	24 53	016	15 06	052	26 10	078	36 19	132	43 06	225	62 13	282	19 44	347
17	25 06	015	15 42	052	26 55	077	36 53	132	42 34	225	61 28	281	19 33	346
18	25 17	014	16 18	051	27 40	077	37 27	132	42 02	224	60 42	280	19 22	345
19	25 28	013	16 53	050	28 24	076	38 02	131	41 30	224	59 57	279	19 10	344
20	25 38	012	17 28	049	29 09	075	38 36	131	40 58	224	59 12	278	18 57	343
21	25 47	011	18 03	048	29 53	074	39 11	131	40 26	224	58 26	278	18 43	343
22	25 55	010	18 37	048	30 38	074	39 46	131	39 54	224	57 41	277	18 29	342
23	26 02	009	19 11	047	31 22	073	40 20	131	39 22	224	56 55	276	18 14	341
24	26 09	008	19 44	046	32 05	072	40 55	130	38 50	223	56 09	275	17 59	340
25	26 15	007	20 17	045	32 49	071	41 31	130	38 19	223	55 23	275	17 43	339
26	26 20	006	20 49	044	33 32	071	42 06	130	37 47	223	54 38	274	17 26	338
27	26 24	005	21 21	043	34 16	070	42 41	130	37 16	223	53 52	273	17 09	337
28	26 28	004	21 52	043	34 59	069	43 16	130	36 45	223	53 06	272	16 51	336
29	26 31	003	22 23	042	35 41	068	43 52	129	36 14	222	52 20	272	16 32	336
	Hamal		◆ALDEBARAN		RIGEL		SIRIUS		◆CANOPUS		Peacock		◆FOMALHAUT	
30	26 32	002	22 54	041	36 24	067	24 53	091	44 28	129	35 43	222	51 34	271
31	26 33	001	23 23	040	37 06	066	25 39	091	45 03	129	35 12	222	50 48	270
32	26 34	000	23 53	039	37 48	065	26 25	090	45 39	129	34 41	222	50 02	270
33	26 33	359	24 21	038	38 30	065	27 11	089	46 15	129	34 11	221	49 16	269
34	26 32	358	24 49	037	39 11	064	27 57	089	46 51	129	33 41	221	48 30	269
35	26 29	357	25 17	036	39 52	063	28 43	088	47 27	128	33 10	221	47 44	268
36	26 26	356	25 44	035	40 33	062	29 29	088	48 03	128	32 40	221	46 58	267
37	26 22	355	26 10	034	41 13	061	30 15	087	48 39	128	32 11	220	46 12	267
38	26 17	354	26 36	033	41 53	060	31 01	086	49 15	128	31 41	220	45 26	266
39	26 12	353	27 01	032	42 33	059	31 47	086	49 51	128	31 11	220	44 41	265
40	26 06	352	27 25	031	43 12	058	32 33	085	50 28	128	30 42	220	43 55	265
41	25 58	351	27 49	030	43 51	057	33 18	084	51 04	128	30 13	219	43 09	264
42	25 50	350	28 11	029	44 29	056	34 04	083	51 40	128	29 44	219	42 23	264
43	25 42	349	28 34	028	45 07	055	34 50	083	52 16	128	29 15	219	41 38	263
44	25 32	348	28 55	027	45 44	054	35 35	082	52 53	128	28 47	218	40 52	263
	◆ALDEBARAN		RIGEL		SIRIUS		◆CANOPUS		ACHERNAR		◆FOMALHAUT		Hamal	
45	29 16	026	46 20	052	36 21	081	53 29	128	68 11	211	40 06	262	25 22	347
46	29 36	025	46 57	051	37 06	081	54 06	128	67 48	212	39 21	262	25 11	346
47	29 56	024	47 32	050	37 51	080	54 42	128	67 23	213	38 36	261	24 59	345
48	30 14	023	48 07	049	38 36	079	55 18	128	66 58	214	37 50	260	24 46	344
49	30 32	022	48 42	048	39 22	078	55 55	128	66 32	214	37 05	260	24 33	343
50	30 49	021	49 15	046	40 06	078	56 31	128	66 06	215	36 20	259	24 19	342
51	31 05	020	49 48	045	40 51	077	57 07	128	65 39	216	35 35	259	24 04	341
52	31 21	019	50 20	044	41 36	076	57 43	128	65 12	217	34 49	258	23 48	340
53	31 35	018	50 52	043	42 20	075	58 20	128	64 44	217	34 05	258	23 32	339
54	31 49	017	51 23	041	43 05	074	58 56	128	64 16	218	33 20	257	23 15	338
55	32 02	016	51 53	040	43 49	073	59 32	128	63 48	219	32 35	257	22 57	337
56	32 14	015	52 22	038	44 33	073	60 08	129	63 19	219	31 50	256	22 39	336
57	32 25	014	52 50	037	45 17	072	60 44	129	62 50	220	31 06	256	22 19	335
58	32 36	012	53 17	036	46 00	071	61 19	129	62 20	220	30 21	255	22 00	334
59	32 45	011	53 43	034	46 43	070	61 55	129	61 50	221	29 37	255	21 39	333
	ALDEBARAN		◆BETELGEUSE		SIRIUS		◆ACRUX		ACHERNAR		◆FOMALHAUT		Hamal	
60	32 54	010	35 41	036	47 26	069	21 29	157	61 20	221	28 53	254	21 18	332
61	33 01	009	36 08	035	48 09	068	21 48	157	60 50	221	28 08	254	20 56	331
62	33 08	008	36 34	034	48 52	067	22 06	156	60 19	222	27 24	253	20 34	330
63	33 14	007	36 59	033	49 34	066	22 25	156	59 49	222	26 40	252	20 11	330
64	33 19	006	37 23	031	50 16	065	22 43	156	59 18	222	25 57	252	19 47	329
65	33 23	004	37 47	030	50 57	064	23 03	155	58 47	223	25 13	251	19 23	328
66	33 26	003	38 09	029	51 38	063	23 22	155	58 15	223	24 30	251	18 58	327
67	33 28	002	38 31	028	52 19	062	23 42	155	57 44	223	23 46	250	18 33	326
68	33 30	001	38 53	027	52 59	061	24 01	154	57 12	224	23 03	250	18 07	325
69	33 30	000	39 13	026	53 39	060	24 22	154	56 40	224	22 20	249	17 40	324
70	33 29	359	39 32	024	54 19	058	24 42	154	56 09	224	21 37	249	17 13	323
71	33 28	358	39 51	023	54 57	057	25 03	153	55 37	224	20 54	248	16 46	323
72	33 26	356	40 08	022	55 36	056	25 23	153	55 05	224	20 12	248	16 17	322
73	33 22	355	40 25	021	56 14	055	25 45	153	54 33	224	19 29	247	15 49	321
74	33 18	354	40 41	019	56 51	053	26 06	152	54 00	225	18 47	247	15 20	320
	BETELGEUSE		◆SIRIUS		Suhail		◆ACRUX		ACHERNAR		◆Diphda		ALDEBARAN	
75	40 56	018	57 27	052	44 42	116	26 27	152	53 28	225	31 05	272	33 13	353
76	41 09	017	58 03	050	45 24	115	26 49	152	52 56	225	30 19	271	33 07	352
77	41 22	016	58 38	049	46 05	115	27 11	151	52 23	225	29 33	271	33 00	351
78	41 34	014	59 12	048	46 47	115	27 33	151	51 51	225	28 47	270	32 52	350
79	41 45	013	59 46	046	47 29	114	27 55	151	51 19	225	28 01	269	32 43	348
80	41 55	012	60 19	044	48 11	114	28 18	150	50 46	225	27 15	269	32 34	347
81	42 04	010	60 50	043	48 53	114	28 41	150	50 14	225	26 29	268	32 23	346
82	42 11	009	61 21	041	49 35	113	29 04	150	49 41	225	25 43	267	32 12	345
83	42 18	008	61 51	039	50 17	113	29 27	150	49 09	225	24 57	267	32 00	344
84	42 24	006	62 19	038	50 59	113	29 50	149	48 37	225	24 11	266	31 47	343
85	42 28	005	62 47	036	51 42	113	30 14	149	48 04	225	23 26	266	31 33	342
86	42 32	004	63 13	034	52 24	112	30 37	149	47 32	225	22 40	265	31 18	341
87	42 34	002	63 38	032	53 07	112	31 01	149	46 59	225	21 54	264	31 02	340
88	42 35	001	64 02	030	53 49	112	31 25	148	46 27	225	21 08	264	30 46	339
89	42 36	000	64 24	028	54 32	112	31 50	148	45 55	225	20 23	263	30 29	338

LHA ♈	Hc	Zn	Hc	Zn	Hc	Zn	Hc	Zn	Hc	Zn	Hc	Zn	Hc	Zn
	PROCYON		REGULUS		◆Suhail		ACRUX		◆ACHERNAR		RIGEL		◆BETELGEUSE	
90	39 21	033	12 35	062	55 15	111	32 14	148	45 22	225	56 37	339	42 35	358
91	39 45	031	13 16	062	55 58	111	32 38	148	44 50	224	56 20	337	42 33	357
92	40 09	030	13 56	061	56 40	111	33 03	147	44 18	224	56 01	336	42 30	356
93	40 32	029	14 36	060	57 23	111	33 28	147	43 46	224	55 42	334	42 26	354
94	40 54	028	15 16	059	58 06	110	33 53	147	43 14	224	55 21	332	42 21	353
95	41 15	027	15 55	059	58 50	110	34 18	147	42 42	224	54 59	331	42 14	352
96	41 35	025	16 34	058	59 33	110	34 43	147	42 10	224	54 36	329	42 07	350
97	41 54	024	17 13	057	60 16	110	35 09	146	41 38	224	54 12	328	41 59	349
98	42 12	023	17 51	056	60 59	110	35 34	146	41 07	224	53 47	326	41 50	348
99	42 30	022	18 29	055	61 42	110	36 00	146	40 35	223	53 21	325	41 39	346
100	42 46	020	19 07	055	62 26	109	36 26	146	40 04	223	52 54	323	41 28	345
101	43 01	019	19 44	054	63 09	109	36 51	146	39 32	223	52 26	322	41 15	344
102	43 16	018	20 21	053	63 53	109	37 17	145	39 01	223	51 57	320	41 02	342
103	43 29	016	20 58	052	64 36	109	37 44	145	38 30	223	51 27	319	40 48	341
104	43 41	015	21 34	051	65 20	109	38 10	145	37 59	222	50 57	318	40 32	340
	PROCYON		REGULUS		◆Gienah		ACRUX		◆ACHERNAR		RIGEL		◆BETELGEUSE	
105	43 53	014	22 10	051	19 32	097	38 36	145	37 28	222	50 25	316	40 16	339
106	44 03	012	22 45	050	20 17	096	39 02	145	36 57	222	49 53	315	39 59	337
107	44 12	011	23 20	049	21 03	096	39 29	145	36 26	222	49 20	314	39 41	336
108	44 20	009	23 54	048	21 49	095	39 56	145	35 56	222	48 47	312	39 22	335
109	44 27	008	24 28	047	22 35	094	40 22	145	35 25	221	48 12	311	39 02	334
110	44 33	007	25 02	046	23 20	094	40 49	144	34 55	221	47 37	310	38 41	333
111	44 38	005	25 35	045	24 06	093	41 16	144	34 25	221	47 02	309	38 19	331
112	44 41	004	26 07	044	24 52	093	41 43	144	33 55	221	46 26	308	37 57	330
113	44 44	002	26 39	044	25 38	092	42 09	144	33 25	220	45 49	307	37 34	329
114	44 45	001	27 11	043	26 24	091	42 36	144	32 55	220	45 12	305	37 10	328
115	44 46	000	27 41	042	27 10	091	43 03	144	32 26	220	44 34	304	36 45	327
116	44 45	358	28 12	041	27 56	090	43 30	144	31 56	220	43 56	303	36 19	326
117	44 43	357	28 41	040	28 42	089	43 58	144	31 27	219	43 18	302	35 53	325
118	44 40	355	29 10	039	29 28	089	44 25	144	30 58	219	42 39	301	35 26	323
119	44 36	354	29 39	038	30 14	088	44 52	144	30 29	219	41 59	300	34 58	322
	REGULUS		◆SPICA		ACRUX		◆CANOPUS		RIGEL		BETELGEUSE		◆PROCYON	
120	30 07	037	13 49	093	45 19	144	69 17	224	41 19	299	34 30	321	44 30	353
121	30 34	036	14 35	092	45 46	144	68 44	225	40 39	298	34 01	320	44 24	351
122	31 01	035	15 20	092	46 13	144	68 11	226	39 58	297	33 31	319	44 16	350
123	31 27	034	16 06	091	46 40	144	67 38	226	39 17	296	33 01	318	44 08	349
124	31 52	033	16 52	091	47 08	144	67 05	227	38 36	296	32 30	317	43 58	347
125	32 16	032	17 38	090	47 35	144	66 31	228	37 54	295	31 59	316	43 47	346
126	32 40	031	18 24	089	48 02	144	65 57	228	37 13	294	31 26	315	43 36	344
127	33 03	030	19 10	089	48 29	144	65 23	229	36 30	293	30 54	314	43 23	343
128	33 26	029	19 56	088	48 56	144	64 48	229	35 48	292	30 21	313	43 09	342
129	33 47	027	20 42	087	49 23	144	64 13	229	35 05	291	29 47	312	42 54	340
130	34 08	026	21 28	087	49 50	144	63 38	230	34 22	290	29 13	311	42 38	339
131	34 28	025	22 14	086	50 17	144	63 03	230	33 39	290	28 38	311	42 21	338
132	34 47	024	23 00	085	50 44	144	62 28	230	32 56	289	28 03	310	42 04	337
133	35 05	023	23 45	085	51 11	144	61 52	231	32 12	288	27 27	309	41 45	335
134	35 23	022	24 31	084	51 37	145	61 17	231	31 28	287	26 51	308	41 25	334
	◆REGULUS		SPICA		◆ACRUX		CANOPUS		◆RIGEL		BETELGEUSE		PROCYON	
135	35 39	021	25 17	083	52 04	145	60 41	231	30 44	286	26 15	307	41 05	333
136	35 55	019	26 03	083	52 30	145	60 05	231	30 00	286	25 38	306	40 43	332
137	36 10	018	26 48	082	52 57	145	59 29	232	29 16	285	25 00	305	40 21	330
138	36 24	017	27 33	081	53 23	145	58 53	232	28 31	284	24 23	304	39 58	329
139	36 37	016	28 19	080	53 49	146	58 17	232	27 46	283	23 45	304	39 34	328
140	36 49	015	29 04	080	54 15	146	57 41	232	27 02	283	23 06	303	39 09	327
141	37 01	014	29 49	079	54 41	146	57 04	232	26 17	282	22 27	302	38 43	326
142	37 11	012	30 34	078	55 06	146	56 28	232	25 32	281	21 48	301	38 17	325
143	37 20	011	31 19	078	55 32	147	55 52	232	24 47	280	21 09	300	37 50	323
144	37 29	010	32 04	077	55 57	147	55 15	232	24 01	280	20 29	300	37 22	322
145	37 36	009	32 49	076	56 22	147	54 39	232	23 16	279	19 49	299	36 54	321
146	37 42	007	33 33	075	56 47	147	54 03	232	22 31	278	19 08	298	36 25	320
147	37 48	006	34 18	075	57 11	148	53 26	232	21 45	278	18 27	297	35 55	319
148	37 52	005	35 02	074	57 36	148	52 50	232	20 59	277	17 46	296	35 24	318
149	37 56	004	35 46	073	58 00	149	52 14	232	20 14	276	17 05	296	34 53	317
	REGULUS		◆SPICA		ACRUX		◆Miaplacidus		CANOPUS		SIRIUS		◆PROCYON	
150	37 58	003	36 30	072	58 24	149	59 40	188	51 37	232	41 57	284	34 21	316
151	38 00	001	37 13	071	58 47	149	59 33	189	51 01	232	41 12	284	33 49	315
152	38 00	000	37 57	070	59 10	150	59 26	189	50 25	232	40 27	283	33 16	314
153	38 00	359	38 40	070	59 33	150	59 19	190	49 48	232	39 42	282	32 43	313
154	37 59	358	39 23	069	59 56	151	59 10	191	49 12	232	38 57	281	32 09	312
155	37 56	356	40 06	068	60 18	151	59 02	191	48 36	232	38 12	281	31 35	311
156	37 53	355	40 48	067	60 40	152	58 53	192	48 00	232	37 27	280	31 00	310
157	37 48	354	41 30	066	61 01	153	58 43	192	47 24	232	36 42	279	30 24	309
158	37 43	353	42 12	065	61 22	153	58 33	193	46 48	231	35 56	278	29 48	308
159	37 36	351	42 54	064	61 43	154	58 22	194	46 12	231	35 11	278	29 12	307
160	37 29	350	43 35	063	62 03	154	58 11	194	45 36	231	34 25	277	28 35	306
161	37 21	349	44 16	062	62 23	155	58 00	195	45 00	231	33 39	276	27 58	305
162	37 11	348	44 56	061	62 42	156	57 48	195	44 25	231	32 54	275	27 20	305
163	37 01	347	45 36	060	63 00	156	57 36	196	43 49	231	32 08	275	26 42	304
164	36 50	345	46 16	059	63 18	157	57 23	196	43 14	230	31 22	274	26 04	303
	ARCTURUS		◆ANTARES		ACRUX		◆CANOPUS		SIRIUS		PROCYON		◆REGULUS	
165	15 20	048	22 13	107	63 36	158	42 38	230	30 36	273	25 25	302	36 38	344
166	15 54	047	22 58	106	63 53	159	42 03	230	29 50	273	24 46	301	36 25	343
167	16 27	046	23 42	105	64 09	160	41 28	230	29 04	272	24 06	300	36 11	342
168	17 00	045	24 26	105	64 25	160	40 53	230	28 19	271	23 27	300	35 56	341
169	17 32	044	25 11	104	64 40	161	40 18	229	27 33	271	22 47	299	35 40	339
170	18 04	043	25 55	104	64 54	162	39 43	229	26 47	270	22 06	298	35 24	338
171	18 35	043	26 40	103	65 08	163	39 08	229	26 01	270	21 25	297	35 06	337
172	19 06	042	27 25	103	65 21	164	38 33	229	25 15	269	20 44	296	34 48	336
173	19 37	041	28 10	102	65 33	165	37 59	229	24 29	268	20 03	296	34 29	335
174	20 07	040	28 55	102	65 44	166	37 24	228	23 43	268	19 21	295	34 09	334
175	20 36	039	29 40	101	65 55	167	36 50	228	22 57	267	18 40	294	33 48	333
176	21 05	038	30 25	100	66 05	168	36 16	228	22 11	266	17 58	293	33 27	332
177	21 33	038	31 10	100	66 14	169	35 42	227	21 25	266	17 15	293	33 04	330
178	22 01	037	31 55	099	66 22	170	35 08	227	20 39	265	16 33	292	32 41	329
179	22 28	036	32 41	099	66 29	171	34 35	227	19 54	265	15 50	291	32 18	328

 LAT 40°S

LHA	Hc	Zn	Hc	Zn	Hc	Zn	Hc	Zn	Hc	Zn	Hc	Zn	Hc	Zn
♈	ARCTURUS		◆ANTARES		RIGIL KENT.		ACRUX		◆CANOPUS		Alphard		◆REGULUS	
180	22 54	035	33 26	098	58 00	144	66 36	172	34 01	227	43 47	302	31 53	327
181	23 20	034	34 12	098	58 27	144	66 41	174	33 28	226	43 08	301	31 28	326
182	23 46	033	34 57	097	58 54	145	66 46	175	32 55	226	42 29	300	31 02	325
183	24 10	032	35 43	096	59 21	145	66 50	176	32 22	226	41 49	299	30 36	324
184	24 35	031	36 29	096	59 47	145	66 53	177	31 49	225	41 08	298	30 08	323
185	24 58	030	37 15	095	60 13	146	66 55	178	31 16	225	40 28	297	29 40	322
186	25 21	029	38 00	095	60 38	146	66 56	179	30 44	225	39 47	296	29 12	321
187	25 43	028	38 46	094	61 04	147	66 56	180	30 11	225	39 05	295	28 43	320
188	26 04	027	39 32	093	61 29	147	66 55	182	29 39	224	38 24	295	28 13	319
189	26 25	026	40 18	093	61 53	148	66 53	183	29 07	224	37 42	294	27 43	318
190	26 45	025	41 04	092	62 17	149	66 50	184	28 35	224	36 59	293	27 12	317
191	27 04	024	41 50	092	62 41	149	66 47	185	28 04	223	36 17	292	26 41	316
192	27 23	023	42 36	091	63 04	150	66 42	186	27 33	223	35 34	291	26 09	316
193	27 41	022	43 22	090	63 27	151	66 37	187	27 01	222	34 51	290	25 36	315
194	27 58	021	44 08	090	63 50	151	66 30	188	26 31	222	34 08	289	25 03	314
	ARCTURUS		◆ANTARES		RIGIL KENT.		◆ACRUX		Suhail		◆REGULUS		Denebola	
195	28 14	020	44 54	089	64 11	152	66 23	190	47 25	246	24 30	313	32 57	339
196	28 30	019	45 40	088	64 33	153	66 15	191	46 43	245	23 56	312	32 40	338
197	28 44	018	46 25	088	64 54	154	66 06	192	46 01	245	23 22	311	32 23	337
198	28 58	017	47 11	087	65 14	154	65 57	193	45 20	245	22 47	310	32 04	336
199	29 11	016	47 57	086	65 33	155	65 46	194	44 38	244	22 12	309	31 45	335
200	29 24	015	48 43	086	65 52	156	65 35	195	43 57	244	21 36	309	31 25	334
201	29 35	014	49 29	085	66 10	157	65 23	196	43 15	244	21 00	308	31 05	333
202	29 46	013	50 15	084	66 28	158	65 10	197	42 34	243	20 23	307	30 44	332
203	29 56	012	51 00	083	66 45	159	64 56	198	41 53	243	19 46	306	30 21	331
204	30 05	011	51 46	083	67 01	160	64 42	198	41 12	243	19 09	305	29 59	330
205	30 13	010	52 32	082	67 16	161	64 27	199	40 32	242	18 31	305	29 35	329
206	30 20	009	53 17	081	67 30	162	64 12	200	39 51	242	17 53	304	29 11	328
207	30 27	008	54 02	080	67 44	163	63 55	201	39 10	242	17 15	303	28 46	327
208	30 32	006	54 48	080	67 57	165	63 39	202	38 30	241	16 36	302	28 21	326
209	30 37	005	55 33	079	68 08	166	63 21	203	37 50	241	15 57	301	27 54	325
	ARCTURUS		◆ANTARES		Peacock		RIGIL KENT.		◆ACRUX		Suhail		◆SPICA	
210	30 41	004	56 18	078	29 26	141	68 19	167	63 03	203	37 09	241	60 07	343
211	30 44	003	57 03	077	29 55	141	68 29	168	62 45	204	36 29	240	59 52	341
212	30 46	002	57 47	076	30 24	141	68 38	170	62 26	205	35 50	240	59 36	339
213	30 47	001	58 32	075	30 53	140	68 46	171	62 06	206	35 10	240	59 19	337
214	30 47	000	59 16	074	31 22	140	68 53	172	61 46	206	34 30	239	59 00	335
215	30 47	359	60 00	073	31 52	140	68 58	173	61 26	207	33 51	239	58 40	333
216	30 45	358	60 44	072	32 22	140	69 03	175	61 05	207	33 12	238	58 19	332
217	30 43	357	61 28	071	32 52	139	69 07	176	60 43	208	32 33	238	57 56	330
218	30 40	355	62 11	070	33 22	139	69 09	178	60 22	209	31 54	238	57 33	328
219	30 36	354	62 54	069	33 52	139	69 11	179	59 59	209	31 15	237	57 08	327
220	30 31	353	63 36	067	34 22	138	69 11	180	59 37	210	30 36	237	56 42	325
221	30 25	352	64 19	066	34 53	138	69 10	182	59 14	210	29 58	237	56 15	323
222	30 18	351	65 01	065	35 24	138	69 08	183	58 51	210	29 20	236	55 47	322
223	30 11	350	65 42	063	35 54	138	69 06	184	58 27	211	28 42	236	55 18	320
224	30 02	349	66 23	062	36 25	138	69 01	186	58 04	211	28 04	235	54 48	319
	Alphecca		Rasalhague		◆Nunki		Peacock		◆RIGIL KENT.		◆SPICA		ARCTURUS	
225	22 47	008	26 21	043	39 53	093	36 56	137	68 56	187	54 18	317	29 53	348
226	22 54	007	26 52	042	40 39	092	37 28	137	68 50	188	53 46	316	29 43	347
227	22 59	006	27 23	041	41 25	092	37 59	137	68 43	190	53 14	315	29 32	346
228	23 04	005	27 53	040	42 11	091	38 31	137	68 35	191	52 41	313	29 20	345
229	23 08	004	28 22	039	42 57	091	39 02	136	68 26	192	52 07	312	29 08	344
230	23 11	004	28 51	038	43 43	090	39 34	136	68 15	193	51 32	311	28 54	343
231	23 13	003	29 19	037	44 29	089	40 06	136	68 04	195	50 57	309	28 40	342
232	23 15	002	29 46	036	45 15	089	40 38	136	67 52	196	50 21	308	28 25	340
233	23 16	001	30 13	035	46 00	088	41 10	136	67 39	197	49 45	307	28 10	339
234	23 16	000	30 39	034	46 46	087	41 42	136	67 25	198	49 08	306	27 53	338
235	23 15	359	31 05	033	47 32	087	42 14	135	67 10	199	48 30	305	27 36	337
236	23 14	358	31 29	032	48 18	086	42 46	135	66 55	200	47 52	303	27 18	336
237	23 12	357	31 53	031	49 04	085	43 19	135	66 39	201	47 13	302	26 59	335
238	23 09	356	32 17	030	49 50	084	43 51	135	66 22	202	46 34	301	26 39	334
239	23 05	355	32 39	029	50 35	084	44 23	135	66 04	203	45 55	300	26 19	333
	Rasalhague		ALTAIR		◆Nunki		Peacock		◆RIGIL KENT.		SPICA		◆ARCTURUS	
240	33 01	028	17 50	061	51 21	083	44 56	135	65 45	204	45 15	299	25 58	332
241	33 22	027	18 30	060	52 07	082	45 29	135	65 26	205	44 35	298	25 37	331
242	33 43	026	19 10	060	52 52	081	46 01	135	65 06	206	43 54	297	25 14	331
243	34 02	025	19 49	059	53 38	081	46 34	135	64 46	207	43 13	296	24 51	330
244	34 21	023	20 29	058	54 23	080	47 07	134	64 25	208	42 32	295	24 28	329
245	34 39	022	21 08	057	55 08	079	47 40	134	64 03	208	41 50	294	24 04	328
246	34 56	021	21 46	057	55 53	078	48 12	134	63 41	209	41 08	293	23 39	327
247	35 12	020	22 24	056	56 38	077	48 45	134	63 19	210	40 25	293	23 13	326
248	35 27	019	23 02	055	57 23	076	49 18	134	62 56	210	39 43	292	22 47	325
249	35 42	018	23 39	054	58 07	075	49 51	134	62 32	211	39 00	291	22 20	324
250	35 55	017	24 16	053	58 52	074	50 24	134	62 08	212	38 17	290	21 53	323
251	36 08	015	24 53	052	59 36	073	50 57	134	61 44	212	37 34	289	21 25	322
252	36 20	014	25 29	051	60 20	072	51 30	134	61 19	213	36 50	288	20 57	321
253	36 31	013	26 05	051	61 03	071	52 03	134	60 54	213	36 06	287	20 28	320
254	36 40	012	26 40	050	61 47	070	52 35	134	60 29	214	35 22	287	19 58	320
	Rasalhague		◆ALTAIR		FOMALHAUT		◆ACHERNAR		RIGIL KENT.		◆SPICA		ARCTURUS	
255	36 49	011	27 15	049	19 00	113	16 08	154	60 03	214	34 38	286	19 28	319
256	36 57	009	27 49	048	19 42	113	16 28	154	59 37	215	33 54	285	18 58	318
257	37 04	008	28 23	047	20 25	112	16 48	153	59 11	215	33 10	284	18 27	317
258	37 10	007	28 57	046	21 08	112	17 09	153	58 44	216	32 25	284	17 55	316
259	37 15	006	29 30	045	21 50	111	17 30	152	58 17	216	31 40	283	17 23	315
260	37 20	005	30 02	044	22 33	111	17 52	152	57 50	216	30 55	282	16 51	315
261	37 23	003	30 34	043	23 16	110	18 13	152	57 23	217	30 10	281	16 18	314
262	37 25	002	31 05	042	24 00	109	18 35	151	56 56	217	29 25	281	15 44	313
263	37 26	001	31 35	041	24 43	109	18 58	151	56 28	217	28 40	280	15 10	312
264	37 26	000	32 05	040	25 27	108	19 20	150	56 00	217	27 55	279	14 36	311
265	37 25	358	32 35	039	26 10	108	19 43	150	55 32	218	27 09	278	14 02	311
266	37 24	357	33 04	038	26 54	107	20 06	150	55 04	218	26 24	278	13 27	310
267	37 21	356	33 32	037	27 38	107	20 30	149	54 36	218	25 38	277	12 51	309
268	37 17	355	33 59	036	28 22	106	20 53	149	54 07	218	24 52	276	12 15	308
269	37 12	353	34 26	035	29 06	106	21 17	148	53 39	219	24 07	276	11 39	308

LHA	Hc	Zn	Hc	Zn	Hc	Zn	Hc	Zn	Hc	Zn	Hc	Zn	Hc	Zn
♈	ALTAIR		Enif		◆FOMALHAUT		ACHERNAR		◆RIGIL KENT.		ANTARES		◆Rasalhague	
270	34 52	034	18 12	059	29 51	105	21 42	148	53 10	219	66 45	299	37 07	352
271	35 17	033	18 51	059	30 35	105	22 06	148	52 41	219	66 04	297	37 00	351
272	35 42	032	19 30	058	31 20	104	22 31	147	52 12	219	65 23	296	36 52	350
273	36 05	031	20 09	057	32 04	104	22 56	147	51 43	219	64 42	295	36 44	349
274	36 29	030	20 47	056	32 49	103	23 21	146	51 14	219	64 00	293	36 34	347
275	36 51	028	21 25	055	33 34	103	23 47	146	50 45	219	63 17	292	36 24	346
276	37 12	027	22 03	054	34 19	102	24 12	146	50 16	219	62 35	291	36 12	345
277	37 33	026	22 40	054	35 04	102	24 38	145	49 47	219	61 51	290	36 00	344
278	37 53	025	23 17	053	35 49	101	25 05	145	49 18	219	61 08	289	35 47	343
279	38 11	024	23 53	052	36 34	100	25 31	145	48 49	219	60 24	288	35 33	342
280	38 29	022	24 29	051	37 19	100	25 58	144	48 19	220	59 40	287	35 18	340
281	38 47	021	25 05	050	38 04	099	26 25	144	47 50	220	58 56	286	35 02	339
282	39 03	020	25 40	049	38 50	099	26 52	144	47 21	220	58 12	285	34 45	338
283	39 18	019	26 14	048	39 35	098	27 19	143	46 52	220	57 27	284	34 27	337
284	39 32	018	26 49	048	40 21	098	27 47	143	46 22	220	56 42	283	34 09	336
	ALTAIR		Enif		◆FOMALHAUT		ACHERNAR		◆RIGIL KENT.		ANTARES		◆Rasalhague	
285	39 46	016	27 22	047	41 06	097	28 14	143	45 53	219	55 58	282	33 50	335
286	39 58	015	27 55	046	41 52	097	28 42	142	45 24	219	55 12	281	33 30	334
287	40 10	014	28 28	045	42 38	096	29 10	142	44 55	219	54 27	280	33 09	333
288	40 20	013	29 00	044	43 23	095	29 39	142	44 26	219	53 42	279	32 47	331
289	40 30	011	29 32	043	44 09	095	30 07	142	43 56	219	52 57	278	32 25	330
290	40 38	010	30 03	042	44 55	094	30 36	141	43 27	219	52 11	278	32 02	329
291	40 46	009	30 33	041	45 41	094	31 05	141	42 58	219	51 25	277	31 38	328
292	40 52	007	31 03	040	46 27	093	31 34	141	42 29	219	50 40	276	31 13	327
293	40 57	006	31 32	039	47 13	093	32 03	140	42 01	219	49 54	275	30 48	326
294	41 02	005	32 01	038	47 58	092	32 33	140	41 32	219	49 08	275	30 22	325
295	41 05	003	32 29	037	48 44	091	33 02	140	41 03	219	48 22	274	29 56	324
296	41 07	002	32 56	036	49 30	091	33 32	140	40 34	219	47 37	273	29 28	323
297	41 08	001	33 23	035	50 16	090	34 02	139	40 06	218	46 51	273	29 01	322
298	41 09	000	33 49	034	51 02	089	34 32	139	39 37	218	46 05	272	28 32	321
299	41 08	358	34 14	033	51 48	089	35 02	139	39 09	218	45 19	271	28 03	320
	Enif		◆FOMALHAUT		ACHERNAR		◆RIGIL KENT.		ANTARES		◆Rasalhague		ALTAIR	
300	34 39	032	52 34	088	35 32	139	38 40	218	44 33	271	27 33	319	41 06	357
301	35 02	031	53 20	087	36 03	138	38 12	218	43 47	270	27 03	318	41 03	356
302	35 25	029	54 06	087	36 33	138	37 44	218	43 01	269	26 32	317	40 59	354
303	35 48	028	54 52	086	37 04	138	37 16	217	42 15	269	26 01	317	40 54	353
304	36 09	027	55 38	085	37 35	138	36 48	217	41 29	268	25 29	316	40 47	352
305	36 30	026	56 23	084	38 06	138	36 20	217	40 43	267	24 56	315	40 40	350
306	36 49	025	57 09	084	38 37	137	35 53	217	39 57	267	24 24	314	40 32	349
307	37 08	024	57 55	083	39 08	137	35 25	217	39 11	266	23 50	313	40 23	348
308	37 26	023	58 40	082	39 40	137	34 58	216	38 25	266	23 16	312	40 13	347
309	37 43	021	59 26	081	40 11	137	34 30	216	37 40	265	22 42	311	40 02	345
310	38 00	020	60 11	080	40 43	137	34 03	216	36 54	264	22 07	310	39 49	344
311	38 15	019	60 57	080	41 14	136	33 36	216	36 08	264	21 32	310	39 36	343
312	38 29	018	61 42	079	41 46	136	33 09	216	35 23	263	20 56	309	39 22	342
313	38 43	016	62 27	078	42 18	136	32 43	215	34 37	263	20 20	308	39 07	340
314	38 56	015	63 12	077	42 50	136	32 16	215	33 51	262	19 44	307	38 51	339
	Enif		◆Diphda		ACHERNAR		◆RIGIL KENT.		ANTARES		Nunki		◆ALTAIR	
315	39 07	014	37 26	082	43 22	136	31 50	215	33 06	262	60 41	288	38 34	338
316	39 18	013	38 12	082	43 54	136	31 24	215	32 20	261	59 57	287	38 16	337
317	39 27	011	38 57	081	44 26	136	30 58	214	31 35	260	59 13	286	37 58	335
318	39 36	010	39 42	080	44 58	136	30 32	214	30 50	260	58 28	285	37 38	334
319	39 44	009	40 28	079	45 30	135	30 06	214	30 05	259	57 44	284	37 18	333
320	39 50	008	41 13	079	46 02	135	29 41	214	29 19	259	56 59	283	36 57	332
321	39 56	006	41 58	078	46 35	135	29 15	213	28 34	258	56 15	282	36 35	331
322	40 01	005	42 43	077	47 07	135	28 50	213	27 49	258	55 30	282	36 12	330
323	40 04	004	43 27	076	47 39	135	28 25	213	27 05	257	54 44	281	35 48	329
324	40 07	003	44 12	075	48 12	135	28 01	212	26 20	257	53 59	280	35 24	327
325	40 08	001	44 56	075	48 44	135	27 36	212	25 35	256	53 14	279	34 59	326
326	40 09	000	45 41	074	49 17	135	27 12	212	24 51	255	52 28	278	34 33	325
327	40 08	359	46 25	073	49 49	135	26 48	211	24 06	255	51 43	277	34 06	324
328	40 07	357	47 08	072	50 21	135	26 24	211	23 22	254	50 57	277	33 39	323
329	40 04	356	47 52	071	50 54	135	26 00	211	22 38	254	50 12	276	33 11	322
	◆Alpheratz		Diphda		◆ACHERNAR		RIGIL KENT.		◆Nunki		ALTAIR		Enif	
330	14 48	029	48 35	070	51 26	135	25 37	211	49 26	275	32 43	321	40 01	355
331	15 10	028	49 18	069	51 59	135	25 14	210	48 40	274	32 13	320	39 56	354
332	15 31	027	50 01	068	52 31	135	24 51	210	47 54	274	31 44	319	39 50	352
333	15 51	026	50 44	067	53 03	135	24 28	210	47 08	273	31 13	318	39 44	351
334	16 11	025	51 26	066	53 36	135	24 05	209	46 22	272	30 42	317	39 36	350
335	16 31	024	52 08	065	54 08	136	23 43	209	45 36	272	30 10	316	39 27	348
336	16 49	024	52 49	064	54 40	136	23 21	208	44 51	271	29 38	315	39 17	347
337	17 07	023	53 30	063	55 12	136	22 59	208	44 05	270	29 06	314	39 07	346
338	17 25	022	54 11	062	55 44	136	22 38	208	43 19	270	28 32	313	38 55	345
339	17 42	021	54 51	061	56 16	136	22 16	207	42 33	269	27 59	312	38 43	343
340	17 58	020	55 31	059	56 48	136	21 55	207	41 47	269	27 24	311	38 29	342
341	18 13	019	56 10	058	57 20	137	21 35	207	41 01	268	26 50	311	38 15	341
342	18 28	018	56 49	057	57 51	137	21 14	206	40 15	267	26 15	310	37 59	340
343	18 42	018	57 27	056	58 23	137	20 54	206	39 29	267	25 39	309	37 43	339
344	18 56	017	58 05	054	58 54	137	20 34	206	38 43	266	25 03	308	37 26	337
	◆Alpheratz		Diphda		◆CANOPUS		RIGIL KENT.		◆Nunki		ALTAIR		Enif	
345	19 09	016	58 42	053	20 12	143	20 14	205	37 57	265	24 26	307	37 08	336
346	19 21	015	59 18	051	20 39	142	19 55	205	37 11	265	23 49	306	36 49	335
347	19 32	014	59 54	050	21 08	142	19 36	204	36 26	264	23 12	305	36 29	334
348	19 43	013	60 29	049	21 36	142	19 17	204	35 40	264	22 34	305	36 08	333
349	19 53	012	61 03	047	22 05	141	18 59	204	34 54	263	21 56	304	35 47	332
350	20 02	011	61 36	045	22 33	141	18 40	203	34 09	263	21 18	303	35 25	331
351	20 11	010	62 08	044	23 03	140	18 22	203	33 23	262	20 39	302	35 02	329
352	20 19	009	62 39	042	23 32	140	18 05	202	32 38	261	20 00	301	34 38	328
353	20 26	008	63 10	040	24 02	140	17 47	202	31 52	261	19 21	301	34 13	327
354	20 32	007	63 39	038	24 31	139	17 30	202	31 07	260	18 41	300	33 48	326
355	20 38	007	64 07	036	25 01	139	17 14	201	30 22	260	18 01	299	33 22	325
356	20 43	006	64 33	035	25 32	139	16 57	201	29 36	259	17 20	298	32 56	324
357	20 47	005	64 59	033	26 02	138	16 41	200	28 51	259	16 40	297	32 28	323
358	20 50	004	65 23	030	26 33	138	16 26	200	28 06	258	15 59	297	32 00	322
359	20 53	003	65 45	028	27 04	137	16 10	199	27 21	257	15 18	296	31 32	321

LHA	Hc	Zn	Hc	Zn	Hc	Zn	Hc	Zn	Hc	Zn	Hc	Zn	Hc	Zn
♈	Alpheratz		Hamal		◆RIGEL		CANOPUS		◆RIGIL KENT.		Peacock		◆Enif	
0	19 55	002	19 09	031	13 59	089	28 19	137	16 52	199	52 32	227	30 16	320
1	19 56	001	19 32	030	14 44	088	28 50	136	16 37	199	51 59	227	29 47	319
2	19 56	000	19 54	029	15 29	087	29 21	136	16 23	198	51 26	227	29 17	318
3	19 56	359	20 16	028	16 15	087	29 53	136	16 09	198	50 53	227	28 47	317
4	19 55	358	20 37	027	17 00	086	30 25	135	15 55	197	50 20	227	28 16	316
5	19 53	357	20 57	026	17 45	085	30 56	135	15 42	197	49 48	227	27 45	316
6	19 50	356	21 17	025	18 30	085	31 29	135	15 29	196	49 15	226	27 13	315
7	19 47	355	21 36	024	19 15	084	32 01	134	15 17	196	48 42	226	26 40	314
8	19 43	354	21 54	023	20 00	083	32 33	134	15 04	195	48 09	226	26 07	313
9	19 38	354	22 12	023	20 45	083	33 06	134	14 52	195	47 36	226	25 34	312
10	19 33	353	22 29	022	21 30	082	33 39	133	14 41	195	47 04	226	25 00	311
11	19 27	352	22 45	021	22 15	081	34 12	133	14 30	194	46 31	226	24 25	310
12	19 20	351	23 00	020	23 00	081	34 45	133	14 19	194	45 58	226	23 50	309
13	19 12	350	23 15	019	23 44	080	35 18	133	14 09	193	45 26	226	23 15	308
14	19 04	349	23 29	018	24 29	079	35 52	132	13 58	193	44 53	226	22 39	307
	◆Hamal		RIGEL		◆CANOPUS		RIGIL KENT.		◆Peacock		FOMALHAUT		Alpheratz	
15	23 43	017	25 13	078	36 25	132	13 49	192	44 21	226	62 43	285	18 55	348
16	23 56	016	25 57	078	36 59	132	13 39	192	43 49	225	61 59	284	18 45	347
17	24 08	015	26 42	077	37 33	131	13 30	191	43 17	225	61 15	283	18 35	346
18	24 19	014	27 26	076	38 07	131	13 22	191	42 44	225	60 31	282	18 24	345
19	24 29	013	28 10	075	38 41	131	13 13	190	42 12	225	59 47	281	18 12	344
20	24 39	012	28 53	075	39 15	131	13 06	190	41 40	225	59 02	280	17 59	344
21	24 48	011	29 37	074	39 50	130	12 58	189	41 09	225	58 17	279	17 46	343
22	24 56	010	30 20	073	40 24	130	12 51	189	40 37	224	57 33	278	17 32	342
23	25 03	009	31 04	072	40 59	130	12 44	188	40 05	224	56 48	278	17 18	341
24	25 10	008	31 47	072	41 34	130	12 38	188	39 34	224	56 03	277	17 03	340
25	25 16	007	32 30	071	42 09	129	12 32	187	39 02	224	55 18	276	16 47	339
26	25 21	006	33 12	070	42 44	129	12 26	187	38 31	224	54 33	275	16 30	338
27	25 25	005	33 55	069	43 19	129	12 21	186	38 00	223	53 48	274	16 13	337
28	25 28	004	34 37	068	43 54	129	12 16	186	37 29	223	53 03	274	15 56	337
29	25 31	003	35 19	067	44 30	129	12 12	185	36 58	223	52 17	273	15 37	336
	◆Hamal		RIGEL		SIRIUS		◆CANOPUS		RIGIL KENT.		◆Peacock		FOMALHAUT	
30	25 32	002	36 00	067	24 55	091	45 05	128	12 08	185	36 27	223	51 32	272
31	25 33	001	36 42	066	25 40	090	45 41	128	12 04	184	35 57	222	50 47	272
32	25 34	000	37 23	065	26 25	090	46 16	128	12 01	184	35 26	222	50 02	271
33	25 33	359	38 04	064	27 10	089	46 52	128	11 58	183	34 56	222	49 16	270
34	25 32	358	38 44	063	27 56	088	47 28	128	11 56	183	34 26	222	48 31	270
35	25 29	357	39 24	062	28 41	088	48 04	128	11 53	182	33 56	221	47 46	269
36	25 26	356	40 04	061	29 26	087	48 40	127	11 52	182	33 26	221	47 00	268
37	25 22	355	40 44	060	30 11	086	49 16	127	11 50	181	32 56	221	46 15	268
38	25 18	354	41 23	059	30 57	086	49 52	127	11 50	181	32 27	220	45 30	267
39	25 12	353	42 01	058	31 42	085	50 28	127	11 49	180	31 57	220	44 45	266
40	25 06	352	42 40	057	32 27	084	51 04	127	11 49	180	31 28	220	44 00	266
41	24 59	351	43 17	056	33 12	083	51 40	127	11 49	179	30 59	220	43 14	265
42	24 51	350	43 55	055	33 57	083	52 16	127	11 50	179	30 30	219	42 29	265
43	24 43	349	44 31	054	34 42	082	52 53	127	11 51	178	30 02	219	41 44	264
44	24 34	348	45 08	053	35 26	081	53 29	127	11 52	178	29 34	219	40 59	264
	◆RIGEL		SIRIUS		CANOPUS		◆RIGIL KENT.		Peacock		◆FOMALHAUT		Hamal	
45	45 44	052	36 11	081	54 05	127	11 54	177	29 05	218	40 14	263	24 23	347
46	46 19	050	36 56	080	54 42	127	11 56	177	28 37	218	39 29	262	24 13	346
47	46 53	049	37 40	079	55 18	127	11 59	176	28 10	218	38 45	262	24 01	345
48	47 27	048	38 25	078	55 54	127	12 02	176	27 42	217	38 00	261	23 49	344
49	48 01	047	39 09	077	56 31	127	12 05	175	27 15	217	37 15	261	23 36	343
50	48 34	046	39 53	077	57 07	127	12 09	175	26 48	217	36 30	260	23 22	342
51	49 06	044	40 37	076	57 44	127	12 13	174	26 21	216	35 46	260	23 07	341
52	49 37	043	41 21	075	58 20	127	12 18	174	25 54	216	35 01	259	22 52	340
53	50 08	042	42 05	074	58 56	127	12 23	173	25 27	216	34 17	258	22 36	339
54	50 37	040	42 48	073	59 32	127	12 28	173	25 01	215	33 33	258	22 19	338
55	51 06	039	43 31	073	60 08	127	12 34	172	24 35	215	32 48	257	22 02	337
56	51 34	038	44 14	072	60 45	127	12 40	172	24 10	214	32 04	257	21 44	336
57	52 02	036	44 57	071	61 21	127	12 47	172	23 44	214	31 20	256	21 25	335
58	52 28	035	45 40	070	61 57	128	12 54	171	23 19	214	30 36	256	21 06	334
59	52 53	033	46 22	069	62 32	128	13 01	171	22 54	213	29 52	255	20 46	333
	ALDEBARAN		◆SIRIUS		Suhail		◆RIGIL KENT.		ACHERNAR		◆Diphda		Hamal	
60	31 55	010	47 04	068	35 04	120	13 08	170	62 05	222	42 14	284	20 25	332
61	32 02	009	47 46	067	35 44	120	13 17	170	61 34	223	41 30	283	20 04	331
62	32 09	008	48 28	066	36 23	119	13 25	169	61 04	223	40 46	282	19 42	331
63	32 14	007	49 09	065	37 03	119	13 34	169	60 33	223	40 01	281	19 19	330
64	32 19	006	49 50	064	37 42	119	13 43	168	60 02	224	39 17	280	18 56	329
65	32 23	004	50 30	063	38 22	118	13 52	168	59 30	224	38 32	280	18 32	328
66	32 26	003	51 10	062	39 02	118	14 02	167	58 59	224	37 48	279	18 08	327
67	32 28	002	51 50	061	39 42	118	14 12	167	58 27	224	37 03	278	17 43	326
68	32 30	001	52 29	060	40 22	117	14 23	166	57 55	225	36 18	277	17 18	325
69	32 30	000	53 08	058	41 03	117	14 34	166	57 23	225	35 33	277	16 52	325
70	32 29	359	53 47	057	41 43	116	14 45	165	56 51	225	34 48	276	16 25	324
71	32 28	358	54 24	056	42 24	116	14 57	165	56 19	225	34 03	275	15 58	323
72	32 26	356	55 02	055	43 05	116	15 09	164	55 47	225	33 18	275	15 30	322
73	32 23	355	55 38	053	43 45	115	15 21	164	55 15	225	32 33	274	15 02	321
74	32 18	354	56 14	052	44 26	115	15 34	163	54 43	225	31 48	273	14 34	320
	BETELGEUSE		◆SIRIUS		Suhail		◆RIGIL KENT.		ACHERNAR		◆Diphda		ALDEBARAN	
75	39 59	018	56 50	051	45 08	115	15 47	163	54 11	226	31 02	273	32 13	353
76	40 12	017	57 24	049	45 49	114	16 00	163	53 38	226	30 17	272	32 08	352
77	40 24	015	57 58	048	46 30	114	16 14	162	53 06	226	29 32	271	32 01	351
78	40 36	014	58 31	046	47 11	114	16 28	162	52 33	226	28 47	271	31 53	350
79	40 46	013	59 04	045	47 53	113	16 43	161	52 01	226	28 01	270	31 45	349
80	40 56	011	59 35	043	48 35	113	16 57	161	51 28	226	27 16	269	31 35	347
81	41 05	010	60 06	042	49 16	113	17 12	160	50 56	226	26 31	269	31 25	346
82	41 12	009	60 35	040	49 58	112	17 28	160	50 24	226	25 46	268	31 14	345
83	41 18	008	61 04	038	50 40	112	17 43	159	49 51	226	25 00	267	31 02	344
84	41 24	006	61 31	037	51 22	112	17 59	159	49 19	226	24 15	267	30 49	343
85	41 28	005	61 58	035	52 04	111	18 16	159	48 46	226	23 30	266	30 36	342
86	41 32	004	62 23	033	52 46	111	18 32	158	48 14	226	22 45	265	30 21	341
87	41 34	002	62 47	031	53 29	111	18 49	158	47 42	226	22 00	265	30 06	340
88	41 35	001	63 10	029	54 11	111	19 07	157	47 09	225	21 15	264	29 50	339
89	41 36	000	63 31	027	54 53	110	19 24	157	46 37	225	20 30	264	29 33	338

LAT 41°S

LHA	Hc	Zn	Hc	Zn	Hc	Zn	Hc	Zn	Hc	Zn	Hc	Zn	Hc	Zn
♈	PROCYON		Alphard		◆Suhail		RIGIL KENT.		◆ACHERNAR		RIGEL		◆BETELGEUSE	
90	38 30	032	34 02	070	55 36	110	19 42	157	46 05	225	55 40	340	41 35	358
91	38 54	031	34 44	069	56 19	110	20 00	156	45 33	225	55 24	338	41 33	357
92	39 17	030	35 26	068	57 01	109	20 19	156	45 01	225	55 06	336	41 30	356
93	39 39	029	36 08	067	57 44	109	20 37	155	44 29	225	54 48	335	41 26	354
94	40 00	027	36 50	066	58 27	109	20 57	155	43 57	225	54 28	333	41 21	353
95	40 21	026	37 31	065	59 10	109	21 16	155	43 25	225	54 07	331	41 15	352
96	40 40	025	38 12	064	59 52	108	21 35	154	42 53	224	53 44	330	41 08	350
97	40 59	024	38 53	064	60 35	108	21 55	154	42 22	224	53 21	328	41 00	349
98	41 17	022	39 33	063	61 19	108	22 15	153	41 50	224	52 57	327	40 51	348
99	41 34	021	40 13	062	62 02	108	22 36	153	41 18	224	52 32	325	40 41	347
100	41 50	020	40 53	061	62 45	108	22 57	153	40 47	224	52 06	324	40 30	345
101	42 04	019	41 32	060	63 28	107	23 17	152	40 16	224	51 38	323	40 18	344
102	42 18	017	42 11	059	64 11	107	23 39	152	39 45	223	51 11	321	40 05	343
103	42 31	016	42 49	058	64 55	107	24 00	152	39 14	223	50 42	320	39 51	341
104	42 43	015	43 27	057	65 38	107	24 22	151	38 43	223	50 12	318	39 36	340
	PROCYON		REGULUS		◆Gienah		RIGIL KENT.		◆ACHERNAR		RIGEL		◆BETELGEUSE	
105	42 54	013	21 32	050	19 39	096	24 44	151	38 12	223	49 42	317	39 20	339
106	43 04	012	22 06	049	20 24	096	25 06	150	37 41	223	49 10	316	39 03	338
107	43 13	011	22 40	049	21 09	095	25 28	150	37 11	222	48 38	315	38 46	336
108	43 21	009	23 14	048	21 54	095	25 51	150	36 40	222	48 06	313	38 27	335
109	43 28	008	23 47	047	22 39	094	26 14	149	36 10	222	47 32	312	38 08	334
110	43 33	007	24 20	046	23 24	093	26 37	149	35 40	222	46 59	311	37 48	333
111	43 38	005	24 52	045	24 09	093	27 00	149	35 10	221	46 24	310	37 27	332
112	43 42	004	25 24	044	24 55	092	27 24	148	34 40	221	45 49	309	37 05	331
113	43 44	002	25 56	043	25 40	091	27 48	148	34 10	221	45 13	307	36 42	329
114	43 45	001	26 26	042	26 25	091	28 12	148	33 41	221	44 37	306	36 19	328
115	43 46	000	26 56	041	27 10	090	28 36	148	33 12	220	44 00	305	35 55	327
116	43 45	358	27 26	040	27 56	089	29 00	147	32 42	220	43 23	304	35 30	326
117	43 43	357	27 55	039	28 41	089	29 25	147	32 13	220	42 45	303	35 04	325
118	43 40	356	28 24	038	29 26	088	29 50	147	31 45	219	42 07	302	34 38	324
119	43 36	354	28 51	037	30 11	087	30 15	146	31 16	219	41 28	301	34 11	323
	REGULUS		◆SPICA		RIGIL KENT.		◆ACHERNAR		RIGEL		BETELGEUSE		◆PROCYON	
120	29 19	037	13 52	093	30 40	146	30 47	219	40 49	300	33 43	322	43 31	353
121	29 45	036	14 37	092	31 05	146	30 19	218	40 10	299	33 15	321	43 25	351
122	30 11	034	15 22	092	31 31	145	29 51	218	39 30	298	32 46	320	43 17	350
123	30 37	033	16 07	091	31 57	145	29 23	218	38 50	297	32 16	319	43 09	349
124	31 01	032	16 53	090	32 23	145	28 55	218	38 10	296	31 46	318	43 00	347
125	31 25	031	17 38	090	32 49	145	28 28	217	37 29	295	31 15	317	42 49	346
126	31 48	030	18 23	089	33 15	144	28 01	217	36 48	294	30 44	316	42 38	345
127	32 11	029	19 09	088	33 41	144	27 34	217	36 07	294	30 12	315	42 25	343
128	32 33	028	19 54	088	34 08	144	27 07	216	35 25	293	29 39	314	42 12	342
129	32 54	027	20 39	087	34 35	144	26 40	216	34 43	292	29 06	313	41 57	341
130	33 14	026	21 24	086	35 02	143	26 14	215	34 01	291	28 33	312	41 42	339
131	33 33	025	22 09	086	35 29	143	25 48	215	33 19	290	27 59	311	41 26	338
132	33 52	024	22 55	085	35 56	143	25 22	215	32 36	289	27 25	310	41 08	337
133	34 10	023	23 40	084	36 23	143	24 56	214	31 53	289	26 50	309	40 50	336
134	34 27	022	24 25	083	36 51	143	24 30	214	31 10	288	26 14	308	40 31	334
	REGULUS		◆SPICA		RIGIL KENT.		◆ACHERNAR		CANOPUS		SIRIUS		◆PROCYON	
135	34 43	020	25 10	083	37 18	142	24 05	214	61 18	233	52 08	300	40 11	333
136	34 59	019	25 54	082	37 46	142	23 40	213	60 42	233	51 29	299	39 50	332
137	35 13	018	26 39	081	38 14	142	23 16	213	60 06	233	50 49	298	39 29	331
138	35 27	017	27 24	081	38 42	142	22 51	213	59 29	233	50 09	297	39 06	330
139	35 39	016	28 09	080	39 10	142	22 27	212	58 53	233	49 28	295	38 43	328
140	35 51	015	28 53	079	39 38	141	22 03	212	58 17	233	48 47	294	38 19	327
141	36 02	013	29 38	078	40 06	141	21 39	211	57 41	233	48 06	293	37 54	326
142	36 12	012	30 22	078	40 35	141	21 16	211	57 04	233	47 24	292	37 28	325
143	36 21	011	31 06	077	41 03	141	20 53	211	56 28	233	46 42	292	37 02	324
144	36 29	010	31 50	076	41 32	141	20 30	210	55 52	233	46 00	291	36 35	323
145	36 37	009	32 34	075	42 00	141	20 07	210	55 15	233	45 17	290	36 07	322
146	36 43	007	33 18	075	42 29	141	19 45	209	54 39	233	44 34	289	35 38	321
147	36 48	006	34 01	074	42 58	140	19 23	209	54 03	233	43 51	288	35 09	319
148	36 53	005	34 45	073	43 27	140	19 01	208	53 26	233	43 08	287	34 40	318
149	36 56	004	35 28	072	43 56	140	18 40	208	52 50	233	42 25	286	34 09	317
	REGULUS		◆SPICA		RIGIL KENT.		◆ACHERNAR		CANOPUS		SIRIUS		◆PROCYON	
150	36 59	002	36 11	071	44 24	140	18 19	208	52 14	233	41 41	285	33 38	316
151	37 00	001	36 54	071	44 54	140	17 58	207	51 37	233	40 57	284	33 07	315
152	37 00	000	37 36	070	45 23	140	17 37	207	51 01	233	40 14	284	32 35	314
153	37 00	359	38 19	069	45 52	140	17 17	206	50 25	233	39 29	283	32 02	313
154	36 59	358	39 01	068	46 21	140	16 57	206	49 49	233	38 45	282	31 29	312
155	36 56	356	39 43	067	46 50	140	16 38	205	49 13	233	38 01	281	30 55	311
156	36 53	355	40 24	066	47 19	140	16 18	205	48 37	233	37 16	281	30 21	310
157	36 49	354	41 06	065	47 49	140	15 59	204	48 01	232	36 32	280	29 46	310
158	36 43	353	41 47	064	48 18	140	15 41	204	47 25	232	35 47	279	29 11	309
159	36 37	351	42 27	063	48 47	140	15 23	204	46 49	232	35 02	278	28 35	308
160	36 30	350	43 07	062	49 16	140	15 05	203	46 13	232	34 18	278	27 59	307
161	36 22	349	43 47	061	49 46	140	14 47	203	45 38	232	33 33	277	27 23	306
162	36 13	348	44 27	060	50 15	140	14 30	202	45 02	232	32 48	276	26 46	305
163	36 03	347	45 06	059	50 44	140	14 13	202	44 27	231	32 03	275	26 09	304
164	35 52	345	45 45	058	51 13	140	13 57	201	43 52	231	31 17	275	25 31	303
	SPICA		◆ANTARES		RIGIL KENT.		ACHERNAR		◆CANOPUS		SIRIUS		◆REGULUS	
165	46 23	057	22 30	106	51 42	140	13 40	201	43 16	231	30 32	274	35 40	344
166	47 01	056	23 14	106	52 11	140	13 25	200	42 41	231	29 47	273	35 27	343
167	47 38	055	23 58	105	52 40	140	13 09	200	42 06	231	29 02	273	35 14	342
168	48 15	054	24 41	104	53 09	140	12 54	199	41 31	230	28 17	272	34 59	341
169	48 52	053	25 25	104	53 38	140	12 39	199	40 56	230	27 31	271	34 44	340
170	49 27	052	26 09	103	54 07	141	12 25	198	40 22	230	26 46	271	34 28	339
171	50 02	050	26 53	103	54 36	141	12 11	198	39 47	230	26 01	270	34 11	337
172	50 37	049	27 38	102	55 05	141	11 57	197	39 13	229	25 16	269	33 53	336
173	51 11	048	28 22	102	55 33	141	11 44	197	38 38	229	24 30	269	33 34	335
174	51 44	046	29 06	101	56 01	141	11 31	196	38 04	229	23 45	268	33 15	334
175	52 16	045	29 51	100	56 30	141	11 19	196	37 30	229	23 00	267	32 55	333
176	52 48	044	30 35	100	56 58	142	11 07	195	36 56	228	22 15	267	32 34	332
177	53 19	042	31 20	099	57 26	142	10 55	195	36 23	228	21 29	266	32 12	331
178	53 49	041	32 05	099	57 54	142	10 44	194	35 49	228	20 44	266	31 50	330
179	54 18	040	32 50	098	58 21	143	10 33	194	35 15	227	19 59	265	31 26	329

LAT 41°S

LAT 41°S

LHA ♈	Hc	Zn	Hc	Zn	Hc	Zn	Hc	Zn	Hc	Zn	Hc	Zn	Hc	Zn
	◆ARCTURUS		ANTARES		◆RIGIL KENT.		ACHERNAR		CANOPUS		◆Suhail		REGULUS	
180	22 05	035	33 34	098	58 49	143	10 23	193	34 42	227	58 23	251	31 03	328
181	22 31	034	34 19	097	59 16	143	10 12	193	34 09	227	57 40	251	30 38	327
182	22 55	033	35 04	096	59 43	144	10 03	192	33 36	227	56 57	251	30 13	326
183	23 20	032	35 49	096	60 10	144	09 54	192	33 03	226	56 15	250	29 47	325
184	23 43	031	36 34	095	60 36	144	09 45	191	32 31	226	55 32	250	29 20	324
185	24 06	030	37 20	095	61 02	145	09 36	190	31 58	226	54 49	250	28 53	323
186	24 28	029	38 05	094	61 28	145	09 28	190	31 26	225	54 07	249	28 25	322
187	24 50	028	38 50	093	61 54	146	09 21	189	30 54	225	53 25	249	27 57	321
188	25 11	027	39 35	093	62 19	146	09 13	189	30 22	225	52 42	249	27 28	320
189	25 31	026	40 20	092	62 44	147	09 07	188	29 50	224	52 00	249	26 58	319
190	25 51	025	41 06	091	63 08	148	09 00	188	29 19	224	51 18	248	26 28	318
191	26 10	024	41 51	091	63 32	148	08 54	187	28 48	224	50 36	248	25 57	317
192	26 28	023	42 36	090	63 56	149	08 49	187	28 16	223	49 54	248	25 26	316
193	26 45	022	43 22	089	64 19	149	08 44	186	27 46	223	49 12	247	24 54	315
194	27 02	021	44 07	089	64 42	150	08 39	186	27 15	222	48 31	247	24 22	314
	◆ARCTURUS		ANTARES		◆Peacock		ACHERNAR		CANOPUS		◆Suhail		REGULUS	
195	27 18	020	44 52	088	23 28	146	08 35	185	26 44	222	47 49	247	23 49	313
196	27 33	019	45 37	087	23 54	146	08 31	185	26 14	222	47 08	246	23 16	312
197	27 47	018	46 23	087	24 19	145	08 27	184	25 44	221	46 26	246	22 42	311
198	28 01	017	47 08	086	24 45	145	08 25	183	25 14	221	45 45	246	22 08	311
199	28 14	016	47 53	085	25 11	145	08 22	183	24 45	221	45 04	245	21 33	310
200	28 26	015	48 38	085	25 37	144	08 20	182	24 16	220	44 23	245	20 58	309
201	28 37	014	49 23	084	26 04	144	08 18	182	23 46	220	43 42	245	20 23	308
202	28 48	013	50 08	083	26 31	144	08 17	181	23 18	219	43 01	244	19 47	307
203	28 57	012	50 53	082	26 58	143	08 16	181	22 49	219	42 20	244	19 11	306
204	29 06	011	51 38	081	27 25	143	08 16	180	22 21	219	41 39	244	18 34	306
205	29 14	010	52 22	081	27 52	143	08 16	180	21 53	218	40 59	243	17 57	305
206	29 21	009	53 07	080	28 20	142	08 16	179	21 25	218	40 19	243	17 20	304
207	29 27	007	53 52	079	28 48	142	08 17	179	20 57	217	39 38	242	16 42	303
208	29 33	006	54 36	078	29 16	142	08 19	178	20 30	217	38 58	242	16 04	302
209	29 37	005	55 20	077	29 44	141	08 20	177	20 03	216	38 18	242	15 26	302
	◆ARCTURUS		ANTARES		◆Peacock		ACHERNAR		CANOPUS		◆Suhail		SPICA	
210	29 41	004	56 04	076	30 13	141	08 23	177	19 36	216	37 39	241	59 09	343
211	29 44	003	56 48	075	30 41	141	08 25	176	19 10	216	36 59	241	58 55	341
212	29 46	002	57 32	074	31 10	140	08 28	176	18 43	215	36 19	241	58 40	339
213	29 47	001	58 16	073	31 39	140	08 32	175	18 18	215	35 40	240	58 23	338
214	29 47	000	58 59	072	32 08	140	08 36	175	17 52	214	35 01	240	58 05	336
215	29 47	359	59 42	071	32 38	139	08 40	174	17 27	214	34 22	239	57 46	334
216	29 46	358	60 25	070	33 07	139	08 45	174	17 02	213	33 43	239	57 26	332
217	29 43	357	61 07	069	33 37	139	08 50	173	16 37	213	33 04	239	57 04	331
218	29 40	355	61 49	068	34 07	139	08 56	173	16 13	212	32 25	238	56 41	329
219	29 36	354	62 31	067	34 37	138	09 02	172	15 49	212	31 47	238	56 18	327
220	29 31	353	63 13	066	35 07	138	09 08	172	15 25	211	31 09	237	55 53	326
221	29 26	352	63 54	064	35 37	138	09 15	171	15 01	211	30 31	237	55 27	324
222	29 19	351	64 34	063	36 08	138	09 22	170	14 38	210	29 53	237	55 00	323
223	29 12	350	65 14	062	36 39	137	09 30	170	14 15	210	29 15	236	54 32	321
224	29 04	349	65 54	060	37 09	137	09 38	169	13 53	209	28 38	236	54 03	320
	Rasalhague		◆Nunki		Peacock		ACHERNAR		◆ACRUX		SPICA		◆ARCTURUS	
225	25 37	043	39 56	092	37 40	137	09 47	169	58 30	213	53 33	318	28 55	348
226	26 08	042	40 41	092	38 11	137	09 56	168	58 06	213	53 03	317	28 45	347
227	26 37	041	41 26	091	38 43	136	10 05	168	57 41	213	52 31	315	28 34	346
228	27 07	040	42 11	090	39 14	136	10 15	167	57 16	214	51 59	314	28 23	345
229	27 35	039	42 57	090	39 45	136	10 25	167	56 51	214	51 26	313	28 10	344
230	28 03	038	43 42	089	40 17	136	10 36	166	56 26	214	50 53	312	27 57	343
231	28 31	037	44 27	088	40 49	136	10 47	166	56 00	215	50 18	310	27 43	342
232	28 58	036	45 13	088	41 20	135	10 58	165	55 34	215	49 44	309	27 29	341
233	29 24	035	45 58	087	41 52	135	11 10	165	55 08	215	49 08	308	27 13	340
234	29 49	034	46 43	086	42 24	135	11 22	164	54 42	215	48 32	307	26 57	339
235	30 14	033	47 28	085	42 56	135	11 34	164	54 16	215	47 56	306	26 40	338
236	30 38	032	48 13	085	43 29	135	11 47	163	53 50	216	47 18	304	26 23	337
237	31 02	031	48 58	084	44 01	135	12 01	163	53 23	216	46 41	303	26 04	336
238	31 25	030	49 43	083	44 33	134	12 14	162	52 57	216	46 03	302	25 45	335
239	31 47	029	50 28	082	45 06	134	12 28	162	52 30	216	45 24	301	25 25	334
	Rasalhague		◆ALTAIR		Peacock		◆ACHERNAR		ACRUX		◆SPICA		ARCTURUS	
240	32 08	028	17 21	061	45 38	134	12 43	161	52 03	216	44 45	300	25 05	333
241	32 29	026	18 00	060	46 11	134	12 58	161	51 37	216	44 06	299	24 44	332
242	32 49	025	18 40	059	46 43	134	13 13	160	51 10	216	43 26	298	24 22	331
243	33 08	024	19 18	059	47 16	134	13 28	160	50 43	217	42 46	297	24 00	330
244	33 26	023	19 57	058	47 49	134	13 44	159	50 16	217	42 05	296	23 36	329
245	33 43	022	20 35	057	48 21	134	14 00	159	49 49	217	41 25	295	23 13	328
246	34 00	021	21 13	056	48 54	134	14 17	158	49 22	217	40 44	294	22 48	327
247	34 16	020	21 50	055	49 27	134	14 34	158	48 54	217	40 02	293	22 23	326
248	34 31	019	22 27	055	50 00	133	14 51	157	48 27	217	39 20	292	21 58	325
249	34 45	018	23 04	054	50 33	133	15 09	157	48 00	217	38 38	292	21 32	324
250	34 58	016	23 40	053	51 06	133	15 27	156	47 33	217	37 56	291	21 05	323
251	35 10	015	24 16	052	51 38	133	15 45	156	47 06	217	37 14	290	20 38	322
252	35 22	014	24 52	051	52 11	133	16 04	155	46 39	217	36 31	289	20 10	322
253	35 32	013	25 27	050	52 44	133	16 23	155	46 11	217	35 48	288	19 41	321
254	35 42	012	26 01	049	53 17	133	16 42	155	45 44	217	35 05	287	19 12	320
	Rasalhague		◆ALTAIR		FOMALHAUT		◆ACHERNAR		RIGIL KENT.		◆SPICA		Alphecca	
255	35 50	010	26 35	048	19 24	113	17 02	154	60 52	215	34 22	287	19 25	340
256	35 58	009	27 09	048	20 05	112	17 22	154	60 26	216	33 38	286	19 09	339
257	36 05	008	27 42	047	20 47	112	17 42	153	60 00	216	32 54	285	18 53	338
258	36 11	007	28 15	046	21 30	111	18 03	153	59 33	216	32 11	284	18 36	337
259	36 16	006	28 47	045	22 12	111	18 24	152	59 06	217	31 27	283	18 18	336
260	36 20	004	29 19	044	22 54	110	18 45	152	58 38	217	30 42	283	17 59	335
261	36 23	003	29 50	043	23 37	110	19 06	151	58 11	218	29 58	282	17 40	334
262	36 25	002	30 20	042	24 20	109	19 28	151	57 43	218	29 14	281	17 20	334
263	36 26	001	30 50	041	25 02	109	19 50	151	57 15	218	28 29	280	17 00	333
264	36 26	000	31 19	040	25 45	108	20 12	150	56 47	218	27 45	280	16 39	332
265	36 25	358	31 48	039	26 29	107	20 35	150	56 19	219	27 00	279	16 17	331
266	36 24	357	32 16	038	27 12	107	20 58	149	55 51	219	26 15	278	15 55	330
267	36 21	356	32 44	037	27 55	106	21 21	149	55 22	219	25 30	278	15 32	329
268	36 17	355	33 10	036	28 39	106	21 45	149	54 54	219	24 46	277	15 08	329
269	36 13	354	33 37	035	29 22	105	22 08	148	54 25	219	24 01	276	14 45	328

LHA ♈	Hc	Zn	Hc	Zn	Hc	Zn	Hc	Zn	Hc	Zn	Hc	Zn	Hc	Zn
	ALTAIR		Enif		◆FOMALHAUT		ACHERNAR		◆RIGIL KENT.		ANTARES		◆Rasalhague	
270	34 02	034	17 41	059	30 06	105	22 32	148	53 56	220	66 15	301	36 07	352
271	34 27	032	18 19	058	30 50	104	22 57	147	53 28	220	65 36	299	36 01	351
272	34 51	031	18 58	057	31 34	104	23 21	147	52 59	220	64 56	298	35 53	350
273	35 14	030	19 36	057	32 18	103	23 46	147	52 30	220	64 16	296	35 45	349
274	35 36	029	20 13	056	33 02	103	24 11	146	52 01	220	63 35	295	35 36	348
275	35 58	028	20 51	055	33 46	102	24 36	146	51 31	220	62 54	294	35 25	346
276	36 19	027	21 28	054	34 31	101	25 02	145	51 02	220	62 12	293	35 14	345
277	36 39	026	22 04	053	35 15	101	25 28	145	50 33	220	61 30	291	35 02	344
278	36 58	025	22 40	052	36 00	100	25 54	145	50 04	220	60 48	290	34 49	343
279	37 16	023	23 16	052	36 44	100	26 20	144	49 35	220	60 05	289	34 36	342
280	37 34	022	23 51	051	37 29	099	26 46	144	49 05	220	59 23	288	34 21	341
281	37 51	021	24 26	050	38 14	099	27 13	144	48 36	220	58 39	287	34 05	339
282	38 06	020	25 01	049	38 59	098	27 40	143	48 07	220	57 56	286	33 49	338
283	38 21	019	25 35	048	39 43	097	28 07	143	47 38	220	57 12	285	33 32	337
284	38 35	017	26 08	047	40 28	097	28 34	143	47 08	220	56 29	284	33 14	336
	ALTAIR		Enif		◆FOMALHAUT		ACHERNAR		◆RIGIL KENT.		ANTARES		◆Rasalhague	
285	38 48	016	26 41	046	41 13	096	29 02	142	46 39	220	55 45	283	32 55	335
286	39 00	015	27 14	045	41 58	096	29 30	142	46 10	220	55 00	282	32 36	334
287	39 12	014	27 45	044	42 43	095	29 58	142	45 41	220	54 16	281	32 15	333
288	39 22	012	28 17	044	43 29	095	30 26	141	45 12	220	53 32	281	31 54	332
289	39 31	011	28 48	043	44 14	094	30 54	141	44 43	220	52 47	280	31 33	331
290	39 39	010	29 18	042	44 59	093	31 23	141	44 14	220	52 02	279	31 10	330
291	39 46	009	29 48	041	45 44	093	31 51	141	43 45	220	51 18	278	30 47	329
292	39 53	007	30 17	040	46 29	092	32 20	140	43 16	220	50 33	277	30 23	328
293	39 58	006	30 46	039	47 15	091	32 49	140	42 47	219	49 48	277	29 58	327
294	40 02	005	31 14	038	48 00	091	33 18	140	42 18	219	49 03	276	29 33	325
295	40 05	003	31 41	037	48 45	090	33 48	139	41 50	219	48 18	275	29 07	324
296	40 07	002	32 08	036	49 30	090	34 17	139	41 21	219	47 33	274	28 40	323
297	40 09	001	32 34	035	50 16	089	34 47	139	40 52	219	46 47	274	28 13	323
298	40 09	000	32 59	033	51 01	088	35 17	139	40 24	219	46 02	273	27 45	322
299	40 08	358	33 24	032	51 46	087	35 47	138	39 56	219	45 17	272	27 17	321
	◆Enif		FOMALHAUT		◆ACHERNAR		RIGIL KENT.		◆ANTARES		Rasalhague		ALTAIR	
300	33 47	031	52 31	087	36 17	138	39 27	218	44 32	272	26 48	320	40 06	357
301	34 11	030	53 17	086	36 47	138	38 59	218	43 46	271	26 18	319	40 03	356
302	34 33	029	54 02	085	37 18	138	38 31	218	43 01	270	25 48	318	39 59	354
303	34 55	028	54 47	085	37 48	137	38 03	218	42 16	270	25 17	317	39 54	353
304	35 15	027	55 32	084	38 19	137	37 36	218	41 31	269	24 46	316	39 48	352
305	35 36	026	56 17	083	38 50	137	37 08	218	40 45	268	24 14	315	39 41	351
306	35 55	025	57 02	082	39 21	137	36 40	217	40 00	268	23 42	314	39 33	349
307	36 13	023	57 47	081	39 52	137	36 13	217	39 15	267	23 09	313	39 24	348
308	36 31	022	58 31	081	40 23	136	35 46	217	38 30	266	22 36	312	39 14	347
309	36 47	021	59 16	080	40 54	136	35 19	217	37 44	266	22 02	312	39 03	345
310	37 03	020	60 00	079	41 26	136	34 52	216	36 59	265	21 28	311	38 52	344
311	37 18	019	60 45	078	41 57	136	34 25	216	36 14	265	20 54	310	38 39	343
312	37 32	017	61 29	077	42 29	136	33 58	216	35 29	264	20 19	309	38 25	342
313	37 45	016	62 13	076	43 01	136	33 32	216	34 44	263	19 43	308	38 11	341
314	37 58	015	62 57	075	43 32	135	33 05	215	33 59	263	19 07	307	37 55	339
	Enif		◆Diphda		ACHERNAR		Miaplacidus		◆RIGIL KENT.		ANTARES		◆ALTAIR	
315	38 09	014	37 18	082	44 04	135	20 43	181	32 39	215	33 14	262	37 39	338
316	38 19	013	38 03	081	44 36	135	20 43	181	32 13	215	32 29	262	37 21	337
317	38 29	011	38 47	080	45 08	135	20 42	180	31 47	215	31 45	261	37 03	336
318	38 37	010	39 32	079	45 40	135	20 42	180	31 21	214	31 00	260	36 44	335
319	38 44	009	40 16	078	46 13	135	20 42	180	30 56	214	30 15	260	36 24	333
320	38 51	008	41 00	078	46 45	135	20 42	179	30 31	214	29 31	259	36 04	332
321	38 56	006	41 45	077	47 17	135	20 43	179	30 05	214	28 46	259	35 42	331
322	39 01	005	42 29	076	47 49	134	20 44	179	29 41	213	28 02	258	35 20	330
323	39 04	004	43 13	075	48 22	134	20 45	178	29 16	213	27 18	258	34 57	329
324	39 07	003	43 56	074	48 54	134	20 47	178	28 51	213	26 34	257	34 33	328
325	39 08	001	44 40	074	49 26	134	20 48	178	28 27	212	25 49	256	34 09	327
326	39 09	000	45 23	073	49 59	134	20 51	177	28 03	212	25 06	256	33 43	326
327	39 08	359	46 06	072	50 31	134	20 53	177	27 39	212	24 22	255	33 18	325
328	39 07	357	46 49	071	51 04	134	20 56	176	27 15	211	23 38	255	32 51	324
329	39 04	356	47 32	070	51 36	134	20 59	176	26 52	211	22 54	254	32 24	322
	◆Diphda		ACHERNAR		◆Miaplacidus		RIGIL KENT.		Nunki		◆ALTAIR		Enif	
330	48 14	069	52 09	134	21 02	176	26 28	211	49 20	276	31 56	321	39 01	355
331	48 56	068	52 41	134	21 05	175	26 05	210	48 35	276	31 27	320	38 56	354
332	49 38	067	53 13	134	21 09	175	25 43	210	47 50	275	30 58	319	38 51	352
333	50 20	066	53 46	134	21 13	175	25 20	210	47 05	274	30 28	318	38 44	351
334	51 01	065	54 18	134	21 18	174	24 58	209	46 19	273	29 58	317	38 37	350
335	51 42	064	54 50	135	21 23	174	24 36	209	45 34	273	29 27	316	38 28	349
336	52 22	063	55 23	135	21 28	173	24 14	209	44 49	272	28 56	316	38 19	347
337	53 02	062	55 55	135	21 33	173	23 52	208	44 04	271	28 24	315	38 09	346
338	53 42	061	56 27	135	21 38	173	23 31	208	43 18	271	27 51	314	37 57	345
339	54 21	059	56 59	135	21 44	172	23 10	208	42 33	270	27 18	313	37 45	344
340	55 00	058	57 31	135	21 50	172	22 49	207	41 48	269	26 45	312	37 32	342
341	55 38	057	58 03	135	21 57	172	22 28	207	41 03	269	26 11	311	37 18	341
342	56 16	056	58 35	136	22 03	171	22 08	206	40 17	268	25 36	310	37 03	340
343	56 53	054	59 06	136	22 10	171	21 48	206	39 32	267	25 01	309	36 47	339
344	57 29	053	59 38	136	22 18	171	21 28	206	38 47	267	24 26	308	36 30	338
	◆Alpheratz		Diphda		◆CANOPUS		RIGIL KENT.		◆Nunki		ALTAIR		Enif	
345	18 11	016	58 05	052	20 59	143	21 09	205	38 02	266	23 50	307	36 13	337
346	18 23	015	58 40	050	21 27	142	20 49	205	37 16	266	23 14	307	35 54	335
347	18 34	014	59 15	049	21 55	142	20 30	205	36 31	265	22 37	306	35 35	334
348	18 45	013	59 49	047	22 23	141	20 12	204	35 46	264	22 00	305	35 15	333
349	18 54	012	60 21	046	22 51	141	19 53	204	35 01	264	21 23	304	34 54	332
350	19 03	011	60 53	044	23 20	141	19 35	203	34 16	263	20 45	303	34 32	331
351	19 12	010	61 24	042	23 49	140	19 18	203	33 31	263	20 07	302	34 10	330
352	19 19	009	61 54	041	24 18	140	19 00	202	32 46	262	19 29	302	33 47	329
353	19 26	008	62 23	039	24 47	139	18 43	202	32 02	261	18 50	301	33 23	328
354	19 33	007	62 51	037	25 17	139	18 26	202	31 17	261	18 11	300	32 58	327
355	19 38	007	63 18	035	25 47	139	18 10	201	30 32	260	17 32	299	32 33	325
356	19 43	006	63 43	033	26 17	138	17 53	201	29 48	260	16 52	298	32 07	324
357	19 47	005	64 08	031	26 47	138	17 38	200	29 03	259	16 12	298	31 40	323
358	19 50	004	64 31	029	27 17	138	17 22	200	28 19	259	15 32	297	31 13	322
359	19 53	003	64 52	027	27 48	137	17 07	199	27 34	258	14 51	296	30 45	321

LHA ♈	Hc	Zn	Hc	Zn	Hc	Zn	Hc	Zn	Hc	Zn	Hc	Zn	Hc	Zn
	Alpheratz		Hamal		◆RIGEL		CANOPUS		◆RIGIL KENT.		Peacock		◆Enif	
0	18 55	002	18 17	031	13 58	089	29 03	136	17 48	199	53 13	228	29 30	321
1	18 56	001	18 40	030	14 42	088	29 33	136	17 34	199	52 40	228	29 01	320
2	18 56	000	19 01	029	15 27	087	30 04	136	17 20	198	52 07	228	28 32	319
3	18 56	359	19 23	028	16 11	086	30 36	135	17 06	198	51 34	227	28 03	318
4	18 55	358	19 43	027	16 56	086	31 07	135	16 53	197	51 01	227	27 32	317
5	18 53	357	20 03	026	17 40	085	31 39	135	16 40	197	50 29	227	27 02	316
6	18 51	356	20 22	025	18 25	084	32 11	134	16 27	196	49 56	227	26 30	315
7	18 47	355	20 41	024	19 09	084	32 43	134	16 14	196	49 23	227	25 58	314
8	18 43	354	20 59	023	19 53	083	33 15	134	16 02	196	48 50	227	25 26	313
9	18 39	354	21 16	022	20 37	082	33 47	133	15 50	195	48 18	227	24 53	312
10	18 33	353	21 33	021	21 22	082	34 20	133	15 39	195	47 45	227	24 20	311
11	18 27	352	21 49	020	22 06	081	34 53	133	15 28	194	47 12	227	23 46	310
12	18 21	351	22 04	020	22 50	080	35 25	132	15 17	194	46 40	227	23 12	310
13	18 13	350	22 18	019	23 33	079	35 58	132	15 07	193	46 07	227	22 38	309
14	18 05	349	22 32	018	24 17	079	36 32	132	14 57	193	45 35	226	22 03	308
	◆Hamal		RIGEL		◆CANOPUS		RIGIL KENT.		◆Peacock		FOMALHAUT		Alpheratz	
15	22 45	017	25 01	078	37 05	131	14 47	192	45 03	226	62 27	287	17 56	348
16	22 58	016	25 44	077	37 39	131	14 38	192	44 31	226	61 44	286	17 47	347
17	23 10	015	26 28	076	38 12	131	14 29	191	43 59	226	61 01	285	17 36	346
18	23 20	014	27 11	076	38 46	131	14 21	191	43 27	226	60 18	284	17 26	345
19	23 31	013	27 54	075	39 20	130	14 13	190	42 55	226	59 34	283	17 14	345
20	23 40	012	28 37	074	39 54	130	14 05	190	42 23	225	58 51	282	17 02	344
21	23 49	011	29 20	073	40 28	130	13 57	189	41 51	225	58 07	281	16 49	343
22	23 57	010	30 03	073	41 03	129	13 50	189	41 19	225	57 23	280	16 35	342
23	24 04	009	30 45	072	41 37	129	13 44	188	40 48	225	56 39	279	16 21	341
24	24 10	008	31 27	071	42 12	129	13 37	188	40 17	225	55 55	278	16 06	340
25	24 16	007	32 10	070	42 47	129	13 31	187	39 45	224	55 11	277	15 51	339
26	24 21	006	32 51	069	43 22	128	13 26	187	39 14	224	54 27	277	15 35	338
27	24 25	005	33 33	068	43 57	128	13 21	186	38 43	224	53 42	276	15 18	338
28	24 28	004	34 14	068	44 32	128	13 16	186	38 12	224	52 58	275	15 00	337
29	24 31	003	34 55	067	45 07	128	13 12	185	37 42	223	52 14	274	14 42	336
	◆Hamal		RIGEL		SIRIUS		◆CANOPUS		RIGIL KENT.		◆Peacock		FOMALHAUT	
30	24 32	002	35 36	066	24 55	090	45 42	128	13 08	185	37 11	223	51 29	274
31	24 33	001	36 17	065	25 40	090	46 17	127	13 04	184	36 41	223	50 45	273
32	24 34	000	36 57	064	26 25	089	46 53	127	13 01	184	36 10	223	50 00	272
33	24 33	359	37 37	063	27 09	088	47 29	127	12 58	183	35 40	222	49 15	271
34	24 32	358	38 17	062	27 54	088	48 04	127	12 55	183	35 10	222	48 31	271
35	24 29	357	38 56	061	28 38	087	48 40	127	12 53	182	34 41	222	47 46	270
36	24 26	356	39 35	060	29 23	086	49 16	126	12 52	182	34 11	222	47 02	269
37	24 23	355	40 13	059	30 07	086	49 52	126	12 50	181	33 41	221	46 17	269
38	24 18	354	40 52	058	30 52	085	50 28	126	12 50	181	33 12	221	45 32	268
39	24 13	353	41 29	057	31 36	084	51 04	126	12 49	180	32 43	221	44 48	267
40	24 07	352	42 07	056	32 20	084	51 40	126	12 49	180	32 14	220	44 03	267
41	24 00	351	42 43	055	33 05	083	52 16	126	12 49	179	31 45	220	43 19	266
42	23 52	350	43 20	054	33 49	082	52 52	126	12 50	179	31 17	220	42 34	266
43	23 44	349	43 56	053	34 33	081	53 28	126	12 51	178	30 48	219	41 50	265
44	23 35	348	44 31	052	35 17	081	54 04	126	12 52	178	30 20	219	41 06	264
	◆RIGEL		SIRIUS		CANOPUS		◆RIGIL KENT.		Peacock		◆FOMALHAUT		Hamal	
45	45 06	051	36 01	080	54 41	125	12 54	177	29 52	219	40 21	264	23 25	347
46	45 40	050	36 45	079	55 17	125	12 56	177	29 25	218	39 37	263	23 14	346
47	46 14	049	37 28	078	55 53	125	12 59	176	28 57	218	38 53	263	23 03	345
48	46 47	047	38 12	078	56 30	125	13 02	176	28 30	218	38 08	262	22 51	344
49	47 20	046	38 56	077	57 06	125	13 05	175	28 02	217	37 24	261	22 38	343
50	47 51	045	39 39	076	57 42	125	13 09	175	27 36	217	36 40	261	22 25	342
51	48 22	044	40 22	075	58 19	125	13 13	174	27 09	217	35 56	260	22 10	341
52	48 53	042	41 05	074	58 55	125	13 18	174	26 42	216	35 12	260	21 56	340
53	49 23	041	41 48	073	59 31	125	13 23	173	26 16	216	34 29	259	21 40	339
54	49 51	040	42 30	073	60 08	126	13 28	173	25 50	215	33 45	259	21 24	338
55	50 19	038	43 13	072	60 44	126	13 34	172	25 24	215	33 01	258	21 07	337
56	50 47	037	43 55	071	61 20	126	13 40	172	24 59	215	32 18	257	20 49	336
57	51 13	036	44 37	070	61 56	126	13 46	171	24 34	214	31 34	257	20 31	335
58	51 38	034	45 19	069	62 32	126	13 53	171	24 09	214	30 51	256	20 12	334
59	52 03	033	46 00	068	63 09	126	14 00	170	23 44	214	30 08	256	19 52	333
	ALDEBARAN		◆SIRIUS		Suhail		◆RIGIL KENT.		ACHERNAR		◆Diphda		Hamal	
60	30 55	010	46 41	067	35 34	120	14 08	170	62 49	224	42 00	285	19 32	333
61	31 03	009	47 22	066	36 13	119	14 16	170	62 18	224	41 16	284	19 11	332
62	31 09	008	48 03	065	36 52	119	14 24	169	61 47	224	40 33	283	18 49	331
63	31 15	007	48 43	064	37 31	118	14 32	169	61 16	225	39 49	282	18 27	330
64	31 20	006	49 23	063	38 11	118	14 41	168	60 44	225	39 06	281	18 05	329
65	31 23	004	50 03	062	38 50	118	14 51	168	60 13	225	38 22	280	17 41	328
66	31 26	003	50 42	061	39 30	117	15 01	167	59 41	225	37 38	280	17 18	327
67	31 28	002	51 20	060	40 10	117	15 11	167	59 09	226	36 54	279	16 53	326
68	31 30	001	51 59	058	40 49	116	15 21	166	58 38	226	36 10	278	16 28	326
69	31 30	000	52 36	057	41 29	116	15 32	166	58 06	226	35 26	277	16 03	325
70	31 30	359	53 14	056	42 10	116	15 43	165	57 34	226	34 42	277	15 37	324
71	31 28	358	53 50	055	42 50	115	15 55	165	57 01	226	33 57	276	15 10	323
72	31 26	357	54 26	054	43 30	115	16 07	164	56 29	226	33 13	275	14 43	322
73	31 23	355	55 02	052	44 11	114	16 19	164	55 57	226	32 28	275	14 15	321
74	31 19	354	55 37	051	44 51	114	16 31	163	55 25	227	31 44	274	13 47	321
	BETELGEUSE		◆SIRIUS		Suhail		◆RIGIL KENT.		ACHERNAR		◆Diphda		ALDEBARAN	
75	39 01	018	56 11	050	45 32	114	16 44	163	54 52	227	30 59	273	31 14	353
76	39 14	016	56 45	048	46 13	113	16 58	162	54 20	227	30 15	272	31 08	352
77	39 27	015	57 18	047	46 54	113	17 11	162	53 47	227	29 30	272	31 01	351
78	39 38	014	57 50	045	47 35	113	17 25	162	53 15	227	28 46	271	30 54	350
79	39 48	013	58 21	044	48 16	112	17 39	161	52 42	227	28 01	270	30 46	349
80	39 57	011	58 51	042	48 58	112	17 54	161	52 10	227	27 17	270	30 37	348
81	40 05	010	59 21	041	49 39	112	18 09	160	51 38	227	26 32	269	30 27	347
82	40 13	009	59 49	039	50 20	111	18 24	160	51 05	227	25 47	268	30 16	345
83	40 19	007	60 17	037	51 02	111	18 40	159	50 33	227	25 03	268	30 04	344
84	40 24	006	60 43	035	51 44	111	18 55	159	50 00	227	24 18	267	29 52	343
85	40 29	005	61 08	034	52 26	110	19 12	159	49 28	226	23 34	266	29 39	342
86	40 32	004	61 32	032	53 07	110	19 28	158	48 56	226	22 49	266	29 25	341
87	40 34	002	61 55	030	53 49	110	19 45	158	48 23	226	22 05	265	29 10	340
88	40 35	001	62 17	028	54 31	109	20 02	157	47 51	226	21 20	265	28 54	339
89	40 36	000	62 37	026	55 14	109	20 19	157	47 19	226	20 36	264	28 38	338

LHA ♈	Hc	Zn	Hc	Zn	Hc	Zn	Hc	Zn	Hc	Zn	Hc	Zn	Hc	Zn
	PROCYON		Alphard		◆Suhail		RIGIL KENT.		◆ACHERNAR		RIGEL		◆BETELGEUSE	
90	37 39	032	33 41	069	55 56	109	20 37	156	46 47	226	54 44	340	40 35	358
91	38 02	031	34 22	068	56 38	108	20 55	156	46 15	226	54 28	339	40 33	357
92	38 25	029	35 03	067	57 20	108	21 13	156	45 43	226	54 11	337	40 30	356
93	38 46	028	35 44	066	58 03	108	21 32	155	45 11	226	53 53	335	40 26	354
94	39 07	027	36 25	066	58 45	107	21 51	155	44 39	225	53 34	334	40 22	353
95	39 27	026	37 06	065	59 28	107	22 10	154	44 07	225	53 14	332	40 16	352
96	39 46	025	37 46	064	60 11	107	22 29	154	43 36	225	52 52	331	40 09	351
97	40 04	023	38 26	063	60 53	106	22 49	154	43 04	225	52 30	329	40 01	349
98	40 21	022	39 05	062	61 36	106	23 09	153	42 33	225	52 07	328	39 52	348
99	40 38	021	39 44	061	62 19	106	23 29	153	42 01	225	51 42	326	39 42	347
100	40 53	020	40 23	060	63 02	106	23 50	152	41 30	224	51 17	325	39 32	345
101	41 08	018	41 01	059	63 45	105	24 11	152	40 59	224	50 51	323	39 20	344
102	41 21	017	41 39	058	64 28	105	24 32	152	40 28	224	50 24	322	39 07	343
103	41 34	016	42 17	057	65 11	105	24 53	151	39 57	224	49 56	321	38 54	342
104	41 45	014	42 54	056	65 54	105	25 14	151	39 26	224	49 27	319	38 39	340
	PROCYON		REGULUS		◆Gienah		RIGIL KENT.		◆ACHERNAR		RIGEL		◆BETELGEUSE	
105	41 56	013	20 53	050	19 45	096	25 36	151	38 56	223	48 57	318	38 24	339
106	42 06	012	21 27	049	20 30	095	25 58	150	38 25	223	48 27	317	38 08	338
107	42 14	010	22 01	048	21 14	095	26 20	150	37 55	223	47 56	315	37 51	337
108	42 22	009	22 34	047	21 58	094	26 43	150	37 25	223	47 24	314	37 33	336
109	42 28	008	23 06	047	22 43	094	27 06	149	36 55	222	46 52	313	37 14	334
110	42 34	006	23 38	046	23 27	093	27 28	149	36 25	222	46 19	312	36 54	333
111	42 38	005	24 10	045	24 12	092	27 52	149	35 55	222	45 45	310	36 34	332
112	42 42	004	24 41	044	24 57	092	28 15	148	35 25	222	45 11	309	36 12	331
113	42 44	002	25 12	043	25 41	091	28 39	148	34 56	221	44 36	308	35 50	330
114	42 45	001	25 42	042	26 26	090	29 02	148	34 26	221	44 01	307	35 28	329
115	42 46	000	26 11	041	27 10	090	29 26	147	33 57	221	43 25	306	35 04	328
116	42 45	358	26 40	040	27 55	089	29 51	147	33 28	220	42 49	305	34 40	326
117	42 43	357	27 09	039	28 39	088	30 15	147	32 59	220	42 12	304	34 15	325
118	42 40	356	27 37	038	29 24	088	30 40	146	32 31	220	41 35	303	33 49	324
119	42 36	354	28 04	037	30 09	087	31 05	146	32 02	219	40 57	302	33 23	323
	REGULUS		◆SPICA		RIGIL KENT.		◆ACHERNAR		RIGEL		BETELGEUSE		◆PROCYON	
120	28 30	036	13 55	093	31 30	146	31 34	219	40 19	301	32 56	322	42 31	353
121	28 56	035	14 39	092	31 55	145	31 06	219	39 41	300	32 28	321	42 25	352
122	29 22	034	15 24	091	32 20	145	30 38	219	39 02	299	32 00	320	42 18	350
123	29 47	033	16 08	091	32 46	145	30 10	218	38 23	298	31 31	319	42 10	349
124	30 11	032	16 53	090	33 12	145	29 43	218	37 43	297	31 01	318	42 01	348
125	30 34	031	17 37	089	33 38	144	29 16	218	37 03	296	30 31	317	41 51	346
126	30 57	030	18 22	089	34 04	144	28 49	217	36 23	295	30 01	316	41 40	345
127	31 19	029	19 07	088	34 30	144	28 22	217	35 42	294	29 29	315	41 28	344
128	31 40	028	19 51	087	34 56	144	27 55	217	35 02	293	28 58	314	41 15	342
129	32 00	027	20 36	087	35 23	143	27 29	216	34 20	292	28 25	313	41 01	341
130	32 20	026	21 20	086	35 50	143	27 03	216	33 39	292	27 53	312	40 46	340
131	32 39	025	22 05	085	36 17	143	26 37	215	32 58	291	27 19	311	40 30	339
132	32 57	024	22 49	084	36 44	143	26 11	215	32 16	290	26 46	310	40 13	337
133	33 15	022	23 33	084	37 11	142	25 45	215	31 34	289	26 12	310	39 55	336
134	33 31	021	24 18	083	37 38	142	25 20	214	30 51	288	25 37	309	39 37	335
	REGULUS		◆SPICA		RIGIL KENT.		◆ACHERNAR		CANOPUS		SIRIUS		◆PROCYON	
135	33 47	020	25 02	082	38 06	142	24 55	214	61 54	234	51 38	301	39 17	334
136	34 02	019	25 46	082	38 33	142	24 30	214	61 17	234	51 00	300	38 57	332
137	34 16	018	26 30	081	39 01	141	24 06	213	60 41	234	50 21	299	38 36	331
138	34 29	017	27 14	080	39 29	141	23 42	213	60 05	234	49 41	298	38 14	330
139	34 42	016	27 58	079	39 57	141	23 18	212	59 29	235	49 02	297	37 51	329
140	34 53	014	28 42	079	40 25	141	22 54	212	58 52	235	48 22	295	37 28	328
141	35 04	013	29 25	078	40 53	141	22 30	212	58 16	235	47 41	294	37 04	327
142	35 14	012	30 09	077	41 21	141	22 07	211	57 40	235	47 00	293	36 39	325
143	35 22	011	30 52	076	41 49	140	21 44	211	57 03	235	46 19	293	36 13	324
144	35 30	010	31 36	076	42 18	140	21 22	210	56 27	235	45 38	292	35 47	323
145	35 37	008	32 19	075	42 46	140	20 59	210	55 50	235	44 56	291	35 20	322
146	35 43	007	33 02	074	43 15	140	20 37	209	55 14	235	44 15	290	34 52	321
147	35 49	006	33 44	073	43 44	140	20 15	209	54 38	235	43 33	289	34 24	320
148	35 53	005	34 27	072	44 13	140	19 54	209	54 02	234	42 50	288	33 55	319
149	35 56	004	35 09	072	44 41	140	19 33	208	53 25	234	42 08	287	33 25	318
	REGULUS		◆SPICA		RIGIL KENT.		◆ACHERNAR		CANOPUS		SIRIUS		◆PROCYON	
150	35 59	002	35 52	071	45 10	140	19 12	208	52 49	234	41 25	286	32 55	317
151	36 00	001	36 34	070	45 39	139	18 51	207	52 13	234	40 42	285	32 24	316
152	36 00	000	37 15	069	46 08	139	18 31	207	51 37	234	39 59	284	31 52	315
153	36 00	359	37 57	068	46 37	139	18 11	206	51 01	234	39 16	284	31 21	314
154	35 59	358	38 38	067	47 07	139	17 51	206	50 25	234	38 32	283	30 48	313
155	35 56	356	39 19	066	47 36	139	17 32	205	49 49	234	37 49	282	30 15	312
156	35 53	355	40 00	065	48 05	139	17 13	205	49 13	233	37 05	281	29 42	311
157	35 49	354	40 40	064	48 34	139	16 54	205	48 37	233	36 21	281	29 08	310
158	35 44	353	41 20	064	49 03	139	16 36	204	48 01	233	35 37	280	28 33	309
159	35 38	352	42 00	063	49 33	139	16 18	204	47 26	233	34 53	279	27 59	308
160	35 31	350	42 39	062	50 02	139	16 00	203	46 50	233	34 09	278	27 23	307
161	35 23	349	43 18	061	50 31	139	15 43	203	46 15	233	33 25	278	26 48	306
162	35 14	348	43 57	060	51 00	139	15 26	202	45 39	232	32 41	277	26 11	305
163	35 04	347	44 35	058	51 30	139	15 09	202	45 04	232	31 57	276	25 35	305
164	34 54	346	45 13	057	51 59	139	14 52	201	44 29	232	31 12	275	24 58	304
	SPICA		◆ANTARES		RIGIL KENT.		ACHERNAR		◆CANOPUS		SIRIUS		◆REGULUS	
165	45 50	056	22 47	106	52 28	139	14 36	201	43 54	232	30 28	275	34 42	345
166	46 27	055	23 30	105	52 57	139	14 21	200	43 19	232	29 43	274	34 30	343
167	47 04	054	24 13	105	53 26	139	14 06	200	42 44	231	28 59	273	34 17	342
168	47 39	053	24 56	104	53 55	139	13 51	199	42 09	231	28 14	273	34 03	341
169	48 15	052	25 39	103	54 24	140	13 36	199	41 35	231	27 30	272	33 48	340
170	48 50	051	26 23	103	54 53	140	13 22	198	41 00	231	26 45	271	33 32	339
171	49 24	049	27 06	102	55 22	140	13 08	198	40 26	230	26 01	271	33 15	338
172	49 57	048	27 50	102	55 51	140	12 55	197	39 51	230	25 16	270	32 58	337
173	50 30	047	28 34	101	56 19	140	12 42	197	39 17	230	24 31	269	32 40	335
174	51 02	046	29 18	100	56 48	140	12 29	196	38 43	229	23 47	269	32 21	334
175	51 34	044	30 01	100	57 16	141	12 17	196	38 10	229	23 02	268	32 01	333
176	52 04	043	30 45	099	57 45	141	12 05	195	37 36	229	22 18	267	31 41	332
177	52 34	042	31 29	099	58 13	141	11 53	195	37 02	229	21 33	267	31 20	331
178	53 04	040	32 14	098	58 41	141	11 42	194	36 29	228	20 49	266	30 58	330
179	53 32	039	32 58	097	59 09	142	11 31	194	35 56	228	20 04	265	30 35	329

LHA	Hc	Zn	Hc	Zn	Hc	Zn	Hc	Zn	Hc	Zn	Hc	Zn	Hc	Zn
♈	◆ARCTURUS		ANTARES		◆RIGIL KENT.		ACHERNAR		CANOPUS		◆Suhail		REGULUS	
180	21 16	034	33 42	097	59 36	142	11 21	193	35 23	228	58 41	253	30 12	328
181	21 41	033	34 26	096	60 04	142	11 11	193	34 50	227	57 59	252	29 48	327
182	22 05	033	35 11	096	60 31	143	11 01	192	34 17	227	57 16	252	29 23	326
183	22 29	032	35 55	095	60 58	143	10 52	192	33 45	227	56 34	252	28 58	325
184	22 52	031	36 39	094	61 24	143	10 44	191	33 12	226	55 52	251	28 32	324
185	23 14	030	37 24	094	61 51	144	10 35	191	32 40	226	55 10	251	28 05	323
186	23 36	029	38 08	093	62 17	144	10 27	190	32 08	226	54 27	251	27 38	322
187	23 57	028	38 53	092	62 43	145	10 20	189	31 36	225	53 45	250	27 10	321
188	24 17	027	39 38	092	63 08	145	10 13	189	31 05	225	53 03	250	26 42	320
189	24 37	026	40 22	091	63 34	146	10 06	188	30 33	225	52 22	250	26 13	319
190	24 56	025	41 07	091	63 58	146	10 00	188	30 02	224	51 40	249	25 43	318
191	25 15	024	41 51	090	64 23	147	09 54	187	29 31	224	50 58	249	25 13	317
192	25 33	023	42 36	089	64 47	148	09 48	187	29 00	224	50 17	249	24 43	316
193	25 50	022	43 20	088	65 11	148	09 43	186	28 29	223	49 35	248	24 12	315
194	26 06	021	44 05	088	65 34	149	09 39	186	27 59	223	48 54	248	23 40	314
	◆ARCTURUS		ANTARES		◆Peacock		ACHERNAR		CANOPUS		◆Suhail		REGULUS	
195	26 21	020	44 50	087	24 18	146	09 34	185	27 29	222	48 12	248	23 08	314
196	26 36	019	45 34	086	24 43	146	09 31	185	26 59	222	47 31	247	22 35	313
197	26 50	018	46 19	086	25 08	145	09 27	184	26 29	222	46 50	247	22 02	312
198	27 04	017	47 03	085	25 34	145	09 24	183	26 00	221	46 09	247	21 29	311
199	27 16	016	47 47	084	26 00	144	09 22	183	25 30	221	45 28	246	20 55	310
200	27 28	015	48 32	083	26 26	144	09 20	182	25 01	220	44 48	246	20 21	309
201	27 39	014	49 16	083	26 52	144	09 18	182	24 32	220	44 07	245	19 46	308
202	27 49	013	50 00	082	27 19	143	09 17	181	24 04	220	43 26	245	19 11	308
203	27 58	012	50 44	081	27 46	143	09 16	181	23 36	219	42 46	245	18 35	307
204	28 07	011	51 28	080	28 13	143	09 16	180	23 08	219	42 06	244	17 59	306
205	28 15	010	52 12	079	28 40	142	09 16	180	22 40	218	41 26	244	17 23	305
206	28 22	008	52 56	079	29 07	142	09 16	179	22 12	218	40 46	244	16 46	304
207	28 28	007	53 40	078	29 35	142	09 17	179	21 45	218	40 06	243	16 09	303
208	28 33	006	54 23	077	30 03	141	09 19	178	21 18	217	39 26	243	15 32	303
209	28 38	005	55 06	076	30 31	141	09 20	177	20 51	217	38 46	242	14 54	302
	◆ARCTURUS		ANTARES		◆Peacock		ACHERNAR		CANOPUS		◆Suhail		SPICA	
210	28 41	004	55 49	075	30 59	141	09 22	177	20 25	216	38 07	242	58 12	344
211	28 44	003	56 32	074	31 27	140	09 25	176	19 58	216	37 28	242	57 58	342
212	28 46	002	57 15	073	31 56	140	09 28	176	19 33	215	36 49	241	57 44	340
213	28 47	001	57 58	072	32 25	140	09 32	175	19 07	215	36 10	241	57 28	338
214	28 47	000	58 40	071	32 54	139	09 35	175	18 42	214	35 31	240	57 11	336
215	28 47	359	59 22	070	33 23	139	09 40	174	18 17	214	34 52	240	56 52	335
216	28 46	358	60 04	069	33 52	139	09 44	174	17 52	213	34 13	240	56 33	333
217	28 43	357	60 45	068	34 22	138	09 50	173	17 27	213	33 35	239	56 12	331
218	28 40	356	61 26	066	34 52	138	09 55	173	17 03	213	32 57	239	55 50	330
219	28 36	354	62 07	065	35 22	138	10 01	172	16 39	212	32 19	238	55 27	328
220	28 32	353	62 47	064	35 52	138	10 07	171	16 16	212	31 41	238	55 03	327
221	28 26	352	63 27	062	36 22	137	10 14	171	15 53	211	31 03	237	54 38	325
222	28 20	351	64 06	061	36 52	137	10 21	170	15 30	211	30 26	237	54 12	324
223	28 13	350	64 45	060	37 23	137	10 29	170	15 07	210	29 48	237	53 45	322
224	28 05	349	65 23	058	37 53	136	10 37	169	14 45	210	29 11	236	53 17	321
	Rasalhague		◆Nunki		Peacock		ACHERNAR		◆ACRUX		SPICA		◆ARCTURUS	
225	24 53	042	39 57	091	38 24	136	10 46	169	59 21	214	52 48	319	27 56	348
226	25 23	041	40 42	091	38 55	136	10 54	168	58 56	214	52 18	318	27 46	347
227	25 52	040	41 27	090	39 26	136	11 04	168	58 31	214	51 48	316	27 36	346
228	26 20	039	42 11	089	39 57	136	11 13	167	58 06	215	51 17	315	27 25	345
229	26 48	038	42 56	089	40 28	135	11 23	167	57 40	215	50 45	314	27 13	344
230	27 16	038	43 40	088	41 00	135	11 34	166	57 15	215	50 13	312	27 00	343
231	27 43	037	44 25	087	41 31	135	11 45	166	56 49	215	49 39	311	26 46	342
232	28 09	036	45 09	087	42 03	135	11 56	165	56 23	216	49 05	310	26 32	341
233	28 34	035	45 54	086	42 35	135	12 08	165	55 57	216	48 31	309	26 17	340
234	28 59	034	46 38	085	43 06	134	12 20	164	55 31	216	47 56	308	26 01	339
235	29 24	033	47 23	084	43 38	134	12 32	164	55 05	216	47 20	306	25 45	338
236	29 47	031	48 07	084	44 10	134	12 45	163	54 38	216	46 44	305	25 28	337
237	30 10	030	48 51	083	44 43	134	12 58	163	54 12	217	46 08	304	25 10	336
238	30 33	029	49 36	082	45 15	134	13 11	162	53 45	217	45 30	303	24 51	335
239	30 54	028	50 20	081	45 47	133	13 25	162	53 18	217	44 53	302	24 32	334
	Rasalhague		◆ALTAIR		Peacock		◆ACHERNAR		ACRUX		◆SPICA		ARCTURUS	
240	31 15	027	16 52	061	46 20	133	13 40	161	52 51	217	44 15	301	24 12	333
241	31 35	026	17 30	060	46 52	133	13 54	161	52 25	217	43 36	300	23 51	332
242	31 54	025	18 09	059	47 25	133	14 09	160	51 58	217	42 57	299	23 30	331
243	32 13	024	18 47	058	47 57	133	14 25	160	51 31	217	42 18	298	23 08	330
244	32 31	023	19 25	058	48 30	133	14 40	159	51 04	217	41 39	297	22 45	329
245	32 48	022	20 02	057	49 02	133	14 56	159	50 36	217	40 59	296	22 22	328
246	33 04	021	20 39	056	49 35	133	15 13	158	50 09	217	40 19	295	21 58	327
247	33 19	020	21 16	055	50 08	133	15 29	158	49 42	218	39 38	294	21 34	326
248	33 34	018	21 52	054	50 41	133	15 47	157	49 15	218	38 57	293	21 09	325
249	33 47	017	22 28	053	51 14	133	16 04	157	48 48	218	38 16	292	20 43	324
250	34 00	016	23 04	053	51 46	132	16 22	156	48 21	218	37 35	291	20 17	324
251	34 12	015	23 39	052	52 19	132	16 40	156	47 54	218	36 53	291	19 50	323
252	34 23	014	24 14	051	52 52	132	16 58	155	47 26	218	36 11	290	19 23	322
253	34 34	013	24 48	050	53 25	132	17 17	155	46 59	217	35 29	289	18 55	321
254	34 43	012	25 22	049	53 58	132	17 36	154	46 32	217	34 47	288	18 26	320
	Rasalhague		◆ALTAIR		FOMALHAUT		◆ACHERNAR		RIGIL KENT.		◆SPICA		Alphecca	
255	34 51	010	25 56	048	19 47	113	17 56	154	61 41	216	34 04	287	18 29	340
256	34 59	009	26 28	047	20 28	112	18 16	153	61 15	217	33 21	286	18 13	339
257	35 05	008	27 01	046	21 09	111	18 36	153	60 48	217	32 39	286	17 57	338
258	35 11	007	27 33	045	21 51	111	18 56	153	60 21	218	31 56	285	17 40	337
259	35 16	006	28 04	044	22 33	110	19 17	152	59 53	218	31 12	284	17 23	336
260	35 20	004	28 35	043	23 15	110	19 38	152	59 26	218	30 29	283	17 04	335
261	35 23	003	29 06	042	23 57	109	19 59	151	58 58	219	29 46	282	16 46	335
262	35 25	002	29 35	041	24 39	109	20 20	151	58 30	219	29 02	282	16 26	334
263	35 26	001	30 05	040	25 21	108	20 42	150	58 02	219	28 18	281	16 06	333
264	35 26	000	30 33	039	26 04	107	21 04	150	57 34	219	27 34	280	15 46	332
265	35 25	358	31 01	038	26 46	107	21 27	150	57 06	220	26 50	279	15 24	331
266	35 24	357	31 29	037	27 29	106	21 50	149	56 37	220	26 06	279	15 03	330
267	35 21	356	31 55	036	28 12	106	22 13	149	56 09	220	25 22	278	14 40	329
268	35 17	355	32 22	035	28 55	105	22 36	148	55 40	220	24 38	277	14 17	329
269	35 13	354	32 47	034	29 38	105	22 59	148	55 11	220	23 54	277	13 54	328

LAT 42°S

LHA	Hc	Zn	Hc	Zn	Hc	Zn	Hc	Zn	Hc	Zn	Hc	Zn	Hc	Zn
♈	ALTAIR		Enif		◆FOMALHAUT		ACHERNAR		◆RIGIL KENT.		ANTARES		◆Rasalhague	
270	33 12	033	17 10	059	30 21	104	23 23	148	54 42	220	65 44	303	35 08	352
271	33 36	032	17 48	058	31 04	104	23 47	147	54 14	221	65 06	301	35 01	351
272	33 59	031	18 25	057	31 48	103	24 12	147	53 45	221	64 27	300	34 54	350
273	34 22	030	19 03	056	32 31	102	24 36	146	53 15	221	63 48	298	34 46	349
274	34 44	029	19 40	056	33 15	102	25 01	146	52 46	221	63 09	297	34 37	348
275	35 05	028	20 16	055	33 59	101	25 26	146	52 17	221	62 29	296	34 27	347
276	35 25	027	20 52	054	34 42	101	25 51	145	51 48	221	61 48	294	34 16	345
277	35 45	025	21 28	053	35 26	100	26 17	145	51 19	221	61 08	293	34 04	344
278	36 03	024	22 04	052	36 10	100	26 43	144	50 49	221	60 26	292	33 52	343
279	36 21	023	22 39	051	36 54	099	27 09	144	50 20	221	59 45	291	33 39	342
280	36 38	022	23 13	050	37 38	098	27 35	144	49 51	221	59 03	290	33 24	341
281	36 55	021	23 47	050	38 22	098	28 01	143	49 22	221	58 21	289	33 09	340
282	37 10	020	24 21	049	39 06	097	28 28	143	48 52	221	57 39	288	32 53	339
283	37 24	018	24 54	048	39 51	097	28 55	143	48 23	221	56 56	287	32 37	337
284	37 38	017	25 27	047	40 35	096	29 22	142	47 54	221	56 13	286	32 19	336
	ALTAIR		Enif		◆FOMALHAUT		ACHERNAR		◆RIGIL KENT.		ANTARES		◆Rasalhague	
285	37 51	016	25 59	046	41 19	095	29 49	142	47 25	221	55 30	285	32 01	335
286	38 02	015	26 31	045	42 04	095	30 17	142	46 56	221	54 47	284	31 42	334
287	38 13	013	27 03	044	42 48	094	30 45	141	46 27	221	54 03	283	31 22	333
288	38 23	012	27 33	043	43 33	094	31 13	141	45 58	221	53 20	282	31 01	332
289	38 32	011	28 04	042	44 17	093	31 41	141	45 29	221	52 36	281	30 40	331
290	38 40	010	28 33	041	45 02	092	32 09	140	45 00	220	51 52	280	30 18	330
291	38 47	008	29 02	040	45 46	092	32 37	140	44 31	220	51 08	279	29 56	329
292	38 53	007	29 31	039	46 31	091	33 06	140	44 02	220	50 24	279	29 32	328
293	38 58	006	29 59	038	47 16	090	33 35	140	43 33	220	49 40	278	29 08	327
294	39 02	005	30 26	037	48 00	090	34 04	139	43 04	220	48 56	277	28 43	326
295	39 05	003	30 53	036	48 45	089	34 33	139	42 36	220	48 12	276	28 18	325
296	39 07	002	31 19	035	49 29	088	35 03	139	42 07	220	47 27	275	27 52	324
297	39 09	001	31 44	034	50 14	088	35 32	138	41 39	220	46 43	275	27 25	323
298	39 09	000	32 09	033	50 58	087	36 02	138	41 11	219	45 59	274	26 58	322
299	39 08	358	32 33	032	51 43	086	36 32	138	40 42	219	45 14	273	26 30	321
	◆Enif		FOMALHAUT		◆ACHERNAR		RIGIL KENT.		◆ANTARES		Rasalhague		ALTAIR	
300	32 56	031	52 27	085	37 02	138	40 14	219	44 30	273	26 02	320	39 06	357
301	33 19	030	53 12	085	37 32	137	39 46	219	43 45	272	25 33	319	39 03	356
302	33 40	029	53 56	084	38 02	137	39 18	219	43 00	271	25 03	318	38 59	354
303	34 02	028	54 40	083	38 32	137	38 51	218	42 16	271	24 33	317	38 54	353
304	34 22	027	55 25	082	39 03	137	38 23	218	41 31	270	24 03	316	38 49	352
305	34 41	025	56 09	082	39 34	136	37 55	218	40 47	269	23 32	315	38 42	351
306	35 00	024	56 53	081	40 04	136	37 28	218	40 02	269	23 00	314	38 34	349
307	35 18	023	57 37	080	40 35	136	37 01	218	39 17	268	22 28	314	38 26	348
308	35 35	022	58 21	079	41 06	136	36 34	217	38 33	267	21 55	313	38 16	347
309	35 51	021	59 04	078	41 38	136	36 07	217	37 48	267	21 22	312	38 05	346
310	36 07	020	59 48	077	42 09	135	35 40	217	37 04	266	20 49	311	37 54	344
311	36 21	018	60 31	076	42 40	135	35 13	217	36 19	265	20 15	310	37 41	343
312	36 35	017	61 14	075	43 12	135	34 47	216	35 35	265	19 41	309	37 28	342
313	36 48	016	61 57	074	43 43	135	34 20	216	34 51	264	19 06	308	37 14	341
314	37 00	015	62 40	073	44 15	135	33 54	216	34 06	263	18 31	308	36 59	340
	Enif		◆Diphda		ACHERNAR		Miaplacidus		◆RIGIL KENT.		ANTARES		◆ALTAIR	
315	37 11	014	37 09	081	44 47	135	21 43	181	33 28	216	33 22	263	36 43	338
316	37 21	012	37 53	080	45 19	134	21 43	181	33 02	215	32 38	262	36 26	337
317	37 30	011	38 36	079	45 50	134	21 42	180	32 36	215	31 54	262	36 08	336
318	37 38	010	39 20	078	46 22	134	21 42	180	32 11	215	31 10	261	35 50	335
319	37 45	009	40 04	078	46 54	134	21 42	180	31 45	214	30 26	260	35 31	334
320	37 51	007	40 47	077	47 27	134	21 42	179	31 20	214	29 42	260	35 10	333
321	37 57	006	41 31	076	47 59	134	21 43	179	30 55	214	28 58	259	34 50	331
322	38 01	005	42 14	075	48 31	134	21 44	179	30 31	214	28 14	259	34 28	330
323	38 05	004	42 57	074	49 03	134	21 45	178	30 06	213	27 30	258	34 05	329
324	38 07	002	43 40	073	49 36	134	21 47	178	29 42	213	26 47	258	33 42	328
325	38 08	001	44 22	073	50 08	133	21 48	177	29 18	213	26 03	257	33 18	327
326	38 09	000	45 05	072	50 40	133	21 50	177	28 54	212	25 20	256	32 54	326
327	38 08	359	45 47	071	51 13	133	21 53	177	28 30	212	24 37	256	32 29	325
328	38 07	357	46 29	070	51 45	133	21 55	176	28 06	212	23 53	255	32 03	324
329	38 04	356	47 11	069	52 18	133	21 58	176	27 43	211	23 10	255	31 36	323
	◆Diphda		ACHERNAR		◆Miaplacidus		RIGIL KENT.		Nunki		◆ALTAIR		Enif	
330	47 52	068	52 50	133	22 02	176	27 20	211	49 13	278	31 09	322	38 01	355
331	48 33	067	53 23	133	22 05	175	26 57	211	48 28	277	30 41	321	37 57	354
332	49 14	066	53 55	133	22 09	175	26 34	210	47 44	276	30 12	320	37 51	352
333	49 55	065	54 27	133	22 13	175	26 12	210	47 00	275	29 43	319	37 45	351
334	50 35	064	55 00	133	22 18	174	25 50	210	46 15	274	29 14	318	37 38	350
335	51 15	063	55 32	133	22 22	174	25 28	209	45 31	274	28 43	317	37 30	349
336	51 54	062	56 04	134	22 27	173	25 06	209	44 46	273	28 13	316	37 20	348
337	52 33	061	56 37	134	22 32	173	24 45	209	44 02	272	27 41	315	37 10	346
338	53 12	059	57 09	134	22 38	173	24 24	208	43 17	272	27 10	314	36 59	345
339	53 50	058	57 41	134	22 44	172	24 03	208	42 33	271	26 37	313	36 47	344
340	54 28	057	58 13	134	22 50	172	23 42	207	41 48	270	26 04	312	36 35	343
341	55 05	056	58 45	134	22 56	172	23 22	207	41 03	270	25 31	311	36 21	342
342	55 42	054	59 17	134	23 03	171	23 02	207	40 19	269	24 57	310	36 06	340
343	56 18	053	59 49	135	23 10	171	22 42	206	39 34	268	24 23	309	35 51	339
344	56 53	052	60 20	135	23 17	171	22 22	206	38 50	268	23 49	309	35 35	338
	◆Alpheratz		Diphda		◆CANOPUS		RIGIL KENT.		◆Nunki		ALTAIR		Enif	
345	17 13	016	57 28	050	21 47	142	22 03	205	38 05	267	23 14	308	35 18	337
346	17 25	015	58 02	049	22 14	142	21 44	205	37 21	266	22 38	307	35 00	336
347	17 36	014	58 35	047	22 42	142	21 25	205	36 36	266	22 02	306	34 41	335
348	17 46	013	59 07	046	23 10	141	21 07	204	35 52	265	21 26	305	34 21	333
349	17 56	012	59 39	044	23 38	141	20 48	204	35 07	265	20 49	304	34 01	332
350	18 04	011	60 10	043	24 06	140	20 31	203	34 23	264	20 12	304	33 40	331
351	18 13	010	60 40	041	24 35	140	20 13	203	33 39	263	19 35	303	33 18	330
352	18 20	009	61 08	039	25 04	140	19 56	203	32 54	263	18 57	302	32 55	329
353	18 27	008	61 36	038	25 33	139	19 39	202	32 10	262	18 19	301	32 32	328
354	18 33	007	62 03	036	26 02	139	19 22	202	31 26	261	17 41	300	32 08	327
355	18 38	006	62 29	034	26 32	138	19 06	201	30 42	261	17 02	300	31 43	326
356	18 43	006	62 53	032	27 01	138	18 50	201	29 58	260	16 23	299	31 18	325
357	18 47	005	63 16	030	27 31	138	18 34	200	29 14	260	15 44	298	30 52	324
358	18 50	004	63 38	028	28 01	137	18 18	200	28 30	259	15 04	297	30 25	323
359	18 53	003	63 59	026	28 32	137	18 03	200	27 46	258	14 25	296	29 58	322

 LAT 43°S

LHA ♈	Hc	Zn	Hc	Zn	Hc	Zn	Hc	Zn	Hc	Zn	Hc	Zn	Hc	Zn
	Alpheratz		Hamal		◆RIGEL		CANOPUS		◆RIGIL KENT.		Peacock		◆Enif	
0	17 55	002	17 26	030	13 56	088	29 46	136	18 45	199	53 53	229	28 43	321
1	17 56	001	17 47	029	14 40	088	30 16	136	18 31	199	53 20	229	28 16	320
2	17 56	000	18 09	029	15 24	087	30 47	135	18 17	198	52 47	228	27 47	319
3	17 56	359	18 29	028	16 07	086	31 18	135	18 03	198	52 14	228	27 18	318
4	17 55	358	18 50	027	16 51	085	31 49	135	17 50	197	51 42	228	26 49	317
5	17 53	357	19 09	026	17 35	085	32 21	134	17 37	197	51 09	228	26 18	316
6	17 51	356	19 28	025	18 19	084	32 52	134	17 24	197	50 36	228	25 48	315
7	17 48	355	19 46	024	19 02	083	33 24	133	17 12	196	50 03	228	25 17	314
8	17 44	355	20 04	023	19 46	083	33 56	133	17 00	196	49 31	228	24 45	313
9	17 39	354	20 21	022	20 29	082	34 28	133	16 48	195	48 58	228	24 13	313
10	17 34	353	20 37	021	21 13	081	35 00	132	16 37	195	48 26	228	23 40	312
11	17 28	352	20 52	020	21 56	080	35 33	132	16 26	194	47 53	228	23 07	311
12	17 21	351	21 07	019	22 39	080	36 06	132	16 16	194	47 21	228	22 34	310
13	17 14	350	21 22	018	23 22	079	36 38	131	16 05	193	46 48	227	22 00	309
14	17 06	349	21 35	017	24 05	078	37 11	131	15 55	193	46 16	227	21 26	308
	◆Hamal		RIGEL		◆CANOPUS		RIGIL KENT.		◆Peacock		FOMALHAUT		Alpheratz	
15	21 48	017	24 48	078	37 45	131	15 46	192	45 44	227	62 09	288	16 57	348
16	22 00	016	25 31	077	38 18	130	15 37	192	45 12	227	61 27	287	16 48	347
17	22 11	015	26 14	076	38 51	130	15 28	191	44 40	227	60 45	286	16 38	346
18	22 22	014	26 56	075	39 25	130	15 20	191	44 08	226	60 03	285	16 27	345
19	22 32	013	27 38	074	39 59	130	15 12	190	43 36	226	59 20	284	16 16	345
20	22 41	012	28 21	074	40 32	129	15 04	190	43 05	226	58 38	283	16 04	344
21	22 50	011	29 03	073	41 06	129	14 57	189	42 33	226	57 55	282	15 52	343
22	22 58	010	29 45	072	41 41	129	14 50	189	42 02	226	57 12	281	15 38	342
23	23 05	009	30 26	071	42 15	128	14 43	188	41 30	225	56 29	281	15 24	341
24	23 11	008	31 08	070	42 49	128	14 37	188	40 59	225	55 46	280	15 10	340
25	23 16	007	31 49	070	43 24	128	14 31	187	40 28	225	55 02	279	14 55	339
26	23 21	006	32 30	069	43 59	128	14 25	187	39 57	225	54 19	278	14 39	338
27	23 25	005	33 11	068	44 33	127	14 20	186	39 26	224	53 36	277	14 22	338
28	23 28	004	33 51	067	45 08	127	14 16	186	38 56	224	52 52	276	14 05	337
29	23 31	003	34 31	066	45 43	127	14 11	185	38 25	224	52 08	276	13 48	336
	◆Hamal		RIGEL		SIRIUS		◆CANOPUS		RIGIL KENT.		◆Peacock		FOMALHAUT	
30	23 33	002	35 11	065	24 56	090	46 18	127	14 07	185	37 55	224	51 25	275
31	23 33	001	35 51	064	25 39	089	46 54	127	14 04	184	37 24	223	50 41	274
32	23 34	000	36 31	063	26 23	089	47 29	126	14 01	184	36 54	223	49 57	273
33	23 33	359	37 10	062	27 07	088	48 04	126	13 58	183	36 24	223	49 13	273
34	23 32	358	37 48	062	27 51	087	48 40	126	13 55	183	35 55	223	48 29	272
35	23 30	357	38 27	061	28 35	087	49 15	126	13 53	182	35 25	222	47 46	271
36	23 27	356	39 05	060	29 19	086	49 51	126	13 52	182	34 56	222	47 02	270
37	23 23	355	39 42	059	30 02	085	50 27	125	13 50	181	34 26	222	46 18	270
38	23 19	354	40 20	058	30 46	084	51 03	125	13 50	181	33 57	221	45 34	269
39	23 13	353	40 57	057	31 30	084	51 38	125	13 49	180	33 28	221	44 50	268
40	23 07	352	41 33	056	32 13	083	52 14	125	13 49	180	33 00	221	44 06	268
41	23 01	351	42 09	054	32 57	082	52 50	125	13 49	179	32 31	220	43 22	267
42	22 53	350	42 44	053	33 40	081	53 27	125	13 50	179	32 03	220	42 39	267
43	22 45	349	43 19	052	34 24	081	54 03	125	13 51	178	31 35	220	41 55	266
44	22 36	348	43 54	051	35 07	080	54 39	124	13 52	178	31 07	219	41 11	265
	◆RIGEL		SIRIUS		CANOPUS		◆RIGIL KENT.		Peacock		◆FOMALHAUT		Hamal	
45	44 28	050	35 50	079	55 15	124	13 54	177	30 39	219	40 27	265	22 27	347
46	45 01	049	36 33	078	55 51	124	13 56	177	30 11	219	39 44	264	22 16	346
47	45 34	048	37 16	078	56 28	124	13 59	176	29 44	218	39 00	263	22 05	345
48	46 06	047	37 59	077	57 04	124	14 02	176	29 17	218	38 16	263	21 53	344
49	46 38	045	38 41	076	57 40	124	14 05	175	28 50	218	37 33	262	21 41	343
50	47 09	044	39 24	075	58 17	124	14 09	175	28 23	217	36 50	262	21 28	342
51	47 39	043	40 06	074	58 53	124	14 13	174	27 57	217	36 06	261	21 14	341
52	48 08	042	40 48	073	59 29	124	14 17	174	27 31	217	35 23	260	20 59	340
53	48 37	040	41 30	073	60 06	124	14 22	173	27 05	216	34 40	260	20 44	339
54	49 05	039	42 12	072	60 42	124	14 27	173	26 39	216	33 56	259	20 28	338
55	49 32	038	42 54	071	61 18	124	14 33	172	26 13	215	33 13	259	20 11	337
56	49 58	036	43 35	070	61 55	124	14 39	172	25 48	215	32 30	258	19 54	336
57	50 24	035	44 16	069	62 31	124	14 45	171	25 23	215	31 48	257	19 36	335
58	50 49	033	44 57	068	63 07	124	14 52	171	24 58	214	31 05	257	19 18	335
59	51 12	032	45 37	067	63 43	125	14 59	170	24 34	214	30 22	256	18 58	334
	ALDEBARAN		◆SIRIUS		Suhail		◆RIGIL KENT.		ACHERNAR		◆Diphda		Hamal	
60	29 56	010	46 18	066	36 04	119	15 07	170	63 32	225	41 44	285	18 39	333
61	30 03	009	46 57	065	36 42	118	15 15	169	63 01	225	41 02	285	18 18	332
62	30 10	008	47 37	064	37 21	118	15 23	169	62 30	226	40 19	284	17 57	331
63	30 15	007	48 16	063	38 00	118	15 31	169	61 58	226	39 36	283	17 35	330
64	30 20	005	48 55	062	38 38	117	15 40	168	61 27	226	38 54	282	17 13	329
65	30 24	004	49 34	061	39 18	117	15 49	168	60 55	226	38 11	281	16 50	328
66	30 26	003	50 12	060	39 57	116	15 59	167	60 23	227	37 28	280	16 27	327
67	30 28	002	50 49	059	40 36	116	16 09	167	59 51	227	36 44	280	16 03	327
68	30 30	001	51 27	057	41 16	116	16 19	166	59 19	227	36 01	279	15 39	326
69	30 30	000	52 03	056	41 55	115	16 30	166	58 47	227	35 18	278	15 14	325
70	30 30	359	52 40	055	42 35	115	16 41	165	58 15	227	34 34	277	14 48	324
71	30 28	358	53 15	054	43 15	114	16 53	165	57 42	227	33 51	277	14 22	323
72	30 26	357	53 50	052	43 55	114	17 04	164	57 10	227	33 07	276	13 55	322
73	30 23	355	54 25	051	44 35	114	17 16	164	56 38	228	32 23	275	13 28	322
74	30 19	354	54 59	050	45 15	113	17 29	163	56 05	228	31 40	274	13 01	321
	BETELGEUSE		◆SIRIUS		Suhail		◆RIGIL KENT.		ACHERNAR		◆Diphda		ALDEBARAN	
75	38 04	017	55 32	048	45 56	113	17 42	163	55 33	228	30 56	274	30 14	353
76	38 17	016	56 04	047	46 36	112	17 55	162	55 01	228	30 12	273	30 09	352
77	38 29	015	56 36	046	47 17	112	18 08	162	54 28	228	29 28	272	30 02	351
78	38 39	014	57 07	044	47 58	112	18 22	161	53 56	228	28 44	272	29 55	350
79	38 49	012	57 37	043	48 39	111	18 36	161	53 23	228	28 00	271	29 47	349
80	38 58	011	58 06	041	49 19	111	18 51	161	52 51	228	27 17	270	29 38	348
81	39 06	010	58 35	039	50 01	111	19 05	160	52 18	228	26 33	270	29 28	347
82	39 13	009	59 02	038	50 42	110	19 20	160	51 46	228	25 49	269	29 18	346
83	39 20	007	59 28	036	51 23	110	19 36	159	51 14	227	25 05	268	29 06	345
84	39 25	006	59 54	034	52 04	109	19 51	159	50 41	227	24 21	268	28 54	343
85	39 29	005	60 18	033	52 46	109	20 07	158	50 09	227	23 37	267	28 41	342
86	39 32	004	60 41	031	53 27	109	20 24	158	49 37	227	22 53	266	28 28	341
87	39 34	002	61 03	029	54 09	108	20 40	158	49 05	227	22 10	266	28 13	340
88	39 35	001	61 24	027	54 51	108	20 57	157	48 32	227	21 26	265	27 58	339
89	39 36	000	61 43	025	55 32	108	21 15	157	48 00	227	20 42	264	27 42	338

LHA ♈	Hc	Zn	Hc	Zn	Hc	Zn	Hc	Zn	Hc	Zn	Hc	Zn	Hc	Zn
	PROCYON		Alphard		◆Suhail		RIGIL KENT.		◆ACHERNAR		RIGEL		◆BETELGEUSE	
90	36 48	031	33 19	068	56 14	107	21 32	156	47 28	227	53 48	341	39 35	358
91	37 11	030	33 59	068	56 56	107	21 50	156	46 56	227	53 32	339	39 33	357
92	37 32	029	34 40	067	57 38	106	22 08	155	46 24	226	53 16	337	39 30	356
93	37 53	028	35 20	066	58 20	106	22 26	155	45 53	226	52 59	336	39 27	355
94	38 13	027	36 00	065	59 03	106	22 45	155	45 21	226	52 40	334	39 22	353
95	38 33	025	36 40	064	59 45	105	23 04	154	44 49	226	52 21	333	39 16	352
96	38 51	024	37 19	063	60 27	105	23 23	154	44 18	226	52 00	331	39 10	351
97	39 09	023	37 58	062	61 10	105	23 43	153	43 46	226	51 38	330	39 02	349
98	39 26	022	38 36	061	61 52	104	24 03	153	43 15	225	51 16	328	38 54	348
99	39 42	021	39 15	060	62 35	104	24 23	153	42 44	225	50 52	327	38 44	347
100	39 56	019	39 53	059	63 17	104	24 43	152	42 13	225	50 28	325	38 34	346
101	40 11	018	40 30	058	64 00	103	25 03	152	41 42	225	50 02	324	38 22	344
102	40 24	017	41 07	057	64 42	103	25 24	151	41 11	225	49 36	323	38 10	343
103	40 36	016	41 44	056	65 25	103	25 45	151	40 40	224	49 09	321	37 57	342
104	40 47	014	42 20	055	66 08	102	26 07	151	40 10	224	48 41	320	37 43	341
	PROCYON		REGULUS		◆Gienah		RIGIL KENT.		◆ACHERNAR		RIGEL		◆BETELGEUSE	
105	40 57	013	20 14	050	19 51	096	26 28	150	39 39	224	48 13	319	37 28	340
106	41 07	012	20 48	049	20 35	095	26 50	150	39 09	224	47 43	317	37 12	338
107	41 15	010	21 20	048	21 19	094	27 12	150	38 39	223	47 13	316	36 55	337
108	41 22	009	21 53	047	22 03	094	27 34	149	38 09	223	46 42	315	36 38	336
109	41 29	008	22 25	046	22 46	093	27 57	149	37 39	223	46 11	314	36 20	335
110	41 34	006	22 56	045	23 30	092	28 20	149	37 09	223	45 39	312	36 01	334
111	41 38	005	23 27	044	24 14	092	28 43	148	36 39	222	45 06	311	35 41	332
112	41 42	004	23 58	044	24 58	091	29 06	148	36 10	222	44 33	310	35 20	331
113	41 44	002	24 28	043	25 42	090	29 29	148	35 41	222	43 59	309	34 58	330
114	41 45	001	24 57	042	26 26	090	29 53	147	35 11	221	43 25	308	34 36	329
115	41 46	000	25 26	041	27 10	089	30 17	147	34 43	221	42 50	307	34 13	328
116	41 45	358	25 54	040	27 53	088	30 41	147	34 14	221	42 14	306	33 50	327
117	41 43	357	26 22	039	28 37	088	31 05	146	33 45	221	41 38	305	33 25	326
118	41 40	356	26 49	038	29 21	087	31 30	146	33 17	220	41 02	304	33 00	325
119	41 37	354	27 16	037	30 05	086	31 54	146	32 48	220	40 25	303	32 35	324
	REGULUS		◆SPICA		RIGIL KENT.		◆ACHERNAR		RIGEL		BETELGEUSE		◆PROCYON	
120	27 42	036	13 57	092	32 19	145	32 20	220	39 48	302	32 08	323	41 32	353
121	28 07	035	14 41	092	32 44	145	31 53	219	39 10	301	31 41	322	41 26	352
122	28 32	034	15 25	091	33 09	145	31 25	219	38 32	300	31 14	320	41 19	350
123	28 56	033	16 09	090	33 35	144	30 57	219	37 54	299	30 45	319	41 11	349
124	29 20	032	16 53	090	34 00	144	30 30	218	37 15	298	30 17	318	41 02	348
125	29 42	031	17 36	089	34 26	144	30 03	218	36 36	297	29 47	317	40 53	346
126	30 05	030	18 20	088	34 52	144	29 36	218	35 57	296	29 17	316	40 42	345
127	30 26	029	19 04	088	35 18	143	29 10	217	35 17	295	28 47	316	40 30	344
128	30 47	028	19 48	087	35 44	143	28 43	217	34 37	294	28 16	315	40 17	343
129	31 07	027	20 32	086	36 11	143	28 17	216	33 57	293	27 44	314	40 04	341
130	31 26	025	21 16	085	36 38	143	27 51	216	33 17	292	27 12	313	39 49	340
131	31 44	024	21 59	085	37 04	142	27 25	216	32 36	291	26 40	312	39 34	339
132	32 02	023	22 43	084	37 31	142	27 00	215	31 55	291	26 07	311	39 18	338
133	32 19	022	23 27	083	37 58	142	26 35	215	31 14	290	25 33	310	39 01	336
134	32 35	021	24 10	083	38 25	142	26 10	215	30 32	289	24 59	309	38 43	335
	REGULUS		◆SPICA		RIGIL KENT.		◆ACHERNAR		CANOPUS		SIRIUS		◆PROCYON	
135	32 51	020	24 54	082	38 53	141	25 45	214	62 28	236	51 07	302	38 24	334
136	33 05	019	25 37	081	39 20	141	25 20	214	61 52	236	50 29	301	38 04	333
137	33 19	018	26 20	080	39 48	141	24 56	213	61 15	236	49 51	300	37 43	332
138	33 32	017	27 03	080	40 15	141	24 32	213	60 39	236	49 13	299	37 22	330
139	33 44	015	27 47	079	40 43	141	24 08	213	60 03	236	48 34	298	37 00	329
140	33 55	014	28 30	078	41 11	140	23 45	212	59 26	236	47 55	297	36 37	328
141	34 05	013	29 12	077	41 39	140	23 21	212	58 50	236	47 16	295	36 14	327
142	34 15	012	29 55	077	42 07	140	22 58	211	58 14	236	46 36	294	35 49	326
143	34 23	011	30 38	076	42 36	140	22 36	211	57 37	236	45 56	293	35 24	325
144	34 31	010	31 20	075	43 04	140	22 13	211	57 01	236	45 16	292	34 59	324
145	34 38	008	32 03	074	43 32	140	21 51	210	56 25	236	44 35	292	34 32	322
146	34 44	007	32 45	073	44 01	139	21 29	210	55 48	236	43 54	291	34 05	321
147	34 49	006	33 27	073	44 29	139	21 08	209	55 12	236	43 13	290	33 37	320
148	34 53	005	34 08	072	44 58	139	20 46	209	54 36	236	42 31	289	33 09	319
149	34 56	004	34 50	071	45 27	139	20 25	208	54 00	235	41 50	288	32 40	318
	REGULUS		◆SPICA		RIGIL KENT.		◆ACHERNAR		CANOPUS		SIRIUS		◆PROCYON	
150	34 59	002	35 31	070	45 56	139	20 05	208	53 24	235	41 08	287	32 11	317
151	35 00	001	36 13	069	46 25	139	19 44	207	52 48	235	40 26	286	31 41	316
152	35 00	000	36 53	068	46 54	139	19 24	207	52 12	235	39 44	285	31 10	315
153	35 00	359	37 34	067	47 23	139	19 05	207	51 36	235	39 01	284	30 39	314
154	34 59	358	38 14	067	47 52	138	18 45	206	51 00	235	38 19	284	30 07	313
155	34 56	356	38 55	066	48 21	138	18 26	206	50 24	235	37 36	283	29 35	312
156	34 53	355	39 34	065	48 50	138	18 07	205	49 48	234	36 53	282	29 02	311
157	34 49	354	40 14	064	49 19	138	17 49	205	49 12	234	36 10	281	28 29	310
158	34 44	353	40 53	063	49 48	138	17 30	204	48 37	234	35 27	280	27 55	309
159	34 38	352	41 32	062	50 18	138	17 13	204	48 01	234	34 44	280	27 21	309
160	34 32	351	42 10	061	50 47	138	16 55	203	47 26	234	34 00	279	26 47	308
161	34 24	349	42 48	060	51 16	138	16 38	203	46 51	233	33 17	278	26 12	307
162	34 15	348	43 26	059	51 45	138	16 21	202	46 16	233	32 33	277	25 36	306
163	34 06	347	44 03	058	52 15	138	16 05	202	45 40	233	31 50	277	25 01	305
164	33 56	346	44 40	057	52 44	138	15 48	201	45 05	233	31 06	276	24 25	304
	SPICA		◆ANTARES		RIGIL KENT.		ACHERNAR		◆CANOPUS		SIRIUS		◆REGULUS	
165	45 17	056	23 03	105	53 13	138	15 33	201	44 31	233	30 23	275	33 44	345
166	45 53	054	23 45	105	53 42	138	15 17	200	43 56	232	29 39	274	33 32	344
167	46 28	053	24 28	104	54 11	138	15 02	200	43 21	232	28 55	274	33 20	342
168	47 03	052	25 10	104	54 41	138	14 47	199	42 47	232	28 11	273	33 06	341
169	47 37	051	25 53	103	55 10	139	14 33	199	42 12	231	27 28	272	32 51	340
170	48 11	050	26 36	102	55 39	139	14 19	198	41 38	231	26 44	272	32 36	339
171	48 44	049	27 19	102	56 08	139	14 05	198	41 04	231	26 00	271	32 20	338
172	49 17	047	28 02	101	56 36	139	13 52	197	40 30	231	25 16	270	32 03	337
173	49 49	046	28 45	101	57 05	139	13 39	197	39 56	230	24 32	270	31 45	336
174	50 20	045	29 28	100	57 34	139	13 27	196	39 22	230	23 48	269	31 27	335
175	50 50	043	30 11	099	58 02	140	13 14	196	38 49	230	23 04	268	31 08	334
176	51 20	042	30 55	099	58 31	140	13 03	195	38 15	229	22 20	268	30 48	332
177	51 49	041	31 38	098	58 59	140	12 51	195	37 42	229	21 37	267	30 27	331
178	52 17	039	32 22	097	59 27	140	12 40	194	37 09	229	20 53	266	30 06	330
179	52 45	038	33 05	097	59 55	141	12 30	194	36 36	229	20 09	266	29 44	329

LAT 43°S

LHA	Hc	Zn	Hc	Zn	Hc	Zn	Hc	Zn	Hc	Zn	Hc	Zn	Hc	Zn
♈	◆ARCTURUS		ANTARES		◆RIGIL KENT.		ACHERNAR		CANOPUS		◆Suhail		REGULUS	
180	20 26	034	33 49	096	60 23	141	12 19	193	36 03	228	58 59	254	29 21	328
181	20 50	033	34 32	096	60 51	141	12 10	193	35 30	228	58 16	254	28 57	327
182	21 14	032	35 16	095	61 18	141	12 00	192	34 58	228	57 34	253	28 33	326
183	21 37	031	36 00	094	61 45	142	11 51	192	34 26	227	56 52	253	28 09	325
184	22 00	030	36 44	094	62 12	142	11 42	191	33 53	227	56 10	253	27 43	324
185	22 22	030	37 27	093	62 39	143	11 34	191	33 22	227	55 28	252	27 17	323
186	22 43	029	38 11	092	63 05	143	11 26	190	32 50	226	54 47	252	26 51	322
187	23 04	028	38 55	092	63 32	144	11 19	189	32 18	226	54 05	252	26 24	321
188	23 24	027	39 39	091	63 57	144	11 12	189	31 47	225	53 23	251	25 56	320
189	23 43	026	40 23	090	64 23	145	11 05	188	31 16	225	52 42	251	25 27	319
190	24 02	025	41 07	090	64 48	145	10 59	188	30 45	225	52 00	251	24 59	318
191	24 20	024	41 51	089	65 13	146	10 53	187	30 14	224	51 19	250	24 29	317
192	24 37	023	42 35	088	65 37	147	10 48	187	29 43	224	50 38	250	23 59	317
193	24 54	022	43 18	088	66 01	147	10 43	186	29 13	224	49 57	249	23 29	316
194	25 10	021	44 02	087	66 25	148	10 38	186	28 43	223	49 16	249	22 58	315
	◆ARCTURUS		ANTARES		◆Peacock		ACHERNAR		CANOPUS		◆Suhail		REGULUS	
195	25 25	020	44 46	086	25 08	146	10 34	185	28 13	223	48 35	249	22 26	314
196	25 39	019	45 30	085	25 32	145	10 30	185	27 43	222	47 54	248	21 55	313
197	25 53	018	46 13	085	25 58	145	10 27	184	27 14	222	47 13	248	21 22	312
198	26 06	017	46 57	084	26 23	144	10 24	184	26 45	222	46 33	248	20 49	311
199	26 18	016	47 41	083	26 49	144	10 22	183	26 16	221	45 52	247	20 16	310
200	26 30	015	48 24	082	27 14	144	10 20	182	25 47	221	45 12	247	19 43	309
201	26 41	014	49 08	081	27 41	143	10 18	182	25 18	220	44 31	246	19 08	309
202	26 50	013	49 51	081	28 07	143	10 17	181	24 50	220	43 51	246	18 34	308
203	27 00	012	50 34	080	28 33	143	10 16	181	24 22	220	43 11	246	17 59	307
204	27 08	010	51 17	079	29 00	142	10 16	180	23 54	219	42 31	245	17 24	306
205	27 16	009	52 00	078	29 27	142	10 16	180	23 27	219	41 52	245	16 48	305
206	27 22	008	52 43	077	29 54	142	10 16	179	22 59	218	41 12	244	16 12	305
207	27 28	007	53 26	076	30 22	141	10 17	179	22 32	218	40 32	244	15 36	304
208	27 33	006	54 09	075	30 49	141	10 18	178	22 06	217	39 53	244	14 59	303
209	27 38	005	54 51	074	31 17	140	10 20	177	21 39	217	39 14	243	14 22	302
	◆ARCTURUS		ANTARES		◆Peacock		ACHERNAR		CANOPUS		◆Suhail		SPICA	
210	27 41	004	55 33	074	31 45	140	10 22	177	21 13	216	38 35	243	57 14	344
211	27 44	003	56 15	073	32 13	140	10 25	176	20 47	216	37 56	242	57 01	342
212	27 46	002	56 57	071	32 42	139	10 28	176	20 21	216	37 17	242	56 47	340
213	27 47	001	57 38	070	33 10	139	10 31	175	19 56	215	36 38	241	56 32	339
214	27 47	000	58 20	069	33 39	139	10 35	175	19 31	215	36 00	241	56 16	337
215	27 47	359	59 00	068	34 08	139	10 39	174	19 06	214	35 22	241	55 58	335
216	27 46	358	59 41	067	34 37	138	10 44	174	18 42	214	34 44	240	55 39	334
217	27 43	357	60 21	066	35 07	138	10 49	173	18 18	213	34 06	240	55 19	332
218	27 40	356	61 01	065	35 36	138	10 55	173	17 54	213	33 28	239	54 58	330
219	27 37	355	61 41	063	36 06	137	11 01	172	17 30	212	32 50	239	54 36	329
220	27 32	353	62 20	062	36 36	137	11 07	171	17 07	212	32 13	238	54 12	327
221	27 27	352	62 58	061	37 06	137	11 14	171	16 44	211	31 35	238	53 48	326
222	27 21	351	63 36	059	37 36	136	11 21	170	16 22	211	30 58	238	53 23	324
223	27 14	350	64 14	058	38 06	136	11 28	170	15 59	210	30 21	237	52 57	323
224	27 06	349	64 50	056	38 36	136	11 36	169	15 37	210	29 44	237	52 30	321
	Rasalhague		◆Nunki		Peacock		ACHERNAR		◆ACRUX		SPICA		◆ARCTURUS	
225	24 08	042	39 58	091	39 07	136	11 44	169	60 10	214	52 02	320	26 57	348
226	24 37	041	40 42	090	39 38	135	11 53	168	59 46	215	51 34	319	26 48	347
227	25 06	040	41 26	089	40 09	135	12 02	168	59 20	215	51 04	317	26 38	346
228	25 34	039	42 10	088	40 40	135	12 12	167	58 55	215	50 34	316	26 27	345
229	26 01	038	42 54	088	41 11	135	12 22	167	58 29	216	50 03	315	26 15	344
230	26 28	037	43 38	087	41 42	134	12 32	166	58 04	216	49 32	313	26 03	343
231	26 54	036	44 22	086	42 13	134	12 43	166	57 38	216	48 59	312	25 49	342
232	27 20	035	45 06	085	42 45	134	12 54	165	57 12	217	48 27	311	25 35	341
233	27 45	034	45 49	085	43 16	134	13 05	165	56 46	217	47 53	310	25 21	340
234	28 09	033	46 33	084	43 48	134	13 17	164	56 19	217	47 19	308	25 05	339
235	28 33	032	47 16	083	44 20	133	13 29	164	55 53	217	46 44	307	24 49	338
236	28 56	031	48 00	083	44 52	133	13 42	163	55 26	217	46 09	306	24 32	337
237	29 19	030	48 43	082	45 24	133	13 55	163	55 00	217	45 33	305	24 15	336
238	29 40	029	49 27	081	45 56	133	14 08	162	54 33	218	44 57	304	23 57	335
239	30 01	028	50 10	080	46 28	133	14 22	162	54 06	218	44 21	303	23 38	334
	Rasalhague		◆ALTAIR		Peacock		◆ACHERNAR		ACRUX		◆SPICA		ARCTURUS	
240	30 22	027	16 22	060	47 00	133	14 36	161	53 39	218	43 44	302	23 18	333
241	30 41	026	17 00	060	47 33	132	14 51	161	53 12	218	43 06	301	22 58	332
242	31 00	025	17 38	059	48 05	132	15 06	160	52 45	218	42 28	300	22 37	331
243	31 18	024	18 15	058	48 38	132	15 21	160	52 18	218	41 50	299	22 16	330
244	31 35	023	18 52	057	49 10	132	15 36	159	51 51	218	41 11	298	21 54	329
245	31 52	022	19 29	056	49 43	132	15 52	159	51 24	218	40 32	297	21 31	328
246	32 08	020	20 06	056	50 16	132	16 08	158	50 57	218	39 53	296	21 08	327
247	32 23	019	20 42	055	50 48	132	16 25	158	50 30	218	39 13	295	20 44	326
248	32 37	018	21 17	054	51 21	132	16 42	157	50 02	218	38 33	294	20 19	326
249	32 50	017	21 52	053	51 54	132	16 59	157	49 35	218	37 53	293	19 54	325
250	33 03	016	22 27	052	52 27	132	17 17	156	49 08	218	37 12	292	19 28	324
251	33 14	015	23 02	051	52 59	131	17 35	156	48 41	218	36 32	291	19 02	323
252	33 25	014	23 36	050	53 32	131	17 53	155	48 14	218	35 51	290	18 35	322
253	33 35	013	24 09	050	54 05	131	18 12	155	47 47	218	35 09	289	18 08	321
254	33 44	011	24 43	049	54 38	131	18 30	154	47 20	218	34 28	289	17 40	320
	Rasalhague		◆ALTAIR		FOMALHAUT		◆ACHERNAR		RIGIL KENT.		◆SPICA		Alphecca	
255	33 52	010	25 15	048	20 10	112	18 50	154	62 29	217	33 46	288	17 33	340
256	34 00	009	25 48	047	20 50	112	19 09	153	62 02	218	33 04	287	17 17	339
257	34 06	008	26 19	046	21 31	111	19 29	153	61 35	218	32 22	286	17 01	338
258	34 12	007	26 51	045	22 12	110	19 49	152	61 08	219	31 40	285	16 45	337
259	34 16	006	27 21	044	22 53	110	20 10	152	60 40	219	30 58	285	16 28	336
260	34 20	004	27 52	043	23 35	109	20 30	152	60 13	219	30 15	284	16 10	336
261	34 23	003	28 21	042	24 16	109	20 51	151	59 45	220	29 32	283	15 51	335
262	34 25	002	28 50	041	24 58	108	21 13	151	59 17	220	28 50	282	15 32	334
263	34 26	001	29 19	040	25 40	108	21 34	150	58 49	220	28 07	281	15 13	333
264	34 26	000	29 47	039	26 22	107	21 56	150	58 20	220	27 24	281	14 53	332
265	34 25	358	30 14	038	27 04	106	22 19	149	57 52	221	26 40	280	14 32	331
266	34 24	357	30 41	037	27 46	106	22 41	149	57 23	221	25 57	279	14 10	330
267	34 21	356	31 07	036	28 28	105	23 04	149	56 54	221	25 14	278	13 48	330
268	34 18	355	31 33	035	29 10	105	23 27	148	56 26	221	24 30	278	13 26	329
269	34 13	354	31 57	034	29 53	104	23 50	148	55 57	221	23 47	277	13 03	328

LHA	Hc	Zn	Hc	Zn	Hc	Zn	Hc	Zn	Hc	Zn	Hc	Zn	Hc	Zn
♈	ALTAIR		Enif		◆FOMALHAUT		ACHERNAR		◆RIGIL KENT.		ANTARES		◆Rasalhague	
270	32 22	033	16 39	058	30 35	104	24 14	147	55 28	221	65 10	305	34 08	353
271	32 45	032	17 16	058	31 18	103	24 38	147	54 59	221	64 34	303	34 02	351
272	33 08	031	17 53	057	32 01	102	25 02	147	54 30	222	63 57	301	33 55	350
273	33 30	030	18 29	056	32 44	102	25 26	146	54 01	222	63 19	300	33 47	349
274	33 51	028	19 06	055	33 27	101	25 51	146	53 31	222	62 41	299	33 38	348
275	34 12	027	19 41	054	34 10	101	26 15	145	53 02	222	62 02	297	33 29	347
276	34 31	026	20 17	054	34 53	100	26 41	145	52 33	222	61 23	296	33 18	346
277	34 50	025	20 52	053	35 36	099	27 06	145	52 04	222	60 43	295	33 07	344
278	35 09	024	21 27	052	36 20	099	27 31	144	51 34	222	60 03	294	32 55	343
279	35 26	023	22 01	051	37 03	098	27 57	144	51 05	222	59 23	292	32 41	342
280	35 43	022	22 35	050	37 47	098	28 23	143	50 36	222	58 42	291	32 28	341
281	35 58	020	23 08	049	38 30	097	28 49	143	50 07	222	58 01	290	32 13	340
282	36 13	019	23 41	048	39 14	096	29 16	143	49 38	222	57 20	289	31 57	339
283	36 27	018	24 14	047	39 57	096	29 43	142	49 08	222	56 38	288	31 41	338
284	36 41	017	24 46	047	40 41	095	30 09	142	48 39	222	55 56	287	31 24	337
	ALTAIR		Enif		◆FOMALHAUT		ACHERNAR		◆RIGIL KENT.		ANTARES		◆Rasalhague	
285	36 53	016	25 18	046	41 25	095	30 37	142	48 10	222	55 14	286	31 06	336
286	37 04	014	25 49	045	42 08	094	31 04	141	47 41	222	54 32	285	30 48	334
287	37 15	013	26 19	044	42 52	093	31 31	141	47 12	221	53 49	284	30 28	333
288	37 24	012	26 49	043	43 36	093	31 59	141	46 43	221	53 07	283	30 08	332
289	37 33	011	27 19	042	44 20	092	32 27	140	46 14	221	52 24	282	29 48	331
290	37 41	010	27 48	041	45 04	091	32 55	140	45 45	221	51 41	281	29 26	330
291	37 48	008	28 16	040	45 48	091	33 23	140	45 16	221	50 58	281	29 04	329
292	37 53	007	28 44	039	46 31	090	33 52	139	44 47	221	50 15	280	28 41	328
293	37 58	006	29 11	038	47 15	089	34 20	139	44 19	221	49 32	279	28 18	327
294	38 02	005	29 38	037	47 59	089	34 49	139	43 50	221	48 48	278	27 54	326
295	38 05	003	30 04	036	48 43	088	35 18	138	43 22	220	48 05	277	27 29	325
296	38 07	002	30 30	035	49 27	087	35 47	138	42 53	220	47 21	277	27 03	324
297	38 09	001	30 54	034	50 11	086	36 17	138	42 25	220	46 37	276	26 37	323
298	38 09	000	31 18	033	50 55	086	36 46	138	41 57	220	45 54	275	26 11	322
299	38 08	358	31 42	032	51 38	085	37 16	137	41 29	220	45 10	274	25 44	321
	◆Enif		FOMALHAUT		◆ACHERNAR		RIGIL KENT.		◆ANTARES		Rasalhague		ALTAIR	
300	32 05	031	52 22	084	37 46	137	41 01	220	44 26	274	25 16	320	38 06	357
301	32 27	030	53 06	083	38 16	137	40 33	219	43 42	273	24 47	319	38 03	356
302	32 48	028	53 49	083	38 46	137	40 05	219	42 59	272	24 19	318	38 00	355
303	33 08	027	54 33	082	39 16	136	39 37	219	42 15	271	23 49	317	37 55	353
304	33 28	026	55 16	081	39 46	136	39 10	219	41 31	271	23 19	317	37 49	352
305	33 47	025	55 59	080	40 17	136	38 42	219	40 47	270	22 49	316	37 43	351
306	34 05	024	56 42	079	40 48	136	38 15	218	40 03	269	22 18	315	37 35	350
307	34 23	023	57 25	078	41 18	135	37 48	218	39 19	269	21 46	314	37 27	348
308	34 39	022	58 08	077	41 49	135	37 21	218	38 35	268	21 15	313	37 17	347
309	34 55	021	58 51	076	42 20	135	36 54	218	37 52	267	20 42	312	37 07	346
310	35 10	019	59 34	075	42 51	135	36 28	217	37 08	267	20 09	311	36 56	345
311	35 24	018	60 16	074	43 23	135	36 01	217	36 24	266	19 36	310	36 44	343
312	35 38	017	60 58	073	43 54	134	35 35	217	35 40	265	19 03	310	36 31	342
313	35 50	016	61 40	072	44 25	134	35 08	217	34 56	265	18 29	309	36 17	341
314	36 02	015	62 22	071	44 57	134	34 42	216	34 13	264	17 54	308	36 03	340
	Enif		◆Diphda		ACHERNAR		Miaplacidus		◆RIGIL KENT.		ANTARES		◆ALTAIR	
315	36 12	013	36 59	080	45 28	134	22 43	181	34 17	216	33 29	264	35 47	339
316	36 22	012	37 42	079	46 00	134	22 43	181	33 51	216	32 46	263	35 31	338
317	36 31	011	38 25	078	46 32	133	22 42	180	33 25	215	32 02	262	35 13	336
318	36 39	010	39 08	078	47 04	133	22 42	180	33 00	215	31 19	262	34 55	335
319	36 46	009	39 51	077	47 36	133	22 42	180	32 35	215	30 35	261	34 37	334
320	36 52	007	40 33	076	48 08	133	22 42	179	32 10	215	29 52	260	34 17	333
321	36 57	006	41 16	075	48 40	133	22 43	179	31 45	214	29 09	260	33 57	332
322	37 01	005	41 58	074	49 12	133	22 44	179	31 20	214	28 26	259	33 36	331
323	37 05	004	42 40	073	49 44	133	22 45	178	30 56	214	27 42	259	33 14	330
324	37 07	002	43 22	073	50 17	133	22 46	178	30 32	213	27 00	258	32 51	329
325	37 08	001	44 04	072	50 49	133	22 48	177	30 08	213	26 17	257	32 28	327
326	37 09	000	44 46	071	51 21	132	22 50	177	29 44	213	25 34	257	32 04	326
327	37 08	359	45 27	070	51 54	132	22 53	177	29 21	212	24 51	256	31 39	325
328	37 07	358	46 08	069	52 26	132	22 55	176	28 57	212	24 09	256	31 14	324
329	37 05	356	46 49	068	52 58	132	22 58	176	28 34	212	23 26	255	30 48	323
	◆Diphda		ACHERNAR		◆Miaplacidus		RIGIL KENT.		Nunki		◆ALTAIR		Enif	
330	47 29	067	53 31	132	23 02	176	28 11	211	49 04	279	30 21	322	37 01	355
331	48 09	066	54 03	132	23 05	175	27 49	211	48 21	278	29 54	321	36 57	354
332	48 49	065	54 36	132	23 09	175	27 26	211	47 37	277	29 26	320	36 52	353
333	49 29	064	55 08	132	23 13	174	27 04	210	46 54	276	28 58	319	36 46	351
334	50 08	063	55 41	132	23 17	174	26 42	210	46 10	276	28 29	318	36 39	350
335	50 47	062	56 13	132	23 22	174	26 20	209	45 26	275	28 00	317	36 31	349
336	51 25	061	56 45	132	23 27	173	25 59	209	44 42	274	27 29	316	36 22	348
337	52 03	059	57 18	133	23 32	173	25 38	209	43 59	273	26 59	315	36 12	347
338	52 41	058	57 50	133	23 37	173	25 17	208	43 15	273	26 28	314	36 01	345
339	53 18	057	58 22	133	23 43	172	24 56	208	42 31	272	25 56	313	35 50	344
340	53 55	056	58 54	133	23 49	172	24 35	208	41 47	271	25 24	313	35 37	343
341	54 31	055	59 27	133	23 55	172	24 15	207	41 03	270	24 51	312	35 24	342
342	55 06	053	59 59	133	24 02	171	23 55	207	40 19	270	24 18	311	35 10	341
343	55 41	052	60 30	133	24 09	171	23 35	206	39 36	269	23 45	310	34 55	339
344	56 15	051	61 02	134	24 16	170	23 16	206	38 52	268	23 11	309	34 39	338
	◆Alpheratz		Diphda		◆CANOPUS		RIGIL KENT.		◆Nunki		ALTAIR		Enif	
345	16 15	015	56 49	049	22 35	142	22 57	206	38 08	268	22 37	308	34 22	337
346	16 27	015	57 22	048	23 02	142	22 38	205	37 24	267	22 02	307	34 05	336
347	16 37	014	57 54	046	23 29	141	22 20	205	36 40	266	21 27	306	33 47	335
348	16 48	013	58 25	045	23 56	141	22 01	204	35 56	266	20 51	305	33 28	334
349	16 57	012	58 56	043	24 24	140	21 43	204	35 13	265	20 15	305	33 08	333
350	17 06	011	59 25	042	24 52	140	21 26	204	34 29	265	19 39	304	32 47	332
351	17 14	010	59 54	040	25 21	140	21 08	203	33 45	264	19 02	303	32 26	330
352	17 21	009	60 22	038	25 49	139	20 51	203	33 02	263	18 25	302	32 04	329
353	17 28	008	60 48	037	26 18	139	20 34	202	32 18	263	17 48	301	31 41	328
354	17 34	007	61 14	035	26 47	138	20 18	202	31 35	262	17 10	301	31 18	327
355	17 39	006	61 39	033	27 16	138	20 01	201	30 51	261	16 33	300	30 54	326
356	17 43	006	62 02	031	27 46	138	19 46	201	30 08	261	15 54	299	30 29	325
357	17 47	005	62 24	029	28 15	137	19 30	201	29 25	260	15 16	298	30 03	324
358	17 51	004	62 45	027	28 45	137	19 15	200	28 41	260	14 37	297	29 37	323
359	17 53	003	63 05	026	29 15	136	19 00	200	27 58	259	13 58	297	29 11	322

LHA	Hc	Zn	Hc	Zn	Hc	Zn	Hc	Zn	Hc	Zn	Hc	Zn	Hc	Zn
♈	Alpheratz		Hamal		◆RIGEL		CANOPUS		◆RIGIL KENT.		Peacock		◆Enif	
0	16 55	002	16 34	030	13 54	088	30 29	136	19 42	199	54 32	230	27 57	321
1	16 56	001	16 55	029	14 37	087	30 59	135	19 28	199	53 59	230	27 29	320
2	16 56	000	17 16	028	15 20	087	31 30	135	19 14	198	53 27	229	27 02	319
3	16 56	359	17 36	028	16 03	086	32 00	134	19 00	198	52 54	229	26 33	318
4	16 55	358	17 56	027	16 46	085	32 31	134	18 47	198	52 21	229	26 04	318
5	16 53	357	18 15	026	17 29	084	33 02	134	18 34	197	51 48	229	25 35	317
6	16 51	356	18 33	025	18 12	084	33 34	133	18 22	197	51 16	229	25 05	316
7	16 48	355	18 51	024	18 55	083	34 05	133	18 10	196	50 43	229	24 35	315
8	16 44	355	19 08	023	19 38	082	34 37	133	17 58	196	50 10	229	24 04	314
9	16 40	354	19 25	022	20 21	082	35 09	132	17 46	195	49 38	229	23 32	313
10	16 34	353	19 41	021	21 03	081	35 41	132	17 35	195	49 06	229	23 00	312
11	16 29	352	19 56	020	21 46	080	36 13	132	17 24	194	48 33	228	22 28	311
12	16 22	351	20 11	019	22 28	079	36 45	131	17 14	194	48 01	228	21 55	310
13	16 15	350	20 25	018	23 11	079	37 18	131	17 04	193	47 29	228	21 22	309
14	16 07	349	20 38	017	23 53	078	37 51	131	16 54	193	46 57	228	20 48	308
	◆Hamal		RIGEL		◆CANOPUS		RIGIL KENT.		◆Peacock		FOMALHAUT		Alpheratz	
15	20 50	016	24 35	077	38 23	130	16 45	192	46 25	228	61 49	290	15 59	348
16	21 02	015	25 17	076	38 57	130	16 36	192	45 53	228	61 08	289	15 50	347
17	21 13	014	25 59	076	39 30	130	16 27	191	45 21	227	60 27	288	15 40	346
18	21 24	014	26 41	075	40 03	129	16 19	191	44 49	227	59 46	287	15 29	346
19	21 34	013	27 22	074	40 37	129	16 11	190	44 18	227	59 05	286	15 18	345
20	21 43	012	28 04	073	41 10	129	16 03	190	43 46	227	58 23	285	15 07	344
21	21 51	011	28 45	072	41 44	128	15 56	189	43 15	227	57 41	284	14 54	343
22	21 58	010	29 26	072	42 18	128	15 49	189	42 43	226	56 59	283	14 41	342
23	22 05	009	30 07	071	42 52	128	15 42	188	42 12	226	56 17	282	14 28	341
24	22 11	008	30 47	070	43 26	127	15 36	188	41 41	226	55 35	281	14 13	340
25	22 17	007	31 28	069	44 00	127	15 30	187	41 10	226	54 53	280	13 58	339
26	22 21	006	32 08	068	44 35	127	15 25	187	40 40	225	54 10	279	13 43	339
27	22 25	005	32 48	067	45 09	127	15 20	186	40 09	225	53 27	279	13 27	338
28	22 28	004	33 27	066	45 44	126	15 15	186	39 38	225	52 45	278	13 10	337
29	22 31	003	34 07	066	46 19	126	15 11	185	39 08	225	52 02	277	12 53	336
	◆Hamal		RIGEL		SIRIUS		◆CANOPUS		RIGIL KENT.		◆Peacock		FOMALHAUT	
30	22 33	002	34 46	065	24 55	090	46 54	126	15 07	185	38 38	224	51 19	276
31	22 33	001	35 25	064	25 38	089	47 29	126	15 04	184	38 08	224	50 36	275
32	22 34	000	36 03	063	26 22	088	48 04	125	15 00	184	37 38	224	49 53	275
33	22 33	359	36 42	062	27 05	087	48 39	125	14 58	183	37 08	223	49 10	274
34	22 32	358	37 19	061	27 48	087	49 15	125	14 55	183	36 38	223	48 27	273
35	22 30	357	37 57	060	28 31	086	49 50	125	14 53	182	36 09	223	47 44	272
36	22 27	356	38 34	059	29 14	085	50 25	125	14 52	182	35 40	222	47 01	272
37	22 23	355	39 11	058	29 57	085	51 01	124	14 50	181	35 11	222	46 17	271
38	22 19	354	39 47	057	30 40	084	51 37	124	14 50	181	34 42	222	45 34	270
39	22 14	353	40 23	056	31 23	083	52 12	124	14 49	180	34 14	221	44 51	269
40	22 08	352	40 59	055	32 06	082	52 48	124	14 49	180	33 45	221	44 08	269
41	22 02	351	41 34	054	32 48	082	53 24	124	14 49	179	33 17	221	43 25	268
42	21 54	350	42 08	053	33 31	081	54 00	124	14 50	179	32 49	220	42 42	267
43	21 46	349	42 42	052	34 14	080	54 36	123	14 51	178	32 21	220	41 59	267
44	21 38	348	43 16	050	34 56	079	55 12	123	14 52	178	31 53	220	41 16	266
	◆RIGEL		SIRIUS		CANOPUS		◆RIGIL KENT.		Peacock		◆FOMALHAUT		Hamal	
45	43 49	049	35 38	078	55 48	123	14 54	177	31 25	219	40 32	265	21 28	347
46	44 21	048	36 21	078	56 25	123	14 56	177	30 58	219	39 49	265	21 18	346
47	44 53	047	37 03	077	57 01	123	14 59	176	30 31	219	39 07	264	21 07	345
48	45 25	046	37 45	076	57 37	123	15 02	176	30 04	218	38 24	264	20 56	344
49	45 55	045	38 26	075	58 13	123	15 05	175	29 38	218	37 41	263	20 44	343
50	46 25	043	39 08	074	58 50	123	15 09	175	29 11	218	36 58	262	20 31	342
51	46 55	042	39 49	073	59 26	123	15 13	174	28 45	217	36 15	262	20 17	341
52	47 23	041	40 31	073	60 02	123	15 17	174	28 19	217	35 33	261	20 03	340
53	47 51	040	41 12	072	60 39	123	15 22	173	27 53	216	34 50	260	19 48	339
54	48 18	038	41 53	071	61 15	123	15 27	173	27 28	216	34 07	260	19 32	338
55	48 44	037	42 33	070	61 51	123	15 33	172	27 02	216	33 25	259	19 16	337
56	49 10	036	43 14	069	62 28	123	15 38	172	26 37	215	32 43	259	18 59	336
57	49 35	034	43 54	068	63 04	123	15 45	171	26 12	215	32 00	258	18 42	336
58	49 58	033	44 34	067	63 40	123	15 51	171	25 48	214	31 18	257	18 23	335
59	50 21	031	45 13	066	64 17	123	15 58	170	25 24	214	30 36	257	18 05	334
	ALDEBARAN		◆SIRIUS		Suhail		◆RIGIL KENT.		ACHERNAR		◆Diphda		Hamal	
60	28 57	010	45 53	065	36 32	118	16 06	170	64 14	226	41 28	286	17 45	333
61	29 04	009	46 32	064	37 10	118	16 13	169	63 43	227	40 46	285	17 25	332
62	29 10	008	47 10	063	37 49	117	16 22	169	63 11	227	40 05	285	17 05	331
63	29 16	006	47 49	062	38 27	117	16 30	168	62 39	227	39 23	284	16 43	330
64	29 20	005	48 26	061	39 06	117	16 39	168	62 08	228	38 41	283	16 22	329
65	29 24	004	49 04	060	39 44	116	16 48	167	61 36	228	37 58	282	15 59	328
66	29 27	003	49 41	059	40 23	116	16 58	167	61 04	228	37 16	281	15 36	328
67	29 29	002	50 18	057	41 02	115	17 07	167	60 32	228	36 34	280	15 13	327
68	29 30	001	50 54	056	41 41	115	17 18	166	59 59	228	35 51	280	14 49	326
69	29 30	000	51 29	055	42 20	114	17 28	166	59 27	228	35 09	279	14 25	325
70	29 30	359	52 05	054	43 00	114	17 39	165	58 55	228	34 26	278	14 00	324
71	29 28	358	52 39	053	43 39	114	17 51	165	58 23	229	33 43	277	13 34	323
72	29 26	357	53 13	051	44 19	113	18 02	164	57 50	229	33 00	277	13 08	322
73	29 23	355	53 47	050	44 59	113	18 14	164	57 18	229	32 18	276	12 41	322
74	29 19	354	54 19	049	45 39	112	18 26	163	56 45	229	31 35	275	12 14	321
	BETELGEUSE		◆SIRIUS		Suhail		◆RIGIL KENT.		ACHERNAR		◆Diphda		ALDEBARAN	
75	37 07	017	54 52	047	46 19	112	18 39	163	56 13	229	30 52	274	29 15	353
76	37 19	016	55 23	046	46 59	111	18 52	162	55 41	229	30 08	274	29 09	352
77	37 31	015	55 54	045	47 39	111	19 05	162	55 08	229	29 25	273	29 03	351
78	37 41	013	56 23	043	48 19	111	19 19	161	54 36	229	28 42	272	28 56	350
79	37 51	012	56 52	042	49 00	110	19 33	161	54 03	229	27 59	272	28 48	349
80	37 59	011	57 21	040	49 40	110	19 47	160	53 31	229	27 16	271	28 39	348
81	38 07	010	57 48	038	50 21	109	20 02	160	52 58	229	26 33	270	28 30	347
82	38 14	009	58 14	037	51 02	109	20 17	160	52 26	229	25 50	269	28 20	346
83	38 20	007	58 40	035	51 43	109	20 32	159	51 54	228	25 07	269	28 09	345
84	38 25	006	59 04	034	52 24	108	20 47	159	51 21	228	24 23	268	27 57	344
85	38 29	005	59 27	032	53 05	108	21 03	158	50 49	228	23 40	267	27 44	343
86	38 32	003	59 49	030	53 46	107	21 19	158	50 17	228	22 57	267	27 31	341
87	38 34	002	60 10	028	54 27	107	21 36	157	49 45	228	22 14	266	27 17	340
88	38 35	001	60 30	026	55 08	107	21 53	157	49 13	228	21 31	265	27 02	339
89	38 36	000	60 49	025	55 50	106	22 10	157	48 41	228	20 48	265	26 46	338

LHA	Hc	Zn	Hc	Zn	Hc	Zn	Hc	Zn	Hc	Zn	Hc	Zn	Hc	Zn
♈	PROCYON		Alphard		◆Suhail		RIGIL KENT.		◆ACHERNAR		RIGEL		◆BETELGEUSE	
90	35 57	031	32 56	068	56 31	106	22 27	156	48 09	228	52 51	341	38 35	358
91	36 19	030	33 36	067	57 13	105	22 45	156	47 37	227	52 36	339	38 33	357
92	36 40	029	34 16	066	57 55	105	23 03	155	47 06	227	52 21	338	38 30	356
93	37 00	028	34 55	065	58 36	105	23 21	155	46 34	227	52 04	336	38 27	355
94	37 20	026	35 34	064	59 18	104	23 39	154	46 02	227	51 46	335	38 22	353
95	37 39	025	36 13	063	60 00	104	23 58	154	45 31	227	51 27	333	38 17	352
96	37 57	024	36 51	062	60 42	103	24 17	154	44 59	227	51 07	332	38 10	351
97	38 14	023	37 29	061	61 24	103	24 36	153	44 28	226	50 46	330	38 03	350
98	38 30	022	38 07	061	62 06	103	24 56	153	43 57	226	50 25	329	37 55	348
99	38 45	020	38 45	060	62 48	102	25 16	152	43 26	226	50 02	327	37 46	347
100	39 00	019	39 22	059	63 30	102	25 36	152	42 55	226	49 38	326	37 35	346
101	39 13	018	39 58	058	64 13	101	25 56	152	42 24	225	49 14	325	37 24	345
102	39 26	017	40 34	056	64 55	101	26 17	151	41 54	225	48 48	323	37 13	343
103	39 38	015	41 10	055	65 37	101	26 38	151	41 23	225	48 22	322	37 00	342
104	39 49	014	41 46	054	66 20	100	26 59	150	40 53	225	47 55	321	36 46	341
	PROCYON		REGULUS		◆Gienah		RIGIL KENT.		◆ACHERNAR		RIGEL		◆BETELGEUSE	
105	39 59	013	19 36	049	19 57	095	27 20	150	40 22	224	47 27	319	36 32	340
106	40 08	011	20 08	049	20 40	095	27 42	150	39 52	224	46 59	318	36 16	339
107	40 16	010	20 40	048	21 23	094	28 04	149	39 22	224	46 30	317	36 00	337
108	40 23	009	21 12	047	22 06	093	28 26	149	38 52	224	46 00	316	35 43	336
109	40 29	008	21 43	046	22 49	093	28 48	149	38 22	223	45 29	314	35 25	335
110	40 35	006	22 14	045	23 33	092	29 11	148	37 53	223	44 58	313	35 07	334
111	40 39	005	22 44	044	24 16	091	29 34	148	37 24	223	44 26	312	34 47	333
112	40 42	004	23 14	043	24 59	091	29 57	148	36 54	223	43 54	311	34 27	332
113	40 44	002	23 43	042	25 42	090	30 20	147	36 25	222	43 21	310	34 06	331
114	40 45	001	24 12	041	26 25	089	30 43	147	35 56	222	42 47	309	33 45	329
115	40 46	000	24 40	040	27 08	089	31 07	147	35 28	222	42 13	307	33 22	328
116	40 45	358	25 08	039	27 51	088	31 31	146	34 59	221	41 39	306	32 59	327
117	40 43	357	25 35	038	28 35	087	31 55	146	34 31	221	41 04	305	32 36	326
118	40 41	356	26 02	038	29 18	086	32 19	146	34 02	221	40 29	304	32 11	325
119	40 37	354	26 28	037	30 01	086	32 44	145	33 34	220	39 53	303	31 46	324
	REGULUS		◆SPICA		RIGIL KENT.		◆ACHERNAR		RIGEL		BETELGEUSE		◆PROCYON	
120	26 53	036	13 59	092	33 08	145	33 07	220	39 16	302	31 20	323	40 32	353
121	27 18	035	14 43	091	33 33	145	32 39	220	38 40	301	30 54	322	40 26	352
122	27 42	034	15 26	091	33 58	144	32 11	219	38 03	300	30 27	321	40 20	351
123	28 06	033	16 09	090	34 24	144	31 44	219	37 25	299	30 00	320	40 12	349
124	28 29	032	16 52	089	34 49	144	31 17	219	36 47	298	29 32	319	40 04	348
125	28 51	030	17 35	089	35 15	143	30 50	218	36 09	297	29 03	318	39 54	347
126	29 12	029	18 18	088	35 40	143	30 24	218	35 31	296	28 34	317	39 44	345
127	29 33	028	19 01	087	36 06	143	29 57	218	34 52	296	28 04	316	39 32	344
128	29 53	027	19 44	086	36 32	143	29 31	217	34 13	295	27 34	315	39 20	343
129	30 13	026	20 28	086	36 59	142	29 05	217	33 33	294	27 03	314	39 07	342
130	30 32	025	21 11	085	37 25	142	28 39	216	32 54	293	26 31	313	38 53	340
131	30 50	024	21 54	084	37 52	142	28 14	216	32 14	292	26 00	312	38 38	339
132	31 07	023	22 36	084	38 18	142	27 49	216	31 34	291	25 27	311	38 22	338
133	31 23	022	23 19	083	38 45	141	27 24	215	30 53	290	24 55	310	38 06	337
134	31 39	021	24 02	082	39 12	141	26 59	215	30 13	289	24 21	309	37 48	335
	REGULUS		◆SPICA		RIGIL KENT.		◆ACHERNAR		CANOPUS		SIRIUS		◆PROCYON	
135	31 54	020	24 45	081	39 39	141	26 34	214	63 01	237	50 34	303	37 30	334
136	32 08	019	25 27	081	40 07	141	26 10	214	62 25	237	49 58	302	37 11	333
137	32 22	017	26 10	080	40 34	140	25 46	214	61 48	237	49 21	301	36 51	332
138	32 34	016	26 52	079	41 02	140	25 22	213	61 12	237	48 44	300	36 30	331
139	32 46	015	27 35	078	41 29	140	24 59	213	60 36	237	48 06	299	36 08	330
140	32 57	014	28 17	078	41 57	140	24 35	212	59 59	237	47 28	297	35 46	328
141	33 07	013	28 59	077	42 25	140	24 12	212	59 23	237	46 50	296	35 23	327
142	33 16	012	29 41	076	42 53	139	23 50	212	58 47	237	46 11	295	35 00	326
143	33 24	011	30 23	075	43 21	139	23 27	211	58 10	237	45 32	294	34 35	325
144	33 32	009	31 04	074	43 49	139	23 05	211	57 34	237	44 52	293	34 10	324
145	33 39	008	31 46	074	44 18	139	22 43	210	56 58	237	44 12	292	33 44	323
146	33 44	007	32 27	073	44 46	139	22 21	210	56 22	237	43 32	291	33 18	322
147	33 49	006	33 08	072	45 15	139	22 00	209	55 45	237	42 52	291	32 51	321
148	33 53	005	33 49	071	45 43	138	21 39	209	55 09	237	42 12	290	32 24	320
149	33 56	004	34 30	070	46 12	138	21 18	209	54 33	237	41 31	289	31 55	319
	REGULUS		◆SPICA		RIGIL KENT.		◆ACHERNAR		CANOPUS		SIRIUS		◆PROCYON	
150	33 59	002	35 11	069	46 41	138	20 58	208	53 57	236	40 50	288	31 27	318
151	34 00	001	35 51	069	47 10	138	20 38	208	53 21	236	40 09	287	30 57	317
152	34 00	000	36 31	068	47 38	138	20 18	207	52 45	236	39 27	286	30 27	316
153	34 00	359	37 11	067	48 07	138	19 58	207	52 10	236	38 46	285	29 57	315
154	33 59	358	37 50	066	48 36	138	19 39	206	51 34	236	38 04	284	29 26	314
155	33 57	356	38 29	065	49 05	138	19 20	206	50 58	236	37 22	284	28 54	313
156	33 53	355	39 08	064	49 35	138	19 01	205	50 23	235	36 40	283	28 22	312
157	33 50	354	39 47	063	50 04	137	18 43	205	49 47	235	35 58	282	27 50	311
158	33 45	353	40 25	062	50 33	137	18 25	204	49 12	235	35 16	281	27 17	310
159	33 39	352	41 03	061	51 02	137	18 07	204	48 36	235	34 33	280	26 44	309
160	33 32	351	41 41	060	51 31	137	17 50	203	48 01	235	33 51	280	26 10	308
161	33 25	349	42 18	059	52 01	137	17 33	203	47 26	234	33 08	279	25 36	307
162	33 17	348	42 55	058	52 30	137	17 17	202	46 51	234	32 25	278	25 01	306
163	33 07	347	43 31	057	52 59	137	17 00	202	46 16	234	31 43	277	24 26	305
164	32 57	346	44 07	056	53 28	137	16 44	201	45 41	234	31 00	277	23 51	304
	SPICA		◆ANTARES		RIGIL KENT.		ACHERNAR		◆CANOPUS		SIRIUS		◆REGULUS	
165	44 42	055	23 19	105	53 58	137	16 29	201	45 07	233	30 17	276	32 47	345
166	45 17	054	24 00	104	54 27	137	16 13	200	44 32	233	29 34	275	32 35	344
167	45 52	052	24 42	104	54 56	137	15 58	200	43 58	233	28 51	274	32 22	343
168	46 26	051	25 24	103	55 25	138	15 44	199	43 23	233	28 08	274	32 09	341
169	46 59	050	26 06	102	55 54	138	15 30	199	42 49	232	27 25	273	31 55	340
170	47 32	049	26 48	102	56 23	138	15 16	198	42 15	232	26 42	272	31 40	339
171	48 04	048	27 31	101	56 52	138	15 02	198	41 41	232	25 59	271	31 24	338
172	48 36	046	28 13	101	57 21	138	14 49	197	41 08	231	25 15	271	31 08	337
173	49 07	045	28 56	100	57 50	138	14 36	197	40 34	231	24 32	270	30 51	336
174	49 37	044	29 38	099	58 19	138	14 24	196	40 00	231	23 49	269	30 33	335
175	50 07	043	30 21	099	58 48	138	14 12	196	39 27	230	23 06	269	30 14	334
176	50 35	041	31 03	098	59 16	139	14 00	195	38 54	230	22 23	268	29 54	333
177	51 03	040	31 46	097	59 45	139	13 49	195	38 21	230	21 40	267	29 34	332
178	51 31	038	32 29	097	60 13	139	13 38	194	37 48	229	20 57	267	29 14	331
179	51 57	037	33 12	096	60 41	139	13 28	194	37 15	229	20 13	266	28 52	330

LHA ♈	Hc	Zn	Hc	Zn	Hc	Zn	Hc	Zn	Hc	Zn	Hc	Zn	Hc	Zn
	◆ARCTURUS		ANTARES		◆RIGIL KENT.		ACHERNAR		CANOPUS		◆Suhail		REGULUS	
180	19 36	034	33 55	096	61 09	140	13 18	193	36 43	229	59 14	256	28 30	329
181	20 00	033	34 38	095	61 37	140	13 08	193	36 10	228	58 32	255	28 07	328
182	20 23	032	35 21	094	62 05	140	12 59	192	35 38	228	57 51	255	27 43	326
183	20 46	031	36 04	094	62 32	141	12 50	192	35 06	228	57 09	255	27 19	325
184	21 08	030	36 47	093	62 59	141	12 41	191	34 34	227	56 27	254	26 54	324
185	21 30	029	37 30	092	63 26	141	12 33	191	34 03	227	55 46	254	26 29	324
186	21 50	028	38 13	092	63 53	142	12 25	190	33 31	227	55 04	253	26 03	323
187	22 11	027	38 56	091	64 20	142	12 18	190	33 00	226	54 23	253	25 37	322
188	22 30	026	39 40	090	64 46	143	12 11	189	32 29	226	53 42	253	25 09	321
189	22 49	026	40 23	089	65 12	143	12 05	188	31 58	226	53 01	252	24 42	320
190	23 07	025	41 06	089	65 37	144	11 59	188	31 27	225	52 20	252	24 14	319
191	23 25	024	41 49	088	66 02	145	11 53	187	30 57	225	51 39	251	23 45	318
192	23 42	023	42 32	087	66 27	145	11 47	187	30 26	224	50 58	251	23 16	317
193	23 58	022	43 15	087	66 51	146	11 43	186	29 56	224	50 17	251	22 46	316
194	24 14	021	43 58	086	67 15	147	11 38	186	29 27	224	49 37	250	22 16	315
	◆ARCTURUS		ANTARES		◆Peacock		ACHERNAR		CANOPUS		◆Suhail		REGULUS	
195	24 28	020	44 41	085	25 57	145	11 34	185	28 57	223	48 56	250	21 45	314
196	24 43	019	45 24	084	26 22	145	11 30	185	28 28	223	48 16	249	21 14	313
197	24 56	018	46 07	084	26 47	145	11 27	184	27 58	222	47 35	249	20 42	312
198	25 09	017	46 50	083	27 12	144	11 24	184	27 29	222	46 55	249	20 10	311
199	25 21	016	47 33	082	27 37	144	11 22	183	27 01	222	46 15	248	19 37	311
200	25 32	015	48 16	081	28 03	143	11 20	182	26 32	221	45 35	248	19 04	310
201	25 42	013	48 58	080	28 29	143	11 18	182	26 04	221	44 55	247	18 31	309
202	25 52	012	49 41	080	28 55	143	11 17	181	25 36	220	44 15	247	17 57	308
203	26 01	011	50 23	079	29 21	142	11 16	181	25 08	220	43 36	246	17 23	307
204	26 09	010	51 05	078	29 47	142	11 16	180	24 41	219	42 56	246	16 48	306
205	26 16	009	51 47	077	30 14	142	11 16	180	24 13	219	42 17	246	16 13	306
206	26 23	008	52 29	076	30 41	141	11 16	179	23 46	218	41 38	245	15 38	305
207	26 29	007	53 11	075	31 08	141	11 17	179	23 20	218	40 58	245	15 03	304
208	26 34	006	53 53	074	31 36	140	11 18	178	22 53	218	40 19	244	14 27	303
209	26 38	005	54 34	073	32 03	140	11 20	177	22 27	217	39 41	244	13 50	302
	◆ARCTURUS		ANTARES		◆Peacock		ACHERNAR		CANOPUS		◆Suhail		SPICA	
210	26 42	004	55 15	072	32 31	140	11 22	177	22 01	217	39 02	243	56 16	344
211	26 44	003	55 56	071	32 59	139	11 25	176	21 36	216	38 23	243	56 04	343
212	26 46	002	56 37	070	33 27	139	11 28	176	21 10	216	37 45	243	55 51	341
213	26 47	001	57 18	069	33 56	139	11 31	175	20 45	215	37 07	242	55 36	339
214	26 47	000	57 58	068	34 24	138	11 35	175	20 20	215	36 29	242	55 20	338
215	26 47	359	58 37	067	34 53	138	11 39	174	19 56	214	35 51	241	55 03	336
216	26 46	358	59 17	066	35 22	138	11 44	174	19 32	214	35 13	241	54 45	334
217	26 44	357	59 56	064	35 51	137	11 49	173	19 08	213	34 35	240	54 26	333
218	26 41	356	60 35	063	36 20	137	11 54	173	18 44	213	33 58	240	54 05	331
219	26 37	355	61 13	062	36 50	137	12 00	172	18 21	212	33 21	239	53 44	330
220	26 33	354	61 51	060	37 19	137	12 06	171	17 58	212	32 44	239	53 22	328
221	26 27	352	62 28	059	37 49	136	12 13	171	17 35	211	32 07	239	52 58	327
222	26 21	351	63 05	058	38 19	136	12 20	170	17 13	211	31 30	238	52 34	325
223	26 14	350	63 41	056	38 49	136	12 27	170	16 51	210	30 54	238	52 09	324
224	26 07	349	64 16	055	39 19	135	12 35	169	16 29	210	30 17	237	51 43	322
	Rasalhague		◆Nunki		Peacock		ACHERNAR		◆ACRUX		SPICA		◆ARCTURUS	
225	23 24	042	39 59	090	39 50	135	12 43	169	61 00	215	51 16	321	25 58	348
226	23 52	041	40 42	089	40 20	135	12 52	168	60 34	216	50 48	319	25 49	347
227	24 20	040	41 25	088	40 51	135	13 01	168	60 09	216	50 20	318	25 39	346
228	24 47	039	42 08	088	41 22	134	13 10	167	59 44	216	49 51	317	25 29	345
229	25 14	038	42 51	087	41 53	134	13 20	167	59 18	217	49 21	315	25 17	344
230	25 40	037	43 34	086	42 24	134	13 30	166	58 52	217	48 50	314	25 05	343
231	26 06	036	44 17	085	42 55	134	13 41	166	58 26	217	48 19	313	24 52	342
232	26 31	035	45 00	085	43 26	133	13 52	165	58 00	217	47 47	312	24 39	341
233	26 55	034	45 43	084	43 58	133	14 03	165	57 33	218	47 14	310	24 24	340
234	27 19	033	46 26	083	44 29	133	14 15	164	57 07	218	46 41	309	24 09	339
235	27 42	032	47 09	082	45 01	133	14 27	163	56 40	218	46 08	308	23 54	338
236	28 05	031	47 52	081	45 33	133	14 39	163	56 14	218	45 33	307	23 37	337
237	28 27	030	48 34	081	46 05	132	14 52	162	55 47	218	44 59	306	23 20	336
238	28 48	029	49 17	080	46 37	132	15 05	162	55 20	218	44 23	305	23 02	335
239	29 08	028	49 59	079	47 09	132	15 19	161	54 53	219	43 48	304	22 44	334
	Rasalhague		◆ALTAIR		Peacock		◆ACHERNAR		ACRUX		◆SPICA		ARCTURUS	
240	29 28	027	15 53	060	47 41	132	15 33	161	54 26	219	43 12	303	22 25	333
241	29 47	026	16 30	059	48 13	132	15 47	160	53 59	219	42 35	302	22 05	332
242	30 05	025	17 07	059	48 45	131	16 02	160	53 32	219	41 58	300	21 45	331
243	30 23	024	17 44	058	49 18	131	16 17	159	53 05	219	41 21	299	21 24	330
244	30 40	022	18 20	057	49 50	131	16 32	159	52 38	219	40 43	298	21 02	329
245	30 56	021	18 56	056	50 23	131	16 48	158	52 11	219	40 05	297	20 40	329
246	31 11	020	19 32	055	50 55	131	17 04	158	51 44	219	39 26	297	20 17	328
247	31 26	019	20 07	054	51 28	131	17 20	157	51 16	219	38 48	296	19 53	327
248	31 40	018	20 42	054	52 01	131	17 37	157	50 49	219	38 08	295	19 30	326
249	31 53	017	21 16	053	52 33	131	17 54	156	50 22	219	37 29	294	19 05	325
250	32 05	016	21 50	052	53 06	131	18 12	156	49 55	219	36 49	293	18 40	324
251	32 16	015	22 24	051	53 39	130	18 29	156	49 28	219	36 10	292	18 14	323
252	32 27	014	22 57	050	54 12	130	18 47	155	49 01	219	35 29	291	17 48	322
253	32 36	012	23 30	049	54 45	130	19 06	155	48 34	219	34 49	290	17 21	321
254	32 45	011	24 03	048	55 17	130	19 24	154	48 07	219	34 08	289	16 54	320
	Rasalhague		◆ALTAIR		FOMALHAUT		◆ACHERNAR		RIGIL KENT.		◆SPICA		Alphecca	
255	32 53	010	24 35	047	20 32	112	19 43	154	63 16	219	33 27	288	16 36	340
256	33 00	009	25 06	046	21 12	111	20 03	153	62 49	219	32 46	288	16 21	339
257	33 07	008	25 37	046	21 53	111	20 22	153	62 22	219	32 05	287	16 06	338
258	33 12	007	26 08	045	22 33	110	20 42	152	61 54	220	31 24	286	15 50	337
259	33 17	005	26 38	044	23 14	110	21 03	152	61 27	220	30 42	285	15 33	337
260	33 20	004	27 08	043	23 54	109	21 23	151	60 59	220	30 00	284	15 15	336
261	33 23	003	27 37	042	24 35	108	21 44	151	60 31	221	29 19	284	14 57	335
262	33 25	002	28 05	041	25 16	108	22 05	150	60 02	221	28 37	283	14 39	334
263	33 26	001	28 33	040	25 58	107	22 26	150	59 34	221	27 54	282	14 19	333
264	33 26	000	29 00	039	26 39	107	22 48	150	59 06	221	27 12	281	13 59	332
265	33 25	358	29 27	038	27 20	106	23 10	149	58 37	222	26 30	280	13 39	331
266	33 24	357	29 53	037	28 02	105	23 32	149	58 08	222	25 47	280	13 18	331
267	33 21	356	30 18	036	28 44	105	23 55	148	57 39	222	25 05	279	12 57	330
268	33 18	355	30 43	035	29 25	104	24 18	148	57 11	222	24 22	278	12 35	329
269	33 14	354	31 07	034	30 07	104	24 41	147	56 42	222	23 39	277	12 12	328

LAT 44°S

LHA ♈	Hc	Zn	Hc	Zn	Hc	Zn	Hc	Zn	Hc	Zn	Hc	Zn	Hc	Zn
	ALTAIR		Enif		◆FOMALHAUT		ACHERNAR		◆RIGIL KENT.		ANTARES		◆Rasalhague	
270	31 31	033	16 07	058	30 49	103	25 04	147	56 13	222	64 36	306	33 09	353
271	31 54	031	16 44	057	31 31	102	25 28	147	55 43	222	64 01	305	33 03	351
272	32 16	030	17 20	057	32 14	102	25 52	146	55 14	223	63 25	303	32 56	350
273	32 38	029	17 56	056	32 56	101	26 16	146	54 45	223	62 48	302	32 48	349
274	32 58	028	18 31	055	33 38	101	26 40	145	54 16	223	62 11	300	32 40	348
275	33 18	027	19 06	054	34 21	100	27 05	145	53 47	223	61 34	299	32 30	347
276	33 38	026	19 41	053	35 03	099	27 30	145	53 17	223	60 56	298	32 20	346
277	33 56	025	20 15	052	35 46	099	27 55	144	52 48	223	60 17	296	32 09	345
278	34 14	024	20 49	052	36 29	098	28 20	144	52 19	223	59 38	295	31 57	343
279	34 31	023	21 23	051	37 11	098	28 46	144	51 50	223	58 59	294	31 44	342
280	34 47	021	21 56	050	37 54	097	29 11	143	51 20	223	58 20	293	31 31	341
281	35 02	020	22 29	049	38 37	096	29 37	143	50 51	223	57 40	292	31 17	340
282	35 17	019	23 01	048	39 20	096	30 04	142	50 22	223	56 59	291	31 01	339
283	35 30	018	23 33	047	40 03	095	30 30	142	49 53	222	56 19	289	30 46	338
284	35 43	017	24 05	046	40 46	094	30 57	142	49 24	222	55 38	288	30 29	337
	ALTAIR		Enif		◆FOMALHAUT		ACHERNAR		◆RIGIL KENT.		ANTARES		◆Rasalhague	
285	35 55	015	24 36	045	41 29	094	31 24	141	48 55	222	54 57	287	30 12	336
286	36 06	014	25 06	044	42 12	093	31 51	141	48 26	222	54 16	286	29 54	335
287	36 16	013	25 36	043	42 55	092	32 18	141	47 57	222	53 34	285	29 35	334
288	36 26	012	26 05	042	43 38	092	32 45	140	47 28	222	52 52	285	29 15	333
289	36 34	011	26 34	042	44 21	091	33 13	140	46 59	222	52 11	284	28 55	332
290	36 42	009	27 02	041	45 05	090	33 41	140	46 30	222	51 29	283	28 34	330
291	36 48	008	27 30	040	45 48	090	34 09	139	46 01	222	50 46	282	28 12	329
292	36 54	007	27 57	039	46 31	089	34 37	139	45 33	222	50 04	281	27 50	328
293	36 59	006	28 24	038	47 14	088	35 06	139	45 04	221	49 22	280	27 27	327
294	37 03	005	28 50	037	47 57	087	35 34	138	44 36	221	48 39	279	27 04	326
295	37 05	003	29 15	036	48 40	087	36 03	138	44 07	221	47 56	278	26 40	325
296	37 07	002	29 40	035	49 23	086	36 32	138	43 39	221	47 14	278	26 15	324
297	37 09	001	30 04	033	50 06	085	37 01	137	43 11	221	46 31	277	25 49	323
298	37 09	000	30 28	032	50 49	084	37 30	137	42 43	221	45 48	276	25 23	322
299	37 08	358	30 51	031	51 32	084	38 00	137	42 15	220	45 05	275	24 57	322
	◆Enif		FOMALHAUT		◆ACHERNAR		RIGIL KENT.		◆ANTARES		Rasalhague		ALTAIR	
300	31 13	030	52 15	083	38 29	137	41 47	220	44 22	275	24 30	321	37 06	357
301	31 34	029	52 58	082	38 59	136	41 19	220	43 39	274	24 02	320	37 03	356
302	31 55	028	53 41	081	39 29	136	40 51	220	42 56	273	23 34	319	37 00	355
303	32 15	027	54 23	080	39 59	136	40 24	219	42 13	272	23 05	318	36 55	353
304	32 34	026	55 06	079	40 29	135	39 56	219	41 30	272	22 36	317	36 50	352
305	32 53	025	55 48	079	41 00	135	39 29	219	40 47	271	22 06	316	36 43	351
306	33 10	024	56 30	078	41 30	135	39 02	219	40 03	270	21 36	315	36 36	350
307	33 27	023	57 12	077	42 01	135	38 35	219	39 20	270	21 05	314	36 28	348
308	33 44	021	57 54	076	42 31	135	38 08	218	38 37	269	20 34	313	36 19	347
309	33 59	020	58 36	075	43 02	134	37 42	218	37 54	268	20 02	312	36 09	346
310	34 14	019	59 18	074	43 33	134	37 15	218	37 11	267	19 30	312	35 58	345
311	34 27	018	59 59	073	44 04	134	36 49	218	36 28	267	18 57	311	35 46	344
312	34 40	017	60 40	072	44 36	134	36 23	217	35 45	266	18 24	310	35 34	342
313	34 52	016	61 21	071	45 07	133	35 57	217	35 02	265	17 51	309	35 20	341
314	35 04	014	62 02	069	45 38	133	35 31	217	34 19	265	17 17	308	35 06	340
	Enif		◆Diphda		ACHERNAR		Miaplacidus		◆RIGIL KENT.		ANTARES		◆ALTAIR	
315	35 14	013	36 48	079	46 10	133	23 43	181	35 05	216	33 36	264	34 51	339
316	35 23	012	37 30	079	46 41	133	23 43	181	34 39	216	32 53	264	34 35	338
317	35 32	011	38 12	078	47 13	133	23 42	180	34 14	216	32 10	263	34 18	337
318	35 40	010	38 55	077	47 45	133	23 42	180	33 49	216	31 27	262	34 01	335
319	35 47	008	39 37	076	48 17	132	23 42	180	33 24	215	30 44	262	33 43	334
320	35 52	007	40 18	075	48 49	132	23 42	179	32 59	215	30 02	261	33 24	333
321	35 57	006	41 00	074	49 20	132	23 43	179	32 35	215	29 19	260	33 04	332
322	36 02	005	41 41	073	49 53	132	23 44	179	32 10	214	28 36	260	32 43	331
323	36 05	004	42 23	073	50 25	132	23 45	178	31 46	214	27 54	259	32 22	330
324	36 07	002	43 04	072	50 57	132	23 46	178	31 22	214	27 12	259	32 00	329
325	36 08	001	43 45	071	51 29	132	23 48	177	30 58	213	26 29	258	31 37	328
326	36 09	000	44 25	070	52 01	132	23 50	177	30 35	213	25 47	257	31 14	327
327	36 08	359	45 06	069	52 34	131	23 53	177	30 11	213	25 05	257	30 50	326
328	36 07	358	45 46	068	53 06	131	23 55	176	29 48	212	24 23	256	30 25	325
329	36 05	356	46 26	067	53 38	131	23 58	176	29 25	212	23 42	255	30 00	324
	◆Diphda		ACHERNAR		◆Miaplacidus		RIGIL KENT.		Nunki		◆ALTAIR		Enif	
330	47 05	066	54 11	131	24 01	176	29 02	212	48 54	280	29 34	323	36 01	355
331	47 45	065	54 43	131	24 05	175	28 40	211	48 12	279	29 07	322	35 57	354
332	48 23	064	55 16	131	24 09	175	28 18	211	47 29	278	28 40	321	35 52	353
333	49 02	063	55 48	131	24 13	174	27 56	210	46 46	277	28 13	320	35 46	351
334	49 40	062	56 21	131	24 17	174	27 34	210	46 04	277	27 44	319	35 40	350
335	50 18	061	56 53	131	24 21	174	27 12	210	45 21	276	27 15	318	35 32	349
336	50 55	060	57 25	131	24 26	173	26 51	209	44 38	275	26 46	317	35 23	348
337	51 32	058	57 58	131	24 31	173	26 30	209	43 55	274	26 16	316	35 14	347
338	52 09	057	58 30	131	24 37	173	26 09	209	43 12	274	25 46	315	35 03	345
339	52 45	056	59 03	132	24 43	172	25 49	208	42 29	273	25 15	314	34 52	344
340	53 20	055	59 35	132	24 49	172	25 28	208	41 45	272	24 43	313	34 40	343
341	53 55	053	60 07	132	24 55	172	25 08	207	41 02	271	24 12	312	34 27	342
342	54 30	052	60 39	132	25 01	171	24 49	207	40 19	271	23 39	311	34 13	341
343	55 04	051	61 11	132	25 08	171	24 29	207	39 36	270	23 06	310	33 59	340
344	55 37	049	61 43	132	25 15	170	24 10	206	38 53	269	22 33	309	33 43	339
	◆Alpheratz		Diphda		◆CANOPUS		RIGIL KENT.		◆Nunki		ALTAIR		Enif	
345	15 18	015	56 09	048	23 22	142	23 51	206	38 10	269	22 00	308	33 27	337
346	15 29	015	56 41	047	23 49	141	23 32	205	37 27	268	21 26	308	33 10	336
347	15 39	014	57 12	045	24 16	141	23 14	205	36 43	267	20 51	307	32 52	335
348	15 49	013	57 42	044	24 43	141	22 56	205	36 00	267	20 16	306	32 34	334
349	15 58	012	58 12	042	25 10	140	22 38	204	35 17	266	19 41	305	32 14	333
350	16 07	011	58 40	041	25 38	140	22 20	204	34 34	265	19 06	304	31 54	332
351	16 15	010	59 08	039	26 06	139	22 03	203	33 51	265	18 30	303	31 34	331
352	16 22	009	59 34	037	26 34	139	21 46	203	33 08	264	17 53	302	31 12	330
353	16 28	008	60 00	036	27 03	138	21 30	202	32 25	263	17 17	302	30 50	329
354	16 34	007	60 25	034	27 32	138	21 13	202	31 43	263	16 40	301	30 27	328
355	16 39	006	60 48	032	28 01	138	20 57	202	31 00	262	16 03	300	30 04	326
356	16 44	006	61 11	030	28 30	137	20 42	201	30 17	261	15 25	299	29 40	325
357	16 48	005	61 32	028	28 59	137	20 26	201	29 34	261	14 47	298	29 15	324
358	16 51	004	61 52	027	29 29	136	20 11	200	28 52	260	14 09	298	28 49	323
359	16 53	003	62 10	025	29 59	136	19 56	200	28 09	260	13 31	297	28 23	322

LHA ♈	Hc	Zn	Hc	Zn	Hc	Zn	Hc	Zn	Hc	Zn	Hc	Zn	Hc	Zn
	Alpheratz		**Hamal**		**◆RIGEL**		**CANOPUS**		**◆RIGIL KENT.**		**Peacock**		**◆Enif**	
0	15 55	002	15 42	030	13 52	088	31 12	135	20 38	199	55 11	231	27 10	322
1	15 56	001	16 03	029	14 34	087	31 42	135	20 24	199	54 38	231	26 43	321
2	15 56	000	16 23	028	15 16	086	32 12	134	20 11	199	54 05	231	26 16	320
3	15 56	359	16 43	027	15 59	086	32 42	134	19 57	198	53 32	230	25 48	319
4	15 55	358	17 02	027	16 41	085	33 13	134	19 44	198	53 00	230	25 20	318
5	15 53	357	17 21	026	17 23	084	33 44	133	19 32	197	52 27	230	24 51	317
6	15 51	356	17 39	025	18 05	083	34 15	133	19 19	197	51 55	230	24 22	316
7	15 48	355	17 56	024	18 48	083	34 46	132	19 07	196	51 22	230	23 52	315
8	15 44	355	18 13	023	19 30	082	35 17	132	18 56	196	50 50	230	23 22	314
9	15 40	354	18 29	022	20 12	081	35 49	132	18 44	195	50 17	230	22 51	313
10	15 35	353	18 45	021	20 53	080	36 21	131	18 33	195	49 45	230	22 20	312
11	15 29	352	19 00	020	21 35	080	36 53	131	18 22	194	49 13	229	21 49	311
12	15 23	351	19 14	019	22 17	079	37 25	131	18 12	194	48 41	229	21 16	310
13	15 16	350	19 28	018	22 59	078	37 57	130	18 02	193	48 08	229	20 44	310
14	15 08	349	19 41	017	23 40	077	38 29	130	17 52	193	47 36	229	20 11	309
	◆Hamal		**RIGEL**		**◆CANOPUS**		**RIGIL KENT.**		**◆Peacock**		**FOMALHAUT**		**Alpheratz**	
15	19 53	016	24 21	077	39 02	130	17 43	192	47 05	229	61 27	292	15 00	348
16	20 04	015	25 03	076	39 35	129	17 34	192	46 33	228	60 48	291	14 51	347
17	20 15	014	25 44	075	40 08	129	17 26	191	46 01	228	60 08	290	14 41	347
18	20 25	013	26 25	074	40 41	129	17 17	191	45 30	228	59 28	289	14 31	346
19	20 35	012	27 05	073	41 14	128	17 10	190	44 58	228	58 48	287	14 20	345
20	20 44	011	27 46	073	41 47	128	17 02	190	44 27	227	58 07	286	14 09	344
21	20 52	011	28 26	072	42 21	128	16 55	189	43 56	227	57 26	285	13 57	343
22	20 59	010	29 07	071	42 55	127	16 48	189	43 25	227	56 45	284	13 44	342
23	21 06	009	29 47	070	43 28	127	16 42	188	42 54	227	56 04	283	13 31	341
24	21 12	008	30 26	069	44 02	127	16 36	188	42 23	226	55 23	283	13 17	340
25	21 17	007	31 06	068	44 36	126	16 30	187	41 52	226	54 41	282	13 02	340
26	21 22	006	31 45	068	45 11	126	16 25	187	41 21	226	54 00	281	12 47	339
27	21 26	005	32 24	067	45 45	126	16 20	186	40 51	226	53 18	280	12 31	338
28	21 29	004	33 03	066	46 19	126	16 15	186	40 21	225	52 36	279	12 15	337
29	21 31	003	33 42	065	46 54	125	16 11	185	39 51	225	51 54	278	11 58	336
	◆Hamal		**RIGEL**		**SIRIUS**		**◆CANOPUS**		**RIGIL KENT.**		**◆Peacock**		**FOMALHAUT**	
30	21 33	002	34 20	064	24 55	089	47 29	125	16 07	185	39 21	225	51 12	277
31	21 33	001	34 58	063	25 37	088	48 03	125	16 03	184	38 51	225	50 30	276
32	21 34	000	35 36	062	26 19	088	48 38	125	16 00	184	38 21	224	49 48	276
33	21 33	359	36 13	061	27 02	087	49 13	124	15 58	183	37 52	224	49 05	275
34	21 32	358	36 50	060	27 44	086	49 49	124	15 55	183	37 22	224	48 23	274
35	21 30	357	37 27	059	28 26	085	50 24	124	15 53	182	36 53	223	47 41	273
36	21 27	356	38 03	058	29 09	085	50 59	124	15 52	182	36 24	223	46 58	273
37	21 24	355	38 39	057	29 51	084	51 34	123	15 50	181	35 55	223	46 16	272
38	21 19	354	39 14	056	30 33	083	52 10	123	15 50	181	35 27	222	45 34	271
39	21 14	353	39 49	055	31 15	082	52 46	123	15 49	180	34 58	222	44 51	270
40	21 09	352	40 24	054	31 57	082	53 21	123	15 49	180	34 30	222	44 09	270
41	21 02	351	40 58	053	32 39	081	53 57	123	15 49	179	34 02	221	43 26	269
42	20 55	350	41 32	052	33 21	080	54 33	122	15 50	179	33 34	221	42 44	268
43	20 47	349	42 05	051	34 03	079	55 09	122	15 51	178	33 06	221	42 01	268
44	20 39	348	42 38	050	34 44	079	55 45	122	15 52	178	32 39	220	41 19	267
	◆RIGEL		**SIRIUS**		**CANOPUS**		**◆RIGIL KENT.**		**Peacock**		**◆FOMALHAUT**		**Hamal**	
45	43 10	049	35 26	078	56 21	122	15 54	177	32 12	220	40 37	266	20 30	347
46	43 41	047	36 07	077	56 57	122	15 56	177	31 45	219	39 54	266	20 20	346
47	44 12	046	36 49	076	57 33	122	15 59	176	31 18	219	39 12	265	20 09	345
48	44 43	045	37 30	075	58 09	121	16 01	176	30 51	219	38 30	264	19 58	344
49	45 12	044	38 11	074	58 45	121	16 05	175	30 25	218	37 48	264	19 46	343
50	45 41	043	38 51	074	59 21	121	16 08	175	29 59	218	37 06	263	19 33	342
51	46 10	041	39 32	073	59 58	121	16 12	174	29 33	218	36 23	262	19 20	341
52	46 37	040	40 12	072	60 34	121	16 17	174	29 07	217	35 41	262	19 06	340
53	47 04	039	40 53	071	61 10	121	16 21	173	28 41	217	35 00	261	18 52	339
54	47 31	038	41 33	070	61 47	121	16 27	173	28 16	216	34 18	261	18 36	338
55	47 56	036	42 12	069	62 23	121	16 32	172	27 51	216	33 36	260	18 21	338
56	48 21	035	42 52	068	62 59	121	16 38	172	27 26	216	32 54	259	18 04	337
57	48 45	033	43 31	067	63 36	121	16 44	171	27 02	215	32 12	259	17 47	336
58	49 08	032	44 10	066	64 12	121	16 51	171	26 37	215	31 31	258	17 29	335
59	49 30	031	44 49	065	64 48	121	16 58	170	26 13	214	30 49	257	17 11	334
	ALDEBARAN		**◆SIRIUS**		**Suhail**		**◆RIGIL KENT.**		**ACHERNAR**		**◆Diphda**		**Hamal**	
60	27 58	010	45 27	064	37 00	118	17 05	170	64 55	228	41 11	287	16 52	333
61	28 05	009	46 05	063	37 38	117	17 12	169	64 23	228	40 30	286	16 32	332
62	28 11	008	46 43	062	38 16	117	17 20	169	63 51	229	39 49	285	16 12	331
63	28 16	006	47 20	061	38 54	116	17 29	168	63 19	229	39 08	284	15 51	330
64	28 20	005	47 57	060	39 32	116	17 38	168	62 48	229	38 27	284	15 30	329
65	28 24	004	48 33	059	40 10	115	17 47	167	62 15	229	37 46	283	15 08	329
66	28 27	003	49 09	058	40 49	115	17 56	167	61 43	229	37 04	282	14 46	328
67	28 29	002	49 45	056	41 27	114	18 06	166	61 11	229	36 23	281	14 23	327
68	28 30	001	50 20	055	42 06	114	18 16	166	60 39	230	35 41	280	13 59	326
69	28 30	000	50 55	054	42 45	114	18 26	165	60 07	230	34 59	280	13 35	325
70	28 30	359	51 29	053	43 24	113	18 37	165	59 34	230	34 17	279	13 11	324
71	28 28	358	52 02	052	44 03	113	18 48	165	59 02	230	33 35	278	12 46	323
72	28 26	357	52 35	050	44 42	112	19 00	164	58 29	230	32 53	277	12 20	323
73	28 23	356	53 08	049	45 21	112	19 12	164	57 57	230	32 11	276	11 54	322
74	28 20	354	53 39	048	46 01	111	19 24	163	57 25	230	31 29	276	11 28	321
	BETELGEUSE		**◆SIRIUS**		**Suhail**		**◆RIGIL KENT.**		**ACHERNAR**		**◆Diphda**		**ALDEBARAN**	
75	36 10	017	54 10	046	46 41	111	19 36	163	56 52	230	30 47	275	28 15	353
76	36 21	016	54 41	045	47 20	110	19 49	162	56 20	230	30 04	274	28 10	352
77	36 33	015	55 10	043	48 00	110	20 02	162	55 47	230	29 22	273	28 04	351
78	36 43	013	55 39	042	48 40	110	20 16	161	55 15	230	28 40	273	27 57	350
79	36 52	012	56 07	041	49 20	109	20 30	161	54 42	230	27 57	272	27 49	349
80	37 01	011	56 34	039	50 00	109	20 44	160	54 10	230	27 15	271	27 41	348
81	37 08	010	57 01	037	50 40	108	20 58	160	53 38	230	26 32	271	27 31	347
82	37 15	008	57 26	036	51 21	108	21 13	159	53 05	230	25 50	270	27 21	346
83	37 20	007	57 50	034	52 01	107	21 28	159	52 33	229	25 08	269	27 11	345
84	37 25	006	58 14	033	52 42	107	21 43	159	52 01	229	24 25	268	26 59	344
85	37 29	005	58 36	031	53 22	106	21 59	158	51 29	229	23 43	268	26 47	343
86	37 32	003	58 57	029	54 03	106	22 15	158	50 57	229	23 00	267	26 34	342
87	37 34	002	59 17	027	54 44	106	22 31	157	50 25	229	22 18	266	26 20	341
88	37 35	001	59 36	026	55 25	105	22 48	157	49 53	229	21 36	266	26 06	340
89	37 36	000	59 54	024	56 06	105	23 05	156	49 21	229	20 53	265	25 51	339

LAT 45°S

LHA ♈	Hc	Zn	Hc	Zn	Hc	Zn	Hc	Zn	Hc	Zn	Hc	Zn	Hc	Zn
	PROCYON		**Alphard**		**◆Suhail**		**RIGIL KENT.**		**◆ACHERNAR**		**RIGEL**		**◆BETELGEUSE**	
90	35 05	031	32 34	067	56 47	104	23 22	156	48 49	228	51 54	341	37 35	358
91	35 27	030	33 13	066	57 28	104	23 39	155	48 18	228	51 40	340	37 33	357
92	35 47	028	33 51	065	58 09	103	23 57	155	47 46	228	51 25	338	37 31	356
93	36 07	027	34 30	065	58 51	103	24 15	155	47 14	228	51 09	337	37 27	355
94	36 26	026	35 08	064	59 32	103	24 33	154	46 43	228	50 52	335	37 23	353
95	36 44	025	35 46	063	60 13	102	24 52	154	46 12	227	50 33	334	37 17	352
96	37 02	024	36 23	062	60 55	102	25 11	153	45 40	227	50 14	332	37 11	351
97	37 18	022	37 01	061	61 36	101	25 30	153	45 09	227	49 54	331	37 04	350
98	37 34	021	37 37	060	62 18	101	25 49	153	44 38	227	49 33	330	36 56	348
99	37 49	020	38 14	059	63 00	100	26 09	152	44 07	227	49 11	328	36 47	347
100	38 03	019	38 50	058	63 42	100	26 29	152	43 37	226	48 48	327	36 37	346
101	38 16	018	39 26	057	64 23	099	26 49	151	43 06	226	48 24	325	36 27	345
102	38 29	016	40 01	056	65 05	099	27 10	151	42 36	226	48 00	324	36 15	344
103	38 40	015	40 36	055	65 47	098	27 30	151	42 05	226	47 35	323	36 03	342
104	38 51	014	41 10	054	66 29	098	27 51	150	41 35	225	47 08	321	35 49	341
	PROCYON		**REGULUS**		**◆Gienah**		**RIGIL KENT.**		**◆ACHERNAR**		**RIGEL**		**◆BETELGEUSE**	
105	39 00	013	18 56	049	20 03	095	28 12	150	41 05	225	46 42	320	35 35	340
106	39 09	011	19 28	048	20 45	094	28 34	149	40 35	225	46 14	319	35 20	339
107	39 17	010	20 00	047	21 27	094	28 55	149	40 05	225	45 46	318	35 05	338
108	39 24	009	20 31	047	22 10	093	29 17	149	39 35	224	45 17	316	34 48	337
109	39 30	007	21 01	046	22 52	092	29 39	148	39 06	224	44 47	315	34 31	335
110	39 35	006	21 31	045	23 34	092	30 02	148	38 37	224	44 17	314	34 13	334
111	39 39	005	22 01	044	24 17	091	30 24	148	38 07	223	43 46	313	33 54	333
112	39 42	004	22 30	043	24 59	090	30 47	147	37 38	223	43 14	312	33 34	332
113	39 44	002	22 59	042	25 42	089	31 10	147	37 09	223	42 42	310	33 14	331
114	39 45	001	23 27	041	26 24	089	31 34	147	36 41	222	42 10	309	32 53	330
115	39 46	000	23 55	040	27 07	088	31 57	146	36 12	222	41 37	308	32 31	329
116	39 45	358	24 22	039	27 49	087	32 21	146	35 44	222	41 03	307	32 09	328
117	39 43	357	24 48	038	28 31	087	32 45	146	35 16	221	40 29	306	31 46	326
118	39 41	356	25 14	037	29 14	086	33 09	145	34 48	221	39 54	305	31 22	325
119	39 37	355	25 39	036	29 56	085	33 33	145	34 20	221	39 19	304	30 58	324
	REGULUS		**◆SPICA**		**RIGIL KENT.**		**◆ACHERNAR**		**RIGEL**		**BETELGEUSE**		**◆PROCYON**	
120	26 04	035	14 01	092	33 57	145	33 52	220	38 44	303	30 32	323	39 33	353
121	26 28	034	14 44	091	34 22	144	33 25	220	38 08	302	30 07	322	39 27	352
122	26 52	033	15 26	090	34 47	144	32 58	220	37 32	301	29 41	321	39 21	351
123	27 15	032	16 09	090	35 12	144	32 31	219	36 55	300	29 14	320	39 13	349
124	27 37	031	16 51	089	35 37	143	32 04	219	36 18	299	28 46	319	39 05	348
125	27 59	030	17 33	088	36 03	143	31 37	219	35 41	298	28 18	318	38 56	347
126	28 20	029	18 16	088	36 28	143	31 11	218	35 04	297	27 50	317	38 46	346
127	28 40	028	18 58	087	36 54	142	30 45	218	34 26	296	27 21	316	38 35	344
128	29 00	027	19 41	086	37 20	142	30 19	218	33 47	295	26 51	315	38 23	343
129	29 19	026	20 23	085	37 46	142	29 53	217	33 09	294	26 21	314	38 10	342
130	29 37	025	21 05	085	38 12	142	29 28	217	32 30	293	25 50	313	37 56	341
131	29 55	024	21 47	084	38 39	141	29 02	216	31 51	293	25 19	312	37 42	339
132	30 12	023	22 30	083	39 05	141	28 37	216	31 12	292	24 48	312	37 27	338
133	30 28	022	23 12	082	39 32	141	28 13	216	30 32	291	24 16	311	37 10	337
134	30 43	021	23 54	082	39 59	141	27 48	215	29 52	290	23 43	310	36 53	336
	REGULUS		**◆SPICA**		**RIGIL KENT.**		**◆ACHERNAR**		**CANOPUS**		**SIRIUS**		**◆PROCYON**	
135	30 58	020	24 36	081	40 26	140	27 24	215	63 33	239	50 01	304	36 36	335
136	31 11	018	25 17	080	40 53	140	27 00	214	62 56	239	49 26	303	36 17	333
137	31 24	017	25 59	079	41 20	140	26 36	214	62 20	239	48 50	302	35 58	332
138	31 37	016	26 41	079	41 48	140	26 12	214	61 44	239	48 14	301	35 37	331
139	31 48	015	27 22	078	42 15	139	25 49	213	61 07	239	47 37	300	35 17	330
140	31 59	014	28 04	077	42 43	139	25 26	213	60 31	239	47 00	298	34 55	329
141	32 08	013	28 45	076	43 11	139	25 03	212	59 55	239	46 22	297	34 33	328
142	32 17	012	29 26	075	43 38	139	24 41	212	59 18	239	45 45	296	34 10	327
143	32 25	010	30 07	075	44 06	139	24 18	211	58 42	239	45 06	295	33 46	325
144	32 33	009	30 48	074	44 34	138	23 56	211	58 06	239	44 28	294	33 22	324
145	32 39	008	31 29	073	45 03	138	23 35	211	57 30	238	43 49	293	32 56	323
146	32 45	007	32 09	072	45 31	138	23 13	210	56 54	238	43 10	292	32 31	322
147	32 50	006	32 50	071	45 59	138	22 52	210	56 18	238	42 31	291	32 05	321
148	32 53	005	33 30	070	46 28	138	22 31	209	55 42	238	41 51	290	31 38	320
149	32 57	004	34 10	070	46 56	138	22 11	209	55 06	238	41 11	290	31 10	319
	REGULUS		**◆SPICA**		**RIGIL KENT.**		**◆ACHERNAR**		**CANOPUS**		**SIRIUS**		**◆PROCYON**	
150	32 59	002	34 49	069	47 25	137	21 51	208	54 30	238	40 31	289	30 42	318
151	33 00	001	35 29	068	47 54	137	21 31	208	53 54	237	39 51	288	30 13	317
152	33 00	000	36 08	067	48 23	137	21 11	207	53 18	237	39 10	287	29 44	316
153	33 00	359	36 47	066	48 52	137	20 52	207	52 43	237	38 30	286	29 15	315
154	32 59	358	37 25	065	49 20	137	20 33	206	52 07	237	37 49	285	28 44	314
155	32 57	357	38 04	064	49 49	137	20 14	206	51 32	237	37 08	284	28 14	313
156	32 54	355	38 42	063	50 19	137	19 56	205	50 56	236	36 26	283	27 42	312
157	32 50	354	39 19	062	50 48	137	19 37	205	50 21	236	35 45	283	27 11	311
158	32 45	353	39 57	061	51 17	137	19 20	205	49 46	236	35 04	282	26 39	310
159	32 40	352	40 34	060	51 46	137	19 02	204	49 11	236	34 22	281	26 06	309
160	32 33	351	41 10	059	52 15	136	18 45	204	48 36	235	33 40	280	25 33	308
161	32 26	350	41 47	058	52 44	136	18 28	203	48 01	235	32 59	279	24 59	307
162	32 18	348	42 22	057	53 14	136	18 12	203	47 26	235	32 17	279	24 26	307
163	32 09	347	42 58	056	53 43	136	17 56	202	46 51	235	31 35	278	23 51	306
164	31 59	346	43 33	055	54 12	136	17 40	202	46 17	234	30 53	277	23 17	305
	SPICA		**◆ANTARES**		**RIGIL KENT.**		**ACHERNAR**		**◆CANOPUS**		**SIRIUS**		**◆REGULUS**	
165	44 07	054	23 34	105	54 41	136	17 25	201	45 42	234	30 11	276	31 49	345
166	44 41	053	24 15	104	55 11	136	17 10	201	45 08	234	29 28	276	31 37	344
167	45 15	052	24 56	103	55 40	136	16 55	200	44 34	234	28 46	275	31 25	343
168	45 48	050	25 38	103	56 09	137	16 40	200	44 00	233	28 04	274	31 12	342
169	46 20	049	26 19	102	56 38	137	16 26	199	43 26	233	27 21	273	30 58	341
170	46 52	048	27 01	101	57 07	137	16 13	199	42 52	233	26 39	273	30 44	339
171	47 23	047	27 42	101	57 36	137	15 59	198	42 18	232	25 57	272	30 29	338
172	47 54	046	28 24	100	58 05	137	15 46	198	41 45	232	25 14	271	30 13	337
173	48 24	044	29 06	099	58 34	137	15 34	197	41 11	232	24 32	271	29 56	336
174	48 53	043	29 48	099	59 03	137	15 22	196	40 38	231	23 49	270	29 38	335
175	49 22	042	30 30	098	59 32	137	15 10	196	40 05	231	23 07	269	29 20	334
176	49 50	040	31 12	097	60 01	138	14 58	195	39 32	231	22 25	268	29 01	333
177	50 17	039	31 54	097	60 29	138	14 47	195	38 59	230	21 42	268	28 41	332
178	50 43	038	32 36	096	60 58	138	14 36	194	38 27	230	21 00	267	28 21	331
179	51 09	036	33 18	096	61 26	138	14 26	194	37 54	230	20 17	266	28 00	330

LAT 45°S

LHA ♈	Hc Zn	Hc Zn	Hc Zn	Hc Zn	Hc Zn	Hc Zn	Hc Zn
	◆ARCTURUS	ANTARES	◆RIGIL KENT.	ACHERNAR	CANOPUS	◆Suhail	REGULUS
180	18 47 034	34 00 095	61 55 138	14 16 193	37 22 229	59 28 257	27 39 329
181	19 10 033	34 43 094	62 23 139	14 07 193	36 50 229	58 47 257	27 16 328
182	19 33 032	35 25 094	62 50 139	13 57 192	36 18 229	58 05 257	26 53 327
183	19 55 031	36 07 093	63 18 139	13 49 192	35 46 228	57 24 256	26 30 326
184	20 16 030	36 50 092	63 46 140	13 40 191	35 15 228	56 43 256	26 06 325
185	20 37 029	37 32 091	64 13 140	13 32 191	34 43 228	56 02 255	25 41 324
186	20 58 028	38 15 091	64 40 141	13 25 190	34 12 227	55 21 255	25 15 323
187	21 17 027	38 57 090	65 07 141	13 17 190	33 41 227	54 40 254	24 50 322
188	21 36 026	39 39 089	65 33 142	13 10 189	33 10 226	53 59 254	24 23 321
189	21 55 025	40 22 089	65 59 142	13 04 188	32 40 226	53 19 253	23 56 320
190	22 13 024	41 04 088	66 25 143	12 58 188	32 09 226	52 38 253	23 28 319
191	22 30 023	41 47 087	66 51 143	12 52 187	31 39 225	51 57 253	23 00 318
192	22 46 022	42 29 086	67 16 144	12 47 187	31 09 225	51 17 252	22 32 317
193	23 02 021	43 11 086	67 41 145	12 42 186	30 39 224	50 37 252	22 03 316
194	23 17 020	43 54 085	68 05 145	12 38 186	30 10 224	49 56 251	21 33 315
	◆ARCTURUS	ANTARES	◆Peacock	ACHERNAR	CANOPUS	◆Suhail	REGULUS
195	23 32 019	44 36 084	26 46 145	12 34 185	29 41 224	49 16 251	21 03 314
196	23 46 018	45 18 083	27 11 145	12 30 185	29 11 223	48 36 250	20 32 314
197	23 59 017	46 00 083	27 35 144	12 27 184	28 43 223	47 56 250	20 01 313
198	24 11 016	46 42 082	28 00 144	12 24 184	28 14 222	47 16 250	19 30 312
199	24 23 015	47 24 081	28 25 143	12 22 183	27 45 222	46 37 249	18 58 311
200	24 34 014	48 06 080	28 51 143	12 20 182	27 17 221	45 57 249	18 26 310
201	24 44 013	48 48 079	29 16 143	12 18 182	26 49 221	45 18 248	17 53 309
202	24 53 012	49 29 078	29 42 142	12 17 181	26 22 221	44 38 248	17 20 308
203	25 02 011	50 11 077	30 08 142	12 16 181	25 54 220	43 59 247	16 47 307
204	25 10 010	50 52 077	30 35 142	12 16 180	25 27 220	43 20 247	16 13 307
205	25 17 009	51 33 076	31 01 141	12 16 180	25 00 219	42 41 246	15 38 306
206	25 24 008	52 14 075	31 28 141	12 16 179	24 33 219	42 02 246	15 04 305
207	25 29 007	52 55 074	31 55 140	12 17 179	24 07 218	41 24 246	14 29 304
208	25 34 006	53 36 073	32 22 140	12 18 178	23 41 218	40 45 245	13 54 303
209	25 38 005	54 16 072	32 49 140	12 20 177	23 15 217	40 07 245	13 18 303
	◆ARCTURUS	ANTARES	◆Peacock	ACHERNAR	CANOPUS	◆Suhail	SPICA
210	25 42 004	54 56 071	33 17 139	12 22 177	22 49 217	39 28 244	55 19 345
211	25 44 003	55 36 070	33 45 139	12 25 176	22 24 216	38 50 244	55 07 343
212	25 46 002	56 16 069	34 12 139	12 28 176	21 59 216	38 12 243	54 54 341
213	25 47 001	56 55 068	34 41 138	12 31 175	21 34 215	37 35 243	54 40 340
214	25 47 000	57 34 066	35 09 138	12 35 175	21 10 215	36 57 242	54 25 338
215	25 47 359	58 13 065	35 37 138	12 39 174	20 45 215	36 19 242	54 08 337
216	25 46 358	58 51 064	36 06 137	12 43 174	20 21 214	35 42 241	53 51 335
217	25 44 357	59 29 063	36 35 137	12 48 173	19 58 214	35 05 241	53 32 333
218	25 41 356	60 07 062	37 04 137	12 54 172	19 35 213	34 28 240	53 13 332
219	25 37 355	60 44 060	37 33 136	12 59 172	19 12 213	33 51 240	52 52 330
220	25 33 354	61 20 059	38 03 136	13 05 171	18 49 212	33 14 240	52 31 329
221	25 28 353	61 56 057	38 32 136	13 12 171	18 27 212	32 38 239	52 08 327
222	25 22 351	62 32 056	39 02 135	13 19 170	18 05 211	32 02 239	51 45 326
223	25 15 350	63 07 054	39 32 135	13 26 170	17 43 211	31 25 238	51 20 324
224	25 08 349	63 41 053	40 02 135	13 34 169	17 21 210	30 49 238	50 55 323
	Rasalhague	◆Nunki	Peacock	ACHERNAR	◆ACRUX	SPICA	◆ARCTURUS
225	22 39 041	39 58 089	40 32 135	13 42 169	61 48 217	50 29 322	25 00 348
226	23 06 040	40 40 088	41 02 134	13 51 168	61 23 217	50 03 320	24 51 347
227	23 34 039	41 23 087	41 33 134	14 00 168	60 57 217	49 35 319	24 41 346
228	24 00 039	42 05 087	42 03 134	14 09 167	60 32 218	49 07 318	24 31 345
229	24 27 038	42 47 086	42 34 133	14 19 167	60 06 218	48 38 316	24 20 344
230	24 52 037	43 30 085	43 05 133	14 29 166	59 40 218	48 08 315	24 08 343
231	25 17 036	44 12 084	43 36 133	14 39 165	59 13 218	47 38 314	23 55 342
232	25 42 035	44 54 084	44 07 133	14 50 165	58 47 218	47 07 312	23 42 341
233	26 05 034	45 36 083	44 39 132	15 01 164	58 20 219	46 35 311	23 28 340
234	26 29 033	46 18 082	45 10 132	15 13 164	57 54 219	46 03 310	23 13 339
235	26 51 032	47 00 081	45 41 132	15 24 163	57 27 219	45 30 309	22 58 338
236	27 13 031	47 42 080	46 13 132	15 37 163	57 00 219	44 57 308	22 42 337
237	27 34 030	48 24 079	46 45 132	15 49 162	56 34 219	44 23 307	22 25 336
238	27 55 029	49 05 079	47 16 131	16 03 162	56 07 219	43 49 306	22 08 335
239	28 15 028	49 47 078	47 48 131	16 16 161	55 40 220	43 14 304	21 50 334
	Rasalhague	◆ALTAIR	Peacock	◆ACHERNAR	ACRUX	◆SPICA	ARCTURUS
240	28 34 027	15 23 060	48 20 131	16 30 161	55 13 220	42 39 303	21 31 333
241	28 53 025	15 59 059	48 52 131	16 44 160	54 46 220	42 03 302	21 12 332
242	29 11 024	16 35 058	49 25 131	16 58 160	54 19 220	41 27 301	20 52 332
243	29 28 023	17 11 058	49 57 130	17 13 159	53 51 220	40 51 300	20 31 331
244	29 44 022	17 47 057	50 29 130	17 28 159	53 24 220	40 14 299	20 10 330
245	30 00 021	18 22 056	51 02 130	17 44 158	52 57 220	39 37 298	19 48 329
246	30 15 020	18 57 055	51 34 130	18 00 158	52 30 220	38 59 297	19 26 328
247	30 29 019	19 32 054	52 07 130	18 16 157	52 03 220	38 21 296	19 03 327
248	30 43 018	20 06 053	52 39 130	18 32 157	51 36 220	37 43 295	18 40 326
249	30 55 017	20 40 052	53 12 130	18 49 156	51 08 220	37 05 294	18 16 325
250	31 07 016	21 13 052	53 45 130	19 06 156	50 41 220	36 26 293	17 51 324
251	31 18 015	21 46 051	54 17 129	19 24 155	50 14 220	35 47 293	17 26 323
252	31 28 013	22 19 050	54 50 129	19 42 155	49 47 220	35 08 292	17 01 322
253	31 38 012	22 51 049	55 23 129	20 00 154	49 20 220	34 28 291	16 34 322
254	31 46 011	23 23 048	55 56 129	20 18 154	48 53 219	33 48 290	16 08 321
	Rasalhague	◆ALTAIR	FOMALHAUT	◆ACHERNAR	RIGIL KENT.	◆SPICA	Alphecca
255	31 54 010	23 54 047	20 54 112	20 37 153	64 03 220	33 08 289	15 40 340
256	32 01 009	24 25 046	21 34 111	20 56 153	63 35 220	32 28 288	15 25 339
257	32 07 008	24 55 045	22 14 110	21 16 153	63 08 221	31 48 287	15 10 338
258	32 12 007	25 25 044	22 53 110	21 35 152	62 40 221	31 07 287	14 54 338
259	32 17 005	25 54 043	23 33 109	21 55 152	62 12 221	30 26 286	14 38 337
260	32 20 004	26 23 042	24 14 108	22 16 151	61 44 222	29 45 285	14 21 336
261	32 23 003	26 52 041	24 54 108	22 36 151	61 16 222	29 04 284	14 03 335
262	32 25 002	27 19 040	25 34 107	22 57 150	60 47 222	28 23 283	13 45 334
263	32 26 001	27 47 039	26 15 107	23 18 150	60 19 222	27 42 282	13 26 333
264	32 26 000	28 13 038	26 56 106	23 40 149	59 50 223	27 00 282	13 06 332
265	32 25 358	28 39 037	27 37 105	24 02 149	59 21 223	26 19 281	12 46 332
266	32 24 357	29 05 036	28 17 105	24 24 149	58 53 223	25 37 280	12 26 331
267	32 21 356	29 30 035	28 59 104	24 46 148	58 24 223	24 55 279	12 05 330
268	32 18 355	29 54 034	29 40 104	25 09 148	57 55 223	24 13 279	11 43 329
269	32 14 354	30 17 033	30 21 103	25 31 147	57 26 223	23 31 278	11 21 328

LHA ♈	Hc Zn	Hc Zn	Hc Zn	Hc Zn	Hc Zn	Hc Zn	Hc Zn
	ALTAIR	Enif	◆FOMALHAUT	ACHERNAR	◆RIGIL KENT.	ANTARES	◆Rasalhague
270	30 40 032	15 35 058	31 02 102	25 54 147	56 57 223	63 59 308	32 09 353
271	31 03 031	16 11 057	31 44 102	26 18 146	56 27 223	63 26 306	32 03 352
272	31 24 030	16 47 056	32 25 101	26 41 146	55 58 224	62 51 305	31 57 350
273	31 45 029	17 22 056	33 07 101	27 05 146	55 29 224	62 16 303	31 49 349
274	32 05 028	17 57 055	33 49 100	27 29 145	55 00 224	61 40 302	31 41 348
275	32 25 027	18 31 054	34 31 099	27 54 145	54 30 224	61 04 301	31 32 347
276	32 44 026	19 05 053	35 13 099	28 18 144	54 01 224	60 27 299	31 22 346
277	33 02 025	19 39 052	35 55 098	28 43 144	53 32 224	59 50 298	31 11 345
278	33 19 023	20 12 051	36 37 097	29 08 144	53 03 224	59 12 297	30 59 344
279	33 35 022	20 45 050	37 19 097	29 34 143	52 33 224	58 34 295	30 47 343
280	33 51 021	21 17 050	38 01 096	29 59 143	52 04 223	57 56 294	30 34 341
281	34 06 020	21 49 049	38 43 095	30 25 142	51 35 223	57 17 293	30 20 340
282	34 20 019	22 21 048	39 25 095	30 51 142	51 06 223	56 38 292	30 05 339
283	34 33 018	22 52 047	40 08 094	31 17 142	50 37 223	55 58 291	29 50 338
284	34 46 016	23 23 046	40 50 093	31 44 141	50 08 223	55 18 290	29 34 337
	ALTAIR	Enif	◆FOMALHAUT	ACHERNAR	◆RIGIL KENT.	ANTARES	◆Rasalhague
285	34 57 015	23 53 045	41 32 093	32 10 141	49 39 223	54 38 289	29 17 336
286	35 08 014	24 23 044	42 15 092	32 37 141	49 10 223	53 58 288	28 59 335
287	35 18 013	24 52 043	42 57 091	33 04 140	48 41 223	53 17 287	28 41 334
288	35 27 012	25 21 042	43 40 091	33 31 140	48 12 223	52 37 286	28 22 333
289	35 35 011	25 49 041	44 22 090	33 59 140	47 43 223	51 56 285	28 02 332
290	35 42 009	26 17 040	45 04 089	34 26 139	47 14 223	51 15 284	27 42 331
291	35 49 008	26 44 039	45 47 089	34 54 139	46 46 222	50 33 283	27 21 330
292	35 54 007	27 10 038	46 29 088	35 22 138	46 17 222	49 52 282	26 59 329
293	35 59 006	27 36 037	47 12 087	35 51 138	45 49 222	49 11 281	26 37 328
294	36 03 004	28 02 036	47 54 086	36 19 138	45 20 222	48 29 280	26 14 327
295	36 06 003	28 27 035	48 36 086	36 48 138	44 52 222	47 47 280	25 50 326
296	36 07 002	28 51 034	49 19 085	37 16 137	44 24 222	47 05 279	25 26 325
297	36 09 001	29 14 033	50 01 084	37 45 137	43 56 221	46 23 278	25 01 324
298	36 09 000	29 37 032	50 43 083	38 14 137	43 28 221	45 41 277	24 36 323
299	36 08 358	29 59 031	51 25 082	38 43 136	43 00 221	44 59 276	24 10 322
	◆Enif	FOMALHAUT	◆ACHERNAR	RIGIL KENT.	◆ANTARES	Rasalhague	ALTAIR
300	30 21 030	52 07 082	39 13 136	42 32 221	44 17 276	23 43 321	36 06 357
301	30 42 029	52 49 081	39 42 136	42 05 220	43 35 275	23 16 320	36 04 356
302	31 02 028	53 31 080	40 12 135	41 37 220	42 52 274	22 48 319	36 00 355
303	31 21 027	54 13 079	40 42 135	41 10 220	42 10 273	22 20 318	35 56 353
304	31 40 026	54 54 078	41 12 135	40 43 220	41 28 273	21 52 317	35 50 352
305	31 58 025	55 36 077	41 42 135	40 16 220	40 45 272	21 23 316	35 44 351
306	32 16 023	56 17 076	42 12 134	39 49 219	40 03 271	20 53 315	35 37 350
307	32 32 022	56 58 075	42 43 134	39 22 219	39 20 270	20 23 314	35 29 349
308	32 48 021	57 39 074	43 13 134	38 55 219	38 38 270	19 52 314	35 20 347
309	33 03 020	58 20 073	43 44 134	38 29 219	37 55 269	19 21 313	35 11 346
310	33 17 019	59 00 072	44 15 133	38 02 218	37 13 268	18 50 312	35 00 345
311	33 30 018	59 40 071	44 46 133	37 36 218	36 31 268	18 18 311	34 49 344
312	33 43 017	60 20 070	45 17 133	37 10 218	35 48 267	17 46 310	34 37 343
313	33 55 015	61 00 069	45 48 133	36 44 217	35 06 266	17 13 309	34 24 342
314	34 05 014	61 40 068	46 19 132	36 19 217	34 24 266	16 40 308	34 10 340
	Enif	◆Diphda	ACHERNAR	Miaplacidus	◆RIGIL KENT.	ANTARES	◆ALTAIR
315	34 16 013	36 36 079	46 50 132	24 43 181	35 53 217	33 41 265	33 55 339
316	34 25 012	37 18 078	47 22 132	24 43 181	35 28 217	32 59 264	33 40 338
317	34 33 011	37 59 077	47 53 132	24 42 180	35 03 216	32 17 264	33 23 337
318	34 41 010	38 41 076	48 25 132	24 42 180	34 38 216	31 35 263	33 06 336
319	34 47 008	39 22 075	48 57 132	24 42 180	34 13 216	30 53 262	32 48 335
320	34 53 007	40 03 074	49 29 131	24 42 179	33 48 215	30 11 262	32 30 334
321	34 58 006	40 43 074	50 00 131	24 43 179	33 24 215	29 29 261	32 11 332
322	35 02 005	41 24 073	50 32 131	24 44 179	33 00 215	28 47 260	31 51 331
323	35 05 004	42 04 072	51 04 131	24 45 178	32 36 214	28 05 260	31 30 330
324	35 07 002	42 45 071	51 36 131	24 46 178	32 12 214	27 23 259	31 09 329
325	35 08 001	43 24 070	52 09 131	24 48 177	31 48 214	26 42 258	30 46 328
326	35 09 000	44 04 069	52 41 131	24 50 177	31 25 213	26 00 258	30 24 327
327	35 08 359	44 44 068	53 13 130	24 53 177	31 02 213	25 19 257	30 00 326
328	35 07 358	45 23 067	53 45 130	24 55 176	30 39 213	24 38 256	29 36 325
329	35 05 356	46 02 066	54 18 130	24 58 176	30 16 212	23 56 256	29 12 324
	◆Diphda	ACHERNAR	◆Miaplacidus	RIGIL KENT.	Nunki	◆ALTAIR	Enif
330	46 40 065	54 50 130	25 01 176	29 53 212	48 44 281	28 46 323	35 02 355
331	47 19 064	55 22 130	25 05 175	29 31 212	48 02 280	28 20 322	34 58 354
332	47 57 063	55 55 130	25 08 175	29 09 211	47 20 279	27 54 321	34 53 353
333	48 34 062	56 27 130	25 12 174	28 47 211	46 38 278	27 27 320	34 47 352
334	49 11 061	57 00 130	25 17 174	28 26 210	45 56 278	26 59 319	34 40 350
335	49 48 060	57 32 130	25 21 174	28 04 210	45 14 277	26 31 318	34 33 349
336	50 25 058	58 05 130	25 26 173	27 43 210	44 32 276	26 02 317	34 24 348
337	51 00 057	58 37 130	25 31 173	27 22 209	43 50 275	25 33 316	34 15 347
338	51 36 056	59 09 130	25 36 173	27 02 209	43 07 274	25 03 315	34 05 346
339	52 11 055	59 42 130	25 42 172	26 42 208	42 25 274	24 33 314	33 54 344
340	52 45 054	60 14 130	25 48 172	26 21 208	41 43 273	24 02 313	33 42 343
341	53 19 052	60 47 130	25 54 171	26 02 208	41 00 272	23 31 312	33 30 342
342	53 53 051	61 19 131	26 01 171	25 42 207	40 18 272	23 00 311	33 16 341
343	54 25 050	61 51 131	26 07 171	25 23 207	39 36 271	22 28 310	33 02 340
344	54 57 048	62 23 131	26 14 170	25 04 206	38 53 270	21 55 310	32 47 339
	◆Alpheratz	Diphda	◆CANOPUS	RIGIL KENT.	◆Nunki	ALTAIR	Enif
345	14 20 015	55 29 047	24 09 142	24 45 206	38 11 269	21 22 309	32 32 338
346	14 31 014	55 59 046	24 35 141	24 26 206	37 28 269	20 49 308	32 15 337
347	14 41 014	56 29 044	25 02 141	24 08 205	36 46 268	20 15 307	31 58 335
348	14 50 013	56 58 043	25 29 140	23 50 205	36 04 267	19 41 306	31 40 334
349	14 59 012	57 27 041	25 56 140	23 33 204	35 21 267	19 07 305	31 21 333
350	15 08 011	57 54 040	26 24 139	23 15 204	34 39 266	18 32 304	31 01 332
351	15 15 010	58 21 038	26 52 139	22 58 203	33 57 265	17 57 304	30 41 331
352	15 22 009	58 46 036	27 20 139	22 42 203	33 14 265	17 21 303	30 20 330
353	15 29 008	59 11 035	27 48 138	22 25 203	32 32 264	16 45 302	29 59 329
354	15 35 007	59 35 033	28 16 138	22 09 202	31 50 263	16 09 301	29 36 328
355	15 40 006	59 57 031	28 45 137	21 53 202	31 08 263	15 33 300	29 14 327
356	15 44 005	60 19 029	29 14 137	21 37 201	30 26 262	14 56 299	28 50 326
357	15 48 005	60 39 028	29 43 136	21 22 201	29 44 261	14 19 299	28 25 325
358	15 51 004	60 58 026	30 12 136	21 07 200	29 02 261	13 41 298	28 01 324
359	15 53 003	61 16 024	30 42 136	20 53 200	28 20 260	13 04 297	27 36 323

 LAT 46°S

LHA	Hc	Zn	Hc	Zn	Hc	Zn	Hc	Zn	Hc	Zn	Hc	Zn	Hc	Zn
♈	Alpheratz		Hamal		◆RIGEL		CANOPUS		◆RIGIL KENT.		Peacock		◆FOMALHAUT	
0	14 55	002	14 50	030	13 49	088	31 54	135	21 35	200	55 48	232	69 35	318
1	14 56	001	15 10	029	14 31	087	32 24	134	21 21	199	55 16	232	69 06	316
2	14 56	000	15 30	028	15 12	086	32 54	134	21 08	199	54 43	232	68 37	314
3	14 56	359	15 50	027	15 54	085	33 24	134	20 54	198	54 10	232	68 06	312
4	14 55	358	16 09	026	16 36	085	33 54	133	20 41	198	53 38	231	67 34	310
5	14 53	357	16 27	025	17 17	084	34 25	133	20 29	197	53 05	231	67 02	308
6	14 51	356	16 44	025	17 58	083	34 55	132	20 17	197	52 33	231	66 29	307
7	14 48	356	17 02	024	18 40	082	35 26	132	20 05	196	52 00	231	65 55	305
8	14 45	355	17 18	023	19 21	082	35 57	132	19 53	196	51 28	231	65 21	303
9	14 40	354	17 34	022	20 02	081	36 29	131	19 42	195	50 56	231	64 45	302
10	14 35	353	17 49	021	20 43	080	37 00	131	19 31	195	50 23	230	64 10	300
11	14 30	352	18 03	020	21 24	079	37 32	130	19 21	194	49 51	230	63 34	299
12	14 24	351	18 17	019	22 05	079	38 03	130	19 10	194	49 19	230	62 57	298
13	14 17	350	18 31	018	22 46	078	38 35	130	19 00	193	48 47	230	62 20	296
14	14 09	349	18 43	017	23 27	077	39 08	129	18 51	193	48 16	230	61 42	295
	Hamal		◆RIGEL		SIRIUS		CANOPUS		◆RIGIL KENT.		Peacock		◆FOMALHAUT	
15	18 55	016	24 07	076	14 31	099	39 40	129	18 42	192	47 44	229	61 04	294
16	19 07	015	24 48	075	15 12	098	40 12	129	18 33	192	47 12	229	60 26	292
17	19 17	014	25 28	075	15 53	098	40 45	128	18 24	191	46 41	229	59 47	291
18	19 27	013	26 08	074	16 34	097	41 18	128	18 16	191	46 10	229	59 08	290
19	19 36	012	26 48	073	17 16	096	41 51	128	18 09	191	45 38	228	58 29	289
20	19 45	011	27 28	072	17 57	096	42 24	127	18 01	190	45 07	228	57 49	288
21	19 53	010	28 07	071	18 39	095	42 57	127	17 54	190	44 36	228	57 09	287
22	20 00	009	28 47	070	19 20	094	43 31	127	17 47	189	44 05	228	56 29	286
23	20 07	009	29 26	070	20 02	094	44 04	126	17 41	189	43 34	227	55 49	285
24	20 12	008	30 05	069	20 43	093	44 38	126	17 35	188	43 04	227	55 09	284
25	20 18	007	30 44	068	21 25	092	45 12	126	17 29	188	42 33	227	54 28	283
26	20 22	006	31 22	067	22 07	091	45 46	125	17 24	187	42 03	227	53 48	282
27	20 26	005	32 00	066	22 48	091	46 20	125	17 19	187	41 33	226	53 07	281
28	20 29	004	32 38	065	23 30	090	46 54	125	17 15	186	41 03	226	52 26	280
29	20 31	003	33 16	064	24 12	089	47 28	124	17 11	185	40 33	226	51 45	279
	Hamal		◆RIGEL		SIRIUS		CANOPUS		◆RIGIL KENT.		Peacock		◆FOMALHAUT	
30	20 33	002	33 53	063	24 53	089	48 03	124	17 07	185	40 03	225	51 04	279
31	20 33	001	34 31	062	25 35	088	48 37	124	17 03	184	39 33	225	50 22	278
32	20 34	000	35 07	062	26 17	087	49 12	124	17 00	184	39 04	225	49 41	277
33	20 33	359	35 44	061	26 58	086	49 47	123	16 57	183	38 35	224	49 00	276
34	20 32	358	36 20	060	27 40	086	50 22	123	16 55	183	38 06	224	48 18	275
35	20 30	357	36 56	059	28 21	085	50 57	123	16 53	182	37 37	224	47 37	274
36	20 27	356	37 31	058	29 03	084	51 32	123	16 52	182	37 08	223	46 55	274
37	20 24	355	38 06	057	29 44	083	52 07	122	16 50	181	36 39	223	46 13	273
38	20 20	354	38 41	056	30 26	083	52 42	122	16 50	181	36 11	223	45 32	272
39	20 15	353	39 15	055	31 07	082	53 18	122	16 49	180	35 43	222	44 50	271
40	20 09	352	39 48	053	31 48	081	53 53	122	16 49	180	35 15	222	44 08	271
41	20 03	351	40 22	052	32 29	080	54 29	121	16 49	179	34 47	222	43 27	270
42	19 56	350	40 54	051	33 10	079	55 04	121	16 50	179	34 19	221	42 45	269
43	19 49	349	41 27	050	33 51	079	55 40	121	16 51	178	33 52	221	42 03	269
44	19 40	348	41 58	049	34 32	078	56 16	121	16 52	178	33 24	221	41 22	268
	RIGEL		SIRIUS		◆CANOPUS		RIGIL KENT.		◆Peacock		FOMALHAUT		◆Hamal	
45	42 30	048	35 13	077	56 52	121	16 54	177	32 58	220	40 40	267	19 31	347
46	43 00	047	35 53	076	57 28	120	16 56	177	32 31	220	39 59	266	19 22	346
47	43 30	046	36 34	075	58 04	120	16 58	176	32 04	219	39 17	266	19 11	345
48	44 00	044	37 14	075	58 40	120	17 01	176	31 38	219	38 35	265	19 00	344
49	44 29	043	37 54	074	59 16	120	17 05	175	31 12	219	37 54	264	18 49	343
50	44 57	042	38 34	073	59 52	120	17 08	175	30 46	218	37 12	264	18 36	342
51	45 25	041	39 14	072	60 28	120	17 12	174	30 20	218	36 31	263	18 23	341
52	45 51	039	39 53	071	61 04	120	17 16	174	29 54	218	35 50	262	18 10	340
53	46 18	038	40 33	070	61 41	119	17 21	173	29 29	217	35 08	262	17 55	340
54	46 43	037	41 12	069	62 17	119	17 26	173	29 04	217	34 27	261	17 41	339
55	47 08	036	41 50	068	62 53	119	17 31	172	28 39	216	33 46	261	17 25	338
56	47 31	034	42 29	067	63 30	119	17 37	172	28 15	216	33 05	260	17 09	337
57	47 54	033	43 07	066	64 06	119	17 43	171	27 51	215	32 24	259	16 52	336
58	48 17	032	43 45	065	64 42	119	17 50	171	27 27	215	31 43	259	16 35	335
59	48 38	030	44 23	064	65 19	119	17 57	170	27 03	215	31 02	258	16 17	334
	◆ALDEBARAN		BETELGEUSE		SIRIUS		◆Suhail		RIGIL KENT.		◆ACHERNAR		Diphda	
60	26 59	010	30 45	034	45 00	063	37 28	117	18 04	170	65 34	230	40 53	288
61	27 06	009	31 08	033	45 37	062	38 05	116	18 11	169	65 02	230	40 13	287
62	27 11	007	31 30	032	46 14	061	38 43	116	18 19	169	64 30	230	39 33	286
63	27 16	006	31 52	030	46 50	060	39 20	115	18 28	168	63 58	230	38 53	285
64	27 21	005	32 12	029	47 26	059	39 58	115	18 36	168	63 26	230	38 12	284
65	27 24	004	32 33	028	48 02	058	40 36	115	18 45	167	62 54	231	37 32	284
66	27 27	003	32 52	027	48 37	057	41 14	114	18 54	167	62 22	231	36 51	283
67	27 29	002	33 11	026	49 11	056	41 52	114	19 04	166	61 50	231	36 11	282
68	27 30	001	33 28	025	49 46	054	42 30	113	19 14	166	61 17	231	35 30	281
69	27 30	000	33 46	024	50 19	053	43 08	113	19 24	165	60 45	231	34 49	280
70	27 30	359	34 02	023	50 52	052	43 47	112	19 35	165	60 12	231	34 08	279
71	27 28	358	34 18	021	51 25	051	44 26	112	19 46	164	59 40	231	33 27	279
72	27 26	357	34 32	020	51 57	049	45 04	111	19 58	164	59 08	231	32 45	278
73	27 23	356	34 46	019	52 28	048	45 43	111	20 09	163	58 35	231	32 04	277
74	27 20	355	35 00	018	52 59	047	46 22	110	20 21	163	58 03	231	31 23	276
	BETELGEUSE		SIRIUS		◆Suhail		RIGIL KENT.		◆ACHERNAR		Diphda		◆ALDEBARAN	
75	35 12	017	53 29	045	47 01	110	20 34	163	57 30	231	30 41	276	27 15	353
76	35 24	016	53 58	044	47 41	109	20 46	162	56 58	231	30 00	275	27 10	352
77	35 34	014	54 27	043	48 20	109	20 59	162	56 25	231	29 18	274	27 04	351
78	35 44	013	54 54	041	48 59	108	21 13	161	55 53	231	28 36	273	26 58	350
79	35 53	012	55 21	040	49 39	108	21 26	161	55 21	231	27 55	273	26 50	349
80	36 02	011	55 47	038	50 19	108	21 40	160	54 48	231	27 13	272	26 42	348
81	36 09	010	56 13	037	50 59	107	21 54	160	54 16	231	26 32	271	26 33	347
82	36 15	008	56 37	035	51 38	107	22 09	159	53 44	231	25 50	270	26 23	346
83	36 21	007	57 00	033	52 18	106	22 24	159	53 12	230	25 08	270	26 13	345
84	36 26	006	57 23	032	52 59	106	22 39	158	52 40	230	24 27	269	26 02	344
85	36 29	005	57 44	030	53 39	105	22 55	158	52 08	230	23 45	268	25 50	343
86	36 32	003	58 05	028	54 19	105	23 10	157	51 36	230	23 03	268	25 37	342
87	36 34	002	58 24	027	54 59	104	23 26	157	51 04	230	22 22	267	25 24	341
88	36 35	001	58 42	025	55 40	104	23 43	157	50 32	230	21 40	266	25 09	340
89	36 36	000	58 59	023	56 20	103	24 00	156	50 00	229	20 58	265	24 55	339

LHA	Hc	Zn	Hc	Zn	Hc	Zn	Hc	Zn	Hc	Zn	Hc	Zn	Hc	Zn
♈	PROCYON		Alphard		◆Suhail		RIGIL KENT.		◆ACHERNAR		◆RIGEL		BETELGEUSE	
90	34 14	030	32 10	067	57 01	103	24 17	156	49 29	229	50 57	342	36 35	358
91	34 34	029	32 48	066	57 42	102	24 34	155	48 57	229	50 44	340	36 33	357
92	34 54	028	33 26	065	58 22	102	24 51	155	48 26	229	50 29	339	36 31	356
93	35 14	027	34 04	064	59 03	101	25 09	154	47 54	229	50 13	337	36 27	355
94	35 32	026	34 41	063	59 44	101	25 27	154	47 23	228	49 57	336	36 23	353
95	35 50	025	35 18	062	60 25	100	25 46	154	46 52	228	49 39	334	36 18	352
96	36 07	023	35 55	061	61 06	100	26 04	153	46 21	228	49 21	333	36 12	351
97	36 23	022	36 31	060	61 47	099	26 23	153	45 50	228	49 01	332	36 05	350
98	36 38	021	37 07	059	62 28	099	26 43	152	45 19	228	48 41	330	35 57	349
99	36 53	020	37 43	058	63 10	098	27 02	152	44 48	227	48 20	329	35 49	347
100	37 06	019	38 18	057	63 51	098	27 22	152	44 18	227	47 58	327	35 39	346
101	37 19	017	38 53	056	64 32	097	27 42	151	43 47	227	47 35	326	35 29	345
102	37 31	016	39 27	055	65 13	097	28 02	151	43 17	227	47 11	325	35 17	344
103	37 42	015	40 01	054	65 55	096	28 22	150	42 47	226	46 47	323	35 05	343
104	37 52	014	40 34	053	66 36	096	28 43	150	42 17	226	46 21	322	34 53	341
	PROCYON		REGULUS		◆Gienah		RIGIL KENT.		◆ACHERNAR		RIGEL		◆BETELGEUSE	
105	38 02	012	18 17	049	20 08	095	29 04	150	41 47	226	45 55	321	34 39	340
106	38 10	011	18 48	048	20 49	094	29 25	149	41 17	225	45 29	319	34 24	339
107	38 18	010	19 19	047	21 31	093	29 47	149	40 48	225	45 01	318	34 09	338
108	38 25	009	19 49	046	22 13	093	30 09	148	40 18	225	44 33	317	33 53	337
109	38 30	007	20 19	045	22 54	092	30 30	148	39 49	225	44 04	316	33 36	336
110	38 35	006	20 49	044	23 36	091	30 53	148	39 20	224	43 35	315	33 19	334
111	38 39	005	21 18	044	24 18	090	31 15	147	38 51	224	43 05	313	33 00	333
112	38 42	004	21 46	043	24 59	090	31 38	147	38 22	224	42 34	312	32 41	332
113	38 44	002	22 14	042	25 41	089	32 00	147	37 53	223	42 03	311	32 22	331
114	38 45	001	22 42	041	26 23	088	32 24	146	37 25	223	41 31	310	32 01	330
115	38 46	000	23 09	040	27 04	088	32 47	146	36 57	223	40 59	309	31 40	329
116	38 45	358	23 35	039	27 46	087	33 10	146	36 28	222	40 27	308	31 18	328
117	38 43	357	24 01	038	28 28	086	33 34	145	36 01	222	39 53	307	30 56	327
118	38 41	356	24 26	037	29 09	085	33 58	145	35 33	222	39 20	306	30 32	326
119	38 37	355	24 51	036	29 51	085	34 22	145	35 05	221	38 46	305	30 09	325
	REGULUS		◆SPICA		RIGIL KENT.		◆ACHERNAR		RIGEL		BETELGEUSE		◆PROCYON	
120	25 15	035	14 03	092	34 46	144	34 38	221	38 11	304	29 44	324	38 33	353
121	25 39	034	14 45	091	35 11	144	34 11	220	37 36	303	29 19	323	38 28	352
122	26 02	033	15 27	090	35 35	144	33 44	220	37 01	302	28 54	322	38 21	351
123	26 24	032	16 08	089	36 00	143	33 17	220	36 25	301	28 27	321	38 14	350
124	26 46	031	16 50	089	36 25	143	32 50	219	35 49	300	28 01	320	38 06	348
125	27 07	030	17 32	088	36 50	143	32 24	219	35 13	299	27 33	319	37 57	347
126	27 28	029	18 13	087	37 16	142	31 58	219	34 36	298	27 06	318	37 48	346
127	27 47	028	18 55	087	37 41	142	31 32	218	33 59	297	26 37	317	37 37	345
128	28 07	027	19 36	086	38 07	142	31 06	218	33 22	296	26 08	316	37 25	343
129	28 25	026	20 18	085	38 33	141	30 41	217	32 44	295	25 39	315	37 13	342
130	28 43	025	20 59	084	38 59	141	30 16	217	32 06	294	25 09	314	37 00	341
131	29 00	024	21 41	084	39 25	141	29 51	217	31 28	293	24 39	313	36 46	340
132	29 16	023	22 22	083	39 52	141	29 26	216	30 49	292	24 08	312	36 31	338
133	29 32	022	23 04	082	40 18	140	29 01	216	30 11	291	23 37	311	36 15	337
134	29 47	020	23 45	081	40 45	140	28 37	215	29 32	291	23 05	310	35 59	336
	◆REGULUS		SPICA		◆RIGIL KENT.		ACHERNAR		◆CANOPUS		SIRIUS		PROCYON	
135	30 01	019	24 26	080	41 12	140	28 13	215	64 03	241	49 27	305	35 41	335
136	30 14	018	25 07	080	41 39	140	27 49	215	63 27	241	48 53	304	35 23	334
137	30 27	017	25 48	079	42 06	139	27 26	214	62 50	241	48 18	303	35 04	333
138	30 39	016	26 29	078	42 33	139	27 02	214	62 14	241	47 43	302	34 45	331
139	30 50	015	27 10	077	43 00	139	26 39	213	61 38	241	47 07	300	34 25	330
140	31 00	014	27 50	077	43 28	139	26 16	213	61 01	240	46 31	299	34 04	329
141	31 10	013	28 31	076	43 56	138	25 54	213	60 25	240	45 54	298	33 42	328
142	31 19	011	29 11	075	44 23	138	25 32	212	59 49	240	45 18	297	33 19	327
143	31 26	010	29 51	074	44 51	138	25 10	212	59 13	240	44 40	296	32 56	326
144	31 34	009	30 31	073	45 19	138	24 48	211	58 37	240	44 03	295	32 33	325
145	31 40	008	31 11	072	45 47	138	24 26	211	58 01	240	43 25	294	32 08	324
146	31 45	007	31 51	072	46 15	137	24 05	210	57 25	240	42 47	293	31 43	323
147	31 50	006	32 30	071	46 44	137	23 44	210	56 49	239	42 08	292	31 18	322
148	31 54	005	33 09	070	47 12	137	23 24	209	56 13	239	41 30	291	30 51	321
149	31 57	003	33 48	069	47 41	137	23 03	209	55 37	239	40 51	290	30 25	319
	◆REGULUS		SPICA		◆RIGIL KENT.		ACHERNAR		◆CANOPUS		SIRIUS		PROCYON	
150	31 59	002	34 27	068	48 09	137	22 43	208	55 01	239	40 11	289	29 57	318
151	32 00	001	35 06	067	48 38	137	22 24	208	54 26	239	39 32	289	29 29	317
152	32 00	000	35 44	066	49 06	136	22 04	208	53 50	238	38 52	288	29 01	316
153	32 00	359	36 22	065	49 35	136	21 45	207	53 15	238	38 13	287	28 32	315
154	31 59	358	37 00	064	50 04	136	21 26	207	52 39	238	37 33	286	28 02	314
155	31 57	357	37 37	063	50 33	136	21 08	206	52 04	238	36 52	285	27 32	313
156	31 54	355	38 14	062	51 02	136	20 50	206	51 29	237	36 12	284	27 02	313
157	31 50	354	38 51	062	51 31	136	20 32	205	50 54	237	35 32	283	26 31	312
158	31 46	353	39 27	061	52 00	136	20 14	205	50 19	237	34 51	283	26 00	311
159	31 40	352	40 04	060	52 29	136	19 57	204	49 44	237	34 10	282	25 28	310
160	31 34	351	40 39	058	52 58	136	19 40	204	49 09	236	33 29	281	24 55	309
161	31 27	350	41 15	057	53 28	136	19 24	203	48 35	236	32 48	280	24 23	308
162	31 19	349	41 50	056	53 57	135	19 07	203	48 00	236	32 07	279	23 50	307
163	31 10	347	42 24	055	54 26	135	18 51	202	47 26	236	31 26	279	23 16	306
164	31 01	346	42 58	054	54 55	135	18 36	202	46 51	235	30 45	278	22 42	305
	SPICA		◆ANTARES		RIGIL KENT.		ACHERNAR		◆CANOPUS		SIRIUS		◆REGULUS	
165	43 32	053	23 49	104	55 24	135	18 21	201	46 17	235	30 04	277	30 51	345
166	44 05	052	24 29	103	55 54	135	18 06	201	45 43	235	29 22	276	30 40	344
167	44 37	051	25 10	103	56 23	135	17 51	200	45 09	234	28 41	275	30 28	343
168	45 09	050	25 50	102	56 52	135	17 37	200	44 35	234	27 59	275	30 15	342
169	45 41	049	26 31	101	57 21	136	17 23	199	44 02	234	27 18	274	30 02	341
170	46 12	047	27 12	101	57 51	136	17 10	199	43 28	233	26 36	273	29 48	340
171	46 42	046	27 53	100	58 20	136	16 56	198	42 55	233	25 54	272	29 33	339
172	47 12	045	28 34	100	58 49	136	16 44	198	42 21	233	25 13	272	29 17	337
173	47 41	044	29 15	099	59 18	136	16 31	197	41 48	232	24 31	271	29 01	336
174	48 09	042	29 56	098	59 47	136	16 19	197	41 15	232	23 49	270	28 44	335
175	48 37	041	30 38	098	60 16	136	16 07	196	40 43	232	23 08	270	28 26	334
176	49 04	040	31 19	097	60 45	136	15 56	196	40 10	231	22 26	269	28 08	333
177	49 30	038	32 01	096	61 13	136	15 45	195	39 37	231	21 44	268	27 48	332
178	49 56	037	32 42	096	61 42	137	15 35	194	39 05	231	21 03	267	27 29	331
179	50 20	036	33 23	095	62 11	137	15 24	194	38 33	230	20 21	267	27 08	330

LHA ♈	Hc	Zn	Hc	Zn	Hc	Zn	Hc	Zn	Hc	Zn	Hc	Zn	Hc	Zn
	◆ARCTURUS		ANTARES		◆Peacock		ACHERNAR		CANOPUS		◆Suhail		REGULUS	
180	17 57	034	34 05	094	22 05	152	15 15	193	38 01	230	59 40	259	26 47	329
181	18 19	033	34 47	093	22 25	151	15 05	193	37 29	230	58 59	259	26 25	328
182	18 42	032	35 28	093	22 46	151	14 56	192	36 58	229	58 19	258	26 03	327
183	19 03	031	36 10	092	23 06	150	14 47	192	36 26	229	57 38	258	25 40	326
184	19 24	030	36 52	091	23 27	150	14 39	191	35 55	228	56 57	257	25 16	325
185	19 45	029	37 33	091	23 48	149	14 31	191	35 24	228	56 16	257	24 52	324
186	20 05	028	38 15	090	24 10	149	14 24	190	34 53	228	55 36	256	24 28	323
187	20 24	027	38 57	089	24 32	148	14 16	190	34 22	227	54 56	256	24 02	322
188	20 43	026	39 38	089	24 54	148	14 10	189	33 52	227	54 15	255	23 36	321
189	21 01	025	40 20	088	25 16	147	14 03	189	33 21	226	53 35	255	23 10	320
190	21 18	024	41 02	087	25 39	147	13 57	188	32 51	226	52 55	254	22 43	319
191	21 35	023	41 43	086	26 01	147	13 52	187	32 21	226	52 15	254	22 16	318
192	21 51	022	42 25	086	26 25	146	13 47	187	31 52	225	51 35	253	21 48	317
193	22 06	021	43 06	085	26 48	146	13 42	186	31 22	225	50 55	253	21 19	316
194	22 21	020	43 48	084	27 12	145	13 37	186	30 53	224	50 15	252	20 50	316
	ARCTURUS		◆ANTARES		Peacock		◆ACHERNAR		CANOPUS		Suhail		◆REGULUS	
195	22 35	019	44 29	083	27 35	145	13 33	185	30 24	224	49 35	252	20 21	315
196	22 49	018	45 10	082	28 00	144	13 30	185	29 55	224	48 56	251	19 51	314
197	23 01	017	45 52	082	28 24	144	13 27	184	29 27	223	48 16	251	19 21	313
198	23 14	016	46 33	081	28 49	144	13 24	184	28 58	223	47 37	251	18 50	312
199	23 25	015	47 14	080	29 13	143	13 22	183	28 30	222	46 58	250	18 19	311
200	23 36	014	47 55	079	29 39	143	13 20	182	28 02	222	46 19	250	17 47	310
201	23 45	013	48 36	078	30 04	142	13 18	182	27 35	221	45 40	249	17 15	309
202	23 55	012	49 17	077	30 30	142	13 17	181	27 07	221	45 01	249	16 43	309
203	24 03	011	49 57	076	30 55	142	13 16	181	26 40	220	44 22	248	16 10	308
204	24 11	010	50 38	075	31 21	141	13 16	180	26 13	220	43 43	248	15 37	307
205	24 18	009	51 18	074	31 48	141	13 16	180	25 46	220	43 05	247	15 03	306
206	24 24	008	51 58	073	32 14	140	13 16	179	25 20	219	42 26	247	14 29	305
207	24 30	007	52 38	073	32 41	140	13 17	179	24 54	219	41 48	246	13 55	304
208	24 35	006	53 17	072	33 08	140	13 18	178	24 28	218	41 10	246	13 21	304
209	24 39	005	53 57	070	33 35	139	13 20	177	24 02	218	40 32	245	12 46	303
	◆ARCTURUS		ANTARES		◆Peacock		ACHERNAR		CANOPUS		◆Suhail		SPICA	
210	24 42	004	54 36	069	34 02	139	13 22	177	23 37	217	39 54	245	54 21	345
211	24 44	003	55 15	068	34 30	139	13 25	176	23 12	217	39 17	244	54 09	343
212	24 46	002	55 53	067	34 57	138	13 27	176	22 47	216	38 39	244	53 57	342
213	24 47	001	56 32	066	35 25	138	13 31	175	22 23	216	38 02	244	53 43	340
214	24 47	000	57 10	065	35 53	137	13 34	175	21 59	215	37 24	243	53 29	339
215	24 47	359	57 47	064	36 22	137	13 38	174	21 35	215	36 47	243	53 13	337
216	24 46	358	58 24	063	36 50	137	13 43	174	21 11	214	36 10	242	52 56	335
217	24 44	357	59 01	061	37 19	136	13 48	173	20 48	214	35 34	242	52 39	334
218	24 41	356	59 37	060	37 48	136	13 53	172	20 25	213	34 57	241	52 20	332
219	24 38	355	60 13	059	38 16	136	13 59	172	20 02	213	34 21	241	52 00	331
220	24 33	354	60 49	057	38 46	135	14 05	171	19 40	212	33 44	240	51 39	329
221	24 28	353	61 23	056	39 15	135	14 11	171	19 18	212	33 08	240	51 18	328
222	24 23	352	61 58	054	39 44	135	14 18	170	18 56	211	32 33	239	50 55	327
223	24 16	351	62 31	053	40 14	134	14 25	170	18 35	211	31 57	239	50 32	325
224	24 09	350	63 04	051	40 44	134	14 33	169	18 13	210	31 21	238	50 07	324
	ANTARES		◆Nunki		Peacock		ACHERNAR		◆CANOPUS		◆SPICA		ARCTURUS	
225	63 36	050	39 56	088	41 14	134	14 41	169	17 53	210	49 42	322	24 01	348
226	64 08	048	40 38	087	41 44	134	14 49	168	17 32	209	49 16	321	23 52	347
227	64 38	046	41 19	086	42 14	133	14 58	168	17 12	209	48 50	320	23 43	346
228	65 08	045	42 01	086	42 45	133	15 07	167	16 52	208	48 22	318	23 33	345
229	65 37	043	42 42	085	43 15	133	15 17	166	16 33	208	47 54	317	23 22	344
230	66 05	041	43 24	084	43 46	132	15 27	166	16 14	207	47 25	316	23 10	343
231	66 32	039	44 05	083	44 17	132	15 37	165	15 55	206	46 56	314	22 58	342
232	66 57	037	44 47	083	44 48	132	15 48	165	15 37	206	46 26	313	22 45	341
233	67 22	035	45 28	082	45 19	132	15 59	164	15 19	205	45 55	312	22 31	340
234	67 45	033	46 09	081	45 50	131	16 10	164	15 01	205	45 24	311	22 17	339
235	68 07	031	46 50	080	46 21	131	16 22	163	14 44	204	44 52	310	22 02	338
236	68 28	028	47 31	079	46 53	131	16 34	163	14 27	204	44 20	309	21 46	337
237	68 47	026	48 12	078	47 24	131	16 47	162	14 10	203	43 47	307	21 30	336
238	69 04	024	48 53	078	47 56	131	17 00	162	13 54	203	43 14	306	21 13	336
239	69 21	021	49 34	077	48 28	130	17 13	161	13 38	202	42 40	305	20 56	335
	Rasalhague		◆ALTAIR		FOMALHAUT		◆ACHERNAR		ACRUX		◆SPICA		ARCTURUS	
240	27 41	026	14 52	060	11 54	121	17 26	161	55 59	221	42 06	304	20 37	334
241	27 59	025	15 28	059	12 30	120	17 40	160	55 32	221	41 31	303	20 19	333
242	28 16	024	16 04	058	13 06	119	17 55	160	55 04	221	40 56	302	19 59	332
243	28 33	023	16 39	057	13 43	119	18 09	159	54 37	221	40 20	301	19 39	331
244	28 49	022	17 14	056	14 20	118	18 24	159	54 10	221	39 44	300	19 18	330
245	29 04	021	17 48	056	14 56	117	18 39	158	53 43	221	39 08	299	18 57	329
246	29 19	020	18 23	055	15 34	117	18 55	158	53 16	221	38 31	298	18 35	328
247	29 32	019	18 57	054	16 11	116	19 11	157	52 49	221	37 54	297	18 13	327
248	29 45	018	19 30	053	16 48	116	19 27	157	52 21	221	37 17	296	17 50	326
249	29 58	017	20 03	052	17 26	115	19 44	156	51 54	221	36 39	295	17 27	325
250	30 09	015	20 36	051	18 04	114	20 01	156	51 27	221	36 02	294	17 03	324
251	30 20	014	21 08	050	18 42	114	20 18	155	51 00	220	35 23	293	16 38	323
252	30 30	013	21 40	049	19 20	113	20 36	155	50 33	220	34 45	292	16 13	323
253	30 39	012	22 11	049	19 59	112	20 54	154	50 06	220	34 06	291	15 47	322
254	30 47	011	22 42	048	20 37	112	21 12	154	49 39	220	33 27	291	15 21	321
	Rasalhague		◆ALTAIR		FOMALHAUT		◆ACHERNAR		ACRUX		◆SPICA		ANTARES	
255	30 55	010	23 13	047	21 16	111	21 31	153	49 12	220	32 48	290	69 29	340
256	31 02	009	23 43	046	21 55	111	21 50	153	48 45	220	32 09	289	69 14	337
257	31 08	008	24 13	045	22 34	110	22 09	152	48 19	220	31 29	288	68 57	335
258	31 13	006	24 42	044	23 13	109	22 28	152	47 52	220	30 50	287	68 38	333
259	31 17	005	25 11	043	23 53	109	22 48	151	47 25	220	30 10	286	68 19	330
260	31 21	004	25 39	042	24 32	108	23 08	151	46 59	220	29 30	285	67 57	328
261	31 23	003	26 06	041	25 12	107	23 29	151	46 32	219	28 49	285	67 35	326
262	31 25	002	26 34	040	25 52	107	23 49	150	46 06	219	28 09	284	67 11	324
263	31 26	001	27 00	039	26 32	106	24 10	150	45 40	219	27 28	283	66 46	322
264	31 26	000	27 26	038	27 12	106	24 31	149	45 13	219	26 48	282	66 20	320
265	31 25	358	27 51	037	27 52	105	24 53	149	44 47	219	26 07	281	65 52	318
266	31 24	357	28 16	036	28 33	104	25 15	148	44 21	219	25 26	281	65 24	316
267	31 22	356	28 41	035	29 13	104	25 37	148	43 55	218	24 45	280	64 55	315
268	31 18	355	29 04	034	29 54	103	25 59	147	43 30	218	24 04	279	64 25	313
269	31 14	354	29 27	033	30 34	102	26 22	147	43 04	218	23 23	278	63 54	311

LAT 46°S

LHA ♈	Hc	Zn	Hc	Zn	Hc	Zn	Hc	Zn	Hc	Zn	Hc	Zn	Hc	Zn
	◆ALTAIR		FOMALHAUT		◆ACHERNAR		RIGIL KENT.		◆SPICA		ANTARES		Rasalhague	
270	29 49	032	31 15	102	26 45	147	57 40	224	22 41	278	63 22	309	31 10	353
271	30 11	031	31 56	101	27 08	146	57 11	225	22 00	277	62 49	308	31 04	352
272	30 32	030	32 37	101	27 31	146	56 41	225	21 19	276	62 16	306	30 57	351
273	30 53	029	33 18	100	27 55	145	56 12	225	20 37	275	61 42	305	30 50	349
274	31 12	028	33 59	099	28 19	145	55 43	225	19 56	275	61 08	303	30 42	348
275	31 31	026	34 40	099	28 43	144	55 14	225	19 14	274	60 33	302	30 33	347
276	31 49	025	35 21	098	29 07	144	54 44	225	18 33	273	59 57	301	30 24	346
277	32 07	024	36 03	097	29 32	144	54 15	225	17 51	272	59 21	299	30 13	345
278	32 24	023	36 44	097	29 57	143	53 46	225	17 09	272	58 45	298	30 02	344
279	32 40	022	37 25	096	30 22	143	53 17	224	16 28	271	58 08	297	29 50	343
280	32 55	021	38 07	095	30 47	142	52 47	224	15 46	270	57 30	296	29 37	342
281	33 09	020	38 48	095	31 12	142	52 18	224	15 04	269	56 53	295	29 24	341
282	33 23	019	39 30	094	31 38	142	51 49	224	14 23	269	56 14	293	29 09	339
283	33 36	017	40 12	093	32 04	141	51 20	224	13 41	268	55 36	292	28 54	338
284	33 48	016	40 53	093	32 30	141	50 51	224	12 59	267	54 57	291	28 39	337
	ALTAIR		Enif		◆FOMALHAUT		ACHERNAR		◆RIGIL KENT.		ANTARES		◆Rasalhague	
285	33 59	015	23 11	045	41 35	092	32 57	141	50 22	224	54 18	290	28 22	336
286	34 10	014	23 40	044	42 16	091	33 23	140	49 53	224	53 39	289	28 05	335
287	34 19	013	24 08	043	42 58	090	33 50	140	49 24	224	53 00	288	27 47	334
288	34 28	012	24 36	042	43 40	090	34 17	139	48 56	224	52 20	287	27 29	333
289	34 36	010	25 04	041	44 22	089	34 44	139	48 27	223	51 40	286	27 09	332
290	34 43	009	25 31	040	45 03	088	35 12	139	47 58	223	51 00	285	26 49	331
291	34 49	008	25 57	039	45 45	088	35 39	138	47 30	223	50 19	284	26 29	330
292	34 55	007	26 23	038	46 26	087	36 07	138	47 01	223	49 39	283	26 08	329
293	34 59	006	26 49	037	47 08	086	36 35	138	46 33	223	48 58	282	25 46	328
294	35 03	004	27 13	036	47 50	085	37 03	137	46 05	223	48 17	281	25 23	327
295	35 06	003	27 37	035	48 31	084	37 32	137	45 37	222	47 37	281	25 00	326
296	35 08	002	28 01	034	49 13	084	38 00	137	45 09	222	46 56	280	24 37	325
297	35 09	001	28 24	033	49 54	083	38 29	136	44 41	222	46 14	279	24 13	324
298	35 09	000	28 46	032	50 35	082	38 58	136	44 13	222	45 33	278	23 48	323
299	35 08	358	29 08	031	51 17	081	39 27	136	43 45	222	44 52	277	23 22	322
	◆Enif		FOMALHAUT		◆ACHERNAR		RIGIL KENT.		◆ANTARES		Rasalhague		ALTAIR	
300	29 29	030	51 58	080	39 56	135	43 18	221	44 10	276	22 57	321	35 06	357
301	29 49	029	52 39	079	40 25	135	42 50	221	43 29	276	22 30	320	35 04	356
302	30 09	028	53 20	079	40 55	135	42 23	221	42 48	275	22 03	319	35 00	355
303	30 28	027	54 00	078	41 24	135	41 56	221	42 06	274	21 36	318	34 56	354
304	30 46	025	54 41	077	41 54	134	41 29	220	41 24	273	21 08	317	34 51	352
305	31 04	024	55 22	076	42 24	134	41 02	220	40 43	273	20 39	316	34 45	351
306	31 20	023	56 02	075	42 54	134	40 35	220	40 01	272	20 10	316	34 38	350
307	31 36	022	56 42	074	43 24	133	40 08	220	39 19	271	19 41	315	34 30	349
308	31 52	021	57 22	073	43 55	133	39 42	219	38 38	270	19 11	314	34 22	348
309	32 06	020	58 02	072	44 25	133	39 16	219	37 56	270	18 41	313	34 12	346
310	32 20	019	58 41	071	44 56	133	38 49	219	37 14	269	18 10	312	34 02	345
311	32 33	018	59 20	069	45 26	132	38 23	218	36 33	268	17 39	311	33 51	344
312	32 45	016	59 59	068	45 57	132	37 58	218	35 51	268	17 07	310	33 39	343
313	32 57	015	60 38	067	46 28	132	37 32	218	35 09	267	16 35	309	33 27	342
314	33 07	014	61 16	066	46 59	132	37 06	218	34 28	266	16 03	309	33 13	341
	◆Enif		Diphda		◆ACHERNAR		Miaplacidus		RIGIL KENT.		◆ANTARES		ALTAIR	
315	33 17	013	36 24	078	47 30	131	25 43	181	36 41	217	33 46	266	32 59	339
316	33 26	012	37 05	077	48 02	131	25 43	181	36 16	217	33 05	265	32 44	338
317	33 34	011	37 45	076	48 33	131	25 42	180	35 51	217	32 23	264	32 28	337
318	33 41	009	38 26	075	49 05	131	25 42	180	35 26	216	31 42	263	32 11	336
319	33 48	008	39 06	074	49 36	131	25 42	180	35 02	216	31 00	263	31 54	335
320	33 53	007	39 46	074	50 08	130	25 42	179	34 37	216	30 19	262	31 36	334
321	33 58	006	40 26	073	50 40	130	25 43	179	34 13	215	29 38	261	31 17	333
322	34 02	005	41 06	072	51 11	130	25 44	179	33 49	215	28 57	261	30 58	332
323	34 05	004	41 45	071	51 43	130	25 45	178	33 25	215	28 16	260	30 38	331
324	34 07	002	42 24	070	52 15	130	25 46	178	33 01	214	27 35	260	30 17	329
325	34 08	001	43 03	069	52 47	130	25 48	177	32 38	214	26 54	259	29 55	328
326	34 09	000	43 42	068	53 19	130	25 50	177	32 15	214	26 13	258	29 33	327
327	34 08	359	44 21	067	53 52	129	25 52	177	31 52	213	25 32	258	29 10	326
328	34 07	358	44 59	066	54 24	129	25 55	176	31 29	213	24 51	257	28 47	325
329	34 05	356	45 37	065	54 56	129	25 58	176	31 07	213	24 11	256	28 23	324
	◆Diphda		ACHERNAR		◆Miaplacidus		RIGIL KENT.		ANTARES		◆ALTAIR		Enif	
330	46 15	064	55 28	129	26 01	176	30 44	212	23 30	256	27 58	323	34 02	355
331	46 52	063	56 01	129	26 04	175	30 22	212	22 50	255	27 33	322	33 58	354
332	47 29	062	56 33	129	26 08	175	30 00	211	22 10	254	27 07	321	33 53	353
333	48 05	061	57 05	129	26 12	174	29 39	211	21 30	254	26 41	320	33 48	352
334	48 42	060	57 38	129	26 16	174	29 17	211	20 50	253	26 14	319	33 41	350
335	49 17	059	58 10	129	26 21	174	28 56	210	20 10	252	25 46	318	33 34	349
336	49 53	057	58 43	129	26 26	173	28 35	210	19 30	252	25 18	317	33 26	348
337	50 28	056	59 15	129	26 31	173	28 15	209	18 51	251	24 50	316	33 17	347
338	51 02	055	59 48	129	26 36	172	27 54	209	18 12	250	24 21	315	33 07	346
339	51 36	054	60 20	129	26 41	172	27 34	209	17 32	250	23 51	314	32 56	345
340	52 09	053	60 53	129	26 47	172	27 14	208	16 53	249	23 21	314	32 45	344
341	52 42	051	61 25	129	26 53	171	26 55	208	16 14	249	22 51	313	32 33	342
342	53 14	050	61 57	129	27 00	171	26 35	207	15 36	248	22 20	312	32 20	341
343	53 46	049	62 30	129	27 07	171	26 16	207	14 57	247	21 49	311	32 06	340
344	54 17	047	63 02	129	27 13	170	25 57	207	14 19	247	21 17	310	31 51	339
	◆Diphda		Acamar		◆CANOPUS		Miaplacidus		RIGIL KENT.		◆Nunki		Enif	
345	54 47	046	47 14	105	24 56	141	27 21	170	25 39	206	38 11	270	31 36	338
346	55 17	045	47 54	104	25 22	141	27 28	170	25 21	206	37 29	269	31 20	337
347	55 46	043	48 35	104	25 49	140	27 36	169	25 03	205	36 48	269	31 03	336
348	56 14	042	49 16	103	26 15	140	27 44	169	24 45	205	36 06	268	30 46	335
349	56 41	040	49 56	102	26 42	140	27 52	168	24 27	205	35 24	267	30 27	333
350	57 08	039	50 37	102	27 09	139	28 00	168	24 10	204	34 43	267	30 08	332
351	57 33	037	51 18	101	27 37	139	28 09	168	23 53	204	34 01	266	29 49	331
352	57 58	035	51 59	101	28 04	138	28 18	167	23 37	203	33 20	265	29 28	330
353	58 21	034	52 40	100	28 32	138	28 27	167	23 20	203	32 38	265	29 07	329
354	58 44	032	53 21	100	29 00	137	28 37	167	23 04	202	31 57	264	28 46	328
355	59 06	030	54 02	099	29 29	137	28 47	166	22 49	202	31 15	263	28 23	327
356	59 26	029	54 43	099	29 57	136	28 57	166	22 33	201	30 34	263	28 00	326
357	59 45	027	55 24	098	30 26	136	29 07	166	22 18	201	29 53	262	27 37	325
358	60 04	025	56 05	097	30 55	136	29 17	165	22 04	201	29 11	261	27 13	324
359	60 21	023	56 47	097	31 25	135	29 28	165	21 49	200	28 30	261	26 48	323

LHA ♈	Hc	Zn	Hc	Zn	Hc	Zn	Hc	Zn	Hc	Zn	Hc	Zn	Hc	Zn
	Alpheratz		Hamal		◆RIGEL		CANOPUS		◆RIGIL KENT.		Peacock		◆FOMALHAUT	
0	13 55	002	13 58	030	13 47	087	32 36	134	22 31	200	56 25	233	68 50	320
1	13 56	001	14 18	029	14 27	087	33 05	134	22 18	199	55 52	233	68 23	317
2	13 56	000	14 37	028	15 08	086	33 35	133	22 04	199	55 20	233	67 54	316
3	13 56	359	14 56	027	15 49	085	34 05	133	21 51	198	54 47	233	67 25	314
4	13 55	358	15 15	026	16 30	084	34 35	133	21 39	198	54 15	232	66 55	312
5	13 53	357	15 33	025	17 10	084	35 05	132	21 26	197	53 42	232	66 24	310
6	13 51	356	15 50	024	17 51	083	35 35	132	21 14	197	53 10	232	65 53	308
7	13 48	356	16 07	024	18 32	082	36 06	131	21 02	196	52 38	232	65 20	307
8	13 45	355	16 23	023	19 12	081	36 37	131	20 51	196	52 05	232	64 47	305
9	13 41	354	16 38	022	19 53	080	37 08	131	20 40	195	51 33	232	64 13	304
10	13 36	353	16 53	021	20 33	080	37 39	130	20 29	195	51 01	231	63 39	302
11	13 30	352	17 07	020	21 13	079	38 10	130	20 19	195	50 29	231	63 04	301
12	13 24	351	17 21	019	21 53	078	38 42	129	20 09	194	49 58	231	62 28	299
13	13 18	350	17 34	018	22 33	077	39 14	129	19 59	194	49 26	231	61 52	298
14	13 10	349	17 46	017	23 13	077	39 45	129	19 49	193	48 54	231	61 16	297
	Hamal		◆RIGEL		SIRIUS		CANOPUS		◆RIGIL KENT.		Peacock		◆FOMALHAUT	
15	17 58	016	23 53	076	14 40	099	40 17	128	19 40	193	48 23	230	60 39	295
16	18 09	015	24 32	075	15 20	098	40 50	128	19 32	192	47 51	230	60 02	294
17	18 19	014	25 12	074	16 01	098	41 22	128	19 23	192	47 20	230	59 24	293
18	18 29	013	25 51	073	16 42	097	41 54	127	19 15	191	46 49	230	58 47	292
19	18 38	012	26 30	072	17 22	096	42 27	127	19 08	191	46 18	229	58 08	291
20	18 46	011	27 09	072	18 03	095	43 00	126	19 00	190	45 47	229	57 30	289
21	18 54	010	27 48	071	18 44	095	43 33	126	18 53	190	45 16	229	56 51	288
22	19 01	009	28 26	070	19 25	094	44 06	126	18 47	189	44 45	228	56 12	287
23	19 07	008	29 05	069	20 05	093	44 39	125	18 40	189	44 15	228	55 33	286
24	19 13	008	29 43	068	20 46	093	45 13	125	18 34	188	43 44	228	54 54	285
25	19 18	007	30 21	067	21 27	092	45 46	125	18 29	188	43 14	228	54 14	284
26	19 22	006	30 58	066	22 08	091	46 20	124	18 24	187	42 44	227	53 34	283
27	19 26	005	31 36	066	22 49	090	46 54	124	18 19	187	42 14	227	52 55	282
28	19 29	004	32 13	065	23 30	090	47 28	124	18 14	186	41 44	227	52 15	282
29	19 31	003	32 50	064	24 11	089	48 02	123	18 10	186	41 14	226	51 34	281
	Hamal		◆RIGEL		SIRIUS		CANOPUS		◆RIGIL KENT.		Peacock		◆FOMALHAUT	
30	19 33	002	33 26	063	24 52	088	48 36	123	18 06	185	40 45	226	50 54	280
31	19 33	001	34 02	062	25 33	087	49 10	123	18 03	185	40 16	226	50 14	279
32	19 34	000	34 38	061	26 13	087	49 45	123	18 00	184	39 46	225	49 33	278
33	19 33	359	35 14	060	26 54	086	50 19	122	17 57	183	39 17	225	48 53	277
34	19 32	358	35 49	059	27 35	085	50 54	122	17 55	183	38 49	225	48 12	276
35	19 30	357	36 24	058	28 16	084	51 29	122	17 53	182	38 20	224	47 31	276
36	19 27	356	36 59	057	28 57	084	52 04	121	17 52	182	37 51	224	46 51	275
37	19 24	355	37 33	056	29 37	083	52 38	121	17 50	181	37 23	224	46 10	274
38	19 20	354	38 06	055	30 18	082	53 14	121	17 50	181	36 55	223	45 29	273
39	19 15	353	38 40	054	30 58	081	53 49	121	17 49	180	36 27	223	44 48	272
40	19 10	352	39 12	053	31 39	080	54 24	120	17 49	180	35 59	223	44 07	272
41	19 04	351	39 45	052	32 19	080	54 59	120	17 49	179	35 32	222	43 26	271
42	18 57	350	40 17	051	32 59	079	55 35	120	17 50	179	35 04	222	42 45	270
43	18 50	349	40 48	050	33 39	078	56 10	120	17 51	178	34 37	221	42 04	269
44	18 42	348	41 19	048	34 19	077	56 46	119	17 52	178	34 10	221	41 24	269
	RIGEL		SIRIUS		◆CANOPUS		RIGIL KENT.		◆Peacock		FOMALHAUT		◆Hamal	
45	41 49	047	34 59	076	57 22	119	17 54	177	33 43	221	40 43	268	18 33	347
46	42 19	046	35 39	076	57 57	119	17 56	177	33 17	220	40 02	267	18 23	346
47	42 48	045	36 18	075	58 33	119	17 58	176	32 50	220	39 21	267	18 13	345
48	43 17	044	36 58	074	59 09	119	18 01	176	32 24	219	38 40	266	18 03	344
49	43 45	043	37 37	073	59 45	118	18 04	175	31 58	219	37 59	265	17 51	343
50	44 12	041	38 16	072	60 21	118	18 08	175	31 33	219	37 19	265	17 39	342
51	44 39	040	38 55	071	60 57	118	18 12	174	31 07	218	36 38	264	17 26	341
52	45 05	039	39 33	070	61 33	118	18 16	174	30 42	218	35 57	263	17 13	341
53	45 30	038	40 12	069	62 09	118	18 21	173	30 17	217	35 17	263	16 59	340
54	45 55	036	40 50	068	62 46	118	18 26	173	29 52	217	34 36	262	16 45	339
55	46 19	035	41 28	067	63 22	118	18 31	172	29 28	217	33 56	261	16 30	338
56	46 42	034	42 05	066	63 58	118	18 37	172	29 03	216	33 15	261	16 14	337
57	47 04	032	42 43	065	64 34	117	18 43	171	28 39	216	32 35	260	15 57	336
58	47 25	031	43 20	064	65 11	117	18 49	171	28 16	215	31 55	259	15 40	335
59	47 46	030	43 56	063	65 47	117	18 56	170	27 52	215	31 14	259	15 23	334
	◆ALDEBARAN		BETELGEUSE		SIRIUS		◆Suhail		RIGIL KENT.		◆ACHERNAR		Diphda	
60	26 00	010	29 55	033	44 33	062	37 55	116	19 03	170	66 12	231	40 34	289
61	26 06	008	30 17	032	45 09	061	38 32	116	19 10	169	65 40	232	39 55	288
62	26 12	007	30 39	031	45 45	060	39 08	115	19 18	169	65 08	232	39 16	287
63	26 17	006	31 00	030	46 20	059	39 46	115	19 26	168	64 36	232	38 36	286
64	26 21	005	31 20	029	46 55	058	40 23	114	19 35	168	64 04	232	37 57	285
65	26 24	004	31 40	028	47 29	057	41 00	114	19 44	167	63 32	232	37 18	284
66	26 27	003	31 58	027	48 03	056	41 38	113	19 53	167	62 59	232	36 38	283
67	26 29	002	32 17	026	48 37	055	42 15	113	20 02	166	62 27	232	35 58	283
68	26 30	001	32 34	025	49 10	053	42 53	112	20 12	166	61 54	232	35 18	282
69	26 30	000	32 51	023	49 43	052	43 31	112	20 23	165	61 22	232	34 38	281
70	26 30	359	33 07	022	50 15	051	44 09	111	20 33	165	60 50	232	33 58	280
71	26 28	358	33 22	021	50 46	050	44 47	111	20 44	164	60 17	232	33 17	279
72	26 26	357	33 36	020	51 17	048	45 26	110	20 55	164	59 45	232	32 37	279
73	26 24	356	33 50	019	51 48	047	46 04	110	21 07	163	59 12	232	31 56	278
74	26 20	355	34 03	018	52 17	046	46 43	109	21 19	163	58 40	232	31 16	277
	BETELGEUSE		SIRIUS		◆Suhail		RIGIL KENT.		◆ACHERNAR		Diphda		◆ALDEBARAN	
75	34 15	017	52 46	044	47 21	109	21 31	162	58 07	232	30 35	276	26 16	353
76	34 26	015	53 14	043	48 00	108	21 43	162	57 35	232	29 54	275	26 11	352
77	34 36	014	53 42	042	48 39	108	21 56	161	57 03	232	29 14	275	26 05	351
78	34 46	013	54 09	040	49 18	107	22 09	161	56 30	232	28 33	274	25 59	350
79	34 55	012	54 35	039	49 57	107	22 23	161	55 58	232	27 52	273	25 51	349
80	35 03	011	55 00	037	50 36	106	22 37	160	55 26	232	27 11	272	25 43	348
81	35 10	009	55 24	036	51 16	106	22 51	160	54 54	232	26 30	272	25 35	347
82	35 16	008	55 48	034	51 55	105	23 05	159	54 22	232	25 49	271	25 25	346
83	35 21	007	56 10	033	52 34	105	23 20	159	53 50	231	25 08	270	25 15	345
84	35 26	006	56 32	031	53 14	104	23 35	158	53 18	231	24 27	269	25 04	344
85	35 30	005	56 52	029	53 54	104	23 50	158	52 46	231	23 46	269	24 52	343
86	35 32	003	57 12	028	54 34	103	24 06	157	52 14	231	23 06	268	24 40	342
87	35 34	002	57 30	026	55 13	103	24 22	157	51 42	231	22 25	267	24 27	341
88	35 35	001	57 48	024	55 53	102	24 38	156	51 11	231	21 44	267	24 13	340
89	35 36	000	58 04	023	56 33	102	24 54	156	50 39	230	21 03	266	23 59	339

LAT 47°S

LHA ♈	Hc	Zn	Hc	Zn	Hc	Zn	Hc	Zn	Hc	Zn	Hc	Zn	Hc	Zn
	PROCYON		Alphard		◆Suhail		RIGIL KENT.		◆ACHERNAR		◆RIGEL		BETELGEUSE	
90	33 22	030	31 46	066	57 13	101	25 11	156	50 07	230	50 00	342	35 35	358
91	33 42	029	32 23	065	57 54	101	25 28	155	49 36	230	49 47	341	35 33	357
92	34 01	028	33 00	064	58 34	100	25 46	155	49 05	230	49 33	339	35 31	356
93	34 20	027	33 37	063	59 14	100	26 03	154	48 34	230	49 18	338	35 28	355
94	34 38	025	34 13	062	59 54	099	26 21	154	48 03	229	49 02	336	35 24	354
95	34 55	024	34 50	061	60 35	099	26 39	153	47 32	229	48 45	335	35 19	352
96	35 11	023	35 25	060	61 15	098	26 58	153	47 01	229	48 27	333	35 13	351
97	35 27	022	36 01	060	61 56	098	27 17	153	46 30	229	48 09	332	35 06	350
98	35 42	021	36 36	059	62 37	097	27 36	152	45 59	228	47 49	331	34 58	349
99	35 56	020	37 11	058	63 17	096	27 55	152	45 29	228	47 28	329	34 50	348
100	36 09	018	37 45	057	63 58	096	28 14	151	44 58	228	47 07	328	34 41	346
101	36 22	017	38 19	055	64 39	095	28 34	151	44 28	227	46 45	327	34 31	345
102	36 33	016	38 52	054	65 19	095	28 54	150	43 58	227	46 22	325	34 20	344
103	36 44	015	39 25	053	66 00	094	29 14	150	43 28	227	45 58	324	34 08	343
104	36 54	013	39 58	052	66 41	093	29 35	150	42 58	227	45 34	323	33 56	342
	PROCYON		REGULUS		◆Gienah		RIGIL KENT.		◆ACHERNAR		RIGEL		◆BETELGEUSE	
105	37 03	012	17 38	049	20 13	094	29 56	149	42 29	226	45 09	321	33 42	340
106	37 11	011	18 08	048	20 53	094	30 17	149	41 59	226	44 43	320	33 28	339
107	37 19	010	18 38	047	21 34	093	30 38	149	41 30	226	44 16	319	33 13	338
108	37 25	008	19 08	046	22 15	092	31 00	148	41 01	225	43 49	318	32 58	337
109	37 31	007	19 37	045	22 56	091	31 21	148	40 31	225	43 21	316	32 42	336
110	37 36	006	20 06	044	23 37	091	31 43	147	40 03	225	42 53	315	32 24	335
111	37 39	005	20 34	043	24 18	090	32 05	147	39 34	224	42 23	314	32 07	334
112	37 42	003	21 02	042	24 59	089	32 28	147	39 05	224	41 54	313	31 48	333
113	37 44	002	21 29	041	25 40	089	32 50	146	38 37	224	41 23	312	31 29	331
114	37 45	001	21 56	041	26 21	088	33 13	146	38 09	223	40 53	311	31 09	330
115	37 46	000	22 22	040	27 01	087	33 36	146	37 41	223	40 21	310	30 48	329
116	37 45	358	22 48	039	27 42	086	34 00	145	37 13	223	39 50	308	30 27	328
117	37 43	357	23 13	038	28 23	086	34 23	145	36 45	222	39 17	307	30 05	327
118	37 41	356	23 38	037	29 04	085	34 47	144	36 18	222	38 45	306	29 43	326
119	37 38	355	24 02	036	29 45	084	35 11	144	35 50	222	38 11	305	29 20	325
	REGULUS		◆SPICA		RIGIL KENT.		◆ACHERNAR		RIGEL		BETELGEUSE		◆PROCYON	
120	24 26	035	14 05	091	35 35	144	35 23	221	37 38	304	28 56	324	37 33	353
121	24 49	034	14 46	091	35 59	143	34 56	221	37 04	303	28 31	323	37 28	352
122	25 11	033	15 27	090	36 23	143	34 30	221	36 29	302	28 07	322	37 22	351
123	25 33	032	16 07	089	36 48	143	34 03	220	35 54	301	27 41	321	37 15	350
124	25 54	031	16 48	088	37 13	142	33 37	220	35 19	300	27 15	320	37 08	348
125	26 15	030	17 29	088	37 38	142	33 11	219	34 44	299	26 48	319	36 59	347
126	26 35	029	18 10	087	38 03	142	32 45	219	34 08	298	26 21	318	36 49	346
127	26 54	028	18 51	086	38 29	142	32 19	219	33 32	297	25 53	317	36 39	345
128	27 13	027	19 32	085	38 54	141	31 54	218	32 55	296	25 25	316	36 28	344
129	27 31	026	20 13	085	39 20	141	31 28	218	32 18	295	24 57	315	36 16	342
130	27 48	025	20 53	084	39 46	141	31 03	217	31 41	295	24 27	314	36 03	341
131	28 05	023	21 34	083	40 12	140	30 39	217	31 04	294	23 58	313	35 49	340
132	28 21	022	22 15	082	40 38	140	30 14	217	30 26	293	23 28	312	35 35	339
133	28 36	021	22 55	082	41 04	140	29 50	216	29 48	292	22 57	311	35 20	338
134	28 51	020	23 35	081	41 31	139	29 26	216	29 10	291	22 26	310	35 04	336
	◆REGULUS		SPICA		◆RIGIL KENT.		ACHERNAR		◆CANOPUS		SIRIUS		PROCYON	
135	29 04	019	24 16	080	41 57	139	29 02	215	64 31	243	48 52	306	34 47	335
136	29 17	018	24 56	079	42 24	139	28 38	215	63 55	242	48 19	305	34 29	334
137	29 30	017	25 36	078	42 51	139	28 15	215	63 19	242	47 45	304	34 11	333
138	29 41	016	26 16	078	43 18	138	27 52	214	62 43	242	47 11	302	33 52	332
139	29 52	015	26 56	077	43 45	138	27 29	214	62 06	242	46 36	301	33 32	331
140	30 02	014	27 36	076	44 13	138	27 07	213	61 30	242	46 01	300	33 12	330
141	30 11	013	28 16	075	44 40	138	26 44	213	60 54	242	45 26	299	32 51	328
142	30 20	011	28 55	074	45 08	137	26 22	212	60 18	242	44 50	298	32 29	327
143	30 27	010	29 34	074	45 36	137	26 01	212	59 42	242	44 13	297	32 07	326
144	30 34	009	30 14	073	46 03	137	25 39	211	59 06	241	43 37	296	31 44	325
145	30 40	008	30 53	072	46 31	137	25 18	211	58 30	241	43 00	295	31 20	324
146	30 46	007	31 31	071	46 59	137	24 57	211	57 54	241	42 23	294	30 55	323
147	30 50	006	32 10	070	47 27	136	24 36	210	57 19	241	41 45	293	30 31	322
148	30 54	005	32 48	069	47 56	136	24 16	210	56 43	241	41 07	292	30 05	321
149	30 57	003	33 26	068	48 24	136	23 56	209	56 07	240	40 29	291	29 39	320
	◆REGULUS		SPICA		◆RIGIL KENT.		ACHERNAR		◆CANOPUS		SIRIUS		PROCYON	
150	30 59	002	34 04	067	48 52	136	23 36	209	55 32	240	39 51	290	29 12	319
151	31 00	001	34 42	067	49 21	136	23 17	208	54 57	240	39 13	289	28 45	318
152	31 00	000	35 19	066	49 50	136	22 57	208	54 21	240	38 34	288	28 17	317
153	31 00	359	35 57	065	50 18	135	22 39	207	53 46	239	37 55	288	27 49	316
154	30 59	358	36 33	064	50 47	135	22 20	207	53 11	239	37 16	287	27 20	315
155	30 57	357	37 10	063	51 16	135	22 02	206	52 36	239	36 37	286	26 51	314
156	30 54	355	37 46	062	51 45	135	21 44	206	52 01	239	35 57	285	26 21	313
157	30 50	354	38 22	061	52 14	135	21 26	205	51 26	238	35 17	284	25 51	312
158	30 46	353	38 58	060	52 43	135	21 09	205	50 51	238	34 38	283	25 20	311
159	30 41	352	39 33	059	53 12	135	20 52	204	50 16	238	33 58	282	24 49	310
160	30 35	351	40 08	058	53 41	135	20 35	204	49 42	237	33 18	282	24 18	309
161	30 28	350	40 42	057	54 10	135	20 19	203	49 08	237	32 38	281	23 46	308
162	30 20	349	41 16	056	54 39	134	20 03	203	48 33	237	31 57	280	23 13	307
163	30 12	348	41 50	055	55 08	134	19 47	202	47 59	236	31 17	279	22 41	306
164	30 03	346	42 23	053	55 38	134	19 32	202	47 25	236	30 37	278	22 08	305
	SPICA		◆ANTARES		RIGIL KENT.		ACHERNAR		◆CANOPUS		SIRIUS		◆REGULUS	
165	42 55	052	24 03	104	56 07	134	19 17	201	46 51	236	29 56	278	29 53	345
166	43 27	051	24 43	103	56 36	134	19 02	201	46 17	236	29 15	277	29 42	344
167	43 59	050	25 23	102	57 05	134	18 47	200	45 44	235	28 35	276	29 30	343
168	44 30	049	26 03	102	57 35	134	18 33	200	45 10	235	27 54	275	29 18	342
169	45 01	048	26 43	101	58 04	134	18 20	199	44 37	235	27 13	274	29 05	341
170	45 31	047	27 23	100	58 33	134	18 06	199	44 03	234	26 32	274	28 51	340
171	46 00	045	28 03	100	59 02	134	17 53	198	43 30	234	25 52	273	28 37	339
172	46 29	044	28 44	099	59 31	135	17 41	198	42 57	233	25 11	272	28 22	338
173	46 57	043	29 24	098	60 01	135	17 29	197	42 25	233	24 30	271	28 06	337
174	47 25	042	30 05	098	60 30	135	17 17	197	41 52	233	23 49	271	27 49	336
175	47 51	040	30 45	097	60 59	135	17 05	196	41 19	232	23 08	270	27 32	334
176	48 18	039	31 26	096	61 28	135	16 54	196	40 47	232	22 27	269	27 14	333
177	48 43	038	32 07	096	61 56	135	16 43	195	40 15	232	21 46	269	26 55	332
178	49 08	036	32 47	095	62 25	135	16 33	195	39 43	231	21 05	268	26 36	331
179	49 31	035	33 28	094	62 54	136	16 23	194	39 11	231	20 24	267	26 15	330

LHA	Hc	Zn	Hc	Zn	Hc	Zn	Hc	Zn	Hc	Zn	Hc	Zn	Hc	Zn
♈	◆ARCTURUS		ANTARES		◆Peacock		ACHERNAR		CANOPUS		◆Suhail		REGULUS	
180	17 07	033	34 09	093	22 58	151	16 13	193	38 39	231	59 51	261	25 56	329
181	17 29	032	34 50	093	23 18	151	16 04	193	38 08	230	59 10	260	25 34	328
182	17 51	032	35 31	092	23 38	150	15 55	192	37 37	230	58 30	260	25 13	327
183	18 12	031	36 12	091	23 58	150	15 46	192	37 05	229	57 50	259	24 50	326
184	18 32	030	36 53	091	24 19	149	15 38	191	36 34	229	57 10	259	24 27	325
185	18 52	029	37 34	090	24 40	149	15 30	191	36 04	229	56 30	258	24 04	324
186	19 12	028	38 14	089	25 01	149	15 23	190	35 33	228	55 50	258	23 39	323
187	19 30	027	38 55	088	25 23	148	15 16	190	35 03	228	55 10	257	23 15	322
188	19 49	026	39 36	088	25 44	148	15 09	189	34 32	227	54 30	257	22 50	321
189	20 06	025	40 17	087	26 06	147	15 03	189	34 03	227	53 50	256	22 24	320
190	20 23	024	40 58	086	26 29	147	14 57	188	33 33	227	53 10	256	21 57	320
191	20 40	023	41 39	085	26 51	146	14 51	187	33 03	226	52 31	255	21 31	319
192	20 55	022	42 20	085	27 14	146	14 46	187	32 34	226	51 51	255	21 03	318
193	21 10	021	43 00	084	27 37	145	14 41	186	32 05	225	51 12	254	20 36	317
194	21 25	020	43 41	083	28 01	145	14 37	186	31 36	225	50 33	254	20 07	316
	ARCTURUS		◆ANTARES		Peacock		◆ACHERNAR		CANOPUS		Suhail		◆REGULUS	
195	21 39	019	44 21	082	28 24	145	14 33	185	31 07	224	49 53	253	19 39	315
196	21 52	018	45 02	081	28 48	144	14 30	185	30 38	224	49 14	253	19 09	314
197	22 04	017	45 42	081	29 12	144	14 27	184	30 10	223	48 35	252	18 40	313
198	22 16	016	46 23	080	29 37	143	14 24	184	29 42	223	47 56	252	18 10	312
199	22 27	015	47 03	079	30 01	143	14 21	183	29 14	223	47 18	251	17 39	311
200	22 37	014	47 43	078	30 26	142	14 20	182	28 47	222	46 39	251	17 08	311
201	22 47	013	48 23	077	30 51	142	14 18	182	28 19	222	46 00	250	16 37	310
202	22 56	012	49 03	076	31 17	142	14 17	181	27 52	221	45 22	250	16 05	309
203	23 04	011	49 42	075	31 42	141	14 16	181	27 26	221	44 44	249	15 33	308
204	23 12	010	50 22	074	32 08	141	14 16	180	26 59	220	44 06	249	15 01	307
205	23 19	009	51 01	073	32 34	140	14 16	180	26 33	220	43 28	248	14 28	306
206	23 25	008	51 40	072	33 00	140	14 16	179	26 07	219	42 50	248	13 55	305
207	23 30	007	52 19	071	33 27	140	14 17	179	25 41	219	42 12	247	13 21	305
208	23 35	006	52 58	070	33 53	139	14 18	178	25 15	218	41 34	247	12 47	304
209	23 39	005	53 36	069	34 20	139	14 20	177	24 50	218	40 57	246	12 13	303
	◆ARCTURUS		ANTARES		◆Peacock		ACHERNAR		CANOPUS		◆Suhail		SPICA	
210	23 42	004	54 14	068	34 47	138	14 22	177	24 25	217	40 19	246	53 23	345
211	23 44	003	54 52	067	35 14	138	14 24	176	24 00	217	39 42	245	53 12	344
212	23 46	002	55 30	066	35 42	138	14 27	176	23 36	216	39 05	245	53 00	342
213	23 47	001	56 07	065	36 09	137	14 31	175	23 11	216	38 28	244	52 47	341
214	23 47	000	56 44	064	36 37	137	14 34	175	22 48	215	37 51	244	52 33	339
215	23 47	359	57 20	062	37 05	137	14 38	174	22 24	215	37 15	243	52 18	338
216	23 46	358	57 56	061	37 34	136	14 43	174	22 01	214	36 38	243	52 02	336
217	23 44	357	58 32	060	38 02	136	14 47	173	21 38	214	36 02	242	51 44	334
218	23 41	356	59 07	059	38 31	136	14 53	172	21 15	213	35 26	242	51 26	333
219	23 38	355	59 41	057	38 59	135	14 58	172	20 53	213	34 50	241	51 07	332
220	23 34	354	60 16	056	39 28	135	15 04	171	20 30	212	34 14	241	50 47	330
221	23 29	353	60 49	054	39 57	135	15 10	171	20 09	212	33 39	240	50 26	329
222	23 23	352	61 22	053	40 27	134	15 17	170	19 47	211	33 03	240	50 05	327
223	23 17	351	61 54	051	40 56	134	15 24	170	19 26	211	32 28	239	49 42	326
224	23 10	350	62 26	050	41 26	134	15 32	169	19 05	210	31 53	239	49 19	324
	ANTARES		◆Nunki		Peacock		ACHERNAR		◆CANOPUS		◆SPICA		ARCTURUS	
225	62 57	048	39 54	087	41 55	133	15 40	169	18 45	210	48 54	323	23 02	349
226	63 27	047	40 34	086	42 25	133	15 48	168	18 25	209	48 29	322	22 54	348
227	63 56	045	41 15	086	42 55	133	15 57	168	18 05	209	48 04	320	22 44	347
228	64 25	043	41 56	085	43 25	132	16 06	167	17 45	208	47 37	319	22 35	346
229	64 53	041	42 37	084	43 56	132	16 15	166	17 26	208	47 10	318	22 24	345
230	65 19	040	43 17	083	44 26	132	16 25	166	17 07	207	46 42	316	22 13	344
231	65 45	038	43 58	082	44 57	131	16 35	165	16 49	207	46 14	315	22 01	343
232	66 09	036	44 39	082	45 27	131	16 46	165	16 31	206	45 45	314	21 48	342
233	66 32	034	45 19	081	45 58	131	16 57	164	16 13	205	45 15	313	21 35	341
234	66 54	032	45 59	080	46 29	131	17 08	164	15 55	205	44 45	312	21 21	340
235	67 15	030	46 40	079	47 00	130	17 19	163	15 38	204	44 14	310	21 06	339
236	67 35	027	47 20	078	47 32	130	17 31	163	15 22	204	43 42	309	20 51	338
237	67 53	025	48 00	077	48 03	130	17 44	162	15 05	203	43 10	308	20 35	337
238	68 09	023	48 39	076	48 34	130	17 56	162	14 49	203	42 38	307	20 19	336
239	68 25	021	49 19	075	49 06	129	18 10	161	14 34	202	42 05	306	20 01	335
	Rasalhague		◆ALTAIR		FOMALHAUT		◆ACHERNAR		ACRUX		◆SPICA		ARCTURUS	
240	26 47	026	14 22	059	12 25	120	18 23	161	56 44	222	41 32	305	19 44	334
241	27 04	025	14 57	059	13 00	120	18 37	160	56 17	222	40 58	304	19 25	333
242	27 21	024	15 32	058	13 36	119	18 51	160	55 50	222	40 24	303	19 06	332
243	27 38	023	16 06	057	14 12	118	19 05	159	55 22	222	39 49	302	18 47	331
244	27 53	022	16 41	056	14 48	118	19 20	159	54 55	222	39 14	301	18 26	330
245	28 08	021	17 14	055	15 24	117	19 35	158	54 28	222	38 39	300	18 06	329
246	28 22	020	17 48	054	16 01	117	19 51	158	54 01	222	38 03	299	17 44	328
247	28 36	019	18 21	054	16 37	116	20 06	157	53 34	222	37 27	298	17 23	327
248	28 48	018	18 54	053	17 14	115	20 23	157	53 07	221	36 50	297	17 00	326
249	29 00	016	19 26	052	17 51	115	20 39	156	52 40	221	36 14	296	16 37	325
250	29 11	015	19 58	051	18 29	114	20 56	156	52 13	221	35 37	295	16 14	325
251	29 22	014	20 30	050	19 06	113	21 13	155	51 46	221	34 59	294	15 50	324
252	29 32	013	21 01	049	19 44	113	21 30	155	51 19	221	34 22	293	15 25	323
253	29 40	012	21 32	048	20 21	112	21 48	154	50 52	221	33 44	292	15 00	322
254	29 49	011	22 02	047	20 59	111	22 06	154	50 25	221	33 06	291	14 35	321
	Rasalhague		◆ALTAIR		FOMALHAUT		◆ACHERNAR		ACRUX		◆SPICA		ANTARES	
255	29 56	010	22 32	046	21 38	111	22 24	153	49 58	221	32 28	290	68 32	341
256	30 02	009	23 01	046	22 16	110	22 43	153	49 31	221	31 49	289	68 18	338
257	30 08	008	23 30	045	22 54	110	23 02	152	49 04	221	31 11	289	68 02	336
258	30 13	006	23 59	044	23 33	109	23 21	152	48 38	221	30 32	288	67 45	334
259	30 17	005	24 27	043	24 12	108	23 41	151	48 11	220	29 53	287	67 26	332
260	30 21	004	24 54	042	24 51	108	24 01	151	47 45	220	29 13	286	67 06	330
261	30 23	003	25 21	041	25 30	107	24 21	150	47 18	220	28 34	285	66 45	327
262	30 25	002	25 48	040	26 09	106	24 41	150	46 52	220	27 54	284	66 22	325
263	30 26	001	26 13	039	26 48	106	25 02	149	46 26	220	27 15	284	65 58	323
264	30 26	000	26 39	038	27 28	105	25 23	149	46 00	220	26 35	283	65 33	321
265	30 25	359	27 04	037	28 07	104	25 44	148	45 34	219	25 55	282	65 07	320
266	30 24	357	27 28	036	28 47	104	26 06	148	45 08	219	25 15	281	64 40	318
267	30 22	356	27 51	035	29 27	103	26 28	148	44 42	219	24 35	280	64 12	316
268	30 19	355	28 14	034	30 07	103	26 50	147	44 17	219	23 54	280	63 43	314
269	30 15	354	28 37	033	30 47	102	27 12	147	43 51	219	23 14	279	63 14	313

LHA	Hc	Zn	Hc	Zn	Hc	Zn	Hc	Zn	Hc	Zn	Hc	Zn	Hc	Zn
♈	◆ALTAIR		FOMALHAUT		◆ACHERNAR		RIGIL KENT.		◆SPICA		ANTARES		Rasalhague	
270	28 58	032	31 27	101	27 35	146	58 22	226	22 33	278	62 43	311	30 10	353
271	29 20	031	32 07	101	27 57	146	57 53	226	21 53	277	62 12	309	30 05	352
272	29 40	029	32 47	100	28 21	145	57 24	226	21 12	276	61 40	308	29 58	351
273	30 00	028	33 28	099	28 44	145	56 54	226	20 31	276	61 07	306	29 51	349
274	30 19	027	34 08	099	29 08	145	56 25	226	19 51	275	60 34	305	29 43	348
275	30 37	026	34 49	098	29 31	144	55 56	226	19 10	274	60 00	304	29 35	347
276	30 55	025	35 29	097	29 56	144	55 27	226	18 29	273	59 26	302	29 25	346
277	31 12	024	36 10	097	30 20	143	54 57	226	17 48	273	58 51	301	29 15	345
278	31 28	023	36 51	096	30 44	143	54 28	225	17 07	272	58 16	300	29 04	344
279	31 44	022	37 31	095	31 09	142	53 59	225	16 26	271	57 40	298	28 53	343
280	31 59	021	38 12	095	31 34	142	53 30	225	15 46	270	57 04	297	28 40	342
281	32 13	020	38 53	094	32 00	142	53 01	225	15 05	270	56 27	296	28 27	341
282	32 26	018	39 34	093	32 25	141	52 32	225	14 24	269	55 50	295	28 13	340
283	32 39	017	40 15	092	32 51	141	52 03	225	13 43	268	55 13	294	27 58	339
284	32 50	016	40 55	092	33 17	140	51 34	225	13 02	268	54 35	292	27 43	337
	ALTAIR		Enif		◆FOMALHAUT		ACHERNAR		◆RIGIL KENT.		ANTARES		◆Rasalhague	
285	33 01	015	22 28	044	41 36	091	33 43	140	51 05	225	53 57	291	27 27	336
286	33 12	014	22 56	043	42 17	090	34 09	140	50 36	225	53 19	290	27 10	335
287	33 21	013	23 24	042	42 58	090	34 36	139	50 08	225	52 40	289	26 53	334
288	33 29	011	23 51	042	43 39	089	35 03	139	49 39	224	52 02	288	26 35	333
289	33 37	010	24 18	041	44 20	088	35 30	139	49 10	224	51 23	287	26 16	332
290	33 44	009	24 45	040	45 01	087	35 57	138	48 42	224	50 43	286	25 57	331
291	33 50	008	25 11	039	45 42	087	36 24	138	48 13	224	50 04	285	25 37	330
292	33 55	007	25 36	038	46 23	086	36 52	138	47 45	224	49 25	284	25 16	329
293	34 00	006	26 00	037	47 03	085	37 19	137	47 17	223	48 45	283	24 55	328
294	34 03	004	26 25	036	47 44	084	37 47	137	46 49	223	48 05	283	24 33	327
295	34 06	003	26 48	035	48 25	083	38 15	136	46 21	223	47 25	282	24 11	326
296	34 08	002	27 11	034	49 05	083	38 44	136	45 53	223	46 45	281	23 48	325
297	34 09	001	27 33	033	49 46	082	39 12	136	45 25	223	46 05	280	23 24	324
298	34 09	000	27 55	032	50 26	081	39 41	135	44 58	222	45 24	279	23 00	323
299	34 08	358	28 16	031	51 07	080	40 09	135	44 30	222	44 44	278	22 35	322
	◆Enif		FOMALHAUT		◆ACHERNAR		RIGIL KENT.		◆ANTARES		Rasalhague		ALTAIR	
300	28 37	029	51 47	079	40 38	135	44 03	222	44 03	277	22 10	321	34 06	357
301	28 57	028	52 27	078	41 07	135	43 35	222	43 23	277	21 44	320	34 04	356
302	29 16	027	53 07	077	41 37	134	43 08	221	42 42	276	21 18	319	34 01	355
303	29 34	026	53 47	076	42 06	134	42 41	221	42 01	275	20 51	319	33 56	354
304	29 52	025	54 27	075	42 36	134	42 14	221	41 20	274	20 23	318	33 51	352
305	30 09	024	55 06	074	43 05	133	41 47	221	40 40	273	19 56	317	33 46	351
306	30 25	023	55 45	073	43 35	133	41 21	220	39 59	273	19 27	316	33 39	350
307	30 41	022	56 24	072	44 05	133	40 54	220	39 18	272	18 59	315	33 32	349
308	30 56	021	57 03	071	44 35	132	40 28	220	38 37	271	18 29	314	33 23	348
309	31 10	020	57 42	070	45 06	132	40 02	220	37 56	270	18 00	313	33 14	347
310	31 23	019	58 20	069	45 36	132	39 36	219	37 15	270	17 30	312	33 04	345
311	31 36	017	58 58	068	46 07	132	39 10	219	36 34	269	16 59	311	32 53	344
312	31 48	016	59 36	067	46 37	131	38 45	219	35 53	268	16 28	311	32 42	343
313	31 59	015	60 14	066	47 08	131	38 19	218	35 12	268	15 57	310	32 30	342
314	32 09	014	60 51	064	47 39	131	37 54	218	34 32	267	15 25	309	32 17	341
	◆Enif		Diphda		◆ACHERNAR		Miaplacidus		RIGIL KENT.		◆ANTARES		ALTAIR	
315	32 19	013	36 11	077	48 10	131	26 43	181	37 29	218	33 51	266	32 03	340
316	32 27	012	36 51	076	48 41	130	26 43	181	37 04	217	33 10	265	31 48	339
317	32 35	011	37 31	075	49 12	130	26 42	180	36 39	217	32 29	265	31 33	337
318	32 42	009	38 10	075	49 43	130	26 42	180	36 14	217	31 48	264	31 17	336
319	32 48	008	38 49	074	50 15	130	26 42	180	35 50	216	31 08	263	31 00	335
320	32 54	007	39 29	073	50 46	130	26 42	179	35 26	216	30 27	263	30 42	334
321	32 58	006	40 08	072	51 18	129	26 43	179	35 02	216	29 46	262	30 24	333
322	33 02	005	40 46	071	51 50	129	26 44	179	34 38	215	29 06	261	30 05	332
323	33 05	004	41 25	070	52 21	129	26 45	178	34 14	215	28 26	261	29 45	331
324	33 07	002	42 03	069	52 53	129	26 46	178	33 51	215	27 45	260	29 25	330
325	33 08	001	42 41	068	53 25	129	26 48	177	33 28	214	27 05	259	29 04	329
326	33 09	000	43 19	067	53 57	128	26 50	177	33 05	214	26 25	259	28 43	328
327	33 08	359	43 57	066	54 29	128	26 52	177	32 42	214	25 45	258	28 20	327
328	33 07	358	44 34	065	55 01	128	26 55	176	32 19	213	25 05	257	27 58	326
329	33 05	356	45 11	064	55 34	128	26 58	176	31 57	213	24 25	257	27 34	325
	◆Diphda		ACHERNAR		◆Miaplacidus		RIGIL KENT.		ANTARES		◆ALTAIR		Enif	
330	45 48	063	56 06	128	27 01	175	31 35	212	23 45	256	27 10	324	33 02	355
331	46 24	062	56 38	128	27 04	175	31 13	212	23 05	255	26 45	323	32 58	354
332	47 00	061	57 10	128	27 08	175	30 52	212	22 26	255	26 20	322	32 54	353
333	47 36	060	57 43	128	27 12	174	30 30	211	21 46	254	25 55	321	32 48	352
334	48 11	059	58 15	128	27 16	174	30 09	211	21 07	253	25 28	320	32 42	351
335	48 46	058	58 47	128	27 20	174	29 48	211	20 28	253	25 01	319	32 35	349
336	49 20	056	59 20	128	27 25	173	29 27	210	19 49	252	24 34	318	32 27	348
337	49 54	055	59 52	128	27 30	173	29 07	210	19 10	251	24 06	317	32 18	347
338	50 27	054	60 25	128	27 35	172	28 47	209	18 31	251	23 38	316	32 09	346
339	51 00	053	60 57	128	27 41	172	28 27	209	17 53	250	23 09	315	31 58	345
340	51 33	052	61 30	128	27 47	172	28 07	209	17 14	249	22 40	314	31 47	344
341	52 04	050	62 02	128	27 53	171	27 48	208	16 36	249	22 10	313	31 36	343
342	52 35	049	62 35	128	27 59	171	27 28	208	15 58	248	21 40	312	31 23	341
343	53 06	048	63 07	128	28 06	171	27 10	207	15 20	248	21 09	311	31 09	340
344	53 36	046	63 39	128	28 13	170	26 51	207	14 43	247	20 38	310	30 55	339
	◆Diphda		Acamar		◆CANOPUS		Miaplacidus		RIGIL KENT.		◆Nunki		Enif	
345	54 05	045	47 29	104	25 43	141	28 20	170	26 33	206	38 10	271	30 40	338
346	54 34	044	48 09	103	26 09	141	28 27	169	26 15	206	37 29	270	30 25	337
347	55 02	042	48 48	102	26 35	140	28 35	169	25 57	206	36 49	269	30 08	336
348	55 29	041	49 28	102	27 01	140	28 43	169	25 39	205	36 08	269	29 51	335
349	55 55	039	50 09	101	27 28	139	28 51	168	25 22	205	35 27	268	29 34	334
350	56 20	038	50 49	101	27 54	139	28 59	168	25 05	204	34 46	267	29 15	333
351	56 45	036	51 29	100	28 22	138	29 08	168	24 48	204	34 05	267	28 56	332
352	57 08	034	52 09	100	28 49	138	29 17	167	24 32	203	33 24	266	28 36	331
353	57 31	033	52 50	099	29 17	137	29 26	167	24 16	203	32 43	265	28 16	329
354	57 53	031	53 30	098	29 44	137	29 35	167	24 00	203	32 03	265	27 55	328
355	58 14	030	54 11	098	30 12	136	29 45	166	23 44	202	31 22	264	27 33	327
356	58 33	028	54 51	097	30 41	136	29 55	166	23 29	202	30 41	263	27 10	326
357	58 52	026	55 32	097	31 09	136	30 05	165	23 14	201	30 01	262	26 47	325
358	59 09	024	56 12	096	31 38	135	30 15	165	23 00	201	29 20	262	26 24	324
359	59 25	023	56 53	095	32 07	135	30 26	165	22 45	200	28 40	261	26 00	323

LAT 48°S

LHA ♈	Hc	Zn	Hc	Zn	Hc	Zn	Hc	Zn	Hc	Zn	Hc	Zn	Hc	Zn
	Alpheratz		Hamal		◆RIGEL		CANOPUS		◆RIGIL KENT.		Peacock		◆FOMALHAUT	
0	12 55	002	13 06	030	13 44	087	33 18	134	23 28	200	57 01	234	68 04	321
1	12 56	001	13 25	029	14 24	086	33 47	133	23 14	199	56 28	234	67 38	319
2	12 56	000	13 45	028	15 04	086	34 16	133	23 01	199	55 55	234	67 11	317
3	12 56	359	14 03	027	15 44	085	34 46	133	22 48	198	55 23	234	66 43	315
4	12 55	358	14 21	026	16 24	084	35 15	132	22 36	198	54 51	234	66 15	313
5	12 53	357	14 38	025	17 04	083	35 45	132	22 23	198	54 18	233	65 45	312
6	12 51	356	14 55	024	17 43	082	36 15	131	22 11	197	53 46	233	65 15	310
7	12 48	356	15 11	023	18 23	082	36 46	131	22 00	197	53 14	233	64 44	308
8	12 45	355	15 27	023	19 03	081	37 16	130	21 49	196	52 42	233	64 12	307
9	12 41	354	15 42	022	19 42	080	37 47	130	21 38	196	52 10	233	63 39	305
10	12 36	353	15 57	021	20 22	079	38 18	130	21 27	195	51 38	232	63 06	304
11	12 31	352	16 11	020	21 01	079	38 49	129	21 17	195	51 07	232	62 32	302
12	12 25	351	16 24	019	21 41	078	39 20	129	21 07	194	50 35	232	61 58	301
13	12 19	350	16 37	018	22 20	077	39 51	128	20 57	194	50 03	232	61 24	299
14	12 11	349	16 49	017	22 59	076	40 23	128	20 48	193	49 32	231	60 48	298
	Hamal		◆RIGEL		SIRIUS		CANOPUS		◆RIGIL KENT.		Peacock		◆FOMALHAUT	
15	17 00	016	23 38	075	14 49	099	40 54	128	20 39	193	49 01	231	60 13	297
16	17 11	015	24 17	075	15 29	098	41 26	127	20 30	192	48 29	231	59 37	296
17	17 21	014	24 55	074	16 09	097	41 58	127	20 22	192	47 58	231	59 00	294
18	17 30	013	25 34	073	16 49	097	42 30	126	20 14	191	47 27	230	58 24	293
19	17 39	012	26 12	072	17 28	096	43 03	126	20 07	191	46 57	230	57 47	292
20	17 47	011	26 50	071	18 08	095	43 35	126	19 59	190	46 26	230	57 09	291
21	17 55	010	27 28	070	18 48	094	44 08	125	19 52	190	45 55	229	56 32	290
22	18 02	009	28 06	069	19 29	094	44 41	125	19 46	189	45 25	229	55 54	289
23	18 08	008	28 43	069	20 09	093	45 14	125	19 40	189	44 54	229	55 16	288
24	18 13	007	29 20	068	20 49	092	45 47	124	19 34	188	44 24	229	54 37	287
25	18 18	006	29 57	067	21 29	091	46 20	124	19 28	188	43 54	228	53 59	286
26	18 23	006	30 34	066	22 09	091	46 54	124	19 23	187	43 24	228	53 20	285
27	18 26	005	31 11	065	22 49	090	47 27	123	19 18	187	42 55	228	52 41	284
28	18 29	004	31 47	064	23 29	089	48 01	123	19 14	186	42 25	227	52 02	283
29	18 31	003	32 23	063	24 09	088	48 34	123	19 10	186	41 56	227	51 23	282
	Hamal		◆RIGEL		SIRIUS		CANOPUS		◆RIGIL KENT.		Peacock		◆FOMALHAUT	
30	18 33	002	32 59	062	24 50	088	49 08	122	19 06	185	41 26	227	50 43	281
31	18 33	001	33 34	061	25 30	087	49 42	122	19 03	185	40 57	226	50 04	280
32	18 34	000	34 09	060	26 10	086	50 17	122	19 00	184	40 28	226	49 24	279
33	18 33	359	34 44	059	26 50	085	50 51	121	18 57	184	40 00	226	48 45	278
34	18 32	358	35 18	058	27 30	085	51 25	121	18 55	183	39 31	225	48 05	277
35	18 30	357	35 52	057	28 10	084	52 00	121	18 53	182	39 03	225	47 25	277
36	18 27	356	36 26	056	28 50	083	52 34	120	18 52	182	38 34	225	46 45	276
37	18 24	355	36 59	055	29 29	082	53 09	120	18 50	181	38 06	224	46 05	275
38	18 20	354	37 32	054	30 09	081	53 44	120	18 49	181	37 38	224	45 25	274
39	18 16	353	38 04	053	30 49	081	54 19	119	18 49	180	37 11	223	44 45	273
40	18 10	352	38 36	052	31 28	080	54 54	119	18 49	180	36 43	223	44 05	273
41	18 05	351	39 07	051	32 08	079	55 29	119	18 49	179	36 16	223	43 25	272
42	17 58	350	39 38	050	32 47	078	56 04	119	18 50	179	35 49	222	42 45	271
43	17 51	349	40 09	049	33 26	077	56 39	118	18 51	178	35 22	222	42 05	270
44	17 43	348	40 39	048	34 06	077	57 15	118	18 52	178	34 55	222	41 24	270
	RIGEL		SIRIUS		◆CANOPUS		RIGIL KENT.		◆Peacock		FOMALHAUT		◆Hamal	
45	41 08	047	34 45	076	57 50	118	18 54	177	34 29	221	40 44	269	17 34	347
46	41 37	045	35 23	075	58 26	118	18 56	177	34 02	221	40 04	268	17 25	346
47	42 06	044	36 02	074	59 01	117	18 58	176	33 36	220	39 24	267	17 15	345
48	42 33	043	36 41	073	59 37	117	19 01	176	33 10	220	38 44	267	17 05	344
49	43 00	042	37 19	072	60 13	117	19 04	175	32 45	219	38 04	266	16 54	343
50	43 27	041	37 57	071	60 49	117	19 08	175	32 19	219	37 24	265	16 42	343
51	43 53	039	38 35	070	61 25	116	19 11	174	31 54	219	36 44	265	16 30	342
52	44 18	038	39 13	069	62 01	116	19 16	174	31 29	218	36 04	264	16 17	341
53	44 42	037	39 50	068	62 37	116	19 20	173	31 04	218	35 24	263	16 03	340
54	45 06	036	40 27	068	63 13	116	19 25	173	30 40	217	34 44	263	15 49	339
55	45 29	034	41 04	067	63 49	116	19 30	172	30 16	217	34 04	262	15 34	338
56	45 52	033	41 41	066	64 25	116	19 36	172	29 52	217	33 25	261	15 19	337
57	46 13	032	42 17	065	65 01	116	19 42	171	29 28	216	32 45	261	15 03	336
58	46 34	030	42 53	064	65 37	115	19 48	171	29 04	216	32 05	260	14 46	335
59	46 54	029	43 29	063	66 14	115	19 55	170	28 41	215	31 26	259	14 29	334
	◆ALDEBARAN		BETELGEUSE		SIRIUS		◆Suhail		RIGIL KENT.		◆ACHERNAR		Diphda	
60	25 01	009	29 05	033	44 05	061	38 21	116	20 02	170	66 49	233	40 14	290
61	25 07	008	29 27	032	44 40	060	38 57	115	20 09	169	66 17	233	39 36	289
62	25 12	007	29 48	031	45 14	059	39 34	115	20 17	169	65 45	233	38 58	288
63	25 17	006	30 08	030	45 49	058	40 10	114	20 25	168	65 12	234	38 20	287
64	25 21	005	30 28	029	46 23	057	40 47	113	20 33	168	64 40	234	37 41	286
65	25 24	004	30 47	028	46 56	056	41 24	113	20 42	167	64 08	234	37 02	285
66	25 27	003	31 05	027	47 29	055	42 01	112	20 51	167	63 35	234	36 23	284
67	25 29	002	31 22	025	48 02	054	42 38	112	21 01	166	63 03	234	35 44	283
68	25 30	001	31 39	024	48 34	052	43 15	111	21 10	166	62 30	234	35 05	282
69	25 30	000	31 56	023	49 06	051	43 53	111	21 21	165	61 58	234	34 26	282
70	25 30	359	32 11	022	49 37	050	44 30	110	21 31	165	61 26	234	33 47	281
71	25 28	358	32 26	021	50 07	049	45 08	110	21 42	164	60 53	234	33 07	280
72	25 26	357	32 40	020	50 37	047	45 46	109	21 53	164	60 21	234	32 28	279
73	25 24	356	32 53	019	51 06	046	46 24	109	22 04	163	59 48	234	31 48	278
74	25 20	355	33 05	018	51 35	045	47 02	108	22 16	163	59 16	234	31 08	278
	BETELGEUSE		SIRIUS		◆Suhail		RIGIL KENT.		◆ACHERNAR		Diphda		◆ALDEBARAN	
75	33 17	016	52 03	044	47 40	108	22 28	162	58 44	234	30 28	277	25 16	354
76	33 28	015	52 30	042	48 18	107	22 40	162	58 11	234	29 48	276	25 11	352
77	33 38	014	52 57	041	48 57	107	22 53	161	57 39	233	29 08	275	25 06	351
78	33 47	013	53 23	039	49 35	106	23 06	161	57 07	233	28 28	274	24 59	350
79	33 56	012	53 48	038	50 14	106	23 19	160	56 35	233	27 48	274	24 52	349
80	34 04	010	54 12	036	50 53	105	23 33	160	56 03	233	27 08	273	24 45	348
81	34 11	009	54 35	035	51 31	105	23 47	159	55 30	233	26 28	272	24 36	347
82	34 17	008	54 58	033	52 10	104	24 01	159	54 58	233	25 48	271	24 27	346
83	34 22	007	55 19	032	52 49	104	24 16	159	54 27	233	25 08	271	24 17	345
84	34 26	006	55 40	030	53 28	103	24 31	158	53 55	232	24 28	270	24 06	344
85	34 30	004	56 00	029	54 07	103	24 46	158	53 23	232	23 48	269	23 55	343
86	34 32	003	56 18	027	54 47	102	25 01	157	52 51	232	23 07	268	23 43	342
87	34 34	002	56 36	025	55 26	101	25 17	157	52 20	232	22 27	268	23 30	341
88	34 35	001	56 53	024	56 05	101	25 33	156	51 48	232	21 47	267	23 17	340
89	34 36	000	57 08	022	56 45	100	25 49	156	51 17	231	21 07	266	23 03	339

LHA ♈	Hc	Zn	Hc	Zn	Hc	Zn	Hc	Zn	Hc	Zn	Hc	Zn	Hc	Zn
	PROCYON		Alphard		◆Suhail		RIGIL KENT.		◆ACHERNAR		◆RIGEL		BETELGEUSE	
90	32 30	030	31 21	066	57 24	100	26 06	155	50 45	231	49 03	343	34 35	358
91	32 49	029	31 58	065	58 04	099	26 23	155	50 14	231	48 50	341	34 33	357
92	33 08	027	32 34	064	58 44	099	26 40	154	49 43	231	48 37	340	34 31	356
93	33 26	026	33 10	063	59 23	098	26 57	154	49 12	230	48 22	338	34 28	355
94	33 44	025	33 45	062	60 03	097	27 15	154	48 41	230	48 07	337	34 24	354
95	34 00	024	34 21	061	60 43	097	27 33	153	48 11	230	47 51	335	34 19	352
96	34 16	023	34 56	060	61 23	096	27 51	153	47 40	230	47 34	334	34 13	351
97	34 31	022	35 30	059	62 03	096	28 10	152	47 09	229	47 15	333	34 07	350
98	34 46	020	36 04	058	62 43	095	28 29	152	46 39	229	46 57	331	34 00	349
99	34 59	019	36 38	057	63 23	094	28 48	151	46 09	229	46 37	330	33 51	348
100	35 12	018	37 12	056	64 03	094	29 07	151	45 38	229	46 16	328	33 42	347
101	35 24	017	37 45	055	64 43	093	29 27	151	45 08	228	45 55	327	33 33	345
102	35 36	016	38 17	054	65 23	092	29 46	150	44 39	228	45 33	326	33 22	344
103	35 46	014	38 49	053	66 03	092	30 06	150	44 09	228	45 10	325	33 11	343
104	35 56	013	39 21	052	66 43	091	30 27	149	43 39	227	44 46	323	32 59	342
	PROCYON		REGULUS		◆Gienah		RIGIL KENT.		◆ACHERNAR		RIGEL		◆BETELGEUSE	
105	36 05	012	16 58	048	20 17	094	30 47	149	43 10	227	44 22	322	32 46	341
106	36 12	011	17 28	048	20 57	093	31 08	149	42 41	227	43 57	321	32 32	340
107	36 20	010	17 57	047	21 37	093	31 29	148	42 11	226	43 31	319	32 18	338
108	36 26	008	18 26	046	22 17	092	31 50	148	41 42	226	43 04	318	32 03	337
109	36 31	007	18 55	045	22 57	091	32 12	147	41 14	226	42 37	317	31 47	336
110	36 36	006	19 23	044	23 37	090	32 34	147	40 45	225	42 10	316	31 30	335
111	36 40	005	19 50	043	24 18	090	32 56	147	40 16	225	41 41	315	31 13	334
112	36 42	003	20 17	042	24 58	089	33 18	146	39 48	225	41 13	314	30 55	333
113	36 44	002	20 44	041	25 38	088	33 40	146	39 20	224	40 43	312	30 36	332
114	36 45	001	21 10	040	26 18	087	34 03	146	38 52	224	40 13	311	30 17	331
115	36 46	000	21 36	039	26 58	087	34 26	145	38 24	224	39 43	310	29 57	330
116	36 45	358	22 01	038	27 38	086	34 49	145	37 57	223	39 12	309	29 36	328
117	36 44	357	22 26	037	28 18	085	35 12	144	37 29	223	38 41	308	29 15	327
118	36 41	356	22 50	036	28 58	084	35 35	144	37 02	223	38 09	307	28 53	326
119	36 38	355	23 14	035	29 38	083	35 59	144	36 35	222	37 36	306	28 30	325
	REGULUS		◆SPICA		RIGIL KENT.		◆ACHERNAR		RIGEL		BETELGEUSE		◆PROCYON	
120	23 37	034	14 06	091	36 23	143	36 08	222	37 04	305	28 07	324	36 34	353
121	23 59	033	14 46	090	36 47	143	35 41	221	36 31	304	27 44	323	36 29	352
122	24 21	032	15 26	090	37 11	143	35 15	221	35 57	303	27 19	322	36 23	351
123	24 42	031	16 06	089	37 36	142	34 49	221	35 23	302	26 54	321	36 16	350
124	25 03	030	16 47	088	38 00	142	34 23	220	34 49	301	26 29	320	36 09	349
125	25 23	029	17 27	087	38 25	142	33 57	220	34 14	300	26 03	319	36 00	347
126	25 42	028	18 07	087	38 50	141	33 31	219	33 39	299	25 36	318	35 51	346
127	26 01	027	18 47	086	39 15	141	33 06	219	33 04	298	25 10	317	35 41	345
128	26 19	026	19 27	085	39 41	141	32 41	219	32 28	297	24 42	316	35 30	344
129	26 37	025	20 07	084	40 06	140	32 16	218	31 52	296	24 14	315	35 19	343
130	26 54	024	20 47	084	40 32	140	31 51	218	31 16	295	23 46	314	35 06	341
131	27 10	023	21 27	083	40 58	140	31 26	217	30 40	294	23 17	313	34 53	340
132	27 25	022	22 06	082	41 24	139	31 02	217	30 03	293	22 47	312	34 39	339
133	27 40	021	22 46	081	41 50	139	30 38	217	29 26	292	22 17	312	34 24	338
134	27 54	020	23 26	080	42 16	139	30 14	216	28 49	292	21 47	311	34 09	337
	◆REGULUS		SPICA		◆RIGIL KENT.		ACHERNAR		◆CANOPUS		SIRIUS		PROCYON	
135	28 08	019	24 05	080	42 43	139	29 51	216	64 58	244	48 17	307	33 52	335
136	28 20	018	24 45	079	43 09	138	29 27	215	64 22	244	47 45	306	33 35	334
137	28 32	017	25 24	078	43 36	138	29 04	215	63 46	244	47 12	305	33 18	333
138	28 44	016	26 03	077	44 03	138	28 42	214	63 10	244	46 38	303	32 59	332
139	28 54	015	26 42	076	44 30	138	28 19	214	62 34	244	46 05	302	32 40	331
140	29 04	013	27 21	076	44 57	137	27 57	214	61 58	244	45 30	301	32 20	330
141	29 13	012	28 00	075	45 24	137	27 35	213	61 22	243	44 56	300	32 00	329
142	29 21	011	28 39	074	45 52	137	27 13	213	60 46	243	44 21	299	31 38	328
143	29 28	010	29 17	073	46 19	137	26 51	212	60 10	243	43 46	298	31 17	327
144	29 35	009	29 55	072	46 47	136	26 30	212	59 34	243	43 10	297	30 54	325
145	29 41	008	30 34	071	47 15	136	26 09	211	58 59	243	42 34	296	30 31	324
146	29 46	007	31 12	070	47 43	136	25 49	211	58 23	242	41 58	295	30 07	323
147	29 51	006	31 49	070	48 11	136	25 28	210	57 47	242	41 21	294	29 43	322
148	29 54	005	32 27	069	48 39	135	25 08	210	57 12	242	40 44	293	29 18	321
149	29 57	003	33 04	068	49 07	135	24 48	209	56 37	242	40 07	292	28 53	320
	◆REGULUS		SPICA		◆RIGIL KENT.		ACHERNAR		◆CANOPUS		SIRIUS		PROCYON	
150	29 59	002	33 41	067	49 35	135	24 29	209	56 01	241	39 30	291	28 27	319
151	30 00	001	34 18	066	50 04	135	24 09	208	55 26	241	38 52	290	28 00	318
152	30 00	000	34 54	065	50 32	135	23 51	208	54 51	241	38 15	289	27 33	317
153	30 00	359	35 31	064	51 01	135	23 32	207	54 16	241	37 37	288	27 06	316
154	29 59	358	36 07	063	51 29	134	23 14	207	53 41	240	36 58	287	26 38	315
155	29 57	357	36 42	062	51 58	134	22 55	206	53 06	240	36 20	286	26 09	314
156	29 54	356	37 18	061	52 27	134	22 38	206	52 32	240	35 41	286	25 40	313
157	29 51	354	37 52	060	52 56	134	22 20	205	51 57	239	35 03	285	25 11	312
158	29 46	353	38 27	059	53 25	134	22 03	205	51 22	239	34 24	284	24 41	311
159	29 41	352	39 01	058	53 54	134	21 46	204	50 48	239	33 45	283	24 11	310
160	29 35	351	39 35	057	54 23	134	21 30	204	50 14	238	33 05	282	23 40	309
161	29 29	350	40 09	056	54 52	134	21 14	203	49 40	238	32 26	281	23 09	309
162	29 21	349	40 42	055	55 21	133	20 58	203	49 06	238	31 47	281	22 37	308
163	29 13	348	41 14	054	55 50	133	20 42	202	48 32	237	31 07	280	22 05	307
164	29 04	347	41 47	053	56 19	133	20 27	202	47 58	237	30 28	279	21 33	306
	SPICA		◆ANTARES		RIGIL KENT.		ACHERNAR		◆CANOPUS		SIRIUS		◆REGULUS	
165	42 18	052	24 17	103	56 48	133	20 12	201	47 24	237	29 48	278	28 55	345
166	42 50	051	24 56	103	57 18	133	19 58	201	46 51	236	29 08	277	28 44	344
167	43 20	049	25 35	102	57 47	133	19 44	200	46 17	236	28 28	277	28 33	343
168	43 51	048	26 15	101	58 16	133	19 30	200	45 44	236	27 48	276	28 21	342
169	44 20	047	26 54	100	58 45	133	19 16	199	45 11	235	27 08	275	28 08	341
170	44 49	046	27 34	100	59 15	133	19 03	199	44 38	235	26 28	274	27 55	340
171	45 18	045	28 13	099	59 44	133	18 50	198	44 05	235	25 48	273	27 41	339
172	45 46	043	28 53	098	60 13	133	18 38	198	43 33	234	25 08	273	27 26	338
173	46 13	042	29 33	098	60 42	133	18 26	197	43 00	234	24 28	272	27 11	337
174	46 40	041	30 12	097	61 11	133	18 14	197	42 28	233	23 48	271	26 55	336
175	47 05	040	30 52	096	61 41	134	18 03	196	41 56	233	23 08	270	26 38	335
176	47 31	038	31 32	096	62 10	134	17 52	196	41 24	233	22 28	270	26 20	334
177	47 55	037	32 12	095	62 39	134	17 41	195	40 52	232	21 47	269	26 02	333
178	48 19	036	32 52	094	63 07	134	17 31	195	40 20	232	21 07	268	25 43	332
179	48 42	034	33 32	094	63 36	134	17 21	194	39 49	232	20 27	267	25 24	331

 LAT 48°S

LHA	Hc	Zn	Hc	Zn	Hc	Zn	Hc	Zn	Hc	Zn	Hc	Zn	Hc	Zn
♈	◆ARCTURUS		ANTARES		◆Peacock		ACHERNAR		CANOPUS		◆Suhail		REGULUS	
180	16 16	033	34 12	093	23 51	151	17 11	194	39 17	231	59 59	262	25 04	330
181	16 38	032	34 52	092	24 10	151	17 02	193	38 46	231	59 20	262	24 43	329
182	16 59	031	35 33	091	24 30	150	16 53	192	38 15	230	58 40	261	24 22	328
183	17 20	030	36 13	091	24 50	150	16 45	192	37 44	230	58 00	261	24 00	327
184	17 40	030	36 53	090	25 10	149	16 37	191	37 14	230	57 21	260	23 38	326
185	18 00	029	37 33	089	25 31	149	16 29	191	36 43	229	56 41	260	23 15	325
186	18 19	028	38 13	088	25 52	148	16 22	190	36 13	229	56 02	259	22 51	324
187	18 37	027	38 53	088	26 13	148	16 15	190	35 43	228	55 22	259	22 27	323
188	18 55	026	39 33	087	26 35	147	16 08	189	35 13	228	54 43	258	22 03	322
189	19 12	025	40 13	086	26 57	147	16 02	189	34 43	227	54 04	257	21 37	321
190	19 28	024	40 53	085	27 19	146	15 56	188	34 14	227	53 25	257	21 12	320
191	19 44	023	41 33	084	27 41	146	15 51	187	33 45	227	52 46	256	20 46	319
192	20 00	022	42 13	084	28 04	146	15 46	187	33 16	226	52 07	256	20 19	318
193	20 14	021	42 53	083	28 27	145	15 41	186	32 47	226	51 28	255	19 52	317
194	20 29	020	43 33	082	28 50	145	15 37	186	32 18	225	50 49	255	19 24	316
	ARCTURUS		◆ANTARES		Peacock		◆ACHERNAR		CANOPUS		Suhail		◆REGULUS	
195	20 42	019	44 13	081	29 13	144	15 33	185	31 50	225	50 10	254	18 56	315
196	20 55	018	44 52	080	29 37	144	15 30	185	31 22	224	49 32	254	18 28	314
197	21 07	017	45 32	080	30 01	143	15 26	184	30 54	224	48 53	253	17 59	313
198	21 18	016	46 11	079	30 25	143	15 24	184	30 26	223	48 15	253	17 29	312
199	21 29	015	46 51	078	30 49	142	15 21	183	29 58	223	47 37	252	16 59	312
200	21 39	014	47 30	077	31 14	142	15 19	182	29 31	223	46 58	252	16 29	311
201	21 49	013	48 09	076	31 38	142	15 18	182	29 04	222	46 20	251	15 59	310
202	21 57	012	48 48	075	32 04	141	15 17	181	28 37	222	45 42	251	15 28	309
203	22 05	011	49 26	074	32 29	141	15 16	181	28 11	221	45 05	250	14 56	308
204	22 13	010	50 05	073	32 54	140	15 16	180	27 45	221	44 27	250	14 24	307
205	22 19	009	50 43	072	33 20	140	15 16	180	27 19	220	43 49	249	13 52	306
206	22 25	008	51 21	071	33 46	140	15 16	179	26 53	220	43 12	249	13 20	306
207	22 31	007	51 59	070	34 12	139	15 17	179	26 27	219	42 35	248	12 47	305
208	22 35	006	52 37	069	34 39	139	15 18	178	26 02	219	41 58	248	12 14	304
209	22 39	005	53 14	068	35 05	138	15 20	177	25 37	218	41 20	247	11 40	303
	◆ARCTURUS		ANTARES		◆Peacock		ACHERNAR		CANOPUS		◆Suhail		SPICA	
210	22 42	004	53 51	067	35 32	138	15 22	177	25 12	218	40 44	246	52 25	346
211	22 45	003	54 28	066	35 59	138	15 24	176	24 48	217	40 07	246	52 14	344
212	22 46	002	55 04	065	36 26	137	15 27	176	24 24	217	39 30	245	52 03	343
213	22 47	001	55 40	063	36 53	137	15 30	175	24 00	216	38 54	245	51 50	341
214	22 47	000	56 16	062	37 21	136	15 34	175	23 36	216	38 18	244	51 37	340
215	22 47	359	56 52	061	37 49	136	15 38	174	23 13	215	37 41	244	51 22	338
216	22 46	358	57 26	060	38 17	136	15 42	174	22 50	215	37 05	243	51 07	337
217	22 44	357	58 01	058	38 45	135	15 47	173	22 27	214	36 30	243	50 50	335
218	22 41	356	58 35	057	39 13	135	15 52	172	22 05	214	35 54	242	50 33	334
219	22 38	355	59 08	056	39 42	135	15 58	172	21 43	213	35 18	242	50 14	332
220	22 34	354	59 41	054	40 10	134	16 03	171	21 21	213	34 43	241	49 55	331
221	22 29	353	60 14	053	40 39	134	16 10	171	21 00	212	34 08	241	49 35	329
222	22 24	352	60 45	052	41 08	134	16 16	170	20 38	212	33 33	240	49 14	328
223	22 18	351	61 16	050	41 37	133	16 23	170	20 18	211	32 58	240	48 52	326
224	22 11	350	61 47	048	42 07	133	16 31	169	19 57	211	32 24	239	48 30	325
	ANTARES		◆Nunki		Peacock		ACHERNAR		◆CANOPUS		◆SPICA		ARCTURUS	
225	62 17	047	39 50	086	42 36	133	16 39	169	19 37	210	48 06	324	22 03	349
226	62 45	045	40 30	086	43 06	132	16 47	168	19 17	209	47 42	322	21 55	348
227	63 14	044	41 10	085	43 36	132	16 55	167	18 57	209	47 17	321	21 46	347
228	63 41	042	41 50	084	44 05	132	17 04	167	18 38	208	46 52	320	21 36	346
229	64 07	040	42 30	083	44 36	131	17 14	166	18 19	208	46 25	318	21 26	345
230	64 32	038	43 10	082	45 06	131	17 23	166	18 01	207	45 58	317	21 15	344
231	64 57	036	43 50	081	45 36	131	17 33	165	17 42	207	45 31	316	21 03	343
232	65 20	034	44 29	081	46 07	130	17 44	165	17 24	206	45 03	315	20 51	342
233	65 42	032	45 09	080	46 37	130	17 54	164	17 07	206	44 34	314	20 38	341
234	66 03	030	45 48	079	47 08	130	18 05	164	16 50	205	44 04	312	20 25	340
235	66 23	028	46 28	078	47 39	130	18 17	163	16 33	204	43 34	311	20 10	339
236	66 41	026	47 07	077	48 10	129	18 29	163	16 16	204	43 04	310	19 55	338
237	66 58	024	47 46	076	48 41	129	18 41	162	16 00	203	42 33	309	19 40	337
238	67 14	022	48 25	075	49 12	129	18 53	162	15 45	203	42 01	308	19 24	336
239	67 28	020	49 04	074	49 44	129	19 06	161	15 29	202	41 29	307	19 07	335
	Rasalhague		◆ALTAIR		FOMALHAUT		◆ACHERNAR		ACRUX		◆SPICA		ARCTURUS	
240	25 53	026	13 52	059	12 55	120	19 20	161	57 28	223	40 57	306	18 50	334
241	26 10	025	14 26	058	13 30	120	19 33	160	57 01	223	40 24	304	18 32	333
242	26 26	024	15 00	058	14 05	119	19 47	159	56 34	223	39 51	303	18 13	332
243	26 42	023	15 34	057	14 40	118	20 01	159	56 07	223	39 17	302	17 54	331
244	26 57	022	16 07	056	15 16	118	20 16	158	55 40	223	38 43	301	17 34	330
245	27 12	021	16 40	055	15 51	117	20 31	158	55 13	223	38 09	300	17 14	329
246	27 26	020	17 13	054	16 27	116	20 46	157	54 46	223	37 34	299	16 53	328
247	27 39	018	17 45	053	17 03	116	21 02	157	54 18	222	36 59	298	16 32	327
248	27 51	017	18 17	052	17 40	115	21 18	156	53 51	222	36 23	297	16 10	327
249	28 03	016	18 49	052	18 16	114	21 34	156	53 24	222	35 47	296	15 48	326
250	28 14	015	19 20	051	18 53	114	21 50	155	52 57	222	35 11	295	15 25	325
251	28 24	014	19 51	050	19 30	113	22 07	155	52 30	222	34 35	295	15 01	324
252	28 33	013	20 22	049	20 07	112	22 24	154	52 03	222	33 58	294	14 37	323
253	28 42	012	20 52	048	20 44	112	22 42	154	51 37	222	33 21	293	14 13	322
254	28 50	011	21 21	047	21 21	111	23 00	153	51 10	222	32 44	292	13 48	321
	Rasalhague		◆ALTAIR		FOMALHAUT		◆ACHERNAR		ACRUX		◆SPICA		ANTARES	
255	28 57	010	21 50	046	21 59	110	23 18	153	50 43	222	32 07	291	67 35	342
256	29 03	009	22 19	045	22 36	110	23 36	152	50 16	222	31 29	290	67 22	339
257	29 09	007	22 47	044	23 14	109	23 55	152	49 50	221	30 51	289	67 07	337
258	29 14	006	23 15	043	23 52	108	24 14	151	49 23	221	30 13	288	66 51	335
259	29 18	005	23 42	042	24 31	108	24 33	151	48 57	221	29 35	287	66 33	333
260	29 21	004	24 09	041	25 09	107	24 53	151	48 30	221	28 57	287	66 14	331
261	29 23	003	24 36	040	25 47	107	25 13	150	48 04	221	28 18	286	65 54	329
262	29 25	002	25 01	039	26 26	106	25 33	150	47 38	221	27 39	285	65 32	327
263	29 26	001	25 27	038	27 04	105	25 53	149	47 12	220	27 00	284	65 10	325
264	29 26	000	25 51	037	27 43	105	26 14	149	46 46	220	26 21	283	64 46	323
265	29 26	359	26 15	036	28 22	104	26 35	148	46 20	220	25 42	282	64 21	321
266	29 24	357	26 39	035	29 01	103	26 57	148	45 54	220	25 03	282	63 55	319
267	29 22	356	27 02	034	29 40	103	27 18	147	45 29	220	24 24	281	63 29	317
268	29 19	355	27 24	033	30 20	102	27 40	147	45 04	219	23 44	280	63 01	316
269	29 15	354	27 46	032	30 59	101	28 02	146	44 38	219	23 04	279	62 33	314

LHA	Hc	Zn	Hc	Zn	Hc	Zn	Hc	Zn	Hc	Zn	Hc	Zn	Hc	Zn
♈	◆ALTAIR		FOMALHAUT		◆ACHERNAR		RIGIL KENT.		◆SPICA		ANTARES		Rasalhague	
270	28 07	031	31 38	101	28 24	146	59 04	227	22 25	278	62 03	312	29 11	353
271	28 28	030	32 18	100	28 47	146	58 34	227	21 45	278	61 33	311	29 05	352
272	28 48	029	32 57	099	29 10	145	58 05	227	21 05	277	61 03	309	28 59	351
273	29 07	028	33 37	099	29 33	145	57 36	227	20 25	276	60 31	308	28 52	350
274	29 26	027	34 17	098	29 56	144	57 07	227	19 45	275	59 59	306	28 45	348
275	29 43	026	34 57	097	30 20	144	56 37	227	19 05	275	59 27	305	28 36	347
276	30 01	025	35 36	097	30 44	143	56 08	227	18 25	274	58 53	304	28 27	346
277	30 17	024	36 16	096	31 08	143	55 39	227	17 45	273	58 20	302	28 17	345
278	30 33	023	36 56	095	31 32	143	55 10	227	17 05	272	57 45	301	28 07	344
279	30 48	022	37 36	094	31 57	142	54 41	226	16 25	272	57 11	300	27 55	343
280	31 03	020	38 16	094	32 22	142	54 12	226	15 45	271	56 36	298	27 43	342
281	31 16	019	38 56	093	32 47	141	53 43	226	15 05	270	56 00	297	27 30	341
282	31 29	018	39 37	092	33 12	141	53 14	226	14 25	269	55 24	296	27 17	340
283	31 41	017	40 17	092	33 37	140	52 45	226	13 44	269	54 48	295	27 03	339
284	31 53	016	40 57	091	34 03	140	52 16	226	13 04	268	54 11	294	26 48	338
	ALTAIR		Enif		◆FOMALHAUT		ACHERNAR		◆RIGIL KENT.		ANTARES		◆Rasalhague	
285	32 03	015	21 45	044	41 37	090	34 29	140	51 47	226	53 35	293	26 32	337
286	32 13	014	22 13	043	42 17	089	34 55	139	51 19	226	52 57	292	26 16	336
287	32 22	012	22 40	042	42 57	089	35 21	139	50 50	225	52 20	291	25 59	335
288	32 31	011	23 06	041	43 37	088	35 48	139	50 21	225	51 42	289	25 41	334
289	32 38	010	23 33	040	44 17	087	36 14	138	49 53	225	51 04	288	25 23	332
290	32 45	009	23 58	039	44 58	086	36 41	138	49 25	225	50 26	287	25 04	331
291	32 51	008	24 24	038	45 38	086	37 08	137	48 56	225	49 48	286	24 45	330
292	32 56	007	24 48	037	46 18	085	37 36	137	48 28	224	49 09	286	24 25	329
293	33 00	005	25 12	036	46 58	084	38 03	137	48 00	224	48 30	285	24 04	328
294	33 03	004	25 36	035	47 37	083	38 31	136	47 32	224	47 51	284	23 43	327
295	33 06	003	25 59	034	48 17	082	38 59	136	47 04	224	47 12	283	23 21	326
296	33 08	002	26 21	033	48 57	081	39 27	136	46 37	224	46 33	282	22 58	325
297	33 09	001	26 43	032	49 37	081	39 55	135	46 09	223	45 54	281	22 35	324
298	33 09	000	27 04	031	50 16	080	40 23	135	45 42	223	45 14	280	22 12	324
299	33 08	358	27 25	030	50 56	079	40 52	135	45 14	223	44 35	279	21 47	323
	◆Enif		FOMALHAUT		◆ACHERNAR		RIGIL KENT.		◆ANTARES		Rasalhague		ALTAIR	
300	27 44	029	51 35	078	41 20	134	44 47	223	43 55	278	21 23	322	33 06	357
301	28 04	028	52 14	077	41 49	134	44 20	222	43 15	278	20 58	321	33 04	356
302	28 22	027	52 53	076	42 18	134	43 53	222	42 35	277	20 32	320	33 01	355
303	28 40	026	53 32	075	42 47	133	43 26	222	41 55	276	20 06	319	32 57	354
304	28 57	025	54 11	074	43 17	133	42 59	222	41 15	275	19 39	318	32 52	353
305	29 14	024	54 49	073	43 46	133	42 33	221	40 35	274	19 12	317	32 46	351
306	29 30	023	55 27	072	44 16	132	42 06	221	39 55	274	18 44	316	32 40	350
307	29 45	022	56 06	071	44 46	132	41 40	221	39 15	273	18 16	315	32 33	349
308	30 00	021	56 43	070	45 16	132	41 14	220	38 35	272	17 48	314	32 25	348
309	30 13	019	57 21	069	45 46	131	40 48	220	37 55	271	17 19	313	32 16	347
310	30 26	018	57 58	068	46 16	131	40 22	220	37 15	271	16 49	312	32 06	346
311	30 39	017	58 35	066	46 46	131	39 57	219	36 35	270	16 19	312	31 56	344
312	30 50	016	59 12	065	47 17	131	39 31	219	35 55	269	15 49	311	31 44	343
313	31 01	015	59 48	064	47 47	130	39 06	219	35 15	268	15 19	310	31 33	342
314	31 11	014	60 24	063	48 18	130	38 41	219	34 34	268	14 48	309	31 20	341
	◆Enif		Diphda		◆ACHERNAR		Miaplacidus		RIGIL KENT.		◆ANTARES		ALTAIR	
315	31 20	013	35 57	076	48 49	130	27 43	181	38 16	218	33 54	267	31 06	340
316	31 28	012	36 36	076	49 19	130	27 43	181	37 51	218	33 14	266	30 52	339
317	31 36	010	37 15	075	49 51	129	27 42	181	37 27	218	32 34	265	30 37	338
318	31 43	009	37 54	074	50 22	129	27 42	180	37 02	217	31 54	265	30 22	337
319	31 49	008	38 32	073	50 53	129	27 42	180	36 38	217	31 14	264	30 05	335
320	31 54	007	39 11	072	51 24	129	27 42	179	36 14	217	30 34	263	29 48	334
321	31 59	006	39 49	071	51 56	128	27 43	179	35 50	216	29 54	263	29 30	333
322	32 02	005	40 26	070	52 27	128	27 44	179	35 27	216	29 15	262	29 12	332
323	32 05	003	41 04	069	52 59	128	27 45	178	35 03	215	28 35	261	28 53	331
324	32 07	002	41 42	068	53 30	128	27 46	178	34 40	215	27 55	261	28 33	330
325	32 08	001	42 19	067	54 02	128	27 48	177	34 17	215	27 16	260	28 13	329
326	32 09	000	42 56	066	54 34	127	27 50	177	33 54	214	26 36	259	27 52	328
327	32 08	359	43 32	065	55 06	127	27 52	177	33 32	214	25 57	259	27 30	327
328	32 07	358	44 09	064	55 38	127	27 55	176	33 10	214	25 18	258	27 08	326
329	32 05	356	44 45	063	56 10	127	27 58	176	32 47	213	24 38	257	26 45	325
	◆Diphda		ACHERNAR		◆Miaplacidus		RIGIL KENT.		ANTARES		◆ALTAIR		Enif	
330	45 20	062	56 42	127	28 01	175	32 26	213	23 59	256	26 22	324	32 02	355
331	45 56	061	57 14	127	28 04	175	32 04	212	23 20	256	25 58	323	31 59	354
332	46 31	060	57 47	127	28 08	175	31 42	212	22 41	255	25 33	322	31 54	353
333	47 05	059	58 19	126	28 11	174	31 21	212	22 03	254	25 08	321	31 49	352
334	47 39	058	58 51	126	28 16	174	31 00	211	21 24	254	24 43	320	31 43	351
335	48 13	057	59 24	126	28 20	173	30 40	211	20 46	253	24 16	319	31 36	350
336	48 46	056	59 56	126	28 25	173	30 19	210	20 07	252	23 50	318	31 28	348
337	49 19	054	60 28	126	28 30	173	29 59	210	19 29	252	23 23	317	31 20	347
338	49 52	053	61 01	126	28 35	172	29 39	210	18 51	251	22 55	316	31 11	346
339	50 24	052	61 33	126	28 40	172	29 19	209	18 13	250	22 27	315	31 01	345
340	50 55	051	62 06	126	28 46	172	29 00	209	17 35	250	21 58	314	30 50	344
341	51 26	049	62 38	126	28 52	171	28 41	208	16 58	249	21 29	313	30 38	343
342	51 56	048	63 11	126	28 58	171	28 22	208	16 20	248	21 00	312	30 26	342
343	52 25	047	63 43	126	29 05	170	28 03	208	15 43	248	20 30	311	30 13	341
344	52 54	045	64 15	126	29 12	170	27 44	207	15 06	247	19 59	310	29 59	339
	◆Diphda		Acamar		◆CANOPUS		Miaplacidus		RIGIL KENT.		◆Nunki		Enif	
345	53 22	044	47 42	102	26 29	141	29 19	170	27 26	207	38 09	272	29 45	338
346	53 50	043	48 21	102	26 55	140	29 26	169	27 08	206	37 29	271	29 29	337
347	54 17	041	49 01	101	27 21	140	29 34	169	26 51	206	36 49	270	29 14	336
348	54 43	040	49 40	101	27 47	139	29 41	169	26 33	205	36 09	269	28 57	335
349	55 08	038	50 20	100	28 13	139	29 50	168	26 16	205	35 28	269	28 40	334
350	55 33	037	50 59	100	28 39	138	29 58	168	26 00	204	34 48	268	28 22	333
351	55 56	035	51 39	099	29 06	138	30 06	167	25 43	204	34 08	267	28 03	332
352	56 19	034	52 19	098	29 33	137	30 15	167	25 27	204	33 28	267	27 44	331
353	56 41	032	52 58	098	30 01	137	30 24	167	25 11	203	32 48	266	27 24	330
354	57 01	030	53 38	097	30 28	137	30 34	166	24 55	203	32 08	265	27 03	329
355	57 21	029	54 18	096	30 56	136	30 43	166	24 40	202	31 28	264	26 42	328
356	57 40	027	54 58	096	31 24	136	30 53	166	24 25	202	30 48	264	26 20	327
357	57 58	025	55 38	095	31 52	135	31 03	165	24 10	201	30 08	263	25 58	326
358	58 14	024	56 18	094	32 20	135	31 13	165	23 56	201	29 28	262	25 35	325
359	58 30	022	56 58	094	32 49	134	31 24	165	23 42	200	28 49	262	25 12	324

LAT 49°S

LHA	Hc	Zn	Hc	Zn	Hc	Zn	Hc	Zn	Hc	Zn	Hc	Zn	Hc	Zn
♈	Alpheratz		Hamal		◆RIGEL		CANOPUS		◆RIGIL KENT.		Peacock		◆FOMALHAUT	
0	11 55	002	12 14	030	13 40	087	33 59	133	24 24	200	57 35	236	67 16	323
1	11 56	001	12 33	029	14 20	086	34 28	133	24 11	200	57 03	235	66 52	321
2	11 56	000	12 51	028	14 59	085	34 57	132	23 58	199	56 30	235	66 26	319
3	11 56	359	13 10	027	15 38	084	35 26	132	23 45	199	55 58	235	66 00	317
4	11 55	358	13 27	026	16 17	084	35 55	132	23 33	198	55 26	235	65 33	315
5	11 54	357	13 44	025	16 56	083	36 25	131	23 21	198	54 54	235	65 04	313
6	11 51	356	14 01	024	17 35	082	36 55	131	23 09	197	54 22	234	64 35	312
7	11 49	356	14 16	023	18 14	081	37 25	130	22 57	197	53 50	234	64 06	310
8	11 45	355	14 32	022	18 53	081	37 55	130	22 46	196	53 18	234	63 35	308
9	11 41	354	14 46	022	19 32	080	38 25	129	22 35	196	52 46	234	63 04	307
10	11 37	353	15 01	021	20 11	079	38 56	129	22 25	195	52 15	233	62 32	305
11	11 32	352	15 14	020	20 49	078	39 26	129	22 15	195	51 43	233	62 00	304
12	11 26	351	15 27	019	21 28	077	39 57	128	22 05	194	51 12	233	61 27	302
13	11 19	350	15 39	018	22 06	077	40 28	128	21 55	194	50 40	233	60 53	301
14	11 12	349	15 51	017	22 44	076	40 59	127	21 46	193	50 09	232	60 19	300
	Hamal		◆RIGEL		SIRIUS		CANOPUS		◆RIGIL KENT.		Peacock		◆FOMALHAUT	
15	16 02	016	23 22	075	14 58	098	41 31	127	21 37	193	49 38	232	59 45	298
16	16 13	015	24 00	074	15 37	098	42 02	127	21 29	192	49 07	232	59 10	297
17	16 23	014	24 38	073	16 16	097	42 34	126	21 21	192	48 36	231	58 35	296
18	16 32	013	25 16	072	16 55	096	43 06	126	21 13	191	48 05	231	57 59	295
19	16 41	012	25 53	072	17 34	095	43 38	125	21 06	191	47 35	231	57 23	294
20	16 48	011	26 30	071	18 14	095	44 10	125	20 58	190	47 04	231	56 47	292
21	16 56	010	27 07	070	18 53	094	44 42	125	20 52	190	46 34	230	56 11	291
22	17 03	009	27 44	069	19 32	093	45 15	124	20 45	189	46 04	230	55 34	290
23	17 09	008	28 21	068	20 11	093	45 48	124	20 39	189	45 34	230	54 57	289
24	17 14	007	28 57	067	20 51	092	46 20	123	20 33	188	45 04	229	54 19	288
25	17 19	006	29 33	066	21 30	091	46 53	123	20 28	188	44 34	229	53 42	287
26	17 23	006	30 09	065	22 09	090	47 26	123	20 23	187	44 04	229	53 04	286
27	17 26	005	30 45	064	22 49	090	48 00	122	20 18	187	43 35	228	52 26	285
28	17 29	004	31 20	064	23 28	089	48 33	122	20 14	186	43 06	228	51 48	284
29	17 31	003	31 55	063	24 08	088	49 06	122	20 10	186	42 36	228	51 10	283
	Hamal		◆RIGEL		SIRIUS		CANOPUS		◆RIGIL KENT.		Peacock		◆FOMALHAUT	
30	17 33	002	32 30	062	24 47	087	49 40	121	20 06	185	42 07	227	50 31	282
31	17 33	001	33 05	061	25 26	086	50 14	121	20 03	185	41 39	227	49 53	281
32	17 34	000	33 39	060	26 05	086	50 47	121	20 00	184	41 10	227	49 14	280
33	17 33	359	34 13	059	26 45	085	51 21	120	19 57	184	40 41	226	48 35	279
34	17 32	358	34 46	058	27 24	084	51 56	120	19 55	183	40 13	226	47 56	279
35	17 30	357	35 19	057	28 03	083	52 30	119	19 53	182	39 45	225	47 17	278
36	17 28	356	35 52	056	28 42	083	53 04	119	19 51	182	39 17	225	46 38	277
37	17 24	355	36 24	055	29 21	082	53 39	119	19 50	181	38 49	225	45 59	276
38	17 21	354	36 56	054	30 00	081	54 13	119	19 49	181	38 22	224	45 20	275
39	17 16	353	37 28	053	30 39	080	54 48	118	19 49	180	37 54	224	44 41	274
40	17 11	352	37 59	052	31 17	079	55 22	118	19 49	180	37 27	224	44 02	274
41	17 05	351	38 29	050	31 56	078	55 57	118	19 49	179	37 00	223	43 22	273
42	16 59	350	39 00	049	32 35	078	56 32	117	19 50	179	36 33	223	42 43	272
43	16 52	349	39 29	048	33 13	077	57 07	117	19 51	178	36 06	222	42 04	271
44	16 44	348	39 58	047	33 51	076	57 42	117	19 52	178	35 40	222	41 24	271
	RIGEL		SIRIUS		◆CANOPUS		RIGIL KENT.		◆Peacock		FOMALHAUT		◆Hamal	
45	40 27	046	34 29	075	58 18	116	19 54	177	35 14	222	40 45	270	16 36	347
46	40 55	045	35 07	074	58 53	116	19 56	177	34 48	221	40 06	269	16 27	346
47	41 22	044	35 45	073	59 28	116	19 58	176	34 22	221	39 26	268	16 17	345
48	41 49	042	36 23	072	60 04	116	20 01	176	33 56	220	38 47	268	16 07	344
49	42 16	041	37 00	071	60 39	115	20 04	175	33 31	220	38 08	267	15 56	344
50	42 41	040	37 37	071	61 15	115	20 07	175	33 06	219	37 28	266	15 45	343
51	43 06	039	38 14	070	61 51	115	20 11	174	32 41	219	36 49	265	15 33	342
52	43 31	038	38 51	069	62 26	115	20 15	174	32 16	219	36 10	265	15 20	341
53	43 54	036	39 28	068	63 02	114	20 20	173	31 52	218	35 31	264	15 07	340
54	44 17	035	40 04	067	63 38	114	20 25	173	31 28	218	34 52	263	14 53	339
55	44 40	034	40 40	066	64 14	114	20 30	172	31 04	217	34 12	263	14 38	338
56	45 01	033	41 16	065	64 50	114	20 35	172	30 40	217	33 33	262	14 23	337
57	45 22	031	41 51	064	65 26	114	20 41	171	30 16	216	32 55	261	14 08	336
58	45 42	030	42 26	063	66 02	113	20 47	171	29 53	216	32 16	260	13 51	335
59	46 01	029	43 01	062	66 38	113	20 54	170	29 30	216	31 37	260	13 35	334
	◆ALDEBARAN		BETELGEUSE		SIRIUS		◆Suhail		RIGIL KENT.		◆ACHERNAR		Diphda	
60	24 01	009	28 15	033	43 36	061	38 46	115	21 01	170	67 24	235	39 54	290
61	24 07	008	28 36	032	44 10	060	39 22	114	21 08	169	66 52	235	39 17	289
62	24 13	007	28 56	031	44 44	058	39 58	114	21 16	169	66 20	235	38 39	288
63	24 17	006	29 16	030	45 17	057	40 34	113	21 24	168	65 47	235	38 02	288
64	24 21	005	29 35	028	45 50	056	41 11	113	21 32	168	65 15	235	37 24	287
65	24 25	004	29 53	027	46 22	055	41 47	112	21 41	167	64 42	235	36 46	286
66	24 27	003	30 11	026	46 54	054	42 24	112	21 50	167	64 10	236	36 08	285
67	24 29	002	30 28	025	47 26	053	43 00	111	21 59	166	63 38	236	35 30	284
68	24 30	001	30 45	024	47 57	052	43 37	111	22 09	166	63 05	236	34 52	283
69	24 30	000	31 00	023	48 28	050	44 14	110	22 19	165	62 33	235	34 14	282
70	24 30	359	31 15	022	48 58	049	44 51	109	22 29	165	62 00	235	33 35	281
71	24 28	358	31 30	021	49 27	048	45 28	109	22 39	164	61 28	235	32 56	281
72	24 27	357	31 43	020	49 56	047	46 05	108	22 50	164	60 55	235	32 18	280
73	24 24	356	31 56	018	50 24	045	46 43	108	23 02	163	60 23	235	31 39	279
74	24 21	355	32 08	017	50 52	044	47 20	107	23 13	163	59 51	235	31 00	278
	BETELGEUSE		SIRIUS		◆Suhail		RIGIL KENT.		◆ACHERNAR		Diphda		◆ALDEBARAN	
75	32 19	016	51 19	043	47 58	107	23 25	162	59 19	235	30 21	277	24 17	354
76	32 30	015	51 45	041	48 36	106	23 37	162	58 46	235	29 42	277	24 12	353
77	32 40	014	52 11	040	49 14	106	23 50	161	58 14	235	29 03	276	24 06	351
78	32 49	013	52 36	038	49 52	105	24 03	161	57 42	235	28 24	275	24 00	350
79	32 57	012	53 00	037	50 30	105	24 16	160	57 10	234	27 44	274	23 53	349
80	33 05	010	53 23	036	51 08	104	24 29	160	56 38	234	27 05	273	23 46	348
81	33 11	009	53 46	034	51 46	103	24 43	159	56 06	234	26 26	273	23 37	347
82	33 17	008	54 07	033	52 24	103	24 57	159	55 34	234	25 46	272	23 29	346
83	33 22	007	54 28	031	53 03	102	25 11	158	55 03	234	25 07	271	23 19	345
84	33 26	006	54 48	029	53 41	102	25 26	158	54 31	233	24 28	270	23 08	344
85	33 30	004	55 07	028	54 20	101	25 41	157	53 59	233	23 48	270	22 57	343
86	33 33	003	55 25	026	54 58	101	25 56	157	53 28	233	23 09	269	22 46	342
87	33 34	002	55 42	025	55 37	100	26 12	157	52 56	233	22 30	268	22 33	341
88	33 35	001	55 58	023	56 16	099	26 28	156	52 25	233	21 50	267	22 20	340
89	33 36	000	56 13	021	56 55	099	26 44	156	51 54	232	21 11	267	22 07	339

LHA	Hc	Zn	Hc	Zn	Hc	Zn	Hc	Zn	Hc	Zn	Hc	Zn	Hc	Zn
♈	PROCYON		Alphard		◆Suhail		RIGIL KENT.		◆ACHERNAR		◆RIGEL		BETELGEUSE	
90	31 38	029	30 56	065	57 34	098	27 00	155	51 23	232	48 05	343	33 35	358
91	31 56	028	31 32	064	58 13	098	27 17	155	50 52	232	47 53	341	33 33	357
92	32 15	027	32 07	063	58 52	097	27 34	154	50 21	232	47 40	340	33 31	356
93	32 32	026	32 42	062	59 31	096	27 51	154	49 50	231	47 27	339	33 28	355
94	32 49	025	33 17	061	60 10	096	28 09	153	49 19	231	47 12	337	33 24	354
95	33 05	024	33 51	060	60 49	095	28 26	153	48 49	231	46 56	336	33 20	353
96	33 21	023	34 25	059	61 28	094	28 45	152	48 18	230	46 40	334	33 14	351
97	33 36	021	34 59	058	62 08	094	29 03	152	47 48	230	46 22	333	33 08	350
98	33 50	020	35 32	057	62 47	093	29 21	152	47 18	230	46 04	332	33 01	349
99	34 03	019	36 05	056	63 26	092	29 40	151	46 48	230	45 45	330	32 53	348
100	34 15	018	36 38	055	64 06	092	29 59	151	46 18	229	45 25	329	32 44	347
101	34 27	017	37 10	054	64 45	091	30 19	150	45 48	229	45 04	328	32 35	346
102	34 38	016	37 41	053	65 24	090	30 38	150	45 19	229	44 43	326	32 24	344
103	34 48	014	38 13	052	66 04	089	30 58	149	44 49	228	44 21	325	32 13	343
104	34 57	013	38 44	051	66 43	089	31 18	149	44 20	228	43 58	324	32 02	342
	PROCYON		REGULUS		◆Gienah		RIGIL KENT.		◆ACHERNAR		RIGEL		◆BETELGEUSE	
105	35 06	012	16 18	048	20 21	094	31 39	149	43 50	228	43 34	323	31 49	341
106	35 14	011	16 47	047	21 00	093	31 59	148	43 21	227	43 10	321	31 36	340
107	35 20	009	17 16	046	21 39	092	32 20	148	42 53	227	42 45	320	31 22	339
108	35 27	008	17 44	046	22 19	091	32 41	147	42 24	227	42 19	319	31 07	338
109	35 32	007	18 12	045	22 58	091	33 02	147	41 55	226	41 53	318	30 52	336
110	35 36	006	18 39	044	23 38	090	33 24	147	41 27	226	41 26	316	30 36	335
111	35 40	005	19 06	043	24 17	089	33 46	146	40 59	226	40 59	315	30 19	334
112	35 43	003	19 33	042	24 56	088	34 08	146	40 31	225	40 31	314	30 01	333
113	35 44	002	19 59	041	25 36	088	34 30	146	40 03	225	40 03	313	29 43	332
114	35 45	001	20 25	040	26 15	087	34 52	145	39 35	225	39 33	312	29 24	331
115	35 46	000	20 50	039	26 54	086	35 15	145	39 07	224	39 04	311	29 05	330
116	35 45	358	21 14	038	27 33	085	35 38	144	38 40	224	38 34	310	28 45	329
117	35 44	357	21 38	037	28 13	084	36 01	144	38 13	223	38 03	309	28 24	328
118	35 41	356	22 02	036	28 52	084	36 24	144	37 46	223	37 32	308	28 03	327
119	35 38	355	22 25	035	29 31	083	36 47	143	37 19	223	37 01	307	27 41	326
	REGULUS		◆SPICA		RIGIL KENT.		◆ACHERNAR		RIGEL		BETELGEUSE		◆PROCYON	
120	22 47	034	14 07	091	37 11	143	36 53	222	36 29	305	27 18	325	35 34	354
121	23 09	033	14 46	090	37 35	143	36 26	222	35 57	304	26 55	324	35 29	352
122	23 30	032	15 26	089	37 59	142	36 00	221	35 24	303	26 32	323	35 24	351
123	23 51	031	16 05	089	38 23	142	35 34	221	34 51	302	26 07	322	35 17	350
124	24 11	030	16 44	088	38 48	142	35 08	221	34 18	301	25 43	321	35 10	349
125	24 31	029	17 24	087	39 12	141	34 43	220	33 44	300	25 17	320	35 02	348
126	24 50	028	18 03	086	39 37	141	34 17	220	33 10	299	24 52	319	34 53	346
127	25 08	027	18 42	085	40 02	141	33 52	219	32 35	299	24 25	318	34 43	345
128	25 26	026	19 21	085	40 27	140	33 27	219	32 01	298	23 59	317	34 33	344
129	25 43	025	20 01	084	40 52	140	33 03	219	31 26	297	23 31	316	34 21	343
130	25 59	024	20 40	083	41 18	140	32 38	218	30 50	296	23 03	315	34 09	342
131	26 15	023	21 19	082	41 43	139	32 14	218	30 15	295	22 35	314	33 57	340
132	26 30	022	21 58	082	42 09	139	31 50	217	29 39	294	22 07	313	33 43	339
133	26 44	021	22 37	081	42 35	139	31 26	217	29 03	293	21 38	312	33 29	338
134	26 58	020	23 15	080	43 01	138	31 03	216	28 26	292	21 08	311	33 13	337
	◆REGULUS		SPICA		◆RIGIL KENT.		ACHERNAR		◆CANOPUS		SIRIUS		PROCYON	
135	27 11	019	23 54	079	43 27	138	30 39	216	65 23	246	47 41	308	32 58	336
136	27 23	018	24 33	078	43 54	138	30 16	216	64 47	246	47 09	307	32 41	335
137	27 35	017	25 11	078	44 20	137	29 54	215	64 11	246	46 37	305	32 24	333
138	27 46	016	25 50	077	44 47	137	29 31	215	63 35	246	46 05	304	32 06	332
139	27 56	014	26 28	076	45 14	137	29 09	214	62 59	246	45 32	303	31 47	331
140	28 05	013	27 06	075	45 41	137	28 47	214	62 23	245	44 59	302	31 28	330
141	28 14	012	27 44	074	46 08	136	28 25	213	61 48	245	44 25	301	31 08	329
142	28 22	011	28 22	073	46 35	136	28 03	213	61 12	245	43 52	300	30 48	328
143	28 29	010	28 59	072	47 03	136	27 42	212	60 36	245	43 17	299	30 26	327
144	28 36	009	29 37	072	47 30	136	27 21	212	60 01	244	42 43	298	30 05	326
145	28 42	008	30 14	071	47 58	135	27 01	211	59 25	244	42 08	297	29 42	325
146	28 47	007	30 51	070	48 25	135	26 40	211	58 50	244	41 32	296	29 19	324
147	28 51	006	31 28	069	48 53	135	26 20	211	58 15	244	40 57	295	28 56	323
148	28 54	005	32 05	068	49 21	135	26 00	210	57 40	243	40 21	294	28 31	322
149	28 57	003	32 41	067	49 49	134	25 41	210	57 04	243	39 45	293	28 07	321
	◆REGULUS		SPICA		◆RIGIL KENT.		ACHERNAR		◆CANOPUS		SIRIUS		PROCYON	
150	28 59	002	33 17	066	50 17	134	25 21	209	56 29	243	39 08	292	27 41	320
151	29 00	001	33 53	065	50 46	134	25 02	209	55 54	242	38 31	291	27 16	319
152	29 00	000	34 29	064	51 14	134	24 43	208	55 20	242	37 54	290	26 49	318
153	29 00	359	35 04	063	51 42	134	24 25	208	54 45	242	37 17	289	26 22	317
154	28 59	358	35 39	062	52 11	133	24 07	207	54 10	241	36 40	288	25 55	316
155	28 57	357	36 14	061	52 40	133	23 49	207	53 36	241	36 03	287	25 27	315
156	28 54	356	36 48	060	53 08	133	23 32	206	53 01	241	35 25	286	24 59	314
157	28 51	354	37 22	059	53 37	133	23 14	206	52 27	240	34 47	285	24 30	313
158	28 47	353	37 56	058	54 06	133	22 58	205	51 53	240	34 09	285	24 01	312
159	28 42	352	38 29	057	54 35	133	22 41	205	51 19	240	33 31	284	23 32	311
160	28 36	351	39 02	056	55 04	133	22 25	204	50 45	239	32 52	283	23 01	310
161	28 30	350	39 35	055	55 33	132	22 09	204	50 11	239	32 14	282	22 31	309
162	28 22	349	40 07	054	56 02	132	21 53	203	49 37	239	31 35	281	22 00	308
163	28 15	348	40 39	053	56 31	132	21 38	203	49 04	238	30 57	280	21 29	307
164	28 06	347	41 10	052	57 00	132	21 23	202	48 30	238	30 18	279	20 57	306
	SPICA		◆ANTARES		RIGIL KENT.		ACHERNAR		◆CANOPUS		SIRIUS		◆REGULUS	
165	41 41	051	24 30	103	57 29	132	21 08	202	47 57	238	29 39	279	27 56	346
166	42 11	050	25 09	102	57 58	132	20 54	201	47 24	237	29 00	278	27 46	345
167	42 41	049	25 47	101	58 27	132	20 40	201	46 51	237	28 21	277	27 35	343
168	43 10	048	26 26	101	58 57	132	20 26	200	46 18	237	27 42	276	27 24	342
169	43 39	046	27 05	100	59 26	132	20 13	199	45 45	236	27 03	275	27 12	341
170	44 07	045	27 44	099	59 55	132	20 00	199	45 12	236	26 24	275	26 59	340
171	44 35	044	28 22	099	60 24	132	19 47	198	44 40	235	25 44	274	26 45	339
172	45 02	043	29 01	098	60 54	132	19 35	198	44 07	235	25 05	273	26 31	338
173	45 28	041	29 40	097	61 23	132	19 23	197	43 35	235	24 26	272	26 15	337
174	45 54	040	30 19	096	61 52	132	19 12	197	43 03	234	23 46	272	26 00	336
175	46 19	039	30 59	096	62 21	132	19 00	196	42 31	234	23 07	271	25 43	335
176	46 43	038	31 38	095	62 50	132	18 49	196	42 00	233	22 28	270	25 26	334
177	47 07	036	32 17	094	63 20	132	18 39	195	41 28	233	21 48	269	25 09	333
178	47 30	035	32 56	094	63 49	133	18 29	195	40 57	233	21 09	269	24 51	332
179	47 52	034	33 36	093	64 18	133	18 19	194	40 26	232	20 30	268	24 32	331

LHA ♈	Hc	Zn	Hc	Zn	Hc	Zn	Hc	Zn	Hc	Zn	Hc	Zn	Hc	Zn
	◆ARCTURUS		ANTARES		◆Peacock		ACHERNAR		CANOPUS		◆Suhail		REGULUS	
180	15 26	033	34 15	092	24 43	151	18 10	194	39 55	232	60 06	264	24 12	330
181	15 47	032	34 54	091	25 02	150	18 01	193	39 24	231	59 27	264	23 52	329
182	16 08	031	35 34	091	25 22	150	17 52	192	38 53	231	58 48	263	23 31	328
183	16 28	030	36 13	090	25 42	149	17 44	192	38 23	231	58 09	262	23 10	327
184	16 48	029	36 52	089	26 02	149	17 36	191	37 52	230	57 30	262	22 48	326
185	17 07	028	37 32	088	26 22	148	17 28	191	37 22	230	56 51	261	22 26	325
186	17 25	028	38 11	088	26 43	148	17 21	190	36 52	229	56 12	261	22 03	324
187	17 43	027	38 50	087	27 04	148	17 14	190	36 23	229	55 34	260	21 39	323
188	18 01	026	39 30	086	27 25	147	17 07	189	35 53	228	54 55	259	21 15	322
189	18 17	025	40 09	085	27 47	147	17 01	189	35 24	228	54 16	259	20 51	321
190	18 34	024	40 48	084	28 09	146	16 56	188	34 55	227	53 38	258	20 26	320
191	18 49	023	41 27	084	28 31	146	16 50	188	34 26	227	52 59	258	20 00	319
192	19 04	022	42 06	083	28 53	145	16 45	187	33 57	227	52 21	257	19 34	318
193	19 18	021	42 45	082	29 16	145	16 41	186	33 28	226	51 42	256	19 08	317
194	19 32	020	43 24	081	29 39	144	16 37	186	33 00	226	51 04	256	18 41	316
	ARCTURUS		◆ANTARES		Peacock		◆ACHERNAR		CANOPUS		Suhail		◆REGULUS	
195	19 45	019	44 03	080	30 02	144	16 33	185	32 32	225	50 26	255	18 13	315
196	19 58	018	44 42	079	30 25	143	16 29	185	32 04	225	49 48	255	17 46	315
197	20 09	017	45 21	079	30 49	143	16 26	184	31 37	224	49 10	254	17 17	314
198	20 21	016	45 59	078	31 12	143	16 24	184	31 09	224	48 32	254	16 49	313
199	20 31	015	46 37	077	31 37	142	16 21	183	30 42	223	47 54	253	16 19	312
200	20 41	014	47 16	076	32 01	142	16 19	182	30 15	223	47 17	253	15 50	311
201	20 50	013	47 54	075	32 25	141	16 18	182	29 49	222	46 39	252	15 20	310
202	20 59	012	48 32	074	32 50	141	16 17	181	29 22	222	46 02	252	14 50	309
203	21 07	011	49 09	073	33 15	140	16 16	181	28 56	221	45 25	251	14 19	308
204	21 14	010	49 47	072	33 40	140	16 16	180	28 30	221	44 47	250	13 48	307
205	21 20	009	50 24	071	34 06	140	16 16	180	28 04	220	44 10	250	13 17	307
206	21 26	008	51 01	070	34 31	139	16 16	179	27 39	220	43 34	249	12 45	306
207	21 31	007	51 38	069	34 57	139	16 17	179	27 14	220	42 57	249	12 13	305
208	21 36	006	52 15	068	35 23	138	16 18	178	26 49	219	42 20	248	11 40	304
209	21 39	005	52 51	067	35 50	138	16 20	177	26 24	219	41 44	248	11 08	303
	◆ARCTURUS		ANTARES		◆Peacock		ACHERNAR		CANOPUS		◆Suhail		SPICA	
210	21 42	004	53 27	066	36 16	137	16 22	177	26 00	218	41 07	247	51 26	346
211	21 45	003	54 03	064	36 43	137	16 24	176	25 36	218	40 31	247	51 16	345
212	21 46	002	54 38	063	37 10	137	16 27	176	25 12	217	39 55	246	51 05	343
213	21 47	001	55 13	062	37 37	136	16 30	175	24 48	216	39 19	246	50 53	342
214	21 47	000	55 48	061	38 04	136	16 34	175	24 25	216	38 43	245	50 40	340
215	21 47	359	56 22	060	38 32	136	16 37	174	24 02	215	38 07	245	50 26	338
216	21 46	358	56 56	058	38 59	135	16 42	173	23 39	215	37 32	244	50 12	337
217	21 44	357	57 29	057	39 27	135	16 46	173	23 17	214	36 57	244	49 56	336
218	21 42	356	58 02	056	39 55	134	16 51	172	22 55	214	36 22	243	49 39	334
219	21 38	355	58 34	054	40 24	134	16 57	172	22 33	213	35 47	242	49 21	333
220	21 34	354	59 06	053	40 52	134	17 03	171	22 12	213	35 12	242	49 03	331
221	21 30	353	59 37	052	41 21	133	17 09	171	21 50	212	34 37	241	48 43	330
222	21 24	352	60 07	050	41 49	133	17 15	170	21 29	212	34 03	241	48 23	328
223	21 18	351	60 37	049	42 18	133	17 22	170	21 09	211	33 28	240	48 02	327
224	21 12	350	61 07	047	42 47	132	17 30	169	20 49	211	32 54	240	47 40	326
	ANTARES		◆Nunki		Peacock		ACHERNAR		◆CANOPUS		◆SPICA		ARCTURUS	
225	61 35	046	39 46	085	43 16	132	17 37	168	20 29	210	47 18	324	21 04	349
226	62 03	044	40 25	085	43 46	132	17 45	168	20 09	210	46 54	323	20 56	348
227	62 30	042	41 04	084	44 15	131	17 54	167	19 50	209	46 30	322	20 48	347
228	62 56	041	41 43	083	44 45	131	18 03	167	19 31	209	46 06	320	20 38	346
229	63 21	039	42 22	082	45 15	131	18 12	166	19 12	208	45 40	319	20 28	345
230	63 45	037	43 01	081	45 45	130	18 21	166	18 54	207	45 14	318	20 18	344
231	64 08	035	43 40	081	46 15	130	18 31	165	18 36	207	44 47	317	20 06	343
232	64 30	033	44 19	080	46 45	130	18 41	165	18 18	206	44 20	315	19 54	342
233	64 51	031	44 58	079	47 16	129	18 52	164	18 01	206	43 52	314	19 42	341
234	65 11	029	45 36	078	47 46	129	19 03	164	17 44	205	43 24	313	19 28	340
235	65 30	027	46 15	077	48 17	129	19 14	163	17 28	205	42 55	312	19 14	339
236	65 47	025	46 53	076	48 48	128	19 26	163	17 11	204	42 25	311	19 00	338
237	66 03	023	47 31	075	49 18	128	19 38	162	16 55	203	41 55	310	18 45	337
238	66 18	021	48 09	074	49 50	128	19 50	161	16 40	203	41 24	308	18 29	336
239	66 32	019	48 47	073	50 21	128	20 03	161	16 25	202	40 53	307	18 13	335
	Rasalhague		◆ALTAIR		FOMALHAUT		◆ACHERNAR		ACRUX		◆SPICA		ARCTURUS	
240	24 59	026	13 21	059	13 25	120	20 16	160	58 12	224	40 22	306	17 56	334
241	25 15	025	13 54	058	13 59	119	20 29	160	57 45	224	39 50	305	17 38	333
242	25 32	024	14 28	057	14 34	119	20 43	159	57 18	224	39 18	304	17 20	332
243	25 47	023	15 01	057	15 08	118	20 57	159	56 51	224	38 45	303	17 02	331
244	26 02	021	15 33	056	15 43	117	21 12	158	56 24	224	38 12	302	16 42	330
245	26 16	020	16 06	055	16 18	117	21 26	158	55 57	224	37 38	301	16 23	329
246	26 29	019	16 38	054	16 54	116	21 41	157	55 29	223	37 04	300	16 02	328
247	26 42	018	17 09	053	17 29	115	21 57	157	55 02	223	36 30	299	15 41	328
248	26 54	017	17 41	052	18 05	115	22 12	156	54 35	223	35 55	298	15 20	327
249	27 05	016	18 12	051	18 41	114	22 28	156	54 08	223	35 20	297	14 58	326
250	27 16	015	18 42	050	19 17	113	22 45	155	53 41	223	34 45	296	14 36	325
251	27 25	014	19 12	050	19 53	113	23 01	155	53 14	223	34 10	295	14 13	324
252	27 35	013	19 42	049	20 29	112	23 18	154	52 48	223	33 34	294	13 49	323
253	27 43	012	20 11	048	21 06	111	23 36	154	52 21	223	32 58	293	13 26	322
254	27 51	011	20 40	047	21 43	111	23 53	153	51 54	223	32 22	292	13 01	321
	Rasalhague		◆ALTAIR		FOMALHAUT		◆ACHERNAR		ACRUX		◆SPICA		ANTARES	
255	27 58	010	21 09	046	22 20	110	24 11	153	51 27	223	31 45	291	66 38	342
256	28 04	008	21 37	045	22 57	109	24 29	152	51 01	222	31 08	291	66 26	340
257	28 09	007	22 04	044	23 34	109	24 48	152	50 34	222	30 31	290	66 12	338
258	28 14	006	22 31	043	24 11	108	25 07	151	50 08	222	29 54	289	65 56	336
259	28 18	005	22 58	042	24 49	107	25 26	151	49 42	222	29 17	288	65 39	334
260	28 21	004	23 24	041	25 26	107	25 45	150	49 15	222	28 39	287	65 21	332
261	28 23	003	23 50	040	26 04	106	26 05	150	48 49	221	28 02	286	65 02	330
262	28 25	002	24 15	039	26 42	105	26 25	149	48 23	221	27 24	285	64 42	328
263	28 26	001	24 39	038	27 20	105	26 45	149	47 57	221	26 46	285	64 20	326
264	28 26	000	25 03	037	27 58	104	27 05	148	47 32	221	26 07	284	63 58	324
265	28 26	359	25 27	036	28 36	103	27 26	148	47 06	221	25 29	283	63 34	322
266	28 24	357	25 50	035	29 15	103	27 47	147	46 40	220	24 51	282	63 09	320
267	28 22	356	26 12	034	29 53	102	28 08	147	46 15	220	24 12	281	62 44	319
268	28 19	355	26 34	033	30 32	101	28 30	147	45 49	220	23 33	280	62 18	317
269	28 15	354	26 55	032	31 10	101	28 52	146	45 24	220	22 55	280	61 50	315

LAT 49°S

LHA ♈	Hc	Zn	Hc	Zn	Hc	Zn	Hc	Zn	Hc	Zn	Hc	Zn	Hc	Zn
	◆ALTAIR		FOMALHAUT		◆ACHERNAR		RIGIL KENT.		◆SPICA		ANTARES		Rasalhague	
270	27 16	031	31 49	100	29 14	146	59 44	228	22 16	279	61 22	314	28 11	353
271	27 36	030	32 28	099	29 36	145	59 15	228	21 37	278	60 53	312	28 06	352
272	27 55	029	33 07	099	29 59	145	58 46	228	20 58	277	60 24	311	28 00	351
273	28 14	028	33 46	098	30 22	144	58 17	228	20 19	276	59 54	309	27 53	350
274	28 32	027	34 25	097	30 45	144	57 47	228	19 40	276	59 23	308	27 46	349
275	28 49	026	35 04	097	31 08	143	57 18	228	19 00	275	58 52	306	27 38	348
276	29 06	025	35 43	096	31 32	143	56 49	228	18 21	274	58 20	305	27 29	346
277	29 22	024	36 22	095	31 56	143	56 20	228	17 42	273	57 47	304	27 19	345
278	29 38	022	37 01	094	32 20	142	55 51	228	17 03	273	57 14	302	27 09	344
279	29 52	021	37 41	094	32 44	142	55 22	227	16 23	272	56 40	301	26 58	343
280	30 06	020	38 20	093	33 08	141	54 53	227	15 44	271	56 07	300	26 46	342
281	30 20	019	38 59	092	33 33	141	54 24	227	15 05	270	55 32	299	26 34	341
282	30 32	018	39 39	091	33 58	140	53 55	227	14 25	270	54 57	297	26 20	340
283	30 44	017	40 18	091	34 23	140	53 26	227	13 46	269	54 22	296	26 07	339
284	30 55	016	40 57	090	34 49	140	52 58	227	13 06	268	53 47	295	25 52	338
	ALTAIR		Enif		◆FOMALHAUT		ACHERNAR		◆RIGIL KENT.		ANTARES		◆Rasalhague	
285	31 05	015	21 02	044	41 37	089	35 14	139	52 29	227	53 11	294	25 37	337
286	31 15	013	21 29	043	42 16	088	35 40	139	52 00	226	52 35	293	25 21	336
287	31 24	012	21 55	042	42 55	088	36 06	138	51 32	226	51 58	292	25 05	335
288	31 32	011	22 21	041	43 35	087	36 32	138	51 03	226	51 22	291	24 48	334
289	31 39	010	22 47	040	44 14	086	36 59	138	50 35	226	50 45	290	24 30	333
290	31 45	009	23 12	039	44 53	085	37 25	137	50 07	226	50 07	289	24 11	332
291	31 51	008	23 36	038	45 32	085	37 52	137	49 39	225	49 30	288	23 52	331
292	31 56	007	24 00	037	46 12	084	38 19	136	49 11	225	48 52	287	23 33	330
293	32 00	005	24 24	036	46 51	083	38 47	136	48 43	225	48 15	286	23 13	329
294	32 03	004	24 47	035	47 30	082	39 14	136	48 15	225	47 37	285	22 52	328
295	32 06	003	25 09	034	48 09	081	39 41	135	47 47	225	46 58	284	22 31	327
296	32 08	002	25 31	033	48 47	080	40 09	135	47 20	224	46 20	283	22 09	326
297	32 09	001	25 52	032	49 26	079	40 37	135	46 52	224	45 42	282	21 46	325
298	32 09	000	26 13	031	50 05	078	41 05	134	46 25	224	45 03	281	21 23	324
299	32 08	358	26 33	030	50 43	078	41 34	134	45 58	224	44 24	280	21 00	323
	◆Enif		FOMALHAUT		◆ACHERNAR		RIGIL KENT.		◆ANTARES		Rasalhague		ALTAIR	
300	26 52	029	51 22	077	42 02	134	45 31	223	43 46	279	20 36	322	32 06	357
301	27 11	028	52 00	076	42 31	133	45 04	223	43 07	279	20 11	321	32 04	356
302	27 29	027	52 38	075	42 59	133	44 37	223	42 28	278	19 46	320	32 01	355
303	27 46	026	53 16	074	43 28	133	44 11	222	41 49	277	19 20	319	31 57	354
304	28 03	025	53 54	073	43 57	132	43 44	222	41 10	276	18 54	318	31 53	353
305	28 19	024	54 31	072	44 27	132	43 18	222	40 30	275	18 28	317	31 47	351
306	28 35	023	55 08	071	44 56	132	42 51	222	39 51	274	18 01	316	31 41	350
307	28 49	021	55 45	070	45 25	131	42 25	221	39 12	274	17 34	315	31 34	349
308	29 03	020	56 22	068	45 55	131	42 00	221	38 33	273	17 06	314	31 26	348
309	29 17	019	56 58	067	46 25	131	41 34	221	37 53	272	16 37	314	31 17	347
310	29 29	018	57 35	066	46 55	130	41 08	220	37 14	271	16 09	313	31 08	346
311	29 41	017	58 10	065	47 25	130	40 43	220	36 35	271	15 39	312	30 58	345
312	29 52	016	58 46	064	47 55	130	40 18	220	35 55	270	15 10	311	30 47	343
313	30 03	015	59 21	062	48 26	129	39 53	219	35 16	269	14 40	310	30 35	342
314	30 13	014	59 56	061	48 56	129	39 28	219	34 37	268	14 10	309	30 23	341
	◆Enif		Diphda		◆ACHERNAR		Miaplacidus		RIGIL KENT.		◆ANTARES		ALTAIR	
315	30 22	013	35 43	076	49 27	129	28 43	181	39 03	219	33 57	268	30 10	340
316	30 30	011	36 21	075	49 57	129	28 43	181	38 38	218	33 18	267	29 56	339
317	30 37	010	36 59	074	50 28	128	28 42	181	38 14	218	32 39	266	29 42	338
318	30 44	009	37 37	073	50 59	128	28 42	180	37 50	218	31 59	265	29 27	337
319	30 50	008	38 14	072	51 30	128	28 42	180	37 26	217	31 20	265	29 11	336
320	30 55	007	38 52	071	52 01	128	28 42	179	37 02	217	30 41	264	28 54	335
321	30 59	006	39 29	070	52 32	127	28 43	179	36 39	217	30 02	263	28 37	333
322	31 03	005	40 06	069	53 04	127	28 44	179	36 15	216	29 23	262	28 19	332
323	31 05	003	40 42	068	53 35	127	28 45	178	35 52	216	28 44	262	28 00	331
324	31 07	002	41 19	067	54 07	127	28 46	178	35 29	215	28 05	261	27 41	330
325	31 09	001	41 55	066	54 38	126	28 48	177	35 06	215	27 26	260	27 21	329
326	31 09	000	42 31	065	55 10	126	28 50	177	34 44	215	26 47	260	27 01	328
327	31 08	359	43 07	064	55 42	126	28 52	177	34 22	214	26 09	259	26 40	327
328	31 07	358	43 42	063	56 14	126	28 55	176	33 59	214	25 30	258	26 18	326
329	31 05	357	44 17	062	56 46	126	28 57	176	33 38	214	24 52	258	25 56	325
	◆Diphda		ACHERNAR		◆Miaplacidus		RIGIL KENT.		ANTARES		◆ALTAIR		Enif	
330	44 52	061	57 18	126	29 00	175	33 16	213	24 13	257	25 33	324	31 03	355
331	45 26	060	57 50	125	29 04	175	32 54	213	23 35	256	25 10	323	30 59	354
332	46 00	059	58 22	125	29 07	175	32 33	212	22 57	256	24 46	322	30 55	353
333	46 34	058	58 54	125	29 11	174	32 12	212	22 19	255	24 22	321	30 49	352
334	47 07	057	59 26	125	29 15	174	31 52	212	21 41	254	23 57	320	30 44	351
335	47 40	056	59 58	125	29 20	173	31 31	211	21 03	253	23 31	319	30 37	350
336	48 12	055	60 31	125	29 24	173	31 11	211	20 25	253	23 05	318	30 29	349
337	48 44	053	61 03	125	29 29	173	30 51	210	19 48	252	22 39	317	30 21	347
338	49 15	052	61 36	125	29 34	172	30 31	210	19 10	251	22 12	316	30 12	346
339	49 46	051	62 08	125	29 40	172	30 11	210	18 33	251	21 44	315	30 03	345
340	50 16	050	62 40	125	29 45	171	29 52	209	17 56	250	21 16	314	29 52	344
341	50 46	048	63 13	124	29 51	171	29 33	209	17 19	249	20 48	313	29 41	343
342	51 15	047	63 45	124	29 58	171	29 14	208	16 42	249	20 19	313	29 29	342
343	51 44	046	64 18	124	30 04	170	28 56	208	16 06	248	19 50	312	29 16	341
344	52 12	044	64 50	125	30 11	170	28 38	207	15 29	247	19 20	311	29 03	340
	◆Diphda		Acamar		◆CANOPUS		Miaplacidus		RIGIL KENT.		◆Nunki		Enif	
345	52 39	043	47 55	101	27 16	140	30 18	170	28 20	207	38 07	273	28 49	339
346	53 05	042	48 33	101	27 41	140	30 25	169	28 02	206	37 27	272	28 34	337
347	53 31	040	49 12	100	28 06	139	30 32	169	27 45	206	36 48	271	28 19	336
348	53 56	039	49 51	100	28 32	139	30 40	168	27 28	206	36 09	270	28 03	335
349	54 21	037	50 30	099	28 58	138	30 48	168	27 11	205	35 29	269	27 46	334
350	54 44	036	51 09	098	29 24	138	30 56	168	26 54	205	34 50	269	27 28	333
351	55 07	034	51 48	098	29 51	138	31 05	167	26 38	204	34 11	268	27 10	332
352	55 29	033	52 27	097	30 17	137	31 14	167	26 22	204	33 31	267	26 51	331
353	55 49	031	53 06	096	30 44	137	31 23	167	26 06	203	32 52	267	26 32	330
354	56 09	030	53 45	096	31 11	136	31 32	166	25 51	203	32 13	266	26 12	329
355	56 28	028	54 24	095	31 39	136	31 41	166	25 35	202	31 34	265	25 51	328
356	56 46	026	55 03	094	32 06	135	31 51	166	25 21	202	30 54	264	25 30	327
357	57 03	025	55 43	094	32 34	135	32 01	165	25 06	201	30 15	264	25 08	326
358	57 19	023	56 22	093	33 02	134	32 11	165	24 52	201	29 36	263	24 46	325
359	57 34	021	57 01	092	33 31	134	32 22	164	24 38	201	28 57	262	24 23	324

LAT 50°S

LHA ♈	Hc	Zn	Hc	Zn	Hc	Zn	Hc	Zn	Hc	Zn	Hc	Zn	Hc	Zn
	Alpheratz		Hamal		♦RIGEL		CANOPUS		♦RIGIL KENT.		Peacock		♦FOMALHAUT	
0	10 55	002	11 21	029	13 37	087	34 40	133	25 21	200	58 08	237	66 28	324
1	10 56	001	11 40	029	14 15	086	35 09	132	25 07	200	57 36	237	66 05	322
2	10 56	000	11 58	028	14 54	085	35 37	132	24 55	199	57 04	236	65 41	320
3	10 56	359	12 16	027	15 32	084	36 06	131	24 42	199	56 32	236	65 16	318
4	10 55	358	12 33	026	16 11	083	36 35	131	24 30	198	56 00	236	64 50	317
5	10 54	357	12 50	025	16 49	083	37 04	131	24 18	198	55 28	236	64 23	315
6	10 52	356	13 06	024	17 27	082	37 34	130	24 06	197	54 56	235	63 55	313
7	10 49	356	13 21	023	18 05	081	38 03	130	23 55	197	54 25	235	63 26	312
8	10 46	355	13 36	022	18 43	080	38 33	129	23 44	196	53 53	235	62 57	310
9	10 42	354	13 51	021	19 21	079	39 03	129	23 33	196	53 21	235	62 27	308
10	10 37	353	14 04	020	19 59	079	39 33	128	23 23	195	52 50	234	61 57	307
11	10 32	352	14 18	020	20 37	078	40 03	128	23 13	195	52 19	234	61 26	305
12	10 27	351	14 30	019	21 14	077	40 34	127	23 03	194	51 47	234	60 54	304
13	10 20	350	14 42	018	21 52	076	41 05	127	22 54	194	51 16	234	60 22	303
14	10 14	349	14 54	017	22 29	075	41 35	127	22 45	193	50 45	233	59 49	301
	Hamal		♦RIGEL		SIRIUS		CANOPUS		♦RIGIL KENT.		Peacock		♦FOMALHAUT	
15	15 05	016	23 07	075	15 07	098	42 06	126	22 36	193	50 14	233	59 16	300
16	15 15	015	23 44	074	15 45	097	42 38	126	22 28	192	49 44	233	58 42	299
17	15 24	014	24 21	073	16 23	097	43 09	125	22 20	192	49 13	232	58 08	297
18	15 33	013	24 57	072	17 02	096	43 41	125	22 12	191	48 43	232	57 34	296
19	15 42	012	25 34	071	17 40	095	44 12	125	22 04	191	48 12	232	56 59	295
20	15 50	011	26 10	070	18 18	094	44 44	124	21 57	190	47 42	231	56 24	294
21	15 57	010	26 47	069	18 57	094	45 16	124	21 51	190	47 12	231	55 48	293
22	16 03	009	27 22	068	19 35	093	45 48	123	21 44	189	46 42	231	55 12	291
23	16 09	008	27 58	068	20 14	092	46 21	123	21 38	189	46 12	230	54 36	290
24	16 14	007	28 34	067	20 52	091	46 53	123	21 33	188	45 43	230	54 00	289
25	16 19	006	29 09	066	21 31	091	47 26	122	21 27	188	45 13	230	53 24	288
26	16 23	005	29 44	065	22 10	090	47 58	122	21 22	187	44 44	229	52 47	287
27	16 26	005	30 19	064	22 48	089	48 31	121	21 18	187	44 14	229	52 10	286
28	16 29	004	30 53	063	23 27	088	49 04	121	21 13	186	43 45	229	51 33	285
29	16 31	003	31 28	062	24 05	088	49 37	121	21 09	186	43 17	228	50 56	284
	Hamal		♦RIGEL		SIRIUS		CANOPUS		♦RIGIL KENT.		Peacock		♦FOMALHAUT	
30	16 33	002	32 02	061	24 44	087	50 11	120	21 06	185	42 48	228	50 18	283
31	16 33	001	32 35	060	25 22	086	50 44	120	21 03	185	42 19	228	49 40	282
32	16 34	000	33 08	059	26 01	085	51 17	119	21 00	184	41 51	227	49 03	281
33	16 33	359	33 41	058	26 39	084	51 51	119	20 57	184	41 23	227	48 25	281
34	16 32	358	34 14	057	27 17	084	52 25	119	20 55	183	40 55	226	47 47	280
35	16 30	357	34 46	056	27 56	083	52 59	118	20 53	183	40 27	226	47 09	279
36	16 28	356	35 18	055	28 34	082	53 33	118	20 51	182	39 59	226	46 31	278
37	16 25	355	35 49	054	29 12	081	54 07	118	20 50	181	39 32	225	45 52	277
38	16 21	354	36 20	053	29 50	080	54 41	117	20 49	181	39 04	225	45 14	276
39	16 17	353	36 51	052	30 28	080	55 15	117	20 49	180	38 37	225	44 36	275
40	16 12	352	37 21	051	31 06	079	55 50	117	20 49	180	38 10	224	43 57	275
41	16 06	351	37 51	050	31 44	078	56 24	116	20 49	179	37 43	224	43 19	274
42	16 00	350	38 20	049	32 21	077	56 59	116	20 50	179	37 17	223	42 40	273
43	15 53	349	38 49	048	32 59	076	57 34	116	20 51	178	36 51	223	42 02	272
44	15 45	348	39 17	047	33 36	075	58 09	115	20 52	178	36 25	222	41 23	271
	RIGEL		SIRIUS		♦CANOPUS		RIGIL KENT.		♦Peacock		FOMALHAUT		♦Hamal	
45	39 45	045	34 13	074	58 44	115	20 54	177	35 58	222	40 45	271	15 37	347
46	40 12	044	34 51	073	59 19	115	20 56	177	35 33	222	40 06	270	15 29	346
47	40 39	043	35 27	073	59 54	114	20 58	176	35 07	221	39 28	269	15 19	345
48	41 05	042	36 04	072	60 29	114	21 01	176	34 42	221	38 49	268	15 09	345
49	41 30	041	36 41	071	61 04	114	21 04	175	34 17	220	38 11	268	14 59	344
50	41 55	040	37 17	070	61 40	113	21 07	175	33 52	220	37 32	267	14 47	343
51	42 19	038	37 53	069	62 15	113	21 11	174	33 27	219	36 54	266	14 36	342
52	42 43	037	38 29	068	62 50	113	21 15	174	33 03	219	36 15	265	14 23	341
53	43 06	036	39 05	067	63 26	113	21 19	173	32 39	219	35 37	265	14 10	340
54	43 28	035	39 40	066	64 02	112	21 24	173	32 15	218	34 58	264	13 57	339
55	43 50	033	40 15	065	64 37	112	21 29	172	31 51	218	34 20	263	13 43	338
56	44 10	032	40 50	064	65 13	112	21 35	172	31 28	217	33 42	263	13 28	337
57	44 30	031	41 24	063	65 49	112	21 41	171	31 04	217	33 03	262	13 13	336
58	44 50	029	41 58	062	66 25	111	21 47	171	30 41	216	32 25	261	12 57	335
59	45 08	028	42 32	061	67 01	111	21 53	170	30 19	216	31 47	260	12 41	334
	♦ALDEBARAN		BETELGEUSE		SIRIUS		♦Suhail		RIGIL KENT.		♦ACHERNAR		Diphda	
60	23 02	009	27 24	032	43 06	060	39 11	114	22 00	170	67 58	237	39 32	291
61	23 08	008	27 45	031	43 39	059	39 47	114	22 07	169	67 25	237	38 56	290
62	23 13	007	28 04	030	44 12	058	40 22	113	22 15	169	66 53	237	38 20	289
63	23 18	006	28 24	029	44 44	057	40 58	112	22 23	168	66 21	237	37 43	288
64	23 22	005	28 42	028	45 16	055	41 33	112	22 31	168	65 48	237	37 07	287
65	23 25	004	29 00	027	45 48	054	42 09	111	22 39	167	65 16	237	36 30	286
66	23 27	003	29 17	026	46 19	053	42 45	111	22 48	167	64 43	237	35 53	286
67	23 29	002	29 34	025	46 49	052	43 21	110	22 57	166	64 11	237	35 15	285
68	23 30	001	29 50	024	47 19	051	43 58	110	23 07	165	63 38	237	34 38	284
69	23 30	000	30 05	023	47 49	050	44 34	109	23 17	165	63 06	237	34 01	283
70	23 30	359	30 20	022	48 18	048	45 10	109	23 27	164	62 34	237	33 23	282
71	23 28	358	30 34	021	48 47	047	45 47	108	23 37	164	62 01	237	32 45	281
72	23 27	357	30 47	019	49 15	046	46 24	107	23 48	164	61 29	237	32 07	280
73	23 24	356	30 59	018	49 42	044	47 01	107	23 59	163	60 57	237	31 29	280
74	23 21	355	31 11	017	50 09	043	47 38	106	24 10	163	60 25	237	30 51	279
	BETELGEUSE		SIRIUS		♦Suhail		RIGIL KENT.		♦ACHERNAR		Diphda		♦ALDEBARAN	
75	31 22	016	50 35	042	48 15	106	24 22	162	59 52	236	30 13	278	23 17	354
76	31 32	015	51 00	040	48 52	105	24 34	162	59 20	236	29 35	277	23 12	353
77	31 42	014	51 25	039	49 29	105	24 47	161	58 48	236	28 56	276	23 07	352
78	31 50	013	51 49	038	50 07	104	24 59	161	58 16	236	28 18	275	23 01	351
79	31 58	011	52 12	036	50 44	103	25 12	160	57 44	236	27 40	275	22 54	349
80	32 06	010	52 34	035	51 22	103	25 26	160	57 13	236	27 01	274	22 47	348
81	32 12	009	52 56	033	51 59	102	25 39	159	56 41	235	26 23	273	22 39	347
82	32 18	008	53 17	032	52 37	102	25 53	159	56 09	235	25 44	272	22 30	346
83	32 23	007	53 37	030	53 15	101	26 07	158	55 38	235	25 06	272	22 21	345
84	32 27	006	53 56	029	53 53	100	26 22	158	55 06	235	24 27	271	22 11	344
85	32 30	004	54 14	027	54 31	100	26 36	157	54 35	234	23 49	270	22 00	343
86	32 33	003	54 31	026	55 09	099	26 52	157	54 03	234	23 10	269	21 49	342
87	32 34	002	54 47	024	55 47	099	27 07	156	53 32	234	22 31	268	21 37	341
88	32 35	001	55 02	022	56 25	098	27 23	156	53 01	234	21 53	268	21 24	340
89	32 36	000	55 17	021	57 03	097	27 38	155	52 30	233	21 14	267	21 11	339

LAT 50°S

LHA ♈	Hc	Zn	Hc	Zn	Hc	Zn	Hc	Zn	Hc	Zn	Hc	Zn	Hc	Zn
	PROCYON		Alphard		♦Suhail		RIGIL KENT.		♦ACHERNAR		♦RIGEL		BETELGEUSE	
90	30 45	029	30 31	064	57 42	097	27 55	155	51 59	233	47 08	343	32 35	358
91	31 04	028	31 05	064	58 20	096	28 11	154	51 28	233	46 56	342	32 34	357
92	31 21	027	31 40	063	58 58	095	28 28	154	50 58	233	46 44	340	32 31	356
93	31 38	026	32 14	062	59 37	095	28 45	154	50 27	232	46 31	339	32 28	355
94	31 55	025	32 48	061	60 15	094	29 02	153	49 57	232	46 16	338	32 25	354
95	32 10	023	33 21	060	60 54	093	29 20	153	49 26	232	46 01	336	32 20	353
96	32 25	022	33 54	059	61 32	093	29 38	152	48 56	231	45 45	335	32 15	351
97	32 40	021	34 27	058	62 11	092	29 56	152	48 26	231	45 28	334	32 09	350
98	32 53	020	34 59	057	62 49	091	30 14	151	47 56	231	45 11	332	32 02	349
99	33 06	019	35 31	056	63 28	090	30 33	151	47 26	230	44 52	331	31 54	348
100	33 18	018	36 03	055	64 06	090	30 52	150	46 57	230	44 33	330	31 46	347
101	33 29	017	36 34	054	64 45	089	31 11	150	46 27	230	44 13	328	31 36	346
102	33 40	015	37 05	053	65 23	088	31 30	150	45 58	229	43 53	327	31 27	345
103	33 50	014	37 36	051	66 02	087	31 50	149	45 29	229	43 31	326	31 16	343
104	33 59	013	38 06	050	66 41	086	32 10	149	45 00	229	43 09	324	31 04	342
	PROCYON		REGULUS		♦Gienah		RIGIL KENT.		♦ACHERNAR		RIGEL		♦BETELGEUSE	
105	34 07	012	15 38	048	20 24	093	32 30	148	44 31	228	42 46	323	30 52	341
106	34 15	011	16 06	047	21 03	092	32 50	148	44 02	228	42 23	322	30 40	340
107	34 21	009	16 34	046	21 41	092	33 11	148	43 33	228	41 59	321	30 26	339
108	34 27	008	17 02	045	22 20	091	33 32	147	43 05	227	41 34	319	30 12	338
109	34 32	007	17 29	044	22 59	090	33 53	147	42 36	227	41 09	318	29 57	337
110	34 36	006	17 56	043	23 37	089	34 14	146	42 08	227	40 43	317	29 41	336
111	34 40	005	18 22	043	24 16	089	34 35	146	41 40	226	40 16	316	29 25	334
112	34 43	003	18 48	042	24 54	088	34 57	145	41 13	226	39 49	315	29 08	333
113	34 44	002	19 14	041	25 33	087	35 19	145	40 45	226	39 21	314	28 50	332
114	34 45	001	19 38	040	26 11	086	35 41	145	40 17	225	38 53	313	28 32	331
115	34 46	000	20 03	039	26 50	086	36 04	144	39 50	225	38 25	311	28 13	330
116	34 45	358	20 27	038	27 28	085	36 26	144	39 23	224	37 55	310	27 54	329
117	34 44	357	20 50	037	28 07	084	36 49	144	38 56	224	37 26	309	27 33	328
118	34 41	356	21 13	036	28 45	083	37 12	143	38 30	224	36 56	308	27 13	327
119	34 38	355	21 36	035	29 23	082	37 35	143	38 03	223	36 25	307	26 51	326
	REGULUS		♦SPICA		RIGIL KENT.		♦ACHERNAR		RIGEL		BETELGEUSE		♦PROCYON	
120	21 57	034	14 08	091	37 59	142	37 37	223	35 54	306	26 29	325	34 35	354
121	22 19	033	14 46	090	38 22	142	37 11	222	35 23	305	26 07	324	34 30	352
122	22 39	032	15 25	089	38 46	142	36 45	222	34 51	304	25 44	323	34 24	351
123	23 00	031	16 03	088	39 10	141	36 19	222	34 19	303	25 20	322	34 18	350
124	23 19	030	16 42	087	39 34	141	35 54	221	33 46	302	24 56	321	34 11	349
125	23 38	029	17 20	087	39 59	141	35 28	221	33 13	301	24 32	320	34 03	348
126	23 57	028	17 59	086	40 23	140	35 03	220	32 40	300	24 07	319	33 55	346
127	24 14	027	18 37	085	40 48	140	34 38	220	32 07	299	23 41	318	33 45	345
128	24 32	026	19 16	084	41 13	140	34 14	219	31 33	298	23 15	317	33 35	344
129	24 48	025	19 54	084	41 38	139	33 49	219	30 59	297	22 48	316	33 24	343
130	25 04	024	20 32	083	42 03	139	33 25	219	30 24	296	22 21	315	33 12	342
131	25 19	023	21 11	082	42 29	139	33 01	218	29 49	295	21 54	314	33 00	341
132	25 34	022	21 49	081	42 54	138	32 38	218	29 14	294	21 26	313	32 47	339
133	25 48	021	22 27	080	43 20	138	32 14	217	28 39	293	20 57	312	32 33	338
134	26 01	020	23 05	080	43 46	138	31 51	217	28 04	293	20 29	311	32 18	337
	♦REGULUS		SPICA		♦RIGIL KENT.		ACHERNAR		♦CANOPUS		SIRIUS		PROCYON	
135	26 14	019	23 43	079	44 12	137	31 28	216	65 46	248	47 03	309	32 03	336
136	26 26	018	24 20	078	44 38	137	31 05	216	65 10	248	46 33	307	31 47	335
137	26 37	016	24 58	077	45 04	137	30 43	215	64 35	248	46 02	306	31 30	334
138	26 48	015	25 36	076	45 31	136	30 20	215	63 59	248	45 31	305	31 13	333
139	26 58	014	26 13	075	45 57	136	29 58	215	63 23	247	44 59	304	30 55	332
140	27 07	013	26 50	075	46 24	136	29 36	214	62 48	247	44 27	303	30 36	330
141	27 15	012	27 27	074	46 51	136	29 15	214	62 12	247	43 54	302	30 17	329
142	27 23	011	28 04	073	47 18	135	28 54	213	61 37	247	43 21	301	29 57	328
143	27 30	010	28 41	072	47 45	135	28 33	213	61 01	246	42 48	300	29 36	327
144	27 37	009	29 18	071	48 13	135	28 12	212	60 26	246	42 14	299	29 15	326
145	27 42	008	29 54	070	48 40	135	27 52	212	59 51	246	41 40	298	28 53	325
146	27 47	007	30 30	069	49 08	134	27 31	211	59 16	245	41 06	296	28 31	324
147	27 51	006	31 06	068	49 35	134	27 12	211	58 41	245	40 31	295	28 08	323
148	27 54	004	31 42	067	50 03	134	26 52	210	58 06	245	39 56	294	27 44	322
149	27 57	003	32 17	067	50 31	134	26 33	210	57 31	244	39 21	293	27 20	321
	♦REGULUS		SPICA		♦RIGIL KENT.		ACHERNAR		♦CANOPUS		SIRIUS		PROCYON	
150	27 59	002	32 53	066	50 59	133	26 14	209	56 56	244	38 45	293	26 56	320
151	28 00	001	33 28	065	51 27	133	25 55	209	56 22	244	38 10	292	26 31	319
152	28 00	000	34 02	064	51 55	133	25 36	208	55 47	243	37 34	291	26 05	318
153	28 00	359	34 37	063	52 24	133	25 18	208	55 13	243	36 58	290	25 39	317
154	27 59	358	35 11	062	52 52	133	25 00	207	54 38	243	36 21	289	25 12	316
155	27 57	357	35 45	061	53 20	132	24 43	207	54 04	242	35 44	288	24 45	315
156	27 55	356	36 18	060	53 49	132	24 25	206	53 30	242	35 08	287	24 18	314
157	27 51	354	36 52	059	54 18	132	24 08	206	52 56	242	34 31	286	23 50	313
158	27 47	353	37 24	058	54 46	132	23 52	205	52 22	241	33 54	285	23 21	312
159	27 42	352	37 57	057	55 15	132	23 35	205	51 48	241	33 16	284	22 52	311
160	27 37	351	38 29	056	55 44	132	23 19	204	51 15	241	32 39	283	22 23	310
161	27 31	350	39 00	055	56 13	131	23 04	204	50 41	240	32 01	283	21 53	309
162	27 24	349	39 32	054	56 42	131	22 48	203	50 08	240	31 24	282	21 23	308
163	27 16	348	40 03	052	57 11	131	22 33	203	49 35	239	30 46	281	20 53	307
164	27 07	347	40 33	051	57 40	131	22 18	202	49 02	239	30 08	280	20 22	306
	SPICA		♦ANTARES		RIGIL KENT.		ACHERNAR		♦CANOPUS		SIRIUS		♦REGULUS	
165	41 03	050	24 43	102	58 09	131	22 04	202	48 29	239	29 30	279	26 58	346
166	41 32	049	25 21	102	58 38	131	21 50	201	47 56	238	28 52	278	26 48	345
167	42 01	048	25 59	101	59 07	131	21 36	201	47 23	238	28 13	278	26 38	344
168	42 30	047	26 37	100	59 36	131	21 23	200	46 50	237	27 35	277	26 27	343
169	42 57	046	27 15	099	60 06	131	21 09	200	46 18	237	26 57	276	26 15	341
170	43 25	044	27 53	099	60 35	131	20 57	199	45 46	237	26 18	275	26 02	340
171	43 51	043	28 31	098	61 04	131	20 44	199	45 14	236	25 40	274	25 49	339
172	44 17	042	29 09	097	61 33	131	20 32	198	44 42	236	25 02	274	25 35	338
173	44 43	041	29 48	097	62 03	131	20 20	197	44 10	235	24 23	273	25 20	337
174	45 08	040	30 26	096	62 32	131	20 09	197	43 38	235	23 45	272	25 05	336
175	45 32	038	31 04	095	63 01	131	19 58	196	43 07	235	23 06	271	24 49	335
176	45 56	037	31 43	094	63 30	131	19 47	196	42 35	234	22 27	270	24 33	334
177	46 18	036	32 21	094	63 59	131	19 37	195	42 04	234	21 49	270	24 15	333
178	46 41	034	33 00	093	64 29	131	19 27	195	41 33	233	21 10	269	23 58	332
179	47 02	033	33 38	092	64 58	131	19 17	194	41 02	233	20 32	268	23 39	331

LAT 50°S

LHA ♈	Hc	Zn	Hc	Zn	Hc	Zn	Hc	Zn	Hc	Zn	Hc	Zn	Hc	Zn
	◆ARCTURUS		ANTARES		◆Peacock		ACHERNAR		CANOPUS		◆Suhail		REGULUS	
180	14 36	033	34 17	091	25 35	151	19 08	194	40 31	232	60 12	266	23 20	330
181	14 57	032	34 55	091	25 54	150	18 59	193	40 01	232	59 33	265	23 01	329
182	15 17	031	35 34	090	26 14	150	18 50	193	39 31	232	58 55	265	22 41	328
183	15 36	030	36 13	089	26 33	149	18 42	192	39 00	231	58 16	264	22 20	327
184	15 56	029	36 51	088	26 53	149	18 34	191	38 31	231	57 38	263	21 59	326
185	16 14	028	37 30	088	27 13	148	18 27	191	38 01	230	57 00	263	21 37	325
186	16 32	027	38 08	087	27 34	148	18 20	190	37 31	230	56 21	262	21 14	324
187	16 50	026	38 47	086	27 55	147	18 13	190	37 02	229	55 43	261	20 51	323
188	17 07	026	39 25	085	28 16	147	18 07	189	36 33	229	55 05	261	20 28	322
189	17 23	025	40 03	084	28 37	146	18 01	189	36 04	228	54 27	260	20 04	321
190	17 39	024	40 42	084	28 58	146	17 55	188	35 35	228	53 49	260	19 40	320
191	17 54	023	41 20	083	29 20	145	17 50	188	35 06	228	53 11	259	19 15	319
192	18 08	022	41 58	082	29 42	145	17 45	187	34 38	227	52 33	258	18 50	318
193	18 22	021	42 36	081	30 05	144	17 40	186	34 10	227	51 56	258	18 24	317
194	18 36	020	43 15	080	30 27	144	17 36	186	33 42	226	51 18	257	17 57	317
	ARCTURUS		◆ANTARES		Peacock		◆ACHERNAR		CANOPUS		Suhail		◆REGULUS	
195	18 48	019	43 52	079	30 50	143	17 32	185	33 14	226	50 41	257	17 31	316
196	19 01	018	44 30	078	31 13	143	17 29	185	32 47	225	50 03	256	17 03	315
197	19 12	017	45 08	078	31 36	143	17 26	184	32 19	225	49 26	255	16 36	314
198	19 23	016	45 46	077	32 00	142	17 23	184	31 52	224	48 48	255	16 08	313
199	19 33	015	46 23	076	32 24	142	17 21	183	31 26	224	48 11	254	15 39	312
200	19 43	014	47 00	075	32 48	141	17 19	182	30 59	223	47 34	254	15 11	311
201	19 52	013	47 38	074	33 12	141	17 18	182	30 33	223	46 57	253	14 41	310
202	20 00	012	48 14	073	33 37	140	17 17	181	30 07	222	46 20	253	14 12	309
203	20 08	011	48 51	072	34 01	140	17 16	181	29 41	222	45 44	252	13 42	309
204	20 15	010	49 28	071	34 26	139	17 16	180	29 15	221	45 07	251	13 11	308
205	20 21	009	50 04	070	34 51	139	17 16	180	28 50	221	44 31	251	12 41	307
206	20 27	008	50 40	069	35 17	139	17 16	179	28 25	220	43 54	250	12 10	306
207	20 32	007	51 16	068	35 42	138	17 17	179	28 00	220	43 18	250	11 38	305
208	20 36	006	51 51	067	36 08	138	17 18	178	27 35	219	42 42	249	11 07	304
209	20 39	005	52 27	066	36 34	137	17 20	177	27 11	219	42 06	249	10 35	303
	◆ARCTURUS		ANTARES		◆Peacock		ACHERNAR		CANOPUS		◆Suhail		SPICA	
210	20 42	004	53 02	064	37 00	137	17 22	177	26 47	218	41 30	248	50 28	346
211	20 45	003	53 36	063	37 27	137	17 24	176	26 23	218	40 54	248	50 18	345
212	20 46	002	54 11	062	37 53	136	17 27	176	26 00	217	40 19	247	50 08	343
213	20 47	001	54 44	061	38 20	136	17 30	175	25 36	217	39 43	246	49 56	342
214	20 47	000	55 18	060	38 47	135	17 33	175	25 13	216	39 08	246	49 44	340
215	20 47	359	55 51	058	39 14	135	17 37	174	24 51	216	38 33	245	49 31	339
216	20 46	358	56 24	057	39 42	135	17 41	173	24 28	215	37 58	245	49 16	337
217	20 44	357	56 56	056	40 09	134	17 46	173	24 06	215	37 23	244	49 01	336
218	20 42	356	57 27	054	40 37	134	17 51	172	23 45	214	36 48	244	48 45	335
219	20 39	355	57 59	053	41 05	133	17 56	172	23 23	214	36 14	243	48 28	333
220	20 35	354	58 29	052	41 33	133	18 02	171	23 02	213	35 40	243	48 10	332
221	20 30	353	58 59	050	42 01	133	18 08	171	22 41	213	35 05	242	47 51	330
222	20 25	352	59 28	049	42 30	132	18 15	170	22 20	212	34 32	241	47 32	329
223	20 19	351	59 57	047	42 58	132	18 21	170	22 00	211	33 58	241	47 12	328
224	20 13	350	60 25	046	43 27	132	18 29	169	21 40	211	33 24	240	46 51	326
	ANTARES		◆Nunki		Peacock		ACHERNAR		◆CANOPUS		◆SPICA		ARCTURUS	
225	60 52	044	39 41	085	43 56	131	18 36	168	21 20	210	46 29	325	20 06	349
226	61 19	043	40 19	084	44 25	131	18 44	168	21 01	210	46 06	324	19 58	348
227	61 45	041	40 57	083	44 55	130	18 52	167	20 42	209	45 43	322	19 49	347
228	62 10	039	41 36	082	45 24	130	19 01	167	20 23	209	45 19	321	19 40	346
229	62 33	038	42 14	081	45 54	130	19 10	166	20 05	208	44 55	320	19 30	345
230	62 56	036	42 52	080	46 23	129	19 19	166	19 47	208	44 29	319	19 20	344
231	63 19	034	43 30	080	46 53	129	19 29	165	19 29	207	44 04	317	19 09	343
232	63 40	032	44 08	079	47 23	129	19 39	165	19 12	206	43 37	316	18 57	342
233	64 00	030	44 46	078	47 53	128	19 50	164	18 55	206	43 10	315	18 45	341
234	64 18	028	45 23	077	48 24	128	20 01	163	18 38	205	42 43	314	18 32	340
235	64 36	026	46 01	076	48 54	128	20 12	163	18 22	205	42 14	313	18 18	339
236	64 53	024	46 38	075	49 24	127	20 23	162	18 06	204	41 46	311	18 04	338
237	65 08	022	47 15	074	49 55	127	20 35	162	17 50	204	41 17	310	17 50	337
238	65 22	020	47 52	073	50 26	127	20 47	161	17 35	203	40 47	309	17 34	336
239	65 35	018	48 29	072	50 57	127	21 00	161	17 20	202	40 17	308	17 18	335
	Rasalhague		◆ALTAIR		FOMALHAUT		◆ACHERNAR		ACRUX		◆SPICA		ARCTURUS	
240	24 05	025	12 50	059	13 55	120	21 13	160	58 55	225	39 46	307	17 02	334
241	24 21	024	13 23	058	14 29	119	21 26	160	58 28	225	39 15	306	16 45	333
242	24 36	023	13 55	057	15 03	118	21 39	159	58 01	225	38 44	305	16 27	332
243	24 51	022	14 28	056	15 37	118	21 53	159	57 34	225	38 12	304	16 09	331
244	25 06	021	14 59	055	16 11	117	22 07	158	57 07	225	37 39	303	15 50	330
245	25 19	020	15 31	055	16 45	116	22 22	158	56 40	225	37 07	302	15 31	330
246	25 32	019	16 02	054	17 20	116	22 37	157	56 13	225	36 34	301	15 11	329
247	25 45	018	16 33	053	17 55	115	22 52	157	55 46	224	36 00	300	14 51	328
248	25 56	017	17 04	052	18 30	114	23 07	156	55 19	224	35 27	299	14 30	327
249	26 07	016	17 34	051	19 05	114	23 23	156	54 52	224	34 53	298	14 09	326
250	26 18	015	18 04	050	19 40	113	23 39	155	54 25	224	34 18	297	13 47	325
251	26 27	014	18 33	049	20 16	112	23 56	155	53 58	224	33 44	296	13 24	324
252	26 36	013	19 02	048	20 52	112	24 12	154	53 31	224	33 09	295	13 01	323
253	26 44	012	19 31	047	21 28	111	24 29	154	53 05	224	32 34	294	12 38	322
254	26 52	011	19 59	047	22 04	110	24 47	153	52 38	224	31 58	293	12 14	321
	Rasalhague		◆ALTAIR		FOMALHAUT		◆ACHERNAR		ACRUX		◆SPICA		ANTARES	
255	26 58	010	20 27	046	22 40	110	25 04	153	52 11	223	31 23	292	65 41	343
256	27 04	008	20 54	045	23 16	109	25 22	152	51 45	223	30 47	291	65 29	341
257	27 10	007	21 21	044	23 53	108	25 41	152	51 19	223	30 11	290	65 16	339
258	27 14	006	21 47	043	24 30	108	25 59	151	50 52	223	29 35	289	65 01	337
259	27 18	005	22 13	042	25 06	107	26 18	151	50 26	223	28 58	288	64 45	335
260	27 21	004	22 39	041	25 43	106	26 37	150	50 00	222	28 21	288	64 28	333
261	27 24	003	23 04	040	26 20	106	26 57	150	49 34	222	27 45	287	64 10	331
262	27 25	002	23 28	039	26 58	105	27 16	149	49 08	222	27 08	286	63 51	329
263	27 26	001	23 52	038	27 35	104	27 36	149	48 42	222	26 30	285	63 30	327
264	27 26	000	24 16	037	28 12	104	27 56	148	48 17	222	25 53	284	63 09	325
265	27 26	359	24 38	036	28 50	103	28 17	148	47 51	221	25 16	283	62 46	323
266	27 24	357	25 01	035	29 28	102	28 38	147	47 26	221	24 38	282	62 23	322
267	27 22	356	25 23	034	30 05	101	28 59	147	47 00	221	24 00	282	61 58	320
268	27 19	355	25 44	033	30 43	101	29 20	146	46 35	221	23 22	281	61 33	318
269	27 16	354	26 04	032	31 21	100	29 42	146	46 10	220	22 44	280	61 07	317

LHA ♈	Hc	Zn	Hc	Zn	Hc	Zn	Hc	Zn	Hc	Zn	Hc	Zn	Hc	Zn
	◆ALTAIR		FOMALHAUT		◆ACHERNAR		RIGIL KENT.		◆SPICA		ANTARES		Rasalhague	
270	26 24	031	31 59	099	30 03	145	60 24	229	22 06	279	60 40	315	27 11	353
271	26 44	030	32 37	099	30 25	145	59 55	229	21 28	278	60 13	314	27 06	352
272	27 03	029	33 15	098	30 48	144	59 26	229	20 50	278	59 44	312	27 01	351
273	27 21	028	33 54	097	31 10	144	58 56	229	20 12	277	59 15	311	26 54	350
274	27 38	027	34 32	097	31 33	143	58 27	229	19 34	276	58 46	309	26 47	349
275	27 55	025	35 10	096	31 56	143	57 58	229	18 55	275	58 15	308	26 39	348
276	28 12	024	35 49	095	32 20	143	57 29	229	18 17	274	57 45	306	26 30	347
277	28 27	023	36 27	094	32 43	142	57 00	229	17 38	274	57 13	305	26 21	345
278	28 42	022	37 06	094	33 07	142	56 31	229	17 00	273	56 41	304	26 11	344
279	28 56	021	37 44	093	33 31	141	56 02	229	16 21	272	56 09	302	26 00	343
280	29 10	020	38 23	092	33 55	141	55 33	228	15 43	271	55 36	301	25 49	342
281	29 23	019	39 01	091	34 20	140	55 04	228	15 04	271	55 03	300	25 37	341
282	29 35	018	39 40	091	34 44	140	54 36	228	14 26	270	54 29	299	25 24	340
283	29 47	017	40 18	090	35 09	140	54 07	228	13 47	269	53 55	297	25 11	339
284	29 57	016	40 57	089	35 34	139	53 38	228	13 08	268	53 21	296	24 57	338
	ALTAIR		Enif		◆FOMALHAUT		ACHERNAR		◆RIGIL KENT.		ANTARES		◆Rasalhague	
285	30 07	014	20 18	044	41 35	088	36 00	139	53 10	228	52 46	295	24 42	337
286	30 17	013	20 45	043	42 14	088	36 25	138	52 41	227	52 11	294	24 26	336
287	30 25	012	21 10	042	42 52	087	36 51	138	52 13	227	51 36	293	24 10	335
288	30 33	011	21 36	041	43 31	086	37 17	138	51 45	227	51 00	292	23 54	334
289	30 40	010	22 01	040	44 09	085	37 43	137	51 17	227	50 24	291	23 36	333
290	30 46	009	22 25	039	44 48	084	38 09	137	50 49	227	49 48	290	23 19	332
291	30 52	008	22 49	038	45 26	083	38 36	136	50 21	226	49 11	289	23 00	331
292	30 56	006	23 12	037	46 04	083	39 03	136	49 53	226	48 35	288	22 41	330
293	31 00	005	23 35	036	46 43	082	39 30	136	49 25	226	47 58	287	22 21	329
294	31 04	004	23 58	035	47 21	081	39 57	135	48 57	226	47 21	286	22 01	328
295	31 06	003	24 19	034	47 59	080	40 24	135	48 30	225	46 44	285	21 40	327
296	31 08	002	24 41	033	48 37	079	40 51	134	48 03	225	46 06	284	21 19	326
297	31 09	001	25 01	032	49 15	078	41 19	134	47 35	225	45 29	283	20 57	325
298	31 09	000	25 21	031	49 52	077	41 47	134	47 08	225	44 51	282	20 35	324
299	31 08	358	25 41	030	50 30	076	42 15	133	46 41	224	44 13	281	20 12	323
	◆Enif		FOMALHAUT		◆ACHERNAR		RIGIL KENT.		◆ANTARES		Rasalhague		ALTAIR	
300	25 59	029	51 07	075	42 43	133	46 14	224	43 35	280	19 48	322	31 07	357
301	26 18	028	51 44	074	43 11	133	45 48	224	42 57	279	19 24	321	31 04	356
302	26 35	027	52 21	073	43 40	132	45 21	223	42 19	279	19 00	320	31 01	355
303	26 52	026	52 58	072	44 09	132	44 55	223	41 41	278	18 35	319	30 58	354
304	27 08	024	53 35	071	44 37	131	44 28	223	41 03	277	18 10	318	30 53	353
305	27 24	023	54 11	070	45 06	131	44 02	223	40 25	276	17 44	317	30 48	352
306	27 39	022	54 48	069	45 35	131	43 36	222	39 46	275	17 17	317	30 42	350
307	27 53	021	55 24	068	46 05	130	43 10	222	39 08	274	16 51	316	30 35	349
308	28 07	020	55 59	067	46 34	130	42 45	222	38 29	274	16 24	315	30 27	348
309	28 20	019	56 35	066	47 04	130	42 19	221	37 51	273	15 56	314	30 19	347
310	28 32	018	57 10	065	47 33	129	41 54	221	37 12	272	15 28	313	30 10	346
311	28 44	017	57 44	063	48 03	129	41 29	221	36 34	271	14 59	312	30 00	345
312	28 55	016	58 19	062	48 33	129	41 04	220	35 55	270	14 31	311	29 49	344
313	29 05	015	58 53	061	49 03	129	40 39	220	35 17	270	14 01	310	29 38	342
314	29 14	014	59 26	060	49 34	128	40 14	220	34 38	269	13 32	309	29 26	341
	◆Enif		Diphda		◆ACHERNAR		Miaplacidus		RIGIL KENT.		◆ANTARES		ALTAIR	
315	29 23	012	35 28	075	50 04	128	29 43	181	39 50	219	33 59	268	29 14	340
316	29 31	011	36 05	074	50 34	128	29 42	181	39 25	219	33 21	267	29 00	339
317	29 38	010	36 42	073	51 05	127	29 42	181	39 01	219	32 42	267	28 46	338
318	29 45	009	37 19	072	51 36	127	29 42	180	38 37	218	32 04	266	28 31	337
319	29 50	008	37 55	071	52 06	127	29 42	180	38 14	218	31 25	265	28 16	336
320	29 55	007	38 32	070	52 37	127	29 42	179	37 50	217	30 47	265	28 00	335
321	29 59	006	39 08	070	53 08	126	29 43	179	37 27	217	30 09	264	27 43	334
322	30 03	005	39 44	069	53 40	126	29 44	179	37 04	217	29 30	263	27 26	333
323	30 05	003	40 20	068	54 11	126	29 45	178	36 41	216	28 52	262	27 08	332
324	30 07	002	40 56	067	54 42	126	29 46	178	36 18	216	28 14	262	26 49	331
325	30 09	001	41 31	066	55 14	125	29 48	177	35 55	216	27 36	261	26 30	330
326	30 09	000	42 06	065	55 45	125	29 50	177	35 33	215	26 58	260	26 10	328
327	30 08	359	42 41	064	56 17	125	29 52	177	35 11	215	26 20	259	25 49	327
328	30 07	358	43 15	063	56 48	125	29 54	176	34 49	214	25 42	259	25 28	326
329	30 05	357	43 49	062	57 20	124	29 57	176	34 27	214	25 04	258	25 07	325
	◆Diphda		ACHERNAR		◆Miaplacidus		RIGIL KENT.		ANTARES		◆ALTAIR		Enif	
330	44 23	060	57 52	124	30 00	175	34 06	214	24 26	257	24 45	324	30 03	355
331	44 56	059	58 24	124	30 03	175	33 45	213	23 49	257	24 22	323	29 59	354
332	45 29	058	58 56	124	30 07	175	33 24	213	23 11	256	23 59	322	29 55	353
333	46 02	057	59 28	124	30 11	174	33 03	212	22 34	255	23 35	321	29 50	352
334	46 34	056	60 00	124	30 15	174	32 43	212	21 57	255	23 10	320	29 44	351
335	47 06	055	60 32	123	30 19	173	32 22	211	21 20	254	22 46	319	29 38	350
336	47 37	054	61 04	123	30 24	173	32 02	211	20 43	253	22 20	318	29 31	349
337	48 08	052	61 37	123	30 29	173	31 42	211	20 06	252	21 54	318	29 23	348
338	48 38	051	62 09	123	30 34	172	31 23	210	19 29	252	21 28	317	29 14	346
339	49 08	050	62 41	123	30 39	172	31 04	210	18 53	251	21 01	316	29 05	345
340	49 37	049	63 14	123	30 45	171	30 45	209	18 16	250	20 34	315	28 54	344
341	50 06	048	63 46	123	30 51	171	30 26	209	17 40	250	20 07	314	28 44	343
342	50 34	046	64 18	123	30 57	171	30 07	208	17 04	249	19 38	313	28 32	342
343	51 02	045	64 51	123	31 03	170	29 49	208	16 28	248	19 10	312	28 20	341
344	51 29	044	65 23	123	31 10	170	29 31	208	15 52	248	18 41	311	28 07	340
	◆Diphda		Acamar		◆CANOPUS		Miaplacidus		RIGIL KENT.		◆Nunki		Enif	
345	51 55	042	48 06	100	28 02	140	31 17	169	29 13	207	38 04	273	27 53	339
346	52 20	041	48 44	100	28 27	140	31 24	169	28 56	207	37 25	273	27 39	338
347	52 45	039	49 22	099	28 52	139	31 31	169	28 39	206	36 47	272	27 24	337
348	53 09	038	50 00	098	29 17	139	31 39	168	28 22	206	36 08	271	27 08	335
349	53 33	037	50 38	098	29 43	138	31 47	168	28 05	205	35 30	270	26 52	334
350	53 55	035	51 17	097	30 09	138	31 55	168	27 49	205	34 51	269	26 35	333
351	54 17	034	51 55	096	30 35	137	32 04	167	27 32	204	34 12	269	26 17	332
352	54 38	032	52 33	096	31 01	137	32 12	167	27 17	204	33 34	268	25 59	331
353	54 58	030	53 12	095	31 28	136	32 21	166	27 01	204	32 55	267	25 40	330
354	55 17	029	53 50	094	31 55	136	32 30	166	26 46	203	32 17	266	25 21	329
355	55 35	027	54 29	094	32 22	135	32 40	166	26 31	203	31 38	266	25 01	328
356	55 52	026	55 07	093	32 49	135	32 49	165	26 16	202	31 00	265	24 40	327
357	56 09	024	55 46	092	33 16	134	32 59	165	26 02	202	30 22	264	24 19	326
358	56 24	022	56 24	091	33 44	134	33 09	165	25 48	201	29 43	263	23 57	325
359	56 38	021	57 03	091	34 12	133	33 19	164	25 34	201	29 05	263	23 35	324

LAT 51°S

LHA ♈	Hc	Zn	Hc	Zn	Hc	Zn	Hc	Zn	Hc	Zn	Hc	Zn	Hc	Zn
	Diphda		◆RIGEL		CANOPUS		Miaplacidus		◆RIGIL KENT.		Peacock		◆FOMALHAUT	
0	55 54	019	13 33	086	35 21	132	34 28	164	26 17	200	58 41	238	65 39	325
1	56 06	017	14 11	086	35 49	132	34 38	163	26 04	200	58 09	238	65 17	323
2	56 16	015	14 48	085	36 17	131	34 49	163	25 51	199	57 37	238	64 54	322
3	56 26	014	15 26	084	36 46	131	35 00	163	25 39	199	57 05	237	64 30	320
4	56 34	012	16 04	083	37 14	130	35 12	162	25 27	198	56 33	237	64 06	318
5	56 41	010	16 41	082	37 43	130	35 23	162	25 15	198	56 01	237	63 40	316
6	56 47	008	17 18	082	38 12	130	35 35	162	25 03	197	55 30	237	63 13	315
7	56 52	007	17 56	081	38 41	129	35 47	161	24 52	197	54 58	236	62 46	313
8	56 56	005	18 33	080	39 11	129	35 59	161	24 41	196	54 27	236	62 18	311
9	56 59	003	19 10	079	39 40	128	36 12	161	24 31	196	53 56	236	61 49	310
10	57 00	001	19 47	078	40 10	128	36 25	160	24 21	195	53 24	236	61 20	308
11	57 01	000	20 24	077	40 40	127	36 37	160	24 11	195	52 53	235	60 50	307
12	57 00	358	21 01	077	41 10	127	36 51	160	24 01	194	52 22	235	60 20	305
13	56 58	356	21 37	076	41 40	126	37 04	159	23 52	194	51 51	235	59 49	304
14	56 55	354	22 14	075	42 11	126	37 17	159	23 43	193	51 21	234	59 17	303
	◆RIGEL		SIRIUS		CANOPUS		◆RIGIL KENT.		Peacock		◆FOMALHAUT		Diphda	
15	22 50	074	15 15	098	42 42	125	23 34	193	50 50	234	58 45	301	56 51	353
16	23 27	073	15 53	097	43 12	125	23 26	192	50 20	234	58 13	300	56 46	351
17	24 03	072	16 30	096	43 43	125	23 18	192	49 49	233	57 40	299	56 39	349
18	24 39	072	17 08	096	44 15	124	23 11	191	49 19	233	57 06	298	56 32	348
19	25 14	071	17 45	095	44 46	124	23 03	191	48 49	233	56 33	296	56 23	346
20	25 50	070	18 23	094	45 17	123	22 56	190	48 19	232	55 59	295	56 13	344
21	26 25	069	19 00	093	45 49	123	22 50	190	47 49	232	55 24	294	56 02	343
22	27 00	068	19 38	093	46 21	122	22 44	189	47 20	232	54 50	293	55 50	341
23	27 35	067	20 16	092	46 53	122	22 38	189	46 50	231	54 15	292	55 38	339
24	28 10	066	20 54	091	47 25	122	22 32	188	46 21	231	53 40	291	55 24	338
25	28 44	065	21 31	090	47 57	121	22 27	188	45 51	231	53 04	290	55 09	336
26	29 18	064	22 09	089	48 29	121	22 22	187	45 22	230	52 28	289	54 53	334
27	29 52	063	22 47	089	49 02	120	22 17	187	44 53	230	51 53	287	54 36	333
28	30 26	062	23 25	088	49 35	120	22 13	186	44 25	229	51 16	286	54 18	331
29	30 59	062	24 02	087	50 07	120	22 09	186	43 56	229	50 40	285	54 00	330
	◆RIGEL		SIRIUS		CANOPUS		◆RIGIL KENT.		Peacock		◆FOMALHAUT		Diphda	
30	31 32	061	24 40	086	50 40	119	22 06	185	43 28	229	50 04	285	53 40	328
31	32 05	060	25 18	086	51 13	119	22 02	185	42 59	228	49 27	284	53 20	327
32	32 37	059	25 55	085	51 46	118	21 59	184	42 31	228	48 50	283	52 59	325
33	33 09	058	26 33	084	52 20	118	21 57	184	42 03	227	48 13	282	52 37	324
34	33 41	057	27 10	083	52 53	118	21 55	183	41 36	227	47 36	281	52 14	322
35	34 12	056	27 48	082	53 27	117	21 53	183	41 08	227	46 59	280	51 51	321
36	34 43	055	28 25	081	54 00	117	21 51	182	40 41	226	46 22	279	51 27	320
37	35 14	054	29 03	081	54 34	116	21 50	181	40 14	226	45 45	278	51 02	318
38	35 44	052	29 40	080	55 08	116	21 49	181	39 46	225	45 07	277	50 36	317
39	36 14	051	30 17	079	55 42	116	21 49	180	39 20	225	44 30	276	50 10	316
40	36 43	050	30 54	078	56 16	115	21 49	180	38 53	225	43 52	276	49 43	314
41	37 12	049	31 31	077	56 50	115	21 49	179	38 27	224	43 14	275	49 16	313
42	37 40	048	32 08	076	57 25	115	21 50	179	38 00	224	42 37	274	48 48	312
43	38 08	047	32 44	075	57 59	114	21 51	178	37 34	223	41 59	273	48 20	310
44	38 36	046	33 21	075	58 34	114	21 52	178	37 08	223	41 21	272	47 51	309
	RIGEL		◆SIRIUS		CANOPUS		◆RIGIL KENT.		Peacock		FOMALHAUT		◆Diphda	
45	39 03	045	33 57	074	59 08	113	21 54	177	36 43	223	40 44	271	47 21	308
46	39 29	044	34 33	073	59 43	113	21 56	177	36 17	222	40 06	271	46 51	307
47	39 55	043	35 09	072	60 18	113	21 58	176	35 52	222	39 28	270	46 21	306
48	40 20	041	35 45	071	60 52	112	22 00	176	35 27	221	38 50	269	45 50	304
49	40 45	040	36 21	070	61 27	112	22 03	175	35 02	221	38 13	268	45 18	303
50	41 09	039	36 56	069	62 03	112	22 07	175	34 38	220	37 35	268	44 47	302
51	41 32	038	37 31	068	62 38	111	22 11	174	34 14	220	36 57	267	44 14	301
52	41 55	037	38 06	067	63 13	111	22 15	174	33 49	219	36 19	266	43 42	300
53	42 17	035	38 41	066	63 48	111	22 19	173	33 26	219	35 42	265	43 09	299
54	42 38	034	39 15	065	64 24	110	22 24	173	33 02	219	35 04	265	42 36	298
55	42 59	033	39 49	064	64 59	110	22 29	172	32 38	218	34 27	264	42 02	297
56	43 19	031	40 23	063	65 34	110	22 34	172	32 15	218	33 49	263	41 28	296
57	43 39	030	40 57	062	66 10	109	22 40	171	31 52	217	33 12	262	40 54	295
58	43 57	029	41 30	061	66 46	109	22 46	170	31 30	217	32 34	262	40 20	294
59	44 15	028	42 03	060	67 21	109	22 52	170	31 07	216	31 57	261	39 45	293
	RIGEL		◆SIRIUS		Suhail		◆RIGIL KENT.		Peacock		FOMALHAUT		◆Diphda	
60	44 32	026	42 35	059	39 35	113	22 59	169	30 45	216	31 20	260	39 10	292
61	44 49	025	43 08	058	40 10	113	23 06	169	30 23	215	30 43	260	38 35	291
62	45 04	024	43 39	057	40 45	112	23 13	168	30 01	215	30 05	259	38 00	290
63	45 19	022	44 11	056	41 20	112	23 21	168	29 40	214	29 28	258	37 24	289
64	45 33	021	44 42	055	41 55	111	23 29	167	29 19	214	28 51	257	36 48	288
65	45 46	019	45 12	053	42 31	110	23 38	167	28 58	213	28 15	257	36 12	287
66	45 58	018	45 42	052	43 06	110	23 46	166	28 37	213	27 38	256	35 36	286
67	46 09	017	46 12	051	43 42	109	23 55	166	28 17	212	27 01	255	35 00	285
68	46 19	015	46 41	050	44 17	109	24 05	165	27 57	212	26 25	255	34 23	284
69	46 29	014	47 10	049	44 53	108	24 14	165	27 37	211	25 49	254	33 47	284
70	46 37	012	47 38	047	45 29	108	24 24	164	27 18	211	25 12	253	33 10	283
71	46 45	011	48 05	046	46 05	107	24 35	164	26 58	210	24 36	253	32 33	282
72	46 52	010	48 32	045	46 41	106	24 45	163	26 39	210	24 00	252	31 56	281
73	46 58	008	48 59	044	47 18	106	24 56	163	26 21	209	23 24	251	31 19	280
74	47 02	007	49 25	042	47 54	105	25 08	162	26 02	209	22 49	251	30 42	279
	RIGEL		◆SIRIUS		Suhail		◆RIGIL KENT.		Peacock		ACHERNAR		◆Diphda	
75	47 06	005	49 50	041	48 30	105	25 19	162	25 44	208	60 25	238	30 04	278
76	47 09	004	50 14	040	49 07	104	25 31	161	25 26	208	59 53	238	29 27	278
77	47 11	002	50 38	038	49 44	103	25 43	161	25 09	207	59 21	237	28 50	277
78	47 12	001	51 01	037	50 20	103	25 56	160	24 52	207	58 49	237	28 12	276
79	47 12	359	51 23	035	50 57	102	26 09	160	24 35	206	58 18	237	27 35	275
80	47 11	358	51 45	034	51 34	102	26 22	159	24 18	206	57 46	237	26 57	274
81	47 10	356	52 06	033	52 11	101	26 35	159	24 02	205	57 14	237	26 19	274
82	47 07	355	52 25	031	52 48	100	26 49	158	23 46	205	56 43	236	25 42	273
83	47 03	354	52 45	030	53 26	100	27 03	158	23 30	204	56 12	236	25 04	272
84	46 58	352	53 03	028	54 03	099	27 17	158	23 15	204	55 40	236	24 26	271
85	46 53	351	53 20	027	54 40	098	27 32	157	23 00	203	55 09	236	23 48	270
86	46 46	349	53 37	025	55 18	098	27 47	157	22 45	203	54 38	235	23 11	270
87	46 39	348	53 52	024	55 55	097	28 02	156	22 31	202	54 07	235	22 33	269
88	46 30	346	54 07	022	56 33	096	28 17	156	22 17	202	53 36	235	21 55	268
89	46 21	345	54 20	020	57 10	096	28 33	155	22 03	201	53 05	234	21 17	267

LHA ♈	Hc	Zn	Hc	Zn	Hc	Zn	Hc	Zn	Hc	Zn	Hc	Zn	Hc	Zn
	PROCYON		◆Suhail		RIGIL KENT.		◆Peacock		ACHERNAR		◆RIGEL		BETELGEUSE	
90	29 53	029	57 48	095	28 49	155	21 50	200	52 35	234	46 11	344	31 35	359
91	30 10	028	58 25	094	29 05	154	21 37	200	52 04	234	45 59	342	31 34	357
92	30 28	027	59 03	094	29 22	154	21 24	199	51 34	234	45 47	341	31 32	356
93	30 44	025	59 41	093	29 39	153	21 12	199	51 03	233	45 35	339	31 29	355
94	31 00	024	60 18	092	29 56	153	21 00	198	50 33	233	45 21	338	31 25	354
95	31 15	023	60 56	091	30 13	152	20 48	198	50 03	233	45 06	337	31 21	353
96	31 30	022	61 34	091	30 31	152	20 37	197	49 33	232	44 51	335	31 15	352
97	31 44	021	62 12	090	30 49	151	20 26	197	49 03	232	44 35	334	31 09	350
98	31 57	020	62 49	089	31 07	151	20 15	196	48 34	232	44 18	333	31 03	349
99	32 09	019	63 27	088	31 25	151	20 05	196	48 04	231	44 00	331	30 55	348
100	32 21	017	64 05	088	31 44	150	19 55	195	47 35	231	43 41	330	30 47	347
101	32 32	016	64 43	087	32 03	150	19 45	194	47 06	231	43 22	329	30 38	346
102	32 42	015	65 20	086	32 22	149	19 36	194	46 37	230	43 02	327	30 29	345
103	32 52	014	65 58	085	32 41	149	19 27	193	46 08	230	42 42	326	30 18	344
104	33 00	013	66 35	084	33 01	148	19 19	193	45 39	230	42 20	325	30 07	342
	PROCYON		REGULUS		◆Gienah		RIGIL KENT.		◆ACHERNAR		RIGEL		◆SIRIUS	
105	33 08	012	14 58	048	20 28	093	33 21	148	45 10	229	41 58	324	55 35	354
106	33 16	010	15 25	047	21 05	092	33 41	148	44 42	229	41 35	322	55 30	352
107	33 22	009	15 53	046	21 43	091	34 01	147	44 13	228	41 12	321	55 24	350
108	33 28	008	16 20	045	22 21	091	34 22	147	43 45	228	40 48	320	55 17	349
109	33 33	007	16 46	044	22 59	090	34 43	146	43 17	228	40 24	319	55 09	347
110	33 37	006	17 12	043	23 36	089	35 04	146	42 49	227	39 58	318	55 00	345
111	33 40	005	17 38	042	24 14	088	35 25	146	42 22	227	39 33	317	54 50	344
112	33 43	003	18 03	041	24 52	087	35 47	145	41 54	227	39 07	315	54 39	342
113	33 44	002	18 28	040	25 30	087	36 08	145	41 27	226	38 40	314	54 26	340
114	33 45	001	18 52	040	26 07	086	36 30	144	41 00	226	38 12	313	54 13	339
115	33 46	000	19 16	039	26 45	085	36 52	144	40 33	225	37 45	312	53 59	337
116	33 45	359	19 39	038	27 22	084	37 15	143	40 06	225	37 16	311	53 44	336
117	33 44	357	20 02	037	28 00	083	37 37	143	39 39	225	36 48	310	53 28	334
118	33 42	356	20 25	036	28 37	083	38 00	143	39 13	224	36 18	309	53 11	333
119	33 39	355	20 46	035	29 15	082	38 23	142	38 47	224	35 49	308	52 53	331
	REGULUS		◆SPICA		RIGIL KENT.		Peacock		◆ACHERNAR		RIGEL		◆SIRIUS	
120	21 08	034	14 08	090	38 46	142	17 53	184	38 21	223	35 19	307	52 34	330
121	21 28	033	14 46	090	39 10	142	17 50	183	37 55	223	34 48	306	52 15	328
122	21 49	032	15 24	089	39 33	141	17 49	182	37 29	222	34 17	305	51 54	327
123	22 08	031	16 01	088	39 57	141	17 47	182	37 04	222	33 46	304	51 33	325
124	22 27	030	16 39	087	40 21	140	17 46	181	36 39	222	33 14	303	51 11	324
125	22 46	029	17 17	086	40 45	140	17 45	181	36 14	221	32 42	302	50 49	322
126	23 04	028	17 54	086	41 09	140	17 45	180	35 49	221	32 10	301	50 25	321
127	23 21	027	18 32	085	41 34	139	17 45	180	35 24	220	31 37	300	50 01	320
128	23 38	026	19 10	084	41 58	139	17 46	179	35 00	220	31 04	299	49 36	318
129	23 54	025	19 47	083	42 23	139	17 46	178	34 36	219	30 31	298	49 11	317
130	24 09	024	20 25	082	42 48	138	17 48	178	34 12	219	29 57	297	48 45	316
131	24 24	023	21 02	082	43 13	138	17 49	177	33 48	219	29 23	296	48 18	314
132	24 38	022	21 39	081	43 39	138	17 51	177	33 25	218	28 49	295	47 51	313
133	24 52	021	22 17	080	44 04	137	17 54	176	33 02	218	28 15	294	47 23	312
134	25 05	020	22 54	079	44 30	137	17 56	176	32 39	217	27 40	293	46 55	311
	◆REGULUS		SPICA		◆RIGIL KENT.		Peacock		ACHERNAR		◆CANOPUS		SIRIUS	
135	25 17	018	23 31	078	44 56	137	17 59	175	32 16	217	66 07	251	46 26	309
136	25 29	017	24 08	077	45 22	136	18 03	174	31 53	216	65 32	250	45 56	308
137	25 40	016	24 44	077	45 48	136	18 07	174	31 31	216	64 56	250	45 26	307
138	25 50	015	25 21	076	46 14	136	18 11	173	31 09	215	64 21	250	44 56	306
139	26 00	014	25 58	075	46 41	135	18 16	173	30 48	215	63 45	249	44 25	305
140	26 09	013	26 34	074	47 07	135	18 21	172	30 26	214	63 10	249	43 54	304
141	26 17	012	27 10	073	47 34	135	18 26	172	30 05	214	62 35	249	43 22	303
142	26 24	011	27 46	072	48 01	135	18 32	171	29 44	213	62 00	248	42 50	301
143	26 31	010	28 22	071	48 28	134	18 38	170	29 23	213	61 25	248	42 18	300
144	26 37	009	28 58	071	48 55	134	18 44	170	29 03	213	60 50	248	41 45	299
145	26 43	008	29 33	070	49 22	134	18 51	169	28 43	212	60 15	247	41 12	298
146	26 47	007	30 09	069	49 49	133	18 58	169	28 23	212	59 40	247	40 39	297
147	26 51	006	30 44	068	50 17	133	19 06	168	28 03	211	59 05	247	40 05	296
148	26 55	004	31 19	067	50 44	133	19 14	168	27 44	211	58 31	246	39 31	295
149	26 57	003	31 53	066	51 12	133	19 22	167	27 25	210	57 56	246	38 57	294
	◆REGULUS		SPICA		◆RIGIL KENT.		Peacock		ACHERNAR		◆CANOPUS		SIRIUS	
150	26 59	002	32 28	065	51 40	132	19 31	166	27 06	210	57 22	245	38 22	293
151	27 00	001	33 02	064	52 08	132	19 40	166	26 47	209	56 48	245	37 47	292
152	27 00	000	33 36	063	52 36	132	19 49	165	26 29	209	56 13	245	37 12	291
153	27 00	359	34 09	062	53 04	132	19 59	165	26 11	208	55 39	244	36 37	290
154	26 59	358	34 42	061	53 32	132	20 09	164	25 54	208	55 05	244	36 01	289
155	26 57	357	35 15	060	54 00	131	20 19	164	25 36	207	54 31	244	35 26	289
156	26 55	356	35 48	059	54 29	131	20 30	163	25 19	207	53 58	243	34 50	288
157	26 52	355	36 20	058	54 57	131	20 41	163	25 02	206	53 24	243	34 14	287
158	26 48	353	36 52	057	55 26	131	20 52	162	24 46	206	52 51	242	33 38	286
159	26 43	352	37 24	056	55 54	131	21 04	161	24 30	205	52 17	242	33 01	285
160	26 38	351	37 55	055	56 23	130	21 16	161	24 14	205	51 44	242	32 25	284
161	26 31	350	38 25	054	56 52	130	21 29	160	23 59	204	51 11	241	31 48	283
162	26 25	349	38 56	053	57 21	130	21 42	160	23 43	203	50 38	241	31 11	282
163	26 17	348	39 26	052	57 50	130	21 55	159	23 28	203	50 05	240	30 34	281
164	26 09	347	39 55	051	58 19	130	22 08	159	23 14	202	49 32	240	29 57	281
	◆SPICA		ANTARES		◆Peacock		ACHERNAR		CANOPUS		◆SIRIUS		REGULUS	
165	40 24	050	24 56	102	22 22	158	23 00	202	48 59	240	29 20	280	26 00	346
166	40 53	048	25 33	101	22 36	158	22 46	201	48 27	239	28 43	279	25 51	345
167	41 21	047	26 10	100	22 51	157	22 32	201	47 54	239	28 05	278	25 40	344
168	41 48	046	26 47	100	23 06	157	22 19	200	47 22	238	27 28	277	25 29	343
169	42 15	045	27 25	099	23 21	156	22 06	200	46 50	238	26 50	276	25 18	342
170	42 42	044	28 02	098	23 36	156	21 53	199	46 18	238	26 13	276	25 06	341
171	43 07	043	28 39	098	23 52	155	21 41	199	45 46	237	25 35	275	24 53	339
172	43 33	041	29 17	097	24 08	155	21 29	198	45 15	237	24 58	274	24 39	338
173	43 57	040	29 54	096	24 24	154	21 18	198	44 43	236	24 20	273	24 25	337
174	44 21	039	30 32	095	24 41	153	21 06	197	44 12	236	23 42	272	24 10	336
175	44 45	038	31 09	095	24 58	153	20 55	197	43 41	235	23 04	272	23 55	335
176	45 07	036	31 47	094	25 15	152	20 45	196	43 10	235	22 27	271	23 38	334
177	45 30	035	32 25	093	25 33	152	20 35	195	42 39	234	21 49	270	23 22	333
178	45 51	034	33 02	092	25 51	151	20 25	195	42 08	234	21 11	269	23 05	332
179	46 11	032	33 40	092	26 09	151	20 15	194	41 38	234	20 33	269	22 47	331

LAT 51°S

LHA ♈	Hc	Zn	Hc	Zn	Hc	Zn	Hc	Zn	Hc	Zn	Hc	Zn	Hc	Zn
	◆SPICA		ANTARES		◆Peacock		ACHERNAR		CANOPUS		◆Suhail		REGULUS	
180	46 31	031	34 18	091	26 28	150	20 06	194	41 08	233	60 15	268	22 28	330
181	46 50	030	34 56	090	26 46	150	19 57	193	40 38	233	59 37	267	22 09	329
182	47 09	028	35 34	089	27 06	149	19 49	193	40 08	232	58 59	266	21 50	328
183	47 26	027	36 11	088	27 25	149	19 41	192	39 38	232	58 22	266	21 29	327
184	47 43	026	36 49	088	27 45	148	19 33	192	39 08	231	57 44	265	21 09	326
185	47 59	024	37 27	087	28 04	148	19 26	191	38 39	231	57 07	264	20 47	325
186	48 14	023	38 04	086	28 25	147	19 19	190	38 10	230	56 29	264	20 26	324
187	48 28	021	38 42	085	28 45	147	19 12	190	37 41	230	55 52	263	20 03	323
188	48 41	020	39 20	084	29 06	146	19 06	189	37 12	229	55 14	262	19 41	322
189	48 54	018	39 57	084	29 27	146	19 00	189	36 43	229	54 37	262	19 17	321
190	49 05	017	40 35	083	29 48	146	18 54	188	36 15	229	53 59	261	18 54	320
191	49 16	015	41 12	082	30 10	145	18 49	188	35 47	228	53 22	260	18 29	320
192	49 26	014	41 49	081	30 31	145	18 44	187	35 19	228	52 45	260	18 05	319
193	49 34	013	42 27	080	30 53	144	18 40	186	34 51	227	52 08	259	17 39	318
194	49 42	011	43 04	079	31 16	144	18 36	186	34 23	227	51 31	258	17 14	317
	◆ARCTURUS		ANTARES		◆Peacock		ACHERNAR		CANOPUS		◆Suhail		Denebola	
195	17 52	019	43 41	078	31 38	143	18 32	185	33 56	226	50 54	258	22 35	341
196	18 03	018	44 18	077	32 01	143	18 29	185	33 29	226	50 17	257	22 22	340
197	18 15	017	44 55	077	32 24	142	18 26	184	33 02	225	49 40	257	22 09	339
198	18 25	016	45 31	076	32 47	142	18 23	184	32 35	225	49 04	256	21 56	338
199	18 35	015	46 08	075	33 11	141	18 21	183	32 09	224	48 27	255	21 41	337
200	18 45	014	46 44	074	33 34	141	18 19	182	31 43	224	47 51	255	21 26	336
201	18 53	013	47 20	073	33 58	140	18 18	182	31 17	223	47 14	254	21 11	335
202	19 01	012	47 56	072	34 23	140	18 17	181	30 51	223	46 38	254	20 55	334
203	19 09	011	48 32	071	34 47	139	18 16	181	30 25	222	46 02	253	20 38	333
204	19 15	010	49 07	070	35 12	139	18 16	180	30 00	222	45 26	252	20 21	332
205	19 22	009	49 43	069	35 36	139	18 16	180	29 35	221	44 50	252	20 03	331
206	19 27	008	50 18	068	36 02	138	18 16	179	29 10	221	44 14	251	19 45	330
207	19 32	007	50 53	067	36 27	138	18 17	179	28 46	220	43 38	251	19 26	329
208	19 36	006	51 27	065	36 52	137	18 18	178	28 22	220	43 03	250	19 06	328
209	19 40	005	52 01	064	37 18	137	18 20	177	27 58	219	42 27	249	18 46	327
	◆ARCTURUS		ANTARES		◆Peacock		ACHERNAR		CANOPUS		◆Suhail		SPICA	
210	19 43	004	52 35	063	37 44	136	18 22	177	27 34	219	41 52	249	49 30	347
211	19 45	003	53 09	062	38 10	136	18 24	176	27 10	218	41 17	248	49 21	345
212	19 46	002	53 42	061	38 36	136	18 27	176	26 47	218	40 42	248	49 10	344
213	19 47	001	54 15	060	39 03	135	18 30	175	26 24	217	40 07	247	48 59	342
214	19 47	000	54 47	058	39 30	135	18 33	175	26 02	217	39 32	247	48 47	341
215	19 47	359	55 19	057	39 57	134	18 37	174	25 39	216	38 58	246	48 34	339
216	19 46	358	55 51	056	40 24	134	18 41	173	25 17	215	38 23	245	48 21	338
217	19 44	357	56 22	055	40 51	134	18 46	173	24 56	215	37 49	245	48 06	336
218	19 42	356	56 52	053	41 18	133	18 50	172	24 34	214	37 15	244	47 51	335
219	19 39	355	57 22	052	41 46	133	18 56	172	24 13	214	36 41	244	47 34	334
220	19 35	354	57 51	050	42 14	132	19 01	171	23 52	213	36 07	243	47 17	332
221	19 31	353	58 20	049	42 42	132	19 07	171	23 31	213	35 33	243	46 59	331
222	19 26	352	58 48	048	43 10	132	19 14	170	23 11	212	35 00	242	46 40	330
223	19 20	351	59 16	046	43 38	131	19 20	169	22 51	212	34 27	241	46 21	328
224	19 14	350	59 43	045	44 07	131	19 27	169	22 32	211	33 54	241	46 01	327
	ANTARES		◆Nunki		Peacock		ACHERNAR		◆CANOPUS		Suhail		◆SPICA	
225	60 09	043	39 35	084	44 35	130	19 35	168	22 12	211	33 21	240	45 40	326
226	60 34	041	40 12	083	45 04	130	19 43	168	21 53	210	32 48	240	45 18	324
227	60 59	040	40 50	082	45 33	130	19 51	167	21 34	209	32 15	239	44 55	323
228	61 23	038	41 27	081	46 02	129	19 59	167	21 16	209	31 43	239	44 32	322
229	61 46	036	42 04	080	46 32	129	20 08	166	20 58	208	31 11	238	44 09	320
230	62 07	035	42 42	080	47 01	129	20 18	166	20 40	208	30 39	237	43 44	319
231	62 28	033	43 19	079	47 31	128	20 27	165	20 23	207	30 07	237	43 19	318
232	62 48	031	43 56	078	48 00	128	20 37	164	20 06	207	29 36	236	42 54	317
233	63 07	029	44 32	077	48 30	128	20 47	164	19 49	206	29 04	236	42 28	316
234	63 25	027	45 09	076	49 00	127	20 58	163	19 33	205	28 33	235	42 01	314
235	63 42	025	45 46	075	49 30	127	21 09	163	19 16	205	28 02	235	41 34	313
236	63 58	024	46 22	074	50 01	127	21 20	162	19 01	204	27 32	234	41 06	312
237	64 12	022	46 58	073	50 31	126	21 32	162	18 45	204	27 01	233	40 37	311
238	64 26	020	47 34	072	51 02	126	21 44	161	18 30	203	26 31	233	40 09	310
239	64 38	018	48 10	071	51 32	126	21 56	161	18 16	203	26 01	232	39 39	309
	◆ANTARES		Nunki		◆FOMALHAUT		ACHERNAR		CANOPUS		◆ACRUX		SPICA	
240	64 48	015	48 46	070	14 25	120	22 09	160	18 01	202	59 37	226	39 10	308
241	64 58	013	49 21	069	14 58	119	22 22	160	17 47	201	59 10	226	38 40	307
242	65 06	011	49 56	068	15 31	118	22 35	159	17 34	201	58 43	226	38 09	305
243	65 12	009	50 31	067	16 04	118	22 49	159	17 21	200	58 16	226	37 38	304
244	65 18	007	51 06	066	16 38	117	23 03	158	17 08	200	57 49	226	37 07	303
245	65 22	005	51 40	065	17 12	116	23 17	157	16 55	199	57 22	226	36 35	302
246	65 24	003	52 14	064	17 46	115	23 32	157	16 43	198	56 55	226	36 03	301
247	65 25	001	52 48	062	18 20	115	23 47	156	16 31	198	56 28	225	35 30	300
248	65 25	358	53 21	061	18 54	114	24 02	156	16 20	197	56 01	225	34 58	299
249	65 23	356	53 54	060	19 29	113	24 18	155	16 09	197	55 34	225	34 25	298
250	65 20	354	54 26	059	20 04	113	24 34	155	15 58	196	55 08	225	33 51	297
251	65 15	352	54 59	058	20 39	112	24 50	154	15 48	195	54 41	225	33 18	296
252	65 10	350	55 30	056	21 14	111	25 06	154	15 38	195	54 14	225	32 44	295
253	65 01	348	56 01	055	21 49	111	25 23	153	15 29	194	53 48	225	32 09	294
254	64 54	346	56 32	054	22 24	110	25 40	153	15 20	194	53 21	224	31 35	294
	ALTAIR		◆FOMALHAUT		ACHERNAR		CANOPUS		◆ACRUX		SPICA		◆ANTARES	
255	19 45	045	23 00	109	25 58	152	15 11	193	52 55	224	31 00	293	64 44	344
256	20 11	044	23 36	109	26 15	152	15 03	192	52 28	224	30 25	292	64 32	342
257	20 38	043	24 11	108	26 33	151	14 55	192	52 02	224	29 50	291	64 20	340
258	21 03	042	24 47	107	26 52	151	14 47	191	51 36	224	29 14	290	64 06	338
259	21 29	042	25 24	106	27 10	150	14 40	191	51 10	224	28 39	289	63 51	336
260	21 53	041	26 00	106	27 29	150	14 34	190	50 44	223	28 03	288	63 35	334
261	22 18	040	26 36	105	27 48	149	14 27	189	50 18	223	27 27	287	63 17	332
262	22 41	039	27 13	104	28 08	149	14 21	189	49 52	223	26 51	286	62 59	330
263	23 05	038	27 49	104	28 27	148	14 16	188	49 27	223	26 15	285	62 40	328
264	23 28	037	28 26	103	28 47	148	14 11	187	49 01	222	25 38	285	62 19	326
265	23 50	036	29 03	102	29 08	147	14 06	187	48 36	222	25 02	284	61 58	325
266	24 12	035	29 40	102	29 28	147	14 02	186	48 11	222	24 25	283	61 35	323
267	24 33	034	30 17	101	29 49	146	13 58	186	47 45	222	23 48	282	61 12	321
268	24 53	033	30 54	100	30 10	146	13 55	185	47 20	221	23 11	281	60 48	319
269	25 13	032	31 31	099	30 31	145	13 52	184	46 56	221	22 34	280	60 23	318

LHA ♈	Hc	Zn	Hc	Zn	Hc	Zn	Hc	Zn	Hc	Zn	Hc	Zn	Hc	Zn
	◆ALTAIR		FOMALHAUT		◆ACHERNAR		CANOPUS		RIGIL KENT.		◆SPICA		ANTARES	
270	25 33	031	32 09	099	30 53	145	13 49	184	61 03	231	21 57	280	59 57	316
271	25 52	029	32 46	098	31 14	145	13 47	183	60 33	231	21 19	279	59 31	315
272	26 10	028	33 23	097	31 36	144	13 45	182	60 04	231	20 42	278	59 04	313
273	26 28	027	34 01	097	31 59	144	13 43	182	59 35	230	20 05	277	58 36	312
274	26 45	026	34 38	096	32 21	143	13 42	181	59 06	230	19 27	276	58 07	310
275	27 01	025	35 16	095	32 44	143	13 42	181	58 37	230	18 50	276	57 38	309
276	27 17	024	35 54	094	33 07	142	13 42	180	58 08	230	18 12	275	57 09	308
277	27 32	023	36 31	094	33 30	142	13 42	179	57 39	230	17 34	274	56 38	306
278	27 47	022	37 09	093	33 54	141	13 42	179	57 10	230	16 57	273	56 08	305
279	28 00	021	37 47	092	34 18	141	13 43	178	56 41	230	16 19	272	55 36	304
280	28 14	020	38 24	091	34 42	140	13 45	177	56 12	230	15 41	272	55 05	302
281	28 26	019	39 02	091	35 06	140	13 47	177	55 44	229	15 03	271	54 33	301
282	28 38	018	39 40	090	35 30	140	13 49	176	55 15	229	14 26	270	54 00	300
283	28 49	017	40 18	089	35 55	139	13 52	176	54 47	229	13 48	269	53 27	299
284	28 59	015	40 55	088	36 20	139	13 55	175	54 18	229	13 10	268	52 54	297
	◆ALTAIR		FOMALHAUT		◆ACHERNAR		CANOPUS		◆RIGIL KENT.		ANTARES		Rasalhague	
285	29 09	014	41 33	087	36 45	138	13 58	174	53 50	229	52 20	296	23 47	337
286	29 18	013	42 11	087	37 10	138	14 02	174	53 22	228	51 46	295	23 32	336
287	29 26	012	42 49	086	37 35	137	14 07	173	52 53	228	51 12	294	23 16	335
288	29 34	011	43 26	085	38 01	137	14 11	173	52 25	228	50 37	293	23 00	334
289	29 41	010	44 04	084	38 27	137	14 16	172	51 57	228	50 02	292	22 43	333
290	29 47	009	44 41	083	38 53	136	14 22	171	51 29	227	49 27	291	22 26	332
291	29 52	008	45 19	082	39 19	136	14 28	171	51 02	227	48 52	290	22 08	331
292	29 57	006	45 56	082	39 46	135	14 34	170	50 34	227	48 16	289	21 49	330
293	30 01	005	46 34	081	40 12	135	14 41	169	50 06	227	47 40	288	21 30	329
294	30 04	004	47 11	080	40 39	135	14 48	169	49 39	226	47 04	287	21 10	328
295	30 06	003	47 48	079	41 06	134	14 56	168	49 12	226	46 28	286	20 50	327
296	30 08	002	48 25	078	41 33	134	15 04	168	48 45	226	45 51	285	20 29	326
297	30 09	001	49 02	077	42 01	133	15 12	167	48 18	226	45 15	284	20 08	325
298	30 09	000	49 38	076	42 28	133	15 21	166	47 51	225	44 38	283	19 46	324
299	30 08	358	50 15	075	42 56	133	15 30	166	47 24	225	44 01	282	19 24	323
	FOMALHAUT		◆ACHERNAR		CANOPUS		◆RIGIL KENT.		ANTARES		Nunki		◆ALTAIR	
300	50 51	074	43 24	132	15 39	165	46 57	225	43 24	281	62 22	327	30 07	357
301	51 28	073	43 52	132	15 49	165	46 31	224	42 47	280	62 01	325	30 04	356
302	52 04	072	44 20	131	15 59	164	46 04	224	42 10	279	61 39	324	30 02	355
303	52 40	071	44 48	131	16 10	163	45 38	224	41 33	279	61 17	322	29 58	354
304	53 15	070	45 17	131	16 21	163	45 12	224	40 55	278	60 53	320	29 53	353
305	53 51	069	45 46	130	16 32	162	44 46	223	40 18	277	60 28	319	29 48	352
306	54 26	068	46 14	130	16 44	162	44 20	223	39 40	276	60 03	317	29 42	350
307	55 01	067	46 43	130	16 56	161	43 55	223	39 03	275	59 37	316	29 36	349
308	55 35	066	47 12	129	17 09	160	43 29	222	38 25	274	59 10	314	29 29	348
309	56 09	065	47 42	129	17 22	160	43 04	222	37 47	274	58 43	313	29 20	347
310	56 43	063	48 11	129	17 35	159	42 39	222	37 10	273	58 15	311	29 12	346
311	57 17	062	48 41	128	17 49	159	42 14	221	36 32	272	57 46	310	29 02	345
312	57 50	061	49 10	128	18 03	158	41 49	221	35 54	271	57 17	308	28 52	344
313	58 23	060	49 40	128	18 17	157	41 25	220	35 17	270	56 47	307	28 41	343
314	58 55	058	50 10	127	18 32	157	41 00	220	34 39	270	56 16	306	28 29	342
	◆FOMALHAUT		ACHERNAR		◆CANOPUS		RIGIL KENT.		◆ANTARES		Nunki		ALTAIR	
315	59 27	057	50 40	127	18 47	156	40 36	220	34 01	269	55 45	304	28 17	340
316	59 58	056	51 11	127	19 02	156	40 12	219	33 23	268	55 14	303	28 04	339
317	60 29	054	51 41	126	19 18	155	39 48	219	32 46	267	54 42	302	27 50	338
318	61 00	053	52 11	126	19 34	154	39 24	219	32 08	267	54 10	300	27 36	337
319	61 29	051	52 42	126	19 50	154	39 01	218	31 30	266	53 37	299	27 21	336
320	61 58	050	53 13	125	20 07	153	38 37	218	30 52	265	53 04	298	27 06	335
321	62 27	048	53 43	125	20 24	153	38 14	218	30 15	264	52 30	297	26 49	334
322	62 55	047	54 14	125	20 42	152	37 51	217	29 37	264	51 56	296	26 32	333
323	63 22	045	54 45	125	20 59	152	37 29	217	29 00	263	51 22	295	26 15	332
324	63 48	043	55 17	124	21 17	151	37 06	216	28 22	262	50 48	294	25 57	331
325	64 13	041	55 48	124	21 36	151	36 44	216	27 45	261	50 13	292	25 38	330
326	64 38	040	56 19	124	21 55	150	36 22	216	27 08	261	49 38	291	25 19	329
327	65 02	038	56 50	124	22 14	149	36 00	215	26 31	260	49 03	290	24 59	328
328	65 24	036	57 22	123	22 33	149	35 38	215	25 53	259	48 27	289	24 38	327
329	65 46	034	57 53	123	22 53	148	35 17	214	25 16	259	47 51	288	24 17	326
	◆FOMALHAUT		ACHERNAR		◆CANOPUS		RIGIL KENT.		◆ANTARES		Nunki		ALTAIR	
330	66 07	032	58 25	123	23 13	148	34 56	214	24 39	258	47 15	287	23 56	325
331	66 26	030	58 57	123	23 33	147	34 35	214	24 03	257	46 39	286	23 34	324
332	66 44	028	59 29	122	23 54	147	34 14	213	23 26	256	46 03	285	23 11	323
333	67 02	026	60 01	122	24 15	146	33 54	213	22 49	256	45 27	285	22 48	322
334	67 17	024	60 33	122	24 36	146	33 33	212	22 13	255	44 50	284	22 24	321
335	67 32	022	61 05	122	24 57	145	33 13	212	21 36	254	44 13	283	22 00	320
336	67 45	019	61 37	122	25 19	144	32 54	211	21 00	254	43 36	282	21 35	319
337	67 57	017	62 09	122	25 41	144	32 34	211	20 24	253	42 59	281	21 10	318
338	68 08	015	62 41	121	26 04	143	32 15	211	19 48	252	42 22	280	20 45	317
339	68 17	013	63 13	121	26 26	143	31 56	210	19 12	251	41 45	279	20 18	316
340	68 24	010	63 45	121	26 49	142	31 37	210	18 36	251	41 08	278	19 52	315
341	68 30	008	64 18	121	27 12	142	31 18	209	18 01	250	40 30	277	19 25	314
342	68 35	006	64 50	121	27 36	141	31 00	209	17 25	249	39 53	277	18 58	313
343	68 38	003	65 23	121	28 00	141	30 42	208	16 50	249	39 15	276	18 30	312
344	68 39	001	65 55	121	28 24	140	30 24	208	16 15	248	38 38	275	18 02	311
	◆Diphda		Acamar		◆CANOPUS		Miaplacidus		RIGIL KENT.		◆Nunki		FOMALHAUT	
345	51 10	041	48 16	099	28 48	140	32 16	169	30 07	207	38 00	274	68 39	358
346	51 35	040	48 53	099	29 12	139	32 23	169	29 49	207	37 22	273	68 37	356
347	51 59	039	49 31	098	29 37	139	32 30	169	29 32	207	36 44	272	68 33	354
348	52 22	037	50 08	097	30 02	138	32 38	168	29 16	206	36 07	272	68 28	351
349	52 44	036	50 46	097	30 27	138	32 46	168	28 59	206	35 29	271	68 22	349
350	53 06	034	51 23	096	30 53	137	32 54	167	28 43	205	34 51	270	68 14	347
351	53 27	033	52 01	095	31 19	137	33 02	167	28 27	205	34 13	269	68 05	344
352	53 47	031	52 39	094	31 45	136	33 11	167	28 11	204	33 36	269	67 54	342
353	54 06	030	53 16	094	32 11	136	33 19	166	27 56	204	32 58	268	67 41	340
354	54 24	028	53 54	093	32 37	135	33 28	166	27 41	203	32 20	267	67 28	338
355	54 42	027	54 32	092	33 04	135	33 38	166	27 26	203	31 43	266	67 13	335
356	54 58	025	55 09	091	33 31	134	33 47	165	27 12	202	31 05	266	66 56	333
357	55 14	023	55 47	091	33 58	134	33 57	165	26 58	202	30 27	265	66 39	331
358	55 28	022	56 25	090	34 25	133	34 07	164	26 44	201	29 50	264	66 20	329
359	55 42	020	57 03	089	34 53	133	34 17	164	26 30	201	29 12	263	66 00	327

LAT 52°S

LHA ♈	Hc	Zn	Hc	Zn	Hc	Zn	Hc	Zn	Hc	Zn	Hc	Zn	Hc	Zn
	Diphda		◆RIGEL		CANOPUS		Miaplacidus		◆RIGIL KENT.		Peacock		◆FOMALHAUT	
0	54 58	018	13 29	086	36 01	132	35 25	164	27 13	201	59 12	240	64 50	327
1	55 09	017	14 06	085	36 29	131	35 36	163	27 00	200	58 40	239	64 29	325
2	55 19	015	14 43	084	36 57	131	35 47	163	26 48	200	58 08	239	64 07	323
3	55 27	013	15 20	084	37 25	130	35 58	162	26 35	199	57 36	239	63 44	321
4	55 35	012	15 56	083	37 53	130	36 09	162	26 24	199	57 05	239	63 20	319
5	55 42	010	16 33	082	38 21	129	36 20	162	26 12	198	56 33	238	62 56	318
6	55 48	008	17 09	081	38 50	129	36 32	161	26 01	198	56 02	238	62 31	316
7	55 53	007	17 46	080	39 19	128	36 44	161	25 50	197	55 31	238	62 05	314
8	55 56	005	18 22	080	39 48	128	36 56	161	25 39	197	55 00	237	61 38	313
9	55 59	003	18 58	079	40 17	127	37 08	160	25 29	196	54 29	237	61 10	311
10	56 00	001	19 35	078	40 46	127	37 21	160	25 19	196	53 58	237	60 42	310
11	56 01	000	20 11	077	41 16	127	37 34	160	25 09	195	53 27	236	60 14	308
12	56 00	358	20 47	076	41 46	126	37 47	159	24 59	195	52 56	236	59 44	307
13	55 58	356	21 22	075	42 16	126	38 00	159	24 50	194	52 26	236	59 15	305
14	55 55	355	21 58	075	42 46	125	38 13	159	24 41	194	51 55	235	58 44	304
	◆RIGEL		SIRIUS		CANOPUS		◆RIGIL KENT.		Peacock		◆FOMALHAUT		Diphda	
15	22 34	074	15 23	098	43 16	125	24 33	193	51 25	235	58 13	303	55 51	353
16	23 09	073	16 00	097	43 47	124	24 25	192	50 55	235	57 42	301	55 46	351
17	23 44	072	16 36	096	44 17	124	24 17	192	50 25	234	57 10	300	55 40	350
18	24 19	071	17 13	095	44 48	123	24 09	191	49 55	234	56 38	299	55 33	348
19	24 54	070	17 50	095	45 19	123	24 02	191	49 25	234	56 06	298	55 25	346
20	25 29	069	18 27	094	45 50	122	23 55	190	48 55	233	55 33	296	55 15	345
21	26 03	068	19 04	093	46 21	122	23 49	190	48 26	233	54 59	295	55 05	343
22	26 38	068	19 41	092	46 53	122	23 43	189	47 57	232	54 26	294	54 54	341
23	27 12	067	20 18	091	47 24	121	23 37	189	47 27	232	53 52	293	54 41	340
24	27 45	066	20 55	091	47 56	121	23 31	188	46 58	232	53 18	292	54 28	338
25	28 19	065	21 31	090	48 28	120	23 26	188	46 29	231	52 43	291	54 14	337
26	28 52	064	22 08	089	49 00	120	23 21	187	46 01	231	52 09	290	53 59	335
27	29 25	063	22 45	088	49 32	119	23 17	187	45 32	231	51 34	289	53 43	333
28	29 58	062	23 22	087	50 04	119	23 13	186	45 03	230	50 59	288	53 26	332
29	30 30	061	23 59	087	50 36	118	23 09	186	44 35	230	50 23	287	53 08	330
	◆RIGEL		SIRIUS		CANOPUS		◆RIGIL KENT.		Peacock		◆FOMALHAUT		Diphda	
30	31 03	060	24 36	086	51 09	118	23 05	185	44 07	229	49 48	286	52 49	329
31	31 34	059	25 13	085	51 42	118	23 02	185	43 39	229	49 12	285	52 30	327
32	32 06	058	25 50	084	52 14	117	22 59	184	43 11	229	48 37	284	52 09	326
33	32 37	057	26 26	083	52 47	117	22 57	184	42 44	228	48 01	283	51 48	325
34	33 08	056	27 03	083	53 20	116	22 55	183	42 16	228	47 25	282	51 26	323
35	33 38	055	27 40	082	53 54	116	22 53	183	41 49	227	46 48	281	51 04	322
36	34 08	054	28 16	081	54 27	116	22 51	182	41 22	227	46 12	280	50 41	320
37	34 38	053	28 53	080	55 00	115	22 50	181	40 55	227	45 36	279	50 17	319
38	35 07	052	29 29	079	55 34	115	22 49	181	40 28	226	44 59	278	49 52	318
39	35 36	051	30 05	078	56 07	114	22 49	180	40 02	226	44 22	277	49 27	316
40	36 05	050	30 41	077	56 41	114	22 49	180	39 35	225	43 46	277	49 01	315
41	36 33	049	31 17	077	57 15	113	22 49	179	39 09	225	43 09	276	48 35	314
42	37 00	048	31 53	076	57 49	113	22 50	179	38 43	224	42 32	275	48 08	312
43	37 27	046	32 29	075	58 23	113	22 51	178	38 18	224	41 55	274	47 40	311
44	37 54	045	33 04	074	58 57	112	22 52	178	37 52	223	41 19	273	47 12	310
	RIGEL		◆SIRIUS		CANOPUS		◆RIGIL KENT.		Peacock		FOMALHAUT		◆Diphda	
45	38 20	044	33 40	073	59 31	112	22 53	177	37 27	223	40 42	272	46 44	309
46	38 45	043	34 15	072	60 06	111	22 55	177	37 02	223	40 05	272	46 15	308
47	39 10	042	34 50	071	60 40	111	22 58	176	36 37	222	39 28	271	45 45	306
48	39 35	041	35 25	070	61 15	111	23 00	176	36 12	222	38 51	270	45 15	305
49	39 59	040	36 00	069	61 49	110	23 03	175	35 48	221	38 14	269	44 45	304
50	40 22	038	36 34	068	62 24	110	23 07	175	35 23	221	37 37	268	44 14	303
51	40 44	037	37 08	067	62 59	110	23 10	174	34 59	220	37 00	268	43 43	302
52	41 06	036	37 42	066	63 33	109	23 14	174	34 36	220	36 23	267	43 12	301
53	41 28	035	38 16	065	64 08	109	23 18	173	34 12	219	35 46	266	42 40	300
54	41 49	034	38 50	064	64 43	108	23 23	173	33 49	219	35 09	265	42 07	299
55	42 09	032	39 23	063	65 18	108	23 28	172	33 26	218	34 33	265	41 35	298
56	42 28	031	39 56	062	65 54	108	23 33	171	33 03	218	33 56	264	41 02	297
57	42 47	030	40 28	061	66 29	107	23 39	171	32 40	218	33 19	263	40 29	296
58	43 05	028	41 01	060	67 04	107	23 45	170	32 18	217	32 43	262	39 55	295
59	43 22	027	41 33	059	67 40	107	23 51	170	31 56	217	32 06	262	39 22	294
	RIGEL		◆SIRIUS		Suhail		◆RIGIL KENT.		Peacock		FOMALHAUT		◆Diphda	
60	43 38	026	42 04	058	39 59	113	23 58	169	31 34	216	31 29	261	38 48	293
61	43 54	025	42 35	057	40 33	112	24 05	169	31 12	216	30 53	260	38 13	292
62	44 09	023	43 06	056	41 07	111	24 12	168	30 51	215	30 17	259	37 39	291
63	44 23	022	43 37	055	41 42	111	24 20	168	30 30	215	29 40	259	37 04	290
64	44 36	020	44 07	054	42 16	110	24 28	167	30 09	214	29 04	258	36 29	289
65	44 49	019	44 36	053	42 51	110	24 36	167	29 48	214	28 28	257	35 54	288
66	45 01	018	45 05	052	43 26	109	24 45	166	29 28	213	27 52	257	35 19	287
67	45 11	016	45 34	050	44 01	108	24 54	166	29 08	213	27 16	256	34 44	286
68	45 21	015	46 02	049	44 36	108	25 03	165	28 48	212	26 41	255	34 08	285
69	45 31	014	46 30	048	45 11	107	25 12	165	28 28	212	26 05	254	33 32	284
70	45 39	012	46 57	047	45 47	107	25 22	164	28 09	211	25 29	254	32 57	283
71	45 46	011	47 24	045	46 22	106	25 32	164	27 50	211	24 54	253	32 21	282
72	45 53	009	47 50	044	46 58	105	25 43	163	27 31	210	24 19	252	31 44	282
73	45 58	008	48 15	043	47 33	105	25 54	163	27 13	210	23 44	252	31 08	281
74	46 03	007	48 40	042	48 09	104	26 05	162	26 55	209	23 09	251	30 32	280
	RIGEL		◆SIRIUS		Suhail		◆RIGIL KENT.		Peacock		ACHERNAR		◆Diphda	
75	46 07	005	49 04	040	48 45	104	26 16	162	26 37	209	60 56	239	29 55	279
76	46 09	004	49 28	039	49 21	103	26 28	161	26 19	208	60 24	239	29 19	278
77	46 11	002	49 51	038	49 57	102	26 40	161	26 02	208	59 53	239	28 42	277
78	46 12	001	50 13	036	50 33	102	26 52	160	25 45	207	59 21	239	28 06	277
79	46 12	359	50 34	035	51 09	101	27 05	160	25 29	207	58 50	238	27 29	276
80	46 11	358	50 55	033	51 46	100	27 18	159	25 12	206	58 18	238	26 52	275
81	46 10	357	51 15	032	52 22	100	27 31	159	24 56	205	57 47	238	26 15	274
82	46 07	355	51 34	030	52 59	099	27 45	158	24 41	205	57 16	238	25 38	273
83	46 03	354	51 52	029	53 35	098	27 58	158	24 25	204	56 45	237	25 01	272
84	45 59	352	52 10	028	54 12	098	28 13	157	24 10	204	56 14	237	24 25	272
85	45 53	351	52 26	026	54 48	097	28 27	157	23 55	203	55 43	237	23 48	271
86	45 47	349	52 42	025	55 25	096	28 42	156	23 41	203	55 12	236	23 11	270
87	45 40	348	52 57	023	56 02	096	28 57	156	23 27	202	54 41	236	22 34	269
88	45 32	347	53 11	021	56 38	095	29 12	155	23 13	202	54 10	236	21 57	269
89	45 23	345	53 24	020	57 15	094	29 27	155	22 59	201	53 40	236	21 20	268

LHA ♈	Hc	Zn	Hc	Zn	Hc	Zn	Hc	Zn	Hc	Zn	Hc	Zn	Hc	Zn
	PROCYON		◆Suhail		RIGIL KENT.		◆Peacock		ACHERNAR		◆RIGEL		BETELGEUSE	
90	29 00	028	57 52	093	29 43	154	22 46	201	53 09	235	45 13	344	30 35	359
91	29 17	027	58 29	093	29 59	154	22 33	200	52 39	235	45 02	342	30 34	357
92	29 34	026	59 06	092	30 16	154	22 21	200	52 09	235	44 51	341	30 32	356
93	29 50	025	59 43	091	30 32	153	22 09	199	51 39	234	44 38	340	30 29	355
94	30 05	024	60 20	090	30 49	153	21 57	198	51 09	234	44 25	338	30 25	354
95	30 20	023	60 57	090	31 06	152	21 45	198	50 39	234	44 11	337	30 21	353
96	30 34	022	61 34	089	31 24	152	21 34	197	50 10	233	43 56	336	30 16	352
97	30 48	021	62 11	088	31 41	151	21 23	197	49 40	233	43 41	334	30 10	350
98	31 00	020	62 48	087	31 59	151	21 13	196	49 11	233	43 24	333	30 04	349
99	31 12	018	63 24	086	32 17	150	21 03	196	48 41	232	43 07	332	29 57	348
100	31 24	017	64 01	086	32 36	150	20 53	195	48 12	232	42 49	330	29 49	347
101	31 34	016	64 38	085	32 54	149	20 43	195	47 43	231	42 31	329	29 40	346
102	31 44	015	65 15	084	33 13	149	20 34	194	47 15	231	42 12	328	29 31	345
103	31 53	014	65 51	083	33 33	149	20 26	193	46 46	231	41 52	327	29 21	344
104	32 02	013	66 28	082	33 52	148	20 17	193	46 17	230	41 31	325	29 10	343
	PROCYON		REGULUS		◆Gienah		RIGIL KENT.		◆ACHERNAR		RIGEL		◆SIRIUS	
105	32 10	012	14 17	048	20 30	092	34 12	148	45 49	230	41 10	324	54 35	354
106	32 17	010	14 44	047	21 07	092	34 31	147	45 21	230	40 48	323	54 30	352
107	32 23	009	15 11	046	21 44	091	34 52	147	44 53	229	40 25	322	54 25	350
108	32 28	008	15 37	045	22 21	090	35 12	146	44 25	229	40 02	321	54 18	349
109	32 33	007	16 03	044	22 58	089	35 33	146	43 57	228	39 38	319	54 11	347
110	32 37	006	16 29	043	23 35	089	35 53	146	43 30	228	39 14	318	54 02	346
111	32 40	004	16 54	042	24 12	088	36 14	145	43 02	228	38 49	317	53 52	344
112	32 43	003	17 18	041	24 49	087	36 36	145	42 35	227	38 24	316	53 42	342
113	32 45	002	17 42	040	25 26	086	36 57	144	42 08	227	37 58	315	53 30	341
114	32 45	001	18 06	039	26 03	085	37 19	144	41 41	226	37 31	314	53 17	339
115	32 46	000	18 29	038	26 39	085	37 41	143	41 15	226	37 04	313	53 04	338
116	32 45	359	18 52	037	27 16	084	38 03	143	40 48	226	36 37	311	52 49	336
117	32 44	357	19 14	036	27 53	083	38 25	143	40 22	225	36 09	310	52 34	335
118	32 42	356	19 36	036	28 29	082	38 48	142	39 56	225	35 41	309	52 18	333
119	32 39	355	19 57	035	29 06	081	39 10	142	39 30	224	35 12	308	52 01	332
	REGULUS		◆SPICA		RIGIL KENT.		Peacock		◆ACHERNAR		RIGEL		◆SIRIUS	
120	20 18	034	14 08	090	39 33	141	18 53	184	39 04	224	34 42	307	51 43	330
121	20 38	033	14 45	089	39 56	141	18 50	183	38 39	223	34 13	306	51 24	329
122	20 57	032	15 22	088	40 20	141	18 49	182	38 13	223	33 43	305	51 04	327
123	21 17	031	15 59	088	40 43	140	18 47	182	37 48	223	33 12	304	50 44	326
124	21 35	030	16 36	087	41 07	140	18 46	181	37 23	222	32 42	303	50 23	324
125	21 53	029	17 13	086	41 31	140	18 45	181	36 59	222	32 11	302	50 01	323
126	22 10	028	17 50	085	41 55	139	18 45	180	36 34	221	31 39	301	49 38	322
127	22 27	027	18 26	084	42 19	139	18 45	180	36 10	221	31 07	300	49 15	320
128	22 44	026	19 03	084	42 43	138	18 46	179	35 46	220	30 35	299	48 51	319
129	22 59	025	19 40	083	43 08	138	18 46	178	35 22	220	30 03	298	48 27	318
130	23 14	024	20 17	082	43 33	138	18 48	178	34 58	219	29 30	297	48 02	316
131	23 29	022	20 53	081	43 58	137	18 49	177	34 35	219	28 57	296	47 36	315
132	23 43	021	21 30	080	44 23	137	18 51	177	34 12	219	28 24	295	47 09	314
133	23 56	020	22 06	080	44 48	137	18 53	176	33 49	218	27 50	294	46 43	313
134	24 08	019	22 42	079	45 14	136	18 56	176	33 26	218	27 17	294	46 15	311
	◆REGULUS		SPICA		◆RIGIL KENT.		Peacock		ACHERNAR		◆CANOPUS		SIRIUS	
135	24 20	018	23 18	078	45 39	136	18 59	175	33 04	217	66 26	253	45 47	310
136	24 32	017	23 54	077	46 05	136	19 03	174	32 42	217	65 51	252	45 19	309
137	24 42	016	24 30	076	46 31	135	19 06	174	32 20	216	65 16	252	44 50	308
138	24 52	015	25 06	075	46 57	135	19 11	173	31 58	216	64 41	252	44 20	307
139	25 01	014	25 42	074	47 23	135	19 15	173	31 37	215	64 06	251	43 51	306
140	25 10	013	26 17	074	47 49	134	19 20	172	31 15	215	63 31	251	43 20	304
141	25 18	012	26 53	073	48 16	134	19 25	172	30 54	214	62 56	250	42 50	303
142	25 25	011	27 28	072	48 42	134	19 31	171	30 34	214	62 21	250	42 19	302
143	25 32	010	28 03	071	49 09	133	19 37	170	30 13	213	61 46	250	41 47	301
144	25 38	009	28 38	070	49 36	133	19 43	170	29 53	213	61 12	249	41 15	300
145	25 43	008	29 12	069	50 03	133	19 50	169	29 33	212	60 37	249	40 43	299
146	25 48	007	29 47	068	50 30	133	19 57	169	29 14	212	60 03	249	40 11	298
147	25 52	005	30 21	067	50 57	132	20 04	168	28 54	211	59 29	248	39 38	297
148	25 55	004	30 55	066	51 25	132	20 12	168	28 35	211	58 54	248	39 05	296
149	25 57	003	31 29	065	51 52	132	20 20	167	28 16	210	58 20	247	38 32	295
	◆REGULUS		SPICA		◆RIGIL KENT.		Peacock		ACHERNAR		◆CANOPUS		SIRIUS	
150	25 59	002	32 02	064	52 20	131	20 29	166	27 58	210	57 46	247	37 58	294
151	26 00	001	32 35	064	52 48	131	20 38	166	27 40	209	57 12	246	37 24	293
152	26 00	000	33 08	063	53 15	131	20 47	165	27 22	209	56 38	246	36 50	292
153	26 00	359	33 41	062	53 43	131	20 57	165	27 04	208	56 05	246	36 16	291
154	25 59	358	34 13	061	54 11	131	21 06	164	26 47	208	55 31	245	35 41	290
155	25 57	357	34 45	060	54 40	130	21 17	164	26 30	207	54 58	245	35 06	289
156	25 55	356	35 17	059	55 08	130	21 27	163	26 13	207	54 24	244	34 31	288
157	25 52	355	35 48	058	55 36	130	21 38	162	25 56	206	53 51	244	33 56	287
158	25 48	354	36 19	057	56 05	130	21 50	162	25 40	206	53 18	244	33 21	286
159	25 43	352	36 50	055	56 33	129	22 01	161	25 24	205	52 45	243	32 45	286
160	25 38	351	37 20	054	57 02	129	22 13	161	25 09	205	52 12	243	32 10	285
161	25 32	350	37 50	053	57 30	129	22 25	160	24 53	204	51 39	242	31 34	284
162	25 26	349	38 19	052	57 59	129	22 38	160	24 38	204	51 06	242	30 58	283
163	25 18	348	38 48	051	58 28	129	22 51	159	24 24	203	50 34	241	30 22	282
164	25 11	347	39 17	050	58 57	129	23 04	159	24 09	203	50 01	241	29 46	281
	◆SPICA		ANTARES		◆Peacock		ACHERNAR		CANOPUS		◆SIRIUS		REGULUS	
165	39 45	049	25 08	101	23 18	158	23 55	202	49 29	241	29 09	280	25 02	346
166	40 13	048	25 44	101	23 32	158	23 42	202	48 57	240	28 33	279	24 53	345
167	40 40	047	26 21	100	23 46	157	23 28	201	48 25	240	27 57	279	24 43	344
168	41 06	046	26 57	099	24 01	156	23 15	200	47 53	239	27 20	278	24 32	343
169	41 33	044	27 34	098	24 16	156	23 02	200	47 22	239	26 43	277	24 21	342
170	41 58	043	28 10	098	24 31	155	22 50	199	46 50	238	26 07	276	24 09	341
171	42 23	042	28 47	097	24 46	155	22 38	199	46 19	238	25 30	275	23 56	340
172	42 47	041	29 24	096	25 02	154	22 26	198	45 47	238	24 53	275	23 43	339
173	43 11	040	30 00	095	25 18	154	22 15	198	45 16	237	24 16	274	23 29	338
174	43 34	038	30 37	095	25 35	153	22 04	197	44 45	237	23 39	273	23 15	337
175	43 57	037	31 14	094	25 52	153	21 53	197	44 15	236	23 02	272	23 00	335
176	44 19	036	31 51	093	26 09	152	21 43	196	43 44	236	22 26	271	22 44	334
177	44 40	034	32 28	092	26 26	152	21 33	196	43 14	235	21 49	271	22 28	333
178	45 01	033	33 05	092	26 44	151	21 23	195	42 43	235	21 12	270	22 11	332
179	45 21	032	33 42	091	27 02	151	21 13	194	42 13	234	20 35	269	21 54	331

 LAT 52°S

LHA ♈	Hc	Zn	Hc	Zn	Hc	Zn	Hc	Zn	Hc	Zn	Hc	Zn	Hc	Zn
	♦SPICA		ANTARES		♦Peacock		ACHERNAR		CANOPUS		♦Suhail		REGULUS	
180	45 40	031	34 18	090	27 20	150	21 04	194	41 43	234	60 16	269	21 36	330
181	45 58	029	34 55	089	27 38	150	20 56	193	41 14	233	59 39	269	21 18	329
182	46 16	028	35 32	089	27 57	149	20 48	193	40 44	233	59 02	268	20 58	328
183	46 33	027	36 09	088	28 16	149	20 40	192	40 15	232	58 26	267	20 39	327
184	46 49	025	36 46	087	28 36	148	20 32	192	39 45	232	57 49	266	20 19	326
185	47 04	024	37 23	086	28 55	148	20 25	191	39 16	232	57 12	266	19 58	325
186	47 19	022	38 00	085	29 15	147	20 18	190	38 48	231	56 35	265	19 37	325
187	47 32	021	38 37	084	29 35	147	20 11	190	38 19	231	55 58	264	19 15	324
188	47 45	020	39 13	084	29 56	146	20 05	189	37 51	230	55 21	264	18 53	323
189	47 57	018	39 50	083	30 16	146	19 59	189	37 22	230	54 45	263	18 30	322
190	48 08	017	40 27	082	30 37	145	19 54	188	36 54	229	54 08	262	18 07	321
191	48 18	015	41 03	081	30 59	145	19 49	188	36 26	229	53 32	262	17 44	320
192	48 27	014	41 40	080	31 20	144	19 44	187	35 59	228	52 55	261	17 19	319
193	48 36	012	42 16	079	31 42	144	19 40	187	35 31	228	52 19	260	16 55	318
194	48 43	011	42 52	078	32 04	143	19 36	186	35 04	227	51 42	260	16 30	317
	♦ARCTURUS		ANTARES		♦Peacock		ACHERNAR		CANOPUS		♦Suhail		Denebola	
195	16 55	019	43 28	077	32 26	143	19 32	185	34 37	227	51 06	259	21 38	341
196	17 06	018	44 04	077	32 49	142	19 29	185	34 11	226	50 30	258	21 26	340
197	17 17	017	44 40	076	33 11	142	19 26	184	33 44	226	49 54	258	21 13	339
198	17 27	016	45 16	075	33 34	141	19 23	184	33 18	225	49 18	257	21 00	338
199	17 37	015	45 51	074	33 57	141	19 21	183	32 52	225	48 42	256	20 46	337
200	17 46	014	46 27	073	34 21	140	19 19	183	32 26	224	48 06	256	20 31	336
201	17 55	013	47 02	072	34 44	140	19 18	182	32 00	224	47 30	255	20 16	335
202	18 03	012	47 37	071	35 08	139	19 17	181	31 35	223	46 54	255	20 01	334
203	18 10	011	48 12	070	35 32	139	19 16	181	31 10	223	46 19	254	19 44	333
204	18 16	010	48 46	069	35 57	139	19 16	180	30 45	222	45 43	253	19 28	332
205	18 22	009	49 20	068	36 21	138	19 16	180	30 20	222	45 08	253	19 10	331
206	18 28	008	49 54	067	36 46	138	19 16	179	29 56	221	44 33	252	18 52	330
207	18 32	007	50 28	065	37 11	137	19 17	179	29 32	221	43 58	252	18 34	330
208	18 36	006	51 02	064	37 36	137	19 18	178	29 08	220	43 23	251	18 15	329
209	18 40	005	51 35	063	38 02	136	19 20	177	28 44	219	42 48	250	17 55	328
	♦ARCTURUS		ANTARES		♦Peacock		ACHERNAR		CANOPUS		♦Suhail		SPICA	
210	18 43	004	52 08	062	38 27	136	19 22	177	28 21	219	42 13	250	48 31	347
211	18 45	003	52 40	061	38 53	135	19 24	176	27 58	218	41 39	249	48 22	345
212	18 46	002	53 12	060	39 19	135	19 26	176	27 35	218	41 04	249	48 13	344
213	18 47	001	53 44	058	39 45	135	19 29	175	27 12	217	40 30	248	48 02	343
214	18 47	000	54 15	057	40 12	134	19 33	174	26 50	217	39 56	247	47 51	341
215	18 47	359	54 46	056	40 38	134	19 37	174	26 28	216	39 22	247	47 38	340
216	18 46	358	55 16	055	41 05	133	19 41	173	26 06	216	38 48	246	47 25	338
217	18 44	357	55 46	053	41 32	133	19 45	173	25 45	215	38 14	246	47 11	337
218	18 42	356	56 16	052	41 59	132	19 50	172	25 24	215	37 40	245	46 56	336
219	18 39	355	56 45	051	42 26	132	19 55	172	25 03	214	37 07	244	46 40	334
220	18 35	354	57 13	049	42 54	132	20 01	171	24 42	214	36 34	244	46 24	333
221	18 31	353	57 40	048	43 22	131	20 06	171	24 22	213	36 01	243	46 07	331
222	18 26	352	58 08	046	43 50	131	20 13	170	24 02	212	35 28	243	45 48	330
223	18 21	351	58 34	045	44 18	130	20 19	169	23 42	212	34 55	242	45 30	329
224	18 15	350	59 00	043	44 46	130	20 26	169	23 23	211	34 22	242	45 10	327
	ANTARES		♦Nunki		Peacock		ACHERNAR		♦CANOPUS		Suhail		♦SPICA	
225	59 25	042	39 28	083	45 14	130	20 34	168	23 04	211	33 50	241	44 50	326
226	59 49	040	40 04	082	45 43	129	20 41	168	22 45	210	33 18	240	44 29	325
227	60 12	039	40 41	081	46 11	129	20 49	167	22 27	210	32 46	240	44 07	324
228	60 35	037	41 18	080	46 40	129	20 58	167	22 09	209	32 14	239	43 45	322
229	60 57	035	41 54	080	47 09	128	21 07	166	21 51	209	31 42	239	43 23	321
230	61 18	034	42 30	079	47 38	128	21 16	165	21 33	208	31 11	238	42 59	320
231	61 38	032	43 06	078	48 07	127	21 25	165	21 16	207	30 40	237	42 34	319
232	61 57	030	43 42	077	48 37	127	21 35	164	20 59	207	30 09	237	42 10	317
233	62 15	028	44 18	076	49 06	127	21 45	164	20 43	206	29 38	236	41 44	316
234	62 32	026	44 54	075	49 36	126	21 55	163	20 27	206	29 07	236	41 19	315
235	62 48	025	45 30	074	50 06	126	22 06	163	20 11	205	28 37	235	40 52	314
236	63 03	023	46 05	073	50 36	126	22 17	162	19 55	204	28 07	234	40 25	313
237	63 16	021	46 40	072	51 06	125	22 29	162	19 40	204	27 37	234	39 58	312
238	63 29	019	47 15	071	51 36	125	22 41	161	19 26	203	27 07	233	39 30	310
239	63 40	017	47 50	070	52 07	125	22 53	161	19 11	203	26 38	233	39 02	309
	♦ANTARES		Nunki		♦FOMALHAUT		ACHERNAR		CANOPUS		♦ACRUX		SPICA	
240	63 50	015	48 25	069	14 54	119	23 05	160	18 57	202	60 19	227	38 33	308
241	63 59	013	48 59	068	15 27	119	23 18	159	18 43	202	59 52	227	38 04	307
242	64 07	011	49 33	067	15 59	118	23 31	159	18 30	201	59 24	227	37 34	306
243	64 13	009	50 07	066	16 32	117	23 45	158	18 17	200	58 57	227	37 04	305
244	64 18	007	50 41	065	17 05	117	23 59	158	18 04	200	58 30	227	36 34	304
245	64 22	005	51 14	064	17 38	116	24 13	157	17 52	199	58 03	227	36 03	303
246	64 24	003	51 47	062	18 11	115	24 27	157	17 40	199	57 37	227	35 31	302
247	64 25	001	52 19	061	18 45	114	24 42	156	17 29	198	57 10	227	35 00	301
248	64 25	358	52 52	060	19 19	114	24 57	156	17 17	197	56 43	226	34 28	300
249	64 23	356	53 23	059	19 53	113	25 12	155	17 07	197	56 16	226	33 56	299
250	64 20	354	53 55	058	20 27	112	25 28	155	16 56	196	55 49	226	33 23	298
251	64 16	352	54 26	056	21 01	112	25 44	154	16 46	195	55 23	226	32 51	297
252	64 10	350	54 56	055	21 35	111	26 00	154	16 36	195	54 56	226	32 18	296
253	64 04	348	55 27	054	22 10	110	26 17	153	16 27	194	54 30	226	31 44	295
254	63 55	346	55 56	053	22 45	110	26 34	153	16 18	194	54 04	225	31 11	294
	ALTAIR		♦FOMALHAUT		ACHERNAR		CANOPUS		♦ACRUX		SPICA		♦ANTARES	
255	19 02	045	23 19	109	26 51	152	16 10	193	53 37	225	30 37	293	63 46	344
256	19 28	044	23 55	108	27 08	152	16 01	192	53 11	225	30 03	292	63 35	342
257	19 54	043	24 30	107	27 26	151	15 54	192	52 45	225	29 28	291	63 23	340
258	20 19	042	25 05	107	27 44	151	15 46	191	52 19	225	28 54	290	63 10	338
259	20 44	041	25 40	106	28 02	150	15 39	191	51 53	224	28 19	289	62 56	336
260	21 08	040	26 16	105	28 21	150	15 33	190	51 27	224	27 44	289	62 41	335
261	21 31	039	26 52	105	28 40	149	15 27	189	51 02	224	27 09	288	62 24	333
262	21 54	038	27 27	104	28 59	149	15 21	189	50 36	224	26 34	287	62 07	331
263	22 17	037	28 03	103	29 18	148	15 15	188	50 11	223	25 58	286	61 48	329
264	22 39	036	28 39	102	29 38	148	15 10	187	49 45	223	25 23	285	61 29	327
265	23 01	035	29 16	102	29 58	147	15 06	187	49 20	223	24 47	284	61 09	326
266	23 22	034	29 52	101	30 18	147	15 02	186	48 55	223	24 11	283	60 47	324
267	23 43	033	30 28	100	30 39	146	14 58	186	48 30	222	23 35	283	60 25	322
268	24 03	032	31 04	100	30 59	146	14 54	185	48 05	222	22 59	282	60 02	321
269	24 22	031	31 41	099	31 20	145	14 51	184	47 41	222	22 23	281	59 38	319

LHA ♈	Hc	Zn	Hc	Zn	Hc	Zn	Hc	Zn	Hc	Zn	Hc	Zn	Hc	Zn
	♦ALTAIR		FOMALHAUT		♦ACHERNAR		CANOPUS		RIGIL KENT.		♦SPICA		ANTARES	
270	24 41	030	32 17	098	31 42	145	14 49	184	61 40	232	21 46	280	59 14	317
271	24 59	029	32 54	097	32 03	144	14 47	183	61 11	232	21 10	279	58 48	316
272	25 17	028	33 31	097	32 25	144	14 45	182	60 42	232	20 34	278	58 22	314
273	25 34	027	34 07	096	32 47	143	14 43	182	60 13	232	19 57	278	57 55	313
274	25 51	026	34 44	095	33 09	143	14 42	181	59 44	232	19 20	277	57 28	312
275	26 07	025	35 21	094	33 32	142	14 42	181	59 15	232	18 44	276	57 00	310
276	26 22	024	35 58	094	33 54	142	14 42	180	58 46	231	18 07	275	56 32	309
277	26 37	023	36 35	093	34 17	141	14 42	179	58 17	231	17 30	274	56 02	307
278	26 51	022	37 12	092	34 41	141	14 42	179	57 48	231	16 53	273	55 33	306
279	27 04	021	37 49	091	35 04	140	14 43	178	57 20	231	16 16	273	55 03	305
280	27 17	020	38 25	091	35 28	140	14 45	177	56 51	231	15 39	272	54 32	303
281	27 29	019	39 02	090	35 51	140	14 47	177	56 22	230	15 02	271	54 01	302
282	27 41	017	39 39	089	36 16	139	14 49	176	55 54	230	14 25	270	53 30	301
283	27 51	016	40 16	088	36 40	139	14 52	176	55 26	230	13 49	270	52 58	300
284	28 02	015	40 53	087	37 04	138	14 55	175	54 57	230	13 12	269	52 26	299
	♦ALTAIR		FOMALHAUT		♦ACHERNAR		CANOPUS		♦RIGIL KENT.		ANTARES		Rasalhague	
285	28 11	014	41 30	087	37 29	138	14 58	174	54 29	230	51 53	297	22 51	337
286	28 20	013	42 07	086	37 54	137	15 02	174	54 01	229	51 20	296	22 37	336
287	28 28	012	42 44	085	38 19	137	15 06	173	53 33	229	50 47	295	22 22	335
288	28 35	011	43 20	084	38 45	136	15 11	172	53 05	229	50 13	294	22 06	334
289	28 42	010	43 57	083	39 10	136	15 16	172	52 37	229	49 39	293	21 49	333
290	28 48	009	44 34	082	39 36	136	15 21	171	52 10	228	49 05	292	21 33	332
291	28 53	007	45 10	081	40 02	135	15 27	171	51 42	228	48 31	291	21 15	331
292	28 57	006	45 47	081	40 28	135	15 33	170	51 15	228	47 56	290	20 57	330
293	29 01	005	46 23	080	40 54	134	15 40	169	50 47	228	47 21	289	20 38	329
294	29 04	004	47 00	079	41 21	134	15 47	169	50 20	227	46 46	288	20 19	328
295	29 06	003	47 36	078	41 48	134	15 54	168	49 53	227	46 11	287	20 00	327
296	29 08	002	48 12	077	42 14	133	16 02	168	49 26	227	45 35	286	19 39	326
297	29 09	001	48 48	076	42 41	133	16 10	167	48 59	226	45 00	285	19 19	325
298	29 09	000	49 24	075	43 09	132	16 19	166	48 32	226	44 24	284	18 57	324
299	29 08	358	49 59	074	43 36	132	16 28	166	48 06	226	43 48	283	18 36	323
	FOMALHAUT		♦ACHERNAR		CANOPUS		♦RIGIL KENT.		ANTARES		Nunki		♦ALTAIR	
300	50 35	073	44 04	131	16 37	165	47 40	226	43 12	282	61 31	328	29 07	357
301	51 10	072	44 31	131	16 47	164	47 13	225	42 36	281	61 12	326	29 05	356
302	51 45	071	44 59	131	16 57	164	46 47	225	42 00	280	60 51	325	29 02	355
303	52 20	070	45 27	130	17 07	163	46 21	225	41 23	279	60 29	323	28 58	354
304	52 54	069	45 56	130	17 18	163	45 55	224	40 47	279	60 06	321	28 54	353
305	53 29	068	46 24	130	17 29	162	45 30	224	40 10	278	59 43	320	28 49	352
306	54 03	067	46 53	129	17 41	161	45 04	224	39 34	277	59 19	318	28 43	351
307	54 36	066	47 21	129	17 53	161	44 39	223	38 57	276	58 54	317	28 37	349
308	55 10	064	47 50	128	18 05	160	44 13	223	38 20	275	58 28	315	28 30	348
309	55 43	063	48 19	128	18 18	160	43 48	223	37 43	274	58 02	314	28 22	347
310	56 16	062	48 48	128	18 31	159	43 24	222	37 06	274	57 35	312	28 13	346
311	56 48	061	49 18	127	18 44	158	42 59	222	36 30	273	57 07	311	28 04	345
312	57 20	059	49 47	127	18 58	158	42 34	221	35 53	272	56 39	309	27 54	344
313	57 52	058	50 17	127	19 12	157	42 10	221	35 16	271	56 10	308	27 44	343
314	58 23	057	50 46	126	19 27	157	41 46	221	34 39	270	55 41	307	27 32	342
	♦FOMALHAUT		ACHERNAR		♦CANOPUS		RIGIL KENT.		♦ANTARES		Nunki		ALTAIR	
315	58 54	056	51 16	126	19 41	156	41 22	220	34 02	270	55 11	305	27 21	341
316	59 24	054	51 46	126	19 57	155	40 58	220	33 25	269	54 41	304	27 08	340
317	59 54	053	52 16	125	20 12	155	40 34	220	32 48	268	54 10	303	26 55	338
318	60 23	051	52 46	125	20 28	154	40 11	219	32 11	267	53 39	302	26 41	337
319	60 51	050	53 17	125	20 44	154	39 48	219	31 34	266	53 07	300	26 26	336
320	61 19	048	53 47	124	21 01	153	39 25	218	30 57	266	52 35	299	26 11	335
321	61 46	047	54 18	124	21 17	153	39 02	218	30 20	265	52 02	298	25 55	334
322	62 13	045	54 48	124	21 35	152	38 39	218	29 44	264	51 30	297	25 39	333
323	62 39	044	55 19	123	21 52	151	38 17	217	29 07	263	50 57	296	25 22	332
324	63 04	042	55 50	123	22 10	151	37 54	217	28 30	263	50 23	295	25 04	331
325	63 28	040	56 21	123	22 28	150	37 32	216	27 54	262	49 49	294	24 46	330
326	63 51	038	56 52	123	22 47	150	37 11	216	27 17	261	49 15	293	24 27	329
327	64 14	037	57 23	122	23 05	149	36 49	216	26 41	260	48 41	291	24 08	328
328	64 35	035	57 54	122	23 24	149	36 28	215	26 04	260	48 07	290	23 48	327
329	64 56	033	58 26	122	23 44	148	36 06	215	25 28	259	47 32	289	23 28	326
	♦FOMALHAUT		ACHERNAR		♦CANOPUS		RIGIL KENT.		♦ANTARES		Nunki		ALTAIR	
330	65 15	031	58 57	122	24 03	148	35 46	214	24 52	258	46 57	288	23 07	325
331	65 34	029	59 29	121	24 23	147	35 25	214	24 16	258	46 22	287	22 45	324
332	65 51	027	60 00	121	24 44	146	35 04	213	23 40	257	45 47	286	22 23	323
333	66 07	025	60 32	121	25 04	146	34 44	213	23 04	256	45 11	285	22 01	322
334	66 22	023	61 04	121	25 25	145	34 24	213	22 28	255	44 35	285	21 38	321
335	66 36	021	61 36	120	25 46	145	34 04	212	21 52	255	44 00	284	21 14	320
336	66 49	019	62 07	120	26 08	144	33 45	212	21 17	254	43 24	283	20 50	319
337	67 00	017	62 39	120	26 30	144	33 25	211	20 41	253	42 47	282	20 26	318
338	67 10	014	63 12	120	26 52	143	33 06	211	20 06	252	42 11	281	20 01	317
339	67 18	012	63 44	120	27 14	143	32 47	210	19 31	252	41 35	280	19 35	316
340	67 25	010	64 16	119	27 37	142	32 29	210	18 56	251	40 58	279	19 09	315
341	67 31	008	64 48	119	27 59	142	32 11	210	18 21	250	40 22	278	18 43	314
342	67 35	005	65 20	119	28 23	141	31 52	209	17 46	250	39 45	277	18 17	313
343	67 38	003	65 52	119	28 46	140	31 35	209	17 12	249	39 09	277	17 49	312
344	67 39	001	66 25	119	29 10	140	31 17	208	16 37	248	38 32	276	17 22	311
	♦Diphda		Acamar		♦CANOPUS		Miaplacidus		RIGIL KENT.		♦Nunki		FOMALHAUT	
345	50 25	041	48 25	098	29 33	139	33 15	169	31 00	208	37 55	275	67 39	359
346	50 48	039	49 02	097	29 58	139	33 22	169	30 43	207	37 18	274	67 37	356
347	51 11	038	49 38	097	30 22	138	33 29	168	30 26	207	36 41	273	67 34	354
348	51 34	036	50 15	096	30 47	138	33 36	168	30 09	206	36 05	272	67 29	352
349	51 55	035	50 52	095	31 12	137	33 44	168	29 53	206	35 28	272	67 23	349
350	52 16	034	51 29	095	31 37	137	33 52	167	29 37	205	34 51	271	67 16	347
351	52 36	032	52 06	094	32 02	136	34 00	167	29 22	205	34 14	270	67 07	345
352	52 55	031	52 42	093	32 28	136	34 09	167	29 06	204	33 37	269	66 56	343
353	53 14	029	53 19	092	32 54	135	34 18	166	28 51	204	33 00	268	66 45	341
354	53 31	028	53 56	092	33 20	135	34 27	166	28 36	203	32 23	268	66 32	339
355	53 48	026	54 33	091	33 46	134	34 36	165	28 22	203	31 46	267	66 18	336
356	54 04	025	55 10	090	34 13	134	34 45	165	28 07	203	31 09	266	66 02	334
357	54 19	023	55 47	089	34 39	133	34 55	165	27 53	202	30 32	265	65 46	332
358	54 33	021	56 24	088	35 06	133	35 05	164	27 40	202	29 56	265	65 28	330
359	54 46	020	57 01	088	35 34	132	35 15	164	27 26	201	29 19	264	65 09	328

LAT 53°S

LHA ♈	Hc	Zn	Hc	Zn	Hc	Zn	Hc	Zn	Hc	Zn	Hc	Zn	Hc	Zn
	Diphda		◆RIGEL		CANOPUS		Miaplacidus		◆RIGIL KENT.		Peacock		◆FOMALHAUT	
0	54 00	018	13 25	086	36 41	131	36 23	163	28 09	201	59 41	241	63 59	328
1	54 11	016	14 01	085	37 08	131	36 33	163	27 57	200	59 10	241	63 39	326
2	54 20	015	14 37	084	37 36	130	36 44	163	27 44	200	58 38	240	63 19	324
3	54 29	013	15 13	083	38 03	130	36 55	162	27 32	199	58 07	240	62 57	322
4	54 37	011	15 49	083	38 31	129	37 06	162	27 20	199	57 36	240	62 35	321
5	54 43	010	16 24	082	38 59	129	37 17	161	27 09	198	57 05	240	62 11	319
6	54 49	008	17 00	081	39 27	128	37 29	161	26 58	198	56 33	239	61 47	317
7	54 53	006	17 36	080	39 56	128	37 41	161	26 47	197	56 02	239	61 22	316
8	54 57	005	18 11	079	40 24	127	37 53	160	26 36	197	55 32	239	60 57	314
9	54 59	003	18 47	078	40 53	127	38 05	160	26 26	196	55 01	238	60 30	313
10	55 00	001	19 22	078	41 22	126	38 17	160	26 16	196	54 30	238	60 03	311
11	55 01	000	19 57	077	41 51	126	38 30	159	26 07	195	54 00	237	59 36	310
12	55 00	358	20 32	076	42 21	125	38 43	159	25 57	195	53 29	237	59 08	308
13	54 58	356	21 07	075	42 50	125	38 56	159	25 48	194	52 59	237	58 39	307
14	54 56	355	21 42	074	43 20	124	39 09	158	25 40	194	52 29	236	58 10	305
	◆RIGEL		SIRIUS		CANOPUS		◆RIGIL KENT.		Peacock		◆FOMALHAUT		Diphda	
15	22 17	073	15 31	097	43 50	124	25 31	193	51 59	236	57 40	304	54 52	353
16	22 51	072	16 07	097	44 20	123	25 23	193	51 29	236	57 10	303	54 47	351
17	23 26	072	16 43	096	44 50	123	25 16	192	50 59	235	56 40	301	54 41	350
18	24 00	071	17 19	095	45 21	123	25 08	192	50 30	235	56 09	300	54 34	348
19	24 34	070	17 55	094	45 51	122	25 01	191	50 00	235	55 37	299	54 26	347
20	25 07	069	18 31	093	46 22	122	24 54	191	49 31	234	55 05	298	54 17	345
21	25 41	068	19 07	093	46 53	121	24 48	190	49 02	234	54 33	297	54 08	343
22	26 14	067	19 43	092	47 24	121	24 42	189	48 33	233	54 01	295	53 57	342
23	26 48	066	20 19	091	47 55	120	24 36	189	48 04	233	53 28	294	53 45	340
24	27 20	065	20 55	090	48 26	120	24 31	188	47 35	233	52 55	293	53 32	339
25	27 53	064	21 31	089	48 57	119	24 26	188	47 06	232	52 22	292	53 19	337
26	28 25	063	22 07	089	49 29	119	24 21	187	46 38	232	51 48	291	53 04	336
27	28 58	062	22 43	088	50 01	118	24 16	187	46 10	231	51 14	290	52 49	334
28	29 29	061	23 19	087	50 33	118	24 12	186	45 42	231	50 40	289	52 33	333
29	30 01	060	23 55	086	51 05	117	24 09	186	45 14	231	50 06	288	52 15	331
	◆RIGEL		SIRIUS		CANOPUS		◆RIGIL KENT.		Peacock		◆FOMALHAUT		Diphda	
30	30 32	060	24 31	085	51 37	117	24 05	185	44 46	230	49 31	287	51 58	330
31	31 03	059	25 07	085	52 09	117	24 02	185	44 18	230	48 57	286	51 39	328
32	31 34	058	25 43	084	52 41	116	23 59	184	43 51	229	48 22	285	51 19	327
33	32 04	057	26 19	083	53 14	116	23 57	184	43 23	229	47 47	284	50 59	325
34	32 34	056	26 55	082	53 46	115	23 55	183	42 56	228	47 12	283	50 38	324
35	33 04	054	27 31	081	54 19	115	23 53	183	42 29	228	46 36	282	50 17	322
36	33 33	053	28 06	080	54 52	114	23 51	182	42 03	228	46 01	281	49 54	321
37	34 02	052	28 42	080	55 25	114	23 50	181	41 36	227	45 26	280	49 31	320
38	34 30	051	29 17	079	55 58	113	23 49	181	41 10	227	44 50	279	49 08	318
39	34 58	050	29 53	078	56 31	113	23 49	180	40 44	226	44 14	278	48 43	317
40	35 26	049	30 28	077	57 05	112	23 49	180	40 18	226	43 38	277	48 18	316
41	35 53	048	31 03	076	57 38	112	23 49	179	39 52	225	43 03	277	47 53	315
42	36 20	047	31 38	075	58 12	112	23 50	179	39 26	225	42 27	276	47 27	313
43	36 46	046	32 13	074	58 45	111	23 51	178	39 01	224	41 51	275	47 00	312
44	37 11	045	32 48	073	59 19	111	23 52	178	38 36	224	41 15	274	46 33	311
	RIGEL		◆SIRIUS		CANOPUS		◆RIGIL KENT.		Peacock		FOMALHAUT		◆Diphda	
45	37 37	044	33 22	072	59 53	110	23 53	177	38 11	224	40 39	273	46 06	310
46	38 01	043	33 56	072	60 27	110	23 55	177	37 46	223	40 03	272	45 38	308
47	38 25	041	34 31	071	61 01	109	23 58	176	37 21	223	39 27	272	45 09	307
48	38 49	040	35 04	070	61 35	109	24 00	176	36 57	222	38 50	271	44 40	306
49	39 12	039	35 38	069	62 09	109	24 03	175	36 33	222	38 14	270	44 11	305
50	39 35	038	36 12	068	62 43	108	24 06	175	36 09	221	37 38	269	43 41	304
51	39 56	037	36 45	067	63 18	108	24 10	174	35 45	221	37 02	268	43 11	303
52	40 18	035	37 18	066	63 52	107	24 14	174	35 21	220	36 26	268	42 40	302
53	40 38	034	37 51	065	64 27	107	24 18	173	34 58	220	35 50	267	42 10	301
54	40 58	033	38 23	064	65 01	106	24 23	172	34 35	219	35 14	266	41 38	299
55	41 18	032	38 56	063	65 36	106	24 28	172	34 12	219	34 38	265	41 07	298
56	41 36	031	39 28	062	66 11	105	24 33	171	33 50	218	34 02	265	40 35	297
57	41 54	029	39 59	061	66 46	105	24 38	171	33 28	218	33 26	264	40 03	296
58	42 12	028	40 31	060	67 20	105	24 44	170	33 05	217	32 50	263	39 30	295
59	42 28	027	41 02	059	67 55	104	24 50	170	32 44	217	32 14	262	38 57	294
	RIGEL		◆SIRIUS		Suhail		◆RIGIL KENT.		Peacock		FOMALHAUT		◆Diphda	
60	42 44	025	41 32	057	40 21	112	24 57	169	32 22	216	31 39	262	38 24	293
61	42 59	024	42 02	056	40 55	111	25 04	169	32 01	216	31 03	261	37 51	292
62	43 14	023	42 32	055	41 29	111	25 11	168	31 40	215	30 27	260	37 17	291
63	43 27	021	43 02	054	42 03	110	25 18	168	31 19	215	29 52	259	36 44	290
64	43 40	020	43 31	053	42 37	109	25 26	167	30 58	214	29 16	259	36 10	289
65	43 52	019	44 00	052	43 11	109	25 34	167	30 38	214	28 41	258	35 36	289
66	44 03	017	44 28	051	43 45	108	25 43	166	30 18	213	28 06	257	35 01	288
67	44 14	016	44 55	050	44 19	107	25 52	166	29 58	213	27 31	256	34 27	287
68	44 23	015	45 23	048	44 54	107	26 01	165	29 39	212	26 56	256	33 52	286
69	44 32	013	45 49	047	45 29	106	26 10	165	29 19	212	26 21	255	33 17	285
70	44 40	012	46 16	046	46 03	106	26 20	164	29 00	211	25 46	254	32 42	284
71	44 47	011	46 41	045	46 38	105	26 30	164	28 42	211	25 11	253	32 07	283
72	44 53	009	47 06	043	47 13	104	26 40	163	28 23	210	24 37	253	31 32	282
73	44 59	008	47 31	042	47 48	104	26 51	163	28 05	210	24 02	252	30 57	281
74	45 03	006	47 55	041	48 23	103	27 02	162	27 47	209	23 28	251	30 21	280
	RIGEL		◆SIRIUS		Suhail		◆RIGIL KENT.		Peacock		ACHERNAR		◆Diphda	
75	45 07	005	48 18	040	48 58	102	27 13	162	27 30	209	61 26	241	29 46	280
76	45 09	004	48 41	038	49 34	102	27 25	161	27 12	208	60 55	241	29 10	279
77	45 11	002	49 03	037	50 09	101	27 37	161	26 55	208	60 23	240	28 34	278
78	45 12	001	49 24	035	50 45	100	27 49	160	26 39	207	59 52	240	27 58	277
79	45 12	359	49 45	034	51 20	100	28 01	160	26 22	207	59 21	240	27 23	276
80	45 12	358	50 05	033	51 56	099	28 14	159	26 06	206	58 49	240	26 47	275
81	45 10	357	50 24	031	52 32	098	28 27	159	25 50	206	58 18	239	26 11	275
82	45 07	355	50 42	030	53 07	098	28 40	158	25 35	205	57 47	239	25 35	274
83	45 04	354	51 00	028	53 43	097	28 54	158	25 20	205	57 16	239	24 59	273
84	44 59	352	51 16	027	54 19	096	29 08	157	25 05	204	56 46	238	24 23	272
85	44 54	351	51 32	025	54 55	096	29 22	157	24 50	204	56 15	238	23 46	271
86	44 48	350	51 48	024	55 31	095	29 37	156	24 36	203	55 44	238	23 10	271
87	44 41	348	52 02	023	56 07	094	29 51	156	24 22	202	55 14	237	22 34	270
88	44 33	347	52 15	021	56 43	093	30 06	155	24 09	202	54 44	237	21 58	269
89	44 25	345	52 28	019	57 19	093	30 22	155	23 55	201	54 13	237	21 22	268

LHA ♈	Hc	Zn	Hc	Zn	Hc	Zn	Hc	Zn	Hc	Zn	Hc	Zn	Hc	Zn
	PROCYON		◆Suhail		RIGIL KENT.		◆Peacock		ACHERNAR		◆RIGEL		BETELGEUSE	
90	28 07	028	57 55	092	30 37	154	23 42	201	53 43	236	44 15	344	29 35	359
91	28 24	027	58 31	091	30 53	154	23 30	200	53 13	236	44 05	343	29 34	357
92	28 40	026	59 07	090	31 09	153	23 17	200	52 43	236	43 54	341	29 32	356
93	28 56	025	59 43	090	31 26	153	23 05	199	52 14	235	43 42	340	29 29	355
94	29 11	024	60 19	089	31 42	152	22 54	199	51 44	235	43 29	339	29 26	354
95	29 25	023	60 55	088	31 59	152	22 42	198	51 14	235	43 16	337	29 21	353
96	29 38	022	61 32	087	32 16	151	22 31	197	50 45	234	43 02	336	29 17	352
97	29 51	021	62 08	086	32 34	151	22 21	197	50 16	234	42 47	335	29 11	351
98	30 04	019	62 44	085	32 51	150	22 10	196	49 47	233	42 31	333	29 05	349
99	30 15	018	63 20	084	33 09	150	22 00	196	49 18	233	42 14	332	28 58	348
100	30 26	017	63 55	083	33 28	150	21 51	195	48 49	233	41 57	331	28 50	347
101	30 37	016	64 31	083	33 46	149	21 42	195	48 20	232	41 39	330	28 42	346
102	30 46	015	65 07	082	34 05	149	21 33	194	47 52	232	41 21	328	28 33	345
103	30 55	014	65 43	081	34 24	148	21 24	193	47 24	232	41 01	327	28 23	344
104	31 03	013	66 18	080	34 43	148	21 16	193	46 55	231	40 41	326	28 13	343
	PROCYON		REGULUS		◆Gienah		RIGIL KENT.		◆ACHERNAR		RIGEL		◆SIRIUS	
105	31 11	011	13 37	047	20 33	092	35 02	147	46 27	231	40 21	325	53 35	354
106	31 18	010	14 03	047	21 09	091	35 22	147	45 59	230	40 00	323	53 31	352
107	31 24	009	14 29	046	21 45	091	35 42	146	45 32	230	39 38	322	53 26	351
108	31 29	008	14 55	045	22 21	090	36 02	146	45 04	230	39 15	321	53 19	349
109	31 33	007	15 20	044	22 57	089	36 22	146	44 37	229	38 53	320	53 12	348
110	31 37	006	15 45	043	23 33	088	36 43	145	44 09	229	38 29	319	53 04	346
111	31 40	004	16 09	042	24 09	087	37 03	145	43 42	228	38 05	318	52 54	344
112	31 43	003	16 33	041	24 45	086	37 24	144	43 16	228	37 40	316	52 44	343
113	31 45	002	16 56	040	25 21	086	37 46	144	42 49	227	37 15	315	52 33	341
114	31 45	001	17 19	039	25 57	085	38 07	143	42 22	227	36 50	314	52 21	340
115	31 46	000	17 42	038	26 33	084	38 29	143	41 56	227	36 23	313	52 08	338
116	31 45	359	18 04	037	27 09	083	38 51	143	41 30	226	35 57	312	51 54	337
117	31 44	357	18 26	036	27 45	082	39 13	142	41 04	226	35 30	311	51 40	335
118	31 42	356	18 47	035	28 21	082	39 35	142	40 38	225	35 02	310	51 24	334
119	31 39	355	19 07	034	28 56	081	39 57	141	40 13	225	34 34	309	51 08	332
	REGULUS		◆SPICA		RIGIL KENT.		Peacock		◆ACHERNAR		RIGEL		◆SIRIUS	
120	19 28	033	14 08	090	40 20	141	19 52	184	39 47	224	34 06	308	50 50	331
121	19 47	032	14 44	089	40 43	141	19 50	183	39 22	224	33 37	307	50 32	329
122	20 06	031	15 20	088	41 06	140	19 49	183	38 57	224	33 08	306	50 14	328
123	20 25	030	15 57	087	41 29	140	19 47	182	38 32	223	32 39	305	49 54	327
124	20 43	029	16 33	087	41 53	139	19 46	181	38 08	223	32 09	304	49 34	325
125	21 00	028	17 09	086	42 16	139	19 45	181	37 43	222	31 38	303	49 13	324
126	21 17	027	17 45	085	42 40	139	19 45	180	37 19	222	31 08	302	48 51	322
127	21 34	026	18 21	084	43 04	138	19 45	180	36 55	221	30 37	301	48 29	321
128	21 49	025	18 56	083	43 28	138	19 46	179	36 31	221	30 06	300	48 06	320
129	22 05	024	19 32	083	43 52	137	19 46	178	36 08	220	29 34	299	47 42	318
130	22 19	023	20 08	082	44 17	137	19 48	178	35 45	220	29 02	298	47 18	317
131	22 33	022	20 44	081	44 42	137	19 49	177	35 22	219	28 30	297	46 53	316
132	22 47	021	21 19	080	45 07	136	19 51	177	34 59	219	27 58	296	46 28	315
133	22 59	020	21 55	079	45 32	136	19 53	176	34 36	219	27 25	295	46 02	313
134	23 12	019	22 30	078	45 57	136	19 56	176	34 14	218	26 52	294	45 35	312
	◆REGULUS		SPICA		◆RIGIL KENT.		Peacock		ACHERNAR		◆CANOPUS		SIRIUS	
135	23 23	018	23 06	077	46 22	135	19 59	175	33 52	218	66 43	255	45 08	311
136	23 34	017	23 41	077	46 48	135	20 02	174	33 30	217	66 08	255	44 41	310
137	23 45	016	24 16	076	47 13	135	20 06	174	33 08	217	65 33	254	44 13	309
138	23 54	015	24 51	075	47 39	134	20 10	173	32 47	216	64 59	254	43 44	307
139	24 03	014	25 26	074	48 05	134	20 15	173	32 26	216	64 24	253	43 15	306
140	24 12	013	26 00	073	48 31	134	20 19	172	32 05	215	63 49	253	42 46	305
141	24 19	012	26 35	072	48 57	133	20 25	171	31 44	215	63 15	252	42 16	304
142	24 26	011	27 09	071	49 24	133	20 30	171	31 24	214	62 41	252	41 46	303
143	24 33	010	27 43	070	49 50	133	20 36	170	31 03	214	62 06	251	41 16	302
144	24 39	009	28 17	070	50 17	132	20 42	170	30 44	213	61 32	251	40 45	301
145	24 44	008	28 51	069	50 44	132	20 49	169	30 24	213	60 58	251	40 14	300
146	24 48	006	29 24	068	51 10	132	20 56	169	30 05	212	60 24	250	39 42	299
147	24 52	005	29 58	067	51 37	131	21 03	168	29 45	212	59 50	250	39 11	298
148	24 55	004	30 31	066	52 05	131	21 11	167	29 27	211	59 16	249	38 38	297
149	24 57	003	31 03	065	52 32	131	21 19	167	29 08	211	58 43	249	38 06	296
	◆REGULUS		SPICA		◆RIGIL KENT.		Peacock		ACHERNAR		◆CANOPUS		SIRIUS	
150	24 59	002	31 36	064	52 59	131	21 27	166	28 50	210	58 09	248	37 33	295
151	25 00	001	32 08	063	53 27	130	21 36	166	28 32	210	57 35	248	37 00	294
152	25 00	000	32 40	062	53 54	130	21 45	165	28 14	209	57 02	247	36 27	293
153	25 00	359	33 12	061	54 22	130	21 54	165	27 57	209	56 29	247	35 54	292
154	24 59	358	33 43	060	54 50	129	22 04	164	27 40	208	55 56	247	35 20	291
155	24 57	357	34 15	059	55 18	129	22 14	163	27 23	208	55 22	246	34 46	290
156	24 55	356	34 45	058	55 46	129	22 25	163	27 06	207	54 49	246	34 12	289
157	24 52	355	35 16	057	56 14	129	22 36	162	26 50	206	54 17	245	33 38	288
158	24 48	354	35 46	056	56 42	128	22 47	162	26 34	206	53 44	245	33 04	287
159	24 44	352	36 16	055	57 11	128	22 58	161	26 18	205	53 11	244	32 29	286
160	24 39	351	36 45	054	57 39	128	23 10	161	26 03	205	52 39	244	31 54	285
161	24 33	350	37 14	053	58 08	128	23 22	160	25 48	204	52 06	243	31 19	284
162	24 27	349	37 42	052	58 36	128	23 34	160	25 33	204	51 34	243	30 44	283
163	24 20	348	38 11	051	59 05	127	23 47	159	25 19	203	51 02	243	30 09	283
164	24 12	347	38 38	049	59 33	127	24 00	158	25 05	203	50 30	242	29 34	282
	◆SPICA		ANTARES		◆Peacock		ACHERNAR		CANOPUS		◆SIRIUS		REGULUS	
165	39 05	048	25 20	101	24 14	158	24 51	202	49 58	242	28 58	281	24 04	346
166	39 32	047	25 55	100	24 27	157	24 37	202	49 27	241	28 23	280	23 55	345
167	39 58	046	26 31	099	24 41	157	24 24	201	48 55	241	27 47	279	23 45	344
168	40 24	045	27 06	099	24 56	156	24 11	201	48 24	240	27 12	278	23 35	343
169	40 49	044	27 42	098	25 10	156	23 59	200	47 52	240	26 36	277	23 24	342
170	41 14	043	28 18	097	25 25	155	23 47	199	47 21	239	26 00	277	23 12	341
171	41 38	041	28 54	096	25 41	155	23 35	199	46 50	239	25 24	276	23 00	340
172	42 02	040	29 30	096	25 56	154	23 23	198	46 19	238	24 48	275	22 47	339
173	42 25	039	30 06	095	26 12	154	23 12	198	45 49	238	24 12	274	22 34	338
174	42 47	038	30 42	094	26 28	153	23 01	197	45 18	237	23 36	273	22 20	337
175	43 09	036	31 18	093	26 45	153	22 50	197	44 48	237	23 00	273	22 05	336
176	43 30	035	31 54	093	27 02	152	22 40	196	44 18	237	22 24	272	21 50	335
177	43 51	034	32 30	092	27 19	151	22 30	196	43 48	236	21 48	271	21 34	334
178	44 10	033	33 06	091	27 36	151	22 21	195	43 18	236	21 12	270	21 18	333
179	44 30	031	33 42	090	27 54	150	22 12	194	42 48	235	20 36	269	21 01	332

LHA	Hc	Zn	Hc	Zn	Hc	Zn	Hc	Zn	Hc	Zn	Hc	Zn	Hc	Zn
♈	◆SPICA		ANTARES		◆Peacock		ACHERNAR		CANOPUS		◆Suhail		REGULUS	
180	44 48	030	34 18	089	28 12	150	22 03	194	42 18	235	60 16	271	20 44	331
181	45 06	029	34 54	089	28 30	149	21 54	193	41 49	234	59 40	270	20 26	330
182	45 23	027	35 30	088	28 49	149	21 46	193	41 20	234	59 04	270	20 07	329
183	45 39	026	36 06	087	29 07	148	21 38	192	40 51	233	58 28	269	19 48	328
184	45 54	025	36 42	086	29 26	148	21 31	192	40 22	233	57 52	268	19 29	327
185	46 09	023	37 18	085	29 46	147	21 24	191	39 54	232	57 15	267	19 09	326
186	46 23	022	37 54	084	30 05	147	21 17	191	39 25	232	56 39	267	18 48	325
187	46 36	021	38 30	084	30 25	146	21 10	190	38 57	231	56 03	266	18 27	324
188	46 48	019	39 06	083	30 45	146	21 04	189	38 29	231	55 27	265	18 05	323
189	47 00	018	39 42	082	31 06	145	20 59	189	38 01	230	54 51	264	17 43	322
190	47 10	016	40 18	081	31 27	145	20 53	188	37 33	230	54 16	264	17 21	321
191	47 20	015	40 53	080	31 47	144	20 48	188	37 06	229	53 40	263	16 58	320
192	47 29	013	41 29	079	32 09	144	20 44	187	36 39	229	53 04	262	16 34	319
193	47 37	012	42 04	078	32 30	143	20 39	187	36 12	228	52 28	262	16 10	318
194	47 44	011	42 40	077	32 52	143	20 35	186	35 45	228	51 52	261	15 46	317
	◆ARCTURUS		ANTARES		◆Peacock		ACHERNAR		CANOPUS		◆Suhail		Denebola	
195	15 58	019	43 15	077	33 14	142	20 32	185	35 18	227	51 17	260	20 41	342
196	16 09	018	43 50	076	33 36	142	20 28	185	34 52	227	50 41	260	20 29	341
197	16 20	017	44 25	075	33 58	141	20 26	184	34 26	226	50 06	259	20 17	340
198	16 30	016	44 59	074	34 21	141	20 23	184	34 00	226	49 30	258	20 04	339
199	16 39	015	45 34	073	34 44	140	20 21	183	33 34	225	48 55	258	19 51	338
200	16 48	014	46 08	072	35 07	140	20 19	183	33 09	225	48 20	257	19 36	337
201	16 56	013	46 43	071	35 30	139	20 18	182	32 43	224	47 45	256	19 22	336
202	17 04	012	47 17	070	35 54	139	20 17	181	32 18	224	47 10	256	19 07	335
203	17 11	011	47 50	069	36 18	139	20 16	181	31 54	223	46 35	255	18 51	334
204	17 17	010	48 24	068	36 42	138	20 16	180	31 29	223	46 00	254	18 34	333
205	17 23	009	48 57	067	37 06	138	20 16	180	31 05	222	45 25	254	18 18	332
206	17 28	008	49 30	065	37 30	137	20 16	179	30 41	221	44 51	253	18 00	331
207	17 33	007	50 03	064	37 55	137	20 17	178	30 17	221	44 16	252	17 42	330
208	17 37	006	50 35	063	38 20	136	20 18	178	29 53	220	43 42	252	17 24	329
209	17 40	005	51 07	062	38 45	136	20 20	177	29 30	220	43 08	251	17 05	328
	◆ARCTURUS		ANTARES		◆Peacock		ACHERNAR		CANOPUS		◆Suhail		SPICA	
210	17 43	004	51 39	061	39 10	135	20 21	177	29 07	219	42 33	251	47 33	347
211	17 45	003	52 10	060	39 36	135	20 24	176	28 44	219	41 59	250	47 24	346
212	17 46	002	52 41	059	40 01	134	20 26	176	28 22	218	41 26	249	47 15	344
213	17 47	001	53 12	057	40 27	134	20 29	175	28 00	218	40 52	249	47 05	343
214	17 47	000	53 42	056	40 53	134	20 33	174	27 38	217	40 18	248	46 54	341
215	17 47	359	54 12	055	41 20	133	20 36	174	27 16	217	39 45	248	46 42	340
216	17 46	358	54 41	054	41 46	133	20 40	173	26 55	216	39 12	247	46 29	339
217	17 44	357	55 10	052	42 13	132	20 45	173	26 34	215	38 38	246	46 16	337
218	17 42	356	55 38	051	42 39	132	20 49	172	26 13	215	38 05	246	46 01	336
219	17 39	355	56 06	050	43 06	131	20 54	172	25 52	214	37 33	245	45 46	335
220	17 36	354	56 33	048	43 34	131	21 00	171	25 32	214	37 00	245	45 30	333
221	17 32	353	57 00	047	44 01	131	21 06	170	25 12	213	36 27	244	45 14	332
222	17 27	352	57 26	045	44 28	130	21 12	170	24 52	213	35 55	243	44 56	331
223	17 21	351	57 51	044	44 56	130	21 18	169	24 33	212	35 23	243	44 38	329
224	17 15	350	58 16	042	45 24	129	21 25	169	24 14	212	34 51	242	44 19	328
	ANTARES		◆Nunki		Peacock		ACHERNAR		◆CANOPUS		Suhail		◆SPICA	
225	58 40	041	39 20	082	45 52	129	21 32	168	23 55	211	34 19	242	44 00	327
226	59 03	039	39 56	081	46 20	128	21 40	168	23 37	210	33 47	241	43 40	325
227	59 25	038	40 32	080	46 49	128	21 48	167	23 19	210	33 16	240	43 19	324
228	59 47	036	41 07	080	47 17	128	21 56	166	23 01	209	32 45	240	42 57	323
229	60 08	034	41 43	079	47 46	127	22 05	166	22 43	209	32 13	239	42 35	322
230	60 28	033	42 18	078	48 15	127	22 14	165	22 26	208	31 43	239	42 13	320
231	60 47	031	42 53	077	48 44	126	22 23	165	22 09	208	31 12	238	41 49	319
232	61 05	029	43 28	076	49 13	126	22 33	164	21 53	207	30 41	237	41 25	318
233	61 22	027	44 03	075	49 42	126	22 43	164	21 37	206	30 11	237	41 01	317
234	61 38	026	44 38	074	50 11	125	22 53	163	21 21	206	29 41	236	40 36	316
235	61 53	024	45 13	073	50 41	125	23 04	163	21 05	205	29 11	235	40 10	314
236	62 07	022	45 47	072	51 10	125	23 15	162	20 50	205	28 42	235	39 44	313
237	62 20	020	46 21	071	51 40	124	23 26	161	20 35	204	28 12	234	39 18	312
238	62 32	018	46 55	070	52 10	124	23 37	161	20 21	203	27 43	234	38 51	311
239	62 43	016	47 29	069	52 40	123	23 49	160	20 06	203	27 14	233	38 23	310
	◆ANTARES		Nunki		◆FOMALHAUT		ACHERNAR		CANOPUS		◆ACRUX		SPICA	
240	62 52	014	48 03	068	15 24	119	24 02	160	19 53	202	60 59	229	37 56	309
241	63 01	012	48 36	067	15 55	118	24 14	159	19 39	202	60 32	228	37 27	308
242	63 08	010	49 09	066	16 27	118	24 27	159	19 26	201	60 05	228	36 58	307
243	63 14	009	49 42	065	16 59	117	24 41	158	19 13	200	59 38	228	36 29	306
244	63 19	007	50 15	064	17 32	116	24 54	158	19 01	200	59 11	228	36 00	305
245	63 22	005	50 47	063	18 04	116	25 08	157	18 49	199	58 44	228	35 30	304
246	63 24	003	51 19	061	18 37	115	25 22	157	18 37	199	58 17	228	35 00	302
247	63 25	001	51 50	060	19 10	114	25 37	156	18 26	198	57 50	228	34 29	301
248	63 25	359	52 21	059	19 43	113	25 52	156	18 15	197	57 24	228	33 58	300
249	63 23	357	52 52	058	20 16	113	26 07	155	18 04	197	56 57	227	33 27	299
250	63 21	355	53 22	057	20 49	112	26 22	154	17 54	196	56 31	227	32 55	298
251	63 17	353	53 52	055	21 23	111	26 38	154	17 44	196	56 04	227	32 23	297
252	63 11	351	54 22	054	21 57	111	26 54	153	17 34	195	55 38	227	31 51	297
253	63 05	349	54 51	053	22 30	110	27 10	153	17 25	194	55 11	227	31 19	296
254	62 57	347	55 19	051	23 04	109	27 27	152	17 16	194	54 45	226	30 46	295
	ALTAIR		◆FOMALHAUT		ACHERNAR		CANOPUS		◆ACRUX		SPICA		◆ANTARES	
255	18 20	045	23 39	108	27 44	152	17 08	193	54 19	226	30 13	294	62 48	345
256	18 45	044	24 13	108	28 01	151	17 00	192	53 53	226	29 40	293	62 38	343
257	19 10	043	24 47	107	28 18	151	16 52	192	53 27	226	29 06	292	62 27	341
258	19 34	042	25 22	106	28 36	150	16 45	191	53 01	226	28 33	291	62 14	339
259	19 58	041	25 57	106	28 54	150	16 38	191	52 36	225	27 59	290	62 01	337
260	20 22	040	26 32	105	29 12	149	16 32	190	52 10	225	27 25	289	61 46	335
261	20 45	039	27 07	104	29 31	149	16 26	189	51 45	225	26 51	288	61 31	334
262	21 07	038	27 42	103	29 50	148	16 20	189	51 19	225	26 16	287	61 14	332
263	21 29	037	28 17	103	30 09	148	16 15	188	50 54	224	25 42	286	60 57	330
264	21 51	036	28 52	102	30 29	147	16 10	188	50 29	224	25 07	286	60 38	328
265	22 12	035	29 27	101	30 48	147	16 05	187	50 04	224	24 32	285	60 19	327
266	22 32	034	30 03	100	31 08	146	16 01	186	49 39	223	23 57	284	59 58	325
267	22 52	033	30 38	100	31 28	146	15 57	186	49 14	223	23 22	283	59 37	323
268	23 12	032	31 14	099	31 49	145	15 54	185	48 49	223	22 47	282	59 15	322
269	23 31	031	31 50	098	32 10	145	15 51	184	48 25	223	22 11	281	58 52	320

LAT 53°S

LHA	Hc	Zn	Hc	Zn	Hc	Zn	Hc	Zn	Hc	Zn	Hc	Zn	Hc	Zn
♈	◆ALTAIR		FOMALHAUT		◆ACHERNAR		CANOPUS		RIGIL KENT.		◆SPICA		ANTARES	
270	23 49	030	32 26	098	32 30	144	15 49	184	62 16	234	21 36	280	58 29	319
271	24 07	029	33 01	097	32 52	144	15 46	183	61 47	233	21 00	280	58 05	317
272	24 24	028	33 37	096	33 13	143	15 45	182	61 18	233	20 25	279	57 40	316
273	24 41	027	34 13	095	33 35	143	15 43	182	60 49	233	19 49	278	57 14	314
274	24 57	026	34 49	094	33 57	142	15 42	181	60 20	233	19 13	277	56 48	313
275	25 12	025	35 25	094	34 19	142	15 42	181	59 52	233	18 37	276	56 21	311
276	25 27	024	36 01	093	34 41	141	15 42	180	59 23	233	18 01	275	55 54	310
277	25 42	023	36 37	092	35 04	141	15 42	179	58 54	233	17 25	275	55 26	309
278	25 55	022	37 13	091	35 27	140	15 42	179	58 25	232	16 49	274	54 57	307
279	26 08	021	37 50	091	35 50	140	15 43	178	57 57	232	16 13	273	54 28	306
280	26 21	020	38 26	090	36 13	140	15 45	177	57 28	232	15 37	272	53 59	305
281	26 32	018	39 02	089	36 37	139	15 47	177	57 00	232	15 01	271	53 29	303
282	26 43	017	39 38	088	37 01	139	15 49	176	56 32	231	14 25	271	52 58	302
283	26 54	016	40 14	087	37 25	138	15 51	176	56 04	231	13 49	270	52 27	301
284	27 04	015	40 50	087	37 49	138	15 54	175	55 36	231	13 13	269	51 56	300
	◆ALTAIR		FOMALHAUT		◆ACHERNAR		CANOPUS		◆RIGIL KENT.		ANTARES		Rasalhague	
285	27 13	014	41 26	086	38 13	137	15 58	174	55 08	231	51 25	299	21 56	338
286	27 21	013	42 02	085	38 38	137	16 02	174	54 40	230	50 53	297	21 42	336
287	27 29	012	42 38	084	39 03	136	16 06	173	54 12	230	50 21	296	21 27	335
288	27 36	011	43 14	083	39 28	136	16 10	172	53 44	230	49 48	295	21 12	334
289	27 42	010	43 50	082	39 53	135	16 15	172	53 17	230	49 15	294	20 56	333
290	27 48	009	44 25	081	40 19	135	16 21	171	52 49	229	48 42	293	20 39	332
291	27 53	007	45 01	081	40 44	135	16 26	171	52 22	229	48 09	292	20 22	331
292	27 58	006	45 37	080	41 10	134	16 32	170	51 55	229	47 35	291	20 05	330
293	28 01	005	46 12	079	41 36	134	16 39	169	51 27	229	47 01	290	19 47	329
294	28 04	004	46 47	078	42 02	133	16 46	169	51 00	228	46 27	289	19 28	328
295	28 06	003	47 23	077	42 29	133	16 53	168	50 34	228	45 53	288	19 09	328
296	28 08	002	47 58	076	42 55	132	17 01	167	50 07	228	45 18	287	18 49	327
297	28 09	001	48 33	075	43 22	132	17 09	167	49 40	227	44 44	286	18 29	326
298	28 09	000	49 07	074	43 49	132	17 17	166	49 14	227	44 09	285	18 09	325
299	28 08	358	49 42	073	44 16	131	17 26	166	48 47	227	43 34	284	17 47	324
	FOMALHAUT		◆ACHERNAR		CANOPUS		◆RIGIL KENT.		ANTARES		Nunki		◆ALTAIR	
300	50 16	072	44 43	131	17 35	165	48 21	226	42 59	283	60 40	329	28 07	357
301	50 51	071	45 11	130	17 45	164	47 55	226	42 24	282	60 21	327	28 05	356
302	51 25	070	45 38	130	17 55	164	47 29	226	41 48	281	60 01	326	28 02	355
303	51 58	069	46 06	130	18 05	163	47 04	225	41 13	280	59 41	324	27 59	354
304	52 32	068	46 34	129	18 15	163	46 38	225	40 37	279	59 19	323	27 54	353
305	53 05	067	47 02	129	18 26	162	46 13	225	40 02	279	58 57	321	27 50	352
306	53 38	065	47 30	128	18 38	161	45 47	224	39 26	278	58 34	319	27 44	351
307	54 11	064	47 59	128	18 50	161	45 22	224	38 50	277	58 10	318	27 38	350
308	54 43	063	48 27	128	19 02	160	44 57	224	38 14	276	57 45	316	27 31	348
309	55 15	062	48 56	127	19 14	160	44 32	223	37 38	275	57 20	315	27 23	347
310	55 47	061	49 25	127	19 27	159	44 08	223	37 02	274	56 54	313	27 15	346
311	56 18	059	49 54	126	19 40	158	43 43	222	36 26	273	56 27	312	27 06	345
312	56 49	058	50 23	126	19 54	158	43 19	222	35 50	273	56 00	311	26 57	344
313	57 20	057	50 52	126	20 08	157	42 55	222	35 14	272	55 33	309	26 46	343
314	57 50	056	51 21	125	20 22	157	42 31	221	34 38	271	55 04	308	26 35	342
	◆FOMALHAUT		ACHERNAR		◆CANOPUS		RIGIL KENT.		◆ANTARES		Nunki		ALTAIR	
315	58 19	054	51 51	125	20 36	156	42 07	221	34 02	270	54 36	307	26 24	341
316	58 48	053	52 21	125	20 51	155	41 44	221	33 26	269	54 06	305	26 12	340
317	59 17	051	52 50	124	21 06	155	41 20	220	32 50	269	53 37	304	25 59	339
318	59 45	050	53 20	124	21 22	154	40 57	220	32 14	268	53 07	303	25 45	338
319	60 12	048	53 50	124	21 38	154	40 34	219	31 38	267	52 36	302	25 31	336
320	60 39	047	54 20	123	21 54	153	40 12	219	31 02	266	52 05	300	25 17	335
321	61 05	045	54 51	123	22 11	152	39 49	219	30 25	266	51 34	299	25 01	334
322	61 30	044	55 21	123	22 28	152	39 27	218	29 50	265	51 02	298	24 45	333
323	61 55	042	55 52	122	22 45	151	39 04	218	29 14	264	50 30	297	24 29	332
324	62 19	041	56 22	122	23 02	151	38 42	217	28 38	263	49 58	296	24 12	331
325	62 42	039	56 53	122	23 20	150	38 21	217	28 02	262	49 25	295	23 54	330
326	63 04	037	57 24	121	23 38	150	37 59	216	27 26	262	48 52	294	23 36	329
327	63 25	035	57 55	121	23 57	149	37 38	216	26 50	261	48 19	293	23 17	328
328	63 46	034	58 26	121	24 16	148	37 17	216	26 15	260	47 45	291	22 58	327
329	64 05	032	58 57	120	24 35	148	36 56	215	25 39	259	47 12	290	22 38	326
	◆FOMALHAUT		ACHERNAR		◆CANOPUS		RIGIL KENT.		◆ANTARES		Nunki		ALTAIR	
330	64 24	030	59 28	120	24 54	147	36 35	215	25 04	259	46 38	289	22 18	325
331	64 41	028	59 59	120	25 14	147	36 14	214	24 28	258	46 03	288	21 57	324
332	64 58	026	60 31	120	25 34	146	35 54	214	23 53	257	45 29	287	21 35	323
333	65 13	024	61 02	119	25 54	146	35 34	213	23 18	256	44 55	286	21 13	322
334	65 27	022	61 34	119	26 14	145	35 14	213	22 43	256	44 20	285	20 51	321
335	65 40	020	62 05	119	26 35	144	34 55	213	22 08	255	43 45	285	20 28	320
336	65 52	018	62 37	118	26 56	144	34 36	212	21 33	254	43 10	284	20 05	319
337	66 02	016	63 09	118	27 18	143	34 17	212	20 58	254	42 35	283	19 41	318
338	66 11	014	63 40	118	27 40	143	33 58	211	20 24	253	41 59	282	19 17	317
339	66 19	012	64 12	118	28 02	142	33 39	211	19 49	252	41 24	281	18 52	316
340	66 26	009	64 44	118	28 24	142	33 21	210	19 15	251	40 49	280	18 27	315
341	66 31	007	65 16	117	28 46	141	33 03	210	18 41	251	40 13	279	18 01	314
342	66 35	005	65 49	117	29 09	141	32 45	209	18 07	250	39 37	278	17 35	314
343	66 38	003	66 21	117	29 32	140	32 27	209	17 33	249	39 01	277	17 09	313
344	66 39	001	66 53	117	29 55	140	32 10	208	17 00	248	38 26	276	16 42	312
	◆Diphda		Acamar		◆CANOPUS		Miaplacidus		RIGIL KENT.		◆Nunki		FOMALHAUT	
345	49 39	040	48 33	097	30 19	139	34 14	169	31 53	208	37 50	276	66 39	359
346	50 02	038	49 09	096	30 43	139	34 21	169	31 36	208	37 14	275	66 37	356
347	50 24	037	49 45	096	31 07	138	34 28	168	31 19	207	36 38	274	66 34	354
348	50 45	036	50 21	095	31 31	137	34 35	168	31 03	207	36 02	273	66 30	352
349	51 06	034	50 57	094	31 56	137	34 43	168	30 47	206	35 26	272	66 24	350
350	51 26	033	51 33	093	32 20	136	34 51	167	30 31	206	34 50	272	66 17	348
351	51 45	031	52 09	093	32 45	136	34 59	167	30 16	205	34 13	271	66 09	346
352	52 04	030	52 45	092	33 11	135	35 07	166	30 01	205	33 37	270	65 59	343
353	52 21	028	53 21	091	33 36	135	35 16	166	29 46	204	33 01	269	65 48	341
354	52 38	027	53 57	090	34 02	134	35 25	166	29 32	204	32 25	268	65 36	339
355	52 54	025	54 33	089	34 28	134	35 34	165	29 17	203	31 49	268	65 23	337
356	53 09	024	55 09	089	34 54	133	35 43	165	29 03	203	31 13	267	65 08	335
357	53 23	022	55 46	088	35 20	133	35 53	164	28 49	202	30 37	266	64 53	333
358	53 37	021	56 22	087	35 47	132	36 03	164	28 35	202	30 01	265	64 36	331
359	53 49	019	56 58	086	36 14	132	36 13	164	28 22	201	29 25	264	64 18	330

LHA	Hc	Zn	Hc	Zn	Hc	Zn	Hc	Zn	Hc	Zn	Hc	Zn	Hc	Zn
♈	Diphda		◆ RIGEL		CANOPUS		Miaplacidus		◆ RIGIL KENT.		Peacock		◆ FOMALHAUT	
0	53 03	017	13 20	086	37 20	131	37 20	163	29 05	201	60 10	243	63 08	329
1	53 13	016	13 56	085	37 47	130	37 31	163	28 53	200	59 38	242	62 49	327
2	53 22	014	14 31	084	38 14	130	37 41	162	28 41	200	59 07	242	62 30	325
3	53 30	013	15 06	083	38 41	129	37 52	162	28 29	199	58 36	242	62 09	323
4	53 38	011	15 41	082	39 09	129	38 03	162	28 17	199	58 05	241	61 48	322
5	53 44	009	16 16	081	39 36	128	38 14	161	28 06	198	57 34	241	61 26	320
6	53 49	008	16 50	081	40 04	128	38 26	161	27 55	198	57 04	241	61 03	319
7	53 54	006	17 25	080	40 32	127	38 37	161	27 44	197	56 33	240	60 39	317
8	53 57	005	18 00	079	41 01	127	38 49	160	27 34	197	56 02	240	60 14	315
9	53 59	003	18 34	078	41 29	126	39 01	160	27 24	196	55 32	239	59 49	314
10	54 00	001	19 09	077	41 57	126	39 14	159	27 14	196	55 02	239	59 24	312
11	54 01	000	19 43	076	42 26	125	39 26	159	27 05	195	54 32	239	58 57	311
12	54 00	358	20 17	076	42 55	125	39 39	159	26 55	195	54 01	238	58 30	309
13	53 59	356	20 51	075	43 24	124	39 52	158	26 47	194	53 32	238	58 03	308
14	53 56	355	21 25	074	43 54	124	40 05	158	26 38	194	53 02	237	57 35	307
	◆ RIGEL		SIRIUS		CANOPUS		◆ RIGIL KENT.		Peacock		◆ FOMALHAUT		Diphda	
15	21 59	073	15 38	097	44 23	123	26 30	193	52 32	237	57 06	305	53 52	353
16	22 33	072	16 13	096	44 53	123	26 22	193	52 03	237	56 37	304	53 48	352
17	23 06	071	16 49	095	45 22	122	26 14	192	51 33	236	56 08	303	53 42	350
18	23 40	070	17 24	095	45 52	122	26 07	192	51 04	236	55 38	301	53 35	348
19	24 13	069	17 59	094	46 22	121	26 00	191	50 35	236	55 08	300	53 28	347
20	24 46	068	18 34	093	46 53	121	25 53	191	50 06	235	54 37	299	53 19	345
21	25 18	068	19 09	092	47 23	120	25 47	190	49 37	235	54 06	298	53 10	344
22	25 51	067	19 45	091	47 54	120	25 41	190	49 08	234	53 34	297	53 00	342
23	26 23	066	20 20	091	48 24	119	25 35	189	48 40	234	53 03	295	52 48	341
24	26 55	065	20 55	090	48 55	119	25 30	188	48 11	233	52 31	294	52 36	339
25	27 27	064	21 30	089	49 26	118	25 25	188	47 43	233	51 58	293	52 23	338
26	27 58	063	22 06	088	49 57	118	25 20	187	47 15	233	51 26	292	52 09	336
27	28 30	062	22 41	087	50 29	117	25 16	187	46 47	232	50 53	291	51 55	335
28	29 01	061	23 16	087	51 00	117	25 12	186	46 19	232	50 20	290	51 39	333
29	29 31	060	23 51	086	51 32	116	25 08	186	45 51	231	49 47	289	51 23	332
	◆ RIGEL		SIRIUS		CANOPUS		◆ RIGIL KENT.		Peacock		FOMALHAUT		◆ Diphda	
30	30 02	059	24 26	085	52 03	116	25 05	185	45 24	231	49 13	288	51 06	330
31	30 32	058	25 01	084	52 35	115	25 02	185	44 57	230	48 40	287	50 48	329
32	31 01	057	25 37	083	53 07	115	24 59	184	44 30	230	48 06	286	50 29	327
33	31 31	056	26 12	082	53 39	114	24 57	184	44 03	230	47 32	285	50 10	326
34	32 00	055	26 46	082	54 11	114	24 54	183	43 36	229	46 58	284	49 50	325
35	32 29	054	27 21	081	54 44	113	24 53	183	43 09	229	46 23	283	49 29	323
36	32 57	053	27 56	080	55 16	113	24 51	182	42 43	228	45 49	282	49 07	322
37	33 25	052	28 31	079	55 49	112	24 50	182	42 17	228	45 14	281	48 45	321
38	33 52	051	29 05	078	56 21	112	24 49	181	41 51	227	44 40	280	48 22	319
39	34 20	050	29 40	077	56 54	112	24 49	180	41 25	227	44 05	279	47 59	318
40	34 46	049	30 14	076	57 27	111	24 49	180	40 59	226	43 30	278	47 35	317
41	35 13	048	30 48	075	58 00	111	24 49	179	40 34	226	42 55	278	47 11	315
42	35 38	047	31 22	075	58 33	110	24 50	179	40 08	226	42 20	277	46 46	314
43	36 04	045	31 56	074	59 06	110	24 51	178	39 43	225	41 45	276	46 20	313
44	36 29	044	32 30	073	59 39	109	24 52	178	39 18	225	41 10	275	45 54	312
	RIGEL		◆ SIRIUS		CANOPUS		◆ RIGIL KENT.		Peacock		FOMALHAUT		◆ Diphda	
45	36 53	043	33 04	072	60 13	109	24 53	177	38 54	224	40 35	274	45 27	310
46	37 17	042	33 37	071	60 46	108	24 55	177	38 29	224	40 00	273	45 00	309
47	37 40	041	34 10	070	61 20	108	24 57	176	38 05	223	39 25	272	44 33	308
48	38 03	040	34 43	069	61 53	107	25 00	176	37 41	223	38 49	272	44 05	307
49	38 25	039	35 16	068	62 27	107	25 03	175	37 17	222	38 14	271	43 36	306
50	38 47	037	35 49	067	63 01	106	25 06	175	36 54	222	37 39	270	43 07	305
51	39 08	036	36 21	066	63 35	106	25 10	174	36 30	221	37 03	269	42 38	304
52	39 29	035	36 53	065	64 09	105	25 13	173	36 07	221	36 28	268	42 09	302
53	39 49	034	37 25	064	64 43	105	25 18	173	35 44	220	35 53	268	41 39	301
54	40 08	033	37 57	063	65 17	104	25 22	172	35 21	220	35 18	267	41 08	300
55	40 27	031	38 28	062	65 51	104	25 27	172	34 59	219	34 43	266	40 38	299
56	40 45	030	38 59	061	66 26	103	25 32	171	34 37	219	34 07	265	40 07	298
57	41 02	029	39 30	060	67 00	103	25 38	171	34 15	218	33 32	264	39 36	297
58	41 19	028	40 00	059	67 34	102	25 43	170	33 53	218	32 57	264	39 04	296
59	41 35	026	40 30	058	68 09	102	25 49	170	33 31	217	32 22	263	38 32	295
	RIGEL		◆ SIRIUS		Suhail		◆ RIGIL KENT.		Peacock		FOMALHAUT		◆ Diphda	
60	41 50	025	41 00	057	40 43	111	25 56	169	33 10	217	31 47	262	38 00	294
61	42 04	024	41 29	056	41 16	110	26 03	169	32 49	216	31 12	261	37 28	293
62	42 18	022	41 58	055	41 49	110	26 10	168	32 28	216	30 37	261	36 55	292
63	42 31	021	42 26	053	42 23	109	26 17	168	32 08	215	30 03	260	36 22	291
64	42 44	020	42 55	052	42 56	108	26 25	167	31 48	215	29 28	259	35 49	290
65	42 55	018	43 22	051	43 30	108	26 33	167	31 28	214	28 53	258	35 16	289
66	43 06	017	43 49	050	44 03	107	26 41	166	31 08	214	28 19	258	34 43	288
67	43 16	016	44 16	049	44 37	107	26 50	166	30 48	213	27 45	257	34 09	287
68	43 25	014	44 43	048	45 11	106	26 59	165	30 29	213	27 10	256	33 36	286
69	43 34	013	45 08	046	45 45	105	27 08	165	30 10	212	26 36	255	33 02	285
70	43 41	012	45 34	045	46 19	105	27 18	164	29 51	212	26 02	255	32 28	285
71	43 48	010	45 58	044	46 53	104	27 28	163	29 33	211	25 28	254	31 53	284
72	43 54	009	46 23	043	47 27	103	27 38	163	29 15	211	24 54	253	31 19	283
73	43 59	008	46 46	041	48 02	103	27 48	162	28 57	210	24 21	252	30 45	282
74	44 04	006	47 09	040	48 36	102	27 59	162	28 39	210	23 47	252	30 10	281
	RIGEL		◆ SIRIUS		Suhail		◆ RIGIL KENT.		Peacock		ACHERNAR		◆ Diphda	
75	44 07	005	47 32	039	49 11	101	28 10	161	28 22	209	61 54	243	29 35	280
76	44 10	004	47 53	038	49 45	101	28 21	161	28 05	209	61 23	242	29 01	279
77	44 11	002	48 15	036	50 20	100	28 33	160	27 48	208	60 52	242	28 26	278
78	44 12	001	48 35	035	50 55	099	28 45	160	27 32	207	60 21	242	27 51	278
79	44 12	359	48 55	033	51 30	099	28 57	159	27 16	207	59 50	241	27 16	277
80	44 12	358	49 14	032	52 05	098	29 10	159	27 00	206	59 19	241	26 41	276
81	44 10	357	49 32	031	52 40	097	29 23	158	26 44	206	58 48	241	26 06	275
82	44 07	355	49 50	029	53 15	096	29 36	158	26 29	205	58 18	240	25 31	274
83	44 04	354	50 07	028	53 50	096	29 49	157	26 14	205	57 47	240	24 55	273
84	44 00	353	50 23	026	54 25	095	30 03	157	26 00	204	57 17	240	24 20	273
85	43 55	351	50 38	025	55 00	094	30 17	156	25 45	204	56 46	239	23 45	272
86	43 49	350	50 53	024	55 35	093	30 31	156	25 31	203	56 16	239	23 10	271
87	43 42	348	51 06	022	56 10	093	30 46	155	25 18	203	55 46	239	22 34	270
88	43 35	347	51 19	021	56 46	092	31 01	155	25 04	202	55 16	238	21 59	269
89	43 27	346	51 31	019	57 21	091	31 16	154	24 51	201	54 46	238	21 24	269

LAT 54°S

LHA	Hc	Zn	Hc	Zn	Hc	Zn	Hc	Zn	Hc	Zn	Hc	Zn	Hc	Zn
♈	PROCYON		◆ Suhail		RIGIL KENT.		◆ Peacock		ACHERNAR		◆ RIGEL		BETELGEUSE	
90	27 14	028	57 56	090	31 31	154	24 38	201	54 16	237	43 18	344	28 35	359
91	27 30	027	58 31	089	31 47	153	24 26	200	53 46	237	43 08	343	28 34	357
92	27 46	026	59 07	089	32 03	153	24 14	200	53 17	237	42 57	342	28 32	356
93	28 01	025	59 42	088	32 19	152	24 02	199	52 47	236	42 46	340	28 29	355
94	28 16	024	60 17	087	32 35	152	23 51	199	52 18	236	42 33	339	28 26	354
95	28 29	023	60 52	086	32 52	152	23 39	198	51 49	236	42 20	338	28 22	353
96	28 43	021	61 27	085	33 09	151	23 29	198	51 20	235	42 07	336	28 17	352
97	28 55	020	62 03	084	33 26	151	23 18	197	50 51	235	41 52	335	28 12	351
98	29 07	019	62 38	083	33 44	150	23 08	196	50 22	234	41 37	334	28 06	350
99	29 18	018	63 13	082	34 01	150	22 58	196	49 53	234	41 21	333	27 59	348
100	29 29	017	63 48	081	34 19	149	22 49	195	49 25	234	41 05	331	27 52	347
101	29 39	016	64 22	080	34 37	149	22 40	195	48 57	233	40 47	330	27 44	346
102	29 48	015	64 57	079	34 56	148	22 31	194	48 29	233	40 29	329	27 35	345
103	29 57	014	65 32	078	35 14	148	22 22	194	48 00	232	40 11	328	27 26	344
104	30 05	012	66 06	077	35 33	147	22 14	193	47 33	232	39 52	326	27 15	343
	PROCYON		REGULUS		◆ Gienah		RIGIL KENT.		◆ ACHERNAR		RIGEL		◆ SIRIUS	
105	30 12	011	12 56	047	20 35	092	35 53	147	47 05	232	39 32	325	52 36	354
106	30 18	010	13 22	046	21 10	091	36 12	146	46 37	231	39 11	324	52 32	352
107	30 24	009	13 47	045	21 45	090	36 32	146	46 10	231	38 50	323	52 26	351
108	30 29	008	14 12	045	22 21	089	36 51	146	45 43	230	38 29	322	52 20	349
109	30 34	007	14 37	044	22 56	088	37 11	145	45 16	230	38 06	320	52 13	348
110	30 38	006	15 01	043	23 31	088	37 32	145	44 49	229	37 44	319	52 05	346
111	30 41	004	15 24	042	24 06	087	37 52	144	44 22	229	37 20	318	51 57	345
112	30 43	003	15 48	041	24 41	086	38 13	144	43 56	229	36 57	317	51 47	343
113	30 45	002	16 10	040	25 17	085	38 34	143	43 29	228	36 32	316	51 36	342
114	30 46	001	16 33	039	25 52	084	38 55	143	43 03	228	36 08	315	51 25	340
115	30 46	000	16 55	038	26 27	084	39 16	142	42 37	227	35 42	314	51 12	339
116	30 45	359	17 16	037	27 02	083	39 38	142	42 11	227	35 16	313	50 59	337
117	30 44	357	17 37	036	27 37	082	40 00	142	41 45	226	34 50	311	50 45	336
118	30 42	356	17 58	035	28 12	081	40 22	141	41 20	226	34 24	310	50 30	334
119	30 39	355	18 18	034	28 46	080	40 44	141	40 55	226	33 57	309	50 14	333
	REGULUS		◆ SPICA		RIGIL KENT.		Peacock		◆ ACHERNAR		RIGEL		◆ SIRIUS	
120	18 37	033	14 08	090	41 06	140	20 52	184	40 30	225	33 29	308	49 58	331
121	18 57	032	14 43	089	41 29	140	20 50	183	40 05	225	33 01	307	49 41	330
122	19 15	031	15 18	088	41 52	140	20 48	183	39 40	224	32 33	306	49 23	329
123	19 33	030	15 54	087	42 15	139	20 47	182	39 16	224	32 04	305	49 04	327
124	19 51	029	16 29	086	42 38	139	20 46	181	38 52	223	31 35	304	48 44	326
125	20 08	028	17 04	085	43 01	138	20 45	181	38 27	223	31 06	303	48 24	324
126	20 24	027	17 39	085	43 25	138	20 45	180	38 04	222	30 36	302	48 03	323
127	20 40	026	18 14	084	43 48	138	20 45	180	37 40	222	30 06	301	47 42	322
128	20 55	025	18 49	083	44 12	137	20 46	179	37 17	221	29 36	300	47 20	320
129	21 10	024	19 24	082	44 36	137	20 46	178	36 53	221	29 05	299	46 57	319
130	21 24	023	19 59	081	45 01	136	20 48	178	36 30	220	28 34	298	46 34	318
131	21 38	022	20 34	080	45 25	136	20 49	177	36 08	220	28 03	297	46 10	317
132	21 51	021	21 09	080	45 50	136	20 51	177	35 45	219	27 32	296	45 45	315
133	22 03	020	21 43	079	46 14	135	20 53	176	35 23	219	27 00	295	45 20	314
134	22 15	019	22 18	078	46 39	135	20 56	175	35 01	218	26 28	294	44 55	313
	◆ REGULUS		SPICA		◆ RIGIL KENT.		Peacock		ACHERNAR		◆ CANOPUS		SIRIUS	
135	22 26	018	22 52	077	47 04	135	20 59	175	34 39	218	66 57	257	44 28	312
136	22 37	017	23 27	076	47 30	134	21 02	174	34 17	217	66 23	257	44 02	311
137	22 47	016	24 01	075	47 55	134	21 06	174	33 56	217	65 49	256	43 35	309
138	22 56	015	24 35	074	48 21	133	21 10	173	33 35	216	65 14	256	43 07	308
139	23 05	014	25 09	074	48 46	133	21 14	173	33 14	216	64 40	255	42 39	307
140	23 13	013	25 43	073	49 12	133	21 19	172	32 54	216	64 06	255	42 11	306
141	23 21	012	26 16	072	49 38	132	21 24	171	32 33	215	63 32	254	41 42	305
142	23 28	011	26 50	071	50 04	132	21 29	171	32 13	214	62 58	254	41 13	304
143	23 34	010	27 23	070	50 30	132	21 35	170	31 53	214	62 24	253	40 44	303
144	23 39	009	27 56	069	50 57	131	21 41	170	31 34	213	61 51	253	40 14	302
145	23 44	008	28 29	068	51 23	131	21 48	169	31 14	213	61 17	252	39 44	301
146	23 49	006	29 01	067	51 50	131	21 55	169	30 55	212	60 44	252	39 13	299
147	23 52	005	29 34	066	52 17	130	22 02	168	30 36	212	60 10	251	38 42	298
148	23 55	004	30 06	065	52 44	130	22 09	167	30 18	211	59 37	251	38 11	297
149	23 57	003	30 38	064	53 11	130	22 17	167	30 00	211	59 03	250	37 40	296
	◆ REGULUS		SPICA		◆ RIGIL KENT.		Peacock		ACHERNAR		◆ CANOPUS		SIRIUS	
150	23 59	002	31 09	063	53 38	129	22 25	166	29 42	210	58 30	250	37 08	295
151	24 00	001	31 41	062	54 05	129	22 34	166	29 24	210	57 57	249	36 36	294
152	24 00	000	32 12	061	54 33	129	22 43	165	29 07	209	57 24	249	36 04	293
153	24 00	359	32 43	060	55 00	129	22 52	164	28 49	209	56 51	248	35 31	292
154	23 59	358	33 13	059	55 28	128	23 02	164	28 33	208	56 19	248	34 59	291
155	23 58	357	33 43	058	55 55	128	23 12	163	28 16	208	55 46	248	34 26	291
156	23 55	356	34 13	057	56 23	128	23 22	163	28 00	207	55 14	247	33 52	290
157	23 52	355	34 43	056	56 51	128	23 33	162	27 44	207	54 41	247	33 19	289
158	23 49	354	35 12	055	57 19	127	23 44	162	27 28	206	54 09	246	32 46	288
159	23 45	353	35 41	054	57 47	127	23 55	161	27 13	206	53 37	246	32 12	287
160	23 40	351	36 09	053	58 16	127	24 06	161	26 57	205	53 05	245	31 38	286
161	23 34	350	36 37	052	58 44	127	24 18	160	26 43	205	52 33	245	31 04	285
162	23 28	349	37 05	051	59 12	126	24 31	159	26 28	204	52 01	244	30 30	284
163	23 21	348	37 32	050	59 41	126	24 43	159	26 14	203	51 29	244	29 56	283
164	23 14	347	37 59	049	60 09	126	24 56	158	26 00	203	50 58	243	29 21	282
	◆ SPICA		ANTARES		◆ Peacock		ACHERNAR		CANOPUS		◆ SIRIUS		REGULUS	
165	38 25	048	25 31	100	25 09	158	25 46	202	50 26	243	28 47	281	23 05	346
166	38 51	047	26 06	100	25 23	157	25 33	202	49 55	242	28 12	281	22 57	345
167	39 17	045	26 40	099	25 36	157	25 20	201	49 24	242	27 37	280	22 47	344
168	39 41	044	27 15	098	25 51	156	25 07	201	48 53	241	27 03	279	22 37	343
169	40 06	043	27 50	097	26 05	156	24 55	200	48 22	241	26 28	278	22 27	342
170	40 30	042	28 25	097	26 20	155	24 43	200	47 51	240	25 53	277	22 16	341
171	40 53	041	29 00	096	26 35	154	24 31	199	47 21	240	25 18	276	22 04	340
172	41 16	040	29 35	095	26 50	154	24 20	199	46 50	239	24 43	275	21 51	339
173	41 38	038	30 11	094	27 06	153	24 09	198	46 20	239	24 08	275	21 38	338
174	42 00	037	30 46	094	27 22	153	23 58	197	45 50	238	23 32	274	21 25	337
175	42 21	036	31 21	093	27 38	152	23 48	197	45 20	238	22 57	273	21 11	336
176	42 41	035	31 56	092	27 55	152	23 38	196	44 50	237	22 22	272	20 56	335
177	43 01	033	32 31	091	28 11	151	23 28	196	44 21	237	21 47	271	20 41	334
178	43 20	032	33 07	090	28 29	151	23 19	195	43 51	236	21 11	271	20 25	333
179	43 38	031	33 42	090	28 46	150	23 10	195	43 22	236	20 36	270	20 08	332

LHA	Hc	Zn	Hc	Zn	Hc	Zn	Hc	Zn	Hc	Zn	Hc	Zn	Hc	Zn
♈	♦SPICA		ANTARES		♦Peacock		ACHERNAR		CANOPUS		♦Suhail		REGULUS	
180	43 56	030	34 17	089	29 04	150	23 01	194	42 53	235	60 14	273	19 51	331
181	44 13	028	34 52	088	29 22	149	22 53	193	42 24	235	59 39	272	19 34	330
182	44 29	027	35 28	087	29 40	149	22 45	193	41 55	234	59 03	271	19 16	329
183	44 45	026	36 03	086	29 58	148	22 37	192	41 27	234	58 28	270	18 58	328
184	45 00	024	36 38	085	30 17	148	22 29	192	40 58	233	57 53	270	18 38	327
185	45 14	023	37 13	085	30 36	147	22 22	191	40 30	233	57 18	269	18 19	326
186	45 27	022	37 48	084	30 56	147	22 16	191	40 02	232	56 42	268	17 59	325
187	45 40	020	38 23	083	31 15	146	22 09	190	39 34	232	56 07	267	17 38	324
188	45 52	019	38 58	082	31 35	145	22 03	189	39 07	231	55 32	267	17 17	323
189	46 03	017	39 33	081	31 55	145	21 58	189	38 39	231	54 57	266	16 56	322
190	46 13	016	40 08	080	32 15	144	21 53	188	38 12	230	54 21	265	16 34	321
191	46 22	015	40 43	079	32 36	144	21 48	188	37 45	230	53 46	264	16 12	320
192	46 31	013	41 17	078	32 57	143	21 43	187	37 18	229	53 11	264	15 49	319
193	46 38	012	41 52	077	33 18	143	21 39	187	36 51	229	52 36	263	15 26	318
194	46 45	010	42 26	077	33 39	142	21 35	186	36 25	228	52 01	262	15 02	317
	♦ARCTURUS		ANTARES		♦Peacock		ACHERNAR		CANOPUS		♦Suhail		Denebola	
195	15 01	018	43 00	076	34 01	142	21 31	185	35 59	228	51 26	261	19 44	342
196	15 12	017	43 34	075	34 23	141	21 28	185	35 33	227	50 52	261	19 33	341
197	15 22	017	44 08	074	34 45	141	21 25	184	35 07	227	50 17	260	19 21	340
198	15 32	016	44 42	073	35 07	140	21 23	184	34 42	226	49 42	259	19 08	339
199	15 41	015	45 16	072	35 30	140	21 21	183	34 16	226	49 07	259	18 55	338
200	15 50	014	45 49	071	35 53	140	21 19	183	33 51	225	48 33	258	18 41	337
201	15 58	013	46 22	070	36 16	139	21 18	182	33 26	225	47 59	257	18 27	336
202	16 05	012	46 55	069	36 39	139	21 17	181	33 02	224	47 24	257	18 12	335
203	16 12	011	47 28	068	37 02	138	21 16	181	32 37	223	46 50	256	17 57	334
204	16 18	010	48 00	067	37 26	138	21 16	180	32 13	223	46 16	255	17 41	333
205	16 24	009	48 33	066	37 50	137	21 16	180	31 49	222	45 42	255	17 25	332
206	16 29	008	49 05	064	38 14	137	21 16	179	31 26	222	45 08	254	17 08	331
207	16 33	007	49 36	063	38 38	136	21 17	178	31 02	221	44 34	253	16 50	330
208	16 37	006	50 08	062	39 03	136	21 18	178	30 39	221	44 00	253	16 32	329
209	16 40	005	50 39	061	39 28	135	21 20	177	30 16	220	43 26	252	16 14	328
	♦ARCTURUS		ANTARES		♦Peacock		ACHERNAR		CANOPUS		♦Suhail		SPICA	
210	16 43	004	51 09	060	39 53	135	21 21	177	29 54	220	42 53	251	46 34	347
211	16 45	003	51 40	059	40 18	134	21 24	176	29 31	219	42 20	251	46 26	346
212	16 46	002	52 10	057	40 43	134	21 26	176	29 09	219	41 46	250	46 17	345
213	16 47	001	52 39	056	41 09	133	21 29	175	28 47	218	41 13	250	46 07	343
214	16 47	000	53 08	055	41 34	133	21 32	174	28 26	217	40 40	249	45 57	342
215	16 47	359	53 37	054	42 00	132	21 36	174	28 04	217	40 07	248	45 45	340
216	16 46	358	54 05	052	42 26	132	21 40	173	27 43	216	39 35	248	45 33	339
217	16 45	357	54 33	051	42 53	132	21 44	173	27 22	216	39 02	247	45 20	338
218	16 42	356	55 00	050	43 19	131	21 49	172	27 02	215	38 30	246	45 06	336
219	16 39	355	55 27	048	43 46	131	21 54	172	26 42	215	37 57	246	44 52	335
220	16 36	354	55 53	047	44 13	130	21 59	171	26 22	214	37 25	245	44 37	334
221	16 32	353	56 18	046	44 40	130	22 05	170	26 02	214	36 53	245	44 21	332
222	16 27	352	56 43	044	45 07	129	22 11	170	25 43	213	36 22	244	44 04	331
223	16 22	351	57 07	043	45 34	129	22 17	169	25 24	212	35 50	243	43 47	330
224	16 16	350	57 31	041	46 02	129	22 24	169	25 05	212	35 19	243	43 28	328
	ANTARES		♦Nunki		Peacock		ACHERNAR		♦CANOPUS		Suhail		♦SPICA	
225	57 54	040	39 12	081	46 29	128	22 31	168	24 47	211	34 47	242	43 10	327
226	58 16	038	39 46	081	46 57	128	22 39	168	24 29	211	34 16	242	42 50	326
227	58 37	037	40 21	080	47 25	127	22 46	167	24 11	210	33 45	241	42 30	325
228	58 58	035	40 56	079	47 53	127	22 55	166	23 53	209	33 15	240	42 09	323
229	59 18	033	41 30	078	48 22	126	23 03	166	23 36	209	32 44	240	41 48	322
230	59 37	032	42 05	077	48 50	126	23 12	165	23 19	208	32 14	239	41 26	321
231	59 55	030	42 39	076	49 19	126	23 21	165	23 03	208	31 44	238	41 04	320
232	60 12	028	43 13	075	49 48	125	23 30	164	22 46	207	31 14	238	40 41	319
233	60 28	027	43 47	074	50 16	125	23 40	164	22 30	207	30 44	237	40 17	317
234	60 44	025	44 21	073	50 46	124	23 50	163	22 15	206	30 14	237	39 53	316
235	60 58	023	44 55	072	51 15	124	24 01	162	21 59	205	29 45	236	39 28	315
236	61 11	021	45 28	071	51 44	124	24 12	162	21 45	205	29 16	235	39 03	314
237	61 24	019	46 01	070	52 14	123	24 23	161	21 30	204	28 47	235	38 37	313
238	61 35	018	46 34	069	52 43	123	24 34	161	21 16	204	28 18	234	38 11	312
239	61 45	016	47 07	068	53 13	122	24 46	160	21 02	203	27 50	233	37 45	311
	♦ANTARES		Nunki		♦FOMALHAUT		ACHERNAR		CANOPUS		♦ACRUX		SPICA	
240	61 54	014	47 40	067	15 53	119	24 58	160	20 48	202	61 38	230	37 18	309
241	62 02	012	48 12	066	16 24	118	25 10	159	20 35	202	61 11	230	36 50	308
242	62 09	010	48 44	065	16 55	117	25 23	159	20 22	201	60 44	230	36 22	307
243	62 14	008	49 16	064	17 26	117	25 36	158	20 09	201	60 17	230	35 54	306
244	62 19	006	49 47	063	17 58	116	25 50	157	19 57	200	59 50	229	35 25	305
245	62 22	004	50 19	061	18 30	115	26 03	157	19 45	199	59 24	229	34 56	304
246	62 24	002	50 49	060	19 02	115	26 17	156	19 34	199	58 57	229	34 27	303
247	62 25	001	51 20	059	19 34	114	26 32	156	19 23	198	58 30	229	33 57	302
248	62 25	359	51 50	058	20 06	113	26 46	155	19 12	198	58 04	229	33 27	301
249	62 24	357	52 20	057	20 39	112	27 01	155	19 01	197	57 37	229	32 57	300
250	62 21	355	52 49	055	21 12	112	27 16	154	18 51	196	57 11	228	32 26	299
251	62 17	353	53 18	054	21 44	111	27 32	154	18 42	196	56 45	228	31 55	298
252	62 12	351	53 46	053	22 17	110	27 47	153	18 32	195	56 18	228	31 24	297
253	62 06	349	54 14	052	22 51	109	28 03	153	18 23	194	55 52	228	30 52	296
254	61 59	347	54 41	050	23 24	109	28 20	152	18 15	194	55 26	227	30 21	295
	ALTAIR		♦FOMALHAUT		ACHERNAR		CANOPUS		♦ACRUX		SPICA		♦ANTARES	
255	17 37	045	23 57	108	28 37	152	18 06	193	55 00	227	29 49	294	61 50	345
256	18 02	044	24 31	107	28 53	151	17 59	193	54 34	227	29 16	293	61 41	343
257	18 26	043	25 05	107	29 11	151	17 51	192	54 09	227	28 44	292	61 30	342
258	18 50	042	25 39	106	29 28	150	17 44	191	53 43	227	28 11	291	61 18	340
259	19 13	041	26 13	105	29 46	149	17 37	191	53 18	226	27 38	290	61 05	338
260	19 36	040	26 47	104	30 04	149	17 31	190	52 52	226	27 05	290	60 52	336
261	19 58	039	27 21	104	30 22	148	17 25	189	52 27	226	26 32	289	60 37	334
262	20 20	038	27 55	103	30 41	148	17 19	189	52 02	225	25 58	288	60 21	333
263	20 41	037	28 30	102	31 00	147	17 14	188	51 37	225	25 24	287	60 04	331
264	21 02	036	29 04	101	31 19	147	17 09	188	51 12	225	24 51	286	59 47	329
265	21 23	035	29 39	101	31 38	146	17 05	187	50 47	225	24 17	285	59 28	328
266	21 43	034	30 14	100	31 58	146	17 01	186	50 22	224	23 43	284	59 09	326
267	22 02	033	30 48	099	32 18	145	16 57	186	49 58	224	23 08	283	58 49	324
268	22 21	032	31 23	098	32 38	145	16 54	185	49 33	224	22 34	282	58 28	323
269	22 39	031	31 58	098	32 58	144	16 51	184	49 09	223	21 59	282	58 06	321

LAT 54°S

LHA	Hc	Zn	Hc	Zn	Hc	Zn	Hc	Zn	Hc	Zn	Hc	Zn	Hc	Zn
♈	♦ALTAIR		FOMALHAUT		♦ACHERNAR		CANOPUS		RIGIL KENT.		♦SPICA		ANTARES	
270	22 57	030	32 33	097	33 19	144	16 48	184	62 51	235	21 25	281	57 44	320
271	23 15	029	33 08	096	33 40	143	16 46	183	62 22	235	20 50	280	57 20	318
272	23 31	028	33 43	095	34 01	143	16 45	183	61 53	235	20 15	279	56 57	317
273	23 47	027	34 18	095	34 23	142	16 43	182	61 25	235	19 40	278	56 32	315
274	24 03	026	34 54	094	34 44	142	16 42	181	60 56	234	19 06	277	56 07	314
275	24 18	025	35 29	093	35 06	141	16 42	181	60 27	234	18 31	277	55 41	312
276	24 32	024	36 04	092	35 28	141	16 42	180	59 59	234	17 55	276	55 15	311
277	24 46	023	36 39	091	35 50	140	16 42	179	59 30	234	17 20	275	54 48	310
278	24 59	021	37 14	091	36 13	140	16 42	179	59 02	234	16 45	274	54 20	308
279	25 12	020	37 50	090	36 36	139	16 43	178	58 33	233	16 10	273	53 52	307
280	25 24	019	38 25	089	36 59	139	16 45	177	58 05	233	15 35	272	53 24	306
281	25 35	018	39 00	088	37 22	139	16 47	177	57 37	233	15 00	272	52 55	304
282	25 46	017	39 36	087	37 46	138	16 49	176	57 09	233	14 24	271	52 26	303
283	25 56	016	40 11	087	38 09	138	16 51	176	56 41	232	13 49	270	51 56	302
284	26 06	015	40 46	086	38 33	137	16 54	175	56 13	232	13 14	269	51 26	301
	♦ALTAIR		FOMALHAUT		♦ACHERNAR		CANOPUS		♦RIGIL KENT.		ANTARES		Rasalhague	
285	26 15	014	41 21	085	38 57	137	16 57	174	55 45	232	50 56	300	21 00	338
286	26 23	013	41 56	084	39 22	136	17 01	174	55 17	232	50 25	298	20 47	337
287	26 30	012	42 31	083	39 46	136	17 05	173	54 50	231	49 54	297	20 32	336
288	26 37	011	43 06	082	40 11	135	17 10	172	54 22	231	49 22	296	20 18	335
289	26 43	010	43 41	081	40 36	135	17 15	172	53 55	231	48 50	295	20 02	334
290	26 49	008	44 16	080	41 01	134	17 20	171	53 28	230	48 18	294	19 46	333
291	26 54	007	44 51	080	41 26	134	17 25	171	53 01	230	47 46	293	19 30	332
292	26 58	006	45 25	079	41 52	133	17 31	170	52 34	230	47 13	292	19 13	331
293	27 01	005	46 00	078	42 17	133	17 38	169	52 07	229	46 40	291	18 55	330
294	27 04	004	46 34	077	42 43	133	17 45	169	51 40	229	46 07	290	18 37	329
295	27 06	003	47 08	076	43 09	132	17 52	168	51 13	229	45 34	289	18 18	328
296	27 08	002	47 43	075	43 35	132	17 59	167	50 47	228	45 01	288	17 59	327
297	27 09	001	48 16	074	44 02	131	18 07	167	50 21	228	44 27	287	17 40	326
298	27 09	000	48 50	073	44 28	131	18 15	166	49 54	228	43 53	286	17 20	325
299	27 08	358	49 24	072	44 55	130	18 24	166	49 28	227	43 19	285	16 59	324
	FOMALHAUT		♦ACHERNAR		CANOPUS		♦RIGIL KENT.		ANTARES		Nunki		♦ALTAIR	
300	49 57	071	45 22	130	18 33	165	49 02	227	42 45	284	59 48	330	27 07	357
301	50 30	070	45 49	130	18 42	164	48 37	227	42 11	283	59 30	328	27 05	356
302	51 03	069	46 16	129	18 52	164	48 11	226	41 36	282	59 11	327	27 02	355
303	51 36	068	46 44	129	19 02	163	47 46	226	41 02	281	58 52	325	26 59	354
304	52 09	066	47 11	128	19 13	162	47 20	226	40 27	280	58 31	324	26 55	353
305	52 41	065	47 39	128	19 24	162	46 55	225	39 52	279	58 10	322	26 50	352
306	53 13	064	48 07	127	19 35	161	46 30	225	39 17	279	57 48	320	26 45	351
307	53 44	063	48 35	127	19 46	161	46 05	225	38 43	278	57 25	319	26 39	350
308	54 16	062	49 03	127	19 58	160	45 40	224	38 08	277	57 01	317	26 32	349
309	54 46	061	49 32	126	20 10	159	45 16	224	37 33	276	56 37	316	26 25	347
310	55 17	059	50 00	126	20 23	159	44 52	223	36 57	275	56 12	315	26 17	346
311	55 47	058	50 29	125	20 36	158	44 27	223	36 22	274	55 47	313	26 08	345
312	56 17	057	50 58	125	20 49	158	44 03	223	35 47	273	55 21	312	25 59	344
313	56 46	056	51 27	125	21 03	157	43 40	222	35 12	273	54 54	310	25 49	343
314	57 15	054	51 56	124	21 17	156	43 16	222	34 37	272	54 27	309	25 38	342
	♦FOMALHAUT		ACHERNAR		♦CANOPUS		RIGIL KENT.		♦ANTARES		Nunki		ALTAIR	
315	57 43	053	52 25	124	21 31	156	42 53	222	34 01	271	53 59	308	25 27	341
316	58 11	052	52 54	124	21 46	155	42 29	221	33 26	270	53 31	306	25 15	340
317	58 39	050	53 24	123	22 01	155	42 06	221	32 51	269	53 03	305	25 03	339
318	59 05	049	53 53	123	22 16	154	41 43	220	32 16	268	52 34	304	24 50	338
319	59 32	047	54 23	122	22 32	153	41 21	220	31 40	268	52 04	303	24 36	337
320	59 57	046	54 53	122	22 48	153	40 58	219	31 05	267	51 34	301	24 22	336
321	60 22	044	55 23	122	23 04	152	40 36	219	30 30	266	51 04	300	24 07	335
322	60 46	043	55 53	121	23 20	152	40 14	219	29 55	265	50 33	299	23 52	334
323	61 10	041	56 23	121	23 37	151	39 52	218	29 20	265	50 02	298	23 36	332
324	61 33	039	56 53	121	23 55	150	39 30	218	28 44	264	49 31	297	23 19	331
325	61 55	038	57 24	120	24 12	150	39 08	217	28 09	263	48 59	296	23 02	330
326	62 16	036	57 54	120	24 30	149	38 47	217	27 34	262	48 27	295	22 44	329
327	62 36	034	58 25	120	24 48	149	38 26	216	27 00	261	47 55	294	22 26	328
328	62 55	032	58 56	119	25 07	148	38 05	216	26 25	261	47 23	292	22 07	327
329	63 14	031	59 26	119	25 25	148	37 45	216	25 50	260	46 50	291	21 48	326
	♦FOMALHAUT		ACHERNAR		♦CANOPUS		RIGIL KENT.		♦ANTARES		Nunki		ALTAIR	
330	63 31	029	59 57	119	25 44	147	37 24	215	25 15	259	46 17	290	21 28	325
331	63 48	027	60 28	118	26 04	146	37 04	215	24 41	258	45 44	289	21 08	324
332	64 03	025	60 59	118	26 23	146	36 44	214	24 06	258	45 11	288	20 47	323
333	64 18	023	61 31	118	26 43	145	36 24	214	23 32	257	44 37	287	20 26	322
334	64 31	021	62 02	117	27 04	145	36 05	213	22 57	256	44 03	286	20 04	321
335	64 43	019	62 33	117	27 24	144	35 45	213	22 23	255	43 29	285	19 42	320
336	64 54	017	63 05	117	27 45	144	35 26	212	21 49	255	42 55	285	19 19	319
337	65 04	015	63 36	116	28 06	143	35 07	212	21 15	254	42 21	284	18 56	319
338	65 13	013	64 08	116	28 27	142	34 49	212	20 41	253	41 47	283	18 32	318
339	65 21	011	64 40	116	28 49	142	34 31	211	20 08	252	41 12	282	18 08	317
340	65 27	009	65 11	116	29 11	141	34 12	211	19 34	252	40 38	281	17 44	316
341	65 32	007	65 43	115	29 33	141	33 55	210	19 01	251	40 03	280	17 19	315
342	65 35	005	66 15	115	29 55	140	33 37	210	18 28	250	39 28	279	16 54	314
343	65 38	003	66 47	115	30 18	140	33 20	209	17 54	249	38 53	278	16 28	313
344	65 39	001	67 19	115	30 41	139	33 03	209	17 21	249	38 18	277	16 02	312
	♦Diphda		Acamar		♦CANOPUS		Miaplacidus		RIGIL KENT.		♦Nunki		FOMALHAUT	
345	48 53	039	48 40	096	31 04	139	35 13	169	32 46	208	37 43	276	65 39	359
346	49 14	038	49 15	095	31 27	138	35 19	169	32 29	208	37 08	276	65 37	357
347	49 36	036	49 50	094	31 51	138	35 27	168	32 13	207	36 33	275	65 34	354
348	49 56	035	50 25	094	32 15	137	35 34	168	31 57	207	35 58	274	65 30	352
349	50 16	034	51 00	093	32 39	136	35 41	167	31 41	206	35 23	273	65 25	350
350	50 35	032	51 36	092	33 04	136	35 49	167	31 25	206	34 48	272	65 18	348
351	50 54	031	52 11	091	33 28	135	35 57	167	31 10	205	34 12	271	65 10	346
352	51 11	029	52 46	090	33 53	135	36 06	166	30 55	205	33 37	271	65 01	344
353	51 28	028	53 21	090	34 18	134	36 14	166	30 40	204	33 02	270	64 51	342
354	51 44	026	53 57	089	34 44	134	36 23	165	30 26	204	32 27	269	64 40	340
355	52 00	025	54 32	088	35 09	133	36 32	165	30 12	203	31 51	268	64 27	338
356	52 14	023	55 07	087	35 35	133	36 41	165	29 58	203	31 16	267	64 13	336
357	52 28	022	55 42	086	36 01	132	36 51	164	29 44	202	30 41	267	63 59	334
358	52 40	020	56 18	085	36 27	132	37 00	164	29 31	202	30 06	266	63 43	332
359	52 52	019	56 53	085	36 54	131	37 10	163	29 18	201	29 31	265	63 26	331

LAT 55°S

LHA ♈	Hc	Zn	Hc	Zn	Hc	Zn	Hc	Zn	Hc	Zn	Hc	Zn	Hc	Zn
	Diphda		◆RIGEL		CANOPUS		Miaplacidus		◆RIGIL KENT.		Peacock		◆FOMALHAUT	
0	52 06	017	13 16	085	37 59	130	38 18	163	30 01	201	60 37	244	62 17	330
1	52 15	015	13 50	085	38 26	130	38 28	163	29 49	201	60 06	244	61 59	328
2	52 24	014	14 24	084	38 52	129	38 38	162	29 37	200	59 35	243	61 40	326
3	52 32	012	14 58	083	39 19	129	38 49	162	29 25	200	59 04	243	61 21	325
4	52 39	011	15 32	082	39 46	128	39 00	161	29 14	199	58 33	243	61 00	323
5	52 45	009	16 07	081	40 13	127	39 11	161	29 03	199	58 03	242	60 39	321
6	52 50	008	16 41	080	40 41	127	39 22	161	28 52	198	57 33	242	60 17	320
7	52 54	006	17 14	079	41 08	126	39 34	160	28 42	198	57 02	241	59 55	318
8	52 57	004	17 48	079	41 36	126	39 46	160	28 31	197	56 32	241	59 31	317
9	52 59	003	18 22	078	42 04	125	39 58	159	28 21	196	56 02	241	59 07	315
10	53 01	001	18 55	077	42 32	125	40 10	159	28 12	196	55 32	240	58 43	314
11	53 01	000	19 29	076	43 00	124	40 22	159	28 02	195	55 02	240	58 17	312
12	53 00	358	20 02	075	43 29	124	40 35	158	27 53	195	54 32	239	57 52	311
13	52 59	357	20 35	074	43 58	123	40 47	158	27 45	194	54 03	239	57 25	309
14	52 56	355	21 08	073	44 26	123	41 00	158	27 36	194	53 33	239	56 58	308
	◆RIGEL		SIRIUS		CANOPUS		◆RIGIL KENT.		Peacock		◆FOMALHAUT		Diphda	
15	21 41	073	15 46	097	44 55	122	27 28	193	53 04	238	56 31	307	52 53	353
16	22 14	072	16 20	096	45 25	122	27 20	193	52 35	238	56 03	305	52 48	352
17	22 47	071	16 54	095	45 54	121	27 13	192	52 06	237	55 35	304	52 43	350
18	23 19	070	17 28	094	46 23	121	27 06	192	51 37	237	55 06	303	52 37	349
19	23 51	069	18 03	094	46 53	120	26 59	191	51 08	237	54 37	301	52 29	347
20	24 23	068	18 37	093	47 23	120	26 52	191	50 40	236	54 07	300	52 21	346
21	24 55	067	19 11	092	47 53	119	26 46	190	50 11	236	53 37	299	52 12	344
22	25 27	066	19 46	091	48 23	119	26 40	190	49 43	235	53 07	298	52 03	343
23	25 58	065	20 20	090	48 53	118	26 35	189	49 15	235	52 36	297	51 52	341
24	26 29	064	20 55	089	49 24	118	26 29	189	48 46	234	52 05	295	51 40	340
25	27 00	063	21 29	089	49 54	117	26 24	188	48 19	234	51 34	294	51 28	338
26	27 31	062	22 04	088	50 25	117	26 20	187	47 51	233	51 03	293	51 14	337
27	28 01	061	22 38	087	50 56	116	26 16	187	47 23	233	50 31	292	51 00	335
28	28 31	060	23 12	086	51 27	116	26 12	186	46 56	233	49 59	291	50 46	334
29	29 01	059	23 47	085	51 58	115	26 08	186	46 29	232	49 27	290	50 30	332
	◆RIGEL		SIRIUS		CANOPUS		◆RIGIL KENT.		Peacock		◆FOMALHAUT		Diphda	
30	29 31	059	24 21	084	52 29	115	26 05	185	46 01	232	48 54	289	50 13	331
31	30 00	058	24 55	084	53 00	114	26 02	185	45 35	231	48 22	288	49 56	329
32	30 29	057	25 29	083	53 32	114	25 59	184	45 08	231	47 49	287	49 38	328
33	30 57	056	26 03	082	54 03	113	25 56	184	44 41	230	47 16	286	49 20	327
34	31 25	054	26 37	081	54 35	113	25 54	183	44 15	230	46 43	285	49 00	325
35	31 53	053	27 11	080	55 07	112	25 53	183	43 49	229	46 09	284	48 41	324
36	32 21	052	27 45	079	55 39	112	25 51	182	43 23	229	45 36	283	48 20	323
37	32 48	051	28 19	078	56 11	111	25 50	182	42 57	228	45 02	282	47 59	321
38	33 14	050	28 53	078	56 43	111	25 49	181	42 31	228	44 29	281	47 37	320
39	33 41	049	29 26	077	57 15	110	25 49	180	42 05	228	43 55	280	47 14	319
40	34 07	048	30 00	076	57 48	110	25 49	180	41 40	227	43 21	279	46 51	317
41	34 32	047	30 33	075	58 20	109	25 49	179	41 15	227	42 47	278	46 28	316
42	34 57	046	31 06	074	58 53	109	25 50	179	40 50	226	42 13	278	46 04	315
43	35 21	045	31 39	073	59 25	108	25 51	178	40 25	226	41 39	277	45 39	314
44	35 46	044	32 12	072	59 58	107	25 52	178	40 01	225	41 05	276	45 14	312
	RIGEL		◆SIRIUS		CANOPUS		◆RIGIL KENT.		Peacock		FOMALHAUT		◆Diphda	
45	36 09	043	32 45	071	60 31	107	25 53	177	39 37	225	40 30	275	44 48	311
46	36 32	042	33 17	070	61 04	106	25 55	177	39 13	224	39 56	274	44 22	310
47	36 55	040	33 49	069	61 37	106	25 57	176	38 49	224	39 22	273	43 55	309
48	37 17	039	34 21	068	62 10	105	26 00	176	38 25	223	38 47	272	43 28	308
49	37 38	038	34 53	067	62 43	105	26 03	175	38 01	223	38 13	272	43 01	307
50	37 59	037	35 25	066	63 17	104	26 06	174	37 38	222	37 38	271	42 33	305
51	38 20	036	35 56	065	63 50	104	26 09	174	37 15	222	37 04	270	42 05	304
52	38 39	035	36 28	064	64 24	103	26 13	173	36 52	221	36 30	269	41 36	303
53	38 59	033	36 58	063	64 57	103	26 17	173	36 30	221	35 55	268	41 07	302
54	39 17	032	37 29	062	65 31	102	26 22	172	36 07	220	35 21	267	40 38	301
55	39 35	031	37 59	061	66 04	102	26 26	172	35 45	220	34 46	267	40 08	300
56	39 53	030	38 29	060	66 38	101	26 31	171	35 23	219	34 12	266	39 38	299
57	40 09	028	38 59	059	67 12	100	26 37	171	35 02	219	33 38	265	39 08	298
58	40 25	027	39 29	058	67 46	100	26 42	170	34 40	218	33 04	264	38 37	297
59	40 41	026	39 58	057	68 20	099	26 48	170	34 19	218	32 29	264	38 06	296
	RIGEL		◆SIRIUS		Suhail		◆RIGIL KENT.		Peacock		FOMALHAUT		◆Diphda	
60	40 55	025	40 26	056	41 04	110	26 55	169	33 58	217	31 55	263	37 35	295
61	41 09	023	40 55	055	41 37	110	27 01	169	33 37	217	31 21	262	37 04	294
62	41 23	022	41 23	054	42 09	109	27 08	168	33 17	216	30 47	261	36 32	293
63	41 35	021	41 50	053	42 42	108	27 16	168	32 57	216	30 13	260	36 00	292
64	41 47	020	42 18	052	43 15	108	27 23	167	32 37	215	29 39	260	35 28	291
65	41 58	018	42 44	050	43 47	107	27 31	166	32 17	215	29 05	259	34 56	290
66	42 09	017	43 11	049	44 20	106	27 39	166	31 58	214	28 32	258	34 24	289
67	42 18	016	43 36	048	44 54	106	27 48	165	31 38	214	27 58	257	33 51	288
68	42 27	014	44 02	047	45 27	105	27 57	165	31 19	213	27 24	257	33 18	287
69	42 35	013	44 27	046	46 00	104	28 06	164	31 01	213	26 51	256	32 45	286
70	42 43	012	44 51	044	46 33	104	28 15	164	30 42	212	26 18	255	32 12	285
71	42 49	010	45 15	043	47 07	103	28 25	163	30 24	212	25 44	254	31 39	284
72	42 55	009	45 38	042	47 41	102	28 35	163	30 06	211	25 11	254	31 05	283
73	43 00	008	46 01	041	48 14	102	28 45	162	29 49	210	24 38	253	30 32	282
74	43 04	006	46 23	039	48 48	101	28 56	162	29 32	210	24 06	252	29 58	282
	RIGEL		◆SIRIUS		Suhail		◆RIGIL KENT.		Peacock		ACHERNAR		◆Diphda	
75	43 07	005	46 45	038	49 22	100	29 07	161	29 14	209	62 21	244	29 24	281
76	43 10	003	47 06	037	49 56	099	29 18	161	28 58	209	61 50	244	28 51	280
77	43 11	002	47 26	036	50 30	099	29 30	160	28 41	208	61 20	244	28 17	279
78	43 12	001	47 46	034	51 04	098	29 41	160	28 25	208	60 49	243	27 43	278
79	43 12	359	48 05	033	51 38	097	29 54	159	28 09	207	60 18	243	27 09	277
80	43 12	358	48 23	031	52 12	097	30 06	159	27 54	207	59 47	243	26 34	276
81	43 10	357	48 41	030	52 46	096	30 19	158	27 38	206	59 17	242	26 00	276
82	43 08	355	48 57	029	53 21	095	30 32	158	27 23	206	58 47	242	25 26	275
83	43 04	354	49 14	027	53 55	094	30 45	157	27 09	205	58 16	241	24 52	274
84	43 00	353	49 29	026	54 29	094	30 58	157	26 54	204	57 46	241	24 17	273
85	42 56	351	49 44	024	55 04	093	31 12	156	26 40	204	57 16	241	23 43	272
86	42 50	350	49 57	023	55 38	092	31 26	156	26 26	203	56 46	240	23 08	271
87	42 44	349	50 11	022	56 12	091	31 40	155	26 13	203	56 16	240	22 34	271
88	42 36	347	50 23	020	56 47	090	31 55	155	26 00	202	55 47	239	22 00	270
89	42 28	346	50 34	019	57 21	090	32 10	154	25 47	202	55 17	239	21 25	269

LHA ♈	Hc	Zn	Hc	Zn	Hc	Zn	Hc	Zn	Hc	Zn	Hc	Zn	Hc	Zn
	PROCYON		◆Suhail		RIGIL KENT.		◆Peacock		ACHERNAR		◆RIGEL		BETELGEUSE	
90	26 21	028	57 56	089	32 25	154	25 34	201	54 48	239	42 20	345	27 35	359
91	26 37	027	58 30	088	32 40	153	25 22	201	54 18	238	42 10	343	27 34	357
92	26 52	026	59 04	087	32 56	153	25 10	200	53 49	238	42 00	342	27 32	356
93	27 07	025	59 39	086	33 12	152	24 59	199	53 20	237	41 49	341	27 29	355
94	27 21	023	60 13	085	33 28	152	24 47	199	52 51	237	41 37	339	27 26	354
95	27 34	022	60 47	084	33 45	151	24 36	198	52 22	237	41 25	338	27 22	353
96	27 47	021	61 22	083	34 01	151	24 26	198	51 54	236	41 12	337	27 18	352
97	27 59	020	61 56	082	34 18	150	24 15	197	51 25	236	40 58	336	27 13	351
98	28 10	019	62 30	081	34 35	150	24 06	197	50 57	235	40 43	334	27 07	350
99	28 21	018	63 04	080	34 53	149	23 56	196	50 28	235	40 28	333	27 00	349
100	28 32	017	63 38	079	35 11	149	23 47	195	50 00	235	40 12	332	26 53	347
101	28 41	016	64 11	078	35 29	148	23 38	195	49 32	234	39 55	331	26 45	346
102	28 50	015	64 45	077	35 47	148	23 29	194	49 04	234	39 38	329	26 37	345
103	28 58	013	65 19	076	36 05	147	23 21	194	48 37	233	39 20	328	26 28	344
104	29 06	012	65 52	075	36 24	147	23 13	193	48 09	233	39 01	327	26 18	343
	PROCYON		REGULUS		◆Gienah		RIGIL KENT.		◆ACHERNAR		RIGEL		◆SIRIUS	
105	29 13	011	12 15	047	20 36	091	36 43	146	47 42	232	38 42	326	51 36	354
106	29 19	010	12 40	046	21 11	091	37 02	146	47 15	232	38 23	324	51 32	353
107	29 25	009	13 05	045	21 45	090	37 21	146	46 48	232	38 02	323	51 27	351
108	29 30	008	13 29	044	22 20	089	37 41	145	46 21	231	37 41	322	51 21	350
109	29 34	007	13 53	043	22 54	088	38 00	145	45 54	231	37 20	321	51 15	348
110	29 38	005	14 17	043	23 28	087	38 21	144	45 27	230	36 58	320	51 07	347
111	29 41	004	14 40	042	24 03	086	38 41	144	45 01	230	36 36	319	50 59	345
112	29 43	003	15 02	041	24 37	086	39 01	143	44 35	229	36 13	317	50 49	344
113	29 45	002	15 24	040	25 11	085	39 22	143	44 09	229	35 49	316	50 39	342
114	29 46	001	15 46	039	25 46	084	39 43	142	43 43	228	35 25	315	50 28	341
115	29 46	000	16 07	038	26 20	083	40 04	142	43 17	228	35 01	314	50 16	339
116	29 45	359	16 28	037	26 54	082	40 25	142	42 52	228	34 36	313	50 04	338
117	29 44	357	16 49	036	27 28	081	40 47	141	42 27	227	34 10	312	49 50	336
118	29 42	356	17 09	035	28 02	080	41 08	141	42 02	227	33 45	311	49 36	335
119	29 40	355	17 28	034	28 36	080	41 30	140	41 37	226	33 18	310	49 21	333
	REGULUS		◆SPICA		RIGIL KENT.		Peacock		◆ACHERNAR		RIGEL		◆SIRIUS	
120	17 47	033	14 07	089	41 52	140	21 52	184	41 12	226	32 52	309	49 05	332
121	18 06	032	14 42	088	42 15	139	21 50	183	40 47	225	32 25	308	48 48	331
122	18 24	031	15 16	088	42 37	139	21 48	183	40 23	225	31 57	307	48 31	329
123	18 41	030	15 50	087	43 00	139	21 47	182	39 59	224	31 29	306	48 13	328
124	18 58	029	16 25	086	43 23	138	21 46	181	39 35	224	31 01	305	47 54	326
125	19 15	028	16 59	085	43 46	138	21 45	181	39 11	223	30 33	304	47 35	325
126	19 31	027	17 33	084	44 09	137	21 45	180	38 48	223	30 04	303	47 15	324
127	19 46	026	18 08	083	44 33	137	21 46	180	38 25	222	29 35	302	46 54	322
128	20 01	025	18 42	083	44 56	136	21 46	179	38 01	222	29 05	301	46 33	321
129	20 15	024	19 16	082	45 20	136	21 46	178	37 39	221	28 36	300	46 11	320
130	20 29	023	19 50	081	45 44	136	21 47	178	37 16	221	28 06	299	45 49	319
131	20 42	022	20 24	080	46 08	135	21 49	177	36 54	220	27 35	298	45 26	317
132	20 55	021	20 58	079	46 32	135	21 51	177	36 31	220	27 05	297	45 02	316
133	21 07	020	21 31	078	46 57	135	21 53	176	36 09	219	26 34	296	44 38	315
134	21 18	019	22 05	078	47 21	134	21 56	175	35 48	219	26 03	295	44 13	314
	◆REGULUS		SPICA		◆RIGIL KENT.		Peacock		ACHERNAR		◆CANOPUS		SIRIUS	
135	21 29	018	22 39	077	47 46	134	21 58	175	35 26	218	67 09	260	43 48	312
136	21 39	017	23 12	076	48 11	133	22 02	174	35 05	218	66 36	259	43 23	311
137	21 49	016	23 45	075	48 36	133	22 05	174	34 44	217	66 02	258	42 57	310
138	21 58	015	24 19	074	49 01	133	22 09	173	34 23	217	65 28	258	42 30	309
139	22 07	014	24 52	073	49 27	132	22 14	173	34 03	216	64 55	257	42 03	308
140	22 15	013	25 24	072	49 52	132	22 18	172	33 42	216	64 21	257	41 36	307
141	22 22	012	25 57	071	50 18	132	22 23	171	33 22	215	63 48	256	41 08	306
142	22 29	011	26 30	070	50 44	131	22 29	171	33 02	215	63 14	256	40 40	304
143	22 35	010	27 02	069	51 10	131	22 34	170	32 43	214	62 41	255	40 11	303
144	22 40	009	27 34	069	51 36	130	22 40	170	32 24	214	62 08	255	39 42	302
145	22 45	007	28 06	068	52 02	130	22 47	169	32 05	213	61 34	254	39 13	301
146	22 49	006	28 38	067	52 29	130	22 53	168	31 46	213	61 01	254	38 43	300
147	22 52	005	29 09	066	52 55	129	23 00	168	31 27	212	60 28	253	38 13	299
148	22 55	004	29 40	065	53 22	129	23 08	167	31 09	212	59 56	252	37 43	298
149	22 58	003	30 11	064	53 49	129	23 16	167	30 51	211	59 23	252	37 13	297
	◆REGULUS		SPICA		◆RIGIL KENT.		Peacock		ACHERNAR		◆CANOPUS		SIRIUS	
150	22 59	002	30 42	063	54 16	128	23 24	166	30 33	211	58 50	251	36 42	296
151	23 00	001	31 13	062	54 43	128	23 32	166	30 16	210	58 18	251	36 11	295
152	23 00	000	31 43	061	55 10	128	23 41	165	29 59	210	57 45	250	35 40	294
153	23 00	359	32 13	060	55 37	127	23 50	164	29 42	209	57 13	250	35 08	293
154	22 59	358	32 42	059	56 04	127	23 59	164	29 25	209	56 40	249	34 36	292
155	22 58	357	33 12	058	56 32	127	24 09	163	29 09	208	56 08	249	34 04	291
156	22 55	356	33 41	057	56 59	127	24 19	163	28 53	207	55 36	248	33 32	290
157	22 53	355	34 09	056	57 27	126	24 30	162	28 37	207	55 04	248	33 00	289
158	22 49	354	34 38	055	57 55	126	24 40	162	28 22	206	54 33	247	32 27	288
159	22 45	353	35 06	054	58 23	126	24 52	161	28 07	206	54 01	247	31 54	287
160	22 40	352	35 33	053	58 51	125	25 03	160	27 52	205	53 29	246	31 21	286
161	22 35	350	36 00	052	59 19	125	25 15	160	27 37	205	52 58	246	30 48	286
162	22 29	349	36 27	050	59 47	125	25 27	159	27 23	204	52 26	245	30 15	285
163	22 22	348	36 53	049	60 15	125	25 39	159	27 09	204	51 55	245	29 42	284
164	22 15	347	37 19	048	60 44	124	25 52	158	26 55	203	51 24	244	29 08	283
	◆SPICA		ANTARES		◆Peacock		ACHERNAR		CANOPUS		◆SIRIUS		REGULUS	
165	37 45	047	25 42	100	26 05	158	26 42	203	50 53	244	28 35	282	22 07	346
166	38 10	046	26 15	099	26 18	157	26 29	202	50 22	243	28 01	281	21 59	345
167	38 34	045	26 49	098	26 32	156	26 16	201	49 52	243	27 27	280	21 50	344
168	38 58	044	27 24	098	26 45	156	26 04	201	49 21	242	26 53	279	21 40	343
169	39 22	043	27 58	097	27 00	155	25 51	200	48 51	242	26 19	278	21 30	342
170	39 45	041	28 32	096	27 14	155	25 40	200	48 21	241	25 45	278	21 19	341
171	40 07	040	29 06	095	27 29	154	25 28	199	47 51	241	25 11	277	21 07	340
172	40 29	039	29 40	095	27 44	154	25 17	199	47 21	240	24 37	276	20 55	339
173	40 51	038	30 15	094	27 59	153	25 06	198	46 51	240	24 02	275	20 43	338
174	41 12	037	30 49	093	28 15	153	24 56	198	46 21	239	23 28	274	20 30	337
175	41 32	035	31 23	092	28 31	152	24 45	197	45 52	239	22 54	273	20 16	336
176	41 51	034	31 58	091	28 47	152	24 35	196	45 22	238	22 19	273	20 02	335
177	42 10	033	32 32	091	29 04	151	24 26	196	44 53	238	21 45	272	19 47	334
178	42 29	032	33 07	090	29 21	150	24 17	195	44 24	237	21 11	271	19 31	333
179	42 47	030	33 41	089	29 38	150	24 08	195	43 55	237	20 36	270	19 16	332

LAT 55°S

LHA	Hc	Zn	Hc	Zn	Hc	Zn	Hc	Zn	Hc	Zn	Hc	Zn	Hc	Zn
♈	◆ SPICA		ANTARES		◆ Peacock		ACHERNAR		CANOPUS		◆ Suhail		REGULUS	
180	43 04	029	34 16	088	29 55	149	23 59	194	43 27	236	60 10	275	18 59	331
181	43 20	028	34 50	087	30 13	149	23 51	194	42 58	236	59 36	274	18 42	330
182	43 36	027	35 24	086	30 31	148	23 43	193	42 30	235	59 01	273	18 25	329
183	43 51	025	35 59	086	30 49	148	23 35	192	42 02	235	58 27	272	18 07	328
184	44 05	024	36 33	085	31 08	147	23 28	192	41 34	234	57 52	271	17 48	327
185	44 19	023	37 07	084	31 26	147	23 21	191	41 06	234	57 18	270	17 29	326
186	44 31	021	37 41	083	31 45	146	23 15	191	40 38	233	56 44	270	17 10	325
187	44 44	020	38 15	082	32 05	146	23 09	190	40 11	232	56 09	269	16 50	324
188	44 55	018	38 49	081	32 24	145	23 03	190	39 44	232	55 35	268	16 29	323
189	45 05	017	39 23	080	32 44	145	22 57	189	39 17	231	55 00	267	16 09	322
190	45 15	016	39 57	079	33 04	144	22 52	188	38 50	231	54 26	266	15 47	321
191	45 24	014	40 31	078	33 24	144	22 47	188	38 23	230	53 52	266	15 26	320
192	45 32	013	41 05	078	33 45	143	22 43	187	37 57	230	53 17	265	15 03	319
193	45 39	012	41 38	077	34 06	143	22 38	187	37 31	229	52 43	264	14 41	318
194	45 46	010	42 12	076	34 27	142	22 35	186	37 05	229	52 09	263	14 18	318
	◆ ARCTURUS		ANTARES		◆ Peacock		ACHERNAR		CANOPUS		◆ Suhail		Denebola	
195	14 04	018	42 45	075	34 48	142	22 31	185	36 39	228	51 35	263	18 47	342
196	14 15	017	43 18	074	35 10	141	22 28	185	36 13	228	51 01	262	18 36	341
197	14 25	016	43 51	073	35 31	141	22 25	184	35 48	227	50 27	261	18 24	340
198	14 34	015	44 24	072	35 53	140	22 23	184	35 23	227	49 53	261	18 12	339
199	14 43	014	44 56	071	36 16	140	22 21	183	34 58	226	49 19	260	18 00	338
200	14 51	014	45 29	070	36 38	139	22 19	183	34 33	226	48 45	259	17 46	337
201	14 59	013	46 01	069	37 01	139	22 18	182	34 09	225	48 11	258	17 32	336
202	15 06	012	46 33	068	37 24	138	22 17	181	33 45	224	47 37	258	17 18	335
203	15 13	011	47 05	067	37 47	138	22 16	181	33 21	224	47 04	257	17 03	334
204	15 19	010	47 36	066	38 10	137	22 16	180	32 57	223	46 30	256	16 48	333
205	15 24	009	48 07	064	38 34	137	22 16	180	32 33	223	45 57	256	16 32	332
206	15 29	008	48 38	063	38 57	136	22 16	179	32 10	222	45 24	255	16 15	331
207	15 34	007	49 09	062	39 21	136	22 17	178	31 47	222	44 50	254	15 58	330
208	15 37	006	49 39	061	39 46	135	22 18	178	31 24	221	44 17	254	15 41	329
209	15 41	005	50 09	060	40 10	135	22 19	177	31 02	221	43 44	253	15 23	328
	◆ ARCTURUS		ANTARES		◆ Peacock		ACHERNAR		CANOPUS		◆ Suhail		SPICA	
210	15 43	004	50 39	059	40 35	134	22 21	177	30 40	220	43 12	252	45 36	348
211	15 45	003	51 08	058	40 59	134	22 23	176	30 18	219	42 39	252	45 28	346
212	15 46	002	51 37	056	41 24	133	22 26	176	29 56	219	42 06	251	45 19	345
213	15 47	001	52 05	055	41 50	133	22 29	175	29 34	218	41 34	250	45 10	344
214	15 47	000	52 33	054	42 15	132	22 32	174	29 13	218	41 01	250	45 00	342
215	15 47	359	53 01	053	42 41	132	22 35	174	28 52	217	40 29	249	44 49	341
216	15 46	358	53 28	051	43 06	131	22 39	173	28 31	217	39 57	248	44 37	339
217	15 45	357	53 55	050	43 32	131	22 44	173	28 11	216	39 25	248	44 25	338
218	15 42	356	54 21	049	43 58	130	22 48	172	27 51	216	38 53	247	44 11	337
219	15 40	355	54 46	047	44 25	130	22 53	171	27 31	215	38 22	247	43 57	335
220	15 36	354	55 11	046	44 51	129	22 58	171	27 11	214	37 50	246	43 43	334
221	15 32	353	55 36	045	45 18	129	23 04	170	26 52	214	37 19	245	43 27	333
222	15 28	352	56 00	043	45 45	129	23 10	170	26 33	213	36 48	245	43 11	331
223	15 23	351	56 23	042	46 12	128	23 16	169	26 14	213	36 17	244	42 55	330
224	15 17	350	56 45	040	46 39	128	23 23	169	25 56	212	35 46	243	42 37	329
	ANTARES		◆ Nunki		Peacock		ACHERNAR		◆ CANOPUS		Suhail		◆ SPICA	
225	57 07	039	39 02	081	47 06	127	23 30	168	25 38	211	35 15	243	42 19	328
226	57 28	037	39 36	080	47 34	127	23 37	167	25 20	211	34 45	242	42 00	326
227	57 49	036	40 10	079	48 01	126	23 45	167	25 03	210	34 14	241	41 41	325
228	58 09	034	40 44	078	48 29	126	23 53	166	24 45	210	33 44	241	41 21	324
229	58 27	032	41 17	077	48 57	125	24 01	166	24 28	209	33 14	240	41 01	323
230	58 46	031	41 51	076	49 25	125	24 10	165	24 12	209	32 44	240	40 39	322
231	59 03	029	42 24	075	49 53	125	24 19	165	23 56	208	32 15	239	40 18	320
232	59 19	028	42 57	074	50 22	124	24 28	164	23 40	207	31 45	238	39 55	319
233	59 35	026	43 30	073	50 50	124	24 38	163	23 24	207	31 16	238	39 33	318
234	59 49	024	44 03	072	51 19	123	24 48	163	23 09	206	30 47	237	39 09	317
235	60 03	022	44 36	071	51 48	123	24 58	162	22 54	206	30 18	236	38 45	316
236	60 15	021	45 08	070	52 17	122	25 09	162	22 39	205	29 50	236	38 21	314
237	60 27	019	45 40	069	52 46	122	25 20	161	22 25	204	29 21	235	37 56	313
238	60 38	017	46 12	068	53 15	122	25 31	161	22 11	204	28 53	235	37 31	312
239	60 47	015	46 44	067	53 44	121	25 42	160	21 57	203	28 25	234	37 05	311
	◆ ANTARES		Nunki		◆ FOMALHAUT		ACHERNAR		CANOPUS		◆ ACRUX		SPICA	
240	60 56	013	47 16	066	16 22	119	25 54	160	21 44	203	62 16	231	36 39	310
241	61 03	012	47 47	065	16 52	118	26 06	159	21 31	202	61 49	231	36 13	309
242	61 10	010	48 18	064	17 23	117	26 19	158	21 18	201	61 22	231	35 46	308
243	61 15	008	48 49	063	17 53	116	26 32	158	21 06	201	60 56	231	35 18	307
244	61 19	006	49 19	062	18 24	116	26 45	157	20 54	200	60 29	231	34 51	306
245	61 22	004	49 49	060	18 55	115	26 58	157	20 42	199	60 02	231	34 23	305
246	61 24	002	50 19	059	19 27	114	27 12	156	20 31	199	59 36	230	33 54	304
247	61 25	001	50 49	058	19 58	113	27 26	156	20 20	198	59 09	230	33 25	303
248	61 25	359	51 18	057	20 30	113	27 41	155	20 09	198	58 43	230	32 56	302
249	61 24	357	51 46	056	21 02	112	27 55	155	19 59	197	58 17	230	32 27	301
250	61 21	355	52 14	054	21 34	111	28 10	154	19 49	196	57 50	230	31 57	300
251	61 17	353	52 42	053	22 06	111	28 25	153	19 39	196	57 24	229	31 27	299
252	61 13	351	53 09	052	22 38	110	28 41	153	19 30	195	56 58	229	30 56	298
253	61 07	349	53 36	051	23 10	109	28 57	152	19 21	195	56 32	229	30 26	297
254	61 00	348	54 03	049	23 43	108	29 13	152	19 13	194	56 06	229	29 55	296
	ALTAIR		◆ FOMALHAUT		ACHERNAR		CANOPUS		◆ ACRUX		SPICA		◆ ANTARES	
255	16 55	044	24 16	108	29 29	151	19 05	193	55 41	228	29 24	295	60 52	346
256	17 18	043	24 49	107	29 46	151	18 57	193	55 15	228	28 52	294	60 43	344
257	17 42	042	25 22	106	30 03	150	18 50	192	54 49	228	28 21	293	60 33	342
258	18 05	042	25 55	105	30 20	150	18 43	191	54 24	228	27 49	292	60 22	340
259	18 27	041	26 28	105	30 38	149	18 36	191	53 59	227	27 17	291	60 10	339
260	18 50	040	27 01	104	30 55	149	18 30	190	53 33	227	26 45	290	59 57	337
261	19 11	039	27 35	103	31 13	148	18 24	189	53 08	227	26 12	289	59 43	335
262	19 33	038	28 08	102	31 32	148	18 19	189	52 43	226	25 40	288	59 28	333
263	19 53	037	28 42	102	31 50	147	18 14	188	52 19	226	25 07	287	59 12	332
264	20 14	036	29 16	101	32 09	147	18 09	188	51 54	226	24 34	286	58 55	330
265	20 34	035	29 50	100	32 28	146	18 04	187	51 29	225	24 01	286	58 37	328
266	20 53	034	30 24	099	32 47	146	18 00	186	51 05	225	23 28	285	58 19	327
267	21 12	033	30 58	099	33 07	145	17 57	186	50 40	225	22 54	284	58 00	325
268	21 30	032	31 32	098	33 27	145	17 54	185	50 16	224	22 21	283	57 40	324
269	21 48	031	32 06	097	33 47	144	17 51	184	49 52	224	21 47	282	57 19	322

LHA	Hc	Zn	Hc	Zn	Hc	Zn	Hc	Zn	Hc	Zn	Hc	Zn	Hc	Zn
♈	◆ ALTAIR		FOMALHAUT		◆ ACHERNAR		CANOPUS		RIGIL KENT.		◆ SPICA		ANTARES	
270	22 05	030	32 40	096	34 07	143	17 48	184	63 25	237	21 13	281	56 58	321
271	22 22	029	33 14	095	34 28	143	17 46	183	62 56	237	20 40	280	56 35	319
272	22 38	028	33 48	095	34 49	142	17 45	183	62 27	236	20 06	279	56 13	318
273	22 54	027	34 23	094	35 10	142	17 43	182	61 59	236	19 32	279	55 49	316
274	23 09	026	34 57	093	35 31	141	17 42	181	61 30	236	18 58	278	55 25	315
275	23 23	024	35 32	092	35 53	141	17 42	181	61 02	236	18 24	277	55 00	313
276	23 37	023	36 06	091	36 15	140	17 42	180	60 33	235	17 49	276	54 35	312
277	23 51	022	36 40	091	36 37	140	17 42	179	60 05	235	17 15	275	54 09	311
278	24 04	021	37 15	090	36 59	139	17 42	179	59 37	235	16 41	274	53 43	309
279	24 16	020	37 49	089	37 21	139	17 43	178	59 08	235	16 06	274	53 16	308
280	24 27	019	38 24	088	37 44	139	17 45	177	58 40	234	15 32	273	52 49	307
281	24 38	018	38 58	087	38 07	138	17 46	177	58 12	234	14 58	272	52 21	306
282	24 49	017	39 32	087	38 30	138	17 49	176	57 45	234	14 23	271	51 53	304
283	24 59	016	40 07	086	38 53	137	17 51	176	57 17	234	13 49	270	51 24	303
284	25 08	015	40 41	085	39 17	137	17 54	175	56 49	233	13 14	269	50 55	302
	◆ ALTAIR		FOMALHAUT		◆ ACHERNAR		CANOPUS		◆ RIGIL KENT.		ANTARES		Rasalhague	
285	25 16	014	41 15	084	39 41	136	17 57	174	56 22	233	50 25	301	20 05	338
286	25 24	013	41 49	083	40 05	136	18 01	174	55 54	233	49 56	300	19 52	337
287	25 32	012	42 24	082	40 29	135	18 05	173	55 27	232	49 26	298	19 38	336
288	25 38	011	42 58	081	40 53	135	18 09	172	55 00	232	48 55	297	19 23	335
289	25 44	009	43 32	080	41 18	134	18 14	172	54 33	232	48 24	296	19 08	334
290	25 49	008	44 05	079	41 42	134	18 19	171	54 06	231	47 53	295	18 53	333
291	25 54	007	44 39	079	42 07	133	18 25	170	53 39	231	47 22	294	18 37	332
292	25 58	006	45 13	078	42 33	133	18 31	170	53 12	231	46 50	293	18 20	331
293	26 02	005	45 46	077	42 58	132	18 37	169	52 45	230	46 19	292	18 03	330
294	26 04	004	46 20	076	43 23	132	18 43	169	52 19	230	45 47	291	17 46	329
295	26 06	003	46 53	075	43 49	131	18 50	168	51 53	230	45 14	290	17 28	328
296	26 08	002	47 26	074	44 15	131	18 58	167	51 26	229	44 42	289	17 09	327
297	26 09	001	47 59	073	44 41	131	19 06	167	51 00	229	44 09	288	16 50	326
298	26 09	000	48 32	072	45 07	130	19 14	166	50 34	229	43 36	287	16 31	325
299	26 08	358	49 05	071	45 34	130	19 22	165	50 09	228	43 03	286	16 11	324
	FOMALHAUT		◆ ACHERNAR		CANOPUS		◆ RIGIL KENT.		ANTARES		Nunki		◆ ALTAIR	
300	49 37	070	46 00	129	19 31	165	49 43	228	42 30	285	58 56	331	26 07	357
301	50 09	069	46 27	129	19 40	164	49 17	228	41 57	284	58 39	329	26 05	356
302	50 41	068	46 54	128	19 50	164	48 52	227	41 23	283	58 21	328	26 02	355
303	51 13	066	47 21	128	20 00	163	48 27	227	40 50	282	58 02	326	25 59	354
304	51 44	065	47 48	127	20 10	162	48 02	227	40 16	281	57 43	324	25 55	353
305	52 15	064	48 16	127	20 21	162	47 37	226	39 42	280	57 22	323	25 51	352
306	52 46	063	48 43	127	20 31	161	47 12	226	39 08	279	57 01	321	25 46	351
307	53 17	062	49 11	126	20 43	160	46 48	225	38 34	278	56 39	320	25 40	350
308	53 47	061	49 39	126	20 54	160	46 23	225	38 00	278	56 17	318	25 33	349
309	54 16	059	50 07	125	21 06	159	45 59	225	37 26	277	55 54	317	25 26	348
310	54 46	058	50 35	125	21 19	159	45 35	224	36 52	276	55 30	316	25 19	346
311	55 15	057	51 03	124	21 32	158	45 11	224	36 17	275	55 05	314	25 10	345
312	55 44	056	51 32	124	21 45	157	44 47	223	35 43	274	54 40	313	25 01	344
313	56 12	054	52 00	124	21 58	157	44 24	223	35 09	273	54 15	311	24 52	343
314	56 40	053	52 29	123	22 12	156	44 00	223	34 34	272	53 49	310	24 41	342
	◆ FOMALHAUT		ACHERNAR		◆ CANOPUS		RIGIL KENT.		◆ ANTARES		Nunki		ALTAIR	
315	57 07	052	52 58	123	22 26	156	43 37	222	34 00	272	53 22	309	24 30	341
316	57 34	050	53 27	122	22 40	155	43 14	222	33 26	271	52 55	307	24 19	340
317	58 00	049	53 56	122	22 55	154	42 51	221	32 51	270	52 28	306	24 07	339
318	58 25	047	54 25	122	23 10	154	42 29	221	32 17	269	52 00	305	23 54	338
319	58 50	046	54 55	121	23 25	153	42 06	220	31 42	268	51 31	304	23 41	337
320	59 15	044	55 24	121	23 41	153	41 44	220	31 08	267	51 02	303	23 27	336
321	59 39	043	55 54	120	23 57	152	41 22	220	30 34	267	50 33	301	23 13	335
322	60 02	041	56 23	120	24 13	151	41 00	219	29 59	266	50 04	300	22 58	334
323	60 24	040	56 53	120	24 30	151	40 39	219	29 25	265	49 34	299	22 42	333
324	60 46	038	57 23	119	24 47	150	40 17	218	28 51	264	49 03	298	22 26	332
325	61 07	037	57 53	119	25 04	150	39 56	218	28 16	264	48 33	297	22 10	331
326	61 27	035	58 23	119	25 22	149	39 35	217	27 42	263	48 02	296	21 53	330
327	61 46	033	58 54	118	25 39	148	39 14	217	27 08	262	47 31	295	21 35	329
328	62 04	031	59 24	118	25 57	148	38 54	217	26 34	261	46 59	293	21 17	328
329	62 22	030	59 55	117	26 16	147	38 33	216	26 00	260	46 28	292	20 58	327
	◆ FOMALHAUT		ACHERNAR		◆ CANOPUS		RIGIL KENT.		◆ ANTARES		Nunki		ALTAIR	
330	62 39	028	60 25	117	26 35	147	38 13	216	25 26	260	45 56	291	20 39	326
331	62 54	026	60 56	117	26 54	146	37 53	215	24 52	259	45 24	290	20 19	325
332	63 09	024	61 27	116	27 13	146	37 33	215	24 19	258	44 51	289	19 59	324
333	63 23	022	61 58	116	27 33	145	37 14	214	23 45	257	44 19	288	19 38	323
334	63 35	021	62 29	116	27 52	144	36 55	214	23 12	257	43 46	287	19 17	322
335	63 47	019	63 00	115	28 13	144	36 36	213	22 38	256	43 13	286	18 56	321
336	63 57	017	63 31	115	28 33	143	36 17	213	22 05	255	42 40	285	18 34	320
337	64 06	015	64 02	115	28 54	143	35 58	212	21 32	254	42 07	284	18 11	319
338	64 15	013	64 33	114	29 15	142	35 40	212	20 59	254	41 33	284	17 48	318
339	64 22	011	65 05	114	29 36	142	35 22	211	20 26	253	41 00	283	17 25	317
340	64 27	009	65 36	114	29 58	141	35 04	211	19 53	252	40 26	282	17 01	316
341	64 32	007	66 08	113	30 19	140	34 46	211	19 20	251	39 52	281	16 37	315
342	64 36	005	66 39	113	30 41	140	34 29	210	18 48	251	39 18	280	16 12	314
343	64 38	003	67 11	113	31 04	139	34 12	210	18 15	250	38 44	279	15 47	313
344	64 39	001	67 43	112	31 26	139	33 55	209	17 43	249	38 10	278	15 22	312
	◆ Diphda		Acamar		◆ CANOPUS		Miaplacidus		RIGIL KENT.		◆ Nunki		FOMALHAUT	
345	48 06	038	48 45	095	31 49	138	36 11	169	33 38	209	37 36	277	64 39	359
346	48 27	037	49 20	094	32 12	138	36 18	168	33 22	208	37 02	276	64 37	357
347	48 47	036	49 54	093	32 35	137	36 25	168	33 06	208	36 28	275	64 35	355
348	49 07	034	50 28	092	32 59	137	36 32	168	32 50	207	35 54	275	64 31	353
349	49 26	033	51 03	092	33 23	136	36 40	167	32 35	207	35 19	274	64 26	351
350	49 44	032	51 37	091	33 47	135	36 48	167	32 19	206	34 45	273	64 20	349
351	50 02	030	52 12	090	34 11	135	36 56	166	32 04	206	34 11	272	64 12	347
352	50 19	029	52 46	089	34 35	134	37 04	166	31 50	205	33 36	271	64 04	345
353	50 35	027	53 20	088	35 00	134	37 12	166	31 35	205	33 02	270	63 54	343
354	50 51	026	53 55	087	35 25	133	37 21	165	31 21	204	32 27	270	63 43	341
355	51 05	024	54 29	087	35 50	133	37 30	165	31 07	204	31 53	269	63 31	339
356	51 19	023	55 04	086	36 15	132	37 39	164	30 53	203	31 19	268	63 18	337
357	51 32	021	55 38	085	36 41	132	37 48	164	30 40	203	30 44	267	63 04	335
358	51 44	020	56 12	084	37 07	131	37 58	164	30 27	202	30 10	266	62 49	333
359	51 55	018	56 46	083	37 33	131	38 08	163	30 14	202	29 35	266	62 34	332

LAT 56°S

LHA ♈	Hc	Zn	Hc	Zn	Hc	Zn	Hc	Zn	Hc	Zn	Hc	Zn	Hc	Zn
	Diphda		♦RIGEL		CANOPUS		Miaplacidus		♦RIGIL KENT.		Peacock		♦FOMALHAUT	
0	51 08	017	13 11	085	38 38	130	39 15	163	30 57	201	61 02	246	61 25	331
1	51 18	015	13 44	084	39 03	129	39 25	162	30 45	201	60 31	245	61 08	329
2	51 26	014	14 17	083	39 30	128	39 35	162	30 33	200	60 01	245	60 50	327
3	51 33	012	14 51	083	39 56	128	39 46	161	30 22	200	59 31	245	60 31	326
4	51 40	011	15 24	082	40 23	127	39 57	161	30 11	199	59 00	244	60 12	324
5	51 45	009	15 57	081	40 49	127	40 08	161	30 00	199	58 30	244	59 52	322
6	51 50	007	16 30	080	41 16	126	40 19	160	29 49	198	58 00	243	59 31	321
7	51 54	006	17 03	079	41 44	126	40 30	160	29 39	198	57 30	243	59 10	319
8	51 57	004	17 36	078	42 11	125	40 42	160	29 29	197	57 00	242	58 47	318
9	51 59	003	18 09	077	42 38	125	40 54	159	29 19	197	56 31	242	58 24	316
10	52 01	001	18 42	077	43 06	124	41 06	159	29 09	196	56 01	242	58 01	315
11	52 01	000	19 14	076	43 34	124	41 18	158	29 00	196	55 32	241	57 37	313
12	52 00	358	19 47	075	44 02	123	41 30	158	28 51	195	55 02	241	57 12	312
13	51 59	357	20 19	074	44 30	122	41 43	158	28 43	195	54 33	240	56 47	310
14	51 56	355	20 51	073	44 59	122	41 56	157	28 35	194	54 04	240	56 21	309
	♦RIGEL		SIRIUS		CANOPUS		♦RIGIL KENT.		Peacock		♦FOMALHAUT		Diphda	
15	21 23	072	15 53	096	45 27	121	28 27	193	53 35	239	55 55	308	51 53	354
16	21 55	071	16 26	096	45 56	121	28 19	193	53 06	239	55 28	306	51 49	352
17	22 27	070	16 59	095	46 25	120	28 12	192	52 38	238	55 01	305	51 44	351
18	22 58	069	17 33	094	46 54	120	28 05	192	52 09	238	54 33	304	51 38	349
19	23 30	069	18 06	093	47 23	119	27 58	191	51 41	238	54 05	303	51 31	347
20	24 01	068	18 40	092	47 52	119	27 51	191	51 13	237	53 37	301	51 23	346
21	24 32	067	19 13	092	48 22	118	27 45	190	50 44	237	53 08	300	51 15	344
22	25 02	066	19 47	091	48 51	118	27 39	190	50 17	236	52 39	299	51 05	343
23	25 33	065	20 20	090	49 21	117	27 34	189	49 49	236	52 09	298	50 55	341
24	26 03	064	20 54	089	49 51	117	27 29	189	49 21	235	51 39	297	50 44	340
25	26 33	063	21 27	088	50 21	116	27 24	188	48 54	235	51 09	295	50 32	339
26	27 03	062	22 01	087	50 51	116	27 19	188	48 26	234	50 39	294	50 19	337
27	27 32	061	22 35	087	51 22	115	27 15	187	47 59	234	50 08	293	50 06	336
28	28 01	060	23 08	086	51 52	115	27 11	186	47 32	233	49 37	292	49 52	334
29	28 30	059	23 41	085	52 23	114	27 08	186	47 05	233	49 06	291	49 37	333
	♦RIGEL		SIRIUS		CANOPUS		♦RIGIL KENT.		Peacock		♦FOMALHAUT		Diphda	
30	28 59	058	24 15	084	52 53	113	27 04	185	46 38	233	48 34	290	49 21	331
31	29 27	057	24 48	083	53 24	113	27 01	185	46 12	232	48 03	289	49 04	330
32	29 55	056	25 21	082	53 55	112	26 59	184	45 45	232	47 31	288	48 47	329
33	30 23	055	25 55	081	54 26	112	26 56	184	45 19	231	46 59	287	48 30	327
34	30 50	054	26 28	081	54 58	111	26 54	183	44 53	231	46 27	286	48 11	326
35	31 17	053	27 01	080	55 29	111	26 53	183	44 27	230	45 54	285	47 52	325
36	31 44	052	27 34	079	56 00	110	26 51	182	44 02	230	45 22	284	47 32	323
37	32 10	051	28 07	078	56 32	110	26 50	182	43 36	229	44 49	283	47 12	322
38	32 36	050	28 39	077	57 03	109	26 49	181	43 11	229	44 17	282	46 51	321
39	33 01	049	29 12	076	57 35	109	26 49	180	42 46	228	43 44	281	46 29	319
40	33 26	048	29 45	075	58 07	108	26 49	180	42 21	228	43 11	280	46 07	318
41	33 51	047	30 17	074	58 39	107	26 49	179	41 56	227	42 38	279	45 44	317
42	34 15	046	30 49	073	59 11	107	26 50	179	41 31	227	42 05	278	45 21	316
43	34 39	044	31 21	072	59 43	106	26 51	178	41 07	226	41 31	278	44 57	314
44	35 02	043	31 53	072	60 15	106	26 52	178	40 43	226	40 58	277	44 33	313
	RIGEL		♦SIRIUS		CANOPUS		♦RIGIL KENT.		Peacock		FOMALHAUT		♦Diphda	
45	35 25	042	32 25	071	60 48	105	26 53	177	40 19	225	40 25	276	44 08	312
46	35 47	041	32 56	070	61 20	105	26 55	177	39 55	225	39 51	275	43 43	311
47	36 09	040	33 28	069	61 53	104	26 57	176	39 32	224	39 18	274	43 17	310
48	36 30	039	33 59	068	62 25	104	27 00	176	39 08	224	38 44	273	42 51	308
49	36 51	038	34 30	067	62 58	103	27 02	175	38 45	223	38 11	272	42 25	307
50	37 11	036	35 01	066	63 31	102	27 05	174	38 22	223	37 37	271	41 58	306
51	37 31	035	35 31	065	64 03	102	27 09	174	38 00	222	37 04	271	41 31	305
52	37 50	034	36 01	064	64 36	101	27 13	173	37 37	222	36 30	270	41 03	304
53	38 08	033	36 31	063	65 09	101	27 17	173	37 15	221	35 57	269	40 35	303
54	38 26	032	37 01	062	65 42	100	27 21	172	36 53	221	35 23	268	40 07	302
55	38 44	030	37 30	061	66 15	099	27 26	172	36 31	220	34 50	267	39 38	301
56	39 00	029	37 59	060	66 49	099	27 31	171	36 10	220	34 16	267	39 09	300
57	39 16	028	38 28	059	67 22	098	27 36	171	35 48	219	33 43	266	38 40	299
58	39 32	027	38 57	058	67 55	097	27 42	170	35 27	219	33 09	265	38 10	298
59	39 47	026	39 25	056	68 28	097	27 47	170	35 06	218	32 36	264	37 40	296
	RIGEL		♦SIRIUS		Suhail		♦RIGIL KENT.		Peacock		FOMALHAUT		♦Diphda	
60	40 01	024	39 53	055	41 25	109	27 54	169	34 46	218	32 02	263	37 10	295
61	40 14	023	40 20	054	41 56	109	28 00	168	34 25	217	31 29	263	36 39	294
62	40 27	022	40 47	053	42 28	108	28 07	168	34 05	217	30 56	262	36 09	293
63	40 39	021	41 14	052	43 00	107	28 14	167	33 45	216	30 23	261	35 38	292
64	40 51	019	41 40	051	43 32	107	28 22	167	33 26	216	29 50	260	35 07	291
65	41 01	018	42 06	050	44 04	106	28 30	166	33 06	215	29 17	259	34 35	291
66	41 11	017	42 31	049	44 37	105	28 38	166	32 47	215	28 44	259	34 04	290
67	41 21	015	42 56	047	45 09	105	28 46	165	32 28	214	28 11	258	33 32	289
68	41 29	014	43 21	046	45 42	104	28 55	165	32 10	213	27 38	257	33 00	288
69	41 37	013	43 44	045	46 14	103	29 04	164	31 51	213	27 05	256	32 28	287
70	41 44	011	44 08	044	46 47	103	29 13	164	31 33	212	26 33	256	31 56	286
71	41 50	010	44 31	043	47 20	102	29 22	163	31 15	212	26 00	255	31 24	285
72	41 56	009	44 53	041	47 53	101	29 32	163	30 58	211	25 28	254	30 51	284
73	42 00	007	45 15	040	48 26	100	29 43	162	30 40	211	24 56	253	30 19	283
74	42 04	006	45 37	039	48 59	100	29 53	162	30 23	210	24 24	253	29 46	282
	RIGEL		♦SIRIUS		Suhail		♦RIGIL KENT.		Peacock		ACHERNAR		♦Diphda	
75	42 07	005	45 57	038	49 32	099	30 04	161	30 07	210	62 47	246	29 13	281
76	42 10	003	46 17	036	50 05	098	30 15	161	29 50	209	62 16	246	28 40	280
77	42 11	002	46 37	035	50 38	097	30 26	160	29 34	209	61 45	245	28 07	280
78	42 12	001	46 56	034	51 12	097	30 38	159	29 18	208	61 15	245	27 34	279
79	42 12	359	47 14	032	51 45	096	30 50	159	29 03	207	60 45	244	27 01	278
80	42 12	358	47 32	031	52 18	095	31 02	158	28 47	207	60 14	244	26 27	277
81	42 10	357	47 48	030	52 52	094	31 14	158	28 32	206	59 44	244	25 54	276
82	42 08	355	48 05	028	53 25	094	31 27	157	28 17	206	59 14	243	25 21	275
83	42 05	354	48 20	027	53 59	093	31 40	157	28 03	205	58 44	243	24 47	274
84	42 01	353	48 35	025	54 32	092	31 53	156	27 49	205	58 15	242	24 14	273
85	41 56	351	48 49	024	55 06	091	32 07	156	27 35	204	57 45	242	23 40	273
86	41 51	350	49 02	023	55 39	090	32 21	155	27 21	204	57 15	242	23 07	272
87	41 45	349	49 15	021	56 13	090	32 35	155	27 08	203	56 46	241	22 33	271
88	41 38	347	49 26	020	56 46	089	32 49	154	26 55	202	56 17	241	22 00	270
89	41 30	346	49 37	018	57 20	088	33 04	154	26 43	202	55 47	240	21 26	269

LHA ♈	Hc	Zn	Hc	Zn	Hc	Zn	Hc	Zn	Hc	Zn	Hc	Zn	Hc	Zn
	SIRIUS		Alphard		♦Suhail		RIGIL KENT.		♦Peacock		ACHERNAR		♦RIGEL	
90	49 47	017	27 47	061	57 53	087	33 19	153	26 30	201	55 18	240	41 22	345
91	49 57	015	28 16	061	58 27	086	33 34	153	26 18	201	54 49	239	41 13	344
92	50 05	014	28 45	060	59 00	085	33 49	152	26 07	200	54 20	239	41 03	342
93	50 13	012	29 14	059	59 34	084	34 05	152	25 55	200	53 52	239	40 52	341
94	50 20	011	29 43	058	60 07	083	34 21	151	25 44	199	53 23	238	40 41	340
95	50 25	009	30 11	057	60 40	083	34 37	151	25 33	198	52 55	238	40 29	338
96	50 30	008	30 39	056	61 14	082	34 54	150	25 23	198	52 26	237	40 16	337
97	50 35	006	31 06	054	61 47	081	35 10	150	25 13	197	51 58	237	40 03	336
98	50 38	005	31 33	053	62 20	080	35 27	149	25 03	197	51 30	236	39 49	335
99	50 40	003	32 00	052	62 53	079	35 44	149	24 54	196	51 02	236	39 34	333
100	50 42	002	32 26	051	63 26	077	36 02	148	24 44	196	50 35	236	39 19	332
101	50 43	000	32 52	050	63 58	076	36 20	148	24 36	195	50 07	235	39 03	331
102	50 42	359	33 18	049	64 31	075	36 37	147	24 27	194	49 39	235	38 46	330
103	50 41	357	33 43	048	65 03	074	36 56	147	24 19	194	49 12	234	38 29	328
104	50 39	356	34 08	047	65 35	073	37 14	147	24 11	193	48 45	234	38 11	327
	Alphard		♦Gienah		RIGIL KENT.		♦Peacock		ACHERNAR		♦RIGEL		SIRIUS	
105	34 32	046	20 38	091	37 33	146	24 04	193	48 18	233	37 53	326	50 36	354
106	34 56	045	21 11	090	37 51	146	23 56	192	47 51	233	37 34	325	50 32	353
107	35 20	044	21 45	089	38 10	145	23 50	191	47 25	232	37 14	324	50 28	351
108	35 43	043	22 18	088	38 30	145	23 43	191	46 58	232	36 54	323	50 22	350
109	36 05	042	22 52	088	38 49	144	23 37	190	46 32	231	36 33	321	50 16	348
110	36 27	040	23 25	087	39 09	144	23 31	190	46 06	231	36 12	320	50 09	347
111	36 49	039	23 59	086	39 29	143	23 26	189	45 40	231	35 50	319	50 01	345
112	37 10	038	24 32	085	39 49	143	23 21	188	45 14	230	35 28	318	49 52	344
113	37 30	037	25 06	084	40 10	142	23 16	188	44 48	230	35 05	317	49 42	342
114	37 50	036	25 39	083	40 30	142	23 11	187	44 23	229	34 42	316	49 32	341
115	38 09	035	26 12	083	40 51	141	23 07	187	43 57	229	34 19	315	49 20	340
116	38 28	033	26 46	082	41 12	141	23 04	186	43 32	228	33 55	314	49 08	338
117	38 46	032	27 19	081	41 33	141	23 00	186	43 07	228	33 30	312	48 55	337
118	39 04	031	27 52	080	41 55	140	22 57	185	42 42	227	33 05	311	48 41	335
119	39 21	030	28 25	079	42 16	140	22 54	184	42 18	227	32 40	310	48 27	334
	REGULUS		♦SPICA		RIGIL KENT.		Peacock		♦ACHERNAR		RIGEL		♦SIRIUS	
120	16 57	033	14 06	089	42 38	139	22 52	184	41 54	226	32 14	309	48 12	332
121	17 15	032	14 40	088	43 00	139	22 50	183	41 29	226	31 48	308	47 56	331
122	17 32	031	15 14	087	43 22	138	22 48	183	41 05	225	31 21	307	47 39	330
123	17 49	030	15 47	087	43 45	138	22 47	182	40 42	225	30 54	306	47 22	328
124	18 06	029	16 20	086	44 07	137	22 46	181	40 18	224	30 27	305	47 04	327
125	18 22	028	16 54	085	44 30	137	22 45	181	39 55	224	29 59	304	46 46	326
126	18 37	027	17 27	084	44 53	137	22 45	180	39 32	223	29 31	303	46 26	324
127	18 52	026	18 01	083	45 16	136	22 45	180	39 09	223	29 03	302	46 07	323
128	19 06	025	18 34	082	45 39	136	22 46	179	38 46	222	28 35	301	45 46	322
129	19 20	024	19 07	081	46 03	135	22 46	178	38 23	222	28 06	300	45 25	321
130	19 34	023	19 40	081	46 27	135	22 47	178	38 01	221	27 37	299	45 04	319
131	19 46	022	20 13	080	46 50	135	22 49	177	37 39	221	27 07	298	44 41	318
132	19 59	021	20 46	079	47 14	134	22 51	177	37 17	220	26 37	297	44 19	317
133	20 10	020	21 19	078	47 39	134	22 53	176	36 56	220	26 07	296	43 55	316
134	20 21	019	21 52	077	48 03	133	22 55	175	36 34	219	25 37	295	43 32	314
	REGULUS		♦SPICA		RIGIL KENT.		♦Peacock		ACHERNAR		CANOPUS		♦SIRIUS	
135	20 32	018	22 25	076	48 27	133	22 58	175	36 13	219	67 19	262	43 07	313
136	20 42	017	22 57	075	48 52	133	23 01	174	35 52	218	66 46	261	42 43	312
137	20 51	016	23 30	074	49 17	132	23 05	174	35 31	218	66 13	261	42 18	311
138	21 00	015	24 02	074	49 42	132	23 09	173	35 11	217	65 40	260	41 52	310
139	21 08	014	24 34	073	50 07	131	23 13	172	34 51	217	65 07	259	41 26	309
140	21 16	013	25 06	072	50 32	131	23 18	172	34 31	216	64 34	259	40 59	307
141	21 23	012	25 38	071	50 58	131	23 23	171	34 11	216	64 01	258	40 33	306
142	21 30	011	26 09	070	51 23	130	23 28	171	33 52	215	63 28	258	40 05	305
143	21 35	009	26 41	069	51 49	130	23 33	170	33 32	215	62 55	257	39 38	304
144	21 41	008	27 12	068	52 15	129	23 39	170	33 13	214	62 23	256	39 10	303
145	21 45	007	27 43	067	52 41	129	23 46	169	32 55	214	61 50	256	38 41	302
146	21 49	006	28 14	066	53 07	129	23 52	168	32 36	213	61 18	255	38 13	301
147	21 53	005	28 44	065	53 33	128	23 59	168	32 18	213	60 45	255	37 44	300
148	21 55	004	29 15	064	53 59	128	24 06	167	32 00	212	60 13	254	37 15	299
149	21 58	003	29 45	063	54 26	128	24 14	167	31 42	212	59 41	254	36 45	298
	REGULUS		♦SPICA		RIGIL KENT.		♦Peacock		ACHERNAR		CANOPUS		♦SIRIUS	
150	21 59	002	30 15	062	54 52	127	24 22	166	31 25	211	59 08	253	36 15	297
151	22 00	001	30 44	061	55 19	127	24 30	165	31 08	210	58 36	252	35 45	296
152	22 00	000	31 13	060	55 46	127	24 39	165	30 51	210	58 05	252	35 15	295
153	22 00	359	31 42	059	56 13	126	24 48	164	30 34	209	57 33	251	34 44	294
154	21 59	358	32 11	058	56 40	126	24 57	164	30 18	209	57 01	251	34 13	293
155	21 58	357	32 40	057	57 07	126	25 07	163	30 02	208	56 29	250	33 42	292
156	21 56	356	33 08	056	57 35	125	25 17	163	29 46	208	55 58	250	33 11	291
157	21 53	355	33 35	055	58 02	125	25 27	162	29 31	207	55 26	249	32 40	290
158	21 49	354	34 03	054	58 30	125	25 37	161	29 15	207	54 55	249	32 08	289
159	21 46	353	34 30	053	58 57	124	25 48	161	29 01	206	54 24	248	31 36	288
160	21 41	352	34 56	052	59 25	124	25 59	160	28 46	206	53 53	248	31 04	287
161	21 36	351	35 23	051	59 53	124	26 11	160	28 32	205	53 22	247	30 32	286
162	21 30	350	35 49	050	60 21	123	26 23	159	28 18	204	52 51	247	30 00	285
163	21 23	348	36 14	049	60 49	123	26 35	159	28 04	204	52 20	246	29 27	284
164	21 16	347	36 39	048	61 17	123	26 47	158	27 50	203	51 50	245	28 55	283
	♦SPICA		ANTARES		♦Peacock		ACHERNAR		CANOPUS		♦SIRIUS		REGULUS	
165	37 04	047	25 52	100	27 00	157	27 37	203	51 19	245	28 22	282	21 09	346
166	37 28	045	26 25	099	27 13	157	27 24	202	50 49	244	27 49	282	21 01	345
167	37 52	044	26 58	098	27 26	156	27 12	202	50 19	244	27 16	281	20 52	344
168	38 15	043	27 31	097	27 40	156	27 00	201	49 49	243	26 43	280	20 43	343
169	38 38	042	28 05	096	27 54	155	26 48	201	49 19	243	26 10	279	20 33	342
170	39 00	041	28 38	096	28 08	155	26 36	200	48 49	242	25 37	278	20 22	341
171	39 21	040	29 11	095	28 23	154	26 25	199	48 19	242	25 04	277	20 11	340
172	39 43	039	29 45	094	28 38	153	26 14	199	47 50	241	24 30	276	19 59	339
173	40 03	037	30 18	093	28 53	153	26 03	198	47 21	241	23 57	276	19 47	338
174	40 23	036	30 52	092	29 08	152	25 53	198	46 51	240	23 24	275	19 34	337
175	40 43	035	31 25	092	29 24	152	25 43	197	46 22	240	22 50	274	19 21	336
176	41 02	034	31 59	091	29 40	151	25 33	197	45 54	239	22 17	273	19 07	335
177	41 20	032	32 32	090	29 56	151	25 24	196	45 25	239	21 43	272	18 53	334
178	41 38	031	33 06	089	30 13	150	25 14	195	44 56	238	21 10	271	18 38	333
179	41 55	030	33 40	088	30 30	150	25 06	195	44 28	237	20 36	270	18 23	332

LHA ♈	Hc	Zn	Hc	Zn	Hc	Zn	Hc	Zn	Hc	Zn	Hc	Zn	Hc	Zn
	◆SPICA		ANTARES		◆Peacock		ACHERNAR		CANOPUS		◆SIRIUS		Alphard	
180	42 11	029	34 13	087	30 47	149	24 57	194	44 00	237	20 02	270	34 00	313
181	42 27	027	34 47	087	31 04	149	24 49	194	43 32	236	19 29	269	33 35	311
182	42 42	026	35 20	086	31 22	148	24 41	193	43 04	236	18 55	268	33 10	310
183	42 56	025	35 54	085	31 40	147	24 34	193	42 36	235	18 22	267	32 44	309
184	43 10	023	36 27	084	31 58	147	24 27	192	42 09	235	17 48	266	32 18	308
185	43 23	022	37 00	083	32 16	146	24 20	191	41 41	234	17 15	266	31 51	307
186	43 35	021	37 34	082	32 35	146	24 14	191	41 14	234	16 41	265	31 24	306
187	43 47	020	38 07	081	32 54	145	24 08	190	40 47	233	16 08	264	30 57	305
188	43 58	018	38 40	080	33 13	145	24 02	190	40 21	233	15 35	263	30 29	304
189	44 08	017	39 13	079	33 33	144	23 56	189	39 54	232	15 01	262	30 02	303
190	44 17	015	39 46	079	33 53	144	23 51	188	39 28	232	14 28	261	29 33	302
191	44 26	014	40 19	078	34 13	143	23 47	188	39 01	231	13 55	261	29 05	301
192	44 34	013	40 51	077	34 33	143	23 42	187	38 35	230	13 22	260	28 36	300
193	44 41	011	41 24	076	34 53	142	23 38	187	38 10	230	12 49	259	28 07	299
194	44 47	010	41 56	075	35 14	142	23 34	186	37 44	229	12 16	258	27 37	298
	SPICA		◆ANTARES		Peacock		◆ACHERNAR		CANOPUS		Suhail		◆Alphard	
195	44 52	009	42 29	074	35 35	141	23 31	186	37 19	229	51 42	264	27 08	297
196	44 57	007	43 01	073	35 56	141	23 28	185	36 54	228	51 08	263	26 38	296
197	45 01	006	43 33	072	36 18	140	23 25	184	36 29	228	50 35	262	26 07	295
198	45 04	004	44 05	071	36 39	140	23 23	184	36 04	227	50 02	262	25 37	294
199	45 06	003	44 36	070	37 01	139	23 21	183	35 39	227	49 29	261	25 06	293
200	45 08	002	45 08	069	37 23	139	23 19	183	35 15	226	48 56	260	24 35	293
201	45 08	000	45 39	068	37 46	138	23 18	182	34 51	226	48 23	260	24 04	292
202	45 08	359	46 10	067	38 08	138	23 17	181	34 27	225	47 50	259	23 33	291
203	45 07	358	46 40	066	38 31	137	23 16	181	34 04	224	47 17	258	23 02	290
204	45 05	356	47 11	065	38 54	137	23 16	180	33 40	224	46 44	257	22 30	289
205	45 02	355	47 41	063	39 17	136	23 16	180	33 17	223	46 11	257	21 58	288
206	44 59	353	48 11	062	39 41	136	23 16	179	32 54	223	45 39	256	21 26	287
207	44 55	352	48 40	061	40 04	135	23 17	178	32 32	222	45 06	255	20 54	286
208	44 50	351	49 10	060	40 28	135	23 18	178	32 09	222	44 34	255	20 22	285
209	44 44	349	49 39	059	40 52	134	23 19	177	31 47	221	44 01	254	19 49	284
	ANTARES		◆Nunki		Peacock		ACHERNAR		◆CANOPUS		Suhail		◆SPICA	
210	50 07	058	30 31	093	41 16	134	23 21	177	31 25	220	43 29	253	44 37	348
211	50 35	057	31 04	092	41 41	133	23 23	176	31 04	220	42 57	253	44 30	347
212	51 03	055	31 38	091	42 05	133	23 26	176	30 42	219	42 25	252	44 21	345
213	51 31	054	32 11	090	42 30	132	23 29	175	30 21	219	41 53	251	44 12	344
214	51 58	053	32 45	089	42 55	132	23 32	174	30 00	218	41 22	251	44 03	342
215	52 24	052	33 18	089	43 20	131	23 35	174	29 40	218	40 50	250	43 52	341
216	52 50	050	33 52	088	43 46	131	23 39	173	29 19	217	40 19	249	43 41	340
217	53 16	049	34 26	087	44 11	130	23 43	173	28 59	216	39 47	249	43 29	338
218	53 41	048	34 59	086	44 37	130	23 48	172	28 40	216	39 16	248	43 16	337
219	54 05	046	35 32	085	45 03	129	23 52	171	28 20	215	38 45	247	43 03	336
220	54 29	045	36 06	084	45 29	129	23 58	171	28 01	215	38 14	247	42 49	335
221	54 53	044	36 39	083	45 55	128	24 03	170	27 42	214	37 44	246	42 34	333
222	55 16	042	37 13	082	46 22	128	24 09	170	27 23	214	37 13	245	42 19	332
223	55 38	041	37 46	082	46 48	127	24 15	169	27 05	213	36 43	245	42 02	331
224	55 59	039	38 19	081	47 15	127	24 22	169	26 47	212	36 12	244	41 46	329
	ANTARES		◆Nunki		Peacock		ACHERNAR		◆CANOPUS		Suhail		◆SPICA	
225	56 20	038	38 52	080	47 42	126	24 29	168	26 29	212	35 42	243	41 28	328
226	56 40	036	39 25	079	48 09	126	24 36	167	26 12	211	35 12	243	41 10	327
227	57 00	035	39 58	078	48 36	125	24 43	167	25 54	211	34 43	242	40 52	326
228	57 19	033	40 31	077	49 04	125	24 51	166	25 37	210	34 13	241	40 32	324
229	57 37	032	41 03	076	49 31	125	24 59	166	25 21	209	33 44	241	40 13	323
230	57 54	030	41 36	075	49 59	124	25 08	165	25 05	209	33 14	240	39 52	322
231	58 10	028	42 08	074	50 27	124	25 17	164	24 49	208	32 45	240	39 31	321
232	58 26	027	42 40	073	50 55	123	25 26	164	24 33	208	32 17	239	39 10	320
233	58 40	025	43 12	072	51 23	123	25 35	163	24 18	207	31 48	238	38 48	318
234	58 54	023	43 44	071	51 51	122	25 45	163	24 02	206	31 20	238	38 25	317
235	59 07	022	44 16	070	52 20	122	25 55	162	23 48	206	30 51	237	38 02	316
236	59 19	020	44 47	069	52 48	121	26 06	162	23 33	205	30 23	236	37 39	315
237	59 30	018	45 19	068	53 17	121	26 16	161	23 19	205	29 56	236	37 15	314
238	59 40	017	45 50	067	53 46	120	26 27	160	23 05	204	29 28	235	36 51	313
239	59 49	015	46 20	066	54 15	120	26 39	160	22 52	203	29 01	234	36 26	312
	◆ANTARES		Nunki		◆FOMALHAUT		ACHERNAR		CANOPUS		◆ACRUX		SPICA	
240	59 57	013	46 51	065	16 50	118	26 50	159	22 39	203	62 53	233	36 00	311
241	60 04	011	47 21	064	17 20	118	27 02	159	22 26	202	62 26	233	35 35	309
242	60 11	010	47 51	063	17 50	117	27 15	158	22 14	201	61 59	233	35 09	308
243	60 16	008	48 21	062	18 20	116	27 27	158	22 02	201	61 33	232	34 42	307
244	60 20	006	48 50	061	18 50	115	27 40	157	21 50	200	61 06	232	34 15	306
245	60 23	004	49 19	059	19 21	115	27 53	157	21 38	200	60 40	232	33 48	305
246	60 24	002	49 48	058	19 51	114	28 07	156	21 27	199	60 13	232	33 21	304
247	60 25	000	50 16	057	20 22	113	28 21	155	21 17	198	59 47	232	32 53	303
248	60 25	359	50 44	056	20 53	112	28 35	155	21 06	198	59 21	231	32 24	302
249	60 24	357	51 12	055	21 24	112	28 49	154	20 56	197	58 55	231	31 56	301
250	60 21	355	51 39	053	21 55	111	29 04	154	20 46	196	58 29	231	31 27	300
251	60 18	353	52 06	052	22 27	110	29 19	153	20 37	196	58 03	231	30 58	299
252	60 13	351	52 32	051	22 58	109	29 34	153	20 28	195	57 37	230	30 28	298
253	60 08	350	52 58	050	23 30	109	29 50	152	20 20	195	57 11	230	29 59	297
254	60 01	348	53 23	048	24 02	108	30 06	152	20 11	194	56 46	230	29 29	296
	ALTAIR		◆FOMALHAUT		ACHERNAR		CANOPUS		◆ACRUX		SPICA		◆ANTARES	
255	16 12	044	24 34	107	30 22	151	20 03	193	56 20	229	28 58	295	59 54	346
256	16 35	043	25 06	106	30 38	150	19 56	193	55 55	229	28 28	294	59 45	344
257	16 58	042	25 38	106	30 55	150	19 49	192	55 29	229	27 57	293	59 36	343
258	17 20	041	26 10	105	31 12	149	19 42	191	55 04	229	27 26	292	59 25	341
259	17 42	040	26 43	104	31 29	149	19 35	191	54 39	228	26 55	291	59 14	339
260	18 03	039	27 16	103	31 46	148	19 29	190	54 14	228	26 24	291	59 01	337
261	18 24	038	27 48	103	32 04	148	19 23	190	53 49	228	25 52	290	58 48	336
262	18 45	037	28 21	102	32 22	147	19 18	189	53 24	227	25 21	289	58 34	334
263	19 05	036	28 54	101	32 40	147	19 13	188	53 00	227	24 49	288	58 19	332
264	19 25	035	29 27	100	32 59	146	19 08	188	52 35	227	24 17	287	58 03	331
265	19 44	034	30 00	100	33 18	146	19 04	187	52 11	226	23 45	286	57 46	329
266	20 03	033	30 33	099	33 37	145	19 00	186	51 47	226	23 12	285	57 29	328
267	20 21	032	31 06	098	33 56	145	18 57	186	51 23	226	22 40	284	57 10	326
268	20 39	031	31 39	097	34 16	144	18 53	185	50 59	225	22 07	283	56 51	325
269	20 56	030	32 13	096	34 35	144	18 51	184	50 35	225	21 34	282	56 31	323

LAT 56°S

LHA ♈	Hc	Zn	Hc	Zn	Hc	Zn	Hc	Zn	Hc	Zn	Hc	Zn	Hc	Zn
	◆ALTAIR		FOMALHAUT		◆ACHERNAR		CANOPUS		RIGIL KENT.		◆SPICA		ANTARES	
270	21 13	029	32 46	096	34 56	143	18 48	184	63 57	239	21 02	282	56 11	322
271	21 29	028	33 20	095	35 16	143	18 46	183	63 28	238	20 29	281	55 50	320
272	21 45	027	33 53	094	35 36	142	18 45	183	63 00	238	19 56	280	55 28	319
273	22 00	026	34 27	093	35 57	142	18 43	182	62 31	238	19 23	279	55 05	317
274	22 15	025	35 00	092	36 18	141	18 42	181	62 03	238	18 49	278	54 42	316
275	22 29	024	35 34	092	36 39	141	18 42	181	61 35	237	18 16	277	54 19	314
276	22 42	023	36 07	091	37 01	140	18 42	180	61 07	237	17 43	276	53 54	313
277	22 55	022	36 41	090	37 22	140	18 42	179	60 38	237	17 09	276	53 30	312
278	23 08	021	37 14	089	37 44	139	18 42	179	60 10	236	16 36	275	53 04	310
279	23 19	020	37 48	088	38 06	138	18 43	178	59 43	236	16 03	274	52 38	309
280	23 31	019	38 21	087	38 29	138	18 45	177	59 15	236	15 29	273	52 12	308
281	23 41	018	38 55	087	38 51	137	18 46	177	58 47	236	14 56	272	51 45	307
282	23 51	017	39 28	086	39 14	137	18 48	176	58 19	235	14 22	271	51 18	305
283	24 01	016	40 02	085	39 37	137	18 51	175	57 52	235	13 48	271	50 51	304
284	24 10	015	40 35	084	40 00	136	18 54	175	57 24	235	13 15	270	50 23	303
	◆ALTAIR		FOMALHAUT		◆ACHERNAR		CANOPUS		◆RIGIL KENT.		ANTARES		Nunki	
285	24 18	014	41 08	083	40 24	136	18 57	174	56 57	234	49 54	302	60 17	358
286	24 26	013	41 42	082	40 47	135	19 00	174	56 30	234	49 26	301	60 15	356
287	24 33	012	42 15	081	41 11	135	19 04	173	56 03	234	48 57	299	60 13	354
288	24 39	010	42 48	080	41 35	134	19 09	172	55 36	233	48 27	298	60 09	352
289	24 45	009	43 21	079	41 59	134	19 13	172	55 09	233	47 57	297	60 04	351
290	24 50	008	43 54	079	42 24	133	19 18	171	54 42	233	47 27	296	59 58	349
291	24 55	007	44 27	078	42 48	133	19 24	170	54 16	232	46 57	295	59 50	347
292	24 59	006	45 00	077	43 13	132	19 30	170	53 49	232	46 27	294	59 42	345
293	25 02	005	45 32	076	43 38	132	19 36	169	53 23	232	45 56	293	59 33	343
294	25 04	004	46 05	075	44 03	131	19 42	169	52 57	231	45 25	292	59 23	342
295	25 06	003	46 37	074	44 29	131	19 49	168	52 31	231	44 53	291	59 12	340
296	25 08	002	47 09	073	44 54	130	19 56	167	52 05	230	44 22	290	59 00	338
297	25 09	001	47 41	072	45 20	130	20 04	167	51 39	230	43 50	289	58 48	337
298	25 09	000	48 13	071	45 46	129	20 12	166	51 14	230	43 18	288	58 34	335
299	25 08	359	48 44	070	46 12	129	20 20	165	50 48	229	42 46	287	58 19	333
	FOMALHAUT		◆ACHERNAR		CANOPUS		◆RIGIL KENT.		ANTARES		Nunki		◆ALTAIR	
300	49 16	069	46 38	128	20 29	165	50 23	229	42 14	286	58 04	332	25 07	357
301	49 47	067	47 04	128	20 38	164	49 57	229	41 42	285	57 47	330	25 05	356
302	50 17	066	47 31	127	20 47	163	49 32	228	41 09	284	57 30	328	25 03	355
303	50 48	065	47 58	127	20 57	163	49 08	228	40 37	283	57 12	327	25 00	354
304	51 18	064	48 24	127	21 07	162	48 43	227	40 04	282	56 54	325	24 56	353
305	51 48	063	48 51	126	21 17	162	48 18	227	39 31	281	56 34	324	24 51	352
306	52 18	062	49 19	126	21 28	161	47 54	227	38 58	280	56 14	322	24 46	351
307	52 48	061	49 46	125	21 39	160	47 29	226	38 25	279	55 53	321	24 41	350
308	53 17	059	50 13	125	21 51	160	47 05	226	37 52	278	55 32	319	24 35	349
309	53 45	058	50 41	124	22 03	159	46 41	225	37 19	277	55 09	318	24 28	348
310	54 14	057	51 09	124	22 15	158	46 18	225	36 45	277	54 47	317	24 20	347
311	54 42	056	51 37	123	22 27	158	45 54	224	36 12	276	54 23	315	24 12	345
312	55 09	054	52 05	123	22 40	157	45 31	224	35 38	275	53 59	314	24 03	344
313	55 36	053	52 33	123	22 53	157	45 07	224	35 05	274	53 35	312	23 54	343
314	56 03	052	53 01	122	23 07	156	44 44	223	34 32	273	53 10	311	23 44	342
	◆FOMALHAUT		ACHERNAR		◆CANOPUS		RIGIL KENT.		◆ANTARES		Nunki		ALTAIR	
315	56 29	050	53 30	122	23 20	155	44 21	223	33 58	272	52 44	310	23 34	341
316	56 55	049	53 58	121	23 34	155	43 59	222	33 24	271	52 18	309	23 23	340
317	57 20	048	54 27	121	23 49	154	43 36	222	32 51	271	51 52	307	23 11	339
318	57 44	046	54 56	120	24 04	154	43 14	221	32 17	270	51 25	306	22 59	338
319	58 08	045	55 25	120	24 19	153	42 52	221	31 44	269	50 57	305	22 46	337
320	58 32	043	55 54	120	24 34	152	42 30	221	31 10	268	50 30	304	22 32	336
321	58 54	042	56 23	119	24 50	152	42 08	220	30 37	267	50 02	302	22 19	335
322	59 16	040	56 53	119	25 06	151	41 47	220	30 03	266	49 33	301	22 04	334
323	59 38	039	57 22	118	25 22	151	41 25	219	29 30	266	49 04	300	21 49	333
324	59 58	037	57 52	118	25 39	150	41 04	219	28 56	265	48 35	299	21 33	332
325	60 18	036	58 22	117	25 56	149	40 43	218	28 23	264	48 05	298	21 17	331
326	60 37	034	58 51	117	26 13	149	40 22	218	27 50	263	47 36	297	21 01	330
327	60 56	032	59 21	117	26 30	148	40 02	217	27 16	262	47 05	296	20 44	329
328	61 13	031	59 51	116	26 48	148	39 42	217	26 43	262	46 35	294	20 26	328
329	61 30	029	60 22	116	27 06	147	39 22	217	26 10	261	46 04	293	20 08	327
	◆FOMALHAUT		ACHERNAR		◆CANOPUS		RIGIL KENT.		◆ANTARES		Nunki		ALTAIR	
330	61 45	027	60 52	115	27 25	146	39 02	216	25 37	260	45 33	292	19 49	326
331	62 00	025	61 22	115	27 43	146	38 42	216	25 04	259	45 02	291	19 30	325
332	62 14	024	61 53	115	28 02	145	38 23	215	24 31	259	44 31	290	19 11	324
333	62 27	022	62 23	114	28 22	145	38 03	215	23 58	258	43 59	289	18 51	323
334	62 39	020	62 54	114	28 41	144	37 44	214	23 25	257	43 28	288	18 30	322
335	62 50	018	63 25	113	29 01	144	37 26	214	22 53	256	42 56	287	18 09	321
336	63 00	016	63 55	113	29 21	143	37 07	213	22 20	255	42 23	286	17 48	320
337	63 08	014	64 26	113	29 41	142	36 49	213	21 48	255	41 51	285	17 26	319
338	63 16	012	64 57	112	30 02	142	36 31	212	21 15	254	41 19	284	17 04	318
339	63 23	010	65 28	112	30 23	141	36 13	212	20 43	253	40 46	283	16 41	317
340	63 28	008	65 59	112	30 44	141	35 55	211	20 11	252	40 13	282	16 18	316
341	63 33	007	66 31	111	31 05	140	35 38	211	19 39	252	39 41	282	15 54	315
342	63 36	005	67 02	111	31 27	140	35 21	210	19 08	251	39 08	281	15 31	314
343	63 38	003	67 33	111	31 49	139	35 04	210	18 36	250	38 35	280	15 06	313
344	63 39	001	68 05	110	32 11	138	34 47	209	18 04	249	38 02	279	14 42	312
	◆Diphda		Acamar		CANOPUS		◆Miaplacidus		RIGIL KENT.		◆Nunki		FOMALHAUT	
345	47 18	038	48 50	094	32 34	138	37 10	169	34 31	209	37 28	278	63 39	359
346	47 39	036	49 23	093	32 56	137	37 17	168	34 15	208	36 55	277	63 37	357
347	47 58	035	49 57	092	33 19	137	37 24	168	33 59	208	36 22	276	63 35	355
348	48 17	034	50 30	091	33 42	136	37 31	167	33 44	207	35 48	275	63 31	353
349	48 35	032	51 04	090	34 06	136	37 38	167	33 28	207	35 15	274	63 27	351
350	48 53	031	51 37	090	34 29	135	37 46	167	33 13	206	34 42	274	63 21	349
351	49 10	030	52 11	089	34 53	134	37 54	166	32 58	206	34 08	273	63 14	347
352	49 26	028	52 44	088	35 17	134	38 02	166	32 44	205	33 35	272	63 06	345
353	49 42	027	53 18	087	35 41	133	38 10	165	32 29	205	33 01	271	62 57	343
354	49 56	025	53 51	086	36 06	133	38 19	165	32 15	204	32 27	270	62 46	341
355	50 10	024	54 25	085	36 31	132	38 28	165	32 02	204	31 54	269	62 35	340
356	50 24	022	54 58	084	36 56	132	38 37	164	31 48	203	31 20	269	62 23	338
357	50 36	021	55 32	083	37 21	131	38 46	164	31 35	203	30 47	268	62 10	336
358	50 48	020	56 05	082	37 46	131	38 55	163	31 22	202	30 13	267	61 56	334
359	50 58	018	56 38	082	38 12	130	39 05	163	31 10	202	29 40	266	61 41	332

LAT 57°S

LHA ♈	Hc	Zn	Hc	Zn	Hc	Zn	Hc	Zn	Hc	Zn	Hc	Zn	Hc	Zn
	Diphda		◆RIGEL		CANOPUS		Miaplacidus		◆RIGIL KENT.		Peacock		◆FOMALHAUT	
0	50 11	016	13 05	085	39 15	129	40 12	162	31 53	202	61 26	247	60 32	332
1	50 20	015	13 38	084	39 41	128	40 22	162	31 41	201	60 56	247	60 16	330
2	50 27	013	14 10	083	40 07	128	40 32	162	31 30	201	60 26	247	59 59	328
3	50 35	012	14 43	082	40 33	127	40 43	161	31 18	200	59 56	246	59 42	327
4	50 41	010	15 15	081	40 59	127	40 53	161	31 07	199	59 26	246	59 23	325
5	50 46	009	15 48	081	41 25	126	41 04	160	30 56	199	58 56	245	59 04	323
6	50 51	007	16 20	080	41 52	126	41 15	160	30 46	198	58 26	245	58 44	322
7	50 55	006	16 52	079	42 18	125	41 27	160	30 36	198	57 57	244	58 24	320
8	50 57	004	17 24	078	42 45	124	41 38	159	30 26	197	57 28	244	58 03	319
9	50 59	003	17 56	077	43 12	124	41 50	159	30 16	197	56 58	243	57 41	317
10	51 01	001	18 28	076	43 39	123	42 02	158	30 07	196	56 29	243	57 18	316
11	51 01	000	18 59	075	44 07	123	42 14	158	29 58	196	56 00	242	56 55	314
12	51 00	358	19 31	074	44 34	122	42 26	158	29 49	195	55 31	242	56 32	313
13	50 59	357	20 02	074	45 02	122	42 38	157	29 41	195	55 02	241	56 07	312
14	50 57	355	20 34	073	45 30	121	42 51	157	29 33	194	54 34	241	55 43	310
	◆RIGEL		SIRIUS		CANOPUS		◆RIGIL KENT.		Peacock		◆FOMALHAUT		Diphda	
15	21 05	072	15 59	096	45 58	121	29 25	194	54 05	241	55 17	309	50 53	354
16	21 36	071	16 32	095	46 26	120	29 17	193	53 37	240	54 52	308	50 49	352
17	22 06	070	17 04	095	46 55	119	29 10	193	53 09	240	54 26	306	50 45	351
18	22 37	069	17 37	094	47 23	119	29 03	192	52 41	239	53 59	305	50 39	349
19	23 07	068	18 09	093	47 52	118	28 57	191	52 13	239	53 32	304	50 32	348
20	23 38	067	18 42	092	48 21	118	28 50	191	51 45	238	53 05	303	50 25	346
21	24 08	066	19 15	091	48 50	117	28 44	190	51 17	238	52 37	301	50 17	345
22	24 37	065	19 47	090	49 19	117	28 39	190	50 49	237	52 09	300	50 08	343
23	25 07	064	20 20	090	49 48	116	28 33	189	50 22	237	51 40	299	49 58	342
24	25 36	063	20 53	089	50 17	116	28 28	189	49 55	236	51 12	298	49 47	340
25	26 05	062	21 25	088	50 47	115	28 23	188	49 28	236	50 43	297	49 36	339
26	26 34	062	21 58	087	51 17	114	28 19	188	49 01	235	50 13	295	49 24	338
27	27 03	061	22 31	086	51 47	114	28 15	187	48 34	235	49 44	294	49 11	336
28	27 31	060	23 03	085	52 16	113	28 11	187	48 07	234	49 14	293	48 57	335
29	27 59	059	23 36	084	52 47	113	28 07	186	47 41	234	48 44	292	48 43	333
	◆RIGEL		SIRIUS		CANOPUS		◆RIGIL KENT.		Peacock		◆FOMALHAUT		Diphda	
30	28 27	058	24 08	084	53 17	112	28 04	185	47 14	233	48 13	291	48 28	332
31	28 54	057	24 41	083	53 47	112	28 01	185	46 48	233	47 43	290	48 12	331
32	29 22	056	25 13	082	54 17	111	27 58	184	46 22	232	47 12	289	47 56	329
33	29 48	055	25 46	081	54 48	111	27 56	184	45 57	232	46 41	288	47 39	328
34	30 15	054	26 18	080	55 19	110	27 54	183	45 31	231	46 10	287	47 21	326
35	30 41	052	26 50	079	55 49	109	27 53	183	45 05	231	45 38	286	47 03	325
36	31 07	051	27 22	078	56 20	109	27 51	182	44 40	230	45 07	285	46 44	324
37	31 32	050	27 54	077	56 51	108	27 50	182	44 15	230	44 35	284	46 24	323
38	31 57	049	28 26	077	57 22	108	27 49	181	43 50	229	44 03	283	46 04	321
39	32 22	048	28 57	076	57 54	107	27 49	180	43 25	229	43 32	282	45 43	320
40	32 46	047	29 29	075	58 25	106	27 49	180	43 01	228	43 00	281	45 22	319
41	33 10	046	30 00	074	58 56	106	27 49	179	42 37	228	42 27	280	45 00	317
42	33 33	045	30 32	073	59 28	105	27 50	179	42 12	227	41 55	279	44 38	316
43	33 56	044	31 03	072	59 59	105	27 50	178	41 48	227	41 23	278	44 15	315
44	34 18	043	31 34	071	60 31	104	27 52	178	41 25	226	40 51	278	43 52	314
	RIGEL		◆SIRIUS		CANOPUS		◆RIGIL KENT.		Peacock		FOMALHAUT		◆Diphda	
45	34 40	042	32 05	070	61 03	104	27 53	177	41 01	226	40 18	277	43 28	313
46	35 02	041	32 35	069	61 34	103	27 55	177	40 38	225	39 46	276	43 04	311
47	35 23	039	33 06	068	62 06	102	27 57	176	40 14	225	39 13	275	42 39	310
48	35 43	038	33 36	067	62 38	102	27 59	175	39 52	224	38 41	274	42 14	309
49	36 03	037	34 06	066	63 10	101	28 02	175	39 29	224	38 08	273	41 48	308
50	36 23	036	34 36	065	63 42	100	28 05	174	39 06	223	37 35	272	41 22	307
51	36 42	035	35 05	064	64 15	100	28 09	174	38 44	223	37 03	271	40 56	306
52	37 00	034	35 34	063	64 47	099	28 12	173	38 22	222	36 30	271	40 29	305
53	37 18	032	36 03	062	65 19	098	28 16	173	38 00	222	35 57	270	40 02	304
54	37 35	031	36 32	061	65 52	098	28 20	172	37 38	221	35 25	269	39 35	302
55	37 52	030	37 01	060	66 24	097	28 25	172	37 17	221	34 52	268	39 07	301
56	38 08	029	37 29	059	66 56	096	28 30	171	36 56	220	34 19	267	38 39	300
57	38 23	028	37 57	058	67 29	096	28 35	171	36 35	220	33 47	266	38 11	299
58	38 38	026	38 24	057	68 01	095	28 41	170	36 14	219	33 14	266	37 42	298
59	38 53	025	38 51	056	68 34	094	28 46	169	35 53	219	32 42	265	37 13	297
	RIGEL		◆SIRIUS		Suhail		◆RIGIL KENT.		Peacock		FOMALHAUT		◆Diphda	
60	39 06	024	39 18	055	41 44	108	28 53	169	35 33	218	32 09	264	36 44	296
61	39 19	023	39 45	054	42 15	108	28 59	168	35 13	218	31 37	263	36 14	295
62	39 31	021	40 11	052	42 46	107	29 06	168	34 53	217	31 04	262	35 45	294
63	39 43	020	40 36	051	43 18	106	29 13	167	34 34	217	30 32	262	35 15	293
64	39 54	019	41 02	050	43 49	106	29 20	167	34 14	216	29 59	261	34 44	292
65	40 04	018	41 27	049	44 21	105	29 28	166	33 55	215	29 27	260	34 14	291
66	40 14	016	41 51	048	44 52	104	29 36	166	33 36	215	28 55	259	33 44	290
67	40 23	015	42 15	047	45 24	104	29 44	165	33 18	214	28 23	258	33 13	289
68	40 31	014	42 39	046	45 56	103	29 53	165	33 00	214	27 51	258	32 42	288
69	40 38	013	43 02	044	46 28	102	30 01	164	32 42	213	27 19	257	32 11	287
70	40 45	011	43 24	043	47 00	101	30 10	164	32 24	213	26 47	256	31 39	286
71	40 51	010	43 46	042	47 32	101	30 20	163	32 06	212	26 16	255	31 08	285
72	40 56	009	44 08	041	48 04	100	30 30	162	31 49	212	25 44	255	30 36	285
73	41 01	007	44 29	039	48 36	099	30 40	162	31 32	211	25 13	254	30 05	284
74	41 05	006	44 50	038	49 08	099	30 50	161	31 15	211	24 41	253	29 33	283
	RIGEL		◆SIRIUS		Suhail		◆RIGIL KENT.		Peacock		ACHERNAR		◆Diphda	
75	41 08	005	45 09	037	49 41	098	31 00	161	30 59	210	63 10	248	29 01	282
76	41 10	003	45 29	036	50 13	097	31 11	160	30 43	209	62 40	247	28 29	281
77	41 11	002	45 47	034	50 45	096	31 22	160	30 27	209	62 10	247	27 57	280
78	41 12	001	46 06	033	51 18	096	31 34	159	30 11	208	61 40	247	27 25	279
79	41 12	359	46 23	032	51 50	095	31 46	159	29 56	208	61 10	246	26 52	278
80	41 12	358	46 40	030	52 23	094	31 58	158	29 41	207	60 40	246	26 20	277
81	41 10	357	46 56	029	52 56	093	32 10	158	29 26	207	60 10	245	25 47	277
82	41 08	355	47 12	028	53 28	092	32 22	157	29 11	206	59 41	245	25 15	276
83	41 05	354	47 26	026	54 01	092	32 35	157	28 57	205	59 11	244	24 42	275
84	41 01	353	47 41	025	54 34	091	32 48	156	28 43	205	58 42	244	24 10	274
85	40 57	352	47 54	024	55 06	090	33 02	156	28 30	204	58 12	243	23 37	273
86	40 52	350	48 07	022	55 39	089	33 15	155	28 16	204	57 43	243	23 05	272
87	40 46	349	48 19	021	56 12	088	33 29	155	28 03	203	57 14	243	22 32	271
88	40 39	348	48 30	019	56 44	087	33 43	154	27 51	203	56 45	242	21 59	271
89	40 32	346	48 40	018	57 17	086	33 58	154	27 38	202	56 16	242	21 27	270

LHA ♈	Hc	Zn	Hc	Zn	Hc	Zn	Hc	Zn	Hc	Zn	Hc	Zn	Hc	Zn
	SIRIUS		Alphard		◆Suhail		RIGIL KENT.		◆Peacock		ACHERNAR		◆RIGEL	
90	48 50	016	27 18	061	57 50	086	34 12	153	27 26	201	55 48	241	40 24	345
91	48 59	015	27 47	060	58 22	085	34 27	153	27 14	201	55 19	241	40 15	344
92	49 07	014	28 15	059	58 55	084	34 42	152	27 03	200	54 51	240	40 06	343
93	49 14	012	28 43	058	59 27	083	34 58	152	26 52	200	54 22	240	39 56	341
94	49 21	011	29 10	057	59 59	082	35 14	151	26 41	199	53 54	239	39 45	340
95	49 26	009	29 37	056	60 32	081	35 29	151	26 30	199	53 26	239	39 33	339
96	49 31	008	30 04	055	61 04	080	35 46	150	26 20	198	52 58	238	39 21	337
97	49 35	006	30 31	054	61 36	079	36 02	150	26 10	197	52 30	238	39 08	336
98	49 38	005	30 57	053	62 08	078	36 19	149	26 00	197	52 03	238	38 55	335
99	49 40	003	31 23	052	62 40	077	36 36	149	25 51	196	51 35	237	38 40	334
100	49 42	002	31 49	051	63 12	076	36 53	148	25 42	196	51 08	237	38 26	333
101	49 43	000	32 14	050	63 43	074	37 10	148	25 34	195	50 41	236	38 10	331
102	49 42	359	32 39	049	64 15	073	37 28	147	25 25	194	50 14	236	37 54	330
103	49 41	357	33 03	048	64 46	072	37 46	147	25 17	194	49 47	235	37 38	329
104	49 39	356	33 27	047	65 17	071	38 04	146	25 10	193	49 20	235	37 21	328
	Alphard		◆Gienah		RIGIL KENT.		◆Peacock		ACHERNAR		◆RIGEL		SIRIUS	
105	33 51	046	20 38	091	38 22	146	25 02	193	48 53	234	37 03	327	49 37	354
106	34 14	044	21 11	090	38 41	145	24 55	192	48 27	234	36 44	325	49 33	353
107	34 36	043	21 44	089	39 00	145	24 48	192	48 01	233	36 26	324	49 29	351
108	34 58	042	22 16	088	39 19	144	24 42	191	47 35	233	36 06	323	49 23	350
109	35 20	041	22 49	087	39 38	144	24 36	190	47 09	232	35 46	322	49 17	349
110	35 41	040	23 22	086	39 57	143	24 30	190	46 43	232	35 26	321	49 10	347
111	36 02	039	23 54	086	40 17	143	24 25	189	46 17	231	35 05	320	49 03	346
112	36 22	038	24 27	085	40 37	142	24 20	189	45 52	231	34 43	318	48 54	344
113	36 42	036	24 59	084	40 57	142	24 15	188	45 27	230	34 22	317	48 45	343
114	37 01	035	25 32	083	41 17	141	24 11	187	45 01	230	33 59	316	48 35	341
115	37 20	034	26 04	082	41 38	141	24 07	187	44 37	229	33 36	315	48 24	340
116	37 38	033	26 37	081	41 58	140	24 03	186	44 12	229	33 13	314	48 12	339
117	37 55	032	27 09	080	42 19	140	24 00	186	43 47	228	32 49	313	48 00	337
118	38 12	031	27 41	079	42 40	140	23 57	185	43 23	228	32 25	312	47 47	336
119	38 29	029	28 13	079	43 02	139	23 54	184	42 59	227	32 01	311	47 33	334
	REGULUS		◆SPICA		RIGIL KENT.		Peacock		◆ACHERNAR		RIGEL		◆SIRIUS	
120	16 06	033	14 05	089	43 23	139	23 52	184	42 35	227	31 36	310	47 19	333
121	16 24	032	14 38	088	43 45	138	23 50	183	42 11	226	31 10	309	47 03	332
122	16 41	031	15 11	087	44 07	138	23 48	183	41 47	226	30 45	308	46 48	330
123	16 57	030	15 43	086	44 29	137	23 47	182	41 24	225	30 19	307	46 31	329
124	17 13	029	16 16	085	44 51	137	23 46	181	41 01	225	29 52	306	46 14	328
125	17 29	028	16 48	085	45 14	136	23 45	181	40 38	224	29 26	305	45 56	326
126	17 44	027	17 21	084	45 36	136	23 45	180	40 15	224	28 58	304	45 38	325
127	17 58	026	17 53	083	45 59	135	23 45	180	39 52	223	28 31	303	45 18	324
128	18 12	025	18 26	082	46 22	135	23 46	179	39 30	223	28 03	302	44 59	322
129	18 25	024	18 58	081	46 45	135	23 46	178	39 08	222	27 35	301	44 39	321
130	18 38	023	19 30	080	47 09	134	23 47	178	38 46	222	27 07	300	44 18	320
131	18 51	022	20 02	079	47 32	134	23 49	177	38 24	221	26 39	299	43 57	319
132	19 03	021	20 35	079	47 56	133	23 51	177	38 03	221	26 10	298	43 35	317
133	19 14	020	21 07	078	48 20	133	23 53	176	37 42	220	25 41	297	43 12	316
134	19 25	019	21 38	077	48 44	133	23 55	175	37 20	220	25 11	296	42 49	315
	REGULUS		◆SPICA		RIGIL KENT.		◆Peacock		ACHERNAR		CANOPUS		◆SIRIUS	
135	19 35	018	22 10	076	49 08	132	23 58	175	37 00	219	67 26	264	42 26	314
136	19 45	017	22 42	075	49 32	132	24 01	174	36 39	219	66 54	264	42 02	313
137	19 54	016	23 13	074	49 57	131	24 05	174	36 19	218	66 21	263	41 38	311
138	20 02	015	23 45	073	50 21	131	24 08	173	35 59	218	65 49	262	41 13	310
139	20 10	014	24 16	072	50 46	130	24 13	172	35 39	217	65 17	262	40 48	309
140	20 18	013	24 47	071	51 11	130	24 17	172	35 19	217	64 44	261	40 23	308
141	20 24	012	25 18	070	51 36	130	24 22	171	35 00	216	64 12	260	39 57	307
142	20 31	010	25 48	069	52 01	129	24 27	171	34 41	216	63 40	260	39 31	306
143	20 36	009	26 19	069	52 27	129	24 32	170	34 22	215	63 08	259	39 04	305
144	20 41	008	26 49	068	52 52	128	24 38	169	34 03	215	62 36	258	38 37	304
145	20 46	007	27 19	067	53 18	128	24 44	169	33 44	214	62 04	258	38 09	303
146	20 50	006	27 49	066	53 44	128	24 51	168	33 26	213	61 32	257	37 42	302
147	20 53	005	28 19	065	54 10	127	24 58	168	33 08	213	61 00	256	37 14	300
148	20 56	004	28 48	064	54 36	127	25 05	167	32 51	212	60 28	256	36 45	299
149	20 58	003	29 18	063	55 02	127	25 12	166	32 33	212	59 57	255	36 17	298
	REGULUS		◆SPICA		RIGIL KENT.		◆Peacock		ACHERNAR		CANOPUS		◆SIRIUS	
150	20 59	002	29 46	062	55 28	126	25 20	166	32 16	211	59 25	255	35 48	297
151	21 00	001	30 15	061	55 55	126	25 28	165	31 59	211	58 54	254	35 19	296
152	21 00	000	30 44	060	56 21	125	25 37	165	31 43	210	58 22	253	34 49	295
153	21 00	359	31 12	059	56 48	125	25 46	164	31 26	210	57 51	253	34 20	294
154	20 59	358	31 39	058	57 15	125	25 55	164	31 10	209	57 20	252	33 50	293
155	20 58	357	32 07	057	57 42	124	26 04	163	30 55	209	56 49	252	33 20	292
156	20 56	356	32 34	056	58 09	124	26 14	162	30 39	208	56 18	251	32 49	291
157	20 53	355	33 01	055	58 36	124	26 24	162	30 24	207	55 47	251	32 19	290
158	20 50	354	33 27	054	59 03	123	26 34	161	30 09	207	55 16	250	31 48	289
159	20 46	353	33 54	053	59 31	123	26 45	161	29 54	206	54 46	249	31 17	289
160	20 42	352	34 19	052	59 58	123	26 56	160	29 40	206	54 15	249	30 46	288
161	20 37	351	34 45	050	60 26	122	27 07	159	29 26	205	53 45	248	30 15	287
162	20 31	350	35 10	049	60 53	122	27 19	159	29 12	205	53 14	248	29 44	286
163	20 25	349	35 34	048	61 21	122	27 31	158	28 59	204	52 44	247	29 12	285
164	20 18	348	35 59	047	61 49	121	27 43	158	28 45	204	52 14	247	28 40	284
	◆SPICA		ANTARES		◆Peacock		ACHERNAR		CANOPUS		◆SIRIUS		REGULUS	
165	36 22	046	26 01	099	27 55	157	28 32	203	51 44	246	28 09	283	20 11	346
166	36 46	045	26 34	098	28 08	157	28 20	202	51 14	245	27 37	282	20 03	345
167	37 08	044	27 06	097	28 21	156	28 08	202	50 45	245	27 05	281	19 54	344
168	37 31	043	27 39	097	28 35	155	27 56	201	50 15	244	26 33	280	19 45	343
169	37 53	042	28 11	096	28 48	155	27 44	201	49 46	244	26 00	279	19 35	342
170	38 14	040	28 44	095	29 03	154	27 32	200	49 16	243	25 28	279	19 25	341
171	38 35	039	29 16	094	29 17	154	27 21	200	48 47	243	24 56	278	19 15	340
172	38 55	038	29 49	093	29 31	153	27 11	199	48 18	242	24 23	277	19 03	339
173	39 15	037	30 21	093	29 46	153	27 00	198	47 50	242	23 51	276	18 51	338
174	39 34	036	30 54	092	30 01	152	26 50	198	47 21	241	23 18	275	18 39	337
175	39 53	034	31 27	091	30 17	152	26 40	197	46 52	241	22 46	274	18 26	336
176	40 12	033	31 59	090	30 33	151	26 30	197	46 24	240	22 13	273	18 13	335
177	40 29	032	32 32	089	30 49	150	26 21	196	45 56	239	21 41	273	17 59	334
178	40 46	031	33 05	088	31 05	150	26 12	196	45 28	239	21 08	272	17 44	333
179	41 03	029	33 37	088	31 21	149	26 04	195	45 00	238	20 35	271	17 29	332

LAT 57°S

LHA ♈	Hc	Zn	Hc	Zn	Hc	Zn	Hc	Zn	Hc	Zn	Hc	Zn	Hc	Zn
	◆SPICA		ANTARES		◆Peacock		ACHERNAR		CANOPUS		◆SIRIUS		Alphard	
180	41 18	028	34 10	087	31 38	149	25 55	194	44 32	238	20 03	270	33 19	313
181	41 34	027	34 43	086	31 55	148	25 47	194	44 05	237	19 30	269	32 55	312
182	41 48	026	35 15	085	32 13	148	25 40	193	43 37	237	18 57	268	32 31	311
183	42 02	024	35 48	084	32 30	147	25 33	193	43 10	236	18 25	267	32 06	310
184	42 15	023	36 20	083	32 48	147	25 26	192	42 43	236	17 52	267	31 40	309
185	42 27	022	36 53	082	33 06	146	25 19	191	42 16	235	17 19	266	31 15	308
186	42 39	021	37 25	081	33 25	145	25 13	191	41 49	234	16 47	265	30 49	307
187	42 50	019	37 57	081	33 43	145	25 07	190	41 23	234	16 14	264	30 22	306
188	43 01	018	38 29	080	34 02	144	25 01	190	40 57	233	15 42	263	29 56	305
189	43 10	017	39 02	079	34 21	144	24 56	189	40 31	233	15 09	263	29 28	304
190	43 19	015	39 33	078	34 41	143	24 51	189	40 05	232	14 37	262	29 01	303
191	43 28	014	40 05	077	35 01	143	24 46	188	39 39	232	14 05	261	28 33	302
192	43 35	013	40 37	076	35 20	142	24 42	187	39 13	231	13 32	260	28 05	301
193	43 42	011	41 09	075	35 41	142	24 38	187	38 48	231	13 00	259	27 37	300
194	43 48	010	41 40	074	36 01	141	24 34	186	38 23	230	12 28	258	27 09	299
	SPICA		◆ANTARES		Peacock		◆ACHERNAR		CANOPUS		Suhail		◆Alphard	
195	43 53	008	42 12	073	36 22	141	24 31	186	37 58	229	51 47	265	26 40	298
196	43 57	007	42 43	072	36 42	140	24 28	185	37 33	229	51 15	264	26 11	297
197	44 01	006	43 14	071	37 03	140	24 25	184	37 09	228	50 42	264	25 41	296
198	44 04	004	43 44	070	37 25	139	24 23	184	36 45	228	50 10	263	25 12	295
199	44 06	003	44 15	069	37 46	139	24 21	183	36 20	227	49 38	262	24 42	294
200	44 08	002	44 45	068	38 08	138	24 19	183	35 57	227	49 05	261	24 12	293
201	44 08	000	45 16	067	38 30	137	24 18	182	35 33	226	48 33	261	23 42	292
202	44 08	359	45 46	066	38 52	137	24 17	181	35 10	225	48 01	260	23 12	291
203	44 07	358	46 15	065	39 15	136	24 16	181	34 46	225	47 29	259	22 41	290
204	44 05	356	46 45	064	39 37	136	24 16	180	34 24	224	46 57	258	22 10	289
205	44 03	355	47 14	063	40 00	135	24 16	180	34 01	224	46 25	258	21 39	288
206	43 59	353	47 43	061	40 23	135	24 16	179	33 38	223	45 53	257	21 08	287
207	43 55	352	48 11	060	40 46	134	24 17	178	33 16	223	45 21	256	20 37	287
208	43 50	351	48 39	059	41 10	134	24 18	178	32 54	222	44 49	256	20 06	286
209	43 45	349	49 07	058	41 33	133	24 19	177	32 32	221	44 18	255	19 34	285
	ANTARES		◆Nunki		Peacock		ACHERNAR		◆CANOPUS		Suhail		◆SPICA	
210	49 35	057	30 33	092	41 57	133	24 21	177	32 11	221	43 46	254	43 38	348
211	50 02	056	31 06	091	42 21	132	24 23	176	31 50	220	43 15	254	43 31	347
212	50 29	054	31 39	090	42 46	132	24 26	175	31 29	220	42 43	253	43 23	345
213	50 55	053	32 11	090	43 10	131	24 28	175	31 08	219	42 12	252	43 15	344
214	51 21	052	32 44	089	43 35	131	24 31	174	30 47	219	41 41	251	43 05	343
215	51 46	051	33 17	088	43 59	130	24 35	174	30 27	218	41 10	251	42 55	341
216	52 11	049	33 49	087	44 24	130	24 39	173	30 07	217	40 39	250	42 45	340
217	52 36	048	34 22	086	44 50	129	24 43	173	29 48	217	40 09	249	42 33	339
218	53 00	047	34 54	085	45 15	129	24 47	172	29 28	216	39 38	249	42 21	337
219	53 24	045	35 27	084	45 40	128	24 52	171	29 09	216	39 08	248	42 08	336
220	53 46	044	35 59	084	46 06	128	24 57	171	28 50	215	38 38	247	41 55	335
221	54 09	043	36 32	083	46 32	127	25 02	170	28 32	214	38 08	247	41 40	334
222	54 31	041	37 04	082	46 58	127	25 08	170	28 13	214	37 38	246	41 25	332
223	54 52	040	37 37	081	47 24	126	25 14	169	27 55	213	37 08	245	41 10	331
224	55 13	038	38 09	080	47 51	126	25 20	168	27 37	213	36 38	245	40 54	330
	ANTARES		◆Nunki		Peacock		ACHERNAR		◆CANOPUS		Suhail		◆SPICA	
225	55 32	037	38 41	079	48 17	125	25 27	168	27 20	212	36 09	244	40 37	329
226	55 52	035	39 13	078	48 44	125	25 34	167	27 03	211	35 40	243	40 20	327
227	56 10	034	39 45	077	49 11	124	25 42	167	26 46	211	35 10	243	40 02	326
228	56 28	032	40 17	076	49 38	124	25 49	166	26 29	210	34 41	242	39 43	325
229	56 45	031	40 48	075	50 05	124	25 57	165	26 13	210	34 13	241	39 24	324
230	57 02	029	41 20	074	50 32	123	26 06	165	25 57	209	33 44	241	39 05	323
231	57 17	028	41 51	073	51 00	123	26 14	164	25 41	208	33 16	240	38 45	321
232	57 32	026	42 22	072	51 27	122	26 23	164	25 26	208	32 47	239	38 24	320
233	57 46	024	42 54	071	51 55	122	26 33	163	25 11	207	32 19	239	38 03	319
234	57 59	023	43 24	070	52 23	121	26 42	163	24 56	207	31 52	238	37 41	318
235	58 11	021	43 55	069	52 51	121	26 52	162	24 42	206	31 24	237	37 19	317
236	58 23	020	44 26	068	53 19	120	27 02	161	24 28	205	30 56	237	36 56	316
237	58 33	018	44 56	067	53 47	120	27 13	161	24 14	205	30 29	236	36 33	314
238	58 43	016	45 26	066	54 16	119	27 24	160	24 00	204	30 02	235	36 10	313
239	58 51	014	45 56	065	54 44	119	27 35	160	23 47	203	29 35	235	35 46	312
	◆ANTARES		Nunki		◆FOMALHAUT		ACHERNAR		CANOPUS		◆ACRUX		SPICA	
240	58 59	013	46 25	064	17 19	118	27 47	159	23 34	203	63 28	235	35 21	311
241	59 06	011	46 54	063	17 48	117	27 58	159	23 22	202	63 02	234	34 56	310
242	59 11	009	47 23	062	18 17	117	28 10	158	23 10	202	62 35	234	34 31	309
243	59 16	008	47 52	061	18 46	116	28 23	157	22 58	201	62 09	234	34 06	308
244	59 20	006	48 20	060	19 16	115	28 35	157	22 46	200	61 43	234	33 40	307
245	59 23	004	48 48	058	19 45	114	28 48	156	22 35	200	61 16	233	33 13	306
246	59 24	002	49 16	057	20 15	114	29 02	156	22 24	199	60 50	233	32 47	305
247	59 25	000	49 43	056	20 45	113	29 15	155	22 14	198	60 24	233	32 20	304
248	59 25	359	50 10	055	21 16	112	29 29	155	22 03	198	59 58	233	31 52	303
249	59 24	357	50 37	054	21 46	111	29 43	154	21 53	197	59 32	232	31 25	302
250	59 22	355	51 03	052	22 16	111	29 58	154	21 44	197	59 06	232	30 57	301
251	59 18	353	51 28	051	22 47	110	30 12	153	21 35	196	58 40	232	30 28	300
252	59 14	352	51 54	050	23 18	109	30 27	152	21 26	195	58 15	231	30 00	299
253	59 09	350	52 18	049	23 49	108	30 43	152	21 18	195	57 49	231	29 31	298
254	59 03	348	52 43	047	24 20	107	30 58	151	21 09	194	57 24	231	29 02	297
	ALTAIR		◆FOMALHAUT		ACHERNAR		CANOPUS		◆ACRUX		SPICA		◆ANTARES	
255	15 28	044	24 51	107	31 14	151	21 02	193	56 59	231	28 33	296	58 56	347
256	15 51	043	25 23	106	31 30	150	20 54	193	56 33	230	28 03	295	58 47	345
257	16 13	042	25 54	105	31 47	150	20 47	192	56 08	230	27 33	294	58 38	343
258	16 35	041	26 26	104	32 03	149	20 40	192	55 43	230	27 03	293	58 28	341
259	16 56	040	26 57	104	32 20	149	20 34	191	55 19	229	26 33	292	58 18	340
260	17 17	039	27 29	103	32 37	148	20 28	190	54 54	229	26 03	291	58 06	338
261	17 37	038	28 01	102	32 55	147	20 23	190	54 29	229	25 32	290	57 53	336
262	17 57	037	28 33	101	33 13	147	20 17	189	54 05	228	25 01	289	57 40	335
263	18 17	036	29 05	101	33 31	146	20 12	188	53 40	228	24 30	288	57 25	333
264	18 36	035	29 37	100	33 49	146	20 08	188	53 16	228	23 59	287	57 10	332
265	18 55	034	30 10	099	34 07	145	20 04	187	52 52	227	23 28	286	56 54	330
266	19 13	033	30 42	098	34 26	145	20 00	186	52 28	227	22 56	285	56 38	328
267	19 30	032	31 14	097	34 45	144	19 56	186	52 04	227	22 25	285	56 20	327
268	19 48	031	31 47	097	35 04	144	19 53	185	51 41	226	21 53	284	56 02	325
269	20 04	030	32 19	096	35 24	143	19 50	184	51 17	226	21 21	283	55 43	324

LHA ♈	Hc	Zn	Hc	Zn	Hc	Zn	Hc	Zn	Hc	Zn	Hc	Zn	Hc	Zn
	◆ALTAIR		FOMALHAUT		◆ACHERNAR		CANOPUS		RIGIL KENT.		◆SPICA		ANTARES	
270	20 21	029	32 52	095	35 43	143	19 48	184	64 27	240	20 49	282	55 24	322
271	20 36	028	33 24	094	36 03	142	19 46	183	63 59	240	20 17	281	55 03	321
272	20 52	027	33 57	093	36 23	142	19 44	183	63 31	240	19 45	280	54 42	320
273	21 06	026	34 30	093	36 44	141	19 43	182	63 03	239	19 13	279	54 21	318
274	21 20	025	35 02	092	37 05	141	19 42	181	62 34	239	18 41	278	53 59	317
275	21 34	024	35 35	091	37 25	140	19 42	181	62 06	239	18 08	278	53 36	315
276	21 47	023	36 08	090	37 47	140	19 42	180	61 39	239	17 36	277	53 13	314
277	22 00	022	36 40	089	38 08	139	19 42	179	61 11	238	17 04	276	52 49	313
278	22 12	021	37 13	088	38 29	138	19 42	179	60 43	238	16 31	275	52 25	311
279	22 23	020	37 46	087	38 51	138	19 43	178	60 15	238	15 58	274	52 00	310
280	22 34	019	38 18	087	39 13	137	19 45	177	59 48	237	15 26	273	51 35	309
281	22 44	018	38 51	086	39 35	137	19 46	177	59 20	237	14 53	272	51 09	308
282	22 54	017	39 23	085	39 58	136	19 48	176	58 53	237	14 21	272	50 43	306
283	23 03	016	39 56	084	40 20	136	19 51	175	58 26	236	13 48	271	50 17	305
284	23 12	015	40 28	083	40 43	135	19 53	175	57 59	236	13 15	270	49 50	304
	◆ALTAIR		FOMALHAUT		◆ACHERNAR		CANOPUS		◆RIGIL KENT.		ANTARES		Nunki	
285	23 20	014	41 01	082	41 06	135	19 57	174	57 32	236	49 22	303	59 17	358
286	23 27	013	41 33	081	41 30	134	20 00	174	57 05	235	48 55	302	59 16	356
287	23 34	011	42 05	080	41 53	134	20 04	173	56 38	235	48 27	300	59 13	354
288	23 40	010	42 38	079	42 17	133	20 08	172	56 11	234	47 58	299	59 09	353
289	23 46	009	43 10	079	42 40	133	20 13	172	55 45	234	47 30	298	59 04	351
290	23 51	008	43 42	078	43 04	132	20 18	171	55 18	234	47 01	297	58 59	349
291	23 55	007	44 13	077	43 29	132	20 23	170	54 52	233	46 31	296	58 52	347
292	23 59	006	44 45	076	43 53	131	20 29	170	54 26	233	46 02	295	58 44	346
293	24 02	005	45 17	075	44 18	131	20 35	169	54 00	233	45 32	294	58 36	344
294	24 05	004	45 48	074	44 42	130	20 41	168	53 34	232	45 02	293	58 26	342
295	24 07	003	46 19	073	45 07	130	20 48	168	53 08	232	44 32	292	58 16	341
296	24 08	002	46 51	072	45 33	129	20 55	167	52 43	231	44 01	291	58 05	339
297	24 09	001	47 21	071	45 58	129	21 02	167	52 17	231	43 31	290	57 52	337
298	24 09	000	47 52	070	46 23	129	21 10	166	51 52	231	43 00	289	57 39	336
299	24 08	359	48 23	069	46 49	128	21 18	165	51 27	230	42 29	288	57 25	334
	FOMALHAUT		◆ACHERNAR		CANOPUS		◆RIGIL KENT.		ANTARES		Nunki		◆ALTAIR	
300	48 53	067	47 15	128	21 27	165	51 02	230	41 57	287	57 11	332	24 07	357
301	49 23	066	47 41	127	21 36	164	50 37	229	41 26	286	56 55	331	24 05	356
302	49 53	065	48 07	127	21 45	163	50 12	229	40 55	285	56 39	329	24 03	355
303	50 22	064	48 33	126	21 54	163	49 48	229	40 23	284	56 22	328	24 00	354
304	50 52	063	49 00	126	22 04	162	49 23	228	39 51	283	56 04	326	23 56	353
305	51 21	062	49 26	125	22 14	161	48 59	228	39 19	282	55 45	325	23 52	352
306	51 49	061	49 53	125	22 25	161	48 35	227	38 47	281	55 26	323	23 47	351
307	52 18	060	50 20	124	22 36	160	48 11	227	38 15	280	55 06	322	23 42	350
308	52 46	058	50 47	124	22 47	160	47 47	226	37 43	279	54 46	320	23 36	349
309	53 13	057	51 14	123	22 59	159	47 23	226	37 10	278	54 25	319	23 29	348
310	53 41	056	51 42	123	23 10	158	47 00	226	36 38	277	54 03	318	23 22	347
311	54 07	055	52 09	122	23 23	158	46 37	225	36 06	276	53 40	316	23 14	346
312	54 34	053	52 37	122	23 35	157	46 14	225	35 33	276	53 17	315	23 06	345
313	55 00	052	53 05	121	23 48	156	45 51	224	35 01	275	52 54	313	22 57	343
314	55 25	051	53 33	121	24 01	156	45 28	224	34 28	274	52 30	312	22 47	342
	◆FOMALHAUT		ACHERNAR		◆CANOPUS		RIGIL KENT.		◆ANTARES		Nunki		ALTAIR	
315	55 50	049	54 01	121	24 15	155	45 05	223	33 55	273	52 05	311	22 37	341
316	56 15	048	54 29	120	24 29	155	44 43	223	33 23	272	51 41	309	22 26	340
317	56 39	047	54 57	120	24 43	154	44 21	223	32 50	271	51 15	308	22 15	339
318	57 02	045	55 26	119	24 57	153	43 59	222	32 17	270	50 49	307	22 03	338
319	57 25	044	55 55	119	25 12	153	43 37	222	31 45	270	50 23	306	21 51	337
320	57 48	042	56 23	118	25 27	152	43 15	221	31 12	269	49 56	305	21 38	336
321	58 09	041	56 52	118	25 43	152	42 54	221	30 39	268	49 29	303	21 24	335
322	58 30	039	57 21	117	25 58	151	42 33	220	30 07	267	49 02	302	21 10	334
323	58 51	038	57 50	117	26 14	150	42 12	220	29 34	266	48 34	301	20 56	333
324	59 10	036	58 19	116	26 31	150	41 51	219	29 01	265	48 06	300	20 41	332
325	59 29	035	58 49	116	26 47	149	41 30	219	28 29	265	47 37	299	20 25	331
326	59 47	033	59 18	116	27 04	149	41 10	218	27 56	264	47 08	298	20 09	330
327	60 05	031	59 48	115	27 21	148	40 49	218	27 24	263	46 39	296	19 52	329
328	60 21	030	60 17	115	27 39	147	40 29	218	26 52	262	46 10	295	19 35	328
329	60 37	028	60 47	114	27 57	147	40 10	217	26 19	261	45 40	294	19 18	327
	◆FOMALHAUT		ACHERNAR		◆CANOPUS		RIGIL KENT.		◆ANTARES		Nunki		ALTAIR	
330	60 52	026	61 17	114	28 15	146	39 50	217	25 47	261	45 10	293	19 00	326
331	61 06	025	61 47	113	28 33	146	39 31	216	25 15	260	44 40	292	18 41	325
332	61 19	023	62 17	113	28 52	145	39 11	216	24 43	259	44 10	291	18 22	324
333	61 31	021	62 47	113	29 11	144	38 53	215	24 11	258	43 39	290	18 03	323
334	61 42	019	63 17	112	29 30	144	38 34	215	23 39	257	43 08	289	17 43	322
335	61 53	017	63 48	112	29 49	143	38 15	214	23 07	257	42 37	288	17 22	321
336	62 02	016	64 18	111	30 09	143	37 57	214	22 35	256	42 06	287	17 02	320
337	62 10	014	64 48	111	30 29	142	37 39	213	22 03	255	41 35	286	16 41	319
338	62 17	012	65 19	110	30 49	141	37 21	213	21 32	254	41 03	285	16 19	318
339	62 24	010	65 50	110	31 10	141	37 04	212	21 00	254	40 32	284	15 57	317
340	62 29	008	66 21	110	31 30	140	36 46	212	20 29	253	40 00	283	15 35	316
341	62 33	006	66 51	109	31 51	140	36 29	211	19 58	252	39 28	282	15 12	315
342	62 36	004	67 22	109	32 13	139	36 13	211	19 27	251	38 56	281	14 49	314
343	62 38	003	67 53	108	32 34	139	35 56	210	18 56	250	38 24	281	14 25	313
344	62 39	001	68 24	108	32 56	138	35 40	210	18 26	250	37 52	280	14 01	313
	◆Diphda		Acamar		CANOPUS		◆Miaplacidus		RIGIL KENT.		◆Nunki		FOMALHAUT	
345	46 31	037	48 53	092	33 18	137	38 09	169	35 24	209	37 20	279	62 39	359
346	46 50	036	49 26	092	33 40	137	38 16	168	35 08	209	36 47	278	62 37	357
347	47 09	034	49 58	091	34 03	136	38 23	168	34 52	208	36 15	277	62 35	355
348	47 27	033	50 31	090	34 25	136	38 30	167	34 37	208	35 43	276	62 32	353
349	47 45	032	51 04	089	34 48	135	38 37	167	34 22	207	35 10	275	62 27	351
350	48 01	030	51 36	088	35 11	135	38 44	166	34 07	207	34 37	274	62 22	349
351	48 18	029	52 09	087	35 35	134	38 52	166	33 52	206	34 05	273	62 15	348
352	48 33	028	52 42	087	35 59	133	39 00	166	33 38	206	33 32	273	62 08	346
353	48 48	026	53 14	086	36 22	133	39 08	165	33 24	205	33 00	272	61 59	344
354	49 02	025	53 47	085	36 46	132	39 17	165	33 10	205	32 27	271	61 49	342
355	49 15	023	54 19	084	37 11	132	39 25	164	32 57	204	31 54	270	61 39	340
356	49 28	022	54 52	083	37 35	131	39 34	164	32 43	204	31 22	269	61 27	338
357	49 40	021	55 24	082	38 00	131	39 43	164	32 30	203	30 49	268	61 15	337
358	49 51	019	55 56	081	38 25	130	39 53	163	32 18	203	30 16	268	61 01	335
359	50 01	018	56 29	080	38 50	129	40 02	163	32 05	202	29 44	267	60 47	333

LAT 58°S

LHA ♈	Hc	Zn	Hc	Zn	Hc	Zn	Hc	Zn	Hc	Zn	Hc	Zn	Hc	Zn
	Diphda		◆RIGEL		CANOPUS		Miaplacidus		◆RIGIL KENT.		Peacock		◆FOMALHAUT	
0	49 13	016	13 00	085	39 53	128	41 09	162	32 49	202	61 48	249	59 39	332
1	49 21	014	13 32	084	40 18	128	41 19	162	32 37	201	61 18	249	59 24	331
2	49 29	013	14 03	083	40 43	127	41 29	161	32 26	201	60 49	248	59 08	329
3	49 36	012	14 35	082	41 09	127	41 40	161	32 15	200	60 19	248	58 51	327
4	49 42	010	15 06	081	41 34	126	41 50	161	32 04	200	59 50	247	58 34	326
5	49 47	009	15 38	080	42 00	125	42 01	160	31 53	199	59 20	247	58 16	324
6	49 51	007	16 09	079	42 26	125	42 12	160	31 43	199	58 51	246	57 57	323
7	49 55	006	16 40	079	42 52	124	42 23	159	31 33	198	58 22	246	57 37	321
8	49 58	004	17 11	078	43 19	124	42 34	159	31 23	198	57 53	245	57 17	320
9	49 59	003	17 42	077	43 45	123	42 46	159	31 14	197	57 25	245	56 56	318
10	50 01	001	18 13	076	44 12	123	42 57	158	31 05	196	56 56	244	56 35	317
11	50 01	000	18 44	075	44 39	122	43 09	158	30 56	196	56 27	244	56 13	315
12	50 00	358	19 15	074	45 06	121	43 21	157	30 47	195	55 59	243	55 50	314
13	49 59	357	19 45	073	45 33	121	43 34	157	30 39	195	55 30	243	55 27	313
14	49 57	355	20 15	072	46 01	120	43 46	157	30 31	194	55 02	242	55 03	311
	◆RIGEL		SIRIUS		CANOPUS		◆RIGIL KENT.		Peacock		◆FOMALHAUT		Diphda	
15	20 46	071	16 06	096	46 28	120	30 23	194	54 34	242	54 39	310	49 54	354
16	21 16	071	16 37	095	46 56	119	30 16	193	54 06	241	54 15	309	49 50	352
17	21 46	070	17 09	094	47 24	119	30 09	193	53 38	241	53 50	307	49 45	351
18	22 15	069	17 41	093	47 52	118	30 02	192	53 11	240	53 24	306	49 40	349
19	22 45	068	18 12	093	48 20	117	29 55	192	52 43	240	52 58	305	49 34	348
20	23 14	067	18 44	092	48 48	117	29 49	191	52 16	239	52 32	304	49 27	347
21	23 43	066	19 16	091	49 17	116	29 43	190	51 49	239	52 05	302	49 19	345
22	24 12	065	19 48	090	49 45	116	29 38	190	51 21	238	51 38	301	49 10	344
23	24 41	064	20 19	089	50 14	115	29 32	189	50 54	238	51 11	300	49 01	342
24	25 09	063	20 51	088	50 43	114	29 27	189	50 28	237	50 43	299	48 51	341
25	25 38	062	21 23	087	51 12	114	29 23	188	50 01	237	50 15	298	48 40	339
26	26 06	061	21 55	087	51 41	113	29 18	188	49 34	236	49 47	297	48 28	338
27	26 33	060	22 27	086	52 10	113	29 14	187	49 08	236	49 18	295	48 16	337
28	27 01	059	22 58	085	52 40	112	29 10	187	48 42	235	48 50	294	48 03	335
29	27 28	058	23 30	084	53 09	112	29 07	186	48 16	235	48 20	293	47 49	334
	◆RIGEL		SIRIUS		CANOPUS		◆RIGIL KENT.		Peacock		◆FOMALHAUT		Diphda	
30	27 55	057	24 01	083	53 39	111	29 04	185	47 50	234	47 51	292	47 35	332
31	28 21	056	24 33	082	54 09	110	29 01	185	47 24	234	47 22	291	47 20	331
32	28 47	055	25 04	081	54 38	110	28 58	184	46 59	233	46 52	290	47 04	330
33	29 13	054	25 36	080	55 08	109	28 56	184	46 33	233	46 22	289	46 48	328
34	29 39	053	26 07	080	55 38	109	28 54	183	46 08	232	45 52	288	46 31	327
35	30 04	052	26 38	079	56 09	108	28 52	183	45 43	232	45 21	287	46 13	326
36	30 29	051	27 10	078	56 39	107	28 51	182	45 18	231	44 51	286	45 55	324
37	30 54	050	27 41	077	57 09	107	28 50	182	44 53	231	44 20	285	45 36	323
38	31 18	049	28 11	076	57 40	106	28 49	181	44 29	230	43 49	284	45 17	322
39	31 41	048	28 42	075	58 10	106	28 49	180	44 05	230	43 19	283	44 57	321
40	32 05	047	29 13	074	58 41	105	28 49	180	43 40	229	42 47	282	44 37	319
41	32 28	046	29 43	073	59 12	104	28 49	179	43 16	229	42 16	281	44 16	318
42	32 50	045	30 14	072	59 43	104	28 50	179	42 53	228	41 45	280	43 54	317
43	33 12	043	30 44	071	60 14	103	28 50	178	42 29	228	41 14	279	43 32	316
44	33 34	042	31 14	070	60 45	102	28 52	178	42 06	227	40 42	278	43 10	314
	RIGEL		◆SIRIUS		CANOPUS		◆RIGIL KENT.		Peacock		FOMALHAUT		◆Diphda	
45	33 55	041	31 44	069	61 16	102	28 53	177	41 43	227	40 11	277	42 47	313
46	34 16	040	32 14	068	61 47	101	28 55	177	41 20	226	39 39	277	42 24	312
47	34 36	039	32 43	067	62 18	100	28 57	176	40 57	226	39 08	276	42 00	311
48	34 56	038	33 12	066	62 49	100	28 59	175	40 34	225	38 36	275	41 36	310
49	35 15	037	33 41	065	63 21	099	29 02	175	40 12	224	38 04	274	41 11	309
50	35 34	036	34 10	064	63 52	098	29 05	174	39 50	224	37 33	273	40 46	308
51	35 52	034	34 39	063	64 24	098	29 08	174	39 28	223	37 01	272	40 21	306
52	36 10	033	35 07	062	64 55	097	29 12	173	39 06	223	36 29	271	39 55	305
53	36 27	032	35 35	061	65 27	096	29 16	173	38 44	222	35 57	270	39 29	304
54	36 44	031	36 03	060	65 58	096	29 20	172	38 23	222	35 25	270	39 02	303
55	37 00	030	36 30	059	66 30	095	29 24	172	38 02	221	34 54	269	38 35	302
56	37 15	029	36 58	058	67 02	094	29 29	171	37 41	221	34 22	268	38 08	301
57	37 30	027	37 24	057	67 34	093	29 34	170	37 21	220	33 50	267	37 41	300
58	37 44	026	37 51	056	68 05	092	29 40	170	37 00	220	33 18	266	37 13	299
59	37 58	025	38 17	055	68 37	092	29 45	169	36 40	219	32 47	265	36 45	298
	RIGEL		◆SIRIUS		Suhail		◆RIGIL KENT.		Peacock		FOMALHAUT		◆Diphda	
60	38 11	024	38 43	054	42 03	108	29 51	169	36 20	219	32 15	265	36 17	297
61	38 24	022	39 09	053	42 33	107	29 58	168	36 00	218	31 43	264	35 48	296
62	38 35	021	39 34	052	43 04	106	30 04	168	35 41	217	31 12	263	35 20	295
63	38 47	020	39 59	051	43 34	106	30 11	167	35 22	217	30 40	262	34 51	294
64	38 57	019	40 23	050	44 05	105	30 19	167	35 03	216	30 09	261	34 22	293
65	39 07	017	40 47	048	44 36	104	30 26	166	34 44	216	29 37	261	33 52	292
66	39 16	016	41 11	047	45 06	103	30 34	166	34 26	215	29 06	260	33 22	291
67	39 25	015	41 34	046	45 37	103	30 42	165	34 07	215	28 35	259	32 53	290
68	39 33	014	41 56	045	46 09	102	30 50	164	33 49	214	28 04	258	32 23	289
69	39 40	012	42 19	044	46 40	101	30 59	164	33 32	214	27 33	257	31 53	288
70	39 46	011	42 40	042	47 11	100	31 08	163	33 14	213	27 02	257	31 22	287
71	39 52	010	43 02	041	47 42	100	31 17	163	32 57	212	26 31	256	30 52	286
72	39 57	009	43 22	040	48 14	099	31 27	162	32 40	212	26 00	255	30 21	285
73	40 01	007	43 43	039	48 45	098	31 37	162	32 23	211	25 29	254	29 50	284
74	40 05	006	44 02	038	49 17	097	31 47	161	32 07	211	24 59	253	29 19	283
	RIGEL		◆SIRIUS		Suhail		◆RIGIL KENT.		Peacock		ACHERNAR		◆Diphda	
75	40 08	005	44 21	036	49 48	097	31 57	161	31 51	210	63 32	250	28 48	282
76	40 10	003	44 40	035	50 20	096	32 08	160	31 35	210	63 02	249	28 17	281
77	40 12	002	44 58	034	50 51	095	32 19	160	31 19	209	62 32	249	27 46	281
78	40 12	001	45 15	032	51 23	094	32 30	159	31 04	209	62 03	248	27 15	280
79	40 12	359	45 32	031	51 55	093	32 41	159	30 49	208	61 33	248	26 43	279
80	40 12	358	45 48	030	52 27	093	32 53	158	30 34	207	61 04	247	26 12	278
81	40 10	357	46 04	029	52 58	092	33 05	157	30 20	207	60 35	247	25 40	277
82	40 08	356	46 18	027	53 30	091	33 18	157	30 05	206	60 05	246	25 09	276
83	40 05	354	46 33	026	54 02	090	33 30	156	29 51	206	59 36	246	24 37	275
84	40 02	353	46 46	024	54 34	089	33 43	156	29 38	205	59 07	245	24 06	274
85	39 58	352	46 59	023	55 05	088	33 56	155	29 24	205	58 39	245	23 34	274
86	39 53	350	47 11	022	55 37	088	34 10	155	29 11	204	58 10	244	23 02	273
87	39 47	349	47 22	020	56 09	087	34 23	154	28 59	203	57 41	244	22 30	272
88	39 41	348	47 33	019	56 41	086	34 37	154	28 46	203	57 13	243	21 59	271
89	39 34	347	47 43	018	57 12	085	34 51	153	28 34	202	56 44	243	21 27	270

LHA ♈	Hc	Zn	Hc	Zn	Hc	Zn	Hc	Zn	Hc	Zn	Hc	Zn	Hc	Zn
	SIRIUS		Alphard		◆Suhail		RIGIL KENT.		◆Peacock		ACHERNAR		◆RIGEL	
90	47 52	016	26 49	061	57 44	084	35 06	153	28 22	202	56 16	242	39 26	345
91	48 01	015	27 16	060	58 16	083	35 20	152	28 10	201	55 48	242	39 17	344
92	48 08	013	27 44	059	58 47	082	35 35	152	27 59	200	55 20	242	39 08	343
93	48 15	012	28 11	058	59 19	081	35 51	151	27 48	200	54 52	241	38 59	342
94	48 22	010	28 37	057	59 50	080	36 06	151	27 37	199	54 24	241	38 48	340
95	48 27	009	29 04	056	60 21	079	36 22	150	27 27	199	53 57	240	38 37	339
96	48 32	008	29 30	055	60 52	078	36 38	150	27 17	198	53 29	240	38 25	338
97	48 35	006	29 56	054	61 23	077	36 54	149	27 07	198	53 02	239	38 13	337
98	48 38	005	30 21	053	61 54	076	37 10	149	26 58	197	52 35	239	38 00	335
99	48 41	003	30 46	051	62 25	075	37 27	148	26 49	196	52 08	238	37 47	334
100	48 42	002	31 11	050	62 56	074	37 44	148	26 40	196	51 41	238	37 32	333
101	48 43	000	31 35	049	63 26	072	38 01	147	26 31	195	51 14	237	37 18	332
102	48 42	359	31 59	048	63 56	071	38 18	147	26 23	195	50 47	237	37 02	330
103	48 41	357	32 22	047	64 26	070	38 36	146	26 15	194	50 21	236	36 46	329
104	48 39	356	32 46	046	64 56	069	38 54	146	26 08	193	49 54	236	36 30	328
	Alphard		◆Gienah		RIGIL KENT.		◆Peacock		ACHERNAR		◆RIGEL		SIRIUS	
105	33 08	045	20 39	090	39 12	145	26 01	193	49 28	235	36 13	327	48 37	355
106	33 31	044	21 11	089	39 30	145	25 54	192	49 02	235	35 55	326	48 33	353
107	33 52	043	21 42	089	39 48	144	25 47	192	48 36	234	35 37	325	48 29	352
108	34 14	042	22 14	088	40 07	144	25 41	191	48 11	234	35 18	323	48 24	350
109	34 35	041	22 46	087	40 26	143	25 35	190	47 45	233	34 59	322	48 18	349
110	34 55	039	23 18	086	40 45	143	25 30	190	47 20	233	34 39	321	48 12	347
111	35 15	038	23 49	085	41 05	142	25 24	189	46 54	232	34 19	320	48 04	346
112	35 35	037	24 21	084	41 24	142	25 19	189	46 29	232	33 58	319	47 56	345
113	35 54	036	24 53	083	41 44	141	25 15	188	46 05	231	33 37	318	47 47	343
114	36 12	035	25 24	082	42 04	141	25 10	187	45 40	231	33 16	317	47 38	342
115	36 30	034	25 56	082	42 24	140	25 06	187	45 15	230	32 54	316	47 27	340
116	36 47	033	26 27	081	42 44	140	25 03	186	44 51	230	32 31	314	47 16	339
117	37 04	031	26 58	080	43 05	139	25 00	186	44 27	229	32 08	313	47 05	338
118	37 21	030	27 30	079	43 26	139	24 57	185	44 03	229	31 45	312	46 52	336
119	37 36	029	28 01	078	43 47	138	24 54	184	43 39	228	31 21	311	46 39	335
	REGULUS		◆SPICA		RIGIL KENT.		Peacock		◆ACHERNAR		RIGEL		◆SIRIUS	
120	15 16	033	14 04	089	44 08	138	24 52	184	43 15	228	30 57	310	46 25	333
121	15 33	032	14 36	088	44 29	138	24 50	183	42 52	227	30 33	309	46 10	332
122	15 49	031	15 07	087	44 51	137	24 48	183	42 29	227	30 08	308	45 55	331
123	16 05	030	15 39	086	45 13	137	24 47	182	42 06	226	29 43	307	45 39	329
124	16 21	029	16 11	085	45 35	136	24 46	181	41 43	226	29 17	306	45 23	328
125	16 35	028	16 43	084	45 57	136	24 45	181	41 20	225	28 51	305	45 06	327
126	16 50	027	17 14	083	46 19	135	24 45	180	40 58	225	28 25	304	44 48	326
127	17 04	026	17 46	083	46 42	135	24 45	180	40 36	224	27 59	303	44 30	324
128	17 18	025	18 17	082	47 04	134	24 46	179	40 14	224	27 32	302	44 11	323
129	17 31	024	18 49	081	47 27	134	24 46	178	39 52	223	27 05	301	43 52	322
130	17 43	023	19 20	080	47 50	133	24 47	178	39 30	222	26 37	300	43 32	321
131	17 55	022	19 51	079	48 13	133	24 49	177	39 09	222	26 10	299	43 11	319
132	18 06	021	20 22	078	48 37	133	24 51	177	38 48	221	25 42	298	42 50	318
133	18 17	020	20 53	077	49 00	132	24 53	176	38 27	221	25 14	297	42 29	317
134	18 28	019	21 24	076	49 24	132	24 55	175	38 06	220	24 45	296	42 07	316
	REGULUS		◆SPICA		RIGIL KENT.		◆Peacock		ACHERNAR		CANOPUS		◆SIRIUS	
135	18 38	018	21 55	075	49 48	131	24 58	175	37 46	220	67 31	267	41 44	314
136	18 47	017	22 26	075	50 12	131	25 01	174	37 26	219	66 59	266	41 21	313
137	18 56	016	22 57	074	50 36	130	25 04	174	37 06	219	66 28	265	40 58	312
138	19 04	015	23 27	073	51 00	130	25 08	173	36 46	218	65 56	264	40 34	311
139	19 12	014	23 57	072	51 25	129	25 12	172	36 26	218	65 24	264	40 10	310
140	19 19	012	24 27	071	51 49	129	25 16	172	36 07	217	64 53	263	39 45	309
141	19 26	011	24 57	070	52 14	129	25 21	171	35 48	217	64 21	262	39 20	308
142	19 32	010	25 27	069	52 39	128	25 26	171	35 29	216	63 50	262	38 55	307
143	19 37	009	25 57	068	53 04	128	25 32	170	35 11	216	63 18	261	38 29	305
144	19 42	008	26 26	067	53 29	127	25 37	169	34 52	215	62 47	260	38 03	304
145	19 46	007	26 55	066	53 55	127	25 43	169	34 34	214	62 16	260	37 37	303
146	19 50	006	27 24	065	54 20	127	25 50	168	34 16	214	61 44	259	37 10	302
147	19 53	005	27 53	064	54 46	126	25 56	168	33 59	213	61 13	258	36 43	301
148	19 56	004	28 22	063	55 11	126	26 03	167	33 41	213	60 42	258	36 16	300
149	19 58	003	28 50	062	55 37	125	26 11	166	33 24	212	60 11	257	35 48	299
	REGULUS		◆SPICA		RIGIL KENT.		◆Peacock		ACHERNAR		CANOPUS		◆SIRIUS	
150	19 59	002	29 18	061	56 03	125	26 18	166	33 07	212	59 40	256	35 20	298
151	20 00	001	29 46	060	56 29	125	26 26	165	32 51	211	59 09	256	34 52	297
152	20 00	000	30 13	059	56 56	124	26 35	165	32 35	211	58 39	255	34 23	296
153	20 00	359	30 40	058	57 22	124	26 43	164	32 19	210	58 08	254	33 55	295
154	19 59	358	31 07	057	57 48	123	26 52	163	32 03	209	57 37	254	33 26	294
155	19 58	357	31 34	056	58 15	123	27 01	163	31 47	209	57 07	253	32 57	293
156	19 56	356	32 00	055	58 42	123	27 11	162	31 32	208	56 37	253	32 27	292
157	19 53	355	32 26	054	59 09	122	27 21	162	31 17	208	56 06	252	31 58	291
158	19 50	354	32 52	053	59 36	122	27 31	161	31 02	207	55 36	251	31 28	290
159	19 46	353	33 17	052	60 03	122	27 41	160	30 48	207	55 06	251	30 58	289
160	19 42	352	33 42	051	60 30	121	27 52	160	30 34	206	54 36	250	30 28	288
161	19 37	351	34 06	050	60 57	121	28 03	159	30 20	205	54 06	250	29 58	287
162	19 32	350	34 30	049	61 24	120	28 15	159	30 07	205	53 36	249	29 27	286
163	19 26	349	34 54	048	61 52	120	28 26	158	29 53	204	53 07	248	28 56	285
164	19 19	348	35 18	047	62 19	120	28 38	158	29 40	204	52 37	248	28 26	284
	◆SPICA		ANTARES		◆Peacock		ACHERNAR		CANOPUS		◆SIRIUS		REGULUS	
165	35 40	046	26 11	099	28 51	157	29 28	203	52 08	247	27 55	284	19 12	347
166	36 03	044	26 42	098	29 03	156	29 15	203	51 39	247	27 24	283	19 05	346
167	36 25	043	27 14	097	29 16	156	29 03	202	51 10	246	26 53	282	18 56	345
168	36 47	042	27 45	096	29 29	155	28 51	201	50 41	245	26 22	281	18 48	343
169	37 08	041	28 17	095	29 43	155	28 40	201	50 12	245	25 50	280	18 38	342
170	37 28	040	28 49	094	29 57	154	28 29	200	49 43	244	25 19	279	18 28	341
171	37 48	039	29 20	094	30 11	154	28 18	200	49 14	244	24 48	278	18 18	340
172	38 08	038	29 52	093	30 25	153	28 07	199	48 46	243	24 16	277	18 07	339
173	38 27	036	30 24	092	30 39	152	27 57	199	48 18	243	23 45	276	17 56	338
174	38 46	035	30 56	091	30 54	152	27 47	198	47 49	242	23 13	276	17 44	337
175	39 04	034	31 27	090	31 10	151	27 37	197	47 21	241	22 41	275	17 31	336
176	39 21	033	31 59	089	31 25	151	27 28	197	46 54	241	22 10	274	17 18	335
177	39 38	032	32 31	089	31 41	150	27 19	196	46 26	240	21 38	273	17 05	334
178	39 55	030	33 03	088	31 57	150	27 10	196	45 58	240	21 07	272	16 51	333
179	40 10	029	33 34	087	32 13	149	27 02	195	45 31	239	20 34	271	16 36	332

LAT 58°S

LHA ♈	Hc Zn	Hc Zn	Hc Zn	Hc Zn	Hc Zn	Hc Zn	Hc Zn
	◆SPICA	ANTARES	◆Peacock	ACHERNAR	CANOPUS	◆SIRIUS	Alphard
180	40 25 028	34 06 086	32 29 148	26 54 194	45 04 239	20 02 270	32 38 313
181	40 40 027	34 38 085	32 46 148	26 46 194	44 37 238	19 31 270	32 15 312
182	40 54 025	35 10 084	33 03 147	26 38 193	44 10 237	18 59 269	31 51 311
183	41 07 024	35 41 083	33 21 147	26 31 193	43 43 237	18 27 268	31 27 310
184	41 20 023	36 13 082	33 38 146	26 24 192	43 17 236	17 55 267	31 03 309
185	41 32 021	36 44 082	33 56 146	26 18 192	42 50 236	17 24 266	30 38 308
186	41 43 020	37 16 081	34 14 145	26 12 191	42 24 235	16 52 265	30 13 307
187	41 54 019	37 47 080	34 32 145	26 06 190	41 58 235	16 20 264	29 47 306
188	42 04 018	38 18 079	34 51 144	26 00 190	41 32 234	15 49 264	29 21 305
189	42 13 016	38 49 078	35 10 143	25 55 189	41 07 233	15 17 263	28 55 304
190	42 21 015	39 20 077	35 29 143	25 50 189	40 41 233	14 45 262	28 29 303
191	42 29 014	39 51 076	35 48 142	25 45 188	40 16 232	14 14 261	28 02 302
192	42 37 012	40 22 075	36 08 142	25 41 187	39 51 232	13 43 260	27 35 301
193	42 43 011	40 53 074	36 27 141	25 37 187	39 26 231	13 11 260	27 07 300
194	42 49 010	41 23 073	36 48 141	25 34 186	39 01 231	12 40 259	26 40 299
	SPICA	◆ANTARES	Peacock	◆ACHERNAR	CANOPUS	Suhail	◆Alphard
195	42 54 008	41 54 072	37 08 140	25 30 186	38 37 230	51 52 266	26 12 298
196	42 58 007	42 24 071	37 28 140	25 27 185	38 13 229	51 20 266	25 44 297
197	43 01 006	42 54 070	37 49 139	25 25 184	37 49 229	50 48 265	25 15 296
198	43 04 004	43 23 069	38 10 139	25 22 184	37 25 228	50 17 264	24 47 295
199	43 06 003	43 53 068	38 31 138	25 21 183	37 01 228	49 45 263	24 18 294
200	43 08 002	44 22 067	38 52 137	25 19 183	36 38 227	49 14 263	23 49 293
201	43 08 000	44 52 066	39 14 137	25 18 182	36 15 227	48 42 262	23 19 292
202	43 08 359	45 20 065	39 36 136	25 17 181	35 52 226	48 11 261	22 50 291
203	43 07 358	45 49 064	39 58 136	25 16 181	35 29 225	47 39 260	22 20 291
204	43 05 356	46 18 063	40 20 135	25 16 180	35 06 225	47 08 260	21 50 290
205	43 03 355	46 46 062	40 43 135	25 16 180	34 44 224	46 37 259	21 20 289
206	43 00 354	47 13 060	41 05 134	25 16 179	34 22 224	46 06 258	20 50 288
207	42 56 352	47 41 059	41 28 134	25 17 178	34 00 223	45 35 257	20 20 287
208	42 51 351	48 08 058	41 51 133	25 18 178	33 39 222	45 04 257	19 49 286
209	42 46 350	48 35 057	42 14 133	25 19 177	33 17 222	44 33 256	19 19 285
	ANTARES	◆Nunki	Peacock	ACHERNAR	◆CANOPUS	Suhail	◆SPICA
210	49 01 056	30 35 091	42 38 132	25 21 177	32 56 221	44 02 255	42 40 348
211	49 28 055	31 07 091	43 01 132	25 23 176	32 35 221	43 31 254	42 33 347
212	49 53 053	31 39 090	43 25 131	25 25 175	32 15 220	43 01 254	42 25 346
213	50 19 052	32 10 089	43 49 131	25 28 175	31 54 219	42 30 253	42 17 344
214	50 43 051	32 42 088	44 14 130	25 31 174	31 34 219	42 00 252	42 08 343
215	51 08 050	33 14 087	44 38 130	25 34 174	31 14 218	41 30 252	41 58 342
216	51 32 048	33 46 086	45 03 129	25 38 173	30 55 218	41 00 251	41 48 340
217	51 55 047	34 17 085	45 27 129	25 42 172	30 36 217	40 30 250	41 37 339
218	52 18 046	34 49 085	45 52 128	25 46 172	30 16 217	40 00 249	41 25 338
219	52 41 044	35 21 084	46 17 128	25 51 171	29 58 216	39 30 249	41 13 337
220	53 03 043	35 52 083	46 43 127	25 56 171	29 39 215	39 00 248	41 00 335
221	53 24 042	36 24 082	47 08 127	26 01 170	29 21 215	38 31 247	40 46 334
222	53 45 040	36 55 081	47 34 126	26 07 169	29 03 214	38 02 247	40 32 333
223	54 06 039	37 27 080	48 00 126	26 13 169	28 45 213	37 33 246	40 17 332
224	54 25 037	37 58 079	48 25 125	26 19 168	28 28 213	37 04 245	40 02 330
	ANTARES	◆Nunki	Peacock	ACHERNAR	◆CANOPUS	Suhail	◆SPICA
225	54 44 036	38 29 078	48 52 125	26 26 168	28 11 212	36 35 245	39 46 329
226	55 03 035	39 00 077	49 18 124	26 33 167	27 54 212	36 06 244	39 29 328
227	55 20 033	39 31 076	49 44 124	26 40 167	27 37 211	35 38 243	39 12 327
228	55 37 032	40 02 075	50 11 123	26 48 166	27 21 210	35 09 243	38 54 325
229	55 54 030	40 33 074	50 38 123	26 55 165	27 05 210	34 41 242	38 36 324
230	56 09 029	41 03 073	51 04 122	27 04 165	26 49 209	34 13 241	38 17 323
231	56 24 027	41 34 072	51 32 122	27 12 164	26 34 209	33 45 241	37 58 322
232	56 38 025	42 04 071	51 59 121	27 21 164	26 19 208	33 18 240	37 38 321
233	56 51 024	42 34 070	52 26 121	27 30 163	26 04 207	32 50 239	37 17 320
234	57 04 022	43 04 069	52 53 120	27 40 162	25 50 207	32 23 239	36 56 318
235	57 15 021	43 33 068	53 21 120	27 49 162	25 36 206	31 56 238	36 35 317
236	57 26 019	44 03 067	53 49 119	27 59 161	25 22 206	31 29 237	36 13 316
237	57 36 017	44 32 066	54 17 119	28 10 161	25 08 205	31 02 237	35 51 315
238	57 45 016	45 01 065	54 45 118	28 20 160	24 55 204	30 36 236	35 28 314
239	57 53 014	45 30 064	55 13 118	28 31 160	24 42 204	30 10 235	35 05 313
	◆ANTARES	Nunki	◆FOMALHAUT	ACHERNAR	CANOPUS	◆ACRUX	SPICA
240	58 00 012	45 58 063	17 47 118	28 43 159	24 30 203	64 03 236	34 42 312
241	58 07 011	46 27 062	18 15 117	28 54 158	24 17 202	63 36 236	34 18 311
242	58 12 009	46 55 061	18 44 116	29 06 158	24 05 202	63 10 236	33 53 309
243	58 17 007	47 22 060	19 12 116	29 18 157	23 54 201	62 44 235	33 29 308
244	58 20 006	47 49 059	19 41 115	29 31 157	23 42 201	62 17 235	33 03 307
245	58 23 004	48 16 057	20 10 114	29 43 156	23 31 200	61 51 235	32 38 306
246	58 25 002	48 43 056	20 39 113	29 56 156	23 21 199	61 25 235	32 12 305
247	58 25 000	49 09 055	21 08 112	30 10 155	23 10 199	61 00 234	31 46 304
248	58 25 359	49 35 054	21 38 112	30 23 154	23 00 198	60 34 234	31 20 303
249	58 24 357	50 01 053	22 07 111	30 37 154	22 51 197	60 08 234	30 53 302
250	58 22 355	50 26 051	22 37 110	30 51 153	22 41 197	59 43 233	30 26 301
251	58 19 354	50 50 050	23 07 109	31 06 153	22 33 196	59 17 233	29 58 300
252	58 15 352	51 15 049	23 37 109	31 21 152	22 24 195	58 52 233	29 31 299
253	58 10 350	51 38 048	24 07 108	31 36 152	22 16 195	58 26 232	29 03 298
254	58 04 349	52 01 046	24 38 107	31 51 151	22 08 194	58 01 232	28 35 297
	ALTAIR	◆FOMALHAUT	ACHERNAR	CANOPUS	◆ACRUX	SPICA	◆ANTARES
255	14 45 044	25 08 106	32 06 150	22 00 194	57 36 232	28 06 296	57 57 347
256	15 07 043	25 39 105	32 22 150	21 53 193	57 11 231	27 38 295	57 50 345
257	15 28 042	26 10 105	32 38 149	21 46 192	56 46 231	27 09 294	57 41 344
258	15 49 041	26 40 104	32 55 149	21 39 192	56 22 231	26 40 293	57 32 342
259	16 10 040	27 11 103	33 11 148	21 33 191	55 57 230	26 10 292	57 21 340
260	16 30 039	27 42 102	33 28 148	21 27 190	55 33 230	25 41 291	57 10 339
261	16 50 038	28 13 102	33 45 147	21 22 190	55 08 230	25 11 290	56 58 337
262	17 10 037	28 45 101	34 03 147	21 16 189	54 44 229	24 41 290	56 45 335
263	17 28 036	29 16 100	34 20 146	21 12 188	54 20 229	24 11 289	56 32 334
264	17 47 035	29 47 099	34 38 145	21 07 188	53 56 229	23 41 288	56 17 332
265	18 05 034	30 19 098	34 56 145	21 03 187	53 32 228	23 11 287	56 02 331
266	18 23 033	30 50 098	35 15 144	20 59 186	53 09 228	22 40 286	55 46 329
267	18 40 032	31 22 097	35 34 144	20 56 186	52 45 228	22 10 285	55 30 328
268	18 56 031	31 53 096	35 52 143	20 53 185	52 22 227	21 39 284	55 12 326
269	19 13 030	32 25 095	36 12 143	20 50 185	51 59 227	21 08 283	54 54 325

LHA ♈	Hc Zn	Hc Zn	Hc Zn	Hc Zn	Hc Zn	Hc Zn	Hc Zn
	◆ALTAIR	FOMALHAUT	◆ACHERNAR	CANOPUS	RIGIL KENT.	◆SPICA	ANTARES
270	19 28 029	32 57 094	36 31 142	20 48 184	64 56 242	20 37 282	54 36 323
271	19 43 028	33 28 093	36 50 142	20 46 183	64 28 242	20 06 281	54 16 322
272	19 58 027	34 00 093	37 10 141	20 44 183	64 00 242	19 35 281	53 56 320
273	20 12 026	34 32 092	37 30 141	20 43 182	63 32 241	19 03 280	53 36 319
274	20 26 025	35 04 091	37 51 140	20 42 181	63 04 241	18 32 279	53 15 318
275	20 39 024	35 35 090	38 11 140	20 42 181	62 37 240	18 00 278	52 53 316
276	20 52 023	36 07 089	38 32 139	20 42 180	62 09 240	17 29 277	52 31 315
277	21 04 022	36 39 088	38 53 138	20 42 179	61 42 240	16 57 276	52 08 314
278	21 16 021	37 11 088	39 14 138	20 42 179	61 14 239	16 26 275	51 45 312
279	21 27 020	37 42 087	39 36 137	20 43 178	60 47 239	15 54 274	51 21 311
280	21 37 019	38 14 086	39 57 137	20 45 177	60 20 239	15 22 274	50 57 310
281	21 47 018	38 46 085	40 19 136	20 46 177	59 52 238	14 50 273	50 32 309
282	21 57 017	39 18 084	40 41 136	20 48 176	59 25 238	14 19 272	50 07 307
283	22 05 016	39 49 083	41 03 135	20 50 175	58 59 238	13 47 271	49 42 306
284	22 14 015	40 21 082	41 26 135	20 53 175	58 32 237	13 15 270	49 16 305
	◆ALTAIR	FOMALHAUT	◆ACHERNAR	CANOPUS	◆RIGIL KENT.	ANTARES	Nunki
285	22 21 014	40 52 081	41 48 134	20 56 174	58 05 237	48 50 304	58 17 358
286	22 29 012	41 24 080	42 11 134	21 00 174	57 39 236	48 23 302	58 16 356
287	22 35 011	41 55 080	42 34 133	21 03 173	57 12 236	47 56 301	58 13 354
288	22 41 010	42 26 079	42 58 133	21 08 172	56 46 236	47 29 300	58 10 353
289	22 47 009	42 57 078	43 21 132	21 12 172	56 20 235	47 01 299	58 05 351
290	22 51 008	43 28 077	43 45 132	21 17 171	55 53 235	46 33 298	58 00 349
291	22 56 007	43 59 076	44 08 131	21 22 170	55 28 234	46 05 297	57 53 348
292	22 59 006	44 30 075	44 32 131	21 28 170	55 02 234	45 36 296	57 46 346
293	23 02 005	45 00 074	44 57 130	21 34 169	54 36 234	45 08 295	57 38 344
294	23 05 004	45 31 073	45 21 130	21 40 168	54 10 233	44 38 294	57 29 343
295	23 07 003	46 01 072	45 46 129	21 46 168	53 45 233	44 09 293	57 19 341
296	23 08 002	46 31 071	46 10 129	21 53 167	53 20 232	43 40 292	57 08 339
297	23 09 001	47 01 070	46 35 128	22 01 166	52 55 232	43 10 290	56 57 338
298	23 09 000	47 31 069	47 00 128	22 08 166	52 30 232	42 40 289	56 44 336
299	23 08 359	48 00 068	47 26 127	22 16 165	52 05 231	42 10 288	56 31 335
	FOMALHAUT	◆ACHERNAR	CANOPUS	◆RIGIL KENT.	ANTARES	Nunki	◆ALTAIR
300	48 30 066	47 51 127	22 25 165	51 40 231	41 40 287	56 17 333	23 07 357
301	48 59 065	48 17 126	22 33 164	51 16 230	41 09 286	56 03 332	23 05 356
302	49 27 064	48 42 126	22 42 163	50 51 230	40 39 286	55 47 330	23 03 355
303	49 56 063	49 08 125	22 52 163	50 27 229	40 08 285	55 31 329	23 00 354
304	50 24 062	49 34 125	23 01 162	50 03 229	39 37 284	55 14 327	22 57 353
305	50 52 061	50 01 124	23 11 161	49 39 229	39 06 283	54 56 326	22 53 352
306	51 20 060	50 27 124	23 22 161	49 15 228	38 35 282	54 38 324	22 48 351
307	51 47 058	50 53 123	23 32 160	48 51 228	38 04 281	54 19 323	22 43 350
308	52 14 057	51 20 123	23 43 159	48 28 227	37 33 280	53 59 321	22 37 349
309	52 40 056	51 47 122	23 55 159	48 05 227	37 01 279	53 39 320	22 30 348
310	53 06 055	52 14 122	24 06 158	47 42 226	36 30 278	53 18 318	22 23 347
311	53 32 053	52 41 121	24 18 158	47 19 226	35 59 277	52 57 317	22 16 346
312	53 58 052	53 08 121	24 31 157	46 56 226	35 27 276	52 35 316	22 08 345
313	54 22 051	53 36 120	24 43 156	46 33 225	34 55 275	52 12 314	21 59 344
314	54 47 050	54 03 120	24 56 156	46 11 225	34 24 274	51 49 313	21 50 343
	◆FOMALHAUT	ACHERNAR	◆CANOPUS	RIGIL KENT.	◆ANTARES	Nunki	ALTAIR
315	55 11 048	54 31 119	25 09 155	45 49 224	33 52 274	51 26 312	21 40 341
316	55 34 047	54 59 119	25 23 154	45 27 224	33 20 273	51 02 310	21 30 340
317	55 57 045	55 27 118	25 37 154	45 05 223	32 48 272	50 38 309	21 19 339
318	56 20 044	55 55 118	25 51 153	44 43 223	32 17 271	50 13 308	21 07 338
319	56 42 043	56 23 117	26 05 153	44 21 222	31 45 270	49 47 307	20 55 337
320	57 03 041	56 51 117	26 20 152	44 00 222	31 13 269	49 22 305	20 43 336
321	57 23 040	57 19 116	26 35 151	43 39 221	30 41 268	48 56 304	20 30 335
322	57 43 038	57 48 116	26 51 151	43 18 221	30 09 268	48 29 303	20 16 334
323	58 03 037	58 17 115	27 06 150	42 57 220	29 38 267	48 02 302	20 02 333
324	58 21 035	58 45 115	27 22 150	42 37 220	29 06 266	47 35 301	19 47 332
325	58 39 034	59 14 115	27 39 149	42 17 219	28 34 265	47 08 300	19 32 331
326	58 57 032	59 43 114	27 55 148	41 56 219	28 03 264	46 40 299	19 17 330
327	59 13 030	60 12 114	28 12 148	41 37 219	27 31 264	46 12 297	19 01 329
328	59 29 029	60 42 113	28 29 147	41 17 218	26 59 263	45 44 296	18 44 328
329	59 44 027	61 11 113	28 47 146	40 57 218	26 28 262	45 15 295	18 27 327
	◆FOMALHAUT	ACHERNAR	◆CANOPUS	RIGIL KENT.	◆ANTARES	Nunki	ALTAIR
330	59 58 025	61 40 112	29 04 146	40 38 217	25 56 261	44 46 294	18 10 326
331	60 11 024	62 10 112	29 22 145	40 19 217	25 25 260	44 17 293	17 52 325
332	60 23 022	62 39 111	29 41 145	40 00 216	24 54 259	43 48 292	17 33 324
333	60 35 020	63 09 111	29 59 144	39 41 216	24 23 259	43 18 291	17 15 323
334	60 46 019	63 39 110	30 18 143	39 23 215	23 51 258	42 48 290	16 55 322
335	60 55 017	64 09 110	30 37 143	39 05 215	23 20 257	42 18 289	16 36 321
336	61 04 015	64 39 109	30 56 142	38 47 214	22 49 256	41 48 288	16 16 320
337	61 12 013	65 09 109	31 16 142	38 29 214	22 19 255	41 18 287	15 55 319
338	61 19 012	65 39 108	31 36 141	38 12 213	21 48 255	40 47 286	15 34 318
339	61 24 010	66 09 108	31 56 140	37 54 213	21 17 254	40 17 285	15 13 317
340	61 29 008	66 39 107	32 16 140	37 37 212	20 47 253	39 46 284	14 51 316
341	61 33 006	67 10 107	32 37 139	37 21 212	20 16 252	39 15 283	14 29 316
342	61 36 004	67 40 106	32 58 139	37 04 211	19 46 252	38 44 282	14 07 315
343	61 38 002	68 11 106	33 19 138	36 48 211	19 16 251	38 13 281	13 44 314
344	61 39 001	68 41 105	33 40 138	36 32 210	18 46 250	37 42 280	13 21 313
	◆Diphda	Acamar	CANOPUS	◆Miaplacidus	RIGIL KENT.	◆Nunki	FOMALHAUT
345	45 43 036	48 55 091	34 02 137	39 08 168	36 16 210	37 10 279	61 39 359
346	46 01 035	49 27 090	34 24 136	39 14 168	36 00 209	36 39 279	61 38 357
347	46 19 034	49 58 090	34 46 136	39 21 168	35 45 209	36 07 278	61 35 355
348	46 37 032	50 30 089	35 08 135	39 28 167	35 30 208	35 36 277	61 32 353
349	46 53 031	51 02 088	35 31 135	39 35 167	35 15 208	35 04 276	61 28 352
350	47 10 030	51 34 087	35 53 134	39 43 166	35 00 207	34 33 275	61 23 350
351	47 25 028	52 06 086	36 16 133	39 50 166	34 46 207	34 01 274	61 17 348
352	47 40 027	52 37 085	36 40 133	39 58 165	34 32 206	33 29 273	61 09 346
353	47 54 026	53 09 084	37 03 132	40 06 165	34 18 205	32 57 272	61 01 344
354	48 08 024	53 40 083	37 27 132	40 15 165	34 05 205	32 26 271	60 52 343
355	48 20 023	54 12 082	37 50 131	40 23 164	33 51 204	31 54 271	60 42 341
356	48 32 022	54 44 082	38 15 131	40 32 164	33 38 204	31 22 270	60 31 339
357	48 44 020	55 15 081	38 39 130	40 41 163	33 25 203	30 50 269	60 20 337
358	48 54 019	55 46 080	39 03 129	40 50 163	33 13 203	30 18 268	60 07 336
359	49 04 017	56 17 079	39 28 129	41 00 163	33 01 202	29 47 267	59 53 334

 LAT 59°S

LHA ♈	Hc	Zn	Hc	Zn	Hc	Zn	Hc	Zn	Hc	Zn	Hc	Zn	Hc	Zn
	Diphda		◆RIGEL		CANOPUS		Miaplacidus		◆RIGIL KENT.		Peacock		◆FOMALHAUT	
0	48 15	016	12 54	084	40 30	128	42 06	162	33 44	202	62 08	251	58 46	333
1	48 23	014	13 25	084	40 54	127	42 16	161	33 33	202	61 39	250	58 31	332
2	48 31	013	13 56	083	41 19	126	42 26	161	33 22	201	61 10	250	58 16	330
3	48 37	011	14 26	082	41 44	126	42 36	161	33 11	200	60 41	249	58 00	328
4	48 43	010	14 57	081	42 09	125	42 47	160	33 00	200	60 12	249	57 44	327
5	48 48	008	15 27	080	42 35	125	42 57	160	32 50	199	59 43	248	57 27	325
6	48 52	007	15 58	079	43 00	124	43 08	159	32 40	199	59 15	248	57 09	324
7	48 55	006	16 28	078	43 26	123	43 19	159	32 30	198	58 46	247	56 50	322
8	48 58	004	16 58	077	43 52	123	43 30	159	32 20	198	58 18	247	56 31	321
9	49 00	003	17 28	077	44 18	122	43 41	158	32 11	197	57 49	246	56 11	319
10	49 01	001	17 58	076	44 44	122	43 53	158	32 02	197	57 21	246	55 51	318
11	49 01	000	18 28	075	45 10	121	44 05	157	31 53	196	56 53	245	55 30	317
12	49 00	358	18 58	074	45 37	120	44 17	157	31 45	196	56 25	245	55 08	315
13	48 59	357	19 28	073	46 03	120	44 29	157	31 37	195	55 57	244	54 46	314
14	48 57	355	19 57	072	46 30	119	44 41	156	31 29	194	55 30	244	54 23	312
	◆RIGEL		SIRIUS		CANOPUS		◆RIGIL KENT.		Peacock		◆FOMALHAUT		Diphda	
15	20 26	071	16 12	096	46 57	119	31 21	194	55 02	243	54 00	311	48 54	354
16	20 56	070	16 42	095	47 25	118	31 14	193	54 35	243	53 37	310	48 50	353
17	21 25	069	17 13	094	47 52	118	31 07	193	54 07	242	53 13	308	48 46	351
18	21 53	068	17 44	093	48 19	117	31 01	192	53 40	241	52 48	307	48 41	350
19	22 22	067	18 15	092	48 47	116	30 54	192	53 13	241	52 24	306	48 35	348
20	22 50	066	18 46	091	49 15	116	30 48	191	52 46	240	51 58	305	48 28	347
21	23 19	065	19 17	091	49 43	115	30 42	191	52 19	240	51 33	303	48 21	345
22	23 47	065	19 48	090	50 11	115	30 37	190	51 52	239	51 07	302	48 13	344
23	24 14	064	20 18	089	50 39	114	30 32	189	51 26	239	50 41	301	48 04	343
24	24 42	063	20 49	088	51 07	113	30 27	189	51 00	238	50 14	300	47 54	341
25	25 09	062	21 20	087	51 36	113	30 22	188	50 33	238	49 47	299	47 44	340
26	25 36	061	21 51	086	52 04	112	30 18	188	50 07	237	49 20	298	47 33	338
27	26 03	060	22 22	085	52 33	112	30 14	187	49 41	237	48 52	296	47 21	337
28	26 30	059	22 53	084	53 02	111	30 10	187	49 16	236	48 24	295	47 09	336
29	26 56	058	23 23	084	53 31	110	30 07	186	48 50	236	47 56	294	46 55	334
	◆RIGEL		SIRIUS		CANOPUS		◆RIGIL KENT.		Peacock		◆FOMALHAUT		Diphda	
30	27 22	057	23 54	083	54 00	110	30 03	186	48 25	235	47 28	293	46 42	333
31	27 48	056	24 25	082	54 29	109	30 01	185	47 59	235	46 59	292	46 27	332
32	28 13	055	24 55	081	54 58	108	29 58	184	47 34	234	46 31	291	46 12	330
33	28 38	054	25 26	080	55 27	108	29 56	184	47 09	234	46 02	290	45 57	329
34	29 03	053	25 56	079	55 57	107	29 54	183	46 44	233	45 33	289	45 40	328
35	29 27	052	26 26	078	56 26	107	29 52	183	46 20	233	45 03	288	45 24	326
36	29 51	051	26 57	077	56 56	106	29 51	182	45 55	232	44 34	287	45 06	325
37	30 15	049	27 27	076	57 26	105	29 50	182	45 31	231	44 04	286	44 48	324
38	30 38	048	27 57	075	57 56	105	29 49	181	45 07	231	43 34	285	44 30	323
39	31 01	047	28 27	075	58 26	104	29 49	180	44 43	230	43 05	284	44 11	321
40	31 24	046	28 56	074	58 56	103	29 49	180	44 19	230	42 34	283	43 51	320
41	31 46	045	29 26	073	59 26	103	29 49	179	43 56	229	42 04	282	43 31	319
42	32 07	044	29 55	072	59 56	102	29 50	179	43 32	229	41 34	281	43 10	318
43	32 29	043	30 24	071	60 26	101	29 50	178	43 09	228	41 04	280	42 49	316
44	32 50	042	30 54	070	60 57	101	29 52	178	42 46	228	40 33	279	42 28	315
	RIGEL		◆SIRIUS		CANOPUS		◆RIGIL KENT.		Peacock		FOMALHAUT		◆Diphda	
45	33 10	041	31 22	069	61 27	100	29 53	177	42 24	227	40 03	278	42 06	314
46	33 30	040	31 51	068	61 58	099	29 55	177	42 01	227	39 32	277	41 43	313
47	33 50	039	32 20	067	62 28	099	29 57	176	41 39	226	39 01	276	41 20	312
48	34 09	037	32 48	066	62 59	098	29 59	175	41 16	226	38 31	276	40 57	310
49	34 27	036	33 16	065	63 29	097	30 02	175	40 54	225	38 00	275	40 33	309
50	34 45	035	33 44	064	64 00	096	30 05	174	40 33	225	37 29	274	40 09	308
51	35 03	034	34 12	063	64 31	096	30 08	174	40 11	224	36 58	273	39 45	307
52	35 20	033	34 39	062	65 01	095	30 11	173	39 50	223	36 27	272	39 20	306
53	35 36	032	35 06	061	65 32	094	30 15	173	39 29	223	35 56	271	38 55	305
54	35 52	031	35 33	060	66 03	093	30 19	172	39 08	222	35 26	270	38 29	304
55	36 08	029	36 00	059	66 34	092	30 24	171	38 47	222	34 55	269	38 03	303
56	36 23	028	36 26	058	67 05	092	30 28	171	38 27	221	34 24	269	37 37	302
57	36 37	027	36 52	057	67 36	091	30 33	170	38 06	221	33 53	268	37 11	301
58	36 51	026	37 17	056	68 07	090	30 39	170	37 46	220	33 22	267	36 44	300
59	37 04	025	37 43	054	68 38	089	30 44	169	37 26	220	32 51	266	36 17	298
	RIGEL		◆SIRIUS		Suhail		◆RIGIL KENT.		Peacock		FOMALHAUT		◆Diphda	
60	37 16	023	38 08	053	42 20	107	30 50	169	37 07	219	32 20	265	35 50	297
61	37 28	022	38 32	052	42 50	106	30 57	168	36 48	218	31 50	264	35 22	296
62	37 39	021	38 57	051	43 20	105	31 03	168	36 28	218	31 19	264	34 54	295
63	37 50	020	39 20	050	43 50	105	31 10	167	36 10	217	30 48	263	34 26	294
64	38 00	018	39 44	049	44 20	104	31 17	166	35 51	217	30 17	262	33 58	293
65	38 10	017	40 07	048	44 50	103	31 24	166	35 33	216	29 47	261	33 30	292
66	38 19	016	40 30	047	45 20	102	31 32	165	35 14	216	29 16	260	33 01	291
67	38 27	015	40 52	045	45 50	102	31 40	165	34 56	215	28 46	260	32 32	290
68	38 34	013	41 14	044	46 20	101	31 48	164	34 39	215	28 16	259	32 03	289
69	38 41	012	41 35	043	46 51	100	31 57	164	34 21	214	27 45	258	31 34	289
70	38 47	011	41 56	042	47 21	099	32 05	163	34 04	213	27 15	257	31 04	288
71	38 53	010	42 16	041	47 52	099	32 15	163	33 47	213	26 45	256	30 35	287
72	38 58	008	42 36	039	48 22	098	32 24	162	33 31	212	26 15	256	30 05	286
73	39 02	007	42 56	038	48 53	097	32 34	162	33 14	212	25 45	255	29 35	285
74	39 05	006	43 14	037	49 24	096	32 43	161	32 58	211	25 16	254	29 05	284
	RIGEL		◆SIRIUS		Suhail		◆RIGIL KENT.		Peacock		ACHERNAR		◆Diphda	
75	39 08	005	43 33	036	49 54	095	32 54	160	32 42	211	63 52	252	28 35	283
76	39 10	003	43 51	034	50 25	095	33 04	160	32 27	210	63 22	251	28 05	282
77	39 12	002	44 08	033	50 56	094	33 15	159	32 12	209	62 53	251	27 35	281
78	39 12	001	44 24	032	51 27	093	33 26	159	31 56	209	62 24	250	27 04	280
79	39 12	359	44 40	031	51 58	092	33 37	158	31 42	208	61 55	250	26 34	279
80	39 12	358	44 56	029	52 29	091	33 49	158	31 27	208	61 26	249	26 03	278
81	39 10	357	45 11	028	53 00	091	34 01	157	31 13	207	60 57	248	25 33	277
82	39 08	356	45 25	027	53 30	090	34 13	157	30 59	207	60 29	248	25 02	277
83	39 06	354	45 38	025	54 01	089	34 25	156	30 45	206	60 00	247	24 31	276
84	39 02	353	45 51	024	54 32	088	34 38	156	30 32	205	59 32	247	24 01	275
85	38 58	352	46 04	023	55 03	087	34 51	155	30 19	205	59 03	246	23 30	274
86	38 54	351	46 15	021	55 34	086	35 04	155	30 06	204	58 35	246	22 59	273
87	38 48	349	46 26	020	56 05	085	35 17	154	29 54	204	58 07	245	22 28	272
88	38 42	348	46 36	019	56 36	084	35 31	153	29 41	203	57 39	245	21 57	271
89	38 35	347	46 46	017	57 06	083	35 45	153	29 29	202	57 11	244	21 26	270

LHA ♈	Hc	Zn	Hc	Zn	Hc	Zn	Hc	Zn	Hc	Zn	Hc	Zn	Hc	Zn
	SIRIUS		Alphard		◆Suhail		RIGIL KENT.		◆Peacock		ACHERNAR		◆RIGEL	
90	46 55	016	26 19	060	57 37	082	35 59	152	29 18	202	56 43	244	38 28	345
91	47 03	014	26 46	059	58 07	081	36 13	152	29 06	201	56 15	243	38 20	344
92	47 10	013	27 12	058	58 38	080	36 28	151	28 55	201	55 48	243	38 11	343
93	47 17	012	27 38	057	59 08	079	36 43	151	28 45	200	55 20	242	38 02	342
94	47 23	010	28 04	056	59 39	078	36 58	150	28 34	199	54 53	242	37 52	341
95	47 28	009	28 30	055	60 09	077	37 14	150	28 24	199	54 26	241	37 41	339
96	47 32	007	28 55	054	60 39	076	37 29	149	28 14	198	53 59	241	37 30	338
97	47 36	006	29 20	053	61 09	075	37 45	149	28 04	198	53 32	240	37 18	337
98	47 39	005	29 44	052	61 39	074	38 01	148	27 55	197	53 05	240	37 06	336
99	47 41	003	30 08	051	62 08	073	38 18	148	27 46	197	52 39	239	36 53	334
100	47 42	002	30 32	050	62 38	072	38 34	147	27 38	196	52 12	239	36 39	333
101	47 43	000	30 56	049	63 07	071	38 51	147	27 29	195	51 46	238	36 25	332
102	47 42	359	31 19	048	63 36	069	39 08	146	27 21	195	51 20	238	36 10	331
103	47 41	357	31 42	047	64 05	068	39 25	146	27 14	194	50 54	237	35 55	330
104	47 40	356	32 04	046	64 34	067	39 43	145	27 06	194	50 28	237	35 39	328
	Alphard		◆Gienah		RIGIL KENT.		◆Peacock		ACHERNAR		◆RIGEL		SIRIUS	
105	32 26	045	20 39	090	40 01	145	26 59	193	50 02	236	35 22	327	47 37	355
106	32 47	044	21 10	089	40 19	144	26 52	192	49 36	236	35 05	326	47 34	353
107	33 08	042	21 41	088	40 37	144	26 46	192	49 11	235	34 48	325	47 30	352
108	33 29	041	22 12	087	40 55	143	26 40	191	48 46	235	34 30	324	47 25	350
109	33 49	040	22 42	086	41 14	143	26 34	191	48 21	234	34 11	323	47 19	349
110	34 09	039	23 13	086	41 33	142	26 29	190	47 56	234	33 52	322	47 13	348
111	34 28	038	23 44	085	41 52	142	26 23	189	47 31	233	33 33	320	47 06	346
112	34 47	037	24 15	084	42 11	141	26 19	189	47 06	233	33 13	319	46 58	345
113	35 05	036	24 45	083	42 31	141	26 14	188	46 42	232	32 53	318	46 50	343
114	35 23	034	25 16	082	42 50	140	26 10	187	46 17	232	32 32	317	46 41	342
115	35 40	033	25 47	081	43 10	140	26 06	187	45 53	231	32 11	316	46 31	341
116	35 57	032	26 17	080	43 30	139	26 03	186	45 29	230	31 49	315	46 20	339
117	36 13	031	26 48	079	43 50	139	25 59	186	45 06	230	31 27	314	46 09	338
118	36 29	030	27 18	078	44 11	138	25 56	185	44 42	229	31 04	313	45 57	337
119	36 44	029	27 48	077	44 32	138	25 54	184	44 19	229	30 42	312	45 45	335
	REGULUS		◆SPICA		RIGIL KENT.		Peacock		◆ACHERNAR		RIGEL		◆SIRIUS	
120	14 25	032	14 02	088	44 52	137	25 52	184	43 56	228	30 18	311	45 31	334
121	14 42	031	14 33	087	45 13	137	25 50	183	43 33	228	29 55	310	45 17	333
122	14 57	030	15 04	087	45 35	136	25 48	183	43 10	227	29 31	309	45 03	331
123	15 13	029	15 35	086	45 56	136	25 47	182	42 47	227	29 06	308	44 48	330
124	15 28	028	16 06	085	46 18	135	25 46	181	42 25	226	28 42	307	44 32	329
125	15 42	027	16 36	084	46 39	135	25 45	181	42 03	226	28 17	305	44 16	327
126	15 56	027	17 07	083	47 01	134	25 45	180	41 41	225	27 51	304	43 59	326
127	16 10	026	17 38	082	47 24	134	25 45	180	41 19	225	27 26	303	43 41	325
128	16 23	025	18 08	081	47 46	134	25 46	179	40 57	224	27 00	302	43 23	324
129	16 36	024	18 39	080	48 08	133	25 46	178	40 36	224	26 34	301	43 04	322
130	16 48	023	19 09	080	48 31	133	25 47	178	40 15	223	26 07	300	42 45	321
131	16 59	022	19 40	079	48 54	132	25 49	177	39 54	222	25 40	299	42 26	320
132	17 10	021	20 10	078	49 17	132	25 50	177	39 33	222	25 13	299	42 05	319
133	17 21	020	20 40	077	49 40	131	25 52	176	39 12	221	24 46	298	41 45	317
134	17 31	019	21 10	076	50 03	131	25 55	175	38 52	221	24 18	297	41 24	316
	REGULUS		◆SPICA		RIGIL KENT.		◆Peacock		ACHERNAR		CANOPUS		◆SIRIUS	
135	17 41	017	21 40	075	50 27	130	25 57	175	38 32	220	67 33	269	41 02	315
136	17 50	016	22 10	074	50 51	130	26 00	174	38 12	220	67 02	268	40 40	314
137	17 58	015	22 39	073	51 14	129	26 04	174	37 52	219	66 32	267	40 18	313
138	18 06	014	23 09	072	51 38	129	26 07	173	37 33	219	66 01	267	39 55	312
139	18 13	013	23 38	071	52 02	129	26 11	172	37 14	218	65 30	266	39 31	310
140	18 20	012	24 08	070	52 27	128	26 16	172	36 55	218	64 59	265	39 08	309
141	18 27	011	24 37	070	52 51	128	26 20	171	36 36	217	64 28	264	38 44	308
142	18 33	010	25 05	069	53 16	127	26 25	170	36 17	216	63 58	264	38 19	307
143	18 38	009	25 34	068	53 40	127	26 31	170	35 59	216	63 27	263	37 54	306
144	18 43	008	26 03	067	54 05	126	26 36	169	35 41	215	62 56	262	37 29	305
145	18 47	007	26 31	066	54 30	126	26 42	169	35 23	215	62 26	261	37 04	304
146	18 50	006	26 59	065	54 55	125	26 48	168	35 06	214	61 55	261	36 38	303
147	18 53	005	27 27	064	55 21	125	26 55	167	34 49	214	61 25	260	36 12	302
148	18 56	004	27 54	063	55 46	125	27 02	167	34 32	213	60 54	259	35 45	301
149	18 58	003	28 22	062	56 11	124	27 09	166	34 15	213	60 24	259	35 19	300
	REGULUS		◆SPICA		RIGIL KENT.		◆Peacock		ACHERNAR		CANOPUS		◆SIRIUS	
150	18 59	002	28 49	061	56 37	124	27 16	166	33 58	212	59 54	258	34 52	299
151	19 00	001	29 16	060	57 03	123	27 24	165	33 42	211	59 23	257	34 24	298
152	19 00	000	29 42	059	57 29	123	27 32	164	33 26	211	58 53	257	33 57	297
153	19 00	359	30 09	058	57 55	122	27 41	164	33 10	210	58 23	256	33 29	296
154	18 59	358	30 35	057	58 21	122	27 50	163	32 55	210	57 53	255	33 01	295
155	18 58	357	31 00	056	58 47	122	27 59	163	32 40	209	57 24	255	32 33	294
156	18 56	356	31 26	055	59 13	121	28 08	162	32 25	209	56 54	254	32 04	293
157	18 54	355	31 51	054	59 40	121	28 18	161	32 10	208	56 24	253	31 36	292
158	18 51	354	32 15	053	60 07	120	28 28	161	31 56	207	55 55	253	31 07	291
159	18 47	353	32 40	052	60 33	120	28 38	160	31 42	207	55 25	252	30 38	290
160	18 43	352	33 04	051	61 00	120	28 49	160	31 28	206	54 56	252	30 09	289
161	18 38	351	33 28	049	61 27	119	28 59	159	31 14	206	54 26	251	29 39	288
162	18 33	350	33 51	048	61 54	119	29 11	159	31 01	205	53 57	250	29 10	287
163	18 27	349	34 14	047	62 21	118	29 22	158	30 48	205	53 28	250	28 40	286
164	18 21	348	34 36	046	62 48	118	29 34	157	30 35	204	52 59	249	28 11	285
	◆SPICA		ANTARES		◆Peacock		ACHERNAR		CANOPUS		◆SIRIUS		REGULUS	
165	34 58	045	26 19	098	29 46	157	30 23	203	52 31	248	27 41	284	18 14	347
166	35 20	044	26 50	097	29 58	156	30 11	203	52 02	248	27 11	283	18 06	346
167	35 41	043	27 21	096	30 11	156	29 59	202	51 33	247	26 40	282	17 58	345
168	36 02	042	27 51	096	30 24	155	29 47	202	51 05	247	26 10	281	17 50	344
169	36 22	041	28 22	095	30 37	154	29 36	201	50 37	246	25 40	280	17 41	343
170	36 42	039	28 53	094	30 50	154	29 25	201	50 08	245	25 09	279	17 31	342
171	37 01	038	29 24	093	31 04	153	29 14	200	49 40	245	24 39	279	17 21	341
172	37 20	037	29 55	092	31 18	153	29 04	199	49 13	244	24 08	278	17 11	340
173	37 39	036	30 26	091	31 33	152	28 54	199	48 45	244	23 38	277	17 00	339
174	37 57	035	30 56	091	31 47	152	28 44	198	48 17	243	23 07	276	16 48	338
175	38 14	034	31 27	090	32 02	151	28 34	198	47 50	242	22 36	275	16 36	337
176	38 31	032	31 58	089	32 17	150	28 25	197	47 22	242	22 05	274	16 24	336
177	38 47	031	32 29	088	32 33	150	28 16	196	46 55	241	21 35	273	16 11	335
178	39 03	030	33 00	087	32 48	149	28 08	196	46 28	241	21 04	272	15 57	334
179	39 18	029	33 31	086	33 04	149	28 00	195	46 01	240	20 33	272	15 43	333

LHA ♈	Hc	Zn	Hc	Zn	Hc	Zn	Hc	Zn	Hc	Zn	Hc	Zn	Hc	Zn
	◆SPICA		ANTARES		◆Peacock		ACHERNAR		CANOPUS		◆SIRIUS		Alphard	
180	39 32	027	34 02	085	33 20	148	27 52	195	45 35	239	20 02	271	31 57	314
181	39 46	026	34 32	084	33 37	148	27 44	194	45 08	239	19 31	270	31 34	313
182	40 00	025	35 03	084	33 54	147	27 37	193	44 42	238	19 00	269	31 11	312
183	40 12	024	35 34	083	34 11	146	27 30	193	44 16	238	18 29	268	30 48	311
184	40 24	022	36 04	082	34 28	146	27 23	192	43 50	237	17 58	267	30 24	310
185	40 36	021	36 35	081	34 45	145	27 17	192	43 24	237	17 27	266	30 00	309
186	40 47	020	37 05	080	35 03	145	27 10	191	42 58	236	16 57	266	29 36	308
187	40 57	019	37 36	079	35 21	144	27 05	190	42 32	235	16 26	265	29 12	307
188	41 06	017	38 06	078	35 39	144	26 59	190	42 07	235	15 55	264	28 47	306
189	41 15	016	38 36	077	35 58	143	26 54	189	41 42	234	15 24	263	28 21	305
190	41 24	015	39 06	076	36 17	142	26 49	189	41 17	234	14 54	262	27 56	304
191	41 31	013	39 36	075	36 36	142	26 45	188	40 52	233	14 23	261	27 30	303
192	41 38	012	40 06	074	36 55	141	26 41	187	40 28	232	13 53	261	27 04	302
193	41 44	011	40 36	073	37 14	141	26 37	187	40 03	232	13 22	260	26 37	301
194	41 50	010	41 05	072	37 34	140	26 33	186	39 39	231	12 52	259	26 10	300
	SPICA		◆ANTARES		Peacock		◆ACHERNAR		CANOPUS		Suhail		◆Alphard	
195	41 54	008	41 35	071	37 54	140	26 30	186	39 15	231	51 55	268	25 43	299
196	41 58	007	42 04	070	38 14	139	26 27	185	38 51	230	51 24	267	25 16	298
197	42 02	006	42 33	069	38 34	139	26 25	184	38 28	229	50 53	266	24 49	297
198	42 04	004	43 02	068	38 55	138	26 22	184	38 04	229	50 22	265	24 21	296
199	42 06	003	43 30	067	39 15	137	26 20	183	37 41	228	49 52	264	23 53	295
200	42 08	002	43 59	066	39 36	137	26 19	183	37 18	228	49 21	264	23 25	294
201	42 08	000	44 27	065	39 58	136	26 18	182	36 56	227	48 50	263	22 56	293
202	42 08	359	44 55	064	40 19	136	26 17	181	36 33	226	48 19	262	22 28	292
203	42 07	358	45 22	063	40 41	135	26 16	181	36 11	226	47 49	261	21 59	291
204	42 05	356	45 50	062	41 03	135	26 16	180	35 49	225	47 18	261	21 30	290
205	42 03	355	46 17	061	41 25	134	26 16	180	35 27	225	46 48	260	21 01	289
206	42 00	354	46 43	060	41 47	134	26 16	179	35 05	224	46 18	259	20 32	288
207	41 56	352	47 10	058	42 09	133	26 17	178	34 44	223	45 47	258	20 02	287
208	41 52	351	47 36	057	42 32	133	26 18	178	34 23	223	45 17	258	19 33	286
209	41 47	350	48 02	056	42 55	132	26 19	177	34 02	222	44 47	257	19 03	285
	ANTARES		◆Nunki		Peacock		ACHERNAR		◆CANOPUS		Suhail		◆SPICA	
210	48 27	055	30 36	091	43 18	132	26 21	177	33 41	222	44 17	256	41 41	348
211	48 52	054	31 07	090	43 41	131	26 23	176	33 21	221	43 47	255	41 34	347
212	49 17	052	31 38	089	44 05	130	26 25	175	33 00	220	43 17	255	41 27	346
213	49 41	051	32 09	088	44 28	130	26 28	175	32 41	220	42 47	254	41 19	345
214	50 05	050	32 40	087	44 52	129	26 31	174	32 21	219	42 18	253	41 11	343
215	50 29	049	33 11	087	45 16	129	26 34	174	32 01	219	41 48	252	41 01	342
216	50 52	047	33 42	086	45 40	128	26 38	173	31 42	218	41 19	252	40 52	341
217	51 14	046	34 12	085	46 04	128	26 42	172	31 23	217	40 49	251	40 41	339
218	51 36	045	34 43	084	46 29	127	26 46	172	31 05	217	40 20	250	40 30	338
219	51 58	044	35 14	083	46 54	127	26 50	171	30 46	216	39 51	250	40 18	337
220	52 19	042	35 44	082	47 18	126	26 55	171	30 28	216	39 22	249	40 06	336
221	52 39	041	36 15	081	47 43	126	27 00	170	30 10	215	38 54	248	39 52	334
222	52 59	039	36 46	080	48 09	125	27 06	169	29 53	214	38 25	247	39 39	333
223	53 19	038	37 16	079	48 34	125	27 12	169	29 35	214	37 57	247	39 25	332
224	53 37	037	37 46	078	49 00	124	27 18	168	29 18	213	37 28	246	39 10	331
	ANTARES		◆Nunki		Peacock		ACHERNAR		◆CANOPUS		Suhail		◆SPICA	
225	53 55	035	38 16	077	49 25	124	27 24	168	29 01	213	37 00	245	38 54	329
226	54 13	034	38 47	077	49 51	123	27 31	167	28 45	212	36 32	245	38 38	328
227	54 30	032	39 17	076	50 17	123	27 38	166	28 29	211	36 04	244	38 22	327
228	54 46	031	39 46	075	50 43	122	27 46	166	28 13	211	35 37	243	38 05	326
229	55 01	029	40 16	074	51 09	121	27 53	165	27 57	210	35 09	243	37 47	325
230	55 16	028	40 46	073	51 36	121	28 02	165	27 42	209	34 42	242	37 29	323
231	55 30	026	41 15	072	52 02	120	28 10	164	27 27	209	34 14	241	37 10	322
232	55 44	025	41 44	071	52 29	120	28 18	163	27 12	208	33 47	241	36 51	321
233	55 56	023	42 13	070	52 56	119	28 27	163	26 58	208	33 21	240	36 32	320
234	56 08	022	42 42	069	53 23	119	28 37	162	26 43	207	32 54	239	36 11	319
235	56 19	020	43 11	068	53 50	118	28 46	162	26 29	206	32 28	238	35 51	318
236	56 29	018	43 39	066	54 17	118	28 56	161	26 16	206	32 01	238	35 30	317
237	56 39	017	44 08	065	54 45	117	29 06	161	26 03	205	31 35	237	35 08	315
238	56 47	015	44 36	064	55 12	117	29 17	160	25 50	204	31 09	236	34 47	314
239	56 55	014	45 03	063	55 40	116	29 27	159	25 37	204	30 44	236	34 24	313
	◆ANTARES		Nunki		◆FOMALHAUT		ACHERNAR		CANOPUS		◆ACRUX		SPICA	
240	57 02	012	45 31	062	18 15	118	29 39	159	25 25	203	64 35	238	34 01	312
241	57 08	010	45 58	061	18 42	117	29 50	158	25 13	203	64 09	238	33 38	311
242	57 13	009	46 25	060	19 10	116	30 02	158	25 01	202	63 43	237	33 15	310
243	57 17	007	46 51	059	19 38	115	30 13	157	24 50	201	63 17	237	32 51	309
244	57 20	005	47 18	058	20 06	114	30 26	156	24 39	201	62 51	237	32 27	308
245	57 23	004	47 44	057	20 34	114	30 38	156	24 28	200	62 25	236	32 02	307
246	57 25	002	48 09	055	21 03	113	30 51	155	24 17	199	62 00	236	31 37	306
247	57 25	000	48 35	054	21 31	112	31 04	155	24 07	199	61 34	236	31 12	305
248	57 25	359	48 59	053	22 00	111	31 17	154	23 57	198	61 08	235	30 47	304
249	57 24	357	49 24	052	22 29	111	31 31	154	23 48	197	60 43	235	30 21	303
250	57 22	355	49 48	050	22 58	110	31 45	153	23 39	197	60 18	235	29 55	302
251	57 19	354	50 12	049	23 27	109	31 59	152	23 30	196	59 53	234	29 28	301
252	57 15	352	50 35	048	23 56	108	32 14	152	23 22	196	59 27	234	29 01	300
253	57 11	351	50 57	047	24 26	107	32 28	151	23 14	195	59 02	234	28 34	299
254	57 05	349	51 20	045	24 55	107	32 43	151	23 06	194	58 38	233	28 07	298
	ALTAIR		◆FOMALHAUT		ACHERNAR		CANOPUS		◆ACRUX		SPICA		◆ANTARES	
255	14 02	044	25 25	106	32 59	150	22 58	194	58 13	233	27 40	297	56 59	347
256	14 23	043	25 55	105	33 14	150	22 51	193	57 48	233	27 12	296	56 51	346
257	14 44	042	26 25	104	33 30	149	22 44	192	57 24	232	26 44	295	56 43	344
258	15 04	041	26 55	103	33 46	148	22 38	192	56 59	232	26 16	294	56 34	342
259	15 24	040	27 25	103	34 02	148	22 32	191	56 35	232	25 47	293	56 25	341
260	15 44	039	27 55	102	34 19	147	22 26	190	56 11	231	25 19	292	56 14	339
261	16 03	038	28 25	101	34 36	147	22 21	190	55 47	231	24 50	291	56 03	338
262	16 22	037	28 55	100	34 53	146	22 16	189	55 23	230	24 21	290	55 51	336
263	16 40	036	29 26	099	35 10	146	22 11	188	54 59	230	23 52	289	55 38	335
264	16 58	035	29 56	099	35 28	145	22 07	188	54 35	230	23 23	288	55 24	333
265	17 15	034	30 27	098	35 45	145	22 03	187	54 12	229	22 53	287	55 10	331
266	17 32	033	30 58	097	36 03	144	21 59	186	53 49	229	22 24	286	54 55	330
267	17 49	032	31 28	096	36 22	143	21 56	186	53 25	229	21 54	285	54 39	329
268	18 05	031	31 59	095	36 40	143	21 53	185	53 02	228	21 24	284	54 22	327
269	18 21	030	32 30	095	36 59	142	21 50	185	52 39	228	20 54	284	54 05	326

LAT 59°S

LHA ♈	Hc	Zn	Hc	Zn	Hc	Zn	Hc	Zn	Hc	Zn	Hc	Zn	Hc	Zn
	◆ALTAIR		FOMALHAUT		◆ACHERNAR		CANOPUS		RIGIL KENT.		◆SPICA		ANTARES	
270	18 36	029	33 01	094	37 18	142	21 48	184	65 23	244	20 24	283	53 47	324
271	18 50	028	33 32	093	37 37	141	21 46	183	64 56	244	19 54	282	53 29	323
272	19 05	027	34 02	092	37 57	141	21 44	183	64 28	243	19 23	281	53 10	321
273	19 18	026	34 33	091	38 17	140	21 43	182	64 00	243	18 53	280	52 50	320
274	19 32	025	35 04	090	38 37	140	21 42	181	63 33	243	18 23	279	52 30	319
275	19 44	024	35 35	089	38 57	139	21 42	181	63 06	242	17 52	278	52 09	317
276	19 57	023	36 06	089	39 17	138	21 42	180	62 38	242	17 21	277	51 48	316
277	20 08	022	36 37	088	39 38	138	21 42	179	62 11	241	16 51	276	51 26	315
278	20 20	021	37 08	087	39 58	137	21 42	179	61 44	241	16 20	276	51 04	313
279	20 30	020	37 39	086	40 20	137	21 43	178	61 17	241	15 49	275	50 41	312
280	20 40	019	38 09	085	40 41	136	21 44	177	60 50	240	15 18	274	50 18	311
281	20 50	018	38 40	084	41 02	136	21 46	177	60 23	240	14 48	273	49 55	309
282	20 59	017	39 11	083	41 24	135	21 48	176	59 57	239	14 17	272	49 31	308
283	21 08	016	39 42	082	41 46	135	21 50	175	59 30	239	13 46	271	49 06	307
284	21 16	014	40 12	081	42 08	134	21 53	175	59 04	239	13 15	270	48 41	306
	◆ALTAIR		FOMALHAUT		◆ACHERNAR		CANOPUS		◆RIGIL KENT.		ANTARES		Nunki	
285	21 23	013	40 43	081	42 30	134	21 56	174	58 37	238	48 16	305	57 17	358
286	21 30	012	41 13	080	42 53	133	21 59	173	58 11	238	47 50	303	57 16	356
287	21 36	011	41 43	079	43 15	133	22 03	173	57 45	237	47 24	302	57 13	355
288	21 42	010	42 14	078	43 38	132	22 07	172	57 19	237	46 58	301	57 10	353
289	21 47	009	42 44	077	44 01	132	22 11	172	56 53	237	46 31	300	57 06	351
290	21 52	008	43 14	076	44 24	131	22 16	171	56 27	236	46 04	299	57 01	350
291	21 56	007	43 44	075	44 48	130	22 21	170	56 02	236	45 37	298	56 55	348
292	22 00	006	44 14	074	45 11	130	22 27	170	55 36	235	45 10	297	56 48	346
293	22 03	005	44 43	073	45 35	129	22 32	169	55 11	235	44 42	296	56 40	345
294	22 05	004	45 13	072	45 59	129	22 39	168	54 46	234	44 14	294	56 32	343
295	22 07	003	45 42	071	46 23	128	22 45	168	54 21	234	43 46	293	56 22	342
296	22 08	002	46 11	070	46 48	128	22 52	167	53 56	234	43 17	292	56 12	340
297	22 09	001	46 40	069	47 12	127	22 59	166	53 31	233	42 49	291	56 01	338
298	22 09	000	47 08	068	47 37	127	23 06	166	53 06	233	42 20	290	55 49	337
299	22 08	359	47 37	067	48 01	126	23 14	165	52 42	232	41 51	289	55 37	335
	FOMALHAUT		◆ACHERNAR		CANOPUS		◆RIGIL KENT.		ANTARES		Nunki		◆ALTAIR	
300	48 05	065	48 26	126	23 22	164	52 18	232	41 21	288	55 24	334	22 07	357
301	48 33	064	48 52	125	23 31	164	51 53	231	40 52	287	55 10	332	22 05	356
302	49 01	063	49 17	125	23 40	163	51 29	231	40 22	286	54 55	331	22 03	355
303	49 28	062	49 42	124	23 49	162	51 05	230	39 53	285	54 39	329	22 00	354
304	49 55	061	50 08	124	23 58	162	50 42	230	39 23	284	54 23	328	21 57	353
305	50 22	060	50 34	123	24 08	161	50 18	230	38 53	283	54 06	326	21 53	352
306	50 49	059	51 00	123	24 18	161	49 55	229	38 23	282	53 49	325	21 49	351
307	51 15	057	51 26	122	24 29	160	49 31	229	37 53	282	53 31	323	21 44	350
308	51 41	056	51 52	122	24 39	159	49 08	228	37 22	281	53 12	322	21 38	349
309	52 06	055	52 18	121	24 50	159	48 45	228	36 52	280	52 53	321	21 32	348
310	52 31	054	52 45	121	25 02	158	48 23	227	36 21	279	52 33	319	21 25	347
311	52 56	052	53 12	120	25 14	157	48 00	227	35 51	278	52 13	318	21 18	346
312	53 20	051	53 38	120	25 26	157	47 38	226	35 20	277	51 52	317	21 10	345
313	53 44	050	54 05	119	25 38	156	47 15	226	34 49	276	51 30	315	21 01	344
314	54 08	049	54 32	119	25 51	155	46 53	225	34 19	275	51 08	314	20 53	343
	◆FOMALHAUT		ACHERNAR		◆CANOPUS		RIGIL KENT.		◆ANTARES		Nunki		ALTAIR	
315	54 31	047	55 00	118	26 04	155	46 31	225	33 48	274	50 46	313	20 43	342
316	54 53	046	55 27	118	26 17	154	46 10	224	33 17	273	50 23	311	20 33	341
317	55 15	044	55 54	117	26 31	154	45 48	224	32 46	273	49 59	310	20 22	340
318	55 36	043	56 22	117	26 44	153	45 27	223	32 15	272	49 35	309	20 11	339
319	55 57	042	56 50	116	26 59	152	45 06	223	31 44	271	49 11	308	20 00	337
320	56 17	040	57 18	116	27 13	152	44 45	223	31 13	270	48 46	306	19 48	336
321	56 37	039	57 45	115	27 28	151	44 24	222	30 43	269	48 21	305	19 35	335
322	56 56	037	58 14	114	27 43	150	44 03	222	30 12	268	47 56	304	19 22	334
323	57 14	036	58 42	114	27 58	150	43 43	221	29 41	267	47 30	303	19 08	333
324	57 32	034	59 10	113	28 14	149	43 23	221	29 10	267	47 04	302	18 54	332
325	57 49	033	59 38	113	28 30	149	43 03	220	28 39	266	46 38	301	18 40	331
326	58 05	031	60 07	112	28 46	148	42 43	220	28 08	265	46 11	299	18 25	330
327	58 21	030	60 36	112	29 03	147	42 23	219	27 38	264	45 44	298	18 09	329
328	58 36	028	61 04	111	29 20	147	42 04	219	27 07	263	45 17	297	17 53	328
329	58 50	026	61 33	111	29 37	146	41 45	218	26 36	262	44 49	296	17 37	327
	◆FOMALHAUT		ACHERNAR		◆CANOPUS		RIGIL KENT.		◆ANTARES		Nunki		ALTAIR	
330	59 03	025	62 02	110	29 54	146	41 26	218	26 06	262	44 21	295	17 20	326
331	59 16	023	62 31	110	30 12	145	41 07	217	25 35	261	43 53	294	17 03	325
332	59 28	021	63 00	109	30 29	144	40 48	217	25 05	260	43 25	293	16 45	324
333	59 39	020	63 29	109	30 48	144	40 30	216	24 34	259	42 56	292	16 27	323
334	59 49	018	63 59	108	31 06	143	40 12	216	24 04	258	42 27	291	16 08	322
335	59 58	016	64 28	108	31 25	142	39 54	215	23 34	257	41 58	290	15 49	321
336	60 06	015	64 57	107	31 44	142	39 36	215	23 04	257	41 29	289	15 29	320
337	60 13	013	65 27	107	32 03	141	39 19	214	22 34	256	41 00	288	15 10	320
338	60 20	011	65 57	106	32 22	141	39 02	214	22 04	255	40 30	287	14 49	319
339	60 25	009	66 26	106	32 42	140	38 45	213	21 34	254	40 01	286	14 29	318
340	60 30	008	66 56	105	33 02	139	38 28	213	21 04	253	39 31	285	14 08	317
341	60 34	006	67 26	105	33 22	139	38 12	212	20 35	253	39 01	284	13 46	316
342	60 36	004	67 56	104	33 43	138	37 55	212	20 05	252	38 31	283	13 24	315
343	60 38	002	68 26	103	34 03	138	37 39	211	19 36	251	38 01	282	13 02	314
344	60 39	001	68 56	103	34 24	137	37 23	211	19 07	250	37 30	281	12 40	313
	◆Diphda		Acamar		CANOPUS		◆Miaplacidus		RIGIL KENT.		◆Nunki		FOMALHAUT	
345	44 54	036	48 56	090	34 46	136	40 07	168	37 08	210	37 00	280	60 39	359
346	45 12	035	49 27	089	35 07	136	40 13	168	36 52	209	36 30	279	60 38	357
347	45 29	033	49 57	088	35 29	135	40 20	167	36 37	209	35 59	278	60 36	355
348	45 46	032	50 28	088	35 50	135	40 27	167	36 23	208	35 28	277	60 33	354
349	46 02	031	50 59	087	36 13	134	40 34	167	36 08	208	34 58	277	60 29	352
350	46 17	029	51 30	086	36 35	133	40 41	166	35 54	207	34 27	276	60 24	350
351	46 32	028	52 01	085	36 57	133	40 49	166	35 40	207	33 56	275	60 18	348
352	46 46	027	52 32	084	37 20	132	40 56	165	35 26	206	33 25	274	60 11	347
353	47 00	025	53 02	083	37 43	132	41 04	165	35 12	206	32 55	273	60 03	345
354	47 13	024	53 33	082	38 06	131	41 13	164	34 59	205	32 24	272	59 55	343
355	47 25	023	54 03	081	38 30	131	41 21	164	34 46	205	31 53	271	59 46	341
356	47 36	021	54 34	080	38 53	130	41 30	164	34 33	204	31 22	270	59 35	340
357	47 47	020	55 04	079	39 17	129	41 38	163	34 20	204	30 51	270	59 24	338
358	47 57	018	55 35	078	39 41	129	41 48	163	34 08	203	30 20	269	59 12	336
359	48 07	017	56 05	077	40 05	128	41 57	162	33 56	203	29 49	268	58 59	335

LAT 60°S

LHA	Hc	Zn	Hc	Zn	Hc	Zn	Hc	Zn	Hc	Zn	Hc	Zn	Hc	Zn
♈	Diphda		◆RIGEL		CANOPUS		Miaplacidus		◆RIGIL KENT.		Peacock		◆FOMALHAUT	
0	47 18	015	12 48	084	41 06	127	43 03	162	34 40	202	62 27	253	57 52	334
1	47 25	014	13 18	083	41 30	126	43 13	161	34 29	202	61 58	252	57 38	332
2	47 32	012	13 48	082	41 54	126	43 23	161	34 18	201	61 30	252	57 24	331
3	47 38	011	14 18	082	42 19	125	43 33	160	34 07	201	61 01	251	57 09	329
4	47 44	010	14 47	081	42 43	124	43 43	160	33 57	200	60 33	250	56 53	328
5	47 48	008	15 17	080	43 08	124	43 53	159	33 46	200	60 05	250	56 37	326
6	47 52	007	15 46	079	43 33	123	44 04	159	33 36	199	59 37	249	56 20	325
7	47 55	005	16 16	078	43 58	123	44 15	159	33 27	198	59 09	249	56 02	323
8	47 58	004	16 45	077	44 24	122	44 26	158	33 17	198	58 41	248	55 44	322
9	48 00	003	17 14	076	44 49	121	44 37	158	33 08	197	58 13	248	55 25	320
10	48 01	001	17 43	075	45 15	121	44 48	157	33 00	197	57 45	247	55 06	319
11	48 01	000	18 12	074	45 41	120	45 00	157	32 51	196	57 18	247	54 46	317
12	48 00	358	18 41	073	46 07	120	45 12	157	32 43	196	56 50	246	54 25	316
13	47 59	357	19 10	073	46 33	119	45 24	156	32 35	195	56 23	245	54 04	315
14	47 57	356	19 38	072	46 59	118	45 36	156	32 27	195	55 56	245	53 43	313
	◆RIGEL		SIRIUS		CANOPUS		◆RIGIL KENT.		Peacock		◆FOMALHAUT		Diphda	
15	20 07	071	16 17	095	47 26	118	32 20	194	55 29	244	53 21	312	47 54	354
16	20 35	070	16 47	094	47 52	117	32 13	193	55 02	244	52 58	311	47 51	353
17	21 03	069	17 17	094	48 19	117	32 06	193	54 35	243	52 35	309	47 47	351
18	21 31	068	17 47	093	48 46	116	31 59	192	54 08	243	52 12	308	47 42	350
19	21 59	067	18 17	092	49 13	115	31 53	192	53 41	242	51 48	307	47 36	348
20	22 26	066	18 47	091	49 40	115	31 47	191	53 15	242	51 24	306	47 30	347
21	22 53	065	19 17	090	50 08	114	31 41	191	52 49	241	50 59	304	47 23	346
22	23 21	064	19 47	089	50 35	113	31 36	190	52 23	240	50 34	303	47 15	344
23	23 47	063	20 17	088	51 03	113	31 31	190	51 57	240	50 09	302	47 06	343
24	24 14	062	20 47	088	51 30	112	31 26	189	51 31	239	49 44	301	46 57	341
25	24 41	061	21 17	087	51 58	112	31 21	188	51 05	239	49 18	300	46 47	340
26	25 07	060	21 47	086	52 26	111	31 17	188	50 39	238	48 51	299	46 37	339
27	25 33	059	22 17	085	52 54	110	31 13	187	50 14	238	48 25	297	46 26	337
28	25 58	058	22 47	084	53 22	110	31 10	187	49 49	237	47 58	296	46 14	336
29	26 24	057	23 16	083	53 51	109	31 06	186	49 23	237	47 31	295	46 01	335
	◆RIGEL		SIRIUS		CANOPUS		◆RIGIL KENT.		Peacock		◆FOMALHAUT		Diphda	
30	26 49	056	23 46	082	54 19	108	31 03	186	48 58	236	47 04	294	45 48	333
31	27 14	055	24 16	081	54 48	108	31 00	185	48 34	236	46 36	293	45 34	332
32	27 38	054	24 46	080	55 16	107	30 58	184	48 09	235	46 09	292	45 20	331
33	28 02	053	25 15	080	55 45	106	30 56	184	47 44	234	45 41	291	45 05	329
34	28 26	052	25 45	079	56 14	106	30 54	183	47 20	234	45 13	290	44 50	328
35	28 50	051	26 14	078	56 43	105	30 52	183	46 56	233	44 44	289	44 34	327
36	29 13	050	26 43	077	57 12	104	30 51	182	46 32	233	44 16	288	44 17	326
37	29 36	049	27 12	076	57 41	104	30 50	182	46 08	232	43 47	287	44 00	324
38	29 58	048	27 41	075	58 10	103	30 49	181	45 44	232	43 19	286	43 42	323
39	30 20	047	28 10	074	58 39	102	30 49	180	45 21	231	42 50	285	43 24	322
40	30 42	046	28 39	073	59 09	102	30 49	180	44 58	231	42 21	284	43 05	321
41	31 03	045	29 08	072	59 38	101	30 49	179	44 35	230	41 51	283	42 46	319
42	31 24	044	29 36	071	60 08	100	30 50	179	44 12	230	41 22	282	42 26	318
43	31 45	043	30 04	070	60 37	100	30 50	178	43 49	229	40 53	281	42 06	317
44	32 05	042	30 33	069	61 07	099	30 52	178	43 26	228	40 23	280	41 45	316
	RIGEL		◆SIRIUS		CANOPUS		◆RIGIL KENT.		Peacock		FOMALHAUT		◆Diphda	
45	32 24	040	31 01	068	61 36	098	30 53	177	43 04	228	39 54	279	41 24	315
46	32 44	039	31 28	067	62 06	097	30 55	176	42 42	227	39 24	278	41 02	313
47	33 02	038	31 56	066	62 36	097	30 57	176	42 20	227	38 54	277	40 40	312
48	33 21	037	32 23	065	63 06	096	30 59	175	41 58	226	38 24	276	40 18	311
49	33 39	036	32 50	064	63 36	095	31 01	175	41 37	226	37 55	275	39 55	310
50	33 56	035	33 17	063	64 06	094	31 04	174	41 15	225	37 25	275	39 32	309
51	34 13	034	33 44	062	64 35	094	31 07	174	40 54	225	36 55	274	39 08	308
52	34 29	032	34 10	061	65 05	093	31 11	173	40 33	224	36 25	273	38 44	307
53	34 45	031	34 37	060	65 35	092	31 15	173	40 12	223	35 55	272	38 20	306
54	35 00	030	35 03	059	66 05	091	31 19	172	39 52	223	35 25	271	37 56	304
55	35 15	029	35 28	058	66 35	090	31 23	171	39 32	222	34 55	270	37 31	303
56	35 30	028	35 53	057	67 05	089	31 28	171	39 11	222	34 25	269	37 05	302
57	35 43	027	36 18	056	67 35	088	31 33	170	38 52	221	33 55	268	36 40	301
58	35 56	025	36 43	055	68 05	088	31 38	170	38 32	221	33 25	268	36 14	300
59	36 09	024	37 08	054	68 35	087	31 43	169	38 13	220	32 55	267	35 48	299
	RIGEL		◆SIRIUS		Suhail		◆RIGIL KENT.		Peacock		FOMALHAUT		◆Diphda	
60	36 21	023	37 32	053	42 37	106	31 49	169	37 53	220	32 25	266	35 22	298
61	36 33	022	37 55	052	43 06	105	31 55	168	37 34	219	31 55	265	34 55	297
62	36 43	021	38 19	051	43 35	104	32 02	167	37 16	218	31 25	264	34 28	296
63	36 54	019	38 42	049	44 04	104	32 08	167	36 57	218	30 55	263	34 01	295
64	37 03	018	39 04	048	44 34	103	32 15	166	36 39	217	30 26	263	33 34	294
65	37 12	017	39 26	047	45 03	102	32 22	166	36 21	217	29 56	262	33 06	293
66	37 21	016	39 48	046	45 32	101	32 30	165	36 03	216	29 26	261	32 39	292
67	37 29	014	40 10	045	46 02	101	32 38	165	35 45	216	28 57	260	32 11	291
68	37 36	013	40 31	044	46 31	100	32 46	164	35 28	215	28 27	259	31 43	290
69	37 42	012	40 51	042	47 01	099	32 54	164	35 11	214	27 58	258	31 14	289
70	37 48	011	41 11	041	47 30	098	33 03	163	34 54	214	27 28	258	30 46	288
71	37 54	010	41 31	040	48 00	097	33 12	162	34 38	213	26 59	257	30 17	287
72	37 58	008	41 50	039	48 30	097	33 21	162	34 21	213	26 30	256	29 49	286
73	38 02	007	42 08	038	49 00	096	33 30	161	34 05	212	26 01	255	29 20	285
74	38 06	006	42 26	036	49 30	095	33 40	161	33 50	211	25 32	254	28 51	284
	RIGEL		◆SIRIUS		Suhail		◆RIGIL KENT.		Peacock		ACHERNAR		◆Diphda	
75	38 08	004	42 44	035	50 00	094	33 50	160	33 34	211	64 10	254	28 22	283
76	38 10	003	43 01	034	50 29	093	34 00	160	33 19	210	63 41	253	27 52	282
77	38 12	002	43 17	033	50 59	093	34 11	159	33 04	210	63 12	252	27 23	282
78	38 12	001	43 33	031	51 29	092	34 22	159	32 49	209	62 44	252	26 54	281
79	38 12	359	43 49	030	51 59	091	34 33	158	32 34	209	62 15	251	26 24	280
80	38 12	358	44 03	029	52 29	090	34 44	157	32 20	208	61 47	251	25 55	279
81	38 10	357	44 18	028	52 59	089	34 56	157	32 06	207	61 19	250	25 25	278
82	38 09	356	44 31	026	53 29	088	35 08	156	31 53	207	60 50	250	24 55	277
83	38 06	354	44 44	025	53 59	087	35 20	156	31 39	206	60 22	249	24 25	276
84	38 03	353	44 56	024	54 29	087	35 32	155	31 26	206	59 54	249	23 55	275
85	37 59	352	45 08	022	54 59	086	35 45	155	31 13	205	59 27	248	23 26	274
86	37 54	351	45 19	021	55 29	085	35 58	154	31 01	204	58 59	247	22 56	274
87	37 49	349	45 30	020	55 59	084	36 11	154	30 48	204	58 31	247	22 26	273
88	37 43	348	45 39	018	56 29	083	36 24	153	30 36	203	58 04	246	21 56	272
89	37 37	347	45 48	017	56 58	082	36 38	153	30 25	203	57 36	246	21 26	271

LHA	Hc	Zn	Hc	Zn	Hc	Zn	Hc	Zn	Hc	Zn	Hc	Zn	Hc	Zn
♈	SIRIUS		Alphard		◆Suhail		RIGIL KENT.		◆Peacock		ACHERNAR		◆RIGEL	
90	45 57	016	25 49	060	57 28	081	36 52	152	30 13	202	57 09	245	37 30	346
91	46 05	014	26 15	059	57 58	080	37 06	152	30 02	201	56 42	245	37 22	344
92	46 12	013	26 40	058	58 27	079	37 21	151	29 51	201	56 15	244	37 14	343
93	46 18	011	27 06	057	58 57	078	37 35	150	29 41	200	55 48	244	37 05	342
94	46 23	010	27 30	056	59 26	077	37 50	150	29 31	200	55 21	243	36 55	341
95	46 28	009	27 55	055	59 55	076	38 05	149	29 21	199	54 54	243	36 45	340
96	46 33	007	28 19	054	60 24	075	38 21	149	29 11	198	54 28	242	36 34	338
97	46 36	006	28 43	053	60 53	073	38 36	148	29 02	198	54 01	241	36 23	337
98	46 39	005	29 07	052	61 21	072	38 52	148	28 53	197	53 35	241	36 11	336
99	46 41	003	29 30	051	61 50	071	39 08	147	28 44	197	53 09	240	35 58	335
100	46 42	002	29 53	050	62 18	070	39 25	147	28 35	196	52 43	240	35 45	334
101	46 43	000	30 16	048	62 46	069	39 41	146	28 27	195	52 17	239	35 32	332
102	46 42	359	30 38	047	63 14	068	39 58	146	28 19	195	51 51	239	35 17	331
103	46 41	358	31 00	046	63 42	066	40 15	145	28 12	194	51 26	238	35 03	330
104	46 40	356	31 22	045	64 09	065	40 32	145	28 05	194	51 00	238	34 47	329
	Alphard		◆Gienah		RIGIL KENT.		◆Peacock		ACHERNAR		◆RIGEL		SIRIUS	
105	31 43	044	20 39	089	40 50	144	27 58	193	50 35	237	34 32	328	46 37	355
106	32 04	043	21 09	089	41 07	144	27 51	192	50 10	237	34 15	327	46 34	353
107	32 24	042	21 38	088	41 25	143	27 45	192	49 45	236	33 59	325	46 30	352
108	32 44	041	22 08	087	41 43	143	27 39	191	49 20	236	33 41	324	46 26	351
109	33 03	040	22 38	086	42 01	142	27 33	191	48 55	235	33 23	323	46 21	349
110	33 22	039	23 08	085	42 20	142	27 28	190	48 31	234	33 05	322	46 15	348
111	33 41	038	23 38	084	42 39	141	27 23	189	48 07	234	32 47	321	46 08	346
112	33 59	036	24 08	083	42 58	141	27 18	189	47 42	233	32 27	320	46 01	345
113	34 16	035	24 38	082	43 17	140	27 14	188	47 18	233	32 08	319	45 52	344
114	34 33	034	25 07	082	43 36	140	27 09	188	46 55	232	31 48	318	45 44	342
115	34 50	033	25 37	081	43 56	139	27 06	187	46 31	232	31 27	316	45 34	341
116	35 06	032	26 07	080	44 15	139	27 02	186	46 07	231	31 06	315	45 24	340
117	35 21	031	26 36	079	44 35	138	26 59	186	45 44	231	30 45	314	45 13	338
118	35 36	029	27 06	078	44 55	138	26 56	185	45 21	230	30 23	313	45 02	337
119	35 51	028	27 35	077	45 16	137	26 54	184	44 58	230	30 01	312	44 50	336
	REGULUS		◆SPICA		RIGIL KENT.		Peacock		◆ACHERNAR		RIGEL		◆SIRIUS	
120	13 34	032	14 00	088	45 36	137	26 51	184	44 35	229	29 39	311	44 37	334
121	13 50	031	14 30	087	45 57	136	26 50	183	44 13	229	29 16	310	44 24	333
122	14 06	030	15 00	086	46 18	136	26 48	183	43 50	228	28 53	309	44 10	332
123	14 21	029	15 30	085	46 39	135	26 47	182	43 28	227	28 30	308	43 56	330
124	14 35	028	16 00	085	47 00	135	26 46	181	43 06	227	28 06	307	43 40	329
125	14 49	027	16 30	084	47 22	134	26 45	181	42 44	226	27 42	306	43 25	328
126	15 03	026	17 00	083	47 43	134	26 45	180	42 23	226	27 17	305	43 09	327
127	15 16	025	17 29	082	48 05	133	26 45	180	42 01	225	26 52	304	42 52	325
128	15 28	024	17 59	081	48 27	133	26 46	179	41 40	225	26 27	303	42 35	324
129	15 41	023	18 29	080	48 49	132	26 46	178	41 19	224	26 02	302	42 17	323
130	15 52	022	18 58	079	49 11	132	26 47	178	40 58	224	25 36	301	41 58	322
131	16 03	021	19 28	078	49 34	131	26 49	177	40 38	223	25 11	300	41 39	320
132	16 14	020	19 57	077	49 56	131	26 50	177	40 17	223	24 44	299	41 20	319
133	16 24	019	20 26	077	50 19	130	26 52	176	39 57	222	24 18	298	41 00	318
134	16 34	018	20 55	076	50 42	130	26 55	175	39 37	221	23 51	297	40 40	317
	REGULUS		◆SPICA		RIGIL KENT.		◆Peacock		ACHERNAR		CANOPUS		◆SIRIUS	
135	16 43	017	21 24	075	51 05	129	26 57	175	39 17	221	67 33	271	40 19	316
136	16 52	016	21 53	074	51 29	129	27 00	174	38 58	220	67 03	271	39 58	315
137	17 00	015	22 22	073	51 52	128	27 03	173	38 39	220	66 33	270	39 37	313
138	17 08	014	22 51	072	52 16	128	27 07	173	38 20	219	66 03	269	39 15	312
139	17 15	013	23 19	071	52 39	127	27 11	172	38 01	219	65 33	268	38 52	311
140	17 22	012	23 47	070	53 03	127	27 15	172	37 42	218	65 03	267	38 29	310
141	17 28	011	24 15	069	53 27	127	27 20	171	37 24	217	64 33	266	38 06	309
142	17 34	010	24 43	068	53 51	126	27 25	170	37 06	217	64 03	266	37 43	308
143	17 39	009	25 11	067	54 16	126	27 30	170	36 48	216	63 33	265	37 19	307
144	17 43	008	25 39	066	54 40	125	27 35	169	36 30	216	63 03	264	36 55	306
145	17 47	007	26 06	065	55 05	125	27 41	169	36 13	215	62 34	263	36 30	304
146	17 51	006	26 33	064	55 30	124	27 47	168	35 55	215	62 04	263	36 05	303
147	17 54	005	27 00	063	55 54	124	27 53	167	35 38	214	61 34	262	35 40	302
148	17 56	004	27 27	062	56 19	123	28 00	167	35 22	214	61 04	261	35 14	301
149	17 58	003	27 53	061	56 45	123	28 07	166	35 05	213	60 35	260	34 49	300
	REGULUS		◆SPICA		RIGIL KENT.		◆Peacock		ACHERNAR		CANOPUS		◆SIRIUS	
150	17 59	002	28 19	060	57 10	122	28 15	166	34 49	212	60 05	260	34 23	299
151	18 00	001	28 45	059	57 35	122	28 22	165	34 33	212	59 36	259	33 56	298
152	18 00	000	29 11	058	58 01	122	28 30	164	34 18	211	59 06	258	33 30	297
153	18 00	359	29 36	057	58 26	121	28 38	164	34 02	211	58 37	258	33 03	296
154	17 59	358	30 01	056	58 52	121	28 47	163	33 47	210	58 08	257	32 36	295
155	17 58	357	30 26	055	59 18	120	28 56	162	33 32	210	57 39	256	32 09	294
156	17 56	356	30 51	054	59 44	120	29 05	162	33 17	209	57 10	256	31 41	293
157	17 54	355	31 15	053	60 10	119	29 15	161	33 03	208	56 41	255	31 13	292
158	17 51	354	31 39	052	60 36	119	29 24	161	32 49	208	56 12	254	30 46	291
159	17 47	353	32 02	051	61 03	118	29 34	160	32 35	207	55 43	254	30 18	290
160	17 43	352	32 26	050	61 29	118	29 45	159	32 21	207	55 14	253	29 49	289
161	17 39	351	32 48	049	61 56	118	29 55	159	32 08	206	54 45	252	29 21	288
162	17 34	350	33 11	048	62 22	117	30 06	158	31 55	205	54 17	252	28 52	287
163	17 28	349	33 33	047	62 49	117	30 18	158	31 42	205	53 49	251	28 24	286
164	17 22	348	33 54	046	63 16	116	30 29	157	31 30	204	53 20	250	27 55	285
	◆SPICA		ANTARES		◆Peacock		ACHERNAR		CANOPUS		◆SIRIUS		REGULUS	
165	34 16	045	26 27	098	30 41	157	31 18	204	52 52	250	27 26	285	17 15	347
166	34 37	043	26 57	097	30 53	156	31 06	203	52 24	249	26 57	284	17 08	346
167	34 57	042	27 27	096	31 05	155	30 54	202	51 56	248	26 27	283	17 01	345
168	35 17	041	27 57	095	31 18	155	30 43	202	51 28	248	25 58	282	16 52	344
169	35 36	040	28 27	094	31 31	154	30 32	201	51 00	247	25 29	281	16 44	343
170	35 56	039	28 57	093	31 44	154	30 21	201	50 33	247	24 59	280	16 35	342
171	36 14	038	29 27	093	31 58	153	30 11	200	50 05	246	24 30	279	16 25	341
172	36 32	037	29 57	092	32 11	152	30 00	200	49 38	245	24 00	278	16 15	340
173	36 50	035	30 27	091	32 26	152	29 51	199	49 11	245	23 30	277	16 04	339
174	37 07	034	30 57	090	32 40	151	29 41	198	48 44	244	23 00	276	15 53	338
175	37 24	033	31 27	089	32 54	151	29 32	198	48 17	243	22 31	275	15 41	337
176	37 40	032	31 57	088	33 09	150	29 23	197	47 50	243	22 01	275	15 29	336
177	37 55	031	32 27	087	33 24	149	29 14	197	47 24	242	21 31	274	15 16	335
178	38 10	029	32 57	086	33 40	149	29 06	196	46 57	242	21 01	273	15 03	334
179	38 25	028	33 27	086	33 55	148	28 57	195	46 31	241	20 31	272	14 50	333

LAT 60°S

LHA ♈	Hc	Zn	Hc	Zn	Hc	Zn	Hc	Zn	Hc	Zn	Hc	Zn	Hc	Zn
	◆SPICA		ANTARES		◆Peacock		ACHERNAR		CANOPUS		◆SIRIUS		Alphard	
180	38 39	027	33 56	085	34 11	148	28 50	195	46 05	240	20 01	271	31 15	314
181	38 52	026	34 26	084	34 27	147	28 42	194	45 39	240	19 31	270	30 53	313
182	39 05	025	34 56	083	34 44	147	28 35	194	45 13	239	19 01	269	30 31	312
183	39 17	023	35 26	082	35 01	146	28 28	193	44 47	239	18 31	268	30 09	311
184	39 29	022	35 55	081	35 17	145	28 22	192	44 22	238	18 01	268	29 46	310
185	39 40	021	36 25	080	35 35	145	28 15	192	43 56	237	17 31	267	29 23	309
186	39 50	020	36 55	079	35 52	144	28 09	191	43 31	237	17 01	266	28 59	308
187	40 00	018	37 24	078	36 10	144	28 04	191	43 06	236	16 31	265	28 36	307
188	40 09	017	37 53	077	36 27	143	27 58	190	42 41	235	16 01	264	28 11	306
189	40 18	016	38 23	076	36 46	143	27 53	189	42 17	235	15 31	263	27 47	305
190	40 25	015	38 52	075	37 04	142	27 49	189	41 52	234	15 02	263	27 22	304
191	40 33	013	39 21	074	37 23	141	27 44	188	41 28	234	14 32	262	26 57	303
192	40 39	012	39 49	073	37 41	141	27 40	188	41 04	233	14 02	261	26 32	302
193	40 45	011	40 18	072	38 00	140	27 36	187	40 40	232	13 33	260	26 06	301
194	40 50	009	40 47	071	38 20	140	27 33	186	40 16	232	13 03	259	25 41	300
	SPICA		◆ANTARES		Peacock		◆ACHERNAR		CANOPUS		Suhail		◆Alphard	
195	40 55	008	41 15	070	38 39	139	27 30	186	39 53	231	51 57	269	25 14	299
196	40 59	007	41 43	069	38 59	139	27 27	185	39 30	231	51 27	268	24 48	298
197	41 02	006	42 11	068	39 19	138	27 24	184	39 07	230	50 57	267	24 21	297
198	41 05	004	42 39	067	39 39	137	27 22	184	38 44	229	50 27	266	23 55	296
199	41 06	003	43 06	066	39 59	137	27 20	183	38 21	229	49 57	266	23 28	295
200	41 08	002	43 34	065	40 20	136	27 19	183	37 58	228	49 27	265	23 00	294
201	41 08	000	44 01	064	40 41	136	27 18	182	37 36	228	48 57	264	22 33	293
202	41 08	359	44 28	063	41 02	135	27 17	181	37 14	227	48 27	263	22 05	292
203	41 07	358	44 54	062	41 23	135	27 16	181	36 52	226	47 57	262	21 37	291
204	41 06	356	45 21	061	41 45	134	27 16	180	36 31	226	47 28	262	21 09	290
205	41 03	355	45 47	060	42 06	134	27 16	180	36 09	225	46 58	261	20 41	289
206	41 00	354	46 13	059	42 28	133	27 16	179	35 48	225	46 28	260	20 13	289
207	40 57	352	46 38	057	42 50	132	27 17	178	35 27	224	45 59	259	19 44	288
208	40 53	351	47 03	056	43 12	132	27 18	178	35 06	223	45 29	259	19 15	287
209	40 48	350	47 28	055	43 35	131	27 19	177	34 46	223	45 00	258	18 47	286
	ANTARES		◆Nunki		Peacock		ACHERNAR		◆CANOPUS		Suhail		◆SPICA	
210	47 52	054	30 37	090	43 57	131	27 21	177	34 26	222	44 31	257	40 42	349
211	48 16	053	31 07	089	44 20	130	27 23	176	34 06	222	44 02	256	40 36	347
212	48 40	052	31 37	089	44 43	130	27 25	175	33 46	221	43 33	256	40 29	346
213	49 03	050	32 07	088	45 06	129	27 28	175	33 26	220	43 04	255	40 21	345
214	49 26	049	32 37	087	45 30	129	27 30	174	33 07	220	42 35	254	40 13	344
215	49 49	048	33 07	086	45 53	128	27 34	174	32 48	219	42 06	253	40 04	342
216	50 11	047	33 37	085	46 17	128	27 37	173	32 29	218	41 37	253	39 55	341
217	50 32	045	34 07	084	46 41	127	27 41	172	32 11	218	41 09	252	39 45	340
218	50 53	044	34 36	083	47 05	126	27 45	172	31 52	217	40 40	251	39 34	338
219	51 14	043	35 06	082	47 29	126	27 50	171	31 34	217	40 12	250	39 23	337
220	51 34	041	35 36	081	47 54	125	27 54	171	31 17	216	39 44	250	39 11	336
221	51 54	040	36 05	080	48 18	125	28 00	170	30 59	215	39 16	249	38 58	335
222	52 13	039	36 35	080	48 43	124	28 05	169	30 42	215	38 48	248	38 45	334
223	52 31	037	37 04	079	49 08	124	28 11	169	30 25	214	38 20	247	38 32	332
224	52 49	036	37 34	078	49 33	123	28 17	168	30 08	214	37 52	247	38 17	331
	ANTARES		◆Nunki		Peacock		ACHERNAR		◆CANOPUS		Suhail		◆SPICA	
225	53 06	034	38 03	077	49 58	123	28 23	167	29 52	213	37 25	246	38 02	330
226	53 23	033	38 32	076	50 23	122	28 30	167	29 36	212	36 57	245	37 47	329
227	53 39	032	39 01	075	50 49	121	28 37	166	29 20	212	36 30	245	37 31	327
228	53 54	030	39 30	074	51 14	121	28 44	166	29 04	211	36 03	244	37 15	326
229	54 09	029	39 59	073	51 40	120	28 51	165	28 49	210	35 36	243	36 58	325
230	54 23	027	40 27	072	52 06	120	28 59	164	28 34	210	35 10	243	36 41	324
231	54 36	026	40 56	071	52 32	119	29 08	164	28 19	209	34 43	242	36 23	323
232	54 49	024	41 24	070	52 59	119	29 16	163	28 05	208	34 17	241	36 04	322
233	55 01	023	41 52	069	53 25	118	29 25	163	27 51	208	33 51	240	35 45	320
234	55 12	021	42 20	068	53 51	118	29 34	162	27 37	207	33 25	240	35 26	319
235	55 23	020	42 48	067	54 18	117	29 43	162	27 23	207	32 59	239	35 06	318
236	55 32	018	43 15	066	54 45	117	29 53	161	27 10	206	32 33	238	34 46	317
237	55 41	016	43 42	065	55 12	116	30 03	160	26 57	205	32 08	238	34 25	316
238	55 49	015	44 09	063	55 39	115	30 13	160	26 44	205	31 42	237	34 04	315
239	55 56	013	44 36	062	56 06	115	30 24	159	26 32	204	31 17	236	33 43	314
	◆ANTARES		Nunki		◆FOMALHAUT		ACHERNAR		CANOPUS		◆ACRUX		SPICA	
240	56 03	012	45 02	061	18 43	117	30 34	159	26 20	203	65 06	240	33 21	313
241	56 09	010	45 29	060	19 09	116	30 46	158	26 08	203	64 40	239	32 59	312
242	56 14	009	45 54	059	19 36	116	30 57	157	25 57	202	64 15	239	32 36	310
243	56 18	007	46 20	058	20 03	115	31 09	157	25 45	201	63 49	239	32 13	309
244	56 21	005	46 45	057	20 31	114	31 21	156	25 35	201	63 23	238	31 50	308
245	56 23	004	47 10	056	20 58	113	31 33	156	25 24	200	62 58	238	31 26	307
246	56 25	002	47 35	054	21 26	113	31 45	155	25 14	200	62 32	238	31 02	306
247	56 25	000	47 59	053	21 54	112	31 58	154	25 04	199	62 07	237	30 38	305
248	56 25	359	48 23	052	22 21	111	32 11	154	24 54	198	61 42	237	30 13	304
249	56 24	357	48 46	051	22 50	110	32 25	153	24 45	198	61 17	237	29 48	303
250	56 22	356	49 09	050	23 18	109	32 38	153	24 36	197	60 52	236	29 23	302
251	56 19	354	49 32	048	23 46	109	32 52	152	24 28	196	60 27	236	28 57	301
252	56 16	352	49 54	047	24 15	108	33 06	152	24 19	196	60 02	236	28 31	300
253	56 11	351	50 16	046	24 43	107	33 21	151	24 12	195	59 37	235	28 05	299
254	56 06	349	50 37	044	25 12	106	33 36	150	24 04	194	59 13	235	27 39	298
	ALTAIR		◆FOMALHAUT		ACHERNAR		CANOPUS		◆ACRUX		SPICA		◆ANTARES	
255	13 18	043	25 41	105	33 50	150	23 57	194	58 48	234	27 12	297	56 00	348
256	13 39	042	26 10	105	34 06	149	23 50	193	58 24	234	26 46	296	55 53	346
257	13 59	042	26 39	104	34 21	149	23 43	192	58 00	234	26 19	295	55 46	344
258	14 19	041	27 08	103	34 37	148	23 37	192	57 36	233	25 51	294	55 37	343
259	14 38	040	27 37	102	34 53	147	23 31	191	57 12	233	25 24	293	55 28	341
260	14 57	039	28 07	101	35 09	147	23 25	190	56 48	232	24 56	292	55 18	340
261	15 15	038	28 36	101	35 26	146	23 20	190	56 24	232	24 28	291	55 07	338
262	15 34	037	29 06	100	35 42	146	23 15	189	56 01	232	24 00	290	54 56	337
263	15 51	036	29 35	099	35 59	145	23 10	189	55 37	231	23 32	289	54 43	335
264	16 09	035	30 05	098	36 17	145	23 06	188	55 14	231	23 04	289	54 30	334
265	16 25	034	30 35	097	36 34	144	23 02	187	54 51	230	22 35	288	54 17	332
266	16 42	033	31 05	096	36 52	144	22 59	187	54 28	230	22 07	287	54 02	331
267	16 58	032	31 35	096	37 10	143	22 55	186	54 05	230	21 38	286	53 47	329
268	17 13	031	32 04	095	37 28	142	22 52	185	53 42	229	21 09	285	53 32	328
269	17 29	030	32 34	094	37 46	142	22 50	185	53 19	229	20 40	284	53 15	326

LHA ♈	Hc	Zn	Hc	Zn	Hc	Zn	Hc	Zn	Hc	Zn	Hc	Zn	Hc	Zn
	◆ALTAIR		FOMALHAUT		◆ACHERNAR		CANOPUS		RIGIL KENT.		◆SPICA		ANTARES	
270	17 43	029	33 04	093	38 05	141	22 48	184	65 49	246	20 11	283	52 58	325
271	17 57	028	33 34	092	38 24	141	22 46	183	65 21	246	19 41	282	52 41	324
272	18 11	027	34 04	091	38 43	140	22 44	183	64 54	245	19 12	281	52 23	322
273	18 24	026	34 34	090	39 02	140	22 43	182	64 27	245	18 42	280	52 04	321
274	18 37	025	35 04	090	39 22	139	22 42	181	64 00	244	18 13	279	51 45	319
275	18 49	024	35 34	089	39 42	138	22 42	181	63 33	244	17 43	279	51 25	318
276	19 01	023	36 04	088	40 02	138	22 42	180	63 06	244	17 14	278	51 05	317
277	19 13	022	36 34	087	40 22	137	22 42	179	62 39	243	16 44	277	50 44	315
278	19 23	021	37 04	086	40 42	137	22 42	179	62 12	243	16 14	276	50 23	314
279	19 34	020	37 34	085	41 03	136	22 43	178	61 46	242	15 44	275	50 01	313
280	19 44	019	38 04	084	41 24	136	22 44	177	61 19	242	15 14	274	49 39	312
281	19 53	018	38 34	083	41 45	135	22 46	177	60 53	241	14 44	273	49 16	310
282	20 02	016	39 03	082	42 06	135	22 48	176	60 26	241	14 14	272	48 53	309
283	20 10	015	39 33	082	42 28	134	22 50	175	60 00	241	13 44	271	48 30	308
284	20 18	014	40 03	081	42 49	134	22 53	175	59 34	240	13 14	271	48 06	307
	◆ALTAIR		FOMALHAUT		◆ACHERNAR		CANOPUS		◆RIGIL KENT.		ANTARES		Nunki	
285	20 25	013	40 32	080	43 11	133	22 56	174	59 08	240	47 41	306	56 18	358
286	20 31	012	41 02	079	43 33	132	22 59	173	58 42	239	47 17	304	56 16	356
287	20 37	011	41 31	078	43 56	132	23 02	173	58 17	239	46 52	303	56 14	355
288	20 43	010	42 00	077	44 18	131	23 06	172	57 51	238	46 27	302	56 11	353
289	20 48	009	42 30	076	44 41	131	23 11	171	57 26	238	46 01	301	56 07	352
290	20 53	008	42 59	075	45 03	130	23 15	171	57 00	237	45 35	300	56 02	350
291	20 57	007	43 28	074	45 26	130	23 20	170	56 35	237	45 09	299	55 56	348
292	21 00	006	43 56	073	45 50	129	23 26	169	56 10	236	44 42	298	55 50	347
293	21 03	005	44 25	072	46 13	129	23 31	169	55 45	236	44 16	296	55 42	345
294	21 05	004	44 53	071	46 36	128	23 37	168	55 20	236	43 49	295	55 34	344
295	21 07	003	45 22	070	47 00	128	23 44	168	54 56	235	43 22	294	55 25	342
296	21 08	002	45 50	069	47 24	127	23 50	167	54 31	235	42 54	293	55 16	340
297	21 09	001	46 17	068	47 48	126	23 57	166	54 07	234	42 26	292	55 05	339
298	21 09	000	46 45	067	48 12	126	24 05	166	53 42	234	41 59	291	54 54	337
299	21 08	359	47 12	066	48 37	125	24 12	165	53 18	233	41 30	290	54 42	336
	FOMALHAUT		◆ACHERNAR		CANOPUS		◆RIGIL KENT.		ANTARES		Nunki		◆ALTAIR	
300	47 40	064	49 01	125	24 20	164	52 54	233	41 02	289	54 30	334	21 07	357
301	48 07	063	49 26	124	24 28	164	52 31	232	40 34	288	54 16	333	21 06	356
302	48 33	062	49 51	124	24 37	163	52 07	232	40 05	287	54 02	331	21 03	355
303	49 00	061	50 16	123	24 46	162	51 43	231	39 36	286	53 48	330	21 01	354
304	49 26	060	50 41	123	24 55	162	51 20	231	39 08	285	53 32	329	20 57	353
305	49 52	059	51 06	122	25 05	161	50 57	230	38 39	284	53 16	327	20 54	352
306	50 17	058	51 32	122	25 15	160	50 34	230	38 09	283	53 00	326	20 49	351
307	50 42	056	51 57	121	25 25	160	50 11	230	37 40	282	52 42	324	20 44	350
308	51 07	055	52 23	121	25 35	159	49 48	229	37 11	281	52 25	323	20 39	349
309	51 31	054	52 49	120	25 46	158	49 25	229	36 41	280	52 06	321	20 33	348
310	51 55	053	53 15	119	25 57	158	49 03	228	36 12	279	51 47	320	20 27	347
311	52 19	051	53 41	119	26 09	157	48 41	228	35 42	279	51 28	319	20 20	346
312	52 42	050	54 08	118	26 21	157	48 19	227	35 12	278	51 08	317	20 12	345
313	53 05	049	54 34	118	26 33	156	47 57	227	34 43	277	50 47	316	20 04	344
314	53 27	048	55 01	117	26 45	155	47 35	226	34 13	276	50 26	315	19 55	343
	◆FOMALHAUT		ACHERNAR		◆CANOPUS		RIGIL KENT.		◆ANTARES		Nunki		ALTAIR	
315	53 49	046	55 27	117	26 58	155	47 14	226	33 43	275	50 05	314	19 46	342
316	54 11	045	55 54	116	27 11	154	46 52	225	33 13	274	49 43	312	19 36	341
317	54 32	043	56 21	116	27 24	153	46 31	225	32 43	273	49 20	311	19 26	340
318	54 52	042	56 48	115	27 38	153	46 10	224	32 13	272	48 57	310	19 16	339
319	55 12	041	57 15	115	27 52	152	45 49	224	31 43	271	48 34	309	19 04	338
320	55 31	039	57 43	114	28 06	151	45 29	223	31 13	271	48 10	307	18 53	337
321	55 50	038	58 10	114	28 20	151	45 08	223	30 43	270	47 46	306	18 41	336
322	56 08	036	58 38	113	28 35	150	44 48	222	30 13	269	47 22	305	18 28	335
323	56 25	035	59 05	112	28 50	150	44 28	222	29 43	268	46 57	304	18 15	334
324	56 42	033	59 33	112	29 06	149	44 08	221	29 13	267	46 32	303	18 01	333
325	56 58	032	60 01	111	29 21	148	43 48	221	28 43	266	46 07	301	17 47	332
326	57 14	030	60 29	111	29 37	148	43 29	220	28 13	265	45 41	300	17 33	331
327	57 29	029	60 57	110	29 53	147	43 10	220	27 43	265	45 15	299	17 18	330
328	57 43	027	61 25	110	30 10	146	42 51	219	27 14	264	44 49	298	17 02	329
329	57 56	026	61 54	109	30 26	146	42 32	219	26 44	263	44 22	297	16 46	328
	◆FOMALHAUT		ACHERNAR		◆CANOPUS		RIGIL KENT.		◆ANTARES		Nunki		ALTAIR	
330	58 09	024	62 22	109	30 43	145	42 13	218	26 14	262	43 55	296	16 30	327
331	58 21	022	62 50	108	31 01	145	41 55	218	25 44	261	43 28	295	16 13	326
332	58 32	021	63 19	107	31 18	144	41 36	217	25 15	260	43 01	294	15 56	325
333	58 42	019	63 48	107	31 36	143	41 18	217	24 45	260	42 33	293	15 38	324
334	58 51	018	64 16	106	31 54	143	41 01	216	24 16	259	42 05	292	15 20	323
335	59 00	016	64 45	106	32 12	142	40 43	216	23 46	258	41 37	291	15 02	322
336	59 08	014	65 14	105	32 31	141	40 26	215	23 17	257	41 09	290	14 43	321
337	59 15	013	65 43	105	32 50	141	40 09	215	22 48	256	40 41	289	14 24	320
338	59 21	011	66 12	104	33 09	140	39 52	214	22 19	255	40 12	288	14 04	319
339	59 26	009	66 41	103	33 28	140	39 35	214	21 50	255	39 44	287	13 44	318
340	59 30	007	67 11	103	33 48	139	39 18	213	21 21	254	39 15	286	13 24	317
341	59 34	006	67 40	102	34 07	138	39 02	212	20 52	253	38 46	285	13 03	316
342	59 36	004	68 09	102	34 27	138	38 46	212	20 24	252	38 17	284	12 42	315
343	59 38	002	68 39	101	34 48	137	38 30	211	19 55	251	37 48	283	12 21	314
344	59 39	001	69 08	100	35 08	137	38 15	211	19 27	251	37 18	282	11 59	313
	◆Diphda		Acamar		CANOPUS		◆Miaplacidus		RIGIL KENT.		◆Nunki		FOMALHAUT	
345	44 05	035	48 55	089	35 29	136	41 05	168	38 00	210	36 49	281	59 39	359
346	44 22	034	49 25	088	35 50	135	41 12	168	37 45	210	36 20	280	59 38	357
347	44 39	033	49 55	087	36 11	135	41 18	167	37 30	209	35 50	279	59 36	355
348	44 55	031	50 25	086	36 32	134	41 25	167	37 15	209	35 20	278	59 33	354
349	45 10	030	50 55	085	36 54	134	41 32	166	37 01	208	34 51	277	59 29	352
350	45 25	029	51 25	085	37 16	133	41 39	166	36 47	208	34 21	276	59 25	350
351	45 39	027	51 55	084	37 38	132	41 47	165	36 33	207	33 51	275	59 19	349
352	45 53	026	52 25	083	38 00	132	41 54	165	36 19	207	33 21	275	59 13	347
353	46 06	025	52 54	082	38 23	131	42 02	165	36 06	206	32 51	274	59 06	345
354	46 18	023	53 24	081	38 45	131	42 10	164	35 53	206	32 21	273	58 57	344
355	46 29	022	53 53	080	39 08	130	42 19	164	35 40	205	31 51	272	58 49	342
356	46 40	021	54 23	079	39 32	129	42 27	163	35 28	204	31 21	271	58 39	340
357	46 51	019	54 52	078	39 55	129	42 36	163	35 15	204	30 51	270	58 28	339
358	47 00	018	55 22	077	40 18	128	42 45	162	35 03	203	30 21	269	58 17	337
359	47 09	017	55 51	076	40 42	127	42 54	162	34 52	203	29 51	268	58 05	335

LAT 61°S

LHA ♈	Hc	Zn	Hc	Zn	Hc	Zn	Hc	Zn	Hc	Zn	Hc	Zn	Hc	Zn
	Diphda		◆RIGEL		CANOPUS		Miaplacidus		◆RIGIL KENT.		Peacock		◆FOMALHAUT	
0	46 20	015	12 42	084	41 42	126	44 00	161	35 35	203	62 44	255	56 58	335
1	46 27	014	13 11	083	42 05	126	44 10	161	35 24	202	62 16	254	56 45	333
2	46 33	012	13 40	082	42 29	125	44 19	160	35 14	201	61 48	253	56 32	331
3	46 39	011	14 09	081	42 53	124	44 29	160	35 03	201	61 20	253	56 17	330
4	46 44	010	14 37	080	43 17	124	44 39	160	34 53	200	60 52	252	56 02	328
5	46 49	008	15 06	080	43 41	123	44 49	159	34 43	200	60 25	252	55 47	327
6	46 53	007	15 35	079	44 06	122	45 00	159	34 33	199	59 57	251	55 31	325
7	46 56	005	16 03	078	44 30	122	45 11	158	34 24	199	59 30	250	55 14	324
8	46 58	004	16 32	077	44 55	121	45 21	158	34 15	198	59 02	250	54 57	323
9	47 00	003	17 00	076	45 20	121	45 32	157	34 06	198	58 35	249	54 39	321
10	47 01	001	17 28	075	45 45	120	45 44	157	33 57	197	58 08	249	54 20	320
11	47 01	000	17 56	074	46 11	119	45 55	157	33 49	196	57 41	248	54 01	318
12	47 00	358	18 24	073	46 36	119	46 07	156	33 41	196	57 14	247	53 42	317
13	46 59	357	18 52	072	47 02	118	46 19	156	33 33	195	56 47	247	53 22	316
14	46 57	356	19 19	071	47 27	117	46 31	155	33 25	195	56 21	246	53 01	314
	◆RIGEL		SIRIUS		CANOPUS		◆RIGIL KENT.		Peacock		◆FOMALHAUT		Diphda	
15	19 47	070	16 23	095	47 53	117	33 18	194	55 54	246	52 40	313	46 55	354
16	20 14	069	16 52	094	48 19	116	33 11	194	55 28	245	52 18	312	46 51	353
17	20 41	069	17 21	093	48 46	116	33 04	193	55 01	244	51 57	310	46 47	351
18	21 08	068	17 50	092	49 12	115	32 58	192	54 35	244	51 34	309	46 43	350
19	21 35	067	18 19	092	49 38	114	32 52	192	54 09	243	51 11	308	46 37	349
20	22 02	066	18 48	091	50 05	114	32 46	191	53 43	243	50 48	307	46 31	347
21	22 28	065	19 17	090	50 32	113	32 40	191	53 17	242	50 25	305	46 25	346
22	22 54	064	19 46	089	50 58	112	32 35	190	52 52	242	50 01	304	46 17	345
23	23 20	063	20 15	088	51 25	112	32 30	190	52 26	241	49 37	303	46 09	343
24	23 46	062	20 44	087	51 52	111	32 25	189	52 01	240	49 12	302	46 00	342
25	24 11	061	21 13	086	52 20	110	32 21	188	51 35	240	48 47	301	45 51	340
26	24 37	060	21 42	085	52 47	110	32 17	188	51 10	239	48 22	300	45 41	339
27	25 02	059	22 11	085	53 14	109	32 13	187	50 45	239	47 57	298	45 30	338
28	25 27	058	22 40	084	53 42	108	32 09	187	50 21	238	47 31	297	45 19	336
29	25 51	057	23 09	083	54 10	108	32 06	186	49 56	238	47 05	296	45 07	335
	◆RIGEL		SIRIUS		CANOPUS		◆RIGIL KENT.		Peacock		◆FOMALHAUT		Diphda	
30	26 15	056	23 38	082	54 37	107	32 03	186	49 31	237	46 39	295	44 54	334
31	26 39	055	24 07	081	55 05	106	32 00	185	49 07	237	46 12	294	44 41	333
32	27 03	054	24 35	080	55 33	106	31 58	184	48 43	236	45 46	293	44 28	331
33	27 26	053	25 04	079	56 01	105	31 56	184	48 19	235	45 19	292	44 13	330
34	27 49	052	25 32	078	56 29	104	31 54	183	47 55	235	44 52	291	43 59	329
35	28 12	051	26 01	077	56 58	104	31 52	183	47 31	234	44 25	290	43 43	327
36	28 34	050	26 29	076	57 26	103	31 51	182	47 08	234	43 57	289	43 27	326
37	28 56	049	26 57	075	57 54	102	31 50	182	46 44	233	43 29	288	43 11	325
38	29 18	048	27 26	074	58 23	101	31 49	181	46 21	233	43 02	287	42 54	324
39	29 39	047	27 53	074	58 51	101	31 49	180	45 58	232	42 34	286	42 36	322
40	30 00	045	28 21	073	59 20	100	31 49	180	45 35	231	42 06	285	42 18	321
41	30 21	044	28 49	072	59 49	099	31 49	179	45 13	231	41 37	284	42 00	320
42	30 41	043	29 17	071	60 17	099	31 50	179	44 50	230	41 09	283	41 41	319
43	31 00	042	29 44	070	60 46	098	31 50	178	44 28	230	40 41	282	41 21	318
44	31 20	041	30 11	069	61 15	097	31 51	178	44 06	229	40 12	281	41 02	316
	RIGEL		◆SIRIUS		CANOPUS		◆RIGIL KENT.		Peacock		FOMALHAUT		◆Diphda	
45	31 39	040	30 38	068	61 44	096	31 53	177	43 44	229	39 44	280	40 41	315
46	31 57	039	31 05	067	62 13	095	31 54	176	43 22	228	39 15	279	40 21	314
47	32 15	038	31 32	066	62 42	095	31 56	176	43 01	227	38 46	278	40 00	313
48	32 33	037	31 58	065	63 11	094	31 59	175	42 39	227	38 17	277	39 38	312
49	32 50	036	32 24	064	63 40	093	32 01	175	42 18	226	37 48	276	39 16	311
50	33 07	034	32 50	063	64 09	092	32 04	174	41 57	226	37 20	275	38 54	309
51	33 23	033	33 16	062	64 38	091	32 07	174	41 37	225	36 51	274	38 31	308
52	33 39	032	33 41	061	65 07	091	32 10	173	41 16	225	36 22	273	38 08	307
53	33 54	031	34 07	060	65 36	090	32 14	172	40 56	224	35 52	273	37 45	306
54	34 08	030	34 31	059	66 05	089	32 18	172	40 36	223	35 23	272	37 21	305
55	34 23	029	34 56	058	66 34	088	32 22	171	40 16	223	34 54	271	36 57	304
56	34 36	027	35 21	056	67 03	087	32 27	171	39 56	222	34 25	270	36 33	303
57	34 50	026	35 45	055	67 32	086	32 32	170	39 37	222	33 56	269	36 09	302
58	35 02	025	36 08	054	68 01	085	32 37	170	39 17	221	33 27	268	35 44	301
59	35 14	024	36 32	053	68 30	084	32 42	169	38 58	221	32 58	267	35 19	300
	RIGEL		◆SIRIUS		Suhail		◆RIGIL KENT.		Peacock		FOMALHAUT		◆Diphda	
60	35 26	023	36 55	052	42 53	105	32 48	168	38 39	220	32 29	266	34 53	299
61	35 37	022	37 18	051	43 21	104	32 54	168	38 21	219	32 00	266	34 28	298
62	35 47	020	37 40	050	43 50	103	33 00	167	38 02	219	31 31	265	34 02	297
63	35 57	019	38 02	049	44 18	103	33 07	167	37 44	218	31 02	264	33 36	296
64	36 06	018	38 24	048	44 46	102	33 14	166	37 26	218	30 33	263	33 09	295
65	36 15	017	38 45	047	45 15	101	33 21	166	37 09	217	30 04	262	32 43	294
66	36 23	016	39 06	045	45 44	100	33 28	165	36 51	217	29 35	261	32 16	293
67	36 31	014	39 27	044	46 12	100	33 36	164	36 34	216	29 07	261	31 49	292
68	36 37	013	39 47	043	46 41	099	33 44	164	36 17	215	28 38	260	31 22	291
69	36 44	012	40 07	042	47 10	098	33 52	163	36 00	215	28 09	259	30 54	290
70	36 49	011	40 26	041	47 39	097	34 00	163	35 44	214	27 41	258	30 27	289
71	36 54	009	40 45	040	48 07	096	34 09	162	35 28	214	27 13	257	29 59	288
72	36 59	008	41 03	038	48 36	096	34 18	162	35 12	213	26 44	256	29 32	287
73	37 03	007	41 21	037	49 05	095	34 27	161	34 56	212	26 16	256	29 04	286
74	37 06	006	41 38	036	49 34	094	34 37	161	34 41	212	25 48	255	28 36	285
	RIGEL		◆SIRIUS		Suhail		◆RIGIL KENT.		Peacock		ACHERNAR		◆Diphda	
75	37 08	004	41 55	035	50 03	093	34 47	160	34 25	211	64 26	256	28 07	284
76	37 10	003	42 11	033	50 32	092	34 57	159	34 10	211	63 58	255	27 39	283
77	37 12	002	42 27	032	51 01	091	35 07	159	33 56	210	63 29	254	27 11	282
78	37 12	001	42 42	031	51 31	091	35 18	158	33 41	209	63 02	254	26 42	281
79	37 12	359	42 57	030	52 00	090	35 28	158	33 27	209	62 34	253	26 14	280
80	37 12	358	43 11	028	52 29	089	35 40	157	33 13	208	62 06	253	25 45	279
81	37 11	357	43 24	027	52 58	088	35 51	157	33 00	208	61 38	252	25 16	278
82	37 09	356	43 37	026	53 27	087	36 03	156	32 46	207	61 11	251	24 47	278
83	37 06	355	43 50	025	53 56	086	36 15	156	32 33	207	60 43	251	24 19	277
84	37 03	353	44 01	023	54 25	085	36 27	155	32 20	206	60 16	250	23 50	276
85	36 59	352	44 13	022	54 54	084	36 39	154	32 08	205	59 48	250	23 21	275
86	36 55	351	44 23	021	55 23	083	36 52	154	31 55	205	59 21	249	22 52	274
87	36 50	350	44 33	019	55 52	082	37 05	153	31 43	204	58 54	248	22 23	273
88	36 45	348	44 42	018	56 20	081	37 18	153	31 32	204	58 27	248	21 54	272
89	36 38	347	44 51	017	56 49	080	37 31	152	31 20	203	58 00	247	21 25	271

LHA ♈	Hc	Zn	Hc	Zn	Hc	Zn	Hc	Zn	Hc	Zn	Hc	Zn	Hc	Zn
	◆SIRIUS		Suhail		◆RIGIL KENT.		Peacock		◆ACHERNAR		Acamar		RIGEL	
90	44 59	015	57 18	079	37 45	152	31 09	202	57 33	247	55 36	286	36 32	346
91	45 06	014	57 46	078	37 59	151	30 58	202	57 07	246	55 08	285	36 24	345
92	45 13	013	58 15	077	38 13	151	30 47	201	56 40	246	54 40	284	36 16	343
93	45 19	011	58 43	076	38 27	150	30 37	200	56 14	245	54 12	283	36 08	342
94	45 24	010	59 11	075	38 42	150	30 27	200	55 48	244	53 43	282	35 58	341
95	45 29	009	59 39	074	38 57	149	30 17	199	55 21	244	53 15	281	35 49	340
96	45 33	007	60 07	073	39 12	148	30 08	199	54 55	243	52 46	280	35 38	339
97	45 36	006	60 35	072	39 27	148	29 59	198	54 29	243	52 17	279	35 27	337
98	45 39	004	61 02	071	39 43	147	29 50	197	54 04	242	51 48	278	35 16	336
99	45 41	003	61 30	069	39 59	147	29 41	197	53 38	242	51 20	277	35 04	335
100	45 42	002	61 57	068	40 15	146	29 33	196	53 12	241	50 51	276	34 51	334
101	45 43	000	62 24	067	40 31	146	29 25	196	52 47	240	50 22	275	34 38	333
102	45 42	359	62 50	066	40 47	145	29 17	195	52 22	240	49 53	274	34 25	332
103	45 41	358	63 17	064	41 04	145	29 10	194	51 57	239	49 24	273	34 11	330
104	45 40	356	63 43	063	41 21	144	29 03	194	51 32	239	48 55	272	33 56	329
	Alphard		◆Gienah		RIGIL KENT.		◆Peacock		ACHERNAR		◆RIGEL		SIRIUS	
105	31 00	044	20 38	089	41 38	144	28 56	193	51 07	238	33 41	328	45 38	355
106	31 20	043	21 07	088	41 55	143	28 50	193	50 42	238	33 25	327	45 35	353
107	31 39	042	21 36	087	42 13	143	28 43	192	50 18	237	33 09	326	45 31	352
108	31 58	040	22 05	086	42 31	142	28 38	191	49 54	237	32 52	325	45 27	351
109	32 17	039	22 34	086	42 49	142	28 32	191	49 29	236	32 35	323	45 22	349
110	32 35	038	23 03	085	43 07	141	28 27	190	49 05	235	32 18	322	45 16	348
111	32 53	037	23 32	084	43 25	141	28 22	189	48 41	235	32 00	321	45 10	347
112	33 10	036	24 01	083	43 44	140	28 17	189	48 18	234	31 41	320	45 03	345
113	33 27	035	24 30	082	44 03	140	28 13	188	47 54	234	31 23	319	44 55	344
114	33 43	034	24 58	081	44 22	139	28 09	188	47 31	233	31 03	318	44 46	343
115	33 59	033	25 27	080	44 41	138	28 05	187	47 08	233	30 44	317	44 37	341
116	34 15	031	25 56	079	45 00	138	28 02	186	46 45	232	30 24	316	44 28	340
117	34 30	030	26 24	078	45 20	137	27 59	186	46 22	232	30 03	315	44 18	339
118	34 44	029	26 53	077	45 40	137	27 56	185	45 59	231	29 42	314	44 07	337
119	34 58	028	27 21	076	45 59	136	27 53	185	45 37	230	29 21	313	43 55	336
	Alphard		◆SPICA		RIGIL KENT.		◆Peacock		ACHERNAR		◆RIGEL		SIRIUS	
120	35 11	027	13 58	088	46 20	136	27 51	184	45 14	230	28 59	312	43 43	335
121	35 24	026	14 27	087	46 40	135	27 50	183	44 52	229	28 37	310	43 30	333
122	35 36	024	14 56	086	47 00	135	27 48	183	44 30	229	28 15	309	43 17	332
123	35 48	023	15 25	085	47 21	134	27 47	182	44 08	228	27 52	308	43 03	331
124	35 59	022	15 54	084	47 42	134	27 46	181	43 47	228	27 30	307	42 49	330
125	36 10	021	16 23	083	48 03	133	27 45	181	43 25	227	27 06	306	42 34	328
126	36 20	020	16 52	082	48 24	133	27 45	180	43 04	226	26 43	305	42 18	327
127	36 29	018	17 21	082	48 46	132	27 45	180	42 43	226	26 19	304	42 02	326
128	36 38	017	17 50	081	49 07	132	27 46	179	42 22	225	25 55	303	41 46	325
129	36 47	016	18 18	080	49 29	131	27 46	178	42 02	225	25 30	302	41 29	323
130	36 54	015	18 47	079	49 51	131	27 47	178	41 41	224	25 05	301	41 11	322
131	37 01	013	19 15	078	50 13	130	27 49	177	41 21	224	24 40	300	40 53	321
132	37 08	012	19 44	077	50 35	130	27 50	176	41 01	223	24 15	299	40 35	320
133	37 14	011	20 12	076	50 58	129	27 52	176	40 41	223	23 50	298	40 16	319
134	37 19	010	20 40	075	51 20	129	27 54	175	40 22	222	23 24	297	39 56	317
	REGULUS		◆SPICA		RIGIL KENT.		◆Peacock		ACHERNAR		CANOPUS		◆SIRIUS	
135	15 46	017	21 08	074	51 43	128	27 57	175	40 03	221	67 30	274	39 36	316
136	15 54	016	21 36	073	52 06	128	28 00	174	39 43	221	67 01	273	39 16	315
137	16 02	015	22 04	072	52 29	127	28 03	173	39 25	220	66 32	272	38 55	314
138	16 10	014	22 32	072	52 52	127	28 07	173	39 06	220	66 03	271	38 34	313
139	16 17	013	22 59	071	53 15	126	28 10	172	38 47	219	65 34	270	38 12	312
140	16 23	012	23 27	070	53 39	126	28 14	172	38 29	219	65 05	269	37 51	311
141	16 29	011	23 54	069	54 03	125	28 19	171	38 11	218	64 36	269	37 28	309
142	16 34	010	24 21	068	54 26	125	28 24	170	37 53	217	64 07	268	37 06	308
143	16 39	009	24 48	067	54 50	124	28 29	170	37 36	217	63 38	267	36 43	307
144	16 44	008	25 14	066	55 14	124	28 34	169	37 19	216	63 09	266	36 19	306
145	16 48	007	25 41	065	55 38	124	28 40	168	37 01	216	62 40	265	35 56	305
146	16 51	006	26 07	064	56 03	123	28 46	168	36 45	215	62 11	264	35 32	304
147	16 54	005	26 33	063	56 27	123	28 52	167	36 28	215	61 42	264	35 08	303
148	16 56	004	26 59	062	56 52	122	28 59	167	36 12	214	61 13	263	34 43	302
149	16 58	003	27 24	061	57 17	122	29 05	166	35 56	213	60 44	262	34 18	301
	REGULUS		◆SPICA		RIGIL KENT.		◆Peacock		ACHERNAR		CANOPUS		◆SIRIUS	
150	16 59	002	27 49	060	57 41	121	29 13	165	35 40	213	60 15	261	33 53	300
151	17 00	001	28 15	059	58 06	121	29 20	165	35 24	212	59 46	261	33 28	299
152	17 00	000	28 39	058	58 31	120	29 28	164	35 09	212	59 18	260	33 02	298
153	17 00	359	29 04	057	58 57	120	29 36	164	34 54	211	58 49	259	32 36	297
154	16 59	358	29 28	056	59 22	119	29 44	163	34 39	210	58 21	258	32 10	296
155	16 58	357	29 52	055	59 47	119	29 53	162	34 24	210	57 52	258	31 44	295
156	16 56	356	30 16	054	60 13	118	30 02	162	34 10	209	57 24	257	31 17	294
157	16 54	355	30 39	053	60 39	118	30 11	161	33 56	209	56 55	256	30 51	293
158	16 51	354	31 02	052	61 04	117	30 21	161	33 42	208	56 27	256	30 24	292
159	16 48	353	31 24	051	61 30	117	30 31	160	33 28	207	55 59	255	29 56	291
160	16 44	352	31 47	050	61 56	116	30 41	159	33 15	207	55 31	254	29 29	290
161	16 40	351	32 09	048	62 23	116	30 51	159	33 02	206	55 03	254	29 02	289
162	16 35	350	32 30	047	62 49	115	31 02	158	32 49	206	54 35	253	28 34	288
163	16 29	349	32 52	046	63 15	115	31 13	157	32 37	205	54 07	252	28 06	287
164	16 23	348	33 12	045	63 41	114	31 24	157	32 25	205	53 40	252	27 39	286
	◆SPICA		ANTARES		◆Peacock		ACHERNAR		CANOPUS		◆SIRIUS		Alphard	
165	33 33	044	26 35	097	31 36	156	32 13	204	53 12	251	27 10	285	34 53	332
166	33 53	043	27 04	096	31 48	156	32 01	203	52 45	250	26 42	284	34 39	331
167	34 12	042	27 33	095	32 00	155	31 50	203	52 18	250	26 14	283	34 25	329
168	34 32	041	28 02	095	32 12	154	31 39	202	51 50	249	25 46	282	34 10	328
169	34 50	040	28 31	094	32 25	154	31 28	202	51 23	248	25 17	281	33 54	327
170	35 09	038	29 00	093	32 38	153	31 17	201	50 56	248	24 49	280	33 38	326
171	35 27	037	29 29	092	32 51	153	31 07	200	50 29	247	24 20	279	33 22	325
172	35 44	036	29 58	091	33 05	152	30 57	200	50 03	246	23 51	279	33 05	324
173	36 01	035	30 27	090	33 18	151	30 47	199	49 36	246	23 22	278	32 47	323
174	36 17	034	30 56	089	33 32	151	30 38	199	49 10	245	22 54	277	32 29	321
175	36 33	033	31 25	088	33 47	150	30 29	198	48 43	244	22 25	276	32 11	320
176	36 49	031	31 54	088	34 01	150	30 20	197	48 17	244	21 56	275	31 52	319
177	37 04	030	32 24	087	34 16	149	30 11	197	47 51	243	21 27	274	31 33	318
178	37 18	029	32 53	086	34 31	149	30 03	196	47 25	243	20 58	273	31 13	317
179	37 32	028	33 22	085	34 46	148	29 55	196	47 00	242	20 29	272	30 53	316

LAT 61°S

LHA ♈	Hc Zn	Hc Zn	Hc Zn	Hc Zn	Hc Zn	Hc Zn	Hc Zn
	SPICA	♦ANTARES	Peacock	♦ACHERNAR	CANOPUS	Suhail	♦Alphard
180	37 45 027	33 51 084	35 02 147	29 48 195	46 34 241	59 09 285	30 33 315
181	37 58 025	34 19 083	35 18 147	29 40 194	46 09 241	58 40 284	30 12 314
182	38 10 024	34 48 082	35 34 146	29 33 194	45 43 240	58 12 283	29 51 313
183	38 22 023	35 17 081	35 50 146	29 27 193	45 18 239	57 44 282	29 29 312
184	38 33 022	35 46 080	36 07 145	29 20 192	44 53 239	57 15 281	29 07 311
185	38 44 021	36 14 079	36 23 144	29 14 192	44 28 238	56 46 280	28 45 310
186	38 54 019	36 43 078	36 41 144	29 08 191	44 04 238	56 18 279	28 22 308
187	39 03 018	37 11 077	36 58 143	29 03 191	43 39 237	55 49 278	27 59 307
188	39 12 017	37 40 077	37 15 143	28 57 190	43 15 236	55 20 277	27 36 306
189	39 20 016	38 08 076	37 33 142	28 52 189	42 51 236	54 51 276	27 12 305
190	39 27 014	38 36 075	37 51 142	28 48 189	42 27 235	54 22 275	26 49 304
191	39 34 013	39 04 074	38 09 141	28 44 188	42 03 234	53 53 274	26 24 303
192	39 41 012	39 32 073	38 28 140	28 40 188	41 40 234	53 24 273	26 00 302
193	39 46 011	40 00 072	38 46 140	28 36 187	41 16 233	52 55 272	25 35 301
194	39 51 009	40 27 071	39 05 139	28 32 186	40 53 233	52 26 271	25 10 300
	SPICA	♦ANTARES	Peacock	♦ACHERNAR	CANOPUS	Suhail	♦Alphard
195	39 55 008	40 54 070	39 24 139	28 29 186	40 30 232	51 57 270	24 45 299
196	39 59 007	41 22 069	39 44 138	28 27 185	40 07 231	51 28 269	24 20 298
197	40 02 005	41 49 068	40 03 137	28 24 185	39 45 231	50 59 269	23 54 297
198	40 05 004	42 15 066	40 23 137	28 22 184	39 22 230	50 30 268	23 28 296
199	40 06 003	42 42 065	40 43 136	28 20 183	39 00 229	50 01 267	23 02 295
200	40 08 002	43 08 064	41 03 136	28 19 183	38 38 229	49 32 266	22 36 295
201	40 08 000	43 34 063	41 24 135	28 18 182	38 16 228	49 03 265	22 09 294
202	40 08 359	44 00 062	41 44 135	28 17 181	37 55 228	48 34 264	21 42 293
203	40 07 358	44 26 061	42 05 134	28 16 181	37 33 227	48 05 264	21 15 292
204	40 06 356	44 51 060	42 26 133	28 16 180	37 12 226	47 36 263	20 48 291
205	40 03 355	45 16 059	42 47 133	28 16 180	36 51 226	47 07 262	20 21 290
206	40 01 354	45 41 058	43 09 132	28 16 179	36 31 225	46 38 261	19 53 289
207	39 57 353	46 05 057	43 30 132	28 17 178	36 10 224	46 10 260	19 26 288
208	39 53 351	46 30 055	43 52 131	28 18 178	35 50 224	45 41 260	18 58 287
209	39 49 350	46 53 054	44 14 131	28 19 177	35 30 223	45 12 259	18 30 286
	ANTARES	♦Nunki	Peacock	ACHERNAR	♦CANOPUS	Suhail	♦SPICA
210	47 17 053	30 37 090	44 36 130	28 21 177	35 10 223	44 44 258	39 43 349
211	47 40 052	31 06 089	44 59 129	28 23 176	34 50 222	44 15 257	39 37 348
212	48 02 051	31 35 088	45 21 129	28 25 175	34 31 221	43 47 256	39 31 346
213	48 25 049	32 04 087	45 44 128	28 27 175	34 12 221	43 19 256	39 23 345
214	48 47 048	32 33 086	46 07 128	28 30 174	33 53 220	42 51 255	39 16 344
215	49 08 047	33 02 085	46 30 127	28 33 173	33 35 219	42 23 254	39 07 342
216	49 29 046	33 31 084	46 53 127	28 37 173	33 16 219	41 55 253	38 58 341
217	49 50 044	34 00 083	47 17 126	28 40 172	32 58 218	41 27 253	38 48 340
218	50 10 043	34 29 083	47 40 126	28 45 172	32 40 218	40 59 252	38 38 339
219	50 30 042	34 58 082	48 04 125	28 49 171	32 23 217	40 32 251	38 27 338
220	50 49 041	35 26 081	48 28 124	28 54 170	32 05 216	40 04 250	38 16 336
221	51 07 039	35 55 080	48 52 124	28 59 170	31 48 216	39 37 250	38 04 335
222	51 25 038	36 24 079	49 16 123	29 04 169	31 31 215	39 10 249	37 51 334
223	51 43 036	36 52 078	49 41 123	29 10 169	31 15 214	38 42 248	37 38 333
224	52 00 035	37 21 077	50 05 122	29 15 168	30 58 214	38 16 248	37 25 331
	ANTARES	♦Nunki	FOMALHAUT	ACHERNAR	♦CANOPUS	Suhail	♦SPICA
225	52 16 034	37 49 076	13 04 129	29 22 167	30 42 213	37 49 247	37 10 330
226	52 32 032	38 17 075	13 27 128	29 28 167	30 26 213	37 22 246	36 56 329
227	52 48 031	38 45 074	13 50 127	29 35 166	30 11 212	36 56 245	36 41 328
228	53 02 029	39 13 073	14 13 127	29 42 166	29 56 211	36 29 245	36 25 327
229	53 16 028	39 41 072	14 37 126	29 49 165	29 41 211	36 03 244	36 09 326
230	53 29 027	40 08 071	15 01 125	29 57 164	29 26 210	35 37 243	35 52 324
231	53 42 025	40 36 070	15 25 124	30 05 164	29 12 209	35 11 242	35 35 323
232	53 54 024	41 03 069	15 49 123	30 13 163	28 57 209	34 45 242	35 17 322
233	54 05 022	41 30 068	16 13 123	30 22 163	28 44 208	34 20 241	34 59 321
234	54 16 021	41 57 067	16 38 122	30 31 162	28 30 207	33 55 240	34 40 320
235	54 26 019	42 23 066	17 03 121	30 40 161	28 17 207	33 29 240	34 21 319
236	54 35 018	42 50 065	17 28 120	30 50 161	28 04 206	33 04 239	34 02 318
237	54 43 016	43 16 064	17 53 119	30 59 160	27 51 206	32 40 238	33 42 316
238	54 51 015	43 42 063	18 18 119	31 09 160	27 39 205	32 15 237	33 22 315
239	54 58 013	44 08 062	18 44 118	31 20 159	27 27 204	31 50 237	33 01 314
	♦ANTARES	Nunki	♦FOMALHAUT	ACHERNAR	CANOPUS	♦ACRUX	SPICA
240	55 04 011	44 33 060	19 10 117	31 30 158	27 15 204	65 36 242	32 40 313
241	55 10 010	44 58 059	19 36 116	31 41 158	27 03 203	65 10 241	32 19 312
242	55 14 008	45 23 058	20 02 115	31 52 157	26 52 202	64 45 241	31 57 311
243	55 18 007	45 48 057	20 28 115	32 04 157	26 41 202	64 19 241	31 35 310
244	55 21 005	46 12 056	20 55 114	32 15 156	26 31 201	63 54 240	31 12 309
245	55 23 004	46 36 055	21 22 113	32 27 155	26 20 200	63 29 240	30 50 308
246	55 25 002	47 00 054	21 49 112	32 40 155	26 10 200	63 04 239	30 26 307
247	55 25 000	47 23 052	22 16 111	32 52 154	26 01 199	62 39 239	30 03 306
248	55 25 359	47 46 051	22 43 111	33 05 154	25 51 198	62 14 239	29 39 305
249	55 24 357	48 08 050	23 10 110	33 18 153	25 42 198	61 49 238	29 15 304
250	55 22 356	48 30 049	23 37 109	33 32 152	25 34 197	61 24 238	28 51 303
251	55 20 354	48 52 047	24 05 108	33 45 152	25 25 196	61 00 237	28 26 302
252	55 16 353	49 13 046	24 33 107	33 59 151	25 17 196	60 35 237	28 01 301
253	55 12 351	49 34 045	25 01 107	34 13 151	25 09 195	60 11 237	27 36 300
254	55 07 349	49 54 044	25 29 106	34 28 150	25 02 194	59 47 236	27 11 299
	Nunki	♦FOMALHAUT	ACHERNAR	CANOPUS	♦ACRUX	SPICA	♦ANTARES
255	50 14 042	25 57 105	34 42 149	24 55 194	59 23 236	26 45 298	55 02 348
256	50 33 041	26 25 104	34 57 149	24 48 193	58 59 235	26 19 297	54 55 346
257	50 52 040	26 53 103	35 12 148	24 42 193	58 35 235	25 53 296	54 48 345
258	51 10 038	27 21 102	35 28 148	24 36 192	58 11 235	25 27 295	54 40 343
259	51 28 037	27 50 102	35 43 147	24 30 191	57 47 234	25 00 294	54 31 342
260	51 45 036	28 18 101	35 59 147	24 24 191	57 24 234	24 33 293	54 22 340
261	52 02 034	28 47 100	36 16 146	24 19 190	57 01 233	24 06 292	54 11 339
262	52 18 033	29 16 099	36 32 145	24 14 189	56 37 233	23 39 291	54 00 337
263	52 34 031	29 44 098	36 49 145	24 10 189	56 14 232	23 12 290	53 49 336
264	52 49 030	30 13 097	37 05 144	24 06 188	55 51 232	22 45 289	53 36 334
265	53 03 029	30 42 097	37 23 144	24 02 187	55 28 232	22 17 288	53 24 333
266	53 16 027	31 11 096	37 40 143	23 58 187	55 06 231	21 49 287	53 10 331
267	53 29 026	31 40 095	37 58 142	23 55 186	54 43 231	21 21 286	52 56 330
268	53 42 024	32 09 094	38 15 142	23 52 185	54 21 230	20 53 285	52 41 328
269	53 53 023	32 38 093	38 33 141	23 50 185	53 58 230	20 25 284	52 25 327

LHA ♈	Hc Zn	Hc Zn	Hc Zn	Hc Zn	Hc Zn	Hc Zn	Hc Zn
	♦ALTAIR	FOMALHAUT	♦ACHERNAR	CANOPUS	♦RIGIL KENT.	SPICA	ANTARES
270	16 51 029	33 07 092	38 52 141	23 48 184	66 12 248	19 57 283	52 09 326
271	17 04 028	33 36 092	39 10 140	23 46 183	65 45 248	19 29 282	51 52 324
272	17 18 027	34 05 091	39 29 140	23 44 183	65 18 247	19 00 282	51 35 323
273	17 30 026	34 34 090	39 48 139	23 43 182	64 51 247	18 32 281	51 17 322
274	17 43 025	35 03 089	40 07 138	23 42 181	64 25 246	18 03 280	50 59 320
275	17 55 024	35 32 088	40 27 138	23 42 181	63 58 246	17 34 279	50 40 319
276	18 06 023	36 02 087	40 46 137	23 42 180	63 32 245	17 05 278	50 21 318
277	18 17 022	36 31 086	41 06 137	23 42 179	63 05 245	16 37 277	50 01 316
278	18 27 020	37 00 085	41 26 136	23 42 179	62 39 244	16 08 276	49 41 315
279	18 37 019	37 29 084	41 46 136	23 43 178	62 13 244	15 39 275	49 20 314
280	18 47 018	37 57 083	42 07 135	23 44 177	61 47 243	15 10 274	48 59 312
281	18 56 017	38 26 083	42 27 135	23 46 177	61 21 243	14 41 273	48 37 311
282	19 04 016	38 55 082	42 48 134	23 48 176	60 55 243	14 12 273	48 15 310
283	19 12 015	39 24 081	43 09 133	23 50 175	60 29 242	13 43 272	47 52 309
284	19 19 014	39 53 080	43 30 133	23 52 175	60 03 242	13 14 271	47 29 308
	♦ALTAIR	FOMALHAUT	♦ACHERNAR	CANOPUS	♦RIGIL KENT.	ANTARES	Nunki
285	19 26 013	40 21 079	43 52 132	23 55 174	59 38 241	47 06 306	55 18 358
286	19 33 012	40 50 078	44 13 132	23 58 173	59 13 241	46 43 305	55 16 356
287	19 39 011	41 18 077	44 35 131	24 02 173	58 47 240	46 19 304	55 14 355
288	19 44 010	41 46 076	44 57 131	24 06 172	58 22 240	45 54 303	55 11 353
289	19 49 009	42 14 075	45 19 130	24 10 171	57 57 239	45 30 302	55 07 352
290	19 53 008	42 42 074	45 42 129	24 15 171	57 32 239	45 05 301	55 03 350
291	19 57 007	43 10 073	46 04 129	24 19 170	57 07 238	44 40 300	54 57 349
292	20 00 006	43 38 072	46 27 128	24 25 169	56 43 238	44 14 298	54 51 347
293	20 03 005	44 06 071	46 50 128	24 30 169	56 18 237	43 49 297	54 44 346
294	20 05 004	44 33 070	47 13 127	24 36 168	55 54 237	43 23 296	54 37 344
295	20 07 003	45 00 069	47 36 127	24 42 167	55 29 236	42 56 295	54 28 342
296	20 08 002	45 27 068	48 00 126	24 49 167	55 05 236	42 30 294	54 19 341
297	20 09 001	45 54 067	48 23 126	24 55 166	54 41 235	42 03 293	54 09 339
298	20 09 000	46 21 066	48 47 125	25 03 165	54 17 235	41 36 292	53 59 338
299	20 08 359	46 47 065	49 11 124	25 10 165	53 54 234	41 09 291	53 47 336
	FOMALHAUT	♦ACHERNAR	CANOPUS	♦RIGIL KENT.	ANTARES	Nunki	♦ALTAIR
300	47 13 063	49 35 124	25 18 164	53 30 234	40 42 290	53 35 335	20 07 358
301	47 39 062	49 59 123	25 26 163	53 07 233	40 15 289	53 23 334	20 06 356
302	48 05 061	50 24 123	25 34 163	52 43 233	39 47 288	53 09 332	20 04 355
303	48 30 060	50 48 122	25 43 162	52 20 232	39 19 287	52 56 331	20 01 354
304	48 55 059	51 13 122	25 52 162	51 57 232	38 51 286	52 41 329	19 58 353
305	49 20 058	51 38 121	26 02 161	51 35 231	38 23 285	52 26 328	19 54 352
306	49 44 057	52 03 121	26 11 160	51 12 231	37 55 284	52 10 326	19 50 351
307	50 08 055	52 28 120	26 21 160	50 49 230	37 27 283	51 54 325	19 45 350
308	50 32 054	52 53 119	26 32 159	50 27 230	36 59 282	51 37 324	19 40 349
309	50 56 053	53 18 119	26 42 158	50 05 229	36 30 281	51 19 322	19 34 348
310	51 19 052	53 44 118	26 53 158	49 43 229	36 02 280	51 01 321	19 28 347
311	51 41 050	54 10 118	27 04 157	49 21 228	35 33 279	50 42 320	19 21 346
312	52 03 049	54 35 117	27 16 156	48 59 228	35 04 278	50 23 318	19 14 345
313	52 25 048	55 01 117	27 28 156	48 38 227	34 35 277	50 04 317	19 06 344
314	52 47 047	55 28 116	27 40 155	48 16 227	34 06 277	49 44 316	18 58 343
	♦FOMALHAUT	ACHERNAR	♦CANOPUS	RIGIL KENT.	♦ANTARES	Nunki	ALTAIR
315	53 07 045	55 54 115	27 52 154	47 55 226	33 37 276	49 23 314	18 49 342
316	53 28 044	56 20 115	28 05 154	47 34 226	33 09 275	49 02 313	18 40 341
317	53 48 043	56 46 114	28 18 153	47 13 225	32 39 274	48 41 312	18 30 340
318	54 07 041	57 13 114	28 31 152	46 53 225	32 10 273	48 19 311	18 20 339
319	54 26 040	57 40 113	28 45 152	46 32 224	31 41 272	47 56 309	18 09 338
320	54 44 038	58 06 113	28 59 151	46 12 224	31 12 271	47 34 308	17 58 337
321	55 02 037	58 33 112	29 13 151	45 52 223	30 43 270	47 11 307	17 46 336
322	55 19 035	59 00 111	29 27 150	45 32 223	30 14 269	46 47 306	17 34 335
323	55 36 034	59 28 111	29 42 149	45 12 222	29 45 269	46 23 305	17 21 334
324	55 52 033	59 55 110	29 57 149	44 53 222	29 16 268	45 59 304	17 08 333
325	56 07 031	60 22 110	30 12 148	44 34 221	28 47 267	45 35 302	16 54 332
326	56 22 030	60 50 109	30 28 147	44 14 221	28 18 266	45 10 301	16 40 331
327	56 36 028	61 17 109	30 43 147	43 56 220	27 49 265	44 45 300	16 26 330
328	56 49 027	61 45 108	31 00 146	43 37 220	27 20 264	44 20 299	16 11 329
329	57 02 025	62 12 107	31 16 145	43 18 219	26 51 263	43 54 298	15 56 328
	♦FOMALHAUT	Diphda	ACHERNAR	♦CANOPUS	RIGIL KENT.	♦ANTARES	Nunki
330	57 14 023	38 16 052	62 40 107	31 33 145	43 00 219	26 22 263	43 29 297
331	57 25 022	38 39 051	63 08 106	31 49 144	42 42 218	25 53 262	43 02 296
332	57 36 020	39 01 050	63 36 106	32 07 144	42 24 218	25 25 261	42 36 295
333	57 45 019	39 23 049	64 04 105	32 24 143	42 06 217	24 56 260	42 10 294
334	57 54 017	39 45 048	64 32 104	32 42 142	41 49 217	24 27 259	41 43 293
335	58 02 015	40 07 047	65 01 104	32 59 142	41 32 216	23 59 258	41 16 292
336	58 10 014	40 28 046	65 29 103	33 18 141	41 15 216	23 30 258	40 49 291
337	58 16 012	40 48 044	65 57 102	33 36 140	40 58 215	23 02 257	40 21 289
338	58 22 011	41 08 043	66 26 102	33 55 140	40 41 215	22 34 256	39 54 288
339	58 27 009	41 28 042	66 54 101	34 14 139	40 25 214	22 06 255	39 26 287
340	58 31 007	41 47 041	67 23 100	34 33 139	40 09 213	21 38 254	38 58 286
341	58 34 006	42 06 040	67 51 100	34 52 138	39 53 213	21 10 253	38 30 286
342	58 37 004	42 24 038	68 20 099	35 12 137	39 37 212	20 42 253	38 02 285
343	58 38 002	42 42 037	68 49 098	35 31 137	39 22 212	20 14 252	37 34 284
344	58 39 001	42 59 036	69 18 098	35 52 136	39 06 211	19 47 251	37 06 283
	♦Diphda	Acamar	♦CANOPUS	RIGIL KENT.	♦ANTARES	Nunki	FOMALHAUT
345	43 16 035	48 54 088	36 12 135	38 51 211	19 19 250	36 37 282	58 39 359
346	43 33 033	49 23 087	36 32 135	38 37 210	18 52 249	36 09 281	58 38 357
347	43 48 032	49 52 086	36 53 134	38 22 210	18 25 248	35 40 280	58 36 356
348	44 03 031	50 21 085	37 14 134	38 08 209	17 58 248	35 11 279	58 33 354
349	44 18 030	50 50 084	37 35 133	37 54 209	17 31 247	34 43 278	58 30 352
350	44 32 028	51 19 083	37 57 132	37 40 208	17 04 246	34 14 277	58 25 351
351	44 46 027	51 47 082	38 18 132	37 26 208	16 38 245	33 45 276	58 20 349
352	44 59 026	52 16 081	38 40 131	37 13 207	16 11 244	33 16 275	58 14 347
353	45 11 024	52 45 080	39 02 131	37 00 206	15 45 244	32 47 274	58 07 346
354	45 23 023	53 14 079	39 24 130	36 47 206	15 19 243	32 18 273	58 00 344
355	45 34 022	53 42 078	39 47 129	36 35 205	14 54 242	31 49 272	57 51 342
356	45 44 020	54 10 077	40 09 129	36 22 205	14 28 241	31 20 272	57 42 341
357	45 54 019	54 39 076	40 32 128	36 10 204	14 03 240	30 51 271	57 32 339
358	46 03 018	55 07 075	40 55 127	35 58 204	13 37 240	30 22 270	57 22 338
359	46 12 016	55 35 074	41 18 127	35 47 203	13 12 239	29 53 269	57 10 336

 LAT 62°S

LHA ♈	Hc	Zn	Hc	Zn	Hc	Zn	Hc	Zn	Hc	Zn	Hc	Zn	Hc	Zn
	Diphda		◆RIGEL		CANOPUS		Miaplacidus		◆RIGIL KENT.		Peacock		◆FOMALHAUT	
0	45 22	015	12 36	084	42 17	125	44 57	161	36 31	203	62 59	257	56 04	335
1	45 29	013	13 04	083	42 40	125	45 06	161	36 20	202	62 31	256	55 52	334
2	45 35	012	13 32	082	43 03	124	45 16	160	36 09	202	62 04	255	55 39	332
3	45 40	011	14 00	081	43 26	124	45 25	160	35 59	201	61 37	255	55 25	331
4	45 45	009	14 27	080	43 50	123	45 35	159	35 49	201	61 10	254	55 11	329
5	45 49	008	14 55	079	44 14	122	45 45	159	35 39	200	60 43	253	54 56	328
6	45 53	007	15 23	078	44 38	122	45 56	158	35 30	199	60 16	253	54 41	326
7	45 56	005	15 50	077	45 02	121	46 06	158	35 21	199	59 49	252	54 25	325
8	45 58	004	16 18	077	45 26	120	46 17	157	35 12	198	59 22	251	54 09	323
9	46 00	003	16 45	076	45 50	120	46 28	157	35 03	198	58 56	251	53 52	322
10	46 01	001	17 12	075	46 15	119	46 39	157	34 54	197	58 29	250	53 34	321
11	46 01	000	17 39	074	46 40	118	46 50	156	34 46	197	58 03	249	53 16	319
12	46 00	358	18 06	073	47 04	118	47 02	156	34 38	196	57 36	249	52 57	318
13	45 59	357	18 33	072	47 29	117	47 13	155	34 31	195	57 10	248	52 38	317
14	45 57	356	19 00	071	47 55	116	47 25	155	34 23	195	56 44	248	52 19	315
	◆RIGEL		SIRIUS		CANOPUS		◆RIGIL KENT.		Peacock		◆FOMALHAUT		Diphda	
15	19 26	070	16 28	095	48 20	116	34 16	194	56 18	247	51 59	314	45 55	354
16	19 53	069	16 56	094	48 45	115	34 09	194	55 52	246	51 38	313	45 52	353
17	20 19	068	17 24	093	49 11	114	34 03	193	55 26	246	51 17	311	45 48	352
18	20 45	067	17 52	092	49 37	114	33 56	193	55 01	245	50 56	310	45 44	350
19	21 11	066	18 20	091	50 02	113	33 50	192	54 35	245	50 34	309	45 39	349
20	21 37	065	18 48	090	50 28	113	33 45	191	54 10	244	50 12	308	45 33	348
21	22 02	064	19 17	089	50 54	112	33 39	191	53 45	243	49 50	306	45 26	346
22	22 27	063	19 45	089	51 21	111	33 34	190	53 20	243	49 27	305	45 19	345
23	22 53	062	20 13	088	51 47	110	33 29	190	52 55	242	49 04	304	45 12	343
24	23 17	061	20 41	087	52 13	110	33 24	189	52 30	242	48 40	303	45 03	342
25	23 42	060	21 09	086	52 40	109	33 20	189	52 05	241	48 16	302	44 54	341
26	24 06	059	21 37	085	53 07	108	33 16	188	51 40	240	47 52	301	44 45	339
27	24 31	058	22 05	084	53 33	108	33 12	187	51 16	240	47 28	299	44 35	338
28	24 54	057	22 33	083	54 00	107	33 09	187	50 52	239	47 03	298	44 24	337
29	25 18	056	23 01	082	54 27	106	33 06	186	50 28	239	46 38	297	44 12	336
	◆RIGEL		SIRIUS		CANOPUS		◆RIGIL KENT.		Peacock		◆FOMALHAUT		Diphda	
30	25 41	055	23 29	081	54 54	106	33 03	186	50 04	238	46 13	296	44 01	334
31	26 04	054	23 57	080	55 22	105	33 00	185	49 40	237	45 48	295	43 48	333
32	26 27	053	24 25	080	55 49	104	32 58	185	49 16	237	45 22	294	43 35	332
33	26 50	052	24 52	079	56 16	104	32 56	184	48 53	236	44 56	293	43 21	330
34	27 12	051	25 20	078	56 44	103	32 54	183	48 29	236	44 30	292	43 07	329
35	27 34	050	25 47	077	57 11	102	32 52	183	48 06	235	44 04	291	42 52	328
36	27 55	049	26 15	076	57 39	101	32 51	182	47 43	235	43 37	290	42 37	327
37	28 16	048	26 42	075	58 06	101	32 50	182	47 20	234	43 11	289	42 21	325
38	28 37	047	27 09	074	58 34	100	32 49	181	46 57	233	42 44	288	42 05	324
39	28 58	046	27 36	073	59 02	099	32 49	180	46 35	233	42 17	287	41 49	323
40	29 18	045	28 03	072	59 30	098	32 49	180	46 12	232	41 50	286	41 31	322
41	29 37	044	28 30	071	59 58	098	32 49	179	45 50	232	41 23	285	41 14	321
42	29 57	043	28 56	070	60 25	097	32 50	179	45 28	231	40 55	284	40 56	319
43	30 16	042	29 23	069	60 53	096	32 50	178	45 06	230	40 28	283	40 37	318
44	30 34	041	29 49	068	61 21	095	32 51	178	44 45	230	40 00	282	40 18	317
	RIGEL		◆SIRIUS		CANOPUS		◆RIGIL KENT.		Peacock		FOMALHAUT		◆Diphda	
45	30 53	040	30 15	067	61 50	094	32 53	177	44 23	229	39 33	281	39 59	316
46	31 10	039	30 41	066	62 18	094	32 54	176	44 02	229	39 05	280	39 39	315
47	31 28	037	31 07	065	62 46	093	32 56	176	43 41	228	38 37	279	39 18	314
48	31 45	036	31 32	064	63 14	092	32 58	175	43 20	228	38 09	278	38 58	312
49	32 01	035	31 57	063	63 42	091	33 01	175	42 59	227	37 42	277	38 37	311
50	32 17	034	32 22	062	64 10	090	33 04	174	42 39	226	37 14	276	38 15	310
51	32 33	033	32 47	061	64 38	089	33 07	174	42 19	226	36 46	275	37 54	309
52	32 48	032	33 12	060	65 07	088	33 10	173	41 59	225	36 17	274	37 32	308
53	33 02	031	33 36	059	65 35	087	33 14	172	41 39	225	35 49	273	37 09	307
54	33 16	029	34 00	058	66 03	087	33 18	172	41 19	224	35 21	272	36 47	306
55	33 30	028	34 24	057	66 31	086	33 22	171	40 59	223	34 53	272	36 24	305
56	33 43	027	34 47	056	66 59	085	33 26	171	40 40	223	34 25	271	36 00	304
57	33 56	026	35 10	055	67 27	084	33 31	170	40 21	222	33 57	270	35 37	302
58	34 08	025	35 33	054	67 55	083	33 36	169	40 02	222	33 29	269	35 13	301
59	34 19	024	35 56	053	68 23	082	33 41	169	39 44	221	33 00	268	34 48	300
	RIGEL		◆SIRIUS		Suhail		◆RIGIL KENT.		Peacock		FOMALHAUT		◆Diphda	
60	34 30	022	36 18	052	43 08	104	33 47	168	39 25	221	32 32	267	34 24	299
61	34 41	021	36 40	050	43 36	103	33 53	168	39 07	220	32 04	266	33 59	298
62	34 51	020	37 01	049	44 03	103	33 59	167	38 49	219	31 36	265	33 34	297
63	35 00	019	37 23	048	44 31	102	34 05	167	38 31	219	31 08	265	33 09	296
64	35 09	018	37 43	047	44 58	101	34 12	166	38 14	218	30 40	264	32 44	295
65	35 17	017	38 04	046	45 26	100	34 19	165	37 56	218	30 12	263	32 18	294
66	35 25	015	38 24	045	45 54	099	34 26	165	37 39	217	29 44	262	31 53	293
67	35 32	014	38 44	044	46 22	099	34 33	164	37 23	216	29 16	261	31 27	292
68	35 39	013	39 03	043	46 50	098	34 41	164	37 06	216	28 48	260	31 00	291
69	35 45	012	39 22	041	47 17	097	34 49	163	36 50	215	28 21	259	30 34	290
70	35 50	010	39 40	040	47 45	096	34 57	163	36 34	215	27 53	259	30 08	289
71	35 55	009	39 58	039	48 14	095	35 06	162	36 18	214	27 25	258	29 41	288
72	35 59	008	40 16	038	48 42	094	35 15	161	36 02	213	26 58	257	29 14	287
73	36 03	007	40 33	037	49 10	094	35 24	161	35 47	213	26 31	256	28 47	286
74	36 06	006	40 49	035	49 38	093	35 33	160	35 31	212	26 03	255	28 20	285
	RIGEL		◆SIRIUS		Suhail		◆RIGIL KENT.		Peacock		ACHERNAR		◆Diphda	
75	36 09	004	41 05	034	50 06	092	35 43	160	35 17	212	64 40	258	27 53	284
76	36 10	003	41 21	033	50 34	091	35 53	159	35 02	211	64 12	257	27 25	284
77	36 12	002	41 36	032	51 02	090	36 03	159	34 48	210	63 45	256	26 58	283
78	36 12	001	41 50	030	51 30	089	36 13	158	34 33	210	63 17	256	26 30	282
79	36 12	359	42 04	029	51 59	088	36 24	157	34 20	209	62 50	255	26 03	281
80	36 12	358	42 18	028	52 27	087	36 35	157	34 06	209	62 23	254	25 35	280
81	36 11	357	42 31	027	52 55	087	36 46	156	33 53	208	61 56	254	25 07	279
82	36 09	356	42 43	025	53 23	086	36 57	156	33 39	207	61 29	253	24 39	278
83	36 07	355	42 55	024	53 51	085	37 09	155	33 27	207	61 02	252	24 11	277
84	36 04	353	43 06	023	54 19	084	37 21	155	33 14	206	60 35	252	23 44	276
85	36 00	352	43 17	022	54 47	083	37 33	154	33 02	206	60 09	251	23 15	275
86	35 56	351	43 27	020	55 15	082	37 46	154	32 50	205	59 42	251	22 47	274
87	35 51	350	43 36	019	55 43	081	37 58	153	32 38	204	59 15	250	22 19	273
88	35 46	348	43 45	018	56 11	080	38 11	152	32 27	204	58 49	249	21 51	273
89	35 40	347	43 54	016	56 38	079	38 24	152	32 15	203	58 23	249	21 23	272

LHA ♈	Hc	Zn	Hc	Zn	Hc	Zn	Hc	Zn	Hc	Zn	Hc	Zn	Hc	Zn
	◆SIRIUS		Suhail		◆RIGIL KENT.		Peacock		◆ACHERNAR		Acamar		RIGEL	
90	44 01	015	57 06	078	38 38	151	32 04	203	57 56	248	55 19	287	35 33	346
91	44 08	014	57 33	077	38 51	151	31 54	202	57 30	248	54 52	286	35 26	345
92	44 14	012	58 01	076	39 05	150	31 43	201	57 04	247	54 25	285	35 19	344
93	44 20	011	58 28	075	39 19	150	31 33	201	56 39	246	53 58	284	35 10	342
94	44 25	010	58 55	073	39 34	149	31 23	200	56 13	246	53 30	283	35 02	341
95	44 30	008	59 22	072	39 48	149	31 14	199	55 47	245	53 03	282	34 52	340
96	44 33	007	59 49	071	40 03	148	31 05	199	55 22	245	52 35	281	34 42	339
97	44 37	006	60 15	070	40 18	147	30 56	198	54 56	244	52 07	280	34 32	338
98	44 39	004	60 42	069	40 33	147	30 47	198	54 31	243	51 40	279	34 21	337
99	44 41	003	61 08	068	40 49	146	30 39	197	54 06	243	51 12	278	34 10	335
100	44 42	002	61 34	066	41 04	146	30 31	196	53 41	242	50 44	277	33 58	334
101	44 43	000	61 59	065	41 20	145	30 23	196	53 16	242	50 16	276	33 45	333
102	44 42	359	62 25	064	41 37	145	30 15	195	52 51	241	49 48	275	33 32	332
103	44 42	358	62 50	063	41 53	144	30 08	195	52 27	240	49 20	274	33 18	331
104	44 40	356	63 15	061	42 09	144	30 01	194	52 02	240	48 52	273	33 04	330
	Alphard		◆Gienah		RIGIL KENT.		◆Peacock		ACHERNAR		◆RIGEL		SIRIUS	
105	30 16	043	20 37	089	42 26	143	29 55	193	51 38	239	32 50	328	44 38	355
106	30 35	042	21 05	088	42 43	143	29 48	193	51 14	239	32 35	327	44 35	354
107	30 54	041	21 33	087	43 00	142	29 42	192	50 50	238	32 19	326	44 32	352
108	31 12	040	22 01	086	43 18	142	29 36	191	50 26	238	32 03	325	44 27	351
109	31 30	039	22 29	085	43 35	141	29 31	191	50 02	237	31 47	324	44 23	350
110	31 48	038	22 57	084	43 53	140	29 26	190	49 39	236	31 30	323	44 17	348
111	32 05	037	23 25	083	44 11	140	29 21	190	49 16	236	31 13	322	44 11	347
112	32 22	036	23 53	082	44 30	139	29 17	189	48 52	235	30 55	321	44 04	346
113	32 38	034	24 21	082	44 48	139	29 12	188	48 29	235	30 37	319	43 57	344
114	32 53	033	24 49	081	45 07	138	29 08	188	48 06	234	30 19	318	43 49	343
115	33 09	032	25 17	080	45 25	138	29 05	187	47 44	234	30 00	317	43 41	342
116	33 23	031	25 44	079	45 44	137	29 01	186	47 21	233	29 40	316	43 31	340
117	33 38	030	26 12	078	46 04	137	28 58	186	46 59	232	29 21	315	43 22	339
118	33 52	029	26 39	077	46 23	136	28 56	185	46 36	232	29 01	314	43 11	338
119	34 05	028	27 07	076	46 43	136	28 53	185	46 14	231	28 40	313	43 00	336
	Alphard		◆SPICA		RIGIL KENT.		◆Peacock		ACHERNAR		◆RIGEL		SIRIUS	
120	34 18	026	13 56	088	47 02	135	28 51	184	45 53	231	28 19	312	42 49	335
121	34 30	025	14 24	087	47 22	135	28 49	183	45 31	230	27 58	311	42 37	334
122	34 42	024	14 52	086	47 43	134	28 48	183	45 09	229	27 37	310	42 24	333
123	34 53	023	15 20	085	48 03	134	28 47	182	44 48	229	27 15	309	42 11	331
124	35 04	022	15 48	084	48 23	133	28 46	181	44 27	228	26 53	308	41 57	330
125	35 14	021	16 16	083	48 44	133	28 45	181	44 06	228	26 31	307	41 43	329
126	35 23	019	16 44	082	49 05	132	28 45	180	43 45	227	26 08	306	41 28	328
127	35 32	018	17 12	081	49 26	132	28 45	180	43 25	227	25 45	305	41 12	326
128	35 41	017	17 40	080	49 47	131	28 46	179	43 04	226	25 21	304	40 57	325
129	35 49	016	18 07	079	50 08	130	28 46	178	42 44	225	24 58	303	40 40	324
130	35 56	015	18 35	079	50 30	130	28 47	178	42 24	225	24 34	302	40 23	323
131	36 03	013	19 03	078	50 52	129	28 48	177	42 04	224	24 10	301	40 06	322
132	36 09	012	19 30	077	51 13	129	28 50	176	41 45	224	23 46	300	39 48	320
133	36 15	011	19 58	076	51 35	128	28 52	176	41 25	223	23 21	299	39 30	319
134	36 20	010	20 25	075	51 58	128	28 54	175	41 06	223	22 56	298	39 12	318
	REGULUS		◆SPICA		RIGIL KENT.		◆Peacock		ACHERNAR		CANOPUS		◆SIRIUS	
135	14 49	017	20 52	074	52 20	127	28 57	175	40 47	222	67 25	276	38 53	317
136	14 57	016	21 19	073	52 42	127	28 59	174	40 29	221	66 57	275	38 33	316
137	15 04	015	21 46	072	53 05	126	29 03	173	40 10	221	66 29	274	38 13	315
138	15 12	014	22 13	071	53 28	126	29 06	173	39 52	220	66 01	273	37 53	313
139	15 18	013	22 39	070	53 51	125	29 10	172	39 34	220	65 32	272	37 32	312
140	15 24	012	23 06	069	54 14	125	29 14	171	39 16	219	65 04	272	37 11	311
141	15 30	011	23 32	068	54 37	124	29 18	171	38 58	218	64 36	271	36 50	310
142	15 35	010	23 58	067	55 00	124	29 23	170	38 41	218	64 08	270	36 28	309
143	15 40	009	24 24	066	55 24	123	29 28	170	38 24	217	63 40	269	36 05	308
144	15 44	008	24 49	065	55 47	123	29 33	169	38 07	217	63 12	268	35 44	307
145	15 48	007	25 15	064	56 11	122	29 39	168	37 50	216	62 44	267	35 21	306
146	15 51	006	25 40	063	56 35	122	29 44	168	37 34	216	62 15	266	34 58	305
147	15 54	005	26 05	062	56 59	121	29 51	167	37 17	215	61 47	266	34 35	304
148	15 56	004	26 30	061	57 23	121	29 57	166	37 01	214	61 19	265	34 11	302
149	15 58	003	26 55	060	57 47	120	30 04	166	36 46	214	60 51	264	33 47	301
	REGULUS		◆SPICA		RIGIL KENT.		◆Peacock		ACHERNAR		CANOPUS		◆SIRIUS	
150	15 59	002	27 19	059	58 12	120	30 11	165	36 30	213	60 23	263	33 23	300
151	16 00	001	27 43	058	58 36	119	30 18	165	36 15	213	59 55	262	32 59	299
152	16 00	000	28 07	057	59 01	119	30 26	164	36 00	212	59 27	262	32 34	298
153	16 00	359	28 31	056	59 26	118	30 34	163	35 45	211	59 00	261	32 09	297
154	16 00	358	28 54	055	59 51	118	30 42	163	35 30	211	58 32	260	31 44	296
155	15 58	357	29 17	054	60 16	117	30 50	162	35 16	210	58 04	259	31 18	295
156	15 57	356	29 40	053	60 41	117	30 59	162	35 02	210	57 36	259	30 53	294
157	15 54	355	30 02	052	61 06	116	31 08	161	34 48	209	57 09	258	30 27	293
158	15 52	354	30 24	051	61 31	116	31 17	160	34 35	208	56 41	257	30 01	292
159	15 48	353	30 46	050	61 57	115	31 27	160	34 22	208	56 14	256	29 35	291
160	15 45	352	31 08	049	62 22	115	31 37	159	34 09	207	55 47	256	29 09	290
161	15 40	351	31 29	048	62 48	114	31 47	158	33 56	207	55 19	255	28 42	289
162	15 36	350	31 49	047	63 14	114	31 58	158	33 43	206	54 52	254	28 15	288
163	15 30	349	32 10	046	63 39	113	32 08	157	33 31	205	54 25	254	27 49	287
164	15 25	348	32 30	045	64 05	113	32 19	157	33 19	205	53 58	253	27 22	286
	◆SPICA		ANTARES		◆Peacock		ACHERNAR		CANOPUS		◆SIRIUS		Alphard	
165	32 50	044	26 42	097	32 31	156	33 07	204	53 31	252	26 55	286	34 01	332
166	33 09	043	27 10	096	32 42	155	32 56	204	53 05	251	26 27	285	33 47	331
167	33 28	041	27 38	095	32 54	155	32 45	203	52 38	251	26 00	284	33 33	330
168	33 46	040	28 06	094	33 06	154	32 34	202	52 11	250	25 33	283	33 19	329
169	34 04	039	28 35	093	33 19	154	32 23	202	51 45	249	25 05	282	33 04	327
170	34 22	038	29 03	092	33 31	153	32 13	201	51 19	249	24 38	281	32 48	326
171	34 39	037	29 31	091	33 44	152	32 03	201	50 52	248	24 10	280	32 32	325
172	34 55	036	29 59	091	33 58	152	31 53	200	50 26	247	23 42	279	32 16	324
173	35 12	035	30 27	090	34 11	151	31 44	199	50 00	247	23 14	278	31 59	323
174	35 27	033	30 55	089	34 25	151	31 35	199	49 35	246	22 46	277	31 42	322
175	35 43	032	31 23	088	34 39	150	31 26	198	49 09	245	22 18	276	31 24	321
176	35 58	031	31 52	087	34 53	149	31 17	198	48 43	245	21 50	275	31 06	320
177	36 12	030	32 20	086	35 07	149	31 09	197	48 18	244	21 22	274	30 48	318
178	36 26	029	32 48	085	35 22	148	31 00	197	47 53	243	20 54	274	30 29	317
179	36 39	028	33 16	084	35 37	148	30 53	196	47 27	243	20 26	273	30 10	316

LHA ♈	Hc	Zn	Hc	Zn	Hc	Zn	Hc	Zn	Hc	Zn	Hc	Zn	Hc	Zn
	SPICA		◆ANTARES		Peacock		◆ACHERNAR		CANOPUS		Suhail		◆Alphard	
180	36 52	026	33 44	083	35 52	147	30 46	195	47 02	242	58 52	286	29 50	315
181	37 04	025	34 12	082	36 08	146	30 38	194	46 38	242	58 25	285	29 30	314
182	37 16	024	34 40	081	36 24	146	30 32	194	46 13	241	57 58	284	29 10	313
183	37 27	023	35 08	081	36 39	145	30 25	193	45 48	240	57 31	283	28 49	312
184	37 37	022	35 35	080	36 56	145	30 19	193	45 24	240	57 03	282	28 28	311
185	37 47	020	36 03	079	37 12	144	30 13	192	45 00	239	56 36	281	28 06	310
186	37 57	019	36 31	078	37 29	143	30 07	191	44 36	238	56 08	280	27 45	309
187	38 06	018	36 58	077	37 46	143	30 02	191	44 12	238	55 40	279	27 23	308
188	38 14	017	37 25	076	38 03	142	29 56	190	43 48	237	55 12	278	27 00	307
189	38 22	015	37 53	075	38 20	142	29 52	190	43 25	236	54 44	277	26 38	306
190	38 29	014	38 20	074	38 38	141	29 47	189	43 01	236	54 16	276	26 15	305
191	38 36	013	38 47	073	38 56	140	29 43	188	42 38	235	53 48	275	25 51	304
192	38 42	012	39 14	072	39 14	140	29 39	188	42 15	235	53 20	274	25 28	303
193	38 47	010	39 40	071	39 32	139	29 35	187	41 52	234	52 52	273	25 04	302
194	38 52	009	40 07	070	39 51	139	29 32	186	41 29	233	52 24	272	24 40	301
	SPICA		◆ANTARES		Peacock		◆ACHERNAR		CANOPUS		Suhail		◆Alphard	
195	38 56	008	40 33	069	40 09	138	29 29	186	41 07	233	51 56	272	24 16	300
196	39 00	007	40 59	068	40 28	137	29 26	185	40 45	232	51 28	271	23 51	299
197	39 03	005	41 25	067	40 47	137	29 24	185	40 23	231	51 00	270	23 26	298
198	39 05	004	41 51	066	41 07	136	29 22	184	40 01	231	50 31	269	23 01	297
199	39 07	003	42 17	065	41 26	136	29 20	183	39 39	230	50 03	268	22 36	296
200	39 08	002	42 42	064	41 46	135	29 19	183	39 17	229	49 35	267	22 10	295
201	39 08	000	43 07	062	42 06	135	29 17	182	38 56	229	49 07	266	21 45	294
202	39 08	359	43 32	061	42 26	134	29 17	181	38 35	228	48 39	265	21 19	293
203	39 07	358	43 56	060	42 47	133	29 16	181	38 14	228	48 11	265	20 53	292
204	39 06	357	44 21	059	43 07	133	29 16	180	37 54	227	47 43	264	20 27	291
205	39 04	355	44 45	058	43 28	132	29 16	180	37 33	226	47 15	263	20 00	290
206	39 01	354	45 09	057	43 49	132	29 16	179	37 13	226	46 47	262	19 34	289
207	38 58	353	45 32	056	44 10	131	29 17	178	36 53	225	46 19	261	19 07	288
208	38 54	351	45 55	055	44 31	130	29 18	178	36 33	224	45 51	261	18 40	287
209	38 49	350	46 18	053	44 53	130	29 19	177	36 13	224	45 23	260	18 13	286
	ANTARES		◆Nunki		Peacock		ACHERNAR		◆CANOPUS		Suhail		◆SPICA	
210	46 40	052	30 36	089	45 15	129	29 21	177	35 54	223	44 56	259	38 44	349
211	47 02	051	31 04	088	45 36	129	29 22	176	35 35	222	44 28	258	38 39	348
212	47 24	050	31 33	087	45 59	128	29 25	175	35 16	222	44 01	257	38 32	346
213	47 45	049	32 01	086	46 21	128	29 27	175	34 57	221	43 33	257	38 25	345
214	48 06	047	32 29	086	46 43	127	29 30	174	34 39	221	43 06	256	38 18	344
215	48 27	046	32 57	085	47 06	126	29 33	173	34 21	220	42 39	255	38 10	343
216	48 47	045	33 25	084	47 29	126	29 36	173	34 03	219	42 11	254	38 01	341
217	49 07	044	33 53	083	47 52	125	29 40	172	33 45	219	41 44	254	37 52	340
218	49 26	042	34 21	082	48 15	125	29 44	172	33 28	218	41 17	253	37 42	339
219	49 45	041	34 49	081	48 38	124	29 48	171	33 10	217	40 51	252	37 32	338
220	50 03	040	35 16	080	49 01	123	29 53	170	32 53	217	40 24	251	37 21	337
221	50 21	038	35 44	079	49 25	123	29 58	170	32 37	216	39 57	250	37 09	335
222	50 38	037	36 12	078	49 49	122	30 03	169	32 20	215	39 31	250	36 57	334
223	50 55	036	36 39	077	50 13	122	30 08	168	32 04	215	39 04	249	36 45	333
224	51 11	034	37 07	076	50 37	121	30 14	168	31 48	214	38 38	248	36 32	332
	ANTARES		◆Nunki		FOMALHAUT		ACHERNAR		◆CANOPUS		Suhail		◆SPICA	
225	51 26	033	37 34	075	13 42	129	30 20	167	31 32	214	38 12	247	36 18	331
226	51 41	032	38 01	074	14 04	128	30 27	167	31 17	213	37 46	247	36 04	329
227	51 56	030	38 28	073	14 26	127	30 33	166	31 02	212	37 20	246	35 50	328
228	52 10	029	38 55	072	14 49	126	30 40	165	30 47	212	36 55	245	35 35	327
229	52 23	027	39 22	071	15 12	126	30 47	165	30 32	211	36 29	245	35 19	326
230	52 36	026	39 48	070	15 35	125	30 55	164	30 18	210	36 04	244	35 03	325
231	52 48	025	40 15	069	15 58	124	31 03	164	30 04	210	35 39	243	34 47	324
232	52 59	023	40 41	068	16 22	123	31 11	163	29 50	209	35 14	242	34 30	322
233	53 10	022	41 07	067	16 45	122	31 19	162	29 36	208	34 49	242	34 12	321
234	53 20	020	41 33	066	17 09	121	31 28	162	29 23	208	34 24	241	33 54	320
235	53 29	019	41 59	065	17 33	121	31 37	161	29 10	207	33 59	240	33 36	319
236	53 38	017	42 24	064	17 58	120	31 46	161	28 58	206	33 35	239	33 18	318
237	53 46	016	42 49	063	18 22	119	31 56	160	28 45	206	33 11	239	32 59	317
238	53 53	014	43 14	062	18 47	118	32 05	159	28 33	205	32 47	238	32 39	316
239	54 00	013	43 39	061	19 12	117	32 16	159	28 21	204	32 23	237	32 19	315
	◆ANTARES		Nunki		◆FOMALHAUT		ACHERNAR		CANOPUS		◆ACRUX		SPICA	
240	54 05	011	44 03	060	19 37	117	32 26	158	28 10	204	66 03	244	31 59	314
241	54 10	010	44 27	058	20 02	116	32 37	157	27 59	203	65 38	243	31 38	313
242	54 15	008	44 51	057	20 28	115	32 48	157	27 48	203	65 13	243	31 17	311
243	54 18	007	45 15	056	20 53	114	32 59	156	27 37	202	64 48	242	30 56	310
244	54 21	005	45 38	055	21 19	113	33 10	156	27 27	201	64 23	242	30 35	309
245	54 23	004	46 01	054	21 45	113	33 22	155	27 17	201	63 58	242	30 13	308
246	54 25	002	46 24	053	22 11	112	33 34	154	27 07	200	63 33	241	29 50	307
247	54 25	000	46 46	052	22 37	111	33 46	154	26 57	199	63 09	241	29 28	306
248	54 25	359	47 08	050	23 04	110	33 59	153	26 48	199	62 44	240	29 05	305
249	54 24	357	47 29	049	23 30	109	34 12	153	26 40	198	62 20	240	28 42	304
250	54 22	356	47 50	048	23 57	109	34 25	152	26 31	197	61 56	239	28 18	303
251	54 20	354	48 11	047	24 24	108	34 38	151	26 23	197	61 31	239	27 54	302
252	54 17	353	48 31	045	24 50	107	34 52	151	26 15	196	61 07	239	27 30	301
253	54 13	351	48 51	044	25 17	106	35 05	150	26 07	195	60 43	238	27 06	300
254	54 08	350	49 10	043	25 45	105	35 19	150	26 00	195	60 19	238	26 42	299
	Nunki		◆FOMALHAUT		ACHERNAR		CANOPUS		◆ACRUX		SPICA		◆ANTARES	
255	49 29	042	26 12	104	35 34	149	25 53	194	59 56	237	26 17	298	54 03	348
256	49 48	040	26 39	104	35 48	148	25 47	193	59 32	237	25 52	297	53 57	347
257	50 06	039	27 07	103	36 03	148	25 40	193	59 09	236	25 27	296	53 50	345
258	50 23	038	27 34	102	36 18	147	25 34	192	58 45	236	25 01	295	53 42	344
259	50 40	036	28 02	101	36 34	147	25 29	191	58 22	235	24 36	294	53 34	342
260	50 57	035	28 29	100	36 49	146	25 23	191	57 59	235	24 10	293	53 25	341
261	51 12	034	28 57	099	37 05	146	25 18	190	57 36	234	23 44	292	53 15	339
262	51 28	032	29 25	099	37 21	145	25 13	189	57 13	234	23 18	291	53 05	338
263	51 42	031	29 53	098	37 37	144	25 09	189	56 50	234	22 51	290	52 54	336
264	51 57	029	30 21	097	37 54	144	25 05	188	56 28	233	22 25	289	52 42	335
265	52 10	028	30 49	096	38 11	143	25 01	187	56 05	233	21 58	288	52 30	333
266	52 23	027	31 17	095	38 28	143	24 58	187	55 43	232	21 31	287	52 17	332
267	52 35	025	31 45	094	38 45	142	24 55	186	55 21	232	21 04	286	52 04	331
268	52 47	024	32 13	093	39 02	141	24 52	185	54 59	231	20 37	286	51 49	329
269	52 58	022	32 41	093	39 20	141	24 50	185	54 37	231	20 10	285	51 35	328

LAT 62°S

LHA ♈	Hc	Zn	Hc	Zn	Hc	Zn	Hc	Zn	Hc	Zn	Hc	Zn	Hc	Zn
	◆ALTAIR		FOMALHAUT		◆ACHERNAR		CANOPUS		◆RIGIL KENT.		SPICA		ANTARES	
270	15 58	028	33 09	092	39 38	140	24 47	184	66 33	250	19 43	284	51 19	326
271	16 11	027	33 37	091	39 56	140	24 46	183	66 07	250	19 15	283	51 03	325
272	16 24	026	34 06	090	40 15	139	24 44	183	65 40	249	18 48	282	50 47	324
273	16 36	025	34 34	089	40 33	138	24 43	182	65 14	249	18 20	281	50 30	322
274	16 48	024	35 02	088	40 52	138	24 42	181	64 48	248	17 53	280	50 13	321
275	16 59	023	35 30	087	41 11	137	24 42	181	64 22	248	17 25	279	49 55	320
276	17 10	022	35 58	086	41 30	137	24 42	180	63 56	247	16 57	278	49 36	318
277	17 21	021	36 26	085	41 49	136	24 42	179	63 30	247	16 29	277	49 17	317
278	17 31	020	36 54	085	42 09	136	24 42	179	63 04	246	16 01	276	48 58	316
279	17 41	019	37 22	084	42 29	135	24 43	178	62 38	246	15 33	276	48 38	315
280	17 50	018	37 50	083	42 49	134	24 44	177	62 13	245	15 05	275	48 18	313
281	17 58	017	38 18	082	43 09	134	24 46	177	61 47	245	14 37	274	47 57	312
282	18 06	016	38 46	081	43 30	133	24 48	176	61 22	244	14 09	273	47 36	311
283	18 14	015	39 14	080	43 50	133	24 50	175	60 56	244	13 41	272	47 14	310
284	18 21	014	39 42	079	44 11	132	24 52	175	60 31	243	13 13	271	46 53	308
	◆ALTAIR		FOMALHAUT		◆ACHERNAR		CANOPUS		◆RIGIL KENT.		ANTARES		Nunki	
285	18 28	013	40 09	078	44 32	132	24 55	174	60 06	243	46 30	307	54 18	358
286	18 34	012	40 37	077	44 53	131	24 58	173	59 41	242	46 08	306	54 16	357
287	18 40	011	41 04	076	45 14	130	25 02	173	59 16	242	45 45	305	54 14	355
288	18 45	010	41 31	075	45 36	130	25 05	172	58 52	241	45 21	304	54 11	353
289	18 50	009	41 58	074	45 58	129	25 09	171	58 27	241	44 58	303	54 08	352
290	18 54	008	42 25	073	46 20	129	25 14	171	58 03	240	44 34	301	54 03	350
291	18 57	007	42 52	072	46 42	128	25 19	170	57 38	240	44 10	300	53 58	349
292	19 01	006	43 19	071	47 04	128	25 24	169	57 14	239	43 45	299	53 53	347
293	19 03	005	43 46	070	47 26	127	25 29	169	56 50	239	43 21	298	53 46	346
294	19 05	004	44 12	069	47 49	126	25 35	168	56 26	238	42 56	297	53 39	344
295	19 07	003	44 38	068	48 12	126	25 41	167	56 02	238	42 31	296	53 31	343
296	19 08	002	45 04	067	48 35	125	25 47	167	55 38	237	42 05	295	53 22	341
297	19 09	001	45 30	066	48 58	125	25 54	166	55 15	237	41 39	294	53 13	340
298	19 09	000	45 56	065	49 21	124	26 01	165	54 51	236	41 14	293	53 03	338
299	19 08	359	46 21	064	49 44	123	26 08	165	54 28	236	40 48	292	52 52	337
	FOMALHAUT		◆ACHERNAR		CANOPUS		◆RIGIL KENT.		ANTARES		Nunki		◆ALTAIR	
300	46 46	062	50 08	123	26 16	164	54 05	235	40 21	291	52 41	336	19 07	358
301	47 11	061	50 32	122	26 23	163	53 42	234	39 55	290	52 29	334	19 06	356
302	47 35	060	50 56	122	26 32	163	53 19	234	39 28	289	52 16	333	19 04	355
303	48 00	059	51 20	121	26 40	162	52 56	233	39 01	288	52 03	331	19 01	354
304	48 24	058	51 44	121	26 49	161	52 34	233	38 35	287	51 49	330	18 58	353
305	48 47	057	52 08	120	26 58	161	52 12	232	38 08	286	51 35	328	18 55	352
306	49 11	056	52 33	119	27 08	160	51 49	232	37 40	285	51 20	327	18 51	351
307	49 34	054	52 57	119	27 17	159	51 27	231	37 13	284	51 04	326	18 46	350
308	49 57	053	53 22	118	27 27	159	51 05	231	36 46	283	50 48	324	18 41	349
309	50 19	052	53 47	118	27 38	158	50 43	230	36 18	282	50 31	323	18 36	348
310	50 41	051	54 12	117	27 48	157	50 22	230	35 51	281	50 14	322	18 30	347
311	51 03	049	54 37	116	27 59	157	50 00	229	35 23	280	49 56	320	18 23	346
312	51 24	048	55 02	116	28 11	156	49 39	229	34 55	279	49 38	319	18 16	345
313	51 45	047	55 28	115	28 22	155	49 18	228	34 27	278	49 19	318	18 09	344
314	52 05	046	55 53	115	28 34	155	48 57	228	33 59	277	49 00	316	18 01	343
	◆FOMALHAUT		ACHERNAR		◆CANOPUS		RIGIL KENT.		◆ANTARES		Nunki		ALTAIR	
315	52 25	044	56 19	114	28 46	154	48 36	227	33 31	276	48 41	315	17 52	342
316	52 44	043	56 45	114	28 59	154	48 16	227	33 03	275	48 21	314	17 43	341
317	53 03	042	57 11	113	29 11	153	47 55	226	32 35	274	48 00	313	17 34	340
318	53 22	040	57 37	112	29 24	152	47 35	226	32 07	274	47 39	311	17 24	339
319	53 40	039	58 03	112	29 38	152	47 15	225	31 39	273	47 18	310	17 13	338
320	53 57	037	58 29	111	29 51	151	46 55	225	31 11	272	46 56	309	17 03	337
321	54 14	036	58 55	111	30 05	150	46 35	224	30 43	271	46 34	308	16 51	336
322	54 30	035	59 22	110	30 19	150	46 16	224	30 14	270	46 12	307	16 39	335
323	54 46	033	59 48	109	30 33	149	45 57	223	29 46	269	45 49	306	16 27	334
324	55 01	032	60 15	109	30 48	148	45 37	223	29 18	268	45 26	304	16 15	333
325	55 16	030	60 42	108	31 03	148	45 18	222	28 50	267	45 03	303	16 01	332
326	55 30	029	61 08	107	31 18	147	45 00	221	28 22	266	44 39	302	15 48	331
327	55 43	027	61 35	107	31 34	146	44 41	221	27 54	266	44 15	301	15 34	330
328	55 55	026	62 02	106	31 49	146	44 23	220	27 26	265	43 50	300	15 20	329
329	56 07	024	62 29	106	32 05	145	44 05	220	26 58	264	43 26	299	15 05	328
	◆FOMALHAUT		Diphda		ACHERNAR		◆CANOPUS		RIGIL KENT.		◆ANTARES		Nunki	
330	56 19	023	37 39	052	62 57	105	32 21	144	43 47	219	26 30	263	43 01	298
331	56 29	021	38 01	051	63 24	104	32 38	144	43 29	219	26 02	262	42 36	297
332	56 39	020	38 23	050	63 51	104	32 55	143	43 11	218	25 34	261	42 11	295
333	56 48	018	38 44	048	64 19	103	33 12	143	42 54	218	25 06	260	41 45	294
334	56 57	017	39 05	047	64 46	102	33 29	142	42 37	217	24 38	260	41 19	293
335	57 04	015	39 25	046	65 14	102	33 46	141	42 20	217	24 11	259	40 53	292
336	57 11	013	39 45	045	65 41	101	34 04	141	42 03	216	23 43	258	40 27	291
337	57 18	012	40 05	044	66 09	100	34 22	140	41 47	216	23 16	257	40 01	290
338	57 23	010	40 24	043	66 37	099	34 40	139	41 30	215	22 48	256	39 34	289
339	57 28	009	40 43	041	67 05	099	34 59	139	41 14	214	22 21	255	39 08	288
340	57 31	007	41 02	040	67 32	098	35 17	138	40 59	214	21 54	255	38 41	287
341	57 34	005	41 20	039	68 00	097	35 36	137	40 43	213	21 27	254	38 14	286
342	57 37	004	41 37	038	68 28	097	35 56	137	40 28	213	21 00	253	37 47	285
343	57 38	002	41 54	036	68 56	096	36 15	136	40 12	212	20 33	252	37 20	284
344	57 39	001	42 11	035	69 24	095	36 35	136	39 57	212	20 06	251	36 52	283
	◆Diphda		Acamar		◆CANOPUS		RIGIL KENT.		◆ANTARES		Nunki		FOMALHAUT	
345	42 27	034	48 51	087	36 54	135	39 43	211	19 39	250	36 25	282	57 39	359
346	42 42	033	49 19	086	37 14	134	39 28	211	19 13	250	35 57	281	57 38	357
347	42 57	032	49 47	085	37 35	134	39 14	210	18 47	249	35 30	280	57 36	356
348	43 12	030	50 15	084	37 55	133	39 00	210	18 20	248	35 02	280	57 34	354
349	43 26	029	50 43	083	38 16	132	38 46	209	17 54	247	34 34	279	57 30	352
350	43 39	028	51 11	082	38 37	132	38 33	208	17 28	246	34 06	278	57 26	351
351	43 52	027	51 39	081	38 58	131	38 19	208	17 03	245	33 38	277	57 21	349
352	44 05	025	52 06	080	39 19	131	38 06	207	16 37	245	33 10	276	57 16	348
353	44 16	024	52 34	079	39 41	130	37 54	207	16 12	244	32 42	275	57 09	346
354	44 27	023	53 02	078	40 02	129	37 41	206	15 47	243	32 14	274	57 02	344
355	44 38	021	53 29	077	40 24	129	37 29	206	15 22	242	31 46	273	56 54	343
356	44 48	020	53 57	076	40 46	128	37 17	205	14 57	241	31 18	272	56 46	341
357	44 57	019	54 24	075	41 09	127	37 05	205	14 32	241	30 50	271	56 36	340
358	45 06	017	54 51	074	41 31	127	36 53	204	14 08	240	30 22	270	56 26	338
359	45 14	016	55 18	073	41 54	126	36 42	203	13 43	239	29 53	270	56 15	337

 LAT 63°S

LHA ♈	Hc	Zn	Hc	Zn	Hc	Zn	Hc	Zn	Hc	Zn	Hc	Zn	Hc	Zn
	Diphda		◆RIGEL		CANOPUS		Miaplacidus		◆RIGIL KENT.		Peacock		◆FOMALHAUT	
0	44 24	014	12 29	084	42 51	125	45 54	161	37 26	203	63 12	258	55 09	336
1	44 30	013	12 56	083	43 14	124	46 03	160	37 15	203	62 45	258	54 58	334
2	44 36	012	13 23	082	43 36	123	46 12	160	37 05	202	62 19	257	54 46	333
3	44 41	011	13 50	081	43 59	123	46 22	159	36 55	201	61 52	256	54 33	331
4	44 46	009	14 17	080	44 22	122	46 31	159	36 45	201	61 26	256	54 19	330
5	44 50	008	14 44	079	44 45	121	46 41	158	36 36	200	60 59	255	54 05	328
6	44 53	006	15 10	078	45 09	121	46 51	158	36 26	200	60 33	254	53 51	327
7	44 56	005	15 37	077	45 32	120	47 02	158	36 17	199	60 07	254	53 36	326
8	44 58	004	16 04	076	45 56	119	47 12	157	36 08	199	59 41	253	53 20	324
9	45 00	002	16 30	075	46 20	119	47 23	157	36 00	198	59 15	252	53 04	323
10	45 01	001	16 56	074	46 44	118	47 34	156	35 52	197	58 49	252	52 47	322
11	45 01	000	17 22	073	47 08	117	47 45	156	35 44	197	58 23	251	52 30	320
12	45 00	358	17 48	073	47 32	117	47 56	155	35 36	196	57 57	250	52 12	319
13	44 59	357	18 14	072	47 56	116	48 08	155	35 28	196	57 32	250	51 54	318
14	44 58	356	18 40	071	48 21	115	48 19	154	35 21	195	57 06	249	51 36	316
	◆RIGEL		SIRIUS		CANOPUS		◆RIGIL KENT.		Peacock		◆FOMALHAUT		Diphda	
15	19 06	070	16 33	094	48 46	115	35 14	195	56 41	248	51 17	315	44 55	354
16	19 31	069	17 00	094	49 10	114	35 07	194	56 16	248	50 57	314	44 52	353
17	19 57	068	17 27	093	49 35	113	35 01	193	55 50	247	50 37	312	44 49	352
18	20 22	067	17 54	092	50 00	113	34 55	193	55 25	247	50 17	311	44 44	350
19	20 47	066	18 21	091	50 25	112	34 49	192	55 00	246	49 56	310	44 40	349
20	21 11	065	18 49	090	50 51	111	34 43	192	54 36	245	49 35	309	44 34	348
21	21 36	064	19 16	089	51 16	111	34 38	191	54 11	245	49 14	307	44 28	346
22	22 00	063	19 43	088	51 42	110	34 33	190	53 46	244	48 52	306	44 21	345
23	22 25	062	20 10	087	52 07	109	34 28	190	53 22	243	48 30	305	44 14	344
24	22 49	061	20 38	086	52 33	109	34 24	189	52 58	243	48 07	304	44 06	342
25	23 12	060	21 05	086	52 59	108	34 19	189	52 34	242	47 44	303	43 58	341
26	23 36	059	21 32	085	53 25	107	34 15	188	52 10	242	47 21	301	43 48	340
27	23 59	058	21 59	084	53 51	106	34 12	188	51 46	241	46 58	300	43 39	339
28	24 22	057	22 26	083	54 17	106	34 08	187	51 22	240	46 34	299	43 29	337
29	24 45	056	22 53	082	54 44	105	34 05	186	50 58	240	46 10	298	43 18	336
	◆RIGEL		SIRIUS		CANOPUS		◆RIGIL KENT.		Peacock		◆FOMALHAUT		Diphda	
30	25 07	055	23 20	081	55 10	104	34 02	186	50 35	239	45 46	297	43 06	335
31	25 29	054	23 47	080	55 36	104	34 00	185	50 12	239	45 22	296	42 54	333
32	25 51	053	24 14	079	56 03	103	33 57	185	49 48	238	44 57	295	42 42	332
33	26 13	052	24 40	078	56 29	102	33 55	184	49 25	237	44 32	294	42 29	331
34	26 34	051	25 07	077	56 56	101	33 54	183	49 03	237	44 07	293	42 15	330
35	26 55	050	25 33	076	57 23	101	33 52	183	48 40	236	43 42	292	42 01	328
36	27 16	049	26 00	075	57 50	100	33 51	182	48 17	235	43 17	291	41 47	327
37	27 36	048	26 26	074	58 17	099	33 50	182	47 55	235	42 51	290	41 32	326
38	27 56	047	26 52	073	58 43	098	33 49	181	47 33	234	42 25	289	41 16	325
39	28 16	046	27 18	073	59 10	097	33 49	180	47 11	234	41 59	287	41 00	323
40	28 35	045	27 44	072	59 37	097	33 49	180	46 49	233	41 33	286	40 44	322
41	28 54	044	28 10	071	60 05	096	33 49	179	46 27	232	41 07	285	40 27	321
42	29 13	043	28 36	070	60 32	095	33 50	179	46 06	232	40 41	284	40 10	320
43	29 31	041	29 01	069	60 59	094	33 50	178	45 44	231	40 14	284	39 52	319
44	29 49	040	29 26	068	61 26	093	33 51	178	45 23	231	39 48	283	39 34	318
	RIGEL		◆SIRIUS		CANOPUS		◆RIGIL KENT.		Peacock		FOMALHAUT		◆Diphda	
45	30 06	039	29 52	067	61 53	093	33 53	177	45 02	230	39 21	282	39 15	316
46	30 23	038	30 16	066	62 20	092	33 54	176	44 41	229	38 55	281	38 56	315
47	30 40	037	30 41	065	62 48	091	33 56	176	44 21	229	38 28	280	38 37	314
48	30 56	036	31 06	064	63 15	090	33 58	175	44 00	228	38 01	279	38 17	313
49	31 12	035	31 30	063	63 42	089	34 01	175	43 40	228	37 34	278	37 57	312
50	31 27	034	31 54	062	64 09	088	34 03	174	43 20	227	37 07	277	37 37	311
51	31 42	033	32 18	061	64 37	087	34 06	173	43 00	226	36 40	276	37 16	310
52	31 57	031	32 41	060	65 04	086	34 10	173	42 41	226	36 13	275	36 55	308
53	32 11	030	33 05	058	65 31	085	34 13	172	42 21	225	35 46	274	36 33	307
54	32 24	029	33 28	057	65 58	084	34 17	172	42 02	225	35 18	273	36 11	306
55	32 37	028	33 51	056	66 25	083	34 21	171	41 43	224	34 51	272	35 49	305
56	32 50	027	34 13	055	66 52	082	34 25	171	41 24	223	34 24	271	35 27	304
57	33 02	026	34 36	054	67 19	081	34 30	170	41 05	223	33 57	270	35 04	303
58	33 13	025	34 57	053	67 46	080	34 35	169	40 47	222	33 29	270	34 41	302
59	33 24	023	35 19	052	68 13	079	34 40	169	40 29	222	33 02	269	34 18	301
	RIGEL		◆SIRIUS		Suhail		◆RIGIL KENT.		Peacock		FOMALHAUT		◆Diphda	
60	33 35	022	35 40	051	43 23	103	34 45	168	40 11	221	32 35	268	33 54	300
61	33 45	021	36 01	050	43 49	102	34 51	168	39 53	220	32 08	267	33 31	299
62	33 54	020	36 22	049	44 16	102	34 57	167	39 35	220	31 41	266	33 07	298
63	34 03	019	36 42	048	44 43	101	35 03	166	39 18	219	31 13	265	32 43	297
64	34 12	018	37 02	047	45 09	100	35 10	166	39 01	219	30 46	264	32 18	296
65	34 20	016	37 22	045	45 36	099	35 17	165	38 44	218	30 19	263	31 53	295
66	34 27	015	37 41	044	46 03	098	35 24	165	38 27	217	29 52	263	31 29	294
67	34 34	014	38 00	043	46 30	098	35 31	164	38 11	217	29 25	262	31 04	293
68	34 40	013	38 18	042	46 57	097	35 39	164	37 55	216	28 58	261	30 38	292
69	34 46	012	38 36	041	47 24	096	35 47	163	37 39	216	28 31	260	30 13	291
70	34 51	010	38 54	040	47 51	095	35 55	162	37 23	215	28 05	259	29 47	290
71	34 56	009	39 11	038	48 18	094	36 03	162	37 07	214	27 38	258	29 22	289
72	35 00	008	39 28	037	48 46	093	36 12	161	36 52	214	27 11	257	28 56	288
73	35 03	007	39 44	036	49 13	092	36 21	161	36 37	213	26 45	257	28 30	287
74	35 06	006	40 00	035	49 40	092	36 30	160	36 22	213	26 18	256	28 04	286
	RIGEL		◆SIRIUS		Suhail		◆RIGIL KENT.		Peacock		ACHERNAR		◆Diphda	
75	35 09	004	40 15	034	50 07	091	36 39	159	36 08	212	64 51	260	27 38	285
76	35 11	003	40 30	032	50 35	090	36 49	159	35 53	211	64 25	259	27 11	284
77	35 12	002	40 45	031	51 02	089	36 59	158	35 39	211	63 58	258	26 45	283
78	35 12	001	40 59	030	51 29	088	37 09	158	35 25	210	63 31	258	26 18	282
79	35 12	359	41 12	029	51 56	087	37 19	157	35 12	210	63 05	257	25 51	281
80	35 12	358	41 25	028	52 23	086	37 30	157	34 59	209	62 38	256	25 25	280
81	35 11	357	41 37	026	52 51	085	37 41	156	34 45	208	62 12	256	24 58	279
82	35 09	356	41 49	025	53 18	084	37 52	155	34 33	208	61 46	255	24 31	278
83	35 07	355	42 00	024	53 45	083	38 04	155	34 20	207	61 19	254	24 04	277
84	35 04	353	42 11	023	54 12	082	38 15	154	34 08	206	60 53	254	23 37	277
85	35 01	352	42 21	021	54 39	081	38 27	154	33 56	206	60 27	253	23 10	276
86	34 57	351	42 31	020	55 06	080	38 39	153	33 44	205	60 01	252	22 43	275
87	34 52	350	42 40	019	55 32	079	38 52	153	33 33	205	59 35	252	22 15	274
88	34 47	349	42 48	017	55 59	078	39 04	152	33 21	204	59 09	251	21 48	273
89	34 41	347	42 56	016	56 26	077	39 17	152	33 10	203	58 44	250	21 21	272

LHA ♈	Hc	Zn	Hc	Zn	Hc	Zn	Hc	Zn	Hc	Zn	Hc	Zn	Hc	Zn
	◆SIRIUS		Suhail		◆RIGIL KENT.		Peacock		◆ACHERNAR		Acamar		RIGEL	
90	43 03	015	56 52	076	39 30	151	33 00	203	58 18	250	55 01	289	34 35	346
91	43 10	013	57 19	075	39 44	150	32 49	202	57 53	249	54 35	287	34 28	345
92	43 16	012	57 45	074	39 57	150	32 39	202	57 27	248	54 09	286	34 21	344
93	43 21	011	58 11	073	40 11	149	32 29	201	57 02	248	53 42	285	34 13	343
94	43 26	010	58 37	072	40 25	149	32 20	200	56 37	247	53 16	284	34 05	341
95	43 30	008	59 03	071	40 39	148	32 10	200	56 12	246	52 50	283	33 56	340
96	43 34	007	59 29	070	40 54	148	32 01	199	55 47	246	52 23	282	33 46	339
97	43 37	006	59 54	068	41 09	147	31 53	198	55 22	245	51 56	281	33 36	338
98	43 39	004	60 19	067	41 23	146	31 44	198	54 57	245	51 30	280	33 26	337
99	43 41	003	60 44	066	41 39	146	31 36	197	54 33	244	51 03	279	33 15	336
100	43 42	002	61 09	065	41 54	145	31 28	197	54 08	243	50 36	278	33 03	334
101	43 43	000	61 34	064	42 10	145	31 20	196	53 44	243	50 09	277	32 51	333
102	43 42	359	61 58	062	42 25	144	31 13	195	53 20	242	49 42	276	32 39	332
103	43 42	358	62 22	061	42 41	144	31 06	195	52 56	242	49 15	275	32 26	331
104	43 40	356	62 45	060	42 58	143	30 59	194	52 32	241	48 48	274	32 13	330
	Alphard		◆Gienah		RIGIL KENT.		◆Peacock		ACHERNAR		◆RIGEL		SIRIUS	
105	29 32	043	20 35	088	43 14	143	30 53	193	52 08	240	31 59	329	43 38	355
106	29 51	042	21 02	087	43 31	142	30 47	193	51 45	240	31 44	328	43 35	354
107	30 09	041	21 29	087	43 48	141	30 41	192	51 21	239	31 29	326	43 32	352
108	30 26	040	21 57	086	44 05	141	30 35	192	50 58	239	31 14	325	43 28	351
109	30 44	039	22 24	085	44 22	140	30 30	191	50 35	238	30 58	324	43 24	350
110	31 00	037	22 51	084	44 39	140	30 25	190	50 12	237	30 42	323	43 18	348
111	31 17	036	23 18	083	44 57	139	30 20	190	49 49	237	30 26	322	43 13	347
112	31 33	035	23 45	082	45 15	139	30 16	189	49 26	236	30 09	321	43 06	346
113	31 48	034	24 12	081	45 33	138	30 12	188	49 04	236	29 51	320	42 59	345
114	32 03	033	24 39	080	45 51	138	30 08	188	48 41	235	29 34	319	42 52	343
115	32 18	032	25 06	079	46 10	137	30 04	187	48 19	234	29 15	318	42 44	342
116	32 32	031	25 32	078	46 28	137	30 01	187	47 57	234	28 57	317	42 35	341
117	32 46	030	25 59	077	46 47	136	29 58	186	47 35	233	28 38	316	42 26	339
118	32 59	028	26 25	076	47 06	135	29 55	185	47 13	233	28 19	314	42 16	338
119	33 12	027	26 52	075	47 25	135	29 53	185	46 52	232	27 59	313	42 05	337
	Alphard		◆SPICA		RIGIL KENT.		◆Peacock		ACHERNAR		◆RIGEL		SIRIUS	
120	33 24	026	13 53	087	47 45	134	29 51	184	46 30	231	27 39	312	41 54	336
121	33 36	025	14 20	086	48 04	134	29 49	183	46 09	231	27 19	311	41 43	334
122	33 47	024	14 47	086	48 24	133	29 48	183	45 48	230	26 58	310	41 31	333
123	33 58	023	15 15	085	48 44	133	29 47	182	45 27	230	26 37	309	41 18	332
124	34 08	021	15 42	084	49 04	132	29 46	181	45 07	229	26 16	308	41 05	331
125	34 18	020	16 09	083	49 24	132	29 45	181	44 46	228	25 54	307	40 51	329
126	34 27	019	16 36	082	49 45	131	29 45	180	44 26	228	25 33	306	40 37	328
127	34 35	018	17 03	081	50 05	131	29 45	180	44 06	227	25 10	305	40 22	327
128	34 43	017	17 30	080	50 26	130	29 46	179	43 46	227	24 48	304	40 07	326
129	34 51	016	17 56	079	50 47	130	29 46	178	43 26	226	24 25	303	39 52	324
130	34 58	014	18 23	078	51 08	129	29 47	178	43 06	226	24 02	302	39 36	323
131	35 04	013	18 50	077	51 29	128	29 48	177	42 47	225	23 39	301	39 19	322
132	35 10	012	19 16	076	51 51	128	29 50	176	42 28	224	23 16	300	39 02	321
133	35 16	011	19 43	075	52 12	127	29 52	176	42 09	224	22 52	299	38 45	320
134	35 21	010	20 09	075	52 34	127	29 54	175	41 50	223	22 28	298	38 27	319
	REGULUS		◆SPICA		RIGIL KENT.		◆Peacock		ACHERNAR		CANOPUS		◆SIRIUS	
135	13 51	017	20 35	074	52 56	126	29 56	175	41 32	223	67 17	279	38 09	317
136	13 59	016	21 01	073	53 18	126	29 59	174	41 14	222	66 50	278	37 50	316
137	14 07	015	21 27	072	53 40	125	30 02	173	40 55	221	66 23	277	37 31	315
138	14 13	014	21 53	071	54 02	125	30 06	173	40 38	221	65 56	276	37 12	314
139	14 20	013	22 19	070	54 25	124	30 09	172	40 20	220	65 29	275	36 52	313
140	14 26	012	22 44	069	54 47	124	30 13	171	40 02	220	65 02	274	36 32	312
141	14 31	011	23 09	068	55 10	123	30 17	171	39 45	219	64 34	273	36 11	311
142	14 36	010	23 34	067	55 33	123	30 22	170	39 28	218	64 07	272	35 50	310
143	14 41	009	23 59	066	55 56	122	30 27	169	39 11	218	63 40	271	35 29	308
144	14 45	008	24 24	065	56 19	122	30 32	169	38 55	217	63 13	270	35 08	307
145	14 49	007	24 49	064	56 42	121	30 37	168	38 38	217	62 45	269	34 46	306
146	14 52	006	25 13	063	57 06	120	30 43	168	38 22	216	62 18	268	34 24	305
147	14 54	005	25 37	062	57 29	120	30 49	167	38 06	215	61 51	267	34 01	304
148	14 57	004	26 01	061	57 53	119	30 55	166	37 51	215	61 24	267	33 39	303
149	14 58	003	26 25	060	58 17	119	31 02	166	37 35	214	60 57	266	33 16	302
	REGULUS		◆SPICA		RIGIL KENT.		◆Peacock		ACHERNAR		CANOPUS		◆SIRIUS	
150	15 00	002	26 48	059	58 41	118	31 09	165	37 20	214	60 29	265	32 52	301
151	15 00	001	27 12	058	59 05	118	31 16	164	37 05	213	60 02	264	32 29	300
152	15 00	000	27 35	057	59 29	117	31 23	164	36 51	212	59 35	263	32 05	299
153	15 00	359	27 57	056	59 53	117	31 31	163	36 36	212	59 08	262	31 41	298
154	15 00	358	28 20	055	60 18	116	31 39	163	36 22	211	58 41	262	31 17	297
155	14 58	357	28 42	054	60 42	116	31 47	162	36 08	211	58 14	261	30 53	296
156	14 57	356	29 04	053	61 07	115	31 56	161	35 54	210	57 48	260	30 28	295
157	14 55	355	29 25	052	61 32	115	32 05	161	35 41	209	57 21	259	30 03	294
158	14 52	354	29 47	051	61 56	114	32 14	160	35 27	209	56 54	259	29 38	293
159	14 49	353	30 08	050	62 21	113	32 23	159	35 15	208	56 27	258	29 13	292
160	14 45	352	30 28	049	62 46	113	32 33	159	35 02	208	56 01	257	28 48	291
161	14 41	351	30 48	048	63 12	112	32 43	158	34 49	207	55 34	256	28 22	290
162	14 37	350	31 08	046	63 37	112	32 53	158	34 37	206	55 08	256	27 56	289
163	14 32	349	31 28	045	64 02	111	33 04	157	34 25	206	54 41	255	27 30	288
164	14 26	348	31 47	044	64 28	111	33 14	156	34 14	205	54 15	254	27 04	287
	◆SPICA		ANTARES		◆Peacock		ACHERNAR		CANOPUS		◆SIRIUS		Alphard	
165	32 06	043	26 49	096	33 26	156	34 02	204	53 49	253	26 38	286	33 07	332
166	32 24	042	27 16	095	33 37	155	33 51	204	53 23	253	26 12	285	32 55	331
167	32 42	041	27 43	094	33 48	155	33 40	203	52 57	252	25 46	284	32 41	330
168	33 00	040	28 10	093	34 00	154	33 29	203	52 31	251	25 19	283	32 27	329
169	33 17	039	28 38	093	34 12	153	33 19	202	52 05	251	24 53	282	32 13	328
170	33 34	038	29 05	092	34 25	153	33 09	201	51 40	250	24 26	281	31 58	327
171	33 51	036	29 32	091	34 37	152	32 59	201	51 14	249	23 59	280	31 43	326
172	34 07	035	29 59	090	34 50	151	32 50	200	50 49	249	23 32	279	31 27	324
173	34 22	034	30 26	089	35 03	151	32 40	200	50 23	248	23 06	279	31 11	323
174	34 37	033	30 54	088	35 17	150	32 32	199	49 58	247	22 39	278	30 55	322
175	34 52	032	31 21	087	35 31	150	32 23	198	49 33	247	22 12	277	30 38	321
176	35 06	031	31 48	086	35 44	149	32 14	198	49 08	246	21 44	276	30 21	320
177	35 20	030	32 15	085	35 59	148	32 06	197	48 44	245	21 17	275	30 03	319
178	35 33	028	32 42	085	36 13	148	31 58	196	48 19	244	20 50	274	29 45	318
179	35 46	027	33 10	084	36 28	147	31 51	196	47 54	244	20 23	273	29 26	317

 LAT 63°S

LHA ♈	Hc	Zn	Hc	Zn	Hc	Zn	Hc	Zn	Hc	Zn	Hc	Zn	Hc	Zn
	SPICA		◆ANTARES		Peacock		◆ACHERNAR		CANOPUS		Suhail		◆Alphard	
180	35 58	026	33 37	083	36 43	147	31 44	195	47 30	243	58 35	288	29 07	316
181	36 10	025	34 04	082	36 58	146	31 37	195	47 06	242	58 09	287	28 48	315
182	36 21	024	34 30	081	37 13	145	31 30	194	46 42	242	57 43	286	28 29	313
183	36 31	022	34 57	080	37 29	145	31 23	193	46 18	241	57 16	285	28 09	312
184	36 42	021	35 24	079	37 44	144	31 17	193	45 54	241	56 50	284	27 48	311
185	36 51	020	35 51	078	38 01	144	31 11	192	45 30	240	56 23	283	27 28	310
186	37 00	019	36 17	077	38 17	143	31 06	191	45 07	239	55 57	282	27 07	309
187	37 09	018	36 44	076	38 33	142	31 01	191	44 44	239	55 30	281	26 46	308
188	37 17	016	37 10	075	38 50	142	30 56	190	44 20	238	55 03	280	26 24	307
189	37 24	015	37 36	074	39 07	141	30 51	190	43 57	237	54 36	279	26 02	306
190	37 31	014	38 03	073	39 24	141	30 46	189	43 35	237	54 09	278	25 40	305
191	37 37	013	38 29	072	39 42	140	30 42	188	43 12	236	53 42	277	25 18	304
192	37 43	011	38 54	071	39 59	139	30 38	188	42 49	235	53 15	276	24 55	303
193	37 48	010	39 20	070	40 17	139	30 35	187	42 27	235	52 48	275	24 32	302
194	37 53	009	39 46	069	40 35	138	30 32	186	42 05	234	52 21	274	24 09	301
	SPICA		◆ANTARES		Peacock		◆ACHERNAR		CANOPUS		Suhail		◆Alphard	
195	37 57	008	40 11	068	40 54	137	30 29	186	41 43	233	51 54	273	23 46	300
196	38 00	007	40 36	067	41 12	137	30 26	185	41 21	233	51 26	272	23 22	299
197	38 03	005	41 01	066	41 31	136	30 24	185	41 00	232	50 59	271	22 58	298
198	38 05	004	41 26	065	41 50	136	30 22	184	40 38	231	50 32	270	22 34	297
199	38 07	003	41 50	064	42 09	135	30 20	183	40 17	231	50 05	269	22 10	296
200	38 08	002	42 15	063	42 28	134	30 19	183	39 56	230	49 37	268	21 45	295
201	38 08	000	42 39	062	42 48	134	30 17	182	39 35	229	49 10	267	21 20	294
202	38 08	359	43 03	061	43 08	133	30 17	181	39 15	229	48 43	267	20 55	293
203	38 07	358	43 26	059	43 28	133	30 16	181	38 54	228	48 16	266	20 30	292
204	38 06	357	43 50	058	43 48	132	30 16	180	38 34	227	47 49	265	20 05	291
205	38 04	355	44 13	057	44 08	131	30 16	180	38 14	227	47 22	264	19 40	290
206	38 01	354	44 35	056	44 28	131	30 16	179	37 55	226	46 55	263	19 14	290
207	37 58	353	44 58	055	44 49	130	30 17	178	37 35	226	46 28	262	18 48	289
208	37 55	352	45 20	054	45 10	130	30 18	178	37 16	225	46 01	262	18 22	288
209	37 50	350	45 42	053	45 31	129	30 19	177	36 57	224	45 34	261	17 56	287
	ANTARES		◆Nunki		Peacock		ACHERNAR		◆CANOPUS		Suhail		◆SPICA	
210	46 03	051	30 35	089	45 52	128	30 20	176	36 38	224	45 07	260	37 45	349
211	46 24	050	31 02	088	46 14	128	30 22	176	36 19	223	44 40	259	37 40	348
212	46 45	049	31 29	087	46 35	127	30 24	175	36 01	222	44 13	258	37 34	347
213	47 05	048	31 57	086	46 57	127	30 27	175	35 42	222	43 47	258	37 27	345
214	47 25	047	32 24	085	47 19	126	30 29	174	35 24	221	43 20	257	37 20	344
215	47 45	045	32 51	084	47 41	125	30 32	173	35 07	220	42 54	256	37 13	343
216	48 04	044	33 18	083	48 03	125	30 36	173	34 49	220	42 27	255	37 04	342
217	48 23	043	33 45	082	48 26	124	30 39	172	34 32	219	42 01	254	36 55	341
218	48 41	042	34 12	081	48 48	124	30 43	171	34 15	218	41 35	254	36 46	339
219	48 59	040	34 39	080	49 11	123	30 47	171	33 58	218	41 09	253	36 36	338
220	49 16	039	35 06	079	49 34	122	30 52	170	33 41	217	40 43	252	36 26	337
221	49 33	038	35 33	078	49 57	122	30 57	170	33 25	216	40 17	251	36 15	336
222	49 50	036	35 59	077	50 20	121	31 02	169	33 09	216	39 51	251	36 03	335
223	50 06	035	36 25	076	50 44	121	31 07	168	32 53	215	39 26	250	35 51	333
224	50 21	034	36 52	075	51 07	120	31 13	168	32 38	215	39 00	249	35 39	332
	ANTARES		◆Nunki		FOMALHAUT		ACHERNAR		◆CANOPUS		Suhail		◆SPICA	
225	50 36	032	37 18	074	14 19	129	31 19	167	32 22	214	38 35	248	35 26	331
226	50 50	031	37 44	073	14 41	128	31 25	166	32 07	213	38 09	247	35 12	330
227	51 04	030	38 10	072	15 03	127	31 31	166	31 52	213	37 44	247	34 59	329
228	51 17	028	38 36	071	15 24	126	31 38	165	31 38	212	37 19	246	34 44	327
229	51 30	027	39 02	070	15 47	125	31 45	165	31 24	211	36 55	245	34 29	326
230	51 42	025	39 28	069	16 09	124	31 53	164	31 10	211	36 30	244	34 14	325
231	51 53	024	39 53	068	16 31	124	32 00	163	30 56	210	36 05	244	33 58	324
232	52 04	023	40 18	067	16 54	123	32 08	163	30 42	209	35 41	243	33 42	323
233	52 14	021	40 43	066	17 17	122	32 16	162	30 29	209	35 17	242	33 25	322
234	52 23	020	41 08	065	17 40	121	32 25	162	30 16	208	34 53	242	33 08	321
235	52 32	018	41 33	064	18 04	120	32 34	161	30 04	207	34 29	241	32 51	320
236	52 40	017	41 57	063	18 27	120	32 43	160	29 51	207	34 05	240	32 33	318
237	52 48	015	42 21	062	18 51	119	32 52	160	29 39	206	33 42	239	32 15	317
238	52 55	014	42 45	061	19 15	118	33 02	159	29 27	205	33 19	239	31 56	316
239	53 01	012	43 09	060	19 39	117	33 11	158	29 16	205	32 55	238	31 37	315
	◆ANTARES		Nunki		◆FOMALHAUT		ACHERNAR		CANOPUS		◆ACRUX		SPICA	
240	53 06	011	43 32	059	20 04	116	33 22	158	29 05	204	66 29	246	31 17	314
241	53 11	009	43 56	058	20 28	115	33 32	157	28 54	203	66 04	245	30 58	313
242	53 15	008	44 18	057	20 53	115	33 43	157	28 43	203	65 39	245	30 38	312
243	53 19	006	44 41	055	21 18	114	33 54	156	28 33	202	65 15	244	30 17	311
244	53 21	005	45 03	054	21 43	113	34 05	155	28 23	201	64 50	244	29 56	310
245	53 23	003	45 25	053	22 08	112	34 16	155	28 13	201	64 26	243	29 35	309
246	53 25	002	45 47	052	22 33	111	34 28	154	28 03	200	64 02	243	29 14	308
247	53 25	000	46 08	051	22 59	111	34 40	154	27 54	199	63 37	242	28 52	307
248	53 25	359	46 29	050	23 24	110	34 52	153	27 45	199	63 13	242	28 30	306
249	53 24	357	46 50	048	23 50	109	35 05	152	27 37	198	62 49	242	28 08	305
250	53 23	356	47 10	047	24 16	108	35 18	152	27 28	197	62 25	241	27 45	304
251	53 20	354	47 29	046	24 42	107	35 31	151	27 20	197	62 02	241	27 22	302
252	53 17	353	47 49	045	25 08	106	35 44	151	27 13	196	61 38	240	26 59	301
253	53 14	351	48 08	043	25 34	106	35 57	150	27 05	195	61 14	240	26 36	300
254	53 09	350	48 26	042	26 00	105	36 11	149	26 58	195	60 51	239	26 12	299
	Nunki		◆FOMALHAUT		ACHERNAR		CANOPUS		◆ACRUX		SPICA		◆ANTARES	
255	48 44	041	26 26	104	36 25	149	26 51	194	60 28	239	25 48	298	53 04	348
256	49 02	040	26 53	103	36 39	148	26 45	193	60 04	238	25 24	297	52 58	347
257	49 19	038	27 20	102	36 54	147	26 39	193	59 41	238	25 00	296	52 52	345
258	49 35	037	27 46	101	37 09	147	26 33	192	59 18	237	24 36	295	52 45	344
259	49 52	036	28 13	101	37 24	146	26 27	191	58 56	237	24 11	295	52 37	343
260	50 07	034	28 40	100	37 39	146	26 22	191	58 33	236	23 46	294	52 28	341
261	50 22	033	29 07	099	37 54	145	26 17	190	58 10	236	23 21	293	52 19	340
262	50 37	032	29 34	098	38 10	144	26 13	189	57 48	235	22 56	292	52 09	338
263	50 51	030	30 01	097	38 26	144	26 08	189	57 25	235	22 30	291	51 59	337
264	51 04	029	30 28	096	38 42	143	26 04	188	57 03	234	22 05	290	51 48	335
265	51 17	027	30 55	095	38 59	143	26 01	187	56 41	234	21 39	289	51 36	334
266	51 29	026	31 22	095	39 15	142	25 57	187	56 19	233	21 13	288	51 24	333
267	51 41	025	31 49	094	39 32	141	25 54	186	55 57	233	20 47	287	51 11	331
268	51 52	023	32 16	093	39 49	141	25 52	185	55 36	232	20 21	286	50 58	330
269	52 02	022	32 43	092	40 06	140	25 49	185	55 14	232	19 55	285	50 44	328

LHA ♈	Hc	Zn	Hc	Zn	Hc	Zn	Hc	Zn	Hc	Zn	Hc	Zn	Hc	Zn
	◆ALTAIR		FOMALHAUT		◆ACHERNAR		CANOPUS		◆RIGIL KENT.		SPICA		ANTARES	
270	15 05	028	33 11	091	40 24	140	25 47	184	66 52	252	19 28	284	50 29	327
271	15 18	027	33 38	090	40 42	139	25 46	183	66 26	252	19 02	283	50 14	326
272	15 30	026	34 05	089	41 00	139	25 44	183	66 01	251	18 35	282	49 59	324
273	15 42	025	34 32	088	41 18	138	25 43	182	65 35	251	18 09	281	49 42	323
274	15 53	024	35 00	087	41 36	137	25 42	181	65 09	250	17 42	280	49 26	322
275	16 04	023	35 27	087	41 55	137	25 42	181	64 44	250	17 15	279	49 09	320
276	16 15	022	35 54	086	42 14	136	25 42	180	64 18	249	16 48	279	48 51	319
277	16 25	021	36 21	085	42 32	136	25 42	179	63 53	249	16 21	278	48 33	318
278	16 35	020	36 48	084	42 52	135	25 42	179	63 27	248	15 54	277	48 15	317
279	16 44	019	37 15	083	43 11	134	25 43	178	63 02	247	15 27	276	47 56	315
280	16 53	018	37 42	082	43 31	134	25 44	177	62 37	247	15 00	275	47 36	314
281	17 01	017	38 09	081	43 50	133	25 46	177	62 12	246	14 33	274	47 16	313
282	17 09	016	38 36	080	44 10	133	25 47	176	61 47	246	14 06	273	46 56	312
283	17 16	015	39 03	079	44 31	132	25 49	175	61 22	245	13 39	272	46 36	310
284	17 23	014	39 30	078	44 51	131	25 52	175	60 58	245	13 11	271	46 15	309
	◆ALTAIR		FOMALHAUT		◆ACHERNAR		CANOPUS		◆RIGIL KENT.		ANTARES		Nunki	
285	17 29	013	39 56	077	45 11	131	25 55	174	60 33	244	45 54	308	53 18	358
286	17 35	012	40 23	076	45 32	130	25 58	173	60 09	244	45 32	307	53 16	357
287	17 41	011	40 49	075	45 53	130	26 01	173	59 44	243	45 10	306	53 14	355
288	17 46	010	41 15	074	46 14	129	26 05	172	59 20	243	44 48	305	53 12	354
289	17 50	009	41 42	073	46 35	128	26 09	171	58 56	242	44 25	303	53 08	352
290	17 54	008	42 08	072	46 57	128	26 13	171	58 32	242	44 02	302	53 04	351
291	17 58	007	42 33	071	47 18	127	26 18	170	58 08	241	43 39	301	52 59	349
292	18 01	006	42 59	070	47 40	127	26 23	169	57 44	240	43 16	300	52 54	348
293	18 03	005	43 25	069	48 02	126	26 28	169	57 21	240	42 52	299	52 48	346
294	18 05	004	43 50	068	48 24	125	26 33	168	56 57	239	42 28	298	52 41	345
295	18 07	003	44 15	067	48 46	125	26 39	167	56 34	239	42 04	297	52 34	343
296	18 08	002	44 40	066	49 09	124	26 45	167	56 11	238	41 39	296	52 25	342
297	18 09	001	45 05	065	49 31	124	26 52	166	55 47	238	41 15	295	52 17	340
298	18 09	000	45 29	064	49 54	123	26 59	165	55 24	237	40 50	294	52 07	339
299	18 08	359	45 54	063	50 17	122	27 06	165	55 02	237	40 25	293	51 57	337
	FOMALHAUT		◆ACHERNAR		CANOPUS		◆RIGIL KENT.		ANTARES		Nunki		◆ALTAIR	
300	46 18	062	50 40	122	27 13	164	54 39	236	40 00	292	51 46	336	18 07	358
301	46 42	060	51 03	121	27 21	163	54 16	236	39 34	291	51 35	335	18 06	357
302	47 05	059	51 27	121	27 29	163	53 54	235	39 09	289	51 23	333	18 04	355
303	47 28	058	51 50	120	27 37	162	53 32	235	38 43	288	51 10	332	18 02	354
304	47 51	057	52 14	119	27 46	161	53 10	234	38 17	287	50 57	330	17 59	353
305	48 14	056	52 38	119	27 55	161	52 48	233	37 51	286	50 43	329	17 55	352
306	48 37	055	53 02	118	28 04	160	52 26	233	37 25	285	50 29	328	17 51	351
307	48 59	053	53 26	118	28 14	159	52 04	232	36 58	285	50 14	326	17 47	350
308	49 20	052	53 50	117	28 23	159	51 43	232	36 32	284	49 59	325	17 42	349
309	49 42	051	54 14	116	28 33	158	51 21	231	36 05	283	49 43	324	17 37	348
310	50 03	050	54 39	116	28 44	157	51 00	231	35 39	282	49 27	322	17 31	347
311	50 23	049	55 03	115	28 55	157	50 39	230	35 12	281	49 10	321	17 25	346
312	50 43	047	55 28	115	29 06	156	50 18	230	34 45	280	48 53	320	17 18	345
313	51 03	046	55 53	114	29 17	155	49 58	229	34 18	279	48 35	319	17 11	344
314	51 23	045	56 18	113	29 28	155	49 37	229	33 51	278	48 17	317	17 03	343
	◆FOMALHAUT		ACHERNAR		◆CANOPUS		RIGIL KENT.		◆ANTARES		Nunki		ALTAIR	
315	51 42	043	56 43	113	29 40	154	49 17	228	33 24	277	47 58	316	16 55	342
316	52 00	042	57 08	112	29 52	153	48 56	228	32 57	276	47 39	315	16 46	341
317	52 18	041	57 33	111	30 05	153	48 36	227	32 30	275	47 19	314	16 37	340
318	52 36	039	57 59	111	30 17	152	48 17	226	32 03	274	46 59	312	16 28	339
319	52 53	038	58 24	110	30 30	151	47 57	226	31 36	273	46 39	311	16 18	338
320	53 09	037	58 50	110	30 43	151	47 37	225	31 09	272	46 18	310	16 07	337
321	53 25	035	59 15	109	30 57	150	47 18	225	30 41	271	45 57	309	15 56	336
322	53 41	034	59 41	108	31 11	149	46 59	224	30 14	271	45 36	308	15 45	335
323	53 56	032	60 07	108	31 25	149	46 40	224	29 47	270	45 14	306	15 33	334
324	54 10	031	60 33	107	31 39	148	46 21	223	29 20	269	44 52	305	15 21	333
325	54 24	030	60 59	106	31 54	147	46 03	223	28 53	268	44 29	304	15 09	332
326	54 37	028	61 25	106	32 08	147	45 44	222	28 25	267	44 07	303	14 56	331
327	54 49	027	61 52	105	32 23	146	45 26	222	27 58	266	43 44	302	14 42	330
328	55 01	025	62 18	104	32 39	145	45 08	221	27 31	265	43 20	301	14 28	329
329	55 13	024	62 45	104	32 54	145	44 50	221	27 04	264	42 57	300	14 14	328
	◆FOMALHAUT		Diphda		ACHERNAR		◆CANOPUS		RIGIL KENT.		◆ANTARES		Nunki	
330	55 23	022	37 02	051	63 11	103	33 10	144	44 33	220	26 37	264	42 33	298
331	55 33	021	37 23	050	63 38	102	33 26	143	44 15	219	26 10	263	42 09	297
332	55 43	019	37 43	049	64 04	102	33 43	143	43 58	219	25 43	262	41 44	296
333	55 51	018	38 04	048	64 31	101	33 59	142	43 41	218	25 16	261	41 20	295
334	55 59	016	38 24	047	64 58	100	34 16	142	43 24	218	24 49	260	40 55	294
335	56 06	015	38 43	046	65 25	099	34 33	141	43 08	217	24 22	259	40 30	293
336	56 13	013	39 03	044	65 51	099	34 50	140	42 52	217	23 55	258	40 05	292
337	56 19	012	39 21	043	66 18	098	35 08	140	42 35	216	23 29	258	39 40	291
338	56 24	010	39 40	042	66 45	097	35 26	139	42 19	216	23 02	257	39 14	290
339	56 28	008	39 58	041	67 12	096	35 44	138	42 04	215	22 36	256	38 49	289
340	56 32	007	40 16	040	67 40	096	36 02	138	41 48	214	22 09	255	38 23	288
341	56 35	005	40 33	039	68 07	095	36 20	137	41 33	214	21 43	254	37 57	287
342	56 37	004	40 49	037	68 34	094	36 39	136	41 18	213	21 17	253	37 31	286
343	56 38	002	41 06	036	69 01	093	36 58	136	41 03	213	20 51	252	37 04	285
344	56 39	001	41 22	035	69 28	092	37 17	135	40 48	212	20 25	252	36 38	284
	◆Diphda		Acamar		◆CANOPUS		RIGIL KENT.		◆ANTARES		Nunki		FOMALHAUT	
345	41 37	034	48 47	086	37 37	134	40 34	212	19 59	251	36 12	283	56 39	359
346	41 52	032	49 14	085	37 56	134	40 20	211	19 34	250	35 45	282	56 38	357
347	42 06	031	49 41	084	38 16	133	40 06	211	19 08	249	35 18	281	56 36	356
348	42 20	030	50 08	083	38 36	132	39 52	210	18 43	248	34 52	280	56 34	354
349	42 33	029	50 35	082	38 56	132	39 39	209	18 17	247	34 25	279	56 31	353
350	42 46	027	51 02	081	39 17	131	39 25	209	17 52	247	33 58	278	56 27	351
351	42 58	026	51 29	080	39 37	131	39 12	208	17 27	246	33 31	277	56 22	350
352	43 10	025	51 55	079	39 58	130	39 00	208	17 03	245	33 04	276	56 17	348
353	43 21	024	52 22	078	40 19	129	38 47	207	16 38	244	32 37	276	56 11	346
354	43 32	022	52 49	077	40 40	129	38 35	207	16 14	243	32 10	275	56 04	345
355	43 42	021	53 15	076	41 02	128	38 23	206	15 49	242	31 42	274	55 57	343
356	43 52	020	53 42	075	41 23	127	38 11	205	15 25	242	31 15	273	55 49	342
357	44 00	018	54 08	074	41 45	127	37 59	205	15 02	241	30 48	272	55 40	340
358	44 09	017	54 34	073	42 07	126	37 48	204	14 38	240	30 21	271	55 30	339
359	44 16	016	55 00	071	42 29	125	37 37	204	14 14	239	29 54	270	55 20	337

 LAT 64°S

LHA	Hc	Zn	Hc	Zn	Hc	Zn	Hc	Zn	Hc	Zn	Hc	Zn	Hc	Zn
♈	Diphda		◆RIGEL		CANOPUS		Miaplacidus		◆RIGIL KENT.		Peacock		◆FOMALHAUT	
0	43 25	014	12 22	083	43 25	124	46 50	160	38 21	203	63 23	260	54 14	336
1	43 32	013	12 48	082	43 47	123	46 59	160	38 11	203	62 57	260	54 03	335
2	43 37	012	13 14	081	44 09	123	47 08	159	38 01	202	62 31	259	53 52	333
3	43 42	010	13 40	081	44 31	122	47 18	159	37 51	202	62 05	258	53 40	332
4	43 47	009	14 06	080	44 54	121	47 27	158	37 41	201	61 40	257	53 27	331
5	43 51	008	14 32	079	45 16	121	47 37	158	37 32	201	61 14	257	53 14	329
6	43 54	006	14 58	078	45 39	120	47 47	158	37 23	200	60 48	256	53 00	328
7	43 56	005	15 24	077	46 02	119	47 57	157	37 14	199	60 23	255	52 46	326
8	43 58	004	15 49	076	46 25	119	48 07	157	37 05	199	59 57	255	52 31	325
9	44 00	002	16 15	075	46 48	118	48 18	156	36 57	198	59 32	254	52 16	324
10	44 01	001	16 40	074	47 11	117	48 29	156	36 49	198	59 07	253	52 00	322
11	44 01	000	17 05	073	47 35	116	48 40	155	36 41	197	58 42	253	51 44	321
12	44 00	358	17 30	072	47 59	116	48 51	155	36 33	196	58 17	252	51 27	320
13	43 59	357	17 55	071	48 22	115	49 02	154	36 26	196	57 52	251	51 10	318
14	43 58	356	18 20	070	48 46	114	49 13	154	36 19	195	57 27	251	50 52	317
	◆RIGEL		SIRIUS		CANOPUS		◆RIGIL KENT.		Peacock		◆FOMALHAUT		Diphda	
15	18 45	069	16 37	094	49 10	114	36 12	195	57 02	250	50 34	316	43 56	354
16	19 09	068	17 03	093	49 34	113	36 06	194	56 38	249	50 15	314	43 53	353
17	19 34	067	17 30	092	49 59	112	35 59	194	56 13	249	49 56	313	43 49	352
18	19 58	067	17 56	091	50 23	112	35 53	193	55 49	248	49 37	312	43 45	351
19	20 22	066	18 22	091	50 47	111	35 48	192	55 24	247	49 17	311	43 41	349
20	20 46	065	18 49	090	51 12	110	35 42	192	55 00	247	48 57	310	43 35	348
21	21 10	064	19 15	089	51 37	110	35 37	191	54 36	246	48 37	308	43 30	347
22	21 33	063	19 41	088	52 02	109	35 32	191	54 12	245	48 16	307	43 23	345
23	21 56	062	20 07	087	52 27	108	35 27	190	53 48	245	47 55	306	43 16	344
24	22 19	061	20 34	086	52 52	107	35 23	189	53 25	244	47 33	305	43 09	343
25	22 42	060	21 00	085	53 17	107	35 19	189	53 01	243	47 12	304	43 01	341
26	23 05	059	21 26	084	53 42	106	35 15	188	52 38	243	46 49	302	42 52	340
27	23 27	058	21 52	083	54 07	105	35 11	188	52 14	242	46 27	301	42 43	339
28	23 49	057	22 18	082	54 33	104	35 08	187	51 51	241	46 05	300	42 33	338
29	24 11	056	22 44	081	54 58	104	35 05	186	51 28	241	45 42	299	42 23	336
	◆RIGEL		SIRIUS		CANOPUS		◆RIGIL KENT.		Peacock		◆FOMALHAUT		Diphda	
30	24 33	055	23 10	081	55 24	103	35 02	186	51 05	240	45 19	298	42 12	335
31	24 54	054	23 36	080	55 50	102	34 59	185	50 42	240	44 55	297	42 01	334
32	25 15	053	24 02	079	56 15	101	34 57	185	50 20	239	44 32	296	41 49	333
33	25 36	052	24 28	078	56 41	101	34 55	184	49 57	238	44 08	295	41 36	331
34	25 56	051	24 53	077	57 07	100	34 54	183	49 35	238	43 44	294	41 24	330
35	26 16	050	25 19	076	57 33	099	34 52	183	49 13	237	43 20	292	41 10	329
36	26 36	048	25 44	075	57 59	098	34 51	182	48 51	236	42 55	291	40 56	328
37	26 56	047	26 10	074	58 25	097	34 50	182	48 29	236	42 31	290	40 42	326
38	27 15	046	26 35	073	58 51	097	34 49	181	48 07	235	42 06	289	40 27	325
39	27 34	045	27 00	072	59 17	096	34 49	180	47 46	235	41 41	288	40 12	324
40	27 52	044	27 25	071	59 44	095	34 49	180	47 25	234	41 16	287	39 56	323
41	28 11	043	27 50	070	60 10	094	34 49	179	47 03	233	40 51	286	39 40	322
42	28 28	042	28 15	069	60 36	093	34 50	179	46 42	233	40 25	285	39 24	320
43	28 46	041	28 39	068	61 02	092	34 50	178	46 22	232	40 00	284	39 07	319
44	29 03	040	29 03	067	61 29	092	34 51	178	46 01	231	39 34	283	38 49	318
	RIGEL		◆SIRIUS		CANOPUS		◆RIGIL KENT.		Peacock		FOMALHAUT		◆Diphda	
45	29 20	039	29 27	066	61 55	091	34 53	177	45 40	231	39 09	282	38 32	317
46	29 36	038	29 51	065	62 21	090	34 54	176	45 20	230	38 43	281	38 13	316
47	29 52	037	30 15	064	62 47	089	34 56	176	45 00	230	38 17	280	37 55	315
48	30 07	036	30 39	063	63 14	088	34 58	175	44 40	229	37 51	279	37 36	314
49	30 23	034	31 02	062	63 40	087	35 00	175	44 20	228	37 25	279	37 17	312
50	30 37	033	31 25	061	64 06	086	35 03	174	44 01	228	36 59	278	36 57	311
51	30 51	032	31 48	060	64 33	085	35 06	173	43 41	227	36 33	277	36 37	310
52	31 05	031	32 11	059	64 59	084	35 09	173	43 22	227	36 07	276	36 17	309
53	31 19	030	32 33	058	65 25	083	35 13	172	43 03	226	35 41	275	35 56	308
54	31 32	029	32 55	057	65 51	082	35 16	172	42 44	225	35 15	274	35 36	307
55	31 44	028	33 17	056	66 17	081	35 20	171	42 26	225	34 48	273	35 14	306
56	31 56	027	33 39	055	66 43	080	35 25	170	42 07	224	34 22	272	34 53	305
57	32 08	025	34 00	054	67 09	079	35 29	170	41 49	223	33 56	271	34 31	304
58	32 19	024	34 21	053	67 35	078	35 34	169	41 31	223	33 30	270	34 09	303
59	32 29	023	34 42	052	68 00	077	35 39	169	41 13	222	33 03	269	33 47	302
	RIGEL		◆SIRIUS		Suhail		◆RIGIL KENT.		Peacock		FOMALHAUT		◆Diphda	
60	32 39	022	35 02	050	43 36	102	35 44	168	40 56	222	32 37	268	33 24	300
61	32 49	021	35 23	049	44 02	101	35 50	167	40 38	221	32 11	268	33 01	299
62	32 58	020	35 42	048	44 27	101	35 56	167	40 21	220	31 44	267	32 38	298
63	33 07	018	36 02	047	44 53	100	36 02	166	40 04	220	31 18	266	32 15	297
64	33 15	017	36 21	046	45 19	099	36 08	166	39 47	219	30 52	265	31 52	296
65	33 22	016	36 40	045	45 45	098	36 15	165	39 31	219	30 26	264	31 28	295
66	33 29	015	36 58	044	46 11	097	36 22	165	39 15	218	30 00	263	31 04	294
67	33 36	014	37 16	043	46 37	096	36 29	164	38 59	217	29 34	262	30 40	293
68	33 42	013	37 34	041	47 04	096	36 36	163	38 43	217	29 07	261	30 16	292
69	33 47	011	37 51	040	47 30	095	36 44	163	38 27	216	28 42	261	29 51	291
70	33 52	010	38 08	039	47 56	094	36 52	162	38 12	215	28 16	260	29 27	290
71	33 57	009	38 24	038	48 22	093	37 00	162	37 57	215	27 50	259	29 02	289
72	34 01	008	38 40	037	48 48	092	37 08	161	37 42	214	27 24	258	28 37	288
73	34 04	007	38 56	036	49 15	091	37 17	160	37 27	214	26 58	257	28 12	287
74	34 07	005	39 11	034	49 41	090	37 26	160	37 13	213	26 33	256	27 47	286
	RIGEL		◆SIRIUS		Suhail		◆RIGIL KENT.		Peacock		ACHERNAR		◆Diphda	
75	34 09	004	39 25	033	50 07	089	37 35	159	36 58	212	65 01	262	27 22	285
76	34 11	003	39 40	032	50 34	089	37 45	159	36 44	212	64 35	261	26 56	285
77	34 12	002	39 53	031	51 00	088	37 54	158	36 31	211	64 09	260	26 31	284
78	34 12	001	40 06	030	51 26	087	38 04	157	36 17	211	63 43	260	26 05	283
79	34 12	359	40 19	028	51 52	086	38 15	157	36 04	210	63 17	259	25 39	282
80	34 12	358	40 31	027	52 19	085	38 25	156	35 51	209	62 52	258	25 14	281
81	34 11	357	40 43	026	52 45	084	38 36	156	35 38	209	62 26	257	24 48	280
82	34 09	356	40 54	025	53 11	083	38 47	155	35 26	208	62 00	257	24 22	279
83	34 07	355	41 05	023	53 37	082	38 58	155	35 13	207	61 35	256	23 56	278
84	34 04	354	41 15	022	54 03	081	39 09	154	35 01	207	61 09	255	23 30	277
85	34 01	352	41 25	021	54 29	080	39 21	153	34 50	206	60 44	255	23 04	276
86	33 57	351	41 34	020	54 55	079	39 33	153	34 38	206	60 19	254	22 37	275
87	33 53	350	41 43	018	55 21	078	39 45	152	34 27	205	59 53	253	22 11	274
88	33 48	349	41 51	017	55 46	077	39 57	152	34 16	204	59 28	253	21 45	273
89	33 43	348	41 58	016	56 12	076	40 10	151	34 05	204	59 03	252	21 19	272

LHA	Hc	Zn	Hc	Zn	Hc	Zn	Hc	Zn	Hc	Zn	Hc	Zn	Hc	Zn
♈	◆SIRIUS		Suhail		◆RIGIL KENT.		Peacock		◆ACHERNAR		Acamar		RIGEL	
90	42 05	015	56 37	075	40 23	151	33 55	203	58 38	251	54 41	290	33 37	346
91	42 11	013	57 03	074	40 36	150	33 45	202	58 13	250	54 16	289	33 30	345
92	42 17	012	57 28	073	40 49	149	33 35	202	57 49	250	53 51	288	33 23	344
93	42 22	011	57 53	071	41 02	149	33 25	201	57 24	249	53 26	287	33 16	343
94	42 27	009	58 18	070	41 16	148	33 16	201	56 59	249	53 01	286	33 08	342
95	42 31	008	58 42	069	41 30	148	33 07	200	56 35	248	52 35	285	32 59	341
96	42 34	007	59 07	068	41 44	147	32 58	199	56 11	247	52 10	283	32 50	339
97	42 37	006	59 31	067	41 59	147	32 50	199	55 47	247	51 44	282	32 41	338
98	42 39	004	59 55	066	42 13	146	32 41	198	55 22	246	51 18	281	32 31	337
99	42 41	003	60 19	064	42 28	145	32 33	197	54 59	245	50 53	280	32 20	336
100	42 42	002	60 43	063	42 43	145	32 26	197	54 35	245	50 27	279	32 09	335
101	42 43	000	61 06	062	42 58	144	32 18	196	54 11	244	50 01	278	31 58	334
102	42 42	359	61 29	061	43 14	144	32 11	195	53 47	243	49 35	278	31 46	332
103	42 42	358	61 52	059	43 30	143	32 04	195	53 24	243	49 09	277	31 33	331
104	42 40	356	62 14	058	43 45	143	31 58	194	53 01	242	48 42	276	31 21	330
	Alphard		◆Gienah		RIGIL KENT.		◆Peacock		ACHERNAR		◆RIGEL		SIRIUS	
105	28 48	043	20 33	088	44 02	142	31 51	194	52 37	242	31 07	329	42 38	355
106	29 06	041	20 59	087	44 18	141	31 45	193	52 14	241	30 54	328	42 36	354
107	29 23	040	21 26	086	44 34	141	31 39	192	51 51	240	30 39	327	42 33	352
108	29 40	039	21 52	085	44 51	140	31 34	192	51 29	240	30 25	326	42 29	351
109	29 57	038	22 18	084	45 08	140	31 29	191	51 06	239	30 10	325	42 25	350
110	30 13	037	22 44	083	45 25	139	31 24	190	50 44	238	29 54	323	42 20	349
111	30 28	036	23 10	082	45 42	139	31 19	190	50 21	238	29 38	322	42 14	347
112	30 44	035	23 36	082	46 00	138	31 15	189	49 59	237	29 22	321	42 08	346
113	30 58	034	24 02	081	46 17	138	31 11	188	49 37	237	29 05	320	42 01	345
114	31 13	033	24 28	080	46 35	137	31 07	188	49 15	236	28 48	319	41 54	343
115	31 27	032	24 54	079	46 53	136	31 04	187	48 53	235	28 31	318	41 47	342
116	31 40	030	25 20	078	47 12	136	31 01	187	48 32	235	28 13	317	41 38	341
117	31 53	029	25 46	077	47 30	135	30 58	186	48 11	234	27 55	316	41 29	340
118	32 06	028	26 11	076	47 49	135	30 55	185	47 49	233	27 37	315	41 20	338
119	32 18	027	26 37	075	48 07	134	30 53	185	47 28	233	27 18	314	41 10	337
	Alphard		◆SPICA		RIGIL KENT.		◆Peacock		ACHERNAR		◆RIGEL		SIRIUS	
120	32 30	026	13 50	087	48 26	134	30 51	184	47 07	232	26 59	313	41 00	336
121	32 41	025	14 16	086	48 46	133	30 49	183	46 47	232	26 39	312	40 49	335
122	32 52	024	14 43	085	49 05	132	30 48	183	46 26	231	26 19	311	40 37	333
123	33 02	022	15 09	084	49 24	132	30 47	182	46 06	230	25 59	310	40 25	332
124	33 12	021	15 35	083	49 44	131	30 46	181	45 46	230	25 39	309	40 12	331
125	33 21	020	16 01	083	50 04	131	30 45	181	45 26	229	25 18	308	39 59	330
126	33 30	019	16 27	082	50 24	130	30 45	180	45 06	229	24 57	306	39 46	329
127	33 38	018	16 53	081	50 44	130	30 45	180	44 46	228	24 36	305	39 32	327
128	33 46	017	17 19	080	51 04	129	30 46	179	44 27	227	24 14	304	39 18	326
129	33 53	015	17 45	079	51 25	129	30 46	178	44 07	227	23 52	303	39 03	325
130	34 00	014	18 11	078	51 45	128	30 47	178	43 48	226	23 30	302	38 47	324
131	34 06	013	18 36	077	52 06	127	30 48	177	43 29	226	23 08	301	38 32	323
132	34 12	012	19 02	076	52 27	127	30 50	176	43 11	225	22 45	300	38 15	321
133	34 17	011	19 27	075	52 48	126	30 52	176	42 52	224	22 23	299	37 59	320
134	34 21	009	19 53	074	53 10	126	30 54	175	42 34	224	22 00	298	37 42	319
	REGULUS		◆SPICA		RIGIL KENT.		◆Peacock		ACHERNAR		CANOPUS		◆SIRIUS	
135	12 54	017	20 18	073	53 31	125	30 56	174	42 16	223	67 07	281	37 24	318
136	13 02	016	20 43	072	53 52	125	30 59	174	41 58	223	66 41	280	37 06	317
137	13 09	015	21 08	071	54 14	124	31 02	173	41 40	222	66 15	279	36 48	316
138	13 15	014	21 33	070	54 36	124	31 05	173	41 23	221	65 49	278	36 30	315
139	13 21	013	21 58	069	54 58	123	31 09	172	41 06	221	65 23	277	36 11	313
140	13 27	012	22 22	068	55 20	122	31 12	171	40 48	220	64 57	276	35 51	312
141	13 32	011	22 47	067	55 42	122	31 17	171	40 32	219	64 30	275	35 32	311
142	13 37	010	23 11	067	56 05	121	31 21	170	40 15	219	64 04	274	35 12	310
143	13 42	009	23 35	066	56 27	121	31 26	169	39 59	218	63 38	273	34 52	309
144	13 46	008	23 59	065	56 50	120	31 31	169	39 42	218	63 12	273	34 31	308
145	13 49	007	24 22	064	57 13	120	31 36	168	39 26	217	62 45	271	34 10	307
146	13 52	006	24 46	063	57 36	119	31 42	167	39 11	216	62 19	270	33 49	306
147	13 55	005	25 09	062	57 59	119	31 47	167	38 55	216	61 53	269	33 27	305
148	13 57	004	25 32	061	58 22	118	31 54	166	38 40	215	61 26	268	33 06	304
149	13 58	003	25 55	060	58 45	117	32 00	166	38 25	215	61 00	268	32 44	303
	REGULUS		◆SPICA		RIGIL KENT.		◆Peacock		ACHERNAR		CANOPUS		◆SIRIUS	
150	14 00	002	26 17	059	59 09	117	32 07	165	38 10	214	60 34	267	32 21	301
151	14 00	001	26 40	058	59 32	116	32 14	164	37 55	213	60 08	266	31 59	300
152	14 00	000	27 02	057	59 56	116	32 21	164	37 41	213	59 41	265	31 36	299
153	14 00	359	27 24	056	60 20	115	32 28	163	37 27	212	59 15	264	31 13	298
154	14 00	358	27 45	054	60 43	115	32 36	162	37 13	212	58 49	263	30 50	297
155	13 58	357	28 06	053	61 07	114	32 44	162	36 59	211	58 23	262	30 26	296
156	13 57	356	28 27	052	61 31	113	32 53	161	36 46	210	57 57	262	30 02	295
157	13 55	355	28 48	051	61 56	113	33 01	161	36 33	210	57 31	261	29 39	294
158	13 52	354	29 08	050	62 20	112	33 10	160	36 20	209	57 05	260	29 15	293
159	13 49	353	29 29	049	62 44	112	33 19	159	36 07	208	56 39	259	28 50	292
160	13 46	352	29 48	048	63 09	111	33 29	159	35 55	208	56 13	259	28 26	291
161	13 42	351	30 08	047	63 33	110	33 39	158	35 43	207	55 48	258	28 01	290
162	13 38	350	30 27	046	63 58	110	33 49	157	35 31	207	55 22	257	27 37	289
163	13 33	349	30 46	045	64 23	109	33 59	157	35 19	206	54 56	256	27 12	288
164	13 27	348	31 04	044	64 48	109	34 09	156	35 08	205	54 31	256	26 47	287
	◆SPICA		ANTARES		◆Peacock		ACHERNAR		CANOPUS		◆SIRIUS		Alphard	
165	31 22	043	26 55	096	34 20	155	34 57	205	54 05	255	26 22	286	32 14	333
166	31 40	042	27 21	095	34 31	155	34 46	204	53 40	254	25 56	286	32 02	331
167	31 57	041	27 47	094	34 43	154	34 35	204	53 15	253	25 31	285	31 49	330
168	32 14	039	28 14	093	34 54	154	34 25	203	52 50	253	25 05	284	31 36	329
169	32 30	038	28 40	092	35 06	153	34 15	202	52 25	252	24 40	283	31 22	328
170	32 47	037	29 06	091	35 18	152	34 05	202	52 00	251	24 14	282	31 08	327
171	33 02	036	29 33	090	35 30	152	33 55	201	51 35	250	23 48	281	30 54	326
172	33 18	035	29 59	089	35 43	151	33 46	200	51 10	250	23 22	280	30 39	325
173	33 32	034	30 25	088	35 56	150	33 37	200	50 46	249	22 56	279	30 23	324
174	33 47	033	30 51	088	36 09	150	33 28	199	50 21	248	22 30	278	30 07	323
175	34 01	032	31 18	087	36 22	149	33 20	199	49 57	248	22 04	277	29 51	321
176	34 14	030	31 44	086	36 36	149	33 12	198	49 32	247	21 38	276	29 35	320
177	34 27	029	32 10	085	36 50	148	33 04	197	49 08	246	21 12	275	29 18	319
178	34 40	028	32 36	084	37 04	147	32 56	197	48 44	246	20 46	274	29 00	318
179	34 52	027	33 02	083	37 18	147	32 49	196	48 20	245	20 20	273	28 42	317

LAT 64°S

LAT 64°S

LHA ♈	Hc Zn	Hc Zn	Hc Zn	Hc Zn	Hc Zn	Hc Zn	Hc Zn
	SPICA	◆ANTARES	Peacock	◆ACHERNAR	CANOPUS	Suhail	◆Alphard
180	35 04 026	33 29 082	37 32 146	32 41 195	47 57 244	58 15 290	28 24 316
181	35 15 025	33 55 081	37 47 146	32 35 195	47 33 243	57 50 288	28 06 315
182	35 26 023	34 21 080	38 02 145	32 28 194	47 10 243	57 25 287	27 47 314
183	35 36 022	34 46 079	38 17 144	32 22 193	46 46 242	57 00 286	27 28 313
184	35 46 021	35 12 078	38 33 144	32 16 193	46 23 241	56 35 285	27 09 312
185	35 55 020	35 38 077	38 49 143	32 10 192	46 00 241	56 09 284	26 49 311
186	36 03 019	36 04 076	39 05 142	32 05 192	45 37 240	55 44 283	26 29 310
187	36 12 017	36 29 075	39 21 142	31 59 191	45 14 239	55 18 282	26 08 309
188	36 19 016	36 54 074	39 37 141	31 55 190	44 52 239	54 52 281	25 48 308
189	36 26 015	37 20 073	39 54 141	31 50 190	44 30 238	54 27 280	25 27 307
190	36 33 014	37 45 072	40 10 140	31 46 189	44 07 237	54 01 279	25 05 306
191	36 39 013	38 10 071	40 28 139	31 42 188	43 45 237	53 35 278	24 44 305
192	36 44 011	38 35 070	40 45 139	31 38 188	43 23 236	53 09 277	24 22 304
193	36 49 010	38 59 069	41 02 138	31 35 187	43 02 235	52 42 276	24 00 303
194	36 53 009	39 24 068	41 20 138	31 31 187	42 40 235	52 16 275	23 38 302
	SPICA	◆ANTARES	Peacock	◆ACHERNAR	CANOPUS	Suhail	◆Alphard
195	36 57 008	39 48 067	41 38 137	31 28 186	42 19 234	51 50 274	23 15 301
196	37 00 006	40 12 066	41 56 136	31 26 185	41 57 233	51 24 273	22 52 300
197	37 03 005	40 36 065	42 14 136	31 24 185	41 36 233	50 57 272	22 29 299
198	37 05 004	41 00 064	42 33 135	31 22 184	41 16 232	50 31 271	22 06 298
199	37 07 003	41 24 063	42 51 134	31 20 183	40 55 231	50 05 270	21 43 297
200	37 08 002	41 47 062	43 10 134	31 19 183	40 35 231	49 39 270	21 19 296
201	37 08 000	42 10 061	43 29 133	31 17 182	40 14 230	49 12 269	20 55 295
202	37 08 359	42 33 060	43 48 133	31 17 182	39 54 229	48 46 268	20 31 294
203	37 07 358	42 55 059	44 08 132	31 16 181	39 34 229	48 20 267	20 07 293
204	37 06 357	43 18 058	44 28 131	31 16 180	39 15 228	47 53 266	19 43 292
205	37 04 355	43 40 056	44 47 131	31 16 180	38 55 227	47 27 265	19 18 291
206	37 02 354	44 02 055	45 07 130	31 16 179	38 36 227	47 01 264	18 54 290
207	36 59 353	44 23 054	45 28 130	31 17 178	38 17 226	46 35 263	18 29 289
208	36 55 352	44 44 053	45 48 129	31 18 178	37 58 225	46 09 263	18 04 288
209	36 51 350	45 05 052	46 09 128	31 19 177	37 39 225	45 43 262	17 39 287
	ANTARES	◆Nunki	Peacock	ACHERNAR	◆CANOPUS	Suhail	◆SPICA
210	45 25 051	30 33 088	46 29 128	31 20 176	37 21 224	45 17 261	36 46 349
211	45 46 049	30 59 087	46 50 127	31 22 176	37 03 223	44 51 260	36 41 348
212	46 05 048	31 26 086	47 11 126	31 24 175	36 45 223	44 25 259	36 36 347
213	46 25 047	31 52 085	47 33 126	31 26 175	36 27 222	43 59 258	36 29 346
214	46 44 046	32 18 084	47 54 125	31 29 174	36 09 221	43 33 258	36 22 344
215	47 03 045	32 44 083	48 15 125	31 32 173	35 52 221	43 08 257	36 15 343
216	47 21 043	33 10 082	48 37 124	31 35 173	35 35 220	42 42 256	36 07 342
217	47 39 042	33 36 081	48 59 123	31 39 172	35 18 220	42 17 255	35 59 341
218	47 56 041	34 02 080	49 21 123	31 43 171	35 02 219	41 51 254	35 50 340
219	48 13 040	34 28 080	49 43 122	31 47 171	34 45 218	41 26 254	35 40 338
220	48 30 038	34 54 079	50 06 121	31 51 170	34 29 218	41 01 253	35 31 337
221	48 46 037	35 20 078	50 28 121	31 56 170	34 13 217	40 36 252	35 20 336
222	49 01 036	35 45 077	50 51 120	32 01 169	33 57 216	40 11 251	35 09 335
223	49 16 034	36 11 076	51 14 120	32 06 168	33 42 216	39 46 251	34 58 334
224	49 31 033	36 36 075	51 37 119	32 11 168	33 27 215	39 21 250	34 46 332
	ANTARES	◆Nunki	FOMALHAUT	ACHERNAR	◆CANOPUS	Suhail	◆SPICA
225	49 45 032	37 02 074	14 57 128	32 17 167	33 12 214	38 57 249	34 33 331
226	49 58 030	37 27 073	15 18 128	32 23 166	32 57 214	38 32 248	34 21 330
227	50 12 029	37 52 072	15 39 127	32 30 166	32 43 213	38 08 247	34 07 329
228	50 24 028	38 17 071	16 00 126	32 36 165	32 29 212	37 43 247	33 53 328
229	50 36 026	38 42 070	16 21 125	32 43 164	32 15 212	37 19 246	33 39 327
230	50 47 025	39 06 069	16 43 124	32 50 164	32 01 211	36 55 245	33 25 326
231	50 58 023	39 31 068	17 05 123	32 58 163	31 48 210	36 32 244	33 09 324
232	51 08 022	39 55 067	17 27 123	33 05 163	31 35 210	36 08 244	32 54 323
233	51 18 021	40 19 066	17 49 122	33 13 162	31 22 209	35 45 243	32 38 322
234	51 27 019	40 43 064	18 11 121	33 22 161	31 09 208	35 21 242	32 22 321
235	51 35 018	41 06 063	18 34 120	33 30 161	30 57 208	34 58 241	32 05 320
236	51 43 016	41 30 062	18 57 119	33 39 160	30 45 207	34 35 241	31 48 319
237	51 50 015	41 53 061	19 20 118	33 48 159	30 33 206	34 12 240	31 30 318
238	51 56 014	42 16 060	19 43 118	33 58 159	30 22 206	33 50 239	31 12 317
239	52 02 012	42 39 059	20 07 117	34 07 158	30 10 205	33 27 238	30 54 316
	◆ANTARES	Nunki	◆FOMALHAUT	ACHERNAR	CANOPUS	◆ACRUX	SPICA
240	52 08 011	43 01 058	20 30 116	34 17 158	29 59 204	66 52 248	30 36 314
241	52 12 009	43 23 057	20 54 115	34 27 157	29 49 204	66 28 247	30 17 313
242	52 16 008	43 45 056	21 18 114	34 38 156	29 38 203	66 04 247	29 57 312
243	52 19 006	44 07 055	21 42 113	34 48 156	29 28 202	65 40 246	29 38 311
244	52 22 005	44 28 053	22 06 113	34 59 155	29 18 202	65 16 246	29 18 310
245	52 24 003	44 49 052	22 30 112	35 10 154	29 09 201	64 52 245	28 58 309
246	52 25 002	45 09 051	22 55 111	35 22 154	29 00 200	64 28 245	28 37 308
247	52 25 000	45 30 050	23 19 110	35 34 153	28 51 200	64 04 244	28 16 307
248	52 25 359	45 50 049	23 44 109	35 46 153	28 42 199	63 41 244	27 55 306
249	52 24 357	46 09 048	24 09 109	35 58 152	28 34 198	63 17 243	27 34 305
250	52 23 356	46 28 046	24 34 108	36 10 151	28 26 198	62 54 243	27 12 304
251	52 21 355	46 47 045	24 59 107	36 23 151	28 18 197	62 30 242	26 50 303
252	52 18 353	47 06 044	25 24 106	36 36 150	28 10 196	62 07 242	26 28 302
253	52 14 352	47 24 043	25 50 105	36 49 150	28 03 196	61 44 241	26 05 301
254	52 10 350	47 41 041	26 15 104	37 03 149	27 56 195	61 21 241	25 43 300
	Nunki	◆FOMALHAUT	ACHERNAR	CANOPUS	◆ACRUX	SPICA	◆ANTARES
255	47 58 040	26 41 103	37 16 148	27 50 194	60 58 240	25 20 299	52 05 349
256	48 15 039	27 06 103	37 30 148	27 43 194	60 35 240	24 57 298	52 00 347
257	48 31 038	27 32 102	37 44 147	27 37 193	60 13 239	24 33 297	51 54 346
258	48 47 036	27 58 101	37 59 146	27 32 192	59 50 239	24 10 296	51 47 344
259	49 03 035	28 24 100	38 14 146	27 26 191	59 28 238	23 46 295	51 39 343
260	49 17 034	28 50 099	38 28 145	27 21 191	59 05 238	23 22 294	51 31 342
261	49 32 032	29 16 098	38 43 145	27 16 190	58 43 237	22 58 293	51 23 340
262	49 45 031	29 42 097	38 59 144	27 12 189	58 21 237	22 34 292	51 14 339
263	49 59 030	30 08 097	39 14 143	27 08 189	57 59 236	22 09 291	51 04 337
264	50 11 028	30 34 096	39 30 143	27 04 188	57 38 236	21 44 290	50 53 336
265	50 24 027	31 00 095	39 46 142	27 00 187	57 16 235	21 20 289	50 42 335
266	50 35 025	31 26 094	40 02 142	26 57 187	56 54 235	20 55 288	50 31 333
267	50 46 024	31 53 093	40 19 141	26 54 186	56 33 234	20 30 287	50 18 332
268	50 57 023	32 19 092	40 36 140	26 51 185	56 12 234	20 04 286	50 06 330
269	51 06 021	32 45 091	40 52 140	26 49 185	55 51 233	19 39 285	49 52 329

LHA ♈	Hc Zn	Hc Zn	Hc Zn	Hc Zn	Hc Zn	Hc Zn	Hc Zn
	◆ALTAIR	FOMALHAUT	◆ACHERNAR	CANOPUS	◆RIGIL KENT.	SPICA	ANTARES
270	14 12 028	33 11 090	41 10 139	26 47 184	67 09 255	19 14 284	49 39 328
271	14 24 027	33 38 090	41 27 139	26 45 183	66 44 254	18 48 283	49 24 326
272	14 36 026	34 04 089	41 44 138	26 44 183	66 19 254	18 23 283	49 10 325
273	14 48 025	34 30 088	42 02 137	26 43 182	65 54 253	17 57 282	48 54 324
274	14 59 024	34 57 087	42 20 137	26 42 181	65 28 252	17 31 281	48 38 322
275	15 09 023	35 23 086	42 38 136	26 42 181	65 03 252	17 05 280	48 22 321
276	15 19 022	35 49 085	42 57 135	26 42 180	64 39 251	16 39 279	48 05 320
277	15 29 021	36 15 084	43 15 135	26 42 179	64 14 250	16 13 278	47 48 319
278	15 38 020	36 41 083	43 34 134	26 42 179	63 49 250	15 47 277	47 31 317
279	15 47 019	37 07 082	43 53 134	26 43 178	63 24 249	15 21 276	47 13 316
280	15 56 018	37 34 081	44 12 133	26 44 177	63 00 249	14 55 275	46 54 315
281	16 04 017	37 59 080	44 31 132	26 46 177	62 35 248	14 29 274	46 35 314
282	16 11 016	38 25 079	44 51 132	26 47 176	62 11 248	14 02 273	46 16 312
283	16 18 015	38 51 078	45 10 131	26 49 175	61 47 247	13 36 272	45 57 311
284	16 25 014	39 17 077	45 30 131	26 52 175	61 23 246	13 10 272	45 37 310
	◆ALTAIR	FOMALHAUT	◆ACHERNAR	CANOPUS	◆RIGIL KENT.	ANTARES	Nunki
285	16 31 013	39 42 076	45 50 130	26 54 174	60 58 246	45 16 309	52 18 358
286	16 37 012	40 08 075	46 11 129	26 57 173	60 35 245	44 56 308	52 16 357
287	16 42 011	40 33 074	46 31 129	27 01 173	60 11 245	44 35 307	52 15 355
288	16 47 010	40 59 073	46 52 128	27 04 172	59 47 244	44 13 305	52 12 354
289	16 51 009	41 24 072	47 12 128	27 08 171	59 23 244	43 52 304	52 09 352
290	16 55 008	41 49 071	47 33 127	27 12 170	59 00 243	43 30 303	52 05 351
291	16 58 007	42 14 070	47 54 126	27 17 170	58 37 242	43 08 302	52 01 349
292	17 01 006	42 38 069	48 16 126	27 21 169	58 13 242	42 45 301	51 55 348
293	17 04 005	43 03 068	48 37 125	27 27 168	57 50 241	42 23 300	51 50 347
294	17 06 004	43 27 067	48 59 125	27 32 168	57 27 241	42 00 299	51 43 345
295	17 07 003	43 51 066	49 20 124	27 38 167	57 04 240	41 36 298	51 36 344
296	17 08 002	44 15 065	49 42 123	27 44 166	56 42 240	41 13 297	51 28 342
297	17 09 001	44 39 064	50 04 123	27 50 166	56 19 239	40 49 295	51 20 341
298	17 09 000	45 02 063	50 27 122	27 57 165	55 56 238	40 26 294	51 11 339
299	17 08 359	45 26 062	50 49 121	28 04 164	55 34 238	40 01 293	51 01 338
	FOMALHAUT	◆ACHERNAR	CANOPUS	◆RIGIL KENT.	ANTARES	Nunki	◆ALTAIR
300	45 49 061	51 11 121	28 11 164	55 12 237	39 37 292	50 51 337	17 07 358
301	46 12 060	51 34 120	28 18 163	54 50 237	39 13 291	50 40 335	17 06 357
302	46 34 058	51 57 120	28 26 162	54 28 236	38 48 290	50 29 334	17 04 355
303	46 56 057	52 20 119	28 34 162	54 06 236	38 23 289	50 17 332	17 02 354
304	47 18 056	52 43 118	28 43 161	53 44 235	37 59 288	50 05 331	16 59 353
305	47 40 055	53 06 118	28 51 160	53 23 235	37 33 287	49 52 330	16 56 352
306	48 01 054	53 29 117	29 00 160	53 02 234	37 08 286	49 38 328	16 52 351
307	48 22 053	53 53 116	29 10 159	52 40 233	36 43 285	49 24 327	16 48 350
308	48 43 051	54 17 116	29 19 158	52 19 233	36 18 284	49 10 326	16 43 349
309	49 04 050	54 40 115	29 29 158	51 58 232	35 52 283	48 55 324	16 38 348
310	49 24 049	55 04 115	29 39 157	51 38 232	35 26 282	48 39 323	16 33 347
311	49 43 048	55 28 114	29 50 156	51 17 231	35 01 281	48 23 322	16 27 346
312	50 02 046	55 52 113	30 00 156	50 57 231	34 35 280	48 07 321	16 20 345
313	50 21 045	56 16 113	30 11 155	50 36 230	34 09 279	47 50 319	16 13 344
314	50 40 044	56 41 112	30 22 154	50 16 230	33 43 279	47 32 318	16 06 343
	◆FOMALHAUT	ACHERNAR	◆CANOPUS	RIGIL KENT.	◆ANTARES	Nunki	ALTAIR
315	50 58 043	57 05 111	30 34 154	49 56 229	33 17 278	47 14 317	15 58 342
316	51 15 041	57 30 111	30 46 153	49 37 228	32 51 277	46 56 316	15 50 341
317	51 32 040	57 54 110	30 58 152	49 17 228	32 25 276	46 38 314	15 41 340
318	51 49 039	58 19 109	31 10 152	48 58 227	31 58 275	46 19 313	15 32 339
319	52 05 037	58 44 109	31 23 151	48 38 227	31 32 274	45 59 312	15 22 338
320	52 21 036	59 09 108	31 36 150	48 19 226	31 06 273	45 39 311	15 12 337
321	52 36 035	59 34 107	31 49 150	48 00 226	30 40 272	45 19 309	15 02 336
322	52 51 033	59 59 107	32 02 149	47 42 225	30 13 271	44 59 308	14 51 335
323	53 05 032	60 25 106	32 16 148	47 23 225	29 47 270	44 38 307	14 39 334
324	53 18 030	60 50 105	32 30 148	47 05 224	29 21 269	44 17 306	14 28 333
325	53 31 029	61 15 105	32 44 147	46 47 223	28 54 268	43 55 305	14 16 332
326	53 44 028	61 41 104	32 58 146	46 29 223	28 28 268	43 34 304	14 03 331
327	53 56 026	62 06 103	33 13 146	46 11 222	28 02 267	43 12 303	13 50 330
328	54 07 025	62 32 102	33 28 145	45 53 222	27 36 266	42 49 301	13 37 329
329	54 18 023	62 58 102	33 43 144	45 36 221	27 09 265	42 27 300	13 23 328
	◆FOMALHAUT	Diphda	ACHERNAR	◆CANOPUS	RIGIL KENT.	◆ANTARES	Nunki
330	54 28 022	36 24 051	63 23 101	33 59 144	45 19 221	26 43 264	42 04 299
331	54 37 020	36 44 049	63 49 100	34 14 143	45 02 220	26 17 263	41 41 298
332	54 46 019	37 04 048	64 15 100	34 30 142	44 45 219	25 51 262	41 18 297
333	54 54 017	37 23 047	64 41 099	34 46 142	44 28 219	25 25 261	40 54 296
334	55 02 016	37 42 046	65 07 098	35 03 141	44 12 218	24 59 261	40 30 295
335	55 08 014	38 01 045	65 33 097	35 20 140	43 56 218	24 33 260	40 06 294
336	55 14 013	38 20 044	65 59 096	35 36 140	43 40 217	24 07 259	39 42 293
337	55 20 011	38 38 043	66 26 096	35 53 139	43 24 217	23 42 258	39 18 292
338	55 25 010	38 55 042	66 52 095	36 11 138	43 08 216	23 16 257	38 53 291
339	55 29 008	39 12 040	67 18 094	36 28 138	42 53 216	22 50 256	38 29 290
340	55 32 007	39 29 039	67 44 093	36 46 137	42 38 215	22 25 255	38 04 289
341	55 35 005	39 46 038	68 10 092	37 04 136	42 23 214	21 59 255	37 39 288
342	55 37 004	40 02 037	68 37 092	37 22 136	42 08 214	21 34 254	37 14 287
343	55 38 002	40 17 036	69 03 091	37 41 135	41 53 213	21 09 253	36 48 286
344	55 39 001	40 32 034	69 29 090	37 59 135	41 39 213	20 44 252	36 23 285
	◆Diphda	Acamar	◆CANOPUS	RIGIL KENT.	◆ANTARES	Nunki	FOMALHAUT
345	40 47 033	48 41 084	38 18 134	41 25 212	20 19 251	35 58 284	55 39 359
346	41 01 032	49 08 083	38 37 133	41 11 212	19 54 250	35 32 283	55 38 357
347	41 15 031	49 34 083	38 57 133	40 57 211	19 29 249	35 06 282	55 36 356
348	41 28 030	50 00 082	39 16 132	40 44 210	19 05 249	34 41 281	55 34 354
349	41 41 028	50 26 081	39 36 131	40 31 210	18 40 248	34 15 280	55 31 353
350	41 53 027	50 52 080	39 56 131	40 18 209	18 16 247	33 49 279	55 28 351
351	42 05 026	51 17 079	40 16 130	40 05 209	17 52 246	33 23 278	55 23 350
352	42 16 025	51 43 078	40 36 129	39 53 208	17 28 245	32 57 277	55 18 348
353	42 26 023	52 09 077	40 57 129	39 40 208	17 04 244	32 31 276	55 13 347
354	42 36 022	52 34 076	41 17 128	39 28 207	16 41 244	32 04 275	55 06 345
355	42 46 021	53 00 074	41 38 127	39 17 206	16 17 243	31 38 274	54 59 344
356	42 55 019	53 25 073	41 59 127	39 05 206	15 54 242	31 12 273	54 52 342
357	43 03 018	53 50 072	42 20 126	38 54 205	15 31 241	30 46 273	54 45 341
358	43 11 017	54 15 071	42 42 125	38 43 205	15 08 240	30 19 272	54 34 339
359	43 19 016	54 40 070	43 03 125	38 32 204	14 45 239	29 53 271	54 25 338

LAT 65°S

LHA	Hc	Zn	Hc	Zn	Hc	Zn	Hc	Zn	Hc	Zn	Hc	Zn	Hc	Zn
♈	Diphda		♦RIGEL		CANOPUS		Miaplacidus		♦RIGIL KENT.		Peacock		♦FOMALHAUT	
0	42 27	014	12 15	083	43 58	123	47 47	160	39 16	204	63 32	262	53 19	337
1	42 33	013	12 40	082	44 20	122	47 55	159	39 06	203	63 07	262	53 09	335
2	42 38	011	13 05	081	44 41	122	48 04	159	38 56	203	62 42	261	52 58	334
3	42 43	010	13 30	080	45 03	121	48 14	159	38 47	202	62 17	260	52 47	333
4	42 47	009	13 55	079	45 24	120	48 23	158	38 37	201	61 52	259	52 35	331
5	42 51	008	14 20	078	45 46	120	48 33	158	38 28	201	61 27	259	52 22	330
6	42 54	006	14 45	078	46 08	119	48 42	157	38 19	200	61 02	258	52 09	328
7	42 57	005	15 10	077	46 31	118	48 52	157	38 10	200	60 37	257	51 56	327
8	42 59	004	15 34	076	46 53	118	49 02	156	38 02	199	60 13	256	51 42	326
9	43 00	002	15 59	075	47 16	117	49 13	156	37 54	198	59 48	256	51 27	324
10	43 01	001	16 23	074	47 38	116	49 23	155	37 46	198	59 23	255	51 12	323
11	43 01	000	16 48	073	48 01	115	49 34	155	37 38	197	58 59	254	50 57	322
12	43 00	358	17 12	072	48 24	115	49 45	154	37 31	197	58 35	253	50 41	320
13	42 59	357	17 36	071	48 47	114	49 56	154	37 24	196	58 10	253	50 25	319
14	42 58	356	18 00	070	49 10	113	50 07	153	37 17	195	57 46	252	50 08	318
	♦RIGEL		SIRIUS		CANOPUS		♦RIGIL KENT.		Peacock		♦FOMALHAUT		Diphda	
15	18 24	069	16 41	094	49 34	113	37 10	195	57 22	251	49 51	317	42 56	355
16	18 47	068	17 07	093	49 57	112	37 04	194	56 58	251	49 33	315	42 53	353
17	19 11	067	17 32	092	50 21	111	36 58	194	56 34	250	49 15	314	42 50	352
18	19 34	066	17 57	091	50 45	110	36 52	193	56 11	249	48 57	313	42 46	351
19	19 57	065	18 23	090	51 08	110	36 46	192	55 47	249	48 38	312	42 42	349
20	20 20	064	18 48	089	51 32	109	36 41	192	55 23	248	48 19	310	42 37	348
21	20 43	063	19 13	088	51 56	108	36 36	191	55 00	247	47 59	309	42 31	347
22	21 05	062	19 39	088	52 20	108	36 31	191	54 37	247	47 39	308	42 25	346
23	21 28	061	20 04	087	52 45	107	36 26	190	54 13	246	47 19	307	42 19	344
24	21 50	060	20 29	086	53 09	106	36 22	189	53 50	245	46 59	306	42 12	343
25	22 12	059	20 55	085	53 33	105	36 18	189	53 27	245	46 38	304	42 04	342
26	22 33	058	21 20	084	53 58	105	36 14	188	53 05	244	46 17	303	41 56	340
27	22 55	057	21 45	083	54 22	104	36 11	188	52 42	243	45 56	302	41 47	339
28	23 16	056	22 10	082	54 47	103	36 07	187	52 19	243	45 34	301	41 38	338
29	23 37	055	22 35	081	55 12	102	36 04	186	51 57	242	45 12	300	41 28	337
	♦RIGEL		SIRIUS		CANOPUS		♦RIGIL KENT.		Peacock		♦FOMALHAUT		Diphda	
30	23 58	054	23 00	080	55 37	101	36 02	186	51 35	241	44 50	299	41 18	335
31	24 18	053	23 25	079	56 02	101	35 59	185	51 12	241	44 28	298	41 07	334
32	24 38	052	23 50	078	56 26	100	35 57	185	50 50	240	44 05	297	40 55	333
33	24 58	051	24 15	077	56 51	099	35 55	184	50 28	239	43 42	296	40 44	332
34	25 18	050	24 40	076	57 17	098	35 53	183	50 07	239	43 19	294	40 31	331
35	25 37	049	25 04	075	57 42	097	35 52	183	49 45	238	42 56	293	40 19	329
36	25 56	048	25 29	074	58 07	097	35 51	182	49 24	237	42 33	292	40 06	328
37	26 15	047	25 53	073	58 32	096	35 50	182	49 02	237	42 09	291	39 52	327
38	26 33	046	26 17	073	58 57	095	35 49	181	48 41	236	41 46	290	39 38	326
39	26 52	045	26 41	072	59 23	094	35 49	180	48 20	235	41 22	289	39 23	324
40	27 09	044	27 05	071	59 48	093	35 49	180	48 00	235	40 58	288	39 08	323
41	27 27	043	27 29	070	60 13	092	35 49	179	47 39	234	40 33	287	38 53	322
42	27 44	042	27 53	069	60 39	092	35 50	179	47 18	234	40 09	286	38 37	321
43	28 00	041	28 16	068	61 04	091	35 50	178	46 58	233	39 45	285	38 21	320
44	28 17	040	28 40	067	61 29	090	35 51	177	46 38	232	39 20	284	38 05	319
	RIGEL		♦SIRIUS		CANOPUS		♦RIGIL KENT.		Peacock		FOMALHAUT		♦Diphda	
45	28 33	039	29 03	066	61 55	089	35 52	177	46 18	232	38 56	283	37 48	317
46	28 48	037	29 26	065	62 20	088	35 54	176	45 58	231	38 31	282	37 30	316
47	29 04	036	29 49	064	62 45	087	35 56	176	45 38	230	38 06	281	37 13	315
48	29 19	035	30 11	063	63 11	086	35 58	175	45 19	230	37 41	280	36 55	314
49	29 33	034	30 34	062	63 36	085	36 00	174	45 00	229	37 16	279	36 36	313
50	29 47	033	30 56	061	64 01	084	36 03	174	44 41	229	36 51	278	36 17	312
51	30 01	032	31 18	059	64 26	083	36 06	173	44 22	228	36 26	277	35 58	311
52	30 14	031	31 40	058	64 51	082	36 09	173	44 03	227	36 01	276	35 39	310
53	30 27	030	32 01	057	65 17	081	36 12	172	43 45	227	35 36	275	35 19	309
54	30 39	029	32 22	056	65 42	080	36 16	171	43 26	226	35 10	275	34 59	307
55	30 51	027	32 43	055	66 06	079	36 20	171	43 08	225	34 45	274	34 39	306
56	31 02	026	33 04	054	66 31	078	36 24	170	42 50	225	34 20	273	34 18	305
57	31 13	025	33 25	053	66 56	077	36 28	170	42 32	224	33 54	272	33 58	304
58	31 24	024	33 45	052	67 21	075	36 33	169	42 15	223	33 29	271	33 37	303
59	31 34	023	34 05	051	67 45	074	36 38	168	41 57	223	33 04	270	33 15	302
	RIGEL		♦SIRIUS		Suhail		♦RIGIL KENT.		Peacock		FOMALHAUT		♦Diphda	
60	31 44	022	34 24	050	43 48	101	36 43	168	41 40	222	32 38	269	32 54	301
61	31 53	021	34 43	049	44 13	101	36 48	167	41 23	222	32 13	268	32 32	300
62	32 01	019	35 02	048	44 38	100	36 54	167	41 07	221	31 48	267	32 10	299
63	32 10	018	35 21	047	45 03	099	37 00	166	40 50	220	31 22	266	31 47	298
64	32 17	017	35 39	045	45 28	098	37 06	165	40 34	220	30 57	265	31 25	297
65	32 25	016	35 57	044	45 53	097	37 13	165	40 18	219	30 32	265	31 02	296
66	32 31	015	36 15	043	46 18	096	37 19	164	40 02	218	30 06	264	30 39	295
67	32 38	014	36 32	042	46 44	095	37 26	164	39 46	218	29 41	263	30 16	294
68	32 43	012	36 49	041	47 09	095	37 34	163	39 31	217	29 16	262	29 53	293
69	32 49	011	37 05	040	47 34	094	37 41	163	39 15	217	28 51	261	29 29	292
70	32 53	010	37 21	039	47 59	093	37 49	162	39 00	216	28 26	260	29 06	291
71	32 57	009	37 37	037	48 25	092	37 57	161	38 46	215	28 01	259	28 42	290
72	33 01	008	37 52	036	48 50	091	38 05	161	38 31	215	27 36	258	28 18	289
73	33 04	007	38 07	035	49 16	090	38 14	160	38 17	214	27 11	258	27 54	288
74	33 07	005	38 21	034	49 41	089	38 22	160	38 03	213	26 47	257	27 30	287
	RIGEL		♦SIRIUS		Suhail		♦RIGIL KENT.		Peacock		ACHERNAR		♦Diphda	
75	33 09	004	38 35	033	50 06	088	38 31	159	37 49	213	65 09	264	27 05	286
76	33 11	003	38 49	032	50 32	087	38 41	158	37 35	212	64 43	263	26 41	285
77	33 12	002	39 02	030	50 57	086	38 50	158	37 22	212	64 18	262	26 17	284
78	33 12	001	39 14	029	51 22	085	39 00	157	37 09	211	63 53	262	25 52	283
79	33 12	359	39 26	028	51 47	085	39 10	157	36 56	210	63 28	261	25 27	282
80	33 12	358	39 38	027	52 13	084	39 20	156	36 43	210	63 03	260	25 02	281
81	33 11	357	39 49	026	52 38	083	39 30	155	36 31	209	62 38	259	24 37	280
82	33 09	356	40 00	024	53 03	082	39 41	155	36 19	208	62 13	258	24 12	279
83	33 07	355	40 10	023	53 28	081	39 52	154	36 07	208	61 48	258	23 47	278
84	33 05	354	40 20	022	53 53	080	40 03	154	35 55	207	61 24	257	23 22	277
85	33 02	352	40 29	021	54 18	079	40 14	153	35 44	206	60 59	256	22 57	277
86	32 58	351	40 38	019	54 43	078	40 26	152	35 32	206	60 34	256	22 32	276
87	32 54	350	40 46	018	55 07	077	40 38	152	35 21	205	60 10	255	22 07	275
88	32 49	349	40 53	017	55 32	075	40 50	151	35 11	205	59 45	254	21 41	274
89	32 44	348	41 00	016	55 56	074	41 02	151	35 00	204	59 21	253	21 16	273

LHA	Hc	Zn	Hc	Zn	Hc	Zn	Hc	Zn	Hc	Zn	Hc	Zn	Hc	Zn
♈	♦SIRIUS		Suhail		♦RIGIL KENT.		Peacock		♦ACHERNAR		Acamar		RIGEL	
90	41 07	014	56 21	073	41 15	150	34 50	203	58 57	253	54 20	291	32 38	347
91	41 13	013	56 45	072	41 28	150	34 40	203	58 33	252	53 56	290	32 32	345
92	41 18	012	57 09	071	41 40	149	34 31	202	58 09	251	53 32	289	32 26	344
93	41 23	011	57 33	070	41 54	148	34 21	201	57 45	251	53 08	288	32 19	343
94	41 28	009	57 57	069	42 07	148	34 12	201	57 21	250	52 44	287	32 11	342
95	41 32	008	58 20	068	42 21	147	34 03	200	56 57	249	52 20	286	32 03	341
96	41 35	007	58 44	067	42 35	147	33 55	199	56 33	249	51 55	285	31 54	340
97	41 37	005	59 07	065	42 49	146	33 46	199	56 10	248	51 31	284	31 45	338
98	41 40	004	59 30	064	43 03	145	33 38	198	55 46	247	51 06	283	31 35	337
99	41 41	003	59 52	063	43 17	145	33 31	198	55 23	247	50 41	282	31 25	336
100	41 42	002	60 15	062	43 32	144	33 23	197	55 00	246	50 16	281	31 15	335
101	41 43	000	60 37	060	43 47	144	33 16	196	54 37	245	49 51	280	31 04	334
102	41 42	359	60 59	059	44 02	143	33 09	196	54 14	245	49 26	279	30 53	333
103	41 42	358	61 21	058	44 17	143	33 02	195	53 51	244	49 01	278	30 41	332
104	41 40	356	61 42	056	44 33	142	32 56	194	53 28	243	48 36	277	30 28	330
	Alphard		♦Gienah		RIGIL KENT.		♦Peacock		ACHERNAR		♦RIGEL		SIRIUS	
105	28 04	042	20 31	088	44 49	141	32 50	194	53 06	243	30 16	329	41 39	355
106	28 21	041	20 56	087	45 05	141	32 44	193	52 43	242	30 03	328	41 36	354
107	28 37	040	21 21	086	45 21	140	32 38	192	52 21	241	29 49	327	41 33	353
108	28 54	039	21 47	085	45 37	140	32 33	192	51 59	241	29 35	326	41 30	351
109	29 09	038	22 12	084	45 53	139	32 28	191	51 36	240	29 21	325	41 25	350
110	29 25	037	22 37	083	46 10	139	32 23	191	51 15	239	29 06	324	41 21	349
111	29 40	036	23 02	082	46 27	138	32 18	190	50 53	239	28 51	323	41 16	348
112	29 54	035	23 27	081	46 44	137	32 14	189	50 31	238	28 35	322	41 10	346
113	30 08	033	23 52	080	47 01	137	32 10	189	50 10	238	28 19	321	41 04	345
114	30 22	032	24 17	079	47 19	136	32 07	188	49 48	237	28 03	319	40 57	344
115	30 36	031	24 42	078	47 37	136	32 03	187	49 27	236	27 46	318	40 49	342
116	30 48	030	25 07	077	47 54	135	32 00	187	49 06	236	27 29	317	40 41	341
117	31 01	029	25 32	076	48 12	134	31 57	186	48 45	235	27 12	316	40 33	340
118	31 13	028	25 56	075	48 31	134	31 55	185	48 25	234	26 54	315	40 24	339
119	31 25	027	26 21	074	48 49	133	31 53	185	48 04	234	26 36	314	40 15	337
	Alphard		♦SPICA		RIGIL KENT.		♦Peacock		ACHERNAR		♦RIGEL		SIRIUS	
120	31 36	026	13 47	087	49 07	133	31 51	184	47 44	233	26 18	313	40 05	336
121	31 47	024	14 12	086	49 26	132	31 49	183	47 24	233	25 59	312	39 54	335
122	31 57	023	14 38	085	49 45	132	31 48	183	47 03	232	25 40	311	39 43	334
123	32 07	022	15 03	084	50 04	131	31 47	182	46 44	231	25 21	310	39 32	333
124	32 16	021	15 28	083	50 23	130	31 46	181	46 24	231	25 01	309	39 20	331
125	32 25	020	15 53	082	50 43	130	31 45	181	46 04	230	24 41	308	39 08	330
126	32 33	019	16 18	081	51 02	129	31 45	180	45 45	229	24 21	307	38 55	329
127	32 41	018	16 43	080	51 22	129	31 45	180	45 26	229	24 01	306	38 41	328
128	32 48	016	17 08	079	51 42	128	31 45	179	45 07	228	23 40	305	38 28	327
129	32 55	015	17 33	079	52 02	128	31 46	178	44 48	228	23 19	304	38 13	325
130	33 02	014	17 58	078	52 22	127	31 47	178	44 30	227	22 58	303	37 59	324
131	33 08	013	18 23	077	52 42	126	31 48	177	44 11	226	22 37	302	37 44	323
132	33 13	012	18 47	076	53 03	126	31 50	176	43 53	226	22 15	301	37 28	322
133	33 18	010	19 12	075	53 23	125	31 51	176	43 35	225	21 53	300	37 12	321
134	33 22	009	19 36	074	53 44	125	31 54	175	43 17	225	21 31	299	36 56	320
	REGULUS		♦SPICA		RIGIL KENT.		♦Peacock		ACHERNAR		CANOPUS		♦SIRIUS	
135	11 57	017	20 00	073	54 05	124	31 56	174	42 59	224	66 54	283	36 40	318
136	12 04	016	20 25	072	54 26	124	31 58	174	42 42	223	66 29	282	36 23	317
137	12 11	015	20 49	071	54 47	123	32 01	173	42 25	223	66 04	281	36 05	316
138	12 17	014	21 13	070	55 09	122	32 05	172	42 08	222	65 40	280	35 47	315
139	12 23	013	21 36	069	55 30	122	32 08	172	41 51	221	65 15	279	35 29	314
140	12 29	012	22 00	068	55 52	121	32 12	171	41 34	221	64 49	278	35 11	313
141	12 34	011	22 23	067	56 14	121	32 16	171	41 18	220	64 24	277	34 52	312
142	12 38	010	22 47	066	56 35	120	32 20	170	41 02	219	63 59	276	34 33	311
143	12 42	009	23 10	065	56 57	119	32 25	169	40 46	219	63 34	275	34 14	309
144	12 46	008	23 33	064	57 20	119	32 30	169	40 30	218	63 09	274	33 54	308
145	12 50	007	23 55	063	57 42	118	32 35	168	40 14	218	62 43	273	33 34	307
146	12 52	006	24 18	062	58 04	118	32 40	167	39 59	217	62 18	272	33 14	306
147	12 55	005	24 40	061	58 27	117	32 46	167	39 44	216	61 53	271	32 53	305
148	12 57	004	25 02	060	58 49	117	32 52	166	39 29	216	61 27	270	32 32	304
149	12 58	003	25 24	059	59 12	116	32 58	165	39 14	215	61 02	269	32 11	303
	REGULUS		♦SPICA		RIGIL KENT.		♦Peacock		ACHERNAR		CANOPUS		♦SIRIUS	
150	13 00	002	25 46	058	59 35	115	33 05	165	39 00	214	60 37	268	31 50	302
151	13 00	001	26 07	057	59 58	115	33 11	164	38 45	214	60 11	268	31 28	301
152	13 00	000	26 28	056	60 21	114	33 18	163	38 31	213	59 46	267	31 06	300
153	13 00	359	26 49	055	60 44	114	33 26	163	38 18	213	59 21	266	30 44	299
154	13 00	358	27 10	054	61 08	113	33 33	162	38 04	212	58 55	265	30 22	298
155	12 59	357	27 30	053	61 31	112	33 41	162	37 51	211	58 30	264	29 59	297
156	12 57	356	27 51	052	61 54	112	33 49	161	37 38	211	58 05	263	29 37	296
157	12 55	355	28 10	051	62 18	111	33 58	160	37 25	210	57 40	262	29 14	295
158	12 53	354	28 30	050	62 42	110	34 07	160	37 12	209	57 15	262	28 51	294
159	12 50	353	28 49	049	63 06	110	34 16	159	37 00	209	56 50	261	28 27	293
160	12 46	352	29 08	048	63 30	109	34 25	158	36 48	208	56 24	260	28 04	292
161	12 43	351	29 27	047	63 54	109	34 34	158	36 36	208	56 00	259	27 40	291
162	12 38	350	29 45	046	64 18	108	34 44	157	36 24	207	55 35	258	27 16	290
163	12 34	349	30 03	045	64 42	107	34 54	156	36 13	206	55 10	258	26 53	289
164	12 29	348	30 21	043	65 06	107	35 04	156	36 02	206	54 45	257	26 28	288
	♦SPICA		ANTARES		♦Peacock		ACHERNAR		CANOPUS		♦SIRIUS		Alphard	
165	30 38	042	27 01	095	35 15	155	35 51	205	54 20	256	26 04	287	31 21	333
166	30 55	041	27 26	094	35 25	155	35 40	204	53 56	255	25 40	286	31 09	332
167	31 11	040	27 51	093	35 37	154	35 30	204	53 31	255	25 16	285	30 57	331
168	31 28	039	28 17	092	35 48	153	35 20	203	53 07	254	24 51	284	30 44	330
169	31 43	038	28 42	092	35 59	153	35 10	203	52 43	253	24 26	283	30 31	328
170	31 59	037	29 07	091	36 11	152	35 01	202	52 18	252	24 02	282	30 18	327
171	32 14	036	29 33	090	36 23	151	34 51	201	51 54	252	23 37	281	30 04	326
172	32 28	035	29 58	089	36 35	151	34 42	201	51 30	251	23 12	280	29 49	325
173	32 43	033	30 23	088	36 48	150	34 33	200	51 06	250	22 47	279	29 35	324
174	32 56	032	30 49	087	37 01	149	34 25	199	50 43	249	22 22	278	29 20	323
175	33 10	031	31 14	086	37 14	149	34 17	199	50 19	249	21 57	278	29 04	322
176	33 23	030	31 39	085	37 27	148	34 09	198	49 55	248	21 32	277	28 48	321
177	33 35	029	32 04	084	37 40	148	34 01	197	49 32	247	21 06	276	28 32	320
178	33 47	028	32 30	083	37 54	147	33 53	197	49 09	247	20 41	275	28 15	319
179	33 59	027	32 55	082	38 08	146	33 46	196	48 45	246	20 16	274	27 58	317

 LAT 65°S

LHA	Hc	Zn	Hc	Zn	Hc	Zn	Hc	Zn	Hc	Zn	Hc	Zn	Hc	Zn
♈	SPICA		♦ANTARES		Peacock		♦ACHERNAR		CANOPUS		Suhail		♦Alphard	
180	34 10	025	33 20	081	38 22	146	33 39	196	48 22	245	57 55	291	27 41	316
181	34 20	024	33 45	080	38 37	145	33 33	195	47 59	244	57 31	290	27 23	315
182	34 31	023	34 10	079	38 51	144	33 26	194	47 37	244	57 07	289	27 05	314
183	34 40	022	34 35	078	39 06	144	33 20	194	47 14	243	56 43	288	26 47	313
184	34 50	021	35 00	078	39 21	143	33 14	193	46 51	242	56 19	287	26 28	312
185	34 58	020	35 24	077	39 36	143	33 09	192	46 29	242	55 54	285	26 10	311
186	35 07	018	35 49	076	39 52	142	33 03	192	46 07	241	55 30	284	25 50	310
187	35 14	017	36 13	075	40 08	141	32 58	191	45 45	240	55 05	283	25 31	309
188	35 21	016	36 38	074	40 24	141	32 54	190	45 23	240	54 40	282	25 11	308
189	35 28	015	37 02	073	40 40	140	32 49	190	45 01	239	54 16	281	24 51	307
190	35 34	014	37 26	072	40 56	139	32 45	189	44 39	238	53 51	280	24 30	306
191	35 40	012	37 50	071	41 13	139	32 41	189	44 18	238	53 26	279	24 10	305
192	35 45	011	38 14	070	41 30	138	32 37	188	43 56	237	53 01	278	23 49	304
193	35 50	010	38 38	069	41 47	138	32 34	187	43 35	236	52 35	277	23 28	303
194	35 54	009	39 01	068	42 04	137	32 31	187	43 14	235	52 10	276	23 06	302
	SPICA		♦ANTARES		Peacock		♦ACHERNAR		CANOPUS		Suhail		♦Alphard	
195	35 58	008	39 25	066	42 21	136	32 28	186	42 54	235	51 45	275	22 45	301
196	36 01	006	39 48	065	42 39	136	32 26	185	42 33	234	51 20	274	22 23	300
197	36 03	005	40 11	064	42 57	135	32 23	185	42 12	233	50 54	273	22 01	299
198	36 05	004	40 33	063	43 15	134	32 21	184	41 52	233	50 29	273	21 38	298
199	36 07	003	40 56	062	43 33	134	32 20	183	41 32	232	50 04	272	21 16	297
200	36 08	001	41 18	061	43 51	133	32 18	183	41 12	231	49 38	271	20 53	296
201	36 08	000	41 40	060	44 10	133	32 17	182	40 52	231	49 13	270	20 30	295
202	36 08	359	42 02	059	44 29	132	32 17	182	40 33	230	48 48	269	20 07	294
203	36 07	358	42 24	058	44 48	131	32 16	181	40 14	229	48 22	268	19 44	293
204	36 06	357	42 45	057	45 07	131	32 16	180	39 54	229	47 57	267	19 20	292
205	36 04	355	43 06	056	45 26	130	32 16	180	39 36	228	47 32	266	18 57	291
206	36 02	354	43 27	054	45 46	129	32 16	179	39 17	227	47 06	265	18 33	290
207	35 59	353	43 47	053	46 05	129	32 17	178	38 58	227	46 41	264	18 09	289
208	35 56	352	44 08	052	46 25	128	32 18	178	38 40	226	46 16	264	17 45	288
209	35 52	351	44 28	051	46 45	127	32 19	177	38 22	225	45 51	263	17 21	287
	ANTARES		♦Nunki		Peacock		ACHERNAR		♦CANOPUS		Suhail		♦SPICA	
210	44 47	050	30 31	087	47 06	127	32 20	176	38 04	225	45 26	262	35 48	349
211	45 06	049	30 56	086	47 26	126	32 22	176	37 46	224	45 01	261	35 43	348
212	45 25	047	31 21	085	47 47	126	32 24	175	37 29	223	44 36	260	35 37	347
213	45 44	046	31 46	085	48 07	125	32 26	174	37 11	223	44 11	259	35 31	346
214	46 02	045	32 12	084	48 28	124	32 29	174	36 54	222	43 46	259	35 25	345
215	46 20	044	32 37	083	48 49	124	32 32	173	36 37	221	43 21	258	35 18	343
216	46 37	043	33 02	082	49 10	123	32 35	173	36 21	221	42 56	257	35 10	342
217	46 54	041	33 27	081	49 32	122	32 38	172	36 04	220	42 32	256	35 02	341
218	47 10	040	33 52	080	49 53	122	32 42	171	35 48	219	42 07	255	34 54	340
219	47 27	039	34 17	079	50 15	121	32 46	171	35 32	219	41 42	255	34 45	339
220	47 42	038	34 42	078	50 37	120	32 50	170	35 16	218	41 18	254	34 35	337
221	47 57	036	35 07	077	50 59	120	32 54	169	35 01	217	40 54	253	34 25	336
222	48 12	035	35 31	076	51 21	119	32 59	169	34 46	217	40 30	252	34 15	335
223	48 27	034	35 56	075	51 43	118	33 05	168	34 31	216	40 06	251	34 04	334
224	48 40	032	36 20	074	52 05	118	33 10	167	34 16	215	39 42	251	33 53	333
	ANTARES		♦Nunki		FOMALHAUT		ACHERNAR		♦CANOPUS		Suhail		♦SPICA	
225	48 54	031	36 45	073	15 34	128	33 16	167	34 01	215	39 18	250	33 41	332
226	49 07	030	37 09	072	15 54	127	33 21	166	33 47	214	38 54	249	33 28	330
227	49 19	028	37 33	071	16 14	126	33 28	166	33 33	213	38 30	248	33 16	329
228	49 31	027	37 57	070	16 35	126	33 34	165	33 19	213	38 07	247	33 03	328
229	49 42	026	38 20	069	16 55	125	33 41	164	33 06	212	37 44	247	32 49	327
230	49 53	024	38 44	068	17 16	124	33 48	164	32 53	211	37 20	246	32 35	326
231	50 03	023	39 07	067	17 38	123	33 55	163	32 40	211	36 57	245	32 21	325
232	50 13	022	39 31	066	17 59	122	34 03	162	32 27	210	36 34	244	32 06	324
233	50 22	020	39 54	065	18 20	121	34 10	162	32 14	209	36 12	244	31 50	323
234	50 30	019	40 16	064	18 42	121	34 19	161	32 02	209	35 49	243	31 35	321
235	50 38	017	40 39	063	19 04	120	34 27	160	31 50	208	35 27	242	31 19	320
236	50 45	016	41 02	062	19 26	119	34 35	160	31 38	207	35 04	241	31 03	319
237	50 52	015	41 24	060	19 48	118	34 44	159	31 27	207	34 42	240	30 46	318
238	50 58	013	41 46	059	20 11	117	34 53	159	31 16	206	34 20	240	30 29	317
239	51 04	012	42 07	058	20 33	116	35 03	158	31 05	205	33 58	239	30 11	316
	♦ANTARES		Nunki		♦FOMALHAUT		ACHERNAR		CANOPUS		♦ACRUX		SPICA	
240	51 09	010	42 29	057	20 56	116	35 12	157	30 54	205	67 14	250	29 53	315
241	51 13	009	42 50	056	21 19	115	35 22	157	30 44	204	66 50	249	29 35	314
242	51 16	008	43 11	055	21 42	114	35 33	156	30 34	203	66 27	249	29 17	313
243	51 19	006	43 31	054	22 06	113	35 43	155	30 24	202	66 03	248	28 58	312
244	51 22	005	43 52	053	22 29	112	35 54	155	30 14	202	65 39	248	28 39	311
245	51 24	003	44 12	051	22 52	111	36 05	154	30 05	201	65 16	247	28 20	310
246	51 25	002	44 31	050	23 16	111	36 16	154	29 56	200	64 53	247	28 00	309
247	51 25	000	44 51	049	23 40	110	36 27	153	29 47	200	64 29	246	27 40	307
248	51 25	359	45 10	048	24 04	109	36 39	152	29 39	199	64 06	246	27 20	306
249	51 24	358	45 28	047	24 28	108	36 51	152	29 31	198	63 43	245	26 59	305
250	51 23	356	45 47	046	24 52	107	37 03	151	29 23	198	63 20	245	26 38	304
251	51 21	355	46 05	044	25 16	106	37 15	150	29 15	197	62 57	244	26 17	303
252	51 18	353	46 22	043	25 41	106	37 28	150	29 08	196	62 35	243	25 56	302
253	51 15	352	46 39	042	26 05	105	37 41	149	29 01	196	62 12	243	25 34	301
254	51 11	350	46 56	041	26 30	104	37 54	149	28 54	195	61 50	242	25 13	300
	Nunki		♦FOMALHAUT		ACHERNAR		CANOPUS		♦ACRUX		SPICA		♦ANTARES	
255	47 12	039	26 54	103	38 07	148	28 48	194	61 27	242	24 51	299	51 06	349
256	47 28	038	27 19	102	38 21	147	28 42	194	61 05	241	24 28	298	51 01	348
257	47 44	037	27 44	101	38 35	147	28 36	193	60 43	241	24 06	297	50 55	346
258	47 59	036	28 09	100	38 49	146	28 30	192	60 21	240	23 43	296	50 49	345
259	48 13	034	28 34	100	39 03	145	28 25	192	59 59	240	23 20	295	50 42	343
260	48 27	033	28 59	099	39 18	145	28 20	191	59 37	239	22 57	294	50 34	342
261	48 41	032	29 24	098	39 32	144	28 15	190	59 15	239	22 34	293	50 26	341
262	48 54	030	29 49	097	39 47	144	28 11	190	58 54	238	22 11	292	50 18	339
263	49 06	029	30 14	096	40 02	143	28 07	189	58 32	238	21 47	291	50 08	338
264	49 18	028	30 40	095	40 18	142	28 03	188	58 11	237	21 24	290	49 58	336
265	49 30	026	31 05	094	40 33	142	28 00	187	57 50	236	21 00	289	49 48	335
266	49 41	025	31 30	093	40 49	141	27 57	187	57 29	236	20 36	289	49 37	334
267	49 51	024	31 55	092	41 05	140	27 54	186	57 08	235	20 12	288	49 25	332
268	50 01	022	32 21	092	41 22	140	27 51	185	56 47	235	19 47	287	49 13	331
269	50 10	021	32 46	091	41 38	139	27 49	185	56 26	234	19 23	286	49 01	330

LHA	Hc	Zn	Hc	Zn	Hc	Zn	Hc	Zn	Hc	Zn	Hc	Zn	Hc	Zn
♈	♦ALTAIR		FOMALHAUT		♦ACHERNAR		CANOPUS		♦RIGIL KENT.		SPICA		ANTARES	
270	13 19	028	33 12	090	41 55	139	27 47	184	67 24	257	18 59	285	48 48	328
271	13 31	027	33 37	089	42 12	138	27 45	183	66 59	256	18 34	284	48 34	327
272	13 42	026	34 02	088	42 29	137	27 44	183	66 35	256	18 09	283	48 20	326
273	13 53	025	34 28	087	42 46	137	27 43	182	66 10	255	17 45	282	48 06	324
274	14 04	024	34 53	086	43 04	136	27 42	181	65 46	254	17 20	281	47 51	323
275	14 14	023	35 18	085	43 21	135	27 42	181	65 21	254	16 55	280	47 35	322
276	14 24	022	35 43	084	43 39	135	27 42	180	64 57	253	16 30	279	47 19	321
277	14 33	021	36 09	083	43 57	134	27 42	179	64 33	252	16 05	278	47 03	319
278	14 42	020	36 34	082	44 15	134	27 42	179	64 09	252	15 40	277	46 46	318
279	14 51	019	36 59	081	44 34	133	27 43	178	63 45	251	15 14	276	46 29	317
280	14 59	018	37 24	080	44 53	132	27 44	177	63 21	251	14 49	275	46 12	316
281	15 06	017	37 49	079	45 11	132	27 45	177	62 57	250	14 24	275	45 54	314
282	15 14	016	38 14	079	45 30	131	27 47	176	62 33	249	13 59	274	45 35	313
283	15 20	015	38 39	078	45 50	130	27 49	175	62 09	249	13 33	273	45 17	312
284	15 27	014	39 03	077	46 09	130	27 51	174	61 46	248	13 08	272	44 58	311
	♦ALTAIR		FOMALHAUT		♦ACHERNAR		CANOPUS		♦RIGIL KENT.		ANTARES		Nunki	
285	15 33	013	39 28	076	46 29	129	27 54	174	61 22	247	44 38	310	51 18	358
286	15 38	012	39 52	075	46 48	129	27 57	173	60 59	247	44 19	308	51 17	357
287	15 43	011	40 17	074	47 08	128	28 00	172	60 36	246	43 59	307	51 15	355
288	15 48	010	40 41	073	47 28	127	28 03	172	60 12	246	43 38	306	51 12	354
289	15 52	009	41 05	072	47 49	127	28 07	171	59 49	245	43 18	305	51 09	352
290	15 55	008	41 29	070	48 09	126	28 11	170	59 26	244	42 57	304	51 06	351
291	15 59	007	41 53	069	48 30	125	28 16	170	59 04	244	42 36	303	51 02	350
292	16 01	006	42 17	068	48 50	125	28 20	169	58 41	243	42 14	302	50 57	348
293	16 04	005	42 40	067	49 11	124	28 25	168	58 18	243	41 52	301	50 51	347
294	16 06	004	43 03	066	49 32	124	28 31	168	57 56	242	41 30	299	50 45	345
295	16 07	003	43 27	065	49 53	123	28 36	167	57 34	241	41 08	298	50 38	344
296	16 08	002	43 49	064	50 15	122	28 42	166	57 11	241	40 46	297	50 31	343
297	16 09	001	44 12	063	50 36	122	28 48	166	56 49	240	40 23	296	50 23	341
298	16 09	000	44 35	062	50 58	121	28 55	165	56 27	240	40 00	295	50 15	340
299	16 08	359	44 57	061	51 20	120	29 01	164	56 05	239	39 37	294	50 06	338
	FOMALHAUT		♦ACHERNAR		CANOPUS		♦RIGIL KENT.		ANTARES		Nunki		♦ALTAIR	
300	45 19	060	51 42	120	29 08	164	55 44	239	39 14	293	49 56	337	16 07	358
301	45 41	059	52 04	119	29 16	163	55 22	238	38 51	292	49 46	336	16 06	357
302	46 02	057	52 26	118	29 23	162	55 01	237	38 27	291	49 35	334	16 04	356
303	46 24	056	52 48	118	29 31	162	54 39	237	38 03	290	49 24	333	16 02	354
304	46 44	055	53 11	117	29 39	161	54 18	236	37 39	289	49 12	332	15 59	353
305	47 05	054	53 33	117	29 48	160	53 57	236	37 15	288	49 00	330	15 56	352
306	47 26	053	53 56	116	29 57	160	53 36	235	36 51	287	48 47	329	15 53	351
307	47 46	052	54 19	115	30 06	159	53 16	235	36 27	286	48 34	328	15 49	350
308	48 05	050	54 42	115	30 15	158	52 55	234	36 02	285	48 20	326	15 44	349
309	48 25	049	55 05	114	30 24	157	52 35	233	35 38	284	48 06	325	15 39	348
310	48 44	048	55 28	113	30 34	157	52 14	233	35 13	283	47 51	324	15 34	347
311	49 02	047	55 52	113	30 44	156	51 54	232	34 48	282	47 36	322	15 28	346
312	49 21	046	56 15	112	30 55	155	51 34	232	34 24	281	47 20	321	15 22	345
313	49 39	044	56 39	111	31 06	155	51 15	231	33 59	280	47 04	320	15 15	344
314	49 56	043	57 03	111	31 16	154	50 55	230	33 34	279	46 47	319	15 08	343
	♦FOMALHAUT		ACHERNAR		♦CANOPUS		RIGIL KENT.		♦ANTARES		Nunki		ALTAIR	
315	50 13	042	57 26	110	31 28	153	50 35	230	33 09	278	46 30	317	15 01	342
316	50 30	040	57 50	109	31 39	153	50 16	229	32 43	277	46 13	316	14 53	341
317	50 46	039	58 14	108	31 51	152	49 57	229	32 18	276	45 55	315	14 44	340
318	51 02	038	58 38	108	32 03	151	49 38	228	31 53	275	45 37	314	14 36	339
319	51 17	036	59 03	107	32 15	151	49 19	228	31 28	274	45 19	313	14 26	338
320	51 32	035	59 27	106	32 28	150	49 01	227	31 03	274	45 00	311	14 17	337
321	51 46	034	59 51	106	32 41	149	48 42	226	30 37	273	44 41	310	14 07	336
322	52 00	032	60 16	105	32 54	149	48 24	226	30 12	272	44 21	309	13 56	335
323	52 14	031	60 40	104	33 07	148	48 06	225	29 47	271	44 01	308	13 45	334
324	52 26	030	61 05	104	33 20	147	47 48	225	29 21	270	43 41	307	13 34	333
325	52 39	028	61 29	103	33 34	147	47 30	224	28 56	269	43 21	306	13 23	332
326	52 50	027	61 54	102	33 48	146	47 12	224	28 30	268	43 00	304	13 11	331
327	53 02	026	62 19	101	34 03	145	46 55	223	28 05	267	42 39	303	12 58	330
328	53 12	024	62 44	101	34 17	145	46 38	222	27 40	266	42 18	302	12 45	329
329	53 22	023	63 09	100	34 32	144	46 21	222	27 15	265	41 56	301	12 32	328
	♦FOMALHAUT		Diphda		ACHERNAR		♦CANOPUS		RIGIL KENT.		♦ANTARES		Nunki	
330	53 32	021	35 45	050	63 34	099	34 47	143	46 04	221	26 49	265	41 34	300
331	53 41	020	36 05	049	63 59	098	35 02	143	45 47	221	26 24	264	41 12	299
332	53 49	018	36 24	048	64 24	097	35 18	142	45 31	220	25 59	263	40 50	298
333	53 57	017	36 42	047	64 49	097	35 33	141	45 15	220	25 34	262	40 27	297
334	54 04	015	37 01	046	65 14	096	35 49	141	44 59	219	25 09	261	40 05	296
335	54 10	014	37 18	044	65 40	095	36 06	140	44 43	218	24 44	260	39 42	295
336	54 16	012	37 36	043	66 05	094	36 22	139	44 27	218	24 19	259	39 18	294
337	54 21	011	37 53	042	66 30	093	36 39	139	44 12	217	23 54	258	38 55	293
338	54 26	010	38 10	041	66 56	093	36 56	138	43 56	217	23 29	258	38 32	292
339	54 29	008	38 27	040	67 21	092	37 13	137	43 41	216	23 04	257	38 08	291
340	54 33	007	38 43	039	67 46	091	37 30	137	43 27	215	22 40	256	37 44	289
341	54 35	005	38 58	037	68 12	090	37 47	136	43 12	215	22 15	255	37 20	288
342	54 37	004	39 13	036	68 37	089	38 05	135	42 58	214	21 51	254	36 56	287
343	54 38	002	39 28	035	69 02	088	38 23	135	42 43	214	21 26	253	36 32	286
344	54 39	001	39 43	034	69 28	087	38 41	134	42 29	213	21 02	252	36 07	285
	♦Diphda		Acamar		♦CANOPUS		RIGIL KENT.		♦ANTARES		Nunki		FOMALHAUT	
345	39 57	033	48 35	083	39 00	133	42 16	213	20 38	251	35 43	285	54 39	359
346	40 10	032	49 00	082	39 18	133	42 02	212	20 14	251	35 18	284	54 38	358
347	40 23	030	49 25	081	39 37	132	41 49	211	19 50	250	34 54	283	54 37	356
348	40 36	029	49 50	080	39 56	131	41 36	211	19 26	249	34 29	282	54 34	355
349	40 48	028	50 15	079	40 15	131	41 23	210	19 03	248	34 04	281	54 32	353
350	40 59	027	50 40	078	40 35	130	41 10	210	18 39	247	33 39	280	54 28	352
351	41 10	025	51 05	077	40 54	129	40 58	209	18 16	246	33 14	279	54 24	350
352	41 21	024	51 30	076	41 14	129	40 45	208	17 53	246	32 49	278	54 20	349
353	41 31	023	51 54	075	41 34	128	40 33	208	17 30	245	32 24	277	54 14	347
354	41 41	022	52 19	074	41 54	127	40 22	207	17 07	244	31 59	276	54 08	346
355	41 50	020	52 43	073	42 14	127	40 10	207	16 44	243	31 33	275	54 02	344
356	41 58	019	53 07	072	42 35	126	39 59	206	16 22	242	31 08	274	53 54	343
357	42 06	018	53 31	071	42 55	125	39 48	206	16 00	241	30 43	273	53 46	341
358	42 14	017	53 55	070	43 16	124	39 37	205	15 38	240	30 17	272	53 38	340
359	42 21	015	54 19	069	43 37	124	39 26	204	15 16	240	29 52	271	53 29	338

LHA ♈	Hc	Zn	Hc	Zn	Hc	Zn	Hc	Zn	Hc	Zn	Hc	Zn	Hc	Zn
	Diphda		♦RIGEL		CANOPUS		Miaplacidus		♦RIGIL KENT.		Peacock		♦FOMALHAUT	
0	41 29	014	12 08	083	44 31	122	48 43	160	40 11	204	63 39	264	52 24	337
1	41 35	013	12 32	082	44 51	122	48 52	159	40 01	204	63 14	264	52 14	336
2	41 40	011	12 56	081	45 12	121	49 00	159	39 51	203	62 50	263	52 04	335
3	41 44	010	13 20	080	45 33	120	49 09	158	39 42	202	62 26	262	51 53	333
4	41 48	009	13 44	079	45 54	119	49 19	158	39 33	202	62 02	261	51 42	332
5	41 52	007	14 08	078	46 16	119	49 28	157	39 24	201	61 38	260	51 30	330
6	41 54	006	14 32	077	46 37	118	49 38	157	39 15	201	61 14	260	51 18	329
7	41 57	005	14 56	076	46 59	117	49 47	156	39 07	200	60 50	259	51 05	328
8	41 59	004	15 20	075	47 20	117	49 57	156	38 59	199	60 26	258	50 52	326
9	42 00	002	15 43	074	47 42	116	50 07	155	38 51	199	60 02	257	50 38	325
10	42 01	001	16 07	074	48 04	115	50 18	155	38 43	198	59 38	256	50 24	324
11	42 01	000	16 30	073	48 27	114	50 28	154	38 36	197	59 15	256	50 10	323
12	42 00	359	16 53	072	48 49	114	50 39	154	38 28	197	58 51	255	49 54	321
13	42 00	357	17 16	071	49 11	113	50 50	153	38 21	196	58 27	254	49 39	320
14	41 58	356	17 39	070	49 34	112	51 01	153	38 15	196	58 04	254	49 23	319
	♦RIGEL		CANOPUS		Miaplacidus		♦RIGIL KENT.		Peacock		♦FOMALHAUT		Diphda	
15	18 02	069	49 56	112	51 12	152	38 08	195	57 41	253	49 07	317	41 56	355
16	18 25	068	50 19	111	51 23	152	38 02	194	57 17	252	48 50	316	41 54	353
17	18 47	067	50 42	110	51 35	151	37 56	194	56 54	251	48 33	315	41 51	352
18	19 10	066	51 05	109	51 47	151	37 50	193	56 31	251	48 15	314	41 47	351
19	19 32	065	51 28	109	51 59	151	37 45	193	56 08	250	47 58	312	41 43	350
20	19 54	064	51 51	108	52 11	150	37 40	192	55 45	249	47 39	311	41 38	348
21	20 16	063	52 15	107	52 23	150	37 35	191	55 23	249	47 21	310	41 33	347
22	20 37	062	52 38	106	52 35	149	37 30	191	55 00	248	47 02	309	41 27	346
23	20 59	061	53 01	106	52 48	149	37 25	190	54 37	247	46 43	308	41 21	345
24	21 20	060	53 25	105	53 01	148	37 21	190	54 15	246	46 23	306	41 14	343
25	21 41	059	53 49	104	53 13	148	37 17	189	53 53	246	46 04	305	41 07	342
26	22 02	058	54 12	103	53 27	147	37 14	188	53 30	245	45 44	304	40 59	341
27	22 22	057	54 36	102	53 40	147	37 10	188	53 08	244	45 23	303	40 51	340
28	22 43	056	55 00	102	53 53	146	37 07	187	52 46	244	45 03	302	40 42	338
29	23 03	055	55 24	101	54 07	146	37 04	187	52 25	243	44 42	301	40 33	337
	♦RIGEL		CANOPUS		Miaplacidus		♦RIGIL KENT.		Peacock		♦FOMALHAUT		Diphda	
30	23 23	054	55 48	100	54 20	146	37 01	186	52 03	242	44 21	300	40 23	336
31	23 42	053	56 12	099	54 34	145	36 59	185	51 41	242	43 59	299	40 13	335
32	24 02	052	56 36	098	54 48	145	36 57	185	51 20	241	43 38	297	40 02	333
33	24 21	051	57 00	098	55 03	144	36 55	184	50 59	240	43 16	296	39 51	332
34	24 39	050	57 24	097	55 17	144	36 53	184	50 37	240	42 54	295	39 39	331
35	24 58	049	57 49	096	55 31	143	36 52	183	50 16	239	42 32	294	39 27	330
36	25 16	048	58 13	095	55 46	143	36 51	182	49 56	238	42 10	293	39 15	329
37	25 34	047	58 37	094	56 01	143	36 50	182	49 35	238	41 47	292	39 02	327
38	25 52	046	59 02	093	56 16	142	36 49	181	49 14	237	41 24	291	38 48	326
39	26 09	045	59 26	092	56 31	142	36 49	180	48 54	236	41 02	290	38 34	325
40	26 26	044	59 50	092	56 46	141	36 49	180	48 34	236	40 39	289	38 20	324
41	26 43	043	60 15	091	57 01	141	36 49	179	48 14	235	40 15	288	38 06	323
42	26 59	041	60 39	090	57 17	140	36 50	179	47 54	234	39 52	287	37 51	321
43	27 15	040	61 04	089	57 32	140	36 50	178	47 34	234	39 29	286	37 35	320
44	27 31	039	61 28	088	57 48	140	36 51	177	47 14	233	39 05	285	37 19	319
	RIGEL		♦SIRIUS		Suhail		♦RIGIL KENT.		Peacock		♦FOMALHAUT		Diphda	
45	27 46	038	28 38	065	38 09	113	36 52	177	46 55	233	38 42	284	37 03	318
46	28 01	037	29 00	064	38 32	112	36 54	176	46 36	232	38 18	283	36 47	317
47	28 15	036	29 22	063	38 54	111	36 56	176	46 16	231	37 54	282	36 30	316
48	28 29	035	29 44	062	39 17	110	36 58	175	45 57	231	37 30	281	36 13	315
49	28 43	034	30 05	061	39 40	109	37 00	174	45 39	230	37 06	280	35 55	313
50	28 57	033	30 26	060	40 03	109	37 02	174	45 20	229	36 42	279	35 37	312
51	29 10	032	30 47	059	40 27	108	37 05	173	45 02	229	36 18	278	35 19	311
52	29 22	031	31 08	058	40 50	107	37 08	173	44 44	228	35 54	277	35 01	310
53	29 34	029	31 29	057	41 13	106	37 11	172	44 25	227	35 29	276	34 42	309
54	29 46	028	31 49	056	41 37	105	37 15	171	44 08	227	35 05	275	34 23	308
55	29 58	027	32 09	055	42 00	105	37 19	171	43 50	226	34 41	274	34 03	307
56	30 08	026	32 29	054	42 24	104	37 23	170	43 32	225	34 16	273	33 44	306
57	30 19	025	32 48	053	42 48	103	37 27	170	43 15	225	33 52	272	33 24	305
58	30 29	024	33 08	052	43 12	102	37 32	169	42 58	224	33 28	272	33 03	304
59	30 39	023	33 27	050	43 35	101	37 36	168	42 41	223	33 03	271	32 43	303
	RIGEL		♦SIRIUS		Suhail		♦RIGIL KENT.		Peacock		♦FOMALHAUT		Diphda	
60	30 48	022	33 45	049	43 59	100	37 42	168	42 25	223	32 39	270	32 22	302
61	30 57	020	34 04	048	44 23	100	37 47	167	42 08	222	32 15	269	32 01	301
62	31 05	019	34 22	047	44 48	099	37 52	167	41 52	222	31 50	268	31 40	299
63	31 13	018	34 39	046	45 12	098	37 58	166	41 36	221	31 26	267	31 19	298
64	31 20	017	34 57	045	45 36	097	38 04	165	41 20	220	31 01	266	30 57	297
65	31 27	016	35 14	044	46 00	096	38 11	165	41 04	220	30 37	265	30 36	296
66	31 33	015	35 31	043	46 24	095	38 17	164	40 49	219	30 13	264	30 14	295
67	31 39	013	35 47	042	46 49	094	38 24	163	40 33	218	29 48	263	29 52	294
68	31 45	012	36 03	040	47 13	093	38 31	163	40 18	218	29 24	263	29 29	293
69	31 50	011	36 19	039	47 37	093	38 38	162	40 04	217	29 00	262	29 07	292
70	31 54	010	36 34	038	48 02	092	38 46	162	39 49	216	28 36	261	28 44	291
71	31 58	009	36 49	037	48 26	091	38 54	161	39 35	216	28 12	260	28 21	290
72	32 02	008	37 03	036	48 51	090	39 02	160	39 20	215	27 48	259	27 58	289
73	32 05	007	37 18	035	49 15	089	39 10	160	39 06	215	27 24	258	27 35	288
74	32 07	005	37 31	034	49 39	088	39 19	159	38 53	214	27 00	257	27 12	287
	RIGEL		♦SIRIUS		Suhail		♦RIGIL KENT.		Peacock		♦FOMALHAUT		Diphda	
75	32 09	004	37 44	032	50 04	087	39 27	159	38 39	213	26 36	256	26 49	286
76	32 11	003	37 57	031	50 28	086	39 36	158	38 26	213	26 13	256	26 25	285
77	32 12	002	38 10	030	50 52	085	39 46	157	38 13	212	25 49	255	26 02	285
78	32 12	001	38 22	029	51 17	084	39 55	157	38 00	211	25 26	254	25 38	284
79	32 12	000	38 33	028	51 41	083	40 05	156	37 48	211	25 02	253	25 14	283
80	32 12	358	38 44	026	52 05	082	40 15	156	37 35	210	24 39	252	24 50	282
81	32 11	357	38 55	025	52 29	081	40 25	155	37 23	209	24 16	251	24 26	281
82	32 10	356	39 05	024	52 54	080	40 35	154	37 11	209	23 53	250	24 02	280
83	32 08	355	39 15	023	53 18	079	40 46	154	37 00	208	23 30	250	23 38	279
84	32 05	354	39 24	022	53 41	078	40 57	153	36 48	207	23 07	249	23 14	278
85	32 02	352	39 33	020	54 05	077	41 08	153	36 37	207	22 44	248	22 50	277
86	31 59	351	39 41	019	54 29	076	41 19	152	36 26	206	22 22	247	22 26	276
87	31 55	350	39 49	018	54 53	075	41 31	151	36 16	206	21 59	246	22 01	275
88	31 50	349	39 56	017	55 16	074	41 42	151	36 05	205	21 37	245	21 37	274
89	31 45	348	40 03	015	55 40	073	41 54	150	35 55	204	21 15	244	21 13	273

LAT 66°S

LHA ♈	Hc	Zn	Hc	Zn	Hc	Zn	Hc	Zn	Hc	Zn	Hc	Zn	Hc	Zn
	♦SIRIUS		Suhail		♦RIGIL KENT.		Peacock		♦ACHERNAR		Acamar		RIGEL	
90	40 09	014	56 03	072	42 07	150	35 45	204	59 14	254	53 57	292	31 40	347
91	40 15	013	56 26	071	42 19	149	35 36	203	58 50	254	53 35	291	31 34	346
92	40 20	012	56 49	070	42 32	148	35 26	202	58 27	253	53 12	290	31 28	344
93	40 24	010	57 12	069	42 45	148	35 17	202	58 04	252	52 49	289	31 21	343
94	40 29	009	57 34	067	42 58	147	35 08	201	57 41	251	52 26	288	31 14	342
95	40 32	008	57 57	066	43 11	147	35 00	200	57 17	251	52 03	287	31 06	341
96	40 35	007	58 19	065	43 25	146	34 51	200	56 54	250	51 39	286	30 58	340
97	40 38	005	58 41	064	43 38	146	34 43	199	56 32	249	51 16	285	30 49	339
98	40 40	004	59 03	063	43 52	145	34 35	198	56 09	249	50 52	284	30 40	337
99	40 41	003	59 24	061	44 06	144	34 28	198	55 46	248	50 28	283	30 31	336
100	40 42	002	59 46	060	44 21	144	34 20	197	55 24	247	50 05	282	30 21	335
101	40 42	000	60 07	059	44 35	143	34 13	196	55 01	247	49 41	281	30 10	334
102	40 42	359	60 27	058	44 50	143	34 07	196	54 39	246	49 17	280	29 59	333
103	40 42	358	60 48	056	45 05	142	34 00	195	54 17	245	48 53	279	29 48	332
104	40 40	357	61 08	055	45 20	141	33 54	195	53 54	245	48 28	278	29 36	331
	Alphard		♦Gienah		RIGIL KENT.		♦Peacock		ACHERNAR		♦RIGEL		SIRIUS	
105	27 19	042	20 28	087	45 35	141	33 48	194	53 32	244	29 24	330	40 39	355
106	27 36	041	20 52	086	45 51	140	33 42	193	53 11	243	29 12	329	40 36	354
107	27 51	040	21 17	085	46 07	140	33 37	193	52 49	243	28 59	327	40 34	353
108	28 07	039	21 41	084	46 22	139	33 31	192	52 27	242	28 45	326	40 30	351
109	28 22	038	22 05	084	46 39	138	33 27	191	52 06	241	28 32	325	40 26	350
110	28 36	036	22 30	083	46 55	138	33 22	191	51 45	241	28 17	324	40 22	349
111	28 51	035	22 54	082	47 11	137	33 18	190	51 23	240	28 03	323	40 17	348
112	29 05	034	23 18	081	47 28	137	33 13	189	51 02	239	27 48	322	40 12	346
113	29 18	033	23 42	080	47 45	136	33 10	189	50 41	239	27 33	321	40 06	345
114	29 31	032	24 06	079	48 02	135	33 06	188	50 21	238	27 17	320	39 59	344
115	29 44	031	24 30	078	48 19	135	33 03	187	50 00	237	27 01	319	39 52	343
116	29 57	030	24 54	077	48 37	134	33 00	187	49 40	237	26 45	318	39 45	341
117	30 08	029	25 17	076	48 54	134	32 57	186	49 19	236	26 28	317	39 37	340
118	30 20	028	25 41	075	49 12	133	32 55	185	48 59	235	26 12	316	39 28	339
119	30 31	026	26 04	074	49 30	132	32 53	185	48 39	235	25 54	314	39 19	338
	Alphard		♦SPICA		RIGIL KENT.		♦Peacock		ACHERNAR		♦RIGEL		SIRIUS	
120	30 42	025	13 43	087	49 48	132	32 51	184	48 19	234	25 37	313	39 10	337
121	30 52	024	14 08	086	50 06	131	32 49	183	48 00	233	25 19	312	39 00	335
122	31 02	023	14 32	085	50 24	131	32 48	183	47 40	233	25 01	311	38 49	334
123	31 11	022	14 56	084	50 43	130	32 47	182	47 21	232	24 42	310	38 38	333
124	31 20	021	15 21	083	51 02	130	32 46	182	47 02	231	24 23	309	38 27	332
125	31 28	020	15 45	082	51 21	129	32 45	181	46 43	231	24 04	308	38 15	331
126	31 36	018	16 09	081	51 40	128	32 45	180	46 24	230	23 45	307	38 03	329
127	31 44	017	16 33	080	51 59	128	32 45	180	46 05	230	23 26	306	37 50	328
128	31 51	016	16 57	079	52 18	127	32 45	179	45 47	229	23 06	305	37 37	327
129	31 57	015	17 21	078	52 38	127	32 46	178	45 28	228	22 46	304	37 24	326
130	32 03	014	17 45	077	52 58	126	32 47	178	45 10	228	22 25	303	37 10	325
131	32 09	013	18 09	076	53 17	125	32 48	177	44 52	227	22 05	302	36 56	323
132	32 14	012	18 32	075	53 37	125	32 50	176	44 35	226	21 44	301	36 41	322
133	32 19	010	18 56	074	53 58	124	32 51	176	44 17	226	21 23	300	36 26	321
134	32 23	009	19 19	073	54 18	124	32 53	175	44 00	225	21 02	299	36 10	320
	Alphard		♦SPICA		RIGIL KENT.		♦Peacock		ACHERNAR		♦CANOPUS		SIRIUS	
135	32 27	008	19 43	073	54 38	123	32 56	174	43 42	224	66 39	286	35 55	319
136	32 30	007	20 06	072	54 59	122	32 58	174	43 25	224	66 15	284	35 38	318
137	32 32	006	20 29	071	55 19	122	33 01	173	43 09	223	65 52	283	35 22	317
138	32 35	005	20 52	070	55 40	121	33 04	172	42 52	223	65 28	282	35 05	316
139	32 36	003	21 15	069	56 01	121	33 07	172	42 36	222	65 04	281	34 48	314
140	32 37	002	21 37	068	56 22	120	33 11	171	42 19	221	64 40	280	34 30	313
141	32 38	001	22 00	067	56 44	119	33 15	170	42 03	221	64 16	279	34 12	312
142	32 38	000	22 22	066	57 05	119	33 19	170	41 48	220	63 52	278	33 54	311
143	32 38	359	22 44	065	57 26	118	33 24	169	41 32	219	63 28	277	33 35	310
144	32 37	357	23 06	064	57 48	117	33 28	168	41 17	219	63 03	276	33 16	309
145	32 36	356	23 28	063	58 10	117	33 33	168	41 02	218	62 39	275	32 57	308
146	32 34	355	23 50	062	58 31	116	33 39	167	40 47	217	62 15	274	32 38	307
147	32 32	354	24 11	061	58 53	116	33 44	167	40 32	217	61 50	273	32 18	306
148	32 29	353	24 32	060	59 15	115	33 50	166	40 17	216	61 26	272	31 58	305
149	32 25	352	24 53	059	59 38	114	33 56	165	40 03	216	61 02	271	31 38	304
	♦SPICA		RIGIL KENT.		♦Peacock		ACHERNAR		♦CANOPUS		SIRIUS		Alphard	
150	25 14	058	60 00	114	34 02	165	39 49	215	60 37	270	31 18	303	32 22	350
151	25 35	057	60 22	113	34 09	164	39 35	214	60 13	269	30 57	302	32 17	349
152	25 55	056	60 45	112	34 16	163	39 21	214	59 48	268	30 36	300	32 13	348
153	26 15	055	61 07	112	34 23	163	39 08	213	59 24	267	30 15	299	32 07	347
154	26 35	054	61 30	111	34 31	162	38 55	212	59 00	267	29 53	298	32 01	346
155	26 54	053	61 53	111	34 38	161	38 42	212	58 35	266	29 32	297	31 55	345
156	27 13	052	62 16	110	34 46	161	38 29	211	58 11	265	29 10	296	31 49	343
157	27 32	051	62 39	109	34 54	160	38 17	210	57 47	264	28 48	295	31 41	342
158	27 51	049	63 02	109	35 03	159	38 04	210	57 22	263	28 26	294	31 34	341
159	28 10	048	63 25	108	35 11	159	37 52	209	56 58	262	28 04	293	31 26	340
160	28 28	047	63 48	107	35 20	158	37 41	209	56 34	262	27 41	292	31 17	339
161	28 45	046	64 12	107	35 30	157	37 29	208	56 10	261	27 19	291	31 08	338
162	29 03	045	64 35	106	35 39	157	37 18	207	55 46	260	26 56	290	30 59	337
163	29 20	044	64 59	105	35 49	156	37 07	207	55 22	259	26 33	289	30 49	335
164	29 37	043	65 22	105	35 59	156	36 56	206	54 58	258	26 10	288	30 38	334
	♦SPICA		ANTARES		♦Peacock		ACHERNAR		♦CANOPUS		SIRIUS		Alphard	
165	29 53	042	27 06	095	36 09	155	36 45	205	54 34	258	25 47	287	30 27	333
166	30 10	041	27 30	094	36 20	154	36 35	205	54 10	257	25 23	286	30 16	332
167	30 25	040	27 54	093	36 30	154	36 25	204	53 47	256	25 00	285	30 05	331
168	30 41	039	28 19	092	36 41	153	36 15	203	53 23	255	24 36	285	29 52	330
169	30 56	038	28 43	091	36 53	152	36 06	203	52 59	254	24 12	284	29 40	329
170	31 11	036	29 08	090	37 04	152	35 56	202	52 36	254	23 49	283	29 27	328
171	31 25	035	29 32	089	37 16	151	35 47	202	52 13	253	23 25	282	29 14	326
172	31 39	034	29 56	088	37 28	150	35 38	201	51 49	252	23 01	281	29 00	325
173	31 52	033	30 21	087	37 40	150	35 30	200	51 26	251	22 37	280	28 46	324
174	32 05	032	30 45	086	37 52	149	35 21	200	51 03	251	22 13	279	28 32	323
175	32 18	031	31 09	085	38 05	148	35 13	199	50 40	250	21 49	278	28 17	322
176	32 31	030	31 34	085	38 18	148	35 06	198	50 17	249	21 24	277	28 02	321
177	32 42	029	31 58	084	38 31	147	34 58	198	49 55	248	21 00	276	27 46	320
178	32 54	027	32 22	083	38 44	147	34 51	197	49 32	248	20 36	275	27 30	319
179	33 05	026	32 46	082	38 58	146	34 44	196	49 09	247	20 12	274	27 14	318

LHA	Hc	Zn	Hc	Zn	Hc	Zn	Hc	Zn	Hc	Zn	Hc	Zn	Hc	Zn
♈	♦SPICA		ANTARES		♦Peacock		ACHERNAR		♦CANOPUS		SIRIUS		Alphard	
180	33 16	025	33 11	081	39 12	145	34 37	196	48 47	246	19 47	273	26 58	317
181	33 26	024	33 35	080	39 26	145	34 31	195	48 25	245	19 23	272	26 41	316
182	33 35	023	33 59	079	39 40	144	34 24	194	48 03	245	18 58	271	26 23	315
183	33 45	022	34 22	078	39 54	143	34 18	194	47 41	244	18 34	271	26 06	314
184	33 53	020	34 46	077	40 09	143	34 13	193	47 19	243	18 10	270	25 48	313
185	34 02	019	35 10	076	40 24	142	34 07	193	46 57	243	17 45	269	25 30	311
186	34 10	018	35 34	075	40 39	141	34 02	192	46 35	242	17 21	268	25 12	310
187	34 17	017	35 57	074	40 54	141	33 57	191	46 14	241	16 57	267	24 53	309
188	34 24	016	36 21	073	41 10	140	33 53	191	45 53	240	16 32	266	24 34	308
189	34 30	015	36 44	072	41 26	139	33 48	190	45 32	240	16 08	265	24 15	307
190	34 36	013	37 07	071	41 42	139	33 44	189	45 11	239	15 44	264	23 55	306
191	34 42	012	37 30	070	41 58	138	33 40	189	44 50	238	15 19	263	23 35	305
192	34 46	011	37 53	069	42 14	138	33 37	188	44 29	238	14 55	262	23 15	304
193	34 51	010	38 15	068	42 31	137	33 34	187	44 08	237	14 31	261	22 55	303
194	34 55	009	38 38	067	42 47	136	33 31	187	43 48	236	14 07	261	22 34	302
	♦SPICA		ANTARES		♦Peacock		ACHERNAR		♦CANOPUS		Suhail		Alphard	
195	34 58	007	39 00	066	43 04	136	33 28	186	43 28	236	51 39	277	22 14	301
196	35 01	006	39 22	065	43 22	135	33 25	185	43 08	235	51 15	276	21 53	300
197	35 04	005	39 44	064	43 39	134	33 23	185	42 48	234	50 50	275	21 31	299
198	35 05	004	40 06	063	43 57	134	33 21	184	42 28	233	50 26	274	21 10	298
199	35 07	003	40 28	061	44 14	133	33 20	183	42 09	233	50 02	273	20 48	297
200	35 08	001	40 49	060	44 32	132	33 18	183	41 49	232	49 37	272	20 27	296
201	35 08	000	41 10	059	44 50	132	33 17	182	41 30	231	49 13	271	20 05	295
202	35 08	359	41 31	058	45 09	131	33 17	182	41 11	231	48 48	270	19 43	294
203	35 07	358	41 52	057	45 27	130	33 16	181	40 52	230	48 24	269	19 20	293
204	35 06	357	42 12	056	45 46	130	33 16	180	40 34	229	48 00	268	18 58	292
205	35 04	355	42 32	055	46 05	129	33 16	180	40 15	229	47 35	267	18 35	291
206	35 02	354	42 52	054	46 24	129	33 16	179	39 57	228	47 11	266	18 12	290
207	35 00	353	43 11	053	46 43	128	33 17	178	39 39	227	46 46	266	17 49	290
208	34 56	352	43 31	051	47 02	127	33 18	178	39 21	227	46 22	265	17 26	289
209	34 53	351	43 49	050	47 22	127	33 19	177	39 04	226	45 58	264	17 03	288
	♦ANTARES		Nunki		♦FOMALHAUT		ACHERNAR		CANOPUS		♦Suhail		SPICA	
210	44 08	049	30 28	087	11 49	141	33 20	176	38 46	225	45 34	263	34 49	350
211	44 26	048	30 52	086	12 05	140	33 22	176	38 29	225	45 09	262	34 44	348
212	44 44	047	31 16	085	12 21	139	33 24	175	38 12	224	44 45	261	34 39	347
213	45 02	046	31 40	084	12 37	138	33 26	174	37 55	223	44 21	260	34 33	346
214	45 19	044	32 05	083	12 53	137	33 28	174	37 39	222	43 57	260	34 27	345
215	45 36	043	32 29	082	13 10	136	33 31	173	37 22	222	43 33	259	34 20	344
216	45 52	042	32 53	081	13 27	136	33 34	172	37 06	221	43 09	258	34 13	342
217	46 09	041	33 17	080	13 44	135	33 38	172	36 50	220	42 45	257	34 05	341
218	46 24	039	33 41	079	14 02	134	33 41	171	36 34	220	42 22	256	33 57	340
219	46 40	038	34 05	078	14 20	133	33 45	171	36 19	219	41 58	255	33 49	339
220	46 54	037	34 29	077	14 38	132	33 49	170	36 04	218	41 34	255	33 40	338
221	47 09	036	34 53	076	14 56	131	33 54	169	35 49	218	41 11	254	33 30	337
222	47 23	034	35 16	075	15 14	130	33 58	169	35 34	217	40 48	253	33 20	335
223	47 36	033	35 40	074	15 33	130	34 03	168	35 19	216	40 24	252	33 10	334
224	47 50	032	36 03	073	15 52	129	34 08	167	35 05	216	40 01	251	32 59	333
	♦ANTARES		Nunki		♦FOMALHAUT		ACHERNAR		CANOPUS		♦Suhail		SPICA	
225	48 02	031	36 27	072	16 11	128	34 14	167	34 51	215	39 38	251	32 48	332
226	48 14	029	36 50	071	16 30	127	34 20	166	34 37	214	39 15	250	32 36	331
227	48 26	028	37 13	070	16 50	126	34 26	165	34 23	214	38 52	249	32 24	330
228	48 37	027	37 36	069	17 10	125	34 32	165	34 10	213	38 30	248	32 11	329
229	48 48	025	37 58	068	17 30	125	34 39	164	33 57	212	38 07	247	31 59	327
230	48 58	024	38 21	067	17 50	124	34 45	163	33 44	212	37 45	247	31 45	326
231	49 08	023	38 43	066	18 10	123	34 52	163	33 31	211	37 22	246	31 31	325
232	49 17	021	39 06	065	18 31	122	35 00	162	33 19	210	37 00	245	31 17	324
233	49 25	020	39 28	064	18 52	121	35 07	162	33 06	210	36 38	244	31 03	323
234	49 33	018	39 50	063	19 13	120	35 15	161	32 55	209	36 16	243	30 48	322
235	49 41	017	40 11	062	19 34	120	35 23	160	32 43	208	35 54	243	30 33	321
236	49 48	016	40 33	061	19 55	119	35 32	160	32 32	207	35 33	242	30 17	320
237	49 54	014	40 54	060	20 17	118	35 40	159	32 20	207	35 11	241	30 01	319
238	50 00	013	41 15	059	20 38	117	35 49	158	32 10	206	34 50	240	29 45	317
239	50 05	012	41 35	058	21 00	116	35 58	158	31 59	205	34 29	240	29 28	316
	♦ANTARES		Nunki		♦FOMALHAUT		ACHERNAR		CANOPUS		♦ACRUX		SPICA	
240	50 09	010	41 56	056	21 22	115	36 08	157	31 49	205	67 33	252	29 11	315
241	50 14	009	42 16	055	21 44	114	36 17	156	31 38	204	67 10	252	28 54	314
242	50 17	007	42 36	054	22 06	114	36 27	156	31 29	203	66 47	251	28 36	313
243	50 20	006	42 56	053	22 29	113	36 37	155	31 19	203	66 24	250	28 18	312
244	50 22	005	43 15	052	22 51	112	36 48	154	31 10	202	66 01	250	28 00	311
245	50 24	003	43 34	051	23 14	111	36 58	154	31 01	201	65 38	249	27 41	310
246	50 25	002	43 53	050	23 37	110	37 09	153	30 52	201	65 15	249	27 22	309
247	50 25	000	44 11	048	24 00	109	37 20	153	30 44	200	64 53	248	27 03	308
248	50 25	359	44 29	047	24 23	108	37 32	152	30 35	199	64 30	248	26 44	307
249	50 24	358	44 47	046	24 46	108	37 43	151	30 27	199	64 08	247	26 24	306
250	50 23	356	45 04	045	25 10	107	37 55	151	30 20	198	63 45	246	26 04	305
251	50 21	355	45 21	044	25 33	106	38 07	150	30 12	197	63 23	246	25 44	304
252	50 19	353	45 38	042	25 57	105	38 20	149	30 05	197	63 01	245	25 24	303
253	50 15	352	45 54	041	26 20	104	38 32	149	29 59	196	62 39	245	25 03	302
254	50 12	351	46 10	040	26 44	103	38 45	148	29 52	195	62 17	244	24 42	301
	♦Nunki		FOMALHAUT		♦ACHERNAR		CANOPUS		♦ACRUX		SPICA		ANTARES	
255	46 26	039	27 08	102	38 58	147	29 46	194	61 55	243	24 21	300	50 07	349
256	46 41	037	27 31	102	39 11	147	29 40	194	61 33	243	24 00	299	50 03	348
257	46 55	036	27 55	101	39 25	146	29 34	193	61 11	242	23 38	298	49 57	346
258	47 10	035	28 19	100	39 38	146	29 29	192	60 50	242	23 16	297	49 51	345
259	47 23	034	28 44	099	39 52	145	29 24	192	60 28	241	22 55	296	49 45	344
260	47 37	032	29 08	098	40 06	144	29 19	191	60 07	241	22 33	295	49 37	342
261	47 49	031	29 32	097	40 21	144	29 14	190	59 46	240	22 10	294	49 30	341
262	48 02	030	29 56	096	40 35	143	29 10	190	59 25	239	21 48	293	49 21	340
263	48 14	028	30 20	095	40 50	142	29 06	189	59 04	239	21 25	292	49 13	338
264	48 25	027	30 45	095	41 05	142	29 03	188	58 43	238	21 03	291	49 03	337
265	48 36	026	31 09	094	41 20	141	28 59	188	58 22	238	20 40	290	48 53	336
266	48 46	024	31 33	093	41 36	140	28 56	187	58 02	237	20 17	289	48 43	334
267	48 56	023	31 58	092	41 51	140	28 53	186	57 41	237	19 53	288	48 32	333
268	49 06	022	32 22	091	42 07	139	28 51	185	57 21	236	19 30	287	48 21	332
269	49 14	020	32 47	090	42 23	139	28 49	185	57 01	236	19 07	286	48 09	330

LAT 66°S

LHA	Hc	Zn	Hc	Zn	Hc	Zn	Hc	Zn	Hc	Zn	Hc	Zn	Hc	Zn
♈	♦Nunki		FOMALHAUT		♦ACHERNAR		CANOPUS		♦ACRUX		SPICA		ANTARES	
270	49 23	019	33 11	089	42 40	138	28 47	184	56 41	235	18 43	285	47 57	329
271	49 30	018	33 35	088	42 56	137	28 45	183	56 21	234	18 20	284	47 44	328
272	49 37	016	34 00	087	43 13	137	28 44	183	56 01	234	17 56	283	47 30	326
273	49 44	015	34 24	086	43 29	136	28 43	182	55 41	233	17 32	282	47 17	325
274	49 50	014	34 48	085	43 46	135	28 42	181	55 22	233	17 08	281	47 02	324
275	49 56	012	35 13	084	44 04	135	28 42	181	55 02	232	16 44	280	46 48	323
276	50 00	011	35 37	084	44 21	134	28 42	180	54 43	232	16 20	279	46 33	321
277	50 05	009	36 01	083	44 39	133	28 42	179	54 24	231	15 56	278	46 17	320
278	50 08	008	36 25	082	44 57	133	28 42	179	54 05	231	15 32	278	46 01	319
279	50 11	007	36 50	081	45 15	132	28 43	178	53 47	230	15 08	277	45 45	318
280	50 14	005	37 14	080	45 33	132	28 44	177	53 28	229	14 43	276	45 28	316
281	50 16	004	37 38	079	45 51	131	28 45	177	53 09	229	14 19	275	45 11	315
282	50 17	002	38 01	078	46 10	130	28 47	176	52 51	228	13 55	274	44 54	314
283	50 18	001	38 25	077	46 28	130	28 49	175	52 33	228	13 30	273	44 36	313
284	50 18	000	38 49	076	46 47	129	28 51	174	52 15	227	13 06	272	44 18	312
	♦ALTAIR		FOMALHAUT		♦ACHERNAR		CANOPUS		♦RIGIL KENT.		ANTARES		Nunki	
285	14 34	013	39 13	075	47 06	128	28 54	174	61 44	249	44 00	310	50 18	358
286	14 39	012	39 36	074	47 25	128	28 56	173	61 22	249	43 41	309	50 17	357
287	14 44	011	39 59	073	47 45	127	28 59	172	60 59	248	43 22	308	50 15	355
288	14 49	010	40 23	072	48 04	126	29 03	172	60 36	247	43 03	307	50 13	354
289	14 52	009	40 46	071	48 24	126	29 07	171	60 14	247	42 43	306	50 10	353
290	14 56	008	41 09	070	48 44	125	29 10	170	59 52	246	42 23	305	50 07	351
291	14 59	007	41 32	069	49 04	125	29 15	170	59 29	245	42 03	304	50 03	350
292	15 02	006	41 54	068	49 24	124	29 19	169	59 07	245	41 42	302	49 58	348
293	15 04	005	42 17	067	49 44	123	29 24	168	58 45	244	41 22	301	49 53	347
294	15 06	004	42 39	065	50 05	123	29 29	168	58 23	244	41 01	300	49 47	346
295	15 07	003	43 01	064	50 26	122	29 35	167	58 02	243	40 39	299	49 41	344
296	15 08	002	43 23	063	50 46	121	29 40	166	57 40	242	40 18	298	49 34	343
297	15 09	001	43 45	062	51 07	121	29 46	165	57 18	242	39 56	297	49 26	342
298	15 09	000	44 06	061	51 28	120	29 53	165	56 57	241	39 34	296	49 18	340
299	15 08	359	44 27	060	51 50	119	29 59	164	56 36	240	39 12	295	49 10	339
	♦FOMALHAUT		ACHERNAR		♦CANOPUS		RIGIL KENT.		♦ANTARES		Nunki		ALTAIR	
300	44 48	059	52 11	119	30 06	163	56 14	240	38 50	294	49 01	337	15 07	358
301	45 09	058	52 32	118	30 13	163	55 53	239	38 28	293	48 51	336	15 06	357
302	45 30	057	52 54	117	30 20	162	55 32	239	38 05	292	48 41	335	15 05	356
303	45 50	055	53 16	117	30 28	161	55 12	238	37 42	291	48 30	333	15 02	355
304	46 10	054	53 38	116	30 36	161	54 51	237	37 20	290	48 19	332	15 00	353
305	46 30	053	54 00	115	30 44	160	54 31	237	36 57	289	48 08	331	14 57	352
306	46 49	052	54 22	115	30 53	159	54 10	236	36 33	288	47 55	330	14 53	351
307	47 08	051	54 44	114	31 02	159	53 50	236	36 10	287	47 43	328	14 50	350
308	47 27	050	55 06	113	31 11	158	53 30	235	35 47	286	47 30	327	14 45	349
309	47 45	048	55 29	113	31 20	157	53 10	234	35 23	285	47 16	326	14 41	348
310	48 03	047	55 51	112	31 29	157	52 50	234	34 59	284	47 02	324	14 36	347
311	48 21	046	56 14	111	31 39	156	52 31	233	34 36	283	46 48	323	14 30	346
312	48 38	045	56 37	110	31 49	155	52 11	233	34 12	282	46 33	322	14 24	345
313	48 55	043	57 00	110	32 00	155	51 52	232	33 48	281	46 18	321	14 18	344
314	49 12	042	57 23	109	32 10	154	51 33	231	33 24	280	46 02	319	14 11	343
	♦FOMALHAUT		ACHERNAR		♦CANOPUS		Miaplacidus		RIGIL KENT.		♦ANTARES		Nunki	
315	49 28	041	57 46	108	32 21	153	45 43	182	51 14	231	33 00	279	45 46	318
316	49 44	040	58 09	108	32 32	152	45 42	181	50 55	230	32 36	278	45 29	317
317	49 59	038	58 33	107	32 44	152	45 42	181	50 36	230	32 11	277	45 13	316
318	50 14	037	58 56	106	32 56	151	45 42	180	50 18	229	31 47	276	44 55	315
319	50 29	036	59 19	105	33 07	150	45 42	180	49 59	228	31 23	275	44 38	313
320	50 43	034	59 43	105	33 20	150	45 42	179	49 41	228	30 59	274	44 20	312
321	50 56	033	60 07	104	33 32	149	45 43	179	49 23	227	30 34	273	44 02	311
322	51 09	032	60 30	103	33 45	148	45 43	178	49 05	227	30 10	272	43 43	310
323	51 22	030	60 54	103	33 58	148	45 44	178	48 48	226	29 45	271	43 24	309
324	51 34	029	61 18	102	34 11	147	45 45	177	48 30	226	29 21	270	43 05	308
325	51 46	028	61 42	101	34 24	146	45 47	177	48 13	225	28 57	270	42 45	306
326	51 57	026	62 06	100	34 38	146	45 48	176	47 56	224	28 32	269	42 26	305
327	52 07	025	62 30	099	34 52	145	45 50	176	47 39	224	28 08	268	42 06	304
328	52 17	024	62 54	099	35 06	144	45 52	175	47 22	223	27 43	267	41 45	303
329	52 27	022	63 18	098	35 20	144	45 54	175	47 05	223	27 19	266	41 25	302
	♦FOMALHAUT		ACHERNAR		♦CANOPUS		Miaplacidus		RIGIL KENT.		♦ANTARES		Nunki	
330	52 36	021	63 42	097	35 35	143	45 56	174	46 49	222	26 55	265	41 04	301
331	52 44	019	64 07	096	35 50	142	45 59	174	46 33	221	26 30	264	40 43	300
332	52 52	018	64 31	095	36 05	142	46 02	173	46 17	221	26 06	263	40 21	299
333	52 59	016	64 55	095	36 20	141	46 05	173	46 01	220	25 42	262	40 00	298
334	53 06	015	65 20	094	36 36	140	46 08	172	45 45	220	25 18	261	39 38	296
335	53 12	014	65 44	093	36 51	139	46 11	172	45 30	219	24 54	261	39 16	295
336	53 17	012	66 08	092	37 07	139	46 15	171	45 14	218	24 30	260	38 54	294
337	53 22	011	66 33	091	37 24	138	46 19	171	44 59	218	24 06	259	38 32	293
338	53 26	009	66 57	090	37 40	137	46 23	170	44 44	217	23 42	258	38 09	292
339	53 30	008	67 21	089	37 57	137	46 27	170	44 30	217	23 18	257	37 47	291
340	53 33	006	67 46	088	38 13	136	46 31	169	44 15	216	22 54	256	37 24	290
341	53 35	005	68 10	087	38 30	135	46 36	169	44 01	215	22 31	255	37 01	289
342	53 37	003	68 35	086	38 48	135	46 41	168	43 47	215	22 07	254	36 38	288
343	53 38	002	68 59	085	39 05	134	46 46	168	43 33	214	21 44	254	36 14	287
344	53 39	001	69 23	084	39 23	133	46 51	167	43 19	214	21 20	253	35 51	286
	Diphda		♦CANOPUS		Miaplacidus		RIGIL KENT.		♦ANTARES		Nunki		♦FOMALHAUT	
345	39 06	032	39 41	133	46 57	167	43 06	213	20 57	252	35 28	285	53 39	359
346	39 19	031	39 59	132	47 02	166	42 53	212	20 34	251	35 04	284	53 38	358
347	39 31	030	40 17	131	47 08	166	42 40	212	20 11	250	34 40	283	53 37	356
348	39 43	029	40 35	131	47 14	165	42 27	211	19 48	249	34 16	282	53 35	355
349	39 55	027	40 54	130	47 21	165	42 14	211	19 25	248	33 53	281	53 32	353
350	40 06	026	41 13	129	47 27	164	42 02	210	19 03	248	33 29	280	53 29	352
351	40 16	025	41 32	129	47 34	164	41 50	210	18 40	247	33 05	279	53 25	350
352	40 26	024	41 51	128	47 41	163	41 38	209	18 18	246	32 40	278	53 21	349
353	40 36	023	42 10	127	47 48	163	41 26	208	17 56	245	32 16	277	53 16	347
354	40 45	021	42 30	126	47 55	162	41 15	208	17 33	244	31 52	277	53 10	346
355	40 54	020	42 50	126	48 03	162	41 04	207	17 12	243	31 28	276	53 04	344
356	41 02	019	43 09	125	48 10	161	40 53	207	16 50	242	31 03	275	52 57	343
357	41 09	018	43 29	124	48 18	161	40 42	206	16 28	242	30 39	274	52 50	342
358	41 16	016	43 50	124	48 26	160	40 31	205	16 07	241	30 15	273	52 42	340
359	41 23	015	44 10	123	48 34	160	40 21	205	15 46	240	29 50	272	52 33	339

LAT 67°S

LHA ♈	Diphda Hc	Zn	◆RIGEL Hc	Zn	CANOPUS Hc	Zn	Miaplacidus Hc	Zn	◆RIGIL KENT. Hc	Zn	Peacock Hc	Zn	◆FOMALHAUT Hc	Zn
0	40 31	014	12 00	083	45 02	121	49 39	159	41 06	204	63 44	266	51 28	338
1	40 36	012	12 24	082	45 22	121	49 47	159	40 56	204	63 20	266	51 19	337
2	40 41	011	12 47	081	45 43	120	49 56	158	40 47	203	62 57	265	51 10	335
3	40 45	010	13 10	080	46 03	119	50 05	158	40 38	203	62 33	264	51 00	334
4	40 49	009	13 33	079	46 23	119	50 14	157	40 29	202	62 10	263	50 49	332
5	40 52	007	13 56	078	46 44	118	50 23	157	40 20	201	61 47	262	50 38	331
6	40 55	006	14 19	077	47 05	117	50 33	156	40 11	201	61 24	261	50 26	330
7	40 57	005	14 42	076	47 26	116	50 42	156	40 03	200	61 01	261	50 14	328
8	40 59	004	15 04	075	47 47	116	50 52	155	39 55	200	60 37	260	50 02	327
9	41 00	002	15 27	074	48 08	115	51 02	155	39 48	199	60 14	259	49 49	326
10	41 01	001	15 49	073	48 29	114	51 12	154	39 40	198	59 51	258	49 36	325
11	41 01	000	16 12	072	48 51	113	51 22	154	39 33	198	59 29	257	49 22	323
12	41 00	359	16 34	071	49 12	113	51 33	153	39 26	197	59 06	257	49 07	322
13	41 00	357	16 56	070	49 34	112	51 43	153	39 19	197	58 43	256	48 53	321
14	40 58	356	17 18	069	49 56	111	51 54	152	39 12	196	58 20	255	48 38	319

LHA ♈	◆RIGEL Hc	Zn	CANOPUS Hc	Zn	Miaplacidus Hc	Zn	◆RIGIL KENT. Hc	Zn	Peacock Hc	Zn	◆FOMALHAUT Hc	Zn	Diphda Hc	Zn
15	17 40	068	50 18	110	52 05	152	39 06	195	57 58	254	48 22	318	40 56	355
16	18 02	068	50 40	110	52 16	151	39 00	195	57 35	254	48 06	317	40 54	353
17	18 23	067	51 02	109	52 27	151	38 54	194	57 13	253	47 50	316	40 51	352
18	18 45	066	51 24	108	52 39	150	38 49	193	56 50	252	47 34	314	40 48	351
19	19 06	065	51 47	107	52 51	150	38 43	193	56 28	251	47 17	313	40 44	350
20	19 27	064	52 09	107	53 02	149	38 38	192	56 06	251	47 00	312	40 39	348
21	19 48	063	52 32	106	53 14	149	38 33	192	55 44	250	46 42	311	40 34	347
22	20 09	062	52 54	105	53 27	149	38 29	191	55 22	249	46 24	310	40 29	346
23	20 29	061	53 17	104	53 39	148	38 24	190	55 00	248	46 06	308	40 23	345
24	20 50	060	53 40	103	53 51	148	38 20	190	54 38	248	45 47	307	40 17	344
25	21 10	059	54 02	103	54 04	147	38 17	189	54 17	247	45 29	306	40 10	342
26	21 30	058	54 25	102	54 17	147	38 13	189	53 55	246	45 10	305	40 02	341
27	21 49	057	54 48	101	54 30	146	38 10	188	53 34	246	44 50	304	39 54	340
28	22 09	056	55 11	100	54 43	146	38 06	187	53 12	245	44 31	303	39 46	339
29	22 28	055	55 34	099	54 56	145	38 04	187	52 51	244	44 11	302	39 37	337

LHA ♈	◆RIGEL Hc	Zn	CANOPUS Hc	Zn	Miaplacidus Hc	Zn	◆RIGIL KENT. Hc	Zn	Peacock Hc	Zn	◆FOMALHAUT Hc	Zn	Diphda Hc	Zn
30	22 47	054	55 58	099	55 10	145	38 01	186	52 30	244	43 51	300	39 28	336
31	23 06	053	56 21	098	55 23	144	37 59	185	52 09	243	43 30	299	39 18	335
32	23 24	052	56 44	097	55 37	144	37 57	185	51 48	242	43 10	298	39 08	334
33	23 43	051	57 07	096	55 51	143	37 55	184	51 28	241	42 49	297	38 58	333
34	24 01	050	57 31	095	56 05	143	37 53	184	51 07	241	42 28	296	38 47	331
35	24 18	048	57 54	094	56 19	142	37 52	183	50 47	240	42 07	295	38 35	330
36	24 36	047	58 17	093	56 34	142	37 51	182	50 27	239	41 46	294	38 23	329
37	24 53	046	58 41	093	56 48	142	37 50	182	50 06	239	41 24	293	38 11	328
38	25 10	045	59 04	092	57 03	141	37 49	181	49 46	238	41 02	292	37 58	327
39	25 26	044	59 28	091	57 18	141	37 49	181	49 27	237	40 41	291	37 45	325
40	25 42	043	59 51	090	57 32	140	37 49	180	49 07	237	40 19	290	37 32	324
41	25 58	042	60 15	089	57 48	140	37 49	179	48 47	236	39 57	289	37 18	323
42	26 14	041	60 38	088	58 03	139	37 50	179	48 28	235	39 34	288	37 03	322
43	26 29	040	61 01	087	58 18	139	37 50	178	48 09	235	39 12	287	36 49	321
44	26 44	039	61 25	086	58 34	138	37 51	177	47 50	234	38 49	286	36 34	320

LHA ♈	RIGEL Hc	Zn	◆SIRIUS Hc	Zn	Suhail Hc	Zn	◆RIGIL KENT. Hc	Zn	Peacock Hc	Zn	◆FOMALHAUT Hc	Zn	Diphda Hc	Zn
45	26 59	038	28 12	065	38 32	112	37 52	177	47 31	233	38 27	285	36 18	318
46	27 13	037	28 34	064	38 54	111	37 54	176	47 12	233	38 04	284	36 03	317
47	27 27	036	28 54	063	39 16	110	37 55	176	46 54	232	37 41	283	35 47	316
48	27 40	035	29 15	062	39 38	109	37 57	175	46 35	231	37 18	282	35 30	315
49	27 53	034	29 36	061	40 00	109	38 00	174	46 17	231	36 55	281	35 14	314
50	28 06	032	29 56	060	40 22	108	38 02	174	45 59	230	36 32	280	34 57	313
51	28 18	031	30 16	058	40 44	107	38 05	173	45 41	229	36 09	279	34 39	312
52	28 30	030	30 36	057	41 07	106	38 08	172	45 23	229	35 46	278	34 22	311
53	28 42	029	30 56	056	41 30	105	38 11	172	45 06	228	35 23	277	34 04	310
54	28 53	028	31 15	055	41 52	105	38 14	171	44 48	227	34 59	276	33 45	309
55	29 04	027	31 34	054	42 15	104	38 18	171	44 31	227	34 36	275	33 27	307
56	29 15	026	31 53	053	42 38	103	38 22	170	44 14	226	34 13	274	33 08	306
57	29 25	025	32 12	052	43 01	102	38 26	169	43 58	225	33 49	273	32 49	305
58	29 34	024	32 30	051	43 24	101	38 31	169	43 41	225	33 26	272	32 30	304
59	29 43	022	32 48	050	43 47	100	38 35	168	43 25	224	33 02	271	32 10	303

LHA ♈	RIGEL Hc	Zn	◆SIRIUS Hc	Zn	Suhail Hc	Zn	◆RIGIL KENT. Hc	Zn	Peacock Hc	Zn	◆FOMALHAUT Hc	Zn	Diphda Hc	Zn
60	29 52	021	33 06	049	44 10	099	38 40	168	43 08	223	32 39	270	31 51	302
61	30 00	020	33 23	048	44 33	099	38 45	167	42 52	223	32 15	269	31 31	301
62	30 08	019	33 41	047	44 56	098	38 51	166	42 36	222	31 52	268	31 11	300
63	30 16	018	33 58	046	45 19	097	38 56	166	42 21	222	31 29	268	30 50	299
64	30 23	017	34 14	045	45 43	096	39 02	165	42 05	221	31 05	267	30 30	298
65	30 29	016	34 30	043	46 06	095	39 08	164	41 50	220	30 42	266	30 09	297
66	30 35	014	34 46	042	46 29	094	39 15	164	41 35	220	30 18	265	29 48	296
67	30 41	013	35 02	041	46 53	093	39 21	163	41 20	219	29 55	264	29 27	295
68	30 46	012	35 17	040	47 16	092	39 28	163	41 06	218	29 32	263	29 05	294
69	30 51	011	35 32	039	47 40	091	39 35	162	40 51	218	29 09	262	28 44	293
70	30 55	010	35 47	038	48 03	091	39 43	161	40 37	217	28 45	261	28 22	292
71	30 59	009	36 01	037	48 26	090	39 50	161	40 23	216	28 22	260	28 00	291
72	31 02	008	36 15	035	48 50	089	39 58	160	40 09	216	27 59	260	27 38	290
73	31 05	006	36 28	034	49 13	088	40 06	160	39 56	215	27 36	259	27 16	289
74	31 07	005	36 41	033	49 37	087	40 15	159	39 42	214	27 13	258	26 54	288

LHA ♈	RIGEL Hc	Zn	◆SIRIUS Hc	Zn	Suhail Hc	Zn	◆RIGIL KENT. Hc	Zn	Peacock Hc	Zn	◆FOMALHAUT Hc	Zn	Diphda Hc	Zn
75	31 09	004	36 54	032	50 00	086	40 23	158	39 29	214	26 50	257	26 32	287
76	31 11	003	37 06	031	50 24	085	40 32	158	39 16	213	26 27	256	26 09	286
77	31 12	002	37 18	030	50 47	084	40 41	157	39 04	212	26 05	255	25 46	285
78	31 12	001	37 29	028	51 10	083	40 50	157	38 51	212	25 42	254	25 24	284
79	31 12	000	37 40	027	51 33	082	41 00	156	38 39	211	25 20	253	25 01	283
80	31 12	358	37 51	026	51 57	081	41 09	155	38 27	210	24 57	253	24 38	282
81	31 11	357	38 01	025	52 20	080	41 19	155	38 15	210	24 35	252	24 15	281
82	31 10	356	38 10	024	52 43	079	41 29	154	38 04	209	24 13	251	23 52	280
83	31 08	355	38 19	022	53 06	078	41 40	153	37 53	208	23 51	250	23 29	279
84	31 05	354	38 28	021	53 29	077	41 50	153	37 41	208	23 29	249	23 06	278
85	31 03	353	38 36	020	53 51	076	42 01	152	37 31	207	23 07	248	22 43	277
86	30 59	351	38 44	019	54 14	075	42 12	152	37 20	206	22 45	247	22 19	276
87	30 56	350	38 52	018	54 37	074	42 23	151	37 10	206	22 24	247	21 56	275
88	30 51	349	38 58	016	54 59	073	42 35	150	37 00	205	22 02	246	21 33	275
89	30 47	348	39 05	015	55 21	072	42 46	150	36 50	205	21 41	245	21 09	274

LHA ♈	◆SIRIUS Hc	Zn	Suhail Hc	Zn	◆RIGIL KENT. Hc	Zn	Peacock Hc	Zn	◆ACHERNAR Hc	Zn	Acamar Hc	Zn	RIGEL Hc	Zn
90	39 11	014	55 44	071	42 58	149	36 40	204	59 29	256	53 34	294	30 42	347
91	39 16	013	56 06	069	43 10	149	36 31	203	59 07	255	53 12	293	30 36	346
92	39 21	011	56 27	068	43 23	148	36 22	203	58 44	254	52 51	292	30 30	345
93	39 25	010	56 49	067	43 35	147	36 13	202	58 21	254	52 29	290	30 24	343
94	39 29	009	57 11	066	43 48	147	36 04	201	57 59	253	52 07	289	30 17	342
95	39 33	008	57 32	065	44 01	146	35 56	201	57 37	252	51 45	288	30 09	341
96	39 36	007	57 53	064	44 14	146	35 48	200	57 14	252	51 22	287	30 02	340
97	39 38	005	58 14	062	44 28	145	35 40	199	56 52	251	51 00	286	29 53	339
98	39 40	004	58 35	061	44 41	144	35 32	199	56 30	250	50 37	285	29 45	338
99	39 41	003	58 55	060	44 55	144	35 25	198	56 08	249	50 14	284	29 36	337
100	39 42	002	59 15	059	45 09	143	35 18	197	55 46	249	49 52	283	29 26	335
101	39 43	000	59 35	057	45 23	143	35 11	197	55 24	248	49 29	282	29 16	334
102	39 42	359	59 55	056	45 37	142	35 04	196	55 03	247	49 06	281	29 06	333
103	39 42	358	60 14	055	45 52	141	34 58	195	54 41	247	48 43	280	28 55	332
104	39 41	357	60 33	053	46 07	141	34 52	195	54 20	246	48 20	279	28 44	331

LHA ♈	Alphard Hc	Zn	◆Gienah Hc	Zn	RIGIL KENT. Hc	Zn	◆Peacock Hc	Zn	ACHERNAR Hc	Zn	◆RIGEL Hc	Zn	SIRIUS Hc	Zn
105	26 35	042	20 25	087	46 22	140	34 46	194	53 58	245	28 32	330	39 39	355
106	26 50	040	20 48	086	46 37	140	34 40	193	53 37	244	28 20	329	39 37	354
107	27 05	039	21 12	085	46 52	139	34 35	193	53 16	244	28 08	328	39 34	353
108	27 20	038	21 35	084	47 08	138	34 30	192	52 55	243	27 55	327	39 31	352
109	27 34	037	21 58	083	47 23	138	34 25	191	52 34	242	27 42	326	39 27	350
110	27 48	036	22 22	082	47 39	137	34 21	191	52 14	242	27 29	324	39 23	349
111	28 02	035	22 45	081	47 55	136	34 17	190	51 53	241	27 15	323	39 18	348
112	28 15	034	23 08	080	48 11	136	34 13	189	51 33	240	27 01	322	39 13	347
113	28 28	033	23 31	079	48 28	135	34 09	189	51 12	240	26 46	321	39 08	345
114	28 40	032	23 54	078	48 44	135	34 06	188	50 52	239	26 31	320	39 01	344
115	28 53	031	24 17	077	49 01	134	34 02	187	50 32	238	26 16	319	38 55	343
116	29 04	030	24 40	076	49 18	133	33 59	187	50 12	238	26 01	318	38 48	342
117	29 16	028	25 03	075	49 35	133	33 57	186	49 52	237	25 45	317	38 40	341
118	29 27	027	25 25	075	49 52	132	33 54	185	49 33	236	25 29	316	38 32	339
119	29 37	026	25 48	074	50 10	132	33 52	185	49 13	236	25 12	315	38 24	338

LHA ♈	Alphard Hc	Zn	◆SPICA Hc	Zn	RIGIL KENT. Hc	Zn	◆Peacock Hc	Zn	ACHERNAR Hc	Zn	◆RIGEL Hc	Zn	SIRIUS Hc	Zn
120	29 47	025	13 40	086	50 28	131	33 50	184	48 54	235	24 55	314	38 15	337
121	29 57	024	14 03	085	50 45	130	33 49	184	48 35	234	24 38	313	38 05	336
122	30 06	023	14 27	084	51 03	130	33 48	183	48 16	234	24 21	312	37 55	334
123	30 15	022	14 50	084	51 21	129	33 47	182	47 57	233	24 03	311	37 45	333
124	30 24	021	15 13	083	51 40	129	33 46	182	47 39	232	23 45	310	37 34	332
125	30 32	019	15 36	082	51 58	128	33 45	181	47 20	232	23 27	309	37 23	331
126	30 39	018	15 59	081	52 17	127	33 45	180	47 02	231	23 09	308	37 11	330
127	30 46	017	16 23	080	52 35	127	33 45	180	46 44	230	22 50	307	36 59	329
128	30 53	016	16 46	079	52 54	126	33 45	179	46 26	230	22 31	306	36 47	327
129	30 59	015	17 09	078	53 13	125	33 46	178	46 08	229	22 12	305	36 34	326
130	31 05	014	17 31	077	53 32	125	33 47	178	45 50	228	21 52	304	36 21	325
131	31 10	013	17 54	076	53 52	124	33 48	177	45 33	228	21 33	303	36 07	324
132	31 15	011	18 17	075	54 11	124	33 49	176	45 16	227	21 13	302	35 53	323
133	31 20	010	18 40	074	54 31	123	33 51	176	44 59	226	20 53	301	35 39	322
134	31 24	009	19 02	073	54 50	122	33 53	175	44 42	226	20 32	300	35 24	321

LHA ♈	Alphard Hc	Zn	◆SPICA Hc	Zn	RIGIL KENT. Hc	Zn	◆Peacock Hc	Zn	ACHERNAR Hc	Zn	◆CANOPUS Hc	Zn	SIRIUS Hc	Zn
135	31 27	008	19 24	072	55 10	122	33 55	174	44 25	225	66 22	288	35 09	319
136	31 30	007	19 47	071	55 30	121	33 58	174	44 09	224	65 59	287	34 54	318
137	31 33	006	20 09	070	55 50	120	34 00	173	43 52	224	65 37	285	34 38	317
138	31 35	004	20 31	069	56 11	120	34 03	172	43 36	223	65 14	284	34 22	316
139	31 36	003	20 53	068	56 31	119	34 07	172	43 20	223	64 51	283	34 05	315
140	31 37	002	21 14	067	56 52	119	34 10	171	43 04	222	64 29	282	33 49	314
141	31 38	001	21 36	066	57 12	118	34 14	170	42 49	221	64 06	281	33 32	313
142	31 38	000	21 57	065	57 33	117	34 18	170	42 33	221	63 43	280	33 14	312
143	31 38	359	22 19	064	57 54	117	34 23	169	42 18	220	63 19	279	32 56	311
144	31 37	357	22 40	063	58 15	116	34 27	168	42 03	219	62 56	278	32 38	309
145	31 36	356	23 00	062	58 36	115	34 32	168	41 49	219	62 33	277	32 20	308
146	31 34	355	23 21	061	58 57	115	34 37	167	41 34	218	62 10	276	32 02	307
147	31 32	354	23 42	060	59 19	114	34 43	166	41 20	217	61 46	275	31 43	306
148	31 29	353	24 02	059	59 40	113	34 48	166	41 06	217	61 23	274	31 24	305
149	31 26	352	24 22	058	60 02	113	34 54	165	40 52	216	61 00	273	31 05	304

LHA ♈	◆SPICA Hc	Zn	RIGIL KENT. Hc	Zn	◆Peacock Hc	Zn	ACHERNAR Hc	Zn	◆CANOPUS Hc	Zn	SIRIUS Hc	Zn	Alphard Hc	Zn
150	24 42	057	60 23	112	35 00	164	40 38	215	60 36	272	30 45	303	31 22	351
151	25 01	056	60 45	111	35 07	164	40 25	215	60 13	271	30 25	302	31 18	349
152	25 21	055	61 07	111	35 13	163	40 11	214	59 49	270	30 05	301	31 14	348
153	25 40	054	61 29	110	35 20	162	39 58	213	59 26	269	29 45	300	31 09	347
154	25 59	053	61 51	109	35 28	162	39 45	213	59 02	268	29 25	299	31 03	346
155	26 18	052	62 13	109	35 35	161	39 33	212	58 39	267	29 04	298	30 57	345
156	26 36	051	62 35	108	35 43	160	39 21	212	58 16	266	28 43	297	30 51	344
157	26 54	050	62 58	107	35 51	160	39 08	211	57 52	266	28 22	296	30 44	342
158	27 12	049	63 20	107	35 59	159	38 56	210	57 29	265	28 01	295	30 37	341
159	27 30	048	63 43	106	36 07	158	38 45	210	57 05	264	27 40	294	30 29	340
160	27 47	047	64 05	105	36 16	158	38 33	209	56 42	263	27 18	293	30 21	339
161	28 04	046	64 28	105	36 25	157	38 22	208	56 19	262	26 56	292	30 12	338
162	28 21	045	64 51	104	36 34	157	38 11	208	55 56	261	26 35	291	30 03	337
163	28 37	044	65 13	103	36 44	156	38 00	207	55 33	261	26 13	290	29 54	336
164	28 53	043	65 36	102	36 53	155	37 50	206	55 09	260	25 51	289	29 44	335

LHA ♈	◆SPICA Hc	Zn	ANTARES Hc	Zn	◆Peacock Hc	Zn	ACHERNAR Hc	Zn	◆CANOPUS Hc	Zn	SIRIUS Hc	Zn	Alphard Hc	Zn
165	29 09	042	27 10	094	37 03	155	37 40	206	54 46	259	25 28	288	29 34	333
166	29 24	041	27 34	093	37 14	154	37 29	205	54 23	258	25 06	287	29 23	332
167	29 39	039	27 57	092	37 24	153	37 20	204	54 01	257	24 43	286	29 12	331
168	29 54	038	28 20	091	37 35	153	37 10	204	53 38	256	24 21	285	29 01	330
169	30 08	037	28 44	090	37 46	152	37 01	203	53 15	256	23 58	284	28 49	329
170	30 22	036	29 07	089	37 57	151	36 52	202	52 52	255	23 35	283	28 36	328
171	30 36	035	29 31	089	38 08	151	36 43	202	52 30	254	23 12	282	28 24	327
172	30 49	034	29 54	088	38 20	150	36 34	201	52 07	253	22 50	281	28 11	326
173	31 02	033	30 18	087	38 32	149	36 26	200	51 45	253	22 26	280	27 57	325
174	31 15	032	30 41	086	38 44	149	36 18	200	51 22	252	22 03	279	27 44	324
175	31 27	031	31 04	085	38 56	148	36 10	199	51 00	251	21 40	278	27 29	322
176	31 38	029	31 28	084	39 08	147	36 03	199	50 38	250	21 17	277	27 15	321
177	31 50	028	31 51	083	39 21	147	35 55	198	50 16	249	20 54	276	27 00	320
178	32 01	027	32 14	082	39 34	146	35 48	197	49 54	249	20 30	275	26 45	319
179	32 11	026	32 37	081	39 47	145	35 41	197	49 32	248	20 07	275	26 30	318

LAT 67°S

LHA	Hc	Zn	Hc	Zn	Hc	Zn	Hc	Zn	Hc	Zn	Hc	Zn	Hc	Zn
♈	◆SPICA		ANTARES		◆Peacock		ACHERNAR		◆CANOPUS		SIRIUS		Alphard	
180	32 21	025	33 01	080	40 01	145	35 35	196	49 11	247	19 44	274	26 14	317
181	32 31	024	33 24	079	40 14	144	35 28	195	48 49	246	19 20	273	25 58	316
182	32 40	023	33 47	078	40 28	143	35 22	195	48 28	246	18 57	272	25 41	315
183	32 49	021	34 09	077	40 42	143	35 17	194	48 06	245	18 33	271	25 24	314
184	32 57	020	34 32	076	40 57	142	35 11	193	47 45	244	18 10	270	25 07	313
185	33 05	019	34 55	075	41 11	142	35 06	193	47 24	244	17 46	269	24 50	312
186	33 12	018	35 18	074	41 26	141	35 01	192	47 03	243	17 23	268	24 32	311
187	33 19	017	35 40	073	41 41	140	34 56	191	46 42	242	17 00	267	24 15	310
188	33 26	016	36 03	072	41 56	140	34 52	191	46 22	241	16 36	266	23 56	309
189	33 32	014	36 25	071	42 11	139	34 47	190	46 01	241	16 13	265	23 38	308
190	33 38	013	36 47	070	42 27	138	34 43	189	45 41	240	15 50	264	23 19	307
191	33 43	012	37 09	069	42 42	138	34 40	189	45 21	239	15 26	264	23 00	306
192	33 48	011	37 31	068	42 58	137	34 36	188	45 01	238	15 03	263	22 41	305
193	33 52	010	37 52	067	43 14	136	34 33	187	44 41	238	14 40	262	22 22	304
194	33 55	009	38 14	066	43 31	136	34 30	187	44 21	237	14 17	261	22 02	303
	◆SPICA		ANTARES		◆Peacock		ACHERNAR		◆CANOPUS		Suhail		Alphard	
195	33 59	007	38 35	065	43 47	135	34 27	186	44 01	236	51 31	278	21 42	302
196	34 01	006	38 56	064	44 04	134	34 25	185	43 42	236	51 08	277	21 22	301
197	34 04	005	39 17	063	44 21	134	34 23	185	43 23	235	50 45	276	21 02	300
198	34 06	004	39 38	062	44 38	133	34 21	184	43 04	234	50 21	275	20 41	299
199	34 07	003	39 59	061	44 55	132	34 20	184	42 45	234	49 58	274	20 21	298
200	34 08	001	40 19	060	45 12	132	34 18	183	42 26	233	49 35	273	20 00	297
201	34 08	000	40 39	059	45 30	131	34 17	182	42 07	232	49 11	272	19 39	296
202	34 08	359	40 59	057	45 48	130	34 16	182	41 49	231	48 48	271	19 18	295
203	34 07	358	41 19	056	46 06	130	34 16	181	41 31	231	48 24	270	18 56	294
204	34 06	357	41 38	055	46 24	129	34 16	180	41 13	230	48 01	269	18 35	293
205	34 05	356	41 57	054	46 42	128	34 16	180	40 55	229	47 37	268	18 13	292
206	34 03	354	42 16	053	47 01	128	34 16	179	40 37	229	47 14	268	17 51	291
207	34 00	353	42 35	052	47 19	127	34 17	178	40 20	228	46 51	267	17 29	290
208	33 57	352	42 53	051	47 38	126	34 17	178	40 02	227	46 27	266	17 07	289
209	33 54	351	43 11	050	47 57	126	34 19	177	39 45	227	46 04	265	16 45	288
	◆ANTARES		Nunki		◆FOMALHAUT		ACHERNAR		CANOPUS		◆Suhail		SPICA	
210	43 28	048	30 24	086	12 36	140	34 20	176	39 28	226	45 40	264	33 50	350
211	43 46	047	30 47	085	12 51	140	34 22	176	39 12	225	45 17	263	33 45	348
212	44 03	046	31 11	084	13 06	139	34 23	175	38 55	224	44 54	262	33 40	347
213	44 20	045	31 34	083	13 22	138	34 26	174	38 39	224	44 31	261	33 35	346
214	44 36	044	31 57	082	13 37	137	34 28	174	38 23	223	44 08	260	33 29	345
215	44 52	042	32 20	081	13 53	136	34 31	173	38 07	222	43 44	260	33 23	344
216	45 08	041	32 43	080	14 10	135	34 34	172	37 51	222	43 21	259	33 16	343
217	45 23	040	33 07	079	14 26	134	34 37	172	37 36	221	42 58	258	33 09	341
218	45 38	039	33 30	079	14 43	134	34 40	171	37 20	220	42 36	257	33 01	340
219	45 52	038	33 52	078	15 00	133	34 44	170	37 05	220	42 13	256	32 53	339
220	46 06	036	34 15	077	15 18	132	34 48	170	36 50	219	41 50	255	32 44	338
221	46 20	035	34 38	076	15 35	131	34 53	169	36 36	218	41 27	255	32 35	337
222	46 33	034	35 01	075	15 53	130	34 57	168	36 22	217	41 05	254	32 26	336
223	46 46	033	35 23	074	16 11	129	35 02	168	36 07	217	40 42	253	32 16	334
224	46 58	031	35 46	073	16 29	129	35 07	167	35 53	216	40 20	252	32 06	333
	◆ANTARES		Nunki		◆FOMALHAUT		ACHERNAR		CANOPUS		◆Suhail		SPICA	
225	47 10	030	36 08	072	16 48	128	35 12	167	35 40	215	39 58	251	31 55	332
226	47 22	029	36 30	071	17 06	127	35 18	166	35 26	215	39 36	250	31 44	331
227	47 33	027	36 52	070	17 25	126	35 24	165	35 13	214	39 14	250	31 32	330
228	47 43	026	37 14	069	17 44	125	35 30	165	35 00	213	38 52	249	31 20	329
229	47 53	025	37 36	067	18 04	124	35 36	164	34 47	213	38 30	248	31 08	328
230	48 03	023	37 57	066	18 23	123	35 43	163	34 35	212	38 08	247	30 55	327
231	48 12	022	38 19	065	18 43	123	35 50	163	34 22	211	37 47	246	30 42	325
232	48 21	021	38 40	064	19 03	122	35 57	162	34 10	211	37 25	246	30 29	324
233	48 29	019	39 01	063	19 23	121	36 04	161	33 59	210	37 04	245	30 15	323
234	48 36	018	39 22	062	19 43	120	36 12	161	33 47	209	36 43	244	30 01	322
235	48 43	017	39 43	061	20 03	119	36 20	160	33 36	208	36 22	243	29 46	321
236	48 50	015	40 03	060	20 24	118	36 28	159	33 25	208	36 01	242	29 31	320
237	48 56	014	40 23	059	20 44	117	36 36	159	33 14	207	35 40	242	29 16	319
238	49 01	013	40 43	058	21 05	117	36 45	158	33 03	206	35 20	241	29 00	318
239	49 06	011	41 03	057	21 26	116	36 54	157	32 53	206	34 59	240	28 44	317
	◆ANTARES		Nunki		◆FOMALHAUT		ACHERNAR		CANOPUS		◆ACRUX		SPICA	
240	49 10	010	41 22	056	21 47	115	37 03	157	32 43	205	67 51	255	28 28	316
241	49 14	009	41 42	055	22 09	114	37 12	156	32 33	204	67 28	254	28 12	315
242	49 17	007	42 01	053	22 30	113	37 22	155	32 24	204	67 05	253	27 55	314
243	49 20	006	42 19	052	22 52	112	37 32	155	32 14	203	66 43	253	27 38	312
244	49 22	005	42 38	051	23 14	112	37 42	154	32 05	202	66 21	252	27 20	311
245	49 24	003	42 56	050	23 36	111	37 52	154	31 57	202	65 58	251	27 02	310
246	49 25	002	43 14	049	23 58	110	38 03	153	31 48	201	65 36	251	26 44	309
247	49 25	000	43 31	048	24 20	109	38 14	152	31 40	200	65 14	250	26 26	308
248	49 25	359	43 48	047	24 42	108	38 25	152	31 32	199	64 52	249	26 08	307
249	49 24	358	44 05	045	25 04	107	38 36	151	31 24	199	64 30	249	25 49	306
250	49 23	356	44 22	044	25 27	106	38 47	150	31 17	198	64 08	248	25 30	305
251	49 21	355	44 38	043	25 49	105	38 59	150	31 10	197	63 47	248	25 11	304
252	49 19	354	44 54	042	26 12	105	39 11	149	31 03	197	63 25	247	24 51	303
253	49 16	352	45 09	041	26 35	104	39 23	148	30 56	196	63 04	246	24 31	302
254	49 13	351	45 24	039	26 57	103	39 36	148	30 50	195	62 42	246	24 11	301
	◆Nunki		FOMALHAUT		◆ACHERNAR		CANOPUS		◆ACRUX		SPICA		ANTARES	
255	45 39	038	27 20	102	39 48	147	30 44	195	62 21	245	23 51	300	49 09	349
256	45 53	037	27 43	101	40 01	146	30 38	194	62 00	245	23 31	299	49 04	348
257	46 07	036	28 06	100	40 14	146	30 33	193	61 39	244	23 10	298	48 59	347
258	46 20	034	28 29	099	40 28	145	30 27	193	61 17	243	22 49	297	48 53	345
259	46 33	033	28 53	098	40 41	144	30 22	192	60 57	243	22 28	296	48 47	344
260	46 46	032	29 16	098	40 55	144	30 18	191	60 36	242	22 07	295	48 40	343
261	46 58	031	29 39	097	41 09	143	30 13	190	60 15	242	21 46	294	48 33	342
262	47 10	029	30 02	096	41 23	143	30 09	190	59 55	241	21 24	293	48 25	340
263	47 21	028	30 26	095	41 37	142	30 05	189	59 34	240	21 03	292	48 17	339
264	47 32	027	30 49	094	41 52	141	30 02	188	59 14	240	20 41	291	48 08	337
265	47 42	025	31 12	093	42 07	141	29 59	188	58 54	239	20 19	290	47 59	336
266	47 52	024	31 36	092	42 22	140	29 56	187	58 34	239	19 57	289	47 49	335
267	48 01	023	31 59	091	42 37	139	29 53	186	58 14	238	19 35	288	47 39	333
268	48 10	021	32 23	090	42 52	139	29 51	186	57 54	237	19 12	287	47 28	332
269	48 18	020	32 46	089	43 08	138	29 48	185	57 34	237	18 50	286	47 17	331

LHA	Hc	Zn	Hc	Zn	Hc	Zn	Hc	Zn	Hc	Zn	Hc	Zn	Hc	Zn
♈	◆Nunki		FOMALHAUT		◆ACHERNAR		CANOPUS		◆ACRUX		SPICA		ANTARES	
270	48 26	019	33 10	088	43 24	137	29 47	184	57 15	236	18 27	285	47 05	329
271	48 33	017	33 33	088	43 40	137	29 45	183	56 55	236	18 05	284	46 53	328
272	48 40	016	33 57	087	43 56	136	29 44	183	56 36	235	17 42	283	46 40	327
273	48 46	015	34 20	086	44 12	135	29 43	182	56 17	235	17 19	283	46 27	326
274	48 52	013	34 43	085	44 29	135	29 42	181	55 58	234	16 56	282	46 14	324
275	48 57	012	35 07	084	44 46	134	29 42	181	55 39	233	16 33	281	46 00	323
276	49 01	011	35 30	083	45 03	133	29 42	180	55 20	233	16 10	280	45 46	322
277	49 05	009	35 53	082	45 20	133	29 42	179	55 01	232	15 47	279	45 31	321
278	49 09	008	36 16	081	45 37	132	29 42	179	54 43	232	15 24	278	45 16	319
279	49 12	007	36 39	080	45 55	131	29 43	178	54 25	231	15 01	277	45 01	318
280	49 14	005	37 02	079	46 12	131	29 44	177	54 07	230	14 37	276	44 45	317
281	49 16	004	37 25	078	46 30	130	29 45	176	53 49	230	14 14	275	44 29	316
282	49 17	002	37 48	077	46 48	129	29 47	176	53 31	229	13 51	274	44 12	315
283	49 18	001	38 11	076	47 06	129	29 49	175	53 13	229	13 27	273	43 55	313
284	49 18	000	38 34	075	47 25	128	29 51	174	52 55	228	13 04	272	43 38	312
	◆ALTAIR		FOMALHAUT		◆ACHERNAR		CANOPUS		◆RIGIL KENT.		ANTARES		Nunki	
285	13 36	013	38 56	074	47 43	128	29 53	174	62 05	251	43 21	311	49 18	358
286	13 41	012	39 19	073	48 02	127	29 56	173	61 43	250	43 03	310	49 17	357
287	13 45	011	39 41	072	48 21	126	29 59	172	61 21	250	42 45	309	49 15	356
288	13 49	010	40 03	071	48 40	126	30 02	172	60 59	249	42 26	308	49 13	354
289	13 53	009	40 26	070	48 59	125	30 06	171	60 37	248	42 08	307	49 10	353
290	13 57	008	40 47	069	49 18	124	30 10	170	60 15	248	41 49	305	49 07	351
291	14 00	007	41 09	068	49 37	124	30 14	170	59 54	247	41 29	304	49 03	350
292	14 02	006	41 31	067	49 57	123	30 18	169	59 32	246	41 10	303	48 59	349
293	14 04	005	41 52	066	50 17	122	30 23	168	59 11	246	40 50	302	48 54	347
294	14 06	004	42 14	065	50 37	122	30 28	167	58 49	245	40 30	301	48 49	346
295	14 07	003	42 35	064	50 57	121	30 33	167	58 28	244	40 10	300	48 43	345
296	14 08	002	42 56	062	51 17	120	30 39	166	58 07	244	39 49	299	48 36	343
297	14 09	001	43 16	061	51 37	120	30 44	165	57 46	243	39 29	298	48 29	342
298	14 09	000	43 37	060	51 58	119	30 50	165	57 25	242	39 08	297	48 22	341
299	14 08	359	43 57	059	52 18	118	30 57	164	57 05	242	38 47	296	48 14	339
	◆FOMALHAUT		ACHERNAR		◆CANOPUS		RIGIL KENT.		◆ANTARES		Nunki		ALTAIR	
300	44 17	058	52 39	117	31 03	163	56 44	241	38 26	295	48 05	338	14 08	358
301	44 37	057	53 00	117	31 10	163	56 24	241	38 04	293	47 56	337	14 06	357
302	44 56	056	53 21	116	31 17	162	56 03	240	37 43	292	47 47	335	14 05	356
303	45 16	055	53 42	115	31 25	161	55 43	239	37 21	291	47 37	334	14 03	355
304	45 35	053	54 03	115	31 33	160	55 23	239	36 59	290	47 26	333	14 00	354
305	45 53	052	54 25	114	31 41	160	55 03	238	36 37	289	47 15	331	13 57	353
306	46 12	051	54 46	113	31 49	159	54 43	237	36 15	288	47 04	330	13 54	351
307	46 30	050	55 08	113	31 57	158	54 23	237	35 52	287	46 52	329	13 50	350
308	46 48	049	55 29	112	32 06	158	54 04	236	35 30	286	46 39	328	13 46	349
309	47 05	048	55 51	111	32 15	157	53 44	236	35 07	285	46 27	326	13 42	348
310	47 22	046	56 13	110	32 24	156	53 25	235	34 45	284	46 13	325	13 37	347
311	47 39	045	56 35	110	32 34	156	53 06	234	34 22	283	46 00	324	13 32	346
312	47 55	044	56 57	109	32 44	155	52 47	234	33 59	282	45 46	323	13 26	345
313	48 11	043	57 19	108	32 54	154	52 28	233	33 36	281	45 31	321	13 20	344
314	48 27	041	57 42	108	33 04	154	52 10	232	33 13	280	45 16	320	13 13	343
	◆FOMALHAUT		ACHERNAR		◆CANOPUS		Miaplacidus		RIGIL KENT.		◆ANTARES		Nunki	
315	48 43	040	58 04	107	33 15	153	46 43	182	51 51	232	32 50	279	45 01	319
316	48 57	039	58 27	106	33 26	152	46 42	181	51 33	231	32 27	279	44 45	318
317	49 12	038	58 49	105	33 37	151	46 42	181	51 15	231	32 04	278	44 29	316
318	49 26	036	59 12	105	33 48	151	46 42	180	50 57	230	31 41	277	44 13	315
319	49 40	035	59 35	104	34 00	150	46 42	180	50 39	229	31 17	276	43 56	314
320	49 53	034	59 57	103	34 11	149	46 42	179	50 21	229	30 54	275	43 39	313
321	50 06	032	60 20	102	34 23	149	46 43	179	50 03	228	30 30	274	43 22	312
322	50 18	031	60 43	102	34 36	148	46 43	178	49 46	228	30 07	273	43 04	311
323	50 30	030	61 06	101	34 48	147	46 44	178	49 29	227	29 44	272	42 46	309
324	50 42	028	61 29	100	35 01	147	46 45	177	49 12	226	29 20	271	42 28	308
325	50 52	027	61 52	099	35 14	146	46 46	177	48 55	226	28 57	270	42 10	307
326	51 03	026	62 16	098	35 27	145	46 48	176	48 38	225	28 33	269	41 51	306
327	51 13	024	62 39	098	35 41	145	46 50	176	48 22	225	28 10	268	41 32	305
328	51 22	023	63 02	097	35 55	144	46 52	175	48 05	224	27 46	267	41 12	304
329	51 31	022	63 25	096	36 08	143	46 54	175	47 49	223	27 23	266	40 53	303
	◆FOMALHAUT		ACHERNAR		◆CANOPUS		Miaplacidus		RIGIL KENT.		◆ANTARES		Nunki	
330	51 40	020	63 49	095	36 23	142	46 56	174	47 33	223	27 00	266	40 33	302
331	51 47	019	64 12	094	36 37	142	46 58	174	47 17	222	26 36	265	40 13	300
332	51 55	018	64 35	093	36 52	141	47 01	173	47 02	221	26 13	264	39 52	299
333	52 02	016	64 59	092	37 07	140	47 04	173	46 46	221	25 50	263	39 32	298
334	52 08	015	65 22	092	37 22	140	47 07	172	46 31	220	25 26	262	39 11	297
335	52 14	013	65 46	091	37 37	139	47 11	172	46 16	220	25 03	261	38 50	296
336	52 19	012	66 09	090	37 52	138	47 14	171	46 01	219	24 40	260	38 29	295
337	52 23	010	66 33	089	38 08	138	47 18	171	45 46	218	24 17	259	38 08	294
338	52 27	009	66 56	088	38 24	137	47 22	170	45 32	218	23 54	258	37 46	293
339	52 31	008	67 19	087	38 40	136	47 26	170	45 18	217	23 31	257	37 25	292
340	52 33	006	67 43	086	38 56	136	47 30	169	45 04	217	23 08	257	37 03	291
341	52 36	005	68 06	085	39 13	135	47 35	169	44 50	216	22 46	256	36 41	290
342	52 37	003	68 30	084	39 30	134	47 40	168	44 36	215	22 23	255	36 19	289
343	52 38	002	68 53	083	39 47	133	47 45	168	44 23	215	22 00	254	35 56	288
344	52 39	000	69 16	082	40 04	133	47 50	167	44 09	214	21 38	253	35 34	287
	Diphda		◆CANOPUS		Miaplacidus		RIGIL KENT.		◆ANTARES		Nunki		◆FOMALHAUT	
345	38 15	032	40 21	132	47 55	167	43 56	214	21 15	252	35 12	286	52 39	359
346	38 27	031	40 38	131	48 01	166	43 43	213	20 53	251	34 49	285	52 38	358
347	38 39	029	40 56	131	48 06	166	43 31	212	20 31	250	34 26	284	52 37	356
348	38 50	028	41 14	130	48 12	165	43 18	212	20 09	250	34 03	283	52 35	355
349	39 01	027	41 32	129	48 19	165	43 06	211	19 47	249	33 40	282	52 33	353
350	39 12	026	41 50	129	48 25	164	42 54	211	19 25	248	33 18	281	52 30	352
351	39 22	025	42 09	128	48 31	164	42 42	210	19 04	247	32 54	280	52 26	350
352	39 31	023	42 27	127	48 38	163	42 30	209	18 42	246	32 31	279	52 22	349
353	39 40	022	42 46	126	48 45	163	42 19	209	18 21	245	32 08	278	52 17	348
354	39 49	021	43 05	126	48 52	162	42 08	208	18 00	244	31 45	277	52 12	346
355	39 57	020	43 24	125	49 00	162	41 57	208	17 38	244	31 22	276	52 06	345
356	40 05	019	43 44	124	49 07	161	41 46	207	17 18	243	30 58	275	52 00	343
357	40 12	017	44 03	124	49 15	161	41 36	206	16 57	242	30 35	274	51 53	342
358	40 19	016	44 23	123	49 23	160	41 26	206	16 36	241	30 12	273	51 45	341
359	40 25	015	44 42	122	49 31	160	41 15	205	16 16	240	29 48	272	51 37	339

LHA ♈	Hc	Zn	Hc	Zn	Hc	Zn	Hc	Zn	Hc	Zn	Hc	Zn	Hc	Zn
	Diphda		◆RIGEL		CANOPUS		Miaplacidus		◆RIGIL KENT.		Peacock		◆FOMALHAUT	
0	39 32	013	11 53	082	45 33	121	50 35	159	42 00	205	63 46	268	50 33	338
1	39 37	012	12 15	082	45 53	120	50 43	158	41 51	204	63 24	268	50 24	337
2	39 42	011	12 37	081	46 12	119	50 52	158	41 42	204	63 01	267	50 15	336
3	39 46	010	12 59	080	46 32	118	51 00	157	41 33	203	62 39	266	50 06	334
4	39 49	008	13 21	079	46 52	118	51 09	157	41 24	202	62 17	265	49 56	333
5	39 53	007	13 43	078	47 12	117	51 18	156	41 16	202	61 54	264	49 45	332
6	39 55	006	14 05	077	47 32	116	51 27	156	41 08	201	61 32	263	49 34	330
7	39 57	005	14 27	076	47 52	115	51 37	155	41 00	200	61 10	262	49 23	329
8	39 59	004	14 49	075	48 12	115	51 46	155	40 52	200	60 47	261	49 11	328
9	40 00	002	15 10	074	48 33	114	51 56	154	40 44	199	60 25	261	48 59	326
10	40 01	001	15 32	073	48 54	113	52 06	154	40 37	199	60 03	260	48 46	325
11	40 01	000	15 53	072	49 14	112	52 16	153	40 30	198	59 41	259	48 33	324
12	40 01	359	16 15	071	49 35	112	52 26	153	40 23	197	59 19	258	48 20	323
13	40 00	357	16 36	070	49 56	111	52 36	152	40 16	197	58 57	257	48 06	321
14	39 58	356	16 57	069	50 17	110	52 47	152	40 10	196	58 35	257	47 52	320
	◆RIGEL		CANOPUS		Miaplacidus		◆RIGIL KENT.		Peacock		◆FOMALHAUT		Diphda	
15	17 18	068	50 38	109	52 58	151	40 04	196	58 13	256	47 37	319	39 57	355
16	17 39	067	51 00	109	53 09	151	39 58	195	57 51	255	47 22	318	39 54	354
17	17 59	066	51 21	108	53 20	150	39 52	194	57 30	254	47 07	316	39 52	352
18	18 20	065	51 42	107	53 31	150	39 47	194	57 08	254	46 51	315	39 48	351
19	18 40	064	52 04	106	53 42	149	39 42	193	56 47	253	46 35	314	39 45	350
20	19 00	063	52 26	105	53 54	149	39 37	192	56 25	252	46 19	313	39 40	349
21	19 20	062	52 47	105	54 06	148	39 32	192	56 04	251	46 02	312	39 36	347
22	19 40	061	53 09	104	54 18	148	39 28	191	55 43	251	45 46	310	39 31	346
23	20 00	060	53 31	103	54 30	147	39 23	191	55 21	250	45 28	309	39 25	345
24	20 19	059	53 53	102	54 42	147	39 20	190	55 00	249	45 11	308	39 19	344
25	20 38	058	54 15	101	54 54	146	39 16	189	54 39	248	44 53	307	39 13	343
26	20 57	057	54 37	100	55 07	146	39 12	189	54 19	248	44 35	306	39 06	341
27	21 16	056	54 59	100	55 19	145	39 09	188	53 58	247	44 16	305	38 58	340
28	21 35	055	55 21	099	55 32	145	39 06	187	53 37	246	43 58	304	38 50	339
29	21 53	054	55 43	098	55 45	144	39 03	187	53 17	245	43 39	302	38 42	338
	◆RIGEL		CANOPUS		Miaplacidus		◆RIGIL KENT.		Peacock		◆FOMALHAUT		Diphda	
30	22 11	053	56 06	097	55 58	144	39 01	186	52 56	245	43 20	301	38 33	336
31	22 29	052	56 28	096	56 12	143	38 58	186	52 36	244	43 01	300	38 24	335
32	22 47	051	56 50	095	56 25	143	38 56	185	52 16	243	42 41	299	38 14	334
33	23 04	050	57 13	095	56 39	142	38 55	184	51 56	243	42 21	298	38 04	333
34	23 21	049	57 35	094	56 53	142	38 53	184	51 36	242	42 01	297	37 54	332
35	23 38	048	57 58	093	57 07	142	38 52	183	51 16	241	41 41	296	37 43	331
36	23 55	047	58 20	092	57 21	141	38 51	182	50 57	240	41 21	295	37 32	329
37	24 11	046	58 43	091	57 35	141	38 50	182	50 37	240	41 00	294	37 20	328
38	24 27	045	59 05	090	57 49	140	38 49	181	50 18	239	40 40	293	37 08	327
39	24 43	044	59 28	089	58 04	140	38 49	181	49 59	238	40 19	292	36 56	326
40	24 59	043	59 50	088	58 18	139	38 49	180	49 40	238	39 58	291	36 43	325
41	25 14	042	60 12	087	58 33	139	38 49	179	49 21	237	39 37	290	36 30	324
42	25 29	041	60 35	086	58 48	138	38 49	179	49 02	236	39 16	289	36 16	322
43	25 43	040	60 57	085	59 03	138	38 50	178	48 43	236	38 54	287	36 02	321
44	25 57	039	61 20	084	59 18	137	38 51	177	48 25	235	38 33	286	35 48	320
	RIGEL		◆SIRIUS		Suhail		◆RIGIL KENT.		Peacock		◆FOMALHAUT		Diphda	
45	26 11	038	27 47	064	38 54	111	38 52	177	48 06	234	38 11	285	35 33	319
46	26 25	037	28 07	063	39 15	110	38 54	176	47 48	234	37 49	284	35 18	318
47	26 38	035	28 27	062	39 36	109	38 55	176	47 30	233	37 28	283	35 03	317
48	26 51	034	28 46	061	39 57	109	38 57	175	47 12	232	37 06	282	34 48	316
49	27 03	033	29 06	060	40 19	108	38 59	174	46 55	232	36 44	281	34 32	314
50	27 15	032	29 25	059	40 40	107	39 02	174	46 37	231	36 22	281	34 16	313
51	27 27	031	29 44	058	41 02	106	39 04	173	46 20	230	35 59	280	33 59	312
52	27 39	030	30 03	057	41 23	105	39 07	172	46 03	230	35 37	279	33 42	311
53	27 50	029	30 22	056	41 45	104	39 10	172	45 46	229	35 15	278	33 25	310
54	28 00	028	30 41	055	42 07	104	39 14	171	45 29	228	34 53	277	33 08	309
55	28 11	027	30 59	054	42 29	103	39 17	171	45 12	227	34 30	276	32 50	308
56	28 20	026	31 17	053	42 51	102	39 21	170	44 56	227	34 08	275	32 32	307
57	28 30	024	31 35	052	43 13	101	39 25	169	44 39	226	33 46	274	32 14	306
58	28 39	023	31 52	051	43 35	100	39 29	169	44 23	225	33 23	273	31 56	305
59	28 48	022	32 09	049	43 57	099	39 34	168	44 07	225	33 01	272	31 37	304
	RIGEL		◆SIRIUS		Suhail		◆RIGIL KENT.		Peacock		◆FOMALHAUT		Diphda	
60	28 56	021	32 26	048	44 19	098	39 39	167	43 52	224	32 38	271	31 19	303
61	29 04	020	32 43	047	44 41	098	39 44	167	43 36	223	32 16	270	31 00	302
62	29 11	019	32 59	046	45 04	097	39 49	166	43 21	223	31 53	269	30 40	301
63	29 19	018	33 15	045	45 26	096	39 55	166	43 05	222	31 31	268	30 21	300
64	29 25	017	33 31	044	45 48	095	40 00	165	42 51	221	31 08	267	30 01	298
65	29 31	015	33 47	043	46 11	094	40 06	164	42 36	221	30 46	266	29 41	297
66	29 37	014	34 02	042	46 33	093	40 12	164	42 21	220	30 23	265	29 21	296
67	29 43	013	34 17	041	46 56	092	40 19	163	42 07	219	30 01	265	29 01	295
68	29 47	012	34 31	040	47 18	091	40 26	162	41 53	219	29 39	264	28 41	294
69	29 52	011	34 45	038	47 41	090	40 32	162	41 39	218	29 16	263	28 20	293
70	29 56	010	34 59	037	48 03	089	40 40	161	41 25	217	28 54	262	27 59	292
71	30 00	009	35 13	036	48 25	089	40 47	161	41 11	217	28 32	261	27 39	291
72	30 03	008	35 26	035	48 48	088	40 55	160	40 58	216	28 10	260	27 18	290
73	30 06	006	35 38	034	49 10	087	41 02	159	40 45	215	27 48	259	26 56	289
74	30 08	005	35 51	033	49 33	086	41 11	159	40 32	215	27 26	258	26 35	288
	RIGEL		◆SIRIUS		Suhail		◆RIGIL KENT.		Peacock		◆FOMALHAUT		Diphda	
75	30 10	004	36 03	032	49 55	085	41 19	158	40 19	214	27 04	257	26 14	287
76	30 11	003	36 14	030	50 18	084	41 27	157	40 07	213	26 42	256	25 52	286
77	30 12	002	36 25	029	50 40	083	41 36	157	39 54	213	26 20	256	25 31	285
78	30 12	001	36 36	028	51 02	082	41 45	156	39 42	212	25 58	255	25 09	284
79	30 12	000	36 47	027	51 24	081	41 54	156	39 30	211	25 37	254	24 47	284
80	30 12	358	36 57	026	51 47	080	42 04	155	39 19	211	25 15	253	24 25	283
81	30 11	357	37 06	025	52 09	079	42 13	154	39 07	210	24 54	252	24 03	282
82	30 10	356	37 15	023	52 31	078	42 23	154	38 56	209	24 32	251	23 41	281
83	30 08	355	37 24	022	52 53	077	42 33	153	38 45	209	24 11	250	23 19	280
84	30 06	354	37 32	021	53 14	076	42 44	152	38 34	208	23 50	249	22 57	279
85	30 03	353	37 40	020	53 36	075	42 54	152	38 24	208	23 29	249	22 35	278
86	30 00	352	37 47	019	53 58	074	43 05	151	38 14	207	23 08	248	22 12	277
87	29 57	350	37 54	017	54 19	072	43 16	151	38 04	206	22 47	247	21 50	276
88	29 53	349	38 01	016	54 41	071	43 27	150	37 54	206	22 27	246	21 28	275
89	29 48	348	38 07	015	55 02	070	43 38	149	37 44	205	22 06	245	21 05	274

LHA ♈	Hc	Zn	Hc	Zn	Hc	Zn	Hc	Zn	Hc	Zn	Hc	Zn	Hc	Zn
	◆SIRIUS		Suhail		◆RIGIL KENT.		Peacock		◆ACHERNAR		Acamar		RIGEL	
90	38 12	014	55 23	069	43 50	149	37 35	204	59 43	258	53 09	295	29 43	347
91	38 17	013	55 44	068	44 02	148	37 26	204	59 21	257	52 49	294	29 38	346
92	38 22	011	56 05	067	44 14	147	37 17	203	58 59	256	52 28	293	29 32	345
93	38 26	010	56 25	066	44 26	147	37 08	202	58 37	255	52 07	292	29 26	344
94	38 30	009	56 46	065	44 38	146	37 00	202	58 16	255	51 46	291	29 20	342
95	38 33	008	57 06	063	44 51	146	36 52	201	57 54	254	51 25	289	29 13	341
96	38 36	006	57 26	062	45 03	145	36 44	200	57 32	253	51 04	288	29 05	340
97	38 38	005	57 45	061	45 16	144	36 36	200	57 11	252	50 42	287	28 57	339
98	38 40	004	58 05	060	45 30	144	36 29	199	56 50	251	50 21	286	28 49	338
99	38 41	003	58 24	059	45 43	143	36 22	198	56 28	251	49 59	285	28 40	337
100	38 42	002	58 43	057	45 57	143	36 15	198	56 07	250	49 38	284	28 31	336
101	38 43	000	59 02	056	46 10	142	36 08	197	55 46	249	49 16	283	28 22	335
102	38 42	359	59 21	055	46 24	141	36 02	196	55 25	249	48 54	282	28 12	333
103	38 42	358	59 39	053	46 39	141	35 56	196	55 04	248	48 32	281	28 02	332
104	38 41	357	59 57	052	46 53	140	35 50	195	54 44	247	48 10	280	27 51	331
	Alphard		◆Gienah		RIGIL KENT.		◆Peacock		ACHERNAR		◆RIGEL		SIRIUS	
105	25 50	041	20 21	086	47 07	139	35 44	194	54 23	246	27 40	330	38 39	355
106	26 04	040	20 44	086	47 22	139	35 39	194	54 02	246	27 29	329	38 37	354
107	26 19	039	21 06	085	47 37	138	35 34	193	53 42	245	27 17	328	38 35	353
108	26 33	038	21 29	084	47 52	138	35 29	192	53 22	244	27 05	327	38 32	352
109	26 46	037	21 51	083	48 07	137	35 24	192	53 02	244	26 53	326	38 28	350
110	27 00	036	22 13	082	48 23	136	35 20	191	52 41	243	26 40	325	38 24	349
111	27 13	035	22 35	081	48 38	136	35 16	190	52 22	242	26 27	324	38 20	348
112	27 25	034	22 58	080	48 54	135	35 12	190	52 02	241	26 13	323	38 15	347
113	27 37	033	23 20	079	49 10	134	35 08	189	51 42	241	25 59	322	38 09	346
114	27 49	031	23 42	078	49 26	134	35 05	188	51 23	240	25 45	320	38 04	344
115	28 01	030	24 04	077	49 43	133	35 02	188	51 03	239	25 31	319	37 57	343
116	28 12	029	24 26	076	49 59	133	34 59	187	50 44	239	25 16	318	37 51	342
117	28 23	028	24 47	075	50 16	132	34 56	186	50 25	238	25 01	317	37 43	341
118	28 33	027	25 09	074	50 32	131	34 54	186	50 06	237	24 45	316	37 36	340
119	28 43	026	25 31	073	50 49	131	34 52	185	49 47	237	24 30	315	37 28	338
	Alphard		◆SPICA		RIGIL KENT.		◆Peacock		ACHERNAR		◆RIGEL		SIRIUS	
120	28 53	025	13 36	086	51 07	130	34 50	184	49 28	236	24 14	314	37 19	337
121	29 02	024	13 58	085	51 24	129	34 49	184	49 10	235	23 57	313	37 10	336
122	29 11	023	14 21	084	51 41	129	34 48	183	48 51	235	23 41	312	37 01	335
123	29 19	021	14 43	083	51 59	128	34 47	182	48 33	234	23 24	311	36 51	334
124	29 28	020	15 05	082	52 17	128	34 46	182	48 15	233	23 07	310	36 41	332
125	29 35	019	15 27	081	52 34	127	34 45	181	47 57	233	22 50	309	36 30	331
126	29 42	018	15 50	080	52 53	126	34 45	180	47 39	232	22 32	308	36 19	330
127	29 49	017	16 12	080	53 11	126	34 45	180	47 22	231	22 14	307	36 08	329
128	29 55	016	16 34	079	53 29	125	34 45	179	47 04	231	21 56	306	35 56	328
129	30 01	015	16 56	078	53 48	124	34 46	178	46 47	230	21 38	305	35 44	327
130	30 07	014	17 18	077	54 06	124	34 47	178	46 30	229	21 19	304	35 32	326
131	30 12	012	17 40	076	54 25	123	34 48	177	46 13	228	21 00	303	35 19	324
132	30 16	011	18 01	075	54 44	122	34 49	176	45 56	228	20 41	302	35 05	323
133	30 21	010	18 23	074	55 03	122	34 51	176	45 40	227	20 22	301	34 52	322
134	30 24	009	18 45	073	55 22	121	34 53	175	45 23	226	20 03	300	34 38	321
	Alphard		◆SPICA		RIGIL KENT.		◆Peacock		ACHERNAR		◆CANOPUS		SIRIUS	
135	30 28	008	19 06	072	55 41	121	34 55	174	45 07	226	66 02	290	34 23	320
136	30 31	007	19 27	071	56 01	120	34 57	174	44 51	225	65 41	289	34 09	319
137	30 33	006	19 48	070	56 20	119	35 00	173	44 35	225	65 20	288	33 54	318
138	30 35	004	20 09	069	56 40	119	35 03	172	44 20	224	64 58	286	33 38	316
139	30 36	003	20 30	068	57 00	118	35 06	172	44 04	223	64 37	285	33 23	315
140	30 37	002	20 51	067	57 20	117	35 10	171	43 49	223	64 15	284	33 07	314
141	30 38	001	21 12	066	57 40	117	35 13	170	43 34	222	63 53	283	32 51	313
142	30 38	000	21 32	065	58 00	116	35 17	170	43 19	221	63 31	282	32 34	312
143	30 38	359	21 52	064	58 20	115	35 21	169	43 04	221	63 09	281	32 17	311
144	30 37	358	22 13	063	58 41	115	35 26	168	42 50	220	62 47	280	32 00	310
145	30 36	356	22 33	062	59 01	114	35 31	168	42 35	219	62 25	279	31 43	309
146	30 34	355	22 52	061	59 22	113	35 36	167	42 21	219	62 03	278	31 25	308
147	30 32	354	23 12	060	59 42	113	35 41	166	42 07	218	61 40	277	31 07	307
148	30 30	353	23 31	059	60 03	112	35 46	166	41 54	217	61 18	276	30 49	306
149	30 27	352	23 50	058	60 24	111	35 52	165	41 40	217	60 56	275	30 31	305
	◆SPICA		RIGIL KENT.		◆Peacock		ACHERNAR		◆CANOPUS		SIRIUS		Alphard	
150	24 09	057	60 45	110	35 58	164	41 27	216	60 33	274	30 12	304	30 23	351
151	24 28	056	61 06	110	36 04	164	41 14	215	60 11	273	29 53	303	30 19	349
152	24 47	055	61 27	109	36 11	163	41 01	215	59 48	272	29 34	301	30 15	348
153	25 05	054	61 49	108	36 18	162	40 48	214	59 26	271	29 15	300	30 10	347
154	25 23	053	62 10	108	36 24	162	40 36	213	59 03	270	28 55	299	30 05	346
155	25 41	052	62 32	107	36 32	161	40 24	213	58 41	269	28 36	298	29 59	345
156	25 58	051	62 53	106	36 39	160	40 12	212	58 18	268	28 16	297	29 53	344
157	26 16	050	63 15	106	36 47	160	40 00	211	57 56	267	27 56	296	29 47	343
158	26 33	049	63 36	105	36 55	159	39 48	211	57 33	266	27 36	295	29 40	342
159	26 49	048	63 58	104	37 03	158	39 37	210	57 11	265	27 15	294	29 33	340
160	27 06	047	64 20	103	37 12	158	39 26	209	56 49	265	26 55	293	29 25	339
161	27 22	046	64 42	103	37 20	157	39 15	209	56 26	264	26 34	292	29 17	338
162	27 38	044	65 04	102	37 29	156	39 04	208	56 04	263	26 13	291	29 08	337
163	27 53	043	65 26	101	37 38	156	38 54	207	55 42	262	25 52	290	28 59	336
164	28 09	042	65 48	100	37 48	155	38 44	207	55 19	261	25 31	289	28 50	335
	◆SPICA		ANTARES		◆Peacock		ACHERNAR		◆CANOPUS		SIRIUS		Alphard	
165	28 24	041	27 14	094	37 57	154	38 34	206	54 57	260	25 10	288	28 40	334
166	28 38	040	27 37	093	38 07	154	38 24	205	54 35	259	24 48	287	28 30	333
167	28 53	039	27 59	092	38 17	153	38 14	205	54 13	259	24 27	286	28 19	331
168	29 07	038	28 22	091	38 28	152	38 05	204	53 51	258	24 05	285	28 08	330
169	29 20	037	28 44	090	38 38	152	37 56	203	53 29	257	23 43	284	27 57	329
170	29 34	036	29 06	089	38 49	151	37 47	203	53 07	256	23 22	283	27 45	328
171	29 47	035	29 29	088	39 00	150	37 39	202	52 46	255	23 00	283	27 33	327
172	29 59	034	29 51	087	39 12	150	37 30	201	52 24	255	22 38	282	27 21	326
173	30 11	032	30 14	086	39 23	149	37 22	201	52 02	254	22 16	281	27 08	325
174	30 23	031	30 36	085	39 35	148	37 14	200	51 41	253	21 54	280	26 55	324
175	30 35	030	30 59	084	39 47	148	37 07	199	51 19	252	21 31	279	26 42	323
176	30 46	029	31 21	083	39 59	147	36 59	199	50 58	251	21 09	278	26 28	322
177	30 57	028	31 43	082	40 11	146	36 52	198	50 37	251	20 47	277	26 14	321
178	31 07	027	32 06	081	40 24	146	36 45	197	50 15	250	20 24	276	25 59	320
179	31 17	026	32 28	080	40 37	145	36 39	197	49 54	249	20 02	275	25 45	318

LHA ♈	Hc Zn	Hc Zn	Hc Zn	Hc Zn	Hc Zn	Hc Zn	Hc Zn
	◆SPICA	ANTARES	◆Peacock	ACHERNAR	◆CANOPUS	SIRIUS	Alphard
180	31 27 025	32 50 079	40 50 144	36 32 196	49 33 248	19 40 274	25 30 317
181	31 36 023	33 12 078	41 03 144	36 26 195	49 13 248	19 17 273	25 14 316
182	31 45 022	33 34 077	41 16 143	36 20 195	48 52 247	18 55 272	24 59 315
183	31 53 021	33 56 076	41 30 142	36 15 194	48 31 246	18 32 271	24 43 314
184	32 01 020	34 18 076	41 44 142	36 09 193	48 11 245	18 10 270	24 26 313
185	32 08 019	34 39 075	41 58 141	36 04 193	47 50 245	17 47 269	24 10 312
186	32 15 018	35 01 074	42 12 140	35 59 192	47 30 244	17 25 268	23 53 311
187	32 22 017	35 22 073	42 27 140	35 55 192	47 10 243	17 02 267	23 36 310
188	32 28 015	35 44 072	42 41 139	35 50 191	46 50 242	16 40 267	23 19 309
189	32 34 014	36 05 070	42 56 138	35 46 190	46 30 242	16 18 266	23 01 308
190	32 39 013	36 26 069	43 11 138	35 43 190	46 11 241	15 55 265	22 43 307
191	32 44 012	36 47 068	43 26 137	35 39 189	45 51 240	15 33 264	22 25 306
192	32 49 011	37 08 067	43 42 136	35 36 188	45 32 239	15 10 263	22 07 305
193	32 53 010	37 29 066	43 57 136	35 33 188	45 12 239	14 48 262	21 48 304
194	32 56 008	37 49 065	44 13 135	35 30 187	44 53 238	14 26 261	21 30 303
	◆SPICA	ANTARES	◆Peacock	ACHERNAR	◆CANOPUS	Suhail	Alphard
195	32 59 007	38 09 064	44 29 134	35 27 186	44 34 237	51 22 279	21 11 302
196	33 02 006	38 30 063	44 45 134	35 25 186	44 16 236	51 00 278	20 52 301
197	33 04 005	38 50 062	45 02 133	35 23 185	43 57 236	50 38 277	20 32 300
198	33 06 004	39 09 061	45 18 132	35 21 184	43 38 235	50 15 276	20 13 299
199	33 07 003	39 29 060	45 35 132	35 19 184	43 20 234	49 53 275	19 53 298
200	33 08 001	39 48 059	45 52 131	35 18 183	43 02 234	49 31 274	19 33 297
201	33 08 000	40 07 058	46 09 130	35 17 182	42 44 233	49 08 273	19 13 296
202	33 08 359	40 26 057	46 26 130	35 16 182	42 26 232	48 46 272	18 52 295
203	33 07 358	40 45 056	46 44 129	35 16 181	42 08 231	48 23 271	18 32 294
204	33 06 357	41 03 055	47 01 128	35 16 180	41 51 231	48 01 270	18 11 293
205	33 05 356	41 22 053	47 19 128	35 16 180	41 34 230	47 38 270	17 51 292
206	33 03 354	41 40 052	47 37 127	35 16 179	41 17 229	47 16 269	17 30 291
207	33 00 353	41 57 051	47 55 126	35 17 178	41 00 229	46 54 268	17 09 290
208	32 58 352	42 15 050	48 13 125	35 17 178	40 43 228	46 31 267	16 48 289
209	32 54 351	42 32 049	48 32 125	35 18 177	40 26 227	46 09 266	16 26 288
	◆ANTARES	Nunki	◆FOMALHAUT	ACHERNAR	CANOPUS	◆Suhail	SPICA
210	42 48 048	30 19 086	13 22 140	35 20 176	40 10 226	45 46 265	32 51 350
211	43 05 047	30 42 085	13 36 139	35 21 176	39 54 226	45 24 264	32 46 349
212	43 21 045	31 04 084	13 51 139	35 23 175	39 38 225	45 02 263	32 42 347
213	43 37 044	31 27 083	14 06 138	35 25 174	39 22 224	44 39 262	32 36 346
214	43 52 043	31 49 082	14 21 137	35 28 174	39 06 224	44 17 261	32 31 345
215	44 07 042	32 11 081	14 37 136	35 30 173	38 51 223	43 55 261	32 25 344
216	44 22 041	32 33 080	14 52 135	35 33 172	38 36 222	43 33 260	32 19 343
217	44 37 039	32 55 079	15 08 134	35 36 172	38 21 221	43 11 259	32 12 342
218	44 51 038	33 17 078	15 25 133	35 40 171	38 06 221	42 49 258	32 04 340
219	45 04 037	33 39 077	15 41 133	35 43 170	37 51 220	42 27 257	31 57 339
220	45 18 036	34 01 076	15 58 132	35 47 170	37 37 219	42 05 256	31 48 338
221	45 31 034	34 23 075	16 15 131	35 51 169	37 23 219	41 43 255	31 40 337
222	45 43 033	34 44 074	16 32 130	35 56 168	37 09 218	41 21 255	31 31 336
223	45 55 032	35 06 073	16 49 129	36 01 168	36 55 217	41 00 254	31 22 335
224	46 07 031	35 27 072	17 07 128	36 05 167	36 42 217	40 38 253	31 12 334
	◆ANTARES	Nunki	◆FOMALHAUT	ACHERNAR	CANOPUS	◆Suhail	SPICA
225	46 18 029	35 49 071	17 24 127	36 11 166	36 29 216	40 17 252	31 02 333
226	46 29 028	36 10 070	17 42 127	36 16 166	36 15 215	39 55 251	30 51 331
227	46 39 027	36 31 069	18 00 126	36 22 165	36 03 214	39 34 250	30 40 330
228	46 49 026	36 52 068	18 19 125	36 28 164	35 50 214	39 13 250	30 29 329
229	46 59 024	37 13 067	18 37 124	36 34 164	35 38 213	38 52 249	30 17 328
230	47 08 023	37 33 066	18 56 123	36 40 163	35 26 212	38 31 248	30 05 327
231	47 16 022	37 54 065	19 15 122	36 47 162	35 14 212	38 10 247	29 53 326
232	47 24 020	38 14 064	19 34 121	36 54 162	35 02 211	37 50 246	29 40 325
233	47 32 019	38 34 063	19 53 121	37 01 161	34 51 210	37 29 246	29 27 324
234	47 39 018	38 54 062	20 13 120	37 08 160	34 39 210	37 09 245	29 13 323
235	47 46 016	39 13 060	20 32 119	37 16 160	34 28 209	36 48 244	28 59 321
236	47 52 015	39 33 059	20 52 118	37 24 159	34 18 208	36 28 243	28 45 320
237	47 58 014	39 52 058	21 12 117	37 32 158	34 07 207	36 08 242	28 31 319
238	48 03 012	40 11 057	21 32 116	37 41 158	33 57 207	35 48 242	28 16 318
239	48 07 011	40 30 056	21 52 115	37 49 157	33 47 206	35 29 241	28 01 317
	ANTARES	◆Nunki	FOMALHAUT	◆ACHERNAR	CANOPUS	◆ACRUX	SPICA
240	48 11 010	40 48 055	22 13 115	37 58 156	33 37 205	68 05 257	27 45 316
241	48 15 008	41 07 054	22 33 114	38 07 156	33 28 205	67 43 256	27 29 315
242	48 18 007	41 25 053	22 54 113	38 16 155	33 19 204	67 22 256	27 13 314
243	48 20 006	41 42 052	23 15 112	38 26 154	33 10 203	67 00 255	26 57 313
244	48 22 004	42 00 050	23 35 111	38 36 154	33 01 202	66 38 254	26 40 312
245	48 24 003	42 17 049	23 56 110	38 46 153	32 52 202	66 17 254	26 23 311
246	48 25 002	42 34 048	24 18 109	38 56 152	32 44 201	65 55 253	26 06 310
247	48 25 000	42 50 047	24 39 108	39 07 152	32 36 200	65 34 252	25 49 309
248	48 25 359	43 07 046	25 00 108	39 17 151	32 29 200	65 12 251	25 31 308
249	48 24 358	43 23 045	25 22 107	39 28 151	32 21 199	64 51 251	25 13 307
250	48 23 356	43 38 043	25 43 106	39 39 150	32 14 198	64 30 250	24 55 306
251	48 22 355	43 54 042	26 05 105	39 51 149	32 07 198	64 09 250	24 37 305
252	48 19 354	44 09 041	26 27 104	40 03 149	32 00 197	63 48 249	24 18 303
253	48 17 352	44 23 040	26 49 103	40 14 148	31 54 196	63 27 248	23 59 302
254	48 13 351	44 37 039	27 10 102	40 26 147	31 48 195	63 06 248	23 40 301
	◆Nunki	FOMALHAUT	◆ACHERNAR	CANOPUS	◆ACRUX	SPICA	ANTARES
255	44 51 037	27 32 101	40 39 147	31 42 195	62 45 247	23 21 300	48 10 350
256	45 05 036	27 55 101	40 51 146	31 36 194	62 25 246	23 01 299	48 05 348
257	45 18 035	28 17 100	41 04 145	31 31 193	62 04 246	22 42 298	48 00 347
258	45 31 034	28 39 099	41 17 145	31 26 193	61 44 245	22 22 297	47 55 346
259	45 43 032	29 01 098	41 30 144	31 21 192	61 23 244	22 02 296	47 49 344
260	45 55 031	29 23 097	41 43 143	31 17 191	61 03 244	21 42 295	47 43 343
261	46 06 030	29 46 096	41 57 143	31 12 191	60 43 243	21 21 294	47 36 342
262	46 17 029	30 08 095	42 11 142	31 08 190	60 23 242	21 01 293	47 29 340
263	46 28 027	30 30 094	42 24 141	31 05 189	60 03 242	20 40 292	47 21 339
264	46 38 026	30 53 093	42 39 141	31 01 188	59 43 241	20 19 292	47 13 338
265	46 48 025	31 15 092	42 53 140	30 58 188	59 24 241	19 58 291	47 04 336
266	46 57 024	31 38 092	43 08 139	30 55 187	59 04 240	19 37 290	46 55 335
267	47 06 022	32 00 091	43 22 139	30 53 186	58 45 239	19 16 289	46 45 334
268	47 14 021	32 23 090	43 37 138	30 50 186	58 26 239	18 54 288	46 35 333
269	47 22 020	32 45 089	43 52 137	30 48 185	58 06 238	18 33 287	46 24 331

LAT 68°S

LHA ♈	Hc Zn	Hc Zn	Hc Zn	Hc Zn	Hc Zn	Hc Zn	Hc Zn
	◆Nunki	FOMALHAUT	◆ACHERNAR	CANOPUS	◆ACRUX	SPICA	ANTARES
270	47 29 018	33 08 088	44 08 137	30 46 184	57 47 238	18 11 286	46 13 330
271	47 36 017	33 30 087	44 23 136	30 45 183	57 28 237	17 50 285	46 02 329
272	47 42 016	33 53 086	44 39 135	30 44 183	57 10 236	17 28 284	45 50 327
273	47 48 014	34 15 085	44 55 135	30 43 182	56 51 236	17 06 283	45 38 326
274	47 53 013	34 37 084	45 11 134	30 42 181	56 32 235	16 44 282	45 25 325
275	47 58 012	35 00 083	45 27 133	30 42 181	56 14 235	16 22 281	45 12 324
276	48 02 010	35 22 082	45 44 133	30 42 180	55 56 234	16 00 280	44 58 323
277	48 06 009	35 44 081	46 00 132	30 42 179	55 38 233	15 38 279	44 44 321
278	48 10 008	36 06 080	46 17 131	30 42 179	55 20 233	15 16 278	44 30 320
279	48 12 006	36 29 079	46 34 131	30 43 178	55 02 232	14 53 277	44 16 319
280	48 15 005	36 51 078	46 51 130	30 44 177	54 44 232	14 31 276	44 01 318
281	48 16 004	37 13 077	47 08 129	30 45 176	54 27 231	14 09 275	43 45 316
282	48 17 002	37 34 076	47 26 129	30 47 176	54 09 230	13 46 274	43 30 315
283	48 18 001	37 56 075	47 43 128	30 48 175	53 52 230	13 24 273	43 14 314
284	48 18 000	38 18 074	48 01 127	30 50 174	53 35 229	13 01 272	42 57 313
	◆ALTAIR	FOMALHAUT	◆ACHERNAR	CANOPUS	◆RIGIL KENT.	ANTARES	Nunki
285	12 37 013	38 39 073	48 19 127	30 53 174	62 24 253	42 41 312	48 18 358
286	12 42 012	39 01 072	48 37 126	30 55 173	62 02 252	42 24 311	48 17 357
287	12 46 011	39 22 071	48 56 125	30 58 172	61 41 251	42 07 310	48 15 356
288	12 50 010	39 43 070	49 14 125	31 02 172	61 20 251	41 49 308	48 13 354
289	12 54 009	40 05 069	49 33 124	31 05 171	60 58 250	41 32 307	48 11 353
290	12 57 008	40 25 068	49 51 123	31 09 170	60 37 249	41 13 306	48 08 352
291	13 00 007	40 46 067	50 10 123	31 13 169	60 16 249	40 55 305	48 04 350
292	13 02 006	41 07 066	50 29 122	31 17 169	59 55 248	40 37 304	48 00 349
293	13 04 005	41 27 065	50 48 121	31 22 168	59 35 247	40 18 303	47 56 348
294	13 06 004	41 48 064	51 08 120	31 26 167	59 14 247	39 59 302	47 51 346
295	13 07 003	42 08 063	51 27 120	31 31 167	58 53 246	39 40 301	47 45 345
296	13 08 002	42 27 062	51 47 119	31 37 166	58 33 245	39 20 300	47 39 344
297	13 09 001	42 47 061	52 06 118	31 42 165	58 13 245	39 01 298	47 32 342
298	13 09 000	43 07 059	52 26 118	31 48 164	57 52 244	38 41 297	47 25 341
299	13 08 359	43 26 058	52 46 117	31 54 164	57 32 243	38 21 296	47 18 340
	◆FOMALHAUT	ACHERNAR	◆CANOPUS	RIGIL KENT.	◆ANTARES	Nunki	ALTAIR
300	43 45 057	53 06 116	32 01 163	57 12 243	38 00 295	47 10 338	13 08 358
301	44 04 056	53 26 116	32 08 162	56 52 242	37 40 294	47 01 337	13 06 357
302	44 22 055	53 47 115	32 14 162	56 33 241	37 19 293	46 52 336	13 05 356
303	44 40 054	54 07 114	32 22 161	56 13 241	36 59 292	46 43 334	13 03 355
304	44 58 053	54 28 113	32 29 160	55 54 240	36 38 291	46 33 333	13 01 354
305	45 16 052	54 49 113	32 37 160	55 34 239	36 17 290	46 22 332	12 58 353
306	45 34 050	55 09 112	32 45 159	55 15 239	35 56 289	46 11 331	12 55 352
307	45 51 049	55 30 111	32 53 158	54 56 238	35 34 288	46 00 329	12 51 351
308	46 08 048	55 51 111	33 02 157	54 37 237	35 13 287	45 49 328	12 47 349
309	46 24 047	56 12 110	33 10 157	54 18 237	34 51 286	45 36 327	12 43 348
310	46 40 046	56 33 109	33 19 156	53 59 236	34 30 285	45 24 326	12 38 347
311	46 56 044	56 55 108	33 29 155	53 41 235	34 08 284	45 11 324	12 33 346
312	47 12 043	57 16 108	33 38 155	53 22 235	33 46 283	44 58 323	12 28 345
313	47 27 042	57 38 107	33 48 154	53 04 234	33 24 282	44 44 322	12 22 344
314	47 42 041	57 59 106	33 58 153	52 46 233	33 02 281	44 30 321	12 16 343
	◆FOMALHAUT	ACHERNAR	◆CANOPUS	Miaplacidus	RIGIL KENT.	◆ANTARES	Nunki
315	47 56 039	58 21 105	34 08 153	47 43 182	52 28 233	32 40 280	44 16 319
316	48 11 038	58 43 105	34 19 152	47 42 181	52 10 232	32 18 279	44 01 318
317	48 24 037	59 04 104	34 29 151	47 42 181	51 52 232	31 56 278	43 46 317
318	48 38 036	59 26 103	34 40 150	47 42 180	51 35 231	31 33 277	43 30 316
319	48 50 034	59 48 102	34 51 150	47 42 180	51 17 230	31 11 276	43 14 315
320	49 03 033	60 10 101	35 03 149	47 42 179	51 00 230	30 49 275	42 58 314
321	49 15 032	60 32 101	35 15 148	47 43 179	50 43 229	30 26 274	42 42 312
322	49 27 031	60 54 100	35 27 148	47 43 178	50 26 228	30 04 273	42 25 311
323	49 38 029	61 16 099	35 39 147	47 44 178	50 09 228	29 41 273	42 08 310
324	49 49 028	61 39 098	35 51 146	47 45 177	49 53 227	29 19 272	41 51 309
325	49 59 027	62 01 097	36 04 146	47 46 177	49 36 227	28 56 271	41 33 308
326	50 09 025	62 23 096	36 16 145	47 48 176	49 20 226	28 34 270	41 15 307
327	50 18 024	62 46 096	36 30 144	47 49 176	49 04 225	28 11 269	40 57 306
328	50 27 023	63 08 095	36 43 143	47 51 175	48 48 225	27 49 268	40 39 304
329	50 35 021	63 30 094	36 56 143	47 53 174	48 33 224	27 27 267	40 20 303
	◆FOMALHAUT	ACHERNAR	◆CANOPUS	Miaplacidus	RIGIL KENT.	◆ANTARES	Nunki
330	50 43 020	63 53 093	37 10 142	47 56 174	48 17 223	27 04 266	40 01 302
331	50 51 019	64 15 092	37 24 141	47 58 173	48 02 223	26 42 265	39 42 301
332	50 58 017	64 38 091	37 38 141	48 01 173	47 46 222	26 19 264	39 23 300
333	51 04 016	65 00 090	37 53 140	48 04 172	47 31 222	25 57 263	39 03 299
334	51 10 014	65 23 089	38 07 139	48 07 172	47 17 221	25 35 262	38 43 298
335	51 15 013	65 45 088	38 22 138	48 10 171	47 02 220	25 12 262	38 23 297
336	51 20 012	66 08 087	38 37 138	48 13 171	46 47 220	24 50 261	38 03 296
337	51 24 010	66 30 086	38 52 137	48 17 170	46 33 219	24 28 260	37 43 295
338	51 28 009	66 53 086	39 08 136	48 21 170	46 19 218	24 06 259	37 22 294
339	51 31 007	67 15 084	39 23 136	48 25 169	46 05 218	23 44 258	37 02 293
340	51 34 006	67 37 083	39 39 135	48 29 169	45 52 217	23 22 257	36 41 292
341	51 36 005	68 00 082	39 55 134	48 34 168	45 38 217	23 00 256	36 20 291
342	51 37 003	68 22 081	40 11 134	48 38 168	45 25 216	22 38 255	35 59 290
343	51 38 002	68 44 080	40 28 133	48 43 167	45 12 215	22 17 254	35 38 289
344	51 39 000	69 06 079	40 44 132	48 48 167	44 59 215	21 55 253	35 16 288
	Diphda	◆CANOPUS	Miaplacidus	RIGIL KENT.	◆ANTARES	Nunki	◆FOMALHAUT
345	37 24 031	41 01 131	48 53 166	44 46 214	21 34 253	34 55 287	51 39 359
346	37 36 030	41 18 131	48 59 166	44 33 214	21 12 252	34 33 286	51 38 358
347	37 47 029	41 35 130	49 05 165	44 21 213	20 51 251	34 11 285	51 37 356
348	37 57 028	41 52 129	49 10 165	44 09 212	20 30 250	33 50 284	51 35 355
349	38 08 027	42 10 129	49 16 164	43 57 212	20 09 249	33 28 283	51 33 353
350	38 18 026	42 27 128	49 23 164	43 45 211	19 48 248	33 06 282	51 30 352
351	38 27 024	42 45 127	49 29 163	43 34 210	19 27 247	32 44 281	51 27 351
352	38 36 023	43 03 126	49 36 163	43 23 210	19 06 246	32 22 280	51 23 349
353	38 45 022	43 22 126	49 42 162	43 12 209	18 46 246	31 59 279	51 18 348
354	38 53 021	43 40 125	49 49 162	43 01 209	18 25 245	31 37 278	51 13 347
355	39 01 020	43 58 124	49 56 161	42 50 208	18 05 244	31 15 277	51 08 345
356	39 08 018	44 17 124	50 04 161	42 40 207	17 45 243	30 53 276	51 02 344
357	39 15 017	44 36 123	50 11 160	42 29 207	17 25 242	30 30 275	50 55 342
358	39 21 016	44 55 122	50 19 160	42 19 206	17 05 241	30 08 274	50 48 341
359	39 27 015	45 14 121	50 27 159	42 10 205	16 46 240	29 45 273	50 41 340

LAT 69°S

LHA	Hc	Zn	Hc	Zn	Hc	Zn	Hc	Zn	Hc	Zn	Hc	Zn	Hc	Zn
♈	Diphda		◆RIGEL		CANOPUS		Miaplacidus		◆RIGIL KENT.		Peacock		◆FOMALHAUT	
0	38 34	013	11 45	082	46 03	120	51 31	158	42 55	205	63 47	270	49 37	339
1	38 39	012	12 06	081	46 22	119	51 39	158	42 45	205	63 25	270	49 29	337
2	38 43	011	12 27	080	46 41	118	51 47	157	42 37	204	63 04	269	49 20	336
3	38 47	010	12 48	079	47 00	117	51 56	157	42 28	203	62 42	268	49 12	335
4	38 50	008	13 09	078	47 19	117	52 04	156	42 20	203	62 21	267	49 02	334
5	38 53	007	13 30	078	47 38	116	52 13	156	42 11	202	61 59	266	48 52	332
6	38 55	006	13 51	077	47 58	115	52 22	155	42 03	201	61 38	265	48 42	331
7	38 57	005	14 12	076	48 17	114	52 31	155	41 56	201	61 17	264	48 31	330
8	38 59	003	14 33	075	48 37	114	52 40	154	41 48	200	60 55	263	48 20	328
9	39 00	002	14 54	074	48 57	113	52 50	154	41 41	200	60 34	262	48 09	327
10	39 01	001	15 14	073	49 17	112	52 59	153	41 34	199	60 13	262	47 57	326
11	39 01	000	15 35	072	49 37	111	53 09	153	41 27	198	59 51	261	47 45	325
12	39 01	359	15 55	071	49 57	111	53 19	152	41 20	198	59 30	260	47 32	323
13	39 00	357	16 15	070	50 17	110	53 29	152	41 14	197	59 09	259	47 19	322
14	38 59	356	16 36	069	50 37	109	53 40	151	41 08	196	58 48	258	47 06	321
	◆RIGEL		CANOPUS		Miaplacidus		◆RIGIL KENT.		Peacock		◆FOMALHAUT		Diphda	
15	16 56	068	50 58	108	53 50	151	41 02	196	58 27	257	46 52	320	38 57	355
16	17 15	067	51 18	107	54 01	150	40 56	195	58 06	257	46 38	318	38 55	354
17	17 35	066	51 39	107	54 12	150	40 51	194	57 45	256	46 23	317	38 52	352
18	17 55	065	51 59	106	54 23	149	40 45	194	57 24	255	46 09	316	38 49	351
19	18 14	064	52 20	105	54 34	149	40 40	193	57 04	254	45 53	315	38 46	350
20	18 33	063	52 41	104	54 45	148	40 35	193	56 43	253	45 38	314	38 42	349
21	18 52	062	53 02	103	54 57	148	40 31	192	56 22	253	45 22	312	38 37	348
22	19 11	061	53 23	102	55 08	147	40 27	191	56 02	252	45 06	311	38 32	346
23	19 30	060	53 44	102	55 20	147	40 22	191	55 41	251	44 50	310	38 27	345
24	19 48	059	54 05	101	55 32	146	40 19	190	55 21	250	44 33	309	38 21	344
25	20 07	058	54 26	100	55 44	146	40 15	189	55 01	250	44 16	308	38 15	343
26	20 25	057	54 47	099	55 56	145	40 12	189	54 41	249	43 59	307	38 09	342
27	20 43	056	55 08	098	56 09	145	40 08	188	54 21	248	43 42	305	38 02	340
28	21 01	055	55 30	097	56 21	144	40 06	187	54 01	247	43 24	304	37 54	339
29	21 18	054	55 51	097	56 34	144	40 03	187	53 41	247	43 06	303	37 46	338
	◆RIGEL		CANOPUS		Miaplacidus		◆RIGIL KENT.		Peacock		◆FOMALHAUT		Diphda	
30	21 35	053	56 12	096	56 47	143	40 00	186	53 21	246	42 48	302	37 38	337
31	21 52	052	56 34	095	57 00	143	39 58	186	53 02	245	42 30	301	37 29	336
32	22 09	051	56 55	094	57 13	142	39 56	185	52 42	244	42 11	300	37 20	334
33	22 26	050	57 17	093	57 26	142	39 54	184	52 23	244	41 53	299	37 11	333
34	22 42	049	57 38	092	57 40	141	39 53	184	52 04	243	41 34	298	37 01	332
35	22 58	048	58 00	091	57 53	141	39 52	183	51 45	242	41 15	297	36 51	331
36	23 14	047	58 21	090	58 07	140	39 51	182	51 26	242	40 55	296	36 40	330
37	23 30	046	58 43	089	58 21	140	39 50	182	51 07	241	40 36	295	36 29	329
38	23 45	045	59 04	088	58 35	139	39 49	181	50 48	240	40 16	293	36 18	327
39	24 00	044	59 26	087	58 49	139	39 49	181	50 30	239	39 56	292	36 06	326
40	24 14	043	59 47	086	59 03	138	39 49	180	50 11	239	39 36	291	35 54	325
41	24 29	042	60 09	085	59 18	138	39 49	179	49 53	238	39 16	290	35 41	324
42	24 43	040	60 30	084	59 32	137	39 49	179	49 35	237	38 56	289	35 28	323
43	24 57	039	60 51	083	59 47	137	39 50	178	49 17	237	38 36	288	35 15	322
44	25 10	038	61 13	082	60 02	136	39 51	177	48 59	236	38 15	287	35 02	321
	RIGEL		◆SIRIUS		Suhail		◆RIGIL KENT.		Peacock		◆FOMALHAUT		Diphda	
45	25 23	037	27 20	064	39 15	110	39 52	177	48 41	235	37 55	286	34 48	319
46	25 36	036	27 39	063	39 35	110	39 53	176	48 23	234	37 34	285	34 34	318
47	25 49	035	27 58	062	39 56	109	39 55	175	48 06	234	37 13	284	34 19	317
48	26 01	034	28 17	061	40 16	108	39 57	175	47 49	233	36 52	283	34 05	316
49	26 13	033	28 36	060	40 37	107	39 59	174	47 32	232	36 31	282	33 49	315
50	26 25	032	28 54	059	40 57	106	40 01	174	47 15	232	36 10	281	33 34	314
51	26 36	031	29 12	058	41 18	105	40 04	173	46 58	231	35 49	280	33 19	313
52	26 47	030	29 30	056	41 39	104	40 07	172	46 41	230	35 28	279	33 03	312
53	26 57	029	29 48	055	41 59	104	40 10	172	46 25	230	35 07	278	32 46	311
54	27 07	028	30 06	054	42 20	103	40 13	171	46 09	229	34 45	277	32 30	310
55	27 17	026	30 23	053	42 41	102	40 16	170	45 52	228	34 24	276	32 13	308
56	27 26	025	30 40	052	43 03	101	40 20	170	45 36	228	34 03	275	31 56	307
57	27 35	024	30 57	051	43 24	100	40 24	169	45 21	227	33 41	274	31 39	306
58	27 44	023	31 14	050	43 45	099	40 28	168	45 05	226	33 20	274	31 22	305
59	27 52	022	31 30	049	44 06	098	40 33	168	44 50	225	32 58	273	31 04	304
	RIGEL		◆SIRIUS		Suhail		◆RIGIL KENT.		Peacock		◆FOMALHAUT		Diphda	
60	28 00	021	31 46	048	44 27	098	40 37	167	44 34	225	32 37	272	30 46	303
61	28 08	020	32 02	047	44 49	097	40 42	167	44 19	224	32 15	271	30 28	302
62	28 15	019	32 18	046	45 10	096	40 47	166	44 04	223	31 54	270	30 10	301
63	28 21	018	32 33	045	45 32	095	40 53	165	43 50	223	31 32	269	29 51	300
64	28 28	016	32 48	044	45 53	094	40 58	165	43 35	222	31 11	268	29 32	299
65	28 34	015	33 03	042	46 14	093	41 04	164	43 21	221	30 49	267	29 13	298
66	28 39	014	33 17	041	46 36	092	41 10	163	43 07	221	30 28	266	28 54	297
67	28 44	013	33 31	040	46 57	091	41 16	163	42 53	220	30 07	265	28 35	296
68	28 49	012	33 45	039	47 19	090	41 23	162	42 39	219	29 45	264	28 16	295
69	28 53	011	33 58	038	47 40	089	41 29	161	42 26	219	29 24	263	27 56	294
70	28 57	010	34 11	037	48 02	088	41 36	161	42 12	218	29 02	262	27 36	293
71	29 00	009	34 24	036	48 23	087	41 44	160	41 59	217	28 41	261	27 16	292
72	29 03	007	34 36	035	48 45	086	41 51	160	41 46	217	28 20	261	26 56	291
73	29 06	006	34 48	033	49 06	085	41 59	159	41 33	216	27 59	260	26 36	290
74	29 08	005	35 00	032	49 28	085	42 06	158	41 21	215	27 38	259	26 16	289
	RIGEL		◆SIRIUS		Suhail		◆RIGIL KENT.		Peacock		◆FOMALHAUT		Diphda	
75	29 10	004	35 11	031	49 49	084	42 14	158	41 09	215	27 16	258	25 56	288
76	29 11	003	35 22	030	50 10	083	42 23	157	40 56	214	26 56	257	25 35	287
77	29 12	002	35 33	029	50 32	082	42 31	156	40 45	213	26 35	256	25 14	286
78	29 12	001	35 43	028	50 53	081	42 40	156	40 33	213	26 14	255	24 54	285
79	29 12	000	35 53	027	51 14	080	42 49	155	40 21	212	25 53	254	24 33	284
80	29 12	358	36 02	025	51 35	079	42 58	155	40 10	211	25 32	253	24 12	283
81	29 11	357	36 11	024	51 56	078	43 07	154	39 59	211	25 12	253	23 51	282
82	29 10	356	36 20	023	52 17	076	43 17	153	39 48	210	24 51	252	23 30	281
83	29 08	355	36 28	022	52 38	075	43 27	153	39 38	209	24 31	251	23 09	280
84	29 06	354	36 36	021	52 59	074	43 37	152	39 27	209	24 11	250	22 48	279
85	29 04	353	36 43	020	53 20	073	43 47	151	39 17	208	23 51	249	22 26	278
86	29 01	352	36 50	018	53 40	072	43 57	151	39 07	207	23 31	248	22 05	277
87	28 57	350	36 57	017	54 00	071	44 08	150	38 57	207	23 11	247	21 44	276
88	28 54	349	37 03	016	54 21	070	44 19	149	38 48	206	22 51	246	21 22	275
89	28 49	348	37 09	015	54 41	069	44 30	149	38 39	205	22 31	246	21 01	274

LHA	Hc	Zn	Hc	Zn	Hc	Zn	Hc	Zn	Hc	Zn	Hc	Zn	Hc	Zn
♈	◆SIRIUS		Suhail		◆RIGIL KENT.		Peacock		◆ACHERNAR		Acamar		RIGEL	
90	37 14	014	55 01	068	44 41	148	38 30	205	59 55	259	52 43	296	28 45	347
91	37 19	012	55 21	067	44 52	148	38 21	204	59 34	258	52 24	295	28 40	346
92	37 23	011	55 40	066	45 04	147	38 12	203	59 13	258	52 04	294	28 34	345
93	37 27	010	56 00	064	45 16	146	38 04	202	58 52	257	51 45	293	28 29	344
94	37 31	009	56 19	063	45 28	146	37 56	202	58 31	256	51 25	292	28 22	343
95	37 34	008	56 38	062	45 40	145	37 48	201	58 10	255	51 05	291	28 16	341
96	37 36	006	56 57	061	45 52	144	37 40	200	57 49	255	50 44	290	28 09	340
97	37 38	005	57 16	060	46 05	144	37 33	200	57 29	254	50 24	288	28 01	339
98	37 40	004	57 34	058	46 18	143	37 26	199	57 08	253	50 04	287	27 53	338
99	37 41	003	57 52	057	46 31	143	37 19	198	56 48	252	49 43	286	27 45	337
100	37 42	001	58 10	056	46 44	142	37 12	198	56 27	251	49 22	285	27 37	336
101	37 43	000	58 28	055	46 57	141	37 06	197	56 07	251	49 02	284	27 28	335
102	37 42	359	58 45	053	47 11	141	36 59	196	55 47	250	48 41	283	27 18	334
103	37 42	358	59 02	052	47 25	140	36 54	196	55 26	249	48 20	282	27 09	333
104	37 41	357	59 19	051	47 39	139	36 48	195	55 06	248	47 59	281	26 59	332
	Alphard		◆Gienah		RIGIL KENT.		◆Peacock		ACHERNAR		◆RIGEL		SIRIUS	
105	25 04	041	20 18	086	47 53	139	36 42	194	54 46	248	26 48	330	37 39	355
106	25 18	040	20 39	085	48 07	138	36 37	194	54 27	247	26 37	329	37 37	354
107	25 32	039	21 00	084	48 21	137	36 32	193	54 07	246	26 26	328	37 35	353
108	25 45	038	21 22	083	48 36	137	36 27	192	53 47	245	26 15	327	37 32	352
109	25 58	037	21 43	082	48 51	136	36 23	192	53 28	245	26 03	326	37 29	351
110	26 11	036	22 04	081	49 06	136	36 19	191	53 08	244	25 51	325	37 25	349
111	26 23	034	22 26	080	49 21	135	36 15	190	52 49	243	25 38	324	37 21	348
112	26 35	033	22 47	079	49 36	134	36 11	190	52 30	243	25 25	323	37 16	347
113	26 47	032	23 08	078	49 52	134	36 08	189	52 11	242	25 12	322	37 11	346
114	26 58	031	23 29	078	50 07	133	36 04	188	51 52	241	24 59	321	37 06	345
115	27 09	030	23 50	077	50 23	132	36 01	188	51 33	240	24 45	320	37 00	343
116	27 20	029	24 11	076	50 39	132	35 59	187	51 15	240	24 31	319	36 54	342
117	27 30	028	24 32	075	50 55	131	35 56	186	50 56	239	24 17	318	36 47	341
118	27 40	027	24 52	074	51 12	130	35 54	186	50 38	238	24 02	317	36 40	340
119	27 49	026	25 13	073	51 28	130	35 52	185	50 20	238	23 47	315	36 32	339
	Alphard		◆SPICA		RIGIL KENT.		◆Peacock		ACHERNAR		◆RIGEL		SIRIUS	
120	27 58	025	13 32	086	51 45	129	35 50	184	50 01	237	23 32	314	36 24	337
121	28 07	023	13 53	085	52 02	128	35 49	184	49 43	236	23 16	313	36 15	336
122	28 16	022	14 14	084	52 18	128	35 47	183	49 26	236	23 01	312	36 07	335
123	28 24	021	14 36	083	52 36	127	35 46	182	49 08	235	22 45	311	35 57	334
124	28 31	020	14 57	082	52 53	127	35 46	182	48 51	234	22 28	310	35 48	333
125	28 38	019	15 18	081	53 10	126	35 45	181	48 33	233	22 12	309	35 38	332
126	28 45	018	15 40	080	53 28	125	35 45	180	48 16	233	21 55	308	35 27	330
127	28 52	017	16 01	079	53 45	125	35 45	180	47 59	232	21 38	307	35 17	329
128	28 58	016	16 22	078	54 03	124	35 45	179	47 42	231	21 21	306	35 05	328
129	29 03	015	16 43	077	54 21	123	35 46	178	47 25	231	21 03	305	34 54	327
130	29 08	013	17 04	076	54 39	123	35 47	178	47 09	230	20 46	304	34 42	326
131	29 13	012	17 25	075	54 57	122	35 48	177	46 52	229	20 28	303	34 30	325
132	29 18	011	17 45	074	55 15	121	35 49	176	46 36	229	20 10	302	34 17	324
133	29 22	010	18 06	073	55 34	121	35 51	175	46 20	228	19 51	301	34 04	323
134	29 25	009	18 27	072	55 53	120	35 53	175	46 04	227	19 33	300	33 51	321
	Alphard		◆SPICA		RIGIL KENT.		◆Peacock		ACHERNAR		◆CANOPUS		SIRIUS	
135	29 28	008	18 47	072	56 11	119	35 55	174	45 49	227	65 41	292	33 37	320
136	29 31	007	19 07	071	56 30	119	35 57	173	45 33	226	65 21	291	33 24	319
137	29 33	005	19 28	070	56 49	118	36 00	173	45 18	225	65 01	290	33 09	318
138	29 35	004	19 48	069	57 08	117	36 02	172	45 03	225	64 40	288	32 55	317
139	29 36	003	20 08	068	57 27	117	36 05	171	44 48	224	64 20	287	32 40	316
140	29 37	002	20 27	067	57 47	116	36 09	171	44 33	223	63 59	286	32 25	315
141	29 38	001	20 47	066	58 06	115	36 12	170	44 18	222	63 39	285	32 09	314
142	29 38	000	21 07	065	58 25	114	36 16	169	44 04	222	63 18	284	31 54	313
143	29 38	359	21 26	064	58 45	114	36 20	169	43 49	221	62 57	283	31 38	311
144	29 37	358	21 45	063	59 05	113	36 25	168	43 35	220	62 36	282	31 21	310
145	29 36	356	22 04	062	59 25	112	36 29	167	43 22	220	62 15	281	31 05	309
146	29 35	355	22 23	061	59 45	112	36 34	167	43 08	219	61 54	280	30 48	308
147	29 33	354	22 42	060	60 05	111	36 39	166	42 54	218	61 32	279	30 31	307
148	29 30	353	23 00	059	60 25	110	36 44	165	42 41	218	61 11	278	30 14	306
149	29 27	352	23 18	058	60 45	110	36 50	165	42 28	217	60 50	276	29 56	305
	◆SPICA		RIGIL KENT.		◆Peacock		ACHERNAR		◆CANOPUS		SIRIUS		Alphard	
150	23 36	057	61 05	109	36 56	164	42 15	216	60 28	275	29 39	304	29 24	351
151	23 54	056	61 26	108	37 02	163	42 03	216	60 07	275	29 21	303	29 20	350
152	24 12	055	61 46	107	37 08	163	41 50	215	59 45	274	29 03	302	29 16	348
153	24 29	054	62 07	107	37 15	162	41 38	214	59 24	273	28 44	301	29 12	347
154	24 47	052	62 27	106	37 21	161	41 26	214	59 02	272	28 26	300	29 07	346
155	25 03	051	62 48	105	37 28	161	41 14	213	58 41	271	28 07	299	29 01	345
156	25 20	050	63 09	104	37 36	160	41 02	212	58 19	270	27 48	298	28 56	344
157	25 37	049	63 30	104	37 43	159	40 51	212	57 58	269	27 29	297	28 50	343
158	25 53	048	63 51	103	37 51	159	40 40	211	57 36	268	27 10	296	28 43	342
159	26 09	047	64 12	102	37 59	158	40 29	210	57 15	267	26 50	295	28 36	341
160	26 24	046	64 33	101	38 07	157	40 18	210	56 54	266	26 31	294	28 29	339
161	26 40	045	64 54	100	38 15	157	40 07	209	56 32	265	26 11	293	28 21	338
162	26 55	044	65 15	100	38 24	156	39 57	208	56 11	264	25 51	292	28 13	337
163	27 10	043	65 36	099	38 33	155	39 47	208	55 49	263	25 31	291	28 04	336
164	27 24	042	65 58	098	38 42	155	39 37	207	55 28	263	25 11	290	27 56	335
	◆SPICA		ANTARES		◆Peacock		ACHERNAR		◆CANOPUS		SIRIUS		Alphard	
165	27 38	041	27 18	093	38 51	154	39 27	206	55 07	262	24 51	289	27 46	334
166	27 52	040	27 39	092	39 01	153	39 18	206	54 45	261	24 30	288	27 37	333
167	28 06	039	28 01	091	39 11	153	39 09	205	54 24	260	24 10	287	27 27	332
168	28 19	038	28 22	090	39 21	152	39 00	204	54 03	259	23 49	286	27 16	331
169	28 32	037	28 44	089	39 31	151	38 51	204	53 42	258	23 28	285	27 06	330
170	28 45	035	29 05	088	39 42	151	38 42	203	53 21	257	23 07	284	26 54	328
171	28 57	034	29 27	087	39 52	150	38 34	202	53 00	257	22 46	283	26 43	327
172	29 09	033	29 48	086	40 03	149	38 26	202	52 39	256	22 26	282	26 31	326
173	29 21	032	30 10	086	40 14	148	38 18	201	52 18	255	22 04	281	26 19	325
174	29 32	031	30 31	085	40 26	148	38 11	200	51 58	254	21 43	280	26 07	324
175	29 43	030	30 52	084	40 37	147	38 03	200	51 37	253	21 22	279	25 54	323
176	29 54	029	31 14	083	40 49	146	37 56	199	51 16	253	21 01	278	25 41	322
177	30 04	028	31 35	082	41 01	146	37 49	198	50 56	252	20 40	277	25 27	321
178	30 14	027	31 56	081	41 13	145	37 43	198	50 36	251	20 18	276	25 14	320
179	30 23	025	32 17	080	41 26	144	37 36	197	50 15	250	19 57	275	25 00	319

LAT 69°S

LHA ♈	Hc	Zn	Hc	Zn	Hc	Zn	Hc	Zn	Hc	Zn	Hc	Zn	Hc	Zn
	♦SPICA		ANTARES		♦Peacock		ACHERNAR		♦CANOPUS		SIRIUS		Alphard	
180	30 32	024	32 39	079	41 38	144	37 30	196	49 55	249	19 35	274	24 45	318
181	30 41	023	33 00	078	41 51	143	37 24	196	49 35	249	19 14	273	24 31	317
182	30 49	022	33 21	077	42 04	142	37 18	195	49 15	248	18 52	272	24 16	316
183	30 57	021	33 41	076	42 17	142	37 13	194	48 55	247	18 31	272	24 01	315
184	31 04	020	34 02	075	42 31	141	37 08	194	48 35	246	18 09	271	23 45	314
185	31 12	019	34 23	074	42 44	140	37 03	193	48 16	246	17 48	270	23 30	312
186	31 18	018	34 44	073	42 58	140	36 58	192	47 56	245	17 26	269	23 14	311
187	31 24	016	35 04	072	43 12	139	36 54	192	47 37	244	17 05	268	22 57	310
188	31 30	015	35 24	071	43 26	138	36 49	191	47 18	243	16 43	267	22 41	309
189	31 36	014	35 45	070	43 41	138	36 45	190	46 58	243	16 22	266	22 24	308
190	31 41	013	36 05	069	43 55	137	36 42	190	46 39	242	16 01	265	22 07	307
191	31 45	012	36 25	068	44 10	136	36 38	189	46 21	241	15 39	264	21 50	306
192	31 50	011	36 45	067	44 25	136	36 35	188	46 02	240	15 18	263	21 32	305
193	31 53	010	37 04	066	44 40	135	36 32	188	45 43	239	14 56	262	21 15	304
194	31 57	008	37 24	065	44 55	134	36 29	187	45 25	239	14 35	261	20 57	303
	♦SPICA		ANTARES		♦Peacock		ACHERNAR		♦CANOPUS		Suhail		Alphard	
195	32 00	007	37 43	064	45 11	134	36 27	186	45 06	238	51 12	280	20 39	302
196	32 02	006	38 02	063	45 27	133	36 25	186	44 48	237	50 51	279	20 21	301
197	32 04	005	38 21	061	45 42	132	36 23	185	44 30	237	50 30	278	20 02	300
198	32 06	004	38 40	060	45 58	132	36 21	184	44 12	236	50 08	277	19 43	299
199	32 07	003	38 59	059	46 15	131	36 19	184	43 55	235	49 47	276	19 25	298
200	32 08	001	39 17	058	46 31	130	36 18	183	43 37	234	49 26	275	19 06	297
201	32 08	000	39 35	057	46 47	129	36 17	182	43 20	234	49 04	274	18 46	296
202	32 08	359	39 53	056	47 04	129	36 16	182	43 03	233	48 43	273	18 27	295
203	32 07	358	40 11	055	47 21	128	36 16	181	42 46	232	48 21	273	18 07	294
204	32 06	357	40 28	054	47 38	127	36 16	180	42 29	231	48 00	272	17 48	293
205	32 05	356	40 46	053	47 55	127	36 16	180	42 12	231	47 38	271	17 28	292
206	32 03	354	41 03	052	48 13	126	36 16	179	41 55	230	47 17	270	17 08	291
207	32 01	353	41 19	050	48 30	125	36 17	178	41 39	229	46 55	269	16 48	290
208	31 58	352	41 36	049	48 48	125	36 17	178	41 23	228	46 34	268	16 28	289
209	31 55	351	41 52	048	49 05	124	36 18	177	41 07	228	46 12	267	16 07	288
	♦ANTARES		Nunki		♦FOMALHAUT		ACHERNAR		CANOPUS		♦Suhail		SPICA	
210	42 08	047	30 15	085	14 08	140	36 20	176	40 51	227	45 51	266	31 51	350
211	42 23	046	30 36	084	14 22	139	36 21	176	40 35	226	45 29	265	31 47	349
212	42 39	045	30 57	083	14 36	138	36 23	175	40 20	226	45 08	264	31 43	348
213	42 54	044	31 19	082	14 50	138	36 25	174	40 05	225	44 47	263	31 38	346
214	43 08	042	31 40	081	15 05	137	36 27	174	39 50	224	44 25	262	31 33	345
215	43 22	041	32 01	080	15 20	136	36 30	173	39 35	223	44 04	262	31 27	344
216	43 36	040	32 22	079	15 35	135	36 33	172	39 20	223	43 43	261	31 21	343
217	43 50	039	32 43	078	15 50	134	36 36	172	39 06	222	43 22	260	31 15	342
218	44 03	038	33 04	077	16 06	133	36 39	171	38 51	221	43 01	259	31 08	341
219	44 16	036	33 25	076	16 22	132	36 43	170	38 37	221	42 39	258	31 00	340
220	44 29	035	33 46	075	16 38	132	36 46	170	38 23	220	42 18	257	30 53	338
221	44 41	034	34 07	074	16 54	131	36 50	169	38 10	219	41 58	256	30 45	337
222	44 53	033	34 27	073	17 10	130	36 55	168	37 56	218	41 37	255	30 36	336
223	45 04	031	34 48	072	17 27	129	36 59	168	37 43	218	41 16	255	30 27	335
224	45 15	030	35 08	071	17 44	128	37 04	167	37 30	217	40 55	254	30 18	334
	♦ANTARES		Nunki		♦FOMALHAUT		ACHERNAR		CANOPUS		♦Suhail		SPICA	
225	45 26	029	35 29	070	18 01	127	37 09	166	37 17	216	40 35	253	30 08	333
226	45 36	028	35 49	069	18 18	126	37 14	166	37 04	216	40 14	252	29 58	332
227	45 46	026	36 09	068	18 35	125	37 20	165	36 52	215	39 54	251	29 48	331
228	45 55	025	36 29	067	18 53	125	37 25	164	36 40	214	39 33	250	29 37	329
229	46 04	024	36 49	066	19 11	124	37 31	163	36 28	213	39 13	250	29 26	328
230	46 13	023	37 08	065	19 29	123	37 38	163	36 16	213	38 53	249	29 15	327
231	46 21	021	37 28	064	19 47	122	37 44	162	36 05	212	38 33	248	29 03	326
232	46 28	020	37 47	063	20 05	121	37 51	161	35 53	211	38 13	247	28 51	325
233	46 35	019	38 06	062	20 24	120	37 58	161	35 42	211	37 54	246	28 38	324
234	46 42	017	38 25	061	20 42	119	38 05	160	35 31	210	37 34	245	28 25	323
235	46 48	016	38 43	060	21 01	119	38 12	159	35 21	209	37 14	245	28 12	322
236	46 54	015	39 02	059	21 20	118	38 20	159	35 11	208	36 55	244	27 59	321
237	46 59	014	39 20	058	21 39	117	38 28	158	35 00	208	36 36	243	27 45	320
238	47 04	012	39 38	056	21 58	116	38 36	157	34 51	207	36 17	242	27 31	319
239	47 08	011	39 56	055	22 18	115	38 44	157	34 41	206	35 58	241	27 16	317
	ANTARES		♦Nunki		FOMALHAUT		♦ACHERNAR		CANOPUS		♦ACRUX		SPICA	
240	47 12	010	40 14	054	22 37	114	38 53	156	34 31	206	68 17	259	27 02	316
241	47 16	008	40 31	053	22 57	113	39 02	155	34 22	205	67 56	259	26 47	315
242	47 18	007	40 48	052	23 17	112	39 11	155	34 13	204	67 35	258	26 32	314
243	47 21	006	41 05	051	23 37	112	39 20	154	34 05	203	67 14	257	26 16	313
244	47 23	004	41 21	050	23 57	111	39 30	153	33 56	203	66 53	256	26 00	312
245	47 24	003	41 38	049	24 17	110	39 39	153	33 48	202	66 32	256	25 44	311
246	47 25	002	41 54	047	24 37	109	39 49	152	33 40	201	66 12	255	25 28	310
247	47 25	000	42 09	046	24 58	108	39 59	151	33 32	201	65 51	254	25 11	309
248	47 25	359	42 25	045	25 18	107	40 10	151	33 25	200	65 30	254	24 54	308
249	47 25	358	42 40	044	25 39	106	40 20	150	33 18	199	65 10	253	24 37	307
250	47 23	356	42 55	043	25 59	105	40 31	149	33 11	198	64 49	252	24 20	306
251	47 22	355	43 09	042	26 20	105	40 42	149	33 04	198	64 29	251	24 03	305
252	47 20	354	43 23	040	26 41	104	40 54	148	32 58	197	64 08	251	23 45	304
253	47 17	352	43 37	039	27 02	103	41 05	147	32 52	196	63 48	250	23 27	303
254	47 14	351	43 50	038	27 23	102	41 17	147	32 46	196	63 28	249	23 09	302
	♦Nunki		FOMALHAUT		♦ACHERNAR		CANOPUS		♦ACRUX		SPICA		ANTARES	
255	44 03	037	27 44	101	41 29	146	32 40	195	63 08	249	22 50	301	47 11	350
256	44 16	036	28 05	100	41 41	145	32 35	194	62 48	248	22 32	300	47 06	348
257	44 29	034	28 26	099	41 53	145	32 29	194	62 28	247	22 13	299	47 02	347
258	44 40	033	28 48	098	42 06	144	32 24	193	62 08	247	21 54	298	46 57	346
259	44 52	032	29 09	097	42 18	143	32 20	192	61 48	246	21 35	297	46 51	345
260	45 03	031	29 30	096	42 31	143	32 16	191	61 29	245	21 16	296	46 45	343
261	45 14	029	29 52	096	42 44	142	32 11	191	61 09	245	20 56	295	46 39	342
262	45 24	028	30 13	095	42 58	141	32 08	190	60 50	244	20 37	294	46 32	341
263	45 34	027	30 35	094	43 11	141	32 04	189	60 31	243	20 17	293	46 25	339
264	45 44	026	30 56	093	43 25	140	32 01	189	60 12	243	19 57	292	46 17	338
265	45 53	024	31 18	092	43 39	139	31 58	188	59 52	242	19 37	291	46 09	337
266	46 02	023	31 39	091	43 53	139	31 55	187	59 34	241	19 17	290	46 00	336
267	46 10	022	32 01	090	44 07	138	31 52	186	59 15	241	18 57	289	45 51	334
268	46 18	021	32 22	089	44 22	137	31 50	186	58 56	240	18 36	288	45 41	333
269	46 25	019	32 44	088	44 36	137	31 48	185	58 37	240	18 16	287	45 31	332

LHA ♈	Hc	Zn	Hc	Zn	Hc	Zn	Hc	Zn	Hc	Zn	Hc	Zn	Hc	Zn
	♦Nunki		FOMALHAUT		♦ACHERNAR		CANOPUS		♦ACRUX		SPICA		ANTARES	
270	46 32	018	33 05	087	44 51	136	31 46	184	58 19	239	17 55	286	45 21	331
271	46 38	017	33 27	086	45 06	135	31 45	184	58 01	238	17 34	285	45 10	329
272	46 44	015	33 48	085	45 21	135	31 44	183	57 42	238	17 13	284	44 59	328
273	46 50	014	34 09	084	45 37	134	31 43	182	57 24	237	16 53	283	44 48	327
274	46 55	013	34 31	083	45 52	133	31 42	181	57 06	236	16 32	282	44 36	326
275	46 59	012	34 52	082	46 08	133	31 42	181	56 48	236	16 11	281	44 23	324
276	47 03	010	35 13	081	46 24	132	31 42	180	56 31	235	15 49	280	44 11	323
277	47 07	009	35 35	080	46 40	131	31 42	179	56 13	235	15 28	279	43 57	322
278	47 10	008	35 56	079	46 56	131	31 42	179	55 56	234	15 07	278	43 44	321
279	47 13	006	36 17	078	47 13	130	31 43	178	55 38	233	14 46	277	43 30	320
280	47 15	005	36 38	077	47 29	129	31 44	177	55 21	233	14 24	276	43 16	318
281	47 16	004	36 59	076	47 46	128	31 45	176	55 04	232	14 03	276	43 02	317
282	47 17	002	37 20	075	48 03	128	31 46	176	54 47	231	13 42	275	42 47	316
283	47 18	001	37 41	074	48 20	127	31 48	175	54 31	231	13 20	274	42 32	315
284	47 18	000	38 01	073	48 37	126	31 50	174	54 14	230	12 59	273	42 16	314
	♦ALTAIR		FOMALHAUT		♦ACHERNAR		CANOPUS		♦RIGIL KENT.		ANTARES		Nunki	
285	11 39	013	38 22	072	48 55	126	31 52	174	62 40	255	42 01	312	47 18	358
286	11 43	012	38 42	071	49 12	125	31 55	173	62 20	254	41 45	311	47 17	357
287	11 47	011	39 02	070	49 30	124	31 58	172	61 59	253	41 28	310	47 16	356
288	11 51	010	39 23	069	49 48	124	32 01	171	61 39	252	41 12	309	47 14	354
289	11 55	009	39 43	068	50 06	123	32 04	171	61 18	252	40 55	308	47 11	353
290	11 58	008	40 03	067	50 24	122	32 08	170	60 58	251	40 38	307	47 08	352
291	12 00	007	40 22	066	50 42	122	32 12	169	60 37	250	40 20	306	47 05	350
292	12 03	006	40 42	065	51 00	121	32 16	169	60 17	249	40 03	305	47 01	349
293	12 05	005	41 01	064	51 19	120	32 20	168	59 57	249	39 45	303	46 57	348
294	12 06	004	41 21	063	51 38	119	32 25	167	59 37	248	39 27	302	46 52	346
295	12 07	003	41 40	062	51 56	119	32 30	166	59 17	247	39 09	301	46 47	345
296	12 08	002	41 59	061	52 15	118	32 35	166	58 57	247	38 50	300	46 41	344
297	12 09	001	42 17	060	52 34	117	32 40	165	58 38	246	38 32	299	46 35	343
298	12 09	000	42 36	059	52 54	117	32 46	164	58 18	245	38 13	298	46 28	341
299	12 08	359	42 54	058	53 13	116	32 52	164	57 59	245	37 54	297	46 21	340
	♦FOMALHAUT		ACHERNAR		♦CANOPUS		RIGIL KENT.		♦ANTARES		Nunki		ALTAIR	
300	43 12	056	53 32	115	32 58	163	57 39	244	37 34	296	46 14	339	12 08	358
301	43 30	055	53 52	114	33 05	162	57 20	243	37 15	295	46 06	337	12 07	357
302	43 47	054	54 11	114	33 11	161	57 01	243	36 55	294	45 57	336	12 05	356
303	44 05	053	54 31	113	33 18	161	56 42	242	36 36	293	45 48	335	12 03	355
304	44 22	052	54 51	112	33 26	160	56 23	241	36 16	292	45 39	334	12 01	354
305	44 39	051	55 11	111	33 33	159	56 04	241	35 56	291	45 29	332	11 58	353
306	44 55	050	55 31	111	33 41	159	55 46	240	35 36	290	45 19	331	11 55	352
307	45 11	048	55 51	110	33 49	158	55 27	239	35 15	289	45 08	330	11 52	351
308	45 27	047	56 12	109	33 57	157	55 09	239	34 55	288	44 57	329	11 48	350
309	45 43	046	56 32	108	34 05	156	54 50	238	34 34	287	44 46	327	11 44	349
310	45 58	045	56 52	108	34 14	156	54 32	237	34 14	286	44 34	326	11 40	348
311	46 13	044	57 13	107	34 23	155	54 14	237	33 53	285	44 22	325	11 35	347
312	46 28	042	57 34	106	34 32	154	53 56	236	33 32	284	44 10	324	11 30	346
313	46 42	041	57 54	105	34 42	154	53 39	235	33 11	283	43 57	323	11 24	345
314	46 56	040	58 15	105	34 51	153	53 21	235	32 50	282	43 43	321	11 18	344
	♦FOMALHAUT		ACHERNAR		♦CANOPUS		Miaplacidus		RIGIL KENT.		♦ANTARES		Nunki	
315	47 10	039	58 36	104	35 01	152	48 43	182	53 04	234	32 29	281	43 30	320
316	47 23	038	58 57	103	35 11	152	48 42	181	52 46	233	32 08	280	43 16	319
317	47 36	036	59 18	102	35 22	151	48 42	181	52 29	233	31 47	279	43 02	318
318	47 49	035	59 39	101	35 32	150	48 42	180	52 12	232	31 25	278	42 47	317
319	48 01	034	60 00	101	35 43	149	48 42	180	51 55	231	31 04	277	42 32	315
320	48 13	033	60 21	100	35 54	149	48 42	179	51 39	231	30 43	276	42 17	314
321	48 24	031	60 42	099	36 06	148	48 43	179	51 22	230	30 21	275	42 01	313
322	48 35	030	61 04	098	36 17	147	48 43	178	51 06	229	30 00	274	41 45	312
323	48 45	029	61 25	097	36 29	147	48 44	178	50 49	229	29 38	273	41 29	311
324	48 55	027	61 46	096	36 41	146	48 45	177	50 33	228	29 17	272	41 13	310
325	49 05	026	62 08	095	36 53	145	48 46	176	50 17	227	28 55	271	40 56	309
326	49 14	025	62 29	095	37 05	144	48 48	176	50 02	227	28 34	270	40 39	307
327	49 23	023	62 51	094	37 18	144	48 49	175	49 46	226	28 12	269	40 22	306
328	49 31	022	63 12	093	37 31	143	48 51	175	49 31	226	27 51	268	40 04	305
329	49 39	021	63 33	092	37 44	142	48 53	174	49 15	225	27 29	267	39 47	304
	♦FOMALHAUT		ACHERNAR		♦CANOPUS		Miaplacidus		RIGIL KENT.		♦ANTARES		Nunki	
330	49 47	019	63 55	091	37 57	142	48 55	174	49 00	224	27 08	267	39 29	303
331	49 54	018	64 16	090	38 11	141	48 58	173	48 45	224	26 47	266	39 11	302
332	50 00	017	64 38	089	38 24	140	49 00	173	48 31	223	26 25	265	38 52	301
333	50 06	015	64 59	088	38 38	139	49 03	172	48 16	222	26 04	264	38 34	300
334	50 12	014	65 21	087	38 52	139	49 06	172	48 02	222	25 42	263	38 15	299
335	50 17	013	65 42	086	39 07	138	49 09	171	47 47	221	25 21	262	37 56	298
336	50 21	011	66 04	085	39 21	137	49 13	171	47 33	220	25 00	261	37 37	296
337	50 25	010	66 25	084	39 36	137	49 16	170	47 20	220	24 39	260	37 17	295
338	50 29	009	66 47	083	39 51	136	49 20	170	47 06	219	24 17	259	36 58	294
339	50 32	007	67 08	082	40 06	135	49 24	169	46 52	219	23 56	258	36 38	293
340	50 34	006	67 29	081	40 21	134	49 28	169	46 39	218	23 35	257	36 18	292
341	50 36	005	67 50	080	40 37	134	49 32	168	46 26	217	23 14	257	35 59	291
342	50 38	003	68 12	079	40 52	133	49 37	168	46 13	217	22 53	256	35 38	290
343	50 38	002	68 33	078	41 08	132	49 42	167	46 00	216	22 33	255	35 18	289
344	50 39	000	68 54	077	41 24	131	49 47	167	45 48	215	22 12	254	34 58	288
	Diphda		♦CANOPUS		Miaplacidus		RIGIL KENT.		♦ANTARES		Nunki		♦FOMALHAUT	
345	36 33	031	41 40	131	49 52	166	45 35	215	21 51	253	34 37	287	50 39	359
346	36 44	030	41 57	130	49 57	165	45 23	214	21 31	252	34 17	286	50 38	358
347	36 54	029	42 13	129	50 03	165	45 11	213	21 10	251	33 56	285	50 37	356
348	37 04	028	42 30	129	50 08	164	45 00	213	20 50	250	33 35	284	50 35	355
349	37 14	026	42 47	128	50 14	164	44 48	212	20 30	249	33 14	283	50 33	354
350	37 23	025	43 04	127	50 20	163	44 37	212	20 10	249	32 53	282	50 31	352
351	37 32	024	43 21	126	50 26	163	44 26	211	19 50	248	32 32	281	50 28	351
352	37 41	023	43 39	126	50 33	162	44 15	210	19 30	247	32 11	280	50 24	350
353	37 49	022	43 56	125	50 39	162	44 04	210	19 10	246	31 50	279	50 20	348
354	37 57	020	44 14	124	50 46	161	43 53	209	18 51	245	31 29	278	50 15	347
355	38 04	019	44 32	123	50 53	161	43 43	208	18 31	244	31 07	277	50 10	345
356	38 11	018	44 50	123	51 00	160	43 33	208	18 12	243	30 46	276	50 04	344
357	38 17	017	45 08	122	51 08	160	43 23	207	17 53	242	30 25	275	49 58	343
358	38 23	016	45 26	121	51 15	159	43 13	206	17 34	241	30 03	275	49 52	341
359	38 29	014	45 45	120	51 23	159	43 04	206	17 15	241	29 42	274	49 45	340

 LAT 70°S

LHA	Hc	Zn	Hc	Zn	Hc	Zn	Hc	Zn	Hc	Zn	Hc	Zn	Hc	Zn
♈	Diphda		◆RIGEL		CANOPUS		Miaplacidus		◆RIGIL KENT.		Peacock		◆FOMALHAUT	
0	37 36	013	11 36	082	46 33	119	52 26	158	43 49	206	63 45	272	48 41	339
2	37 44	011	12 17	080	47 09	117	52 42	157	43 31	204	63 04	271	48 26	337
4	37 51	008	12 57	078	47 46	116	52 59	156	43 15	203	62 23	269	48 08	334
6	37 56	006	13 37	076	48 23	114	53 16	155	42 59	202	61 42	267	47 50	331
8	37 59	003	14 17	074	49 01	113	53 34	154	42 44	200	61 01	265	47 29	329
10	38 01	001	14 56	072	49 39	111	53 53	152	42 30	199	60 20	263	47 07	326
12	38 01	359	15 35	071	50 17	109	54 12	151	42 17	198	59 40	262	46 44	324
14	37 59	356	16 14	069	50 56	108	54 32	150	42 05	197	58 59	260	46 19	321
16	37 55	354	16 52	067	51 35	106	54 53	149	41 54	195	58 19	258	45 53	319
18	37 50	351	17 29	065	52 15	104	55 14	148	41 44	194	57 39	257	45 25	317
20	37 43	349	18 06	063	52 55	103	55 36	147	41 34	193	56 59	255	44 56	314
22	37 34	347	18 42	061	53 35	101	55 58	146	41 25	191	56 20	253	44 26	312
24	37 24	344	19 17	059	54 15	099	56 21	145	41 18	190	55 41	252	43 55	310
26	37 12	342	19 52	057	54 56	098	56 45	144	41 11	189	55 02	250	43 23	307
28	36 58	339	20 26	055	55 37	096	57 09	143	41 05	188	54 23	249	42 50	305
	RIGEL		◆SIRIUS		Suhail		◆RIGIL KENT.		Peacock		◆FOMALHAUT		Diphda	
30	20 59	053	22 04	078	34 59	122	41 00	186	53 45	247	42 16	303	36 43	337
32	21 31	051	22 44	076	35 34	120	40 56	185	53 08	246	41 41	301	36 26	335
34	22 03	049	23 23	074	36 10	119	40 53	184	52 30	244	41 06	298	36 08	332
36	22 33	046	24 03	072	36 46	117	40 51	182	51 54	243	40 29	296	35 48	330
38	23 02	044	24 41	070	37 23	115	40 49	181	51 18	241	39 52	294	35 27	328
40	23 30	042	25 20	068	38 00	114	40 49	180	50 42	240	39 14	292	35 04	326
42	23 57	040	25 58	066	38 38	112	40 49	179	50 07	238	38 36	290	34 40	323
44	24 23	038	26 35	064	39 16	110	40 51	177	49 32	237	37 57	288	34 15	321
46	24 48	036	27 12	062	39 55	109	40 53	176	48 58	235	37 18	286	33 49	319
48	25 11	034	27 47	060	40 34	107	40 57	175	48 24	234	36 38	284	33 21	317
50	25 33	032	28 23	058	41 13	105	41 01	173	47 52	233	35 58	282	32 52	314
52	25 54	029	28 57	056	41 53	104	41 06	172	47 19	231	35 18	280	32 22	312
54	26 14	027	29 31	054	42 33	102	41 12	171	46 48	230	34 37	278	31 52	310
56	26 32	025	30 03	052	43 13	100	41 19	170	46 17	228	33 57	276	31 20	308
58	26 49	023	30 35	050	43 54	098	41 27	168	45 46	227	33 16	274	30 47	306
	RIGEL		◆SIRIUS		Suhail		◆RIGIL KENT.		Peacock		◆FOMALHAUT		Diphda	
60	27 04	021	31 06	047	44 35	097	41 36	167	45 17	226	32 35	272	30 13	304
62	27 18	019	31 36	045	45 16	095	41 45	166	44 48	224	31 54	270	29 38	302
64	27 30	016	32 04	043	45 57	093	41 56	164	44 20	223	31 13	268	29 03	299
66	27 41	014	32 32	041	46 38	091	42 07	163	43 52	221	30 32	267	28 27	297
68	27 50	012	32 58	039	47 19	089	42 20	162	43 25	220	29 51	265	27 50	295
70	27 58	010	33 23	036	48 00	087	42 33	161	42 59	219	29 10	263	27 13	293
72	28 04	007	33 47	034	48 41	085	42 47	159	42 34	217	28 29	261	26 35	291
74	28 08	005	34 09	032	49 21	083	43 02	158	42 10	216	27 49	259	25 56	289
76	28 11	003	34 30	030	50 02	081	43 18	157	41 46	214	27 09	257	25 17	287
78	28 12	001	34 50	027	50 43	079	43 35	155	41 23	213	26 29	256	24 38	285
80	28 12	358	35 08	025	51 23	077	43 52	154	41 01	212	25 49	254	23 58	283
82	28 10	356	35 25	023	52 03	075	44 10	153	40 40	210	25 10	252	23 18	281
84	28 07	354	35 40	020	52 42	073	44 29	152	40 20	209	24 31	250	22 38	280
86	28 01	352	35 53	018	53 21	071	44 49	150	40 00	208	23 53	249	21 57	278
88	27 55	349	36 05	016	54 00	069	45 10	149	39 42	206	23 15	247	21 17	276
	◆SIRIUS		Suhail		◆RIGIL KENT.		Peacock		FOMALHAUT		◆ACHERNAR		RIGEL	
90	36 16	013	54 38	066	45 32	148	39 24	205	22 37	245	60 05	261	27 46	347
92	36 24	011	55 15	064	45 54	146	39 07	203	22 00	243	59 25	259	27 36	345
94	36 31	009	55 51	062	46 17	145	38 51	202	21 24	242	58 45	258	27 25	343
96	36 37	006	56 27	059	46 41	144	38 36	201	20 48	240	58 05	256	27 12	341
98	36 40	004	57 02	057	47 06	143	38 22	199	20 13	238	57 25	254	26 58	338
100	36 42	001	57 36	055	47 31	141	38 09	198	19 39	236	56 46	253	26 42	336
102	36 42	359	58 09	052	47 57	140	37 57	197	19 05	235	56 06	251	26 24	334
104	36 41	357	58 41	049	48 24	139	37 46	195	18 32	233	55 28	250	26 06	332
106	36 38	354	59 11	047	48 51	137	37 35	194	17 59	231	54 49	248	25 46	330
108	36 33	352	59 41	044	49 20	136	37 26	193	17 28	229	54 12	247	25 24	327
110	36 26	350	60 09	041	49 48	135	37 18	191	16 57	228	53 34	245	25 01	325
112	36 18	347	60 35	039	50 18	133	37 10	190	16 27	226	52 57	244	24 37	323
114	36 08	345	61 00	036	50 48	132	37 04	188	15 58	224	52 20	242	24 12	321
116	35 56	342	61 23	033	51 19	131	36 58	187	15 30	222	51 44	241	23 46	319
118	35 43	340	61 44	030	51 50	129	36 54	186	15 03	221	51 09	239	23 18	317
	Suhail		◆SPICA		RIGIL KENT.		Peacock		◆FOMALHAUT		ACHERNAR		◆SIRIUS	
120	62 04	027	13 27	086	52 22	128	36 50	184	14 37	219	50 34	238	35 28	338
122	62 21	024	14 08	084	52 55	127	36 47	183	14 11	217	49 59	236	35 12	335
124	62 37	021	14 49	082	53 28	125	36 46	182	13 47	215	49 25	235	34 54	333
126	62 50	018	15 29	080	54 02	124	36 45	180	13 24	214	48 52	234	34 35	331
128	63 02	014	16 09	078	54 36	123	36 45	179	13 02	212	48 19	232	34 14	329
130	63 11	011	16 49	076	55 11	121	36 47	177	12 41	210	47 47	231	33 52	326
132	63 18	008	17 29	074	55 46	120	36 49	176	12 20	208	47 16	229	33 29	324
134	63 22	005	18 08	072	56 22	119	36 52	175	12 01	207	46 45	228	33 04	322
136	63 24	002	18 47	070	56 58	117	36 57	173	11 44	205	46 15	227	32 38	320
138	63 24	358	19 26	068	57 35	116	37 02	172	11 27	203	45 45	225	32 11	317
140	63 22	355	20 04	066	58 12	114	37 08	171	11 11	201	45 16	224	31 42	315
142	63 17	352	20 41	064	58 50	113	37 15	169	10 57	200	44 48	222	31 13	313
144	63 10	349	21 17	062	59 28	112	37 23	168	10 44	198	44 21	221	30 42	311
146	63 01	345	21 53	060	60 06	110	37 32	167	10 32	196	43 54	220	30 11	309
148	62 50	342	22 29	058	60 45	109	37 42	165	10 21	194	43 28	218	29 38	307
	◆SPICA		ANTARES		◆Peacock		FOMALHAUT		ACHERNAR		◆CANOPUS		Suhail	
150	23 03	056	22 18	106	37 53	164	10 11	193	43 03	217	60 22	277	62 36	339
152	23 37	054	22 57	104	38 05	162	10 03	191	42 39	216	59 41	275	62 20	336
154	24 10	052	23 37	103	38 18	161	09 56	189	42 16	214	59 00	273	62 03	333
156	24 42	050	24 17	101	38 32	160	09 50	187	41 53	213	58 19	271	61 43	330
158	25 13	048	24 58	099	38 47	158	09 45	186	41 31	212	57 38	269	61 22	327
160	25 43	046	25 38	097	39 02	157	09 42	184	41 10	210	56 57	268	60 59	324
162	26 12	044	26 19	095	39 19	156	09 40	182	40 50	209	56 16	266	60 34	321
164	26 39	042	27 00	093	39 36	154	09 39	180	40 30	207	55 35	264	60 07	318
166	27 06	039	27 41	092	39 54	153	09 39	179	40 12	206	54 54	262	59 39	316
168	27 32	037	28 22	090	40 14	151	09 41	177	39 54	205	54 14	261	59 10	313
170	27 56	035	29 03	088	40 34	150	09 44	175	39 38	203	53 33	259	58 39	310
172	28 19	033	29 44	086	40 55	149	09 48	173	39 22	202	52 53	257	58 07	308
174	28 41	031	30 25	084	41 16	147	09 54	171	39 07	201	52 13	255	57 35	305
176	29 01	029	31 06	082	41 39	146	10 00	170	38 53	199	51 34	254	57 01	303
178	29 20	026	31 46	080	42 02	145	10 08	168	38 40	198	50 54	252	56 26	300

LHA	Hc	Zn	Hc	Zn	Hc	Zn	Hc	Zn	Hc	Zn	Hc	Zn	Hc	Zn
♈	SPICA		◆ANTARES		Nunki		FOMALHAUT		◆ACHERNAR		CANOPUS		◆Suhail	
180	29 37	024	32 27	078	20 06	112	10 17	166	38 28	197	50 16	251	55 50	298
182	29 53	022	33 07	076	20 44	110	10 28	164	38 16	195	49 37	249	55 13	296
184	30 08	020	33 46	074	21 23	108	10 39	163	38 06	194	48 59	247	54 36	293
186	30 21	017	34 26	072	22 02	107	10 52	161	37 57	192	48 21	246	53 58	291
188	30 32	015	35 04	070	22 41	105	11 06	159	37 48	191	47 44	244	53 19	289
190	30 42	013	35 43	068	23 21	103	11 21	157	37 41	190	47 07	243	52 40	287
192	30 51	011	36 21	066	24 01	101	11 38	156	37 34	188	46 31	241	52 01	285
194	30 57	008	36 58	064	24 42	099	11 55	154	37 29	187	45 56	240	51 21	283
196	31 03	006	37 34	062	25 22	098	12 14	152	37 24	186	45 20	238	50 41	281
198	31 06	004	38 10	060	26 03	096	12 33	150	37 21	184	44 46	237	50 00	279
200	31 08	001	38 45	058	26 44	094	12 54	149	37 18	183	44 12	235	49 19	277
202	31 08	359	39 19	055	27 25	092	13 16	147	37 16	182	43 39	234	48 39	275
204	31 07	357	39 53	053	28 06	090	13 39	145	37 16	180	43 06	232	47 58	273
206	31 03	355	40 25	051	28 47	088	14 03	143	37 16	179	42 34	231	47 17	271
208	30 59	352	40 56	049	29 28	086	14 28	142	37 17	178	42 02	229	46 36	269
	◆ANTARES		Nunki		◆FOMALHAUT		ACHERNAR		CANOPUS		◆Suhail		SPICA	
210	41 27	046	30 09	084	14 54	140	37 20	176	41 32	228	45 55	267	30 52	350
212	41 56	044	30 50	082	15 21	138	37 23	175	41 02	226	45 14	265	30 44	348
214	42 24	042	31 30	081	15 49	136	37 27	173	40 32	225	44 33	263	30 35	345
216	42 50	039	32 11	079	16 17	135	37 32	172	40 04	223	43 52	262	30 24	343
218	43 16	037	32 51	077	16 47	133	37 38	171	39 36	222	43 12	260	30 11	341
220	43 40	035	33 31	075	17 17	131	37 45	169	39 09	220	42 31	258	29 57	339
222	44 02	032	34 10	073	17 48	130	37 53	168	38 43	219	41 51	256	29 41	336
224	44 23	030	34 49	071	18 21	128	38 02	167	38 18	217	41 12	255	29 24	334
226	44 43	027	35 27	069	18 53	126	38 12	165	37 53	216	40 32	253	29 05	332
228	45 01	025	36 05	066	19 27	124	38 23	164	37 29	215	39 53	251	28 45	330
230	45 17	022	36 43	064	20 01	123	38 35	163	37 07	213	39 14	249	28 24	328
232	45 32	020	37 19	062	20 36	121	38 48	161	36 45	212	38 36	248	28 01	325
234	45 45	017	37 55	060	21 12	119	39 01	160	36 23	210	37 58	246	27 37	323
236	45 56	015	38 30	058	21 48	117	39 16	159	36 03	209	37 21	244	27 12	321
238	46 05	012	39 05	056	22 25	116	39 31	157	35 44	207	36 44	243	26 46	319
	ANTARES		◆Nunki		FOMALHAUT		◆ACHERNAR		CANOPUS		◆ACRUX		SPICA	
240	46 13	009	39 38	054	23 02	114	39 48	156	35 26	206	68 27	262	26 18	317
242	46 19	007	40 11	051	23 40	112	40 05	154	35 08	204	67 47	260	25 50	315
244	46 23	004	40 42	049	24 18	110	40 23	153	34 52	203	67 06	259	25 20	313
246	46 25	002	41 13	047	24 57	108	40 42	152	34 36	202	66 26	257	24 49	310
248	46 25	359	41 42	045	25 36	107	41 02	150	34 21	200	65 46	256	24 17	308
250	46 24	356	42 10	042	26 15	105	41 23	149	34 08	199	65 07	254	23 45	306
252	46 20	354	42 37	040	26 55	103	41 44	148	33 55	197	64 27	253	23 11	304
254	46 15	351	43 03	037	27 35	101	42 07	146	33 43	196	63 48	251	22 37	302
256	46 08	349	43 27	035	28 15	100	42 30	145	33 33	194	63 09	250	22 02	300
258	45 59	346	43 50	033	28 56	098	42 54	144	33 23	193	62 31	248	21 26	298
260	45 48	344	44 12	030	29 37	096	43 19	142	33 14	191	61 53	247	20 49	296
262	45 35	341	44 31	028	30 18	094	43 44	141	33 07	190	61 15	246	20 12	294
264	45 21	338	44 50	025	30 59	092	44 11	139	33 00	189	60 38	244	19 35	292
266	45 05	336	45 06	023	31 40	090	44 38	138	32 54	187	60 01	243	18 56	290
268	44 48	333	45 21	020	32 21	088	45 05	137	32 50	186	59 25	242	18 17	288
	◆Nunki		FOMALHAUT		◆ACHERNAR		CANOPUS		◆ACRUX		RIGIL KENT.		ANTARES	
270	45 35	018	33 02	086	45 34	135	32 46	184	58 49	240	68 00	269	44 29	331
272	45 46	015	33 43	085	46 03	134	32 44	183	58 14	239	67 19	267	44 08	329
274	45 56	013	34 23	083	46 33	133	32 42	181	57 39	238	66 38	266	43 46	326
276	46 04	010	35 04	081	47 04	131	32 42	180	57 04	236	65 57	264	43 22	324
278	46 11	007	35 44	079	47 35	130	32 42	179	56 30	235	65 16	262	42 57	321
280	46 15	005	36 25	077	48 07	128	32 44	177	55 57	234	64 36	261	42 31	319
282	46 17	002	37 04	075	48 39	127	32 46	176	55 24	233	63 55	259	42 03	317
284	46 18	000	37 44	073	49 13	125	32 50	174	54 52	231	63 15	257	41 35	314
286	46 17	357	38 23	071	49 46	124	32 55	173	54 20	230	62 35	256	41 05	312
288	46 14	354	39 01	069	50 21	123	33 00	171	53 49	229	61 56	254	40 34	310
290	46 09	352	39 39	067	50 55	121	33 07	170	53 18	227	61 16	253	40 02	307
292	46 02	349	40 16	064	51 31	120	33 15	168	52 48	226	60 37	251	39 29	305
294	45 54	347	40 53	062	52 07	118	33 23	167	52 19	225	59 59	250	38 55	303
296	45 44	344	41 29	060	52 43	117	33 33	166	51 50	224	59 20	248	38 20	301
298	45 32	342	42 04	058	53 20	115	33 44	164	51 22	222	58 43	247	37 44	299
	◆FOMALHAUT		ACHERNAR		◆CANOPUS		ACRUX		RIGIL KENT.		◆ANTARES		Nunki	
300	42 39	056	53 57	114	33 56	163	50 55	221	58 05	245	37 08	297	45 18	339
302	43 12	053	54 35	112	34 08	161	50 28	220	57 28	244	36 31	295	45 02	337
304	43 44	051	55 13	111	34 22	160	50 02	219	56 51	243	35 53	292	44 45	334
306	44 16	049	55 52	109	34 37	158	49 37	218	56 15	241	35 15	290	44 26	332
308	44 46	047	56 31	108	34 52	157	49 12	216	55 39	240	34 36	288	44 06	329
310	45 15	044	57 10	106	35 09	155	48 48	215	55 04	238	33 57	286	43 44	327
312	45 43	042	57 49	105	35 26	154	48 25	214	54 30	237	33 18	284	43 21	324
314	46 10	039	58 29	103	35 45	153	48 03	213	53 55	236	32 38	282	42 56	322
316	46 35	037	59 09	101	36 04	151	47 41	211	53 22	234	31 57	280	42 30	320
318	46 59	034	59 50	100	36 24	150	47 20	210	52 49	233	31 17	278	42 03	317
320	47 22	032	60 30	098	36 45	148	47 00	209	52 16	232	30 36	277	41 35	315
322	47 43	029	61 11	096	37 07	147	46 41	208	51 44	230	29 55	275	41 05	313
324	48 02	027	61 52	094	37 30	145	46 22	206	51 13	229	29 14	273	40 34	310
326	48 20	024	62 33	093	37 54	144	46 04	205	50 42	228	28 33	271	40 02	308
328	48 36	022	63 14	091	38 19	142	45 47	204	50 12	226	27 52	269	39 30	306
	◆FOMALHAUT		Diphda		◆CANOPUS		ACRUX		RIGIL KENT.		◆ANTARES		Nunki	
330	48 50	019	32 28	047	38 44	141	45 31	203	49 43	225	27 11	267	38 56	304
332	49 03	016	32 57	045	39 10	140	45 15	201	49 14	224	26 30	265	38 21	301
334	49 13	014	33 26	043	39 37	138	45 01	200	48 46	222	25 50	263	37 46	299
336	49 22	011	33 53	041	40 05	137	44 47	199	48 19	221	25 09	262	37 10	297
338	49 29	009	34 20	039	40 34	135	44 34	198	47 52	220	24 28	260	36 33	295
340	49 34	006	34 45	036	41 03	134	44 22	196	47 26	219	23 48	258	35 55	293
342	49 38	003	35 08	034	41 33	132	44 11	195	47 01	217	23 08	256	35 17	291
344	49 39	000	35 31	032	42 04	131	44 01	194	46 37	216	22 28	254	34 39	289
346	49 38	358	35 51	030	42 35	129	43 51	193	46 13	215	21 49	252	34 00	287
348	49 36	355	36 11	027	43 07	128	43 42	192	45 50	213	21 10	251	33 20	285
350	49 31	352	36 29	025	43 40	126	43 35	190	45 28	212	20 32	249	32 40	283
352	49 25	350	36 45	023	44 13	125	43 28	189	45 06	211	19 54	247	32 00	281
354	49 17	347	37 00	020	44 47	123	43 22	188	44 46	210	19 16	245	31 20	279
356	49 07	344	37 14	018	45 22	122	43 17	187	44 26	208	18 39	244	30 39	277
358	48 55	342	37 26	015	45 57	120	43 12	185	44 07	207	18 03	242	29 58	275

LAT 71°S

LHA	Hc	Zn	Hc	Zn	Hc	Zn	Hc	Zn	Hc	Zn	Hc	Zn	Hc	Zn
♈	Diphda		♦RIGEL		CANOPUS		Miaplacidus		♦RIGIL KENT.		Peacock		♦FOMALHAUT	
0	36 37	013	11 28	082	47 01	118	53 22	157	44 43	206	63 42	274	47 45	340
2	36 45	011	12 07	080	47 36	116	53 37	156	44 26	205	63 03	273	47 30	337
4	36 51	008	12 45	078	48 11	115	53 53	155	44 10	203	62 24	271	47 14	334
6	36 56	006	13 23	076	48 47	113	54 10	154	43 55	202	61 45	269	46 57	332
8	36 59	003	14 01	074	49 23	111	54 28	153	43 41	201	61 06	267	46 38	329
10	37 01	001	14 38	072	50 00	110	54 46	152	43 27	200	60 27	265	46 17	327
12	37 01	359	15 15	070	50 36	108	55 05	151	43 14	198	59 48	263	45 55	325
14	36 59	356	15 52	068	51 14	107	55 24	150	43 03	197	59 09	262	45 32	322
16	36 55	354	16 28	066	51 51	105	55 44	149	42 52	196	58 30	260	45 07	320
18	36 50	351	17 03	064	52 29	103	56 05	148	42 42	194	57 52	258	44 41	317
20	36 44	349	17 38	062	53 07	102	56 26	146	42 33	193	57 14	256	44 14	315
22	36 36	347	18 13	060	53 46	100	56 48	145	42 24	192	56 36	255	43 46	313
24	36 26	344	18 46	058	54 24	098	57 10	144	42 17	190	55 59	253	43 17	310
26	36 15	342	19 19	056	55 03	096	57 33	143	42 10	189	55 21	252	42 47	308
28	36 02	340	19 51	054	55 42	094	57 57	142	42 04	188	54 44	250	42 15	306
	RIGEL		♦SIRIUS		Suhail		♦RIGIL KENT.		Peacock		♦FOMALHAUT		Diphda	
30	20 23	052	21 51	078	35 31	121	42 00	186	54 08	248	41 43	304	35 48	337
32	20 53	050	22 29	076	36 04	120	41 56	185	53 32	247	41 10	301	35 32	335
34	21 23	048	23 07	074	36 38	118	41 53	184	52 56	245	40 37	299	35 15	333
36	21 51	046	23 44	072	37 13	116	41 51	182	52 21	244	40 02	297	34 56	330
38	22 19	044	24 21	070	37 48	115	41 49	181	51 46	242	39 27	295	34 36	328
40	22 46	042	24 57	068	38 24	113	41 49	180	51 12	241	38 51	293	34 15	326
42	23 11	040	25 33	066	39 00	111	41 49	179	50 38	239	38 15	291	33 52	324
44	23 36	038	26 09	064	39 37	110	41 51	177	50 04	238	37 38	289	33 29	321
46	23 59	036	26 43	062	40 14	108	41 53	176	49 32	236	37 01	287	33 04	319
48	24 21	034	27 17	060	40 51	106	41 56	175	48 59	235	36 23	285	32 37	317
50	24 42	031	27 51	058	41 29	104	42 00	173	48 28	233	35 45	283	32 10	315
52	25 02	029	28 23	056	42 07	103	42 05	172	47 57	232	35 07	281	31 42	313
54	25 21	027	28 55	053	42 45	101	42 11	171	47 26	231	34 29	279	31 13	310
56	25 38	025	29 26	051	43 24	099	42 18	169	46 56	229	33 50	277	30 43	308
58	25 53	023	29 56	049	44 02	097	42 26	168	46 27	228	33 11	275	30 12	306
	RIGEL		♦SIRIUS		Suhail		♦RIGIL KENT.		Peacock		♦FOMALHAUT		Diphda	
60	26 08	021	30 25	047	44 41	096	42 34	167	45 58	226	32 32	273	29 40	304
62	26 21	018	30 53	045	45 20	094	42 44	165	45 31	225	31 53	271	29 07	302
64	26 32	016	31 20	043	45 59	092	42 54	164	45 03	223	31 14	269	28 33	300
66	26 43	014	31 46	041	46 38	090	43 05	163	44 37	222	30 35	267	27 59	298
68	26 51	012	32 11	038	47 17	088	43 17	162	44 11	221	29 56	265	27 24	296
70	26 59	010	32 35	036	47 56	086	43 30	160	43 46	219	29 17	263	26 49	294
72	27 04	007	32 57	034	48 35	084	43 43	159	43 22	218	28 38	262	26 13	292
74	27 08	005	33 18	032	49 14	082	43 58	158	42 58	216	28 00	260	25 36	290
76	27 11	003	33 38	029	49 52	080	44 13	156	42 35	215	27 22	258	24 59	288
78	27 12	001	33 57	027	50 31	078	44 29	155	42 13	214	26 43	256	24 22	286
80	27 12	358	34 14	025	51 09	076	44 46	154	41 52	212	26 06	254	23 44	284
82	27 10	356	34 29	023	51 47	074	45 04	152	41 32	211	25 28	253	23 06	282
84	27 07	354	34 44	020	52 24	072	45 22	151	41 12	209	24 51	251	22 28	280
86	27 02	352	34 56	018	53 01	070	45 41	150	40 53	208	24 14	249	21 49	278
88	26 56	350	35 08	016	53 37	067	46 01	148	40 36	207	23 38	247	21 10	276
	♦SIRIUS		Suhail		♦RIGIL KENT.		Peacock		FOMALHAUT		♦ACHERNAR		RIGEL	
90	35 17	013	54 13	065	46 22	147	40 19	205	23 02	245	60 14	263	26 48	347
92	35 25	011	54 48	063	46 44	146	40 02	204	22 27	244	59 35	261	26 38	345
94	35 32	009	55 23	061	47 06	144	39 47	202	21 52	242	58 57	259	26 28	343
96	35 37	006	55 56	058	47 29	143	39 33	201	21 18	240	58 18	258	26 16	341
98	35 40	004	56 29	056	47 53	142	39 19	200	20 45	238	57 40	256	26 02	339
100	35 42	001	57 01	053	48 18	141	39 06	198	20 12	237	57 02	254	25 47	336
102	35 42	359	57 32	051	48 43	139	38 54	197	19 39	235	56 25	253	25 31	334
104	35 41	357	58 01	048	49 09	138	38 44	195	19 08	233	55 48	251	25 13	332
106	35 38	354	58 30	046	49 35	137	38 34	194	18 37	231	55 11	250	24 54	330
108	35 33	352	58 57	043	50 02	135	38 25	193	18 07	230	54 35	248	24 34	328
110	35 27	350	59 23	040	50 30	134	38 16	191	17 38	228	53 59	246	24 12	326
112	35 19	347	59 48	038	50 59	132	38 09	190	17 09	226	53 23	245	23 49	323
114	35 10	345	60 11	035	51 28	131	38 03	189	16 41	224	52 48	243	23 26	321
116	34 59	343	60 32	032	51 58	130	37 58	187	16 14	223	52 13	242	23 01	319
118	34 47	340	60 52	029	52 28	128	37 53	186	15 48	221	51 39	240	22 34	317
	Suhail		♦SPICA		RIGIL KENT.		Peacock		♦FOMALHAUT		ACHERNAR		♦SIRIUS	
120	61 10	026	13 22	085	52 59	127	37 50	184	15 23	219	51 05	239	34 33	338
122	61 26	023	14 01	083	53 30	126	37 47	183	14 59	217	50 32	237	34 18	336
124	61 41	020	14 40	082	54 02	124	37 46	182	14 36	216	49 59	236	34 01	333
126	61 53	017	15 19	080	54 35	123	37 45	180	14 14	214	49 27	235	33 43	331
128	62 03	014	15 57	078	55 08	122	37 45	179	13 53	212	48 56	233	33 23	329
130	62 12	011	16 35	076	55 41	120	37 47	177	13 32	210	48 25	232	33 02	327
132	62 18	008	17 13	074	56 15	119	37 49	176	13 13	209	47 54	230	32 40	324
134	62 22	005	17 50	072	56 50	117	37 52	175	12 55	207	47 25	229	32 17	322
136	62 24	001	18 27	070	57 25	116	37 56	173	12 38	205	46 55	227	31 52	320
138	62 24	358	19 03	068	58 00	114	38 01	172	12 22	203	46 27	226	31 27	318
140	62 22	355	19 39	066	58 36	113	38 07	171	12 07	201	45 59	225	31 00	316
142	62 18	352	20 15	064	59 12	111	38 14	169	11 53	200	45 32	223	30 32	313
144	62 12	349	20 49	062	59 49	110	38 22	168	11 41	198	45 06	222	30 03	311
146	62 03	346	21 24	060	60 26	108	38 31	166	11 29	196	44 40	220	29 33	309
148	61 52	343	21 57	058	61 03	107	38 40	165	11 19	194	44 15	219	29 02	307
	♦SPICA		ANTARES		♦Peacock		FOMALHAUT		ACHERNAR		♦CANOPUS		Suhail	
150	22 30	056	22 34	106	38 51	164	11 10	193	43 51	218	60 13	279	61 40	340
152	23 02	054	23 12	104	39 02	162	11 02	191	43 28	216	59 34	277	61 26	337
154	23 33	052	23 50	102	39 15	161	10 55	189	43 05	215	58 56	275	61 09	334
156	24 03	050	24 28	100	39 28	159	10 49	187	42 43	213	58 17	273	60 51	331
158	24 32	048	25 07	099	39 42	158	10 45	186	42 22	212	57 38	271	60 31	328
160	25 01	046	25 46	097	39 57	157	10 42	184	42 02	211	56 59	269	60 10	325
162	25 28	043	26 24	095	40 13	155	10 40	182	41 42	209	56 20	267	59 47	322
164	25 55	041	27 03	093	40 30	154	10 39	180	41 24	208	55 41	265	59 22	320
166	26 20	039	27 42	091	40 48	152	10 39	179	41 06	206	55 02	264	58 56	317
168	26 44	037	28 21	089	41 06	151	10 41	177	40 49	205	54 23	262	58 29	314
170	27 07	035	29 01	087	41 26	150	10 44	175	40 33	204	53 44	260	58 00	312
172	27 29	033	29 40	085	41 46	148	10 48	173	40 17	202	53 06	258	57 30	309
174	27 49	031	30 18	083	42 07	147	10 53	171	40 03	201	52 28	257	56 59	307
176	28 08	028	30 57	081	42 28	145	10 59	170	39 49	200	51 50	255	56 27	304
178	28 26	026	31 36	080	42 51	144	11 07	168	39 37	198	51 12	253	55 55	302

LHA	Hc	Zn	Hc	Zn	Hc	Zn	Hc	Zn	Hc	Zn	Hc	Zn	Hc	Zn
♈	SPICA		♦ANTARES		Nunki		FOMALHAUT		♦ACHERNAR		CANOPUS		♦Suhail	
180	28 43	024	32 14	078	20 28	112	11 16	166	39 25	197	50 35	252	55 21	299
182	28 58	022	32 52	076	21 05	110	11 26	164	39 14	195	49 58	250	54 47	297
184	29 11	019	33 30	074	21 41	108	11 37	163	39 04	194	49 22	248	54 11	295
186	29 24	017	34 07	072	22 19	106	11 49	161	38 55	193	48 45	247	53 36	292
188	29 35	015	34 44	070	22 57	104	12 02	159	38 47	191	48 10	245	52 59	290
190	29 44	013	35 20	067	23 34	103	12 17	157	38 40	190	47 35	244	52 22	288
192	29 52	010	35 56	065	24 13	101	12 32	156	38 34	189	47 00	242	51 45	286
194	29 58	008	36 31	063	24 51	099	12 49	154	38 28	187	46 25	241	51 07	284
196	30 03	006	37 06	061	25 30	097	13 07	152	38 24	186	45 52	239	50 29	282
198	30 06	004	37 39	059	26 09	095	13 26	150	38 21	184	45 19	237	49 51	280
200	30 08	001	38 13	057	26 48	093	13 46	149	38 18	183	44 46	236	49 12	278
202	30 08	359	38 45	055	27 27	091	14 06	147	38 16	182	44 14	234	48 33	276
204	30 07	357	39 16	052	28 06	090	14 28	145	38 16	180	43 42	233	47 54	274
206	30 04	355	39 47	050	28 45	088	14 51	143	38 16	179	43 12	231	47 15	272
208	29 59	352	40 16	048	29 24	086	15 15	142	38 17	178	42 41	230	46 36	270
	♦ANTARES		Nunki		♦FOMALHAUT		ACHERNAR		CANOPUS		♦Suhail		SPICA	
210	40 45	046	30 03	084	15 40	140	38 19	176	42 12	228	45 57	268	29 53	350
212	41 12	043	30 42	082	16 05	138	38 23	175	41 43	227	45 18	266	29 46	348
214	41 39	041	31 20	080	16 32	136	38 27	173	41 15	225	44 39	264	29 37	346
216	42 04	039	31 58	078	16 59	135	38 32	172	40 47	224	44 00	263	29 26	343
218	42 27	036	32 37	076	17 28	133	38 37	171	40 21	222	43 22	261	29 14	341
220	42 50	034	33 14	074	17 57	131	38 44	169	39 55	221	42 43	259	29 01	339
222	43 11	032	33 52	072	18 27	129	38 52	168	39 30	219	42 05	257	28 46	337
224	43 31	029	34 29	070	18 57	128	39 01	166	39 05	218	41 27	255	28 30	334
226	43 49	027	35 05	068	19 29	126	39 10	165	38 41	216	40 49	254	28 12	332
228	44 06	024	35 41	066	20 01	124	39 21	164	38 19	215	40 12	252	27 54	330
230	44 21	022	36 16	064	20 33	122	39 32	162	37 57	214	39 35	250	27 33	328
232	44 35	019	36 51	062	21 07	120	39 44	161	37 35	212	38 59	249	27 12	326
234	44 47	017	37 25	059	21 41	119	39 58	160	37 15	211	38 22	247	26 49	323
236	44 58	014	37 58	057	22 15	117	40 12	158	36 56	209	37 47	245	26 25	321
238	45 07	012	38 31	055	22 50	115	40 27	157	36 37	208	37 12	244	26 00	319
	ANTARES		♦Nunki		FOMALHAUT		♦ACHERNAR		CANOPUS		♦ACRUX		SPICA	
240	45 14	009	39 02	053	23 26	113	40 42	155	36 19	206	68 34	264	25 34	317
242	45 19	007	39 33	051	24 02	112	40 59	154	36 03	205	67 55	263	25 07	315
244	45 23	004	40 03	048	24 38	110	41 17	153	35 47	203	67 17	261	24 39	313
246	45 25	002	40 32	046	25 15	108	41 35	151	35 32	202	66 38	259	24 10	311
248	45 25	359	40 59	044	25 53	106	41 54	150	35 18	200	66 00	258	23 40	309
250	45 24	357	41 26	042	26 30	104	42 14	149	35 05	199	65 22	256	23 09	307
252	45 20	354	41 51	039	27 08	103	42 35	147	34 52	197	64 44	255	22 37	305
254	45 15	351	42 15	037	27 47	101	42 57	146	34 41	196	64 06	253	22 05	303
256	45 09	349	42 38	035	28 25	099	43 19	144	34 31	195	63 29	252	21 31	301
258	45 00	346	42 59	032	29 04	097	43 42	143	34 21	193	62 52	250	20 57	298
260	44 50	344	43 20	030	29 43	095	44 06	142	34 13	192	62 16	249	20 23	296
262	44 39	341	43 38	027	30 22	093	44 31	140	34 06	190	61 39	247	19 48	294
264	44 25	339	43 55	025	31 01	092	44 56	139	33 59	189	61 03	246	19 12	293
266	44 11	336	44 11	022	31 40	090	45 22	137	33 54	187	60 28	245	18 35	291
268	43 54	334	44 25	020	32 19	088	45 49	136	33 49	186	59 53	243	17 59	289
	♦Nunki		FOMALHAUT		♦ACHERNAR		CANOPUS		♦ACRUX		RIGIL KENT.		ANTARES	
270	44 38	017	32 58	086	46 16	135	33 46	184	59 18	242	68 00	272	43 36	331
272	44 48	015	33 37	084	46 45	133	33 44	183	58 44	240	67 21	270	43 17	329
274	44 58	012	34 15	082	47 13	132	33 42	181	58 10	239	66 41	268	42 56	327
276	45 05	010	34 54	080	47 43	130	33 42	180	57 37	238	66 02	266	42 34	324
278	45 11	007	35 32	078	48 13	129	33 42	179	57 04	236	65 23	264	42 10	322
280	45 15	005	36 10	076	48 44	127	33 44	177	56 32	235	64 45	263	41 46	320
282	45 18	002	36 48	074	49 15	126	33 46	176	56 00	234	64 06	261	41 20	317
284	45 18	000	37 26	072	49 47	125	33 50	174	55 29	232	63 28	259	40 53	315
286	45 17	357	38 02	070	50 19	123	33 54	173	54 58	231	62 49	258	40 24	313
288	45 14	355	38 39	068	50 52	122	34 00	171	54 28	230	62 11	256	39 55	310
290	45 10	352	39 15	066	51 26	120	34 06	170	53 59	229	61 33	254	39 25	308
292	45 03	349	39 50	064	52 00	119	34 13	168	53 30	227	60 56	253	38 54	306
294	44 55	347	40 25	061	52 35	117	34 22	167	53 01	226	60 19	251	38 22	304
296	44 46	344	40 59	059	53 10	116	34 31	165	52 33	225	59 42	250	37 49	302
298	44 35	342	41 32	057	53 45	114	34 41	164	52 06	223	59 05	248	37 15	299
	♦FOMALHAUT		ACHERNAR		♦CANOPUS		ACRUX		RIGIL KENT.		♦ANTARES		Nunki	
300	42 04	055	54 21	113	34 53	162	51 40	222	58 29	247	36 41	297	44 22	339
302	42 36	053	54 57	111	35 05	161	51 14	221	57 54	245	36 06	295	44 07	337
304	43 07	050	55 34	109	35 18	160	50 49	220	57 18	244	35 30	293	43 51	335
306	43 36	048	56 11	108	35 32	158	50 24	218	56 43	242	34 54	291	43 33	332
308	44 05	046	56 48	106	35 47	157	50 00	217	56 09	241	34 17	289	43 14	330
310	44 32	043	57 26	105	36 03	155	49 37	216	55 35	240	33 40	287	42 54	327
312	44 58	041	58 04	103	36 20	154	49 15	214	55 02	238	33 02	285	42 32	325
314	45 23	039	58 42	101	36 38	152	48 53	213	54 29	237	32 24	283	42 09	322
316	45 47	036	59 20	100	36 56	151	48 32	212	53 56	235	31 46	281	41 45	320
318	46 10	034	59 59	098	37 16	149	48 12	211	53 24	234	31 08	279	41 19	318
320	46 31	031	60 38	096	37 36	148	47 52	209	52 53	233	30 29	277	40 52	315
322	46 50	029	61 17	094	37 58	146	47 34	208	52 22	231	29 50	275	40 24	313
324	47 08	026	61 56	093	38 20	145	47 16	207	51 52	230	29 11	273	39 55	311
326	47 25	024	62 35	091	38 42	143	46 58	206	51 22	229	28 32	271	39 25	309
328	47 40	021	63 14	089	39 06	142	46 42	204	50 53	227	27 53	269	38 54	306
	♦FOMALHAUT		Diphda		♦CANOPUS		ACRUX		RIGIL KENT.		♦ANTARES		Nunki	
330	47 53	019	31 47	047	39 31	141	46 26	203	50 25	226	27 14	268	38 22	304
332	48 05	016	32 15	045	39 56	139	46 11	202	49 57	225	26 35	266	37 50	302
334	48 15	014	32 42	043	40 22	138	45 57	201	49 30	223	25 56	264	37 16	300
336	48 23	011	33 08	040	40 48	136	45 44	199	49 04	222	25 17	262	36 42	298
338	48 30	008	33 33	038	41 16	135	45 31	198	48 38	221	24 39	260	36 07	296
340	48 35	006	33 56	036	41 44	133	45 20	197	48 13	219	24 00	258	35 32	294
342	48 38	003	34 18	034	42 13	132	45 09	195	47 49	218	23 22	256	34 56	292
344	48 39	000	34 39	031	42 43	130	44 59	194	47 25	217	22 45	255	34 19	290
346	48 38	358	34 59	029	43 13	129	44 50	193	47 02	215	22 07	253	33 42	288
348	48 36	355	35 18	027	43 44	127	44 41	192	46 40	214	21 30	251	33 04	285
350	48 32	353	35 34	025	44 15	126	44 34	190	46 18	213	20 53	249	32 27	283
352	48 26	350	35 50	022	44 47	124	44 27	189	45 58	211	20 17	247	31 48	282
354	48 18	347	36 04	020	45 20	122	44 21	188	45 38	210	19 41	246	31 10	280
356	48 09	345	36 17	018	45 53	121	44 16	187	45 19	209	19 06	244	30 31	278
358	47 58	342	36 28	015	46 27	119	44 12	185	45 00	207	18 31	242	29 53	276

LHA ♈	Hc	Zn	Hc	Zn	Hc	Zn	Hc	Zn	Hc	Zn	Hc	Zn	Hc	Zn
	Diphda		◆RIGEL		CANOPUS		Miaplacidus		◆RIGIL KENT.		Peacock		◆FOMALHAUT	
0	35 39	013	11 19	082	47 29	117	54 17	157	45 36	207	63 36	277	46 48	340
2	35 46	010	11 56	080	48 02	115	54 32	156	45 20	205	62 59	274	46 35	337
4	35 52	008	12 32	078	48 36	114	54 48	154	45 05	204	62 22	273	46 20	335
6	35 56	006	13 08	076	49 10	112	55 04	153	44 50	203	61 45	271	46 04	332
8	35 59	003	13 44	074	49 44	110	55 21	152	44 37	201	61 08	269	45 46	330
10	36 01	001	14 20	072	50 19	109	55 39	151	44 24	200	60 31	267	45 27	328
12	36 01	359	14 55	070	50 55	107	55 57	150	44 11	199	59 54	265	45 06	325
14	35 59	356	15 29	068	51 30	105	56 16	149	44 00	197	59 17	263	44 44	323
16	35 56	354	16 04	066	52 06	104	56 35	148	43 50	196	58 40	261	44 21	320
18	35 51	352	16 37	064	52 42	102	56 55	147	43 40	195	58 04	260	43 57	318
20	35 45	349	17 10	062	53 19	100	57 16	146	43 31	193	57 27	258	43 32	316
22	35 37	347	17 43	060	53 55	098	57 37	144	43 23	192	56 51	256	43 05	313
24	35 28	345	18 15	058	54 32	097	57 59	143	43 16	191	56 15	255	42 38	311
26	35 18	342	18 46	056	55 09	095	58 21	142	43 09	189	55 40	253	42 09	309
28	35 06	340	19 16	054	55 46	093	58 44	141	43 04	188	55 04	251	41 40	307
	RIGEL		◆SIRIUS		Suhail		◆RIGIL KENT.		Peacock		◆FOMALHAUT		Diphda	
30	19 46	052	21 38	077	36 02	121	42 59	187	54 29	250	41 10	304	34 52	338
32	20 15	050	22 14	075	36 34	119	42 55	185	53 55	248	40 39	302	34 37	335
34	20 43	048	22 50	073	37 06	117	42 53	184	53 20	247	40 07	300	34 21	333
36	21 10	046	23 25	071	37 40	116	42 50	183	52 47	245	39 34	298	34 04	331
38	21 36	044	24 00	069	38 13	114	42 49	181	52 13	243	39 01	296	33 45	329
40	22 01	042	24 34	067	38 47	112	42 49	180	51 40	242	38 28	294	33 25	326
42	22 25	040	25 08	065	39 22	111	42 49	179	51 08	240	37 53	292	33 04	324
44	22 48	038	25 42	063	39 57	109	42 51	177	50 36	239	37 19	289	32 41	322
46	23 10	035	26 15	061	40 32	107	42 53	176	50 04	237	36 43	287	32 18	320
48	23 31	033	26 47	059	41 08	105	42 56	175	49 33	236	36 08	285	31 53	317
50	23 51	031	27 18	057	41 43	104	43 00	173	49 03	234	35 32	283	31 28	315
52	24 10	029	27 49	055	42 20	102	43 05	172	48 33	233	34 56	281	31 01	313
54	24 27	027	28 19	053	42 56	100	43 10	171	48 04	231	34 19	279	30 34	311
56	24 43	025	28 48	051	43 33	098	43 17	169	47 35	230	33 43	277	30 05	309
58	24 58	023	29 17	049	44 09	096	43 24	168	47 07	228	33 06	275	29 36	307
	RIGEL		◆SIRIUS		Suhail		◆RIGIL KENT.		Peacock		◆FOMALHAUT		Diphda	
60	25 12	020	29 44	047	44 46	095	43 33	167	46 40	227	32 29	274	29 06	305
62	25 24	018	30 11	044	45 23	093	43 42	165	46 13	226	31 52	272	28 35	302
64	25 35	016	30 36	042	46 00	091	43 51	164	45 47	224	31 15	270	28 03	300
66	25 44	014	31 01	040	46 37	089	44 02	163	45 21	223	30 38	268	27 31	298
68	25 53	012	31 24	038	47 15	087	44 14	161	44 56	221	30 01	266	26 58	296
70	25 59	009	31 46	036	47 51	085	44 26	160	44 32	220	29 24	264	26 24	294
72	26 05	007	32 07	034	48 28	083	44 39	159	44 09	218	28 47	262	25 50	292
74	26 09	005	32 27	031	49 05	081	44 53	157	43 46	217	28 10	260	25 16	290
76	26 11	003	32 46	029	49 42	079	45 08	156	43 24	216	27 34	258	24 41	288
78	26 12	001	33 03	027	50 18	077	45 23	155	43 03	214	26 58	257	24 05	286
80	26 12	358	33 19	025	50 54	075	45 40	153	42 43	213	26 22	255	23 30	284
82	26 10	356	33 34	022	51 29	073	45 57	152	42 23	211	25 46	253	22 53	282
84	26 07	354	33 47	020	52 05	071	46 15	151	42 04	210	25 11	251	22 17	280
86	26 03	352	33 59	018	52 39	068	46 33	149	41 46	208	24 36	249	21 41	278
88	25 57	350	34 10	015	53 14	066	46 52	148	41 29	207	24 01	248	21 04	276
	◆SIRIUS		Suhail		◆RIGIL KENT.		Peacock		FOMALHAUT		◆ACHERNAR		RIGEL	
90	34 19	013	53 47	064	47 13	147	41 13	206	23 27	246	60 20	264	25 49	347
92	34 27	011	54 20	062	47 33	145	40 57	204	22 54	244	59 44	263	25 40	345
94	34 33	008	54 53	059	47 55	144	40 42	203	22 21	242	59 07	261	25 30	343
96	34 37	006	55 24	057	48 17	142	40 28	201	21 48	240	58 30	259	25 19	341
98	34 41	004	55 55	055	48 40	141	40 15	200	21 16	239	57 54	258	25 06	339
100	34 42	001	56 24	052	49 04	140	40 03	199	20 45	237	57 18	256	24 52	337
102	34 42	359	56 53	050	49 28	138	39 52	197	20 14	235	56 42	254	24 36	334
104	34 41	357	57 21	047	49 53	137	39 41	196	19 44	233	56 07	253	24 20	332
106	34 38	354	57 48	045	50 18	136	39 32	194	19 14	232	55 31	251	24 02	330
108	34 34	352	58 13	042	50 45	134	39 23	193	18 46	230	54 56	249	23 43	328
110	34 28	350	58 37	039	51 11	133	39 15	191	18 18	228	54 22	248	23 23	326
112	34 21	347	59 00	037	51 39	132	39 08	190	17 51	226	53 48	246	23 01	324
114	34 12	345	59 21	034	52 07	130	39 02	189	17 24	225	53 14	245	22 39	322
116	34 02	343	59 41	031	52 36	129	38 57	187	16 59	223	52 41	243	22 15	319
118	33 50	341	59 59	028	53 05	127	38 53	186	16 34	221	52 08	242	21 50	317
	Suhail		◆SPICA		RIGIL KENT.		Peacock		◆FOMALHAUT		ACHERNAR		◆SIRIUS	
120	60 16	025	13 17	085	53 35	126	38 50	184	16 10	219	51 36	240	33 37	338
122	60 31	022	13 54	083	54 05	125	38 47	183	15 47	217	51 04	239	33 23	336
124	60 44	019	14 31	081	54 36	123	38 46	182	15 25	216	50 32	237	33 07	334
126	60 56	017	15 08	079	55 07	122	38 45	180	15 04	214	50 02	236	32 50	331
128	61 05	014	15 44	077	55 39	120	38 45	179	14 43	212	49 31	234	32 32	329
130	61 13	011	16 20	075	56 11	119	38 47	177	14 24	210	49 01	233	32 12	327
132	61 19	008	16 56	074	56 44	117	38 49	176	14 06	209	48 32	231	31 51	325
134	61 23	004	17 31	072	57 17	116	38 52	175	13 49	207	48 04	230	31 29	323
136	61 25	001	18 06	070	57 50	114	38 56	173	13 32	205	47 36	228	31 06	320
138	61 24	358	18 41	068	58 24	113	39 01	172	13 17	203	47 08	227	30 42	318
140	61 22	355	19 15	066	58 59	111	39 06	170	13 03	202	46 42	225	30 17	316
142	61 18	352	19 48	064	59 33	110	39 13	169	12 50	200	46 16	224	29 50	314
144	61 13	349	20 21	062	60 08	108	39 21	168	12 38	198	45 50	222	29 23	312
146	61 05	346	20 53	060	60 44	107	39 29	166	12 27	196	45 26	221	28 55	310
148	60 55	343	21 25	058	61 19	105	39 38	165	12 17	194	45 02	220	28 26	307
	◆SPICA		ANTARES		◆Peacock		FOMALHAUT		ACHERNAR		◆CANOPUS		Suhail	
150	21 56	056	22 50	105	39 48	163	12 08	193	44 38	218	60 03	281	60 44	340
152	22 26	054	23 26	104	40 00	162	12 01	191	44 16	217	59 26	279	60 30	337
154	22 56	051	24 02	102	40 11	161	11 54	189	43 54	215	58 50	277	60 15	335
156	23 24	049	24 39	100	40 24	159	11 49	187	43 33	214	58 13	275	59 58	332
158	23 52	047	25 15	098	40 38	158	11 45	186	43 13	212	57 36	273	59 40	329
160	24 19	045	25 52	096	40 52	156	11 42	184	42 53	211	56 59	271	59 20	326
162	24 45	043	26 29	094	41 08	155	11 40	182	42 34	210	56 22	269	58 59	323
164	25 09	041	27 06	092	41 24	153	11 39	180	42 16	208	55 45	267	58 36	321
166	25 33	039	27 43	091	41 41	152	11 39	179	41 59	207	55 08	265	58 12	318
168	25 56	037	28 20	089	41 59	151	11 41	177	41 43	205	54 31	263	57 46	315
170	26 17	035	28 57	087	42 17	149	11 44	175	41 27	204	53 54	261	57 20	313
172	26 38	032	29 34	085	42 37	148	11 47	173	41 13	203	53 17	260	56 52	310
174	26 57	030	30 11	083	42 57	146	11 52	171	40 59	201	52 41	258	56 23	308
176	27 15	028	30 48	081	43 18	145	11 58	170	40 46	200	52 05	256	55 53	305
178	27 32	026	31 24	079	43 39	144	12 06	168	40 34	198	51 29	255	55 23	303

LAT 72°S

LHA ♈	Hc	Zn	Hc	Zn	Hc	Zn	Hc	Zn	Hc	Zn	Hc	Zn	Hc	Zn
	SPICA		◆ANTARES		Nunki		FOMALHAUT		◆ACHERNAR		CANOPUS		◆Suhail	
180	27 48	024	32 01	077	20 50	111	12 14	166	40 22	197	50 53	253	54 51	301
182	28 02	021	32 37	075	21 25	109	12 23	164	40 12	196	50 18	251	54 19	298
184	28 15	019	33 12	073	22 00	108	12 34	163	40 02	194	49 43	250	53 46	296
186	28 26	017	33 48	071	22 35	106	12 46	161	39 54	193	49 09	248	53 12	294
188	28 37	015	34 22	069	23 11	104	12 58	159	39 46	191	48 34	246	52 38	291
190	28 45	013	34 57	067	23 47	102	13 12	157	39 39	190	48 01	245	52 03	289
192	28 53	010	35 31	065	24 24	100	13 27	155	39 33	189	47 27	243	51 28	287
194	28 59	008	36 04	063	25 00	098	13 43	154	39 28	187	46 55	241	50 52	285
196	29 03	006	36 36	060	25 37	097	14 00	152	39 24	186	46 22	240	50 16	283
198	29 06	004	37 08	058	26 14	095	14 18	150	39 20	184	45 50	238	49 40	281
200	29 08	001	37 39	056	26 51	093	14 37	148	39 18	183	45 19	237	49 03	279
202	29 08	359	38 10	054	27 28	091	14 57	147	39 16	182	44 48	235	48 27	277
204	29 07	357	38 39	052	28 05	089	15 17	145	39 16	180	44 18	234	47 50	275
206	29 04	355	39 08	050	28 42	087	15 39	143	39 16	179	43 49	232	47 13	273
208	29 00	352	39 36	047	29 19	085	16 02	141	39 17	177	43 20	231	46 36	271
	◆ANTARES		Nunki		◆FOMALHAUT		ACHERNAR		CANOPUS		◆Suhail		SPICA	
210	40 03	045	29 56	083	16 25	140	39 19	176	42 51	229	45 59	269	28 54	350
212	40 28	043	30 33	081	16 50	138	39 22	175	42 24	228	45 22	267	28 47	348
214	40 53	041	31 09	079	17 15	136	39 26	173	41 57	226	44 45	265	28 39	346
216	41 17	038	31 46	077	17 41	134	39 31	172	41 30	224	44 08	264	28 29	343
218	41 39	036	32 22	075	18 08	133	39 37	170	41 05	223	43 31	262	28 18	341
220	42 00	034	32 57	073	18 36	131	39 43	169	40 40	221	42 54	260	28 05	339
222	42 20	031	33 33	071	19 04	129	39 51	168	40 16	220	42 18	258	27 51	337
224	42 38	029	34 08	069	19 34	127	39 59	166	39 52	218	41 42	256	27 36	335
226	42 56	026	34 42	067	20 03	125	40 08	165	39 30	217	41 06	255	27 19	332
228	43 11	024	35 16	065	20 34	124	40 18	163	39 08	215	40 30	253	27 02	330
230	43 26	021	35 49	063	21 05	122	40 29	162	38 47	214	39 55	251	26 43	328
232	43 39	019	36 22	061	21 37	120	40 41	161	38 26	212	39 20	249	26 22	326
234	43 50	017	36 54	059	22 09	118	40 54	159	38 07	211	38 46	248	26 01	324
236	44 00	014	37 26	057	22 42	117	41 07	158	37 48	210	38 12	246	25 39	322
238	44 08	012	37 56	054	23 16	115	41 22	156	37 30	208	37 38	244	25 15	320
	ANTARES		◆Nunki		FOMALHAUT		◆ACHERNAR		CANOPUS		◆ACRUX		SPICA	
240	44 15	009	38 26	052	23 49	113	41 37	155	37 13	207	68 39	267	24 50	317
242	44 20	007	38 55	050	24 24	111	41 53	154	36 57	205	68 02	265	24 25	315
244	44 23	004	39 23	048	24 59	109	42 10	152	36 42	204	67 25	263	23 58	313
246	44 25	002	39 50	046	25 34	108	42 27	151	36 27	202	66 48	262	23 31	311
248	44 25	359	40 16	043	26 09	106	42 46	149	36 14	201	66 11	260	23 02	309
250	44 24	357	40 41	041	26 45	104	43 05	148	36 01	199	65 35	258	22 33	307
252	44 21	354	41 04	039	27 21	102	43 25	147	35 50	198	64 59	257	22 03	305
254	44 16	352	41 27	036	27 58	100	43 46	145	35 39	196	64 23	255	21 32	303
256	44 10	349	41 48	034	28 34	098	44 07	144	35 29	195	63 47	254	21 01	301
258	44 02	347	42 09	032	29 11	097	44 30	142	35 20	193	63 12	252	20 29	299
260	43 53	344	42 27	029	29 48	095	44 53	141	35 12	192	62 36	251	19 56	297
262	43 42	342	42 45	027	30 25	093	45 16	140	35 05	190	62 02	249	19 22	295
264	43 29	339	43 01	024	31 02	091	45 41	138	34 59	189	61 27	248	18 49	293
266	43 15	337	43 15	022	31 39	089	46 06	137	34 53	187	60 53	246	18 14	291
268	43 00	334	43 29	020	32 16	087	46 32	135	34 49	186	60 19	245	17 39	289
	◆Nunki		FOMALHAUT		◆ACHERNAR		CANOPUS		◆ACRUX		RIGIL KENT.		ANTARES	
270	43 40	017	32 53	085	46 58	134	34 46	184	59 46	243	67 57	274	42 43	332
272	43 50	015	33 30	083	47 25	132	34 43	183	59 13	242	67 19	272	42 25	330
274	43 59	012	34 07	081	47 53	131	34 42	181	58 41	240	66 42	270	42 06	327
276	44 06	010	34 43	079	48 21	129	34 42	180	58 08	239	66 05	268	41 45	325
278	44 12	007	35 20	077	48 50	128	34 42	178	57 37	238	65 28	267	41 23	322
280	44 15	005	35 56	075	49 20	127	34 44	177	57 06	236	64 51	265	41 00	320
282	44 18	002	36 31	073	49 50	125	34 46	176	56 35	235	64 14	263	40 35	318
284	44 18	000	37 07	071	50 21	124	34 49	174	56 05	234	63 38	261	40 10	316
286	44 17	357	37 42	069	50 52	122	34 54	173	55 35	232	63 01	260	39 43	313
288	44 15	355	38 16	067	51 23	121	34 59	171	55 06	231	62 25	258	39 16	311
290	44 10	352	38 50	065	51 56	119	35 05	170	54 38	230	61 49	256	38 47	309
292	44 04	350	39 23	063	52 28	118	35 12	168	54 10	228	61 13	255	38 18	307
294	43 57	347	39 56	061	53 01	116	35 20	167	53 43	227	60 37	253	37 48	304
296	43 48	345	40 28	059	53 35	114	35 29	165	53 16	226	60 02	251	37 17	302
298	43 37	342	40 59	056	54 09	113	35 39	164	52 50	224	59 27	250	36 45	300
	◆FOMALHAUT		ACHERNAR		◆CANOPUS		ACRUX		RIGIL KENT.		◆ANTARES		Nunki	
300	41 30	054	54 43	111	35 50	162	52 24	223	58 52	248	36 13	298	43 25	340
302	41 59	052	55 18	110	36 02	161	51 59	222	58 18	247	35 40	296	43 12	337
304	42 28	050	55 53	108	36 14	159	51 35	220	57 44	245	35 06	294	42 57	335
306	42 56	047	56 28	106	36 28	158	51 11	219	57 11	244	34 32	292	42 40	332
308	43 23	045	57 04	105	36 42	156	50 48	218	56 38	242	33 57	290	42 23	330
310	43 48	043	57 40	103	36 58	155	50 26	216	56 05	241	33 22	288	42 03	328
312	44 13	040	58 16	101	37 14	153	50 04	215	55 33	239	32 47	286	41 43	325
314	44 36	038	58 53	100	37 31	152	49 43	214	55 01	238	32 11	284	41 21	323
316	44 59	036	59 29	098	37 49	150	49 23	213	54 30	237	31 34	282	40 53	321
318	45 20	033	60 06	096	38 07	149	49 03	211	53 59	235	30 58	280	40 34	318
320	45 39	031	60 43	094	38 27	147	48 44	210	53 29	234	30 21	278	40 09	316
322	45 58	028	61 20	093	38 47	146	48 26	209	52 59	232	29 45	276	39 43	314
324	46 15	026	61 57	091	39 09	144	48 09	207	52 30	231	29 08	274	39 16	312
326	46 30	023	62 34	089	39 30	143	47 52	206	52 02	230	28 31	272	38 47	309
328	46 44	021	63 11	087	39 53	141	47 36	205	51 34	228	27 53	270	38 18	307
	◆FOMALHAUT		Diphda		◆CANOPUS		ACRUX		RIGIL KENT.		◆ANTARES		Nunki	
330	46 56	018	31 06	047	40 17	140	47 21	203	51 06	227	27 16	268	37 48	305
332	47 07	016	31 32	044	40 41	138	47 07	202	50 40	225	26 39	266	37 17	303
334	47 17	013	31 58	042	41 06	137	46 53	201	50 14	224	26 02	264	36 46	301
336	47 24	011	32 22	040	41 31	135	46 40	200	49 48	223	25 26	262	36 14	298
338	47 31	008	32 45	038	41 58	134	46 28	198	49 23	221	24 49	261	35 41	296
340	47 35	006	33 07	036	42 25	132	46 17	197	48 59	220	24 12	259	35 07	294
342	47 38	003	33 28	033	42 53	131	46 07	196	48 36	219	23 36	257	34 33	292
344	47 39	000	33 48	031	43 21	129	45 57	194	48 13	217	23 00	255	33 59	290
346	47 38	358	34 07	029	43 50	128	45 48	193	47 51	216	22 25	253	33 24	288
348	47 36	355	34 24	027	44 19	126	45 40	192	47 29	215	21 49	251	32 48	286
350	47 32	353	34 40	024	44 50	125	45 33	191	47 09	213	21 14	250	32 12	284
352	47 27	350	34 54	022	45 20	123	45 26	189	46 49	212	20 40	248	31 36	282
354	47 20	348	35 08	020	45 52	122	45 21	188	46 30	211	20 06	246	31 00	280
356	47 11	345	35 19	017	46 24	120	45 16	187	46 11	209	19 32	244	30 23	278
358	47 00	342	35 30	015	46 56	118	45 12	186	45 53	208	18 59	242	29 46	276

LAT 73°S

LHA ♈	Hc	Zn	Hc	Zn	Hc	Zn	Hc	Zn	Hc	Zn	Hc	Zn	Hc	Zn
	Diphda		◆RIGEL		CANOPUS		Miaplacidus		◆RIGIL KENT.		Peacock		◆FOMALHAUT	
0	34 40	013	11 11	081	47 55	116	55 12	156	46 30	207	63 28	278	45 52	340
2	34 47	010	11 45	080	48 27	114	55 27	155	46 15	206	62 53	276	45 40	338
4	34 53	008	12 20	078	48 59	113	55 42	154	46 00	204	62 18	274	45 26	335
6	34 57	006	12 54	076	49 30	111	55 57	153	45 46	203	61 43	272	45 10	333
8	34 59	003	13 28	074	50 05	109	56 14	152	45 32	202	61 08	270	44 54	331
10	35 01	001	14 01	072	50 38	108	56 31	150	45 20	200	60 33	269	44 36	328
12	35 01	359	14 34	070	51 12	106	56 49	149	45 08	199	59 58	267	44 17	326
14	34 59	356	15 07	068	51 46	104	57 07	148	44 57	197	59 23	265	43 56	323
16	34 56	354	15 39	066	52 20	102	57 26	147	44 47	196	58 48	263	43 35	321
18	34 52	352	16 11	064	52 54	101	57 45	146	44 38	195	58 14	261	43 12	319
20	34 46	349	16 42	062	53 29	099	58 05	145	44 29	193	57 39	260	42 48	316
22	34 39	347	17 13	060	54 03	097	58 26	144	44 22	192	57 05	258	42 24	314
24	34 30	345	17 43	058	54 38	095	58 47	142	44 15	191	56 30	256	41 58	312
26	34 20	342	18 12	056	55 13	093	59 08	141	44 09	189	55 56	254	41 31	310
28	34 09	340	18 41	054	55 48	092	59 31	140	44 03	188	55 23	253	41 04	307
	RIGEL		◆SIRIUS		Suhail		◆RIGIL KENT.		Peacock		◆FOMALHAUT		Diphda	
30	19 09	052	21 25	077	36 32	120	43 59	187	54 49	251	40 36	305	33 57	338
32	19 36	050	21 59	075	37 03	118	43 55	185	54 16	249	40 06	303	33 43	336
34	20 02	048	22 32	073	37 34	117	43 52	184	53 44	248	39 37	301	33 28	333
36	20 28	046	23 06	071	38 05	115	43 50	183	53 11	246	39 06	299	33 11	331
38	20 52	044	23 39	069	38 37	113	43 49	181	52 40	245	38 35	296	32 54	329
40	21 16	041	24 11	067	39 10	112	43 49	180	52 08	243	38 03	294	32 35	327
42	21 39	039	24 43	065	39 42	110	43 49	179	51 37	241	37 31	292	32 15	324
44	22 01	037	25 15	063	40 16	108	43 51	177	51 06	240	36 58	290	31 54	322
46	22 21	035	25 46	061	40 49	106	43 53	176	50 36	238	36 25	288	31 32	320
48	22 41	033	26 16	059	41 23	105	43 56	174	50 07	237	35 52	286	31 09	318
50	23 00	031	26 46	057	41 57	103	44 00	173	49 38	235	35 18	284	30 45	316
52	23 17	029	27 15	055	42 31	101	44 04	172	49 09	234	34 44	282	30 20	314
54	23 33	027	27 43	053	43 06	099	44 10	170	48 41	232	34 09	280	29 54	311
56	23 49	025	28 10	050	43 41	097	44 16	169	48 13	231	33 35	278	29 27	309
58	24 03	022	28 37	048	44 16	095	44 23	168	47 47	229	33 00	276	29 00	307
	RIGEL		◆SIRIUS		Suhail		◆RIGIL KENT.		Peacock		◆FOMALHAUT		Diphda	
60	24 15	020	29 03	046	44 51	094	44 31	166	47 20	228	32 25	274	28 31	305
62	24 27	018	29 28	044	45 26	092	44 40	165	46 55	226	31 50	272	28 02	303
64	24 37	016	29 52	042	46 01	090	44 49	164	46 30	225	31 15	270	27 33	301
66	24 46	014	30 15	040	46 36	088	44 59	162	46 05	223	30 40	268	27 02	299
68	24 54	012	30 37	038	47 11	086	45 10	161	45 41	222	30 05	267	26 31	297
70	25 00	009	30 57	035	47 46	084	45 22	160	45 18	220	29 30	265	26 00	295
72	25 05	007	31 17	033	48 21	082	45 35	158	44 56	219	28 55	263	25 27	293
74	25 09	005	31 36	031	48 55	080	45 48	157	44 34	218	28 20	261	24 55	291
76	25 11	003	31 53	029	49 30	078	46 02	155	44 13	216	27 45	259	24 22	289
78	25 12	001	32 10	027	50 04	076	46 17	154	43 53	215	27 11	257	23 48	287
80	25 12	358	32 25	024	50 38	074	46 33	153	43 33	213	26 37	255	23 15	285
82	25 10	356	32 38	022	51 11	072	46 49	151	43 14	212	26 03	253	22 40	283
84	25 08	354	32 51	020	51 44	069	47 07	150	42 56	210	25 30	252	22 06	281
86	25 03	352	33 02	017	52 17	067	47 25	149	42 39	209	24 57	250	21 32	279
88	24 58	350	33 12	015	52 49	065	47 43	147	42 23	207	24 24	248	20 57	277
	◆SIRIUS		Suhail		◆RIGIL KENT.		Peacock		FOMALHAUT		◆ACHERNAR		RIGEL	
90	33 20	013	53 20	063	48 02	146	42 07	206	23 52	246	60 25	266	24 51	348
92	33 28	011	53 51	060	48 22	145	41 52	205	23 20	244	59 50	264	24 42	345
94	33 33	008	54 21	058	48 43	143	41 38	203	22 48	243	59 15	263	24 33	343
96	33 38	006	54 51	056	49 04	142	41 24	202	22 17	241	58 41	261	24 22	341
98	33 41	004	55 19	053	49 27	140	41 12	200	21 47	239	58 06	259	24 10	339
100	33 42	001	55 47	051	49 49	139	41 00	199	21 17	237	57 32	257	23 57	337
102	33 42	359	56 14	048	50 13	138	40 49	197	20 48	235	56 58	256	23 42	335
104	33 41	357	56 40	046	50 36	136	40 39	196	20 19	234	56 24	254	23 27	332
106	33 39	355	57 04	043	51 01	135	40 30	194	19 52	232	55 50	252	23 10	330
108	33 34	352	57 28	041	51 26	133	40 22	193	19 24	230	55 17	251	22 52	328
110	33 29	350	57 50	038	51 52	132	40 14	192	18 58	228	54 44	249	22 33	326
112	33 22	348	58 11	036	52 18	130	40 07	190	18 32	227	54 11	247	22 13	324
114	33 14	345	58 31	033	52 45	129	40 02	189	18 07	225	53 39	246	21 52	322
116	33 04	343	58 49	030	53 13	128	39 57	187	17 43	223	53 07	244	21 29	320
118	32 54	341	59 06	027	53 41	126	39 53	186	17 19	221	52 36	243	21 06	318
	Suhail		◆SPICA		RIGIL KENT.		Peacock		◆FOMALHAUT		ACHERNAR		◆SIRIUS	
120	59 21	025	13 12	085	54 09	125	39 49	185	16 56	219	52 05	241	32 41	339
122	59 35	022	13 47	083	54 38	123	39 47	183	16 34	218	51 35	240	32 28	336
124	59 47	019	14 22	081	55 08	122	39 46	182	16 13	216	51 05	238	32 13	334
126	59 58	016	14 56	079	55 38	120	39 45	180	15 53	214	50 35	237	31 57	332
128	60 07	013	15 31	077	56 08	119	39 45	179	15 34	212	50 06	235	31 40	330
130	60 14	010	16 05	075	56 39	117	39 47	177	15 16	211	49 38	234	31 22	327
132	60 19	007	16 39	073	57 11	116	39 49	176	14 58	209	49 10	232	31 02	325
134	60 23	004	17 12	071	57 42	114	39 52	175	14 42	207	48 42	231	30 41	323
136	60 25	001	17 45	069	58 14	113	39 55	173	14 27	205	48 15	229	30 20	321
138	60 24	358	18 18	067	58 47	111	40 00	172	14 12	203	47 49	228	29 57	319
140	60 23	356	18 50	065	59 20	110	40 05	170	13 59	202	47 24	226	29 33	316
142	60 19	353	19 21	063	59 53	108	40 12	169	13 46	200	46 59	225	29 09	314
144	60 14	350	19 52	061	60 26	107	40 19	167	13 35	198	46 34	223	28 43	312
146	60 06	347	20 23	059	61 00	105	40 27	166	13 24	196	46 11	222	28 17	310
148	59 58	344	20 53	057	61 34	103	40 36	165	13 15	195	45 48	220	27 49	308
	◆SPICA		ANTARES		◆Peacock		FOMALHAUT		ACHERNAR		◆CANOPUS		Suhail	
150	21 22	055	23 06	105	40 46	163	13 07	193	45 25	219	59 51	282	59 47	341
152	21 50	053	23 40	103	40 57	162	13 00	191	45 04	217	59 17	280	59 35	338
154	22 18	051	24 14	101	41 08	160	12 53	189	44 43	216	58 42	278	59 21	335
156	22 45	049	24 49	099	41 20	159	12 48	187	44 23	214	58 07	276	59 05	333
158	23 11	047	25 24	098	41 33	157	12 44	186	44 03	213	57 32	274	58 48	330
160	23 36	045	25 58	096	41 47	156	12 42	184	43 44	212	56 57	272	58 30	327
162	24 01	043	26 33	094	42 02	154	12 40	182	43 26	210	56 22	270	58 10	324
164	24 24	041	27 08	092	42 17	153	12 39	180	43 09	209	55 47	268	57 49	322
166	24 46	039	27 44	090	42 34	152	12 39	179	42 53	207	55 12	267	57 27	319
168	25 08	036	28 19	088	42 51	150	12 41	177	42 37	206	54 37	265	57 03	316
170	25 28	034	28 54	086	43 09	149	12 43	175	42 22	204	54 02	263	56 38	314
172	25 47	032	29 29	084	43 27	147	12 47	173	42 08	203	53 27	261	56 13	311
174	26 05	030	30 03	082	43 47	146	12 52	171	41 55	202	52 53	259	55 46	309
176	26 22	028	30 38	080	44 07	144	12 57	170	41 42	200	52 18	258	55 18	306
178	26 38	026	31 13	078	44 27	143	13 04	168	41 31	199	51 44	256	54 49	304

LHA ♈	Hc	Zn	Hc	Zn	Hc	Zn	Hc	Zn	Hc	Zn	Hc	Zn	Hc	Zn
	SPICA		◆ANTARES		Nunki		FOMALHAUT		◆ACHERNAR		CANOPUS		◆Suhail	
180	26 53	023	31 47	076	21 12	111	13 12	166	41 20	197	51 10	254	54 20	302
182	27 06	021	32 21	074	21 45	109	13 21	164	41 10	196	50 37	252	53 50	299
184	27 18	019	32 54	072	22 18	107	13 31	163	41 01	194	50 04	251	53 19	297
186	27 29	017	33 28	070	22 52	105	13 42	161	40 52	193	49 31	249	52 47	295
188	27 38	015	34 00	068	23 26	104	13 54	159	40 45	192	48 58	247	52 15	293
190	27 47	012	34 33	066	24 00	102	14 07	157	40 38	190	48 26	246	51 43	290
192	27 54	010	35 05	064	24 34	100	14 22	155	40 32	189	47 54	244	51 10	288
194	27 59	008	35 36	062	25 09	098	14 37	154	40 27	187	47 23	242	50 36	286
196	28 03	006	36 07	060	25 44	096	14 53	152	40 23	186	46 52	241	50 02	284
198	28 06	004	36 37	058	26 19	094	15 10	150	40 20	185	46 22	239	49 28	282
200	28 08	001	37 06	056	26 54	092	15 28	148	40 18	183	45 52	238	48 54	280
202	28 08	359	37 34	053	27 29	090	15 47	147	40 16	182	45 22	236	48 19	278
204	28 07	357	38 02	051	28 04	088	16 06	145	40 16	180	44 53	234	47 44	276
206	28 04	355	38 29	049	28 39	087	16 27	143	40 16	179	44 25	233	47 09	274
208	28 00	352	38 55	047	29 14	085	16 49	141	40 17	177	43 58	231	46 34	272
	◆ANTARES		Nunki		◆FOMALHAUT		ACHERNAR		CANOPUS		◆Suhail		SPICA	
210	39 20	045	29 49	083	17 11	139	40 19	176	43 30	230	45 59	270	27 55	350
212	39 44	042	30 23	081	17 34	138	40 22	175	43 04	228	45 24	268	27 48	348
214	40 07	040	30 58	079	17 58	136	40 26	173	42 38	227	44 49	266	27 41	346
216	40 29	038	31 32	077	18 23	134	40 30	172	42 13	225	44 14	264	27 31	344
218	40 50	035	32 06	075	18 49	132	40 36	170	41 48	224	43 39	263	27 21	341
220	41 10	033	32 40	073	19 15	131	40 42	169	41 25	222	43 04	261	27 09	339
222	41 28	031	33 13	071	19 42	129	40 49	168	41 01	221	42 30	259	26 56	337
224	41 46	028	33 46	069	20 10	127	40 57	166	40 39	219	41 55	257	26 42	335
226	42 02	026	34 19	067	20 38	125	41 06	165	40 17	217	41 21	255	26 26	333
228	42 16	024	34 51	065	21 07	123	41 16	163	39 56	216	40 48	254	26 09	331
230	42 30	021	35 22	062	21 37	122	41 26	162	39 36	214	40 14	252	25 52	328
232	42 42	019	35 53	060	22 07	120	41 38	160	39 17	213	39 41	250	25 33	326
234	42 52	016	36 23	058	22 38	118	41 50	159	38 58	211	39 08	248	25 12	324
236	43 01	014	36 52	056	23 09	116	42 03	158	38 40	210	38 36	247	24 51	322
238	43 09	011	37 21	054	23 41	114	42 17	156	38 23	208	38 04	245	24 29	320
	ANTARES		◆Nunki		FOMALHAUT		◆ACHERNAR		CANOPUS		◆ACRUX		SPICA	
240	43 15	009	37 49	052	24 13	113	42 31	155	38 07	207	68 40	270	24 06	318
242	43 20	006	38 16	049	24 45	111	42 47	153	37 51	205	68 05	268	23 42	316
244	43 23	004	38 42	047	25 18	109	43 03	152	37 37	204	67 30	266	23 17	314
246	43 25	002	39 08	045	25 52	107	43 20	150	37 23	202	66 55	264	22 51	311
248	43 25	359	39 32	043	26 25	105	43 37	149	37 10	201	66 21	262	22 24	309
250	43 24	357	39 55	040	26 59	103	43 56	148	36 58	199	65 46	261	21 57	307
252	43 21	354	40 17	038	27 34	102	44 15	146	36 47	198	65 11	259	21 29	305
254	43 17	352	40 39	036	28 08	100	44 35	145	36 36	196	64 37	257	21 00	303
256	43 11	349	40 59	034	28 43	098	44 56	143	36 27	195	64 03	256	20 30	301
258	43 04	347	41 17	031	29 18	096	45 17	142	36 18	193	63 29	254	20 00	299
260	42 55	344	41 35	029	29 53	094	45 39	140	36 11	192	62 55	252	19 29	297
262	42 45	342	41 51	026	30 28	092	46 02	139	36 04	190	62 22	251	18 57	295
264	42 33	340	42 06	024	31 03	090	46 25	137	35 58	189	61 49	249	18 25	293
266	42 20	337	42 20	022	31 38	088	46 49	136	35 53	187	61 16	248	17 53	291
268	42 06	335	42 32	019	32 13	087	47 14	134	35 49	186	60 44	246	17 20	289
	◆Nunki		FOMALHAUT		◆ACHERNAR		CANOPUS		◆ACRUX		RIGIL KENT.		ANTARES	
270	42 43	017	32 48	085	47 39	133	35 46	184	60 12	245	67 51	277	41 50	332
272	42 52	014	33 23	083	48 05	132	35 43	183	59 41	243	67 16	275	41 33	330
274	43 00	012	33 57	081	48 32	130	35 42	181	59 09	242	66 41	273	41 15	328
276	43 07	010	34 32	079	48 59	129	35 42	180	58 39	240	66 06	271	40 56	325
278	43 12	007	35 06	077	49 27	127	35 42	178	58 08	239	65 31	269	40 35	323
280	43 16	005	35 40	075	49 55	126	35 43	177	57 38	238	64 56	267	40 14	321
282	43 18	002	36 14	073	50 24	124	35 46	175	57 09	236	64 21	265	39 51	318
284	43 18	000	36 47	071	50 53	123	35 49	174	56 40	235	63 46	263	39 27	316
286	43 17	357	37 20	068	51 23	121	35 53	173	56 12	233	63 11	261	39 02	314
288	43 15	355	37 52	066	51 53	119	35 58	171	55 44	232	62 36	260	38 36	312
290	43 11	352	38 24	064	52 24	118	36 04	170	55 16	231	62 02	258	38 10	309
292	43 05	350	38 55	062	52 55	116	36 11	168	54 50	229	61 28	256	37 42	307
294	42 58	347	39 26	060	53 27	115	36 19	167	54 23	228	60 54	255	37 14	305
296	42 50	345	39 56	058	53 59	113	36 27	165	53 57	227	60 20	253	36 45	303
298	42 40	343	40 26	056	54 32	112	36 37	164	53 32	225	59 47	251	36 15	301
	◆FOMALHAUT		ACHERNAR		◆CANOPUS		ACRUX		RIGIL KENT.		◆ANTARES		Nunki	
300	40 54	053	55 04	110	36 47	162	53 08	224	59 14	250	35 44	299	42 29	340
302	41 22	051	55 37	108	36 58	161	52 44	222	58 41	248	35 13	296	42 16	338
304	41 49	049	56 11	107	37 10	159	52 20	221	58 08	247	34 42	294	42 02	335
306	42 15	047	56 45	105	37 23	158	51 58	220	57 36	245	34 09	292	41 47	333
308	42 40	044	57 19	103	37 37	156	51 35	218	57 05	244	33 37	290	41 30	331
310	43 04	042	57 53	102	37 52	155	51 14	217	56 33	242	33 04	288	41 13	328
312	43 27	040	58 27	100	38 07	153	50 53	216	56 03	241	32 30	286	40 53	326
314	43 49	037	59 02	098	38 24	152	50 33	214	55 32	239	31 56	284	40 33	324
316	44 10	035	59 37	096	38 41	150	50 13	213	55 02	238	31 22	282	40 12	321
318	44 29	033	60 12	094	38 59	149	49 54	212	54 33	236	30 48	280	39 49	319
320	44 48	030	60 47	093	39 17	147	49 36	211	54 04	235	30 13	278	39 26	317
322	45 05	028	61 22	091	39 37	146	49 19	209	53 35	233	29 38	276	39 01	314
324	45 20	025	61 57	089	39 57	144	49 02	208	53 08	232	29 03	274	38 36	312
326	45 35	023	62 32	087	40 18	142	48 46	207	52 40	231	28 28	272	38 09	310
328	45 48	021	63 07	085	40 40	141	48 31	205	52 13	229	27 53	271	37 42	308
	◆FOMALHAUT		Diphda		◆CANOPUS		ACRUX		RIGIL KENT.		◆ANTARES		Nunki	
330	46 00	018	30 24	046	41 02	139	48 16	204	51 47	228	27 18	269	37 14	306
332	46 10	016	30 49	044	41 26	138	48 02	203	51 21	226	26 43	267	36 45	303
334	46 18	013	31 13	042	41 49	136	47 49	201	50 56	225	26 08	265	36 15	301
336	46 25	011	31 36	040	42 14	135	47 37	200	50 32	224	25 33	263	35 45	299
338	46 31	008	31 58	037	42 39	133	47 25	199	50 08	222	24 58	261	35 14	297
340	46 35	005	32 19	035	43 05	132	47 14	197	49 45	221	24 24	259	34 42	295
342	46 38	003	32 38	033	43 32	130	47 04	196	49 22	219	23 50	257	34 10	293
344	46 39	000	32 57	031	43 59	129	46 55	195	49 00	218	23 15	255	33 38	291
346	46 38	358	33 14	029	44 26	127	46 46	193	48 39	217	22 42	254	33 05	289
348	46 36	355	33 30	026	44 55	126	46 39	192	48 19	215	22 08	252	32 31	287
350	46 33	353	33 45	024	45 23	124	46 32	191	47 59	214	21 35	250	31 57	285
352	46 28	350	33 59	022	45 53	122	46 25	190	47 40	212	21 02	248	31 23	283
354	46 21	348	34 11	019	46 23	121	46 20	188	47 21	211	20 30	246	30 49	281
356	46 13	345	34 22	017	46 53	119	46 15	187	47 03	210	19 58	244	30 14	279
358	46 03	343	34 32	015	47 24	118	46 12	186	46 46	208	19 27	243	29 40	277

LHA ♈	Hc	Zn	Hc	Zn	Hc	Zn	Hc	Zn	Hc	Zn	Hc	Zn	Hc	Zn
	Diphda		◆RIGEL		CANOPUS		Miaplacidus		◆RIGIL KENT.		Peacock		◆FOMALHAUT	
0	33 41	012	11 02	081	48 21	115	56 07	156	47 23	207	63 18	280	44 56	341
2	33 48	010	11 34	079	48 51	113	56 21	154	47 09	206	62 46	278	44 44	338
4	33 53	008	12 07	077	49 22	112	56 35	153	46 54	205	62 13	276	44 31	336
6	33 57	006	12 39	075	49 53	110	56 51	152	46 41	203	61 40	274	44 17	333
8	34 00	003	13 11	073	50 24	108	57 06	151	46 28	202	61 07	272	44 01	331
10	34 01	001	13 42	072	50 56	106	57 23	150	46 16	201	60 34	270	43 45	329
12	34 01	359	14 13	070	51 28	105	57 40	148	46 05	199	60 01	268	43 27	326
14	33 59	356	14 44	068	52 00	103	57 57	147	45 55	198	59 28	267	43 08	324
16	33 56	354	15 14	066	52 32	101	58 16	146	45 45	196	58 55	265	42 48	322
18	33 52	352	15 44	064	53 05	099	58 34	145	45 36	195	58 22	263	42 27	319
20	33 47	349	16 14	062	53 37	098	58 54	144	45 28	194	57 49	261	42 05	317
22	33 40	347	16 43	060	54 10	096	59 13	143	45 20	192	57 16	259	41 42	315
24	33 32	345	17 11	058	54 43	094	59 34	141	45 14	191	56 44	258	41 18	312
26	33 23	343	17 38	056	55 16	092	59 55	140	45 08	189	56 12	256	40 53	310
28	33 13	340	18 05	054	55 49	090	60 16	139	45 03	188	55 40	254	40 27	308
	RIGEL		◆SIRIUS		Suhail		◆RIGIL KENT.		Peacock		◆FOMALHAUT		Diphda	
30	18 32	052	21 11	077	37 02	119	44 58	187	55 08	252	40 01	306	33 01	338
32	18 57	049	21 43	075	37 31	118	44 55	185	54 37	251	39 34	304	32 48	336
34	19 22	047	22 15	073	38 00	116	44 52	184	54 06	249	39 06	301	32 34	334
36	19 46	045	22 46	071	38 30	114	44 50	183	53 35	247	38 37	299	32 19	331
38	20 09	043	23 17	069	39 01	113	44 49	181	53 05	246	38 08	297	32 02	329
40	20 31	041	23 48	067	39 31	111	44 49	180	52 35	244	37 38	295	31 45	327
42	20 52	039	24 18	065	40 02	109	44 49	178	52 05	243	37 08	293	31 26	325
44	21 13	037	24 47	063	40 34	107	44 51	177	51 36	241	36 37	291	31 07	323
46	21 32	035	25 16	060	41 06	105	44 53	176	51 07	239	36 06	289	30 46	320
48	21 51	033	25 45	058	41 38	104	44 56	174	50 39	238	35 35	287	30 24	318
50	22 08	031	26 13	056	42 10	102	44 59	173	50 11	236	35 03	285	30 02	316
52	22 25	029	26 40	054	42 42	100	45 04	172	49 44	235	34 31	283	29 39	314
54	22 40	026	27 06	052	43 15	098	45 09	170	49 17	233	33 58	281	29 14	312
56	22 54	024	27 32	050	43 48	096	45 15	169	48 51	232	33 26	279	28 49	310
58	23 07	022	27 57	048	44 21	094	45 22	167	48 25	230	32 53	277	28 23	308
	RIGEL		◆SIRIUS		Suhail		◆RIGIL KENT.		Peacock		◆FOMALHAUT		Diphda	
60	23 19	020	28 21	046	44 54	093	45 29	166	48 00	229	32 20	275	27 57	305
62	23 30	018	28 45	044	45 27	091	45 37	165	47 36	227	31 47	273	27 30	303
64	23 39	016	29 07	042	46 00	089	45 47	163	47 12	226	31 14	271	27 02	301
66	23 48	014	29 28	039	46 33	087	45 56	162	46 48	224	30 41	269	26 33	299
68	23 55	011	29 49	037	47 06	085	46 07	161	46 26	223	30 08	267	26 04	297
70	24 01	009	30 08	035	47 39	083	46 18	159	46 04	221	29 35	265	25 34	295
72	24 06	007	30 27	033	48 12	081	46 31	158	45 42	220	29 02	263	25 04	293
74	24 09	005	30 44	031	48 44	079	46 43	156	45 22	218	28 29	261	24 33	291
76	24 11	003	31 01	028	49 16	077	46 57	155	45 01	217	27 57	260	24 02	289
78	24 12	001	31 16	026	49 48	075	47 11	154	44 42	215	27 24	258	23 31	287
80	24 12	358	31 30	024	50 20	073	47 26	152	44 23	214	26 52	256	22 59	285
82	24 11	356	31 43	022	50 52	070	47 42	151	44 05	212	26 20	254	22 27	283
84	24 08	354	31 54	020	51 23	068	47 58	149	43 48	211	25 48	252	21 55	281
86	24 04	352	32 05	017	51 53	066	48 16	148	43 31	209	25 17	250	21 22	279
88	23 59	350	32 14	015	52 23	064	48 33	147	43 16	208	24 46	248	20 49	277
	◆SIRIUS		Suhail		◆RIGIL KENT.		Peacock		FOMALHAUT		◆ACHERNAR		RIGEL	
90	32 22	013	52 52	062	48 52	145	43 01	206	24 16	247	60 28	268	23 52	348
92	32 29	010	53 21	059	49 11	144	42 46	205	23 45	245	59 55	266	23 44	345
94	32 34	008	53 49	057	49 31	142	42 33	203	23 16	243	59 22	264	23 35	343
96	32 38	006	54 17	055	49 51	141	42 20	202	22 47	241	58 49	262	23 25	341
98	32 41	004	54 43	052	50 12	140	42 08	201	22 18	239	58 17	261	23 14	339
100	32 42	001	55 09	050	50 34	138	41 57	199	21 50	238	57 44	259	23 02	337
102	32 42	359	55 34	047	50 57	137	41 46	198	21 22	236	57 12	257	22 48	335
104	32 41	357	55 58	045	51 19	135	41 37	196	20 55	234	56 40	255	22 33	333
106	32 39	355	56 20	042	51 43	134	41 28	195	20 29	232	56 08	254	22 18	330
108	32 35	352	56 42	040	52 07	132	41 20	193	20 03	230	55 36	252	22 01	328
110	32 30	350	57 03	037	52 32	131	41 13	192	19 38	229	55 05	250	21 43	326
112	32 24	348	57 22	035	52 57	130	41 06	190	19 13	227	54 34	249	21 24	324
114	32 16	346	57 40	032	53 23	128	41 01	189	18 49	225	54 03	247	21 04	322
116	32 07	343	57 57	029	53 49	127	40 56	187	18 26	223	53 33	245	20 43	320
118	31 57	341	58 13	027	54 16	125	40 52	186	18 04	221	53 03	244	20 22	318
	Suhail		◆SPICA		RIGIL KENT.		Peacock		◆FOMALHAUT		ACHERNAR		◆SIRIUS	
120	58 27	024	13 07	085	54 43	124	40 49	185	17 43	220	52 34	242	31 45	339
122	58 39	021	13 40	083	55 11	122	40 47	183	17 22	218	52 04	241	31 33	337
124	58 51	018	14 12	081	55 39	121	40 46	182	17 02	216	51 36	239	31 19	334
126	59 00	016	14 45	079	56 08	119	40 45	180	16 43	214	51 08	238	31 04	332
128	59 08	013	15 17	077	56 37	118	40 45	179	16 25	212	50 40	236	30 48	330
130	59 15	010	15 49	075	57 06	116	40 47	177	16 07	211	50 13	234	30 31	328
132	59 20	007	16 21	073	57 36	115	40 48	176	15 51	209	49 46	233	30 13	325
134	59 23	004	16 53	071	58 06	113	40 51	174	15 36	207	49 20	231	29 54	323
136	59 25	001	17 24	069	58 37	111	40 55	173	15 21	205	48 54	230	29 33	321
138	59 25	359	17 54	067	59 08	110	40 59	172	15 07	204	48 29	228	29 12	319
140	59 23	356	18 25	065	59 39	108	41 05	170	14 54	202	48 05	227	28 50	317
142	59 19	353	18 54	063	60 11	107	41 11	169	14 43	200	47 41	225	28 27	315
144	59 15	350	19 23	061	60 43	105	41 18	167	14 32	198	47 18	224	28 03	313
146	59 08	347	19 52	059	61 15	103	41 25	166	14 22	196	46 55	222	27 38	310
148	59 00	344	20 20	057	61 47	102	41 34	164	14 13	195	46 33	221	27 12	308
	◆SPICA		ANTARES		◆Peacock		FOMALHAUT		ACHERNAR		◆CANOPUS		Suhail	
150	20 48	055	23 21	105	41 43	163	14 05	193	46 12	219	59 37	284	58 50	342
152	21 14	053	23 54	103	41 53	161	13 59	191	45 51	218	59 05	282	58 39	339
154	21 40	051	24 26	101	42 04	160	13 53	189	45 31	216	58 32	280	58 26	336
156	22 06	049	24 59	099	42 16	158	13 48	187	45 12	215	58 00	278	58 12	333
158	22 30	047	25 31	097	42 29	157	13 44	186	44 53	214	57 27	276	57 56	331
160	22 54	045	26 04	095	42 42	156	13 41	184	44 35	212	56 54	274	57 39	328
162	23 16	042	26 37	093	42 56	154	13 40	182	44 18	211	56 21	272	57 21	325
164	23 38	040	27 10	091	43 11	153	13 39	180	44 02	209	55 48	270	57 02	323
166	23 59	038	27 43	089	43 26	151	13 39	179	43 46	208	55 15	268	56 41	320
168	24 19	036	28 16	088	43 43	150	13 41	177	43 31	206	54 42	266	56 19	317
170	24 38	034	28 49	086	44 00	148	13 43	175	43 17	205	54 09	264	55 56	315
172	24 56	032	29 22	084	44 18	147	13 46	173	43 03	203	53 36	262	55 33	312
174	25 13	030	29 55	082	44 36	145	13 51	171	42 50	202	53 03	261	55 08	310
176	25 29	028	30 28	080	44 55	144	13 56	170	42 39	200	52 31	259	54 42	308
178	25 44	025	31 00	078	45 15	142	14 03	168	42 27	199	51 58	257	54 15	305

LAT 74°S

LHA ♈	Hc	Zn	Hc	Zn	Hc	Zn	Hc	Zn	Hc	Zn	Hc	Zn	Hc	Zn
	SPICA		◆ANTARES		Nunki		FOMALHAUT		◆ACHERNAR		CANOPUS		◆Suhail	
180	25 58	023	31 32	076	21 33	111	14 10	166	42 17	198	51 26	255	53 48	303
182	26 10	021	32 04	074	22 04	109	14 19	164	42 07	196	50 54	254	53 20	301
184	26 21	019	32 36	072	22 35	107	14 28	162	41 59	195	50 23	252	52 51	298
186	26 32	017	33 07	070	23 07	105	14 39	161	41 51	193	49 52	250	52 22	296
188	26 40	015	33 38	068	23 39	103	14 50	159	41 43	192	49 21	248	51 52	294
190	26 48	012	34 08	066	24 12	101	15 03	157	41 37	190	48 50	247	51 21	292
192	26 55	010	34 38	063	24 44	099	15 16	155	41 32	189	48 20	245	50 50	289
194	27 00	008	35 07	061	25 17	098	15 30	153	41 27	187	47 50	243	50 19	287
196	27 04	006	35 36	059	25 50	096	15 46	152	41 23	186	47 21	242	49 47	285
198	27 06	004	36 04	057	26 23	094	16 02	150	41 20	185	46 52	240	49 15	283
200	27 08	001	36 32	055	26 56	092	16 19	148	41 18	183	46 23	238	48 43	281
202	27 08	359	36 58	053	27 29	090	16 37	146	41 16	182	45 55	237	48 10	279
204	27 07	357	37 24	051	28 02	088	16 55	145	41 16	180	45 28	235	47 37	277
206	27 05	355	37 49	048	28 35	086	17 15	143	41 16	179	45 01	234	47 04	275
208	27 01	353	38 14	046	29 08	084	17 35	141	41 17	177	44 35	232	46 31	273
	◆ANTARES		Nunki		◆FOMALHAUT		ACHERNAR		CANOPUS		◆Suhail		SPICA	
210	38 37	044	29 41	082	17 57	139	41 19	176	44 09	230	45 58	271	26 56	350
212	39 00	042	30 13	080	18 19	137	41 22	175	43 44	229	45 25	269	26 50	348
214	39 21	039	30 46	078	18 41	136	41 25	173	43 19	227	44 52	267	26 42	346
216	39 42	037	31 18	076	19 05	134	41 30	172	42 55	226	44 19	265	26 34	344
218	40 01	035	31 50	074	19 29	132	41 35	170	42 32	224	43 46	264	26 24	342
220	40 20	033	32 22	072	19 54	130	41 41	169	42 09	223	43 13	262	26 13	339
222	40 37	030	32 53	070	20 20	128	41 48	167	41 47	221	42 41	260	26 01	337
224	40 53	028	33 24	068	20 46	127	41 55	166	41 26	220	42 08	258	25 47	335
226	41 08	026	33 54	066	21 13	125	42 04	164	41 05	218	41 36	256	25 33	333
228	41 21	023	34 24	064	21 40	123	42 13	163	40 45	216	41 04	254	25 17	331
230	41 34	021	34 54	062	22 08	121	42 23	162	40 25	215	40 32	253	25 00	329
232	41 45	018	35 23	060	22 37	119	42 34	160	40 07	213	40 01	251	24 43	326
234	41 55	016	35 51	058	23 06	118	42 46	159	39 49	212	39 30	249	24 24	324
236	42 03	014	36 19	055	23 35	116	42 58	157	39 32	210	38 59	247	24 04	322
238	42 10	011	36 45	053	24 05	114	43 11	156	39 16	209	38 29	246	23 43	320
	ANTARES		◆Nunki		FOMALHAUT		◆ACHERNAR		CANOPUS		◆ACRUX		SPICA	
240	42 16	009	37 12	051	24 36	112	43 25	154	39 00	207	68 39	272	23 22	318
242	42 20	006	37 37	049	25 06	110	43 40	153	38 45	206	68 06	270	22 59	316
244	42 23	004	38 01	047	25 38	109	43 56	151	38 31	204	67 33	268	22 36	314
246	42 25	002	38 25	044	26 09	107	44 12	150	38 18	203	67 00	266	22 11	312
248	42 25	359	38 48	042	26 41	105	44 29	148	38 06	201	66 27	265	21 46	310
250	42 24	357	39 09	040	27 13	103	44 46	147	37 54	200	65 54	263	21 20	308
252	42 21	354	39 30	038	27 45	101	45 05	146	37 44	198	65 22	261	20 54	306
254	42 17	352	39 50	035	28 18	099	45 24	144	37 34	197	64 49	259	20 27	304
256	42 12	349	40 08	033	28 51	097	45 44	143	37 25	195	64 17	258	19 59	301
258	42 05	347	40 26	031	29 24	095	46 04	141	37 17	194	63 44	256	19 30	299
260	41 57	345	40 42	028	29 57	094	46 25	140	37 09	192	63 13	254	19 01	297
262	41 48	342	40 57	026	30 30	092	46 47	138	37 03	191	62 41	253	18 32	295
264	41 37	340	41 11	024	31 03	090	47 09	137	36 57	189	62 09	251	18 01	293
266	41 25	337	41 24	021	31 36	088	47 32	135	36 52	188	61 38	250	17 31	291
268	41 12	335	41 35	019	32 09	086	47 56	134	36 48	186	61 07	248	17 00	289
	◆Nunki		FOMALHAUT		◆ACHERNAR		CANOPUS		◆ACRUX		RIGIL KENT.		ANTARES	
270	41 45	017	32 42	084	48 20	132	36 45	185	60 37	246	67 43	279	40 57	333
272	41 54	014	33 15	082	48 45	131	36 43	183	60 07	245	67 10	277	40 41	330
274	42 02	012	33 47	080	49 10	129	36 42	181	59 37	243	66 37	275	40 24	328
276	42 08	009	34 20	078	49 36	128	36 42	180	59 08	242	66 04	273	40 06	326
278	42 12	007	34 52	076	50 03	126	36 43	178	58 39	240	65 31	271	39 47	323
280	42 16	005	35 24	074	50 30	125	36 43	177	58 10	239	64 58	269	39 27	321
282	42 18	002	35 55	072	50 57	123	36 46	175	57 42	238	64 25	267	39 06	319
284	42 18	000	36 27	070	51 25	121	36 49	174	57 14	236	63 52	265	38 43	317
286	42 17	357	36 58	068	51 53	120	36 53	172	56 47	235	63 19	263	38 20	314
288	42 15	355	37 28	066	52 22	118	36 57	171	56 20	233	62 46	262	37 56	312
290	42 11	352	37 58	064	52 52	117	37 03	169	55 54	232	62 14	260	37 31	310
292	42 06	350	38 27	061	53 21	115	37 10	168	55 28	230	61 41	258	37 06	308
294	42 00	348	38 56	059	53 52	114	37 17	166	55 03	229	61 09	256	36 39	306
296	41 52	345	39 24	057	54 22	112	37 25	165	54 38	228	60 37	255	36 12	303
298	41 43	343	39 51	055	54 53	110	37 34	163	54 14	226	60 05	253	35 44	301
	◆FOMALHAUT		ACHERNAR		◆CANOPUS		ACRUX		RIGIL KENT.		◆ANTARES		Nunki	
300	40 18	053	55 24	109	37 44	162	53 51	225	59 33	252	35 15	299	41 33	340
302	40 44	051	55 56	107	37 55	160	53 28	223	59 02	250	34 46	297	41 21	338
304	41 09	048	56 27	105	38 06	159	53 05	222	58 31	248	34 16	295	41 08	336
306	41 34	046	56 59	103	38 19	157	52 43	221	58 01	247	33 46	293	40 54	333
308	41 57	044	57 32	102	38 32	156	52 22	219	57 31	245	33 16	291	40 38	331
310	42 19	042	58 04	100	38 46	154	52 01	218	57 01	244	32 44	289	40 21	329
312	42 41	039	58 37	098	39 01	153	51 41	217	56 31	242	32 13	287	40 04	326
314	43 01	037	59 10	096	39 16	151	51 22	215	56 02	241	31 41	285	39 45	324
316	43 21	035	59 43	095	39 33	150	51 03	214	55 34	239	31 09	283	39 25	322
318	43 39	032	60 16	093	39 50	148	50 45	212	55 06	238	30 37	281	39 04	319
320	43 56	030	60 49	091	40 08	147	50 28	211	54 38	236	30 04	279	38 42	317
322	44 12	027	61 22	089	40 26	145	50 11	210	54 11	235	29 31	277	38 19	315
324	44 26	025	61 55	087	40 46	143	49 55	208	53 44	233	28 58	275	37 55	313
326	44 40	023	62 28	085	41 06	142	49 40	207	53 18	232	28 25	273	37 30	311
328	44 52	020	63 01	083	41 26	140	49 25	206	52 52	230	27 52	271	37 05	308
	◆FOMALHAUT		Diphda		◆CANOPUS		ACRUX		RIGIL KENT.		◆ANTARES		Nunki	
330	45 02	018	29 43	046	41 48	139	49 11	204	52 27	229	27 19	269	36 38	306
332	45 12	015	30 06	044	42 10	137	48 58	203	52 03	227	26 46	267	36 11	304
334	45 20	013	30 28	041	42 33	136	48 45	202	51 39	226	26 13	265	35 44	302
336	45 26	010	30 50	039	42 56	134	48 33	200	51 15	224	25 40	263	35 15	300
338	45 32	008	31 10	037	43 20	133	48 22	199	50 52	223	25 08	262	34 46	298
340	45 36	005	31 29	035	43 45	131	48 12	198	50 30	222	24 35	260	34 17	296
342	45 38	003	31 48	033	44 10	129	48 02	196	50 09	220	24 02	258	33 47	293
344	45 39	000	32 05	030	44 36	128	47 53	195	49 48	219	23 30	256	33 16	291
346	45 38	358	32 21	028	45 02	126	47 45	194	49 27	217	22 58	254	32 45	289
348	45 37	355	32 36	026	45 29	125	47 37	192	49 07	216	22 27	252	32 14	287
350	45 33	353	32 50	024	45 57	123	47 31	191	48 48	214	21 55	250	31 42	285
352	45 28	350	33 03	021	46 25	121	47 25	190	48 30	213	21 24	248	31 10	283
354	45 22	348	33 14	019	46 53	120	47 19	188	48 12	212	20 54	247	30 37	281
356	45 15	346	33 25	017	47 22	118	47 15	187	47 55	210	20 24	245	30 05	279
358	45 06	343	33 34	015	47 51	117	47 11	186	47 39	209	19 54	243	29 32	277

LAT 75°S

LAT 75°S

LHA ♈	Hc	Zn	Hc	Zn	Hc	Zn	Hc	Zn	Hc	Zn	Hc	Zn	Hc	Zn
	◆Diphda		◆CANOPUS		ACRUX		RIGIL KENT.		◆ANTARES		Peacock		FOMALHAUT	
0	32 43	012	48 46	114	48 08	184	48 17	208	19 54	241	63 06	282	43 59	341
2	32 49	010	49 14	112	48 06	183	48 02	207	19 27	240	62 36	280	43 48	339
4	32 54	008	49 43	110	48 05	182	47 49	205	19 00	238	62 05	278	43 36	336
6	32 57	005	50 13	109	48 04	180	47 36	204	18 34	236	61 34	276	43 23	334
8	33 00	003	50 42	107	48 04	179	47 24	202	18 09	234	61 03	274	43 09	331
10	33 01	001	51 12	105	48 05	178	47 12	201	17 44	232	60 32	272	42 53	329
12	33 01	359	51 42	103	48 07	176	47 02	199	17 19	230	60 01	270	42 37	327
14	32 59	356	52 12	102	48 09	175	46 52	198	16 56	229	59 30	268	42 19	324
16	32 57	354	52 43	100	48 12	174	46 42	197	16 33	227	58 59	266	42 01	322
18	32 53	352	53 14	098	48 16	172	46 34	195	16 11	225	58 28	265	41 41	320
20	32 48	350	53 44	096	48 21	171	46 26	194	15 49	223	57 58	263	41 21	318
22	32 42	347	54 15	094	48 26	170	46 19	192	15 28	221	57 27	261	40 59	315
24	32 34	345	54 46	092	48 32	168	46 13	191	15 08	219	56 56	259	40 37	313
26	32 26	343	55 17	091	48 39	167	46 07	190	14 49	218	56 26	257	40 14	311
28	32 16	341	55 48	089	48 46	165	46 02	188	14 30	216	55 56	256	39 50	309
	RIGEL		◆SIRIUS		Suhail		◆RIGIL KENT.		Peacock		◆FOMALHAUT		Diphda	
30	17 54	051	20 57	076	37 31	119	45 58	187	55 26	254	39 25	306	32 05	338
32	18 18	049	21 27	074	37 58	117	45 55	185	54 56	252	39 00	304	31 53	336
34	18 41	047	21 56	072	38 26	115	45 52	184	54 27	250	38 34	302	31 40	334
36	19 04	045	22 26	070	38 55	114	45 50	183	53 57	249	38 07	300	31 26	332
38	19 25	043	22 55	068	39 23	112	45 49	181	53 29	247	37 40	298	31 11	330
40	19 46	041	23 24	066	39 52	110	45 49	180	53 00	245	37 13	296	30 54	327
42	20 06	039	23 52	064	40 22	108	45 49	178	52 32	244	36 44	294	30 37	325
44	20 25	037	24 19	062	40 51	106	45 51	177	52 05	242	36 16	292	30 19	323
46	20 43	035	24 47	060	41 21	105	45 53	176	51 37	241	35 47	289	30 00	321
48	21 00	033	25 13	058	41 51	103	45 55	174	51 10	239	35 17	287	29 40	319
50	21 17	031	25 39	056	42 22	101	45 59	173	50 44	237	34 47	285	29 19	316
52	21 32	028	26 05	054	42 52	099	46 03	171	50 18	236	34 17	283	28 57	314
54	21 46	026	26 29	052	43 23	097	46 08	170	49 53	234	33 47	281	28 34	312
56	21 59	024	26 53	050	43 54	095	46 14	169	49 28	233	33 16	279	28 11	310
58	22 12	022	27 17	048	44 25	093	46 20	167	49 03	231	32 46	277	27 47	308
	RIGEL		◆SIRIUS		Suhail		◆RIGIL KENT.		Peacock		◆FOMALHAUT		ACHERNAR	
60	22 23	020	27 39	045	44 56	092	46 27	166	48 40	229	32 15	275	67 52	303
62	22 33	018	28 01	043	45 27	090	46 35	164	48 16	228	31 44	273	67 25	301
64	22 42	016	28 22	041	45 58	088	46 44	163	47 53	226	31 13	272	66 58	298
66	22 49	014	28 42	039	46 29	086	46 53	162	47 31	225	30 42	270	66 31	296
68	22 56	011	29 01	037	47 00	084	47 04	160	47 10	223	30 11	268	66 02	293
70	23 02	009	29 19	035	47 31	082	47 14	159	46 49	222	29 40	266	65 34	291
72	23 06	007	29 36	033	48 01	080	47 26	157	46 28	220	29 09	264	65 04	289
74	23 09	005	29 53	030	48 32	078	47 38	156	46 08	219	28 38	262	64 35	286
76	23 11	003	30 08	028	49 02	076	47 51	155	45 49	217	28 07	260	64 05	284
78	23 12	001	30 22	026	49 32	074	48 05	153	45 31	216	27 37	258	63 34	282
80	23 12	358	30 35	024	50 02	071	48 19	152	45 13	214	27 06	256	63 04	280
82	23 11	356	30 47	022	50 31	069	48 34	150	44 56	213	26 36	254	62 33	278
84	23 08	354	30 58	019	51 00	067	48 50	149	44 39	211	26 07	253	62 02	276
86	23 04	352	31 07	017	51 28	065	49 06	147	44 24	210	25 37	251	61 31	274
88	23 00	350	31 16	015	51 56	063	49 23	146	44 09	208	25 08	249	61 00	272
	◆SIRIUS		Suhail		◆RIGIL KENT.		Peacock		◆FOMALHAUT		ACHERNAR		RIGEL	
90	31 23	013	52 23	060	49 41	145	43 54	207	24 39	247	60 29	270	22 53	348
92	31 30	010	52 50	058	49 59	143	43 41	205	24 11	245	59 58	268	22 46	346
94	31 35	008	53 16	056	50 18	142	43 28	204	23 43	243	59 27	266	22 38	343
96	31 38	006	53 41	054	50 38	140	43 16	202	23 15	242	58 56	264	22 29	341
98	31 41	004	54 06	051	50 58	139	43 04	201	22 48	240	58 26	262	22 18	339
100	31 42	001	54 30	049	51 19	137	42 53	199	22 22	238	57 55	260	22 06	337
102	31 42	359	54 53	046	51 40	136	42 44	198	21 56	236	57 24	259	21 54	335
104	31 41	357	55 15	044	52 02	134	42 34	196	21 30	234	56 54	257	21 40	333
106	31 39	355	55 36	041	52 24	133	42 26	195	21 05	232	56 24	255	21 25	331
108	31 36	352	55 56	039	52 47	132	42 18	193	20 41	231	55 54	253	21 10	329
110	31 31	350	56 15	036	53 11	130	42 12	192	20 17	229	55 24	252	20 53	326
112	31 25	348	56 33	034	53 35	129	42 05	191	19 54	227	54 55	250	20 35	324
114	31 18	346	56 49	031	53 59	127	42 00	189	19 32	225	54 26	248	20 17	322
116	31 10	343	57 05	029	54 24	126	41 56	188	19 10	223	53 57	247	19 57	320
118	31 00	341	57 19	026	54 50	124	41 52	186	18 49	222	53 29	245	19 37	318
	Suhail		◆SPICA		RIGIL KENT.		Peacock		◆FOMALHAUT		ACHERNAR		◆SIRIUS	
120	57 32	023	13 01	084	55 16	123	41 49	185	18 29	220	53 01	243	30 50	339
122	57 43	021	13 32	082	55 42	121	41 47	183	18 09	218	52 33	242	30 38	337
124	57 54	018	14 03	081	56 09	119	41 46	182	17 51	216	52 06	240	30 25	335
126	58 02	015	14 33	079	56 36	118	41 45	180	17 33	214	51 39	239	30 11	332
128	58 10	012	15 04	077	57 04	116	41 45	179	17 15	213	51 13	237	29 56	330
130	58 16	010	15 34	075	57 32	115	41 46	177	16 59	211	50 47	235	29 40	328
132	58 20	007	16 03	073	58 00	113	41 48	176	16 44	209	50 22	234	29 23	326
134	58 23	004	16 33	071	58 29	112	41 51	174	16 29	207	49 57	232	29 05	324
136	58 25	001	17 02	069	58 58	110	41 54	173	16 15	205	49 33	231	28 46	321
138	58 25	359	17 31	067	59 28	108	41 59	171	16 02	204	49 09	229	28 27	319
140	58 23	356	17 59	065	59 57	107	42 04	170	15 50	202	48 46	228	28 06	317
142	58 20	353	18 27	063	60 27	105	42 10	168	15 39	200	48 23	226	27 44	315
144	58 15	350	18 54	061	60 57	103	42 16	167	15 29	198	48 01	225	27 22	313
146	58 09	347	19 21	059	61 28	101	42 24	166	15 20	196	47 39	223	26 59	311
148	58 02	345	19 47	057	61 58	100	42 32	164	15 11	195	47 18	222	26 35	309
	SPICA		◆ANTARES		Peacock		FOMALHAUT		◆ACHERNAR		CANOPUS		◆Suhail	
150	20 13	055	23 36	104	42 41	163	15 04	193	46 58	220	59 22	286	57 53	342
152	20 38	053	24 07	102	42 50	161	14 57	191	46 38	219	58 52	284	57 43	339
154	21 02	051	24 37	100	43 01	160	14 52	189	46 19	217	58 21	281	57 31	337
156	21 26	048	25 08	099	43 12	158	14 47	187	46 01	216	57 51	279	57 18	334
158	21 49	046	25 38	097	43 24	157	14 44	186	45 43	214	57 20	277	57 04	331
160	22 11	044	26 09	095	43 37	155	14 41	184	45 26	213	56 49	275	56 48	329
162	22 32	042	26 40	093	43 50	154	14 40	182	45 10	211	56 18	273	56 32	326
164	22 53	040	27 11	091	44 04	152	14 39	180	44 54	210	55 47	271	56 14	324
166	23 12	038	27 42	089	44 19	151	14 39	179	44 39	208	55 16	269	55 55	321
168	23 31	036	28 14	087	44 34	149	14 41	177	44 25	207	54 45	267	55 35	318
170	23 49	034	28 44	085	44 51	148	14 43	175	44 11	205	54 14	266	55 14	316
172	24 05	032	29 15	083	45 08	146	14 46	173	43 58	204	53 43	264	54 52	314
174	24 21	030	29 46	081	45 25	145	14 50	171	43 46	202	53 12	262	54 29	311
176	24 36	027	30 17	079	45 43	143	14 55	170	43 35	201	52 42	260	54 05	309
178	24 50	025	30 47	077	46 02	142	15 02	168	43 24	199	52 11	258	53 40	306

LHA ♈	Hc	Zn	Hc	Zn	Hc	Zn	Hc	Zn	Hc	Zn	Hc	Zn	Hc	Zn
	SPICA		◆ANTARES		Nunki		FOMALHAUT		◆ACHERNAR		CANOPUS		◆Suhail	
180	25 02	023	31 17	075	21 54	110	15 09	166	43 14	198	51 41	256	53 15	304
182	25 14	021	31 47	073	22 23	108	15 17	164	43 05	196	51 11	255	52 49	302
184	25 25	019	32 17	071	22 53	106	15 26	162	42 57	195	50 41	253	52 22	299
186	25 34	017	32 46	069	23 23	105	15 35	161	42 49	193	50 11	251	51 55	297
188	25 42	014	33 15	067	23 53	103	15 46	159	42 42	192	49 42	249	51 27	295
190	25 50	012	33 43	065	24 23	101	15 58	157	42 36	191	49 13	248	50 58	293
192	25 56	010	34 11	063	24 54	099	16 11	155	42 31	189	48 45	246	50 30	291
194	26 00	008	34 38	061	25 25	097	16 24	153	42 26	188	48 17	244	50 00	288
196	26 04	006	35 05	059	25 55	095	16 38	152	42 23	186	47 49	243	49 31	286
198	26 07	004	35 31	057	26 26	093	16 54	150	42 20	185	47 21	241	49 01	284
200	26 08	001	35 57	054	26 57	091	17 10	148	42 18	183	46 54	239	48 30	282
202	26 08	359	36 22	052	27 28	089	17 27	146	42 16	182	46 28	238	48 00	280
204	26 07	357	36 46	050	28 00	087	17 44	144	42 16	180	46 02	236	47 29	278
206	26 05	355	37 09	048	28 31	085	18 03	143	42 16	179	45 36	234	46 59	276
208	26 01	353	37 32	046	29 01	084	18 22	141	42 17	177	45 11	233	46 28	274
	◆ANTARES		Nunki		FOMALHAUT		◆ACHERNAR		CANOPUS		◆Suhail		SPICA	
210	37 54	043	29 32	082	18 42	139	42 19	176	44 47	231	45 57	272	25 57	350
212	38 15	041	30 03	080	19 03	137	42 21	174	44 23	230	45 26	270	25 51	348
214	38 35	039	30 33	078	19 24	135	42 25	173	44 00	228	44 54	268	25 44	346
216	38 54	037	31 04	076	19 46	134	42 29	172	43 37	226	44 23	266	25 36	344
218	39 12	034	31 33	074	20 09	132	42 34	170	43 14	225	43 52	265	25 27	342
220	39 29	032	32 03	072	20 33	130	42 40	169	42 53	223	43 22	263	25 17	340
222	39 45	030	32 32	069	20 57	128	42 46	167	42 32	222	42 51	261	25 05	337
224	40 00	028	33 01	067	21 21	126	42 54	166	42 12	220	42 20	259	24 53	335
226	40 14	025	33 30	065	21 47	125	43 02	164	41 52	219	41 50	257	24 39	333
228	40 26	023	33 58	063	22 13	123	43 11	163	41 33	217	41 20	255	24 25	331
230	40 38	021	34 25	061	22 39	121	43 20	161	41 15	215	40 50	253	24 09	329
232	40 48	018	34 52	059	23 06	119	43 31	160	40 57	214	40 20	252	23 53	327
234	40 57	016	35 18	057	23 33	117	43 42	158	40 40	212	39 51	250	23 35	325
236	41 05	013	35 44	055	24 01	115	43 54	157	40 24	211	39 22	248	23 17	322
238	41 11	011	36 09	053	24 29	114	44 06	155	40 08	209	38 53	246	22 57	320
	ANTARES		◆Nunki		FOMALHAUT		◆ACHERNAR		CANOPUS		◆ACRUX		SPICA	
240	41 17	009	36 34	050	24 58	112	44 19	154	39 53	208	68 36	275	22 37	318
242	41 21	006	36 57	048	25 27	110	44 33	152	39 39	206	68 05	273	22 16	316
244	41 24	004	37 20	046	25 56	108	44 48	151	39 26	205	67 34	271	21 54	314
246	41 25	002	37 42	044	26 26	106	45 04	149	39 14	203	67 03	269	21 31	312
248	41 25	359	38 03	042	26 56	104	45 20	148	39 02	201	66 32	267	21 08	310
250	41 24	357	38 23	039	27 26	102	45 37	146	38 51	200	66 01	265	20 44	308
252	41 22	354	38 42	037	27 57	101	45 54	145	38 41	198	65 30	263	20 19	306
254	41 18	352	39 01	035	28 27	099	46 12	143	38 31	197	64 59	261	19 53	304
256	41 13	350	39 18	033	28 58	097	46 31	142	38 23	195	64 28	260	19 27	302
258	41 07	347	39 34	030	29 29	095	46 50	140	38 15	194	63 58	258	19 01	300
260	40 59	345	39 49	028	30 00	093	47 11	139	38 08	192	63 28	256	18 33	298
262	40 51	342	40 03	026	30 31	091	47 31	137	38 02	191	62 58	255	18 06	296
264	40 41	340	40 16	023	31 02	089	47 53	136	37 56	189	62 28	253	17 37	294
266	40 29	338	40 28	021	31 33	087	48 14	134	37 52	188	61 58	251	17 09	292
268	40 17	335	40 39	019	32 04	085	48 37	133	37 48	186	61 29	250	16 40	290
	◆Nunki		FOMALHAUT		◆ACHERNAR		CANOPUS		◆ACRUX		RIGIL KENT.		ANTARES	
270	40 48	016	32 35	083	49 00	131	37 45	185	61 00	248	67 32	281	40 04	333
272	40 56	014	33 06	081	49 24	130	37 43	183	60 31	247	67 02	279	39 49	331
274	41 03	012	33 36	079	49 48	128	37 42	182	60 03	245	66 31	277	39 33	329
276	41 09	009	34 07	077	50 12	127	37 42	180	59 35	243	66 00	275	39 17	326
278	41 13	007	34 37	075	50 38	125	37 42	178	59 07	242	65 29	273	38 59	324
280	41 16	004	35 07	073	51 03	124	37 43	177	58 40	240	64 58	271	38 40	322
282	41 18	002	35 36	071	51 29	122	37 45	175	58 13	239	64 27	269	38 20	319
284	41 18	000	36 06	069	51 56	120	37 48	174	57 47	237	63 56	267	38 00	317
286	41 17	357	36 35	067	52 23	119	37 52	172	57 21	236	63 25	265	37 38	315
288	41 15	355	37 03	065	52 50	117	37 57	171	56 56	234	62 54	264	37 16	313
290	41 12	353	37 31	063	53 18	116	38 02	169	56 31	233	62 23	262	36 53	311
292	41 07	350	37 58	061	53 46	114	38 08	168	56 06	232	61 52	260	36 29	308
294	41 01	348	38 25	059	54 15	112	38 15	166	55 42	230	61 22	258	36 04	306
296	40 54	345	38 51	056	54 44	111	38 23	165	55 18	229	60 52	257	35 38	304
298	40 46	343	39 17	054	55 13	109	38 32	163	54 55	227	60 22	255	35 12	302
	◆FOMALHAUT		ACHERNAR		◆CANOPUS		ACRUX		RIGIL KENT.		◆ANTARES		Nunki	
300	39 42	052	55 43	107	38 41	162	54 33	226	59 52	253	34 46	300	40 36	341
302	40 06	050	56 12	105	38 51	160	54 11	224	59 22	251	34 19	298	40 25	338
304	40 29	048	56 42	104	39 02	158	53 49	223	58 53	250	33 51	296	40 13	336
306	40 52	045	57 13	102	39 14	157	53 29	222	58 24	248	33 23	294	40 00	334
308	41 13	043	57 43	100	39 27	155	53 08	220	57 55	247	32 54	292	39 46	331
310	41 34	041	58 14	098	39 40	154	52 49	219	57 27	245	32 25	289	39 30	329
312	41 54	039	58 45	097	39 54	152	52 29	217	56 59	243	31 55	287	39 14	327
314	42 13	036	59 16	095	40 09	151	52 11	216	56 31	242	31 26	285	38 56	324
316	42 31	034	59 47	093	40 24	149	51 53	215	56 04	240	30 55	283	38 38	322
318	42 48	032	60 18	091	40 41	148	51 36	213	55 37	239	30 25	281	38 18	320
320	43 04	029	60 49	089	40 58	146	51 19	212	55 11	237	29 55	279	37 58	318
322	43 18	027	61 20	087	41 15	145	51 03	210	54 45	236	29 24	277	37 36	316
324	43 32	025	61 51	085	41 34	143	50 48	209	54 20	234	28 53	275	37 14	313
326	43 44	022	62 22	083	41 53	141	50 33	208	53 55	233	28 22	274	36 51	311
328	43 55	020	62 52	081	42 12	140	50 19	206	53 30	231	27 51	272	36 27	309
	◆FOMALHAUT		ACHERNAR		◆CANOPUS		ACRUX		RIGIL KENT.		◆ANTARES		Nunki	
330	44 05	017	63 23	079	42 33	138	50 05	205	53 06	230	27 20	270	36 03	307
332	44 14	015	63 53	077	42 54	137	49 53	204	52 43	228	26 49	268	35 38	305
334	44 21	013	64 23	075	43 15	135	49 41	202	52 20	227	26 18	266	35 12	302
336	44 27	010	64 53	072	43 38	133	49 29	201	51 58	225	25 47	264	34 45	300
338	44 32	008	65 22	070	44 00	132	49 19	199	51 36	224	25 16	262	34 18	298
340	44 36	005	65 51	068	44 24	130	49 09	198	51 15	222	24 45	260	33 50	296
342	44 38	003	66 20	065	44 48	129	48 59	197	50 54	221	24 15	258	33 22	294
344	44 38	000	66 48	063	45 12	127	48 51	195	50 34	219	23 45	256	32 54	292
346	44 38	358	67 15	060	45 37	125	48 43	194	50 15	218	23 15	254	32 25	290
348	44 37	356	67 42	058	46 03	124	48 36	193	49 56	217	22 45	253	31 55	288
350	44 34	353	68 08	055	46 29	122	48 29	191	49 38	215	22 15	251	31 26	286
352	44 29	351	68 33	052	46 55	121	48 24	190	49 20	214	21 46	249	30 56	284
354	44 24	348	68 57	050	47 22	119	48 19	189	49 03	212	21 17	247	30 25	282
356	44 17	346	69 20	047	47 50	117	48 15	187	48 47	211	20 49	245	29 55	280
358	44 08	343	69 42	044	48 18	116	48 11	186	48 31	209	20 21	243	29 24	278

LAT 76°S

LHA ♈	Hc	Zn	Hc	Zn	Hc	Zn	Hc	Zn	Hc	Zn	Hc	Zn	Hc	Zn
	♦Diphda		♦CANOPUS		ACRUX		RIGIL KENT.		♦ANTARES		Peacock		FOMALHAUT	
0	31 44	012	49 10	113	49 08	185	49 09	209	20 22	242	62 52	284	43 02	341
2	31 50	010	49 37	111	49 06	183	48 56	207	19 57	240	62 24	282	42 52	339
4	31 54	008	50 04	109	49 05	182	48 43	206	19 32	238	61 56	280	42 41	337
6	31 58	005	50 31	108	49 04	180	48 31	204	19 07	236	61 27	278	42 29	334
8	32 00	003	50 59	106	49 04	179	48 19	203	18 44	234	60 58	276	42 16	332
10	32 01	001	51 27	104	49 05	178	48 08	201	18 20	233	60 29	274	42 02	330
12	32 01	359	51 55	102	49 07	176	47 58	200	17 58	231	60 00	272	41 47	327
14	31 59	356	52 24	100	49 09	175	47 49	198	17 35	229	59 31	270	41 30	325
16	31 57	354	52 53	099	49 12	173	47 40	197	17 14	227	59 02	268	41 13	323
18	31 54	352	53 21	097	49 16	172	47 32	196	16 53	225	58 33	266	40 55	320
20	31 49	350	53 50	095	49 20	171	47 24	194	16 33	223	58 04	264	40 36	318
22	31 43	347	54 19	093	49 25	169	47 17	193	16 13	222	57 36	262	40 16	316
24	31 36	345	54 48	091	49 31	168	47 11	191	15 54	220	57 07	261	39 56	314
26	31 28	343	55 17	089	49 37	167	47 06	190	15 36	218	56 38	259	39 34	311
28	31 19	341	55 46	087	49 44	165	47 01	188	15 19	216	56 10	257	39 12	309
	RIGEL		♦SIRIUS		Suhail		♦RIGIL KENT.		Peacock		♦FOMALHAUT		Diphda	
30	17 16	051	20 42	076	38 00	118	46 58	187	55 42	255	38 50	307	31 09	339
32	17 39	049	21 10	074	38 25	116	46 54	186	55 14	253	38 26	305	30 58	336
34	18 00	047	21 38	072	38 52	115	46 52	184	54 46	252	38 02	303	30 46	334
36	18 21	045	22 05	070	39 18	113	46 50	183	54 19	250	37 37	301	30 33	332
38	18 41	043	22 32	068	39 45	111	46 49	181	53 51	248	37 12	298	30 19	330
40	19 01	041	22 59	066	40 12	109	46 49	180	53 25	247	36 46	296	30 04	328
42	19 19	039	23 25	064	40 40	107	46 49	178	52 58	245	36 20	294	29 48	325
44	19 37	037	23 51	062	41 08	106	46 50	177	52 32	243	35 53	292	29 31	323
46	19 54	035	24 16	060	41 36	104	46 52	176	52 06	242	35 26	290	29 13	321
48	20 10	032	24 41	058	42 04	102	46 55	174	51 41	240	34 59	288	28 54	319
50	20 25	030	25 05	056	42 33	100	46 58	173	51 16	238	34 31	286	28 35	317
52	20 39	028	25 29	053	43 01	098	47 02	171	50 52	237	34 03	284	28 15	315
54	20 52	026	25 52	051	43 30	096	47 07	170	50 28	235	33 35	282	27 54	313
56	21 05	024	26 14	049	43 59	094	47 12	168	50 04	234	33 06	280	27 32	310
58	21 16	022	26 36	047	44 28	093	47 19	167	49 41	232	32 38	278	27 10	308
	RIGEL		♦SIRIUS		Suhail		♦RIGIL KENT.		Peacock		♦FOMALHAUT		ACHERNAR	
60	21 26	020	26 57	045	44 57	091	47 26	166	49 18	230	32 09	276	67 18	305
62	21 36	018	27 17	043	45 26	089	47 33	164	48 56	229	31 40	274	66 54	303
64	21 44	016	27 37	041	45 55	087	47 41	163	48 35	227	31 11	272	66 29	300
66	21 51	013	27 55	039	46 24	085	47 50	161	48 14	226	30 42	270	66 04	298
68	21 57	011	28 13	037	46 53	083	48 00	160	47 53	224	30 13	268	65 38	295
70	22 02	009	28 30	034	47 22	081	48 10	158	47 33	223	29 44	266	65 11	293
72	22 07	007	28 46	032	47 50	079	48 21	157	47 14	221	29 15	264	64 44	291
74	22 10	005	29 01	030	48 19	077	48 33	156	46 55	219	28 46	262	64 17	288
76	22 12	003	29 15	028	48 47	075	48 45	154	46 37	218	28 17	261	63 49	286
78	22 12	001	29 28	026	49 14	072	48 58	153	46 19	216	27 49	259	63 21	284
80	22 12	358	29 40	024	49 42	070	49 12	151	46 02	215	27 20	257	62 53	282
82	22 11	356	29 51	021	50 09	068	49 26	150	45 46	213	26 52	255	62 24	280
84	22 08	354	30 01	019	50 36	066	49 41	148	45 31	212	26 24	253	61 56	278
86	22 05	352	30 10	017	51 02	064	49 57	147	45 16	210	25 57	251	61 27	276
88	22 00	350	30 18	015	51 28	062	50 13	145	45 01	209	25 29	249	60 58	274
	♦SIRIUS		Suhail		♦RIGIL KENT.		Peacock		♦FOMALHAUT		ACHERNAR		RIGEL	
90	30 25	012	51 53	059	50 30	144	44 48	207	25 02	247	60 29	272	21 55	348
92	30 31	010	52 18	057	50 47	142	44 35	206	24 36	246	60 00	270	21 48	346
94	30 35	008	52 42	055	51 05	141	44 23	204	24 10	244	59 31	268	21 40	344
96	30 39	006	53 05	052	51 24	140	44 11	203	23 44	242	59 02	266	21 32	341
98	30 41	004	53 28	050	51 43	138	44 00	201	23 18	240	58 33	264	21 22	339
100	30 42	001	53 50	048	52 03	137	43 50	200	22 53	238	58 04	262	21 11	337
102	30 42	359	54 11	045	52 23	135	43 41	198	22 29	236	57 35	260	20 59	335
104	30 41	357	54 31	043	52 44	134	43 32	197	22 05	235	57 07	258	20 47	333
106	30 39	355	54 50	040	53 05	132	43 24	195	21 42	233	56 38	257	20 33	331
108	30 36	352	55 09	038	53 27	131	43 17	194	21 19	231	56 10	255	20 18	329
110	30 32	350	55 26	035	53 49	129	43 10	192	20 57	229	55 42	253	20 03	327
112	30 26	348	55 42	033	54 12	127	43 04	191	20 35	227	55 15	251	19 47	325
114	30 20	346	55 58	030	54 35	126	42 59	189	20 14	225	54 47	250	19 29	323
116	30 12	344	56 12	028	54 59	124	42 55	188	19 54	224	54 20	248	19 11	320
118	30 03	341	56 25	025	55 23	123	42 52	186	19 34	222	53 54	246	18 52	318
	Suhail		♦SPICA		RIGIL KENT.		Peacock		♦FOMALHAUT		ACHERNAR		♦SIRIUS	
120	56 37	023	12 55	084	55 48	121	42 49	185	19 15	220	53 27	245	29 53	339
122	56 47	020	13 24	082	56 13	120	42 47	183	18 56	218	53 01	243	29 43	337
124	56 56	017	13 53	080	56 38	118	42 46	182	18 39	216	52 35	241	29 31	335
126	57 04	015	14 21	078	57 04	117	42 45	180	18 22	215	52 10	240	29 18	333
128	57 11	012	14 50	076	57 30	115	42 45	179	18 06	213	51 45	238	29 04	330
130	57 17	009	15 18	074	57 56	113	42 46	177	17 51	211	51 21	236	28 49	328
132	57 21	007	15 45	072	58 23	112	42 48	176	17 36	209	50 57	235	28 34	326
134	57 23	004	16 13	070	58 50	110	42 51	174	17 22	207	50 33	233	28 17	324
136	57 25	001	16 40	068	59 18	108	42 54	173	17 09	206	50 10	232	27 59	322
138	57 25	359	17 07	066	59 46	107	42 58	171	16 57	204	49 48	230	27 41	320
140	57 23	356	17 33	064	60 13	105	43 03	170	16 46	202	49 26	229	27 22	318
142	57 20	353	17 59	062	60 42	103	43 08	168	16 35	200	49 04	227	27 02	315
144	57 16	351	18 25	060	61 10	101	43 15	167	16 26	198	48 43	225	26 41	313
146	57 11	348	18 50	058	61 39	100	43 22	165	16 17	197	48 23	224	26 20	311
148	57 04	345	19 14	056	62 07	098	43 29	164	16 09	195	48 03	222	25 57	309
	SPICA		♦ANTARES		Peacock		FOMALHAUT		♦ACHERNAR		CANOPUS		♦Suhail	
150	19 38	054	23 51	104	43 38	162	16 02	193	47 44	221	59 05	287	56 56	343
152	20 01	052	24 19	102	43 47	161	15 56	191	47 25	219	58 37	285	56 47	340
154	20 24	050	24 48	100	43 57	159	15 51	189	47 07	218	58 09	283	56 36	337
156	20 46	048	25 16	098	44 08	158	15 47	188	46 50	216	57 40	281	56 24	335
158	21 07	046	25 45	096	44 19	156	15 44	186	46 33	215	57 12	279	56 11	332
160	21 28	044	26 14	094	44 31	155	15 41	184	46 16	213	56 43	277	55 57	329
162	21 48	042	26 43	092	44 44	153	15 40	182	46 01	212	56 14	275	55 42	327
164	22 07	040	27 12	090	44 57	152	15 39	180	45 46	210	55 45	273	55 25	324
166	22 25	038	27 41	088	45 11	150	15 39	179	45 32	209	55 16	271	55 08	322
168	22 42	036	28 10	086	45 26	149	15 40	177	45 18	207	54 47	269	54 49	319
170	22 59	034	28 39	085	45 41	147	15 43	175	45 05	206	54 18	267	54 30	317
172	23 14	031	29 08	083	45 57	146	15 46	173	44 53	204	53 49	265	54 10	315
174	23 29	029	29 37	081	46 14	144	15 50	171	44 42	203	53 20	263	53 49	312
176	23 43	027	30 05	079	46 31	143	15 54	169	44 31	201	52 51	261	53 27	310
178	23 55	025	30 34	077	46 49	141	16 00	168	44 21	200	52 23	260	53 04	307

LHA ♈	Hc	Zn	Hc	Zn	Hc	Zn	Hc	Zn	Hc	Zn	Hc	Zn	Hc	Zn
	SPICA		♦ANTARES		Nunki		FOMALHAUT		♦ACHERNAR		CANOPUS		♦Suhail	
180	24 07	023	31 02	075	22 14	110	16 07	166	44 11	198	51 54	258	52 41	305
182	24 18	021	31 29	073	22 42	108	16 14	164	44 03	197	51 26	256	52 17	303
184	24 28	019	31 57	071	23 10	106	16 23	162	43 55	195	50 58	254	51 52	301
186	24 37	016	32 24	068	23 38	104	16 32	160	43 47	194	50 30	252	51 27	298
188	24 44	014	32 51	066	24 06	102	16 42	159	43 41	192	50 03	251	51 01	296
190	24 51	012	33 17	064	24 34	100	16 53	157	43 35	191	49 35	249	50 35	294
192	24 56	010	33 43	062	25 03	098	17 05	155	43 30	189	49 09	247	50 08	292
194	25 01	008	34 09	060	25 32	097	17 18	153	43 26	188	48 42	245	49 41	290
196	25 04	006	34 34	058	26 01	095	17 31	151	43 22	186	48 16	244	49 13	287
198	25 07	004	34 58	056	26 30	093	17 45	150	43 20	185	47 50	242	48 45	285
200	25 08	001	35 22	054	26 59	091	18 01	148	43 18	183	47 25	240	48 17	283
202	25 08	359	35 45	052	27 28	089	18 16	146	43 16	182	47 00	239	47 49	281
204	25 07	357	36 07	049	27 57	087	18 33	144	43 16	180	46 35	237	47 20	279
206	25 05	355	36 29	047	28 26	085	18 50	142	43 16	179	46 11	235	46 52	277
208	25 02	353	36 50	045	28 54	083	19 08	141	43 17	177	45 47	234	46 23	275
	♦ANTARES		Nunki		FOMALHAUT		♦ACHERNAR		CANOPUS		♦Suhail		SPICA	
210	37 10	043	29 23	081	19 27	139	43 19	176	45 24	232	45 54	273	24 58	351
212	37 29	041	29 52	079	19 47	137	43 21	174	45 01	230	45 25	271	24 52	348
214	37 48	038	30 20	077	20 07	135	43 24	173	44 39	229	44 56	269	24 46	346
216	38 05	036	30 48	075	20 28	133	43 28	171	44 18	227	44 27	267	24 38	344
218	38 22	034	31 16	073	20 49	132	43 33	170	43 57	226	43 58	265	24 30	342
220	38 38	032	31 44	071	21 11	130	43 39	168	43 36	224	43 29	264	24 20	340
222	38 53	029	32 11	069	21 34	128	43 45	167	43 17	222	43 00	262	24 10	338
224	39 06	027	32 38	067	21 57	126	43 52	165	42 57	221	42 31	260	23 58	335
226	39 19	025	33 04	065	22 21	124	43 59	164	42 39	219	42 03	258	23 46	333
228	39 31	023	33 30	063	22 45	122	44 08	162	42 21	218	41 35	256	23 32	331
230	39 41	020	33 56	061	23 10	121	44 17	161	42 03	216	41 07	254	23 18	329
232	39 51	018	34 21	058	23 35	119	44 27	159	41 47	214	40 39	252	23 02	327
234	39 59	016	34 46	056	24 01	117	44 37	158	41 31	213	40 11	251	22 46	325
236	40 06	013	35 09	054	24 27	115	44 49	156	41 15	211	39 44	249	22 29	323
238	40 13	011	35 33	052	24 53	113	45 01	155	41 00	210	39 17	247	22 11	321
	ANTARES		♦Nunki		FOMALHAUT		♦ACHERNAR		CANOPUS		♦ACRUX		SPICA	
240	40 17	009	35 55	050	25 20	111	45 13	153	40 46	208	68 30	277	21 52	319
242	40 21	006	36 17	048	25 47	109	45 26	152	40 33	206	68 01	275	21 32	316
244	40 24	004	36 38	046	26 15	108	45 40	150	40 21	205	67 32	273	21 12	314
246	40 25	001	36 58	043	26 43	106	45 55	149	40 09	203	67 03	271	20 51	312
248	40 25	359	37 18	041	27 11	104	46 10	147	39 58	202	66 34	269	20 29	310
250	40 24	357	37 37	039	27 39	102	46 26	146	39 47	200	66 05	267	20 07	308
252	40 22	354	37 55	037	28 07	100	46 43	144	39 38	199	65 36	265	19 44	306
254	40 19	352	38 11	034	28 36	098	47 00	143	39 29	197	65 07	264	19 20	304
256	40 14	350	38 27	032	29 05	096	47 18	141	39 21	196	64 38	262	18 55	302
258	40 08	347	38 42	030	29 34	094	47 37	140	39 13	194	64 09	260	18 31	300
260	40 01	345	38 56	028	30 03	092	47 56	138	39 07	192	63 41	258	18 05	298
262	39 53	343	39 09	025	30 32	090	48 15	137	39 01	191	63 13	257	17 39	296
264	39 44	340	39 21	023	31 01	089	48 35	135	38 56	189	62 44	255	17 13	294
266	39 34	338	39 32	021	31 30	087	48 56	134	38 51	188	62 17	253	16 46	292
268	39 22	336	39 42	018	31 59	085	49 17	132	38 48	186	61 49	251	16 19	290
	♦Nunki		FOMALHAUT		♦ACHERNAR		CANOPUS		♦ACRUX		RIGIL KENT.		ANTARES	
270	39 50	016	32 28	083	49 39	130	38 45	185	61 22	250	67 19	284	39 10	333
272	39 58	014	32 56	081	50 02	129	38 43	183	60 54	248	66 51	282	38 57	331
274	40 04	011	33 25	079	50 24	127	38 42	182	60 28	247	66 22	279	38 42	329
276	40 09	009	33 53	077	50 48	126	38 42	180	60 01	245	65 53	277	38 27	327
278	40 13	007	34 21	075	51 12	124	38 42	178	59 35	243	65 25	275	38 10	324
280	40 16	004	34 49	073	51 36	123	38 43	177	59 09	242	64 56	273	37 53	322
282	40 18	002	35 17	071	52 01	121	38 45	175	58 44	240	64 27	271	37 35	320
284	40 18	000	35 44	068	52 26	119	38 48	174	58 19	239	63 58	269	37 15	318
286	40 17	357	36 11	066	52 51	118	38 51	172	57 54	237	63 29	267	36 56	316
288	40 15	355	36 37	064	53 17	116	38 56	171	57 30	236	63 00	266	36 35	313
290	40 12	353	37 03	062	53 43	114	39 01	169	57 06	234	62 31	264	36 13	311
292	40 08	350	37 29	060	54 10	113	39 07	168	56 43	233	62 02	262	35 51	309
294	40 03	348	37 53	058	54 37	111	39 13	166	56 20	231	61 33	260	35 28	307
296	39 56	346	38 18	056	55 04	109	39 21	164	55 58	230	61 05	258	35 05	305
298	39 48	343	38 41	054	55 32	108	39 29	163	55 36	228	60 36	257	34 40	303
	♦FOMALHAUT		ACHERNAR		♦CANOPUS		ACRUX		RIGIL KENT.		♦ANTARES		Nunki	
300	39 05	051	56 00	106	39 38	161	55 14	227	60 08	255	34 16	300	39 39	341
302	39 27	049	56 28	104	39 48	160	54 53	225	59 40	253	33 50	298	39 29	339
304	39 49	047	56 56	102	39 58	158	54 33	224	59 13	251	33 25	296	39 18	336
306	40 09	045	57 24	100	40 09	157	54 13	222	58 45	250	32 58	294	39 06	334
308	40 30	043	57 53	099	40 21	155	53 54	221	58 18	248	32 32	292	38 53	332
310	40 49	040	58 22	097	40 34	153	53 35	220	57 51	246	32 05	290	38 39	329
312	41 07	038	58 51	095	40 47	152	53 17	218	57 25	245	31 37	288	38 23	327
314	41 25	036	59 20	093	41 01	150	52 59	217	56 59	243	31 09	286	38 07	325
316	41 41	034	59 49	091	41 16	149	52 42	215	56 33	242	30 41	284	37 50	323
318	41 57	031	60 18	089	41 31	147	52 26	214	56 08	240	30 13	282	37 32	320
320	42 11	029	60 47	087	41 47	146	52 10	212	55 43	238	29 44	280	37 13	318
322	42 25	027	61 16	085	42 04	144	51 55	211	55 18	237	29 16	278	36 53	316
324	42 37	024	61 45	083	42 21	142	51 40	210	54 54	235	28 47	276	36 33	314
326	42 49	022	62 13	081	42 39	141	51 26	208	54 31	234	28 18	274	36 11	312
328	42 59	020	62 42	079	42 58	139	51 13	207	54 08	232	27 49	272	35 49	309
	♦FOMALHAUT		ACHERNAR		♦CANOPUS		ACRUX		RIGIL KENT.		♦ANTARES		Nunki	
330	43 08	017	63 10	077	43 17	138	51 00	205	53 45	231	27 20	270	35 27	307
332	43 16	015	63 38	075	43 37	136	50 48	204	53 23	229	26 51	268	35 03	305
334	43 23	012	64 06	073	43 58	134	50 36	203	53 01	228	26 22	266	34 39	303
336	43 28	010	64 34	070	44 19	133	50 25	201	52 40	226	25 53	264	34 15	301
338	43 33	008	65 01	068	44 40	131	50 15	200	52 19	225	25 24	262	33 49	299
340	43 36	005	65 28	066	45 02	130	50 06	198	51 59	223	24 56	261	33 24	297
342	43 38	003	65 54	063	45 25	128	49 57	197	51 39	222	24 27	259	32 58	295
344	43 39	000	66 19	061	45 48	126	49 49	196	51 20	220	23 59	257	32 31	293
346	43 39	358	66 44	058	46 12	125	49 41	194	51 02	219	23 31	255	32 04	291
348	43 37	356	67 09	056	46 36	123	49 34	193	50 44	217	23 03	253	31 37	289
350	43 34	353	67 32	053	47 01	121	49 28	191	50 27	216	22 35	251	31 09	286
352	43 30	351	67 55	050	47 26	120	49 23	190	50 10	214	22 08	249	30 41	284
354	43 25	348	68 17	048	47 51	118	49 18	189	49 54	213	21 41	247	30 13	282
356	43 18	346	68 38	045	48 17	116	49 14	187	49 38	211	21 14	245	29 44	280
358	43 11	344	68 58	042	48 43	115	49 11	186	49 24	210	20 48	244	29 16	278

LHA ♈	Hc	Zn	Hc	Zn	Hc	Zn	Hc	Zn	Hc	Zn	Hc	Zn	Hc	Zn
	♦Diphda		♦CANOPUS		ACRUX		RIGIL KENT.		♦ANTARES		Peacock		FOMALHAUT	
0	30 46	012	49 32	112	50 08	185	50 02	209	20 50	242	62 37	286	42 05	342
2	30 51	010	49 58	110	50 06	183	49 49	208	20 27	240	62 11	284	41 56	339
4	30 55	008	50 23	108	50 05	182	49 37	206	20 03	238	61 44	282	41 46	337
6	30 58	005	50 49	106	50 04	180	49 25	205	19 41	237	61 18	280	41 35	335
8	31 00	003	51 15	105	50 04	179	49 14	203	19 18	235	60 51	278	41 23	332
10	31 01	001	51 41	103	50 05	178	49 04	202	18 57	233	60 24	276	41 10	330
12	31 01	359	52 08	101	50 07	176	48 54	200	18 35	231	59 57	274	40 56	328
14	31 00	356	52 34	099	50 09	175	48 45	199	18 15	229	59 30	272	40 41	325
16	30 57	354	53 01	097	50 12	173	48 37	197	17 55	227	59 03	270	40 25	323
18	30 54	352	53 28	095	50 15	172	48 29	196	17 35	225	58 36	268	40 09	321
20	30 50	350	53 55	093	50 19	171	48 22	194	17 16	224	58 10	266	39 51	319
22	30 45	348	54 22	092	50 24	169	48 16	193	16 58	222	57 43	264	39 33	316
24	30 38	345	54 49	090	50 29	168	48 10	191	16 40	220	57 16	262	39 14	314
26	30 31	343	55 16	088	50 35	166	48 05	190	16 23	218	56 49	260	38 54	312
28	30 23	341	55 42	086	50 42	165	48 01	189	16 07	216	56 23	258	38 34	310
	RIGEL		♦SIRIUS		Suhail		♦RIGIL KENT.		Peacock		♦FOMALHAUT		Diphda	
30	16 39	051	20 27	075	38 28	117	47 57	187	55 56	257	38 13	308	30 13	339
32	16 59	049	20 53	073	38 52	116	47 54	186	55 30	255	37 51	306	30 03	337
34	17 19	047	21 19	071	39 16	114	47 52	184	55 04	253	37 29	303	29 52	334
36	17 38	045	21 44	069	39 41	112	47 50	183	54 38	251	37 06	301	29 40	332
38	17 57	043	22 10	067	40 06	110	47 49	181	54 13	250	36 43	299	29 27	330
40	18 15	041	22 34	065	40 32	108	47 49	180	53 48	248	36 19	297	29 13	328
42	18 32	038	22 59	063	40 58	107	47 49	178	53 23	246	35 55	295	28 58	326
44	18 49	036	23 23	061	41 24	105	47 50	177	52 58	244	35 30	293	28 43	324
46	19 04	034	23 46	059	41 50	103	47 52	176	52 34	243	35 05	291	28 26	321
48	19 19	032	24 09	057	42 16	101	47 55	174	52 10	241	34 40	289	28 09	319
50	19 33	030	24 31	055	42 43	099	47 58	173	51 47	239	34 14	287	27 51	317
52	19 46	028	24 53	053	43 09	097	48 02	171	51 24	238	33 48	285	27 32	315
54	19 58	026	25 15	051	43 36	095	48 06	170	51 01	236	33 22	283	27 13	313
56	20 10	024	25 35	049	44 03	093	48 11	168	50 39	235	32 55	281	26 53	311
58	20 20	022	25 55	047	44 30	092	48 17	167	50 17	233	32 29	279	26 32	309
	RIGEL		♦SIRIUS		Suhail		♦RIGIL KENT.		Peacock		♦FOMALHAUT		ACHERNAR	
60	20 30	020	26 15	045	44 57	090	48 24	165	49 56	231	32 02	277	66 42	307
62	20 38	018	26 33	043	45 24	088	48 31	164	49 35	230	31 35	275	66 21	305
64	20 46	015	26 51	041	45 51	086	48 39	162	49 15	228	31 08	273	65 58	302
66	20 53	013	27 08	038	46 18	084	48 47	161	48 55	226	30 41	271	65 35	300
68	20 58	011	27 25	036	46 45	082	48 56	159	48 36	225	30 14	269	65 11	297
70	21 03	009	27 40	034	47 11	080	49 06	158	48 17	223	29 47	267	64 47	295
72	21 07	007	27 55	032	47 38	078	49 16	157	47 59	222	29 20	265	64 22	293
74	21 10	005	28 09	030	48 04	076	49 28	155	47 41	220	28 54	263	63 57	290
76	21 12	003	28 22	028	48 30	073	49 39	154	47 24	219	28 27	261	63 31	288
78	21 12	001	28 34	025	48 56	071	49 52	152	47 08	217	28 00	259	63 06	286
80	21 12	358	28 45	023	49 21	069	50 04	151	46 52	215	27 34	257	62 40	284
82	21 11	356	28 55	021	49 46	067	50 18	149	46 36	214	27 08	255	62 13	282
84	21 09	354	29 04	019	50 11	065	50 32	148	46 21	212	26 42	254	61 47	279
86	21 06	352	29 13	017	50 35	063	50 47	146	46 07	211	26 16	252	61 20	277
88	21 01	350	29 20	015	50 59	060	51 02	145	45 54	209	25 50	250	60 53	275
	♦SIRIUS		Suhail		♦RIGIL KENT.		Peacock		♦FOMALHAUT		ACHERNAR		RIGEL	
90	29 26	012	51 22	058	51 18	143	45 41	208	25 25	248	60 26	273	20 56	348
92	29 32	010	51 45	056	51 34	142	45 29	206	25 00	246	59 59	271	20 50	346
94	29 36	008	52 07	054	51 52	140	45 17	205	24 36	244	59 32	269	20 43	344
96	29 39	006	52 28	051	52 09	139	45 06	203	24 12	242	59 05	267	20 35	342
98	29 41	004	52 49	049	52 27	137	44 56	202	23 48	240	58 38	266	20 26	339
100	29 42	001	53 09	047	52 46	136	44 46	200	23 25	239	58 11	264	20 16	337
102	29 42	359	53 28	044	53 05	134	44 38	198	23 02	237	57 45	262	20 05	335
104	29 42	357	53 47	042	53 25	133	44 29	197	22 40	235	57 18	260	19 53	333
106	29 40	355	54 04	040	53 45	131	44 22	195	22 18	233	56 52	258	19 41	331
108	29 37	353	54 21	037	54 05	129	44 15	194	21 56	231	56 25	256	19 27	329
110	29 33	350	54 37	035	54 26	128	44 09	192	21 36	229	55 59	255	19 13	327
112	29 28	348	54 52	032	54 48	126	44 03	191	21 15	228	55 33	253	18 58	325
114	29 21	346	55 06	030	55 10	125	43 59	189	20 56	226	55 08	251	18 42	323
116	29 14	344	55 19	027	55 32	123	43 55	188	20 37	224	54 42	249	18 25	321
118	29 06	342	55 30	025	55 55	122	43 51	186	20 18	222	54 17	248	18 07	319
	Suhail		♦SPICA		RIGIL KENT.		Peacock		♦FOMALHAUT		ACHERNAR		♦SIRIUS	
120	55 41	022	12 49	084	56 18	120	43 49	185	20 01	220	53 52	246	28 57	339
122	55 51	020	13 16	082	56 42	118	43 47	183	19 44	218	53 28	244	28 47	337
124	55 59	017	13 42	080	57 06	117	43 46	182	19 27	217	53 04	243	28 36	335
126	56 06	014	14 09	078	57 30	115	43 45	180	19 11	215	52 40	241	28 25	333
128	56 12	012	14 35	076	57 55	113	43 45	179	18 56	213	52 16	239	28 12	331
130	56 17	009	15 01	074	58 19	112	43 46	177	18 42	211	51 53	238	27 58	329
132	56 21	006	15 27	072	58 45	110	43 48	176	18 28	209	51 31	236	27 44	326
134	56 23	004	15 53	070	59 10	108	43 50	174	18 15	208	51 09	234	27 28	324
136	56 25	001	16 18	068	59 36	107	43 53	173	18 03	206	50 47	233	27 12	322
138	56 25	359	16 43	066	60 02	105	43 57	171	17 52	204	50 26	231	26 55	320
140	56 23	356	17 07	064	60 28	103	44 02	170	17 41	202	50 05	229	26 37	318
142	56 21	353	17 31	062	60 54	101	44 07	168	17 32	200	49 45	228	26 19	316
144	56 17	351	17 55	060	61 21	100	44 13	167	17 23	198	49 25	226	26 00	314
146	56 12	348	18 18	058	61 48	098	44 20	165	17 15	197	49 06	225	25 40	312
148	56 06	346	18 41	056	62 14	096	44 27	164	17 07	195	48 47	223	25 19	309
	SPICA		♦ANTARES		Peacock		FOMALHAUT		♦ACHERNAR		CANOPUS		♦Suhail	
150	19 03	054	24 05	103	44 35	162	17 01	193	48 29	222	58 46	289	55 59	343
152	19 25	052	24 31	101	44 44	160	16 55	191	48 11	220	58 20	287	55 50	340
154	19 45	050	24 58	100	44 53	159	16 50	189	47 54	218	57 54	285	55 41	338
156	20 06	048	25 25	098	45 03	157	16 46	188	47 38	217	57 28	282	55 30	335
158	20 26	046	25 51	096	45 14	156	16 43	186	47 22	215	57 02	280	55 18	333
160	20 45	044	26 18	094	45 25	154	16 41	184	47 07	214	56 35	278	55 05	330
162	21 03	042	26 45	092	45 37	153	16 40	182	46 52	212	56 08	276	54 51	328
164	21 20	040	27 12	090	45 50	151	16 39	180	46 38	211	55 41	274	54 36	325
166	21 37	038	27 39	088	46 03	150	16 39	178	46 24	209	55 14	272	54 20	323
168	21 53	035	28 06	086	46 17	148	16 40	177	46 11	208	54 47	270	54 04	320
170	22 09	033	28 33	084	46 32	147	16 42	175	45 59	206	54 20	268	53 46	318
172	22 23	031	29 00	082	46 47	145	16 45	173	45 48	205	53 53	266	53 27	316
174	22 36	029	29 26	080	47 02	143	16 49	171	45 37	203	53 27	265	53 08	313
176	22 49	027	29 53	078	47 19	142	16 53	169	45 27	201	53 00	263	52 48	311
178	23 01	025	30 19	076	47 36	140	16 59	168	45 17	200	52 33	261	52 27	308

LAT 77°S

LHA ♈	Hc	Zn	Hc	Zn	Hc	Zn	Hc	Zn	Hc	Zn	Hc	Zn	Hc	Zn
	SPICA		♦ANTARES		Nunki		FOMALHAUT		♦ACHERNAR		CANOPUS		♦Suhail	
180	23 12	023	30 45	074	22 34	109	17 05	166	45 08	198	52 06	259	52 06	306
182	23 22	021	31 11	072	23 00	108	17 12	164	45 00	197	51 40	257	51 44	304
184	23 31	018	31 37	070	23 26	106	17 20	162	44 52	195	51 14	255	51 21	302
186	23 39	016	32 02	068	23 52	104	17 29	160	44 46	194	50 48	254	50 58	299
188	23 46	014	32 27	066	24 18	102	17 38	159	44 40	192	50 22	252	50 34	297
190	23 52	012	32 51	064	24 45	100	17 48	157	44 34	191	49 57	250	50 10	295
192	23 57	010	33 15	062	25 11	098	17 59	155	44 29	189	49 31	248	49 45	293
194	24 02	008	33 39	060	25 38	096	18 11	153	44 25	188	49 06	246	49 20	291
196	24 05	006	34 02	057	26 05	094	18 24	151	44 22	186	48 42	245	48 55	289
198	24 07	003	34 24	055	26 32	092	18 37	149	44 19	185	48 18	243	48 29	286
200	24 08	001	34 46	053	26 59	090	18 51	148	44 17	183	47 54	241	48 03	284
202	24 08	359	35 07	051	27 26	088	19 06	146	44 16	182	47 30	240	47 37	282
204	24 07	357	35 28	049	27 53	086	19 22	144	44 16	180	47 07	238	47 10	280
206	24 05	355	35 48	047	28 20	084	19 38	142	44 16	179	46 45	236	46 43	278
208	24 02	353	36 07	045	28 47	082	19 55	140	44 17	177	46 22	235	46 17	276
	♦ANTARES		Nunki		FOMALHAUT		♦ACHERNAR		CANOPUS		♦Suhail		SPICA	
210	36 26	042	29 13	080	20 12	139	44 19	176	46 01	233	45 50	274	23 58	351
212	36 44	040	29 40	078	20 30	137	44 21	174	45 39	231	45 23	272	23 53	348
214	37 01	038	30 06	076	20 49	135	44 24	173	45 19	230	44 56	270	23 48	346
216	37 17	036	30 32	074	21 09	133	44 28	171	44 58	228	44 29	268	23 41	344
218	37 32	034	30 58	072	21 29	131	44 32	170	44 39	226	44 02	266	23 33	342
220	37 47	031	31 24	070	21 49	129	44 37	168	44 19	225	43 35	265	23 24	340
222	38 00	029	31 49	068	22 10	128	44 43	167	44 01	223	43 08	263	23 14	338
224	38 13	027	32 14	066	22 32	126	44 50	165	43 43	221	42 41	261	23 04	336
226	38 25	024	32 39	064	22 54	124	44 57	164	43 25	220	42 15	259	22 52	334
228	38 35	022	33 03	062	23 17	122	45 05	162	43 08	218	41 49	257	22 40	331
230	38 45	020	33 26	060	23 40	120	45 14	161	42 52	216	41 22	255	22 26	329
232	38 54	018	33 49	058	24 04	118	45 23	159	42 36	215	40 56	253	22 12	327
234	39 01	015	34 12	056	24 28	116	45 33	158	42 21	213	40 31	251	21 57	325
236	39 08	013	34 34	054	24 52	115	45 44	156	42 06	212	40 05	250	21 41	323
238	39 14	011	34 56	052	25 17	113	45 55	155	41 53	210	39 40	248	21 24	321
	ANTARES		♦Nunki		FOMALHAUT		♦ACHERNAR		CANOPUS		♦ACRUX		SPICA	
240	39 18	008	35 16	049	25 42	111	46 07	153	41 39	208	68 21	280	21 07	319
242	39 21	006	35 37	047	26 07	109	46 19	151	41 27	207	67 54	278	20 49	317
244	39 24	004	35 56	045	26 33	107	46 33	150	41 15	205	67 27	276	20 30	315
246	39 25	001	36 15	043	26 59	105	46 46	148	41 04	204	67 00	274	20 10	313
248	39 25	359	36 33	041	27 25	103	47 01	147	40 53	202	66 33	272	19 50	311
250	39 24	357	36 50	038	27 51	101	47 16	145	40 43	200	66 06	270	19 29	308
252	39 22	355	37 06	036	28 18	100	47 32	144	40 34	199	65 39	268	19 08	306
254	39 19	352	37 22	034	28 44	098	47 48	142	40 26	197	65 12	266	18 46	304
256	39 15	350	37 36	032	29 11	096	48 05	141	40 18	196	64 46	264	18 23	302
258	39 10	348	37 50	030	29 38	094	48 22	139	40 11	194	64 19	262	18 00	300
260	39 03	345	38 03	027	30 05	092	48 40	137	40 05	193	63 52	260	17 37	298
262	38 56	343	38 15	025	30 32	090	48 59	136	40 00	191	63 26	258	17 13	296
264	38 48	341	38 26	023	30 59	088	49 18	134	39 55	189	62 59	257	16 49	294
266	38 38	338	38 36	020	31 26	086	49 37	133	39 51	188	62 33	255	16 24	292
268	38 28	336	38 45	018	31 53	084	49 57	131	39 47	186	62 07	253	15 59	290
	♦Nunki		FOMALHAUT		♦ACHERNAR		CANOPUS		♦ACRUX		RIGIL KENT.		ANTARES	
270	38 53	016	32 20	082	50 18	130	39 45	185	61 41	252	67 04	286	38 16	334
272	39 00	014	32 46	080	50 39	128	39 43	183	61 16	250	66 37	284	38 04	332
274	39 05	011	33 13	078	51 00	126	39 42	182	60 51	248	66 11	282	37 51	329
276	39 10	009	33 39	076	51 22	125	39 42	180	60 26	247	65 45	280	37 36	327
278	39 14	007	34 05	074	51 45	123	39 42	178	60 01	245	65 18	277	37 21	325
280	39 16	004	34 31	072	52 08	121	39 43	177	59 37	243	64 51	275	37 05	323
282	39 18	002	34 57	070	52 31	120	39 45	175	59 13	242	64 24	273	36 48	320
284	39 18	000	35 22	068	52 54	118	39 48	174	58 49	240	63 57	271	36 31	318
286	39 17	357	35 47	066	53 18	116	39 51	172	58 26	239	63 30	269	36 13	316
288	39 16	355	36 11	064	53 43	115	39 55	171	58 03	237	63 03	267	35 53	314
290	39 13	353	36 35	062	54 08	113	40 00	169	57 41	235	62 36	266	35 34	312
292	39 09	350	36 58	059	54 32	111	40 05	167	57 19	234	62 09	264	35 13	310
294	39 04	348	37 21	057	54 58	110	40 12	166	56 57	232	61 43	262	34 52	307
296	38 58	346	37 44	055	55 23	108	40 19	164	56 36	231	61 16	260	34 30	305
298	38 51	344	38 06	053	55 49	106	40 26	163	56 15	229	60 49	258	34 08	303
	♦FOMALHAUT		ACHERNAR		♦CANOPUS		ACRUX		RIGIL KENT.		♦ANTARES		Nunki	
300	38 27	051	56 15	104	40 35	161	55 55	228	60 23	256	33 45	301	38 43	341
302	38 48	049	56 41	103	40 44	159	55 35	226	59 57	255	33 22	299	38 33	339
304	39 07	046	57 08	101	40 54	158	55 16	225	59 31	253	32 58	297	38 23	337
306	39 27	044	57 34	099	41 04	156	54 57	223	59 05	251	32 33	295	38 12	334
308	39 45	042	58 01	097	41 15	155	54 39	222	58 40	250	32 09	293	38 00	332
310	40 03	040	58 28	095	41 27	153	54 21	220	58 15	248	31 44	291	37 47	330
312	40 20	038	58 55	093	41 40	151	54 04	219	57 50	246	31 18	289	37 33	328
314	40 36	035	59 22	091	41 53	150	53 47	218	57 25	245	30 52	287	37 18	325
316	40 51	033	59 49	089	42 07	148	53 31	216	57 01	243	30 26	285	37 02	323
318	41 05	031	60 16	087	42 21	147	53 15	215	56 37	241	30 00	283	36 45	321
320	41 19	028	60 43	085	42 37	145	53 00	213	56 14	240	29 34	281	36 28	319
322	41 31	026	61 10	083	42 52	143	52 46	212	55 51	238	29 07	279	36 10	317
324	41 42	024	61 36	081	43 09	142	52 32	210	55 28	237	28 40	277	35 51	314
326	41 53	022	62 03	079	43 26	140	52 19	209	55 06	235	28 14	275	35 31	312
328	42 02	019	62 30	077	43 43	139	52 06	207	54 44	233	27 47	273	35 11	310
	♦FOMALHAUT		ACHERNAR		♦CANOPUS		ACRUX		RIGIL KENT.		♦ANTARES		Nunki	
330	42 11	017	62 56	075	44 01	137	51 54	206	54 22	232	27 20	271	34 50	308
332	42 18	015	63 22	073	44 20	135	51 42	205	54 01	230	26 53	269	34 28	306
334	42 24	012	63 47	071	44 39	134	51 31	203	53 41	229	26 26	267	34 06	304
336	42 29	010	64 13	068	44 59	132	51 21	202	53 21	227	25 59	265	33 43	302
338	42 33	007	64 37	066	45 19	130	51 12	200	53 01	226	25 32	263	33 20	299
340	42 36	005	65 02	064	45 40	129	51 03	199	52 42	224	25 05	261	32 56	297
342	42 38	003	65 26	061	46 02	127	50 54	197	52 24	223	24 39	259	32 32	295
344	42 39	000	65 49	059	46 23	125	50 46	196	52 06	221	24 12	257	32 08	293
346	42 39	358	66 12	056	46 45	124	50 39	195	51 48	220	23 46	255	31 43	291
348	42 37	356	66 34	054	47 08	122	50 33	193	51 31	218	23 20	253	31 17	289
350	42 34	353	66 56	051	47 31	120	50 27	192	51 15	217	22 54	251	30 52	287
352	42 31	351	67 16	049	47 55	119	50 22	190	50 59	215	22 29	250	30 26	285
354	42 26	349	67 36	046	48 19	117	50 17	189	50 44	214	22 04	248	29 59	283
356	42 20	346	67 55	043	48 43	115	50 14	187	50 29	212	21 39	246	29 33	281
358	42 13	344	68 13	040	49 07	113	50 10	186	50 15	211	21 14	244	29 07	279

LAT 78°S

LHA	Hc	Zn	Hc	Zn	Hc	Zn	Hc	Zn	Hc	Zn	Hc	Zn	Hc	Zn
♈	◆Diphda		◆CANOPUS		ACRUX		RIGIL KENT.		◆ANTARES		Peacock		FOMALHAUT	
0	29 47	012	49 54	111	51 08	185	50 54	210	21 18	242	62 19	288	41 08	342
2	29 52	010	50 18	109	51 06	183	50 42	208	20 56	241	61 55	286	41 00	340
4	29 55	008	50 41	107	51 05	182	50 31	207	20 35	239	61 31	284	40 51	337
6	29 58	005	51 05	105	51 04	180	50 20	205	20 14	237	61 07	282	40 41	335
8	30 00	003	51 29	103	51 04	179	50 10	204	19 53	235	60 42	279	40 30	333
10	30 01	001	51 54	102	51 05	178	50 00	202	19 33	233	60 17	277	40 18	330
12	30 01	359	52 18	100	51 07	176	49 51	201	19 13	231	59 53	275	40 05	328
14	30 00	357	52 43	098	51 09	175	49 42	199	18 54	229	59 28	273	39 52	326
16	29 58	354	53 08	096	51 11	173	49 34	198	18 35	227	59 03	271	39 37	324
18	29 55	352	53 33	094	51 15	172	49 27	196	18 17	226	58 38	269	39 22	321
20	29 51	350	53 57	092	51 18	170	49 20	195	18 00	224	58 13	267	39 06	319
22	29 46	348	54 22	090	51 23	169	49 14	193	17 43	222	57 48	266	38 49	317
24	29 40	346	54 47	088	51 28	167	49 09	192	17 26	220	57 23	264	38 32	315
26	29 34	343	55 12	086	51 34	166	49 04	190	17 11	218	56 59	262	38 14	313
28	29 26	341	55 37	084	51 40	165	49 00	189	16 55	216	56 34	260	37 55	310
	RIGEL		◆SIRIUS		Suhail		◆RIGIL KENT.		Peacock		◆FOMALHAUT		Diphda	
30	16 01	051	20 12	075	38 55	117	48 57	187	56 09	258	37 36	308	29 17	339
32	16 20	049	20 36	073	39 17	115	48 54	186	55 45	256	37 16	306	29 08	337
34	16 38	046	21 00	071	39 40	113	48 52	184	55 21	254	36 56	304	28 58	335
36	16 56	044	21 23	069	40 03	111	48 50	183	54 57	253	36 35	302	28 47	333
38	17 13	042	21 46	067	40 27	109	48 49	181	54 33	251	36 13	300	28 35	330
40	17 29	040	22 09	065	40 50	108	48 49	180	54 10	249	35 52	298	28 22	328
42	17 45	038	22 32	063	41 14	106	48 49	178	53 47	247	35 29	296	28 09	326
44	18 00	036	22 54	061	41 38	104	48 50	177	53 24	246	35 07	294	27 54	324
46	18 15	034	23 15	059	42 03	102	48 52	175	53 01	244	34 44	291	27 39	322
48	18 28	032	23 36	057	42 27	100	48 54	174	52 39	242	34 20	289	27 23	320
50	18 41	030	23 57	055	42 52	098	48 57	172	52 17	241	33 57	287	27 07	318
52	18 53	028	24 17	053	43 17	096	49 01	171	51 56	239	33 33	285	26 50	315
54	19 04	026	24 37	051	43 41	094	49 05	169	51 34	237	33 08	283	26 32	313
56	19 15	024	24 56	049	44 06	092	49 10	168	51 14	236	32 44	281	26 13	311
58	19 24	022	25 14	046	44 31	091	49 15	167	50 53	234	32 19	279	25 54	309
	RIGEL		◆SIRIUS		Suhail		◆RIGIL KENT.		Peacock		◆FOMALHAUT		ACHERNAR	
60	19 33	020	25 32	044	44 56	089	49 22	165	50 33	232	31 55	277	66 05	309
62	19 41	017	25 49	042	45 21	087	49 28	164	50 14	231	31 30	275	65 46	306
64	19 48	015	26 05	040	45 46	085	49 36	162	49 55	229	31 05	273	65 25	304
66	19 54	013	26 21	038	46 11	083	49 44	161	49 36	227	30 40	271	65 04	302
68	20 00	011	26 36	036	46 35	081	49 52	159	49 18	226	30 15	269	64 43	299
70	20 04	009	26 50	034	47 00	079	50 02	158	49 00	224	29 50	267	64 21	297
72	20 07	007	27 04	032	47 24	077	50 11	156	48 43	222	29 25	266	63 58	294
74	20 10	005	27 17	030	47 49	074	50 22	155	48 27	221	29 01	264	63 35	292
76	20 12	003	27 29	027	48 12	072	50 33	153	48 11	219	28 36	262	63 12	290
78	20 12	001	27 40	025	48 36	070	50 44	152	47 55	218	28 11	260	62 48	288
80	20 12	358	27 50	023	48 59	068	50 57	150	47 40	216	27 47	258	62 24	286
82	20 11	356	27 59	021	49 22	066	51 09	149	47 26	214	27 22	256	62 00	283
84	20 09	354	28 08	019	49 45	064	51 23	147	47 12	213	26 58	254	61 36	281
86	20 06	352	28 15	017	50 07	062	51 36	145	46 59	211	26 34	252	61 11	279
88	20 02	350	28 22	014	50 29	059	51 51	144	46 46	210	26 11	250	60 47	277
	◆SIRIUS		Suhail		◆RIGIL KENT.		Peacock		◆FOMALHAUT		ACHERNAR		RIGEL	
90	28 28	012	50 50	057	52 06	142	46 34	208	25 48	248	60 22	275	19 58	348
92	28 32	010	51 11	055	52 21	141	46 23	207	25 25	246	59 57	273	19 52	346
94	28 36	008	51 31	053	52 37	139	46 12	205	25 02	245	59 32	271	19 45	344
96	28 39	006	51 50	050	52 54	138	46 01	203	24 39	243	59 07	269	19 38	342
98	28 41	004	52 09	048	53 11	136	45 52	202	24 17	241	58 42	267	19 30	340
100	28 42	001	52 28	046	53 28	135	45 43	200	23 56	239	58 17	265	19 20	337
102	28 42	359	52 45	043	53 46	133	45 34	199	23 35	237	57 52	263	19 10	335
104	28 42	357	53 02	041	54 05	132	45 27	197	23 14	235	57 28	261	19 00	333
106	28 40	355	53 18	039	54 24	130	45 20	196	22 54	233	57 03	260	18 48	331
108	28 37	353	53 33	036	54 43	128	45 13	194	22 34	232	56 39	258	18 36	329
110	28 33	350	53 47	034	55 03	127	45 07	193	22 15	230	56 14	256	18 22	327
112	28 29	348	54 01	031	55 23	125	45 02	191	21 56	228	55 50	254	18 09	325
114	28 23	346	54 13	029	55 43	124	44 58	190	21 38	226	55 26	252	17 54	323
116	28 17	344	54 25	027	56 04	122	44 54	188	21 20	224	55 03	251	17 38	321
118	28 09	342	54 36	024	56 26	120	44 51	186	21 03	222	54 39	249	17 22	319
	Suhail		◆SPICA		RIGIL KENT.		Peacock		◆FOMALHAUT		ACHERNAR		◆SIRIUS	
120	54 45	022	12 43	084	56 48	119	44 48	185	20 46	221	54 16	247	28 01	340
122	54 54	019	13 07	082	57 10	117	44 47	183	20 30	219	53 53	245	27 52	337
124	55 02	017	13 32	080	57 32	115	44 46	182	20 15	217	53 31	244	27 42	335
126	55 08	014	13 56	078	57 55	114	44 45	180	20 01	215	53 09	242	27 31	333
128	55 14	011	14 21	076	58 18	112	44 45	179	19 47	213	52 47	240	27 19	331
130	55 18	009	14 45	074	58 41	110	44 46	177	19 33	211	52 25	239	27 07	329
132	55 21	006	15 09	072	59 04	109	44 48	176	19 21	210	52 04	237	26 54	327
134	55 24	004	15 32	070	59 28	107	44 50	174	19 09	208	51 43	235	26 40	324
136	55 25	001	15 55	068	59 52	105	44 53	173	18 57	206	51 23	234	26 25	322
138	55 25	359	16 18	066	60 16	103	44 57	171	18 47	204	51 03	232	26 09	320
140	55 23	356	16 41	064	60 41	101	45 01	169	18 37	202	50 44	230	25 53	318
142	55 21	354	17 03	062	61 05	100	45 06	168	18 28	200	50 25	229	25 36	316
144	55 18	351	17 25	060	61 30	098	45 11	166	18 20	199	50 06	227	25 18	314
146	55 13	348	17 46	058	61 55	096	45 17	165	18 12	197	49 48	226	25 00	312
148	55 08	346	18 07	056	62 20	094	45 24	163	18 05	195	49 31	224	24 41	310
	SPICA		◆ANTARES		Peacock		FOMALHAUT		◆ACHERNAR		CANOPUS		◆Suhail	
150	18 28	054	24 19	103	45 32	162	17 59	193	49 14	222	58 26	290	55 01	343
152	18 47	052	24 43	101	45 40	160	17 54	191	48 57	221	58 02	288	54 54	341
154	19 07	050	25 08	099	45 49	159	17 50	189	48 41	219	57 39	286	54 45	338
156	19 25	048	25 32	097	45 58	157	17 46	188	48 26	218	57 14	284	54 35	336
158	19 44	046	25 57	095	46 08	155	17 43	186	48 11	216	56 50	282	54 24	333
160	20 01	043	26 22	093	46 19	154	17 41	184	47 56	214	56 26	280	54 13	331
162	20 18	041	26 47	091	46 30	152	17 39	182	47 42	213	56 01	278	54 00	328
164	20 34	039	27 12	089	46 42	151	17 39	180	47 29	211	55 36	276	53 47	326
166	20 50	037	27 37	087	46 55	149	17 39	178	47 17	210	55 11	274	53 32	324
168	21 04	035	28 02	085	47 08	148	17 40	177	47 05	208	54 46	272	53 17	321
170	21 18	033	28 26	083	47 21	146	17 42	175	46 53	207	54 21	270	53 01	319
172	21 32	031	28 51	081	47 36	144	17 45	173	46 42	205	53 56	268	52 44	316
174	21 44	029	29 16	079	47 50	143	17 48	171	46 32	203	53 32	266	52 27	314
176	21 56	027	29 40	077	48 06	141	17 52	169	46 22	202	53 07	264	52 08	312
178	22 07	025	30 05	075	48 22	140	17 57	168	46 13	200	52 42	262	51 50	309

LHA	Hc	Zn	Hc	Zn	Hc	Zn	Hc	Zn	Hc	Zn	Hc	Zn	Hc	Zn
♈	SPICA		◆ANTARES		Nunki		FOMALHAUT		◆ACHERNAR		CANOPUS		◆Suhail	
180	22 17	023	30 29	073	22 54	109	18 03	166	46 05	199	52 17	260	51 30	307
182	22 26	020	30 52	071	23 18	107	18 10	164	45 57	197	51 53	258	51 10	305
184	22 34	018	31 16	069	23 42	105	18 17	162	45 50	196	51 28	257	50 49	303
186	22 41	016	31 39	067	24 06	103	18 25	160	45 44	194	51 04	255	50 28	300
188	22 48	014	32 02	065	24 30	101	18 34	158	45 38	193	50 40	253	50 06	298
190	22 54	012	32 24	063	24 55	099	18 43	157	45 33	191	50 17	251	49 44	296
192	22 58	010	32 46	061	25 20	098	18 54	155	45 29	190	49 53	249	49 21	294
194	23 02	008	33 08	059	25 44	096	19 05	153	45 25	188	49 30	248	48 58	292
196	23 05	006	33 29	057	26 09	094	19 16	151	45 22	186	49 07	246	48 35	290
198	23 07	003	33 50	055	26 34	092	19 29	149	45 19	185	48 44	244	48 11	288
200	23 08	001	34 10	053	26 59	090	19 42	147	45 17	183	48 22	242	47 47	285
202	23 08	359	34 29	051	27 24	088	19 56	146	45 16	182	48 00	241	47 23	283
204	23 07	357	34 48	048	27 49	086	20 10	144	45 16	180	47 39	239	46 59	281
206	23 05	355	35 07	046	28 14	084	20 25	142	45 16	179	47 18	237	46 34	279
208	23 03	353	35 24	044	28 39	082	20 41	140	45 17	177	46 57	235	46 10	277
	◆ANTARES		Nunki		FOMALHAUT		◆ACHERNAR		CANOPUS		◆Suhail		SPICA	
210	35 41	042	29 03	080	20 57	138	45 18	176	46 37	234	45 45	275	22 59	351
212	35 58	040	29 28	078	21 14	136	45 21	174	46 17	232	45 20	273	22 55	349
214	36 13	038	29 52	076	21 31	135	45 23	173	45 57	230	44 55	271	22 49	346
216	36 28	035	30 16	074	21 50	133	45 27	171	45 38	229	44 30	269	22 43	344
218	36 42	033	30 40	072	22 08	131	45 31	170	45 20	227	44 05	267	22 36	342
220	36 55	031	31 03	070	22 27	129	45 36	168	45 02	225	43 40	265	22 28	340
222	37 08	029	31 27	068	22 47	127	45 42	166	44 44	224	43 15	264	22 19	338
224	37 19	026	31 50	066	23 07	125	45 48	165	44 27	222	42 51	262	22 09	336
226	37 30	024	32 12	064	23 28	124	45 55	163	44 11	220	42 26	260	21 58	334
228	37 40	022	32 34	062	23 49	122	46 02	162	43 55	219	42 02	258	21 47	332
230	37 49	020	32 56	059	24 10	120	46 10	160	43 40	217	41 37	256	21 35	329
232	37 57	017	33 17	057	24 32	118	46 19	159	43 25	215	41 13	254	21 21	327
234	38 03	015	33 38	055	24 54	116	46 28	157	43 11	214	40 49	252	21 08	325
236	38 10	013	33 58	053	25 17	114	46 38	156	42 57	212	40 26	250	20 53	323
238	38 15	011	34 18	051	25 40	112	46 49	154	42 44	211	40 02	249	20 38	321
	ANTARES		◆Nunki		FOMALHAUT		◆ACHERNAR		CANOPUS		◆ACRUX		SPICA	
240	38 19	008	34 37	049	26 03	110	47 00	152	42 32	209	68 09	282	20 22	319
242	38 22	006	34 56	047	26 26	109	47 12	151	42 20	207	67 45	280	20 05	317
244	38 24	004	35 13	045	26 50	107	47 24	149	42 09	206	67 20	278	19 48	315
246	38 25	001	35 31	042	27 14	105	47 37	148	41 59	204	66 55	276	19 30	313
248	38 25	359	35 47	040	27 38	103	47 51	146	41 49	202	66 31	274	19 11	311
250	38 24	357	36 03	038	28 03	101	48 05	145	41 40	201	66 06	272	18 52	309
252	38 22	355	36 18	036	28 27	099	48 20	143	41 31	199	65 41	270	18 32	307
254	38 20	352	36 32	034	28 52	097	48 35	141	41 23	198	65 16	268	18 12	305
256	38 16	350	36 45	031	29 17	095	48 51	140	41 16	196	64 51	266	17 51	303
258	38 11	348	36 58	029	29 42	093	49 07	138	41 10	194	64 26	264	17 30	301
260	38 05	345	37 10	027	30 07	091	49 24	137	41 04	193	64 01	262	17 08	299
262	37 59	343	37 21	025	30 32	089	49 41	135	40 58	191	63 37	260	16 46	297
264	37 51	341	37 30	022	30 57	087	49 59	133	40 54	190	63 12	259	16 24	295
266	37 42	339	37 40	020	31 21	085	50 18	132	40 50	188	62 48	257	16 01	293
268	37 33	336	37 48	018	31 46	083	50 36	130	40 47	186	62 23	255	15 38	291
	◆Nunki		FOMALHAUT		◆ACHERNAR		CANOPUS		◆ACRUX		RIGIL KENT.		ANTARES	
270	37 55	016	32 11	081	50 56	129	40 45	185	61 59	253	66 46	288	37 22	334
272	38 01	013	32 36	079	51 15	127	40 43	183	61 36	252	66 22	286	37 11	332
274	38 06	011	33 00	077	51 36	125	40 42	182	61 12	250	65 58	284	36 59	330
276	38 11	009	33 24	075	51 56	124	40 42	180	60 49	248	65 34	282	36 46	328
278	38 14	007	33 48	073	52 17	122	40 42	178	60 26	247	65 09	280	36 32	325
280	38 16	004	34 12	071	52 38	120	40 43	177	60 03	245	64 44	277	36 17	323
282	38 18	002	34 36	069	53 00	119	40 45	175	59 41	243	64 20	275	36 02	321
284	38 18	000	34 59	067	53 22	117	40 47	174	59 19	242	63 55	273	35 46	319
286	38 18	357	35 22	065	53 45	115	40 50	172	58 57	240	63 30	271	35 29	317
288	38 16	355	35 44	063	54 07	113	40 54	170	58 35	238	63 05	269	35 12	314
290	38 13	353	36 06	061	54 30	112	40 59	169	58 14	237	62 40	268	34 53	312
292	38 10	351	36 28	059	54 54	110	41 04	167	57 54	235	62 15	266	34 35	310
294	38 05	348	36 49	057	55 17	108	41 10	166	57 33	234	61 50	264	34 15	308
296	38 00	346	37 09	055	55 41	106	41 16	164	57 13	232	61 25	262	33 55	306
298	37 53	344	37 29	052	56 05	105	41 24	162	56 54	231	61 01	260	33 35	304
	◆FOMALHAUT		ACHERNAR		◆CANOPUS		ACRUX		RIGIL KENT.		◆ANTARES		Nunki	
300	37 49	050	56 29	103	41 31	161	56 35	229	60 36	258	33 14	302	37 46	341
302	38 08	048	56 54	101	41 40	159	56 16	227	60 12	256	32 52	299	37 37	339
304	38 26	046	57 18	099	41 49	158	55 58	226	59 48	255	32 30	297	37 28	337
306	38 44	044	57 43	097	41 59	156	55 40	224	59 24	253	32 08	295	37 18	335
308	39 00	042	58 08	095	42 10	154	55 23	223	59 00	251	31 45	293	37 07	332
310	39 17	039	58 33	094	42 21	153	55 06	221	58 37	249	31 22	291	36 55	330
312	39 32	037	58 58	092	42 32	151	54 50	220	58 13	248	30 59	289	36 42	328
314	39 47	035	59 23	090	42 45	149	54 34	218	57 50	246	30 35	287	36 28	326
316	40 01	033	59 47	088	42 58	148	54 19	217	57 28	244	30 11	285	36 14	324
318	40 14	030	60 12	086	43 11	146	54 05	215	57 05	243	29 47	283	35 59	321
320	40 26	028	60 37	084	43 26	145	53 50	214	56 43	241	29 23	281	35 43	319
322	40 37	026	61 02	082	43 40	143	53 37	212	56 22	239	28 58	279	35 26	317
324	40 47	024	61 27	080	43 56	141	53 24	211	56 01	238	28 33	277	35 09	315
326	40 57	021	61 51	077	44 12	140	53 11	209	55 40	236	28 08	275	34 51	313
328	41 05	019	62 15	075	44 28	138	52 59	208	55 19	235	27 44	273	34 32	311
	◆FOMALHAUT		ACHERNAR		◆CANOPUS		ACRUX		RIGIL KENT.		◆ANTARES		Nunki	
330	41 13	017	62 39	073	44 45	136	52 48	207	54 59	233	27 19	271	34 13	308
332	41 20	014	63 03	071	45 03	135	52 37	205	54 39	231	26 54	269	33 53	306
334	41 25	012	63 26	069	45 21	133	52 27	204	54 20	230	26 29	267	33 33	304
336	41 30	010	63 49	066	45 39	131	52 17	202	54 01	228	26 04	265	33 12	302
338	41 34	007	64 12	064	45 58	130	52 08	201	53 43	227	25 39	263	32 50	300
340	41 37	005	64 34	062	46 17	128	51 59	199	53 25	225	25 14	261	32 29	298
342	41 38	003	64 56	059	46 37	126	51 51	198	53 08	223	24 50	260	32 06	296
344	41 39	000	65 17	057	46 58	125	51 44	196	52 51	222	24 25	258	31 44	294
346	41 39	358	65 38	054	47 18	123	51 37	195	52 34	220	24 01	256	31 21	292
348	41 37	356	65 58	052	47 40	121	51 31	193	52 18	219	23 37	254	30 57	290
350	41 35	353	66 17	049	48 01	119	51 26	192	52 03	217	23 13	252	30 34	288
352	41 32	351	66 36	047	48 23	118	51 21	191	51 48	216	22 49	250	30 10	286
354	41 27	349	66 54	044	48 45	116	51 17	189	51 34	214	22 26	248	29 46	284
356	41 22	346	67 11	042	49 08	114	51 13	188	51 20	213	22 03	246	29 21	282
358	41 16	344	67 27	039	49 31	112	51 10	186	51 07	211	21 41	244	28 57	280

 LAT 79°S

LHA	Hc	Zn	Hc	Zn	Hc	Zn	Hc	Zn	Hc	Zn	Hc	Zn	Hc	Zn
♈	◆Diphda		◆CANOPUS		ACRUX		RIGIL KENT.		◆ANTARES		Peacock		FOMALHAUT	
0	28 48	012	50 15	110	52 07	185	51 46	210	21 46	243	62 00	290	40 11	342
2	28 52	010	50 36	108	52 06	183	51 35	209	21 26	241	61 38	288	40 04	340
4	28 56	007	50 58	106	52 05	182	51 24	207	21 06	239	61 16	285	39 55	338
6	28 58	005	51 20	104	52 04	180	51 14	206	20 46	237	60 54	283	39 46	335
8	29 00	003	51 43	102	52 04	179	51 04	204	20 27	235	60 31	281	39 36	333
10	29 01	001	52 05	100	52 05	177	50 55	203	20 09	233	60 09	279	39 26	331
12	29 01	359	52 28	098	52 06	176	50 47	201	19 51	231	59 46	277	39 14	329
14	29 00	357	52 50	097	52 08	175	50 39	200	19 33	230	59 23	275	39 02	326
16	28 58	354	53 13	095	52 11	173	50 31	198	19 16	228	59 01	273	38 49	324
18	28 55	352	53 36	093	52 14	172	50 25	197	18 59	226	58 38	271	38 35	322
20	28 52	350	53 59	091	52 18	170	50 18	195	18 43	224	58 15	269	38 20	320
22	28 47	348	54 22	089	52 22	169	50 13	193	18 27	222	57 52	267	38 05	318
24	28 42	346	54 45	087	52 27	167	50 08	192	18 12	220	57 29	265	37 50	315
26	28 36	344	55 08	085	52 32	166	50 03	190	17 58	218	57 06	263	37 33	313
28	28 29	341	55 30	083	52 38	164	49 59	189	17 44	217	56 44	261	37 16	311
	RIGEL		◆SIRIUS		Suhail		◆RIGIL KENT.		Peacock		◆FOMALHAUT		Diphda	
30	15 22	050	19 56	075	39 22	116	49 56	187	56 21	260	36 59	309	28 21	339
32	15 40	048	20 18	073	39 42	114	49 54	186	55 59	258	36 41	307	28 13	337
34	15 57	046	20 40	071	40 03	112	49 51	184	55 36	256	36 22	305	28 04	335
36	16 13	044	21 02	069	40 25	110	49 50	183	55 14	254	36 03	303	27 53	333
38	16 28	042	21 23	067	40 46	109	49 49	181	54 52	252	35 43	300	27 43	331
40	16 43	040	21 44	065	41 08	107	49 49	180	54 31	250	35 23	298	27 31	328
42	16 58	038	22 04	063	41 30	105	49 49	178	54 09	249	35 03	296	27 19	326
44	17 12	036	22 24	061	41 52	103	49 50	177	53 48	247	34 42	294	27 06	324
46	17 25	034	22 44	059	42 15	101	49 52	175	53 27	245	34 21	292	26 52	322
48	17 37	032	23 03	056	42 37	099	49 54	174	53 06	243	34 00	290	26 38	320
50	17 49	030	23 22	054	43 00	097	49 57	172	52 46	242	33 38	288	26 23	318
52	18 00	028	23 41	052	43 23	095	50 00	171	52 26	240	33 16	286	26 07	316
54	18 10	026	23 58	050	43 46	093	50 04	169	52 06	238	32 54	284	25 51	314
56	18 20	024	24 16	048	44 08	092	50 09	168	51 47	237	32 32	282	25 34	312
58	18 29	022	24 33	046	44 31	090	50 14	166	51 28	235	32 09	280	25 16	310
	RIGEL		◆SIRIUS		Suhail		◆RIGIL KENT.		Peacock		◆FOMALHAUT		ACHERNAR	
60	18 37	019	24 49	044	44 54	088	50 20	165	51 10	233	31 47	278	65 27	311
62	18 44	017	25 04	042	45 17	086	50 26	163	50 52	232	31 24	276	65 09	308
64	18 50	015	25 19	040	45 40	084	50 33	162	50 34	230	31 01	274	64 51	306
66	18 56	013	25 34	038	46 03	082	50 40	160	50 17	228	30 38	272	64 32	303
68	19 01	011	25 47	036	46 25	080	50 48	159	50 00	227	30 16	270	64 13	301
70	19 05	009	26 00	034	46 48	078	50 57	157	49 43	225	29 53	268	63 53	299
72	19 08	007	26 13	031	47 10	075	51 06	156	49 27	223	29 30	266	63 32	296
74	19 10	005	26 24	029	47 32	073	51 16	154	49 12	222	29 07	264	63 12	294
76	19 12	003	26 35	027	47 54	071	51 26	152	48 57	220	28 44	262	62 51	292
78	19 12	001	26 45	025	48 15	069	51 37	151	48 43	218	28 22	260	62 29	290
80	19 12	359	26 55	023	48 36	067	51 48	149	48 29	217	27 59	258	62 08	287
82	19 11	356	27 03	021	48 57	065	52 00	148	48 15	215	27 37	256	61 46	285
84	19 09	354	27 11	019	49 18	063	52 13	146	48 02	214	27 15	255	61 23	283
86	19 07	352	27 18	016	49 38	061	52 26	145	47 50	212	26 53	253	61 01	281
88	19 03	350	27 24	014	49 58	058	52 39	143	47 38	210	26 31	251	60 38	279
	◆SIRIUS		Suhail		◆RIGIL KENT.		Peacock		◆FOMALHAUT		ACHERNAR		RIGEL	
90	27 29	012	50 17	056	52 53	142	47 27	209	26 09	249	60 16	277	18 59	348
92	27 33	010	50 36	054	53 08	140	47 16	207	25 48	247	59 53	275	18 54	346
94	27 37	008	50 54	052	53 23	138	47 06	206	25 27	245	59 30	273	18 48	344
96	27 40	006	51 12	049	53 38	137	46 56	204	25 07	243	59 07	271	18 41	342
98	27 41	003	51 29	047	53 54	135	46 47	202	24 47	241	58 44	269	18 33	340
100	27 42	001	51 45	045	54 10	134	46 39	201	24 27	239	58 21	267	18 25	338
102	27 42	359	52 01	043	54 27	132	46 31	199	24 07	238	57 59	265	18 16	336
104	27 42	357	52 16	040	54 44	131	46 24	198	23 48	236	57 36	263	18 06	333
106	27 40	355	52 31	038	55 02	129	46 17	196	23 29	234	57 13	261	17 55	331
108	27 38	353	52 44	036	55 20	127	46 11	194	23 11	232	56 51	259	17 44	329
110	27 34	351	52 57	033	55 38	126	46 06	193	22 53	230	56 28	257	17 32	327
112	27 30	348	53 09	031	55 57	124	46 01	191	22 36	228	56 06	256	17 19	325
114	27 25	346	53 21	028	56 16	122	45 57	190	22 19	226	55 44	254	17 06	323
116	27 19	344	53 31	026	56 36	121	45 53	188	22 03	224	55 22	252	16 52	321
118	27 12	342	53 41	023	56 56	119	45 51	187	21 47	223	55 00	250	16 37	319
	Suhail		◆SPICA		RIGIL KENT.		Peacock		◆FOMALHAUT		ACHERNAR		◆SIRIUS	
120	53 49	021	12 36	084	57 16	117	45 48	185	21 32	221	54 39	248	27 05	340
122	53 57	019	12 59	082	57 36	116	45 47	183	21 17	219	54 18	247	26 57	338
124	54 04	016	13 21	080	57 57	114	45 46	182	21 03	217	53 57	245	26 47	335
126	54 10	014	13 44	078	58 18	112	45 45	180	20 50	215	53 36	243	26 38	333
128	54 15	011	14 06	076	58 39	110	45 45	179	20 37	213	53 16	241	26 27	331
130	54 19	009	14 28	074	59 01	109	45 46	177	20 24	212	52 56	240	26 16	329
132	54 22	006	14 50	072	59 23	107	45 48	176	20 13	210	52 36	238	26 03	327
134	54 24	004	15 11	070	59 45	105	45 50	174	20 02	208	52 17	236	25 51	325
136	54 25	001	15 33	068	60 07	103	45 52	172	19 51	206	51 58	235	25 37	323
138	54 25	359	15 54	066	60 29	102	45 56	171	19 42	204	51 40	233	25 23	321
140	54 24	356	16 14	064	60 52	100	46 00	169	19 33	202	51 22	231	25 08	319
142	54 22	354	16 35	062	61 14	098	46 04	168	19 24	201	51 04	230	24 52	316
144	54 19	351	16 55	060	61 37	096	46 10	166	19 17	199	50 47	228	24 36	314
146	54 15	349	17 14	058	62 00	094	46 15	165	19 10	197	50 30	226	24 20	312
148	54 10	346	17 33	056	62 23	092	46 22	163	19 03	195	50 14	225	24 03	310
	SPICA		◆ANTARES		Peacock		FOMALHAUT		◆ACHERNAR		CANOPUS		◆Suhail	
150	17 52	053	24 32	102	46 29	161	18 58	193	49 58	223	58 04	292	54 04	344
152	18 10	051	24 54	101	46 36	160	18 53	191	49 42	221	57 43	290	53 57	341
154	18 28	049	25 17	099	46 45	158	18 49	189	49 27	220	57 21	288	53 49	339
156	18 45	047	25 39	097	46 53	157	18 45	188	49 13	218	56 59	285	53 40	336
158	19 02	045	26 02	095	47 03	155	18 43	186	48 59	217	56 37	283	53 31	334
160	19 17	043	26 25	093	47 13	153	18 41	184	48 46	215	56 15	281	53 20	332
162	19 33	041	26 48	091	47 23	152	18 39	182	48 33	213	55 52	279	53 09	329
164	19 48	039	27 11	089	47 34	150	18 39	180	48 20	212	55 29	277	52 57	327
166	20 02	037	27 34	087	47 46	149	18 39	178	48 09	210	55 07	275	52 44	324
168	20 15	035	27 57	085	47 58	147	18 40	177	47 57	209	54 44	273	52 30	322
170	20 28	033	28 19	083	48 11	145	18 42	175	47 47	207	54 21	271	52 16	320
172	20 40	031	28 42	081	48 24	144	18 44	173	47 37	205	53 58	269	52 00	317
174	20 51	029	29 05	079	48 38	142	18 47	171	47 27	204	53 35	267	51 45	315
176	21 02	027	29 27	077	48 52	141	18 51	169	47 18	202	53 12	265	51 28	313
178	21 12	025	29 49	075	49 07	139	18 56	167	47 10	201	52 50	263	51 11	310

LHA	Hc	Zn	Hc	Zn	Hc	Zn	Hc	Zn	Hc	Zn	Hc	Zn	Hc	Zn
♈	SPICA		◆ANTARES		Nunki		FOMALHAUT		◆ACHERNAR		CANOPUS		◆Suhail	
180	21 21	022	30 11	073	23 14	109	19 01	166	47 02	199	52 27	262	50 53	308
182	21 29	020	30 33	071	23 35	107	19 07	164	46 55	198	52 04	260	50 35	306
184	21 37	018	30 54	069	23 57	105	19 14	162	46 48	196	51 42	258	50 16	304
186	21 44	016	31 16	067	24 20	103	19 21	160	46 42	194	51 20	256	49 57	302
188	21 50	014	31 36	065	24 42	101	19 30	158	46 37	193	50 57	254	49 37	299
190	21 55	012	31 57	063	25 05	099	19 38	156	46 32	191	50 35	252	49 17	297
192	21 59	010	32 17	061	25 27	097	19 48	155	46 28	190	50 14	250	48 56	295
194	22 03	008	32 37	058	25 50	095	19 58	153	46 24	188	49 52	249	48 35	293
196	22 05	006	32 56	056	26 13	093	20 09	151	46 21	187	49 31	247	48 14	291
198	22 07	003	33 15	054	26 36	091	20 20	149	46 19	185	49 10	245	47 53	289
200	22 08	001	33 33	052	26 59	089	20 32	147	46 17	183	48 50	243	47 31	287
202	22 08	359	33 51	050	27 22	087	20 45	145	46 16	182	48 29	242	47 09	284
204	22 07	357	34 08	048	27 44	085	20 58	144	46 16	180	48 09	240	46 47	282
206	22 06	355	34 25	046	28 07	083	21 12	142	46 16	179	47 50	238	46 24	280
208	22 03	353	34 41	044	28 30	081	21 27	140	46 17	177	47 31	236	46 01	278
	◆ANTARES		Nunki		FOMALHAUT		◆ACHERNAR		CANOPUS		◆Suhail		SPICA	
210	34 57	041	28 52	079	21 42	138	46 18	176	47 12	235	45 39	276	22 00	351
212	35 11	039	29 15	077	21 57	136	46 20	174	46 53	233	45 16	274	21 56	349
214	35 26	037	29 37	075	22 13	134	46 23	172	46 35	231	44 53	272	21 51	346
216	35 39	035	29 59	073	22 30	132	46 26	171	46 18	229	44 30	270	21 45	344
218	35 52	033	30 21	071	22 47	131	46 30	169	46 00	228	44 07	268	21 39	342
220	36 04	031	30 42	069	23 05	129	46 35	168	45 44	226	43 44	266	21 31	340
222	36 15	028	31 04	067	23 23	127	46 40	166	45 28	224	43 22	265	21 23	338
224	36 25	026	31 25	065	23 42	125	46 46	165	45 12	223	42 59	263	21 14	336
226	36 35	024	31 45	063	24 00	123	46 52	163	44 56	221	42 36	261	21 04	334
228	36 44	022	32 06	061	24 20	121	46 59	161	44 42	219	42 14	259	20 54	332
230	36 52	019	32 25	059	24 40	119	47 07	160	44 28	218	41 51	257	20 43	330
232	36 59	017	32 45	057	25 00	118	47 15	158	44 14	216	41 29	255	20 31	328
234	37 06	015	33 04	055	25 20	116	47 23	157	44 01	214	41 07	253	20 18	326
236	37 11	013	33 22	053	25 41	114	47 33	155	43 48	213	40 45	251	20 05	323
238	37 16	010	33 40	050	26 02	112	47 43	154	43 36	211	40 24	249	19 51	321
	ANTARES		◆Nunki		FOMALHAUT		◆ACHERNAR		CANOPUS		◆ACRUX		SPICA	
240	37 19	008	33 57	048	26 24	110	47 53	152	43 24	209	67 55	285	19 36	319
242	37 22	006	34 14	046	26 45	108	48 04	150	43 13	208	67 33	282	19 21	317
244	37 24	004	34 30	044	27 07	106	48 16	149	43 03	206	67 11	280	19 05	315
246	37 25	001	34 46	042	27 29	104	48 28	147	42 53	204	66 48	278	18 49	313
248	37 25	359	35 01	040	27 51	102	48 41	146	42 44	203	66 25	276	18 32	311
250	37 24	357	35 15	038	28 14	100	48 54	144	42 36	201	66 03	274	18 14	309
252	37 23	355	35 29	035	28 36	098	49 08	142	42 28	199	65 40	272	17 56	307
254	37 20	352	35 42	033	28 59	097	49 22	141	42 20	198	65 17	270	17 38	305
256	37 17	350	35 54	031	29 22	095	49 36	139	42 14	196	64 54	268	17 19	303
258	37 12	348	36 05	029	29 45	093	49 52	138	42 08	195	64 31	266	16 59	301
260	37 07	346	36 16	027	30 08	091	50 07	136	42 02	193	64 08	264	16 39	299
262	37 01	343	36 26	024	30 31	089	50 24	134	41 57	191	63 46	262	16 19	297
264	36 54	341	36 35	022	30 53	087	50 40	133	41 53	190	63 23	261	15 59	295
266	36 46	339	36 43	020	31 16	085	50 57	131	41 50	188	63 00	259	15 38	293
268	36 38	337	36 51	018	31 39	083	51 15	129	41 47	186	62 38	257	15 16	291
	◆Nunki		FOMALHAUT		◆ACHERNAR		CANOPUS		◆ACRUX		RIGIL KENT.		ANTARES	
270	36 57	015	32 02	081	51 33	128	41 44	185	62 16	255	66 26	290	36 28	335
272	37 03	013	32 24	079	51 51	126	41 43	183	61 54	253	66 04	288	36 18	332
274	37 08	011	32 47	077	52 10	124	41 42	182	61 32	252	65 42	286	36 07	330
276	37 11	009	33 09	075	52 29	123	41 42	180	61 10	250	65 20	284	35 55	328
278	37 14	006	33 31	073	52 48	121	41 42	178	60 49	248	64 58	282	35 43	326
280	37 17	004	33 53	071	53 08	119	41 43	177	60 28	246	64 35	280	35 29	324
282	37 18	002	34 14	069	53 28	117	41 45	175	60 07	245	64 13	277	35 15	321
284	37 18	000	34 35	067	53 49	116	41 47	173	59 46	243	63 50	275	35 01	319
286	37 18	357	34 56	064	54 10	114	41 50	172	59 26	241	63 27	273	34 45	317
288	37 16	355	35 16	062	54 31	112	41 53	170	59 06	240	63 04	271	34 30	315
290	37 14	353	35 37	060	54 52	110	41 58	169	58 46	238	62 41	269	34 13	313
292	37 11	351	35 56	058	55 14	109	42 02	167	58 27	237	62 19	268	33 56	311
294	37 06	348	36 15	056	55 35	107	42 08	165	58 08	235	61 56	266	33 38	308
296	37 01	346	36 34	054	55 57	105	42 14	164	57 50	233	61 33	264	33 20	306
298	36 55	344	36 52	052	56 20	103	42 21	162	57 32	232	61 10	262	33 01	304
	◆FOMALHAUT		ACHERNAR		◆CANOPUS		ACRUX		RIGIL KENT.		◆ANTARES		Nunki	
300	37 10	050	56 42	101	42 28	160	57 14	230	60 48	260	32 42	302	36 49	342
302	37 27	048	57 04	100	42 36	159	56 56	229	60 25	258	32 23	300	36 41	339
304	37 44	045	57 27	098	42 45	157	56 39	227	60 03	256	32 03	298	36 33	337
306	38 00	043	57 50	096	42 54	156	56 23	225	59 41	255	31 42	296	36 23	335
308	38 15	041	58 13	094	43 04	154	56 07	224	59 19	253	31 21	294	36 13	333
310	38 30	039	58 36	092	43 14	152	55 51	222	58 57	251	31 00	292	36 03	331
312	38 44	037	58 58	090	43 25	151	55 36	221	58 35	249	30 39	290	35 51	328
314	38 57	034	59 21	088	43 36	149	55 21	219	58 14	247	30 17	288	35 39	326
316	39 10	032	59 44	086	43 48	147	55 07	218	57 53	246	29 55	286	35 25	324
318	39 22	030	60 07	084	44 01	146	54 53	216	57 32	244	29 33	284	35 12	322
320	39 33	028	60 30	082	44 14	144	54 40	215	57 12	242	29 11	282	34 57	320
322	39 43	025	60 52	080	44 28	142	54 27	213	56 52	241	28 48	280	34 42	318
324	39 52	023	61 15	078	44 42	141	54 15	212	56 32	239	28 26	278	34 26	315
326	40 01	021	61 37	076	44 57	139	54 03	210	56 13	237	28 03	276	34 10	313
328	40 09	019	61 59	073	45 12	137	53 52	209	55 53	236	27 40	274	33 53	311
	◆FOMALHAUT		ACHERNAR		◆CANOPUS		ACRUX		RIGIL KENT.		◆ANTARES		Nunki	
330	40 16	016	62 21	071	45 28	136	53 41	207	55 35	234	27 17	272	33 35	309
332	40 22	014	62 43	069	45 44	134	53 31	206	55 16	232	26 54	270	33 17	307
334	40 27	012	63 04	067	46 01	132	53 21	204	54 58	231	26 31	268	32 59	305
336	40 31	010	63 25	065	46 18	131	53 12	203	54 41	229	26 09	266	32 40	303
338	40 34	007	63 45	062	46 36	129	53 04	201	54 24	228	25 46	264	32 20	301
340	40 37	005	64 05	060	46 54	127	52 56	200	54 07	226	25 23	262	32 00	298
342	40 38	003	64 25	058	47 12	125	52 48	198	53 51	224	25 00	260	31 40	296
344	40 39	000	64 44	055	47 31	124	52 42	197	53 35	223	24 38	258	31 19	294
346	40 39	358	65 02	053	47 51	122	52 35	195	53 20	221	24 16	256	30 58	292
348	40 37	356	65 20	050	48 10	120	52 30	194	53 05	220	23 53	254	30 37	290
350	40 35	354	65 37	048	48 30	118	52 24	192	52 51	218	23 32	252	30 15	288
352	40 32	351	65 54	045	48 50	117	52 20	191	52 37	217	23 10	250	29 53	286
354	40 28	349	66 10	043	49 11	115	52 16	189	52 23	215	22 48	248	29 31	284
356	40 23	347	66 25	040	49 32	113	52 13	188	52 10	213	22 27	247	29 09	282
358	40 18	344	66 39	037	49 53	111	52 10	186	51 58	212	22 06	245	28 47	280

LAT 80°S

LHA	Hc	Zn	Hc	Zn	Hc	Zn	Hc	Zn	Hc	Zn	Hc	Zn	Hc	Zn
♈	ACHERNAR		◆SIRIUS		Suhail		ACRUX		◆RIGIL KENT.		Peacock		◆FOMALHAUT	
0	66 03	033	14 31	104	35 47	142	53 07	185	52 38	211	61 38	292	39 14	342
2	66 14	031	14 52	102	36 00	141	53 06	183	52 27	209	61 19	289	39 07	340
4	66 24	028	15 12	100	36 13	139	53 05	182	52 17	208	60 59	287	39 00	338
6	66 34	025	15 33	098	36 27	137	53 04	180	52 08	206	60 39	285	38 52	336
8	66 42	023	15 53	096	36 41	135	53 04	179	51 59	205	60 19	283	38 43	333
10	66 50	020	16 14	094	36 56	133	53 05	177	51 51	203	59 58	281	38 33	331
12	66 56	017	16 35	092	37 12	132	53 06	176	51 43	202	59 38	279	38 23	329
14	67 02	014	16 56	090	37 28	130	53 08	174	51 35	200	59 17	277	38 12	327
16	67 07	012	17 17	088	37 44	128	53 10	173	51 28	198	58 57	275	38 00	325
18	67 11	009	17 37	086	38 00	126	53 13	171	51 22	197	58 36	273	37 47	322
20	67 13	006	17 58	084	38 17	124	53 17	170	51 16	195	58 15	271	37 34	320
22	67 15	003	18 19	082	38 35	123	53 21	168	51 11	194	57 54	269	37 21	318
24	67 16	001	18 39	080	38 52	121	53 25	167	51 06	192	57 33	267	37 07	316
26	67 15	358	19 00	078	39 11	119	53 30	165	51 02	191	57 12	265	36 52	314
28	67 14	355	19 20	076	39 29	117	53 36	164	50 59	189	56 52	263	36 37	312
	◆SIRIUS		Suhail		ACRUX		◆RIGIL KENT.		Peacock		◆FOMALHAUT		ACHERNAR	
30	19 40	074	39 48	115	53 42	162	50 56	188	56 31	261	36 21	309	67 12	352
32	20 00	072	40 07	113	53 48	161	50 53	186	56 11	259	36 04	307	67 08	349
34	20 20	070	40 26	112	53 55	159	50 51	184	55 50	257	35 48	305	67 04	347
36	20 40	068	40 45	110	54 03	158	50 50	183	55 30	255	35 30	303	66 59	344
38	20 59	066	41 05	108	54 11	156	50 49	181	55 10	254	35 13	301	66 52	341
40	21 18	064	41 25	106	54 20	155	50 49	180	54 50	252	34 55	299	66 45	338
42	21 36	062	41 45	104	54 29	153	50 49	178	54 30	250	34 36	297	66 37	336
44	21 55	060	42 05	102	54 39	152	50 50	177	54 11	248	34 18	295	66 28	333
46	22 13	058	42 26	100	54 49	150	50 52	175	53 52	246	33 58	293	66 18	330
48	22 30	056	42 46	098	54 59	149	50 54	174	53 33	245	33 39	291	66 07	328
50	22 47	054	43 07	096	55 11	147	50 56	172	53 14	243	33 19	289	65 56	325
52	23 04	052	43 28	094	55 22	145	50 59	171	52 56	241	33 00	287	65 43	322
54	23 20	050	43 49	093	55 34	144	51 03	169	52 37	239	32 40	285	65 30	320
56	23 36	048	44 09	091	55 47	142	51 07	167	52 20	238	32 19	283	65 17	317
58	23 51	046	44 30	089	56 00	141	51 12	166	52 02	236	31 59	281	65 02	315
	◆SIRIUS		Suhail		ACRUX		◆RIGIL KENT.		Peacock		◆FOMALHAUT		ACHERNAR	
60	24 06	044	44 51	087	56 13	139	51 17	164	51 45	234	31 38	279	64 47	312
62	24 20	042	45 12	085	56 27	138	51 23	163	51 28	232	31 18	277	64 31	310
64	24 33	040	45 33	083	56 41	136	51 30	161	51 12	231	30 57	275	64 15	307
66	24 46	037	45 53	081	56 56	134	51 37	160	50 56	229	30 36	273	63 58	305
68	24 59	035	46 14	079	57 11	133	51 44	158	50 41	227	30 15	271	63 41	303
70	25 10	033	46 34	076	57 26	131	51 52	157	50 25	226	29 54	269	63 23	300
72	25 22	031	46 54	074	57 42	130	52 01	155	50 11	224	29 34	267	63 05	298
74	25 32	029	47 14	072	57 59	128	52 10	153	49 57	222	29 13	265	62 46	296
76	25 42	027	47 34	070	58 15	126	52 19	152	49 43	221	28 52	263	62 28	294
78	25 51	025	47 53	068	58 32	125	52 29	150	49 29	219	28 32	261	62 08	291
80	25 59	023	48 13	066	58 49	123	52 40	149	49 16	217	28 11	259	61 49	289
82	26 07	021	48 31	064	59 07	121	52 51	147	49 04	216	27 51	257	61 29	287
84	26 14	018	48 50	062	59 25	120	53 02	146	48 52	214	27 30	255	61 09	285
86	26 20	016	49 08	060	59 43	118	53 14	144	48 41	212	27 10	253	60 49	283
88	26 26	014	49 26	057	60 02	116	53 27	142	48 30	211	26 51	251	60 28	281
	◆SIRIUS		Suhail		ACRUX		◆RIGIL KENT.		Peacock		◆FOMALHAUT		ACHERNAR	
90	26 30	012	49 43	055	60 21	115	53 40	141	48 19	209	26 31	249	60 08	279
92	26 34	010	50 00	053	60 40	113	53 53	139	48 09	208	26 12	247	59 47	276
94	26 37	008	50 17	051	60 59	111	54 07	138	48 00	206	25 52	245	59 26	274
96	26 40	006	50 32	049	61 19	109	54 21	136	47 51	204	25 34	244	59 05	272
98	26 42	003	50 48	046	61 38	108	54 36	134	47 43	203	25 15	242	58 45	270
100	26 42	001	51 03	044	61 58	106	54 51	133	47 35	201	24 57	240	58 24	268
102	26 42	359	51 17	042	62 18	104	55 07	131	47 28	200	24 39	238	58 03	267
104	26 42	357	51 30	039	62 39	102	55 23	129	47 21	198	24 22	236	57 42	265
106	26 40	355	51 43	037	62 59	100	55 39	128	47 15	196	24 05	234	57 22	263
108	26 38	353	51 55	035	63 20	099	55 56	126	47 09	195	23 48	232	57 01	261
110	26 35	351	52 07	032	63 40	097	56 13	124	47 04	193	23 32	230	56 40	259
112	26 31	348	52 18	030	64 01	095	56 30	123	47 00	191	23 16	228	56 20	257
114	26 27	346	52 28	028	64 22	093	56 48	121	46 56	190	23 00	227	56 00	255
116	26 21	344	52 37	025	64 43	091	57 06	119	46 53	188	22 46	225	55 40	253
118	26 15	342	52 46	023	65 04	089	57 24	118	46 50	187	22 31	223	55 20	252
	Suhail		SPICA		◆RIGIL KENT.		Peacock		◆FOMALHAUT		ACHERNAR		◆SIRIUS	
120	52 53	021	12 29	083	57 43	116	46 48	185	22 17	221	55 00	250	26 09	340
122	53 00	018	12 50	081	58 01	114	46 46	183	22 04	219	54 41	248	26 01	338
124	53 06	016	13 10	079	58 21	112	46 45	182	21 51	217	54 22	246	25 53	336
126	53 12	013	13 31	077	58 40	111	46 45	180	21 39	215	54 03	244	25 44	334
128	53 16	011	13 51	075	59 00	109	46 45	179	21 27	214	53 44	243	25 34	331
130	53 19	008	14 11	073	59 19	107	46 46	177	21 15	212	53 26	241	25 24	329
132	53 22	006	14 31	071	59 39	105	46 47	175	21 05	210	53 08	239	25 13	327
134	53 24	004	14 50	069	60 00	103	46 49	174	20 55	208	52 50	237	25 01	325
136	53 25	001	15 10	067	60 20	102	46 52	172	20 45	206	52 32	236	24 49	323
138	53 25	359	15 29	065	60 40	100	46 55	171	20 36	204	52 15	234	24 36	321
140	53 24	356	15 48	063	61 01	098	46 59	169	20 28	202	51 59	232	24 23	319
142	53 22	354	16 06	061	61 22	096	47 03	167	20 20	201	51 42	231	24 09	317
144	53 19	351	16 24	059	61 42	094	47 08	166	20 13	199	51 26	229	23 54	315
146	53 16	349	16 42	057	62 03	092	47 13	164	20 07	197	51 11	227	23 39	313
148	53 11	347	16 59	055	62 24	090	47 19	163	20 01	195	50 56	226	23 24	311
	SPICA		◆ANTARES		Peacock		FOMALHAUT		◆ACHERNAR		CANOPUS		◆Suhail	
150	17 16	053	24 44	102	47 26	161	19 56	193	50 41	224	57 41	293	53 06	344
152	17 33	051	25 05	100	47 33	159	19 52	191	50 27	222	57 22	291	53 00	342
154	17 49	049	25 25	098	47 40	158	19 48	190	50 13	221	57 02	289	52 53	339
156	18 04	047	25 46	096	47 48	156	19 45	188	50 00	219	56 42	287	52 45	337
158	18 19	045	26 07	094	47 57	155	19 42	186	49 47	217	56 22	285	52 37	335
160	18 34	043	26 28	092	48 06	153	19 41	184	49 35	216	56 02	283	52 27	332
162	18 48	041	26 49	090	48 16	151	19 39	182	49 23	214	55 42	281	52 17	330
164	19 01	039	27 09	088	48 26	150	19 39	181	49 11	212	55 21	279	52 06	327
166	19 14	037	27 30	086	48 37	148	19 39	178	49 00	211	55 01	277	51 55	325
168	19 26	035	27 51	084	48 48	146	19 40	177	48 50	209	54 40	275	51 43	323
170	19 38	033	28 12	082	49 00	145	19 42	175	48 40	208	54 19	273	51 30	320
172	19 48	031	28 32	080	49 12	143	19 44	173	48 31	206	53 58	271	51 16	318
174	19 59	029	28 53	078	49 25	141	19 47	171	48 22	204	53 37	269	51 02	316
176	20 08	026	29 13	076	49 38	140	19 50	169	48 13	203	53 17	267	50 47	314
178	20 17	024	29 33	074	49 52	138	19 55	167	48 06	201	52 56	265	50 32	311

LHA	Hc	Zn	Hc	Zn	Hc	Zn	Hc	Zn	Hc	Zn	Hc	Zn	Hc	Zn
♈	SPICA		◆ANTARES		Peacock		FOMALHAUT		◆ACHERNAR		CANOPUS		◆Suhail	
180	20 26	022	29 53	072	50 06	136	19 59	166	47 58	199	52 35	263	50 16	309
182	20 33	020	30 13	070	50 21	135	20 05	164	47 52	198	52 14	261	49 59	307
184	20 40	018	30 32	068	50 36	133	20 11	162	47 46	196	51 54	259	49 42	305
186	20 46	016	30 52	066	50 51	131	20 18	160	47 40	195	51 33	257	49 25	303
188	20 51	014	31 11	064	51 07	130	20 25	158	47 35	193	51 13	255	49 07	300
190	20 56	012	31 29	062	51 23	128	20 33	156	47 31	191	50 53	253	48 49	298
192	21 00	010	31 47	060	51 40	126	20 42	154	47 27	190	50 33	252	48 31	296
194	21 03	008	32 05	058	51 57	125	20 51	153	47 24	188	50 14	250	48 12	294
196	21 06	006	32 23	056	52 14	123	21 01	151	47 21	187	49 54	248	47 52	292
198	21 07	003	32 40	054	52 32	121	21 12	149	47 19	185	49 35	246	47 33	290
200	21 08	001	32 56	052	52 50	119	21 23	147	47 17	183	49 16	244	47 13	288
202	21 08	359	33 12	050	53 08	118	21 34	145	47 16	182	48 58	243	46 53	285
204	21 07	357	33 28	047	53 27	116	21 47	143	47 16	180	48 39	241	46 33	283
206	21 06	355	33 43	045	53 46	114	21 59	141	47 16	179	48 21	239	46 13	281
208	21 04	353	33 57	043	54 05	112	22 13	140	47 17	177	48 03	237	45 52	279
	◆ANTARES		Nunki		◆FOMALHAUT		ACHERNAR		CANOPUS		◆Suhail		SPICA	
210	34 11	041	28 41	079	22 26	138	47 18	176	47 46	235	45 32	277	21 01	351
212	34 25	039	29 01	077	22 41	136	47 20	174	47 29	234	45 11	275	20 57	349
214	34 38	037	29 22	075	22 55	134	47 22	172	47 13	232	44 50	273	20 53	347
216	34 50	034	29 42	073	23 11	132	47 25	171	46 56	230	44 29	271	20 47	344
218	35 01	032	30 01	071	23 26	130	47 29	169	46 41	228	44 08	269	20 41	342
220	35 12	030	30 21	069	23 42	128	47 33	168	46 25	227	43 48	267	20 35	340
222	35 22	028	30 40	067	23 59	127	47 38	166	46 10	225	43 27	265	20 27	338
224	35 32	026	30 59	065	24 16	125	47 43	164	45 56	223	43 06	264	20 19	336
226	35 40	024	31 18	063	24 33	123	47 49	163	45 42	222	42 45	262	20 11	334
228	35 48	021	31 36	060	24 51	121	47 56	161	45 28	220	42 25	260	20 01	332
230	35 55	019	31 54	058	25 09	119	48 03	160	45 15	218	42 04	258	19 51	330
232	36 02	017	32 12	056	25 27	117	48 10	158	45 02	217	41 44	256	19 40	328
234	36 08	015	32 29	054	25 46	115	48 19	156	44 50	215	41 24	254	19 29	326
236	36 12	013	32 45	052	26 05	113	48 27	155	44 38	213	41 04	252	19 17	324
238	36 17	010	33 02	050	26 24	111	48 36	153	44 27	211	40 44	250	19 04	322
	ANTARES		◆Nunki		FOMALHAUT		◆ACHERNAR		CANOPUS		◆ACRUX		SPICA	
240	36 20	008	33 17	048	26 44	109	48 46	151	44 17	210	67 39	287	18 51	320
242	36 22	006	33 33	046	27 04	108	48 56	150	44 07	208	67 19	285	18 37	317
244	36 24	004	33 47	044	27 23	106	49 07	148	43 57	206	66 59	283	18 23	315
246	36 25	001	34 01	041	27 44	104	49 18	147	43 48	205	66 38	280	18 08	313
248	36 25	359	34 15	039	28 04	102	49 30	145	43 39	203	66 18	278	17 52	311
250	36 25	357	34 28	037	28 24	100	49 42	143	43 32	201	65 57	276	17 36	309
252	36 23	355	34 40	035	28 45	098	49 55	142	43 24	200	65 36	274	17 20	307
254	36 21	353	34 52	033	29 06	096	50 08	140	43 17	198	65 16	272	17 03	305
256	36 18	350	35 02	031	29 26	094	50 22	138	43 11	196	64 55	270	16 46	303
258	36 14	348	35 13	028	29 47	092	50 36	137	43 06	195	64 34	268	16 28	301
260	36 09	346	35 22	026	30 08	090	50 50	135	43 01	193	64 13	266	16 10	299
262	36 04	344	35 31	024	30 29	088	51 05	133	42 56	192	63 52	264	15 52	297
264	35 57	341	35 39	022	30 50	086	51 20	132	42 52	190	63 32	263	15 33	295
266	35 50	339	35 47	020	31 11	084	51 36	130	42 49	188	63 11	261	15 14	293
268	35 42	337	35 53	017	31 31	082	51 52	128	42 46	187	62 51	259	14 55	291
	◆Nunki		FOMALHAUT		◆ACHERNAR		CANOPUS		◆ACRUX		RIGIL KENT.		ANTARES	
270	35 59	015	31 52	080	52 09	127	42 44	185	62 30	257	66 04	293	35 34	335
272	36 04	013	32 12	078	52 26	125	42 43	183	62 10	255	65 44	290	35 25	333
274	36 09	011	32 33	076	52 43	123	42 42	182	61 50	253	65 25	288	35 15	330
276	36 12	009	32 53	074	53 01	121	42 42	180	61 30	252	65 05	286	35 04	328
278	36 15	006	33 13	072	53 19	120	42 42	178	61 10	250	64 45	284	34 53	326
280	36 17	004	33 32	070	53 37	118	42 43	177	60 51	248	64 24	282	34 41	324
282	36 18	002	33 52	068	53 55	116	42 44	175	60 32	246	64 04	280	34 28	322
284	36 18	000	34 11	066	54 14	114	42 46	173	60 13	245	63 43	277	34 15	320
286	36 18	357	34 30	064	54 33	113	42 49	172	59 54	243	63 23	275	34 01	317
288	36 16	355	34 48	062	54 53	111	42 52	170	59 36	241	63 02	273	33 47	315
290	36 14	353	35 07	060	55 12	109	42 56	168	59 17	240	62 41	271	33 32	313
292	36 11	351	35 24	058	55 32	107	43 01	167	59 00	238	62 20	269	33 17	311
294	36 08	349	35 42	055	55 52	105	43 06	165	58 42	236	61 59	267	33 01	309
296	36 03	346	35 59	053	56 12	104	43 11	163	58 25	235	61 39	266	32 44	307
298	35 58	344	36 15	051	56 33	102	43 18	162	58 08	233	61 18	264	32 27	305
	◆FOMALHAUT		ACHERNAR		◆CANOPUS		ACRUX		RIGIL KENT.		◆ANTARES		Nunki	
300	36 31	049	56 53	100	43 25	160	57 52	231	60 57	262	32 10	303	35 52	342
302	36 47	047	57 14	098	43 32	158	57 36	230	60 37	260	31 52	301	35 45	340
304	37 02	045	57 34	096	43 40	157	57 20	228	60 16	258	31 34	299	35 37	338
306	37 16	043	57 55	094	43 48	155	57 05	227	59 56	256	31 16	296	35 29	335
308	37 30	040	58 16	092	43 57	153	56 50	225	59 36	254	30 57	294	35 20	333
310	37 43	038	58 37	090	44 07	152	56 35	223	59 16	253	30 38	292	35 10	331
312	37 56	036	58 58	088	44 17	150	56 21	222	58 56	251	30 18	290	35 00	329
314	38 08	034	59 18	086	44 28	148	56 07	220	58 36	249	29 59	288	34 49	327
316	38 19	032	59 39	084	44 39	147	55 54	219	58 17	247	29 39	286	34 37	324
318	38 30	030	60 00	082	44 50	145	55 41	217	57 58	245	29 19	284	34 24	322
320	38 39	027	60 20	080	45 03	143	55 29	215	57 39	244	28 58	282	34 11	320
322	38 49	025	60 41	078	45 15	142	55 17	214	57 21	242	28 38	280	33 58	318
324	38 57	023	61 01	076	45 29	140	55 06	212	57 02	240	28 17	278	33 43	316
326	39 05	021	61 21	074	45 42	138	54 55	211	56 44	239	27 57	276	33 29	314
328	39 12	018	61 41	072	45 56	137	54 44	209	56 27	237	27 36	274	33 13	312
	◆FOMALHAUT		ACHERNAR		◆CANOPUS		ACRUX		RIGIL KENT.		◆ANTARES		Nunki	
330	39 18	016	62 01	070	46 11	135	54 34	208	56 09	235	27 15	272	32 57	309
332	39 23	014	62 20	067	46 26	133	54 25	206	55 52	234	26 54	270	32 41	307
334	39 28	012	62 39	065	46 41	131	54 16	205	55 36	232	26 33	268	32 24	305
336	39 32	009	62 58	063	46 57	130	54 08	203	55 20	230	26 13	266	32 07	303
338	39 35	007	63 16	061	47 13	128	54 00	202	55 04	229	25 52	264	31 49	301
340	39 37	005	63 34	058	47 30	126	53 52	200	54 48	227	25 31	262	31 31	299
342	39 38	003	63 52	056	47 47	124	53 45	199	54 33	225	25 11	260	31 13	297
344	39 39	000	64 09	053	48 04	123	53 39	197	54 19	224	24 50	259	30 54	295
346	39 39	358	64 25	051	48 22	121	53 33	196	54 05	222	24 30	257	30 35	293
348	39 38	356	64 41	049	48 40	119	53 28	194	53 51	221	24 10	255	30 16	291
350	39 36	354	64 56	046	48 58	117	53 23	193	53 37	219	23 50	253	29 56	289
352	39 33	351	65 11	044	49 17	116	53 19	191	53 25	217	23 30	251	29 36	287
354	39 29	349	65 25	041	49 36	114	53 15	189	53 12	216	23 10	249	29 16	285
356	39 25	347	65 39	039	49 55	112	53 12	188	53 00	214	22 51	247	28 56	283
358	39 20	345	65 51	036	50 14	110	53 09	186	52 49	213	22 32	245	28 36	281

LHA ♈	Hc	Zn	Hc	Zn	Hc	Zn	Hc	Zn	Hc	Zn	Hc	Zn	Hc	Zn
	ACHERNAR		◆SIRIUS		Suhail		ACRUX		◆RIGIL KENT.		Peacock		◆FOMALHAUT	
0	65 13	032	14 46	104	36 34	142	54 07	185	53 29	212	61 15	293	38 17	343
2	65 22	030	15 04	102	36 46	140	54 06	184	53 20	210	60 58	291	38 11	340
4	65 31	027	15 22	100	36 58	138	54 05	182	53 10	208	60 40	289	38 04	338
6	65 39	024	15 41	098	37 11	136	54 04	180	53 02	207	60 23	287	37 57	336
8	65 47	022	16 00	096	37 24	135	54 04	179	52 53	205	60 05	285	37 49	334
10	65 53	019	16 18	094	37 37	133	54 05	177	52 46	204	59 46	283	37 40	332
12	65 59	017	16 37	092	37 51	131	54 06	176	52 38	202	59 28	280	37 31	329
14	66 04	014	16 56	090	38 06	129	54 08	174	52 32	200	59 09	278	37 21	327
16	66 08	011	17 15	088	38 20	127	54 10	173	52 25	199	58 51	276	37 11	325
18	66 11	009	17 33	086	38 36	126	54 13	171	52 19	197	58 32	274	37 00	323
20	66 14	006	17 52	084	38 51	124	54 16	170	52 14	196	58 13	272	36 48	321
22	66 15	003	18 11	082	39 07	122	54 19	168	52 09	194	57 55	270	36 36	319
24	66 16	001	18 29	080	39 23	120	54 23	167	52 05	192	57 36	268	36 23	316
26	66 15	358	18 48	078	39 39	118	54 28	165	52 01	191	57 17	266	36 10	314
28	66 14	355	19 06	076	39 56	116	54 33	163	51 58	189	56 58	264	35 56	312
	◆SIRIUS		Suhail		ACRUX		◆RIGIL KENT.		Peacock		◆FOMALHAUT		ACHERNAR	
30	19 24	074	40 13	115	54 39	162	51 55	188	56 40	263	35 42	310	66 12	352
32	19 42	072	40 30	113	54 45	160	51 53	186	56 21	261	35 28	308	66 09	350
34	20 00	070	40 47	111	54 51	159	51 51	185	56 03	259	35 13	306	66 06	347
36	20 17	068	41 05	109	54 58	157	51 50	183	55 44	257	34 57	304	66 01	344
38	20 35	066	41 23	107	55 06	156	51 49	181	55 26	255	34 42	302	65 56	342
40	20 52	064	41 41	105	55 14	154	51 49	180	55 08	253	34 25	300	65 49	339
42	21 08	062	41 59	103	55 22	152	51 49	178	54 50	251	34 09	297	65 42	337
44	21 25	060	42 18	101	55 31	151	51 50	177	54 32	249	33 52	295	65 34	334
46	21 41	058	42 36	099	55 41	149	51 51	175	54 15	248	33 35	293	65 26	331
48	21 56	056	42 55	097	55 50	148	51 53	174	53 58	246	33 18	291	65 16	329
50	22 12	054	43 13	095	56 01	146	51 56	172	53 41	244	33 00	289	65 06	326
52	22 27	052	43 32	094	56 11	145	51 58	170	53 24	242	32 42	287	64 56	324
54	22 41	050	43 51	092	56 22	143	52 02	169	53 08	241	32 24	285	64 44	321
56	22 55	048	44 10	090	56 34	141	52 06	167	52 51	239	32 06	283	64 32	319
58	23 09	046	44 28	088	56 46	140	52 10	166	52 35	237	31 48	281	64 19	316
	◆SIRIUS		Suhail		ACRUX		◆RIGIL KENT.		Peacock		◆FOMALHAUT		ACHERNAR	
60	23 22	043	44 47	086	56 58	138	52 15	164	52 20	235	31 29	279	64 06	314
62	23 35	041	45 06	084	57 11	137	52 20	162	52 05	234	31 11	277	63 52	311
64	23 47	039	45 24	082	57 24	135	52 26	161	51 50	232	30 52	275	63 38	309
66	23 59	037	45 43	080	57 37	133	52 33	159	51 35	230	30 33	273	63 23	307
68	24 10	035	46 01	078	57 51	132	52 40	158	51 21	228	30 14	271	63 08	304
70	24 20	033	46 20	075	58 06	130	52 47	156	51 07	227	29 56	269	62 52	302
72	24 30	031	46 38	073	58 20	128	52 55	154	50 54	225	29 37	267	62 36	300
74	24 39	029	46 55	071	58 35	127	53 03	153	50 41	223	29 18	265	62 20	298
76	24 48	027	47 13	069	58 50	125	53 12	151	50 28	222	28 59	263	62 03	295
78	24 56	025	47 31	067	59 06	123	53 21	150	50 16	220	28 41	261	61 46	293
80	25 04	023	47 48	065	59 22	122	53 31	148	50 04	218	28 22	259	61 28	291
82	25 11	020	48 05	063	59 38	120	53 41	146	49 52	216	28 04	257	61 11	289
84	25 17	018	48 21	061	59 54	118	53 52	145	49 42	215	27 46	256	60 53	287
86	25 22	016	48 37	059	60 11	116	54 03	143	49 31	213	27 28	254	60 35	284
88	25 27	014	48 53	056	60 28	115	54 14	142	49 21	211	27 10	252	60 16	282
	◆SIRIUS		Suhail		ACRUX		◆RIGIL KENT.		Peacock		◆FOMALHAUT		ACHERNAR	
90	25 32	012	49 09	054	60 45	113	54 26	140	49 11	210	26 52	250	59 58	280
92	25 35	010	49 24	052	61 02	111	54 38	138	49 02	208	26 34	248	59 39	278
94	25 38	008	49 38	050	61 20	109	54 51	137	48 54	206	26 17	246	59 21	276
96	25 40	006	49 52	048	61 38	108	55 04	135	48 46	205	26 00	244	59 02	274
98	25 42	003	50 06	045	61 56	106	55 18	133	48 38	203	25 43	242	58 43	272
100	25 42	001	50 19	043	62 14	104	55 32	132	48 31	202	25 27	240	58 25	270
102	25 42	359	50 32	041	62 32	102	55 46	130	48 24	200	25 11	238	58 06	268
104	25 42	357	50 44	039	62 50	100	56 00	128	48 18	198	24 55	236	57 47	266
106	25 41	355	50 55	036	63 09	098	56 15	127	48 12	197	24 40	234	57 28	264
108	25 39	353	51 06	034	63 28	097	56 30	125	48 07	195	24 25	233	57 10	262
110	25 36	351	51 16	032	63 46	095	56 46	123	48 03	193	24 10	231	56 51	260
112	25 32	349	51 26	029	64 05	093	57 02	121	47 59	192	23 55	229	56 33	259
114	25 28	346	51 35	027	64 24	091	57 18	120	47 55	190	23 42	227	56 14	257
116	25 24	344	51 43	025	64 43	089	57 34	118	47 52	188	23 28	225	55 56	255
118	25 18	342	51 50	022	65 01	087	57 51	116	47 50	187	23 15	223	55 38	253
	Suhail		SPICA		◆RIGIL KENT.		Peacock		◆FOMALHAUT		ACHERNAR		◆SIRIUS	
120	51 57	020	12 22	083	58 08	114	47 48	185	23 02	221	55 20	251	25 12	340
122	52 03	018	12 41	081	58 25	113	47 46	184	22 50	219	55 03	249	25 05	338
124	52 08	015	12 59	079	58 43	111	47 45	182	22 39	218	54 45	248	24 58	336
126	52 13	013	13 17	077	59 00	109	47 45	180	22 27	216	54 28	246	24 50	334
128	52 17	011	13 36	075	59 18	107	47 45	179	22 17	214	54 11	244	24 42	332
130	52 20	008	13 54	073	59 36	105	47 46	177	22 06	212	53 54	242	24 32	330
132	52 22	006	14 12	071	59 54	104	47 47	175	21 57	210	53 38	240	24 23	327
134	52 24	004	14 29	069	60 13	102	47 49	174	21 48	208	53 22	239	24 12	325
136	52 25	001	14 47	067	60 31	100	47 51	172	21 39	206	53 06	237	24 01	323
138	52 25	359	15 04	065	60 50	098	47 54	170	21 31	204	52 50	235	23 50	321
140	52 24	356	15 21	063	61 08	096	47 58	169	21 23	203	52 35	233	23 38	319
142	52 22	354	15 37	061	61 27	094	48 01	167	21 17	201	52 20	232	23 25	317
144	52 20	352	15 54	059	61 46	092	48 06	166	21 10	199	52 05	230	23 12	315
146	52 17	349	16 09	057	62 05	090	48 11	164	21 04	197	51 51	228	22 59	313
148	52 13	347	16 25	055	62 23	088	48 16	162	20 59	195	51 37	227	22 45	311
	SPICA		◆ANTARES		Peacock		FOMALHAUT		◆ACHERNAR		CANOPUS		◆Suhail	
150	16 40	053	24 57	102	48 22	161	20 55	193	51 24	225	57 17	295	52 08	345
152	16 55	051	25 15	100	48 29	159	20 50	191	51 11	223	56 59	293	52 03	342
154	17 09	049	25 34	098	48 36	157	20 47	190	50 58	221	56 42	291	51 57	340
156	17 23	047	25 52	096	48 43	156	20 44	188	50 46	220	56 24	288	51 50	337
158	17 37	045	26 11	094	48 51	154	20 42	186	50 34	218	56 06	286	51 42	335
160	17 50	043	26 30	092	49 00	152	20 40	184	50 23	216	55 48	284	51 34	333
162	18 02	041	26 49	090	49 09	151	20 39	182	50 12	215	55 30	282	51 25	330
164	18 14	039	27 07	088	49 18	149	20 39	180	50 02	213	55 11	280	51 16	328
166	18 26	037	27 26	086	49 28	147	20 39	178	49 52	211	54 53	278	51 05	326
168	18 37	035	27 45	084	49 38	146	20 40	177	49 42	210	54 34	276	50 55	324
170	18 47	033	28 03	082	49 49	144	20 41	175	49 33	208	54 16	274	50 43	321
172	18 57	030	28 22	080	50 00	142	20 43	173	49 24	206	53 57	272	50 31	319
174	19 06	028	28 40	078	50 12	141	20 46	171	49 16	205	53 38	270	50 18	317
176	19 15	026	28 59	076	50 24	139	20 49	169	49 09	203	53 19	268	50 05	315
178	19 23	024	29 17	074	50 37	137	20 53	167	49 02	201	53 01	266	49 52	312

LAT 81°S

LHA ♈	Hc	Zn	Hc	Zn	Hc	Zn	Hc	Zn	Hc	Zn	Hc	Zn	Hc	Zn
	SPICA		◆ANTARES		Peacock		FOMALHAUT		◆ACHERNAR		CANOPUS		◆Suhail	
180	19 30	022	29 35	072	50 49	136	20 57	165	48 55	200	52 42	264	49 38	310
182	19 37	020	29 52	070	51 03	134	21 02	164	48 49	198	52 23	262	49 23	308
184	19 43	018	30 10	068	51 16	132	21 08	162	48 43	197	52 05	260	49 08	306
186	19 48	016	30 27	066	51 31	130	21 14	160	48 38	195	51 46	258	48 52	303
188	19 53	014	30 44	064	51 45	129	21 21	158	48 34	193	51 28	256	48 37	301
190	19 57	012	31 01	062	52 00	127	21 28	156	48 29	192	51 10	255	48 20	299
192	20 01	010	31 17	059	52 15	125	21 36	154	48 26	190	50 52	253	48 04	297
194	20 04	008	31 33	057	52 30	124	21 45	152	48 23	188	50 34	251	47 47	295
196	20 06	005	31 49	055	52 46	122	21 54	151	48 20	187	50 16	249	47 30	293
198	20 07	003	32 04	053	53 02	120	22 03	149	48 18	185	49 59	247	47 12	291
200	20 08	001	32 19	051	53 19	118	22 13	147	48 17	184	49 42	245	46 55	289
202	20 08	359	32 33	049	53 35	116	22 24	145	48 16	182	49 25	244	46 37	287
204	20 07	357	32 47	047	53 52	115	22 35	143	48 16	180	49 08	242	46 19	284
206	20 06	355	33 01	045	54 10	113	22 46	141	48 16	179	48 52	240	46 00	282
208	20 04	353	33 14	043	54 27	111	22 58	139	48 17	177	48 36	238	45 42	280
	◆ANTARES		Nunki		◆FOMALHAUT		ACHERNAR		CANOPUS		◆Suhail		SPICA	
210	33 26	041	28 29	078	23 11	137	48 18	175	48 20	236	45 23	278	20 02	351
212	33 38	038	28 47	076	23 24	136	48 20	174	48 04	235	45 05	276	19 58	349
214	33 49	036	29 06	074	23 37	134	48 22	172	47 49	233	44 46	274	19 54	347
216	34 00	034	29 24	072	23 51	132	48 25	171	47 34	231	44 27	272	19 50	345
218	34 10	032	29 41	070	24 05	130	48 28	169	47 20	229	44 09	270	19 44	342
220	34 20	030	29 59	068	24 19	128	48 32	167	47 06	228	43 50	268	19 38	340
222	34 29	028	30 16	066	24 34	126	48 36	166	46 52	226	43 31	266	19 32	338
224	34 37	025	30 33	064	24 50	124	48 41	164	46 39	224	43 12	264	19 24	336
226	34 45	023	30 50	062	25 05	122	48 47	162	46 26	222	42 54	263	19 17	334
228	34 52	021	31 06	060	25 21	120	48 53	161	46 14	221	42 35	261	19 08	332
230	34 59	019	31 22	058	25 38	119	48 59	159	46 02	219	42 17	259	18 59	330
232	35 04	017	31 38	056	25 54	117	49 06	157	45 50	217	41 58	257	18 49	328
234	35 10	015	31 53	054	26 11	115	49 13	156	45 39	215	41 40	255	18 39	326
236	35 14	012	32 08	052	26 28	113	49 21	154	45 28	214	41 22	253	18 28	324
238	35 18	010	32 23	049	26 46	111	49 30	153	45 18	212	41 04	251	18 17	322
	ANTARES		◆Nunki		FOMALHAUT		◆ACHERNAR		CANOPUS		◆ACRUX		SPICA	
240	35 21	008	32 37	047	27 04	109	49 39	151	45 09	210	67 20	289	18 05	320
242	35 23	006	32 50	045	27 21	107	49 48	149	44 59	209	67 03	287	17 53	318
244	35 24	004	33 04	043	27 39	105	49 58	148	44 51	207	66 45	285	17 40	316
246	35 25	001	33 16	041	27 58	103	50 08	146	44 42	205	66 26	283	17 26	314
248	35 25	359	33 28	039	28 16	101	50 19	144	44 35	204	66 08	281	17 13	312
250	35 25	357	33 40	037	28 34	099	50 30	143	44 27	202	65 49	279	16 58	310
252	35 23	355	33 51	035	28 53	097	50 42	141	44 21	200	65 31	276	16 44	307
254	35 21	353	34 01	032	29 12	095	50 54	139	44 14	198	65 12	274	16 29	305
256	35 18	350	34 11	030	29 30	093	51 06	138	44 09	197	64 53	272	16 13	303
258	35 15	348	34 20	028	29 49	091	51 19	136	44 04	195	64 35	270	15 57	301
260	35 11	346	34 28	026	30 08	090	51 32	134	43 59	193	64 16	268	15 41	299
262	35 06	344	34 36	024	30 27	088	51 46	132	43 55	192	63 57	266	15 25	297
264	35 00	342	34 44	022	30 45	086	52 00	131	43 51	190	63 38	265	15 08	295
266	34 54	339	34 50	019	31 04	084	52 14	129	43 48	188	63 20	263	14 51	293
268	34 47	337	34 56	017	31 23	082	52 29	127	43 46	187	63 01	261	14 33	291
	◆Nunki		FOMALHAUT		◆ACHERNAR		CANOPUS		◆ACRUX		RIGIL KENT.		ANTARES	
270	35 01	015	31 41	080	52 44	126	43 44	185	62 43	259	65 40	295	34 40	335
272	35 06	013	32 00	078	53 00	124	43 43	183	62 24	257	65 23	292	34 31	333
274	35 10	011	32 18	076	53 15	122	43 42	182	62 06	255	65 05	290	34 23	331
276	35 13	008	32 36	073	53 32	120	43 42	180	61 48	253	64 47	288	34 13	329
278	35 15	006	32 54	071	53 48	119	43 42	178	61 30	252	64 29	286	34 03	326
280	35 17	004	33 12	069	54 04	117	43 43	177	61 12	250	64 11	284	33 52	324
282	35 18	002	33 29	067	54 21	115	43 44	175	60 55	248	63 53	282	33 41	322
284	35 18	000	33 46	065	54 39	113	43 46	173	60 38	246	63 35	279	33 29	320
286	35 18	358	34 03	063	54 56	111	43 48	172	60 21	245	63 16	277	33 17	318
288	35 17	355	34 20	061	55 13	110	43 51	170	60 04	243	62 57	275	33 04	316
290	35 15	353	34 36	059	55 31	108	43 55	168	59 47	241	62 39	273	32 51	314
292	35 12	351	34 52	057	55 49	106	43 59	167	59 31	239	62 20	271	32 37	312
294	35 09	349	35 07	055	56 07	104	44 04	165	59 15	238	62 01	269	32 23	309
296	35 05	347	35 23	053	56 26	102	44 09	163	58 59	236	61 42	267	32 08	307
298	35 00	344	35 37	051	56 44	100	44 15	161	58 44	234	61 24	265	31 53	305
	◆FOMALHAUT		ACHERNAR		◆CANOPUS		ACRUX		RIGIL KENT.		◆ANTARES		Nunki	
300	35 52	049	57 03	098	44 21	160	58 29	233	61 05	264	31 37	303	34 55	342
302	36 05	046	57 21	096	44 28	158	58 14	231	60 46	262	31 22	301	34 49	340
304	36 19	044	57 40	095	44 35	156	57 59	229	60 28	260	31 05	299	34 42	338
306	36 32	042	57 59	093	44 43	155	57 45	228	60 09	258	30 49	297	34 34	336
308	36 44	040	58 17	091	44 51	153	57 32	226	59 51	256	30 32	295	34 26	333
310	36 56	038	58 36	089	45 00	151	57 18	224	59 33	254	30 15	293	34 18	331
312	37 07	036	58 55	087	45 09	150	57 05	223	59 15	252	29 57	291	34 08	329
314	37 18	033	59 14	085	45 19	148	56 53	221	58 57	251	29 40	289	33 58	327
316	37 28	031	59 32	083	45 29	146	56 41	220	58 40	249	29 22	287	33 48	325
318	37 37	029	59 51	081	45 40	145	56 29	218	58 22	247	29 04	285	33 37	323
320	37 46	027	60 09	078	45 51	143	56 18	216	58 05	245	28 45	283	33 25	321
322	37 54	025	60 28	076	46 02	141	56 07	215	57 48	243	28 27	281	33 13	318
324	38 02	023	60 46	074	46 14	139	55 56	213	57 31	242	28 09	279	33 00	316
326	38 09	020	61 04	072	46 27	138	55 46	212	57 15	240	27 50	277	32 47	314
328	38 15	018	61 22	070	46 40	136	55 37	210	56 59	238	27 31	275	32 33	312
	◆FOMALHAUT		ACHERNAR		◆CANOPUS		ACRUX		RIGIL KENT.		◆ANTARES		Nunki	
330	38 20	016	61 39	068	46 53	134	55 27	208	56 43	237	27 13	273	32 19	310
332	38 25	014	61 56	066	47 06	132	55 19	207	56 28	235	26 54	271	32 04	308
334	38 29	011	62 13	063	47 21	131	55 10	205	56 12	233	26 35	269	31 49	306
336	38 33	009	62 30	061	47 35	129	55 03	204	55 58	231	26 16	267	31 34	304
338	38 35	007	62 46	059	47 50	127	54 55	202	55 43	230	25 57	265	31 18	302
340	38 37	005	63 02	057	48 05	125	54 49	201	55 29	228	25 39	263	31 02	300
342	38 38	003	63 17	054	48 20	124	54 42	199	55 15	226	25 20	261	30 46	297
344	38 39	000	63 32	052	48 36	122	54 36	197	55 02	225	25 02	259	30 29	295
346	38 39	358	63 47	049	48 52	120	54 31	196	54 49	223	24 43	257	30 12	293
348	38 38	356	64 01	047	49 09	118	54 26	194	54 36	221	24 25	255	29 54	291
350	38 36	354	64 14	045	49 25	116	54 22	193	54 24	220	24 07	253	29 37	289
352	38 34	351	64 27	042	49 42	115	54 18	191	54 12	218	23 49	251	29 19	287
354	38 30	349	64 40	040	50 00	113	54 14	190	54 01	217	23 32	249	29 01	285
356	38 27	347	64 51	037	50 17	111	54 11	188	53 50	215	23 14	247	28 43	283
358	38 22	345	65 02	035	50 35	109	54 09	187	53 39	213	22 57	245	28 24	281

 LAT 82°S

LHA ♈	ACHERNAR Hc	Zn	◆SIRIUS Hc	Zn	Suhail Hc	Zn	ACRUX Hc	Zn	◆RIGIL KENT. Hc	Zn	Peacock Hc	Zn	◆FOMALHAUT Hc	Zn
0	64 22	031	15 00	103	37 21	141	55 07	185	54 20	212	60 51	295	37 19	343
2	64 30	029	15 16	101	37 32	140	55 05	184	54 11	211	60 36	293	37 14	341
4	64 37	026	15 32	100	37 43	138	55 05	182	54 03	209	60 20	291	37 08	338
6	64 44	024	15 49	098	37 54	136	55 04	180	53 55	207	60 04	288	37 02	336
8	64 51	021	16 06	096	38 06	134	55 04	179	53 48	206	59 49	286	36 55	334
10	64 56	018	16 22	094	38 18	132	55 05	177	53 41	204	59 32	284	36 47	332
12	65 01	016	16 39	092	38 31	130	55 06	176	53 34	203	59 16	282	36 39	330
14	65 06	013	16 56	090	38 43	129	55 07	174	53 28	201	59 00	280	36 31	328
16	65 09	011	17 12	088	38 57	127	55 09	173	53 22	199	58 43	278	36 21	326
18	65 12	008	17 29	086	39 10	125	55 12	171	53 17	198	58 27	276	36 12	323
20	65 14	006	17 46	084	39 24	123	55 15	169	53 12	196	58 10	274	36 02	321
22	65 15	003	18 02	082	39 38	121	55 18	168	53 07	194	57 53	272	35 51	319
24	65 16	000	18 19	080	39 53	119	55 22	166	53 04	193	57 37	270	35 40	317
26	65 15	358	18 35	078	40 07	117	55 26	165	53 00	191	57 20	268	35 28	315
28	65 14	355	18 51	076	40 22	116	55 31	163	52 57	190	57 03	266	35 16	313

LHA ♈	◆SIRIUS Hc	Zn	Suhail Hc	Zn	ACRUX Hc	Zn	◆RIGIL KENT. Hc	Zn	Peacock Hc	Zn	◆FOMALHAUT Hc	Zn	ACHERNAR Hc	Zn
30	19 07	074	40 37	114	55 36	161	52 55	188	56 47	264	35 03	311	65 13	353
32	19 23	072	40 53	112	55 41	160	52 52	186	56 30	262	34 51	308	65 10	350
34	19 39	070	41 08	110	55 47	158	52 51	185	56 14	260	34 37	306	65 07	348
36	19 55	068	41 24	108	55 54	157	52 50	183	55 57	258	34 24	304	65 03	345
38	20 10	066	41 40	106	56 00	155	52 49	181	55 41	256	34 10	302	64 58	342
40	20 25	064	41 56	104	56 08	153	52 49	180	55 25	255	33 55	300	64 53	340
42	20 40	062	42 12	102	56 15	152	52 49	178	55 09	253	33 41	298	64 47	337
44	20 54	060	42 29	100	56 23	150	52 50	177	54 53	251	33 26	296	64 40	335
46	21 09	058	42 45	098	56 32	149	52 51	175	54 37	249	33 11	294	64 33	332
48	21 23	055	43 02	096	56 41	147	52 53	173	54 22	247	32 56	292	64 25	330
50	21 36	053	43 19	095	56 50	145	52 55	172	54 06	245	32 40	290	64 16	327
52	21 49	051	43 35	093	57 00	144	52 58	170	53 51	243	32 24	288	64 07	325
54	22 02	049	43 52	091	57 10	142	53 01	169	53 36	242	32 08	286	63 57	323
56	22 15	047	44 09	089	57 20	140	53 04	167	53 22	240	31 52	284	63 46	320
58	22 27	045	44 25	087	57 31	139	53 08	165	53 08	238	31 36	282	63 35	318

LHA ♈	◆SIRIUS Hc	Zn	Suhail Hc	Zn	ACRUX Hc	Zn	◆RIGIL KENT. Hc	Zn	Peacock Hc	Zn	◆FOMALHAUT Hc	Zn	ACHERNAR Hc	Zn
60	22 38	043	44 42	085	57 42	137	53 13	164	52 54	236	31 19	280	63 24	315
62	22 50	041	44 59	083	57 54	135	53 18	162	52 40	235	31 03	278	63 12	313
64	23 00	039	45 15	081	58 06	134	53 23	160	52 26	233	30 46	276	63 00	311
66	23 11	037	45 31	079	58 18	132	53 29	159	52 13	231	30 30	274	62 47	308
68	23 20	035	45 48	077	58 31	130	53 35	157	52 00	229	30 13	272	62 33	306
70	23 30	033	46 04	074	58 44	129	53 42	156	51 48	228	29 56	270	62 20	304
72	23 39	031	46 20	072	58 57	127	53 49	154	51 36	226	29 39	268	62 06	301
74	23 47	029	46 36	070	59 10	125	53 56	152	51 24	224	29 23	266	61 51	299
76	23 55	027	46 51	068	59 24	124	54 04	151	51 13	222	29 06	264	61 36	297
78	24 02	024	47 07	066	59 38	122	54 13	149	51 01	221	28 50	262	61 21	295
80	24 08	022	47 22	064	59 52	120	54 22	147	50 51	219	28 33	260	61 06	293
82	24 14	020	47 37	062	60 07	118	54 31	146	50 40	217	28 17	258	60 50	290
84	24 20	018	47 51	060	60 22	117	54 41	144	50 31	215	28 00	256	60 35	288
86	24 25	016	48 06	058	60 37	115	54 51	142	50 21	214	27 44	254	60 19	286
88	24 29	014	48 20	055	60 52	113	55 01	141	50 12	212	27 28	252	60 03	284

LHA ♈	◆SIRIUS Hc	Zn	Suhail Hc	Zn	ACRUX Hc	Zn	◆RIGIL KENT. Hc	Zn	Peacock Hc	Zn	◆FOMALHAUT Hc	Zn	ACHERNAR Hc	Zn
90	24 33	012	48 33	053	61 07	111	55 12	139	50 03	210	27 13	250	59 46	282
92	24 36	010	48 46	051	61 23	110	55 23	137	49 55	209	26 57	248	59 30	280
94	24 39	008	48 59	049	61 39	108	55 34	136	49 47	207	26 41	246	59 13	278
96	24 40	006	49 12	047	61 55	106	55 46	134	49 40	205	26 26	244	58 57	276
98	24 42	003	49 23	045	62 11	104	55 58	132	49 33	204	26 11	243	58 40	274
100	24 42	001	49 35	042	62 27	102	56 11	131	49 27	202	25 57	241	58 24	272
102	24 42	359	49 46	040	62 44	100	56 24	129	49 21	200	25 42	239	58 07	270
104	24 42	357	49 56	038	63 00	098	56 37	127	49 15	199	25 28	237	57 50	268
106	24 41	355	50 06	036	63 17	096	56 50	125	49 10	197	25 14	235	57 34	266
108	24 39	353	50 16	033	63 33	095	57 04	124	49 05	195	25 01	233	57 17	264
110	24 37	351	50 25	031	63 50	093	57 18	122	49 01	194	24 48	231	57 00	262
112	24 34	349	50 33	029	64 07	091	57 33	120	48 57	192	24 35	229	56 44	260
114	24 30	347	50 41	027	64 23	089	57 47	118	48 54	190	24 22	227	56 27	258
116	24 26	344	50 48	024	64 40	087	58 02	117	48 52	189	24 10	225	56 11	256
118	24 21	342	50 55	022	64 57	085	58 17	115	48 49	187	23 59	223	55 55	254

LHA ♈	Suhail Hc	Zn	SPICA Hc	Zn	◆RIGIL KENT. Hc	Zn	Peacock Hc	Zn	◆FOMALHAUT Hc	Zn	ACHERNAR Hc	Zn	◆SIRIUS Hc	Zn
120	51 01	020	12 15	083	58 32	113	48 48	185	23 47	222	55 39	253	24 16	340
122	51 06	017	12 31	081	58 48	111	48 46	184	23 36	220	55 23	251	24 10	338
124	51 11	015	12 48	079	59 03	109	48 45	182	23 26	218	55 07	249	24 03	336
126	51 15	013	13 04	077	59 19	108	48 45	180	23 16	216	54 52	247	23 56	334
128	51 18	010	13 20	075	59 35	106	48 45	179	23 06	214	54 37	245	23 49	332
130	51 21	008	13 36	073	59 51	104	48 46	177	22 57	212	54 22	243	23 41	330
132	51 23	006	13 52	071	60 08	102	48 47	175	22 49	210	54 07	242	23 32	328
134	51 24	003	14 08	069	60 24	100	48 49	174	22 40	208	53 52	240	23 23	326
136	51 25	001	14 23	067	60 41	098	48 51	172	22 33	207	53 38	238	23 13	324
138	51 25	359	14 38	065	60 57	096	48 53	170	22 26	205	53 24	236	23 03	321
140	51 24	356	14 53	063	61 14	094	48 56	169	22 19	203	53 10	234	22 52	319
142	51 23	354	15 08	061	61 30	092	49 00	167	22 13	201	52 57	233	22 41	317
144	51 21	352	15 23	059	61 47	090	49 04	165	22 07	199	52 44	231	22 30	315
146	51 18	349	15 37	057	62 04	088	49 08	164	22 02	197	52 31	229	22 18	313
148	51 14	347	15 51	055	62 21	086	49 13	162	21 57	195	52 18	227	22 05	311

LHA ♈	SPICA Hc	Zn	◆ANTARES Hc	Zn	Peacock Hc	Zn	FOMALHAUT Hc	Zn	◆ACHERNAR Hc	Zn	CANOPUS Hc	Zn	◆Suhail Hc	Zn
150	16 04	053	25 08	101	49 19	160	21 53	193	52 06	226	56 51	296	51 10	345
152	16 17	051	25 25	099	49 25	159	21 49	192	51 55	224	56 36	294	51 06	343
154	16 30	049	25 41	097	49 31	157	21 46	190	51 43	222	56 20	292	51 00	340
156	16 42	047	25 58	095	49 38	155	21 44	188	51 32	221	56 05	290	50 54	338
158	16 54	045	26 15	093	49 45	154	21 42	186	51 21	219	55 49	288	50 48	336
160	17 06	043	26 31	091	49 53	152	21 40	184	51 11	217	55 33	286	50 41	333
162	17 17	041	26 48	089	50 01	150	21 39	182	51 01	215	55 17	284	50 33	331
164	17 27	038	27 05	087	50 09	148	21 39	180	50 52	214	55 00	281	50 24	329
166	17 38	036	27 21	085	50 18	147	21 39	178	50 43	212	54 44	279	50 16	327
168	17 47	034	27 38	083	50 28	145	21 40	177	50 34	210	54 27	277	50 06	324
170	17 56	032	27 55	081	50 37	143	21 41	175	50 26	209	54 11	275	49 56	322
172	18 05	030	28 11	079	50 48	142	21 43	173	50 18	207	53 54	273	49 45	320
174	18 13	028	28 27	077	50 58	140	21 45	171	50 11	205	53 37	271	49 34	318
176	18 21	026	28 44	075	51 09	138	21 48	169	50 04	204	53 21	269	49 23	315
178	18 28	024	29 00	073	51 20	136	21 52	167	49 57	202	53 04	267	49 11	313

LHA ♈	SPICA Hc	Zn	◆ANTARES Hc	Zn	Peacock Hc	Zn	FOMALHAUT Hc	Zn	◆ACHERNAR Hc	Zn	CANOPUS Hc	Zn	◆Suhail Hc	Zn
180	18 34	022	29 16	071	51 32	135	21 56	165	49 51	200	52 47	265	48 59	311
182	18 40	020	29 31	069	51 44	133	22 00	163	49 46	199	52 31	263	48 46	309
184	18 46	018	29 47	067	51 56	131	22 05	162	49 41	197	52 14	262	48 33	307
186	18 51	016	30 02	065	52 09	130	22 11	160	49 36	195	51 58	260	48 19	304
188	18 55	014	30 17	063	52 22	128	22 17	158	49 32	194	51 41	258	48 05	302
190	18 59	012	30 32	061	52 36	126	22 23	156	49 28	192	51 25	256	47 51	300
192	19 02	010	30 46	059	52 49	124	22 30	154	49 25	190	51 09	254	47 36	298
194	19 04	008	31 01	057	53 03	122	22 38	152	49 22	189	50 53	252	47 21	296
196	19 06	005	31 14	055	53 17	121	22 46	150	49 20	187	50 37	250	47 06	294
198	19 07	003	31 28	053	53 32	119	22 54	148	49 18	185	50 22	248	46 51	292
200	19 08	001	31 41	051	53 47	117	23 03	147	49 17	184	50 06	246	46 35	290
202	19 08	359	31 54	049	54 02	115	23 13	145	49 16	182	49 51	245	46 19	288
204	19 08	357	32 06	046	54 17	113	23 23	143	49 16	180	49 36	243	46 03	285
206	19 06	355	32 18	044	54 32	112	23 33	141	49 16	179	49 21	241	45 47	283
208	19 05	353	32 29	042	54 48	110	23 44	139	49 17	177	49 07	239	45 31	281

LHA ♈	◆ANTARES Hc	Zn	Nunki Hc	Zn	◆FOMALHAUT Hc	Zn	ACHERNAR Hc	Zn	CANOPUS Hc	Zn	◆Suhail Hc	Zn	SPICA Hc	Zn
210	32 40	040	28 17	078	23 55	137	49 18	175	48 53	237	45 14	279	19 02	351
212	32 51	038	28 33	076	24 06	135	49 19	174	48 39	236	44 58	277	18 59	349
214	33 01	036	28 49	074	24 18	133	49 21	172	48 25	234	44 41	275	18 56	347
216	33 10	034	29 05	072	24 31	131	49 24	170	48 12	232	44 24	273	18 52	345
218	33 19	032	29 21	070	24 43	130	49 27	169	47 59	230	44 08	271	18 47	343
220	33 28	029	29 36	068	24 56	128	49 30	167	47 46	228	43 51	269	18 42	341
222	33 36	027	29 52	066	25 10	126	49 34	165	47 34	227	43 34	267	18 36	338
224	33 43	025	30 07	064	25 23	124	49 39	164	47 22	225	43 18	265	18 29	336
226	33 50	023	30 21	061	25 37	122	49 44	162	47 10	223	43 01	263	18 23	334
228	33 56	021	30 36	059	25 52	120	49 49	160	46 59	221	42 45	261	18 15	332
230	34 02	019	30 50	057	26 06	118	49 55	159	46 48	220	42 28	260	18 07	330
232	34 07	017	31 04	055	26 21	116	50 01	157	46 38	218	42 12	258	17 58	328
234	34 11	014	31 18	053	26 36	114	50 08	155	46 28	216	41 55	256	17 49	326
236	34 15	012	31 31	051	26 52	112	50 15	154	46 18	214	41 39	254	17 40	324
238	34 18	010	31 44	049	27 07	110	50 23	152	46 09	213	41 23	252	17 30	322

LHA ♈	ANTARES Hc	Zn	◆Nunki Hc	Zn	FOMALHAUT Hc	Zn	◆ACHERNAR Hc	Zn	CANOPUS Hc	Zn	◆ACRUX Hc	Zn	SPICA Hc	Zn
240	34 21	008	31 56	047	27 23	109	50 31	150	46 00	211	67 00	291	17 19	320
242	34 23	006	32 08	045	27 39	107	50 39	149	45 52	209	66 44	289	17 08	318
244	34 24	004	32 20	043	27 55	105	50 48	147	45 44	207	66 28	287	16 57	316
246	34 25	001	32 31	041	28 11	103	50 58	145	45 37	206	66 12	285	16 45	314
248	34 25	359	32 41	038	28 27	101	51 07	144	45 30	204	65 56	283	16 33	312
250	34 25	357	32 51	036	28 44	099	51 17	142	45 23	202	65 39	281	16 20	310
252	34 24	355	33 01	034	29 00	097	51 28	140	45 17	200	65 23	279	16 07	308
254	34 22	353	33 10	032	29 17	095	51 39	138	45 11	199	65 06	277	15 54	306
256	34 19	351	33 19	030	29 34	093	51 50	137	45 06	197	64 50	275	15 40	304
258	34 16	348	33 27	028	29 50	091	52 02	135	45 02	195	64 33	273	15 26	302
260	34 13	346	33 34	026	30 07	089	52 14	133	44 57	194	64 16	271	15 12	300
262	34 08	344	33 41	024	30 24	087	52 26	131	44 54	192	64 00	269	14 57	298
264	34 03	342	33 48	021	30 40	085	52 39	130	44 50	190	63 43	267	14 42	296
266	33 58	340	33 54	019	30 57	083	52 52	128	44 48	188	63 26	265	14 27	294
268	33 52	338	33 59	017	31 14	081	53 05	126	44 46	187	63 10	263	14 11	292

LHA ♈	◆Nunki Hc	Zn	FOMALHAUT Hc	Zn	◆ACHERNAR Hc	Zn	CANOPUS Hc	Zn	◆ACRUX Hc	Zn	RIGIL KENT. Hc	Zn	ANTARES Hc	Zn
270	34 03	015	31 30	079	53 19	124	44 44	185	62 53	261	65 14	297	33 45	335
272	34 07	013	31 46	077	53 33	123	44 43	183	62 37	259	64 59	294	33 38	333
274	34 11	011	32 03	075	53 47	121	44 42	182	62 20	257	64 43	292	33 30	331
276	34 13	008	32 19	073	54 01	119	44 42	180	62 04	255	64 28	290	33 22	329
278	34 16	006	32 34	071	54 16	117	44 42	178	61 48	253	64 12	288	33 13	327
280	34 17	004	32 50	069	54 31	116	44 43	177	61 32	252	63 56	286	33 03	325
282	34 18	002	33 06	067	54 46	114	44 44	175	61 17	250	63 40	284	32 54	323
284	34 18	000	33 21	065	55 02	112	44 46	173	61 01	248	63 24	281	32 43	320
286	34 18	358	33 36	063	55 17	110	44 48	171	60 46	246	63 07	279	32 32	318
288	34 17	355	33 51	061	55 33	108	44 51	170	60 30	244	62 51	277	32 21	316
290	34 15	353	34 05	058	55 49	106	44 54	168	60 15	243	62 34	275	32 09	314
292	34 13	351	34 19	056	56 05	104	44 57	166	60 01	241	62 17	273	31 57	312
294	34 10	349	34 33	054	56 21	103	45 02	165	59 46	239	62 01	271	31 44	310
296	34 06	347	34 46	052	56 38	101	45 06	163	59 32	237	61 44	269	31 31	308
298	34 02	345	34 59	050	56 54	099	45 12	161	59 18	236	61 27	267	31 18	306

LHA ♈	◆FOMALHAUT Hc	Zn	ACHERNAR Hc	Zn	◆CANOPUS Hc	Zn	ACRUX Hc	Zn	RIGIL KENT. Hc	Zn	◆ANTARES Hc	Zn	Nunki Hc	Zn
300	35 12	048	57 11	097	45 17	159	59 05	234	61 11	265	31 04	304	33 57	342
302	35 24	046	57 27	095	45 23	158	58 51	232	60 54	263	30 50	302	33 52	340
304	35 36	044	57 44	093	45 30	156	58 38	231	60 37	261	30 36	300	33 46	338
306	35 47	042	58 00	091	45 37	154	58 25	229	60 21	260	30 21	298	33 40	336
308	35 58	040	58 17	089	45 44	153	58 13	227	60 05	258	30 06	295	33 33	334
310	36 08	037	58 34	087	45 52	151	58 01	226	59 48	256	29 51	293	33 25	332
312	36 18	035	58 51	085	46 01	149	57 49	224	59 32	254	29 36	291	33 17	329
314	36 28	033	59 07	083	46 09	147	57 38	222	59 16	252	29 20	289	33 08	327
316	36 36	031	59 24	081	46 19	146	57 27	221	59 00	250	29 04	287	32 59	325
318	36 45	029	59 40	079	46 28	144	57 16	219	58 45	248	28 48	285	32 49	323
320	36 52	027	59 56	077	46 38	142	57 06	217	58 29	247	28 32	283	32 39	321
322	37 00	024	60 13	075	46 49	140	56 56	216	58 14	245	28 16	281	32 28	319
324	37 06	022	60 29	073	47 00	139	56 46	214	57 59	243	27 59	279	32 17	317
326	37 12	020	60 44	070	47 11	137	56 37	212	57 44	241	27 43	277	32 05	315
328	37 18	018	61 00	068	47 22	135	56 28	211	57 30	240	27 26	275	31 53	313

LHA ♈	◆FOMALHAUT Hc	Zn	ACHERNAR Hc	Zn	◆CANOPUS Hc	Zn	ACRUX Hc	Zn	RIGIL KENT. Hc	Zn	◆ANTARES Hc	Zn	Nunki Hc	Zn
330	37 23	016	61 16	066	47 34	133	56 20	209	57 16	238	27 09	273	31 40	310
332	37 27	014	61 31	064	47 47	132	56 12	208	57 02	236	26 53	271	31 27	308
334	37 30	011	61 46	062	47 59	130	56 05	206	56 48	234	26 36	269	31 14	306
336	37 33	009	62 00	059	48 12	128	55 58	204	56 34	233	26 19	267	31 01	304
338	37 36	007	62 14	057	48 26	126	55 51	203	56 21	231	26 03	265	30 47	302
340	37 37	005	62 28	055	48 39	124	55 45	201	56 09	229	25 46	263	30 32	300
342	37 38	003	62 42	053	48 53	123	55 39	200	55 56	227	25 29	261	30 18	298
344	37 39	000	62 55	050	49 07	121	55 33	198	55 44	226	25 13	259	30 03	296
346	37 39	358	63 07	048	49 22	119	55 29	196	55 32	224	24 57	257	29 48	294
348	37 38	356	63 19	046	49 37	117	55 24	195	55 21	222	24 40	256	29 32	292
350	37 36	354	63 31	043	49 52	115	55 20	193	55 10	221	24 24	254	29 17	290
352	37 34	352	63 42	041	50 07	113	55 16	192	54 59	219	24 08	252	29 01	288
354	37 31	349	63 53	038	50 22	112	55 13	190	54 49	217	23 53	250	28 45	286
356	37 28	347	64 03	036	50 38	110	55 11	188	54 39	216	23 37	248	28 29	284
358	37 24	345	64 13	034	50 54	108	55 09	187	54 29	214	23 22	246	28 12	282

LHA ♈	Hc	Zn	Hc	Zn	Hc	Zn	Hc	Zn	Hc	Zn	Hc	Zn	Hc	Zn
	ACHERNAR		◆SIRIUS		Suhail		ACRUX		◆RIGIL KENT.		Peacock		◆FOMALHAUT	
0	63 30	030	15 14	103	38 08	141	56 07	185	55 10	213	60 25	297	36 22	343
2	63 37	028	15 28	101	38 17	139	56 05	184	55 03	211	60 12	294	36 18	341
4	63 43	025	15 42	099	38 27	137	56 05	182	54 55	210	59 58	292	36 13	339
6	63 49	023	15 57	097	38 37	135	56 04	180	54 48	208	59 45	290	36 07	337
8	63 55	020	16 11	095	38 47	134	56 04	179	54 41	206	59 31	288	36 01	334
10	63 59	018	16 26	093	38 58	132	56 05	177	54 35	205	59 17	286	35 54	332
12	64 04	015	16 40	091	39 09	130	56 06	176	54 29	203	59 03	284	35 47	330
14	64 07	013	16 55	089	39 21	128	56 07	174	54 24	201	58 48	282	35 40	328
16	64 10	010	17 10	087	39 32	126	56 09	172	54 19	200	58 34	280	35 32	326
18	64 12	008	17 24	085	39 44	124	56 11	171	54 14	198	58 20	278	35 23	324
20	64 14	005	17 39	083	39 56	122	56 14	169	54 09	196	58 05	276	35 15	322
22	64 15	003	17 53	081	40 09	120	56 17	167	54 06	195	57 51	274	35 05	320
24	64 16	000	18 08	079	40 22	119	56 20	166	54 02	193	57 36	272	34 56	317
26	64 16	358	18 22	077	40 35	117	56 24	164	53 59	191	57 21	270	34 46	315
28	64 15	355	18 36	075	40 48	115	56 28	163	53 56	190	57 07	268	34 35	313
	◆SIRIUS		Suhail		ACRUX		◆RIGIL KENT.		Peacock		◆FOMALHAUT		ACHERNAR	
30	18 50	073	41 01	113	56 32	161	53 54	188	56 52	266	34 24	311	64 13	353
32	19 04	071	41 15	111	56 37	159	53 52	186	56 38	264	34 13	309	64 11	351
34	19 18	069	41 28	109	56 43	158	53 51	185	56 23	262	34 02	307	64 08	348
36	19 32	067	41 42	107	56 49	156	53 50	183	56 09	260	33 50	305	64 05	346
38	19 45	065	41 56	105	56 55	154	53 49	181	55 54	258	33 38	303	64 01	343
40	19 58	063	42 11	103	57 01	153	53 49	180	55 40	256	33 25	301	63 57	341
42	20 11	061	42 25	101	57 08	151	53 49	178	55 26	254	33 12	299	63 51	338
44	20 24	059	42 39	099	57 15	149	53 50	177	55 12	252	32 59	297	63 46	336
46	20 36	057	42 54	098	57 23	148	53 51	175	54 58	250	32 46	295	63 39	333
48	20 48	055	43 08	096	57 31	146	53 52	173	54 44	248	32 33	292	63 33	331
50	21 00	053	43 23	094	57 39	144	53 54	172	54 31	247	32 19	290	63 25	328
52	21 12	051	43 37	092	57 48	143	53 57	170	54 18	245	32 05	288	63 17	326
54	21 23	049	43 52	090	57 57	141	53 59	168	54 04	243	31 51	286	63 09	324
56	21 34	047	44 07	088	58 06	139	54 03	167	53 51	241	31 37	284	63 00	321
58	21 44	045	44 21	086	58 16	138	54 06	165	53 39	239	31 23	282	62 51	319
	◆SIRIUS		Suhail		ACRUX		◆RIGIL KENT.		Peacock		◆FOMALHAUT		ACHERNAR	
60	21 55	043	44 36	084	58 26	136	54 10	163	53 26	237	31 09	280	62 41	317
62	22 04	041	44 50	082	58 36	134	54 15	162	53 14	236	30 54	278	62 31	314
64	22 14	039	45 05	080	58 47	133	54 19	160	53 02	234	30 40	276	62 20	312
66	22 23	037	45 19	078	58 58	131	54 25	158	52 51	232	30 25	274	62 09	310
68	22 31	035	45 33	076	59 09	129	54 30	157	52 39	230	30 11	272	61 57	307
70	22 39	033	45 47	073	59 21	127	54 36	155	52 28	228	29 56	270	61 46	305
72	22 47	030	46 01	071	59 32	126	54 43	153	52 17	227	29 41	268	61 34	303
74	22 54	028	46 15	069	59 44	124	54 49	152	52 07	225	29 27	266	61 21	301
76	23 01	026	46 29	067	59 57	122	54 57	150	51 57	223	29 12	264	61 08	299
78	23 07	024	46 42	065	60 09	120	55 04	148	51 47	221	28 58	262	60 55	296
80	23 13	022	46 55	063	60 22	119	55 12	147	51 37	220	28 43	260	60 42	294
82	23 18	020	47 08	061	60 35	117	55 20	145	51 28	218	28 29	259	60 29	292
84	23 23	018	47 21	059	60 48	115	55 29	143	51 19	216	28 15	257	60 15	290
86	23 27	016	47 33	057	61 01	113	55 38	141	51 11	214	28 00	255	60 01	288
88	23 31	014	47 45	055	61 15	111	55 47	140	51 03	213	27 46	253	59 47	286
	◆SIRIUS		Suhail		ACRUX		◆RIGIL KENT.		Peacock		◆FOMALHAUT		ACHERNAR	
90	23 34	012	47 57	052	61 28	110	55 57	138	50 55	211	27 33	251	59 33	284
92	23 37	010	48 08	050	61 42	108	56 07	136	50 48	209	27 19	249	59 19	282
94	23 39	008	48 19	048	61 56	106	56 17	135	50 41	208	27 05	247	59 04	279
96	23 41	005	48 30	046	62 10	104	56 27	133	50 34	206	26 52	245	58 50	277
98	23 42	003	48 40	044	62 25	102	56 38	131	50 28	204	26 39	243	58 35	275
100	23 42	001	48 50	042	62 39	100	56 49	129	50 22	202	26 26	241	58 21	273
102	23 43	359	49 00	039	62 53	098	57 01	128	50 17	201	26 13	239	58 06	271
104	23 42	357	49 09	037	63 08	096	57 13	126	50 12	199	26 01	237	57 52	269
106	23 41	355	49 18	035	63 23	094	57 25	124	50 07	197	25 49	235	57 37	267
108	23 39	353	49 26	033	63 37	093	57 37	122	50 03	196	25 37	233	57 22	265
110	23 37	351	49 33	031	63 52	091	57 49	121	49 59	194	25 25	231	57 08	263
112	23 35	349	49 40	028	64 06	089	58 02	119	49 56	192	25 14	230	56 53	262
114	23 32	347	49 47	026	64 21	087	58 15	117	49 53	190	25 03	228	56 39	260
116	23 28	345	49 53	024	64 36	085	58 28	115	49 51	189	24 52	226	56 25	258
118	23 24	342	49 59	022	64 50	082	58 41	113	49 49	187	24 42	224	56 10	256
	Suhail		SPICA		◆RIGIL KENT.		Peacock		◆FOMALHAUT		ACHERNAR		◆SIRIUS	
120	50 04	019	12 07	083	58 55	111	49 47	185	24 32	222	55 56	254	23 19	340
122	50 09	017	12 22	081	59 09	110	49 46	184	24 23	220	55 42	252	23 14	338
124	50 13	015	12 36	079	59 23	108	49 45	182	24 13	218	55 28	250	23 08	336
126	50 16	012	12 50	077	59 37	106	49 45	180	24 04	216	55 15	248	23 02	334
128	50 19	010	13 04	075	59 51	104	49 45	179	23 56	214	55 01	246	22 56	332
130	50 21	008	13 18	073	60 05	102	49 46	177	23 48	212	54 48	245	22 49	330
132	50 23	006	13 32	071	60 19	100	49 47	175	23 40	211	54 35	243	22 41	328
134	50 24	003	13 46	069	60 34	098	49 48	173	23 33	209	54 22	241	22 33	326
136	50 25	001	14 00	067	60 48	096	49 50	172	23 26	207	54 09	239	22 25	324
138	50 25	359	14 13	065	61 03	094	49 52	170	23 20	205	53 57	237	22 16	322
140	50 24	357	14 26	063	61 17	092	49 55	168	23 14	203	53 45	236	22 07	320
142	50 23	354	14 39	061	61 32	091	49 58	167	23 09	201	53 33	234	21 57	318
144	50 21	352	14 51	059	61 47	089	50 02	165	23 04	199	53 21	232	21 47	316
146	50 19	350	15 04	057	62 01	087	50 06	163	22 59	197	53 10	230	21 36	313
148	50 16	347	15 16	055	62 16	084	50 10	162	22 55	195	52 59	228	21 26	311
	SPICA		◆ANTARES		Peacock		FOMALHAUT		◆ACHERNAR		CANOPUS		◆Suhail	
150	15 28	053	25 20	101	50 15	160	22 51	194	52 48	227	56 24	298	50 12	345
152	15 39	051	25 34	099	50 20	158	22 48	192	52 37	225	56 10	295	50 08	343
154	15 50	048	25 49	097	50 26	156	22 45	190	52 27	223	55 57	293	50 04	341
156	16 01	046	26 03	095	50 32	155	22 43	188	52 17	221	55 44	291	49 59	338
158	16 11	044	26 18	093	50 39	153	22 41	186	52 08	220	55 30	289	49 53	336
160	16 21	042	26 32	091	50 45	151	22 40	184	51 59	218	55 16	287	49 47	334
162	16 31	040	26 47	089	50 53	150	22 39	182	51 50	216	55 02	285	49 40	332
164	16 40	038	27 02	087	51 00	148	22 39	180	51 41	214	54 48	283	49 33	329
166	16 49	036	27 16	085	51 08	146	22 39	178	51 33	213	54 33	281	49 25	327
168	16 58	034	27 31	083	51 17	144	22 40	177	51 26	211	54 19	279	49 17	325
170	17 06	032	27 45	081	51 25	143	22 41	175	51 18	209	54 04	277	49 08	323
172	17 13	030	28 00	079	51 34	141	22 42	173	51 11	208	53 50	275	48 59	321
174	17 20	028	28 14	077	51 44	139	22 45	171	51 05	206	53 35	273	48 50	318
176	17 27	026	28 28	075	51 53	137	22 47	169	50 59	204	53 21	271	48 40	316
178	17 33	024	28 42	073	52 04	136	22 50	167	50 53	202	53 06	269	48 30	314

LAT 83°S

LHA ♈	Hc	Zn	Hc	Zn	Hc	Zn	Hc	Zn	Hc	Zn	Hc	Zn	Hc	Zn
	SPICA		◆ANTARES		Peacock		FOMALHAUT		◆ACHERNAR		CANOPUS		◆Suhail	
180	17 39	022	28 56	071	52 14	134	22 54	165	50 48	201	52 51	267	48 19	312
182	17 44	020	29 10	069	52 25	132	22 58	163	50 43	199	52 37	265	48 08	310
184	17 49	018	29 23	067	52 36	130	23 02	161	50 38	197	52 22	263	47 56	307
186	17 53	016	29 37	065	52 47	129	23 07	160	50 34	196	52 08	261	47 45	305
188	17 57	014	29 50	063	52 59	127	23 12	158	50 30	194	51 53	259	47 33	303
190	18 00	012	30 03	060	53 10	125	23 18	156	50 27	192	51 39	257	47 20	301
192	18 03	010	30 15	058	53 22	123	23 24	154	50 24	190	51 25	255	47 07	299
194	18 05	007	30 28	056	53 35	121	23 31	152	50 22	189	51 11	253	46 55	297
196	18 06	005	30 40	054	53 47	119	23 38	150	50 20	187	50 57	251	46 41	295
198	18 07	003	30 51	052	54 00	118	23 45	148	50 18	185	50 43	249	46 28	293
200	18 08	001	31 03	050	54 13	116	23 53	146	50 17	184	50 30	248	46 14	291
202	18 08	359	31 14	048	54 27	114	24 02	144	50 16	182	50 16	246	46 01	289
204	18 08	357	31 24	046	54 40	112	24 10	143	50 16	180	50 03	244	45 47	286
206	18 07	355	31 35	044	54 54	110	24 19	141	50 16	179	49 50	242	45 33	284
208	18 05	353	31 45	042	55 08	108	24 29	139	50 16	177	49 37	240	45 18	282
	◆ANTARES		Nunki		◆FOMALHAUT		ACHERNAR		CANOPUS		◆Suhail		SPICA	
210	31 54	040	28 04	077	24 39	137	50 17	175	49 25	238	45 04	280	18 03	351
212	32 03	038	28 18	075	24 49	135	50 19	174	49 12	236	44 50	278	18 01	349
214	32 12	035	28 32	073	24 59	133	50 21	172	49 00	235	44 35	276	17 57	347
216	32 20	033	28 46	071	25 10	131	50 23	170	48 48	233	44 21	274	17 54	345
218	32 28	031	29 00	069	25 21	129	50 26	168	48 37	231	44 06	272	17 50	343
220	32 36	029	29 13	067	25 33	127	50 29	167	48 26	229	43 51	270	17 45	341
222	32 42	027	29 27	065	25 45	125	50 32	165	48 15	227	43 37	268	17 40	339
224	32 49	025	29 40	063	25 57	123	50 36	163	48 04	226	43 22	266	17 34	337
226	32 55	023	29 53	061	26 09	122	50 41	162	47 54	224	43 08	264	17 28	334
228	33 00	021	30 05	059	26 22	120	50 46	160	47 44	222	42 53	262	17 22	332
230	33 05	019	30 18	057	26 34	118	50 51	158	47 34	220	42 39	260	17 15	330
232	33 09	016	30 30	055	26 47	116	50 56	157	47 25	218	42 24	259	17 07	328
234	33 13	014	30 42	053	27 01	114	51 02	155	47 16	217	42 10	257	17 00	326
236	33 17	012	30 53	051	27 14	112	51 09	153	47 08	215	41 56	255	16 51	324
238	33 19	010	31 04	049	27 28	110	51 16	151	46 59	213	41 42	253	16 42	322
	ANTARES		◆Nunki		FOMALHAUT		◆ACHERNAR		CANOPUS		◆ACRUX		SPICA	
240	33 22	008	31 15	046	27 42	108	51 23	150	46 52	211	66 37	294	16 33	320
242	33 23	006	31 25	044	27 56	106	51 30	148	46 44	210	66 23	291	16 24	318
244	33 25	004	31 35	042	28 10	104	51 38	146	46 37	208	66 09	289	16 14	316
246	33 25	001	31 45	040	28 24	102	51 47	145	46 31	206	65 55	287	16 03	314
248	33 25	359	31 54	038	28 38	100	51 55	143	46 24	204	65 41	285	15 53	312
250	33 25	357	32 03	036	28 53	098	52 04	141	46 18	203	65 27	283	15 42	310
252	33 24	355	32 11	034	29 07	096	52 14	139	46 13	201	65 13	281	15 30	308
254	33 22	353	32 19	032	29 22	094	52 23	138	46 08	199	64 58	279	15 19	306
256	33 20	351	32 27	030	29 36	092	52 33	136	46 03	197	64 44	277	15 07	304
258	33 17	348	32 34	028	29 51	090	52 44	134	45 59	196	64 29	275	14 54	302
260	33 14	346	32 40	025	30 06	088	52 54	132	45 56	194	64 15	273	14 42	300
262	33 11	344	32 46	023	30 20	086	53 05	130	45 52	192	64 00	271	14 29	298
264	33 06	342	32 52	021	30 35	084	53 17	129	45 49	190	63 46	269	14 16	296
266	33 02	340	32 57	019	30 49	082	53 28	127	45 47	189	63 31	267	14 03	294
268	32 56	338	33 01	017	31 04	080	53 40	125	45 45	187	63 16	265	13 49	292
	◆Nunki		FOMALHAUT		◆ACHERNAR		CANOPUS		◆ACRUX		RIGIL KENT.		ANTARES	
270	33 05	015	31 18	078	53 52	123	45 44	185	63 02	263	64 46	299	32 50	336
272	33 09	013	31 33	076	54 05	122	45 42	183	62 47	261	64 33	296	32 44	334
274	33 12	010	31 47	074	54 17	120	45 42	182	62 33	259	64 20	294	32 37	331
276	33 14	008	32 01	072	54 30	118	45 42	180	62 19	257	64 06	292	32 30	329
278	33 16	006	32 15	070	54 43	116	45 42	178	62 04	255	63 53	290	32 23	327
280	33 17	004	32 28	068	54 56	114	45 42	176	61 50	253	63 39	288	32 14	325
282	33 18	002	32 42	066	55 10	112	45 44	175	61 36	251	63 25	285	32 06	323
284	33 18	000	32 55	064	55 23	111	45 45	173	61 23	250	63 11	283	31 57	321
286	33 18	358	33 08	062	55 37	109	45 47	171	61 09	248	62 56	281	31 47	319
288	33 17	355	33 21	060	55 51	107	45 50	170	60 56	246	62 42	279	31 38	317
290	33 16	353	33 33	058	56 05	105	45 52	168	60 42	244	62 28	277	31 27	315
292	33 14	351	33 46	056	56 19	103	45 56	166	60 29	242	62 13	275	31 17	312
294	33 11	349	33 57	054	56 33	101	45 59	164	60 16	241	61 58	273	31 06	310
296	33 08	347	34 09	052	56 48	099	46 04	163	60 04	239	61 44	271	30 54	308
298	33 04	345	34 20	050	57 02	097	46 08	161	59 51	237	61 29	269	30 43	306
	◆FOMALHAUT		ACHERNAR		◆CANOPUS		ACRUX		RIGIL KENT.		◆ANTARES		Nunki	
300	34 31	048	57 17	095	46 13	159	59 39	235	61 15	267	30 31	304	33 00	343
302	34 42	045	57 31	093	46 19	157	59 27	234	61 00	265	30 19	302	32 56	340
304	34 52	043	57 46	091	46 25	156	59 16	232	60 45	263	30 06	300	32 50	338
306	35 02	041	58 01	089	46 31	154	59 04	230	60 31	261	29 53	298	32 45	336
308	35 11	039	58 15	087	46 37	152	58 53	228	60 17	259	29 40	296	32 39	334
310	35 20	037	58 30	085	46 44	150	58 42	227	60 02	258	29 27	294	32 32	332
312	35 29	035	58 44	083	46 52	149	58 32	225	59 48	256	29 14	292	32 25	330
314	35 37	033	58 59	081	47 00	147	58 22	223	59 34	254	29 00	290	32 17	328
316	35 45	031	59 13	079	47 08	145	58 12	222	59 20	252	28 46	288	32 09	326
318	35 52	028	59 28	077	47 16	143	58 02	220	59 06	250	28 32	286	32 01	323
320	35 59	026	59 42	075	47 25	141	57 53	218	58 52	248	28 18	284	31 52	321
322	36 05	024	59 56	073	47 35	140	57 44	217	58 39	246	28 04	282	31 43	319
324	36 11	022	60 10	071	47 44	138	57 36	215	58 26	245	27 49	280	31 33	317
326	36 16	020	60 24	069	47 54	136	57 28	213	58 13	243	27 35	278	31 23	315
328	36 21	018	60 37	067	48 05	134	57 20	212	58 00	241	27 20	276	31 12	313
	◆FOMALHAUT		ACHERNAR		◆CANOPUS		ACRUX		RIGIL KENT.		◆ANTARES		Nunki	
330	36 25	016	60 50	064	48 15	133	57 12	210	57 47	239	27 06	274	31 01	311
332	36 28	013	61 03	062	48 26	131	57 05	208	57 35	237	26 51	272	30 50	309
334	36 32	011	61 16	060	48 37	129	56 58	207	57 22	236	26 36	270	30 38	307
336	36 34	009	61 29	058	48 49	127	56 52	205	57 10	234	26 22	268	30 27	305
338	36 36	007	61 41	056	49 01	125	56 46	203	56 59	232	26 07	266	30 14	303
340	36 38	005	61 53	053	49 13	123	56 40	202	56 47	230	25 53	264	30 02	301
342	36 39	003	62 05	051	49 25	122	56 35	200	56 36	229	25 38	262	29 49	299
344	36 39	000	62 16	049	49 38	120	56 30	198	56 25	227	25 24	260	29 36	296
346	36 39	358	62 27	047	49 50	118	56 26	197	56 15	225	25 09	258	29 23	294
348	36 38	356	62 37	044	50 03	116	56 22	195	56 05	223	24 55	256	29 10	292
350	36 37	354	62 47	042	50 17	114	56 18	193	55 55	222	24 41	254	28 56	290
352	36 35	352	62 56	040	50 30	112	56 15	192	55 45	220	24 27	252	28 42	288
354	36 33	350	63 06	037	50 44	110	56 12	190	55 36	218	24 13	250	28 28	286
356	36 30	347	63 14	035	50 58	109	56 10	189	55 27	217	24 00	248	28 14	284
358	36 26	345	63 22	032	51 11	107	56 08	187	55 19	215	23 46	246	28 00	282

LAT 84°S

LHA ♈	Hc	Zn	Hc	Zn	Hc	Zn	Hc	Zn	Hc	Zn	Hc	Zn	Hc	Zn
	ACHERNAR		◆SIRIUS		Suhail		ACRUX		◆RIGIL KENT.		Peacock		◆FOMALHAUT	
0	62 38	029	15 27	103	38 54	140	57 06	185	56 00	214	59 57	298	35 25	343
2	62 43	027	15 39	101	39 02	139	57 05	184	55 54	212	59 46	296	35 21	341
4	62 49	024	15 52	099	39 11	137	57 05	182	55 47	210	59 35	294	35 17	339
6	62 54	022	16 04	097	39 19	135	57 04	180	55 41	209	59 23	292	35 12	337
8	62 58	020	16 17	095	39 28	133	57 04	179	55 35	207	59 12	290	35 07	335
10	63 02	017	16 29	093	39 38	131	57 05	177	55 29	205	59 00	287	35 01	333
12	63 06	015	16 42	091	39 47	129	57 06	175	55 24	204	58 48	285	34 55	331
14	63 09	012	16 54	089	39 57	127	57 07	174	55 19	202	58 36	283	34 49	328
16	63 11	010	17 07	087	40 07	125	57 08	172	55 15	200	58 23	281	34 42	326
18	63 13	008	17 19	085	40 18	124	57 10	170	55 11	199	58 11	279	34 35	324
20	63 14	005	17 32	083	40 28	122	57 12	169	55 07	197	57 59	277	34 27	322
22	63 15	003	17 44	081	40 39	120	57 15	167	55 04	195	57 46	275	34 20	320
24	63 16	000	17 57	079	40 50	118	57 18	165	55 00	193	57 34	273	34 11	318
26	63 16	358	18 09	077	41 01	116	57 21	164	54 58	192	57 21	271	34 03	316
28	63 15	356	18 21	075	41 13	114	57 25	162	54 55	190	57 08	269	33 54	314
	◆SIRIUS		Suhail		ACRUX		◆RIGIL KENT.		Peacock		◆FOMALHAUT		ACHERNAR	
30	18 33	073	41 24	112	57 29	160	54 53	188	56 56	267	33 45	312	63 14	353
32	18 45	071	41 36	110	57 33	159	54 52	187	56 43	265	33 35	310	63 12	351
34	18 57	069	41 48	108	57 38	157	54 50	185	56 31	263	33 25	307	63 10	348
36	19 08	067	42 00	106	57 43	155	54 50	183	56 18	261	33 15	305	63 07	346
38	19 20	065	42 12	104	57 49	154	54 49	182	56 06	259	33 05	303	63 04	344
40	19 31	063	42 24	102	57 54	152	54 49	180	55 54	257	32 54	301	63 00	341
42	19 42	061	42 36	101	58 00	150	54 49	178	55 42	255	32 43	299	62 56	339
44	19 53	059	42 49	099	58 07	149	54 50	176	55 30	254	32 32	297	62 51	337
46	20 04	057	43 01	097	58 13	147	54 51	175	55 18	252	32 21	295	62 46	334
48	20 14	055	43 14	095	58 20	145	54 52	173	55 06	250	32 10	293	62 40	332
50	20 24	053	43 26	093	58 28	143	54 54	171	54 54	248	31 58	291	62 34	330
52	20 34	051	43 39	091	58 35	142	54 56	170	54 43	246	31 46	289	62 27	327
54	20 44	049	43 51	089	58 43	140	54 58	168	54 31	244	31 34	287	62 20	325
56	20 53	047	44 04	087	58 52	138	55 01	166	54 20	242	31 22	285	62 13	323
58	21 02	045	44 16	085	59 00	137	55 04	165	54 09	240	31 10	283	62 05	320
	◆SIRIUS		Suhail		ACRUX		◆RIGIL KENT.		Peacock		◆FOMALHAUT		ACHERNAR	
60	21 11	043	44 29	083	59 09	135	55 08	163	53 58	239	30 58	281	61 57	318
62	21 19	041	44 41	081	59 18	133	55 12	161	53 48	237	30 45	279	61 48	316
64	21 27	039	44 53	079	59 27	131	55 16	159	53 37	235	30 33	277	61 39	313
66	21 34	036	45 06	077	59 37	130	55 20	158	53 27	233	30 21	275	61 30	311
68	21 42	034	45 18	075	59 46	128	55 25	156	53 17	231	30 08	273	61 20	309
70	21 49	032	45 30	072	59 56	126	55 30	154	53 07	229	29 55	271	61 10	307
72	21 55	030	45 42	070	60 07	124	55 36	153	52 58	228	29 43	269	61 00	305
74	22 01	028	45 53	068	60 17	122	55 42	151	52 49	226	29 30	267	60 50	302
76	22 07	026	46 05	066	60 28	121	55 48	149	52 40	224	29 18	265	60 39	300
78	22 12	024	46 16	064	60 39	119	55 55	147	52 31	222	29 05	263	60 28	298
80	22 17	022	46 28	062	60 50	117	56 02	146	52 23	221	28 53	261	60 17	296
82	22 22	020	46 39	060	61 01	115	56 09	144	52 15	219	28 41	259	60 05	294
84	22 26	018	46 49	058	61 13	113	56 17	142	52 07	217	28 28	257	59 54	292
86	22 29	016	47 00	056	61 24	112	56 24	141	52 00	215	28 16	255	59 42	289
88	22 33	014	47 10	054	61 36	110	56 32	139	51 53	213	28 04	253	59 30	287
	◆SIRIUS		Suhail		ACRUX		◆RIGIL KENT.		Peacock		◆FOMALHAUT		ACHERNAR	
90	22 35	012	47 20	052	61 48	108	56 41	137	51 46	212	27 52	251	59 18	285
92	22 38	010	47 30	049	62 00	106	56 50	135	51 40	210	27 40	249	59 06	283
94	22 40	008	47 39	047	62 12	104	56 59	133	51 34	208	27 29	247	58 54	281
96	22 41	005	47 48	045	62 24	102	57 08	132	51 28	206	27 17	245	58 41	279
98	22 42	003	47 57	043	62 36	100	57 17	130	51 23	205	27 06	243	58 29	277
100	22 42	001	48 05	041	62 49	098	57 27	128	51 18	203	26 55	241	58 17	275
102	22 43	359	48 13	039	63 01	096	57 37	126	51 13	201	26 44	240	58 04	273
104	22 42	357	48 21	036	63 14	094	57 47	125	51 09	199	26 33	238	57 51	271
106	22 41	355	48 28	034	63 26	092	57 58	123	51 05	198	26 23	236	57 39	269
108	22 40	353	48 35	032	63 39	091	58 08	121	51 01	196	26 12	234	57 26	267
110	22 38	351	48 41	030	63 51	089	58 19	119	50 58	194	26 02	232	57 14	265
112	22 36	349	48 48	028	64 04	087	58 30	117	50 55	192	25 53	230	57 01	263
114	22 33	347	48 53	026	64 16	084	58 41	115	50 52	191	25 43	228	56 49	261
116	22 30	345	48 58	023	64 29	082	58 53	114	50 50	189	25 34	226	56 37	259
118	22 27	343	49 03	021	64 41	080	59 04	112	50 48	187	25 25	224	56 24	257
	Suhail		SPICA		◆RIGIL KENT.		Peacock		◆FOMALHAUT		ACHERNAR		◆SIRIUS	
120	49 07	019	11 59	082	59 16	110	50 47	185	25 17	222	56 12	255	22 23	341
122	49 11	017	12 12	080	59 28	108	50 46	184	25 08	220	56 00	253	22 18	338
124	49 15	014	12 24	078	59 40	106	50 45	182	25 00	218	55 48	252	22 14	336
126	49 17	012	12 36	076	59 52	104	50 45	180	24 53	216	55 36	250	22 08	334
128	49 20	010	12 48	074	60 04	102	50 45	179	24 46	215	55 25	248	22 03	332
130	49 22	008	13 00	072	60 17	100	50 46	177	24 39	213	55 13	246	21 57	330
132	49 23	006	13 12	070	60 29	098	50 47	175	24 32	211	55 02	244	21 50	328
134	49 24	003	13 24	068	60 41	097	50 48	173	24 26	209	54 51	242	21 43	326
136	49 25	001	13 36	066	60 54	095	50 50	172	24 20	207	54 40	240	21 36	324
138	49 25	359	13 47	064	61 06	093	50 52	170	24 14	205	54 29	239	21 29	322
140	49 24	357	13 58	062	61 19	091	50 54	168	24 09	203	54 18	237	21 21	320
142	49 23	354	14 09	060	61 32	089	50 57	166	24 05	201	54 08	235	21 12	318
144	49 22	352	14 20	058	61 44	087	51 00	165	24 00	199	53 58	233	21 04	316
146	49 20	350	14 31	056	61 57	085	51 03	163	23 56	197	53 48	231	20 55	314
148	49 17	348	14 41	054	62 09	083	51 07	161	23 53	196	53 38	229	20 46	312
	SPICA		◆ANTARES		Peacock		FOMALHAUT		◆ACHERNAR		CANOPUS		◆Suhail	
150	14 51	052	25 31	100	51 11	159	23 50	194	53 29	228	55 55	299	49 14	345
152	15 01	050	25 43	098	51 16	158	23 47	192	53 20	226	55 44	297	49 11	343
154	15 10	048	25 55	096	51 21	156	23 44	190	53 11	224	55 33	295	49 07	341
156	15 20	046	26 08	094	51 26	154	23 43	188	53 02	222	55 21	293	49 03	339
158	15 28	044	26 20	092	51 32	152	23 41	186	52 54	220	55 10	290	48 58	337
160	15 37	042	26 33	090	51 38	151	23 40	184	52 46	219	54 58	288	48 53	334
162	15 45	040	26 46	088	51 44	149	23 39	182	52 38	217	54 46	286	48 47	332
164	15 53	038	26 58	086	51 51	147	23 39	180	52 31	215	54 34	284	48 41	330
166	16 01	036	27 11	084	51 58	145	23 39	178	52 24	213	54 21	282	48 35	328
168	16 08	034	27 23	082	52 05	144	23 40	177	52 17	212	54 09	280	48 28	326
170	16 15	032	27 35	080	52 13	142	23 41	175	52 11	210	53 57	278	48 21	323
172	16 21	030	27 48	078	52 21	140	23 42	173	52 04	208	53 44	276	48 13	321
174	16 27	028	28 00	076	52 29	138	23 44	171	51 59	206	53 32	274	48 05	319
176	16 33	026	28 12	074	52 37	136	23 46	169	51 53	205	53 19	272	47 56	317
178	16 38	024	28 24	072	52 46	135	23 49	167	51 48	203	53 07	270	47 48	315

LHA ♈	Hc	Zn	Hc	Zn	Hc	Zn	Hc	Zn	Hc	Zn	Hc	Zn	Hc	Zn
	SPICA		◆ANTARES		Peacock		FOMALHAUT		◆ACHERNAR		CANOPUS		◆Suhail	
180	16 43	022	28 36	070	52 55	133	23 52	165	51 44	201	52 54	268	47 39	313
182	16 48	020	28 48	068	53 04	131	23 55	163	51 39	199	52 42	266	47 29	310
184	16 52	018	28 59	066	53 14	129	23 59	161	51 35	198	52 29	264	47 20	308
186	16 55	016	29 11	064	53 24	127	24 03	159	51 32	196	52 17	262	47 10	306
188	16 58	014	29 22	062	53 34	126	24 08	158	51 28	194	52 04	260	46 59	304
190	17 01	012	29 33	060	53 44	124	24 13	156	51 25	192	51 52	258	46 49	302
192	17 03	009	29 44	058	53 55	122	24 18	154	51 23	191	51 40	256	46 38	300
194	17 05	007	29 54	056	54 06	120	24 24	152	51 21	189	51 28	254	46 27	298
196	17 07	005	30 04	054	54 16	118	24 30	150	51 19	187	51 16	252	46 16	296
198	17 08	003	30 14	052	54 28	116	24 36	148	51 18	186	51 04	251	46 04	294
200	17 08	001	30 24	050	54 39	115	24 43	146	51 17	184	50 52	249	45 53	292
202	17 08	359	30 33	048	54 50	113	24 50	144	51 16	182	50 40	247	45 41	289
204	17 08	357	30 43	046	55 02	111	24 58	142	51 16	180	50 29	245	45 29	287
206	17 07	355	30 51	043	55 14	109	25 06	140	51 16	179	50 18	243	45 17	285
208	17 06	353	31 00	041	55 26	107	25 14	138	51 16	177	50 07	241	45 05	283
	◆ANTARES		Nunki		◆FOMALHAUT		ACHERNAR		CANOPUS		◆Suhail		SPICA	
210	31 08	039	27 50	077	25 22	137	51 17	175	49 56	239	44 53	281	17 04	351
212	31 16	037	28 02	075	25 31	135	51 18	173	49 45	237	44 40	279	17 02	349
214	31 23	035	28 14	073	25 40	133	51 20	172	49 35	236	44 28	277	16 59	347
216	31 30	033	28 26	071	25 49	131	51 22	170	49 24	234	44 16	275	16 56	345
218	31 37	031	28 38	069	25 59	129	51 24	168	49 14	232	44 03	273	16 52	343
220	31 43	029	28 50	067	26 09	127	51 27	166	49 05	230	43 51	271	16 49	341
222	31 49	027	29 01	065	26 19	125	51 30	165	48 55	228	43 38	269	16 44	339
224	31 54	025	29 12	063	26 30	123	51 34	163	48 46	226	43 25	267	16 39	337
226	31 59	023	29 23	060	26 40	121	51 38	161	48 37	225	43 13	265	16 34	335
228	32 04	020	29 34	058	26 51	119	51 42	160	48 28	223	43 00	263	16 29	333
230	32 08	018	29 45	056	27 02	117	51 46	158	48 20	221	42 48	261	16 23	331
232	32 12	016	29 55	054	27 13	115	51 51	156	48 12	219	42 36	259	16 16	328
234	32 15	014	30 05	052	27 25	113	51 57	154	48 04	217	42 23	257	16 10	326
236	32 18	012	30 15	050	27 36	111	52 02	153	47 57	216	42 11	256	16 02	324
238	32 20	010	30 24	048	27 48	109	52 08	151	47 49	214	41 59	254	15 55	322
	ANTARES		◆Nunki		FOMALHAUT		◆ACHERNAR		CANOPUS		◆ACRUX		SPICA	
240	32 22	008	30 33	046	28 00	108	52 14	149	47 43	212	66 11	296	15 47	320
242	32 24	006	30 42	044	28 12	106	52 21	147	47 36	210	66 00	294	15 39	318
244	32 25	003	30 51	042	28 24	104	52 28	146	47 30	208	65 48	291	15 30	316
246	32 25	001	30 59	040	28 36	102	52 35	144	47 24	207	65 37	289	15 22	314
248	32 25	359	31 07	038	28 49	100	52 43	142	47 19	205	65 25	287	15 12	312
250	32 25	357	31 14	036	29 01	098	52 51	140	47 14	203	65 13	285	15 03	310
252	32 24	355	31 21	034	29 14	096	52 59	138	47 09	201	65 01	283	14 53	308
254	32 23	353	31 28	031	29 26	094	53 07	137	47 05	199	64 48	281	14 43	306
256	32 21	351	31 35	029	29 39	092	53 16	135	47 01	198	64 36	279	14 33	304
258	32 19	349	31 41	027	29 51	090	53 25	133	46 57	196	64 24	277	14 23	302
260	32 16	347	31 46	025	30 04	088	53 34	131	46 54	194	64 11	275	14 12	300
262	32 13	344	31 51	023	30 16	086	53 44	129	46 51	192	63 59	273	14 01	298
264	32 09	342	31 56	021	30 29	084	53 54	128	46 48	191	63 46	271	13 50	296
266	32 05	340	32 00	019	30 41	082	54 04	126	46 46	189	63 33	269	13 38	294
268	32 01	338	32 04	017	30 54	080	54 14	124	46 45	187	63 21	267	13 27	292
	◆Nunki		FOMALHAUT		◆ACHERNAR		CANOPUS		◆ACRUX		RIGIL KENT.		ANTARES	
270	32 07	015	31 06	078	54 25	122	46 43	185	63 08	265	64 17	300	31 56	336
272	32 10	012	31 18	076	54 35	120	46 42	183	62 56	263	64 06	298	31 50	334
274	32 13	010	31 30	074	54 46	118	46 42	182	62 44	261	63 55	296	31 45	332
276	32 15	008	31 42	072	54 57	117	46 42	180	62 31	259	63 43	294	31 39	330
278	32 16	006	31 54	070	55 09	115	46 42	178	62 19	257	63 32	292	31 32	328
280	32 17	004	32 06	068	55 20	113	46 42	176	62 07	255	63 20	290	31 25	325
282	32 18	002	32 17	066	55 32	111	46 43	175	61 55	253	63 08	287	31 18	323
284	32 18	000	32 28	064	55 44	109	46 45	173	61 43	251	62 56	285	31 10	321
286	32 18	358	32 40	061	55 55	107	46 46	171	61 31	250	62 44	283	31 02	319
288	32 17	355	32 50	059	56 08	105	46 49	169	61 19	248	62 31	281	30 54	317
290	32 16	353	33 01	057	56 20	103	46 51	168	61 08	246	62 19	279	30 45	315
292	32 14	351	33 12	055	56 32	102	46 54	166	60 56	244	62 07	277	30 36	313
294	32 12	349	33 22	053	56 44	100	46 57	164	60 45	242	61 54	275	30 27	311
296	32 10	347	33 32	051	56 57	098	47 01	162	60 34	240	61 42	273	30 17	309
298	32 06	345	33 41	049	57 09	096	47 05	160	60 23	239	61 29	271	30 07	307
	◆FOMALHAUT		ACHERNAR		◆CANOPUS		ACRUX		RIGIL KENT.		◆ANTARES		Nunki	
300	33 51	047	57 22	094	47 09	159	60 13	237	61 17	269	29 57	305	32 03	343
302	34 00	045	57 34	092	47 14	157	60 02	235	61 04	267	29 46	303	31 59	341
304	34 08	043	57 47	090	47 19	155	59 52	233	60 52	265	29 36	301	31 55	339
306	34 17	041	57 59	088	47 25	153	59 42	231	60 39	263	29 25	299	31 50	336
308	34 25	039	58 12	086	47 30	152	59 33	230	60 27	261	29 14	296	31 45	334
310	34 32	037	58 24	084	47 36	150	59 23	228	60 14	259	29 02	294	31 39	332
312	34 40	034	58 37	082	47 43	148	59 14	226	60 02	257	28 51	292	31 33	330
314	34 47	032	58 49	080	47 50	146	59 05	224	59 50	255	28 39	290	31 27	328
316	34 53	030	59 01	078	47 57	144	58 56	223	59 38	254	28 27	288	31 20	326
318	34 59	028	59 14	076	48 04	143	58 48	221	59 26	252	28 15	286	31 12	324
320	35 05	026	59 26	073	48 12	141	58 40	219	59 14	250	28 03	284	31 05	322
322	35 10	024	59 38	071	48 20	139	58 32	218	59 02	248	27 51	282	30 57	320
324	35 15	022	59 49	069	48 29	137	58 25	216	58 51	246	27 39	280	30 49	318
326	35 19	020	60 01	067	48 37	135	58 17	214	58 39	244	27 26	278	30 40	316
328	35 23	017	60 13	065	48 46	134	58 11	212	58 28	242	27 14	276	30 31	313
	◆FOMALHAUT		ACHERNAR		◆CANOPUS		ACRUX		RIGIL KENT.		◆ANTARES		Nunki	
330	35 27	015	60 24	063	48 55	132	58 04	211	58 17	241	27 02	274	30 22	311
332	35 30	013	60 35	061	49 05	130	57 58	209	58 06	239	26 49	272	30 12	309
334	35 33	011	60 46	059	49 15	128	57 52	207	57 56	237	26 36	270	30 02	307
336	35 35	009	60 56	056	49 25	126	57 46	206	57 45	235	26 24	268	29 52	305
338	35 37	007	61 07	054	49 35	124	57 41	204	57 35	233	26 11	266	29 42	303
340	35 38	005	61 17	052	49 45	122	57 36	202	57 25	232	25 59	264	29 31	301
342	35 39	003	61 26	050	49 56	121	57 32	201	57 15	230	25 46	262	29 20	299
344	35 39	000	61 36	047	50 07	119	57 27	199	57 06	228	25 34	260	29 09	297
346	35 39	358	61 45	045	50 18	117	57 23	197	56 57	226	25 22	258	28 58	295
348	35 38	356	61 53	043	50 29	115	57 20	196	56 48	224	25 09	256	28 47	293
350	35 37	354	62 02	041	50 41	113	57 17	194	56 39	223	24 57	254	28 35	291
352	35 36	352	62 10	038	50 52	111	57 14	192	56 31	221	24 45	253	28 23	289
354	35 33	350	62 17	036	51 04	109	57 11	190	56 23	219	24 33	251	28 11	287
356	35 31	348	62 25	034	51 16	107	57 09	189	56 15	217	24 22	249	27 59	285
358	35 28	345	62 31	031	51 28	106	57 08	187	56 08	216	24 10	247	27 47	283

 LAT 85°S

LHA	Hc	Zn	Hc	Zn	Hc	Zn	Hc	Zn	Hc	Zn	Hc	Zn	Hc	Zn
♈	ACHERNAR		◆SIRIUS		Suhail		ACRUX		◆RIGIL KENT.		Peacock		◆FOMALHAUT	
0	61 45	028	15 40	103	39 40	140	58 06	186	56 50	215	59 29	300	34 27	343
2	61 50	026	15 51	101	39 47	138	58 05	184	56 44	213	59 19	297	34 24	341
4	61 54	024	16 01	099	39 54	136	58 05	182	56 39	211	59 10	295	34 21	339
6	61 58	021	16 11	097	40 02	134	58 04	180	56 33	209	59 00	293	34 17	337
8	62 02	019	16 22	095	40 09	132	58 04	179	56 28	208	58 51	291	34 12	335
10	62 05	017	16 32	093	40 17	130	58 05	177	56 24	206	58 41	289	34 08	333
12	62 08	014	16 43	091	40 25	129	58 05	175	56 19	204	58 31	287	34 03	331
14	62 10	012	16 53	089	40 33	127	58 06	174	56 15	202	58 21	285	33 58	329
16	62 12	010	17 04	087	40 42	125	58 08	172	56 11	201	58 11	283	33 52	327
18	62 14	007	17 14	085	40 51	123	58 09	170	56 08	199	58 01	281	33 46	325
20	62 15	005	17 24	083	40 59	121	58 11	168	56 04	197	57 50	279	33 40	323
22	62 15	003	17 35	081	41 08	119	58 14	167	56 01	195	57 40	277	33 33	320
24	62 16	000	17 45	079	41 18	117	58 16	165	55 59	194	57 29	275	33 27	318
26	62 16	358	17 55	077	41 27	115	58 19	163	55 56	192	57 19	273	33 20	316
28	62 15	356	18 05	075	41 37	113	58 22	162	55 54	190	57 09	271	33 12	314
	◆SIRIUS		Suhail		ACRUX		◆RIGIL KENT.		Peacock		◆FOMALHAUT		ACHERNAR	
30	18 15	073	41 46	111	58 26	160	55 53	189	56 58	269	33 05	312	62 14	353
32	18 25	071	41 56	109	58 29	158	55 51	187	56 48	267	32 57	310	62 13	351
34	18 35	069	42 06	107	58 33	156	55 50	185	56 37	265	32 49	308	62 11	349
36	18 45	067	42 16	105	58 38	155	55 49	183	56 27	263	32 40	306	62 09	347
38	18 54	065	42 26	104	58 42	153	55 49	182	56 16	261	32 32	304	62 06	344
40	19 04	063	42 36	102	58 47	151	55 49	180	56 06	259	32 23	302	62 03	342
42	19 13	061	42 47	100	58 52	149	55 49	178	55 56	257	32 14	300	61 59	340
44	19 22	059	42 57	098	58 58	148	55 50	176	55 46	255	32 05	298	61 56	337
46	19 31	057	43 07	096	59 03	146	55 50	175	55 36	253	31 55	296	61 51	335
48	19 39	055	43 18	094	59 09	144	55 52	173	55 26	251	31 46	294	61 47	333
50	19 48	053	43 28	092	59 16	142	55 53	171	55 16	249	31 36	292	61 42	330
52	19 56	050	43 39	090	59 22	141	55 55	169	55 06	247	31 26	290	61 36	328
54	20 04	048	43 49	088	59 29	139	55 57	168	54 57	245	31 16	288	61 31	326
56	20 12	046	44 00	086	59 36	137	55 59	166	54 47	244	31 06	286	61 25	324
58	20 19	044	44 10	084	59 43	135	56 02	164	54 38	242	30 56	284	61 18	321
	◆SIRIUS		Suhail		ACRUX		◆RIGIL KENT.		Peacock		◆FOMALHAUT		ACHERNAR	
60	20 26	042	44 20	082	59 51	134	56 05	162	54 29	240	30 46	282	61 12	319
62	20 33	040	44 31	080	59 58	132	56 08	161	54 20	238	30 36	279	61 05	317
64	20 40	038	44 41	078	60 06	130	56 12	159	54 11	236	30 25	277	60 57	315
66	20 46	036	44 51	076	60 14	128	56 16	157	54 03	234	30 15	275	60 50	313
68	20 52	034	45 01	074	60 23	126	56 20	155	53 54	232	30 05	273	60 42	310
70	20 58	032	45 11	072	60 31	125	56 24	154	53 46	231	29 54	271	60 34	308
72	21 03	030	45 21	069	60 40	123	56 29	152	53 38	229	29 44	269	60 26	306
74	21 08	028	45 31	067	60 49	121	56 34	150	53 30	227	29 33	268	60 17	304
76	21 13	026	45 40	065	60 58	119	56 40	148	53 23	225	29 23	266	60 08	302
78	21 18	024	45 50	063	61 07	117	56 45	147	53 15	223	29 12	264	59 59	299
80	21 22	022	45 59	061	61 16	115	56 51	145	53 08	221	29 02	262	59 50	297
82	21 25	020	46 08	059	61 26	114	56 57	143	53 02	220	28 52	260	59 41	295
84	21 29	018	46 17	057	61 36	112	57 04	141	52 55	218	28 41	258	59 31	293
86	21 32	016	46 26	055	61 45	110	57 10	140	52 49	216	28 31	256	59 21	291
88	21 34	014	46 34	053	61 55	108	57 17	138	52 43	214	28 21	254	59 12	289
	◆SIRIUS		Suhail		ACRUX		◆RIGIL KENT.		Peacock		◆FOMALHAUT		ACHERNAR	
90	21 37	012	46 42	051	62 05	106	57 24	136	52 37	212	28 11	252	59 02	287
92	21 39	010	46 50	049	62 15	104	57 32	134	52 32	211	28 01	250	58 52	285
94	21 40	007	46 58	046	62 25	102	57 39	132	52 27	209	27 52	248	58 41	283
96	21 41	005	47 06	044	62 36	100	57 47	131	52 22	207	27 42	246	58 31	281
98	21 42	003	47 13	042	62 46	098	57 55	129	52 17	205	27 33	244	58 21	279
100	21 42	001	47 20	040	62 56	096	58 04	127	52 13	203	27 23	242	58 10	277
102	21 43	359	47 26	038	63 07	094	58 12	125	52 09	202	27 14	240	58 00	275
104	21 42	357	47 32	036	63 17	092	58 21	123	52 05	200	27 05	238	57 50	273
106	21 41	355	47 38	034	63 28	090	58 30	121	52 02	198	26 56	236	57 39	271
108	21 40	353	47 44	032	63 38	088	58 39	120	51 59	196	26 48	234	57 29	269
110	21 39	351	47 49	029	63 49	086	58 48	118	51 56	194	26 39	232	57 18	267
112	21 37	349	47 54	027	63 59	084	58 57	116	51 53	193	26 31	230	57 08	265
114	21 35	347	47 59	025	64 09	082	59 07	114	51 51	191	26 23	228	56 57	263
116	21 32	345	48 03	023	64 20	080	59 16	112	51 49	189	26 16	226	56 47	261
118	21 29	343	48 07	021	64 30	078	59 26	110	51 48	187	26 08	224	56 37	259
	◆Suhail		SPICA		◆ANTARES		Peacock		FOMALHAUT		◆ACHERNAR		SIRIUS	
120	48 11	019	11 51	082	23 20	129	51 47	186	26 01	223	56 27	257	21 26	341
122	48 14	016	12 02	080	23 28	127	51 46	184	25 54	221	56 16	255	21 22	339
124	48 16	014	12 12	078	23 36	125	51 45	182	25 47	219	56 06	253	21 19	337
126	48 19	012	12 22	076	23 45	123	51 45	180	25 41	217	55 56	251	21 14	335
128	48 21	010	12 32	074	23 54	121	51 45	179	25 35	215	55 47	249	21 10	332
130	48 22	008	12 42	072	24 03	119	51 46	177	25 29	213	55 37	247	21 04	330
132	48 24	005	12 52	070	24 12	117	51 46	175	25 24	211	55 27	245	20 59	328
134	48 24	003	13 02	068	24 21	115	51 47	173	25 18	209	55 18	243	20 54	326
136	48 25	001	13 12	066	24 31	113	51 49	171	25 13	207	55 09	242	20 48	324
138	48 25	359	13 21	064	24 41	112	51 51	170	25 09	205	54 59	240	20 41	322
140	48 24	357	13 30	062	24 50	110	51 53	168	25 04	203	54 51	238	20 35	320
142	48 23	354	13 40	060	25 00	108	51 55	166	25 00	201	54 42	236	20 28	318
144	48 22	352	13 49	058	25 10	106	51 58	164	24 57	199	54 33	234	20 21	316
146	48 21	350	13 57	056	25 20	104	52 01	163	24 54	198	54 25	232	20 13	314
148	48 19	348	14 06	054	25 31	102	52 04	161	24 51	196	54 17	230	20 06	312
	SPICA		◆ANTARES		Peacock		FOMALHAUT		◆ACHERNAR		CANOPUS		◆Suhail	
150	14 14	052	25 41	100	52 08	159	24 48	194	54 09	229	55 26	300	48 16	346
152	14 22	050	25 51	098	52 11	157	24 46	192	54 01	227	55 16	298	48 14	344
154	14 30	048	26 02	096	52 16	155	24 44	190	53 53	225	55 07	296	48 10	341
156	14 38	046	26 12	094	52 20	154	24 42	188	53 46	223	54 58	294	48 07	339
158	14 45	044	26 23	092	52 25	152	24 41	186	53 39	221	54 48	292	48 03	337
160	14 52	042	26 33	090	52 30	150	24 40	184	53 32	220	54 38	290	47 59	335
162	14 59	040	26 43	088	52 35	148	24 39	182	53 26	218	54 28	288	47 54	333
164	15 06	038	26 54	086	52 41	146	24 39	180	53 20	216	54 18	286	47 49	331
166	15 12	036	27 04	084	52 47	145	24 39	178	53 14	214	54 08	284	47 44	328
168	15 18	034	27 15	082	52 53	143	24 40	177	53 08	212	53 58	281	47 38	326
170	15 24	032	27 25	080	53 00	141	24 40	175	53 02	211	53 48	279	47 32	324
172	15 29	030	27 35	078	53 06	139	24 41	173	52 57	209	53 37	277	47 26	322
174	15 34	028	27 45	076	53 13	137	24 43	171	52 52	207	53 27	275	47 19	320
176	15 39	026	27 56	074	53 20	136	24 45	169	52 48	205	53 16	273	47 12	318
178	15 43	024	28 06	072	53 28	134	24 47	167	52 43	203	53 06	271	47 05	316

LHA	Hc	Zn	Hc	Zn	Hc	Zn	Hc	Zn	Hc	Zn	Hc	Zn	Hc	Zn
♈	SPICA		◆ANTARES		Peacock		FOMALHAUT		◆ACHERNAR		CANOPUS		◆Suhail	
180	15 47	022	28 15	070	53 36	132	24 50	165	52 39	202	52 56	269	46 58	313
182	15 51	020	28 25	068	53 43	130	24 52	163	52 36	200	52 45	267	46 50	311
184	15 54	018	28 35	066	53 52	128	24 56	161	52 32	198	52 35	265	46 42	309
186	15 57	016	28 44	064	54 00	126	24 59	159	52 29	196	52 24	263	46 34	307
188	16 00	014	28 53	062	54 08	125	25 03	157	52 26	194	52 14	261	46 25	305
190	16 02	011	29 03	060	54 17	123	25 07	155	52 24	193	52 04	260	46 17	303
192	16 04	009	29 12	057	54 26	121	25 12	154	52 22	191	51 53	258	46 08	301
194	16 06	007	29 20	055	54 35	119	25 17	152	52 20	189	51 43	256	45 59	299
196	16 07	005	29 29	053	54 44	117	25 22	150	52 19	187	51 33	254	45 49	297
198	16 08	003	29 37	051	54 54	115	25 27	148	52 17	186	51 23	252	45 40	295
200	16 08	001	29 45	049	55 03	113	25 33	146	52 16	184	51 13	250	45 30	292
202	16 08	359	29 53	047	55 13	111	25 39	144	52 16	182	51 03	248	45 21	290
204	16 08	357	30 00	045	55 23	109	25 45	142	52 16	180	50 54	246	45 11	288
206	16 07	355	30 08	043	55 33	108	25 52	140	52 16	179	50 44	244	45 01	286
208	16 06	353	30 15	041	55 43	106	25 59	138	52 16	177	50 35	242	44 51	284
	◆ANTARES		Nunki		FOMALHAUT		◆ACHERNAR		CANOPUS		◆Suhail		SPICA	
210	30 21	039	27 36	076	26 06	136	52 17	175	50 26	240	44 41	282	16 05	351
212	30 28	037	27 46	074	26 13	134	52 18	173	50 17	238	44 30	280	16 03	349
214	30 34	035	27 56	072	26 21	132	52 19	171	50 08	237	44 20	278	16 01	347
216	30 40	033	28 06	070	26 29	130	52 21	170	49 59	235	44 10	276	15 58	345
218	30 45	031	28 16	068	26 37	128	52 23	168	49 51	233	43 59	274	15 55	343
220	30 50	029	28 25	066	26 45	127	52 26	166	49 43	231	43 49	272	15 52	341
222	30 55	026	28 35	064	26 53	125	52 28	164	49 35	229	43 38	270	15 48	339
224	31 00	024	28 44	062	27 02	123	52 31	163	49 27	227	43 28	268	15 44	337
226	31 04	022	28 53	060	27 11	121	52 34	161	49 19	225	43 17	266	15 40	335
228	31 08	020	29 02	058	27 20	119	52 38	159	49 12	224	43 07	264	15 35	333
230	31 11	018	29 11	056	27 29	117	52 42	157	49 05	222	42 57	262	15 30	331
232	31 14	016	29 20	054	27 39	115	52 46	156	48 58	220	42 46	260	15 25	329
234	31 17	014	29 28	052	27 48	113	52 51	154	48 52	218	42 36	258	15 20	327
236	31 19	012	29 36	050	27 58	111	52 55	152	48 45	216	42 26	256	15 14	325
238	31 21	010	29 44	048	28 08	109	53 00	150	48 39	214	42 16	254	15 07	323
	ANTARES		Nunki		◆FOMALHAUT		ACHERNAR		◆CANOPUS		ACRUX		◆RIGIL KENT.	
240	31 23	008	29 52	046	28 18	107	53 06	148	48 33	213	65 44	298	65 27	336
242	31 24	006	29 59	044	28 28	105	53 11	147	48 28	211	65 35	296	65 23	334
244	31 25	003	30 06	041	28 38	103	53 17	145	48 23	209	65 26	293	65 18	331
246	31 25	001	30 13	039	28 48	101	53 23	143	48 18	207	65 16	291	65 13	329
248	31 25	359	30 19	037	28 59	099	53 30	141	48 13	205	65 06	289	65 07	327
250	31 25	357	30 25	035	29 09	097	53 37	139	48 09	203	64 56	287	65 01	325
252	31 24	355	30 31	033	29 19	095	53 43	138	48 05	202	64 46	285	64 55	322
254	31 23	353	30 37	031	29 30	093	53 51	136	48 01	200	64 36	283	64 48	320
256	31 22	351	30 42	029	29 40	091	53 58	134	47 58	198	64 26	281	64 42	318
258	31 20	349	30 47	027	29 51	089	54 06	132	47 55	196	64 15	279	64 34	315
260	31 18	347	30 52	025	30 01	087	54 14	130	47 52	194	64 05	277	64 27	313
262	31 15	345	30 56	023	30 11	085	54 22	128	47 50	193	63 55	275	64 19	311
264	31 12	342	31 00	021	30 22	083	54 30	127	47 47	191	63 44	273	64 11	309
266	31 09	340	31 03	019	30 32	081	54 38	125	47 46	189	63 34	271	64 03	306
268	31 05	338	31 06	016	30 43	079	54 47	123	47 44	187	63 23	269	63 54	304
	Nunki		◆FOMALHAUT		ACHERNAR		◆CANOPUS		ACRUX		◆RIGIL KENT.		ANTARES	
270	31 09	014	30 53	077	54 56	121	47 43	185	63 13	267	63 46	302	31 01	336
272	31 12	012	31 03	075	55 05	119	47 42	184	63 02	265	63 37	300	30 57	334
274	31 14	010	31 13	073	55 14	117	47 42	182	62 52	263	63 27	298	30 52	332
276	31 15	008	31 23	071	55 24	115	47 42	180	62 42	261	63 18	296	30 47	330
278	31 17	006	31 33	069	55 33	113	47 42	178	62 31	259	63 09	293	30 41	328
280	31 18	004	31 43	067	55 43	112	47 42	176	62 21	257	62 59	291	30 36	326
282	31 18	002	31 52	065	55 53	110	47 43	175	62 11	255	62 49	289	30 30	324
284	31 18	000	32 01	063	56 03	108	47 44	173	62 01	253	62 39	287	30 23	322
286	31 18	358	32 11	061	56 13	106	47 46	171	61 51	251	62 29	285	30 17	320
288	31 17	356	32 20	059	56 23	104	47 48	169	61 41	249	62 19	283	30 10	317
290	31 16	353	32 29	057	56 33	102	47 50	167	61 31	248	62 09	281	30 02	315
292	31 15	351	32 37	055	56 43	100	47 52	166	61 22	246	61 58	279	29 55	313
294	31 13	349	32 46	053	56 53	098	47 55	164	61 12	244	61 48	277	29 47	311
296	31 11	347	32 54	051	57 04	096	47 58	162	61 03	242	61 38	275	29 39	309
298	31 09	345	33 02	049	57 14	094	48 01	160	60 54	240	61 27	273	29 31	307
	◆FOMALHAUT		ACHERNAR		◆CANOPUS		ACRUX		◆RIGIL KENT.		ANTARES		Nunki	
300	33 10	047	57 25	092	48 05	158	60 45	238	61 17	271	29 23	305	31 06	343
302	33 17	044	57 35	090	48 09	156	60 36	236	61 06	269	29 14	303	31 02	341
304	33 24	042	57 46	088	48 13	155	60 28	235	60 56	267	29 05	301	30 59	339
306	33 31	040	57 56	086	48 18	153	60 19	233	60 45	265	28 56	299	30 55	337
308	33 38	038	58 07	084	48 23	151	60 11	231	60 35	263	28 47	297	30 51	335
310	33 44	036	58 17	082	48 28	149	60 03	229	60 25	261	28 37	295	30 46	333
312	33 50	034	58 27	080	48 34	147	59 55	227	60 14	259	28 28	293	30 41	330
314	33 56	032	58 37	078	48 39	146	59 47	226	60 04	257	28 18	291	30 36	328
316	34 01	030	58 48	076	48 45	144	59 40	224	59 54	255	28 08	289	30 30	326
318	34 06	028	58 58	074	48 52	142	59 33	222	59 44	253	27 58	287	30 24	324
320	34 11	026	59 08	072	48 58	140	59 26	220	59 34	251	27 48	285	30 18	322
322	34 15	024	59 18	070	49 05	138	59 19	219	59 24	249	27 38	283	30 11	320
324	34 19	021	59 27	068	49 12	136	59 13	217	59 14	248	27 28	281	30 04	318
326	34 23	019	59 37	066	49 20	135	59 07	215	59 05	246	27 18	279	29 57	316
328	34 26	017	59 46	063	49 27	133	59 01	213	58 55	244	27 07	277	29 50	314
	◆FOMALHAUT		ACHERNAR		◆CANOPUS		ACRUX		◆RIGIL KENT.		ANTARES		Nunki	
330	34 29	015	59 56	061	49 35	131	58 55	212	58 46	242	26 57	275	29 42	312
332	34 32	013	60 05	059	49 43	129	58 50	210	58 37	240	26 46	273	29 34	310
334	34 34	011	60 14	057	49 51	127	58 45	208	58 28	238	26 36	271	29 26	308
336	34 36	009	60 22	055	50 00	125	58 40	206	58 19	236	26 25	269	29 18	306
338	34 37	007	60 31	053	50 08	123	58 36	205	58 10	235	26 15	267	29 09	304
340	34 38	005	60 39	051	50 17	121	58 32	203	58 02	233	26 05	265	29 00	302
342	34 39	002	60 47	048	50 26	120	58 28	201	57 54	231	25 54	263	28 51	299
344	34 39	000	60 55	046	50 35	118	58 24	199	57 46	229	25 44	261	28 42	297
346	34 39	358	61 02	044	50 45	116	58 21	198	57 38	227	25 34	259	28 33	295
348	34 38	356	61 09	042	50 54	114	58 18	196	57 30	226	25 23	257	28 23	293
350	34 37	354	61 16	039	51 04	112	58 15	194	57 23	224	25 13	255	28 13	291
352	34 36	352	61 22	037	51 14	110	58 13	192	57 16	222	25 03	253	28 04	289
354	34 34	350	61 29	035	51 23	108	58 10	191	57 09	220	24 53	251	27 54	287
356	34 32	348	61 34	033	51 33	106	58 09	189	57 02	218	24 43	249	27 44	285
358	34 30	346	61 40	030	51 44	104	58 07	187	56 56	217	24 34	247	27 33	283

LHA	Hc	Zn	Hc	Zn	Hc	Zn	Hc	Zn	Hc	Zn	Hc	Zn	Hc	Zn
♈	ACHERNAR		◆SIRIUS		Suhail		ACRUX		◆RIGIL KENT.		Peacock		◆FOMALHAUT	
0	60 52	027	15 53	102	40 26	139	59 06	186	57 39	216	58 58	301	33 30	344
2	60 56	025	16 02	100	40 31	137	59 05	184	57 34	214	58 51	299	33 27	342
4	60 59	023	16 10	098	40 37	136	59 04	182	57 30	212	58 44	297	33 24	340
6	61 02	021	16 18	096	40 43	134	59 04	181	57 25	210	58 36	295	33 21	337
8	61 05	018	16 27	094	40 49	132	59 04	179	57 21	208	58 28	293	33 18	335
10	61 07	016	16 35	092	40 56	130	59 05	177	57 17	207	58 21	291	33 14	333
12	61 09	014	16 43	090	41 02	128	59 05	175	57 14	205	58 13	288	33 10	331
14	61 11	012	16 52	088	41 09	126	59 06	173	57 10	203	58 05	286	33 06	329
16	61 13	009	17 00	086	41 16	124	59 07	172	57 07	201	57 57	284	33 02	327
18	61 14	007	17 08	084	41 23	122	59 08	170	57 04	199	57 49	282	32 57	325
20	61 15	005	17 17	082	41 30	120	59 10	168	57 02	198	57 40	280	32 52	323
22	61 15	003	17 25	080	41 37	118	59 12	166	56 59	196	57 32	278	32 47	321
24	61 16	000	17 33	078	41 45	116	59 14	165	56 57	194	57 24	276	32 42	319
26	61 16	358	17 41	076	41 52	114	59 16	163	56 55	192	57 15	274	32 36	317
28	61 15	356	17 49	074	42 00	112	59 19	161	56 53	191	57 07	272	32 30	315
	◆SIRIUS		Suhail		ACRUX		◆RIGIL KENT.		Peacock		◆FOMALHAUT		ACHERNAR	
30	17 57	072	42 08	110	59 22	159	56 52	189	56 59	270	32 24	313	61 14	354
32	18 05	070	42 16	109	59 25	158	56 51	187	56 50	268	32 18	311	61 13	351
34	18 13	068	42 24	107	59 28	156	56 50	185	56 42	266	32 11	308	61 12	349
36	18 21	066	42 32	105	59 32	154	56 49	183	56 34	264	32 05	306	61 10	347
38	18 29	064	42 40	103	59 36	152	56 49	182	56 25	262	31 58	304	61 08	345
40	18 36	062	42 48	101	59 40	150	56 49	180	56 17	260	31 51	302	61 06	342
42	18 43	060	42 56	099	59 44	149	56 49	178	56 09	258	31 44	300	61 03	340
44	18 51	058	43 05	097	59 48	147	56 49	176	56 01	256	31 36	298	61 00	338
46	18 58	056	43 13	095	59 53	145	56 50	174	55 53	254	31 29	296	60 57	336
48	19 04	054	43 21	093	59 58	143	56 51	173	55 45	253	31 21	294	60 53	334
50	19 11	052	43 30	091	60 03	141	56 52	171	55 37	251	31 14	292	60 49	331
52	19 18	050	43 38	089	60 08	140	56 54	169	55 29	249	31 06	290	60 45	329
54	19 24	048	43 46	087	60 14	138	56 55	167	55 21	247	30 58	288	60 41	327
56	19 30	046	43 55	085	60 20	136	56 57	166	55 13	245	30 50	286	60 36	325
58	19 36	044	44 03	083	60 25	134	57 00	164	55 06	243	30 42	284	60 31	323
	◆SIRIUS		Suhail		ACRUX		◆RIGIL KENT.		Peacock		◆FOMALHAUT		ACHERNAR	
60	19 42	042	44 11	081	60 32	132	57 02	162	54 58	241	30 34	282	60 26	320
62	19 47	040	44 19	079	60 38	131	57 05	160	54 51	239	30 26	280	60 20	318
64	19 53	038	44 28	077	60 44	129	57 08	158	54 44	237	30 17	278	60 15	316
66	19 58	036	44 36	075	60 51	127	57 11	157	54 37	235	30 09	276	60 09	314
68	20 02	034	44 44	073	60 58	125	57 14	155	54 30	233	30 01	274	60 03	312
70	20 07	032	44 52	071	61 05	123	57 18	153	54 24	232	29 52	272	59 56	310
72	20 11	030	45 00	069	61 12	121	57 22	151	54 17	230	29 44	270	59 50	307
74	20 15	028	45 07	066	61 19	119	57 26	149	54 11	228	29 36	268	59 43	305
76	20 19	026	45 15	064	61 26	118	57 30	148	54 05	226	29 27	266	59 36	303
78	20 23	024	45 22	062	61 34	116	57 35	146	53 59	224	29 19	264	59 29	301
80	20 26	022	45 30	060	61 41	114	57 40	144	53 53	222	29 11	262	59 22	299
82	20 29	020	45 37	058	61 49	112	57 45	142	53 48	220	29 02	260	59 14	297
84	20 32	018	45 44	056	61 57	110	57 50	140	53 42	219	28 54	258	59 07	295
86	20 34	016	45 51	054	62 05	108	57 56	138	53 37	217	28 46	256	58 59	293
88	20 36	014	45 58	052	62 13	106	58 01	137	53 32	215	28 38	254	58 51	290
	◆SIRIUS		Suhail		ACRUX		◆RIGIL KENT.		Peacock		◆FOMALHAUT		ACHERNAR	
90	20 38	011	46 04	050	62 21	104	58 07	135	53 28	213	28 30	252	58 43	288
92	20 39	009	46 10	048	62 29	102	58 13	133	53 23	211	28 22	250	58 35	286
94	20 41	007	46 16	046	62 37	100	58 19	131	53 19	209	28 14	248	58 27	284
96	20 42	005	46 22	044	62 45	098	58 26	129	53 15	208	28 06	246	58 19	282
98	20 42	003	46 28	042	62 54	096	58 32	127	53 11	206	27 59	244	58 11	280
100	20 42	001	46 33	039	63 02	094	58 39	126	53 08	204	27 51	242	58 03	278
102	20 43	359	46 39	037	63 10	092	58 46	124	53 04	202	27 44	240	57 54	276
104	20 42	357	46 44	035	63 19	090	58 53	122	53 01	200	27 37	238	57 46	274
106	20 42	355	46 48	033	63 27	088	59 00	120	52 59	198	27 30	237	57 38	272
108	20 41	353	46 53	031	63 36	086	59 07	118	52 56	197	27 23	235	57 29	270
110	20 40	351	46 57	029	63 44	084	59 15	116	52 54	195	27 16	233	57 21	268
112	20 38	349	47 01	027	63 52	082	59 22	114	52 52	193	27 09	231	57 13	266
114	20 36	347	47 04	025	64 00	080	59 30	112	52 50	191	27 03	229	57 04	264
116	20 34	345	47 08	022	64 09	078	59 38	111	52 49	189	26 57	227	56 56	262
118	20 32	343	47 11	020	64 17	076	59 46	109	52 47	188	26 51	225	56 48	260
	◆Suhail		SPICA		◆ANTARES		Peacock		FOMALHAUT		◆ACHERNAR		SIRIUS	
120	47 14	018	11 43	082	23 57	129	52 46	186	26 45	223	56 40	258	20 29	341
122	47 16	016	11 51	080	24 04	127	52 46	184	26 40	221	56 31	256	20 27	339
124	47 18	014	12 00	078	24 11	125	52 45	182	26 34	219	56 23	254	20 23	337
126	47 20	012	12 08	076	24 18	123	52 45	180	26 29	217	56 15	252	20 20	335
128	47 22	010	12 16	074	24 25	121	52 45	178	26 24	215	56 07	251	20 16	333
130	47 23	007	12 24	072	24 32	119	52 46	177	26 19	213	55 59	249	20 12	331
132	47 24	005	12 32	070	24 40	117	52 46	175	26 15	211	55 52	247	20 08	329
134	47 24	003	12 40	068	24 47	115	52 47	173	26 11	209	55 44	245	20 04	327
136	47 25	001	12 47	066	24 55	113	52 48	171	26 07	207	55 37	243	19 59	324
138	47 25	359	12 55	064	25 02	111	52 50	169	26 03	205	55 29	241	19 54	322
140	47 24	357	13 02	062	25 10	109	52 51	168	26 00	203	55 22	239	19 49	320
142	47 24	355	13 10	060	25 18	107	52 53	166	25 56	202	55 15	237	19 43	318
144	47 23	352	13 17	058	25 26	105	52 55	164	25 53	200	55 08	235	19 37	316
146	47 22	350	13 24	056	25 34	103	52 58	162	25 51	198	55 01	233	19 32	314
148	47 20	348	13 31	054	25 43	101	53 01	160	25 48	196	54 54	232	19 26	312
	SPICA		◆ANTARES		Peacock		FOMALHAUT		◆ACHERNAR		CANOPUS		◆Suhail	
150	13 37	052	25 51	099	53 03	159	25 46	194	54 48	230	54 55	301	47 18	346
152	13 44	050	25 59	097	53 07	157	25 44	192	54 42	228	54 48	299	47 16	344
154	13 50	048	26 07	095	53 10	155	25 43	190	54 36	226	54 40	297	47 13	342
156	13 56	046	26 16	093	53 14	153	25 41	188	54 30	224	54 33	295	47 11	340
158	14 02	044	26 24	091	53 18	151	25 40	186	54 24	222	54 25	293	47 08	337
160	14 08	042	26 33	089	53 22	149	25 40	184	54 18	220	54 17	291	47 04	335
162	14 13	040	26 41	087	53 26	148	25 39	182	54 13	219	54 09	289	47 01	333
164	14 19	038	26 49	085	53 31	146	25 39	180	54 08	217	54 01	287	46 57	331
166	14 24	036	26 58	083	53 36	144	25 39	178	54 03	215	53 53	285	46 53	329
168	14 28	034	27 06	081	53 41	142	25 39	176	53 58	213	53 45	283	46 48	327
170	14 33	032	27 14	079	53 46	140	25 40	175	53 54	211	53 37	281	46 43	325
172	14 37	030	27 22	077	53 51	138	25 41	173	53 50	209	53 29	279	46 38	323
174	14 41	028	27 30	075	53 57	136	25 42	171	53 46	208	53 21	277	46 33	320
176	14 45	026	27 38	073	54 03	135	25 44	169	53 42	206	53 12	275	46 28	318
178	14 48	024	27 46	071	54 09	133	25 45	167	53 38	204	53 04	273	46 22	316

LHA	Hc	Zn	Hc	Zn	Hc	Zn	Hc	Zn	Hc	Zn	Hc	Zn	Hc	Zn
♈	SPICA		◆ANTARES		Peacock		FOMALHAUT		◆ACHERNAR		CANOPUS		◆Suhail	
180	14 52	022	27 54	069	54 15	131	25 48	165	53 35	202	52 56	271	46 16	314
182	14 55	020	28 02	067	54 22	129	25 50	163	53 32	200	52 47	269	46 10	312
184	14 57	018	28 10	065	54 28	127	25 52	161	53 29	198	52 39	267	46 04	310
186	15 00	015	28 17	063	54 35	125	25 55	159	53 27	197	52 30	265	45 57	308
188	15 02	013	28 25	061	54 42	123	25 58	157	53 25	195	52 22	263	45 51	306
190	15 03	011	28 32	059	54 49	121	26 02	155	53 23	193	52 14	261	45 44	304
192	15 05	009	28 39	057	54 56	120	26 05	153	53 21	191	52 06	259	45 37	302
194	15 06	007	28 46	055	55 04	118	26 09	151	53 19	189	51 57	257	45 29	300
196	15 07	005	28 53	053	55 11	116	26 13	149	53 18	188	51 49	255	45 22	298
198	15 08	003	28 59	051	55 19	114	26 18	148	53 17	186	51 41	253	45 15	295
200	15 08	001	29 06	049	55 26	112	26 22	146	53 16	184	51 33	251	45 07	293
202	15 08	359	29 12	047	55 34	110	26 27	144	53 16	182	51 25	249	44 59	291
204	15 08	357	29 18	045	55 42	108	26 32	142	53 16	180	51 18	247	44 51	289
206	15 07	355	29 24	043	55 50	106	26 38	140	53 16	179	51 10	245	44 43	287
208	15 06	353	29 29	041	55 58	104	26 43	138	53 16	177	51 02	243	44 35	285
	◆ANTARES		Nunki		FOMALHAUT		◆ACHERNAR		CANOPUS		◆Suhail		SPICA	
210	29 35	039	27 21	076	26 49	136	53 17	175	50 55	241	44 27	283	15 05	351
212	29 40	037	27 30	074	26 55	134	53 18	173	50 48	239	44 19	281	15 04	349
214	29 45	034	27 38	072	27 01	132	53 19	171	50 41	238	44 11	279	15 02	347
216	29 49	032	27 45	070	27 07	130	53 20	169	50 34	236	44 03	277	15 00	345
218	29 54	030	27 53	068	27 14	128	53 22	168	50 27	234	43 54	275	14 58	343
220	29 58	028	28 01	066	27 20	126	53 24	166	50 20	232	43 46	273	14 55	341
222	30 01	026	28 08	064	27 27	124	53 26	164	50 14	230	43 38	271	14 52	339
224	30 05	024	28 16	062	27 34	122	53 28	162	50 07	228	43 29	269	14 49	337
226	30 08	022	28 23	060	27 41	120	53 31	160	50 01	226	43 21	267	14 46	335
228	30 11	020	28 30	057	27 49	118	53 34	159	49 55	224	43 12	265	14 42	333
230	30 14	018	28 37	055	27 56	116	53 37	157	49 49	222	43 04	263	14 38	331
232	30 16	016	28 44	053	28 04	114	53 41	155	49 44	221	42 56	261	14 34	329
234	30 19	014	28 51	051	28 11	112	53 44	153	49 39	219	42 48	259	14 29	327
236	30 20	012	28 57	049	28 19	110	53 48	151	49 33	217	42 39	257	14 25	325
238	30 22	010	29 03	047	28 27	108	53 52	149	49 29	215	42 31	255	14 20	323
	ANTARES		Nunki		◆FOMALHAUT		ACHERNAR		◆CANOPUS		ACRUX		◆RIGIL KENT.	
240	30 23	008	29 09	045	28 35	106	53 57	148	49 24	213	65 16	300	64 32	337
242	30 24	005	29 15	043	28 43	105	54 01	146	49 19	211	65 08	298	64 29	335
244	30 25	003	29 21	041	28 51	103	54 06	144	49 15	209	65 01	295	64 25	332
246	30 25	001	29 26	039	28 59	101	54 11	142	49 11	208	64 53	293	64 21	330
248	30 25	359	29 31	037	29 08	099	54 16	140	49 07	206	64 45	291	64 17	328
250	30 25	357	29 36	035	29 16	097	54 22	138	49 04	204	64 38	289	64 12	326
252	30 24	355	29 41	033	29 24	095	54 27	137	49 01	202	64 30	287	64 07	323
254	30 24	353	29 45	031	29 33	093	54 33	135	48 58	200	64 22	285	64 02	321
256	30 22	351	29 50	029	29 41	091	54 39	133	48 55	198	64 13	283	63 57	319
258	30 21	349	29 54	027	29 49	089	54 46	131	48 52	197	64 05	281	63 51	317
260	30 19	347	29 57	025	29 58	087	54 52	129	48 50	195	63 57	279	63 45	315
262	30 17	345	30 01	023	30 06	085	54 58	127	48 48	193	63 49	277	63 39	312
264	30 15	343	30 04	020	30 14	083	55 05	125	48 46	191	63 40	275	63 33	310
266	30 12	341	30 06	018	30 23	081	55 12	124	48 45	189	63 32	273	63 26	308
268	30 09	338	30 09	016	30 31	079	55 19	122	48 44	187	63 24	271	63 20	306
	Nunki		◆FOMALHAUT		ACHERNAR		◆CANOPUS		ACRUX		◆RIGIL KENT.		ANTARES	
270	30 11	014	30 39	077	55 26	120	48 43	185	63 15	269	63 13	304	30 06	336
272	30 13	012	30 47	075	55 34	118	48 42	184	63 07	267	63 06	302	30 02	334
274	30 15	010	30 55	073	55 41	116	48 42	182	62 59	265	62 59	300	29 59	332
276	30 16	008	31 03	071	55 49	114	48 42	180	62 50	263	62 51	297	29 55	330
278	30 17	006	31 11	069	55 56	112	48 42	178	62 42	261	62 44	295	29 50	328
280	30 18	004	31 19	066	56 04	110	48 42	176	62 34	259	62 36	293	29 46	326
282	30 18	002	31 26	064	56 12	108	48 43	174	62 26	257	62 28	291	29 41	324
284	30 18	000	31 34	062	56 20	106	48 44	173	62 17	255	62 21	289	29 36	322
286	30 18	358	31 41	060	56 28	104	48 45	171	62 09	253	62 13	287	29 31	320
288	30 18	356	31 48	058	56 36	102	48 46	169	62 01	251	62 05	285	29 25	318
290	30 17	353	31 56	056	56 45	100	48 48	167	61 54	249	61 56	283	29 20	316
292	30 16	351	32 02	054	56 53	099	48 50	165	61 46	247	61 48	281	29 14	314
294	30 14	349	32 09	052	57 01	097	48 52	163	61 38	245	61 40	279	29 07	312
296	30 13	347	32 16	050	57 09	095	48 55	162	61 31	244	61 32	277	29 01	310
298	30 11	345	32 22	048	57 18	093	48 58	160	61 23	242	61 23	275	28 55	308
	◆FOMALHAUT		ACHERNAR		◆CANOPUS		ACRUX		◆RIGIL KENT.		ANTARES		Nunki	
300	32 28	046	57 26	091	49 01	158	61 16	240	61 15	273	28 48	306	30 08	343
302	32 34	044	57 35	089	49 04	156	61 09	238	61 07	271	28 41	304	30 06	341
304	32 40	042	57 43	087	49 08	154	61 02	236	60 58	269	28 34	302	30 03	339
306	32 45	040	57 51	085	49 11	152	60 55	234	60 50	267	28 27	299	30 00	337
308	32 50	038	58 00	083	49 15	150	60 48	232	60 42	265	28 19	297	29 56	335
310	32 55	036	58 08	081	49 20	149	60 42	231	60 33	263	28 12	295	29 53	333
312	33 00	034	58 16	079	49 24	147	60 35	229	60 25	261	28 04	293	29 49	331
314	33 05	032	58 24	076	49 29	145	60 29	227	60 17	259	27 56	291	29 44	329
316	33 09	030	58 32	074	49 34	143	60 23	225	60 08	257	27 49	289	29 40	327
318	33 13	027	58 40	072	49 39	141	60 17	223	60 00	255	27 41	287	29 35	325
320	33 17	025	58 48	070	49 44	139	60 11	221	59 52	253	27 33	285	29 30	322
322	33 20	023	58 56	068	49 50	137	60 06	220	59 44	251	27 25	283	29 25	320
324	33 23	021	59 04	066	49 55	136	60 01	218	59 36	249	27 16	281	29 20	318
326	33 26	019	59 11	064	50 02	134	59 56	216	59 29	247	27 08	279	29 14	316
328	33 29	017	59 19	062	50 08	132	59 51	214	59 21	245	27 00	277	29 08	314
	◆FOMALHAUT		ACHERNAR		◆CANOPUS		ACRUX		◆RIGIL KENT.		ANTARES		Nunki	
330	33 31	015	59 26	060	50 14	130	59 46	212	59 14	243	26 52	275	29 02	312
332	33 33	013	59 33	058	50 20	128	59 42	211	59 06	242	26 43	273	28 56	310
334	33 35	011	59 40	056	50 27	126	59 38	209	58 59	240	26 35	271	28 49	308
336	33 36	009	59 47	053	50 34	124	59 34	207	58 52	238	26 26	269	28 42	306
338	33 37	007	59 54	051	50 41	122	59 30	205	58 45	236	26 18	267	28 36	304
340	33 38	005	60 00	049	50 48	120	59 27	203	58 38	234	26 10	265	28 28	302
342	33 39	002	60 06	047	50 55	119	59 24	202	58 31	232	26 01	263	28 21	300
344	33 39	000	60 13	045	51 03	117	59 21	200	58 25	230	25 53	261	28 14	298
346	33 39	358	60 18	043	51 10	115	59 18	198	58 18	229	25 45	259	28 07	296
348	33 38	356	60 24	040	51 18	113	59 15	196	58 12	227	25 37	257	27 59	294
350	33 38	354	60 29	038	51 26	111	59 13	195	58 06	225	25 29	255	27 51	292
352	33 37	352	60 34	036	51 34	109	59 11	193	58 00	223	25 20	253	27 43	290
354	33 35	350	60 39	034	51 42	107	59 09	191	57 55	221	25 12	251	27 35	288
356	33 34	348	60 44	032	51 50	105	59 08	189	57 49	219	25 05	249	27 27	286
358	33 32	346	60 48	030	51 58	103	59 07	188	57 44	218	24 57	247	27 19	284

 LAT 87°S

LHA ♈	Hc	Zn	Hc	Zn	Hc	Zn	Hc	Zn	Hc	Zn	Hc	Zn	Hc	Zn
	ACHERNAR		◆SIRIUS		Suhail		ACRUX		◆RIGIL KENT.		Peacock		◆FOMALHAUT	
0	59 58	026	16 06	102	41 11	139	60 05	186	58 27	217	58 27	302	32 32	344
2	60 01	024	16 12	100	41 15	137	60 05	184	58 24	215	58 21	300	32 30	342
4	60 03	022	16 19	098	41 20	135	60 04	182	58 20	213	58 16	298	32 28	340
6	60 06	020	16 25	096	41 24	133	60 04	181	58 17	211	58 10	296	32 26	338
8	60 08	018	16 31	094	41 29	131	60 04	179	58 14	209	58 05	294	32 23	336
10	60 10	016	16 37	092	41 34	129	60 05	177	58 11	207	57 59	292	32 21	334
12	60 11	013	16 44	090	41 39	127	60 05	175	58 08	206	57 53	290	32 18	331
14	60 13	011	16 50	088	41 44	125	60 06	173	58 05	204	57 47	288	32 15	329
16	60 14	009	16 56	086	41 49	123	60 06	171	58 03	202	57 41	286	32 11	327
18	60 15	007	17 02	084	41 54	121	60 08	170	58 01	200	57 35	284	32 08	325
20	60 15	005	17 09	082	42 00	119	60 09	168	57 59	198	57 29	282	32 04	323
22	60 16	003	17 15	080	42 05	117	60 10	166	57 57	196	57 23	280	32 00	321
24	60 16	000	17 21	078	42 11	115	60 12	164	57 55	194	57 16	278	31 56	319
26	60 16	358	17 27	076	42 17	114	60 14	162	57 54	193	57 10	276	31 52	317
28	60 15	356	17 33	074	42 22	112	60 16	161	57 52	191	57 04	274	31 48	315
	◆SIRIUS		Suhail		ACRUX		◆RIGIL KENT.		Peacock		◆FOMALHAUT		ACHERNAR	
30	17 39	072	42 28	110	60 18	159	57 51	189	56 58	272	31 43	313	60 15	354
32	17 45	070	42 34	108	60 20	157	57 50	187	56 51	270	31 39	311	60 14	352
34	17 51	068	42 40	106	60 23	155	57 50	185	56 45	268	31 34	309	60 13	350
36	17 57	066	42 46	104	60 25	153	57 49	183	56 39	266	31 29	307	60 12	347
38	18 02	064	42 52	102	60 28	151	57 49	182	56 33	264	31 24	305	60 10	345
40	18 08	062	42 59	100	60 31	150	57 49	180	56 26	262	31 19	303	60 08	343
42	18 14	060	43 05	098	60 35	148	57 49	178	56 20	260	31 13	301	60 07	341
44	18 19	058	43 11	096	60 38	146	57 49	176	56 14	258	31 08	299	60 04	339
46	18 24	056	43 17	094	60 42	144	57 50	174	56 08	256	31 02	297	60 02	336
48	18 29	054	43 24	092	60 46	142	57 51	173	56 02	254	30 57	295	59 59	334
50	18 34	052	43 30	090	60 49	140	57 51	171	55 56	252	30 51	293	59 57	332
52	18 39	050	43 36	088	60 54	138	57 53	169	55 50	250	30 45	291	59 53	330
54	18 44	048	43 42	086	60 58	137	57 54	167	55 44	248	30 39	289	59 50	328
56	18 49	046	43 49	084	61 02	135	57 55	165	55 38	246	30 33	287	59 47	326
58	18 53	044	43 55	082	61 07	133	57 57	163	55 33	244	30 27	285	59 43	324
	◆SIRIUS		Suhail		ACRUX		◆RIGIL KENT.		Peacock		◆FOMALHAUT		ACHERNAR	
60	18 57	042	44 01	080	61 11	131	57 59	161	55 27	242	30 21	283	59 39	321
62	19 01	040	44 07	078	61 16	129	58 01	160	55 21	240	30 15	281	59 35	319
64	19 05	038	44 13	076	61 21	127	58 03	158	55 16	238	30 09	279	59 31	317
66	19 09	036	44 19	074	61 26	125	58 06	156	55 11	237	30 02	277	59 27	315
68	19 13	034	44 25	072	61 31	123	58 09	154	55 06	235	29 56	275	59 22	313
70	19 16	032	44 31	070	61 37	122	58 11	152	55 01	233	29 50	273	59 18	311
72	19 19	030	44 37	068	61 42	120	58 14	150	54 56	231	29 44	271	59 13	309
74	19 22	028	44 43	066	61 48	118	58 18	149	54 51	229	29 37	269	59 08	307
76	19 25	026	44 49	064	61 53	116	58 21	147	54 46	227	29 31	267	59 03	304
78	19 28	024	44 54	061	61 59	114	58 24	145	54 42	225	29 25	265	58 57	302
80	19 30	022	45 00	059	62 05	112	58 28	143	54 37	223	29 19	263	58 52	300
82	19 32	020	45 05	057	62 11	110	58 32	141	54 33	221	29 12	261	58 47	298
84	19 34	018	45 10	055	62 16	108	58 36	139	54 29	219	29 06	259	58 41	296
86	19 36	015	45 15	053	62 22	106	58 40	137	54 25	218	29 00	257	58 35	294
88	19 38	013	45 20	051	62 29	104	58 45	136	54 21	216	28 54	255	58 30	292
	◆SIRIUS		Suhail		ACRUX		◆RIGIL KENT.		Peacock		◆FOMALHAUT		ACHERNAR	
90	19 39	011	45 25	049	62 35	102	58 49	134	54 18	214	28 48	253	58 24	290
92	19 40	009	45 30	047	62 41	100	58 54	132	54 14	212	28 42	251	58 18	288
94	19 41	007	45 34	045	62 47	098	58 58	130	54 11	210	28 36	249	58 12	286
96	19 42	005	45 39	043	62 53	096	59 03	128	54 08	208	28 30	247	58 06	284
98	19 42	003	45 43	041	62 59	094	59 08	126	54 05	206	28 25	245	58 00	282
100	19 42	001	45 47	039	63 06	092	59 13	124	54 02	204	28 19	243	57 53	280
102	19 43	359	45 51	037	63 12	091	59 19	122	54 00	203	28 13	241	57 47	278
104	19 42	357	45 54	035	63 18	089	59 24	120	53 58	201	28 08	239	57 41	276
106	19 42	355	45 58	033	63 25	086	59 29	119	53 55	199	28 03	237	57 35	274
108	19 41	353	46 01	030	63 31	084	59 35	117	53 54	197	27 57	235	57 28	272
110	19 40	351	46 04	028	63 37	082	59 41	115	53 52	195	27 52	233	57 22	270
112	19 39	349	46 07	026	63 43	080	59 46	113	53 50	193	27 47	231	57 16	268
114	19 38	347	46 10	024	63 49	078	59 52	111	53 49	191	27 43	229	57 10	266
116	19 36	345	46 12	022	63 56	076	59 58	109	53 48	190	27 38	227	57 03	264
118	19 35	343	46 14	020	64 02	074	60 04	107	53 47	188	27 33	225	56 57	262
	◆Suhail		SPICA		◆ANTARES		Peacock		FOMALHAUT		◆ACHERNAR		SIRIUS	
120	46 17	018	11 35	082	24 35	128	53 46	186	27 29	223	56 51	260	19 33	341
122	46 18	016	11 41	080	24 40	126	53 45	184	27 25	221	56 45	258	19 31	339
124	46 20	014	11 47	078	24 45	124	53 45	182	27 21	219	56 39	256	19 28	337
126	46 21	012	11 53	076	24 50	122	53 45	180	27 17	217	56 33	254	19 26	335
128	46 22	009	11 59	074	24 56	121	53 45	178	27 13	215	56 27	252	19 23	333
130	46 23	007	12 05	072	25 01	119	53 45	177	27 09	213	56 21	250	19 20	331
132	46 24	005	12 11	070	25 07	117	53 46	175	27 06	211	56 15	248	19 17	329
134	46 24	003	12 17	068	25 12	115	53 47	173	27 03	210	56 09	246	19 13	327
136	46 25	001	12 23	066	25 18	113	53 47	171	27 00	208	56 03	244	19 10	325
138	46 25	359	12 28	064	25 24	111	53 49	169	26 57	206	55 58	242	19 06	323
140	46 24	357	12 34	062	25 30	109	53 50	167	26 55	204	55 52	240	19 02	321
142	46 24	355	12 39	060	25 36	107	53 51	165	26 52	202	55 47	238	18 58	319
144	46 23	353	12 45	058	25 42	105	53 53	164	26 50	200	55 41	237	18 54	317
146	46 22	350	12 50	056	25 48	103	53 55	162	26 48	198	55 36	235	18 50	315
148	46 21	348	12 55	054	25 54	101	53 57	160	26 46	196	55 31	233	18 45	313
	SPICA		◆ANTARES		Peacock		FOMALHAUT		◆ACHERNAR		CANOPUS		◆Suhail	
150	13 00	052	26 00	099	53 59	158	26 44	194	55 26	231	54 23	303	46 20	346
152	13 05	050	26 06	097	54 02	156	26 43	192	55 22	229	54 18	301	46 18	344
154	13 10	048	26 13	095	54 04	154	26 42	190	55 17	227	54 12	298	46 16	342
156	13 14	046	26 19	093	54 07	152	26 41	188	55 12	225	54 07	296	46 14	340
158	13 19	044	26 25	091	54 10	151	26 40	186	55 08	223	54 01	294	46 12	338
160	13 23	042	26 31	089	54 13	149	26 39	184	55 04	221	53 55	292	46 10	336
162	13 27	040	26 38	087	54 17	147	26 39	182	55 00	219	53 49	290	46 07	334
164	13 31	038	26 44	085	54 20	145	26 39	180	54 56	218	53 43	288	46 04	332
166	13 35	036	26 50	083	54 24	143	26 39	178	54 52	216	53 37	286	46 01	329
168	13 38	034	26 56	081	54 28	141	26 39	176	54 48	214	53 31	284	45 58	327
170	13 42	032	27 03	079	54 32	139	26 40	174	54 45	212	53 25	282	45 54	325
172	13 45	030	27 09	077	54 36	137	26 40	173	54 42	210	53 19	280	45 51	323
174	13 48	028	27 15	075	54 40	135	26 41	171	54 39	208	53 13	278	45 47	321
176	13 51	026	27 21	073	54 45	134	26 43	169	54 36	206	53 07	276	45 43	319
178	13 53	023	27 27	071	54 49	132	26 44	167	54 33	204	53 00	274	45 38	317

LHA ♈	Hc	Zn	Hc	Zn	Hc	Zn	Hc	Zn	Hc	Zn	Hc	Zn	Hc	Zn
	SPICA		◆ANTARES		Peacock		FOMALHAUT		◆ACHERNAR		CANOPUS		◆Suhail	
180	13 56	021	27 33	069	54 54	130	26 45	165	54 31	203	52 54	272	45 34	315
182	13 58	019	27 39	067	54 59	128	26 47	163	54 28	201	52 48	270	45 30	313
184	14 00	017	27 44	065	55 04	126	26 49	161	54 26	199	52 42	268	45 25	311
186	14 02	015	27 50	063	55 09	124	26 51	159	54 24	197	52 35	266	45 20	309
188	14 03	013	27 55	061	55 14	122	26 54	157	54 22	195	52 29	264	45 15	307
190	14 05	011	28 01	059	55 20	120	26 56	155	54 21	193	52 23	262	45 10	305
192	14 06	009	28 06	057	55 25	118	26 59	153	54 20	191	52 17	260	45 05	303
194	14 07	007	28 11	055	55 31	116	27 02	151	54 18	190	52 10	258	44 59	300
196	14 07	005	28 16	052	55 36	114	27 05	149	54 18	188	52 04	256	44 54	298
198	14 08	003	28 21	050	55 42	113	27 08	147	54 17	186	51 58	254	44 48	296
200	14 08	001	28 26	048	55 48	111	27 12	145	54 16	184	51 52	252	44 43	294
202	14 08	359	28 31	046	55 54	109	27 15	143	54 16	182	51 46	250	44 37	292
204	14 08	357	28 35	044	56 00	107	27 19	141	54 16	180	51 40	248	44 31	290
206	14 07	355	28 39	042	56 06	105	27 23	139	54 16	179	51 35	246	44 25	288
208	14 07	353	28 44	040	56 12	103	27 27	137	54 16	177	51 29	244	44 19	286
	◆ANTARES		Nunki		FOMALHAUT		◆ACHERNAR		CANOPUS		◆Suhail		SPICA	
210	28 48	038	27 06	075	27 32	136	54 17	175	51 23	242	44 13	284	14 06	351
212	28 51	036	27 12	073	27 36	134	54 17	173	51 18	241	44 07	282	14 05	349
214	28 55	034	27 18	071	27 41	132	54 18	171	51 12	239	44 01	280	14 04	347
216	28 58	032	27 24	069	27 46	130	54 19	169	51 07	237	43 55	278	14 02	345
218	29 02	030	27 30	067	27 51	128	54 20	167	51 02	235	43 48	276	14 00	343
220	29 05	028	27 36	065	27 56	126	54 22	166	50 57	233	43 42	274	13 58	341
222	29 08	026	27 42	063	28 01	124	54 24	164	50 52	231	43 36	272	13 56	339
224	29 10	024	27 47	061	28 06	122	54 25	162	50 47	229	43 30	270	13 54	337
226	29 13	022	27 53	059	28 11	120	54 27	160	50 42	227	43 23	268	13 51	335
228	29 15	020	27 58	057	28 17	118	54 30	158	50 38	225	43 17	266	13 49	333
230	29 17	018	28 03	055	28 23	116	54 32	156	50 33	223	43 11	264	13 46	331
232	29 19	016	28 08	053	28 28	114	54 35	154	50 29	221	43 05	262	13 43	329
234	29 20	014	28 13	051	28 34	112	54 38	153	50 25	219	42 58	260	13 39	327
236	29 22	012	28 18	049	28 40	110	54 41	151	50 21	218	42 52	258	13 36	325
238	29 23	010	28 23	047	28 46	108	54 44	149	50 17	216	42 46	256	13 32	323
	ANTARES		Nunki		◆FOMALHAUT		ACHERNAR		◆CANOPUS		ACRUX		◆RIGIL KENT.	
240	29 24	007	28 27	045	28 52	106	54 47	147	50 14	214	64 45	301	63 37	338
242	29 24	005	28 31	043	28 58	104	54 51	145	50 10	212	64 40	299	63 34	336
244	29 25	003	28 36	041	29 04	102	54 54	143	50 07	210	64 34	297	63 32	333
246	29 25	001	28 40	039	29 10	100	54 58	141	50 04	208	64 29	295	63 29	331
248	29 25	359	28 43	037	29 16	098	55 02	139	50 01	206	64 23	293	63 25	329
250	29 25	357	28 47	035	29 23	096	55 06	137	49 59	204	64 17	291	63 22	327
252	29 25	355	28 51	033	29 29	094	55 11	136	49 56	202	64 11	289	63 19	325
254	29 24	353	28 54	031	29 35	092	55 15	134	49 54	201	64 05	287	63 15	323
256	29 23	351	28 57	028	29 41	090	55 20	132	49 52	199	63 59	285	63 11	320
258	29 22	349	29 00	026	29 48	088	55 24	130	49 50	197	63 53	283	63 07	318
260	29 21	347	29 03	024	29 54	086	55 29	128	49 48	195	63 47	281	63 03	316
262	29 19	345	29 05	022	30 00	084	55 34	126	49 47	193	63 41	279	62 58	314
264	29 17	343	29 07	020	30 06	082	55 39	124	49 45	191	63 34	277	62 54	312
266	29 15	341	29 09	018	30 13	080	55 45	122	49 44	189	63 28	275	62 49	310
268	29 13	339	29 11	016	30 19	078	55 50	120	49 43	187	63 22	273	62 44	308
	Nunki		◆FOMALHAUT		ACHERNAR		◆CANOPUS		ACRUX		◆RIGIL KENT.		ANTARES	
270	29 13	014	30 25	076	55 56	118	49 42	186	63 16	271	62 39	305	29 11	337
272	29 14	012	30 31	074	56 01	117	49 42	184	63 09	269	62 34	303	29 08	335
274	29 16	010	30 37	072	56 07	115	49 42	182	63 03	267	62 28	301	29 06	333
276	29 17	008	30 43	070	56 13	113	49 42	180	62 57	265	62 23	299	29 03	331
278	29 17	006	30 49	068	56 18	111	49 42	178	62 51	263	62 17	297	28 59	328
280	29 18	004	30 55	066	56 24	109	49 42	176	62 44	261	62 12	295	28 56	326
282	29 18	002	31 00	064	56 30	107	49 43	174	62 38	259	62 06	293	28 52	324
284	29 18	000	31 06	062	56 36	105	49 43	172	62 32	257	62 00	291	28 49	322
286	29 18	358	31 11	060	56 42	103	49 44	171	62 26	255	61 54	289	28 45	320
288	29 18	356	31 17	058	56 49	101	49 45	169	62 20	253	61 48	287	28 41	318
290	29 17	354	31 22	056	56 55	099	49 47	167	62 14	251	61 42	285	28 36	316
292	29 16	352	31 27	054	57 01	097	49 48	165	62 08	249	61 36	283	28 32	314
294	29 15	349	31 32	052	57 07	095	49 50	163	62 02	247	61 30	281	28 27	312
296	29 14	347	31 37	050	57 13	093	49 52	161	61 56	245	61 24	278	28 23	310
298	29 13	345	31 42	048	57 20	091	49 54	159	61 51	243	61 18	276	28 18	308
	◆FOMALHAUT		ACHERNAR		◆CANOPUS		ACRUX		◆RIGIL KENT.		ANTARES		Nunki	
300	31 46	046	57 26	089	49 56	157	61 45	241	61 11	274	28 13	306	29 11	343
302	31 51	044	57 32	087	49 59	156	61 40	240	61 05	272	28 08	304	29 09	341
304	31 55	042	57 39	085	50 01	154	61 34	238	60 59	270	28 02	302	29 07	339
306	31 59	039	57 45	083	50 04	152	61 29	236	60 52	268	27 57	300	29 04	337
308	32 03	037	57 51	081	50 07	150	61 24	234	60 46	266	27 51	298	29 02	335
310	32 07	035	57 57	079	50 11	148	61 19	232	60 40	264	27 46	296	28 59	333
312	32 10	033	58 03	077	50 14	146	61 14	230	60 34	262	27 40	294	28 56	331
314	32 14	031	58 09	075	50 18	144	61 09	228	60 27	261	27 34	292	28 53	329
316	32 17	029	58 15	073	50 21	142	61 05	226	60 21	259	27 29	290	28 50	327
318	32 20	027	58 21	071	50 25	140	61 00	224	60 15	257	27 23	288	28 46	325
320	32 22	025	58 27	069	50 29	139	60 56	223	60 09	255	27 17	286	28 42	323
322	32 25	023	58 33	067	50 34	137	60 52	221	60 03	253	27 10	284	28 39	321
324	32 27	021	58 39	065	50 38	135	60 48	219	59 57	251	27 04	282	28 35	319
326	32 30	019	58 45	063	50 43	133	60 44	217	59 51	249	26 58	280	28 30	317
328	32 31	017	58 50	060	50 47	131	60 40	215	59 45	247	26 52	278	28 26	315
	◆FOMALHAUT		ACHERNAR		◆CANOPUS		ACRUX		◆RIGIL KENT.		ANTARES		Nunki	
330	32 33	015	58 55	058	50 52	129	60 37	213	59 40	245	26 46	276	28 21	313
332	32 35	013	59 01	056	50 57	127	60 33	212	59 34	243	26 39	274	28 17	311
334	32 36	011	59 06	054	51 02	125	60 30	210	59 28	241	26 33	272	28 12	309
336	32 37	009	59 11	052	51 07	123	60 27	208	59 23	239	26 27	270	28 07	307
338	32 38	007	59 16	050	51 13	121	60 24	206	59 18	237	26 21	268	28 02	304
340	32 38	004	59 21	048	51 18	119	60 22	204	59 12	235	26 14	266	27 57	302
342	32 39	002	59 25	046	51 23	117	60 19	202	59 07	234	26 08	264	27 51	300
344	32 39	000	59 30	044	51 29	115	60 17	201	59 02	232	26 02	262	27 46	298
346	32 39	358	59 34	042	51 35	114	60 15	199	58 57	230	25 56	260	27 40	296
348	32 39	356	59 38	039	51 41	112	60 13	197	58 53	228	25 50	258	27 34	294
350	32 38	354	59 42	037	51 46	110	60 11	195	58 48	226	25 43	256	27 29	292
352	32 37	352	59 45	035	51 52	108	60 10	193	58 44	224	25 37	254	27 23	290
354	32 36	350	59 49	033	51 58	106	60 08	191	58 39	222	25 31	252	27 17	288
356	32 35	348	59 52	031	52 05	104	60 07	190	58 35	220	25 25	250	27 11	286
358	32 34	346	59 55	029	52 11	102	60 06	188	58 31	219	25 20	248	27 05	284

LHA ♈	Hc	Zn	Hc	Zn	Hc	Zn	Hc	Zn	Hc	Zn	Hc	Zn	Hc	Zn
	ACHERNAR		◆SIRIUS		Suhail		ACRUX		◆RIGIL KENT.		Peacock		◆FOMALHAUT	
0	59 04	026	16 19	102	41 56	138	61 05	186	59 15	218	57 54	304	31 34	344
2	59 06	024	16 23	100	41 59	136	61 05	184	59 13	216	57 50	302	31 33	342
4	59 08	022	16 27	098	42 02	134	61 04	182	59 10	214	57 47	300	31 32	340
6	59 09	019	16 31	096	42 05	132	61 04	181	59 08	212	57 43	298	31 30	338
8	59 11	017	16 35	094	42 08	130	61 04	179	59 06	210	57 39	296	31 29	336
10	59 12	015	16 39	092	42 11	128	61 04	177	59 04	208	57 36	294	31 27	334
12	59 13	013	16 44	090	42 15	126	61 05	175	59 02	206	57 32	291	31 25	332
14	59 14	011	16 48	088	42 18	124	61 05	173	59 00	204	57 28	289	31 23	330
16	59 14	009	16 52	086	42 22	123	61 06	171	58 59	202	57 24	287	31 21	328
18	59 15	007	16 56	084	42 25	121	61 07	169	58 57	201	57 20	285	31 18	326
20	59 15	005	17 00	082	42 29	119	61 07	167	58 56	199	57 16	283	31 16	324
22	59 16	003	17 04	080	42 33	117	61 08	166	58 54	197	57 12	281	31 13	322
24	59 16	000	17 08	078	42 36	115	61 09	164	58 53	195	57 08	279	31 11	320
26	59 16	358	17 13	076	42 40	113	61 11	162	58 52	193	57 03	277	31 08	318
28	59 15	356	17 17	074	42 44	111	61 12	160	58 51	191	56 59	275	31 05	316
	◆SIRIUS		Suhail		ACRUX		◆RIGIL KENT.		Peacock		◆FOMALHAUT		ACHERNAR	
30	17 21	072	42 48	109	61 14	158	58 51	189	56 55	273	31 02	314	59 15	354
32	17 25	070	42 52	107	61 15	156	58 50	187	56 51	271	30 59	311	59 15	352
34	17 28	068	42 56	105	61 17	154	58 49	185	56 47	269	30 56	309	59 14	350
36	17 32	066	43 00	103	61 19	152	58 49	184	56 43	267	30 53	307	59 13	348
38	17 36	064	43 04	101	61 21	151	58 49	182	56 38	265	30 49	305	59 12	346
40	17 40	062	43 08	099	61 23	149	58 49	180	56 34	263	30 46	303	59 11	343
42	17 43	060	43 12	097	61 25	147	58 49	178	56 30	261	30 42	301	59 10	341
44	17 47	058	43 17	095	61 28	145	58 49	176	56 26	259	30 39	299	59 08	339
46	17 51	056	43 21	093	61 30	143	58 50	174	56 22	257	30 35	297	59 07	337
48	17 54	054	43 25	091	61 33	141	58 50	172	56 18	255	30 31	295	59 05	335
50	17 57	052	43 29	089	61 35	139	58 51	170	56 14	253	30 27	293	59 03	333
52	18 01	050	43 33	087	61 38	137	58 51	169	56 10	251	30 24	291	59 01	331
54	18 04	048	43 37	085	61 41	135	58 52	167	56 06	249	30 20	289	58 59	329
56	18 07	046	43 42	083	61 44	133	58 53	165	56 02	248	30 16	287	58 57	327
58	18 10	044	43 46	081	61 47	132	58 55	163	55 58	246	30 12	285	58 55	325
	◆SIRIUS		Suhail		ACRUX		◆RIGIL KENT.		Peacock		◆FOMALHAUT		ACHERNAR	
60	18 13	042	43 50	079	61 50	130	58 56	161	55 54	244	30 08	283	58 52	322
62	18 15	040	43 54	077	61 53	128	58 57	159	55 51	242	30 03	281	58 49	320
64	18 18	038	43 58	075	61 57	126	58 59	157	55 47	240	29 59	279	58 47	318
66	18 20	036	44 02	073	62 00	124	59 01	155	55 43	238	29 55	277	58 44	316
68	18 23	034	44 06	071	62 04	122	59 02	153	55 40	236	29 51	275	58 41	314
70	18 25	032	44 10	069	62 07	120	59 04	152	55 36	234	29 47	273	58 38	312
72	18 27	030	44 14	067	62 11	118	59 06	150	55 33	232	29 43	271	58 35	310
74	18 29	028	44 18	065	62 15	116	59 09	148	55 30	230	29 38	269	58 31	308
76	18 31	026	44 21	063	62 19	114	59 11	146	55 27	228	29 34	267	58 28	306
78	18 33	023	44 25	061	62 22	112	59 13	144	55 24	226	29 30	265	58 25	304
80	18 34	021	44 29	059	62 26	110	59 16	142	55 21	224	29 26	263	58 21	302
82	18 36	019	44 32	057	62 30	108	59 18	140	55 18	222	29 22	261	58 18	300
84	18 37	017	44 36	055	62 34	106	59 21	138	55 15	220	29 18	259	58 14	298
86	18 38	015	44 39	052	62 38	104	59 24	136	55 12	218	29 14	257	58 10	296
88	18 39	013	44 42	050	62 42	102	59 27	134	55 10	217	29 09	255	58 06	294
	◆SIRIUS		Suhail		ACRUX		◆RIGIL KENT.		Peacock		◆FOMALHAUT		ACHERNAR	
90	18 40	011	44 46	048	62 47	100	59 30	132	55 07	215	29 05	253	58 02	291
92	18 41	009	44 49	046	62 51	098	59 33	131	55 05	213	29 01	251	57 59	289
94	18 42	007	44 52	044	62 55	096	59 36	129	55 03	211	28 58	249	57 55	287
96	18 42	005	44 54	042	62 59	095	59 40	127	55 01	209	28 54	247	57 51	285
98	18 42	003	44 57	040	63 03	093	59 43	125	54 59	207	28 50	245	57 47	283
100	18 43	001	45 00	038	63 07	091	59 47	123	54 57	205	28 46	243	57 42	281
102	18 43	359	45 02	036	63 12	089	59 50	121	54 55	203	28 42	241	57 38	279
104	18 42	357	45 05	034	63 16	087	59 54	119	54 54	201	28 39	239	57 34	277
106	18 42	355	45 07	032	63 20	085	59 57	117	54 52	199	28 35	237	57 30	275
108	18 42	353	45 09	030	63 24	082	60 01	115	54 51	197	28 32	235	57 26	273
110	18 41	351	45 11	028	63 28	080	60 05	113	54 50	196	28 28	233	57 22	271
112	18 40	349	45 13	026	63 32	078	60 09	111	54 49	194	28 25	231	57 17	269
114	18 40	347	45 15	024	63 36	076	60 13	109	54 48	192	28 22	229	57 13	267
116	18 39	345	45 17	022	63 40	074	60 17	107	54 47	190	28 19	228	57 09	265
118	18 37	343	45 18	020	63 44	072	60 21	105	54 46	188	28 15	226	57 05	263
	◆Suhail		SPICA		◆ANTARES		Peacock		FOMALHAUT		◆ACHERNAR		SIRIUS	
120	45 19	018	11 26	082	25 12	128	54 46	186	28 13	224	57 01	261	18 36	341
122	45 21	015	11 30	080	25 15	126	54 45	184	28 10	222	56 57	259	18 35	339
124	45 22	013	11 34	078	25 19	124	54 45	182	28 07	220	56 53	257	18 33	337
126	45 22	011	11 38	076	25 22	122	54 45	180	28 04	218	56 48	255	18 31	335
128	45 23	009	11 42	074	25 26	120	54 45	178	28 02	216	56 44	253	18 30	333
130	45 24	007	11 46	072	25 29	118	54 45	176	27 59	214	56 40	251	18 28	331
132	45 24	005	11 50	070	25 33	116	54 46	175	27 57	212	56 36	249	18 25	329
134	45 25	003	11 54	068	25 37	114	54 46	173	27 55	210	56 33	248	18 23	327
136	45 25	001	11 58	066	25 41	112	54 47	171	27 53	208	56 29	246	18 21	325
138	45 25	359	12 02	064	25 45	110	54 47	169	27 51	206	56 25	244	18 18	323
140	45 25	357	12 06	062	25 49	108	54 48	167	27 49	204	56 21	242	18 16	321
142	45 24	355	12 09	060	25 53	106	54 49	165	27 48	202	56 18	240	18 13	319
144	45 24	353	12 13	058	25 57	104	54 50	163	27 46	200	56 14	238	18 10	317
146	45 23	351	12 16	056	26 01	102	54 52	161	27 45	198	56 11	236	18 07	315
148	45 22	349	12 20	054	26 05	100	54 53	159	27 44	196	56 07	234	18 04	313
	SPICA		◆ANTARES		Peacock		FOMALHAUT		◆ACHERNAR		CANOPUS		◆Suhail	
150	12 23	052	26 09	098	54 55	157	27 43	194	56 04	232	53 50	304	45 22	347
152	12 26	050	26 13	096	54 56	156	27 42	192	56 01	230	53 47	302	45 20	344
154	12 29	048	26 17	094	54 58	154	27 41	190	55 57	228	53 43	300	45 19	342
156	12 32	046	26 22	092	55 00	152	27 40	188	55 54	226	53 39	298	45 18	340
158	12 35	044	26 26	090	55 02	150	27 40	186	55 51	224	53 36	296	45 16	338
160	12 38	042	26 30	088	55 04	148	27 39	184	55 48	222	53 32	294	45 15	336
162	12 41	039	26 34	086	55 07	146	27 39	182	55 46	220	53 28	292	45 13	334
164	12 44	037	26 38	084	55 09	144	27 39	180	55 43	218	53 24	289	45 11	332
166	12 46	035	26 42	082	55 11	142	27 39	178	55 40	217	53 20	287	45 09	330
168	12 48	033	26 47	080	55 14	140	27 39	176	55 38	215	53 16	285	45 07	328
170	12 51	031	26 51	078	55 17	138	27 40	174	55 36	213	53 12	283	45 05	326
172	12 53	029	26 55	076	55 20	136	27 40	172	55 33	211	53 08	281	45 02	324
174	12 55	027	26 59	074	55 23	134	27 41	171	55 31	209	53 04	279	45 00	322
176	12 57	025	27 03	072	55 26	133	27 41	169	55 29	207	53 00	277	44 57	320
178	12 58	023	27 07	070	55 29	131	27 42	167	55 28	205	52 55	275	44 54	318

LHA ♈	Hc	Zn	Hc	Zn	Hc	Zn	Hc	Zn	Hc	Zn	Hc	Zn	Hc	Zn
	SPICA		◆ANTARES		Peacock		FOMALHAUT		◆ACHERNAR		CANOPUS		◆Suhail	
180	13 00	021	27 11	068	55 32	129	27 43	165	55 26	203	52 51	273	44 51	316
182	13 01	019	27 15	066	55 35	127	27 44	163	55 24	201	52 47	271	44 48	314
184	13 03	017	27 18	064	55 39	125	27 46	161	55 23	199	52 43	269	44 45	312
186	13 04	015	27 22	062	55 42	123	27 47	159	55 22	197	52 39	267	44 42	309
188	13 05	013	27 26	060	55 46	121	27 49	157	55 20	196	52 35	265	44 39	307
190	13 06	011	27 29	058	55 49	119	27 51	155	55 19	194	52 30	263	44 36	305
192	13 07	009	27 33	056	55 53	117	27 52	153	55 18	192	52 26	261	44 32	303
194	13 07	007	27 36	054	55 57	115	27 54	151	55 18	190	52 22	259	44 29	301
196	13 08	005	27 40	052	56 01	113	27 56	149	55 17	188	52 18	257	44 25	299
198	13 08	003	27 43	050	56 05	111	27 59	147	55 16	186	52 14	255	44 21	297
200	13 08	001	27 46	048	56 08	109	28 01	145	55 16	184	52 10	253	44 18	295
202	13 08	359	27 49	046	56 12	107	28 03	143	55 16	182	52 06	251	44 14	293
204	13 08	357	27 52	044	56 16	105	28 06	141	55 16	180	52 02	250	44 10	291
206	13 08	355	27 55	042	56 21	103	28 09	139	55 16	178	51 58	248	44 06	289
208	13 07	353	27 58	040	56 25	101	28 12	137	55 16	177	51 54	246	44 02	287
	◆ANTARES		Nunki		FOMALHAUT		◆ACHERNAR		CANOPUS		◆Suhail		SPICA	
210	28 00	038	26 51	075	28 14	135	55 16	175	51 50	244	43 58	285	13 07	351
212	28 03	036	26 55	073	28 17	133	55 17	173	51 47	242	43 54	283	13 06	349
214	28 05	034	26 59	071	28 21	131	55 17	171	51 43	240	43 50	281	13 05	347
216	28 08	032	27 03	069	28 24	129	55 18	169	51 39	238	43 46	279	13 04	345
218	28 10	030	27 07	067	28 27	127	55 19	167	51 36	236	43 41	277	13 03	343
220	28 12	028	27 10	065	28 30	125	55 20	165	51 33	234	43 37	275	13 02	341
222	28 14	026	27 14	063	28 34	123	55 21	163	51 29	232	43 33	273	13 00	339
224	28 15	024	27 18	061	28 37	121	55 22	161	51 26	230	43 29	271	12 59	337
226	28 17	022	27 21	059	28 41	119	55 24	159	51 23	228	43 25	269	12 57	335
228	28 18	020	27 25	057	28 45	117	55 25	158	51 20	226	43 21	267	12 55	333
230	28 20	018	27 28	055	28 48	115	55 27	156	51 17	224	43 16	265	12 53	331
232	28 21	016	27 32	053	28 52	113	55 29	154	51 14	222	43 12	263	12 51	329
234	28 22	014	27 35	051	28 56	111	55 31	152	51 11	220	43 08	261	12 49	327
236	28 23	011	27 38	048	29 00	109	55 33	150	51 09	218	43 04	259	12 47	325
238	28 24	009	27 41	046	29 04	107	55 35	148	51 06	216	43 00	257	12 44	323
	ANTARES		Nunki		◆FOMALHAUT		ACHERNAR		◆CANOPUS		ACRUX		◆RIGIL KENT.	
240	28 24	007	27 44	044	29 08	105	55 37	146	51 04	214	64 13	303	62 41	338
242	28 25	005	27 47	042	29 12	103	55 40	144	51 01	213	64 09	301	62 39	336
244	28 25	003	27 50	040	29 16	101	55 42	142	50 59	211	64 06	299	62 38	334
246	28 25	001	27 53	038	29 20	099	55 45	140	50 57	209	64 02	297	62 36	332
248	28 25	359	27 55	036	29 24	097	55 47	138	50 55	207	63 58	295	62 34	330
250	28 25	357	27 58	034	29 29	095	55 50	137	50 53	205	63 55	293	62 32	328
252	28 25	355	28 00	032	29 33	093	55 53	135	50 51	203	63 51	291	62 29	326
254	28 24	353	28 02	030	29 37	091	55 56	133	50 50	201	63 47	289	62 27	324
256	28 24	351	28 04	028	29 41	089	55 59	131	50 48	199	63 43	287	62 24	322
258	28 23	349	28 06	026	29 45	087	56 03	129	50 47	197	63 39	285	62 22	319
260	28 22	347	28 08	024	29 50	085	56 06	127	50 46	195	63 35	283	62 19	317
262	28 21	345	28 10	022	29 54	083	56 09	125	50 45	193	63 31	281	62 16	315
264	28 20	343	28 11	020	29 58	081	56 13	123	50 44	191	63 26	279	62 13	313
266	28 19	341	28 12	018	30 02	079	56 16	121	50 43	190	63 22	277	62 10	311
268	28 17	339	28 14	016	30 06	077	56 20	119	50 43	188	63 18	275	62 07	309
	Nunki		◆FOMALHAUT		ACHERNAR		◆CANOPUS		ACRUX		◆RIGIL KENT.		ANTARES	
270	28 15	014	30 10	075	56 24	117	50 42	186	63 14	273	62 03	307	28 16	337
272	28 16	012	30 14	073	56 27	115	50 42	184	63 10	271	62 00	305	28 14	335
274	28 16	010	30 18	071	56 31	113	50 42	182	63 06	269	61 57	303	28 12	333
276	28 17	008	30 22	069	56 35	111	50 42	180	63 01	267	61 53	301	28 10	331
278	28 18	006	30 26	067	56 39	109	50 42	178	62 57	265	61 49	299	28 08	329
280	28 18	004	30 30	065	56 43	107	50 42	176	62 53	263	61 46	297	28 06	327
282	28 18	002	30 34	063	56 47	105	50 42	174	62 49	261	61 42	295	28 04	325
284	28 18	000	30 37	061	56 51	103	50 43	172	62 45	259	61 38	292	28 01	323
286	28 18	358	30 41	059	56 55	101	50 43	170	62 41	257	61 34	290	27 58	321
288	28 18	356	30 45	057	56 59	099	50 44	168	62 37	255	61 30	288	27 56	319
290	28 17	354	30 48	055	57 03	097	50 45	167	62 33	253	61 26	286	27 53	317
292	28 17	352	30 51	053	57 07	095	50 46	165	62 29	251	61 22	284	27 50	315
294	28 16	350	30 55	051	57 12	093	50 47	163	62 25	249	61 18	282	27 47	313
296	28 15	348	30 58	049	57 16	091	50 49	161	62 21	247	61 14	280	27 44	311
298	28 14	345	31 01	047	57 20	089	50 50	159	62 17	245	61 10	278	27 41	308
	◆FOMALHAUT		ACHERNAR		◆CANOPUS		ACRUX		◆RIGIL KENT.		ANTARES		Nunki	
300	31 04	045	57 24	087	50 52	157	62 13	243	61 06	276	27 37	306	28 13	343
302	31 07	043	57 28	085	50 53	155	62 10	241	61 02	274	27 34	304	28 12	341
304	31 10	041	57 33	083	50 55	153	62 06	239	60 57	272	27 30	302	28 11	339
306	31 13	039	57 37	081	50 57	151	62 02	237	60 53	270	27 27	300	28 09	337
308	31 15	037	57 41	079	50 59	149	61 59	235	60 49	268	27 23	298	28 07	335
310	31 18	035	57 45	077	51 01	147	61 55	233	60 45	266	27 19	296	28 06	333
312	31 20	033	57 49	075	51 04	145	61 52	232	60 41	264	27 16	294	28 04	331
314	31 22	031	57 53	073	51 06	144	61 49	230	60 36	262	27 12	292	28 02	329
316	31 24	029	57 57	071	51 09	142	61 46	228	60 32	260	27 08	290	27 59	327
318	31 26	027	58 01	069	51 11	140	61 43	226	60 28	258	27 04	288	27 57	325
320	31 28	025	58 05	067	51 14	138	61 40	224	60 24	256	27 00	286	27 55	323
322	31 30	023	58 09	065	51 17	136	61 37	222	60 20	254	26 56	284	27 52	321
324	31 31	021	58 12	063	51 20	134	61 34	220	60 16	252	26 52	282	27 49	319
326	31 33	019	58 16	061	51 23	132	61 31	218	60 12	250	26 48	280	27 47	317
328	31 34	017	58 20	059	51 26	130	61 29	216	60 08	248	26 44	278	27 44	315
	◆FOMALHAUT		ACHERNAR		◆CANOPUS		ACRUX		◆RIGIL KENT.		ANTARES		Nunki	
330	31 35	015	58 23	057	51 29	128	61 27	214	60 04	247	26 39	276	27 41	313
332	31 36	013	58 27	055	51 33	126	61 24	212	60 00	245	26 35	274	27 37	311
334	31 37	011	58 30	053	51 36	124	61 22	211	59 57	243	26 31	272	27 34	309
336	31 38	009	58 34	051	51 40	122	61 20	209	59 53	241	26 27	270	27 31	307
338	31 38	006	58 37	049	51 43	120	61 18	207	59 49	239	26 23	268	27 28	305
340	31 39	004	58 40	047	51 47	118	61 16	205	59 46	237	26 19	266	27 24	303
342	31 39	002	58 43	045	51 51	116	61 14	203	59 42	235	26 14	264	27 21	301
344	31 39	000	58 46	043	51 54	114	61 13	201	59 39	233	26 10	262	27 17	299
346	31 39	358	58 48	040	51 58	112	61 11	199	59 36	231	26 06	260	27 13	297
348	31 39	356	58 51	038	52 02	110	61 10	197	59 32	229	26 02	258	27 09	295
350	31 38	354	58 54	036	52 06	108	61 09	196	59 29	227	25 58	256	27 06	293
352	31 38	352	58 56	034	52 10	106	61 08	194	59 26	225	25 54	254	27 02	291
354	31 37	350	58 58	032	52 14	104	61 07	192	59 23	223	25 50	252	26 58	289
356	31 36	348	59 01	030	52 18	103	61 06	190	59 21	221	25 46	250	26 54	287
358	31 35	346	59 03	028	52 22	101	61 06	188	59 18	220	25 42	248	26 50	285

LAT 89°S

LHA	Hc	Zn	Hc	Zn	Hc	Zn	Hc	Zn	Hc	Zn	Hc	Zn	Hc	Zn
♈	ACHERNAR		◆SIRIUS		Suhail		ACRUX		◆RIGIL KENT.		Peacock		◆FOMALHAUT	
0	58 10	025	16 31	102	42 41	138	62 05	186	60 02	219	57 20	305	30 37	344
2	58 11	023	16 33	100	42 42	136	62 04	184	60 01	217	57 18	303	30 36	342
4	58 12	021	16 35	098	42 44	134	62 04	182	60 00	215	57 17	301	30 35	340
6	58 13	019	16 37	096	42 45	132	62 04	181	59 59	213	57 15	299	30 35	338
8	58 13	017	16 39	094	42 47	130	62 04	179	59 58	211	57 13	297	30 34	336
10	58 14	015	16 41	092	42 48	128	62 04	177	59 57	209	57 11	295	30 33	334
12	58 14	013	16 43	090	42 50	126	62 05	175	59 56	207	57 09	293	30 32	332
14	58 15	011	16 45	088	42 52	124	62 05	173	59 55	205	57 07	291	30 31	330
16	58 15	009	16 47	086	42 54	122	62 05	171	59 54	203	57 05	289	30 30	328
18	58 15	007	16 49	084	42 55	120	62 05	169	59 53	201	57 03	287	30 29	326
20	58 16	005	16 52	082	42 57	118	62 06	167	59 52	199	57 01	285	30 28	324
22	58 16	002	16 54	080	42 59	116	62 06	165	59 52	197	56 59	283	30 26	322
24	58 16	000	16 56	078	43 01	114	62 07	163	59 51	195	56 57	281	30 25	320
26	58 16	358	16 58	076	43 03	112	62 08	161	59 51	193	56 55	279	30 24	318
28	58 16	356	17 00	074	43 05	110	62 08	159	59 50	191	56 53	277	30 22	316
	◆SIRIUS		Suhail		ACRUX		◆RIGIL KENT.		Peacock		◆FOMALHAUT		ACHERNAR	
30	17 02	072	43 07	108	62 09	157	59 50	190	56 51	275	30 21	314	58 15	354
32	17 04	070	43 09	106	62 10	155	59 49	188	56 49	273	30 19	312	58 15	352
34	17 06	068	43 11	104	62 11	153	59 49	186	56 47	271	30 18	310	58 15	350
36	17 08	066	43 13	102	62 12	152	59 49	184	56 45	269	30 16	308	58 14	348
38	17 09	064	43 15	100	62 13	150	59 49	182	56 43	267	30 14	306	58 14	346
40	17 11	061	43 17	098	62 14	148	59 49	180	56 40	265	30 13	304	58 13	344
42	17 13	059	43 19	096	62 15	146	59 49	178	56 38	263	30 11	302	58 13	342
44	17 15	057	43 21	094	62 16	144	59 49	176	56 36	261	30 09	300	58 12	340
46	17 17	055	43 23	092	62 18	142	59 49	174	56 34	259	30 07	298	58 11	338
48	17 18	053	43 25	090	62 19	140	59 49	172	56 32	257	30 05	296	58 11	336
50	17 20	051	43 27	088	62 20	138	59 50	170	56 30	255	30 03	294	58 10	334
52	17 22	049	43 30	086	62 22	136	59 50	168	56 28	253	30 01	292	58 09	332
54	17 23	047	43 32	084	62 23	134	59 51	166	56 26	251	30 00	290	58 08	330
56	17 25	045	43 34	082	62 25	132	59 51	164	56 24	249	29 58	288	58 07	328
58	17 26	043	43 36	080	62 26	130	59 52	162	56 22	247	29 56	286	58 05	325
	◆SIRIUS		Suhail		ACRUX		◆RIGIL KENT.		Peacock		◆FOMALHAUT		ACHERNAR	
60	17 28	041	43 38	078	62 28	128	59 52	160	56 20	245	29 54	284	58 04	323
62	17 29	039	43 40	076	62 30	126	59 53	158	56 18	243	29 51	282	58 03	321
64	17 30	037	43 42	074	62 31	124	59 54	157	56 17	241	29 49	280	58 02	319
66	17 31	035	43 44	072	62 33	122	59 55	155	56 15	239	29 47	278	58 00	317
68	17 33	033	43 46	070	62 35	120	59 56	153	56 13	237	29 45	276	57 59	315
70	17 34	031	43 48	068	62 37	118	59 57	151	56 11	235	29 43	274	57 57	313
72	17 35	029	43 50	066	62 38	116	59 58	149	56 10	233	29 41	272	57 56	311
74	17 36	027	43 52	064	62 40	114	59 59	147	56 08	231	29 39	270	57 54	309
76	17 37	025	43 53	062	62 42	112	60 00	145	56 06	229	29 37	268	57 52	307
78	17 38	023	43 55	060	62 44	110	60 01	143	56 05	227	29 35	266	57 51	305
80	17 38	021	43 57	058	62 46	108	60 03	141	56 03	225	29 33	264	57 49	303
82	17 39	019	43 59	056	62 48	106	60 04	139	56 02	223	29 31	262	57 47	301
84	17 40	017	44 01	054	62 50	104	60 05	137	56 00	221	29 29	260	57 45	299
86	17 40	015	44 02	052	62 52	102	60 07	135	55 59	219	29 27	258	57 44	297
88	17 41	013	44 04	050	62 54	101	60 08	133	55 58	217	29 24	256	57 42	295
	◆SIRIUS		Suhail		ACRUX		◆RIGIL KENT.		Peacock		◆FOMALHAUT		ACHERNAR	
90	17 41	011	44 05	048	62 56	099	60 10	131	55 56	215	29 22	254	57 40	293
92	17 42	009	44 07	046	62 58	097	60 12	129	55 55	213	29 20	252	57 38	291
94	17 42	007	44 08	044	63 01	095	60 13	127	55 54	212	29 18	250	57 36	289
96	17 42	005	44 10	042	63 03	093	60 15	125	55 53	210	29 17	248	57 34	287
98	17 42	003	44 11	040	63 05	091	60 17	123	55 52	208	29 15	246	57 32	285
100	17 43	001	44 13	038	63 07	089	60 18	121	55 51	206	29 13	244	57 30	283
102	17 43	359	44 14	036	63 09	087	60 20	119	55 50	204	29 11	242	57 28	281
104	17 42	357	44 15	033	63 11	085	60 22	117	55 49	202	29 09	240	57 26	279
106	17 42	355	44 16	031	63 13	083	60 24	115	55 49	200	29 07	238	57 24	277
108	17 42	353	44 17	029	63 15	081	60 26	113	55 48	198	29 05	236	57 22	275
110	17 42	351	44 18	027	63 17	079	60 28	111	55 47	196	29 04	234	57 19	273
112	17 41	349	44 19	025	63 19	076	60 30	109	55 47	194	29 02	232	57 17	271
114	17 41	347	44 20	023	63 21	074	60 32	108	55 46	192	29 00	230	57 15	269
116	17 41	345	44 21	021	63 23	072	60 34	106	55 46	190	28 59	228	57 13	267
118	17 40	343	44 21	019	63 25	070	60 36	104	55 46	188	28 57	226	57 11	265
	◆Suhail		SPICA		◆ANTARES		Peacock		FOMALHAUT		◆ACHERNAR		SIRIUS	
120	44 22	017	11 17	081	25 49	128	55 45	186	28 56	224	57 09	263	17 39	341
122	44 23	015	11 19	079	25 50	126	55 45	184	28 54	222	57 07	261	17 39	339
124	44 23	013	11 21	077	25 52	124	55 45	182	28 53	220	57 05	259	17 38	337
126	44 24	011	11 23	075	25 54	122	55 45	180	28 52	218	57 03	257	17 37	335
128	44 24	009	11 25	073	25 56	120	55 45	178	28 50	216	57 01	255	17 36	333
130	44 24	007	11 27	071	25 58	118	55 45	176	28 49	214	56 59	253	17 35	331
132	44 25	005	11 29	069	25 59	116	55 45	174	28 48	212	56 57	251	17 34	329
134	44 25	003	11 31	067	26 01	114	55 46	173	28 47	210	56 55	249	17 33	327
136	44 25	001	11 33	065	26 03	112	55 46	171	28 46	208	56 53	247	17 32	325
138	44 25	359	11 35	063	26 05	110	55 46	169	28 45	206	56 51	245	17 31	323
140	44 25	357	11 37	061	26 07	108	55 47	167	28 44	204	56 49	243	17 29	321
142	44 24	355	11 39	059	26 09	106	55 47	165	28 43	202	56 47	241	17 28	319
144	44 24	353	11 41	057	26 11	104	55 48	163	28 43	200	56 45	239	17 27	317
146	44 24	351	11 42	055	26 13	102	55 49	161	28 42	198	56 44	237	17 25	315
148	44 24	349	11 44	053	26 15	100	55 49	159	28 41	196	56 42	235	17 24	313
	SPICA		◆ANTARES		Peacock		FOMALHAUT		◆ACHERNAR		CANOPUS		◆Suhail	
150	11 46	051	26 17	098	55 50	157	28 41	194	56 40	233	53 16	305	44 23	347
152	11 47	049	26 19	096	55 51	155	28 40	192	56 39	231	53 15	303	44 23	345
154	11 49	047	26 22	094	55 52	153	28 40	190	56 37	229	53 13	301	44 22	343
156	11 50	045	26 24	092	55 53	151	28 40	188	56 35	227	53 11	299	44 21	341
158	11 52	043	26 26	090	55 54	149	28 39	186	56 34	225	53 09	297	44 21	339
160	11 53	041	26 28	088	55 55	147	28 39	184	56 32	223	53 07	295	44 20	337
162	11 55	039	26 30	086	55 56	145	28 39	182	56 31	221	53 05	293	44 19	335
164	11 56	037	26 32	084	55 57	143	28 39	180	56 30	219	53 03	291	44 18	333
166	11 57	035	26 34	082	55 59	141	28 39	178	56 28	217	53 01	289	44 17	330
168	11 58	033	26 36	080	56 00	139	28 39	176	56 27	215	52 59	287	44 16	328
170	11 59	031	26 38	078	56 01	137	28 39	174	56 26	214	52 57	285	44 15	326
172	12 00	029	26 40	076	56 03	135	28 39	172	56 25	212	52 55	283	44 14	324
174	12 01	027	26 42	074	56 04	133	28 40	170	56 24	210	52 53	281	44 12	322
176	12 02	025	26 44	072	56 06	131	28 40	168	56 23	208	52 51	279	44 11	320
178	12 03	023	26 46	070	56 07	129	28 41	166	56 22	206	52 49	277	44 10	318

LHA	Hc	Zn	Hc	Zn	Hc	Zn	Hc	Zn	Hc	Zn	Hc	Zn	Hc	Zn
♈	SPICA		◆ANTARES		Peacock		FOMALHAUT		◆ACHERNAR		CANOPUS		◆Suhail	
180	12 04	021	26 48	068	56 09	128	28 41	164	56 21	204	52 47	275	44 08	316
182	12 05	019	26 50	066	56 11	126	28 42	163	56 20	202	52 45	273	44 07	314
184	12 05	017	26 52	064	56 12	124	28 42	161	56 19	200	52 43	271	44 05	312
186	12 06	015	26 54	062	56 14	122	28 43	159	56 19	198	52 41	269	44 04	310
188	12 07	013	26 56	060	56 16	120	28 44	157	56 18	196	52 39	267	44 02	308
190	12 07	011	26 58	058	56 18	118	28 45	155	56 18	194	52 37	265	44 00	306
192	12 07	009	26 59	056	56 20	116	28 46	153	56 17	192	52 35	263	43 59	304
194	12 08	007	27 01	054	56 22	114	28 47	151	56 17	190	52 33	261	43 57	302
196	12 08	005	27 03	052	56 24	112	28 48	149	56 16	188	52 30	259	43 55	300
198	12 08	003	27 04	050	56 26	110	28 49	147	56 16	186	52 28	257	43 53	298
200	12 08	001	27 06	048	56 28	108	28 50	145	56 16	184	52 26	255	43 52	296
202	12 08	359	27 07	046	56 30	106	28 51	143	56 16	182	52 24	253	43 50	294
204	12 08	357	27 09	044	56 32	104	28 53	141	56 16	180	52 22	251	43 48	292
206	12 08	355	27 10	042	56 34	102	28 54	139	56 16	178	52 20	249	43 46	290
208	12 08	353	27 12	040	56 36	100	28 55	137	56 16	176	52 18	247	43 44	288
	◆ANTARES		Nunki		FOMALHAUT		◆ACHERNAR		CANOPUS		◆Suhail		SPICA	
210	27 13	038	26 35	074	28 57	135	56 16	175	52 17	245	43 42	286	12 07	351
212	27 14	036	26 37	072	28 58	133	56 16	173	52 15	243	43 40	284	12 07	349
214	27 15	034	26 39	070	29 00	131	56 17	171	52 13	241	43 38	282	12 07	347
216	27 16	032	26 41	068	29 02	129	56 17	169	52 11	239	43 36	280	12 06	345
218	27 18	030	26 43	066	29 03	127	56 17	167	52 09	237	43 34	278	12 06	343
220	27 19	028	26 45	064	29 05	125	56 18	165	52 08	235	43 32	276	12 05	341
222	27 19	025	26 46	062	29 07	123	56 18	163	52 06	233	43 29	274	12 04	339
224	27 20	023	26 48	060	29 08	121	56 19	161	52 04	231	43 27	272	12 03	337
226	27 21	021	26 50	058	29 10	119	56 20	159	52 03	229	43 25	270	12 03	335
228	27 22	019	26 52	056	29 12	117	56 21	157	52 01	227	43 23	268	12 02	333
230	27 23	017	26 54	054	29 14	115	56 21	155	52 00	225	43 21	266	12 01	331
232	27 23	015	26 55	052	29 16	113	56 22	153	51 58	223	43 19	264	12 00	329
234	27 24	013	26 57	050	29 18	111	56 23	151	51 57	221	43 17	262	11 59	327
236	27 24	011	26 58	048	29 20	109	56 24	149	51 55	219	43 15	260	11 57	325
238	27 24	009	27 00	046	29 22	107	56 26	147	51 54	217	43 13	258	11 56	323
	ANTARES		Nunki		◆FOMALHAUT		ACHERNAR		◆CANOPUS		ACRUX		◆RIGIL KENT.	
240	27 25	007	27 01	044	29 24	105	56 27	145	51 53	215	63 39	305	61 45	339
242	27 25	005	27 03	042	29 26	103	56 28	143	51 52	213	63 38	303	61 44	337
244	27 25	003	27 04	040	29 28	101	56 29	141	51 50	211	63 36	301	61 43	335
246	27 25	001	27 06	038	29 30	099	56 31	139	51 49	209	63 34	299	61 43	333
248	27 25	359	27 07	036	29 32	097	56 32	137	51 48	207	63 32	297	61 42	331
250	27 25	357	27 08	034	29 34	095	56 33	135	51 47	205	63 30	295	61 41	329
252	27 25	355	27 09	032	29 36	093	56 35	134	51 47	203	63 28	293	61 39	327
254	27 25	353	27 10	030	29 38	091	56 36	132	51 46	201	63 26	291	61 38	325
256	27 25	351	27 11	028	29 40	089	56 38	130	51 45	200	63 24	289	61 37	323
258	27 24	349	27 12	026	29 42	087	56 40	128	51 44	198	63 22	287	61 36	321
260	27 24	347	27 13	024	29 45	085	56 41	126	51 44	196	63 20	285	61 34	319
262	27 23	345	27 14	022	29 47	083	56 43	124	51 43	194	63 18	283	61 33	317
264	27 23	343	27 15	020	29 49	081	56 45	122	51 43	192	63 16	281	61 31	315
266	27 22	341	27 15	018	29 51	079	56 47	120	51 42	190	63 14	279	61 30	312
268	27 21	339	27 16	016	29 53	077	56 48	118	51 42	188	63 12	277	61 28	310
	Nunki		◆FOMALHAUT		ACHERNAR		◆CANOPUS		ACRUX		◆RIGIL KENT.		ANTARES	
270	27 16	014	29 55	075	56 50	116	51 42	186	63 10	275	61 27	308	27 21	337
272	27 17	012	29 57	073	56 52	114	51 42	184	63 08	273	61 25	306	27 20	335
274	27 17	010	29 59	071	56 54	112	51 42	182	63 06	271	61 23	304	27 19	333
276	27 18	008	30 01	069	56 56	110	51 42	180	63 04	269	61 22	302	27 18	331
278	27 18	006	30 03	067	56 58	108	51 42	178	63 02	267	61 20	300	27 17	329
280	27 18	004	30 05	065	57 00	106	51 42	176	63 00	265	61 18	298	27 16	327
282	27 18	002	30 07	063	57 02	104	51 42	174	62 58	263	61 16	296	27 15	325
284	27 18	000	30 08	061	57 04	102	51 42	172	62 56	261	61 14	294	27 13	323
286	27 18	358	30 10	059	57 06	100	51 42	170	62 53	259	61 12	292	27 12	321
288	27 18	356	30 12	057	57 08	098	51 43	168	62 51	257	61 10	290	27 11	319
290	27 18	354	30 14	055	57 10	096	51 43	166	62 49	255	61 08	288	27 09	317
292	27 18	352	30 15	053	57 12	094	51 44	164	62 47	253	61 06	286	27 08	315
294	27 17	350	30 17	051	57 15	092	51 44	162	62 45	251	61 04	284	27 06	313
296	27 17	348	30 19	049	57 17	090	51 45	160	62 43	249	61 02	282	27 05	311
298	27 16	346	30 20	047	57 19	088	51 46	158	62 41	247	61 00	280	27 03	309
	◆FOMALHAUT		ACHERNAR		◆CANOPUS		ACRUX		◆RIGIL KENT.		ANTARES		Nunki	
300	30 22	045	57 21	086	51 47	156	62 40	245	60 58	278	27 01	307	27 16	344
302	30 23	043	57 23	084	51 48	155	62 38	243	60 56	276	27 00	305	27 15	342
304	30 25	041	57 25	082	51 48	153	62 36	241	60 54	274	26 58	303	27 14	340
306	30 26	039	57 27	080	51 49	151	62 34	239	60 52	272	26 56	301	27 14	338
308	30 27	037	57 29	078	51 51	149	62 32	237	60 50	270	26 54	299	27 13	336
310	30 28	035	57 31	076	51 52	147	62 31	235	60 48	268	26 53	297	27 12	334
312	30 30	033	57 33	074	51 53	145	62 29	233	60 46	266	26 51	295	27 11	331
314	30 31	031	57 35	072	51 54	143	62 27	231	60 44	264	26 49	293	27 10	329
316	30 32	029	57 37	070	51 55	141	62 26	229	60 42	262	26 47	291	27 09	327
318	30 33	027	57 39	068	51 57	139	62 24	227	60 39	260	26 45	289	27 08	325
320	30 34	025	57 41	066	51 58	137	62 23	225	60 37	258	26 43	287	27 06	323
322	30 34	023	57 43	064	52 00	135	62 21	223	60 35	256	26 41	285	27 05	321
324	30 35	021	57 45	062	52 01	133	62 20	221	60 33	254	26 39	283	27 04	319
326	30 36	019	57 47	060	52 03	131	62 18	219	60 31	252	26 37	281	27 02	317
328	30 36	017	57 48	058	52 04	129	62 17	217	60 29	250	26 35	279	27 01	315
	◆FOMALHAUT		ACHERNAR		◆CANOPUS		ACRUX		◆RIGIL KENT.		ANTARES		Nunki	
330	30 37	014	57 50	056	52 06	127	62 16	215	60 27	248	26 33	277	27 00	313
332	30 38	012	57 52	054	52 08	125	62 15	213	60 25	246	26 31	275	26 58	311
334	30 38	010	57 53	052	52 09	123	62 13	212	60 24	244	26 28	273	26 56	309
336	30 38	008	57 55	050	52 11	121	62 12	210	60 22	242	26 26	271	26 55	307
338	30 39	006	57 57	048	52 13	119	62 11	208	60 20	240	26 24	269	26 53	305
340	30 39	004	57 58	045	52 15	117	62 10	206	60 18	238	26 22	267	26 51	303
342	30 39	002	58 00	043	52 17	115	62 10	204	60 16	236	26 20	265	26 50	301
344	30 39	000	58 01	041	52 19	113	62 09	202	60 15	234	26 18	263	26 48	299
346	30 39	358	58 02	039	52 20	111	62 08	200	60 13	232	26 16	261	26 46	297
348	30 39	356	58 04	037	52 22	109	62 07	198	60 11	230	26 14	259	26 44	295
350	30 39	354	58 05	035	52 24	107	62 07	196	60 10	228	26 12	257	26 42	293
352	30 38	352	58 06	033	52 26	105	62 06	194	60 08	227	26 10	255	26 40	291
354	30 38	350	58 07	031	52 28	103	62 06	192	60 07	225	26 08	253	26 38	289
356	30 38	348	58 08	029	52 31	101	62 05	190	60 05	223	26 06	251	26 36	287
358	30 37	346	58 09	027	52 33	099	62 05	188	60 04	221	26 04	249	26 34	285

a. GHA ♈ AT 00^h ON THE FIRST DAY OF EACH MONTH

Year	Jan. 1	Feb. 1	Mar. 1	Apr. 1	May 1	June 1	July 1	Aug. 1	Sept. 1	Oct. 1	Nov. 1	Dec. 1
	° ′	° ′	° ′	° ′	° ′	° ′	° ′	° ′	° ′	° ′	° ′	° ′
1991	100 09	130 42	158 18	188 51	218 26	248 59	278 33	309 06	339 40	9 14	39 47	69 21
1992	99 55	130 28	159 03	189 36	219 10	249 44	279 18	309 51	340 25	9 59	40 32	70 06
1993	100 39	131 13	158 49	189 22	218 56	249 29	279 04	309 37	340 10	9 44	40 18	69 52
1994	100 25	130 58	158 34	189 08	218 42	249 15	278 49	309 23	339 56	9 30	40 03	69 37
1995	100 11	130 44	158 20	188 53	218 27	249 01	278 35	309 08	339 41	9 16	39 49	69 23
1996	99 56	130 30	159 05	189 38	219 12	249 45	279 20	309 53	340 26	10 00	40 34	70 08
1997	100 41	131 14	158 50	189 24	218 58	249 31	279 05	309 38	340 12	9 46	40 19	69 53
1998	100 27	131 00	158 36	189 09	218 43	249 17	278 51	309 24	339 57	9 31	40 05	69 39
1999	100 12	130 46	158 21	188 55	218 29	249 02	278 36	309 10	339 43	9 17	39 50	69 25

b. INCREMENT OF GHA ♈ FOR DAYS AND HOURS

Day	**1**	**2**	**3**	**4**	**5**	**6**	**7**	**8**	**9**	**10**	**11**	**12**	**13**	**14**	**15**	**16**
h	° ′	° ′	° ′	° ′	° ′	° ′	° ′	° ′	° ′	° ′	° ′	° ′	° ′	° ′	° ′	° ′
00	0 00	0 59	1 58	2 57	3 57	4 56	5 55	6 54	7 53	8 52	9 51	10 51	11 50	12 49	13 48	14 47
01	15 02	16 02	17 01	18 00	18 59	19 58	20 57	21 56	22 56	23 55	24 54	25 53	26 52	27 51	28 50	29 50
02	30 05	31 04	32 03	33 02	34 01	35 01	36 00	36 59	37 58	38 57	39 56	40 55	41 55	42 54	43 53	44 52
03	45 07	46 07	47 06	48 05	49 04	50 03	51 02	52 01	53 01	54 00	54 59	55 58	56 57	57 56	58 55	59 54
04	60 10	61 09	62 08	63 07	64 06	65 06	66 05	67 04	68 03	69 02	70 01	71 00	72 00	72 59	73 58	74 57
05	75 12	76 11	77 11	78 10	79 09	80 08	81 07	82 06	83 05	84 05	85 04	86 03	87 02	88 01	89 00	89 59
06	90 15	91 14	92 13	93 12	94 11	95 10	96 10	97 09	98 08	99 07	100 06	101 05	102 04	103 04	104 03	105 02
07	105 17	106 16	107 16	108 15	109 14	110 13	111 12	112 11	113 10	114 09	115 09	116 08	117 07	118 06	119 05	120 04
08	120 20	121 19	122 18	123 17	124 16	125 15	126 15	127 14	128 13	129 12	130 11	131 10	132 09	133 09	134 08	135 07
09	135 22	136 21	137 20	138 20	139 19	140 18	141 17	142 16	143 15	144 14	145 14	146 13	147 12	148 11	149 10	150 09
10	150 25	151 24	152 23	153 22	154 21	155 20	156 19	157 19	158 18	159 17	160 16	161 15	162 14	163 13	164 13	165 12
11	165 27	166 26	167 25	168 25	169 24	170 23	171 22	172 21	173 20	174 19	175 18	176 18	177 17	178 16	179 15	180 14
12	180 30	181 29	182 28	183 27	184 26	185 25	186 24	187 24	188 23	189 22	190 21	191 20	192 19	193 18	194 18	195 17
13	195 32	196 31	197 30	198 29	199 29	200 28	201 27	202 26	203 25	204 24	205 23	206 23	207 22	208 21	209 20	210 19
14	210 34	211 34	212 33	213 32	214 31	215 30	216 29	217 28	218 28	219 27	220 26	221 25	222 24	223 23	224 22	225 22
15	225 37	226 36	227 35	228 34	229 34	230 33	231 32	232 31	233 30	234 29	235 28	236 27	237 27	238 26	239 25	240 24
16	240 39	241 39	242 38	243 37	244 36	245 35	246 34	247 33	248 33	249 32	250 31	251 30	252 29	253 28	254 27	255 27
17	255 42	256 41	257 40	258 39	259 38	260 38	261 37	262 36	263 35	264 34	265 33	266 32	267 32	268 31	269 30	270 29
18	270 44	271 43	272 43	273 42	274 41	275 40	276 39	277 38	278 37	279 37	280 36	281 35	282 34	283 33	284 32	285 31
19	285 47	286 46	287 45	288 44	289 43	290 43	291 42	292 41	293 40	294 39	295 38	296 37	297 36	298 36	299 35	300 34
20	300 49	301 48	302 48	303 47	304 46	305 45	306 44	307 43	308 42	309 42	310 41	311 40	312 39	313 38	314 37	315 36
21	315 52	316 51	317 50	318 49	319 48	320 47	321 47	322 46	323 45	324 44	325 43	326 42	327 41	328 41	329 40	330 39
22	330 54	331 53	332 52	333 52	334 51	335 50	336 49	337 48	338 47	339 46	340 46	341 45	342 44	343 43	344 42	345 41
23	345 57	346 56	347 55	348 54	349 53	350 52	351 52	352 51	353 50	354 49	355 48	356 47	357 46	358 45	359 45	0 44

Day	**17**	**18**	**19**	**20**	**21**	**22**	**23**	**24**	**25**	**26**	**27**	**28**	**29**	**30**	**31**	**32**
h	° ′	° ′	° ′	° ′	° ′	° ′	° ′	° ′	° ′	° ′	° ′	° ′	° ′	° ′	° ′	° ′
00	15 46	16 45	17 44	18 44	19 43	20 42	21 41	22 40	23 39	24 38	25 38	26 37	27 36	28 35	29 34	30 33
01	30 49	31 48	32 47	33 46	34 45	35 44	36 44	37 43	38 42	39 41	40 40	41 39	42 38	43 37	44 37	45 36
02	45 51	46 50	47 49	48 49	49 48	50 47	51 46	52 45	53 44	54 43	55 43	56 42	57 41	58 40	59 39	60 38
03	60 54	61 53	62 52	63 51	64 50	65 49	66 48	67 48	68 47	69 46	70 45	71 44	72 43	73 42	74 42	75 41
04	75 56	76 55	77 54	78 53	79 53	80 52	81 51	82 50	83 49	84 48	85 47	86 47	87 46	88 45	89 44	90 43
05	90 59	91 58	92 57	93 56	94 55	95 54	96 53	97 53	98 52	99 51	100 50	101 49	102 48	103 47	104 46	105 46
06	106 01	107 00	107 59	108 58	109 58	110 57	111 56	112 55	113 54	114 53	115 52	116 52	117 51	118 50	119 49	120 48
07	121 03	122 03	123 02	124 01	125 00	125 59	126 58	127 57	128 57	129 56	130 55	131 54	132 53	133 52	134 51	135 51
08	136 06	137 05	138 04	139 03	140 02	141 02	142 01	143 00	143 59	144 58	145 57	146 56	147 56	148 55	149 54	150 53
09	151 08	152 08	153 07	154 06	155 05	156 04	157 03	158 02	159 02	160 01	161 00	161 59	162 58	163 57	164 56	165 55
10	166 11	167 10	168 09	169 08	170 07	171 07	172 06	173 05	174 04	175 03	176 02	177 01	178 01	179 00	179 59	180 58
11	181 13	182 12	183 12	184 11	185 10	186 09	187 08	188 07	189 06	190 06	191 05	192 04	193 03	194 02	195 01	196 00
12	196 16	197 15	198 14	199 13	200 12	201 11	202 11	203 10	204 09	205 08	206 07	207 06	208 05	209 05	210 04	211 03
13	211 18	212 17	213 17	214 16	215 15	216 14	217 13	218 12	219 11	220 11	221 10	222 09	223 08	224 07	225 06	226 05
14	226 21	227 20	228 19	229 18	230 17	231 16	232 16	233 15	234 14	235 13	236 12	237 11	238 10	239 10	240 09	241 08
15	241 23	242 22	243 21	244 21	245 20	246 19	247 18	248 17	249 16	250 15	251 15	252 14	253 13	254 12	255 11	256 10
16	256 26	257 25	258 24	259 23	260 22	261 21	262 20	263 20	264 19	265 18	266 17	267 16	268 15	269 14	270 14	271 13
17	271 28	272 27	273 26	274 26	275 25	276 24	277 23	278 22	279 21	280 20	281 19	282 19	283 18	284 17	285 16	286 15
18	286 31	287 30	288 29	289 28	290 27	291 26	292 25	293 25	294 24	295 23	296 22	297 21	298 20	299 19	300 19	301 18
19	301 33	302 32	303 31	304 30	305 30	306 29	307 28	308 27	309 26	310 25	311 24	312 24	313 23	314 22	315 21	316 20
20	316 36	317 35	318 34	319 33	320 32	321 31	322 30	323 29	324 29	325 28	326 27	327 26	328 25	329 24	330 23	331 23
21	331 38	332 37	333 36	334 35	335 35	336 34	337 33	338 32	339 31	340 30	341 29	342 28	343 28	344 27	345 26	346 25
22	346 40	347 40	348 39	349 38	350 37	351 36	352 35	353 34	354 34	355 33	356 32	357 31	358 30	359 29	0 28	1 28
23	1 43	2 42	3 41	4 40	5 39	6 39	7 38	8 37	9 36	10 35	11 34	12 33	13 33	14 32	15 31	16 30

c. INCREMENT OF GHA ϒ FOR MINUTES AND SECONDS

m	00ˢ	04ˢ	08ˢ	12ˢ	16ˢ	20ˢ	24ˢ	28ˢ	m	32ˢ	36ˢ	40ˢ	44ˢ	48ˢ	52ˢ	56ˢ	60ˢ	m
	° ′	° ′	° ′	° ′	° ′	° ′	° ′	° ′		° ′	° ′	° ′	° ′	° ′	° ′	° ′	° ′	
00	0 00	0 01	0 02	0 03	0 04	0 05	0 06	0 07	00	0 08	0 09	0 10	0 11	0 12	0 13	0 14	0 15	00
01	0 15	0 16	0 17	0 18	0 19	0 20	0 21	0 22	01	0 23	0 24	0 25	0 26	0 27	0 28	0 29	0 30	01
02	0 30	0 31	0 32	0 33	0 34	0 35	0 36	0 37	02	0 38	0 39	0 40	0 41	0 42	0 43	0 44	0 45	02
03	0 45	0 46	0 47	0 48	0 49	0 50	0 51	0 52	03	0 53	0 54	0 55	0 56	0 57	0 58	0 59	1 00	03
04	1 00	1 01	1 02	1 03	1 04	1 05	1 06	1 07	04	1 08	1 09	1 10	1 11	1 12	1 13	1 14	1 15	04
05	1 15	1 16	1 17	1 18	1 19	1 20	1 21	1 22	05	1 23	1 24	1 25	1 26	1 27	1 28	1 29	1 30	05
06	1 30	1 31	1 32	1 33	1 34	1 35	1 36	1 37	06	1 38	1 39	1 40	1 41	1 42	1 43	1 44	1 45	06
07	1 45	1 46	1 47	1 48	1 49	1 50	1 51	1 52	07	1 53	1 54	1 55	1 56	1 57	1 58	1 59	2 00	07
08	2 00	2 01	2 02	2 03	2 04	2 05	2 06	2 07	08	2 08	2 09	2 10	2 11	2 12	2 13	2 14	2 15	08
09	2 15	2 16	2 17	2 18	2 19	2 20	2 21	2 22	09	2 23	2 24	2 25	2 26	2 27	2 28	2 29	2 30	09
10	2 30	2 31	2 32	2 33	2 34	2 35	2 36	2 37	10	2 38	2 39	2 40	2 41	2 42	2 43	2 44	2 45	10
11	2 45	2 46	2 47	2 48	2 49	2 50	2 51	2 52	11	2 53	2 54	2 55	2 56	2 57	2 58	2 59	3 00	11
12	3 00	3 01	3 02	3 04	3 05	3 06	3 07	3 08	12	3 09	3 10	3 11	3 12	3 13	3 14	3 15	3 16	12
13	3 16	3 17	3 18	3 19	3 20	3 21	3 22	3 23	13	3 24	3 25	3 26	3 27	3 28	3 29	3 30	3 31	13
14	3 31	3 32	3 33	3 34	3 35	3 36	3 37	3 38	14	3 39	3 40	3 41	3 42	3 43	3 44	3 45	3 46	14
15	3 46	3 47	3 48	3 49	3 50	3 51	3 52	3 53	15	3 54	3 55	3 56	3 57	3 58	3 59	4 00	4 01	15
16	4 01	4 02	4 03	4 04	4 05	4 06	4 07	4 08	16	4 09	4 10	4 11	4 12	4 13	4 14	4 15	4 16	16
17	4 16	4 17	4 18	4 19	4 20	4 21	4 22	4 23	17	4 24	4 25	4 26	4 27	4 28	4 29	4 30	4 31	17
18	4 31	4 32	4 33	4 34	4 35	4 36	4 37	4 38	18	4 39	4 40	4 41	4 42	4 43	4 44	4 45	4 46	18
19	4 46	4 47	4 48	4 49	4 50	4 51	4 52	4 53	19	4 54	4 55	4 56	4 57	4 58	4 59	5 00	5 01	19
20	5 01	5 02	5 03	5 04	5 05	5 06	5 07	5 08	20	5 09	5 10	5 11	5 12	5 13	5 14	5 15	5 16	20
21	5 16	5 17	5 18	5 19	5 20	5 21	5 22	5 23	21	5 24	5 25	5 26	5 27	5 28	5 29	5 30	5 31	21
22	5 31	5 32	5 33	5 34	5 35	5 36	5 37	5 38	22	5 39	5 40	5 41	5 42	5 43	5 44	5 45	5 46	22
23	5 46	5 47	5 48	5 49	5 50	5 51	5 52	5 53	23	5 54	5 55	5 56	5 57	5 58	5 59	6 00	6 01	23
24	6 01	6 02	6 03	6 04	6 05	6 06	6 07	6 08	24	6 09	6 10	6 11	6 12	6 13	6 14	6 15	6 16	24
25	6 16	6 17	6 18	6 19	6 20	6 21	6 22	6 23	25	6 24	6 25	6 26	6 27	6 28	6 29	6 30	6 31	25
26	6 31	6 32	6 33	6 34	6 35	6 36	6 37	6 38	26	6 39	6 40	6 41	6 42	6 43	6 44	6 45	6 46	26
27	6 46	6 47	6 48	6 49	6 50	6 51	6 52	6 53	27	6 54	6 55	6 56	6 57	6 58	6 59	7 00	7 01	27
28	7 01	7 02	7 03	7 04	7 05	7 06	7 07	7 08	28	7 09	7 10	7 11	7 12	7 13	7 14	7 15	7 16	28
29	7 16	7 17	7 18	7 19	7 20	7 21	7 22	7 23	29	7 24	7 25	7 26	7 27	7 28	7 29	7 30	7 31	29
30	7 31	7 32	7 33	7 34	7 35	7 36	7 37	7 38	30	7 39	7 40	7 41	7 42	7 43	7 44	7 45	7 46	30
31	7 46	7 47	7 48	7 49	7 50	7 51	7 52	7 53	31	7 54	7 55	7 56	7 57	7 58	7 59	8 00	8 01	31
32	8 01	8 02	8 03	8 04	8 05	8 06	8 07	8 08	32	8 09	8 10	8 11	8 12	8 13	8 14	8 15	8 16	32
33	8 16	8 17	8 18	8 19	8 20	8 21	8 22	8 23	33	8 24	8 25	8 26	8 27	8 28	8 29	8 30	8 31	33
34	8 31	8 32	8 33	8 34	8 35	8 36	8 37	8 38	34	8 39	8 40	8 41	8 42	8 43	8 44	8 45	8 46	34
35	8 46	8 47	8 48	8 49	8 50	8 51	8 52	8 53	35	8 54	8 55	8 56	8 57	8 58	8 59	9 00	9 01	35
36	9 01	9 02	9 03	9 04	9 05	9 06	9 07	9 08	36	9 10	9 11	9 12	9 13	9 14	9 15	9 16	9 17	36
37	9 17	9 18	9 19	9 20	9 21	9 22	9 23	9 24	37	9 25	9 26	9 27	9 28	9 29	9 30	9 31	9 32	37
38	9 32	9 33	9 34	9 35	9 36	9 37	9 38	9 39	38	9 40	9 41	9 42	9 43	9 44	9 45	9 46	9 47	38
39	9 47	9 48	9 49	9 50	9 51	9 52	9 53	9 54	39	9 55	9 56	9 57	9 58	9 59	10 00	10 01	10 02	39
40	10 02	10 03	10 04	10 05	10 06	10 07	10 08	10 09	40	10 10	10 11	10 12	10 13	10 14	10 15	10 16	10 17	40
41	10 17	10 18	10 19	10 20	10 21	10 22	10 23	10 24	41	10 25	10 26	10 27	10 28	10 29	10 30	10 31	10 32	41
42	10 32	10 33	10 34	10 35	10 36	10 37	10 38	10 39	42	10 40	10 41	10 42	10 43	10 44	10 45	10 46	10 47	42
43	10 47	10 48	10 49	10 50	10 51	10 52	10 53	10 54	43	10 55	10 56	10 57	10 58	10 59	11 00	11 01	11 02	43
44	11 02	11 03	11 04	11 05	11 06	11 07	11 08	11 09	44	11 10	11 11	11 12	11 13	11 14	11 15	11 16	11 17	44
45	11 17	11 18	11 19	11 20	11 21	11 22	11 23	11 24	45	11 25	11 26	11 27	11 28	11 29	11 30	11 31	11 32	45
46	11 32	11 33	11 34	11 35	11 36	11 37	11 38	11 39	46	11 40	11 41	11 42	11 43	11 44	11 45	11 46	11 47	46
47	11 47	11 48	11 49	11 50	11 51	11 52	11 53	11 54	47	11 55	11 56	11 57	11 58	11 59	12 00	12 01	12 02	47
48	12 02	12 03	12 04	12 05	12 06	12 07	12 08	12 09	48	12 10	12 11	12 12	12 13	12 14	12 15	12 16	12 17	48
49	12 17	12 18	12 19	12 20	12 21	12 22	12 23	12 24	49	12 25	12 26	12 27	12 28	12 29	12 30	12 31	12 32	49
50	12 32	12 33	12 34	12 35	12 36	12 37	12 38	12 39	50	12 40	12 41	12 42	12 43	12 44	12 45	12 46	12 47	50
51	12 47	12 48	12 49	12 50	12 51	12 52	12 53	12 54	51	12 55	12 56	12 57	12 58	12 59	13 00	13 01	13 02	51
52	13 02	13 03	13 04	13 05	13 06	13 07	13 08	13 09	52	13 10	13 11	13 12	13 13	13 14	13 15	13 16	13 17	52
53	13 17	13 18	13 19	13 20	13 21	13 22	13 23	13 24	53	13 25	13 26	13 27	13 28	13 29	13 30	13 31	13 32	53
54	13 32	13 33	13 34	13 35	13 36	13 37	13 38	13 39	54	13 40	13 41	13 42	13 43	13 44	13 45	13 46	13 47	54
55	13 47	13 48	13 49	13 50	13 51	13 52	13 53	13 54	55	13 55	13 56	13 57	13 58	13 59	14 00	14 01	14 02	55
56	14 02	14 03	14 04	14 05	14 06	14 07	14 08	14 09	56	14 10	14 11	14 12	14 13	14 14	14 15	14 16	14 17	56
57	14 17	14 18	14 19	14 20	14 21	14 22	14 23	14 24	57	14 25	14 26	14 27	14 28	14 29	14 30	14 31	14 32	57
58	14 32	14 33	14 34	14 35	14 36	14 37	14 38	14 39	58	14 40	14 41	14 42	14 43	14 44	14 45	14 46	14 47	58
59	14 47	14 48	14 49	14 50	14 51	14 52	14 53	14 54	59	14 55	14 56	14 57	14 58	14 59	15 00	15 01	15 02	59

Example. The value of GHAϒ for 1994 June 16 at $21^h\ 33^m\ 07^s$ UT is **(a)** 249° 15′ + **(b)** 330° 39′ + **(c)** 8° 18′ = 228° 12′.

LHA ♈	North latitudes N 89°	N 80°	N 70°	N 60°	N 50°	N 40°	N 20°	0°	South latitudes S 20°	S 40°	S 50°	S 60°	S 70°	S 80°	S 89°	LHA ♈
								1991								
°	′ °	′ °	′ °	′ °	′ °	′ °	′ °	′ °	′ °	′ °	′ °	′ °	′ °	′ °	′ °	°
0	**1** 180	**1** 200	**1** 220	**2** 230	**2** 240	**2** 240	**3** 250	**3** 250	**3** 250	**2** 240	**2** 240	**2** 230	**1** 220	**1** 200	**1** 180	**0**
30	**1** 210	**1** 230	**2** 240	**2** 240	**2** 250	**2** 250	**3** 250	**3** 250	**2** 250	**2** 240	**1** 230	**1** 220	**1** 200	**1** 170	**1** 150	**30**
60	**1** 240	**1** 250	**2** 250	**2** 260	**2** 260	**3** 260	**3** 260	**2** 260	**2** 260	**1** 250	**1** 240	**1** 220	**0** —	**1** 130	**1** 120	**60**
90	**1** 270	**1** 270	**2** 270	**2** 270	**2** 270	**3** 270	**3** 270	**2** 270	**2** 270	**1** 270	**1** 270	**0** —	**0** —	**1** 090	**1** 090	**90**
120	**1** 300	**1** 290	**2** 290	**2** 290	**2** 280	**3** 280	**3** 280	**3** 280	**2** 290	**1** 290	**1** 300	**1** 320	**1** 000	**1** 040	**1** 060	**120**
150	**1** 330	**1** 320	**2** 310	**2** 300	**2** 300	**2** 290	**3** 290	**3** 290	**2** 290	**2** 300	**2** 310	**1** 320	**1** 340	**1** 000	**1** 030	**150**
180	**1** 000	**1** 340	**1** 320	**2** 310	**2** 300	**2** 300	**3** 290	**3** 290	**3** 290	**2** 300	**2** 300	**2** 310	**1** 320	**1** 340	**1** 000	**180**
210	**1** 030	**1** 010	**1** 340	**1** 320	**1** 310	**2** 300	**2** 290	**3** 290	**3** 290	**2** 290	**2** 290	**2** 300	**2** 300	**1** 310	**1** 330	**210**
240	**1** 060	**1** 050	**0** —	**1** 320	**1** 300	**1** 290	**2** 280	**2** 280	**3** 280	**3** 280	**2** 280	**2** 280	**2** 290	**1** 290	**1** 300	**240**
270	**1** 090	**1** 090	**0** —	**0** —	**1** 270	**1** 270	**2** 270	**2** 270	**3** 270	**3** 270	**2** 270	**2** 270	**2** 270	**1** 270	**1** 270	**270**
300	**1** 120	**1** 140	**1** 180	**1** 220	**1** 240	**1** 250	**2** 250	**3** 260	**3** 260	**3** 260	**2** 260	**2** 250	**2** 250	**1** 250	**1** 240	**300**
330	**1** 150	**1** 180	**1** 200	**1** 220	**2** 230	**2** 240	**2** 250	**3** 250	**3** 250	**2** 250	**2** 240	**2** 240	**2** 230	**1** 220	**1** 210	**330**
360	**1** 180	**1** 200	**1** 220	**2** 230	**2** 240	**2** 240	**3** 250	**3** 250	**3** 250	**2** 240	**2** 240	**2** 230	**1** 220	**1** 200	**1** 180	**360**
								1992								
0	**1** 180	**1** 200	**1** 220	**1** 230	**1** 240	**1** 240	**2** 250	**2** 250	**2** 250	**1** 240	**1** 240	**1** 230	**1** 220	**1** 200	**1** 180	**0**
30	**1** 210	**1** 230	**1** 240	**1** 240	**1** 240	**2** 250	**2** 250	**2** 250	**2** 250	**1** 240	**1** 230	**1** 220	**1** 200	**1** 170	**1** 150	**30**
60	**1** 240	**1** 250	**1** 250	**1** 250	**2** 260	**2** 260	**2** 260	**2** 260	**1** 250	**1** 250	**1** 240	**0** —	**0** —	**0** —	**1** 120	**60**
90	**1** 270	**1** 270	**1** 270	**1** 270	**2** 270	**2** 270	**2** 270	**2** 270	**1** 270	**1** 270	**1** 270	**0** —	**0** —	**0** —	**1** 090	**90**
120	**1** 300	**1** 290	**1** 290	**1** 280	**2** 280	**2** 280	**2** 280	**2** 280	**1** 280	**1** 290	**1** 300	**0** —	**0** —	**0** —	**1** 060	**120**
150	**1** 330	**1** 310	**1** 300	**1** 300	**1** 290	**2** 290	**2** 290	**2** 290	**2** 290	**1** 300	**1** 310	**1** 320	**1** 340	**1** 010	**1** 030	**150**
180	**1** 000	**1** 340	**1** 320	**1** 310	**1** 300	**1** 300	**2** 290	**2** 290	**2** 290	**1** 300	**1** 300	**1** 310	**1** 320	**1** 340	**1** 000	**180**
210	**1** 030	**1** 010	**1** 340	**1** 320	**1** 310	**1** 300	**2** 290	**2** 290	**2** 290	**2** 290	**1** 300	**1** 300	**1** 300	**1** 310	**1** 330	**210**
240	**1** 060	**0** —	**0** —	**0** —	**1** 300	**1** 290	**1** 290	**2** 280	**2** 280	**2** 280	**2** 280	**1** 290	**1** 290	**1** 290	**1** 300	**240**
270	**1** 090	**0** —	**0** —	**0** —	**1** 270	**1** 270	**1** 270	**2** 270	**2** 270	**2** 270	**2** 270	**1** 270	**1** 270	**1** 270	**1** 270	**270**
300	**1** 120	**0** —	**0** —	**0** —	**1** 240	**1** 250	**1** 260	**2** 260	**2** 260	**2** 260	**2** 260	**1** 260	**1** 250	**1** 250	**1** 240	**300**
330	**1** 150	**1** 170	**1** 200	**1** 220	**1** 230	**1** 240	**2** 250	**2** 250	**2** 250	**2** 250	**1** 250	**1** 240	**1** 240	**1** 230	**1** 210	**330**
360	**1** 180	**1** 200	**1** 220	**1** 230	**1** 240	**1** 240	**2** 250	**2** 250	**2** 250	**1** 240	**1** 240	**1** 230	**1** 220	**1** 200	**1** 180	**360**
								1993								
0	**0** —	**0** —	**0** —	**1** 230	**1** 230	**1** 240	**1** 240	**1** 250	**1** 250	**1** 240	**1** 240	**1** 230	**1** 220	**0** —	**0** —	**0**
30	**0** —	**0** —	**1** 230	**1** 240	**1** 240	**1** 240	**1** 250	**1** 250	**1** 250	**1** 240	**1** 230	**0** —	**0** —	**0** —	**0** —	**30**
60	**0** —	**1** 240	**1** 250	**1** 250	**1** 250	**1** 250	**1** 260	**1** 250	**1** 250	**1** 240	**0** —	**0** —	**0** —	**0** —	**0** —	**60**
90	**0** —	**1** 260	**1** 260	**1** 270	**1** 270	**1** 270	**1** 270	**1** 270	**1** 270	**0** —	**0** —	**0** —	**0** —	**0** —	**0** —	**90**
120	**0** —	**1** 290	**1** 280	**1** 280	**1** 280	**1** 280	**1** 280	**1** 280	**1** 280	**0** —	**0** —	**0** —	**0** —	**0** —	**0** —	**120**
150	**0** —	**1** 310	**1** 300	**1** 290	**1** 290	**1** 290	**1** 290	**1** 290	**1** 290	**1** 300	**0** —	**0** —	**0** —	**0** —	**0** —	**150**
180	**0** —	**0** —	**1** 320	**1** 310	**1** 300	**1** 300	**1** 290	**1** 290	**1** 300	**1** 300	**1** 310	**1** 310	**0** —	**0** —	**0** —	**180**
210	**0** —	**0** —	**0** —	**0** —	**1** 310	**1** 300	**1** 290	**1** 290	**1** 290	**1** 300	**1** 300	**1** 300	**1** 310	**0** —	**0** —	**210**
240	**0** —	**0** —	**0** —	**0** —	**0** —	**1** 300	**1** 290	**1** 290	**1** 280	**1** 290	**1** 290	**1** 290	**1** 290	**1** 300	**0** —	**240**
270	**0** —	**0** —	**0** —	**0** —	**0** —	**0** —	**1** 270	**1** 270	**1** 270	**1** 270	**1** 270	**1** 270	**1** 280	**1** 280	**0** —	**270**
300	**0** —	**0** —	**0** —	**0** —	**0** —	**0** —	**1** 260	**1** 260	**1** 260	**1** 260	**1** 260	**1** 260	**1** 260	**1** 250	**0** —	**300**
330	**0** —	**0** —	**0** —	**0** —	**0** —	**1** 240	**1** 250	**1** 250	**1** 250	**1** 250	**1** 250	**1** 250	**1** 240	**1** 230	**0** —	**330**
360	**0** —	**0** —	**0** —	**1** 230	**1** 230	**1** 240	**1** 240	**1** 250	**1** 250	**1** 240	**1** 240	**1** 230	**1** 220	**0** —	**0** —	**360**

There is no correction for **1994** or **1995**

The above table gives the correction to be applied to a position line or a fix, *deduced from the tables in this volume,* for the effects of precession and nutation. Each entry consists of a group of four figures of which the first (in bold type) is the distance, in nautical miles, which the position line or fix is to be moved, and the group of three figures is the direction (true bearing). The table is entered firstly by the year, then by choosing the column nearest the latitude and finally the entry nearest the LHA♈ of observation; no interpolation is necessary.

Example. In 1992 a fix is obtained in latitude N 23° when LHA♈ is 71°. Entering the table with the year 1992, latitude N 20° and LHA♈ 60° gives **2**′ 260° which indicates that the fix is to be transferred 2 miles in true bearing 260°.

LHA ♈	North latitudes								South latitudes							LHA ♈
	N 89°	N 80°	N 70°	N 60°	N 50°	N 40°	N 20°	0°	S 20°	S 40°	S 50°	S 60°	S 70°	S 80°	S 89°	
								1996								
°	′ °	′ °	′ °	′ °	′ °	′ °	′ °	′ °	′ °	′ °	′ °	′ °	′ °	′ °	′ °	°
0	1 020	1 030	1 050	1 050	1 060	1 060	1 070	1 070	1 060	1 060	1 050	1 040	1 030	1 010	1 350	**0**
30	1 050	1 060	1 060	1 070	1 070	1 070	1 070	1 070	1 070	1 060	1 050	0 —	0 —	0 —	1 310	**30**
60	1 080	1 080	1 080	1 080	1 080	1 080	1 080	1 080	1 080	1 080	0 —	0 —	0 —	0 —	1 280	**60**
90	1 110	1 100	1 100	1 100	1 100	1 100	1 100	1 100	1 100	1 110	0 —	0 —	0 —	0 —	1 250	**90**
120	1 140	1 120	1 120	1 110	1 110	1 110	1 110	1 110	1 110	1 120	1 130	0 —	0 —	0 —	1 220	**120**
150	1 160	1 150	1 140	1 130	1 120	1 120	1 110	1 110	1 120	1 120	1 130	1 140	1 150	1 170	1 190	**150**
180	1 190	1 170	1 150	1 140	1 130	1 120	1 120	1 110	1 110	1 120	1 120	1 130	1 130	1 150	1 160	**180**
210	1 230	0 —	0 —	0 —	1 130	1 120	1 110	1 110	1 110	1 110	1 110	1 110	1 120	1 120	1 130	**210**
240	1 260	0 —	0 —	0 —	0 —	1 100	1 100	1 100	1 100	1 100	1 100	1 100	1 100	1 100	1 100	**240**
270	1 290	0 —	0 —	0 —	0 —	1 070	1 080	1 080	1 080	1 080	1 080	1 080	1 080	1 080	1 070	**270**
300	1 320	0 —	0 —	0 —	1 050	1 060	1 070	1 070	1 070	1 070	1 070	1 070	1 060	1 060	1 040	**300**
330	1 350	1 010	1 030	1 040	1 050	1 060	1 060	1 070	1 070	1 060	1 060	1 050	1 040	1 030	1 020	**330**
360	1 020	1 030	1 050	1 050	1 060	1 060	1 070	1 070	1 060	1 060	1 050	1 040	1 030	1 010	1 350	**360**
								1997								
0	1 010	1 030	1 040	1 050	2 060	2 060	2 070	2 070	2 060	2 060	1 050	1 040	1 030	1 010	1 350	**0**
30	1 040	1 050	1 060	2 070	2 070	2 070	2 070	2 070	2 070	1 060	1 050	1 040	1 010	1 340	1 320	**30**
60	1 070	1 080	1 080	2 080	2 080	2 080	2 080	2 080	2 080	1 070	1 070	0 —	0 —	1 300	1 290	**60**
90	1 100	1 100	1 100	2 100	2 100	2 090	2 090	2 090	2 100	1 100	1 110	0 —	0 —	1 250	1 260	**90**
120	1 130	1 120	1 110	2 110	2 110	2 110	2 110	2 110	2 110	1 120	1 130	1 140	1 170	1 210	1 230	**120**
150	1 160	1 140	1 130	1 120	2 120	2 120	2 110	2 110	2 120	2 120	1 130	1 140	1 150	1 180	1 200	**150**
180	1 190	1 170	1 150	1 140	1 130	2 120	2 120	2 110	2 110	2 120	2 120	1 130	1 140	1 150	1 170	**180**
210	1 220	1 200	1 170	1 140	1 130	1 120	2 110	2 110	2 110	2 110	2 110	2 110	1 120	1 130	1 140	**210**
240	1 250	1 240	0 —	0 —	1 110	1 110	2 100	2 100	2 100	2 100	2 100	2 100	1 100	1 100	1 110	**240**
270	1 280	1 290	0 —	0 —	1 070	1 080	2 080	2 090	2 090	2 090	2 080	2 080	1 080	1 080	1 080	**270**
300	1 310	1 330	1 010	1 040	1 050	1 060	2 070	2 070	2 070	2 070	2 070	2 070	1 070	1 060	1 050	**300**
330	1 340	1 000	1 030	1 040	1 050	2 060	2 060	2 070	2 070	2 060	2 060	1 060	1 050	1 040	1 020	**330**
360	1 010	1 030	1 040	1 050	2 060	2 060	2 070	2 070	2 060	2 060	1 050	1 040	1 030	1 010	1 350	**360**
								1998								
0	1 010	1 030	2 040	2 050	2 060	2 060	3 070	3 070	3 060	2 060	2 050	2 050	1 030	1 020	1 350	**0**
30	1 040	1 050	2 060	2 060	2 070	3 070	3 070	3 070	2 070	2 060	1 050	1 040	1 010	1 350	1 320	**30**
60	1 070	2 070	2 080	2 080	2 080	3 080	3 080	3 080	2 080	1 070	1 060	1 040	0 —	1 310	1 290	**60**
90	1 100	2 100	2 090	2 090	3 090	3 090	3 090	3 090	2 090	1 100	1 100	0 —	0 —	1 260	1 260	**90**
120	1 130	1 120	2 110	2 110	2 110	3 110	3 100	3 100	2 110	2 120	1 130	1 140	1 180	1 210	1 230	**120**
150	1 160	1 140	2 130	2 120	2 120	2 110	3 110	3 110	3 110	2 120	2 130	1 140	1 160	1 180	1 200	**150**
180	1 190	1 160	1 150	2 130	2 130	2 120	3 120	3 110	3 110	2 120	2 120	2 130	2 140	1 150	1 170	**180**
210	1 220	1 190	1 170	1 140	1 130	2 120	2 110	3 110	3 110	3 110	2 110	2 120	2 120	1 130	1 140	**210**
240	1 250	1 230	0 —	1 140	1 120	1 110	2 100	3 100	3 100	3 100	2 100	2 100	2 100	2 110	1 110	**240**
270	1 280	1 280	0 —	0 —	1 080	1 080	2 090	3 090	3 090	3 090	3 090	2 090	2 090	2 080	1 080	**270**
300	1 310	1 330	1 000	1 040	1 050	2 060	2 070	3 080	3 080	3 070	2 070	2 070	2 070	1 060	1 050	**300**
330	1 340	1 000	1 020	1 040	2 050	2 060	3 070	3 070	3 070	2 070	2 060	2 060	2 050	1 040	1 020	**330**
360	1 010	1 030	2 040	2 050	2 060	2 060	3 070	3 070	3 060	2 060	2 050	2 050	1 030	1 020	1 350	**360**
								1999								
0	1 010	2 030	2 040	2 050	3 060	3 060	3 070	4 070	3 060	3 060	2 050	2 050	2 040	1 020	1 000	**0**
30	1 040	2 050	2 060	3 060	3 070	3 070	4 070	3 070	3 070	2 060	2 050	2 040	1 020	1 350	1 330	**30**
60	1 070	2 070	2 080	3 080	3 080	3 080	4 080	3 080	3 080	2 070	1 060	1 040	1 350	1 310	1 300	**60**
90	1 100	2 090	2 090	3 090	3 090	3 090	4 090	3 090	3 090	2 090	1 100	0 —	0 —	1 260	1 270	**90**
120	1 120	2 120	2 110	3 110	3 110	3 100	4 100	3 100	3 110	2 110	1 120	1 140	1 180	1 220	1 230	**120**
150	1 150	2 140	2 130	3 120	3 120	3 110	4 110	4 110	3 110	2 120	2 130	2 140	1 160	1 180	1 200	**150**
180	1 180	1 160	2 140	2 130	2 130	3 120	3 120	4 110	3 110	3 120	3 120	2 130	2 140	2 150	1 170	**180**
210	1 210	1 190	1 160	2 140	2 130	2 120	3 110	3 110	4 110	3 110	3 110	3 120	2 120	2 130	1 140	**210**
240	1 240	1 230	1 190	1 140	1 120	2 110	3 100	3 100	4 100	3 100	3 100	3 100	2 100	2 110	1 110	**240**
270	1 270	1 280	0 —	0 —	1 080	2 090	3 090	3 090	4 090	3 090	3 090	3 090	2 090	2 090	1 080	**270**
300	1 310	1 320	1 000	1 040	1 060	2 070	3 070	3 080	4 080	3 080	3 070	3 070	2 070	2 060	1 060	**300**
330	1 340	1 000	1 020	2 040	2 050	2 060	3 070	4 070	4 070	3 070	3 060	3 060	2 050	2 040	1 030	**330**
360	1 010	2 030	2 040	2 050	3 060	3 060	3 070	4 070	3 060	3 060	2 050	2 050	2 040	1 020	1 000	**360**

Example. In 1999 a position line is obtained in latitude S 52° when LHA♈ is 327°. Entering the table with the year 1999, latitude S 50°, and LHA♈ 330° gives 3′ 060° which indicates that the position line is to be transferred 3 miles in true bearing 060°.

LHA ♈	*Q*	LHA ♈	*Q*	LHA ♈	*Q*	LHA ♈	*Q*	LHA ♈	*Q*	LHA ♈	*Q*	LHA ♈	*Q*	LHA ♈	*Q*
° ′	′	° ′	′	° ′	′	° ′	′	° ′	′	° ′	′	° ′	′	° ′	′
358 46		86 00		120 50		153 30		227 46		281 46		314 17		349 32	
	−36		−29		− 4		+21		+44		+19		− 6		−31
0 50		87 38		122 07		154 56		233 14		283 10		315 33		351 16	
	−37		−28		− 3		+22		+43		+18		− 7		−32
3 01		89 14		123 23		156 23		237 18		284 32		316 50		353 02	
	−38		−27		− 2		+23		+42		+17		− 8		−33
5 20		90 49		124 39		157 51		240 42		285 55		318 07		354 53	
	−39		−26		− 1		+24		+41		+16		− 9		−34
7 48		92 21		125 54		159 21		243 42		287 16		319 24		356 47	
	−40		−25		0		+25		+40		+15		−10		−35
10 29		93 52		127 11		160 53		246 25		288 36		320 42		358 46	
	−41		−24		+ 1		+26		+39		+14		−11		−36
13 27		95 21		128 27		162 26		248 55		289 56		322 00		0 50	
	−42		−23		+ 2		+27		+38		+13		−12		−37
16 49		96 49		129 43		164 01		251 15		291 16		323 19		3 01	
	−43		−22		+ 3		+28		+37		+12		−13		−38
20 50		98 15		130 59		165 37		253 28		292 35		324 38		5 20	
	−44		−21		+ 4		+29		+36		+11		−14		−39
26 13		99 40		132 16		167 17		255 33		293 53		325 57		7 48	
	−45		−20		+ 5		+30		+35		+10		−15		−40
47 38		101 05		133 32		168 58		257 34		295 11		327 18		10 29	
	−44		−19		+ 6		+31		+34		+ 9		−16		−41
53 01		102 28		134 49		170 43		259 29		296 28		328 39		13 27	
	−43		−18		+ 7		+32		+33		+ 8		−17		−42
57 02		103 51		136 05		172 31		261 20		297 46		330 00		16 49	
	−42		−17		+ 8		+33		+32		+ 7		−18		−43
60 24		105 12		137 23		174 22		263 08		299 02		331 23		20 50	
	−41		−16		+ 9		+34		+31		+ 6		−19		−44
63 22		106 33		138 40		176 17		264 53		300 19		332 46		26 13	
	−40		−15		+10		+35		+30		+ 5		−20		−45
66 03		107 54		139 58		178 18		266 34		301 35		334 11		47 38	
	−39		−14		+11		+36		+29		+ 4		−21		−44
68 31		109 13		141 16		180 23		268 14		302 52		335 36		53 01	
	−38		−13		+12		+37		+28		+ 3		−22		−43
70 50		110 32		142 35		182 36		269 50		304 08		337 02		57 02	
	−37		−12		+13		+38		+27		+ 2		−23		−42
73 01		111 51		143 55		184 56		271 25		305 24		338 30		60 24	
	−36		−11		+14		+39		+26		+ 1		−24		−41
75 05		113 09		145 15		187 26		272 58		306 40		339 59		63 22	
	−35		−10		+15		+40		+25		0		−25		−40
77 04		114 27		146 35		190 09		274 30		307 57		341 30		66 03	
	−34		− 9		+16		+41		+24		− 1		−26		−39
78 58		115 44		147 56		193 09		276 00		309 12		343 02		68 31	
	−33		− 8		+17		+42		+23		− 2		−27		−38
80 49		117 01		149 19		196 33		277 28		310 28		344 37		70 50	
	−32		− 7		+18		+43		+22		− 3		−28		−37
82 35		118 18		150 41		200 37		278 55		311 44		346 13		73 01	
	−31		− 6		+19		+44		+21		− 4		−29		−36
84 19		119 34		152 05		206 05		280 21		313 01		347 51		75 05	
	−30		− 5		+20		+45		+20		− 5		−30		−35
86 00		120 50		153 30		227 46		281 46		314 17		349 32		77 04	

The above table, which does *not* include refraction, gives the quantity *Q* to be applied to the corrected sextant altitude of *Polaris* to give the latitude of the observer. In critical cases ascend.

Polaris: Mag. 2.1, SHA 323° 04′, Dec N 89° 14′.7

TABLE 7 — AZIMUTH OF *POLARIS*

LHA ♈	Latitude 0°	30°	50°	55°	60°	65°	70°	LHA ♈	Latitude 0°	30°	50°	55°	60°	65°	70°
°	°	°	°	°	°	°	°	°	°	°	°	°	°	°	°
0	0·5	0·5	0·7	0·8	0·9	1·1	1·4	180	359·5	359·5	359·3	359·2	359·1	359·0	358·7
10	0·3	0·4	0·5	0·6	0·7	0·8	1·0	190	359·7	359·6	359·5	359·4	359·3	359·2	359·0
20	0·2	0·3	0·3	0·4	0·4	0·5	0·7	200	359·8	359·7	359·7	359·6	359·6	359·5	359·4
30	0·1	0·1	0·1	0·2	0·2	0·2	0·3	210	359·9	359·9	359·9	359·8	359·8	359·8	359·7
40	0·0	0·0	359·9	359·9	359·9	359·9	359·9	220	0·0	0·0	0·1	0·1	0·1	0·1	0·1
50	359·8	359·8	359·7	359·7	359·7	359·6	359·5	230	0·2	0·2	0·3	0·3	0·3	0·4	0·5
60	359·7	359·7	359·5	359·5	359·4	359·3	359·1	240	0·3	0·3	0·5	0·5	0·6	0·7	0·8
70	359·6	359·5	359·4	359·3	359·2	359·0	358·8	250	0·4	0·5	0·6	0·7	0·8	1·0	1·2
80	359·5	359·4	359·2	359·1	359·0	358·8	358·5	260	0·5	0·6	0·8	0·9	1·0	1·2	1·5
90	359·4	359·3	359·1	358·9	358·8	358·5	358·2	270	0·6	0·7	0·9	1·0	1·2	1·4	1·7
100	359·3	359·2	358·9	358·8	358·6	358·4	358·0	280	0·7	0·8	1·0	1·2	1·3	1·6	1·9
110	359·3	359·2	358·9	358·7	358·5	358·3	357·9	290	0·7	0·8	1·1	1·3	1·4	1·7	2·1
120	359·3	359·1	358·8	358·7	358·5	358·2	357·8	300	0·7	0·9	1·2	1·3	1·5	1·8	2·2
130	359·2	359·1	358·8	358·7	358·5	358·2	357·8	310	0·8	0·9	1·2	1·3	1·5	1·8	2·2
140	359·3	359·2	358·9	358·7	358·5	358·3	357·9	320	0·7	0·9	1·1	1·3	1·5	1·8	2·2
150	359·3	359·2	358·9	358·8	358·6	358·4	358·0	330	0·7	0·8	1·1	1·2	1·4	1·7	2·1
160	359·4	359·3	359·0	358·9	358·8	358·5	358·2	340	0·6	0·7	1·0	1·1	1·3	1·5	1·9
170	359·4	359·4	359·2	359·1	358·9	358·7	358·4	350	0·6	0·6	0·9	1·0	1·1	1·3	1·7
180	359·5	359·5	359·3	359·2	359·1	359·0	358·7	360	0·5	0·5	0·7	0·8	0·9	1·1	1·4

When Cassiopeia is left (right), *Polaris* is west (east).

TABLE 8—REFRACTION

TO BE *SUBTRACTED* FROM SEXTANT ALTITUDE

(a) Height in thousands of feet

R	0	5	10	15	20	25	30	35	40	45	50	55	*R*
′	°	°	°	°	°	°	°	°	°	°	°	°	′
	90	90	90	90	90	90	90	90	90	90	90	90	
0													0
	63	59	55	51	46	41	36	31	26	20	17	13	
1													1
	33	29	26	22	19	16	14	11	10	10	10	10	
2													2
	21	19	16	14	12	10	10	10					
3													3
	16	14	12	10	10								
4													4
	12	11	10										
5													5
	10	10											

(b) Height in thousands of metres

R	0	1	2	3	4	5	6	7	8	9	10	11	12	13	14	15	16	17	18	19	*R*
′	°	°	°	°	°	°	°	°	°	°	°	°	°	°	°	°	°	°	°	°	′
	90	90	90	90	90	90	90	90	90	90	90	90	90	90	90	90	90	90	90	90	
0																					0
	63	61	58	55	53	50	46	43	40	37	33	30	26	23	20	17	15	13	11	10	
1																					1
	33	31	28	26	24	21	19	17	16	14	12	11	10	10	10	10	10	10	10		
2																					2
	21	20	18	16	15	13	12	11	10	10	10	10									
3																					3
	16	14	13	12	11	10	10	10													
4																					4
	12	11	10	10	10																
5																					5
	10	10																			

Choose the column appropriate to height, in units of 1000 feet in table 8(a) or in units of 1000 metres in table 8(b), and find the range of altitude in which the sextant altitude lies; the corresponding value of *R* is the refraction to be subtracted from the sextant altitude.

TABLE 9—CORIOLIS (*Z*) CORRECTION

STANDARD DOME REFRACTION

To be *subtracted* from sextant altitude when using sextant suspension in a perspex dome.

Alt.	Refn.	Alt.	Refn.
°	′	°	′
10	8	50	4
20	7	60	4
30	6	70	3
40	5	80	3

This table must not be used if a calibration table is fitted to the dome, or if a flat glass plate is provided, or for non-standard domes.

Ground speed knots	Latitude 0°	10°	20°	30°	40°	50°	60°	70°	80°	90°	Ground speed knots
	′	′	′	′	′	′	′	′	′	′	
50	0	0	0	1	1	1	1	1	1	1	**50**
100	0	0	1	1	2	2	2	2	3	3	**100**
150	0	1	1	2	3	3	3	4	4	4	**150**
200	0	1	2	3	3	4	5	5	5	5	**200**
250	0	1	2	3	4	5	6	6	6	7	**250**
300	0	1	3	4	5	6	7	7	8	8	**300**
350	0	2	3	5	6	7	8	9	9	9	**350**
400	0	2	4	5	7	8	9	10	10	10	**400**
450	0	2	4	6	8	9	10	11	12	12	**450**
500	0	2	4	7	8	10	11	12	13	13	**500**
550	0	3	5	7	9	11	12	14	14	14	**550**
600	0	3	5	8	10	12	14	15	16	16	**600**
650	0	3	6	9	11	13	15	16	17	17	**650**
700	0	3	6	9	12	14	16	17	18	18	**700**
750	0	3	7	10	13	15	17	18	19	20	**750**
800	0	4	7	10	13	16	18	20	21	21	**800**
850	0	4	8	11	14	17	19	21	22	22	**850**
900	0	4	8	12	15	18	20	22	23	24	**900**

Apply by moving the position line a distance *Z* to starboard (right) of the track in northern latitudes, and to port (left) in southern latitudes.

BUBBLE SEXTANT ERROR

Sextant No.

Alt.	Corr.
°	′